Table of atomic weights to four significant figures*

(*Scaled to the relative atomic mass of $^{12}C = 12$ exactly*)
Values quoted in this table are reliable to ± 1 or better in the fourth significant figure except for the five elements for which the larger indicated uncertainties apply. Each element that has neither a stable isotope nor a characteristic natural isotopic composition is represented in this table by one of that element's commonly known radioisotopes identified by its mass number (in superscript preceding the chemical symbol) and its relative atomic mass, in the Atomic Weight column.

Atomic Number	Name	Symbol	Atomic Weight	Atomic Number	Name	Symbol	Atomic Weight
1	Hydrogen	H	1.008	53	Iodine	I	126.9
2	Helium	He	4.003	54	Xenon	Xe	131.3
3	Lithium	Li	6.941 ± 2	55	Caesium	Cs	132.9
4	Beryllium	Be	9.012	56	Barium	Ba	137.3
5	Boron	B	10.81	57	Lanthanum	La	138.9
6	Carbon	C	12.01	58	Cerium	Ce	140.1
7	Nitrogen	N	14.01	59	Praseodymium	Pr	140.9
8	Oxygen	O	16.00	60	Neodymium	Nd	144.2
9	Fluorine	F	19.00	61	Promethium	^{145}Pm	144.9
10	Neon	Ne	20.18	62	Samarium	Sm	150.4
11	Sodium	Na	22.99	63	Europium	Eu	152.0
12	Magnesium	Mg	24.30	64	Gadolinium	Gd	157.2
13	Aluminium	Al	26.98	65	Terbium	Tb	158.9
14	Silicon	Si	28.09	66	Dysprosium	Dy	162.5
15	Phosphorus	P	30.97	67	Holmium	Ho	164.9
16	Sulfur	S	32.07	68	Erbium	Er	167.3
17	Chlorine	Cl	35.45	69	Thulium	Tm	168.9
18	Argon	Ar	39.95	70	Ytterbium	Yb	173.0
19	Potassium	K	39.10	71	Lutetium	Lu	175.0
20	Calcium	Ca	40.08	72	Hafnium	Hf	178.5
21	Scandium	Sc	44.96	73	Tantalum	Ta	180.9
22	Titanium	Ti	47.88 ± 3	74	Wolfram (Tungsten)	W	183.8
23	Vanadium	V	50.94	75	Rhenium	Re	186.2
24	Chromium	Cr	52.00	76	Osmium	Os	190.2
25	Manganese	Mn	54.94	77	Iridium	Ir	192.2
26	Iron	Fe	55.85	78	Platinum	Pt	195.1
27	Cobalt	Co	58.93	79	Gold	Au	197.0
28	Nickel	Ni	58.69	80	Mercury	Hg	200.6
29	Copper	Cu	63.55	81	Thallium	Tl	204.4
30	Zinc	Zn	65.39 ± 2	82	Lead	Pb	207.2
31	Gallium	Ga	69.72	83	Bismuth	Bi	209.0
32	Germanium	Ge	72.61 ± 2	84	Polonium	^{210}Po	210.0
33	Arsenic	As	74.92	85	Astatine	^{210}At	210.0
34	Selenium	Se	78.96 ± 3	86	Radon	^{222}Rn	222.0
35	Bromine	Br	79.90	87	Francium	^{223}Fr	223.0
36	Krypton	Kr	83.80	88	Radium	^{226}Ra	226.0
37	Rubidium	Rb	85.47	89	Actinium	^{227}Ac	227.0
38	Strontium	Sr	87.62	90	Thorium	Th	232.0
39	Yttrium	Y	88.91	91	Protactinium	Pa	231.0
40	Zirconium	Zr	91.22	92	Uranium	U	238.0
41	Niobium	Nb	92.91	93	Neptunium	^{237}Np	237.0
42	Molybdenum	Mo	95.94	94	Plutonium	^{239}Pu	239.1
43	Technetium	^{99}Tc	98.91	95	Americium	^{243}Am	243.1
44	Ruthenium	Ru	101.1	96	Curium	^{247}Cm	247.1
45	Rhodium	Rh	102.9	97	Berkelium	^{247}Bk	247.1
46	Palladium	Pd	106.4	98	Californium	^{252}Cf	252.1
47	Silver	Ag	107.9	99	Einsteinium	^{252}Es	252.1
48	Cadmium	Cd	112.4	100	Fermium	^{257}Fm	257.1
49	Indium	In	114.8	101	Mendelevium	^{256}Md	256.1
50	Tin	Sn	118.7	102	Nobelium	^{259}No	259.1
51	Antimony	Sb	121.8	103	Lawrencium	^{260}Lr	260.1
52	Tellurium	Te	127.6				

Reproduced with permission from the International Union of Pure and Applied Chemistry (IUPAC) in consultation with the IUPAC Commission on Atomic Weights and Isotopic Abundances. Taken from *Pure Applied Chemistry*, Volume 60, No.6 (1988), pp. 841-854.

Hazardous Chemicals
Desk Reference

Hazardous Chemicals Desk Reference

Second Edition

Richard J. Lewis, Sr.

VNR VAN NOSTRAND REINHOLD
New York

Disclaimer

Extreme care has been taken in preparation of this work.
However, neither the publisher nor the author shall be
held responsible or liable for any damages resulting in
connection with or arising from the use of any of the
information in this book.

Copyright © 1991 by Van Nostrand Reinhold

Library of Congress Catalog Card Number 90–48602
ISBN 0–442–00497–4

All rights reserved. Certain portions of this work © 1987 by
Van Nostrand Reinhold. No part of this work covered by the copyright
hereon may be reproduced or used in any form or by any means—graphic,
electronic, or mechanical, including photocopying, recording, taping,
or informational storage and retrieval systems—without written permission
of the publisher.

Manufactured in the United States of America

Published by Van Nostrand Reinhold
115 Fifth Avenue
New York, NY 10003

Chapman and Hall
2–6 Boundary Row
London, SE 1 8HN

Thomas Nelson Australia
102 Dodds Street
South Melbourne 3205
Victoria, Australia

Nelson Canada
1120 Birchmount Road
Scarborough, Ontario M1K 5G4, Canada

16 15 14 13 12 11 10 9 8 7 6 5 4 3 2

Library of Congress Cataloging-in-Publication Data
Lewis, Richard J., Sr.
    Hazardous materials desk reference / by Richard J. Lewis, Sr.—
2nd ed.
        p. cm.
    Includes index.
        ISBN 0–442—00497–4
        1. Hazardous substances—Handbooks, manuals, etc.   I. Title.
T55.3.H3L49   1990
604.7—dc20   90–48602
                                                  CIP

To Scott Allen Ross
for many stimulating discussions

Contents

Preface

This second edition of the *Hazardous Chemicals Desk Reference* fills the need for a reference work of moderate size which would serve the information needs of many who must work with and evaluate the hazard of chemicals.

Over 5,500 chemical entries are provided. More than 1,000 new entries were added based upon their importance in industry, their toxicity, or fire and explosion hazard. About 400 entries present in the first edition were removed to provide space for more interesting substances. The basic data were extracted from *Dangerous Properties of Industrial Materials*, 7th Edition with extensive updating of information. Citations to toxicity data and other less relevant material are absent from this reference but will be found in *Dangerous Properties of Industrial Materials*, 7th Edition.

Several data fields were completely updated:

OSHA Standards reflect the revised standards which became effective on September 1, 1989.
ACGIH, MAK, and DOT entries were updated through 1990 published changes.
All carcinogenic data were updated and Safety Profile statements rewritten.
Additional synonyms were added.

Two cross-indices are provided to permit rapid location of a material if either a Chemical Abstract Service (CAS) number or a synonym for the material is the point of entry.

RICHARD J. LEWIS, SR.

Introduction

This condensation of information on potentially hazardous materials includes drugs, food additives, preservatives, ores, pesticides, dyes, detergents, lubricants, soaps, plastics, extracts from plant and animal sources, plants and animals which are toxic by contact or consumption, and industrial intermediates and waste products from production processes. Some of the information refers to materials of undefined composition. The chemicals included are assumed to exhibit the reported toxic effect in their pure state unless otherwise noted. However, even in the case of a supposedly "pure" chemical, there is usually some degree of uncertainty about its exact composition and the impurities which may be present. This possibility must be considered in attempting to interpret the data presented since a contaminant could cause the toxic effects observed. Some radioactive materials are included but the effect reported is the chemically produced effect rather than the radiation effect.

Tradename products representing compounded or formulated proprietary mixtures available as commercial products are excluded from this compilation. These exclusions are necessary because of difficulties in assessing the contribution of each component of a mixture to that material's total toxicity and because a product's formulation is often changed by varying the components, their concentration, or their purity. Some commercial product tradenames are included as synonyms, particularly when they represent a single active chemical entity or a well-defined mixture of relatively constant composition.

For each entry the following data are provided when available: the entry number, Hazard Rating, entry name, CAS number, DOT number, molecular formula, molecular weight, a description of the material and its physical properties, and synonyms. Following, where available, are IARC reviews, NTP Carcinogenesis Testing Program results, EPA Extremely Hazardous Substances List notations, the EPA Genetic Toxicology Program inclusion, and presence on the Community Right-To-Know List. The presence of the material on the update of the EPA TSCA Inventory of chemicals in use in the United States is noted. The next grouping consists of the U.S. Occupational Safety and Health Administration's (OSHA) Permissible Exposure Levels, the American Conference of Governmental Industrial Hygienists' (ACGIH) Threshold Limit Values (TLV's), German Research Society's (MAK) values, National Institute for Occupational Safety and Health (NIOSH) Recommended Exposure Levels, and U.S. Department of Transportation (DOT) classifications. Each entry concludes with a Safety Profile which discusses the toxic and other hazards of the entry.

1. *Entry Number* identifies each entry by a unique number consisting of three letters and three numbers, for example, AAA123. The first letter of the entry number indicates the alphabetical position of the entry. Numbers beginning with "A" are assigned to entries indexed with the A's. The two cross-indexes use the entry number as a locator.

2. *Entry Name* The name of each material is selected, where possible, to be the most commonly used designation. Certain entries with identical synonyms are grouped together because of similar toxicity data.

3. *Hazard Rating (HR:)* was assigned to each material in the form of a number 1, 2, or 3 that briefly identifies the level of the toxicity or hazard. The letter "D" is used where the data available are insufficient to indicate a relative rating. In most cases a "D" rating is assigned when only in vitro mutagenic or experimental reproductive data are available. Ratings are assigned on the basis of low (1), medium (2), or high (3) toxic, fire, explosive, or reactivity hazard.

The number "3" indicates an LD50 below 400 mg/kg or an LC50 below 100 ppm; or that the material is explosive, highly flammable, or highly reactive.

The number "2" indicates an LD50 of 400–4,000 mg/kg or an LC50 of 100–500 ppm; or that the material is flammable or reactive.

The number "1" indicates an LD50 of 4,000–40,000 mg/kg or an LC50 of 500–4,000 ppm; or that the material is combustible.

4. *Chemical Abstracts Service Registry Number (CAS)* is a numeric designation assigned by the American Chemical Society's Chemical Abstracts Service and uniquely identifies a specific chemical compound. This number allows one to conclusively identify a material regardless of the name or naming system used.

5. *DOT:* indicates a four digit hazard code assigned by the U.S. Department of Transportation. This code is recognized internationally and is in agreement with the United Nations coding system. The code is used on transport documents, labels, and placards. It is also used to determine the regulations for shipping the material.

6. *Molecular formula (mf:)* or atomic formula (*af:*) designates the elemental composition of the material and is structured according to the Hill System (see *Journal of the American Chemical Society*, 22(8): 478–494, 1900) in which carbon and hydrogen (if present) are listed first, followed by the other elemental symbols in alphabetical order. The formula for compounds that do not contain carbon are ordered strictly alphabetically by element symbol. Compounds such as salts or those containing waters of hydration have molecular formulas incorporating the CAS dot-disconnect convention. In this convention, the components are listed individually and separated by a period. The individual components of the formula are given in order of decreasing carbon atom count, and the component ratios given. A lower case "x" indicates that the ratio is unknown. A lower case "n" indicates a repeating, polymerlike structure. The formula is obtained from one of the cited references or a chemical reference text, or derived from the name of the material.

7. *Molecular Weight (mw:)* or atomic weight (*aw:*) is calculated from the molecular formula using standard elemental molecular weights (carbon = 12.01).

8. *Properties (PROP:)* are selected to be useful in evaluating the hazard of a material and determining its proper storage and use procedures. A definition of the material is included where necessary. The physical description of the material may include the form, color, and odor to aid in positive identification. When available, the boiling point, melting point, density, vapor pressure, vapor density, and refractive index are given. The flash point, autoignition temperature, and lower and upper explosive limits are included to aid in fire protection and control. An indication is given of the solubility or miscibility of the material in water and common solvents.

9. *Synonyms* for the entry name are listed alphabetically. Synonyms include other chemical names, common or generic names, foreign names (with the language in parentheses), or codes. Some synonyms consist in whole or in part of registered trademarks. These trademarks are not identified as such. The reader is cautioned that some synonyms, particularly common names, may be ambiguous and refer to more than one material.

10. *CONSENSUS REPORTS*. Five types of information are listed: (a) International Agency for Research on Cancer (IARC) monograph reviews, (b) National Toxicology Program (NTP) reports, (c) EPA Extremely Hazardous Substances List, (d) Community Right-To-Know List, (e) Genetic Toxicology Program, and (f) TSCA status.

a. Cancer Reviews. The U.N. International Agency for Research on Cancer (IARC) monographs examine information on suspected environmental carcinogens, and provide summaries of available data with appropriate references. These reviews contain synonyms, physical and chemical properties, uses and occurrence, and biological data relevant to the evaluation of carcinogenic risk to humans. The over 44 monographs in the series contain an evaluation of approximately 900 materials. They are coded IMEMDT followed by volume, page, and year. Single copies of the individual monographs (specify volume number) can be ordered from WHO Publications Centre USA, 49 Sheridan Avenue, Albany, New York 12210, telephone (518) 436-9686.

The format of the IARC data line is as follows. The entry ''IARC Cancer Review:'' indicates that the carcinogenicity data pertaining to a compound has been reviewed by an IARC committee. The committee's conclusions are summarized in three words. The first word indicates whether the data pertains to humans or to animals. The next two words indicate the degree of carcinogenic risk as defined by IARC.

For experimental animals the evidence of carcinogenicity is assessed by IARC and judged to fall into one of four groups defined as follows:

(1) Sufficient Evidence of carcinogenicity is provided when there is an increased incidence of malignant tumors: (a) in multiple species or strains; or (b) in multiple experiments (preferably with different routes of administration or using different dose levels); or (c) to an unusual degree with regard to the incidence, site, or type of tumor, or age at onset. Additional evidence may be provided by data on dose-response effects.

(2) Limited Evidence of carcinogenicity is available when the data suggest a carcinogenic effect but are limited because: (a) the studies involve a single species, strain, or experiment; (b) the experiments are restricted by inadequate dosage levels, inadequate duration of exposure to the agent, inadequate period of follow-up, poor survival, too few animals, or inadequate reporting; or (c) the neoplasms produced often occur spontaneously and, in the past, have been difficult to classify as malignant by histological criteria alone (e.g., lung adenomas and adenocarcinomas, and liver tumors in certain strains of mice).

(3) Inadequate Evidence is available when, because of major qualitative or quantitative limitations, the studies cannot be interpreted as showing either the presence or absence of a carcinogenic effect.

(4) No Evidence applies when several adequate studies are available which show that within the limitations of the tests used, the chemical is not carcinogenic.

It should be noted that the categories Sufficient Evidence and Limited Evidence refer only to the strength of the experimental evidence that these chemicals are carcinogenic and not to the extent of their carcinogenic activity nor to the mechanism involved. The classification of any chemical may change as new information becomes available.

The evidence for carcinogenicity from studies in humans is assessed by the IARC committees and judged to fall into one of four groups defined as follows:

(1) Sufficient Evidence of carcinogenicity indicates that there is a causal relationship between the exposure and human cancer.

(2) Limited Evidence of carcinogenicity indicates that a causal relationship is credible, but that alternative explanations, such as chance, bias, or confounding, could not adequately be excluded.

(3) Inadequate Evidence, which applies to both positive and negative evidence, indicates that one of two conditions prevailed: (a) there are few pertinent data; or (b) the available studies, while showing evidence of association, do not exclude chance, bias, or confounding.

(4) No Evidence applies when several adequate studies are available which do not show evidence of carcinogenicity.

This cancer review reflects only the conclusion of the IARC committee based on the data available for the committee's evaluation. Some substances previously reviewed by IARC may be reexamined as additional data become available. These substances will contain multiple IARC review lines, each of which is referenced to the applicable IARC monograph volume.

b. NTP Status. This entry indicates that the material has been tested by the National Toxicology Program (NTP) under its Carcinogenesis Testing Program. These entries are also identified as National Cancer Institute (NCI), which reported the studies before the NCI Carcinogenesis Testing Program was absorbed by NTP. Testing is reported in NTP reports indicated by NTPTR*, or in the journal *Cancer.* (J. B. Lippincott Co., E. Washington Sq., Philadelphia, PA 19105) V.1-1948-indicated by CANCAR. To obtain additional information about NTP, the Carcinogenesis Testing Program, or the status of a particular material under test, contact the Toxicology Information and Scientific Evaluation Group, NTP/TRTP/NIEHS, Mail Drop 18-01, P.O. Box 12233, Research Triangle Park, NC 27709.

c. EPA Extremely Hazardous Substances List. This list was developed by the U.S. Environmental Protection Agency (EPA) as required by the Superfund Amendments and Reauthorization Act of 1986 (SARA). Title III Section 304 requires notification by facilities of a release of certain extremely hazardous substances. These 402 substances were listed by EPA in the *Federal Register* November 17, 1986.

d. Community Right to Know List. This list was developed by the EPA as required by the Superfund Amendments and Reauthorization Act of 1986 (SARA). Title III, Sections 311–312 require manufacturing facilities to prepare Material Safety Data Sheets and notify local authorities of the presence of listed chemicals. Both specific chemicals and classes of chemicals are covered by these Sections.

e. EPA Genetic Toxicology Program. (GENE-TOX). This status line indicates that the material has had genetic effects reported in the literature during the period 1969–1979. The test protocol in the literature is evaluated by an EPA expert panel on mutations and the positive or negative genetic effect of the substance is reported. To obtain additional information about this program, contact GENE-TOX program, USEPA, 401 M Street, SW, TS796, Washington, DC 20460, Telephone (202) 382-3513.

f. EPA TSCA Status Line. This line indicates that the material appears on the chemical inventory prepared by the Environmental Protection Agency in accordance with provisions of the Toxic Substances Control Act (TSCA). Materials reported in the inventory include those that are produced commercially in or imported into this country. The reader should note, however, that materials already regulated by EPA under FIFRA and by the Food and Drug Administration under the Food, Drug, and Cosmetic Act, as amended, are not included in the TSCA inventory. Similarly, alcohol, tobacco, and explosive materials are not regulated under TSCA. TSCA regulations should be consulted for an exact definition of reporting requirements. For additional information about TSCA, contact EPA, Office of Toxic Substances, Washington, D.C. 20402. Specific questions about the inventory can be directed to the EPA Office of Industry Assistance, telephone (800) 424-9065.

11. *Standards and Recommendations*. This section contains regulations by agencies of the United States Government or recommendations by expert groups. "OSHA" refers to standards promulgated under Section 6 of the Occupational Safety and Health Act of 1970. "EPA" refers to Worker Protection Standards for Agricultural Pesticides promulgated by the Environmental Protection Agency under the Federal Insecticide, Fungicide, and Rodenticide Act (FIFRA). "DOT" refers to materials regulated for shipment by the Department of Transportation. Because of frequent changes to and litigation of Federal regulations, it is recommended that the reader contact the applicable agency for information about the current standards for a particular material. Omission of a material or regulatory notation from this edition does not imply any relief from regulatory responsibility.

a. OSHA Air Contaminant Standards. The values given are for the revised standards which were published in January 13, 1989 and take effect on September 1, 1989 through December 31, 1992. These are noted with the entry "OSHA PEL:" followed by "TWA" or "CL" meaning either time-weighted avarage or ceiling value, respectively, to which workers can be exposed for a normal, 8-hour day, 40-hour work week without ill effects. For some materials, TWA, CL, and Pk (peak) values are given in the standard. In those cases, all three are listed. Finally, some entries may be followed by the designation "(skin)." This designation indicates that the compound may be absorbed by the skin and, even though the air concentration may be below the standard, significant additional exposure through the skin may be possible.

b. ACGIH Threshold Limit Values. The American Conference of Governmental Industrial Hygienists (ACGIH) Threshold Limit Values are noted with the entry "ACGIH TLV:" followed by "TWA" or "CL" meaning either time-weighted average or ceiling value, respectively, to which workers can be exposed for a normal 8-hour day, 40-hour work week without ill effects. The notation "CL" indicates a ceiling limit which must not be exceeded. The notation "skin" indicates that the material penetrates intact skin, and skin contact should be avoided even though the TLV concentration is not exceeded. STEL indicates a short-term exposure limit, usually a 15-minute time-weighted average, which should not be exceeded. Biological Exposure Indices (*BEI:*) are, according to the ACGIH, set to provide a warning level ". . . of biological response to the chemical, or warning levels of that chemical or its metabolic product(s) in tissues, fluids, or exhaled air of exposed workers . . ."

The latest annual TLV list is contained in the publication *Threshold Limit Values and Biological Exposure Indices*. This publication should be consulted for future trends in recommendations. The ACGIH TLV's are adopted in whole or in part by many countries and local administrative agencies throughout the world. As a result, these recommendations have a major effect on the control of workplace contaminant concentrations. The ACGIH may be contacted for additional information at 6500 Glenway Ave., Cincinnati, Ohio 45211, USA.

c. DFG MAK. These lines contain the German Research Society's Maximum Allowable Concentration values. Those materials which are classified as to workplace hazard potential by the German Research Society are noted on this line. The MAK values are also revised annually and discussions of materials under consideration for MAK assignment are included in the annual publication together with the current values. *BAT:* indicates Biological Tolerance Value for a Working Material which is defined as, ". . . the maximum permissible quantity of a chemical compound, its metabolites, or any deviation from the norm of biological parameters induced by these substances in exposed humans." *TRK:* values are Technical Guiding Concentrations for workplace control of carcinogens. For additional information, write to Deutsche Forschungsgemeinschaft (German Research Society), Kennedyallee 40, D-5300 Bonn 2, Federal Republic of Germany. The publication *Maximum Concentrations at the Workplace and Biological Tolerance Values for Working Materials* can be obtained from Verlag Chemie GmbH, Buchauslieferung,

P.O. Box 1260/1280, D-6940 Weinheim, Federal Republic of Germany, or VCH Publishers, Suite 909, 220 East 23rd Street, New York, N.Y. 10010.

d. NIOSH REL. This line indicates that a NIOSH criteria document recommending a certain occupational exposure has been published for this compound or for a class of compounds to which this material belongs. These documents contain extensive data, analysis, and references. The more recent publications can be obtained from the National Institute for Occupational Safety and Health, U.S. Department of Health and Human Services, 4676 Columbia Pkwy., Cincinnati, Ohio 45226.

e. DOT Classification. This is the hazard classification according to the U.S. Department of Transportation (DOT) or the International Maritime Organization (IMO). This classification gives an indication of the hazards expected in transportation, and serves as a guide to the development of proper labels, placards, and shipping instructions. The basic hazard classes include compressed gases, flammables, oxidizers, corrosives, explosives, radioactive materials, and poisons. Although a material may be designated by only one hazard class, additional hazards may be indicated by adding labels or by using other means as directed by DOT. Many materials are regulated under general headings such as "pesticides" or "combustible liquids" as defined in the regulations. These are not noted here, as their specific concentration or properties must be known for proper classification. Special regulations may govern shipment by air. This information should serve *only as a guide*, since the regulation of transported materials is carefully controlled in most countries by federal and local agencies. Because of frequent changes to regulations, it is recommended that the reader contact the applicable agency for information about the current standards for a particular material. United States transportation regulations are found in 40 CFR, Parts 100 to 189. Contact the U.S. Department of Transportation, Materials Transportation Bureau, Washington, D.C. 20590.

12. *SAFETY PROFILE*. These are text summaries of the hazardous properties of the entry. For toxicity, human effects are identified either by "human" or more specifically by "man, woman, child, or infant." The affected human target organs and specific effects are reported when available. The word "experimental" indicates that the reported effects resulted from a controlled exposure of laboratory animals to the substance. Information is provided on irritant, mutagenic, acute toxic effects, and chronic carcinogenic, reproductive, and toxic effects.

Carcinogenic potential is denoted by the words confirmed, suspected, or questionable based on experimental evidence and the opinion of expert review groups. The OSHA, IARC, ACGIH, and DFG MAK decision schedules are not related or synchronized. Thus an entry may have had a recent review by only one group. The most stringent classification of any regulation or expert group is taken as governing.

Confirmed carcinogen indicates those substances which are capable of causing cancer in exposed humans. An entry was given this designation if it had one or more of the following data items present:

a. An OSHA regulated carcinogen.
b. An ACGIH assignment as a human or animal carcinogen.
c. A DFG MAK assignment as a confirmed human or animal carcinogen.
d. An IARC assignment of human or animal sufficient evidence of carcinogenicity or higher.
e. NTP Fourth Annual Report on Carcinogens.

Suspected carcinogen indicates that these substances may be capable of causing cancer in exposed humans. The evidence is suggestive, but not sufficient to convince expert review committees. Some entries have not yet had expert review, but contain experimental reports of carcinogenic activity. In particular, an entry is included if it has positive reports of carcinogenic endpoint in two species. As more studies are published, many suspected carcinogens will have their carcinogenicity confirmed. On the other hand, some will be judged noncarcinogenic in the future. An entry was given this designation if it had one or more of the following data items present:

a. An ACGIH assignment of suspected carcinogen.
b. A DFG MAK assignment of suspected carcinogen.
c. An IARC assignment of human or animal limited evidence.
d. Two animal studies reported positive carcinogenic endpoint in different species.

Questionable carcinogens have minimal published evidence of possible carcinogenic activity. The reported endpoint is often neoplastic growth with no spread or invasion characteristic of carcinogenic pathology. An even weaker endpoint is that of equivocal tumorigenic agent (ETA). Reports are assigned this designation when the study was defective. The report may have lacked control animals, may have used very small sample size, often lack complete pathology reporting, or suffer many other study design defects. Many of these were designed for other than carcinogenic evaluation, and the reported carcinogenic effect is a byproduct of the study, not the goal. The data are presented because some of these substances may be carcinogens. There are insufficient data to affirm or deny the possibility. An entry was given this designation if it had one or more of the following data items present:

a. An IARC assignment of inadequate or no evidence.
b. A single human report of carcinogenicity.
c. A single experimental carcinogenic report, or duplicate reports in the same species.
d. One or more experimental neoplastic or equivocal tumorigenic agent report.

For flammable, combustible, or reactive materials the fire and explosion hazards are briefly summarized. Materials which are incompatible with the entry substance are listed here. Where feasible, fire-fighting materials and methods are discussed.

A material with a flash point of 100°F or less is considered flammable and dangerous; if the flash point is from 100 to 200°F, the substance is combustible and of moderate hazard; if it is above 200°F, the substance is combustible and of low fire hazard.

Disaster hazards comments are incorporated, where available, to alert readers to the dangers that may be encountered on entering storage premises during a fire or other emergency. The presence of water, steam, acid fumes, or powerful vibrations can cause the decomposition of many materials into dangerous compounds. High temperatures (such as those resulting from a fire) are of particular concern since these can cause many otherwise mild chemicals to emit highly toxic gases or vapors such as NO_x, SO_x, acids, and so forth, or evolve toxic elemental vapors of antimony, arsenic, mercury, and the like.

Acknowledgments

Thanks to Grace, my wife, for her constant help and advice in all steps in producing this work. I extend thanks to Dr. Mark Licker, Executive Editor, Chemistry and Industrial Safety, for encouragement. My best to Alberta Gordon and Louise Kurtz for their expert professional advice and assistance in converting the manuscript to this volume.

Key to Abbreviations

abs – absolute
ACGIH – American Conference of Governmental Industrial Hygienists
alc – alcohol
alk – alkaline
amorph – amorphous
anhyd – anhydrous
approx – approximately
aq – aqueous
atm – atmosphere
autoign – autoignition
aw – atomic weight
af – atomic formula
bp – boiling point
CAS – Chemical Abstracts Service
cc – cubic centimeter
CC – closed cup
CC – ceiling concentration
COC – Cleveland open cup
compd(s) – compound(s)
conc – concentration, concentrated
cryst, crys – crystal(s), crystalline
d – density
D – day(s)
decomp, dec – decomposition
deliq – deliquescent
dil – dilute
DOT – U.S. Department of Transportation
EPA – U.S. Environmental Protection Agency
eth – ether
(F) – Fahrenheit
FCC – Food Chemical Codex
FDA – U.S. Food and Drug Administration
flash p – flash point
fp – freezing point
g, gm – gram
gran – granular, granules
H, hr – hour(s)
HR: – hazard rating
IARC – International Agency for Research on Cancer
insol – insoluble
kg – kilogram (one thousand grams)
L,l – liter

lel – lower explosive level
liq – liquid
M – minute(s)
m^3 – cubic meter
mf – molecular formula
mg – milligram
μ, u – micron
misc – miscible
mL, ml – milliliter
mm – millimeter
mod – moderately
mp – melting point
mppcf – million particles per cubic foot
mw – molecular weight
NIOSH – National Institute for Occupational Safety
  and Health
ng – nanogram
NTP – National Toxicology Program
OC – open cup
org – organic
OSHA – Occupational Safety and Health Administration
PEL – permissible exposure level
petr – petroleum
pg – picogram (one trillionth of a gram)
Pk – peak concentration
pmole – picomole
ppb – parts per billion (v/v)
pph – parts per hundred (v/v) (percent)
ppm – parts per million (v/v)
ppt – parts per trillion (v/v)
PROP – properties
refr—refractive
S,sec – second(s)
slt, sltly – slightly
sol – soluble
soln – solution
subl – sublimes
TCC – Tag closed cup
tech – technical
temp – temperature
TLV – Threshold Limit Value
TOC – Tag open cup

TWA – time weighted average
U, unk – unknown, unreported
μ, u – micron
uel – upper explosive limits
μg, ug – microgram
ULC, ulc – Underwriters Laboratory Classification
USDA – U.S. Department of Agriculture
vac – vacuum
vap – vapor
vap d – vapor density
vap press – vapor pressure

vol – volume
visc – viscosity
W – week(s)
Y – year(s)
% – percent(age)
> – greater than
< – less than
<= – equal to or less than
=> – equal to or greater than
° – degrees of temperature in Celsius (Centigrade)
F, °F – temperature in Fahrenheit

I. General Chemical Entries

A

AAC250   CAS: 8021-27-0   **HR: 2**
ABIES ALBA OIL

PROP: Colorless to pale-yellow oil from the steam distillation of the crushed cones of *Abies Alba Mill* (FCTXAV 12,807,74).

SYNS: OIL of ABIES ALBA * OIL of FUR * OIL of SILVER FIR * OIL of SILVER PINE * SILVER FIR NEEDLE OIL * SILVER FIR OIL * SILVER PINE OIL * TEMPLIN OIL

CONSENSUS REPORTS: Reported in EPA TSCA Inventory.

SAFETY PROFILE: A skin irritant. When heated to decomposition it emits acrid smoke and irritating fumes.

AAD000   CAS: 1393-62-0   **HR: 3**
ABRIN

PROP: Yellowish-white powder. Sol in solns of sodium chloride, usually with turbidity. Incubation at 60° for 30 min fails to remove toxic effect, but at 80°, most of the toxicity is lost.

SYNS: ABRINS * AGGLUTININ * CRAB'S EYES * INDIAN LICORICE SEED * JUMBLE BEAD * PRAYER BEAD * TOXALBUMIN

SAFETY PROFILE: A deadly poison to humans by ingestion. Poison by ingestion, intravenous, and intraperitoneal route. Mutation data reported. When heated to decomposition it emits acrid fumes and irritating smoke. Note: Do not confuse with abrine.

AAE250   CAS: 827-61-2   **HR: 3**
ACECLIDINE
mf: $C_9H_{15}NO_2$ mw: 169.25

SYNS: 3-ACETOXYQUINUCLIDINE GLAUCOSTAT * 3-QUINUCLIDINOL ACETATE

SAFETY PROFILE: Poison by ingestion, subcutaneous, and intravenous routes. When heated to decomposition it emits toxic fumes of NO_x.

AAF750   CAS: 3598-37-6   **HR: 3**
ACEPROMAZINE MALEATE
mf: $C_{19}H_{22}N_2OS \cdot C_4H_4O_4$ mw: 442.57

SYNS: 2-ACETYL-10-(3-(DIMETHYLAMINO) PROPYL)PHENOTHIAZINE, MALEATE * ACETYL- PROMAZINE MALEATE (1:1) * ATRAVET * 10-(3-(DIMETHYLAMINO)PROPYL)PHENOTHIA- ZIN-2-YL METHYL KETONE MALEATE (1:1) * MA- LEATE ACIDE de l'ACETYL-3-DIMETHYLAMINO-3- PROPYL-10-PHENOTHIAZINE (FRENCH) * NOTENSIL * PREGICIL * SOPRONTIN

SAFETY PROFILE: Poison by ingestion, subcutaneous, and intravenous routes. When heated to decomposition it emits highly toxic fumes of NO_x and SO_x.

AAG000   CAS: 105-57-7   **HR: 2**
ACETAL
DOT: UN 1088
mf: $C_6H_{14}O_2$ mw: 118.20

PROP: Colorless, volatile liquid; agreeable odor, nutty after-taste. Bp: 102.7°, flash p: −5°F (CC), lel: 1.65%, uel: 10.4%, d: 0.831, autoign temp: 446°F, vap press: 10 mm @ 8.0°, vap d: 4.08, mp: −100°. Sltly sol in water, misc in alc and ether.

SYNS: ACETAAL (DUTCH) * ACETAL DIETHYLI- QUE (FRENCH) * ACETALE (ITALIAN) * 1,1-DIA- ETHOXY-AETHAN (GERMAN) * DIAETHYLACETAL (GERMAN) * 1,1-DIETHOXY-ETHAAN (DUTCH) * 1,1-DIETHOXYETHANE * DIETHYL ACETAL * 1,1-DIETOSSIETANO (ITALIAN) * ETHYLIDENE DIETHYL ETHER * USAF DO-45

CONSENSUS REPORTS: Reported in EPA TSCA Inventory.

DOT Classification: Label: Flammable Liquid.

SAFETY PROFILE: Moderately toxic by ingestion and intraperitoneal routes. A skin and eye irritant. A narcotic. Dangerous fire hazard when exposed to heat or flame; can react vigorously with oxidizing materials. Forms heat-sensitive explosive peroxides on contact with air. When heated to decomposition it emits acrid smoke and fumes.

AAG250   CAS: 75-07-0   **HR: 3**
ACETALDEHYDE
DOT: UN 1089
mf: C_2H_4O mw: 44.06

PROP: Colorless, fuming liquid; pungent, fruity odor. Mp: −123.5°, bp: 20.8°, lel: 4.0%, uel: 57%, flash p: −36°F (CC), d: 0.804 @ 0°/20°, autoign temp: 347°F, vap d: 1.52. Misc in water, alc, and ether.

SYNS: ACETALDEHYD (GERMAN) * ACETIC ALDEHYDE * ALDEHYDE ACETIQUE (FRENCH) * ALDEIDE ACETICA (ITALIAN) * ETHANAL * ETHYL ALDEHYDE * FEMA No. 2003 * NCIC56326 * OCTOWY ALDEHYD (POLISH) * RCRA WASTE NUMBER U001

CONSENSUS REPORTS: IARC Cancer Review: GROUP 2B IMEMDT 7,77,87; Animal Sufficient Evidence IMEMDT 36,101,85; Human Inadequate Evidence IMEMDT 36,101,85. On Community Right-To-Know List. Reported in EPA TSCA Inventory. EPA Genetic Toxicology Program.

OSHA PEL: (Transitional: TWA 200 ppm) TWA 100 ppm; STEL 150 ppm
ACGIH TLV: TWA 100 ppm; STEL 150 ppm
DFG MAK: 50 ppm (90 mg/m^3), Suspected Carcinogen.
DOT Classification: Flammable Liquid; LABEL: Flammable Liquid

SAFETY PROFILE: Suspected carcinogen with experimental carcinogenic and tumorigenic data. Poison by intratracheal and intravenous routes. A human systemic irritant by inhalation. A narcotic. Human mutation data reported. An experimental teratogen. Other experimental reproductive effects. A skin and severe eye irritant. A common air contaminant. Highly flammable liquid. Mixtures of 30-60 percent of the vapor in air ignite above 100°. It can react violently with acid anhydrides; alcohols; ketones; phenols; NH$_3$; HCN; H$_2$S; halogens; P; isocyanates; strong alkalies; and amines. Reactions with cobalt chloride; mercury(II) chlorate; or mercury(II) perchlorate form sensitive, explosive products. Polymerizes violently in the presence of traces of metals or acids. Reaction with oxygen may lead to detonation. When heated to decomposition it emits acrid smoke and fumes.

AAG500   CAS: 75-39-8   **HR: 2**
ACETALDEHYDE AMMONIA
DOT: UN 1841
mf: C$_2$H$_4$O • H$_3$N   mw: 61.10

PROP: White, crystalline solid. Bp: 110°, mp: 97°. Very sol in water, alc; sltly sol in ether.

SYNS: ACETALDEHYDE, AMINE SALT * ALDEHYDE AMMONIA * 1-AMINOETHANOL * α-AMINOETHYL ALCOHOL

DOT Classification: ORM-A; Label: None.

SAFETY PROFILE: It readily decomposes into acetaldehyde and ammonia when heated, causing the hazards of these substances. Moderate fire and explosion hazard when exposed to heat or flame. Moderately dangerous when heated to decomposition it emits NH$_3$ and NO$_x$ toxic fumes; can react with oxidizing materials.

AAH000   CAS: 16568-02-8   **HR: 3**
ACETALDEHYDE-N-METHYL-N-FORMYLHYDRAZONE
mf: C$_4$H$_8$N$_2$O   mw: 100.14

SYNS: ACETALDEHYDE-N-FORMYL-N-METHYLHYDRAZONE * ETHYLIDENE GYROMITRIN * GYROMITRIN * N-METHYL-N-FORMYL HYDRAZONE of ACETALDEHYDE

CONSENSUS REPORTS: IARC Cancer Review: GROUP 3 IMEMDT 7,56,87; Animal Limited Evidence IMEMDT 7,391,87. EPA Genetic Toxicology Program.

SAFETY PROFILE: Poison via ingestion and possibly other routes. Questionable carcinogen with experimental carcinogenic and tumorigenic data. When heated to decomposition it emits toxic fumes of NO$_x$.

AAH250   CAS: 107-29-9   **HR: 3**
ACETALDEHYDE OXIME
DOT: UN 2332
mf: C$_2$H$_5$NO   mw: 59.08

PROP: A water-sol, crystalline material; sol in alc, ether. Mp: (α) 46.5°, mp: (β) 12°, d: 0.966, bp: 114.5°, flash p: ≤ 72°F.

SYNS: ACETALDOXIME * ALDOXIME * ETHANAL OXIME * ETHYLIDENEHYDROXYLAMINE * USAF AM-5

CONSENSUS REPORTS: Reported in EPA TSCA Inventory.

DOT Classification: Label: Flammable Liquid.

SAFETY PROFILE: Poison via intraperitoneal route. A dangerous fire hazard with a flash point is at room temperature. When heated to decomposition it emits toxic fumes of NO$_x$.

AAH750 CAS: 107-89-1 **HR: 3**
ACETALDOL

DOT: UN 2839

mf: $C_4H_8O_2$ mw: 88.12

PROP: Clear, white-to-yellow syrupy liquid. Bp: 83° @ 20 mm, flash p: 150°F (OC), d: 1.11, autoign temp: 482°F, vap d: 3.04.

SYNS: ALDOL * 3-BUTANOLAL * 3-HYDROXY-BUTANAL * β-HYDROXYBUTYRALDEHYDE * 3-HYDROXYBUTYRALDEHYDE * OXYBUTANAL * OXYBUTYRIC ALDEHYDE

CONSENSUS REPORTS: Reported in EPA TSCA Inventory.

DOT Classification: Poison B; Label: Poison.

SAFETY PROFILE: Poison via skin contact. Moderately toxic by ingestion. A skin and eye irritant. Moderate fire hazard when exposed to heat or flame; emits crotonaldehyde and water when heated. Can react with oxidizing materials.

AAI000 CAS: 60-35-5 **HR: 3**
ACETAMIDE
mf: C_2H_5NO mw: 59.08

PROP: Colorless crystals; mousey odor. Mp: 81°, bp: 221.2°, d: 1.159 @ 20°/4°, vap press: 1 mm @ 65°. Decomp in hot water.

SYNS: ACETIC ACID AMIDE * ACETIMIDIC ACID * AMID KYSELINY OCTOVE * ETHANAMIDE * METHANECARBOXAMIDE * NCI-C02108

CONSENSUS REPORTS: IARC Cancer Review: GROUP 2B IMEMDT 7,56,87; Animal Sufficient Evidence IMEMDT 7,389,87. On Community Right-To-Know List. Reported in EPA TSCA Inventory.

DFG MAK: Suspected Carcinogen.

SAFETY PROFILE: Suspected carcinogen with experimental carcinogenic and neoplastigenic data. Experimental teratogenic data. Moderately toxic by intraperitoneal and possibly other routes. Other experimental reproductive effects. Mutation data reported. When heated to decomposition it emits toxic fumes of NO_x.

AAI250 CAS: 59-66-5 **HR: 3**
5-ACETAMIDE-1,3,4-THIADIAZOLE-2-SULFONAMIDE
mf: $C_4H_6N_4O_3S_2$ mw: 222.26

SYNS: 2-ACETAMIDO-5-SULFONAMIDO-1,3,4-THIADIAZOLE * ACETAMIDOTHIADIAZOLE-SULFONAMIDE * ACETAMOX * ACETAZOL-AMID * ACETAZOLAMIDE * ACETAZOLEAMIDE * ACETOZALAMIDE * 2-ACETYLAMINO-1,3,-4-THIADIAZOLE-5-SULFONAMIDE * N-(5-(AMI-NOSULFONYL)-1,3,4-THIADIAZOL-2-YL)ACETAMIDE * CARBONIC ANHYDRASE INHIBITOR NO. 6063 * CIDAMEX * DEFILTRAN * DEHYDRATIN * DIACARB * DIAKARB * DIAMOX * DIDOC * DILURAN * DIURAMID * DIURETICUM-HOLZINGER * DIUTAZOL * DONMOX * EDEMOX * EUMICTON * FONURIT * GLAUPAX * GLUPAX * MUIRAMID * NATRIONEX * NEPHR-AMIDE * PHONURIT * N-(5-SULFAMOYL-1,-3,4-THIADIAZOL-2-YL)ACETAMIDE * VETAMOX

CONSENSUS REPORTS: Reported in EPA TSCA Inventory.

SAFETY PROFILE: Poison by subcutaneous and intravenous routes. Moderately toxic by intraperitoneal route. Experimental teratogenic and reproductive effects. When heated to decomposition it emits very toxic fumes of NO_x and SO_x. A carbonic anhydrase inhibitor and diuretic used to treat glaucoma.

AAL750 CAS: 531-82-8 **HR: 3**
2-ACETAMIDO-4-(5-NITRO-2-FURYL)THIAZOLE
mf: $C_9H_7N_3O_4S$ mw: 253.25

SYNS: 2-ACETAMINO-4-(5-NITRO-2-FURYL)THIAZOLE * 2-ACETYLAMINO-4-(5-NITRO-2-FURYL) THIA-ZOLE * N-(4-(5-NITRO-2-FURANYL)-2-THIAZOLYL)-ACETAMIDE * N-(4-(5-NITRO-2-FURYL)-2-THIAZOLYL)ACETAMIDE * N-(4-(5-NITRO-2-FURYL)THIAZOL-2-YL)ACETAMIDE

CONSENSUS REPORTS: IARC Cancer Review: GROUP 2B IMEMDT 7,56,87; Animal Sufficient Evidence IMEMDT 1,181,72; IMEMDT 7,185,74

SAFETY PROFILE: Suspected carcinogen with experimental carcinogenic, tumorigenic, and neoplastigenic data. Mutation data reported. When heated to decomposition it emits very toxic fumes of SO_x and NO_x.

AAQ250 CAS: 2832-40-8 **HR: 3**
ACETAMINE YELLOW CG
mf: $C_{15}H_{15}N_3O_2$ mw: 269.33

SYNS: ACTIOQUINONE LIGHT YELLOW * AMACEL YELLOW G * CALCOSYN YELLOW GC * CELLITON FAST YELLOW G * C.I. 11855 * CIBACET YELLOW GBA * C.I. DISPERSE YELLOW 3 * HISPERSE YELLOW G * N-(4-((2-HYDROXY-5-METHYLPHENYL)AZO)PHENYL)ACETAMIDE * 4'-((6-HYDROXY-m-TOLYL)AZO)ACETANILIDE * INTRASPERSE YELLOW GBA EXTRA * MICROSETILE YELLOW GR * NACELAN FAST YELLOW CG * NCI-C53781 * YELLOW Z

CONSENSUS REPORTS: Community Right-To-Know List. Reported in EPA TSCA Inventory. IARC Cancer Review: Animal Inadequate Evidence IMEMDT 8,97,75; NTP Carcinogenesis Bioassay (feed); Clear Evidence: mouse, rat NTPTR* NTP-TR-222,82

SAFETY PROFILE: Suspected carcinogen with experimental tumorigenic and carcinogenic data. An allergen. Mutation data reported. When heated to decomposition it emits toxic fumes of NO_x.

AAQ500   CAS: 103-84-4   **HR: 3**
ACETANILIDE
mf: C_8H_9NO mw: 135.18

PROP: White, shining, crystalline scales. Mp: 113.5°, bp: 305°, flash p: 345°F (OC), d: 1.2105 @ 4°/4°, autoign temp: 1004°F, vap press: 1 mm @ 114.0°, vap d: 4.65. Somewhat sol in water, alc and ether.

SYNS: ACETAMIDOBENZENE * ACETANIL * ACETIC ACID ANILIDE * ACETOANILIDE * ACETYLAMINOBENZENE * ACETYLANILINE * N-ACETYLANILINE * AN * ANTIFEBRIN * PHENALGENE * N-PHENYLACETAMIDE * USAF EK-3

CONSENSUS REPORTS: Reported in EPA TSCA Inventory. EPA Genetic Toxicology Program.

SAFETY PROFILE: A human poison by an unspecified route. Poison by ingestion and intravenous routes. Moderately toxic by intraperitoneal route. Human systemic effects by ingestion: hallucinations and distorted perceptions, sleepiness, constipation, cyanosis, respiratory stimulation, kidney damage, methemoglobinemia-carboxhemoglobinemia and decreased body temperature. When heated to decomposition it emits toxic fumes of NO_x. Combustible when exposed to heat or flame.

AAS250   CAS: 5421-48-7   **HR: 3**
(ACETATO)(DIETHOXYPHOSPHINYL)MERCURY
mf: $C_6H_{13}HgO_5P$ mw: 396.75

SYN: (DIETHOXY-PHOSPHINYL)MERCURY ACETATE

CONSENSUS REPORTS: Mercury and its compounds are on the Community Right-To-Know List.

OSHA PEL: (Transitional: CL 1 mg/10m³) CL 0.1 mg(Hg)/m³ (skin)
ACGIH TLV: TWA 0.1 mg(Hg)/m³ (skin)
NIOSH REL: TWA 0.05 mg(Hg)/m³.

SAFETY PROFILE: Poison by intraperitoneal route. When heated to decomposition it emits very toxic fumes of Hg and PO_x.

AAS500   CAS: 21450-81-7   **HR: 3**
(ACETATO)(2,3,5,6-TETRAMETHYL-PHENYL)MERCURY
mf: $C_{12}H_{16}HgO_2$ mw: 392.87

SYN: (2,3,5,6-TETRAMETHYLPHENYL)MERCURY ACETATE

CONSENSUS REPORTS: Mercury and its compounds are on the Community Right-To-Know List.

OSHA PEL: (Transitional: CL 1 mg/10m³) CL 0.1 mg(Hg)/m³ (skin)
ACGIH TLV: TWA 0.1 mg(Hg)m³ (skin)
NIOSH REL: TWA 0.05 mg(Hg)/m³

SAFETY PROFILE: Poison by intravenous route. When heated to decomposition it emits toxic fumes of Hg.

AAS750   CAS: 1424-27-7   **HR: D**
ACETAZOLAMIDE SODIUM
mf: $C_4H_5N_4O_3S_2 \cdot Na$ mw: 244.24

SYNS: ACETAZOLAMIDE SODIUM SALT * SODIUM ACETAZOLAMIDE

SAFETY PROFILE: An experimental teratogen. Other experimental reproductive effects. When heated to decomposition it emits very toxic fumes of NO_x, Na_2O, and SO_x.

AAT250   CAS: 64-19-7   **HR: 3**
ACETIC ACID

DOT: UN 2789/UN 2790
mf: $C_2H_4O_2$ mw: 60.06

PROP: Clear, colorless liquid; pungent odor. Mp: 16.7°, bp: 118.1°, flash p: 109°F (CC), lel: 5.4%, uel: 16.0% @ 212°F, d: 1.049 @ 20°/4°, autoign temp: 869°F, vap press: 11.4 mm @ 20°, vap d: 2.07. Misc in water, alc, and ether.

SYNS: ACETIC ACID (aqueous solution) (DOT) * ACETIC ACID, GLACIAL (DOT) * ACIDE ACETIQUE (FRENCH) * ACIDO ACETICO (ITALIAN) * AZIJNZUUR (DUTCH) * ESSIGSAEURE (GERMAN) * ETHANOIC ACID * ETHYLIC ACID * FEMA No. 2006 * GLACIAL ACETIC ACID * METHANECARBOXYLIC ACID * OCTOWY KWAS (POLISH) * VINEGAR ACID

CONSENSUS REPORTS: Reported in EPA TSCA Inventory.

OSHA PEL: TWA 10 ppm
ACGIH TLV: TWA 10 ppm; STEL 15 ppm
DFG MAK: 10 ppm (25 mg/m^3)
DOT Classification: Corrosive, Flammable Liquid; Label: Corrosive, Flammable Liquid.

SAFETY PROFILE: A human poison by an unspecified route. Moderately toxic by various routes. A severe eye and skin irritant. Can cause burns, lachrymation, and conjunctivitis. Human systemic effects by ingestion: changes in the esophagus, ulceration or bleeding from the small and large intestines. Human systemic irritant effects and mucous membrane irritant. Experimental reproductive effects. Mutation data reported. A common air contaminant. A combustible liquid. Moderate fire and explosion hazard when exposed to heat or flame; can react vigorously with oxidizing materials. To fight fire, use CO_2, dry chemical, alcohol foam, foam and mist. When heated to decomposition it emits irritating fumes.

Potentially explosive reaction with 5-azidotetrazole; bromine pentafluoride; chromium trioxide; hydrogen peroxide; potassium permanganate; sodium peroxide; and phosphorus trichloride. Potentially violent reactions with acetaldehyde and acetic anhydride. Ignites on contact with potassium-tert-butoxide. Incompatible with chromic acid; nitric acid; 2-amino-ethanol; NH_4NO_3; ClF_3; chlorosulfonic acid; (O_3 + diallyl methyl carbinol); ethylenediamine; ethylene imine; (HNO_3 + acetone); oleum; $HClO_4$; permanganates; $P(OCN)_3$; KOH; NaOH; n-xylene.

AAU000   CAS: 150-84-5   **HR: 1**
ACETIC ACID, CITRONELLYL ESTER
mf: $C_{12}H_{22}O_2$ mw: 198.34

PROP: Found in oils of Citronella Ceylon, Geranium, and about 20 other oils (FCTXAV 11,1011,73). Colorless liquid; fruity odor. D: 0.883-0.893, refr index: 1.440-1.450, flash p: +212°F. Sol in alc and fixed oils; insol in glycerin, propylene glycol, and water @229°.

SYNS: ACETIC ACID-3,7-DIMETHYL-6-OCTEN-1-YL ESTER * CITRONELLYL ACETATE (FCC) * 2,6-DIMETHYL-2-OCTEN-8-OL ACETATE * 3,7-DIMETHYL-6-OCTEN-1-YL ACETATE * FEMA No. 2311

CONSENSUS REPORTS: Reported in EPA TSCA Inventory.

SAFETY PROFILE: Mildly toxic by ingestion. A human skin irritant. Combustible liquid. When heated to decomposition it emits acrid smoke and irritating fumes.

AAU250   CAS: 18461-55-7   **HR: 3**
ACETIC ACID-4,6-DINITRO-o-CRESYL ESTER
mf: $C_9H_8N_2O_6$ mw: 240.19

SYNS: DNOK-ACETAT (CZECH) * 4,6-DINITRO-o-KRESYLESTER KYSELINY OCTOVE (CZECH)

NIOSH REL: (Dinitro ortho-Cresyl) TWA 0.2 mg/m^3

SAFETY PROFILE: Poison by ingestion and intraperitoneal routes. A skin and severe eye irritant. When heated to decomposition it emits toxic fumes of NO_x.

AAW000   CAS: 56856-83-8   **HR: 3**
ACETIC ACID METHYLNITROSAMINOMETHYL ESTER
mf: $C_4H_8N_2O_3$ mw: 132.14

SYNS: α-ACETOXY DIMETHYLNITROSAMINE * N-α-ACETOXYMETHYL-N-METHYLNITROSAMINE * ACETOXYMETHYL-METHYL-NITROSAMIN (GERMAN) * ACETOXYMETHYL METHYLNITROSAMINE * 1-ACETOXY-N-NITROSODIMETHYLAMINE * AMMN * ANN (GERMAN) * DMN-OAC * MAMN * METHYL(ACETOXYMETHYL)NITROSAMINE * N-NITROSO-N-(ACETOXY)METHYL-N-METHYLAMINE * N-NITROSO-N-METHYL-N-ACETOXYMETHYLAMINE

SAFETY PROFILE: Suspected carcinogen with experimental carcinogenic, neoplastigenic, and tumorigenic data. Poison by ingestion, subcutaneous, intravenous, and intraperitoneal routes. Experimental teratogenic data. Human mutation data reported. When heated to decomposition it emits toxic fumes of NO_x.

AAW500 CAS: 1118-39-4 **HR: 1**
ACETIC ACID MYRCENYL ESTER
mf: $C_{12}H_{20}O_2$ mw: 196.32

SYNS: ACETIC ACID-2-METHYL-6-METHYLENE-7-OCTEN-2-YL ESTER * 3-METHYLENE-7-METHYL-1-OCTEN-7-YL ACETATE * 2-METHYL-6-METHYLENE-7-OCTEN-2-OL ACETATE * 2-METHYL-6-METHYLENE-7-OCTEN-2-YL ACETATE * MYRCENYL ACETATE

CONSENSUS REPORTS: Reported in EPA TSCA Inventory.

SAFETY PROFILE: Mildly toxic by ingestion. A skin irritant. When heated to decomposition it emits acrid smoke and irritating fumes.

AAX175 CAS: 9003-22-9 **HR: 1**
ACETIC ACID, VINYL ESTER,
POLYMER with CHLOROETHYLENE
mf: $(C_4H_6O_2 \cdot C_2H_3Cl)n$

SYNS: ACETIC ACID ETHENYL ESTER POLYMER with CHLORETHENE (9CI) * A 15 (POLYMER) * BAKELITE LP 70 * BAKELITE VLFV * BAKELITE VMCC * BAKELITE VYNS * BREON 351 * CHLOROETHYLENEVINYL ACETATE POLYMER * CORVIC 236581 * DENKALAC 61 * DIAMOND SHAMROCK 744 * EXON 450 * EXON 454 * GEON 135 * HOSTAFLEX VP 150 * LEUCOVYL PA 1302 * NORVINYL P 6 * OPALON 400 * PLIOVAC AO * POLYVINYL CHLORIDE-POLYVINYL ACETATE * PVC CORDO * RHODOPAS 6000 * SARPIFAN HP 1 * SCONATEX * SOLVIC 523KC * SUMILIT PCX * TENNUS 0565 * TYGON * VAGD * VINNOL H 10/60 * VINYL ACETATE-VINYL CHLORIDE COPOLYMER * VINYL ACETATE-VINYL CHLORIDE POLYMER * VINYL CHLORIDE-VINYL ACETATE POLYMER * VINYLITE VYDR 21 * VLVF * VMCC * VYNW

CONSENSUS REPORTS: IARC Cancer Review: Animal Limited Evidence IMEMDT 19,377,79. Reported in EPA TSCA Inventory.

SAFETY PROFILE: Suspected carcinogen with experimental tumorigenic data. When heated to decomposition it emits toxic fumes of HCl.

AAX250 CAS: 9003-20-7 **HR: 3**
ACETIC ACID VINYL ESTER
POLYMERS
mf: $(C_4H_6O_2)_n$

PROP: Clear, water-white solid resin. Sol in benzene, acetone; insol in water.

SYNS: ACETIC ACID ETHENYL ESTER HOMOPOLYMER * ASAHISOL 1527 * ASB 516 * AYAA * AYAF * BAKELITE AYAA * BAKELITE LP 90 * BASCOREZ * BOND CH 18 * BOOKSAVER * BORDEN 2123 * CEVIAN A 678 * D 50 * DANFIRM * DARATAK * DCA 70 * DUVILAX BD 20 * ELMER'S GLUE ALL * EP 1463 * FORMVAR 1285 * GELVA CSV 16 * GOHSENYL E 50 Y * KURARE OM 100 * LEMAC 1000 * MERCKOGEN 6000 * MOVINYL 114 * NATIONAL 120-1207 * POLYVINYL ACETATE (FCC) * PROTEX (POLYMER) * RHODOPAS M * SOVIOL * SP 60 ESTER * TOABOND 40H * UCAR 130 * VA 0112 * VINAC B 7 * VINYL ACETATE HOMOPOLYMER * VINYL ACETATE POLYMER * VINYL ACETATE RESIN * VINYL PRODUCTS R 10688 * WINACET D

CONSENSUS REPORTS: IARC Cancer Review: Animal Inadequate Evidence IMEMDT 19,341,79. Reported in EPA TSCA Inventory.

SAFETY PROFILE: When heated to decomposition it emits acrid smoke and irritating fumes.

AAX500 CAS: 108-24-7 **HR: 2**
ACETIC ANHYDRIDE
DOT: UN 1715
mf: $C_4H_6O_3$ mw: 102.10

PROP: Colorless, very mobile, strongly refractive liquid; very strong acetic odor. Mp: $-73.1°$, bp: $140°$, flash p: $129°F$ (CC), d: 1.082 @ $20°/4°$, lel: 2.9%, uel: 10.3%, autoign temp: $734°F$, vap press: 10 mm @ $36.0°$, vap d: 3.52. Somewhat sol in cold water; decomp in hot water and hot alc; misc in alc and ether.

SYNS: ACETIC ACID, ANHYDRIDE * ACETIC OXIDE * ACETYL ANHYDRIDE * ACETYL ETHER * ACETYL OXIDE * ANHYDRIDE ACETIQUE (FRENCH) * ANIDRIDE ACETICA (ITALIAN) * AZIJNZUURANHYDRIDE (DUTCH) * ESSIGSAEUREANHYDRID (GERMAN) * ETHANOIC ANHYDRATE * OCTOWY BEZWODNIK (POLISH)

CONSENSUS REPORTS: Reported in EPA TSCA Inventory.

OSHA PEL: CL 5 ppm
ACGIH TLV: CL 5 ppm
DFG MAK: 5 ppm (20 mg/m^3)
DOT Classification: IMO: Corrosive Material; Label: Corrosive, Flammable Liquid.

SAFETY PROFILE: Moderately toxic by inhalation, ingestion, and skin contact. A skin and severe eye irritant. Moderate fire and explosion hazard when exposed to heat or flame. Potentially explosive reactions with barium peroxide; boric acid; chromium trioxide; 1,3-diphenyltriazene; hydrochloric acid + water; hypochlorous acid; nitric acid; perchloric acid + water; peroxyacetic acid; potassium permanganate; tetrafluoroboric acid; 4-toluenesulfonic acid + water; and acetic acid + water. Reactions with ethanol + sodium hydrogen sulfate; and hydrogen peroxide form explosive products. Reactions with ammonium nitrate + hexamethylenetetraminium acetate + nitric acid form as products the military explosives RDX and HMX. Reacts violently with N-tert-butylphthalimic acid + tetrafluoroboric acid, chromic acid, glycerol + phosphoryl chloride; and metal nitrates (e.g., copper or sodium nitrates). Incompatible with 2-aminoethanol; aniline; chlorosulfonic acid; (CrO_3 + acetic acid); ethylenediamine; ethyleneimine; glycerol; oleum; HF; permanganates; NaOH; Na_2O_2; H_2SO_4; water; N_2O_2; (glycerol + phosphoryl chloride). When heated to decomposition it emits toxic fumes; can react vigorously with oxidizing materials, will react violently on contact with water or steam. To fight fire, use CO_2, dry chemical, water mist, alcohol foam.

AAX750　　　　CAS: 93-29-8　　　　**HR: 2**
ACETISOEUGENOL
mf: $C_{12}H_{14}O_3$　　　mw: 206.26

PROP: White crystals; clove odor. Flash p: 153°F. Sol in alc, chloroform, ether; insol in water.

SYNS: 4-ACETOXY-3-METHOXY-1-PROPENYLBENZENE * ACETYLISOEUGENOL * FEMA No. 2470 * ISOEUGENOL ACETATE * ISOEUGENYL ACETATE (FCC) * 2-METHOXY-4-PROPENYLPHENYL ACETATE

CONSENSUS REPORTS: Reported in EPA TSCA Inventory.

SAFETY PROFILE: Moderately toxic by ingestion. Combustible liquid. When heated to decomposition it emits acrid smoke and irritating fumes.

AAY000　　　　CAS: 102-01-2　　　　**HR: 3**
ACETOACETANILIDE
mf: $C_{10}H_{11}NO_2$　　　mw: 177.22

PROP: White, crystalline solid. Mp: 85°, bp: decomp, flash p: 365°F (COC), d: 1.260 @ 20°, vap press: 0.01 mm @ 20°.

SYNS: ACETOACETAMIDOBENZENE * ACETOACETIC ACID ANILIDE * ACETOACETIC ANILIDE * ((ACETOACETYL)AMINO)BENZENE * ACETOACETYLANILINE * ACETYLACETANILIDE * α-ACETYLACETANILIDE * N-(ACETYLACETYL)ANILINE * β-KETOBUTYRANILIDE * N-PHENYLACETOACETAMIDE * USAF EK-1239

CONSENSUS REPORTS: Reported in EPA TSCA Inventory.

SAFETY PROFILE: Poison by intraperitoneal route. A weak allergen. Combustible when exposed to heat or flame. When heated to decomposition it emits toxic NO_x fumes. To fight fire, use alcohol foam, water mist, CO_2, dry chemical.

ABA000　　　　CAS: 93-68-5　　　　**HR: 2**
ACETOACET-o-TOLUIDIDE
mf: $C_{11}H_{13}NO_2$　　　mw: 191.25

PROP: Crystals. Mp: 106°, bp: decomp, d: 1.300 @ 20°, vap press: 0.01 mm @ 20°, flash p: 320°F (COC).

SYNS: 2-ACETOACETYLAMINOTOLUENE * ACETOACETYL-2-METHYLANILIDE * 2'-METHYLACETOACETANILIDE

CONSENSUS REPORTS: Reported in EPA TSCA Inventory.

SAFETY PROFILE: Moderately toxic by ingestion. When heated to decomposition it emits toxic fumes of NO_x.

ABA500　　　　CAS: 92-15-9　　　　**HR: 2**
ACETOACETYL-o-ANISIDINE
mf: $C_{11}H_{13}NO_3$　　　mw: 207.25

PROP: Crystals. Mp: 86.6°, flash p: 325°F (OC), d: 1.132 @ 86.6°/20°, vap d: 7.0.

SYNS: o-ACETOACETANISIDE * ACETOACET-o-ANISIDIN (CZECH) * ACETOACETIC ACID-o-ANISIDIDE * 2-ACETOACETYLAMINOANISOLE * ACETOACETYL-o-ANISIDE * ACETOACETYL-o-ANISINE * o-METHOXYACETOACETANILIDE * 2-METHOXYACETOACETANILIDE * 2'-METHOXYACETOACETANILIDE

CONSENSUS REPORTS: Reported in EPA TSCA Inventory.

SAFETY PROFILE: Moderately toxic by ingestion. A skin and eye irritant. When heated to decomposition it emits toxic fumes of NO_x. Combustible when exposed to heat or flame or oxidizing materials. To fight fire, use CO_2, mist, dry chemicals.

ABB500   CAS: 513-86-0   **HR: 3**
ACETOIN
DOT: UN 2621
mf: $C_4H_8O_2$ mw: 88.12

PROP: Sltly yellow liquid or crystalline solid; buttery odor. D: 1.016, bp: 147-148°, refr index: 1.417, mp: 15°, flash p: 106°F. Misc with water, alc, propylene glycol; insol in vegetable oil.

SYNS: ACETYL METHYL CARBINOL * 2-BUTANOL-3-ONE * DIMETHYLKETOL * FEMA No. 2008 * 3-HYDROXY-2-BUTANONE * 1-HYDROXYETHYL METHYL KETONE * γ-HYDROXY-β-OXOBUTANE

CONSENSUS REPORTS: Reported in EPA TSCA Inventory.

DOT Classification: IMO: Flammable Liquid; Label: Flammable Liquid.

SAFETY PROFILE: Experimental reproductive effects. Mildly toxic by subcutaneous route. A moderate skin irritant. Flammable liquid. When heated to decomposition it emits acrid smoke and fumes.

ABC000   CAS: 116-09-6   **HR: 2**
ACETOL (1)
mf: $C_3H_6O_2$ mw: 74.09

PROP: Colorless liquid. D: 1.084 @ 20°/4°, mp: −7°, bp: 145°-146° decomp; misc in water, alc and ether.

SYNS: HYDROXYACETONE * 1-HYDROXY-2-PROPANONE

CONSENSUS REPORTS: Reported in EPA TSCA Inventory.

SAFETY PROFILE: An experimental teratogen. Moderately toxic by ingestion. Mutation data reported. An allergen. Implicated in aplastic anemia. A 10 gram dose may be fatal to an adult. Skin contact, inhalation, or ingestion can cause asthma, sneezing, irritation of eyes and nose, hives and eczema. Combustible when exposed to heat or flame. When heated to decomposition it emits acrid smoke and fumes.

ABC250   CAS: 828-00-2   **HR: 3**
ACETOMETHOXANE
mf: $C_8H_{14}O_4$ mw: 174.22

PROP: Yellow to amber, clear liquid. Sol in water and org solvents. D: 1.068-1.075 @ 25/25; bp: 66-68° @ 3 mm; fp: < −25°.

SYNS: ACETIC ACID-2,6-DIMETHYL-m-DIOXAN-4-YL ESTER * ACETOMETHOXAN * 6-ACETOXY-2,4-DI-METHYL-m-DIOXANE * DDOA * DIMETHOXANE * 2,6-DIMETHYL-m-DIOXAN-4-OL ACETATE * 2,6-DIMETHYL-m-DIOXAN-4-YL ACETATE * DIOXIN (bactericide) (OBS.) * G1V GARD DXN * NCI-C56213

CONSENSUS REPORTS: IARC Cancer Review: GROUP 3 IMEMDT 7,56,87; Animal Limited Evidence IMEMDT 15,177,77. NTP Fourth Annual Report On Carcinogens, 1984.

SAFETY PROFILE: Confirmed carcinogen with experimental carcinogenic data. Moderately toxic by ingestion. When heated to decomposition it emits acrid smoke and fumes.

ABC500   CAS: 93-08-3   **HR: 2**
2′-ACETONAPHTHONE
mf: $C_{12}H_{10}O$ mw: 170.22

PROP: White crystalline solid; orange blossom odor. Flash p: 264°F. Sol in fixed oils; sltly sol in propylene glycol; insol in glycerin.

SYNS: β-ACETONAPHTHALENE * ACETONAPH-THONE * β-ACETONAPHTHONE * 2-ACETONAPH-THONE * β-ACETYLNAPHTHALENE * 2-ACETYL-NAPHTHALENE * FEMA No. 2723 * METHYL-β-NA-PHTHYL KETONE (FCC) * METHYL-2-NAPHTHYL KETONE * β-METHYL NAPHTHYL KETONE * 1-(2-NAPHTHALENYL)ETHANONE * β-NAPHTHYL METHYL KETONE * 2-NAPHTHYL METHYL KETONE * ORANGE CRYSTALS

CONSENSUS REPORTS: Reported in EPA TSCA Inventory.

SAFETY PROFILE: Moderately toxic by ingestion. A human skin irritant. Combustible liquid. When heated to decomposition it emits acrid smoke and fumes.

ABC750   CAS: 67-64-1   **HR: 2**
ACETONE
DOT: UN 1090
mf: C_3H_6O mw: 58.09

PROP: Colorless liquid; fragrant mint-like odor. Mp: −94.6°, bp: 56.48°, refr index: 1.356, flash p: 0°F (CC), lel: 2.6%, uel: 12.8%, d: 0.7972 @ 15°, autoign temp: (color) 869°F, vap press: 400 mm @ 39.5°, vap d: 2.00. Misc in water, alc, and ether.

SYNS: ACETON (GERMAN, DUTCH, POLISH)
* DIMETHYLFORMALDEHYDE * DIMETHYLKETAL
* DIMETHYL KETONE * FEMA No. 3326 * KE-
TONE PROPANE * β-KETOPROPANE * METHYL
KETONE * PROPANONE * 2-PROPANONE
* PYROACETIC ACID * PYROACETIC ETHER
* RCRA WASTE NUMBER U002

CONSENSUS REPORTS: On Community Right-To-Know List. Reported in EPA TSCA Inventory.

OSHA PEL: (Transitional: TWA 1000 ppm) TWA 750 ppm; STEL 1000 ppm
ACGIH TLV: TWA 750 ppm; STEL 1000 ppm
DFG MAK: 1000 ppm (2400 mg/m^3)
NIOSH REL: (Ketones) TWA 590 mg/m^3
DOT Classification: Flammable Liquid; Label: Flammable Liquid.

SAFETY PROFILE: Moderately toxic by various routes. A skin and severe eye irritant. Human systemic effects by inhalation: changes in EEG, changes in carbohydrate metabolism, nasal effects, conjunctiva irritation, respiratory system effects, nausea and vomiting, and muscle weakness. Human systemic effects by ingestion: coma, kidney damage, and metabolic changes. Narcotic in high concentration. In industry, no injurious effects have been reported other than skin irritation resulting from its defatting action, or headache from prolonged inhalation. A common air contaminant. Highly flammable liquid. Dangerous disaster hazard due to fire and explosion hazard; can react vigorously with oxidizing materials.

   Potentially explosive reaction with nitric acid + sulfuric acid; bromine trifluoride; nitrosyl chloride + platinum; nitrosyl perchlorate; chromyl chloride; thiotrithiazyl perchlorate; and 2,4,6-trichloro-1,3,5-triazine + water. Reacts to form explosive peroxide products with 2-methyl-1,3-butadiene; hydrogen peroxide; and peroxomonosulfuric acid. Ignites on contact with activated carbon; chromium trioxide; dioxygen difluoride + carbon dioxide; and potassium-tert-butoxide. Reacts violently with bromoform; chloroform + alkalies; bromine; and sulfur dichloride. Incompatible with CrO;

(nitric + acetic acid); NOCl; nitryl perchlorate; permonosulfuric acid; NaOBr; (sulfuric acid + potassium dichromate); (thio-diglycol + hydrogen peroxide); trichloromelamine; air; HNO$_3$; chloroform; and H$_2$SO$_4$. To fight fire, use CO$_2$, dry chemical, alcohol foam.

ABD000   CAS: 57-15-8   **HR: 3**
ACETONE CHLOROFORM
mf: C$_4$H$_7$Cl$_3$O   mw: 177.46

PROP: Crystals, camphor odor. Mp: 97°, bp: 167°.

SYNS: ANHYDROUS CHLOROBUTANOL
* CHLORBUTANOL * CHLORBUTOL
* CHLORETONE * CHLOROBUTANOL
* CLORTRAN * HCP * METHAFORM
* SEDAFORM * β,β,β-TRICHLORO-tert-BUTYL ALCO-
HOL * TRICHLORO-tert-BUTYL ALCOHOL * tert-TRI-
CHLOROBUTYL ALCOHOL * 1,1,1-TRICHLORO-2-
METHYL-2-PROPANOL

CONSENSUS REPORTS: Reported in EPA TSCA Inventory.

SAFETY PROFILE: Poison by ingestion. Moderately toxic by parenteral route. A narcotic. A skin and eye irritant. Mutation data reported. Dangerous; can react with oxidizing materials. Combustible when exposed to heat or flame. When heated to decomposition it emits toxic fumes of Cl$^-$.

ABD750   **HR: 3**
ACETONE OIL

PROP: (a) Standard: light, lemon-yellow. (b) Refined: almost water white. (c) Heavy: dark, orange-yellow. Bp: (a) 75-160°, (c) 80-225°. D: (a) 0.826-0.830, (b) 0.812, (c) 0.885-0.865.

DOT Classification: Flammable Liquid; Label: Flammable Liquid.

SAFETY PROFILE: Dangerous fire and explosion hazard when exposed to heat or flame. Can react vigorously with oxidizing materials. Some carcinogenic activity. To fight fire, use CO$_2$, dry chemical.

ABE000   **HR: 3**
ACETONE PEROXIDE

PROP: Liquid or absorbed on cornstarch. The trimeric form is crystalline. Mp: 97°.

SAFETY PROFILE: Severe skin and eye irritant. Flammable by spontaneous chemical reaction; can react vigorously with reducing materials. The trimeric form is shock-sensitive and static-electricity-sensitive and may detonate.

ABE250   CAS: 110-20-3   **HR: 3**
ACETONE SEMICARBAZONE
mf: $C_4H_9N_3O$ mw: 115.16

PROP: Mp: 190-199° (decomp). Sol in cold water; sltly sol in cold alc; insol in ether.

CONSENSUS REPORTS: Reported in EPA TSCA Inventory.

SAFETY PROFILE: Poison by intravenous route. When heated to decomposition it emits toxic fumes of NO_x.

ABE500   CAS: 75-05-8   **HR: 3**
ACETONITRILE
DOT: UN 1648
mf: C_2H_3N mw: 41.06

PROP: Colorless liquid, aromatic odor. Mp: −45°, bp: 81.1°, flash p: 42°F (COC), d: 0.7868 @ 20°/20°, vap d: 1.42, vap press: 100 mm @ 27°, lel: 4.4%, uel: 16%, autoign temp: 975°F. Misc in water, alc, and ether.

SYNS: ACETONITRIL (GERMAN, DUTCH) * CYANO-METHANE * CYANURE de METHYL (FRENCH) * ETHANENITRILE * ETHYL NITRILE * METHA-NECARBONITRILE * METHYL CYANIDE * NCI-C60822 * RCRA WASTE NUMBER U003 * USAF EK-488

CONSENSUS REPORTS: On Community Right-To-Know List. Reported in EPA TSCA Inventory.

OSHA PEL: TWA 40 ppm; STEL 60 ppm
ACGIH TLV: TWA 40 ppm; STEL 60 ppm (skin)
DFG MAK: 40 ppm (70 mg/m³)
NIOSH REL: TWA 34 mg/m³
DOT Classification: Flammable Liquid; Label: Flammable Liquid and Poison.

SAFETY PROFILE: Poison by ingestion and intraperitoneal routes. Moderately toxic by several routes. A skin and severe eye irritant. Human systemic effects by ingestion: convulsions, nausea or vomiting, and metabolic acidosis. Human respiratory system effects by inhalation. Experimental reproductive effects. An experimental teraacitogen. Mutation data reported. Dangerous fire hazard when exposed to heat, flame, or oxidizers. When heated to decomposi-

tion it emits highly toxic fumes of CN^- and NO_x. Potentially explosive reaction with lanthanide perchlorates and nitrogen-fluorine compounds. Exothermic reaction with sulfuric acid at 53°C. Will react with water, steam, and acids to produce toxic and flammable vapors. Incompatible with oleum; chlorosulfonic acid; perchlorates; nitrating agents; indium; dinitrogen tetraoxide; N-fluoro compounds (i.e., perfluorourea + acetonitrile); HNO_3; SO_3. To fight fire use foam, CO_2, dry chemical.

ABF500   CAS: 117-52-2   **HR: 3**
3-(α-ACETONYLFURFURYL)-4-HYDROXYCOUMARIN
mf: $C_{17}H_{14}O_5$ mw: 298.31

PROP: White powder; practically insol in water, sol in alcohols. Mp: 124°.

SYNS: CUMAFURYL (GERMAN) * COUMAFURYL * FOUMARIN * 3-(α-FURYL-β-ACETYLAETHYL)-4-HYDROXYCUMARIN (GERMAN) * 3-(1-FURYL-3-ACETYLETHYL)-4-HYDROXYCOUMARIN * KRUMKIL * RATAFIN * RAT-A-WAY

SAFETY PROFILE: Poison by ingestion and possibly other routes.

ABG000   CAS: 5714-00-1   **HR: 3**
ACETOPHENAZINE
mf: $C_{23}H_{29}N_3O_2S \cdot 2C_4H_4O_4$ mw: 643.77

SYNS: ACETOPHENAZINE MALEATE * 2-ACETYL-10-(3-(4-(β-HYDROXYETHYL)PIPERAZINYL)PROPYL)-PHENOTHIAZINE * 1-(2-HYDROXYETHYL)-4-(3-(2-ACETYL-10-PHENOTHIAZYL)PROPYL)PIPERAZINE * 1-(10-(3-(4-(2-HYDROXYETHYL)-1-PIPERAZINYL)-PROPYL)-10H-PHENOTHIAZIN-2-YL)ETHANONE * 10-(3-(4-(2-HYDROXYETHYL)-1-PIPERAZINYL)PROPYL) PHENOTHIAZIN-2-YLMETHYL KETONE * 10-(3-(4-(2-HYDROXYETHYL)-1-PIPERAZINYL)PROPYL)PHE-NOTHIAZIN-2-YL METHYL KETONE DIMALEATE * SCH 6673 * TINDAL

SAFETY PROFILE: Poison by ingestion, intraperitoneal, and intravenous routes. Severe eye irritant.

ABG750   CAS: 62-44-2   **HR: 3**
p-ACETOPHENETIDIDE
mf: $C_{10}H_{13}NO_2$ mw: 179.24

SYNS: 1-ACETAMIDO-4-ETHOXYBENZENE * ACETO-p-PHENALIDE * p-ACETOPHENETIDE

* ACETO-p-PHENETIDIDE * ACETO-4-PHENETIDINE
* ACET-p-PHENALIDE * ACET-p-PHENETIDIN
* ACETOPHENETIDIN * p-ACETPHENETIDIN
* ACETOPHENETIDINE * ACETOPHENETIN
* ACETPHENETIDIN * ACETYLPHENETIDIN
* N-ACETYL-p-PHENETIDINE * ACHROCIDIN
* ANAPAC * APC * ASA COMPOUND
* BROMO SELTZER * BUFF-A-COMP * CITRA-
FORT * CODEMPIRAL * COMMOTIONAL
* CONTRADOL * CORICIDIN * CORIFORTE
* CORYBAN-D * DAPRISAL * DARVON COM-
POUND * DASIKON * EMPIRIN COMPOUND
* 4-ETHOXYACETANILIDE * p-ETHOXYACETANI-
LIDE * N-(4-ETHOXYPHENYL)ACETAMIDE
* N-p-ETHOXYPHENYLACETAMIDE * FENACETINA
* FIORINAL * MELABON * PARACETOPHENETI-
DIN * PERCOBARB * PERCODAN * p-PHEN-
ACETIN * RCRA WASTE NUMBER U187 * SINUTAB
* TETRACYDIN * XARIL * ZACTIRIN COM-
POUND

CONSENSUS REPORTS: IARC Cancer Review: GROUP 2A IMEMDT 7,310,87; Animal Inadequate Evidence IMEMDT 13,141,77; Human Limited Evidence IMEMDT 13,141,77; Animal Limited Evidence IMEMDT 24,135,80; Human Limited Evidence IMEMDT 24,135,80. NTP Fourth Annual Report On Carcinogens, 1984. Reported in EPA TSCA Inventory.

SAFETY PROFILE: Confirmed carcinogen producing tumors of the kidney and bladder. Experimental teratogenic data. A human poison by an unspecified route. Poison by intravenous and possibly other routes. Moderately toxic by several routes. Human systemic effects by ingestion: cyanosis, liver damage, and methemoglobinemia-carboxhemoglobinemia. Mutation data reported. Experimental reproductive effects. Chronic effects consist of weight loss, insomnia, shortness of breath, weakness and often aplastic anemia. When heated to decomposition it emits toxic fumes of NO_x.

ABH000   CAS: 98-86-2   HR: 3
ACETOPHENONE
mf: C_8H_8O mw: 120.16

PROP: Colorless liquid or plates; sweet, pungent odor. Mp: 19.7°, bp: 202.3°, flash p: 180°F (OC), d: 1.026 @ 20°/4°, vap d: 4.14, vap press: 1 mm @ 15°, autoign temp: 1060°F. Very sol in propylene glycol and fixed oils; sol in alc, chloroform, and ether; sltly sol in water; insol in glycerin.

SYNS: ACETYLBENZENE * BENZOYL METHIDE
* DYMEX * FEMA No. 2009 * HYPNONE
* KETONE METHYL PHENYL * METHYL PHENYL
KETONE * 1-PHENYLETHANONE * PHENYL
METHYL KETONE * USAF EK-496

CONSENSUS REPORTS: Reported in EPA TSCA Inventory.

SAFETY PROFILE: Poison by intraperitoneal and subcutaneous routes. Moderately toxic by ingestion. A skin and severe eye irritant. Mutation data reported. Narcotic in high concentration. A hypnotic. Combustible liquid. To fight fire use foam, CO_2, dry chemical. When heated to decomposition it emits acrid smoke and fumes.

ABH500   CAS: 61-00-7   HR: 3
ACETOPROMAZINE
mf: $C_{19}H_{22}N_2OS$ mw: 326.49

SYNS: ACEPROMAZINA * ACEPROMAZINE
* ACEPROMIZINA * ACETAZINE * ACETHYL-
PROMAZIN * 3-ACETYL-10-(3-DIMETHYLAM-
INOPROPYL)PHENOTHIAZINE * ACETYL-
PROMAZINE * ANATRAN * ANERGAN
* ATRAVET * ATSETOZIN * AY-57,062
* AZEPROMAZINE * 1522 CB * 10-(3-
DIMETHYLAMINOPROPYL)PHENOTHIAZINE-3-
ETHYLONE * 1-(10-(3-(DIMETHYLAMINO)PRO-
PYL)-10H-PHENOTHIAZIN-2-YL)ETHANONE
* 10-(3-DIMETHYLAMINOPROPYL)PHENOTHI-
AZIN-3-YLMETHYL KETONE * LISERGAN
* NOTENQUIL * NOTENSIL * NOTESIL
* PLEGECYL * PLEGICIN * PLIVAPHEN
* SOPRINTIN * SOPRONTIN * SOPROTIN
* SV-1522 * VETRANQUIL * WY-1172

SAFETY PROFILE: Poison by ingestion, intravenous, and subcutaneous routes. When heated to decomposition it emits toxic fumes of SO_x and NO_x. An animal tranquilizer.

ABJ250   CAS: 103-89-9   HR: 2
p-ACETOTOLUIDIDE
mf: $C_9H_{11}NO$ mw: 149.21

PROP: Crystals. Bp: 307°, flash p: 335°F (CC), d: 1.212, vap d: 5.14, mp: 153°.

SYNS: p-ACETAMIDOTOLUENE * p-ACETOTOLUIDE
* 4-ACETOTOLUIDE * 4-(ACETYLAMINO)TOLUENE
* ACETYL-p-TOLUIDINE * N-ACETYL-p-TOLUIDIDE
* p-METHYLACETANILIDE * 4-METHYLACETANI-
LIDE * 4'-METHYLACETANILIDE

CONSENSUS REPORTS: Reported in EPA TSCA Inventory.

SAFETY PROFILE: Moderately toxic by ingestion. Combustible. When heated to decomposition it emits toxic fumes of NO_x. To fight fire use water, foam, CO_2, dry chemical.

ABL000   CAS: 6098-44-8   **HR: 3**
N-ACETOXY-N-ACETYL-2-AMINOFLUORENE
mf: $C_{17}H_{15}NO_3$ mw: 281.33

SYNS: ACETIC ACID (N-ACETYL-N-(2-FLUORENYL) AMINO) ESTER * N-ACETOXY-2-ACETAMIDO-FLUORENE * N-ACETOXY-2-ACETYLAMINO-FLUORENE * N-ACETOXY-2-FLUORENYLACETAMIDE * N-(FLUOREN-2-YL)ACETOHYDROXAMIC ACET-AMIDE

CONSENSUS REPORTS: EPA Genetic Toxicology Program.

SAFETY PROFILE: Questionable carcinogen with experimental tumorigenic and neoplastigenic data. Human mutation data reported. When heated to decomposition it emits toxic fumes of NO_x.

ABL500   CAS: 3061-65-2   **HR: 3**
2-ACETOXYACRYLONITRILE
mf: $C_5H_5NO_2$ mw: 111.11

SYNS: α-ACETOXYACRYLONITRILE * α-CYANOVI-NYL ACETATE

CONSENSUS REPORTS: Cyanide and its compounds are on the Community Right-To-Know List.

SAFETY PROFILE: Poison by inhalation, ingestion, and skin contact. A skin irritant. When heated to decomposition it emits toxic fumes of NO_x.

ABM250   CAS: 1515-76-0   **HR: 3**
1-ACETOXY-1,3-BUTADIENE
mf: $C_6H_8O_2$ mw: 112.14

SYN: ACETIC ACID-1,3-BUTADIENYL ESTER

SAFETY PROFILE: Poison by inhalation. Moderately toxic by other routes. A skin irritant. Mutation data reported. When heated to decomposition it emits acrid smoke.

ABN000   CAS: 2885-39-4   **HR: 3**
ACETOXYCYCLOHEXIMIDE
mf: $C_{17}H_{25}NO_6$ mw: 339.43

SYNS: ACETYLOXYCYCLOHEXIMIDE * 3-(2-(5-ACE-TOXY-3,5-DIMETHYL-2-OXOCYCLOHEXYL)-2-HYDROXYETHYL)GLUTARIMIDE * AXM * E-73 ACETATE * NSC 32743 * STREPTOVITACIN E 73

SAFETY PROFILE: Deadly poison by ingestion, intravenous, intraperitoneal, and subcutaneous routes. Human mutation data reported. When heated to decomposition it emits toxic fumes, such as NO_x.

ABO000   CAS: 60-31-1   **HR: 3**
2-ACETOXYETHYLTRIMETHYLAMMO-NIUM CHLORIDE
mf: $C_7H_{16}NO_2 \bullet Cl$ mw: 181.69

SYNS: ACECOLINE * ACETYLCHOLINE CHLORIDE * ACETYLCHOLINE HYDROCHLORIDE * ACETYL-CHOLINIUM CHLORIDE * 2-(ACETYLOXY)-N,N,N-TRIMETHYLETHANAMINIUM CHLORIDE * ACH CHLORIDE * ARTEROCOLINE * CHOLINE CHLORIDE ACETATE * (2-HYDROXYETHYL)-TRIMETHYLAMMONIUM CHLORIDE ACETATE * OVISOT * TL 1505

CONSENSUS REPORTS: Reported in EPA TSCA Inventory.

SAFETY PROFILE: Poison by subcutaneous, intravenous, intraperitoneal, and parenteral routes. Moderately toxic by ingestion. When heated to decomposition it emits very toxic fumes of NO_x and Cl^-. A cholinergic agent.

ABQ000   CAS: 6283-24-5   **HR: 3**
p-(ACETOXYMERCURI)ANILINE
mf: $C_8H_9HgNO_2$ mw: 351.77

PROP: Colorless crystals, insol in water. Mp: 167°.

SYNS: (ACETATO)(p-AMINOPHENYL)MERCURY * p-AMINOPHENYLMERCURIC ACETATE

CONSENSUS REPORTS: Reported in EPA TSCA Inventory.

OSHA PEL: (Transitional: CL 1 mg/10m^3) CL 0.1 mg(Hg)/m^3 (skin)
ACGIH TLV: TWA 0.1 mg(Hg)/m^3 (skin)
NIOSH REL: TWA 0.05 mg(Hg)/m^3

SAFETY PROFILE: Poison by intravenous routes. When heated to decomposition it emits very toxic fumes of NO_x and Hg.

ABQ250   CAS: 54481-45-7   **HR: 3**
2-(ACETOXYMERCURI)-4-NITROANILINE
mf: $C_8H_8HgN_2O_4$ mw: 396.77

SYN: ACETATO(2-AMINO-5-NITROPHENYL)MERCURY

OSHA PEL: (Transitional: CL 1 mg/10m^3) CL
   0.1 mg(Hg)/m^3 (skin)
ACGIH TLV: TWA 0.1 mg(Hg)/m^3 (skin)
NIOSH REL: TWA 0.05 mg(Hg)/m^3

SAFETY PROFILE: Poison by intraperitoneal
route. When heated to decomposition it emits
very toxic fumes of Hg and NO$_x$.

ABS750   CAS: 830-03-5   **HR: 3**
p-ACETOXYNITROBENZENE
mf: $C_8H_7NO_4$ mw: 181.16

SYNS: p-NITROPHENOL ACETATE * p-NITROPHE-
NYL ACETATE * 4-NITROPHENYL ACETATE

CONSENSUS REPORTS: Reported in EPA
TSCA Inventory.

SAFETY PROFILE: Poison by intravenous
route. When heated to decomposition it emits
toxic fumes of NO$_x$.

ABT750   CAS: 53198-41-7   **HR: 3**
1-ACETOXY-N-NITROSODIPROPYLAMINE
mf: $C_8H_{16}N_2O_3$ mw: 188.26

SYNS: ACETIC ACID-1-(PROPYLNITROSAMINO)-
PROPYL ESTER * N-(α-ACETOXY)PROPYL-N-N-PRO-
PYLNITROSAMINE * 1-(PROPYLNITROSAMINO)
PROPYL ACETATE

SAFETY PROFILE: Moderately toxic by subcutaneous route. Mutation data reported. Questionable carcinogen with experimental carcinogenic data. When heated to decomposition it
emits toxic fumes of NO$_x$.

ABU000   CAS: 51-98-9   **HR: 3**
17-ACETOXY-19-NOR-17-α-PREGN-4-EN-20-YN-3-ONE
mf: $C_{22}H_{28}O_3$ mw: 340.50

SYNS: 17-β-ACETOXY-19-NOR-17-α-PREGN-4-EN-20-YN-
3-ONE * (17-α)-17-(ACETYLOXY)-19-NORPREGN-4-
EN-20-YN-3-ONE * 17-ACETYLOXY(17-α)-19-
NORPREGN-4-ESTREN-17-β-OL-ACETATE-3-ONE
* 17-ENT * 17-α-ETHINYL-19-NORTESTOSTERONE
ACETATE * 17-α-ETHINYL-19-NORTESTOS-

TERONE-17-β-ACETATE * 17-α-ETHYNYL-17-β-
ACETOXY-19-NORANDROST-4-EN-3-ONE * 17-α-
ETHYNYL-17-HYDROXYESTR-4-EN-3-ONE ACE-
TATE * 17-α-ETHYNYL-19-NORTESTOSTERONE
ACETATE * 17-HYDROXY-19-NOR-17-α-PREGN-
4-EN-20-YN-3-ONE ACETATE * 17-β-HYDROXY-19-
NOR-17-α-PREGN-4-EN-20-YN-3-ONE ACETATE
* NORETHINDRONE-17-ACETATE * 19-NORETHIS-
TERONE ACETATE * 19-NORETHYNYLTESTOSTERONE
ACETATE * NORETHYSTERONE ACETATE
* NORLUTATE * NORLUTINE ACETATE
* ORLUTATE

CONSENSUS REPORTS: IARC Cancer Review: Animal Limited Evidence IMEMDT
21,441,79; Animal Sufficient Evidence
IMEMDT 6,179,74. EPA Genetic Toxicology
Program.

SAFETY PROFILE: Suspected carcinogen with
experimental tumorigenic data. Human reproductive effects by ingestion and implant routes:
menstrual cycle changes, postpartum effects and
changes in fertility. A human teratogen by an
unspecified route with developmental abnormalities of the urogenital system. Experimental reproductive effects. Mutation data reported.
When heated to decomposition it emits acrid
smoke and irritating fumes. Used in the treatment of menstrual disorders and uterine bleeding.

ABU500   CAS: 62-38-4   **HR: 3**
ACETOXYPHENYLMERCURY

DOT: UN 1674
mf: $C_8H_8HgO_2$ mw: 336.75

PROP: Lustrous crystals, sltly sol in water.
Mp: 149°.

SYNS: ACETATE PHENYLMERCURIQUE (FRENCH)
* (ACETATO)PHENYLMERCURY * ACETIC ACID,
PHENYLMERCURY DERIV. * (ACETOXYMER-
CURI)BENZENE * AGROSAN * ALGIMYCIN
* ANTIMUCIN WDR * BUFEN * CEKUSIL
* CELMER * CERESAN * CONTRA CREME
* DYANACIDE * FEMMA * FENYLMERCURIACE-
TAT (CZECH) * FMA * FUNGITOX OR * GALLO-
TOX * HL-331 * HONG KIEN * HOSTAQUICK
* KWIKSAN * LEYTOSAN * LIQUIPHENE
* MERCURIPHENYL ACETATE * NORFORMS
* NYLMERATE * OCTAN FENYLRTUTNATY (CZECH)
* PAMISAN * PHENMAD * PHENOMERCURIC
ACETATE * PHENYLMERCURIACETATE * PHENYL
MERCURIC ACETATE * PHENYLMERCURY ACETATE

* PHENYLQUECKSILBERACETAT (GERMAN)
* PHIX * PMA * PMAC * PMACETATE
* PMAL * PMAS * PURASAN-SC-10 * PURA-
TURF 10 * QUICKSAN * RCRA WASTE NUMBER
P092 * SANITIZED SPG * SC-110 * SEEDTOX
* SPOR-KIL * TAG * TAG FUNGICIDE
* ZIARNIK

CONSENSUS REPORTS: EPA Extremely Hazardous Substances List. Reported in EPA TSCA Inventory. EPA Genetic Toxicology Program. Mercury and its compounds are on the Community Right-To-Know List.

OSHA PEL: (Transitional: CL 1 mg/10m^3) CL 0.1 mg(Hg)/m^3 (skin)
ACGIH TLV: TWA 0.1 mg(Hg)/m^3 (skin)
NIOSH REL: TWA 0.05 mg(Hg)/m^3
DOT Classification: IMO: Poison B; Label: Poison.

SAFETY PROFILE: Poison by ingestion, intravenous, intraperitoneal, subcutaneous, and possibly other routes. Experimental teratogenic and reproductive effects. Mutation data reported. An eye and severe human skin irritant. When heated to decomposition it emits toxic fumes of Hg.

ABV250   CAS: 17427-00-8   **HR: 3**
3-ACETOXYPHENYLTRIMETHYLAMMO-NIUM IODIDE
mf: $C_{11}H_{16}NO_2 \cdot I$ mw: 321.18

SYN: NU 2017

SAFETY PROFILE: A poison via subcutaneous and intravenous routes. When heated to decomposition it emits very toxic fumes of NO$_x$ and I$^-$.

ABW750   CAS: 1907-13-7   **HR: 3**
ACETOXYTRIETHYLSTANNANE
mf: $C_8H_{18}O_2Sn$ mw: 264.95

SYNS: ACETOXYTRIETHYLTIN * TIN TRIAETHYL-ZINNACETAT (GERMAN) * TRIETHYLTIN ACETATE

OSHA PEL: TWA 0.1 mg(Sn)/m^3 (skin)
ACGIH TLV: TWA 0.1 mg(Sn)/m^3 (skin) (Proposed: TWA 0.1 mg(Sn)/m^3; STEL 0.2 mg(Sn)/m^3 (skin))
NIOSH REL: (Organotin Compounds) TWA 0.1 mg(Sn)/m^3

SAFETY PROFILE: Poison by ingestion and intravenous routes. When heated to decomposition it emits acrid smoke and irritating fumes.

ABX000   CAS: 2897-46-3   **HR: 3**
ACETOXYTRIHEXYLSTANNANE
mf: $C_{20}H_{42}O_2Sn$ mw: 433.31

SYNS: ACETOXYTRIHEXYLTIN * TRIHEXYLTIN ACETATE * TRI-N-HEXYLZINNACETAT (GERMAN)

OSHA PEL: TWA 0.1 mg(Sn)/m^3 (skin)
ACGIH TLV: TWA 0.1 mg(Sn)/m^3 (skin) (Proposed: TWA 0.1 mg(Sn)/m^3; STEL 0.2 mg(Sn)/m^3 (skin))
NIOSH REL: (Organotin Compounds) TWA 0.1 mg(Sn)/m^3

SAFETY PROFILE: Poison by intravenous route. Moderately toxic by ingestion. When heated to decomposition it emits acrid smoke and fumes.

ABX250   CAS: 900-95-8   **HR: 3**
ACETOXYTRIPHENYLSTANNANE
mf: $C_{20}H_{18}O_2Sn$ mw: 409.07

PROP: Practically insol, crystalline solid. Mp: 120°.

SYNS: ACETATE de TRIPHENYL-ETAIN (FRENCH)
* ACETATO di STAGNO TRIFENILE (ITALIAN)
* ACETATOTRIPHENYLSTANNANE * ACETOXY-TRI-PHENYL-STANNAN (GERMAN) * ACETOXY-TRIPHEN-YLSTANNANE * ACETOXYTRIPHENYLTIN
* (ACETYLOXY)TRIPHENYL-STANNANE (9CI)
* BATASAN * BRESTAN * ENT 25,208
* FENOLOVO ACETATE * FENTIN ACETAAT
(DUTCH) * FENTIN ACETAT (GERMAN) * FENTIN
ACETATE * FENTINE ACETATE (FRENCH) * FINTIN
ACETATO (ITALIAN) * GC 6936 * HOE-2824
* LIROMATIN * LIROSTANOL * PHENTIN ACE-TATE * PHENTINOACETATE * SUZU * TINES-TAN * TINESTAN 60 WP * TIN TRIPHENYL ACE-TATE * TPTA * TPZA * TRIFENYLTINACETAAT
(DUTCH) * TRIPHENYLACETO STANNANE
* TRIPHENYLTIN ACETATE * TRIPHENYL-ZINNACETAT (GERMAN) * TUBOTIN
* VP 1940

CONSENSUS REPORTS: EPA Extremely Hazardous Substances List. Reported in EPA TSCA Inventory.

OSHA PEL: TWA 0.1 mg(Sn)/m^3 (skin)
ACGIH TLV: TWA 0.1 mg(Sn)/m^3 (skin) (Proposed: TWA 0.1 mg(Sn)/m^3; STEL 0.2 mg(Sn)/m^3 (skin))
NIOSH REL: (Organotin Compounds) TWA 0.1 mg(Sn)/m^3

SAFETY PROFILE: Poison by ingestion, skin contact, intraperitoneal, intravenous, and subcutaneous routes. Questionable carcinogen with experimental neoplastigenic data. Experimental teratogenic and reproductive effects. A fungicide and algicide used as a wood preservative. When heated to decomposition it emits acrid smoke and fumes.

ABX500   CAS: 97-44-9   **HR: 3**
ACETPHENARSINE
mf: $C_8H_{10}AsNO_5$ mw: 275.11

PROP: Crystalline material, sltly water sol. Decomp @ 240-250°.

SYNS: 3-ACETAMIDO-4-HYDROXY-PHENYLARSONIC ACID * ACETARSOL * ACETARSONE * 3-ACETYLAMINO-4-HYDROXYPHENYLARSONIC ACID * (3-(ACETYLAMINO)-4-HYDROXYPHENYL) ARSONINE (9CI) * N-ACETYL-4-HYDROXY-m-ARSANILIC ACID * AMARSAN * AMOEBAL * ARSPHEN * ARSONIC ACID * DEVEGAN * DISPARICIDA * DYNARSAN * EHRLICH 594 * 190 F * F 190 * FOURNEAU 190 * GINARSOL * GOYL * GYNOPLIX * KHAROPHEN * KUBARSOL * LIMARSOL MALAGRIDE * MEXYL * MONARGAN * NILACID * ORALCID * ORARSAN * OSARSAL * OSARSOLE * OSVARSAN * PALLICID * PAROXYL * SPIROCID * SPIROZID * STOVARSAL * STOVARSOL * STOVARSOLAN * SVC * VAGISEPT * VAGOFLOR

CONSENSUS REPORTS: Arsenic and its compounds are on the Community Right-To-Know List.

OSHA PEL: TWA 500 $\mu g(As)/m^3$
ACGIH TLV: TWA 0.2 $mg(As)/m^3$

SAFETY PROFILE: Poison by ingestion and intravenous routes. Human systemic effects by ingestion: respiratory system, endocrine system, dermatitis and fever. Human systemic effects by intravaginal route: hallucinations, distorted perceptions, convulsions, nausea or vomiting, decreased urine volume and fever. Mutation data reported. When heated to decomposition it emits very toxic fumes of NO_x and As.

ABX750   CAS: 123-54-6   **HR: 3**
ACETYL ACETONE

DOT: UN 2310
mf: $C_5H_8O_2$ mw: 100.13

PROP: Colorless to sltly yellow liquid; pleasant odor. Mp: −23.2°, bp: 139° @ 746 mm, flash p: 105°F (OC), d: 0.952-0.962, refr index: 1.402, vap d: 3.45, autoign temp: 644°F. Misc in alc, ether, chloroform, acetone, glacial acetic acid, and propylene glycol; insol in glycerin and water.

SYNS: ACETOACETONE * DIACETYLMETHANE * FEMA No. 2841 * PENTANEDIONE * 2,4-PENTANEDIONE (FCC)

CONSENSUS REPORTS: Reported in EPA TSCA Inventory.

DOT Classification: IMO: Flammable or Combustible Liquid; Label: Flammable Liquid.

SAFETY PROFILE: Moderately toxic via ingestion, intraperitoneal and inhalation routes. A skin and severe eye irritant. Flammable liquid when exposed to heat or flame. Incompatible with oxidizing materials. To fight fire, use alcohol foam, CO_2, dry chemical.

ABY000   CAS: 28322-02-3   **HR: 3**
4-ACETYLAMINOFLUORENE
mf: $C_{15}H_{13}NO$ mw: 223.29

SYNS: 4-ACETYLAMINOFLUOREN (GERMAN) * N-FLUOREN-4-YLACETAMIDE * N-4-FLUORENYL-ACETAMIDE

CONSENSUS REPORTS: EPA Genetic Toxicology Program.

SAFETY PROFILE: Poison by intraperitoneal route. Questionable carcinogen with experimental tumorigenic data. Mutation data reported. When heated to decomposition it emits toxic fumes of NO_x.

ABY900   CAS: 140-40-9   **HR: 3**
2-ACETYLAMINO-5-NITROTHIAZOLE
mf: $C_5H_5N_3O_3S$ mw: 187.19

PROP: Needles from alc, elongated plates from acetic acid. Mp: 264-265°. The commercial product may be yellow. Sol in aq solns of NaOH and NH_3 with deep orange color.

SYNS: ACETAMIDO-5-NITROTHIAZOLE * ACINITRAZOLE * AMINITROZOLE * ENHEPTIN-A * GYNOFON * N-(5-NITRO-2-THIAZOLYL)-ACETAMIDE * PLEOCIDE * TRICHORAD * TRICHORAL * TRITHEOM

SAFETY PROFILE: Poison by ingestion. Mutation data reported. When heated to decomposition it emits toxic fumes of SO_x and NO_x.

ACB250   CAS: 460-07-1   **HR: 3**
1-ACETYLAZIRIDINE
mf: C_4H_7NO mw: 85.12

SYN: ACETYLETHYLENEIMINE

CONSENSUS REPORTS: Reported in EPA TSCA Inventory. EPA Genetic Toxicology Program.

SAFETY PROFILE: Poison by intraperitoneal route. Questionable carcinogen with experimental tumorigenic and neoplastigenic data. When heated to decomposition it emits toxic fumes of NO_x.

ACC250   CAS: 644-31-5   **HR: 3**
ACETYL BENZOYL PEROXIDE (solid)
DOT: UN 2081
mf: $C_9H_8O_4$ mw: 180.17

PROP: White crystals. Sol in oils, alc, ether and chloroform. Mp: 36-37°, bp: 130° @ 19 mm.

DOT Classification: Forbidden.

SAFETY PROFILE: Poison by inhalation and ingestion. Severe irritant. A powerful oxidizing agent which is corrosive to the skin and mucous membranes. Dangerous; shock or heat will cause detonation with evolution of toxic fumes; will react with water or steam to produce heat; can react vigorously with reducing materials. Flammable by spontaneous chemical reaction. To fight fire use CO_2 or dry chemical. When heated to decomposition it emits acrid smoke and fumes.

ACC500   CAS: 644-31-5   **HR: 3**
ACETYL BENZOYLPEROXIDE (solution)
DOT: UN 2081

PROP: Solution contains not over 40% acetyl benzoyl peroxide (FEREAC 41,15972,76).

DOT Classification: Organic Peroxide; Label: Organic Peroxide.

SAFETY PROFILE: Highly irritating to skin, eyes, and mucous membranes. When heated to decomposition it emits acrid smoke and fumes.

ACD000   CAS: 4463-22-3   **HR: 3**
N-ACETYL-4-BIPHENYLHY-DROXYLAMINE
mf: $C_{14}H_{13}NO_2$ mw: 227.28

SYNS: 4-BIPHENYLACETHYDROXAMIC ACID * N-HYDROXY-AABP * N-HYDROXY-4-ACETAMIDO-BIPHENYL * N-4-(N-HYDROXYACETAMIDO)BI-PHENYL * N-HYDROXY-4-ACETAMIDODIPHENYL * N-HYDROXY-4-ACETYLAMINOBIPHENYL * N-HYDROXY-N-4-BIPHENYLACETAMIDE

SAFETY PROFILE: Questionable carcinogen with experimental carcinogenic and tumorigenic data. Human mutation data reported. When heated to decomposition it emits toxic fumes of NO_x.

ACD250   CAS: 3733-45-7   **HR: 3**
N-(N-ACETYL-3-(p-(BIS(2-CHLOROETHYL)AMINO)PHENYL)-ALANYL-3-PHENYLALANINE ETHYL ESTER

SYN: ETHYL ESTER of N-ACETYL-dl-SARCOLYSYL-l-PHENYLALANINE

SAFETY PROFILE: Poison by ingestion and intramuscular routes. When heated to decomposition it emits very toxic fumes of Cl^- and NO_x.

ACD750   CAS: 506-96-7   **HR: 3**
ACETYL BROMIDE
DOT: UN 1716
mf: C_2H_3BrO mw: 122.96

PROP: Colorless, fuming liquid; turns yellow in air. Mp: −96.5°, bp: 76.7°, d: 1.52 @ 9.5°/4°. Decomp in water and alc; misc in benzene, ether, and chloroform.

CONSENSUS REPORTS: Reported in EPA TSCA Inventory.

DOT Classification: Corrosive Material; Label: Corrosive.

SAFETY PROFILE: Poison by inhalation, ingestion, skin contact, and intraperitoneal routes. Violent reaction on contact with water, steam, methanol, or ethanol produces toxic and reactive HBr. When heated to decomposition it emits highly corrosive and toxic fumes of carbonyl bromide and bromine. To fight fire, use dry chemical, CO_2.

ACE000   CAS: 77-66-7   **HR: 2**
1-ACETYL-3-(2-BROMO-2-ETHYLBUTYRYL)UREA
mf: $C_9H_{15}BrN_2O_3$ mw: 279.17

SYNS: ABASIN * ABSIN * ACECARBROMAL
* ACETCARBROMAL * ACETKARBROMAL
* ACETYL ADALIN * N-((ACETYLAMINO)CARBO-
NYL)-2-BROMO-2-ETHYLBUTANAMIDE * ACETYL-
BROMODIETHYLACETYLCARBAMIDE * N-ACETYL-
N-BROMODIETHYLACETYLCARBAMIDE
* N-ACETYL-N-BROMODIETHYLACETYLUREA
* N-ACETYL-N'-α-BROMO-α-ETHYLBUTYRYL-
CARBAMIDE * 1-ACETYL-3-(α-BROMO-α-ETH-
YLBUTYRYL)UREA * ACETYLCARBROMAL
* ADITYL * CARBASED * DAROLON * IBA-
TRAN * PAXAREL * SEDAMYL * SEDMYNOL
* SEDTRAN

CONSENSUS REPORTS: Reported in EPA
TSCA Inventory.

SAFETY PROFILE: Moderately toxic by injection. Human systemic effects by ingestion: toxic
psychosis. When heated to decomposition it
emits very toxic fumes of Br^- and NO_x. A
sedative.

ACE500   CAS: 2813-95-8   **HR: 3**
o-ACETYL-2-sec-BUTYL-4,6-
DINITROPHENOL
mf: $C_{12}H_{14}N_2O_6$ mw: 282.28

SYNS: ACETIC ACID-(2,4-DINITRO-6-sec-BUTYLPHE-
NYL) ESTER * ACETIC ACID-(4,6-DINITRO-2-sec-BU-
TYLPHENYL) ESTER * ARETIT * 2-sec-BUTYL-4,6-
DINITROPHENYLACETATE * 2,4-DINITRO-6-sec-BU-
TYLFENYLESTER KYSELINY OCTOVE (CZECH)
* 6-sec-BUTYL-2,4-DINITROPHENYLACETATE
* 2,4-DINITRO-6-sek.BUTYL-PHENYLACETAT (GERMAN)
* 4,6-DINITRO-2-sec-BUTYLPHENYL ACETATE
* DINOSEB-ACETATE * HOE 2904 * β-(2-
HYDROXY-3,5-DINITROPHENYL)BUTANE ACE-
TATE * IVOSIT * 2-(1-METHYLPROPYL)-4,6-
DINITROPHENYL ACETATE * PHENOTAN

SAFETY PROFILE: Poison by ingestion. A skin
and eye irritant. When heated to decomposition
it emits toxic fumes of NO_x. A herbicide.

ACF000   CAS: 36573-63-4   **HR: 3**
3'-o-ACETYLCALOTROPIN
mf: $C_{31}H_{42}O_{10}$ mw: 574.73

PROP: A glycoside isolated *Asclepius cuns-
suica* (ARZNAD 28,1095,78).

SYN: ASCLEPIN

SAFETY PROFILE: Poison by ingestion and
intraperitoneal routes. When heated to decom-

position it emits acrid smoke and irritating
fumes.

ACF750   CAS: 75-36-5   **HR: 3**
ACETYL CHLORIDE
DOT: UN 1717
mf: C_2H_3ClO mw: 78.50

PROP: Colorless, fuming liquid. Mp: $-112°$,
bp: $51°-52°$, flash p: $40°F$ (CC), autoign temp:
$734°F$, d: 1.1051 @ $20°/4°$, vap d: 2.70. lel:
5%. Decomp in water and alc; misc in benzene,
ether, and chloroform.

SYNS: ACETIC ACID CHLORIDE * ACETIC CHLO-
RIDE * ETHANOYL CHLORIDE * RCRA WASTE
NUMBER U006

CONSENSUS REPORTS: Reported in EPA
TSCA Inventory.

DOT Classification: Flammable Liquid; Label:
   Flammable Liquid, Corrosive.

SAFETY PROFILE: Poison by inhalation and
ingestion. A human systemic irritant by inhalation. Violent hydrolysis reaction with water or
steam produces heat, acetic acid, HCl and other
corrosive chlorides. May decompose during
preparation. Dangerous fire hazard when exposed to heat or flame. Explosion hazard by
spontaneous chemical reaction, with dimethyl
sulfoxide or ethanol. Also incompatible with
PCl_3. When heated to decomposition it emits
highly toxic fumes of phosgene and Cl^-. To
fight fire, use CO_2 or dry chemical.

ACH000   CAS: 616-91-1   **HR: 3**
N-ACETYL-I-CYSTEINE
mf: $C_5H_9NO_3S$ mw: 163.21

SYNS: 1-α-ACETAMIDO-β-MERCAPTOPROPIONIC ACID
* ACETEIN * N-ACETYL-1-CYSTEINE (9CI)
* ACETYLCYSTEINE * N-ACETYLCYSTEINE
* N-ACETYL-N-CYSTEINE * N-ACETYL-3-
MERCAPTOALANINE * AIRBRON * BRONCHOLY-
SIN * FLUIMUCETIN * FLUIMUCIL * FLUMICIL
* INSPIR * MERCAPTURIC ACID * (R)-MERCAP-
TURIC ACID * MUCOLYTICUM * MUCOLYTICUM
LAPPE * MUCOMYST * MUCOSOLVIN * NAC
* NAC-TB * NSC 111180 * PARVOLEX
* RESPAIRE

CONSENSUS REPORTS: Reported in EPA
TSCA Inventory.

SAFETY PROFILE: Poison by intraperitoneal
route. Moderately toxic by other routes. Muta-

tion data reported. When heated to decomposition it emits very toxic fumes of NO_x and SO_x.

ACH500   CAS: 1111-39-3   **HR: 3**
ACETYLDIGITOXIN-a
mf: $C_{43}H_{66}O_{14}$ mw: 807.09

SYNS: α-ACETYLDIGITOXIN * ACYLANID

SAFETY PROFILE: Poison by ingestion and intravenous routes. When heated to decomposition it emits acrid smoke and fumes.

ACI000   CAS: 5511-98-8   **HR: 3**
ACETYLDIGOXIN-a
mf: $C_{43}H_{66}O_{15}$ mw: 823.09

SYNS: α-ACETYLDIGOXIN * DIGORID A
* DIGOXIGENIN + ZUCKERKETTE WIE BIE ACETYL-DIGITOXIN A (GERMAN)

SAFETY PROFILE: Deadly poison by ingestion, intravenous, and intraduodenal routes. When heated to decomposition it emits acrid smoke and fumes.

ACI250   CAS: 5355-48-6   **HR: 3**
ACETYLDIGOXIN-b
mf: $C_{43}H_{66}O_{15}$ mw: 823.09

SYNS: β-ACETYLDIGOXIN * DIGORID B
* DIGOXIGENIN + ZUCKERKETTE WIE BEI ACETYL-DIGITOXIN-α (GERMAN) * HEXAMETHYL-ENEIMINE-3,5-DINITROBENZOATE

SAFETY PROFILE: Deadly poison by ingestion, intravenous, and intraduodenal routes. When heated to decomposition it emits acrid smoke and fumes.

ACI750   CAS: 74-86-2   **HR: 3**
ACETYLENE

DOT: UN 1001
mf: C_2H_2 mw: 26.04

PROP: Colorless gas, garlic-like odor. Flammable. Bp: $-84.0°$ (sublimes), lel: 2.5%, uel: 82%, mp: $-81.8°$, flash p: 0°F (CC), d: 1.173 g/L @ 0°, autoign temp: 581°F, vap press: 40 atm @ 16.8°, vap d: 0.91; d: (liquid) 0.613 @ $-80°$. D: (solid) 0.730 @ $-85°$. Quite sol in water; very sol in alc; almost misc in ether.

SYNS: ACETYLEN * ACETYLENE, dissolved (DOT)
* ETHINE * ETHYNE * NARCYLEN

CONSENSUS REPORTS: Reported in EPA TSCA Inventory.

OSHA PEL: CL 2500 ppm
ACGIH TLV: Simple asphyxiant.
NIOSH REL: CL 2500 ppm
DOT Classification: Forbidden; Flammable Gas; Label: Flammable Gas.

SAFETY PROFILE: Mildly toxic by inhalation. Human systemic effects by inhalation: headache and dyspnea. Narcotic in high concentration. In general industrial practice, acetylene does not constitute a serious toxic hazard. It is a very dangerous fire hazard when exposed to heat, flame, or oxidizers. Moderate explosion hazard when exposed to heat or flame or by spontaneous chemical reaction. At high pressures and moderate temperatures, and in the absence of air, acetylene has been known to decompose explosively. Reacts with copper to form the explosive copper acetylide. Incompatible with brass; copper salts; copper carbide; powdered Co; Hg; Hg salts; K; Ag and Ag salts; RbH; CsH; halogens; HNO_3; NaH; oxidants. Acetylene + halide + UV can explode. Molten K ignites in C_2H_2 and then explodes. C_2H_2 reacts vigorously with trifluoromethyl hypofluorite. With O_2, C_2H_2 can detonate very powerfully. When ignited, it burns with an intensely hot flame; can react vigorously with oxidizing materials.

When mixed with O_2 in proportions of 40% or more, acetylene acts as a narcotic and has been used in anesthesia. Acetylene acts as a simple asphyxiant by diluting the O_2 in the air to a level which will not support life. However, the presence of impurities in commercial acetylene may result in the production of symptoms before an asphyxiant concentration is reached. Thus: 10% in air produces slight intoxication, 20% produces staggering gait, 30% produces general incoordination, 33% leads to unconsciousness in 7 minutes, up to 80% produces complete anesthesia, increased blood pressure, narcosis and stimulated respiration.

Dizziness, headache, mild gastric symptoms, and (in high concentration) semi-asphyxia and brief loss of consciousness have all been reported. To fight fire, use CO_2, water spray, or dry chemical. Stop flow of gas.

ACJ000   **HR: 3**
ACETYLENE CHLORIDE

PROP: A gas. Bp: $-31°$, vap d: 2.0, mp: $-126°$.

SYN: CHLOROETHYNE

SAFETY PROFILE: Dangerous fire hazard by spontaneous chemical reaction. Spontaneously flammable in air. Shock will explode it. When heated to decomposition it emits highly toxic fumes of phosgene; can react vigorously with oxidizing materials.

ACJ250   CAS: 543-21-5   **HR: 3**
ACETYLENEDICARBOXAMIDE
mf: $C_4H_4N_2O_2$ mw: 112.10

PROP: Produced by *Str. reticuli var. Aquamyceticus* and is identical to Cellocidin.

SYNS: ACETYLENEDICARBOXYLIC ACID DIAMIDE * AQUAMYCIN * 2-BUTYNEDIAMIDE * CELLOCIDIN * LENAMYCIN

CONSENSUS REPORTS: Reported in EPA TSCA Inventory.

SAFETY PROFILE: Poison by intravenous and intraperitoneal routes. When heated to decomposition it emits toxic fumes of NO_x.

ACJ500   CAS: 928-04-1   **HR: 3**
ACETYLENEDICARBOXYLIC ACID MONOPOTASSIUM SALT
mf: $C_4HO_4 \cdot K$ mw: 152.15

SYNS: U-4783 * MONOPOTASSIUM SALT of ACETYLENEDICARBOXYLIC ACID

CONSENSUS REPORTS: Reported in EPA TSCA Inventory.

SAFETY PROFILE: Poison by ingestion, intravenous, and intraperitoneal routes. When heated to decomposition it emits acrid smoke and fumes of KO_x.

ACK000   CAS: 156-60-5   **HR: 2**
trans-ACETYLENE DICHLORIDE
mf: $C_2H_2Cl_2$ mw: 96.94

PROP: Colorless liquid, pleasant odor. Mp: −50°, bp: 48°, flash p: 36°F, autoign temp: 860°F, lel: 9.7%, uel: 12.8%, d: 1.2743 @ 25°/4°, vap press: 400 mm @ 30.8°, vap d: 3.34.

SYNS: trans-DICHLOROETHYLENE * trans-1,2-DICHLOROETHYLENE (MAK) * RCRA WASTE NUMBER U079

CONSENSUS REPORTS: Reported in EPA TSCA Inventory.

DFG MAK: 200 ppm (790 mg/m³)

SAFETY PROFILE: Mildly toxic by inhalation and other routes. Human systemic effects by inhalation: sleep, hallucinations and distorted perceptions. Mutation data reported. Exposure to high vapor concentration can cause nausea, vomiting, weakness, tremor and cramps. Recovery is usually prompt following removal from exposure. Dermatitis may result from de-fatting action on skin. Dangerous fire hazard when exposed to heat, flame or oxidizers. Moderate explosion hazard in the form of vapor when exposed to flame. Violent reaction with difluoromethylene dihypofluorite. Forms shock-sensitive explosive mixtures with dinitrogen tetraoxide. Reaction with solid caustic alkalies or their concentrated solutions produces chloracetylene gas which ignites spontaneously in air. Reacts violently with N_2O_4; KOH; Na; NaOH. Moderate explosion hazard in the form of vapor when exposed to flame. Can react vigorously with oxidizing materials. To fight fire, use water spray, foam, CO_2, dry chemical. When heated to decomposition it emits toxic fumes of Cl^-.

ACK250   CAS: 79-27-6   **HR: 3**
ACETYLENE TETRABROMIDE
DOT: UN 2504
mf: $C_2H_2Br_4$ mw: 345.68

PROP: Colorless to yellow liquid. Bp: 151° @ 54 mm, fp: −1°, d: 2.9638 @ 20°/4°, autoign temp: 635°F.

SYNS: MUTHMANN'S LIQUID * TBE * 1,1,2,2-TETRABROMAETHAN (GERMAN) * TETRABROMOACETYLENE * 1,1,2,2-TETRABROMOETANO (ITALIAN) * S-TETRABROMOETHANE * 1,1,2,2-TETRABROMOETHANE * 1,1,2,2-TETRABROOMETHAAN (DUTCH)

CONSENSUS REPORTS: Reported in EPA TSCA Inventory. EPA Genetic Toxicology Program.

OSHA PEL: TWA 1 ppm
ACGIH TLV: TWA 1 ppm
DFG MAK: 1 ppm (14 mg/m³)
DOT Classification: ORM-A; Label: None.

SAFETY PROFILE: Poison by inhalation and ingestion. An eye and skin irritant and a narcotic. Questionable carcinogen with experimental neoplastigenic data. Mutation data reported. When heated it emits highly toxic fumes of carbonyl bromide and Br^-.

ACL750 CAS: 88-29-9 **HR: 3**
ACETYL ETHYL TETRAMETHYL TETRALIN
mf: $C_{18}H_{26}O$ mw: 258.44

PROP: White crystals.

SYNS: 7-ACETYL-1,1,4,4-TETRAMETHYL-1,2,3,4-TETRAHYDRONAPHTHALENE * ACETYLETHYL TETRAMETHYLTETRALIN * 6-ACETYL-1,1,4,4-TETRAMETHYL-7-ETHYL-1,2,3,4,-TETRALIN * AETT * ETHANONE-1-(3-ETHYL-5,6,7,8-TETRAHYDRO-5,5,8,8-TETRAMETHYL-2-NAPHTHALENYL)(9CI) * 3'-ETHYL-5',6',7',8'-TETRAHYDRO-5',5',8'-TETRAMETHYL-2'-ACETONAPHTHONE * 1-(3-ETHYL-5,6,7,8-TETRAHYDRO-5,5,8,8-TETRAMETHYL-2-NAPHTHALENYL)-ETHANONE * MUSK 36A * POLYCYCLIC MUSK * VERSALIDE

CONSENSUS REPORTS: Reported in EPA TSCA Inventory.

SAFETY PROFILE: Poison by ingestion and intraperitoneal routes. Moderately toxic by other routes, especially by skin contact and subcutaneous routes. A skin irritant. Exposure causes blue coloration of internal organs and central nervous system effects, i.e., hyperexcitability, tremors, lack of coordination, hunched back and loss of weight. It is slowly metabolized and excreted via feces. Symptoms persist for 90 days after exposure. Severity of symptoms seems proportional to length of exposure. It is freely absorbed via human skin. When heated to decomposition it emits acrid smoke and fumes.

ACM000 CAS: 557-99-3 **HR: 3**
ACETYL FLUORIDE
mf: C_2H_3FO mw: 62.05

PROP: D: 1.002 @ 15°/4°; mp: −60°, bp: 20.8°. Sltly sol in alc, ether, acetone, and benzene.

SYN: METHYLCARBONYL FLUORIDE

CONSENSUS REPORTS: Reported in EPA TSCA Inventory.

OSHA PEL: TWA 2.5 mg(F)/m^3
ACGIH TLV: TWA 2.5 mg(F)/m^3

SAFETY PROFILE: Poison by inhalation. When heated to decomposition it emits toxic fumes of F$^-$.

ACM750 CAS: 1068-57-1 **HR: 3**
ACETYL HYDRAZIDE
mf: $C_2H_6N_2O$ mw: 74.10

SYNS: ACETHYDRAZIDE * ACETOHYDRAZIDE * N-ACETYLHYDRAZINE * ENT 61,241 * ETHANEHYDRAZONIC ACID * MONOACETYLHYDRAZINE

CONSENSUS REPORTS: Reported in EPA TSCA Inventory.

SAFETY PROFILE: Poison by ingestion and intraperitoneal routes. Mutation data reported. Exposure can cause hemolysis and liver damage. When heated to decomposition it emits toxic fumes of NO$_x$.

ACO000 **HR: 3**
ACETYLIDES

SAFETY PROFILE: Severe explosion hazard when shocked or exposed to heat. Acetylides are very sensitive to shock, friction, and heat. They explode readily and are one of the few commercial explosives which contain no O_2 or N_2 and therefore produce no gas. The explosion simply results from the large amount of heat instantaneously produced. Acetylides are used for detonating compositions, or in combination with lead azide in detonating rivets where the acetylides reduce the flash point of the more insensitive azides. They are in a class with the fulminates and the azides as primary detonants. Because these materials are so sensitive to shock and temperature, they must be handled with extreme care. They must be kept cool, and should be kept wet if they are to be stored. Metal powders, such as finely divided Cu or Ag, should not be stored or kept with acetylene or acetylides since it is possible for them to react with these metal powders to form very sensitive acetylides which, while they are not dangerous in themselves, can cause enough of a flash to ignite a possibly explosive mixture of gases and thus cause an explosion in a warehouse or storage area. Examples of commercially used acetylides are silver acetylide and copper acetylide.

ACO500 CAS: 507-02-8 **HR: 3**
ACETYL IODIDE

DOT: UN 1898
mf: C_2H_3IO mw: 169.95

PROP: Brown, transparent, fuming liquid. Bp: 108°, d: 2.067 @ 20°/4°, decomp in water and alc; sol in ether.

CONSENSUS REPORTS: Reported in EPA TSCA Inventory.

DOT Classification: Corrosive Material; Label: Corrosive.

SAFETY PROFILE: A toxic, corrosive material. Reacts with water or steam to produce toxic and corrosive fumes. Dangerous to use. When heated to decomposition it emits toxic fumes of I^-.

ACP000   CAS: 39293-24-8   HR: 3
ACETYLKIDAMYCIN
mf: $C_{46}H_{58}N_2O_{13}$ mw: 847.06

SAFETY PROFILE: Poison by intravenous and intraperitoneal routes. Moderately toxic by ingestion. Human mutation data reported. When heated to decomposition it emits toxic fumes of NO_x.

ACP500   CAS: 63938-24-9   HR: 3
1-ACETYLLYSERGIC ACID DIETHYLAMIDE BITARTRATE
mf: $C_{22}H_{27}N_3O_2 \cdot 2C_4H_4O_6$ mw: 661.68

SYN: 1-ACETYL-9,10-DIDEHYDRO-N,N-DIETHYL-6-METHYLERGOLINE-8-β-CARBOXAMIDE BITARTRATE

SAFETY PROFILE: Deadly poison by intravenous route. Human systemic effects by ingestion of very small amounts: EEG changes, hallucinations, distorted perceptions and changes in psychophysiological test scores. When heated to decomposition it emits toxic fumes of NO_x.

ACP750   CAS: 50485-03-5   HR: 3
d-1-ACETYL LYSERGIC ACID MONOETHYLAMIDE
mf: $C_{20}H_{23}N_3O_2$ mw: 337.46

SYNS: 1-ACETYL-9,10-DIDEHYDRO-N-ETHYL-6-METHYLERGOLINE-8-β-CARBOXAMIDE * 1-ACETYLLYSERGIC ACID ETHYLAMIDE

SAFETY PROFILE: Poison by ingestion and intravenous routes. Ingesting very small amounts produce psychotropic effects in humans. When heated to decomposition it emits toxic fumes of NO_x.

ACR300   CAS: 83-63-6   HR: 3
N-ACETYL-N-(2-METHYL-4-((2-METHYLPHENYL)AZO)PHENYL)ACETAMIDE
mf: $C_{18}H_{19}N_3O_2$ mw: 309.40

SYNS: DERMAGAN * DERMAGEN * DIACETAZOTOL * DIACETOTOLUIDE * o-DIACETOTO-

LUIDIDE, 4''-(o-TOLYLAZO)-(8CI) * DIACETYLAMINOAZOTOLUENE * N,N-DIACETYL-o-TOLYLAZO-o-TOLUIDINE * DIAMAZO * DIMAZON * EPIDERMOL * EPITHELONE * GRANULIN * PELLIDOL * PELLIDOLE * PERIPHERMIN * 4-o-TOLYLAZO-o-DIACETOTOLUIDE * 4'-(o-TOLYLAZO)-o-DIACETOTOLUIDIDE

CONSENSUS REPORTS: IARC Cancer Review: GROUP 3 IMEMDT 7,56,87; Animal Inadequate Evidence IMEMDT 8,113,75

SAFETY PROFILE: Questionable carcinogen. When heated to decomposition it emits dangerous and toxic fumes of NO_x.

ACR400   CAS: 28895-91-2   HR: 3
ACETYLMETHYLNITROSOUREA
mf: $C_4H_7N_3O_3$ mw: 145.14

SYNS: ACETYL-METHYL-NITROSO-HARNSTOFF (GERMAN) * N'-ACETYL-METHYLNITROSOUREA * N-METHYL-N-NITROSO-N'-ACETYLUREA * 1-METHYL-1-NITROSOACETYLUREA

SAFETY PROFILE: Poison by ingestion. Questionable carcinogen with experimental tumorigenic data. Mutation data reported. When heated to decomposition it emits toxic fumes of NO_x.

ACS750   CAS: 591-09-3   HR: 3
ACETYL NITRATE
mf: $C_2H_3NO_4$ mw: 105.06

PROP: Colorless, fuming, mobile liquid. Bp: 22° @ 70 mm; d: 1.24 @ 15°/4°.

SYN: ACETIC ACID, ANHYDRIDE with NITRIC ACID (1:1)

SAFETY PROFILE: Corrosive to the eye. Violently unstable. Reacts explosively with ethyl-3,4-dihydroxybenzenesulfonate + oleum, HgO, and other active oxides. Solutions may explode violently above 60°C and the pure material explodes above 100°C. When heated to decomposition it emits toxic fumes of NO_x and/or explodes.

ACV500   CAS: 110-22-5   HR: 3
ACETYL PEROXIDE
mf: $C_4H_6O_4$ mw: 118.04

PROP: Solid or colorless crystals or liquid. Sltly sol in cold water, decomp. D: 1.18, mp: 30°, bp: 63° @ 21 mm.

SYN: DIACETYL PEROXIDE (MAK)

CONSENSUS REPORTS: Reported in EPA TSCA Inventory.

DFG MAK: Strong Skin Effects.
DOT Classification: Forbidden.

SAFETY PROFILE: Severe skin and eye irritant. Questionable carcinogen with experimental tumorigenic data. Dangerous fire hazard by spontaneous chemical reaction. A powerful oxidizing agent; can cause ignition of organic materials on contact. Severe explosion hazard when shocked or exposed to heat. It may explode spontaneously in storage and should be used as soon as prepared. It will react with water or steam to produce heat; can react vigorously with reducing materials; emits toxic fumes on contact with acid or acid fumes. To fight fire use CO_2, dry chemical.

Storage and Handling: Must be kept below 27° and not warmed over 30°. Do not add to hot materials. Do not add accelerator to this material. Store in original container with vented cap. Avoid bodily contact. This material is nearly always stored and handled as a 25% solution in an inert solvent.

ADA725 CAS: 50-78-2 **HR: 3**
ACETYLSALICYLIC ACID
mf: $C_9H_8O_4$ mw: 180.17

PROP: Colorless needles. Mp: 135°. Very sltly sol in alc, sol in benzene. Solubility in water = 1% @ 37°, in ether = 5% @ 20°.

SYNS: AC 5230 * ACENTERINE * ACESAL * ACETAL * ACETICYL * ACETILSALICILICO * ACETILUM ACIDULATUM * ACETISAL * ACETOL * ACETONYL * ACETOPHEN * ACETOSAL * ACETOSALIC ACID * ACETOSALIN * 2-ACETOXYBENZOIC ACID * o-ACETOXY-BENZOIC ACID * ACETYLIN * 2-(ACETYLOXY)BENZOIC ACID * ACETYLSAL * ACETYLSALICYL-SAURE (GERMAN) * ACIDE ACETYLSALICYLIQUE (FRENCH) * ACIDO o-ACETIL-BENZOICO (ITALIAN) * ACIDO ACETILSALICILICO (ITALIAN) * ACIDUM ACETYLSALICYLICUM * ACIMETTEN * ACISAL * ACYLPYRIN * ASA * A.S.A. * A.S.A. EMPIRIN * ASAGRAN * ASATARD * ASPALON * ASPERGUM * ASPIRDROPS * ASPIRIN * ASPIRINE * ASPRO * ASTERIC * BENASPIR * BIALPIRINIA * CAPRIN * o-CARBOXYPHENYL ACETATE * COLFARIT * CONTRHEUMA RETARD * CRYSTAR * DELGESIC * DOLEAN pH 8 * DURAMAX * ECM * ECOTRIN * EMPIRIN

* ENDYDOL * ENTERICIN * ENTEROPHEN * ENTEROSARINE * ENTROPHEN * EXTREN * GLOBOID * HELICON * IDRAGIN * MEASURIN * NEURONIKA * NOVID * POLOPIRYNA * RHEUMIN TABLETTEN * RHODINE * SALACETIN * SALCETOGEN * SALETIN * SOLPYRON * XAXA

CONSENSUS REPORTS: EPA Genetic Toxicology Program. Reported in EPA TSCA Inventory.

OSHA PEL: TWA 5 mg/m^3
ACGIH TLV: TWA 5 mg/m^3

SAFETY PROFILE: Poison by ingestion, intraperitoneal, and possibly other routes. Moderately toxic by several routes. Human systemic effects by ingestion: general anesthetic, nausea or vomiting, liver damage, hematuria and kidney damage, dehydration, metabolic effects, changes in the blood, respiratory stimulation, sputum, effects on the joints, tinnitus, bleeding from large intestine, constipation, and other gastrointestinal effects. Implicated in aplastic anemia. A 10 grams dose to an adult may be fatal. Human reproductive effects by ingestion and possibly other routes: menstrual cycle changes, parturition, various effects on newborn including apgar score, developmental abnormalities of the cardiovascular and respiratory systems. Experimental animal reproductive effects. Human mutation data reported. A human teratogen. An allergen; skin contact, inhalation, or ingestion can cause asthma, sneezing, irritation of eyes and nose, hives and eczema. Combustible when exposed to heat or flame. When heated to decomposition it emits acrid smoke and fumes.

ADC750 CAS: 584-26-9 **HR: 3**
1-ACETYL-2-THIOHYDANTOIN
mf: $C_5H_6N_2O_2S$ mw: 158.19

PROP: Insol in water and ether; sltly sol in alc. Mp: 175-176°.

SYNS: USAF B-7 * USAF BE-0405

CONSENSUS REPORTS: Reported in EPA TSCA Inventory.

SAFETY PROFILE: Poison by intravenous and intraperitoneal routes. When heated to decomposition it emits very toxic fumes such as SO_x and NO_x.

ADD750   CAS: 77-89-4   **HR: 2**
ACETYL TRIETHYL CITRATE
mf: $C_{14}H_{22}O_8$ mw: 318.36

SYNS: CITRIC ACID, ACETYL TRIETHYL ESTER
* TRICARBALLYLIC ACID-β-ACETOXYTRIBUTYL ES-
TER * TRIETHYL ACETYLCITRATE

CONSENSUS REPORTS: Reported in EPA
TSCA Inventory.

SAFETY PROFILE: Moderately toxic by intraperitoneal route. Mildly toxic by ingestion.
When heated to decomposition it emits acrid
smoke and fumes.

ADF250   CAS: 12788-93-1   **HR: 3**
ACID BUTYL PHOSPHATE

DOT: UN 1718
mf: $C_4H_{10}O_4P$ mw: 153.1

PROP: Water-white liquid; sol in alc, acetone
and toluene; insol in water, petroleum, and
naphtha. D: 1.120-1.125 @ 25°/40°, flash p:
230°F (COC).

SYNS: n-BUTYL ACID PHOSPHATE * BUTYL PHOS-
PHORIC ACID

DOT Classification: Corrosive Material; Label:
Corrosive.

SAFETY PROFILE: Toxic and corrosive. When
heated to decomposition it emits highly toxic
fumes of PO_x. Combustible when exposed to
heat or flame.

ADF500   **HR: 3**
ACID CARBOYS, EMPTY

SAFETY PROFILE: *Warning:* These containers
may contain concentrated vapors or even some
liquid acid remaining from their original contents. Therefore, they can give rise to all the
hazards of their original contents.

ADI250   **HR: 3**
ACONITUM CARMICHAELI

SAFETY PROFILE: Poison by intraperitoneal
and subcutaneous routes. Moderately toxic by
intravenous route. Mildly toxic by ingestion.
When heated to decomposition it emits acrid
smoke and fumes.

ADJ500   CAS: 260-94-6   **HR: 3**
ACRIDINE

DOT: UN 2713
mf: $C_{13}H_9N$ mw: 179.23

PROP: Small, colorless needles. Mp: 110.5°,
bp: 346°, d: 1.005 @ 19.7°/4°, vap press: 1
mm @ 129.4°. Sltly sol in hot water; sol in
alc, ether, and CS_2.

SYNS: 9-AZAANTHRACENE * 10-AZAANTHRACENE
* BENZO(b)QUINOLINE * 2,3-BENZOQUINOLINE
* DIBENZO(b,e)PYRIDINE

OSHA PEL: TWA 0.2 mg/m^3

CONSENSUS REPORTS: Reported in EPA
TSCA Inventory.

DOT Classification: IMO: Flammable Solid;
Label: Flammable Solid.

SAFETY PROFILE: Poison by subcutaneous
and intravenous routes. Mutation data reported.
A skin, eye, and mucous membrane irritant.
When heated to decomposition it emits toxic
fumes of NO_x.

ADR000   CAS: 107-02-8   **HR: 3**
ACROLEIN

DOT: UN 1092
mf: C_3H_4O mw: 56.07

PROP: Colorless or yellowish liquid; disagreeable, choking odor. Sol in water, alc, and ether.
Mp: −87.7°, bp: 52.5°, flash p: <0°F, d: 0.841
@ 20°/4°, autoign temp: unstable (455°F), lel:
2.8%, uel: 31%, vap d: 1.94.

SYNS: ACQUINITE * ACRALDEHYDE * ACRO-
LEINA (ITALIAN) * ACROLEINE (DUTCH, FRENCH)
* ACRYLALDEHYD (GERMAN) * ACRYLALDEHYDE
* ACRYLIC ALDEHYDE * AKROLEIN (CZECH)
* ALDEIDE ACRILICA (ITALIAN) * ALDEHYDE
ACRYLIQUE (FRENCH) * ALLYL ALDEHYDE
* AKROLEINA (POLISH) * AQUALINE * BIOCIDE
* CROLEAN * ETHYLENE ALDEHYDE * MAGNA-
CIDE H * NSC 8819 * PROPENAL (CZECH)
* 2-PROPENAL * PROP-2-EN-1-AL
* 2-PROPEN-1-ONE * PROPYLENE ALDEHYDE
* RCRA WASTE NUMBER P003 * SLIMICIDE

CONSENSUS REPORTS: IARC Cancer Review: GROUP 3 IMEMDT 7,78,87; Animal
Inadequate Evidence IMEMDT 36,133,85;
IMEMDT 19,479,79; Human Inadequate Evidence IMEMDT 36,133,85. Community Right-
To-Know List. EPA Extremely Hazardous Substances List. Reported in EPA TSCA Inventory.

OSHA PEL: (Transitional: TWA 0.1 ppm)
TWA 0.1 ppm; STEL 0.3 ppm

ACGIH TLV: TWA 0.1 ppm; STEL 0.3 ppm
DFG MAK: 0.1 ppm (0.25 mg/m^3)
DOT Classification: Flammable Liquid; Label: Flammable Liquid and Poison.

SAFETY PROFILE: Human poison by inhalation and intradermal route. Poison experimentally by most routes. Severe eye and skin irritant. Human systemic irritant and pulmonary system effects by inhalation include: lacrimation, delayed hypersensitivity with multiple organ involvement, and respiratory system damage. Human mutation data reported. Experimental reproductive effects. Questionable carcinogen. Dangerous fire hazard when exposed to heat, flame, or oxidizers. An explosion hazard. Incompatible with amines; SO$_2$; metal salts; oxidants; (light + heat). Violent polymerization reaction on contact with strong acid; strong base; weak acid conditions (e.g., nitrous fumes; sulfur dioxide; carbon dioxide); thiourea; or dimethylamine. When heated to decomposition it emits highly toxic fumes; can react vigorously with oxidizing materials. To fight fire, use CO$_2$, dry chemical or alcohol foam.

ADR500   CAS: 100-73-2   HR: 2
ACROLEIN DIMER
DOT: UN 2607
mf: C$_6$H$_8$O$_2$ mw: 112.14

PROP: Liquid, sol in water. D: 1.0775 (20°), bp: 151.3°, fp: −100°, flash p: 118°F (OC).

SYNS: 3,4-DIHYDRO-2H-PYRAN-2-CARBOXALDEHYDE * 2-FORMYL-3,4-DIHYDRO-2H-PYRAN * PYRAN ALDEHYDE

DOT Classification: DOT-IMO: Flammable or Combustible Liquid; Label: Flammable Liquid.

SAFETY PROFILE: Mildly toxic by ingestion. A skin and severe eye irritant. Flammable when exposed to heat, flame or powerful oxidizing agents. To fight fire, use alcohol foam and multipurpose dry chemical. When heated to decomposition it emits acrid smoke and fumes.

ADS250   CAS: 79-06-1   HR: 3
ACRYLAMIDE
DOT: UN 2074
mf: C$_3$H$_5$NO   mw: 71.09

PROP: White, crystalline solid. Very sol in water, alc, and ether. Mp: 84.5 ± 0.3°, bp:

125° @ 25 mm, d: 1.122 @ 30°, vap press: 1.6 mm @ 84.5°, vap d: 2.45.

SYNS: ACRYLIC AMIDE * AKRYLAMID (CZECH) * ETHYLENECARBOXAMIDE * PROPENAMIDE * 2-PROPENAMIDE * RCRA WASTE NUMBER U007

CONSENSUS REPORTS: IARC Cancer Review: GROUP 2B IMEMDT 7,56,87; Animal Sufficient Evidence IMEMDT 39,41,86. EPA Extremely Hazardous Substances List. Community Right-To-Know List. Reported in EPA TSCA Inventory.

OSHA PEL: (Transitional: TWA 0.3 mg/m^3 (skin)) TWA 0.03 mg/m^3 (skin)
ACGIH TLV: Suspected Human Carcinogen, TWA 0.03 mg/m^3 (skin)
DFG MAK: Animal Carcinogen, Suspected Human Carcinogen.
NIOSH REL: TWA 0.3 mg/m^3
DOT Classification: IMO: Poison B; Label: St. Andrews Cross.

SAFETY PROFILE: Confirmed carcinogen with experimental carcinogenic and neoplastigenic data. Poison by ingestion, skin contact, intravenous, intraperitoneal, and possibly other routes. Experimental reproductive effects. Mutation data reported. A skin and eye irritant. Intoxication from it has caused a peripheral neuropathy, erythema and peeling of the palms. In industry, intoxication is mainly via dermal route, next via inhalation and last via ingestion. Time of onset varied from 1-24 months to 8 years. Symptoms were, via dermal route, a numbness, tingling and touch tenderness. In a couple of weeks, coldness of extremities; later, excessive sweating, bluish-red and peeling of palms, marked fatigue and limb-weakness. It is dangerous because it can be absorbed through the unbroken skin. From animal experiments it seems to be a central nervous system toxin. Adult rats fed an average of 30 mg/kg for 14 days were all partially paralyzed and had reduced their food consumption by 50 percent. Polymerizes violently at its melting point. When heated to decomposition it emits acrid fumes and NO$_x$.

ADS750   CAS: 79-10-7   HR: 3
ACRYLIC ACID
DOT: UN 2218
mf: C$_3$H$_4$O$_2$ mw: 72.07

PROP: Liquid, acrid odor. Misc in water, benzene, alc, chloroform, ether, and acetone. Mp:

13°, bp: 141°, d: 1.062, vap press: 10 mm @ 39.9°, flash p: 130°F (OC), vap d: 2.45.

SYNS: ACROLEIC ACID * ACRYLIC ACID (ACGIH,DOT,OSHA) * ACRYLIC ACID, inhibited (DOT) * ACRYLIC ACID, GLACIAL * ETHYLENECARBOX-YLIC ACID * GLACIAL ACRYLIC ACID * KYSE-LINA AKRYLOVA * PROPENE ACID * PROPENOIC ACID * 2-PROPENOIC ACID (9CI) * RCRA WASTE NUMBER U008 * VINYLFORMIC ACID

CONSENSUS REPORTS: IARC Cancer Review: Human Inadequate Evidence IMEMDT 19,47,79. Community Right-To-Know List. Reported in EPA TSCA Inventory.

OSHA PEL: TWA 10 ppm (skin)
ACGIH TLV: TWA 10 ppm (Proposed: 2 ppm (skin))
DOT Classification: IMO: Corrosive Material; Label: Corrosive, Flammable Liquid.

SAFETY PROFILE: Poison by ingestion, skin contact, intraperitoneal, and possibly other routes. Experimental teratogenic and reproductive effects. A severe skin and eye irritant. Questionable carcinogen with experimental carcinogenic and tumorigenic data. Corrosive. Exothermic polymerization at room temperature may become explosive if confined. A fire hazard when exposed to heat or flame.

ADX500   CAS: 107-13-1   **HR: 3**
ACRYLONITRILE
DOT: UN 1093
mf: C_3H_3N mw: 53.07

PROP: Colorless, mobile liquid; mild odor. Sol in water. Mp: −82°, bp: 77.3°, fp: −83°, flash p: 30°F (TCC), lel: 3.1%, uel: 17%, d: 0.806 @ 20°/4°, autoign temp: 898°F, vap press: 100 mm @ 22.8°, vap d: 1.83, flash p: (of 5% aq sol): <50°F.

SYNS: ACRYLNITRIL (GERMAN, DUTCH) * ACRYL-ONITRILE MONOMER * AKRYLONITRYL (POLISH) * CARBACRYL * CIANURO DI VINILE (ITALIAN) * CYANOETHYLENE * CYANURE de VINYLE (FRENCH) * ENT 54 * FUMIGRAIN * NITRILE ACRILICO (ITALIAN) * NITRILE ACRYLIQUE (FRENCH) * PROPENENITRILE * 2-PROPENENITRILE * RCRA WASTE NUMBER U009 * TL 314 * VENTOX * VINYL CYANIDE

CONSENSUS REPORTS: IARC Cancer Review: Human Limited Evidence IMEMDT 19,73,79; Animal Limited Evidence IMEMDT 19,73,79. NTP Fourth Annual Report On Carcinogens, 1984. Community Right-To-Know List. EPA Extremely Hazardous Substances List. Reported in EPA TSCA Inventory.

OSHA PEL: TWA 2 ppm; CL 10 ppm/15M; Cancer Hazard.
ACGIH TLV: Suspected Human Carcinogen, TWA 2 ppm (skin).
DFG TRK: 3 ppm (7 mg/m^3), Animal Carcinogen, Suspected Human Carcinogen.
NIOSH REL: TWA 1 ppm; CL 10 ppm/15M
DOT Classification: Flammable Liquid and Poison.

SAFETY PROFILE: Confirmed human carcinogen with experimental carcinogenic, neoplastigenic, and tumorigenic data. Poison by inhalation, ingestion, skin contact, and other routes. Human systemic irritant, somnolence, general anesthesia, cyanosis and diarrhea by inhalation and skin contact. Human mutation data reported. Experimental teratogenic and reproductive effects. Dangerous fire hazard when exposed to heat, flame, or oxidizers. Moderate explosion hazard when exposed to flame. Can react vigorously with oxidizing materials (see also CYANIDES).

Acrylonitrile closely resembles hydrocyanic acid in its toxic action. By inhibiting the respiratory enzymes of tissue, it renders the tissue cells incapable of oxygen absorption. Poisoning is acute; there is little evidence of cumulative action on repeated exposure. Exposure to low concentration is followed by flushing of the face and increased salivation; further exposure results in irritation of the eyes and nose, photophobia, deepened respiration, and, if exposure continues, shallow respiration, nausea, vomiting, weakness, an oppressive feeling in the chest, and occasionally headache and diarrhea are other complaints. Several cases of mild jaundice accompanied by mild anemia and leucocytosis have been reported. Urinalysis is generally negative, except for an increase in bile pigment. Serum and bile thiocyanates are raised. Unstable and easily oxidized. Explosive polymerization may occur on storage with silver nitrate. Potentially explosive reactions with benzyltrimethylammonium hydroxide + pyrrole; tetrahydrocarbazole + benzyltrimethylammonium hydroxide. Violent reactions with strong acids (e.g., nitric or sulfuric); strong bases; azoisobutyronitrile; dibenzoyl peroxide; di-tert-butylperoxide; or bromine. Incompatible with AgNO$_3$ and amines. To fight fire use CO$_2$, dry chemical or alcohol foam. When heated to decomposition it emits toxic fumes of NO$_x$ and CN$^-$.

ADY500   CAS: 9003-54-7   **HR: 3**
ACRYLONITRILE POLYMER with STYRENE
mf: (C$_8$H$_8$ • C$_3$H$_3$N)$_x$

SYNS: ACRILAFIL * ACRYLONITRILE-STYRENE COPOLYMER * ACRYLONITRILE-STYRENE POLYMER * ACRYLONITRILE-STYRENE RESIN * ACS * AS 61CL * BAKELITE RMD 4511 * CEVIAN HL * DIALUX * ESTYRENE AS * KOSTIL * LITAC * LURAN * LUSTRAN * POLYSTYRENE-ACRYLONITRILE * 2-PROPENENITRILE POLYMER with ETHENYLBENZENE * REXENE 106 * SANREX * SN 20 * STYRENE-ACRYLONITRILE COPOLYMER * STYREN-ACRYLONITRILE-POLYMER * TERULAN KP 2540 * TYRIL

CONSENSUS REPORTS: Reported in EPA TSCA Inventory. Cyanide and its compounds are on the Community Right-To-Know List.

SAFETY PROFILE: Moderately to highly toxic by ingestion. When heated to decomposition it emits toxic fumes of NO$_x$ and CN$^-$.

AEA000   **HR: 3**
ACTINIC RADIATION

SAFETY PROFILE: Outdoor workers, such as fishermen, sailors, soldiers and farmers, show a high incidence of skin cancer. The commonest acute manifestation of actinic radiation effects on skin is sunburn.

AEB750   CAS: 102488-99-3   **HR: 3**
ACTINOMYCIN L

SYN: ACTINOMYCIN 2104L

CONSENSUS REPORTS: IARC Cancer Review: Animal Sufficient Evidence IMEMDT 10,29,76

SAFETY PROFILE: Confirmed carcinogen with experimental neoplastigenic data.

AEC000   CAS: 12623-78-8   **HR: 3**
ACTINOMYCIN S

SYN: ACTINOMYCIN 1048A

CONSENSUS REPORTS: IARC Cancer Review: Animal Sufficient Evidence IMEMDT 10,29,76

SAFETY PROFILE: Confirmed carcinogen with experimental neoplastigenic data.

AEN000   CAS: 628-94-4   **HR: 3**
ADIPAMIDE
mf: C$_6$H$_{12}$N$_2$O$_2$ mw: 144.20

PROP: Crystals. Mp: 220°. Sol in alc.

SYNS: ADIPIC ACID DIAMIDE * ADIPIC DIAMIDE * 1,4-BUTANEDICARBOXAMIDE * HEXANEDIAMIDE (9CI) * NCI-C02095

CONSENSUS REPORTS: Reported in EPA TSCA Inventory.

SAFETY PROFILE: Moderately toxic by ingestion. Questionable carcinogen with experimental carcinogenic data. When heated to decomposition it emits toxic fumes of NO$_x$.

AEN250   CAS: 124-04-9   **HR: 3**
ADIPIC ACID

DOT: NA 9077
mf: C$_6$H$_{10}$O$_4$ mw: 146.16

PROP: White monoclinic prisms. Mp: 152°, flash p: 385°F (CC), d: 1.360 @ 25°/4°, vap press: 1 mm @ 159.5°, vap d: 5.04, autoign temp: 788°F, bp: 337.5°. Very sol in alc. Sol in acetone, water = 1.4% @15°; 0.6% @ 15° in ether.

SYNS: ACIFLOCTIN * ACINETTEN * ADILACTETTEN * ADIPINIC ACID * 1,4-BUTANEDICARBOXYLIC ACID * FEMA No. 2011 * 1,6-HEXANEDIOIC ACID * KYSELINA ADIPOVA (CZECH) * MOLTEN ADIPIC ACID

CONSENSUS REPORTS: Reported in EPA TSCA Inventory.

DOT Classification: ORM-E; Label: None.

SAFETY PROFILE: Poison by intraperitoneal route. Moderately toxic by other routes. A severe eye irritant. Combustible when exposed to heat or flame; can react with oxidizing materials. When heated to decomposition it emits acrid smoke and fumes.

AER250   CAS: 111-69-3   **HR: 3**
ADIPONITRILE

DOT: UN 2205
mf: C$_6$H$_8$N$_2$ mw: 108.16

PROP: Water-white liquid, practically odorless. Mp: 2.3°, bp: 295°, flash p: 199.4°F (OC), d: 0.965 @ 20°/4°, vap d: 3.73.

SYNS: ADIPIC ACID DINITRILE * ADIPIC ACID NITRILE * ADIPODINITRILE * 1,4-DICYANOBUTANE * HEXANEDINITRILE * HEXANEDIOIC ACID DINITRILE * NITRILE ADIPICO (ITALIAN) * TETRAMETHYLENE CYANIDE

CONSENSUS REPORTS: EPA Extremely Hazardous Substances List. Reported in EPA TSCA Inventory. Cyanide and its compounds are on the Community Right-To-Know List.

NIOSH REL: TWA 18 mg/m^3
DOT Classification: IMO: Poison B; Label: St. Andrews Cross.

SAFETY PROFILE: Poison by inhalation, ingestion, subcutaneous, and intraperitoneal routes. It is toxic since the nitrile group will behave as a cyanide when ingested or absorbed in the body. It produces disturbances of the respiration and circulation, irritation of the stomach and intestines, and loss of weight. Its low vapor pressure at room temperature makes exposure to harmful concentrations of its vapors unlikely if handled with reasonable care in well ventilated areas. Flammable when exposed to heat or flame. When heated to decomposition it emits toxic fumes of CN$^-$. Can react with oxidizing materials. To fight fire, use foam, CO$_2$, dry chemical.

AES250   CAS: 150-05-0   **HR: 3**
d-ADRENALINE
mf: $C_9H_{13}NO_3$ mw: 183.23

PROP: Light brown or nearly white crystals. Mp: 211-212°. Very sltly sol in water, alc, 1:1 chloroform, and ether.

SYNS: l-(+)-ADRENALINE * d-EPINEPHRINE

SAFETY PROFILE: Poison by subcutaneous and intravenous routes. Can cause contact dermatitis. Usually the symptoms are of short duration and clear up spontaneously. Combustible when heated. Upon decomposition it emits toxic fumes of NO$_x$.

AES750   CAS: 23214-92-8   **HR: 3**
ADRIAMYCIN
mf: $C_{27}H_{29}NO_{11} \cdot ClH$ mw: 543.57

PROP: Isolated from cultures of *Streptomyces peucetius var. Caesius.*

SYNS: ADM * ADRIAMYCIN-HCl * ADRIAMYCIN SEMIQUINONE * ADRIBLASTINA * DOXORUBICIN * DX * F.I 106 * 14-HYDROXYDAUNO-MYCIN * 14'-HYDROXYDAUNOMYCIN * 14-HYDROXYDAUNORUBICINE * KW-125 * NCI-C01514 * NSC-123127

CONSENSUS REPORTS: IARC Cancer Review: Animal Inadequate Evidence IMEMDT 10,43,76. NTP Fourth Annual Report On Carcinogens, 1984.

SAFETY PROFILE: Confirmed carcinogen with experimental carcinogenic, neoplastigenic, and tumorigenic data. Poison by intraperitoneal, subcutaneous, parenteral, and intravenous routes. Experimental teratogenic and reproductive effects. Human mutation data reported. Human systemic effects by intravenous route: cardiac myopathy including infarction, nausea or vomiting, and effects on the hair. When heated to decomposition it emits very toxic fumes of NO$_x$ and HCl.

AET750   CAS: 1402-68-2   **HR: 3**
AFLATOXIN

CONSENSUS REPORTS: IARC Cancer Review: GROUP 1 IMEMDT 7,83,87; Human Limited Evidence IMEMDT 10,51,76. NTP Fourth Annual Report On Carcinogens, 1984.

SAFETY PROFILE: Confirmed human carcinogen with experimental tumorigenic data. Human poison by ingestion. Moderately toxic by other routes. Experimental teratogenic and reproductive effects. Mutation data reported.

AEU250   CAS: 1162-65-8   **HR: 3**
AFLATOXIN B1
mf: $C_{17}H_{12}O_6$ mw: 312.29

PROP: A metabolite of *Aspergillus flavus link ex fries* (12VXA5 8,24,68). A crystalline material. Mp: 268°.

SYNS: AFBI * AFLATOXIN B

CONSENSUS REPORTS: IARC Cancer Review: Animal Sufficient Evidence IMEMDT 10,51,76; 1,145,72. NTP Fourth Annual Report On Carcinogens, 1984. EPA Genetic Toxicology Program.

SAFETY PROFILE: Confirmed human carcinogen with experimental tumorigenic, neoplastigenic, and carcinogenic data. Acute poison by ingestion, intraperitoneal, and possibly other routes. Experimental teratogenic and reproductive effects. Mutation data reported. When heated to decomposition it emits acrid smoke.

AEU750   CAS: 7220-81-7   **HR: 3**
AFLATOXIN B2
mf: $C_{17}H_{14}O_6$ mw: 314.31

SYN: DIHYDROAFLATOXIN B1

CONSENSUS REPORTS: IARC Cancer Review: Animal Sufficient Evidence IMEMDT 10,51,76; Animal Limited Evidence IMEMDT 1,145,72. NTP Fourth Annual Report On Carcinogens, 1984.

SAFETY PROFILE: Confirmed human carcinogen with experimental tumorigenic data. Poison by ingestion. Mutation data reported. When heated to decomposition it emits acrid smoke and fumes.

AEV000   CAS: 1165-39-5   **HR: 3**
AFLATOXIN G1
mf: $C_{17}H_{12}O_7$ mw: 328.29

PROP: A metabolite of *Aspergillus flavus link ex fries*.

CONSENSUS REPORTS: IARC Cancer Review: Animal Sufficient Evidence IMEMDT 10,51,76. NTP Fourth Annual Report On Carcinogens, 1984.

SAFETY PROFILE: Confirmed human carcinogen with experimental carcinogenic and neoplastigenic data. Poison by ingestion and intraperitoneal routes. A suspected human carcinogenic. Mutation data reported. When heated to decomposition it emits acrid smoke and irritating fumes.

AEV250   **HR: 3**
AFLATOXIN G1 mixed with AFLATOXIN B1
CONSENSUS REPORTS: NTP Fourth Annual Report On Carcinogens, 1984.

SAFETY PROFILE: Confirmed human carcinogen with experimental carcinogenic, neoplastigenic, and tumorigenic data.

AEV500   CAS: 7241-98-7   **HR: 3**
AFLATOXIN G2
mf: $C_{17}H_{14}O_7$ mw: 330.31

CONSENSUS REPORTS: IARC Cancer Review: Animal Inadequate Evidence IMEMDT 1,145,72. NTP Fourth Annual Report On Carcinogens, 1984. EPA Genetic Toxicology Program.

SAFETY PROFILE: Confirmed carcinogen. Acute poison by ingestion. Mutation data reported. When heated to decomposition it emits acrid smoke and irritating fumes.

AEW000   CAS: 6795-23-9   **HR: 3**
AFLATOXIN M1
mf: $C_{17}H_{12}O_7$ mw: 328.29

SYN: 4-HYDROXYAFLATOXIN B1

CONSENSUS REPORTS: IARC Cancer Review: Animal Sufficient Evidence IMEMDT 10,51,76

CONSENSUS REPORTS: EPA Genetic Toxicology Program.

SAFETY PROFILE: Confirmed carcinogen with experimental tumorigenic data. Poison by ingestion. Mutation data reported. When heated to decomposition it emits acrid smoke and irritating fumes.

AEW500   CAS: 29611-03-8   **HR: 3**
AFLATOXIN Ro
mf: $C_{17}H_{14}O_6$ mw: 314.31

SYN: AFLATOXICOL

SAFETY PROFILE: Suspected carcinogen with experimental carcinogenic data. Mutation data reported. When heated to decomposition it emits acrid smoke and irritating fumes.

AEX250   CAS: 9002-18-0   **HR: 1**
AGAR

PROP: Extracted from the red algae *Rhodopyceae*. Unground: in thin, translucent, membranous pieces; ground: pale buff powder. Sol in boiling water; insol in cold water and organic solvents.

SYNS: AGAR-AGAR * AGAR AGAR FLAKE
* AGAR-AGAR GUM * BENGAL GELATIN
* BENGAL ISINGLASS * CEYLON ISINGLASS
* CHINESE ISINGLASS * DIGENEA SIMPLEX MUCILAGE * GELOSE * JAPAN AGAR * JAPAN ISINGLASS * LAYOR CARANG * NCI-C50475

CONSENSUS REPORTS: NTP Carcinogenesis Bioassay (feed); No Evidence: mouse, rat NTPTR* NTP-TR-230,82. Reported in EPA TSCA Inventory.

SAFETY PROFILE: Mildly toxic by ingestion. When heated to decomposition it emits acrid smoke and fumes.

AFG250   HR: 2
AIR, compressed

PROP: Bluish, mobile liquid. O_2 + N_2. Bp: $-189°$ (liq); flash p: none; autoign temp: none.

DOT Classification: Nonflammable Gas; Label: Nonflammable Gas.

SAFETY PROFILE: Liquid air can cause tissue damage due to low temperature. Personnel exposed to compressed air may develop caisson disease (the bends, the chokes) if decompression is too rapid. Moderate explosion hazard when containers under pressure are shocked or exposed to heat or flame or flammable materials; i.e., ethyl ether; hydrocarbons; or charcoal, which have been in contact with liquid air may explode very easily. Ordinary oxidation is greatly accelerated in compressed air. Moderately dangerous disaster hazard; can react vigorously with reducing materials.

AFI850   CAS: 70536-17-3   HR: 3
ALBUMIN MACRO AGGREGATES

SYNS: ALBUMIN * MAA

SAFETY PROFILE: Poison by intravenous route. When heated to decomposition it emits acrid smoke and irritating fumes.

AFJ000   HR: 3
ALCOHOL, DENATURED

PROP: Liquid. Composed of alcohol and denaturants.

SYN: DENATURED SPIRITS

SAFETY PROFILE: Potentially poisonous by ingestion. Toxicity depends upon alcohols in question, generally ethanol with methanol as a denaturant. Dangerous fire hazard; can react vigorously with oxidizing materials. Moderate explosion hazard.

AFJ250   HR: 3
ALCOHOLS, N.O.S.

CONSENSUS REPORTS: A generic term applied to a series of compounds, the simplest of which has the general formula $C_nH_{2n+1}OH$. (See also specific compound.)

SAFETY PROFILE: No general statement can be made due to wide variations in toxic effects. Dangerous fire hazard when exposed to heat or flame. Can react violently in contact with $(H_2O + H_2SO_4)$; HOCl; Cl_2; isocyanates; $LiAl_4$;

N_2O_4; $HClO_4$; H_2SO_5 (Caro's acid); $Ba(ClO_4)_2$, $(CH_2)_2O$; acetaldehyde; diethyl aluminum bromide; hexamethylene diisocyanate; triisobutyl aluminum.

AFJ800   HR: 2
ALDEHYDES

PROP: A class of chemicals with the general formula R • CHO, and characterized by an unsaturated carbonyl group (C=O).

SAFETY PROFILE: Aldehydes are widely used in many industrial processes. The US production of acetaldehyde in 1982 was 281,000 tons. The world production of acrolein in 1975 was 59,000 tons. They occur in nature and are gaseous byproducts of incomplete combustion of wood and coal, in exhaust from gasoline and diesel engines, industrial waste gases and fumes, tobacco smoke and wood fires. Formaldehyde and acetaldehyde are carcinogens. Many of the aldehydes are mutagens. They are reactive compounds participating in oxidation, reduction, addition and polymerization reactions. All the aldehydes possess anesthetic properties, but this is obscured by their highly irritating action on the eyes and mucous membranes of the respiratory tract. The lower aldehydes, very soluble in water, act chiefly on the eyes and tissues of the upper respiratory tract. The higher aldehydes, less soluble in water, tend to penetrate more deeply into the respiratory system and may affect the lungs. Some higher aldehydes and also the aromatic aldehydes may exhibit much lower toxicity.

AFK250   CAS: 309-00-2   HR: 3
ALDRIN

DOT: UN 2761/NA 2762
mf: $C_{12}H_8Cl_6$ mw: 364.90

PROP: Crystals. Mp: 104-105°. Insol in water; sol in aromatics, esters, ketones, paraffins, and halogenated solvents.

SYNS: ALDREX * ALDREX 30 * ALDRIN, cast solid (DOT) * ALDRINE (FRENCH) * ALDRITE * ALDROSOL * ALTOX * COMPOUND 118 * DRINOX * ENT 15,949 * HEXACHLOROHEXA-HYDRO-endo-exo-DIMETHANONAPHTHALENE * 1,2,3,4,10,10-HEXACHLORO-1,4,4a,5,8,8a-HEXAHYDRO-1,-4,5,8-DIMETHANONAPHTHALENE * 1,2,3,4,10,10-HEXACHLORO-1,4,4a,5,8,8a-HEXAHYDRO-exo-1,4,-endo-5,8-DIMETHANONAPHTHALENE * 1,2,3,4,10,10-HEXA-

CHLORO-1,4,4a,5,8,8a-HEXAHYDRO-1,4-endo-exo-5, 8-DI-METHANONAPHTHALENE * HHDN * NCI-C00044 * OCTALENE * RCRA WASTE NUMBER P004 * SEEDRIN

CONSENSUS REPORTS: IARC Cancer Review: Human Inadequate Evidence IMEMDT 5,25,74; Animal Inadequate Evidence IMEMDT 5,25,74; NCI Carcinogenesis Bioassay (feed); Clear Evidence: mouse NCITR* NCI-CG-TR-21,78; Inadequate Studies: rat NCITR* NCI-CG-TR-21,78. EPA Genetic Toxicology Program. EPA Extremely Hazardous Substances List. Community Right-To-Know List.

OSHA PEL: TWA 0.25 mg/m^3 (skin)
ACGIH TLV: TWA 0.25 mg/m^3/
DFG MAK: 0.25 mg/m^3
NIOSH REL: (Aldrin) Reduce to lowest detectable level.
DOT Classification: Poison B; ORM-A.

SAFETY PROFILE: Poison by ingestion, skin contact, intravenous, intraperitoneal and other routes. Human systemic effects by ingestion: excitement, tremors and nausea or vomiting. Experimental teratogenic and reproductive effects. Continued acute exposure causes liver damage. Human mutation data reported. Questionable carcinogen with experimental carcinogenic, neoplastigenic, and tumorigenic data. When heated to decomposition it emits toxic fumes of Cl$^-$.

AFK750 HR: 1
ALFALFA MEAL

SAFETY PROFILE: An allergen. Skin contact may cause dermatitis. Flammable when exposed to heat or flame; by spontaneous chemical reaction. Avoid moisture content extremes. Fires may smolder for 72 hours before becoming noticeable.

AFL000 CAS: 9005-32-7 HR: 2
ALGINIC ACID

PROP: Extracted from brown seaweeds. White to yellow white fibrous powder; odorless and tasteless. Sol in alkaline solutions; insol in organic solvents.

SYNS: KELACID * LANDALGINE * NORGINE * PLOYMANNURONIC ACID * SAZZIO

CONSENSUS REPORTS: Reported in EPA TSCA Inventory.

SAFETY PROFILE: Moderately toxic by intraperitoneal route. When heated to decomposition it emits acrid smoke and irritating fumes.

AFM250 HR: 3
ALIPHATIC and AROMATIC EPOXIDES

SAFETY PROFILE: Suspected carcinogen with experimental tumors of the skin, lung, and blood-forming tissues.

AFM500 HR: D
ALKALIES

PROP: A term loosely applied to the hydroxides and carbonates of the alkali metals and alkaline earth metals, as well as the bicarbonate and hydroxide of ammonium. They can neutralize acids, change the color of indicators, and impart a soapy taste and feel to aq solns.

SAFETY PROFILE: Variable toxicity. As a group, they constitute the commonest causes of contact dermatitis. Systemically ammonia is most troublesome.

AFM750 HR: 3
ALKALOID SALTS

SYN: ALKALOIDS

SAFETY PROFILE: Nearly all alkaloid salts are poisonous. Some are also allergens. Dangerous; when heated to decomposition they emits highly toxic fumes.

AFN250 HR: D
ALKANES

PROP: All colorless neutral liquids with light aromatic odors. (n-pentane, n-hexane, n-heptane, n-octane)

SAFETY PROFILE: Hexane can cause neuropathy with chronic exposure. Other alkanes or mixtures may have the same effect. Many are dangerous fire hazards when exposed to flame, heat or oxidizers.

AFP750 CAS: 8023-53-8 HR: 3
ALKYL(C$_8$C$_{18}$)DIMETHYL-3,4-DICHLOROBENZYLAMMONIUM CHLORIDE

SYNS: ALKYL(C$_8$H$_{17}$ to C$_{18}$H$_{37}$) DIMETHYL-3,4-DICHLOROBENZYL AMMONIUM CHLORIDE * DICHLOROBENZALKONIUM CHLORIDE * TETROSAN

SAFETY PROFILE: A deadly poison by ingestion. Poison by intravenous route. Moderately toxic ingestion. A severe eye irritant. Can cause liver and kidney damage. A moderate allergen. Mutation data reported. When heated to decomposition it emits very toxic fumes of NO_x, NH_3, and Cl^-.

AFR250   CAS: 584-79-2   **HR: 3**
ALLETHRIN

DOT: UN 2902
mf: $C_{19}H_{26}O_3$ mw: 302.45

PROP: A viscous liquid.

SYNS: (+)-ALLELRETHONYL (+)-cis,trans-CHRY-SANTHEMATE * d-ALLETHRIN * ALLETHRIN I * ALLYL CINERIN * ALLYL HOMOLOG of CINERIN I * d,l-2-ALLYL-4-HYDROXY-3-METHYL-2-CYCLOPEN-TEN-1-ONE-d,l-CHRYSANTHEMUMMONOCARBOXYLATE * 3-ALLYL-4-KETO-2-METHYLCYCLOPENTENYL CHRYSANTHEMUMMONOCARBOXYLATE * 3-ALLYL-2-METHYL-4-OXO-2-CYCLOPENTEN-1-YL CHRYSANTHE-MATE * dl-3-ALLYL-2-METHYL-4-OXOCYCLOPENT-2-ENYL-dl-cis trans CHRYSANTHEMATE * ALLYLRE-THRONYL dl-cis-trans-CHRYSANTHEMATE * BIOALLE-THRIN * CINERIN I ALLYL HOMOLOG * ENT 17,510 * EXTHRIN * FDA 1446 * FMC 249 * NECARBOXYLIC ACID * NIA 249 * PALLE-THRINE * PYNAMIN * PYNAMIN-FORTE * PYRESIN * PYRESYN * SYNTHETIC PY-RETHRINS

CONSENSUS REPORTS: Reported in EPA TSCA Inventory. EPA Genetic Toxicology Program.

DOT Classification: ORM-A; Label: None.

SAFETY PROFILE: Poison by ingestion, intravenous, intracerebral, and intraperitoneal routes. Moderately toxic by ingestion. An allergen. An insecticide. It can cause liver and kidney damage by all routes of entry into the body. Lung congestion may occur due to exposure. Local contact may cause contact dermatitis. Inhalation may cause asthma, coughing, wheezing, running nose and eyes. Mutation data reported. Slight fire hazard. When heated to decomposition it emits acrid fumes.

AFR500   CAS: 34624-48-1   **HR: 3**
(+)-cis-ALLETHRIN
mf: $C_{19}H_{26}O_3$ mw: 302.45

SYN: (+)-(Z)-2,2-DIMETHYL-3-(2-METHYLPROPENYL)-CYCLOPROPANECARBOXYLIC ACID ESTER with 2-AL-LYL-4-HYDROXY-3-METHYL-2-CYCLOPENTEN-ONE

SAFETY PROFILE: Poison by ingestion and inhalation. When heated to decomposition it emits acrid smoke and irritating fumes.

AFS000   **HR: 3**
ALLETHRIN RACEMIC MIXTURE

SYN: 4-HYDROXY-3-METHYL-2-CYCLOPENTEN-1-ONE, cis- mixed with trans-2,2-DIMETHYL-3-(2-METHYL-PROPENYL)CYCLOPROPANECARBOXYLIC ACID ESTER with 2-ALLYL-4-HYDROXY-3-METHYL-2-CYCLOPEN-TEN-1-ONE (1:4)

SAFETY PROFILE: Poison by inhalation. Moderately toxic by ingestion. When heated to decomposition it emits toxic fumes of Cl^-.

AFU750   CAS: 591-87-7   **HR: 3**
ALLYL ACETATE

DOT: UN 2333
mf: $C_5H_8O_2$ mw: 100.13

PROP: Liquid, vap d: 3.45, bp: 104°, d: 0.928. Flash p: 72°F. Insol in water.

SYNS: ACETIC ACID ALLYL ESTER * ACETIC ACID-2-PROPENYL ESTER * 3-ACETOXYPROPENE

CONSENSUS REPORTS: Reported in EPA TSCA Inventory.

DOT Classification: IMO: Label: Flammable Liquid and Poison.

SAFETY PROFILE: Poison by ingestion. Moderately toxic by inhalation and skin contact. A skin and eye irritant. When heated to decomposition it emits acrid smoke and irritating fumes. Dangerous fire hazard.

AFV500   CAS: 107-18-6   **HR: 3**
ALLYL ALCOHOL

DOT: UN 1098
mf: C_3H_6O mw: 58.09

PROP: Limpid liquid; pungent odor. Mp: −129°, bp: 96°-97°, lel: 2.5%, uel: 18%, flash p: 70°F (CC), d: 0.854 @ 20°/4°, autoign temp: 713°F, vap press: 10 mm @ 10.5°, vap d: 2.00. Misc in water, alc, and ether.

SYNS: ALCOOL ALLILCO (ITALIAN) * ALCOOL AL-LYLIQUE (FRENCH) * ALLILOWY ALKOHOL (POLISH)

* ALLYL AL * ALLYLALKOHOL (GERMAN)
* ALLYLIC ALCOHOL * 3-HYDROXYPROPENE
* ORVINYLCARBINOL * PROPENOL * PROPEN-1-OL-3 * 1-PROPEN-3-OL * 2-PROPEN-1-OL
* PROPENYL ALCOHOL * 2-PROPENYL ALCOHOL * RCRA WASTE NUMBER P005
* SHELL UNDRAUTTED A * VINYLCARBINOL
* WEED DRENCH

CONSENSUS REPORTS: EPA Extremely Hazardous Substances List. Reported in EPA TSCA Inventory.

OSHA PEL: (Transitional: TWA 2 ppm (skin))
  TWA 2 ppm; STEL 4 ppm (skin)
ACGIH TLV: TWA 2 ppm; STEL 4 ppm (skin)
DFG MAK 2 ppm (5 mg/m^3)
DOT Classification: Flammable Liquid; Label: Poison and Flammable Liquid.

SAFETY PROFILE: Poison by inhalation, ingestion, skin contact, subcutaneous, intraperitoneal, and possibly other routes. A skin, severe eye (human), and systemic irritant. Mutation data reported. Dangerous fire and explosion hazard when exposed to heat or flame or oxidizers. Exposive or violent reaction with sulfuric acid; alkali + 2,4,6-trichloro-1,3,5-triazine; or 2,4,6-tris(bromoamino)-1,3,5-triazine. Reaction with carbon tetrachloride produces explosively unstable halogenated C_4 epoxides. Incompatible with chlorosulfonic acid; HNO_3; H_2SO_4; oleum; NaOH; diallyl phosphite; PCl_3; tri-n-bromo-melamine. When heated to decomposition it emits acrid smoke and fumes. To fight fire, use CO_2, alcohol foam, dry chemical.

AFW000   CAS: 107-11-9   **HR: 3**
ALLYLAMINE

DOT: UN 2334
mf: C_3H_7N mw: 57.11

PROP: Colorless liquid, burning taste, sharp odor. Bp: 56.5°, d: 0.761 @ 20°/4°, flash p: −20°F, autoign temp: 705°F, vap d: 2.00, lel: 2.2%, uel: 22%. Misc in water, alc, and ether.

SYNS: 3-AMINOPROPENE * 3-AMINOPROPYLENE
* MONOALLYLAMINE * 2-PROPENAMINE
* 2-PROPEN-1-AMINE

CONSENSUS REPORTS: EPA Extremely Hazardous Substances List. Reported in EPA TSCA Inventory.

DOT Classification: IMO: Flammable Liquid; Label: Flammable Liquid and Poison.

SAFETY PROFILE: Poison by inhalation, ingestion, intraperitoneal and skin contact. Human systemic effects by inhalation: lacrimation and lung effects. A systemic irritant. Mutation data reported. A severe eye and skin irritant. Extraordinary precautions against fumes are advised. Dangerous fire and explosion hazard when exposed to heat, flame or oxidizers. Highly reactive. When heated to decomposition it emits toxic fumes of NO_x. To fight fire, use alcohol foam, CO_2, dry chemical.

AFW750   CAS: 140-67-0   **HR: 3**
p-ALLYLANISOLE
mf: $C_{10}H_{12}O$ mw: 148.22

PROP: Isolated from rind of *Persea Gratissima Garth,* and from Oil of Estragon; found in oils of Russian Anise, Basil, Fennel, Turpentine, and others (FCTXAV 14,601,76). Colorless to sltly yellow liquid; anise odor. D: 0.960-0.968, refr index: 1.519-1.524, flash p: 178°F. Sol in alc; insol in water.

SYNS: 4-ALLYL-1-METHOXYBENZENE * CHAVICOL METHYL ETHER * ESDRAGOL * ISOANETHOLE
* p-METHOXYALLYLBENZENE * 1-METHOXY-4-(2-PROPENYL)BENZENE * METHYL CHAVICOL
* NCI-C60946 * TARRAGON

CONSENSUS REPORTS: Reported in EPA TSCA Inventory.

SAFETY PROFILE: Moderate acute toxicity by many routes. A skin irritant. Questionable carcinogen with experimental carcinogenic and neoplastigenic data. Mutation data reported. Combustible liquid. When heated to decomposition it emits acrid smoke and irritating fumes. A spice used in foods, liqueurs, and perfumes.

AFY000   CAS: 106-95-6   **HR: 3**
ALLYL BROMIDE

DOT: UN 1099
mf: C_3H_5Br mw: 120.99

PROP: Colorless liquid, pungent odor. Mp: −119°, bp: 71.3°, flash p: 30°F, d: 1.3980 @ 20°/4°, autoign temp: 563°F, vap d: 4.17, lel: 4.4%, uel: 7.3%. Insol in water.

SYNS: BROMALLYLENE * 3-BROMOPROPENE
* 3-BROMOPROPYLENE

CONSENSUS REPORTS: Reported in EPA TSCA Inventory.

DOT Classification: DOT-IMO: Flammable Liquid; Label: Flammable Liquid and Poison.

SAFETY PROFILE: Poison by ingestion and intraperitoneal routes. Mildly toxic by inhalation. Human mutation data reported. Dangerous fire and explosion hazard when exposed to heat, flame or oxidizers. When heated to decomposition it emits toxic fumes of Br^-. To fight fire, use alcohol foam, water spray or mist, CO_2, dry chemical.

AGA500   CAS: 123-68-2   **HR: 3**
ALLYL CAPROATE
mf: $C_9H_{16}O_2$ mw: 156.25

PROP: Bp: 186-188°. Insol in water; sol in alc and ether.

SYNS: ALLYL HEXANOATE (FCC) * FEMA No. 2032 * 2-PROPENYL-N-HEXANOATE

CONSENSUS REPORTS: Reported in EPA TSCA Inventory.

SAFETY PROFILE: Poison by ingestion and skin contact. Mutation data reported. An irritant to human skin. When heated to decomposition it emits acrid smoke and irritating fumes.

AGB250   CAS: 107-05-1   **HR: 3**
ALLYL CHLORIDE
DOT: UN 1100
mf: C_3H_5Cl mw: 76.53

PROP: Colorless liquid. Mp: −136.4°, bp: 44.6°, d: 0.938 @ 20°/4°, flash p: −25°F, lel: 2.9%, uel: 11.2%, autoign temp: 905°F, vap d: 2.64. Solubility = <0.1 in water.

SYNS: ALLILE (CLORURO DI) (ITALIAN) * ALLYL-CHLORID (GERMAN) * ALLYLE (CHLORURE D') (FRENCH) * CHLORALLYLENE * CHLOROALLYLENE * 3-CHLOROPRENE * 3-CHLORO-1-PROPENE * 3-CHLOROPROPENE * 1-CHLORO PROPENE-2 * 1-CHLORO-2-PROPENE * α-CHLOROPROPYLENE * 3-CHLORO-1-PROPYLENE * 3-CHLOROPROPYLENE * 3-CHLORPROPEN (GERMAN) * NCI-C04615 * 2-PROPENYL CHLORIDE

CONSENSUS REPORTS: IARC Cancer Review: Animal Inadequate Evidence IMEMDT 36,39,85; NCI Carcinogenesis Bioassay (gavage); No Evidence: rat NCITR* NCI-CG-TR-73,78; Clear Evidence: mouse NCITR* NCI-CG-TR-73,78. Reported in EPA TSCA Inventory. EPA Genetic Toxicology Program. Community Right-To-Know List.

OSHA PEL: (Transitional: TWA 1 ppm) TWA 1 ppm; STEL 2 ppm
ACGIH TLV: TWA 1 ppm; STEL 2 ppm
DFG MAK: 1 ppm (3 mg/m^3), Suspected Carcinogen.
NIOSH REL: TWA 1 ppm; CL 3 ppm/15M
DOT Classification: Flammable Liquid; Label: Flammable Liquid.

SAFETY PROFILE: Suspected carcinogen with experimental tumorigenic data. Poison by ingestion, intraperitoneal, and intravenous routes. Moderately toxic by inhalation and skin contact. Experimental teratogenic and reproductive effects. Human mutation data reported. A skin, eye, and mucous membrane irritant. Chronic exposure may cause liver and kidney damage. The vapors of allyl chloride are quite irritating to the eyes, nose, and throat. Contact of the liquid with the skin, in addition to local vasoconstriction and numbness, may lead to rapid absorption and distribution through the body. If remedial measures are not taken promptly, such contact may result in burns and internal injuries. Inhalation may cause headache, dizziness, and in high concentration, loss of consciousness; however, even in low concentration, its odor in most cases is irritating enough to give warning of its presence. Concentration of the vapors high enough to cause serious effects, including damage to the lungs, especially on repeated exposure, may not be intolerable. Consequently, the warning characteristics should never be disregarded. In general, precautions should be taken AT ALL TIMES to avoid spillage and accumulation of noticeable concentration of the vapors in the atmosphere. Acute exposure in experimental animals has resulted in marked inflammation of lungs, irritation of skin, and swelling of the kidneys. Chronically exposed animals have shown degenerative changes in the liver and kidneys. Reported human exposures have been principally cases of irritation of the eyes, skin and respiratory tract, sometimes accompanied by aches and pains in the bones. Liver and kidney injury is possible.

Dangerous fire and explosion hazard when exposed to heat, flame or oxidizers. Vigorous or explosive reaction above −70°C with alkyl aluminum chlorides (e.g., trichlorotriethyl dialuminum; ethyl aluminum dichloride; or diethyl aluminum chloride) + aromatic hydrocarbons (e.g., benzene or toluene). Violently exothermic polymerization reaction with Lewis acids (e.g., aluminum chloride; boron trifluoride; or sulfuric acid) and metals (e.g., aluminum; magnesium; zinc; or galvanized metals). Incompatible with HNO_3; ethylene imine; ethylenediamine; chlorosulfonic acid; oleum; NaOH. To fight fire, use CO_2, alcohol foam, dry chemical.

Storage and Handling: Keep cool, away from heat sources. Maintain good ventilation. Work in a fume hood or with closed system if possible; otherwise, use adequate ventilation so that the odor of allyl chloride does not persist. If it should be necessary to enter an area in which the odor of allyl chloride is at all noticeable, use a gas mask equipped with an ''organic vapor'' canister. Do not disregard the warning odor or eye irritation of allyl chloride.

AGB500   CAS: 2937-50-0   HR: 3
ALLYL CHLOROCARBONATE
DOT: UN 1722
mf: $C_4H_5ClO_2$ mw: 120.54

PROP: Liquid. Bp: 106-114°, flash p: 88°F (CC), d: 1.14, vap d: 4.2.

SYNS: ALLYL CHLOROFORMATE (DOT)

CONSENSUS REPORTS: Reported in EPA TSCA Inventory.

DOT Classification: Flammable Liquid; Label: Flammable Liquid; IMO: Corrosive Material; Label: Corrosive, Flammable Liquid.

SAFETY PROFILE: Poison by inhalation and ingestion. Corrosive. Dangerous when exposed to heat, open flame (or sparks), or powerful oxidizers. Can react with oxidizing materials. To fight fire, use alcohol foam, spray or mist, dry chemical. When heated to decomposition it emits toxic fumes of Cl^-.

AGC000   CAS: 1866-31-5   HR: 2
ALLYL CINNAMATE
mf: $C_{12}H_{12}O_2$ mw: 188.24

PROP: Colorless to light yellow liquid; cherry odor. D: 1.052 @ 25°/25°; bp: 150-152° @ 15 mm. Insol in water; sol in alc; very sol in ether.

SYNS: ALLYL-3-PHENYLACRYLATE * PROPENYL CINNAMATE * VINYL CARBINYL CINNAMATE

CONSENSUS REPORTS: Reported in EPA TSCA Inventory.

SAFETY PROFILE: Moderately toxic by ingestion. Human skin irritant. When heated to decomposition it emits acrid smoke and irritating fumes.

AGC500   CAS: 2705-87-5   HR: 3
ALLYL CYCLOHEXANEPROPIONATE
mf: $C_{12}H_{20}O_2$ mw: 196.32

PROP: Colorless liquid; pineapple odor. D: 0.945-0.950, refr index: 1.457-1.463, flash p: +212°F. Misc in alc, chloroform, ether; insol in glycerin and water.

SYNS: 3-ALLYLCYCLOHEXYL PROPIONATE
* ALLYL HEXAHYDROPHENYLPROPIONATE
* FEMA No. 2026

CONSENSUS REPORTS: Reported in EPA TSCA Inventory.

SAFETY PROFILE: Poison by ingestion. When heated to decomposition it emits acrid smoke and irritating fumes. Combustible liquid.

AGE250   CAS: 93-15-2   HR: 3
4-ALLYL-1,2-DIMETHOXYBENZENE
mf: $C_{11}H_{14}O_2$ mw: 178.25

PROP: Colorless to pale yellow liquid; clove, carnation odor. D: 1.032-1.036, refr index: 1.532, flash p: 212°F. Sol in fixed oils; insol in glycerin, propylene glycol.

SYNS: 1-ALLYL-3,4-DIMETHOXYBENZENE * 4-ALLYLVERATROLE * 1,2-DIMETHOXY-4-ALLYLBENZENE * 1-(3,4-DIMETHOXYPHENYL)-2-PROPENE * ENT 21,040 * 1,3,4-EUGENOL METHYL ETHER * EUGENYL METHYL ETHER * FEMA No. 2475 * METHYL EUGENOL (FCC) * VERATROLE METHYL ETHER

CONSENSUS REPORTS: Reported in EPA TSCA Inventory.

SAFETY PROFILE: Poison by intravenous route. Moderately toxic by ingestion and intraperitoneal routes. A skin irritant. Mutation data reported. Combustible liquid. When heated to decomposition it emits acrid smoke and irritating fumes. Some other alkenylbenzenes have carcinogenic activity.

AGG500   HR: 3
ALLYL FLUORIDE
mf: C_3H_5F mw: 60.07

PROP: Colorless gas. Bp: −10°.

SYN: 3-FLUOROPROPENE

SAFETY PROFILE: Poison by inhalation and ingestion. A strong irritant. When heated to decomposition it emits highly toxic fumes of F^-. Incompatible with water or steam to produce toxic and corrosive fumes.

AGH000   CAS: 1838-59-1   **HR: 3**
ALLYL FORMATE

DOT: UN 2336
mf: $C_4H_6O_2$ mw: 86.10

PROP: Liquid, sltly water-sol, sol in organic solvents. D: 0.948 @ 18°/4°, bp: 83°, flash p: $<-50°F$.

SYNS: FORMIC ACID, ALLYL ESTER   *   3-PROPENYL METHANOATE

CONSENSUS REPORTS: Reported in EPA TSCA Inventory.

DOT Classification: IMO: Flammable Liquid; Label: Flammable Liquid and Poison.

SAFETY PROFILE: Poison by ingestion. Mildly toxic by inhalation. Very flammable and reactive. Dangerous fire hazard. When heated to decomposition it yields irritating smoke and fumes.

AGH150   CAS: 106-92-3   **HR: 3**
ALLYL GLYCIDYL ETHER

DOT: UN 2219
mf: $C_6H_{10}O_2$ mw: 114.16

PROP: Bp: 153.9°, fp: $-100°$ (forms glass), flash p: 135°F (OC), d: 0.9698 @ 20°/4°, vap press: 21.59 mm @ 60°, vap d: 3.94.

SYNS: AGE   *   ALLIL-GLICIDIL-ETERE (ITALIAN)
*   1-ALLILOSSI-2,3 EPOSSIPROPANO (ITALIAN)
*   ALLYL-2,3-EPOXYPROPYL ETHER   *   ALLYLGLYCI-
DAETHER (GERMAN)   *   1-ALLYLOXY-2,3-EPOXY-PRO-
PAAN (DUTCH)   *   1-ALLYLOXY-2,3-EPOXYPROPAN
(GERMAN)   *   1-(ALLYLOXY)-2,3-EPOXYPROPANE
*   NCI-C56666   *   OXYDE d'ALLYLE et de GLYCIDYLE
(FRENCH)   *   ((2-PROPENYLOXY)METHYL)OXIRANE

CONSENSUS REPORTS: Reported in EPA TSCA Inventory.

OSHA PEL: (Transitional: CL 10 ppm) TWA 5 ppm; STEL 10 ppm
ACGIH TLV: TWA 5 ppm; STEL 10 ppm (skin)
NIOSH REL: (Glycidyl Ethers) CL 45 mg/m³/ 15M
DFG MAK: 10 ppm (45 mg/m³)

DOT Classification: Flammable or Combustible Liquid; Label: Flammable Liquid.

SAFETY PROFILE: Poison by ingestion. Moderately toxic by inhalation and skin contact. Mutation data reported. A severe skin and eye irritant. Can cause central nervous system depression and pulmonary edema. Flammable when exposed to heat or flame; can react with oxidizing materials. To fight fire, use foam, CO_2, dry chemical. When heated to decomposition it emits acrid smoke and irritating fumes.

AGH250   CAS: 142-19-8   **HR: 2**
ALLYL HEPTANOATE

mf: $C_{10}H_{18}O_2$ mw: 170.28

PROP: Colorless to pale yellow liquid; fruity, sweet, pineapple odor. D: 0.880, refr index: 1.426, flash p: 154°F.

SYNS: ALLYL ENANTHATE   *   ALLYL HEPTOATE
*   ALLYL HEPTYLATE   *   FEMA No. 2031   *   2-PRO-
PENYL HEPTANOATE

CONSENSUS REPORTS: Reported in EPA TSCA Inventory.

SAFETY PROFILE: Moderately toxic by ingestion and skin contact. A human skin irritant. Combustible liquid. When heated to decomposition it emits acrid smoke and irritating fumes.

AGH750   **HR: 3**
ALLYL HYDROPEROXIDE
mf: $C_3H_6O_2$ mw: 74.1

SAFETY PROFILE: Highly toxic. A potentially explosive liquid. Unstable to heat, light, and solid alkalies. Mixtures with sand are impact sensitive. Upon decomposition it emits acrid smoke and fumes.

AGI250   **HR: 3**
ALLYL IODIDE
mf: C_3H_5I mw: 168.0

PROP: Yellow liquid, pungent odor. Mp: $-99°$, bp: 103.1°, d: 1.825 @ 20°/4°, vap d: 5.8.

SAFETY PROFILE: Poison by inhalation and ingestion. A powerful irritant. When heated to decomposition it emits highly toxic fumes of I^-. Moderately flammable. Incompatible with oxidizing materials. To fight fire, use water, foam, CO_2, dry chemical.

AGI500   CAS: 79-78-7   **HR: 2**
ALLYL-α-IONONE
mf: $C_{16}H_{24}O$ mw: 232.40

PROP: Colorless to yellow liquid; fruity, woody odor. D: 0.928-0.935, refr index: 1.503-1.507, flash p: +212°F. Sol in alc; insol in water @ 265°.

SYNS: CETONE V * FEMA No. 2033 * 1-(2,6,6-TRIMETHYL-2-CYCLOHEXEN-1-YL)-1,6-HEPTADIEN-3-ONE

CONSENSUS REPORTS: Reported in EPA TSCA Inventory.

SAFETY PROFILE: A skin irritant. Combustible liquid. When heated to decomposition it emits acrid smoke and irritating fumes.

AGJ250   CAS: 57-06-7   **HR: 3**
ALLYL ISOTHIOCYANATE

DOT: UN 1545
mf: C_4H_5NS mw: 99.16

PROP: Colorless to pale yellow liquid; irritating odor with mustard taste. Mp: −80°, bp: 150.7°, flash p: 115°F, d: 1.013-1.016 @ 25°/25°, vap press: 10 mm @ 38.3°, vap d: 3.41, refr index: 1.527-1.531. Misc with alc, carbon disulfide, and ether.

SYNS: AITC * ALLYL ISORHODANIDE * ALLYL ISOSULFOCYANATE * ALLYL ISOTHIOCYANATE, stabilized (DOT) * ALLYL MUSTARD OIL * ALLYLSENFOEL (GERMAN) * ALLYL SEVENOLUM * ALLYL THIOCARBONIMIDE * ARTIFICIAL MUSTARD OIL * CARBOSPOL * FEMA No. 2034 * ISOTHIOCYANATE d'ALLYLE (FRENCH) * 3-ISOTHIOCYANATO-1-PROPENE * MUSTARD OIL * NCI-C50464 * OIL of MUSTARD, ARTIFICIAL * OLEUM SINAPIS VOLATILE * 2-PROPENYL ISOTHIOCYANATE * REDSKIN * SENF OEL (GERMAN) * SYNTHETIC MUSTARD OIL * VOLATILE OIL of MUSTARD

CONSENSUS REPORTS: IARC Cancer Review: Animal Limited Evidence IMEMDT 36,55,85; NTP Carcinogenesis Bioassay (gavage); No Evidence: mouse NTPTR* NTP-TR-234,82; Clear Evidence: rat NTPTR* NTP-TR-234,82. Reported in EPA TSCA Inventory.

DOT Classification: Poison B; Label: Flammable Liquid and Poison

SAFETY PROFILE: Suspected carcinogen with experimental neoplastigenic, and tumorigenic data. Poison by ingestion, skin contact, intravenous, subcutaneous, and intraperitoneal routes. Experimental teratogenic and reproductive effects. An allergen. May cause contact dermatitis. Mutation data reported. Combustible liquid. Highly reactive. When heated to decomposition (above 250°) or on contact with acid or acid fumes it emits highly toxic fumes of CN^-, SO_x and NO_x. To fight fire, use foam, CO_2, dry chemical.

AGJ500   **HR: 3**
ALLYL MERCAPTAN
mf: C_3H_6S mw: 74.15

PROP: Water-white liquid with a strong garlic odor, darkens on standing. D: 0.925 @ 23°/4°, bp: 68°, flash p: 14°F.

SYN: 2-PROPENE-1-THIOL

SAFETY PROFILE: Poison by inhalation and ingestion. Strong irritant to skin and mucous membranes. When heated to decomposition it emits highly toxic fumes of SO_x. Very dangerous fire hazard. To fight fire, use water mist or spray, alcohol foam, CO_2, or dry chemical.

AGM500   CAS: 4230-97-1   **HR: 2**
ALLYL OCTANOATE
mf: $C_{11}H_{20}O_2$ mw: 184.31

PROP: Colorless liquid; fruity odor. D: 0.8550.861, refr index: 1.425, flash p: +151°F. Sol in alc, fixed oils; sltly sol in propylene glycol; insol in glycerin and water @ 260°.

SYNS: ALLYL CAPRYLATE * FEMA No. 2037 * OCTANOIC ACID ALLYL ESTER * OCTANOIC ACID-2-PROPENYL ESTER

CONSENSUS REPORTS: Reported in EPA TSCA Inventory.

SAFETY PROFILE: Moderately toxic by ingestion. A skin irritant. When heated to decomposition it emits acrid smoke and irritating fumes.

AGQ750   CAS: 7493-74-5   **HR: 2**
ALLYL PHENOXYACETATE
mf: $C_{11}H_{12}O_3$ mw: 192.23

PROP: Colorless to light yellow liquid; heavy fruit odor.

SYN: ACETATE P.A.

CONSENSUS REPORTS: Reported in EPA TSCA Inventory.

SAFETY PROFILE: Moderately toxic by ingestion and skin contact. When heated to decomposition it emits acrid smoke and irritating fumes.

AGR500   CAS: 2179-59-1   **HR: 1**
ALLYL PROPYL DISULFIDE
mf: $C_6H_{12}S_2$ mw: 148.30

PROP: Liquid, pungent odor.
OSHA PEL: (Transitional: TWA 2 PPM) TWA 2 ppm; STEL 3 ppm
ACGIH TLV: TWA 2 ppm; STEL 3 ppm
DFG MAK: 2 ppm (12 mg/m^3)

SAFETY PROFILE: A powerful irritant. Moderately flammable by exposure to heat, flame or oxidizers. When heated to decomposition it emits highly toxic SO_x. To fight fire, use foam, CO_2, dry chemical.

AGU250   CAS: 107-37-9   **HR: 3**
ALLYL TRICHLOROSILANE
DOT: UN 1724
mf: $C_3H_5Cl_3Si$ mw: 175.52

PROP: Colorless liquid, pungent, irritating odor. Bp: 117.5°, d: 1.217 @ 27°, flash p: 95°F (COC).

SYN: TRICHLOROALLYLSILANE

CONSENSUS REPORTS: Reported in EPA TSCA Inventory.

DOT Classification: IMO: Corrosive Material; Label: Corrosive, Flammable Liquid.

SAFETY PROFILE: Poison by intravenous route. Corrosive. When heated to decomposition it emits toxic Cl^-. A dangerous fire hazard. To fight fire, use foam, mist, spray, dry chemical.

AGX000   CAS: 7429-90-5   **HR: 3**
ALUMINUM
DOT: UN 1309/UN 1383/UN 1396
af: Al   aw: 26.98

PROP: A silvery ductile metal. Mp: 660°, bp: 2450°, d: 2.702, vap press 1 mm @ 1284°. Sol in HCl, H_2SO_4, and alkalies.

SYNS: A OO * AD1M * ALAUN (GERMAN) * ALUMINA FIBRE * ALUMINUM FLAKE * ALUMINUM DEHYDRATED * ALUMINUM, METALLIC, POWDER (DOT) * ALUMINUM POWDER * ALUMINUM POWDER, UNCOATED, NONPYROPHORIC (DOT) * C.I. 77000 * EMANAY * ATOMIZED ALUMINUM POWDER * JISC 3108 * JISC 3110 * METANA ALUMINUM PASTE * NORAL INK GRADE ALUMINUM * PAP-1

CONSENSUS REPORTS: Community Right-To-Know List (fume or dust). Reported in EPA TSCA Inventory.

OSHA PEL: Total Dust: TWA 15 mg/m^3; Respirable Fraction: TWA 5 mg/m^3; Pyro Powders and Welding Fumes: 5 mg/m^3; Soluble Salts and Alkyls: 2 mg/m^3.
ACGIH TLV: Metal and Oxide: TWA 10 mg/m^3 (dust); Pyro Powders and Welding Fumes: TWA 5 mg/m^3; Soluble Salts and Alkyls) TWA 2 mg/m^3
DFG MAK: 6 mg/m^3; BAT: 170 μg/L in urine at end of shift.
DOT Classification: Label: Flammable Solid; Label: Spontaneously Combustible (pyrophoric); IMO: Flammable Solid; Label: Dangerous When Wet (non-pyrophoric).

SAFETY PROFILE: Although aluminum is not generally regarded as an industrial poison, inhalation of finely divided powder has been reported to cause pulmonary fibrosis. It is a reactive metal and the greatest industrial hazards are with chemical reactions. As with other metals the powder and dust are the most dangerous forms. Dust is moderately flammable/explosive by heat, flame, or chemical reaction with powerful oxidizers. To fight fire, use special mixtures of dry chemical.

Powdered aluminum undergoes the following dangerous interactions: explosive reaction after a delay period with $KClO_4$ + $Ba(NO_3)_2$ + KNO_3 + H_2O; also with $Ba(NO_3)_2$ + KNO_3 + sulfur + vegetable adhesives + H_2O. Mixtures with powdered AgCl; NH_4NO_3 or NH_4NO_3 + $Ca(NO_3)_2$ + formamide + H_2O are powerful explosives. Mixture with ammonium peroxodisulfate + water is explosive. Violent or explosive 'thermite' reaction when heated with metal oxides; oxosalts (nitrates, sulfates); or sulfides; and with hot copper oxide worked with an iron or steel tool. Potentially explosive reaction with CCl_4 during ball milling operations. Many violent or explosive reactions with the following halocarbons have occurred in industry: bromomethane; bromotrifluoromethane; CCl_4; chlorodifluoromethane; chloroform; chloromethane; chloromethane + 2-methylpropane; dichlorodifluoromethane; 1,2-dichloroethane; dichloromethane; 1,2-dichloropropane; 1,2-difluorotetrafluoroethane; fluorotrichloroethane; hexachloroethane + alcohol; polytrifluoroethylene oils and greases; tetrachloroethylene; tetrafluoromethane; 1,1,1-trichloroethane; trichloroethylene; 1,1,2-trichlorotrifluoroethane; and trichlorotrifluoroethane-dichlorobenzene. Potentially explosive reaction with chloroform amidinium nitrate. Ignites on contact with vapors of $AsCl_3$; SCl_2; Se_2Cl_2; and PCl_5. Reacts violently on heating with Sb or As. Ignites on heating in $SbCl_3$ vapor. Ignites on contact with barium peroxide. Potentially violent reaction with sodium acetylide. Mixture with sodium peroxide may ignite or react violently. Spontaneously ignites in CS_2 vapor. Halogens: ignites in chlorine gas, foil reacts vigorously with liquid Br_2, violent reaction with $H_2O + I_2$. Violent reaction with hydrochloric acid; hydrofluoric acid; and hydrogen chloride gas. Violent reaction with disulfur dibromide. Violent reaction with the non-metals phosphorus; sulfur; and selenium. Violent reaction or ignition with the interhalogens: bromine pentafluoride; chlorine fluoride; iodine chloride; iodine pentafluoride; and iodine heptafluoride. Burns when heated in CO_2. Ignites on contact with O_2 and mixtures with $O_2 + H_2O$ ignite and react violently. Mixture with picric acid + water ignites after a delay period. Explosive reaction above 800°C with sodium sulfate. Violent reaction with sulfur when heated. Exothermic reaction with iron powder + water releases explosive hydrogen gas.

Aluminum powder also forms sensitive explosive mixtures with oxidants such as: liquid Cl_2 and other halogens; N_2O_4; tetranitromethane; bromates; iodates; $NaClO_3$; $KClO_3$; and other chlorates; $NaNO_3$; aqueous nitrates; $KClO_4$ and other perchlorate salts; nitryl fluoride; ammonium peroxodisulfate; sodium peroxide; zinc peroxide; and other peroxides; red phosphorus; and powdered polytetrafluoroethylene (PTFE).

Bulk aluminum may undergo the following dangerous interactions: exothermic reaction with butanol; methanol; 2-propanol; or other alcohols; sodium hydroxide to release explosive hydrogen gas. Reaction with diborane forms pyrophoric product. Ignition on contact with niobium oxide + sulfur. Explosive reaction with molten metal oxides; oxosalts (nitrates, sulfates); sulfides; and sodium carbonate. Reaction with arsenic trioxide + sodium arsenate + sodium hydroxide produces the toxic arsine gas.

Violent reaction with chlorine trifluoride. Incandescent reaction with formic acid. Potentially violent alloy formation with palladium, platinum at mp of Al, 600°C. Vigorous dissolution reaction in methanol + carbon tetrachloride. Vigorous amalgamation reaction with mercury (II) salts + moisture. Violent reaction with molten silicon steels. Violent exothermic reaction above 600°C with sodium diuranate.

AGX250   HR: 1
ALUMINUM AMMONIUM SULFATE
mf: $Al_2(SO_4)_3(NH_4)_2SO_4 \cdot 24H_2O$ mw: 906

PROP: Colorless crystals; odorless with sweet taste. D: 1.645, mp: 94.5°, bp: loses 20 waters @ 120°. Sol in water, glycerin; insol in alc.

SAFETY PROFILE: Irritating if inhaled or ingested. Upon decomposition it emits toxic fumes of NO_x and SO_x.

AGX500   HR: 3
ALUMINUM BOROHYDRIDE
mf: AlB_3H_{12} mw: 71.53

PROP: Liquid. Bp: 44.5°, mp: −64.5°, vap press: 400 mm @ 28.1°.

SYN: ALUMINUM TETRAHYDROBORATE

SAFETY PROFILE: Dangerous by spontaneous chemical reaction; ignites spontaneously in air, particularly in moist air. Explodes in O_2 at temperatures as low as 20°. An explosive range of 5 to 90%. Incompatible with water; steam; oxidizing materials; acid; acid fumes; will react with water or steam to produce heat, H_2, or toxic fumes. To fight fire, use CO_2, dry chemical.

AGX750   CAS: 7727-15-3   HR: 2
ALUMINUM BROMIDE
DOT: UN 1725/UN 2580
mf: $AlBr_3$ mw: 266.71

PROP: White to yellow-red lumps. Mp: 97.5°, bp: 263.3° @ 748 mm, d: 3.2, vap press: 1 mm @ 81.3°.

SYNS: ALUMINUM BROMIDE, anhydrous * ALUMINUM BROMIDE, solution (DOT) * ALUMINUM TRIBROMIDE * TRIBROMOALUMINUM

CONSENSUS REPORTS: Reported in EPA TSCA Inventory.

ACGIH TLV: TWA 2 mg(Al)/m^3
DOT Classification: Label: Corrosive.

SAFETY PROFILE: A toxic, corrosive material. Mixtures with sodium or potassium explode violently upon impact. When heated to decomposition it emits toxic fumes of Br$^-$. Do not add H$_2$O to anhydrous material. Hydrolysis can be violent.

AGY750   CAS: 7446-70-0   **HR: 3**
ALUMINUM CHLORIDE

DOT: UN 1726/UN 2581
mf: AlCl$_3$ mw: 133.33

PROP: White hexagonal deliquescent crystals. D: 2.44, mp: 194° @ 5.2 atm, bp: subl @ 181°, vap press: 1 mm @ 100.0°. Violently sol in water; sol in alc and ether.

SYNS: ALLUMINIO(CLORURO DI) (ITALIAN) * ALUMINUMCHLORID (GERMAN) * ALUMINUM CHLORIDE (1:3) * ALUMINUM CHLORIDE, anhydrous (DOT) * ALUMINUM CHLORIDE, solution (DOT) * ALUMINUM TRICHLORIDE * CHLORURE d'ALUMINUM (FRENCH) * PEARSALL * TRICHLOROALUMINUM

CONSENSUS REPORTS: Reported in EPA TSCA Inventory.

ACGIH TLV: TWA 2 mg(Al)/m^3
DOT Classification: IMO: Corrosive Material; Label: Corrosive.

SAFETY PROFILE: Moderately toxic by ingestion. Mutation data reported. Experimental teratogenic and reproductive effects. Other experimental reproductive effects. The dust is an irritant by ingestion, inhalation, and skin contact. Highly exothermic polymerization reactions with alkenes. Incompatible with nitrobenzenes or nitrobenzene + phenol. Highly exothermic reaction with water or steam produces toxic fumes of HCl.

AHA250   CAS: 7047-84-9   **HR: 3**
ALUMINUM DEXTRAN
mf: C$_{18}$H$_{37}$AlO$_4$ mw: 344.48

PROP: Powder. A complex containing aluminum and dextran, a chain of molecular weight 2,500, corresponding to a chain of 15 anhydroglucose units.

SYN: ALUMINUM MONOSTEARATE * STEARIC ACID ALUMINUM DIHYDROXIDE SALT

ACGIH TLV: TWA 2 mg(Al)/m^3

SAFETY PROFILE: Questionable carcinogen with experimental tumorigenic data. When heated to decomposition it emits acrid smoke and fumes.

AHA750   **HR: 3**
ALUMINUM ETHYLATE
mf: Al(OC$_2$H$_5$)$_3$ mw: 162.15

PROP: Liquid. Decomp by H$_2$O. Bp: 200° @ 6-8 mm; mp: 140°.

SAFETY PROFILE: Strong irritant to skin, eyes, and mucous membranes by inhalation.

AHB500   CAS: 7784-21-6   **HR: 3**
ALUMINUM HYDRIDE

DOT: UN 2463
mf: AlH$_3$ mw: 30.01

PROP: Colorless powder.

SYNS: ALANE * ALUMINUM TRIHYDRIDE * α-ALUMINUM TRIHYDRIDE

CONSENSUS REPORTS: Reported in EPA TSCA Inventory.

ACGIH TLV: TWA 2 mg(Al)/m^3
DOT Classification: Label: Flammable Solid and Dangerous When Wet; IMO: Flammable Solid; Label: Spontaneously Combustible.

SAFETY PROFILE: Little is known about the toxicity of AlH$_3$, however hydrides of some metals (such as AsH$_3$) are extremely toxic. Dangerous fire hazard. An unstable material which is spontaneously flammable in air or O$_2$. Evolves explosive H$_2$ upon contact with moisture. Severe explosion hazard by chemical reaction wherein H$_2$ gas is produced, also in contact with methyl ethers contaminated by CO$_2$. Mixtures with tetrazole derivatives are explosive. Reacts with oxidizing materials. On contact with acid or acid fumes, it can emit toxic fumes.

AHC500   **HR: 3**
ALUMINUM IODIDE
mf: AlI$_3$ mw: 407.7

PROP: White leaflets. Mp: 191°, bp: 360°, d: 3.98 @ 25°, vap press: 1 mm @ 178.0° (sublimes).

SAFETY PROFILE: Incompatible with water.

AHD250   **HR: 3**
ALUMINUM MAGNESIUM PHOSPHIDE
mf: Mg$_3$AlP$_3$ mw: 192.8

SYN: MAGNESIUM ALUMINUM PHOSPHIDE (DOT)

ACGIH TLV: TWA 2 mg(Al)/m^3
DOT Classification: Label: Flammable Solid and Dangerous When Wet.

SAFETY PROFILE: A poison. Dangerous fire hazard. Evolves spontaneously flammable PH$_3$ in contact with water.

AHD500   HR: 3
ALUMINUM METHYL
mf: Al(CH$_3$)$_3$ mw: 72.07

PROP: Colorless liquid. Bp: 130°; mp: 0°.

SAFETY PROFILE: Related alkyl aluminum compounds are poisonous and strong irritants. Very flammable by spontaneous chemical reaction with air. Incompatible with water; halogenated hydrocarbons; and oxidizing materials. When heated to decomposition it emits toxic fumes. To fight fire, do not use water, foam, or halogenated extinguishing agents. Use dry chemical.

AHD750   CAS: 13473-90-0   HR: 3
ALUMINUM(III) NITRATE (1:3)
DOT: UN 1438
mf: N$_3$O$_9$ • Al   mw: 213.01

PROP: White crystals.

SYNS: ALUMINUM NITRATE (DOT) * ALUMINUM TRINITRATE * NITRIC ACID, ALUMIUM SALT * NITRIC ACID, ALUMINUM(3+) SALT

CONSENSUS REPORTS: Reported in EPA TSCA Inventory.

ACGIH TLV: TWA 2 mg(Al)/m^3
DOT Classification: Label: Oxidizer.

SAFETY PROFILE: A poison. A severe eye and mild skin irritant. A powerful oxidizer. When heated to decomposition it emits toxic NO$_x$. A nitrating agent.

AHE250   CAS: 1344-28-1   HR: 2
ALUMINUM OXIDE (2:3)
mf: Al$_2$O$_3$ mw: 101.96

PROP: White powder. Mp: 2050°, bp: 2977°, d: 3.5-4.0, vap press: 1 mm @ 2158°.

SYNS: A 1 (SORBENT) * A1-0109 P * ABRAREX * ACTIVATED ALUMINUM OXIDE * ALCOA F 1

* ALMITE * ALON * ALUMINA * α-ALUMINA (OSHA) * β-ALUMINA * γ-ALUMINA * ALUMINUM OXIDE * α-ALUMINUM OXIDE * β-ALUMINUM OXIDE * γ-ALUMINUM OXIDE * ALUMINUM SESQUIOXIDE * ALUMITE * ALUNDUM * BROCKMANN, ALUMINUM OXIDE * CAB-O-GRIP * COMPALOX * DIALUMINUM TRIOXIDE * DISPAL * DOTMENT 324 * FASERTON * G 2 (OXIDE) * KHP 2 * LUCALOX * MICROGRIT WCA * PS 1 * RC 172DBM

CONSENSUS REPORTS: Community Right-To-Know List. Reported in EPA TSCA Inventory.

OSHA PEL: Total Dust: (Transitional: TWA 5 mg/m^3) TWA 10 mg/m^3; Respirable Fraction: TWA 5 mg/m^3
ACGIH TLV: TWA (nuisance particulate) 10 mg/m^3 of total dust (when toxic impurities are not present, e.g., quartz < 1%).
DFG MAK: 6 mg/m^3 (fume)

SAFETY PROFILE: Inhalation of finely divided particles may cause lung damage (Shaver's disease). Questionable carcinogen with experimental neoplastigenic and tumorigenic data by implantation. Incompatible with hot chlorinated rubber. Exothermic reaction above 200°C with halocarbon vapors produces toxic HCl and phosgene.

AHE750   CAS: 20859-73-8   HR: 3
ALUMINUM PHOSPHIDE
DOT: UN 1397
mf: AlP   mw: 57.95

PROP: Dark gray or dark yellow crystals. D: 2.85 @ 25°/4°. Mp: >1000°.

SYNS: AIP * AL-PHOS * ALUMINUM FOSFIDE (DUTCH) * ALUMINUM MONOPHOSPHIDE * CELPHIDE * CELPHOS * DELICIA * DETIA GAS EX-B * FOSFURI di ALLUMINIO (ITALIAN) * FUMITOXIN * PHOSPHURES d'ALUMIUM (FRENCH) * RCRA WASTE NUMBER P006

CONSENSUS REPORTS: EPA Extremely Hazardous Substances List. Reported in EPA TSCA Inventory.

ACGIH TLV: TWA 2 mg(Al)/m^3
DOT Classification: Label: Flammable Solid and Dangerous When Wet; IMO: Flammable Solid; Label: Dangerous When Wet and Poison.

SAFETY PROFILE: A human poison by inhalation and ingestion. Dangerous; in contact with water, steam or alkali it slowly yields PH_3, which is spontaneously flammable in air. Explosive reaction on contact with mineral acids produces phosphine. When heated to decomposition it yields toxic PO_x.

AHF000   **HR: 3**
ALUMINUM PICRATE
mf: $Al(C_6H_2O(NO_2)_3)_3$ mw: 711.3

PROP: A solid.

SAFETY PROFILE: A poison. A powerful irritant. Very flammable by reaction with reducing materials. Severe explosion hazard when shocked or exposed to heat. When heated to decomposition it emits highly toxic fumes of NO_x and explodes.

AHG000   CAS: 11138-49-1   **HR: 2**
ALUMINUM SODIUM OXIDE
DOT: UN 1819/UN 2812
mf: $NaAlO_2$ mw: 82.0

PROP: White, hygroscopic powder. Mp: 1650°.

SYNS: β-ALUMINA * β''-ALUMINA * NALCO 680 * SODIUM ALUMINATE, solid (DOT) * SODIUM ALUMINUM OXIDE * SODIUM POLYALUMINATE

ACGIH TLV: TWA 2 mg(Al)/m^3
DOT Classification: ORM-B; Label: None, solid; Corrosive Material; Label: Corrosive, solution.

SAFETY PROFILE: Moderate irritant to skin, eyes and mucous membranes. A corrosive substance. When heated to decomposition it emits toxic fumes of Na_2O.

AHG500   **HR: 2**
ALUMINUM SODIUM SULFATE
mf: $NaAl(SO_4)_2 \cdot 12 H_2O$ mw: 458.29

PROP: Colorless crystals. Mp: 61°; d: 1.675. Anhydrous: sol in alc; sltly sol in water. Dodecahydrate: sol in water and alc.

SYNS: SODA ALUM * SODIUM ALUMINUM SULFATE

SAFETY PROFILE: A weak sensitizer. A general-purpose food additive. Local contact may cause contact dermatitis. An irritant. When heated to decomposition it emits toxic fumes of SO_x and Na_2O.

AHG750   CAS: 10043-01-3   **HR: 2**
ALUMINUM SULFATE (2:3)
DOT: UN 1760/UN 9078
mf: $O_{12}S_3 \cdot 2Al$ mw: 342.14

PROP: White powder; sweet taste. Mp: decomp @ 770°, d: 2.71. Solubility in water = 36.4% @ 20°.

SYNS: ALUM * ALUMINUM TRISULFATE * CAKE ALUM * DIALUMINUM SULPHATE * DIALUMINUM TRISULFATE * SULFURIC ACID, ALUMINUM SALT (3:2)

CONSENSUS REPORTS: Reported in EPA TSCA Inventory.

ACGIH TLV: TWA 2 mg(Al)/m^3
DOT Classification: ORM-E; Label: None, solid; ORM-B; Label: None, solution.

SAFETY PROFILE: Moderately toxic by ingestion and intraperitoneal routes. Experimental reproductive effects. Hydrolyzes to form sulfuric acid which irritates tissue, especially lungs. When heated to decomposition it emits toxic fumes of SO_x.

AHG875   CAS: 16962-07-5   **HR: 3**
ALUMINUM TETRAHYDROBORATE
DOT: UN 2870
mf: AlB_3H_{12} mw: 71.51

DOT Classification: Flammable Solid; Label: Spontaneously Combustible, Danger When Wet

SAFETY PROFILE: A poison. Spontaneously flammable in air. Explodes in oxygen with traces of water. Incompatible with alkenes and water.

AHH000   **HR: 3**
ALUMINUM THALLIUM SULFATE
mf: $AlTl(SO_4)_2 \cdot 12H_2O$ mw: 639.6

PROP: Cubic, octagonal, colorless crystals. Mp: 91°; d: 2.32 @ 20°/4°.

CONSENSUS REPORTS: Thallium and its compounds are on the Community Right-To-Know List.

SAFETY PROFILE: A poison.

AHH750   **HR: 3**
ALUMINUM TRIPROPYL
mf: $Al(C_3H_7)_3$ mw: 156.24

PROP: Liquid.

SYN: TRIPROPYL ALUMINUM

SAFETY PROFILE: Related alkyl aluminum compounds are poisons. Very flammable by spontaneous reaction with air. Incompatible with halogenated hydrocarbons. Hydrolyzes to evolve flammable vapor. To fight fire, do not use water, foam or halogenated extinguishing agents. Use dry chemical or a special powder extinguisher.

AHI750 HR: 3
AMATOL

PROP: A high explosive. Composition: NH_4NO_3, 80%; and TNT, 20%; d: 1.47.

SAFETY PROFILE: Moderately toxic by inhalation and ingestion routes. An allergen. May cause contact dermatitis. Dangerous fire hazard. An explosive by shock, spontaneous chemical reaction, or exposure to flame. Decomposition emits highly toxic fumes.

AHJ000 CAS: 9000-02-6 HR: 1
AMBERGRIS TINCTURE

PROP: Concretion from intestine of sperm whale, composed mostly of cholesterol.

SYNS: AMBER * AMBRA * GRAY AMBER

SAFETY PROFILE: A mild skin irritant. When heated to decomposition it emits acrid smoke and irritating fumes.

AHJ750 CAS: 52645-53-1 HR: 3
AMBUSH
mf: $C_{21}H_{20}Cl_2O_3$ mw: 391.31

SYNS: AI3-29158 * BW-21-Z * ECTIBAN * EXMIN * FMC 33297 * FMC 41655 * ICI-PP 557 * KESTREL (Pesticide) * NDRC-143 * NIA 33297 * OUTFLANK * OUTFLANK-STOCKADE * PERMETHRIN (USDA) * PERMETRIN (HUNGARIAN) * PERMETRINA (PORTUGUESE) * 3-PHENOXYBEN-ZYL (±)-3-(2,2-DICHLOROVINYL)-2,2-DIMETHYLCYCLO-PROPANECARBOXYLATE * (3-PHENOXYPHENYL)-METHYL-3-(2,2-DICHLORETHENYL)-2,2-DIMETH-YLCYCLOPROPANECARBOXYLATE * POUNCE * PP 557 * S-3151 * SBP-1513 * TALCORD * WL 43479

SAFETY PROFILE: Poison by inhalation, intravenous, and intracerebral routes. Moderately toxic by ingestion. Experimental reproductive effects. Mutation data reported. A skin irritant.

When heated to decomposition it emits toxic fumes of Cl⁻.

AHK000 HR: 3
AMERICIUM
af: Am aw: 243

PROP: A silvery, somewhat malleable radioactive metal. Mp: 994°; bp: 2607°; d: 13.67 @ 20°.

SAFETY PROFILE: A poison. Bone-seeking, long-lived radioactive element. Flammable; see POWDERED METALS. In a disaster, this highly toxic radioactive material can be disseminated over a wide area, causing a long-lived inhalation hazard which is difficult to remove from surfaces or from the body once it enters.

AHK250 HR: 3
AMERICIUM TRICHLORIDE
mf: $AmCl_3$ mw: 349.4

SAFETY PROFILE: Due to its alpha particle radioactivity, it can cause radiolysis and build pressure in sealed containers and eventually explode.

AHK500 CAS: 50-07-7 HR: 3
AMETYCIN
mf: $C_{15}H_{18}N_4O_5$ mw: 334.37

SYNS: 7-AMINO-9-α-METHOXYMITOSANE * MIT-C * MITO-C * MITOCIN-C * MITOMY-CIN * MITOMYCIN-C * MITOMYCINUM * MMC * MUTAMYCIN * MUTAMYCIN (MITOMYCIN for INJECTION) * MYTOMYCIN * NCI-C04706 * NSC 26980 * RCRA WASTE NUM-BER U010

CONSENSUS REPORTS: IARC Cancer Review: GROUP 2B IMEMDT 7,56,87; Animal Sufficient Evidence IMEMDT 10,171,76; NCI Carcinogenesis Studies (ipr); Clear Evidence: rat RRCRBU 52,1,75; No Evidence: mouse RRCRBU 52,1,75. EPA Extremely Hazardous Substances List. EPA Genetic Toxicology Program. Reported in EPA TSCA Inventory.

SAFETY PROFILE: Suspected carcinogen with experimental carcinogenic and neoplastigenic data. Poison by ingestion, subcutaneous, intravenous, and intraperitoneal routes. Human systemic effects by intravenous route: dyspnea and lung fibrosis. Experimental teratogenic and reproductive effects. Human mutation data reported. When heated to decomposition it emits toxic fumes of NO_x.

AHL750   HR: B
AMIDES

PROP: Organic compounds containing the structural group $-CONH_2$, and closely related to the organic acids with the grouping $-COOH$. Common examples are: acetamide (CH_3CONH_2) and urea $(CO(NH_2)_2)$.

SAFETY PROFILE: Most of the saturated amides have low toxicity, but the unsaturated and N-substituted amides are irritants and may be absorbed via skin contact. Can cause injury to the liver, kidney, and brain.

AHP750   HR: D
AMINES

PROP: A large group of organic compounds containing nitrogen and considered as derived from ammonia (NH_3) by replacement of one or more hydrogen atoms by an organic radical.

SAFETY PROFILE: Variable toxicity; some are poisons, some are only slightly toxic. Many are skin irritants and some are sensitizers.

AHR250   CAS: 613-89-8   HR: 3
2-AMINOACETOPHENONE
mf: C_8H_9NO mw: 135.18

PROP: Yellow, oily liquid. Bp: 251° (slt decomp); insol in water; sol in alc and ether.

SYNS: φ-AMINOACETOPHENONE * PHENACYL-AMINE

SAFETY PROFILE: Experimental teratogenic effects. Questionable carcinogen with experimental carcinogenic, neoplastigenic, and tumorigenic data. When heated to decomposition it emits toxic fumes of NO_x.

AHR500   CAS: 99-03-6   HR: 2
3'-AMINOACETOPHENONE
mf: C_8H_9NO mw: 135.18

PROP: Yellow, oily liquid. Bp: 251°; (slt decomp); insol in water; sol in alc and ether.

SYNS: m-ACETYLANILINE * 3-ACETYLANILINE * β-AMINOACETOPHENONE * m-AMINOACETOPHE-NONE * m-AMINOACETYLBENZENE

CONSENSUS REPORTS: Reported in EPA TSCA Inventory.

SAFETY PROFILE: Moderately toxic by ingestion. Mildly toxic by skin contact. An eye irritant. Mutation data reported. When heated to decomposition it emits toxic fumes of NO_x.

AHR750   CAS: 99-92-3   HR: 3
p-AMINO ACETOPHENONE
mf: C_8H_9NO mw: 135.18

PROP: Crystalline. Mp: 106°, bp: 293-295°. Sol in hot water, alc, and ether.

SYNS: 4-ACETYLANILINE * 4'-AMINOACETOPHE-NONE * p-AMINOACETYLBENZENE * USAF EK-631

CONSENSUS REPORTS: Reported in EPA TSCA Inventory.

SAFETY PROFILE: Poison by ingestion and intraperitoneal routes. When heated to decomposition it emits toxic fumes of NO_x.

AHS500   CAS: 90-45-9   HR: 3
9-AMINOACRIDINE
mf: $C_{13}H_{10}N_2$ mw: 194.25

SYNS: 9AA * 9-ACRIDINAMINE * AMINACRINE * 5-AMINOACRIDINE * IZOACRIDINA * MONAC-RIN

CONSENSUS REPORTS: EPA Genetic Toxicology Program.

SAFETY PROFILE: Poison by intraperitoneal and subcutaneous routes. Mutation data reported. When heated to decomposition it emits toxic fumes of NO_x.

AHT850   CAS: 53222-25-6   HR: 3
6-AMINO-4-((3-AMINO-4-(((4-((1-METHYLPYRIDINIUM-4-YL)AMINO)PHENYL)AMINO)CARBONYL)PHENYL)AMINO)-1-METHYLQUINOLINIUM),DIIODIDE
mf: $C_{29}H_{29}N_6O \cdot 2I$ mw: 731.44

SAFETY PROFILE: Poison by intraperitoneal route. Mutagenic data reported. When heated to decomposition it emits very toxic fumes of NO_x and I^-.

AIA750   CAS: 82-45-1   HR: 3
1-AMINOANTHRAQUINONE
mf: $C_{14}H_9NO_2$ mw: 223.24

PROP: Red needles. Mp: 256°, bp: subl. Insol in water; sol in HCl, alc, benzene, ether, and chloroform.

SYNS: 1-AMINO-9,10-ANTHRACENEDIONE * 1-AMI-NOANTHRACHINON (CZECH) * α-AMINOANTHRAQUI-NONE * 1-AMINO-9,10-ANTHRAQUINONE * α-AN-THRAQUINONYLAMINE * C.I. 37275 * DIAZO FAST RED AL

CONSENSUS REPORTS: Reported in EPA TSCA Inventory.

SAFETY PROFILE: Moderately toxic by intraperitoneal route. An eye irritant. Questionable carcinogen with experimental tumorigenic data. Mutation data reported. When heated to decomposition it emits toxic NO_x.

AIB000   CAS: 117-79-3   HR: 3
2-AMINOANTHRAQUINONE
mf: $C_{14}H_9NO_2$ mw: 223.24

PROP: Red needles from alc. Mp: 302°, bp: subl. Insol in water and ether; sol in alc and benzene.

SYNS: 2-AMINO-9,10-ANTHRACENEDIONE * 2-AMINO-9,10-ANTRAQUINONE * β-AMINO-ANTHRAQUINONE * β-ANTHRAQUINONYLAMINE * NCI-C01876

CONSENSUS REPORTS: IARC Cancer Review: GROUP 3 IMEMDT 7,56,87; Animal Limited Evidence IMEMDT 27,191,82. NTP Fourth Annual Report On Carcinogens, 1984. NCI Carcinogenesis Bioassay (feed); Clear Evidence: mouse, rat NCITR* NCI-CG-TR-144,78. Community Right-To-Know List. Reported in EPA TSCA Inventory.

SAFETY PROFILE: Confirmed carcinogen with experimental carcinogenic, neoplastigenic, and tumorigenic data. Moderately toxic via intraperitoneal route. Mutation data reported. When heated to decomposition it emits toxic NO_x.

AIC250   CAS: 97-56-3   HR: 3
2-AMINO-5-AZOTOLUENE
mf: $C_{14}H_{15}N_3$ mw: 225.32

SYNS: AAT * o-AAT * o-AMIDOAZOTOLUOL (GERMAN) * AMINOAZOTOLUENE (indicator) * o-AMINOAZOTOLUENE (MAK) * 4'-AMINO-2,3'-AZOTOLUENE * o-AMINOAZOTOLUENO (SPANISH) * 4'-AMINO-2:3'-AZOTOLUENE * o-AMINOAZOTO-LUOL * 4-AMINO-2',3-DIMETHYLAZOBENZENE * 4'-AMINO-2,3'-DIMETHYLAZOBENZENE * o-AT * BRASILAZINA OIL YELLOW R * BUTTER YELLOW * C.I. 11160 * C.I. 11160B * C.I. SOLVENT YELLOW 3 * 2',3-DIMETHYL-4-AMINOAZO-BENZENE * FAST GARNET GBC BASE * FAST OIL YELLOW * FAST YELLOW AT * FAST YELLOW B * HIDACO OIL YELLOW * 2-METHYL-4-((2-METHYLPHENYL)AZO)BENZENAMINE * OAAT

* OIL YELLOW * OIL YELLOW 21 * OIL YELLOW 2681 * OIL YELLOW AT * OIL YELLOW A * OIL YELLOW C * OIL YELLOW I * OIL YELLOW 2R * OIL YELLOW T * ORGANOL YELLOW 25 * SOMALIA YELLOW R * SUDAN YELLOW RRA * o-TOLUENEAZO-o-TOLUIDINE * o-TOLUOL-AZO-o-TOLUIDIN (GERMAN) * 5-(o-TOLYLAZO)-2-AMINOTO-LUENE * 4-(o-TOLYLAZO)-o-TOLUIDINE * TULA-BASE FAST GARNET GB * TULABASE FAST GARNET GBC * WAXAKOL YELLOW NL

CONSENSUS REPORTS: IARC Cancer Review: GROUP 2B IMEMDT 7,56,87; Animal Sufficient Evidence IMEMDT 8,61,75. Community Right-To-Know List. Reported in EPA TSCA Inventory. EPA Genetic Toxicology Program.

DFG MAK: Animal Carcinogen; Suspected Human Carcinogen.

SAFETY PROFILE: Confirmed carcinogen with experimental carcinogenic, neoplastigenic, and tumorigenic data. Poison by ingestion. Moderately toxic by subcutaneous route. Human mutation data reported. Experimental reproductive effects. When heated to decomposition it emits toxic fumes of NO_x.

AID500   CAS: 98-16-8   HR: 3
m-AMINOBENZAL FLUORIDE

DOT: UN 2948
mf: $C_7H_6F_3N$ mw: 161.14

PROP: Colorless liquid with aniline-like odor. Mp: 3°, bp: 189°, d: 1.303 @ 15.5°/15.5°, vap d: 5.56.

SYNS: m-AMINOBENZOTRIFLUORIDE * 3-AMINO-BENZOTRIFLUORIDE * m-(TRIFLUOROMETHYL)-ANILINE * 3-(TRIFLUOROMETHYL)ANILINE * 3-(TRIFLUOROMETHYL)BENZENAMINE * USAF MA-4

CONSENSUS REPORTS: EPA Extremely Hazardous Substances List. Reported in EPA TSCA Inventory.

DOT Classification: IMO: Poison B; Label: Poison.

SAFETY PROFILE: Poison by inhalation, ingestion, and intraperitoneal routes. May be moderately toxic by other routes. When heated to decomposition it emits very toxic fumes of F^- and NO_x.

AIF500   CAS: 137-07-5   **HR: 3**
2-AMINOBENZENETHIOL
mf: C_6H_7NS mw: 125.20

PROP: Liquid. Mp: 23°, bp: 227.2°, flash p: 175°F, d: 1.168, vap d: 4.3.

SYNS: o-AMINOTHIOPHENOL * 2-AMINOTHIOPHE-NOL * o-MERCAPTOANILINE * USAF EK-4376

CONSENSUS REPORTS: Reported in EPA TSCA Inventory.

SAFETY PROFILE: Poison by intraperitoneal route. Moderately toxic by ingestion. Moderately flammable. Can react with oxidizing materials. To fight fire use water, foam, CO_2, mist or spray, dry chemical.

AIG000   CAS: 934-32-7   **HR: 3**
2-AMINOBENZIMIDAZOLE
mf: $C_7H_7N_3$ mw: 133.17

PROP: Aqueous leaflets. Mp: 222-224°; sol in water, alkalies, alc, acetone; very sltly sol in ether.

SYN: USAF EK-4037

CONSENSUS REPORTS: Reported in EPA TSCA Inventory.

SAFETY PROFILE: Poison by intravenous and intraperitoneal routes. Moderately toxic by ingestion. An experimental teratogen. Mutation data reported. When heated to decomposition it emits toxic NO_x.

AIH600   CAS: 150-13-0   **HR: 2**
p-AMINOBENZOIC ACID
mf: $C_7H_7NO_2$ mw: 137.15

PROP: Yellowish to red crystals. Mp: 187°. Sol in water, alc, and ether.

SYNS: γ-AMINOBENZOIC ACID * 4-AMINOBENZOIC ACID * 1-AMINO-4-CARBOXYBENZENE * ANTI-CHROMOTRICHIA FACTOR * BACTERIAL VITAMIN H1 * 4-CARBOXYANILINE * p-CARBOXYPHENYLAMINE * CHROMOTRICHIA FACTOR * PABA * TRICHO-CHROMOGENIC FACTOR * VITAMIN H

CONSENSUS REPORTS: IARC Cancer Review: Animal Inadequate Evidence IMEMDT 16,249,78. Reported in EPA TSCA Inventory.

SAFETY PROFILE: Moderately toxic by ingestion and intravenous routes. Ingesting large doses can cause nausea, vomiting, skin rash, methemoglobinemia and possibly toxic hepatitis. Experimental reproductive effects. Mutation data reported. Combustible. When heated to decomposition it emits toxic fumes of NO_x. A topical sunscreen.

AIL500   CAS: 63917-76-0   **HR: 3**
p-AMINOBENZOIC ACID-3-(β-DIETHYLAMINO)ETHOXY)PROPYL ESTER
mf: $C_{16}H_{26}N_2O_3$ mw: 294.44

SAFETY PROFILE: Poison by intravenous and subcutaneous routes. When heated to decomposition it emits toxic fumes of NO_x.

AIL750   CAS: 59-46-1   **HR: 3**
p-AMINOBENZOIC ACID-2-DIETHYLAMINOETHYL ESTER
mf: $C_{13}H_{20}N_2O_2$ mw: 236.35

SYNS: ALLOCAINE * 4-AMINOBENZOIC ACID DI-ETHYLAMINOETHYL ESTER * p-AMINO BENZOYLDIETHYLAMINOETHANOL * DIETHYL-AMINOETHYL-p-AMINOBENZOATE * β-DI-ETHYLAMINOETHYL-4-AMINOBENZOATE * 2-DIETHYLAMINOETHYL-p-AMINOBENZOATE * GEROVITAL * JENACAINE * NEOCAINE * NISSOCAINE * NOROCAINE * NOVOCAINE * PROCAINE * PROCAINE, BASE * SCUROCAINE * SPINOCAINE

SAFETY PROFILE: Poison by ingestion, intraperitoneal, intravenous, and subcutaneous routes. Moderately toxic by parenteral route. Human systemic effects by intramuscular route: lack of muscular control, rigidity and possibly catalepsy. When heated to decomposition it emits toxic fumes of NO_x. Used as a local anesthetic.

AIR250   CAS: 1137-41-3   **HR: 3**
p-AMINOBENZOPHENONE
mf: $C_{13}H_{11}NO$ mw: 197.25

PROP: Leaflets from alc. Mp: 124°. Very sltly sol in cold water, very sol in alc.

SYN: USAF A-233.

CONSENSUS REPORTS: Reported in EPA TSCA Inventory.

SAFETY PROFILE: Poison by intraperitoneal route. When heated to decomposition it emits toxic fumes of NO_x.

AIS600 CAS: 4570-41-6 **HR: 3**
2-AMINOBENZOXAZOLE
mf: $C_7H_6N_2O$ mw: 134.15
SAFETY PROFILE: Poison by intravenous and intraperitoneal routes. Moderately toxic by ingestion. When heated to decomposition it emits toxic fumes of NO_x.

AIT000 CAS: 5892-15-9 **HR: 3**
AMINOBENZOYLDIBUTYLAMINOPRO-PANOL HYDROCHLORIDE
mf: $C_{18}H_{30}N_2O_2 \cdot ClH$ mw: 342.96
SYN: p-AMINOBENZOIC ACID-3-(DIBUTYL-AMINO)PROPYL ESTER, HYDROCHLORIDE

SAFETY PROFILE: Poison by subcutaneous, intraperitoneal, and possibly other routes. When heated to decomposition it emits very toxic fumes of HCl and NO_x.

AIT250 CAS: 51-05-8 **HR: 3**
p-AMINOBENZOYLDIETHYLAMINO-ETHANOL HYDROCHLORIDE
mf: $C_{13}H_{20}N_2O_2 \cdot ClH$ mw: 272.81
SYNS: ALLOCAINE * 4-AMINOBENZOIC ACID 2-(DI-ETHYLAMINO)ETHYL ESTER, HYDROCHLORIDE * p-AMINOBENZOIC ACID-2-DIETHYLAMINOETHYL ES-TER, HYDROCHLORIDE * AMINOCAINE * ANADO-LOR * ANESTIL * ANESTHESOL * ATOXICO-CAINE * BERNOCAINE * CETAIN * CHLOROCAINE * 2-DIETHYLAMINOETHYL-p-AMINOBENZOATE HYDROCHLORIDE * DIETHYL-AMINOETHANOL-4-AMINOBENZOATE HYDROCHLORIDE * DUGERASE * ETHOCAINE * IROCAINE * ISOCAINE-ASID * ISOCAINE-HEISLER * JUVOCAINE * KEROCAINE * LACTOCAINE * NAUCAINE * NEOCAINE * NOVOCAIN-CHLORHYDRAT (GERMAN) * NOVOCAIN HYDROCHLORID (GERMAN) * NOVOCAINE HYDROCHLORIDE * PARACAIN * PLANOCAINE * PROCAINE HYDROCHLORIDE * SCUROCAINE * SEVICAINE * SYNCAINE * TOPOKAIN * WESTOCAINE

CONSENSUS REPORTS: Reported in EPA TSCA Inventory. EPA Genetic Toxicology Program.
SAFETY PROFILE: Poison by ingestion, subcutaneous, intravenous, intraperitoneal, intraspinal, parenteral, and possibly other routes. May have human reproductive effects. When heated to decomposition it emits very toxic fumes of HCl and NO_x. Used as a local anesthetic.

AIT750 CAS: 532-62-7 **HR: 3**
p-AMINOBENZOYLDIMETHYLAMINO-1,2-DIMETHYLPROPANOL HYDROCHLORIDE
mf: $C_{14}H_{22}N_2O_2 \cdot ClH$ mw: 286.84

SYNS: p-AMINOBENZOIC ACID 3-(DIMETHYLAMINO)-1,2-DIMETHYLPROPYL ESTER, HYDROCHLORIDE * BUTAMIN * 3-DIMETHYLAMINO-1,2-DIMETHYL-PROPYL p-AMINOBENZOATE HYDROCHLORIDE * 4-(DIMETHYLAMINO)-3-METHYL-2-BUTANOL 4-AMINOBENZOATE (ester) HYDROCHLORIDE * 3-DI-METHYL-1,2-DIMETHYLPROPYL p-AMINOBENZOATE HYDROCHLORIDE * TOTOCAINE HYDROCHLORIDE * TUTOCAINE HYDROCHLORIDE

SAFETY PROFILE: Poison by subcutaneous, intravenous, intraperitoneal, and intraspinal routes. When heated to decomposition it emits very toxic fumes of HCl and NO_x. Used as a surface and infiltration anesthetic.

AIV500 CAS: 69-53-4 **HR: 3**
AMINOBENZYLPENICILLIN
mf: $C_{16}H_{19}N_3O_4S$ mw: 349.44
SYNS: ACILLIN * ADOBACILLIN * ALPEN * AMBLOSIN * AMCILL * AMFIPEN * d-(−)-α-AMINOBENZYLPENICILLIN * d-(−)-α-AMINOPENICIL-LIN * 6-(d(−)-α-AMINOPHENYLACETAMIDO)PENI-CILLANIC ACID * (AMINOPHENYLMETHYL)-PENICIL-LIN * AMIPENIX S * AMPERIL * AMPI-BOL * AMPICILLIN (USDA) * d-AMPICILLIN * d-(−)-AMPICILLIN * AMPICILLIN A * AMPICILLIN ACID * AMPICILLIN ANHYDRATE * AMPICIN * AMPIKEL * AMPIMED * AMPIPENIN * AMPLISOM * AMPLITAL * AMPY-PENYL * AUSTRAPEN * AY-6108 * BINOTAL * BONAPICILLIN * BRITACIL * BRL * BRL 1341 * COPHARCILIN * CYMBI * DIVERCILLIN * DOKTACILLIN * GRAM-PENIL * GUICITRINA * GUICITRINE * LIFE-AMPIL * MARISILAN * NSC-528986 * NUVAPEN * OMNIPEN * P-50 * PENBRISTOL * PENBRITIN * PENBRITIN PAEDIATRIC * PENBRITIN SYRUP * PENBROCK * PENICLINE * PENTREX * PENTREXL * PFIZERPEN A * POLYCILLIN * PONECIL * PRINCIPEN * QIDAMP * RO-AMPEN * SEMICILLIN * SK-AMPICILLIN * SYNPENIN * TOKIOCILLIN * TOLOMOL * TOTACILLIN * TOTALCICLINA * TOTAPEN * ULTRABION * ULTRABRON * VICCILLIN * VICCILLIN S * VICILLIN * WY-5103

CONSENSUS REPORTS: EPA Genetic Toxicology Program.

SAFETY PROFILE: Poison by intracerebral and other unspecified routes. Moderately toxic by intraperitoneal route. Human systemic effects by ingestion: fever, angranulocytosis, and other blood effects. Experimental reproductive effects. Mutation data reported. When heated to decomposition it emits very toxic fumes of NO_x and SO_x.

AIX000   CAS: 1031-47-6   **HR: 3**
5-AMINO-1-BIS(DIMETHYLAMIDE) PHOSPHORYL-3-PHENYL-1,2,4-TRIAZOLE
mf: $C_{12}H_{19}N_6OP$ mw: 294.34

SYNS: 5-AMINO-1-BIS(DIMETHYLAMIDO)PHOSPHORYL-3-PHENYL-1,2,4-TRIAZOLE * 5-AMINO-1-(BIS(DIMETHYLAMINO)PHOSPHINYL)-3-PHENYL-1,2,4-TRIAZOLE * 5-AMINO-3-FENIL-1-BIS(-DIMETILAMINO)-FOSFORIL-1,2,4-TRIAZOLO (ITALIAN) * 5-AMINO-3-FENYL-1-BIS-(DIMETHYL-AMINO)-FOSFORYL-1,2,4-TRIAZOOL (DUTCH) * 5-AMINO-3-PHENYL-1-BIS (DIMETHYL-AMINO)-PHOSPHORYLE-1,2,4-TRIAZOLE (FRENCH) * 5-AMINO-3-PHENYL-1-BIS(DIMETHYLAMINO)-PHOSPHORYL-1H-1,2,4-TRIAZOL (GERMAN) * 5-AMINO-3-PHENYL-1,2,4-TRIAZOLE-1-YL-N,N,N′,N′-TETRAMETHYLPHOSPHODIAMIDE * 5-AMINO-3-PHENYL-1,2,4-TRIAZOLYL-1-BIS(DIMETHYLAMIDO)PHEOSPHATE * 5-AMINO-3-PHENYL-1,2,4-TRIAZOLYL-N,N,N′N′-TETRAMETHYL-PHOSPHONAMIDE * p-(5-AMINO-3-PHENYL-1H-1,2,4-TRIAZOL-1-YL)-N,N,N′-TETRAMETHYL PHOSPHONIC DIAMIDE * BIS(DIMETHYLAMINO)-3-AMINO-5-PHENYLTRIAZOLYL PHOSPHINE OXIDE * ENT 27,223 * NIAGARA 5943 * 3-PHENYL-5-AMINO-1,2,4-TRIAZOLYL-(1)-(N,N′-TETRAMETHYL) DIAMIDOPHOSPHONATE * TRIAMIFOS (GERMAN, DUTCH, ITALIAN) * TRIAMIPHOS * TRIAMPHOS * WEPSIN * WEPSYN * WEPSYN 155 * WP 155

CONSENSUS REPORTS: EPA Extremely Hazardous Substances List.

SAFETY PROFILE: Poison by ingestion, skin contact, intraperitoneal, and possibly other routes. Experimental teratogenic and reproductive effects. Mutation data reported. When heated to decomposition it emits very toxic fumes of PO_x and NO_x.

AIX250   CAS: 56-18-8   **HR: 3**
AMINOBIS(PROPYLAMINE)

DOT: UN 2269
mf: $C_6H_{17}N_3$ mw: 131.26

SYNS: BIS-(3-AMINOPROPYL)AMINE * 3,3-DIAMINODIPROPYLAMINE * 3,3′-DIAMINODIPROPYLAMINE * DIPROPYLENETRIAMINE * IMINOBIS(PROPYLAMINE) * 3,3′-IMINOBIS(PROPYLAMINE) * INITIATING EXPLOSIVE IMINOBISPROPYLAMINE (DOT)

CONSENSUS REPORTS: Reported in EPA TSCA Inventory.

DOT Classification: Corrosive Material; Label: Corrosive.

SAFETY PROFILE: Poison by skin contact. Moderately toxic by ingestion. A skin and severe eye irritant. When heated to decomposition it emits toxic fumes of NO_x. An explosive.

AJA250   CAS: 96-20-8   **HR: 3**
2-AMINOBUTAN-1-OL
mf: $C_4H_{11}NO$ mw: 89.16

PROP: Water-white liquid. Mp: $-2°$, bp: 178°, flash p: 165°F (OC), d: 0.944 @ 20°/20°, vap d: 3.06.

SYNS: 2-AMINO-1-BUTANOL * 2-AMINO-n-BUTYL ALCOHOL * BUTANOL-2-AMINE

CONSENSUS REPORTS: Reported in EPA TSCA Inventory.

SAFETY PROFILE: Poison by intravenous and intraperitoneal routes. Moderately toxic by ingestion. Moderately flammable when exposed to heat, flame or oxidizing materials. To fight fire use water spray, alcohol foam, dry chemical. When heated to decomposition it yields NO_x.

AJB250   CAS: 118-68-3   **HR: 3**
3-(2-AMINOBUTYL)INDOLE ACETATE
mf: $C_{12}H_{16}N_2 \cdot C_2H_4O_2$ mw: 248.36

SYNS: α-ETHYLTRYPTAMINE ACETATE * dl-α-ETHYLTRYPTAMINE ACETATE * ETRYPTAMINE ACETATE * INDOLE-3-(2-AMINOBUTYL) ACETATE

SAFETY PROFILE: Poison by ingestion, intraperitoneal, and intravenous routes. When heated to decomposition it emits toxic fumes of NO_x.

AJC500   CAS: 34562-99-7   **HR: 3**
γ-AMINOBUTYRIC ACID CETYL ESTER
mf: $C_{20}H_{41}NO_2$ mw: 327.62

SYNS: CETYL-γ-AMINOBUTYRATE * CETYL GABA

SAFETY PROFILE: Poison by intravenous and intraperitoneal routes. When heated to decomposition it emits toxic fumes of NO_x.

AJD750   CAS: 26148-68-5   **HR: 3**
AMINO-α-CARBOLINE
mf: $C_{11}H_9N_3$ mw: 183.2

SYNS: 2-AMINO-α-CARBOLINE * 2-AMINO-9H-PYRIDO(2,3-B)INDOLE

CONSENSUS REPORTS: Cancer Review: GROUP 2B IMEMDT 7,56,87; Animal Sufficient Evidence IMEMDT 40,245,86

SAFETY PROFILE: Suspected carcinogen with experimental carcinogenic data. Human mutation data reported. When heated to decomposition it emits toxic fumes of NO_x.

AJP750   CAS: 109-55-7   **HR: 2**
1-AMINO-3-DIMETHYLAMINOPROPANE
mf: $C_5H_{14}N_2$ mw: 102.21

PROP: Colorless liquid. Mp: $< -70°$, bp: 123°, flash p: 100°F (OC), d: 0.8100 @ 30°, vap press: 10 mm @ 30°, vap d: 3.52.

SYNS: N,N-DIMETHYL-N-(3-AMINOPROPYL)-AMINE * 3-(DIMETHYLAMINO)PROPYLAMINE * N,N-DIMETHYL-1,3-DIAMINOPROPANE * N,N-DI-METHYL-1,3-PROPANEDIAMINE * N,N-DIMETHYL-1,3-PROPYLENEDIAMINE

CONSENSUS REPORTS: Reported in EPA TSCA Inventory.

SAFETY PROFILE: Moderately toxic by ingestion. A skin and eye irritant. Very flammable when exposed to heat, flame or oxidizers. Reaction with 1,2-dichloroethane produces explosive acetylene gas. This and other amines ignite on contact with cellulose nitrate of high surface area. To fight fire, use alcohol foam, CO_2, dry chemical. When heated to decomposition it emits toxic fumes of NO_x.

AJQ675   CAS: 77500-04-0   **HR: 3**
2-AMINO-3,8-DIMETHYLIMIDAZO
(4,5-f)QUINOXALINE
mf: $C_{11}H_{11}N_5$ mw: 213.27

SYNS: 2-AMINO-3,8-DIMETHYL-3H-IMIDAZO(4,5-f)-QUINOXALINE * 3,8-DIMETHYL-3H-IMIDAZO(4,5-f)-QUINOXALIN-2-AMINE

CONSENSUS REPORTS: IARC Cancer Review: GROUP 3 IMEMDT 7,56,87; Animal Inadequate Evidence IMEMDT 40,283,86

SAFETY PROFILE: Suspected carcinogen with experimental carcinogenic data. Mutation data reported. When heated to decomposition it emits toxic fumes of NO_x.

AJR500   CAS: 68808-54-8   **HR: 3**
3-AMINO-1,4-DIMETHYL-5H-
PYRIDO(4,3-b)INDOLE ACETATE
mf: $C_{13}H_{13}N_3 \cdot C_2H_4O_2$ mw: 271.35

SYNS: 1,4-DIMETHYL-5H-PYRIDO(4,3-b)INDOL-3-AMINE ACETATE * 1,4-DIMETHYL-5H-PYRIDO(4,3-b)INDOL-3-AMINE MONOACETATE * TRP-P-1 (ACETATE)

SAFETY PROFILE: Suspected carcinogen with experimental carcinogenic data. Mutation data reported. When heated to decomposition it emits toxic fumes of NO_x.

AJS100   CAS: 92-67-1   **HR: 3**
4-AMINODIPHENYL
mf: $C_{12}H_{11}N$ mw: 169.24

PROP: Colorless crystals. Mp. 53°, bp: 302°, d: 1.160 @ 20°/20°, autoign temp: 842°F.

SYNS: p-AMINOBIPHENYL * 4-AMINOBIPHENYL * 4-AMINODIFENIL (SPANISH) * p-AMINODIPHENYL * BIPHENYLAMINE * 4-BIPHENYLAMINE * (1,1'-BIPHENYL)-4-AMINE * p-BIPHENYLAMINE * PARAAMINODIPHENYL * p-PHENYLANILINE * XENYLAMIN (CZECH) * XENYLAMINE

CONSENSUS REPORTS: IARC Cancer Review: GROUP 1 IMEMDT 7,91,87; Human Limited Evidence IMEMDT 1,74,72; Animal Sufficient Evidence IMEMDT 1,74,72; Human Sufficient Evidence IMEMDT 28,151,82. NTP Fourth Annual Report On Carcinogens, 1984. Reported in EPA TSCA Inventory. EPA Genetic Toxicology Program. Community Right-To-Know List.

OSHA PEL: Cancer Suspect Agent
ACGIH TLV: Confirmed Human Carcinogen.
DFG MAK: Human Carcinogen.

SAFETY PROFILE: Confirmed human carcinogen with experimental carcinogenic and tumorigenic data. Poison by ingestion and intraperitoneal routes. Human mutation data reported. An irritant. Effects resemble those of benzidine.

Slight to moderate fire hazard when exposed to heat, flames (sparks), or powerful oxidizers. To fight fire, use water spray, mist, dry chemical. When heated to decomposition it emits toxic fumes of NO_x.

AJT250   CAS: 60-23-1   **HR: 3**
2-AMINOETHANETHIOL
mf: C_2H_7NS mw: 77.16

SYNS: 2-AMINOETHYL MERCAPTAN * BECAPTAN * CISTEAMINA (ITALIAN) * CYCTEINAMINE * CYSTEAMIDE * CYSTEAMINE * DECARBOXY-CYSTEINE * LAMBRATEN * MEA * MECRA-MINE * MERCAMINE * MERCAPTAMINE * β-MERCAPTOETHYLAMINE * (2-MERCAPTO-ETHYL)AMINE * THIOETHANOLAMINE

CONSENSUS REPORTS: EPA Genetic Toxicology Program.

SAFETY PROFILE: Poison by intravenous, subcutaneous, and intraperitoneal routes. Moderately toxic by ingestion. Experimental reproductive effects. Mutation data reported. When heated to decomposition it emits very toxic fumes of SO_x and NO_x.

AJU250   CAS: 929-06-6   **HR: 2**
2-AMINOETHOXYETHANOL

DOT: NA 1760
mf: $C_4H_{11}NO_2$ mw: 105.16

SYNS: 2-(2-AMINOETHOXY)ETHANOL * DIGLYCO-LAMINE

CONSENSUS REPORTS: Reported in EPA TSCA Inventory.

DOT Classification: Corrosive Material; Label: Corrosive.

SAFETY PROFILE: Moderately toxic by skin contact. Mildly toxic by ingestion. Severe eye and skin irritant. Corrosive and a powerful irritant. When heated to decomposition it emits toxic fumes of NO_x.

AJV000   CAS: 132-32-1   **HR: 3**
3-AMINO-9-ETHYLCARBAZOLE
mf: $C_{14}H_{14}N_2$ mw: 210.30

PROP: In cancer bioassay both free amine and hydrochloride salt used NCITR* NCI-CG-TR-93,78.

SYN: 3-AMINO-N-ETHYLCARBAZOLE

CONSENSUS REPORTS: Reported in EPA TSCA Inventory.

DFG MAK: Suspected Carcinogen.

SAFETY PROFILE: Suspected carcinogen with experimental carcinogenic data. Poison by ingestion and intraperitoneal routes. When heated to decomposition it emits toxic fumes of NO_x.

AJV250   CAS: 6109-97-3   **HR: 3**
3-AMINO-9-ETHYLCARBAZOLEHY-DROCHLORIDE
mf: $C_{14}H_{14}N_2 \cdot ClH$ mw: 246.76

PROP: In cancer bioassay both free amine and hydrochloride salt used NCITR* NCI-CG-TR-93,78.

SYN: NCI-C03043

CONSENSUS REPORTS: NCI Carcinogenesis Bioassay Completed; Results Positive: mouse, rat NCITR* NCI-CG-TR-93,78.

SAFETY PROFILE: Suspected carcinogen with experimental carcinogenic data. Poison by ingestion. Mutation data reported. When heated to decomposition it emits very toxic fumes of NO_x and HCl.

AKB000   CAS: 140-31-8   **HR: 3**
N-AMINOETHYLPIPERAZINE

DOT: UN 2815
mf: $C_6H_{15}N_3$ mw: 129.24

PROP: Light-colored liquid. D: 0.9852 @ 20°/20°, mp: −19°, bp: 220.4°, flash p: 200°F (OC), vap d: 4.4.

SYNS: AMINOETHYLPIPERAZINE * N-(β-AMINO-ETHYL)PIPERAZINE * N-(2-AMINOETHYL)PIPERAZINE * 1-(2-AMINOETHYL)PIPERAZINE * USAF DO-46

CONSENSUS REPORTS: Reported in EPA TSCA Inventory.

DOT Classification: Corrosive Material; Label: Corrosive.

SAFETY PROFILE: Poison by intraperitoneal routes. Moderately toxic by ingestion and skin contact. A skin and eye irritant. Mutation data reported. Moderately flammable when exposed to heat, flame or sparks, powerful oxidizers. To fight fire, use alcohol foam. When heated to decomposition it emits toxic fumes of NO_x.

AKE250   CAS: 116-85-8   **HR: 3**
1-AMINO-4-HYDROXYANTHRA-
QUINONE
mf: $C_{14}H_9NO_3$ mw: 239.24

PROP: Red-violet powder. Mp: 207-208°. Sol in water, HCl, alc, ether, and benzene.

SYNS: 1-AMINO-4-OXYANTHRAQUINONE (RUSSIAN) * C.I. 60710 * 1-HYDROXY-4-AMINOANTHRAQUI-NONE * 4-HYDROXY-1-ANTHRAQUINONYLAMINE

CONSENSUS REPORTS: Reported in EPA TSCA Inventory.

SAFETY PROFILE: Poison by intraperitoneal route. Mutation data reported. When heated to decomposition it emits toxic fumes of NO_x.

AKF000   CAS: 536-25-4   **HR: 3**
3-AMINO-4-HYDROXYBENZOIC ACID
METHYL ESTER
mf: $C_8H_9NO_3$ mw: 167.18

SYNS: AMINOBENZ * ORTHOCAINE * ORTHO-DERM * ORTHOFORM

CONSENSUS REPORTS: Reported in EPA TSCA Inventory.

SAFETY PROFILE: Poison by intraperitoneal route. Moderately toxic by ingestion. When heated to decomposition it emits toxic fumes of NO_x.

AKP750   CAS: 82-28-0   **HR: 3**
1-AMINO-2-METHYLANTHRAQUINONE
mf: $C_{15}H_{11}NO_2$ mw: 237.27

SYNS: ACETATE FAST ORANGE R * ACETOQUI-NONE LIGHT ORANGE JL * 1-AMINO-2-METHYL-9,10-ANTHRACENEDIONE * ARTISIL ORANGE 3RP * CELLITON ORANGE R * C.I. 60700 * C.I. DIS-PERSE ORANGE 11 * CILLA ORANGE R * DIS-PERSE ORANGE * DURANOL ORANGE G * 2-METHYL-1-ANTHRAQUINONYLAMINE * MICRO-SETILE ORANGE RA * NCI-C01901 * NYLO-QUINONE ORANGE JR * PERLITON ORANGE 3R * SERISOL ORANGE YL * SUPRACET ORANGE R

CONSENSUS REPORTS: IARC Cancer Review: GROUP 3 IMEMDT 7,56,87; Animal Limited Evidence IMEMDT 27,199,82. NTP Fourth Annual Report On Carcinogens, 1984. NCI Carcinogenesis Bioassay (feed); Clear Evidence: mouse, rat NCITR* NCI-CG-TR-111,78. Community Right-To-Know List. Reported in EPA TSCA Inventory.

SAFETY PROFILE: Confirmed carcinogen with experimental carcinogenic, neoplastigenic, and tumorigenic data. When heated to decomposition it emits toxic fumes of NO_x.

AKS250   CAS: 67730-11-4   **HR: 3**
2-AMINO-6-METHYLDIPYRIDO
(1,2-a:3′,2′-d)IMIDAZOLE
mf: $C_{11}H_{10}N_4$ mw: 198.25

SYNS: GLU-P-I * 6-ME-GLU-P-2 * 6-METHYL DIPYRIDO(1,2-a:3′,2′-d)IMIDAZOL-2-AMINE

CONSENSUS REPORTS: IARC Cancer Review: GROUP 2B IMEMDT 7,56,87; Animal Sufficient Evidence IMEMDT 40,223,86

SAFETY PROFILE: Suspected carcinogen with experimental carcinogenic data. Human mutation data reported. When heated to decomposition it emits toxic fumes of NO_x.

AKT600   CAS: 76180-96-6   **HR: 3**
2-AMINO-3-METHYLIMIDAZO(4,5-f)
QUINOLINE
mf: $C_{11}H_{10}N_4$ mw: 198.25

CONSENSUS REPORTS: IARC Cancer Review: GROUP 2B IMEMDT 7,56,87; Animal Sufficient Evidence IMEMDT 40,261,86

SAFETY PROFILE: Suspected carcinogen with experimental carcinogenic and tumorigenic data. Mutation data reported. When heated to decomposition it emits toxic fumes of NO_x.

ALC500   CAS: 1824-81-3   **HR: 3**
2-AMINO-6-METHYLPYRIDINE
mf: $C_6H_8N_2$ mw: 108.16

PROP: Mp: 43.7°, bp: 214.4, vap d: 3.73.

CONSENSUS REPORTS: Reported in EPA TSCA Inventory.

SAFETY PROFILE: Poison by intravenous route. When heated to decomposition it emits toxic fumes of NO_x.

ALD500   CAS: 62450-07-1   **HR: 3**
3-AMINO-1-METHYL-5H-PYRIDO(4,3-b)
INDOLE
mf: $C_{12}H_{11}N_3$ mw: 197.26

SYNS: 3-AMINO-1-METHYL-γ-CARBOLINE * 1-METHYL-3-AMINO-5H-PYRIDO(4,3-b)INDOLE * TRP-P-2 * TRYPTOPHAN P2

CONSENSUS REPORTS: IARC Cancer Review: GROUP 2B IMEMDT 7,56,87; Animal

Sufficient Evidence IMEMDT 31,255,83. EPA Genetic Toxicology Program.

SAFETY PROFILE: Suspected carcinogen with experimental carcinogenic and neoplastigenic data. Mutation data reported. When heated to decomposition it emits toxic fumes of NO_x.

ALD750   CAS: 68006-83-7   **HR: 3**
2-AMINO-3-METHYL-9H-PYRIDO(2,3-b) INDOLE
mf: $C_{12}H_{11}N_3$ mw: 197.2

SYN: 2-AMINO-3-METHYL-α-CARBOLINE

CONSENSUS REPORTS: IARC Cancer Review: GROUP 2B IMEMDT 7,56,87; Animal Sufficient Evidence IMEMDT 40,253,86

SAFETY PROFILE: Suspected carcinogen with experimental carcinogenic data. Mutation data reported. When heated to decomposition it emits toxic fumes of NO_x.

ALJ750   CAS: 118-46-7   **HR: 3**
8-AMINO-2-NAPHTHOL
mf: $C_{10}H_9NO$ mw: 159.20

PROP: Crystals from benzene or ligroin. Mp: 95-97° (decomp); sol in hot water, alkali and HCl.

CONSENSUS REPORTS: Reported in EPA TSCA Inventory.

SAFETY PROFILE: Poison by intravenous route. When heated to decomposition it emits toxic NO_x.

ALL750   CAS: 5307-14-2   **HR: 3**
4-AMINO-2-NITROANILINE
mf: $C_6H_7N_3O_2$ mw: 153.16

SYNS: C.I. 76070   *   C.I. OXIDATION BASE 22   *   1,4-DIAMINO-2-NITROBENZENE   *   DURAFUR BROWN   *   DURAFUR BROWN 2R   *   DYE GS   *   FOURAMIEN 2R   *   FOURRINE 36   *   FOURRINE BROWN 2R   *   NCI-C02222   *   2NDB   *   2-NITRO-1,4-BENZENEDIAMINE   *   2-NITRO-1,4-DIAMINOBENZENE   *   NITRO-p-PHENYLENEDIAMINE   *   o-NITRO-p-PHE-NYLENEDIAMINE (MAK)   *   2-NITRO-1,4-PHENYLENE-DIAMINE   *   2-NITRO-p-PHENYLENEDIAMINE   *   2-NP   *   2-NPPD   *   2-N-p-PDA   *   OXIDATION BASE 22   *   URSOL BROWN RR   *   ZOBA BROWN RR

CONSENSUS REPORTS: IARC Cancer Review: Animal Inadequate Evidence IMEMDT 16,73,78. NCI Carcinogenesis Bioassay (feed);

No Evidence: rat NCITR* NCI-CG-TR-169,79; Clear Evidence: mouse NCITR* NCI-CG-TR-169,79. Reported in EPA TSCA Inventory. EPA Genetic Toxicology Program.

DFG MAK: Suspected Carcinogen.

SAFETY PROFILE: Suspected carcinogen with experimental carcinogenic and neoplastigenic data. Poison by intraperitoneal route. Moderately toxic by ingestion. Experimental teratogenic and reproductive effects. Mutation data reported. When heated to decomposition it emits toxic fumes of NO_x.

ALO750   CAS: 2871-01-4   **HR: D**
2-((4-AMINO-2-NITROPHENYL) AMINO)ETHANOL
mf: $C_8H_{11}N_3O_3$ mw: 197.22

SYNS: HC RED NO. 3   *   NCI-C54922

CONSENSUS REPORTS: NTP Carcinogenesis Studies (gavage); Equivocal Evidence: mouse NTPTR* NTP-TR-281,86; No Evidence: rat NTPTR* NTP-TR-281,86. Reported in EPA TSCA Inventory.

SAFETY PROFILE: Questionable carcinogen with experimental carcinogenic data. Mutation data. When heated to decomposition it emits toxic fumes of NO_x.

ALT000   CAS: 95-55-6   **HR: 3**
2-AMINOPHENOL

DOT: UN 2512
mf: C_7H_7NO mw: 109.14

PROP: Colorless needles. Mp: 173°; bp: subl. Sol in water, alc; very sol in ether.

SYNS: 2-AMINO-1-HYDROXYBENZENE   *   o-AMINOPHENOL   *   BASF URSOL 3GA   *   BENZOFUR GG   *   C.I. 76520   *   C.I. OXI-DATION BASE 17   *   FOURAMINE OP   *   o-HY-DROXYANILINE   *   2-HYDROXYANILINE   *   NAKO YELLOW EGA   *   PARADONE OLIVE GREEN B   *   PELAGOL 3GA   *   PELAGOL GREY GG   *   ZOBA 3GA

CONSENSUS REPORTS: Reported in EPA TSCA Inventory.

SAFETY PROFILE: Poison by ingestion, intraperitoneal and subcutaneous routes. Moderately toxic by an unspecified route. Mutation data reported. Experimental reproductive effects. An eye irritant. When heated to decomposition it emits toxic NO_x.

ALT500   CAS: 591-27-5   **HR: 3**
m-AMINOPHENOL
DOT: UN 2521
mf: C_6H_7NO mw: 109.14

PROP: Prisms from toluene. Mp: 123°. Sol in water, alc; sltly sol in ether.

SYNS: m-AMINOFENOL (CZECH) * 3-AMINO-1-HY-DROXYBENZENE * 3-AMINOPHENOL * m-AMINO-PHENOL (DOT) * BASF URSOL EG * C.I. 76545 * C.I. OXIDATION BASE 7 * FOURAMINE EG * FOURRINE 65 * FOURRINE EG * FURRO EG * FUTRAMINE EG * 3-HYDROXYANILINE * NAKO TEG * PELAGOL EG * RENAL EG * TERTRAL EG * URSOL EG * ZOBA EG

CONSENSUS REPORTS: Reported in EPA TSCA Inventory. EPA Genetic Toxicology Program.

DOT Classification: IMO: Poison B; Label: St. Andrews Cross.

SAFETY PROFILE: Poison by ingestion, subcutaneous, intraperitoneal and possibly other routes. Experimental teratogenic and reproductive effects. Mutation data reported. A skin and eye irritant. When heated to decomposition it emits toxic fumes of NO_x.

ALW750   CAS: 144-14-9   **HR: 3**
N-β-(p-AMINOPHENYL)
ETHYLNORMEPERIDINE
mf: $C_{22}H_{28}N_2O_2$ mw: 352.52

SYNS: 1-(p-AMINOPHENETHYL)-4-PHENYLISONIPE-COTIC ACID, ETHYL ESTER * 1-(p-AMINOPHEN-ETHYL)-4-PHENYLPIPERIDINE-4-CARBOXYLIC ACID ETHYL ESTER * N-(β-(p-AMINOPHENYL)ETHYL)-4-PHENYL-4-CARBETHOXYPIPERIDINE * ETHYL-1-(p-AMINOPHENETHYL)-4-PHENYLISONIPECOTATE

SAFETY PROFILE: Poison by ingestion, subcutaneous, intravenous, and intraperitoneal routes. When heated to decomposition it emits toxic fumes of NO_x.

ALX250   CAS: 13425-22-4   **HR: 3**
2-AMINO-5-PHENYL-OXAZOLINE
FORMATE
mf: $C_9H_{10}N_2O • C_4H_4O_4$ mw: 278.29

SYNS: AMINOREXFUMARATE * MENOCIL

SAFETY PROFILE: Poison by ingestion. Human central nervous system effects by ingestion.

When heated to decomposition it emits toxic fumes of NO_x.

ALZ000   CAS: 51249-05-9   **HR: 2**
AMINOPHON
mf: $C_{18}H_{37}NO_3P$ mw: 346.53

SYNS: 1-(BUTYLAMINO)CYCLOHEXYLPHOSPHONIC ACID DIBUTYL ESTER * O,O-DIBUTYL-1-BUTYL-AMINO-CYCLOHEXYLPHOSPHONATE

SAFETY PROFILE: Moderately toxic by several routes. When heated to decomposition it emits very toxic fumes of PO_x and NO_x.

AMA750   **HR: 2**
3-AMINOPROPANOL
mf: C_3H_9NO mw: 75.11

PROP: Colorless liquid, fishy odor. Bp: 168° @ 500 mm, flash p: 175°F (TOC), fp: 12.4°, d: 0.9786 @ 30°, vap press: 2.1 mm @ 60°, vap d: 2.59.

SAFETY PROFILE: Moderately toxic by skin contact and ingestion. An irritant. Moderately flammable. Incompatible with oxidizing materials. To fight fire, use foam, CO_2, dry chemical.

AMB000   CAS: 138-61-4   **HR: 3**
AMINOPROPANOL PYROCATECHOL-
HYDROCHLORIDE
mf: $C_9H_{13}NO_3 • ClH$ mw: 219.69

SYNS: 3,4-DIHYDROXYNOREPHEDRINE HYDROCHLO-RIDE * 3,4-DIHYDROXYPHENYLAMINOPROPANOL HYDROCHLORIDE * 3,4-DIHYDROXYPHENYLPROPA-NOLAMINE HYDROCHLORIDE * ISO-ADRENALINE HYDROCHLORIDE * α-METHYL-NORADRENALINE HYDROCHLORIDE * NORHOMOEPI-NEPHRINE HYDROCHLORIDE

SAFETY PROFILE: Poison by subcutaneous and intravenous routes. When heated to decomposition it emits very toxic fumes of NO_x and Cl^-.

AMB250   **HR: 3**
2-AMINO PROPIONITRILE
mf: $C_3H_6N_2$ mw: 70.1

CONSENSUS REPORTS: Cyanide and its compounds are on the Community Right To Know List.

SAFETY PROFILE: A poison and dangerous fire hazard. Can explode in storage. Upon decomposition it emits toxic fumes of CN^- and NO_x.

AMB500   CAS: 151-18-8   **HR: 3**
3-AMINOPROPIONITRILE
mf: $C_3H_6N_2$ mw: 70.11

PROP: Liquid, amine odor. Bp: 185°.

SYNS: β-AMINOPROPIONITRILE * BAPN
* β-CYANOETHYLAMINE

CONSENSUS REPORTS: EPA Genetic Toxicology Program. Reported in EPA TSCA Inventory. Cyanide and its compounds are on the Community Right-To-Know List.

SAFETY PROFILE: Experimental teratogenic and reproductive effects. Nitriles usually have cyanide-like effects. Easily oxidized and unstable. A storage hazard; it polymerizes to an explosive yellow solid. When heated to decomposition it emits toxic fumes of CN^- and NO_x.

AMD750   CAS: 4985-85-7   **HR: 3**
AMINOPROPYLDIETHANOLAMINE
DOT: NA 1760
mf: $C_7H_{18}N_2O_2$ mw: 162.27

CONSENSUS REPORTS: Reported in EPA TSCA Inventory.

DOT Classification: Corrosive Material; Label: Corrosive.

SAFETY PROFILE: A corrosive. A powerful skin, eye and mucous membrane irritant. When heated to decomposition it emits toxic fumes of NO_x.

AME500   CAS: 299-26-3   **HR: 3**
3-(2-AMINOPROPYL)INDOLE
mf: $C_{11}H_{14}N_2$ mw: 174.27

SYNS: INDOPAN * α-METHYL-β-INDOLAETHYL-
AMINE (GERMAN) * α-METHYL-β-INDOLEETHYL-
AMINE * α-METHYLTRYPTAMINE

SAFETY PROFILE: Poison by ingestion and intraperitoneal routes. Moderately toxic by subcutaneous route. Human psychotropic effects by ingestion. An experimental teratogen. When heated to decomposition it emits toxic fumes of NO_x.

AME750   CAS: 18237-15-5   **HR: 3**
3-(γ-AMINOPROPYL)
INDOLEHYDROCHLORIDE
mf: $C_{11}H_{14}N_2 \cdot ClH$ mw: 210.73

SYNS: HOMOTRYPTAMINE HYDROCHLORIDE
* INDOLE-3-PROPYLAMINE HYDROCHLORIDE
* γ-3-INDOLYLPROPYLAMINE HYDROCHLORIDE

SAFETY PROFILE: Poison by intravenous and intraperitoneal routes. When heated to decomposition it emits very toxic fumes of HCl and NO_x.

AMF250   CAS: 123-00-2   **HR: 3**
4-AMINOPROPYLMORPHOLINE
DOT: UN 1760
mf: $C_7H_{16}N_2O$ mw: 144.25

PROP: Liquid. Mp: $-15°$, bp: 224.7°, flash p: 220°F (OC), d: 0.9872 @ 20°/20°, vap press: 0.06 mm @ 20°, vap d: 4.97.

SYN: N-AMINOPROPYLMORPHOLINE (DOT)

CONSENSUS REPORTS: Reported in EPA TSCA Inventory.

DOT Classification: Corrosive Material; Label: Corrosive.

SAFETY PROFILE: A corrosive material. Moderately toxic by several routes. A severe skin and eye irritant. Combustible. Can react with oxidizing materials. To fight fire, use alcohol foam, dry chemical. When heated to decomposition it emits toxic fumes of NO_x.

AMG750   CAS: 54-62-6   **HR: 3**
AMINOPTERIDINE
mf: $C_{19}H_{20}N_8O_5$ mw: 440.47

PROP: Yellow needles, sol in sodium hydroxide soln.

SYNS: 4-AMINO-4-DEOXYPTEROYLGLUTAMATE
* 4-AMINO-PGA * AMINOPTERIN * 4-AMINOP-
TEROYLGLUTAMIC ACID * APGA * ENT 26,079
* FOLIC ACID, 4-AMINO- * NSC 739

CONSENSUS REPORTS: EPA Extremely Hazardous Substances List.

SAFETY PROFILE: Poison by ingestion and intraperitoneal routes. Human and experimental teratogenic data. Mutation data reported. Human systemic effects by ingestion: gastrointestinal. Questionable carcinogen with experimental tumorigenic data. When heated to decomposition it emits toxic fumes of NO_x.

AMI000   CAS: 504-29-0   **HR: 3**
2-AMINOPYRIDINE

DOT: UN 2671
mf: $C_5H_6N_2$ mw: 94.13

PROP: White powder or crystals. Mp: 58.1, bp: 210.6°. Sol in water and ether; very sol in alc; sltly sol in ligroin.

SYNS: α-AMINOPYRIDINE * AMINO-2-PYRIDINE * o-AMINOPYRIDINE * α-PYRIDINAMINE * α-PYRIDYLAMINE

CONSENSUS REPORTS: Reported in EPA TSCA Inventory.

OSHA PEL: TWA 0.5 ppm
ACGIH TLV: TWA 0.5 ppm
DFG MAK: 0.5 ppm (2 mg/m³)
DOT Classification: IMO: Poison B; Label: Poison.

SAFETY PROFILE: Poison by inhalation, subcutaneous, intravenous, and intraperitoneal routes. Toxic effects resemble strychnine poisoning. Human systemic effects by inhalation: somnolence, convulsions, and antipsychotic effects. Human central nervous system effects by inhalation. When heated to decomposition it emits highly toxic fumes of NO_x.

AMI250   CAS: 462-08-8   **HR: 3**
3-AMINOPYRIDINE

DOT: UN 2671
mf: $C_5H_6N_2$ mw: 94.13

PROP: Leaflets from benzene or ligroin. Mp: 64°; bp: 251°. Very sol in water, alc, ether; insol in ligroin.

SYNS: AMINO-3-PYRIDINE * m-AMINOPYRIDINE (DOT) * 3-PYRIDINAMINE * 3-PYRIDYLAMINE

CONSENSUS REPORTS: Reported in EPA TSCA Inventory.

DOT Classification: IMO: Poison B; Label: Poison.

SAFETY PROFILE: Poison by ingestion, intraperitoneal, subcutaneous, and intravenous routes. When heated to decomposition it emits toxic fumes of NO_x.

AMK500   CAS: 68-89-3   **HR: 3**
AMINOPYRINE SODIUM SULFONATE
mf: $C_{13}H_{17}N_3O_4S \cdot Na$ mw: 334.38

SYNS: (ANTIPYRINYLMETHYLAMINO)METHANE-SULFONIC ACID SODIUM SALT * METHYLAMINO-ANTIPYRINE SODIUM METHANESULFONATE * 4-METHYLAMINO-1,5-DIMETHYL-2-PHENYL-3-PYRAZOLONE SODIUM METHANESULFONATE * METHYLAMINOPHENYLDIMETHYLPYRA-ZOLONE METHANESULFONATE SODIUM * 1-PHENYL-2,3-DIMETHYL-5-PYRAZOLONE-4-METHYLAMINOMETHANESULFONATESODIUM * 1-PHENYL-2,3-DIMETHYLPYRAZOLONE-(5)-4-METHYLAMINOMETHANESULFONICACID SODIUM * PHENYL DIMETHYL PYRAZOLON METHYL AMINO-METHANE SODIUM SULFONATE * 4-SODIUM METH-ANESULFONATE METHYLAMINE-ANTIPYRINE * SODIUM METHYLAMINOANTIPYRINE METHANESUL-FONATE * SODIUM-4-METHYLAMINO-1,5-DIMETHYL-2-PHENYL-3-PYRAZOLONE 4-METHANESULFONATE * SODIUM NORAMIDOPYRINE METHANESULFONATE * SODIUM-1-PHENYL-2,3-DIMETHYL-4-METHYLAMINO-PYRAZOLON-N-METHANESULFONATE * SODIUM-1-PHENYL-2,3-DIMETHYL-5-PYRAZOLONE-4-METHYL-AMINO METHANESULFONATE * SODIUM PHENYLDI-METHYLPYRAZOLONMETHYLAMINOMETHANE SULFONATE

SAFETY PROFILE: Poison by subcutaneous route. Moderately toxic by several other routes. An experimental teratogen. Human mutation data reported. Questionable carcinogen with experimental neoplastigenic data. When heated to decomposition it emits very toxic fumes of NO_x, Na_2O and SO_x.

AMQ500   CAS: 23757-42-8   **HR: 3**
4-AMINO-2,2,5,5-TETRAKIS (TRIFLUOROMETHYL)-3-IMIDAZOLINE
mf: $C_7H_3F_{12}N_3$ mw: 357.13

SYNS: 5-AMINO-2,2,4,4-TETRAKIS(TRIFLUORO-METHYL)IMIDAZOLIDINE * EXP 338

SAFETY PROFILE: Poison by ingestion, intraperitoneal, and intravenous routes. When heated to decomposition it emits very toxic fumes of F^- and NO_x.

AMT500   CAS: 139-13-9   **HR: 3**
AMINOTRIACETIC ACID
mf: $C_6H_9NO_6$ mw: 191.16

SYNS: N,N-BIS(CARBOXYMETHYL)GLYCINE * NCI-C02766 * NITRILOTRIACETIC ACID * TRIGLYCINE * TRIGLYCOLLAMIC ACID * VERSENE NTA ACID

CONSENSUS REPORTS: NTP Fourth Annual Report On Carcinogens, 1984. NCI Carcinogenesis Bioassay (feed); Clear Evidence: mouse, rat NCITR* NCI-CG-TR-6,77. Reported in EPA TSCA Inventory. Community Right-To-Know List.

SAFETY PROFILE: Confirmed carcinogen with experimental carcinogenic and neoplastigenic data. Poison by intraperitoneal route. Moderately toxic by ingestion. When heated to decomposition it emits toxic fumes of NO_x.

AMW000   CAS: 2432-99-7   **HR: 3**
11-AMINOUNDECANOIC ACID
mf: $C_{11}H_{23}NO_2$ mw: 201.35

SYNS: AMINOUNDECANOIC ACID * 11-AMINOUN-DECYLIC ACID * NCI-C50613

CONSENSUS REPORTS: IARC Cancer Review: GROUP 3 IMEMDT 7,56,87; Animal Limited Evidence IMEMDT 39,239,86. NTP Carcinogenesis Bioassay (feed): Clear Evidence: mouse, rat NTPTR* NTP-TR-216,82. Reported in EPA TSCA Inventory.

SAFETY PROFILE: Questionable carcinogen with experimental carcinogenic and neoplastigenic data. Mutation data reported. When heated to decomposition it emits toxic fumes of NO_x.

AMX750   CAS: 57-43-2   **HR: 3**
AMITAL
mf: $C_{11}H_{18}N_2O_3$ mw: 226.31

PROP: Slightly bitter crystals.

SYNS: AMOBARBITAL * AMYLBARBITONE * AMYLOBARBITAL * AMYLOBARBITONE * AMYTAL * 5-ETHYL-5-ISOAMYLBARBITURIC ACID * 5-ETHYL-5-ISOAMYLMALONYL UREA * ETHYLISOPENTYLBARBITURIC ACID * 5-ETHYL-5-ISOPENTYLBARBITURIC ACID * 5-ETHYL-5-(3-METHYLBUTYL)BARBITURIC ACID * ISOAMYLETHYLBARBITURIC ACID * 5-ISOAMYL-5-ETHYLBARBITURIC ACID * NSC 10815

SAFETY PROFILE: A human poison by ingestion. An experimental poison by ingestion, intravenous, intraperitoneal, and subcutaneous routes. When heated to decomposition it emits toxic fumes of NO_x.

AMY050   CAS: 61-82-5   **HR: 3**
AMITROLE
mf: $C_2H_4N_4$ mw: 84.10

SYNS: AMEROL * AMINOTRIAZOLE * 2-AMINO-TRIAZOLE * 3-AMINOTRIAZOLE * 3-AMINO-s-TRIAZOLE * 3-AMINO-1,2,4-TRIAZOLE * 2-AMINO-1,3,4-TRIAZOLE * 3-AMINO-1H-1,2,4-TRIAZOLE * AMINOTRIAZOLE (plant regulator) * AMINO TRIA-ZOLE WEEDKILLER 90 * AMINOTRIAZOL-SPRITZPUL-VER * AMITOL * AMITRIL * AMITRIL T.L. * AMITROL * AMITROL 90 * AMITROL-T * AMIZOL * AT * ATA * AT LIQUID * AZAPLANT * AZOLAN * AZOLE * CAMPA-PRIM A 1544 * CYTROL * DIUROL * DOMATOL * ELMASIL * EMISOL * ENT 25,445 * FENA-MINE * FENAVAR * HERBICIDE TOTAL * HERBIZOLE * KLEER-LOT * ORGA-414 * RADOXONE TL * RAMIZOL * RCRA WASTE NUMBER U011 * SIMAZOL * SOLUTION CONCEN-TREE T271 * TRIAZOLAMINE * 1H-1,2,4-TRI-AZOL-3-AMINE * USAF XR-22 * VOROX * WEEDAR ADS * WEEDAZIN * WEEDAZOL * WEEDEX GRANULAT * WEEDOCLOR * X-ALL LIQUID

CONSENSUS REPORTS: IARC Cancer Review: GROUP 2B IMEMDT 7,92,87; Human Inadequate Evidence 41,293,86; IMEMDT 7,31,74; Animal Sufficient Evidence IMEMDT 7,31,74; IMEMDT 41,293,86. NTP Fourth Annual Report On Carcinogens, 1984. Reported in EPA TSCA Inventory. EPA Genetic Toxicology Program.

OSHA PEL: TWA 0.2 mg/m^3
ACGIH TLV: TWA 0.2 mg/m^3
DFG MAK: 0.2 mg/m^3

SAFETY PROFILE: Confirmed carcinogen with experimental carcinogenic, tumorigenic, and neoplastigenic data. Poison by intraperitoneal route. Moderately toxic by ingestion. Experimental teratogenic and reproductive effects. Mutation data reported. When heated to decomposition it emits toxic fumes of NO_x. An herbicide and plant growth regulator.

AMY500   CAS: 7664-41-7   **HR: 3**
AMMONIA

DOT: UN 1005/UN 2073/UN 2672
mf: H_3N mw: 17.04

PROP: Colorless gas, extremely pungent odor, liquefied by compression. Mp: −77.7°, bp: −33.35°, lel: 16%, uel: 25%, d: 0.771 g/liter @ 0°, 0.817 g/liter @ −79°, autoign temp: 1204°F, vap press: 10 atm @ 25.7°, vap d: 0.6. Very sol in water; moderately sol in alc.

SYNS: AMMONIAC (FRENCH) * AMMONIACA (ITALIAN) * AMMONIA GAS * AMMONIAK (GERMAN) * AMONIAK (POLISH) * ANHYDROUS AMMONIA * SPIRIT of HARTSHORN

CONSENSUS REPORTS: EPA Extremely Hazardous Substances List. Community Right-To-Know List. Reported in EPA TSCA Inventory.

OSHA PEL: TWA 35 ppm
ACGIH TLV: TWA 25 ppm; STEL 35 ppm
DFG MAK: 50 ppm (35 mg/m^3)
NIOSH REL: CL 50 ppm
DOT Classification: Nonflammable Gas; Label: Nonflammable Gas.

SAFETY PROFILE: A human poison by an unspecified route. Poison experimentally by inhalation, ingestion, and possibly other routes. An eye, mucous membrane, and systemic irritant by inhalation. Mutation data reported. A common air contaminant. Difficult to ignite. Explosion hazard when exposed to flame or in a fire. NH_3 + air in a fire can detonate. Potentially violent or explosive reactions on contact with interhalogens (e.g., bromine pentafluoride; chlorine trifluoride); 1,2-dichloroethane (with liquid NH_3); boron halides; chloroformamidnium nitrate; ethylene oxide (polymerization reaction); magnesium perchlorate; nitrogen trichloride; oxygen + platinum; or strong oxidants (e.g., potassium chlorate; nitryl chloride; chromyl chloride; dichlorine oxide; chromium trioxide; trioxygen difluoride; nitric acid; hydrogen peroxide; tetramethylammonium amide; thiocarbonyl azide thiocyanate; sulfinyl chloride; thiotriazyl chloride; ammonium peroxodisulfate; fluorine; nitrogen oxide; dinitrogen tetraoxide; and liquid oxygen). Forms sensitive explosive mixtures with air + hydrocarbons; 1-chloro-2,4-dinitrobenzene; 2-,or 4-chloronitrobenzene (above 160°C/30 bar); ethanol + silver nitrate; germanium derivatives; stibine; and chlorine. Reaction with silver chloride; silver nitrate; silver azide; and silver oxide form the explosive silver nitride. Reactions with chlorine azide; bromine; iodine; iodine + potassium; heavy metals and their compounds (e.g. gold(III) chloride; mercury; and potassium thallium amide ammoniate); tellurium halides (e.g., tellurium tetrabromide; and tellurium tetrachloride)and pentaborane(9) give explosive products. Incompatible in contact with Ag; acetaldehyde; acrolein; B; BI_3; halogens; $HClO_3$; ClO;

chlorites; chlorosilane; (ethylene dichloride + liquid ammonia); Au; hexachloromelamine; (hydrazine + alkali metals); HBr; HOCl; $Mg(ClO_4)_2$; N_2O_4; NCl_3; NF_3; OF_2; P_2O_5; P_2O_3; picric acid; (K + AsH_3); (K + PH_3); (K + $NaNO_2$); potassium ferricyanide; potassium mercuric cyanide; (Na + CO); Sb; S; SCl_2; tellurium hydropentachloride; trichloromelamine; NO_2Cl; SbH_3; tetramethylammonium amide; $SOCl_2$; thiotrithiazylchloride. Incandescent reaction when heated with calcium. Emits toxic fumes of NH_3 and NO_x when exposed to heat. To fight fire stop flow of gas.

ANA000 CAS: 631-61-8 **HR: 3**
AMMONIUM ACETATE
DOT: NA 9079
mf: $C_2H_4O_2 \cdot H_3N$ mw: 77.10
PROP: Crystals. Mp: 114°, d: 1.07.

SYN: ACETIC ACID, AMMONIUM SALT

CONSENSUS REPORTS: Reported in EPA TSCA Inventory.

DOT Classification: ORM-E; Label: None.

SAFETY PROFILE: Poison by intravenous route. Moderately toxic by intraperitoneal routes. When heated to decomposition it emits toxic fumes of NO_x and NH_3.

ANA750 **HR: 3**
AMMONIUM AZIDE
mf: NH_4N_3 mw: 60.1
PROP: Colorless plates. Mp: 160°, bp: explodes, d: 1.346, vap press: 1 mm @ 59.2° (sublimes).

SAFETY PROFILE: Poison by inhalation and ingestion. Moderately flammable. Unstable. Explosion hazard upon rapid heating.

ANB250 CAS: 1066-33-7 **HR: 3**
AMMONIUM BICARBONATE (1:1)
DOT: NA 9081
mf: $HCO_3 \cdot H_4N$ mw: 79.1
PROP: Hard, colorless to white crystals; faint ammonia odor, stable at room temp, volatile. Decomp @ 60°, mp: 107.5° (rapid heating). D: 1.586. Sol in water; insol in alc.

SYNS: ACID AMMONIUM CARBONATE * AMMONIUM CARBONATE * AMMONIUM HYDROGEN CARBONATE * CARBONIC ACID, MONOAMMONIUM SALT * MONOAMMONIUM CARBONATE

CONSENSUS REPORTS: Reported in EPA TSCA Inventory.

DOT Classification: ORM-E; Label: None.

SAFETY PROFILE: Poison by intravenous route. When heated to decomposition it emits toxic fumes of NO_x and NH_3.

ANB500   CAS: 7789-09-5   HR: 3
AMMONIUM BICHROMATE

DOT: UN 1439
mf: $Cr_2H_8N_2O_7$ mw: 252.10

PROP: Red crystals. Mp: decomp, d: 2.936.

SYNS: AMMONIO (DICROMATO DI) (ITALIAN)
* AMMONIUMBICHROMAAT (DUTCH) * AMMONI-
UMDICHROMAT (GERMAN) * AMMONIUMDICHRO-
MAAT (DUTCH) * AMMONIUM DICHROMATE
* AMMONIUM DICHROMATE(VI) * BICHROMATE
d'AMMONIUM (FRENCH)

CONSENSUS REPORTS: Reported in EPA TSCA Inventory. Chromium and its compounds are on the Community Right-To-Know List.

OSHA PEL: CL 0.1 mg(CrO_3)/m^3
ACGIH TLV: TWA 0.05 mg(Cr)/m^3
NIOSH REL: TWA 25 μg(Cr(VI))/m^3; CL 50 μg/m^3/15M
DOT Classification: Oxidizer; Label: Oxidizer.

SAFETY PROFILE: Poison by inhalation, ingestion, skin contact, and subcutaneous routes. An unstable oxidizer. Moderately flammable; reacts with reducing agents.

ANC000   HR: 3
AMMONIUM BROMATE
mf: NH_4BrO_3 mw: 145.96

PROP: Colorless crystals. Very sol in water. Mp: explodes.

SAFETY PROFILE: An unstable, explosive oxidizing material. Severe explosion hazard.

ANC250   CAS: 12124-97-9   HR: D
AMMONIUM BROMIDE
mf: BrH_4N mw: 97.96

PROP: Colorless, cubic, sltly hygroscopic crystals. Mp: subl @ 452°, bp: 235° in vac, d: 2.429, vap press: 1 mm @ 198.3°.

SYN: HYDROBROMIC ACID MONOAMMONIATE

CONSENSUS REPORTS: Reported in EPA TSCA Inventory.

SAFETY PROFILE: Mutation data reported. When heated to decomposition it emits very toxic fumes of NO_x, Br^-, and NH_3. Incompatible with BrF_3; IF_7; K.

AND250   HR: 3
AMMONIUM CADMIUM CHLORIDE
mf: $4NH_4Cl \cdot CdCl_2$ mw: 397.3

PROP: Colorless, rhombic crystals. D: 2.01; sol in water.

CONSENSUS REPORTS: Cadmium and its compounds are on the Community Right-To-Know List.

OSHA PEL: TWA 0.1 mg(Cd)/m^3; CL 0.6 mg(Cd)/m^3 (fume)
ACGIH TLV: TWA 0.05 mg(Cd)/m^3 (Proposed: TWA 0.01 mg(Cd)/m^3 (dust), Human Carcinogen); BEI: 10 μg/g creatinine in urine; 10 μg/L in blood.
DFG BAT: Blood: 1.5 μg/dL; Urine: 15 μg/dL; Suspected Carcinogen.
NIOSH REL: (Cadmium) Reduce to lowest feasible level

SAFETY PROFILE: Confirmed human carcinogen. A poison. When heated to decomposition it emits toxic fumes of NH_3, NO_x, and Cl^-.

AND500   HR: 3
AMMONIUM CALCIUM ARSENATE
mf: $NH_4CaAsO_4 \cdot 6H_2O$ mw: 305.1

PROP: Colorless crystals. Mp: 140° (decomp), d: 1.905 @ 15°. Slightly sol in cold water; sol in hot water, NH_4Cl, and NH_4OH.

CONSENSUS REPORTS: Arsenic and its compounds are on the Community Right-To-Know List.

SAFETY PROFILE: A poison.

AND750   CAS: 1111-78-0   HR: 3
AMMONIUM CARBAMATE

DOT: NA 9083
mf: $CH_3NO_2 \cdot H_3N$ mw: 78.09

PROP: White, crystalline, rhombic powder; sol in water and alc; ammonia odor. Sublimates at 60°.

SYN: AMMONIUM AMINOFORMATE

CONSENSUS REPORTS: Reported in EPA TSCA Inventory.

DOT Classification: ORM-A; Label: None

SAFETY PROFILE: Poison by intravenous route.

ANE000   CAS: 506-87-6   HR: 3
AMMONIUM CARBONATE
DOT: NA 9084
mf: $(NH_4)_2CO_3$ mw: 96.09

PROP: Colorless crystals. Decomposes on standing to ammonium bicarbonate. Sltly sol in water.

SYNS: AMMONIUMCARBONAT (GERMAN) * CARBONIC ACID, AMMONIUM SALT * CARBONIC ACID, DIAMMONIUM SALT * DIAMMONIUM CARBONATE

CONSENSUS REPORTS: Reported in EPA TSCA Inventory.

DOT Classification: ORM-A; Label: None

SAFETY PROFILE: Poison by subcutaneous and intravenous routes. When heated to decomposition it emits toxic fumes of NO_x and NH_3.

ANE250   CAS: 10192-29-7   HR: 3
AMMONIUM CHLORATE
mf: ClH_3NO_3 mw: 100.49

PROP: White crystals or mass.

DOT Classification: Forbidden

SAFETY PROFILE: A powerful oxidizer. Moderately flammable due to spontaneous chemical reaction. Explosion hazard due to shock, chemical reaction, or exposure to heat. A storage hazard; it may explode at room temperature. Explodes when heated to 100°C. When contaminated it is very sensitive. Solution in water may explode if heated or dried. When heated to decomposition it emits highly toxic fumes of Cl^- and NO_x. Incompatible with reducing materials; BrF_3; BrF_5.

ANE500   CAS: 12125-02-9   HR: 2
AMMONIUM CHLORIDE
DOT: NA 9085
mf: $H_4N \cdot Cl$ mw: 53.50

PROP: White crystals; salty taste. Bp: 520°, mp: 337.8°, d: 1.520, vap press: 1 mm @ 160.4° (sublimes). Sol in water, alc, and glycerin.

SYNS: AMMONIUMCHLORID (GERMAN) * AMMONIUM MURIATE * CHLORID AMONNY (CZECH) * SAL AMMONIA * SAL AMMONIAC

CONSENSUS REPORTS: Reported in EPA TSCA Inventory.

OSHA PEL: (Fume) TWA 10 mg/m^3; STEL 20 mg/m^3
ACGIH TLV: TWA 10 mg/m^3; STEL 20 mg/m^3
DOT Classification: ORM-E.

SAFETY PROFILE: Poison by subcutaneous, intravenous, and intramuscular routes. Moderately toxic by other routes. A severe eye irritant. Explosive reaction with potassium chlorate or bromine trifluoride. Violent reaction (ignition) with bromine pentafluoride; NH_4; NO_3; and IF_7. Reaction with hydrogen cyanide may give the explosive nitrogen trichloride. When heated to decomposition it emits very toxic fumes of NO_x, Cl^-, and NH_3.

ANF000   CAS: 19168-23-1   HR: 3
AMMONIUM CHLOROPALLADATE(IV)
mf: $Cl_6H_8N_2Pd$ mw: 355.20

PROP: Red-brown crystals. D: 2.418, mp: decomp.

SYNS: AMMONIUM HEXACHLOROPALLADATE * DIAMMONIUM HEXACHLOROPALLADATE

CONSENSUS REPORTS: Reported in EPA TSCA Inventory.

SAFETY PROFILE: A poison skin irritant. When heated to decomposition it emits very toxic fumes of NO_x, Cl^-, and NH_3.

ANF250   CAS: 16919-58-7   HR: 3
AMMONIUM CHLOROPLATINATE
mf: $Cl_6Pt \cdot 2H_4N$ mw: 443.89

PROP: Cubic, yellow crystals. D: 3.065, mp: decomposes.

SYNS: AMMONIUM HEXACHLOROPLATINATE(IV) * AMMONIUM PLATINIC CHLORIDE * DIAMMONIUM HEXACHLOROPLATINATE (2-) * PLATINIC AMMONIUM CHLORIDE

CONSENSUS REPORTS: Reported in EPA TSCA Inventory.

OSHA PEL: TWA 0.002 mg(Pt)/m^3
ACGIH TLV: TWA 0.002 mg(Pt)/m^3

SAFETY PROFILE: Poison by inhalation and ingestion. Human pulmonary system effects by inhalation. An explosively unstable compound. Incompatible with KOH (boiling with alkali yields a product which, after drying, will explode @ 205° or if mixed with combustibles). When heated to decomposition it emits very toxic fumes of Cl^-, NO_x, and NH_3.

ANF500   HR: 3
AMMONIUM CHROMATE
mf: $(NH_4)_2CrO_4$ mw: 152.1

PROP: Yellow, crystalline material. Mp: decomp @ 180°, d: 1.91 @ 12°. Sol in cold water.

CONSENSUS REPORTS: Chromium and its compounds are on the Community Right-To-Know List.

SAFETY PROFILE: A poison. A powerful oxidizer. An explosion hazard when shocked or heated. When heated to decomposition it emits toxic fumes of NH_3 and NO_x. Incompatible with reducing agents.

ANF800   CAS: 3012-65-5   HR: 2
AMMONIUM CITRATE
DOT: NA 9087
mf: $C_6H_8O_7 \cdot xH_3N$ mw: 311.42

PROP: Granules or crystals. D: 1.48. Sol in water; sltly sol in alc.

SYNS: AMMONIUM CITRATE, DIBASIC (DOT) * CITRIC ACID, AMMONIUM SALT * DIAMMONIUM CITRATE

CONSENSUS REPORTS: Reported in EPA TSCA Inventory.

DOT Classification: ORM-E.

SAFETY PROFILE: Experimental poison by intravenous route. A skin and eye irritant. When heated to decomposition it emits acrid smoke and irritating fumes.

ANG000   HR: 3
AMMONIUM CYANIDE
mf: NH_4CN mw: 44.1

PROP: Solid, white powder or crystals. Mp: 36° (decomp), bp: sublimes @ 40°, d: 1.002 @ 100°, vap press: 400 ppm @ 20.5°. Very sol in water, alc; decomp in hot water.

CONSENSUS REPORTS: Cyanide and its compounds are on the Community Right-To-Know List.

SAFETY PROFILE: A poison. When heated to decomposition it emits toxic CN^-, NH_3, NO_x.

ANG250   HR: 3
AMMONIUM DIFLUORIDE mixed with HYDROCHLORIC ACID

SYN: WHITE ACID (DOT)

DOT Classification: Corrosive Material; Label: Corrosive.

SAFETY PROFILE: A corrosive. Poison by inhalation, ingestion, and skin contact. When heated to decomposition it emits very toxic fumes of F^-, HF, and HCl.

ANH000   CAS: 13826-83-0   HR: 3
AMMONIUM FLUOBORATE
DOT: NA 9088
mf: NH_4BF_4 mw: 104.9

PROP: White, rhombic crystals. D: 1.871 @ 15°, mp: subl, sol in NH_4OH, and water.

SYNS: AMMONIUM BOROFLUORIDE * AMMONIUM FLUOROBORATE * AMMONIUM TETRAFLUORO-BORATE * AMMONIUM TETRAFLUOROBORATE(1-)

CONSENSUS REPORTS: Reported in EPA TSCA Inventory.

NIOSH REL: TWA 2.5 mg(F)/m^3
DOT Classification: ORM-B; Label: None

SAFETY PROFILE: A poison; strong irritant. When heated to decomposition it emits very toxic fumes of F^- and NO_x, and NH_3.

ANH250   CAS: 12125-01-8   HR: 3
AMMONIUM FLUORIDE
DOT: UN 2505
mf: $H_4N \cdot F$ mw: 37.05

PROP: Colorless crystals. Mp: subl; d: 1.009 @ 25°.

SYNS: AMMONIUM FLUORURE (FRENCH) * NEUTRAL AMMONIUM FLUORIDE

CONSENSUS REPORTS: Reported in EPA TSCA Inventory.

OSHA PEL: TWA 2.5 mg(F)/m^3
NIOSH REL: TWA 2.5 mg(F)/m^3
DOT Classification: ORM-B; Label: None.

SAFETY PROFILE: Poison by subcutaneous and intraperitoneal routes. When heated to decomposition it emits very toxic fumes of F^-, NO_x and NH_3. Incompatible with ClF_3.

ANI250 CAS: 16962-40-6 **HR: 3**
AMMONIUM HEXAFLUOROTITANATE
mf: $F_6Ti \cdot H_4N_2$ mw: 193.96

CONSENSUS REPORTS: Reported in EPA
TSCA Inventory.

NIOSH REL: TWA 2.5 mg(F)/m^3

SAFETY PROFILE: Poison by intravenous
route. When heated to decomposition it emits
very toxic fumes of F$^-$ and NO$_x$.

ANI500 CAS: 13815-31-1 **HR: 3**
AMMONIUM HEXAFLUOROVANADATE
mf: $F_6H_{12}N_3V$ mw: 219.09

SYN: HEXAFLUORO VANADATE (3-) TRIAMMONIUM
SALT

OSHA PEL: TWA 2.5 mg(F)/m^3
NIOSH REL: (Vanadium Compounds) CL 0.05
mg(V)/m^3/15M

SAFETY PROFILE: Poison by intravenous
route. When heated to decomposition it emits
very toxic NH$_3$, NO$_x$, VO$_x$ and fluorides.

ANJ000 CAS: 1341-49-7 **HR: 3**
AMMONIUM HYDROGEN FLUORIDE

DOT: UN 1727/UN 2817
mf: F_2H_5N mw: 57.06

PROP: White crystals. D: 1.51, mp: 124.6°.
Will etch glass. Water-sol.

SYNS: AMMONIUM BIFLUORIDE * AMMONIUM
HYDROGEN FLUORIDE, solid

CONSENSUS REPORTS: Reported in EPA
TSCA Inventory.

NIOSH REL: TWA 2.5 mg(F)/m^3
DOT Classification: ORM-B; Label: None.

SAFETY PROFILE: Caustic poison and strong
irritant by all routes. When heated to decomposition it emits very toxic fumes of F$^-$, NO$_x$ and
NH$_3$.

ANJ250 CAS: 1341-49-7 **HR: 3**
**AMMONIUM HYDROGEN FLUORIDE
(solution)**

DOT: UN 2817

SYN: AMMONIUM HYDROGEN FLUORIDE, solution
(DOT)

CONSENSUS REPORTS: Reported in EPA
TSCA Inventory.

NIOSH REL: TWA 2.5 mg(F)/m^3
DOT Classification: Corrosive Material; Label:
Corrosive.

SAFETY PROFILE: Caustic poison and strong
irritant by all routes. When heated to decomposition it emits very toxic fumes of HF, F$^-$, and
NO$_x$.

ANJ750 CAS: 12124-99-1 **HR: 3**
AMMONIUM HYDROSULFIDE

DOT: NA 2683
mf: NH$_4$HS mw: 51.11

PROP: Powder or crystals. Mp: 118° (150 atm);
d: 1.17; vap press: 400 mm @ 21.8°.

SYNS: AMMONIUM BISULFIDE * AMMONIUM HY-
DROGEN SULFIDE * AMMONIUM HYDROSULFIDE, so-
lution (DOT) * AMMONIUM MERCAPTAN * AMMO-
NIUM SULFHYDRATE * MONOAMMONIUM SULFIDE
* SIRNIK AMONNY * TRUE AMMONIUM SULFIDE

CONSENSUS REPORTS: Reported in EPA
TSCA Inventory.

DOT Classification: ORM-A; Label: None

SAFETY PROFILE: Poison by ingestion, skin
contact, subcutaneous, intravenous, intradermal, parenteral, and intraperitoneal routes. Pyroforic in air. When heated to decomposition
it emits very toxic fumes of SO$_x$, NO$_x$, and
NH$_3$. Incompatible with zinc.

ANK000 CAS: 12124-99-1 **HR: 3**
AMMONIUM HYDROSULFIDE (solution)

DOT: NA 2683

SYN: AMMONIUM HYDROSULFIDE, solution (DOT)

CONSENSUS REPORTS: Reported in EPA
TSCA Inventory.

DOT Classification: ORM-A; Label: None.

SAFETY PROFILE: Poison due to easily liberated H$_2$S. When heated to decomposition it emits
very toxic fumes of NO$_x$, SO$_x$, NH$_3$, and H$_2$S.

ANK250 CAS: 1336-21-6 **HR: 3**
AMMONIUM HYDROXIDE

DOT: NA 2672
mf: $H_4N \cdot HO$ mw: 35.06

PROP: Clear, colorless liquid solution of ammonia; very pungent odor. D: 0.90, mp: −77°.

Sol in water. Soln contains not more than 44% ammonia.

SYNS: AMMONIA, solution (DOT) * AQUA AMMO-
NIA * AQUEOUS AMMONIA

CONSENSUS REPORTS: Reported in EPA TSCA Inventory.

NIOSH REL: CL 50 ppm
DOT Classification: Corrosive Material; Label: Corrosive.

SAFETY PROFILE: A human poison by inges-
tion. An experimental poison by inhalation and ingestion. A severe eye irritant. Human systemic eye and other systemic irritant effects by inhalation. Mutation data reported. Incompatible with acrolein; nitromethane; acrylic acid; chlorosul-
fonic acid; dimethyl sulfate; halogens; (Au + aqua regia); HCl; HF; HNO_3; oleum; β-propiolactone; propylene oxide; $AgNO_3$; Ag_2O; (Ag_2O + C_2H_5OH); $AgMnO_4$; H_2SO_4. Dangerous; liquid can inflict burns. Use with adequate ventila-
tion. When heated to decomposition it emits NH_3 and NO_x.

ANK750 HR: 3
AMMONIUM IODATE
mf: H_4INO_3 mw: 192.94

PROP: Colorless crystals. D: 3.309 @ 21°, mp: 150° (decomp). Sltly sol in cold water; insol in hot water.

SAFETY PROFILE: A powerful, unstable oxi-
dizer. When heated to decomposition it emits very toxic fumes of I^- and NO_x. Has detonated upon contact with a scoop, possibly due to contamination by ammonium periodate.

ANL000 HR: 2
AMMONIUM IODIDE
mf: NH_4I mw: 145

PROP: Colorless, hygroscopic crystals. Mp: subl @ 551°, bp: 220° (vacuo), d: 2.514 @ 25°, vap press: 1 mm @ 210.9°.

SAFETY PROFILE: Moderately toxic. Incompatible with BrF_3; IF_7; K. When heated to decomposition it emits toxic fumes of I^-, NH_3, and NO_x.

ANL750 HR: 3
AMMONIUM MAGNESIUM ARSENATE
mf: $NH_4MgAsO_4 \cdot 6H_2O$ mw: 289.4

PROP: Colorless crystals. Mp: decomp: d: 1.932 @ 15°. Very sltly water-sol.

CONSENSUS REPORTS: Arsenic and its compounds are on the Community Right-To-Know List.

SAFETY PROFILE: When heated to decompo-
sition it emits very toxic fumes of As, NH_3, and NO_x.

ANM000 HR: 3
AMMONIUM MAGNESIUM CHROMATE
mf: $(NH_4)_2CrO_4 \cdot MgCrO_4 \cdot 6H_2O$ mw: 400.5

PROP: Yellow crystals. Mp: decomp, d: 1.84. Very water sol.

CONSENSUS REPORTS: Chromium and its compounds are on the Community Right-To-Know List.

SAFETY PROFILE: A poison. Moderately flammable; can explode. Incompatible with reducing agents. When heated to decomposition it can emit toxic fumes of NH_3 and NO_x.

ANM500 CAS: 5421-46-5 HR: 3
AMMONIUM MERCAPTOACETATE
mf: $C_2H_3O_2S \cdot H_3N$ mw: 108.15

PROP: Colorless liquid; strong skunk-like odor.

SYNS: AMMONIUM THIOGLYCOLATE * AMMO-
NIUM THIOGLYCOLLATE * THIOGLYCOLLIC ACID, AMMONIUM SALT * USAF MO-2

CONSENSUS REPORTS: Reported in EPA TSCA Inventory.

SAFETY PROFILE: Poison by intraperitoneal route. An allergen; can cause contact dermatitis. Emits hydrogen sulfide. When heated to decomposition it emits very toxic NO_x, SO_x, and NH_3.

ANM750 CAS: 13106-76-8 HR: 3
AMMONIUM MOLYBDATE
mf: $MoO_4 \cdot 2H_4N$ mw: 196.04

SYNS: AMMONIUM PARAMOLYBDATE * DIAMMO-
NIUM MOLYBDATE * MOLYBDIC ACID DIAMMO-
NIUM SALT

CONSENSUS REPORTS: Reported in EPA TSCA Inventory.

OSHA PEL: TWA 5 mg(Mo)/m³
ACGIH TLV: TWA 5 mg(Mo)/m³

SAFETY PROFILE: Poison by ingestion and intraperitoneal route. Moderately toxic by other routes. An irritant. When heated to decomposition it emits toxic fumes of NH_3 and NO_x.

ANN000   CAS: 6484-52-2   **HR: 3**
AMMONIUM(I) NITRATE(1:1)
DOT: NA 1942/UN 0222/UN 2426
mf: $HNO_3 \cdot H_3N$ mw: 80.06

PROP: Colorless crystals. Mp: 169.6°, d: 1.725
@ 25°, bp: decomp >210°. Solubility: 192/100
@ 20°.

SYNS: AMMONIUM NITRATE * AMMONIUM NI-
TRATE (DOT) * NITRIC ACID, AMMONIUM SALT

CONSENSUS REPORTS: Community Right-
To-Know List. Reported in EPA TSCA Inventory.

DOT Classification: Oxidizer; Label: Oxidizer.

SAFETY PROFILE: A powerful oxidizer and
an allergen. A relatively stable explosive which
has, however, caused many industrial explosions. Violent or explosive spontaneous reactions with acetic anhydride + nitric acid; ammonium sulfate + potassium; copper iron(II)
sulfide; sawdust; urea; barium nitrate; hot water;
and ammonium chloride + water + zinc. Forms
heat- or shock-sensitive explosive mixtures with
acetic acid; aluminum + calcium nitrate +
formamide (a blasting explosive); ammonia;
charcoal + metal oxides (e.g., rust; copper oxide; zinc oxide above 80°C); chloride salts (e.g.,
ammonium chloride; calcium chloride; iron(III)
chloride; and aluminum chloride); cyanoguanidine; fertilizers (e.g., super phosphate + organic
materials above 90°C); hydrocarbon oils; powdered metals (e.g., aluminum; antimony; bismuth; cadmium; chromium; cobalt; copper; iron;
lead; magnesium; manganese; nickel; tin; zinc;
brass; stainless steel; titanium; and potassium);
nonmetals (e.g., charcoal; and phosphorus); organic fuels (e.g., wax; oils; and stearates); potassium permanganate; sugar; sulfur; and trini-
troanisole. Reaction with alkali metals (e.g.,
sodium) forms an explosive product. Ignites on
contact with ammonium dichromate; potassium
dichromate; potassium chromate; barium chloride; sodium chloride; potassium nitrate; and
chromium (VI) salts. Can ignite when mixed
with acetic acid. Use water in large amounts
to fight fire. It is important that the mass of
materials be kept cool and that burning be extinguished promptly. Ventilate well. May explode
under confinement and high temperatures. When
heated to decomposition it emits highly toxic
fumes of NO_x. Can react vigorously with reducing materials. Incompatible with; (NH_4Cl +

heat); (C + heat); organic matter; P; NaOCl;
$NaClO_4$. Occasional explosions in presence of
oil; ($NH_4)_2SO_4$ with K or Na will explode.

ANO500   CAS: 135-20-6   **HR: 3**
**AMMONIUM-N-
NITROSOPHENYLHYDROXYLAMINE**
mf: $C_6H_6N_2O_2 \cdot H_4N$ mw: 156.19

SYNS: CUPFERRON * N-HYDROXY-N-NITROSO-
BENZENAMINE, AMMONIUM SALT * KUPFERRON
(CZECH) * NCI-C03258 * N-NITROSOFENYLHY-
DROXYLAMIN AMONNY (CZECH) * N-NITROSO-
PHENYLHYDROXYLAMIN AMMONIUM SALZ
(GERMAN) * N-NITROSOPHENYLHYDROXYLAMINE
AMMONIUM SALT

CONSENSUS REPORTS: NTP Fourth Annual
Report On Carcinogens, 1984. NCI Carcinogen-
esis Bioassay (feed); Clear Evidence: mouse,
rat NCITR* NCI-CG-TR-100,78. Reported in
EPA TSCA Inventory. Community Right-To-
Know List.

SAFETY PROFILE: Confirmed carcinogen with
experimental carcinogenic and tumorigenic
data. Poison by intravenous route. Powerful eye
irritant. Solutions with thorium salts are unstable
explosives above 15°C. Solutions with titanium
or zirconium salts are unstable explosives above
40°C. When heated to decomposition it emits
very toxic NH_3 and NO_x.

ANO750   **HR: 3**
AMMONIUM OXALATE
mf: ($NH_4)_2C_2O_4 \cdot H_2O$ mw: 142.12

PROP: Colorless crystals. Mp: decomp; d: 1.50.
Sltly sol in water.

SAFETY PROFILE: A poison. Can react violently with (NaOCl + ammonium acetate).
When heated to decomposition it can emit toxic
fumes of NH_3 and NO_x.

ANP250   **HR: 3**
AMMONIUM PERCHLORATE
mf: NH_4ClO_4 mw: 117.50

PROP: White crystals. Mp: decomp; d: 1.95.

SAFETY PROFILE: Easily ignited by friction.
Can explode when mixed with sugar; charcoal;
or on contact with hot copper pipes. Can be
sensitized by nitryl perchlorate; KIO_4; $KMnO_4$;
metals (as cocrystallized impurities). When contaminated by powdered carbon; ferrocene; S;

organic matter; powdered metals; it becomes impact sensitive. When heated to decomposition it emits toxic fumes of NH_3, Cl^-, and NO_x.

ANP500   HR: 3
AMMONIUM PERCHLORYL AMIDE
mf: $H_5N_2O_3Cl$ mw: 116.6

PROP: Mp: 80°.

SAFETY PROFILE: A shock-sensitive explosive. May detonate @ 80°. When heated to decomposition it emits very toxic fumes of NH_3, NO_x, and Cl^-.

ANP625   CAS: 3825-26-1   HR: 3
AMMONIUM PERFLUOROOCTANOATE
mf: $C_8F_{15}O_2 \cdot H_4N$ mw: 431.13

SYNS: AMMONIUM PENTADECAFLUOROOCTANATE * AMMONIUM PERFLUOROCAPRILATE * AMMONIUM PERFLUOROCAPRYLATE * APFO * FC-143 * PERFLUOROAMMONIUM OCTANOATE

ACGIH TLV: (Proposed: 0.1 mg/m^3)

SAFETY PROFILE: Poison by inhalation. Moderately toxic by ingestion. An eye and skin irritant. Experimental reproductive effects. When heated to decomposition it emits toxic fumes of F^- and NH_3.

ANP750   HR: 3
AMMONIUM-m-PERIODATE
mf: NH_4IO_4 mw: 209

PROP: Colorless crystals. Mp: explodes; d: 3.056.

SAFETY PROFILE: A contact explosive. Heat, impact, and touch as from a scoop or an abrasive impact may cause explosion. When heated to decomposition it can emit toxic fumes of NH_3, NO_x, and I^-.

ANQ000   HR: 3
AMMONIUM PERMANGANATE
mf: NH_4MnO_4 mw: 137.0

PROP: Crystalline solid. Mp: explodes; d: 2.208 @ 10°.

CONSENSUS REPORTS: Manganese compounds are on the Community Right-To-Know List.

SAFETY PROFILE: A powerful oxidizer. Moderately flammable by chemical reaction with reducing agents. Explosive when shocked or warmed to 60°. Can be exploded by percussion. When heated to decomposition it emits toxic fumes of NO_x and NH_3. Incompatible with reducing material; friction.

ANQ250   HR: 3
AMMONIUM PEROXO BORATE
mf: $BH_4NO_3 \cdot 1/2H_2O$ mw: 85.86

PROP: White crystals. Mp: decomp; sltly sol in water.

SAFETY PROFILE: Potentially explosive by heat, friction, or impact. When heated to decomposition it emits toxic fumes of NO_x and NH_3.

ANQ750   HR: 3
AMMONIUM PEROXY CHROMATE
mf: $(NH_4)_3CrO_2$ mw: 234.1

PROP: Red-brown crystals. Mp: decomp @ 40°, bp: explodes @ 50°.

CONSENSUS REPORTS: Chromium and its compounds are on the Community Right-To-Know List.

SAFETY PROFILE: A poison. Moderately flammable by chemical reaction with reducing agents. A powerful oxidizer. Moderately explosive when heated. When heated to decomposition it emits toxic fumes of NO_x and NH_3.

ANR000   CAS: 7727-54-0   HR: 3
AMMONIUM PERSULFATE
DOT: UN 1444
mf: $O_8S_2 \cdot 2H_4N$ mw: 228.22

PROP: White crystals. Mp: decomp @ 120°, d: 1.982.

SYNS: AMMONIUM PEROXYDISULFATE * AMMONIUM PERSULFATE (DOT) * PERSULFATE d'AMMONIUM (FRENCH)

CONSENSUS REPORTS: Reported in EPA TSCA Inventory.

DOT Classification: IMO: Oxidizer; Label: Oxidizer.

SAFETY PROFILE: Poison by intravenous and intraperitoneal routes. Moderately toxic by ingestion. A powerful oxidizer which can react vigorously with reducing agents. Releases oxygen when heated. Mixtures with sodium peroxide are explosives sensitive to friction, heating above 75°C, or contact with CO_2 or water. Mixtures with (powdered aluminum + water) or (zinc + ammonia) are explosive. Violent reaction with iron or solutions of ammonia + silver salts. Solution with sulfuric acid is a strong oxidizing cleaning solution. When heated to decomposition it emits toxic fumes of SO_x, NH_3, and NO_x.

ANR500   CAS: 7783-28-0   **HR: 2**
AMMONIUM PHOSPHATE DIBASIC
mf: $H_6N_2 \cdot H_3O_4P$ mw: 132.08

PROP: White crystals or powder; salty taste. D: 1.619, mp: 155° (decomp). Sol in water; insol in alc.

SYNS: AMMONIUM PHOSPHATE * DIAMMONIUM HYDROGEN PHOSPHATE * DIBASIC AMMONIUM PHOSPHATE * SECONDARY AMMONIUM PHOSPHATE

CONSENSUS REPORTS: Reported in EPA TSCA Inventory.

SAFETY PROFILE: Low to moderate toxicity. When heated to decomposition it emits very toxic fumes of PO_x, NO_x, and NH_3.

ANR750   CAS: 7772-76-1   **HR: 2**
AMMONIUM PHOSPHATE, MONOBASIC
mf: $NH_4H_2PO_4$ mw: 115

PROP: Brilliant white crystals or powder. D: 1.803 @ 19°; mp: 190°. Sol in water.

SAFETY PROFILE: Incompatible with NaOCl.

ANS000   **HR: 3**
AMMONIUM PHOSPHIDE
mf: $P(NH_4)_3$ mw: 85.07

SAFETY PROFILE: Poison by inhalation and ingestion. When heated to decomposition it emits toxic fumes of PO_x, NO_x, and NH_3.

ANS250   CAS: 51503-61-8   **HR: 3**
AMMONIUM PHOSPHITE
mf: H_6NO_3P mw: 99.04

SYN: AMMONIUM ORTHOPHOSPHITE

SAFETY PROFILE: Poison by inhalation. When heated to decomposition it emits very toxic fumes of NO_x, NH_3, and PO_x.

ANS500   CAS: 131-74-8   **HR: 3**
AMMONIUM PICRATE

DOT: UN 0004
mf: $C_6H_3N_3O_7 \cdot H_3N$ mw: 246.16

PROP: Red or yellow, rhombic crystals. D: 1.719, mp: decomp; bp: expl @ 423°. Solubility: 1.1/100 @ 20°.

SYNS: AMMONIUM PICRONITRATE * OBELINE PICRATE * PICRATE of AMMONIA (DOT) * PICRIC ACID, AMMONIUM SALT

CONSENSUS REPORTS: Reported in EPA TSCA Inventory.

DOT Classification: Class A Explosive; Label: Explosive A.

SAFETY PROFILE: An allergen. Moderately irritating to skin, eyes, and mucous membranes. Moderately flammable by spontaneous chemical reaction. A powerful oxidizer which reacts vigorously with reducing materials. Dangerous explosive when shocked or heated. The presence of trace metals increases its heat sensitivity. When heated to decomposition it emits highly toxic fumes of NO_x.

ANS750   CAS: 131-74-8   **HR: 3**
AMMONIUM PICRATE (wet)

DOT: UN 1310

PROP: Bright yellow crystals; bitter taste. Compound contains 10% or more water.

SYNS: AMMONIUM PICRATE, wet with 10% or more water (DOT) * AMMONIUM PICRATE, wet with 10% or more water, over 16 oz. in one outside packaging (DOT) * EXPLOSIVE D * OBELINE PICRATE * PHENOL, 2,4,6-TRINITRO-, AMMONIUM SALT (9CI) * PICRATOL * PICRIC ACID, AMMONIUM SALT * RCRA WASTE NUMBER P009

CONSENSUS REPORTS: Reported in EPA TSCA Inventory.

DOT Classification: Flammable Solid; Label: Flammable Solid.

SAFETY PROFILE: An explosive. A flammable material that is a powerful oxidizer. When heated to decomposition it emits fumes of NO_x; explodes.

ANU650   CAS: 7773-06-0   **HR: 2**
AMMONIUM SULFAMATE

DOT: UN 9089
mf: $H_2NO_3S \cdot H_4N$ mw: 114.14

PROP: Deliquescent, crystalline material (white crystalline solid). Bp: 160° (decomp), mp: 131°.

SYNS: AMCIDE * AMICIDE * AMMAT
* AMMATE * AMMONIUM AMIDOSULFONATE
* AMMONIUM AMIDOSULPHATE * AMMONIUM-
SALZ der AMIDOSULFONSAURE (GERMAN) * AMMO-
NIUM SULPHAMATE * AMS * IKURIN * MONO-
AMMONIUM SULFAMATE * SULFAMATE
* SULFAMIC ACID, MONOAMMONIUM SALT
* SULFAMINSAURE (GERMAN)

CONSENSUS REPORTS: Reported in EPA
TSCA Inventory.

OSHA PEL: (Transitional: TWA Total Dust:
15 mg/m^3; Respirable Fraction: 5 mg/m^3)
TWA 10 mg/m^3; Respirable Fraction: 5
mg/m^3
ACGIH TLV: TWA 10 mg/m^3
DFG MAK: 15 mg/m^3
DOT Classification: ORM-E; Label: None

SAFETY PROFILE: Moderately toxic by ingestion and intraperitoneal routes. Somewhat explosive when heated or by spontaneous chemical
reaction in a hot acid solution. A powerful oxidizer. When heated to decomposition it emits
very toxic fumes of NH_3, NO_x, and SO_x.

ANU750   CAS: 7783-20-2   **HR: 2**
AMMONIUM SULFATE (2:1)
DOT: UN 2506
mf: $H_8N_2O_4S$ mw: 132.16

PROP: White crystals. Mp: > 280° (decomp);
d: 1.77. Sol in water; insol in alc.

SYNS: AMMONIUM SULPHATE * DIAMMONIUM
SULFATE * SULFURIC ACID, DIAMMONIUM SALT

CONSENSUS REPORTS: Community Right-
To-Know List. Reported in EPA TSCA Inventory.

DOT Classification: ORM-B; Label: None.

SAFETY PROFILE: Moderately toxic by several routes. Incandescent reaction on heating
with potassium chlorate. Reaction with sodium
hypochlorite gives the unstable explosive nitrogen trichloride. Incompatible with (K +
NH_4NO_3); KNO_2; (NaK + NH_4NO_3). When
heated to decomposition it emits very toxic
fumes of NO_x, NH_3, and SO_x.

ANW750   CAS: 1762-95-4   **HR: 3**
AMMONIUM THIOCYANATE
DOT: NA 9092
mf: $CNS \cdot H_4N$ mw: 76.13

PROP: Colorless solid or deliquescent crystals.
Mp: 149.6°, bp: decomp @ 170°, d: 1.305.

SYNS: AMMONIUM RHODANATE * AMMONIUM
RHODANIDE * AMMONIUM SULFOCYANATE
* AMMONIUM SULFOCYANIDE * AMTHIO
* RHODANID * RHODANIDE * TRANS-AID
* USAF EK-P-433 * WEEDAZOL TL

CONSENSUS REPORTS: Reported in EPA
TSCA Inventory. EPA Genetic Toxicology Program.

DOT Classification: ORM-E; Label: None.

SAFETY PROFILE: Poison by ingestion. Moderately toxic by other routes. Human systemic
effects by ingestion: hallucinations and distorted
perceptions, nausea or vomiting, and other gastrointestinal effects. When heated to decomposition it emits toxic fumes of NH_3, NO_x, SO_x,
and CN^-. Incompatible with $KClO_3$ and mixtures with $Pb(NO_3)_2$.

ANX750   **HR: 3**
AMMONIUM TRICHLOROACETATE
mf: $NH_4O_2CCCl_3$ mw: 180.6

SAFETY PROFILE: Poison by inhalation and
ingestion. A powerful irritant. When heated to
decomposition or on contact with acid or acid
fumes it emits toxic fumes of Cl^-, NH_3, and
NO_x. Incompatible with water or steam.

ANY250   CAS: 7803-55-6   **HR: 3**
AMMONIUM VANADATE
DOT: UN 2859
mf: $O_3V \cdot H_4N$ mw: 116.99

PROP: Colorless to yellow crystals. Mp: 200°
(decomp), d: 2.326.

SYNS: AMMONIUM METAVANADATE (DOT)
* RCRA WASTE NUMBER P119 * VANADIC ACID,
AMMONIUM SALT

CONSENSUS REPORTS: Reported in EPA
TSCA Inventory. EPA Genetic Toxicology Program.

NIOSH REL: (Vanadium, Compound) CL 0.05
mg(V)/m^3/15M
DOT Classification: IMO: Poison B; Label:
Poison.

SAFETY PROFILE: Poison by ingestion, subcutaneous, intravenous, intratracheal, and intraperitoneal routes. Experimental teratogenic and
reproductive effects. Mutation data reported.

When heated to decomposition it emits toxic fumes of NH_3, VO_x, and NO_x.

AOA100 CAS: 61336-70-7 **HR: 2**
AMOXICILLIN TRIHYDDRATE
mf: $C_{16}H_{19}N_3O_5S \cdot 3H_2O$ mw: 419.50

SYNS: α-AMINO-p-HYDROXYBENZYLPENICILLIN
TRIHYDRATE * (2S-(2-α,5-α,6-β(S*)))-6((AMINO-
(4-HYEROXYPHENYL)ACETYL)AMINO)-3,3-DI-
METHYL-7-OXO-4-THIA-1-AZABICYCLO(3.2.0)HEPTANE-2-
CARBOXYLIC ACID TRIHYDRATE * BRL 2333
TRIHYDRATE

SAFETY PROFILE: Moderately toxic. Experimental teratogenic and reproductive effects. When heated to decomposition it emits toxic fumes of SO_x and NO_x.

AOA750 CAS: 2706-50-5 **HR: 3**
AMPHETAMINE HYDROCHLORIDE
mf: $C_9H_{13}N \cdot ClH$ mw: 171.69

SYNS: dl-α-METHYL-PHENETHYLAMINE HYDROCHLO-
RIDE * dl-β-PHENYLISOPROPYLAMINE HYDROCHLO-
RIDE

SAFETY PROFILE: Poison by subcutaneous, intravenous, and intraperitoneal routes. When heated to decomposition it emits very toxic fumes of HCl and NO_x.

AOC500 CAS: 1397-89-3 **HR: 3**
AMPHOTERICIN B
mf: $C_{47}H_{73}NO_{17}$ mw: 924.21

SYNS: AMPHOMORONAL * AMPHOTERICINE B
* FUNGILIN * FUNGISONE

SAFETY PROFILE: Poison by intravenous and intraperitoneal routes. Mutation data reported. When heated to decomposition it emits toxic fumes of NO_x.

AOD125 CAS: 7177-48-2 **HR: D**
AMPICILLIN TRIHYDRATE
mf: $C_{16}H_{19}N_3O_4S \cdot 3H_2O$ mw: 403.50

SYNS: AMCAP * AMCILL * AMINOBENZYLPENI-
CILLIN TRIHYDRATE * α-AMINOBENZYLPENICILLIN
TRIHYDRATE * AMPERIL * AMPICHEL
* AMPIKEL * AMPINOVA * AMPLIN * ANCIL-
LIN * CYMBI * DIVERCILLIN * LIFEAMPIL
* MOREPEN * NCI-C56086 * PEN A * PENSYN
* POLYCILLIN * PRINCILLIN * RO-AMPEN
* TRAFARBIOT * UKOPEN * VIDOPEN

SAFETY PROFILE: Experimental teratogenic and reproductive effects. When heated to decomposition it emits toxic fumes of SO_x and NO_x.

AOD725 CAS: 628-63-7 **HR: 2**
n-AMYL ACETATE

DOT: UN 1104
mf: $C_7H_{14}O_2$ mw: 130.21

PROP: Colorless liquid; pear or banana-like odor. Mp: −78.5°, bp: 148° @ 737 mm, ULC: 55-60, lel: 1.1%, uel: 7.5%, flash p: 77°F (CC), d: 0.879 @ 20°/20°, autoign temp: 714°F, vap d: 4.5. Very sltly sol in water; misc in alc and ether.

SYNS: ACETATE d'AMYLE (FRENCH) * ACETIC
ACID, AMYL ESTER * AMYL ACETATE (DOT)
* AMYL ACETIC ESTER * AMYL ACETIC ETHER
* AMYLAZETAT (GERMAN) * BIRNENOEL
* OCTAN AMYLU (POLISH) * PEAR OIL * PENT-
ACETATE * 1-PENTANOL ACETATE * PENTYL
ACETATE * n-PENTYL ACETATE * 1-PENTYL
ACETATE * PRIMARY AMYL ACETATE

CONSENSUS REPORTS: Reported in EPA TSCA Inventory.

OSHA PEL: TWA 100 ppm
ACGIH TLV: TWA 100 ppm
DOT Classification: Flammable or Combustible Liquid; Label: Flammable Liquid.

SAFETY PROFILE: Moderately toxic by intraperitoneal route. Human systemic effects by inhalation: conjunctiva irritation, headache, and somnolence. A human eye irritant. Apparently more toxic than butyl acetate. Chronic toxicity is of a low order. Dangerous fire hazard when exposed to heat or flame; can react with oxidizing materials. Moderately explosive in the form of vapor when exposed to flame. To fight fire, use alcohol foam, dry chemical. When heated to decomposition it emits acrid smoke and irritating fumes.

AOD735 CAS: 626-38-0 **HR: 2**
sec-AMYL ACETATE

DOT: UN 1104
mf: $C_7H_{14}O_2$ mw: 130.21

PROP: Colorless liquid. Bp: 120°, flash p: 73.4°F (CC), d: 0.862-0.866 @ 20°/20°, vap d: 4.48, lel: 1.1%, uel: 7.5%. Sltly sol in water; misc in alc and ether.

SYNS: 2-ACETOXYPENTANE * 1-METHYLBUTYL ACETATE * 2-PENTANOL, ACETATE * 2-PENTYL ACETATE

CONSENSUS REPORTS: Reported in EPA TSCA Inventory.

OSHA PEL: TWA 125 ppm
ACGIH TLV: TWA 125 ppm
DOT Classification: Flammable Liquid; Label: Flammable Liquid.

SAFETY PROFILE: Mildly toxic by inhalation. Human systemic effects by inhalation: conjunctiva irritation. Dangerous fire hazard when exposed to heat or flame; can react with oxidizing materials. Moderately explosive in the form of vapor when exposed to heat or flame. To fight fire, use alcohol foam, dry chemical. When heated to decomposition it emits acrid smoke and irritating fumes.

AOD750 HR: 2
AMYL ACETATE (mixed isomers)
PROP: Colorless liquid, pear-like odor. Mp: −78.5°, bp: 148° @ 737 mm, ULC: 55-60, lel: 1.1%, uel: 7.5%, flash p: 77°F (CC), d: 0.879 @ 20°/20°, autoign temp: 714°F, vap d: 4.5.

SYN: ACETIC ACID, AMYL ESTER

DFG MAK: 100 ppm (525 mg/m^3)

SAFETY PROFILE: A skin irritant. Mildly toxic by ingestion. Dangerous fire hazard; can react with oxidizing materials. Moderately explosive in the form of vapor when exposed to flame. To fight fire, use alcohol foam, dry chemical. When heated to decomposition it emits acrid smoke and irritating fumes.

AOE000 HR: 2
AMYL ALCOHOL
mf: $C_5H_{12}O$ mw: 88.1

PROP: Clear liquid. Mp: −79°, bp: 137.8°, flash p: 91°F (CC), d: 0.8168 @ 20°/20°, ULC: 40, lel: 1.2%, uel: 10% @ 212°F, vap press: 1 mm @ 13.6°, 10 mm @ 44.9°, vap d: 3.04. Sol in water; misc in alc and ether.

SYNS: ALCOOL AMYLIQUE (FRENCH) * N-AMYL ALCOHOL * AMYL ALCOHOL, NORMAL * N-AMYLALKOHOL (CZECH) * N-BUTYLCARBINOL * PENTANOL-1 * N-PENTANOL * PENTAN-1-OL * PENTASOL * PENTYL ALCOHOL * PRIMARY AMYL ALCOHOL

CONSENSUS REPORTS: Reported in EPA TSCA Inventory.

SAFETY PROFILE: Moderately toxic by ingestion and skin contact. An eye and upper respiratory irritant by inhalation. A severe skin and eye irritant. Questionable carcinogen with experimental tumorigenic data. Ingestion can cause headache, nausea, vomiting, delirium, and methemoglobin formation. Extremely flammable if exposed to heat, flame or powerful oxidizers. Moderately explosive when exposed to flame. Incompatible with oxidizing materials, hydrogen trisulfide. To fight fire use alcohol foam, dry chemical.

AOF250 CAS: 63905-98-6 HR: 3
4-AMYL-N-BENZOHYDRYLPYRIDINIUM BROMIDE
mf: $C_{22}H_{26}N \cdot Br$ mw: 384.40

SYN: B-45

SAFETY PROFILE: Poison by ingestion, intraperitoneal, subcutaneous, and intravenous routes. When heated to decomposition it emits very toxic fumes of NO_x and Br^-.

AOF750 HR: 3
d-AMYL BROMIDE
mf: $CH_3(CH_2)_4Br$ mw: 151.1

PROP: Colorless liquid. Bp: 120°, flash p: 90°F, fp: < −30°, d: 1.211 @ 25°/25°.

SAFETY PROFILE: Poison by intraperitoneal route. It can cause liver damage, is narcotic in high concentrations, and is a local irritant. Extremely flammable. To fight fire, use alcohol foam, water mist or spray, dry chemical. When heated to decomposition it emits very toxic bromides. Incompatible with oxidizing materials.

AOG500 CAS: 122-40-7 HR: 2
α-AMYL CINNAMALDEHYDE
mf: $C_{14}H_{18}O$ mw: 202.32

PROP: Yellow liquid; floral jasmine odor. D: 0.963, refr index: 1.554, bp: 174-175° @ 20 mm. Sol in fixed oils; insol in glycerin and propylene glycol

SYNS: α-AMYL CINNAMIC ALDEHYDE * α-AMYL-β-PHENYLACROLEIN * FEMA No. 2061 * JASMINALDEHYDE * α-PENTYL-CINNAMALDEHYDE

CONSENSUS REPORTS: Reported in EPA TSCA Inventory.

SAFETY PROFILE: Moderately toxic by ingestion. A mild skin irritant. When heated to decomposition it emits acrid smoke and irritating fumes.

AOG600 HR: 1
AMYL CINNAMATE
mf: $C_{14}H_{18}O_2$ mw: 218.28

PROP: Colorless to pale yellow liquid; slt cocoa odor. D: 0.992-0.997, refr index: 1.535, flash p: +212°F. Sol in fixed oils; sltly sol in propylene glycol; insol in glycerin @ 310°.

SYNS: FEMA No. 2063 * ISOAMYL CINNAMATE * ISOAMYL 3-PENTYL PROPENATE

SAFETY PROFILE: Combustible liquid. When heated to decomposition it emits acrid smoke and irritating fumes.

AOI750 CAS: 513-35-9 HR: 2
α,η-AMYLENE
DOT: UN 2371/UN 2460
mf: C_5H_{10} mw: 70.15

PROP: Liquid, disagreeable odor. Mp: −124° bp: 30.1°, lel: 1.6%, uel: 8.7%, flash p: 0°F (OC), d: 0.643, vap d: 2.42, autoign temp: 527°F.

SYNS: β-ISOAMYLENE * 2-METHYL-2-BUTENE * TRIMETHYLETHYLENE

CONSENSUS REPORTS: Reported in EPA TSCA Inventory.

DOT Classification: Flammable Liquid; Label: Flammable Liquid.

SAFETY PROFILE: Moderately toxic by ingestion and inhalation. Narcotic in high concentration. A simple asphyxiant. Extremely flammable. Moderately explosive when exposed to heat, flame, or powerful oxidizers. To fight fire, use alcohol foam, spray, mist, dry chemical. When heated to decomposition it emits acrid smoke and irritating fumes.

AOJ000 HR: 2
AMYLENES, MIXED
DOT: UN 1108/UN 2371/UN 2460
mf: C_5H_{10} mw: 70.58

PROP: Water-white liquid. Bp: 32.2°, flash p: 0°F, d: 0.66 @ 20°.

CONSENSUS REPORTS: Reported in EPA TSCA Inventory.

DOT Classification: Flammable Liquid; Label: Flammable Liquid

SAFETY PROFILE: Moderately toxic. Very flammable; reacts with heat, flame and oxidizing materials. To fight fire, use foam, CO_2, dry chemical.

AOJ500 CAS: 638-49-3 HR: 2
n-AMYL FORMATE
DOT: UN 1109
mf: $C_6H_{12}O_2$ mw: 116.18

PROP: Clear liquid. D: 0.902, 0.893 @ 15°/4°, mp: −73.5°, bp: 130.4°, flash p: 80°F. Very sltly sol in water; misc in alc and ether.

SYNS: AMYL FORMATE (DOT) * PENTYL FORMATE * m-PENTYL FORMATE

CONSENSUS REPORTS: Reported in EPA TSCA Inventory.

DOT Classification: Flammable Liquid; Label: Flammable Liquid.

SAFETY PROFILE: A moderate irritant by ingestion and skin contact. Dangerously flammable; reacts vigorously with heat, flame, oxidizing materials. To fight fire, use foam, CO_2, dry chemical.

AOK250 HR: 1
AMYL LACTATE
mf: $C_8H_{16}O_3$ mw: 160.2

PROP: Colorless liquid. Bp: 210°; flash p: 175°F; d: 0.960 @ 20°.

SAFETY PROFILE: An irritant by inhalation and ingestion. Moderately flammable. Incompatible with heat, flame, oxidizing materials. To fight fire, use foam, CO_2, dry chemical.

AOK500 HR: 2
AMYL LAURATE
mf: $C_5H_{11}O_2C(CH_2)_{10}CH_3$ mw: 270.44

PROP: Bp: 290°, flash p: 300°F, d: 0.86.

SAFETY PROFILE: It may defat skin and cause contact dermatitis. Combustible. Incompatible with oxidizing materials. To fight fire, use CO_2, dry chemical.

AOK750   CAS: 105-30-6   **HR: 2**
AMYL METHYL ALCOHOL
mf: $C_6H_{14}O$ mw: 102.20

PROP: Liquid. Bp: 130°, flash p: 114°F (CC), d: 0.804, vap d: 3.52.

SYNS: 1,3-DIMETHYL BUTANOL * ISOHEXYL AL-COHOL * ISOPROPYL DIMETHYL CARBINOL * METHYLAMYL ALCOHOL * METHYL ISOBUTYL CARBINOL * 2-METHYLPENTANOL-1 * 2-METHYL-2-PROPYLETHANOL

CONSENSUS REPORTS: Reported in EPA TSCA Inventory.

SAFETY PROFILE: Moderately toxic by ingestion and skin contact. A skin irritant. Human systemic irritant by inhalation. Moderately flammable; can react with oxidizing materials. To fight fire, use CO_2, dry chemical. When heated to decomposition it emits smoke and acrid fumes.

AOL000   CAS: 13256-07-0   **HR: 3**
n-AMYL-N-METHYLNITROSAMINE
mf: $C_6H_{14}N_2O$ mw: 130.22

SYNS: AMN * METHYLAMYLNITROSAMIN (GER-MAN) * METHYLAMYLNITROSAMINE * METHYL-N-AMYLNITROSAMINE * N-METHYL-N-NITROSOPEN-TYLAMINE * METHYL-N-PENTYLNITROSAMINE * N-NITROSO-N-METHYL-N-AMYLAMINE * NITRO-SOMETHYL-N-PENTYLAMINE

CONSENSUS REPORTS: EPA Genetic Toxicology Program.

SAFETY PROFILE: Suspected carcinogen with experimental carcinogenic, neoplastigenic, and tumorigenic data. Poison by ingestion, subcutaneous, and intraperitoneal routes. Mutation data reported. When heated to decomposition it emits toxic NO_x.

AOL250   CAS: 1002-16-0   **HR: 2**
AMYL NITRATE
DOT: UN 1112
mf: $C_5H_{11}NO_3$ mw: 133.17

PROP: Liquid. Bp: 145°, flash p: 125°F (OC), d: 0.99.

SYN: NITRATE D'AMYLE (FRENCH)

DOT Classification: Flammable or Combustible Liquid; Label:Flammable Liquid

SAFETY PROFILE: Moderately toxic by inhalation. Moderately flammable when exposed to heat or flame or by spontaneous chemical reaction. An oxidizing agent. When heated to decomposition it emits toxic fumes of NO_x.

AOL500   CAS: 463-04-7   **HR: 2**
n-AMYL NITRITE
DOT: UN 1113
mf: $C_5H_{11}NO_2$ mw: 117.17

PROP: Clear, yellowish liquid; peculiar, ethereal, fruity odor and pungent, aromatic taste. Bp: 96°-99°, d: 0.8528 @ 20°/4°, autoign temp: 408°F, vap d: 4.0.

SYNS: AMYL NITRITE (DOT) * 1-NITROPENTANE * NITROUS ACID, PENTYL ESTER * PENTYL NI-TRITE

CONSENSUS REPORTS: Reported in EPA TSCA Inventory.

DOT Classification: Flammable Liquid; Label: Flammable Liquid.

SAFETY PROFILE: Moderately toxic by inhalation and ingestion. Causes flushing of skin, rapid pulse, headache, and fall in blood pressure. Mutation data reported. Moderately flammable when exposed to heat or flame or by spontaneous chemical reaction. To fight fire, use alcohol foam. An oxidizing material. Vapors explode when heated. It will react with oxidizing or reducing materials. When heated to decomposition it emits toxic fumes of NO_x.

AOM000   CAS: 644-26-8   **HR: 3**
AMYLOCAINE
mf: $C_{14}H_{21}NO_2$ mw: 235.36

SYNS: AMYLEINE * 1-(DIMETHYLAMINO)-2-METHYL-2-BUTANOL BENZOATE (ESTER) * STOVAINE

SAFETY PROFILE: Poison by intravenous, subcutaneous, and intraperitoneal routes. When heated to decomposition it emits toxic fumes of NO_x.

AOM250   CAS: 14938-35-3   **HR: 3**
4-n-AMYLPHENOL
mf: $C_{11}H_{16}O$ mw: 164.27

PROP: Liquid. Bp: 342°, vap d: 5.66, flash p: 219°F (OC), d: 0.966.

SYN: p-PENTYLPHENOL

CONSENSUS REPORTS: Reported in EPA TSCA Inventory.

SAFETY PROFILE: Questionable carcinogen with experimental neoplastigenic data. Moderately flammable. To fight fire, use foam, CO_2, dry chemical. When heated to decomposition it emits acrid smoke and irritating fumes.

AOM500   HR: 3
2-sec-AMYLPHENOL

PROP: Clear, straw-colored liquid. D: 0.955-0.971 @ 30°/30°, bp: 235-250°, flash p: 200°F. Very sltly sol in water, sol in oils and organic solvents.

SYN: o-(sec-PENTYL) PHENOL

SAFETY PROFILE: Poison by intravenous route. Questionable carcinogen with experimental neoplastigenic data by skin contact. Moderately flammable when exposed to heat or flame. To fight fire, use foam, fog, dry chemical, water mist or spray, multi-purpose dry chemical. When heated to decomposition it emits acrid smoke and irritating fumes.

AON350   HR: 2
AMYL PROPIONATE
mf: $C_8H_{16}O_2$ mw: 144.21

PROP: Colorless liquid; fruity, apricot-pineapple odor. D: 0.866, refr index: 1.405-1.409, flash p: 106°F. Sol in alc, fixed oils; insol in glycerine, propylene glycol, water @ 160°.

SYNS: FEMA No. 2082 * ISOAMYL PROPIONATE

SAFETY PROFILE: Combustible liquid. When heated to decomposition it emits acrid smoke and irritating fumes.

AON500   CAS: 32446-40-5   HR: 3
n-AMYL THIOCYANATE
mf: $C_6H_{11}NS$ mw: 129.24

PROP: Pale yellow oil. D: 0.905, bp: 197°. Insol in water; sol in alc and ether.

SYN: THIOCYANIC ACID, AMYL ESTER

SAFETY PROFILE: Poison by subcutaneous and intraperitoneal routes. When heated to decomposition it emits toxic fumes of NO_x and SO_x.

AON750   CAS: 64-43-7   HR: 3
AMYTAL SODIUM
mf: $C_{11}H_{17}N_2O_3 \cdot Na$ mw: 248.29

SYNS: 5-ETHYL-5-ISOPENTYLBARBITURIC ACID SODIUM SALT * 5-ETHYL-5-(3-METHYLBUTYL)-

BARBITURIC ACID SODIUM DERIVATIVE * 5-ISOAMYL-5-ETHYLBARBITURIC ACID, SODIUM DERIVATIVE * SODIUM AMYLOBARBITONE * SODIUM ETHYLISOAMYLBARBITURATE * SODIUM ISOAMYLETHYL BARBITURATE

SAFETY PROFILE: Poison by ingestion, subcutaneous, intravenous, parenteral, intraperitoneal and rectal routes. When heated to decomposition it emits toxic NO_x and Na_2O.

AOP250   CAS: 2270-40-8   HR: 3
ANGUIDIN
mf: $C_{18}H_{26}O_7$ mw: 354.44

SYNS: ANG 66 * ANGUIDINE * DAS * 4-β,15-DIACETOXY-3-α-HYDROXY-12,13-EPOXYTRICHOTHEC-9-ENE * DIACETOXYSCIRPENOL * 4,15-DIACETOXYSCIRPEN-3-OL * DIAZETOXYSKIRPENOL (GERMAN) * (3-α,4-β)-12,13-EPOXY-4,15-DIACETATE-TRICHOTHEC-9-ENE-3,4,15-TRIOL * 12,13-EPOXY-4-β,15-DIAZETOXY-3-α-HYDROXY-TRICHOTHEC-9-ENE * MM 4462 * NSC-141537

CONSENSUS REPORTS: EPA Genetic Toxicology Program.

SAFETY PROFILE: A deadly poison by ingestion, inhalation, intravenous, intraperitoneal and subcutaneous routes. Human systemic effects by intraperitoneal route: muscle weakness, nausea or vomiting, and fever. Mutation data reported. A skin irritant. When heated to decomposition it emits acrid smoke and fumes.

AOP500   HR: 2
ANHYDRIDES

PROP: Chemical compounds derived from acids by elimination of a molecule of water. Thus, sulfur trioxide (SO_3) is the anhydride of sulfuric acid (H_2SO_4); carbon dioxide (CO_2) is the anhydride of carbonic acid (H_2CO_3); phthalic acid ($C_6H_4(CO_2H)_2$) minus water gives phthalic anhydride ($C_6H_4(CO_2)O$). This term should not be confused with anhydrous, meaning without water.

SAFETY PROFILE: Anhydrides are acidic and react with bases in tissue. Thus, they tend to attack and irritate tissue.

AOQ000   CAS: 62-53-3   HR: 3
ANILINE
DOT: UN 1547
mf: C_6H_7N mw: 93.14

PROP: Colorless, oily liquid; characteristic odor. Bp: 184.4°, lel: 1.3%, ULC: 20-25, flash p: 158°F (CC), fp: −6.2°, d: 1.02 @ 20°/4°, autoign temp: 1139°F, vap press: 1 mm @ 34.8°, vap d: 3.22.

SYNS: AMINOBENZENE * AMINOPHEN * ANILIN (CZECH) * ANILINA (ITALIAN, POLISH) * ANILINE OIL * BENZENAMINE * BLUE OIL * C.I. 76000 * HUILE d'ANILINE (FRENCH) * NCI-C03736 * PHENYLAMINE

CONSENSUS REPORTS: IARC Cancer Review: Animal Inadequate Evidence IMEMDT 4,27,74; Human No Evidence IMEMDT 4, 27,74. EPA Extremely Hazardous Substances List. Community Right-To-Know List. Reported in EPA TSCA Inventory.

OSHA PEL: (Transitional: TWA 5 ppm (skin)) TWA 2 ppm (skin)
ACGIH TLV: TWA 2 ppm (skin) (Proposed: BEI: 50 mg/L total p-aminophenol in urine at end of shift.
DFG MAK: 2 ppm (8 mg/m^3), Suspected Carcinogen; BAT: 1 mg/L in urine at end of shift.
DOT Classification: Label: Poison B.

SAFETY PROFILE: Suspected carcinogen with experimental neoplastigenic data. A human poison by an unspecified route. Poison experimentally by most routes including inhalation and ingestion. A skin and severe eye irritant, and a mild sensitizer. In the body, aniline causes formation of methemoglobin, resulting in prolonged anoxemia and depression of the central nervous system; less acute exposure causes hemolysis of the red blood cells, followed by stimulation of the bone marrow. The liver may be affected with resulting jaundice. Long-term exposure to aniline dye manufacture has been associated with malignant bladder growths. A common air contaminant. Moderately flammable when exposed to heat or flame. To fight fire, use alcohol foam, CO_2, dry chemical. It can react vigorously with oxidizing materials. When heated to decomposition it emits highly toxic fumes of NO_x. Spontaneously explosive reactions occur with benzenediazonium-2-carboxylate; dibenzoyl peroxide; fluorine nitrate; nitrosyl perchlorate; red fuming nitric acid; peroxodisulfuric acid; and tetranitromethane. Violent reactions with boron trichloride; peroxyformic acid; diisopropyl peroxydicarbonate; fluorine; trichloronitromethane (145°C); acetic anhydride; chlorosulfonic acid; hexachloromelamine; (HNO_3 + N_2O_4 + H_2SO_4); (nitrobenzene + glycerin); oleum; ($HCHO$ + $HClO_4$); perchromates; K_2O_2; β-propiolactone; $AgClO_4$; Na_2O_2; H_2SO_4; trichloromelamine; acids; peroxydisulfuric acid; FO_3Cl; diisopropyl peroxy-dicarbonate; n-haloimides; and trichloronitromethane. Ignites on contact with sodium peroxide + water. Forms heat or shock sensitive explosive mixtures with anilinium chloride (detonates at 240°C/7.6 bar); nitromethane; hydrogen peroxide; 1-chloro-2,3-epoxypropane; and peroxomonosulfuric acid. Reactions with perchloryl fluoride; perchloric acid; and ozone form explosive products.

AOQ500   HR: D
ANILINE DYES

SAFETY PROFILE: The finished dyes are generally very much less toxic than many of the intermediates occurring or used in the manufacture of the dyes. Some of the aniline dyes cause local irritating effects to the eyes, mucous membranes and skin; the basic dyes are believed to be more irritating than the acid dyes. Allergic responses to aniline dyes have been known to occur. When heated to decomposition it emits toxic fumes of NO_x, possibly SO_x.

AOR000   HR: 3
ANILINE OIL DRUMS, EMPTY

SAFETY PROFILE: Combustible if full of vapors, such drums may ignite under the proper conditions. A dangerous disaster hazard if many drums are involved. They emit highly toxic fumes of aniline.

AOR500   CAS: 548-62-9   HR: 3
ANILINE VIOLET
mf: $C_{25}H_{30}N_3 \cdot Cl$ mw: 408.03

SYNS: AIZEN CRYSTAL VIOLET EXTRA PURE * GENTIAN VIOLET * HEXAMETHYL-p-ROSANILINE HYDROCHLORIDE * HEXAMETHYL VIOLET * METHYLROSANILINE CHLORIDE * NCI-C55969

CONSENSUS REPORTS: Reported in EPA TSCA Inventory.

SAFETY PROFILE: Poison by ingestion, intravenous, intraperitoneal, and intraduodenal routes. Experimental reproductive effects. A human skin irritant. Human mutation data reported. Questionable carcinogen with experimental carcinogenic data. When heated to decomposition it emits very toxic fumes of NO_x, and Cl^-.

AOR750   CAS: 122-98-5   HR: 3
2-ANILINOETHANOL
mf: $C_8H_{11}NO$ mw: 137.20

PROP: D: 1.1, bp: 268°, flash p: 305°F (OC).

SYNS: N-(2-HYDROXYETHYL)PHENYLAMINE * PHENYL ETHANOLAMINE * N-PHENYLETHANO-LAMINE * 2-(PHENYLAMINO)ETHANOL

CONSENSUS REPORTS: Reported in EPA TSCA Inventory.

SAFETY PROFILE: Poison by skin contact, intraperitoneal, and intravenous routes. Moderately toxic by ingestion. A skin and severe eye irritant. Combustible when exposed to heat or flame. To fight fire, use dry chemical, water mist. When heated to decomposition it emits toxic fumes of NO_x.

AOT250   HR: 3
ANILITE

SAFETY PROFILE: A highly explosive mixture composed of liquid NO_2 and carbon disulfide or gasoline. Extremely sensitive to shock.

AOT500   CAS: 123-11-5   HR: 2
p-ANISALDEHYDE
mf: $C_8H_8O_2$ mw: 136.15

PROP: Colorless oil; hawthorn odor. D: 1.123 @ 20°/4°, refr index: 1.571-1.574, mp: 2.5°, bp: 247-248°, flash p: 250°F. Misc in alc, ether, fixed oils; sol in propylene glycol; insol in glycerin, water.

SYNS: ANISIC ALDEHYDE * FEMA No. 2670 * 4-METHOXYBENZALDEHYDE * p-METHOXY-BENZALDEHYDE (FCC)

CONSENSUS REPORTS: Reported in EPA TSCA Inventory.

SAFETY PROFILE: Moderately toxic by ingestion. A skin irritant. Mutation data reported. Combustible liquid. When heated to decomposition it emits acrid smoke and irritating fumes.

AOU250   CAS: 8007-70-3   HR: 2
ANISE OIL

PROP: Consists of (80-90%) of Anethole. Small quantities of methyl chavicol, p-methoxyaceto-phenone and other materials also. Found in the dried ripe fruit of *Impinella anisum L.* (FCTXAV 11,855,73). D: 0.978-0.988 @ 25°/25°.

SYNS: ANISEED OIL * ANIS OEL (GERMAN) * OIL of ANISE * STAR ANISE OIL

CONSENSUS REPORTS: Reported in EPA TSCA Inventory.

SAFETY PROFILE: Moderately toxic by ingestion. A weak sensitizer. May cause contact dermatitis. Combustible liquid. When heated to decomposition it emits acrid smoke and irritating fumes.

AOU500   CAS: 586-38-9   HR: 3
m-ANISIC ACID
mf: $C_8H_8O_3$ mw: 152.15

PROP: Needles from aq solns. Mp: 107-109°, bp: 170-172° @ 10 mm. Sol in hot water, alc, and ether.

SYNS: m-METHOXYBENZOIC ACID * 3-METH-OXYBENZOIC ACID

CONSENSUS REPORTS: Reported in EPA TSCA Inventory.

SAFETY PROFILE: Poison by intraperitoneal route. When heated to decomposition it emits acrid smoke and irritating fumes.

AOV000   CAS: 94-30-4   HR: 2
p-ANISIC ACID, ETHYL ESTER
mf: $C_{10}H_{12}O_3$ mw: 180.21

PROP: Colorless liquid; fruity, anise odor. D: 1.103 @ 25/25, refr index: 1.522-1.526, mp: 7-8°, bp: 269-270°, flash p: +212°F. Sol in alc and ether; sltly sol in water.

SYNS: ETHYL ANISATE * ETHYL-p-ANISATE (FCC) * ETHYL-4-METHOXYBENZOATE * ETHYL-p-METH-OXYBENZOATE * FEMA No. 2420

CONSENSUS REPORTS: Reported in EPA TSCA Inventory.

SAFETY PROFILE: Moderately toxic by ingestion. Combustible liquid. When heated to decomposition it emits acrid smoke and irritating fumes.

AOV500   CAS: 3290-99-1   HR: 3
p-ANISIC ACID, HYDRAZIDE
mf: $C_8H_{10}N_2O_2$ mw: 166.20

SYNS: ANISIC ACID HYDRAZIDE * ANISIC HYDRA-
ZIDE * ANISOYLHYDRAZINE * p-ANISOYLHYDRA-
ZINE * p-METHOXYBENZOIC ACID HYDRAZIDE
* 4-METHOXYBENZOIC ACID HYDRAZIDE
* p-METHOXYBENZOIC HYDRAZIDE
* 4-METHOXYBENZOYL HYDRAZIDE * (p-METH-
OXYBENZOYL)HYDRAZINE * 4-METHOXYBENZOYL-
HYDRAZINE

SAFETY PROFILE: Poison by intravenous
route. Questionable carcinogen with experimental neoplastigenic data. When heated to decomposition it emits toxic fumes of NO_x.

AOV900   CAS: 90-04-0   HR: 2
o-ANISIDINE

DOT: UN 2431
mf: C_7H_9NO mw: 123.17

SYNS: o-AMINOANISOLE * 2-AMINOANISOLE
* 1-AMINO-2-METHOXYBENZENE * 2-ANISIDINE
* o-ANISYLAMINE * 2-METHOXY-1-AMINOBENZENE
* o-METHOXYANILINE * 2-METHOXY-BENZEN-
AMINE (9CI) * o-METHOXYPHENYLAMINE

CONSENSUS REPORTS: IARC Cancer Re-
view: GROUP 2B IMEMDT 7,56,87; Human
Limited Evidence IMEMDT 27,63,82. EPA Ge-
netic Toxicology Program. Reported in EPA
TSCA Inventory. Community Right-To-Know
List.

OSHA PEL: TWA 0.5 mg/m^3
ACGIH TLV: TWA 0.5 mg/m^3 (skin)
DFG MAK: 0.1 ppm (0.5 mg/m^3)
DOT Classification: DOT-IMO: Poison B; La-
bel: St. Andrews Cross.

SAFETY PROFILE: Suspected carcinogen.
Moderately toxic by ingestion. Mutation data
reported. When heated to decomposition it emits
toxic fumes of NO_x.

AOW000   CAS: 104-94-9   HR: 2
p-ANISIDINE
mf: C_7H_9NO mw: 123.16

PROP: Plates from aq soln. D: 1.089 @ 55°/
55°, mp: 57.2°, bp: 243°, vap d: 4.28. Sol in
hot water, alc, and ether.

SYNS: p-AMINOANISOLE * 4-AMINOANISOLE
* 1-AMINO-4-METHOXYBENZENE * 4-ANISIDINE
* p-ANISYLAMINE * p-METHOXYANILINE
* 4-METHOXYANILINE * 4-METHOXYBENZEN-
AMINE * 4-METHOXYBENZENEAMINE
* p-METHOXYPHENYLAMINE

CONSENSUS REPORTS: IARC Cancer Review: Human Inadequate Evidence IMEMDT
27,63,82. Community Right-To-Know List.
Reported in EPA TSCA Inventory.

OSHA PEL: TWA 0.5 mg/m^3
ACGIH TLV: TWA 0.5 mg/m^3 (skin)
DFG MAK: 0.1 ppm (0.5 mg/m^3)

SAFETY PROFILE: Moderately toxic by several routes. A mild sensitizer. May cause a contact dermatitis. When heated to decomposition
it evolves toxic fumes of NO_x.

AOX250   CAS: 134-29-2   HR: 3
o-ANISIDINE HYDROCHLORIDE
mf: $C_7H_9NO \cdot ClH$ mw: 159.63

SYNS: C.I. 37115 * 2-METHOXYANILINE HYDRO-
CHLORIDE * NCI-C03747

CONSENSUS REPORTS: IARC Cancer Review: Animal Sufficient Evidence IMEMDT
27,63,82. NTP Fourth Annual Report On Carcinogens, 1984. NCI Carcinogenesis Bioassay
(feed); Clear Evidence: mouse, rat NCITR*
NCI-CG-TR-89,78. Community Right-To-
Know List.

SAFETY PROFILE: Confirmed carcinogen with
experimental carcinogenic, neoplastigenic, and
tumorigenic data. Mutation data reported. When
heated to decomposition it emits very toxic
fumes of NO_x and HCl.

AOX500   CAS: 20265-97-8   HR: 3
p-ANISIDINE HYDROCHLORIDE
mf: $C_7H_9NO \cdot ClH$ mw: 159.63

SYN: NCI-C03758

CONSENSUS REPORTS: IARC Cancer Review: Animal Inadequate Evidence IMEMDT
27,63,82; NCI Carcinogenesis Bioassay (feed);
No Evidence: mouse NCITR* NCI-CG-TR-
116,78; Inadequate Studies: rat NCITR* NCI-
CG-TR-116,78. Reported in EPA TSCA Inventory.

SAFETY PROFILE: Questionable carcinogen
with experimental carcinogenic and tumorigenic
data. When heated to decomposition it emits
very toxic fumes of NO_x and HCl.

AOX750   CAS: 100-66-3   HR: 2
ANISOLE

DOT: UN 2222
mf: C_7H_8O mw: 108.15

PROP: Mobile liquid, clear straw color; phenol, anise odor. Vapor d: 3.72, mp: $-37.3°$, bp: 153.8°, flash p: 125°F (COC), d: 0.983-0.988, refr index: 1.513-1.518, vap press: 10 mm @ 42.2°, autoign temp: 887°F. Insol in water; sol in alc and ether.

SYNS: FEMA No. 2097 * METHOXYBENZENE * METHYL PHENYL ETHER * PHENYL METHYL ETHER

CONSENSUS REPORTS: Reported in EPA TSCA Inventory.

DOT Classification: Flammable or Combustible Liquid; Label: Flammable Liquid

SAFETY PROFILE: Moderately toxic by ingestion. A skin irritant. Combustible liquid. To fight fire, use foam, CO_2, dry chemical. When heated to decomposition it emits acrid fumes.

AOY250   CAS: 100-07-2   HR: 3
ANISOYL CHLORIDE
DOT: UN 1729
mf: $C_8H_7ClO_2$ mw: 170.60

PROP: Needle-like crystals, insol in water, sol in ether and acetone. Mp: 22°, bp: 262-263°, (slt decomp).

SYNS: p-ANISYL CHLORIDE * METHOXYBENZOYL CHLORIDE

CONSENSUS REPORTS: Reported in EPA TSCA Inventory.

DOT Classification: Corrosive Material; Label: Corrosive.

SAFETY PROFILE: Corrosive to skin, eyes, mucous membranes, and other tissue. Evolves HCl by hydrolysis. A storage hazard; can explode spontaneously at room temperature. When heated to decomposition it emits toxic fumes of Cl^- and may explode.

AOY400   HR: 1
ANISYL ACETATE
mf: $C_{10}H_{12}O_3$ mw: 180.20

PROP: Colorless to slt yellow liquid; fruity, balsamic odor. D: 1.104, refr index: 1.511-1.516, flash p: +210°F. Sol in alc, most oils; insol in glycerin and propylene glycol.

SYNS: FEMA No. 2098 * p-METHOXYBENZYL ACETATE

SAFETY PROFILE: Combustible liquid. When heated to decomposition it emits acrid smoke and irritating fumes.

APE100   HR: 2
ANNATTO EXTRACT
PROP: From solvent extraction of *Bixa orellana* L. seeds (JAPMA8 49,218,60). Yellow red solutions or powder.

SYNS: ACHIOTE * BIXA ORELLANA

SAFETY PROFILE: Moderately toxic by intraperitoneal route. Human systemic effects by skin contact. When heated to decomposition it emits acrid smoke and irritating fumes.

APE750   CAS: 191-26-4   HR: 3
ANTHANTHRENE
mf: $C_{22}H_{12}$ mw: 276.34

SYNS: ANTHANTHREN (GERMAN) * ANTHRANTHRENE * DIBENZO-(drf,mno)CHRYSENE * DIBENZO(cd,mk)PYRENE

CONSENSUS REPORTS: IARC Cancer Review: GROUP 3 IMEMDT 7,56,87; Animal Limited Evidence IMEMDT 32,95,83

SAFETY PROFILE: Questionable carcinogen with experimental carcinogenic and tumorigenic data. Mutation data reported. A polycyclic hydrocarbon found in polluted air. When heated to decomposition it emits acrid fumes.

APG000   CAS: 613-13-8   HR: 3
2-ANTHRACENAMINE
mf: $C_{14}H_{11}N$ mw: 193.26

PROP: Yellow leaflets from alc. Mp: 238°; bp: subl @ 93° @ 9 mm. Insol in water; sltly sol in alc and ether.

SYNS: β-AMINOANTHRACENE * 2-AMINOANTHRACENE * 2-ANTHRACYLAMINE * 2-ANTHRAMINE * 2-ANTHRYLAMINE

SAFETY PROFILE: Suspected carcinogen with experimental carcinogenic and tumorigenic data. Mutation data reported. When heated to decomposition it emits toxic fumes of NO_x.

APG500   CAS: 120-12-7   HR: 3
ANTHRACENE
mf: $C_{14}H_{10}$ mw: 178.24

PROP: Colorless crystals, violet fluorescence. Mp: 217°, lel: 0.6%, flash p: 250°F (CC), d: 1.24 @ 27°/4°, autoign temp: 1004°F, vap press:

1 mm @ 145.0°, (sublimes), vap d: 6.15, bp: 339.9°. Insol in water. Solubility in alc @ 1.9/100 @ 20°; in ether 12.2/100 @ 20°.

SYNS: ANTHRACEN (GERMAN) * ANTHRACIN * GREEN OIL * PARANAPHTHALENE * TETRA OLIVE N2G

CONSENSUS REPORTS: IARC Cancer Review: Animal No Evidence IMEMDT 32,105, 83.Reported in EPA TSCA Inventory. Community Right-To-Know List.

OSHA PEL: TWA 0.2 mg/m^3

SAFETY PROFILE: A skin irritant and allergen. Questionable carcinogen with experimental neoplastigenic and tumorigenic data. Mutation data reported. Combustible when exposed to heat, flame, or oxidizing materials. Moderately explosive when exposed to flame; Ca(OCl)$_2$; chromic acid. To fight fire, use water, foam, CO$_2$, water spray or mist, dry chemical. Explodes on contact with fluorine.

APH250   CAS: 480-22-8   **HR: 3**
1,8,9-ANTHRACENETRIOL
mf: C$_{14}$H$_{10}$O$_3$ mw: 226.24

PROP: Yellow powder. Mp: 178-180°. Insol in water; sol in fat, hot alc, benzene, and dilute alkalies.

SYNS: ANTHRALIN * 1,8,9-ANTHRATRIOL * DIHYDROXYANTHRANOL * 1,8-DIHYDROXY-ANTHRANOL * 1,8-DIHYDROXY-9-ANTHRANOL * 1,8-DIHYDROXY-9-ANTHRONE * DIOXYANTHRANOL * 1,8,9-TRIHYDROXYANTHRACENE

CONSENSUS REPORTS: IARC Cancer Review: GROUP 3 IMEMDT 7,56,87; Animal Limited Evidence IMEMDT 13,75,77

SAFETY PROFILE: Questionable carcinogen with experimental neoplastigenic and tumorigenic data. Mutation data reported. Skin contact can cause folliculitis. Absorption can cause kidney damage and intestinal disturbances. Combustible when heated. When heated to decomposition it emits acrid smoke and irritating fumes.

API500   CAS: 118-92-3   **HR: 3**
ANTHRANILIC ACID
mf: C$_7$H$_7$NO$_2$ mw: 137.15

PROP: Needle-like crystals. Mp: 146°, bp: subl, d: 1.412 @ 20°. Solubility: in water = 0.35/100 @ 14°, in 90% alc = 10.7/100 @ 10°, in ether = 16/100 @ 70°.

SYNS: o-AMIDOBENZOIC ACID * o-AMINOBENZOIC ACID * 2-AMINOBENZOIC ACID * 1-AMINO-2-CARBOXYBENZENE * CARBOXYANILINE * o-CARBOXYANILINE * 2-CARBOXYANILINE * NCI-C01730 * VITAMIN L

CONSENSUS REPORTS: IARC Cancer Review: Animal Inadequate Evidence IMEMDT 16,265,78. NTP Carcinogenesis Bioassay (feed): No Evidence: mouse, rat NCITR* NCI-TR-36,78. Reported in EPA TSCA Inventory.

SAFETY PROFILE: Experimental reproductive effects. Moderately toxic by intraperitoneal route. Questionable carcinogen with experimental tumorigenic data. Combustible. When heated to decomposition it emits toxic fumes of NO$_x$.

API750   CAS: 87-29-6   **HR: 3**
ANTHRANILIC ACID, CINNAMYL ESTER
mf: C$_{16}$H$_{15}$NO$_2$ mw: 253.32

PROP: Reddish yellow powder; balsamic odor. Mp: 60°, flash p: +212°F. Sol in alc, chloroform, ether; insol in water.

SYNS: 2-AMINOBENZOIC ACID-3-PHENYL-2-PROPENYL ESTER * CINNAMYL ALCOHOL ANTHRANILATE * CINNAMYL-2-AMINOBENZOATE * CINNAMYL-o-AMINOBENZOATE * CINNAMYL ANTHRANILATE (FCC) * FEMA No. 2295 * NCI-C03510 * 3-PHENYL-2-PROPENYLANTHRANILATE * 3-PHENYL-2-PROPEN-1-YL ANTHRANILATE

CONSENSUS REPORTS: IARC Cancer Review: GROUP 3 IMEMDT 7,56,87; Animal Limited Evidence IMEMDT 31,133,83; Animal Inadequate Evidence IMEMDT 16,287,78; NCI Carcinogenesis Bioassay (feed); Clear Evidence: mouse, rat NCITR* NCI-CG-TR-196,80. Reported in EPA TSCA Inventory.

SAFETY PROFILE: Suspected carcinogen with experimental carcinogenic and neoplastigenic data. Combustible liquid. When heated to decomposition it emits toxic fumes of NO$_x$.

APJ250   CAS: 134-20-3   **HR: 3**
ANTHRANILIC ACID, METHYL ESTER
mf: C$_8$H$_9$NO$_2$ mw: 151.18

PROP: Plates from alc or colorless liquid; grape odor. D: 1.161-1.169, mp: 23.8°, bp: 225-230° @ 15 mm, flash p: 219°F. Very sol in water, propylene glycol, hot abs alc (23/100); insol in ether, chloroform, glycerin.

SYNS: o-AMINOBENZOIC ACID METHYL ESTER * 2-AMINOBENZOIC ACID METHYL ESTER * 2-CARBOMETHOXYANILINE * o-CARBOMETHOXYANILINE * FEMA No. 2682 * 2-(METHOXYCARBONYL)ANILINE * METHYL 2-AMINOBENZOATE * METHYL o-AMINOBENZOATE * METHYL ANTHRANILATE (FCC) * METHYLESTER KYSELINY ANTHRANILOVE * NEROLI OIL, ARTIFICAL

CONSENSUS REPORTS: Reported in EPA TSCA Inventory.

SAFETY PROFILE: Moderately toxic by ingestion. Experimental reproductive effects. A skin irritant. Combustible liquid. When heated to decomposition it emits toxic fumes of NO_x.

APJ500   CAS: 133-18-6   **HR: 1**
ANTHRANILIC ACID, PHENETHYL ESTER
mf: $C_{15}H_{15}NO_2$ mw: 241.31
PROP: White to yellow crystals; grape odor.
SYNS: BENZYLCARBINYL ANTHRANILATE * β-PHENETHYL-o-AMINOBENZOATE * 2-PHENYLETHYL-o-AMINOBENZOATE * PHENETHYL ANTHRANILATE * 2-PHENYLETHYL ANTHRANILATE

CONSENSUS REPORTS: Reported in EPA TSCA Inventory.

SAFETY PROFILE: A skin irritant. When heated to decomposition it emits toxic fumes of NO_x.

APK250   CAS: 84-65-1   **HR: 2**
ANTHRAQUINONE
mf: $C_{14}H_8O_2$ mw: 208.22
PROP: Yellow crystals. Mp: 286°, bp: 376.9°, flash p: 365°F (CC), d: 1.438, vap press: 1 mm @ 190.0°, vap d: 7.16. Insol in water; very sltly sol in ether. Solubility in alc = 0.05/100 @ 18°, in hot alc = 2.25/100.
SYNS: 9,10-ANTHRACENEDIONE * ANTHRADIONE * 9,10-ANTHRAQUINONE * 9,10-DIOXOANTHRACENE

CONSENSUS REPORTS: Reported in EPA TSCA Inventory.

SAFETY PROFILE: Moderately toxic by several routes. A mild allergen. Combustible when exposed to heat or flame. To fight fire, use water, foam, CO_2, water spray or mist, dry chemical. When heated to decomposition it emits acrid smoke and irritating fumes.

AQB000   CAS: 31282-04-9   **HR: 3**
ANTIHELMYCIN
mf: $C_{20}H_{37}N_3O_{13}$ mw: 527.60
SYNS: HYGROMIX-8 * HYGROMYCIN B (USDA)

SAFETY PROFILE: Poison by intraperitoneal route. When heated to decomposition it emits toxic fumes of NO_x.

AQB750   CAS: 7440-36-0   **HR: 3**
ANTIMONY
DOT: UN 2871
af: Sb   aw: 121.75
PROP: Silvery or gray, lustrous metal. Mp: 630°, bp: 1635°, d: 6.684 @ 25°, vap press: 1 mm @ 886°. Insol in water; sol in hot concentrated H_2SO_4.
SYNS: ANTIMONY BLACK * ANTIMONY REGULUS * ANTYMON (POLISH) * C.I. 77050 * STIBIUM

CONSENSUS REPORTS: Antimony and its compounds are on the Community Right-To-Know List. Reported in EPA TSCA Inventory.
OSHA PEL: TWA 0.5 mg(Sb)/m³
ACGIH TLV: TWA 0.5 mg(Sb)/m³
DFG MAK: 0.5 mg(Sb)/m³
NIOSH REL: TWA 0.5 mg(Sb)/m³
DOT Classification: Poison B; LABEL: St. Andrews Cross.

SAFETY PROFILE: An experimental poison by intraperitoneal route. Questionable carcinogen with experimental carcinogenic data. Moderate fire and explosion hazard in the forms of dust and vapor, when exposed to heat or flame. When heated or on contact with acid it emits toxic fumes of SbH_3. Electrolysis of acid sulfides and stirred Sb halide yields explosive Sb. It can react violently with NH_4NO_3; halogens; BrN_3; BrF_3; $HClO_3$; ClO; ClF_3; HNO_3; KNO_3; $KMnO_4$; K_2O_2; $NaNO_3$; oxidants.

AQC500   CAS: 10025-91-9   **HR: 3**
ANTIMONY(III) CHLORIDE
DOT: UN 1733
mf: Cl_3Sb mw: 228.10
PROP: Colorless, rhombic, deliq crystals. D: 3.06, mp: 73.4°, bp: 220°, vap press: 1 mm @ 49.2° (subl). Sol in water @ 20°; sol in alc, benzene, and chloroform.
SYNS: ANTIMOINE (TRICHLORURE d') * ANTIMONIO (TRICLORURO di) * ANTIMONOUS CHLO-

RIDE * ANTIMONOUS CHLORIDE (DOT) * ANTI-MONTRICHLORID * ANTIMONY BUTTER * ANTIMONY CHLORIDE * ANTIMONY CHLORIDE (DOT) * ANTIMONY TRICHLORIDE * ANTIMONY TRI-CHLORIDE, liquid (DOT) * ANTIMONY TRICHLORIDE, solid (DOT) * ANTIMONY TRICHLORIDE, solution (DOT) * ANTIMOONTRICHLRIDE * BUTTER of ANTIMONY * CHLORID ANTIMONITY * CHLORURE ANTIMONIEUX * C.I. 77056 * STIBINE, TRICHLORO- * TRICHLOROSTIBINE * TRICHLORURE d'ANTIMOINE

CONSENSUS REPORTS: Reported in EPA TSCA Inventory. Antimony and its compounds are on the Community Right-To-Know List.

OSHA PEL: TWA 500 μg(Sb)/m^3
NIOSH REL: TWA 0.5 mg(Sb)/m^3
DOT Classification: Corrosive Material; Label: Corrosive.

SAFETY PROFILE: Moderately toxic by ingestion. Human pulmonary system effects by inhalation. Corrosive by vigorous reaction with moisture, generating heat and hydrogen chloride gas (a strong irritant) which can cause pulmonary edema when inhaled. Systemic effects can be caused by the antimony. Mutation data reported. When heated to decomposition it emits very toxic fumes of chlorine and antimony. It can react violently with aluminum; potassium; sodium.

AQD000   CAS: 7647-18-9   HR: 3
ANTIMONY(V) CHLORIDE
DOT: UN 1730/UN 1731
mf: Cl$_5$Sb   mw: 299.01

PROP: Red-yellow oil, liquid, offensive odor. Mp: 2.8°, bp: 140°, d: 2.336, vap press: 1 mm @ 22.7°. Decomp in water; sol in HCl, HBr, and CS$_2$.

SYNS: ANTIMONIC CHLORIDE * ANTIMONIO (PENTACLORURO DI) (ITALIAN) * ANTIMONPENTACHLORID (GERMAN) * ANTIMONY PENTACHLORIDE * ANTIMONY PENTACHLORIDE (DOT) * ANTIMONY PERCHLORIDE * ANTIMOONPENTACHLORIDE (DUTCH) * BUTTER of ANTIMONY * PENTACHLOROANTIMONY * PENTACHLORURE D'ANTIMOINE (FRENCH) * PERCHLORURE D'ANTIMOINE (FRENCH)

CONSENSUS REPORTS: Reported in EPA TSCA Inventory. Antimony and its compounds are on the Community Right-To-Know List.

OSHA PEL: TWA 500 μg(Sb)/m^3
NIOSH REL: TWA 0.5 mg(Sb)/m^3
DOT Classification: Corrosive Material; Label: Corrosive.

SAFETY PROFILE: Poison by ingestion. Corrosive. Mutation data reported. When heated to decomposition it emits very toxic fumes of Cl$^-$ and Sb.

AQD500   HR: 3
ANTIMONY COMPOUNDS
CONSENSUS REPORTS: On Community Right-To-Know List.

SAFETY PROFILE: Most antimony compounds are poisons by ingestion, inhalation, and intraperitoneal routes. Locally antimony compounds irritate the skin and mucous membranes. (Sb^{+++} and hot HClO$_3$) can form an explosive mixture.

AQE000   CAS: 7783-56-4   HR: 3
ANTIMONY(III) FLUORIDE (1:3)
DOT: NA 1549
mf: F$_3$Sb   mw: 178.75

PROP: Colorless, rhombic, deliq crystals. Mp: 292°, bp: 376° (subl), d: 4.379 @ 20.9°. Sol in water @ 20°.

SYNS: ANTIMOINE FLUORURE (FRENCH) * ANTIMONOUS FLUORIDE * ANTIMONY TRIFLUORIDE * TRIFLUOROANTIMONY

CONSENSUS REPORTS: Reported in EPA TSCA Inventory. Antimony and its compounds are on the Community Right-To-Know List.

NIOSH REL: TWA 0.5 mg(Sb)/m^3
DOT Classification: Corrosive Material; Label: Corrosive

SAFETY PROFILE: Poison by subcutaneous route. Corrosive to skin and eyes. When heated to decomposition it emits very toxic fumes of F$^-$ and Sb.

AQE250   CAS: 58164-88-8   HR: 3
ANTIMONY LACTATE
mf: C$_9$H$_{15}$O$_9$ • Sb   mw: 388.99

PROP: Tan-colored mass, water-sol.

SYNS: ANTIMONY LACTATE, solid (DOT) * LACTIC ACID, ANTIMONY SALT

CONSENSUS REPORTS: Reported in EPA TSCA Inventory. Antimony and its compounds are on the Community Right-To-Know List.

NIOSH REL: TWA 0.5 mg(Sb)/m^3
DOT Classification: ORM-A; Label: None.

SAFETY PROFILE: When heated to decomposition it emits toxic fumes of Sb.

AQE750 HR: 3
ANTIMONY NITRIDE
mf: NSb   mw: 135.76

CONSENSUS REPORTS: Antimony and its compounds are on the Community Right-To-Know List.

SAFETY PROFILE: Explosively decomposes upon warming (in vacuo). When heated to decomposition it emits very toxic fumes of Sb, NO$_x$, and NH$_3$.

AQF000 CAS: 1309-64-4 HR: 3
ANTIMONY OXIDE
mf: O$_3$Sb$_2$ mw: 291.50

PROP: White cubes. D: 5.2, mp: 650°, bp: 1550° subl. Very sltly sol in water; sol in KOH and HCl.

SYNS: ANTIMONIOUS OXIDE * ANTIMONY PEROX-
IDE * ANTIMONY SESQUIOXIDE * ANTIMONY TRI-
OXIDE (MAK) * ANTIMONY WHITE * C.I. PIG-
MENT WHITE 11 * DECHLORANE-A-O * DI-
ANTIMONY TRIOXIDE * FLOWERS of ANTIMONY
* NCI-C55152

CONSENSUS REPORTS: Reported in EPA TSCA Inventory. Antimony and its compounds are on the Community Right-To-Know List.

OSHA PEL: TWA 0.5 mg(Sb)/m^3
ACGIH TLV: TWA 0.5 mg(Sb)/m^3; Suspected Carcinogen
DFG MAK: Animal Carcinogen, Suspected Human Carcinogen.
NIOSH REL: TWA 0.5 mg(Sb)/m^3

SAFETY PROFILE: Confirmed carcinogen with experimental carcinogenic and neoplastigenic data. Poison by intravenous and subcutaneous routes. Moderately toxic by other routes. Experimental reproductive effects. Mutation data reported. When heated to decomposition it emits toxic Sb fumes. Incompatible with chlorinated rubber and heat of 216° and with BrF$_3$.

AQF250 CAS: 7783-70-2 HR: 3
ANTIMONY(V) PENTAFLUORIDE
DOT: UN 1732
mf: F$_5$Sb   mw: 216.75

PROP: Oily, colorless liquid. Very reactive. Mp: 7.0°, bp: 149.5°, d: (liq) 2.99 @ 23°. Sol in water and KF.

SYNS: ANTIMONY FLUORIDE * ANTIMONY(V)
FLUORIDE * PENTAFLUOROANTIMONY

CONSENSUS REPORTS: Reported in EPA TSCA Inventory. Antimony and its compounds are on the Community Right-To-Know List. EPA Extremely Hazardous Substances List.
NIOSH REL: TWA 0.5 mg(Sb)/m^3
DOT Classification: Corrosive Material; Label: Corrosive.

SAFETY PROFILE: A very reactive, corrosive liquid to skin, eyes, mucous membranes. Violent reaction with phosphates. When heated to decomposition it emits very toxic fumes of F$^-$ and Sb.

AQG250 CAS: 28300-74-5 HR: 3
ANTIMONY POTASSIUM TARTRATE
DOT: UN 1551
mf: C$_4$H$_4$O$_7$Sb • K   mw: 324.93

PROP: Colorless crystals to white powder. D: 2.607, mp: loses H$_2$O @ 100°.

SYNS: ANTIMONYL POTASSIUM TARTRATE
* POTASSIUM ANTIMONYL-d-TARTRATE * POTAS-
SIUM ANTIMONY TARTRATE * EMETIQUE (FRENCH)
* ENT 50,434 * POTASSIUM ANTIMONYL TARTRATE
* TARTAR EMETIC * TARTARIZED ANTIMONY
* TARTRATE ANTIMONIO-POTASSIQUE (FRENCH)
* TARTRATED ANTIMONY

CONSENSUS REPORTS: Antimony and its compounds are on the Community Right-To-Know List.

OSHA PEL: TWA 500 µg(Sb)/m^3
NIOSH REL: TWA 0.5 mg(Sb)/m^3
DOT Classification: ORM-A; Label: None.

SAFETY PROFILE: Poison by ingestion, subcutaneous, intravenous, intramuscular, and intraperitoneal routes. Large doses cause severe liver damage. Used medicinally, the therapeutic dose is close to the toxic dose. Upon decomposition it emits toxic fumes of K$_2$O and Sb.

AQI750 CAS: 34521-09-0 HR: 3
ANTIMONY SODIUM TARTRATE
mf: C$_4$H$_4$O$_7$Sb • Na   mw: 308.82

SYNS: SODIUM ANTIMONYL TARTRATE * ANTI-
MONY SODIUM OXIDE-l-(+)-TARTRATE * NATRIUM-

ANTIMONYLTARTRAT (GERMAN) * SODIUM ANTI-
MONY TARTRATE

CONSENSUS REPORTS: Antimony and its compounds are on the Community Right-To-Know List.

OSHA PEL: TWA 500 $\mu g(Sb)/m^3$
NIOSH REL: TWA 0.5 mg(Sb)/m^3

SAFETY PROFILE: Poison by subcutaneous, intravenous, and intraperitoneal routes. Human mutation data reported. When heated to decomposition it emits toxic fumes of Sb.

AQK500   **HR: 3**
ANTIMONY TRIETHYL
mf: $Sb(C_2H_5)_3$ mw: 209.0

PROP: Liquid, water-insol. D: 1.324 @ 16°, mp: −29°, bp: 159.5°.

CONSENSUS REPORTS: Antimony and its compounds are on the Community Right-To-Know List.

SAFETY PROFILE: Alkyl metal compounds are often highly toxic. Dangerous fire hazard by spontaneous chemical reaction. Explodes in air; water; carbon tetrachloride; other halogenated hydrocarbons; dimethyl formamide; triethyl borine. When heated to decomposition it emits highly toxic fumes of Sb.

AQK750   **HR: 3**
ANTIMONY TRIIODIDE
mf: SbI_3 mw: 502.5

PROP: Red-to-yellow crystals. Mp: 170°, bp: 401°, d: 4.768 @ 22°, vap press: 1 mm @ 163.6°.

CONSENSUS REPORTS: Antimony and its compounds are on the Community Right-To-Know List.

SAFETY PROFILE: Poison by ingestion. Incompatible with sodium; potassium. When heated to decomposition it emits highly toxic Sb fumes and I$^-$.

AQL000   **HR: 3**
ANTIMONY TRIMETHYL
mf: $Sb(CH_3)_3$ mw: 166.9

PROP: Liquid, sltly sol in water. Bp: 80.6°, d: 1.523 @ 15°

SYN: TRIMETHYL STIBINE

CONSENSUS REPORTS: Antimony and its compounds are on the Community Right-To-Know List.

SAFETY PROFILE: Toxic. Dangerous fire hazard by spontaneous reaction in air. Explodes in water. When heated to decomposition it emits highly toxic fumes of antimony. Incompatible with oxidizing materials; halogenated hydrocarbons.

AQL500   CAS: 1345-04-6   **HR: 3**
ANTIMONY TRISULFIDE

DOT: NA 1325
mf: S_3Sb_2 mw: 339.68

PROP: Red-to-black crystals. Mp: 546°, d: 4.64, bp: ca. 1150°. Sol in H_2SO_4, solubility in water = 0.002/100 @ 20° (decomp).

SYNS: ANTIMONOUS SULFIDE * ANTIMONY GLANCE * ANTIMONY ORANGE * ANTIMONY SULFIDE * C.I. 77060 * CRIMSON ANTIMONY * NEEDLE ANTIMONY

CONSENSUS REPORTS: Reported in EPA TSCA Inventory. Antimony and its compounds are on the Community Right-To-Know List.

OSHA PEL: TWA 500 $\mu g(Sb)/m^3$
NIOSH REL: TWA 0.5 mg(Sb)/m^3
DOT Classification: ORM-A; Label: None.

SAFETY PROFILE: Toxic by intraperitoneal route. Human blood and gastrointestinal system effects by inhalation. Spontaneously flammable when exposed to strong oxidizers. Flammable when exposed to heat or flame. Moderately explosive by spontaneous reaction with chlorates; perchlorates; ClO; thallic oxide. When heated to decomposition or on contact with acid or acid fumes it emits highly toxic fumes of oxides of sulfur and antimony. Will react with water or steam to produce toxic and flammable vapors.

AQL750   **HR: 3**
ANTIMONY TRITELLURIDE
mf: Sb_2Te_3 mw: 626.4

PROP: Gray powder. Mp: 629°; d: 6.50 @ 13°.

SYN: ANTIMONY TELLURIDE

CONSENSUS REPORTS: Antimony and its compounds are on the Community Right-To-Know List.

SAFETY PROFILE: Probably a poison. Flammable by spontaneous reaction with strong oxidizers. Moderately explosive by chemical reaction in contact with chlorates and perchlorates. When heated to decomposition or on contact with acid or acid fumes it emits highly toxic fumes of Sb and tellurium. Incompatible with water or steam and oxidizing materials.

AQN000   CAS: 60-80-0   **HR: 3**
ANTIPYRINE
mf: $C_{11}H_{12}N_2O$ mw: 188.23

PROP: Fine, white crystals. Mp: 113°, bp: 319° @ 174 mm, d: 1.19. Very sol in water and alc; sltly sol in ether.

SYNS: DIMETHYLOXYQUINAZINE * 2,3-DI-METHYL-1-PHENYL-3-PYRAZOLIN-5-ONE * 2,3-DI-METHYL-1-PHENYL-5-PYRAZOLONE * OXYDI-METHYLQUINAZINE * PHENAZONE (PHARMACEUTICAL) * 1-PHENYL-2,3-DIMETHYL-PYRAZOLE-5-ONE * 1-PHENYL-2,3-DI-METHYL-5-PYRAZOLONE

CONSENSUS REPORTS: Reported in EPA TSCA Inventory.

SAFETY PROFILE: A human poison by an unspecified route. Moderately toxic via ingestion, subcutaneous, and intravenous routes. Questionable carcinogen with experimental tumorigenic data. When heated to decomposition it emits toxic fumes of NO_x.

AQN635   CAS: 86-88-4   **HR: 3**
ANTU

DOT: UN 1651
mf: $C_{11}H_{10}N_2S$ mw: 202.29

PROP: Crystals; bitter taste. Mp: 198°. Sltly sol in hot alc.

SYNS: ALPHANAPHTHYL THIOUREA * ALPHANA-PHTYL THIOUREE (FRENCH) * ALRATO * ANTU-RAT * CHEMICAL 109 * DIRAX * KILL KANTZ * KRYSID * 1-NAFTIL-TIOUREA (ITALIAN) * 1-NAFTYLTHIOUREUM (DUTCH) * 1-NAPHTHAL-ENYLTHIOUREA * α-NAPHTHALTHIOHARNSTOFF (GERMAN) * α-NAPHTHOTHIOUREA * α-NAPH-THYLTHIOCARBAMIDE * 1-NAPHTHYL-THIO-HARNSTOFF (GERMAN) * 1-NAPHTHYL THIOUREA (MAK) * α-NAPHTHYLTHIOUREA * 1-(1-NAPH-THYL)-2-THIOUREA * N-(1-NAPHTHYL)-2-THIOUREA * α-NAPHTHYLTHIOUREA (DOT) * 1-NAPHTHYL-THIOUREE (FRENCH) * NAPHTOX

* RATTRACK * RCRA WASTE NUMBER P072 * SMEESANA * U-5227 * USAF EK-P-5976

CONSENSUS REPORTS: IARC Cancer Review: Animal Inadequate Evidence IMEMDT 30,347,83. Reported in EPA TSCA Inventory. EPA Extremely Hazardous Substances List. EPA Genetic Toxicology Program.

OSHA PEL: TWA 0.3 mg/m³
ACGIH TLV: TWA 0.3 mg/m³
DFG MAK: 0.3 mg/m³
DOT Classification: Poison B; Label: Poison.

SAFETY PROFILE: Poison by ingestion and intraperitoneal routes. Moderately toxic to humans by an unspecified route. Questionable carcinogen with experimental tumorigenic data. Mutagenic data. A rodenticide used extensively. Death is caused by pulmonary edema. Chronic toxicity has been known to cause dermatitis and a decrease in the white blood cells. When heated to decomposition it emits toxic fumes of NO_x and SO_x.

AQO750   **HR: 3**
APOCODEINE
mf: $C_{18}H_{19}NO_2$ mw: 281.34

PROP: White, crystalline solid. Mp: 124°.

SAFETY PROFILE: Poison by inhalation and ingestion. A weak sensitizer and may cause contact dermatitis. When heated to decomposition it emits highly toxic fumes of NO_x.

AQP000   CAS: 1937-37-7   **HR: 3**
APOMINE BLACK GX
mf: $C_{34}H_{25}N_9O_7S_2 \cdot 2Na$ mw: 781.78

SYNS: AHCO DIRECT BLACK GX * AIREDALE BLACK ED * AIZEN DIRECT DEEP BLACK GH * AMANIL BLACK GL * APOMINE BLACK GX * ATLANTIC BLACK BD * ATUL DIRECT BLACK E * AZINE DEEP BLACK EW * AZOCARD BLACK EW * AZOMINE BLACK EWO * BELAMINE BLACK GX * BENCIDAL BLACK E * BENZAMIL BLACK E * BENZO DEEP BLACK E * BENZOFORM BLACK BCN-CF * BLACK 2EMBL * BRASILAMINA BLACK GN * CALCOMINE BLACK * CARBIDE BLACK E * CERN PRIMA 38 * CHLORAMINE BLACK C * CHLORAZOL BLACK E (biological stain) * CHLORAZOL BLACK EA * CHLORAZOL BLACK EN * CHROME LEATHER BLACK EM * C.I. 30235 * C.I. DIRECT BLACK 38 * COIR DEEP BLACK C * COLUMBIA BLACK EP * DIACOTTON DEEP

BLACK * DIAMINE DEEP BLACK EC * DI-
APHTAMINE BLACK V * DIAZINE BLACK E
* DIAZOL BLACK 2V * DIPHENYL DEEP BLACK G
* DIRECT BLACK A * DIRECT BLACK META
* ENIANIL BLACK CN * ERIE BLACK B * FENA-
MIN BLACK E * FIBRE BLACK VF * FIXANOL
BLACK E * FORMALINE BLACK C * FORMIC
BLACK C * HISPAMIN BLACK EF * INTERCHEM
DIRECT BLACK Z * KAYAKU DIRECT DEEP BLACK
EX * LURAZOL BLACK BA * META BLACK
* MITSUI DIRECT BLACK EX * NCI-C54557
* NIPPON DEEP BLACK * PAPER BLACK BA
* PARAMINE BLACK B * PEERAMINE BLACK E
* PHENO BLACK EP * PONTAMINE BLACK E
* SANDOPEL BLACK EX * SERISTAN BLACK B
* TELON FAST BLACK E * TETRAZO DEEP
BLACK G * TERTRODIRECT BLACK E * TETRO-
DIRECT BLACK EFD * UNION BLACK EM
* VONDACEL BLACK N

CONSENSUS REPORTS: IARC Cancer Review: Animal Sufficient Evidence IMEMDT
29,295,82, Human Limited Evidence IMEMDT
29,295,82. NTP Fourth Annual Report On Carcinogens, 1984. Reported in EPA TSCA Inventory. NTP Carcinogenesis Bioassay (feed):
Clear Evidence: rat NCICTR* NCI-TR-108,78;
No Evidence: mouse NCICTR NCI-TR-108,78.
On Community-Right-To-Know List.

SAFETY PROFILE: Confirmed carcinogen with
carcinogenic and tumorigenic data. Mutation
data reported. When heated to decomposition
it emits very toxic fumes of NO_x, Na_2O, and
SO_2.

AQQ500   CAS: 9000-01-5   **HR: 2**
ARABIC GUM
mw: 240,000

PROP: A gum from the stems and branches
of *Acacia senegal (L.)* Willd. or of *Acacia* (Fam.
Leguminosae). Sol in water; insol in alc.

SYNS: ACACIA * ACACIA DEALBATA GUM
* ACACIA GUM * ACACIA SENEGAL * ACACIA
SYRUP * AUSTRALIAN GUM * GUM ARABIC
* GUM OVALINE * GUM SENEGAL * INDIAN
GUM * NCI-C50748 * SENEGAL GUM * STAR-
SOL NO. 1 * WATTLE GUM

CONSENSUS REPORTS: NTP Carcinogen-
esis Bioassay (feed); No Evidence: mouse, rat
NTPTR* NTP-TR-227,82. Reported in EPA
TSCA Inventory.

SAFETY PROFILE: Inhalation or ingestion has
produced hives, eczema, and angiodema. A
weak allergen. Combustible. When heated to
decomposition it emits acrid smoke.

AQT500   CAS: 39300-45-3   **HR: 3**
ARATHANE
mf: $C_{18}H_{24}N_2O_6$ mw: 364.44
PROP: Liquid.

SYNS: CAPRYLDINITROPHENYL CROTONATE
* 2-CAPRYL-4,6-DINITROPHENYL CROTONATE
* CROTONATE de 2,4-DINITRO 6-(1-METHYL-
HEPTYL)-PHENYLE (FRENCH) * 4,6-DINITRO-2-CA-
PRYLPHENYL CROTONATE * 4,6-DINITRO-2-(2-CA-
PRYL)PHENYL CROTONATE * DINITRO(1-METHYL-
HEPTYL)PHENYL CROTONATE * 2,4-DINITRO-6-(1-
METHYLHEPTYL)PHENYL CROTONATE * 2,4-DINI-
TRO-6-(2-OCTYL)PHENYL CROTONATE * ENT 24727
* (6-(1-METHYL-HEPTYL)-2,4-DINITRO-FENYL)-CROTO-
NAAT (DUTCH) * (6-(1-METHYL-HEPTYL)-2,3-DINI-
TRO-PHENYL)-CROTONAT (GERMAN) * 2-(1-METHYL-
HEPTYL)-4,6-DINITROPHENYL CROTONATE * (6-(1-
METIL-EPITL)-2,4-DINITRO-FENIL)-CROTONATO
(ITALIAN)

SAFETY PROFILE: Poison by ingestion and
intravenous routes. Questionable carcinogen
with experimental neoplastigenic data. When
heated to decomposition it emits toxic fumes
of NO_x.

AQT750   CAS: 63-75-2   **HR: 3**
ARECOLINE
mf: $C_8H_{13}NO_2$ mw: 155.22
PROP: Oily liquid. Bp: 209°.

SYNS: ARECAIDINE METHYL ESTER * ARECOLINE
BASE * METHYL-1,2,5,6-TETRAHYDRO-1-METHYLNI-
COTINATE * N-METHYL-Δ-TETRAHYDRONICOTINIC
ACID METHYL ESTER * N-METHYLTETRAHYDRO-
PYRIDINE-β-CARBOXYLIC ACID METHYL ESTER
* 1,2,5,6-TETRAHYDRO-1-METHYLNICOTINIC ACID,
METHYL ESTER

CONSENSUS REPORTS: IARC Cancer Review: Animal Inadequate Evidence IMEMDT
37,141,85

SAFETY PROFILE: Poison by inhalation, ingestion, and subcutaneous routes. Questionable
carcinogen with experimental neoplastigenic
data. It mimics the action of acetylcholine a
neuro transmitter, and is a parasympathetic nervous system stimulant. Its action on the central
nervous system can cause tremors. A mutagen.

It is easily nitrosated to several nitrosamines. It is the major alkaloid found in betel quid. Combustible, can react with oxidizing materials. When heated to decomposition it emits highly toxic fumes of NO_x.

AQW000   CAS: 1119-34-2   **HR: 2**
l-ARGININE MONOHYDROCHLORIDE
mf: $C_6H_{14}N_4O_2 \cdot ClH$ mw: 210.70

PROP: White crystalline powder; odorless. Mp: 222-235° (decomp). Very sol in water; sltly sol in alc.

SYNS: ARGAMINE * ARGININE HYDROCHLORIDE * l-ARGININE HYDROCHLORIDE * ARGININE MO-NOHYDROCHLORIDE * ARGIVENE * DETOX-ARGIN * l-HYDROCHLORIDE ARGININE * LEVAR-GIN * MINOPHAGEN A * R-GENE

CONSENSUS REPORTS: Reported in EPA TSCA Inventory.

SAFETY PROFILE: Moderately toxic by intraperitoneal route. Mildly toxic by ingestion. An experimental teratogen. When heated to decomposition it emits very toxic fumes of NO_x and HCl.

AQW250   CAS: 7440-37-1   **HR: 1**
ARGON
af: Ar   aw: 39.94

DOT: UN 1006/UN 1951

PROP: Colorless, inert gas. Mp: $-189.2°$, bp: $-185.7°$, d: 1.784 g/L @ 0°, 1.40 @ $-186°$, 1.65 @ $-233°$. Solubility in water 3.36 mL/100 g @ 20°.

CONSENSUS REPORTS: Reported in EPA TSCA Inventory.

DOT Classification: Nonflammable Gas; Label: Nonflammable Gas.

SAFETY PROFILE: A simple asphyxiant gas. As an inert gas, it has no specific inherent dangerous properties. Gases of this type have no specific toxicity effect, but they act by excluding O_2 from the lungs. The effect of simple asphyxiant gases is proportional to the extent to which they diminish the amount (partial pressure) of O_2 in the air that is breathed. The oxygen may be diminished to 0.75% of its normal percentage in air before appreciable symptoms develop, and this in turn requires the presence of a simple asphyxiant in a concentration of 33% in the mixture of air and gas. When the simple asphyxi-

ant reaches a concentration of 50%, marked symptoms can be produced. A concentration of 75% is fatal in a matter of minutes. The first symptoms produced by simple asphyxiant gases such as argon are rapid respirations and air hunger. Mental alertness is diminished and muscular coordination is impaired. Later, judgment becomes faulty and all sensations are depressed. Emotional instability often results and fatigue occurs rapidly. As the asphyxia progresses, there may be nausea and vomiting, prostration, and loss of consciousness, and finally, convulsions, deep coma and death.

AQY250   CAS: 313-67-7   **HR: 3**
ARISTOLOCHINE
mf: $C_{17}H_{11}NO_7$ mw: 341.29

PROP: From alcoholic extract of *Aristolochia indico* (CNCRA6 42,35,64).

SYNS: ARISTOLOCHIC ACID * BIRTHWORT * 8-METHOXY-6-NITROPHENANTHOL-(3,4-d)-1,3-DIOX-OLE-5-CARBOXYLIC ACID * NSC-50413

SAFETY PROFILE: Confirmed carcinogen. Poison by intravenous route. When heated to decomposition it emits toxic fumes of NO_x.
    From International Register of Potentially Toxic Chemicals: April 1982. Vol 5 No. 1: The Ministry of Health of the Federal Republic of Germany has withdrawn from the national market drugs containing aristolochic acid. The decision resulted from the demonstration of a carcinogenic potential in a three-month ingestion toxicity study undertaken in rats. Aristolochic acid is claimed to promote phagocytosis and to have immunostimulant activity. A growth-inhibiting effect on experimentally induced tumors has been described, but this effect has not been shown to have any clinical relevance. Extracts of species of *Aristolochiacea* have traditionally been used as a bitter, and a broad range of therapeutic effects has been claimed.

AQY500   **HR: 3**
ARNICA

PROP: An alcoholic infusion

SYNS: MOUNTAIN TOBACCO * WOLFSBANE

SAFETY PROFILE: Poison by inhalation and ingestion. A moderate irritant and allergen. It can cause gastroenteritis, nervous disturbances, and collapse. May cause contact dermatitis. Combustible when exposed to heat or flame. Incompatible with oxidizing materials.

AQY750   HR: 3
AROMATIC AMINES

PROP: Amines which contain one or more rings of unsaturated or cyclic HC, such as benzene. There are vast numbers of such amines. The term is largely due to the characteristic odor.

SAFETY PROFILE: Many of these aromatic amines are recognized as carcinogenic to the human bladder, ureter, and renal pelvis, intestines, lung, liver, and prostate.

AQZ000   HR: 2
AROMATIC SPIRITS of AMMONIA

PROP: Colorless liquid, suffocating odor of ammonia. Composition: 10% by weight of NH_3 in alcohol.

SAFETY PROFILE: A dangerous fire hazard due to its alcohol content. Moderately explosive. When heated, it emits toxic fumes of ammonia. Incompatible with oxidizing materials.

ARA250   CAS: 98-50-0   HR: 3
ARSANILIC ACID
mf: $C_6H_8AsNO_3$ mw: 217.06

PROP: Needles from aq solns. Mp: 232°, bp: decomp, $-H_2O$ @ 15°. Very sol in hot water, alc; insol in ether and benzene.

SYNS: 4-AMINOBENZENEARSONIC ACID * p-AMINOBENZENEARSONIC ACID * AMINOPHENYLARSINE ACID * p-AMINOPHENYLARSINE ACID * p-AMINOPHENYLARSINIC ACID * 4-AMINOPHENYLARSONIC ACID * p-AMINOPHENYLARSONIC ACID * p-ANILINEARSONIC ACID * ANTOXYLIC ACID * 4-ARSANILIC ACID * p-ARSANILIC ACID * ATOXYLIC ACID

CONSENSUS REPORTS: IARC Cancer Review: Animal Inadequate Evidence IMEMDT 23,39,80. Reported in EPA TSCA Inventory. Arsenic and its compounds are on the Community Right-To-Know List.

OSHA PEL: TWA 0.01 mg(As)/m^3
ACGIH TLV: TWA 0.2 mg(As)/m^3

SAFETY PROFILE: Poison by ingestion, intravenous, and intraperitoneal routes. Flammable, decomposes with heat to yield flammable vapors. When heated to decomposition or on

contact with acid or acid fumes emits highly toxic fumes of As and NO_x.

ARA500   CAS: 127-85-5   HR: 3
ARSANILIC ACID, MONOSODIUM SALT

DOT: UN 2473
mf: $C_6H_7AsNO_3 \cdot Na$ mw: 239.05

PROP: Tetrahydrate: white, odorless, crystalline powder; faint salty taste. Sol in water; somewhat sol in alc.

SYNS: (4-AMINOPHENYL)ARSONIC ACID SODIUM SALT * ARSANILIC ACID SODIUM SALT * ATOXYL * NCI-C61176 * SODIUM AMINARSONATE * SODIUM-p-AMINOBENZENEARSONATE * SODIUM AMINOPHENOL ARSONATE * SODIUM-p-AMINOPHENYLARSONATE * SODIUM ANILARSONATE * SODIUM-ANILINE ARSONATE * SODIUM ARSANILATE * SODIUM-p-ARSANILATE * SODIUM ARSONILATE

CONSENSUS REPORTS: Arsenic and its compounds are on the Community Right-To-Know List.

OSHA PEL: TWA 0.01 mg(As)/m^3
ACGIH TLV: TWA 0.2 mg(As)/m^3
DOT Classification: Poison B; Label: St. Andrews Cross

SAFETY PROFILE: Poison by subcutaneous route. Can cause blindness. When heated to decomposition it emits very toxic fumes of As and NO_x.

ARA750   CAS: 7440-38-2   HR: 3
ARSENIC

DOT: UN 1558
af: As   aw: 74.92

PROP: Silvery to black, brittle, crystalline and amorphous metalloid. Mp: 814° @ 36 atm, bp: sublimes @ 612°, d: black crystals 5.724 @ 14°; black amorphous 4.7, vap press: 1 mm @ 372° (sublimes). Insol in water; sol in HNO_3.

SYNS: ARSEN (GERMAN, POLISH) * ARSENICALS * ARSENIC-75 * ARSENIC BLACK * COLLOIDAL ARSENIC * GREY ARSENIC * METALLIC ARSENIC

CONSENSUS REPORTS: IARC Cancer Review: GROUP 1 IMEMDT 7,100,87; Human Sufficient Evidence IMEMDT 23,39,80; Human Inadequate Evidence IMEMDT 2,48,73. NTP Fourth Annual Report On Carcinogens, 1984.

Reported in EPA TSCA Inventory. Arsenic and its compounds are on the Community Right-To-Know List.

OSHA PEL: TWA 0.01 mg(As)/m^3; Cancer Hazard

ACGIH TLV: TWA 0.2 mg(As)/m^3

DFG TRK: 0.2 mg/m^3 calculated as arsenic in that portion of dust that can possibly be inhaled.

NIOSH REL: CL 2 μg(As)/m^3

DOT Classification: Poison B; Label: Poison.

SAFETY PROFILE: Confirmed human carcinogen producing liver tumors. Poison by subcutaneous, intramuscular, and intraperitoneal routes. Human systemic skin and gastrointestinal effects by ingestion. Experimental teratogenic and reproductive data. Mutation data reported. Flammable in the form of dust when exposed to heat or flame or by chemical reaction with powerful oxidizers such as bromates; chlorates; iodates; peroxides; lithium; NCl_3; KNO_3; $KMnO_4$; Rb_2C_2; $AgNO_4$; $NOCl$; IF_5; CrO_3; ClF_3; ClO; BrF_3; BrF_5; BrN_3; RbC_3BCH; CsC_3BCH. Slightly explosive in the form of dust when exposed to flame. When heated or on contact with acid or acid fumes, it emits highly toxic fumes; can react vigorously on contact with oxidizing materials. Incompatible with bromine azide; dirubidium acetylide; halogens; palladium; zinc; platinum; NCl_3; $AgNO_3$; CrO_3; Na_2O_2; hexafluoro isopropylideneamino lithium.

ARB000   CAS: 10102-53-1   HR: 3
m-ARSENIC ACID

mf: $AsHO_3$ mw: 123.93

SYN: METAARSENIC ACID

CONSENSUS REPORTS: Reported in EPA TSCA Inventory. Arsenic and its compounds are on the Community Right-To-Know List.

OSHA PEL: TWA 0.01 mg(As)/m^3
ACGIH TLV: TWA 0.2 mg(As)/m^3
DFG MAK: Human Carcinogen.
NIOSH REL: CL 2 μg(As)/m^3/15M

SAFETY PROFILE: Confirmed human carcinogen. When heated to decomposition it emits toxic fumes of arsenic.

ARB250   CAS: 7778-39-4   HR: 3
o-ARSENIC ACID

DOT: UN 1553/UN 1554
mf: AsH_3O_4 mw: 141.95

SYNS: ACIDE ARSENIQUE LIQUIDE (FRENCH)
* ARSENATE * ARSENIC ACID, liquid (DOT)
* ARSENIC ACID, solid (DOT) * DESICCANT
L-10 * HI-YIELD DESSICANT H-10 * ORTHOARSENIC ACID * RCRA WASTE NUMBER P010
* ZOTOX * ZOTOX CRAB GRASS KILLER

CONSENSUS REPORTS: Reported in EPA TSCA Inventory. Arsenic and its compounds are on the Community Right-To-Know List.

OSHA PEL: TWA 0.01 mg(As)/m^3
ACGIH TLV: TWA 0.2 mg(As)/m^3
DFG MAK: Human Carcinogen.
NIOSH REL: CL 2 μg(As)/m^3/15M
DOT Classification: Poison B; Label: Poison.

SAFETY PROFILE: Confirmed human carcinogen. Poison by ingestion. Experimental teratogenic and reproductive effects. Human mutation data reported. When heated to decomposition it emits toxic fumes of arsenic.

ARB750   CAS: 7778-44-1   HR: 3
ARSENIC ACID, CALCIUM SALT (2:3)

DOT: UN 1573
mf: $As_2O_8 \cdot 3Ca$ mw: 398.08

PROP: Colorless, amorphous powder. D: 3.620. Solubility in water = 0.013/100 @ 25°.

SYNS: ARSENIATE de CALCIUM (FRENCH) * CALCIUMARSENAT * CALCIUM ARSENATE (MAK)
* CALCIUM ORTHOARSENATE * KALZIUMARSENIAT (GERMAN) * TRICALCIUMARSENAT (GERMAN)
* TRICALCIUM ARSENATE

CONSENSUS REPORTS: IARC Cancer Review: Human Sufficient Evidence IMEMDT 23,39,80; Animal No Evidence IMEMDT 2,48,73; Animal Inadequate Evidence IMEMDT 23,39,80. Reported in EPA TSCA Inventory. Arsenic and its compounds are on the Community Right-To-Know List. EPA Extremely Hazardous Substances List.

OSHA PEL: TWA 0.01 mg(As)/m^3; Cancer Hazard
ACGIH TLV: TWA 0.2 mg(As)/m^3
DFG MAK: Human Carcinogen.
NIOSH REL: CL 2 μg(As)/m^3/15M
DOT Classification: Poison B; Label: Poison.

SAFETY PROFILE: Confirmed human carcinogen. Poison by ingestion. Moderately toxic by skin contact. When heated to decomposition it emits toxic fumes of arsenic.

ARC000   CAS: 7778-43-0   **HR: 3**
ARSENIC ACID, DISODIUM SALT
mf: $Na_2HAsO_4 \cdot 7H_2O$ mw: 312.01

PROP: Colorless powder, effloresces. D: 1.88, mp: $-7H_2O$ @ 130°, bp: decomp @ 150°. Solubility in water = 61/100 @ 15°, sol in glycerol.

SYNS: DISODIUM ARSENATE * DISODIUM ARSENIC ACID * DISODIUM HYDROGEN ARSENATE * DISODIUM HYDROGEN ORTHOARSENATE * DISODIUM MONOHYDROGEN ARSENATE * SODIUM ACID ARSENATE * SODIUM ARSENATE * SODIUM ARSENATE DIBASIC, anhydrous

CONSENSUS REPORTS: Reported in EPA TSCA Inventory. Arsenic and its compounds are on the Community Right-To-Know List.

OSHA PEL: TWA 0.5 mg(As)/m³: Cancer Hazard
ACGIH TLV: TWA 0.2 mg(As)/m³
NIOSH REL: CL 2 µg(As)/m³/15M
DFG MAK: Human Carcinogen.

SAFETY PROFILE: Confirmed human carcinogen. Poison by intraperitoneal route. Human mutation data reported. When heated to decomposition it emits toxic fumes of arsenic.

ARC250   CAS: 10048-95-0   **HR: 3**
ARSENIC ACID, DISODIUM SALT, HEPTAHYDRATE
mf: $AsHO_4 \cdot 2Na \cdot 7H_2O$ mw: 312.05

SYNS: DISODIUM ARSENATE, HEPTAHYDRATE * SODIUM ACID ARSENATE, HEPTAHYDRATE * SODIUM ARSENATE, DIBASIC, HEPTAHYDRATE * SODIUM ARSENATE HEPTAHYDRATE

CONSENSUS REPORTS: Arsenic and its compounds are on the Community Right-To-Know List.

OSHA PEL: TWA 0.01 mg(As)/m³
ACGIH TLV: TWA 0.2 mg(As)/m³
NIOSH REL: CL 2 µg(As)/m³/15M
DFG MAK: Human Carcinogen.

SAFETY PROFILE: Confirmed human carcinogen. Poison by subcutaneous route. An experimental teratogen. When heated to decomposition it emits toxic fumes of arsenic.

ARC500   CAS: 7774-41-6   **HR: 3**
o-ARSENIC ACID, HEMIHYDRATE
mf: $AsH_3O_4 \cdot 1/2H_2O$ mw: 150.96

PROP: White, translucent crystals. Mp: 35.5°, bp: $-H_2O$ @ 160°, d: 2.0-2.5.

SYNS: ARSENIC ACID, solid (DOT) * ORTHOARSENIC ACID HEMIHYDRATE

CONSENSUS REPORTS: Arsenic and its compounds are on the Community Right-To-Know List.

OSHA PEL: TWA 0.01 mg(As)/m³
ACGIH TLV: TWA 0.2 mg(As)/m³
NIOSH REL: CL 2 µg(As)/m³/15M
DFG MAK: Human Carcinogen.
DOT Classification: Poison B; Label: Poison.

SAFETY PROFILE: Confirmed human carcinogen. Poison by intravenous route. When heated to decomposition it emits toxic fumes of arsenic.

ARC750   CAS: 7645-25-2   **HR: 3**
ARSENIC ACID, LEAD SALT
DOT: UN 1617
mf: $AsH_3O_4 \cdot 7Pb$ mw: 1592.28

SYNS: ARSENIATE de PLOMB (FRENCH) * LEAD ARSENATE

CONSENSUS REPORTS: Arsenic compounds and Lead compounds are on the Community Right-To-Know List.

OSHA PEL: TWA 0.01 mg(As)/m³; Cancer Hazard
ACGIH TLV: TWA 0.15 mg(Pb)/m³
NIOSH REL: TWA 0.10 mg(Pb)/m³; CL 2 µg(As)/m³/15M
DFG MAK: Human Carcinogen.
DOT Classification: Poison B; Label: Poison

SAFETY PROFILE: Confirmed human carcinogen. Poison by ingestion. When heated to decomposition it emits very toxic fumes of lead and arsenic.

ARD000   CAS: 10103-50-1   **HR: 3**
ARSENIC ACID, MAGNESIUM SALT
DOT: UN 1622
mf: $AsH_3O_4 \cdot 7Mg$ mw: 312.12

PROP: Monoclinic, white crystals. D: 2.60-2.61.

SYNS: ARSENIATE de MAGNESIUM (FRENCH) * MAGNESIUM ARSENATE * MAGNESIUM ARSENATE PHOSPHOR

CONSENSUS REPORTS: Reported in EPA TSCA Inventory. Arsenic and its compounds are on the Community Right-To-Know List.

OSHA PEL: TWA 0.01 mg(As)/m³
ACGIH TLV: TWA 0.2 mg(As)/m³

DFG MAK: Human Carcinogen.
NIOSH REL: CL 2 μg(As)/m^3/15M
DOT Classification: Poison B; Label: Poison.

SAFETY PROFILE: Confirmed human carcinogen. Poison by ingestion. When heated to decomposition it emits toxic fumes of arsenic.

ARD250   CAS: 7784-41-0   **HR: 3**
ARSENIC ACID, MONOPOTASSIUM SALT
DOT: UN 1677
mf: AsH$_2$O$_4$ • K   mw: 180.04

SYNS: MONOPOTASSIUM ARSENATE * MONOPOTASSIUM DIHYDROGEN ARSENATE * MACQUER'S SALT * POTASSIUM ACID ARSENATE * POTASSIUM ARSENATE * POTASSIUM DIHYDROGEN ARSENATE * POTASSIUM HYDROGEN ARSENATE

CONSENSUS REPORTS: IARC Cancer Review: Human Sufficient Evidence IMEMDT 23,39,80. Reported in EPA TSCA Inventory. Arsenic and its compounds are on the Community Right-To-Know List.

OSHA PEL: TWA 0.01 mg(As)/m^3, Cancer Hazard
ACGIH TLV: TWA 0.2 mg(As)/m^3
NIOSH REL: CL 2 μg(As)/m^3/15M
DOT Classification: Poison B; Label: Poison.

SAFETY PROFILE: Confirmed human carcinogen. Mutation data reported. When heated to decomposition it emits toxic fumes of arsenic.

ARD500   CAS: 15120-17-9   **HR: 3**
ARSENIC ACID, MONOSODIUM SALT
mf: AsO$_3$ • Na   mw: 145.91

SYNS: ARSENIC ACID, SODIUM SALT (9CI) * SODIUM ARSENATE * SODIUM METAARSENATE * SODIUM MONOHYDROGEN ARSENATE

CONSENSUS REPORTS: Arsenic and its compounds are on the Community Right-To-Know List.

OSHA PEL: TWA 0.5 mg(As)/m^3: Cancer Hazard
ACGIH TLV: TWA 0.2 mg(As)/m^3
NIOSH REL: CL 2 μg(As)/m^3/15M
DFG MAK: Human Carcinogen.

SAFETY PROFILE: Confirmed human carcinogen. A poison. Mutation data reported. When heated to decomposition it emits toxic fumes of arsenic.

ARD600   CAS: 10103-60-3   **HR: 3**
ARSENIC ACID, MONOSODIUM SALT
mf: AsH$_2$O$_4$ • Na   mw: 163.93

SYNS: MONOSODIUM ARSENATE * SODIUM ARSENATE * SODIUM DIHYDROGEN ARSENATE * SODIUM DIHYDROGEN ORTHOARSENATE

OSHA PEL: TWA 0.5 mg(As)/m^3: Cancer Hazard
ACGIH TLV: TWA 0.2 mg(As)/m^3
NIOSH REL: CL 2 μg(As)/m^3/15M
DFG MAK: Human Carcinogen.

CONSENSUS REPORTS: Arsenic and its compounds are on the Community Right-To-Know List.

SAFETY PROFILE: Confirmed human carcinogen. Poison by intravenous route. When heated to decomposition it emits toxic fumes of arsenic.

ARD750   CAS: 7631-89-2   **HR: 3**
ARSENIC ACID, SODIUM SALT
DOT: UN 1685
mf: AsH$_3$O$_4$ • 7Na   mw: 202.94

SYNS: SODIUM ARSENATE (DOT) * SODIUM ORTHOARSENATE

CONSENSUS REPORTS: IARC Cancer Review: Human Sufficient Evidence IMEMDT 23,39,80; Animal Inadequate Evidence IMEMDT 2,48,73; IMEMDT 23,39,80. Reported in EPA TSCA Inventory. Arsenic and its compounds are on the Community Right-To-Know List.

OSHA PEL: TWA 0.01 mg(As)/m^3; Cancer Hazard
ACGIH TLV: TWA 0.2 mg(As)/m^3
NIOSH REL: CL 2 μg(As)/m^3/15M
DOT Classification: Poison B; Label: Poison.

SAFETY PROFILE: Confirmed human carcinogen with experimental tumorigenic data. Poison by ingestion, intravenous, and intraperitoneal routes. Experimental teratogenic and reproductive data. Mutation data reported. When heated to decomposition it emits toxic fumes of As and Na$_2$O.

ARE000   CAS: 64070-83-3   **HR: 3**
ARSENIC(V) ACID, TRISODIUM SALT, HEPTAHYDRATE (1:3:7)
mf: AsO$_4$ • 3Na • 7H$_2$O   mw: 334.03

SYN: TRISODIUM ARSENATE, HEPTAHYDRATE

CONSENSUS REPORTS: Arsenic and its compounds are on the Community Right-To-Know List.

OSHA PEL: TWA 0.01 mg(As)/m^3; Cancer Hazard
ACGIH TLV: TWA 0.2 mg(As)g/m^3
NIOSH REL: CL 2 μg(As)/m^3/15M
DFG MAK: Human Carcinogen.

SAFETY PROFILE: Confirmed human carcinogen with experimental carcinogenic data. Poison by intraperitoneal route. When heated to decomposition it emits toxic fumes of arsenic.

ARE250   CAS: 8028-75-9   HR: 3
ARSENICAL DIP

DOT: NA 1557

SYNS: ARSENICAL DIP, liquid (DOT)   *   SHEEP DIP

CONSENSUS REPORTS: Arsenic and its compounds are on the Community Right-To-Know List.

OSHA PEL: TWA 0.01 mg(As)/m^3; Cancer Hazard
ACGIH TLV: TWA 0.2 mg(As)/m^3
NIOSH REL: CL 2 μg(As)/m^3/15M
DOT Classification: Poison B; Label: Poison.

SAFETY PROFILE: A poison.

ARE500   CAS: 8028-73-7   HR: 3
ARSENICAL DUST

DOT: UN 1562

SYNS: ARSENICAL FLUE DUST   *   FLUE DUST, AR-SENIC containing

CONSENSUS REPORTS: Reported in EPA TSCA Inventory. Arsenic and its compounds are on the Community Right-To-Know List.

OSHA PEL: TWA 0.5 mg(As)/m^3
ACGIH TLV: TWA 0.2 mg(As)/m^3
NIOSH REL: CL 2 μg(As)/m^3/15M
DOT Classification: Poison B; Label: Poison.

SAFETY PROFILE: A poison. Questionable carcinogen with experimental tumorigenic data.

ARE750   HR: 3
ARSENICAL FLUE DUST (DOT)

OSHA PEL: TWA 0.01 mg(As)/m^3
ACGIH TLV: TWA 0.2 mg(As)/m^3

NIOSH REL: CL 2 μg(As)/m^3/15M
DFG MAK: Human Carcinogen.
DOT Classification: Poison B; Label: Poison.

SAFETY PROFILE: Confirmed human carcinogen. Poison by inhalation and ingestion.

ARF000   HR: 3
ARSENIC BISULFIDE
mf: As$_2$S$_2$ mw: 214

PROP: Red-brown crystals. Bp: 565°, mp: (β) 307°, d: (α) 3.506 @ 19°, (β) 3.254 @ 19°.

SYN: REALGAR

CONSENSUS REPORTS: Arsenic and its compounds are on the Community Right-To-Know List.

SAFETY PROFILE: A poison. Flammable in the form of dust when exposed to heat or flame. Explosion hazard when intimately mixed with powerful oxidizers such as Cl$_2$; KNO$_3$; chlorates. It will react with water or steam to produce toxic and flammable vapors.

ARF250   CAS: 7784-33-0   HR: 3
ARSENIC(III) BROMIDE

DOT: UN 1555
mf: AsBr$_3$ mw: 314.65

PROP: Colorless, rhombic crystals. Mp: 32.8°, bp: 220.0°, vap press: 1 mm @ 41.8°, d: 3.3972 @ 25°, (liq), 3.3282.

SYNS: ARSENIC TRIBROMIDE   *   ARSENOUS BROM-IDE   *   ARSENOUS TRIBROMIDE   *   TRIBROMOAR-SINE

CONSENSUS REPORTS: Reported in EPA TSCA Inventory. Arsenic and its compounds are on the Community Right-To-Know List.

OSHA PEL: TWA 0.01 mg(As)/m^3
ACGIH TLV: TWA 0.2 mg(As)/m^3
NIOSH REL: CL 2 μg(As)/m^3/15M
DOT Classification: Poison B; Label: Poison.

SAFETY PROFILE: A poison. When heated to decomposition it emits very toxic fumes of As and Br$^-$.

ARF500   CAS: 7784-34-1   HR: 3
ARSENIC CHLORIDE

DOT: UN 1560
mf: AsCl$_3$ mw: 181.28

PROP: Colorless, oily liquid. D: 2.15 @ 25°, mp: −16°, bp: 130°. Decomp in water and by UV light; misc in chloroform, CCl_4, ether, iodine, P, S, alkali iodides, oils and fats. Vap d: 6.25, vap press: 10 mm @ 23.5°.

SYNS: ARSENIC BUTTER * ARSENIC(III) CHLORIDE * ARSENIOUS CHLORIDE * ARSENOUS CHLORIDE * ARSENOUS TRICHLORIDE (9CI) * CHLORURE d'ARSENIC (FRENCH) * CHLORURE ARSENIEUX (FRENCH) * FUMING LIQUID ARSENIC * TRICHLO-ROARSINE * TRICHLORURE d'ARSENIC (FRENCH)

CONSENSUS REPORTS: Reported in EPA TSCA Inventory. Arsenic and its compounds are on the Community Right-To-Know List. EPA Extremely Hazardous Substances List.

OSHA PEL: TWA 0.01 mg(As)/m^3
ACGIH TLV: TWA 0.2 mg(As)/m^3
NIOSH REL: CL 2 μg(As)/m^3/15M
DOT Classification: Poison B; Label: Poison.

SAFETY PROFILE: A poison via inhalation. Very poisonous; fumes in air. Mutation data reported. When heated to decomposition it emits very toxic fumes of As and Cl$^−$. Highly reactive. Explodes with Na; K; Al on impact.

ARF750
ARSENIC COMPOUNDS
HR: 3

SYN: ARSENICALS

CONSENSUS REPORTS: Arsenic and its compounds are on the Community Right-To-Know List.

OSHA PEL: Inorganic: TWA 0.01 mg(As)/m^3; Cancer Hazard; Organic: TWA 0.5 mg(As)/m^3
ACGIH TLV: TWA 0.2 mg(As)/m^3
NIOSH REL: CL 2 μg(As)/m^3/15M

SAFETY PROFILE: Inorganic compounds are confirmed human carcinogens producing tumors of the mouth, esophagus, larynx, bladder, and para nasal sinus. A recognized carcinogen of the skin, lungs, and liver. Used as insecticides, herbicides, silvicides, defoliants, desiccants and rodenticides. Poisoning from arsenic compounds may be acute or chronic. Acute poisoning usually results from swallowing arsenic compounds; chronic poisoning from either swallowing or inhaling. Acute allergic reactions to arsenic compounds used in medical therapy have been fairly common, the type and severity of reaction depending upon the compound. Inorganic arsenicals are more toxic than organics. Trivalent is more toxic than pentavalent. Acute arsenic poisoning (from ingestion) results in marked irritation of the stomach and intestines with nausea, vomiting, and diarrhea. In severe cases, the vomitus and stools are bloody and the patient goes into collapse and shock with weak, rapid pulse, cold sweats, coma, and death. Chronic arsenic poisoning, whether through ingestion or inhalation, may manifest itself in many different ways. There may be disturbances of the digestive system such as loss of appetite, cramps, nausea, constipation, or diarrhea. Liver damage may occur, resulting in jaundice. Disturbances of the blood, kidneys, and nervous system are not infrequent. Arsenic can cause a variety of skin abnormalities including itching, pigmentation, and even cancerous changes. A characteristic of arsenic poisoning is the great variety of symptoms that can be produced. Dangerous; when heated to decomposition, or when metallic arsenic contacts acids or acid fumes, or when water solutions of arsenicals are in contact with active metals such as Fe; Al; Zn; they emits highly toxic fumes of arsenic.

In treating acute poisoning from ingestion BAL (dimercaptol) is of questionable effectiveness for acute and chronic poisoning with trivalent arsenicals, such as arsenic trioxide, arsine, and arsenites. It is of no value for pentavalent arsenicals, such as cacodylic acid, methanearsonic acid, sodium, cacodylate, MSMA, DSMA, arsanilic acid, arsenic acid, and arsenates. Vomiting and gastric lavage are the preferred emergency treatments for acute arsenical poisoning. Modern medical treatment of arsenical poisoning uses exchange transfusion and dialysis (A. E. De Palma, *J. Occup Med.*, Vol. 11,582-587 (1969). Note: Arsenic compounds are common air contaminants.

ARG000
ARSENIC DIETHYL
HR: 3

mf: [As(C$_2$H$_5$)$_2$]   mw: 266.2

PROP: Liquid or oil. Bp: 185°-190°, d: about 1.

CONSENSUS REPORTS: Arsenic and its compounds are on the Community Right-To-Know List.

SAFETY PROFILE: A poison. A dangerous fire hazard by spontaneous chemical reaction. Dangerous when heated. Incompatible with oxidizing materials.

ARG250   HR: 3
ARSENIC DIMETHYL
mf: [As(CH$_3$)$_2$]   mw: 210.0

PROP: Colorless to yellow oily liquid. Mp: −6°; bp: 186°; d: 1.15.

CONSENSUS REPORTS: Arsenic and its compounds are on the Community Right-To-Know List.

SAFETY PROFILE: Poison by inhalation and ingestion. Flammable. Evolves dangerous fumes of arsenic when heated.

ARG500   HR: 3
ARSENIC HEMISELENIDE
mf: As$_2$Se   mw: 228.78

CONSENSUS REPORTS: Arsenic compounds and its compounds as well as selenium and its compounds are on the Community Right-To-Know List.

OSHA PEL: TWA 0.01 mg(As)/m^3; Cancer Hazard; TWA 0.2 mg(Se)/m^3
ACGIH TLV: TWA 0.2 mg(As)/m^3; TWA 0.2 mg(Se)/m^3
DFG TRK: 0.2 mg/m^3 calculated as arsenic in that portion of dust that can possibly be inhaled; 0.1 mg(Se)/m^3
NIOSH REL: CL 2 μg(As)/m^3

SAFETY PROFILE: When heated to decomposition it emits fumes of As and Se. Incompatible with oxidizing materials. When heated to decomposition it emits highly toxic fumes of Se and arsenic.

ARG750   CAS: 7784-45-4   HR: 3
ARSENIC IODIDE

DOT: NA 1557
mf: AsI$_3$ mw: 455.62

PROP: Red hexagonal crystals. Mp: 141.8°; bp: 403°; d: 4.38 @ 13°. Solubility: in water = 6/100 @ 25°, in CS$_2$ = 5.2/100.

SYNS: ARSENIC TRIIODIDE * ARSENOUS IODIDE * ARSENOUS TRIIODIDE (9CI) * TRI-IODOARSINE

CONSENSUS REPORTS: Reported in EPA TSCA Inventory. Arsenic and its compounds are on the Community Right-To-Know List.

OSHA PEL: TWA 0.01 mg(As)/m^3
ACGIH TLV: TWA 0.2 mg(As)/m^3
NIOSH REL: CL 2 μg(As)/m^3/15M
DOT Classification: Poison B; Label: Poison.

SAFETY PROFILE: A poison. Can form a shock sensitive compound with sodium or potassium. When heated to decomposition it emits very toxic fumes of I$^-$ and arsenic.

ARH250   HR: 3
ARSENIC PENTASULFIDE
mf: As$_2$S$_5$ mw: 310.2

PROP: Brownish-yellow glassy, amorphous, highly refractive mass. Mp: 500° (sublimes).

CONSENSUS REPORTS: Arsenic and its compounds are on the Community Right-To-Know List.

OSHA PEL: TWA 0.01 mg(As)/m^3
ACGIH TLV: TWA 0.2 mg(As)/m^3
NIOSH REL: CL 2 μg(As)/m^3/15M

SAFETY PROFILE: Flammable in the form of dust when exposed to heat or flame. Explosive when intimately mixed with powerful oxidizers, such as Cl$_2$; KNO$_3$; chlorates. Will react with water and steam to produce toxic and flammable vapors. Incompatible with water, steam, and strong oxidizers.

ARH500   CAS: 1303-28-2   HR: 3
ARSENIC PENTOXIDE

DOT: UN 1559
mf: As$_2$O$_5$ mw: 229.84

PROP: White, amorphous, deliquescent solid. Mp: decomp @ 800°, d: 4.32. Sol in alc. Solubility in water = 65.8/100 @ 20°.

SYNS: ANHYDRIDE ARSENIQUE (FRENCH) * ARSENIC ACID * ARSENIC ACID ANHYDRIDE * ARSENIC ANHYDRIDE * ARSENIC OXIDE * ARSENIC(V) OXIDE * DIARSENIC PENTOXIDE

CONSENSUS REPORTS: IARC Cancer Review: Human Sufficient Evidence IMEMDT 23,39,80. Reported in EPA TSCA Inventory. Arsenic and its compounds are on the Community Right-To-Know List. EPA Extremely Hazardous Substances List.

OSHA PEL: TWA 0.01 mg(As)/m^3
ACGIH TLV: TWA 0.2 mg(As)/m^3
DFG MAK: Human Carcinogen.
NIOSH REL: CL 2 μg(As)/m^3/15M
DOT Classification: Poison B; Label: Poison.

SAFETY PROFILE: Confirmed human carcinogen. Poison by ingestion and intravenous routes. Mutation data reported. Reacts vigorously with Rb_2C_2. When heated to decomposition it emits toxic fumes of arsenic.

ARH750   HR: 3
ARSENIC PHOSPHIDE
mf: AsP   mw: 105.9

PROP: Brown to red powder. Mp: sublimes with decomp.

CONSENSUS REPORTS: Arsenic and its compounds are on the Community Right-To-Know List.

SAFETY PROFILE: Flammable by spontaneous chemical reaction. Phosphine is liberated upon contact with moisture. Dangerous when heated; see PHOSPHOROUS and ARSENIC COMPOUNDS. Incompatible with water or steam; oxidizing materials.

ARI000   CAS: 1303-33-9   HR: 3
ARSENIC SULFIDE
DOT: NA 1557
mf: As_2S_3 mw: 246.04

PROP: Yellow or red crystals. Bp: 707°, d: 3.43; mp: 312°. Insol in water; sol in alkalies.

SYNS: ARSENIC SESQUISULFIDE * ARSENIC SULFIDE YELLOW * ARSENIC SULPHIDE * ARSENIC TRISULFIDE * ARSENIC YELLOW * ARSENIOUS SULPHIDE * ARSENOUS SULFIDE * C.I. 77086 * DIARSENIC TRISULFIDE * KING'S YELLOW * ORPIMENT

CONSENSUS REPORTS: IARC Cancer Review: Human Sufficient Evidence IMEMDT 23,39,80. Reported in EPA TSCA Inventory. Arsenic and its compounds are on the Community Right-To-Know List.

OSHA PEL: TWA 0.01 mg(As)/m³; Cancer Hazard
ACGIH TLV: TWA 0.2 mg(As)/m³
NIOSH REL: CL 2 μg(As)/m³/15M
DOT Classification: Poison B; Label: Poison.

SAFETY PROFILE: Confirmed human carcinogen with experimental tumorigenic data. A poison. Reacts violently with H_2O_2; $(KNO_3 + S)$. When heated to decomposition or contact with acid or acid fumes it emits highly toxic fumes of SO_2, H_2S, and As. Reacts with water or steam to emit toxic and flammable vapors.

ARI250   CAS: 7784-35-2   HR: 3
ARSENIC TRIFLUORIDE
mf: AsF_3 mw: 131.92

PROP: Colorless liquid. D: 3.01, mp: −5.95, bp: 51°, vap press: 100 mm @ 13.2°, 400 mm @ 41.5°. Insol in water; sol in alc, benzene, and mercury.

SYNS: ARSENIC FLUORIDE * ARSENOUS FLUORIDE * TRIFLUOROARSINE

CONSENSUS REPORTS: Reported in EPA TSCA Inventory. Arsenic and its compounds are on the Community Right-To-Know List.

OSHA PEL: TWA 0.01 mg(As)/m³; Cancer Hazard
ACGIH TLV: TWA 0.2 mg(As)/m³
NIOSH REL: CL 2 μg(As)/m³/15M

SAFETY PROFILE: Confirmed human carcinogen. A poison by inhalation. Strong reaction with P_2O_3. When heated to decomposition it emits very toxic fumes of As and F^-.

ARI500   CAS: 8012-54-2   HR: 3
ARSENIC TRIIODIDE mixed with MERCURIC IODIDE
DOT: NA 2810

SYNS: ARSENIOUS and MERCURIC IODIDE, solution (DOT) * DONOVAN'S SOLUTION

CONSENSUS REPORTS: Arsenic compounds and Mercury compounds are on the Community Right-To-Know List.

ACGIH TLV: TWA 0.1 mg(Hg)/m³ (skin)
NIOSH REL: TWA 0.05 mg(Hg)/m³; CL 2 μg(As)/m³/15M
DOT Classification: Poison B; Label: Poison.

SAFETY PROFILE: A poison. When heated to decomposition it emits very toxic fumes of Hg, As, and I^-.

ARI750   CAS: 1327-53-3   HR: 3
ARSENIC TRIOXIDE
DOT: UN 1561
mf: As_4O_6 mw: 395.68

PROP: Colorless, rhombic crystals (dimer, claudetite). D: 4.15, mp: 278°, bp: 460°. Solubility in water = 1.82/100 @ 20°; sol in alc. Cubes: Colorless. D: 3.865, mp: 309°. Solubility in water = 1.2/100 @ 20°.

SYNS: ACIDE ARSENIEUX (FRENCH) * ANHYDRIDE ARSENIEUX (FRENCH) * ARSENIC BLANC (FRENCH)

* ARSENIC OXIDE * ARSENIC(III) OXIDE
* ARSENIC SESQUIOXIDE * ARSENIGEN SAURE
(GERMAN) * ARSENIOUS ACID (MAK) * ARSENI-
OUS OXIDE * ARSENIOUS TRIOXIDE * ARSENOUS
ACID * ARSENOUS ACID ANHYDRIDE * ARSE-
NOUS ANHYDRIDE * ARSENOUS OXIDE * ARSE-
NOUS OXIDE ANHYDRIDE * CRUDE ARSENIC
* DIARSENIC TRIOXIDE * WHITE ARSENIC

CONSENSUS REPORTS: IARC Cancer Re-
view: GROUP 1 IMEMDT 7,100,87; Human
Limited Evidence IMEMDT 2,48,73; Human
Sufficient Evidence IMEMDT 23,39,80; Ani-
mal Inadequate Evidence IMEMDT 2,48,73;
IMEMDT 23,39,80. NTP Fourth Annual Report
On Carcinogens, 1984. Reported in EPA TSCA
Inventory. Arsenic and its compounds are on
the Community Right-To-Know List. EPA Ex-
tremely Hazardous Substances List.

OSHA PEL: TWA 0.01 mg(As)/m^3: Cancer
    Hazard
ACGIH TLV: Production: Suspected Human
    Carcinogen
DFG MAK: Human Carcinogen.
NIOSH REL: CL 2 μg(As)/m^3/15M
DOT Classification: Poison B; Label: Poison.

SAFETY PROFILE: Confirmed human carcinogen with experimental neoplastigenic and tumorigenic data. Poison by ingestion, subcutaneous, intradermal, intravenous, and possibly
other routes. Human gastrointestinal effects by
ingestion. Mutation data reported. Reacts vigorously with Rb_2C_2; CIF_3; F_2; Hg; OF_2; $NaClO_3$.

ARJ000   HR: 3
ARSENIC TRIOXIDE mixed with
SELENIUM DIOXIDE (1:1)

SYN: SELENIUM DIOXIDE mixed with ARSENIC TRIOX-
IDE (1:1)

CONSENSUS REPORTS: Arsenic and its
compounds, as well as selenium and its compounds, are on the Community Right-To-Know
List.

OSHA PEL: TWA 0.01 mg(As)/m^3; Cancer
    Hazard
ACGIH TLV: TWA 0.2 mg(As)/m^3; Suspected
    Carcinogen; 0.2 mg(Se)/m^3
DFG MAK: 0.1 mg(Se)/m^3

SAFETY PROFILE: Confirmed human carcinogen with experimental tumorigenic data. When
heated to decomposition it emits very toxic
fumes of As and Se.

ARJ250   HR: 3
ARSENIDES
CONSENSUS REPORTS: Arsenic and its
compounds are on the Community Right-To-
Know List.

SAFETY PROFILE: Compounds of arsenic and
hydrogen or metals, (i.e., transitional, alkaline
earth, or rare-earth). These materials are dangerous because they readily emit very toxic arsine
and arsenic fumes when exposed to heat, moisture, acids, and acid fumes.

ARJ500   CAS: 14060-38-9   HR: 3
ARSENIOUS ACID SODIUM SALT
mf: $AsH_3O_3 \cdot 7Na$ mw: 286.88

PROP: Colorless or grayish-white powder. D:
1.87.

SYNS: ARSONIC ACID, SODIUM SALT (9CI)
* ARSENIOUS ACID, SODIUM SALT POLYMERS
* NATRIUMARSENIT (GERMAN) * SODIUM OR-
THOARSENITE

CONSENSUS REPORTS: Arsenic and its
compounds are on the Community Right-To-
Know List.

OSHA PEL: TWA 0.01 mg(As)/m^3; Cancer
    Hazard
ACGIH TLV: TWA 0.2 mg(As)/m^3
NIOSH REL: CL 2 μg(As)/m^3/15M

SAFETY PROFILE: Confirmed human carcinogen. Poison by intraperitoneal and subcutaneous
routes. Moderately toxic by ingestion. When
heated to decomposition it emits toxic fumes
of arsenic.

ARJ750   CAS: 1303-18-0   HR: 3
ARSENOPYRITE
mf: AsFeS   mw: 162.83

SYNS: ARSENOMARCASITE * MISPICKEL

CONSENSUS REPORTS: Arsenic and its
compounds are on the Community Right-To-
Know List.

OSHA PEL: TWA 0.01 mg(As)/m^3
ACGIH TLV: TWA 0.2 mg(As)/m^3
NIOSH REL: CL 2 μg(As)/m^3/15M

SAFETY PROFILE: Poison by intravenous
route. When heated to decomposition it emits
very toxic fumes of As and SO_x.

ARK250   CAS: 7784-42-1   **HR: 3**
ARSINE
DOT: UN 2188
mf: AsH$_3$ mw: 77.95

PROP: Colorless gas, mild garlic odor. D: 2.695 g/L; bp: −62.5°; vap d: 2.66; mp: −116°. Solubility in water = 28 mg/100 @ 20°. Sol in benzene and chloroform.

SYNS: ARSENIC HYDRIDE * ARSENIC TRIHYDRIDE * ARSENIURETTED HYDROGEN * ARSENOUS HYDRIDE * ARSENOWODOR (POLISH) * ARSENWASSERSTOFF (GERMAN) * HYDROGEN ARSENIDE

CONSENSUS REPORTS: IARC Cancer Review: Human Sufficient Evidence IMEMDT 23,39,80. Reported in EPA TSCA Inventory. Arsenic and its compounds are on the Community Right-To-Know List. EPA Extremely Hazardous Substances List.

OSHA PEL: TWA 0.05 ppm
ACGIH TLV: TWA 0.05 ppm
DFG MAK: 0.05 ppm (0.2 mg/m^3)
NIOSH REL: CL 2 μg(As)/m^3/15M
DOT Classification: Poison A; Label: Poison Gas and Flammable Gas.

SAFETY PROFILE: Confirmed human carcinogen. Poison by inhalation. Human red blood cell, gastrointestinal system, central nervous system, and other systemic effects by inhalation. Flammable when exposed to flame. Moderately explosive when exposed to Cl$_2$; HNO$_3$; (K + NH$_3$); open flame; or powerful shock. Dangerous, more toxic than its oxidation product. When heated to decomposition it emits highly toxic fumes of arsenic.

ARK500   **HR: 3**
ARSINE BORON TRIBROMIDE
mf: AsH$_3$ • BBr$_3$ mw: 328.6

CONSENSUS REPORTS: Arsenic and its compounds are on the Community Right-To-Know List.

SAFETY PROFILE: A poison. A highly unstable compound. Ignites in air. When heated to decomposition it emits very toxic fumes of As and Br$^-$.

ARL250   CAS: 8022-37-5   **HR: 2**
ARTEMISIA OIL

PROP: Chief constituent is Thujone, and found in the plant *Artemisia absinthium* L. (FCTXAV 13,681,75).

SYNS: ABSINTHIUM * ARTEMISIA OIL (WORMWOOD) * OIL, ARTEMISIA

CONSENSUS REPORTS: Reported in EPA TSCA Inventory.

SAFETY PROFILE: Moderately toxic by ingestion. An allergen. Habitual users develop, ''absinthism'' with tremors, vertigo, vomiting, and hallucinations. May cause a contact dermatitis. When heated to decomposition it emits acrid smoke and irritating fumes.

ARM000   CAS: 13425-94-0   **HR: 3**
ASALIN

SYN: ETHYL ESTER of N-ACETYL-dl-SARCOSYLYL-dl-VALINE

SAFETY PROFILE: Poison by ingestion, intramuscular, rectal, and intraperitoneal routes.

ARM250   CAS: 1332-21-4   **HR: 3**
ASBESTOS
DOT: UN 2212/UN 2590

SYNS: AMIANTHUS * AMOSITE (OBS.) * AMPHIBOLE * ASBEST (GERMAN) * ASBESTOS FIBER * FIBROUS GRUNERITE * NCI-C08991 * SERPENTINE

CONSENSUS REPORTS: IARC Cancer Review: GROUP 1 IMEMDT 7,106,87; Human Sufficient Evidence IMEMDT 2,17,73; IMEMDT 14,11,77; Animal Sufficient Evidence IMEMDT 2,17,73; IMEMDT 14,11,77. NTP Fourth Annual Report On Carcinogens, 1984. Reported in EPA TSCA Inventory. On Community Right-To-Know List. EPA Genetic Toxicology Program.

OSHA PEL: TWA 2 million fb/m^3; CL 10 million fb/m^3; Cancer Hazard
ACGIH TLV: Human Carcinogen, TWA 2 fb/cc
DFG TRK: (Fine dust particles which are able to reach the alveolar area of the lung) crocidolite: 0.05 × 10^6 fibers/m^3 (0.025 mg/m^3) (definition of fiber: length greater than 5 μm; diameter less than 3 μm; length/diameter greater than 3:1, equivalent to 1 fiber/cc); chrysotile, amosite, anthophyllite, tremolite, actinolite: 1 × 10^6 fibers/m^3 (0.05 mg/m^3) applicable when there is more than 2.5% asbestos in

the dust; 2.0 mg/m^3 applicable when there is less than or equal to 2.5 wt % asbestos in fine dust.
NIOSH REL: TWA 100,000 fb/m^3 over 5 μm in length
DOT Classification: ORM-C.

SAFETY PROFILE: Confirmed human carcinogen producing lung tumors. Experimental neoplastigenic and tumorigenic data. Human pulmonary system effects by inhalation. Usually at least 4 to 7 years of exposure are required before serious lung damage (fibrosis) results. A common air contaminant.

ARM260   CAS: 77536-66-4   **HR: 3**
ASBESTOS, ACTINOLITE

SYNS: ACTINOLITE ASBESTOS * ASBESTOS (ACGIH)

CONSENSUS REPORTS: IARC Cancer Review: GROUP 1 IMEMDT 7,106,87; Animal Sufficient Evidence IMEMDT 14,11,77.

OSHA PEL: TWA 2 million fb/m^3; CL 10 million fb/m^3; Cancer Hazard
ACGIH TLV: Human Carcinogen, TWA 2 fb/cc
DFG TRK: (Fine dust particles which are able to reach the alveolar area of the lung) 1 × 10^6 fibers/m^3 (0.05 mg/m^3) applicable when there is more than 2.5% asbestos in the dust.
NIOSH REL: TWA 100,000 fb/m^3 over 5 μm in length

SAFETY PROFILE: Confirmed human carcinogen.

ARM262   CAS: 12172-73-5   **HR: 3**
ASBESTOS, AMOSITE

SYNS: AMOSITE ASBESTOS * ASBESTOS (ACGIH)
* MYSORITE * NCI-C60253A

CONSENSUS REPORTS: IARC Cancer Review: GROUP 1 IMEMDT 7,106,87; Animal Sufficient Evidence IMEMDT 2,17,73; IMEMDT 14,11,77; Human Sufficient Evidence IMEMDT 2,17,73; IMEMDT 14,11,77. NTP Fourth Annual Report On Carcinogens, 1984. NTP Carcinogenesis Studies (feed); No Evidence: hamster NTPTR* NTP-TR-249,83. EPA Genetic Toxicology Program.

OSHA PEL: TWA 2 million fb/m^3; CL 10 million fb/m^3; Cancer Hazard
ACGIH TLV: Human Carcinogen, TWA 0.5 fb/cc

DFG TRK: (Fine dust particles which are able to reach the alveolar area of the lung) 1 × 10^6 fibers/m^3 (0.05 mg/m^3) applicable when there is more than 2.5% asbestos in the dust.
NIOSH REL: TWA 100,000 fb/m^3 over 5 μm in length

SAFETY PROFILE: Confirmed human carcinogen with experimental carcinogenic, neoplastigenic, and tumorigenic data. Mutation data reported.

ARM264   CAS: 77536-67-5   **HR: 3**
ASBESTOS, ANTHOPHYLITE

SYNS: ANTHOPHYLITE * ASBESTOS (ACGIH)
* AZBOLEN ASBESTOS * FERROANTHOPHYLLITE

CONSENSUS REPORTS: IARC Cancer Review: Animal Sufficient Evidence IMEMDT 2,17,73; IMEMDT 14,11,77; Human Sufficient Evidence IMEMDT 14,11,77. NTP Fourth Annual Report On Carcinogens, 1984. EPA Genetic Toxicology Program.

OSHA PEL: TWA 2 million fb/m^3; CL 10 million fb/m^3; Cancer Hazard
ACGIH TLV: Human Carcinogen, TWA 2 fb/cc
DFG TRK: (Fine dust particles which are able to reach the alveolar area of the lung) 1 × 10^6 fibers/m^3 (0.05 mg/m^3) applicable when there is more than 2.5% asbestos in the dust
NIOSH REL: TWA 100,000 fb/m^3 over 5 μm in length

SAFETY PROFILE: Confirmed human carcinogen with experimental carcinogenic, neoplastigenic, and tumorigenic data. Mutation data reported.

ARM266   CAS: 17068-78-9   **HR: 1**
ASBESTOS, ANTHOPHYLLITE

SYNS: AZBLLEN ASBESTOS * 16 F

SAFETY PROFILE: Confirmed carcinogen with experimental tumorigenic data.

ARM268   CAS: 12001-29-5   **HR: 3**
ASBESTOS, CHRYSOTILE

DOT: UN 2590

SYNS: 7-45 ASBESTOS * ASBESTOS (ACGIH)
* ASBESTOS, WHITE (DOT) * AVIBEST C
* CALIDRIA RG 100 * CALIDRIA RG 144
* CALIDRIA RG 600 * CASSIAR AK
* CHRYSOTILE ASBESTOS * CHRYSOTILE
(DOT) * HOOKER NO. 1 CHRYSOTILE ASBESTOS

* METAXITE * NCI-C61223A * PLASTIBEST 20
* SERPENTINE * SERPENTINE CHRYSOTILE
* SYLODEX * WHITE ASBESTOS

CONSENSUS REPORTS: IARC Cancer Review: Human Sufficient Evidence IMEMDT 2,17,73; Animal Sufficient Evidence IMEMDT 2,17,73. NTP Fourth Annual Report On Carcinogens, 1984. NTP Carcinogenesis Studies (feed); Some Evidence: rat NTPTR* NTP-TR-295,85. EPA Genetic Toxicology Program.

OSHA PEL: TWA 2 million fb/m^3; CL 10 million fb/m^3:; Cancer Hazard
ACGIH TLV: Human Carcinogen, TWA 2 fb/cc
DFG TRK: (Fine dust particles which are able to reach the alveolar area of the lung) 1×10^6 fibers/m^3 (0.05 mg/m^3) applicable when there is more than 2.5% asbestos in the dust.
NIOSH REL: TWA 100,000 fb/m^3 over 5 μm in length

SAFETY PROFILE: Confirmed human carcinogen producing tumors of the lung. Human mutation data reported. Poison by intraperitoneal route. Human systemic effects by inhalation: lung fibrosis, dyspnea, and cough.

ARM275 CAS: 12001-28-4 HR: 3
ASBESTOS, CROCIDOLITE
DOT: UN 2212

SYNS: AMORPHOUS CROCIDOLITE ASBESTOS
* ASBESTOS (ACGIH) * BLUE ASBESTOS (DOT)
* CROCIDOLITE ASBESTOS * CROCIDOLITE (DOT)
* FIBROUS CROCIDOLITE ASBESTOS * KROKYDO-
LITH (GERMAN) * NCI C09007

CONSENSUS REPORTS: IARC Cancer Review: Animal Sufficient Evidence IMEMDT 14,11,77, IMEMDT 2,17,73; Human Sufficient Evidence IMEMDT 14,11,77. NTP Fourth Annual Report On Carcinogens, 1984. EPA Genetic Toxicology Program.

OSHA PEL: TWA 2 million fb/m^3; CL 10 million fb/m^3; Cancer Hazard
ACGIH TLV: Human Carcinogen, TWA 0.2 fb/cc
DFG TRK: (Fine dust particles which are able to reach the alveolar area of the lung) crocidolite: 0.05×10^6 fibers/m^3 (0.025 mg/m^3) (definition of fiber: length greater than 5 μm; diameter less than 3 μm; length/diameter greater than 3:1, equivalent to 1 fiber/cc).

NIOSH REL: TWA 100,000 fb/m^3 over 5 μm in length
DOT Classification: ORM-C; LABEL: none

SAFETY PROFILE: Confirmed human carcinogen with experimental carcinogenic, neoplastigenic, and tumorigenic data by inhalation. Human mutation data reported.

ARM280 CAS: 77536-68-6 HR: 3
ASBESTOS, TREMOLITE

SYNS: ASBESTOS (ACGIH) * FIBROUS TREMO-
LITE * NCI-C08991 * TREMOLITE ASBESTOS

CONSENSUS REPORTS: IARC Cancer Review: Human Sufficient Evidence IMEMDT 14,11,77; Animal Sufficient Evidence IMEMDT 14,11,77. NTP Fourth Annual Report On Carcinogens, 1984.

OSHA PEL: TWA 2 million fb/m^3; CL 10 million fb/m^3; Cancer Hazard
ACGIH TLV: Human Carcinogen, TWA 2 fb/cc
DFG TRK: (Fine dust particles which are able to reach the alveolar area of the lung) 1×10^6 fibers/m^3 (0.05 mg/m^3) applicable when there is more than 2.5% asbestos in the dust.
NIOSH REL: TWA 100,000 fb/m^3 over 5 μm in length

SAFETY PROFILE: Confirmed human carcinogen with experimental tumorigenic and neoplastigenic data.

ARM500 CAS: 512-85-6 HR: 3
ASCARIDOLE
mf: C$_{10}$H$_{16}$O$_2$ mw: 168.26

PROP: Colorless unstable liquid. Mp: 3.3°, bp: 40° @ 2 mm; 115° @ 15 mm, d: 1.011 @ 13°/15°.

SYNS: ASCARISIN * 1,4-PEROXIDO-p-MENTHENE-2

DOT Classification: Forbidden.

SAFETY PROFILE: Poison by ingestion. Questionable carcinogen with experimental neoplastigenic and tumorigenic data. Flammable by spontaneous chemical reaction. An oxidizer. Explodes when heated >130° or when exposed to organic acids. Dangerous; heating emits toxic fumes and may explode; reacts with reducing materials.

ARN000   CAS: 50-81-7   **HR: 2**
l-ASCORBIC ACID
mf: $C_6H_8O_6$ mw: 176.14

PROP: White crystals. Mp: 192°, flash p: +210°F. Sol in water; sltly sol in alc; insol in ether, chloroform, benzene, petroleum ether, fixed oils and fats.

SYNS: ASCORBIC ACID * l(+)-ASCORBIC ACID * ASCORBUTINA * CEVITAMIC ACID * CEVITA-MIN * FEMA No. 2109 * 3-KETO-l-GULOFURANO-LACTONE * l-3-KETOTHREOHEXURONIC ACID LAC-TONE * NATRASCORB INJECTABLE * NCI-C54808 * 3-OXO-l-GULOFURANOLACTONE * VITACIN * VITAMIN C * VITAMISIN * VITASCORBOL * XITIX * l-XYLOASCORBIC ACID

CONSENSUS REPORTS: NTP Carcinogenesis Bioassay (feed); No Evidence: mouse, rat NTPTR* NTP-TR-247,83; NTPTR* NTP-TR-214,82. Reported in EPA TSCA Inventory.

SAFETY PROFILE: Moderately toxic. Human blood systemic effects by intravenous route. Mutation data reported. Combustible liquid. When heated to decomposition it emits acrid smoke and irritating fumes.

ARN125   CAS: 134-03-2   **HR: D**
ASCORBIC ACID SODIUM SALT
mf: $C_6H_8O_6 \cdot Na$ mw: 199.13

PROP: Minute white to yellow crystals; odorless. Decomp at 218°. Freely sol in water; very sltly sol in alc; insol in chloroform, ether.

SYNS: l-ASCORBIC ACID SODIUM SALT * ASCOR-BICIN * ASCORBIN * CEBITATE * CENOLATE * ISKIA-C * MONOSODIUM ASCORBATE * NATRASCORB * NATRI-C * SODASCORBATE * SODIUM ASCORBATE (FCC) * SODIUM-l-AS-CORBATE * VITAMIN C * VITAMIN C SODIUM

CONSENSUS REPORTS: Reported in EPA TSCA Inventory.

SAFETY PROFILE: Human mutation data reported. When heated to decomposition it emits toxic fumes of Na_2O.

ARN800   CAS: 9015-68-3   **HR: 3**
l-ASPARAGINASE

SYNS: ASPARAGINASE * l-ASPARAGINASE X * l-ASPARAGINASI (ITALIAN) * l-ASPARAGINE AMI-DOHYDROLASE * LEUCOGEN * NSC-109229

SAFETY PROFILE: Human (child) systemic effects by intramuscular route. Questionable carcinogen with experimental neoplastigenic data.

ARN825   CAS: 22839-47-0   **HR: 1**
ASPARTAME
mf: $C_{14}H_{18}N_2O_5$ mw: 294.34

PROP: White crystalline powder; odorless with a sweet taste. Sltly sol in water, alc.

SYNS: 3-AMINO-N-(α-CARBOXYPHENETHYL) SUCCINAMIC ACID N-METHYL ESTER, stereoisomer * ASPARTYLPHENYLALANINE METHYL ESTER * N-l-α-ASPARTYL-l-PHENYLALANINE 1-METHYL ESTER (9CI) * CANDEREL * DIPEPTIDE SWEET-ENER * EQUAL * METHYL ASPARTYLPHENYLA-LANATE * 1-METHYL N-l-α-ASPARTYL-l-PHE-NYLALANINE * NUTRASWEET * SWEET DIPEP-TIDE

SAFETY PROFILE: Human systemic effects by ingestion: allergic dermatitis. Experimental reproductive effects. When heated to decomposition it emits toxic fumes of NO_x.

ARO500   CAS: 8052-42-4   **HR: 3**
ASPHALT

DOT: NA 1999

PROP: Black or dark brown mass. Bp: <470°, flash p: 400+°F (CC), d: 0.95−1.1, autoign temp: 905°F.

SYNS: ASPHALTUM * BITUMEN (MAK) * JU-DEAN PITCH * MINERAL PITCH * PETROLEUM PITCH * ROAD ASPHALT (DOT) * ROAD TAR (DOT)

CONSENSUS REPORTS: IARC Cancer Review: Human Inadequate Evidence IMEMDT 35,39,85. Reported in EPA TSCA Inventory.

ACGIH TLV: TWA 5 mg/m^3
DFG MAK: Suspected Carcinogen.
NIOSH REL: CL 5 mg/m^3/15M
DOT Classification: ORM-C; Label: None.

SAFETY PROFILE: Suspected carcinogen with experimental carcinogenic and tumorigenic data. A moderate irritant. May contain carcinogenic components. Combustible when exposed to heat or flame. To fight fire, use foam, CO_2, or dry chemical.

ARO750   CAS: 8052-42-4   **HR: 3**
ASPHALT (CUT BACK)

DOT: UN 1999

PROP: A liquid petroleum product, solubility of residue from distillation in carbon tetrachloride = 99.5%. Flash p: < 50°F.

SYNS: ROAD ASPHALT (DOT) * ROAD TAR, liquid (DOT)

DOT Classification: Flammable liquid. Label: Flammable liquid.

SAFETY PROFILE: Contains carcinogenic components. A dangerous fire hazard when exposed to heat or flame. To fight fire use dry chemical, water mist, fog. When heated to decomposition it emits smoke and irritating, acrid fumes.

ARP250   CAS: 8003-03-0   HR: 2
ASPIRIN, PHENACETIN and CAFFEINE
mf: $C_{10}H_{13}NO_2 \cdot C_9H_8O_4 \cdot C_8H_{10}N_4O_2$
mw: 553.63

PROP: Composed of 50% aspirin, 46% phenacetin, and 4% caffeine (NCIMR* NIH-71-E-2144)

SYNS: 2-(ACETYLOXY)BENZOIC ACID, mixed with 3,7-DIHYDRO-1,3,7-TRIMETHYL-1H-PURINE-2,6-DIONE and N-(4-ETHOXYPHENYL)ACETAMIDE * APC (pharmaceutical) * ASCOPHEN * CITRAMON * EMPIRIN COMPOUND * NCI-C02697 * OSCOPHEN * THOMAPYRIN

CONSENSUS REPORTS: NCI Carcinogenesis Bioassay (feed); Inadequate Studies: mouse, rat NCITR* NCI-CG-TR-67,78

SAFETY PROFILE: Moderately toxic by ingestion. Questionable carcinogen with experimental tumorigenic data. When heated to decomposition it emits toxic fumes of NO_x.

ARQ250   CAS: 83-89-6   HR: 3
ATABRINE
mf: $C_{23}H_{30}ClN_3O$ mw: 400.01

PROP: Bright yellow crystals. Mp: decomp @ 248°.

SYNS: 6-CHLORO-9-((4-(DIETHYL AMINO)-1-METHYL BUTYL)AMINO)-2-METHOXYACRIDINE * 3-CHLORO-7-METHOXY-9-(1-METHYL-4-DIETHYLAMINOBUTYL-AMINO)ACRIDINE * 2-METHOXY-6-CHLORO-9-DI-ETHYLAMINOPENTYLAMINOACRIDINE * QUINACRINE

SAFETY PROFILE: Poison by subcutaneous route. Moderately toxic by ingestion. Mutation

data reported. Has been implicated in aplastic anemia. When heated to decomposition, it emits very toxic fumes of Cl^- and NO_x.

ARQ725   CAS: 1912-24-9   HR: 3
ATRAZINE
mf: $C_8H_{14}ClN_5$ mw: 215.72

PROP: Crystals. Mp: 171-174°. Solubility at 25°: in water, 70 ppm; ether, 12,000 ppm; chloroform, 52,000 ppm; methanol, 18,000 ppm.

SYNS: A 361 * AATREX * AATREX 4L * AATREX NINE-O * AATREX 80W * 2-AETHYLAMINO-4-CHLOR-6-ISOPROPYL-AMINO-1,3,5-TRIAZIN (GERMAN) * 2-AETHYL-AMINO-4-ISOPROPYLAMINO-6-CHLOR-1,3,5-TRIAZIN (GERMAN) * AKTIKON * AKTIKON PK * AKTINIT A * AKTINIT PK * ARGEZIN * ATAZINAX * ATRANEX * ATRASINE * ATRATOL A * ATRAZIN * ATRED * ATREX * CANDEX * CEKUZINA-T * 2-CHLORO-4-ETHYLAMINEISOPROPYLAMINE-s-TRIAZINE * 1-CHLORO-3-ETHYLAMINO-5-ISO-PROPYLAMINO-s-TRIAZINE * 1-CHLORO-3-ETHYLAMINO-5-ISOPROPYLAMINO-2,4,6-TRI-AZINE * 2-CHLORO-4-ETHYLAMINO-6-ISOPRO-PYLAMINO-s-TRIAZINE * 2-CHLORO-4-ETHYLAMINO-6-ISOPROPYLAMINO-1,3,5-TRI-AZINE * 6-CHLORO-N-ETHYL-N'-(1-METH-YLETHYL)-1,3,5-TRIAZINE-2,4-DIAMINE (9CI) * 2-CHLORO-4-(2-PROPYLAMINO)-6-ETHYLAMINO-s-TRIAZINE * CRISATRINA * CRISAZINE * CYAZIN * FARMCO ATRAZINE * FENAMIN * FENAMINE * FENATROL * G 30027 * GEIGY 30,027 * GESAPRIM * GESOPRIM * GRIFFEX * HUNGAZIN * HUNGAZIN PK * INAKOR * OLEOGESAPRIM * PRIMATOL * PRIMAZE * RADAZIN * RADIZINE * SHELL ATRAZINE HERBICIDE * STRAZINE * TRIAZINE A 1294 * VECTAL * VECTAL SC * WEEDEX A * WONUK * ZEAZIN * ZEAZINE

CONSENSUS REPORTS: EPA Genetic Toxicology Program. Reported in EPA TSCA Inventory.

OSHA PEL: TWA 5 mg/m^3
ACGIH TLV: TWA 5 mg/m^3
DFG MAK: 2 mg/m^3

SAFETY PROFILE: Poison by intraperitoneal route. Moderately toxic by ingestion. Mildly toxic by inhalation and skin contact. Human mutation data reported. Experimental reproductive effects. A skin and severe eye irritant. Questionable carcinogen with experimental tumorigenic data. When heated to decomposition it emits toxic fumes of Cl^- and NO_x.

ARR000   CAS: 51-55-8   **HR: 3**
ATROPINE
mf: $C_{17}H_{23}NO_3$ mw: 289.41

PROP: Colorless crystalline alkaloid.

SYNS: ATROPIN (GERMAN) * dl-HYOSCYAMINE * 2-PHENYLHYDRACRYLIC ACID-3-α-TROPANYL ESTER * β-PHENYL-γ-OXYPROPIONSAEURE-TROPYL-ESTER (GERMAN) * 1-α-H,5-α-H-TROPAN-3-α-OL (±)-TROPATE (ESTER) * dl-TROPANYL-2-HYDROXY-1-PHENYL-PROPIONATE * TROPIC ACID-3-α-TROPANYL ESTER * TROPIC ACID, ESTER with TROPINE * TROPINE TROPATE * dl-TROPYLTROPATE * (+,-)-TROPYL TROPATE

CONSENSUS REPORTS: Reported in EPA TSCA Inventory.

SAFETY PROFILE: Poison by ingestion, subcutaneous, intravenous, and intraperitoneal routes. An alkaloid. When heated to decomposition it emits toxic fumes of NO_x.

ARR250   CAS: 2472-17-5   **HR: 3**
ATROPINE SULFATE (1:1)
mf: $C_{17}H_{23}NO_3 \cdot H_2O_4S$ mw: 387.49

SAFETY PROFILE: Poison by intravenous route. Mutation data reported. When heated to decomposition it emits very toxic fumes of NO_x and SO_x.

ARR500   CAS: 55-48-1   **HR: 3**
ATROPINE SULFATE (2:1)
mf: $C_{34}H_{46}N_2O_6 \cdot H_2O_4S$ mw: 676.90

SYNS: ATROPIN SIRAN (CZECH) * ATROPINSULFAT (GERMAN) * SULFATE D'ATROPINE (FRENCH) * 1-α-H,5-α-H-TROPAN-3-α-OL (±)-TROPATE (ESTER), SULFATE (2:1) SALT * dl-TROPANYL-2-HYDROXY-1-HENYLPROPIONATE SULFATE

CONSENSUS REPORTS: Reported in EPA TSCA Inventory.

SAFETY PROFILE: Poison by subcutaneous, intravenous, and intraperitoneal routes. Moderately toxic by ingestion. Human (child) pulmonary system effects by ingestion. When heated to decomposition it emits very toxic fumes of NO_x and SO_x.

ARS000   CAS: 60748-45-0   **HR: 3**
ATX II

PROP: A polypeptide isolated from the sea anemone, *Anemonia sulcata* (TOXIA6 16, 561,78)

SYN: SEA ANEMONE TOXIN II

SAFETY PROFILE: A deadly poison by intravenous and parenteral routes.

ARS750   **HR: 3**
AUREMETINE

PROP: Percentage composition: 28% emetine, 16% auramine and 56% iodine (AJTMAQ 10,249,30)

SAFETY PROFILE: Poison by ingestion. When heated to decomposition it emits very toxic fumes of I^- and NO_x.

ART250   CAS: 12192-57-3   **HR: 3**
1-AUROTHIO-d-GLUCOPYRANOSE
mf: $C_6H_{11}O_5S \cdot Au$ mw: 392.20

SYNS: AUREOTAN * AUROMYOSE * AUROTAN * AUROTHIOGLUCOSE * AURUMINE * AU-THRON * BRENOL * (d-GLUCOPYRANOSYLTHIO) GOLD * (1-d-GLUCOSYLTHIO)GOLD * GLYSANOL B * GOLD THIOGLUCOSE * GTG * ORONOL * ROMOSOL * SOLGANAL * SOLGANAL B * (1-THIO-d-GLUCOPYRANOSATO)GOLD * 1-THIO-GLUCOPYRANOSE, MONOGOLD(1+) SALT * THIO-GLUCOSE d'OR (FRENCH)

CONSENSUS REPORTS: IARC Cancer Review: GROUP 1 IMEMDT 7,56,87; Animal Limited Evidence IMEMDT 13,39,77

SAFETY PROFILE: Confirmed carcinogen with experimental carcinogenic and neoplastigenic data. A deadly human poison by an unspecified route. An experimental poison by intramuscular route. Moderately toxic by subcutaneous and intravenous routes. Experimental teratogenic and reproductive effects. When heated to decomposition it emits very toxic fumes of SO_x. Used to treat rheumatoid arthritis.

ARW250   CAS: 75-80-9   **HR: 3**
AVERTIN
mf: $C_2H_3Br_3O$ mw: 282.78

PROP: Crystals, ethereal odor, aromatic taste, sltly water-sol, sol in alc and organic solvents. Mp: 70-82°, bp: 92-93° @ 10 mm.

SYNS: 2,2,2-TRIBROMOETHANOL * 2,2,2-TRIBRO-
MOETHYL ALCOHOL

CONSENSUS REPORTS: Reported in EPA
TSCA Inventory.

SAFETY PROFILE: Poison by intravenous
route. Moderately toxic by ingestion and other
routes. Dangerous when heated; see also
BROMIDES.

ASA500   CAS: 115-02-6   **HR: 3**
AZASERINE
mf: $C_5H_7N_3O_4$ mw: 173.15

PROP: Produced by the strain *Streptomyces fra-
gilis* (85ERAY 2,1249,78)

SYNS: AZASERIN * l-AZASERINE * AZS
* CI-337 * CL 337 * CN-15,757 * DIAZOACE-
TATE (ESTER)-l-SERINE * l-DIAZOACETATE (ESTER)
SERINE * DIAZO-ACETIC ACID ESTER with SERINE
* o-DIAZOACETYL-l-SERINE * NSC-742 * P-165
* RCRA WASTE NUMBER U015 * l-SERINE DIAZ-
OACETATE * l-SERINE DIAZOACETATE (ester)

CONSENSUS REPORTS: IARC Cancer Review: GROUP 2B IMEMDT 7,56,87; Animal
Limited Evidence IMEMDT 10,73,76. EPA Genetic Toxicology Program.

SAFETY PROFILE: Suspected carcinogen with
experimental carcinogenic, neoplastigenic, and
tumorigenic data. Poison by ingestion, intraperitoneal, and subcutaneous routes. Experimental
teratogenic and reproductive effects. Human
mutation data reported. When heated to decomposition it emits toxic fumes of NO_x.

ASB250   CAS: 446-86-6   **HR: 3**
AZATHIOPRINE
mf: $C_9H_7N_7O_2S$ mw: 277.29

SYNS: AZANIN * AZATIOPRIN * AZOTHIOPRINE
* BW 57-322 * CCUCOL * IMURAN * IMUREK
* IMUREL * METHYLNITROIMIDAZOLYLMER-
CAPTOPURINE * 6-(1'-METHYL-4'-NITRO-5'-IMIDAZO-
LYL)-MERCAPTOPURINE * 6-(METHYL-p-NITRO-5-IMI-
DAZOLYL)-THIOPURINE * 6-((1-METHYL-4-NITRO-
IMIDAZOL-5-YL)THIO)PURINE * 6-(1-METHYL-p-
NITRO-5-IMIDAZOLYL)-THIOPURINE * 6-(1-METHYL-4-
NITROIMIDAZOL-5-YLTHIO)PURINE * 6-((1-METHYL-4-
NITRO-1H-IMIDAZOL-5-YL)THIO)-1H-PURINE * NCI-
C03474 * NSC-39084 * RORASUL

CONSENSUS REPORTS: IARC Cancer Review: GROUP 1 IMEMDT 7,119,87 Human
Sufficient Evidence IMEMDT 26,47,81; Animal Limited Evidence IMEMDT 26,47,81. NTP
Fourth Annual Report On Carcinogens, 1984.
NCI Carcinogenesis Studies (ipr); No Evidence:
rat CANCAR 40,1935,77; Clear Evidence:
mouse CANCAR 40,1935,77. EPA Genetic
Toxicology Program.

SAFETY PROFILE: Confirmed human carcinogen producing bladder tumors and leukemia.
Poison by subcutaneous, intradermal, and intraperitoneal routes. Moderately toxic by ingestion. Other human systemic effects by ingestion
and unspecified routes: anemia, bone marrow
abnormalities, hair effects, and metabolic effects. Experimental teratogenic and reproductive effects. Human mutation data reported.
When heated to decomposition it emits very
toxic fumes of NO_x and SO_x. An immunosup-
pressant.

ASC750   **HR: D**
AZIDES

SAFETY PROFILE: Variable toxicity. Many
azides are poisonous, cause a fall in blood pressure and some inhibit enzyme action, thus resembling nitrites and cyanides. An azide is a
compound of hydrogen or a metal ion and the
monovalent $-N_3$ radical. All of its salts and
the acid are unstable and some decompose explosively; although lead azide, which is one
of the most important azides, is not very sensitive. Dangerous; shock and heat will explode
it. When heated to decomposition it emits highly
toxic fumes. If exposed to CS_2, it forms violently
explosive salts. Organic azides are sensitized
by metal salts or traces of strong acid. (See
also specific compound).

ASH500   CAS: 86-50-0   **HR: 3**
AZINPHOS METHYL

DOT: NA 2783
mf: $C_{10}H_{12}N_3O_3PS_2$ mw: 317.34

PROP: Crystals or brown, waxy solid. D: 1.44,
mp: 74°. Sltly sol in water; sol in organic solvents.

SYNS: AZINFOS-METHYL (DUTCH) * AZINPHOS-
METILE (ITALIAN) * AZINPHOS METHYL, liquid (DOT)
* BAY 9027 * BAYER 17147 * BENZO-

TRIAZINEDITHIOPHOSPHORIC ACID DIMETHOXY ESTER * BENZOTRIAZINE derivative of a METHYL DITHIOPHOSPHATE * CARFENE * COTNION METHYL * CRYSTHION 2L * CRYSTHYON * DBD * S-(3,4-DIHYDRO-4-OXO-BENZO(α)(1,2,3)TRIAZIN-3-YLMETHYL)-O,O-DIMETHYL PHOSPHORODITHIOATE * S-(3,4-DIHYDRO-4-OXO-1,2,3-BENZOTRIAZIN-3-YLMETHYL)- O,O-DIMETHYL PHOSPHORODITHIOATE * O,O-DIMETHYL-S-(BENZAZIMINOMETHYL) DITHIOPHOSPHATE * O,O-DIMETHYL-S-(1,2,3-BENZOTRIAZINYL-4-KETO) METHYL PHOSPHORODITHIOATE * O,O-DIMETHYL-S-(3,4-DIHYDRO-4-KETO-1,2,3-BENZOTRIAZINYL-3-METHYL) DITHIOPHOSPHATE * DIMETHYLDITHIO-PHOSPHORIC-ACID N-METHYLBENZAZIMIDE ESTER * O,O-DIMETHYL-S-(4-OXO-3H-1,2,3-BENZOTRIZIANE-3-METHYL)PHOSPHORODITHIOATE * O,O-DIMETH-YL-S-(4-OXOBENZOTRIAZINO-3-METHYL)PHOSPHORO-DITHIOATE * O,O-DIMETHYL-S-(4-OXO-1,2,3-BENZO-TRIAZINO(3)-METHYL) THIOTHIONOPHOSPHATE * O,O-DIMETHYL-S-((4-OXO-3H-1,2,3-BENZOTRIAZIN-3-YL)-METHYL)-DITHIOFOSFAAT (DUTCH) * O,O-DI-METHYL-S-((4-OXO-3H-1,2,3-BENZOTRIAZIN-3-YL)-METHYL)-DITHIOPHOSPHAT (GERMAN) * O,O-DI-METHYL-S-4-OXO-1,2,3-BENZOTRIAZIN-3(4H)-YL-METHYL PHOSPHORODITHIOATE * O,O-DIMETIL-S-((4-OXO-3H-1,2,3-BENZOTRIAZIN-3-IL)-METIL)-DITIO-FOSFATO (ITALIAN) * ENT 23,233 * GOTHNION * GUSATHION * GUTHION (DOT) * GUTHION, liquid (DOT) * 3-(MERCAPTOMETHYL)-1,2,3-BENZO-TRIAZIN-4(3H)-ONE-O,O-DIMETHYL PHOSPHORO-DITHIOATE * 3-(MERCAPTOMETHYL)-1,2,3-BENZO-TRIAZIN-4(3H)-ONE-O,O-DIMETHYL PHOSPHORODI-THIOATE-S-ESTER * METHYLAZINPHOS * N-METHYLBENZAZIMIDE, DIMETHYLDITHIOPHOS-PHORIC ACID ESTER * METHYL GUTHION * METILTRIAZOTION * NCI-C00066

CONSENSUS REPORTS: NCI Carcinogenesis Bioassay (feed); Inadequate Studies: rat NCITR* NCI-CG-TR-69,78; No Evidence: mouse NCITR* NCI-CG-TR-69,78. EPA Genetic Toxicology Program. EPA Extremely Hazardous Substances List.

OSHA PEL: TWA 0.2 mg/m^3 (skin)
ACGIH TLV: TWA 0.2 mg/m^3 (skin)
DFG MAK: 0.2 mg/m^3
DOT Classification: Poison B; Label: Poison, liquid mixture.

SAFETY PROFILE: Poison by inhalation, ingestion, skin contact, intravenous, intraperitoneal, and possibly other routes. Experimental teratogenic and reproductive effects. Questionable carcinogen with experimental tumorigenic data. Human mutation data reported. When heated to decomposition it emits very toxic fumes of PO$_x$, SO$_x$, and NO$_x$.

ASH750   CAS: 671-51-2   **HR: 3**
AZIRIDINE CARBOXYLIC ACID ETHYL ESTER
mf: C$_5$H$_9$NO$_2$ mw: 115.15

SYNS: N-CARBETHOXYETHYLENIMINE * N-(ETHOXYCARBONYL)AZIRIDINE * N-ETH-OXYCARBONYLETHYLENEIMINE * ETHOXY-CARBONYL-1-ETHYLENIMINE * ETHYL AZIRI-DINECARBOXYLATE * ETHYL-1-AZIRIDINECAR-BOXYLATE * ETHYL AZIRIDINOCARBOXYLATE * ETHYL-1-AZIRIDINYLCARBOXYLATE * ETHYL AZIRIDINYLFORMATE

SAFETY PROFILE: Poison by intravenous and intraperitoneal routes. Mutation data reported. When heated to decomposition it emits toxic fumes of NO$_x$.

ASI000   CAS: 1072-52-2   **HR: 3**
1-AZIRIDINE ETHANOL
mf: C$_4$H$_9$NO   mw: 87.14

SYNS: 2-(1-AZIRIDINYL)ETHANOL * β-HYDROXY-1-ETHYLAZIRIDINE * 2-HYDROXY-1-ETHYLAZIRI-DINE * N-(β-HYDROXYETHYL)AZIRIDINE * N-(2-HYDROXYETHYL)AZIRIDINE * N-HYDROXYETHYL ETHYLENE IMINE * N-(2-HYDROXYETHYL)ETHYL-ENIMINE * 1-(2-HYDROXYETHYL)ETHYLENIMINE

CONSENSUS REPORTS: IARC Cancer Review: GROUP 3 IMEMDT 7,56,87; Animal Limited Evidence IMEMDT 9,47,75. Reported in EPA TSCA Inventory.

SAFETY PROFILE: Poison by ingestion, skin contact, and intravenous routes. A skin and eye irritant. Questionable carcinogen with experimental neoplastigenic data. When heated to decomposition it emits toxic fumes of NO$_x$.

ASL250   CAS: 103-33-3   **HR: 3**
AZOBENZENE
mf: C$_{12}$H$_{10}$N$_2$ mw: 182.23

PROP: Orange, monoclinic crystals. Mp: 68°, bp: 297°, d: 1.203 @ 20°/4°, vap press: 1 mm @ 103.5°. Insol in water. Solubility in alc =

4.2/100 @ 20° in ether (ligroin) = 12/100 @ 20°.

SYNS: AZOBENZEEN (DUTCH) * AZOBENZIDE * AZOBENZOL * AZOBISBENZENE * AZODIBENZENE * AZODIBENZENEAZOFUME * BENZENE-AZOBENZENE * DIAZOBENZENE * DIPHENYLDIAZENE * 1,2-DIPHENYLDIAZENE * DIPHENYLDIIMIDE * ENT 14,611 * NCI-C02926 * USAF-EK-704

CONSENSUS REPORTS: IARC Cancer Review: GROUP 3 IMEMDT 7,56,87; Animal Limited Evidence IMEMDT 8,75,75; NCI Carcinogenesis Bioassay (feed); Clear Evidence: rat NCITR* NCI-CG-TR-154,79; No Evidence: mouse NCITR* NCI-CG-TR-154,79. Reported in EPA TSCA Inventory.

SAFETY PROFILE: Moderately toxic by ingestion and possibly other routes. Questionable carcinogen with experimental carcinogenic, neoplastigenic, and tumorigenic data. When heated to decomposition it emits toxic fumes of NO_x.

ASL750   CAS: 78-67-1   HR: 3
AZOBISISOBUTYLONITRILE
DOT: UN 2952
mf: $C_8H_{12}N_4$ mw: 164.24

SYNS: ACETO AZIB * AIBN * α,α'-AZOBISISO-BUTYLONITRILE * AZOBISISOBUTYRONITRILE * 2,2'-AZOBIS(ISOBUTYRONITRILE) * 2,2'-AZOBIS(2-METHYLPROPIONITRILE) * AZODIISOBUTYRO-NITRILE * α,α'-AZODIISOBUTYRONITRILE * 2,2'-AZODIISOBUTYRONITRILE * AZODIISO-BUTYRONITRILE (DOT) * 2,2'-DICYANO-2,2'-AZOPROPANE * POLYZOLE AZDN * POROFOR 57 * VAZO 64

CONSENSUS REPORTS: Cyanide and its compounds are on the Community Right-To-Know List. Reported in EPA TSCA Inventory.

DOT Classification: IMO: Flammable Solid; Label: Flammable Solid.

SAFETY PROFILE: Poison by intraperitoneal route. Moderately toxic by ingestion. Easily oxidized, unstable. Violent exothermic decomposition when heated. Solution in acetone may decompose explosively. Explodes when heated with heptane. When heated to decomposition

it emits toxic fumes of NO_x and CN^-. A free radical generator.

ASM270   CAS: 123-77-3   HR: D
AZODICARBAMIDE
mf: $C_2H_4N_4O_2$ mw: 116.10

PROP: Orange-red powder. Decomp @ 180°-200°. Very sltly sol in hot water; insol in alc. Decomp in hot HCl.

SYNS: 1,1'-AZOBISCARBAMIDE * AZOBISCARBONAMIDE * AZOBISCARBOXAMIDE * 1,1'-AZO-BIS(FORMAMIDE) * AZODICARBOAMIDE * AZODI-CARBONAMIDE * AZODICARBOXAMIDE * AZODICARBOXYLIC ACID DIAMIDE * $\Delta(1,1')$-BIUREA * CELOSEN AZ * ChKhZ 21 * ChKhZ 21R * DIAZENEDICARBOXAMIDE * GENITRON AC * GENITRON AC 2 * GENITRON AC 4 * KEMPORE * KEMPORE 125 * KEMPORE R 125 * LUCEL ADA * NCI-C55981 * NITROPORE * PINHOLE AK 2 * POROFOR 505 * POROFOR ADC/R * POROFOR ChKhZ 21 * POROFOR ChKhZ 21R * UNIFOAM AZ * UNIFORM AZ * YUNIHOMU AZ

CONSENSUS REPORTS: Reported in EPA TSCA Inventory.

SAFETY PROFILE: Mutation data reported. When heated to decomposition it emits toxic fumes of NO_x.

ASM300   CAS: 123-77-3   HR: 3
AZODICARBONAMIDE
mf: $C_2H_4N_4O_2$ mw: 116.08

PROP: Yellow to orange-red crystalline powder. Mp: above 180° (decomp). Sltly sol in dimethyl sulfoxide; insol in water, organic solvents.

SAFETY PROFILE: Flammable solid. When heated to decomposition emits toxic fumes of NO_x.

ASN000   HR: 3
"AZODRIN"
mf: $C_6H_{14}O_5NP$ mw: 211.2

PROP: Reddish-brown solid, mild ester odor. Bp: 125°.

SYN: MONOCROTOPHOS

SAFETY PROFILE: Poison by ingestion and skin contact. A dangerous fire hazard. When heated to decomposition it evolves highly toxic fumes of NO_x and PO_x.

ASN250   CAS: 821-14-7   **HR: 3**
AZO ETHANE
mf: $C_4H_{10}N_2$ mw: 86.16

SYN: AZOAETHAN (GERMAN)

SAFETY PROFILE: Moderate acute toxicity. An experimental teratogen. Questionable carcinogen with experimental carcinogenic and tumorigenic data. When heated to decomposition it emits toxic fumes of NO_x. An unstable, dangerously explosive material in concentrated state.

ASO750   CAS: 495-48-7   **HR: 3**
AZOXYBENZENE
mf: $C_{12}H_{10}N_2O$ mw: 198.23

PROP: Yellow, rhombic crystals. D: 1.248 @ 20°/20°; mp: 36°; bp: decomp. Insol in water, solubility in alc = 11.4/100 @ 15°, solubility in ether (ligroin) = 43.5/100 @ 15°.

SYNS: AZOBENZENE OXIDE * AZOSSIBENZENE (ITALIAN) * AZOXYBENZEEN (DUTCH) * AZOXYBENZIDE * AZOXYBENZOL (GERMAN) * AZOXYDIBENZENE * ORDINARY AZOXYBENZENE

CONSENSUS REPORTS: Reported in EPA TSCA Inventory.

SAFETY PROFILE: Poison by subcutaneous route. Moderately toxic by ingestion, skin contact, and other routes. A skin and eye irritant. Combustible. When heated to decomposition it emits toxic fumes of NO_x.

ASP000   CAS: 16301-26-1   **HR: 3**
AZOXYETHANE
mf: $C_4H_{10}N_2O$ mw: 102.16

SYNS: AZOXYAETHAN (GERMAN) * DIETHYLDIAZENE-1-OXIDE

SAFETY PROFILE: Poison by subcutaneous and intravenous routes. An experimental transplacental teratogen. Questionable carcinogen with experimental carcinogenic and tumorigenic data. When heated to decomposition it emits toxic fumes of NO_x.

ASP250   CAS: 25843-45-2   **HR: 3**
AZOXYMETHANE
mf: $C_2H_6N_2O$ mw: 74.10

SYN: AOM

SAFETY PROFILE: Suspected carcinogen with experimental carcinogenic and tumorigenic data. Poison by subcutaneous route. Mutation data reported. When heated to decomposition it emits toxic fumes of NO_x.

B

BAB750   CAS: 1395-21-7   **HR: 3**
BACILLUS SUBTILIS BPN

PROP: A commercial raw proteolytic enzyme used in laundry detergents.

SYNS: BACILLOMYCIN (8CI, 9CI) * BACILLOMYCIN R * FUNGOCIN * SUBTILISINS (ACGIH) * SUBTILISINS BPN

OSHA PEL: CL 0.00006 mg/m^3
ACGIH TLV: CL 0.00006 mg/m^3

SAFETY PROFILE: A poison via intraperitoneal route. A severe eye irritant. When heated to decomposition it emits toxic fumes of NO$_x$.

BAC000   CAS: 9014-01-1   **HR: 3**
BACILLUS SUBTILIS CARLSBERG

PROP: A commercial raw proteolytic enzyme used in laundry detergents.

SYNS: ALCALASE * ALK-ENZYME * BACILLO-PEPTIDASE A * BACILLOPEPTIDASE B * BIOPRASE * COLISTINASE * E.C. 3.4.4.16 * E.C. 3.4.21.14 * MAXATASE * NAGARSE * SUBTILISIN (9CI, ACGIH) * SUBTILISIN CARLSBURG * SUBTILISIN NOVO * SUBTILOPEPTIDASE A * SUBTILOPEPTIDASE B * SUBTILOPEPTIDASE BPN' * SUBTILOPEPTIDASE C * THERMOASE PC-10

CONSENSUS REPORTS: Reported in EPA TSCA Inventory.

ACGIH TLV: CL 0.00006 mg/m^3

SAFETY PROFILE: Moderately toxic by ingestion. An eye irritant. When heated to decomposition it emits toxic fumes of NO$_x$.

BAC250   CAS: 1405-87-4   **HR: 3**
BACITRACIN

PROP: White to pale buff, hygroscopic powder; odorless or slt odor. Freely sol in water, alc, methanol, and glacial acetic acid; insol in acetone, chloroform, and ether. When heated to decomposition it emits acrid smoke and irritating fumes.

SYNS: AYFIVIN * BACIGUENT * BACI-JEL * BACILIQUIN * BACITEK OINTMENT * FORTRACIN * PARENTRACIN * PENITRACIN * TOPITRACIN * USAF CB-7 * ZUTRACIN

CONSENSUS REPORTS: Reported in EPA TSCA Inventory.

SAFETY PROFILE: A poison by intraperitoneal and intravenous routes. Moderately toxic by ingestion and subcutaneous routes. Mutation data reported.

BAD250   **HR: 1**
BAGASSE DUST

SAFETY PROFILE: A nuisance dust from the fibrous residue of cane sugar manufacture. Inhalation can cause bronchial asthma, sneezing, rhinorrhea, pneumonitis, etc. Fire and explosion hazard when exposed to heat, flame, or oxidizers.

BAD625   CAS: 10309-37-2   **HR: 3**
BAKUCHIOL
mf: C$_{18}$H$_{24}$O   mw: 256.42

SAFETY PROFILE: Poison by intravenous and intraperitoneal routes. Moderately toxic by ingestion. When heated to decomposition it emits acrid smoke and fumes.

BAD750   CAS: 59-52-9   **HR: 3**
BAL
mf: C$_3$H$_8$OS$_2$ mw: 124.23

PROP: Viscous, oily liquid; pungent odor, bp: 140° @ 40 mm, vap d: 4.3, d: 1.2385 @ 25°/4°.

SYNS: BRITISH ANTILEWISITE * DICAPTOL * DIMERCAPROL PROPANOL * DIMERCAPTOL * 2,3-DIMERCAPTOL-1-PROPANOL * DIMERCAPTO-PROPANOL * 2,3-DIMERCAPTOPROPANOL * 2,3-DIMERCAPTOPROPAN-1-OL * DITHIOGLYCEROL * 1,2-DITHIOGLYCEROL * 2,3-DITHIOPROPANOL * SULFACTIN * USAF ME-1

CONSENSUS REPORTS: EPA Genetic Toxicology Program. Reported in EPA TSCA Inventory.

SAFETY PROFILE: Poison via intramuscular, parenteral, intraperitoneal, and intravenous routes. Experimental teratogenic and reproductive effects. Human systemic effects by intramuscular route: hemorrhage and dermatitis. Human blood and systemic skin effects by intramuscular route. It causes redness and swelling when applied locally to the skin, but does not produce blisters or ulcers. Intensely irritating to eyes and mucous membranes. Systemic symptoms are caused by injection. When heated to decomposition, it emits toxic fumes of SO_x. Used as an antidote to arsenic, gold and mercury poisoning.

BAE750   HR: 1
BALSAM of PERU

PROP: Dark brown, viscid liquid; vanilla odor. Sol in fixed oils; sltly sol in propylene glycol; insol in glycerin. Extracted from *Myroxylon pereirae Klotzsch*.

SYNS: BALSAM PERU OIL (FCC) * PERUVIAN BALSAM

SAFETY PROFILE: A mild allergen. Combustible when heated. When heated to decomposition it emits acrid smoke and irritating fumes.

BAG250   CAS: 144-02-5   HR: 3
BARBITAL SODIUM
mf: $C_8H_{11}N_2O_3 \cdot Na$ mw: 206.20

PROP: Bitter crystals or powder.

SYNS: BARBITAL Na * BARBITAL SOLUBLE
* BARBITONE SODIUM * DIETHYLBARBITURATE
MONOSODIUM * 5,5-DIETHYLBARBITURIC ACID SODIUM deriv. * DIETHYLMALONYLUREA SODIUM
* EMBINAL * MEDINAL * NATRINAL
* NATRIUMBARBITALS (GERMAN) * NERVOSETON
* 2,4,6(1H,3H,5H)-PYRIMIDINETRIONE, 5,5-DIETHYL-,
MONOSODIUM SALT (9CI) * SODIUM BARBITAL
* SODIUM BARBITONE * SODIUM DIETHYLBARBITURATE * SODIUM-5,5-DIETHYLBARBITURATE
* SODIUM ETHYLBARBITAL * SODIUM MALONYLUREA * SODIUM VERONAL * SOLUBLE BARBITAL
* SOPRINAL * THYALONE * VERONAL SODIUM

SAFETY PROFILE: Poison by ingestion, subcutaneous, intravenous, and intraperitoneal routes. Large doses cause marked depression (sometimes preceded by excitation), prolonged coma, and death. Allergic skin reactions may occur on contact. Implicated in development of aplastic anemia. A truly habit-forming drug. Experimental teratogenic and reproductive effects. Mutation data reported. Combustible. When heated to decomposition it emits toxic fumes of NO_x and Na_2O.

BAG500   HR: 3
BARBITURATES

SYNS: BARBITAL * BARBITAL SODIUM
* BARBITONE

SAFETY PROFILE: Salts or derivatives of barbituric acid are central nervous system depressants, and are used as hypnotics, sedatives and anesthetics. Usually administered orally. They are strongly habit forming. Several compounds including amo-, seco-, and pentabarbital are restricted chemicals. Their use can cause a reaction called barbiturism which is marked by chills, headache, fever, and cutaneous eruptions.

BAH250   CAS: 7440-39-3   HR: 3
BARIUM

DOT: UN 1399/UN 1400/UN 1854
af: Ba   aw: 137.36

PROP: Silver-white, sltly lustrous, somewhat malleable metal. Mp: 725°, bp: 1640°, d: 3.5 @ 20°, vap press: 10 mm @ 1049°.

CONSENSUS REPORTS: Reported in EPA TSCA Inventory. Community Right-To-Know List.

DOT Classification: Flammable Solid; Label: Dangerous When Wet, Spontaneously Combustible, pyrophoric.

SAFETY PROFILE: Water and stomach acids solubilize barium salts and can cause poisoning. Symptoms are vomiting, colic, diarrhea, slow irregular pulse, transient hypertension, and convulsive tremors and muscular paralysis. Death may occur from a few hours to a few days. Half-life of barium in bone has been estimated at 50 days. Dust is dangerous and explosive when exposed to heat, flame, or chemical reaction. Violent or explosive reaction with water, CCl_4, fluorotrichloromethane, trichloroethylene, and C_2Cl_4. Incompatible with acids, $C_2Cl_3F_3$, $C_2H_2FCl_3$, C_2HCl_3 and water, 1,1,2-trichloro trifluoro ethane, and fluorotrichloroethane. The powder may ignite or explode in air or other oxidizing gases.

BAI000   CAS: 18810-58-7   HR: 3
BARIUM AZIDE
DOT: UN 0224
mf: BaN_6 mw: 221.40

PROP: Monoclinic prisms. Mp: $-N_2$ @ about 120°, bp: explodes, d: 2.936.

SYN: BARIUM AZIDE, dry or containing less than 50% water (DOT)

CONSENSUS REPORTS: Reported in EPA TSCA Inventory. Barium and its compounds are on the Community Right-To-Know List.

OSHA PEL: TWA 0.5 mg(Ba)/m^3
ACGIH TLV: TWA 0.5 mg/(Ba)m^3
DOT Classification: Class A Explosive; Label: Explosive A and Poison.

SAFETY PROFILE: A poison. Moderate explosion hazard when shocked or heated to 275°. Spontaneously flammable in air. Very unstable. When heated to decomposition it emits toxic fumes of NO_x.

BAI500   **HR: 3**
BARIUM BENZOATE
mf: $Ba(C_7H_5O_2)_2 \cdot 2H_2O$ mw: 415.61

PROP: White, nacreous leaflets. Mp: loses $2H_2O$ @ 100°.

CONSENSUS REPORTS: Barium and its compounds are on the Community Right-To-Know List.

SAFETY PROFILE: Deadly poison.

BAI750   CAS: 13967-90-3   **HR: 3**
BARIUM BROMATE
mf: $Ba(BrO_3)_2 \cdot H_2O$ mw: 411.21

PROP: White crystals or crystalline powder, mp: decomp @ 260°, d: 3.99 @ 18°.

CONSENSUS REPORTS: Barium and its compounds are on the Community Right-To-Know List.

DOT Classification: Oxidizer; Label: Oxidizer and Poison

SAFETY PROFILE: Very toxic. Fire hazard by chemical reaction with easily oxidized materials. Explodes at 300°. Mixtures with sulfur are unstable storage hazards; igniting immediately at 91°C and after a 2-11 day delay period at room temperature. Incompatible with Al, As, C, Cu, metal sulfides, organic matter, P, and reducing materials. When heated to decomposition, it emits toxic fumes of Br^-.

BAJ000   **HR: 3**
BARIUM CARBIDE
mf: BaC_2 mw: 161.4

PROP: Gray crystals. D: 3.75.

CONSENSUS REPORTS: Barium and its compounds are on the Community Right-To-Know List.

SAFETY PROFILE: A poison. A fire and explosion hazard by chemical reaction with moisture to form acetylene. Incompatible with Se, S, and H_2O. To fight fire, use CO_2, dry chemical.

BAJ500   CAS: 13477-00-4   **HR: 3**
BARIUM CHLORATE
DOT: UN 1445
mf: $Cl_2O_6 \cdot Ba$ mw: 304.24

PROP: Colorless prisms or white powder. Mp: loses H_2O @ 414°, d: 3.18.

SYN: CHLORIC ACID, BARIUM SALT

CONSENSUS REPORTS: Reported in EPA TSCA Inventory. Barium and its compounds are on the Community Right-To-Know List.

OSHA PEL: TWA 0.5 mg(Ba)/m^3
ACGIH TLV: TWA 0.5 mg(Ba)/m^3
DOT Classification: Oxidizer; Label: Oxidizer.

SAFETY PROFILE: A poison. For fire and explosion hazards, see CHLORATES. Incompatible with Al, As, C, charcoal, Cu, MnO_2, metal sulfides, S_4N_4, organic matter, P, and S.

BAK250   CAS: 10294-40-3   **HR: 3**
BARIUM CHROMATE(VI)
mf: $Ba \cdot CrH_2O_4$ mw: 255.36

PROP: Heavy, yellow, crystalline powder. D: 4.498 @ 15°.

SYNS: BARIUM CHROMATE (1:1) * BARIUM CHROMATE OXIDE * BARYTA YELLOW * CHROMIC ACID, BARIUM SALT (1:1) * C.I. 77103 * C.I. PIGMENT YELLOW 31 * LEMON CHROME * LEMON YELLOW * PERMANENT YELLOW * STEINBUHL YELLOW * ULTRAMARINE YELLOW

CONSENSUS REPORTS: IARC Cancer Review: GROUP 1 IMEMDT 7,165,87, Animal Inadequate Evidence IMEMDT 2,100,73; Human Sufficient Evidence IMEMDT 23,205,80. Reported in EPA TSCA Inventory. Barium and its compounds are on the Community Right-To-Know List.

OSHA PEL: TWA 0.1 mg (C_3O_3)m^3; 0.5 mg(Ba)/m^3

ACGIH TLV: TWA 0.5 mg(Ba)/m³; 0.05 mg(Cr)/m³; Confirmed Human Carcinogen
NIOSH REL: TWA 0.001 mg(Cr(VI))/m³

SAFETY PROFILE: Confirmed human carcinogen. A poison. Mutation data reported. Reacts vigorously with reducing materials. Used in pyrotechnics and as an explosive initiator.

BAK500 HR: 3
BARIUM COMPOUNDS (soluble)

CONSENSUS REPORTS: Barium and its compounds are on the Community Right-To-Know List.

OSHA PEL: Soluble Compounds: TWA 0.5 mg(Ba)/m³
ACGIH TLV: Soluble Compounds:TWA 0.5 mg/m³
DFG MAK: Soluble Compounds: 0.5 mg/m³
DOT Classification: Some barium compounds are flammable or explosive.

SAFETY PROFILE: The chromate is a human carcinogen. The soluble barium salts, such as the chloride and sulfide, are poisonous when ingested. The insoluble sulfate used in radiography is not acutely toxic. Few cases of industrial systemic poisoning have been reported, but one investigator describes a fatal case of poisoning attributed to barium oxide, the symptoms being severe abdominal pain with vomiting, dyspnoea, rapid pulse, paralysis of the arm and leg, and eventually cyanosis and death. The same investigator produced paralysis in animals with barium oxide and carbonate. The usual result of exposure to the sulfide, oxide, and carbonate is irritation of the eyes, nose, and throat, and of the skin, producing dermatitis. The salts mentioned are somewhat caustic.

BAK750 CAS: 542-62-1 HR: 3
BARIUM CYANIDE

DOT: UN 1565
mf: C_2BaN_2 mw: 189.38

PROP: White, crystalline powder.

SYNS: BARIUM CYANIDE, solid (DOT) * BARIUM DICYANIDE * RCRA WASTE NUMBER P013

CONSENSUS REPORTS: Reported in EPA TSCA Inventory. Cyanide and its compounds, as well as barium and its compounds, are on the Community Right-To-Know List.

OSHA PEL: TWA 0.5 mg(Ba)/m³
ACGIH TLV: TWA 0.5 mg(Ba)/m³
DOT Classification: Poison B; Label: Poison.

SAFETY PROFILE: A deadly poison. When heated to decomposition it emits toxic fumes of CN^-.

BAL000 HR: 3
BARIUM CYANOPLATINITE
mf: $BaPt(CN)_4 \cdot 4H_2O$ mw: 508.6

PROP: (a) Monoclinic, yellow crystals; (b) rhombic crystals. Mp: loses $2H_2O$ @ 100°; d: (a) 2.076, (b) 2.085.

CONSENSUS REPORTS: Cyanide and its compounds, as well as barium and its compounds, are on the Community Right-To-Know List.

OSHA PEL: TWA 5 mg(CN)/m³
ACGIH TLV: TWA 5 mg(CN)/m³ (skin)
DFG MAK: 5 mg/m³
NIOSH REL: (Cyanide) CL 5 mg(CN)/m³/10M

SAFETY PROFILE: A poison. When heated to decomposition it emits highly toxic fumes of CN^- and NO_x.

BAL500 HR: 3
BARIUM DICHROMATE
mf: $BaCr_2O_7$ mw: 353.38

PROP: Brownish-red, crystalline masses.

SYN: BARIUM BICHROMATE

CONSENSUS REPORTS: Barium and its compounds, as well as chromium and its compounds, are on the Community Right-To-Know List.

SAFETY PROFILE: A poison. Some chromates are carcinogenic. A moderate fire hazard by chemical reaction with easily oxidized materials. A powerful oxidizer. Incompatible with reducing materials.

BAM250 CAS: 13477-09-3 HR: 3
BARIUM HYDRIDE
mf: BaH_2 mw: 139.38

PROP: Gray crystals or lumps. Mp: decomp @ 675°, bp: 1400°, d: 4.21 @ 0°.

CONSENSUS REPORTS: Barium and its compounds are on the Community Right-To-Know List.

SAFETY PROFILE: A poison. Rapidly decomposed by water and acids. In powder form, it ignites spontaneously in air and reacts vigorously with water. Coarser material ignites when heated in oxygen. It is incompatible with water; acids; and metal halogenates. A dangerous fire hazard because moisture may cause it to ignite. To fight fire, use dry chemical, graphite, CO_2.

BAM750 HR: 3
BARIUM HYPOPHOSPHITE
mf: $Ba(H_2PO_2)_2 \cdot H_2O$ mw: 285.38

PROP: Crystalline powder. Mp: decomp, d: 2.90 @ 17°.

CONSENSUS REPORTS: Barium and its compounds are on the Community Right-To-Know List.

SAFETY PROFILE: A poison. When heated to decomposition it emits highly toxic fumes of PO_x. Incompatible with $KClO_3$. When heated to decomposition it emits toxic fumes of PO_x.

BAN000 HR: 3
BARIUM IODATE
mf: $Ba(IO_3)_2$ mw: 487.20

PROP: White, crystalline powder. Mp: decomp, d: 4.998.

CONSENSUS REPORTS: Barium and its compounds are on the Community Right-To-Know List.

SAFETY PROFILE: A poison. A powerful oxidizer. Incompatible with Al; As; C; Cu; metal sulfides; organic matter. When heated to decomposition it emits toxic fumes of I^-.

BAN250 CAS: 10022-31-8 HR: 3
BARIUM(II) NITRATE (1:2)
DOT: UN 1446
mf: $N_2O_6 \cdot Ba$ mw: 261.36

PROP: Lustrous crystals. Mp: 592°, bp: decomp, d: 3.24 @ 23°.

SYNS: BARIUM DINITRATE * BARIUM NITRATE (DOT) * DUSICNAN BARNATY (CZECH) * NITRATE de BARYUM (FRENCH) * NITRIC ACID, BARIUM SALT

CONSENSUS REPORTS: Reported in EPA TSCA Inventory. Barium and its compounds are on the Community Right-To-Know List.

OSHA PEL: TWA 0.5 mg(Ba)/m³
ACGIH TLV: TWA: 0.5 mg(Ba)/m³

DOT Classification: Oxidizer; Label: Oxidizer; DOT-IMO: Oxidizer; Label: Oxidizer and Poison.

SAFETY PROFILE: A poison via ingestion, subcutaneous, parenteral, and intravenous routes. An irritant to skin and eyes. When heated to decomposition it emits very toxic fumes of NO_x. An oxidizer. Mixtures with finely divided aluminum-magnesium alloys are easily ignitable and extremely sensitive to friction or impact. Such mixtures are used in chemical photoflash applications. Incompatible with (Mg + BaO_2 + Zn), Al and Mg alloys. When heated to decomposition it emits toxic fumes of NO_x.

BAN500 CAS: 12047-79-9 HR: 3
BARIUM NITRIDE
mf: Ba_3N_2 mw: 440.10

PROP: Colorless crystals. Bp: 1000° (vac), d: 4.783 @ 25°/4°.

CONSENSUS REPORTS: Barium and its compounds are on the Community Right-To-Know List.

SAFETY PROFILE: A poison. Flammable by spontaneous chemical reaction with water to liberate explosive ammonia gas. Dangerous; explodes upon heating and by spontaneous chemical reaction, liberating NH_3 vapor which can form explosive mixtures with air. Violent reaction with air or moisture.

BAO000 CAS: 1304-28-5 HR: 3
BARIUM OXIDE
DOT: UN 1884
mf: BaO   mw: 153.34

PROP: White to yellowish-white powder. Mp: 1923°, bp: 2000° (approx), d: 5.72.

SYNS: BARIUM MONOXIDE * BARIUM PROTOXIDE * BARYTA * CALCINED BARYTA * OXYDE de BARYUM (FRENCH)

CONSENSUS REPORTS: Reported in EPA TSCA Inventory. Barium and its compounds are on the Community Right-To-Know List.

OSHA PEL: TWA 0.5 mg(Ba)/m³
ACGIH TLV: TWA 0.5 mg(Ba)/m³
DOT Classification: ORM-B; Label: None; DOT-IMO: Poison B; Label: St. Andrews Cross.

SAFETY PROFILE: A poison via subcutaneous route. Combustible by spontaneous chemical reaction; produces heat on contact with water or steam. Incompatible with H_2S, hydroxyl-amine, N_2O_4, triuranium octaoxide, and SO_3.

BAO250   CAS: 1304-29-6   **HR: 3**
BARIUM PEROXIDE
DOT: UN 1449
mf: BaO_2 mw: 169.34

PROP: Grayish-white powder. Mp: 450°, bp: -0 @ 800°, d: 4.96.

SYNS: BARIO (PEROSSIDO di) (ITALIAN) * BARIUM BINOXIDE * BARIUM DIOXIDE * BARIUMPEROXID (GERMAN) * BARIUMPEROXYDE (DUTCH) * BARIUM SUPEROXIDE * DIOXYDE de BARYUM (FRENCH) * PEROXYDE de BARYUM (FRENCH)

CONSENSUS REPORTS: Reported in EPA TSCA Inventory. Barium and its compounds are on the Community Right-To-Know List.
OSHA PEL: TWA 0.5 mg(Ba)/m³
ACGIH TLV: TWA 0.5 mg(Ba)/m³
DOT Classification: Oxidizer; Label: Oxidizer and Poison.

SAFETY PROFILE: A poison via subcutaneous route. A powerful oxidizer. Explodes on contact with acetic anhydride. Ignites when mixed with calcium-silicon alloys, powdered aluminum, powdered magnesium, water + organic compounds. Mixtures with propane react violently when heated. The powder ignites when heated to 265°C with selenium. Wood ignites with friction from the peroxide. Incompatible with H_2S, water, peroxy formic acid, hydroxylamine solution, mixture of (Mg + Zn + $Ba(NO_3)_2$), and organic matter.

BAP000   CAS: 7727-43-7   **HR: 2**
BARIUM SULFATE
mf: $O_4S \cdot Ba$ mw: 233.40

PROP: White, heavy, odorless powder. D: 4.50 @ 15°, mp: 1580°. Insol in water or dilute acids.

SYNS: ACTYBARYTE * ARTIFICIAL BARITE * ARTIFICIAL HEAVY SPAR * BAKONTAL * BARIDOL * BARITE * BARITOP * BAROSPERSE * BAROTRAST * BARYTA WHITE * BARYTES * BAYRITES * BLANC FIXE * C.I. 77120 * C.I. PIGMENT WHITE 21 * CITOBARYUM * COLONATRAST * ENAMEL WHITE * ESOPHOTRAST * EWEISS * E-Z-PAQUE * FINEMEAL * LACTOBARYT * LIQUIBARINE * MACROPAQUE * NEOBAR * ORATRAST * PERMANENT WHITE * PRECIPITATED BARIUM SULPHATE * RAYBAR * REDI-FLOW * SOLBAR

* SULFURIC ACID, BARIUM SALT (1:1) * SUPRAMIKE * TRAVAD * UNIBARYT

CONSENSUS REPORTS: Reported in EPA TSCA Inventory. Barium and its compounds are on the Community Right-To-Know List.

OSHA PEL: (Transitional: Total Dust: TWA 15 mg/m³; Respirable Fraction: 5 mg/m³) Total Dust: TWA 10 mg/m³; Respirable Fraction: 5 mg/m³
ACGIH TLV: TWA (nuisance particulate) 10 mg/m³ of total dust (when toxic impurities are not present, e.g., quartz < 1%).

SAFETY PROFILE: Questionable carcinogen with experimental tumorigenic data. A relatively insoluble salt used as an opaque medium in radiography. Soluble impurities can lead to toxic reactions. Heating with aluminum can produce an explosion. Incompatible with Al or P. When heated to decomposition it emits toxic fumes of SO_x.

BAP250   CAS: 21109-95-5   **HR: 3**
BARIUM SULFIDE
mf: BaS   mw: 169.4

PROP: Cubic, colorless crystals. D: 4.25 @ 15°, mp: 1200°.

CONSENSUS REPORTS: Barium and its compounds are on the Community Right-To-Know List.

SAFETY PROFILE: A poison. Flammable by spontaneous chemical reaction; air, moisture, or acid fumes may cause it to ignite. To fight fire, use CO_2, dry chemical. Reacts violently with phosphorous (V) oxide. Mixtures with lead dioxide, potassium chlorate, or potassium nitrite explode when heated. Incompatible with Cl_2O, $Ca(NO_3)_2$, $Sr(NO_3)_2$, $Ca(ClO_3)_2$, $Sr(ClO_3)_2$, $(ClO_3)_2$.

BAR250   CAS: 8015-73-4   **HR: 2**
BASIL OIL

PROP: Contains about 55% methyl chavicol and 35% of alcohols calculated as lenatoal and other compounds found in the leaves of *Ocimum resilium* L. a pale yellow liquid; floral, spicy odor. Sol in fixed oils, propylene glycol; insol in glycerin.

SYNS: BASIL OIL, EUROPEAN TYPE (FCC) * OCIMUM BASILICUM OIL * OIL of BASIL

CONSENSUS REPORTS: Reported in EPA TSCA Inventory.

SAFETY PROFILE: Moderately toxic by ingestion. A skin and eye irritant. When heated to decomposition it emits acrid smoke and irritating fumes.

BAT500   HR: 2
BAY OIL

PROP: Consists mainly of eugenol and chavicol (55-65%), major portion of balance consists of terpenes (alpha-pinene, myrcene, and dipentene) small quantities of citro, nerol, cineol, and other terpenoids have also been found. Yellow or brown liquid; aromatic odor, pungent, spicy taste. Sol in alc and glacial acetic acid.

SYNS: BAY LEAF OIL * BOIS d'INDE * LAUREL LEAF OIL * MYRCIA OIL * MYRICIA OIL * OIL of BAY * OIL of MYRCIA

CONSENSUS REPORTS: Reported in EPA TSCA Inventory.

SAFETY PROFILE: Moderately toxic by ingestion. When heated to decomposition it emits acrid smoke.

BAT750   CAS: 14816-18-3   HR: 3
BAYTHION
mf: $C_{12}H_{15}N_2O_3PS$ mw: 298.32

SYNS: B 77488 * BAY 5621 * BAY 77488 * BAYRE 77488 * BENZOYL CYANIDE-o-(DIETHOXY-PHOSPHINOTHIOYL)OXIME * O,O-DIAETHYL-o-(α-CYANBENZYLIDEN-AMINO)-THIONPHOSPHAT (GERMAN) * O,O-DIAETHYL-o-(α-CYANO-BENZYLIDENAMINO)-MONOTHIOPHOSPHAT (GERMAN) * α-(((DIETHOXY-PHOSPHINOTHIOYL)OXY)IMINO)BENZENEACETONITRILE * (DIETHOXY-THIOPHOSPHORYLOXYIMINO)-PHENYL ACETONITRILE * O,O-DIETHYL PHOSPHOROTHIOATE, o-ESTER with PHENYLGLYOXYLONITRILE OXIME * ENT 27,488 * 4-ETHOXY-7-PHENYL-3,5-DIOXA-6-AZA-4-PHOSPHAOCT-6-ENE-8-NITRILE 4 SULFIDE * PHENYLGLYOXYLONITRILE OXIME-O, O-DIETHYL PHOSPHOROTHIOATE * PHOXIME * PHOXIN * SEBACIL * VALEXONE * VOLATON

CONSENSUS REPORTS: Cyanide and its compounds are on the Community Right-To-Know List.

SAFETY PROFILE: Poison by ingestion. An experimental teratogen. When heated to decomposition it emits very toxic fumes of CN^-, NO_x, PO_x, and SO_x.

BAU000   CAS: 8012-89-3   HR: 1
BEESWAX

PROP: Yellow to brownish-yellow, soft to brittle wax. Mp: 62-65°, d: 0.95-0.96. Sol in chloroform, ether, fixed oils; sltly sol in alc

SYNS: BEESWAX, WHITE * BEESWAX, YELLOW

SAFETY PROFILE: A mild allergen. Combustible when heated.

BAU500   HR: 3
BELLADONNA

PROP: An extract from the deadly nightshade plant. The alkaloids atropine and belladonnine are derivatives.

SYN: DEADLY NIGHTSHADE

SAFETY PROFILE: A deadly poison. Local contact may cause a contact dermatitis. A poisonous constituent of some berries and plants, and of some folk remedies.

BAU750   CAS: 147-24-0   HR: 3
BENADRYL HYDROCHLORIDE
mf: $C_{17}H_{21}NO \cdot ClH$ mw: 291.85

SYNS: AMBENYL * BAX * BENA * BENADRYL * BENDYLATE * BENOCTEN * BENZEHIST * BENZHYDRAMINE HYDROCHLORIDE * 2-(BENZHYDRYLOXY)-N,N-DIMETHYLETHYLAMINE-HYDROCHLORIDE * DABYLEN * DIFENHYDRAMINE HYDROCHLORIDE * DIMETHYLAMINE BENZHYDRYL ESTER HYDROCHLORIDE * β-DIMETHYLAMINOETHYL BENZHYDRYL ETHER HYDROCHLORIDE * DIPHENYLHYDRAMINE HYDROCHLORIDE * 2-(DIPHENYLMETHOXY)-N,N-DIMETHYLETHANAMINE HYDROCHLORIDE * 2-DIPHENYLMETHOXY-N,N-DIMETHYLETHYLAMINE HYDROCHLORIDE * DOLESTAN * ELDADRYL * FELBEN * FENYLHIST * HALBMOND * α-HYDROXYDIPHENYLMETHANE-β-DIMETHYLAMINOETHYL ETHER HYDROCHLORIDE * NCI-C56075 * ROHYDRA * SK-DIPHENHYDRAMINE * VALDRENE * WEHYDRYL

SAFETY PROFILE: Poison by ingestion, intramuscular, subcutaneous, intravenous, and intraperitoneal routes. Experimental teratogenic and reproductive effects. When heated to decomposition it emits very toxic fumes of NO_x and HCl.

BAV575   CAS: 17804-35-2   **HR: 3**
BENOMYL
mf: $C_{14}H_{18}N_4O_3$ mw: 290.36

SYNS: ARILATE * BBC * BENLATE 50
* BENOMYL 50W * BNM * 1-(BUTYLCARBA-
MOYL)-2-BENZIMIDAZOLECARBAMIC ACID, METHYL
ESTER * 1-(BUTYLCARBAMOYL)-2-BENZIMIDAZOL-
METHYLCARBAMAT (GERMAN) * 1-(N-BUTYLCAR-
BAMOYL)-2-(METHOXY-CARBOXAMIDO)-BENZIMIDAZOL
(GERMAN) * DU PONT 1991 * FUNDASOL
* FUNGICIDE 1991 * MBC * METHYL-1-(BUTYL-
CARBAMOYL)-2-BENZIMIDAZOLYLCARBAMATE
* TERSAN 1991

CONSENSUS REPORTS: Reported in EPA
TSCA Inventory. EPA Genetic Toxicology Program.

OSHA PEL: (Transitional: Total Dust: TWA
15 mg/m³; Respirable Fraction: 5 mg/m³) Total Dust: TWA 10 mg/m³; Respirable Fraction: 5 mg/m³
ACGIH TLV: TWA 10 mg/m³

SAFETY PROFILE: Poison by ingestion.
Mildly toxic by inhalation. Experimental teratogenic and reproductive effects. Human mutation
data reported. A human skin irritant. When
heated to decomposition it emits toxic fumes
of NO_x.

BAV750   CAS: 1302-78-9   **HR: 1**
BENTONITE

PROP: A clay containing appreciable amounts
of the clay mineral montmorillonite; light yellow
or green, cream, pink, gray to black solid. Insol
in water and common organic solvents.

SYNS: ALBAGEL PREMIUM USP 4444 * BENTONITE
2073 * BENTONITE MAGMA * HI-JEL * IMVITE
I.G.B.A. * MAGBOND * MONTMORILLONITE
* PANTHER CREEK BENTONITE * SOUTHERN BEN-
TONITE * TIXOTON * VOLCLAY * VOLCLAY
BENTONITE BC * WILKINITE

CONSENSUS REPORTS: Reported in EPA
TSCA Inventory.

SAFETY PROFILE: Poison by intravenous
route causing blood clotting. Questionable carcinogen with experimental tumorigenic data.

BAW250   CAS: 205-99-2   **HR: 3**
BENZ(e)ACEPHENANTHRYLENE
mf: $C_{20}H_{12}$ mw: 252.32

PROP: Mp: 168°.

SYNS: 3,4-BENZ(e)ACEPHENANTHRYLENE * 2,3-
BENZFLUORANTHENE * 3,4-BENZFLUORANTHENE
* BENZO(b)FLUORANTHENE * BENZO(e)FLUORAN-
THENE * 2,3-BENZOFLUORANTHENE * 3,4-BENZO-
FLUORANTHENE * 2,3-BENZOFLUORANTHRENE
* B(b)F

CONSENSUS REPORTS: IARC Cancer Re-
view: GROUP 2B IMEMDT 7,56,87, Animal
Sufficient Evidence IMEMDT 32,147,83;
IMEMDT 3,69,73. NTP Fourth Annual Report
On Carcinogens, 1984. EPA Genetic Toxicol-
ogy Program.

SAFETY PROFILE: Confirmed carcinogen with
experimental carcinogenic and tumorigenic
data. Mutation data reported. When heated to
decomposition it emits acrid smoke and irritating
fumes.

BAW500   CAS: 71-79-4   **HR: 3**
BENZACINE HYDROCHLORIDE
mf: $C_{18}H_{21}O_3ClH$ mw: 403.28

SYNS: BENZACIN * BENZACINE * BENZACIN
HYDROCHLORIDE * DIMETHYLAMINOETHYL BENZI-
LATE, HYDROCHLORIDE * β-DIMETHYLAMINO-
ETHYL BENZILATE HYDROCHLORIDE * 2-(DI-
METHYLAMINO)ETHYL BENZILATE HYDROCHLO-
RIDE * DIMETHYLAMINOETHYL BENZYLATE
HYDROCHLORIDE * DIMETHYLAMINOETHYL DI-
PHENYLHYDROXYACETATE HYDROCHLORIDE
* HK-141

SAFETY PROFILE: Poison by ingestion, intravenous, and intraperitoneal routes. When
heated to decomposition it emits toxic fumes
of HCl.

BAW750   CAS: 225-51-4   **HR: 3**
BENZ(c)ACRIDINE
mf: $C_{17}H_{11}N$ mw: 229.29
PROP: Mp: 108°.

SYNS: 12-AZABENZ(a)ANTHRACENE * B(c)AC
* 3,4-BENZACRIDINE * 7,8-BENZACRIDINE
(FRENCH) * 3,4-BENZOACRIDINE * α-CHRYSIDINE
* α-NAPHTHACRIDINE * RCRA WASTE NUMBER
U016

CONSENSUS REPORTS: IARC Cancer Review: GROUP 3 IMEMDT 7,56,87; Animal
Sufficient Evidence IMEMDT 3,241,73; Animal Limited Evidence IMEMDT 32,129,83

SAFETY PROFILE: Questionable carcinogen
with experimental neoplastigenic and tumori-

genic data. Mutation data reported. When heated to decomposition it emits toxic fumes of NO_x.

BAY300   CAS: 98-87-3   **HR: 3**
BENZAL CHLORIDE

DOT: UN 1886
mf: $C_7H_6Cl_2$ mw: 161.03

PROP: Very refractive liquid. Mp: $-16°$, bp: 214°, d: 1.29.

SYNS: BENZYL DICHLORIDE * BENZYLENE CHLO-
RIDE * BENZYLIDENE CHLORIDE (DOT) * CHLO-
RURE de BENZYLIDENE (FRENCH) * α,α-DICHLORO-
TOLUENE * RCRA WASTE NUMBER U017

CONSENSUS REPORTS: IARC Cancer Review: Human Inadequate Evidence IMEMDT 29,65,82; Animal Limited Evidence IMEMDT 29,65,82. Reported in EPA TSCA Inventory. EPA Genetic Toxicology Program. EPA Extremely Hazardous Substances List. Community Right-To-Know List.

DFG MAK: Suspected Carcinogen.
DOT Classification: Poison B; Label: Poison.

SAFETY PROFILE: Suspected carcinogen with experimental carcinogenic and neoplastigenic data. Poison by inhalation. Moderately toxic by ingestion. A suspected human carcinogenic. A strong irritant and lachrymator. Causes central nervous system depression. Mutation data reported. When heated to decomposition it emits toxic fumes of Cl^-.

BAY500   CAS: 100-52-7   **HR: 3**
BENZALDEHYDE

DOT: UN 1989
mf: C_7H_6O mw: 106.13

PROP: Colorless liquid; burning taste with bitter almond odor. Mp: $-26°$, bp: 179°, flash p: 148°F, d: 1.041, autoign temp: 377°F, vap press: 1 mm @ 26.2°, vap d: 3.65, refr index: 1.544. Sltly sol in water; misc in alc, ether, oils.

SYNS: ALMOND ARTIFICIAL ESSENTIAL OIL
* ARTIFICIAL ALMOND OIL * BENZENECARBAL-
DEHYDE * BENZENECARBONAL * BENZOIC AL-
DEHYDE * FEMA No. 2127 * NCI-C56133

CONSENSUS REPORTS: EPA Genetic Toxicology Program. Reported in EPA TSCA Inventory.

DOT Classification: Combustible Liquid; Label: None.

SAFETY PROFILE: Poison by ingestion and intraperitoneal routes. Moderately toxic by subcutaneous route. An allergen. Acts as a feeble local anesthetic. Local contact may cause contact dermatitis. Causes central nervous system depression in small doses and convulsions in larger doses. A skin irritant. Mutation data reported. Combustible liquid. To fight fire, use water (may be used as a blanket), alcohol, foam, dry chemical. A strong reducing agent. Reacts violently with peroxyformic acid and other oxidizers.

BAY750   CAS: 633-03-4   **HR: 3**
BENZALDEHYDE GREEN
mf: $C_{27}H_{33}N_2 \cdot HO_4S$ mw: 482.69

PROP: Bright green crystals.

SYNS: ADC BRILLIANT GREEN CRYSTALS * AIZEN
DIAMOND GREEN GH * ANILINE GREEN * ASTRA
DIAMOND GREEN GX * AVON GREEN A-4379
* BASIC BRIGHT GREEN * BRILLIANT GREEN SUL-
FATE * CALCOZINE BRILLIANT GREEN G * C.I.
42040 * C.I. BASIC GREEN 1, SULFATE (1:1)
* DEORLENE GREEN JJO * DIAMOND GREEN G
* EMERALD GREEN * ETHYL GREEN * FAST
GREEN JJO * HIDACO BRILLIANT GREEN * MALA-
CHITE GREEN G * MITSUI BRILLIANT GREEN G
* TERTROPHENE BRILLIANT GREEN G * TOKYO
ANILINE BRILLIANT GREEN

CONSENSUS REPORTS: Reported in EPA TSCA Inventory.

SAFETY PROFILE: Poison by intraperitoneal and intravenous routes. A mild human skin irritant. When heated to decomposition it emits very toxic fumes of NO_x, NH_3 and SO_x.

BBA000   CAS: 1708-39-0   **HR: 2**
BENZAL GLYCERYL ACETAL
mf: $C_{10}H_{12}O_3$ mw: 180.22

PROP: Colorless to pale yellow liquid; mild almond odor. D: 1.183-1.193, refr index: 1.535-1.541, flash p: 165°F.

SYNS: BENZALDEHYDE GLYCERYL ACETAL (FCC)
* BENZYLIDENE GLYCEROL * BUTYL PHENYL AC-
ETATE * FEMA No. 2209 * 2-PHENYL-m-DIOXAN-5-
OL

SAFETY PROFILE: Moderately toxic by ingestion and intraperitoneal routes. Mildly toxic by skin contact. Combustible liquid. When heated to decomposition it emits acrid smoke and irritating fumes.

BBC250   CAS: 56-55-3   **HR: 3**
BENZ(a)ANTHRACENE
mf: $C_{18}H_{12}$ mw: 228.30

PROP: Colorless leaflets or plates. Bp: 400°, mp: 160°.

SYNS: BA * BENZANTHRACENE * 1,2-BENZAN-THRACENE * 1,2-BENZ(a)ANTHRACENE * 1,2-BEN-ZANTHRAZEN (GERMAN) * BENZANTHRENE * 1,2-BENZANTHRENE * BENZOANTHRACENE * 1,2-BENZOANTHRACENE * BENZO(a)ANTHRA-CENE * BENZO(a)PHENANTHRENE * BENZO(b)PHENANTHRENE * 2,3-BENZOPHENANTHRENE * 2,3-BENZPHENANTHRENE * NAPHTHANTHRA-CENE * RCRA WASTE NUMBER U018 * TETRA-PHENE

CONSENSUS REPORTS: IARC Cancer Review: GROUP 2A IMEMDT 7,56,87, Animal Sufficient Evidence IMEMDT 32,135,83; IMEMDT 3,45,73. NTP Fourth Annual Report On Carcinogens, 1984. EPA Genetic Toxicology Program. Reported in EPA TSCA Inventory.

SAFETY PROFILE: Confirmed carcinogen with experimental carcinogenic, neoplastigenic, tumorigenic data by skin contact and other routes. Poison by intravenous route. Human mutation data reported. It is found in oils, waxes, smoke, food, drugs. When heated to decomposition it emits acrid smoke and irritating fumes.

BBD250   CAS: 60967-88-6   **HR: 3**
BENZ(a)ANTHRACENE-1,2-DIHYDRODIOL
mf: $C_{18}H_{14}O_2$ mw: 262.32

SYNS: BA-1,2-DIHYDRODIOL * trans-1,2-DIHY-DROXY-1,2-DIHYDROBENZ(a)ANTHRACENE

CONSENSUS REPORTS: EPA Genetic Toxicology Program.

SAFETY PROFILE: Questionable carcinogen with experimental tumorigenic data by skin contact. Mutation data reported. When heated to decomposition it emits acrid smoke and irritating fumes.

BBD500   CAS: 60967-89-7   **HR: 3**
BENZ(a)ANTHRACENE-3,4-DIHYDRODIOL
mf: $C_{18}H_{14}O_2$ mw: 262.32

SYNS: BA-3,4-DIHYDRODIOL * trans-3,4-DIHYDRO-3,4-DIHYDROXYBENZO(a)ANTHRACENE * trans-3,4-DIHYDROXY-3,4-DIHYDROBENZ(a)ANTHRACENE

CONSENSUS REPORTS: EPA Genetic Toxicology Program.

SAFETY PROFILE: Questionable carcinogen with experimental neoplastigenic data by skin contact. Mutation data reported. When heated to decomposition it emits acrid smoke and irritating fumes.

BBE250   CAS: 3719-37-7   **HR: 3**
BENZ(a)ANTHRACENE-5,6-DIHYDRODIOL
mf: $C_{18}H_{14}O_2$ mw: 262.32

SYNS: BA-5,6-DIHYDRODIOL * BA-5,6-trans-DIHY-DRODIOL * BENZ(a)ANTHRACENE-5,6-trans-DIHYDRO-DIOL * trans-5,6-DIHYDROXY-5,6-DIHYDROBENZ(a)AN-THRACENE

CONSENSUS REPORTS: EPA Genetic Toxicology Program.

SAFETY PROFILE: Questionable carcinogen with experimental tumorigenic and neoplastigenic data by skin contact. Mutation data reported. When heated to decomposition it emits acrid smoke and irritating fumes.

BBE750   CAS: 34501-24-1   **HR: 3**
trans-BENZ(a)ANTHRACENE-8,9-DIHYDRODIOL
mf: $C_{18}H_{14}O_2$ mw: 262.32

SYNS: BA-8,9-DIHYDRODIOL * trans-8,9-DIHY-DROXY-8,9-DIHYDROBENZ(a)ANTHRACENE

CONSENSUS REPORTS: EPA Genetic Toxicology Program.

SAFETY PROFILE: Questionable carcinogen with experimental tumorigenic and neoplastigenic data by skin contact. Mutation data reported. When heated to decomposition it emits acrid smoke and irritating fumes.

BBF000   CAS: 60967-90-0   **HR: 3**
BENZ(a)ANTHRACENE-10,11-DIHYDRODIOL
mf: $C_{18}H_{14}O_2$ mw: 262.32

SYNS: BA-10,11-DIHYDRODIOL * trans-10,11-DIHY-DROXY-10,11-DIHYDROBENZ(a)ANTHRACENE

CONSENSUS REPORTS: EPA Genetic Toxicology Program.

SAFETY PROFILE: Questionable carcinogen with experimental tumorigenic data by skin contact. When heated to decomposition it emits acrid smoke and irritating fumes.

BBH250   CAS: 16110-13-7   **HR: 3**
BENZ(a)ANTHRACENE-7-METHANOL
mf: $C_{19}H_{14}O$ mw: 258.33

SYNS: 7-HMBA * 7-HYDROXYMETHYLBENZ(a)AN-THRACENE * 10-HYDROXYMETHYL-1,2-BENZAN-THRACENE

CONSENSUS REPORTS: EPA Genetic Toxicology Program.

SAFETY PROFILE: Questionable carcinogen with experimental tumorigenic data. Mutation data reported. When heated to decomposition it emits acrid smoke and irritating fumes.

BBJ750   CAS: 59-97-2   **HR: 3**
BENZAZOLINE HYDROCHLORIDE
mf: $C_{10}H_{12}N_2 \cdot ClH$ mw: 196.70

SYNS: ARTERODY * BENZYLIMIDAZOLINE HY-DROCHLORIDE * 2-BENZYL-2-IMIDAZOLINE MONO-HYDROCHLORIDE * IMIDALINE HYDROCHLORIDE * PRISCOL * PRISCOLINE HYDROCHLORIDE * TOLAVAD * TOLAZOLINE CHLORIDE * TOLA-ZOLINE HYDROCHLORIDE * TOLPAL

CONSENSUS REPORTS: Reported in EPA TSCA Inventory.

SAFETY PROFILE: Poison by ingestion, intravenous, and intraperitoneal routes. Human systemic effects by intravenous route: change in heart rate, unspecified vascular effects, and sweating. When heated to decomposition it emits very toxic fumes of NO_x and HCl.

BBK000   CAS: 300-62-9   **HR: 3**
BENZEDRINE
mf: $C_9H_{13}N$ mw: 135.23

PROP: Liquid. Bp: 200°, flash p: < 212°F (OC), d: 0.931, vap d: 4.65.

SYNS: ACTEDRON * ADIPAN * ALLODENE * dl-AMPHETAMINE * ANOREXIDE * (±)-BEN-ZEDRINE * dl-BENZEDRINE * DEOXY-NOREPHEDRINE * (±)-DESOXYNOREPHEDRINE * racemic-DESOXYNOR-EPHEDRINE * ELASTONON * ISOAMYCIN * ISOMYN * MECODRIN * α-METHYLBENZENEETHANEAMINE * dl-α-METHYLPHENETHYLAMINE * (±)-α-METHYL-PHENETHYLAMINE * NOREPHEDRANE * NOVYDRINE * ORTEDRINE * PHENEDRINE

* dl-1-PHENYL-2-AMINOPROPANE * PROFAMINA * PROPISAMINE * PSYCHEDRINE * RAPHET-AMINE * SIMPATEDRIN * SYMPAMINE * SYMPATEDRINE * WECKAMINE

CONSENSUS REPORTS: Reported in EPA TSCA Inventory. EPA Extremely Hazardous Substances List.

SAFETY PROFILE: A deadly human poison by an unspecified route. An experimental poison by ingestion, subcutaneous, intraperitoneal, intravenous, and possibly other routes. Experimental reproductive effects. Mutation data reported. A central nervous system stimulant. Overdoses cause hyperactivity, restlessness, insomnia, rapid pulse, rise in blood pressure, dilated pupils, dryness of the throat. Combustible when exposed to heat, flame or oxidizers. When heated to decomposition it emits toxic fumes of NO_x. To fight fire, use CO_2, dry chemical, alcohol foam, water mist, fog.

BBK250   CAS: 156-31-0   **HR: 3**
BENZEDRINE SULFATE
mf: $C_{18}H_{26}N_2 \cdot H_2O_4S$ mw: 368.54

SYNS: AMITRENE * AMPHOIDS S * AMPHORDS S * BAR-TIME * DIAMPHETAMINE SULFATE * KLINE * dl-α-METHYLPHENETHYLAMINE SUL-FATE * PHENETHYLAMINE, α-METHYL-, SULFATE (2:1) * 1-PHENYL-2-AMINOPROPANE SULFATE

SAFETY PROFILE: A poison via ingestion, parenteral, intramuscular, intraperitoneal, and subcutaneous routes. When heated to decomposition it emits very toxic fumes of SO_x and NO_x.

BBK500   CAS: 51-63-8   **HR: 3**
d-BENZEDRINE SULFATE
mf: $C_{18}H_{26}N_2 \cdot H_2O_4S$ mw: 368.54

SYNS: ACEDRON * ADJUDETS * ADRIXINE * AFATIN * ALBEMAP * ALGO-DEX * AMDEX * d-AMFETASUL * AMITRENE * AMPHAETEX * AMPHEDRINE * AMPHEREX * (+)-AMPHET-AMINE SULFATE * d-AMPHETAMINE SULFATE * AMPHETASUL * AMPHEX * AMPTREREX * AMSUSTAIN * APETAIN * ARDEX * BETA-FEDRINA * BETAFEDRINE * d-BETAPHEDRINE * CARRTIME * CRADEX * DADEX * DADOX d-CITRAMINE * DAMS * DAS * DELLIPSOIDS * DEPHADREN * DESOXYN * DEXAIME * DEXALINE * DEXALME * DEXALONE * DEXAMED * DEXAMINE * DEXAMPHAMINE

* DEXAMPHETAMINE * DEXAMPHETAMINE SUL-
FATE * DEXAMYL * DEXEDRINA * DEXE-
DRINE SULFATE * DEXIES * DEXTROAMPHET-
AMINE SULFATE * DEXTRO-α-METHYLPHENETHYL-
AMINE SULFATE * DEXTRO-1-PHENYL-2-AMINOPRO-
PANE SULFATE * DEXTRO-β-PHENYLISOPROPYL-
AMINE SULFATE * OBESEDRIN * FASTBALLS
* HEARTS * (S)-α-METHYL-BENZENEETHANAMINE
SULFATE (2:1) * d-α-METHYLPHENETHYLAMINE
SULFATE * OBESONIL * ORANGES * PELLCAFS
* PELLCAP * PELLCAPS * PERKE * PHENO-
PROMIN * d-1-PHENYL-2-AMINOPROPANE SULFATE
* d-β-PHENYLISOPROPYLAMINE SULFATE
* PHETADEX * POMADEX * PRO-DEXTER
* PSYCHODRINE * RECORDATI * REVIDEX
* SIMPAMINA-D * SYMPAMIN * SYMPAMINA-D
* TEMPODEX * TUPHETAMINE * TYDEX
* ZAMINE

SAFETY PROFILE: Poison by ingestion, intraperitoneal, subcutaneous, and intravenous
routes. A human teratogen which causes developmental abnormalities of the central nervous
system. Experimental reproductive effects including other teratogenic effects. A habit-forming stimulant. When heated to decomposition
it emits very toxic fumes of SO_x and NO_x.

BBK750   CAS: 51-62-7   **HR: 3**
I-BENZEDRINE SULFATE
mf: $C_{18}H_{26}N_2 \cdot H_2O_4S$ mw: 368.54

SYNS: (−)-AMPHETAMINE SULFATE * 1-AMPHET-
AMINE SULFATE * LEVEDRINE * 1-1-PHENYL-2-
AMINOPROPANE SULFATE

SAFETY PROFILE: A poison via subcutaneous
and intraperitoneal routes. When heated to decomposition it emits very toxic fumes of SO_x
and NO_x.

BBL000   CAS: 142-04-1   **HR: 3**
BENZENAMINE HYDROCHLORIDE
DOT: UN 1548
mf: $C_6H_7N \cdot ClH$ mw: 129.60

PROP: Crystals. Vap d: 4.46, d: 1.22, mp:
198°, bp: 245°, flash p: 380°F (OC).

SYNS: ANILINE CHLORIDE * ANILINE HYDRO-
CHLORIDE (DOT) * "ANILINE SALT" * ANILINIUM
CHLORIDE * CHLORHYDRATE d'ANILINE (FRENCH)
* CHLORID ANILINU (CZECH) * NCI-C03736
* PHENYLAMINE HYDROCHLORIDE * SUL ANILI-
NOVA (CZECH) * USAF EK-442

CONSENSUS REPORTS: IARC Cancer Review: Animal Limited Evidence IMEMDT
27,39,82. NCI Carcinogenesis Bioassay Completed; Results Positive: rat NCITR* NCI-CG-
TR-130,78; Results Negative: Mouse NCITR*
NCI-CG-TR-130,78. Reported in EPA TSCA
Inventory. EPA Genetic Toxicology Program.

DOT Classification: DOT-IMO: Poison B; Label: St. Andrews Cross.

SAFETY PROFILE: Suspected carcinogen with
experimental carcinogenic and tumorigenic
data. Poison by intraperitoneal route. Moderately toxic by ingestion. Experimental teratogenic effects. Human mutation data reported.
A skin and eye irritant. Combustible when exposed to heat or flame. When heated to decomposition or on contact with acid or acid fumes,
it emits highly toxic fumes of aniline and chlorine compounds. Reacts explosively with aniline
at 240°C/7.6 bar. Can react vigorously with
oxidizing materials. To fight fire, use water,
CO_2, water mist or spray, dry chemical.

BBL250   CAS: 71-43-2   **HR: 3**
BENZENE
DOT: UN 1114
mf: C_6H_6 mw: 78.12

PROP: Clear, colorless liquid. Mp: 5.51°, bp:
80.093°-80.094°, flash p: 12°F (CC), d: 0.8794
@ 20°, autoign temp: 1044°F, lel: 1.4%, uel:
8.0%, vap press: 100 mm @ 26.1°, vap d:
2.77, ULC: 95-100.

SYNS: (6)ANNULENE * BENZEEN (DUTCH)
* BENZEN (POLISH) * BENZIN (OBS.) * BENZINE
(OBS.) * BENZOL (DOT) * BENZOLE * BENZO-
LENE * BENZOLO (ITALIAN) * BICARBURET of
HYDROGEN * CARBON OIL * COAL NAPHTHA
* CYCLOHEXATRIENE * FENZEN (CZECH)
* MINERAL NAPHTHA * MOTOR BENZOL
* NCI-C55276 * NITRATION BENZENE * PHENE
* PHENYL HYDRIDE * PYROBENZOL * PYRO-
BENZOLE * RCRA WASTE NUMBER U019

CONSENSUS REPORTS: IARC Cancer Re-
view: GROUP 1 IMEMDT 7,120,87, Human
Limited Evidence IMEMDT 7,203,74; Animal
Inadequate Evidence IMEMDT 7,203,74; IARC
Cancer Review: Animal Limited Evidence
IMEMDT 29,93,82; Human Sufficient Evi-
dence IMEMDT 29,93,82. NTP Fourth Annual
Report On Carcinogens, 1984. NTP Carcino-
genesis Studies (gavage); Clear Evidence:

mouse, rat NTPTR* NTP-TR-289,86. EPA Genetic Toxicology Program. Reported in EPA TSCA Inventory. On Community Right-To-Know List.rm

OSHA PEL: (Transitional: TWA 10 ppm; CL 25 ppm; Pk 50 ppm/10M) TWA 1 ppm; STEL 5 ppm; Pk 5 ppm/15M/8H; Cancer Hazard

ACGIH TLV: TWA 10 ppm; Suspected Human Carcinogen; BEI: 50 mg(total phenol)/L in urine at end of shift recommended as a mean value.

DFG TRK: 5 ppm (16 mg/m^3) Human Carcinogen.

NIOSH REL: TWA 0.32 mg/m^3; CL 3.2 mg/m^3/15M

DOT Classification: Flammable Liquid; Label: Flammable Liquid.

SAFETY PROFILE: Confirmed human carcinogen producing myeloid leukemia, Hodgkin's disease, and lymphomas by inhalation. Experimental carcinogenic, neoplastigenic, and tumorigenic data. A human poison by inhalation. An experimental poison by skin contact, intraperitoneal, intravenous, and possibly other routes. Moderately toxic by ingestion and subcutaneous routes. A severe eye and moderate skin irritant. Human systemic effects by inhalation and ingestion: euphoria, somnolence, changes in REM sleep, changes in motor activity, nausea or vomiting, reduced number of blood platelets, other unspecified blood effects, dermatitis, and fever. Experimental teratogenic and reproductive effects. Human mutation data reported. A narcotic. In industry, inhalation is the primary route of chronic benzene poisoning. Poisoning by skin contact has been reported. Recent (1987) research indicates that effects are seen at less than 1 ppm. Exposures needed to be reduced to 0.1 ppm before no toxic effects were observed. Elimination is chiefly through the lungs. A common air contaminant.

A dangerous fire hazard when exposed to heat or flame. Explodes on contact with diborane, bromine pentafluoride, permanganic acid, peroxomonosulfuric acid, and peroxodisulfuric acid. Forms sensitive, explosive mixtures with iodine pentafluoride, silver perchlorate, nitryl perchlorate, nitric acid, liquid oxygen, ozone, arsenic pentafluoride + potassium methoxide (explodes above 30°C). Ignites on contact with sodium peroxide + water, dioxygenyl tetrafluoroborate, iodine heptafluoride, and dioxygen difluoride. Vigorous or incandescent reaction with hydrogen + Raney nickel (above 210°C), uranium hexafluoride, and bromine trifluoride. Can react vigorously with oxidizing materials, such as Cl_2, CrO_3, O_2, $NClO_4$, O_3, perchlorates, $(AlCl_3 + FClO_4)$, $(H_2SO_4 + \text{permanganates})$, K_2O_2, $(AgClO_4 + \text{acetic acid})$, Na_2O_2. Moderate explosion hazard when exposed to heat or flame. Use with adequate ventilation. To fight fire, use foam, CO_2, dry chemical.

Poisoning occurs most commonly via inhalation of the vapor, although benzene can penetrate the skin and cause poisoning. Locally, benzene has a comparatively strong irritating effect, producing erythema and burning, and, in more severe cases, edema and even blistering. Exposure to high concentrations of the vapor (3000 ppm or higher) may result from failure of equipment or spillage. Such exposure, while rare in industry, may cause acute poisoning, characterized by the narcotic action of benzene on the central nervous system. The anesthetic action of benzene is similar to that of other anesthetic gases, consisting of a preliminary stage of excitation followed by depression and, if exposure is continued, death through respiratory failure. The chronic, rather than the acute form, of benzene poisoning is important in industry. It is a recognized leukemogen. There is no specific blood picture occurring in cases of chronic benzol poisoning. The bone marrow may be hypoplastic, normal, or hyperplastic, the changes reflected in the peripheral blood. Anemia, leucopenia, macrocytosis, reticulocytosis, thrombocytopenia, high color index, and prolonged bleeding time may be present. Cases of myeloid leukemia have been reported. For the worker, repeated blood examinations are necessary, including hemoglobin determinations, white and red cell counts, and differential smears. Where a worker shows a progressive drop in either red or white cells, or where the white count remains low, 5,000/mm$_3$ or the red count <4.0 million/mm$_3$, on two successive monthly examinations, the worker should be immediately removed from benzene exposure. Elimination is chiefly through the lungs, when fresh air is breathed. The portion that is absorbed is oxidized, and the oxidation products are combined with sulfuric and glycuronic acids and eliminated in the urine. This may be used as a diagnostic sign. Benzene has a definite cumulative action, and exposure to a relatively high concentration is not serious from the point of view of causing damage to the blood-forming system, provided the exposure is not repeated. In acute poisoning, the worker becomes confused and dizzy, complains of tightening of the leg muscles and of pressure over the forehead, then passes into a stage of excitement. If allowed to remain exposed, he quickly becomes stupefied and lapses into coma. In non-fatal cases, recovery is usually complete with no permanent disability. In chronic poisoning the onset is slow, with the symptoms vague: fatigue, headache, dizziness, nausea and loss of appetite, loss of weight and weakness are common complaints in early cases. Later, pallor, nosebleeds, bleeding gums, menorrhagia, petechiae and purpura may develop. There is great individual variation in the signs and symptoms of chronic benzene poisoning.

BBL500   CAS: 122-78-1   HR: 2
BENZENEACETALDEHYDE
mf: C_8H_8O mw: 120.16

PROP: Oily, colorless liquid which polymerizes and grows more viscous on standing; odor similar to lilac and hyacinth. Has been crystallized, mp: 33-34°, d:(25/25) 1.023-1.030, refr index: 1.525-1.545, bp: (10) 78°, n (20/D) 1.524-1.528, flash p: 154°F. Sltly sol in water; sol in alc, ether, propylene glycol. One part is sol in two parts of 80% alc forming a clear solution.

SYNS: FEMA No. 2874 * HYACINTHIN * PAA * PHENYLACETALDEHYDE (FCC) * PHENYLACETIC ALDEHYDE * PHENYLETHANAL * α-TOLUALDE-HYDE * α-TOLUIC ALDEHYDE

CONSENSUS REPORTS: Reported in EPA TSCA Inventory.

SAFETY PROFILE: Moderately toxic by ingestion. Human skin irritant. Combustible liquid. When heated to decomposition it emits acrid smoke and irritating fumes.

BBP000   CAS: 123-61-5   HR: 3
BENZENE-1,3-DIISOCYANATE
mf: $C_8H_4N_2O_2$ mw: 160.14

SYNS: 1,3-DIISOCYANATOBENZENE * NACCONATE 400 * m-PHENYLENE DIISOCYANATE * m-PHENYLENE ISOCYANATE

CONSENSUS REPORTS: Reported in EPA TSCA Inventory. Cyanide and its compounds are on the Community Right-To-Know List.

NIOSH REL: TWA (Diisocyanates) 0.005 ppm; CL 0.02 ppm/10M

SAFETY PROFILE: Deadly poison by intravenous route. When heated to decomposition it emits toxic fumes of NO_x and CN^-.

BBP250   CAS: 623-26-7   HR: 2
p-BENZENEDINITRILE
mf: $C_8H_4N_2$ mw: 128.14

PROP: Crystals, vap d: 4.42.

SYNS: 4-CYANOBENZONITRILE * p-DICYANOBEN-ZENE * 1,4-DICYANOBENZENE * NITRIL KYSE-LINY TEREFTALOVE (CZECH) * p-PDN * p-PHTHALODINITRILE * TEREFTALODINITRIL (CZECH) * TEREPHTHALONITRILE

CONSENSUS REPORTS: Reported in EPA TSCA Inventory. Cyanide and its compounds are on the Community Right-To-Know List.

SAFETY PROFILE: Moderately toxic by intraperitoneal route. Slightly toxic by ingestion. A skin and eye irritant. When heated to decomposition it emits toxic fumes of CN^- and NO_x.

BBP750   CAS: 608-73-1   HR: 3
BENZENE HEXACHLORIDE
mf: $C_6H_6Cl_6$ mw: 290.82

PROP: Technical grade contains 68.7% α-BHC, 6.5% β-BHC, and 13.5% γ-BHC. White, crystalline powder. Mp: 113°, vap press: 0.0317 mm @ 20°.

SYNS: BHC (USDA) * COMPOUND-666 * DBH * ENT 8,601 * GAMMEXANE * HCCH * HEXA * HEXACHLOR * HEXACHLORAN * HEXACHLO-ROCYCLOHEXANE * 1,2,3,4,5,6-HEXACHLOROCYCLO-HEXANE * HEXYLAN * LINDANE

CONSENSUS REPORTS: IARC Cancer Review: Animal Sufficient Evidence IMEMDT 5,47,74

SAFETY PROFILE: Confirmed carcinogen with experimental carcinogenic, neoplastigenic, and tumorigenic data by ingestion and skin contact. Poison by ingestion and subcutaneous routes. Moderately toxic by skin contact. Human systemic effects by inhalation: headache, nausea or vomiting, and fever. Implicated in aplastic anemia. Experimental reproductive effects. Mutation data reported. Lindane is more toxic than DDT or dieldrin. When heated to decomposition it emits highly toxic fumes of phosgene, HCl and Cl⁻. Potentially violent reaction with dimethylformamide + iron. When heated to decomposition it emits highly toxic fumes of phosgene, HCl and Cl⁻.

A toxic organochlorine which is persistent in the environment and accumulates in mammalian tissue. For cattle, the oral LD50 <= 100 mg/kg. The various isomers have different actions; the γ (lindane) and α isomers are central nervous system stimulants, the principal symptom being convulsions. The β and Δ isomers are central nervous system depressants. The use of thermal vaporizers with lindane has caused acute poisoning by inhalation.

The dangerous acute dose of the technical mixture has been estimated at about 30 grams and the dangerous dose of lindane at about 7 to 15 grams. However, as already mentioned, a single dose of 45 mg (or approximately 0.65 mg/kg) of lindane caused convulsions. Lindane shows a marked difference in toxicity to different species. Its toxic effect on laboratory animals compares favorably with that of DDT, but for several domestic animals, notably calves, lindane is more toxic than DDT or dieldrin. On a chronic systemic basis the α, β and γ isomers are experimental carcinogens. Has been implicated in aplastic anemia.

Dermatitis and perhaps other manifestations based on sensitivity represent a sort of chronic, though probably not systemic intoxication, which has been observed in humans.

The signs and symptoms of confirmed acute poisoning in humans have paralleled those of experimental animals. These signs and symptoms are: excitation, hyperirritability, loss of equilibrium, clonic-tonic convulsions, and later depression.

There is some evidence that the pulmonary edema and vascular collapse may be of neurogenic origin also. The symptoms in animals systemically poisoned by the γ isomer alone are essentially similar to those caused by mixtures, although the onset may be earlier. Workers acutely exposed to high air concentrations of lindane and its decomposition products show headache, nausea, and irritation of eyes, nose, and throat.

In rare instances, urticaria has followed exposure to lindane vapor. Unlike the signs and symptoms already mentioned, this allergic manifestation occurs only in susceptible individuals, and usually only after a period of sensitization.

BBQ000   CAS: 319-84-6   **HR: 3**
BENZENE HEXACHLORIDE-α-isomer
mf: $C_6H_6Cl_6$ mw: 290.82

SYNS: α-BENZENEHEXACHLORIDE * α-BHC * ENT 9,232 * α-HCH * α-HEXACHLORANE * HEXACHLORCYCLOHEXAN (GERMAN) * α-HEXA-CHLOROCYCLOHEXANE * α-1,2,3,4,5,6-HEXACHLO-ROCYCLOHEXANE (MAK) * 1-α,2-α,3-β,4-α,5-β,6-β-HEXACHLOROCYCLOHEXANE * α-LINDANE

CONSENSUS REPORTS: IARC Cancer Review: Animal Sufficient Evidence IMEMDT 20,195,79; IMEMDT 5,47,74. NTP Fourth Annual Report On Carcinogens, 1984. EPA Genetic Toxicology Program. Reported in EPA TSCA Inventory.

DFG MAK: 0.5 mg/m³

SAFETY PROFILE: Confirmed carcinogen with experimental carcinogenic, tumorigenic, and neoplastigenic data. Poison by ingestion. Mutation data reported. When heated to decomposition it emits toxic fumes of Cl⁻.

BBQ500   CAS: 58-89-9   **HR: 3**
BENZENE HEXACHLORIDE-γ isomer
DOT: NA 2761
mf: $C_6H_6Cl_6$ mw: 290.82

SYNS: AALINDAN * AFICIDE * AGRISOL G-20 * AGROCIDE * AGRONEXIT * AMEISENATOD * AMEISENMITTEL MERCK * APARSIN * APHTIRIA * APLIDAL * ARBITEX * BBH * BEN-HEX * BENTOX 10 * γ-BENZENE HEXACHLORIDE * BEXOL * BHC * γ-BHC * CELANEX * CHLORESENE * CODECHINE * DBH * DETMOL-EXTRAKT * DETOX 25 * DEVORAN * DOL GRANULE * DRILL TOX-SPEZIAL AGLUKON * ENT 7,796 * ENTOMOXAN * EXAGAMA * FORLIN * GALLOGAMA * GAMACID * GAMAPHEX * GAMENE * GAMISO * GAMMA-COL * GAMMAHEXA * GAMMAHEX-ANE * GAMMALIN * GAMMOPAZ * HCCH * HCH * γ-HCH * HECLOTOX * HEXACHLO-RAN * γ-HEXACHLORAN * γ-HEXACHLORANE * γ-HEXACHLOROBENZENE * 1-α,2-α,3-β,4-α,5-α,6-β-HEXACHLOROCYCLOHEXANE * γ-HEXACHLORO-CYCLOHEXANE (MAK) * 1,2,3,4,5,6-HEXACHLOROCY-CLOHEXANE, γ-ISOMER * HEXATOX * HEXICIDE * HGI * INEXIT * ISOTOX * JACUTIN * KOKOTINE * KWELL * LENDINE * LENTOX * LIDENAL * LINDAGRAIN * LINDANE (ACGIH, DOT, USDA) * LINTOX * MILBOL 49 * MSZY-COL * NCI-C00204 * NEO-SCABICIDOL * NEXIT

* NOVIGAM * OVADZIAK * PEDRACZAK
* QUELLADA * RCRA WASTE NUMBER U129
* SANG gamma * STREUNEX * TAP 85
* VITON

CONSENSUS REPORTS: IARC Cancer Review: Animal Sufficient Evidence IMEMDT 5,47,74; IMEMDT 20,195,79. NTP Fourth Annual Report On Carcinogens, 1984. NCI Carcinogenesis Bioassay (feed); No Evidence: mouse, rat NCITR* NCI-CG-TR-14,77. EPA Extremely Hazardous Substances List. EPA Genetic Toxicology Program. Community Right-To-Know List. Reported in EPA TSCA Inventory.

OSHA PEL: TWA 0.5 mg/m^3 (skin)
ACGIH TLV: TWA 0.5 mg/m^3 (skin)
DFG MAK: 0.5 mg/m^3
DOT Classification: ORM-A; Label: None.

SAFETY PROFILE: Confirmed carcinogen with experimental carcinogenic neoplastigenic data. A human systemic poison by ingestion. Also a poison by ingestion, skin contact, intraperitoneal, intravenous, and intramuscular routes. Human systemic effects by ingestion: convulsions, dyspnea, and cyanosis. Experimental teratogenic and reproductive effects. Mutation data reported. When heated to decomposition it emits toxic fumes of Cl$^-$, HCl, and phosgene.

BBQ750   **HR: 3**
BENZENEHEXACHLORIDE
(mixed isomers)

PROP: Technical BHC contains about 64% α, 10% β, 13% γ, 9% Δ and 1% ϵ isomers of 1,2,3,4,5,6-hexachlorocyclohexane.

SYNS: BENZAHEX * BENZEX * DOL
* DOLMIX * FBHC * FHCH * 1,2,3,4,5,6-
HEXACHLOROCYCLOHEXANE (mixture of isomers)
* HEXYCLAN * KOTOL * SOPROCIDE
* TECHNICAL BHC * TECHNICAL HCH

CONSENSUS REPORTS: IARC Cancer Review: Animal Sufficient Evidence IMEMDT 5,47,74; IMEMDT 20,195,79. NTP Fourth Annual Report On Carcinogens, 1984.

SAFETY PROFILE: Confirmed carcinogen with experimental tumorigenic and neoplastigenic data. Poison by inhalation and ingestion. Human systemic effects by an unspecified route: convulsions. Potentially dangerous reaction with DMF in presence of Fe, or CCl$_4$. When heated to decomposition it emits highly toxic fumes of Cl$^-$, HCl, and phosgene.

BBR000   CAS: 319-85-7   **HR: 3**
trans-α-BENZENEHEXACHLORIDE
mf: C$_6$H$_6$Cl$_6$ mw: 290.82

SYNS: β-BENZENEHEXACHLORIDE * β-BHC
* ENT 9,233 * β-HCH * β-HEXACHLOROBENZENE
* 1-α,2-β,3-α,4-β,5-α,6-β-HEXACHLOROCYCLOHEXANE
* β-HEXACHLOROCYCLOHEXANE * β-1,2,3,4,5,6-
HEXACHLOROCYCLOHEXANE (MAK) * β-ISOMER
* β-LINDANE

CONSENSUS REPORTS: IARC Cancer Review: Animal Sufficient Evidence IMEMDT 5,47,74; Animal Limited Evidence IMEMDT 20,195,79. NTP Fourth Annual Report On Carcinogens, 1984. Reported in EPA TSCA Inventory.

DFG MAK: 0.5 mg/m^3

SAFETY PROFILE: Confirmed carcinogen with experimental neoplastigenic data. Mildly toxic by ingestion. When heated to decomposition it emits very toxic fumes of Cl$^-$, HCl, and phosgene.

BBS250   CAS: 98-11-3   **HR: 3**
BENZENESULFONIC ACID
mf: C$_6$H$_6$O$_3$S   mw: 158.18

PROP: Deliquescent plates or tablets. Mp: 43-44°.

SYN: PHENYLSULFONIC ACID

CONSENSUS REPORTS: Reported in EPA TSCA Inventory.

SAFETY PROFILE: Poison by ingestion, skin contact and probably inhalation. A skin irritant.

BBS750   CAS: 98-09-9   **HR: 3**
BENZENESULFONYL CHLORIDE

DOT: UN 2225
mf: C$_6$H$_5$ClO$_2$S   mw: 176.62

SYNS: BENZENE SULFONCHLORIDE * BENZENE-
SULFONIC (ACID) CHLORIDE * BENZENE SULPHONYL
CHLORIDE (DOT) * BENZENOSULFOCHLOREK
(POLISH) * BENZENOSULPHOCHLORIDE * BSC-
REFINE D * RCRA WASTE NUMBER U020

CONSENSUS REPORTS: Reported in EPA TSCA Inventory.

DOT Classification: IMO: Corrosive Material; Label: Corrosive.

SAFETY PROFILE: Poison by intraperitoneal route. A dangerous storage hazard. It may explode in a sealed bottle. Explosive reaction with dimethyl sulfoxide. Reacts vigorously with methyl formamide. When heated to decomposition it emits toxic fumes of Cl^- and SO_x.

BBT250   CAS: 368-43-4   **HR: 3**
BENZENESULPHONYL FLUORIDE
mf: $C_6H_5FO_2S$ mw: 160.17

PROP: Clear liquid. Bp: 209°, fp: −5°, flash p: 196°F, d: 1.329, vap press: 8 mm @ 80°, vap d: 5.52.

CONSENSUS REPORTS: Reported in EPA TSCA Inventory.

SAFETY PROFILE: A poison by intraperitoneal routes. Slightly irritating to skin. Flammable when exposed to heat or flame. It can react vigorously with oxidizing materials. To fight fire, use water, foam, CO_2, water spray or mist, dry chemical. When heated to decomposition it emits toxic fumes of F^- and SO_x.

BBU250   CAS: 533-73-3   **HR: 3**
1,2,4-BENZENETRIOL
mf: $C_6H_6O_3$ mw: 126.12

SYNS: HYDROXYHYDROQUINONE * HYDROXY-QUINOL * OXYHYDROCHINON (GERMAN) * OXYHYDROQUINONE * 1,2,4-TRIHYDROXYBENZENE

CONSENSUS REPORTS: EPA Genetic Toxicology Program. Reported in EPA TSCA Inventory.

SAFETY PROFILE: Poison by subcutaneous and intraperitoneal routes. Human mutation data reported. When heated to decomposition it emits acrid smoke and irritating fumes.

BBV000   CAS: 52-49-3   **HR: 3**
BENZHEXOL HYDROCHLORIDE
mf: $C_{20}H_{31}NO \cdot ClH$ mw: 337.98

SYNS: APARKAN * ARTANE * ARTANE HYDROCHLORIDE * ARTANE TRIHEXYPHENIDYL * BENZHEXOL CHLORIDE * CYCLODOL * α-CYCLOHEXYL-α-PHENYL-1-PIPERIDINEPROPANOL HYDROCHLORIDE * PACITANE * PARALEST * PARGITAN * PARKINSAN * PARKOPAN * PERAGIT * 1-PHENYL-1-CYCLOHEXYL-3-PIPERIDYL-1-PROPANOL HYDROCHLORIDE * PIPANOL * 3-(1-PIPERIDYL)-1-CYCLOHEXYL-1-PHENYL-1-PROPANOL HYDROCHLORIDE * ROMPARKIN * SEDRENA * TREMIN * TRI-

ESIFENIDILE * TRIEXIFENIDILA * TRIPHEDINON * TRIPHENIDYL * TRIHEXYLPHENIDYL HYDROCHLORIDE * TSIKLODOL

SAFETY PROFILE: Poison by ingestion, intraperitoneal, intravenous, and subcutaneous routes. An anticholinergic agent which causes human psychotropic effects. When heated to decomposition it emits very toxic fumes of NO_x and HCl.

BBV250   CAS: 613-94-5   **HR: 3**
BENZHYDRAZIDE
mf: $C_7H_8N_2O$ mw: 136.17

SYNS: BENZOHYDRAZIDE * BENZOHYDRAZINE * BENZOIC HYDRAZIDE * BENZOYL HYDRAZIDE

CONSENSUS REPORTS: Reported in EPA TSCA Inventory.

SAFETY PROFILE: Poison by subcutaneous and intraperitoneal routes. Questionable carcinogen with experimental carcinogenic and neoplastigenic data. Violent reaction with benzeneseleninic acid. When heated to decomposition it emits toxic fumes of NO_x.

BBV500   CAS: 58-73-1   **HR: 3**
BENZHYDRYL
mf: $C_{17}H_{21}NO$ mw: 255.39

SYNS: ALERYL * ALLEDRYL * ALLERGAN B * ALLERGEVAL * ALLERGICAL * ALLERGIN * ALLERGINA * ALLERGIVAL * AMIDRYL * ANTISTOMINUM * ANTOMIN * AUTOMIN * BAGAODRYL * BARAMINE * BENA * BENACHLOR * BENADON * BENADRIN * BENADRYL * BEN-ALLERGIN * BENAPON * BENODIN * BENODINE * BENYLAN * BENZANTINE * BENZHYDRAMINE * BENZHYDRAMINUM * BENZHYDRIL * o-BENZHYDRYLDIMETHYLAMINOETHANOL * 2-(BENZHYDRYLOXY)-N,N-DIMETHYLETHYLAMINE * 2-(BENZOHYDRYLOXY)-N,N-DIMETHYLETHYLAMINE * BETRAMIN * DABYLEN * DEBENDRIN * DERMISTINE * DERMODRIN * DESENTOL * DIABENYL * DIABYLEN * DIBONDRIN * DIFEDRYL * DIFENHYDRAMIN * DIFENIDRAMINA (ITALIAN) * DIHIDRAL * DIMEDROL * DIMEDRYL * β-DIMETHYLAMINO-AETHYL-BENZHYDRYL-AETHER (GERMAN) * β-DIMETHYLAMINOETHANOL DIPHENYLMETHYL ETHER * α-(2-DIMETHYLAMINOETHOXY)DIPHENYLMETHANE * β-DIMETHYLAMINOETHYLBENZHYDRYLETHER * DIPHANTINE * DIPHENYLHYDRAMINE * 2-(DIPHENYLME-

THOXY)-N,N-DIMETHYLETHYLAMINE * DRYISTAN
* DRYLISTAN * DYLAMON * ETANAUTINE
* HISTAXIN * HYADRINE * IBIODRAL
* MEDIDRYL * MEPHADRYL * NAUSEN
* PROBEDRYL * RESTAMIN * RESTAMINE
* RIGIDIL * RIGIDYL * S51 * SYNTEDRIL
* SYNTODRIL * VENA

CONSENSUS REPORTS: Reported in EPA TSCA Inventory.

SAFETY PROFILE: Deadly human poison by an unspecified route. Poison by ingestion, intravenous, intraperitoneal, and subcutaneous routes. Experimental reproductive effects. Human mutation data reported. When heated to decomposition it emits toxic fumes of NO_x.

BBW500   CAS: 132-69-4   HR: 3
BENZIDAMINE HYDROCHLORIDE
mf: $C_{19}H_{23}N_3O \cdot ClH$ mw: 345.91

SYNS: AF 864 * BENALGIN * BENZINDAMINE HYDROCHLORIDE * BENZYDAMINE HYDROCHLORIDE * 1-BENZYL-3-γ-DIMETHYLAMINOPROPOXY-1H-INDAZOLE HYDROCHLORIDE * 1-BENZYL-3-(3-(DI-METHYLAMINO)PROPOXY)-1H-INDAZOLE HYDROCHLORIDE * BENZYRIN * DIFFLAM * N,N-DI-METHYL-3((1-PHENYLMETHYL)-1H-INDAZOL-3-YL)OXY)-1-PROPANAMINE HYDROCHLORIDE * DORINAMIN * ENZAMIN * EPIROTIN * IMOTRYL * INDOLIN * RIRILIM * RIRIPEN * SALYZORON * TAMAS * TANTUM * VERAX

SAFETY PROFILE: Poison by intraperitoneal, subcutaneous, and intravenous routes. Moderately toxic by ingestion. Experimental teratogenic and reproductive effects. An eye irritant. A nonsteroidal anti-inflammatory analgesic. When heated to decomposition it emits very toxic fumes of HCl and NO_x.

BBX000   CAS: 92-87-5   HR: 3
BENZIDINE
DOT: UN 1885
mf: $C_{12}H_{12}N_2$ mw: 184.26

PROP: Grayish-yellow, crystalline powder; white or sltly reddish crystals, powder, or leaf. Mp: 127.5-128.7° @ 740 mm, bp: 401.7°, d: 1.250 @ 20°/4°.

SYNS: BENZIDIN (CZECH) * BENZIDINA (ITALIAN) * BENZYDYNA (POLISH) * p,p-BIANILINE * 4,4'-BIANILINE * (1,1'-BIPHENYL)-4,4'-DIAMINE

(9CI) * 4,4'-BIPHENYLDIAMINE * 4,4'-BIPHENYL-ENEDIAMINE * C.I. 37225 * C.I. AZOIC DIAZO COMPONENT 112 * p,p'-DIAMINOBIPHENYL * 4,4'-DIAMINOBIPHENYL * 4,4'-DIAMINO-1,1'-BI-PHENYL * p-DIAMINODIPHENYL * 4,4'-DIAMINO-DIPHENYL * p,p'-DIANILINE * 4,4'-DIPHENYLENE-DIAMINE * FAST CORINTH BASE B * NCI-C03361 * RCRA WASTE NUMBER U021

CONSENSUS REPORTS: IARC Cancer Review: Human Limited Evidence IMEMDT 1,80,72; Human Sufficient Evidence IMEMDT 29,149,82; Animal Sufficient Evidence IMEMDT 1,80,72; IMEMDT 29,149,82. NTP Fourth Annual Report On Carcinogens, 1984. EPA Genetic Toxicology Program. Community Right-To-Know List. Reported in EPA TSCA Inventory.

OSHA: Cancer Suspect Agent
ACGIH TLV: Confirmed Human Carcinogen
DFG MAK: Human Carcinogen.
DOT Classification: Poison B; Label: Poison.

SAFETY PROFILE: Confirmed human carcinogen producing bladder tumors. Experimental carcinogenic and tumorigenic data. Poison by ingestion and intraperitoneal routes. Human mutation data reported. Can cause damage to blood, including hemolysis and bone marrow depression. On ingestion causes nausea and vomiting which may be followed by liver and kidney damage. Any exposure is considered extremely hazardous. When heated to decomposition it emits highly toxic fumes of NO_x.

BBX750   CAS: 531-85-1   HR: 3
BENZIDINE HYDROCHLORIDE
mf: $C_{12}H_{12}N_2 \cdot 2ClH$ mw: 257.18

SYNS: (1,1'-BIPHENYL)-4,4'-DIAMINE, DIHYDROCHLO-RIDE * DIHIDROCLORURO de BENZIDINA (SPANISH)

CONSENSUS REPORTS: Reported in EPA TSCA Inventory. EPA Genetic Toxicology Program.

SAFETY PROFILE: Suspected carcinogen with experimental carcinogenic and tumorigenic data. Human mutation data reported. When heated to decomposition it emits very toxic fumes of HCl and NO_x.

BBY000   CAS: 531-86-2   HR: 3
BENZIDINE SULFATE
mf: $C_{12}H_{12}N_2 \cdot H_2O_4S$ mw: 282.34

SYN: (1,1'-BIPHENYL)-4,4'-DIAMINE SULFATE (1:1)

OSHA: Carcinogen

SAFETY PROFILE: Confirmed human carcinogen with experimental carcinogenic data. When heated to decomposition it emits toxic fumes of SO_x and NO_x.

BBY300   HR: 2
BENZIDINE SULPHATE
and HYDRAZINE-BENZENE
mf: $C_6H_8N_2 \cdot C_{12}H_{12}N_2 \cdot H_2O_4S$ mw: 390.50

SYN: HYDRAZINE-BENZENE and BENZIDINE SULFATE

SAFETY PROFILE: Suspected carcinogen with experimental carcinogenic data. When heated to decomposition it emits toxic fumes of NO_x and SO_x.

BCA000   CAS: 57-37-4   HR: 3
BENZILIC ACID-β-DIETHYLAMINO-
ETHYL ESTER HYDROCHLORIDE
mf: $C_{20}H_{25}NO_3 \cdot ClH$ mw: 363.92

SYNS: ACTOZINE * AMIOYL * AMISYL * AMITAKON * AMIZIL HYDROCHLORIDE * ARCADINE * AY-5406 * BENACTIZINE HYDROCHLORIDE * BENACTYZIN (CZECH) * BENACTYZINE CHLORIDE * BENACTYZINE HYDROCHLORIDE * BENAKTIN * BENZILATE DU DIETHYLAMINO-ETHANOL CHLORHYDRATE (FRENCH) * CAFRON * CEDAD * CEVANOL * DESTENDO * β-DIETHYLAMINOETHYL BENZILATE HYDROCHLORIDE * 2-DIETHYLAMINOETHYL BENZILATE HYDROCHLORIDE * 2-DIETHYLAMINOETHYL DIPHENYLGLYCOLATE HYDROCHLORIDE * 2-(DIFENYL-HYDROXYACETOXY)ETHYL-DIETHYLAMMONIUMCHLORID (CZECH) * DIPHENYLGLYCOLLIC ACID-2-(DIETHYLAMINO) ETHYL ESTER HYDROCHLORIDE * FOBEX * IBIOTYZIL * KATRON * LEUCIDIL * NERVACTON * NERVATIL * NEURAKTIL * NEUROBENZIL * NEUROLEPTONE * NUTINAL * PARASAN * PARPON * PHOBEX * PROCALM * STOIKON * SUAVITIL * TRANQUILLIN * VALLADAN * WIN 5606

CONSENSUS REPORTS: Reported in EPA TSCA Inventory.

SAFETY PROFILE: Poison by ingestion, intraperitoneal, subcutaneous, intradermal, and intravenous routes. Human systemic effects by ingestion of very small amounts: toxic phycho-

sis. Experimental reproductive effects. When heated to decomposition it emits very toxic fumes of NO_x and HCl.

BCB750   CAS: 51-17-2   HR: 3
BENZIMIDAZOLE
mf: $C_7H_6N_2$ mw: 118.15

PROP: Tabular crystals sol in alcohol, sparingly sol in water. Mp: 170.5°, bp: >360°.

SYNS: 3-AZAINDOLE * AZINDOLE * 1H-BENZIMIDAZOLE (9CI) * o-BENZIMIDAZOLE * BENZIMINAZOLE * 1,3-BENZODIAZOLE * BENZOIMIDAZOLE * BZI * 1,3-DIAZAINDENE * N,N'-METHENYL-o-PHENYLENEDIAMINE * NSC 759

CONSENSUS REPORTS: Reported in EPA TSCA Inventory.

SAFETY PROFILE: Poison by intravenous and intraperitoneal routes. Moderately toxic by ingestion. Mutation data reported. When heated to decomposition it emits highly toxic fumes of NO_x.

BCC250   CAS: 6898-43-7   HR: 3
BENZIMIDAZOLE METHYLENE
MUSTARD
mf: $C_{14}H_{19}Cl_2N_3 \cdot ClH$ mw: 336.72

SYNS: BENZIMIDAZOLE MUSTARD * 2-(BIS(2-CHLOROETHYL)AMINOMETHYL)-5,5-DIMETHYLBENZIMIDAZOLE HYDROCHLORIDE * 2-(DI-2-CHLOROETHYL)AMINOMETHYL-5,6-DIMETHYLBENZIMIDAZOLE * NSC-23892

SAFETY PROFILE: Questionable carcinogen with experimental carcinogenic data. When heated to decomposition it emits very toxic fumes HCl and NO_x.

BCC500   CAS: 583-39-1   HR: 3
2-BENZIMIDAZOLETHIOL
mf: $C_7H_6N_2S$ mw: 150.21

SYNS: ANTIEGENE MB * ANTIOXIDANT MB (CZECH) * AOMB * ASM MB * 2-MERCAPTOBENZIMIDAZOLE * MERCAPTOBENZOIMIDAZOLE * 2-MERCAPTOBENZOIMIDAZOLE * MERKAPTOBENZIMIDAZOL (CZECH) * NCI-C60980 * o-PHENYLENETHIOUREA * USAF EK-6540 * USAF XF-21

CONSENSUS REPORTS: Reported in EPA TSCA Inventory.

SAFETY PROFILE: Poison by intraperitoneal and intravenous routes. Moderately toxic by ingestion. Skin and eye irritant. When heated to decomposition it emits toxic fumes of SO_x and NO_x.

BCE500   CAS: 81-07-2   HR: 3
1,2-BENZISOTHIAZOL-3(2H)-ONE-1,1-DIOXIDE
mf: $C_7H_5NO_3S$ mw: 183.19

PROP: White crystals or powder; odorless with sweet taste. Mp: 228° (decomp), bp: sublimes. Sol in water, alc, chloroform, and ether.

SYNS: ANHYDRO-o-SULFAMINE BENZOIC ACID * 3-BENZISOTHIAZOLINONE-1,1-DIOXIDE * o-BENZOIC SULPHIMIDE * o-BENZOSULFIMIDE * BENZOSULPHIMIDE * BENZO-2-SULPHIMIDE * o-BENZOYL SULFIMIDE * o-BENZOYL SULPHIMIDE * 1,2-DIHYDRO-2-KETOBENZISOSULFONAZOLE * 1,2-DIHYDRO-2-KETOBENZISOSULPHONAZOLE * 2,3-DIHYDRO-3-OXOBENZISOSULFONAZOLE * 2,3-DIHYDRO-3-OXOBENZISOSULPHONAZOLE * GARANTOSE * GLUCID * GLUSIDE * HERMESETAS * 3-HYDROXYBENZISOTHIAZOL-S,S-DIOXIDE * INSOLUBLE SACCHARINE * KANDISET * NATREEN * RCRA WASTE NUMBER U202 * SACARINA * SACCAHARIMIDE * SACCHARINA * SACCHARIN ACID * SACCHARINE * SACCHARINOL * SACCHARINOSE * SACCHAROL * SAXIN * SUCRE EDULCOR * SUCRETTE * o-SULFOBENZIMIDE * o-SULFOBENZOIC ACID IMIDE * 2-SULPHOBENZOIC IMIDE * SYKOSE * SYNCAL * ZAHARINA

CONSENSUS REPORTS: IARC Cancer Review: GROUP 2B IMEMDT 7,334,87, Human Inadequate Evidence IMEMDT 22,111,80; Animal Sufficient Evidence IMEMDT 22,111,80. EPA Genetic Toxicology Program. Reported in EPA TSCA Inventory. Community Right-To-Know List.

SAFETY PROFILE: Suspected carcinogen with experimental neoplastigenic and tumorigenic data. Mild acute toxicity by ingestion. Experimental teratogenic and reproductive effects. Mutation data reported. When heated to decomposition it emits toxic NO_x and SO_x.

BCG500   CAS: 214-17-5   HR: 3
BENZO(b)CHRYSENE
mf: $C_{22}H_{14}$ mw: 278.36

SYNS: 2,3-BENZOCHRYSENE * 3,4-BENZOTETRACENE * 3,4-BENZOTETRAPHENE * BENZO(c)TETRAPHENE * 1,2:6,7-DIBENZOPHENANTHRENE * 2,3:7,8-DIBENZOPHENANTHRENE * DIBENZO-2,3,7,8-PHENANTHRENE

SAFETY PROFILE: Questionable carcinogen with experimental neoplastigenic data by skin contact. Mutation data reported. When heated to decomposition it emits acrid smoke and irritating fumes.

BCH750   CAS: 10085-81-1   HR: 3
BENZOCTAMINE HYDROCHLORIDE
mf: $C_{18}H_{19}N \cdot ClH$ mw: 285.84

SYNS: BA 30,803 * 1-METHYLAMINOMETHYLDIBENZO(b,c)BICYCLO(2,2,2)OCTADIENE HYDROCHLORIDE * N-METHYLETHANOANTHRACENE-9-(10H)-METHYLAMINE HYDROCHLORIDE * TACITIN

SAFETY PROFILE: Poison by intravenous route. Moderately toxic by ingestion. Experimental teratogenic effects. A sedative and muscle relaxant. When heated to decomposition it emits very toxic fumes of NO_x and HCl^-.

BCI500   CAS: 135-87-5   HR: 3
BENZODIOXANE HYDROCHLORIDE
mf: $C_{14}H_{19}NO_2 \cdot ClH$ mw: 269.80

SYNS: BENODAINE HYDROCHLORIDE * 1-(1,4-BENZODIOXAN-2-YLMETHYL)PIPERIDINEHYDROCHLORIDE * F 933 * FOURNEAU 933 * 2-PIPERIDINOMETHYL-1,4-BENZODIOXAN HYDROCHLORIDE * 2-(1-PIPERIDYLMETHYL)-1,4-BENZODIOXAN HYDROCHLORIDE * PIPEROXANE HYDROCHLORIDE

SAFETY PROFILE: Poison by intraperitoneal and intravenous routes. Moderately toxic by ingestion and subcutaneous routes. Experimental reproductive effects. When heated to decomposition it emits very toxic fumes of NO_x and HCl.

BCJ000   CAS: 5208-87-7   HR: 3
1,3-BENZODIOXOLE-5-(2-PROPEN-1-OL)
mf: $C_{10}H_{10}O_3$ mw: 178.20

SYNS: 1'-HYDROXYSAFROLE * 1,2-METHYLENEDIOXY-4-(1-HYDROXYALLYL)BENZENE * α-VINYLPIPERONYL ALCOHOL

SAFETY PROFILE: Suspected carcinogen with experimental carcinogenic, neoplastigenic, and tumorigenic data. Human mutation data reported. When heated to decomposition it emits acrid smoke and irritating fumes.

BCJ500   CAS: 205-82-3   **HR: 3**
BENZO(j)FLUORANTHENE
mf: $C_{20}H_{12}$ mw: 252.32

SYNS: 10,11-BENZFLUORANTHENE * BENZ(j)FLUO-
ROANTHRENE * BENZO(1)FLUORANTHENE
* 7,8-BENZOFLUORANTHENE * B(j)F * DIBEN-
ZO(a,jk)FLUORENE

CONSENSUS REPORTS: IARC Cancer Re-
view: GROUP 2B IMEMDT 7,56,87; Animal
Limited Evidence IMEMDT 3,82,73; Animal
Sufficient Evidence IMEMDT 32,155,83

SAFETY PROFILE: Suspected carcinogen with
experimental carcinogenic, neoplastigenic, and
tumorigenic data. Mutation data reported. When
heated to decomposition it emits acrid smoke
and irritating fumes.

BCJ750   CAS: 207-08-9   **HR: 3**
BENZO(k)FLUORANTHENE
mf: $C_{20}H_{12}$ mw: 252.32

SYNS: 8,9-BENZOFLUORANTHENE * 11,12-BENZO-
FLUORANTHENE * 11,12-BENZO(k)FLUORANTHENE
* 2,3,1′,8′-BINAPHTHYLENE * DIBENZO(b,jk)
FLUORENE

CONSENSUS REPORTS: IARC Cancer Review: Animal Sufficient Evidence IMEMDT
32,163,83

SAFETY PROFILE: Confirmed carcinogen with
experimental tumorigenic data. Mutation data
reported. When heated to decomposition it emits
acrid smoke and irritating fumes.

BCL250   CAS: 23844-24-8   **HR: 3**
BENZOGUANAMINE
mf: $C_{22}H_{32}N_2O_5$ mw: 404.56

PROP: Crystals. Mp: 227°, d: 1.4.

SYNS: 2-ACETOXY-3-DIETHYLCARBAMYL-9,10-DI-
METHOXY-1,2,3,4,6,7-HEXAHYDRO-11B-BENZO(a)
QUINOLIZINE * BENZOCHINAMIDE * BENZO-
QUINAMIDE * BENZQUINAMIDE * BENZ-
QUINAMIDU (POLISH) * BZQ * P 2647 * QUAN-
TRIL * QUANTRYL

SAFETY PROFILE: Poison by ingestion, intraperitoneal, and intravenous routes. When heated
to decomposition it emits toxic fumes of NO_x.

BCL750   CAS: 65-85-0   **HR: 3**
BENZOIC ACID
mf: $C_7H_6O_2$ mw: 122.13

PROP: White crystalline powder. Mp: 121.7°,
bp: 249°, flash p: 250°F (CC), d: 1.316, autoign
temp: 1060°F, vap press: 1 mm @ 96.0° (sublimes), vap d: 4.21. Moderately sol in water;
sol in alc, ether, chloroform, fixed oils.

SYNS: ACIDE BENZOIQUE (FRENCH) * BENZENE-
CARBOXYLIC ACID * BENZENEFORMIC ACID
* BENZENEMETHANOIC ACID * BENZOATE
* BENZOESAEURE (GERMAN) * BENZOIC ACID
(DOT) * CARBOXYBENZENE * DRACYLIC ACID
* KYSELINA BENZOOVA (CZECH) * PHENYL CAR-
BOXYLIC ACID * PHENYLFORMIC ACID * RE-
TARDER BA * RETARDEX * SALVO LIQUID
* SALVO POWDER * TENN-PLAS

CONSENSUS REPORTS: Reported in EPA
TSCA Inventory. EPA Genetic Toxicology Program.

DOT Classification: ORM-E; Label: None.

SAFETY PROFILE: Poison by subcutaneous
route. Moderately toxic by ingestion and intraperitoneal routes. Human systemic effects by
inhalation: dyspnea and allergic dermatitis. Severe eye irritant. A human skin irritant. Combustible when exposed to heat or flame; can
react with oxidizing materials. The powder
burns rapidly in oxygen. To fight fire, use water,
CO_2, water spray or mist, dry chemical. When
heated to decomposition it emits acrid smoke
and irritating fumes.

BCM000   CAS: 120-51-4   **HR: 2**
BENZOIC ACID, BENZYL ESTER
mf: $C_{14}H_{12}O_2$ mw: 212.26

PROP: Found in Peru and Tolu Balsams, in
Ylang-Ylang and in about 20 other essential
oils. Colorless oily liquid; slt aromatic odor.
Mp: 21°, bp: 324°, flash p: 298°F (CC), d: 1.116,
refr index: 1.568, vap d: 7.3, autoign temp:
898°F. Misc with alc, chloroform, ether; insol
in glycerin, water.

SYNS: ASCABIN * ASCABIOL * BENYLATE
* BENZOIC ACID, PHENYLMETHYL ESTER * BEN-
ZYL ALCOHOL BENZOIC ESTER * BENZYL BENZENE-
CARBOXYLATE * BENZYL BENZOATE (FCC)
* BENZYLETS * BENZYL PHENYLFORMATE
* COLEBENZ * FEMA No. 2138 * NOVOSCABIN
* PERUSCABIN * SCABANCA * VANZOATE
* VENZONATE

CONSENSUS REPORTS: Reported in EPA
TSCA Inventory.

SAFETY PROFILE: Moderately toxic by ingestion and skin contact. Combustible liquid. Can react with oxidizing materials. To fight fire, use CO_2, water spray or mist, dry chemical. When heated to decomposition it emits acrid and irritating fumes and smoke.

BCP250   CAS: 119-53-9   **HR: D**
BENZOIN
mf: $C_{14}H_{12}O_2$ mw: 212.26

SYNS: BENZOYLPHENYLCARBINOL * BITTER AL-
MOND OIL CAMPHOR * α-HYDROXYBENZYL PHENYL
KETONE * α-HYDROXY-α-PHENYLACETOPHENONE
* 2-HYDROXY-2-PHENYLACETOPHENONE * NCI-
C50011

NCI Carcinogenesis Bioassay (feed); No Evidence: mouse, rat NCITR* NCI-CG-TR-204,80. Reported in EPA TSCA Inventory.

SAFETY PROFILE: Mutation data reported. When heated to decomposition it emits acrid smoke and irritating fumes.

BCQ250   CAS: 100-47-0   **HR: 3**
BENZONITRILE
DOT: UN 2224
mf: C_7H_5N mw: 103.13

PROP: Transparent, colorless oil; almond-like odor. D: 1.246 @ 20°/4°, bp: 191°, mp: −12.8°.

SYNS: BENZENENITRILE * BENZOIC ACID NITRILE
* BENZONITRILE (DOT) * CYANOBENZENE
* PHENYL CYANIDE

CONSENSUS REPORTS: Reported in EPA TSCA Inventory. Cyanide and its compounds are on the Community Right-To-Know List.

DOT Classification: Combustible Liquid; Label: None; DOT-IMO: Poison B; Label: Poison.

SAFETY PROFILE: Poison by intraperitoneal and subcutaneous routes. Moderately toxic by ingestion, inhalation, and skin contact. A skin irritant. Combustible liquid. When heated to decomposition it emits toxic fumes of CN^- and NO_x.

BCQ500   CAS: 189-55-9   **HR: 3**
BENZO(rst)PENTAPHENE
mf: $C_{24}H_{14}$ mw: 302.38

PROP: Green-yellow needles. Mp: 280-282°.

SYNS: DB(a,i)P * DIBENZO(a,i)PYRENE * DI-
BENZO(b,h)PYRENE * 1,2,7,8-DIBENZOPYRENE
* 3,4:9,10-DIBENZOPYRENE * DIBENZ(a,i)PYRENE
* 1,2:7,8-DIBENZPYRENE * 3,4:9,10-DIBENZPYRENE
* RCRA WASTE NUMBER U064

CONSENSUS REPORTS: IARC Cancer Review: GROUP 2B IMEMDT 7,56,87; Animal Sufficient Evidence IMEMDT 3,215,73; IMEMDT 32,337,83. NTP Fourth Annual Report On Carcinogens, 1984. EPA Genetic Toxicology Program.

SAFETY PROFILE: Confirmed with experimental neoplastigenic and tumorigenic data. Mutation data reported. When heated to decomposition it emits acrid smoke and irritating fumes.

BCR750   CAS: 195-19-7   **HR: 3**
BENZO(c)PHENANTHRENE
mf: $C_{18}H_{12}$ mw: 228.30

SYNS: 3,4-BENZOPHENANTHRENE * 3,4-BENZPHE-
NANTHRENE * TETRAHELICENE

CONSENSUS REPORTS: IARC Cancer Review: GROUP 3 IMEMDT 7,56,87, Animal Inadequate Evidence IMEMDT 32,205,83

SAFETY PROFILE: Questionable carcinogen with experimental tumorigenic data. Mutation data reported. When heated to decomposition it emits acrid and irritating fumes.

BCS250   CAS: 119-61-9   **HR: 3**
BENZOPHENONE
mf: $C_{13}H_{10}O$ mw: 182.23

PROP: Rhombic, white crystals; persistent rose-like odor. mp (α): 49°, mp (β): 26°, mp (γ): 47°, bp: 305.4°, d (α): 1.0976 @ 50°/50°, d (β): 1.108 @ 23°/40°, vap press: 1 mm @ 108.2. Sol in fixed oils; sltly sol in propylene glycol; insol in glycerol.

SYNS: BENZOYLBENZENE * DIPHENYL KETONE
* DIPHENYLMETHANONE * FEMA No. 2134
* α-OXODIPHENYLMETHANE * PHENYL KETONE

CONSENSUS REPORTS: Reported in EPA TSCA Inventory.

SAFETY PROFILE: Moderately toxic by ingestion and intraperitoneal routes. Combustible when heated. Incompatible with oxidizers.

When heated to decomposition it emits acrid and irritating fumes.

BCS750 CAS: 50-32-8 **HR: 3**
BENZO(a)PYRENE
mf: $C_{20}H_{12}$ mw: 252.32

PROP: Yellow crystals. Mp: 179°, bp: 312° @ 10 mm. Insol in water; sol in benzene, toluene, and xylene.

SYNS: BENZO(d,e,f)CHRYSENE * 3,4-BENZOPIRENE (ITALIAN) * 3,4-BENZOPYRENE * 6,7-BENZOPYRENE * 3,4-BENZPYREN (GERMAN) * BENZ(a)PYRENE * 3,4-BENZ(a)PYRENE * 3,4-BENZYPYRENE

CONSENSUS REPORTS: IARC Cancer Review: GROUP 2A IMEMDT 7,56,87; Animal Sufficient Evidence IMEMDT 32,211,83; IMEMDT 3,91,73. Reported in EPA TSCA Inventory.

OSHA PEL: TWA 0.2 mg/m^3

SAFETY PROFILE: Suspected carcinogen with experimental carcinogenic, neoplastigenic, and tumorigenic data. A poison via subcutaneous, intraperitoneal and intrarenal routes. Experimental teratogenic and reproductive effects. Human mutation data reported. A common air contaminant of water, food, and smoke. When heated to decomposition it emits acrid smoke and fumes.

BCT000 CAS: 192-97-2 **HR: 3**
BENZO(e)PYRENE
mf: $C_{20}H_{12}$ mw: 252.32

SYNS: 1,2-BENZOPYRENE * 4,5-BENZOPYRENE * 1,2-BENZPYRENE * B(e)P

CONSENSUS REPORTS: IARC Cancer Review: GROUP 3 IMEMDT 7,56,87; Animal Inadequate Evidence IMEMDT 32,225,83; Animal Limited Evidence IMEMDT 3,137,73. EPA Genetic Toxicology Program.

SAFETY PROFILE: Questionable carcinogen with experimental tumorigenic data. Experimental teratogenic and reproductive effects. Human mutation data reported. When heated to decomposition it emits acrid smoke and irritating fumes.

BCU000 CAS: 60268-85-1 **HR: 3**
anti-BENZO(a)PYRENE-7,8-
DIHYDRODIOL-9,10-OXIDE
mf: $C_{20}H_{14}O_3$ mw: 302.34

SYNS: BENZO(a)PYRENE-7,8-DIHYDRODIOL-9,10-EPOXIDE (anti) * BP-7,8-DIHYDRODIOL-9,10-EPOXIDE (anti) * anti-BP-7,8-DIHYDRODIOL-9,10-OXIDE

SAFETY PROFILE: Questionable carcinogen with experimental tumorigenic data by skin contact. Human mutation data reported. When heated to decomposition it emits acrid smoke and irritating fumes.

BCV250 CAS: 21247-98-3 **HR: 3**
BENZO(a)PYRENE-6-METHANOL
mf: $C_{21}H_{14}O$ mw: 282.35

SYN: 6-HYDROXYMETHYLBENZO(a)PYRENE

CONSENSUS REPORTS: EPA Genetic Toxicology Program.

SAFETY PROFILE: Suspected carcinogen with experimental carcinogenic, neoplastigenic, and tumorigenic data. Mutation data reported. When heated to decomposition it emits acrid smoke and fumes.

BCX000 CAS: 56892-30-9 **HR: 3**
BENZO(a)PYREN-2-OL
mf: $C_{20}H_{12}O$ mw: 268.32

SYN: 2-HYDROXYBENZO(a)PYRENE

CONSENSUS REPORTS: EPA Genetic Toxicology Program.

SAFETY PROFILE: Questionable carcinogen with experimental tumorigenic and neoplastigenic data. Human mutation data reported. When heated to decomposition it emits acrid smoke and irritating fumes.

BCX250 CAS: 13345-21-6 **HR: 3**
BENZO(a)PYREN-3-OL
mf: $C_{20}H_{12}O$ mw: 268.32

SYNS: BP-3-HYDROXY * 3-HYDROXYBENZO(a)PYRENE * 8-HYDROXY-3,4-BENZPYRENE

CONSENSUS REPORTS: EPA Genetic Toxicology Program.

SAFETY PROFILE: Questionable carcinogen with experimental tumorigenic and neoplastigenic data by skin contact. Human mutation data reported. When heated to decomposition it emits acrid smoke and irritating fumes.

BCY000 CAS: 37994-82-4 **HR: 3**
BENZO(a)PYREN-7-OL
mf: $C_{20}H_{12}O$ mw: 268.32

SYN: 7-HYDROXYBENZO(a)PYRENE

CONSENSUS REPORTS: EPA Genetic Toxicology Program.

SAFETY PROFILE: Questionable carcinogen with experimental tumorigenic and neoplastigenic data by skin contact. Human mutation data reported. When heated to decomposition it emits acrid smoke and fumes.

BDC250   CAS: 583-63-1   **HR: 3**
o-BENZOQUINONE
DOT: UN 2587
mf: $C_6H_4O_2$ mw: 108.10

SYNS: 1,2-BENZOQUINONE * BENZOQUINONE (DOT) * 3,5-CYCLOHEXADIENE-1,2-DIONE * o-QUINONE

DOT Classification: Poison B; Label: Poison.

SAFETY PROFILE: A poison. Mutation data reported. When heated to decomposition it emits acrid smoke and irritating fumes.

BDC750   CAS: 800-24-8   **HR: 3**
BENZOQUINONE AZIRIDINE
mf: $C_{16}H_{22}N_2O_6$ mw: 338.40

SYNS: A-139 * AZIRIDYL BENZOQUINONE * BAYER A 139 * BAYER R39 SOLUBLE * 2,5-BIS (1-AZIRIDINYL)-3,6-BIS(2-METHOXYETHOXY)-p-BENZOQUINONE * 2,5-BIS(1-AZIRIDINYL)-3,6-BIS(2-METHOXYETHOXY)-2,5-CYCLOHEXADIENE-1,4-DIONE * 2,5-BISMETHOXYETHOXY-3,6-BISETHYLENEIMINO-1,4-BENZOQUINONE * 3,6-BIS(β-METHOXYETHOXY)-2,5-BIS(ETHYLENIMINO)-p-BENZOQUINONE * 3,6-BIS (β-METHOXYETHOXY)-2,5-BIS(ETHYLENEIMINO)-p-BENZOQUINONE * E 39 SOLUBLE * NSC-17262

CONSENSUS REPORTS: IARC Cancer Review: GROUP 3 IMEMDT 7,56,87; Animal Limited Evidence IMEMDT 9,51,75. EPA Genetic Toxicology Program.

SAFETY PROFILE: Deadly poison by intravenous route. Questionable carcinogen with experimental carcinogenic data. Mutation data reported. When heated to decomposition it emits toxic fumes of NO_x.

BDD000   CAS: 495-73-8   **HR: 3**
1,4-BENZOQUINONE-N′-
BENZOYLHYDRAZONE OXIME
mf: $C_{13}H_{11}N_3O_2$ mw: 241.27

SYNS: BAYER 15080 * BENCHINOX * BENGUINOX * BENQUINOX * BENZOIC ACID(4-(HYDROXYIMINO)-2,5-CYCLOHEXADIEN-1-YLIDENE) HYDRAZIDE * p-BENZOQUINONE OXIME BENZOYLHYDRAZONE * CEREDON * CERELINE * CERENOX * CHINONOXIM-BENZOYLHYDRAZON (GERMAN) * CHINONOXIME-BENZOYLHYDRAZONE * COBH * GBH * LERENOX * QGH * QUINONE OXIME BENZOYLHYDRAZONE * TILLANTOX * TSERENOX

SAFETY PROFILE: Poison by ingestion and possibly other routes. When heated to decomposition it emits toxic NO_x.

BDE250   CAS: 91-33-8   **HR: 3**
BENZOTHIAZIDE
mf: $C_{15}H_{14}ClN_3O_4S_3$ mw: 431.95

SYNS: AQUATAG * 3-((BENZYLTHIO)METHYL)-6-CHLORO-1,2,4-BENZOTHIADIAZINE-7-SULFONAMIDE-1,1-DIOXIDE * 3-BENZYLTHIOMETHYL-6-CHLORO-2H-1,2,4-BENZOTHIADIAZINE-7-SULFONAMIDE-1,1-DIOXIDE * 3-BENZYLTHIOMETHYL-6-CHLORO-7-SULFAMOYL-1,2,4-BENZOTHIADIAZINE-1,1-DIOXIDE * 3-BENZYL-THIOMETHYL-6-CHLORO-7-SULFAMYL-1,2,4-BENZOTHIADIAZINE-1,1-DIOXIDE * 3-BENZYLTHIOMETHYL-6-CHLORO-7-SULFAMYL-2H-1,2,4-BENZOTHIADIAZINE-1,1-DIOXIDE * 6-CHLORO-3-(((PHENYLMETHYL)THIO)METHYL)-2H-1,2,4-BENZOTHIADIAZINE-7-SULFONAMIDE DIOXIDE * EDEMEX * EXNA * EXOSALT * FOVANE * FREEURIL * NACLEX * P 1393 * PFIZER 1393 * URESE

SAFETY PROFILE: Poison by intravenous route. A diuretic and antihypertensive agent. When heated to decomposition it emits very toxic fumes of SO_x, NO_x, and Cl^-.

BDE500   CAS: 95-16-9   **HR: 3**
BENZOTHIAZOLE
mf: C_7H_5NS mw: 135.19

PROP: Liquid, odor of quinoline, sltly water-sol. D: 1.246 @ 20°/4°, bp: 228° @ 765 mm.

SYNS: BENZOSULFONAZOLE * O-2857 * 1-THIA-3-AZAINDENE * USAF EK-4812

CONSENSUS REPORTS: Reported in EPA TSCA Inventory.

SAFETY PROFILE: Poison by intraperitoneal, intravenous, and possibly other routes. When heated to decomposition it emits very toxic fumes of SO_x, CN^-, and NO_x.

BDE750   CAS: 120-78-5   **HR: 3**
BENZOTHIAZOLE DISULFIDE
mf: $C_{14}H_8N_2S_4$ mw: 332.48

PROP: Cream to light yellow powder; mp: 175°, d: 1.5.

SYNS: ALTAX * BENZOTHIAZOLYL DISULFIDE * 2-BENZOTHIAZOLYL DISULFIDE * BIS(BENZO-THIAZOLYL)DISULFIDE * BIS(2-BENZOTHIAZYL) DISULFIDE * DI-2-BENZOTHIAZOLYLDISULFIDE * DIBENZOTHIAZYL DISULFIDE * 2,2'-DIBENZOTHI-AZYLDISULFIDE * DIBENZOYLTHIAZYL DISULFIDE * DIBENZTHIAZYL DISULFIDE * 2,2'-DITHIOBIS-(BENZOTHIAZOLE) * DWUSIARCZEK DWUBENZOTI-AZYLU (POLISH) * MBTS * MBTS RUBBER ACCEL-ERATOR * 2-MERCAPTOBENZOTHIAZOLEDISULFIDE * 2-MERCAPTOBENZOTHIAZYLDISULFIDE * ROYAL MBTS * THIOFIDE * USAF B-33 * USAF CY-5 * USAF EK-5432 * VULKACIT DM * VULKACIT DM/MGC

CONSENSUS REPORTS: Reported in EPA TSCA Inventory.

SAFETY PROFILE: Poison by intravenous and intraperitoneal routes. Mildly toxic by ingestion. Experimental teratogenic and reproductive effects. Questionable carcinogen with experimental tumorigenic data. Mutation data reported. When heated to decomposition it emits very toxic fumes of SO_x and NO_x.

BDF000   CAS: 149-30-4   **HR: 3**
2-BENZOTHIAZOLETHIOL
mf: $C_7H_5NS_2$ mw: 167.25

PROP: Light yellow powder. Mp: 170°, d: 1.42 @ 25°.

SYNS: CAPTAX * MBT * MERCAPTOBENZO-THIAZOLE * 2-MERCAPTOBENZOTHIAZOLE * 2-MERKAPTOBENZOTIAZOL (POLISH) * NCI-C56519 * PENNAC MBT POWDER * ROKON * ROTAX * SULFADENE * USAF GY-3 * USAF XR-29

CONSENSUS REPORTS: NTP Carcinogenesis Studies (gavage); Some Evidence rat NTPTR* NTP-TR-332,88: (gavage); Equivocal Evidence: mouse NTPTR* NTP-TR-332,88. Reported in EPA TSCA Inventory.

SAFETY PROFILE: Suspected carcinogen with experimental carcinogenic and tumorigenic data. Poison by intraperitoneal routes. Moderately toxic by ingestion. Experimental terato-

genic and reproductive effects. Incompatible with oxidizers. When heated to decomposition or on contact with acids or acid fumes it emits toxic SO_x and NO_x.

BDG000   CAS: 102-77-2   **HR: 3**
2-BENZOTHIAZOLYL-N-MORPHOLINOSULFIDE
mf: $C_{11}H_{12}N_2OS_2$ mw: 252.37

SYNS: AMAX * 2-BENZOTHIAZOLYLSULFENYL MORPHOLINE * 4-(2-BENZOTHIAZOLYLTHIO)MOR-PHOLINE * 2-(MORPHOLINOTHIO)BENZOTHIAZOLE * MORPHOLINYLMERCAPTOBENZOTHIAZOLE * 2-(4-MORPHOLINYLTHIO)BENZOTHIAZOLE * N-(OXYDIETHYLENE)BENZOTHIAZOLE-2-SULFENA-MIDE * SANTOCURE MOR * SULFENAMIDE M * USAF CY-7 * VULCAFOR BSM

CONSENSUS REPORTS: Reported in EPA TSCA Inventory.

SAFETY PROFILE: Poison by intraperitoneal route. Moderately toxic by ingestion. Questionable carcinogen with experimental neoplastigenic data. Experimental teratogenic effects. Mutation data reported. When heated to decomposition it emits very toxic fumes of NO_x and SO_x.

BDH250   CAS: 95-14-7   **HR: 3**
1H-BENZOTRIAZOLE
mf: $C_6H_5N_3$ mw: 119.14

PROP: Needle-like crystals. Mp: 100°, bp: 204° @ 15 mm.

SYNS: 1,2,-AMINOZOPHENYLENE * AZIMIDOBEN-ZENE * AZIMINOBENZENE * BENZENE AZIMIDE * BENZISOTRIAZOLE * 1,2,3-BENZOTRIAZOLE * COBRATEC #99 * 2,3-DIAZAINDOLE * NCI-C03521 * NSC-3058 * 1,2,3-TRIAZAINDENE * U-6233

CONSENSUS REPORTS: NCI Carcinogenesis Bioassay (feed); Inadequate Studies: mouse, rat NCITR* NCI-CG-TR-88,78. Reported in EPA TSCA Inventory.

SAFETY PROFILE: Poison by intravenous route. Moderately toxic by ingestion and intraperitoneal routes. Questionable carcinogen with experimental tumorigenic data. Mutation data reported. May detonate at 220°C or during vacuum distillation. When heated to decomposition it emits toxic fumes of NO_x.

BDH500   CAS: 98-08-8   **HR: 3**
BENZOTRIFLUORIDE

DOT: UN 2338
mf: $C_7H_5F_3$ mw: 146.12

PROP: Water-white liquid, aromatic odor. Mp: $-29.1°$, bp: $104°$, flash p: $54°F$ (CC), d: 1.197 @ $15.5°/15.5°$, vap d: 5.04, vap press: 11 mm @ $0°$.

SYNS: BENZENYL FLUORIDE * BENZYLIDYNE FLUORIDE * PHENYLFLUOROFORM * (TRIFLUO-ROMETHYL)BENZENE * α,α,α-TRIFLUOROTOLUENE * φ-TRIFLUOROTOLUENE * USAF MA-16

CONSENSUS REPORTS: Reported in EPA TSCA Inventory.

DOT Classification: Flammable Liquid; Label: Flammable Liquid.

SAFETY PROFILE: Poison by intraperitoneal route. Moderately toxic by subcutaneous route. Dangerous fire hazard. To fight fire, use water, foam, CO_2, spray mist, dry chemical. When heated to decomposition it emits toxic fumes of F^-. Incompatible with oxidizing materials.

BDH750   CAS: 215-58-7   **HR: 3**
BENZO(b)TRIPHENYLENE
mf: $C_{22}H_{14}$ mw: 278.36

PROP: Clear plates or leaflets. Mp: $267°$.

SYNS: DB(a,c)A * DIBENZ(a,c)ANTHRACENE * 1,2:3,4-DIBENZANTHRACENE * DIBENZO(a,c)ANTHRACENE * 1,2:3,4-DIBENZOANTHRACENE

CONSENSUS REPORTS: EPA Genetic Toxicology Program. IARC Cancer Review: GROUP 3 IMEMDT 7,56,87; Animal Limited Evidence IMEMDT 32,289,83

SAFETY PROFILE: Questionable carcinogen with experimental tumorigenic data. Human mutation data reported. When heated to decomposition it emits acrid smoke and irritating fumes.

BDJ250   CAS: 2310-17-0   **HR: 3**
S-((3-BENZOXAZOLINYL-6-CHLORO-2-OXO)METHYL) O,O-DIETHYLPHOS-PHORODITHIOATE
mf: $C_{12}H_{15}ClNO_4PS_2$ mw: 367.82

SYNS: AZOFENE * BENZOPHOSPHATE * BENZPHOS * CHIPMAN 11974 * S-(6-CHLORO-3-(MERCAPTOMETHYL)-2-BENZOXAZOLINONE)-O,O-DIETHYL PHOSPHORODITHIOATE * 3-(6-CHLORO-2-OXOBENZOXAZOLIN-3-YL)METHYL-O,O-DIETHYL PHOS-PHOROTHIOLOTHIONATE * O,O-DIAETHYL-S-(6-CHLOR-2-OXO-BEN(b)-1,3-OXALIN-3-YL)-METHYL-DITHIOPHOSPHAT (GERMAN) * O,O-DIETHYL-S-((6-CHLOR-2-OXO-BENZOXAZOLIN-3-YL)-METHYL)-DITHIO FOSFAAT (DUTCH) * O,O-DIETHYL-S-(6-CHLOROBENZOX-AZOLINYL-3-METHYL)DITHIOPHOSPHATE * O,O-DIETHYL-S-((6-CHLORO-2-OXOBENZOXAZOLIN-3-YL)METHYL) PHOSPHORODITHIOATE * O,O-DI-ETHYL-S-(6-CHLORO-2-OXO-BENZOXAZOLIN-3-YL)METHYL-PHOSPHORO THIOLOTHIONATE * 3-DI-ETHYLDITHIOPHOSPHORYLMETHYL-6-CHLOROBENZ-OXAZOLONE-2 * O,O-DIETIL-S-((6-CLORO-2-OXO-BENZOSSAZOLIN-3-IL)-METIL)-DITIOFOS-FATO (ITALIAN) * ENT 27,163 * FOZALON * NIA-9241 * NIAGARA 9241 * NPH-1091 * PHASOLON * PHOSALON * PHOSALONE * PHOZALON * RHODIA RP 11974 * RUBITOX * ZOLON * ZOLONE * ZOLONE PM * ZOOLON

CONSENSUS REPORTS: EPA: Farm Worker Field Reentry.

SAFETY PROFILE: Poison by ingestion, skin contact, and possibly other routes. A cholinesterase inhibitor. When heated to decomposition it emits very toxic fumes of Cl^-, NO_x, PO_x, and SO_x.

BDL750   CAS: 582-61-6   **HR: 3**
BENZOYL AZIDE
mf: $C_7H_5N_3O$ mw: 147.14

SYN: BENZAZIDE, BENZOIC ACID AZIDE

DOT Classification: Forbidden.

SAFETY PROFILE: May explode when heated above $120°C$.

BDM500   CAS: 98-88-4   **HR: 3**
BENZOYL CHLORIDE

DOT: UN 1736
mf: C_7H_5ClO mw: 140.57

PROP: Colorless, fuming, pungent liquid; decomposes in water. Mp: $-0.5°$, bp: $197°$, flash p: $162°F$ (CC), d: 1.2187 @ $15°/15°$, vap press: 1 mm @ $32.1°$, vap d: 4.88.

SYNS: BENZENECARBONYL CHLORIDE * BENZOIC ACID, CHLORIDE * BENZOYL CHLORIDE (DOT) * α-CHLOROBENZALDEHYDE

CONSENSUS REPORTS: IARC Cancer Review: GROUP 3 IMEMDT 7,56,87, Human Inadequate Evidence IMEMDT 29,83,82; Animal Inadequate Evidence IMEMDT 29,83,82. Community Right-To-Know List. Reported in EPA TSCA Inventory. EPA Genetic Toxicology Program.

DOT Classification: Corrosive Material; Label: Corrosive.

SAFETY PROFILE: Questionable carcinogen with experimental tumorigenic data by skin contact. Human systemic effects by inhalation: unspecified effects on olfaction, and respiratory systems. Corrosive effects on the skin, eyes, and mucous membranes by inhalation. Flammable when exposed to heat or flame. Will react with water or steam to produce heat and toxic and corrosive fumes. Violent or explosive reaction with dimethyl sulfoxide, and aluminum chloride + naphthalene. To fight fire, use alcohol foam, CO_2, dry chemical. Incompatible with dimethyl sulfoxide, (NaN_3 + KOH), water, steam, and oxidizers. When heated to decomposition it emits toxic fumes of Cl^-.

BDR750   CAS: 4342-36-3   **HR: 3**
BENZOYLOXYTRIBUTYLSTANNANE
mf: $C_{19}H_{32}O_2Sn$ mw: 411.20

SYNS: TRIBUTYLTIN BENZOATE * TRI-N-BUTYL-ZINN BENZOATE (GERMAN)

CONSENSUS REPORTS: Reported in EPA TSCA Inventory.

OSHA PEL: TWA 0.1 mg(Sn)/m^3 (skin)
ACGIH TLV: TWA 0.1 mg(Sn)/m^3 (skin) (Proposed: TWA 0.1 mg(Sn)/m^3; STEL 0.2 mg(Sn)/m^3 (skin))
NIOSH REL: (Organotin Compounds) TWA 0.1 mg(Sn)/m^3

SAFETY PROFILE: Poison by ingestion and intravenous routes. Moderately toxic by subcutaneous route. When heated to decomposition it emits acrid smoke and irritating fumes.

BDS000   CAS: 94-36-0   **HR: 3**
BENZOYL PEROXIDE

DOT: UN 2085/UN 2086/UN 2087/UN 2088/UN 2089/UN 2090
mf: $C_{14}H_{10}O_4$ mw: 242.24

PROP: White, granular, tasteless, odorless powder. Mp: 103-106° (decomp), bp: decom-poses explosively, autoign temp: 176°F. Sol in benzene, acetone, chloroform; sltly sol in alc; insol in water.

SYNS: ACETOXYL * ACNEGEL * AZTEC BPO * BENOXYL * BENZAC * BENZAKNEW * BENZOIC ACID, PEROXIDE * BENZOPEROXIDE * BENZOYL * BENZOYLPEROXID (GERMAN) * BENZOYLPEROXYDE (DUTCH) * BENZOYL SUPEROXIDE * BZF-60 * CADET * CADOX * CLEARASIL BENZOYL PEROXIDE LOTION * CLEARASIL BP ACNE TREATMENT * CUTICURA ACNE CREAM * DEBROXIDE * DIBENZOYLPEROXID (GERMAN) * DIBENZOYL PEROXIDE (MAK) * DIBENZOYLPEROXYDE (DUTCH) * DIPHENYLGLYOXAL PEROXIDE * DRY AND CLEAR * EPICLEAR * FOSTEX * GAROX * INCIDOL * LOROXIDE * LUCIDOL * LUPERCO * LUPEROX FL * NAYPER B and BO * NOROX BZP-250 * NOVADELOX * OXY-5 * OXY-10 * OXYLITE * OXY WASH * PANOXYL * PEROSSIDO di BENZOILE(ITALIAN) * PEROXYDE de BENZOYLE (FRENCH) * PERSADOX * QUINOLOR COMPOUND * SULFOXYL * SUPEROX * THERADERM * TOPEX * VANOXIDE * XERAC

CONSENSUS REPORTS: IARC Cancer Review: GROUP 3 IMEMDT 7,56,87, Animal Inadequate Evidence IMEMDT 36,267,85; Human Inadequate Evidence IMEMDT 36,267,85. Reported in EPA TSCA Inventory. EPA Genetic Toxicology Program. Community Right-To-Know List.

OSHA PEL: TWA 5 mg/m^3
ACGIH TLV: TWA 5 mg/m^3
DFG MAK: 5 mg/m^3
NIOSH REL: TWA 5 mg/m^3
DOT Classification: Organic Peroxide; Label: Organic Peroxide.

SAFETY PROFILE: Poison by ingestion and intraperitoneal routes. Can cause dermatitis, asthmatic effects, testicular atrophy, and vasodilation. An allergen and eye irritant. Human mutation data reported. Questionable carcinogen with experimental tumorigenic data. Moderate fire hazard by spontaneous chemical reaction in contact with reducing agents. It ignites readily and burns rapidly. A powerful oxidizer. Dangerous explosion hazard; may explode spontaneously, when heated to above melting point, or when overheated under confinement. It is moderately sensitive to heat, shock, friction or contact with combustible materials. Explosive decomposition above the mp (103°) forms flammable products.

Explosive or violent reaction on contact with N,N-dimethylaniline; aniline; dimethyl sulfide; lithium tetrahydroaluminate; and N-bromosuccinimide + 4-toluic acid. Mixture with carbon tetrachloride + ethylene explodes at elevated temperatures and pressures. Reacts violently in contact with various organic or inorganic acids; alcohols; amines; metallic naphthenates; as well as with polymerization accelerators; i.e., dimethylaniline; and (CCl_4 + C_2H_4). Violent reaction with charcoal when heated above 50°. Decomposition produces dense white smoke of benzoic acid; phenyl benzoate; terphenyls; biphenyls; benzene and carbon dioxide. Vigorous reaction leading to ignition with methylmethacrylate; and vinyl acetate + ethyl acetate. To fight fire, use water spray, foam. All precautions must be taken to guard against fire and explosion hazards. Keep in a cool place; out of the direct rays of the sun; away from sparks, open flames, and other sources of heat; avoid shock, rough handling, friction from grinding, etc. Isolated storage is required; keep away from possible contact with acids; alcohols; ethers; or other reducing agents or polymerization catalysts such as dimethylaniline. Complete instructions on storage and handling available from manufacturer.

BDS250   HR: 2
BENZOYL PEROXIDE, WET

PROP: A paste or wetted granular material containing at least 30% water. Autoign temp 176°F.

SAFETY PROFILE: Moderate fire hazard by chemical reaction with reducing agents; a powerful oxidizer. Mixed with a large surplus of water (i.e., 30%), this material is relatively safe. It is most dangerous when it contains very little water (1% or less). To fight fire, use water, foam or spray. Care must be taken to prevent drying out of wet material.

BDX000   CAS: 140-11-4   HR: 3
BENZYL ACETATE
mf: $C_9H_{10}O_2$ mw: 150.19

PROP: Colorless liquid; sweet, floral fruity odor. Mp: −51.5°, bp: 213.5°, flash p: 216°F (CC), d: 1.06, autoign temp: 862°F, vap press: 1 mm @ 45°, vap d: 5.1, refr index: 1.501. Sol in alc, most fixed oils, propylene glycol; insol in glycerin and water @ 214°.

SYNS: ACETIC ACID BENZYL ESTER * ACETIC ACID PHENYLMETHYL ESTER * α-ACETOXYTOLUENE * BENZYL ETHANOATE * FEMA No. 2135 * NCI-C06508

CONSENSUS REPORTS: IARC Cancer Review: GROUP 3 IMEMDT 7,56,87; Animal Limited Evidence IMEMDT 40,109,86. NTP Carcinogenesis Studies (gavage); Some Evidence: mouse, rat NTPTR* NTP-TR-250,86. Reported in EPA TSCA Inventory.

SAFETY PROFILE: A poison by inhalation. Moderately toxic by ingestion and subcutaneous routes. Human systemic effects by inhalation: an antipsychotic, unspecified respiratory and urinary system effects. Questionable carcinogen with experimental tumorigenic data. Combustible liquid. To fight fire, use alcohol foam, CO_2. When heated to decomposition it emits irritating fumes.

BDX500   CAS: 100-51-6   HR: 3
BENZYL ALCOHOL
mf: C_7H_8O mw: 108.15

PROP: Found in jasmine, hyacinth, ylang-ylang oils and at least two dozen other essential oils. Water-white liquid; faint, aromatic odor, sharp burning taste. Mp: −15.3°, bp: 205.7°, flash p: 213°F (CC), d: 1.042, autoign temp: 817°F, vap press: 1 mm @ 58.0°, vap d: 3.72, refr index: 1.540. Misc with alc, chloroform, ether, water @ 206°(decomp).

SYNS: BENZAL ALCOHOL * BENZENECARBINOL * BENZENEMETHANOL * BENZOYL ALCOHOL * FEMA No. 2137 * HYDROXYTOLUENE * α-HYDROXYTOLUENE * NCI-C06111 * PHENOLCARBINOL * PHENYLCARBINOL * PHENYLMETHANOL * PHENYLMETHYL ALCOHOL * α-TOLUENOL

CONSENSUS REPORTS: EPA Genetic Toxicology Program. Reported in EPA TSCA Inventory.

SAFETY PROFILE: Poison by ingestion, intraperitoneal, intravenous, parenteral routes. Moderately toxic by inhalation, skin contact and subcutaneous routes. A moderate skin and severe eye irritant. Combustible liquid. Mixtures with sulfuric acid decompose explosively at 180°. Exothermic polymerization is catalyzed by HBr + iron when heated above 100°. To fight fire, use alcohol foam, CO_2, dry chemical. When heated to decomposition it emits acrid smoke and fumes.

BDY669 CAS: 61-33-6 **HR: 3**
BENZYL-6-AMINOPENICILLINIC ACID
mf: $C_{16}H_{18}N_2O_4S$ mw: 334.42

SYNS: ABBOCILLIN * (5R,6R)-BENXYLPENICILLIN * BENZOPENICILLIN * BENZYLPENICILLIN * BENZYLPENICILLIN G * BENZYLPENICILLINIC ACID * CILLORAL * CILOPEN * COMPOCILLIN G * COSMOPEN * DROPCILLIN * FREE BENZYLPENICILLIN * PENICILLIN G * GALOFAK * GELACILLIN * LIQUACILLIN * PHENYLACETAMIDOPENICILLANIC ACID * (PHENYLMETHYL) PENICILLINIC ACID * PRADUPEN * SPECILLINE G

CONSENSUS REPORTS: EPA Genetic Toxicology Program.

SAFETY PROFILE: Poison by ingestion, intravenous, intracerebral, intraspinal, subcutaneous and possibly other routes. Human (child) systemic effects by parenteral route: changes in cochlear (inner ear) structure or function, convulsions, and dyspnea. Questionable carcinogen with experimental tumorigenic data. Mutation data reported. When heated to decomposition it emits very toxic fumes of NO_x and SO_x.

BEA500 CAS: 36226-64-9 **HR: 3**
BENZYLBARBITAL
mf: $C_{13}H_{14}N_2O_3$ mw: 246.29

SYNS: 5-BENZYL-5-ETHYLBARBITURIC ACID * ETHYLBENZYLBARBITURIC ACID * 5-ETHYL-5-(PHENYLMETHYL)-2,4,6(1H,3H,5H)-PYRIMIDINETRIONE (9CI)

SAFETY PROFILE: Poison by ingestion, intraperitoneal, and subcutaneous routes. When heated to decomposition it emits toxic fumes of NO_x. An hypnotic agent.

BEC000 CAS: 100-39-0 **HR: 2**
BENZYL BROMIDE
DOT: UN 1737
mf: C_7H_7Br mw: 171.05

PROP: Clear, refractive liquid; pleasant odor, lachrymator, insol in water. Mp: −4.0°, bp: 198°, d: 1.438 @ 22°/0°, vap d: 5.8.

SYNS: (BROMOMETHYL)BENZENE * p-(BROMOMETHYL)NITROBENZENE * BROMOPHENYLMETHANE * φ-BROMOTOLUENE * α-BROMOTOLUENE (DOT)

CONSENSUS REPORTS: Reported in EPA TSCA Inventory.

DOT Classification: Corrosive Material; Label: Corrosive.

SAFETY PROFILE: Intensely irritating and corrosive to skin, eyes, and mucous membranes. Large doses cause central nervous system depression. Mutation data reported. Reaction with molecular sieve produces toxic hydrogen bromide gas.

BEC500 CAS: 85-68-7 **HR: 3**
BENZYL BUTYL PHTHALATE
mf: $C_{19}H_{20}O_4$ mw: 312.39

PROP: Clear, oily liquid. Mp: $< -35°$, bp: 370°, flash p: 390°F, d: 1.116 @ 25°/25°, vap d: 10.8.

SYNS: BBP * 1,2-BENZENEDICARBOXYLIC ACID, BUTYL PHENYLMETHYL ESTER * BUTYL BENZYL PHTHALATE * n-BUTYL BENZYL PHTHALATE * NCI-C54375 * PALATINOL BB * SANTICIZER 160 * SICOL 160 * UNIMOLL BB

CONSENSUS REPORTS: IARC Cancer Review: GROUP 3 IMEMDT 7,56,87, Animal Inadequate Evidence IMEMDT 29,193,82; NTP Carcinogenesis Bioassay (feed); No Evidence: mouse NTPTR* NTP-TR-213,82; Clear Evidence: rat NTPTR* NTP-TR-213,82. Reported in EPA TSCA Inventory. Community Right-To-Know List.

SAFETY PROFILE: Questionable carcinogen with experimental carcinogenic data. Moderately toxic by ingestion and intraperitoneal routes. Experimental reproductive effects. Combustible when exposed to heat or flame; can react with oxidizers. To fight fire, use spray or mist, CO_2, dry chemical. When heated to decomposition it emits acrid smoke and irritating fumes.

BED000 CAS: 103-37-7 **HR: 2**
BENZYL n-BUTYRATE
mf: $C_{11}H_{14}O_2$ mw: 178.25

PROP: Colorless liquid; floral plum-like odor. D: 1.006, refr index: 1.492, flash p: +212°F. Sol in fixed oils; insol in glycerin, propylene glycol, water @ 239°.

SYNS: BENZYL n-BUTANOATE * FEMA No. 2140

CONSENSUS REPORTS: Reported in EPA TSCA Inventory.

SAFETY PROFILE: Moderately toxic by ingestion. Combustible liquid. When heated to decomposition it emits acrid smoke and irritating fumes.

BEE375   CAS: 100-44-7   HR: 3
BENZYL CHLORIDE

DOT: UN 1738
mf: C_7H_7Cl mw: 126.59

PROP: Colorless liquid, very refractive; irritating, unpleasant odor. Mp: −43°, bp: 179°, lel: 1.1%, flash p: 153°F, d: 1.1026 @ 18/4°, autoign temp: 1085°F, vap d: 4.36.

SYNS: BENZILE (CLORURO di) (ITALIAN) * BENZYLCHLORID (GERMAN) * BENZYLE (CHLORURE de) (FRENCH) * CHLOROMETHYLBENZENE * CHLOROPHENYLMETHANE * α-CHLOROTOLUENE * φ-CHLOROTOLUENE * α-CHLORTOLUOL (GERMAN) * CHLORURE de BENZYLE (FRENCH) * NCI-C06360 * RCRA WASTE NUMBER P028 * TOLYL CHLORIDE

CONSENSUS REPORTS: IARC Cancer Review: Animal Limited Evidence IMEMDT 29,49,82; Animal Sufficient Evidence IMEMDT 11,217,76; Human Inadequate Evidence IMEMDT 29,49,82. EPA Genetic Toxicology Program. Community Right-To-Know List. Reported in EPA TSCA Inventory. EPA Extremely Hazardous Substances List.

OSHA PEL: TWA 1 ppm
ACGIH TLV: TWA 1 ppm
DFG MAK: 1 ppm (5 mg/m^3); Suspected Carcinogen.
NIOSH REL: (Benzyl Chloride) CL 5 mg/m^3/ 15M
DOT Classification: Corrosive Material; Label: Corrosive.

SAFETY PROFILE: Suspected carcinogen with experimental carcinogenic and tumorigenic data. Poison by inhalation. Moderately toxic by ingestion and subcutaneous routes. Experimental reproductive effects. Human mutation data reported. A corrosive irritant to skin, eyes, and mucous membranes. Flammable and moderately explosive when exposed to heat or flame. Can react vigorously with oxidizing materials. May explode during distillation. The decomposition rate can reach explosive violence in presence of metals such as iron. Catalytic impurities (e.g., aluminum, iron, rust) or sodium acetate + pyridine + iron (at 115°C) may cause violent polymerization reactions. Will react with water or steam to produce toxic and corrosive fumes.

Incompatible with dimethyl sulfoxide. When heated to decomposition it emits toxic fumes of Cl$^-$.

BEF500   CAS: 501-53-1   HR: 3
BENZYL CHLOROFORMATE

DOT: UN 1739
mf: $C_8H_7ClO_2$ mw: 170.60

PROP: Colorless to pale yellow liquid, odor of phosgene.

SYNS: BENZYLCARBONYL CHLORIDE * BENZYL CHLOROCARBONATE (DOT) * BENZYL CHLOROFORMATE (DOT) * BENZYLOXYCARBONYL CHLORIDE * BZCF * CARBOBENZOXY CHLORIDE * CARBOBENZYLOXY CHLORIDE * CHLOROFORMIC ACID BENZYL ESTER

CONSENSUS REPORTS: Reported in EPA TSCA Inventory.

DOT Classification: Corrosive Material; Label: Corrosive.

SAFETY PROFILE: Poison by ingestion and inhalation routes. A powerful corrosive irritant. Thermally unstable. Will react with water or steam to produce toxic and corrosive fumes and heat. Iron salts catalyze the explosive decomposition of the ester. When heated to decomposition it emits toxic fumes of Cl$^-$ and phosgene.

BEG750   CAS: 103-41-3   HR: 2
BENZYL CINNAMATE
mf: $C_{16}H_{14}O_2$ mw: 238.30

PROP: Found in balsams of Peru, Tolu, Styrax, Copaiba and others. White crystals; aromatic odor. Mp: 39°, bp: 350.0°, vap press: 1 mm @ 173.8°, flash p: +212°F. Sol in fixed oils; insol in glycerin, propylene glycol.

SYNS: BENZYL ALCOHOL CINNAMIC ESTER * BENZYL γ-PHENYLACRYLATE * CINNAMEIN * trans-CINNAMIC ACID BENZYL ESTER * FEMA No. 2142 * 3-PHENYL-2-PROPENOIC ACID PHENYLMETHYL ESTER (9CI)

CONSENSUS REPORTS: Reported in EPA TSCA Inventory.

SAFETY PROFILE: Moderately toxic by ingestion. A mild allergen and skin irritant. Combustible liquid. When heated to decomposition it emits acrid smoke and irritating fumes.

BEM000   CAS: 139-07-1   **HR: 2**
**BENZYLDIMETHYLDODECYLAMMO-
NIUM CHLORIDE**
mf: $C_{21}H_{38}N \cdot Cl$ mw: 340.05

SYN: DODECYL DIMETHYL BENZYLAMMONIUM CHLORIDE

SAFETY PROFILE: A skin and eye irritant. When heated to decomposition it emits very toxic fumes of NO_x, NH_3, and Cl^-.

BEM750   CAS: 525-02-0   **HR: 3**
**1-BENZYL-2,5-DIMETHYL SEROTONIN
HYDROCHLORIDE**
mf: $C_{19}H_{22}N_2O \cdot ClH$ mw: 330.89

SYNS: 3-(2-AMINOETHYL)-1-BENZYL-5-METHOXY-2-METHYLINDOLE HYDROCHLORIDE * BAS * BENANSERIN HYDROCHLORIDE * BENZYL ANTISEROTONIN * 1-BENZYL-2-METHYL-3-(2-AMINO-ETHYL)-5-METHOXYINDOLE HYDROCHLORIDE * 1-BENZYL-2-METHYL-5-METHOXYTRYPTAMINE HYDROCHLORIDE * SEROTONIN BENZYL ANALOG * WOOLLEY'S ANTISEROTONIN

SAFETY PROFILE: Poison by intraperitoneal route. A serotonin antagonist which causes psychotropic effects in humans. Experimental reproductive effects. When heated to decomposition it emits very toxic fumes of HCl and NO_x.

BEN000   CAS: 121-54-0   **HR: 3**
**BENZYLDIMETHYL(2-(2-(p-(1,1,3,3-
TETRAMETHYLBUTYL)PHENOXY)
ETHOXY)ETHYL) AMMONIUM
CHLORIDE**
mf: $C_{27}H_{42}NO_2 \cdot Cl$ mw: 448.15

PROP: Colorless crystals. Sol in water.

SYNS: ANTI-GERM 77 * ANTISEPTOL * BENZETHONIUM CHLORIDE * BENZETONIUM CHLORIDE * BENZYLDIMETHYL-p-(1,1,3,3-TETRAMETHYLBUTYL)PHENOXYETHOXY-ETHYLAMMONIUM CHLORIDE * BZT * DIAPP * DIISOBUTYLPHENOXYETHOXY-ETHYLDIMETHYL BENZYL AMMONIUM CHLORIDE * DISILYN * HYAMINE * HYAMINE 1622 * NCI-C61494 * p-tert-OCTYLPHENOXYETHOXY-ETHYLDIMETHYLBENZYL AMMONIUM CHLORIDE * PHEMERIDE * PHEMEROL CHLORIDE * PHEMITHYN * POLYMINE D * QUATRACHLOR * SOLAMINE

CONSENSUS REPORTS: Reported in EPA TSCA Inventory.

SAFETY PROFILE: Poison by ingestion, subcutaneous, intraperitoneal, and intravenous routes. A severe eye irritant. Questionable carcinogen with experimental neoplastigenic data. When heated to decomposition it emits very toxic fumes of Cl^-, NH_3, and NO_x. A topical anti-infective agent.

BEO250   CAS: 103-50-4   **HR: 2**
BENZYL ETHER
mf: $C_{14}H_{14}O$ mw: 198.28

PROP: Colorless to pale yellow liquid. Mp: 5°, bp: 298°, flash p: 275°F (CC), d: 1.039, vap d: 6.84, refr index: 1.557.

SYNS: BENZYL OXIDE (CZECH) * DIBENZYLETHER (CZECH) * FEMA No. 2371

CONSENSUS REPORTS: Reported in EPA TSCA Inventory.

SAFETY PROFILE: Moderately toxic by ingestion. Vapors are probably narcotic in high concentration. A skin and eye irritant. Combustible when exposed to heat or flame; can react with oxidizing materials. Moderate explosion hazard by spontaneous chemical reaction. To fight fire, use CO_2, dry chemical.

BEO750   CAS: 2016-63-9   **HR: 3**
**8'-BENZYL-7(2-(ETHYL(2-
HYDROXYETHYL)AMINO)ETHYL)
THEOPHYLLINE HYDROCHLORIDE**
mf: $C_{20}H_{27}N_5O_3 \cdot ClH$ mw: 421.98

SYNS: BAMIFYLLINE HYDROCHLORIDE * BAMIPHYLLINE HYDROCHLORIDE * BAX 2793Z * BENZETAMOPHYLLINE HYDROCHLORIDE * 8-BENZYL-7-(N-ETHYL-N-(β-HYDROXYETHYL)AMINO-ETHYL)THEOPHYLLINE HYDROCHLORIDE * 8102 CB HYDROCHLORIDE * 7-(N-(β-HYDROXYETHYL)-N-ETHYL)-AMINOETHYL-8-BENZYL-THEOPHYLLINE * TRENTADIL HYDROCHLORIDE

SAFETY PROFILE: Poison by ingestion, intraperitoneal, and intravenous routes. When heated to decomposition it emits toxic fumes of HCl and NO_x. A bronchodilator.

BEP250   CAS: 104-57-4   **HR: 2**
BENZYL FORMATE
mf: $C_8H_8O_2$ mw: 136.16

SYNS: BENZYL ALCOHOL FORMATE * BENZYL METHANOATE

CONSENSUS REPORTS: Reported in EPA TSCA Inventory.

SAFETY PROFILE: Moderately toxic by ingestion and skin contact. Probably narcotic in high concentrations. When heated to decomposition it emits acrid, irritating fumes.

BEP500   CAS: 10453-86-8   **HR: 3**
5-BENZYL-3-FURYL METHYL(±)-cis,trans-CHRYSANTHEMATE
mf: $C_{22}H_{26}O_3$ mw: 338.48

SYNS: BENZOFUROLINE * BENZYFUROLINE * (5-BENZYL-3-FURYL) METHYL-2,2-DIMETHYL-3-(2-METHYLPROPENYL)-CYCLOPROPANECARBOXYLATE * CHRYSON * CHRYSRON * DIMETHYL-3-(2-METHYL-1-PROPENYL)CYCLOPROPANECARBOXYLATE * ENT 27,474 * FMC 17370 * FOR-SYN * NIA 17170 * NRDC 104 * NSC 195022 * OMS-1206 * PREMGARD * PYNOSECT * PYRETHERM * RESMETHRIN * RESMETRINA (PORTUGUESE) * SBP-1382 * S.B. PENICK 1382 * SYNTHRIN

CONSENSUS REPORTS: EPA Genetic Toxicology Program.

SAFETY PROFILE: Poison by inhalation, ingestion, and intravenous routes. Moderately toxic by skin contact. When heated to decomposition it emits acrid and irritating fumes.

BEU250   CAS: 622-78-6   **HR: 3**
BENZYL-ISOTHIOCYANATE
mf: C_8H_7NS mw: 149.22

PROP: Orange-red, crystalline solid. Mp: 41°, bp: 230°, d: 1.125

SYNS: BENZYL MUSTARD OIL * BENZYLSENFOEL (GERMAN) * ISOTHIOCYANIC ACID BENZYL ESTER

CONSENSUS REPORTS: Reported in EPA TSCA Inventory.

SAFETY PROFILE: Poison by intraperitoneal and subcutaneous routes. Intensely irritating. Mutation data reported. Moderate fire hazard via heat, flame, and oxidizers. To fight fire, use water, spray, foam, dry chemical. When heated to decomposition it emits very toxic NO_x and SO_x.

BEU500   CAS: 538-28-3   **HR: 3**
BENZYLISOTHIOUREA HYDROCHLORIDE
mf: $C_8H_{10}N_2S \cdot ClH$ mw: 202.72

SYNS: BENZYLISOTHIOURONIUM CHLORIDE * 2-BENZYLISOTHIOURONIUM CHLORIDE * 2-BENZYL-2-THIO-PSEUDOUREA HYDROCHLORIDE * BENZYL THIOPSEUDOUREA HYDROCHLORIDE * BENZYLTHIURONIUM CHLORIDE * S-BENZYLTHIURONIUM CHLORIDE * BTKH * ISOTHIOURONIUM CHLORIDE, BENZYL * 2-THIO-2-BENZYL-PSEUDOUREA HYDROCHLORIDE * TL 944 * USAF EK-2124

CONSENSUS REPORTS: Reported in EPA TSCA Inventory.

SAFETY PROFILE: Poison by ingestion, intraperitoneal, subcutaneous, and intravenous routes. When heated to decomposition it emits very toxic fumes of HCl, SO_x, and NO_x.

BFD250   CAS: 69-57-8   **HR: 3**
BENZYL PENICILLINIC ACID SODIUM SALT
mf: $C_{16}H_{17}N_2O_4S \cdot Na$ mw: 356.40

SYNS: AMERICAN PENICILLIN * BENZYLPENICILLIN SODIUM * CRYSTAPEN * MYCOFARM * NOVOCILLIN * PEN-A-BRASIVE * PENICILLIN-G, MONOSODIUM SALT * PENICILLIN G, SODIUM * PENICILLIN G, SODIUM SALT * PENILARYN * PENZYLPENICILLIN SODIUM SALT * SODIUM BENZYLPENICILLIN * SODIUM BENZYLPENICILLIN G * SODIUM BENZYLPENICILLINATE * SODIUM PENICILLIN * SODIUM PENICILLIN G * SODIUM PENICILLIN II * VETICILLIN

CONSENSUS REPORTS: EPA Genetic Toxicology Program.

SAFETY PROFILE: Poison by intracerebral, parenteral, and intramuscular routes. Moderately toxic via intravenous route. Mildly toxic by ingestion. Experimental teratogenic and reproductive effects. Questionable carcinogen with experimental tumorigenic data. When heated to decomposition it emits very toxic fumes of NO_x, Na_2O, and SO_x. An antibiotic.

BFD400   **HR: 1**
BENZYL PHENYLACETATE
mf: $C_{15}H_{14}O_2$ mw: 226.27

PROP: Colorless liquid; sweet, floral odor with honey undertone. D: 1.095-1.099, refr index: 1.553-1.558, flash p: +212°F. Sol in alc, chloroform, ether.

SYN: FEMA No. 2149

SAFETY PROFILE: Combustible liquid. When heated to decomposition it emits acrid smoke and irritating fumes.

BFJ750   CAS: 118-58-1   **HR: 2**
BENZYL SALICYLATE
mf: $C_{14}H_{12}O_3$ mw: 228.26

PROP: Thick colorless liquid, pleasant odor. Bp: 208° @ 26 mm, d: 1.175 @ 20°, refr index: 1.579. Sol in fixed oils; insol in glycerin, propylene glycol.

SYNS: BENZYL-o-HYDROXYBENZOATE * FEMA No. 2151

CONSENSUS REPORTS: Reported in EPA TSCA Inventory.

SAFETY PROFILE: Moderately toxic by ingestion. Combustible when exposed to heat or flame. When heated to decomposition it emits acrid smoke and irritating fumes. Incompatible with oxidizing materials.

BFL000   CAS: 3012-37-1   **HR: 3**
BENZYL THIOCYANATE
mf: C_8H_7NS mw: 149.22

PROP: Orange-red, crystalline solid. Mp: 41°; bp: 230°; d: 1.125.

SYNS: BENZYL MUSTARD OIL * PHENYLMETHYL ESTER THIOCYANIC ACID (9CI) * SOLVAT 14 * α-THIOCYANATOTOLUENE * TROPEOLIN

CONSENSUS REPORTS: Reported in EPA TSCA Inventory.

SAFETY PROFILE: Poison by subcutaneous and intraperitoneal routes. When heated to decomposition it emits very toxic fumes of NO_x, SO_x, and CN^-.

BFL250   CAS: 98-07-7   **HR: 3**
BENZYL TRICHLORIDE

DOT: UN 2226
mf: $C_7H_5Cl_3$ mw: 195.47

PROP: Clear, colorless to yellowish liquid; penetrating odor. Mp: −5°, bp: 221°, d: 1.38 @ 15.5°/15.5°, vap d: 6.77.

SYNS: BENZENYL CHLORIDE * BENZENYL TRI-CHLORIDE * BENZOIC TRICHLORIDE * BENZOTRI-CHLORIDE (DOT, MAK) * BENZYLIDYNE CHLORIDE * CHLORURE de BENZENYLE (FRENCH) * PHENYL CHLOROFORM * PHENYLTRICHLOROMETHANE * RCRA WASTE NUMBER U023 * TOLUENE TRI-

CHLORIDE * TRICHLOORMETHYLBENZEEN (DUTCH) * TRICHLORMETHYLBENZOL (GERMAN) * TRI-CHLOROMETHYLBENZENE * 1-(TRICHLOROMETHYL) BENZENE * TRICHLOROPHENYLMETHANE * α,α,α-TRICHLOROTOLUENE * φ,φ,φ-TRICHLORO-TOLUENE * TRICLOROMETILBENZENE (ITALIAN) * TRICLOROTOLUENE (ITALIAN)

CONSENSUS REPORTS: IARC Cancer Review: Human Limited Evidence IMEMDT 29,73,82; Animal Sufficient Evidence IMEMDT 29,73,82. NTP Fourth Annual Report On Carcinogens, 1984. EPA Genetic Toxicology Program. EPA Extremely Hazardous Substances List. Reported in EPA TSCA Inventory.

DFG MAK: Suspected Carcinogen.
DOT Classification: IMO: Corrosive Material; Label: Corrosive.

SAFETY PROFILE: Confirmed carcinogen with experimental carcinogenic data by skin contact and neoplastiginic data by inhalation. Experimental poison by inhalation. Corrosive to the skin, eyes, and mucous membranes. Large doses can cause central nervous system depression. Mutation data reported. When heated to decomposition it emits toxic fumes of Cl^-.

BFM750   CAS: 4525-46-6   **HR: 3**
BENZYL TRIMETHYL AMMONIUM IODIDE
mf: $C_{10}H_{16}N \cdot I$ mw: 277.17

SYNS: BENZYLDIMETHYLAMINE METHIODIDE * PHENMETHYL TRIMETHYLAMMONIUM IODIDE

CONSENSUS REPORTS: Reported in EPA TSCA Inventory.

SAFETY PROFILE: Poison by intraperitoneal and intravenous routes. When heated to decomposition it emits very toxic fumes of NO_x, NH_3, and I^-.

BFN500   CAS: 2086-83-1   **HR: 3**
BERBERINE
mf: $C_{20}H_{18}NO_4$ mw: 336.39

PROP: White to yellow crystals. Mp (anhyd): 145°.

SYNS: BERBERIN * 9,10-DIMETHOXY-2,3-(METHYL-ENEDIOXY)-7,8,13,13A-TETRAHYDROBERBINIUM

SAFETY PROFILE: An alkaloid poison by ingestion and subcutaneous routes. In humans, toxic doses lower the body temperature, increase peristalsis, and cause death by central paralysis. Mutation data reported. Should carry a poison label. Should never be ingested without the advice of a physician. Should not be handled excessively since it may be absorbed through the skin and have a toxic effect upon the body. An antimalarial agent. When heated to decomposition it emits highly toxic fumes of NO_x.

BFN750   CAS: 69352-97-2   **HR: 3**
BERBERINE SULFATE TRIHYDRATE
mf: $C_{40}H_{36}N_2O_8 \cdot O_4S \cdot 3H_2O$ mw: 822.90

SYNS: 5,6-DIHYDRO-9,10-DIMETHOXYBENZO(g)-1,3-BENZODIOXOLO(5,6-a)QUINOLIZINIUM SULFATE TRIHYDRATE * 7,8,13,13A-TETRADEHYDRO-9,10-DIMETHOXY-2,3-(METHYLENEDIOXY)BERBINIUM SULFATE TRIHYDRATE * UMBELLATINE SULFATE TRIHYDRATE

SAFETY PROFILE: Poison by intraperitoneal and subcutaneous routes. When heated to decomposition it emits very toxic SO_x and NO_x.

BFO000   CAS: 8007-75-8   **HR: 1**
BERGAMOT OIL RECTIFIED

PROP: Yellow-green liquid; agreeable odor. *Composition:* 1-linalyl acetate, 1-linalool, d-limonene, dipentene, bergaptene. By rectification of bergamot oil expressed, under vacuum, to remove completely the furocoumarins and other related nonvolatile residues; found in the fruit of citrus *Bergamia risso et poiteau (Fam. rutaceae).* D: 0.875-0.880 @ 25°/25°. Misc with alc, glacial acetic acid; sol in fixed oils; insol in glycerin, propylene glycol.

SYNS: BERGAMOTTE OEL (GERMAN) * OIL of BERGAMOT, COLDPRESSED * OIL of BERGAMOT, RECTIFIED

CONSENSUS REPORTS: Reported in EPA TSCA Inventory.

SAFETY PROFILE: Mildly toxic by ingestion. A mild skin irritant and allergen. Combustible. When heated to decomposition it emits acrid smoke and irritating fumes.

BFO250   CAS: 12161-82-9   **HR: 3**
BERTRANDITE
mf: $H_{10}O_9Si_2 \cdot H_2O \cdot Be_4$ mw: 264.34

SYN: BERYLLIUM SILICATE HYDRATE

CONSENSUS REPORTS: IARC Cancer Review: Animal Sufficient Evidence IMEMDT

1,17,72; Animal Inadequate Evidence IMEMDT 23,143,80. Reported in EPA TSCA Inventory. Beryllium and its compounds are on the Community Right-To-Know List.

OSHA PEL: (Transitional: TWA 0.002 mg(Be)/m^3; CL 0.005; Pk 0.025/30M/8H) TWA 0.002 mg(Be)/m^3; STEL 0.005 mg(Be)/m^3/30M; CL 0.025 mg(Be)/m^3
ACGIH TLV: TWA 0.002 mg(Be)/m^3, Suspected Human Carcinogen.
NIOSH REL: CL not to exceed 0.0005 mg (Be)/m^3

SAFETY PROFILE: Confirmed carcinogen. When heated to decomposition it emits very toxic fumes of BeO.

BFO500   CAS: 1302-52-9   **HR: 3**
BERYL
mf: $Al_2O_{18}Si_6 \cdot 3Be$ mw: 537.53

PROP: Green, blue, yellow or white crystals. D: 2.63-2.91.

SYNS: BERYL ORE * BERYLLIUM ALUMINOSILICATE * BERYLLIUM ALUMINUM SILICATE

CONSENSUS REPORTS: IARC Cancer Review: GROUP 2A IMEMDT 7,127,87; Animal Sufficient Evidence IMEMDT 23,143,80; IMEMDT 1,17,72. NTP Fourth Annual Report On Carcinogens, 1984. Reported in EPA TSCA Inventory. Beryllium and its compounds are on the Community Right-To-Know List.

OSHA PEL: (Transitional: TWA 0.002 mg(Be)/m^3; CL 0.005; Pk 0.025/30M/8H) TWA 0.002 mg(Be)/m^3; STEL 0.005 mg(Be)/m^3/30M; CL 0.025 mg(Be)/m^3
ACGIH TLV: TWA 0.002 mg(Be)/m^3, Suspected Human Carcinogen.
NIOSH REL: CL not to exceed 0.0005 mg (Be)/m^3

SAFETY PROFILE: Confirmed carcinogen with experimental carcinogenic, neoplastigenic, and tumorigenic data. When heated to decomposition it emits toxic fumes of BeO.

BFO750   CAS: 7440-41-7   **HR: 3**
BERYLLIUM
DOT: UN 1567
af: Be   aw: 9.01

PROP: A grayish-white, hard, light metal. Mp: 1278°, bp: 2970°, d: 1.85.

SYNS: BERYLLIUM-9 * BERYLLIUM, metal powder (DOT) * GLUCINUM * RCRA WASTE NUMBER P015

CONSENSUS REPORTS: IARC Cancer Review: GROUP 2A IMEMDT 7,127,87; Human Limited Evidence IMEMDT 23,143,80; Animal Sufficient Evidence IMEMDT 23,143,80; IMEMDT 1,17,72. NTP Fourth Annual Report On Carcinogens, 1984. Beryllium and its compounds are on the Community Right-To-Know List. Reported in EPA TSCA Inventory.

OSHA PEL: (Transitional: TWA 0.002 mg(Be)/m^3; CL 0.005; Pk 0.025/30M/8H) TWA 0.002 mg(Be)/m^3; STEL 0.005 mg(Be)/m^3/30M; CL 0.025 mg(Be)/m^3
ACGIH TLV: TWA 0.002 mg/m^3, Suspected Human Carcinogen.
DFG TRK: Animal Carcinogen, Suspected Human Carcinogen. Grinding of beryllium metal and alloys: 0.005 mg/m^3 calculated as Be in that portion of dust that can possibly be inhaled; other Be compounds: 0.002 mg/m^3 calculated as Be in that portion of dust that can possibly be inhaled
NIOSH REL: CL not to exceed 0.0005 mg (Be)/m^3
DOT Classification: Poison B, Flammable Solid Powder and Poison (metal).

SAFETY PROFILE: Confirmed carcinogen with experimental carcinogenic, neoplastigenic, and tumorigenic data. A deadly poison by intravenous route. Human systemic effects by inhalation: lung fibrosis, dyspnea, and weight loss. Human mutation data reported. A moderate fire hazard in the form of dust or powder, or when exposed to flame or by spontaneous chemical reaction. Slight explosion hazard in the form of powder or dust. Incompatible with halocarbons. Reacts incandescently with fluorine or chlorine. Mixtures of the powder with CCl_4 or trichloroethylene will flash or spark on impact. When heated to decomposition in air it emits very toxic fumes of BeO. Reacts with Li and P.

BFP000 CAS: 543-81-7 **HR: 3**
BERYLLIUM ACETATE
mf: $C_4H_6O_4 \cdot Be$ mw: 127.11

PROP: Plates. Mp: decomp @ 300°.

SYN: BERYLLIUM ACETATE, NORMAL

CONSENSUS REPORTS: IARC Cancer Review: Animal Inadequate Evidence IMEMDT 23,143,80. Beryllium and its compounds are on the Community Right-To-Know List.

OSHA PEL: (Transitional: TWA 0.002 mg(Be)/m^3; CL 0.005; Pk 0.025/30M/8H) TWA 0.002 mg(Be)/m^3; STEL 0.005 mg(Be)/m^3/30M; CL 0.025 mg(Be)/m^3
ACGIH TLV: TWA 0.002 mg(Be)/m^3, Suspected Human Carcinogen.
DFG MAK: Animal Carcinogen, Suspected Human Carcinogen.
NIOSH REL: CL not to exceed 0.0005 mg (Be)/m^3

SAFETY PROFILE: Confirmed carcinogen. Poison by intraperitoneal route. When heated to decomposition it emits toxic fumes of BeO.

BFP250 CAS: 12770-50-2 **HR: 3**
BERYLLIUM ALUMINUM ALLOY
PROP: Alloy is 62% beryllium and 38% aluminum.

SYNS: ALUMINUM ALLOY, Al,Be * ALUMINUM BERYLLIUM ALLOY

CONSENSUS REPORTS: IARC Cancer Review: GROUP 2A IMEMDT 7,127,87; Animal Sufficient Evidence IMEMDT 23,143,80. NTP Fourth Annual Report On Carcinogens, 1984. Beryllium and its compounds are on the Community Right-To-Know List.

OSHA PEL: (Transitional: TWA 0.002 mg(Be)/m^3; CL 0.005; Pk 0.025/30M/8H) TWA 0.002 mg(Be)/m^3; STEL 0.005 mg(Be)/m^3/30M; CL 0.025 mg(Be)/m^3
ACGIH TLV: TWA 0.002 mg(Be)/m^3, Suspected Human Carcinogen.
DFG MAK: Animal Carcinogen, Suspected Human Carcinogen.
NIOSH REL: CL not to exceed 0.0005 mg (Be)/m^3

SAFETY PROFILE: Confirmed carcinogen with experimental carcinogenic and tumorigenic data. When heated to decomposition it emits very toxic BeO.

BFP500 CAS: 66104-24-3 **HR: 3**
BERYLLIUM CARBONATE
mf: $C_2H_2Be_3O_8$ mw: 181.07

SYNS: BERYLLIUM CARBONATE, BASIC * BERYLLIUMOXIDE CARBONATE * BIS(CARBONATO(2-)) DIHYDROXYTRIBERYLLIUM

CONSENSUS REPORTS: IARC Cancer Review: GROUP 2A IMEMDT 7,127,87; Animal Sufficient Evidence IMEMDT 23,143,80. Reported in EPA TSCA Inventory. Beryllium and its compounds are on the Community Right-To-Know List.

OSHA PEL: (Transitional: TWA 0.002 mg(Be)/m^3; CL 0.005; Pk 0.025/30M/8H) TWA 0.002 mg(Be)/m^3; STEL 0.005 mg(Be)/m^3/30M; CL 0.025 mg(Be)/m^3

ACGIH TLV: TWA 0.002 mg(Be)/m^3, Suspected Human Carcinogen.

DFG MAK: Animal Carcinogen, Suspected Human Carcinogen.

NIOSH REL: CL not to exceed 0.0005 mg (Be)/m^3

SAFETY PROFILE: Confirmed carcinogen. When heated to decomposition it emits toxic BeO dust.

BFP750   CAS: 13106-47-3   HR: 3
BERYLLIUM CARBONATE (1:1)
mf: $CO_3 \cdot Be$ mw: 69.02

SYN: CARBONIC ACID BERYLLIUM SALT (1:1)

CONSENSUS REPORTS: Reported in EPA TSCA Inventory. Beryllium and its compounds are on the Community Right-To-Know List.

OSHA PEL: (Transitional: TWA 0.002 mg(Be)/m^3; CL 0.005; Pk 0.025/30M/8H) TWA 0.002 mg(Be)/m^3; STEL 0.005 mg(Be)/m^3/30M; CL 0.025 mg(Be)/m^3

ACGIH TLV: TWA 0.002 mg(Be)/m^3, Suspected Human Carcinogen.

DFG MAK: 50 ppm (90 mg/m^3)

NIOSH REL: CL not to exceed 0.0005 mg (Be)/m^3

SAFETY PROFILE: Confirmed carcinogen. Poison by intraperitoneal route. When heated to decomposition it emits highly toxic fumes of BeO.

BFQ000   CAS: 7787-47-5   HR: 3
BERYLLIUM CHLORIDE

DOT: NA 1566
mf: $BeCl_2$ mw: 79.91

PROP: Colorless, deliquescent needles. Mp: 440°, bp: 520°, d: 1.899 @ 25°, vap press: 1 mm @ 291° (sublimes).

SYN: BERYLLIUM DICHLORIDE

CONSENSUS REPORTS: IARC Cancer Review: GROUP 2A IMEMDT 7,127,87; Animal Sufficient Evidence IMEMDT 23,143,80. NTP Fourth Annual Report On Carcinogens, 1984. EPA Genetic Toxicology Program. Reported in EPA TSCA Inventory. Beryllium and its compounds are on the Community Right-To-Know List.

OSHA PEL: (Transitional: TWA 0.002 mg(Be)/m^3; CL 0.005; Pk 0.025/30M/8H) TWA 0.002 mg(Be)/m^3; STEL 0.005 mg(Be)/m^3/30M; CL 0.025 mg(Be)/m^3

ACGIH TLV: TWA 0.002 mg(Be)/m^3, Suspected Human Carcinogen.

DFG MAK: Animal Carcinogen, Suspected Human Carcinogen.

NIOSH REL: CL not to exceed 0.0005 mg (Be)/m^3

DOT Classification: Poison B; Label: Poison.

SAFETY PROFILE: Confirmed carcinogen with experimental tumorigenic data. Poison by ingestion and intraperitoneal routes. Experimental reproductive effects. Mutation data reported. When heated to decomposition it emits very toxic fumes of BeO and Cl$^-$.

BFQ250   CAS: 13466-27-8   HR: 3
BERYLLIUM CHLORIDE TETRAHYDRATE
mf: $BeCl_2 \cdot 4H_2O$ mw: 151.99

CONSENSUS REPORTS: Beryllium and its compounds are on the Community Right-To-Know List.

OSHA PEL: (Transitional: TWA 0.002 mg(Be)/m^3; CL 0.005; Pk 0.025/30M/8H) TWA 0.002 mg(Be)/m^3; STEL 0.005 mg(Be)/m^3/30M; CL 0.025 mg(Be)/m^3

ACGIH TLV: TWA 0.002 mg(Be)/m^3, Suspected Human Carcinogen.

DFG MAK: Animal Carcinogen, Suspected Human Carcinogen.

NIOSH REL: CL not to exceed 0.0005 mg (Be)/m^3

SAFETY PROFILE: Confirmed carcinogen. Poison by intraperitoneal route. When heated to decomposition it emits very toxic Cl$^-$ and BeO.

BFQ500   HR: 3
BERYLLIUM COMPOUNDS

CONSENSUS REPORTS: Beryllium and its compounds are on the Community Right-To-Know List.

OSHA PEL: (Transitional: TWA 0.002 mg(Be)/m^3; CL 0.005; Pk 0.025/30M/8H) TWA 0.002 mg(Be)/m^3; STEL 0.005 mg(Be)/m^3/30M; CL 0.025 mg(Be)/m^3

ACGIH TLV: TWA 0.002 mg/m^3, Suspected Human Carcinogen.

DFG TRK: Animal Carcinogen, Suspected Human Carcinogen. Grinding of beryllium metal and alloys: 0.005 mg/m^3 calculated as Be in that portion of dust that can possibly be inhaled; other Be compounds: 0.002 mg/m^3 calculated as Be in that portion of dust that can possibly be inhaled

SAFETY PROFILE: Confirmed carcinogens. Beryllium compounds can enter the body through inhalation of dusts and fumes, and may act locally on the skin. Even alloys of low beryllium content have been shown to be dangerous. In industry, inhalation of the dust can cause severe lung damage with symptoms appearing within months. Effects have been reported in persons living near processing plants and in families of beryllium workers. The fluoride, ammonium fluoride, sulfate, oxide, and hydroxide occur during extraction from beryllium ore. Exposure to the oxide may occur in processing of beryllium alloys and beryllium ceramics.

The extraction of Be from its ore is attended by exposure to acid salts of the metal, particularly the fluoride (BeF$_2$), the ammonium fluoride and the sulfate (BeSO$_4$) and also to beryllium oxide (BeO), and hydroxide [Be(OH)$_2$]. Exposure to the oxide also occurs in the casting of Be alloys and in operations with beryllia ceramics. In the manufacture of fluorescent powders, lamps and sign tubes there may be exposure to Be carbonate and to more complex salts, such as ZnMnBe silicate. Exposure to Be compounds encountered in the extraction of the metal or its oxide from the ore, particularly the halide salts, has been attended, in certain individuals, by the development of dermatitis of an edematous and papulovesicular type, chronic skin ulcers, rhinitis, nasopharyngitis, epistaxis, bronchitis and in severe cases, by the development of an acute pneumonitis, with cough, scanty sputum, low-grade fever, rales, dyspnea and substernal pain. Radiographs show diffuse haziness throughout both lungs, followed by the appearance of soft, ill-defined opacities. The condition occurs while the worker is exposed, sometimes within 1 or 2 months of starting work, and recovery occurs within 2 months, as a rule, though radiographic changes sometimes persist for longer periods. Occasionally, recovery may not occur and lung fibrosis results. In severe cases of pneumonitis the patient may die. Necropsies have revealed diffuse pulmonary edema, hemorrhagic extravasation, large numbers of plasma cells and a relative absence of polymorphonuclear infiltration. On the basis of experimental work with animals, certain investigators are of the opinion that the acute upper and lower respiratory effects are due chiefly to the acid radical present in the dust or fume, but this view has little support. A delayed form of lung disease, characterized by the occurrence of granulomatous areas in the lung tissue, has been reported in workers manufacturing fluorescent powders, lamps and sign tubes, casting beryllium master alloys, and in the production of beryllium from beryl ore. Symptoms can start during exposure, but they might be delayed up to 5 years or more after leaving work. The commonest symptoms are coughing, shortness of breath, loss of appetite, loss of weight, and fatigue. Rales are usually present in the bases and axillae, and the red cell count is frequently elevated. Cyanosis is common and the pulse and respiratory rates are often increased. Radiographically, three stages of the disease are described: (1) a diffuse, uniform granular shadowing extending throughout both lung fields; (2) a diffuse reticular pattern on the granular background; (3) the appearance of distinct nodules scattered through the lungs, with some enlargement and blurring of the hilar shadows. The intensity of the shadowing is usually greater in the middle third of the lung fields. The prognosis is poor. Clinical improvement may occur gradually over a period of several years, but there appears to be little tendency for the radiographic shadowing to clear. In certain cases, the disease has progressed gradually for some months or years, with death resulting from respiratory and cardiac failure. In several instances necropsies have shown the presence of a diffuse fibrosis with coarse strands of hyalinized collagen between the alveoli and, in some places, replacing them. The hyalinized areas contained granulomatous foci, the alveolar walls are thickened and fibrosed, the blood vessels being engorged and dilated. In some cases the hilar lymph nodes show granulomatous change and fibrosis. Granulomatous change has also been noted in the liver and hyaline fibrosis in the spleen. Two cases of delayed lung disease coming to autopsy have presented papular lesions on the dorsum of the hands; on the biopsy these showed "sarcoid-like" lesions with central necrosis.

Several cases have been reported in which localized granulomatous lesions developed following penetrating wounds caused by splinters of glass from broken fluorescent light tubes. Several weeks or months following the accident, swellings were noted in the injured areas and excision revealed granulomatous tumors, which in one case was shown to contain beryllium.

There is no specific treatment, but temporary remissions have been produced by ACTH and cortisone.

BFQ750 CAS: 12010-12-7 **HR: 3**
BERYLLIUM COMPOUND with NIOBIUM (12:1)
mf: $Be_{12}Nb$ mw: 201.03

CONSENSUS REPORTS: Beryllium and its compounds are on the Community Right-To-Know List.

OSHA PEL: (Transitional: TWA 0.002 mg(Be)/m^3; CL 0.005; Pk 0.025/30M/8H) TWA 0.002 mg(Be)/m^3; STEL 0.005 mg(Be)/m^3/30M; CL 0.025 mg(Be)/m^3
ACGIH TLV: TWA 0.002 mg(Be)/m^3, Suspected Human Carcinogen.
NIOSH REL: (Beryllium) CL not to exceed 0.005 mg(Be)/m^3

SAFETY PROFILE: Confirmed carcinogen with experimental tumorigenic data. When heated to decomposition in air it emits very toxic fumes of BeO.

BFR000 CAS: 12232-67-6 **HR: 3**
BERYLLIUM COMPOUND with TITANIUM (12:1)
mf: $Be_{12}Ti$ mw: 156.02

SYN: TITANIUM compounded with BERYLLIUM (1:12)

CONSENSUS REPORTS: Beryllium and its compounds are on the Community Right-To-Know List.

OSHA PEL: (Transitional: TWA 0.002 mg(Be)/m^3; CL 0.005; Pk 0.025/30M/8H) TWA 0.002 mg(Be)/m^3; STEL 0.005 mg(Be)/m^3/30M; CL 0.025 mg(Be)/m^3
ACGIH TLV: TWA 0.002 mg(Be)/m^3, Suspected Human Carcinogen.

NIOSH REL: (Beryllium) CL not to exceed 0.0005 mg(Be)/m^3

SAFETY PROFILE: Confirmed carcinogen with experimental tumorigenic data. When heated to decomposition it emits very toxic fumes of BeO.

BFR250 CAS: 12400-16-7 **HR: 3**
BERYLLIUM COMPOUND with VANADIUM (12:1)
mf: $Be_{12}V$ mw: 159.06

SYN: TITANIUM compounded with BERYLLIUM (1:12)

CONSENSUS REPORTS: Beryllium and its compounds are on the Community Right-To-Know List.

OSHA PEL: (Transitional: TWA 0.002 mg(Be)/m^3; CL 0.005; Pk 0.025/30M/8H) TWA 0.002 mg(Be)/m^3; STEL 0.005 mg(Be)/m^3/30M; CL 0.025 mg(Be)/m^3
ACGIH TLV: TWA 0.002 mg(Be)/m^3, Suspected Human Carcinogen
NIOSH REL: (Beryllium) CL not to exceed 0.0005 mg(Be)/m^3; (REL to Vanadium) 1.0 mg(V)/m^3

SAFETY PROFILE: Confirmed carcinogen with experimental tumorigenic data. When heated to decomposition it emits very toxic fumes of BeO and VO_x.

BFR500 CAS: 7787-49-7 **HR: 3**
BERYLLIUM FLUORIDE
DOT: NA 1566
mf: BeF_2 mw: 47.01

PROP: Amorphous, colorless mass. Mp: 800°, d: 1.986 @ 25°.

SYN: BERYLLIUM DIFLUORIDE

CONSENSUS REPORTS: IARC Cancer Review: GROUP 2A IMEMDT 7,127,87; Animal Sufficient Evidence IMEMDT 23,143,80. NTP Fourth Annual Report On Carcinogens, 1984. Beryllium and its compounds are on the Community Right-To-Know List. Reported in EPA TSCA Inventory.

OSHA PEL: (Transitional: TWA 0.002 mg(Be)/m^3; CL 0.005; Pk 0.025/30M/8H) TWA 0.002 mg(Be)/m^3; STEL 0.005 mg(Be)/m^3/30M; CL 0.025 mg(Be)/m^3
ACGIH TLV: TWA 0.002 mg(Be)/m^3, Suspected Human Carcinogen; 2.5 mg(F)/m^3

NIOSH REL: CL not to exceed 0.0005 mg (Be)/m^3

DOT Classification: Poison B; Label: Poison.

SAFETY PROFILE: Confirmed carcinogen with experimental carcinogenic and tumorigenic data by inhalation. Poison by ingestion, subcutaneous, intravenous, and intraperitoneal routes. Incompatible with Mg. When heated to decomposition, it emits very toxic fumes of BeO and F$^-$.

BFR750 CAS: 7787-52-2 HR: 3
BERYLLIUM HYDRIDE
mf: BeH$_2$ mw: 11.03

PROP: White solid.

CONSENSUS REPORTS: Beryllium and its compounds are on the Community Right-To-Know List.

SAFETY PROFILE: Confirmed carcinogen. A dangerous fire hazard. When heated to 220°C it liberates explosive hydrogen gas. Reacts violently with methanol, water, and dilute acids. When heated to decomposition it emits toxic fumes of BeO.

BFS000 CAS: 13598-15-7 HR: 3
BERYLLIUM HYDROGEN PHOSPHATE (1:1)
mf: BeHO$_4$P   mw: 104.99

SYNS: BERYLLIUM PHOSPHATE * PHOSPHORIC ACID, BERYLLIUM SALT (1:1) * PHOSPHOROUS ACID, BERYLLIUM SALT

CONSENSUS REPORTS: IARC Cancer Review: GROUP 2A IMEMDT 7,127,87; Animal Sufficient Evidence IMEMDT 23,143,80; IMEMDT 1,17,72. NTP Fourth Annual Report On Carcinogens, 1984. Beryllium and its compounds are on the Community Right-To-Know List.

OSHA PEL: (Transitional: TWA 0.002 mg(Be)/m^3; CL 0.005; Pk 0.025/30M/8H) TWA 0.002 mg(Be)/m^3; STEL 0.005 mg(Be)/m^3/30M; CL 0.025 mg(Be)/m^3

ACGIH TLV: TWA 0.002 mg(Be)/m^3, Suspected Human Carcinogen.

NIOSH REL: CL not to exceed 0.0005 mg (Be)/m^3

SAFETY PROFILE: Confirmed carcinogen with experimental carcinogenic and tumorigenic data. Poison by intravenous route. When heated to decomposition it emits very toxic fumes of BeO and PO$_x$.

BFS250 CAS: 13327-32-7 HR: 3
BERYLLIUM HYDROXIDE
mf: H$_2$O$_2$ • Be   mw: 43.03

PROP: Amorphous powder or crystals. Mp: decomp @ 138°, d(cr): 1.909.

SYNS: BERYLLIUM DIHYDROXIDE * BERYLLIUM HYDRATE

CONSENSUS REPORTS: IARC Cancer Review: GROUP 2A IMEMDT 7,127,87; Animal Sufficient Evidence IMEMDT 23,143,80. NTP Fourth Annual Report On Carcinogens, 1984. Beryllium and its compounds are on the Community Right-To-Know List. Reported in EPA TSCA Inventory.

OSHA PEL: (Transitional: TWA 0.002 mg(Be)/m^3; CL 0.005; Pk 0.025/30M/8H) TWA 0.002 mg(Be)/m^3; STEL 0.005 mg(Be)/m^3/30M; CL 0.025 mg(Be)/m^3

ACGIH TLV: TWA 0.002 mg(Be)/m^3, Suspected Human Carcinogen.

NIOSH REL: CL not to exceed 0.0005 mg (Be)/m^3

SAFETY PROFILE: Confirmed carcinogen with experimental carcinogenic and tumorigenic data. Poison by intravenous route. When heated to decomposition it emits very toxic fumes of BeO.

BFS750 HR: 3
BERYLLIUM MANGANESE ZINC SILICATE

SYNS: MANGANESE ZINC BERYLLIUM SILICATE * ZINC MANGANESE BERYLLIUM SILICATE

CONSENSUS REPORTS: Beryllium, manganese, zinc, and their compounds are on the Community Right-To-Know List.

OSHA PEL: (Transitional: TWA 0.002 mg(Be)/m^3; CL 0.005; Pk 0.025/30M/8H) TWA 0.002 mg(Be)/m^3; STEL 0.005 mg(Be)/m^3/30M; CL 0.025 mg(Be)/m^3

ACGIH TLV: TWA 0.002 mg(Be)/m^3, Suspected Human Carcinogen; TWA 5 mg (Mn)/m^3

NIOSH REL: (Beryllium) CL Not to exceed 0.0005 mg(Be)/m^3

SAFETY PROFILE: Confirmed carcinogen with experimental tumorigenic data. When heated to decomposition it emits very toxic fumes of BeO and ZnO.

BFT000   CAS: 13597-99-4   HR: 3
BERYLLIUM NITRATE
mf: BeN_2O_6 mw: 133.03

DOT: UN 2464

PROP: White-yellowish crystals, deliquescent. Mp: 60°, bp: decomp @ 100-200°.

SYNS: BERYLLIUM DINITRATE * NITRIC ACID, BE-RYLLIUM SALT

CONSENSUS REPORTS: Beryllium and its compounds are on the Community Right-To-Know List.

OSHA PEL: (Transitional: TWA 0.002 mg(Be)/m³; CL 0.005; Pk 0.025/30M/8H) TWA 0.002 mg(Be)/m³; STEL 0.005 mg(Be)/m³/30M; CL 0.025 mg(Be)/m³
ACGIH TLV: TWA 0.002 mg(Be)m³, Suspected Human Carcinogen.
NIOSH REL: CL not to exceed 0.0005 mg (Be)/m³
DOT Classification: Label: Oxidizer and Poison.

SAFETY PROFILE: Confirmed carcinogen. Poison by intraperitoneal and subcutaneous routes. Experimental reproductive effects. When heated to decomposition it emits very toxic fumes of BeO and NO_x.

BFT250   CAS: 1304-56-9   HR: 3
BERYLLIUM OXIDE
mf: BeO   mw: 25.01

PROP: White, amorphous powder. Mp: 2530° ± 30°, bp: 3900° (approx), d: 3.025.

SYNS: BERYLLIA * BERYLLIUM MONOXIDE * THERMALOX

CONSENSUS REPORTS: IARC Cancer Review: GROUP 2A IMEMDT 7,127,87; Animal Sufficient Evidence IMEMDT 1,17,72; IMEMDT 23,143,80. NTP Fourth Annual Report On Carcinogens, 1984. Beryllium and its compounds are on the Community Right-To-Know List. Reported in EPA TSCA Inventory.

OSHA PEL: (Transitional: TWA 0.002 mg(Be)/m³; CL 0.005; Pk 0.025/30M/8H) TWA 0.002 mg(Be)/m³; STEL 0.005 mg(Be)/m³/30M; CL 0.025 mg(Be)/m³

ACGIH TLV: TWA 0.002 mg(Be)/m³, Suspected Human Carcinogen.
NIOSH REL: (Beryllium) CL not to exceed 0.0005 mg(Be)/m³

SAFETY PROFILE: Confirmed carcinogen with experimental tumorigenic data. Experimental reproductive data. Incompatible with (Mg + heat). When heated to decomposition it emits very toxic fumes of BeO.

BFT500   CAS: 19049-40-2   HR: 3
BERYLLIUM OXYACETATE
mf: $C_{12}H_{18}Be_4O_{13}$ mw: 406.34

SYNS: BERYLLIUM ACETATE, BASIC * BERYLLIUM OXIDE ACETATE * HEXAKIS(μ-ACETATO-O: O'))-μ(⁴)-OXOTETRABERYLLIUM * HEXAKIS(μ-ACETATO)-μ(⁴)-OXOTETRABERYLLIUM

CONSENSUS REPORTS: Beryllium and its compounds are on the Community Right-To-Know List.

OSHA PEL: (Transitional: TWA 0.002 mg(Be)/m³; CL 0.005; Pk 0.025/30M/8H) TWA 0.002 mg(Be)/m³; STEL 0.005 mg(Be)/m³/30M; CL 0.025 mg(Be)/m³
ACGIH TLV: TWA 0.002 mg(Be)/m³, Suspected Human Carcinogen.
NIOSH REL: CL not to exceed 0.0005 mg (Be)/m³

SAFETY PROFILE: Confirmed carcinogen. When heated to decomposition it emits toxic fumes BeO.

BFT750   CAS: 63990-88-5   HR: 3
BERYLLIUM OXYFLUORIDE
mf: BeF_2O_2 mw: 79.01

CONSENSUS REPORTS: Beryllium and its compounds are on the Community Right-To-Know List.

OSHA PEL: (Transitional: TWA 0.002 mg(Be)/m³; CL 0.005; Pk 0.025/30M/8H) TWA 0.002 mg(Be)/m³; STEL 0.005 mg(Be)/m³/30M; CL 0.025 mg(Be)/m³
ACGIH TLV: TWA 0.002 mg(Be)/m³, Suspected Human Carcinogen; 2.5 mg(F)/m³
NIOSH REL: CL not to exceed 0.0005 mg (Be)/m³

SAFETY PROFILE: Confirmed carcinogen. Poison by ingestion, subcutaneous, intravenous, and intraperitoneal routes. When heated to decomposition it emits very toxic fumes of BeO and F^-.

BFU000   CAS: 13597-95-0   **HR: 3**
BERYLLIUM PERCHLORATE
mf: $Be(ClO_4)_2$ mw: 207.91

PROP: Very hygroscopic crystals, sol in water: 148.6 g/100 mL.
OSHA PEL: (Transitional: TWA 0.002 mg(Be)/m³; CL 0.005; Pk 0.025/30M/8H) TWA 0.002 mg(Be)/m³; STEL 0.005 mg(Be)/m³/30M; CL 0.025 mg(Be)/m³
ACGIH TLV: TWA 0.002 mg(Be)/m³, Suspected Human Carcinogen.
NIOSH REL: CL not to exceed 0.0005 mg (Be)/m³

SAFETY PROFILE: Confirmed carcinogen. A powerful oxidant used in propellant and igniter systems. When heated to decomposition it emits toxic fumes of Cl⁻ and BeO.

BFU250   CAS: 13510-49-1   **HR: 3**
BERYLLIUM SULFATE (1:1)
mf: $O_4S \cdot Be$ mw: 105.07

PROP: Crystals. Mp: 550-600° (decomp), d: 2.443.

SYN: SULFURIC ACID, BERYLLIUM SALT (1:1)

CONSENSUS REPORTS: IARC Cancer Review: GROUP 2A IMEMDT 7,127,87; Animal Sufficient Evidence IMEMDT 23,143,80. NTP Fourth Annual Report On Carcinogens, 1984. Beryllium and its compounds are on the Community Right-To-Know List. Reported in EPA TSCA Inventory.

OSHA PEL: (Transitional: TWA 0.002 mg(Be)/m³; CL 0.005; Pk 0.025/30M/8H) TWA 0.002 mg(Be)/m³; STEL 0.005 mg(Be)/m³/30M; CL 0.025 mg(Be)/m³
ACGIH TLV: TWA 0.002 mg(Be)/m³, Suspected Human Carcinogen.
NIOSH REL: CL not to exceed 0.0005 mg (Be)/m³

SAFETY PROFILE: Confirmed carcinogen with experimental tumorigenic data. Acute poison by inhalation, ingestion, intraperitoneal, subcutaneous, intravenous, and intratracheal routes. Mutation data reported. When heated to decomposition it emits very toxic fumes of SO_x and BeO.

BFU500   CAS: 7787-56-6   **HR: 3**
BERYLLIUM SULFATE
TETRAHYDRATE (1:1:4)
mf: $O_4S \cdot Be \cdot 4H_2O$ mw: 177.15

SYNS: BERYLLIUM SULPHATE TETRAHYDRATE
∗ SULFURIC ACID, BERYLLIUM SALT (1:1), TETRA-HYDRATE

CONSENSUS REPORTS: IARC Cancer Review: GROUP 2A IMEMDT 7,127,87; Animal Sufficient Evidence IMEMDT 23,143,80; IMEMDT 1,17,72. NTP Fourth Annual Report On Carcinogens, 1984. Beryllium and its compounds are on the Community Right-To-Know List.

OSHA PEL: (Transitional: TWA 0.002 mg(Be)/m³; CL 0.005; Pk 0.025/30M/8H) TWA 0.002 mg(Be)/m³; STEL 0.005 mg(Be)/m³/30M; CL 0.025 mg(Be)/m³
ACGIH TLV: TWA 0.002 mg(Be)/m³, Suspected Human Carcinogen.
NIOSH REL: CL not to exceed 0.0005 mg (Be)/m³

SAFETY PROFILE: Confirmed carcinogen with experimental carcinogenic data by inhalation. Deadly poison by subcutaneous and intravenous routes. Human mutation data reported. When heated to decomposition it emits very toxic fumes of BeO and SO_x.

BFU750   **HR: 3**
BERYLLIUM TETRAHYDROBORATE
mf: B_2BeH_8 mw: 38.70

CONSENSUS REPORTS: Beryllium and its compounds are on the Community Right-To-Know List.

OSHA PEL: (Transitional: TWA 0.002 mg(Be)/m³; CL 0.005; Pk 0.025/30M/8H) TWA 0.002 mg(Be)/m³; STEL 0.005 mg(Be)/m³/30M; CL 0.025 mg(Be)/m³
ACGIH TLV: TWA 0.002 mg(Be)/m³, Suspected Human Carcinogen.
NIOSH REL: CL not to exceed 0.0005 mg (Be)/m³

SAFETY PROFILE: Confirmed carcinogen. Ignites and then explodes in air or on contact with water. Upon decomposition it emits toxic fumes of BeO and BO_x.

BFV000   **HR: 3**
BERYLLIUM TETRAHYDROBORA-
TETRIMETHYLAMINE
mf: $C_3H_{17}B_2BeN$ mw: 97.78

CONSENSUS REPORTS: Beryllium and its compounds are on the Community Right-To-Know List.

OSHA PEL: (Transitional: TWA 0.002 mg(Be)/m^3; CL 0.005; Pk 0.025/30M/8H) TWA 0.002 mg(Be)/m^3; STEL 0.005 mg(Be)/m^3/30M; CL 0.025 mg(Be)/m^3

ACGIH TLV: TWA 0.002 mg(Be)/m^3, Suspected Human Carcinogen.

NIOSH REL: CL not to exceed 0.0005 mg (Be)/m^3

SAFETY PROFILE: Confirmed carcinogen. It will ignite in contact with air or water. When heated to decomposition it emits toxic fumes of BeO, BO$_x$ and NO$_x$.

BFV250   CAS: 39413-47-3   **HR: 3**
BERYLLIUM ZINC SILICATE
mf: O$_2$Si • Zn • Be   mw: 134.47

SYN: ZINC BERYLLIUM SILICATE

CONSENSUS REPORTS: IARC Cancer Review: GROUP 2A IMEMDT 7,127,87; Animal Sufficient Evidence IMEMDT 23,143,80; IMEMDT 1,17,72. NTP Fourth Annual Report On Carcinogens, 1984. Beryllium and its compounds, as well as zinc and its compounds, are on the Community Right-To-Know List.

OSHA PEL: (Transitional: TWA 0.002 mg(Be)/m^3; CL 0.005; Pk 0.025/30M/8H) TWA 0.002 mg(Be)/m^3; STEL 0.005 mg(Be)/m^3/30M; CL 0.025 mg(Be)/m^3

ACGIH TLV: TWA 0.002 mg(Be)/m^3, Suspected Human Carcinogen.

NIOSH REL: CL not to exceed 0.0005 mg (Be)/m^3

SAFETY PROFILE: Confirmed carcinogen with experimental tumorigenic data. When heated to decomposition it emits toxic fumes of BeO and ZnO.

BFV750   CAS: 378-44-9   **HR: D**
BETAMETHASONE
mf: C$_{22}$H$_{29}$FO$_5$ mw: 392.51

SYNS: BETNELAN * BETSOLAN * CELESTONE * 9-α-FLUORO-16-β-METHYLPREDNISOLONE * 9-α-FLUORO-16-β-METHYL- 1,4-PREGNADIENE-11-β,17-α,21-TRIOL-3,20-DIONE * 9-FLUORO-11-β,17,21-TRIHYDROXY-16-β-METHYLPREGNA-1,4-DIENE-3,20-DIONE * 9-α-FLUORO-11-β,17,21-TRIHYDROXY-16-β-METHYLPREGNA-1,4-DIENE- 3,20-DIONE * 16-β-METHYL-1,4-PREGNADIENE-9-α-FLUORO-11-β,17-α,21-TRIOL- 3,20-DIONE * NSC-39470 * Sch 4831

SAFETY PROFILE: Experimental reproductive effects. When heated to decomposition it emits toxic fumes of F$^-$.

BFW000   CAS: 39323-48-3   **HR: 3**
BETEL NUT

PROP: Mottled brown with fawn color. Extract of 50 grams sun-dried betel nut in 100 mL boiling water.

SYNS: ARECA CATECHU * ARECA CATECHU Linn., fruit extract * ARECA CATECHU Linn., nut extract * BN * PINANG * POOGIPHALAM, nut extract * SUPARI, nut extract

CONSENSUS REPORTS: IARC Cancer Review: Animal Limited Evidence IMEMDT 37,141,85

SAFETY PROFILE: Suspected carcinogen with experimental carcinogenic and neoplastigenic data. Moderately toxic by intraperitoneal route. Experimental teratogenic and reproductive effects. When heated to decomposition it emits toxic fumes of NO$_x$.

BFW125   **HR: 3**
BETEL QUID EXTRACT

SAFETY PROFILE: Suspected carcinogen with experimental carcinogenic and tumorigenic data by skin contact. Human mutation data reported.

BFW135   **HR: 3**
BETEL TOBACCO EXTRACT

SYN: JAFFNA TOBACCO

CONSENSUS REPORTS: IARC Cancer Review: Human Sufficient Evidence IMEMDT 37,141,85; Animal Limited Evidence IMEMDT 37,141,85

SAFETY PROFILE: Confirmed human carcinogen. Human mutation data reported.

BFW500   CAS: 319-86-8   **HR: 2**
Δ-BHC
mf: C$_6$H$_6$Cl$_6$ mw: 290.82

SYNS: Δ-BENZENEHEXACHLORIDE * ENT 9,234 * 1-α,2-α,3-α,4-β,5-α,6-β-HEXACHLOROCYCLOHEXANE * Δ-HEXACHLOROCYCLOHEXANE * Δ-1,2,3,4,5,6-HEXACHLOROCYCLOHEXANE * Δ-LINDANE

CONSENSUS REPORTS: Reported in EPA TSCA Inventory.

SAFETY PROFILE: Moderately toxic by ingestion. When heated to decomposition it emits toxic fumes of Cl^-.

BFW750　　　CAS: 128-37-0　　　**HR: 2**
BHT (food grade)
mf: $C_{15}H_{24}O$　　　mw: 220.39

PROP: White, crystalline solid; faint characteristic odor. Bp: 265°, fp: 68°, flash p: 260°F (TOC), d: 1.048 @ 20°/4°, vap d: 7.6. Sol in alc; insol in water and propylene glycol.

SYNS: ADVASTAB 401 * AGIDOL * ANTIOXI-DANT DBPC * ANTIOXIDANT 29 * AO 29 * AO 4K * 2,6-BIS(1,1-DIMETHYLETHYL)-4-METHYL-PHENOL * BUKS * BUTYLATED HYDROXY-TOLUENE * BUTYLHYDROXYTOLUENE * CAO 1 * CAO 3 * CATALIN CAO-3 * CHEMANOX 11 * DBMP * DBPC (technical grade) * DIBUTYLATED HYDROXYTOLUENE * 2,6-DI-tert-BUTYL-p-CRESOL (OSHA, ACGIH) * 2,6-DI-tert-BUTYL-1-HYDROXY-4-METHYLBENZENE * 3,5-DI-tert-BUTYL-4-HYDROXY-TOLUENE * 2,6-DI-terc. BUTYL-p-KRESOL (CZECH) * 2,6-DI-tert-BUTYL-p-METHYLPHENOL * 2,6-DI-tert-BUTYL-4-METHYLPHENOL * FEMA No. 2184 * 4-HYDROXY-3,5-DI-tert-BUTYLTOLUENE * IMPRUVOL * IONOL * IONOL (antioxidant) * 4-METHYL-2,6-DI-terc. BUTYLFENOL (CZECH) * METHYL DI-tert-BUTYLPHENOL * 4-METHYL-2,6-DI-tert-BUTYLPHENOL * NCI-C03598 * NONOX TBC * PARABAR 441 * SUSTANE * TENOX BHT * TOPANOL * VANLUBE PCX

CONSENSUS REPORTS: IARC Cancer Review: GROUP 3 IMEMDT 7,56,87; Animal Limited Evidence IMEDT 40,161,86. NCI Carcinogenesis Bioassay Completed; (feed): No Evidence: mouse,rat NCITR* NCI-CG-TR-150,79. Reported in EPA TSCA Inventory. EPA Genetic Toxicology Program.

OSHA PEL: TLV 10 mg/m^3
ACGIH TLV: TLV 10 mg/m^3

SAFETY PROFILE: Poison by intraperitoneal and intravenous routes. Moderately toxic by ingestion. Experimental reproductive effects. A human skin irritant. A skin and eye irritant. Questionable carcinogen with experimental carcinogenic and neoplastigenic data. Combustible when exposed to heat or flame. It can react with oxidizing materials. To fight fire, use CO_2, dry chemical. When heated to decomposition it emits acrid smoke and fumes.

BFX000　　　CAS: 613-35-4　　　**HR: 3**
4′,4‴-BIACETANILIDE
mf: $C_{16}H_{16}N_2O_2$　　　mw: 268.34

PROP: Mp: 329°.

SYNS: N,N′-(1,1′-BIPHENYL)-4,4′-DIYLBIS-ACETAMIDE 4′,4‴-BIACETANILIDE * N,N′-4,4′-BIPHENYLYLENE-BISACETAMIDE * 4,4′-DIACETYLAMINOBIPHENYL * N,N′-DIACETYL BENZIDINE * 4,4′-DIACETYLBEN-ZIDINE

CONSENSUS REPORTS: IARC Cancer Review: GROUP 2B IMEMDT 7,56,87, Animal Sufficient Evidence IMEMDT 16,293,78. Reported in EPA TSCA Inventory.

SAFETY PROFILE: Suspected carcinogen with experimental carcinogenic, neoplastigenic, and tumorigenic data. Mutation data reported. When heated to decomposition it emits toxic fumes of NO_x.

BFX250　　　CAS: 2130-56-5　　　**HR: 3**
5,5′-BIANTHRANILIC ACID
mf: $C_{14}H_{12}N_2O_4$　　　mw: 272.28

SYNS: 3,3′-BENZIDINEDICARBOXYLIC ACID * 4,4′-DIAMINO-3,3′-BIPHENYLDICARBOXYLIC ACID * 4,4′-DIAMINOBIPHENYL-3,3′-DICARBOXYLIC ACID * 3,3′-DICARBOXYBENZIDINE * KWAS BENZYDY-NODWUKAROKSYLOWY (POLISH)

CONSENSUS REPORTS: Reported in EPA TSCA Inventory.

SAFETY PROFILE: Questionable carcinogen with experimental tumorigenic data. Mutation data reported. When heated to decomposition it emits toxic fumes of NO_x.

BFX500　　　CAS: 103-29-7　　　**HR: 3**
BIBENZYL
mf: $C_{14}H_{14}$　　　mw: 182.28

PROP: Flash p: 264°F, autoign temp: 896°F, d: 1.0, vap d: 6.29, bp: 285°.

SYNS: DIBENZYL * 1,2-DIPHENYLETHANE

CONSENSUS REPORTS: Reported in EPA TSCA Inventory.

SAFETY PROFILE: Poison by intravenous route. Moderately toxic by intraperitoneal route.

Combustible. To fight fire, use water, spray, mist, alcohol foam, dry chemical. When heated to decomposition it emits acrid smoke and fumes.

BGA750   CAS: 1464-53-5   **HR: 3**
1,1′-BI(ETHYLENE OXIDE)
mf: $C_4H_6O_2$ mw: 86.10

PROP: Colorless liquid. Bp: 142°, mp: 19°, d: 1.113 @ 18°/4°.

SYNS: BIOXIRANE * 2,2′-BIOXIRANE * BUTA-DIENDIOXYD (GERMAN) * BUTADIENE DIEPOXIDE * 1,3-BUTADIENE DIEPOXIDE * BUTADIENE DIOX-IDE * BUTANE DIEPOXIDE * DEB * DIEPOXY-BUTANE * 2,4-DIEPOXYBUTANE * 1,2:3,4-DI-EPOXYBUTANE * DIOXYBUTADIENE * ENT 26,592 * ERYTHRITOL ANHYDRIDE * RCRA WASTE NUM-BER U085

CONSENSUS REPORTS: NTP Fourth Annual Report On Carcinogens, 1984. EPA Extremely Hazardous Substances List. EPA Genetic Toxi-cology Program. Community Right-To-Know List. Reported in EPA TSCA Inventory.

SAFETY PROFILE: Confirmed carcinogen with experimental tumorigenic data. Poison by inges-tion, inhalation, skin contact and intraperitoneal routes. Human mutation data reported. A severe skin and eye irritant. When heated to decomposi-tion it emits acrid smoke and irritating fumes.

BGC250   CAS: 69382-20-3   **HR: 3**
BINDON ETHYL ETHER
mf: $C_{20}H_{14}O_3$ mw: 302.34

SYNS: BINDON ATHYLATHER * 2-(3-ETHOXY-1-IN-DANYLIDENE)-1,3-DINDANDIONE

SAFETY PROFILE: Poison by intraperitoneal route. An experimental teratogen. Other experi-mental reproductive effects. When heated to de-composition it emits acrid smoke and irritating fumes.

BGE000   CAS: 92-52-4   **HR: 3**
BIPHENYL
mf: $C_{12}H_{10}$ mw: 154.22

PROP: White scales, pleasant odor. Mp: 70°, bp: 255°, flash p: 235°F (CC), d: 0.991 @ 75°/4°, autoign temp: 1004°F, vap d: 5.31, lel: 0.6% @ 232°, uel: 5.8% @ 331°F.

SYNS: BIBENZENE * 1,1′-BIPHENYL * DIPHE-NYL (OSHA) * LEMONENE * PHENADOR-X * PHENYLBENZENE * PHPH * XENENE

CONSENSUS REPORTS: EPA Genetic Toxi-cology Program. Reported in EPA TSCA Inven-tory. Community Right-To-Know List.

OSHA PEL: TWA 0.2 ppm
ACGIH TLV: TWA 0.2 ppm
DFG MAK: 0.2 ppm (1 mg/m³)

SAFETY PROFILE: Poison by intravenous route. Moderately toxic by ingestion. A power-ful irritant by inhalation in humans. Human sys-temic effects by inhalation of very small amounts: flaccid paralysis, nausea or vomiting, and other unspecified gastrointestinal effects. Questionable carcinogen with experimental tu-morigenic and neoplastigenic data. Mutation data reported. Combustible when exposed to heat or flame; can react with oxidizing materials. To fight fire, use CO_2, dry chemical, water spray, mist, fog. When heated to decomposition it emits acrid smoke and fumes.

BGJ500   CAS: 92-69-3   **HR: 3**
4-BIPHENYLOL
mf: $C_{12}H_{10}O$ mw: 170.22

SYNS: p-HYDROXYBIPHENYL * 4-HYDROXYBIPHE-NYL * p-HYDROXYDIPHENYL * 4-HYDROXYDI-PHENYL * PARAXENOL * p-PHENYLPHENOL * 4-PHENYLPHENOL

CONSENSUS REPORTS: Reported in EPA TSCA Inventory.

SAFETY PROFILE: Acute poison by intraperi-toneal route. Questionable carcinogen with ex-perimental carcinogenic and tumorigenic data. When heated to decomposition it emits acrid, irritating fumes.

BGJ750   CAS: 132-27-4   **HR: 3**
2-BIPHENYLOL, SODIUM SALT
mf: $C_{12}H_9O \cdot Na$ mw: 192.20

SYNS: BACTROL * D.C.S. * DORVICIDE A * DOWICIDE * 2-HYDROXYDIPHENYL SODIUM * MIL-DU-RID * MYSTOX WFA * NATRIPHENE * OPP-Na * ORPHENOL * o-PHENYLPHENOL SO-DIUM SALT * 2-PHENYLPHENOL SODIUM SALT * PREVENTOL-ON * SODIUM-2-HYDROXYDIPHENYL * SODIUM-o-PHENYLPHENATE * SODIUM-2-PHE-NYLPHENATE * SODIUM-o-PHENYLPHENOLATE * SODIUM-o-PHENYLPHENOXIDE * SOPP * STOMOLD B * TOPANE

CONSENSUS REPORTS: IARC Cancer Re-view: GROUP 2B IMEMDT 7,56,87, Animal

Limited Evidence IMEMDT 30,329,83. Reported in EPA TSCA Inventory.

SAFETY PROFILE: Suspected carcinogen with experimental carcinogenic, neoplastigenic, and tumorigenic data. Moderately toxic by ingestion. Experimental teratogenic and reproductive effects. A human skin irritant. A severe skin irritant to experimental animals. When heated to decomposition it emits toxic fumes of Na_2O.

BGO750   CAS: 8001-88-5   **HR: 2**
BIRCH TAR OIL

PROP: Brown liquid; leather-like odor. D: 0.886-0.950. Found in the tar of the bark and wood of *Betula pendula* Roth (Fam. *Betulaceae*) and prepared by steam distillation of the tar obtained by dry distillation of the bark and wood. Sol in fixed oils; insol in glycerin, mineral oil, and propylene glycol.

SYN: BIRCH TAR OIL, RECTIFIED (FCC)

CONSENSUS REPORTS: Reported in EPA TSCA Inventory.

SAFETY PROFILE: A skin irritant. Moderately irritating to eyes and mucous membranes. A mild allergen. Combustible when exposed to heat or flame; can react with oxidizing materials.

BGP250   CAS: 304-28-9   **HR: 3**
2,7-BIS(ACETAMIDO)FLUORENE
mf: $C_{17}H_{16}N_2O_2$ mw: 280.35

SYNS: 2,7-DIACETAMIDOFLUORENE * 2,7-DIACE-TYLAMINOFLUORENE * 2,7-FAA * 2,7-FLUORE-NYLBISACETAMIDE * N,N'-FLUOREN-2,7-YLBIS-ACETAMIDE * N,N'-FLUOREN-2,7-YLENEBISACET-AMIDE * N,N'-2,7-FLUORENYLENEBISACETAMIDE * N,N'-(FLUOREN-2,7-YLENE)BIS(ACETYLAMINE) * N,N'-2,7-FLUORENYLENEDIACETAMIDE

CONSENSUS REPORTS: EPA Genetic Toxicology Program.

SAFETY PROFILE: Suspected carcinogen with experimental carcinogenic, neoplastigenic, and tumorigenic data. Mutation data reported. When heated to decomposition it emits toxic fumes of NO_x.

BGQ750   CAS: 14024-64-7   **HR: 3**
BIS(ACETYLACETONATO) TITANIUM OXIDE
mf: $C_{10}H_{14}O_5Ti$ mw: 262.14

SYNS: BIS(2,4-PENTANEDIONATO)TITANIUM OXIDE * TITANIUM ACETONYL ACETONATE * TITANIUM OXIDE BIS(ACETYLACETONATE) * TITANIUM, OXOBIS(2,4-PENTANEDIONATO-O,O') * TITANYL BIS(ACETYLACETONATE)

SAFETY PROFILE: Moderate toxic by intraperitoneal route. Questionable carcinogen with experimental tumorigenic data. When heated to decomposition it emits acrid smoke and irritating fumes.

BGT000   CAS: 28434-86-8   **HR: 3**
BIS(4-AMINO-3-CHLOROPHENYL) ETHER
mf: $C_{12}H_{10}Cl_2N_2O$ mw: 269.14

SYNS: 3,3'-DICHLOR-4,4'-DIAMINO-DIPHENYLAETHER (GERMAN) * 3,3'-DICHLORO-4,4'-DIAMINODIPHENYL ETHER * 4,4'-OXYBIS(2-CHLOROANILINE) * 4,4'-OXYBIS(2-CHLORO-BENZENAMINE)

CONSENSUS REPORTS: IARC Cancer Review: GROUP 2B IMEMDT 7,56,87; Animal Sufficient Evidence IMEMDT 16,309,78

SAFETY PROFILE: Suspected carcinogen with experimental carcinogenic data. When heated to decomposition it emits toxic fumes of Cl^- and NO_x.

BGU750   CAS: 105-83-9   **HR: 3**
BIS(γ-AMINOPROPYL) METHYLAMINE
mf: $C_7H_{19}N_3$ mw: 145.29

PROP: Liquid, completely miscible in water. D: 0.9307 @ 20°/20°, bp: 240.6°, fp: −29.6°, flash p: 220°F.

SYNS: BIS(φ-AMINOPROPYL)METHYLAMINE * BIS(3-AMINOPROPYL)METHYLAMINE * N,N-BIS(γ-AMINOPROPYL)METHYLAMINE * N,N-BIS(3-AMINO-PROPYL)METHYLAMINE * 3,7'-DIAMINO-N-METHYL-DIPROPYLAMINE * METHYLBIS(3-AMINOPROPYL)-AMINE

CONSENSUS REPORTS: Reported in EPA TSCA Inventory.

SAFETY PROFILE: Poison by inhalation and skin contact. Moderately toxic by ingestion. A skin and severe eye irritant. Combustible when exposed to heat or flame. To fight fire, use foam, fog, dry chemical. When heated to decomposition it emits toxic fumes of NO_x.

BGV000   CAS: 7209-38-3   HR: 3
1,4-BIS(AMINOPROPYL)PIPERAZINE

DOT: NA 1760
mf: $C_{10}H_{24}N_4$ mw: 200.38

SYN: BIS(AMINOPROPYL)PIPERAZINE (DOT)

CONSENSUS REPORTS: Reported in EPA TSCA Inventory.

DOT Classification: Corrosive Material; Label: Corrosive.

SAFETY PROFILE: Poison by intravenous route. A corrosive material and a powerful irritant to skin, eyes, and mucous membranes. When heated to decomposition it emits toxic fumes of NO_x.

BGY000   CAS: 1078-79-1   HR: 3
BIS(1-AZIRIDINYL)(2-METHYL-3-THIAZOLIDINYL)PHOSPHINE OXIDE
mf: $C_8H_{16}N_3OPS$ mw: 233.30

SYNS: IMIPHOS * MARCOPHANE * MARKOFANE

SAFETY PROFILE: Poison by ingestion and intraperitoneal routes. When heated to decomposition it emits very toxic fumes of SO_x, PO_x, and NO_x.

BHA750   CAS: 155-04-4   HR: 3
BIS(2-BENZOTHIAZOLYLTHIO)ZINC
mf: $C_{14}H_8N_2S_4 \cdot Zn$ mw: 397.85

SYNS: 2-BENZOTHIAZOLETHIOL, ZINC SALT (2:1)
* BIS(MERCAPTOBENZOTHIAZOLATO)ZINC
* HERMAT Zn-MBT * 2-MERCAPTOBENZOTHIAZOLE ZINC SALT * OXAF * PENNAC ZT * TISPERSE MB-58 * USAF GY-7 * VULKACIT ZM * ZENITE * ZENITE SPECIAL * ZETAX * ZINC-2-BENZOTHIAZOLETHIOLATE * ZINC BENZOTHIAZOLYL MERCAPTIDE * ZINC BENZOTHIAZOL-2-YLTHIOLATE * ZINC BENZOTHIAZYL-2-MERCAPTIDE * ZINC MERCAPTOBENZOTHIAZOLATE * ZINC-2-MERCAPTOBENZOTHIAZOLE * ZINC MERCAPTOBENZOTHIAZOLE SALT * ZMBT * ZnMB

CONSENSUS REPORTS: Reported in EPA TSCA Inventory. Zinc compounds are on the Community Right-To-Know List.

SAFETY PROFILE: Poison by intraperitoneal route. Moderately toxic by ingestion and subcutaneous routes. Questionable carcinogen with experimental carcinogenic data. When heated to decomposition it emits very toxic fumes of SO_x, NO_x, and ZnO.

BHB000   CAS: 64092-23-5   HR: 3
BIS(2-BENZOYLBENZOATO)BIS(3-(1-METHYL-2-PYRROLIDINYL)PYRIDINE) NICKEL TRIHYDRATE
mf: $C_{48}H_{46}N_4NiO_6 \cdot 3H_2O$ mw: 887.75

SYN: NICOTINE, COMPOUND, with NICKEL(II)-o-BENZOYL BENZOATE TRIHYDRATE (2:1)

CONSENSUS REPORTS: Nickel and its compounds are on The Community Right-To-Know List.

OSHA PEL: (Transitional: TWA 1 mg/m^3) TWA 0.1 mg (Ni)/m^3
ACGIH TLV: TWA 0.1 mg (Ni)/m^3; (Proposed: TWA 0.05 mg(Ni)/m^3; Human Carcinogen)
NIOSH REL: (Inorganic Nickel) TWA 0.015 mg(Ni)/m^3

SAFETY PROFILE: Suspected carcinogen. Poison by ingestion and intraperitoneal routes. When heated to decomposition it emits toxic fumes of NO_x.

BHB750   CAS: 4420-79-5   HR: 3
2,5-BIS(BIS-(2-CHLOROETHYL)AMINOMETHYL)HYDROQUINONE
mf: $C_{16}H_{24}Cl_4N_2O_2$ mw: 418.22

SYNS: HYDROQUINONE MUSTARD * NSC 18321 * WEATHERBEE MUSTARD

SAFETY PROFILE: Deadly poison by intravenous and intraperitoneal routes. A powerful irritant. Questionable carcinogen with experimental carcinogenic data. When heated to decomposition it emits highly toxic fumes of NO_x and Cl^-.

BHD250   CAS: 3785-34-0   HR: 3
1,2-BIS(BROMOACETOXY)ETHANE
mf: $C_6H_8Br_2O_4$ mw: 303.96

SYNS: BROMOACETIC ACID ETHYLENE ESTER * ETHYLENE BIS(BROMOACETATE) * ETHYLENE BROMOACETATE * ETHYLENE GLYCOL BIS(BROMOACETATE) * PANDUROL * S 13

SAFETY PROFILE: Poison by intraperitoneal and intravenous routes. When heated to decomposition it emits toxic fumes of Br^-.

BHK250   CAS: 15546-16-4   HR: 3
BIS(BUTOXYMALEOYLOXY)DIBUTYL-STANNANE
mf: $C_{24}H_{40}O_8Sn$ mw: 575.33

SYNS: DI-N-BUTYLTIN DI(MONOBUTYL)MALEATE * DI-N-BUTYL-ZINN-DI(MONOBUTYL)MALEINAT (GER-MAN)

CONSENSUS REPORTS: Reported in EPA TSCA Inventory.

OSHA PEL: TWA 0.1 mg(Sn)/m^3 (skin)
ACGIH TLV: TWA 0.1 mg(Sn)/m^3 (skin) (Proposed: TWA 0.1 mg(Sn)/m^3; STEL 0.2 mg(Sn)/m^3 (skin))
NIOSH REL: (Organotin Compounds) TWA 0.1 mg(Sn)/m^3

SAFETY PROFILE: Poison by ingestion. When heated to decomposition it emits acrid smoke and irritating fumes.

BHK500   CAS: 29575-02-8   HR: 2
BIS(BUTOXYMALEOYLOXY)DIOCTYL-STANNANE
mf: $C_{32}H_{56}O_8Sn$ mw: 687.57

SYNS: DI-N-OCTYLTIN BIS(BUTYL MALEATE) * DI-N-OCTYLTIN DIMONOBUTYLMALEATE * DI-N-OCTYLZINN-DIMONOBUTYLMALEINAT (GER-MAN)

CONSENSUS REPORTS: Reported in EPA TSCA Inventory.

OSHA PEL: TWA 0.1 mg(Sn)/m^3 (skin)
ACGIH TLV: TWA 0.1 mg(Sn)m^3 (skin) (Proposed: TWA 0.1 mg(Sn)/m^3; STEL 0.2 mg(Sn)/m^3 (skin))
NIOSH REL: (Organotin Compounds) TWA 0.1 mg(Sn)/m^3

SAFETY PROFILE: Moderately toxic by ingestion. When heated to decomposition it emits acrid smoke and irritating fumes.

BHM000   CAS: 111-17-1   HR: 3
BIS(2-CARBOXYETHYL) SULFIDE
mf: $C_6H_{10}O_4S$ mw: 178.22

PROP: Very sol in alc, hot water, acetate; sltly sol in water. Mp: 134°.

SYNS: DIETHYL SULFIDE-2,2'-DICARBOXYLIC ACID * KYSELINA-β,β'-THIODIPROPIONOVA (CZECH) * TDPA * 2-(2,3,5,6-TETRAMETHYLPHENOXY)PROPIONIC AICD * 4-THIAHEPTANEDIOIC ACID * THIODIPROPIONIC ACID * β,β'-THIODIPROPIONIC ACID * 3,3'-THIODIPROPIONIC ACID * TYOX A

CONSENSUS REPORTS: Reported in EPA TSCA Inventory.

SAFETY PROFILE: A poison by intraperitoneal and intravenous routes. Moderately toxic by ingestion. A skin and eye irritant. When heated to decomposition it emits toxic fumes of SO_x.

BHM750   CAS: 94-17-7   HR: 2
BIS(p-CHLOROBENZOYL) PEROXIDE
DOT: UN 2113/UN 2114/UN 2115
mf: $C_{14}H_8Cl_2O_4$ mw: 311.12

PROP: A white, granular material. Insol in water, sol in organic solvents.

SYNS: CADPX PS * p-CHLOROBENZOYL PEROXIDE * p-CHLOROBENZOYL PEROXIDE (DOT) * p,p'-DICHLOROBENZOYL PEROXIDE * DI-(4-CHLOROBENZOYL) PEROXIDE

CONSENSUS REPORTS: Reported in EPA TSCA Inventory.

DOT Classification: Organic Peroxide; Label: Organic Peroxide.

SAFETY PROFILE: Moderately toxic by intraperitoneal route. Probably an irritant to skin and mucous membranes. Dangerous fire hazard; a powerful oxidizer. Store in a cool place away from fire hazards, sparks, open flames, and out of the direct rays of the sun. Dangerous explosion hazard; this material may explode by heat (over 38°) or contamination. Any contaminant which acts as an accelerator to the polymerization or decomposition of this material can cause an explosion. Heat or contact with certain fumes or mists can cause it to explode. To fight small fires, use CO_2 or foam extinguishers. Water spray or mist may also be used. Dry chemical is effective. When heated to decomposition it emits toxic fumes of Cl$^-$.

BHN000   CAS: 366-93-8   HR: 2
trans-N,N'-BIS(2-CHLOROBENZYL)-1,4-CYCLOHEXANEBIS(METHYLAMINE) DIHYDROCHLORIDE
mf: $C_{22}H_{28}Cl_2N_2 \cdot 2ClH$ mw: 464.34

SYNS: AY 9944 * trans-1,4-BIS(2-DICHLOROBENZYLAMINOETHYL)CYCLOHEXANE DICHLORHYDRATE (FRENCH) * trans-N,N'-(1,4-CYCLOHEXYLENEDIMETHYLENE)BIS(2-CHLOROBENZYLAMINE) DIHYDROCHLORIDE

SAFETY PROFILE: Moderately toxic by ingestion. Experimental teratogenic and reproductive effects. Inhibits cholesterol synthesis. When heated to decomposition it emits very toxic fumes of NO_x and Cl^-.

BHN500   CAS: 3374-04-7   HR: 3
N,N-BIS(β-CHLOROETHYL)-dl-ALANINE HYDROCHLORIDE
mf: $C_7H_{13}Cl_2NO_2 \cdot ClH$ mw: 250.57

SYNS: ALANINE MUSTARD * NSC 17663

SAFETY PROFILE: Deadly poison by intracerebral and intravenous routes. Human systemic effects by an unspecified route: bone marrow changes. When heated to decomposition it emits very toxic fumes of Cl^-, NO_x and HCl.

BHN750   CAS: 334-22-5   HR: 3
BIS-β-CHLOROETHYLAMINE
mf: $C_4H_9Cl_2N$ mw: 142.04

SYNS: N,N-BIS-(β-CHLORAETHYL)-AMIN (GERMAN) * NH-LOST * NOR-NITROGEN MUSTARD * NSC-10873

SAFETY PROFILE: Poison by intraperitoneal, subcutaneous, and intravenous routes. Human mutation data reported. When heated to decomposition it emits very toxic NO_x and Cl^-.

BHO250   CAS: 821-48-7   HR: 3
BIS(2-CHLOROETHYL)AMINE HYDROCHLORIDE
mf: $C_4H_9Cl_2N \cdot ClH$ mw: 178.50

SYNS: BIS(β-CHLOROETHYL)AMINE HYDROCHLORIDE * N,N-BIS-(2-CHLOROETHYL)AMINE HYDROCHLORIDE * BIS(2-CHLOROETHYL)AMMONIUM CHLORIDE * 2-CHLORO-N-(2-CHLOROETHYL)ETHANAMINE HYDROCHLORIDE * β,β'-DICHLORODIETHYLAMINE HYDROCHLORIDE * 2,2'-DICHLORO DIETHYLAMINE HYDROCHLORIDE * DI-2-CHLOROETHYLAMINE HYDROCHLORIDE * LEO 72a * NC 26 * NOR-HN2 * NOR-HN2 HYDROCHLORIDE * NOR-LOST HYDROCHLORID (GERMAN) * NORNITROGEN MUSTARD HYDROCHLORIDE * NSC 10873 * SK 555 * TL 161

CONSENSUS REPORTS: EPA Genetic Toxicology Program. Reported in EPA TSCA Inventory.

SAFETY PROFILE: A poison by inhalation, intraperitoneal, intramuscular, and subcutaneous routes. An experimental teratogen. Human mutation data reported. When heated to decomposition it emits toxic fumes of NH_3, NO_x and Cl^-.

BHO500   CAS: 1215-16-3   HR: 3
4'-(BIS(2-CHLOROETHYL)AMINO) ACETANILIDE
mf: $C_{12}H_{16}Cl_2N_2O$ mw: 275.20

SYN: p-ACETYLAMINOPHENYL DERIVATIVE of NITROGEN MUSTARD

SAFETY PROFILE: Poison via intraperitoneal route. An experimental teratogen. When heated to decomposition it emits very toxic fumes of Cl^- and NO_x.

BHP750   CAS: 1492-93-9   HR: 3
4'-(BIS(2-CHLOROETHYL)AMINO)-2-FLUORO ACETANILIDE
mf: $C_{12}H_{15}Cl_2FN_2O$ mw: 293.19

SYN: p-FLUOROACETYLAMINOPHENYL DERIVATIVE of NITROGEN MUSTARD

SAFETY PROFILE: Poison by intraperitoneal route. An experimental teratogen. When heated to decomposition it emits very toxic fumes of Cl^-, F^-, and NO_x.

BHT250   CAS: 342-95-0   HR: 3
3-(o-(BIS-(β-CHLOROETHYL)AMINO)PHENYL)-dl-ALANINE
mf: $C_{13}H_{18}Cl_2N_2O_2$ mw: 305.23

SYNS: CB 1729 * o-DI-2-CHLOROETHYLAMINO-dl-PHENYLALANINE * FDA 0109 * MEROPHAN * o-MEROPHAN * NSC-57199 * ORTHOPHENYL-ALANINE MUSTARD * (±)-o-PHENYLALANINE MUSTARD * o-PHENYLALANINE MUSTARD * o-dl-SARCOLYSIN

SAFETY PROFILE: Deadly poison by intraperitoneal, subcutaneous, and intravenous routes. When heated to decomposition it emits very toxic fumes of Cl^- and NO_x.

BHT750   CAS: 531-76-0   HR: 3
dl-3-(p-(BIS(2-CHLOROETHYL)AMINO) PHENYL)ALANINE
mf: $C_{13}H_{18}Cl_2N_2O_2$ mw: 305.23

SYNS: 4-(BIS(2-CHLOROETHYL)AMINO)-dl-PHENYL-ALANINE * 3-(p-(BIS(2-CHLOROETHYL)AMINO)PHENYL)ALANINE * CB-3307 * p-DI-(2-CHLORAETHYL)-AMINO-dl-PHENYL-ALANIN (GERMAN) * p-DI(2-CHLOROETHYL)AMINO-dl-PHENYLALANINE * MERFALAN * MERPHALAN * o-MERPHALAN * NCI-C04944 * NSC-14210 * PHENYLALANIN-LOST (GERMAN) * dl-PHENYLALANINE MUSTARD

* SAKOLYSIN (GERMAN) * SARCOCLORIN
* dl-SARCOLYSIN * dl-SARCOLYSINE

CONSENSUS REPORTS: IARC Cancer Review: GROUP 2B IMEMDT 7,56,87; Animal Limited Evidence IMEMDT 9,167,75; NCI Carcinogenesis Studies (ipr); Clear Evidence: mouse CANCAR 40,1935,77; No Evidence: rat CANCAR 40,1935,77

SAFETY PROFILE: Suspected carcinogen with experimental tumorigenic data. A poison by ingestion, intraperitoneal, intravenous, and intracerebral routes. Other experimental reproductive effects. Mutation data reported. An antineoplastic agent. When heated to decomposition it emits very toxic fumes of Cl^- and NO_x.

BHU750   CAS: 4213-32-5   **HR: 3**
3-(p-(BIS(β-CHLOROETHYL)
AMINO)PHENYL)-d-ALANINE
HYDROCHLORIDE
mf: $C_{13}H_{18}Cl_2N_2O_2 \cdot ClH$ mw: 341.69

SYNS: 4-(BIS(2-CHLOROETHYL)AMINO)-d-PHENYLALANINE MONOHYDROCHLORIDE * NSC-35051
* PHENYLALANINE MUSTARD * d-PHENYLALANINE MUSTARD

SAFETY PROFILE: An intravenous poison. When heated to decomposition it emits very toxic fumes of Cl^-, NO_x and HCl.

BHV000   CAS: 1465-26-5   **HR: 3**
3-(p-(BIS(β-CHLOROETHYL)
AMINO)PHENYL)-dl-ALANINE
HYDROCHLORIDE
mf: $C_{13}H_{18}Cl_2N_2O_2 \cdot ClH$ mw: 341.69

SYNS: ALKERAN (RUSSIAN) * 4-(BIS(2-CHLOROETHYL)AMINO)-dl-PHENYLALANINE MONOHYDROCHLORIDE * CB 3008 * MELPHALAN (RUSSIAN)
* MERPHALAN HYDROCHLORIDE * NCS-14210
* dl-PHENYLALANINE MUSTARD HYDROCHLORIDE
* dl-SARCOLYSINE HYDROCHLORIDE * SARCOLYSIN HYDROCHLORIDE * SARKOKLORIN
* SKI 21739

SAFETY PROFILE: Deadly poison by intravenous route. Human mutation data reported. When heated to decomposition it emits very toxic fumes of Cl^- and NO_x.

BHV250   CAS: 3223-07-2   **HR: 3**
l-3-(p-(BIS(2-CHLOROETHYL)AMINO)
PHENYL)ALANINE MONO-
HYDROCHLORIDE
mf: $C_{13}H_{18}Cl_2N_2O_2 \cdot ClH$ mw: 341.69

SYNS: ALANINE NITROGEN MUSTARD * CB 3025
* MELPHALAN HYDROCHLORIDE * NSC-8806
* l-PHENYLALANINE MUSTARD HYDROCHLORIDE
* l-SARCOLYSINE HYDROCHLORIDE

SAFETY PROFILE: Deadly poison by intravenous route. Human systemic effects by ingestion: nausea and vomiting. Questionable carcinogen with experimental carcinogenic data. When heated to decomposition it emits very toxic fumes of Cl^-, NO_x, and HCl.

BHX250   CAS: 35849-41-3   **HR: 3**
l-3-(p-(BIS(2-CHLOROETHYL)AMINO)
PHENYL)-N-FORMYLALANINE-3
mf: $C_{14}H_{18}Cl_2N_2O_3$ mw: 333.24

SYN: N-FORMYL-l-p-SARCOLYSIN

SAFETY PROFILE: Poison by intraperitoneal route. Moderately toxic by ingestion. Human gastrointestinal effects by ingestion. When heated to decomposition it emits very toxic fumes of Cl^- and NO_x.

BHY500   CAS: 857-95-4   **HR: 3**
o-(4-(BIS(2-CHLOROETHYL)AMINO)
PHENYL-dl-TYROSINE
mf: $C_{19}H_{22}Cl_2N_2O_3$ mw: 397.26

SYN: PHENTYRIN

SAFETY PROFILE: A poison by ingestion, intravenous, and intraperitoneal routes. When heated to decomposition it emits very toxic fumes of Cl^- and NO_x.

BIA000   CAS: 4213-40-5   **HR: 3**
o-(4-BIS(β-CHLOROETHYL)AMINO-o-
TOLYLAZO)BENZOIC ACID
mf: $C_{18}H_{19}Cl_2N_3O_2$ mw: 380.30

SYN: NSC-16498

SAFETY PROFILE: Poison by intraperitoneal and intravenous routes. Mutation data reported. When heated to decomposition it emits very toxic fumes of Cl^- and NO_x.

BIA250   CAS: 66-75-1   **HR: 3**
5-(BIS(2-CHLOROETHYL)AMINO)
URACIL
mf: $C_8H_{11}Cl_2N_3O_2$ mw: 252.12

SYNS: AMINOURACIL MUSTARD * 5-(BIS(2-CHLOROETHYL)AMINO)-2,4(1H,3H)PYRIMIDINEDIONE
* 5-N,N-BIS(2-CHLOROETHYL)AMINOURACIL

* CB-4835 * CHLORETHAMINACIL * DEMETHYL-DOPAN * DESMETHYLDOPAN * 5-(DI-(β-CHLORO-ETHYL)AMINO)URACIL * 5-(DI-2-CHLOROETHYL) AMINOURACIL * 2,6-DIHYDROXY-5-BIS(2-CHLORO-ETHYL)AMINOPYRAMIDINE * ENT 50,439 * NCI-C04820 * NORDOPAN * NSC-34462 * RCRA WASTE NUMBER U237 * SK-19849 * U-8344 * URACILLOST * URACILMOSTAZA * URACIL MUSTARD * URAMUSTIN * URAMUSTINE

CONSENSUS REPORTS: IARC Cancer Review: GROUP 2B IMEMDT 7,370,87; Animal Sufficient Evidence IMEMDT 9,235,75; NCI Carcinogenesis Studies (ipr); Clear Evidence: mouse, rat RRCRBU 52,1,75. EPA Genetic Toxicology Program.

SAFETY PROFILE: Suspected carcinogen with experimental carcinogenic and neoplastigenic data. A deadly poison by ingestion and intraperitoneal routes. Mutation data reported. When heated to decomposition it emits very toxic fumes of Cl^- and NO_x.

BIA750   CAS: 55-51-6   HR: 3
N,N-BIS(2-CHLOROETHYL)BENZYL-AMINE
mf: $C_{11}H_{15}Cl_2N$ mw: 232.17

SYNS: BENZYLBIS(β-CHLOROETHYL)AMINE * BENZYL NORMECHLORETHAMINE * N,N-BIS(2-CHLOROETHYL)BENZENEMETHANAMINE * BIS(2-CHLOROETHYL)BENZYLAMINE * DCBA * DI-(2-CHLOROETHYL)BENZYLAMINE * TL 965

CONSENSUS REPORTS: EPA Genetic Toxicology Program.

SAFETY PROFILE: Poison by ingestion and possibly other routes. Mutation data reported. When heated to decomposition it emits very toxic fumes of Cl^- and NO_x.

BID250   CAS: 538-07-8   HR: 3
BIS(2-CHLOROETHYL)ETHYLAMINE
mf: $C_6H_{13}Cl_2N$ mw: 170.10

SYNS: 2,2'-DICHLOROTRIETHYLAMINE * ETHYL-BIS(β-CHLOROETHYL)AMINE * ETHYLBIS(2-CHLORO-ETHYL)AMINE * ETHYL-S * HN1 * TL 329 * TL 1149

CONSENSUS REPORTS: Reported in EPA TSCA Inventory. EPA Extremely Hazardous Substances List.

SAFETY PROFILE: Deadly poison by inhalation, skin contact, ingestion, intravenous, subcutaneous, and intraperitoneal routes. When heated to decomposition it emits very toxic fumes of Cl^- and NO_x.

BID750   CAS: 111-91-1   HR: 3
BIS(β-CHLOROETHYL)FORMAL
mf: $C_5H_{10}Cl_2O_2$ mw: 173.05

PROP: Liquid. Bp: 217.5°, flash p: 230°F (OC), d: 1.23, vap d: 5.9.

SYNS: BIS(2-CHLOROETHOXY)METHANE * BIS(2-CHLOROETHYL)FORMAL * DICHLOROETHYL FORMAL * DI-2-CHLOROETHYL FORMAL * FORMAL-DEHYDE BIS(β-CHLOROETHYL) ACETAL * 1,1'-(METHYLENEBIS(OXY)BIS(2-CHLOROETHANE) * RCRA WASTE NUMBER U024

CONSENSUS REPORTS: Reported in EPA TSCA Inventory.

SAFETY PROFILE: Poison by ingestion, inhalation, and skin contact. A skin and eye irritant. Combustible when exposed to heat or flame. Incompatible with oxidizers. To fight fire, use alcohol foam, foam, CO_2, dry chemical. When heated to decomposition it emits toxic fumes of Cl^-.

BIE250   CAS: 51-75-2   HR: 3
BIS(β-CHLOROETHYL)METHYLAMINE
mf: $C_5H_{11}Cl_2N$ mw: 156.07

PROP: Dark liquid. Mp: 1° @ 10 mm, d: 1.09 @ 25°, vap press: 0.17 mm @ 25°, vap d: 5.9.

SYNS: BIS(2-CHLOROETHYL)METHYLAMINE * N,N-BIS(2-CHLOROETHYL)METHYLAMINE * CARYOLYSIN * CHLORMETHINE * CLORAMIN * DICHLOR AMINE * DICHLOREN (GERMAN) * β,β'-DICHLORODIETHYL-N-METHYLAMINE * DI(2-CHLOROETHYL)METHYLAMINE * 2,2'-DI-CHLORO-N-METHYLDIETHYLAMINE * EMBICHIN * ENT 25,294 * HN2 * MBA * MECHLORETH-AMINE * N-METHYL-BIS-CHLORAETHYLAMIN (GER-MAN) * METHYLBIS(β-CHLOROETHYL)AMINE * N-METHYL-BIS(β-CHLOROETHYL)AMINE * N-METHYL-BIS(2-CHLOROETHYL)AMINE (MAK) * N-METHYL-2,2'-DICHLORODIETHYLAMINE * METHYLDI(2-CHLOROETHYL)AMINE * N-METHYL-LOST * MUSTARGEN * MUSTINE * MUTAGEN * NITROGEN MUSTARD * N-LOST (GERMAN) * NSC 762 * TL 146

CONSENSUS REPORTS: EPA Genetic Toxicology Program. Reported in EPA TSCA Inventory. EPA Extremely Hazardous Substances List. Community Right-To-Know List.

DFG MAK: Human Carcinogen.

SAFETY PROFILE: Confirmed human carcinogen producing skin tumors by skin contact. Experimental carcinogenic, tumorigenic, and neoplastigenic data. A deadly poison by inhalation, ingestion, skin contact, and most other routes. Experimental teratogenic and reproductive effects. A powerful skin and eye irritant. Human mutation data reported. It has been used as a blistering agent in chemical warfare. When heated to decomposition it emits very toxic fumes of Cl^- and NO_x.

BIE500   CAS: 55-86-7   **HR: 3**
BIS(2-CHLOROETHYL)METHYLAMINE HYDROCHLORIDE
mf: $C_5H_{11}Cl_2N \cdot ClH$ mw: 192.53

SYNS: ANTIMIT * AZOTOYPERITE * C 6866 * CAROLYSINE * CARYOLYSINE * CARYOLYSINE HYDROCHLORIDE * CHLORAMIN * CHLORAMINE * CHLORAMIN HYDROCHLORIDE * CHLORETHAMINE * CHLORETHAZINE * CHLORMETHINE HYDROCHLORIDE * CHLORMETHINUM * 2-CHLORO-N-(2-CHLOROETHYL)-N-METHYLETHANAMINE HYDROCHLORIDE * DEMA * DICHLOREN * DICHLOREN HYDROCHLORIDE * β,β′-DICHLORODIETHYL-N-METHYLAMINE HYDROCHLORIDE * DI(2-CHLOROETHYL)METHYLAMINE HYDROCHLORIDE * 1,5-DICHLORO-3-METHYL-3-AZAPENTANE HYDROCHLORIDE * 2,2′-DICHLORO-N-METHYLDIETHYLAMINE HYDROCHLORIDE * DIMITAN * EMBECHINE * EMBICHIN * EMBICHIN HYDROCHLORIDE * EMBIKHINE * ERASOL * ERASOL HYDROCHLORIDE * ERASOL-IDO * HN2.HCl * HN2 HYDROCHLORIDE * KLORAMIN * N-LOST * MBA HYDROCHLORIDE * MEBICHLORAMINE * MECHLORETHAMINE HYDROCHLORIDE * MERCHLORETHANAMINE * METHYLBIS(β-CHLOROETHYL)AMINE HYDROCHLORIDE * N-METHYL-BIS-β-CHLORETHYLAMINE HYDROCHLORIDE * METHYLBIS(2-CHLOROETHYL)AMINE HYDROCHLORIDE * N-METHYLBIS (2-CHLOROETHYL)AMINE HYDROCHLORIDE * N-METHYL-2,2′-DICHLORODIETHYLAMINE HYDROCHLORIDE * N-METHYL-DI-2-CHLOROETHYLAMINE HYDROCHLORIDE * METHYLDI(β-CHLOROETHYL)AMINE HYDROCHLORIDE * METHYLDI(2-CHLOROETHYL)AMINE HYDROCHLORIDE * MITO-XINE * N-MUSTARD (GERMAN) * MUSTARGEN * MUSTARGEN HYDROCHLORIDE * MUSTINE HYDROCHLOR * MUSTINE HYDROCHLORIDE * NCI-C56382 * NITOL * NITOL "TAKEDA" * NITROGEN MUSTARD HYDROCHLORIDE * NITROGRANULOGEN * NITROGRANULOGEN HYDROCHLORIDE * NSC 762 * NSC-762 HYDROCHLORIDE * PLIVA * STICKSTOFFLOST * ZAGREB

CONSENSUS REPORTS: IARC Cancer Review: GROUP 2A IMEMDT 7,269,87; Animal Sufficient Evidence IMEMDT 9,193,75. NTP Fourth Annual Report On Carcinogens, 1984. EPA Genetic Toxicology Program.

SAFETY PROFILE: Confirmed carcinogen with experimental carcinogenic, neoplastigenic, and tumorigenic data. Deadly poison by ingestion, intravenous, subcutaneous, intraperitoneal, and parenteral routes. Experimental teratogenic and reproductive effects. Human systemic effects by intravenous route: nausea or vomiting, reduction in the number of white blood cells and blood platelates. Other experimental reproductive effects. Human mutation data reported.

BIF250   CAS: 494-03-1   **HR: 3**
N,N-BIS(2-CHLOROETHYL)-2-NAPHTHYLAMINE
mf: $C_{14}H_{15}Cl_2N$ mw: 268.20

SYNS: 2-BIS(2-CHLOROETHYL)AMINONAPHTHALENE * BIS(2-CHLOROETHYL)-β-NAPHTHYLAMINE * CHLORNAFTINA * CHLORNAPHAZIN * CHLORNAPHTHIN * CHLORONAFTINA * CHLORONAPHTHINE * CLORNAPHAZINE * DICHLOROETHYL-β-NAPHTHYLAMINE * DI(2-CHLOROETHYL)-β-NAPHTHYLAMINE * N,N-DI(2-CHLOROETHYL)-β-NAPHTHYLAMINE * 2-N,N-DI(2-CHLOROETHYL)NAPHTHYLAMINE * ERYSAN * NAPHTHYLAMINE MUSTARD * β-NAPHTHYL-BIS-(β-CHLOROETHYL)AMINE * 2-NAPHTHYLBIS(2-CHLOROETHYL)AMINE * β-NAPHTHYL-DI-(2-CHLOROETHYL)AMINE * NSC-62209 * R48 * RCRA WASTE NUMBER U026

CONSENSUS REPORTS: IARC Cancer Review: GROUP 1 IMEMDT 7,130,87; Animal Sufficient Evidence IMEMDT 4,119,74; Human Sufficient Evidence IMEMDT 4,119,74. NTP Fourth Annual Report On Carcinogens, 1984. EPA Genetic Toxicology Program.

SAFETY PROFILE: Confirmed human carcinogen producing bladder tumors. Human and experimental carcinogenic data. Moderately toxic by intraperitoneal route. When heated to decomposition it emits very toxic fumes of Cl^- and NO_x.

BIF750 CAS: 154-93-8 **HR: 3**
N,N′-BIS(2-CHLOROETHYL)-N-NITROSOUREA
mf: $C_5H_9Cl_2N_3O_2$ mw: 214.07

SYNS: BCNU * BiCNU * BIS(2-CHLOROETHYL) NITROSOUREA * 1,3-BIS(β-CHLOROETHYL)-1-NITRO-SOUREA * 1,3-BIS-(2-CHLOROETHYL)-1-NITRO-SOUREA * BISCHLOROETHYLNITROSOUREA * CARMUBRIS * CARMUSTIN * CARMUSTINE * FDA 0345 * NCI-C04773 * NITRUMON * NSC-409962 * SK 27702 * SRI 1720

CONSENSUS REPORTS: IARC Cancer Review: GROUP 2A IMEMDT 7,150,87; Human Limited Evidence IMEMDT 26,79,81; Animal Sufficient Evidence IMEMDT 26,79,81. NTP Fourth Annual Report On Carcinogens, 1984. NCI Carcinogenesis Studies (ipr); Some Evidence: rat CANCAR 40,1935,77; Clear Evidence: mouse CANCAR 40,1935,77. EPA Genetic Toxicology Program.

SAFETY PROFILE: Confirmed carcinogen with experimental carcinogenic and tumorigenic data. A human poison by parenteral route. An experimental poison by ingestion, intravenous, intraperitoneal, parenteral, and subcutaneous routes. Human systemic effects by parenteral, intravenous, and possibly other routes: nausea or vomiting, reduced white blood cell and blood platelet counts, bone marrow damage and potentially fatal respiratory system effects including lung fibrosis, dyspnea, and cyanosis. Experimental teratogenic and reproductive effects. Human mutation data reported. When heated to decomposition it emits very toxic fumes of Cl^- and NO_x.

BIH250 CAS: 505-60-2 **HR: 3**
BIS(2-CHLOROETHYL)SULFIDE
mf: $C_4H_8Cl_2S$ mw: 159.08

PROP: Colorless (if pure), to light yellow, oily liquid. Bp: 228°, fp: 14.4°, flash p: 221°F, d: 1.2741 @ 20°/4°, vap d: 5.4, vap press: 0.09 mm @ 30°.

SYNS: BIS(β-CHLOROETHYL)SULFIDE * BIS(2-CHLOROETHYL)SULPHIDE * 1-CHLORO-2-(β-CHLORO-ETHYLTHIO)ETHANE * β,β-DICHLOR-ETHYL-SUL-PHIDE * 2,2′-DICHLORODIETHYL SULFIDE * DI-2-CHLOROETHYL SULFIDE * β,β′-DICHLORO-ETHYL SULFIDE * 2,2′-DICHLOROETHYL SULPHIDE (MAK) * DISTILLED MUSTARD * KAMPSTOFF "LOST" * MUSTARD GAS * MUSTARD HD * MUSTARD VAPOR * SCHWEFEL-LOST * S-LOST * S MUSTARD * SULFUR MUSTARD * SULFUR MUSTARD GAS * SULPHUR MUSTARD GAS * 1,1′-THIOBIS(2-CHLOROETHANE) * YELLOW CROSS LIQUID * YPERITE

CONSENSUS REPORTS: IARC Cancer Review: GROUP 1 IMEMDT 7,259,87; Animal Sufficient Evidence IMEMDT 9,181,75; Human Limited Evidence IMEMDT 9,181,75. NTP Fourth Annual Report On Carcinogens, 1984. EPA Extremely Hazardous Substances List. Community Right-To-Know List. EPA Genetic Toxicology Program. Reported in EPA TSCA Inventory.

DFG MAK: Human Carcinogen.

SAFETY PROFILE: Confirmed human carcinogen with experimental carcinogenic, neoplastigenic, and tumorigenic data. A human poison by inhalation and subcutaneous routes. An experimental poison by inhalation, skin contact, subcutaneous, and intravenous routes. A severe human skin and eye irritant. Human mutation data reported. A military blistering gas. Strongly effects the skin, eyes, lungs, and gastric system. Pulmonary lesions are often fatal. It penetrates the skin deeply and injures blood vessels. Minute amounts can cause inflammation. Secondary infections are common. Combustible when exposed to heat or flame; can be ignited by a large explosive charge. It will react with water or steam to produce toxic and corrosive fumes. Vigorous reaction with oxidizing materials. Incompatible with bleaching powder. To fight fire, use water, foam, CO_2, dry chemical. Dangerous; when heated to decomposition or on contact with acid or acid fumes it emits highly toxic fumes of SO_x and Cl^-.

BIH500 CAS: 471-03-4 **HR: 3**
BIS(2-CHLOROETHYL)SULFONE
mf: $C_4H_8Cl_2O_2S$ mw: 191.08

SYNS: BIS(β-CHLOROETHYL)SULFONE * MUSTARD GAS SULFONE * MUSTARD SULFONE * YPERITE SULFONE

SAFETY PROFILE: A poison via intravenous and subcutaneous routes. Moderately toxic via inhalation. When heated to decomposition it emits very toxic fumes of Cl$^-$ and SO$_x$.

heated to decomposition it emits toxic fumes of Cl$^-$.

BII250   CAS: 108-60-1   HR: 3
BIS(2-CHLOROISOPROPYL) ETHER

DOT: UN 2490
mf: C$_6$H$_{12}$Cl$_2$O   mw: 171.08

PROP: Colorless liquid. Bp: 187.8°, fp: >−20°, flash p: 185°F (OC), d: 1.11 @ 25°/25°, vap d: 6.0, vap press: 0.10 mm @ 20°.

SYNS: BIS(2-CHLORO-1-METHYLETHYL) ETHER * (2-CHLORO-1-METHYLETHYL) ETHER * DICHLORODIISOPROPYL ETHER * DICHLOROISOPROPYL ETHER (DOT) * 2,2′-DICHLOROISOPROPYL ETHER * NCI-C50044 * RCRA WASTE NUMBER U027

CONSENSUS REPORTS: IARC Cancer Review: GROUP 3 IMEMDT 7,56,87, Animal Limited Evidence IMEMDT 41,149,86. NCI Carcinogenesis Bioassay (gavage); No Evidence: rat NCITR* NCI-CG-TR-191,79. Community Right-To-Know List. Reported in EPA TSCA Inventory.

DOT Classification: Corrosive Material; Label: Corrosive; IMO: Poison B; Label: Poison.

SAFETY PROFILE: Poison by ingestion. Moderately toxic by skin contact and inhalation. An eye irritant. Questionable carcinogen. Mutation data reported. A corrosive material. Moderate fire hazard when exposed to heat, flame, or powerful oxidizers. Incompatible with oxidizing materials. To fight fire, use water to blanket fire, or use foam, CO$_2$, dry chemical. When heated to decomposition it emits highly toxic fumes of Cl$^-$.

BIJ250   CAS: 13483-18-6   HR: 3
BIS-1,2-(CHLOROMETHOXY)ETHANE
mf: C$_4$H$_8$Cl$_2$O$_2$ mw: 159.02

PROP: Viscous liquid. Bp: 99-100° @ 22 mm, d: 1.2879 @ 14°/15°.

SYN: ETHYLENE GLYCOL BIS(CHLOROMETHYL)ETHER

CONSENSUS REPORTS: IARC Cancer Review: GROUP 3 IMEMDT 7,56,87; Animal Sufficient Evidence IMEMDT 15,31,77. Reported in EPA TSCA Inventory. Glycol ethers are on the Community Right-To-Know List.

SAFETY PROFILE: Questionable carcinogen with experimental neoplastigenic data. When

BIJ500   CAS: 56894-91-8   HR: 2
1,4-BIS(CHLOROMETHOXYMETHYL) BENZENE
mf: C$_{10}$H$_{12}$Cl$_2$O$_2$ mw: 235.12

SYN: BIS-1,4-(CHLOROMETHOXY)-p-XYLENE

CONSENSUS REPORTS: IARC Cancer Review: GROUP 3 IMEMDT 7,56,87; Animal Sufficient Evidence IMEMDT 15,37,77

SAFETY PROFILE: Questionable carcinogen with experimental neoplastigenic and tumorigenic data. When heated to decomposition it emits toxic fumes of Cl$^-$.

BIK000   CAS: 542-88-1   HR: 3
BIS(CHLOROMETHYL) ETHER

DOT: UN 2249
mf: C$_2$H$_4$Cl$_2$O   mw: 114.96

PROP: Volatile liquid. Bp: 105°, d: 1.315 @ 20°, vap d: 4.0. flash p: <19°.

SYNS: BCME * BIS-CME * CHLORO(CHLOROMETHOXY)METHANE * DICHLORDIMETHYLAETHER (GERMAN) * sym-DICHLORODIMETHYL ETHER (DOT) * sym-DICHLOROMETHYL ETHER * DIMETHYL-1,1′-DICHLOROETHER * OXYBIS(CHLOROMETHANE) * RCRA WASTE NUMBER P016

CONSENSUS REPORTS: IARC Cancer Review: GROUP 1 IMEMDT 7,131,87; Animal Sufficient Evidence IMEMDT 4,231,74; Human Sufficient Evidence IMEMDT 4,231,74. NTP Fourth Annual Report On Carcinogens, 1984. Community Right-To-Know List. EPA Extremely Hazardous Substances List. Reported in EPA TSCA Inventory.

OSHA: Cancer Suspect Agent
ACGIH TLV: TWA 0.001 ppm; Confirmed Human Carcinogen.
DFG MAK: Human Carcinogen.
DOT Classification: Poison B; Label: Flammable Liquid and Poison

SAFETY PROFILE: Confirmed human carcinogen with experimental carcinogenic, neoplastigenic, and tumorigenic data. Poison by inhalation, ingestion, and skin contact. Human systemic effects by inhalation: irritation of the conjunctiva, unspecified nasal and respiratory effects. Human mutation data reported. A dangerous fire hazard. When heated to decomposition it emits very toxic fumes of Cl^-.

BIK250   CAS: 534-07-6   HR: 3
BIS(CHLOROMETHYL)KETONE

DOT: UN 2649
mf: $C_3H_4Cl_2O$ mw: 126.97

PROP: Crystals. Mp: 45°, bp: 173°, d: 1.3826 @ 46°/4°, vap d: 4.38.

SYNS: sym-DICHLOROACETONE * α,α'-DICHLORO-ACETONE * α,γ-DICHLOROACETONE * 1,3-DI-CHLOROACETONE * 1,3-DICHLOROACETONE (DOT) * 1,3-DICHLORO-2-PROPANONE

CONSENSUS REPORTS: EPA Genetic Toxicology Program. Reported in EPA TSCA Inventory. EPA Extremely Hazardous Substances List.

DOT Classification: Poison B; Label: Poison.

SAFETY PROFILE: Poison by inhalation. Mutation data reported. A sytemic irritant by ingestion and inhalation routes. Dangerous; when heated to decomposition emits highly toxic fumes of Cl^-.

BIM250   CAS: 55-56-1   HR: 2
1,6-BIS(5-(p-CHLOROPHENYL)BIGUANI-DINO)HEXANE
mf: $C_{22}H_{30}Cl_2N_{10}$ mw: 505.52

SYNS: 1,6-BIS(p-CHLOROPHENYLDIGUANIDO)HEXANE * CHLORHEXIDIN (CZECH) * CHLORHEXIDINE * 1,6-DI(4'-CHLOROPHENYLDIGUANIDO)HEXANE * 1,1'-HEXAMETHYLENEBIS(5-(p-CHLOROPHENYL)BI-GUANIDE * HIBITANE * NOLVASAN * ROTERSEPT * STERIDO

SAFETY PROFILE: Mildly toxic by ingestion. Experimental reproductive effects. A human skin irritant. Mutation data reported. When heated to decomposition it emits very toxic fumes of Cl^- and NO_x.

BIM500   CAS: 72-54-8   HR: 3
1,1-BIS(4-CHLOROPHENYL)-2,2-DICHLOROETHANE

DOT: NA 2761
mf: $C_{14}H_{10}Cl_4$ mw: 320.04

PROP: Crystalline solid. Mp: 110°, vap d: 11.

SYNS: 1,1-BIS(p-CHLOROPHENYL)-2,2-DICHLORO-ETHANE * 2,2-BIS(p-CHLOROPHENYL)-1,1-DICHLORO-ETHANE * 2,2-BIS(4-CHLOROPHENYL)-1,1-DICHLORO-ETHANE * DDD * p,p'-DDD * 1,1-DICHLOOR-2,2-BIS(4-CHLOOR FENYL)-ETHAAN (DUTCH) * 1,1-DICHLOR-2,2-BIS(4-CHLOR-PHENYL)-AETHAN (GERMAN) * 1,1-DICHLORO-2,2-BIS(p-CHLOROPHENYL)ETHANE * 1,1-DICHLORO-2,2-BIS(4-CHLOROPHENYL)-ETHANE (FRENCH) * 1,1-DICHLORO-2,2-BIS(p-CHLOROPHE-NYL)ETHANE (DOT) * 1,1-DICHLORO-2,2-BIS(PARA-CHLOROPHENYL)ETHANE (DOT) * 1,1-DICHLORO-2,2-DI(4-CHLOROPHENYL)ETHANE * DICHLORODI-PHENYL DICHLOROETHANE * p,p'-DICHLORODIPHEN-YLDICHLOROETHANE * 1,1-DICLORO-2,2-BIS(4-CLORO-FENIL)-ETANO (ITALIAN) * DILENE * ENT 4,225 * ME-1700 * NCI-C00475 * RCRA WASTE NUMBER U060 * RHOTHANE * RHO-THANE D-3 * ROTHANE * TDE * p,p'-TDE * TDE (DOT) * TETRACHLORODIPHENYLETHANE

CONSENSUS REPORTS: IARC Cancer Review: Animal Sufficient Evidence IMEMDT 5,83,74. NCI Carcinogenesis Bioassay (feed); Clear Evidence: rat NCITR* NCI-CG-TR-131,78; No Evidence: mouse NCITR* NCI-CG-TR-131,78. EPA Genetic Toxicology Program.

DOT Classification: ORM-A; Label: None.

SAFETY PROFILE: Confirmed carcinogen with experimental carcinogenic, neoplastigenic, and tumorigenic data. Poison by ingestion. Moderately toxic by skin contact. Mutation data reported. An insecticide. When heated to decomposition it emits toxic fumes of Cl^-.

BIM750   CAS: 72-55-9   HR: 3
2,2-BIS(p-CHLOROPHENYL)-1,1-DICHLOROETHYLENE
mf: $C_{14}H_8Cl_4$ mw: 318.02

SYNS: DDE * p,p'-DDE * DDT DEHYDROCHLO-RIDE * 1,1-DICHLORO-2,2-BIS(p-CHLOROPHENYL)ETHYLENE * p,p'-DICHLORODIPHENYL DICHLORO-ETHYLENE * 1,1'-DICHLOROETHENYLIDENE)BIS(4-CHLOROBENZENE) * NCI-C00555

CONSENSUS REPORTS: IARC Cancer Review: Animal Limited Evidence IMEMDT 5,83,74. NCI Carcinogenesis Bioassay (feed); Clear Evidence: mouse NCITR* NCI-CG-TR-131,78; No Evidence: rat NCITR* NCI-CG-TR-131,78. EPA Genetic Toxicology Program.

SAFETY PROFILE: Suspected carcinogen with experimental carcinogenic and neoplastigenic data. Poison by ingestion. Experimental reproductive effects. Mutation data reported. An insecticide. When heated to decomposition it emits very toxic fumes of Cl$^-$.

BIO750   CAS: 115-32-2   **HR: 3**
1,1-BIS(p-CHLOROPHENYL)-2,2,2-TRICHLOROETHANOL

DOT: NA 2761
mf: $C_{14}H_9Cl_5O$ mw: 370.48

PROP: Material used in cancer bioassay was 40-60% pure NCITR* NCI-CG-TR-90,78.

SYNS: ACARIN * 1,1-BIS(CHLOROPHENYL)-2,2,2-TRICHLOROETHANOL * 1,1-BIS(4-CHLOROPHENYL)-2,2,2-TRICHLOROETHANOL * CARBAX * CEKUDIFOL * 4-CHLORO-α-(4-CHLOROPHENYL)-α-(TRICHLOROMETHYL)BENZENEMETHANOL * CPCA * DECOFOL * DICHLOROKELTHANE * DI-(p-CHLOROPHENYL)TRICHLOROMETHYLCARBINOL * 4,4'-DICHLORO-α-(TRICHLOROMETHYL)BENZHYDROL * DICOFOL * DTMC * ENT 23,648 * FW 293 * HIFOL * KELTANE * p,p'-KELTHANE * KELTHANE (DOT) * KELTHANE DUST BASE * KELTHANETHANOL * MILBOL * MITIGAN * NCI-C00486 * 2,2,2-TRICHLOOR-1,1-BIS(4-CHLOOR FENYL)-ETHANOL (DUTCH) * 1,1,1-TRICHLOR-2,2-BIS(4-CHLORPHENYL)-AETHANOL (GERMAN) * 2,2,2-TRICHLOR-1,1-BIS(4-CHLOR-PHENYL)-AETHANOL (GERMAN) * 2,2,2-TRICHLORO-1,1-BIS(4-CHLOROPHENYL)-ETHANOL (FRENCH) * 2,2,2-TRICHLORO-1,1-BIS(4-CLORO-FENIL)-ETANOLO (ITALIAN) * 2,2,2-TRICHLORO-1,1-DI-(4-CHLOROPHENYL)ETHANOL

CONSENSUS REPORTS: IARC Cancer Review: GROUP 3 IMEMDT 7,56,87; Animal Limited Evidence IMEMDT 30,87,83. NCI Carcinogenesis Bioassay (feed); Clear Evidence: mouse NCITR* NCI-CG-TR-90,78; No Evidence: rat NCITR* NCI-CG-TR-90,78. Community Right-To-Know List.

DOT Classification: ORM-E; Label: None.

SAFETY PROFILE: Poison by ingestion and skin contact. Moderately toxic by intraperitoneal route. Human mutation data reported. Questionable carcinogen with experimental carcinogenic data. When heated to decomposition it emits toxic fumes of Cl$^-$.

BIQ250   CAS: 40334-69-8   **HR: 3**
BIS(2-CHLOROVINYL)CHLOROARSINE
mf: $C_4H_4AsCl_3$ mw: 233.35

SYN: LEWISITE II

OSHA PEL: TWA 0.5 mg(As)/m^3

CONSENSUS REPORTS: Arsenic and its compounds are on The Community Right-To-Know List.

SAFETY PROFILE: A poison by skin contact and subcutaneous routes. When heated to decomposition it emits very toxic fumes of As and Cl$^-$.

BIQ500   CAS: 111-94-4   **HR: 3**
BIS(β-CYANOETHYL)AMINE
mf: $C_6H_9N_3$ mw: 123.18

PROP: Liquid. Mp: $-5.5°$, bp: 173° @ 10 mm, d: 1.0165 @ 30°, vap d: 3.3.

SYNS: BBCE * BIS-(2-CYANOETHYL)AMINE * N,N-BIS(2-CYANOETHYL)AMINE * 2-CYANO-N-(2-CYANOETHYL)ETHANAMINE * DI(2-CIANOETIL) AMMINA (ITALIAN) * DI-(2-CYANOETHYL)AMINE * 2,2'-DICYANODIETHYLAMINE * IDPN * 3,3'-IMINOBISPROPANENITRILE * IMINO-β,β'-DIPROPIONITRILE * β,β-IMINODIPROPIONITRILE * β,β'-IMINODIPROPIONITRILE * 3,3'-IMINODIPROPIONITRILE * 2341 I.S. * USAF A-8564

CONSENSUS REPORTS: Reported in EPA TSCA Inventory. Cyanide and its compounds are on the Community Right-To-Know List.

SAFETY PROFILE: A poison via intraperitoneal route. Moderately toxic by ingestion and skin contact. Experimental teratogenic and reproductive effects. An eye irritant. A storage hazard, may explode in a sealed container. When heated to decomposition it emits toxic fumes of NO$_x$ and CN$^-$.

BIS250   CAS: 38780-36-8   **HR: 3**
cis-BIS(CYCLOPENTYLAMMINE)PLATINUM(II)
mf: $C_{10}H_{22}Cl_2N_2Pt$ mw: 436.33

SYNS: cis-DICHLOROBIS(CYCLOPENTYLAMMINE) PLATINUM(II) * cis-DICYCLOPENTYLAMMINEDICHLOROPLATINUM(II)

SAFETY PROFILE: Moderately toxic by intraperitoneal and possibly other routes. Questionable carcinogen with experimental carcinogenic data. Mutation data reported. When heated to decomposition it emits very toxic fumes of Cl$^-$ and NO$_x$.

BIS500   CAS: 3465-75-6   **HR: 3**
BIS(DECANOYLOXY)DI-n-
BUTYLSTANNANE
mf: $C_{28}H_{56}O_4Sn$ mw: 575.53

SYN: BIS(DECANOYLOXY)DI-N-BUTYLTIN

OSHA PEL: TWA 0.1 mg(Sn)/m^3 (skin)
ACGIH TLV: TWA 0.1 mg(Sn)/m^3 (skin) (Proposed: TWA 0.1 mg(Sn)/m^3; STEL 0.2 mg(Sn)/m^3 (skin))
NIOSH REL: (Organotin Compounds) TWA 0.1 mg(Sn)/m^3

SAFETY PROFILE: Poison by ingestion. When heated to decomposition it emits acrid and irritant fumes.

BIW750   CAS: 13927-77-0   **HR: 3**
BIS(DIBUTYLDITHIOCARBAMATO)
NICKEL
mf: $C_{18}H_{36}N_2S_4 \cdot Ni$ mw: 467.51

SYNS: DIBUTYLDITHIOCARBAMIC ACID, NICKEL SALT * NICKEL DIBUTYLDITHIOCARBAMATE * UV CHEK AM 104 * VANGUARD N

CONSENSUS REPORTS: Reported in EPA TSCA Inventory. Nickel and its compounds are on The Community Right-To-Know List.

NIOSH REL: TWA 0.015 mg(Ni)/m^3

SAFETY PROFILE: Questionable carcinogen with experimental tumorigenic data. When heated to decomposition it emits very toxic fumes of SO$_x$ and NO$_x$.

BIX000   CAS: 136-23-2   **HR: 3**
BIS(DIBUTYLDITHIOCARBAMATO)ZINC
mf: $C_{18}H_{38}N_2S_4Zn$ mw: 476.19

PROP: White powder. Mp: 104-108°; d: 1.24 @ 20°/20°.

SYNS: ACETO ZDBD * BUTAZATE * BUTAZATE 50-D * BUTYL ZIMATE * BUTYL ZIRAM * DIBUTYLDITHIO-CARBAMIC ACID ZINC COMPLEX * DIBUTYLDITHIOCARBAMIC ACID ZINC SALT * USAF GY-5 * VULCACURE * VULKACIT LDB/C * ZINC-BIBUTYLDITHIOCARBAMATE * ZINC-DIBUTYLDITHIOCARBAMATE * ZINC-N,N-DIBUTYL-DITHIOCARBAMATE

CONSENSUS REPORTS: Reported in EPA TSCA Inventory. Zinc and its compounds are on the Community Right-To-Know List.

SAFETY PROFILE: Poison by intraperitoneal route. Questionable carcinogen with experimental tumorigenic data. When heated to decomposition it emits very toxic fumes of NO$_x$, ZnO and SO$_x$.

BIX500   CAS: 15442-77-0   **HR: 3**
BIS(3,4-DICHLOROBENZOATO)NICKEL
mf: $C_{14}H_6Cl_4NiO_4$ mw: 438.71

CONSENSUS REPORTS: Nickel and its compounds are on The Community Right-To-Know List.

OSHA PEL: (Transitional: TWA 1 mg/m^3) TWA 0.1 mg (Ni)/m^3
ACGIH TLV: TWA 0.1 mg(Ni)/m^3; (Proposed: TWA 0.05 mg(Ni)/m^3; Human Carcinogen)
NIOSH REL: (Inorganic Nickel) TWA 0.015 mg(Ni)/m^3

SAFETY PROFILE: Suspected carcinogen. Poison by intravenous route. When heated to decomposition it emits toxic fumes of Cl$^-$.

BIX750   CAS: 80-43-3   **HR: 2**
BIS(2,4-DICHLORO BENZOYL)
PEROXIDE
mf: $C_{14}H_6Cl_4O_4$ mw: 380.01

CONSENSUS REPORTS: Reported in EPA TSCA Inventory.

DOT Classification: Organic Peroxide; Label: Organic Peroxide.

SAFETY PROFILE: Moderately toxic by inhalation. Explosion Hazard: Pure compound is extremely shock sensitive and decomposes rapidly @ 80°. When heated to decomposition it emits toxic fumes of Cl$^-$.

BJB500   CAS: 14239-68-0   **HR: 3**
BIS(DIETHYLDITHIOCARBAMATO)
CADMIUM
mf: $C_{10}H_{20}CdN_2S_4$ mw: 408.96

SYNS: CADMIUM DIETHYL DITHIOCARBAMATE * ETHYL CADMATE * ETHYL TUADS

CONSENSUS REPORTS: Reported in EPA TSCA Inventory. Cadmium and its compounds are on the Community Right-To-Know List.

OSHA PEL: TWA 0.1 mg(Cd)/m^3; CL 0.6 mg(Cd)/m^3 (fume)
ACGIH TLV: TWA 0.05 mg(Cd)/m^3 (Proposed: TWA 0.01 mg(Cd)/m^3 (dust), Human Car-

cinogen); BEI: 10 μg/g creatinine in urine; 10 μg/L in blood.

DFG BAT: Blood 1.5 μg/dL; Urine 15 μg/dL, Suspected Carcinogen.

NIOSH REL: (Cadmium) Reduce to lowest feasible level

SAFETY PROFILE: Confirmed human carcinogen with experimental tumorigenic data. Mutation data reported. When heated to decomposition it emits very toxic fumes of NO_x and SO_x.

BJB750   CAS: 14239-51-1   HR: 3
BIS(DIETHYLDITHIOCARBAMATO) MERCURY

mf: $C_{10}H_{20}HgN_2S_4$ mw: 497.15

CONSENSUS REPORTS: Mercury and its compounds are on the Community Right-To-Know List.

OSHA PEL: (Transitional: CL 1 mg/10m^3) CL 0.1 mg(Hg)/m^3 (skin)
ACGIH TLV: TWA 0.1 mg(Hg)/m^3
NIOSH REL: TWA 0.05 mg(Hg)/m^3

SAFETY PROFILE: Poison by intravenous and intraperitoneal routes. When heated to decomposition it emits very toxic fumes of NO_x, SO_x, and Hg.

BJC000   CAS: 14324-55-1   HR: 3
BIS(DIETHYLDITHIOCARBAMATO)ZINC

mf: $C_{10}H_{22}N_2S_4 \cdot Zn$ mw: 363.95

PROP: White powder. D: 1.47 @ 20°/20°.

SYNS: DIETHYLDITHIOCARBAMIC ACID ZINC SALT * ETHAZATE * ETHYL CYMATE * ETHYL ZIMATE * ETHYL ZIRUM * VULCACURE * VULKACIT LDA * ZINC DIETHYLDITHIOCARBAMATE * ZINC-N,N-DIETHYLDITHIOCARBAMATE

CONSENSUS REPORTS: Reported in EPA TSCA Inventory. Zinc and its compounds are on the Community Right-To-Know List.

SAFETY PROFILE: Poison by intraperitoneal route. Moderately toxic by ingestion and subcutaneous routes. Severe irritant to eyes, nose, and throat. Questionable carcinogen with experimental carcinogenic and tumorigenic data. Mutation data reported. When heated to decomposition it emits very toxic fumes of NO_x and SO_x.

BJC250   CAS: 738-99-8   HR: 3
1,4-BIS(N,N'-DIETHYLENE PHOSPHAMIDE)PIPERAZINE

mf: $C_{12}H_{24}N_6O_2P_2$ mw: 346.36

SYNS: 1,4-BIS(BIS(1-AZIRIDINYL)PHOSPHINYL)PIPERAZINE * DIPIN * DIPINE * ENT 50,107 * 1,4-PIPERAZINEDIYLBIS(BIS(1-AZIRIDINYL)PHOSPHINE OXIDE * TETRAETHYLENEIMIDEPIPERAZINE-N,N'-DIPHOSPHORIC ACID

SAFETY PROFILE: Poison by ingestion, intraperitoneal, and subcutaneous routes. Human mutation data reported. When heated to decomposition it emits very toxic fumes of PO_x and NO_x.

BJD000   CAS: 34491-12-8   HR: 3
BIS(DIETHYLTHIO)CHLORO METHYL PHOSPHONATE

mf: $C_5H_{12}ClOPS_2$ mw: 218.71

SYNS: CHEMAGRO 5461 * CHEMAGRO R-5461 * S,S-DIETHYL(CHLOROMETHYL)PHOSPHONODITHIOATE * ENT 27,267 * R-5461

SAFETY PROFILE: Poison by ingestion, skin contact, and intraperitoneal routes. When heated to decomposition it emits very toxic fumes of SO_x, PO_x and Cl^-.

BJE750   CAS: 115-26-4   HR: 3
BIS(DIMETHYLAMIDO)FLUORO PHOSPHATE

mf: $C_4H_{12}FN_2OP$ mw: 154.15

SYNS: BFP * BFPO * BIS(DIMETHYLAMINO)FLUOROPHOSPHATE * BISDIMETHYLAMINOFLUOROPHOSPHINE OXIDE * BIS(DIMETHYLAMIDO)PHOSPHORYL FLUORIDE * CR 409 * DIFO * DIMEFOX * DMF * ENT 19,109 * FLUOPHOSPHORIC ACID DI(DIMETHYLAMIDE) * FLUORURE de N,N,N',N'-TETRAMETHYLE PHOSPHORO-DIAMIDE (FRENCH) * HANANE * PESTOX IV * PESTOX XIV * PESTOX 14 * T-2002 * TERRA-SYSTAM * TERRA-SYTAM * TERRASYTUM * N,N,N',N'-TETRAMETHYL-DIAMIDO-FOSFORZUUR-FLUORIDE (DUTCH) * TETRAMETHYLDIAMIDOPHOSPHORIC FLUORIDE * N,N,N',N'-TETRAMETHYL-DIAMIDO-PHOSPHORSAEURE-FLUORID (GERMAN) * TETRAMETHYL-PHOSPHORODIAMIDIC FLUORIDE * N,N,N,N-TETRAMETHYLPHOSPHORODIAMIDIC FLUORIDE * N,N,N',N'-TETRAMETIL-FOSFORODIAMMIDO-FLUORURO (ITALIAN) * TETRA SYTAM * TL 792 * WACKER S 14/10

CONSENSUS REPORTS: EPA Extremely Hazardous Substances List.

SAFETY PROFILE: Poison by ingestion, skin contact, intraperitoneal, subcutaneous, and intravenous routes. When heated to decomposition it emits very toxic fumes of F^-, NO_x and PO_x.

BJF000   CAS: 494-38-2   **HR: 3**
3,6-BIS(DIMETHYLAMINO)ACRIDINE
mf: $C_{17}H_{19}N_3$ mw: 265.39

SYNS: ACRIDINE ORANGE * ACRIDINE ORANGE FREE BASE * BASIC ORANGE 3RN * 2,8-BISDI-METHYLAMINOACRIDINE * BRILLIANT ACRIDINE ORANGE E * C.I. 46005 * C.I. NO. 46005:1 * C.I. BASIC ORANGE 14 * C.I. SOLVENT OR-ANGE 15 * 3,6-DI(DIMETHYLAMINO)ACRIDINE * EUCHRYSINE * RHODULINE ORANGE * SOL-VENT ORANGE 15 * N,N,N'-TETRAMETHYL-3,6-ACRI-DINEDIAMINE * WAXOLINE ORANGE A

CONSENSUS REPORTS: IARC Cancer Review: GROUP 3 IMEMDT 7,56,87; Animal Inadequate Evidence IMEMDT 16,145,78

SAFETY PROFILE: Poison by subcutaneous route. Questionable carcinogen with experimental tumorigenic and carcinogenic data. Mutation data reported. When heated to decomposition it emits toxic fumes of NO_x.

BJI000   CAS: 541-19-5   **HR: 3**
BIS(β-DIMETHYLAMINOETHYL)SUC-CINATE BIS(METHYLIODIDE)
mf: $C_{14}H_{30}N_2O_4 \cdot 2I$ mw: 517.92

SYNS: ASCURON * CELOCURINE * CHOLINE IODIDE SUCCINATE (2:1) * CURACIT * DIACETYL-CHOLINE DIIODIDE * DITILIN IODIDE * SUCCINIC ACID BIS(β-DIMETHYLAMINOETHYL) ESTER BISMETH-IODIDE * SUCCINIC ACID, DIESTER with CHOLINE IODIDE * SUCCINYLDICHOLINE IODIDE * o,o-SUC-CINYLDICHOLINE IODIDE * SUXAMETHONIUM IO-DIDE

SAFETY PROFILE: Poison by intravenous and intraperitoneal routes. When heated to decomposition it emits very toxic fumes of NO_x and I^-.

BJI250   CAS: 61-73-4   **HR: 3**
3,7-BIS(DIMETHYL AMINO)PHENAZA THIONIUM CHLORIDE
mf: $C_{16}H_{18}N_3S \cdot Cl$ mw: 319.88

SYNS: AIZEN METHYLENE BLUE BH * BASIC BLUE 9 * 3,7-BIS(DIMETHYLAMINO)PHENOTHIAZIN-5-IUM CHLORIDE * CALCOZINE BLUE ZF * CHROMOS-MON * C.I. BASIC BLUE 9 * C.I. 52 015 (CZECH) * D&C BLUE NUMBER 1 * EXTERNAL BLUE 1 * HIDACO METHYLENE BLUE SALT FREE * LEATHER PURE BLUE HB * METHYLENE BLUE * METHYLENE BLUE A * METHYLENE BLUE BB * METHYLENE BLUE BB ZINC FREE * METHYLENE BLUE CHLORIDE * METHYLENE BLUE CHLORIDE (BIOLOGICAL STAIN) * METHYLENE BLUE D * METHYLENE BLUE I (MEDICINAL) * METHYLENE BLUE (MEDICINAL) * METHYLENE BLUE NF (MEDICI-NAL) * METHYLENE BLUE POLYCHROME * METHYLENE BLUE USP (MEDICINAL) * METHY-LENE BLUE USP XII (MEDICINAL) * METHYLENIUM CERULEUM * METHYLTHIONINE CHLORIDE * METHYLTHIONIUM CHLORIDE * MITSUI METHY-LENE BLUE * MODR METHYLENOVA (CZECH) * SANDOCRYL BLUE BRL * SCHULTZ NO. 1038 * SWISS BLUE * TETRAMETHYLTHIONINE CHLO-RIDE * YAMAMOTO METHYLENE BLUE B

CONSENSUS REPORTS: EPA Genetic Toxicology Program. Reported in EPA TSCA Inventory.

SAFETY PROFILE: Poison by ingestion, intraperitoneal, intravenous, and subcutaneous routes. Mutation data reported. When heated to decomposition it emits very toxic fumes of NO_x, SO_x, and Cl^-.

BJK500   CAS: 137-30-4   **HR: 3**
BIS(DIMETHYLDITHIOCARBA-MATO)ZINC
mf: $C_6H_{12}N_2S_4 \cdot Zn$ mw: 305.81

PROP: White powder. Mp: 248-250°; d: 1.65 @ 20°/20°.

SYNS: AAPROTECT * AAVOLEX * AAZIRA * ACCELERATOR L * ACETO ZDED * ACETO ZDMD * ALCOBAM ZM * AMYL ZIMATE * ANTENE * BIS(DIMETHYLCARBAMODITHIOA-TO-S,S')ZINC * BIS(DIMETHYLDITHIOCARBAMATE de ZINC) (FRENCH) * BIS(N,N-DIMETIL-DITIOCAR-BAMMATO) DI ZINCO (ITALIAN) * CARBAMIC ACID, DIMETHYLDITHIO-, ZINC SALT (2:) * CARBAZINC * CIRAM * CORONA COROZATE * COROZATE * CUMAN * CUMAN L * CYMATE * DIMETHYLCARBAMODITHIOIC ACID, ZINC COM-PLEX * DIMETHYLCARBAMODITHIOIC ACID, ZINC SALT * DIMETHYLDITHIOCARBAMATE ZINC SALT * DIMETHYLDITHIOCARBAMIC ACID, ZINC SALT * DRUPINA 90 * EPTAC 1 * ENT 988

* FUCLASIN * FUCLASIN ULTRA * FUKLASIN * FUNGOSTOP * HERMAT ZDM * HEX-AZIR * KARBAM WHITE * METHASAN * METHAZATE * METHYL ZIMATE * METHYL ZINEB * METHYL ZIRAM * MEXENE * MEZ-ENE * MILBAM * MILBAN * MOLURAME * MYCRONIL * NCI-C50442 * ORCHARD BRAND ZIRAM * POMARSOL Z FORTE * PRO-DARAM * RHODIACID * SOXINAL PZ * SOXI-NOL PZ * TRICARBAMIX Z * TSIMAT * TSIRAM (RUSSIAN) * USAF P-2 * VANCIDE MZ-96 * VULCACURE * VULKACITE L * Z 75 * ZARLATE * Z-C SPRAY * ZERLATE * ZI-MATE * ZIMATE METHYL * ZINC BIS(DIMETHYL-DITHIOCARBAMATE) * ZINC BIS(DIMETHYLDITHIO-CARBAMOYL)DISULPHIDE * ZINC BIS(DIMETH-YLTHIOCARBAMOYL)DISULFIDE * ZINC DIMETHYL-DITHIOCARBAMATE * ZINC N,N-DIMETHYLDITHIO-CARBAMATE * ZINCMATE * ZINK-BIS(N,N-DI-METHYL-DITHIOCARBAMAAT) (DUTCH) * ZINK-BIS(N,N-DIMETHYL-DITHIOCARBAMAT) (GERMAN) * ZINKCARBAMATE * ZINK-(N,N-DIMETHYL-DITHIOCARBAMAT) (GERMAN) * ZIRAM * ZIRAM TECHNICAL * ZIRAMVIS * ZIRASAN * ZIRBERK * ZIREX 90 * ZIRIDE * ZIRTHANE * ZITOX

CONSENSUS REPORTS: IARC Cancer Review: GROUP 3 IMEMDT 7,56,87, Animal Inadequate Evidence IMEMDT 12,259,76; NTP Carcinogenesis Bioassay (feed); Clear Evidence: mouse, rat NTPTR* NTP-TR-238,83. EPA Genetic Toxicology Program. Reported in EPA TSCA Inventory. Zinc and its compounds are on the Community Right-To-Know List.

SAFETY PROFILE: Poison by ingestion, intraperitoneal, and intravenous routes. Moderately toxic by inhalation. Questionable carcinogen with experimental carcinogenic and tumorigenic data. Mutation data reported. Severe irritant to eyes, nose, and throat. When heated to decomposition it emits very toxic fumes of NO_x and SO_x.

BJK750   CAS: 4636-83-3   HR: 3
1,1'-BIS(3,5-DIMETHYLMORPHOLINO-CARBONYLMETHYL)-4,4'-BIPYRIDYNIUM DICHLORIDE
mf: $C_{26}H_{36}N_4O_4 \cdot 2Cl$ mw: 539.56

SYNS: 1,1'-BIS(3,5-DIMETHYLMORPHOLINOCAR-BONYLMETHYL)-4,4'-BIPYRIDINIUM-DICHLORID (GER-MAN) * 1,1'-BIS(2-(3,5-DIMETHYL-4-MORPHO-LINYL)-2-OXOETHYL)-4,4'-BIPYRIDINIUM DICHLORIDE * CEROXONE * MORFAMQUAT * MORFOXONE * MORPHANQUAT DICHLORIDE * PP 745

SAFETY PROFILE: Poison by ingestion. When heated to decomposition it emits very toxic fumes of Cl^- and NO_x.

BJL600   CAS: 97-74-5   HR: 3
BIS(DIMETHYLTHIOCARBAMOYL)SULFIDE
mf: $C_6H_{12}N_2S_3$ mw: 208.38

SYNS: ACETO TMTM * BIS(DIMETHYLTHIOCARBA-MYL) MONOSULFIDE * CARBAMIC ACID, DIMETHYL-DITHIO-, ANHYDROSULFIDE * MONEX * MONO-THIURAD * MONOTHIURAM * PENNAC MS * TETRAMETHYLTHIURAMMONIUM SULFIDE * TETRAMETHYLTHIURAM MONOSULFIDE * TETRAMETHYLTHIURAMONOSULFIDE * TETRA-METHYLTHIURAM SULFIDE * TETRAMETHYL-TRITHIO CARBAMIC ANHYDRIDE * 1,1'-THIOBIS (N,N-DIMETHYLTHIO)FORMAMIDE * THIONEX *. THIONEX RUBBER ACCELERATOR * TMTM * TMTMS * UNADS * USAF B-32 * USAF EK-P-6255 * VULKACIT THIURAM MS/C

CONSENSUS REPORTS: Reported in EPA TSCA Inventory.

SAFETY PROFILE: Poison by ingestion and intraperitoneal routes. Questionable carcinogen with experimental tumorigenic data. Mutation data reported. Experimental reproductive effects. When heated to decomposition it emits very toxic fumes of NO_x and SO_x.

BJN250   CAS: 2386-90-5   HR: 3
BIS(2,3-EPOXYCYCLOPENTYL) ETHER
mf: $C_{10}H_{14}O_3$ mw: 182.24

SYNS: EP-205 * ERR 4205 * 2,2'-OXYBIS-6-OXA-BICYCLO-(3.1.0)HEXANE

CONSENSUS REPORTS: EPA Genetic Toxicology Program. Reported in EPA TSCA Inventory.

SAFETY PROFILE: Moderately toxic by ingestion. A systemic irritant by skin contact and ingestion. Questionable carcinogen with experimental carcinogenic and neoplastigenic data. When heated to decomposition it emits acrid smoke and irritating fumes.

BJP000   CAS: 122-34-9   **HR: 3**
2,4-BIS(ETHYLAMINO)-6-CHLORO-s-TRIAZINE
mf: $C_7H_{12}ClN_5$ mw: 201.69

SYNS: A 2079 * AKTINIT S * AQUAZINE
* BATAZINA * 2,4-BIS(AETHYLAMINO)-6-CHLOR-
1,3,5-TRIAZIN (GERMAN) * BITEMOL * BITEMOL S
50 * CAT (HERBICIDE) * CDT * CEKUSAN
* CEKUZINA-S * CET * 1-CHLORO-3,5-BISETHY-
LAMINO-2,4,6-TRIAZINE * 2-CHLORO-4,6-BIS(ETH-
YLAMINO)-s-TRIAZINE * 2-CHLORO-4,6-BIS(ETHYL-
AMINO)-1,3,5-TRIAZINE * FRAMED * G 27692
* GEIGY 27,692 * GESARAN * GESATOP
* GESATOP 50 * H 1803 * HERBAZIN * HER-
BAZIN 50 * HERBEX * HERBOXY * HUNGAZIN
DT * PREMAZINE * PRIMATOL S * PRINCEP
* PRINTOP * RADOCON * RADOKOR * SIMADEX * SIMANEX * SIMAZIN * SIMA-
ZINE (USDA) * SIMAZINE 80W * SYMAZINE
* TAFAZINE * TAFAZINE 50-W * TAPHAZINE
* TRIAZINE A 384 * W 6658 * ZEAPUR

CONSENSUS REPORTS: EPA Genetic Toxi-cology Program. Reported in EPA TSCA Inven-tory.

SAFETY PROFILE: Poison by intravenous route. Questionable carcinogen with experimen-tal tumorigenic data. A skin and eye irritant. Mutation data reported. When heated to decom-position it emits very toxic fumes of Cl⁻ and NO_x.

BJP500   CAS: 6708-69-6   **HR: 3**
2,6-BIS(ETHYLEN-IMINO)-4-AMINO-s-TRIAZINE
mf: $C_7H_{10}N_6$ mw: 178.23

SAFETY PROFILE: A poison by intraperitoneal and intravenous routes. When heated to decom-position it emits toxic fumes of NO_x.

BJQ250   CAS: 2781-10-4   **HR: 3**
BIS(2-ETHYLHEXANOYLOXY)DIBUTYL STANNANE
mf: $C_{24}H_{48}O_4Sn$ mw: 519.41

SYNS: DIBUTYLBIS((2-ETHYLHEXANOYL)OXY)-STAN-
NANE * DIBUTYLBIS((2-ETHYL-1-OXOHEXYL)OXY)-
STANNANE (9CI) * DIBUTYLTIN BIS(α-ETHYLHEXA-
NOATE) * DIBUTYLTIN BIS(2-ETHYLHEXANOATE)
* DIBUTYLTIN DI(2-ETHYLHEXANOATE) * DI-n-BU-
TYLTIN DI-2-ETHYLHEXANOATE * DIBUTYLTIN DI(2-
ETHYLHEXOATE)

CONSENSUS REPORTS: Reported in EPA TSCA Inventory.

OSHA PEL: TWA 0.1 mg(Sn)/m³ (skin)
ACGIH TLV: TWA 0.1 mg(Sn)/m³ (skin) (Pro-posed: TWA 0.1 mg(Sn)/m³; STEL 0.2 mg(Sn)/m³ (skin))
NIOSH REL: (Organotin Compounds) TWA 0.1 mg(Sn)/m³

SAFETY PROFILE: Poison by ingestion and intravenous route. When heated to decomposi-tion it emits acrid smoke and irritating fumes.

BJR750   CAS: 298-07-7   **HR: 3**
BIS(2-ETHYLHEXYL)PHOSPHATE
DOT: NA 1902
mf: $C_{16}H_{35}O_4P$ mw: 322.48

SYNS: BIS(2-ETHYLHEXYL)HYDROGEN PHOSPHATE
* BIS(2-ETHYLHEXYL)ORTHOPHOSPHORIC ACID
* BIS(2-ETHYLHEXYL)PHOSPHORIC ACID * DEHPA
EXTRACTANT * DI-(2-ETHYLHEXYL)PHOSPHATE
* DI-2(ETHYLHEXYL)PHOSPHORIC ACID * DI-(2-
ETHYLHEXYL)PHOSPHORIC ACID (DOT) * 2-ETHYL-1-
HEXANOL HYDROGEN PHOSPHATE * HDEHP

CONSENSUS REPORTS: Reported in EPA TSCA Inventory.

DOT Classification: Corrosive Material; Label: Corrosive.

SAFETY PROFILE: Poison by intraperitoneal route. A corrosive material. A skin and eye irritant. When heated to decomposition it emits toxic fumes of PO_x.

BJS250   CAS: 122-62-3   **HR: 2**
BIS(2-ETHYLHEXYL)SEBACATE
mf: $C_{26}H_{50}O_4$ mw: 426.76

PROP: Light, clear liquid; mild odor. Bp: 248° @ 9 mm, fp: −55°, flash p: 410°F, d: 0.913 @ 25°/25°, vap d: 14.7.

SYNS: BISOFLEX DOS * DECANEDIOIC ACID,
BIS(2-ETHYLHEXYL) ESTER * DI(2-ETHYLHEXYL)SE-
BACATE * DIOCTYL SEBACATE * DOS
* 2-ETHYLHEXYL SEBACATE * MONOPLEX DOS
* OCTOIL S * OCTYL SEBACATE * PX 438
* STALFLEX DOS * UNIFLEX DOS

CONSENSUS REPORTS: Reported in EPA TSCA Inventory.

SAFETY PROFILE: Moderately toxic by inges-tion and intravenous routes. Combustible when exposed to heat or flame; can react with oxidizing materials. To fight fire, use foam, CO_2, dry chemical. When heated to decomposition it emits acrid and irritating fumes.

BJT250 CAS: 2440-45-1 HR: 3
BIS(ETHYLMERCURI)PHOSPHATE
mf: $C_4H_{11}Hg_2O_4P$ mw: 555.30

PROP: Solid

SYNS: ETHYLMERCURIC PHOSPHATE * ETHYL-MERCURY PHOSPHATE * LIGNASAN FUNGICIDE * LIGNASAN-X * NEW IMPROVED CERESAN * NEW IMPROVED GRANOSAN

CONSENSUS REPORTS: Mercury and its compounds are on the Community Right-To-Know List.

OSHA PEL: (Transitional: CL 1 mg/10m^3) TWA 0.01 mg(Hg)/m^3; STEL 0.03 mg/m^3 (skin)
ACGIH TLV: TWA 0.01 mg(Hg)/m^3; STEL 0.03 mg(Hg)/m^3
NIOSH REL: TWA 0.05 mg(Hg)/m^3

SAFETY PROFILE: Poison by ingestion and subcutaneous routes. When heated to decomposition it emits very toxic fumes of Hg and PO$_x$.

BJU000 CAS: 502-55-6 HR: 3
BISETHYL XANTHOGEN DISULFIDE
mf: $C_6H_{10}O_2S_4$ mw: 242.40

SYNS: AULIGEN * BEK * BEXIDE * BEXT * BIETHYLXANTHOGENTRISULFIDE * BIS(ETHYLXANTHIC)DISULFIDE * DEX * DIETHYLDITHIO BIS(THIONOFORMATE) * DIETHYL DIXANTHOGEN * DIETHYL XANTHOGENATE * DIETHYLXANTHOGEN DISULFIDE * DITHIOBIS(THIOFORMIC ACID)-o,o-DIETHYL ESTER * DIXANTHOGEN * ETHYL XANTHOGEN DISULFIDE * EXD * K PREPARATION * THIOPEROXYDICARBONIC ACID DIETHYL ESTER

CONSENSUS REPORTS: Reported in EPA TSCA Inventory.

SAFETY PROFILE: Poison by ingestion and intraperitoneal routes. Moderately toxic by skin contact and possibly other routes. When heated to decomposition it emits highly toxic fumes of SO$_x$.

BJU250 CAS: 1851-71-4 HR: 3
BIS(ETHYLXANTHOGEN)
TETRASULFIDE
mf: $C_6H_{10}O_2S_6$ mw: 306.52

SYN: TETRASULFIDE, BIS(ETHOXYTHIOCARBONYL)

SAFETY PROFILE: Poison by ingestion. When heated to decomposition it emits toxic fumes such as SO$_x$.

BJY000 CAS: 14873-10-0 HR: 3
BIS(l-HISTIDINE)COBALT
mf: $C_{12}H_{14}N_6O_5 \cdot Co$ mw: 365.25

SYNS: α-AMINOIMIDAZOLE-4-PROPIONIC ACID, COBALT(2+) SALT * BIS(l-HISTIDINATO)COBALT * COBALT-HISTIDINE * KOBALT HISTIDIN (GERMAN)

CONSENSUS REPORTS: Cobalt and its compounds are on the Community Right-To-Know List.

SAFETY PROFILE: Poison by intraperitoneal and intravenous routes. When heated to decomposition it emits toxic fumes of NO$_x$.

BJY825 CAS: 2614-76-8 HR: 3
2,2-BIS(HYDROPEROXY)PROPANE
DOT: UN 2178
mf: $C_3H_8O_4$ mw: 108.09

DOT Classification: Organic Peroxide; Label:Organic Peroxide

SAFETY PROFILE: Ignites or explodes when heated. When heated to decomposition it emits acrid smoke and fumes.

BJZ000 CAS: 66-76-2 HR: 3
BISHYDROXYCOUMARIN
mf: $C_{19}H_{12}O_6$ mw: 336.31

PROP: Very small crystals, slight pleasant odor, bitter taste, sol in alkali. Mp: 287-293°.

SYNS: ACADYL * ACAVYL * ANTITROMBOSIN * BARACOUMIN * BHC * BIS(4-HYDROXY-COUMARIN-3-YL)METHANE * CUMA * CUMID * DICOUMARIN * DICOUMAROL * DICUMAN * DICUMARINE * DI-(4-HYDROXY-3-COUMARINYL) METHANE * DI-4-HYDROXY-3,3'-METHYLENEDICOUMARIN * DUFALONE * KUMORAN * MELITOXIN * 3,3'-METHYLEEN-BIS(4-HYDROXY-CUMARINE) (DUTCH) * 3,3'-METHYLEN-BIS(4-HYDROXY-CUMARIN) (GERMAN) * 3,3'-METHYLENEBIS(4-HYDROXY-1,2-BENZOPYRONE) * 3,3'-METHYLENEBIS-(4-HYDROXYCOUMARIN) * 3,3'-METHYLENE-BIS(4-HYDROXYCOUMARINE) (FRENCH) * 3,3'-METILEN-BIS(4-IDROSSI-CUMARINA) (ITALIAN) * TEMPARIN * TROMBOSAN

CONSENSUS REPORTS: Reported in EPA TSCA Inventory.

SAFETY PROFILE: Poison by ingestion, subcutaneous, intravenous, and intraperitoneal routes. Human reproductive effects by ingestion and possibly other routes: fetal death, unspecified developmental abnormalities, stillbirth, and unspecified neonatal effects. An anticoagulant. Excessive doses can cause hemorrhages. When heated to decomposition it emits acrid smoke and fumes.

BKA000   CAS: 548-00-5   HR: 3
BIS(4-HYDROXY-3-COUMARIN) ACETIC ACID ETHYL ESTER
mf: $C_{22}H_{16}O_8$ mw: 408.38

SYNS: BIS-3,3'-(4-HYDROXYCOUMARINYL)ACETIC ACID ETHYL ESTER * BIS-(4-HYDROXY-3-COUMARINYL)ETHYL ACETATE * BIS(4-HYDROXY-2-OXO-2H-1-BENZOPYRAN-3-YL)ACETIC ACID ETHYL ESTER * BOEA * B.O.E.A. * 3,3'-(CARBOXYMETHYLENE)BIS(4-HYDROXYCOUMARIN) ETHYL ESTER * DICUMACYL * ETHYL BISCOUMACETATE * ETHYL BIS(4-HYDROXYCOUMARINYL) ACETATE * ETHYL BIS(4-HYDROXY-3-COUMARINYL)ACETATE * ETHYLDICOUMAROL * ETHYLDICOUMAROL ACETATE * ETHYL-4,4'-DIHYDROXYDICOUMARINYL-3,3'-ACETATE * NEODICOUMARIN * NEODICOUMAROL * NEODICUMARINUM * PELENTAN * STABILENE * TROMBARIN * TROMBIL * TROMBOLYSAN * TROMEXAN * TROMEXAN ETHYL ACETATE

SAFETY PROFILE: Poison by intraperitoneal route. Moderately toxic by ingestion and subcutaneous routes. Human reproductive effects by ingestion: developmental abnormalities of the cardiovascular system, stillbirth, and unspecified neonatal effects. An anticoagulant. When heated to decomposition it emits acrid and irritating fumes.

BKD750   CAS: 64036-91-5   HR: 3
BIS(2-HYDROXYETHYL)-2-(2-CHLORO ETHYL THIO)ETHYL SULFONIUM) CHLORIDE
mf: $C_8H_{18}ClO_2S_2 \cdot Cl$ mw: 281.28

SYNS: β-CHLOROETHYL-β-(BIS(β-HYDROXYETHYL) SULFONIUM)ETHYL SULFIDE CHLORIDE * 2-(2-CHLOROETHYL)THIOETHYLBIS(2-HYDROXYETHYL)-CHLORIDE

SAFETY PROFILE: A poison by skin contact, subcutaneous, and intravenous routes. When heated to decomposition it emits very toxic fumes of Cl^- and SO_x.

BKE500   CAS: 120-40-1   HR: 2
N,N-BIS(2-HYDROXYETHYL)DODECAN AMIDE
mf: $C_{16}H_{33}NO_3$ mw: 287.50

PROP: Solid. Mp: 36°

SYNS: BIS(2-HYDROXYETHYL)LAURAMIDE * N,N-BIS(HYDROXYETHYL)LAURAMIDE * N,N-BIS(β-HYDROXYETHYL)LAURAMIDE * N,N-BIS(2-HYDROXYETHYL)LAURAMIDE * CLINDROL 101CG * CLINDROL SUPERAMIDE 100L * COCO DIETHANOLAMIDE * COCONUT OIL AMIDE of DIETHANOLAMINE * COMPERLAN LD * CONDENSATE PL * CRILLON L.D.E. * DIETHANOLLAURAMIDE * N,N-DIETHANOLLAURAMIDE * N,N-DIETHANOL-LAURIC ACID AMIDE * EMID 6511 * EMID 6541 * ETHYLAN MLD * HETAMIDE ML * LAURAMIDE DEA * LAURIC ACID DIETHANOLAMIDE * LAURIC DIETHANOLAMIDE * LAUROYL DIETHANOLAMIDE * LAURYL DIETHANOLAMIDE * LDA * LDE * MONAMID 150-LW * NCI-C55323 * NINOL AA-62 EXTRA * NINOL 4821 * NINOL AA62 * ONYXOL 345 * REWOMID DLMS * RICHAMIDE 6310 * ROLAMID CD * STANDAMIDD LD * STEINAMID DL 203 S * SUPER AMIDE L-9A * SYNOTOL L-60 * UNAMIDE J-56 * VARAMID ML 1

CONSENSUS REPORTS: Reported in EPA TSCA Inventory.

SAFETY PROFILE: Moderately toxic by ingestion. When heated to decomposition it emits toxic fumes of NO_x.

BKF250   CAS: 2784-94-3   HR: 3
N',N'-BIS(2-HYDROXYETHYL)-N-METHYL-2-NITRO-p-PHENYLENEDI-AMINE
mf: $C_{11}H_{17}N_3O_4$ mw: 255.31

SYNS: HC BLUE 1 * NCI-C04159

CONSENSUS REPORTS: NTP Carcinogenesis Studies (feed); Some Evidence: rat NTPTR* NTP-TR-271,85;(feed); Clear Evidence: mouse NTPTR* NTP-TR-271,85. Reported in EPA TSCA Inventory.

SAFETY PROFILE: Suspected carcinogen with experimental carcinogenic data. Mutation data reported. When heated to decomposition it emits toxic fumes of NO_x.

BKH500   CAS: 794-93-4   **HR: 3**
BIS(HYDROXYMETHYL)FURATRIZINE
mf: $C_{11}H_{11}N_5O_5$ mw: 293.27

SYNS: 3-BIS(HYDROXYMETHYL)AMINO-6-(5-NITRO-2-FURYLETHENYL)-1,2,4-TRIAZINE * DHNT * 3-DI-(HYDROXYMETHYL)AMINO-6-(5-NITRO-2-FURYLETH-NEYL)-1,2,4-TRIAZINE * 3-DI(HYDROXYMETHYL)AMINO-6-(2-(5-NITRO-2-FURYL)VINYL)-1,2,4-TRIAZINE * DIHYDROXYMETHYL FURATRIZINE * FURATONE * FURATONE-S * N-(6-(5-NITROFURFURYLIDENE-METHYL)-1,2,4-TRIAZIN-3-YL)IMINODIMETHANOL * 6-(5-NITRO-2-FURYLVINYL)-3-(DIHYDROXYDIMETHY-LAMINO)-1,2,4-TRIAZENE * N-(6-(2-(5-NITRO-2-FU-RYL)VINYL)-1,2,4-TRIAZIN-3-YL)IMINODIMETHANOL * ((6-(2-(5-NITRO-2-FURYL)VINYL)-as-TRIAZIN-3-YL)IMINO)DIMETHANOL * PANFURAN-S

CONSENSUS REPORTS: IARC Cancer Review: GROUP 2B IMEMDT 7,56,87; Animal Sufficient Evidence IMEMDT 24,77,80

SAFETY PROFILE: Suspected carcinogen with experimental carcinogenic and tumorigenic data. Moderately toxic by ingestion, intraperitoneal, and subcutaneous route. Mutation data reported. An antibacterial agent. When heated to decomposition it emits toxic fumes of NO_x.

BKK750   CAS: 26401-97-8   **HR: 2**
BIS(ISOOCTYLOXYCARBONYLMETHYL-THIO)DIOCTYL STANNANE
mf: $C_{36}H_{72}O_4S_2Sn$ mw: 751.89

SYNS: ADVASTAB 17 MO * BIS(MERCAPTOACE-TATE)DIOCTYL-TIN BIS(ISOOCTYL) ESTER * DIISOOC-TYL ((DIOCTYLSTANNYLENE)DITHIO)DIACETATE * DIOCTYLTIN BIS(ISOOCTYL MERCAPTOACETATE) * DIOCTYLTIN-S,S'-BIS(ISOOCTYL MERCAPTOACE-TATE) * DIOCTYLTIN BIS(ISOOCTYL THIOGLYCO-LATE) * DIOCTYL-TIN BIS(ISOOCTYLTHIOGLYCOL-LATE) * DI-n-OCTYLTIN DIISOOCTYL THIO-GLYCOLATE * DI-n-OCTYL-ZINN-DI-ISOOCTYLTHIO-GLYKOLAT (GERMAN) * DOTG * THERMOLITE 831

CONSENSUS REPORTS: Reported in EPA TSCA Inventory.

OSHA PEL: TWA 0.1 mg(Sn)/m^3 (skin)
ACGIH TLV: TWA 0.1 mg(Sn)/m^3 (skin) (Proposed: TWA 0.1 mg(Sn)/m^3; STEL 0.2 mg(Sn)/m^3 (skin))
NIOSH REL: (Organotin Compounds) TWA 0.1 mg(Sn)/m^3

SAFETY PROFILE: Moderately toxic by ingestion and possibly other routes. An experimental teratogen. When heated to decomposition it emits toxic fumes of SO_x.

BKL250   CAS: 7287-19-6   **HR: 2**
2,4-BIS(ISOPROPYLAMINO)-6-METHYLMERCAPTO-s-TRIAZINE
mf: $C_{10}H_{19}N_5S$ mw: 241.40

SYNS: 4,6-BIS(ISOPROPYLAMINO)-2-METHYLMER-CAPTO-s-TRIAZINE * 2,4-BIS(ISOPROPYLAMINO)-6-METHYLTHIO-s-TRIAZINE * 2,4-BIS(ISOPROPYL-AMINO)-6-METHYLTHIO-1,3,5-TRIAZINE * N,N'-BIS(1-METHYLETHYL)-6-METHYL-THIO-1,3,5-TRIAZINE-2,4-DI-AMINE * CAPAROL * G 34161 * GESAGARD * MERKAZIN * 2-METHYLMERCAPTO-4,6-BIS(ISO-PROPYLAMINO)-s-TRIAZINE * 2-METHYLTHIO-4,6-BIS(ISOPROPYLAMINO)-s-TRIAZINE * POLISIN * PRIMATOL Q * PROMETREX * PROMETRIN * PROMETRYN * PROMETRYNE (USDA) * SE-LEKTIN * SESAGARD

SAFETY PROFILE: Moderately toxic by ingestion. An eye irritant. An herbicide. When heated to decomposition it emits very toxic fumes of NO_x and SO_x.

BKL750   CAS: 3006-93-7   **HR: 3**
1,3-BISMALEIMIDO BENZENE
mf: $C_{14}H_8N_2O_4$ mw: 268.24

SYNS: 1,3-DIMALEIMIDOBENZENE * HVA 2 * HVA-2 CURING AGENT * M-PHDM * N,N'-(m-PHENYLENE)BISMALEIMIDE * 1,1'-(m-PHENYLENE)BIS-1H-PYROLE-2,5-DIONE (9CI) * N,N'-(m-PHENYL-ENEDIMALEIMIDE)

CONSENSUS REPORTS: Reported in EPA TSCA Inventory.

SAFETY PROFILE: Poison by ingestion and intraperitoneal routes. When heated to decomposition it emits toxic fumes of NO_x.

BKM500   CAS: 1187-00-4   **HR: 3**
BIS(METHANE SULFONYL)-d-MANNITOL
mf: $C_8H_{18}O_{10}S_2$ mw: 338.38

SYNS: 1,6-BIS-o-METHYLSULFONYL-d-MANNITOL * CB 2511 * 1,6-DIMESYL-d-MANNITOL * 1,6-DI-METHANESULFONATE-d-MANNITOL * 1,6-DIMETH-ANE-SULFONOXY-d-MANNITOL * 1,6-DIMETHANE-SULFHONOXY-1,6-DIDEOXY-d-MANNITOL * DMM * d-MANNITOL BUSULFAN * MANNITOL MYLERAN * MANNOGRANOL * MM * NSC-37538

CONSENSUS REPORTS: EPA Genetic Toxicology Program.

SAFETY PROFILE: Poison by intravenous route. Moderately toxic by intraperitoneal route. Mildly toxic by ingestion. Questionable carcinogen with experimental neoplastigenic data. Mutation data reported. When heated to decomposition it emits toxic fumes of SO_x.

BKU500   CAS: 103-34-4   HR: 3
N,N′-BISMORPHOLINE DISULFIDE
mf: $C_8H_{16}N_2O_2S_2$ mw: 236.38

PROP: Tan to gray powder. Mp: 122° min, d: 1.36 @ 25°.

SYNS: ACCEL R * BISMORPHOLINO DISULFIDE * DIMORPHOLINE DISULFIDE * DIMORPHOLINO DISULFIDE * DITHIOBISMORPHOLINE * 4,4′-DITHIOBIS(MORPHOLINE) * N,N-DITHIODIMORPHOLINE * 4,4′-DITHIODIMORPHOLINE * 4,4′-DITHIOMORPHOLINE * MORPHOLINE DISULFIDE * MORPHOLINODISULFIDE * SULFASAN * SULFASAN R POWDER * USAF B-17 * USAF EK-T-6645

CONSENSUS REPORTS: Reported in EPA TSCA Inventory.

SAFETY PROFILE: Poison by intraperitoneal and intravenous routes. Moderately toxic by ingestion. Mutation data reported. When heated to decomposition it emits very toxic fumes of NO_x and SO_x.

BKU750   CAS: 7440-69-9   HR: 3
BISMUTH
of: Bi   aw: 208.98

PROP: Hexagonal silver-white or reddish metallic crystals. Mp: 271.3°, bp: 1420-1560°, d: 9.80, vap press: 1 mm @ 1021°.

SYN: BISMUTH-209

CONSENSUS REPORTS: Reported in EPA TSCA Inventory.

SAFETY PROFILE: Poisonous to humans. Flammable when exposed to flame. Reaction with $[Bi(OH)_3 + Al(OH)_3]$, coprecipitated and H_2 reduced produces a spontaneously flammable product. Moderately dangerous, can react with acid or acid fumes to emit toxic fumes. Incompatible with Al; BrF_3; acids; NOF; NH_4NO_3; $HClO_3$; Cl_2; IF_5; HNO_3; $HClO_4$.

BKV250   CAS: 12001-47-7   HR: 3
BISMUTH ARSPHENAMINE SULFONATE
mf: $C_{21}H_{24}As_3Bi_2N_3O_{12}S_3 \cdot 3Na$ mw: 1318.35

SYNS: BISMARSEN * SULFARSPHENAMINE BISMUTH

CONSENSUS REPORTS: Arsenic and its compounds are on the Community Right-To-Know List.

OSHA PEL: TWA 0.5 mg(As)/m^3

SAFETY PROFILE: A poison by intraperitoneal and intramuscular routes. When heated to decomposition it emits very toxic fumes of Na_2O, NO_x, SO_x, As, and Bi.

BKV750   HR: 3
BISMUTH COMPOUNDS

SAFETY PROFILE: Bismuth and its salts can cause kidney damage, although the degree of such damage is usually mild. Large doses can be fatal. Industrially it is considered one of the less toxic of the heavy metals, although intoxication has occurred from its use in medicine. The similarity between the pharmacologic and toxic behavior of lead and bismuth has been pointed out in the literature. Like lead, bismuth may be liberated from tissue deposits during periods of acidosis. Serious and sometimes fatal poisoning may occur from the injection of large doses into closed cavities and from extensive application to burns. Death of animals from bismuth nephritis following injections of soluble salts occurs within several hours to 24 days, the time being generally inversely proportional to the dose, and it appears to be in the order of 5-10 times higher than the dose by slow intravenous injection for rabbits. It is stated that the administration of bismuth should be stopped when gingivitis appears, for otherwise serious ulcerative stomatitis is likely to result. Other toxic results may develop, such as malaise, albuminuria, diarrhea, skin reactions and sometimes serious exodermatitis. Industrial bismuth poisoning has not been reported, although bismuth absorbed in industrial cases may complicate a diagnosis of plumbism, since the dark line in the gums, which is often present in lead poisoning, is also produced by bismuth. All bismuth compounds do not have equal toxicity.

Treatment and Antidotes: Personnel showing some of the symptoms noted above which might indicate that they were absorbing too much bismuth into the body should be removed from exposure as soon as possible. Get medical advice. Personnel should be cautioned against careless handling of these materials.

BKW000   CAS: 21260-46-8   **HR: 3**
BISMUTH DIMETHYL DITHIO CARBAMATE
mf: $C_9H_{18}N_3S_6 \cdot Bi$ mw: 569.64

SYNS: BISMATE * TRIS(DIMETHYLDITHIOCAR-BAMATO)BISMUTH

CONSENSUS REPORTS: Reported in EPA TSCA Inventory.

SAFETY PROFILE: Questionable carcinogen with experimental tumorigenic data. When heated to decomposition it emits very toxic fumes of SO_x and NO_x.

BKW250   CAS: 10361-44-1   **HR: 3**
BISMUTH NITRATE
mf: BiN_3O_9 mw: 395.01

PROP: Triclinic, colorless, slightly hygroscopic crystals. Bp: $-5H_2O$ @ 80°, d: 2.83, mp: 30° (decomp).

SYN: NITRIC ACID, BISMUTH(3+) SALT

CONSENSUS REPORTS: Reported in EPA TSCA Inventory.

SAFETY PROFILE: Poison by intravenous route. Moderately toxic by intraperitoneal route. When heated to decomposition it emits toxic fumes of Bi and NO_x.

BKW750   CAS: 7787-62-4   **HR: 3**
BISMUTH PENTAFLUORIDE
mf: BiF_5 mw: 303.98

PROP: Crystals. Sublimes @ 550°.

SAFETY PROFILE: An irritant poison via ingestion and inhalation routes. Decomposes vigorously and sometimes ignites on contact with moisture to yield O_3 and bismuth trifluoride. Very dangerous. Incompatible with water and petrolatum above 50°; and acids. Reacts violently with water and petrolatum above 50°; and acids at room temperature, liberating much heat and ozone. When heated to decomposition it emits highly toxic fumes of F^-.

BKX750   CAS: 150-49-2   **HR: 3**
BISMUTH SODIUM THIOGLYCOLLATE
mf: $C_6H_6BiNa_3O_6S_3$ mw: 548.25

SYNS: BISTRIMATE * MERCAPTOACETIC ACID, SODIUM-BISMUTH SALT * SODIUM BISMUTH THIO-GLYCOLATE * SODIUM BISMUTH THIOGLYCOLLATE * THIOBISMOL

SAFETY PROFILE: Systemic toxic effects in children: somnolence, nausea or vomiting, kidney damage, and decreased urine volume. Poison by intramuscular route. When heated to decomposition it emits very toxic fumes of SO_x and Na_2O.

BLB250   CAS: 4731-77-5   **HR: 3**
BIS(OCTANOYLOXY)DI-n-BUTYL STANNANE
mf: $C_{24}H_{48}O_4Sn$ mw: 519.41

SYNS: BIS(OCTANOYLOXY)DI-n-BUTYLTIN * DIBUTYLBIS(OCTANOYLOXY)STANNANE * DIBUTYLBIS((1-OXOOCTYL)OXY)STANNANE * DIBUTYLTIN DICAPRYLATE * DIBUTYLTIN DIOC-TANOATE * DIBUTYLTIN DIOCTATE * DIBUTYL-TIN OCTANOATE * KAPRYLAN DI-N-BUTYLCINICITY (CZECH)

OSHA PEL: TWA 0.1 $mg(Sn)/m^3$ (skin)
ACGIH TLV: TWA 0.1 $mg(Sn)/m^3$ (skin) (Proposed: TWA 0.1 $mg(Sn)/m^3$; STEL 0.2 $mg(Sn)/m^3$ (skin))
NIOSH REL: (Organotin Compounds) TWA 0.1 $mg(Sn)/m^3$

SAFETY PROFILE: Poison by ingestion. A severe skin and eye irritant. When heated to decomposition it emits acrid and irritating fumes.

BLC000   CAS: 868-18-8   **HR: 2**
BISODIUM TARTRATE
mf: $C_4H_4O_6 \cdot 2Na$ mw: 194.06

PROP: Transparent crystals; colorless and odorless. Sol in water.

SYNS: 2,3-DIHYDROXY-(R-(R*,R*))-BUTANEDIOIC ACID DISODIUM SALT (9CI) * DISODIUM TARTRATE * DISODIUM l-(+)-TARTRATE * SODIUM TARTRATE (FCC) * SODIUM l-(+)-TARTRATE

CONSENSUS REPORTS: Reported in EPA TSCA Inventory.

SAFETY PROFILE: Moderately toxic by ingestion. When heated to decomposition it emits acrid smoke and irritating fumes.

BLC250   CAS: 10380-28-6   **HR: 3**
BIS(8-OXYQUINOLINE)COPPER
mf: $C_{18}H_{12}CuN_2O_2$ mw: 351.86

PROP: Yellow-green powder.

SYNS: BIOQUIN * BIOQUIN 1 * BIS(8-QUINO-
LINATO)COPPER * BIS(8-QUINOLINOLATO)COPPER
* BIS(8-QUINOLINOLATO-N(^{1}),O(^{8}))-COPPER
* CELLU-QUIN * COPPER-8 * COPPER HY-
DROXYQUINOLATE * COPPER-8-HYDROXYQUINO-
LATE * COPPER-8-HYDROXYQUINOLINATE
* COPPER-8-HYDROXYQUINOLINE * COPPER OXI-
NATE * COPPER (2+) OXINATE * COPPER OXINE
* COPPER OXYQUINOLATE * COPPER OXYQUINO-
LINE * COPPER QUINOLATE * COPPER-8-QUINO-
LATE * COPPER-8-QUINOLINOL * COPPER
QUINOLINOLATE * COPPER-8-QUINOLINOLATE
* CUNILATE * CUNILATE 2472 * CUPRIC-8-HY-
DROXYQUINOLATE * CUPRIC-8-QUINOLINOLATE
* DOKIRIN * FRUITDO * 8-HYDROXYQUINOLINE
COPPER COMPLEX * MILMER * OXIME COPPER
* OXINE COPPER * OXINE CUIVRE * OXYQUIN-
OLINOLEATE de CUIVRE (FRENCH) * QUINONDO

CONSENSUS REPORTS: IARC Cancer Review: GROUP 3 IMEMDT 7,56,87, Animal
Inadequate Evidence IMEMDT 15,103,77. Reported in EPA TSCA Inventory. Copper and
its compounds are on the Community Right-
To-Know List.

SAFETY PROFILE: Poison by intraperitoneal
route. Questionable carcinogen with experimental tumorigenic data. Mutation data reported.
When heated to decomposition it emits toxic
fumes of NO_x.

BLD000   CAS: 42310-84-9   **HR: 3**
BISPENTAFLUOROSULFUR OXIDE
mf: $F_{10}OS_2$ mw: 270.12

SYN: SULFUR FLUORIDE OXIDE

OSHA PEL: TWA 2.5 mg(F)/m^3
NIOSH REL: TWA 2.5 mg(F)/m^3

SAFETY PROFILE: Poison by inhalation.
When heated to decomposition it emits very
toxic fumes of F^- and SO_x.

BLD500   CAS: 80-05-7   **HR: 3**
BISPHENOL A
mf: $C_{15}H_{16}O_2$ mw: 228.31

PROP: White flakes, mild phenolic odor. Insol
in water; sol in alcohol and dilute alkalies; sltly
sol in CCl_4.

SYNS: BISFEROL A (GERMAN) * 2,2-BIS-4′-HY-
DROXYFENYLPROPAN (CZECH) * BIS(4-HYDROXY-
PHENYL) DIMETHYLMETHANE * BIS(4-HYDROXY-
PHENYL)PROPANE * 2,2-BIS(p-HYDROXYPHENYL)
PROPANE * 2,2-BIS(4-HYDROXYPHENYL)PROPANE
* DIAN * p,p′-DIHYDROXYDIPHENYLDIMETHYL-
METHANE * 4,4′-DIHYDROXYDIPHENYLDIMETHYL-
METHANE * p,p′-DIHYDROXYDIPHENYLPROPANE
* 2,2-(4,4′-DIHYDROXYDIPHENYL)PROPANE * 4,4′-
DIHYDROXYDIPHENYLPROPANE * 4,4′-DIHY-
DROXYDIPHENYL-2,2-PROPANE * 4,4′-DIHYDROXY-
2,2-DIPHENYLPROPANE * β-DI-p-HYDROXY-
PHENYLPROPANE * 2,2-DI(4-HYDROXYPHENYL)PRO-
PANE * DIMETHYL BIS(p-HYDROXYPHENYL)METH-
ANE * DIMETHYLMETHYLENE-p,p′-DIPHENOL
* 2,2-DI(4-PHENYLOL)PROPANE * p,p′-ISOPRO-
PYLIDENEBISPHENOL * 4,4′-ISOPROPYLIDENEBIS-
PHENOL * p,p′-ISOPROPYLIDENEDIPHENOL
* NCI-C50635

CONSENSUS REPORTS: NTP Carcinogen-
esis Bioassay (feed); Inadequate Studies: mouse,
rat NTPTR* NTP-TR-215,82. Community
Right-To-Know List. Reported in EPA TSCA
Inventory.

SAFETY PROFILE: Poison by intraperitoneal
route. Moderately toxic by ingestion and skin
contact. Experimental teratogenic and reproductive effects. A skin and eye irritant. When heated
to decomposition it emits acrid and irritating
fumes.

BLD750   CAS: 1675-54-3   **HR: 3**
BISPHENOL A DIGLYCIDYL ETHER
mf: $C_{21}H_{24}O_4$ mw: 340.45

SYNS: 2,2-BIS(4-(2,3-EPOXYPROPYLOXY)PHENYL)PRO-
PANE * BIS(4-GLYCIDYLOXYPHENYL)DIMETHY-
AMETHANE * 2,2-BIS(p-GLYCIDYLOXYPHE-
NYL)PROPANE * BIS(4-HYDROXYPHENYL)DI-
METHYLMETHANE DIGLYCIDYL ETHER * 2,2-BIS(4-
HYDROXYPHENYL)PROPANE, DIGLYCIDYL ETHER
* 2,2-BIS (p-HYDROXYPHENYL)PROPANE, DIGLYCIDYL
ETHER * D.E.R. 332 * DIGLYCIDYL BISPHENOL
A ETHER * DIGLYCIDYL ETHER of 2,2-BIS(p-HY-
DROXYPHENYL)PROPANE * DIGLYCIDYL ETHER of
2,2-BIS(4-HYDROXYPHENYL)PROPANE * DIGLYCIDYL
ETHER of BISPHENOL A * DIGLYCIDYL ETHER of
4,4′-ISOPROPYLIDENEDIPHENOL * 4,4′-DIHYDROXY-
DIPHENYLDIMETHYLMETHANE DIGLYCIDYL ETHER
* p,p′-DIHYDROXYDIPHENYLDIMETHYLMETHANE DI-
GLYCIDYL ETHER * EPI-REZ 508 * EPI-REZ 510
* EPON 828 * EPOXIDE A * ERL-2774 * 4,4′-

ISOPROPYLIDENEDIPHENOL DIGLYCIDYL ETHER * 2,2'-((1-METHYLETHYLIDENE)BIS(4,1-PHENYLENE-OXYMETHYLENE))BISOXIRANE

CONSENSUS REPORTS: EPA Genetic Toxicology Program. Reported in EPA TSCA Inventory.

SAFETY PROFILE: Poison by skin contact. Mildly toxic by ingestion. Mutation data reported. A skin and severe eye irritant. Questionable carcinogen with experimental carcinogenic and tumorigenic data. When heated to decomposition it emits acrid and irritating fumes.

BLE500 CAS: 74-31-7 **HR: 3**
1,4-BIS(PHENYL AMINO)BENZENE
mf: $C_{18}H_{16}N_2$ mw: 260.36
PROP: A solid. D: 1.20, vap d: 9.0.

SYNS: AGERITE * AGERITEDPPD * N,N'-DIFE-NYL-p-FENYLENDIAMIN (CZECH) * N,N'-DIPHENYL-p-PHENYLENEDIAMINE * DIPHENYL-p-PHENYLENEDI-AMINE * DPPD * FLEXAMINE G * JZF * NONOX DPPD * p-PHENYLAMINODIPHENYL-AMINE * 4-PHENYLAMINODIPHENYLAMINE * USAF GY-2

CONSENSUS REPORTS: Reported in EPA TSCA Inventory.

SAFETY PROFILE: Poison by intraperitoneal route. Moderately toxic by ingestion. A weak allergen. Experimental teratogenic and reproductive effects. An eye irritant. Questionable carcinogen with experimental tumorigenic data. Mutation data reported. Combustible when exposed to heat or flame; can react with oxidizing materials. When heated to decomposition it emits toxic fumes of NO_x.

BLG500 CAS: 1113-14-0 **HR: 3**
trans-1,2-BIS(n-PROPYLSULFONYL)
ETHYLENE
mf: $C_8H_{16}O_4S_2$ mw: 240.36

SYNS: B-1843 * C-272 * CHEMAGRO B-1843 * VANCIDE PA * VANCIDE PA DISPERSION

SAFETY PROFILE: Poison by ingestion and intravenous routes. When heated to decomposition it emits toxic fumes of SO_x.

BLK000 CAS: 128-80-3 **HR: 2**
1,4-BIS(p-TOLYLAMINO)ANTHRA-
QUINONE
mf: $C_{28}H_{22}N_2O_2$ mw: 418.52

SYNS: ALIZARINE CYANINE GREEN BASE * AMA-PLAST GREEN OZ * ARLOSOL GREEN B * BIS-1,4-p-TOLYLAMINOANTHRCHINON (CZECH) * C-GREEN 10 * C.I. 61565 * C.I. SOLVENT GREEN 3 * CYANINE GREEN G BASE * D&C GREEN NO. 6 * 1,4-DI-p-TOLUIDINOANTHRAQUINONE * FAT SOLUBLE GREEN ANTHRAQUINONE * 11091 GREEN * GREENNO.2 * MICRO-LEXGREEN5B * NITRO FAST GREEN GB * ORGANOL FAST GREEN J * QUINIZARINE GREEN BASE * SUDAN GREEN 4B * TOYO ORIENTAL OIL BLUE G * WAXOLINE GREEN

CONSENSUS REPORTS: Reported in EPA TSCA Inventory.

SAFETY PROFILE: Moderately toxic by ingestion. An eye irritant. When heated to decomposition it emits toxic fumes of NO_x.

BLL750 CAS: 56-35-9 **HR: 3**
BIS(TRIBUTYL TIN)OXIDE
mf: $C_{24}H_{54}OSn_2$ mw: 596.16

SYNS: BIOMET TBTO * BIS-(TRI-N-BUTYLCIN)OXID (CZECH) * BIS(TRIBUTYLOXIDE) of TIN * BIS(TRI-BUTYLSTANNYL)OXIDE * BIS(TRI-N-BUTYLZINN)-OXYD (GERMAN) * BTO * BUTINOX * C-Sn-9 * ENT 24,979 * HEXABUTYLDISTANNOXANE * HEXABUTYLDITIN * KYSLICNIK TRI-N-BUTYL-CINICITY (CZECH) * L.S. 3394 * OTBE (FRENCH) * OXYBIS(TRIBUTYLTIN) * OXYDE de TRIBUTYLE-TAIN * TBOT * TBTO * TRI-n-BUTYL-STAN-NANE OXIDE * TRIBUTYLTIN OXIDE

CONSENSUS REPORTS: Reported in EPA TSCA Inventory.

OSHA PEL: TWA 0.1 mg(Sn)/m^3 (skin)
ACGIH TLV: TWA 0.1 mg(Sn)/m^3 (skin) (Proposed: TWA 0.1 mg(Sn)/m^3; STEL 0.2 mg(Sn)/m^3 (skin))
NIOSH REL: (Organotin Compounds) TWA 0.1 mg(Sn)/m^3

SAFETY PROFILE: A poison by ingestion, intraperitoneal, and intravenous routes. Moderately toxic by skin contact. Mutation data reported. A severe eye irritant. When heated to decomposition it emits acrid and irritating fumes.

BLM750 CAS: 2532-50-5 **HR: 3**
BIS(TRICHLORO METHYL)TRISULFIDE
mf: $C_2Cl_6S_3$ mw: 332.90

SYNS: BISTRICHLOROMETHYLTRISULFID (CZECH)
* TRITHIOBIS(TRICHLOROMETHANE)

SAFETY PROFILE: Poison by intravenous route. Moderately toxic by ingestion. A skin and eye irritant. When heated to decomposition it emits very toxic fumes of Cl^- and SO_x.

BLN250   CAS: 57-52-3   **HR: 3**
BIS(TRIETHYLTIN) SULFATE
mf: $C_{12}H_{30}O_4SSn_2$ mw: 507.86

SYNS:TRIAETHYLZINNSULFAT(GERMAN) * TRI-ETHYLHYDROXY-STANNANE SULFATE (2:1) (8CI)
* TRIETHYLHYDROXYTIN SULFATE * TRIETHYLTIN SULPHATE

OSHA PEL: TWA 0.1 mg(Sn)/m³ (skin)
ACGIH TLV: TWA 0.1 mg(Sn)/m³ (skin) (Proposed: TWA 0.1 mg(Sn)/m³; STEL 0.2 mg(Sn)/m³ (skin))
NIOSH REL: (Organotin Compounds) TWA 0.1 mg(Sn)/m³

SAFETY PROFILE: Poison by ingestion, intraperitoneal, subcutaneous, intravenous, and parenteral routes. When heated to decomposition it emits toxic fumes of SO_x.

BLR750   CAS: 28930-30-5   **HR: 2**
BIS(TRINITROPHENYL)SULFIDE
mf: $C_{12}H_4N_6O_{12}S$ mw: 456.28

SYNS: HEXANITRODIPHENYLSULFIDE * PICRYL SULFIDE

SAFETY PROFILE: Moderately toxic by ingestion. This material is a powerful explosive and has an added military advantage in that its explosive gases contain irritating and very toxic SO_x.

BLS250   CAS: 14264-16-5   **HR: 3**
BIS(TRIPHENYLPHOSPHINE)DICHLORONICKEL
mf: $C_{24}H_{54}P_2 \cdot Cl_2Ni$ mw: 534.33

SYNS: BIS(TRI-N-BUTYLPHOSPHINE)DICHLORONICKEL
* TRIBUTYL-PHOSPHINE compd. with NICKELCHLORIDE (2:1)

CONSENSUS REPORTS: Reported in EPA TSCA Inventory. Nickel and its compounds are on the Community Right-To-Know List.

OSHA PEL: (Transitional: TWA 1 mg/m³) TWA 0.1 mg (Ni)/m³

ACGIH TLV: TWA 0.1 mg(Ni)/m³; (Proposed: TWA 0.05 mg(Ni)/m³; Human Carcinogen)
NIOSH REL: TWA 0.015 mg/(Ni)/m³

SAFETY PROFILE: Suspected carcinogen. Poison by intravenous route. When heated to decomposition it emits very toxic fumes of Cl^- and PO_x.

BLS500   CAS: 15709-62-3   **HR: 3**
BIS(TRIPHENYL PHOSPHINE)NICKEL DITHIOCYANATE
mf: $C_{38}H_{30}N_2NiP_2S_2$ mw: 699.47

SYN: NICKEL BISTRIPHENYLPHOSPHINE DITHIOCYANATE

CONSENSUS REPORTS: Nickel and its compounds are on the Community Right-To-Know List.

NIOSH REL: TWA 0.015 mg(Ni)/m³

SAFETY PROFILE: Poison by intravenous route. When heated to decomposition it emits very toxic fumes of SO_x, PO_x, NO_x, and CN^-.

BLS750   CAS: 1624-02-8   **HR: 2**
BIS(TRIPHENYL SILYL)CHROMATE
mf: $C_{36}H_{30}CrO_4Si_2$ mw: 634.84

SYN: CHROMIC ACID, BIS(TRIPHENYLSILYL) ESTER

CONSENSUS REPORTS: Reported in EPA TSCA Inventory. Chromium and its compounds are on the Community Right-To-Know List.

OSHA PEL: CL 0.1 mg(Cr)₃)/m³
ACGIH TLV: TWA 0.05 mg(Cr)/m³
NIOSH REL: (Chromium(VI) TWA 0.025 mg(Cr(VI))/m³; CL 0.05/15M

SAFETY PROFILE: Moderately toxic by ingestion and skin contact. When heated to decomposition it emits toxic fumes of CrO_3 particulates.

BLT250   CAS: 77-80-5   **HR: 3**
BIS(TRIPHENYL TIN)SULFIDE
mf: $C_{36}H_{30}SSn_2$ mw: 732.10

SYN: 1,1,1,3,3,3-HEXAPHENYLDISTANNTHIANE

OSHA PEL: TWA 0.1 mg(Sn)/m³ (skin)
ACGIH TLV: TWA 0.1 mg(Sn)/m³ (skin) (Proposed: TWA 0.1 mg(Sn)/m³; STEL 0.2 mg(Sn)/m³ (skin))
NIOSH REL: (Organotin Compounds) TWA 0.1 mg(Sn)/m³

SAFETY PROFILE: A poison via intravenous route. Moderately toxic by ingestion. When heated to decomposition it emits toxic fumes of SO_x.

BLU000   CAS: 13356-08-6   **HR: 2**
BIS(TRIS(β,β-DIMETHYLPHENETHYL) TIN)OXIDE
mf: $C_{60}H_{78}OSn_2$ mw: 1052.76

SYNS: BENDEX * BIS(TRIS(2-METHYL-2-PHENYL-PROPYL)TIN)OXIDE * DI(TRI-(2,2-DIMETHYL-2-PHE-NYLETHYL)TIN)OXIDE * ENT 27,738 * FENBUTA-TIN OXIDE * HEXAKIS(β,β-DIMETHYLPHENETHYL) DISTANNOXANE * HEXAKIS(2-METHYL-2-PHENYL-PROPYL)DISTANNOXANE * SD 14114 * SHELL SD-14114 * TORQUE * VENDEX

OSHA PEL: TWA 0.1 mg(Sn)/m^3 (skin)
ACGIH TLV: TWA 0.1 mg(Sn)/m^3 (skin) (Proposed: TWA 0.1 mg(Sn)/m^3; STEL 0.2 mg(Sn)/m^3 (skin))
NIOSH REL: (Organotin Compounds) TWA 0.1 mg(Sn)/m^3

SAFETY PROFILE: Moderately toxic by ingestion and skin contact. When heated to decomposition it emits acrid smoke and irritating fumes.

BLV000   CAS: 5169-78-8   **HR: 3**
BITIODIN
mf: $C_{15}H_{17}NS_2$ mw: 275.45

SYNS: AT 327 * CR/662 * 3-(DI-2-THIENYLMETH-YLENE)-1-METHYLPIPERIDINE * 1-METHYL-3-PIPE-RIDYLIDENEDI(2-THIENYL)METHANE * TIPEDINE * TIPEPIDINE

SAFETY PROFILE: A poison via subcutaneous, intraperitoneal, intravenous, and intramuscular routes. Moderately toxic by ingestion. An antitussive. When heated to decomposition it emits very toxic fumes of NO_x and SO_x.

BLV250   CAS: 13394-86-0   **HR: 3**
(m,o'-BITOLYL)-4-AMINE
mf: $C_{14}H_{15}N$ mw: 197.30

SYNS: 2',3-DIMETHYL-4-AMINOBIPHENYL * 3,2'-DIMETHYL-4-AMINOBIPHENYL * 3,2'-DIMETHYL-4-AMINODIPHENYL * 3,2'-DIMETHYL-4-BIPHENYL-AMINE * 3,2'-DMAB

CONSENSUS REPORTS: EPA Genetic Toxicology Program.

SAFETY PROFILE: Suspected carcinogen with experimental carcinogenic and tumorigenic data. Moderately toxic by intraperitoneal route. Mutation data reported. When heated to decomposition it emits toxic fumes of NO_x.

BLV500   CAS: 8013-76-1   **HR: 3**
BITTER ALMOND OIL

PROP: Volatile oil from dried ripe kernels of bitter almonds or from other kernels containing amygdalin, such as apricots, cherries, plums, and especially peaches. Colorless liquid; strong almond odor. Bp: 179°, d: 1.045-1.070 @ 15°. Sltly sol in water; sol in fixed oils and propylene glycol; insol in glycerin.

SYNS: ALMOND OIL BITTER, FFPA (FCC) * OIL, BITTER ALMOND

CONSENSUS REPORTS: Reported in EPA TSCA Inventory.

SAFETY PROFILE: A human poison by ingestion. Moderately toxic by skin contact. A skin irritant. When heated to decomposition it emits toxic fumes of CN^-.

BLV750   CAS: 68916-04-1   **HR: 1**
BITTER ORANGE OIL

PROP: Main constituent is d-limonene. Pale yellow liquid, bitter taste. D: 0.842-0.848 @ 25°/25°. Very sltly sol in water; miscible with absolute alc; sol in 4 vols alc, in 1 vol glacial acetic acid. Keep well closed, cool and protected from light.

SAFETY PROFILE: A skin irritant. When heated to decomposition it emits acrid smoke and irritating fumes.

BLW250   CAS: 8006-82-4   **HR: 1**
BLACK PEPPER OIL

PROP: From steam distillation of dried fruit of *Piper nigrum L.* (Fam. *Piperaceae*). Main constituents include α- and β-pinene, β-caryophyllene, l-limonene, d-hydrocarveol, piperidine and piperrine. A colorless to greenish liquid; odor and taste of pepper. Sol in fixed oils, mineral oil, propylene glycol; sltly sol in glycerin.

CONSENSUS REPORTS: Reported in EPA TSCA Inventory.

SAFETY PROFILE: A moderate skin irritant. Mutation data reported. When heated to decomposition it emits acrid smoke and irritating fumes.

BLW750   CAS: 21725-46-2   **HR: 3**
BLADEX
mf: $C_9H_{13}ClN_6$ mw: 240.73

PROP: A white, crystalline material. Mp: 167°.

SYNS: BLADEX 80WP * 2-CHLORO-4-(1-CYANO-1-METHYLETHYLAMINO)-6-ETHYLAMINO-1,3,5-TRIAZINE * 2-CHLORO-4-ETHYLAMINO-6-(1-CYANO-1-METHYL) ETHYLAMINO-s-TRIAZINE * 2-((4-CHLORO-6-(ETHYLAMINO)-1,3,5-TRIAZIN-2-YL)AMINO)-2-METHYL-PROPANENITRILE * 2-(4-CHLORO-6-ETHYLAMINO-s-TRIAZINE-2-YLAMINO)-2-METHYL-PROPIONITRILE * 2-((4-CHLORO-6-(ETHYLAMINO)-s-TRIAZIN-2-YL) AMINO)-2-METHYLPROPIONITRILE * 2-(4-CHLORO-6-ETHYLAMINO-1,3,5-TRIAZINE-2-YLAMINO)-2-METHYL-PROPIONITRILE * CYANAZINE * DW3418 * FORTROL * PAYZE * SD 15418 * WL 19805

CONSENSUS REPORTS: EPA Genetic Toxicology Program. Cyanide and its compounds are on the Community Right-To-Know List.

SAFETY PROFILE: Poison by ingestion. Moderately toxic by skin contact. An experimental teratogen. Mutation data reported. An herbicide. When heated to decomposition it emits very toxic fumes of Cl^-, NO_x, and CN^-.

BLY000   CAS: 11056-06-7   **HR: 3**
BLEOMYCIN

PROP: A group of related glycopeptide antibiotics isolated from *Streptomyces verticillus* (12VXA5 9, 171,76).

SYNS: BLEO * BLENOXANE * BLEOCIN * BLM

CONSENSUS REPORTS: IARC Cancer Review: GROUP 2B IMEMDT 7,134,87, Human Inadequate Evidence IMEMDT 26,97,81. EPA Genetic Toxicology Program.

SAFETY PROFILE: Suspected carcinogen. A human poison by intravenous route; moderately toxic to humans by intramuscular route. Poison experimentally by intravenous and intraperitoneal routes. Human systemic effects by ingestion and intramuscular routes: dyspnea and fibrosing alveolitis (lung). Experimental reproductive effects. Human mutation data reported. When heated to decomposition it emits toxic fumes of NO_x.

BLY250   CAS: 11116-31-7   **HR: 3**
BLEOMYCIN A2

SYN: ZHENGGUANGMYCIN A2 (CHINESE)

SAFETY PROFILE: Poison by intravenous and intraperitoneal routes. Noted for adverse pulmonary effects in humans. Mutation data reported. When heated to decomposition it emits toxic fumes of NO_x.

BMA750   CAS: 8001-85-2   **HR: 2**
BONE OIL

PROP: Product of destructive distillation of bones in preparation of bone charcoal containing nitrogenous compounds such as pyridine, aniline, methylamine, and pyrrole.

SYNS: ANIMAL OIL * DIPPEL'S OIL * OIL of HARTSHORN

CONSENSUS REPORTS: Reported in EPA TSCA Inventory.

SAFETY PROFILE: Moderately toxic by ingestion. When heated to decomposition it emits toxic fumes of NO_x.

BMB500   CAS: 6569-51-3   **HR: 3**
BORAZINE
mf: $B_3H_6N_3$ mw: 80.5

SYN: BORAZOLE

PROP: Colorless liquid. Mp: −58°, bp: 53°, d: 0.824 @ 0°.

SAFETY PROFILE: A powerful irritant to skin, eyes, and mucous membranes. May explode spontaneously when stored in the light. Reacts with water to form toxic and flammable boron hydrides. A dangerous fire hazard. When heated to decomposition it emits toxic fumes of NO_x.

BMB750   **HR: 3**
BORDEAUX ARSENITE

SYN: BORDEAU ARSENITE, liquid or solid (DOT)

CONSENSUS REPORTS: Arsenic and its compounds, as well as copper and its compounds, are on the Community Right-To-Know List.

OSHA PEL: TWA 0.01 mg(As)/m^3
NIOSH REL: CL 0.002 mg(As)/m^3/15M
DOT Classification: Poison B; Label: Poison, liquid or solid.

SAFETY PROFILE: A poison. When heated to decomposition it emits toxic fumes of As.

BMC000 CAS: 10043-35-3 **HR: 3**
BORIC ACID
mf: BH₃O₃ mw: 61.84

PROP: White crystals or powder. Mp: 185° (decomp), −1.5H₂O @ 300°, d: 1.435 @ 15°.

SYNS: BORACIC ACID * BOROFAX * BOR-SAURE (GERMAN) * NCI-C56417 * ORTHOBORIC ACID * THREE ELEPHANT

CONSENSUS REPORTS: Reported in EPA TSCA Inventory.

SAFETY PROFILE: A human poison by ingestion and possibly other routes. Moderately toxic by skin contact and subcutaneous routes in humans. Poison experimentally by inhalation and subcutaneous routes. Moderately toxic experimentally by intraperitoneal and intravenous routes. Human systemic effects by an unspecified route: wakefulness, anorexia, nausea and vomiting. Ingestion or absorption by other routes may also cause diarrhea, abdominal cramps, erythematous lesions on skin and mucous membranes, circulatory collapse, tachycardia, cyanosis, delirium, convulsions, and coma. Death has occurred from ingestion of less than 5 grams in infants, and from 5 to 20 grams in adults. Chronic exposure may result in borism (dry skin, eruptions, and gastrointestinal disturbances). Experimental reproductive effects. Mutation data reported. A human skin irritant. Incompatible with K; (CH₃CO)₂O.

BMC250 CAS: 34099-73-5 **HR: 3**
BORIC ACID, ETHYL ESTER
DOT: UN 1176
mf: C₂H₇BO₃ mw: 89.90

PROP: Colorless liquid, mild odor, decomp in water. Bp: 120°, flash p: 52°F (CC), d: 0.864 @ 26.5°, vap d: 5.04.

SYN: ETHYL BORATE (DOT)

DOT Classification: Flammable Liquid; Label: Flammable Liquid.

SAFETY PROFILE: A severe eye irritant. Dangerous fire hazard when exposed to heat or flame; will react with water or steam to produce flammable vapors. Incompatible with oxidizers, heat, or open flame. To fight fire, use CO₂, dry chemical.

BMD000 CAS: 507-70-0 **HR: 2**
BORNEOL
DOT: UN 1312
mf: C₁₀H₁₈O mw: 154.28

PROP: Hexagonal crystals, peppery odor and burning taste. Mp: 208°, bp: 212°, flash p: 150°F, d: 1.01 @ 20°/4°, vap d: 5.31.

SYNS: BAROS CAMPHOR * BHIMSAIM CAMPHOR * BORNEO CAMPHOR * trans-BORNEOL * BOR-NEOL (DOT) * BORNYL ALCOHOL * 2-CAMPHA-NOL * DRYOBALANOPS CAMPHOR * 2-HYDROXY-CAMPHANE * MALAYAN CAMPHOR * SUMATRA CAMPHOR * endo-1,7,7-TRIMETHYL-BICYCLO(2.2.1) HEPTAN-2-OL

CONSENSUS REPORTS: Reported in EPA TSCA Inventory.

DOT Classification: Flammable Solid; Label: None.

SAFETY PROFILE: Moderately toxic by ingestion. Mutation data reported. A mild irritant. Flammable when exposed to heat or flame; can react with oxidizing materials. To fight fire, use water, CO₂, water spray, dry chemical. When heated to decomposition it emits acrid smoke and fumes.

BMD100 CAS: 76-49-3 **HR: 1**
BORNYL ACETATE
mf: C₁₂H₂₀O₂ mw: 196.29

PROP: Colorless liquid or white crystalline solid; sweet, piney odor. D: 0.981-0.985, refr index: 1.462. flash p: 192°F. Sol in alc, fixed oils; sltly sol in water; insol in glycerin, propylene glycol @ 226°.

SYNS: l-BORNYL ACETATE * FEMA No. 2159

SAFETY PROFILE: Combustible liquid. When heated to decomposition it emits acrid smoke and irritating fumes.

BMD500 CAS: 7440-42-8 **HR: 3**
BORON
af: B aw: 10.81

PROP: Monoclinic crystals, yellow or brown amorphous powder. Mp: 2300°, bp: 2550°, d: 3.33 @ 20°.

CONSENSUS REPORTS: Reported in EPA TSCA Inventory.

SAFETY PROFILE: A poison by ingestion. A relatively inert metal except in the form of powder or when exposed to highly oxidizing agents. Flammable in the form of dust when exposed to air or by chemical reaction. An explosion hazard in the form of dust which ignites on contact with air. Reacts explosively when ground with lead fluoride or silver fluoride. Ignites in contact with gaseous chlorine or fluorine at room temperature. Incompatible with NH_3, Br_2, BrF_3, Cs_2C_2, Cl_2, CuO, HIO_3, PbO_2, HNO_3, NO, NOF, N_2O, $KClO_3$, KNO_3, Rb_2C_2, S, BrF_5, IF_5, metal fluorides, inter halogens, nitryl fluoride (FNO_2), OF_2, KNO_2, NO_x, Na_2O_2, PbO, and air.

BME500   HR: 3
BORON COMPOUNDS

SAFETY PROFILE: Very toxic and therefore considered an industrial poison. Used in medicine as sodium borate, boric acid, or borax, which is a common cleanser. Fatal poisoning of children has been caused by the accidental substitution of boric acid for powdered milk. The medical literature reveals instances of accidental poisoning due to boric acid; ingestion of borates or boric acid; and presumably absorption of boric acid from wounds and burns. The fatal dose of orally ingested boric acid for an adult is somewhat greater than 15 to 20 grams and for an infant, 5 to 6 grams. Boron is one of a group of elements, such as Pb, Mn, As, which affects the central nervous system. Boron poisoning causes depression of the circulation, persistent vomiting and diarrhea, followed by profound shock and coma. The temperature becomes subnormal and a scarletina-form rash may cover the entire body. Containers of boric acid should be plainly labeled and should differ radically from those which contain powdered milk, particularly in institutions such as hospitals.

BMG000   CAS: 1303-86-2   HR: 2
BORON OXIDE
mf: B_2O_3 mw: 69.62

PROP: Vitreous, colorless crystals. Mp: 450° (approx), bp: 1860°, d: 2.46.

SYNS: BORIC ANHYDRIDE * BORON SESQUIOXIDE
* BORON TRIOXIDE * FUSED BORIC ACID

CONSENSUS REPORTS: Reported in EPA TSCA Inventory.

OSHA PEL: (Transitional: Total Dust: TWA 15 mg/m³; Respirable Fraction: TWA 5 mg/ m³) Total Dust: TWA 10 mg/m³; Respirable Fraction: TWA 5 mg/m³
ACGIH TLV: TWA 10 mg/m³
DFG MAK: 15 mg/m³

SAFETY PROFILE: Moderately toxic by ingestion and intraperitoneal routes. An eye and skin irritant. A pesticide. Mixed with CaO and put into fused $CaCl_2$, the mixture incandesces.

BMG250   HR: 3
BORON PHOSPHIDE
mf: BP   mw: 41.79

PROP: Maroon powder. Mp: 200°.

SAFETY PROFILE: A poison. Ignites @ 200°. Deflagrates with fused alkali nitrates. Incompatible with HNO_3; oxidants; i.e., nitrates. When heated to decomposition it emits toxic fumes of PO_x.

BMG400   CAS: 10294-33-4   HR: 3
BORON TRIBROMIDE
DOT: UN 2692
mf: BBr_3 mw: 250.54

PROP: Colorless, fuming liquid. Mp: −45°, bp: 91.7°, d: 2.650 @ 0°, vap press: 40 mm @ 14.0°, 100 mm @ 33.5°.

SYNS: BORON BROMIDE * TRONA

CONSENSUS REPORTS: Reported in EPA TSCA Inventory.

OSHA PEL: CL 1 ppm
ACGIH TLV: CL 1 ppm
DOT Classification: Corrosive Material; Label: Corrosive.

SAFETY PROFILE: A poison. Corrosive. A skin, eye, and mucous membrane irritant. Dangerous; may explode when heated. This and other boron halides react with water or steam to produce toxic and corrosive fumes and may explode. Incompatible with K or Na. When heated to decomposition it emits toxic fumes of Br^-.

BMG500   CAS: 10294-34-5   HR: 3
BORON TRICHLORIDE
DOT: UN 1741
mf: BCl_3 mw: 117.16

PROP: Colorless, fuming liquid. Pungent, irritating odor. Mp: −107°, bp: 12.5°, d: 1.434 @ 0°, vap press: 1 atm @ 12.7°, vap d: 4.03.

SYNS: BORON CHLORIDE * CHLORURE de BORE (FRENCH)

CONSENSUS REPORTS: Reported in EPA TSCA Inventory. EPA Extremely Hazardous Substances List.

DOT Classification: Corrosive Material; Label: Corrosive; IMO: Nonflammable Gas; Label: Nonflammable Gas, Corrosive.

SAFETY PROFILE: A poison by inhalation and probably other routes. An irritant to skin, eyes, and mucous membranes. Reacts with water or steam to produce heat, toxic and corrosive fumes. Violent reaction with aniline or phosphine. Incompatible with hexafluorisopropylidene amino lithium, NO_2, grease, organic matter, O_2. When heated to decomposition it emits toxic fumes of Cl^-.

BMG700   CAS: 7637-07-2   **HR: 3**
BORON TRIFLUORIDE

DOT: UN 1008
mf: BF_3 mw: 67.81

PROP: Colorless gas; pungent, irritating odor. Mp: $-126.8°$, bp: $-99.9°$, d: 2.99 g/L.

SYNS: BORON FLUORIDE * FLUORURE de BORE (FRENCH)

CONSENSUS REPORTS: Reported in EPA TSCA Inventory. EPA Extremely Hazardous Substances List.

OSHA PEL: CL 1 ppm
ACGIH TLV: CL 1 ppm
DFG MAK: 1 ppm (3 mg/m^3)
NIOSH REL: (Boron Trifluoride) No Exposure Limit
DOT Classification: Nonflammable Gas; Label: Nonflammable Gas and Poison; IMO: Poison A; Label: Poison Gas.

SAFETY PROFILE: A poison by inhalation. A strong irritant. Dangerous; when heated to decomposition or upon contact with water or steam, will produce toxic and corrosive fumes of F^-. Incompatible with alkali metals, alkaline earth metals (except Mg), alkyl nitrates, CaO.

BMG750   CAS: 753-53-7   **HR: 3**
BORON TRIFLUORIDE-ACETIC ACID COMPLEX

DOT: UN 1742

SYN: BORON TRIFLUORIDE-ACETIC ACID COMPLEX (DOT)

CONSENSUS REPORTS: Reported in EPA TSCA Inventory.

DOT Classification: Corrosive Material; Label: Corrosive.

SAFETY PROFILE: A very corrosive material. When heated to decomposition it emits very toxic fumes of F^-, B oxides.

BMH000   CAS: 353-42-4   **HR: 3**
BORON TRIFLUORIDE-DIMETHYL ETHER

DOT: UN 2965
mf: $C_2H_6O \cdot BF_3$ mw: 113.89

SYN: BORON TRIFLUORIDE DIMETHYL ETHERATE (DOT)

CONSENSUS REPORTS: Reported in EPA TSCA Inventory. EPA Extremely Hazardous Substances List.

DOT Classification: IMO: Flammable Solid; Label: Dangerous When Wet, Flammable Liquid, Corrosive.

SAFETY PROFILE: Poison by inhalation. When heated to decomposition it emits toxic fumes of F^-.

BMH500   CAS: 13517-10-7   **HR: 3**
BORON TRIIODIDE
mf: BI_3 mw: 391.52

PROP: Colorless, hygroscopic plates. Mp: $43°$, bp: $210°$, d: 3.35 @ $50°$.

SAFETY PROFILE: A poison. Reacts violently with water. Incandescent reaction with red or white phosphorous. Exothermic reaction with ammonia. Incompatible with ethers; carbohydrates; POCl. When heated to decomposition it emits toxic fumes of I^-.

BML000   **HR: 3**
BRACKEN FERN, DRIED

SYNS: 1-CYCLOHEXENE-1-CARBOXYLIC ACID, 3,4,5 * PTERIDIUM AQUILINUM * PTERIS AQUALINA * S. EGRELTRI ATUNUN (TURKISH)

CONSENSUS REPORTS: IARC Cancer Review: Human Inadequate Evidence IMEMDT

40,47,86; Animal Sufficient Evidence IMEMDT 40,47,86

SAFETY PROFILE: Confirmed carcinogen with experimental carcinogenic, neoplastigenic, and tumorigenic data. Experimental teratogenic and reproductive effects. Mutation data reported.

BML500   CAS: 58-82-2   **HR: D**
BRADYKININ
mf: $C_{50}H_{73}N_{15}O_{11}$ mw: 1060.38

SYNS: BK * BRADYKININ (SYNTHETIC) * BRS 640 * KALLIDIN * PRS 640 * SYNTHETIC BRADYKININ

SAFETY PROFILE: Experimental teratogenic and reproductive effects. Mutation data reported. When heated to decomposition it emits toxic fumes of NO_x.

BMM500   CAS: 2580-78-1   **HR: 3**
BRILLIANT BLUE R
mf: $C_{22}H_{16}N_2O_{11}S_3 \cdot 2Na$ mw: 626.56

SYNS: CAVALITE BRILLIANT BLUE R * C.I. 61200 * C.I. REACTIVE BLUE 19 * C.I. REACTIVE BLUE 19, DISODIUM SALT * REACTIVE BLUE 19 * REMALAN BRILLIANT BLUE R * REMAZOL BRILLIANT BLUE R

CONSENSUS REPORTS: Reported in EPA TSCA Inventory.

SAFETY PROFILE: Questionable carcinogen with experimental tumorigenic data. When heated to decomposition it emits very toxic fumes of Na_2O, NO_x and SO_x.

BMM650   CAS: 314-40-9   **HR: 2**
BROMACIL
mf: $C_9H_{13}BrN_2O_2$ mw: 261.15

SYNS: BOREA * BROMAZIL * 5-BROMO-3-sec-BUTYL-6-METHYLURACIL * 5-BROMO-6-METHYL-3-(1-METHYLPROPYL)-2,4(1H,3H)-PYRIMIDINEDIONE * 5-BROMO-6-METHYL-3-(1-METHYLPROPYL)URACIL * 3-sek.BUTYL-5-BROM-6-METHYLURACIL (GERMAN) * CYNOGAN * DU PONT HERBICIDE 976 * EEREX GRANULAR WEED KILLER * EEREX WATER SOLUBLE CONCENTRATE WEED KILLER * HERBICIDE 976 * HYVAR * HYVAREX * HYVAR X * HYVAR X BROMACIL * HYVAR X WEED KILLER * KROVAR II * NALKIL * URAGAN * URAGON * UROX B WATER

SOLUBLE CONCENTRATE WEED KILLER * UROX HX GRANULAR WEED KILLER

CONSENSUS REPORTS: EPA Genetic Toxicology Program.

OSHA PEL: TWA 1 ppm
ACGIH TLV: TWA 1 ppm

SAFETY PROFILE: Moderately toxic by ingestion. An experimental teratogen. Mutation data reported. An herbicide. When heated to decomposition it emits very toxic fumes of Br^- and NO_x.

BMN250   CAS: 13977-28-1   **HR: 3**
BROMADRYL
mf: $C_{18}H_{22}BrNO \cdot ClH$ mw: 384.78

SYNS: 2-(1-(4-BROMODIPHENYL)ETHOXY)-N,N-DI-METHYLETHYLAMINE HYDROCHLORIDE * p-BROMO-α-METHYLBENZHYDRYL-2-DIMETHYL-AMINOETHYL ETHER HYDROCHLORIDE * 2-((p-BROMO-α-METHYL-α-PHENYLBENZYL)OXY)-N,N-DIMETHYLETHYLAMINE HYDROCHLORIDE * 2-(1-(4-BROMOPHENYL)-1-PHENYLETHOXY)-N,N-DI-METHYLETHANAMINE HYDROCHLORIDE * 1-(p-BRO-MOPHENYL)-1-PHENYL-1-(2-DIMETHYLAMINOETH-OXY)ETHANE HYDROCHLORIDE * (2-(1-p-BRO-MOPHENYL-1-PHENYLETHOXY)ETHYL)DIMETHYL-ETHYLAMINE HYDROCHLORIDE * β-DIMETHYL-AMINOETHYL-p-BROMO-α-METHYLBENZHYDRYL ETHER HYDROCHLORIDE * EMBRAMINE HYDRO-CHLORIDE * MEBROPHENHYDRAMINE * ME-BROPHENHYDRAMINE HYDROCHLORIDE * MEBRYL

SAFETY PROFILE: Poison by ingestion and intravenous routes. An antihistaminic agent. When heated to decomposition it emits very toxic fumes of HCl, Br^-, and NO_x.

BMN500   **HR: 2**
BROMATES

SAFETY PROFILE: Generally considered to be more toxic than chlorates; causes central nervous system paralysis. They may form methemoglobin, but less actively than chlorates. Flammable in the form of gas, vapor, or dust by chemical reaction with (powdered metals + acids), Al, As, CaH_2, C, Cu, powdered metals, metal sulfides, organic matter, PH_4I, P, SrH, S, (H_2SO_4 + metals). When heated to decomposition they emit toxic fumes of Br^-; can react with reducing materials.

BMO000 CAS: 9001-00-7 **HR: 3**
BROMELAIN

PROP: From pineapples *Ananas comosus* and *Ananas bracteatus* L. White to tan amorphous powder. Sol in water; insol in alc, chloroform, ether.

SYNS: ANANASE * BROMELAINS * BROMELIN * E.C. 3.4.4.24 * EXTRANASE * INFLAMEN * PLANT PROTEASE CONCENTRATE * TRAUMA-NASE

CONSENSUS REPORTS: Reported in EPA TSCA Inventory.

SAFETY PROFILE: A poison via intraperitoneal and intravenous routes. When heated to decomposition it emits acrid smoke and fumes.

BMO750 **HR: 3**
BROMIDES

SAFETY PROFILE: The most common inorganic bromides are Na, K, NH_4, Ca and Mg bromides. Methyl and ethyl bromides are among the most common organic bromides. The inorganic bromides produce depression, emaciation, and, in severe cases, psychosis and mental deterioration. Bromide rashes (bromoderma), especially of the face and resembling acne and furunculosis, often occur when bromide inhalation or administration is prolonged. Organic bromides, such as methyl bromide and ethyl bromide, are volatile liquids of relatively high toxicity. When strongly heated they emit highly toxic fumes of Br^-.

BMO825 **HR: D**
BROMINATED VEGETABLE (SOYBEAN) OIL

SYN: VEGETABLE (SOYBEAN) OIL, brominated

SAFETY PROFILE: Experimental reproductive effects. When heated to decomposition it emits toxic fumes of Br^-.

BMP000 CAS: 7726-95-6 **HR: 3**
BROMINE

DOT: UN 1744
mf: Br_2 mw: 159.82

PROP: Rhombic crystals or dark red liquid. Fp: $-7.3°$, bp: $58.73°$, d: 2.928 @ $59°$, 3.12 @ $20°$, vap press: 175 mm @ $21°$, 1 atm @ $58.2°$, vap d: 5.5.

SYNS: BROM (GERMAN) * BROME (FRENCH) * BROMINE, solution (DOT) * BROMO (ITALIAN) * BROOM (DUTCH)

CONSENSUS REPORTS: Reported in EPA TSCA Inventory. EPA Genetic Toxicology Program.

OSHA PEL: (Transitional: TWA 0.1 ppm) TWA 0.1 ppm; STEL 0.3 ppm
ACGIH TLV: TWA 0.1 ppm; STEL 0.3 ppm
DFG MAK: 0.1 ppm (0.7 mg/m^3)
DOT Classification: Corrosive Material; Label: Corrosive; IMO: Corrosive Material; Label: Corrosive and Poison.

SAFETY PROFILE: A human poison by ingestion and moderately toxic by inhalation. A poison by ingestion and inhalation experimentally. Corrosive. The action of bromine is essentially the same as that of chlorine, irritating the mucous membranes of the eyes and upper respiratory tract. Severe exposure may result in pulmonary edema. Usually, however, the irritant qualities of the chemical force the worker to leave the exposure area before serious poisoning can result. Chronic exposure is similar to the therapeutic ingestion of excessive bromides. Regular physical examinations should be made of people who work with bromine or bromides. Flammable in the form of liquid or vapor by spontaneous chemical reaction with reducing materials. A very powerful oxidizer. Highly dangerous; when heated it emits highly toxic fumes; will react with water or steam to produce toxic and corrosive fumes. Reacts explosively with diethylzinc, germane, disilane, dimethylformamide, hydrogen, isobutyrophenone, metal azides (particularly silver or sodium azide), potassium, silane and homologs, praseodymium, antimony, trimethylamine, ammonia.

BMP250 **HR: 3**
BROMINE AZIDE
mf: BrN_3 mw: 121.93

PROP: Crystals or red liquid. Mp: $45°$, bp: explodes.

SYN: BROMOAZIDE

SAFETY PROFILE: A poison. Can explode spontaneously. The solid, liquid and vapor are shock-sensitive explosives. Concentrated solutions in organic solvents may explode. Moderate fire hazard in the form of vapor by chemical reaction. A powerful oxidant. Moderately explosive when exposed to heat. The liquid explodes on contact with arsenic, sodium, silver foil, or phosphorous. Incompatible with Sb, ethyl ether, Ag, metals. When heated to decomposition it emits highly toxic fumes of Br^- and explodes.

BMP500   CAS: 21255-83-4   HR: 3
BROMINE DIOXIDE
mf: BrO_2 mw: 111.91

PROP: Light yellow crystals. Mp: $0°$ (decomp).

CONSENSUS REPORTS: EPA Extremely Hazardous Substances List.

SAFETY PROFILE: Very unstable material. Flammable in the form of vapor by chemical reaction with reducing agents. Potentially explosive if heated rapidly. A strong oxidant. Reaction with water, steam, or reducing materials produces toxic and corrosive fumes. Must be stored at low temperatures. When heated to decomposition it emits toxic fumes of Br^-.

BMP750   CAS: 13863-59-7   HR: 3
BROMINE FLUORIDE
mf: BrF   mw: 98.91

SAFETY PROFILE: A poison and powerful irritant. Very reactive. Ignites on contact with H_2. Incompatible with organic matter; water. When heated to decomposition it emits toxic fumes of F^- and Br^-.

BMQ000   CAS: 7789-30-2   HR: 3
BROMINE PENTAFLUORIDE
DOT: UN 1745
mf: BrF_5 mw: 174.91

PROP: Colorless fuming liquid. Mp: -61.3, bp: 40.5, d: 2.466 @ 25°, vap d: 6.05.
OSHA PEL: TWA 0.1 ppm
ACGIH TLV: TWA 0.1 ppm
NIOSH REL: (Inorganic Fluorides) TWA 2.5 mg(F)/m^3
DOT Classification: Oxidizer; Label: Oxidizer; IMO: Oxidizer; Label: Oxidizer, Poison, Corrosive.

SAFETY PROFILE: A poisonous, corrosive and extremely reactive gas. It is a powerful oxidizer. Will react with water or steam to produce toxic and corrosive fumes. The liquefied gas reacts violently with many organic compounds and some inorganic compounds. Explodes or ignites on contact with hydrogen-containing materials (i.e., acetic acid, ammonia, benzene, ethanol, hydrogen, hydrogen sulfide, methane, cork, grease, paper, wax, chloromethane). Reacts violently and may ignite on contact with acids, halogens, non-metals, metal halides, metals, oxides, concentrated nitric or sulfuric acids, aluminum powder, ammonium chloride, antimony, arsenic, arsenic pentoxide, barium, bismuth, boron powder, boron trioxide, calcium oxide, carbon monoxide, charcoal, chlorine, chromium, chromium trioxide, cobalt powder, iodine, iodine pentoxide, iridium powder, iron powder, lithium powder, manganese, magnesium oxide, molybdenum, molybdenum trioxide, nickel powder, red phosphorous, phosphorous pentoxide, potassium iodide, rhodium powder, selenium, sulfur, sulfur dioxide, tellurium, tungsten, tungsten trioxide, water, zinc. When heated to decomposition it emits very toxic fumes of F^- and Br^-.

BMQ325   CAS: 7787-71-5   HR: 3
BROMINE TRIFLUORIDE
DOT: UN 1746
mf: BrF_3 mw: 136.91

PROP: Colorless, fuming liquid. Mp: 8.8°, bp: 127°, d: 2.84.

CONSENSUS REPORTS: Reported in EPA TSCA Inventory.

OSHA PEL: TWA 2.5 mg(F)/m^3
ACGIH TLV: TWA 2.5 mg(F)/m^3
NIOSH REL: (Inorganic Fluorides) TWA 2.5 mg(F)/m^3
DOT Classification: Oxidizer; Label: Oxidizer and Poison; IMO: Oxidizer; Label: Oxidizer, Poison, Corrosive.

SAFETY PROFILE: Poisonous and corrosive. Very reactive; a powerful oxidizer. Explosive or violent reaction with organic materials, water, acetone, ammonium halides, antimony, antimony trichloride oxide, arsenic, benzene, boron, bromine, carbon, carbon monoxide, carbon tetrachloride, carbon tetraiodide, chloromethane, cobalt, ether, halogens, iodine, powdered molybdenum, niobium, 2-pentanone, phosphorus, potassium hexachloroplatinate, pyridine, silicon, silicone grease, sulfur, tantalum, tin dichloride, titanium, toluene, vanadium, uranium, uranium hexafluoride. Incompatible with Sb_2O_3, $BaCl_2$, Bi_2O_5, $CdCl_2$, $CaCl_2$, CsCl,

LiCl, MnIO$_3$, metals, Nb$_2$O$_5$, PtBr$_4$, PtCl$_4$, (Pt + KFO), KBr, KCl, KI, RhBr$_4$, RbCl, AgCl, NaBr, NaCl, NaI, Ta$_2$O$_5$, Sn, W, UO$_x$, rubber, plastics. The product of reaction with pyridine ignites when dry. When heated to decomposition it emits toxic fumes of F$^-$ and Br$^-$. Very dangerous.

BMR750   CAS: 79-08-3   **HR: 3**
α-BROMOACETIC ACID

DOT: UN 1938
mf: C$_2$H$_3$O$_2$Br   mw: 138.04

PROP: Hygroscopic crystals, sol in water and alc. D: 1.93, mp: 50°, bp: 208°.

SYNS: ACIDE BROMACETIQUE (FRENCH) * BRO-MOACETIC ACID, solid (DOT) * BROMOACETIC ACID, solution (DOT) * BROMOETHANIOC ACID * α-BRO-MOETHANIOC ACID * MONOBROMESSIGSAEURE (GERMAN) * MONOBROMOACETIC ACID * TO NTU

CONSENSUS REPORTS: Reported in EPA TSCA Inventory.

DOT Classification: Corrosive Material; Label: Corrosive, solid; Corrosive Material; Label: Corrosive Solution.

SAFETY PROFILE: Poison by ingestion, intraperitoneal, and intravenous routes. Irritating and corrosive to skin and mucous membranes. Mutation data reported. When heated to decomposition it emits toxic fumes of Br$^-$.

BMS500   **HR: 3**
BROMOACETYLENE
mf: C$_2$HBr   mw: 104.9

PROP: Gas. Bp: −2°, vap d: 4.684

SYNS: BROMACETYLENE * BROMOETHYNE

SAFETY PROFILE: Toxicity is probably similar to dibromoacetylene. A dangerous fire hazard by spontaneous chemical reaction. A spontaneously flammable gas. Highly explosive. May explode or ignite on contact with air. Incompatible with oxidizing materials, even when solid at −196°. When heated to decomposition it burns and emits toxic fumes of Br$^-$.

BMV750   CAS: 61-75-6   **HR: 3**
(o-BROMOBENZYL)ETHYLDIMETHYL-AMMONIUM-p-TOLUENESULFONATE
mf: C$_{11}$H$_{17}$BrN • C$_7$H$_7$O$_3$S   mw: 414.40

SYNS: ASL-603 * BRETYLAN * BRETYLATE * BRETYLIUM-p-TOLUENESULFONATE * BRETY-LIUM TOSYLATE * BRETYLOL * 2-BROMO-N-ETHYL-N,N-DIMETHYLBENZENEMETHANAMINIUM 4-METHYLBENZENESULFONATE * DARENTHIN * N-ETHYL-N-o-BROMOBENZYL-N,N-DIMETHYLAMMO-NIUM TOSYLATE * ORNID

SAFETY PROFILE: A poison by ingestion, intraperitoneal, subcutaneous, intravenous, and intramuscular routes. An anti-adrenergic agent and antiarrhythmic cardiac depressant. When heated to decomposition it emits very toxic fumes of SO$_x$, NH$_3$, NO$_x$, and Br$^-$.

BMW250   CAS: 5798-79-8   **HR: 3**
BROMOBENZYLNITRILE

DOT: UN 1694
mf: C$_8$H$_6$BrN   mw: 196.06

PROP: Pure: Yellowish-white crystals. Tech: brown, oily liquid with pungent odor of sour fruit; mp: 29°, bp: 242°, fp: 25.5°, flash p: none, d: 1.5160 @ 20°, vap d: 6.8, vap press: 0.011 mm @ 20°.

SYNS: BBC * BBN * BROMBENZYL CYANIDE * α-BROMOBENZYL CYANIDE * α-BROMOBENZYL-NITRILE * α-BROMOPHENYLACETONITRILE * α-BROMO-α-TOLUNITRILE * CA * CAMITE

CONSENSUS REPORTS: Cyanide and its compounds are on the Community Right-To-Know List.

DOT Classification: IMO: Poison B; Label: Poison.

SAFETY PROFILE: Poison by ingestion. Moderately toxic to humans by inhalation. When heated to decomposition it emits very toxic fumes of NO$_x$, Br$^-$, and CN$^-$.

BMX500   CAS: 109-65-9   **HR: 2**
1-BROMOBUTANE

DOT: UN 1126
mf: C$_4$H$_9$Br   mw: 137.04

PROP: Colorless to pale straw-colored liquid. Mp: −112.4°, bp: 101.4°, flash p: 65°F (OC), d: 1.274 @ 25°/25°, autoign temp: 509°F, vap d: 4.72, lel: 2.8% @ 212°F, uel: 6.6% @ 212°F.

SYNS: BUTYL BROMIDE (DOT) * BUTYL BROMIDE, NORMAL (DOT) * N-BUTYL BROMIDE

CONSENSUS REPORTS: EPA Genetic Toxicology Program. Reported in EPA TSCA Inventory.

DOT Classification: Flammable Liquid; Label: Flammable Liquid.

SAFETY PROFILE: Moderately toxic by intraperitoneal route. Mildly toxic by inhalation. Dangerous fire hazard when exposed to heat, flame or oxidizers. Violent reaction with bromobenzene + sodium above 30°C. Can react with oxidizing materials. To fight fire, use CO_2, dry chemical, mist or spray.

BMX750   CAS: 78-76-2   **HR: 3**
2-BROMOBUTANE

DOT: UN 2339
mf: C_4H_9Br mw: 137.04

PROP: Colorless liquid, fp: $< -50°$, bp: 91.4°, flash p: 70°F, d: 1.257 @ 25°/25°.

SYNS: sec-BUTYL BROMIDE * METHYLETHYLBRO-MOMETHANE

CONSENSUS REPORTS: EPA Genetic Toxicology Program. Reported in EPA TSCA Inventory.

DOT Classification: Flammable Liquid; Label: Flammable Liquid.

SAFETY PROFILE: Narcotic in high concentrations. Questionable carcinogen with experimental neoplastigenic data. Dangerous fire hazard when exposed to heat or flame. When heated to decomposition it emits toxic fumes of Br^-; can react with oxidizing materials. To fight fire, use water, spray or mist, foam, CO_2, dry chemical.

BNA250   CAS: 353-59-3   **HR: 1**
BROMOCHLORODIFLUOROMETHANE

DOT: UN 1974
mf: $CBrClF_2$ mw: 165.37

PROP: Colorless gas.

SYNS: CHLORODIFLUOROBROMOMETHANE (DOT)
* CHLORODIFLUOROMONOBROMOMETHANE
* HALON 1211

CONSENSUS REPORTS: Reported in EPA TSCA Inventory.

DOT Classification: Nonflammable Gas; Label: Nonflammable Gas.

SAFETY PROFILE: Mutation data reported. An asphyxiant. When heated to decomposition it emits very toxic fumes of Br^-, Cl^-, and F^-.

BNA750   CAS: 41198-08-7   **HR: 3**
O-(4-BROMO-2-CHLOROPHENYL)-O-ETHYL-S-PROPYL PHOSPHORO-THIOATE
mf: $C_{11}H_{15}BrClO_3PS$ mw: 373.65

SYNS: CGA 15324 * CURACRON * POLYCRON
* PROFENOFOS * SELECRON

SAFETY PROFILE: Poison by ingestion and skin contact. When heated to decomposition it emits very toxic SO_x, PO_x, Br^-, and Cl^-.

BNA825   CAS: 109-70-6   **HR: 2**
1-BROMO-3-CHLOROPROPANE

DOT: UN 2688
mf: C_3H_6BrCl mw: 157.45

SYNS: 3-BROMOPROPYL CHLORIDE * 1,3-CHBP
* omega-CHLOROBROMOPROPANE * 1-CHLORO-3-BROMOPROPANE (DOT) * 3-CHLOROPROPYL BRO-MIDE * TRIMETHYLENE BROMIDE CHLORIDE
* TRIMETHYLENE CHLOROBROMIDE

CONSENSUS REPORTS: Reported in EPA TSCA Inventory.

DOT Classification: Poison B; Label: St. Andrews Cross.

SAFETY PROFILE: Moderately toxic by ingestion. When heated to decomposition it emits toxic fumes of Cl^- and Br^-.

BNC750   CAS: 59-14-3   **HR: 2**
5-BROMO-2′-DEOXY URIDINE
mf: $C_9H_{11}BrN_2O_5$ mw: 307.13

SYNS: BDU * 5-BDU * BROMODEOXYURIDINE
* 5-BROMODEOXYURIDINE * 5-BROMO-2-DEOXYU-RIDINE * 5-BROMODESOXYURIDINE * BROMOUR-ACIL DEOXYRIBOSIDE * 5-BROMOURACIL DEOXYRI-BOSIDE * 5-BROMOURACIL-2-DEOXYRIBOSIDE
* BROXURIDINE * BRUDR * BUDR * 5-BUDR

CONSENSUS REPORTS: EPA Genetic Toxicology Program.

SAFETY PROFILE: Moderately toxic by subcutaneous, intravenous, intraperitoneal, and possibly other routes. Mildly toxic by ingestion. Experimental teratogenic and reproductive effects. Human mutation data reported. When heated to decomposition it emits very toxic fumes of Br^- and NO_x.

BND500   CAS: 75-27-4   HR: 3
BROMODICHLOROMETHANE
mf: $CHBrCl_2$ mw: 163.83

PROP: Colorless liquid. Bp: 89.2−90.6°, d: 1.971 @ 25°/25°.

SYNS: BDCM * DICHLOROBROMOMETHANE * NCI-C55243

CONSENSUS REPORTS: NTP Carcinogenesis Studies (gavage): Clear Evidence: rat, mouse NTPTR* NTP-TR-321,87. EPA Genetic Toxicology Program. Community Right-To-Know List. Reported in EPA TSCA Inventory.

SAFETY PROFILE: Suspected carcinogen with experimental carcinogenic data. Moderately toxic by ingestion. Human mutation data reported. When heated to decomposition it emits very toxic fumes of Br^- and Cl^-.

BNG750   CAS: 776-74-9   HR: 3
BROMODIPHENYLMETHANE
DOT: UN 1770
mf: $C_{13}H_{11}Br$ mw: 247.15

PROP: Solid, decomp in hot water, sol in alc, very sol in benzene. Mp: 45°; bp: 193° @ 26 mm.

SYNS: DIPHENYLMETHYL BROMIDE (DOT)
* DIPHENYL METHYL BROMIDE, solid (DOT)
* DIPHENYL METHYL BROMIDE, solution (DOT)

CONSENSUS REPORTS: Reported in EPA TSCA Inventory.

DOT Classification: Corrosive Material; Label: Corrosive, Solid or Solution.

SAFETY PROFILE: A corrosive poison. When heated to decomposition it emits toxic fumes of Br^-.

BNH500   CAS: 17372-87-1   HR: 3
BROMOEOSINE
mf: $C_{20}H_8Br_4O_5 \cdot 2Na$ mw: 693.90

SYNS: AIZEN EOSINE GH * BROMO ACID
* BROMOFLUORESCEIC ACID * BROMO FLUORESCEIN * BRONZE BROMO * CERTIQUAL EOSINE
* C.I. 45380 * D&C RED NO. 22 * DISODIUM EOSIN * EOSINE * EOSINE SODIUM SALT
* EOSINE YELLOWISH * EOSIN GELBLICH (GER-

MAN) * FENAZO EOSINE XG * HIDACID DIBROMO FLUORESCEIN * IRGALITE BRONZE RED CL
* PHLOXINE TONER B * PHLOX RED TONER X-1354
* PURE EOSINE YY * 11445 RED * SODIUM EOSINATE * SYMULER EOSIN TONER * 2,4,5,7-TETRA-BROMO-9-o-CARBOXYPHENYL-6-HYDROXY-3-ISOXAN-THONE, DISODIUM SALT * 2,4,5,7-TETRABROMO-3,6-FLUORANDIOL * TETRABROMOFLUORESCEIN
* 2',4',5',7'-TETRABROMOFLUORESCEIN DISODIUM SALT * TETRABROMOFLUORESCEIN S * TETRA-BROMOFLUORESCEIN SOLUBLE * 2-(2,4,5,7-TETRA-BROMO-6-HYDROXY-3-OXO-3H-XANTHENE-9-YL)BEN-ZOIC ACID, DISODIUM SALT * TOYO EOSINE G
* 1903 YELLOW PINK

CONSENSUS REPORTS: IARC Cancer Review: Animal Inadequate Evidence IMEMDT 15,183,77. EPA Genetic Toxicology Program. Reported in EPA TSCA Inventory.

SAFETY PROFILE: Poison by intravenous route. Moderately toxic by ingestion, subcutaneous, and intraperitoneal routes. Questionable carcinogen with experimental tumorigenic data. When heated to decomposition it emits very toxic fumes of Br^- and Na_2O.

BNI000   CAS: 3132-64-7   HR: 3
3-BROMO-1,2-EPOXYPROPANE
DOT: UN 2558
mf: C_3H_5BrO mw: 136.99

PROP: Flash p: < 22°.

SYNS: EPIBROMHYDRIN * EPIBROMOHYDRIN (DOT) * EPIBROMOHYDRINE

CONSENSUS REPORTS: EPA Genetic Toxicology Program. Reported in EPA TSCA Inventory.

DOT Classification: Poison B; Label: Flammable Liquid and Poison.

SAFETY PROFILE: Poison by intraperitoneal route. Human mutation data reported. A dangerous fire hazard when exposed to heat or flame. When heated to decomposition it emits toxic fumes of Br^-.

BNI500   CAS: 540-51-2   HR: 3
2-BROMO ETHANOL
mf: C_2H_5BrO mw: 124.98

SYNS: BE * BROMOETHANOL * ETHYLENEBRO-MOHYDRIN * GLYCOL BROMOHYDRIN

CONSENSUS REPORTS: EPA Genetic Toxicology Program. Reported in EPA TSCA Inventory.

SAFETY PROFILE: Poison by intraperitoneal route. Questionable carcinogen with experimental neoplastigenic and tumorigenic data. Mutation data reported. When heated to decomposition it emits toxic fumes of Br⁻.

BNK000   CAS: 77-65-6   **HR: 3**
2-BROMO-2-ETHYLBUTYRYLUREA
mf: $C_7H_{13}BrN_2O_2$ mw: 237.13

SYNS: ADALIN * ADDISOMNOL * N-(AMINO-CARBONYL)-2-BROMO-2-ETHYLBUTANAMIDE * BRO-MACETOCARBAMIDE * BROMADAL * BROMADEL * BROMODIETHYLACETYLCARBAMIDE * BRO-MODIETHYLACETYLUREA * (α-BROMO-α-ETHYL-BUTYRYL)CARBAMIDE * (α-BROMO-α-ETHYLBUTY-RYL)UREA * 1-BROMO-ETHYL-BUTYRYL-UREA * 2-BROMO-2-ETHYLBUTYRLUREA * CARBOMAL * DIACID * DORMITURIN * FYDALIN * HOGGAR * KARBROMAL * KARTRYL * NCI-C03805 * NENESIN * NYCTAL * PAR-KOSED * PELIDORM * PIANADALIN * PLANA-DALIN * TILDIN * URADAL

CONSENSUS REPORTS: NCI Carcinogenesis Bioassay (feed); No Evidence: mouse, rat NCITR* NCI-CG-TR-173,79. Reported in EPA TSCA Inventory.

SAFETY PROFILE: Poison by ingestion, subcutaneous, and possibly other routes. Moderately toxic via intravenous and intraperitoneal routes. Mutation data reported. A sedative, hypnotic and central nervous system depressant. When heated to decomposition it emits very toxic fumes of NO_x and Br⁻.

BNK250   **HR: 3**
2-BROMO ETHYL ETHYL ETHER
mf: C_4H_9BrO mw: 155

PROP: Liquid. Vap d: 5.25, flash p: 5°.

SAFETY PROFILE: An insecticide. A dangerous fire hazard when exposed to heat or flame.

BNL000   CAS: 75-25-2   **HR: 3**
BROMOFORM

DOT: UN 2515
mf: CHBr₃   mw: 252.75

PROP: Colorless liquid or hexagonal crystals. Mp: 6-7°, bp: 149.5°, flash p: none, d: 2.890 @ 20°/4°.

SYNS: BROMOFORME (FRENCH) * BROMOFORMIO (ITALIAN) * METHENYL TRIBROMIDE * NCI-C55130 * RCRA WASTE NUMBER U225 * TRIBROM-METHAAN (DUTCH) * TRIBROMMETHAN (GERMAN) * TRIBROMOMETAN (ITALIAN) * TRIBROMOMETH-ANE

CONSENSUS REPORTS: Reported in EPA TSCA Inventory. Community Right-To-Know List.

OSHA PEL: TWA 0.5 ppm (skin)
ACGIH TLV: TWA 0.5 ppm (skin)
DOT Classification: Poison B; Label: St. Andrews Cross.

SAFETY PROFILE: A human poison by ingestion. Moderately toxic via intraperitoneal and subcutaneous routes. Human mutation data reported. A lachrymator. Questionable carcinogen with experimental neoplastigenic data. It can damage the liver to a serious degree and cause death. It has anesthetic properties similar to those of chloroform, but is not sufficiently volatile for inhalation purposes and is far too toxic for human use. As a sedative and antitussive its medicinal application has resulted in numerous poisonings. Inhalation of small amounts causes irritation, provoking the flow of tears and saliva, and reddening of the face. Abuse can lead to addiction and serious consequences. Explosive reaction with crown ethers or potassium hydroxide. Violent reaction with acetone or bases. Incompatible with Li or NaK alloys. When heated to decomposition it emits highly toxic fumes of Br⁻.

BNM250   CAS: 478-84-2   **HR: 3**
2-BROMO-d-LYSERGIC ACID DIETHYLAMIDE
mf: $C_{20}H_{26}BrN_3O$ mw: 404.40

SYNS: BOL * BOL-148 * d-2-BROM-DIETHYLAM-IDE of LYSERGIC ACID * BROM LSD * BROMLYS-ERGAMIDE * 2-BROM-d-LYSERGIC ACID DIETHYL-AMINE * 2-BROMO-9,10-DIDEHYDRO-N,N-DIETHYL-6-METHYLERGOLINE-8-β-CARBOXAMIDE * BRO-MOLYSERGIDE * 9,10-DIDEHYDRO-N,N-DIETHYL-2-BROMO-6-METHYLERGOLINE-8-β-CARBOXAMIDE * USAF SZ-1

SAFETY PROFILE: Poison by intraperitoneal and intravenous routes. Experimental teratogenic and reproductive effects. Human systemic effects by ingestion: dilation or the arteries or veins. Many lysergic acid derivatives have central nervous system effects. When heated to decomposition it emits very toxic fumes such as Br^- and NO_x.

BNO750 CAS: 24961-39-5 **HR: 3**
7-BROMO METHYL BENZ(a) ANTHRACENE
mf: $C_{19}H_{13}Br$ mw: 321.23

SYNS: 7-BMBA * ICR 498

CONSENSUS REPORTS: EPA Genetic Toxicology Program.

SAFETY PROFILE: A deadly poison by intravenous route. Questionable carcinogen with experimental carcinogenic, neoplastigenic, and tumorigenic data. Human mutation data reported. When heated to decomposition it emits toxic fumes of Br^-.

BNP250 CAS: 107-82-4 **HR: 2**
1-BROMO-3-METHYL BUTANE
DOT: UN 2341
mf: $C_5H_{11}Br$ mw: 151.05

PROP: Colorless liquid. D: 1.210, mp: $-112°$, bp: 120-121°, flash p: 21°. Sltly sol in water; misc with alc and ether.

SYNS: ISOAMYL BROMIDE * ISOPENTYL BROMIDE * 3-METHYLBUTYL BROMIDE

CONSENSUS REPORTS: EPA Genetic Toxicology Program.

DOT Classification: Label: Flammable Liquid.

SAFETY PROFILE: Moderately toxic by intraperitoneal route. Dangerous fire hazard when exposed to heat or flame. When heated to decomposition it emits toxic fumes of Br^-.

BNP750 CAS: 496-67-3 **HR: 2**
2-BROMO-3-METHYLBUTYRYLUREA
mf: $C_6H_{11}BrN_2O_2$ mw: 223.10

SYNS: ABROVAL * ALLUVAL * ALURAL * N-(AMINOCARBONYL)-2-BROMO-3-METHYLBUTANAMIDE * BROMARAL * BROMCARBAMIDE * BROMISOVAL * BROMISOVALERYLUREA * α-BROMISOVALERYLUREA * BROMISOVALUM * BROMIZOVAL * BROMOCARBAMIDE * α-BROMO-β-DIMETHYLPROPANOYLUREA * α-BROMOISO-

VALERIC ACID UREIDE * α-BROMOISOVALEROYLUREA * (α-BROMOISOVALERYL)UREA * BROMOVAL * BROMOVALEROCARBAMIDE * BROMOVALERYLUREA * BROMOXIL * BROMURAL * BROMUVAN * BROMVALERYLUREA * BROMVALETONE * BROMVALETONUM * BROMVALUREA * BROMYL * BROVALIN * BROVALUREA * BROVARIN * BVU * CALMOTIN * DIAGRABROMYL * DIBROLUUR * DORMIGENE * ISOBROMYL * ISOVAL * MONOBROMOISOVALERYLUREA * 2-MONOBROMOISOVALERYLUREA * PIVADORM * PIVADORN * SOMNUROL * UPIOL * UVALERAL

CONSENSUS REPORTS: Reported in EPA TSCA Inventory.

SAFETY PROFILE: Moderately toxic by ingestion and possibly other routes. Human systemic effects by ingestion: nausea or vomiting, and coma. A sedative and hypnotic agent. When heated to decomposition it emits very toxic fumes of Br^- and NO_x.

BNR750 CAS: 78-77-3 **HR: 3**
1-BROMO-2-METHYL PROPANE
DOT: UN 2342
mf: C_4H_9Br mw: 137.04

PROP: Flash p: 22°C.

SYNS: 1-BUTYL BROMIDE * i-BUTYL BROMIDE * ISOBUTYL BROMIDE

CONSENSUS REPORTS: EPA Genetic Toxicology Program. Reported in EPA TSCA Inventory.

DOT Classification: IMO: Flammable or Combustible Liquid; Label: Flammable Liquid.

SAFETY PROFILE: Questionable carcinogen with experimental neoplastigenic data. Moderately toxic by intraperitoneal route. A dangerous fire hazard when exposed to heat or flame. When heated to decomposition it emits toxic fumes of Br^-.

BNT250 CAS: 52-51-7 **HR: 3**
2-BROMO-2-NITRO-1,3-PROPANEDIOL
mf: $C_3H_6BrNO_4$ mw: 200.01

SYNS: 2-BROMO-2-NITROPANE-1,3-DIOL * 2-BROMO-2-NITROPROPAN-1,3-DIOL * β-BROMO-β-NITROTRIMETHYLENEGLYCOL * BRONOCOT * BRONOPOL * BRONOSOL

CONSENSUS REPORTS: Reported in EPA TSCA Inventory.

SAFETY PROFILE: Poison by ingestion, subcutaneous, intravenous, and intraperitoneal routes. Moderately toxic by skin contact. An eye and human skin irritant. An antiseptic. When heated to decomposition it emits very toxic fumes of NO_x and Br^-.

BNU500   CAS: 107-81-3   HR: 3
2-BROMOPENTANE

DOT: UN 2343
mf: $C_5H_{11}Br$ mw: 151.07

PROP: Colorless to yellow liquid, strong odor, bp: 120°, fp: $< -30°$, d: 1.211 @ 25°/25°, flash p: 90°F.

CONSENSUS REPORTS: Reported in EPA TSCA Inventory.

DOT Classification: Flammable Liquid; Label: Flammable Liquid.

SAFETY PROFILE: Poison by intraperitoneal route. A local irritant and narcotic in high concentration. Ingestion can cause liver damage. A dangerous fire hazard when exposed to heat or flame. When heated to decomposition it emits toxic fumes of Br^-.

BNV250   HR: 2
BROMO PHENOLS
mf: $HO(C_6H_4)Br$ mw: 173

PROP: (m-) Crystals; insol in water; sol in alc, ether, and alkalis. (p-) Crystals; sltly sol in water; sol in alc, ether, chloroform, and glacial acetic acid. (o-) Yellow to oily, red liquid; unpleasant odor; insol in water; sol in alc, ether, and chloroform. D: (p-) 1.840 (15°), 1.5875 (80°); (o-) 1.5. Mp: (m-) 33°; (p-) 64°; (o-) 6°. Bp: (m-) 236°; (p-) 238°; (o-) 194°.

SAFETY PROFILE: Moderately toxic by subcutaneous route. Dangerous in a fire. When heated to decomposition it emits toxic fumes of Br^-.

BNV750   CAS: 16532-79-9   HR: 3
4-BROMOPHENYLACETONITRILE

DOT: UN 1694
mf: C_8H_6BrN mw: 196.06

SYNS: 4-BROMOBENZENEACETONITRILE * p-BRO-MOBENZYL CYANIDE * 4-BROMOBENZYLCYANIDE * p-BROMOPHENYLACETONITRILE * 2-(4-BROMO-PHENYL)ACETONITRILE

CONSENSUS REPORTS: Reported in EPA TSCA Inventory. Cyanide and its compounds are on the Community Right-To-Know List.

DOT Classification: Poison B; Label: Poison.

SAFETY PROFILE: Poison by intravenous route. When heated to decomposition it emits very toxic fumes of Br^-, NO_x and CN^-.

BNW500   CAS: 1808-12-4   HR: 3
BROMO PHENYL HYDRAMINE HYDROCHLORIDE
mf: $C_{17}H_{20}BrNO \cdot ClH$ mw: 370.75

SYNS: β-(p-BROMOBENZHYDRYLOXY)ETHYLDIMETH-YLAMINE HYDROCHLORIDE * 2-(4-BROMOBENZO-HYDRYLOXY)ETHYLDIMETHYLAMINE HYDRO-CHLORIDE

SAFETY PROFILE: Poison by ingestion, intravenous, and intraperitoneal routes. When heated to decomposition it emits very toxic fumes of Br^-, NO_x, and HCl.

BNX750   CAS: 106-94-5   HR: 2
1-BROMOPROPANE

DOT: UN 2344
mf: C_3H_7Br mw: 123.01

PROP: Liquid. Mp: $-110°$, bp: 70.9°, d: 1.353 @ 20°/4°, autoign temp: 914°F, flash p: $<22°$, lel: 4.6%.

SYNS: 1-BROMOPROPANE (DOT) * PROPYL BRO-MIDE

CONSENSUS REPORTS: Reported in EPA TSCA Inventory.

DOT Classification: Flammable or Combustible Liquid; Label: Flammable Liquid.

SAFETY PROFILE: Moderately toxic by ingestion and intraperitoneal routes. Mildly toxic by inhalation. Experimental reproductive effects. Mutation data reported. Dangerous fire hazard when heated or exposed to flame or oxidizers. To fight fire, use water, foam, CO_2, dry chemical. When heated to decomposition it emits toxic fumes of Br^-.

BNZ000   CAS: 598-31-2   HR: 3
BROMO-2-PROPANONE

DOT: UN 1569
mf: C_3H_5BrO mw: 136.99

SYNS: ACETONYL BROMIDE * ACETYL METHYL BROMIDE * BROMOACETONE * BROMOACETONE (DOT) * BROMOACETONE, liquid (DOT) * BROMO-METHYL METHYL KETONE * 1-BROMO-2-PROPA-NONE * MONOBROMOACETONE * RCRA WASTE NUMBER P017

DOT Classification: Poison A; Label: Poison Gas; Poison B; Label: Flammable Liquid and Poison.

SAFETY PROFILE: A poisonous gas. Moderately toxic to humans by inhalation. When heated to decomposition it emits toxic fumes of Br^-.

BOB250   CAS: 590-92-1   **HR: 3**
3-BROMOPROPIONIC ACID
mf: $C_3H_5BrO_2$ mw: 152.99

SYN: β-BROMOPROPIONIC ACID

CONSENSUS REPORTS: Reported in EPA TSCA Inventory.

SAFETY PROFILE: Moderately toxic by intraperitoneal route. Questionable carcinogen with experimental tumorigenic data. Mutation data reported. When heated to decomposition it emits toxic fumes of Br^-.

BOD500   CAS: 41287-56-3   **HR: 3**
2-(6-(5-BROMO-2-PYRIDYL OXY)HEXYL)AMINOETHANE THIOL HYDROCHLORIDE
mf: $C_{13}H_{21}BrN_2OS \cdot ClH$ mw: 369.79

SAFETY PROFILE: Poison by ingestion and intraperitoneal route. When heated to decomposition it emits very toxic Br^-, Cl^-, SO_x, and NO_x.

BOE750   CAS: 13465-73-1   **HR: 3**
BROMO SILANE
mf: BrH_3Si mw: 111.02

DOT Classification: Forbidden

SAFETY PROFILE: Ignites spontaneously upon exposure to air. When heated to decomposition it emits toxic fumes of Br^-.

BOF500   CAS: 128-08-5   **HR: 3**
N-BROMO SUCCINIMIDE
mf: $C_4H_4BrNO_2$ mw: 178.00

PROP: White to pale buff, fine, crystalline powder with faint odor of bromine. Mp: 173-175°, d: 2.098.

SYNS: 1-BROMO-2,5-PYRROLIDINEDIONE * N-BRO-MOSUCCIMIDE * SUCCINBROMIMIDE * SUCCINI-BROMIMIDE

CONSENSUS REPORTS: Reported in EPA TSCA Inventory.

SAFETY PROFILE: Poison by intraperitoneal route. An irritating poison to skin, eyes, and mucous membranes. Reacts explosively with aniline, diallyl sulfide, and hydrazine hydrate. Explosive reaction with propiononitrile after heating to 105°C for 24 hours. Violent reaction with dibenzoyl peroxide + 4-toluic acid. When heated to decomposition it emits toxic fumes of Br^- and NO_x.

BOH750   CAS: 75-62-7   **HR: 3**
BROMOTRICHLOROMETHANE
mf: $CBrCl_3$ mw: 198.27

PROP: Colorless liquid; bp: 103.8-105.1°; d: 1.997 @ 25°/25°.

CONSENSUS REPORTS: Reported in EPA TSCA Inventory.

SAFETY PROFILE: Poison by ingestion and intraperitoneal routes. Narcotic in high concentration. Mutation data reported. Incompatible with ethylene. When heated to decomposition it emits very toxic fumes of Cl^- and Br^-.

BOJ000   CAS: 598-73-2   **HR: 3**
BROMO TRIFLUOROETHYLENE

DOT: UN 2419
mf: BrF_3C_2 mw: 160.94

PROP: Bp: −3°.

SYNS: BROMOTRIFLUOROETHENE * TRIFLUORO-BROMOETHYLENE * TRIFLUOROVINYLBROMIDE

DOT Classification: Flammable Gas; Label: Flammable Gas.

SAFETY PROFILE: A poison. Flammable gas or liquid. Ignites spontaneously in air. Incompatible with powerful oxidizers, O_2. When heated to decomposition it emits highly toxic fumes of Br^-, F^-, and $COCF_2$.

BOL750   CAS: 357-57-3   **HR: 3**
BRUCINE

DOT: UN 1570
mf: $C_{23}H_{26}N_2O_4$ mw: 394.51

PROP: Monoclinic prisms. Mp: 178°. An alkaloid extracted from *Strychnos* seeds (WQCHM* 4,-,74).

SYNS: BRUCIN (GERMAN) * BRUCINA (ITALIAN) * (−)-BRUCINE * BRUCINE, solid (DOT) * BRUCINE ALKALOID * DIMETHOXY STRYCHNINE (DOT) * 2,3-DIMETHOXYSTRYCHNINE * 10,11-DIMETHY-STRYCHNINE * RCRA WASTE NUMBER P018

CONSENSUS REPORTS: Reported in EPA TSCA Inventory.

DOT Classification: Poison B; Label: Poison.

SAFETY PROFILE: A poison by subcutaneous, intravenous, and intraperitoneal routes. An alkaloid-like strychnine, but one-sixth as toxic. When heated to decomposition it emits toxic fumes of NO_x.

BOM000   CAS: 60723-51-5   **HR: 3**
BRUCINE METHIODIDE
mf: $C_{23}H_{26}N_2O_4 \cdot CH_3I$ mw: 536.45

SYNS: BRUCINE IODOMETHYLATE * BRUCINE IODOMETHYLE (FRENCH)

SAFETY PROFILE: A poison via intravenous route. When heated to decomposition it emits very toxic fumes of NO_x and I^-.

BOM750   CAS: 15537-73-2   **HR: 3**
BUFORMIN HYDROCHLORIDE
mf: $C_6H_{15}N_5 \cdot 7ClH$ mw: 412.48

SYNS: BUFONAMIN * DIABRIN * INSULAMIN

SAFETY PROFILE: A poison by ingestion. When heated to decomposition it emits very toxic fumes of HCl and NO_x.

BON000   CAS: 471-95-4   **HR: 3**
BUFOTALINE
mf: $C_{26}H_{36}O_6$ mw: 444.62

SYNS: BUFOTALIN * 3-β,14,16-β-TRIHYDROXY-5-β-BUFA-20,22-DIENOLIDE-16-ACETATE

SAFETY PROFILE: A deadly poison by ingestion, subcutaneous, and intravenous routes. When heated to decomposition it emits acrid and irritating fumes.

BON750   CAS: 27262-46-0   **HR: 3**
BUPICAINE HYDROCHLORIDE (+)
mf: $C_{18}H_{28}N_2O \cdot ClH$ mw: 324.94

SYN: 1-BUTYL-2′,6′-PIPECOLOXYLIDIDE HYDROCHLORIDE (+)

SAFETY PROFILE: Poison by ingestion, subcutaneous, intravenous, parenteral, and intratracheal routes. When heated to decomposition it emits very toxic fumes of HCl and NO_x.

BOO750   CAS: 149-16-6   **HR: 3**
BUTACAINE
mf: $C_{18}H_{30}N_2O_2$ mw: 306.50

PROP: Colorless, odorless powder. Mp: 98-100°.

SYNS: 3-(p-AMINOBENZOXY)-1-DI-n-BUTYLAMINOPROPANE * p-AMINOBENZOYLDIBUTYLAMINOPROPANOL * BUTYN * 3-(DIBUTYLAMINO)-1-PROPANOL-p-AMINOBENZOATE * 3-DIBUTYLAMINOPROPYL-p-AMINOBENZOATE

SAFETY PROFILE: A poison via subcutaneous and intravenous routes. A weak allergen. Combustible. When heated to decomposition it emits toxic fumes of NO_x.

BOP000   CAS: 149-15-5   **HR: 3**
BUTACAINE SULFATE
mf: $C_{36}H_{60}N_4O_4 \cdot H_2O_4S$ mw: 711.08

SYNS: 3-(p-AMINOBENZOXY)-1-DI-n-BUTYLAMINOPROPANE SULFATE * p-AMINOBENZOYLDIBUTYLAMINOPROPANOL SULFATE * BUTELLINE * BUTYN SULFATE * 3-(DIBUTYLAMINO)-1-PROPANOL-p-AMINOBENZOATE (ESTER) SULFATE (2:1) * 3-DIBUTYLAMINO-1-PROPANOL-4-AMINOBENZOATE (ESTER) SULFATE (SALT) (2:1) * DIBUTYLAMINOPROPYL-p-AMINOBENZOATE SULFATE * 3′-DIBUTYLAMINOPROPYL-4-AMINOBENZOATE SULFATE

SAFETY PROFILE: A poison by ingestion, subcutaneous, intravenous, and intraperitoneal routes. A topical anesthetic. When heated to decomposition it emits very toxic fumes of SO_x and NO_2.

BOP500   CAS: 106-99-0   **HR: 3**
1,3-BUTADIENE
mf: C_4H_6 mw: 54.10

PROP: Colorless gas; mild aromatic odor. Very reactive. Bp: −4.5°, mp: −113°, fp: −108.9°, flash p: −105°F, lel: 2.0%, uel: 11.5%, d: 0.621 @ 2 0°/4°, autoign temp: 788°F, vap d: 1.87, vap press: 1840 mm @ 21°.

SYNS: BIETHYLENE * BIVINYL * BUTADIEEN (DUTCH) * BUTA-1,3-DIEEN (DUTCH) * BUTADIEN

(POLISH) * BUTA-1,3-DIEN (GERMAN) * BUTA-1,3-DIENE * α-γ-BUTADIENE * DIVINYL * ERYTHRENE * NCI-C50602 * PYRROLYLENE * VINYLETHYLENE

CONSENSUS REPORTS: IARC Cancer Review: GROUP 2B IMEMDT 7,136,87; Human Inadequate Evidence IMEMDT 39,155,86; Animal Sufficient Evidence IMEMDT 39,155,86. NTP Carcinogenesis Studies (inhalation); Clear Evidence: mouse NTPTR* NTP-TR-288,84. Reported in EPA TSCA Inventory. Community Right-To-Know List.

OSHA PEL: TWA 1000 ppm
ACGIH TLV: TWA 10 ppm; Suspected Human Carcinogen
DFG MAK: Processing after polymerization and loading: 15 ppm; Others: 5 ppm; Animal Carcinogen, Suspected Human Carcinogen.
NIOSH REL: Reduce to lowest feasible level
DOT Classification: Flammable Gas; Label: Flammable Gas.

SAFETY PROFILE: Confirmed carcinogen with experimental carcinogenic and neoplastigenic data. Experimental reproductive effects. Mutation data reported. Inhalation of high concentrations can cause unconsciousness and death. Human systemic effects by inhalation: cough, hallucinations, distorted perceptions, changes in the visual field and other unspecified eye effects. The vapors are irritating to eyes and mucous membranes. If spilled on skin or clothing, it can cause burns or frost bite (due to rapid vaporization). Chronic systemic poisoning in humans has not been reported. Dangerous fire hazard when exposed to heat, flame, or powerful oxidizers. Upon exposure to air it forms explosive peroxides sensitive to heat, shock, or heating above 27°C. May decompose explosively when heated above 200°C/1.0 kbar. Explodes on contact with aluminum tetrahydroborate. Potentially explosive reaction with NO_x + O_2, ethanol + iodine + mercury oxide (at 35°C), ClO_2, crotonaldehyde (above 180°C), buten-3-yne (with heat and pressure). Reaction with sodium nitrite forms a spontaneously flammable product. Exothermic reaction with boron trifluoride etherate + phenol. To fight fire, stop flow of gas. When heated to decomposition it emits acrid smoke and fumes.

BOP750   CAS: 30031-64-2   **HR: 3**
I-BUTADIENE DIEPOXIDE
mf: $C_4H_6O_2$ mw: 86.10

SYNS: (S-(R*,R*))-2,2'-BIOXIRANE * 1-DIEPOXYBUTANE * 1-1,2:3,4-DIEPOXYBUTANE * (2S,3S)-DIEPOXYBUTANE * (2S,3S)-1,2:3,4-DIEPOXYBUTANE * NSC-32606

CONSENSUS REPORTS: IARC Cancer Review: GROUP 2B IMEMDT 7,56,87; Animal Sufficient Evidence IMEMDT 11,115,76. EPA Genetic Toxicology Program.

SAFETY PROFILE: Suspected carcinogen with experimental neoplastigenic data. Poison by intraperitoneal route. Mutation data reported. When heated to decomposition it emits acrid and irritating fumes.

BOQ750   CAS: 125-88-2   **HR: 3**
BUTALBITAL SODIUM
mf: $C_{10}H_{14}N_2O_3 \cdot Na$ mw: 233.25

SYNS: APROBARBITAL SODIUM * APROBARBITONE SODIUM * SODIUM-5-ALLYL-5-ISOPROPYLBARBITURATE

SAFETY PROFILE: A poison via intraperitoneal and subcutaneous routes. When heated to decomposition it emits toxic fumes of NO_x and Na_2O.

BOR000   CAS: 1142-70-7   **HR: 3**
BUTALLYLONAL
mf: $C_{11}H_{15}BrN_2O_3$ mw: 303.19

SYNS: 5-(2-BROMOALLYL)-5-sec-BUTYLBARBITURIC ACID * 5-(2'-BROMOALLYL)-5-(1'-METHYL-N-PROPYL)BARBITURIC ACID * BUTYLALYLONAL * 5-sec-BUTYL-5-(β-BROMOALLYL)BARBITURIC ACID * PERNOCTON * PERNOSTON * 2,4,6(1H,3H,5H)-PYRIMIDINETRIONE, 5-(2-BROMO-2-PROPENYL)-5-(1-METHYLPROPYL)-(9CI) * SONBUTAL

SAFETY PROFILE: Poison by ingestion, intravenous, intraperitoneal, and subcutaneous routes. A central nervous system depressant (hypnotic) by ingestion. When heated to decomposition it emits very toxic fumes of Br^- and NO_x.

BOR250   CAS: 3486-86-0   **HR: 3**
BUTALLYLONAL SODIUM
mf: $C_{11}H_{14}BrN_2O_3 \cdot Na$ mw: 325.17

SYNS: sec-BUTYL-BROM-ALLYL BARBITURIC ACID SODIUM SALT * SODIUM-5-(2-BROMOALLYL)-5-sec-BUTYLBARBITURATE

SAFETY PROFILE: Poison by ingestion and intraperitoneal routes. When heated to decomposition it emits very toxic fumes of Br^- and NO_x.

BOR500   CAS: 106-97-8   **HR: 1**
BUTANE
DOT: UN 1011/UN 1075
mf: C_4H_{10} mw: 58.14

PROP: Colorless gas; faint disagreeable odor. Bp: $-0.5°$, fp: $-138°$, lel: 1.9%, uel: 8.5%, flash p: $-76°F$ (CC), d: 0.599, autoign temp: 761°F, vap press: 2 atm @ 18.8°, vap d: 2.046.

SYNS: n-BUTANE (DOT) * BUTANEN (DUTCH) * BUTANI (ITALIAN) * DIETHYL * METHYLE-THYLMETHANE

CONSENSUS REPORTS: Reported in EPA TSCA Inventory.

OSHA PEL: TWA 800 ppm
ACGIH TLV: TWA 800 ppm
DFG MAK: 1000 ppm (2350 mg/m^3)
DOT Classification: Flammable Gas; Label: Flammable Gas.

SAFETY PROFILE: Mildly toxic via inhalation. Causes drowsiness. An asphyxiant. Very dangerous fire hazard when exposed to heat, flame, or oxidizers. Highly explosive when exposed to flame, or when mixed with [$Ni(CO)_4 + O_2$]. To fight fire, stop flow of gas. When heated to decomposition it emits acrid smoke and fumes.

BOR750   CAS: 590-88-5   **HR: 2**
1,3-BUTANEDIAMINE
mf: $C_4H_{12}N_2$ mw: 88.18

PROP: Liquid. Bp: 142-150°, flash p: 125°F, d: 0.85, vap d: 3.04.

SYN: 1,3-DIAMINOBUTANE

SAFETY PROFILE: Moderately toxic by ingestion and skin contact. Severe skin and eye irritant. Moderate fire hazard when exposed to heat or flame. To fight fire, use alcohol foam, foam, CO_2, dry chemical. Incompatible with oxidizing materials. When heated to decomposition it emits toxic fumes of NO_x.

BOS250   CAS: 584-03-2   **HR: 2**
1,2-BUTANEDIOL
mf: $C_4H_{10}O_2$ mw: 90.14

PROP: D: 1.0, vap d: 3.1, bp: 194°, flash p: 194°F.

SYN: 1,2-BUTYLENE GLYCOL

CONSENSUS REPORTS: Reported in EPA TSCA Inventory.

SAFETY PROFILE: Moderately toxic by ingestion. Flammable when exposed to heat or flame. To fight fire, use alcohol foam. When heated to decomposition it emits acrid and irritating fumes.

BOS500   CAS: 107-88-0   **HR: 1**
1,3-BUTANEDIOL
mf: $C_4H_{10}O_2$ mw: 90.14

PROP: Viscous liquid. Bp: 207.5°, fp: $<-50°$, flash p: 250°F, d: 1.006 @ 20°/20°, autoign temp: 741°F, vap press: 0.06 mm @ 20°, vap d: 3.2.

SYNS: 1,3-BUTANDIOL (GERMAN) * BUTANE-1,3-DIOL * β-BUTYLENE GLYCOL * 1,3-BUTYLENE GLYCOL (FCC) * 1,3-DIHYDROXYBUTANE * METHYLTRIMETHYLENE GLYCOL

CONSENSUS REPORTS: Reported in EPA TSCA Inventory.

SAFETY PROFILE: Mildly toxic by ingestion and subcutaneous routes. An eye irritant. Combustible when exposed to heat or flame. Incompatible with oxidizing materials. To fight fire, use foam, alcohol foam, CO_2, dry chemical. When heated to decomposition it emits acrid smoke and irritating fumes.

BOS750   CAS: 110-63-4   **HR: 2**
1,4-BUTANEDIOL
mf: $C_4H_{10}O_2$ mw: 90.14

PROP: Nearly odorless, colorless, viscid liquid. Bp: 228°, fp: 20.9°, flash p: 250°F (OC), d: 1.0154 @ 25°/4°, vap d: 3.1.

SYNS: BUTANE-1,4-DIOL * 1,4-BUTYLENE GLYCOL * 1,4-DIHYDROXYBUTANE * 1,4-TETRAMETHYLENE GLYCOL

CONSENSUS REPORTS: Reported in EPA TSCA Inventory.

SAFETY PROFILE: Moderately toxic by ingestion and intraperitoneal routes. Combustible when exposed to heat or flame. To fight fire, use alcohol foam, mist, foam, CO_2, dry chemical. Incompatible with oxidizing materials. When heated to decomposition it emits acrid smoke and fumes.

BOT000   CAS: 513-85-9   **HR: 1**
2,3-BUTANEDIOL
mf: $C_4H_{10}O_2$ mw: 90.14

PROP: Colorless liquid or solid. Bp: 180°, fp: 19°, flash p: 185°F (TOC), d: 1.0095 @ 20°/20°, autoign temp: 756°F, vap press: 0.17 mm @ 20°, vap d: 3.1.

SYNS: 2,3-BUTYLENE GLYCOL * 2,3-DIHYDROXY-BUTANE * DIMETHYLENE GLYCOL

CONSENSUS REPORTS: Reported in EPA TSCA Inventory.

SAFETY PROFILE: Mildly toxic by ingestion. Flammable when exposed to heat or flame. Incompatible with oxidizing materials. To fight fire, use alcohol foam, CO_2, dry chemical. When heated to decomposition it emits acrid smoke and fumes.

BOT250   CAS: 55-98-1   **HR: 3**
1,4-BUTANEDIOL DIMETHYL SULFONATE
mf: $C_6H_{14}O_6S_2$ mw: 246.32

PROP: White crystals. Mp: 114-118°.

SYNS: 1,4-BIS(METHANESULFONOXY)BUTANE * (1,4-BIS(METHANESULFONYLOXY)BUTANE) * BISULFAN * BISULPHANE * 1,4-BUTANEDIOL DIMETHANESULPHONATE * BUZULFAN * C.B. 2041 * CITOSULFAN * 1,4-DIMESYLOXYBUTANE * 1,4-DIMETHANESULFONOXYBUTANE * 1,4-DI-(METHANESULFONYLOXY)BUTANE * 1,4-DIMETH-ANESULPHONYLOXYBUTANE * 1,4-DIMETHYLSUL-FONOXYBUTANE * GT41 * GT 2041 * LEUCO-SULFAN * MABLIN * METHANESULFONIC ACID TETRAMETHYLENE ESTER * MIELUCIN * MI-SULBAN * MITOSTAN * MYELOLEUKON * MYLERAN * NCI-C01592 * NSC-750 * SULPHABUTIN * TETRAMETHYLENE BIS(METHANESULFONATE) * TETRAMETHYLENE DIMETHANE SULFONATE * X 149

CONSENSUS REPORTS: IARC Cancer Review: GROUP 1 IMEMDT 7,137,87; Animal Inadequate Evidence IMEMDT 4,247,74; Human Inadequate Evidence IMEMDT 4,247,74. NTP Fourth Annual Report On Carcinogens, 1984. EPA Genetic Toxicology Program.

SAFETY PROFILE: Confirmed carcinogen producing leukemia, kidney, and uterine tumors. Experimental neoplastigenic and tumorigenic data. Poison by ingestion, subcutaneous, intraperitoneal, intravenous, and possibly other routes. Ingestion by pregnant women can cause cancer of the reproductive system of the fetus including the uterus. Human teratogenic effects by ingestion and possibly other routes include developmental abnormalities of the eye, ear, craniofacial area including the nose and tongue, gastrointestinal system, endocrine system, urogenital system, and other unspecified areas. Other human reproductive effects by ingestion and possibly other routes include: impotence; changes in the uterus, cervix, and vagina; and menstrual cycle disorders. Experimental reproductive effects. Human mutation data reported. When heated to decomposition it emits toxic fumes of SO_x.

BOT500   CAS: 431-03-8   **HR: 3**
2,3-BUTANEDIONE

DOT: UN 2346
mf: $C_4H_6O_2$ mw: 86.10

PROP: Greenish-yellow liquid; strong odor. Bp: 88°, flash p: 80°F, d: 0.9904 @ 15°/15°, refr index: 1.393-1.397, vap d: 3.00.Misc in alc, fixed oils, propylene glycol; sol in glycerin, water.

SYNS: BIACETYL * DIACETYL (FCC) * 2,3-DIKE-TOBUTANE * DIMETHYL DIKETONE * DIMETHYL-GLYOXAL * FEMA No. 2370

CONSENSUS REPORTS: Reported in EPA TSCA Inventory.

DOT Classification: Flammable Liquid; Label: Flammable Liquid.

SAFETY PROFILE: A poison by intraperitoneal route. Moderately toxic by ingestion. A skin irritant. Human mutation data reported. Flammable liquid. Dangerous fire hazard when exposed to heat or flame. To fight fire, use alcohol foam, CO_2, dry chemical. When heated to decomposition it emits acrid smoke and fumes.

BOU250   CAS: 1633-83-6   **HR: 3**
BUTANE SULTONE
mf: $C_4H_8O_3S$ mw: 136.18

SYNS: BUTANESULFONE * Δ-BUTANE SULTONE * 1,4-BUTANESULTONE (MAK) * 1,4-BUTYLENE SULFONE * Δ-VALEROSULTONE

CONSENSUS REPORTS: EPA Genetic Toxicology Program. Reported in EPA TSCA Inventory.

DFG MAK: Suspected Carcinogen.

SAFETY PROFILE: Suspected carcinogen with experimental tumorigenic data. Poison by subcutaneous, intravenous, and intraperitoneal routes. Moderately toxic by ingestion. Human mutation data reported. When heated to decomposition it emits toxic fumes of SO_x.

BOV000   CAS: 96-48-0   **HR: 3**
4-BUTANOLIDE
mf: $C_4H_6O_2$ mw: 86.10

PROP: Colorless liquid; mild caramel odor. Mp: $-44°$, bp: 206°, flash p: 209°F (OC), d: 1.124 @ 25°/4°, refr index: 1.434-1.454 @ 25°, vap d: 3.0.

SYNS: γ-6480 * γ-BL * BLO * BLON * BUTYRIC ACID LACTONE * γ-BUTYROLACTONE (FCC) * α-BUTYROLACTONE * BUTYRYL LACTONE * 4-DEOXYTETRONIC ACID * DIHYDRO-2(3H)-FURANONE * FEMA No. 3291 * 4-HYDROXY-BUTANOIC ACID LACTONE * γ-HYDROXYBUTYRIC ACID CYCLIC ESTER * 4-HYDROXYBUTYRIC ACID γ-LACTONE * γ-HYDROXYBUTYROLACTONE * NCI-C55878 * TETRAHYDRO-2-FURANONE

CONSENSUS REPORTS: IARC Cancer Review: GROUP 3 IMEMDT 7,56,87; Animal No Evidence IMEMDT 11,231,76. EPA Genetic Toxicology Program. Reported in EPA TSCA Inventory.

SAFETY PROFILE: Moderately toxic by ingestion, intravenous, and intraperitoneal routes. Experimental reproductive effects. Questionable carcinogen with experimental tumorigenic data by skin contact. Mutation data reported. Less acutely toxic than β-propiolactone. Combustible when exposed to heat or flame; can react with oxidizing materials. To fight fire, use foam, alcohol foam, CO_2, dry chemical. Potentially explosive reaction with butanol + 2,4-dichlorophenol + sodium hydroxide. When heated to decomposition it emits acrid and irritating fumes.

BOV750   CAS: 129-18-0   **HR: 3**
BUTAZOLIDINE SODIUM
mf: $C_{19}H_{20}N_2O_2 \cdot Na$ mw: 331.40

SYNS: 4-BUTYL-1,2-DIPHENYL-3,5-PYRAZOLIDINE-DIONE SODIUM SALT * 3,5-DIOXO-1,2-DIPHENYL-4-N-BUTYLPYRAZOLIDIN SODIUM * DIPHENYLDIOXOBU-TYLPYRAZOLIDINE-BUTAZOLIDINE-SODIUM * PHENYLBUTAZONE SODIUM * SODIUM BU-TAZOLIDINE * SODIUM PHENYLBUTAZONE * SODIUM SALT of PHENYLBUTAZONE

SAFETY PROFILE: A human poison by ingestion. Human systemic effects by ingestion: respiratory system damage, agranulocytosis, and dermatitis. An experimental poison via subcutaneous, intravenous, and intraperitoneal routes. An anti-inflammatory drug. When heated to decomposition it emits toxic fumes of NO_x and Na_2O.

BOW250   CAS: 25167-67-3   **HR: 3**
1-BUTENE
DOT: UN 1012
mf: C_4H_8 mw: 56.11

PROP: A colorless, flammable gas; sltly aromatic odor. Bp: $-6.3°$, fp: $-185.3°$, lel: 1.6%, uel: 9.3% flash p: $-80°$ ($-112°F$), d: 0.668 @ 0°/1°, vap d: 1.93, vap press: 3480 mm @ 21°, autoign temp: 723°F.

SYNS: BUTYLENE * α-BUTYLENE

CONSENSUS REPORTS: Reported in EPA TSCA Inventory.

DOT Classification: Flammable Gas; Label: Flammable Gas.

SAFETY PROFILE: A simple asphyxiant. Very dangerous fire hazard when exposed to heat, flame, or oxidizers. To fight fire, stop flow of gas. Moderately explosive when exposed to flame. Mixtures with aluminum tetrahydroborate explode after an induction period. When heated to decomposition it emits acrid smoke and fumes.

BOW500   **HR: 3**
cis-2-BUTENE
mf: C_4H_8 mw: 56.11

PROP: Colorless, flammable gas; sltly aromatic odor. Bp: 1°, fp: $-139°$, flash p: $-100°F$, d: 0.627 @ 15.5°/15.5°, vap press: 1410 mm @ 21°, autoign temp: 615°F, lel: 1.7%, uel: 9.0%, vap d: 1.9.

SYNS: DIMETHYLETHYLENE * PSEUDO-BUTYLENE

SAFETY PROFILE: A simple asphyxiant. Very dangerous fire hazard when exposed to heat or flame. Very likely to explode. Incompatible with oxidizing materials. To fight fire, stop flow of gas. When heated to decomposition it emits acrid smoke and fumes.

BOW750   **HR: 3**
trans-2-BUTENE
mf: C_4H_8 mw: 56.11

PROP: A colorless, flammable gas; sltly aromatic odor. Bp: 2.5°, fp: −105.6°, flash p: −100°F, d: 0.613 @ 15.5°/15.5° vap d: 1.95, vap press: 1592 mm @ 21°, autoign temp: 615 F, lel: 1.8%, uel: 9.7%, vap d: 1.9.

CONSENSUS REPORTS: EPA Extremely Hazardous Substances List.

SAFETY PROFILE: A simple asphyxiant. Very dangerous fire hazard when exposed to heat or flame. Very likely to explode. To fight fire, stop flow of gas. Incompatible with oxidizing materials. When heated to decomposition it emits acrid smoke and fumes.

BOY000   CAS: 6117-91-5   **HR: 2**
2-BUTEN-1-OL
mf: C_4H_8O mw: 72.12

PROP: Colorless liquid. Mp: <30°, bp: 118°, flash p: 92°F, d: 0.8726 @ 0°/4°, vap d: 2.49.

SYNS: 2-BUTENOL * 2-BUTENYL ALCOHOL * CROTONYL ALCOHOL * CROTYL ALCOHOL

CONSENSUS REPORTS: Reported in EPA TSCA Inventory.

SAFETY PROFILE: Moderately toxic by ingestion and skin contact. Mildly toxic by inhalation. Mutation data reported. Dangerous fire hazard when exposed to heat or flame; can react with oxidizing materials. To fight fire, use alcohol foam, CO_2, dry chemical. When heated to decomposition it emits acrid smoke and fumes.

BOY500   CAS: 78-94-4   **HR: 3**
3-BUTEN-2-ONE
mf: C_4H_6O mw: 70.10

DOT: UN 1251

PROP: Colorless liquid, powerfully irritating odor. Bp: 81.4°, flash p: 20°F (CC), d: 0.8393 @ 25°/4°, vap d: 2.41.

SYNS: ACETYL ETHYLENE * 3-BUTENE-2-ONE * METHYLENE ACETONE * METHYL-VINYL-CE-

TONE (FRENCH) * METHYLVINYLKETON (GERMAN) * METHYL VINYL KETONE * γ-OXO-α-BUTYLENE * VINYL METHYL KETONE

CONSENSUS REPORTS: Reported in EPA TSCA Inventory. EPA Extremely Hazardous Substances List.

DOT Classification: Flammable Liquid; Label: Flammable Liquid.

SAFETY PROFILE: Poison by ingestion, inhalation, and intraperitoneal routes. A severe irritant to skin, eyes, and mucous membranes. A lachrymator. Mutation data reported. Dangerous fire hazard when exposed to heat, flame or oxidizers. To fight fire, use CO_2, dry chemical. When heated to decomposition it emits acrid smoke and fumes.

BPE109   CAS: 689-97-4   **HR: 3**
BUTEN-3-YNE
mf: C_4H_4 mw: 52.08

PROP: Flash p: < −5°, lel: 2%, uel: 100% d: 0.68 @ 1.7 atm, vap d: 1.8, bp: 11°.

SYN: VINYL ACETYLENE

SAFETY PROFILE: Forms explosive peroxides with air or oxygen. Very exothermic decomposition when heated. Reacts explosively when heated with 1,3-butadiene or oxygen. Reacts with silver nitrate to form the explosive silver buten-3-ynide. When heated to decomposition it emits acrid smoke and irritating fumes.

BPF000   CAS: 125-40-6   **HR: 3**
BUTISOL
mf: $C_{10}H_{16}N_2O_3$ mw: 212.28

SYNS: BUTABARB * BUTABARBITAL * BUTABARBITONE * BUTATAB * BUTATAL * BUTICAPS * BUTRATE * 5-sec-BUTYL-5-ETHYLBARBITURIC ACID * 5-sec-BUTYL-5-ETHYLMALONYL UREA * 5-ETHYL-5-(1-METHYLPROPYL)BARBITURATE * 5-ETHYL-5-(1-METHYLPROPYL)BARBITURIC ACID * 5-ETHYL-5-(1-METHYLPROPYL)-2,4,6(1H,3H,5H)-PYRIMIDINETRIONE (9CI) * MEDARSED * NILOX * SECBUBARBITAL * SECBUTABARBITAL * SECBUTOBARBITONE * UNICELLES

SAFETY PROFILE: Poison by ingestion, intravenous, intraperitoneal, and subcutaneous routes. A central nervous system depressant. When heated to decomposition it emits toxic fumes of NO_x.

BPF500   CAS: 77-28-1   **HR: 3**
BUTOBARBITAL
mf: $C_{10}H_{16}N_2O_3$ mw: 212.28

SYNS: BUDORM * BUTETHAL * BUTOBARBI-
TONE * BUTOBARBITURAL * 5-BUTYL-5-ETHYL-
BARBITURIC ACID * 5-BUTYL-5-ETHYL-2,4,-
6(1H,3H,5H)-PYRIMIDINETRIONE (9CI) * 5-ETHYL-5-N-
BUTYLBARBITURIC ACID * ETOVAL * HYPERBU-
TAL * LONGANOCT * MEONAL * MONODORM
* NEONAL * SONERILE * SONERYL

SAFETY PROFILE: A poison by ingestion, intraperitoneal, subcutaneous, and intravenous
routes. Experimental teratogenic and reproductive effects. Human systemic effects by ingestion: changes in motor activity, coma, and nausea or vomiting. A central nervous system
depressant. When heated to decomposition it
emits toxic fumes of NO_x.

BPG000   CAS: 126-22-7   **HR: 3**
BUTONATE
mf: $C_8H_{14}Cl_3O_5P$ mw: 327.54

SYNS: BUTANOIC ACID 2,2,2-TRICHLORO-1-(DIME-
THOXYPHOSPHINYL)ETHYL ESTER * BUTILCHLORO-
FOS * DIMETHOXY-2,2,2-TRICHLORO-1-N-BUTYRY-
LOXY-ETHYLPHOSPHINE OXIDE * O,O-DIMETHYL-(1-
BUTYRYLOXY-2,2,2-TRICHLOROETHYL) PHOSPHONATE
* O,O-DIMETHYL 2,2,2-TRICHLORO-1-(N-BUTYRYL-
OXY)ETHYLPHOSPHONATE * ENT 20,852 * F-139
* T-113 * TRIBUFON

SAFETY PROFILE: Poison by ingestion. Moderately toxic by skin contact, intraperitoneal,
subcutaneous and possibly other routes. Mutation data reported. When heated to decomposition it emits highly toxic fumes of PO_x and
Cl^-.

BPG250   CAS: 6365-83-9   **HR: 3**
BUTOPHEN
mf: $C_{10}H_{12}N_2O_5 \cdot H_3N$ mw: 257.28

SYNS: 2-sec-BUTYL-4,6-DINITROPHENOL AMMONIUM
SALT * CHEMOX SELECTIVE * 4,6-DINITRO-2-sec.
BUTYLFENOLATE AMMONY (CZECH) * 4,6-DINITRO-
o-sec-BUTYLPHENOL AMMONIUM SALT * 4,6-DINI-
TRO-2-sec-BUTYLPHENOL AMMONIUM SALT * DINO-
SEB (AMINE) * DNBP AMMONIUM SALT * DOW
SELECTIVE * 2-(1-METHYL-N-PROPYL) 4,6-DINI-
TROPHENOL AMMONIUM SALT * SELECTIVE
* SINOX W

SAFETY PROFILE: A poison by ingestion and
skin contact. A severe eye irritant. When heated

to decomposition it emits very toxic fumes of
NH_3 and NO_x.

BPJ850   CAS: 111-76-2   **HR: 3**
2-BUTOXYETHANOL

DOT: UN 2369
mf: $C_6H_{14}O_2$ mw: 118.20

PROP: Clear, mobile liquid; pleasant odor. Bp:
168.4-170.2°, fp: −74.8°, flash p: 160°F
(COC), d: 0.9012 @ 20°/20°, vap press: 300
mm @ 140°.

SYNS: BUCS * BUTOKSYETYLOWY ALKOHOL
(POLISH) * 2-BUTOSSI-ETANOLO (ITALIAN)
* 2-BUTOXY-AETHANOL (GERMAN) * BUTOXYETH-
ANOL * 2-BUTOXY-1-ETHANOL * n-BUTOXYETHA-
NOL * BUTYL CELLOSOLVE * o-BUTYL ETHYL-
ENE GLYCOL * BUTYL GLYCOL * BUTYLGLYCOL
(FRENCH, GERMAN) * BUTYL OXITOL * DOW-
ANOL EB * EKTASOLVE EB * ETHYLENE GLYCOL-
n-BUTYL ETHER * ETHYLENE GLYCOL MONOBUTYL
ETHER (MAK, DOT) * GAFCOL EB * GLYCOL BU-
TYL ETHER * GLYCOL ETHER EB * GLYCOL
ETHER EB ACETATE * GLYCOL MONOBUTYL ETHER
* JEFFERSOL EB * MONOBUTYL GLYCOL ETHER
* 3-OXA-1-HEPTANOL * POLY-SOLV EB

CONSENSUS REPORTS: Reported in EPA
TSCA Inventory. Glycol ethers are on the Community Right-To-Know List.

OSHA PEL: (Transitional: TWA 50 ppm (skin))
  TWA 25 ppm (skin)
ACGIH TLV: TWA 25 ppm (skin)
DFG MAK: 20 ppm (100 mg/m^3)
DOT Classification: Poison B; Label: St. An-
  drews Cross, Flammable Liquid.

SAFETY PROFILE: Poison via ingestion, skin
contact, intraperitoneal, and intravenous routes.
Moderately toxic via inhalation and subcutaneous routes. Human systemic effects by inhalation: nausea or vomiting, headache, nose tumors, unspecified eye effects. Experimental
teratogenic and reproductive effects. A skin and
eye irritant. Flammable liquid when exposed
to heat or flame. To fight fire, use foam, CO_2,
and dry chemical. Incompatible with oxidizing
materials, heat, and flame. When heated to decomposition it emits acrid smoke and irritating
fumes.

BPK250   CAS: 78-51-3   **HR: 3**
2-BUTOXYETHANOL PHOSPHATE
mf: $C_{18}H_{39}O_7P$ mw: 398.54

PROP: Light-colored liquid, butyl-like odor. Mp: −70°; bp: 200-230° @ 4 mm, flash p: 435°F, d: 1.02 @ 20°/20°, vap press: 0.03 mm @ 150°, vap d: 13.8.

SYNS: KP 140 * KRONITEX KP-140 * PHOSFLEX T-BEP * TBEP * TRI(2-BUTOXYETHANOL PHOSPHATE) * TRIBUTOXYETHYL PHOSPHATE * TRI(2-BUTOXYETHYL) PHOSPHATE * TRIBUTYL CELLOSOLVE PHOSPHATE * TRIS(2-BUTOXYETHYL) ESTER PHOSPHORIC ACID * TRIS(2-BUTOXYETHYL) PHOSPHATE

CONSENSUS REPORTS: Reported in EPA TSCA Inventory.

SAFETY PROFILE: A poison by intravenous route. Moderately toxic by ingestion. Combustible when exposed to heat or flame. Dangerous; can react with oxidizing materials. To fight fire, use water, foam, CO_2, dry chemical. When heated to decomposition it emits toxic fumes of PO_x.

BPL250   CAS: 112-56-1   **HR: 3**
2-(2-BUTOXY ETHOXY)ETHYL THIOCYANATE
mf: $C_9H_{17}NO_2S$ mw: 203.33

SYNS: 2-(2-(BUTOXY)ETHOXY)ETHYL THIOCYANIC ACID ESTER * BUTOXYRHODANODIETHYL ETHER * β-BUTOXY-β′-THIOCYANODIETHYL ETHER * 2-BUTOXY-2′-THIOCYANODIETHYL ETHER * 1-BUTOXY-2-(2-THIOCYANOETHOXY)ETHANE * 1-BUTOXY-2-(2-THIOCYANATOETHYXY)ENTHANE * BUTYL CARBITOL RHODANATE * BUTYL CARBITOL THIOCYANATE * ENT 6 * ETHANOL-2-(2-BUTOXYETHOXY) THIOCYANATE * LETHANE * LETHANE 384 * LETHANE 384 REGULAR

SAFETY PROFILE: A poison by ingestion, skin contact, intraperitoneal, subcutaneous, and intravenous routes. Moderately toxic by an unspecified route. High concentrations can cause central nervous system depression. An insecticide. When heated to decomposition it emits very toxic fumes of SO_x, NO_x, and CN^-.

BPL500   CAS: 124-16-3   **HR: 2**
1-BUTOXY ETHOXY-2-PROPANOL
mf: $C_9H_{20}O_3$ mw: 176.29

PROP: Sol in water. D: 0.9310 @ 20°/20°, bp: 230.3°, fp: −90°, flash p: 250°F (OC).

SYN: 1-(2-BUTOXYETHOXY)-2-PROPANOL

CONSENSUS REPORTS: Reported in EPA TSCA Inventory.

SAFETY PROFILE: Moderately toxic by ingestion and skin contact. A skin and eye irritant. Combustible when exposed to heat or flame. To fight fire, use alcohol foam, dry chemical, spray, or mist. When heated to decomposition it emits acrid and irritating fumes.

BPM000   CAS: 112-07-2   **HR: 2**
2-BUTOXYETHYL ACETATE
mf: $C_8H_{16}O_3$ mw: 160.24

PROP: Colorless liquid; fruity odor. Bp: 192.3°, d: 0.9424 @ 20°/20°, fp: −63.5°, flash p: 190°F. Sol in hydrocarbons and organic solvents; insol in water.

SYNS: 2-BUTOXYETHANOL ACETATE * 2-BUTOXYETHYL ESTER ACETIC ACID * BUTYL CELLOSOLVE ACETATE * EKTASOLVE EB ACETATE * ETHYLENE GLYCOL MONOBUTYL ETHER ACETATE (MAK) * GLYCOL MONOBUTYL ETHERACETATE

CONSENSUS REPORTS: Reported in EPA ISCA Inventory. Glycol ethers are on the Community Right-To-Know List.

DFG MAK: 20 ppm (135 mg/m^3)

SAFETY PROFILE: Moderately toxic by ingestion and skin contact. Mild skin irritant. Flammable when exposed to heat, flame, or oxidizers. To fight fire, use alcohol foam. When heated to decomposition it emits acrid smoke and irritating fumes.

BPP750   CAS: 2438-72-4   **HR: 2**
p-BUTOXY PHENYL ACETO-HYDROXAMIC ACID
mf: $C_{12}H_{17}NO_3$ mw: 223.30

SYNS: BUFEXAMIC ACID * 4-BUTOXYPHENYLACETOHYDROXAMIC ACID * CP 1044 J3 * DROXAROL * DROXARYL * FLOGICID * FLOGOCID N PLASTIGEL * J3 * PARFENAC * PARFENAL

CONSENSUS REPORTS: EPA Genetic Toxicology Program.

SAFETY PROFILE: Moderately toxic by ingestion and intraperitoneal routes. Experimental teratogenic and reproductive effects. Mutation data reported. When heated to decomposition it emits toxic fumes of NO_x.

BPR500   CAS: 536-43-6   **HR: 3**
4'-BUTOXY-3-PIPERIDINO
PROPIOPHENONE HYDROCHLORIDE
mf: $C_{18}H_{27}NO_2 \cdot ClH$ mw: 325.92

SYNS: 1-(2-(4-BUTOXYBENZOYL)ETHYL)PIPERIDINE
HYDROCHLORIDE * 4-n-BUTOXY-β-(1-PIPERIDYL)PRO-
PIOPHENONE HYDROCHLORIDE * DICLONIA
* DYCLOCAINUM * DYCLONE HYDROCHLORIDE
* DYCLONINE HYDROCLORIDE * DYCLOTHANE
* P-267 * S 154

SAFETY PROFILE: A poison by ingestion, intraperitoneal, subcutaneous, and intravenous
routes. A skin and eye irritant. When heated
to decomposition it emits very toxic fumes of
HCl and NO_x.

BPU000   CAS: 16227-10-4   **HR: 3**
BUTRIZOL
mf: $C_6H_{11}N_3$ mw: 125.20

SYNS: BT * 4-N-BUTYL-4H-1,2,4-TRIAZOLE
* 4-BUTYL-s-TRIAZOLE * DITHANE R-24
* INDAR * RH-124

SAFETY PROFILE: A poison by ingestion and
skin contact. When heated to decomposition it
emits toxic fumes of NO_x.

BPU750   CAS: 123-86-4   **HR: 2**
n-BUTYL ACETATE
DOT: UN 1123
mf: $C_6H_{12}O_2$ mw: 116.18

PROP: Colorless liquid; strong fruity odor. Bp:
126°, fp: −73.5°, ULC: 50-60, lel: 1.4%, uel:
7.5°, flash p: 72°F, d: 0.88 @ 20°/20°, refr
index: 1.393-1.396, autoign temp: 797°F, vap
press: 15 mm @ 25°. Misc with alc, ether,
and propylene glycol; sltly sol in water.

SYNS: ACETATE de BUTYLE (FRENCH) * ACETIC
ACID n-BUTYL ESTER * BUTILE (ACETATI di) (ITAL-
IAN) * BUTYLACETAT (GERMAN) * BUTYL ACE-
TATE * 1-BUTYL ACETATE * BUTYLACETATEN
(DUTCH) * BUTYLE(ACETATEde)(FRENCH) * BU-
TYL ETHANOATE * FEMA No. 2174 * OCTAN
n-BUTYLU (POLISH)

CONSENSUS REPORTS: Reported in EPA
TSCA Inventory.

OSHA PEL: (Transitional: TWA 150 ppm)
TWA 150 ppm; STEL 200 ppm

ACGIH TLV: TWA 150 ppm; STEL 200 ppm
DFG MAK: 200 ppm (950 mg/m^3)
DOT Classification: Flammable Liquid; Label:
Flammable Liquid.

SAFETY PROFILE: Moderately toxic by intraperitoneal route. Mildly toxic by inhalation and
ingestion. An experimental teratogen. A skin
and severe eye irritant. Human systemic effects
by inhalation: conjunctiva irritation, unspecified
nasal and respiratory system effects. A mild
allergen. High concentrations are irritating to
eyes and respiratory tract and cause narcosis.
Evidence of chronic systemic toxicity is incon-
clusive. Flammable liquid. Moderately explosive when exposed to flame. Ignites on contact
with potassium-tert-butoxide. To fight fire, use
alcohol foam, CO_2, dry chemical. When heated
to decomposition it emits acrid and irritating
fumes.

BPV000   CAS: 105-46-4   **HR: 2**
sec-BUTYL ACETATE
DOT: UN 1123
mf: $C_6H_{12}O_2$ mw: 116.18

PROP: Colorless liquid, mild odor. Bp: 112°,
flash p: 18°, d: 0.862-0.866 @ 20°/20°, vap
d: 4.00. lel: 1.3%, uel: 7.5%.

SYNS: ACETATE de BUTYLE SECONDAIRE (FRENCH)
* ACETIC ACID-2-BUTOXY ESTER * ACETIC ACID-1-
METHYLPROPYL ESTER (9CI) * 2-BUTANOL ACETATE
* 2-BUTYL ACETATE * sec-BUTYL ALCOHOL ACE-
TATE

CONSENSUS REPORTS: Reported in EPA
TSCA Inventory.

OSHA PEL: TWA 200 ppm
ACGIH TLV: TWA 200 ppm
DFG MAK: 200 ppm (950 mg/m^3)
DOT Classification: Flammable Liquid; Label:
Flammable Liquid.

SAFETY PROFILE: An irritant and allergen.
Flammable. To fight fire, use alcohol foam,
CO_2, dry chemical. When heated to decomposition it emits acrid and irritating fumes.

BPV100   CAS: 540-88-5   **HR: 3**
tert-BUTYL ACETATE
DOT: UN 1123
mf: $C_6H_{12}O_2$ mw: 116.18

SYNS: ACETIC ACID-tert-BUTYL ESTER * ACETIC ACID-1,1-DIMETHYLETHYL ESTER * TEXACO LEAD APPRECIATOR * TLA

CONSENSUS REPORTS: Reported in EPA TSCA Inventory.

OSHA PEL: TWA 200 ppm
ACGIH TLV: TWA 200 ppm
DFG MAK: 200 ppm (950 mg/m^3)
DOT Classification: IMO: Flammable Liquid; Label: Flammable Liquid.

SAFETY PROFILE: Poison by inhalation and ingestion. Flammable. To fight fire, use alcohol foam, CO_2, dry chemical. When heated to decomposition it emits acrid smoke and irritating fumes.

BPV250   CAS: 591-60-6   **HR: 1**
BUTYL ACETOACETATE
mf: $C_8H_{14}O_3$ mw: 158.22

PROP: Bp: 214°, flash p: 185°F, d: 0.96, vap d: 5.55.

SYNS: ACETOACETIC ACID BUTYL ESTER * 3-OXO-BUTANOIC ACID BUTYL ESTER

CONSENSUS REPORTS: Reported in EPA TSCA Inventory.

SAFETY PROFILE: Mildly toxic by ingestion. An eye irritant. Flammable. To fight fire, use alcohol foam, CO_2, dry chemical. When heated to decomposition it emits acrid and irritating fumes.

BPW100   CAS: 141-32-2   **HR: 2**
n-BUTYL ACRYLATE
DOT: UN 2348
mf: $C_7H_{12}O_2$ mw: 128.19

PROP: Water-white, extremely reactive monomer. Bp: 69° @ 50 mm, fp: −64.6°, flash p: 120°F (OC), d: 0.89 @ 25°/25°, vap press: 10 mm @ 35.5°, vap d: 4.42.

SYNS: ACRYLIC ACID BUTYL ESTER * ACRYLIC ACID n-BUTYL ESTER (MAK) * BUTYL ACRYLATE * BUTYLACRYLATE, INHIBITED (DOT) * BUTYL-2-PROPENOATE

CONSENSUS REPORTS: IARC Cancer Review: GROUP 3 IMEMDT 7,56,87, Animal Inadequate Evidence IMEMDT 39,67,86. Re-

ported in EPA TSCA Inventory. Community Right-To-Know List.

OSHA PEL: TWA 10 ppm
ACGIH TLV: TWA 10 ppm
DFG MAK: 10 ppm (55 mg/m^3)
DOT Classification: Flammable or Combustible Liquid; Label: Flammable Liquid.

SAFETY PROFILE: Moderately toxic by ingestion, inhalation, skin contact, and intraperitoneal routes. Experimental reproductive effects. A skin and eye irritant. Flammable when exposed to heat or flame. To fight fire, use foam, CO_2, dry chemical. Incompatible with oxidizing materials. When heated to decomposition it emits acrid and irritating fumes.

BPW500   CAS: 71-36-3   **HR: 3**
n-BUTYL ALCOHOL
DOT: UN 1120
mf: $C_4H_{10}O$ mw: 74.14

PROP: Colorless liquid; vinous odor. Bp: 117.5°, ULC: 40, lel: 1.4%, uel: 11.2%, fp: −88.9°, flash p: 95-100°F, d: 0.80978 @ 20°/4°, autoign temp: 689°F, vap press: 5.5 mm @ 20°, vap d: 2.55. Misc in alc, ether, and organic solvents; sltly sol in water.

SYNS: ALCOOL BUTYLIQUE (FRENCH) * BUTANOL (FRENCH) * 1-BUTANOL * n-BUTANOL * BUTAN-1-OL * BUTANOL (DOT) * BUTANOLEN (DUTCH) * BUTANOLO (ITALIAN) * BUTYL ALCOHOL (DOT) * BUTYL HYDROXIDE * BUTYLOWY ALKOHOL (POLISH) * BUTYRIC or NORMAL PRIMARY BUTYL ALCOHOL * CCS 203 * FEMA No. 2178 * 1-HYDROXYBUTANE * METHYLOLPROPANE * PROPYLCARBINOL * PROPYLMETHANOL * RCRA WASTE NUMBER U031

CONSENSUS REPORTS: Community Right-To-Know List. EPA Genetic Toxicology Program. Reported in EPA TSCA Inventory.

OSHA PEL: (Transitional: TWA 100 ppm) CL 50 ppm (skin)
ACGIH TLV: CL 50 ppm (skin)
DFG MAK: 100 ppm (300 mg/m^3)
DOT Classification: Flammable or Combustible Liquid; Label: Flammable Liquid.

SAFETY PROFILE: A poison by intravenous route. Moderately toxic by skin contact, ingestion, subcutaneous, intraperitoneal, and possibly other routes. Human systemic effects by inhalation: conjunctiva irritation, unspecified respiratory system, and nasal effects. A skin and severe eye irritant. Though animal experiments have shown the butyl alcohols to possess toxic properties, they have produced few cases of poisoning in industry probably because of their low volatility. The use of normal butyl alcohol is reported to have resulted in irritation of the eyes, with corneal inflammation, slight headache and dizziness, slight irritation of the nose and throat, and dermatitis about the fingernails and along the side of the fingers. Keratitis has also been reported. Mutation data reported. Flammable liquid. Moderately explosive when exposed to flame. Incompatible with Al, chromium trioxide, oxidizing materials. To fight fire, use water spray, alcohol foam, CO_2, dry chemical. When heated to decomposition it emits acrid smoke and fumes.

BPW750   CAS: 78-92-2   HR: 3
sec-BUTYL ALCOHOL

DOT: UN 1120
mf: $C_4H_{10}O$ mw: 74.14

PROP: Colorless liquid. Mp: −89°, bp: 99.5°, flash p: 14°, d: 0.808 @ 20°/4°, autoign temp: 763°F, vap press: 10 mm @ 20°, vap d: 2.55, lel: 1.7% @ 212°F, uel: 9.8% @ 212°F.

SYNS: ALCOOL BUTYLIQUE SECONDAIRE (FRENCH) * BUTAN-2-OL * sec-BUTANOL (DOT) * 2-BUTA-NOL * BUTANOL SECONDAIRE (FRENCH) * 2-BU-TYL ALCOHOL * BUTYLENE HYDRATE * CCS 301 * ETHYLMETHYL CARBINOL * 2-HYDROXYBUTANE * METHYLETHYLCARBINOL * S.B.A.

CONSENSUS REPORTS: Community Right-To-Know List. Reported in EPA TSCA Inventory.

OSHA PEL: (Transitional: TWA 150 ppm) TWA 100 ppm
ACGIH TLV: TWA 100 ppm
DFG MAK: 100 ppm (300 mg/m³)
DOT Classification: Flammable or Combustible Liquid; Label: Flammable Liquid.

SAFETY PROFILE: Poison by intravenous and intraperitoneal routes. Mildly toxic by ingestion. An eye irritant. Mutation data reported. Dangerous fire hazard when exposed to heat or flame. Auto-oxidizes to an explosive peroxide. Ignites on contact with chromium trioxide. To fight fire, use water spray, alcohol foam, CO_2, dry chemical. Incompatible with oxidizing materi-

als. When heated to decomposition it emits acrid smoke and fumes.

BPX000   CAS: 75-65-0   HR: 2
tert-BUTYL ALCOHOL

DOT: UN 1120
mf: $C_4H_{10}O$ mw: 74.14

PROP: Colorless liquid or rhombic prisms or plates. Mp: 25.3°, bp: 82.8°, flash p: 50°F (CC), d: 0.7887 @ 20°/4°, autoign temp: 896°F, vap press: 40 mm @ 24.5°, vap d: 2.55, lel: 2.4%, uel: 8.0%.

SYNS: ALCOOL BUTYLIQUE TERTIAIRE (FRENCH) * tert-BUTANOL * BUTANOL TERTIAIRE (FRENCH) * tert-BUTYL HYDROXIDE * 1,1-DIMETHYLETHA-NOL * 2-METHYL-2-PROPANOL * NCI-C55367 * TRIMETHYLCARBINOL

CONSENSUS REPORTS: Community Right-To-Know List. Reported in EPA TSCA Inventory. EPA Genetic Toxicology Program.

OSHA PEL: (Transitional: TWA 100 ppm) TWA 100 ppm; STEL 150 ppm
ACGIH TLV: TWA 100 ppm; STEL 150 ppm
DFG MAK: 100 ppm (300 mg/m³)
DOT Classification: Flammable Liquid; Label: Flammable Liquid.

SAFETY PROFILE: Moderately toxic by ingestion, intravenous, and intraperitoneal routes. Experimental reproductive effects. Mutation data reported. Dangerous fire hazard when exposed to heat or flame. Moderately explosive in the form of vapor when exposed to flame. Ignites on contact with potassium-sodium alloys. To fight fire, use alcohol foam, CO_2, dry chemical. Incompatible with oxidizing materials, H_2O_2.

BPX750   CAS: 109-73-9   HR: 3
n-BUTYLAMINE

DOT: UN 1125
mf: $C_4H_{11}N$ mw: 73.16

PROP: Liquid, ammonia-like odor. Mp: −50°, bp: 77°, flash p: 10°F (OC), 10°F (CC), d: 0.74-0.76 @ 20°/20°, autoign temp: 594°F, vap d: 2.52, lel: 1.7%, uel: 9.8%.

SYNS: 1-AMINO-BUTAAN (DUTCH) * 1-AMINOBU-TAN (GERMAN) * 1-AMINOBUTANE * 1-BUTAN-AMINE * n-BUTILAMINA (ITALIAN) * n-BUTYLA-MIN (GERMAN) * MONO-n-BUTYLAMINE * NOR-VALAMINE

CONSENSUS REPORTS: Reported in EPA TSCA Inventory.

OSHA PEL: CL 5 ppm (skin)
ACGIH TLV: CL 5 ppm
DFG MAK: 5 ppm (15 mg/m^3)
DOT Classification: Flammable Liquid; Label: Flammable Liquid.

SAFETY PROFILE: Poison by ingestion, skin contact, and intravenous routes. Moderately toxic by inhalation, intraperitoneal, parenteral, and possibly other routes. A severe skin irritant. Questionable carcinogen with experimental tumorigenic data. Mutation data reported. Dangerous fire hazard when exposed to heat, flame or oxidizing materials. To fight fire, use alcohol foam, CO_2, dry chemical. Explodes on contact with perchloryl fluoride. When heated to decomposition it emits toxic fumes of NO_x.

BPY000 CAS: 13952-84-6 **HR: 3**
sec-BUTYLAMINE

DOT: UN 1125
mf: $C_4H_{11}N$ mw: 73.16

PROP: Liquid. Mp: −104°, bp: 63°, flash p: 15°F, d: 0.724 @ 20°.

SYNS: 2-AB * 2-AMINOBUTANE * BUTAFUME * 2-BUTANAMINE * DECCOTANE * FRUCOTE * 1-METHYLPROPYLAMINE * TUTANE

CONSENSUS REPORTS: Reported in EPA TSCA Inventory.

DFG MAK: 5 ppm (15 mg/m^3)
DOT Classification: Flammable Liquid; Label: Flammable Liquid.

SAFETY PROFILE: A poison by ingestion. A powerful irritant. Moderately toxic by skin contact. Dangerous fire hazard when exposed to heat or flame. To fight fire, use alcohol foam, water spray or mist, dry chemical. Incompatible with oxidizing materials. When heated to decomposition it emits toxic fumes of NO_x.

BPY250 CAS: 75-64-9 **HR: 3**
tert-BUTYLAMINE

DOT: UN 1125
mf: $C_4H_{11}N$ mw: 73.16

PROP: Colorless liquid. Mp: −67.5°, bp: 44-46°, d: 0.700 @ 15°, lel: 1.7% @ 212°F, uel: 8.9% @ 212°F, vap d: 2.5, autoign temp: 716°F.

SYNS: 2-AMINOISOBUTANE * 2-AMINO-2-METHYL-PROPANE * BUTYLAMINE, tertiary * 1,1-DIMETH-YLETHYLAMINE * TRIMETHYLAMINOMETHANE

CONSENSUS REPORTS: Reported in EPA TSCA Inventory.

DFG MAK: 5 ppm (15 mg/m^3)
DOT Classification: Flammable Liquid; Label: Flammable Liquid.

SAFETY PROFILE: Poison by ingestion. Very dangerous fire hazard when exposed to heat or flame. Very exothermic reaction with 2,2-dibromo-1,3-dimethylcyclopropanoic acid. To fight fire, use alcohol foam. When heated to decomposition it emits toxic fumes of NO_x.

BQA010 CAS: 94-24-6 **HR: 3**
p-(BUTYLAMINO)BENZOIC ACID-2-(DIMETHYLAMINO)ETHYL ESTER
mf: $C_{15}H_{24}N_2O_2$ mw: 264.41

SYNS: AMETHOCAINE * ANETAIN * p-BUTYL-AMINOBENZOYL-2-DIMETHYLAMINOETHANOL * CONTRALGIN * DICAIN * DICAINE * DI-KAIN * DIMETHYLAMINOETHYL-p-BUTYL-AMINO-BENZOATE * 2-DIMETHYLAMINOETHYL-p-BUTYL-AMINOBENZOATE * FISSUCAIN * INTERCAIN * LANDOCAINE * LAUDOCAINE * MEDICAINE * MEDIHALER-TETRACAINE * MEETHOBALM * METRASPRAY * MUCAESTHIN * NIPHANOID * PANTOCAINE * PONTOCAINE * REXOCAINE * TETRACAINE * UROMUCAESTHIN

SAFETY PROFILE: A human poison by parenteral route with systemic effects including: muscle contractions, coma, and cyanosis. A poison experimentally by intravenous, parenteral, intratracheal, intraperitoneal, and subcutaneous routes. Human mutation data reported. A local anesthetic. When heated to decomposition it emits toxic fumes of NO_x.

BQC000 CAS: 111-75-1 **HR: 2**
2-BUTYLAMINOETHANOL
mf: $C_6H_{15}NO$ mw: 117.22

PROP: Liquid. Bp: 192°, flash p: 170°F (OC), d: 0.89, vap d: 4.03.

SYN: 2-n-BUYTLAMINOETHANOL

CONSENSUS REPORTS: Reported in EPA TSCA Inventory.

SAFETY PROFILE: Moderately toxic by ingestion and intraperitoneal routes. A skin and severe eye irritant. Flammable when exposed to heat or flame. To fight fire, use alcohol foam, foam, CO_2, dry chemical. Incompatible with oxidizing materials. When heated to decomposition it emits toxic fumes of NO_x.

BQD250   CAS: 3775-90-4   **HR: 3**
tert-BUTYL AMINO ETHYL METHACRYLATE
mf: $C_{10}H_{19}NO_2$ mw: 185.30

PROP: Liquid; bp: 100-105°; d: 0.914. flash p: 205°F (OC).

SYNS: AGEFLEX FM-4   *   2-(tert-BUTYLAMINO)E-
THYL METHACRYLATE

CONSENSUS REPORTS: Reported in EPA TSCA Inventory.

SAFETY PROFILE: Poison by intraperitoneal route. Combustible when exposed to heat or flame. To fight fire, use alcohol foam, water spray or mist, dry chemical. When heated to decomposition it emits toxic fumes of NO_x.

BQF500   CAS: 18559-94-9   **HR: 3**
α′-((tert-BUTYL AMINO)METHYL)-4-HYDROXY-m-XYLENE-α,α′-DIOL
mf: $C_{13}H_{21}NO_3$ mw: 239.35

SYNS: AEORLIN   *   AH 3365   *   ALBUTEROL
*   BRONCOVALEAS   *   2-(tert-BUTYLAMINO)-1-(4-HY-
DROXY-3-HYDROXYMETHYLPHENYL)ETHANOL
*   α-1-((tert-BUTYLAMINO)METHYL)-4-HYDROXY-m-XY-
LENE-α-DIOL   *   α-1-(((1,1-DIMETHYLETHYL)AMINO)
METHYL)-4-HYDROXY-1,3-BENZENEDIMETHANOL
*   4-HYDROXY-3-HYDROXYMETHYL-α-((tert-BUTYL-
AMINO)METHYL)BENZYL ALCOHOL   *   PROVENTIL
*   SALBUTAMOL   *   SOLBUTAMOL   *   SULTANOL
*   VENETLIN   *   VENTOLIN

SAFETY PROFILE: A poison by intraperitoneal and intravenous routes. Moderately toxic by ingestion and subcutaneous routes. Human cardiovascular system effects by intravenous route including arrythmias, change in heart rate and plasma or blood volume. Human (child) behavioral and cardiac effects by ingestion including tremors, excitement and change in heart rate. Human maternal effects of the uterus, cervix and vagina by ingestion. An experimental teratogen. A bronchodilator. When heated to decomposition it emits toxic fumes of NO_x.

BQH250   CAS: 528-97-2   **HR: 3**
p-BUTYLAMINO SALICYLIC ACID-2-(DIETHYLAMINO)ETHYL ESTER HYDROCHLORIDE
mf: $C_{17}H_{28}N_2O_3 \cdot ClH$ mw: 344.93

SYNS: BRONCHIOCAIN   *   BRONCHOCAIN
*   BRONCHOCAINE   *   4-(BUTYLAMINO)SALICYLIC
ACID 2-(DIETHYLAMINO)ETHYL ESTER HYDROCHLO-
RIDE   *   4-(BUTYLAMINO)-SALICYLIC ACID 2-(DIETH-
YLAMINO)ETHYL ESTER MONOHYDROCHLORIDE
*   C 4208   *   HCl SALZ DES p,N,N-BUTYLAMINOSALI-
CYLSAEUREDIAETHYLAMINOAETHYLESTER (GERMAN)
*   PARAESIN   *   PHENOCAINE   *   S 650   *   SALI-
CYL-DIAETHYL (GERMAN)   *   WOFACAIN A

SAFETY PROFILE: A poison via intraperitoneal, subcutaneous, and intravenous routes. A severe eye irritant. When heated to decomposition, it emits very toxic fumes of NO_x and HCl.

BQH850   CAS: 1126-78-9   **HR: 3**
N-BUTYLANILINE
DOT: UN 2738
mf: $C_{10}H_{15}N$ mw: 149.26

SYNS: N-(n-BUTYL)ANILINE   *   N-n-BUTYLANILINE
(DOT)   *   N-BUTYLBENZENAMINE (9CI)   *   4-(PHENY-
LAMINO)BUTANE

CONSENSUS REPORTS: Reported in EPA TSCA Inventory.

DOT Classification: Poison B; Label: Poison.

SAFETY PROFILE: Poison by an unspecified route. Moderately toxic by skin contact and ingestion. A severe skin and eye irritant. When heated to decomposition it emits toxic fumes of NO_x.

BQI000   CAS: 25013-16-5   **HR: 3**
BUTYLATED HYDROXYANISOLE
mf: $C_{11}H_{16}O_2$ mw: 180.27

PROP: White waxy solid; faint characteristic odor. Mp: 48-63°. Sol in alc and propylene glycol; insol in water.

SYNS: ANTRANCINE 12   *   BHA (FCC)   *   BUTYL-
HYDROXYANISOLE   *   tert-BUTYLHYDROXYANISOLE
*   tert-BUTYL-4-HYDROXYANISOLE   *   2(3)-tert-BUTYL-
4-HYDROXYANISOLE   *   BUTYLOHYDROKSYANIZOL
(POLISH)   *   EMBANOX   *   FEMA No. 2183
*   NIPANTIOX 1-F   *   PREMERGE PLUS   *   SUS-
TANE   *   SUSTANE 1-F   *   TENOX BHA   *   VERTAC

CONSENSUS REPORTS: IARC Cancer Review: GROUP 2B IMEMDT 7,56,87; Animal Sufficient Evidence IMEMDT 40,123,86. Reported in EPA TSCA Inventory. EPA Genetic Toxicology Program.

SAFETY PROFILE: Suspected carcinogen with experimental carcinogenic, neoplastigenic, and tumorigenic data. Moderately toxic by ingestion and intraperitoneal routes. Experimental reproductive effects. Mutation data reported. When heated to decomposition it emits acrid and irritating fumes.

BQI250   CAS: 1070-19-5   **HR: 3**
tert-BUTYL AZIDO FORMATE
mf: $C_5H_9N_3O_2$ mw: 143.1

SYNS: tert-BUTOXY CARBONYL AZIDE * tert-BUTYLOXYCARBONYL AZIDE * CARBONAZIDIC ACID, 1,1-DIMETHYLETHYL ESTER

CONSENSUS REPORTS: Reported in EPA TSCA Inventory.

DOT Classification: Forbidden.

An unstable shock- and heat-sensitive explosive. It may explode above 100°C and ignites at 143°C. When heated to decomposition it emits toxic fumes of NO_x.

BQI750   CAS: 104-51-8   **HR: 1**
n-BUTYLBENZENE
DOT: UN 2709
mf: $C_{10}H_{14}$ mw: 134.24

PROP: Colorless liquid. Mp: −81.2°, bp: 182.1°, fp: −88.2°, flash p: 160°F (TOC), d: 0.8601 @ 20°/4°, vap press: 1 mm @ 22.7°, autoign temp: 774°F, lel: 0.8%, uel: 5.8%, vap d: 4.6.

SYN: 1-PHENYLBUTANE

CONSENSUS REPORTS: Reported in EPA TSCA Inventory.

DOT Classification: Flammable or Combustible Liquid; Label: Flammable Liquid.

SAFETY PROFILE: Mildly toxic by ingestion. Flammable when exposed to heat or flame. To fight fire, use alcohol foam, CO_2, dry chemical. Incompatible with oxidizing materials. When heated to decomposition it emits acrid and irritating fumes.

BQJ000   CAS: 135-98-8   **HR: 2**
sec-BUTYLBENZENE
DOT: UN 2709
mf: $C_{10}H_{14}$ mw: 134.24

PROP: Colorless liquid. Mp: −82.7°, bp: 173.5°, fp: −75.8°, flash p: 126°F (TOC), d: 0.8621 @ 20°, vap press: 1 mm @ 18.6°, vap d: 4.62, autoign temp: 788°F, lel: 0.8%, uel: 6.9%.

SYN: 2-PHENYLBUTANE

CONSENSUS REPORTS: Reported in EPA TSCA Inventory.

DOT Classification: Flammable or Combustible Liquid; Label: Flammable Liquid.

SAFETY PROFILE: Moderately toxic by ingestion. Flammable when exposed to heat or flame. To fight fire, use foam, CO_2, dry chemical, water spray or mist. Incompatible with oxidizing materials. When heated to decomposition it emits acrid smoke and fumes.

BQJ250   CAS: 98-06-6   **HR: 1**
tert-BUTYLBENZENE
DOT: UN 2709
mf: $C_{10}H_{14}$ mw: 134.24

PROP: Colorless liquid. Bp: 168.2°, fp: −58°, flash p: 140°F (TOC), d: 0.8665 @ 20°, vap press: 1 mm @ 13.0°, vap d: 4.62, autoign temp: 842°F, lel: 0.7% @ 212°F, uel: 5.7% @ 212°F.

SYNS: 2-METHYL-2-PHENYLPROPANE * PSEUDO-BUTYLBENZENE * TRIMETHYLPHENYLMETHANE

CONSENSUS REPORTS: Reported in EPA TSCA Inventory.

DOT Classification: Flammable or Combustible Liquid; Label: Flammable Liquid.

SAFETY PROFILE: Mildly toxic by ingestion. Flammable when exposed to heat or flame. To fight fire, use foam, CO_2, dry chemical, water spray, fog, mist. Incompatible with oxidizing materials. When heated to decomposition it emits acrid smoke and fumes.

BQK000   CAS: 14255-87-9   **HR: 2**
5-BUTYL-2-BENZIMIDAZOLECARBAMIC ACID METHYL ESTER
mf: $C_{13}H_{17}N_3O_2$ mw: 247.33

SYNS: N-(BUTYL-5-BENZIMIDAZOLYL)-2-CARBAMATE de METHYLE (FRENCH) * (4-BUTYL-1H-BENZIMID-AZOL-2-YL)-CARBAMIC ACID METHYL ESTER * 5-BUTYL-2-(CARBOMETHOXYAMINO)BENZIMID-AZOLE * HELMATAC * METHYL-5-BUTYL-2-BENZIMIDAZOLECARBAMATE * PARBENDAZOLE * PBDZ * SKF 29044 * VERMINUM * WORM GUARD

SAFETY PROFILE: Moderately toxic by ingestion. Experimental teratogenic and reproductive effects. Human mutation data reported. An anthelminthic agent. When heated to decomposition it emits toxic fumes of NO_x.

BQK250   CAS: 136-60-7   **HR: 2**
BUTYL BENZOATE
mf: $C_{11}H_{14}O_2$ mw: 178.25

PROP: Liquid. Mp: $-21.5°$, bp: $250°$, flash p: $225°F$ (OC), d: 1.0073 @ $20°/20°$, vap press: <0.01 mm @ $20°$, vap d: 6.15.

SYNS: ANTHRAPOLE AZ * BENZOIC ACID-n-BUTYL ESTER * n-BUTYL BENZOATE * DAI CARI XBN

CONSENSUS REPORTS: Reported in EPA TSCA Inventory.

SAFETY PROFILE: Moderately toxic by skin contact. Mildly toxic by ingestion. Severe skin irritant and moderate eye irritant. Combustible when exposed to heat or flame; can react with oxidizing materials. To fight fire, use CO_2, dry chemical, water mist, fog, spray. When heated to decomposition it emits acrid and irritating fumes.

BQK500   CAS: 98-73-7   **HR: 2**
p-tert-BUTYL BENZOIC ACID
mf: $C_{11}H_{14}O_2$ mw: 178.25

PROP: Colorless, fine, crystalline powder. Mp: 166.3°, d: 1.142 @ $20°/4°$.

SYN: TBBA

CONSENSUS REPORTS: Reported in EPA TSCA Inventory.

SAFETY PROFILE: Moderately toxic by ingestion. Experimental reproductive effects. An irritant. Combustible when exposed to heat or flame. Incompatible with oxidizing materials. To fight fire, use foam, CO_2, dry chemical. When heated to decomposition it emits acrid smoke and irritating fumes.

BQL000   CAS: 1190-53-0   **HR: 3**
N-BUTYLBIGUANIDE HYDROCHLORIDE
mf: $C_6H_{15}N_5 \cdot ClH$ mw: 193.72

SYNS: ANDERE * BIFORON * BIGUNAL * BUFONAMIN * BUFORMIN HYDROCHLORIDE * BULBONIN * 1-BUTYLBIGUANIDE HYDROCHLO-RIDE * N-BUTYLIMIDODICARBONIMIDIC DIAMIDE MONOHYDROCHLORIDE (9CI) * 1-BUTYLDIGUANIDE HYDROCHLORIDE * DIABRIN * DIBETOS * GLIBUTIDE * GLIPORAL * INSULAMIN * KREBON * PANFORMIN * SILUBIN * SIN-DIATIL * TIDEMOL * ZIAVETINE

SAFETY PROFILE: A poison via ingestion, intravenous, and intraperitoneal routes. When heated to decomposition it emits very toxic fumes of HCl and NO_x.

BQM000   CAS: 102-79-4   **HR: 2**
N-BUTYL-N,N-BIS(HYDROXY ETHYL) AMINE
mf: $C_8H_{19}NO_2$ mw: 161.28

PROP: Liquid. Bp: $262°$, flash p: $245°F$ (OC), d: 0.97, vap d: 5.55.

SYNS: N-BUTYLDIETHANOLAMINE * N-BUTYL-2,2'-IMINODIETHANOL

CONSENSUS REPORTS: Reported in EPA TSCA Inventory.

SAFETY PROFILE: Mildly toxic via ingestion. No chronic effects data. A skin and severe eye irritant. Combustible when exposed to heat or flame. To fight fire, use alcohol foam, foam, CO_2, dry chemical. Incompatible with oxidizing materials. When heated to decomposition it emits toxic fumes of NO_x.

BQM250   CAS: 507-19-7   **HR: 3**
tert-BUTYL BROMIDE
DOT: UN 2342
mf: C_4H_9Br mw: 137.04

PROP: Colorless liquid. Mp: $-20°$, bp: $73.3°$, fp: $-18°$, d: 1.215 @ $25°/25°$.

SYNS: 2-BROMOISOBUTANE * 2-BROMO-2-METH-YLPROPANE (DOT) * TRIMETHYLBROMOMETHANE

CONSENSUS REPORTS: EPA Genetic Toxicology Program. Reported in EPA TSCA Inventory.

DOT Classification: Flammable or Combustible Liquid; Label: Flammable Liquid.

SAFETY PROFILE: Moderately toxic by intraperitoneal route. Questionable carcinogen with experimental neoplastigenic data. When heated to decomposition it emits toxic fumes of Br^-.

BQM500 CAS: 109-21-7 **HR: 2**
n-BUTYL n-BUTANOATE
mf: $C_8H_{16}O_2$ mw: 144.24

PROP: Colorless liquid; pineapple odor. Bp: 166°, flash p: 128°F (OC), d: 0.67-0.871, refr index: 1.405, vap d: 5.0. Misc with alc, ether, vegetable oils; sltly sol in propylene glycol, water.

SYNS: BUTYL BUTYRATE (FCC) * n-BUTYL BUTY-RATE * n-BUTYL n-BUTYRATE * FEMA No. 2186

CONSENSUS REPORTS: Reported in EPA TSCA Inventory.

SAFETY PROFILE: Moderately toxic via intraperitoneal route. Mildly toxic by ingestion. Moderately irritating to eyes, skin, and mucous membranes by inhalation. Narcotic in high concentrations. Combustible liquid. To fight fire, use alcohol foam, foam, CO_2, dry chemical. Incompatible with oxidizing materials. When heated to decomposition it emits acrid and irritating fumes.

BQP000 CAS: 7492-70-8 **HR: 1**
BUTYL BUTYROLLACTATE
mf: $C_{11}H_{20}O_4$ mw: 216.28

PROP: Colorless liquid; butter, creamlike odor. D: 0.970, refr index: 1.420, flash p: +212°F. Misc with alc, fixed oils; sol in propylene glycol; insol in water.

SYNS: BUTANOIC ACID-2-BUTOXY-1-METHYL-2-OXO-ETHYL ESTER (9CI) * BUTYL BUTYRYL LACTATE * BUTYRIC ACID ESTER with BUTYL LACTATE * FEMA No. 2190 * LACTIC ACID, BUTYL ESTER, BUTYRATE

CONSENSUS REPORTS: Reported in EPA TSCA Inventory.

SAFETY PROFILE: A skin irritant. Combustible liquid. When heated to decomposition it emits acrid smoke and irritating fumes.

BQP250 CAS: 592-35-8 **HR: 3**
BUTYL CARBAMATE
mf: $C_5H_{11}NO_2$ mw: 117.17

SYNS: CARBAMIC ACID, BUTYL ESTER * USAF EL-101 * USAF FO-1

CONSENSUS REPORTS: Reported in EPA TSCA Inventory.

SAFETY PROFILE: A poison via intraperitoneal route. Moderately toxic via subcutaneous route. Experimental teratogenic and reproductive effects. Questionable carcinogen with experimental neoplastigenic data. Mutation data reported. When heated to decomposition it emits toxic fumes of NO_x.

BQP500 CAS: 124-17-4 **HR: 2**
BUTYL CARBITOL ACETATE
mf: $C_{10}H_{20}O_4$ mw: 204.30

PROP: Colorless liquid. fp: −32.2°, bp: 247°, flash p: 240°F (OC), d: 0.981 @ 20°/20°, autoign temp: 570°F, vap press: 0.01 mm @ 20°.

SYNS: 2-(2-BUTOXYETHOXY)ETHANOL ACETATE * 2-(2-BUTOXYETHOXY)ETHYL ACETATE * DIETH-YLENE GLYCOL BUTYL ETHER ACETATE * DIGLY-COL MONOBUTYL ETHER ACETATE * EKTASOLVE DB ACETATE * GLYCOL ETHER DB ACEATATE

CONSENSUS REPORTS: Reported in EPA TSCA Inventory. Glycol ethers are on the Community Right-To-Know List.

SAFETY PROFILE: Moderately toxic by ingestion. Mild skin and eye irritant. Combustible when exposed to heat or flame. To fight fire, use foam, CO_2, dry chemical. Incompatible with oxidizing materials; heat; flame. When heated to decomposition it emits acrid and irritating fumes.

BQP750 CAS: 85-70-1 **HR: 2**
BUTYL CARBOBUTOXYMETHYL PHTHALATE
mf: $C_{18}H_{24}O_6$ mw: 336.42

SYNS: BUTYL PHTHALATE BUTYL GLYCOLATE * BUTYL PHTHALYL BUTYL GLYCOLATE * DIBU-TYL-o-(o-CARBOXYBENZOYL) GLYCOLATE * DIBU-TYL-o-CARBOXYBENZOYLOXYACETATE * SANTICI-ZIER B-16

CONSENSUS REPORTS: Reported in EPA TSCA Inventory.

SAFETY PROFILE: Mildly toxic via intraperitoneal route. Experimental teratogenic and reproductive effects. Mutation data reported. An eye irritant. When heated to decomposition it emits acrid and irritating fumes.

BQQ250   CAS: 38252-74-3   **HR: 3**
N-BUTYL-(3-CARBOXY PROPYL)NITROSAMINE
mf: $C_8H_{16}N_2O_3$ mw: 188.26

SYNS: BCPN * 4-(BUTYLNITROSOAMINO)BUTA-NOIC ACID * N-NITROSO-N-BUTYL-N-(3-CARBOXY-PROPYL)AMINE

CONSENSUS REPORTS: IARC Cancer Review: Animal Limited Evidence IMEMDT 17,51,78. NTP Fourth Annual Report On Carcinogens, 1984. EPA Genetic Toxicology Program.

SAFETY PROFILE: Confirmed carcinogen with experimental carcinogenic and tumorigenic data. Mutation data reported. When heated to decomposition it emits toxic fumes of NO_x.

BQQ750   CAS: 109-69-3   **HR: 2**
n-BUTYL CHLORIDE
DOT: UN 1127
mf: C_4H_9Cl mw: 92.58

PROP: Colorless liquid. Mp: $-123.1°$, bp: 78°, lel: 1.9%, uel: 10.1%, flash p: 15°F (OC), d: 0.884, autoign temp: 860°F, vap d: 3.20.

SYNS: BUTYL CHLORIDE (DOT) * 1-CHLOROBU-TANE (DOT) * CHLORURE de BUTYLE (FRENCH) * NCI-C06155 * N-PROPYLCARBINYL CHLORIDE

CONSENSUS REPORTS: NTP Carcinogenesis Studies (gavage); No Evidence: mouse, rat NTPTR* NTP-TR-312,86. EPA Genetic Toxicology Program. Reported in EPA TSCA Inventory.

DOT Classification: Flammable Liquid; Label: Flammable Liquid.

SAFETY PROFILE: Moderately toxic by ingestion and possible other routes. Mutation data reported. Skin and eye irritant. Dangerous fire hazard when exposed to heat or flame. Moderately explosive when exposed to flame. When heated to decomposition it emits highly toxic fumes of phosgene and Cl^-. To fight fire, use foam, CO_2, dry chemical. Incompatible with oxidizing materials.

BQR000   CAS: 507-20-0   **HR: 3**
tert-BUTYL CHLORIDE
mf: C_4H_9Cl mw: 92.58

PROP: Flash p: 32°F, d: 0.87, vap d: 3.2, bp: 51°.

SYNS: 2-CHLOROISOBUTANE * 2-CHLORO-2-METH-YLPROPANE * TRIMETHYLCHLOROMETHANE

CONSENSUS REPORTS: Reported in EPA TSCA Inventory.

SAFETY PROFILE: Questionable carcinogen with experimental neoplastigenic data. Dangerous fire hazard when exposed to heat, flame (sparks), and oxidizers. To fight fire, use water, spray, fog, alcohol foam, dry chemical. When heated to decomposition it emits toxic fumes of Cl^-.

BQV000   CAS: 1189-85-1   **HR: 3**
tert-BUTYL CHROMATE
mf: $C_8H_{18}CrO_4$ mw: 230.26

SYN: CHROMIC ACID, DI-tert-BUTYL ESTER

CONSENSUS REPORTS: Chromium and its compounds are on the Community Right-To-Know List.

OSHA PEL: CL 0.1 mg(CrO_3)/m^3 (skin)
ACGIH TLV: CL 0.1 mg(CrO_3)/m^3 (skin)
NIOSH REL: CL 0.001 Mg(Cr(VI))/m^3

SAFETY PROFILE: A very flammable mixture. When heated to decomposition it emits acrid and irritating fumes.

BQV750   CAS: 2409-55-4   **HR: 3**
2-tert-BUTYL-p-CRESOL
mf: $C_{11}H_{16}O$ mw: 164.27

PROP: Clear liquid, sol in organic solvents and aqueous potassium hydroxide. Fp: 23.1°, bp: 244°, d: 0.922, flash p: 116°F.

SYNS: 2-tert-BUTYL-p-KRESOL (CZECH) * 2-tert-BU-TYL-4-METHYLPHENOL

CONSENSUS REPORTS: Reported in EPA TSCA Inventory.

SAFETY PROFILE: A poison by intraperitoneal and intravenous routes. Moderately toxic by ingestion and skin contact. Questionable carcinogen with experimental neoplastigenic data. A severe skin and eye irritant. Mutation data reported. Flammable when exposed to heat, flame, or oxidizers. To fight fire, use alcohol foam, foam, water spray, fog, dry chemical. When heated to decomposition it emits acrid and irritating fumes.

BQW750 CAS: 10108-56-2 **HR: 3**
N-BUTYL CYCLOHEXYL AMINE
mf: $C_{10}H_{21}N$ mw: 155.32

PROP: Flash p: 200°F (OC), d: 0.8, bp: 210°.

SAFETY PROFILE: A poison by ingestion. Moderately toxic by skin contact. A skin irritant. Combustible when exposed to heat or flame. To fight fire, use alcohol foam. When heated to decomposition it emits toxic fumes of NO_x.

BQX000 CAS: 61925-70-0 **HR: 3**
N-(4-tert-BUTYL CYCLOHEXYL)-3,3-DIPHENYL PROPYLAMINE HYDROCHLORIDE
mf: $C_{25}H_{25}N \cdot ClH$ mw: 375.97

SYN: MG 18037

SAFETY PROFILE: A poison by intraperitoneal route. Moderately toxic by ingestion. When heated to decomposition it emits very toxic fumes of HCl and NO_x.

BQZ000 CAS: 94-80-4 **HR: 3**
BUTYL DICHLOROPHENOXYACETATE
mf: $C_{12}H_{14}Cl_2O_3$ mw: 277.16

SYNS: BUTYL 2,4-D * BUTYL (2,4-DICHLOROPHE-NOXY)ACETATE * 2,4-D BUTYL ESTER * BUTYL ESTER 2,4-D * (2,4-DICHLOROPHENOXY)ACETIC ACID, BUTYL ESTER * ESSO HERBICIDE 10 * FERNESTA * LIRONOX * SHELL 40

CONSENSUS REPORTS: IARC Cancer Review: Animal Inadequate Evidence IMEMDT 15,111,77

SAFETY PROFILE: Poison by an unspecified route. Moderately toxic by ingestion and possibly other routes. Experimental teratogenic and reproductive effects. Questionable carcinogen. An herbicide. When heated to decomposition it emits toxic fumes of Cl^-.

BRC500 CAS: 51003-83-9 **HR: 3**
2-n-BUTYL-3-DIMETHYLAMINO-5,6-METHYLENEDIOXYINDENE HYDROCHLORIDE
mf: $C_{16}H_{21}NO_2 \cdot ClH$ mw: 295.84

SYNS: 6-BUTYL-5-DIMETHYLAMINO-5H-INDENO(5,6-d)-1,3-DIOXOLE HYDROCHLORIDE * bu-MDI

SAFETY PROFILE: A poison by intraperitoneal and intravenous routes. When heated to decomposition it emits very toxic fumes of NO_x and HCl.

BRE500 CAS: 88-85-7 **HR: 3**
2-sec-BUTYL-4,6-DINITROPHENOL
mf: $C_{10}H_{12}N_2O_5$ mw: 240.24

PROP: Crystals. Vap d: 7.73.

SYNS: ARETIT * BASANITE * BNP 30 * BU-TAPHENE * CALDON * CHEMOX GENERAL * CHEMOX P.E. * DINITRO * DINITRO-3 * 4,6-DINITRO-2-sec.BUTYLFENOL (CZECH) * 2,4-DI-NITRO-6-sec-BUTYLPHENOL * 4,6-DINITRO-o-sec-BU-TYLPHENOL * 4,6-DINITRO-2-sec-BUTYLPHENOL * DINITROBUTYLPHENOL * 4,6-DINITRO-2-(1-METHYL-N-PROPYL)PHENOL * 2,4-DINITRO-6-(1-METHYL-PROPYL)PHENOL (FRENCH) * DINOSEB * DINOSEBE (FRENCH) * DN 289 * DNBP * DNOSBP * DNSBP * DOW GENERAL * DOW GENERAL WEED KILLER * DOW SELECTIVE WEED KILLER * ELGETOL * ELGETOL 318 * ENT 1,122 * GEBUTOX * HEL-FIRE * KILOSEB * 6-(1-METHYL-PROPYL)-2,4-DINITROFENOL (DUTCH) * 2-(1-METHYLPROPYL)-4,6-DINITROPHENOL * 6-(1-METIL-PROPIL)-2,4-DINITRO-FENOLO (ITALIAN) * NITROPONE C * PHENOTAN * PREMERGE * PREMERGE 3 * RCRA WASTE NUMBER P020 * SINOX GENERAL * SPARIC * SPURGE * SUBITEX * UNICROP DNBP * VERTAC DINITRO WEED KILLER * VERTAC GENERAL WEED KILLER * VERTAC SELECTIVE WEED KILLER

CONSENSUS REPORTS: EPA Genetic Toxicology Program. EPA Extremely Hazardous Substances List.

SAFETY PROFILE: A poison by ingestion, inhalation, skin contact, subcutaneous, intraperitoneal, and possibly other routes. Experimental teratogenic and reproductive effects. A severe eye irritant. Questionable carcinogen with experimental tumorigenic data. Mutation data reported. An herbicide. When heated to decomposition it emits toxic fumes of NO_x.

BRF500 CAS: 50-33-9 **HR: 3**
4-BUTYL-1,2-DIPHENYL-3,5-DIOXO PYRAZOLIDINE
mf: $C_{19}H_{20}N_2O_2$ mw: 308.41

SYNS: ALINDOR * ALKABUTAZONA * ALQOVE-
RIN * ANERVAL * ANPUZONE * ANTADOL
* ANUSPIRAMIN * ARTIZIN * ARTRIZONE
* ARTROPAN * AZDID * AZOBUTYL * AZO-
LID * BENZONE * BETAZED * BIZOLIN 200
* B.T.Z. * BUSONE * BUTACOMPREN
* BUTACOTE * BUTADION * BUTADIONA
* BUTAGESIC * BUTALAN * BUTALGINA
* BUTALIDON * BUTALUY * BUTAPHEN
* BUTAPIRAZOL * BUTAPYRAZOLE * BUTAREC-
BON * BUTARTRIL * BUTARTRINA * BUTA-
ZINA * BUTAZOLIDIN * BUTAZONA * BUTA-
ZONE * BUTE * BUTIDIONA * BUTIWAS-
SIMPLE * BUTONE * BUTOZ * 4-BUTYL-1,2-DI-
PHENYLPYRAZOLIDINE-3,5-DIONE * 4-BUTYL-1,2-DI-
PHENYL-3,5-PYRAZOLIDINEDIONE * BUTYLPYRIN
* BUVETZONE * BUZON * CHEMBUTAZONE
* FA-192 * DIGIBUTINA * DIOSSIDONE
* 3,5-DIOXO-1,2-DIPHENYL-4-N-BUTYLPYRAZOLIDENE
* 3,5-DIOXO-1,2-DIPHENYL-4-N-BUTYL-PYRAZOLIDIN
* 3,5-DIOXO-1,2-DIPHENYL-4-N-BUTYL-PYRAZOLIDINE
* DIOZOL * DIPHEBUZOL * DIPHENYLBU-
TAZONE * 1,2-DIPHENYL-4-BUTYL-3,5-DIOXOPY-
RAZOLIDINE * 1,2-DIPHENYL-4-BUTYL-3,5-PYRAZO-
LIDINEDIONE * 1,2-DIPHENYL-3,5-DIOXO-4-BUTYLPY-
RAZOLIDINE * 1,2-DIPHENYL-2,3-DIOXO-4-N-BUTYL-
PYRAZOLINE * ECOBUTAZONE * ELMEDAL
* EQUI BUTE * ERIBUTAZONE * ESTEVE
* FBZ * FEBUZINA * FENARTIL * FENIBUTA-
SAN * FENIBUTAZONA * FENIBUTOL * FENIL-
BUTAZONA * FENILBUTINE * FENILIDINA
* FENOTONE * FENYLBUTAZON * FLEXAZONE
* IA-BUT * INTALBUT * INTRABUTAZONE
* IPSOFLAME * KADOL * LINGEL * MAL-
GESIC * MEPHABUTAZONE * MERIZONE
* NADAZONE * NADOZONE * NCI-C56531
* NEO-ZOLINE * NOVOPHENYL * PBZ
* PHEBUZIN * PHEBUZINE * PHEN-BUTA-VET
* PHENBUTAZOL * PHENOPYRINE * PHENYLBU-
TAZ * PHENYLBUTAZON (GERMAN) * PHENYLBU-
TAZONE * PHENYLBUTAZONUM * PHENYL-MO-
BUZON * PIRARREUMOL 'B'' * PRAECIRHEUMIN
* PYRABUTOL * PYRAZOLIDIN * RECTOFASA
* REUDO * REUDOX * REUMASYL * REUMA-
ZIN * REUMAZOL * REUPOLAR * ROBIZON-V
* RUBATONE * R-3-ZON * SCANBUTAZONE
* SCHEMERGIN * SHIGRODIN * TAZONE
* TETNOR * TEVCODYNE * THERAZONE
* TICINIL * TODALGIL * USAF GE-15
* UZONE * VAC-10 * WESCOZONE * ZOLA-
PHEN * ZOLIDINUM * ZORANE

CONSENSUS REPORTS: IARC Cancer Review: GROUP 3 IMEMDT 7,316,87; Human

Inadequate Evidence IMEMDT 13,183,77.
EPA Genetic Toxicology Program. Reported
in EPA TSCA Inventory.

SAFETY PROFILE: Suspected human carcinogen producing leukemia. A human poison by
parenteral route. An experimental poison by ingestion, intraperitoneal, subcutaneous, intravenous and intramuscular routes. Human systemic
effects by ingestion and possibly other routes:
fever, blood pressure increase, other unspecified
vascular effects, damage to kidney tubules and
glomeruli, decreased urine volume, blood in
the urine, reduction in the number of white blood
cells, and agranulocytosis. Experimental teratogenic and reproductive effects. Human mutation
data reported. An eye irritant. An anti-inflammatory agent. When heated to decomposition
it emits toxic fumes of NO_x.

BRH750   CAS: 142-96-1   **HR: 1**
n-BUTYL ETHER

DOT: UN 1149
mf: $C_8H_{18}O$ mw: 130.26

PROP: Colorless liquid. Mp: $-95°$, bp: 142°,
flash p: 77°F, d: 0.769 @ 20°/20°, autoign temp:
382°F, vap d: 4.48, lel: 1.5%, uel: 7.6%.

SYNS: 1-BUTOXYBUTANE * BUTYL ETHER (DOT)
* DI-n-BUTYL ETHER (DOT) * DIBUTYL OXIDE
* ETHER BUTYLIQUE (FRENCH) * 1,1'-OXYBIS(BU-
TANE)

CONSENSUS REPORTS: Reported in EPA
TSCA Inventory.

DOT Classification: Flammable Liquid; Label:
Flammable Liquid; IMO: Flammable or Combustible Liquid; Label: Flammable Liquid.

SAFETY PROFILE: Mildly toxic by inhalation,
ingestion, and skin contact. Human systemic
effects by inhalation: conjunctiva irritation and
unspecified nasal effects. An experimental skin
and human eye irritant. Dangerous fire hazard
when exposed to heat, flame or oxidizers. Incompatible with NCl_3 and oxidizing materials.
To fight fire, use alcohol foam, dry chemical.
When heated to decomposition it emits acrid
smoke and fumes.

BRI000   CAS: 123-05-7   **HR: 2**
BUTYL ETHYL ACETALDEHYDE

DOT: UN 1191
mf: $C_8H_{16}O$ mw: 128.24

PROP: Bp: 163.4°, flash p: 125°F (OC). autoign temp: 387°F, d: 0.8205, vap press: 1.8 mm @ 20°, vap d: 4.42.

SYNS: ETHYLBUTYLACETALDEHYDE * α-ETHYL-CAPROALDEHYDE * 2-ETHYLHEXALDEHYDE * ETHYLHEXALDEHYDE (DOT) * 2-ETHYLHEXANAL * β-PROPYL-α-ETHYLACROLEIN

CONSENSUS REPORTS: Reported in EPA TSCA Inventory.

DOT Classification: Flammable or Combustible Liquid; Label: Flammable Liquid.

SAFETY PROFILE: Moderately toxic by ingestion and intraperitoneal routes. Mildly toxic by inhalation and skin contact. An eye and severe skin irritant. Dangerous fire hazard; spontaneously flammable in air. To fight fire, use foam, CO_2, dry chemical, water spray, mist, fog. Incompatible with oxidizing materials. When heated to decomposition it emits acrid and irritating fumes.

BRI250 CAS: 149-57-5 **HR: 2**
BUTYL ETHYL ACETIC ACID
mf: $C_8H_{16}O_2$ mw: 144.24

PROP: Flash p: 260°F (OC).

SYNS: α-ETHYLCAPROIC ACID * 2-ETHYLHEXA-NOIC ACID * 2-ETHYLHEXOIC ACID

CONSENSUS REPORTS: Reported in EPA TSCA Inventory.

SAFETY PROFILE: Moderately toxic by ingestion and skin contact. A skin and severe eye irritant. Combustible when exposed to heat or flame. When heated to decomposition, it emits acrid and irritating fumes.

BRJ750 CAS: 2425-74-3 **HR: 3**
tert-BUTYL FORMAMIDE
mf: $C_5H_{11}NO$ mw: 101.17

CONSENSUS REPORTS: Reported in EPA TSCA Inventory.

SAFETY PROFILE: A poison by intravenous route. When heated to decomposition it emits toxic fumes of NO_x.

BRK000 CAS: 592-84-7 **HR: 2**
n-BUTYL FORMATE
DOT: UN 1128
mf: $C_5H_{10}O_2$ mw: 102.15

PROP: Colorless liquid. Mp: −90°, bp: 106.0°, flash p: 64°F (CC), d: 0.911, autoign temp: 612°F, vap press: 40 mm @ 31.6°, vap d: 3.52, lel: 1.7%, uel: 8%.

SYNS: BUTYLESTER KYSELINY MRAVENCI * BUTYL FORMATE (DOT)

CONSENSUS REPORTS: Reported in EPA TSCA Inventory.

DOT Classification: Flammable Liquid; Label: Flammable Liquid.

SAFETY PROFILE: Moderately toxic by ingestion. Mildly toxic by inhalation. Human systemic effects by inhalation: muscle contractions and spasticity, conjunctiva irritation, and unspecified respiratory changes. An irritant and narcotic in high concentrations. Dangerous fire hazard when exposed to heat or flame. To fight fire, use alcohol foam, foam, CO_2, dry chemical. Incompatible with oxidizing materials. When heated to decomposition it emits acrid and irritating fumes.

BRK750 CAS: 2426-08-6 **HR: 3**
n-BUTYL GLYCIDYL ETHER
mf: $C_7H_{14}O_2$ mw: 130.21

SYNS: AGEFLEX BGE * BGE * 2,3-EPOXYPRO-PYL BUTYL ETHER * GLYCIDYL BUTYL ETHER

CONSENSUS REPORTS: Reported in EPA TSCA Inventory.

OSHA PEL: (Transitional: TWA 50 ppm) TWA 25 ppm
ACGIH TLV: TWA 25 ppm
DFG MAK: Suspected Carcinogen.
NIOSH REL: (Glycidyl Ethers) CL 30 mg/m^3/ 15M

SAFETY PROFILE: Moderately toxic by ingestion, skin contact, and intraperitoneal routes. Mildly toxic by inhalation. An experimental teratogen. Mutation data reported. A skin and severe eye irritant. When heated to decomposition it emits acrid and irritating fumes.

BRM250 CAS: 75-91-2 **HR: 3**
tert-BUTYLHYDROPEROXIDE
DOT: UN 2092/UN 2093/UN 2094
mf: $C_4H_{10}O_2$ mw: 90.14

PROP: Water-white liquid, sltly sol in water, very sol in esters and alc. Flash p: 80°F or above, fp: −35°, d: 0.860, vap d: 2.07.

SYNS: terc. BUTYLHYDROPEROXID (CZECH)
* CADOX TBH * 1,1-DIMETHYLETHYL HYDROPER-
OXIDE * HYDROPEROXYDE de BUTYLE TERTIAIRE
(FRENCH) * 2-HYDROPEROXY-2-METHYLPROPANE
* PERBUTYL H * TBHP-70 * TRIGONOX A-75
(CZECH)

CONSENSUS REPORTS: EPA Genetic Toxicology Program. Reported in EPA TSCA Inventory.

DFG MAK: Moderate skin effects.
DOT Classification: Organic Peroxide.

SAFETY PROFILE: A poison by ingestion and
inhalation. A severe skin and eye irritant. Mutation data reported. At highest dosage levels,
symptoms noted were severe depression, incoordination, and cyanosis. Death was due to respiratory arrest. Very dangerous fire hazard when
exposed to heat or flame, or by spontaneous
chemical reaction such as with reducing materials. Moderately explosive; may explode during
distillation. Violent reaction with traces of acid.
Concentrated solutions may ignite spontaneously on contact with molecular sieve. Mixtures
with transition metal salts may react vigorously
and release oxygen. Forms an unstable solution
with 1,2-dichloroethane. To fight fire, use alcohol foam, CO_2, dry chemical. When heated to
decomposition it emits acrid smoke and fumes.

BRM500   CAS: 1948-33-0   **HR: 3**
tert-BUTYLHYDROQUINONE
mf: $C_{10}H_{14}O_2$ mw: 166.24

PROP: White crystalline solid; characteristic
odor. Mp: 126.5-128.5°. Sol in alc, ether; insol
in water.

SYNS: MONO-tert-BUTYLHYDROQUINONE
* MTBHQ * SUSTANE * TBHQ (FCC) * TENOX
TBHQ

CONSENSUS REPORTS: Reported in EPA
TSCA Inventory.

SAFETY PROFILE: Poison by intraperitoneal
route. Moderately toxic by ingestion. Mutation
data reported. When heated to decomposition
it emits acrid smoke and irritating fumes.

BRQ350   **HR: 2**
BUTYL ISOBUTYRATE
mf: $C_8H_{16}O_2$ mw: 44.44

PROP: Colorless liquid; apple-pineapple odor.
D: 0.859-0.864, refr index: 1.401, flash p:

113°F. Misc with alc, ether, fixed oils; insol
in glycerin, propylene glycol, water @ 166°.

SYN: FEMA No. 2188

SAFETY PROFILE: Combustible liquid. When
heated to decomposition it emits acrid smoke
and irritating fumes.

BRQ500   CAS: 111-36-4   **HR: 3**
n-BUTYL ISOCYANATE
DOT: UN 2485
mf: C_5H_9NO mw: 99.15

PROP: Colorless liquid. Bp: 115°, d: 0.880
@ 20°/4°.

SYNS: BIC * ISOCYANIC ACID, BUTYL ESTER

CONSENSUS REPORTS: Reported in EPA
TSCA Inventory.

DOT Classification: Flammable Liquid; Label:
Flammable Liquid and Poison.

SAFETY PROFILE: A poison by ingestion and
intravenous routes. Mildly toxic by inhalation.
A powerful irritant to eyes, skin, and mucous
membranes. Flammable liquid.

BRR250   CAS: 30026-92-7   **HR: 3**
tert-BUTYL ISOPROPYL BENZENE
HYDROPEROXIDE
DOT: NA 2091
mf: $C_{13}H_{20}O_2$ mw: 208.33

PROP: Crystals.

SYN: tert-BUTYL ISOPROPYL BENZENE HYDROPEROX-
IDE (DOT)

DOT Classification: Label: Organic Peroxide.

SAFETY PROFILE: Powerful irritant. Dangerous fire hazard when exposed to heat or flame
or by chemical reaction. Incompatible with oxidizing or reducing materials. When heated to
decomposition it emits acrid smoke and fumes.

BRR500   CAS: 74926-97-9   **HR: 3**
2-sec-BUTYL-6-ISOPROPYLPHENOL
mf: $C_{13}H_{20}O$ mw: 192.2

SAFETY PROFILE: Poison by intravenous
route. When heated to decomposition it emits
acrid smoke and irritating fumes.

BRR600   CAS: 138-22-7   **HR: 3**
n-BUTYL LACTATE
mf: $C_7H_{14}O_3$ mw: 146.21

PROP: Liquid. Sltly sol in water; misc in alc and ether. Mp: $-43°$, bp: $188°$, flash p: $160°F$ (OC), d: 0.968, autoign temp: $720°F$, vap d: 5.04, vap press: 0.4 mm @ $20°$.

SYNS: BUTYL α-HYDROXYPROPIONATE * BUTYL LACTATE * 2-HYDROXYPROPANOIC ACID, BUTYL ESTER * LACTIC ACID, BUTYL ESTER

CONSENSUS REPORTS: Reported in EPA TSCA Inventory.

OSHA PEL: TWA 5 ppm
ACGIH TLV: TWA 5 ppm

SAFETY PROFILE: Poison by intraperitoneal route. A skin irritant. Toxic concentration in air for humans is about 4 ppm. Flammable when exposed to heat or flame; can react with oxidizing materials. To fight fire, use alcohol foam, foam, CO_2, dry chemical. When heated to decomposition it emits acrid smoke and irritating fumes.

BRR900 CAS: 109-79-5 **HR: 3**
n-BUTYL MERCAPTAN
DOT: UN 2347
mf: $C_4H_{10}S$ mw: 90.20

PROP: Colorless liquid, skunk-like odor. Mp: $-116°$, bp: $98°$, d: 0.8365 @ $25°/4°$, flash p: $35°F$, vap d: 3.1.

SYNS: BUTANETHIOL * n-BUTANETHIOL * BUTYL MERCAPTAN * BUTYL MERCAPTAN (DOT) * NCI-C60866

CONSENSUS REPORTS: Reported in EPA TSCA Inventory.

OSHA PEL: (Transitional: TWA 10 ppm) TWA 0.5 ppm
ACGIH TLV: TWA 0.5 ppm
DFG MAK: 0.5 ppm (1.5 mg/m^3)
NIOSH REL: (n-Alkane Mono Thiols) CL 0.5 ppm/15M
DOT Classification: Flammable Liquid; Label: Flammable Liquid.

SAFETY PROFILE: Poison by intraperitoneal route. Moderately toxic by inhalation and ingestion. An eye irritant. Dangerous fire hazard by exposure to heat, flame, sparks, or powerful oxidizers. Reacts violently with HNO_3. Incompatible with acids, acid fumes, oxidizing materials, heat, flame, sparks. To fight fire, use alcohol

foam. When heated to decomposition it emits toxic SO_x.

BRS750 CAS: 543-63-5 **HR: 3**
n-BUTYLMERCURIC CHLORIDE
mf: C_4H_9ClHg mw: 293.17

SYN: BMC

CONSENSUS REPORTS: Mercury and its compounds are on the Community Right-To-Know List.

OSHA PEL: (Transitional: CL 1 mg/10m^3) TWA 0.01 mg(Hg)/m^3; STEL 0.03 mg/m^3 (skin)
ACGIH TLV: TWA 0.01 mg/(Hg)/m^3; STEL 0.03 mg(Hg)/m^3
NIOSH REL: TWA 0.05 mg(Hg)/m^3

SAFETY PROFILE: A poison by subcutaneous route. Mutation data reported. When heated to decomposition it emits very toxic fumes of Cl^- and Hg.

BRT000 CAS: 532-34-3 **HR: 2**
n-BUTYL MESITYL OXIDE OXALATE
mf: $C_{12}H_{18}O_4$ mw: 226.30

PROP: Yellow to reddish liquid. Bp: $113°$, d: 1.052-1.060 @ $25°/25°$, flash p: $315°F$.

SYNS: BMOO * BUTOPYRONOXYL * BUTYL-3,4-DIHYDRO-2,2-DIMETHYL-4-OXO-2H-PYRAN-6-CARBOXYLATE * n-BUTYL ESTER of 3,4-DIHYDRO-2,2-DIMETHYL-4-OXO-2H-PYRAN-6-CARBOXYLIC ACID * n-BUTYLMESITYLOXID OXALATE * 2-CARBO-n-BUTOXY-6,6-DIMETHYL-5,6-DIHYDRO-1,4-PYRONE * 3,4-DIHYDRO-2,2-DIMETHYL-4-OXO-2H-PYRAN-6-CARBOXYLIC ACID-n-BUTYL ESTER * DIHDYROPYRONE * α,α-DIMETHYL-α'-CARBOBUTOXY-DIHYDRO-γ-PYRONE * 2,2-DIMETHYL-6-CARBOBUTOXY-2,3-DIHYDRO-4-PYRONE * ENT 9 * INDALONE

CONSENSUS REPORTS: Reported in EPA TSCA Inventory.

SAFETY PROFILE: Moderately toxic by ingestion. Produces liver necrosis in experimental animals. A mild skin irritant. Combustible when exposed to heat or flame. When heated to decomposition it emits acrid and irritant fumes.

BRU500 CAS: 83-66-9 **HR: 3**
6-tert-BUTYL-3-METHYL-2,4-DINITRO ANISOLE
mf: $C_{12}H_{16}N_2O_5$ mw: 268.30

SYNS: 2,6-DINITRO-3-METHOXY-4-tert-BUTYLTOLUENE
* MUSK AMBRETTE

CONSENSUS REPORTS: Reported in EPA TSCA Inventory.

SAFETY PROFILE: A poison by ingestion. Mutation data reported. A skin irritant. When heated to decomposition it emits toxic fumes of NO_x.

BRV500   CAS: 544-16-1   **HR: 3**
n-BUTYL NITRITE
DOT: UN 2351
mf: $C_4H_9NO_2$ mw: 103.14

PROP: Oily liquid, characteristic odor, misc in alc and ether. Bp: 75°, d: 0.9114 @ 0°/4°, vap d: 3.5, flash p: 10°.

SYNS: BUTYL NITRITE (DOT) * NBN * NCI-C56553 * NITROUS ACID-n-BUTYL ESTER

CONSENSUS REPORTS: Reported in EPA TSCA Inventory.

DOT Classification: Flammable Liquid; Label: Flammable Liquid.

SAFETY PROFILE: A poison by ingestion and intraperitoneal routes. Mildly toxic by inhalation. An irritant. Human systemic effects by ingestion: methemoglobinemia-carboxhemoglobinemia. Resembles amyl nitrite in causing fall in blood pressure, headache, pulse throbbing, and weakness. Mutation data reported. Flammable when exposed to heat or flame or by spontaneous chemical reaction. When heated to decomposition it emits toxic fumes of NO_x.

BRV750   CAS: 924-43-6   **HR: 2**
sec-BUTYL NITRITE
mf: $C_4H_9NO_2$ mw: 103.14

PROP: Liquid. Bp: 68°, d: 0.8981 @ 0°/4°, vap d: 3.5.

SYNS: NITROUS ACID-sec-BUTYL ESTER * NITROUS ACID-1-METHYL PROPYL ESTER

CONSENSUS REPORTS: Reported in EPA TSCA Inventory.

SAFETY PROFILE: Moderately toxic by ingestion, inhalation, and intraperitoneal routes. Mutation data reported. Flammable when exposed to heat or flame or by spontaneous chemical reaction. An oxidizer. Potentially explosive. To fight fire, use water, spray, foam, dry chemical. When heated to decomposition it emits toxic fumes of NO_x.

BRX500   CAS: 56986-36-8   **HR: 3**
BUTYLNITROSOAMINOMETHYL ACETATE
mf: $C_7H_{14}N_2O_3$ mw: 174.23

SYNS: ACETOXYMETHYLBUTYLNITROSAMINE
* N-(ACETOXY)METHYL-N,N-BUTYLNITROSAMINE
* BAMN * BUTYL ACETOXYMETHYLNITROSAMINE
* N-BUTYL-N-(ACETOXYMETHYL)NITROSAMINE
* N-NITROSO-N-(1-ACETOXYMETHYL)BUTYLAMINE

CONSENSUS REPORTS: EPA Genetic Toxicology Program.

SAFETY PROFILE: Moderately toxic by ingestion. Questionable carcinogen with experimental carcinogenic and tumorigenic data. Mutation data reported. When heated to decomposition it emits toxic fumes of NO_x.

BRY250   CAS: 16339-05-2   **HR: 3**
N-BUTYL-N-NITROSO AMYL AMINE
mf: $C_9H_{20}N_2O$ mw: 172.31

SYNS: BUTYLAMYLNITROSAMIN (GERMAN)
* N-BUTYL-N-NITROSOPENTYLAMINE * N-BUTYL-N-PENTYLINITROSAMINE * N-NITROSO-N-BUTYLPENTYLAMINE * N-NITROSO-N-BUTYL-N-PENTYLAMINE

SAFETY PROFILE: Moderately toxic by subcutaneous route. Questionable carcinogen with experimental tumorigenic data. When heated to decomposition it emits toxic fumes of NO_x.

BRY500   CAS: 924-16-3   **HR: 3**
n-BUTYL-N-NITROSO-1-BUTAMINE
mf: $C_8H_{18}N_2O$ mw: 158.28

PROP: Pale yellow liquid. Bp: 235°.

SYNS: DBN * DBNA * DI-n-BUTYLNITROSAMIN (GERMAN) * DIBUTYLNITROSOAMINE * DI-n-BUTYLNITROSAMINE * N,N-DI-n-BUTYLNITROSAMINE * N,N-DIBUTYLNITROSOAMINE * NDBA * N-NITROSODIBUTYLAMINE * N-NITROSODI-n-BUTYLAMINE (MAK) * RCRA WASTE NUMBER U172

CONSENSUS REPORTS: IARC Cancer Review: GROUP 2B IMEMDT 7,56,87; Animal Sufficient Evidence IMEMDT 28,151,82; IMEMDT 17,51,78; IMEMDT 4,197,74; Human Limited Evidence IMEMDT 17,51,78. NTP Fourth Annual Report On Carcinogens,

1984. Community Right-To-Know List. EPA Genetic Toxicology Program. Reported in EPA TSCA Inventory.

DFG MAK: Animal Carcinogen, Suspected Human Carcinogen.

SAFETY PROFILE: Confirmed carcinogen with experimental carcinogenic, tumorigenic, and neoplastigenic data. Moderately toxic by ingestion, subcutaneous and intraperitoneal routes. Experimental teratogenic and reproductive effects. Human mutation data reported. When heated to decomposition it emits toxic fumes of NO_x.

BRZ000   CAS: 6558-78-7   **HR: 3**
N-BUTYL-N-NITROSO ETHYL CARBAMATE
mf: $C_7H_{14}N_2O_3$ mw: 174.23

SYNS: N-BUTYL-N-NITROSOURETHAN * 1-BUTYL-1-NITROSOURETHAN * TL 478

CONSENSUS REPORTS: EPA Genetic Toxicology Program.

SAFETY PROFILE: A poison by inhalation. Moderately toxic by ingestion. Experimental teratogenic data. Questionable carcinogen with experimental neoplastigenic and tumorigenic data. Mutation data reported. When heated to decomposition it emits toxic fumes of NO_x.

BSA250   CAS: 869-01-2   **HR: 3**
n-BUTYLNITROSOUREA
mf: $C_5H_{11}N_3O_2$ mw: 145.19

SYNS: BNU * BUTYLNITROSOHARNSTOFF (GERMAN) * N-n-BUTYL-N-NITROSOUREA * 1-BUTYL-1-NITROSOUREA * N-NITROSOBUTYLUREA

CONSENSUS REPORTS: EPA Genetic Toxicology Program.

SAFETY PROFILE: Suspected carcinogen with experimental carcinogenic and tumorigenic data. An poison by ingestion. Moderately toxic by subcutaneous route. Experimental teratogenic and reproductive effects. Mutation data reported. When heated to decomposition it emits toxic fumes of NO_x.

BSA500   CAS: 3913-02-8   **HR: 1**
2-BUTYL-1-OCTANOL
mf: $C_{12}H_{26}O$ mw: 186.38

PROP: Liquid. Mp: −80°, flash p: 230°F(OC), bp: 253.3°, d: 0.8355 @ 20°/20°, vap d: 6.42.

SYN: 2-BUTYLOCTYL ALCOHOL

CONSENSUS REPORTS: Reported in EPA TSCA Inventory.

SAFETY PROFILE: Mildly toxic by ingestion. A skin and eye irritant. Combustible when exposed to heat or flame. Incompatible with oxidizing materials. To fight fire, use CO_2, dry chemical. When heated to decomposition it emits acrid and irritating fumes.

BSB000   **HR: 1**
BUTYL OLEATE
PROP: Liquid. Bp: 173°, flash p: 356°F(OC), d: 0.873, vap d: 11.3.

SYN: (Z)-9-OCTADECENOIC ACID BUTYL ESTER

SAFETY PROFILE: A skin irritant. Combustible when exposed to heat or flame. To fight fire, use CO_2, dry chemical. Incompatible with oxidizing materials. When heated to decomposition it emits acrid smoke and irritating fumes.

BSB500   CAS: 61734-89-2   **HR: 3**
N-BUTYL-N-(2-OXOBUTYL) NITROSAMINE
mf: $C_8H_{16}N_2O_2$ mw: 172.26

SYN: N-NITROSO-N-(2-OXOBUTYL)BUTYLAMINE

CONSENSUS REPORTS: EPA Genetic Toxicology Program.

SAFETY PROFILE: Questionable carcinogen with experimental tumorigenic data. Mutation data reported. When heated to decomposition it emits toxic fumes of NO_x.

BSC250   CAS: 107-71-1   **HR: 3**
tert-BUTYL PERACETATE
DOT: UN 2096
mf: $C_6H_{12}O_3$ mw: 132.18

PROP: Clear, colorless, benzene solution; insol in water; sol in organic solvents. D: 0.923, vap press: 50 mm @ 26°, flash p: <80°F (COC).

SYNS: tert-BUTYL PEROXYACETATE * tert-BUTYL PEROXYACETATE, more than 76% in solution (DOT)
* ETHANEPEROXOIC ACID-1,1-DIMETHYLETHYL ESTER
* LUPERSOL 70

CONSENSUS REPORTS: Reported in EPA TSCA Inventory.

DFG MAK: Moderate skin irritant.
DOT Classification: Forbidden.

SAFETY PROFILE: Moderately toxic by ingestion. Mildly toxic by inhalation. Moderate skin and eye irritant. A shock and heat sensitive explosive. Dangerous fire hazard when exposed to heat, flame, reducing agents. To fight fire, use dry chemical, alcohol foam, spray and mist. When heated to decomposition it emits acrid smoke and fumes.

BSC500   CAS: 614-45-9   **HR: 3**
tert-BUTYL PERBENZOATE
DOT: UN 2890/UN 2097
mf: $C_{11}H_{14}O_3$ mw: 194.25

PROP: Colorless to slight yellow liquid, mild aromatic odor. Bp: 112° (decomp), flash p: 19°, fp: 8°, vap press: 0.33 mm @ 50°, d: 1.0. Insol in water; sol in organic solvents.

SYNS: tert-BUTYLPERBENZOAN (CZECH) * tert-BUTYL PEROXY BENZOATE * tert-BUTYL PEROXYBEN-ZOATE, technical pure or in concentration of more than 75% (DOT) * ESPEROX 10 * NOVOX * PERBEN-ZOATE de BUTYLE TERTIAIRE (FRENCH) * TRI-GONOX C

CONSENSUS REPORTS: Reported in EPA TSCA Inventory.

DOT Classification: Organic Peroxide; Label: Organic Peroxide.

SAFETY PROFILE: Moderately toxic by ingestion. A skin and eye irritant. Questionable carcinogen with experimental tumorigenic data. Mutation data reported. Potentially explosive when heated above 115°C. Explosive reaction on contact with organic matter or copper(I) bromide + limonene. When heated to decomposition it emits acrid smoke and fumes.

BSC750   CAS: 110-05-4   **HR: 2**
tert-BUTYL PEROXIDE
DOT: UN 2102
mf: $C_8H_{18}O_2$ mw: 146.26

PROP: Clear, water white liquid. Mp: −40°, bp: 80° @ 284 mm, flash p: 65°F (OC), d: 0.79, vap press: 19.51 mm @ 20°, vap d: 5.03.

SYNS: CADOX * DI-tert-BUTYLPEROXID (GERMAN) * DI-tert-BUTYL PEROXIDE (MAK) * DI-tert-BUTYL PEROXYDE (DUTCH) * DTBP * PEROSSIDO di BU-TILE TERZIARIO (ITALIAN) * PEROXYDE de BUTYLE TERTIAIRE (FRENCH) * (TRIBUTYL)PEROXIDE

CONSENSUS REPORTS: Reported in EPA ISCA Inventory

DFG MAK: Mild skin irritant.
DOT Classification: Organic Peroxide; Label: Organic Peroxide, Flammable Liquid.

SAFETY PROFILE: Moderately toxic by intraperitoneal route. A powerful irritant by ingestion and inhalation. A mild skin and eye irritant. Questionable carcinogen with experimental tumorigenic data. Warning: Water may not work to fight fire. When heated to decomposition it emits acrid smoke and fumes.

BSD000   CAS: 19910-65-7   **HR: 2**
sec-BUTYL PEROXYDICARBONATE
DOT: UN 2150/UN 2151
mf: $C_{10}H_{18}O_6$ mw: 234.28

SYNS: DI-sec-BUTYL PEROXYDICARBONATE * DI-sec-BUTYL PEROXYDICARBONATE, not more than 52% in solution (DOT) * DI-sec-BUTYL PEROXYDICAR-BONATE, technically pure (DOT)

CONSENSUS REPORTS: Reported in EPA TSCA Inventory.

DOT Classification: Organic Peroxide; Label: Organic Peroxide.

SAFETY PROFILE: Moderately toxic by skin contact. When heated to decomposition it emits acrid smoke and irritating fumes.

BSD250   CAS: 927-07-1   **HR: 3**
tert-BUTYL PEROXYPIVALATE
DOT: UN 2110
mf: $C_9H_{18}O_3$ mw: 174.27

PROP: Colorless liquid, insol in water and ethylene glycol, sol in most organic solvents. D: 0.854 @ 25°/25°, fp: < 19°, flash p: > 155°F (OC), rapid decomp @ 21°.

SYNS: ESPEROX 31M * TRIGONOZ 25-C75

CONSENSUS REPORTS: Reported in EPA TSCA Inventory.

DOT Classification: Organic Peroxide; Label: Organic Peroxide

SAFETY PROFILE: Mildly toxic by ingestion. Moderately flammable by heat, flame (sparks), oxidizers. Can explode on heating. To fight fire, use water, fog, mist, alcohol foam, dry chemical. When heated to decomposition it emits acrid smoke and fumes.

BSD500 CAS: 3180-09-4 **HR: 3**
2-n-BUTYLPHENOL

DOT: UN 2228/UN 2229
mf: $C_{10}H_{14}O$ mw: 150.24

SYNS: o-BUTYLPHENOL, liquid (DOT) * o-BUTYL-PHENOL, solid (DOT)

DOT Classification: Poison B; Label: St. Andrews Cross.

SAFETY PROFILE: Questionable carcinogen with experimental neoplastigenic data. When heated to decomposition it emits acrid smoke and irritating fumes.

BSD750 CAS: 1638-22-8 **HR: 3**
4-n-BUTYLPHENOL

DOT: UN 2228/UN 2229
mf: $C_{10}H_{14}O$ mw: 150.24

CONSENSUS REPORTS: Reported in EPA TSCA Inventory.

DOT Classification: Poison B; Label: St. Andrews Cross.

SAFETY PROFILE: A poison. Questionable carcinogen with experimental tumorigenic data. When heated to decomposition it emits acrid smoke and irritating fumes.

BSE000 CAS: 89-72-5 **HR: 3**
o-sec-BUTYLPHENOL
mf: $C_{10}H_{14}O$ mw: 150.24

PROP: Colorless liquid. Bp: 226-228° @ 25 mm, fp: 12°, flash p: 225°F, d: 0.981 @ 25°/25°.

SYN: 2-sec.-BUTYLFENOL (CZECH)

CONSENSUS REPORTS: Reported in EPA TSCA Inventory.

OSHA PEL: TWA 5 ppm (skin)
ACGIH TLV: TWA 5 ppm (skin)

SAFETY PROFILE: A poison by intraperitoneal and intravenous routes. Moderately toxic by ingestion and skin contact. A severe skin and eye

irritant. Combustible when exposed to heat or flame. To fight fire, use foam, spray, CO_2, dry chemical. When heated to decomposition it emits acrid and irritating fumes.

BSE250 CAS: 99-71-8 **HR: 3**
p-sec-BUTYL PHENOL
mf: $(CH_3CHC_2H_5)C_6H_4OH$ mw: 150.2

PROP: Nearly white flakes. Bp: 135.4-136.5° @ 25 mm, fp: 51°, flash p: 240°F, d: 0.963 @ 60°/60°.

SYN: 4-sec BUTYL PHENOL

CONSENSUS REPORTS: Reported in EPA TSCA Inventory.

SAFETY PROFILE: Poison by intravenous and intraperitoneal routes. Moderately toxic by ingestion. Combustible when exposed to heat or flame. When heated to decomposition it emits toxic fumes. To fight fire, use foam, CO_2, dry chemical. Incompatible with oxidizing materials.

BSE500 CAS: 98-54-4 **HR: 3**
4-tert-BUTYLPHENOL
mf: $C_{10}H_{14}O$ mw: 150.24

PROP: Crystals or practically white flakes. Bp: 238°, fp: 97°, d: 0.9081 @ 114°/4°, vap press: 1 mm @ 70.0°, vap d: 5.1.

SYNS: p-tert-BUTYLFENOL (CZECH) * BUTYLPHEN * p-tert-BUTYLPHENOL (MAK) * 4-(1,1-DIMETHYL-ETHYL)PHENOL * 1-HYDROXY-4-tert-BUTYLBENZENE * UCAR BUTYLPHENOL 4-T

CONSENSUS REPORTS: Reported in EPA TSCA Inventory.

DFG MAK: 0.08 ppm (0.5 mg/m³)

SAFETY PROFILE: Poison by intraperitoneal route. Moderately toxic by skin contact and ingestion. A skin and severe eye irritant. Questionable carcinogen with experimental neoplastigenic data. Combustible when exposed to heat or flame; can react with oxidizing materials. To fight fire, use foam, CO_2, dry chemical. When heated to decomposition it emits acrid and irritating fumes.

BSF250 CAS: 61005-12-7 **HR: 3**
o-sec-BUTYLPHENYL CARBAMATE
mf: $C_{12}H_{17}NO_2$ mw: 207.30

SAFETY PROFILE: A poison by ingestion. When heated to decomposition it emits toxic fumes of NO_x.

BSF750   CAS: 1126-79-0   **HR: 2**
BUTYL PHENYL ETHER
mf: $C_{10}H_{14}O$ mw: 150.24

PROP: Flash p: 180°F (OC), d: 0.9, vap d: 5.2, bp: 210°.

SYN: BUTOXYPHENYL

CONSENSUS REPORTS: Reported in EPA TSCA Inventory.

SAFETY PROFILE: Moderately toxic by ingestion. When heated to decomposition it emits acrid and irritating fumes.

BSH250   CAS: 78-48-8   **HR: 3**
BUTYL PHOSPHOROTRITHIOATE
mf: $C_{12}H_{27}OPS_3$ mw: 314.54

PROP: Liquid. Bp: 150° @ 0.3 mm. Insol in water; sol in aliphatic, aromatic, and chlorinated hydrocarbons.

SYNS: B-1,776   *   BUTIFOS   *   BUTIPHOS
*   CHEMAGRO 1,776   *   CHEMAGRO B-1776   *   DEF
*   DEF DEFOLIANT   *   DE-GREEN   *   E-Z-OFF D
*   FOS-FALL "A"   *   ORTHO PHOSPHATE DEFOLIANT
*   S,S,S-TRIBUTYL PHOSPHOROTRITHIOATE
*   S,S,S-TRIBUTYL TRITHIOPHOSPHATE

CONSENSUS REPORTS: Reported in EPA TSCA Inventory.

SAFETY PROFILE: A poison by ingestion, skin contact, intraperitoneal, and possibly other routes. Experimental reproductive effects. Animal experiments show an anti-cholinesterase effect.

BSI000   CAS: 536-69-6   **HR: 3**
5-BUTYL PICOLINIC ACID
mf: $C_{10}H_{13}NO_2$ mw: 179.24

SYNS: 5-BUTYL-2-PYRIDINECARBOXYLIC ACID
*   FUSARIC ACID   *   FUSARINIC ACID

CONSENSUS REPORTS: EPA Genetic Toxicology Program. Reported in EPA TSCA Inventory.

SAFETY PROFILE: A poison by ingestion, intraperitoneal, and intravenous routes. When

heated to decomposition it emits toxic fumes of NO_x.

BSI250   CAS: 2180-92-9   **HR: 3**
1-BUTYL-2',6'-PIPECOLOXYLIDIDE
mf: $C_{18}H_{28}N_2O$ mw: 288.48

SYNS: BUPIVACAINE   *   dl-BUPIVACAINE

SAFETY PROFILE: A poison by subcutaneous, intraperitoneal, intratracheal, parenteral, and intravenous routes. Human systemic effects by intravenous route: changes in regional blood flow rates and euphoria. When heated to decomposition it emits toxic fumes of NO_x.

BSJ500   CAS: 590-01-2   **HR: 1**
BUTYL PROPANOATE

DOT: UN 1914
mf: $C_7H_{14}O_2$ mw: 130.2

PROP: Water-white liquid, apple-like odor. Mp: −89.6°, bp: 145.4°, flash p: 90°F, d: 0.875 @ 20°, autoign temp: 800°F, vap d: 4.49.

SYNS: BUTYL PROPIONATE   *   n-BUTYL PROPIONATE   *   PROPANOIC ACID BUTYLESTER (9CI)

DOT Classification: Flammable or Combustible Liquid; Label:Flammable Liquid

SAFETY PROFILE: Mildly toxic by ingestion. A skin irritant. Dangerously flammable when exposed to heat or flame. To fight fire, use foam, CO_2, dry chemical. Incompatible with oxidizing materials.

BSK000   CAS: 98-29-3   **HR: 3**
4-tert-BUTYLPYROCATECHOL
mf: $C_{10}H_{14}O_2$ mw: 166.24

PROP: Fp: 52°, flash p: 265°F, bp: 285°, d: 1.049 @ 60°/25°.

SYNS: 4-tert-BUTYLCATECHOL   *   p-tert-BUTYLPYRO-
CATECHOL   *   4-tert-BUTYLPYROKATECHIN (CZECH)
*   4-(1,1-DIMETHYLETHYL)-1,2-BENZENEDIOL
*   SYNOX TBC

CONSENSUS REPORTS: Reported in EPA TSCA Inventory.

SAFETY PROFILE: A poison by intravenous route. Moderately toxic by ingestion and skin absorption. A severe skin and eye irritant. Combustible when exposed to heat or flame. To fight fire, use CO_2, dry chemical, fog, mist.

When heated to decomposition it emits acrid and irritating fumes.

BSL500   CAS: 2273-43-0   **HR: 3**
BUTYL STANNOIC ACID
mf: $C_4H_{10}O_2Sn$ mw: 208.83

SYN: BUTYLHYDROXYOXOSTANNANE

CONSENSUS REPORTS: Reported in EPA TSCA Inventory.

OSHA PEL: TWA 0.1 mg(Sn)/m^3 (skin)
ACGIH TLV: TWA 0.1 mg(Sn)/m^3 (skin) (Proposed: TWA 0.1 mg(Sn)/m^3; STEL 0.2 mg(Sn)/m^3 (skin))
NIOSH REL: (Organotin Compounds) TWA 0.1 mg(Sn)/m^3

SAFETY PROFILE: A poison by intravenous route. When heated to decomposition it emits acrid smoke and irritating fumes.

BSM000   CAS: 339-43-5   **HR: 3**
1-BUTYL-3-SULFANILYL UREA
mf: $C_{11}H_{17}N_3O_3S$ mw: 271.37

SYNS: ALENTIN * N-(4-AMINOBENZENESULFO-NYL)-N'-BUTYLUREA * 4-AMINO-N-((BUTYL-AMINO)CARBONYL)BENZENESULFONAMIDE * AMINOPHENUROBUTANE * BUCARBAN * BU-CROL * BUKARBAN * BURCOL * BUTISUL-FINA * N'-(BUTYLCARBAMOYL)SULFANILAMIDE * N^1-(BUTYLCARBAMOYL)SULFANILAMIDE * N-BUTYLSULFANILYUREA * CARBUTAMID * CARBUTAMIDE * CICLORAL * DIABORAL * EMEDAN * GLUCIDORAL * GLUCOFREN * GLYBUTAMIDE * INBUTON * INVENOL * NADISAN * NADIZAN * NORBORAL * ORANIL * ORANYL * ORASULIN * N^1-SUL-FANILYL-N^2-BUTYLCARBAMIDE * N^1-SULFANILYL-N^2-BUTYLUREA * N-SULFANILYL-N'BUTYLUREE (FRENCH) * U 6987

CONSENSUS REPORTS: Reported in EPA TSCA Inventory.

SAFETY PROFILE: A poison by intraperitoneal route. Moderately toxic by subcutaneous route. An experimental teratogen. When heated to decomposition it emits very toxic fumes of NO$_x$ and SO$_x$.

BSN500   CAS: 628-83-1   **HR: 3**
n-BUTYL THIOCYANATE
mf: C_5H_9NS mw: 115.21

SYNS: n-BUTYL RHODANATE * BUTYRHODANID (GERMAN) * 1-THIOCYANOBUTANE

CONSENSUS REPORTS: Reported in EPA TSCA Inventory.

SAFETY PROFILE: A poison by ingestion and subcutaneous routes. When heated to decomposition it emits very toxic fumes of NO$_x$ and SO$_x$.

BSO500   CAS: 1516-32-1   **HR: 3**
n-BUTYL THIOUREA
mf: $C_5H_{12}N_2S$ mw: 132.25

SYN: USAF D-5

CONSENSUS REPORTS: Reported in EPA TSCA Inventory.

SAFETY PROFILE: A poison by ingestion and intraperitoneal routes. When heated to decomposition it emits very toxic fumes of NO$_x$ and SO$_x$.

BSP250   CAS: 5593-70-4   **HR: 3**
BUTYL TITANATE
mf: $C_{16}H_{36}O_4 \cdot Ti$ mw: 340.42

PROP: Colorless to light yellow liquid, odor of butanol. Mp: $-55°$, bp: $312°$, flash p: 170°F, vap d: 11.5.

SYN: TETRABUTYLTITANATE (CZECH)

CONSENSUS REPORTS: Reported in EPA TSCA Inventory.

SAFETY PROFILE: A poison by intravenous route. Moderately toxic by ingestion. Flammable when exposed to heat or flame. To fight fire, use water, spray, foam, dry chemical. Incompatible with oxidizing materials. When heated to decomposition it emits acrid and irritating fumes.

BSP500   CAS: 98-51-1   **HR: 2**
p-tert-BUTYLTOLUENE
mf: $C_{11}H_{16}$ mw: 148.27

PROP: Colorless liquid.

SYNS: p-METHYL-tert-BUTYLBENZENE * 1-METH-YL-4-tert-BUTYLBENZENE * TBT

CONSENSUS REPORTS: Reported in EPA TSCA Inventory.

OSHA PEL: (Transitional: TWA 10 ppm) TWA 10 ppm; STEL 20 ppm

ACGIH TLV: TWA 10 ppm; STEL 20 ppm. DFG MAK: 10 ppm (60 mg/m^3)

SAFETY PROFILE: Moderately toxic by inhalation and ingestion. A skin and human eye irritant. Human systemic effects by inhalation: nausea or vomiting, conjunctiva irritation, unspecified effects on the sense of taste. Inhalation of vapors causes irritation of lungs and depression of central nervous system. Prolonged exposure may result in damage to liver and kidneys. Flammable when exposed to heat or flame. Incompatible with oxidizing materials. When heated to decomposition it emits acrid smoke and fumes.

BSQ000   CAS: 64-77-7   **HR: 3**
1-BUTYL-3-(p-TOLYL SULFONYL)UREA
mf: C$_{12}$H$_{18}$N$_2$O$_3$S   mw: 270.38

SYNS: AGLICID * ARKOZAL * ARTOSIN * ARTOZIN * BUTAMID * N-((BUTYLAMI-NO)CARBONYL)-4-METHYLBENZENESULFONAMIDE * 1-BUTYL-3-(p-METHYLPHENYLSULFONYL)UREA * n-BUTYL-N'-p-TOLUENESULFONYLUREA * N-n-BUTYL-N'-TOSYLUREA * 1-BUTYL-3-TOSYL-UREA * BZ 55 * D 860 * DIABEN * DIABET-AMID * DIABETOL * DIABUTON * DOLIPOL * DRABET * HLS 831 * IPOGLICONE * MO-BENOL * NCI-CO1763 * ORABET * ORALIN * OREZAN * ORINASE * ORINAZ * OTERBEN * RASTINON * SK-TOLBUTAMIDE * N-(SULFO-NYL-p-METHYLBENZENE)-N'-N-BUTYLUREA * TOL-BUSAL * TOLBUTAMID * TOLBUTAMIDE * 1-p-TOLUENESULFONYL-3-BUTYLUREA * TO-LUINA * TOLUMID * TOLUVAN * N-(p-TOLYL-SULFONYL)-N'-BUTYLCARBAMIDE * 3-(p-TOLYL-4-SULFONYL)-1-BUTYLUREA * TOLYLSULFONYLBU-TYLUREA * WILLBUTAMIDE

CONSENSUS REPORTS: NCI Carcinogenesis Bioassay (feed); No Evidence: mouse, rat NCITR* NCI-CG-TR-31,77. Reported in EPA TSCA Inventory. EPA Genetic Toxicology Program.

SAFETY PROFILE: Moderately toxic by ingestion and several other routes. A human teratogen. Human reproductive effects by ingestion and possibly other routes: stillbirth, developmental abnormalities of the cardiovascular (circulatory) system and urogenital system, and unspecified neonatal effects. Other experimental teratogenic and reproductive effects. Mutation data reported. Implicated in aplastic anemia.

When heated to decomposition it emits very toxic fumes of NO$_x$ and SO$_x$.

BSQ750   CAS: 93-79-8   **HR: 2**
BUTYL-2,4,5-TRICHLOROPHENOXY-ACETATE
mf: C$_{12}$H$_{13}$Cl$_3$O$_3$ mw: 311.60

SYNS: ARBORICID * BUTYL-2,4,5-T * BUTY-LATE-2,4,5-T * N-BUTYLESTER KYSELINI-2,4,5-TRI-CHLORFENOXYOCTOVE (CZECH) * N-BUTYL (2,4,5-TRICHLOROPHENOXY)ACETATE * FLOMORE * KILEX 3 * KRZEWOTOKS * 2,4,5-T-N-BUTYL ESTER * TORMONA * 2,4,5-TRICHLOROPHENOX-YACETIC ACID, BUTYL ESTER * TRIOXONE * U46KW

CONSENSUS REPORTS: EPA Genetic Toxicology Program.

SAFETY PROFILE: Moderately toxic by ingestion. Experimental teratogenic and reproductive effects. A skin and eye irritant. Mutation data reported. When heated to decomposition it emits toxic fumes of Cl$^-$.

BSR000   CAS: 7521-80-4   **HR: 3**
BUTYL TRICHLORO SILANE

DOT: UN 1747
mf: C$_4$H$_9$Cl$_3$Si   mw: 191.57

PROP: Liquid, vap d: 6.4, flash p: 130°F (OC), d: 1.2.

CONSENSUS REPORTS: Reported in EPA TSCA Inventory.

DOT Classification: Corrosive Material; Label: Corrosive.

SAFETY PROFILE: A corrosive poison. Flammable by heat, flame (sparks), oxidizers. To fight fire, use water to blanket fire, fog, mist, dry chemical, alcohol foam. Reacts with water or steam to produce heat and toxic and corrosive fumes. When heated to decomposition it emits highly toxic fumes of Cl$^-$.

BSS250   CAS: 592-31-4   **HR: 2**
N-BUTYLUREA
mf: C$_5$H$_{12}$N$_2$O   mw: 116.19

SYN: NCI-CO2131

CONSENSUS REPORTS: Reported in EPA TSCA Inventory. EPA Genetic Toxicology Program.

SAFETY PROFILE: Moderately toxic by parenteral route. Mutation data reported. When heated to decomposition it emits toxic fumes of NO_x.

BSS500   **HR: 3**
1-BUTYLUREA and SODIUM NITRITE (2:1)

SAFETY PROFILE: Suspected carcinogen with experimental carcinogenic, neoplastigenic, and tumorigenic data. When heated to decomposition it emits toxic fumes of NO_x. See also NITRITES.

BST500   CAS: 110-65-6   **HR: 3**
2-BUTYNE-1,4-DIOL

DOT: UN 2716
mf: $C_4H_6O_2$ mw: 86.10

PROP: Straw to amber crystals. Mp: 57.5°, bp: 194° @ 100 mm.

SYN: 1,4-BUTYNEDIOL (DOT)

CONSENSUS REPORTS: Reported in EPA TSCA Inventory.

DOT Classification: Flammable Solid; Label: Flammable Solid.

SAFETY PROFILE: A poison by ingestion. A skin sensitizer upon long or repeated contact. Moderately explosive. When heated to decomposition it emits acrid smoke and fumes and may explode. Explosive reaction with traces of alkalies, alkali earth hydroxides, halide salts, strong acids, mercury salts + strong acids.

BST750   **HR: 3**
2-BUTYNE-1-THIOL

SAFETY PROFILE: Forms an explosive polymer on exposure to air. Store at −20° in the presence of a stabilizer under nitrogen. When heated to decomposition it emits toxic fumes of SO_x.

BSU250   CAS: 123-72-8   **HR: 3**
n-BUTYRALDEHYDE

DOT: UN 1129
mf: C_4H_8O mw: 72.12

PROP: Colorless, mobile liquid; pungent, nutty odor. Mp: −100°, bp: 74.7°, flash p: 20°F (CC), (−6°), d: 0.902 @ 20°/4°, autoign temp: 446°F, lel: 2.5%, uel: 12.5%, vap d: 2.5, D: 0.797-0.802. Sol in water; misc with ether @ 74.8°.

SYNS: ALDEHYDE BUTYRIQUE (FRENCH) * ALDEIDE BUTIRRICA (ITALIAN) * BUTAL * BUTALDEHYDE * BUTALYDE * BUTANAL * n-BUTANAL (CZECH) * BUTYRAL * BUTYRALDEHYD (GERMAN) * BUTYRALDEHYDE (CZECH) * n-BUTYL ALDEHYDE * BUTYRIC ALDEHYDE * FEMA No. 2219 * NCI-C56291

CONSENSUS REPORTS: Community Right-To-Know List. Reported in EPA TSCA Inventory.

DOT Classification: Flammable Liquid; Label: Flammable Liquid.

SAFETY PROFILE: Moderately toxic by ingestion, inhalation, skin contact, intraperitoneal, and subcutaneous routes. Severe skin and eye irritant. Human immunological effects by inhalation: delayed hypersensitivity. Highly flammable liquid. To fight fire, use foam, CO_2, dry chemical. Incompatible with oxidizing materials. Reacts vigorously with chlorosulfonic acid, HNO_3, oleum, H_2SO_4. When heated to decomposition it emits acrid smoke and fumes.

BSU500   CAS: 110-69-0   **HR: 3**
m-BUTYRALDEHYDE OXIME

DOT: UN 2840
mf: C_4H_9NO mw: 87.14

PROP: Liquid. Mp: −29.5°, bp: 152°, flash p: 136°F (CC), d: 0.923, vap d: 3.01.

SYNS: BUTANAL OXIME * BUTYRALDOXIME (DOT) * N-BUTYRALDOXIME * SKINO #1 * TROYKYD ANTI-SKIN BTO * USAF AM-6

CONSENSUS REPORTS: Reported in EPA TSCA Inventory.

DOT Classification: Flammable or Combustible Liquid; Label: Flammable Liquid.

SAFETY PROFILE: A poison by intraperitoneal route. Flammable when exposed to heat or flame. To fight fire, use alcohol foam, dry chemical. Highly explosive. Can explode during vacuum distillation. Incompatible with oxidizing materials, metallic impurities. When heated to decomposition it emits toxic fumes of NO_x.

BSW000   CAS: 107-92-6   **HR: 2**
n-BUTYRIC ACID

DOT: UN 2820
mf: $C_4H_8O_2$ mw: 88.12

PROP: Colorless liquid; strong, rancid butter odor. Mp: −7.9°, bp: 163.5°, flash p: 161°F, fp: −5.5°, d: 0.9590 @ 20°/20°, refr index: 1.397, autoign temp: 846°F, vap press: 0.43 mm @ 20°, vap d: 3.04, lel: 2.0%, uel: 10.0%.

SYNS: BUTANOIC ACID * BUTTERSAEURE (GERMAN) * ETHYLACETIC ACID * FEMA No. 2221 * 1-PROPANECARBOXYLIC ACID * PROPYLFORMIC ACID

CONSENSUS REPORTS: Reported in EPA TSCA Inventory.

DOT Classification: Corrosive Material; Label: Corrosive.

SAFETY PROFILE: Moderately toxic by ingestion, skin contact, subcutaneous, intraperitoneal, and intravenous routes. Human mutation data reported. Severe skin and eye irritant. A corrosive material. Combustible liquid. Could react with oxidizing materials. Incandescent reaction with chromium trioxide above 100°. To fight fire, use alcohol foam, CO_2, dry chemical. When heated to decomposition it emits acrid smoke and irritating fumes.

BSW500   CAS: 539-90-2   **HR: 1**
BUTYRIC ACID ISOBUTYL ESTER
mf: $C_8H_{16}O_2$ mw: 144.24

PROP: Colorless liquid; apple-pineapple odor. D: 0.858-0.863, refr index: 1.402. Sol in alc, fixed oils; sltly sol in water; insol in glycerin.

SYNS: FEMA No. 2187 * ISOBUTYL BUTANOATE * ISOBUTYL BUTYRATE (FCC) * 2-METHYLPROPYL BUTYRATE

CONSENSUS REPORTS: Reported in EPA TSCA Inventory.

SAFETY PROFILE: Mildly toxic by ingestion and intraduodenal routes. A skin irritant. When heated to decomposition it emits acrid smoke and irritating fumes.

BSX000   CAS: 3068-88-0   **HR: 3**
β-BUTYROLACTONE
mf: $C_4H_6O_2$ mw: 86.10

SYNS: 3-HYDROXYBUTANOIC ACID-β-LACTONE * HYDROXYBUTYRIC ACID LACTONE * 3-HYDROXYBUTYRIC ACID LACTONE * 4-METHYL-2-OXETANONE

CONSENSUS REPORTS: IARC Cancer Review: GROUP 2B IMEMDT 7,56,87; Animal Sufficient Evidence IMEMDT 11,225,76. Reported in EPA TSCA Inventory.

SAFETY PROFILE: Suspected carcinogen with experimental carcinogenic, neoplastigenic, and tumorigenic data. Mildly toxic by ingestion. A moderate skin irritant. Mutation data reported. When heated to decomposition it emits acrid and irritating fumes.

BSX250   CAS: 109-74-0   **HR: 3**
BUTYRONITRILE
DOT: UN 2411
mf: C_4H_7N mw: 69.12

PROP: Colorless liquid. D: 0.796 @ 15°, mp: −112.6°, bp: 117°, flash p: 79°F (OC). Sltly sol in water; sol in alc and ether.

SYNS: BUTANENITRILE * n-BUTANENITRILE * BUTYRIC ACID NITRILE * BUTYRONITRILE (DOT) * 1-CYANOPROPANE * PROPYL CYANIDE

CONSENSUS REPORTS: Reported in EPA TSCA Inventory. Cyanide and its compounds are on the Community Right-To-Know List.

NIOSH REL: TWA 22 mg/m[3]
DOT Classification: Flammable Liquid; Label: Flammable Liquid and Poison.

SAFETY PROFILE: A poison by ingestion, skin contact, intraperitoneal, and subcutaneous routes. Moderately toxic by inhalation. A skin irritant. Dangerous fire hazard when exposed to heat, flame, or oxidizers. To fight fire, use alcohol foam. When heated to decomposition it emits toxic fumes of NO_x and CN^-.

BSY000   CAS: 10431-86-4   **HR: 3**
1-n-BUTYRYLAZIRIDINE
mf: $C_6H_{11}NO$ mw: 113.18

SYNS: 1-BUTYRYLAZIRIDINE * BUTYRYLETHYLENEIMINE * BUTYRYLETHYLENIMINE * 1-(1-OXOBUTYL)AZIRIDINE

SAFETY PROFILE: Moderately toxic by intraperitoneal route. Questionable carcinogen with experimental tumorigenic data. Mutation data reported. When heated to decomposition it emits toxic fumes of NO_x.

BSY250
BUTYRYL CHLORIDE
HR: 3

mf: C$_4$H$_7$ClO   mw: 106.51

PROP: Clear, colorless liquid with sharp odor. Mp: −89°, bp: 101°, d: 1.028 @ 20°/20°, vap d: 3.67, flash p: <21°.

SAFETY PROFILE: A poisonous irritant to skin, eyes, and mucous membranes. A dangerous fire hazard when exposed to heat or flame. Reaction with water, steam, or oxidizing materials produces toxic and corrosive fumes. When heated to decomposition it emits highly toxic fumes of Cl$^-$.

BTA250   CAS: 8065-36-9   **HR: 3**
BUX-TEN

PROP: A low-melting, amber solid; very sol in xylene, ethanol; nearly insol in water. Mp: 26.4°.

SYNS: BUFENCARB * BUX * METALKAMATE * METHYLCARBAMIC ACID-m-(1-METHYL)BUTYL)-PHENYL ESTER mixed with CARBAMIC ACID, METHYL-m-(1-ETHYLPROPYL)PHENYL ESTER (3:1) * ORTHO 5353

SAFETY PROFILE: A poison by ingestion and skin contact. When heated to decomposition it emits toxic fumes of NO$_x$.

BTA500   CAS: 60452-14-4   **HR: 3**
BZL

mf: C$_{20}$H$_{27}$N$_5$O$_3$ • ClH   mw: 421.98

SYN: 7-(β-DIETHYLAMINOETHYL)-8-(α-HYDROXYBENZYL) THEOPHYLLINE HYDROCHLORIDE

SAFETY PROFILE: Poison by ingestion, intraperitoneal, and intravenous routes. When heated to decomposition it emits very toxic fumes of HCl and NO$_x$.

C

CAC250   **HR: 3**
CACODYL SULFIDE
mf: $((CH_3)_2As)_2S$ mw: 242

PROP: Oily liquid, sltly sol in water. Bp: 211°.

SYN: DICACODYL SULFIDE

CONSENSUS REPORTS: Arsenic and its compounds are on the Community Right-To-Know List.

SAFETY PROFILE: Poison by most routes. Dangerous fire hazard when exposed to heat or by spontaneous chemical reaction, i.e., in air. Vigorous reaction with oxidizing materials. When heated to decomposition it emits toxic fumes of As.

CAD000   CAS: 7440-43-9   **HR: 3**
CADMIUM
af: Cd   aw: 112.40

PROP: Hexagonal crystals, silver-white, malleable metal. Mp: 320.9°, bp: 767 ± 2°, d: 8.642, vap press: 1 mm @ 394°.

SYNS: C.I. 77180 * COLLOIDAL CADMIUM * KADMIUM (GERMAN)

CONSENSUS REPORTS: IARC Cancer Review: GROUP 2A IMEMDT 7,139,87; Animal Sufficient Evidence IMEMDT 11,39,76; IMEMDT 2,74,73. NTP Fourth Annual Report On Carcinogens, 1984. Cadmium and its compounds are on the Community Right-To-Know List. Reported in EPA TSCA Inventory. EPA Genetic Toxicology Program.

OSHA PEL: Fume: TWA 0.1 mg(Cd)/m³; CL 0.6 mg(Cd)/m³; Dust: TWA 0.2 mg(Cd)/m³; CL 0.6 mg(Cd)/m³
ACGIH TLV: Dust and Salts: TWA 0.05 mg(Cd)/m³ (Proposed: TWA 0.01 mg(Cd)/m³ (dust), Human Carcinogen); BEI: 10 μg/g creatinine in urine; 10 μg/L in blood.
DFG BAT: Blood 1.5 μg/dL; Urine 15 μg/dL. MAK: Suspected Carcinogen.
NIOSH REL: (Cadmium) Reduce to lowest feasible level

SAFETY PROFILE: Confirmed human carcinogen with experimental carcinogenic, neoplastigenic, and tumorigenic data. A human poison by inhalation and possibly other routes. Poison experimentally by ingestion, inhalation, intraperitoneal, subcutaneous, intramuscular, and intravenous routes. In humans inhalation causes an excess of protein in the urine. Experimental teratogenic and reproductive effects. Mutation data reported. The dust ignites spontaneously in air and is flammable and explosive when exposed to heat, flame, or by chemical reaction with oxidizing agents, metals, HN_3, Zn, Se, and Te. Explodes on contact with hydrazoic acid. Violent or explosive reaction when heated with ammonium nitrate. Vigorous reaction when heated with nitryl fluoride. When heated to a high temperature it emits toxic fumes of Cd.

CAD250   CAS: 543-90-8   **HR: 3**
CADMIUM(II) ACETATE
DOT: NA 2570
mf: $C_2H_4O_2 \cdot 1/2Cd$ mw: 116.25

PROP: Monoclinic, colorless crystals; odor of acetic acid. Mp: 256°, bp: decomp, d: 2.341.

SYNS: ACETIC ACID, CADMIUM SALT * BIS(ACETOXY)CADMIUM * CADMIUM ACETATE (DOT) * CADMIUM DIACETATE * C.I. 77185

CONSENSUS REPORTS: Reported in EPA TSCA Inventory. EPA Genetic Toxicology Program. Cadmium and its compounds are on the Community Right-To-Know List.

OSHA PEL: TWA 0.2 mg(Cd)/m³; CL 0.6 mg(Cd)/m³ (dust)
ACGIH TLV: TWA 0.05 mg(Cd)/m³ (Proposed: TWA 0.01 mg(Cd)/m³ (dust), Human Carcinogen); BEI: 10 μg/g creatinine in urine; 10 μg/L in blood.
NIOSH REL: (Cadmium) Reduce to lowest feasible level

SAFETY PROFILE: Confirmed human carcinogen. Poison by intraperitoneal route. Experimental teratogenic and reproductive effects. Human mutation data reported. When heated to decomposition it emits toxic fumes of Cd.

CAD325   CAS: 22750-53-4   **HR: 3**
CADMIUM AMIDE
mf: CdH_4N_2 mw: 144.46

SYN: CADMIUM DIAMIDE

CONSENSUS REPORTS: Cadmium compounds are on the Community Right-To-Know List.

OSHA PEL: TWA 0.2 mg(Cd)/m^3; CL 0.6 mg(Cd)/m^3 (dust)

ACGIH TLV: TWA 0.05 mg(Cd)/m^3 (Proposed: TWA 0.01 mg(Cd)/m^3 (dust), Human Carcinogen); BEI: 10 μg/g creatinine in urine; 10 μg/L in blood.

NIOSH REL: (Cadmium) Reduce to lowest feasible level

SAFETY PROFILE: Confirmed human carcinogen. May explode if heated. Reacts violently with water. When heated to decomposition it emits toxic fumes of Cd and NO$_x$.

CAD350   CAS: 14215-29-3   **HR: 3**
CADMIUM AZIDE
mf: CdN$_6$ mw: 196.45

SYN: CADMIUM DIAZIDE

CONSENSUS REPORTS: Cadmium compounds are on the Community Right-To-Know List.

OSHA PEL: TWA 0.2 mg(Cd)/m^3; CL 0.6 mg(Cd)/m^3 (dust)

ACGIH TLV: TWA 0.05 mg(Cd)/m^3 (Proposed: TWA 0.01 mg(Cd)/m^3 (dust), Human Carcinogen); BEI: 10 μg/g creatinine in urine; 10 μg/L in blood.

NIOSH REL: (Cadmium) Reduce to lowest feasible level

SAFETY PROFILE: Confirmed human carcinogen. The dry solid is an unstable heat- and friction-sensitive explosive. When heated to decomposition it emits toxic fumes of NO$_x$ and Cd.

CAD500   CAS: 7495-93-4   **HR: 3**
CADMIUM BIS(2-ETHYLHEXYL) PHOSPHITE
mf: C$_{32}$H$_{68}$O$_6$P$_2$ • Cd   mw: 723.34

SYN: BIS(2-ETHYLHEXYL) ESTER PHOSPHORUS ACID CADMIUM SALT

CONSENSUS REPORTS: Cadmium and its compounds are on the Community Right-To-Know List.

OSHA PEL: TWA 0.2 mg(Cd)/m^3; CL 0.6 mg(Cd)/m^3 (dust)

ACGIH TLV: TWA 0.05 mg(Cd)/m^3 (Proposed: TWA 0.01 mg(Cd)/m^3 (dust), Human Carci-

nogen); BEI: 10 μg/g creatinine in urine; 10 μg/L in blood.

NIOSH REL: (Cadmium) Reduce to lowest feasible level

SAFETY PROFILE: Confirmed human carcinogen. Poison by intraperitoneal route. When heated to decomposition it emits toxic fumes of PO$_x$ and Cd.

CAD750   CAS: 2191-10-8   **HR: 3**
CADMIUM CAPRYLATE
mf: C$_{16}$H$_{30}$O$_4$ • Cd   mw: 398.86

SYN: OCTANOIC ACID, CADMIUM SALT (2:1)

CONSENSUS REPORTS: Reported in EPA TSCA Inventory. Cadmium and its compounds are on the Community Right-To-Know List.

OSHA PEL: TWA 0.2 mg(Cd)/m^3; CL 0.6 mg(Cd)/m^3 (dust)

ACGIH TLV: TWA 0.05 mg(Cd)/m^3 (Proposed: TWA 0.01 mg(Cd)/m^3 (dust), Human Carcinogen); BEI: 10 μg/g creatinine in urine; 10 μg/L in blood.

NIOSH REL: (Cadmium) Reduce to lowest feasible level

SAFETY PROFILE: Confirmed human carcinogen. Poison by ingestion and intratracheal routes. When heated to decomposition it emits toxic fumes of Cd.

CAE000   **HR: 3**
CADMIUM CHLORATE
mf: CdCl$_2$O$_6$ mw: 279.31

PROP: Colorless, deliquescent prisms. Mp: 80°, d: 2.28 @ 18°.

CONSENSUS REPORTS: Cadmium and its compounds are on the Community Right-To-Know List.

OSHA PEL: TWA 0.2 mg(Cd)/m^3; CL 0.6 mg(Cd)/m^3 (dust)

ACGIH TLV: TWA 0.05 mg(Cd)/m^3 (Proposed: TWA 0.01 mg(Cd)/m^3 (dust), Human Carcinogen); BEI: 10 μg/g creatinine in urine; 10 μg/L in blood.

NIOSH REL: (Cadmium) Reduce to lowest feasible level

SAFETY PROFILE: Confirmed human carcinogen. A powerful oxidizing agent. Flammable by chemical reaction with reducing agents. Moderate explosion hazard when shocked or exposed to heat. Violent or explosive reaction with sulfides (e.g., copper(II) sulfide (explodes), antimony(II) sulfide, arsenic(III) sulfide, tin(II) sulfide, tin(IV) sulfide. When heated to decomposition it emits toxic fumes of Cd and Cl^-.

CAE250   CAS: 10108-64-2   **HR: 3**
CADMIUM CHLORIDE
mf: $CdCl_2$ mw: 183.30

DOT: UN 2570

PROP: Hexagonal, colorless crystals. Mp: 568°, d: 4.047 @ 25°, vap press: 10 mm @ 656°, bp: 960°.

SYNS: CADDY * CADMIUM DICHLORIDE * KADMIUMCHLORID (GERMAN) * VI-CAD

CONSENSUS REPORTS: IARC Cancer Review: GROUP 2A IMEMDT 7,139,87; Animal Sufficient Evidence IMEMDT 11,39,76; IMEMDT 2,74,73. NTP Fourth Annual Report On Carcinogens, 1984. EPA Genetic Toxicology Program. Cadmium and its compounds are on the Community Right-To-Know List. Reported in EPA TSCA Inventory.

OSHA PEL: TWA 0.2 mg(Cd)/m^3; CL 0.6 mg(Cd)/m^3 (dust)
ACGIH TLV: TWA 0.05 mg(Cd)/m^3 (Proposed: TWA 0.01 mg(Cd)/m^3 (dust), Human Carcinogen); BEI: 10 µg/g creatinine in urine; 10 µg/L in blood.
DFG MAK: Animal Carcinogen, Suspected Human Carcinogen.
NIOSH REL: (Cadmium) Reduce to lowest feasible level
DOT Classification: ORM-E; Label: None.

SAFETY PROFILE: Confirmed human carcinogen with experimental carcinogenic, and tumorigenic data. Poison by ingestion, inhalation, skin contact, intraperitoneal, subcutaneous, intravenous and possibly other routes. Experimental teratogenic and reproductive effects. Human mutation data reported. Reacts violently with BrF_3 and K. When heated to decomposition it emits very toxic fumes of Cd and Cl^-.

CAE375   CAS: 72589-96-9   **HR: 3**
CADMIUM CHLORIDE, DIHYDRATE
mf: $CdCl_2 \cdot 2H_2O$ mw: 219.34

CONSENSUS REPORTS: Cadmium and its compounds are on the Community Right-To-Know List.

OSHA PEL: TWA 0.2 mg(Cd)/m^3; CL 0.6 mg(Cd)/m^3 (dust)
ACGIH TLV: TWA 0.05 mg(Cd)/m^3 (Proposed: TWA 0.01 mg(Cd)/m^3 (dust), Human Carcinogen); BEI: 10 µg/g creatinine in urine; 10 µg/L in blood.
DFG MAK: Animal Carcinogen, Suspected Human Carcinogen.
NIOSH REL: (Cadmium) Reduce to lowest feasible level

SAFETY PROFILE: Confirmed human carcinogen with experimental tumorigenic data. Experimental reproductive effects. When heated to decomposition it emits toxic fumes of Cl^- and Cd.

CAE425   CAS: 7790-78-5   **HR: 3**
CADMIUM CHLORIDE, HYDRATE (2:5)
mf: $CdCl_2 \cdot 5/2H_2O$ mw: 228.35

CONSENSUS REPORTS: Cadmium and its compounds are on the Community Right-To-Know List.

OSHA PEL: TWA 0.2 mg(Cd)/m^3; CL 0.6 mg(Cd)/m^3 (dust)
ACGIH TLV: TWA 0.05 mg(Cd)/m^3 (Proposed: TWA 0.01 mg(Cd)/m^3 (dust), Human Carcinogen); BEI: 10 µg/g creatinine in urine; 10 µg/L in blood.
DFG MAK: Animal Carcinogen, Suspected Human Carcinogen.
NIOSH REL: (Cadmium) Reduce to lowest feasible level

SAFETY PROFILE: Confirmed human carcinogen. Poison by intraperitoneal route. Experimental reproductive effects. Human mutation data reported. When heated to decomposition it emits toxic fumes of Cl^- and Cd.

CAE500   CAS: 35658-65-2   **HR: 3**
CADMIUM CHLORIDE, MONOHYDRATE
mf: $CdCl_2 \cdot H_2O$ mw: 201.32

CONSENSUS REPORTS: Cadmium and its compounds are on the Community Right-To-Know List.

OSHA PEL: TWA 0.2 mg(Cd)/m^3; CL 0.6 mg(Cd)/m^3 (dust)
ACGIH TLV: TWA 0.05 mg(Cd)/m^3 (Proposed: TWA 0.01 mg(Cd)/m^3 (dust), Human Carcinogen); BEI: 10 µg/g creatinine in urine; 10 µg/L in blood.
DFG MAK: Animal Carcinogen, Suspected Human Carcinogen.

NIOSH REL: (Cadmium) Reduce to lowest feasible level

SAFETY PROFILE: Confirmed human carcinogen with experimental carcinogenic and tumorigenic data. When heated to decomposition it emits very toxic fumes of Cd and Cl$^-$.

CAE750 HR: 3
CADMIUM COMPOUNDS

OSHA PEL: Fume: TWA 0.1 mg(Cd)/m^3; CL 0.6 mg(Cd)/m^3; Dust: TWA 0.2 mg(Cd)/m^3; CL 0.6 mg(Cd)/m^3
ACGIH TLV: Dust and Salts: TWA 0.05 mg(Cd)/m^3 (Proposed: TWA 0.01 mg(Cd)/m^3 (dust), Human Carcinogen); BEI: 10 μg/g creatinine in urine; 10 μg/L in blood.
DFG BAT: Blood 1.5 μg/dL; Urine 15 μg/dL. MAK: Suspected Carcinogen.
NIOSH REL: (Cadmium) Reduce to lowest feasible level

SAFETY PROFILE: Confirmed human carcinogen producing lung tumors. Poison by ingestion. The irritating and emetic action is so violent, however, that little of the cadmium has time to be absorbed and fatal poisoning rarely ensues. Experimental carcinogens and teratogens. Cases of human poisoning have been reported from ingestion of food or beverages prepared or stored in cadmium-plated containers. Inhalation of fumes or dusts affects the respiratory tract and the kidneys. Brief exposure to high concentrations may result in pulmonary edema and death. Fatal concentrations may be breathed without sufficient discomfort to warn a worker to leave the exposure. Cadmium oxide fumes can cause metal fume fever resembling that caused by zinc oxide fumes. When heated to decomposition they emit toxic fumes of Cd.

CAF500 HR: 3
CADMIUM DICYANIDE
mf: C$_2$CdN$_2$ mw: 164.44

CONSENSUS REPORTS: Cadmium and its compounds and Cyanide and its compounds are on the Community Right-To-Know List.

OSHA PEL: TWA 0.2 mg(Cd)/m^3; CL 0.6 mg(Cd)/m^3 (dust)
ACGIH TLV: TWA 0.05 mg(Cd)/m^3 (Proposed: TWA 0.01 mg(Cd)/m^3 (dust), Human Carcinogen); BEI: 10 μg/g creatinine in urine; 10 μg/L in blood.

NIOSH REL: (Cadmium) Reduce to lowest feasible level

SAFETY PROFILE: Confirmed human carcinogen. A poison. Incompatible with magnesium. When heated to decomposition it emits toxic fumes of Cd and CN$^-$.

CAF750 CAS: 15954-91-3 HR: 3
CADMIUM(II) EDTA COMPLEX

SYN: (ETHYLENEDINITRILO)TETRAACETIC ACID CADMIUM(II) COMPLEX

CONSENSUS REPORTS: Cadmium and its compounds are on the Community Right-To-Know List.

OSHA PEL: TWA 0.2 mg(Cd)/m^3; CL 0.6 mg(Cd)/m^3 (dust)
ACGIH TLV: TWA 0.05 mg(Cd)/m^3 (Proposed: TWA 0.01 mg(Cd)/m^3 (dust), Human Carcinogen); BEI: 10 μg/g creatinine in urine; 10 μg/L in blood.
NIOSH REL: (Cadmium) Reduce to lowest feasible level

SAFETY PROFILE: Confirmed human carcinogen. Poison by intraperitoneal route. When heated to decomposition it emits toxic fumes of NO$_x$ and Cd.

CAG000 CAS: 14486-19-2 HR: 3
CADMIUM FLUOBORATE
mf: B$_2$CdF$_8$ mw: 286.02

SYNS: CADMIUM FLUOROBORATE * TL 1026

CONSENSUS REPORTS: Reported in EPA TSCA Inventory. Cadmium and its compounds are on the Community Right-To-Know List.

OSHA PEL: TWA 0.2 mg(Cd)/m^3; CL 0.6 mg(Cd)/m^3 (dust); 2.5 mg(F)/m^3
ACGIH TLV: TWA 0.05 mg(Cd)/m^3 (Proposed: TWA 0.01 mg(Cd)/m^3 (dust), Human Carcinogen); BEI: 10 μg/g creatinine in urine; 10 μg/L in blood.
NIOSH REL: (Cadmium) Reduce to lowest feasible level

SAFETY PROFILE: Confirmed human carcinogen. Poison by ingestion and inhalation. When heated to decomposition it emits very toxic fumes of Cd and F$^-$.

CAG250 CAS: 7790-79-6 HR: 3
CADMIUM FLUORIDE
mf: CdF$_2$ mw: 150.40

PROP: Cubic, white crystals. Mp: 1100°, bp: 1758°, d: 6.64, vap press: 1 mm @ 1112°.

SYN: CADMIUM FLUORURE (FRENCH)

CONSENSUS REPORTS: Reported in EPA TSCA Inventory. Cadmium and its compounds are on the Community Right-To-Know List.

OSHA PEL: TWA 0.2 mg(Cd)/m^3; CL 0.6 mg(Cd)/m^3 (dust); 2.5 mg(F)/m^3
ACGIH TLV: TWA 0.05 mg(Cd)/m^3 (Proposed: TWA 0.01 mg(Cd)/m^3 (dust), Human Carcinogen); BEI: 10 μg/g creatinine in urine; 10 μg/L in blood.
NIOSH REL: (Cadmium) Reduce to lowest feasible level

SAFETY PROFILE: Confirmed human carcinogen. Poison by subcutaneous route. Violent reaction with K. When heated to decomposition it emits very toxic fumes of Cd and F$^-$.

CAG500   CAS: 17010-21-8   **HR: 3**
CADMIUM FLUOSILICATE
mf: CdF$_6$Si   mw: 254.49

PROP: Hexagonal, colorless crystals.

SYN: TL 1070

CONSENSUS REPORTS: Cadmium and its compounds are on the Community Right-To-Know List.

OSHA PEL: TWA 0.2 mg(Cd)/m^3; CL 0.6 mg(Cd)/m^3 (dust); 2.5 mg(F)/m^3
ACGIH TLV: TWA 0.05 mg(Cd)/m^3 (Proposed: TWA 0.01 mg(Cd)/m^3 (dust), Human Carcinogen); BEI: 10 μg/g creatinine in urine; 10 μg/L in blood.
NIOSH REL: (Cadmium) Reduce to lowest feasible level

SAFETY PROFILE: Confirmed human carcinogen. Poison by ingestion and inhalation. When heated to decomposition it emits very toxic fumes of Cd and F$^-$.

CAG750   CAS: 16039-55-7   **HR: 3**
CADMIUM LACTATE
mf: C$_6$H$_{10}$O$_6$•Cd   mw: 290.56

PROP: Needles.

SYN: LACTIC ACID, CADMIUM SALT

CONSENSUS REPORTS: Cadmium and its compounds are on the Community Right-To-Know List.

OSHA PEL: TWA 0.2 mg(Cd)/m^3; CL 0.6 mg(Cd)/m^3 (dust)
ACGIH TLV: TWA 0.05 mg(Cd)/m^3 (Proposed: TWA 0.01 mg(Cd)/m^3 (dust), Human Carcinogen); BEI: 10 μg/g creatinine in urine; 10 μg/L in blood.
NIOSH REL: (Cadmium) Reduce to lowest feasible level

SAFETY PROFILE: Confirmed human carcinogen. A poison. When heated to decomposition it emits toxic fumes of Cd.

CAH000   CAS: 10325-94-7   **HR: 3**
CADMIUM NITRATE
mf: CdN$_2$O$_6$ mw: 236.42

PROP: White, prismatic needles; hygroscopic. Mp: 350°.

SYNS: CADMIUM DINITRATE   *   NITRIC ACID, CADMIUM SALT

CONSENSUS REPORTS: Reported in EPA TSCA Inventory. EPA Genetic Toxicology Program. Cadmium and its compounds are on the Community Right-To-Know List.

OSHA PEL: TWA 0.2 mg(Cd)/m^3; CL 0.6 mg(Cd)/m^3 (dust)
ACGIH TLV: TWA 0.05 mg(Cd)/m^3 (Proposed: TWA 0.01 mg(Cd)/m^3 (dust), Human Carcinogen); BEI: 10 μg/g creatinine in urine; 10 μg/L in blood.
NIOSH REL: (Cadmium) Reduce to lowest feasible level

SAFETY PROFILE: Confirmed human carcinogen. Poison by ingestion and possibly other routes. Moderately toxic by inhalation. Mutation data reported. When heated to decomposition it emits very toxic fumes of Cd and NO$_x$.

CAH250   CAS: 10022-68-1   **HR: 3**
**CADMIUM(II) NITRATE
TETRAHYDRATE (1:2:4)**
mf: N$_2$O$_6$•Cd•4H$_2$O   mw: 308.50

SYNS: DUSICNAN KADEMNATY (CZECH)   *   NITRIC ACID, CADMIUM SALT, TETRAHYDRATE

CONSENSUS REPORTS: Cadmium and its compounds are on the Community Right-To-Know List.

OSHA PEL: TWA 0.2 mg(Cd)/m^3; CL 0.6 mg(Cd)/m^3 (dust)
ACGIH TLV: TWA 0.05 mg(Cd)/m^3 (Proposed: TWA 0.01 mg(Cd)/m^3 (dust), Human Car-

cinogen); BEI: 10 μg/g creatinine in urine; 10 μg/L in blood.

NIOSH REL: (Cadmium) Reduce to lowest feasible level

SAFETY PROFILE: Confirmed human carcinogen. Poison by ingestion. A severe skin and moderate eye irritant. Mutation data reported. When heated to decomposition it emits very toxic fumes of Cd and NO_x.

CAH500   CAS: 1306-19-0   HR: 3
CADMIUM OXIDE
mf: CdO   mw: 128.40

PROP: (1) Amorphous, brown crystals; (2) cubic, brown crystals. Mp (1): <1426°, mp (2): decomp @ 950°, bp: 1559°, d (1): 6.95, d (2): 8.15, vap press: 1 mm @ 1000°.

SYNS: KADMU TLENEK (POLISH)   *   NCI-C02551

CONSENSUS REPORTS: IARC Cancer Review: GROUP 2A IMEMDT 7,139,87; Human Inadequate Evidence IMEMDT 2,74,73; IMEMDT 11,39,76; Animal Sufficient Evidence IMEMDT 11,39,76; IMEMDT 2,74,73. Reported in EPA TSCA Inventory. EPA Extremely Hazardous Substances List. Cadmium and its compounds are on the Community Right-To-Know List.

OSHA PEL: TWA 0.2 mg(Cd)/m³; CL 0.6 mg(Cd)/m³ (dust)
ACGIH TLV: TWA 0.05 mg(Cd)/m³ (Proposed: TWA 0.01 mg(Cd)/m³ (dust), Human Carcinogen); BEI: 10 μg/g creatinine in urine; 10 μg/L in blood.
DFG MAK: Suspected Carcinogen.
NIOSH REL: (Cadmium) Reduce to lowest feasible level

SAFETY PROFILE: Confirmed human carcinogen with experimental neoplastigenic data. Poison by ingestion, inhalation, and intraperitoneal routes. Experimental reproductive effects. Human systemic effects by inhalation include: change in the sense of smell, change in heart rate, blood pressure increase, an excess of protein in the urine and other kidney or bladder changes. Mixtures with magnesium explode when heated. When heated to decomposition it emits toxic fumes of Cd.

CAH750   HR: 3
CADMIUM OXIDE FUME

SYN: CADMIUM FUME

CONSENSUS REPORTS: Reported in EPA TSCA Inventory. Cadmium and its compounds are on the Community Right-To-Know List.

OSHA PEL: TWA 0.1 mg(Cd)/m³; CL 0.3 mg(Cd)/m³
ACGIH TLV: TWA 0.05 mg(Cd)/m³ (Proposed: TWA 0.01 mg(Cd)/m³ (dust), Human Carcinogen); BEI: 10 μg/g creatinine in urine; 10 μg/L in blood.
NIOSH REL: (Cadmium) Reduce to lowest feasible level

SAFETY PROFILE: Confirmed human carcinogen. Poison by inhalation. Moderately toxic to humans by inhalation. Human pulmonary system effects by inhalation including: coughing, difficult breathing, and cyanosis. A strong irritant via inhalation. When heated to decomposition it emits toxic fumes of Cd.

CAI000   CAS: 13477-17-3   HR: 3
CADMIUM PHOSPHATE
mf: $Cd_3O_8P_2 \cdot 4H_2O$ mw: 599.22

PROP: Amorphous or colorless crystals. Mp: 1500°.

SYN: TL 1182

CONSENSUS REPORTS: Reported in EPA TSCA Inventory. Cadmium and its compounds are on the Community Right-To-Know List.

OSHA PEL: TWA 0.2 mg(Cd)/m³; CL 0.6 mg(Cd)/m³ (dust)
ACGIH TLV: TWA 0.05 mg(Cd)/m³ (Proposed: TWA 0.01 mg(Cd)/m³ (dust), Human Carcinogen); BEI: 10 μg/g creatinine in urine; 10 μg/L in blood.
NIOSH REL: (Cadmium) Reduce to lowest feasible level

SAFETY PROFILE: Confirmed human carcinogen. Poison by inhalation. When heated to decomposition it emits toxic fumes of Cd and PO_x.

CAI125   CAS: 12014-28-7   HR: 3
CADMIUM PHOSPHIDE
mf: Cd_3P_2 mw: 399.18

CONSENSUS REPORTS: Cadmium compounds are on the Community Right-To-Know List.

OSHA PEL: Fume: TWA 0.1 mg(Cd)/m³; CL 0.6 mg(Cd)/m³; Dust: TWA 0.2 mg(Cd)/m³; CL 0.6 mg(Cd)/m³

ACGIH TLV: Dust and Salts: TWA 0.05 mg(Cd)/m^3 (Proposed: TWA 0.01 mg(Cd)/ m^3 (dust), Human Carcinogen); BEI: 10 μg/ g creatinine in urine; 10 μg/L in blood.

DFG BAT: Blood 1.5 μg/dL; Urine 15 μg/dL. MAK: Suspected Carcinogen.

NIOSH REL: (Cadmium) Reduce to lowest feasible level

SAFETY PROFILE: Confirmed carcinogen. Explosive reaction with concentrated nitric acid. When heated to decomposition it emits toxic fumes of PO$_x$ and Cd.

CAI250   HR: 3
CADMIUM PROPIONATE
mf: C$_6$H$_{10}$CdO$_5$ mw: 258.55

CONSENSUS REPORTS: Cadmium and its compounds are on the Community Right-To-Know List.

OSHA PEL: Fume: TWA 0.1 mg(Cd)/m^3; CL 0.6 mg(Cd)/m^3; Dust: TWA 0.2 mg(Cd)/m^3; CL 0.6 mg(Cd)/m^3

ACGIH TLV: Dust and Salts: TWA 0.05 mg(Cd)/m^3 (Proposed: TWA 0.01 mg(Cd)/ m^3 (dust), Human Carcinogen); BEI: 10 μg/ g creatinine in urine; 10 μg/L in blood.

DFG BAT: Blood 1.5 μg/dL; Urine 15 μg/dL. MAK: Suspected Carcinogen.

NIOSH REL: (Cadmium) Reduce to lowest feasible level

SAFETY PROFILE: Confirmed carcinogen. The salt has exploded. Incompatible with 3-pentanone vapor. When heated to decomposition it emits toxic fumes of Cd.

CAI500   HR: 3
CADMIUM SELENIDE
mf: CdSe   mw: 191.36

PROP: Preparative hazard.

CONSENSUS REPORTS: Cadmium and its compounds as well as Selenium and its compounds are on the Community Right-To-Know List.

OSHA PEL: Fume: TWA 0.1 mg(Cd)/m^3; CL 0.6 mg(Cd)/m^3; Dust: TWA 0.2 mg(Cd)/m^3; CL 0.6 mg(Cd)/m^3

ACGIH TLV: Dust and Salts: TWA 0.05 mg(Cd)/m^3 (Proposed: TWA 0.01 mg(Cd)/ m^3 (dust), Human Carcinogen); BEI: 10 μg/ g creatinine in urine; 10 μg/L in blood.

DFG BAT: Blood 1.5 μg/dL; Urine 15 μg/dL. MAK: Suspected Carcinogen; 0.1 mg(Se)/ m^3.

NIOSH REL: (Cadmium) Reduce to lowest feasible level

SAFETY PROFILE: Confirmed carcinogen. Selenium compounds are considered to be poisons. When heated to decomposition it emits toxic fumes of Cd and Se.

CAI750   CAS: 141-00-4   HR: 3
CADMIUM SUCCINATE
mf: C$_4$H$_4$O$_4$•Cd   mw: 228.48

SYNS: CADMINATE   *   SUCCINIC ACID, CADMIUM SALT (1:1)

CONSENSUS REPORTS: Reported in EPA TSCA Inventory. Cadmium and its compounds are on the Community Right-To-Know List.

OSHA PEL: Fume: TWA 0.1 mg(Cd)/m^3; CL 0.6 mg(Cd)/m^3; Dust: TWA 0.2 mg(Cd)/m^3; CL 0.6 mg(Cd)/m^3

ACGIH TLV: Dust and Salts: TWA 0.05 mg(Cd)/m^3 (Proposed: TWA 0.01 mg(Cd)/ m^3 (dust), Human Carcinogen); BEI: 10 μg/ g creatinine in urine; 10 μg/L in blood.

DFG BAT: Blood 1.5 μg/dL; Urine 15 μg/dL. MAK: Suspected Carcinogen.

NIOSH REL: (Cadmium) Reduce to lowest feasible level

SAFETY PROFILE: Confirmed carcinogen. Poison by ingestion and intraperitoneal routes. Moderately toxic by ingestion. When heated to decomposition it emits toxic fumes of Cd.

CAJ000   CAS: 10124-36-4   HR: 3
CADMIUM SULFATE (1:1)
mf: O$_4$S•Cd   mw: 208.46

PROP: Rhombic, white crystals. Mp: 1000°, d: 4.691.

SYNS: CADMIUM SULFATE   *   CADMIUM SULPHATE * SULFURIC ACID, CADMIUM(2+) SALT   *   SULPHURIC ACID, CADMIUM SALT (1:1)

CONSENSUS REPORTS: IARC Cancer Review: GROUP 2A IMEMDT 7,139,87; Animal Sufficient Evidence IMEMDT 11,39,76; IMEMDT 2,74,73. Reported in EPA TSCA Inventory. EPA Genetic Toxicology Program. Cadmium and its compounds are on the Community Right-To-Know List.

OSHA PEL: TWA 0.2 mg(Cd)/m^3; CL 0.6 mg(Cd)/m^3 (dust)
ACGIH TLV: TWA 0.05 mg(Cd)/m^3 (Proposed: TWA 0.01 mg(Cd)/m^3 (dust), Human Carcinogen); BEI: 10 μg/g creatinine in urine; 10 μg/L in blood.
DFG MAK: Suspected Carcinogen.
NIOSH REL: (Cadmium) Reduce to lowest feasible level

SAFETY PROFILE: Confirmed human carcinogen with experimental carcinogenic data. Poison by ingestion, subcutaneous, and intraperitoneal routes. Experimental teratogenic and reproductive effects. Mutation data reported. When heated to decomposition it emits very toxic fumes of Cd and SO$_x$.

CAJ250   CAS: 7790-84-3   HR: 3
CADMIUM SULFATE (1:1)
HYDRATE (3:8)
mf: O$_4$S • Cd • 8/3H$_2$O   mw: 256.51

SYNS: CADMIUM SULFATE OCTAHYDRATE * SULFURIC ACID, CADMIUM SALT, HYDRATE

CONSENSUS REPORTS: IARC Cancer Review: Animal Sufficient Evidence IMEMDT 2,74,73. Cadmium and its compounds are on the Community Right-To-Know List.

OSHA PEL: TWA 0.2 mg(Cd)/m^3; CL 0.6 mg(Cd)/m^3 (dust)
ACGIH TLV: TWA 0.05 mg(Cd)/m^3 (Proposed: TWA 0.01 mg(Cd)/m^3 (dust), Human Carcinogen); BEI: 10 μg/g creatinine in urine; 10 μg/L in blood.
NIOSH REL: (Cadmium) Reduce to lowest feasible level

SAFETY PROFILE: Confirmed human carcinogen with experimental tumorigenic and neoplastigenic data. Experimental teratogenic and reproductive effects. Mutation data reported. When heated to decomposition it emits very toxic fumes of Cd and SO$_x$.

CAJ500   CAS: 13477-21-9   HR: 3
CADMIUM SULFATE TETRAHYDRATE
mf: O$_4$S • Cd • 4H$_2$O   mw: 280.54

SYN: SULFURIC ACID, CADMIUM SALT, TETRAHYDRATE

CONSENSUS REPORTS: Cadmium and its compounds are on the Community Right-To-Know List.

OSHA PEL: TWA 0.2 mg(Cd)/m^3; CL 0.6 mg(Cd)/m^3 (dust)
ACGIH TLV: TWA 0.05 mg(Cd)/m^3 (Proposed: TWA 0.01 mg(Cd)/m^3 (dust), Human Carcinogen); BEI: 10 μg/g creatinine in urine; 10 μg/L in blood.
NIOSH REL: (Cadmium) Reduce to lowest feasible level

SAFETY PROFILE: Confirmed human carcinogen with experimental neoplastigenic data. When heated to decomposition it emits very toxic fumes of Cd and SO$_x$.

CAJ750   CAS: 1306-23-6   HR: 3
CADMIUM SULFIDE
mf: CdS   mw: 144.46

PROP: Hexagonal, yellow-orange crystals. Mp: 1750 @ 100 atm, bp: subl in N$_2$, d: 4.82.

SYNS: AURORA YELLOW * CADMIUM GOLDEN 366 * CADMIUM LEMON YELLOW 527 * CADMIUM ORANGE * CADMIUM PRIMROSE 819 * CADMIUM SULPHIDE * CADMIUM YELLOW * CADMOPUR YELLOW * CAPSEBON * C.I. 77199 * C.I. PIGMENT ORANGE 20 * C.I. PIGMENT YELLOW 37 * FERRO YELLOW * GREENOCKITE * NCI-C02711

CONSENSUS REPORTS: IARC Cancer Review: GROUP 2A IMEMDT 7,139,87; Animal Sufficient Evidence IMEMDT 11,39,76; IMEMDT 2,74,73. NTP Fourth Annual Report On Carcinogens, 1984. EPA Genetic Toxicology Program. Cadmium and its compounds are on the Community Right-To-Know List. Reported in EPA TSCA Inventory.

OSHA PEL: TWA 0.2 mg(Cd)/m^3; CL 0.6 mg(Cd)/m^3 (dust)
ACGIH TLV: TWA 0.05 mg(Cd)/m^3 (Proposed: TWA 0.01 mg(Cd)/m^3 (dust), Human Carcinogen); BEI: 10 μg/g creatinine in urine; 10 μg/L in blood.
DFG MAK: Suspected Carcinogen.
NIOSH REL: (Cadmium) Reduce to lowest feasible level

SAFETY PROFILE: Confirmed human carcinogen with experimental carcinogenic and tumorigenic data. Moderately toxic by ingestion and inhalation. Human mutation data reported. When heated to decomposition it emits very toxic fumes of Cd and SO$_x$.

CAK000   **HR: 3**
CADMIUM THERMOVACUUM AEROSOL

SYN: AEROSOL of THERMOVACUUM CADMIUM

CONSENSUS REPORTS: Cadmium and its compounds are on the Community Right-To-Know List.

OSHA PEL: TWA 0.2 mg(Cd)/m^3; CL 0.6 mg(Cd)/m^3 (dust)

ACGIH TLV: TWA 0.05 mg(Cd)/m^3 (Proposed: TWA 0.01 mg(Cd)/m^3 (dust), Human Carcinogen); BEI: 10 μg/g creatinine in urine; 10 μg/L in blood.

NIOSH REL: (Cadmium) Reduce to lowest feasible level

SAFETY PROFILE: Confirmed human carcinogen. Moderately toxic by an unspecified route. When heated to decomposition it emits very toxic fumes of Cd.

CAK250   CAS: 73419-42-8   **HR: 3**
CADMIUM-THIONEIN
mf: C$_{18}$H$_{30}$N$_6$O$_4$S$_2$ • Cd   mw: 571.06

PROP: Cadmium(II) is bound to the protein thioneine from rat or rabbit liver.

CONSENSUS REPORTS: Cadmium and its compounds are on the Community Right-To-Know List.

OSHA PEL: TWA 0.2 mg(Cd)/m^3; CL 0.6 mg(Cd)/m^3 (dust)

ACGIH TLV: TWA 0.05 mg(Cd)/m^3 (Proposed: TWA 0.01 mg(Cd)/m^3 (dust), Human Carcinogen); BEI: 10 μg/g creatinine in urine; 10 μg/L in blood.

NIOSH REL: (Cadmium) Reduce to lowest feasible level

SAFETY PROFILE: Confirmed human carcinogen. Deadly poison by intravenous route. When heated to decomposition it emits very toxic fumes of NO$_x$, SO$_x$, and Cd.

CAK500   CAS: 58-08-2   **HR: 3**
CAFFEINE
mf: C$_8$H$_{10}$N$_4$O$_2$ mw: 194.22

PROP: White, fleecy masses; odorless with bitter taste. Mp: 236.8°. Sol in water, alc, chloroform, ether.

SYNS: CAFFEIN * COFFEIN (GERMAN) * COF-FEINE * 3,7-DIHYDRO-1,3,7-TRIMETHYL-1H-PURINE-2,6-DIONE * ELDIATRIC C * FEMA No. 2224

* GUARANINE * KOFFEIN (GERMAN) * METHYL-THEOBROMIDE * 1-METHYLTHEOBROMINE * 7-METHYLTHEOPHYLLINE * NCI-C02733 * NO-DOZ * ORGANEX * THEIN * THEINE * 1,3,7-TRIMETHYL-2,6-DIOXOPURINE * 1,3,7-TRI-METHYLXANTHINE

CONSENSUS REPORTS: Reported in EPA TSCA Inventory. EPA Genetic Toxicology Program.

SAFETY PROFILE: A human poison by ingestion. An experimental poison by ingestion, subcutaneous, intraperitoneal, intramuscular, rectal, and intravenous routes. Human systemic effects by ingestion, intravenous, and intramuscular routes include: ataxia, blood pressure elevation, convulsions or effect on seizure threshold, diarrhea, distorted perceptions, hallucinations, hypermotility, muscle contraction or spasticity, somnolence (general depressed activity), nausea or vomiting, toxic psychosis, tremors. A human teratogen causing developmental abnormalities of the craniofacial and musculoskeletal systems, pregnancy termination (abortion) and stillbirth. Human maternal effects include an unspecified effect on labor or childbirth. Human mutation data reported. An experimental teratogen. Questionable carcinogen with experimental carcinogenic data. Large doses (above 1.0 gram) cause palpitation, excitement, insomnia, dizziness, headache, and vomiting. Continued excessive use of caffeine in tea or coffee may lead to digestive disturbances, constipation, palpitations, shortness of breath, and depressed mental states. It is also implicated in cardiac disorders under those conditions. When heated to decomposition it emits toxic fumes of NO$_x$.

CAK750   CAS: 5743-18-0   **HR: 3**
CAFFEINE HYDROBROMIDE
mf: C$_8$H$_{10}$N$_4$O$_2$ • BrH   mw: 275.14

SYNS: CAFFEINE BROMIDE * 3,7-DIHYDRO-1,3,7-TRIMETHYL-1H-PURINE-2,6-DIONE MONOHYDROBRO-MIDE

SAFETY PROFILE: Poison by ingestion, subcutaneous, and intravenous routes. When heated to decomposition it emits very toxic fumes of NO$_x$ and HBr.

CAL000   CAS: 470-82-6   **HR: 3**
CAJEPUTOL
mf: C$_{10}$H$_{18}$O   mw: 154.28

PROP: Colorless liquid characteristic odor with pungent, cooling taste. D: 0.921-0.924, refr index: 1.455-1.460, flash p: 122°F. Sol in alc, fixed oils, glycerin, and propylene glycol.

SYNS: 1,8-CINEOL * CINEOLE * 1,8-CINEOLE * 1,8-EPOXY-p-MENTHANE * EUCALYPTOL (FCC) * EUCALYPTOLE * FEMA No. 2465 * LIMONENE OXIDE * NCI-C56575 * 1,8-OXIDO-p-MENTHANE * 1,3,3-TRIMETHYL-2-OXABICYCLO(2.2.2)OCTANE

CONSENSUS REPORTS: Reported in EPA TSCA Inventory.

SAFETY PROFILE: Poison by subcutaneous and intramuscular routes. Moderately toxic by ingestion. Experimental reproductive effects. Combustible liquid. When heated to decomposition it emits acrid smoke and fumes.

CAL250   CAS: 7440-70-2   **HR: 2**
CALCIUM

DOT: NA 1401/UN 11401/UN 1855
af: Ca   aw: 40.08

PROP: Silver-white, soft metal. Mp: 842°, bp: 1484°, d: 1.54 @ 20°, vap press: 10 mm @ 983°.

SYNS: CALCICAT * CALCIUM, non-pyrophoric (DOT) * CALCIUM, pyrophoric (DOT) * CALCIUM, METAL, CRYSTALLINE (DOT) * CALCIUM, METAL (DOT)

CONSENSUS REPORTS: Reported in EPA TSCA Inventory.

DOT Classification: Flammable Solid; Label: Flammable Solid and Dangerous When Wet; IMO: Flammable Solid; Label: Spontaneously Combustible.

SAFETY PROFILE: Flammable when heated or in intimate contact with moisture or acids. Moderate explosion hazard in intimate contact with very powerful oxidizing agents. Reacts with moisture or acids to liberate large quantities of hydrogen; can develop explosive pressure in containers. To fight fire, use special mixtures of dry chemical. Violent reaction with water may evolve explosive hydrogen gas. Potentially explosive reaction with alkali metal hydroxides or carbonates, dinitrogen tetraoxide, lead chloride + heat, phosphorus(V) oxide + heat, sulfur + heat. Molten calcium reacts explosively with asbestos cement. Hypergolic reaction with chlorine fluorides (e.g., chlorine trifluoride, chlorine pentafluoride). Ignition on contact with halogens (e.g., fluorine, chlorine), sulfur + vanadium(V) oxide. Violent reaction with mercury (at 390°C), silicon (above 1050°C), sodium + mixed oxides + heat. Incompatible with air.

CAL500   CAS: 64046-96-4   **HR: 3**
CALCIUM ACETARSONE
mf: $C_8H_{10}AsNO_5 \cdot 7Ca$ mw: 555.67

SYN: N-ACETYL-4-HYDROXY-m-ARSANILIC ACID, CALCIUM SALT

CONSENSUS REPORTS: Arsenic and its compounds are on the Community Right-To-Know List.

OSHA PEL: TWA 0.5 mg(As)/m^3
ACGIH TLV: TWA 0.2 mg(As)/m^3

SAFETY PROFILE: Poison by ingestion. When heated to decomposition it emits very toxic fumes of As and NO$_x$.

CAL750   CAS: 62-54-4   **HR: 3**
CALCIUM ACETATE
mf: $C_4H_6O_4 \cdot Ca$ mw: 158.18

PROP: Fine white, bulky powder. Sol in water; sltly sol in alc.

SYNS: ACETATE of LIME * BROWN ACETATE * CALCIUM DIACETATE * GRAY ACETATE * LIME ACETATE * LIME PYROLIGNITE * SORBO-CALCIAN * SORBO-CALCION * TELTOZAN * VINEGAR SALTS

CONSENSUS REPORTS: Reported in EPA TSCA Inventory.

SAFETY PROFILE: Poison by intravenous route. Mutation data reported. When heated to decomposition it emits acrid smoke and fumes.

CAM500   CAS: 27152-57-4   **HR: 3**
CALCIUM ARSENITE

DOT: NA 1574
mf: $As2O_6 \cdot 3Ca$ mw: 366.08

PROP: White, granular powder.

SYNS: ARSENIOUS ACID, CALCIUM SALT * CALCIUM ARSENITE, solid (DOT) * MONOCALCIUM ARSENITE

CONSENSUS REPORTS: Arsenic and its compounds are on the Community Right-To-Know List.

OSHA PEL: TWA 0.01 mg(As)/m^3
ACGIH TLV: TWA 0.2 mg(As)/m^3

NIOSH REL: (Inorganic Arsenic) CL 0.002 mg(As)/m^3/15M

DOT Classification: Poison B; Label: Poison.

SAFETY PROFILE: A poison by inhalation and ingestion. When heated to decomposition it emits toxic fumes of As.

CAM680　　　　　　　　HR: 2
CALCIUM BENZOATE
mf: $C_{14}H_{10}O_4 \cdot 3H_2O$　　　mw: 374.26

PROP: Orthorhombic crystals or powder. D: 1.44. Sol in water.

SAFETY PROFILE: Combustible when exposed to heat or flame. When heated to decomposition it emits acrid smoke and irritating fumes.

CAM750　　CAS: 6485-34-3　　HR: D
CALCIUM-o-BENZOSULFIMIDE
mf: $C_{14}H_{10}N_2O_6S_2 \cdot Ca$　　mw: 406.46

PROP: White, crystalline powder; odorless or faint aromatic odor; sol in water.

SYNS: 1,2-BENZISOTHIAZOL-3(2H)-ONE-1,1-DIOXIDE, CALCIUM SALT * CALCIUM-o-BENZOSULPHIMIDE * CALCIUM-2-BENZOSULPHIMIDE * CALCIUM SACCHARIN * CALCIUM SACCHARINA * CALCIUM SACCHARINATE * DARAMIN * SACCHARIN CALCIUM * SULPHOBENZOIC IMIDE CALCIUM SALT

CONSENSUS REPORTS: Reported in EPA TSCA Inventory.

SAFETY PROFILE: Mutagenic data reported. When heated to decomposition it emits toxic fumes of SO_x and NO_x.

CAN000　　CAS: 13780-03-5　　HR: 3
CALCIUM BISULFITE (solution)
DOT: UN 1923/UN 2693

PROP: Colorless or sltly yellowish liquid, strong sulfur dioxide odor. D: 1.06.

SYNS: CALCIUM BISULFITE, solution (DOT) * CALCIUM HYDROGEN SULFITE, solution (DOT)

CONSENSUS REPORTS: Reported in EPA TSCA Inventory.

DOT Classification: Corrosive Material; Label: Corrosive.

SAFETY PROFILE: A poison via ingestion. Strong irritant via skin contact, ingestion, and inhalation. When heated to decomposition it emits toxic fumes of SO_x.

CAN750　　CAS: 75-20-7　　HR: 3
CALCIUM CARBIDE
DOT: UN 1402
mf: C_2Ca　　mw: 64.10

PROP: Rhombic, gray crystals. Mp: approx 2300°, d: 2.222.

SYN: CALCIUM ACETYLIDE

CONSENSUS REPORTS: Reported in EPA TSCA Inventory.

DOT Classification: Flammable Solid; Label: Flammable Solid and Dangerous When Wet.

SAFETY PROFILE: Reaction on contact with moisture forms explosive acetylene gas. Flammable on contact with moisture, acid or acid fumes; evolves heat or flammable vapors. Moderate explosion hazard. Incandescent reaction with Cl_2 (245°C), Br_2 (350°C), I_2 (305°C), HCl gas + heat, PbF_2, Mg + heat. Incompatible with Se, (KOH + Cl_2), $AgNO_3$, Na_2O_2, $SnCl_2$, S, water. Mixtures with iron(III) chloride, iron(III) oxide, tin(II) chloride are easily ignited and burn fiercely. Vigorous reaction with methanol after an induction period. Addition to silver nitrate solutions precipitates the dangerously explosive silver acetylide. Copper salt solutions behave similarly.

CAO000　　CAS: 1317-65-3　　HR: 2
CALCIUM CARBONATE
mf: $CO_3 \cdot Ca$　　mw: 100.09

PROP: White microcrystalline powder. Mp: 825° (α), 1339° (β) @ 102.5 atm; d: 2.7-2.95. Found in nature as the minerals limestone, marble, aragonite, calcite, and vaterite. Odorless, tasteless powder or crystals. Two crystalline forms are of commercial importance: Aragonite, orthorhombic, mp: 825° (decomp), d: 2.83, formed at temperatures above 30°; calcite, hexagonal-rhombohedral, mp: 1339° (102.5 atm), d: 2.711, formed at temperatures below 30°. At about 825° it decomposes into CaO and CO_2. Practically insol in water, alc; sol in dilute acids.

SYNS: AGRICULTURAL LIMESTONE * AGSTONE * ARAGONITE * ATOMIT * BELL MINE PULVERIZED LIMESTONE * CALCITE * CARBONIC ACID, CALCIUM SALT (1:1) * CHALK * DOLOMITE * FRANKLIN * LIMESTONE (FCC) * LITHO-

GRAPHIC STONE * MARBLE * NATURAL CALCIUM CARBONATE * PORTLAND STONE * SOHNHOFEN STONE * VATERITE

CONSENSUS REPORTS: Reported in EPA TSCA Inventory.

OSHA PEL: Total Dust: 15 mg/m^3; Respirable Fraction: 5 mg/m^3

ACGIH TLV: TWA (nuisance particulate) 10 mg/m^3 of total dust (when toxic impurities are not present, e.g., quartz $<$ 1%).

SAFETY PROFILE: A severe eye and moderate skin irritant. Ignites on contact with F_2. Incompatible with acids, alum, ammonium salts, (Mg + H_2). Calcium carbonate is a common air contaminant.

CAO500   CAS: 10137-74-3   HR: 2
CALCIUM CHLORATE

DOT: UN 1452/UN 2429
mf: $Cl_2O_6 \cdot Ca$ mw: 206.98

PROP: Monoclinic, white-yellowish, deliquescent crystals. Mp: loses H_2O @ $>100°$, d: 2.711.

SYNS: CALCIUM CHLORATE, aqueous solution (DOT) * CHLORATE de CALCIUM (FRENCH)

DOT Classification: Oxidizer; Label: Oxidizer.

SAFETY PROFILE: Moderately toxic by ingestion and intraperitoneal routes. A powerful oxidant. Incompatible with Al, As, C, Cu, charcoal, MnO_2, metal sulfides, S, dibasic organic acids, organic matter, P. When heated to decomposition it emits toxic fumes of Cl^-.

CAO750   CAS: 10043-52-4   HR: 2
CALCIUM CHLORIDE
mf: $CaCl_2$ mw: 110.98

PROP: Cubic, colorless, deliquescent crystals. Mp: 772°, bp: $>1600°$, d: 2.512 @ 25°. Sol in water and alc.

SYNS: CALCIUM CHLORIDE, anhydrous * CALPLUS * CALTAC * DOWFLAKE * LIQUIDOW * PELADOW * SNOMELT * SUPERFLAKE ANHYDROUS

CONSENSUS REPORTS: Reported in EPA TSCA Inventory. EPA Genetic Toxicology Program.

SAFETY PROFILE: Moderately toxic by ingestion. Poison by intravenous, intramuscular,

intraperitoneal, and subcutaneous routes. Questionable carcinogen with experimental tumorigenic data. Mutation data reported. Reacts violently with (B_2O_3 + CaO), BrF_3. Reaction with zinc releases explosive hydrogen gas. Catalyzes exothermic polymerization of methyl vinyl ether. Exothermic reaction with water. When heated to decomposition it emits toxic fumes of Cl^-.

CAP000   CAS: 14674-72-7   HR: 3
CALCIUM CHLORITE

DOT: UN 1453
mf: $CaCl_2O_4$ mw: 174.98

PROP: White solid.

DOT Classification: Oxidizer; Label: Oxidizer.

SAFETY PROFILE: A strong oxidizer. Ignites on contact with potassium thiocyanate. Reaction with Cl_2 yields explosive ClO_2. When heated to decomposition it emits toxic fumes of Cl^-.

CAP500   CAS: 13765-19-0   HR: 3
CALCIUM CHROMATE

DOT: NA 9096
mf: $CrO_4 \cdot Ca$ mw: 156.08

PROP: Monoclinic prisms; yellow color.

SYNS: CALCIUM CHROMATE (VI) * CALCIUM CHROME YELLOW * CALCIUM CHROMIUM OXIDE ($CaCrO_4$) * CALCIUM MONOCHROMATE * CHROMIC ACID, CALCIUM SALT (1:1) * C.I. 77223 * C.I. PIGMENT YELLOW 33 * GELBIN * RCRA WASTE NUMBER U032 * YELLOW ULTRAMARINE

CONSENSUS REPORTS: IARC Cancer Review: GROUP 1 IMEMDT 7,165,87; Animal Sufficient Evidence IMEMDT 2,100,73; IMEMDT 23,205,80; Human Sufficient Evidence IMEMDT 23,205,80. NTP Fourth Annual Report On Carcinogens, 1984. Reported in EPA TSCA Inventory. EPA Genetic Toxicology Program. Chromium and its compounds are on the Community Right-To-Know List.

OSHA PEL: CL 0.1 mg(CrO_3)/m^3

ACGIH TLV: TWA 0.05 mg(Cr)/m^3; Confirmed Human Carcinogen; (Proposed: TWA 0.001; Suspected Human Carcinogen)

DFG TRK: 0.1 mg/m^3 calculated as CrO_3 in that portion of dust that can possibly be inhaled; 0.2 mg/m^3 arc-welding by hand; others

0.1 mg/m^3. Animal Carcinogen, Suspected Human Carcinogen.
NIOSH REL: TWA 0.001 mg(Cr(VI))/m^3
DOT Classification: ORM-E; Label: None.

SAFETY PROFILE: Confirmed human carcinogen with experimental carcinogenic, neoplastigenic, and tumorigenic data. Experimental reproductive effects. Mutation data reported. A powerful oxidizer. Mixture with boron burns violently if ignited.

CAP750   CAS: 8012-75-7   **HR: 3**
CALCIUM CHROMATE(VI) DIHYDRATE
mf: CrO$_4$•Ca•2H$_2$O   mw: 192.12

SYNS: CALCIUM CHROME YELLOW * CHROMIC ACID, CALCIUM SALT (1:1), DIHYDRATE * C.I. 77223 * C.I. PIGMENT YELLOW 33 * GELBIN YELLOW ULTRAMARINE * PIGMENT YELLOW 33 * STEINBUHL YELLOW

CONSENSUS REPORTS: IARC Cancer Review: Animal Sufficient Evidence IMEMDT 2,100,72. Chromium and its compounds are on the Community Right-To-Know List.

OSHA PEL: CL 0.1 mg(CrO^3)/m^3
ACGIH TLV: TWA 0.05 mg(Cr)/m^3; Confirmed Human Carcinogen
NIOSH REL: (Chromium(VI)) TWA 0.001 mg(Cr(VI))/m^3

SAFETY PROFILE: Confirmed human carcinogen with experimental tumorigenic and carcinogenic data. Poison by ingestion and implant routes. Mutation data reported. A powerful oxidizer.

CAQ000   **HR: D**
CALCIUM COMPOUNDS

SAFETY PROFILE: The fumes evolved by burning calcium in air are composed of calcium oxide (quick lime) which is an irritant to the skin, eyes, and mucous membranes. Generally speaking, calcium compounds should be considered toxic only when they contain toxic components (such as arsenic, etc.) or as calcium oxide or hydroxide. Calcium compounds are common air contaminants.

CAQ250   CAS: 156-62-7   **HR: 3**
CALCIUM CYANAMIDE
mf: CN$_2$•Ca   mw: 80.11
DOT: UN 1403

PROP: Hexagonal, rhombohedral, colorless crystals. Mp: 1300°, subl > 1500°. Compound not hydrated; compound contains more than 0.1% calcium.

SYNS: AERO-CYANAMID * AERO CYANAMID GRANULAR * AERO CYANAMID SPECIAL GRADE * ALZODEF * CALCIUM CARBIMIDE * CALCIUM CYANAMID * CCC * CYANAMIDE * CYANAMIDE, CALCIUM SALT (1:1) * CYANAMIDE CALCIQUE (FRENCH) * CYANAMID GRANULAR * CYANAMID SPECIAL GRADE * CY-L 500 * LIME-NITROGEN (DOT) * NCI-C02937 * NITROGEN LIME * NITROLIME * USAF CY-2

CONSENSUS REPORTS: NCI Carcinogenesis Bioassay (feed); No Evidence: mouse, rat NCITR* NCI-CG-TR-163,79. Community Right-To-Know List. Reported in EPA TSCA Inventory.

OSHA PEL: TWA 0.5 mg/m^3
ACGIH TLV: TWA 0.5 mg/m^3
DFG MAK: 1 mg/m^3
DOT Classification: ORM-C; Label: None; IMO: Flammable Solid; Label: Dangerous When Wet.

SAFETY PROFILE: Poison by ingestion, inhalation, skin contact, intravenous, and intraperitoneal routes. Moderately toxic to humans by ingestion. Questionable carcinogen with experimental tumorigenic data. Mutation data reported. The fatal dose, by ingestion, is probably around 20 to 30 grams for an adult. It does not have a cyanide effect. Calcium cyanamide is not believed to have a cumulative action. Flammable. Reaction with water forms the explosive acetylene gas. When heated to decomposition it emits toxic fumes of NO$_x$ and CN$^-$.

CAQ500   CAS: 592-01-8   **HR: 3**
CALCIUM CYANIDE
DOT: UN 1575
mf: C$_2$CaN$_2$ mw: 92.12

PROP: Rhombohedral crystals or white powder. Mp: decomp > 350°.

SYNS: CALCIUM CYANIDE, solid (DOT) * CALCYANIDE * CYANOGAS * CYANURE de CALCIUM (FRENCH) * RCRA WASTE NUMBER P021

CONSENSUS REPORTS: Cyanide and its compounds are on the Community Right-To-Know List. Reported in EPA TSCA Inventory.

OSHA PEL: TWA 5 mg(CN)/m^3
ACGIH TLV: TWA 5 mg(CN)/m^3 (skin)
DFG MAK: 5 mg/m^3
NIOSH REL: (Cyanide) CL 5 mg(CN)/m^3/10M
DOT Classification: Poison B; Label: Poison.

SAFETY PROFILE: A deadly poison by ingestion and probably other routes. When heated to decomposition it emits toxic fumes of NO$_x$ and CN$^-$.

CAQ750   CAS: 592-01-8   **HR: 3**
CALCIUM CYANIDE (mixture)

DOT: UN 1575

SYN: CALCIUM CYANIDE MIXTURE, solid (DOT)

CONSENSUS REPORTS: Cyanide and its compounds are on the Community Right-To-Know List.

OSHA PEL: TWA 5 mg(CN)/m^3
ACGIH TLV: TWA 5 mg(CN)/m^3 (skin)
DFG MAK: 5 mg/m^3
NIOSH REL: (Cyanide) CL 5 mg(CN)/m^3/10M
DOT Classification: Poison B; Label: Poison.

SAFETY PROFILE: A poison. When heated to decomposition it emits toxic fumes of NO$_x$ and CN$^-$.

CAR000   CAS: 139-06-0   **HR: 3**
CALCIUM CYCLOHEXYLSULPHAMATE
mf: C$_{12}$H$_{24}$N$_2$O$_6$S$_2$•Ca   mw: 396.58

PROP: White, crystalline powder; almost odorless; freely sol in water; practically insol in alc, benzene, chloroform, and ether.

SYNS: CALCIUM CYCLAMATE * CALCIUM CYCLOHEXANESULFAMATE * CALCIUM CYCLOHEXANE SULPHAMATE * CALCIUM CYCLOHEXYLSULFAMATE * CYCLAMATE CALCIUM * CYCLAMATE, CALCIUM SALT * CYCLAN * CYCLOHEXANESULFAMIC ACID, CALCIUM SALT * CYCLOHEXYLSULPHAMIC ACID, CALCIUM SALT * CYLAN * DIETIL * KALZIUMZYKLAMATE (GERMAN) * SUCARYL CALCIUM

CONSENSUS REPORTS: IARC Cancer Review: GROUP 3 IMEMDT 7,178,87; Animal Limited Evidence IMEMDT 22,55,80. Reported in EPA TSCA Inventory. EPA Genetic Toxicology Program.

SAFETY PROFILE: Poison by ingestion and intravenous routes. Experimental reproductive effects. Questionable carcinogen with experimental tumorigenic and neoplastigenic data. Human mutation data reported. When heated to decomposition it emits very toxic fumes of SO$_x$ and NO$_x$.

CAS250   CAS: 544-17-2   **HR: 3**
CALCIUM FORMATE
mf: C$_2$H$_2$O$_4$•Ca   mw: 130.12

SYNS: FORMIC ACID, CALCIUM SALT * MRAVENCAN VAPENATY (CZECH)

CONSENSUS REPORTS: Reported in EPA TSCA Inventory.

SAFETY PROFILE: Poison by intravenous route. Moderately toxic by ingestion. An eye irritant. When heated to decomposition it emits acrid smoke and fumes.

CAS750   CAS: 299-28-5   **HR: 2**
CALCIUM GLUCONATE
mf: C$_{12}$H$_{22}$O$_{14}$•Ca   mw: 430.42

PROP: White, fluffy powder or granules; odorless and tasteless. Sol in hot water; less sol in cold water; insol in alc, acetic acid, and other organic solvents. Mp: loses H$_2$O @ 120°.

SYN: GLUCONATE de CALCIUM (FRENCH)

CONSENSUS REPORTS: Reported in EPA TSCA Inventory.

SAFETY PROFILE: Moderately toxic by subcutaneous, intraperitoneal, and intravenous routes. Human systemic effects in infants by intramuscular route: dermatitis and fever. When heated to decomposition it emits acrid smoke and fumes.

CAT250   CAS: 1305-62-0   **HR: 2**
CALCIUM HYDROXIDE
mf: CaH$_2$O$_2$ mw: 74.10

PROP: Rhombic, trigonal, colorless crystals or white power; sltly bitter taste. Mp: loses H$_2$O @ 580°, bp: decomp, d: 2.343. Sol in water and glycerin; insol in alc.

SYNS: BELL MINE * CALCIUM HYDRATE * HYDRATED LIME * KEMIKAL * LIME WATER * SLAKED LIME

CONSENSUS REPORTS: Reported in EPA TSCA Inventory.

OSHA PEL: TWA 5 mg/m^3
ACGIH TLV: TWA 5 mg/m^3

SAFETY PROFILE: Mildly toxic by ingestion. A severe eye irritant. A skin, mucous membrane and respiratory system irritant. Mutation data reported. Causes dermatitis. Dust is considered to be a significant industrial hazard. A common air contaminant. Violent reaction with maleic anhydride, nitroethane, nitromethane, nitroparaffins, nitropropane, phosphorus. Reaction with polychlorinated phenols + potassium nitrate forms extremely toxic products.

CAU000   CAS: 10124-37-5   **HR: 3**
CALCIUM(II) NITRATE (1:2)
DOT: UN 1454
mf: $N_2O_6 \cdot Ca$ mw: 164.10

SYN: CALCIUM NITRATE (DOT)

CONSENSUS REPORTS: Reported in EPA TSCA Inventory.

DOT Classification: Oxidizer; Label: Oxidizer.

SAFETY PROFILE: An irritant. A strong oxidant. Forms powerfully explosive mixtures with aluminum + ammonium nitrate + formamide + water, ammonium nitrate + hydrocarbon oils, ammonium nitrate + water-soluble fuels, and organic materials. When heated to decomposition it emits toxic fumes of NO_x.

CAU500   CAS: 1305-78-8   **HR: 3**
CALCIUM OXIDE
DOT: UN 1910
mf: CaO   mw: 56.08

PROP: Cubic, white crystals. Mp: 2580°, d: 3.37, bp: 2850°. Sol in water and glycerin; insol in alc.

SYNS: BURNT LIME * CALCIA * CALX * LIME * LIME, BURNED * LIME, UNSLAKED (DOT) * OXYDE de CALCIUM (FRENCH) * QUICKLIME (DOT) * WAPNIOWY TLENEK (POLISH)

CONSENSUS REPORTS: Reported in EPA TSCA Inventory.

OSHA PEL: TWA 5 mg/m^3
ACGIH TLV: TWA 2 mg/m^3
DFG MAK: 5 mg/m^3
DOT Classification: ORM-B; Label: None.

SAFETY PROFILE: A caustic and irritating material. A common air contaminant. A powerful caustic to living tissue. The powdered oxide may react explosively with water. Mixtures with ethanol may ignite if heated and thus can cause an air-vapor explosion. Violent reaction with (B_2O_3 + $CaCl_2$), interhalogens (e.g., BF_3, ClF_3), F_2, HF, P_2O_5 + heat, water. Incandescent reaction with liquid HF. Incompatible with phosphorus(V) oxide.

CAU750   CAS: 137-08-6   **HR: 2**
CALCIUM-d-PANTOTHENATE
mf: $C_{19}H_{34}N_2O_{10} \cdot Ca$ mw: 490.63

PROP: White, sltly hygroscopic powder; odorless; bitter taste. Mp: 170-172°, decomp @ 195-196°. Sol in water and glycerin; insol in alc, chloroform, and ether.

SYNS: CALCIUM d(+)-N-(α,γ-DIHYDROXY-β,β-DIMETHYLBUTYRYL)-β-ALANINATE * CALCIUM PANTHOTHENATE (FCC) * CALCIUM PANTOTHENATE * d-CALCIUM PANTOTHENATE * CALPANATE * DEXTRO CALCIUM PANTOTHENATE * N-(2,4-DIHYDROXY-3,3-DIMETHYLBUTYRYL)-β-ALANINE CALCIUM * PANCAL * PANTHOJECT * PANTHOLIN * PANTOTHENATE CALCIUM * PANTOTHENIC ACID, CALCIUM SALT * (+)-PANTOTHENIC ACID, CALCIUM SALT * VITAMIN B-5

CONSENSUS REPORTS: Reported in EPA TSCA Inventory.

SAFETY PROFILE: Moderately toxic by intraperitoneal, subcutaneous, and intravenous routes. Mildly toxic by ingestion. A vitamin. When heated to decomposition it emits toxic fumes of NO_x.

CAU780   **HR: 1**
CALCIUM PANTOTHENATE, CALCIUM CHLORIDE DOUBLE SALT
mf: $C_{19}H_{34}N_2O_{10} \cdot Ca_2Cl_2$ mw: 601.61

PROP: White, sltly hygroscopic powder; odorless with bitter taste. Sol in water and glycerin; insol in alc, chloroform, and ether.

SAFETY PROFILE: Moderately toxic by intraperitoneal, subcutaneous, and intravenous routes. Mildly toxic by ingestion. A vitamin. When heated to decomposition it emits toxic fumes of NO_x.

CAV000   CAS: 7563-42-0   **HR: 3**
CALCIUM PENTOBARBITAL
mf: $C_{11}H_{18}N_2O_3 \cdot 7Ca$ mw: 506.87

SYNS: CALCIUM NEMBUTAL * INSOM-RAPIDO * NEMBUTAL CALCIUM * PENTOBARBITAL CAL-

CIUM * 2,4,6(1H,3H,5H)-PYRIMIDINETRIONE, 5-
ETHYL-5-(1-METHYLBUTYL)-, CALCIUM SALT (9CI)
* RAVONA * REPOCAL * SCHLAFEN

SAFETY PROFILE: Poison by ingestion and intravenous routes. When heated to decomposition it emits toxic fumes of NO_x.

CAV250 CAS: 10118-76-0 HR: 3
CALCIUM PERMANGANATE
DOT: UN 1456
mf: $Mn_2O_8 \cdot Ca$ mw: 277.96

PROP: Violet, deliquescent crystals. Mp: decomp, d: 2.4.

SYN: KALIUMPERMANGANAT (GERMAN)

CONSENSUS REPORTS: Manganese and its compounds are on the Community Right-To-Know List.

OSHA PEL: CL 5 mg(Mn)/m^3
ACGIH TLV: TWA 5 mg(Mn)/m^3
DOT Classification: Oxidizer; Label: Oxidizer.

SAFETY PROFILE: Poison by intravenous route. A strong oxidant. May explode on contact with acetic acid or acetic anhydride. Ignites on contact with cellulose. Incompatible with hydrogen peroxide.

CAV500 CAS: 1305-79-9 HR: 3
CALCIUM PEROXIDE
DOT: UN 1457
mf: CaO_2 mw: 72.08

PROP: Yellow crystals or powder or white crystals, decomposes in air. Mp: decomp @ 275°. Insol in water; sol in acids, forming hydrogen peroxide.

SYNS: CALCIUM DIOXIDE * CALCIUM SUPEROXIDE

CONSENSUS REPORTS: Reported in EPA TSCA Inventory.

DOT Classification: Oxidizer; Label: Oxidizer.

SAFETY PROFILE: Irritating in concentrated form. Will react with moisture to form slaked lime. Flammable if hot and mixed with finely divided combustible material. Mixtures with oxidizable materials can also be ignited by grinding and are explosion hazards. A strong alkali. An oxidizer. Mixtures with polysulfide polymers may ignite.

CAW100 CAS: 7757-93-9 HR: 1
CALCIUM PHOSPHATE, DIBASIC
mf: $CaHPO_4 \cdot 2H_2O$ mw: 172.09

PROP: White powder. Sol in dilute acid; insol in water, alc.

SYN: DICALCIUM PHOSPHATE

SAFETY PROFILE: Skin and eye irritant. A nuisance dust.

CAW120 CAS: 12167-74-7 HR: 1
CALCIUM PHOSPHATE, TRIBASIC
mf: $10CaO \cdot 3P_2O_5 \cdot H_2O$ mw: 1004.64

PROP: White powder. Sol in dilute HCl; insol in water, alc.

SYNS: PERCIPITATED CALCIUM PHOSPHATE * TRICALCIUM PHOSPHATE

SAFETY PROFILE: Skin and eye irritant. A nuisance dust.

CAW250 CAS: 1305-99-3 HR: 3
CALCIUM PHOSPHIDE
DOT: UN 1360
mf: Ca_3P_2 mw: 182.18

PROP: Red crystals. Mp: >1600°, d: 2.238 @ 25°.

SYN: TRICALCIUM DIPHOSPHIDE

CONSENSUS REPORTS: Reported in EPA TSCA Inventory.

DOT Classification: Flammable Solid; Label: Flammable Solid and Dangerous When Wet.

SAFETY PROFILE: Highly toxic due to phosphide which in presence of moisture emits phosphine. The phosphine may ignite spontaneously in air. Incandescent reaction with oxygen at 300°C. Incompatible with dichlorine oxide. When heated to decomposition it emits toxic fumes of PO_x.

CAW500 CAS: 9007-13-0 HR: 1
CALCIUM RESINATE
DOT: UN 1313/UN 1314
mf: $Ca(C_{44}H_{62}O_4)_2$ mw: 1349.50

PROP: Yellowish-white, amorphous powder or lumps.

SYNS: CALCIUM RESINATE, fused (DOT) * CALCIUM RESINATE, technically pure (DOT) * LIMED ROSIN

CONSENSUS REPORTS: Reported in EPA TSCA Inventory.

DOT Classification: Flammable Solid; Label: Flammable Solid; Flammable Solid; Label: Flammable Solid, fused.

SAFETY PROFILE: Flammable solid when heated; can react with oxidizing materials. When heated to decomposition it emits acrid smoke and fumes.

CAW850   CAS: 1344-95-2   **HR: 1**
CALCIUM SILICATE

PROP: Varying proportions of CaO and SiO_2. White powder. Insol in water.
OSHA PEL: Total Dust: 15 mg/m^3; Respirable Fraction: 5 mg/m^3
ACGIH TLV: TWA (nuisance particulate) 10 mg/m^3 of total dust (when toxic impurities are not present, e.g., quartz $<$ 1%).

SAFETY PROFILE: A nuisance dust.

CAX250   CAS: 16925-39-6   **HR: 3**
CALCIUM SILICOFLUORIDE
mf: CaF$_6$Si   mw: 182.17

PROP: White, crystalline powder. D: 2.662 @ 17.5°.

SYNS: CALCIUM FLUOSILICATE * CALCIUM HEX-AFLUOROSILICATE

CONSENSUS REPORTS: Reported in EPA TSCA Inventory.

OSHA PEL: TWA 2.5 mg(F)/m^3
NIOSH REL: (Inorganic Fluorides) TWA 2.5 mg(F)/m^3

SAFETY PROFILE: Poison by ingestion and subcutaneous routes. When heated to decomposition it emits toxic fumes of F$^-$.

CAX500   CAS: 7778-18-9   **HR: 3**
CALCIUM SULFATE
mf: CaSO$_4$ mw: 136.14

PROP: Pure anhydrous, white powder or odorless crystals. D: 2.964; mp: 1450°.

SYNS: GYPSUM * PLASTER of PARIS

OSHA PEL: Total Dust: 15 mg/m^3; Respirable Fraction: 5 mg/m^3
ACGIH TLV: TWA (nuisance particulate) 10 mg/m^3 of total dust (when toxic impurities are not present, e.g., quartz $<$ 1%)7

SAFETY PROFILE: A nuisance dust. Reacts violently with aluminum when heated. Mixtures with diazomethane react exothermically and eventually explode. Mixtures with phosphorus ignite at high temperatures. When heated to decomposition it emits toxic fumes of SO$_x$.

CAX750   CAS: 10101-41-4   **HR: 1**
CALCIUM(II) SULFATE DIHYDRATE (1:1:2)
mf: O$_4$S•Ca•2H$_2$O   mw: 172.18

PROP: Colorless crystals. D: 2.32, mp: 128° (loses 1.5H$_2$O), bp: 163° (loses 2H$_2$O).

SYNS: ALABASTER * ANNALINE * C.I. 77231 * C.I. PIGMENT WHITE 25 * GYPSUM * GYPSUM STONE * LAND PLASTER * LIGHT SPAR * MAGNESIA WHITE * MINERAL WHITE * NATIVE CALCIUM SULFATE * PRECIPITATED CALCIUM SULFATE * SATINITE * SATIN SPAR * SULFURIC ACID, CALCIUM(2+) SALT, DIHYDRATE * TERRA ALBA

OSHA PEL: Total Dust: 15 mg/m^3; Respirable Fraction: 5 mg/m^3
ACGIH TLV: TWA (nuisance particulate) 10 mg/m^3 of total dust (when toxic impurities are not present, e.g., quartz $<$ 1%)

SAFETY PROFILE: Human systemic effects by inhalation: fibrosing alveolitis (growth of fibrous tissue in the lung); unspecified respiratory system effects and unspecified effects on the nose. Questionable carcinogen with experimental carcinogenic data. Long considered a nuisance dust (depending on silica content). When heated to decomposition it emits toxic fumes of SO$_x$.

CAY000   CAS: 20548-54-3   **HR: 3**
CALCIUM SULFIDE
mf: CaS   mw: 72.14

PROP: Cubic, colorless crystals. Bp: decomp, d: 218 @ 15°.

SYNS: CALCIC LIVER of SULFUR * HEPAR CALCIS * OLDHAMITE

SAFETY PROFILE: A poison via inhalation. Reacts violently with chromyl chloride, lead dioxide, potassium chlorate (mild explosion), potassium nitrate (violent explosion). Incompatible with oxidants. When heated to decomposition it emits toxic fumes of SO$_x$.

CAY250   CAS: 2092-16-2   **HR: 3**
CALCIUM THIOCYANATE
mf: $C_2N_2S_2 \cdot Ca$ mw: 156.24

PROP: White, deliquescent crystals.

SYNS: CALCIUM RHODANID (GERMAN) * THIO-CYANIC ACID, CALCIUM SALT (2:1)

CONSENSUS REPORTS: Reported in EPA TSCA Inventory.

SAFETY PROFILE: Poison by ingestion and intravenous routes. When heated to decomposition it emits toxic fumes of NO_x and SO_x.

CAY500   CAS: 12111-24-9   **HR: 2**
CALCIUM TRISODIUM DIETHYLENE TRIAMINE PENTAACETATE
mf: $C_{14}H_{18}N_3O_{10} \cdot CaNa_3$ mw: 497.40

SYNS: Ba 2797 * CALCIUM CHEL-330 * CAL-CIUM-DTPA * CALCIUM TRISODIUM CHEL 330 * CALCIUM TRISODIUM DTPA * CALCIUM TRISO-DIUM PENTETATE * CALCIUM TRISODIUM SALT of DIETHYLENETRIAMINEPENTAACETIC ACID * DI-ETHYLENETRIAMINE PENTAACETIC ACID, CALCIUM TRISODIUM SALT * DITRIPENTAT * DTPA CAL-CIUM TRISODIUM SALT * PENTACIN * PENTACINE * PENTETATE TRISODIUM CALCIUM * PENTHAMIL

CONSENSUS REPORTS: Reported in EPA TSCA Inventory.

SAFETY PROFILE: Moderately toxic by intra-peritoneal route. Experimental teratogenic and reproductive effects. Mutation data reported. When heated to decomposition it emits toxic fumes of Na_2O and NO_x.

CBA500   CAS: 79-92-5   **HR: 1**
CAMPHENE

DOT: UN 9011
mf: $C_{10}H_{16}$ mw: 136.26

PROP: Colorless cubic crystals; oily odor. Mp: 50-51°, bp: 159°, d: 0.842 @ 54°/4°, refr index: 1.452 @ 55°. Sol in alc; misc in fixed oils; insol in water.

SYN: FEMA No. 2229

CONSENSUS REPORTS: Reported in EPA TSCA Inventory.

DOT Classification: ORM-A; Label: None.

SAFETY PROFILE: Mutation data reported. Combustible; yields flammable vapors when heated and can react with oxidizing materials. To fight fire, use water spray, foam, fog, CO_2. When heated to decomposition it emits acrid smoke and irritating fumes.

CBA750   CAS: 76-22-2   **HR: 3**
CAMPHOR

DOT: UN 2717
mf: $C_{10}H_{16}O$ mw: 152.26

PROP: White, transparent, crystalline masses; penetrating odor; pungent, aromatic taste. Mp: 180°, bp: 204°, lel: 0.6%, uel: 3.5%, flash p: 150°F (CC), d: 0.992 @ 25°/4°, autoign temp: 871°F, vap d: 5.24.

SYNS: 2-BORNANONE * 2-CAMPHANONE * CAMPHOR, synthetic (ACGIH, DOT) * CAMPHOR-NATURAL * FORMOSA CAMPHOR * GUM CAM-PHOR * HUILE de CAMPHRE (FRENCH) * JAPAN CAMPHOR * KAMPFER (GERMAN) * 2-KETO-1,7,7-TRIMETHYLNORCAMPHANE * LAUREL CAMPHOR * MATRICARIA CAMPHOR * 2-OXOBORNANE * 1,7,7-TRIMETHYLBICYCLO(2.2.1)-2-HEPTANONE * 1,7,7-TRIMETHYLNORCAMPHOR

CONSENSUS REPORTS: Reported in EPA TSCA Inventory.

OSHA PEL: TWA 2 mg/m^3
ACGIH TLV: TWA 2 ppm; STEL 3 ppm
DFG MAK: 2 ppm (13 mg/m^3)
DOT Classification: Flammable Solid; Label: Flammable Solid.

SAFETY PROFILE: A human poison by inges-tion, and possibly other routes. An experimental poison by inhalation, subcutaneous and intra-peritoneal routes. A local irritant. Ingestion causes nausea, vomiting, dizziness, excitation, and convulsions. Mutation data reported. Used as a topical anti-infective and anti-itching agent. Flammable when exposed to heat or flame; can react with oxidizing materials. Vapor is explo-sive when exposed to heat or flame or CrO_3. To fight fire, use foam, carbon dioxide, dry chemical.

CBB250   CAS: 464-49-3   **HR: 3**
(1R,4R)-(+)-CAMPHOR
mf: $C_{10}H_{16}O$ mw: 152.26

SYNS: ALCANFOR * (+)-2-BORNANONE * d-2-BORNANONE * d-2-CAMPHANONE * (+)-CAMPHOR * d-CAMPHOR * d-(+)-CAMPHOR * CAMPHOR USP * JAPANESE CAMPHOR * (1R)-1,7,7-TRIMETHYL-BICYCLO(2.2.1)HEPTAN-2-ONE

CONSENSUS REPORTS: Reported in EPA TSCA Inventory.

SAFETY PROFILE: Poison by intraperitoneal route. Moderately toxic by ingestion, subcutaneous and intravenous routes. A skin irritant. When heated to decomposition it emits acrid and irritating fumes.

CBB500   CAS: 8008-51-3   **HR: 3**
CAMPHOR OIL

DOT: UN 1130

PROP: Colorless or yellowish, oily, fragrant liquid. Bp: 175-200°, flash p: 117°F (CC), d: 0.875-0.900 @ 20°/20°. Insol in water; sol in chloroform, ether, oils, and in approx 3 vols alc. Found in the trees and bark of *Cinnamomum carphora sieb* (*Fam. Lauraceae*) and prepared by fractional distillation of crude camphor oil after the camphor has been crystallized out; a white, viscous liquid with cineole as the principal ingredient along with monoterpenes.

SYNS: CAMPHOR OIL, RECTIFIED * CAMPHOR OIL WHITE * CAMPHOR OIL YELLOW * FORMOSA CAMPHOR OIL * FORMOSE OIL of CAMPHOR * JAPANESE CAMPHOR OIL * JAPANESE, OIL of CAMPHOR * LIGHT CAMPHOR OIL * LIGHT OIL of CAMPHOR * LIQUID CAMPHOR * OIL of CAMPHOR RECTIFIED * OIL CAMPHOR SASSAFRASSY * OIL of CAMPHOR WHITE * WHITE CAMPHOR OIL * WHITE OIL of CAMPHOR

CONSENSUS REPORTS: Reported in EPA TSCA Inventory.

DOT Classification: Combustible Liquid; Label: None; Flammable or Combustible Liquid; Label: Flammable Liquid.

SAFETY PROFILE: A human poison by ingestion. Human systemic effects by ingestion: convulsions, tremors, and unspecified respiratory system effects. A skin irritant. Flammable when exposed to heat or flame; can react with oxidizing materials. To fight fire, use foam, CO_2, dry chemical, mist, fog.

CBC250   CAS: 1403-17-4   **HR: 3**
CANDICIDIN
mf: $C_{63}H_{85}N_{21}O_{19}$ mw: 1440.69

SYNS: CANDEPTIN * CANDIMON * LEVORIN * VANOBID

SAFETY PROFILE: Poison by intraperitoneal and subcutaneous routes. Moderately toxic by ingestion. Experimental teratogenic and reproductive effects. When heated to decomposition it emits toxic fumes of NO_x.

CBE250   **HR: 2**
CANTHARIDES
mf: $C_{10}H_{12}O_4$ mw: 196.15

PROP: Brown to black powder or scales. Mp: 218°, bp: subl @ 90°.

SYNS: BLISTERING BEETLES * BLISTERING FLIES * SPANISH FLY

SAFETY PROFILE: Strong irritant via skin contact, ingestion, inhalation, and contact with eyes. An allergen. Can cause conjunctivitis, keratitis, blepharitis, slight swelling of cornea and inflammation of iris. It is often mistakenly used as an aphrodisiac, but it is much too dangerous and irritating a material for this purpose. When heated to decomposition it emits acrid smoke and fumes.

CBF250   CAS: 302-22-7   **HR: 3**
CAP
mf: $C_{23}H_{29}ClO_4$ mw: 404.97

SYNS: 17-ACETOXY-6-CHLORO-6-DEHYDROPROGESTERONE * 17-α-ACETOXY-6-CHLORO-6-DEHYDROPROGESTERONE * 17-α-ACETOXY-6-CHLORO-6,7-DEHYDROPROGESTERONE * 17-α-ACETOXY-6-CHLORO-4,6-PREGNADIENE-3,20-DIONE * 17-α-ACETOXY-6-CHLOROPREGNA-4,6-DIENE-3,20-DIONE * 17-(ACETYLOXY)-6-CHLOROPREGNA-4,6-DIENE-3,20-DIONE * CHLORMADINON ACETATE * CHLORMADINONE ACETATE * CHLORMADINONU (POLISH) * 6-CHLORO-17-α-ACETOXY-4,6-PREGNADIENE-3,20-DIONE * Δ⁶-6-CHLORO-17-α-ACETOXYPROGESTERONE * 6-CHLORO-Δ⁶-17-ACETOXYPROGESTERONE * 6-CHLORO-Δ⁶-(17-α)ACETOXYPROGESTERONE * 6-CHLORO-Δ⁶-DEHYDRO-17-ACETOXYPROGESTERONE * 6-CHLORO-6-DEHYDRO-17-α-ACETOXYPROGESTERONE * 6-CHLORO-6-DEHYDRO-17-α-HYDROXYPROGESTERONE ACETATE * 6-CHLORO-17-α-HYDROXYPREGNA-4,6-DIENE-3,20-DIONE ACETATE * 6-CHLORO-17-α-HYDROXY-Δ⁶-PROGESTERONE ACETATE * CHLOROMADINONE ACETATE * 6-CHLORO-Δ⁴,⁶-PREGNADIENE-17-α-OL-3,20-DIONE-17-ACETATE * 6-CHLORO-PREGNA-4,6-DIEN-17-α-OL-3,20-DIONE ACETATE * CLORDION * CMA * C-QUENS * 6-DEHYDRO-6-CHLORO-17-α-ACETOXYPROGESTERONE * LORMIN * LUTINYL * NSC-92338 * RS 1280 * SKEDULE * ST 155

CONSENSUS REPORTS: IARC Cancer Review: Animal Limited Evidence IMEMDT 21,365,79; Animal Sufficient Evidence IMEMDT 6,149,74.

SAFETY PROFILE: Suspected carcinogen with experimental carcinogenic and tumorigenic data. Moderately toxic by intraperitoneal route. Human maternal and reproductive effects by ingestion, intramuscular, and possibly other routes: ovary, uterus, cervix, vagina, and fallopian tube changes; menstrual cycle changes or disorders; changes in fertility; and other unspecified female effects. A human teratogen which causes developmental abnormalities of the endocrine system in the fetus. Experimental teratogenic and reproductive effects. When heated to decomposition it emits toxic fumes of Cl⁻.

CBF700   CAS: 105-60-2   HR: 3
CAPROLACTAM
mf: $C_6H_{11}NO$ mw: 113.18

PROP: White crystals. Mp: 69°, vap press: 6 mm @ 120°.

SYNS: AMINOCAPROIC LACTAM * 6-AMINOHEXA-NOIC ACID CYCLIC LACTAM * 2-AZACYCLOHEPTA-NONE * 6-CAPROLACTAM * omega-CAPROLACTAM (MAK) * CAPROLATTAME (FRENCH) * CYCLOHEX-ANONE ISO-OXIME * EPSYLON KAPROLAKTAM (PO-LISH) * HEXAHYDRO-2-AZEPINONE * HEXAHY-DRO-2H-AZEPIN-2-ONE * 6-HEXANELACTAM * HEXANONE ISOXIME * HEXANONISOXIM (GERMAN) * 1,6-HEXOLACTAM * e-KAPROLAK-TAM (CZECH) * 2-KETOHEXAMETHYLENIMINE * NCI-C50646 * 2-OXOHEXAMETHYLENIMINE * 2-PERHYDROAZEPINONE

CONSENSUS REPORTS: IARC Cancer Review: GROUP 4 IMEMDT 7,56,87; Animal No Evidence IMEMDT 39,247,86. Reported in EPA TSCA Inventory.

OSHA PEL: Dust: 1 mg/m³; STEL 3 mg/m³; Vapor: 5 ppm; STEL 10 ppm
ACGIH TLV: Dust: 1 mg/m³; STEL 3 mg/m³; Vapor: 5 ppm; STEL 10 ppm; (Proposed: TWA Dust: 1 mg/m³; 5 ppm (vapor and aerosol)
DFG MAK: 25 mg/m³

SAFETY PROFILE: Moderately toxic by ingestion, skin contact, intraperitoneal, and subcutaneous routes. Human systemic effects by inhalation: cough. Experimental reproductive effects. Human mutation data reported. A skin and eye

irritant. Potentially explosive reaction with acetic acid + dinitrogen trioxide. When heated to decomposition it emits toxic fumes of NO_x.

CBF800   CAS: 2425-06-1   HR: 3
CAPTAFOL
mf: $C_{10}H_9Cl_4NO_2S$ mw: 349.06

SYNS: CAPTOFOL * DIFOLATAN * DIFOSAN * FOLCID * ORTHO 5865 * SANSPOR * SUL-FONIMIDE * SULPHEIMIDE * N-(1,1,2,2-TETRA-CHLORAETHYLTHIO)CYCLOHEX-4-EN-1,4-DIACARBOXI-MID (GERMAN) * N-(1,1,2,2-TETRACHLORAETHYL-THIO)TETRAHYDROPHTHALAMID (GERMAN) * N-1,1,2,2-TETRACHLOROETHYLMERCAPTO-4-CY-CLOHEXENE-1,2-CARBOXIMIDE * N-((1,1,2,2-TET-RACHLOROETHYL)SULFENYL)-cis-4-CYCLOHEXENE-1,2-DICARBOXIMIDE * N-(1,1,2,2-TETRACHLORO-ETHYLTHIO)-4-CYCLOHEXENE-1,2-DICARBOXIMIDE

CONSENSUS REPORTS: EPA Genetic Toxicology Program.

OSHA PEL: TWA 0.1 mg/m³
ACGIH TLV: TWA 0.1 mg/m³

SAFETY PROFILE: Poison by intraperitoneal route. Moderately toxic by ingestion. Other experimental reproductive effects. Questionable carcinogen with experimental carcinogenic data. Experimental teratogenic effects. Mutation data reported. A fungicide. When heated to decomposition it emits very toxic fumes of Cl⁻, NO_x, and SO_x.

CBG000   CAS: 133-06-2   HR: 3
CAPTAN
DOT: NA 9099
mf: $C_9H_8Cl_3NO_2S$ mw: 300.59

PROP: Odorless crystals. Insol in water; sol in benzene and chloroform.

SYNS: AACAPTAN * AGROSOL S * AGROX 2-WAY and 3-WAY * AMERCIDE * BANGTON * BEAN SEED PROTECTANT * CAPTAF * CAP-TANCAPTENEET 26,538 * CAPTANE * CAPTAN-STREPTOMYCIN 7.5-0.1 POTATO SEED PIECE PROTEC-TANT * CAPTEX * ENT 26,538 * ESSO FUNGI-CIDE 406 * FLIT 406 * FUNGUS BAN TYPE II * GLYODEX 3722 * GRANOX PPM * GUSTAFSON CAPTAN 30-DD * HEXACAP * KAPTAN * LE CAPTANE (FRENCH) * MALIPUR * MERPAN * MICRO-CHECK 12 * NCI-C00077 * NERACID * ORTHOCIDE * OSOCIDE * SR406 * STAUF-FER CAPTAN * 3a,4,7,7a-TETRAHYDRO-

N-(TRICHLOROMETHANESULPHENYL)PHTHALIMIDE
* 3a,4,7,7a-TETRAHYDRO-2-((TRICHLOROMETHYL)
THIO)-1H-ISOINDOLE-1,3(2H)-DIONE * 1,2,3,6-TETRA-
HYDRO-N-(TRICHLOROMETHYLTHIO)PHTHALIMIDE
* N-(TRICHLOR-METHYLTHIO)-PHTHALIMID (GERMAN)
* N-TRICHLOROMETHYLMERCAPTO-4-CYCLOHEXENE-
1,2-DICARBOXIMIDE * N-(TRICHLOROMETHYLMER-
CAPTO)-Δ⁴-TETRAHYDROPHTHALIMIDE * N-TRI-
CHLOROMETHYLTHIOCYCLOHEX-4-ENE-1,2-DICARBOXI-
MIDE * N-TRICHLOROMETHYLTHIO-cis-Δ⁴-CYCLO-
HEXENE-1,2-DICARBOXIMIDE * N-((TRICHLORO-
METHYL)THIO)-4-CYCLOHEXENE-1,2-DICARBOXIMIDE
* TRICHLOROMETHYLTHIO-1,2,5,6-TETRAHYDRO-
PHTHALAMIDE * N-((TRICHLOROMETHYL)THIO)
TETRAHYDROPHTHALIMIDE * N-TRICHLORO-
METHYLTHIO-3A,4,7,7A-TETRAHYDROPHTHALIMIDE
* VANCIDE 89 * VANGARD K * VANICIDE
* VONDCAPTAN

CONSENSUS REPORTS: IARC Cancer Review: GROUP 3 IMEMDT 7,56,87; Animal Limited Evidence IMEMDT 30,295,83. NCI Carcinogenesis Bioassay (feed); Clear Evidence: mouse NCITR* NCI-CG-TR-15,77; No Evidence: rat NCITR* NCI-CG-TR-15,77. EPA Genetic Toxicology Program. Community Right-To-Know List. Reported in EPA TSCA Inventory.

OSHA PEL: TWA 5 mg/m³
ACGIH TLV: TWA 5 mg/m³
DOT Classification: ORM-E; Label; None.

SAFETY PROFILE: Moderately toxic to humans by ingestion. Moderately toxic experimentally by ingestion, inhalation, and intraperitoneal routes. Experimental teratogenic and reproductive effects. Questionable carcinogen with experimental tumorigenic and neoplastigenic data. Human mutation data reported. When heated to decomposition it emits toxic fumes of Cl⁻, SO_x, and NO_x.

CBG125   CAS: 8028-89-5   HR: D
CARAMEL

PROP: Dark brown to black liquid or solid; burnt sugar odor, pleasant bitter taste. Sol in water (colloidal).

SYN: CARAMEL COLOR

CONSENSUS REPORTS: Reported in EPA TSCA Inventory.

SAFETY PROFILE: Mutation data reported. When heated to decomposition it emits acrid smoke and irritating fumes.

CBG500   CAS: 8000-42-8   HR: 2
CARAWAY OIL

PROP: The main constituent of caraway oil is 1-carvone; found in the fruits of Carum carvi L. (Fam. Umbelliferae). Colorless liquid; odor and taste of caraway.

SYNS: KUEMMEL OIL (GERMAN) * OIL of CARAWAY

CONSENSUS REPORTS: Reported in EPA TSCA Inventory.

SAFETY PROFILE: Moderately toxic by ingestion and skin contact. A skin irritant. Mutation data reported. When heated to decomposition it emits acrid smoke and irritating fumes.

CBH750   HR: 3
CARBAMATES

PROP: Compounds based upon carbamic acid, NH_2COOH. Used only in the form of its numerous salts and derivatives.

SAFETY PROFILE: Many carbamates are poisons or moderately toxic, and some are carcinogenic, teratogenic, or mutagenic. They are used as insecticides, fungicides, herbicides, and as accelerators in the vulcanization of rubber. There is little data on persistence or breakdown in the environment.

The N-alkylcarbamates and thiocarbamates can react with nitrite under mildly acid conditions to form N-nitroso compounds. Nitrite is found in soils, in human saliva and in cured meats. N-nitrosodimethylamine is formed by soil microorganisms from Thiram. Other N-nitroso compounds could similarly be formed from other carbamate pesticides. However, the extent of the reaction of carbamates and nitrite in man is not known. The N-nitrosodialkylamines formed from dialkylthiocarbamate pesticides and nitrite are potent animal carcinogens and mutagens. The N-nitroso derivatives of several N-alkylcarbamates produce cancers in experimental animals at small doses.

Carbaryl, semicarbazide hydrochloride, n-propyl carbamate, Maneb, Zineb, Ferbam and Thiram are experimental teratogens.

Many of the carbamates have central nervous system effects. Carbaryl and Zectran are acetylcholinesterase inhibitors.

Ethylenethiourea, which produces thyroid carcinomas in rats and liver cell tumors in mice by ingestion, is formed from ethylenebisdithiocarbamates such as Maneb and Zineb by metabolic processes and cooking.

CBI250   CAS: 120-02-5   **HR: 3**
4-CARBAMIDOPHENYL BIS(CARBOXY-METHYLTHIO)ARSENITE
mf: $C_{11}H_{13}AsN_2O_5S_2$ mw: 392.30

SYNS: 2,2'-((4-((AMINOCARBONYL)AMINO)PHE-NYL)ARSINIDENE)BIS(THIO)BISACETIC ACID * BIS (CARBOXYMETHYLMERCAPTO)(p-UREIDOPHENYL) ARSINE * BIS(CARBOXYMETHYLTHIO)(p-UREIDO-PHENYL)ARSINE * (p-CARBAMOYLAMINO)PHENY-LARSINOBIS(2-THIO-ACETIC ACID) * CC 914 * C.C. No. 914 * MERCAPTOACETIC ACID, DIESTER with DITHIO-p-UREIDOBENZENEARSONOUS ACID * PHENYL UREA-p-DI(CARBOXYMETHYL) THIOAR-SENITE * THIOCARBARSONE * (p-UREIDOPHENYL-ARSYLENEDITHIO)DIACETIC ACID

CONSENSUS REPORTS: Arsenic and its compounds are on the Community Right-To-Know List.

OSHA PEL: TWA 0.5 mg(As)/m^3
ACGIH TLV: TWA 0.2 mg(As)/m^3

SAFETY PROFILE: Poison by intraperitoneal and intravenous routes. Moderately toxic by ingestion. When heated to decomposition it emits very toxic fumes of As and SO$_x$.

CBJ000   CAS: 121-59-5   **HR: 3**
N-CARBAMOYLARSANILIC ACID
mf: $C_7H_9AsN_2O_4$ mw: 260.10

PROP: White, nearly odorless powder; slt acid taste; sol in alc and water. Mp: 174°.

SYNS: AMABEVAN * AMEBAN * AMEBARSONE * AMIBIARSON * AMINARSON * AMINARSONE * AMINOARSON * (4-((AMINOCARBONYL)AMINO) PHENYL)ARSONIC ACID * ARSAMBIDE * p-AR-SONOPHENYLUREA * p-CARBAMIDOBENZENEAR-SONIC ACID * p-CARBAMINO PHENYL ARSONIC ACID * CARBAMINOPHENYL-p-ARSONIC ACID * 4-CAR-BAMYLAMINOPHENYLARSONIC ACID * N-CAR-BAMYL ARSANILIC ACID * CARBARSONE (USDA) * CARBASONE * FENARSONE * HISTOCARB * LEUCARSONE * p-UREIDOBENZENEARSONIC ACID * 4-UREIDO-1-PHENYLARSONIC ACID

CONSENSUS REPORTS: Arsenic and its compounds are on the Community Right-To-Know List.

OSHA PEL: TWA 10 μmg/m^3
ACGIH TLV: TWA 0.2 mg(As)/m^3

SAFETY PROFILE: Poison by ingestion. Moderately toxic by intraperitoneal route. Questionable carcinogen with experimental tumorigenic data. When heated to decomposition it emits very toxic fumes of As and NO$_x$.

CBJ750   CAS: 618-25-7   **HR: 3**
N-(CARBAMOYLMETHYL)ARSANILIC ACID
mf: $C_8H_{11}AsN_2O_4$ mw: 274.13

PROP: White, crystalline powder.

SYNS: (4-((2-AMINO-2-OXOETHYL)AMINO)PHENYL)AR-SONIC ACID * 4-ARSONOPHENYLGLYCINAMIDE * p-((CARBAMOYLMETHYL)AMINO)-BENXENEARSONIC ACID * SODIUM-N-PHENYLGLYCINAMIDE-p-ARSO-NATE * TRYPARSAMIDE

CONSENSUS REPORTS: Arsenic and its compounds are on the Community Right-To-Know List.

OSHA PEL: TWA 0.5 mg(As)/m^3
ACGIH TLV: TWA 0.2 mg(As)/m^3

SAFETY PROFILE: Poison by ingestion and intramuscular route. Moderately toxic by intravenous route. When heated to decomposition it emits very toxic fumes of As and NO$_x$.

CBM500   CAS: 116-06-3   **HR: 3**
CARBANOLATE
mf: $C_7H_{14}N_2O_2S$ mw: 190.29

PROP: A solid material.

SYNS: ALDECARB * ALDICARB (USDA) * ALDICARBE (FRENCH) * AMBUSH * ENT 27,093 * 2-METHYL-2-(METHYLTHIO)PROPANAL-O-((METHYL-AMINO)CARBONYL)OXIME * 2-METHYL-2-(METHYL-THIO)PROPIONALDEHYDE OXIME * 2-METHYL-2-(METHYLTHIO)PROPIONALDEHYDE-O-(METHYLCAR-BAMOYL)OXIME * 2-METHYL-2-METHYLTHIO-PROPIONALDEHYD-O-(N-METHYL-CARBAMOYL)-OXIM (GERMAN) * 2-METIL-2-TIOMETIL-PROPIONALDEID-O-(N-METIL-CARBAMOIL)-OSSIMA (ITALIAN) * NCI-C08640 * OMS-771 * RCRA WASTE NUMBER P070 * TEMIC * TEMIK * TEMIK G10 * UC-21149

CONSENSUS REPORTS: NCI Carcinogenesis Bioassay (feed); No Evidence: mouse, rat NCITR* NCI-CG-TR-136,79. Reported in EPA TSCA Inventory. EPA Extremely Hazardous Substances List.

SAFETY PROFILE: Deadly poison by ingestion, skin contact, subcutaneous and possibly other routes. Human mutation data reported. A powerful systemic poison. In 1985 over 150 people in California exhibited toxic effects from eating watermelons contaminated with aldicarb. When heated to decomposition it emits very toxic fumes of NO_x and SO_x.

CBM750   CAS: 63-25-2   **HR: 3**
CARBARYL
mf: $C_{12}H_{11}NO_2$ mw: 201.24

PROP: White crystals. Mp: 142°, d: 1.232 @ 20°/20°.

SYNS: CARBATOX-60 * CRAG SEVIN * ENT 23,969 * EXPERIMENTAL INSECTICIDE 7744 * KARBARYL (POLISH) * N-METHYLCARBAMATE de 1-NAPHTYLE (FRENCH) * METHYLCARBAMATE-1-NAPHTHALENOL * METHYLCARBAMATE-1-NAPH-THOL * METHYLCARBAMIC ACID-1-NAPHTHYL ES-TER * N-METHYL-1-NAFTYL-CARBAMAAT (DUTCH) * N-METHYL-1-NAPHTHYL-CARBAMAT (GERMAN) * N-METHYL-α-NAPHTHYLCARBAMATE * N-METHYL-1-NAPHTHYL CARBAMATE * N-METHYL-α-NAPHTHYLURETHAN * N-METIL-1-NAFTIL-CARBAM-MATO (ITALIAN) * α-NAFTYL-N-METHYLKARBAMAT (CZECH) * 1-NAPHTHOL-N-METHYLCARBAMATE * 1-NAPHTHYL METHYLCARBAMATE * α-NAPH-THYL N-METHYLCARBAMATE * 1-NAPHTHYL-N-METHYLCARBAMATE * SEVIN

CONSENSUS REPORTS: IARC Cancer Review: GROUP 3 IMEMDT 7,56,87; Animal Inadequate Evidence IMEMDT 12,37,76. Community Right-To-Know List.

OSHA PEL: TWA 5 mg/m^3
ACGIH TLV: TWA 5 mg/m^3
DFG MAK: 5 mg/m^3
NIOSH REL: TWA 5 mg/m^3
DOT Classification: ORM-A; Label: None.

SAFETY PROFILE: Poison by ingestion, intravenous, intraperitoneal, and possibly other routes. Human systemic effects by ingestion: sensory change involving peripheral nerve, muscle weakness. Experimental teratogenic and reproductive effects. Questionable carcinogen with experimental carcinogenic and tumorigenic data. Human mutation data reported. An eye and severe skin irritant. Absorbed by all routes, although skin absorption is slow. No accumulation in tissue. Symptoms include blurred vision, headache, stomach ache, vomiting. Symptoms

similar to but less severe than those due to parathion. A reversible cholinesterase inhibitor. When heated to decomposition it emits toxic fumes of NO_x.

CBN000   CAS: 86-74-8   **HR: 3**
CARBAZOLE
mf: $C_{12}H_9N$ mw: 167.22

PROP: White crystals. Mp: 244.8°, bp: 354.8°, d: 1.10 @ 18°/4°, vap press: 400 mm @ 323.0°.

SYNS: 9-AZAFLUORENE * 9H-CARBAZOLE * DIBENZOPYRROLE * DIBENZO(b,d)PYRROLE * DIPHENYLENEIMINE * DIPHENYLENIMIDE * DIPHENYLENIMINE * USAF EK-600

CONSENSUS REPORTS: IARC Cancer Review: GROUP 3 IMEMDT 7,56,87; Animal Limited Evidence IMEMDT 32,239,83. Reported in EPA TSCA Inventory.

SAFETY PROFILE: Poison by intraperitoneal route. Questionable carcinogen. Moderately toxic by ingestion. A pesticide. When heated to decomposition it emits toxic fumes of NO_x.

CBN375   **HR: 3**
CARBENDAZIM and SODIUM NITRITE (5:1)

SYNS: METHYL-2-BENZIMIDAZOLE CARBAMATE and SODIUM NITRITE * SODIUM NITRITE and CARBENDA-ZIM (1:5) * SODIUM NITRITE and METHYL-2-BENZIMI-DAZOLE CARBAMATE

SAFETY PROFILE: Suspected carcinogen with experimental carcinogenic data. When heated to decomposition it emits toxic fumes of Na_2O and NO_x.

CBQ750   CAS: 112-15-2   **HR: 2**
CARBITOL ACETATE
mf: $C_8H_{16}O_4$ mw: 176.24

PROP: Liquid. Bp: 217.4°, fp: −25°, flash p: 230°F (OC), d: 1.0114 @ 20°/20°, vap press: 0.05 mm @ 20°, vap d: 6.07.

SYNS: DIETHYLENE GLYCOL MONOETHYL ETHER ACETATE * DIGLYCOL MONOETHYL ETHER ACE-TATE * EKTASOLVE de ACETATE * 2-(2-ETHOXYE-THOXY)ETHANOL ACETATE * GLYCOL ETHER de ACETATE

CONSENSUS REPORTS: Reported in EPA TSCA Inventory. Glycol ether compounds are on the Community Right-To-Know List.

SAFETY PROFILE: Moderately toxic by ingestion. A skin and eye irritant. Combustible when exposed to heat; can react with oxidizing materials. To fight fire, use alcohol foam, water, CO_2, dry chemical. When heated to decomposition it emits acrid smoke and fumes.

CBR000   CAS: 111-90-0   **HR: 2**
CARBITOL CELLOSOLVE
mf: $C_6H_{14}O_3$ mw: 134.20

PROP: Colorless liquid, mild pleasant odor. Bp: 201.9°, flash p: 201°F (OC), d: 0.9902 @ 20°/4°, vap d: 4.62.

SYNS: APV * CARBITOL * CARBITOL SOLVENT * DIETHYLENE GLYCOL ETHYL ETHER * DIETHYLENE GLYCOL MONOETHYL ETHER * DIGLYCOL MONOETHYL ETHER * DIOXITOL * DOWANOL * DOWANOL DE * ETHOXY DIGLYCOL * 2-(2-ETHOXYETHOXY)ETHANOL * ETHYL CARBITOL * ETHYL DIETHYLENE GLYCOL * ETHYLENE DIGLYCOL MONOETHYL ETHER * LOSUNGSMITTEL APV * MONOETHYL ETHER of DIETHYLENE GLYCOL * POLY-SOLV * SOLVOSOL

CONSENSUS REPORTS: Reported in EPA TSCA Inventory. Glycol ether compounds are on the Community Right-To-Know List.

SAFETY PROFILE: Moderately toxic by ingestion, intravenous, intraperitoneal, and possibly other routes. Mildly toxic by skin contact. A skin and eye irritant. Experimental reproductive effects. Combustible when exposed to heat; can react with oxidizing materials. To fight fire, use alcohol foam, CO_2, dry chemical. When heated to decomposition it emits acrid smoke and irritating fumes.

CBS275   CAS: 1563-66-2   **HR: 3**
CARBOFURAN
DOT: NA 2757
mf: $C_{12}H_{15}NO_3$ mw: 221.28

PROP: White, crystalline solid; odorless. Mp: 105-152°, d: 1.180 @ 20°/20°, vap press: 2 × 10^{-5} mm @ 33°. Sltly sol in water.

SYNS: BAY 70143 * CURATERR * D 1221 * 2,3-DIHYDRO-2,2-DIMETHYLBENZOFURANYL-7-N-METHYLCARBAMATE * 2,3-DIHYDRO-2,2-DIMETHYL-7-BENZOFURANYL METHYLCARBAMATE * 2,2-DIMETHYL-7-COUMARANYL N-METHYLCARBAMATE * 2,2-DIMETHYL-2,3-DIHYDROBENZOFURAN-7-YL ESTER, METHYLCARBAMIC ACID * 2,2-DIMETHYL-

2,3-DIHYDRO-7-BENZOFURANYL-N-METHYLCARBAMATE * ENT 27,164 * FMC 10242 * FURADAN * FURODAN * METHYL CARBAMIC ACID 2,3-DIHYDRO-2,2-DIMETHYL-7-BENZOFURANYL ESTER * NIA 10242 * NIAGRA 10242 * YALTOX

CONSENSUS REPORTS: EPA Extremely Hazardous Substances List. Reported in EPA TSCA Inventory. EPA Genetic Toxicology Program.

OSHA PEL: TWA 0.1 mg/m^3
ACGIH TLV: TWA 0.1 mg/m^3
DOT Classification: Poison B; Label: Poison; Poison B; Label: Poison, liquid.

SAFETY PROFILE: Poison by inhalation, ingestion, skin contact, intravenous, and possibly other routes. Experimental teratogenic and reproductive effects. Human mutation data reported. A When heated to decomposition it emits toxic fumes of NO_x.

CBS500   CAS: 497-18-7   **HR: 3**
CARBOHYDRAZIDE
mf: CH_6N_4O mw: 90.11

SYNS: 4-AMINOSEMICARBAZIDE * CARBAZIC ACID HYDRAZIDE * CARBAZIDE * CARBAZIDE (DOT) * CARBODIHYDRAZIDE * CARBONIC ACID DIHYDRAZIDE * CARBONIC DIHYDRAZIDE * CARBONOHYDRAZIDE * CARBONYLDIHYDRAZINE * 1,3-DIAMINOUREA

CONSENSUS REPORTS: Reported in EPA TSCA Inventory.

DOT Classification: Forbidden.

SAFETY PROFILE: Poison by intravenous and intraperitoneal routes. Explodes when heated. Reacts with nitrous acid to form the explosive carbonic diazide. When heated to decomposition it emits toxic fumes of NO_x.

CBS750   CAS: 63042-08-0   **HR: 3**
4'-CARBOMETHOXY-2,3'-DIMETHYLAZOBENZENE
mf: $C_{16}H_{16}N_2O_3$ mw: 284.34

SYNS: 4'-CARBOMETHOXY-2,3'-DIMETHYLAZOBENZOL * CARBONIC ACID METHYL-4-(o-TOLYLAZO)-o-TOLYL ESTER * 2,3'-DIMETHYLAZOBENZENE-4'-METHYLCARBONATE

CONSENSUS REPORTS: NTP Fourth Annual Report On Carcinogens, 1984.

SAFETY PROFILE: Confirmed with experimental tumorigenic data. When heated to decomposition it emits toxic fumes of NO_x.

CBT250   CAS: 4564-87-8   HR: 3
CARBOMYCIN
mf: $C_{42}H_{67}NO_{16}$ mw: 842.10

SYNS: CARBOMYCIN A * DELTAMYCIN A
* 9-DEOXY-12,13-EPOXY-9-OXOLEUCOMYCIN V 3-ACE-
TATE 4^B-(3-METHYLBUTANOATE) * M-4209
* MAGNAMYCIN * MAGNAMYCIN A

SAFETY PROFILE: Poison by subcutaneous route. Moderately toxic by intravenous and intramuscular routes. When heated to decomposition it emits toxic fumes of NO_x.

CBT500   CAS: 7440-44-0   HR: 1
CARBON
af: C   aw: 12.01

PROP: Black crystals, powder or diamond form. Mp: 3652-3697° (sublimes), bp: approx 4200°, d (amorphous): 1.8-2.1, d (graphite): 2.25, d (diamond): 3.51, vap press: 1 mm @ 3586°.

SYNS: BLACK PEARLS * CARBONE (ITALIAN)
* CHARCOAL BLACK * C.I. 77266 * COLUMBIAN
CARBON * GRAPHITE (MAK) * GRAPHITE, NATU-
RAL (ACGIH) * GRAPHITE, SYNTHETIC * PURIFIED
CHARCOAL

CONSENSUS REPORTS: Reported in EPA TSCA Inventory.

OSHA PEL: (Natural graphite) (Transitional: TWA 50 mppcf) TWA 2.5 mg/m^3; (Synthetic graphite) (Transitional: TWA Total Dust: 15 mg/m^3; Respirable Fraction: 5 mg/m^3) TWA Total Dust: 10 mg/m^3; Respirable Fraction: 5 mg/m^3
ACGIH TLV: (Proposed: 2.5 mg/m^3 (respirable))
DFG MAK: 6 mg/m^3
DOT Classification: Flammable Solid; Label: Spontaneously Combustible

SAFETY PROFILE: Moderately toxic by intravenous route. Experimental reproductive effects. It can cause a dust irritation, particularly to the eyes and mucous membranes. Combustible when exposed to heat. Dust is explosive when exposed to heat or flame or oxides, peroxides, oxosalts, halogens, interhalogens, O_2, (NH_4NO_3 + heat), (NH_4ClO_4 @ 240°), bro-

mates, $Ca(OCl)_2$, chlorates, (Cl_2 + $Cr(OCl)_2$), ClO, iodates, IO_5, $(Pb(NO_3)_2$, $HgNO_3$, HNO_3, (oils + air), (K + air), Na_2S, $Zn(NO_3)_2$. Incompatible with air, metals, oxidants, unsaturated oils.

CBT750   CAS: 1333-86-4   HR: 1
CARBON BLACK

PROP: A generic term applied to a family of high-purity colloidal carbons commercially produced by carefully controlled pyrolysis of gaseous or liquid hydrocarbons. Carbon blacks, including commercial colloidal carbons such as furnace blacks, lamp blacks and acetylene blacks, usually contain less than several tenths percent of extractable organic matter and less than one percent ash.

SYNS: ACETYLENE BLACK * CHANNEL BLACK
* FURNACE BLACK * LAMP BLACK

CONSENSUS REPORTS: IARC Cancer Review: GROUP 3 IMEMDT 7,142,87; Human Inadequate Evidence IMEMDT 33,35,84; Animal Inadequate Evidence IMEMDT 33,35,84

OSHA PEL: TWA 3.5 mg/m^3
ACGIH TLV: TWA 3.5 mg/m^3
NIOSH REL: TWA 3.5 mg/m^3

SAFETY PROFILE: Mildly toxic by ingestion, inhalation, and skin contact. Questionable carcinogen. A nuisance dust in high concentrations. While it is true that the tiny particulates of carbon black contain some molecules of carcinogenic materials, the carcinogens are apparently held tightly and are not eluted by hot or cold water, gastric juices, or blood plasma.

CBU250   CAS: 124-38-9   HR: 1
CARBON DIOXIDE
DOT: UN 1013/UN 1845/UN 2187
mf: CO_2 mw: 44.01

PROP: Colorless, odorless gas. Mp: sublimes @ −78.5° (−56.6° @ 5.2 atm), vap d: 1.53.

SYNS: ANHYDRIDE CARBONIQUE (FRENCH)
* CARBONIC ACID GAS * CARBONIC ANHYDRIDE
* KOHLENDIOXYD (GERMAN) * KOHLENSAURE
(GERMAN)

CONSENSUS REPORTS: Reported in EPA TSCA Inventory.

OSHA PEL: (Transitional: TWA 5000 ppm) TWA 10,000 ppm; STEL 30,000 ppm

ACGIH TLV: TWA 5000 ppm; STEL 30,000 ppm

DFG MAK: 5000 ppm (9000 mg/m^3)

NIOSH REL: TWA 10000 ppm; CL 30000 ppm/ 10M

DOT Classification: Nonflammable Gas; Label: Nonflammable Gas; ORM-A; Label: None.

SAFETY PROFILE: An asphyxiant. Experimental teratogenic and reproductive effects. Contact of carbon dioxide snow with the skin can cause burns. Dusts of magnesium, zirconium, titanium, and some magnesium-aluminum alloys ignite and then explode in CO_2 atmospheres. Dusts of aluminum, chromium, and manganese ignite and then explode when heated in CO_2. Several bulk metals will burn in CO_2. Reacts vigorously with (Al + Na$_2$O$_2$), Cs$_2$O, Mg(C$_2$H$_5$)$_2$, Li, (Mg + Na$_2$O$_2$), K, KHC, Na, Na$_2$C$_2$, NaK, Ti. CO_2 fire extinguishers can produce highly incendiary sparks of 5-15 mJ at 10-20 KV by electrostatic discharge. Incompatible with acrylaldehyde, aziridine, metal acetylides, sodium peroxide.

CBU500   CAS: 124-38-9   **HR: 1**
CARBON DIOXIDE (liquefied)

DOT: UN 1013/UN 1845/UN 2187

SYNS: CARBON DIOXIDE, liquefied (DOT) * CARBON DIOXIDE, refrigerated liquid (DOT)

CONSENSUS REPORTS: Reported in EPA TSCA Inventory.

NIOSH REL: TWA 10000 ppm; CL 30000 ppm/ 10M

DOT Classification: Nonflammable Gas; Label: Nonflammable Gas.

SAFETY PROFILE: See CARBON DIOXIDE.

CBV000   CAS: 53569-62-3   **HR: 2**
CARBON DIOXIDE mixed with NITROUS OXIDE

DOT: UN 1015
mf: $CO_2 \cdot N_2O$ mw: 88.03

SYNS: CARBON DIOXIDE, mixture with NITROGEN OXIDE (N$_2$O) * CARBON DIOXIDE-NITROUS OXIDE mixture (DOT)

NIOSH REL: (Carbon Dioxide) TWA 10000 ppm; CL 30000 ppm/10M

NIOSH REL: (N$_2$O as Anesthetic Agent) TWA 25 ppm/1H

DOT Classification: Nonflammable Gas; Label: Nonflammable Gas.

SAFETY PROFILE: An anesthetic mixture. Combustible. An oxidizing mixture. Can react with reducing materials.

CBV500   CAS: 75-15-0   **HR: 3**
CARBON DISULFIDE

DOT: UN 1131
mf: CS$_2$ mw: 76.13

PROP: Clear, colorless liquid; nearly odorless when pure. Mp: −110.8°, bp: 46.5°, lel: 1.3%, uel: 50%, flash p: −22°F (CC), d: 1.261 @ 20°/20°, autoign temp: 257°F, vap press: 400 mm @ 28°, vap d: 2.64.

SYNS: CARBON BISULFIDE (DOT) * CARBON BISULPHIDE * CARBON DISULPHIDE * CARBONE (SUFURE de) (FRENCH) * CARBONIO (SOLFURO di) (ITALIAN) * CARBON SULFIDE * CARBON SULPHIDE (DOT) * DITHIOCARBONIC ANHYDRIDE * KOHLENDISULFID (SCHWEFELKOHLENSTOFF) (GERMAN) * KOOLSTOFDISULFIDE (ZWAVELKOOLSTOF) (DUTCH) * NCI-C04591 * RCRA WASTE NUMBER P022 * SCHWEFELKOHLENSTOFF (GERMAN) * SOLFURO di CARBONIO (ITALIAN) * SULPHOCARBONIC ANHYDRIDE * WEEVILTOX * WEGLA DWUSIARCZEK (POLISH)

CONSENSUS REPORTS: Reported in EPA TSCA Inventory. EPA Genetic Toxicology Program. Community Right-To-Know List. EPA Extremely Hazardous Substances List.

OSHA PEL: (Transitional: TWA 20 ppm; CL 30 ppm; PK 100 ppm/30 min) TWA 4 ppm; STEL 12 (skin)

ACGIH TLV: TWA 10 ppm (skin); BEI: 5 mg(2-thiothiazolidine-4-carboxylic acid (TTCA))/ g creatinine in urine.

DFG MAK: 10 ppm (30 mg/m^3); BAT: 8 mg/ L of 4-thio-4-thiazolidine carboxylic acid (TTCA) at end of shift.

NIOSH REL: TWA 1 ppm; CL 10 ppm/15M

DOT Classification: Flammable Liquid; Label: Flammable Liquid; IMO: Flammable Liquid; Label: Flammable Liquid, Poison.

SAFETY PROFILE: A human poison by ingestion and possibly other routes. Mildly toxic to humans by inhalation. An experimental poison by intraperitoneal route. Human reproductive effects on spermatogenesis by inhalation. Experimental teratogenic and reproductive effects. Human mutation data reported. The main toxic effect is on the central nervous system, acting as a narcotic and anesthetic in acute poisoning with death following from respiratory failure. In chronic poisoning, the effect on the nervous system is one of central and peripheral damage which may be permanent if the damage has been severe.

Flammable liquid. A dangerous fire hazard when exposed to heat, flame, sparks, friction, or oxidizing materials. Severe explosion hazard when exposed to heat or flame. Ignition and potentially explosive reaction when heated in contact with rust or iron. Mixtures with sodium or potassium-sodium alloys are powerful, shock-sensitive explosives. Explodes on contact with permanganic acid. Potentially explosive reaction with nitrogen oxide, chlorine (catalyzed by iron). Mixtures with dinitrogen tetraoxide are heat-, spark- and shock-sensitive explosives. Reacts with metal azides to produce shock- and heat-sensitive, explosive metal azidodithioformates. Aluminum powder ignites in CS_2 vapor. The vapor ignites on contact with fluorine. Reacts violently with azides, CsN_3, ClO, ethylamine diamine, ethylene imine, $Pb(N_3)_2$, LiN_3, (H_2SO_4 + permanganates), KN_3, RbN_3, NaN_3, phenylcopper-triphenylphosphine complexes. Incompatible with air, metals, oxidants. To fight fire, use water, CO_2, dry chemical, fog, mist. When heated to decomposition it emits highly toxic fumes of SO_x.

CBW750   CAS: 630-08-0   **HR: 3**
CARBON MONOXIDE

DOT: UN 1016/NA 9202
mf: CO   mw: 28.01

PROP: Colorless, odorless gas. Mp: $-207°$, bp: $-191.3°$, lel: 12.5%, uel: 74.2%, d: (gas) 1.250 g/L @ 0°, (liquid) 0.793, autoign temp: 1128°F.

SYNS: CARBONE (OXYDE de) (FRENCH) * CAR-
BONIC OXIDE * CARBONIO (OSSIDO di) (ITALIAN)
* CARBON MONOXIDE, CRYOGENIC liquid (DOT)
* CARBON OXIDE (CO) * EXHAUST GAS
* FLUE GAS * KOHLENMONOXID (GERMAN)
* KOHLENOXYD (GERMAN) * KOOLMONOXYDE
(DUTCH) * OXYDE de CARBONE (FRENCH)
* WEGLA TLENEK (POLISH)

CONSENSUS REPORTS: Reported in EPA TSCA Inventory.

OSHA PEL: (Transitional: TWA 50 ppm) TWA 35; CL 200 ppm
ACGIH TLV: TWA 50 ppm; STEL 400 ppm; BEI: less than 8% carboxyhemoglobin in blood at end of shift; less than 40 ppm CO in end-exhaled air at end of shift.
DFG MAK: 30 ppm (33 mg/m^3); BAT: 5% in blood at end of shift.
NIOSH REL: TWA 35 ppm; CL 200 ppm
DOT Classification: Flammable Gas; Label: Flammable Gas; Flammable Gas; Label: Flammable Gas and Poison Gas.

SAFETY PROFILE: Mildly toxic by inhalation in humans but has caused many fatalities. Experimental teratogenic and reproductive effects. Human systemic effects by inhalation: changes in psychophysiological tests and methemoglobinemia-carboxhemoglobinemia. Can cause asphyxiations by preventing hemoglobin from binding oxygen. After being removed from exposure, the half-life of its elimination from the blood is one hour. Chronic exposure effects can occur at lower concentrations. A common air contaminant. Acute cases of poisoning resulting from brief exposures to high concentrations seldom result in any permanent disability if recovery takes place. Chronic effects as the result of repeated exposure to lower concentrations have been described, particularly in the Scandinavian literature. Auditory disturbances and contraction of the visual fields have been demonstrated. Glycosuria does occur, and heart irregularities have been reported. Other workers have found that where the poisoning has been relatively long and severe, cerebral congestion and edema may occur, resulting in long-lasting mental or nervous damage. Repeated exposure to low concentration of the gas, up to 100 ppm in air, is generally believed to cause no signs of poisoning or permanent damage. Industrially, sequelae are rare, as exposure, though often severe, is usually brief. It is a common air contaminant.

A dangerous fire hazard when exposed to flame. Severe explosion hazard when exposed to heat or flame. Violent or explosive reaction on contact with bromine trifluoride, bromine pentafluoride, chlorine dioxide, or peroxodisulfuryl difluoride. Mixture of liquid CO with liquid O_2 is explosive. Reacts with sodium or potassium to form explosive products sensitive to shock, heat, or contact with water. Mixture with copper powder + copper(II) perchlorate + water forms an explosive complex. Mixture of liquid CO with liquid dinitrogen oxide is a rocket propellant combination. Ignites on warming with iodine heptafluoride. Ignites on contact with cesium oxide + water. Potentially explosive reaction with iron(III) oxide between 0-150°C. Exothermic reaction with ClF_3, (Li + H_2O), NF_3, OF_2, (K + O_2), Ag_2O, (Na + NH_3). To fight fire, stop flow of gas.

CBX109   CAS: 1885-14-9   **HR: 3**
CARBONOCHLORIDIC ACID PHENYL ESTER

DOT: UN 2746
mf: $C_7H_5ClO_2$ mw: 156.57

SYNS: CHLOROFORMIC ACID PHENYL ESTER * FENYLESTER KYSELINY CHLORMRAVENCI (CZECH) * PHENYL CHLOROCARBONATE * PHENYL CHLOROFORMATE * PHENYLCHLOROFORMATE (DOT)

CONSENSUS REPORTS: Reported in EPA TSCA Inventory.

DOT Classification: Poison B; Label: Corrosive and Poison.

SAFETY PROFILE: Poison by inhalation. Moderately toxic by ingestion and skin contact. A corrosive skin and eye irritant. When heated to decomposition it emits toxic fumes of Cl^-.

CBX250   **HR: 2**
CARBON REMOVER (liquid)

PROP: Flash p: <80°F.

DOT Classification: Flammable Liquid; Label: Flammable Liquid.

SAFETY PROFILE: Dangerous fire hazard when exposed to heat or flame; can react with oxidizing materials. To fight fire, use CO_2, dry chemical.

CBX750   CAS: 558-13-4   **HR: 3**
CARBON TETRABROMIDE

DOT: UN 2516
mf: CBr_4 mw: 331.65

PROP: Colorless, monoclinic tablets. Mp: (α) 48.4°, (β) 90.1°, bp: 189.5°, d: 3.42, vap press: 40 mm @ 96.3°.

SYNS: CARBON BROMIDE * TETRABROMIDE METHANE * TETRABROMOMETHANE

CONSENSUS REPORTS: Reported in EPA TSCA Inventory.

OSHA PEL: TWA 0.1 ppm; STEL 0.3 ppm
ACGIH TLV: TWA 0.1 ppm; STEL 0.3 ppm
DOT Classification: Poison B; Label: St. Andrews Cross.

SAFETY PROFILE: Poison by subcutaneous and intravenous routes. Moderately toxic by ingestion. Narcotic in high concentration. Mixture with Li particles is an impact-sensitive explosive. Explodes on contact with hexacylcohexyldilead. When heated to decomposition it emits toxic fumes of Br^-.

CBY000   CAS: 56-23-5   **HR: 3**
CARBON TETRACHLORIDE

DOT: UN 1846
mf: CCl_4 mw: 153.81

PROP: Colorless liquid; heavy, ethereal odor. Mp: −22.6°, bp: 76.8°, fp: −22.9°, flash p: none, d: 1.597 @ 20°, vap press: 100 mm @ 23.0°.

SYNS: BENZINOFORM * CARBONA * CARBON CHLORIDE * CARBON TET * CZTEROCHLOREK WEGLA (POLISH) * ENT 4,705 * FASCIOLIN * FLUKOIDS * METHANE TETRACHLORIDE * NECATORINA * NECATORINE * PERCHLOROMETHANE * R 10 * RCRA WASTE NUMBER U211 * TETRACHLOORKOOLSTOF (DUTCH) * TETRACHLOORMETAAN * TETRACHLORKOHLENSTOFF, TETRA (GERMAN) * TETRACHLORMETHAN (GERMAN) * TETRACHLOROCARBON * TETRACHLOROMETHANE * TETRACHLORURE de CARBONE (FRENCH) * TETRACLOROMETANO (ITALIAN) * TETRACLORURO di CARBONIO (ITALIAN) * TETRAFINOL * TETRAFORM * TETRASOL * UNIVERM * VERMOESTRICID

CONSENSUS REPORTS: IARC Cancer Review: GROUP 2B IMEMDT 7,143,87; Animal Sufficient Evidence IMEMDT 20,371,79; IMEMDT 1,53,72; Human Inadequate Evidence IMEMDT 1,53,72; Human Limited Evidence IMEMDT 20,371,79. NTP Fourth Annual Report On Carcinogens, 1984. Community Right-To-Know List. EPA Genetic Toxicology Program. Reported in EPA TSCA Inventory.

OSHA PEL: (Transitional: TWA 10 ppm; CL 25 ppm; PK 200 ppm/5 min) TWA 2 ppm
ACGIH TLV: TWA 5 ppm; STEL 30 (skin); Suspected Human Carcinogen

DFG MAK: 10 ppm (65 mg/m^3); BEI: 1.6 mL/ m^3 in alveolar air 1 hour after exposure; Suspected Carcinogen.
NIOSH REL: CL 2 ppm/60M
DOT Classification: ORM-A; Label: None; Poison B; Label: Poison.

SAFETY PROFILE: Confirmed carcinogen with experimental carcinogenic, neoplastigenic, and tumorigenic data. A human poison by ingestion and possibly other routes. Poison by subcutaneous and intravenous routes. Mildly toxic by inhalation. Human systemic effects by inhalation and ingestion: nausea or vomiting, pupillary constriction, coma, antipsychotic effects, tremors, somnolence, anorexia, unspecified respiratory system and gastrointestinal system effects. Experimental teratogenic and reproductive effects. An eye and skin irritant. Damages liver, kidneys, and lungs. Mutation data reported. A narcotic. Individual susceptibility varies widely. Contact dermatitis can result from skin contact.

Carbon tetrachloride has a narcotic action resembling that of chloroform, though not as strong. Following exposure to high concentrations, the victim may become unconscious, and if exposure is not terminated, death can follow from respiratory failure. The after-effects following recovery from narcosis are more serious than those of delayed chloroform poisoning, usually taking the form of damage to the kidneys, liver, and lungs. Exposure to lower concentrations, insufficient to produce unconsciousness, usually results in severe gastrointestinal upset and may progress to serious kidney and hepatic damage. The kidney lesion is an acute nephrosis; the liver involvement consists of an acute degeneration of the central portions of the lobules. When recovery takes place, there may be no permanent disability. Marked variation in individual susceptibility to carbon tetrachloride exists; some persons appear to be unaffected by exposures which seriously poison their fellow workers. Alcoholism and previous liver and kidney damage seem to render the individual more susceptible. Concentrations on the order of 1000 to 1500 ppm are sufficient to cause symptoms if exposure continues for several hours. Repeated daily exposure to such concentration may result in poisoning.

Though the common form of poisoning following industrial exposure is usually one of gastrointestinal upset, which may be followed by renal damage, other cases have been reported in which the central nervous system has been effected resulting in the production of polyneuritis, narrowing of the visual fields, and other neurological changes. Prolonged exposure to small amounts of carbon tetrachloride has also been reported as causing cirrhosis of the liver.

Locally, a dermatitis may be produced following long or repeated contact with the liquid. The skin oils are removed and the skin becomes red, cracked and dry. The effect of carbon tetrachloride on the eyes either as a vapor or as a liquid, is one of irritation with lacrimation and burning.

Industrial poisoning is usually acute with malaise, headache, nausea, dizziness, and confusion which may be followed by stupor and sometimes loss of consciousness. Symptoms of liver and kidney damage may follow later with development of dark urine, sometimes jaundice and liver enlargement, followed by scanty urine, albumenuria and renal casts; uremia may develop and cause death. Where exposure has been less acute, the symptoms are usually headache, dizziness, nausea, vomiting, epigastric distress, loss of appetite, and fatigue. Visual disturbances (blind spots, spots before the eyes, a visual "haze" and restriction of the visual fields), secondary anemia, and occasionally a slight jaundice may occur. Dermatitis may be noticed on the exposed parts.

Forms impact-sensitive explosive mixtures with particulates of many metals, e.g., aluminum (when ball milled or heated to 152° in a closed container); barium (bulk metal also reacts violently); beryllium; potassium (200 times more shock-sensitive than mercury fulminate); potassium-sodium alloy (more sensitive than potassium); lithium; sodium; zinc (burns readily). Also forms explosive mixtures with chlorine trifluoride; calcium hypochlorite (heat sensitive); calcium disilicide (friction and pressure sensitive); triethyldialuminum trichloride (heat sensitive); decaborane(14) (impact sensitive); dinitrogen tetraoxide. Violent or explosive reaction on contact with fluorine. Forms explosive mixtures with ethylene between 25-105° and 30-80 bar. Potentially explosive reaction on contact with boranes. 9:1 mixtures of methanol and CCl_4 react exothermically with aluminum, magnesium, or zinc. Potentially dangerous reaction with dimethyl formamide; 1,2,3,4,5,6-hexachlorocyclohexane; or dimethylacetamide when iron is present as a catalyst. CCl_4 has caused explosions when used as a fire extinguisher on wax and uranium fires. Incompatible with aluminum trichloride, dibenzoyl peroxide, potassium-tert-butoxide. Vigorous exothermic reaction with allyl alcohol; $Al(C_2H_5)_3$; (benzoyl peroxide + C_2H_4); BrF_3; diborane; disilane; liquid O_2; Pu; ($AgClO_4$ + HCl); potassium-tert-butoxide; tetraethylenepentamine; tetrasilane; trisilane; Zr. When heated to decomposition it emits toxic fumes of Cl^- and phosgene. It has been banned from household use by the FDA.

CBY250   CAS: 75-73-0   **HR: 2**
CARBON TETRAFLUORIDE
DOT: UN 1982
mf: CF_4 mw: 88.01

PROP: Colorless gas. Mp: $-184°$, bp: $-127.7°$, d: 1.96 @ $-184°$.

SYNS: ARCTON O   *   CARBON FLUORIDE   *   FC 14   *   FREON 14   *   HALOCARBON 14   *   HALON 14   *   PERFLUOROMETHANE   *   R 14   *   R 14 (REFRIGERANT)   *   TETRAFLUOROMETHANE (DOT)

CONSENSUS REPORTS: Reported in EPA TSCA Inventory.

DOT Classification: Nonflammable Gas; Label: Nonflammable Gas.

SAFETY PROFILE: Mildly toxic by inhalation. Less chronically toxic than carbon tetrachloride. Violent reaction with Al. When heated to decomposition it emits toxic fumes of F^-.

CBY750   CAS: 75-46-7   **HR: 2**
CARBON TRIFLUORIDE
DOT: UN 1984
mf: CHF_3 mw: 70.02

PROP: Colorless, odorless gas. Mp: $-163°$, bp: $-82.2°$, d: 1.52 (liquid) @ $-100°$.

SYNS: ARCTON   *   FLUOROFORM   *   FLUORYL   *   FREON 23   *   FREON F-23   *   GENETRON-23   *   HALOCARBON 23   *   METHYL TRIFLUORIDE   *   R 23   *   TRIFLUOROMETHANE (DOT)

CONSENSUS REPORTS: EPA Genetic Toxicology Program. Reported in EPA TSCA Inventory.

DOT Classification: Nonflammable Gas; Label: Nonflammable Gas.

SAFETY PROFILE: Narcotic in high concentration. A mild respiratory irritant. Mutation data

reported. When heated to decomposition it emits toxic fumes of F^-.

CCA500   CAS: 353-50-4   **HR: 3**
CARBONYL FLUORIDE
DOT: UN 2417
mf: CF_2O mw: 66.01

PROP: Colorless gas; pungent. Hygroscopic, mp: $-114°$, bp: $-83°$, d: 1.139 @ $-114°$.

SYNS: CARBON DIFLUORIDE OXIDE   *   CARBON FLUORIDE OXIDE   *   CARBONIC DIFLUORIDE   *   CARBON OXYFLUORIDE   *   CARBONYL DIFLUORIDE   *   DIFLUOROFORMALDEHYDE   *   FLUOPHOSGENE   *   FLUOROFORMYL FLUORIDE   *   FLUOROPHOSGENE   *   RCRA WASTE NUMBER U033

CONSENSUS REPORTS: Reported in EPA TSCA Inventory.

OSHA PEL: TWA 2 ppm; STEL 5 ppm
ACGIH TLV: TWA 2 ppm; STEL 5 ppm
DOT Classification: Poison A; Label: Poison Gas.

SAFETY PROFILE: A poison. Moderately toxic by inhalation. A powerful irritant. Hydrolyzes instantly to form HF on contact with moisture. Incompatible with hexafluoroisopropylideneamino-lithium. When heated to decomposition it emits toxic fumes of CO and F^-.

CCB609   **HR: 3**
CARBONYLS
PROP: The (CO) group with a metal (M). They may exist as dimeric acetylene derivatives (MOC≡COM) or as salts of hexahydroxybenzene.

SAFETY PROFILE: Most carbonyls are highly toxic. The toxicity of carbonyls depends in part, but not always entirely, on their ready decomposition which releases carbon monoxide. Symptoms are due in part to carbon monoxide and in part to the direct irritating action of the carbonyl. Many carbonyl metals ignite spontaneously in air, some with a delay period. Others are moderate fire and explosion hazards when exposed to heat or flame. Carbonyls of alkali metals are potentially explosive. Hypergolic reaction with dinitrogen tetraoxide. They react with water or steam to produce toxic and flammable vapors; can react vigorously with oxidizing materials. When heated to decomposition they emit highly toxic fumes of carbon monoxide.

CCC000   CAS: 463-58-1   HR: 3
CARBONYL SULFIDE
DOT: UN 2204
mf: COS   mw: 60.07

PROP: Gas or liquid. Mp: $-138°$, bp: $49.9°$, lel: 12%, uel: 28.5%, d: liq 1.24 @ $-87°$, vap d: 2.1.

SYNS: CARBON OXIDE SULFIDE * CARBON OXY-SULFIDE * CARBONYL SULFIDE-^{32}S * OXYCAR-BON SULFIDE

CONSENSUS REPORTS: Community Right-To-Know List. Reported in EPA TSCA Inventory.

DOT Classification: Poison A; Label: Poison Gas and Flammable Gas.

SAFETY PROFILE: Poison by intraperitoneal route. Mildly toxic by inhalation. Narcotic in high concentration. An irritant. May liberate highly toxic hydrogen sulfide upon decomposition. A very dangerous fire hazard and moderate explosion hazard when exposed to heat or flame. Can react vigorously with oxidizing materials. To fight fire, stop flow of gas or use CO_2, dry chemical or water spray. When heated to decomposition it emits toxic fumes of CO.

CCC500   CAS: 5234-68-4   HR: 3
CARBOXINE
mf: $C_{12}H_{13}NO_2S$ mw: 235.32

SYNS: 5-CARBOXANILIDO-2,3-DIHYDRO-6-METHYL-1,4-OXATHIIN * CARBOXIN (USDA) * D 735 * DCMO * 2,3-DIHYDRO-5-CARBOXANILIDO-6-METHYL-1,4-OXATHIIN * 5,6-DIHYDRO-2-METHYL-3-CARBOXANILIDO-1,4-OXATHIIN (GERMAN) * 2,3-DI-HYDRO-6-METHYL-1,4-OXATHIIN-5-CARBOXANILIDE * 5,6-DIHYDRO-2-METHYL-1,4-OXATHIIN-3-CARBOXA-NILIDE * 5,6-DIHYDRO-2-METHYL-N-PHENYL-1,4-OXATHIIN-3-CARBOXAMIDE * F 735 * FLO PRO V SEED PROTECTANT * VITAVAX

SAFETY PROFILE: Poison by ingestion. Moderately toxic by skin contact and possibly other routes. When heated to decomposition it emits very toxic fumes of NO_x and SO_x.

CCE500   CAS: 493-52-7   HR: 3
2-CARBOXY-4′-(DIMETHYLAMINO) AZOBENZENE
mf: $C_{15}H_{15}N_3O_2$ mw: 269.33

PROP: Shiny violet crystals.

SYNS: C.I. 13020 * C.I. ACID RED 2 * p-(DI-METHYLAMINO)AZOBENZENE-o-CARBOXYLIC ACID * 4′-DIMETHYLAMINOAZOBENZENE-2-CARBOXYLIC ACID * o-((p-(DIMETHYLAMINO)PHENYL)AZO)BEN-ZOIC ACID * 2-((4-DIMETHYLAMINO)PHENYLAZO) BENZOIC ACID * METHYL RED

CONSENSUS REPORTS: IARC Cancer Review: GROUP 3 IMEMDT 7,56,87; Animal Inadequate Evidence IMEMDT 8,161,75. Reported in EPA TSCA Inventory. EPA Genetic Toxicology Program.

SAFETY PROFILE: Questionable carcinogen with experimental tumorigenic data. Mutation data reported. When heated to decomposition it emits toxic fumes of NO_x.

CCG500   CAS: 36568-91-9   HR: 3
(4-(CARBOXY METHOXY)-3-CHLORO-PHENYL)(5,5-DIETHYL-2,4,6(1H,3H,5H)-PYRIMIDINETRIONATO-O^2-MERCURY, MONOSODIUM SALT
mf: $C_{16}H_{18}ClHgN_2O_6 \cdot Na$ mw: 593.39

SYNS: MERBAPHEN * NOVASUROL

CONSENSUS REPORTS: Mercury and its compounds are on the Community Right-To-Know List.

OSHA PEL: (Transitional: CL 1 mg/10m^3) CL 0.1 mg(Hg)/m^3 (skin)
ACGIH TLV: TWA 0.1 mg(Hg)/m^3 (skin)
NIOSH REL: (Inorganic Mercury) TWA 0.05 mg(Hg)/m^3

SAFETY PROFILE: Poison by intravenous route. When heated to decomposition it emits very toxic fumes of Cl$^-$, NO_x, and Hg.

CCJ625   CAS: 8000-66-6   HR: D
CARDAMON OIL

PROP: From the seed of *Elettaria acrdamomun* (L.) Maton (Fam. *Zingiberazeae*). Colorless liquid; aromatic penetrating odor of cardamom, pungent taste. Misc with alc.

SYNS: CARDAMON * OIL of CARDAMON

CONSENSUS REPORTS: Reported in EPA TSCA Inventory.

SAFETY PROFILE: Mutation data reported. When heated to decomposition it emits acrid smoke and fumes.

CCK000   CAS: 3599-32-4   **HR: 3**
CARDIO-GREEN
mf: $C_{43}H_{48}N_2O_6S_2 \cdot Na$ mw: 776.04

SYN: ICG

CONSENSUS REPORTS: Reported in EPA TSCA Inventory.

SAFETY PROFILE: Poison by intraperitoneal and intravenous routes. When heated to decomposition it emits very toxic fumes of SO_x, Na_2O, and NO_x.

CCK250   CAS: 959-24-0   **HR: 3**
β-CARDONE
mf: $C_{12}H_{20}N_2O_3S \cdot ClH$ mw: 308.86

SYNS: 4′-(1-HYDROXY-2-(ISOPROPYLAMINO)ETHYL) METHANESULFOANILIDE HYDROCHLORIDE * 4′-(1-HYDROXY-2-ISOPROPYLAMINO)ETHYL)METHANESUL-FONANILIDE MONOHYDROCHLORIDE * 4-(2-ISOPRO-PYLAMINE-1-HYDROXYETHYL)METHANESULFOANILIDE HYDROCHLORIDE * 4-(2-ISOPROPYLAMINO-1-HYDROXYAETHYL)METHANESULFONALID HYDRO-CHLORID (GERMAN) * ISOPROPYLAMINOHYDROXY-ETHYLMETHANESULFONALIDE HYDROCHLORIDE * N-ISOPROPYL-β-(4-METHANESULFONAMIDOPHENYL) ETHANOLAMINE HYDROCHLORIDE * MEAD JOHN-SON 1999 * MJ 1999 * MJ 1999 HYDROCHLORIDE * SOTACOR * SOTALEX * SOTALOL * SO-TALOL HYDROCHLORIDE

SAFETY PROFILE: A human poison by ingestion. Poison experimentally by intravenous and intraperitoneal routes. Moderately toxic by ingestion. Human systemic effects by ingestion: excitement, dyspnea, and convulsions. When heated to decomposition it emits very toxic fumes of HCl, SO_x, and NO_x.

CCL250   CAS: 9000-07-1   **HR: 3**
CARRAGEEN

PROP: A sulfated polysaccharide. Dried plant of seaweed *Chondrus crispus, Chondrus ocellatus, Eucheuma cottonil, Eucheuma spinosum, Gigartina acicularis, Gigartina pistillata, Gigartina radula, Gigartina stellata*. Yellow-white when powdered. Sol in water @ 80°; insol in organic solvents. Dried, bleached *Chondrus crispus* containing salts of sulfated polygalactose esters.

SYNS: 3,6-ANHYDRO-d-GALACTAN * AUBYGEL GS * AUBYGUM DM * BURTONITE-V-40-E * CARA-STAY * CARASTAY G * CARRAGEENAN (FCC)

* CARRAGEENAN GUM * CARRAGHEANIN * CARRAGHEEN * CARRAGHEENAN * CHON-DRUS * CHONDRUS EXTRACT * COLLOID 775 * COREINE * EUCHEUMA SPINOSUM GUM * FLANOGEN ELA * GALOZONE * GELCARIN * GELCARIN HMR * GELOZONE * GENU * GENUGEL * GENUGEL CJ * GENUGOL RLV * GENUVISCO J * GUM CARRAGEENAN * GUM CHON 2 * GUM CHROND * IRISH GUM * IRISH MOSS EXTRACT * IRISH MOSS GELOSE * KILLEEN * LYGOMME CDS * PEARLPUSS * PELLUGEL * PENCOGEL * PIG-WRACK * SATIAGEL GS 350 * SATIAGUM 3 * SATIAGUM STANDARD * SEAKEM CARRAGEENIN * SEATREM * SELF ROCK MOSS * VISCARIN

CONSENSUS REPORTS: IARC Cancer Review: GROUP 3 IMEMDT 7,56,87; Animal Limited Evidence IMEMDT 10,181,76. Reported in EPA TSCA Inventory.

SAFETY PROFILE: Poison by intravenous route. Questionable carcinogen with experimental neoplastigenic and tumorigenic data. When heated to decomposition it emits acrid smoke and fumes.

CCL500   **HR: 3**
CARRAGEENAN, DEGRADED

CONSENSUS REPORTS: IARC Cancer Review: Animal Sufficient Evidence IMEMDT 31,79,83.

SAFETY PROFILE: Confirmed carcinogen with experimental carcinogenic, neoplastigenic, and tumorigenic data. When heated to decomposition it emits toxic fumes of SO_x.

CCL750   CAS: 8015-88-1   **HR: 1**
CARROT SEED OIL

PROP: Distilled from the seeds of *Daucus carota* L. (Fam. *Umbelliferae*). Light yellow to amber liquid; aromatic odor. Sol in fixed oils, mineral oil; insol in glycerin, propylene glycol.

CONSENSUS REPORTS: Reported in EPA TSCA Inventory.

SAFETY PROFILE: A skin irritant. When heated to decomposition it emits acrid smoke and irritating fumes.

CCM000   CAS: 499-75-2   **HR: 3**
CARVACROL
mf: $C_{10}H_{14}O$ mw: 150.24

PROP: Colorless to pale yellow liquid; spicy thymol odor. D: 0.974-0.980, refr index: 1.521-1.526, flash p: 212.°F. Sol in alc, ether; insol in water.

SYNS: 2-p-CYMENOL * FEMA No. 2245 * 2-HYDROXY-p-CYMENE * ISOPROPYL-o-CRESOL * 5-ISOPROPYL-2-METHYLPHENOL * ISOTHYMOL * 2-METHYL-5-ISOPROPYLPHENOL * o-THYMOL

CONSENSUS REPORTS: Reported in EPA TSCA Inventory.

SAFETY PROFILE: Poison by ingestion and subcutaneous route. Moderately toxic by skin contact. A severe skin irritant. Combustible liquid. When heated to decomposition it emits acrid smoke and irritating fumes.

CCM100   CAS: 2244-16-8   HR: 3
d-CARVONE
mf: $C_{10}H_{14}O$ mw: 150.24

PROP: Colorless liquid; caraway odor. D: 0.956-0.960, refr index: 1.96-1.499. Sol in propylene glycol, fixed oils; misc in alc; insol in glycerin.

SYNS: (+)-CARVONE * d(+)-CARVONE * (S)-CARVONE * (S)-(+)-CARVONE * FEMA No. 2249 * d-p-MENTHA-6,8,(9)-DIEN-2-ONE * d-1-METHYL-4-ISOPROPENYL-6-CYCLOHEXEN-2-ONE * (S)-2-METHYL-5-(1-METHYLETHENYL)-2-CYCLOHEXEN-1-ONE

CONSENSUS REPORTS: Reported in EPA TSCA Inventory.

SAFETY PROFILE: Poison by ingestion and skin contact. A skin irritant. When heated to decomposition it emits acrid smoke and irritating fumes.

CCM120   CAS: 6485-40-1   HR: 2
l(−)-CARVONE
mf: $C_{10}H_{14}O$ mw: 150.22

PROP: Colorless liquid; spearmint odor. D: 0.956-0.960, refr index: 1.495-1.499. Sol in propylene glycol, fixed oils; misc in alc; insol in glycerin.

SYNS: (−)-CARVONE * 1-CARVONE * (R)-CARVONE * FEMA No. 2249 * 1-6,8(9)-p-MENTHADIEN-2-ONE * (R)-(−)-p-MENTHA-6,8-DIEN-2-ONE * 1-1-METHYL-4-ISOPROPENYL-6-CYCLOHEXEN-2-ONE * (R)-2-METHYL-5-(1-METHYLETHENYL)-2-CYCLOHEXEN-1-ONE (9CI)

SAFETY PROFILE: Moderately toxic by ingestion. When heated to decomposition it emits acrid smoke and irritating fumes.

CCN000   CAS: 87-44-5   HR: 1
CARYOPHYLLENE
mf: $C_{15}H_{26}$ mw: 206.41

PROP: Found in oil of clove, cinnamon leaves and copaiba balsam and in minor quantities in various other essential oils, especially lavender; prepared by isolation from clove leaf oil, clove stem oil, cinnamon leaf oil or pine oil fractions. Colorless to sltly yellow oily liquid; clove odor. D: 0.897-0.910, refr index: 1.498-1.504, flash p: 206°F. Sol in alc, ether; insol in water.

SYNS: β-CARYOPHYLLENE (FCC) * FEMA No. 2252 * 8-METHYLENE-4,11,11-(TRIMETHYL)BICYCLO(7.2.0) UNDEC-4-ENE

CONSENSUS REPORTS: Reported in EPA TSCA Inventory.

SAFETY PROFILE: A skin irritant. Combustible liquid. When heated to decomposition it emits acrid smoke and irritating fumes.

CCO750   CAS: 8007-80-5   HR: 3
CASSIA OIL

PROP: Chief constituent is cinnamic aldehyde, found in the leaves and twigs of *Cinnamomum cassia blume*. Yellow liquid; cinnamon odor, spicy burning taste. Sol in fixed oils, propylene glycol; insol in glycerin, mineral oil.

SYNS: ARTIFICIAL CINNAMON OIL * CINNAMON BARK OIL * CINNAMON BARK OIL, CEYLON TYPE (FCC) * CINNAMON OIL * KASSIA OEL (GERMAN) * OIL of CASSIA * OIL of CHINESE CINNAMON * OIL of CINNAMON * OIL of CINNAMON, CEYLON * OILS, CINNAMON

CONSENSUS REPORTS: Reported in EPA TSCA Inventory.

SAFETY PROFILE: Poison by skin contact. Moderately toxic by ingestion and intraperitoneal routes. A human skin irritant. Mutation data reported. When heated to decomposition it emits acrid smoke and irritating fumes.

CCP000   HR: 3
CASTOR BEAN

PROP: An annual which may grow higher than 15 feet. The large, lobed leaves may be 3 feet across. The spiny seed pods grow in clusters and contain plump seeds that are white with brown or black mottling. The seeds have a pleasant taste.

SYNS: AFRICAN COFFEE TREE * CASTOR BEANS (DOT) * CASTOR FLAKE (DOT) * CASTOR MEAL (DOT) * CASTOR OIL PLANT * CASTOR POMACE (DOT) * HIGUERETA (CUBA, PUERTO RICO) * HIGUERILLA (MEXICO) * KOLI (HAWAII) * LA'AU-'AILA (HAWAII) * MAN'S MOTHERWORT * MEXICO WEED * PA'AILA (HAWAII) * PALMA CHRISTI (HAITI) * RICIN (HAITI) * RICINO (PUERTO RICO) * RICINUS COMMUNIS * STEADFAST * WONDER TREE

DOT Classification: ORM-C; Label: None.

SAFETY PROFILE: Deadly poison by ingestion in humans. The seeds contain the deadly poison ricin, a plant lectin (toxalbumin) which inhibits protein synthesis in the intestinal wall. Ingestion of the seeds can cause after a delay period of several hours: nausea, vomiting, diarrhea and intestinal dysfunction. There may be massive fluid and electrolyte loss. Ingestion of as few as 2 seeds could be fatal. A potent allergen. When heated to decomposition it emits toxic fumes of NO_x.

CCP250   CAS: 8001-79-4   HR: 2
CASTOR OIL

PROP: From seeds of *Ricinus communis* L. (Fam. *Euphorbiaceae*). A colorless to pale yellow, viscous liquid; bland taste, characteristic odor. Mp: $-12°$, bp: 313°, flash p: 445°F (CC), d: 0.96, autoign temp: 840°F. Sol in alc; misc in abs alc, glacial acetic acid, chloroform, and ether.

SYNS: AROMATIC CASTOR OIL * CASTOR OIL AROMATIC * COSMETOL * CRYSTAL O * GOLD BOND * NCI-C55163 * NEOLOID * OIL of PALMA CHRISTI * PHORBYOL * RICINUS OIL * RICIRUS OIL * TANGANTANGAN OIL

CONSENSUS REPORTS: Reported in EPA TSCA Inventory.

SAFETY PROFILE: Moderately toxic by ingestion. An allergen. An eye irritant. Combustible when exposed to heat. Spontaneous heating may occur. To fight fire, use CO_2, dry chemical, fog, mist.

CCP500   CAS: 535-89-7   HR: 3
CASTRIX
mf: $C_7H_{10}ClN_3$ mw: 171.65

PROP: Sltly water-sol crystals.

SYNS: 2-CHLOOR-4-DIMETHYLAMINO-6-METHYL-PYRIMIDINE (DUTCH) * 2-CHLOR-4-DIMETHYLAMINO-6-METHYLPYRIMIDIN (GERMAN) * 2-CHLORO-4-DIMETHYLAMINO-6-METHYL-PYRIMIDINE * 2-CHLORO-4-METHYL-6-DIMETHYLAMINOPYRIMIDINE * 2-CLORO-4-DIMETILAMINO-6-METIL-PIRIMIDINA (ITALIAN) * CRIMIDIN (GERMAN) * CRIMIDINA (ITALIAN) * CRIMIDINE * W 491

CONSENSUS REPORTS: EPA Extremely Hazardous Substances List.

SAFETY PROFILE: Deadly poison by ingestion, intraperitoneal, and possibly other routes. Can cause central nervous system damage and convulsions. Intensely poisonous to mammals. A pesticide. When heated to decomposition it emits very toxic fumes of Cl^- and NO_x.

CCP850   CAS: 120-80-9   HR: 3
CATECHOL
mf: $C_6H_6O_2$ mw: 110.12

PROP: Colorless crystals. Mp: 105°, bp: 246°, flash p: 261°F (CC), d: 1.341 @ 15°, vap press: 10 mm @ 118.3°, vap d: 3.79. Sol in water, chloroform, and benzene; very sol in alc and ether.

SYNS: o-BENZENEDIOL * 1,2-BENZENEDIOL * CATECHIN * C.I. 76500 * C.I. OXIDATION BASE 26 * o-DIHYDROXYBENZENE * 1,2-DIHYDROXYBENZENE * o-DIOXYBENZENE * o-DIPHENOL * DURAFUR DEVELOPER C * FOURAMINE PCH * FOURRINE 68 * o-HYDROQUINONE * o-HYDROXYPHENOL * 2-HYDROXYPHENOL * NCI-C55856 * OXYPHENIC ACID * PELAGOL GREY C * o-PHENYLENEDIOL * PYROCATECHIN * PYROCATECHINIC ACID * PYROCATECHOL * PYROCATECHUIC ACID

CONSENSUS REPORTS: IARC Cancer Review: GROUP 3 IMEMDT 7,56,87; Animal Inadequate Evidence IMEMDT 15,155,77. EPA Extremely Hazardous Substances List. Reported in EPA TSCA Inventory. EPA Genetic Toxicology Program.

OSHA PEL: TWA 5 ppm (skin)
ACGIH TLV: TWA 5 ppm

SAFETY PROFILE: Poison by ingestion, subcutaneous, intraperitoneal, intravenous, parenteral, and possibly other routes. Moderately toxic by skin contact. Experimental reproductive effects. Can cause dermatitis on skin contact. An allergen. Human mutation data reported. Questionable carcinogen. Systemic effects similar to phenol. Combustible when exposed to heat or flame; can react vigorously with oxidizing materials. Hypergolic reaction with concentrated nitric acid. To fight fire, use water, CO_2, dry chemical. When heated to decomposition it emits acrid smoke and irritating fumes.

CCQ500   CAS: 8007-20-3   **HR: 2**
CEDAR LEAF OIL

PROP: Constituent is d-α-thujone, found in leaves of *Thuja occidentalis* L. (Fam. *Cupressaaceae*). Yellowish, volatile oil; strong sage odor. D: 0.910-0.920. Sol in fixed oils, mineral oil, propylene glycol; insol in glycerin.

SYNS: OIL of ARBOR VITAE * OIL of CEDAR LEAF * OIL of THUJA * OIL of WHITE CEDAR * OILS, CEDAR LEAF * OIL THUJA * THUJA OIL * WHITE CEDAR OIL

CONSENSUS REPORTS: Reported in EPA TSCA Inventory.

SAFETY PROFILE: Moderately toxic by ingestion. A skin irritant. Ingestion of large quantities causes hypertension, bradycardia, tachypnea, convulsions, death. When heated to decomposition it emits acrid smoke and fumes.

CCS500   CAS: 35607-66-0   **HR: 2**
CEFOXITIN
mf: $C_{16}H_{17}N_3O_7S_2$ mw: 427.48

SYNS: CEPHOXITIN * CFX * REPHOXITIN

SAFETY PROFILE: Mildly toxic by subcutaneous and intravenous routes. Human systemic effects by intravenous route: reduction in the white blood cell count. When heated to decomposition it emits very toxic fumes of NO_x and SO_x.

CCT250   CAS: 9005-81-6   **HR: 3**
CELLOPHANE
mf: $(C_6H_{10}O_5)_n$

SYN: VISKING CELLOPHANE

CONSENSUS REPORTS: Reported in EPA TSCA Inventory.

SAFETY PROFILE: Questionable carcinogen with experimental tumorigenic data by implant. When heated to decomposition it emits acrid smoke and irritating fumes.

CCU000   CAS: 8050-88-2   **HR: 1**
"CELLULOID"

DOT: UN 2000/UN 2002

PROP: Clear or colored cellulose nitrate. D: 1.35-1.60.

SYNS: CELLULOID, in blocks, rods, rolls, sheets, tubes (DOT) * CELLULOID SCRAP (DOT)

DOT Classification: Flammable Solid; Label: Flammable Solid; Flammable Solid; Label: Spontaneously Combustible (Scrap).

SAFETY PROFILE: Mildly toxic. Flammable when exposed to heat or flame. Can react with oxidizing materials. When heated to decomposition it emits acrid smoke and fumes.

CCU150   CAS: 9004-34-6   **HR: 1**
CELLULOSE, POWDERED

PROP: Fine white fibrous particles from treatment of bleached cellulose from wood or cotton. Insol in water and most organic solvents.
OSHA PEL: Total Dust: 15 mg/m^3; Respirable Fraction: 5 mg/m^3
ACGIH TLV: TWA (nuisance particulate) 10 mg/m^3 of total dust (when toxic impurities are not present, e.g., quartz < 1%).

SAFETY PROFILE: A nuisance dust. When heated to decomposition it emits acrid smoke and irritating fumes.

CCU250   CAS: 9004-70-0   **HR: 3**
CELLULOSE TETRANITRATE

DOT: UN 0340/UN 0342/UN 1324/UN 2555/ UN 2556/NA 2059/UN 2060
mf: $C_{12}H_{16}(ONO_2)_4O_6$ mw: 504.3

PROP: White, amorphous solid. D: 1.66, flash p: 55°F.

SYNS: CELLOIDIN * CELLULOSE NITRATE * COLLODION COTTON * COLLOXYLIN * GUNCOTTON * NITROCELLULOSE * NITROCOTTON * PYRALIN * PYROXYLIN * PYROXYLIN PLASTICS (DOT) * PYROXYLIN PLASTIC SCRAP (DOT) * SOLUBLE GUN COTTON * XYLOIDIN

CONSENSUS REPORTS: Reported in EPA TSCA Inventory.

DOT Classification: Flammable Solid; Label: Flammable Solid.

SAFETY PROFILE: Flammable solid. Highly dangerous fire hazard in the dry state when exposed to heat, flame, or powerful oxidizers. When wet with 35% of denatured ethanol it is about as hazardous as ethanol alone or gasoline. Dry cellulose tetranitrate burns rapidly with intense heat and ignites easily. Moderately dangerous explosion hazard. To fight fire, use copious volumes of water, alcohol foam. CO_2 is effective in extinguishing fires of nitrocellulose solvents.

CCV000   HR: 3
CEMENT, leather

DOT Classification: Flammable Liquid; Label: Flammable Liquid.

SAFETY PROFILE: Flammable when exposed to heat or flame.

CCV250   HR: 3
CEMENT (liquid)

SYN: CEMENT, adhesive (DOT)

DOT Classification: Combustible Liquid; Label: None; Flammable Liquid; Label: Flammable Liquid; IMO: Flammable or Combustible Liquid; Label: Flammable Liquid.

SAFETY PROFILE: Combustible when exposed to heat or flame.

CCV750   HR: 2
CEMENT (pyroxylin)

SYN: CEMENT, PYROXYLIN (DOT)

DOT Classification: Flammable Liquid; Label: Flammable Liquid.

SAFETY PROFILE: Dangerous fire hazard when exposed to heat or flame; can react with oxidizing materials.

CCW000   HR: 2
CEMENT (roofing liquid)

SYN: CEMENT, ROOFING, liquid (DOT)

DOT Classification: Flammable Liquid; Label: Flammable Liquid.

SAFETY PROFILE: Dangerous fire hazard when exposed to heat or flame; can react with oxidizing materials.

CCW250   HR: 2
CEMENT (rubber)

PROP: Flash p: 50°F or less.

SYN: CEMENT, RUBBER (DOT)

DOT Classification: Flammable Liquid; Label: Flammable Liquid.

SAFETY PROFILE: May contain benzene or other toxic solvents. Dangerous fire hazard when exposed to heat or flame; can react with oxidizing materials.

CCX000   CAS: 123-03-5   HR: 3
CEPACOL CHLORIDE
mf: $C_{21}H_{38}N \cdot Cl$ mw: 340.05

SYNS: ACETOQUAT CPC * AKTIVEX * AMMONYX CPC * BIOSEPT * CEEPRYN * CEEPRYN CHLORIDE * CEPRIM * CETAMIUM * CETYLPYRIDINIUM CHLORIDE * N-CETYLPYRIDINIUM CHLORIDE * 1-CETYLPYRIDINIUM CHLORIDE * DOBENDAN * HEXADECYLPYRIDINIUM CHLORIDE * n-HEXADECYLPYRIDINIUM CHLORIDE * 1-HEXADECYLPYRIDINIUM CHLORIDE * INTEXSAN CPC * PRISTACIN * PYRISEPT * QUATERNARIO CPC

CONSENSUS REPORTS: Reported in EPA TSCA Inventory.

SAFETY PROFILE: Poison by ingestion, intraperitoneal, subcutaneous, and intravenous routes. Moderately toxic by skin contact. A skin and eye irritant. When heated to decomposition it emits very toxic fumes of NO_x and Cl^-.

CCX500   CAS: 21593-23-7   HR: 2
CEPHAPIRIN
mf: $C_{17}H_{17}N_3O_6S_2$ mw: 423.49

SYNS: CEFAPIRIN (GERMAN) * 3-(HYDROXYMETHYL)-8-OXO-7-(2-(4-PYRIDYLTHIO)ACETAMIDO)-5-THIA-1-AZABICYCLO(4.2.0)OCT-2-ENE-2-CARBOXYLIC ACID, ACETATE (ESTER)

SAFETY PROFILE: Moderately toxic by intravenous route. When heated to decomposition it emits very toxic fumes of NO_x and SO_x.

CCY000   CAS: 1306-38-3   HR: 2
CERIC OXIDE
mf: CeO_2 mw: 172.12

SYN: CERIUM DIOXIDE

CONSENSUS REPORTS: Reported in EPA TSCA Inventory.

SAFETY PROFILE: Moderately toxic by ingestion.

CCY250   CAS: 7440-45-1   HR: 2
CERIUM

DOT: UN 1333
af: Ce   aw: 140.13

PROP: Cubic or hexagonal, steel gray crystals. Mp: 815°, bp: 3257°, d: (cubic form): 6.90, hexagonal form 6.75.

CONSENSUS REPORTS: Reported in EPA TSCA inventory.

DOT Classification: Flammable Solid; Label: Flammable Solid

SAFETY PROFILE: Cerium resembles aluminum in its pharmacological action as well as in its chemical properties. The insoluble salts such as the oxalates are stated to be nontoxic even in large doses. It is used to prevent vomiting in pregnancy. The average dose is from 0.05 to 0.5 grams.
　　The effect on the central nervous system of the rare-earth metals following inhalation may preclude welding operations with these materials to any large extent. Cerium is stated to produce polycythemia but is useless in the treatment of anemia owing to its toxic effects. The salts of cerium increase the blood coagulation rate. A strong reducing agent. Moderate fire hazard; ignites spontaneously in air at 150-180°. Moderate explosion hazard in the form of dust when exposed to flame. The metal or its alloys spark with friction. Many alloys are pyrophoric in air. Explosive reaction with zinc. Very exothermic reaction with antimony or bismuth. Ignites when heated in atmospheres of CO_2 + N_2; Cl_2; or Br_2. Violent reaction when heated with phosphorus (400°C); silicon (1400°C).

CCY500   CAS: 537-00-8   HR: 3
CERIUM ACETATE
mf: $C_6H_9O_6 \cdot Ce$ mw: 317.27

SYNS: CERIUM TRIACETATE ＊ CEROUS ACETATE

CONSENSUS REPORTS: Reported in EPA TSCA Inventory.

SAFETY PROFILE: Human central nervous system effects. When heated to decomposition it emits acrid and irritating fumes.

CCY699   HR: 3
CERIUM AZIDE

SAFETY PROFILE: An explosive. Upon decomposition it emits toxic fumes of NO_x.

CCY750   CAS: 7790-86-5   HR: 3
CERIUM CHLORIDE
mf: $CeCl_3$ mw: 246.47

PROP: Colorless, deliquescent crystals. Mp: 848°, bp: 1727°, d: 3.92.

SYNS: CERIUM(III) CHLORIDE ＊ CERIUM TRICHLORIDE ＊ CEROUS CHLORIDE

CONSENSUS REPORTS: Reported in EPA TSCA Inventory. EPA Genetic Toxicology Program.

SAFETY PROFILE: Poison by intravenous, intraperitoneal, and subcutaneous routes. Moderately toxic by ingestion. When heated to decomposition it emits toxic fumes of Cl^-.

CCZ000   CAS: 512-24-3   HR: 3
CERIUM CITRATE
mf: $C_6H_8O_7 \cdot Ce$ mw: 332.26

SYNS: CERIUM(III) CITRATE ＊ CEROUS CITRATE ＊ 2-HYDROXY-1,2,3-PROPANETRISCARBOXYLIC ACID CERIUM(3+) SALT (1:1) (9CI)

SAFETY PROFILE: Poison by intraperitoneal route. Experimental reproductive effects. When heated to decomposition it emits acrid and irritating fumes.

CDA250   HR: 2
CERIUM COMPOUNDS

PROP: Compounds of cerium and the other rare earth elements are generally of low toxicity. The greatest exposures are likely to be during manufacture of cerium. Exposed workers have experienced sensitivity to heat, itching, and skin lesions. Large doses to experimental animals have caused writhing, ataxia (loss of muscle coordination), labored respiration, sedation, hypotension, and death by cardiovascular collapse. The chloride, bromide, nitrate, bromate, and perchlorate salts are water soluble and thus are more likely to cause systemic effects when ingested. The sulfates, iodides, and iodates are less water soluble. Oxides, oxalates, sulfides, carbonates, fluorides, and phosphates are insoluble. The salts of cerium increase the blood coagulation rate. Cerium tartrate has been found to produce a direct injurious action on the hearts of small animals. Cerium oxalate has been used to suppress motion sickness and to suppress vomiting during pregnancy (by ingestion of 1 gram/24 hours). The toxicity of cerium compounds may be taken to be that of cerium, except when the anion has a toxicity of its own.

CDA500   CAS: 15158-67-5   **HR: 3**
CERIUM EDETATE

SAFETY PROFILE: Poison by intraperitoneal route. When heated to decomposition it emits acrid smoke and irritating fumes.

CDA750   CAS: 7758-88-5   **HR: 3**
CERIUM FLUORIDE
mf: CeF_3 mw: 197.12

PROP: White, hexagonal crystals. D: 6.16, mp: 1430°, bp: 2327°. Insol in water; sol in H_2SO_4.

SYNS: CERIUM FLUORURE (FRENCH) * CERIUM TRIFLUORIDE * CEROUS FLUORIDE

CONSENSUS REPORTS: Reported in EPA TSCA Inventory.

OSHA PEL: TWA 2.5 mg(F)/m^3
ACGIH TLV: TWA 2.5 mg(F)/m^3
NIOSH REL: (Inorganic Fluorides) TWA 2.5 mg(F)/m^3.

SAFETY PROFILE: A poison. When heated to decomposition it emits toxic fumes of F^-.

CDB250   CAS: 10294-41-4   **HR: 3**
CERIUM(III) NITRATE, HEXAHYDRATE (1:3:6)
mf: $N_3O_9 \cdot Ce \cdot 6H_2O$ mw: 434.27

SYNS: CERIUM NITRATE, HEXAHYDRATE * CERIUM TRINITRATE HEXAHYDRATE * CEROUS NITRATE HEXAHYDRATE * NITRIC ACID, CERIUM(3+) SALT, HEXAHYDRATE

SAFETY PROFILE: Poison by intraperitoneal and intravenous routes. Moderately toxic by ingestion. When heated to decomposition it emits toxic fumes of NO_x.

CDB500   **HR: 3**
CERIUM(III) TETRAHYDROALUMINATE

PROP: Decomp @ −80°C.

SAFETY PROFILE: A dangerous fire hazard. Ignites spontaneously in air. Unstable.

CDC000   CAS: 7440-46-2   **HR: 3**
CESIUM
af: Cs   aw: 132.91

DOT: UN 1383/UN 1407

PROP: Hexagonal crystals, silver-white, ductile metal or possibly a silvery liquid. Mp: 28.5°, bp: 705°, d: 1.873, vap press: 1 mm @ 279°.

SYNS: CESIUM-133 * CESIUM METAL (DOT) * CESIUM, POWDERED (DOT)

CONSENSUS REPORTS: Reported in EPA TSCA Inventory.

DOT Classification: Flammable Solid; Label: Flammable Solid and Dangerous When Wet; IMO: Flammable Solid; Label: Spontaneously Combustible (powdered).

SAFETY PROFILE: Cesium is quite similar to potassium in its elemental state. It has been shown, however, to have pronounced physiological action in experimentation with animals. Hyper-irritability, including marked spasms, has been shown to follow the administration of cesium in amounts equal to the potassium content of the diet. It has been found that replacing the potassium in the diet of rats with cesium caused death after 10-17 days. Ignites spontaneously in air. Violent reaction with water, moisture or steam releases hydrogen gas which explodes. Violent reaction with acids, halogens, and other oxidizing materials. Incandescent reaction with non-metals (e.g., sulfur, phosphorus).

CDC500   CAS: 7787-69-1   **HR: 2**
CESIUM BROMIDE
mf: BrCs   mw: 212.82

CONSENSUS REPORTS: Reported in EPA TSCA Inventory.

SAFETY PROFILE: Moderately toxic by intraperitoneal route. When heated to decomposition it emits toxic fumes of Br^-.

CDC699   **HR: 1**
CESIUM BROMOXENATE

SAFETY PROFILE: Solution in water is extremely unstable. When heated to decomposition it emits toxic fumes of Br^-.

CDC750   CAS: 534-17-8   **HR: 2**
CESIUM CARBONATE
mf: $CO_3 \cdot 2Cs$ mw: 325.83

SYNS: CARBONIC ACID, DICESIUM SALT * DICE-SIUM CARBONATE

CONSENSUS REPORTS: EPA Genetic Toxicology Program. Reported in EPA TSCA Inventory.

SAFETY PROFILE: Moderately toxic by ingestion. Mutation data reported. When heated to decomposition it emits acrid smoke and fumes.

CDD000   CAS: 7647-17-8   **HR: 2**
CESIUM CHLORIDE
mf: ClCs   mw: 168.36

PROP: D: 3.99, mp: 646°, bp: 1303°.

SYNS: CESIUM MONOCHLORIDE * DICESIUM DI-CHLORIDE * TRICESIUM TRICHLORIDE

CONSENSUS REPORTS: Reported in EPA TSCA Inventory. EPA Genetic Toxicology Program.

SAFETY PROFILE: Moderately toxic by ingestion and intraperitoneal routes. Mutation data reported. Reacts violently with BF_3. When heated to decomposition it emits toxic fumes of Cl^-.

CDD500   CAS: 13400-13-0   **HR: 3**
CESIUM FLUORIDE
mf: CsF   mw: 151.91

SYNS: CESIUM MONOFLUORIDE * DICESIUM DI-FLUORIDE * TRICESIUM TRIFLUORIDE

CONSENSUS REPORTS: Reported in EPA TSCA Inventory.

OSHA PEL: TWA 2.5 mg(F)m^3
ACGIH TLV: TWA 2.5 mg(F)/m^3
NIOSH REL: (Inorganic Fluorides) TWA 2.5 mg(F)/m^3.

SAFETY PROFILE: A poison. Incompatible with benzenediazonium tetrafluoroborate; difluoroamine. When heated to decomposition it emits toxic fumes of F^-.

CDD750   CAS: 21351-79-1   **HR: 3**
CESIUM HYDROXIDE

DOT: UN 2681/UN 2682
mf: CsHO   mw: 149.92

PROP: Colorless to yellowish, very deliquescent crystals. Mp: 272.3°, d: 3.675.

SYNS: CESIUM HYDRATE * CESIUM HYDROXIDE DIMER * CESIUM HYDROXIDE, solid (DOT) * CESIUM HYDROXIDE, solution (DOT)

CONSENSUS REPORTS: Reported in EPA TSCA Inventory.

OSHA PEL: TWA 2 mg/m^3
ACGIH TLV: TWA 2 mg/m^3

DOT Classification: Corrosive; IMO: Corrosive Material; Label: Corrosive.

SAFETY PROFILE: Poison by intraperitoneal route. Moderately toxic by ingestion. A powerful caustic. A corrosive skin and eye irritant.

CDE000   CAS: 7789-17-5   **HR: 2**
CESIUM IODIDE
mf: CsI   mw: 259.81

CONSENSUS REPORTS: Reported in EPA TSCA Inventory.

SAFETY PROFILE: Moderately toxic by ingestion and intraperitoneal route. When heated to decomposition, it emits toxic fumes of I^-.

CDE250   CAS: 7789-18-6   **HR: 2**
CESIUM(I) NITRATE (1:1)
mf: $NO_3 \cdot Cs$ mw: 194.92

DOT: UN 1451

PROP: Colorless, hexagonal or cubic, glittering crystalline powder. Mp: 414°, bp: decomp, d: 3.685, 2.71 @ 500° (liq).

SYNS: CESIUM NITRATE (DOT) * NITRIC ACID, CESIUM SALT

CONSENSUS REPORTS: Reported in EPA TSCA Inventory. EPA Genetic Toxicology Program.

DOT Classification: Oxidizer; Label: Oxidizer.

SAFETY PROFILE: Moderately toxic by ingestion and intraperitoneal routes. Mutation data reported. When heated to decomposition it emits toxic fumes of NO_x.

CDF750   CAS: 6004-24-6   **HR: 3**
CETYLPYRIDINIUM CHLORIDE MONOHYDRATE
mf: $C_{21}H_{38}N \cdot Cl \cdot H_2O$ mw: 358.07

SYNS: CEEPRYN * CEPACOL * 1-HEXADECYL-
PYRIDINIUM CHLORIDE MONOHYDRATE

SAFETY PROFILE: Poison by ingestion, intraperitoneal, intravenous, and subcutaneous
routes. When heated to decomposition it emits
very toxic fumes of Cl^- and NO_x.

CDH250 CAS: 520-36-5 **HR: 1**
CHAMOMILE
mf: $C_{15}H_{10}O_5$ mw: 270.25

PROP: Blue liquid, turning brownish-yellow.
Composed of amyl and butyl esters of angelic
and tiglic acids, butyric acid, etc. D: 0.905-
0.915 @ 15°/15°.

SYNS: APIGENIN * APIGENINE * APIGENOL
* C.I. NATURAL YELLOW 1 * 5,7-DIHYDROXY-2-
(4-HYDROXYPHENYL)-4H-1-BENZOPYRAN-4-ONE
* 2-(p-HYDROXYPHENYL)-5,7-DIHYDROXYCHROMONE
* PELARGIDENON 1449 * 4',5,7-TRIHYDROXY-
FLAVONE * VERSULIN

SAFETY PROFILE: Mutation data reported. A
mild allergen. When heated to decomposition
it emits acrid smoke and irritating fumes.

CDH500 CAS: 8002-66-2 **HR: 1**
CHAMOMILE OIL

PROP: By steam distillation of the flowers and
stalks of *Matrilaria chamomilla* L. Blue-yellow-
ish-brown liquid; strong odor and bitter aromatic
taste. Composed of amyl and butyl esters of
angelic, tiglic acids, and butyric acid. D: 0.905-
0.915 @ 15°/15°. Sol in fixed oils, propylene
glycol; insol in mineral oil, glycerin.

SYNS: CAMOMILE OIL GERMAN * CHAMOMILE-
GERMAN OIL * GERMAN CHAMOMILE OIL
* HUNGARIAN CHAMOMILE OIL

CONSENSUS REPORTS: Reported in EPA
TSCA Inventory.

SAFETY PROFILE: A mild allergen. A skin
irritant. When heated to decomposition it emits
acrid and irritating fumes.

CDH750 CAS: 8015-92-7 **HR: 1**
CHAMOMILE OIL (ROMAN)

PROP: Obtained by the steam distillation of
the dried flowers of *Anthemis nobilis* L. Blue
liquid, turning brownish-yellow; strong aromatic odor. Composition: Amyl and butyl esters
of angelic and tiglic acids, butyric acid, etc.

D: 0.905-0.915 @ 15°/15°. Sol in fixed oils,
mineral oil, propylene glycol; insol in glycerin.

SYN: CAMOMILE OIL, ENGLISH TYPE (FCC)

CONSENSUS REPORTS: Reported in EPA
TSCA Inventory.

SAFETY PROFILE: A mild allergen. A skin
irritant. Combustible when heated. When heated
to decomposition it emits acrid smoke and irritating fumes.

CDI000 CAS: 64365-11-3 **HR: 1**
CHARCOAL, ACTIVATED
af: C aw: 12.01

DOT: UN 1362

PROP: Black porous solid, coarse granules or
powder. Insol in water, organic solvents.

SYNS: ACTIVATED CARBON * CARBON, ACTI-
VATED * CARBORAFFIN * CARBORAFINE
* KARBORAFIN * NUCHAR 722

DOT Classification: Flammable Solid; Label:
Spontaneously Combustible.

SAFETY PROFILE: It can cause a dust irritation, particularly to the eyes and mucous membranes. Combustible when exposed to heat. Dust
is flammable and explosive when exposed to
heat or flame or oxides.

CDI250 CAS: 16291-96-6 **HR: 2**
CHARCOAL (BRIQUETTES)

DOT: NA 1361

PROP: Black amorphous solid. Composition:
carbon + impurities. Mw: 12.0, mp: >3500°,
bp: 4200°, d: 3.51.

SYN: CHARCOAL

CONSENSUS REPORTS: Reported in EPA
TSCA Inventory.

DOT Classification: Flammable Solid; Label:
Flammable Solid.

SAFETY PROFILE: Carbon itself has no toxic
action, but it contains impurities that may be
toxic. Fire hazard: reacts with liquid air,
$Ba(ClO_3)_2$, BrF_5, ClO, $Ca(ClO_3)_2$, ClF_2, F_2,
H_2O_2, $Mg(ClO_3)_2$, $(O_2 + wood)$, perchlorates,
peroxides, (P + air), K + $KClO_3$, KNO_3, RuO_4,
$AgNO_3$, $NaClO_3$, $(AgCl + NaO_2)$, S, (S +
$NaNO_3$), $Zn(ClO_3)_2$. Heats spontaneously, particularly when wet, freshly calcined, or tightly

packed, and it can ignite and burn. Slight explosion hazard when exposed to heat or flame. To fight fire, use water, mist, foam or dry chemical. When heated to decomposition it emits acrid smoke and fumes.

CDI500   CAS: 16291-96-6   **HR: 2**
CHARCOAL SCREENINGS, MADE from "PINON" WOOD (DOT)

DOT: NA 1361

DOT Classification: Flammable Solid; Label: Flammable Solid.

SAFETY PROFILE: A flammable solid.

CDJ000   **HR: 2**
CHARCOAL (SHELL)

SYN: CHARCOAL, SHELL (DOT)

DOT Classification: Flammable Solid; Label: Flammable Solid.

SAFETY PROFILE: A flammable solid.

CDJ500   CAS: 16291-96-6   **HR: 2**
CHARCOAL (wood, ground, crushed, granulated or pulverized)

DOT: NA 1361

DOT Classification: Flammable Solid; Label: Flammable Solid.

SAFETY PROFILE: A flammable solid.

CDK000   CAS: 16291-96-6   **HR: 2**
CHARCOAL WOOD SCREENINGS, OTHER THAN "PINON" WOOD SCREENINGS (DOT)

DOT Classification: Flammable Solid; Label: Flammable Solid.

SAFETY PROFILE: A flammable solid.

CDL000   CAS: 476-32-4   **HR: 3**
CHELIDONINE
mf: $C_{20}H_{19}NO_5$ mw: 353.40

PROP: White, crystalline powder. Mp: 135-136°.

SAFETY PROFILE: Poison by intravenous and subcutaneous routes. A central nervous system depressant causing sleepiness, depression, slowing of the pulse, and, in large doses, coma

and circulatory failures. Combustible when exposed to heat or flame. When heated to decomposition it emits toxic fumes of NO_x.

CDL500   CAS: 8006-99-3   **HR: 3**
CHENOPODIUM OIL

PROP: American wormseed. Ingredients are ascaridol, cymene, camphor and saponins. Colorless or pale yellow liquid, characteristic disagreeable odor and taste. Composition: 60-70% ascaridol. D: 0.950-0.980 @ 25°/25°. Insol in water; sol in 8 vols 70% alc; sltly sol in glacial acetic acid. Keep well closed, cool, and protected from light.

SYNS: OIL of AMERICAN WORMSEED * OIL of CHENOPODIUM

SAFETY PROFILE: Poison by ingestion. Moderately toxic by skin contact. A skin irritant. When heated to decomposition it emits acrid smoke and irritating fumes.

CDM000   **HR: 3**
CHERRY LAUREL OIL

PROP: Volatile oil from leaves of *Prunus laurocerasus L., Rosacene.* Pale yellow liquid, odor and taste similar to oil of bitter almond. D: 1.054-1.066 @ 20°/20°. Sltly sol in water; sol in 2 vols 70% alc, benzene, chloroform, and ether.

SAFETY PROFILE: Very poisonous. Hydrogen cyanide component is responsible for highly toxic properties. Keep well closed, cool, and protected from light. When heated to decomposition it emits toxic fumes of CN^-.

CDM250   CAS: 1401-55-4   **HR: 3**
CHESTNUT TANNIN

SYNS: CASTANEA SATIVA MILL TANNIN * TANNIN from CHESTNUT

SAFETY PROFILE: Poison by subcutaneous, intramuscular, intravenous, and intraperitoneal routes. Questionable carcinogen with experimental tumorigenic data. When heated to decomposition it emits acrid and irritating fumes.

CDN000   CAS: 53-19-0   **HR: 3**
CHLODITHANE
mf: $C_{14}H_{10}Cl_4$ mw: 320.04

SYNS: CHLODITAN * 1-CHLORO-2-(2,2-DICHLORO-1-(4-CHLOROPHENYL)ETHYL)BENZENE * 2-(o-CHLOROPHENYL)-2-(p-CHLOROPHENYL)-1,1-DICHLOROETHANE * o,p'-DDD * 2,4'-DDD * 1,1-DICHLORO-2,2-BIS(2,4'-DICHLOROPHENYL)ETHANE * 1,1-DICHLORO-2-(o-CHLOROPHENYL)-2-(p-CHLOROPHENYL)ETHANE * o,p'-DICHLORODIPHENYLDICHLOROETHANE * 2,4'-DICHLOROPHENYLDICHLOROETHANE * MITOTANE * NCI-C04933 * NSC 38721 * o,p-TDE * o,p'-TDE

CONSENSUS REPORTS: NCI Carcinogenesis Studies (ipr); Equivocal Evidence: mouse, rat CANCAR 40,1935,77. EPA Genetic Toxicology Program.

SAFETY PROFILE: Human systemic effects by ingestion: somnolence, blood pressure depression, diarrhea, nausea or vomiting, normocytic anemia (decrease in the number of red blood cells), and pigmented or nucleated red blood cells. Experimental teratogenic and reproductive effects. Questionable carcinogen with experimental carcinogenic and tumorigenic data. Mutation data reported. When heated to decomposition it emits toxic fumes of Cl$^-$.

CDN200   CAS: 78-95-5   HR: 3
CHLORACETONE
DOT: UN 1695
mf: C_3H_5ClO mw: 92.53

PROP: Colorless liquid, pungent odor. Mp: −44.5°, bp: 119°, d: 1.162.

SYNS: ACETONYL CHLORIDE * A-STOFF * CHLORACETONE * CHLOROACETONE * CHLOROACETONE, stabilized (DOT) * CHLOROPROPANONE * 1-CHLORO-2-PROPANONE * MONOCHLORACETONE * MONOCHLOROACETONE * MONOCHLOROACETONE, inhibited (DOT) * MONOCHLOROACETONE, stabilized (DOT) * MONOCHLOROACETONE, unstabilized (DOT) * TONITE

CONSENSUS REPORTS: Reported in EPA TSCA Inventory.

ACGIH TLV: CL 1 ppm (skin)
DOT Classification: Forbidden, unstabilized; Irritating Material; LABEL: Irritant; POISON B; LABEL: Poison, stabilized

SAFETY PROFILE: Poison by inhalation, ingestion, and skin contact. Mutation data reported. A lachrymator poison gas. Flammable when exposed to heat or flame, or oxidizers.

Old material can explode. When heated to decomposition it emits highly toxic fumes.

CDN500   CAS: 107-14-2   HR: 3
CHLORACETONITRILE
DOT: UN 2668
mf: C_2H_2ClN mw: 75.50

SYNS: CHLOROACETONITRILE (DOT) * α-CHLOROACETONITRILE * 2-CHLOROACETONITRILE * CHLOROMETHYL CYANIDE * MONOCHLOROACETONITRILE * MONOCHLOROMETHYL CYANIDE * USAF KF-5

CONSENSUS REPORTS: Reported in EPA TSCA Inventory. Cyanide and its compounds are on the Community Right-To-Know List.

DOT Classification: Poison B; Label: Flammable Liquid and Poison.

SAFETY PROFILE: Poison by ingestion, skin contact, and intraperitoneal route. Moderately toxic by inhalation. A skin irritant. Human mutation data reported. Questionable carcinogen with experimental tumorigenic data. When heated to decomposition it emits very toxic fumes of Cl$^-$, NO$_x$, and CN$^-$.

CDO000   CAS: 302-17-0   HR: 3
CHLORAL HYDRATE
mf: $C_2HCl_3O \cdot H_2O$ mw: 165.40

PROP: Transparent, colorless crystals; aromatic, penetrating, sltly acrid odor and sltly bitter, caustic taste. Mp: 52°, bp: 97.5°, d: 1.9.

SYNS: AQUACHLORAL * Bi 3411 * CHLORALDURAT * DORMAL * FELSULES * HYDRAL * HYDRAL de CHLORAL * KESSODRATE * LORINAL * NOCTEC * NORTEC * NYCOTON * PHALDRONE * RECTULES * SK-CHORAL HYDRATE * SOMNI SED * SOMNOS * SONTEC * TOSYL * TRAWOTOX * TRICHLORACETALDEHYD-HYDRAT (GERMAN) * TRICHLOROACETALDEHYDE HYDRATE * TRICHLOROACETALDEHYDE MONOHYDRATE * 2,2,2-TRICHLORO-1,1-ETHANEDIOL

CONSENSUS REPORTS: Reported in EPA TSCA Inventory. EPA Genetic Toxicology Program.

SAFETY PROFILE: A human poison by ingestion and possibly other routes. Poison experimentally by ingestion, intravenous, and rectal routes. Moderately toxic by subcutaneous, parenteral, and intraperitoneal routes. Human systemic effects by ingestion: general anesthetic; cardiac arrythmias, blood pressure depression. Human mutation data reported. Questionable carcinogen with experimental carcinogenic and tumorigenic data by skin contact. A sedative, anesthetic, and narcotic. Combustible when exposed to heat or flame. When heated to decomposition it emits toxic fumes of Cl^-.

CDO250   CAS: 95-06-7   HR: 3
2-CHLORALLYL DIETHYLDITHIO-CARBAMATE
mf: $C_8H_{14}ClNS_2$ mw: 223.80

PROP: Amber liquid. Bp: 129° @ 1 mm.

SYNS: CDEC * CHLORALLYL DIETHYLDITHIOCAR-BAMATE * 2-CHLOROALLYL DIETHYLDITHIOCARBA-MATE * 2-CHLOROALLYL-N,N-DIETHYLDITHIOCAR-BAMATE * 2-CHLORO-2-PROPENE-1-THIOL DIETH-YLDITHIOCARBAMATE * 2-CHLORO-2-PRO-PENYL DIETHYLCARBAMODITHIOATE * CP 4572 * DIETHYLCARBAMODITHIOIC ACID 2-CHLORO-2-PRO-PENYL ESTER * DIETHYLDITHIOCARBAMIC ACID-2-CHLOROALLYL ESTER * NCI-C00453 * SULFAL-LATE * THIOALLATE * VEGADEX * VEGADEX SUPER

CONSENSUS REPORTS: IARC Cancer Review: GROUP 2B IMEMDT 7,56,87; Animal Sufficient Evidence IMEMDT 30,283,83. NTP Fourth Annual Report On Carcinogens, 1984. NCI Carcinogenesis Bioassay (feed); Clear Evidence: mouse, rat NCITR* NCI-CG-TR-115,78. EPA Genetic Toxicology Program.

SAFETY PROFILE: Confirmed with experimental carcinogenic data. Moderately toxic by ingestion and skin contact. Mutation data reported. An herbicide. When heated to decomposition it emits very toxic fumes of Cl^-, NO_x, and SO_x.

CDO500   CAS: 305-03-3   HR: 3
CHLORAMBUCIL
mf: $C_{14}H_{19}Cl_2NO_2$ mw: 304.24

SYNS: AMBOCHLORIN * AMBOCLORIN * 4-(BIS(2-CHLOROETHYL)AMINO)BENZENEBUTANOIC ACID * γ-(p-BIS(2-CHLOROETHYL)AMINOPHENYL) BUTYRIC ACID * 4-(p-(BIS(2-CHLOROETHYL)AMINO) PHENYL)BUTYRIC ACID * 4-(p-BIS(β-CHLOROETH-YL)AMINOPHENYL)BUTYRIC ACID * CB 1348 * CHLORAMINOPHEN * CHLORAMINOPHENE * CHLOROAMBUCIL * CHLOROBUTIN * CHLOROBUTINE * N,N-DI-2-CHLOROETHYL-γ-p-

AMINOPHENYLBUTYRIC ACID * p-(N,N-DI-2-CHLOROETHYL)AMINOPHENYL BUTYRIC ACID * p-N,N-DI-(β-CHLOROETHYL)AMINOPHENYL BU-TYRIC ACID * γ-(p-DI(2-CHLOROETHYL)AMINOPHE-NYL)BUTYRIC ACID * ECLORIL * ELCORIL * LEUKERAN * LEUKERSAN * LEUKORAN * LINFOLIZIN * LINFOLYSIN * NCI-C03485 * NSC-3088 * PHENYLBUTYRIC ACID NITROGEN MUSTARD * RCRA WASTE NUMBER U035

CONSENSUS REPORTS: IARC Cancer Review: GROUP 1 IMEMDT 7,144,87; Human Inadequate Evidence IMEMDT 9,125,75; Human Limited Evidence IMEMDT 26,115,81; Animal Limited Evidence IMEMDT 26,115,81; Animal Sufficient Evidence IMEMDT 9, 125,75. NTP Fourth Annual Report On Carcinogens, 1984. EPA Genetic Toxicology Program.

SAFETY PROFILE: Confirmed carcinogen producing leukemia. Experimental carcinogenic and neoplastigenic data. Poison by ingestion, intravenous, intraperitoneal, and subcutaneous routes. Human respiratory system effects by ingestion: cough, dyspnea, and interstitial fibrosis. Human reproductive effects by ingestion and possibly other routes: changes in spermatogenesis; menstrual cycle changes or disorders; and teratogenic effects of the fetal urogenital system. Experimental teratogenic and reproductive effects. Human mutation data reported. An antineoplastic agent. When heated to decomposition it emits very toxic fumes of Cl^- and NO_x.

CDP000   CAS: 127-65-1   HR: 3
CHLORAMINE T
mf: $C_7H_8ClNO_2S \cdot Na$ mw: 228.66

SYNS: ACTI-CHLORE * AKTIVIN * ANEXOL * BENZENESULFONAMIDE, N-CHLORO-4-METHYL-, SO-DIUM SALT (9CI) * BERKENDYL * CHLORALONE * CHLORASAN * CHLORASEPTINE * CHLORA-ZAN * CHLORAZENE * CHLORAZONE * CHLO-ROZONE * CHLORSEPTOL * CLORINA * CLO-ROSAN * DESINFECT * EUCLORINA * GANSIL * GYNECLORINA * HALAMID * HELIOGEN * KLORAMIN * KLORAMINE-T * MULTICHLOR * SODIUM CHLORAMINE T * SODIUM p-TOLUENE-SULFONYLCHLORAMIDE * SODIUM TOSYLCHLO-RAMIDE * TAMPULES * TOCHLORINE * TOL-AMINE * TOSYLCHLORAMIDE SODIUM

CONSENSUS REPORTS: Reported in EPA TSCA Inventory.

SAFETY PROFILE: Poison by parenteral and intravenous routes. Human mutagenic data reported. When heated to decomposition it emits toxic fumes of Cl^-, SO_x, Na_2O, and NO_x.

CDP250   CAS: 56-75-7   HR: 3
CHLORAMPHENICOL

mf: $C_{11}H_{12}Cl_2N_2O_5$ mw: 323.15

PROP: Crystalline. Mp: 151°. Sltly sol in water.

SYNS: ALFICETYN * AMBOFEN * AMPHENICOL * AMPHICOL * AMSECLOR * ANACETIN * AQUAMYCETIN * AUSTRACIL * AUSTRACOL * BIOCETIN * BIOPHENICOL * CAF * CAM * CAP * CATILAN * CHEMICETIN * CHEMICETINA * CHLOMIN * CHLOMYCOL * CHLORAMEX * CHLORAMFICIN * CHLORAMFILIN * d-CHLORAMPHENICOL * d-threo-CHLORAMPHENICOL * CHLORAMSAAR * CHLORASOL * CHLORA-TABS * CHLORICOL * CHLORNITROMYCIN * CHLOROCAPS * CHLOROCID * CHLOROCIDIN C TETRAN * CHLOROCOL * CHLOROJECT L * CHLOROMAX * CHLOROMYCETIN * CHLORONITRIN * CHLOROPTIC * CHLOROVULES * CIDOCETINE * CIPLAMYCETIN * CLORAMIDINA * CLOROAMFENICOLO (ITALIAN) * CLOROMISAN * CLOROSINTEX * COMYCETIN * CPH * CYLPHENICOL * DESPHEN * DETREOMYCINE * DEXTROMYCETIN * d-(−)-threo-2-DICHLOROACETAMIDO-1-p-NITROPHENYL-1,3-PROPANEDIOL * d-threo-N-DICHLOROACETYL-1-p-NITROPHENYL-2-AMINO-1,3-PROPANEDIOL * d-(−)-threo-2,2-DICHLORO-N-(β-HYDROXY-α-(HYDROXYMETHYL))-p-NITROPHENETHYL-ACETAMIDE * d-(−)-2,2-DICHLORO-N-(β-HYDROXY-α-(HYDROXYMETHYL)-p-NITROPHENYLETHYL)ACETAMIDE * d-threo-N-(1,1'-DIHYDROXY-1-p-NITROPHENYLISOPROPYL)DICHLOROACETAMIDE * DOCTAMICINA * ECONOCHLOR * EMBACETIN * EMETREN * ENICOL * ENTEROMYCETIN * ERBAPLAST * ERTILEN * FARMICETINA * FENICOL * GLOBENICOL * GLOROUS * HALOMYCETIN * HORTFENICOL * I 337A * INTRAMYCETIN * ISMICETINA * ISOPHENICOL * ISOPTO FENICOL * KAMAVER * KEMICETINE * LEUKOMYAN * LEVOMYCETIN * LOROMISIN * MASTIPHEN * MEDIAMYCETINE * MICOCHLORINE * MICROCETINA * MYCHEL * MYCINOL * NCI-C55709 * d-(−)-threo-1-p-NITROPHENYL-2-DICHLORACETAMIDO-1,3-PROPANEDIOL * d-threo-1-(p-NITROPHENYL)-2-(DICHLOROACETYLAMINO)-1,3-PROPANEDIOL * NORIMYCIN V * NOVOCHLOROCAP * NOVOMYCETIN * NOVOPHENICOL * NSC 3069 * OFTALENT * OLEOMYCETIN * OPTHOCHLOR * OTOPHEN * PANTOVERNIL * PARAXIN * PETNAMYCETIN * QUEMICETINA * RIVOMYCIN * ROMPHENIL * SEPTICOL * SINTOMICETINA * STANOMYCETIN * SYNTHOMYCINE * TEVCOCIN * TIFOMYCINE * TREOMICETINA * U-6062 * UNIMYCETIN * VETICOL

CONSENSUS REPORTS: IARC Cancer Review: GROUP 2B IMEMDT 7,145,87; Human Limited Evidence IMEMDT 10,85,76. Reported in EPA TSCA Inventory. EPA Genetic Toxicology Program.

SAFETY PROFILE: Suspected human carcinogen producing leukemia, aplastic anemia, and other bone marrow changes. Experimental tumorigenic data. Poison by intraperitoneal, intravenous, and subcutaneous routes. Moderately toxic by ingestion. Human systemic effects by an unknown route: changes in plasma or blood volume, unspecified liver effects, and hemorrhaging. Experimental teratogenic and reproductive effects. Human mutation data reported. An antibiotic. When heated to decomposition it emits very toxic fumes of NO_x and Cl^-.

CDP700   CAS: 530-43-8   HR: 2
CHLORAMPHENICOL PALMITATE

mf: $C_{27}H_{42}Cl_2N_2O_6$ mw: 561.61

SYNS: CAP-P * CAP-PALMITATE * CHLORAMPHENICOL MONOPALMITATE * DETREOPAL * α-ESTER PALMITIC ACID with D-threo-(−)-2,2-DICHLORO-N-(β-HYDROXY-α-(HYDROXYMETHYL)-p-NITROPHENETHYL)ACETAMIDE

SAFETY PROFILE: Moderately toxic by oral route. An experimental teratogen. An antibiotic. When heated to decomposition it emits very toxic fumes of NO_x and Cl^-.

CDQ000   HR: 3
CHLORATES

PROP: Chlorates are a combination of a metal or hydrogen and $^-ClO_3$ monovalent radical. They are crystalline and somewhat deliquescent.

SAFETY PROFILE: The principal toxic effects of chlorates are the production of methemoglobin in the blood and destruction of red blood corpuscles. The latter may lead to irritation of the kidneys. Damage to heart muscle has been reported.

Dangerous fire hazard in contact with flammable matter. When contaminated with oxidizable materials, they are particularly sensitive to friction, heat, and shock. They are powerful oxidizing agents and can undergo violent reactions with reducing materials. Dangerous explosion hazard when shocked, exposed to heat, or rubbed, particularly when contaminated with sugar; charcoal; shellac; sulfur; starch; sawdust; sulfuric acid; ammonium compounds; cyanides; phosphorous or antimony sulfide; Al; (metals + acids); As_2S_3; CaH_2; MnO_2; metal sulfides; organic acids; powdered metals; Hg_3P_4; PHI_4; SCN; (S + Cu); Se; NaH_2PO_2; SrH; SO_2. Chlorates when mixed with combustible materials may form explosive mixtures. For instance, potassium chlorate, when mixed with sulfur or with other combustible substances explodes on friction. Pure chlorates which have been spilled on the floor, or mixed with small amounts of impurities, become very sensitive to shock and friction. Water is considered the best agent for fighting fires involving chlorates. When heated to decomposition they can emit toxic fumes of Cl^- and explode.

CDR000   CAS: 129-71-5   HR: 3
CHLORCYCLIZINE DIHYDROCHLORIDE
mf: $C_{18}H_{21}ClN_2 \cdot 2ClH$ mw: 373.78

SYNS: AH 289 * 1-(4-CHLOROBENZHYDRYL)-4-METHYLPIPERAZINE DIHYDROCHLORIDE * 1-(p-CHLORO-α-PHENYLBENZYL)-4-METHYL-PIPERAZINE DIHYDROCHLORIDE * DI-PARALENE-2-HYDROCHLORIDE * HISTANTINE DIHYDROCHLORIDE * N-METHYL-N'-(4-CHLOROBENZHYDRYL)PIPERAZINE DIHYDROCHLORIDE * PERAZIL * PERAZIL DIHYDROCHLORIDE * TRIHISTAN

SAFETY PROFILE: Poison by ingestion, subcutaneous, intravenous, and intraperitoneal routes. An antihistamine. When heated to decomposition it emits very toxic fumes of Cl^- and NO_x.

CDR250   CAS: 14362-31-3   HR: 3
CHLORCYCLIZINE HYDROCHLORIDE
mf: $C_{18}H_{21}ClN_2 \cdot xClH$ mw: 556.08

SYNS: AH-289 HYDROCHLORIDE * CHLORCYCLIZINIUM CHLORIDE * 1-(p-CHLOROBENZHYDRYL)-4-METHYLPIPERAZINE HYDROCHLORIDE * DIPARALENE HYDROCHLORIDE * ERAMIDE

SAFETY PROFILE: Poison by ingestion, subcutaneous, intravenous, and intraperitoneal routes. An experimental teratogen. Other experimental reproductive effects. When heated to decompo-

sition it emits very toxic fumes of HCl and NO_x.

CDR750   CAS: 57-74-9   HR: 3
CHLORDANE
DOT: UN 2762
mf: $C_{10}H_6Cl_8$ mw: 409.76

PROP: Colorless to amber; odorless, viscous liquid. Bp: 175°, d: 1.57-1.63 @ 15.5°/15.5°.

SYNS: ASPON-CHLORDANE * BELT * CD 68 * CHLOORDAAN (DUTCH) * CHLORDAN * γ-CHLORDAN * CHLORDANE, liquid (DOT) * CHLORINDAN * CHLOR KIL * CHLORODANE * CHLORTOX * CLORDAN (ITALIAN) * CORODANE * CORTILAN-NEU * DICHLOROCHLORDENE * DOWCHLOR * ENT 9,932 * ENT 25,552-X * HCS 3260 * KYPCHLOR * M 140 * M 410 * NCI-C00099 * NIRAN * 1,2,4,5,6,7,8,8-OCTA-CHLOOR-3a,4,7,7a-TETRAHYDRO-4,7-endo-METHANO-INDAAN (DUTCH) * OCTACHLOR * OCTACHLORO-DIHYDRODICYCLOPENTADIENE * 1,2,4,5,6,7,8,8-OC-TACHLORO-2,3,3a,4,7,7a-HEXAHYDRO-4,7-METHANO-INDENE * 1,2,4,5,6,7,8,8-OCTACHLORO-2,3,3a,4,7,7a-HEXAHYDRO-4,7-METHANO-1H-INDENE * 1,2,4,5,6,7,8,8-OCTACHLORO-3a,4,7,7a-HEXAHYDRO-4,7-METHYLENE INDANE * OCTA-CHLORO-4,7-METHANOHYDROINDANE * OCTA-CHLORO-4,7-METHANOTETRAHYDROINDANE * 1,2,4,5,6,7,8,8-OCTACHLORO-4,7-METHANO-3a,4,7,7a-TETRAHYDROINDANE * 1,2,4,5,6,7,8,8-OCTACHLORO-3a,4,7,7a-TETRAHYDRO-4,7-METHANOINDAN * 1,2,4,5,6,7,8,8-OCTACHLORO-3a,4,7,7a-TETRAHYDRO-4,7-METHANOINDANE * 1,2,4,5,6,7,10,10-OCTACHLORO-4,7,8,9-TETRAHYDRO-4,7-METHYLENEINDANE * 1,2,4,5,6,7,8,8-OCTACHLOR-3a,4,7,7a-TETRAHYDRO-4,7-endo-METHANO-INDAN (GERMAN) * OCTA-KLOR * OKTATERR * ORTHO-KLOR * 1,2,4,5,6,7,8,8-OTTOCHLORO-3A,4,7,7A-TETRAIDRO-4,7-endo-METANO-INDANO (ITALIAN) * RCRA WASTE NUMBER U036 * SD 5532 * SHELL SD-5532 * SYNKLOR * TAT CHLOR 4 * TOPICHLOR 20 * TOPICLOR * TOPICLOR 20 * TOXICHLOR * VELSICOL 1068

CONSENSUS REPORTS: IARC Cancer Review: GROUP 3 IMEMDT 7,146,87; Human Inadequate Evidence IMEMDT 20,45,79; Animal Sufficient Evidence IMEMDT 20,45,79. NCI Carcinogenesis Bioassay (feed); Clear Evidence: mouse NCITR* NCI-CG-TR-8,77; No Evidence: rat NCITR* NCI-CG-TR-8,77. EPA Genetic Toxicology Program. Community Right-To-Know List. EPA Extremely Hazardous Substances List.

OSHA PEL: TWA 0.5 mg/m^3 (skin)
ACGIH TLV: TWA 0.5 mg/m^3 (skin)
DFG MAK: Suspected Carcinogen.
DOT Classification: Combustible Liquid; Label: None; Flammable Liquid; Label: Flammable Liquid.

SAFETY PROFILE: Suspected carcinogen with experimental carcinogenic data. Poison to humans by ingestion and possibly other routes. An experimental poison by ingestion, inhalation, intravenous, and intraperitoneal routes. Moderately toxic by skin contact. Human systemic effects by ingestion or skin contact: tremors, convulsions, excitement, ataxia (loss of muscle coordination), and gastritis. Experimental teratogenic and reproductive effects. Human mutation data reported. Combustible liquid. It is no longer permitted for use as a termiticide in homes.

A central nervous system stimulant whose exact mode of action is unknown, but it may involve microsomal enzyme stimulation. Animals poisoned by this and related compounds show an extremely marked loss of appetite and neurological symptoms. The fatal dose to humans is unknown. It has been estimated to be between 6 to 60 grams (0.2 to 2 ounces). One person receiving an accidental skin application of 25% solution (amounting to something over 30 grams of technical chlordane) developed symptoms within about 40 minutes and died, apparently of respiratory failure, before medical attention was obtained. In two patients, death followed exposure to low ingestion doses of chlordane (2-4 grams). On microscopic examination, both patients showed severe chronic fatty degeneration of the liver, characteristic of chronic alcoholism. Although these two fatalities cannot be attributed exclusively to chlordane, they are entirely consistent with previous observations that the toxicity of other chlorinated hydrocarbons is much enhanced in the presence of chronic liver damage. The dangerous chronic dose in humans is unknown.

Experimental animals exposed to repeated small doses exhibit hyperexcitability, tremors, and convulsions, and those which survive long enough show marked anorexia and loss of weight. Symptoms in animals frequently occur within an hour of the administration of a large dose, but death often is delayed for several days depending on the dosage and route of administration. In any event, symptoms are of longer duration with chlordane than with DDT under similar conditions.

Laboratory analyses on poisoned animals are essentially normal, except that the insecticide is found in tissues by means of bioassay. A method for specific, quantitative chemical analysis for chlordane is now available using small amounts of subcutaneous fat. Chronically poisoned animals show degenerative changes in the liver and kidney tubules.

When heated to decomposition it emits toxic fumes of Cl$^-$.

CDS000   CAS: 115-28-6   HR: 1
CHLORENDIC ACID
mf: $C_9H_4Cl_6O_4$ mw:388.83

SYNS: 1,4,5,6,7,7-HEXACHLORO-5-NORBORNENE-2,3-DICARBOXYLIC ACID * KYSELINA 3,6-ENDOMETHYLEN-3,4,5,6,7,7-HEXACHLOR-Δ^4-TETRAHYDROFTALOVA (CZECH) * KYSELINA HET (CZECH) * NCI-C55072

CONSENSUS REPORTS: NTP Carcinogenesis Studies (feed): Clear Evidence: mouse,rat NTPTR* NTP-TR-304,87. Reported in EPA TSCA Inventory.

SAFETY PROFILE: Suspected carcinogen with experimental carcinogenic data. A severe eye and mild skin irritant. When heated to decomposition it emits toxic fumes of Cl$^-$.

CDS125   CAS: 16672-87-0   HR: 2
CHLORETHEPHON
mf: $C_2H_6ClO_3P$ mw: 144.50

PROP: Very hygroscopic needles from benzene. Mp: 74-75°. Freely sol in water, methanol, acetone, ethylene glycol, propylene glycol; sltly sol in benzene, toluene; practically insol in petr ether.

SYNS: AMCHEM 68-250 * BROMOFLOR * CAMPOSAN * CEP * 2-CEPA * CEPHA * CEPHA 10LS * 2-CHLORAETHYL-PHOSPHONSAEURE (GERMAN) * 2-CHLORETHYLPHOSPHONIC ACID * 2-CHLOROETHANEPHOSPHONIC ACID * ETHEFON * ETHEL * ETHEPHON * ETHEVERSE * ETHREL * FLORDIMEX * FLOREL * G 996 * KAMPOSAN * ROLL-FRUCT * TOMATHREL

CONSENSUS REPORTS: EPA Genetic Toxicology Program.

SAFETY PROFILE: Moderately toxic by ingestion. Mildly toxic by skin contact. A plant growth regulator. Caution: Spray formulations are quite acidic, about pH 1.0. May be irritating to exposed skin and eyes, or if inhaled. When heated to decomposition it emits toxic fumes of Cl⁻ and PO$_x$.

CDS750   CAS: 470-90-6   **HR: 3**
CHLORFENVINFOS
mf: C$_{12}$H$_{14}$Cl$_3$O$_4$P   mw: 359.58

SYNS: APACHLOR * BIRLANE * C-10015 * CFV * CGA 26351 * CHLOFENVINPHOS * O-2-CHLOOR-1-(2,4-DICHLOOR-FENYL)-VINYL-O,O-DI-ETHYLFOSFAAT (DUTCH) * O-2-CHLOR-1-(2,4-DI-CHLOR-PHENYL)-VINYL-O,O-DIAETHYLPHOSPHAT (GERMAN) * CHLORFENVINFOS * CHLORFENVINPHOS * 2-CHLORO-1-(2,4-DICHLOROPHENYL)VINYL DIETHYL PHOSPHATE * β-2-CHLORO-1-(2',4'-DICHLOROPHE-NYL) VINYL DIETHYLPHOSPHATE * CHLOROFEN-VINPHOS * CHLORPHENVINFOS * CHLORPHEN-VINPHOS * O-2-CLORO-1-(2,4-DICLORO-FENIL)-VINYL-O,O-DIETILFOSFATO (ITALIAN) * COMPOUND 4072 * CVP * DERMATON * O,O-DIAETHYL-O-1-(4,5-DICHLORPHENYL)-2-CHLOR-VINYL-PHOSPHAT (GERMAN) * 2,4-DICHLORO-α-(CHLOROMETHYLENE)BEN-ZYL ALCOHOL DIETHYL PHOSPHATE * O,O-DIETHYL-O-(2-CHLORO-1-(2',4'-DICHLOROPHENYL)VINYL) PHOS-PHATE * ENT 24,969 * GC 4072 * OMS 1328 * PHOSPHATE de O,O-DIETHYLE et de O-2-CHLORO-1-(2,4-DICHLOROPHENYL) VINYLE (FRENCH) * SAPE-CRON * SHELL 4072 * STELADONE * SUPONA * SUPONE * UNITOX * VINYLPHATE

CONSENSUS REPORTS: EPA Extremely Hazardous Substances List.

SAFETY PROFILE: Poison by ingestion, skin contact, intraperitoneal, subcutaneous, and intravenous routes. Human systemic effects by skin contact: unspecified blood system effects. Mutation data reported. A cholinesterase inhibitor. An insecticide. When heated to decomposition it emits very toxic fumes of Cl⁻ and PO$_x$.

CDT750   CAS: 96-24-2   **HR: 3**
CHLORHYDRIN

DOT: UN 2689
mf: C$_3$H$_7$ClO$_2$ mw: 110.55

PROP: Colorless liquid. Bp: 213° decomp, d: 1.326.

SYNS: α-CHLORHYDRIN * CHLORODEOXYGLY-CEROL * 1-CHLORO-2,3-DIHYDROXYPROPANE * 3-CHLORO-1,2-DIHYDROXYPROPANE * α-CHLO-ROHYDRIN * 1-CHLOROPROPANE-2,3-DIOL * 1-CHLORO-2,3-PROPANEDIOL * 3-CHLOROPRO-PANE-1,2-DIOL * 3-CHLORO-1,2-PROPANEDIOL * 3-CHLOROPROPYLENE GYLCOL * β,β'-DIHY-DROXYISOPROPYL CHLORIDE * 2,3-DIHYDROXYPRO-PYL CHLORIDE * EPIBLOC * GLYCERIN-α-MONO-CHLORHYDRIN * GLYCEROL CHLOROHYDRIN * GLYCEROL-α-CHLOROHYDRIN * GLYCEROL-α-MONOCHLOROHYDRIN (DOT) * GLYCERYL-α-CHLO-ROHYDRIN * MONOCHLORHYDRIN * MONOCHLO-ROHYDRIN * α-MONOCHLOROHYDRIN * U-5897

CONSENSUS REPORTS: Reported in EPA TSCA Inventory. EPA Genetic Toxicology Program.

DOT Classification: Poison B; Label: St. Andrews Cross.

SAFETY PROFILE: Poison by ingestion and intraperitoneal routes. Moderately toxic by inhalation. Experimental reproductive effects. An eye irritant. Questionable carcinogen with experimental tumorigenic data. Mutation data reported. A chemosterilant for rodents. Combustible when exposed to heat or flame. Reaction with perchloric acid forms a sensitive explosive product more powerful than glyceryl nitrate. When heated to decomposition it emits toxic fumes of Cl⁻.

CDU000   CAS: 7790-93-4   **HR: 3**
CHLORIC ACID

DOT: UN 2626
mf: ClHO$_3$ mw: 84.46

PROP: Colorless solution. Mp: <−20°, bp: decomp @ 40°, d: 1.282 @ 14.2°.

SYN: CHLORIC ACID, solution, containing not more than 10% acid (DOT)

CONSENSUS REPORTS: Reported in EPA TSCA Inventory.

DOT Classification: Oxidizer; Label: Oxidizer and Poison; Oxidizer; Label: Oxidizer, solution.

SAFETY PROFILE: A poison. A strong irritant by ingestion and inhalation. Dangerous fire hazard; ignites organic matter upon contact. A very powerful oxidizing agent. Violent or explosive reaction with oxidizable materials. Aqueous solutions decompose explosively during evaporation. Solutions greater than 40% are unstable. Reacts violently with NH$_3$, Sb, Sb$_2$S$_3$, As$_2$S$_3$, Bi, CuS, PHI$_4$, SnS$_2$, SnS. Reaction with cellulose causes ignition after a delay period. Dangerous reaction with metal sulfides and metal chlorides (e.g., incandescent reaction with antimony trisulfide, arsenic trisulfide, tin(II)sulfide, tin(IV) sulfide, explosion on contact with copper sulfide). Reaction with metals (e.g., antimony, bismuth, iron) forms explosive products. When heated to decomposition it emits toxic fumes of Cl^-.

CDU250   HR: D
CHLORIDES

SAFETY PROFILE: Varies widely. Sodium chloride (table salt) has very low toxicity, while carbonyl chloride (phosgene) is lethal in small doses. Therefore, see specific entries. When heated to decomposition or on contact with acids or acid fumes, they evolve highly toxic chloride fumes. Some organic chlorides decompose to yield phosgene.

CDV100   CAS: 8001-35-2   HR: 3
CHLORINATED CAMPHENE

DOT: NA 2761
mf: $C_{10}H_{10}Cl_8$ mw: 413.80

PROP: Yellow, waxy solid; pleasant piney odor. Mp: 65-90°. Almost insol in water; very sol in aromatic hydrocarbons.

SYNS: AGRICIDE MAGGOT KILLER (F) * ALLTEX * ALLTOX * ATTAC 6 * ATTAC 6-3 * CAMPHECHLOR * CAMPHOCHLOR * CAMPHOCLOR * CAMPHOFENE HUILEUX * CHEM-PHENE * CHLOROCAMPHENE * CLOR CHEM T-590 * COMPOUND 3956 * CRESTOXO * CRISTOXO 90 * ENT 9,735 * ESTONOX * FASCO-TERPENE * GENIPHENE * GY-PHENE * HERCULES 3956 * HERCULES TOXAPHENE * KAMFOCHLOR * M 5055 * MELIPAX * MOTOX * NCI-C00259 * OCTACHLOROCAMPHENE * PCC * PENPHENE * PHENACIDE * PHENATOX * POLYCHLORCAMPHENE * POLYCHLORINATED CAMPHENES * POLYCHLOROCAMPHENE * RCRA WASTE NUMBER P123 * STROBANE-T-90 * SYNTHETIC 3956 * TOXADUST * TOXAFEEN (DUTCH) * TOXAKIL * TOXAPHEN (GERMAN) * TOXAPHENE * TOXON 63 * TOXYPHEN * VERTAC 90% * VERTAC TOXAPHENE 90

CONSENSUS REPORTS: IARC Cancer Review: GROUP 2B IMEMDT 7,56,87; Human Limited Evidence IMEMDT 20,327,79; Animal Sufficient Evidence IMEMDT 20,327,79. NTP Fourth Annual Report On Carcinogens, 1984.

NCI Carcinogenesis Bioassay (feed); Clear Evidence: mouse, rat NCITR* NCI-CG-TR-37,79.

OSHA PEL: (Transitional: TWA 0.5 mg/m^3 (skin)) TWA 0.5 mg/m^3; STEL 1 mg/m^3 (skin)
ACGIH TLV: TWA 0.5 mg/m^3; STEL 1 mg/m^3 (skin)
DFG MAK: 0.5 mg/m^3
DOT Classification: ORM-A; Label: None.

SAFETY PROFILE: Confirmed carcinogen with experimental carcinogenic and tumorigenic data. Human poison by ingestion and possibly other routes. Experimental poison by ingestion, intraperitoneal, and possibly other routes. Moderately toxic experimentally by inhalation and skin contact. May be a human carcinogenic. Human systemic effects by ingestion and skin contact: somnolence, convulsions or effect on seizure threshold, coma and allergic skin dermatitis. Experimental teratogenic and reproductive effects. A skin irritant; absorbed through the skin. Human mutation data reported. Liver injury has been reported. Lethal amounts of toxaphene can enter the body through the mouth, lungs, and skin. Systemic absorption of the insecticide is increased by the presence of digestible oils, and liquid preparations of the insecticide which penetrate the skin more readily than do dusts and wettable powders.

A toxic mixture of organochlorine pesticides stored to some extent in body fat. It resembles chlordane and, to some extent, camphor in its physiological action. It causes diffuse stimulation of the brain and spinal cord resulting in generalized convulsions of a tonic or clonic character. Death usually results from respiratory failure. Detoxification appears to occur in the liver. The lethal ingestion dose for man is estimated to be 2-7 grams, a toxicity of about four times that of DDT. At least seven human deaths have been reported due to toxaphene, all in children. Two families have been made ill by eating vegetables containing a large residue of toxaphene. When heated to decomposition it emits toxic fumes of Cl^-.

CDV175   CAS: 55720-99-5   HR: 3
CHLORINATED DIPHENYL OXIDE
mf: $C_{12}H_4Cl_6O$ mw: 376.86

PROP: Light yellow, very viscous liquid. Bp: 230-260° @ 8 mm, d: 1.60 @ 20°/60°, autoign temp: 1148°F, vap d: 13.0.

SYNS: HEXACHLORO DIPHENYL OXIDE * HEXA-
CHLOROPHENYL ETHER * PHENYL ETHER HEXA-
CHLORO

OSHA PEL: TWA 0.5 mg/m^3
ACGIH TLV: TWA 0.5 mg/m^3
DFG MAK: 0.5 mg/m^3

SAFETY PROFILE: Poison by ingestion and
probably by inhalation. Combustible when exposed to heat, flame, or oxidizing materials.
To fight fire, use water spray, fog, foam, dry
chemical, CO_2. When heated to decomposition
it emits toxic fumes of Cl^-.

CDV250   HR: 2
CHLORINATED HYDROCARBONS, ALIPHATIC

SYN: ALIPHATIC CHLORINATED HYDROCARBONS
* CHLORINATED HC, ALIPHATIC

SAFETY PROFILE: Suspected carcinogen with
experimental tumors of the liver, lung, skin,
and blood-forming tissues. The substitution of
a chlorine (or other halogen) atom for a hydrogen
greatly increases the anesthetic action of the
aliphatic hydrocarbons and increases the range
of their systemic effects. In many cases, the
chlorine derivative is quite toxic. In general,
the unsaturated chlorine derivatives are more
narcotic but less toxic than the saturated derivatives. In the saturated group, the narcotic effect
is proportional to the number of chlorine atoms.
This relationship is not true for toxicity.

In dealing with these chlorinated hydrocarbons, it must be remembered that a toxic action
may result from repeated exposure to concentrations which are too low to produce a narcotic
effect, and which, consequently, are too low
to give warning of danger. Individual susceptibility varies widely. Certain workmen may be
seriously affected by concentrations that seem
to have no effect on fellow employees in the
same exposure.

In general reactivity decreases with greater
substitution of halogen for hydrogen atoms. Halogenated (e.g., fluorine, chlorine, or bromine
containing) acetylene compounds are unstable
and should be treated as explosives. Lightly
substituted haloalkanes are highly flammable
and can react with divalent light metals to form
dangerously reactive products. Lightly substituted haloalkenes are highly flammable, peroxi-
dizable, and may polymerize violently. When
heated to decomposition they emit highly toxic

fumes of phosgene. They may react violently
with Al; liquid O_2; K; and Na.

CDV500   HR: 3
CHLORINATED HYDROCARBONS, AROMATIC

SYN: CHLORINATED HC AROMATIC

SAFETY PROFILE: In most instances, it is difficult to predict the toxicity of these compounds.
However, in the case of most aromatic chlorine
compounds, their toxicity is usually no greater,
and frequently is less, than that of the corresponding aromatic hydrocarbons with the nota-
ble exception of naphthalene and the various
biphenyls. They can react with oxidizing materials. React violently with Al, liquid O_2, K, or
Na. When heated to decomposition it emits toxic
fumes of Cl^-.

CDV625   CAS: 56641-03-3   HR: 3
CHLORINATED POLYETHER POLYURETHAN

PROP: Polymer formed from toluene diisocyanate and 1,4-butanediol and cured with 4,4'-
methylenebis(o-chloroaniline).
mf: $(C_{13}H_{12}Cl_2N_2 \cdot C_9H_6N_2O_2 \cdot (C_4H_8O)_nH_2O)_x$

SYNS: OSTAMER * POLYURETHANE Y-238
* Y-238

CONSENSUS REPORTS: IARC Cancer Review: Animal Sufficient Evidence IMEMDT
19,303,79.

SAFETY PROFILE: Confirmed carcinogen with
experimental tumorigenic data. When heated
to decomposition it emits toxic fumes of Cl^-
and NO_x.

CDV750   CAS: 7782-50-5   HR: 3
CHLORINE

DOT: UN 1017
mf: Cl_2 mw: 70.90

PROP: Greenish-yellow gas, liquid, or rhombic
crystals. Mp: $-101°$, bp: $-34.5°$, d: (liquid)
1.47 @ 0° (3.65 atm), vap press: 4800 mm
@ 20°, vap d: 2.49. Sol in water.

SYNS: BERTHOLITE * CHLOOR (DUTCH)
* CHLOR (GERMAN) * CHLORE (FRENCH)
* CHLORINE MOL. * CLORO (ITALIAN) * MO-
LECULAR CHLORINE

CONSENSUS REPORTS: Reported in EPA TSCA Inventory. Community Right-To-Know List. EPA Extremely Hazardous Substances List.

OSHA PEL: (Transitional: TWA CL 1 ppm) TWA 0.5 ppm; STEL 1 ppm
ACGIH TLV: TWA 0.5 ppm; STEL 1 ppm
DFG MAK: 0.5 ppm (1.5 mg/m^3)
NIOSH REL: CL 0.5 ppm/15M
DOT Classification: Nonflammable Gas; Label: Nonflammable Gas and Poison; Poison A; Label: Poison Gas.

SAFETY PROFILE: Moderately toxic to humans by inhalation. Very irritating by inhalation. Human mutation data reported. Human respiratory system effects by inhalation: changes in the trachea or bronchi, emphysema, chronic pulmonary edema or congestion. A strong irritant to eyes and mucous membranes. Chlorine is extremely irritating to the mucous membranes of the eyes and the respiratory tract at 3 ppm. Combines with moisture to liberate O_2 and forms HCl. Both these substances, if present in quantity, cause inflammation of the tissues with which they come in contact. A concentration of 3.5 ppm produces a detectable odor; 15 ppm causes immediate irritation of the throat. Concentrations of 50 ppm are dangerous for even short exposures; 1000 ppm may be fatal, even when exposure is brief. Because of its intensely irritating properties, severe industrial exposure seldom occurs, as the worker is forced to leave the exposure area before he can be seriously affected. In cases where this is impossible, the initial irritation of the eyes and mucous membranes of the nose and throat is followed by coughing, a feeling of suffocation, and later, pain and a feeling of constriction in the chest. If exposure has been severe, pulmonary edema may follow with rales being heard over the chest. It is a common air contaminant.

Explodes on contact with acetylene + heat or UV light, air + ethylene, molten aluminum, ammonia, amidosulfuric acid, antimony trichloride + tetramethyl silane (at 100°), benzene + light, biuret, bromine pentafluoride + heat, tert-butanol, butyl rubber + naphtha, carbon disulfide + iron catalyst, chlorinated pyridine + iron powder, 3-chloropropyne, cobalt(II) chloride + methanol, diborane, dibutyl phthalate (at 118°), dichloro(methyl)arsine (in a sealed container), diethyl ether, dimethyl phosphoramidiate, dioxygen difluoride, disilyl oxide, 4,4'-dithiodimorpholine, ethane over activated carbon (at 350°), fluorine + sparks, gasoline, glycerol (above 70° in a sealed container), hexachlorodisilane (above 300°), hydrocarbon oils or waxes, iron(III) chloride + monomers (e.g., styrene), methane over mercury oxide, methanol, methanol + tetrapyridine cobalt(II) chloride, naphtha + sodium hydroxide, nitrogen triiodide, oxygen difluoride, white phosphorus (in liquid Cl_2), phosphorus compounds, polypropylene + zinc oxide, propane (at 300°), silicones when heated in a sealed container [e.g., polydimethyl siloxane (above 88°), polymethyl trifluoropropylsiloxane (above 68°)], stibine, synthetic rubber (in liquid Cl_2), tetraselenium tetranitride, trimethyl thionophosphate. Explosive products are formed on reaction with alkylthiouronium salts, amidosulfuric acid, acidic ammonium chloride solutions, aziridine, bis(2,4-dinitrophenyl)disulfide, cyanuric acid, phenyl magnesium bromide. Mixtures with ethylene are explosives initiated by light, heat, or by the presence or mercury, mercury oxide, silver oxide, lead oxide (at 100°). Mixtures with hydrogen are explosives initiated by sparks, light, heating to over 280°, or the presence of yellow mercuric oxide or nitrogen trichloride. Mixtures with hydrogen and other gases (e.g., air, hydrogen chloride, oxygen) are also explosive.

Ignition or explosive reaction with metals (e.g., aluminum, antimony powder, bismuth powder, brass, calcium powder, copper, germanium, iron, manganese, potassium, tin, vanadium powder). Reaction with some metals requires moist Cl_2 or heat. Ignites with diethyl zinc (on contact), polyisobutylene (at 130°), metal acetylides, metal carbides, metal hydrides (e.g., potassium hydride, sodium hydride, copper hydride), metal phosphides (e.g., copper(II) phosphide), methane + oxygen, hydrazine, hydroxylamine, calcium nitride, non-metals (e.g., boron, active carbon, silicon, phosphorus), non-metal hydrides (e.g., arsine, phosphine, silane), steel (above 200° or as low as 50° when impurities are present), sulfides (e.g., arsenic disulfide, boron trisulfide, mercuric sulfide), trialkyl boranes.

Violent reaction with alcohols, N-aryl sulfinamides, dimethyl formamide, polychlorobiphenyl, sodium hydroxide, hydrochloric acid + dinitroanilines. Incandescent reaction when warmed with cesium oxide (above 150°), tellurium, arsenic, tungsten dioxide. Potentially dangerous reaction with hydrocarbons + Lewis acids releases toxic and reactive HCl gas.

Can react to cause fires or explosions upon contact with turpentine, illuminating gas, polypropylene, rubber, sulfamic acid and many other chemicals.

CDW000 CAS: 13973-88-1 **HR: 3**
CHLORINE AZIDE
mf: ClN$_3$ mw: 77.47

PROP: An explosive gas.

SYN: CHLOR(O)AZIDE

DOT Classification: Forbidden

SAFETY PROFILE: Strong irritant by inhalation. An extremely unstable explosive. Reacts with liquid ammonia to form an explosive liquid. Explosive reaction with 1,3-butadiene, C$_2$H$_6$, C$_2$H$_4$, CH$_4$, C$_3$H$_8$, phophorus, silver azide, sodium. Reacts with water or steam to produce toxic and corrosive fumes of HCl. Has been used as an initiator in chemical gas lasers. When heated to decomposition it emits toxic fumes of Cl$^-$ and NO$_x$.

CDW450 CAS: 10049-04-4 **HR: 3**
CHLORINE DIOXIDE
DOT: NA 9191
mf: ClO$_2$ mw: 67.45

PROP: Red-yellow gas or orange-red crystals. Mp: $-59°$, bp: 9.9° @ 731 mm (explodes), d: 3.09 g/L @ 11°.

SYNS: ALCIDE * ANTHIUM DIOXCIDE * CHLORINE DIOXIDE, not hydrated (DOT) * CHLORINE PEROXIDE * CHLORINE OXIDE * CHLORINE(IV) OXIDE * CHLOROPEROXYL * CHLORYL RADICAL * DOXCIDE 50

CONSENSUS REPORTS: Reported in EPA TSCA Inventory. Community Right-To-Know List.

OSHA PEL: (Transitional: TWA 0.1 ppm) TWA 0.1 ppm; STEL 0.3 ppm
ACGIH TLV: TWA 0.1 ppm; STEL 0.3 ppm
DFG MAK: 0.1 ppm (0.3 mg/m^3)
DOT Classification: Forbidden (not hydrated); Oxidizer; Label: Oxidizer and Poison (hydrated, frozen).

SAFETY PROFILE: Moderately toxic by inhalation. Experimental reproductive effects. Muta-

tion data reported. An eye irritant. A powerful explosive sensitive to spark, impact, sunlight, or heating rapidly to 100°C. A powerful oxidizer. Concentrations of greater than 10% in air are explosive. Explodes on mixing with carbon monoxide, hydrocarbons (e.g., butadiene, ethane, ethylene, methane, propane), fluoramines (e.g., difluoramine, trifluoramine). Mixtures with hydrogen explode with sparking or or contact with platinum. Explodes on contact with mercury, potassium hydroxide, phosphorus pentachloride + chlorine. Ignites or explodes on contact with non-metals (e.g., phosphorus, sulfur, sugar). Reacts violently with F$_2$, NHF$_2$. Reacts with water or steam to produce toxic and corrosive fumes of HCl. When heated to decomposition it emits toxic fumes of Cl$^-$.

CDX250 CAS: 13637-63-3 **HR: 3**
CHLORINE PENTAFLUORIDE
mf: ClF$_5$ mw: 130.45

DOT: UN 2548

SYNS: CHLORINE FLUORIDE (ClF$_5$) * CHLORINE PENTAFLUORIDE (DOT)

OSHA PEL: TWA 2.5 mg(F)/m^3
ACGIH TLV: TWA 2.5 mg(F)/m^3
NIOSH REL: (Inorganic Fluorides) TWA 2.5 mg(F)/m^3
DOT Classification: Poison A: Label: Poison Gas, Oxidizer, Corrosive.

SAFETY PROFILE: Poison by inhalation. A corrosive material. Vigorous reaction in contact with water or anhydrous nitric acid. Violent reaction on contact with metals. When heated to decomposition it emits very toxic fumes of Cl$^-$ and F$^-$.

CDX750 CAS: 7790-91-2 **HR: 3**
CHLORINE TRIFLUORIDE
DOT: UN 1749
mf: ClF$_3$ mw: 92.45

PROP: Colorless gas to yellow liquid, sweet odor, mp: $-83°$, bp: 11.8°, d: 1.77 @ 13°.

SYNS: CHLORINE FLUORIDE * CHLOROTRIFLUORIDE * TRIFLUORURE de CHLORE (FRENCH)

CONSENSUS REPORTS: Reported in EPA TSCA Inventory.

OSHA PEL: CL 0.1 ppm
ACGIH TLV: CL 0.1 ppm
DFG MAK: 0.1 ppm (0.4 mg/m^3)

DOT Classification: Oxidizer; Label: Oxidizer and Poison; Poison A; Label: Poison Gas, Oxidizer, Corrosive.

SAFETY PROFILE: Human poison by inhalation. An eye irritant. Spontaneously flammable. A powerful oxidant which may react violently with oxidizable materials. A rocket propellant.

Explosive reaction with water, bis(trifluoromethyl)sulfide or -disulfide, polychlorotrifluoroethylene, trifluoromethanesulfenyl chloride, and other hydrogen containing materials (e.g., ammonia, coal gas, hydrogen, hydrogen sulfide, methane, acetic acid, benzene, ether, cotton, paper, wood). Forms shock-sensitive explosive mixtures with highly chlorinated compounds (e.g., carbon tetrachloride), nitroaryl compounds (e.g., trinitrotoluene, hexanitrobiphenyl, hexanitrodiphenyl amine, hexanitrodiphenyl sulfide, hexanitrodiphenyl ether). Reaction with ammonium fluoride or ammonium hydrogen fluoride forms explosive gaseous products.

Ignition on contact with boron-containing materials, iodine, finely divided refractory materials (e.g., asbestos, glass wool, sand, tungsten carbide), fluorinated polymers (with flowing trifluoride).

Violent reaction with acids (e.g., nitric or sulfuric), chromium trioxide, ruthenium, selenium tetrafluoride (above 106°C), metals, metal oxides, metal salts, non-metals, non-metal salts, organic matter, glass wool, acetic acid, Al, Sb, As, Cu, Ir, Fe, Pb, Mg, Mo, Os, P, K, Rh, Se, Si, Ag, Na, S, Te, Sn, W, Zn, oxides, CO, graphite, HgI_2, HNO_3, K_2CO_3, KI, rubber, AgN_3, $AgNO_3$, NaOH, V_2P_5, WO_3. Incompatible with fuels, nitro compounds. When heated to decomposition or in reaction with water or steam it emits toxic fumes of F^- and Cl^-.

CDY250   HR: 3
CHLORITES

SAFETY PROFILE: Many chlorite salts are heat- and impact-sensitive explosives. The metal salts are powerful oxidants. They are much less stable than the analogous chlorates. React violently with NH_3, organic matter, or metals.

CDY500   CAS: 107-20-0   HR: 3
CHLOROACETALDEHYDE

DOT: UN 2232
mf: C_2H_3ClO mw: 78.50

PROP: Clear, colorless liquid; pungent odor. Bp: 90.0°-100.1° (40% soln), fp: −16.3° (40% soln), flash p: 190°F, d: 1.19 @ 25°/25° (40% soln), vap press: 100 mm @ 45° (40% soln).

SYNS: 2-CHLOROACETALDEHYDE * CHLOROACETALDEHYDE MONOMER * 2-CHLORO-1-ETHANAL * MONOCHLOROACETALDEHYDE * RCRA WASTE NUMBER P023

CONSENSUS REPORTS: Reported in EPA TSCA Inventory. EPA Genetic Toxicology Program.

OSHA PEL: CL 1 ppm
ACGIH TLV: CL 1 ppm
DFG MAK: 1 ppm (3 mg/m^3)
DOT Classification: Poison B; Label: Poison.

SAFETY PROFILE: Poison by ingestion, skin contact, and intraperitoneal routes. Mutation data reported. Combustible when exposed to heat or flame. Reacts with oxidizing materials. To fight fire, use water, foam, CO_2, dry chemical. When heated to decomposition it emits toxic fumes of Cl^-.

CDY850   CAS: 79-07-2   HR: 3
2-CHLORO ACETAMIDE
mf: C_2H_4ClNO mw: 93.52

PROP: Crystals, moderately water-sol, mp: 120°, bp: 225° (decomp).

SYNS: CHLORACETAMID (GERMAN) * CHLOROACETAMIDE * α-CHLOROACETAMIDE * 2-CHLOROETHANAMIDE * USAF DO-29

CONSENSUS REPORTS: Reported in EPA TSCA Inventory.

SAFETY PROFILE: Poison by ingestion, intravenous, and intraperitoneal routes. When heated to decomposition it emits very toxic Cl^- and NO_x.

CEA000   CAS: 79-11-8   HR: 3
CHLOROACETIC ACID
mf: $C_2H_3ClO_2$ mw: 94.50

DOT: UN 1750/UN 1751

PROP: Colorless crystals. Mp: (α) 63°, (β) 56°, (τ) 50°, bp: 189°, flash p: 259°F, d: 1.58 @ 20°/20°, vap d: 3.26.

SYNS: ACIDE CHLORACETIQUE (FRENCH) * ACIDE MONOCHLORACETIQUE (FRENCH) * ACIDOMONO-

CLOROACETICO (ITALIAN) * CHLORACETIC ACID * α-CHLOROACETIC ACID * CHLOROACETIC ACID, liquid (DOT) * CHLOROACETIC ACID, solid (DOT) * CHLOROETHANOIC ACID * MCA * MONO-CHLOORAZIJNZUUR (DUTCH) * MONOCHLORACETIC ACID * MONOCHLORESSIGSAEURE (GERMAN) * MONOCHLOROACETIC ACID * MONOCHLORO-ETHANOIC ACID * NCI-C60231

CONSENSUS REPORTS: Reported in EPA TSCA Inventory. EPA Genetic Toxicology Program. EPA Extremely Hazardous Substances List. Community Right-To-Know List.

DOT Classification: Corrosive Material; Label: Corrosive, liquid, solution or solid.

SAFETY PROFILE: Poison by ingestion, inhalation, subcutaneous, and intravenous route. A corrosive skin, eye, and mucous membrane irritant. Questionable carcinogen with experimental tumorigenic data. Mutation data reported. Combustible liquid when exposed to heat or flame. To fight fire, use water spray, fog, mist, dry chemical, foam. When heated to decomposition it emits toxic fumes of Cl^-.

CEA750   CAS: 532-27-4   HR: 3
α-CHLOROACETOPHENONE
DOT: UN 1697
mf: C_8H_7ClO mw: 154.60

SYNS: CAF * CAP * CHEMICAL MACE * 1-CHLOROACETOPHENONE * omega-CHLOROACE-TOPHENONE * CHLOROACETOPHENONE, gas, liquid or solid (DOT) * CHLOROMETHYL PHENYL KETONE * 2-CHLORO-1-PHENYLETHANONE * CN * MACE (lachrymator) * NCI-C55107 * PHENACYL CHLORIDE * PHENYLCHLOROMETHYLKETONE

CONSENSUS REPORTS: Reported in EPA TSCA Inventory. Community Right-To-Know List.

OSHA PEL: TWA 0.05 ppm
ACGIH TLV: TWA 0.05 ppm
DOT Classification: Irritating Material; Label: Irritant; Poison B; Label: Poison.

SAFETY PROFILE: A human poison by inhalation. An experimental poison by ingestion, inhalation, intraperitoneal, and intravenous routes. Human systemic effects by inhalation: lacrimation, conjunctiva irritation, and unspecified eye effects, cough, and dyspnea. A severe eye and moderate skin irritant. Questionable carcinogen with experimental neoplastigenic data by skin

contact. A riot control agent. When heated to decomposition it emits toxic fumes of Cl^-.

CEB250   CAS: 99-91-2   HR: 3
p-CHLOROACETOPHENONE
mf: C_8H_7ClO mw: 154.60

PROP: Pale straw-colored liquid or white crystals; fragrant, non-persistent odor. Mp: 56°, bp: 237-247°, fp: 59°, d: 1.19 @ 25°/25°, vap press: 0.012 mm @ 0°, vap d: 5.2, flash p: 244°F.

SYNS: 4'-CHLOROACETOPHENONE * 4-CHLORO-ACETOPHENONE * 1-(4-CHLOROPHENYL)ETHANONE * USAF DO-1

CONSENSUS REPORTS: Reported in EPA TSCA Inventory.

SAFETY PROFILE: Poison by intraperitoneal route. Moderately toxic by ingestion. A powerful irritant and lachrymator. Human systemic effects by inhalation: lachrimation and unspecified effects on the eye and sense of smell. Combustible when exposed to heat or flame. To fight fire, use water, foam, alcohol foam, dry chemical. When heated to decomposition or on contact with water or steam it emits toxic fumes of Cl^-.

CEC000   CAS: 140-49-8   HR: 2
4'-CHLOROACETYL ACETANILIDE
mf: $C_{10}H_{10}ClNO_2$ mw: 211.66

SYNS: p-ACETAMIDOPHENACYL CHLORIDE * p-(ACETYLAMINO)PHENACYL CHLORIDE * 4'-(CHLOROACETYL)ACETANILIDE * NCI-C03770

CONSENSUS REPORTS: NCI Carcinogenesis Bioassay (feed); No Evidence: mouse, rat NCITR* NCI- CG-TR-177,79. Reported in EPA TSCA Inventory.

SAFETY PROFILE: Moderately toxic by ingestion. Mutation data reported. When heated to decomposition it emits very toxic fumes of Cl^- and NO_x.

CEC250   CAS: 79-04-9   HR: 3
CHLOROACETYL CHLORIDE
DOT: UN 1752
mf: $C_2H_2Cl_2O$ mw: 112.94

PROP: Water-white or sltly yellow liquid. Bp: 105-106°, fp: −22.5°, flash p: none, d: 1.495 @ 0°.

SYNS: CHLORACETYL CHLORIDE * CHLORO-
ACETIC ACID CHLORIDE * CHLOROACETIC CHLO-
RIDE * CHLORURE de CHLORACETYLE (FRENCH)
* MONOCHLOROACETYL CHLORIDE

CONSENSUS REPORTS: Reported in EPA
TSCA Inventory.

OSHA PEL: TWA 0.05 ppm
ACGIH TLV: TWA 0.05 ppm (Proposed: TWA
  0.05 ppm; STEL: 0.15 ppm)
DOT Classification: Corrosive Material; Label;
  Corrosive.

SAFETY PROFILE: Poison by ingestion and
intravenous routes. Mildly toxic by inhalation.
Corrosive. A lacrymator. When heated to decomposition it emits toxic fumes of Cl⁻.

CEH250   CAS: 95-85-2   HR: 3
p-CHLORO-o-AMINOPHENOL
mf: C_6H_6ClNO mw: 143.58

DOT: UN 2673

SYN: 2-AMINO-4-CHLOROPHENOL (DOT)

CONSENSUS REPORTS: Reported in EPA
TSCA Inventory.

DOT Classification: Poison B; Label: Poison.

SAFETY PROFILE: A poison. Moderately toxic
by ingestion. When heated to decomposition
it emits very toxic fumes of Cl⁻ and NO_x.

CEH670   CAS: 95-51-2   HR: 3
2-CHLOROANILINE
DOT: UN 2018/UN 2019
mf: C_6H_6ClN mw: 127.58

PROP: Liquid. Bp: 208.84°, mp: −1.94°, d:
1.2114, n (20/D) 1.5895. Practically insol in
water; sol in most organic solvents, also in acids.

SYNS: 1-AMINO-2-CHLOROBENZENE * o-CHLOR-
ANILINE * o-CHLOROANILINE * o-CHLOROAN-
ILINE, liquid (DOT) * o-CHLOROANILINE, solid (DOT)
* 2-CHLORO-BENZENAMINE (9CI) * FAST YELLOW
GC BASE

CONSENSUS REPORTS: EPA Genetic Toxicology Program. Reported in EPA TSCA Inventory.

DOT Classification: Poison B; Label: Poison.

SAFETY PROFILE: Poison by skin contact,
ingestion, and subcutaneous routes. Mutation

data reported. When heated to decomposition
it emits toxic fumes of Cl⁻ and NO_x.

CEH675   CAS: 108-42-9   HR: 3
3-CHLOROANILINE
DOT: UN 2018/UN 2019
mf: C_6H_6ClN mw: 127.58

PROP: Liquid. Bp: 230.5°, mp: −10.4°, d:
1.2150, n (20/D) 1.5931. Practically insol in
water; sol in most common organic solvents.

SYNS: m-AMINOCHLOROBENZENE * 1-AMINO-3-
CHLOROBENZENE * 3-CHLOORANILINEN (DUTCH)
* m-CHLORANILINE * m-CHLOROANILINE
* 3-CHLOROANILINE (ITALIAN) * m-CHLOROANI-
LINE, liquid (DOT) * m-CHLOROANILINE, solid (DOT)
* 3-CHLOROBENZENAMINE * m-CHLOROPHENYL-
AMINE * 3-CHLOROPHENYLAMINE * FAST OR-
ANGE GC BASE * ORANGE GC BASE

CONSENSUS REPORTS: EPA Genetic Toxicology Program. Reported in EPA TSCA Inventory.

DOT Classification: Poison B; Label: Poison.

SAFETY PROFILE: Poison by ingestion, skin
contact, subcutaneous, and intravenous routes.
Mutation data reported. When heated to decomposition it emits toxic fumes of Cl⁻ and NO_x.

CEH680   CAS: 106-47-8   HR: 3
4-CHLOROANILINE
DOT: UN 2018/UN 2019
mf: C_6H_6ClN mw: 127.58

PROP: Orthorhombic crystals from alcohol or
petr ether. Mp: 72.5°, bp: 232°, d: 1.169. Sol
in hot water; freely sol in alc, ether, acetone,
carbon disulfide.

SYNS: 1-AMINO-4-CHLOROBENZENE * 4-CHLOR-
ANILIN (CZECH) * p-CHLORANILINE * p-CHLORO-
ANILINE * p-CHLOROANILINE, liquid (DOT) * p-
CHLOROANILINE, solid (DOT) * 4-CHLOROBENZEN-
AMINE * 4-CHLORO BENZENEAMINE * 4-CHLO-
ROPHENYLAMINE * NCI-C02039 * RCRA WASTE
NUMBER P024

CONSENSUS REPORTS: EPA Genetic Toxicology Program. Reported in EPA TSCA Inventory.

DOT Classification: Poison B; Label: Poison.

SAFETY PROFILE: Poison by ingestion, inhalation, skin contact, subcutaneous, and intrave-

nous routes. Moderately toxic by inhalation and intraperitoneal routes. A skin and severe eye irritant. Questionable carcinogen with experimental neoplastigenic and tumorigenic data. Mutation data reported. When heated to decomposition it emits toxic fumes of Cl^- and NO_x.

CEI000   CAS: 82-44-0   HR: 3
1-CHLOROANTHRAQUINONE
mf: $C_{14}H_8ClO_2$ mw: 243.67

SYNS: 1-CHLORANTHRACHINON (CZECH) * 1-CHLORO-9,10-ANTHRACENEDIONE * α-CHLOROAN-THRAQUINONE * 1-CHLORO-9,10-ANTHRAQUINONE * α-MONOCHLOROANTHRAQUINONE

CONSENSUS REPORTS: Reported in EPA TSCA Inventory.

SAFETY PROFILE: Poison by intravenous route. Mildly toxic by ingestion. A skin and eye irritant. When heated to decomposition it emits very toxic fumes of Cl^- and NO_x.

CEI500   CAS: 89-98-5   HR: 3
o-CHLOROBENZALDEHYDE
mf: C_7H_5ClO mw: 140.57

SYNS: o-CHLOORBENZALDEHYDE (DUTCH) * 2-CHLOORBENZALDEHYDE (DUTCH) * 2-CHLOR-BENZALDEHYD (GERMAN) * 2-CHLOROBENZALDE-HYDE * 2-CLOROBENZALDEIDE (ITALIAN) * o-CHLOROBENZENECARBOXALDEHYDE * USAF M-7

CONSENSUS REPORTS: Reported in EPA TSCA Inventory.

SAFETY PROFILE: Poison by intraperitoneal and intravenous routes. When heated to decomposition it emits toxic fumes of Cl^-.

CEJ125   CAS: 108-90-7   HR: 2
CHLOROBENZENE
DOT: UN 1134
mf: C_6H_5Cl mw: 112.56

PROP: Clear, colorless liquid. Bp: 131.7°, lel: 1.3%, uel: 7.1%, @ 150°, mp: −45°, flash p: 85°F (CC), d: 1.113 @ 15.5°/15.5°, autoign temp: 1180°F, vap press: 10 mm @ 22.2°, vap d: 3.88.

SYNS: BENZENE CHLORIDE * CHLOORBENZEEN (DUTCH) * CHLORBENZENE * CHLORBENZOL * CHLORBENZEN (POLISH) * CHLOROBENZOL (DOT) * CLOROBENZENE (ITALIAN) * MCB

* MONOCHLOORBENZEEN (DUTCH) * MONO-CHLORBENZENE * MONOCHLORBENZOL (GERMAN) * MONOCHLOROBENZENE * MONOCLOROBENZENE (ITALIAN) * NCI-C54886 * PHENYL CHLORIDE * RCRA WASTE NUMBER U037

CONSENSUS REPORTS: NTP Carcinogenesis Studies (gavage); Some Evidence: rat NTPTR* NTP-TR-261,85; No Evidence: mouse NTPTR* NTP-TR-261,85. Reported in EPA TSCA Inventory. Community Right-To-Know List.

OSHA PEL: TWA 75 ppm
ACGIH TLV: TWA 75 ppm (Proposed: 10 ppm)
DFG MAK: 50 ppm (230 mg/m^3)
DOT Classification: Flammable or Combustible Liquid; Label: Flammable Liquid.

SAFETY PROFILE: Poison by ingestion. Moderately toxic by intraperitoneal route. Experimental teratogenic and reproductive effects. Mutation data reported. Strong narcotic with slight irritant qualities. Dichlorobenzols are strongly narcotic. Little is known of the effects of repeated exposures at lower concentrations, but it may cause kidney and liver damage. The industrial illnesses reported may possibly be due to nitrobenzol. Dangerous fire hazard when exposed to heat or flame. Moderate explosion hazard when exposed to heat or flame. Potentially explosive reaction with powdered sodium or phosphorus trichloride + sodium. Violent reaction with $AgClO_4$ or dimethyl sulfoxide. Reacts vigorously with oxidizers. To fight fire, use foam, CO_2, dry chemical, water to blanket fire. Associated with EPA Superfund sites.

CEM000   CAS: 873-32-5   HR: 3
o-CHLOROBENZONITRILE
mf: C_7H_4ClN mw: 137.57

PROP: (o- and p-). Crystals.

SYNS: o-CHLORBENZONITRIL (CZECH) * NITRIL KYSELINY-o-CHLORBENZOOVE (CZECH)

CONSENSUS REPORTS: Reported in EPA TSCA Inventory. Cyanide and its compounds are on the Community Right-To-Know List.

SAFETY PROFILE: Poison by intraperitoneal route. Moderately toxic by ingestion. An eye irritant. When heated to decomposition or on contact with water, steam, acid or acid fumes it emits toxic fumes of Cl^- and CN^-.

CEM825 CAS: 98-56-6 **HR: 1**
p-CHLOROBENZOTRIFLUORIDE

DOT: UN 2234
mf: $C_7H_4ClF_3$ mw: 180.56

SYNS: (p-CHLOROPHENYL)TRIFLUOROMETHANE
* p-CHLOROTRIFLUOROMETHYLBENZENE * 4-
CHLOROTRIFLUOROMETHYLBENZENE * 1-CHLORO-
4-(TRIMETHYL)-BENZENE (9CI) * α,α,α-TRIFLUORO-4-
CHLOROTOLUENE * p-TRIFLUOROMETHYLPHENYL
CHLORIDE * p-(TRIFLUOROMETHYL)CHLOROBEN-
ZENE

CONSENSUS REPORTS: Reported in EPA
TSCA Inventory.

DOT Classification: Flammable or Combustible Liquid; Label: Flammable Liquid.

SAFETY PROFILE: Mildly toxic by ingestion
and inhalation. Human mutation data reported.
Flammable. Strongly exothermic reaction with
sodium dimethylsulfinate. When heated to decomposition it emits toxic fumes of F^- and
Cl^-.

CEP000 CAS: 103-17-3 **HR: 2**
**p-CHLOROBENZYL-p-CHLOROPHENYL
SULFIDE**

mf: $C_{13}H_{10}Cl_2S$ mw: 269.19

PROP: Crystals, almond-like odor, insol in water, sol in most organic solvents. Mp: 75-76°,
d: 1.4210 @ 25°/4°, vap press: 1.21 × 10^{-5}
mm @ 30°.

SYNS: CHLOORBENZIDE (DUTCH) * (4-CHLOOR-
BENZYL)-(4-CHLOOR-FENYL)-SULFIDE (DUTCH)
* CHLORBENSID (GERMAN) * CHLORBENSIDE
* CHLORBENXIDE * CHLORBENZIDE * (4-
CHLOR-BENZYL)-(4-CHLOR-PHENYL)-SULFID (GERMAN)
* p-CHLOROBENZYL-p-CHLOROPHENYL SULPHIDE
* 4-CHLOROBENZYL-4-CHLOROPHENYL SULPHIDE
* 1-CHLORO-4-(((4-CHLOROPHENYL)METHYL)THIO)
BENZENE * CHLOROCIDE * CHLOROPARACIDE
* 4-CHLOROPHENYL-4'-CHLOROBENZYL SULFIDE
* CHLOROSULFACIDE * CHLORPARACIDE
* CHLORSULPHACIDE * (4-CLORO-BENZIL)-(4-
CLORO-FENIL)-SOLFURO (ITALIAN) * p,p'-DICHLORO-
DIPHENYL SULFIDE * ENT 20,696 * HRS 860
* METOX * MITOX * RD 2195 * SULFURE de
4-CHLOROBENZYLE et de 4-CHLOROPHENYLE (FRENCH)

SAFETY PROFILE: Moderately toxic by ingestion and possibly other routes. Has caused liver
and kidney injury and skin irritation in experi-

mental animals. When heated to decomposition
it emits toxic fumes of Cl^- and SO_x.

CEQ600 CAS: 2698-41-1 **HR: 3**
**o-CHLOROBENZYLIDENE
MALONONITRILE**

mf: $C_{10}H_5ClN_2$ mw: 188.62

PROP: White crystals. Mp: 95°, bp: 313°.

SYNS: 2-CHLOROBENZAL MALONONITRILE
* o-CHLOROBENZAL MALONONITRILE * o-CHLORO-
BENZYLIDENE MALONITRILE * 2-CHLOROBENZYLI-
DENE MALONONITRILE * 2-CHLOROBMN * CS
* β,β-DICYANO-o-CHLOROSTYRENE * NCI-C55118
* PROPANEDINITRILE((2-CHLOROPHENYL)METHY-
LENE) * USAF KF-11

CONSENSUS REPORTS: Reported in EPA
TSCA Inventory. Cyanide and its compounds
are on the Community Right-To-Know. List.

OSHA PEL: (Transitional: TWA 0.05 ppm) CL
0.05 ppm (skin)
ACGIH TLV: CL 0.05 ppm (skin)
OSHA PEL: TWA 0.05 ppm

SAFETY PROFILE: Poison by ingestion, intraperitoneal, and intravenous routes. Moderately
toxic by inhalation. Human systemic effects by
inhalation: conjuntiva irritation, cough, and unspecified respiratory system effects. A human
skin and eye irritant. Human exposure data suggest relatively low systematic toxicity, but intense irritation of eyes, skin, and mucous membranes. Mutation data reported. A tear-gas used
for riot control. When heated to decomposition
it emits very toxic fumes of Cl^-, NO_x, and
CN^-.

CES650 CAS: 74-97-5 **HR: 3**
CHLOROBROMOMETHANE

DOT: UN 1887
mf: CH_2BrCl mw: 129.39

PROP: Clear, colorless liquid; sweet odor. Bp:
67.8°, fp: −88°, flash p: none, d: 1.930 @
25°/25°, vap d: 4.46.

SYNS: BROMOCHLOROMETHANE * HALON 1011
* METHYLENE CHLOROBROMIDE * MIL-B-4394-B
* MONO-CHLORO-MONO-BROMO-METHANE

CONSENSUS REPORTS: Reported in EPA
TSCA Inventory.

OSHA PEL: TWA 200 ppm
ACGIH TLV: TWA 200 ppm

DFG MAK: 200 ppm (1050 mg/m^3)

DOT Classification: Poison B; Label: St. Andrews Cross.

SAFETY PROFILE: A poison. Mildly toxic by ingestion and inhalation. Mutation data reported. This material has a narcotic action of moderate intensity, although of prolonged duration. Animals exposed for several weeks to 1000 ppm had blood bromide levels as high as 350 mg/100 g. Therefore, until further data are available, it should be considered at least as toxic as carbon tetrachloride and more than minimal exposure to its vapors should be avoided. Dangerous; when heated to decomposition it emits highly toxic fumes of Br$^-$ and Cl$^-$.

CET000   CAS: 73926-87-1   **HR: 3**
trans-CHLORO(2-(3-BROMOPROPION-
AMIDO)CYCLOHEXYL)MERCURY
mf: C$_9$H$_{15}$BrClHgNO   mw: 469.20

SYNS: 3-BROMO-N-(2-CHLOROMERCURICYCLO-
HEXYL)PROPIONAMIDE * MERCURY (E)-CHLORO(2-
(3-BROMOPROPIONAMIDO)CYCLOHEXYL)

CONSENSUS REPORTS: Mercury and its compounds are on the Community Right-To-Know List.

OSHA PEL: (Transitional: CL 1 mg/10m^3) TWA 0.01 mg(Hg)/m^3; STEL 0.03 mg/m^3 (skin)
ACGIH TLV: TWA 0.1 mg(Hg)/m^3 (skin)
NIOSH REL: (Inorganic Mercury) TWA: 0.05 mg(Hg)/m^3

SAFETY PROFILE: Poison by intravenous route. When heated to decomposition it emits very toxic fumes of Br$^-$, Cl$^-$, NO$_x$, and Hg.

CET250   CAS: 627-22-5   **HR: 3**
1-CHLOROBUTADIENE
mf: C$_4$H$_5$Cl   mw: 88.54

PROP: Colorless liquid. Bp: 59.4°, d: 0.9583, flash p: −4°F, lel: 4.0%, uel: 20.0%, vap d: 3.0.

SYN: 1-CHLORO-1,3-BUTADIENE

CONSENSUS REPORTS: EPA Genetic Toxicology Program.

SAFETY PROFILE: Mutation data reported. Probably a poison by ingestion, subcutaneous, and intravenous routes. Dangerous fire hazard when exposed to heat or flame. To fight fire, use alcohol foam. When heated to decomposition it emits toxic fumes of Cl$^-$.

CEU250   CAS: 78-86-4   **HR: 3**
2-CHLOROBUTANE
DOT: UN 1127
mf: C$_4$H$_9$Cl   mw: 92.58

PROP: Flash p: 14°F, d: 0.87, vap d: 3.2, bp: 68.50.

SYN: sec-BUTYL CHLORIDE

CONSENSUS REPORTS: Reported in EPA TSCA Inventory.

DOT Classification: Flammable Liquid; Label: Flammable Liquid.

SAFETY PROFILE: Mildly toxic by ingestion, inhalation, and skin contact. Questionable carcinogen with experimental neoplastigenic data. Dangerous fire hazard when exposed to heat, open flame (sparks), or oxidizers. To fight fire, use water, water spray, fog, mist, dry chemical, alcohol foam. When heated to decomposition it emits toxic fumes of Cl$^-$.

CEU500   CAS: 928-51-8   **HR: 3**
4-CHLORO-1-BUTANOL
mf: C$_4$H$_9$ClO   mw: 108.58

SYNS: 4-CHLORBUTAN-1-OL (GERMAN) * 4-
CHLORO-1-BUTANE-OL * 4-CHLOROBUTANOL
* TETRAMETHYLENE CHLOROHYDRIN

CONSENSUS REPORTS: Reported in EPA TSCA Inventory.

SAFETY PROFILE: Moderately toxic by ingestion. Questionable carcinogen with experimental neoplastigenic data. Mutation data reported. When heated to decomposition it emits toxic fumes of Cl$^-$.

CEW500   CAS: 101-27-9   **HR: 3**
m-CHLORO CARBANILIC ACID-4-
CHLORO-2-BUTYNYL ESTER
mf: C$_{11}$H$_9$Cl$_2$NO$_2$ mw: 258.11

SYNS: A-980 * BARBAMATE * BARBAN
* BARBANE * 2-BUTYNYL-4-CHLORO-m-CHLORO-
CARBANILATE * C-847 * CARBIN * CARBYNE
* CARYNE * CBN * (4-CHLOOR-BUT-2-YN-YL)-
N-(3-CHLOOR-FENYL)-CARBAMAAT (DUTCH)
* (4-CHLOR-BUT-2-IN-YL)-N-(3-CHLOR-PHENYL)-CAR-
BAMAT (GERMAN) * CHLORINAT * CHLORO-2-
BUTYNYL-m-CHLOROCARBAMATE * 4-CHLOROBUT-

2-YNYL-m-CHLOROCARBANILATE * 4-CHLORO-2-BUTYNYL-m-CHLOROCARBANILATE * 4-CHLORO-BUT-2-YNYL-3-CHLOROPHENYLCARBAMATE * 4-CHLORO-2-BUTYNYL-N-(3-CHLOROPHENYL)CARBA-MATE * N-(3-CHLORO PHENYL) CARBAMATE de 4-CHLORO 2-BUTYNYLE (FRENCH) * (3-CHLORO-PHENYL)CARBAMIC ACID 4-CHLORO-2-BUTYNYL ESTER * (4-CLORO-BUT-2-IN-IL)-N-(3-CLORO-FENIL)-CAR-BAMMATO (ITALIAN) * CS-847 * FISONS B25 * NEOBAN * S-847

CONSENSUS REPORTS: EPA Genetic Toxicology Program.

SAFETY PROFILE: Poison by ingestion, inhalation, and possibly other routes. Mildly toxic by skin contact. Mutation data reported. An herbicide. When heated to decomposition it emits very toxic fumes of Cl⁻ and NO_x.

CFA500   CAS: 126-85-2   HR: 3
2-CHLORO-N-(2-CHLOROETHYL)-N-METHYL ETHANAMINE-N-OXIDE
mf: $C_5H_{11}Cl_2NO$ mw: 172.07

SYNS: 2,2'-DICHLORO-N-METHYLDIETHYLAMINE-N-OXIDE * DIETHYLAMINE, 2,2'-DICHLORO-N-METHYL-, OXIDE * HN2 AMINE OXIDE * HN2 OXIDE MUSTARD * MBAO * MECHLORETHAMINE OXIDE * METHYL-BIS(β-CHLOROETHYL)AMINE OXIDE * METHYLBIS(β-CHLOROETHYL)AMINE N-OXIDE * N-METHYL-DI-2-CHLOROETHYLAMINE-N-OXIDE * MITOMEN * MITOMIN * NITRO-GEN MUSTARD OXIDE * NITROGEN MUSTARD N-OX-IDE * NITROMIN * NMO * N-OXYD-LOST * N-OXYD-MUSTARD * NSC 10107 * OXY-NH2

SAFETY PROFILE: Poison by intravenous and intraperitoneal routes. Experimental reproductive effects. Questionable carcinogen with experimental carcinogenic and tumorigenic data. Mutation data reported. When heated to decomposition it emits toxic fumes of Cl⁻ and NO_x.

CFA750   CAS: 302-70-5   HR: 3
2-CHLORO-N-(2-CHLOROETHYL)-N-METHYLETHANAMINE-N-OXIDE HYDROCHLORIDE
mf: $C_5H_{11}Cl_2NO \cdot ClH$ mw: 208.53

SYNS: CHLORMETHINE-N-OXIDE HYDROCHLORIDE * 2,2'-DICHLORO-N-METHYLDIETHYLAMINE N-OXIDE HYDROCHLORIDE * HN₂ OXIDE HYDROCHLORIDE * MBAO HYDROCHLORIDE * MECHLORETHAMINE OXIDE HYDROCHLORIDE * METHYL-BIS-(β-CHLORA-ETHYL)-AMIN-N-OXYD-HYDROCHLORID (GERMAN)

* METHYLBIS(β-CHLOROETHYL)AMINE-N-OXIDE HY-DROCHLORIDE * N-METHYLBIS(2-CHLOROETHYL) AMINE-N-OXIDE HYDROCHLORIDE * N-METHYL-2,2'-DICHLORODIETHYLAMINE-N-OXIDE HYDROCHLORIDE * METHYLDI(2-CHLOROETHYL)AMINE-N-OXIDE HY-DROCHLORIDE * MITOMEN * MUSTRON * NITROGEN MUSTARD OXIDE * NITROGEN MUS-TARD-N-OXIDE * NITROGEN MUSTARD-N-OXIDE HY-DROCHLORIDE * NITROMIM * NITROMIN HYDRO-CHLORIDE * N-OXYD-LOST * NSC-10107 * OSSIAMINA * OSSICHLORIN * OXYAMINE * SK-598 * XA 2

CONSENSUS REPORTS: IARC Cancer Review: GROUP 2B IMEMDT 7,56,87; Animal Sufficient Evidence IMEMDT 9,209,75; EPA Genetic Toxicology Program.

SAFETY PROFILE: Suspected carcinogen with experimental tumorigenic data. Poison by ingestion, subcutaneous, intravenous, and intraperitoneal routes. Human systemic effects by an unspecified route: convulsions and unspecified changes in bone marrow. Mutation data reported. An antineoplastic agent. When heated to decomposition it emits toxic fumes of Cl⁻ and NO_x.

CFE250   CAS: 59-50-7   HR: 3
4-CHLORO-m-CRESOL
mf: C_7H_7ClO mw: 142.59

PROP: Odorless crystals (when pure). Somewhat sol in water, very sol in organic solvents. Mp: 66°, bp: 235°.

SYNS: APTAL * BAKTOL * BAKTOLAN * CANDASETPIC * p-CHLOR-m-CRESOL * CHLO-ROCRESOL * p-CHLOROCRESOL * p-CHLORO-m-CRESOL * 6-CHLORO-m-CRESOL * 2-CHLORO-HY-DROXYTOLUENE * 6-CHLORO-3-HYDROXYTOLUENE * 4-CHLORO-3-METHYLPHENOL * 3-METHYL-4-CHLOROPHENOL * OTTAFACT * PARMETOL * PAROL * PCMC * PREVENTOL CMK * RASCHIT * RASEN-ANICON * RCRA WASTE NUMBER U039

CONSENSUS REPORTS: Reported in EPA TSCA Inventory. Chlorophenol compounds are on the Community Right-To-Know List.

SAFETY PROFILE: Poison by intraperitoneal and subcutaneous routes. Moderately toxic by ingestion. An allergen. Incompatible with sodium hydroxide. When heated to decomposition it emits toxic fumes of Cl⁻ and phosgene.

CFF500   CAS: 82-93-9   **HR: 3**
CHLOROCYCLINE
mf: $C_{18}H_{21}ClN_2$ mw: 300.86

SYNS: CHLORCYCLINE * CHLORCYCLIZINE
* 1-(4-CHLOROBENZHYDRYL)-4-METHYLPIPERAZINE
* CHLOROCYCLIZINE * 1-(p-CHLORO-α-PHENYL-
BENZYL)-4-METHYLPIPERAZINE * DI-PARALEN
* DIPARALENE * HISTANTIN * HISTANTINE

SAFETY PROFILE: Unspecified human reproductive effects. Experimental teratogenic and
reproductive effects. Mutation data reported.
When heated to decomposition it emits very
toxic fumes of Cl⁻ and NO_x.

CFK000   CAS: 93-71-0   **HR: 3**
2-CHLORO-N,N-DIALLYLACETAMIDE
mf: $C_8H_{12}ClNO$ mw: 173.66

PROP: Amber liquid. Bp: 74° @ 0.3 mm. Sltly
sol in H_2O; sol in alc, hexane, and xylene.

SYNS: ALIDOCHLOR * ALLIDOCHLOR * CDAA
* CDAAT * α-CHLORO-N,N-DIALLYLACETAMIDE
* 2-CHLORO-N,N-DI-2-PROPENYLACETAMIDE
* CP 6,343 * DIALLYLCHLOROACETAMIDE
* N,N-DIALLYLCHLOROACETAMIDE * N,N-DIAL-
LYL-α-CHLOROACETAMIDE * N,N-DIALLYL-2-CHLO-
ROACETAMIDE * NCI-CO4035 * RADOX
* RANDOX * RANTOX T

CONSENSUS REPORTS: Reported in EPA
TSCA Inventory.

SAFETY PROFILE: Poison by skin contact.
Moderately toxic by ingestion. An herbicide.
When heated to decomposition it emits very
toxic fumes of Cl⁻ and NO_x.

CFK125   CAS: 95-83-0   **HR: 3**
4-CHLORO-1,2-DIAMINOBENZENE
mf: $C_6H_7ClN_2$ mw: 142.60

SYNS: p-CHLORO-o-PHENYLENEDIAMINE * 4-
CHLORO-o-PHENYLENEDIAMINE * 4-CHLORO-1,2-
PHENYLENEDIAMINE * 4-Cl-o-PD * 1,2-DIAMINO-4-
CHLOROBENZENE * 3,4-DIAMINOCHLOROBENZENE
* 3,4-DIAMINO-1-CHLOROBENZENE * NCI-C03292
* URSOL OLIVE 6G

CONSENSUS REPORTS: IARC Cancer Re-
view: GROUP 2B IMEMDT 7,56,87; Human
Limited Evidence IMEMDT 27,81,82; Animal
Sufficient Evidence IMEMDT 27,81,82. NTP
Fourth Annual Report On Carcinogens, 1984.
NCI Carcinogenesis Bioassay (feed); Clear Evi-

dence: mouse, rat NCITR* NCI-CG-TR-63,78.
Reported in EPA TSCA Inventory.

SAFETY PROFILE: Confirmed with experimental carcinogenic and neoplastigenic data.
Human mutation data reported. When heated
to decomposition it emits toxic fumes of Cl⁻
and NO_x.

CFK500   CAS: 124-48-1   **HR: 2**
CHLORODIBROMOMETHANE
mf: $CHBr_2Cl$ mw: 208.29

PROP: Colorless to pale yellow, heavy liquid.
Bp: 118-122°, fp: < −20°, d: 2.440 @ 25°/
25°.

SYNS: CDBM * DIBROMOCHLOROMETHANE
* NCI-C55254

CONSENSUS REPORTS: NTP Carcinogen-
esis Studies (gavage); Some Evidence: mouse
NTPTR* NTP-TR-282,86; No Evidence: rat
NTPTR* NTP-TR-282,85; Reported in EPA
TSCA Inventory.

SAFETY PROFILE: Moderately toxic by ingestion. Human mutation data reported. Compounds of this type are generally irritating and
narcotic. When heated to decomposition it emits
toxic fumes of Cl⁻ and Br⁻.

CFX000   CAS: 15972-60-8   **HR: 2**
2-CHLORO-2′,6′-DIETHYL-
N-(METHOXYMETHYL)ACETANILIDE
mf: $C_{14}H_{20}ClNO_2$ mw: 269.80

SYNS: ALACHLOR (USDA) * ALANEX * ALO-
CHLOR * CHLORESSIGSAEURE-N-(METHOXY-
METHYL)-2,6-DIAETHYLANILID (GERMAN) * 2-CHLO-
RO-N-(2,6-DIETHYLPHENYL)-N-(METHOXYMETHYL)
ACETAMIDE * CP 50144 * LASSO * LAZO
* METACHLOR * METHACHLOR * PILLARZO

CONSENSUS REPORTS: EPA Genetic Toxicology Program.

SAFETY PROFILE: Moderately toxic by ingestion, skin contact, and possibly other routes.
Human mutation data reported. When heated
to decomposition it emits very toxic fumes of
Cl⁻ and NO_x.

CFX250   CAS: 75-68-3   **HR: 1**
1-CHLORO-1,1-DIFLUOROETHANE
DOT: UN 2517
mf: $C_2H_3ClF_2$ mw: 100.50

PROP: Gas. Mp: −131°, bp: −9.5°, d: 1.19, lel: 9.0%; uel: 14.8%.

SYNS: CHLORODIFLUOROETHANE (DOT) * α-CHLOROETHYLIDENE FLUORIDE * 1,1-DIFLUORO-1-CHLOROETHANE * DIFLUOROMONOCHLOROETHANE (DOT) * FC142b * FLUOROCARBON FC142b * FREON 142 * FREON 142b * GENETRON 101 * GENETRON 142b

CONSENSUS REPORTS: Reported in EPA TSCA Inventory.

DOT Classification: Flammable Gas; Label: Flammable Gas.

SAFETY PROFILE: Very mildly toxic by inhalation. Mutation data reported. A very dangerous fire hazard when exposed to heat, flame or oxidizing materials. To fight fire, stop flow of gas. Can react vigorously with oxidizing materials. When heated to decomposition it emits toxic fumes of F⁻ and Cl⁻.

CFX500   CAS: 75-45-6   **HR: 1**
CHLORODIFLUOROMETHANE

DOT: UN 1018
mf: $CHClF_2$ mw: 86.47

PROP: Gas. D: 3.87 air @ 0°, mp: −146°, bp: −40.8°, autoign temp: 1170°F.

SYNS: ALGOFRENE TYPE 6 * ARCTON 4 * DIFLUOROCHLOROMETHANE * DIFLUOROMONO-CHLOROMETHANE * ELECTRO-CF 22 * ESKIMON 22 * F 22 * FLUOROCARBON-22 * FREON * FREON 22 * FRIGEN * GENETRON 22 * ISCEON 22 * ISOTRON 22 * MONOCHLORODI-FLUOROMETHANE * PROPELLANT 22 * R 22 (DOT) * REFRIGERANT 22 * UCON 22/HALOCARBON 22

CONSENSUS REPORTS: IARC Cancer Review: GROUP 3 IMEMDT 7,149,87; Human Inadequate Evidence IMEMDT 41,237,86; Animal Limited Evidence IMEMDT 41,237,86. Reported in EPA TSCA Inventory. EPA Genetic Toxicology Program.

OSHA PEL: TWA 1000 ppm
ACGIH TLV: TWA 1000 ppm
DFG MAK: 500 ppm (1800 mg/m³)
DOT Classification: Nonflammable Gas; Label: Nonflammable Gas.

SAFETY PROFILE: Mildly toxic by inhalation. Experimental reproductive effects. Mutation data reported. An asphyxiant in high concentrations. At elevated pressures, 50% mixtures with air are combustible although ignition is difficult. When heated to decomposition it emits toxic fumes of F⁻ and Cl⁻.

CFY000   CAS: 58-93-5   **HR: 3**
6-CHLORO-3,4-DIHYDRO-2H-1,2,4-BENZOTHIADIAZINE-7-SULFONAMIDE-1,1-DIOXIDE
mf: $C_7H_8ClN_3O_4S_2$ mw: 297.75

SYNS: AQUARILLS * AQUARIUS * BREMIL * 6-CHLORO-3,4-DIHYDRO-7-SULFAMOYL-2H-1,2,4-BENZOTHIADIAZINE-1,1-DIOXIDE * 6-CHLORO-7-SULFAMOYL-3,4-DIHYDRO-2H-1,2,4-BENZOTHIADIA-ZINE-1,1-DIOXIDE * CHLOROSULTHIADIL * CHLORSULFONAMIDO DIHYDROBENZOTHIADIA-ZINE DIOXIDE * CHLORZIDE * CIDREX * DICHLOROSAL * DICHLOTIAZID * DICHLO-TRIDE * DICLOTRIDE * 3,4-DIHYDRO-6-CHLORO-7-SULFAMYL-1,2,4-BENZOTHIADIAZINE-1,1-DIOXIDE * DIHYDROCHLOROTHIAZID * DIHYDROCHLO-ROTHIAZIDE * 3,4-DIHYDROCHLOROTHIAZIDE * DIHYDROXYCHLOROTHIAZIDUM * DIREMA * DISALUNIL * DRENOL * DYAZIDE * ESI-DREX * ESIDRIX * FLUVIN * HCTZ * HCZ * HIDRIL * HIDROCHLORTIAZID * HIDRORONOL * HIDROTIAZIDA * HYDRO-AQUIL * HYDRO-CHLORTHIAZID * HYDRODIURETIC * HYDRO-DIURIL * HYDROSALURIC * HYDROTHIDE * HYPOTHIAZIDE * IDROTIAZIDE * IVAUGAN * JEN-DIRIL * MASCHITT * MEGADIURIL * NCI-C55925 * NEFRIX * NEO-CODEMA * NEOFLUMEN * ORETIC * PANURIN * RO-HYDRAZIDE * SU 5879 * THIARETIC * THIURETIC * THLARETIC * URODIAZIN * VETIDREX * ZIDE

CONSENSUS REPORTS: Reported in EPA TSCA Inventory. EPA Genetic Toxicology Program.

SAFETY PROFILE: Poison by intraperitoneal and intravenous routes. Moderately toxic by ingestion and subcutaneous routes. Experimental reproductive effects. Mutation data reported. A diuretic. When heated to decomposition it emits very toxic fumes of SO_x, Cl⁻, and NO_x.

CGB000   CAS: 55299-24-6   **HR: 2**
7-CHLORO-1,3-DIHYDRO-5-PHENYL-1-TRIMETHYLSILYL-2H-1,4-BENZODIAZEPIN-2-ONE
mf: $C_{18}H_{18}ClOSi$ mw: 313.90

SYNS: ST 720 (FRENCH) * TRIMETHYL SILYL-1-CHLORO-7-DIHYDRO-1,3-PHENYL-5,2H-BENZODIAZE-PINE-1,4-ONE-2 (FRENCH)

SAFETY PROFILE: Moderately toxic by ingestion and intraperitoneal routes. When heated to decomposition it emits toxic fumes of Cl$^-$.

CGB500   CAS: 1779-25-5   HR: 1
CHLORO DIISOBUTYL ALUMINUM
mf: $C_8H_{18}AlCl$ mw: 176.69

SYNS: ALLUMINIO DIISOBUTIL-MONOCLORURO (ITALIAN) * BIS(ISOBUTYL)ALUMINUM CHLORIDE * CHLOROBIS(2-METHYLPROPYL)ALUMINUM * DIISOBUTYLALUMINUM CHLORIDE * DIISOBUTYLALUMINUM MONOCHLORIDE * DIISOBUTYL-CHLOROALUMINUM

CONSENSUS REPORTS: Reported in EPA TSCA Inventory.

ACGIH TLV: TWA 2 mg(Al)/m^3

SAFETY PROFILE: Mildly toxic by inhalation. Ignites spontaneously in air. When heated to decomposition it emits toxic fumes of Cl$^-$.

CGD250   CAS: 2491-76-1   HR: 3
p-CHLORO DIMETHYLAMINOAZO-BENZENE
mf: $C_{14}H_{14}ClN_3$ mw: 259.76

SYNS: 4'-CHLORO-4-DIMETHYLAMINOAZOBENZENE * N,N-DIMETHYL-p-((p-CHLOROPHENYL)AZO)ANILINE

SAFETY PROFILE: Moderately toxic by subcutaneous route. Questionable carcinogen with experimental neoplastigenic data. Experimental teratogenic effects. When heated to decomposition it emits very toxic fumes of Cl$^-$ and NO$_x$.

CGL750   CAS: 25567-67-3   HR: 3
CHLORODINITROBENZENE
DOT: UN 1577
mf: $C_6H_3ClN_2O_4$ mw: 202.56

SYNS: CHLORODINITROBENZENE (DOT) * CHLORODINITRO BENZENE (mixed isomers) * DINITROCHLOROBENZENE * DINITROCHLOROBENZENE (DOT)

DOT Classification: Poison B; Label: Poison.

SAFETY PROFILE: A poison. Mutation data reported. Potentially explosive. When heated to decomposition it emits very toxic fumes of Cl$^-$ and NO$_x$.

CGM000   CAS: 97-00-7   HR: 3
1-CHLORO-2,4-DINITROBENZENE
mf: $C_6H_3ClN_2O_4$ mw: 202.56

PROP: Yellow rhombic crystals, insol in water. mp(α): 53.4°, mp(β): 43°, mp(γ): 27°, bp: 315°, lel: 2.0%, uel: 22%, flash p: 382°F (CC), d(α): 1.687 @ 22°, d(β):1.680 @ 20°/4°, vap d: 6.98.

SYNS: 1-CHLOOR-2,4-DINITROBENZEEN (DUTCH) * 1-CHLOR-2,4-DINITROBENZENE * 4-CHLORO-1,3-DINITROBENZENE * 6-CHLORO-1,3-DINITROBENZENE * 1-CHLORO-2,4-DINITROBENZOL (GERMAN) * 1-CLORO-2,4-DINITROBENZENE (ITALIAN) * 2,4-DINITROCHLOROBENZENE * 1,3-DINITRO-4-CHLOROBENZENE * 2,4-DINITRO-1-CHLOROBENZENE * DINITROCHLOROBENZOL * DINITROCHLOROBENZOL (DOT) * DNCB

CONSENSUS REPORTS: Reported in EPA TSCA Inventory.

DOT Classification: Poison B; Label: Poison.

SAFETY PROFILE: Poison by by skin contact and intraperitoneal routes. Moderately toxic by by ingestion. A human skin irritant. Acts as a primary irritant as well as a sensitizer of skin. An allergen. Mutation data reported. Combustible when exposed to heat or flame. A moderate explosion hazard when exposed to flame, sparks, heated to 150°, or when shocked in a sealed container. Explosive reaction with ammonia at 170°C/40 bar. To fight fire use CO$_2$, dry chemical. Reacts violently with hydrazine sulfate or hydrazine hydrate.

CGN000   CAS: 712-48-1   HR: 3
CHLORODIPHENYLARSINE
DOT: UN 1699
mf: $C_{12}H_{10}AsCl$ mw: 264.59

PROP: Colorless crystals when pure, technical product is dark brown liquid. Bp: 333° (decomp), fp: 44°, d: 1.363 @ 40° (solid): 1.358 @ 45° (liquid), vap press: 0.00049 mm @ 20°, vap d: 9.15.

SYNS: BLUE CROSS * CLARK I * DA * DIPHENYLARSINOUS CHLORIDE * DIPHENYL-CHLOORARSINE (DUTCH) * DIPHENYLCHLOROARSINE (DOT) * SNEEZING GAS

CONSENSUS REPORTS: Arsenic and its compounds are on the Community Right-To-Know List.

OSHA PEL: TWA 0.5 mg(As)/m^3

DOT Classification: Poison B; Label: Poison.

SAFETY PROFILE: A human poison by inhalation. Poison experimentally by inhalation and skin contact. A powerfully irritating military poison. Exposure yields cold-like symptoms, plus headache, vomiting and nausea. A non-persistent gas. Decontamination is by use of chlorine or caustic soda in confined spaces. When heated to decomposition it emits toxic fumes of As and Cl$^-$.

CGO500 CAS: 115-96-8 **HR: 3**
2-CHLOROETHANOL PHOSPHATE
mf: $C_6H_{12}Cl_3O_4P$ mw: 285.50

PROP: Flash p: 421°F (COC), boiling range: 210-220° @ 20 mm d: 1.425 @ 20°/20°, autoign temp: 1115°F, vap press: 0.5 mm @ 145°.

SYNS: CELLUFLEX * FYROL CEF * NCI-C60128 * NIAX FLAME RETARDANT 3 CF * TRICHLOR-ETHYL PHOSPHATE * TRI-β-CHLOROETHYL PHOS-PHATE * TRI(2-CHLOROETHYL)PHOSPHATE * TRIS(2-CHLOROETHYL)ESTER PHOSPHORIC ACID * TRIS(β-CHLOROETHYL) PHOSPHATE * TRIS(2-CHLOROETHYL) PHOSPHATE

CONSENSUS REPORTS: Reported in EPA TSCA Inventory. EPA Genetic Toxicology Program.

SAFETY PROFILE: Poison by intraperitoneal route. Moderately toxic by ingestion. Experimental reproductive effects. A skin and eye irritant. Combustible when exposed to heat or flame. When heated to decomposition it emits very toxic fumes of PO$_x$ and Cl$^-$.

CGV250 CAS: 13010-47-4 **HR: 3**
1-(2-CHLOROETHYL)-3-CYCLOHEXYL-1-NITROSOUREA
mf: $C_9H_{16}ClN_3O_2$ mw: 233.73

SYNS: BELUSTINE * CCNU * CECENU * CEENU * CHLOROETHYLCYCLOHEXYLNITROSO-UREA * N-(2-CHLOROETHYL)-N'-CYCLOHEXYL-N-NITROSOUREA * ((CHLORO-2-ETHYL)-1-CYCLO-HEXYL-3-NITROSOUREA * CINU * (CLORO-2-ETIL)-1-CICLOESIL-3-NITROSOUREA (ITALIAN) * ICIG 1109 * LOMUSTINE * NCI-C04740 * NSC-79037 * RB 1509 * SRI 2200

CONSENSUS REPORTS: IARC Cancer Review: GROUP 2A IMEMDT 7,150,87; Human Limited Evidence IMEMDT 26,137,81; Animal Sufficient Evidence IMEMDT 26,137,81. NTP Fourth Annual Report On Carcinogens, 1984. NCI Carcinogenesis Studies (ipr); Clear Evidence: mouse CANCAR 40,1935,77; No Evidence: rat CANCAR 40,1935,77. EPA Genetic Toxicology Program.

SAFETY PROFILE: Confirmed carcinogen with experimental carcinogenic and tumorigenic data. Poison by ingestion, intraperitoneal, subcutaneous, intravenous, and possibly other routes. Human systemic effects by ingestion: anorexia, nausea or vomiting, leukopenia (decrease in the white blood cell count), and thrombocytopenia (decrease in the number of blood platelets). Experimental teratogenic and reproductive effects. Human mutation data reported. When heated to decomposition it emits very toxic fumes of Cl$^-$ and NO$_x$.

CGV500 CAS: 100-35-6 **HR: 3**
N-(2-CHLORO ETHYL)DIETHYLAMINE
mf: $C_6H_{14}ClN$ mw: 135.66

SYNS: (2-CHLOROETHYL)DIETHYLAMINE * β-CHLOROTRIETHYLAMINE * 2-CHLOROTRIETHY-LAMINE * 2-(DIETHYLAMINO)CHLOROETHANE * DIETHYLAMINOETHYL CHLORIDE * β-(DIETHYL-AMINO)ETHYL CHLORIDE * N-DIETHYLAMINO-ETHYL CHLORIDE * 2-(DIETHYLAMINO)ETHYL CHLO-RIDE * DIETHYL(2-CHLOROETHYL)AMINE

CONSENSUS REPORTS: EPA Genetic Toxicology Program.

SAFETY PROFILE: Poison by ingestion, skin contact and possibly other routes. A severe eye and moderate skin irritant. When heated to decomposition it emits very toxic fumes of Cl$^-$ and NO$_x$.

CGW000 CAS: 107-99-3 **HR: 3**
N-(2-CHLOROETHYL)DIMETHYLAMINE
mf: $C_4H_{10}ClN$ mw: 107.60

PROP: Liquid. Vap d: 3.72.

SYNS: CHLORO(DIMETHYLAMINO)ETHANE * β-CHLOROETHYLDIMETHYLAMINE * (2-CHLORO-ETHYL)DIMETHYLAMINE * DIMETHYLAMINOETHYL CHLORIDE * β-(DIMETHYLAMINO)ETHYL CHLORIDE * 2-DIMETHYLAMINOETHYLCHLORIDE * DIMETHYL(2-CHLOROETHYL)AMINE * HN 1 * NITROGEN HALF MUSTARD

CONSENSUS REPORTS: EPA Genetic Toxicology Program. Reported in EPA TSCA Inventory.

SAFETY PROFILE: Poison by an unspecified route. A systemic irritant. Mutation data reported. When heated to decomposition it emits highly toxic fumes of Cl⁻ and NO_x.

CGY750   CAS: 693-07-2   HR: 3
CHLOROETHYL ETHYL SULFIDE
mf: C_4H_9ClS mw: 124.64

SYNS: 2-CHLOROETHYL ETHYL SULFIDE * 2-CHLOROETHYL ETHYL THIOETHER * 1-CHLORO-2-(ETHYLTHIO)ETHANE * ETHYL-β-CHLOROETHYL SULFIDE * ETHYL-2-CHLOROETHYL SULFIDE * β-ETHYLMERKAPTOETHYLCHLORID (CZECH) * 2-(ETHYLTHIO)CHLOROETHANE * 2-ETHYLTHIO-ETHYL CHLORIDE * HALF-MUSTARD GAS * h-MG

CONSENSUS REPORTS: Reported in EPA TSCA Inventory. EPA Genetic Toxicology Program.

SAFETY PROFILE: Poison by ingestion and subcutaneous routes. Mutation data reported. A severe skin and eye irritant. When heated to decomposition it emits very toxic fumes of Cl⁻ and SO_x.

CHC500   CAS: 107-27-7   HR: 3
CHLOROETHYL MERCURY
mf: C_2H_5ClHg mw: 265.11

PROP: Silvery, irridescent leaflets. Mp: 192.5°.

SYNS: CERESAN * EMC * ETHYLMERCURIC CHLORIDE * ETHYLMERCURY CHLORIDE * GANOZAN * GRANOSAN

CONSENSUS REPORTS: Mercury and its compounds are on the Community Right-To-Know List.

OSHA PEL: (Transitional: CL 1 mg/10m³) TWA 0.01 mg(Hg)/m³; STEL 0.03 mg/m³ (skin)
ACGIH TLV: TWA 0.01 mg(Hg)/m³; STEL 0.03 mg(Hg)/m³
NIOSH REL: TWA 0.05 mg(Hg)/m³

SAFETY PROFILE: Poison by ingestion, inhalation, skin contact, subcutaneous, intraperitoneal, and possibly other routes. An experimental teratogen. Human mutation data reported. When heated to decomposition it emits very toxic fumes of Cl⁻ and Hg.

CHC750   CAS: 3570-58-9   HR: 3
2-CHLOROETHYL METHANESULFONATE
mf: $C_3H_7ClO_3S$ mw: 158.61

SYNS: CB 1506 * β-CHLOROETHYLMETHANESUL-FONATE * CHLOROETHYL METHANESULPHONATE * CHLOROMETHANE SULFONATE d'ETHYLE (FRENCH) * METHANESULFONIC ACID CHLOROETHYL ESTER * NSC 18016

CONSENSUS REPORTS: EPA Genetic Toxicology Program.

SAFETY PROFILE: Poison by ingestion, intravenous, and intraperitoneal routes. Experimental reproductive effects. Mutation data reported. When heated to decomposition it emits very toxic fumes of Cl⁻ and SO_x.

CHD250   CAS: 13909-09-6   HR: 3
1-(2-CHLOROETHYL)-3-(4-METHYL-CYCLOHEXYL)-1-NITROSOUREA
mf: $C_{10}H_{18}ClN_3O_2$ mw: 247.76

SYNS: 1-(2-CHLOROETHYL)-3-(trans-4-METHYL-CYCLO-HEXYL)-1-NITROSOUREA * N-(2-CHLOROETHYL)-N'-(trans-4-METHYLCYCLOHEXYL)-N-NITROSOUREA * ME-CCNU * METHYL-CCNU * trans-METHYL-CCNU * METHYL-LOMUSTINE * NCI-C04955 * NSC-95441 * SEMUSTINE

CONSENSUS REPORTS: NCI Carcinogenesis Studies (ipr); Clear Evidence: rat CANCAR 40,1935,77; No Evidence: mouse CANCAR 40,1935,77.

SAFETY PROFILE: Suspected human carcinogen producing leukemia. Experimental carcinogenic and tumorigenic data. Poison by ingestion, intraperitoneal, intravenous, and other possibly other routes. Mutation data reported. Human systemic effects by ingestion: nausea or vomiting, damage to kidney tubules and glomeruli, and hematuria (blood in the urine). When heated to decomposition it emits very toxic fumes of Cl⁻ and NO_x.

CHF500   CAS: 6296-45-3   HR: 3
2-CHLOROETHYL-N-NITROSOURETHANE
mf: $C_5H_9ClN_2O_3$ mw: 180.61

SYNS: N-(2-CHLOROETHYL)-N-NITROSOETHYLCARBA-MATE * N-(β-CHLOROETHYL)-N-NITROSOURETHAN * ETHYL-N-(β-CHLOROETHYL)-N-NITROSOCARBAMATE * TL 154

SAFETY PROFILE: Poison by inhalation, ingestion, and intraperitoneal routes. Questionable carcinogen with experimental tumorigenic data. Mutation data reported. Many N-nitroso compounds are carcinogens. When heated to decomposition it emits very toxic fumes of Cl^- and NO_x.

CHG000   CAS: 113-18-8   **HR: 3**
1-CHLORO-3-ETHYL-1-PENTEN-4-YN-3-OL
mf: C_7H_9ClO mw: 144.61

SYNS: A 71 * AETHYL-CHLORVYNOL * ALVI-NOL * ARVYNOL * β-CHLOROVINYL ETHYLETHY-NYL CARBINOL * 3-(β-CHLOROVINYL)-1-PENTYN-3-OL * ETCHLORVINOLO * ETHCHLOROVYNOL * ETHCHLORVINYL * ETHCLORVYNOL * ETHO-CHLORVYNOL * ETHYL-β-CHLOROVINYLETHYNYL CARBINOL * ETHYLCHLORVYNOL * NORMONSON * NORMOSAN * NORMOSON * NOSTEL * PLACIDIL * PLACIDYL * ROERIDORM * SERENIL * SERENSIL

SAFETY PROFILE: Poison by ingestion, subcutaneous, intraperitoneal, and intravenous routes. Human systemic effects by ingestion: general anesthesia and thrombocytopenia (reduction in the number of blood platelets). Human effects on newborn by an unspecified route: drug dependency and apgar score (condition of newborn). Experimental teratogenic and reproductive effects. When heated to decomposition it emits toxic fumes of Cl^-.

CHI250   CAS: 110-75-8   **HR: 3**
2-CHLOROETHYL VINYL ETHER
mf: C_4H_7ClO mw: 106.56

PROP: Liquid. Bp: 109° @ 740 mm, d: 1.0525, flash p: 80°F (OC), mp: −70.3°.

SYNS: 2-CHLORETHYL VINYL ETHER * (2-CHLO-ROETHOXY)ETHENE * RCRA WASTE NUMBER U042 * VINYL-β-CHLOROETHYL ETHER * VINYL-2-CHLO-ROETHYL ETHER

CONSENSUS REPORTS: Reported in EPA TSCA Inventory.

SAFETY PROFILE: Poison by ingestion. Moderately toxic by inhalation and skin contact. A severe eye and skin irritant. Dangerous fire hazard when exposed to heat, flame, or oxidizers. Potentially explosive. May form dangerous peroxides on exposure to air. To fight fire, use alcohol foam, dry chemical. When heated to decomposition it emits toxic fumes of Cl^-.

CHI900   CAS: 593-70-4   **HR: 3**
CHLOROFLUOROMETHANE
mf: CH_2ClF mw: 68.48

SYNS: CFC 31 * CHLOROFLUOROMETHANE * FC 31 * FREON 31 * MONOCHLOROMONO-FLUOROMETHANE * R 31 * R 31 (refrigerant)

CONSENSUS REPORTS: IARC Cancer Review: GROUP 3 IMEMDT 7,56,87; Animal Limited Evidence IMEMDT 41,229,86.

DFG MAK: Animal Carcinogen, Suspected Human Carcinogen.

SAFETY PROFILE: Confirmed carcinogen with experimental carcinogenic data. Moderately toxic by inhalation. Mutation data reported. When heated to decomposition it emits very toxic fumes of Cl^- and F^-.

CHJ500   CAS: 67-66-3   **HR: 3**
CHLOROFORM

DOT: UN 1888
mf: $CHCl_3$ mw: 119.37

PROP: Colorless liquid; heavy, ethereal odor. Mp: −63.5°, bp: 61.26°, fp: −63.5°, flash p: none, d: 1.49845 @ 15°, vap press: 100 mm @ 10.4°, vap d: 4.12.

SYNS: CHLOROFORME (FRENCH) * CLOROFORMIO (ITALIAN) * FORMYL TRICHLORIDE * METHANE TRICHLORIDE * METHENYL TRICHLORIDE * METHYL TRICHLORIDE * NCI-C02686 * R 20 (REFRIGERANT) * RCRA WASTE NUMBER U044 * TCM * TRICHLOORMETHAAN (DUTCH) * TRICHLORMETHAN (CZECH) * TRICHLOROFORM * TRICHLOROMETHANE * TRICLOROMETANO (ITALIAN)

CONSENSUS REPORTS: IARC Cancer Review: GROUP 2B IMEMDT 7,152,87; Animal Limited Evidence IMEMDT 1,61,72; Human Limited Evidence IMEMDT 20,401,79; Animal Sufficient Evidence IMEMDT 20,401,79. NTP Fourth Annual Report On Carcinogens, 1984. NCI Carcinogenesis Bioassay (gavage); Clear Evidence: mouse, rat NCITR* NCI-CG-TR,1976. EPA Genetic Toxicology Program. EPA Extremely Hazardous Substances List. Community Right-To-Know List. Reported in EPA TSCA Inventory.

OSHA PEL: (Transitional: CL 50 ppm) TWA 2 ppm
ACGIH TLV: TWA 10 ppm; Suspected Human Carcinogen

DFG MAK: Suspected Carcinogen.

NIOSH REL: (Waste Anesthetic Gases and Vapors) CL 2 ppm/1H; (Chloroform) CL 2 ppm/60M

DOT Classification: ORM-A: Label: None; IMO: Poison B; Label: Poison.

SAFETY PROFILE: Confirmed carcinogen with experimental carcinogenic, neoplastigenic, and tumorigenic data. A human poison by ingestion and inhalation. An experimental poison by ingestion and intravenous routes. Moderately toxic experimentally by intraperitoneal and subcutaneous routes. Human systemic effects by inhalation: hallucinations and distorted perceptions, nausea, vomiting, and other unspecified gastrointestinal effects. Human mutation data reported. Experimental teratogenic and reproductive effects.

Inhalation of the concentrated vapor causes dilation of the pupils with reduced reaction to light, as well as reduced intraocular pressure (experimental). In the initial stages there is a feeling of warmth of the face and body, then an irritation of the mucous membranes, conjunctiva, and skin; followed by excitation, loss of reflexes, sensation, and consciousness. Prolonged inhalation will bring on paralysis accompanied by cardiac respiratory failure and finally death.

Chloroform has been widely used as an anesthetic. However, due to its toxic effects, this use is being abandoned. Concentrations of 68,000-82,000 ppm in air can kill most animals in a few minutes. 14,000 ppm may cause death after an exposure of from 30 to 60 minutes. 5,000-6,000 ppm can be tolerated by animals for 1 hour without serious disturbances. The maximum concentration tolerated for several hours or for prolonged exposure with slight symptoms is 2,000-2,500 ppm. Prolonged administration as an anesthetic may lead to such serious effects as profound toxemia and damage to the liver, heart, and kidneys. Experimental prolonged but light anesthesia in dogs produces a typical hepatitis.

Explosive reaction with sodium + methanol or sodium methoxide + methanol. Mixtures with sodium or potassium are impact-sensitive explosives. Reacts violently with acetone + alkali (e.g., sodium hydroxide, potassium hydroxide, or calcium hydroxide), Al, disilane, Li, Mg, methanol + alkali, nitrogen tetroxide, perchloric acid + phosphorus pentoxide, potassium-tert-butoxide, sodium methylate, NaK. Incompatible with dinitrogen tetraoxide, fluorine, metals, or triisopropylphosphine. Nonflammable. When heated to decomposition it emits toxic fumes of Cl^-.

CHJ750   CAS: 54-31-9   HR: 3

4-CHLORO-N-FURFURYL-5-SULFAMOYLANTHRANILIC ACID

mf: $C_{12}H_{11}ClN_2O_5S$ mw: 330.76

SYNS: AISEMIDE * ALUZINE * 5-(AMINOSULFO-NYL)-4-CHLORO-2-((2-FURNAYLMETHYL)AMINO)BEN-ZOIC ACID * BERONALD * CHLOR-N-(2-FURYL-METHYL)-5-SULFAMYLANTHRANILSAEURE (GERMAN) * 4-CHLORO-N-(2-FURYLMETHYL)-5-SULFAMOYLAN-THRANILIC ACID * DESDEMIN * DIURAL * DRYPTAL * ERROLON * EUTENSIN * FRUSEMIDE * FRUSEMIN * FRUSID * FULSIX * FULUVAMIDE * FURANTHRIL * FURANTHRYL * FURANTRIL * FURESIS * FUROSEDON * FUROSEMID * FUROSEMIDE * FUROSEMIDE "MITA" * FURSEMID * FURSE-MIDE * FUSID * HYDRO-RAPID * KATLEX * LASEX * LASIX * LB 502 * LOWPSTRON * MACASIROOL * NICOROL * NCI-C55936 * PREFEMIN * PROFEMIN * RADONNA * ROSEMIDE * SALIX * SEGURIL * TRANSIT * TROFURIT * UREX * UROSEMIDE

CONSENSUS REPORTS: EPA Genetic Toxicology Program.

SAFETY PROFILE: Poison by intravenous route. Moderately toxic by ingestion, intraperitoneal and possibly other routes. Human systemic effects by intravenous route: change in the sensitivity of the ear to sound, tinnitus, unspecified effects on the heart, constriction of the arteries, and a decrease in urine volume. Ingestion can damage the liver. Experimental teratogenic and reproductive effects. Human mutation data reported. When heated to decomposition it emits very toxic fumes of Cl^-, NO_x, and SO_x.

CHM000   CAS: 615-67-8   HR: 3

CHLOROHYDROQUINONE

mf: $C_6H_5ClO_2$ mw: 144.56

CONSENSUS REPORTS: Reported in EPA TSCA Inventory.

SAFETY PROFILE: Poison by ingestion and intraperitoneal routes. Moderately toxic by skin contact. When heated to decomposition it emits toxic fumes of Cl^-.

CHP250   CAS: 303-47-9   **HR: 3**
(−)-N-((5-CHLORO-8-HYDROXY-3-METHYL-1-OXO-7-ISOCHROMANYL)CARBONYL)-3-PHENYLALANINE
mf: $C_{29}H_{18}ClNO_6$ mw: 403.84

PROP: Crystals. Mp: 169°.

SYNS: (R)N-((5-CHLORO-3,4-DIHYDRO-8-HYDROXY-3-METHYL-1-OXO-1H-2-BENZOPYRAN-7-YL)PHENYLALANINE * NCI-C56586 * OCHRATOXIN A

CONSENSUS REPORTS: IARC Cancer Review: GROUP 3 IMEMDT 7,271,87; Animal Limited Evidence IMEMDT 31,191,83; Animal Inadequate Evidence IMEMDT 10,191,76; Human Inadequate Evidence IMEMDT 31,191,83.

SAFETY PROFILE: Poison by ingestion, intraperitoneal, intravenous, and subcutaneous routes. Experimental teratogenic and reproductive effects. Questionable carcinogen with experimental carcinogenic, neoplastigenic, tumorigenic data. Mutation data reported. When heated to decomposition it emits very toxic fumes of Cl^- and NO_x.

CHP500   CAS: 5160-02-1   **HR: 3**
5-CHLORO-2-((2-HYDROXY-1-NAPHTHYL)AZO)-p-TOLUENE SULFONIC ACID, BARIUM SALT
mf: $C_{17}H_{12}ClN_2O_4S \cdot 1/2Ba$ mw: 444.49

SYNS: BRIGHT RED * BRILLIANT RED * BRILLIANT SCARLET * BRILLIANT TONER Z * BRONZE RED RO * BRONZE SCARLET * 5-CHLORO-2-((2-HYDROXY-1-NAPHTHALENYL)AZO)-4-METHYLBENZENE SULFONIC ACID, BARIUM SALT (2:1) * 5-CHLORO-2-((2-HYDROXY-1-NAPHTHALENYL)AZO)-4-METHYLBENZENE SULPHONIC ACID, BARIUM SALT * 1-(4-CHLORO-o-SULFO-5-TOLYLAZO)-2-NAPHTHOL,BARIUM SALT * C.I. PIGMENT RED * COSMETIC CORAL RED KO BLUISH * DAINICHI LAKE RED C * D&C RED No. 9 * DESERT RED * ELJON LAKE RED C * HAMILTON RED * HELIO RED TONER LCLL * IRGALITE RED CBN * ISOL LAKE RED LCS 12527 * LAKE RED C * LATEXOL SCARLET R * LD RUBBER RED 16913 * LUTETIA RED CLN * MICROTEX LAKE RED CR * MOHICAN RED A-8008 * NCI-C53792 * No. 3 CONC. SCARLET * PARIDINE RED LCL * PIGMENT RED CD * POTOMAC RED * RECOLITE RED LAKE C * 1860 RED * RED SCARLET * SANYO LAKE RED C * SEGNALE RED LC * SICO LAKE RED 2L * SUPEROL RED C RT-265 * SYMULER LAKE RED C * TERMOSOLIDO RED LCG * TEXAN RED TONER D * TONER LAKE RED C * TRANSPARENT BRONZE SCARLET * VULCAFIX SCARLET R * VULCAN RED LC * VULCOL FAST RED L * WAYNE RED X-2486

CONSENSUS REPORTS: IARC Cancer Review: GROUP 3 IMEMDT 7,56,87; Animal Inadequate Evidence IMEMDT 8,107,75; NTP Carcinogenesis Bioassay (feed); Clear Evidence: rat NTPTR* NTP-TR-225,82; No Evidence: mouse NTPTR* NTP-TR-225,82. Reported in EPA TSCA Inventory.

SAFETY PROFILE: Questionable carcinogen with experimental carcinogenic and tumorigenic data. Mutation data reported. When heated to decomposition it emits very toxic fumes of SO_x, NO_x, and Cl^-.

CHR500   CAS: 130-26-7   **HR: 3**
5-CHLORO-7-IODO-8-QUINOLINOL
mf: C_9H_5ClINO mw: 305.50

SYNS: ALCHLOQUIN * AMEBIL * AMOENOL * BACTOL * BARQUINOL * BUDOFORM * CHINOFORM * 5-CHLOR-7-JOD-8-8HYDROXY-CHINOLIN (GERMAN) * 5-CHLORO-8-HYDROXY-7-IODOQUINOLINE * 5-CHLORO-7-IODO-8-HYDROXY QUINOLINE * CHLOROIODOQUINE * CLIOQUINOL * CLIQUINOL * ECZECIDIN * EMAFORM * ENTERO-BIO FORM * ENTEROQUINOL * ENTEROSEPTOL * ENTERO-VIOFORM * ENTEROZOL * ENTERUM LOCORTEN * ENTROKIN * HI-ENTEROL * HYDRIODIDE-ENTROL * IODENTEROL * IODOCHLORHYDROXYQUINOL * IODOCHLORHYDROXYQUINOLINE * 7-IODO-5-CHLORO-8-HYDROXYQUINOLINE * 7-IODO-5-CHLOROXINE * IODOENTEROL * NIOFORM * QUINAMBICIDE * ROMETIN * VIOFORM * VIOFORM N.N.R.

CONSENSUS REPORTS: Reported in EPA TSCA Inventory. EPA Genetic Toxicology Program.

SAFETY PROFILE: Poison by ingestion. Moderately toxic by intraperitoneal route. Human systemic effects by ingestion: change in central nervous system electrical function, optic nerve damage, and changes in vision. Experimental teratogenic and reproductive effects. Human mutation data reported. When heated to decomposition it emits very toxic fumes of Cl^-, I^-, and NO_x.

CHS500   CAS: 1918-16-7   **HR: 3**
2-CHLORO-N-ISOPROPYLACETANILIDE
mf: $C_{11}H_{14}ClNO$ mw: 211.71

SYNS: BEXTON * CHLORESSIGSAEURE-N-ISOPRO-
PYLANILID (GERMAN) * α-CHLORO-N-ISOPROPYL-
ACETANILIDE * 2-CHLORO-N-ISOPROPYL-N-PHENYL-
ACETAMIDE * 2-CHLORO-N-(1-METHYLETHYL)-N-
PHENYLACETAMIDE * CP 31393 * N-ISO-
PROPYL-α-CHLOROACETANILIDE * N-ISOPROPYL-
2-CHLOROACETANILIDE * NITICID * PROPA-
CHLOR * PROPACHLORE * RAMROD
* SATECID

CONSENSUS REPORTS: EPA Genetic Toxicology Program.

SAFETY PROFILE: Poison by ingestion, skin
contact, and possibly other routes. Mutation data
reported. A selective herbicide. When heated
to decomposition it emits very toxic fumes of
Cl^- and NO_x.

CHU500   CAS: 59-85-8   HR: 3
p-CHLOROMERCURIC BENZOIC ACID
mf: $C_7H_5ClHgO_2$ mw: 357.16

SYNS: (p-CARBOXYPHENYL)CHLOROMERCURY
* p-(CHLOROMERCURI)BENZOIC ACID * USAF D-3

CONSENSUS REPORTS: Reported in EPA
TSCA Inventory. Mercury and its compounds
are on the Community Right-To-Know List.

OSHA PEL: (Transitional: CL 1 mg/10m³) CL
  0.1 mg(Hg)/m³ (skin)
ACGIH TLV: TWA 0.1 mg(Hg)/m³ (skin)
NIOSH REL: (Inorganic Mercury) TWA: 0.05
  mg(Hg)/m³

SAFETY PROFILE: Poison by intraperitoneal
route. When heated to decomposition it emits
very toxic fumes of Cl^- and Hg.

CHX750   HR: 3
CHLOROMETHANE mixed with
DICHLOROMETHANE

SYN: METHYL CHLORIDE-METHYLENE CHLORIDE
MIXTURE (DOT)

DOT Classification: Flammable Gas; Label:
  Flammable Gas.

SAFETY PROFILE: Flammable when exposed
to heat or flame. When heated to decomposition
it emits toxic fumes of Cl^-.

CHY000   CAS: 3518-65-8   HR: 3
CHLOROMETHANE SULFONYL
CHLORIDE
mf: $CH_2Cl_2O_2S$ mw: 148.99

SYNS: CHLORID KYSELINY CHLORMETHANSULFO-
NOVE (CZECH) * CHLORMETHANSULFOCHLORID
(CZECH)

SAFETY PROFILE: Poison by ingestion. A severe skin and eye irritant. When heated to decomposition it emits very toxic fumes of Cl^-
and SO_x.

CHY250   CAS: 148-65-2   HR: 3
CHLOROMETHAPYRILENE
mf: $C_{14}H_{18}ClN_3S$ mw: 295.86

SYNS: CHLOROPYRILENE * CHLOROTHEN
* 2-((5-CHLORO-2-THENYL)(2-DIMETHYLAMINOETHYL)
AMINO)PYRIDINE * CHLOROTHENYLPYRAMINE
* N,N-DIMETHYL-N'-(2-PYRIDYL)-N'-(5-CHLORO-2-
THENYL)ETHYLENEDIAMINE * ETHYLENEDIAMINE,
N-(5-CHLORO-2-THENYL)-N',N'-DIMETHYL-N-2-PYRIDYL-
* NCI-C60559 * PYRITHEN * TAGATHEN
* 2-THENYLAMINE, 5-CHLORO-N-(2-(DIMETHYL-
AMINO)ETHYL)-N-2-PYRIDYL-

SAFETY PROFILE: Poison by intraperitoneal
route. When heated to decomposition it emits
very toxic fumes of Cl^-, NO_x, and SO_x.

CIF250   CAS: 1199-85-5   HR: 3
p-CHLORO-N-METHYLAMPHETAMINE
mf: $C_{10}H_{14}ClN$ mw: 183.70

SYNS: p-CHLORO-N-α-DIMETHYLPHENETHYLAMINE
* d-1-p-CHLORO-METHYLAMPHETAMINE (FRENCH)
* CMA * pCMA * RO 4-6861 * S-33

SAFETY PROFILE: Poison by ingestion, intravenous, intraperitoneal and subcutaneous
routes. When heated to decomposition it emits
very toxic fumes of Cl^- and NO_x.

CIG250   CAS: 6325-54-8   HR: 3
7-CHLOROMETHYL BENZ(a)
ANTHRACENE
mf: $C_{19}H_{13}Cl$ mw: 276.77

SYN: ICR 451

SAFETY PROFILE: Poison by intravenous
route. Questionable carcinogen with experimental neoplastigenic data. Mutation data reported.
When heated to decomposition it emits toxic
Cl^-.

CIH000   CAS: 49852-84-8   HR: 3
6-CHLOROMETHYL BENZO(a)PYRENE
mf: $C_{21}H_{13}Cl$ mw: 300.79

SAFETY PROFILE: Questionable carcinogen with experimental carcinogenic and neoplastigenic data. Mutation data reported. When heated to decomposition it emits very toxic fumes of Cl^-.

CIK500 CAS: 77944-89-9 HR: 3
2-CHLORO-6-METHYLCARBANILIC ACID-2-(PYRROLIDINYL)ETHYL ESTER HYDROCHLORIDE
mf: $C_{14}H_{19}ClN_2O_2 \cdot ClH$ mw: 319.26

SYNS: C 3067 * 2-(PYRROLIDINYL)ETHYL-2-CHLORO-6-METHYLCARBANILATE HYDROCHLORIDE

SAFETY PROFILE: Poison by intraperitoneal, subcutaneous, and intravenous routes. An eye irritant. When heated to decomposition it emits very toxic fumes of NO_x and Cl^-.

CIK750 CAS: 321-54-0 HR: 3
3-CHLORO-4-METHYL-7-COUMARINYL DIETHYLPHOSPHATE
mf: $C_{14}H_{16}ClO_6P$ mw: 346.72

SYNS: COROXON * COUMAPHOS-O-ANALOG * COUMAPHOS OXYGEN ANALOG (USDA) * O,O-DI(2-CHLOROETHYL)-7-(3-CHLORO-4-METHYLCOUMARINYL)PHOSPHATE * O,O-DIETHYL-O-(3-CHLORO-4-METHYLCOUMARIN-7-YL)PHOSPHATE * DIETHYL-3-CHLORO-4-METHYL-7-COUMARINYL PHOSPHATE * PHOSPHORIC ACID, DIETHYL ESTER, with 3-CHLORO-7-HYDROXY-4-METHYLCOUMARIN

SAFETY PROFILE: Deadly poison by ingestion. When heated to decomposition it emits very toxic fumes of PO_x and Cl^-.

CIM000 CAS: 3188-13-4 HR: 3
CHLOROMETHYL ETHYL ETHER
DOT: UN 2354
mf: C_3H_7ClO mw: 94.54

PROP: Flash p: $<-2.2°F$.

SYNS: CHLOROMETHOXY ETHANE * ETHOXY CHLOROMETHANE * ETHOXY METHYL CHLORIDE

CONSENSUS REPORTS: Reported in EPA TSCA Inventory.

DOT Classification: Flammable Liquid and Poison; Label: Flammable Liquid and Poison.

SAFETY PROFILE: A poison by inhalation and ingestion. A very dangerous fire and explosion hazard when exposed to heat or flame.

CIN750 CAS: 13345-62-5 HR: 3
7-CHLOROMETHYL-12-METHYL BENZ(a)ANTHRACENE
mf: $C_{20}H_{15}Cl$ mw: 290.80

SYN: IRC 453

CONSENSUS REPORTS: EPA Genetic Toxicology Program.

SAFETY PROFILE: Poison by intravenous route. Questionable carcinogen with experimental neoplastigenic data. Mutation data reported. When heated to decomposition it emits toxic Cl^-.

CIO250 CAS: 107-30-2 HR: 3
CHLOROMETHYL METHYL ETHER
DOT: UN 1239
mf: C_2H_5ClO mw: 80.52

PROP: Flash p: $<73.4°F$.

SYNS: CHLORDIMETHYLETHER (CZECH) * CMME * DIMETHYLCHLOROETHER * ETHER METHYLIQUE MONOCHLORE (FRENCH) * METHYLCHLOROMETHYL ETHER * METHYLCHLOROMETHYL ETHER (DOT) * METHYL CHLOROMETHYL ETHER, anhydrous (DOT) * MONOCHLORODIMETHYL ETHER (MAK) * RCRA WASTE NUMBER U046

CONSENSUS REPORTS: IARC Cancer Review: GROUP 1 IMEMDT 7,131,87; Animal Sufficient Evidence IMEMDT 4,239,74; Human Limited Evidence IMEMDT 4,239,74. NTP Fourth Annual Report On Carcinogens, 1984. EPA Genetic Toxicology Program. Reported in EPA TSCA Inventory. Community Right-To-Know List. EPA Extremely Hazardous Substances List.

OSHA: Carcinogen
ACGIH TLV: Suspected Human Carcinogen.
DFG MAK: Human Carcinogen.
DOT Classification: Flammable Liquid; Label: Flammable Liquid, Poison, anhydrous; IMO: Flammable Liquid; Label: Flammable Liquid.

SAFETY PROFILE: Confirmed human carcinogen with experimental carcinogenic, tumorigenic, and neoplastigenic data. Poison by inhalation. Moderately toxic by ingestion. Human mutation data reported. A very dangerous fire hazard when exposed to heat or flame. To fight fire, use alcohol foam, water, CO_2, or dry chemical. Reaction with divalent metals forms a very reactive product. When heated to decomposition it emits toxic fumes of Cl^-.

CIP500   CAS: 3688-85-5   HR: 2
4-CHLORO-N-METHYL-3-(METHYLSUL-FAMOYL)BENZAMIDE
mf: $C_9H_{11}ClN_2O_3S$ mw: 262.73

SYNS: 4-CHLORO-N-METHYL-3-((METHYLAMINO)SUL-FONYL)BENZAMIDE * C.I. 456 * CN-36337 * D 1593 * DIAPAMIDE * THIAMIZIDE * TIAMIZID * TIAMIZIDE * VECTREN

SAFETY PROFILE: Moderately toxic by ingestion and intraperitoneal routes. When heated to decomposition it emits very toxic SO_x, NO_x, and Cl^-.

CIQ500   CAS: 16339-16-5   HR: 3
2-CHLORO-N-METHYL-N-NITROSOETHYLAMINE
mf: $C_3H_7ClN_2O$ mw: 122.57

SYNS: 2-CHLORO-2-METHYL-N-NITROSOETHANAMINE * METHYL-2-CHLORAETHYLNITROSAMIN (GERMAN) * METHYL(2-CHLOROETHYL)NITROSAMINE * N-NITROSOMETHYL-2-CHLOROETHYLAMINE

SAFETY PROFILE: Poison by ingestion, intravenous, and possibly other routes. Questionable carcinogen with experimental tumorigenic data. Many nitrosamine compounds are carcinogens. When heated to decomposition it emits very toxic fumes of Cl^- and NO_x.

CIR250   CAS: 94-74-6   HR: 3
(4-CHLORO-2-METHYLPHENOXY) ACETIC ACID
mf: $C_9H_9ClO_3$ mw: 200.63

SYNS: AGRITOX * AGROXONE * ANICON KOMBI * ANICON M * BH MCPA * BORDERMASTER * BROMINAL M & PLUS * B-SELEKTONON M * CHIPTOX * 4-CHLORO-o-CRESOXYACETIC ACID * 4-CHLORO-o-TOLOXYACETIC ACID * ((4-CHLORO-o-TOLYL)OXY)ACETIC ACID * CHWASTOX * CORNOX-M * DED-WEED * DICOPUR-M * DICOTEX * DOW MCP AMINE WEED KILLER * EMCEPAN * EMPAL * HEDAPUR M 52 * HERBICIDE M * HORMOTUHO * 4K-2M * KILSEM * KREZONE * LEGUMEX DB * LEUNA M * LEYSPRAY * LINORMONE * M 40 * 2M-4C * MCP * MCPA * MEPHANAC * METAXON * METHOXONE * 2-METHYL-4-CHLOROPHENOXYACETIC ACID * 2-METHYL-4-CHLORPHENOXYESSIGSAEURE (GERMAN) * 2M-4KH * NETAZOL * OKULTIN M * PHENOXYLENE SUPER * RAPHONE * RAZOL DOCK KILLER * RHOMENE * RHONOX * SEPPIC MMD

* SHAMROX * SOVIET TECHNICAL HERBICIDE 2M-4C * TRASAN * U 46 M-FLUID * USTINEX * VACATE * VERDONE * VESAKONTUHO MCPA * WEEDAR MCPA CONCENTRATE * WEEDONE MCPA ESTER * WEED-RHAP * ZELAN

CONSENSUS REPORTS: IARC Cancer Review: GROUP 2B IMEMDT 7,156,87; Animal Inadequate Evidence IMEMDT 30,255,83; Human Inadequate Evidence IMEMDT 30,255,83; Human Limited Evidence IMEMDT 41,357,86. Reported in EPA TSCA Inventory. EPA Genetic Toxicology Program.

SAFETY PROFILE: Suspected carcinogen. Poison by subcutaneous and intravenous routes. Moderately toxic by ingestion. Human systemic effects by ingestion: blood pressure decrease and coma. Experimental teratogenic and reproductive effects. Mutation data reported. An herbicide. When heated to decomposition it emits toxic fumes of Cl^-.

CIR500   CAS: 93-65-2   HR: 2
4-CHLORO-2-METHYLPHENOXY-α-PROPIONIC ACID
mf: $C_{10}H_{11}ClO_3$ mw: 214.66

SYNS: ACIDE 2-(4-CHLORO-2-METHYL-PHENOXY)PRO-PIONIQUE (FRENCH) * ACIDO 2-(4-CLORO-2-METIL-FENOSSI)-PROPIONICO (ITALIAN) * BH MECOPROP * CHIPCO TURF HERBICIDE MCPP * 2-(4-CHLOOR-2-METHYL-FENOXY)-PROPIONZUUR (DUTCH) * 2-(4-CHLOR-2-METHYL-PHENOXY)-PROPIONSAEURE (GERMAN) * 2-(4-CHLORO-2-METHYLPHENOXY)PROPIONIC ACID * (+)-α-(4-CHLORO-2-METHYLPHENOXY) PROPIONIC ACID * 2-(4-CHLOROPHENOXY-2-METHYL) PROPIONIC ACID * 2-(p-CHLORO-o-TOLYLOXY)PRO-PIONIC ACID * CMPP * COMPITOX * FBC CMPP * HEDONAL MCPP * ISO-CORNOX * KILPROP * LIRANOX * 2M-4CP * MCPP * 2-MCPP * MCPP 2,4-D * MCPP-D-4 * MCPP-K-4 * MECOMEC * MECOPEOP * MECOPER * MECOPEX * MECOPROP * MECOTURF * MECPROP * MEPRO * METHOXONE * α-(2-METHYL-4-CHLOROPHENOXY)PROPIONIC ACID * 2-(2-METHYL-4-CHLOROPHENOXY)PROPIONIC ACID * 2-METHYL-4-CHLOROPHENOXY-α-PROPIONIC ACID * 2-(2-METHYL-4-CHLORPHENOXY)-PROPIONSAEURE (GERMAN) * 2M 4KHP * N.B. MECOPROP * PROPAL * PROPONEX-PLUS * RANKOTEX * RUNCATEX * RD 4593 * U 46 * U 46 KV-ESTER * U 46 KV-FLUID * VI-PAR * VI-PEX

CONSENSUS REPORTS: IARC Cancer Review: GROUP 2B IMEMDT 7,156,87; Human

Limited Evidence IMEMDT 41,357,86. EPA Genetic Toxicology Program. Reported in EPA TSCA Inventory.

SAFETY PROFILE: Suspected carcinogen. Moderately toxic by ingestion, skin contact, intraperitoneal, and possibly other routes. Experimental teratogenic and reproductive effects. Mutation data reported. An herbicide. When heated to decomposition it emits toxic fumes of Cl^-.

CIR750   CAS: 22316-47-8   **HR: 3**
7-CHLORO-1-METHYL-5-PHENYL-1H-1,5-BENZODIAZEPINE-2,4(3H,5H)-DIONE
mf: $C_{16}H_{13}ClN_2O_2$ mw: 300.76

SYNS: CHLOREPIN * CLOBAZAM * CLOREPIN * FRISIUM * H-4723 * HR 376 * LM-2717 * 1-PHENYL-5-METHYL-8-CHLORO-1,2,4,5-TETRAHY-DRO-2,4-DIOXO-3H-1,5-BENZODIAZEPINE * RU-4723 * URBANYL

SAFETY PROFILE: Poison by ingestion and intraperitoneal routes. Moderately toxic by subcutaneous route. Human systemic effects by ingestion: wakefulness, withdrawal, nausea and vomiting. Experimental teratogenic and reproductive effects. A tranquilizer. When heated to decomposition it emits very toxic fumes of NO_x and Cl^-.

CIS750   CAS: 2058-52-8   **HR: 3**
2-CHLORO-11-(4-METHYLPIPER-AZINO)DIBENZO(b,f)(1,4)THIAZEPINE
mf: $C_{18}H_{18}ClN_3S$ mw: 343.90

SYNS: 2-CHLORO-11-(4-METHYL-1-PIPERAZINYL)DI-BENZO(b,f)(1,4)THIAZEPINE * DIBENZOTHIAZEPINE

SAFETY PROFILE: Poison by ingestion. Experimental reproductive effects. When heated to decomposition it emits very toxic fumes of Cl^-, NO_x, and SO_x.

CIU750   CAS: 563-47-3   **HR: 3**
3-CHLORO-2-METHYLPROPENE

DOT: UN 2554
mf: C_4H_7Cl mw: 90.56

PROP: Colorless, volatile liquid, disagreeable odor. Bp: 72.17°, lel: 2.3%, uel: 9.3%, fp: $<-80°$, d: 0.9257 @ 20°/4°, vap press: 101.7 mm @ 20°, vap d: 3.12, flash p: $-10°$. Misc in alc, ether.

SYNS: 3-CHLOR-2-METHYL-PROP-1-EN (GERMAN) * γ-CHLOROISOBUTYLENE * 3-CHLORO-2-METHYL-1-PROPENE * CHLORURE de METHALLYLE (FRENCH) * 3-CLORO-2-METIL-PROP-1-ENE (ITALIAN) * CLORURO di METALLILE (ITALIAN) * ISOBUTENYL CHLORIDE * METHALLYL CHLORIDE * α-METH-ALLYL CHLORIDE * 2-METHYL-ALLYLCHLORID (GERMAN) * β-METHYLALLYL CHLORIDE * 2-METHYLALLYL CHLORIDE * METHYL ALLYL CHLORIDE (DOT) * NCI-C54820

CONSENSUS REPORTS: NTP Carcinogenesis Studies (gavage); Clear Evidence: mouse, rat NTPTR* NTP-TR-300,86. Reported in EPA TSCA Inventory.

DOT Classification: IMO: Flammable Liquid; Label: Flammable Liquid.

SAFETY PROFILE: Suspected carcinogen with experimental carcinogenic, neoplastigenic, and tumorigenic data. Mildly toxic to humans by inhalation. An irritant. Human mutation data reported. Very dangerous fire hazard when exposed to heat, flame, or oxidizers. Moderately explosive when exposed to heat or flame. Can react vigorously with oxidizing materials. To fight fire, use alcohol foam, CO_2, dry chemical. When heated to decomposition it emits toxic fumes of Cl^-.

CIV000   CAS: 6959-48-4   **HR: 3**
3-(CHLOROMETHYL) PYRIDINE HYDROCHLORIDE
mf: $C_6H_6ClN \cdot ClH$ mw: 164.04

SYN: NCI-C03838

CONSENSUS REPORTS: NCI Carcinogenesis Bioassay (gavage); Clear Evidence: mouse, rat NCITR* NCI-CG-TR-95,78. EPA Genetic Toxicology Program.

SAFETY PROFILE: Suspected carcinogen with experimental carcinogenic data. Poison by ingestion. Mutation data reported. When heated to decomposition it emits very toxic fumes of NO_x and Cl^-.

CIW250   **HR: 3**
5'-CHLORO-2-(METHYL(2-(PYRROL-IDINYL)ETHYL)AMINO)-o-ACETOTO-LUIDIDE DIHYDROCHLORIDE

SYNS: C 5420 * 5'-CHLORO-2-(METHYL(2-(PYRROLI-DINYL)ETHYL)AMINO)-O-ACETOTOLUIDIDE DIHYDRO-CHLORIDE

SAFETY PROFILE: Poison by intraperitoneal and subcutaneous routes. An eye irritant. When heated to decomposition it emits very toxic fumes of Cl$^-$ and NO$_x$.

CIY250   CAS: 617-88-9   **HR: 3**
2-CHLOROMETHYLTHIOPHENE
mf: C_5H_5ClS mw: 132.61

SAFETY PROFILE: Flammable when exposed to heat or flame. A storage hazard. It decomposes at room temperature to release hydrogen chloride, and may explode in a sealed container. Highly explosive when shocked, exposed to heat, or by spontaneous chemical reaction. Can react vigorously with oxidizing materials.

CJB250   CAS: 121-73-3   **HR: 3**
1-CHLORO-3-NITROBENZENE

DOT: UN 1578
mf: $C_6H_4ClNO_2$ mw: 157.56

PROP: Yellowish crystals. Mp: 46°, flash p: 103°, bp: 236°, d: 1.534 @ 20°/4°.

SYNS: CHLORO-m-NITROBENZENE * m-CHLORONI-TROBENZENE * m-CHLORONITROBENZENE (DOT) * m-NITROCHLOROBENZENE * m-NITROCHLORO-BENZENE, solid (DOT)

CONSENSUS REPORTS: Reported in EPA TSCA Inventory.

DOT Classification: Poison B; Label: Poison.

SAFETY PROFILE: Poison by ingestion and inhalation. It forms methemoglobin in the body and gives rise to cyanosis and blood changes. Its effects are cumulative and analogous to those of nitrobenzene. The para compound is thought to be somewhat less toxic than the ortho compound. Chemically, it is probably converted in the body to chloroaniline, which is also poisonous. In industry, it is the dust of this material that is most often the source of intoxication. Dangerous fire hazard when exposed to heat or flame. It can react with oxidizing materials. When heated to decomposition it emits toxic fumes of Cl$^-$, NO$_x$, and phosgene.

CJB750   CAS: 88-73-3   **HR: 3**
CHLORO-o-NITROBENZENE

DOT: UN 1578
mf: $C_6H_4ClNO_2$ mw: 157.56

PROP: Yellow crystals. Mp: 32-33°, bp: 245-246°, d: 1.348, flash p: 123°.

SYNS: o-CHLORONITROBENZENE * o-CHLORONI-TROBENZENE (DOT) * 1-CHLORO-2-NITROBENZENE * 2-CHLORONITROBENZENE * 2-CHLORO-1-NITRO-BENZENE * o-NITROCHLOROBENZENE * o-NITRO-CHLOROBENZENE, liquid (DOT) * ONCB

CONSENSUS REPORTS: Reported in EPA TSCA Inventory.

DOT Classification: Poison B; Label: Poison.

SAFETY PROFILE: Poison by ingestion and probably inhalation. Questionable carcinogen with experimental carcinogenic and neoplastigenic data. Combustible when exposed to heat or flame. To fight fire, use water, foam. Potentially explosive reaction with ammonia at 160°C/30 bar. When heated to decomposition it emits toxic fumes of Cl$^-$, NO$_x$, and phosgene.

CJD750   CAS: 2425-66-3   **HR: 3**
CHLORONITROPROPANE
mf: $C_3H_6ClNO_2$ mw: 123.55

SYNS: CHLORONITROPROPAN (POLISH) * 1-CHLORO-2-NITROPROPANE

SAFETY PROFILE: Poison by ingestion and skin contact. Moderately toxic by inhalation. When heated to decomposition it emits very toxic fumes of Cl$^-$ and NO$_x$.

CJE000   CAS: 600-25-9   **HR: 3**
1-CHLORO-1-NITROPROPANE
mf: $C_3H_6ClNO_2$ mw: 123.55

PROP: Liquid. Bp: 139.5°, flash p: 144°F (OC), d: 1.209 @ 20°/20°, vap d: 4.26.

SYNS: CHLORONITROPROPANE * KORAX * LANSTAN

OSHA PEL: (Transitional: TWA 20 ppm) TWA 2 ppm
ACGIH TLV: TWA 2 ppm
DFG MAK: 20 ppm (100 mg/m^3)

SAFETY PROFILE: Poison by ingestion, subcutaneous, and possibly other routes. Moderately toxic by inhalation. Causes injury to kidneys, liver, and cardiovascular system. Combustible when exposed to heat, flame (sparks), and oxidizers. Moderately explosive when exposed to heat. To fight fire, use alcohol foam, water, CO$_2$, or dry chemical. When heated to decomposition it emits toxic fumes of Cl$^-$ and NO$_x$.

CJE250   CAS: 594-71-8   **HR: 3**
2-CHLORO-2-NITROPROPANE
mf: $C_3H_6ClNO_2$ mw: 123.55

PROP: Liquid. Bp: 132°, flash p: 135°F (OC), d: 1.193 @ 20°/20°, vap d: 4.26.

SAFETY PROFILE: Poison by subcutaneous route. Moderately toxic by ingestion. Mildly toxic by inhalation. Flammable when exposed to heat, flame, or oxidizers. Explodes on rapid heating. When heated to decomposition it emits toxic fumes of Cl^- and NO_x.

CJI500   CAS: 76-15-3   **HR: 1**
CHLOROPENTAFLUOROETHANE

DOT: UN 1020
mf: C_2ClF_5 mw: 154.47

PROP: Colorless gas. Bp: −39.3°, mp: −38°, d: 1.5678 @ -42°. Insol in water; sol in alc and ether.

SYNS: F-115 * FLUOROCARBON-115 * FREON 115 * GENETRON 115 * HALOCARBON 115 * MONOCHLOROPENTAFLUOROETHANE (DOT)

CONSENSUS REPORTS: Reported in EPA TSCA Inventory.

OSHA PEL: TWA 1000 ppm
ACGIH TLV: TWA 1000 ppm
DOT Classification: Nonflammable Gas; Label: Nonflammable Gas.

SAFETY PROFILE: Mildly toxic by inhalation. A nonflammable gas. When heated to decomposition it emits toxic fumes of F^- and Cl^-.

CJJ000   CAS: 3691-35-8   **HR: 3**
CHLOROPHACINONE
mf: $C_{23}H_{15}ClO_3$ mw: 374.83

SYNS: AFNOR * CAID * CHLOORFACINON (DUTCH) * 2(2-(4-CHLOOR-FENYL-2-FENYL)-ACETYL)-INDAAN-1,3-DION (DUTCH) * CHLORFACINON (GERMAN) * 2-(α-p-CHLOROPHENYLACETYL)INDANE-1,3-DIONE * 2-((p-CHLOROPHENYL)PHENYLACETYL)-1,3-INDANDIONE * 2(2-(4-CHLOROPHENYL)-2-PHENYL-ACETYL)INDAN-1,3-DIONE * 2-((4-CHLOROPHENYL)PHENYLACETYL)-1H-INDENE-1,3(2H)-DIONE * CHLORPHACINON (ITALIAN) * 2(2-(4-CHLOR-PHENYL-2-PHE-NYL)ACETYL)INDAN-1,3-DION (GERMAN) * ((4-CHLORPHENYL)-1-PHENYL)-ACETYL-1,3-INDANDION (GERMAN) * 1-(4-CHLORPHENYL)-1-PHENYL-ACETYL-INDAN-1,3-DION (GERMAN) * 2(2-(4-CLORO-FENIL-2-FENIL)-ACETIL)INDAN-1,3-DIONE (ITALIAN) * DELTA * DRAT * LIPHADIONE * LM 91 * MICROZUL

* MURIOL * 2-(2-PHENYL-2-(4-CHLOROPHENYL) ACETYL)-1,3-INDANDIONE * QUICK * RAMUCIDE * RANAC * RATOMET * RAVIAC * ROZOL * TOPITOX

CONSENSUS REPORTS: EPA Extremely Hazardous Substances List.

SAFETY PROFILE: Poison by ingestion and skin contact. A pesticide. When heated to decomposition it emits toxic fumes of Cl^-.

CJJ250   CAS: 6164-98-3   **HR: 3**
CHLOROPHENAMIDINE
mf: $C_{10}H_{13}ClN_2$ mw: 196.70

SYNS: ACARON * BERMAT * C 8514 * CARZOL * CDM * CHLORDIMEFORM * CHLORFENAMIDINE * N'-(4-CHLORO-2-METHYL-PHENYL)-N,N-DIMETHYLMETHANIMIDAMIDE * CHLOROPHENAMADIN * N'-(4-CHLORO-o-TOLYL)-N,N-DIMETHYLFORMAMIDINE * CHLORPHENAMI-DINE * N'-(4-CHLOR-o-TOLYL)-N,N-DIMETHYLFOR-MAMIDIN (GERMAN) * CIBA 8514 * N,N-DI-METHYL-N'-(2-METHYL-4-CHLOROPHENYL)-FORMA-MIDINE * N,N-DIMETHYL-N'-(2-METHYL-4-CHLOR-PHENYL)-FORMADIN (GERMAN) * ENT 27,335 * ENT 27,567 * EP-333 * FUNDAL * FUNDAL 500 * FUNDEX * GALECRON * N'-(2-METHYL-4-CHLOROPHENYL)-N,N-DIMETHYLFORMAMIDINE * N'-(2-METHYL-4-CHLORPHENYL)-FORMAMIDIN-HYDROCHLORID (GERMAN) * NSC 190935 * RS 141 * SCHERING 36268 * SN 36268 * SPANON * SPANONE

CONSENSUS REPORTS: IARC Cancer Review: GROUP 3 IMEMDT 7,56,87. EPA Genetic Toxicology Program.

SAFETY PROFILE: Poison by ingestion, skin contact, and intraperitoneal routes. Experimental reproductive effects. Human mutation data reported. An eye and skin irritant. Questionable carcinogen with experimental carcinogenic data. When heated to decomposition it emits very toxic fumes of NO_x and Cl^-.

CJK250   CAS: 95-57-8   **HR: 3**
2-CHLOROPHENOL

DOT: UN 2020/UN 2021
mf: C_6H_5ClO mw: 128.56

PROP: Light amber liquid. Bp: 174.5°, fp: 7°, d: 1.256 @ 25°/25°, flash p: 147°F, vap press: 1 mm @ 12.1°. Sltly water sol; very sol in alc, ether, and alkali.

SYNS: o-CHLOROPHENOL * o-CHLOROPHENOL, liquid (DOT) * o-CHLOROPHENOL, solid (DOT) * o-CHLORPHENOL (GERMAN) * RCRA WASTE NUMBER U048

CONSENSUS REPORTS: Reported in EPA TSCA Inventory. Chlorophenol compounds are on the Community Right-To-Know List.

DOT Classification: Poison B; Label: St. Andrews Cross.

SAFETY PROFILE: Poison by ingestion, intraperitoneal, intravenous, and subcutaneous routes. Experimental reproductive effects. Questionable carcinogen with experimental tumorigenic data. Flammable when exposed to heat, flame, or oxidizers. To fight fire, use alcohol foam. When heated to decomposition it emits toxic fumes of Cl^-.

CJK500　　CAS: 108-43-0　　HR: 3
3-CHLOROPHENOL

DOT: UN 2020/UN 2021
mf: C_6H_5ClO　　mw: 128.56

PROP: Crystals. Mp: 33.5°, bp: 214°, d: 1.245 @ 45°/4°, vap press: 1 mm @ 44.2°, flash p: >112°.

SYNS: m-CHLOROPHENOL * m-CHLOROPHENOL, liquid (DOT) * m-CHLOROPHENOL, solid (DOT)

CONSENSUS REPORTS: Reported in EPA TSCA Inventory. Chlorophenol compounds are on the Community Right-To-Know List.

DOT Classification: Poison B; Label: St. Andrews Cross.

SAFETY PROFILE: Poison by intraperitoneal route. Moderately toxic by ingestion and subcutaneous routes. Questionable carcinogen with experimental tumorigenic data by skin contact. When heated to decomposition it emits toxic fumes of Cl^-.

CJK750　　CAS: 106-48-9　　HR: 3
4-CHLOROPHENOL

mf: C_6H_5ClO　　mw: 128.56

DOT: UN 2020/UN 2021

PROP: Needle-like, white to straw colored crystals; unpleasant odor. fp: 42.8°, flash p: 250°F, d: 1.246 @ 60°/25°, vap press: 1 mm @ 49.8°, mp: 43.5°, bp: 220°, sltly water sol; very sol in alc, chloroform, and ether.

SYNS: p-CHLORFENOL (CZECH) * p-CHLOROPHENOL * p-CHLOROPHENOL, liquid (DOT) * p-CHLOROPHENOL, solid (DOT)

CONSENSUS REPORTS: Reported in EPA TSCA Inventory. Chlorophenol compounds are on the Community Right-To-Know List.

DOT Classification: IMO: Poison B; Label: St. Andrews Cross.

SAFETY PROFILE: Poison by ingestion and intraperitoneal routes. Moderately toxic by skin contact and subcutaneous routes. A severe skin and eye irritant. Mutation data reported. Combustible when exposed to heat or flame. To fight fire, use water, spray, mist, fog, foam, dry chemical. When heated to decomposition it emits toxic fumes of Cl^-.

CJL000　　　　　　　　　　HR: 3
CHLOROPHENOLS

CONSENSUS REPORTS: Chlorophenol compounds are on the Community Right-To-Know List.

SAFETY PROFILE: Many are suspected experimental carcinogens. Most are strong eye and skin irritants. They are systemic irritants by inhalation, ingestion, and skin contact. Generally mutagenic.

　　Trichlorophenols are generally poisons and may be carcinogens. They may contain 2,3,7,8-tetrachlorodibenzo-p-dioxin (TCDD) as a contaminant. Some trichlorophenols are used as herbicides (e.g., 2,4,5-T and silvex). Human exposure may cause chloracne, liver dysfunction, muscle weakness, and prophyria.

　　Pentachlorophenol is a poison by several routes. Human exposure causes increased respiration, fever, tachycardia, muscle weakness, and cardiac failure. Many toxic effects are due to impurities in commercial grade material. A teratogen and mutagen. Pentachlorophenol and 2,4,6-trichlorophenol may interfere with mitochondrial oxidative phosphorylation. When heated to decomposition they emit toxic fumes of Cl^-.

CJM250　　CAS: 58-39-9　　HR: 3
4-(3-(2-CHLOROPHENOTHIAZIN-10-YL)PROPYL)-1-PIPERAZINEETHANOL
mf: $C_{21}H_{26}ClN_3OS$　　mw: 404.01

SYNS: 2-CHLORO-10-3-(1-(2-HYDROXYETHYL)-4-PIPERAZINYL)PROPYL PHENOTHIAZINE * DECENTAN

* ETAPERAZIN * ETAPERAZINE * ETHAPERAZINE * FENTAZIN * 1-(2-HYDROXYETHYL)-4-(3-(2-CHLORO-10-PHENOTHIAZINYL)PROPYL)PIPERAZINE * γ-(4-(β-HYDROXYETHYL)PIPERAZIN-1-YL)PROPYL-2-CHLOROPHENOTHIAZINE * 1′,1-(2-IDROSSIETIL)4-(3-(2-CLORO-10-FENOTIAZIL)PROPILPIPERAZINA (ITALIAN) * PERFENAZINA (ITALIAN) * PERPHENAZIN * PERPHENAZINE * TRIFARON * TRILAFON

SAFETY PROFILE: Poison by ingestion, intravenous, subcutaneous, intraperitoneal, and intramuscular routes. Human systemic effects by intramuscular route: muscle spasms. Experimental teratogenic and reproductive effects. Human mutation data reported. When heated to decomposition it emits very toxic fumes of SO_x, NO_x, and Cl^-.

CJN000   CAS: 122-88-3   HR: 2
p-CHLOROPHENOXYACETIC ACID
mf: $C_8H_7ClO_3$ mw: 186.60

PROP: Mp: 159°.

SYNS: (4-CHLOROPHENOXY)ACETIC ACID * p-CHLOROPHENOXYACETIC ACID * 4-CP * CPA * MARKS 4-CPA * PCPA * SURE-SET * TOMATO HOLD * TOMATO FIX CONCENTRATE * TOMATOTONE

CONSENSUS REPORTS: Reported in EPA TSCA Inventory.

SAFETY PROFILE: Moderately toxic by ingestion and intraperitoneal routes. Human mutation data reported. When heated to decomposition it emits toxic fumes of Cl^-.

CJO250   CAS: 43121-43-3   HR: 3
1-(4-CHLOROPHENOXY)-3,3-DIMETHYL-1-(1,2,4-TRIAZOL-1-YL)-2-BUTAN-2-ONE
mf: $C_{14}H_{16}ClN_3O_2$ mw: 293.78

SYNS: AMIRAL * BAY 6681 F * BAYLETON * BAY-MEB-6447 * 1-((tert-BUTYLCARBONYL-4-CHLOROPHENOXY)METHYL)-1H-1,2,4-TRIAZOLE * 1-(4-CHLOROPHENOXY)-3,3-DIMETHYL-1-(1H-1,2,4-TRIAZOL-1-YL)-2-BUTANONE * MEB 6447 * TRIADIMEFON

SAFETY PROFILE: Poison by ingestion. When heated to decomposition it emits very toxic fumes of Cl^- and NO_x.

CJR500   CAS: 80-38-6   HR: 2
4-CHLOROPHENYL BENZENE-SULFONATE
mf: $C_{12}H_9ClO_3S$ mw: 268.72

PROP: Colorless crystals. Mp: 62°. Insol in water; sol in organic solvents.

SYNS: ARACID * BENZENESULFONATE de 4-CHLOROPHENYLE (FRENCH) * BENZENESULFONIC ACID, 4-CHLOROPHENYL ESTER * (4-CHLOOR-FENYL)-BENZEEN-SULFONAAT (DUTCH) * p-CHLOROFENYLESTER KYSELINY BENZENSULFONOVE (CZECH) * p-CHLOROPHENYL BENZENESULFONATE * p-CHLOROPHENYL BENZENESULPHONATE * 4-CHLOROPHENYL BENZENESULPHONATE * (4-CHLOR-PHENYL)-BENZOLSULFONAT (GERMAN) * (4-CLORO-FENIL)-BENZOL-SOLFONATO (ITALIAN) * CPB * CPBS * ENT 4,585 * FENIZON (FRENCH) * FENSON * GC 928 * MURVESCO * PCBS * PCI * PCPBS * TRIFENSON

SAFETY PROFILE: Moderately toxic by ingestion and possibly other routes. An eye and skin irritant. An acaricide. When heated to decomposition it emits toxic fumes of Cl^- and SO_x.

CJR809   CAS: 1982-36-1   HR: 3
1-(p-CHLORO-α-PHENYLBENZYL) HEXAHYDRO-4-METHYL-1H-1,4-DIAZEPINE DIHYDROCHLORIDE
mf: $C_{19}H_{23}ClN_2 \cdot 2ClH$ mw: 387.81

SYNS: HOMOCHLORCYCLIZINE DIHYDROCHLORIDE * HOMOCHLOROCYCLIZINE DIHYDROCHLORIDE * SA 97 DIHYDROCHLORIDE

SAFETY PROFILE: Poison by ingestion, subcutaneous, intravenous, and intraperitoneal routes. When heated to decomposition it emits very toxic fumes of Cl^- and NO_x.

CJR909   CAS: 68-88-2   HR: 3
1-(p-CHLORO-α-PHENYLBENZYL)-4-(2-((2-HYDROXYETHOXY)ETHYL) PIPERAZINE
mf: $C_{21}H_{27}ClN_2O_2$ mw: 374.95

SYNS: ATARA * ATARAX * ATARAXOID * ATARAZOID * ATAZINA * ATERAX * 1-(p-CHLOROBENZHYDRYL)-4-(2-(2-HYDROXYETHOXY)ETHYL)DIETHYLENEDIAMINE * 1-(p-CHLOROBENZHYDRYL)-4-(2-(2-HYDROXYETHOXY)ETHYL)PIPERAZINE * N-(4-CHLOROBENZHYDRYL)-N′-(HYDROXYETHOXYETHYL)PIPERAZINE * 1-(p-CHLORODIPHENYLMETHYL)-4-(2-(2-HYDROXYETHOXY)ETHYL)PIPERAZINE * 2-(2-(4-(p-CHLORO-α-PHENYLBENZYL)-1-PIPERAZINYL)ETHOXY)ETHANOL * DEINAIT * EQUIPOISE * FENAROL * HYCHOTINE * HYDROXINE * HYDROXYCINE

* HYDROXYZINE * IDROSSIZINA * NEO-CALMA
* NEUROZINA * NP 212 * PAMAZONE
* PARENTERAL * PAXISTIL * PLACIDOL
* PLAXIDOL * TRAN-Q * TRAQUIZINE
* UCB 492 * U.CB 4492 * VESPARAZ-WIRKSTOFF

SAFETY PROFILE: Poison by intravenous and intraperitoneal routes. Moderately toxic by ingestion. Experimental teratogenic and reproductive effects. When heated to decomposition it emits very toxic fumes of Cl⁻ and NO_x.

CJT750   CAS: 80-33-1   HR: 3
4-CHLOROPHENYL-4-CHLOROBEN-ZENESULFONATE
mf: $C_{12}H_8Cl_2O_3S$ mw: 303.16

SYNS: ACARICYDOL E 20 * C-854 * C 1,006
* CCS * CHLOORFENSON (DUTCH) * (4-CHLOOR-FENYL)-4-CHLOOR-BENZEEN-SULFONAAT (DUTCH)
* CHLOREFENIZON (FRENCH) * CHLORFENSON
* CHLORFENSONE * 4-CHLOROBENZENESULFO-NATE de 4-CHLOROPHENYLE (FRENCH) * p-CHLORO-BENZENESULFONIC ACID-p-CHLOROPHENYL ESTER
* CHLOROFENIZON * p-CHLOROPHENYL-p-CHLO-ROBENZENE SULFONATE * 4-CHLOROPHENYL-4-CHLOROBENZENESULPHONATE * 4-CHLORPHENYL-4'-CHLORBENZOLSULFONAT (GERMAN) * (4-CHLOR-PHENYL)-4-CHLOR-BENZOL-SULFONATE (GERMAN)
* (4-CLORO-FENIL)-4-CLORO-VENZOL-SOLFONATO (ITALIAN) * COROTRAN * CPCBS * D 854
* DIFENSON * ENT 16,358 * EPHIRSULPHONATE
* ESTER SULFONATE * ESTONMITE * ETHER-SULFONATE * GENITE 883 * K 6451 * LETH-ALAIRE G-58 * MITICIDE K-101 * NIAGARATRAN
* ONEX * ORTHOTRAN * OTRACID * OVA-TRAN * OVEX * OVOCHLOR * OVOTOX
* OVOTRAN * PCPCBS * SAPPILAN * SAPPI-RAN * TRICHLORFENSON (OBS.)

SAFETY PROFILE: Moderately toxic by ingestion and possibly other routes. Questionable carcinogen with experimental tumorigenic data. A pesticide. When heated to decomposition it emits very toxic fumes of Cl⁻ and SO_x.

CJV250   CAS: 35367-38-5   HR: 2
1-(4-CHLOROPHENYL)-3-(2,6-DIFLUOROBENZOYL)UREA
mf: $C_{14}H_9ClF_2N_2O_2$ mw: 310.70

SYNS: N-(((4-CHLOROPHENYL)AMINO)CARBONYL)-2,6-DIFLUOROBENZAMIDE * DIFLUBENZURON
* DIFLURON * DIMILIN * DU 112307 * ENT 29,054 * OMS 1804 * PDD 6040I * PH 60-40

* PHILIPS-DUPHAR PH 60-40 * TH 6040 * THOMP-SON-HAYWARD TH6040

CONSENSUS REPORTS: Reported in EPA TSCA Inventory. EPA Genetic Toxicology Program.

SAFETY PROFILE: Moderately toxic by skin contact. Mildly toxic by ingestion and possibly other routes. Mutation data reported. When heated to decomposition it emits very toxic fumes of Cl⁻, F⁻, and NO_x.

CJX750   CAS: 150-68-5   HR: 3
3-(p-CHLOROPHENYL)-1,1-DIMETHYLUREA
mf: $C_9H_{11}ClN_2O$ mw: 198.67

PROP: Crystals; slight odor. Mp: 171°, vap press: 0.002 mm @ 100°. Nearly water insol,

SYNS: 3-(4-CHLOOR-FENYL)-1,1-DIMETHYLUREUM (DUTCH) * CHLORFENIDIM * N-(p-CHLOROPHE-NYL)-N',N'-DIMETHYLUREA * N'-(4-CHLOROPHE-NYL)-N,N-DIMETHYLUREA * 1-(p-CHLOROPHENYL)-3,3-DIMETHYLUREA * 3-(4-CHLOROPHENYL)-1,1-DI-METHYLUREA * 1-(4-CHLORO PHENYL)-3,3-DIMETH-YLUREE (FRENCH) * 3-(4-CHLOR-PHENYL)-1,1-DI-METHYL-HARNSTOFF (GERMAN) * 3-(4-CLORO-FENIL)-1,1-DIMETIL-UREA (ITALIAN) * CMU
* N,N-DIMETHYL-N'-(4-CHLOROPHENYL)UREA
* 1,1-DIMETHYL-3-(p-CHLOROPHENYL)UREA
* HERBICIDES, MONURON * KARMEX MONURON HERBICIDE * KARMEX W. MONURON HERBICIDE
* LIROBETAREX * MONUREX * MONURON
* MONUROX * MONURUON * MONUURON
* NCI-C02846 * TELVAR * TELVAR MONURON WEEDKILLER * USAF P-8 * USAF XR-41

CONSENSUS REPORTS: IARC Cancer Review: GROUP 3 IMEMDT 7,56,87; Animal Sufficient Evidence IMEMDT 12,167,76. Reported in EPA TSCA Inventory. EPA Genetic Toxicology Program.

SAFETY PROFILE: Moderately toxic by ingestion, intraperitoneal, and possibly other routes. Experimental teratogenic and reproductive effects. Questionable carcinogen with experimental carcinogenic data. Mutation data reported. An herbicide. When heated to decomposition it emits very toxic fumes of NO_x, and Cl⁻.

CJY120   CAS: 5131-60-2   HR: 3
4-CHLORO-m-PHENYLENEDIAMINE
mf: $C_6H_7ClN_2$ mw: 142.60

PROP: Mp: 90°.

SYNS: C.I. 76027 * 4-CHLORO-1,3-BENZENEDI-AMINE * 1-CHLORO-2,4-DIAMINOBENZENE * 4-CHLOROPHENE-1,3-DIAMINE * 4-CHLOROPHE-NYLENE-1,3-DIAMINE * 4-CHLORO-1,3-PHENYLENEDI-AMINE * 4-Cl-M-PD * NCI-C03305

CONSENSUS REPORTS: IARC Cancer Review: GROUP 3 IMEMDT 7,56,87; Animal Inadequate Evidence IMEMDT 27,81,82. NCI Carcinogenesis Bioassay (feed); Clear Evidence: mouse, rat NCITR* NCI-CG-TR-85,78. Reported in EPA TSCA Inventory.

SAFETY PROFILE: Suspected carcinogen with experimental carcinogenic, neoplastigenic, and tumorigenic data. Mutation data reported. When heated to decomposition it emits toxic fumes of Cl^- and NO_x.

CKB000   CAS: 104-12-1   **HR: 3**
p-CHLOROPHENYL ISOCYANATE
mf: C_7H_4ClNO mw: 153.57

PROP: White solid, sol in organic solvents. Mp: 31°, bp: 204°, flash p: 230°F.

SYNS: p-CHLORFENYLISOKYANAT (CZECH) * ISOCYANIC ACID-p-CHLOROPHENYL ESTER * PCPI

CONSENSUS REPORTS: Reported in EPA TSCA Inventory.

SAFETY PROFILE: Poison by inhalation. Moderately toxic by ingestion. Unspecified human systemic effects. A severe eye and moderate skin irritant. Combustible when exposed to heat or flame. Dangerous, can explode on distillation. When heated to decomposition it emits toxic fumes of Cl^-, CN^- and NO_x.

CKB250   CAS: 500-92-5   **HR: 3**
1-(p-CHLOROPHENYL)-5-ISOPROPYLBIGUANIDE
mf: $C_{11}H_{16}ClN_5$ mw: 253.77

PROP: White powder. Mp: 244°.

SYNS: BIGUMAL * CHLORGUANIDE * CHLORO-GUANIDE * PALUDRINE * PROGUANIL

SAFETY PROFILE: Poison by ingestion, intravenous, and intraperitoneal routes. Experimental reproductive effects. When heated to decomposition it emits toxic fumes of Cl^- and NO_x.

CKC000   CAS: 101-21-3   **HR: 3**
N-3-CHLOROPHENYLISOPROPYL-CARBAMATE
mf: $C_{10}H_{12}ClNO_2$ mw: 213.68

PROP: Light brown, crystalline solid; faint characteristic odor. Mp: 41°, bp: 247° (decomp).

SYNS: BEET-KLEEN * BUD-NIP * N-(3-CHLOOR-FENYL)-ISOPROPYL CARBAMAAT (DUTCH) * CHLOR-IFC * CHLOR-IPC * m-CHLOROCARBANILIC ACID, ISOPROPYL ESTER * 3-CHLOROCARBANILIC ACID, ISOPROPYL ESTER * N-(3-CHLORO PHENYL) CARBA-MATE D'ISOPROPYLE (FRENCH) * N-(3-CHLOROPHE-NYL)CARBAMIC ACID, ISOPROPYL ESTER * (3-CHLO-ROPHENYL)CARBAMIC ACID, 1-METHYLETHYL ESTER * CHLOROPROPHAM * N-(3-CHLOR-PHENYL)-ISO-PROPYL-CARBAMAT (GERMAN) * CHLORPROPHAM * CHLORPROPHAME (FRENCH) * CICP * CI-IPC * CIPC * N-(3-CLORO-FENIL)-ISOPROPIL-CARBAM-MATO (ITALIAN) * ELBANIL * ENT 18,060 * FASCO WY-HOE * FURLOE * FURLOE 4EC * ISOPROPYL-m-CHLOROCARBANILATE * ISOPRO-PYL-3-CHLOROCARBANILATE * ISOPROPYL-3-CHLO-ROPHENYLCARBAMATE * ISOPROPYL-N-(3-CHLORO-PHENYL)CARBAMATE * o-ISOPROPYL-N-(3-CHLORO-PHENYL)CARBAMATE * ISOPROPYL-N-(3-CHLORPHE-NYL)-CARBAMAT (GERMAN) * JACK WILSON CHLORO 51 (oil) * LIRO CIPC * METOXON * NEXOVAL * PREVENOL * PREVENOL 56 * PREVENTOL * PREVENTOL 56 * PREWEED * SPROUT NIP * SPROUT-NIP EC * SPUD-NIC * SPUD-NIE * STOPGERME-S * TATERPEX * TRIHERBICIDE CIPC * UNICROP CIPC * Y 3

CONSENSUS REPORTS: IARC Cancer Review: GROUP 3 IMEMDT 7,56,87; Animal Inadequate Evidence IMEMDT 12,55,76. EPA Genetic Toxicology Program.

SAFETY PROFILE: Moderately toxic by ingestion, intraperitoneal, and possibly other routes. Questionable carcinogen with experimental neoplastigenic and teratogenic data. Human mutation data reported. An herbicide. When heated to decomposition it emits highly toxic fumes of Cl^-, NO_x, and phosgene.

CKD500   CAS: 1746-81-2   **HR: 3**
3-(4-CHLOROPHENYL)-1-METHOXY-1-METHYLUREA
mf: $C_9H_{11}ClN_2O_2$ mw: 214.67

SYNS: AFESIN * ARESIN * AREZIN * ARE-ZINE * ARRESIN * 3-(4-CHLORPHENYL)-1-METH-OXY-1-METHYLHARNSTOFF (GERMAN) * N-(4-CHLO-ROPHENYL)-N'-METHOXY-N-METHYLUREA * N'-(4-

CHLOROPHENYL)-N-METHOXY-N-METHYLUREA
* HOE 2747 * MONOLINURON * PREMALIN

CONSENSUS REPORTS: Reported in EPA TSCA Inventory.

SAFETY PROFILE: Moderately toxic by ingestion. Experimental teratogenic and reproductive effects. When heated to decomposition it emits very toxic fumes of Cl⁻ and NO$_x$.

CKD750   CAS: 1867-66-9   HR: 3
2-(o-CHLOROPHENYL)-2-(METHYL-AMINO)CYCLOHEXANONE HYDROCHLORIDE
mf: C$_{13}$H$_{16}$ClNO • ClH   mw: 274.21

SYNS: CI 581 * CL 369 * CN-52,372-2
* KETAJECT * KETALAR * KETAMINE
* KETAMINE HYDROCHLORIDE * KETANEST
* KETASET * KETAVET * KETOLAR * VETALAR

CONSENSUS REPORTS: EPA Genetic Toxicology Program.

SAFETY PROFILE: Poison by intramuscular, intraperitoneal, and intravenous routes. Moderately toxic by ingestion. Human systemic effects by intravenous and possibly other routes: analgesia, coma, hallucinations and distorted perceptions. An experimental teratogen. An anesthetic. When heated to decomposition it emits very toxic fumes of Cl⁻ and NO$_x$.

CKF000   CAS: 3942-54-9   HR: 3
o-CHLOROPHENYL METHYLCARBAMATE
mf: C$_8$H$_8$ClNO$_2$ mw: 185.62

SYNS: 2-CHLOROPHENYL-N-METHYLCARBAMATE
* CPMC * ETROFOL * HOPCIDE

CONSENSUS REPORTS: EPA Genetic Toxicology Program. Reported in EPA TSCA Inventory.

SAFETY PROFILE: A poison by ingestion and possibly other routes. An insecticide. When heated to decomposition it emits very toxic fumes of Cl⁻ and NO$_x$.

CKF750   CAS: 3766-60-7   HR: 2
3-(p-CHLOROPHENYL)-1-METHYL-1-(1-METHYL-2-PROPYNYL)UREA
mf: C$_{12}$H$_{13}$ClN$_2$O   mw: 236.72

SYNS: ARISAN * BUTURON * BUTYRON
* N'-(4-CHLOROPHENYL)-N-ISOBUTINYL-N-METHYL-

UREA * N'-(4-CHLOROPHENYL)-N-METHYL-N-(1-METHYL-2-PROPYNYL)-UREA * N-(4-CHLORPHE-NYL)-N'-METHYL-N'-ISOBUTINYLHARNSTOFF (GERMAN)
* 3-(4-CHLORPHENYL)-1-METHYL-1-ISOBUTINYL-HARNSTOFF (GERMAN) * EPTAPUR * H 95

CONSENSUS REPORTS: EPA Genetic Toxicology Program.

SAFETY PROFILE: Moderately toxic by ingestion and intraperitoneal routes. Experimental teratogenic and reproductive effects. An herbicide. When heated to decomposition it emits very toxic fumes of Cl⁻ and NO$_x$.

CKM000   CAS: 116-29-0   HR: 2
p-CHLOROPHENYL-2,4,5-TRICHLOROPHENYL SULFONE
mf: C$_{12}$H$_6$Cl$_4$O$_2$S   mw: 356.04

PROP: Crystals. Mp: 147°. Nearly water-insol.

SYNS: AKARITOX * AREDION * 4-CHLOROPHE-NYL-2,4,5-TRICHLOROPHENYL SULFONE * p-CHLO-ROPHENYL-2,4,5-TRICHLOROPHENYL SULPHONE
* DUPHAR * ENT 23,737 * FMC 5488 * MITION
* NIA 5488 * POLACARITOX * ROZTOZOL
* SULFONE-2,4,4',5-TETRACHLORODIPHENYL
* TEDION * TEDION V-18 * 2,4,4',5-TETRA-CHLOOR-DIFENYL-SULFON (DUTCH) * 2,4,4',5-TET-RACHLOR-DIPHENYL-SULFON (GERMAN) * 2,4,4',5-TETRACHLORODIPHENYL SULFONE * 2,4,5,4'-TETRA-CHLORODIPHENYLSULPHONE * 2,4,4',5-TETRA-CLORO-DIFENIL-SOLFONE (ITALIAN) * TETRADI-CHLONE * TETRADIFON * TETRADIPHON
* TETRAFIDON * 1,2,4-TRICHLORO-5-((4-CHLORO-PHENYL)SULFONYL)-BENZENE * V-18

CONSENSUS REPORTS: EPA Genetic Toxicology Program.

SAFETY PROFILE: Moderately toxic by ingestion. Mildly toxic by skin contact. An experimental teratogen. Used to control worms in crops. When heated to decomposition it emits highly toxic fumes of Cl⁻ and SO$_x$.

CKM250   CAS: 26571-79-9   HR: 3
CHLOROPHENYLTRICHLOROSILANE
DOT: UN 1753
mf: C$_6$H$_4$Cl$_4$Si   mw: 245.99

PROP: Colorless to pale yellow liquid, readily hydrolyzed by moisture with the liberation of HCl (a mixture of 3 isomers). Bp: 230°, d: 1.439 @ 25°/25°, flash p: 255°F (COC).

DOT Classification: Corrosive Material; Label: Corrosive.

SAFETY PROFILE: A poison irritant by ingestion and inhalation. A corrosive irritant to the skin, eyes, and mucous membranes. Combustible when exposed to heat or flame. In contact with water it readily hydrolyzes to HCl and evolves heat. When heated to decomposition it emits toxic fumes of Cl$^-$.

CKN000   CAS: 1406-65-1   **HR: 3**
CHLOROPHYLL

PROP: Dark green solution.

SYNS: BIOPHYLL * CHLOROPHYL, GREEN * C.I. 1956 * DAROTOL * DEODOPHYLL * E 140 * GREEN CHLOROPHYL * L-GRUEN No. 1 (GERMAN) * No. 1249 * No. 1403 * No. 75810

SAFETY PROFILE: Poison by intravenous and intraperitoneal routes. When heated to decomposition it emits toxic fumes of NO$_x$.

CKN500   CAS: 76-06-2   **HR: 3**
CHLOROPICRIN

DOT: UN 1580/UN 1583
mf: CCl$_3$NO$_2$ mw: 164.37

PROP: Sltly oily, colorless liquid. D: 1.651 @ 22.8°/4°, mp: −64°, bp: 112.3 @ 766 mm, vap press: 40 mm @ 33.80, vap d: 6.69. Sol in water, alc, and ether.

SYNS: ACQUINITE * CHLOORPIKRINE (DUTCH) * CHLOR-O-PIC * CHLOROPICRIN, ABSORBED (DOT) * CHLOROPICRIN, liquid (DOT) * CHLOROPICRINE (FRENCH) * CHLORPIKRIN (GERMAN) * CLOROPICRINA (ITALIAN) * DOJYOPICRIN * DOLOCHLOR * LARVACIDE * MICROLYSIN * NCI-C00533 * NITROCHLOROFORM * NITROTRICHLOROMETHANE * PIC-CLOR * PICFUME * PICRIDE * PROFUME A * PS * TRICHLOORNITROMETHAAN (DUTCH) * TRICHLORNITROMETHAN (GERMAN) * TRICHLORONITROMETHANE * TRICLOR * TRICLORO-NITRO-METANO (ITALIAN)

CONSENSUS REPORTS: NCI Carcinogenesis Bioassay (gavage); No Evidence: mouse NCITR* NCI-GC-TR-65,78. Reported in EPA TSCA Inventory.

OSHA PEL: TWA 0.1 ppm
ACGIH TLV: TWA 0.1 ppm
DFG MAK: 0.1 ppm (0.7 mg/m^3)
DOT Classification: Poison B; Label: Poison, Liquid and Absorbed.

SAFETY PROFILE: Poison by ingestion, intravenous, and intraperitoneal routes. Moderately toxic by inhalation. Human systemic effects by inhalation: lacrimation, conjunctiva irritation, and pulmonary changes. Mutation data reported. A powerful irritant that affects all body surfaces. It causes lachrymation, vomiting, bronchitis, pulmonary edema, irritation to gastrointestinal and respiratory tracts. Questionable carcinogen with experimental tumorigenic data. An additional toxic effect is its reaction with SH-groups in hemoglobin thus interfering with oxygen transport. Photochemical transformation of chloropicrin into phosgene (carboxy chloride, COCl$_2$) has been reported. A concentration of 1 ppm causes a smarting pain in the eyes and therefore in itself constitutes a good warning of exposure. Inhalation causes vomiting, probably due to swallowing saliva in which small amounts of chloropicrin have dissolved. Its primary lethal effect is to produce lung injury and it is a difficult gas to protect oneself against because it is chemically inert and does not react with the usual chemicals used in gas masks. Four ppm is sufficient to render a worker unfit for action and 20 ppm, when breathed from 1 to 2 minutes, causes definite bronchial or pulmonary lesions. Industrially it is used as a warning agent in commercial fumigants. It is more toxic than chlorine but less so than phosgene.

Above a critical volume it can be shocked into detonation. Mixtures with 3-bromopropyne are shock- and heat-sensitive explosives. Violent reaction with aniline + heat, alcoholic sodium hydroxide, sodium methoxide, propargyl bromide. When heated to decomposition it emits very toxic fumes of Cl$^-$ and NO$_x$.

Used for insect and rodent control in grain elevators and bins and as a soil fumigant and fungicide.

CKN510   CAS: 76-06-2   **HR: 3**
CHLOROPICRIN MIXTURE (flammable)

DOT: NA 2929
mf: CCl$_3$NO$_2$ mw: 164.37

SYNS: CHLOROPICRINE * TRICHLORONITROMETHANE (flammable mixture)

CONSENSUS REPORTS: Reported in EPA TSCA Inventory.

DOT Classification: Poison B; Label: Poison and Flammable Liquid.

SAFETY PROFILE: A poison. Moderately toxic by inhalation. When heated to decomposition it emits very toxic fumes of NO_x and Cl^-.

CKO750   CAS: 16941-12-1   HR: 3
CHLOROPLATINIC ACID

DOT: UN 2507
mf: $Cl_6Pt \cdot 2H$ mw: 409.81

PROP: Brownish-yellow, very deliquescent, crystalline mass; easily sol in water and alc. D: 2.431, mp: 60°.

SYNS: CHLOROPLATINIC(IV) ACID * DIHYDROGEN HEXACHLOROPLATINATE * DIHYDROGEN HEXACHLOROPLATINATE(2-) * HEXACHLOROPLA-TINIC ACID * HEXACHLOROPLATINIC(IV) ACID * HEXACHLOROPLATININIC(4+) ACID, HYDROGEN- * HYDROGEN HEXACHLOROPLATINATE(4+) * PLATINIC CHLORIDE

CONSENSUS REPORTS: Reported in EPA TSCA Inventory. EPA Genetic Toxicology Program.

OSHA PEL: TWA 0.002 mg(Pt)/m³
ACGIH TLV: TWA 0.002 mg(Pt)/m³
DOT Classification: Corrosive Material; Label: Corrosive.

SAFETY PROFILE: Poison by intravenous and intraperitoneal routes. Mutation data reported. Incompatible with BrF_3. When heated to decomposition it emits toxic fumes of Cl^-.

CKP750   CAS: 540-54-5   HR: 2
1-CHLOROPROPANE

DOT: UN 1278
mf: C_3H_7Cl mw: 78.55

PROP: Colorless liquid, chloroform-like odor. Mp: $-122.8°$, bp: 47.2°, lel: 2.6%, uel: 11.1%, flash p: $<0°F$, d: 0.890, vap d: 2.71, autoign temp: 968°F.

SYN: N-PROPYL CHLORIDE

CONSENSUS REPORTS: Reported in EPA TSCA Inventory.

DOT Classification: Flammable Liquid; Label: Flammable Liquid.

SAFETY PROFILE: A moderately poisonous irritant to skin, eyes, and mucous membranes. Narcotic in high concentrations. Dangerous fire hazard when exposed to heat, flame, or oxidiz-ers. Moderately explosive when exposed to flame. Keep away from heat and open flame; can react vigorously with oxidizing materials. To fight fire use CO_2, dry chemical. When heated to decomposition it emits toxic fumes of Cl^-.

CKR500   CAS: 78-89-7   HR: 3
2-CHLORO-1-PROPANOL

DOT: UN 2611
mf: C_3H_7ClO mw: 94.55

PROP: Colorless liquid, mild non-residual odor. Bp: 133.5°, flash p: 125°F (CC), d: 1.103 @ 20°, vap d: 3.26.

SYNS: 2-CHLOROPROPYL ALCOHOL * PROPYLENE-CHLOROHYDRIN

CONSENSUS REPORTS: Reported in EPA TSCA Inventory.

DOT Classification: Poison B; Label: Flamma-ble Liquid and Poison.

SAFETY PROFILE: Poison by ingestion. Mod-erately toxic by inhalation and skin contact. A skin and severe eye irritant. Flammable when exposed to heat, flame, or powerful oxidizers. To fight fire, use alcohol foam, CO_2, dry chemi-cal. When heated to decomposition it emits toxic fumes of Cl^-.

CKS000   CAS: 557-98-2   HR: 1
2-CHLORO-1-PROPENE

DOT: UN 2456
mf: C_3H_5Cl mw: 76.53

PROP: Colorless liq. Bp: 22.65°, fp: $-137.4°$, d: 0.918 @ 9°, flash p: $-4°$. lel: 4.5%; uel: 16%.

SYN: 2-CHLOROPROPENE (DOT)

CONSENSUS REPORTS: Reported in EPA TSCA Inventory.

DOT Classification: Flammable Liquid; Label: Flammable Liquid.

SAFETY PROFILE: Mildy toxic by inhalation. Mutation data reported. Very dangerous fire haz-ard when exposed to heat, flame, sparks, or powerful oxidizers. To fight fire, use water, spray, mist, fog, dry chemical, alcohol foam. When heated to decomposition it emits toxic fumes of Cl^-.

CKT000   CAS: 17639-93-9   **HR: 3**
2-CHLOROPROPIONIC ACID METHYL ESTER

DOT: UN 2933
mf: $C_4H_7ClO_2$ mw: 122.56

SYN: METHYL-2-CHLOROPROPIONATE (DOT)

CONSENSUS REPORTS: Reported in EPA TSCA Inventory.

DOT Classification: Flammable or Combustible Liquid; Label: Flammable Liquid.

SAFETY PROFILE: Poison by intraperitoneal route. Flammable when exposed to heat of flame. When heated to decomposition it emits toxic fumes of Cl^-.

CKT500   CAS: 6285-05-8   **HR: 3**
p-CHLOROPROPIOPHENONE
mf: C_9H_9ClO mw: 168.63

SYN: USAF EK-5296

CONSENSUS REPORTS: Reported in EPA TSCA Inventory.

SAFETY PROFILE: Poison by intraperitoneal and intravenous routes. When heated to decomposition it emits toxic fumes of Cl^-.

CKW000   CAS: 109-09-1   **HR: 3**
2-CHLOROPYRIDINE

DOT: UN 2822
mf: C_5H_4ClN mw: 113.55

PROP: Colorless, oily liquid. Bp: 170°, d: 1.205 @ 25°, vap press: 1 mm @ 13.3°, vap d: 3.93.

SYNS: o-CHLOROPYRIDINE * α-CHLOROPYRIDINE

CONSENSUS REPORTS: Reported in EPA TSCA Inventory.

DOT Classification: Poison B; Label: Poison.

SAFETY PROFILE: Poison by ingestion, inhalation, skin contact, and intraperitoneal routes. Combustible when exposed to heat or flame. Can react with oxidizing materials. When heated to decomposition it emits very toxic fumes of Cl^-, NO_x, and phosgene.

CLE250   **HR: 3**
CHLOROSILANES

PROP: Compounds of Si, Cl, and H where the total number of atoms of Cl and H add up to 4. SiH_xCl_{4-x}.

SAFETY PROFILE: Poison by ingestion and inhalation, and a poisonous irritant to skin, eyes, and mucous membranes. Toxicity is based on HCl which is formed upon hydrolysis of a chlorosilane. Self-ignites in air. With a little ammonia, it forms a self-igniting product. They react with water or steam to produce heat and toxic and corrosive fumes of HCl. When heated to decomposition they emit highly toxic fumes of Cl^-.

CLE750   CAS: 2039-87-4   **HR: 1**
o-CHLOROSTYRENE
mf: C_8H_7Cl mw: 138.60

CONSENSUS REPORTS: Reported in EPA TSCA Inventory.

OSHA PEL: TWA 50 ppm; STEL: 75 ppm
ACGIH TLV: TWA 50 ppm; STEL: 75 ppm

SAFETY PROFILE: Mildly toxic by ingestion and skin contact. A skin and eye irritant. When heated to decomposition it emits toxic fumes of Cl^-.

CLG500   CAS: 7790-94-5   **HR: 3**
CHLOROSULFURIC ACID

DOT: UN 1754
mf: $ClHO_3S$ mw: 116.52

PROP: Clear to cloudy, colorless to pale yellow liquid; sharp odor. Mp: −80°, bp: 151.0°, d: 1.766 @ 18°, vap press: 1 mm @ 32°, vap d: 4.02.

SYNS: CHLOROSULFONIC ACID (DOT) * MONOCHLOROSULFURIC ACID * SULFONIC ACID, MONOCHLORIDE * SULFURIC CHLOROHYDRIN

CONSENSUS REPORTS: Reported in EPA TSCA Inventory.

DOT Classification: Corrosive Material; Label: Corrosive.

SAFETY PROFILE: A poison irritant. Chlorosulfonic acid is corrosive, can cause severe acid burns; and is very irritating to the eyes, lungs and mucous membranes. It can cause acute toxic effects either in the liquid or vapor state. Inhalation of concentrated vapor may cause loss of consciousness with serious damage to lung tissue. Contact of liquid with the eyes can cause severe burns if not immediately and completely removed. It also causes severe skin burns due to its highly corrosive action. Upon ingestion it will irritate the mouth, esophagus, and stomach to a serious degree and on contact with skin cause dermatitis. It may cause conjunctivitis even in the vapor form. If spilled on a person, remove all contaminated clothing, wash contaminated skin with a lot of water, followed by baking soda solution. Irrigate eyes with warm water for 15 minutes. Consult a physician.

Vent stored drums two times per month to control pressure of H_2 produced by action of acid on metal of drum. Decomposes explosively on contact with water, alcohol, and acids. Explosive reaction with phosphorus. Violent reaction with silver nitrate. Potentially violent reaction with sulfuric acid or diphenyl ether. Incompatible with acetic acid, acetic anhydride, acetonitrile, acrolein, acrylic acid, acrylonitrile, allyl alcohol, allyl chloride, 2-amino ethanol, ammonium hydroxide, aniline, n-butyraldehyde, creosote oil, cresol, cumene, dichloroethyl ether, diethylene glycol monomethyl ether, diisobutylene, diisopropylether, epichloro hydrin, ethyl acetate, ethyl acrylate, ethylene chlorohydrin, ethylene cyanohydrin, ethylene diamine, ethylene glycol, ethylene glycol monoethyl ether acetate, ethylene imine, glyoxal, HCl, HF, H_2O_2, isoprene, mesityl oxide, metal powders, methyl ethyl ketone, HNO_3, 2-nitropropane, β-propiolactone, propylene oxide, pyridene, NaOH, sulfolane, styrene monomer, vinyl acetate, vinylidene chloride, water, organic matter, combustibles. Dangerous. To fight fire, avoid water, use dry chemicals. When heated to decomposition it emits toxic fumes of Cl^- and SO_x.

CLH000   CAS: 63938-10-3   **HR: 1**
CHLOROTETRAFLUOROETHANE
DOT: UN 1021
mf: C_2HClF_4 mw: 136.48
PROP: Colorless gas.

SYN: MONOCHLOROTETRAFLUOROETHANE (DOT)

DOT Classification: Nonflammable Gas; Label: Nonflammable Gas.

SAFETY PROFILE: Probably acts as a simple asphyxiant. When heated to decomposition it emits highly toxic fumes of F^- and Cl^-.

CLJ750   CAS: 2812-73-9   **HR: 3**
CHLOROTHIOFORMIC ACID ETHYL ESTER
DOT: UN 2826
mf: C_3H_5ClOS mw: 124.59

SYN: ETHYL CHLOROTHIOFORMATE (DOT)

DOT Classification: Corrosive Material; Label: Corrosive, Flammable Liquid.

SAFETY PROFILE: Probably a poison by inhalation and ingestion. A corrosive irritant to skin, eyes, and mucous membranes. Flammable when exposed to heat or flame. When heated to decomposition it emits very toxic fumes of Cl^- and SO_x.

CLK100   CAS: 95-49-8   **HR: 2**
o-CHLOROTOLUENE
DOT: UN 2238
mf: C_7H_7Cl mw: 126.59

PROP: Liquid. Bp: 158.97°, d: (20/4) 1.0826, mp: -35.59°. Volatile with steam. Sltly sol in water; freely sol in alc, benzene, chloroform, ether.

SYNS: 2-CHLORO-1-METHYLBENZENE (9CI) * o-CHLOROTOLUENE * 2-CHLOROTOLUENE * HALSO 99 * 1-METHYL-2-CHLOROBENZENE * 2-METHYLCHLOROBENZENE * o-TOLYL CHLORIDE

CONSENSUS REPORTS: Reported in EPA TSCA Inventory.

OSHA PEL: TWA 50 ppm
ACGIH TLV: TWA 50 ppm
DOT Classification: Flammable or Combustible Liquid; Label: Flammable Liquid.

SAFETY PROFILE: Moderately toxic by unspecified routes. Flammable when exposed to heat or flame. When heated to decomposition it emits toxic fumes of Cl^-.

CLK220   CAS: 95-69-2   **HR: 3**
4-CHLORO-o-TOLUIDINE
mf: C_7H_8ClN mw: 141.61

SYNS: AMARTHOL FAST RED TR BASE * 2-AMINO-5-CHLOROTOLUENE * AZOENE FAST RED TR BASE * AZOGENE FAST RED TR * AZOIC DIAZO COMPONENT 11 BASE * BRENTAMINE FAST RED TR BASE * 5-CHLORO-2-AMINOTOLUENE * 4-CHLORO-2-METHYLANILINE * 4-CHLORO-6-METHYLANILINE * 4-CHLORO-2-METHYLBENZENEAMINE * 4-CHLORO-2-TOLUIDINE * DAITO RED BASE TR * DEVAL RED K * DEVAL RED TR * DIAZO FAST RED TRA * FAST RED BASE TR * FAST RED 5CT BASE * FAST RED TR * FAST RED TR11 * FAST RED TR BASE * FAST RED TRO BASE

* KAKO RED TR BASE * KAMBAMINE RED TR
* 2-METHYL-4-CHLOROANILINE * MITSUI RED TR
BASE * RED BASE CIBA IX * RED BASE IRGA IX
* RED BASE NTR * RED TR BASE * SANYO FAST
RED TR BASE * TULABASE FAST RED TR

CONSENSUS REPORTS: IARC Cancer Review: GROUP 2B IMEMDT 7,56,87; Human Inadequate Evidence IMEMDT 16,277,78; Animal Sufficient Evidence IMEMDT 30,61,83. Reported in EPA TSCA Inventory.

DFG MAK: Human Carcinogen.

SAFETY PROFILE: Confirmed carcinogen. Poison by ingestion and subcutaneous routes. Human mutation data reported. In the presence of Copper(II) chloride catalyst decomposition occurs above 239°C. When heated to decomposition it emits toxic fumes of Cl^- and NO_x.

CLK225   CAS: 95-79-4   HR: 3
5-CHLORO-o-TOLUIDINE
mf: C_7H_8ClN mw: 141.61

PROP: Solid. Bp: 241°, mp: 29°.

SYNS: ACCO FAST RED KB BASE * 1-AMINO-3-CHLORO-6-METHYLBENZENE * 2-AMINO-4-CHLORO-TOLUENE * ANSIBASE RED KB * AZOENE FAST RED KB BASE * AZOIC DIAZO COMPONENT 32 * 4-CHLORO-2-AMINOTOLUENE * 3-CHLORO-6-METHYLANILINE * 5-CHLORO-2-METHYLANILINE * FAST RED KB AMINE * FAST RED KB BASE * FAST RED KB SALT * FAST RED KB SALT SUPRA * FAST RED KBS SALT * GENAZO RED KB SOLN * HILTONIL FAST RED KB BASE * LAKE RED KB BASE * METROGEN RED FORMER KB SOLN * NAPHTHOSOL FAST RED KB BASE * NCI-C02051 * PHARMAZOID RED KB * RED KB BASE * SPECTROLENE RED KB * STABLE RED KB BASE

CONSENSUS REPORTS: NTP Carcinogenesis Bioassay (feed): Clear Evidence: mouse NCITR* NCI-TR-187,79; (feed): Inadequate Studies: rat NCITR* NCI-TR-187,79. Reported in EPA TSCA Inventory. EPA Genetic Toxicology Program.

DFG MAK: Suspected Carcinogen.

SAFETY PROFILE: Suspected carcinogen with experimental carcinogenic and tumorigenic data. Moderately toxic by ingestion. When heated to decomposition it emits very toxic fumes of Cl^- and NO_x.

CLK235   CAS: 3165-93-3   HR: 3
4-CHLORO-2-TOLUIDINE HYDROCHLORIDE
DOT: UN 1579
mf: $C_7H_8ClN \cdot ClH$ mw: 178.07

SYNS: AMARTHOL FAST RED TR BASE * AMARTHOL FAST RED TR SALT * 2-AMINO-5-CHLORO-TOLUENE HYDROCHLORIDE * AZANIL RED SALT TRD * AZOENE FAST RED TR SALT * AZOGENE FAST RED TR * AZOIC DIAZO COMPONENT 11 BASE * BRENTAMINE FAST RED TR SALT * CHLORHYDRATE de 4-CHLOROORTHOTOLUIDINE (FRENCH) * 5-CHLORO-2-AMINOTOLUENE HYDROCHLORIDE * 4-CHLORO-2-METHYLANILINE HYDROCHLORIDE * 4-CHLORO-6-METHYLANILINE HYDROCHLORIDE * 4-CHLORO-2-METHYLBENZENEAMINE HYDROCHLORIDE * 4-CHLORO-o-TOLUIDINE HYDROCHLORIDE * 4-CHLORO-o-TOLUIDINE HYDROCHLORIDE (DOT) * C.I. 37085 * C.I. AZOIC DIAZO COMPONENT 11 * DAITO RED SALT TR * DEVOL RED K * DEVOL RED TA SALT * DEVOL RED TR * DIAZO FAST RED TR * DIAZO FAST RED TRA * FAST RED 5CT SALT * FAST RED SALT TR * FAST RED SALT TRA * FAST RED SALT TRN * FAST RED TR SALT * HINDASOL RED TR SALT * KROMON GREEN B * 2-METHYL-4-CHLOROANILINE HYDROCHLORIDE * NATASOL FAST RED TR SALT * NCI-C02368 * NEUTROSEL RED TRVA * OFNA-PERL SALT RRA * RCRA WASTE NUMBER U049 * RED BASE CIBA IX * RED BASE IRGA IX * RED SALT CIBA IX * RED SALT IRGA IX * RED TRS SALT * SANYO FAST RED SALT TR

CONSENSUS REPORTS: IARC Cancer Review: Animal Inadequate Evidence, Human Inadequate Evidence IMEMDT 16,277,78. NCI Carcinogenesis Bioassay (Feed); Clear Evidence: Mouse; No Evidence: Rat NCITR* NCI-CG-TR-165,79. Reported in EPA TSCA Inventory.

DOT Classification: Poison B; Label: Poison; DOT-IMO: Poison B; Label: St. Andrews Cross.

SAFETY PROFILE: Suspected carcinogen with experimental carcinogenic data. Moderately toxic by intraperitoneal route. When heated to decomposition it emits toxic fumes of NO_x and Cl^-.

CLO750   CAS: 569-57-3   HR: 3
CHLOROTRIANISENE
mf: $C_{23}H_{21}ClO_3$ mw: 380.89

SYNS: ANISENE * CHLORESTROLO * 1,1',1''-(1-CHLORO-1-ETHENYL-2-YLIDENE)-TRIS(4-METHOXYBENZENE) * CHLOROTRIANIZEN * CHLOROTRISIN * CHLOROTRIS(p-METHOXYPHENYL)ETHYLENE * CHLRTRIANISEN * CLORESTROLO * CLOROTRISIN * CTA * HORMONISENE * KHLORTRIANIZEN * MERBENTUL * METACE * NSC-10108 * RIANIL * TACE * TACE-FN * TRI-p-ANISYLCHLOROETHYLENE * TRIS(p-METHOXYPHENYL)CHLOROETHYLENE

CONSENSUS REPORTS: IARC Cancer Review: Animal Inadequate Evidence IMEMDT 21,139,79; Human Limited Evidence IMEMDT 21,139,79.

SAFETY PROFILE: Suspected human carcinogen with experimental tumorigenic data. Human reproductive effects by ingestion: changes in fertility. Used in cancer treatment. When heated to decomposition it emits very toxic fumes of Cl^-.

CLP500   CAS: 1461-22-9   HR: 3
CHLOROTRIBUTYLSTANNANE
mf: $C_{12}H_{27}ClSn$ mw: 325.53

SYNS: CHLORID TRI-n-BUTYLCINICITY (CZECH) * TRIBUTYLCHLOROSTANNANE * TRI-n-BUTYLTIN CHLORIDE * TRI-n-BUTYLZINN-CHLORID (GERMAN)

CONSENSUS REPORTS: Reported in EPA TSCA Inventory.

OSHA PEL: TWA 0.1 mg(Sn)/m³ (skin)
ACGIH TLV: TWA 0.1 mg(Sn)/m³ (skin) (Proposed: TWA 0.1 mg(Sn)/m³; STEL 0.2 mg(Sn)/m³ (skin))
NIOSH REL: (Organotin Compounds) TWA 0.1 mg(Sn)/m³

SAFETY PROFILE: Poison by ingestion and skin contact. Moderately toxic by an unspecified route. A severe eye irritant. Tributyl tin compounds are extremely toxic to marine life. When heated to decomposition it emits toxic fumes of Cl^-.

CLP750   CAS: 1929-82-4   HR: 3
2-CHLORO-6-(TRICHLOROMETHYL) PYRIDINE
mf: $C_6H_3Cl_4N$ mw: 230.90

SYNS: DOWCO-163 * NITRAPYRIN (ACGIH) * N-SERVE NITROGEN STABILIZER

CONSENSUS REPORTS: NCI Carcinogenesis Studies (ipr); Clear Evidence: mouse, rat

RRCRBU 52,1,75. Reported in EPA TSCA Inventory.

OSHA PEL: Total Dust: 15 mg/m³; Respirable Fraction: 5 mg/m³
ACGIH TLV: TWA 10 mg/m³; STEL 20 mg/m³

SAFETY PROFILE: Poison by ingestion. Moderately toxic by skin contact and possibly other routes. When heated to decomposition it emits very toxic fumes of Cl^- and NO_x.

CLQ500   CAS: 15529-90-5   HR: 3
CHLORO(TRIETHYLPHOSPHINE)GOLD
mf: $C_6H_{15}AuClP$ mw: 350.60

SYNS: SK&F 36914 * TRIETHYLPHOSPHINEAUROUS CHLORIDE

SAFETY PROFILE: Poison by ingestion. Experimental teratogenic and reproductive effects. Human mutation data reported. When heated to decomposition it emits very toxic fumes of Cl^- and PO_x.

CLQ750   CAS: 79-38-9   HR: 3
CHLOROTRIFLUOROETHYLENE
DOT: UN 1082
mf: C_2ClF_3 mw: 116.47

PROP: A gas. Lel: 24%, uel: 40.3%, flash p: $-18°F$.

SYNS: 1-CHLORO-1,2,2-TRIFLUOROETHYLENE * 2-CHLORO-1,1,2-TRIFLUOROETHYLENE * CHLORTRIFLUORAETHYLEN (GERMAN) * CTFE * DAIFLON * FLUOROPLAST 3 * GENETRON 1113 * MONOCHLOROTRIFLUOROETHYLENE * TRIFLUOROCHLOROETHYLENE (DOT) * 1,1,2-TRIFLUORO-2-CHLOROETHYLENE * TRIFLUOROMONOCHLOROETHYLENE * TRIFLUOROVINYL CHLORIDE * TRITHENE

CONSENSUS REPORTS: Reported in EPA TSCA Inventory.

DOT Classification: Flammable Gas; Label: Flammable Gas.

SAFETY PROFILE: Poison by ingestion and intraperitoneal routes. Moderately toxic by inhalation. Very dangerous fire hazard when exposed to heat, flames (sparks), or oxidizers. To fight fire, stop flow of gas. Violent reaction when mixed with $(Br_2 + O_2)$ or $(ClF_3 + water)$. Potentially explosive polymerization reaction with ethylene. Incompatible with 1,1-dichloroethylene, oxygen. When heated to decomposition it emits toxic fumes of F^- and Cl^-.

CLR250   CAS: 75-72-9   HR: 1
CHLOROTRIFLUOROMETHANE

DOT: UN 1022

mf: $CClF_3$ mw: 104.46

PROP: Colorless gas, ethereal odor. Mp: $-181°$, bp: $-80°$.

SYNS: ARCTON 3 * F 13 * FREON 13 * GENETRON 13 * HALOCARBON 13/UCON 13 * MONOCHLOROTRIFLUOROMETHANE (DOT) * R 13 * TRIFLUOROCHLOROMETHANE (DOT) * TRIFLUOROMETHYL CHLORIDE * TRIFLUORO-MONOCHLOROCARBON

CONSENSUS REPORTS: Reported in EPA TSCA Inventory.

DOT Classification: Nonflammable Gas; Label: Nonflammable Gas.

SAFETY PROFILE: A mild irritant. Narcotic in high concentrations. Reacts violently with Al. When heated to decomposition it emits highly toxic fumes of F^- and Cl^-.

CLS250   CAS: 17230-87-4   HR: 3
4-(4-(4-CHLORO-α,α,α-TRIFLUORO-m-TOLYL)-4-HYDROXYPIPERIDINO)BU-TYROPHENONE-4'-FLUOROHYDRO-CHLORIDE
mf: $C_{22}H_{22}ClF_4NO_2 \cdot ClH$ mw: 480.36

SYNS: CLOFLUPEROL HYDROCHLORIDE * R 9298 * SEPERIDOL * SEPEROL

SAFETY PROFILE: Poison by ingestion, subcutaneous, and intravenous routes. When heated to decomposition it emits very toxic fumes of Cl^-, F^- and NO_x.

CLT000   CAS: 1066-45-1   HR: 3
CHLOROTRIMETHYLSTANNANE
mf: C_3H_9ClSn mw: 199.26

PROP: Mp: 37°.

SYNS: CHLOROTRIMETHYLTIN * TRIMETHYL-CHLOROSTANNANE * TRIMETHYLCHLOROTIN * TRIMETHYLSTANNYL CHLORIDE * TRIMETHYL-TIN CHLORIDE

CONSENSUS REPORTS: EPA Extremely Hazardous Substances List. Reported in EPA TSCA Inventory.

OSHA PEL: TWA 0.1 mg(Sn)/m³ (skin)
ACGIH TLV: TWA 0.1 mg(Sn)/m³ (skin) (Proposed: TWA 0.1 mg(Sn)/m³; STEL 0.2 mg(Sn)/m³ (skin))
NIOSH REL: (Organotin Compounds) TWA 0.1 mg(Sn)/m³

SAFETY PROFILE: A deadly poison by intravenous route. When heated to decomposition it emits toxic fumes of Cl^-.

CLT250   CAS: 1943-16-4   HR: 3
CHLOROTRINITROMETHANE
mf: $CClN_3O_6$ mw: 185.49

SAFETY PROFILE: Poison by intraperitoneal route. Mildly toxic by inhalation. Potentially explosive. When heated to decomposition it emits very toxic fumes of Cl^- and NO_x.

CLU000   CAS: 639-58-7   HR: 3
CHLOROTRIPHENYLSTANNANE
mf: $C_{18}H_{15}ClSn$ mw: 385.47

PROP: Colorless crystals, insol in water, sol in organic solvents. Mp: 106°, bp: 240° @ 13.5 mm.

SYNS: AQUATIN * BRESTANOL * CHLOROTRI-PHENYLTIN * FENTIN CHLORIDE * GC 8993 * GENERAL CHEMICALS 8993 * HOE 2872 * LS 4442 * TINMATE * TPTC * TRIPHENYL-CHLOROSTANNANE * TRIPHENYLCHLOROTIN * TRIPHENYLTIN CHLORIDE

CONSENSUS REPORTS: Reported in EPA TSCA Inventory. EPA Extremely Hazardous Substances List.

OSHA PEL: TWA 0.1 mg(Sn)/m³ (skin)
ACGIH TLV: TWA 0.1 mg(Sn)/m³ (skin) (Proposed: TWA 0.1 mg(Sn)/m³; STEL 0.2 mg(Sn)/m³ (skin))
NIOSH REL: (Organotin Compounds) TWA 0.1 mg(Sn)/m³

SAFETY PROFILE: Poison by ingestion and intravenous routes. Experimental reproductive effects. An insect chemosterilant. When heated to decomposition it emits toxic fumes of Cl^-.

CLU250   CAS: 2279-76-7   HR: 3
CHLOROTRIPROPYLSTANNANE
mf: $C_9H_{21}ClSn$ mw: 283.44

PROP: Colorless liquid. Sol in organic solvents. D: 1.2678 @ 28°, mp: $-23.5°$.

SYNS: TRIPROPYLTIN CHLORIDE * TRI-n-PROPYL-TIN CHLORIDE

OSHA PEL: TWA 0.1 mg(Sn)/m^3 (skin)
ACGIH TLV: TWA 0.1 mg(Sn)/m^3 (skin) (Proposed: TWA 0.1 mg(Sn)/m^3; STEL 0.2 mg(Sn)/m^3 (skin))
NIOSH REL: (Organotin Compounds) TWA 0.1 mg(Sn)/m^3

SAFETY PROFILE: Poison by intravenous route. When heated to decomposition it emits toxic fumes of Cl$^-$.

CLU500   CAS: 10008-90-9   **HR: 3**
CHLORO(TRIVINYL)STANNANE
mf: C$_6$H$_9$ClSn   mw: 235.29

SYN: TRIVINYLTIN CHLORIDE

OSHA PEL: TWA 0.1 mg(Sn)/m^3 (skin)
ACGIH TLV: TWA 0.1 mg(Sn)/m^3 (skin) (Proposed: TWA 0.1 mg(Sn)/m^3; STEL 0.2 mg(Sn)/m^3 (skin))
NIOSH REL: TWA (Organotin Compounds) 0.1 mg(Sn)/m^3

SAFETY PROFILE: Poison by intravenous route. When heated to decomposition it emits toxic fumes of Cl$^-$.

CLV000   CAS: 541-25-3   **HR: 3**
CHLOROVINYLARSINE DICHLORIDE
mf: C$_2$H$_2$AsCl$_3$ mw: 207.31

PROP: Liquid, faint odor of geranium. Bp: 190° decomp, fp: −13°, d: 1.888 @ 20°/4°, vap press: 0.4 mm @ 20°, vap d: 7.15.

SYNS: (2-CHLOROETHENYL) ARSONOUS DICHLORIDE * β-CHLOROVINYLBICHLOROARSINE * 2-CHLORO-VINYLDICHLOROARSINE * (2-CHLOROVINYL)DI-CHLOROARSINE * DICHLORO(2-CHLOROVINYL)AR-SINE * LEWISITE * LEWISITE (ARSENIC COMPOUND)

CONSENSUS REPORTS: Reported in EPA TSCA Inventory. EPA Genetic Toxicology Program. EPA Extremely Hazardous Substances List. Arsenic and its compounds are on the Community Right-To-Know List.

SAFETY PROFILE: A human poison by inhalation and skin contact. Poison experimentally by inhalation, skin contact, subcutaneous, intraperitoneal and intravenous routes. A blistering type military poison. Lewisite is absorbed through skin, as little as 2 mL on the skin can cause death. Has a delayed action similar to distilled mustard gas. This gas exhibits a systemic poisoning effect on humans. When heated to decomposition it emits toxic fumes of Cl$^-$ and As.

CLW000   CAS: 88-04-0   **HR: 3**
4-CHLORO-3,5-XYLENOL
mf: C$_8$H$_9$ClO   mw: 156.62

PROP: Crystals, phenolic odor, sltly water-sol. Mp: 115.5°, bp: 246°.

SYNS: BENZYTOL * 4-CHLORO-3,5-DIMETHYLPHE-NOL * CHLORO-XYLENOL * p-CHLORO-m-XYL-ENOL * DESSON * DETTOL * ESPADOL * HUSEPT EXTRA * OTTASEPT * OTTASEPT EX-TRA * PCMX * RBA 777

CONSENSUS REPORTS: Reported in EPA TSCA Inventory. Chlorophenols are on the Community Right-To-Know List.

SAFETY PROFILE: Poison by intraperitoneal route. Moderately toxic by ingestion. An experimental teratogen. An antimicrobial agent. When heated to decomposition it emits toxic fumes of Cl$^-$.

CLW250   CAS: 50892-23-4   **HR: 3**
(4-CHLORO-6-(2,3-XYLIDINO)-2-PYRIMIDINYLTHIO)ACETIC ACID
mf: C$_{14}$H$_{14}$ClN$_3$O$_2$S   mw: 323.82

SYNS: ((4-CHLORO-6-((2,3-DIMETHYLPHENYL)AMINO)-2-PYRIMIDINYL)THIO)ACETIC ACID * WY-14,643

SAFETY PROFILE: Suspected carcinogen with experimental carcinogenic and tumorigenic data. Mutation data reported. When heated to decomposition it emits very toxic fumes of Cl$^-$, NO$_x$, and SO$_x$.

CLW500   CAS: 65089-17-0   **HR: 3**
2-((4-CHLORO-6-(2,3-XYLIDINO)-2-PYRIMIDINYL)THIO)-N-(2-HYDROXYETHYL)ACETAMIDE
mf: C$_{16}$H$_{19}$ClN$_4$O$_2$S   mw: 366.90

SYNS: BR-931 * PIRINIXIL

SAFETY PROFILE: Suspected carcinogen with experimental carcinogenic data. Mutation data reported. When heated to decomposition it emits very toxic fumes of Cl$^-$, NO$_x$, and SO$_x$.

CLZ000   CAS: 4611-02-3   **HR: 3**
CHLORPROETHAZINE HYDROCHLORIDE
mf: C$_{19}$H$_{23}$ClN$_2$S • ClH   mw: 383.41

SYN: 2-CHLORO-10-(3'-DIETHYLAMINOPROPYL)PHENO-
THIAZINE HYDROCHLORIDE

SAFETY PROFILE: Poison by ingestion, intravenous, intraperitoneal, and subcutaneous
routes. When heated to decomposition it emits
very toxic fumes of Cl^-, NO_x and SO_x.

CMA100   CAS: 2921-88-2   HR: 3
CHLORPYRIFOS
DOT: NA 2783
mf: $C_9H_{11}Cl_3NO_3PS$ mw: 350.59

SYNS: BRODAN * O,O-DIAETHYL-O-3,5,6-TRI-
CHLOR-2-PYRIDYLMONOTHIOPHOSPHAT (GERMAN)
* O,O-DIETHYL-O-3,5,6-TRICHLORO-2-PYRIDYL PHOS-
PHOROTHIOATE * DOWCO 179 * DURSBAN
* DURSBAN F * ENT 27,311 * ERADEX
* LORSBAN * OMS-0971 * PYRINEX * 3,5,6-
TRICHLORO-2-PYRIDINOL-O-ESTER with O,O-DIETHYL
PHOSPHOROTHIOATE

CONSENSUS REPORTS: EPA Genetic Toxicology Program.

OSHA PEL: TWA 0.2 mg/m³ (skin)
ACGIH TLV: TWA 0.2 mg/m³ (skin)
DOT Classification: ORM-A; Label: None.

SAFETY PROFILE: Poison by ingestion, skin
contact, inhalation, subcutaneous, and possibly
other routes. An experimental teratogen. Mutation data reported. When heated to decomposition it emits very toxic fumes of Cl^-, NO_x,
PO_x, and SO_x.

CMA250   CAS: 5598-13-0   HR: 3
CHLORPYRIFOS-METHYL
mf: $C_7H_7Cl_3NO_3PS$ mw: 322.53

SYNS: O,O-DIMETHYL-O-(3,5,6-TRICHLORO-2-PYRI-
DYL)PHOSPHOROTHIOATE * DOWCO 217 * DURS-
BAN METHYL * ENT 27,520 * METHYL CHLOR-
PYRIFOS * METHYL DURSBAN * NOLTRAN
* ENT 27520 * NSC 60380 * OMS-1155
* RELDAN * ZERTELL

SAFETY PROFILE: Poison by ingestion. A skin
irritant. A pesticide. When heated to decomposition it emits very toxic fumes of Cl^-, NO_x,
PO_x, and SO_x.

CMA750   CAS: 57-62-5   HR: 3
CHLORTETRACYCLINE
mf: $C_{22}H_{23}ClN_2O_8$ mw: 478.92

PROP: Golden yellow crystals. Mp: 168-169°.
Sltly sol in water; very sol in aq soln pH 7.65;
freely sol in the "cellosolves," dioxane, "Carbitol;" sol in methanol, ethanol, butanol, acetone, ethyl acetate, and benzene; insol in ether
and petroleum ether.

SYNS: ACRONIZE * AUREOCINA * AUREOMY-
CIN * AUREOMYCIN A-377 * AUREOMYKOIN
* BIOMITSIN * BIOMYCIN * 7-CHLORO-4-(DI-
METHYLAMINO)-1,4,4a,5,5a,6,11,12a-OCTAHYDRO-2-
NAPHTHACENECARBOXAMIDE * 7-CHLOROTETRA-
CYCLINE * CHRYSOMYKINE * CTC * DUOMY-
CIN * FLAMYCIN

CONSENSUS REPORTS: Reported in EPA
TSCA Inventory.

SAFETY PROFILE: Poison by intravenous,
subcutaneous, intracerebral, and intraperitoneal
routes. Moderately toxic by ingestion. Experimental reproductive effects. When heated to decomposition it emits toxic fumes of Cl^- and
NO_x.

CMC750   CAS: 67-97-0   HR: 3
CHOLECALCIFEROL
mf: $C_{27}H_{44}O$ mw: 384.71

PROP: White crystals; odorless. Insol in water;
sol in alc, chloroform, fatty oils.

SYNS: COLECALCIFEROL * 7-DEHYDROCHOLES-
TROL, ACTIVATED * DELSTEROL * DEPARAL
* D3-VIGANTOL * OLEOVITAMIN D3 * RICKE-
TON * 9,10-SECOCHOLESTA-5,7,10(19)-TRIEN-3-β-OL
* TRIVITAN * VIGORSAN * VITAMIN D3
* VITINC DAN-DEE-3

CONSENSUS REPORTS: Reported in EPA
TSCA Inventory.

SAFETY PROFILE: Poison by ingestion. An
experimental teratogen. When heated to decomposition it emits acrid smoke and irritating
fumes.

CMD750   CAS: 57-88-5   HR: 3
CHOLESTEROL
mf: $C_{27}H_{46}O$ mw: 386.73

PROP: White or faint yellow, pearly leaflets.
Mp: 148.5°, bp: 360° decomp.

SYNS: CHOLEST-5-EN-3-β-OL * Δ⁵-CHOLESTEN-3-β-
OL * 5-CHOLESTEN-3-β-OL * 5:6-CHOLESTEN-3-β-
OL * CHOLESTERIN * CHOLESTEROL BASE H
* CHOLESTERYL ALCOHOL * CHOLESTRIN

* CHOLESTROL * CORDULAN * DUSOLINE
* DUSORAN * DYTHOL * HYDROCERIN
* 3-β-HYDROXYCHOLEST-5-ENE * KATHRO
* LANOL * NIMCO CHOLESTEROL BASE H
* PROVITAMIN D * SUPER HARTOLAN * TEGO-
LAN

CONSENSUS REPORTS: IARC Cancer Review: GROUP 3 IMEMDT 7,161,87; Human Inadequate Evidence IMEMDT 31,95,83; Animal Inadequate Evidence IMEMDT 10,99,76. Reported in EPA TSCA Inventory. EPA Genetic Toxicology Program.

SAFETY PROFILE: Experimental teratogenic and reproductive effects. Questionable carcinogen with experimental carcinogenic and tumorigenic data. Mutation data reported. Used in pharmaceutical and dermal preparations as an emulsifying agent. When heated to decomposition it emits acrid smoke and irritating fumes.

CME250   CAS: 3546-10-9   **HR: 3**
CHOLESTERYL-p-BIS(2-CHLORO-ETHYL)AMINO PHENYLACETATE
mf: $C_{39}H_{59}Cl_2NO_2$ mw: 644.89

SYNS: (p-BIS(2-CHLOROETHYL)AMINO)PHENYL)ACE-
TATE CHOLESTEROL * (p-(BIS(2-CHLOROETHYL)
AMINO)PHENYL)ACETIC ACID CHOLESTEROL ESTER
* (4-(BIS(2-CHLOROETHYL)AMINO)PHENYL)ACETIC
ACID CHOLESTERYL ESTER * 5-CHOLESTEN-3-β-OL
3-(p-(BIS(2-CHLOROETHYL)AMINO)PHENYL)ACETATE
* FENESTERIN * FENESTRIN * NCI-C01558
* NSC 104469 * PHENESTERINE * PHENESTRIN

CONSENSUS REPORTS: NCI Carcinogenesis Bioassay (gavage); Clear Evidence: mouse, rat NCITR* NCI-CG-TR-60,78.

SAFETY PROFILE: Suspected carcinogen with experimental carcinogenic and neoplastigenic data. When heated to decomposition it emits very toxic fumes of Cl^- and NO_x.

CMF750   CAS: 67-48-1   **HR: 3**
CHOLINE HYDROCHLORIDE
mf: $C_5H_{14}NO \cdot Cl$ mw: 139.65

PROP: Colorless to white hygroscopic crystals; slt odor of trimethylamine. Sol in water and alc.

SYNS: BIOCOLINA * CHLORIDE de CHOLINE
(FRENCH) * CHOLINE CHLORHYDRATE * CHO-
LINE CHLORIDE (FCC) * CHOLINIUM CHLORIDE

* HEPACHOLINE * (2-HYDROXYETHYL)TRIMETH-
YLAMMONIUM CHLORIDE * LIPOTRIL

CONSENSUS REPORTS: Reported in EPA TSCA Inventory.

SAFETY PROFILE: Poison by intraperitoneal and intravenous routes. Moderately toxic experimentally by ingestion and subcutaneous routes. Mutation data reported. A lipotropic agent which induces the reduction in fats contained in the liver. When heated to decomposition it emits toxic fumes of Cl^-, SO_x and NO_x.

CMG250   CAS: 306-40-1   **HR: 3**
CHOLINE SUCCINATE (2:1) (ESTER)
mf: $C_{14}H_{30}N_2O_4$ mw: 290.46

SYNS: ANECTINE * CHOLINE SUCCINATE (ester)
* DIACETYLCHOLINE * DICHOLINE SUCCINATE
* 2,2'-((1,4-DIOXO-1,4-BUTANEDIYL)BIS(OXY))BIS-
(N,N,N-TRIMETHYLETHANAMINIUM * DITILIN
* DITILINE * QUELICIN * SUCCINIC ACID DIES-
TER with CHOLINE * SUCCINOCHOLINE * SUC-
CINOYLCHOLINE * SUCCINYLBISCHOLINE * SUC-
CINYLDICHOLINE * SUXAMETHONIUM * SUXE-
METHONIUM

SAFETY PROFILE: Poison by ingestion, intraperitoneal, intravenous,and subcutaneous routes. Human systemic effects by intravenous route: changes in the trachea or bronchi. When heated to decomposition it emits toxic fumes of NO_x.

CMH000   CAS: 1066-30-4   **HR: 3**
CHROMIC ACETATE
DOT: NA 9101
mf: $C_6H_9O_6 \cdot Cr$ mw: 229.15

PROP: Gray, green powder or bluish-green pasty mass.

SYNS: CHROMIC ACETATE(III) * CHROMIUM ACE-
TATE * CHROMIUM(III) ACETATE * CHROMIUM
TRIACETATE

CONSENSUS REPORTS: IARC Cancer Review: GROUP 3 IMEMDT 7,165,87; Animal Inadequate Evidence IMEMDT 2,100,73; IMEMDT 23,205,80. Chromium and its compounds are on the Community Right-To-Know List. Reported in EPA TSCA Inventory.

OSHA PEL: TWA 0.5 mg(Cr)/m^3
ACGIH TLV: TWA 0.5 mg(Cr)/m^3
DOT Classification: ORM-E; Label: None.

SAFETY PROFILE: Moderately toxic by intravenous route. Questionable carcinogen with experimental tumorigenic data. Human mutation data reported. When heated to decomposition it emits acrid smoke and irritating fumes.

CMH250   CAS: 7738-94-5   HR: 3
CHROMIC ACID
mf: CrH_2O_4 mw: 118.02

SYNS: ACIDE CHROMIQUE (FRENCH) * CHROMIC(VI) ACID

CONSENSUS REPORTS: Reported in EPA TSCA Inventory. Chromium and its compounds are on the Community Right-To-Know List.

OSHA PEL: (Transitional: CL 1 mg/10m^3) CL 0.1 mg(CrO_3)/m^3
ACGIH TLV: TWA 0.05 mg(Cr)/m^3, Confirmed Human Carcinogen.
DFG MAK: Animal Carcinogen, Suspected Human Carcinogen.
NIOSH REL: TWA 0.025 mg(Cr(VI))/m^3; CL 0.05/15M

SAFETY PROFILE: Poison by subcutaneous route. Mutation data reported. A powerful oxidizer. A storage hazard; it may burst a sealed container due to carbon dioxide release. Potentially explosive reactions with oxidizable materials. May ignite on contact with acetone or alcohols. When heated to decomposition it emits acrid smoke and irritating fumes.

CMH500   HR: 3
CHROMIC ACID (mixture)
mf: CrO_3 mw: 99.98

PROP: Dark red crystals. Mp: 196°, d: 2.70, decomp @ 250° to Cr_2O_3 + O_2, a powerful oxidizer. Water-sol.

SYN: CHROMIC ACID MIXTURE, DRY (DOT)

CONSENSUS REPORTS: Chromium and its compounds are on the Community Right-To-Know List.

OSHA PEL: (Transitional: CL 1 mg/10m^3) CL 0.1 mg(CrO_3)/m^3
ACGIH TLV: TWA 0.05 mg(Cr)/m^3, Confirmed Human Carcinogen.
DFG MAK: Animal Carcinogen, Suspected Human Carcinogen.
NIOSH REL: TWA 0.025 mg(Cr(VI))/m^3; CL 0.05/15M
DOT Classification: Oxidizer; Label: Oxidizer.

SAFETY PROFILE: A poison. A powerful irritant of skin, eyes, and mucous membranes. Can cause a dermatitis, bronchoasthma, "chrome holes," damage to the eyes. Dangerously reactive. Incompatible with acetic acid, acetic anhydride, tetrahydronaphthalene, acetone, alcohols, alkali metals, ammonia, arsenic, bromine penta fluoride, butyric acid, n,n-dimethylformamide, hydrogen sulfide, peroxyformic acid, phosphorus, potassium hexacyanoferrate, pyridine, selenium, sodium, sulfur and many other materials.

CMH750   CAS: 1308-14-1   HR: 3
CHROMIC ACID (solution)
DOT: UN 1755
mf: CrH_3O_3 mw: 103.03

SYN: CHROMIC(III) HYDROXIDE

CONSENSUS REPORTS: Reported in EPA TSCA Inventory. Chromium and its compounds are on the Community Right-To-Know List.

OSHA PEL: (Transitional: CL 1 mg/10m^3) CL 0.1 mg(CrO_3)/m^3
ACGIH TLV: TWA 0.05 mg(Cr)/m^3
DFG MAK: Animal Carcinogen, Suspected Human Carcinogen.
NIOSH REL: TWA 0.025 mg(Cr(VI))/m^3; CL 0.05 mg/m^3/15M
DOT Classification: Corrosive Material; Label: Corrosive

SAFETY PROFILE: A poison by many routes. Dangerously reactive.

CMI250   CAS: 24613-89-6   HR: 3
CHROMIC CHROMATE
mf: Cr_3O_{12} • 2Cr   mw: 452.00

SYNS: CHROMIC ACID, CHROMIUM(3+) SALT (3:2) * CHROMIUM CHROMATE (MAK)

CONSENSUS REPORTS: IARC Cancer Review: Animal Sufficient Evidence IMEMDT 2,100,73. Reported in EPA TSCA Inventory. Chromium and its compounds are on the Community Right-To-Know List.

OSHA PEL: (Transitional: CL 1 mg/10m^3) CL 0.1 mg(CrO_3)/m^3
ACGIH TLV: TWA 0.05 mg(Cr)/m^3
DFG MAK: Animal Carcinogen, Suspected Human Carcinogen.
NIOSH REL: TWA 0.001 mg(Cr(VI))/m^3

SAFETY PROFILE: Confirmed carcinogen with experimental carcinogenic and neoplastigenic data. Very powerful oxidizer.

CMI500   CAS: 1308-31-2   **HR: 3**
CHROMITE (MINERAL)
mf: Cr_2FeO_4 mw: 223.85

SYNS: CHROME ORE * CHROMITE * CHROMITE ORE * IRON CHROMITE

CONSENSUS REPORTS: IARC Cancer Review: GROUP 3 IMEMDT 7,165,87; Animal Inadequate Evidence IMEMDT 23,205,80. Chromium and its compounds are on the Community Right-To-Know List.

OSHA PEL: TWA 0.5 mg(Cr)/m^3
ACGIH TLV: TWA 0.05 mg/m^3 (ore processing); Confirmed Human Carcinogen (ore processing)

SAFETY PROFILE: Confirmed human carcinogen during ore processing. Human mutation data reported.

CMI750   CAS: 7440-47-3   **HR: 3**
CHROMIUM
af: Cr   aw: 52.00

SYN: CHROME

CONSENSUS REPORTS: IARC Cancer Review: GROUP 3 IMEMDT 7,165,87; Animal Inadequate Evidence IMEMDT 23,205,80. NTP Fourth Annual Report On Carcinogens, 1984. Chromium and its compounds are on the Community Right-To-Know List. Reported in EPA TSCA Inventory.

OSHA PEL: TWA 1 mg/m^3
ACGIH TLV: TWA 0.5 mg/m^3

SAFETY PROFILE: Confirmed human carcinogen with experimental tumorigenic data. Human poison by ingestion with gastrointestinal effects. Powder will explode spontaneously in air. Ignites and is potentially explosive in atmospheres of carbon dioxide. Violent or explosive reaction when heated with ammonium nitrate. May ignite or react violently with bromine pentafluoride. Incandescent reaction with nitrogen oxide or sulfur dioxide. Incompatible with oxidants.

CMJ250   CAS: 10025-73-7   **HR: 3**
CHROMIUM CHLORIDE
mf: Cl_3Cr mw: 158.35
PROP: Bp: 1300° (subl).

SYNS: CHROMIC CHLORIDE * CHROMIUM(III) CHLORIDE (1:3) * CHROMIUM CHLORIDE, anhydrous * CHROMIUM TRICHLORIDE * C.I. 77295 * PURATRONIC CHROMIUM CHLORIDE * TRICHLOROCHROMIUM

CONSENSUS REPORTS: Reported in EPA TSCA Inventory. EPA Genetic Toxicology Program. Chromium and its compounds are on the Community Right-To-Know List. EPA Extremely Hazardous Substances List.

OSHA PEL: TWA 0.5 mg(Cr)/m^3
ACGIH TLV: TWA 0.5 mg(Cr)/m^3

SAFETY PROFILE: Poison by skin contact, inhalation, intramuscular, intravenous, and intraperitoneal routes. Moderately toxic by subcutaneous route. Experimental teratogenic and reproductive effects. Human mutation data reported. Reacts violently with lithium under nitrogen atmosphere. When heated to decomposition it emits toxic fumes of Cl$^-$.

CMJ500   **HR: 3**
CHROMIUM COMPOUNDS

CONSENSUS REPORTS: Chromium and its compounds are on the Community Right-To-Know List.

SAFETY PROFILE: Chromate salts are suspected human carcinogens producing tumors of the lungs, nasal cavity, and paranasal sinus. Chromic acid and its salts have a corrosive action on the skin and mucous membranes. The lesions are confined to the exposed parts, affecting chiefly the skin of the hands and forearms and the mucous membranes of the nasal septum. The characteristic lesion is a deep, penetrating ulcer, which, for the most part, does not tend to suppurate, and which is slow in healing. Small ulcers, about the size of a matchhead, may be found, chiefly around the base of the nails, on the knuckles, dorsum of the hands and forearms. These ulcers tend to be clean and progress slowly. They are frequently painless, even though quite deep. They heal slowly and leave scars. On the mucous membranes of the nasal septum, the ulcers are usually accompanied by purulent discharge and crusting. If exposure continues, perforation of the nasal septum may result but produces no deformity of the nose. Hexavalent compounds are more toxic than the trivalent. Eczematous dermatitis due to trivalent chromium compounds has been reported.

CMJ900   CAS: 1308-38-9   **HR: 3**
CHROMIUM(III) OXIDE (2:3)
mf: Cr_2O_3 mw: 152.00

SYNS: ANADOMIS GREEN * ANIDRIDE CROMIQUE
(FRENCH) * CASALIS GREEN * CHROME GREEN
* CHROME OCHER * CHROME OXIDE * CHROME
OXIDE GREEN * CHROMIA * CHROMIC ACID
* CHROMIC ACID GREEN * CHROMIC OXIDE
* CHROMIUM OXIDE * CHROMIUM(III) OXIDE
* CHROMIUM(3+) OXIDE * CHROMIUM SESQUI-
OXIDE * CHROMIUM(3+) TRIOXIDE * C.I. 77288
* C.I. No. 77278 * C.I. PIGMENT GREEN 17
* DICHROMIUM TRIOXIDE * 11661 GREEN
* GREEN CHROME OXIDE * GREEN CHROMIC
OXIDE * GREEN CINNABAR * GREEN ROUGE
* GUIGNER'S GREEN * LEAF GREEN * LEVANOX
GREEN GA * OIL GREEN * OXIDE of CHROMIUM
* ULTRAMARINE GREEN

CONSENSUS REPORTS: IARC Cancer Review: GROUP 3 IMEMDT 7,165,87; Animal Inadequate Evidence IMEMDT 23,205,80. Reported in EPA TSCA Inventory. Chromium and its compounds are on the Community Right-To-Know List.

OSHA PEL: TWA 0.5 mg(Cr)/m^3
ACGIH TLV: TWA 0.5 mg(Cr)/m^3
DFG MAK: Suspected Carcinogen.

SAFETY PROFILE: Suspected carcinogen with experimental tumorigenic data. Mutation data reported. Probably a severe eye, skin, and mucous membrane irritant. A powerful oxidizer. Reacts violently with CLF_3.

CMK000   CAS: 1333-82-0   **HR: 3**
CHROMIUM(VI) OXIDE (1:3)

DOT: UN 1463/NA 1463/UN 1755
mf: CrO_3 mw: 100.00

PROP: Red, rhombic, deliquescent crystals. D: 2.70, mp: 196°, bp: decomp, sol: 61.7 g/100 cc @ 0°, 67.45 g/100 cc @ 100°.

SYNS: ANHYDRIDE CHROMIQUE (FRENCH)
* ANIDRIDE CROMICA (ITALIAN) * CHROME (TRI-
OXYDE de) (FRENCH) * CHROMIC ACID * CHRO-
MIC(VI) ACID * CHROMIC ACID, solid (DOT)
* CHROMIC ACID, solution (DOT) * CHROMIC ANHY-
DRIDE (DOT) * CHROMIC TRIOXIDE (DOT)
* CHROMIUM OXIDE * CHROMIUM(VI) OXIDE
* CHROMIUM TRIOXIDE * CHROMIUM(6+) TRI-
OXIDE * CHROMIUM TRIOXIDE, anhydrous (DOT)

* CHROMO (TRIOSSIDO di) (ITALIAN) * CHROM-
SAEUREANHYDRID (GERMAN) * CHROMTRIOXID
(GERMAN) * CHROOMTRIOXYDE (DUTCH)
* CHROOMZUURANHYDRIDE (DUTCH) * MONO-
CHROMIUM OXIDE) * MONOCHROMIUM TRIOXIDE
* PURATRONIC CHROMIUM TRIOXIDE

CONSENSUS REPORTS: IARC Cancer Review: GROUP 1 IMEMDT 7,165,87; Animal Sufficient Evidence IMEMDT 23,205,80. NTP Fourth Annual Report On Carcinogens, 1984. EPA Genetic Toxicology Program. Chromium and its compounds are on the Community Right-To-Know List. Reported in EPA TSCA Inventory.

OSHA PEL: CL 0.1 mg(CrO_3)/m^3
ACGIH TLV: TWA 0.05 mg(Cr)/m^3; Confirmed Human Carcinogen
DFG MAK: 0.1 mg/m^3, Suspected Carcinogen.
NIOSH REL: TWA 0.025 mg(Cr(VI))/m^3; CL 0.05/15M
DOT Classification: Oxidizer; Label: Oxidizer, solid; Corrosive Material; Label: Corrosive, solution; Oxidizer; Label: Oxidizer, Corrosive, anhydrous, solid.

SAFETY PROFILE: Confirmed human carcinogen producing nasal and lung tumors. Experimental carcinogenic and tumorigenic data. Poison by ingestion, intraperitoneal, and subcutaneous routes. Experimental carcinogenic data. Experimental teratogenic and reproductive effects. Human mutation data reported. Corrosive. Probably a severe eye, skin, and mucous membrane irritant.

A powerful oxidizer. Explosive reaction with acetaldehyde, acetic acid + heat, acetic anhydride + heat, benzaldehyde, benzene, benzylthylaniline, butyraldehyde, 1,3-dimethyl-hexahydropyrimidone, diethyl ether, ethylacetate, isopropylacetate, methyl dioxane, pelargonic acid, pentyl acetate, phosphorus + heat, propionaldehyde and other organic materials or solvents. Forms a friction- and heat-sensitive explosive mixture with potassium hexacyanoferrate. Ignites on contact with alcohols, acetic anhydride + tetrahydronaphthalene, acetone, butanol, chromium(II) sulfide, cyclohexanol, dimethyl formamide, ethanol, ethylene glycol, methanol, 2-propanol, pyridine. Violent reaction with acetic anhydride + 3-methylphenol (above 75°C), acetylene, bromine pentafluoride, glycerol, hexamethylphosphoramide, peroxyformic acid, selenium, sodium amide. Incandescent reaction with alkali metals (e.g., sodium, potassium), ammonia, arsenic, butyric acid (above 100°C), chlorine trifluoride, hydrogen sulfide + heat, sodium + heat, and sulfur. Incompatible with N,N-dimethylformamide.

CMK400   CAS: 37224-57-0   HR: 2
CHROMIUM POTASSIUM ZINC OXIDE

SYNS: POTASSIUM ZINC CHROMATE   *   ZINC POTASSIUM CHROMATE

CONSENSUS REPORTS: IARC Cancer Review: Human Sufficient Evidence IMEMDT 23,205,80; Animal Sufficient Evidence IMEMDT 2,100,73. Chromium and its compounds, as well as zinc and its compounds, are on the Community Right-To-Know List.

OSHA PEL: (Transitional: 1 mg(CrO_3)/10m^3) CL 0.1 mg(CrO_3)/m^3
ACGIH TLV: TWA 0.01 mg(Cr)/m^3; Confirmed Human Carcinogen
DFG MAK: Human Carcinogen.
NIOSH REL: (Chromium (VI)) TWA 0.001 mg(Cr(VI))/m^3

SAFETY PROFILE: Confirmed carcinogen with experimental carcinogenic data. When heated to decomposition it emits toxic fumes of Cr$^-$ and Zn$^-$.

CMK500   CAS: 15930-94-6   HR: 3
CHROMIUM(6+)ZINC OXIDE HYDRATE (1:2:6:1)
mf: $CrO_4 \cdot H_2O_2 \cdot Zn_2 \cdot H_2O$ mw: 298.78

SYNS: BUTTERCUP YELLOW   *   CHROMIC ACID, ZINC SALT (1:2)   *   ZINC CHROMATE HYDROXIDE   *   ZINC CHROMATE(VI) HYDROXIDE   *   ZINC HYDROXYCHROMATE   *   ZINC YELLOW

CONSENSUS REPORTS: IARC Cancer Review: Human Sufficient Evidence IMEMDT 23,205,80; Animal Sufficient Evidence IMEMDT 2,100,73. Chromium and its compounds, as well as zinc and its compounds, are on the Community Right-To-Know List.

OSHA PEL: (Transitional: 1 mg(CrO_3)/10m^3) CL 0.1 mg(CrO_3)/m^3
ACGIH TLV: TWA 0.01 mg(Cr)/m^3; Confirmed Human Carcinogen
DFG MAK: Human Carcinogen.

NIOSH REL: (Chromium (VI)) TWA 0.001 mg(Cr(VI))/m^3

SAFETY PROFILE: Confirmed human carcinogen. Mutation data reported. When heated to decomposition it emits toxic fumes of ZnO.

CML125   CAS: 14977-61-8   HR: 3
CHROMYL CHLORIDE

DOT: UN 1758
mf: Cl_2CrO_2 mw: 154.90

PROP: Dark red liquid, musty burning odor. Mp: −96.5°, bp: 115.7°, d: 1.9145 @ 25°/4°, vap press: 20 mm @ 20°.

SYNS: CHLORURE de CHROMYLE (FRENCH)   *   CHROMIC OXYCHLORIDE   *   CHROMIUM CHLORIDE OXIDE   *   CHROMIUM DICHLORIDE DIOXIDE   *   CHROMIUM DIOXIDE DICHLORIDE   *   CHROMIUM-(VI) DIOXYCHLORIDE   *   CHROMIUM OXYCHLORIDE   *   CHROMOXYCHLORID (GERMAN)   *   CHROMYL-CHLORID (GERMAN)   *   CHROOMOXYLCHLORIDE (DUTCH)   *   CROMILE, CLORURO di (ITALIAN)   *   CROMO, OSSICLORURO di (ITALIAN)   *   DICHLORODIOXOCHROMIUM   *   DIOXODICHLOROCHROMIUM   *   OXYCHLORURE CHROMIQUE (FRENCH)

CONSENSUS REPORTS: Reported in EPA TSCA Inventory. Chromium and its compounds are on the Community Right-To-Know List.

OSHA PEL: TWA 0.5 mg(Cr)/m^3
ACGIH TLV: TWA 0.025 ppm
DFG MAK: Suspected Carcinogen.
NIOSH REL: TWA 0.001 mg(Cr(VI))/m^3
DOT Classification: Corrosive Material; Label: Corrosive.

SAFETY PROFILE: Probably a poison by various routes. Mutation data reported. Corrosive. A strong irritant. Hydrolyzes to form chromic and hydrochloric acids. A strong oxidizer and chlorinating agent. Violent reaction with water. Reacts violently with alcohol, ether, acetone, turpentine. Ignites or explodes on contact with non-metal halides (e.g., disulfur dichloride, phosphorus trichloride, and phosphorus tribromide), non-metal hydrides (e.g., hydrogen sulfide, and hydrogen phosphide), flowers of sulfur, moist phosphorus, sodium azide, and urea. During preparation can violently explode. Incompatible with ammonia, disulfur dichloride, organic solvents, phosphorus, phosphorus trichloride, sodium azide, and sulfur. When heated to decomposition it emits toxic fumes of Cl$^-$.

CML750   CAS: 491-59-8   **HR: 3**
CHRYSAROBIN
mf: $C_{15}H_{12}O_3$ mw: 240.27

PROP: Brownish to orange-yellow crystals.

SYNS: CHRYSOPHANIC ACID ANTHRANOL
* 3-METHYL-1,8,9-ANTHRACENETRIOL * 3-METH-
YLANTHRALIN * 1,8,9-TRIHYDROXY-3-METHYLAN-
THRACENE

SAFETY PROFILE: Poison by intraperitoneal
route. Mutation data reported. An irritant and
an allergen. Combustible when exposed to heat
or flame. When heated to decomposition it emits
acrid smoke and fumes.

CML810   CAS: 218-01-9   **HR: 3**
CHRYSENE
mf: $C_{18}H_{12}$ mw: 228.30

PROP: Occurs in coal tar. Is formed during
distillation of coal, in very small amount during
distillation or pyrolysis of many fats and oils.
Orthorhombic bipyramidal plates from benzene.
D: 1.274, mp: 254°. Sublimes easily in vacuum,
bp: 448°. Sltly sol in alc, ether, carbon bisulfide,
and glacial acetic acid; moderately sol in boiling
benzene; insol in water. Chrysene is generally
only sltly sol in cold organic solvents, but fairly
sol in these solvents when hot, including glacial
acetic acid.

SYNS: 1,2-BENZOPHENANTHRENE * BENZO(a)PHE-
NANTHRENE * 1,2-BENZPHENANTHRENE * BENZ-
(a)PHENANTHRENE * 1,2,5,6-DIBENZONAPHTHALENE
* RCRA WASTE NUMBER U050

CONSENSUS REPORTS: IARC Cancer Re-
view: GROUP 1 IMEMDT 7,56,87; Animal
Limited Evidence IMEMDT 32,247,83; Animal
Sufficient Evidence IMEMDT 3,159,73. EPA
Genetic Toxicology Program. Reported in EPA
TSCA Inventory.

OSHA PEL: 0.2 mg/m^3
ACGIH TLV: Suspected Human Carcinogen
DFG MAK: Animal Carcinogen, Suspected Hu-
   man Carcinogen.
NIOSH REL: (Chrysene) To be controlled as
   a carcinogen.

SAFETY PROFILE: Confirmed carcinogen with
experimental carcinogenic, neoplastigenic, and
tumorigenic data by skin contact. Human mutation data reported. When heated to decomposition it emits acrid smoke and fumes.

CMN000   CAS: 505-75-9   **HR: 3**
CICUTOXIN
mf: $C_{17}H_{22}O_2$ mw: 258.39

SYN: 8,10,12-HEPTADECATRIENE-4,6-DIYNE-1,14-DIOL,
(E,E,E)-(-)-

SAFETY PROFILE: Poison by ingestion, intravenous, and possibly other routes. When
heated to decomposition it emits acrid smoke
and irritating fumes.

CMO000   CAS: 2602-46-2   **HR: 3**
**C.I. DIRECT BLUE 6, TETRASODIUM
SALT**
mf: $C_{32}H_{20}N_6O_{14}S_4 \cdot 4Na$ mw: 920.69

PROP: A dye.

SYNS: AIREDALE BLUE 2BD * AIZEN DIRECT BLUE
2BH * AMANIL BLUE 2BX * ATLANTIC BLUE 2B
* ATUL DIRECT BLUE 2B * AZOCARD BLUE 2B
* AZOMINE BLUE 2B * BELAMINE BLUE 2B
* BENCIDAL BLUE 2B * BENZANIL BLUE 2B
* BENZO BLUE GS * BLUE 2B * BRASILAMINA
BLUE 2B * CALCOMINE BLUE 2B * CHLORAMINE
BLUE 2B * CHLORAZOL BLUE B * CHROME
LEATHER BLUE 2B * C.I. 22610 * CRESOTINE
BLUE 2B * DIACOTTON BLUE BB * DIAMINE BLUE
2B * DIAPHTAMINE BLUE BB * DIAZINE BLUE 2B
* DIAZOL BLUE 2B * DIPHENYL BLUE 2B
* DIRECT BLUE 6 * ENIANIL BLUE 2BN * FENA-
MIN BLUE 2B * FIXANOL BLUE 2B * HISPAMIN
BLUE 2B * INDIGO BLUE 2B * KAYAKU DIRECT
* MITSUI DIRECT BLUE 2BN * NAPHTAMINE BLUE
2B * NB2B * NCI-C54579 * NIAGARA BLUE 2B
* NIPPON BLUE BB * PARAMINE BLUE 2B
* PHENAMINE BLUE BB * PHENO BLUE 2B
* PONTAMINE BLUE BB * SODIUM DIPHENYL-4,4'-
BIS-AZO-2''-8''-AMINO-1''-NAPHTHOL-3'',6 '' DISULPHO-
NATE * TERTRODIRECT BLUE 2B * VONDACEL
BLUE 2B

CONSENSUS REPORTS: IARC Cancer Re-
view: Human Limited Evidence IMEMDT
29,311,82; Animal Sufficient Evidence
IMEMDT 29,311,82. NTP Fourth Annual Re-
port On Carcinogens, 1984. NCI Carcinogenesis
Bioassay (feed); Clear Evidence: rat NCITR*
NCI-CG-TR-108,78; No Evidence: mouse
NCITR* NCI-CG-TR-108,78. Reported in EPA
TSCA Inventory. Community Right-To-Know
List.

SAFETY PROFILE: Confirmed carcinogen with
experimental carcinogenic, neoplastigenic, and

tumorigenic data. Experimental teratogenic and reproductive effects. Mutation data reported. When heated to decomposition it emits very toxic fumes of NO_x, Na_2O and SO_x.

CMO250   CAS: 72-57-1   **HR: 3**
C.I. DIRECT BLUE 14, TETRASODIUM SALT
mf: $C_{34}H_{28}N_6O_{14}S_4 \cdot 4Na$ mw: 964.88

SYNS: AMANIL SKY BLUE * AMIDINE BLUE 4B * AZIDINE BLUE 3B * AZURRO DIRETTO 3B * BENCIDAL BLUE 3B * BENZAMINE BLUE * BENZO BLUE * BLEU DIAMINE * BLUE EMB * BRASILAMINA BLUE 3B * CENTRALINE BLUE 3B * CHLORAMINE BLUE * CHLORAZOL BLUE 3B * CHROME LEATHER BLUE 3B * C.I. 23850 * C.I. DIRECT BLUE 14 * CONGOBLAU 3B * CONGO BLUE * CRESOTINE BLUE 3B * DI-AMINE BLUE 3B * DIANILBLAU * DIANIL BLUE * DIAZINE BLUE 3B * DIPHENYL BLUE 3B * DIRECT BLUE 14 * HISPAMIN BLUE 3BX * NAPHTAMINE BLUE 2B * NAPHTHYLAMINE BLUE * NCI-C61289 * NIAGARA BLUE * ORION BLUE 3B * PARAMINE BLUE 3B * PARKIBLEU * PARKIPAN * PONTAMINE BLUE 3BX * PYRA-ZOL BLUE 3B * PYROTROPBLAU * RCRA WASTE NUMBER U236 * RENOLBLAU 3B * SODIUM DITOLYLDIAZOBIS-8-AMINO-1-NAPHTHOL-3,6-DISUL-FONATE * SODIUM DITOLYLDIAZOBIS-8-AMINO-1-NAPHTHOL-3,6-DISULPHONATE * TB * TRIANOL DIRECT BLUE 3B * TRIPAN BLUE * TRYPANBLAU (GERMAN) * TRYPAN BLUE * TRYPAN BLUE SODIUM SALT

CONSENSUS REPORTS: IARC Cancer Review: GROUP 2B IMEMDT 7,56,87; Animal Sufficient Evidence IMEMDT 8,267,75. EPA Genetic Toxicology Program. Reported in EPA TSCA Inventory.

SAFETY PROFILE: Suspected carcinogen with experimental carcinogenic, neoplastigenic, and tumorigenic data. Poison by intraperitoneal, intravenous, and subcutaneous routes. Experimental teratogenic and reproductive effects. Mutation data reported. When heated to decomposition it emits very toxic fumes of NO_x, Na_2O and SO_x.

CMO750   CAS: 16071-86-6   **HR: 3**
C.I. DIRECT BROWN
mf: $C_{31}H_{20}N_6O_9S \cdot Cu \cdot 2Na$ mw: 762.15

SYNS: AIZEN PRIMULA BROWN BRLH * AMANIL SUPRA BROWN LBL * ATLANTIC RESIN FAST BROWN BRL * BENZAMIL SUPRA BROWN BRLL * CALCO-DUR BROWN BRL * CHLORAMINE FAST BROWN BRL * CHROME LEATHER BROWN BRLL * C.I. 30145 * DERMA FAST BROWN W-GL * DIPHENYL FAST BROWN BRL * DIRECT BROWN 95 * NCI-C54568 * SATURN BROWN LBR * SOLAR BROWN PL * TETRAMINE FAST BROWN BRS

CONSENSUS REPORTS: IARC Cancer Review: Animal Limited Evidence IMEMDT 29,321,82; Human Limited Evidence IMEMDT 29,321,82; NCI Carcinogenesis Bioassay (feed); Clear Evidence: rat NCITR* NCI-CG-TR-108,78; No Evidence: mouse NCITR* NCI-CG-TR-108,78. Reported in EPA TSCA Inventory. Community Right-To-Know List.

SAFETY PROFILE: Suspected carcinogen with experimental carcinogenic and neoplastigenic data. Mutation data reported. When heated to decomposition it emits very toxic fumes of Na_2O, SO_x, and NO_x.

CMP500   CAS: 4553-89-3   **HR: 3**
C.I. FOOD BROWN 3, DISODIUM SALT
mf: $C_{27}H_{20}N_4O_9S_2 \cdot 2Na$ mw: 654.61

SYNS: C.I. 20285 * 2,4-DIHYDROXY-3,5-DI(4-SUL-PHO-1-NAPHTHYLAZO)BENZYL ALCOHOL, DISODIUM SALT

SAFETY PROFILE: Poison by intraperitoneal route. When heated to decomposition it emits very toxic fumes of NO_x, Na_2O, and SO_x.

CMP800   **HR: 3**
CIGARETTE REFINED TAR

SYNS: TAR, from tobacco * CIGARETTE TAR * COLOMBIAN BLACK TOBACCO CIGARETTE REFINED TAR * TOBACCO REFINED TAR * TOBACCO TAR * U.S. BLENDED LIGHT TOBACCO CIGARETTE RE-FINED TAR

SAFETY PROFILE: Suspected carcinogen with experimental carcinogenic data. Mutation data reported.

CMP969   CAS: 104-55-2   **HR: 3**
CINNAMALDEHYDE
mf: C_9H_8O mw: 132.17

PROP: Found in Ceylon and Chinese cinnamon oils. Yellowish, oily liquid; strong odor of cinnamon. D: 1.048-1.052, mp: −7.5°, bp: 246.0° (some decomp), d: 1.048-1.052 @ 25°/25°, refr index: 1.619-1.623, flash p: 248°F. Very sltly sol in water; misc with alc, ether, chloroform, fixed oils.

SYNS: BENZYLIDENEACETALDEHYDE * CASSIA ALDEHYDE * CINNAMAL * CINNAMYL ALDEHYDE * CINNIMIC ALDEHYDE * FEMA No. 2286 * NCI-C56111 * PHENYLACROLEIN * 3-PHENYL-ACROLEIN * 3-PHENYLPROPENAL * 3-PHENYL-2-PROPENAL * ZIMTALDEHYDE

CONSENSUS REPORTS: Reported in EPA TSCA Inventory.

SAFETY PROFILE: Poison by intravenous and parenteral routes. Moderately toxic by ingestion and intraperitoneal routes. A severe human skin irritant. Mutation data reported. Combustible liquid. May ignite after a delay period in contact with NaOH. When heated to decomposition it emits acrid smoke and fumes.

CMP975   CAS: 621-82-9   HR: 3
CINNAMIC ACID
mf: $C_9H_8O_2$ mw: 148.17

PROP: Occurs free and partly esterified in storax, balsam Peru or Tolu, oil of cinnamon, coca leaves. White monoclinic crystals; honey floral odor. D: (4/4) 1.2475, mp: 133°, bp: 300°, flash p: +212°F. One gram dissolves in about 2000 mL water at 25° (more sol in hot water), in 6 mL alc, 5 mL methanol, 12 mL chloroform. Freely sol in benzene, ether, acetone, glacial acetic acid, carbon disulfide, fixed oils.

SYNS: FEMA No. 2288 * PHENYLACRYLIC ACID * tert-β-PHENYLACRYLIC ACID * 3-PHENYLA-CRYLIC ACID * 3-PHENYLPROPENOIC ACID * 3-PHENYL-2-PROPENOIC ACID * ZIMTSAEURE (GERMAN)

CONSENSUS REPORTS: Reported in EPA TSCA Inventory.

SAFETY PROFILE: Poison by intravenous and intraperitoneal routes. Moderately toxic by ingestion. A skin irritant. Combustible liquid. When heated to decomposition it emits acrid smoke and fumes.

CMQ730   CAS: 103-54-8   HR: 2
CINNAMYL ACETATE
mf: $C_{11}H_{12}O_2$ mw: 176.23

PROP: Colorless liquid; sweet floral odor. D: 1.047-1.051, refr index: 1.539-1.543, flash p: 244°F. Misc with chloroform, ether, fixed oils; insol in glycerin, water @ 264°.

SYNS: ACETIC ACID, CINNAMYL ESTER * FEMA No. 2293 * γ-PHENYLALLYL ACETATE * 3-PHE-NYL-2-PROPEN-1-YL ACETATE

CONSENSUS REPORTS: Reported in EPA TSCA Inventory.

SAFETY PROFILE: Moderately toxic by ingestion and intraperitoneal routes. Combustible liquid. When heated to decomposition it emits acrid smoke and fumes.

CMQ740   CAS: 104-54-1   HR: 2
CINNAMYL ALCOHOL
mf: $C_9H_{10}O$ mw: 134.19

PROP: Occurs (in the esterified form) in storax and in balsam Peru, cinnamon leaves, hyacinth oil. Needles or crystalline mass; odor of hyacinth. Mp: 33°, d: 1.0397, bp: 250.0°, n (20/D) 1.58190. Sol in water, glycerol, propylene glycol. Freely sol in alc, ether, other common organic solvents.

SYNS: CINNAMIC ALCOHOL * CINNAMYL ALCO-HOL, SYNTHETIC * FEMA No. 2294 * γ-PHENYLAL-LYL ALCOHOL * 3-PHENYLALLYL ALCOHOL * 3-PHENYL-2-PROPEN-1-OL * STYRONE * STYRYL CARBINOL

CONSENSUS REPORTS: Reported in EPA TSCA Inventory.

SAFETY PROFILE: Moderately toxic by ingestion. A skin irritant. When heated to decomposition it emits acrid smoke and fumes.

CMR250   CAS: 64043-53-4   HR: 3
d-CINNAMYLEPHEDRINE HYDROCHLORIDE
mf: $C_{19}H_{23}NO \cdot ClH$ mw: 317.89

SYN: CINNAMYLEPHEDRINE HYDROCHLORIDE, DEX-TRO

SAFETY PROFILE: Poison by subcutaneous route. Human systemic effects by very small amounts administered intradermally: unspecified skin effects. When heated to decomposition it emits very toxic fumes of HCl and NO_x.

CMR500   CAS: 104-65-4   HR: 2
CINNAMYL FORMATE
mf: $C_{10}H_{10}O_2$ mw: 162.20

PROP: Colorless liquid; balsamic odor. D: 1.077-1.082, refr index: 1.550-1.556, flash p:

212°F. Misc with alc, chloroform, ether, fixed oils; insol in water @ 250°.

SYNS: CINNAMYL ALCOHOL, FORMATE * CINNA-MYL METHANOATE * FEMA No. 2299 * FORMIC ACID, CINNAMYL ESTER * 3-PHENYL-2-PROPEN-1-YL FORMATE

CONSENSUS REPORTS: Reported in EPA TSCA Inventory.

SAFETY PROFILE: Moderately toxic by ingestion. Combustible liquid. When heated to decomposition it emits acrid smoke and irritating fumes.

CMR800   HR: 1
CINNAMYL ISOVALERATE
mf: $C_{14}H_{18}O_2$ mw: 218.30

PROP: Colorless to sltly yellow liquid; spicy, floral, fruity odor. D: 0.991-0.996, refr index: 1.518-1.524, flash p: +212°F. Misc in alc, chloroform, ether, most oils; insol in glycerin, propylene glycol, water @ 313°.

SYN: FEMA No. 2302

SAFETY PROFILE: Combustible liquid. When heated to decomposition it emits acrid smoke and irritating fumes.

CMR850   HR: 1
CINNAMYL PROPIONATE
mf: $C_{12}H_{14}O_2$ mw: 190.24

PROP: Colorless to pale yellow liquid; spicy, fruity, balsamic odor. D: 1.029-1.033, refr index: 1.523-1.537, flash p: +212°F. Misc in alc, chloroform, ether, most oils; insol in glycerin, propylene glycol, water @ 289°.

SYN: FEMA No. 2301

SAFETY PROFILE: Combustible liquid. When heated to decomposition it emits acrid smoke and irritating fumes.

CMS750   CAS: 77-92-9   HR: 3
CITRIC ACID
mf: $C_6H_8O_7$ mw: 192.14

PROP: Colorless, odorless crystals (crystals are monoclinic holohedra and crystallize from hot concd aq soln); acid taste. Mp: 153° (anhydrous form), bp: decomp; d: 1.665, flash p: +212°F. Sol in water, alc, ether.

SYNS: ACILETTEN * CITRETTEN * CITRIC ACID, anhydrous * CITRO * FEMA No. 2306

* 2-HYDROXY-1,2,3-PROPANETRICARBOXYLIC ACID
* β-HYDROXYTRICARBALLYLIC ACID * KYSELINA CITRONOVA (CZECH)

CONSENSUS REPORTS: Reported in EPA TSCA Inventory.

SAFETY PROFILE: Poison by intravenous route. Moderately toxic by subcutaneous and intraperitoneal routes. Mildly toxic by ingestion. A severe eye and moderate skin irritant. An irritating organic acid, some allergenic properties. Combustible liquid. Potentially explosive reaction with metal nitrates. When heated to decomposition it emits acrid smoke and fumes.

CMS845   CAS: 106-23-0   HR: 2
CITRONELLAL
mf: $C_{10}H_{18}O$ mw: 154.25

PROP: Colorless to sltly yellow liquid; intense lemon-citronella-rose odor. D: 0.850-0.860, refr index: 1.446-1.456, flash p: 170°F. Sol in alc, most oils; sltly sol in propylene glycol; insol glycerin, water.

SYNS: 3,7-DIMETHYL-6-OCTENAL * FEMA No. 2307

SAFETY PROFILE: Combustible liquid. When heated to decomposition it emits acrid smoke and irritating fumes.

CMS850   CAS: 107-75-5   HR: 1
CITRONELLAL HYDRATE
mf: $C_{10}H_{20}O_2$ mw: 172.30

PROP: Colorless liquid; sweet, floral, lily odor. D: 0.918-0.923, refr index: 1.447-1.450, flash p: +212°F. Sol in fixed oils, propylene glycol; insol in glycerin.

SYNS: CYCLALIA * CYCLOSIA * 3,7-DIMETHYL-7-HYDROXYOCTANAL * FEMA No. 2583 * FIXOL * HYDROXYCITRONELLAL (FCC) * 7-HYDROXYCITRONELLAL * 7-HYDROXY-3,7-DIMETHYLOCTAN-1-AL * 7-HYDROXY-3,7-DIMETHYL OCTANAL * LAURINE * LILYL ALDEHYDE * MUSUET SYNTHETIC * MUSUETTINE PRINCIPLE * PHIXIA

CONSENSUS REPORTS: Reported in EPA TSCA Inventory.

SAFETY PROFILE: A skin irritant. Combustible liquid. When heated to decomposition it emits acrid smoke and irritating fumes.

CMT250   CAS: 106-22-9   HR: 3
CITRONELLOL
mf: $C_{10}H_{20}O$ mw: 156.30

PROP: Colorless oily liquid; rose odor. D: 0.850-0.860, refr index: 1.454-1.462, flash p: 215°F. Sol in fixed oils, propylene glycol; sltly sol in water; insol in glycerin @ 225°.

SYNS: CEPHROL * 2,6-DIMETHYL-2-OCTEN-8-OL * 3,7-DIMETHYL-6-OCTEN-1-OL * FEMA No. 2309 * FEMA No. 2980 * RHODINOL * RODINOL

CONSENSUS REPORTS: Reported in EPA TSCA Inventory.

SAFETY PROFILE: Poison by intravenous route. Moderately toxic by ingestion, skin contact, and intramuscular routes. Combustible liquid. When heated to decomposition it emits acrid smoke and irritating fumes.

CMT600 HR: 1
CITRONELLYL BUTYRATE
mf: $C_{14}H_{26}O_2$ mw: 226.36

PROP: Colorless liquid; strong, fruity-rosy odor. D: 0.873-0.883; refr index: 1.444-1.448, flash p: +212°F. Misc in alc, ether, chloroform, most oils; insol water @ 245°.

SYNS: 3,7-DIMETHYL-6-OCTEN-1-YL BUTYRATE * FEMA No. 2312

SAFETY PROFILE: Combustible liquid. When heated to decomposition it emits acrid smoke and irritating fumes.

CMT750 CAS: 105-85-1 HR: 1
CITRONELLYL FORMATE
mf: $C_{11}H_{20}O_2$ mw: 184.31

PROP: Colorless liquid; strong, fruity odor. D: 0.890-0.93, refr index: 1.443-1.452, flash p: 197°F. Sol in alc, fixed oils; sltly sol in propylene glycol; insol in glycerin, water @ 235°.

SYNS: 3,7-DIMETHYL-6-OCTEN-1-OL FORMATE * 2,6-DIMETHYL-2-OCTEN-8-YL FORMATE * 3,7-DIMETHYL-6-OCTEN-1-YL FORMATE * FEMA No. 2314 * FORMIC ACID, CITRONELLYL ESTER * FORMIC ACID-3,7-DIMETHYL-6-OCTEN-1-YL ESTER

CONSENSUS REPORTS: Reported in EPA TSCA Inventory.

SAFETY PROFILE: Mildly toxic by ingestion. A human skin irritant. Combustible liquid. When heated to decomposition it emits acrid smoke and irritating fumes.

CMT900 HR: 1
CITRONELLYL ISOBUTYRATE
mf: $C_{14}H_{26}O_2$ mw: 226.36

PROP: Colorless liquid; rosy-fruity odor. D: 0.870-0.880, refr index: 1.440-1.448, flash p: +212°F. Misc in alc, chloroform, ether, most oils; insol in water @ 249°.

SYNS: 3,7-DIMETHYL-6-OCTEN-1-YL ISOBUTYRATE * FEMA No. 2313

SAFETY PROFILE: Combustible liquid. When heated to decomposition it emits acrid smoke and irritating fumes.

CMU100 HR: 1
CITRONELLYL PROPIONATE
mf: $C_{13}H_{24}O_2$ mw: 212.33

PROP: Colorless liquid; fruity-rosy odor. D: 0.877-0.886, refr index: 1.443-1.449, flash p: +212°F. Misc in alc, most oils; insol in water @ 242°.

SYN: FEMA No. 2316

SAFETY PROFILE: Combustible liquid. When heated to decomposition it emits acrid smoke and irritating fumes.

CMW000 CAS: 1622-61-3 HR: 2
CLOAZEPAM
mf: $C_{15}H_{10}ClN_3O_3$ mw: 315.73

SYNS: 5-(o-CHLOROPHENYL)-1,3-DIHYDRO-7-NITRO-2H-1,4-BENZODIAZEPIN-2-ONE * 5-(o-CHLOROPHE-NYL)-7-NITRO-1H-1,4-BENZODIAZEPIN-2(3H)-ONE * CLOAZEPAM * CLONAZEPAM * RIVOTRIL * RO 4-8180 * RO 5-4023

CONSENSUS REPORTS: EPA Genetic Toxicology Program.

SAFETY PROFILE: Moderately toxic by ingestion. Experimental teratogenic and reproductive effects. An anticonvulsant. When heated to decomposition it emits very toxic fumes of Cl^- and NO_x.

CMX700 CAS: 50-41-9 HR: 3
racemic-CLOMIPHENE CITRATE
mf: $C_{26}H_{28}ClNO \cdot C_6H_8O_7$ mw: 598.14

SYNS: CHLORAMIPHENE * CHLORAMIPHENE CITRATE * 2-CHLORO-1-(p-(β-DIETHYLAMINOETHOXY) PHENYL)-1,2-DIPHENYLETHYLENE * 2-(p-(2-CHLORO-1,2-DIPHENYL VINYL)PHENOXY)TRIETHYLAMINE CITRATE (1:1) * CLOMID * CLOMIFEN CITRATE

* CLOMIFENO * CLOMIPHENE CITRATE * CLO-MIPHENE DIHYDROGEN CITRATE * CLOMIPHENE-R
* CLOMIPHINE * CLOMIVID * CLOMPHID
* 1-(p-(β-DIETHYLAMINO ETHOXY)PHENYL)-1,2-DIPHE-NYL-2-CHLOROETHYLENE CITRATE * DYNERIC
* GENOZYM * IKACLOMIN * MER-41
* MRL 41 * NSC 35770 * OMIFIN

CONSENSUS REPORTS: IARC Cancer Review: GROUP 3 IMEMDT 7,172,87; Human Inadequate Evidence IMEMDT 21,551,79; Animal Inadequate Evidence IMEMDT 21,551,79. EPA Genetic Toxicology Program.

SAFETY PROFILE: Moderately toxic by ingestion and intraperitoneal routes. Human reproductive effects by ingestion: changes in spermatogenesis and effects on testes, epididymis and sperm duct. Human teratogenic effects by an unspecified route: developmental abnormalities of the eye and ear. Experimental reproductive effects. Questionable carcinogen with experimental tumorigenic data. Used as a drug to induce ovulation and for the treatment of oligospermia. When heated to decomposition it emits very toxic fumes of Cl^- and NO_x.

CMX850   CAS: 2971-90-6   HR: 1
CLOPIDOL
mf: $C_7H_7Cl_2NO$ mw: 192.05

SYNS: COCCIDIOSTAT C * COYDEN * 3,5-DI-CHLORO-2,6-DIMETHYL-4-PYRIDINOL * LERBEK
* METHYLCHLOROPINDOL * METHYLCHLORPIN-DOL * METILCLORPINDOL

OSHA PEL: Total Dust: 15 mg/m³; Respirable Fraction: 5 mg/m³
ACGIH TLV: TWA 10 mg/m³

SAFETY PROFILE: A nuisance dust. When heated to decomposition it emits very toxic fumes of Cl^- and NO_x.

CMX880   CAS: 40665-92-7   HR: D
CLOPROSTENOL
mf: $C_{22}H_{29}ClO_6$ mw: 424.96

SYNS: ESTRUMATE * ICI 80996 * racemic-ICI 80,996 * I.C.I. LTD. COMPOUND NUMBER 80996

SAFETY PROFILE: Experimental reproductive effects. When heated to decomposition it emits toxic fumes of Cl^-.

CMY000   CAS: 17780-75-5   HR: 3
CLORGYLINE HYDROCHLORIDE
mf: $C_{13}H_{15}Cl_2NO \cdot ClH$ mw: 308.65

SYN: N-METHYL-N-PROPARGYL-3-(2,4-DICHLOROPHE-NOXY)PROPYLAMINE HYDROCHLORIDE

SAFETY PROFILE: Poison by ingestion, intraperitoneal, intravenous, and subcutaneous routes. When heated to decomposition it emits very toxic fumes of HCl and NO_x.

CMY100   CAS: 8015-97-2   HR: 2
CLOVE LEAF OIL MADAGASCAR

PROP: From steam distillation of leaves of *Eugenis caryophyllata* Thunberg (*Eugenia aromatica* L. Baill.) (Fam. *Myrtaceae*). Pale yellow liquid. Ref. index: 1.527-1.538 ZBJ000 20°. Sol in propylene glycol, fixed oils; insol in glycerin, mineral oil.

SYNS: CLOVE LEAF OIL * OILS, CLOVE LEAF

CONSENSUS REPORTS: Reported in EPA TSCA Inventory.

SAFETY PROFILE: Moderately toxic by ingestion and skin contact. A severe skin irritant. When heated to decomposition it emits acrid smoke and fumes.

CMY625   HR: 3
COAL CONVERSION MATERIALS, SRC-II HEAVY DISTILLATE

SYN: SRC-II HEAVY DISTILLATE

SAFETY PROFILE: Suspected carcinogen with experimental carcinogenic data by skin contact.

CMY635   HR: 2
COAL DUST

SYN: ANTHRACITE PARTICLES * COAL FACINGS
* COAL, GROUND BITUMINOUS (DOT) * COAL-MILLED * COAL SLAG-MILLED * SEA COAL

OSHA PEL: (Transitional: Respirable Quartz Fraction less than 5% SiO_2: TWA 2.4 mg/m³; Respirable Quartz Fraction greater than or equal to 5% SiO_2: 10 mg/m³) Respirable Quartz Fraction less than 5% SiO_2: TWA 2 mg/m³; Respirable Quartz Fraction greater than or equal to 5% SiO_2: 0.1 mg/m³
ACGIH TLV: TWA 2 mg/m³
DOT Classification: Flammable Solid; Label: Flammable Solid.

SAFETY PROFILE: Questionable carcinogen with experimental tumorigenic data. Variable toxicity depending upon SiO_2 content. Moderately flammable when exposed to heat, flame, or chemical reaction with oxidizers. Slightly explosive when exposed to flame.

CMY800 CAS: 8007-45-2 **HR: 3**
COAL TAR

DOT: UN 1999

SYNS: CARBO-CORT * CRUDE COAL TAR * ESTAR * IMPERVOTAR * LAV * LAVATAR * PIXALBOL * PIX CARBONIS * POLYTAR BATH * SUPERTAH * SYNTAR * TAR * TAR, COAL * TAR, liquid (DOT) * ZETAR

CONSENSUS REPORTS: IARC Cancer Review: GROUP 1 IMEMDT 7,175,87; Animal Sufficient Evidence IMEMDT 34,65,84; IMEMDT 35,83,85; IMEMDT 3,22,73; Human Sufficient Evidence IMEMDT 34,65,84; IMEMDT 3,22,73; Human Limited Evidence IMEMDT 35,83,85. NTP Fourth Annual Report On Carcinogens, 1984. Reported in EPA TSCA Inventory.

OSHA PEL: TWA 0.2 mg/m^3; Carcinogen
DFG MAK: Human Carcinogen.
NIOSH REL: TWA 0.1 mg/m^3
DOT Classification: Flammable or Combustible Liquid; Label: Flammable Liquid.

SAFETY PROFILE: Confirmed human carcinogen with experimental carcinogenic and tumorigenic data. Mutation data reported. A human and experimental skin irritant. When heated to decomposition it emits acrid smoke and irritating fumes.

CMY825 CAS: 8001-58-9 **HR: 3**
COAL TAR CREOSOTE

DOT: UN 1136

SYNS: AWPA #1 * BRICK OIL * COAL TAR OIL * COAL TAR OIL (DOT) * CREOSOTE * CREOSOTE, from COAL TAR * CREOSOTE OIL * CREOSOTE P1 * CREOSOTUM * CRESYLIC CREOSOTE * HEAVY OIL * LIQUID PITCH OIL * NAPHTHALENE OIL * PRESERV-O-SOTE * RCRA WASTE NUMBER U051 * TAR OIL * WASH OIL

CONSENSUS REPORTS: IARC Cancer Review: GROUP 2A IMEMDT 7,177,87; Animal Sufficient Evidence, Human Limited Evidence IMEMDT 35,83,85; Animal Sufficient Evicence IMEMDT 3,22,73. NTP Fourth Annual Report On Carcinogens, 1984. Reported in EPA TSCA Inventory.

NIOSH REL: TWA 0.1 mg/m^3 CHE fraction
DOT Classification: Flammable or Combustible Liquid; Label: Flammable Liquid.

SAFETY PROFILE: Confirmed carcinogen with experimental carcinogenic data. Moderately toxic by ingestion. Experimental reproductive effects. Mutation data reported. When heated to decomposition it emits acrid smoke and fumes.

CMY840 **HR: 3**
COAL TAR DYE, liquid (DOT)

SAFETY PROFILE: A corrosive material. Many of the coal tar dyes are quite harmless and are permitted for foods, drugs, and cosmetics. Some of them may be allergens or carcinogens. When heated to decomposition they emit acrid smoke and fumes.

CMZ100 CAS: 65996-93-2 **HR: 3**
COAL TAR PITCH VOLATILES

SYNS: PITCH * PITCH, COAL TAR

CONSENSUS REPORTS: IARC Cancer Review: GROUP 1 IMEMDT 7,174,87; Animal Sufficient Evidence, Human Sufficient Evidence IMEMDT 35,83,85; Human Sufficient Evidence IMEMDT 3,22,73. Reported in EPA TSCA Inventory.

OSHA PEL: TWA 0.2 mg/m^3; Carcinogen
ACGIH TLV: TWA 0.2 mg/m^3 (volatile), Confirmed Human Carcinogen
NIOSH REL: TWA 0.1 mg/m^3 CHE fraction

SAFETY PROFILE: Confirmed carcinogen with experimental carcinogenic and neoplastigenic data by skin contact. When heated to decomposition it emits acrid smoke and fumes.

CNA250 CAS: 7440-48-4 **HR: 3**
COBALT
af: Co aw: 58.93

PROP: Gray, hard, magnetic, ductile, somewhat malleable metal. Exists in two allotropic forms. At room temperature, the hexagonal form is more stable than the cubic form; both forms can exist at room temperature. Stable in air or toward water at ordinary temperatures. D 8.92, mp 1493°, bp about 3100°, Brinell hardness: 125, latent heat of fusion 62 cal/g, latent heat of vaporization 1500 cal/g, specific heat (15-100°): 0.1056 cal/g/°C. Readily sol in dil HNO$_3$;

very slowly attacked by HCl or cold H_2SO_4. The hydrated salts of cobalt are red, and the sol salts form red solns which become blue on adding concd HCl.

SYNS: AQUACAT * C.I. 77320 * COBALT-59 * KOBALT (GERMAN, POLISH) * NCI-C60311 * SUPER COBALT

CONSENSUS REPORTS: Reported in EPA TSCA Inventory. Cobalt and its compounds are on the Community Right-To-Know List.

OSHA PEL: (Transitional: TWA 0.1 mg/m³) TWA 0.05 mg/m³
ACGIH TLV: (metal, dust, and fume) TWA 0.05 mg(Co)/m³
DFG TRK: 0.5 mg/m³ calculated as cobalt in that portion of dust that can possibly be inhaled in the production of cobalt powder and catalysts; hard metal (tungsten carbide) and magnet production (processing of powder, machine pressing, and mechanical processing of unsintered articles); others 0.1 mg/m³ calculated as cobalt in that portion of dust that can possibly be inhaled. Animal Carcinogen, Suspected Human Carcinogen.

SAFETY PROFILE: Confirmed carcinogen with experimental neoplastigenic and tumorigenic data. Poison by intravenous, intratracheal, and intraperitoneal routes. Moderately toxic by ingestion. Inhalation of the dust may cause pulmonary damage. The powder may cause dermatitis. Ingestion of soluble salts produces nausea and vomiting by local irritation. Powdered cobalt ignites spontaneously in air. Flammable when exposed to heat or flame. Explosive reaction with hydrazinium nitrate, ammonium nitrate + heat, and 1,3,4,7-tetramethylisoindole (at 390°C). Ignites on contact with bromine pentafluoride. Incandescent reaction with acetylene or nitryl fluoride.

CNA500   CAS: 6147-53-1   **HR: 2**
COBALT ACETATE TETRAHYDRATE
mf: $C_4H_6O_4 \cdot Co \cdot 4H_2O$ mw: 249.11

SYNS: ACETIC ACID, COBALT(2+) SALT, TETRAHYDRATE * COBALT DIACETATE TETRAHYDRATE * COBALTOUS ACETATE TETRAHYDRATE * OCTAN KOBALTNATY (CZECH)

CONSENSUS REPORTS: Cobalt and its compounds are on the Community Right-To-Know List.

SAFETY PROFILE: Moderately toxic by ingestion. A skin and eye irritant. Human mutation data reported. When heated to decomposition it emits acrid smoke and irritating fumes.

CNA750   CAS: 11114-92-4   **HR: 3**
COBALT ALLOY, Co, Cr

SYNS: CHROMIUM-COBALT ALLOY * COBALT-CHROMIUM ALLOY * DIN 2.4602 * DIN 2.4964 * HASTELLOY C * HAYNES STELLITE 21 * HEV-4 * VITALLIUM * ZIMALLOY

CONSENSUS REPORTS: IARC Cancer Review: Animal Limited Evidence IMEMDT 23,205,80. NTP Fourth Annual Report On Carcinogens, 1984. Cobalt and its compounds, as well as chromium and its compounds, are on the Community Right-To-Know List.

OSHA PEL: TWA 1 mg(Cr)/m³; 0.1 mg(Co)/m³ (fume and dust)
ACGIH TLV: TWA 0.5 mg(Cr)/m³

SAFETY PROFILE: Confirmed carcinogen with experimental tumorigenic data. Violent reaction with molten Li.

CNB500   CAS: 10210-68-1   **HR: 3**
COBALT CARBONYL
mf: $C_8Co_2O_8$ mw: 341.94

PROP: Orange platelets. D: 1.87, mp: 51°, decomp above 52°. Decomp on exposure to air. Insol in water; sol in organic solvents.

SYNS: COBALT OCTACARBONYL * COBALT TETRACARBONYL * COBALT TETRACARBONYL DIMER * DI-mu-CARBONYLHEXACARBONYLDICOBALT * DICOBALT CARBONYL * DICOBALT OCTACARBONYL * OCTACARBONYLDICOBALT

CONSENSUS REPORTS: Reported in EPA TSCA Inventory. Cobalt and its compounds are on the Community Right-To-Know List. EPA Extremely Hazardous Substances List.

OSHA PEL: TWA 0.1 mg(Co)/m³
ACGIH TLV: TWA 0.1 mg(Co)/m³

SAFETY PROFILE: Poison by ingestion and inhalation. Decomposes in air to form a product which ignites spontaneously in air. When heated to decomposition it emits acrid smoke and fumes.

CNB599   CAS: 7646-79-9   **HR: 3**
COBALT(II) CHLORIDE
mf: Cl_2Co mw: 129.83

PROP: Blue powder. Mp: 724°, bp: 1049°, d: 3.348.

SYNS: COBALT DICHLORIDE * COBALT MURIATE * COBALTOUS CHLORIDE * COBALTOUS DICHLO-RIDE * KOBALT CHLORID (GERMAN)

CONSENSUS REPORTS: Reported in EPA TSCA Inventory. EPA Genetic Toxicology Program. Cobalt and its compounds are on the Community Right-To-Know List.

SAFETY PROFILE: Poison experimentally by ingestion, skin contact, intraperitoneal, intravenous, and subcutaneous routes. Moderately toxic to humans by ingestion. Human systemic effects by ingestion: anorexia, goiter (increased thyroid size), and weight loss. Experimental teratogenic and reproductive effects. Human mutation data reported. Questionable carcinogen with experimental carcinogenic data. Incompatible with metals (e.g., sodium and potassium). When heated to decomposition it emits toxic fumes of Cl^-.

CNB850   HR: 3
COBALT COMPOUNDS

CONSENSUS REPORTS: Cobalt and its compounds are on the Community Right-To-Know List.

DFG TRK: 0.5 mg/m^3 calculated as cobalt in that portion of dust that can possibly be inhaled in the production of cobalt powder and catalysts; hard metal (tungsten carbide) and magnet production (processing of powder, machine pressing, and mechanical processing of unsintered articles); others 0.1 mg/m^3 calculated as cobalt in that portion of dust that can possibly be inhaled. Animal Carcinogen, Suspected Human Carcinogen.

SAFETY PROFILE: Confirmed carcinogen with experimental neoplastigenic and tumorigenic data. Cobalt has a low toxicity by ingestion. Ingestion of soluble salts produces nausea and vomiting by local irritation. In animals, administration of cobalt salts produces an increase in the total red cell mass of the blood. In humans, a single case of poisoning with liver and kidney damage has been attributed to cobalt. Locally, cobalt has been shown to produce dermatitis and investigators have been able to demonstrate a hypersensitivity of the skin to cobalt. There have been reports of hematologic, digestive, and pulmonary changes in humans.

CNC230   CAS: 16842-03-8   HR: 3
COBALT HYDROCARBONYL
mf: C_4HCoO_4 mw: 171.98

CONSENSUS REPORTS: Cobalt and its compounds are on the Community Right-To-Know List.

OSHA PEL: TWA 0.1 mg(Co)/m^3
ACGIH TLV: TWA 0.1 mg(Co)/m^3

SAFETY PROFILE: Poison by inhalation.

CNE125   CAS: 10124-43-3   HR: 3
COBALT(II) SULFATE (1:1)
mf: $O_4S \cdot Co$ mw: 154.99

PROP: Red to lavender dimorphic, orthorhombic crystals. D: 3.71. Stable to 708°. Dissolves slowly in boiling water.

SYNS: COBALTOUS SULFATE * COBALT SULFATE * COBALT SULFATE (1:1) * COBALT (2+) SULFATE * COBALT(II) SULPHATE * SULFURIC ACID, COBALT(2+) SALT (1:1)

CONSENSUS REPORTS: Cobalt and its compounds are on the Community Right-To-Know List. EPA Genetic Toxicology Program. Reported in EPA TSCA Inventory.

SAFETY PROFILE: Poison by intravenous and intraperitoneal routes. Moderately toxic by ingestion. When heated to decomposition it emits toxic fumes of SO_x.

CNF250   CAS: 104-61-0   HR: 2
COCONUT ALDEHYDE
mf: $C_9H_{16}O_2$ mw: 156.25

PROP: Colorless to sltly yellow liquid; coconut odor. D: 0.958-0.966, refr index: 1.446-1.450, flash p: +212°F. Sol in alc, fixed oils, propylene glycol; insol in water.

SYNS: ALDEHYDE C-18 * γ-N-AMYLBUTYROLAC-TONE * FEMA No. 2781 * 4-HYDROXYNONANOIC ACID, γ-LACTONE * γ-NONALACTONE (FCC) * 1,4-NONALOLIDE * PRUNOLIDE

CONSENSUS REPORTS: Reported in EPA TSCA Inventory.

SAFETY PROFILE: Moderately toxic by ingestion. A skin irritant. Combustible liquid. When heated to decomposition it emits acrid smoke and irritating fumes.

CNH000 CAS: 9004-70-0 **HR: 2**
COLLODION
$C_{12}H_{16}O_6(NO_3)_4C_{13}H_{17}O_7(NO_3)_3$ mw: 975

PROP: Soln of nitrated cellulose in ether + alcohol. Flash p: <0°F.

CONSENSUS REPORTS: Reported in EPA TSCA Inventory.

DOT Classification: Flammable Liquid; Label: Flammable Liquid.

SAFETY PROFILE: Very dangerous fire hazard when exposed to heat or flame. To fight fire, use alcohol foam. When heated to decomposition it emits toxic fumes of NO_x.

CNH792 CAS: 8001-61-4 **HR: 2**
COPAIBA OIL

PROP: From steam distillation of South American *Copaifera* L. (Fam. *Leguminosae*) balsam. Colorless to yellow liquid; characteristic odor, aromatic, slightly bitter taste. D: 0.880-0.907; ref. index: 1.493-1.500 @ 20°. Sol in alc, fixed oils, mineral oil.

SYNS: BALSAM CAPTIVI * BALSAMS, COPAIBA * COPAIBA BALSAM * COPAIBA OLEORESIN * JESUIT'S BALSAM

CONSENSUS REPORTS: Reported in EPA TSCA Inventory.

SAFETY PROFILE: Mildly toxic by ingestion. Large doses cause vomiting and diarrhea. Can also cause dermatitis and kidney damage. When heated to decomposition it emits acrid smoke and irritating fumes.

CNI000 CAS: 7440-50-8 **HR: 3**
COPPER
af: Cu aw: 63.54

PROP: A metal with a distinct reddish color. Mp: 1083°, bp: 2324°, d: 8.92, vap press: 1 mm @ 1628°.

SYNS: ALLBRI NATURAL COPPER * ANAC 110 * ARWOOD COPPER * BRONZE POWDER * CDA 101 * CDA 102 * CDA 110 * CDA 122 * C.I. 77400 * C.I. PIGMENT METAL 2 * COPPER-AIRBORNE * COPPER BRONZE * COPPER-MILLED * COPPER SLAG-AIRBORNE * COPPER SLAG-MILLED * 1721 GOLD * GOLD BRONZE * KAFAR COPPER * M1 (COPPER) * M2 (COPPER) * OFHC Cu * RANEY COPPER

CONSENSUS REPORTS: Reported in EPA TSCA Inventory. Copper and its compounds are on the Community Right-To-Know List.

OSHA PEL: TWA (dust, mist) 1 mg(Cu)/m³; (fume) 0.1 mg/m³
ACGIH TLV: TWA (dust, mist) 1 mg(Cu)/m³; (fume) 0.2 mg/m³
DFG MAK: (dust) 1 mg/m³; (fume) 0.1 mg/m³

SAFETY PROFILE: Questionable carcinogen with experimental tumorigenic data. Experimental teratogenic and reproductive effects. Human systemic effects by ingestion: nausea and vomiting. Liquid copper explodes on contact with water. Potentially explosive reaction with actylenic compounds, 3-bromopropyne, ethylene oxide, lead azide, and ammonium nitrate. Ignites on contact with chlorine, chlorine trifluoride, fluorine (above 121°), and hydrazinium nitrate (above 70°). Reacts violently with C_2H_2, bromates, chlorates, iodates, $(Cl_2 + OF_2)$, dimethyl sulfoxide + trichloroacetic acid, ethylene oxide, H_2O_2, hydrazine mononitrate, hydrazoic acid, $H_2S + air$, $Pb(N_3)_2$, K_2O_2, NaN_3, Na_2O_2, sulfuric acid. Incandescent reaction with potassium dioxide. Incompatible with 1-bromo-2-propyne.

CNI250 CAS: 142-71-2 **HR: 3**
COPPER ACETATE
DOT: NA 9106
mf: $C_4H_6O_4 \cdot Cu$ mw: 181.64

PROP: Greenish-blue powder or small crystals.

SYNS: ACETATE de CUIVRE (FRENCH) * ACETIC ACID, CUPRIC SALT * COPPER(2+) ACETATE * COPPER(II) ACETATE * COPPER DIACETATE * COPPER(2+) DIACETATE * CRYSTALLIZED VERDIGRIS * CRYSTALS of VENUS * CUPRIC ACETATE * CUPRIC DIACETATE * NEUTRAL VERDIGRIS * OCTAN MEDNATY (CZECH)

CONSENSUS REPORTS: Reported in EPA TSCA Inventory. Copper and its compounds are on the Community Right-To-Know List.

DOT Classification: ORM-E; Label: None.

SAFETY PROFILE: Poison by subcutaneous and intraperitoneal routes. Moderately toxic by ingestion. Experimental reproductive effects. When heated to decomposition it emits acrid smoke and irritating fumes.

CNI500   CAS: 12540-13-5   **HR: 3**
COPPER(II) ACETYLIDE
mf: C_2Cu mw: 87.56

PROP: A black or brown solid.

CONSENSUS REPORTS: Copper and its compounds are on the Community Right-To-Know List.

DOT Classification: Forbidden

SAFETY PROFILE: Ignites and then explodes when heated to 100°C. Much more sensitive to impact, friction, and heat than copper(I) acetylide (the red-brown form).

CNI750   CAS: 11133-98-5   **HR: 3**
COPPER ALLOY, Cu, Be

SYN: COPPER-BERYLLIUM ALLOY

CONSENSUS REPORTS: IARC Cancer Review: Animal Inadequate Evidence IMEMDT 23,143,80. Copper and its compounds, as well as beryllium and its compounds, are on the Community Right-To-Know List.

OSHA PEL: (Transitional: TWA 0.002 mg(Be)/m^3; CL 0.005; Pk 0.025/30M/8H) TWA 0.002 mg(Be)/m^3; STEL 0.005 mg(Be)/m^3/30M; CL 0.025 mg(Be)/m^3
ACGIH TLV: TWA 0.002 mg(Be)/m^3, Suspected Human Carcinogen.
NIOSH REL: CL not to exceed 0.0005 mg(Be)/m^3

SAFETY PROFILE: Confirmed carcinogen. Cases of berylliosis have been reported from exposure to so called low beryllium alloys. When heated to decomposition it emits very toxic fumes of BeO.

CNI900   CAS: 16102-92-4   **HR: 3**
COPPER ARSENATE HYDROXIDE
mf: $AsCu_2HO_5$ mw: 283.01

PROP: A green solid.

SYNS: COPPER ARSENATE (BASIC)   *   CUPROUS ARSENATE, BASIC

CONSENSUS REPORTS: Copper and its compounds as well as arsenic and its compounds are on the Community Right-To-Know List.

NIOSH REL: CL 2 μg/m^3/15M

SAFETY PROFILE: A poison by various routes. When heated to decomposition it emits toxic fumes of As.

CNK500   CAS: 1344-67-8   **HR: 3**
COPPER(II) CHLORIDE (1:2)

DOT: UN 2802
mf: Cl_2Cu mw: 134.44

PROP: Yellowish-brown, hygroscopic powder. Mp: 498°, d: 3.054.

SYNS: COPPER CHLORIDE (DOT)   *   CUPRIC CHLORIDE   *   KIRTICOPPER

CONSENSUS REPORTS: Copper and its compounds are on the Community Right-To-Know List. Reported in EPA TSCA Inventory.

DOT Classification: ORM-B; Label: None.

SAFETY PROFILE: Poison by ingestion, subcutaneous, intravenous, and intraperitoneal routes. Mutation data reported. Can react violently with K and Na. When heated to decomposition it emits toxic fumes of Cl^-.

CNK700   CAS: 55158-44-6   **HR: 3**
COPPER-COBALT-BERYLLIUM

SYNS: COPPER ALLOY, Cu, Be, Co   *   BERYLLIUM-COPPER-COBALT ALLOY

CONSENSUS REPORTS: IARC Cancer Review: Animal Inadequate Evidence IMEMDT 23,143,80. Copper, cobalt, and beryllium and their compounds are on the Community Right-To-Know List.

OSHA PEL: (Transitional: TWA 0.002 mg(Be)/m^3; CL 0.005; Pk 0.025/30M/8H) TWA 0.002 mg(Be)/m^3; STEL 0.005 mg(Be)/m^3/30M; CL 0.025 mg(Be)/m^3
ACGIH TLV: TWA 0.002 mg(Be)/m^3, Suspected Human Carcinogen
NIOSH REL: (To Beryllium) CL not to exceed 0.0005 mg(Be)/m^3

SAFETY PROFILE: Confirmed carcinogen. When heated to decomposition it emits very toxic fumes of BeO.

CNK750   **HR: 3**
COPPER COMPOUNDS

CONSENSUS REPORTS: Copper and its compounds are on the Community Right-To-Know List.

SAFETY PROFILE: As the sublimed oxide, copper may be responsible for one form of metal fume fever. In animals, inhalation of copper dust has caused hemolysis of the red blood cells, deposition of hemofuscin in the liver and pancreas, and injury to the lung cells; injection of the dust has caused cirrhosis of the liver and pancreas, and a condition closely resembling hemochromatosis, or bronzed diabetes. However, considerable trial exposure to copper compounds has not resulted in such disease. As regards local effect, copper chloride and sulfate have been reported as causing irritation of the skin and conjunctivae which may be on an allergic basis. Cuprous oxide is irritating to the eyes and upper respiratory tract. Discoloration of the skin is often seen in persons handling copper, but this does not indicate any actual injury. There is an excess of cancer cases in the copper smelting industry. In humans the ingestion of a large quantity of copper sulfate has caused vomiting, gastric pain, dizziness, exhaustion, anemia, cramps, convulsions, shock, coma and death. Symptoms attributed to damage to the nervous system and kidney have been recorded, jaundice has been observed and, in some cases, the liver has been enlarged. Deaths have been reported to have occurred following the ingestion of as little as 27 grams of the salt, while other victims have recovered after having taken up to 120 grams. Many copper-containing compounds are used as fungicides. Many copper salts form highly unstable acetylides. Those formed in basic solutions from (Cu^+ salts + C_2H_2) are less stable than those formed from Cu^{++} salts. (copper salts + hydrazine) react strongly, and with nitro-methane are explosive.

CNL000   CAS: 544-92-3   **HR: 3**
COPPER CYANIDE
mf: CCuN   mw: 89.56

PROP: Monoclinic, white prisms. Mp: 473° in N_2, bp: decomp, d: 2.92.

SYNS: CUPRICIN * CUPROUS CYANIDE * RCRA WASTE NUMBER P029

CONSENSUS REPORTS: Reported in EPA TSCA Inventory. Cyanide and its compounds, as well as copper and its compounds, are on the Community Right-To-Know List.

DOT Classification: Poison B; Label: Poison.

SAFETY PROFILE: A poison. Reacts violently with magnesium. When heated to decomposition it emits very toxic CN^- and NO_x.

CNL250   CAS: 14763-77-0   **HR: 3**
COPPER(II) CYANIDE
DOT: UN 1587
mf: C_2CuN_2 mw: 115.58

PROP: Yellowish-green powder. Mp: decomp before melting.

SYNS: COPPER CYANIDE (DOT) * COPPER CYANAMIDE * CUPRIC CYANIDE (DOT) * CYANURE de CUIVRE (FRENCH)

CONSENSUS REPORTS: Copper and its compounds, as well as cyanide and its compounds, are on the Community Right-To-Know List.

DOT Classification: Poison B; Label: Poison.

SAFETY PROFILE: Poison by intraperitoneal route. Incompatible with magnesium. When heated to decomposition it emits toxic fumes of NO_x and CN^-.

CNM750   CAS: 3251-23-8   **HR: 2**
COPPER(II) NITRATE
DOT: UN 1479
mf: CuN_2O_6 mw: 187.55

PROP: Large, blue-green, deliquescent, orthorhombic crystals. D: 2.047, sublimes at 150-225°, mp: 255-256°. Sol in water, ethyl acetate, dioxane; dissolves in and reacts vigorously with ether.

SYNS: COPPER DINITRATE * COPPER(2+) NITRATE * CUPRIC DINITRATE * CUPRIC NITRATE (DOT)

CONSENSUS REPORTS: Copper and its compounds are on the Community Right-To-Know List. Reported in EPA TSCA Inventory.

DOT Classification: Oxidizer; Label: Oxidizer.

SAFETY PROFILE: Moderately toxic by ingestion. A severe eye and skin irritant. Potentially explosive reaction above 220°C with ammonium or potassium hexacyanoferrate(II). Reaction with ammonia + potassium amide gives explosive product. Violent reaction with acetic anhydride. May ignite on prolonged contact with paper. Concentrated solutions may ignite in contact with tin foil. Used as a fungicide, herbicide, and as a catalyst component in solid rocket fuel. When heated to decomposition it emits toxic fumes of NO_x.

CNN500 CAS: 10290-12-7 **HR: 3**
COPPER ORTHOARSENITE

DOT: UN 1586
mf: $AsCuHO_3$ mw: 187.47

PROP: Yellowish-green powder. Mp: decomp.

SYNS: ACID COPPER ARSENITE * AIR-FLO GREEN * ARSONIC ACID, COPPER(2+) SALT (1:1) (9CI) * COPPER ARSENITE, solid (DOT) * CUPRIC ARSENITE * CUPRIC GREEN * SCHEELES GREEN * SCHEELE'S MINERAL * SWEDISH GREEN

CONSENSUS REPORTS: Arsenic and its compounds, as well as copper and its compounds, are on the Community Right-To-Know List.

OSHA PEL: TWA 0.01 mg(As)/m^3
ACGIH TLV: TWA 0.2 mg(As)/m^3
NIOSH REL: CL 0.002 mg(As)/m^3/15M
DOT Classification: Poison B; Label: Poison.

SAFETY PROFILE: Poison. When heated to decomposition it emits toxic fumes of As.

CNP250 CAS: 7758-98-7 **HR: 3**
COPPER(II) SULFATE (1:1)

DOT: NA 9109
mf: $O_4S \cdot Cu$ mw: 159.60

PROP: Blue crystals or blue, crystalline granules or powder. D: 2.284.

SYNS: BCS COPPER FUNGICIDE * BLUE COPPER * BLUE STONE * BLUE VITRIOL * COPPER MONOSULFATE * COPPER SULFATE * CP BASIC SULFATE * CUPRIC SULFATE * KUPPERSULFAT (GERMAN) * ROMAN VITRIOL * SULFATE de CUIVRE (FRENCH) * SULFURIC ACID, COPPER(2+) SALT (1:1) * TNCS 53 * TRINAGLE

CONSENSUS REPORTS: Copper and its compounds are on the Community Right-To-Know List. Reported in EPA TSCA Inventory. EPA Genetic Toxicology Program.

DOT Classification: ORM-E; label: None.

SAFETY PROFILE: A human poison by ingestion. An experimental poison by ingestion, subcutaneous, parenteral, intravenous, and intraperitoneal routes. Human systemic effects by ingestion: gastritis, diarrhea, nausea or vomiting, damage to kidney tubules, and hemolysis. Questionable carcinogen with experimental tumorigenic data. Mutation data reported. Reacts violently with hydroxylamine, magnesium.

When heated to decomposition it emits toxic fumes of SO_x.

CNR000 CAS: 8001-31-8 **HR: 2**
COPRA (OIL)

DOT: UN 1363

PROP: From the kernel of the fruit of the coconut palm *Cocos nucifera*. Fatty solid or liquid; sweet, nutty taste.Mp: 21-27°.

SYNS: COCONUT BUTTER * COCONUT MEAL PELLETS, containing 6-13% moisture and no more than 10% residual fat (DOT) * COCONUT OIL (FCC) * COCONUT PALM OIL * COPRA (DOT) * COPRA PELLETS (DOT) * FREE COCONUT OIL

CONSENSUS REPORTS: Reported in EPA TSCA Inventory.

DOT Classification: ORM-C; Label: None; Flammable Solid; Label: None.

SAFETY PROFILE: Flammable solid when exposed to heat or flame. May spontaneously heat and ignite if stored wet and hot.

CNR735 CAS: 8008-52-4 **HR: 2**
CORIANDER OIL

PROP: From steam distillation of ripe fruit of *Coriandrum sativum* L. (Fam. *Umbelliferae*). Colorless liquid; characteristic odor and taste. D: 0.863-0.875, refr index: 1.462 @ 20°.

SYNS: OIL of CORIANDER * OILS, CORIANDER

CONSENSUS REPORTS: Reported in EPA TSCA Inventory.

SAFETY PROFILE: Moderately toxic by ingestion. Mutation data reported. A skin irritant. When heated to decomposition it emits acrid smoke and fumes.

CNS000 CAS: 8001-30-7 **HR: 1**
CORN OIL

PROP: Light yellow, clear, oily liquid; faint characteristic odor. Mp: −10°, flash p: 490°F (CC), d: 0.92, autoign temp: 740°F. From wet milling of *Zea mays* (85DIA2 2,70,77).

CONSENSUS REPORTS: Reported in EPA TSCA Inventory.

SAFETY PROFILE: Human skin irritant. An experimental teratogen. May be an allergen. Combustible liquid when exposed to heat or

flame. Dangerous spontaneous heating may occur during storage if leaks impregnate rags,
waste, etc. To fight fire, use CO_2, dry chemical.

CNT250   HR: 3
CORUNDUM FUME

PROP: Half finely divided alumina, half silica.

SAFETY PROFILE: Poison by intratracheal
route.

CNT750   HR: 2
COTTON DUST

ACGIH TLV: TWA 0.2 mg/m^3 (raw dust)
DFG MAK: 1.5 mg/m^3 (raw cotton)
NIOSH REL: CL 0.200 mg/m^3 lint-free

SAFETY PROFILE: Human pulmonary effects.
Causes a mild febrile condition of the lungs
resembling metal fume fever. Coarser grades
of cotton contain more dust than the finer varieties, and therefore constitute a greater hazard.
It is considered an inert dust and indeed it is,
within the meaning of the term. However, it
can cause some illness, due to the allergens or
fungi in the cotton or on the dust. Workers in
processing rooms may develop conjunctivitis
or blepharitis from the burned products of the
gassing of the double yarn. It is a mild allergen.
Inhalation may produce bronchial asthma,
sneezing and eczema in sensitized persons.
Moderate fire and explosion hazard when exposed to heat or flame; can react with oxidizing
materials.

CNU000   CAS: 8001-29-4   HR: 3
COTTONSEED OIL (unhydrogenated)

PROP: Oily, pale yellow, nearly odorless liquid
from seeds of species of *Gossypium hirsutum*.
Flash p: 486°F (CC), fp: 0° to 5°, d: 0.915-
0.921 @ 25°/25°, autoign temp: 650°F.

SYNS: DEODORIZED WINTERIZED COTTONSEED OIL
* NCI-C50168

SAFETY PROFILE: Questionable carcinogen
with experimental tumorigenic data. Experimental teratogenic and reproductive effects. An
allergen. Combustible liquid when exposed to
heat or flame. However, if allowed to impregnate rags or oily waste, it can become a dangerous hazard due to spontaneous heating. To fight
fire, use CO_2, dry chemical.

CNU750   CAS: 56-72-4   HR: 3
COUMAPHOS

DOT: UN 2783
mf: $C_{14}H_{16}ClO_5PS$ mw: 362.78

SYNS: AGRIDIP * ASUNTHOL * BAYER 21/199
* BAYMIX 50 * 3-CHLORO-7-HYDROXY-4-METHYL-
COUMARIN-O,O-DIETHYL PHOSPHOROTHIOATE
* 3-CHLORO-7-HYDROXY-4-METHYL-COUMARIN-O-ES-
TER with O,O-DIETHYL PHOSPHOROTHIOATE
* O-3-CHLORO-4-METHYL-7-COUMARINYL-O,O-DI-
ETHYL PHOSPHOROTHIOATE * 3-CHLORO-4-METHYL-
7-COUMARINYL DIETHYL PHOSPHOROTHIOATE
* 3-CHLORO-4-METHYL-7-HYDROXYCOUMARIN DI-
ETHYL THIOPHOSPHORIC ACID ESTER * 3-CHLORO-4-
METHYLUMBELLIFERONE-O-ESTER with O,O-DIETHYL
PHOSPHOROTHIOATE * CUMAFOS (DUTCH)
* O,O-DIAETHYL-O-(3-CHLOR-4-METHYL-CUMARIN-7-
YL)-MONOTHIOPHOSPHAT (GERMAN) * O,O-DI-
ETHYL-O-(3-CHLOOR-4-METHYL-CUMARIN-7-YL)MONO-
THIOFOSFAAT (DUTCH) * O,O-DIETHYL-O-(3-
CHLORO-4-METHYL-7-COUMARINYL)PHOSPHORO-
THIOATE * O,O-DIETHYL-O-(3-CHLORO-4-METHYL-
COUMARINYL-7) THIOPHOSPHATE * O,O-DI-
ETHYL-O-(3-CHLORO-4-METHYL-2-OXO-2H-BENZOPY-
RAN-7-YL)PHOSPHOROTHIOATE * O,O-DIETHYL-3-
CHLORO-4-METHYL-7-UMBELLIFERONE THIOPHOS-
PHATE * O,O-DIETHYL-O-(3-CHLORO-4-
METHYLUMBELLIFERYL)PHOSPHOROTHIOATE
* DIETHYL-3-CHLORO-4-METHYLUMBELLIFERYL
THIONOPHOSPHATE * DIETHYL THIOPHOSPHORIC
ACIDESTER of 3-CHLORO-4-METHYL-7-HYDROXYCOU-
MARIN * O,O-DIETIL-O-(3-CLORO-4-METIL-CUMARIN-
7-IL-MONOTIOFOSFATO) (ITALIAN) * DIOLICE
* ENT 17,956 * MELDONE * NCI-C08662
* THIOPHOSPHATE de O,O-DIETHYLE et de O-
(3-CHLORO-4-METHYL-7-COUMARINYLE) (FRENCH)

CONSENSUS REPORTS: NCI Carcinogen-
esis Bioassay (feed); No Evidence: mouse, rat
NCITR* NCI-CG-TR-96,79. EPA Extremely
Hazardous Substances List.

DOT Classification: Poison B; Label: Poison;
Poison B; Label: Poison, liquid.

SAFETY PROFILE: Poison by ingestion, skin
contact, inhalation, intraperitoneal, and possibly
other routes. Mutation data reported. When
heated to decomposition, it emits very toxic
fumes of SO_x, PO_x, and Cl^-.

CNV000   CAS: 91-64-5   HR: 3
COUMARIN
mf: $C_9H_6O_2$ mw: 146.15

PROP: Crystals; fragrant, pleasant odor; burning taste. Mp: 70°, bp: 291.0°, vap press: 1 mm @ 106.0°.

SYNS: 2H-1-BENZOPYRAN-2-ONE * 1,2-BENZOPY-RONE * cis-o-COUMARINIC ACID LACTONE * COUMARINIC ANHYDRIDE * o-HYDROXYCIN-NAMIC ACID LACTONE * o-HYDROXYZIMTSAURE-LACTON (GERMAN) * NCI-C60297 * 2-OXO-1,2-BENZOPYRAN * RATTEX * TONKA BEAN CAM-PHOR

CONSENSUS REPORTS: IARC Cancer Review: GROUP 3 IMEMDT 7,56,87; Animal Limited Evidence IMEMDT 10,113,76. EPA Genetic Toxicology Program. Reported in EPA TSCA Inventory.

SAFETY PROFILE: Poison by ingestion, intraperitoneal, and subcutaneous routes. Questionable carcinogen with experimental tumorigenic data. Mutation data reported. Experimental teratogenic effects. Combustible when exposed to heat or flame. When heated to decomposition it emits acrid smoke and fumes.

CNW000   CAS: 136-78-7   **HR: 2**
CRAG HERBICIDE
mf: $C_8H_7Cl_2O_5S \cdot Na$ mw: 309.10

SYNS: CRAG HERBICIDE 1 * CRAG SESONE * 2,4-DES-Na * 2,4-DES-NATRIUM (GERMAN) * 2-(2,4-DICHLOROPHENOXY)ETHANOL HYDROGEN SULFATE SODIUM SALT * 2,4-DICHLOROPHENOXY-ETHYL SULFATE, SODIUM SALT * DISUL * DISUL-Na * DISUL-SODIUM * NATRIUM-2,4-DICHLORPHE-NOXYATHYLSULFAT (GERMAN) * SES * SESONE (ACGIH) * SODIUM-2-(2,4-DICHLOROPHENOXY)ETHYL SULFATE * SODIUM-2,4-DICHLOROPHENOXYETHYL SULPHATE * SODIUM-2,4-DICHLOROPHENYL CELLO-SOLVE SULFATE

OSHA PEL: (Transitional: TWA Total Dust: 15 mg/m^3; Respirable Fraction: 5 mg/m^3) TWA Total Dust: 10 mg/m^3; Respirable Fraction: 5 mg/m^3
ACGIH TLV: TWA 10 mg/m^3

SAFETY PROFILE: Moderately toxic by ingestion. Strong solutions are skin irritants. An herbicide. When heated to decomposition it emits very toxic fumes of Cl^-, Na_2O, and SO_x.

CNW500   CAS: 1319-77-3   **HR: 3**
CRESOL
DOT: UN 2022/UN 2076
mf: C_7H_8O mw: 108.15

PROP: Description (U.S.P. XVI): mixture of isomeric cresols obtained from coal tar, colorless or yellowish to brown-yellow or pinkish liquid, phenolic odor. Mp: 10.9-35.5°, bp: 191-203°, flash p: 178°F, d: 1.030-1.038 @ 25°/25°, vap press: 1 mm @ 38-53°, vap d: 3.72.

SYNS: ACEDE CRESYLIQUE (FRENCH) * BACILLOL * CRESOLI (ITALIAN) * CRESYLIC ACID * HY-DROXYTOLUOLE (GERMAN) * KRESOLE (GERMAN) * KRESOLEN (DUTCH) * KREZOL (POLISH) * RCRA WASTE NUMBER U052 * TEKRESOL * ar-TOLUENOL

CONSENSUS REPORTS: Community Right-To-Know List.

OSHA PEL: TWA 5 ppm (skin)
ACGIH TLV: TWA 5 ppm
DFG MAK: (all isomers) 5 ppm (22 mg/m^3)
NIOSH REL: TWA 10 mg/m^3
DOT Classification: Corrosive Material; Label: Corrosive; Poison B; Label: Poison.

SAFETY PROFILE: Moderately toxic by ingestion and skin contact. Corrosive to skin and mucous membranes. Systemic poisoning has rarely been reported, but it is possible that absorption may result in damage to the kidneys, liver, and nervous system. The main hazard accompanying its use in industry lies in severe chemical burns and dermatitis. Flammable when exposed to heat or flame; can react vigorously with oxidizing materials. Slightly explosive in the form of vapor when exposed to heat or flame. Explosive Range: 1.35% @ 300°F. Reacts violently with HNO_3, oleum, or chlorosulfonic acid. When heated to decomposition it emits highly toxic and irritating fumes. To fight fire, use foam, CO_2, dry chemical.

CNW750   CAS: 108-39-4   **HR: 3**
m-CRESOL
DOT: UN 2076
mf: C_7H_8O mw: 108.15

PROP: Colorless to yellowish liquid, phenolic odor. Mp: 10.9° bp: 202.8°, lel: 1.1% @ 302°F, flash p: 202°F, d: 1.034 @ 20°/4°, autoign temp: 1038°F, vap press: 1 mm @ 52.0°, vap d: 3.72.

SYNS: 3-CRESOL * m-CRESYLIC ACID * 1-HY-DROXY-3-METHYLBENZENE * m-HYDROXYTOLUENE * m-KRESOL * m-METHYLPHENOL * 3-METHYL-PHENOL * m-OXYTOLUENE * RCRA WASTE NUM-BER U052 * m-TOLUOL

CONSENSUS REPORTS: Community Right-To-Know List. Reported in EPA TSCA Inventory. EPA Genetic Toxicology Program.

OSHA PEL: TWA 5 ppm (skin)
ACGIH TLV: TWA 5 ppm
NIOSH REL: TWA 10 mg/m^3
DOT Classification: Poison B; Label: Poison.

SAFETY PROFILE: Poison by ingestion, intravenous, intraperitoneal, and subcutaneous routes. Moderately toxic by skin contact. Severe eye and skin irritant. Human mutation data reported. Questionable carcinogen with experimental neoplastigenic data. Flammable when exposed to heat or flame. Moderately explosive in the form of vapor when exposed to heat or flame.

CNX000   CAS: 95-48-7   HR: 3
o-CRESOL

DOT: UN 2076
mf: C$_7$H$_8$O   mw: 108.15

PROP: Crystals or liquid darkening with exposure to air and light. Mp: 30.8°, bp: 190.8°, flash p: 178°F, d: 1.047 @ 20°/4°, autoign temp: 1110°F, vap press: 1 mm @ 38.2°, vap d: 3.72, lel: 1.4% @ 300°F.

SYNS: 2-CRESOL * o-CRESYLIC ACID * 1-HY-DROXY-2-METHYLBENZENE * o-HYDROXYTOLUENE * o-KRESOL (GERMAN) * o-METHYLPHENOL * 2-METHYLPHENOL * ORTHOCRESOL * o-OXY-TOLUENE * RCRA WASTE NUMBER U052 * o-TOL-UOL

CONSENSUS REPORTS: EPA Extremely Hazardous Substances List. Community Right-To-Know List. EPA Genetic Toxicology Program. Reported in EPA TSCA Inventory.

OSHA PEL: TWA 5 ppm (skin)
ACGIH TLV: TWA 5 ppm
NIOSH REL: TWA 10 mg/m^3
DOT Classification: Poison B; Label: Poison.

SAFETY PROFILE: Poison by ingestion, inhalation, subcutaneous, intravenous, and intraperitoneal routes. Moderately toxic by skin contact. A severe eye and skin irritant. Human mutation data reported. Questionable carcinogen with experimental neoplastigenic data. Flammable when exposed to heat, flame, or oxidants. To fight fire, water may be used to blanket fire; foam, fog, mist, dry chemical.

CNX250   CAS: 106-44-5   HR: 3
p-CRESOL

DOT: UN 2076
mf: C$_7$H$_8$O   mw: 108.15

PROP: Found in a score of essential oils, including ylang-ylang and oil of jasmine. Crystals, phenolic odor. Mp: 35.5°, bp: 201.8°, lel: 1.1% @ 302°F, flash p: 202°F, d: 1.0341 @ 20°/4°, autoign temp: 1038°F, vap press 1 mm @ 53.0°, vap d: 3.72.

SYNS: 4-CRESOL * p-CRESYLIC ACID * 1-HY-DROXY-4-METHYLBENZENE * p-HYDROXYTOLUENE * 4-HYDROXYTOLUENE * p-KRESOL * 1-METHYL-4-HYDROXYBENZENE * p-METHYL-PHENOL * 4-METHYLPHENOL * p-OXYTOLUENE * PARAMETHYL PHENOL * RCRA WASTE NUM-BER U052 * p-TOLUOL * p-TOLYL ALCOHOL

CONSENSUS REPORTS: Community Right-To-Know List. Reported in EPA TSCA Inventory. EPA Genetic Toxicology Program.

OSHA PEL: TWA 5 ppm (skin)
ACGIH TLV: TWA 5 ppm
NIOSH REL: TWA 10 mg/m^3
DOT Classification: Poison B; Label: Poison.

SAFETY PROFILE: Poison by ingestion, skin contact, subcutaneous, intravenous, and intraperitoneal routes. A severe skin and eye irritant. Questionable carcinogen with experimental neoplastigenic data by itself and with 7,12-dimethyl benz(a)anthracene. Combustible when exposed to heat or flame. Moderately explosive in the form of vapor when exposed to heat or flame. To fight fire, use CO$_2$, dry chemical, alcohol foam.

COB250   CAS: 4170-30-3   HR: 3
CROTONALDEHYDE

DOT: UN 1143
mf: C$_4$H$_6$O   mw: 70.09

PROP: Water-white, mobile liquid; pungent suffocating odor. Bp: 104°, fp: −76.0°, lel: 2.1%, uel: 15.5%, flash p: 55°F, d: 0.853 @ 20°/20°, vap d: 2.41, autoign temp: 405°F.

SYNS: 2-BUTENAL * CROTONIC ALDEHYDE * KROTONALDEHYD (CZECH) * β-METHYL ACRO-LEIN * RCRA WASTE NUMBER U053

CONSENSUS REPORTS: EPA Extremely Hazardous Substances List. Reported in EPA TSCA Inventory.

OSHA PEL: TWA 2 ppm
DFG MAK: Suspected Carcinogen.
DOT Classification: Flammable Liquid; Label:
   Flammable Liquid and Poison.

SAFETY PROFILE: Suspected carcinogen with experimental carcinogenic data. Poison by ingestion and inhalation. Mutation data reported. An eye, skin, and mucous membrane irritant. A lachrymating material which can cause corneal burns and is very dangerous to the eyes. Caution: Keep away from heat and open flame. Keep container closed. Use with adequate ventilation. Extremely irritating to eyes, skin, mucous membranes. When necessary, the lacrimatory effect of the vapors may be counteracted by ammonia fumes. Dangerous fire hazard when exposed to heat or flame; can react with oxidizing materials. To fight fire, use alcohol foam, CO_2, dry chemical. Reacts violently with 1,3-butadiene. Violent hypergolic reaction with concentrated nitric acid. When heated to decomposition it emits acrid smoke and fumes.

COB260   CAS: 123-73-9   **HR: 3**
(E)-CROTONALDEHYDE
mf: C_4H_6O mw: 70.10

PROP: Water-white, mobile liquid; pungent, suffocating odor. Bp: 104°, fp: −76.0°, lel: 2.1%, uel: 15.5%, flash p: 55°F, d: 0.853 @ 20°/20°, vap d: 2.41, autoign temp: 450°F.

SYNS: ALDEHYDE CROTONIQUE (FRENCH) * trans-2-BUTENAL * (E)-2-BUTENAL * CROTONAL * CROTONALDEHYDE * CROTONIC ALDEHYDE * 1,2-ETHANEDIOL DIPROPANOATE (9CI) * ETHYLENE GLYCOL DIPROPIONATE (8CI) * ETHYLENE PROPIONATE * β-METHYL ACROLEIN * NCI-C56279 * PROPYLENE ALDEHYDE * RCRA WASTE NUMBER U053 * TOPANEL

CONSENSUS REPORTS: Reported in EPA TSCA Inventory.

OSHA PEL: TWA 2 ppm
ACGIH TLV: TWA 2 ppm
DFG MAK: Suspected Carcinogen.

SAFETY PROFILE: A poison by ingestion, subcutaneous, and intraperitoneal routes. Mutation data reported. A lachrymating material which is very dangerous to the eyes. Human respiratory system irritant by inhalation. Can cause corneal burns and is irritating to the skin. In case of contact, immediately flush the skin or eyes with water for at least 15 minutes and get medical attention. Dangerous fire hazard when exposed to heat or flame. To fight fire, use alcohol foam, CO_2, dry chemical. Incompatible with 1,3-butadiene and oxidizing materials. When heated to decomposition it emits acrid smoke and fumes.

COB500   CAS: 3724-65-0   **HR: 3**
CROTONIC ACID
DOT: UN 2823
mf: $C_4H_6O_2$ mw: 86.10

PROP: Colorless, needle-like crystals. Bp: 185°, mp: 72°, flash p: 190°F (COC), d: 1.018 @ 15°/4°, vap press: 0.19 mm @ 20°, vap d: 2.97.

SYNS: α-BUTENOIC ACID * 2-BUTENOIC ACID * CROTONIC ACID, solid * α-CROTONIC ACID * 3-METHYLACRYLIC ACID * β-METHYLACRYLIC ACID

CONSENSUS REPORTS: Reported in EPA TSCA Inventory.

DOT Classification: Corrosive Material; Label: Corrosive.

SAFETY PROFILE: Poison by intraperitoneal route. Moderately toxic by ingestion, skin contact, and subcutaneous routes. A powerful corrosive and irritant. Flammable when exposed to heat or flame; can react with oxidizing materials. To fight fire, use alcohol Foam, CO_2, dry chemical. When heated to decomposition it emits acrid smoke and irritating fumes.

COB750   CAS: 623-70-1   **HR: 2**
α-CROTONIC ACID ETHYL ESTER
DOT: UN 1862
mf: $C_6H_{10}O_2$ mw: 114.16

PROP: Colorless, monoclinic prisms or water-white liquid; pungent odor. Mp: 45° (solid). Bp: 209° (solid), 139° (liquid), flash p: 36.0°F, d: 0.9207 @ 20°/20°, vap d: 3.93.

SYNS: CROTONATE d'ETHYLE (FRENCH) * (E)-CROTONIC ACID, ETHYL ESTER * ETHYLCROTONATE

CONSENSUS REPORTS: Reported in EPA TSCA Inventory.

DOT Classification: Flammable Liquid; Label: Flammable Liquid.

SAFETY PROFILE: Moderately toxic by ingestion and probably by inhalation. A skin, mucous membrane, and severe eye irritant. Very dangerous fire hazard when exposed to heat or flame; can react vigorously with oxidizing materials. To fight fire, use foam, CO_2, or dry chemical. When heated to decomposition it emits acrid smoke and fumes.

COC500   CAS: 503-17-3   HR: 2
CROTONYLENE

DOT: UN 1144
mf: CH_3CCCH_3 mw: 54.09

PROP: Liquid. Bp: 27°, flash p: $< -4°F$, lel: 1.4%, d: 0.688 @ 25°, vap d: 1.91.

SYNS: 2-BUTYNE * DIMETHYLACETYLENE

CONSENSUS REPORTS: Reported in EPA TSCA Inventory.

DOT Classification: Flammable Liquid; Label: Flammable Liquid.

SAFETY PROFILE: A simple asphyxiant. Very dangerous fire hazard when exposed to heat or flame; can react with oxidizing materials. Moderately explosive in the form of vapor when exposed to heat or flame. To fight fire, use foam, CO_2, dry chemicals.

COD750   CAS: 68308-34-9   HR: 3
CRUDE SHALE OILS

DOT: UN 1288

SYNS: BLUE OIL * GREEN OIL * RAW SHALE OIL * SHALE OIL (DOT) * UNFINISHED LUBRICATING OIL

CONSENSUS REPORTS: IARC Cancer Review: GROUP 1 IMEMDT 7,339,87; Human Sufficient Evidence IMEMDT 35,161,85; Animal Limited Evidence IMEMDT 35,161,85; Animal Sufficient Evidence IMEMDT 3,22, 73.

DOT Classification: Flammable or Combustible Liquid; Label: Flammable Liquid.

SAFETY PROFILE: Confirmed human carcinogen with experimental carcinogenic, neoplastigenic, and tumorigenic data. Mildly toxic by ingestion, skin contact, and intraperitoneal routes. A skin irritant. Mutation data reported. Flammable when exposed to heat and flame. When heated to decomposition it emits acrid smoke and fumes.

COD850   CAS: 299-86-5   HR: 3
CRUFORMATE
mf: $C_{12}H_{19}ClNO_3P$ mw: 291.74

SYNS: AMIDOFOS * AMIDOPHOS * o-(4-tert BUTYL-2-CHLOOR-FENYL)-o-METHYL-FOSFORZUUR-N-METHYL-AMIDE (DUTCH) * 4-tert-BUTYL-2-CHLORO PHENYL METHYL METHYL PHOSPHORAMIDATE * o-(4-tert-BUTYL-2-CHLOR-PHENYL)-o-METHYL-PHOSPHORSAEURE-N-METHYL AMID (GERMAN) * 4-tert.-BUTYL 2-CHLOROPHENYL METHYLPHOSPHORAMIDATE de METHYLE (FRENCH) * o-(4-terz.-BUTIL-2-CLORO-FENIL)-o-METIL-FOSFORAMMIDE (ITALIAN) * CRUFOMATE * CRUFOMATE A * DOWCC 132 * ENT 25,602-X * MONTREL * o-METHYL-o-2-CHLORO-4-tert-BUTYLPHENYL-N-METHYLAMIDOPHOSPHATE * RUELENE * RUELENE DRENCH * RUELENE 25E * RULENE

OSHA PEL: TWA 5 mg/m^3
ACGIH TLV: TWA 5 mg/m^3

SAFETY PROFILE: A poison by ingestion and inhalation. Moderately toxic via skin contact and possibly other routes. Experimental reproductive effects. When heated to decomposition it emits very toxic fumes of PO_x, NO_x, and Cl^-.

COE000   CAS: 1309-32-6   HR: 3
CRYPTOHALITE

DOT: UN 2854
mf: $F_6Si \cdot 2H_4N$ mw: 178.19

PROP: Mp: subl, d: 2.01.

SYNS: AMMONIUM FLUOSILICATE * AMMONIUM HEXAFLUOROSILICATE * AMMONIUM SILICOFLUORIDE (DOT) * DIAMMONIUM HEXAFLUOROSILICATE * FLUOSILICATE de AMMONIUM (FRENCH)

NIOSH REL: TWA 2.5 mg(F)/m^3
DOT Classification: ORM-B; Label: None; Poison B; Label: St. Andrews Cross.

SAFETY PROFILE: Poison by ingestion and subcutaneous routes. When heated to decomposition it emits very toxic fumes of F^-, NH_3, and NO_x.

COE500   CAS: 122-03-2   HR: 2
CUMALDEHYDE
mf: $C_{10}H_{12}O$ mw: 148.22

PROP: Found in at least 50 essential oils such as cumin, eucalyptus species, cinnamon, boldo and rue, and as main constituent of oil of *Pectis papposa harn* and *gray*. Colorless to pale yellow liquid; pungent odor of cumin. D: 0.976-0.980, refr index: 1.529-1.534, flash p: 199°F. Sol in alc, ether; insol in water.

SYNS: p-CUMIC ALDEHYDE * CUMINALDEHYDE * CUMINIC ALDEHYDE (FCC) * CUMINYL ALDE-HYDE * FEMA No. 2341 * p-ISOPROPYLBENZALDE-HYDE * 4-ISOPROPYLBENZALDEHYDE * p-ISOPROPYLBENZENECARBOXALDEHYDE * 4-(1-METHYLETHYL)-BENZALDEHYDE (9CI)

CONSENSUS REPORTS: Reported in EPA TSCA Inventory.

SAFETY PROFILE: Moderately toxic by ingestion and skin contact. A skin irritant. Combustible liquid. When heated to decomposition it emits acrid smoke and irritating fumes.

COE750   CAS: 98-82-8   HR: 2
CUMENE
DOT: UN 1918
mf: C_9H_{12} mw: 120.21

PROP: Colorless liquid. Mp: $-96.0°$, bp: 152°, flash p: 111°F, d: 0.864 @ 20°/4°, vap press: 10 mm @ 38.3°, autoign temp: 795°F, lel: 0.9%, uel: 6.5%, vap d: 4.1.

SYNS: BENZENE ISOPROPYL * CUMEEN (DUTCH) * CUM * 2-FENILPROPANO (ITALIAN) * 2-FENYL-PROPAAN (DUTCH) * ISOPROPYLBENZEEN (DUTCH) * ISOPROPILBENZENE (ITALIAN) * ISOPROPYL BEN-ZENE * ISOPROPYLBENZOL * ISOPROPYL-BENZOL (GERMAN) * 2-PHENYLPROPANE * RCRA WASTE NUMBER U055

CONSENSUS REPORTS: Community Right-To-Know List. Reported in EPA TSCA Inventory. EPA Genetic Toxicology Program.

OSHA PEL: TWA 50 ppm (skin)
ACGIH TLV: TWA 50 ppm (skin)
DFG MAK: 50 ppm (245 mg/m^3)
DOT Classification: Flammable or Combustible Liquid; Label: Flammable Liquid.

SAFETY PROFILE: Moderately toxic by ingestion. Mildly toxic by inhalation and skin contact. Human systemic effects by inhalation: an antipsychotic, unspecified changes in the sense of smell and respiratory system. An eye and skin irritant. Potential narcotic action. Central nervous system depressant. There is no apparent difference between the toxicity of natural cumene or that derived from petroleum. Flammable when exposed to heat or flame; can react with oxidizing materials. Violent reaction with HNO_3, oleum, chlorosulfonic acid. To fight fire, use foam, CO_2, dry chemical.

COF000   CAS: 93-53-8   HR: 2
CUMENE ALDEHYDE
mf: $C_9H_{10}O$ mw: 134.19

PROP: Colorless liquid; floral odor. D: 0.998-1.006, refr index: 1.515-1.520, flash p: 156°F. Sol in fixed oils; sltly sol in propylene glycol; insol in glycerin.

SYNS: FEMA No. 2886 * α-FORMYLETHYLBENZENE * HYACINTHAL * HYDRATROP ALDEHYDE * HYDRATROPIC ALDEHYDE * α-METHYL PHENYL-ACETALDEHYDE * α-METHYL-α-TOLUIC ALDEHYDE * 2-PHENYLPROPANAL * α-PHENYLPROPIONAL-DEHYDE * 2-PHENYLPROPIONALDEHYDE (FCC)

CONSENSUS REPORTS: Reported in EPA TSCA Inventory.

SAFETY PROFILE: Moderately toxic by ingestion. Combustible liquid. When heated to decomposition it emits acrid smoke and irritating fumes.

COF325   CAS: 8014-13-9   HR: 2
CUMIN OIL
PROP: From steam distillation of *Cuminum cyminum* L. Light yellow to brown liquid; strong odor. D: 0.905-0.925, refr index: 1.501 @ 20°. Sol in fixed oils, mineral oil; very sol in glycerin, propylene glycol.

SYNS: CUMMIN * OILS, CUMIN

CONSENSUS REPORTS: Reported in EPA TSCA Inventory.

SAFETY PROFILE: Moderately toxic by ingestion and skin contact. A skin irritant. Mutation data reported. When heated to decomposition it emits acrid smoke and irritating fumes.

COF500   CAS: 12002-03-8   HR: 3
CUPRIC ACETOARSENITE
DOT: UN 1585
mf: $C_4H_6As_6Cu_4O_{16}$ mw: 1013.78

PROP: Emerald green powder.

SYNS: (ACETATO)(TRIMETAARSENITO)DICOPPER * ACETOARSENITE de CUIVRE (FRENCH) * BASLE GREEN * C.I. 77410 * C.I. PIGMENT GREEN 21 (9CI) * COPPER ACETOARSENITE (DOT) * COPPER

ACETOARSENITE, solid (DOT) * EMERALD GREEN * ENT 884 * FRENCH GREEN * GENUINE PARIS GREEN * IMPERIAL GREEN * KING'S GREEN * MEADOW GREEN * MINERAL GREEN * MITIS GREEN * MOSS GREEN * MOUNTAIN GREEN * NEUWIED GREEN * NEW GREEN * ORTHO P-G BAIT * PARIS GREEN * PARROT GREEN * PATENT GREEN * POWDER GREEN * SCHWEINFURTERGRUN * SCHWEINFURT GREEN * SOWBUG & CUTWORM BAIT * SWEDISH GREEN * VIENNA GREEN

CONSENSUS REPORTS: Arsenic and its compounds as well as copper and its compounds are on the Community Right-To-Know List.

OSHA PEL: TWA 0.5 mg(As)/m^3

DOT Classification: Poison B; Label: Poison.

SAFETY PROFILE: Poison by ingestion and and possibly other routes. An insecticide. When heated to decomposition it emits very toxic fumes of As.

COG000 CAS: 8024-37-1 **HR: 2**
CURCUMIN

SYNS: C.I. 75300 * CURCUMA OIL * CURCU-MINE * NCI-C60015 * TURMERIC OIL * TURMERIC OLEORESIN

CONSENSUS REPORTS: Reported in EPA TSCA Inventory.

SAFETY PROFILE: Moderately toxic by intraperitoneal route. A skin irritant. Mutation data reported. When heated to decomposition it emits acrid smoke and irritating fumes.

COH000 **HR: 3**
CUTTING OILS

SAFETY PROFILE: Often carcinogenic. The cause of ''cutting oil'' dermatitis is generally due to an insoluble oil. However it can occasionally be caused by a soluble oil. Many have looked for a causative factor other than the oil itself. Bacteria have frequently been blamed, although insoluble oils are usually sterile while the soluble oils may contain bacteria. The metal slivers which occur in these oils after use have also been blamed as well as the sulfur, chlorine, and inhibitors which they contain. The oil itself can plug the pores and form boils. Combustible when exposed to heat or flame.

COH250 CAS: 140-87-4 **HR: 3**
CYANACETIC ACID HYDRAZIDE
mf: C$_3$H$_5$N$_3$O mw: 99.11

PROP: Mp: 115°, a solid.

SYNS: AB-42 * ARMAZAL * CIANAZIL * CYACETACID * CYACETACIDE * CYACETAZID * CYACETAZIDE * CYANACETHYDRAZIDE * CYANACETIC ACID HYDRAZIDE * CYANACETO-HYDRAZIDE * CYANACETYLHYDRAZIDE * CYAN-AZIDE * CYANIZIDE * CYANOACETHYDRAZIDE * CYANOACETIC ACID HYDRAZIDE * CYANO-ACETOHYDRAZIDE * α-CYANOACETOHYDRAZIDE * CYANOACETYLHYDRAZIDE * CYANOETHYDRA-ZIDE * CYAZID * CYAZIDE * DICTYCIDE * DICTYZIDE * HELMOX * HIDACIAN * HIDACIANN * KYANACETHYDRAZID * LEAN-DIN * MACKREAZID * MALONITRILE HYDRAZIDE * MALONONITRILE HYDRAZIDE * NEOHYDRAZID * REACID * REAZID * REAZIDE * TSIAZID * USAF KF-18

CONSENSUS REPORTS: Cyanide and its compounds are on the Community Right To Know List.

SAFETY PROFILE: Poison by ingestion and intraperitoneal route. When heated to decomposition it emits toxic fumes of NO$_x$ and CN$^-$.

COH500 CAS: 420-04-2 **HR: 3**
CYANAMIDE
mf: CH$_2$N$_2$ mw: 42.05

PROP: Deliquescent crystals. Mp: 45°, bp: 260°, flash p: 285°F, d: 1.282, vap d: 1.45.

SYNS: AMIDOCYANOGEN * CARBAMONITRILE * CARBIMIDE * CYANOAMINE * CYANOGEN-AMIDE * CYANOGEN NITRIDE * HYDROGEN CYANAMIDE * USAF EK-1995

CONSENSUS REPORTS: Reported in EPA TSCA Inventory. Cyanide and its compounds are on the Community Right-To-Know List.

OSHA PEL: TWA 2 mg/m^3
ACGIH TLV: TWA 2 mg/m^3

SAFETY PROFILE: Poison by ingestion, inhalation, and intraperitoneal route. Moderately toxic by skin contact. A severe eye irritant. Combustible when exposed to heat or flame. To fight fire, use CO$_2$, dry chemical. Thermally unstable. Contact with moisture (water), acids, or alkalies may cause a violent reaction above 40°. Concentrated aqueous solutions may undergo explosive polymerization. Mixture with 1,2-phenylenediamine salts may cause explosive polymerization. When heated to decomposition or on contact with acid or acid fumes, it emits toxic fumes of CN^- and NO_x.

COI500   CAS: 57-12-5   **HR: 3**
CYANIDE
mf: CN^- mw: 26.02

SYNS: CARBON NITRIDE ION (CN^{1-}) * CYANIDE ANION * CYANURE (FRENCH) * ISOCYANIDE * RCRA WASTE NUMBER P030

CONSENSUS REPORTS: Cyanide and its compounds are on the Community Right-To-Know List.

OSHA PEL: TWA 5 mg(CN)/m^3
ACGIH TLV: TWA 5 mg/m^3 (skin)
DFG MAK: 5 mg/m^3
NIOSH REL: TWA CL 5 mg/m^3/10M

SAFETY PROFILE: Very poisonous by most routes. Cyanide directly stimulates the chemoreceptors of the carotid and aortic bodies with a resultant hyperpnea (increase in the depth and rate of respiration). Cardiac irregularities are often noted, but the heart invariably outlasts the respirations. Death is due to respiratory arrest of central origin. It can occur within seconds or minutes of the inhalation of high concentrations of HCN gas. Because of slower absorption, death may be more delayed after the ingestion of cyanide salts, but the critical events still occur within the first hour. Two other sources of cyanide have been responsible for human poisoning: the naturally occurring amygdalin and the drug nitroprusside.

Amygdalin is a cyanogenic glycoside found in apricot, peach, and similar fruit pits and in sweet almonds (Sayre and Kaymakcalan, 1941). It is a chemical combination of glucose, benzaldehyde, and cyanide from which the latter can be released by the action of β-glucosidase or emulsion. Although these enzymes are not found in mammalian tissues, the human intestinal microflora appears to possess these or similar enzymes capable of effecting cyanide release resulting in human poisoning. For this reason, amygdalin may be as much as 40 times more toxic by the oral route as compared with intravenous injection. Amygdalin is the major ingredient of Laetrile, and this alleged anticancer drug has also been responsible for human cyanide poisoning.

An ethical drug that may also cause cyanide poisoning in overdose is the potent vascular smooth muscle relaxant sodium nitroprusside. Although nitroprusside is related chemically to ferricyanide, unlike the latter it penetrates into erythrocytes and reacts with hemoglobin to release its cyanide (Smith and Kruszyna, 1974). Fortunately, the therapeutic margin for nitroprusside appears to be quite large.

Cyanide is commonly found in certain rat and pest poisons, silver and metal polishes, photographic solutions, and fumigating products. Compounds such as potassium cyanide can also be readily purchased from chemical stores. Cyanide is readily absorbed from all routes, including the skin, mucous membranes, and by inhalation, although alkali salts of cyanide are toxic only when ingested. Death may occur with ingestion of even small amounts of sodium or potassium cyanide and can occur within minutes or hours depending on route of exposure. Inhalation of toxic fumes represents a potentially rapidly fatal type of exposure. A blood cyanide level of greater than 0.2 μg/mL is considered toxic. Lethal cases have usually had levels above 1 μg/mL.

Clinically, cyanide poisoning is reported to produce a bitter, almond odor on the breath of the patient; however, only a small proportion of the population is genetically able to discern this characteristic odor. Typically, cyanide has a bitter, burning taste, and following poisoning, symptoms of salivation, nausea without vomiting, anxiety, confusion, vertigo, giddiness, lower jaw stiffness, convulsions, opisthotonos, paralysis, coma, cardiac arrhythmias, and transient respiratory stimulation followed by respiratory failure may occur. Bradycardia is a common finding, but in most cases heartbeat usually outlasts respiration (Wexler et al., 1947). A prolonged expiratory phase is considered to be characteristic of cyanide poisoning. (Casarett and Doull's, "Toxicology, The Basic Science of Poisons" 2nd ed. Doull, Klaassen and Amdur (eds). Macmillan Pub. Co. Inc. New York, NY).

The volatile cyanides resemble HCN physiologically, inhibiting tissue oxidation and causing death through asphyxia. Cyanogen is probably as toxic as HCN; the nitriles are generally considered somewhat less toxic, probably because of their lower volatility. The non-volatile cyanide salts appear to be relatively non-toxic systemically, so long as they are not ingested and care is taken to prevent the formation of HCN. Workers, such as electroplaters and picklers who are daily exposed to cyanide solutions may develop a ''cyanide'' rash, characterized by itching, and by macular, papular, and vesicular eruptions. Frequently there is secondary infection. Exposure to small amounts of cyanide compounds over long periods of time is reported to cause loss of appetite, headache, weakness, nausea, dizziness, and symptoms of irritation of the upper respiratory tract and eyes.

Flammable by chemical reaction with heat, moisture, acid. Many cyanides evolve HCN rather easily. This is a flammable gas and is highly toxic. Carbon dioxide from the air is sufficiently acidic to liberate HCN from cyanide solutions. Reaction with hypochlorite solutions may be violent at pH 10.0 - 10.3. Explodes if melted with nitrites or chlorates at about 450°. Violent reaction with F_2; Mg; nitrates; HNO_3; nitrites. Metal cyanides are easily oxidized and may be thermally unstable. N-cyano derivatives may be reactive or unstable. Many organic nitriles can be very reactive under the right conditions. When heated to decomposition or on contact with acid, acid fumes, water or steam, it emits toxic and flammable vapors of CN^-.

CON500   CAS: 353-18-4   **HR: 3**
2-CYANO-2′-FLUORODIETHYL ETHER
mf: C_5H_8FNO mw: 117.14

SYNS: 2-CYANOETHYL-2′-FLUOROETHYLETHER
* 2-FLUORO-2′-CYANODIETHYL ETHER

CONSENSUS REPORTS: Cyanide and its compounds are on the Community Right To Know List.

SAFETY PROFILE: Poison by intraperitoneal and subcutaneous routes. When heated to decomposition it emits toxic F^-, NO_x, and CN^-.

COO000   CAS: 460-19-5   **HR: 3**
CYANOGEN

DOT: UN 1026
mf: C_2N_2 mw: 52.04

PROP: Colorless gas, pungent odor. Mp: −34.4°, bp: −21.0°, d: 0.866 @ 17°/4°, lel: 6.6%, uel: 32%, vap d: 1.8.

SYNS: CARBON NITRIDE * CYANOGENE (FRENCH)
* CYANOGEN GAS (DOT) * DICYANOGEN

* ETHANEDINITRILE * NITRILOACETONITRILE
* OXALIC ACID DINITRILE * OXALONITRILE
* OXALYL CYANIDE * PRUSSITE * RCRA WASTE NUMBER P031

CONSENSUS REPORTS: Reported in EPA TSCA Inventory. Cyanide and its compounds are on the Community Right-To-Know List.

OSHA PEL: TWA 10 ppm
ACGIH TLV: TWA 10 ppm
DFG MAK: 10 ppm (22 mg/m^3)
DOT Classification: Poison A; Label: Flammable Gas and Poison Gas.

SAFETY PROFILE: A poison by subcutaneous and possibly other routes. Moderately toxic by inhalation. Human systemic effects by inhalation: damage to the olfactory nerves, and irritation of the conjunctiva. A systemic irritant by inhalation and subcutaneous routes. A human eye irritant. Very dangerous fire hazard when exposed to heat, flames (sparks), or oxidizers. To fight fire, stop flow of gas. Potentially explosive reaction with powerful oxidants (e.g., dichlorine oxide, fluorine, oxygen, ozone). When heated to decomposition or on contact with acid, acid fumes, water or steam, will react to produce highly toxic fumes of NO_x and CN^-.

COO500   CAS: 506-68-3   **HR: 3**
CYANOGEN BROMIDE

DOT: UN 1889
mf: CBrN   mw: 105.93

PROP: Colorless needles. Mp: 52°, bp: 61.6°, d: 2.015 @ 20°/4°, vap press: 100 mm @ 22.6°.

SYNS: BROMINE CYANIDE * BROMOCYAN
* BROMOCYANOGEN * BROMURE de CYANOGEN (FRENCH) * CAMPILIT * CYANOBROMIDE
* CYANOGEN MONOBROMIDE * RCRA WASTE NUMBER U246 * TL 822

CONSENSUS REPORTS: Cyanide and its compounds are on the Community Right-To-Know List. EPA Extremely Hazardous Substances List. Reported in EPA TSCA Inventory.

DOT Classification: Poison B; Label: Poison.

SAFETY PROFILE: A human and experimental poison by inhalation. When heated to decomposition it emits very toxic fumes of CN^- and Br^-. Possibly unstable.

COO750 CAS: 506-77-4 **HR: 3**
CYANOGEN CHLORIDE
DOT: UN 1589
mf: CClN   mw: 61.47

PROP: Colorless liquid or gas; lachrymatory and irritating odor. Mp: −6.5°, bp: 13.1°, d: 1.218 @ 4°/4°, vap press: 1010 mm @ 20°, vap d: 1.98.

SYNS: CHLORCYAN * CHLORINE CYANIDE * CHLOROCYAN * CHLOROCYANIDE * CHLORO-CYANOGEN * CHLORURE de CYANOGENE (FRENCH) * CYANOGEN CHLORIDE, containing less than 0.9% water (DOT) * CYANOGEN CHLORIDE, inhibited (DOT) * RCRA WASTE NUMBER P033

CONSENSUS REPORTS: Cyanide and its compounds are on the Community Right-To-Know List. Reported in EPA TSCA Inventory.

OSHA PEL: CL 0.3 ppm
ACGIH TLV: CL 0.3 ppm
DOT Classification: Poison A; Label: Flammable Gas and Poison Gas.

SAFETY PROFILE: Poison by ingestion, subcutaneous. and possibly other routes. Toxic by inhalation. Human systemic effects by inhalation: lacrimation, conjunctiva irritation, and chronic pulmonary edema or congestion. A primary irritant. A severe human eye irritant. An insecticide. Flammable when exposed to heat or flame. When heated to decomposition or on contact with water or steam, it will react to produce highly toxic and corrosive fumes of Cl⁻, CN⁻, and NO$_x$.

COU000 CAS: 14901-08-7 **HR: 3**
CYCASIN
mf: C$_8$H$_{16}$N$_2$O$_7$ mw: 252.26

SYNS: CYCAS REVOLUTA GLUCOSIDE * CYKA-ZINE * β-d-GLUCOSYLOXYAZOXYMETHANE * METHYLAZOXYMETHANOL GLUCOSIDE * METH-YLAZOXYMETHANOL-β-d-GLUCOSIDE * (METHYL-ONN-AZOXY)METHYL-β-d-GLUCOPYRANOSIDE

CONSENSUS REPORTS: EPA Genetic Toxicology Program. IARC Cancer Review: GROUP 2B IMEMDT 7,56,87; Human Inadequate Evidence IMEMDT 10,121,76; Animal Sufficient Evidence IMEMDT 10,121,76; Animal Sufficient Evidence IMEMDT 1,157,72. NTP Fourth Annual Report On Carcinogens, 1984.

SAFETY PROFILE: Confirmed carcinogen with experimental carcinogenic and tumorigenic data. A poison by ingestion. Mutation data reported. When heated to decomposition it emits toxic fumes of NO$_x$.

COU500 CAS: 103-95-7 **HR: 2**
CYCLAMEN ALDEHYDE
mf: C$_{13}$H$_{18}$O   mw: 190.31

PROP: Colorless liquid; strong, floral odor. D: 0.946-0.952, refr index: 1.503-1.508. Sol in fixed oils; insol in propylene glycol, glycerin.

SYNS: ALDEHYDE B * CYCLAMAL * FEMA No. 2743 * p-ISOPROPYL-α-METHYLHYDROCINNAMIC ALDEHYDE * p-ISOPROPYL-α-METHYLPHENYLPROPYL ALDEHYDE * α-METHYL-p-ISOPROPYLHYDROCIN-NAMALDEHYDE * 2-METHYL-3-(p-ISOPROPYLPHENYL)PROPIONALDEHYDE

CONSENSUS REPORTS: Reported in EPA TSCA Inventory.

SAFETY PROFILE: Moderately toxic by ingestion. A human skin irritant. When heated to decomposition it emits acrid smoke and irritating fumes.

COW000 CAS: 287-23-0 **HR: 1**
CYCLOBUTANE
DOT: UN 2601
mf: C$_4$H$_8$ mw: 56.12

PROP: A gas. Mp: −50°, bp: 12.9°, flash p: <50°F(CC), d: 0.708 @ 11°, vap d: 1.93, lel: 1.8%.

SYN: TETRAMETHYLENE

DOT Classification: Flammable Gas; Label: Flammable Gas

SAFETY PROFILE: May be a simple asphyxiant. Very dangerous fire hazard when exposed to heat or flame; can react with oxidizing materials. To fight fire, stop flow of gas; CO$_2$, dry chemicals, or water spray. When heated to decomposition it emits acrid smoke and fumes.

COW250 **HR: 1**
CYCLOBUTENE
mf: C$_4$H$_6$ mw: 54.09

PROP: Gas. Bp: 2.4°; d: 0.733 @ 0°/4°; flash p: <15°F.

SYN: CYCLOBUTYLENE

SAFETY PROFILE: May be a simple asphyxiant. Dangerous fire hazard when exposed to heat or flame; can react with oxidizing materials. When heated to decomposition it emits acrid smoke and fumes.

COX500   CAS: 291-64-5   HR: 2
CYCLOHEPTANE
DOT: UN 2241
mf: C_7H_{14} mw: 98.19

PROP: An oil. Mp: $-12°$, bp: $117°$, flash p: $59°F$, d: 0.8099 @ $20°/4°$, vap d: 3.3.

SYN: SUBERANE

DOT Classification: Flammable Liquid; Label: Flammable Liquid

SAFETY PROFILE: Dangerous fire hazard when exposed to heat or flame; can react with oxidizing materials. To fight fire, use foam, CO_2, dry chemicals.

COY000   CAS: 544-25-2   HR: 3
1,3,5-CYCLOHEPTATRIENE
DOT: UN 2603
mf: C_7H_8 mw: 92.14

PROP: Flash p: $39.2°F$.

SYNS: CYCLOHEPTATRIENE (DOT) * TROPILIDENE

DOT Classification: Flammable Liquid; Label: Flammable Liquid and Poison.

SAFETY PROFILE: Poison by ingestion and skin contact. Mutation data reported. A very dangerous fire hazard when exposed to heat, flame, or oxidizers. Potentially violent reaction with nitrogen monoxide. When heated to decomposition it emits acrid smoke and fumes.

CPB000   CAS: 110-82-7   HR: 3
CYCLOHEXANE
DOT: UN 1145
mf: C_6H_{12} mw: 84.18

PROP: Colorless, mobile liquid; pungent odor. Mp: $6.5°$, bp: $80.7°$, fp: $4.6°$, flash: p: $1.4°F$, ULC: 90-95, lel: 1.3%, uel: 8.4%, d: 0.7791 @ $20°/4°$, autoign temp: $473°F$, vap press: 100 mm @ $60.8°$, vap d: 2.90.

SYNS: CICLOESANO (ITALIAN) * CYCLOHEXAAN (DUTCH) * CYCLOHEXAN (GERMAN) * CYKLO-

HEKSAN (POLISH) * HEXAHYDROBENZENE * HEXAMETHYLENE * HEXANAPHTHENE * RCRA WASTE NUMBER U056

CONSENSUS REPORTS: Community Right-To-Know List.

OSHA PEL: TWA 300 ppm
ACGIH TLV: TWA 300 ppm
DFG MAK: 300 ppm (1050 mg/m^3)
DOT Classification: Flammable Liquid; Label: Flammable Liquid.

SAFETY PROFILE: Poison by intravenous route. Moderately toxic by ingestion. A systemic irritant by inhalation and ingestion. A skin irritant. Mutation data reported. Flammable liquid. Dangerous fire hazard when exposed to heat or flame; can react with oxidizing materials. Moderate explosion hazard in the form of vapor when exposed to flame. When mixed hot with liquid dinitrogen tetraoxide an explosion resulted. To fight fire, use foam, CO_2, dry chemical, spray, fog. When heated to decomposition it emits acrid smoke and fumes.

CPB625   CAS: 1569-69-3   HR: 2
CYCLOHEXANETHIOL
mf: $C_6H_{12}S$ mw: 116.24

SYNS: CYKLOHEXANTHIOL * CYKLOHEXYLMER-KATPAN (CZECH)

CONSENSUS REPORTS: Reported in EPA TSCA Inventory.

NIOSH REL: CL 0.5 ppm/15M

SAFETY PROFILE: An eye and severe skin irritant. When heated to decomposition it emits toxic fumes of SO_x.

CPB750   CAS: 108-93-0   HR: 3
CYCLOHEXANOL
mf: $C_6H_{12}O$ mw: 100.16

PROP: Colorless needles or viscous liquid; hygroscopic, camphor-like odor. Mp: $24°$, bp: $161.5°$, flash p: $154°F$ (CC), d: 0.9449 @ $25°/4°$, vap press: 1 mm @ $21.0°$, vap d: 3.45, autoign temp: $572°F$.

SYNS: ADRONAL * ANOL * CICLOESANOLO (ITALIAN) * CYCLOHEXYL ALCOHOL * CYKLO-HEKSANOL (POLISH) * HEXAHYDROPHENOL * HEXALIN * HYDRALIN * HYDROPHENOL * HYDROXYCYCLOHEXANE * NAXOL

CONSENSUS REPORTS: EPA Genetic Toxicology Program. Reported in EPA TSCA Inventory.

OSHA PEL: TWA 50 ppm (skin)
ACGIH TLV: TWA 50 ppm (skin)
DFG MAK: 50 ppm (200 mg/m³)

SAFETY PROFILE: Poison by intravenous route. Moderately toxic by ingestion, subcutaneous and intramuscular routes. Mildly toxic by skin contact. Human systemic effects by inhalation: conjunctiva irritation, and changes in the olfactory and respiratory systems. Has caused damage to kidneys, liver and blood vessels in experimental animals. Experimental reproductive effects. Human mutation data reported. A severe eye irritant. Narcotic-like action. Has caused liver, kidney, vascular injury in experimental animals. Flammable when exposed to heat or flame; can react with oxidizing materials. Ignites on contact with chromium trioxide. Violent reaction with HNO_3. Incompatible with oxidants. To fight fire, use alcohol foam, foam, CO_2, dry chemical. When heated to decomposition it emits acrid smoke and fumes.

CPC000   CAS: 108-94-1   **HR: 2**
CYCLOHEXANONE

DOT: UN 1915
mf: $C_6H_{10}O$ mw: 98.16

PROP: Colorless liquid; acetone-like odor. Mp: −45.0°, bp: 115.6°. ULC: 35-40, lel: 1.1% @ 100°, flash p: 111°F, d: 0.9478 @ 20°/4°, autoign temp: 788°F, vap press: 10 mm @ 38.7°, vap d: 3.4.

SYNS: CICLOESANONE (ITALIAN) * CYCLOHEXANON (DUTCH) * CYKLOHEKSANON (POLISH) * HEXANON * KETOHEXAMETHYLENE * NADONE * NCI-C55005 * PIMELIC KETONE * RCRA WASTE NUMBER U057 * SEXTONE

CONSENSUS REPORTS: Reported in EPA TSCA Inventory.

OSHA PEL: (Transitional: TWA 50 ppm) TWA 25 ppm (skin)
ACGIH TLV: TWA 25 ppm (skin)
DFG MAK: 50 ppm (200 mg/m³)
NIOSH REL: TWA 100 mg/m³
DOT Classification: Flammable or Combustible Liquid; Label: Flammable Liquid.

SAFETY PROFILE: Moderately toxic by ingestion, inhalation, subcutaneous, intravenous, and intraperitoneal routes. A skin and severe eye irritant. Human systemic effects by inhalation: changes in the sense of smell, conjunctiva irritation, and unspecified respiratory system changes. Human irritant by inhalation. Mild narcotic properties have also been ascribed to it. Human mutation data reported. Experimental reproductive effects. Flammable when exposed to heat or flame; can react vigorously with oxidizing materials. Slight explosion hazard in its vapor form, when exposed to flame. Explosive reaction with nitric acid at 75°C. Reaction with hydrogen peroxide + nitric acid forms an explosive peroxide. To fight fire, use alcohol foam, dry chemical or CO_2. When heated to decomposition it emits acrid smoke and irritating fumes.

CPC250   **HR: 2**
CYCLOHEXANONE-D
mf: C_6H_8O mw: 96.12

PROP: Liquid. Bp: 155.5°, flash p: 93°F (CC), vap d: 3.31, vap press: 4 mm @ 20°.

SAFETY PROFILE: Skin contact can cause a dermatitis. Irritating to eyes, skin and mucous membranes. Can damage the liver and kidneys. Dangerous fire hazard when exposed to flame and heat; can react with oxidizing materials. To fight fire, use CO_2, dry chemical.

CPC500   **HR: 3**
CYCLOHEXANONE PEROXIDE and BIS(1-HYDROXYCYCLOHEXYL) PEROXIDE MIXTURE

SYN: 1-((1-HYDROPEROXYCYCLOHEXYL)DIOXY) CYCLOHEXENOL with BIS(1-HYDROXYCYCLOHEXYL) PEROXIDE

DOT Classification: Organic Peroxide; Label: Organic Peroxide.

SAFETY PROFILE: A powerful oxidizer. When heated to decomposition it emits acrid smoke and irritating fumes.

CPC579   CAS: 110-83-8   **HR: 2**
CYCLOHEXENE

DOT: UN 2256
mf: C_6H_{10} mw: 82.15

PROP: Colorless liquid. Bp: 83°, fp: −103.7°, flash p: < 21.2°F, d: 0.8102 @ 20°/4°, vap press: 160 mm @ 38°, autoign temp: 590°F, vap d: 2.8, lel: 1.2%.

SYNS: BENZENETETRAHYDRIDE * CYKLOHEKSEN (POLISH) * 1,2,3,4-TETRAHYDROBENZENE

CONSENSUS REPORTS: Reported in EPA TSCA Inventory.

OSHA PEL: 300 ppm
ACGIH TLV: 300 ppm
DFG MAK: 300 ppm (1015 mg/m^3)
DOT Classification: Flammable Liquid.

SAFETY PROFILE: Moderately toxic by inhalation and ingestion. A very dangerous fire hazard when exposed to flame; can react with oxidizers. Dangerous; keep away from heat and open flame. To fight fire, use foam, CO_2, dry chemical.

CPD000   CAS: 286-20-4   **HR: 3**
CYCLOHEXENE OXIDE
mf: $C_6H_{10}O$ mw: 98.16

PROP: Clear liquid. Bp: 129.5°, flash p: 81°F, d: 0.9678 @ 25°/4°, vap d: 3.5.

SYNS: CCHO * CYCLOHEXANE OXIDE * CYCLOHEXENE EPOXIDE * CYCLOHEXENE-1-OXIDE * 1,2-CYCLOHEXENE OXIDE * CYCLOHEXYLENE OXIDE * 1,2-EPOXYCYCLOHEXANE * 7-OXABICYCLO(4.1.0)HEPTANE * TETRAMETHYLENE-OXIRANE

CONSENSUS REPORTS: Reported in EPA TSCA Inventory. EPA Genetic Toxicology Program.

SAFETY PROFILE: Moderately toxic by ingestion, skin contact, intraperitoneal, and intramuscular routes. Mildly toxic by inhalation. Questionable carcinogen with experimental tumorigenic data. Mutation data reported. A dangerous fire hazard when exposed to heat or flame. When heated to decomposition it emits acrid smoke and irritant fumes.

CPD250   CAS: 930-68-7   **HR: 3**
2-CYCLOHEXEN-1-ONE
mf: C_6H_8O mw: 96.14

SYN: CYCLOHEXENONE

CONSENSUS REPORTS: Reported in EPA TSCA Inventory.

SAFETY PROFILE: A poison by ingestion, inhalation, intraperitoneal, and skin routes. When heated to decomposition it emits acrid smoke and irritant fumes.

CPD750   CAS: 100-40-3   **HR: 3**
CYCLOHEXENYLETHYLENE
mf: C_8H_{12} mw: 108.20

PROP: Liquid. Bp: 128°, fp: −109°, flash p: 60°F (TOC), d: 0.832 @ 20°/4°, autoign temp: 517°F, vap press: 25.8 mm @ 38°, vap d: 3.76.

SYNS: BUTADIENE DIMER * 4-ETHENYL-1-CYCLOHEXENE * NCI-C54999 * 1,2,3,4-TETRAHYDROSTYRENE * 1-VINYLCYCLOHEXENE-3 * 1-VINYL-CYCLOHEX-3-ENE * 4-VINYLCYCLOHEXENE-1 * 4-VINYL-1-CYCLOHEXENE

CONSENSUS REPORTS: IARC Cancer Review: GROUP 3 IMEMDT 7,56,87; Animal Inadequate Evidence IMEMDT 11,277,76; Animal Limited Evidence IMEMDT 39,181,86. NTP Carcinogenesis Studies (gavage); Clear Evidence: mouse NTPTR* NTP-TR-303,86; Inadequate Studies: rat NTPTR* NTP-TR-303,86. Reported in EPA TSCA Inventory.

SAFETY PROFILE: Moderately toxic by ingestion and inhalation. Mildly toxic by skin contact. Questionable carcinogen with experimental carcinogenic, neoplastigenic, and tumorigenic data. Dangerous fire hazard when exposed to heat, flame or oxidizers. Can react with oxidizers. To fight fire, use foam, CO_2, dry chemical.

CPE500   CAS: 10137-69-6   **HR: 2**
CYCLOHEXENYL TRICHLOROSILANE
DOT: UN 1762
mf: $C_6H_9Cl_3Si$ mw: 215.59

PROP: Colorless, fuming liquid; HCl odor. Bp: 202°; d: 1.263 @ 25°/25°; flash p: 200°F (COC).

CONSENSUS REPORTS: Reported in EPA TSCA Inventory.

DOT Classification: Corrosive Material; Label: Corrosive.

SAFETY PROFILE: Moderately toxic by ingestion and skin contact. A skin irritant. A corrosive material. It fumes in moist air releasing HCl. Flammable when exposed to heat or flame. When heated to decomposition it emits toxic fumes of Cl$^-$.

CPF000   CAS: 622-45-7   **HR: 2**
CYCLOHEXYL ACETATE
DOT: UN 2243
mf: $C_8H_{14}O_2$ mw: 142.22

PROP: Pale yellow liquid; fruity odor. Bp: 177°, d: 0.996, vap d: 4.9, flash p: 136°F, autoign temp: 633°F.

SYNS: CYCLOHEXANOL ACETATE * CYCLOHEX-ANOLAZETAT (GERMAN) * CYCLOHEXANYL ACETATE

CONSENSUS REPORTS: Reported in EPA TSCA Inventory.

DOT Classification: Flammable or Combustible Liquid; Label: Flammable Liquid.

SAFETY PROFILE: Moderately toxic by subcutaneous route. Mildly toxic by ingestion and skin contact. Human systemic effects by inhalation: conjunctiva irritation and unspecified respiratory system changes. A systemic irritant to humans. Flammable when exposed to heat or flame. When heated to decomposition it emits acrid smoke and irritating fumes.

CPF500　　CAS: 108-91-8　　HR: 3
CYCLOHEXYLAMINE
DOT: UN 2357
mf: $C_6H_{13}N$　　mw: 99.20

PROP: Liquid; strong, fishy odor. Mp: $-17.7°$, bp: 134.5°, flash p: 69.8°F, d: 0.865 @ 25°/25°, autoign temp: 560°F, vap d. 3.42.

SYNS: AMINOCYCLOHEXANE * AMINOHEXAHY-DROBENZENE * CHA * CYCLOHEXANAMINE * HEXAHYDROANILINE * HEXAHYDROBENZEN-AMINE

CONSENSUS REPORTS: IARC Cancer Review: GROUP 3 IMEMDT 7,178,87; Animal No Evidence IMEMDT 22,55,80. EPA Extremely Hazardous Substances List. EPA Genetic Toxicology Program. Reported in EPA TSCA Inventory.

OSHA PEL: TWA 10 ppm
ACGIH TLV: TWA 10 ppm
DFG MAK: 10 ppm (40 mg/m³)
DOT Classification: Flammable Liquid; Label: Flammable Liquid, Corrosive; Flammable or Combustible Liquid; Label: Flammable, Corrosive.

SAFETY PROFILE: A poison by ingestion, skin contact, and intraperitoneal routes. Moderately toxic by subcutaneous and parenteral routes. Experimental teratogenic and reproductive effects. Severe human skin irritant. Can cause dermatitis and convulsions. Human mutation data reported. Questionable carcinogen. Flammable liquid. Dangerous fire hazard when exposed to heat, flame, or oxidizers. To fight fire, use alcohol foam, CO_2, dry chemical. When heated to decomposition it emits toxic fumes of NO_x.

CPK500　　CAS: 131-89-5　　HR: 3
2-CYCLOHEXYL-4,6-DINITROPHENOL
DOT: NA 9026
mf: $C_{12}H_{14}N_2O_5$　　mw: 266.28

PROP: Crystals.

SYNS: 6-CICLOESIL-2,4-DINITR-FENOLO (ITALIAN) * 2-CYCLOHEXYL-4,6-DINITROFENOL (DUTCH) * 6-CYCLOHEXYL-2,4-DINITROPHENOL * DINEX * DINITROCYCLOHEXYLPHENOL * DINITRO-o-CY-CLOHEXYLPHENOL * 2,4-DINITRO-6-CYCLOHEXYL-PHENOL * 4,6-DINITRO-o-CYCLOHEXYLPHENOL * DINITROCYCLOHEXYLPHENOL (DOT) * DN DRY MIX NO. 1 * DN DUST NO. 12 * DNOCHP * DOWSPRAY 17 * DRY MIX NO. 1 * ENT 157 * PEDINEX (FRENCH) * RCRA WASTE NUMBER P034 * SN 46

DOT Classification: ORM-A; Label: None.

SAFETY PROFILE: A poison by ingestion, intraperitoneal, intravenous, subcutaneous, and possibly other routes. Moderately toxic by skin contact. A skin irritant. Fire hazard. Can react with oxidizers. When heated to decomposition it emits toxic fumes of NO_x.

CPN500　　CAS: 3173-53-3　　HR: 3
CYCLOHEXYL ISOCYANATE
DOT: UN 2488
mf: $C_7H_{11}NO$　　mw: 125.19

SYNS: ISOCYANIC ACID, CYCLOHEXYL ESTER * NSC 87419

CONSENSUS REPORTS: Reported in EPA TSCA Inventory.

DOT Classification: Poison B; Label: Flammable Liquid and Poison.

SAFETY PROFILE: Poison by intravenous and intraperitoneal routes. Mutation data reported. Flammable when exposed to heat or flame. When heated to decomposition it emits toxic fumes of NO_x.

CPQ625   CAS: 100-88-9   HR: 3
N-CYCLOHEXYLSULPHAMIC ACID
mf: $C_6H_{13}NO_3S$ mw: 179.26

PROP: Crystals; sweet-sour taste. Mp: 169-170°. Fairly strong acid. Very sparingly soluble in water. Slowly hydrolyzed by hot water.

SYNS: CYCLAMATE * CYCLAMIC ACID * CYCLOHEXANESULPHAMIC ACID * CYCLOHEXYLAMIDOSULPHURIC ACID * CYCLOHEXYLAMINESULPHONIC ACID * CYCLOHEXYLSULFAMIC ACID (9CI) * CYCLOHEXYLSULPHAMIC ACID * HEXAMIC ACID * SUCARYL * SUCARYL ACID

CONSENSUS REPORTS: Reported in EPA TSCA Inventory.

SAFETY PROFILE: Suspected human carcinogen producing bladder tumors. Poison by intravenous route. Mildly toxic by ingestion. When heated to decomposition it emits toxic fumes of SO_x and NO_x.

CPR250   CAS: 98-12-4   HR: 3
CYCLOHEXYLTRICHLOROSILANE
DOT: UN 1763
mf: $C_6H_{11}Cl_3Si$ mw: 217.61

CONSENSUS REPORTS: Reported in EPA TSCA Inventory.

DOT Classification: Corrosive Material; Label: Corrosive.

SAFETY PROFILE: A highly toxic and corrosive material. When heated to decomposition it emits toxic fumes of Cl^-.

CPR800   CAS: 121-82-4   HR: 3
CYCLONITE
DOT: UN 0072/UN 0118
mf: $C_3H_6N_6O_6$ mw: 222.15

PROP: White, crystalline powder. Mp: 202°.

SYNS: CYCLOTRIMETHYLENENITRAMINE * CYCLOTRIMETHYLENETRINITRAMINE * CYCLOTRIMETHYLENETRINITRAMINE, containing at least 10%-25% water (DOT) * CYCLOTRIMETHYLENETRINITRAMINE, desensitized (DOT) * ESAIDRO-1,3,5-TRINITRO-1,3,5-TRIAZINA (ITALIAN) * HEKSOGEN (POLISH) * HEXAHYDRO-1,3,5-TRINITRO-1,3,5-TRIAZIN (GERMAN) * HEXAHYDRO-1,3,5-TRINITRO-s-TRIAZINE * HEXAHYDRO-1,3,5-TRINITRO-1,3,5-TRIAZINE * HEXOGEEN (DUTCH) * HEXOGEN (explosive) * HEXOGEN

5W * HEXOLITE * HEXOLITE, dry or containing, by weight, less than 15% water (DOT) * PBX(AF) 108 * RDX * TRIMETHYLEENTRINITRAMINE (DUTCH) * TRIMETHYLENETRINITRAMINE * sym-TRIMETHYLENETRINITRAMINE * TRINITROCYCLOTRIMETHYLENE TRIAMINE * 1,3,5-TRINITRO-1,3,5-TRIAZACYCLOHEXANE

CONSENSUS REPORTS: Reported in EPA TSCA Inventory.

OSHA PEL: TWA 1.5 mg/m³ (skin)
ACGIH TLV: TWA 1.5 mg/m³ (skin)
DOT Classification: Class A Explosive; Label: Explosive A, Corrosive.

SAFETY PROFILE: Poison by ingestion, intraperitoneal, and intravenous routes. Experimental reproductive effects. A corrosive irritant to skin, eyes, and mucous membranes. Cases of epileptiform convulsions have been reported from exposure. It is one of the most powerful high explosives in use today. Has more shattering power than TNT and is often mixed with TNT as a bursting charge for aerial bombs, mines, and torpedoes. It is easily initiated by mercury fulminate which may be used as a booster. When heated to decomposition it emits toxic fumes of NO_x.

CPS000   CAS: 115-25-3   HR: 1
CYCLOOCTAFLUOROBUTANE
DOT: UN 1976
mf: C_4F_8 mw: 200.03

PROP: Colorless, odorless gas. Bp: -6.04°, mp: -41.4°, d (liquid): 1.513 @ -70°F.

SYNS: FC-C 318 * FREON C-318 * HALOCARBON C-138 * OCTAFLUOROCYCLOBUTANE (DOT) * PERFLUOROCYCLOBUTANE * PROPELLANT C318 * R-C 318

CONSENSUS REPORTS: EPA Genetic Toxicology Program. Reported in EPA TSCA Inventory.

DOT Classification: Nonflammable Gas; Label: Nonflammable Gas.

SAFETY PROFILE: Mildly toxic by ingestion and inhalation. Can cause slight transient effects at high concentrations. No anesthesia or central nervous system effects. Nonflammable Gas. Mutation data reported. When heated to decomposition it emits highly toxic fumes of F^-.

CPS500   CAS: 629-20-9   **HR: 3**
1,3,5,7-CYCLOOCTATETRAENE

DOT: UN 2538
mf: C_8H_8 mw: 104.15

PROP: Liquid. Mp: $-7°$, bp: 140.6°, fp: $-4.7°$, vap press: 7.9 mm @ 25°, flash p: <71.6°F.

DOT Classification: Flammable Liquid; Label: Flammable Liquid

SAFETY PROFILE: May be a simple asphyxiant. A dangerous fire hazard when exposed to heat or flame; can react with oxidizing materials. To fight fire, use spray, mist, fog, foam, dry chemicals. Reaction with oxygen gives explosive peroxide byproducts. When heated to decomposition it emits acrid smoke and fumes.

CPU500   CAS: 542-92-7   **HR: 3**
1,3-CYCLOPENTADIENE
mf: C_5H_6 mw: 66.11

PROP: Colorless liquid. Mp: $-85°$, bp: 42.5°, d: 0.80475 @ 19°/4°. flash p: 77°F.

SYNS: CYCLOPENTADIENE * PENTOLE * PYRO-PENTYLENE * R-PENTINE

CONSENSUS REPORTS: Reported in EPA TSCA Inventory.

OSHA PEL: TWA 75 ppm
ACGIH TLV: TWA 75 ppm
DFG MAK: 75 ppm (200 mg/m³)

SAFETY PROFILE: Probably moderately toxic by inhalation. A dangerous fire hazard when exposed to heat or flame; can react with oxidizing materials. Moderate explosion hazard in the form of gas when exposed to heat or by chemical reaction. It decomposes violently at high temperatures and pressures. Dimerization is highly exothermic. Explosive reaction with fuming nitric acid; dinitrogen tetroxide; sulfuric acid. Reaction with nitrogen oxide + oxygen forms an explosive product. Reaction with oxygen forms a flame-sensitive explosive product. Ignites on contact with oxygen + ozone. Reacts vigorously on contact with potassium hydroxide. Incompatible with oxides of nitrogen, sulfuric acid. When heated to decomposition it emits acrid smoke and fumes.

CPV000   CAS: 12079-65-1   **HR: 3**
CYCLOPENTADIENYLMANGANESE TRICARBONYL
mf: $C_8H_5MnO_3$ mw: 204.07

SYNS: MANGANESE CYCLOPENTADIENYL TRICAR-BONYL * MCT

CONSENSUS REPORTS: Reported in EPA TSCA Inventory. Manganese and its compounds are on the Community Right-To-Know List.

OSHA PEL: TWA 0.1 mg(Mn)/m³ (skin)
ACGIH TLV: TWA 0.1 mg(Mn)/m³

SAFETY PROFILE: A poison by ingestion, inhalation, intraperitoneal, and intravenous routes. A mild narcotic which can damage kidneys. When heated to decomposition it emits acrid smoke and irritating fumes.

CPV750   CAS: 287-92-3   **HR: 1**
CYCLOPENTANE

DOT: UN 1146
mf: C_5H_{10} mw: 70.15

PROP: Colorless liquid. Bp: 49.3°, fp: $-93.7°$, flash p: 19.4°F, autoign temp: 716°F, d: 0.745 @ 20°/4°, vap press: 400 mm @ 31.0°, vap d: 2.42.

SYN: PENTAMETHYLENE

CONSENSUS REPORTS: Reported in EPA TSCA Inventory.

OSHA PEL: TWA 600 ppm
ACGIH TLV: TWA 600 ppm
DOT Classification: Flammable Liquid; Label: Flammable Liquid.

SAFETY PROFILE: Mildly toxic by ingestion and inhalation. High concentrations have narcotic action. A very dangerous fire hazard when exposed to heat or flame; can react with oxidizers. To fight fire, use foam, CO_2, dry chemical. When heated to decomposition it emits acrid smoke and fumes.

CPW500   CAS: 120-92-3   **HR: 2**
CYCLOPENTANONE

DOT: UN 2245
mf: C_5H_8O mw: 84.13

PROP: Liquid. Mp: $-58.2°$, bp: 130.6°, flash p: 79°F, d: 0.9509 @ 18°/4°, vap d: 2.3.

SYNS: ADIPIC KETONE * DUMASIN * KETOCY-CLOPENTANE * KETOPENTAMETHYLENE

CONSENSUS REPORTS: Reported in EPA TSCA Inventory.

DOT Classification: Flammable or Combustible Liquid; Label: Flammable Liquid.

SAFETY PROFILE: Moderately toxic by intra-peritoneal and subcutaneous routes. A skin and severe eye irritant. Dangerous fire hazard when exposed to heat or flame; can react with oxidizers. To fight fire, use alcohol foam, foam, CO_2, dry chemical. Potentially explosive reaction with hydrogen peroxide + nitric acid. When heated to decomposition it emits acrid smoke and fumes.

CPX750   CAS: 142-29-0   HR: 2
CYCLOPENTENE
DOT: UN 2246
mf: C_5H_8 mw: 68.13

PROP: Liquid. Mp: −93.3°, bp: 44.242°, fp: −135.2°, flash p: −20°F, d: 0.77199 @ 20°.

CONSENSUS REPORTS: Reported in EPA TSCA Inventory.

DOT Classification: Flammable Liquid; Label: Flammable Liquid.

SAFETY PROFILE: Moderately toxic by ingestion and skin contact. A very dangerous fire hazard when exposed to flame or heat; can react with oxidizing materials. Keep away from heat and open flame. To fight fire, use foam, CO_2, dry chemical.

CQC675   CAS: 6055-19-2   HR: 3
CYCLOPHOSPHAMIDE HYDRATE
mf: $C_7H_{15}Cl_2N_2O_2P \cdot H_2O$ mw: 279.13

SYNS: 1-BIS(2-CHLOROETHYL)AMINO-1-OXO-2-AZA-5-OXAPHOSPHORIDINE MONOHYDRATE * 2-(BIS(2-CHLOROETHYL)AMINO)-1-OXA-3-AZA-2-PHOSPHOCY-CLOHEXANE 2-OXIDE MONOHYDRATE * (BIS-(CHLORO-2-ETHYL)AMINO)-2-TETRAHYDRO-3,4,5,6-OX-AZAPHOSPHORINE-1,3,2-OXIDE-2-MONOHYDRATE * BIS(2-CHLOROETHYL)PHOSPHORAMIDE CYCLIC PRO-PANOLAMIDE ESTER MONOHYDRATE * N,N-BIS(β-CHLOROETHYL)-N′,O-PROPYLENEPHOSPHORIC ACID ESTER AMINE MONOHYDRATE * N,N-BIS(2-CHLORO-ETHYL)TETRAHYDRO-2H-1,3,2-OXAPHOSPHORIN-2-AMINE-2-OXIDE MONOHYDRATE * N,N-BIS(β-CHLOROETHYL)-N′,O-TRIMETHYLENEPHOSPHORIC ACID ESTER DIAMIDE MONOHYDRATE * CB-4564 * CLAFEN * CYCLIC N′,O-PROPYLENE ESTER of N,N-BIS(2-CHLOROETHYL)PHOSPHORODIAMIDIC ACID MONOHYDRATE * CYCLOPHOSPHAMIDE MONOHY-DRATE * CYCLOPHOSPHAMIDUM * CYCLO-PHOSPHAN * CYCLOPHOSPHANE * CYCLOPHOS-PHANUM * CYTOPHOSPHAN * CYTOXAN * 2-(DI(2-CHLOROETHYL)AMINO)-1-OXA-3-AZA-2-

PHOSPHACYCLOHEXANE-2-OXIDE MONOHYDRATE * N,N-DI(2-CHLOROETHYL)AMINO-N,O-PROPYLENE PHOSPHORIC ACID ESTER DI AMIDE MONOHYDRATE * ENDOXANA * ENDOXAN-ASTA * ENDOXAN MONOHYDRATE * ENDOXAN R * ENDUXAN * GENOXAL * MITOXAN * NSC 26271 * PRO-CYTOX * SEMDOXAN * SENDOXAN * SEN-DUXAN

CONSENSUS REPORTS: IARC Cancer Review: Animal Sufficient Evidence IMEMDT 9,135,75; Human Limited Evidence IMEMDT 9,135,75; Human Sufficient Evidence IMEMDT 26,165,81.

SAFETY PROFILE: Confirmed human carcinogen. Poison by ingestion and intravenous routes. Experimental reproductive effects. Mutation data reported. When heated to decomposition it emits toxic fumes of Cl^-, PO_x, and NO_x.

CQC750   HR: 3
CYCLOPHOSPHAMIDE and MNU (1 : 2)
PROP: A combination of these two drugs is used in chemotherapy to combat far advanced malignant tumors.

SYN: MNU and CYCLOPHOSPHAMIDE (2:1)

SAFETY PROFILE: Human systemic effects by intravenous route: nausea or vomiting and bone marrow changes. When heated to decomposition it emits very toxic fumes of Cl^-, NO_x, and PO_x.

CQD750   CAS: 75-19-4   HR: 3
CYCLOPROPANE
DOT: UN 1027
mf: C_3H_6 mw: 42.09

PROP: Colorless gas. Mp: −126.6°, bp: −33.5°, lel: 2.4%, uel: 10.4%, d: 1.879 g/L @ 0°, autoign temp: 932°F. A minor constituent of MAPP gas.

SYNS: CYCLOPROPANE, liquefied (DOT) * TRI-METHYLENE

CONSENSUS REPORTS: Reported in EPA TSCA Inventory.

DOT Classification: Flammable Gas; Label: Flammable Gas.

SAFETY PROFILE: Mutation data reported. High concentrations are narcotic. Human reproductive effects. Very dangerous fire hazard when exposed to heat or flame; can react with oxidizing materials. Explosion Hazard: Moderate in the form of vapor when exposed to heat or flame. To fight fire, stop flow of gas, then use CO_2, dry chemical or water spray. When heated to decomposition it emits acrid smoke and fumes.

CQH250   CAS: 2691-41-0   **HR: 3**
**CYCLOTETRAMETHYLENE
TETRANITRAMINE**

DOT: UN 0226
mf: $C_4H_8N_8O_8$ mw: 296.20

SYNS: CYCLOTETRAMETHYLENE TETRANITRAMINE, dry (DOT) * HMX (DOT) * beta HMY * HW 4 * LX 14-0 * OCTOGEN * OKTOGEN * TETRAMETHYLENETETRANITRAMINE * 1,3,5,7-TETRANITROPERHYDRO-1,3,5,7-TETRAZOCINE

CONSENSUS REPORTS: Reported in EPA TSCA Inventory.

DOT Classification: Forbidden, Dry; Class A Explosive; Label: Explosive A, wet.

SAFETY PROFILE: A poison by ingestion and intravenous routes. An explosive. Decomposes violently at 279°C. When heated to decomposition it emits toxic fumes of NO_x.

CQH650   CAS: 13121-70-5   **HR: 3**
CYHEXATIN
mf: $C_{18}H_{34}OSn$ mw: 385.21

SYNS: DOWCO-213 * ENT 27,395-X * M 3180 * PLICTRAN * PLYCTRAN * TCTH * TRICYCLOHEXYLHYDROXYSTANNANE * TRICYCLOHEXYLHYDROXYTIN * TRICYCLOHEXYLTIN HYDROXIDE * TRICYCLOHEXYLZINNHYDROXID (GERMAN)

OSHA PEL: TWA 5 mg/m^3
ACGIH TLV: TWA 5 mg/m^3
NIOSH REL: (Organotin Compounds) TWA 0.1 mg(Sn)/m^3.

SAFETY PROFILE: Poison by ingestion, inhalation, and intraperitoneal routes. Moderately toxic by skin contact. Experimental reproductive effects. When heated to decomposition it emits acrid smoke and irritating fumes.

CQI000   CAS: 99-87-6   **HR: 1**
p-CYMENE

DOT: UN 2046
mf: $C_{10}H_{14}$ mw: 134.24

PROP: Colorless to pale yellow liquid; odorless. Mp: −68.2°, bp: 176°, lel: 0.7%, @ 100°, ULC:

30-35, flash p: 117°F (CC), d: 0.853, refr index: 1.489, autoign temp: 817°F, vap d: 4.62, vap press: 1 mm @ 17.3°, flash p: (technical) 127°F, uel (technical): 5.6%. Found in nearly 100 volatile oils including lemongrass, sage, thyme, coriander, star anise, and cinnamon. Sol in alc, ether, acetone, benzene.

SYNS: CAMPHOGEN * CYMENE * CYMOL * DOLCYMENE * FEMA No. 2356 * 4-ISOPROPYL-1-METHYLBENZENE * p-ISOPROPYLTOLUENE * p-METHYL-CUMENE * p-METHYLISOPROPYL BENZENE * 1-METHYL-4-ISOPROPYLBENZENE * PARACYMENE * PARACYMOL

CONSENSUS REPORTS: Reported in EPA TSCA Inventory.

DOT Classification: Flammable or Combustible Liquid; Label: Flammable Liquid.

SAFETY PROFILE: Mildly toxic by ingestion. Humans sustain central nervous system effects at low doses. Mutation data reported. A skin irritant. Flammable or combustible liquid. Explosion Hazard: Slight in the form of vapor. To fight fire, use foam, CO_2, dry chemical. When heated to decomposition it emits acrid smoke and fumes.

CQJ250   CAS: 2759-71-9   **HR: 3**
CYPROMID
mf: $C_{10}H_9Cl_2NO$ mw: 230.10

SYNS: CIPROMID * CLOBBER * 3,4'-DICHLOROCYCLOPROPANECARBOXANILIDE * N-(3,4-DICHLOROPHENYL)CYCLOPROPANECARBOXAMIDE

SAFETY PROFILE: A poison by ingestion. Moderately toxic by skin contact. An herbicide. When heated to decomposition it emits very toxic fumes of HCl and NO_x.

CQK000   CAS: 52-90-4   **HR: 2**
l-CYSTEINE
mf: $C_3H_7NO_2S$ mw: 121.17

PROP: An amino acid derived from cystine, occurring naturally in the l-form, which will be considered here. Colorless crystals; sol in water, ammonium hydroxide, and acetic acid; insol in ether, acetone, benzene, carbon disulfide, and carbon tetrachloride.

SYNS: CYSTEIN * CYSTEINE * l-(+)-CYSTEINE * HALF-CYSTEINE * HALF-CYSTINE * β-MERCAPTOALANINE * THIOSERINE

CONSENSUS REPORTS: Reported in EPA TSCA Inventory. EPA Genetic Toxicology Program.

SAFETY PROFILE: Moderately toxic by ingestion, intraperitoneal, and subcutaneous routes. Experimental reproductive effects. Human mutation data reported. When heated to decomposition it emits very toxic fumes of SO_x and NO_x.

CQK250   CAS: 52-89-1   HR: 2
l-CYSTEIN HYDROCHLORIDE
mf: $C_3H_7NO_2S \cdot ClH$ mw: 157.63

PROP: White crystalline powder; characteristic acetic taste. Mp: 175° (decomp). Sol in water, alc.

SYNS: CYSTEINE CHLORHYDRATE * CYSTEINE HYDROCHLORIDE * l-CYSTEINE HYDROCHLORIDE * l-CYSTEINE MONOHYDROCHLORIDE (FCC)

CONSENSUS REPORTS: Reported in EPA TSCA Inventory.

SAFETY PROFILE: Moderately toxic by intraperitoneal, intravenous, and possibly other routes. Mutation data reported. When heated to decomposition it emits very toxic fumes of NO_x, SO_x, and Cl^-.

CQK325   CAS: 56-89-3   HR: D
l-CYSTINE
mf: $C_6H_{12}N_2O_4S_2$ mw: 240.30

PROP: Naturally occurring levorotatory form. Colorless to white hexagonal tablets from water. Decomp 260-261°. Sltly sol in water, alc. d-Cystine: Crystals. Sltly sol in water. dl-Cystine, the synthetic racemic form: Crystals. Sltly sol in water. meso-Cystine, the internally compensated form: Crystals. Sltly sol in water.

SYNS: CYSTEINE DISULFIDE * CYSTIN * (−)-CYSTINE * CYSTINE ACID * DICYSTEINE * β,β'-DITHIODIALANINE * GELUCYSTINE * OXIDIZED l-CYSTEINE

SAFETY PROFILE: Experimental reproductive effects. When heated to decomposition it emits toxic fumes of PO_x and SO_x.

CQL250   CAS: 115-93-5   HR: 3
CYTHIOATE
mf: $C_8H_{12}NO_5PS_2$ mw: 297.30

SYNS: O,O-DIMETHYL-O,p-SULFAMOYLPHENYL PHOSPHOROTHIOATE * ENT 25,640 * p-HYDROXYBENZENESULFONAMIDE-O-ESTER with O,O-DIMETHYL PHOSPHOROTHIOATE

CONSENSUS REPORTS: Reported in EPA TSCA Inventory.

SAFETY PROFILE: Poison by ingestion. An insecticide. When heated to decomposition it emits very toxic fumes of NO_x, PO_x, and SO_x.

CQM250   CAS: 36011-19-5   HR: 3
CYTOCHALASIN E
mf: $C_{28}H_{32}NO_7$ mw: 494.61

PROP: Food storage mold metabolite of *Aspergillus clavatus*.

SAFETY PROFILE: A poison by ingestion, intraperitoneal, and parenteral routes. When heated to decomposition it emits toxic fumes of NO_x.

CQN000   CAS: 4465-94-5   HR: 3
CYTOXAL ALCOHOL
mf: $C_7H_{17}Cl_2N_2O_3P \cdot C_6H_{13}N$ mw: 378.33

SYNS: 2-(BIS(2-CHLOROETHYL)AMINO)TETRAHYDROOXAZAPHOSPHORINE CYCLOHEXYLAMINE SALT * N,N-BIS(2-CHLOROETHYL)-N'-(3-HYDROXYPROPYL)PHOSPHORODIAMIDATE, CYCLOHEXYLAMMONIUM SALT * N,N-BIS(2-CHLOROETHYL)-N'-3-PHOSPHORODIAMIDIC ACID HYDROXYLPROPYLCYCLOHEXYLAMINE SALT * CYTOXYL ALCOHOL CYCLOHEXYLAMMONIUM SALT * NCI-C04922 * NSC-52695

CONSENSUS REPORTS: NCI Carcinogenesis Studies (ipr); Clear Evidence: mouse, rat RRCRBU 52,1,75.

SAFETY PROFILE: Suspected carcinogen with experimental carcinogenic and tumorigenic data. Poison by intravenous route. Moderately toxic by ingestion and subcutaneous routes. Experimental teratogenic and reproductive effects. Human mutation data reported. When heated to decomposition it emits very toxic fumes of NO_x, NH_3, PO_x, and Cl^-.

D

DAA800   CAS: 94-75-7   **HR: 3**
2,4-D
DOT: NA 2765
mf: $C_8H_6Cl_2O_3$ mw: 221.04

PROP: White powder. Mp: 141°; bp: 160° @ 0.4 mm; vap d: 7.63.

SYNS: ACIDE-2,4-DICHLORO PHENOXYACETIQUE (FRENCH) * ACIDO (2,4-DICLORO-FENOSSI)-ACETICO (ITALIAN) * AGROTECT * AMIDOX * AMOXONE * AQUA-KLEEN * BH 2,4-D * CHIPCO TURF HERBICIDE ''D'' * CHLOROXONE * CROP RIDER * CROTILIN * D 50 * DACAMINE * 2,4-D ACID * DEBROUSSAILLANT 600 * DECAMINE * DED-WEED * DED-WEED LV-69 * DESORMONE * (2,4-DICHLOOR-FENOXY)-AZIJNZUUR (DUTCH) * DICHLOROPHENOXYACETIC ACID * 2,4-DICHLOROPHENOXYACETIC ACID (DOT) * 2,4-DICHLORPHENOXYACETIC ACID * (2,4-DICHLOR-PHENOXY)-ESSIGSAEURE (GERMAN) * DICOPUR * DICOTOX * DINOXOL * DMA-4 * DORMONE * 2,4-DWUCHLOROFENOKSYOCTOSY KWAS (POLISH) * EMULSAMINE BK * EMULSAMINE E-3 * ENT 8,538 * ENVERT 171 * ENVERT DT * ESTERON * ESTERON 99 * ESTERON 76 BE * ESTERON BRUSH KILLER * ESTERON 99 CONCENTRATE * ESTERONE FOUR * ESTERON 44 WEED KILLER * FARMCO * FERNESTA * FERNIMINE * FERNOXONE * FOREDEX 75 * FORMOLA 40 * HEDONAL (The herbicide) * HERBIDAL * IPANER * KROTILINE * LAWN-KEEP * MACRONDRAY * MIRACLE * MONOSAN * MOXONE * NETAGRONE 600 * NSC 423 * PENNAMINE * PHENOX * PIELIK * PLANOTOX * PLANTGARD * RCRA WASTE NUMBER U240 * RHODIA * SALVO * SPRITZHORMIN/2,4-D * SUPER D WEEDONE * SUPERORMONE CONCENTRE * TRANSAMINE * TRIBUTON * TRINOXOL * U 46 * U 46DP * U-5043 * VERGEMASTER * VERTON D * VIDON 638 * VISKO-RHAP DRIFT HERBICIDES * WEED-AG-BAR * WEEDAR-64 * WEED-B-GON * WEEDEZ WONDER BAR * WEEDONE LV4 * WEED TOX * WEEDTROL

CONSENSUS REPORTS: IARC Cancer Review: GROUP 2B IMEMDT 7,156,87; Human Limited Evidence IMEMDT 41,357,86; Animal Inadequate Evidence IMEMDT 15,111,77; Human Inadequate Evidence IMEMDT 15,111,77.

EPA Genetic Toxicology Program. Reported in EPA TSCA Inventory. Community Right-To-Know List.

OSHA PEL: TWA 10 mg/m³
ACGIH TLV: TWA 10 mg/m³
DFG MAK: 10 mg/m³
DOT Classification: ORM-A; Label: None.

SAFETY PROFILE: Suspected human carcinogen. Experimental teratogenic and reproductive effects. Poison by ingestion, intravenous, and intraperitoneal routes. Moderately toxic by skin contact. Human systemic effects by ingestion: somnolence, convulsions, coma, and nausea or vomiting. Can cause liver and kidney injury. A skin and severe eye irritant. Human mutation data reported. When heated to decomposition it emits toxic fumes of Cl⁻.

DAB020   CAS: 2307-55-3   **HR: 3**
2,4-D AMMONIUM SALT
mf: $C_8H_6Cl_2O_3 \cdot H_3N$ mw: 238.08

CONSENSUS REPORTS: IARC Cancer Review: Animal Inadequate Evidence IMEMDT 15,111,77

SAFETY PROFILE: A poison by intraperitoneal route. Questionable carcinogen with experimental tumorigenic data. When heated to decomposition it emits very toxic fumes of Cl⁻, NO_x, and NH_3.

DAB600   CAS: 4342-03-4   **HR: 3**
DACARBAZINE
mf: $C_6H_{10}N_6O$ mw: 182.22

SYNS: DETICENE * DIC * (DIMETHYL-TRIAZENO)IMIDAZOLECARBOXAMIDE * 4-(DIMETHYLTRIAZENO)IMIDAZOLE-5-CARBOXAMIDE * 4-(3,3-DIMETHYL-1-TRIAZENO)IMIDAZOLE-5-CARBOXAMIDE * 4-(5)-(3,3-DIMETHYL-1-TRIAZENO)IMIDAZOLE-5(4)-CARBOXAMIDE * 5-(DIMETHYLTRIAZENO)IMIDAZOLE-4-CARBOXAMIDE * 5-(3,3-DIMETHYLTRIAZENO)IMIDAZOLE-4-CARBOXAMIDE * 5-(3,3-DIMETHYL-1-TRIAZENO)IMIDAZOLE-4-CARBOXAMIDE * 5-(3,3-DIMETHYL-1-TRIAZENYL)-1H-IMIDAZOLE-4-CARBOXAMIDE * DTIC * DTIC-DOME * NCI-C04717 * NSC-45388

CONSENSUS REPORTS: IARC Cancer Review: GROUP 2B IMEMDT 7,184,87; Human Limited Evidence IMEMDT 26,203,81; Animal Sufficient Evidence IMEMDT 26,203,81. NCI Carcinogenesis Studies (ipr); Clear Evidence: mouse, rat RRCRBU 52,1,75. EPA Genetic Toxicology Program.

SAFETY PROFILE: Suspected carcinogen with experimental carcinogenic and tumorigenic data. Poison by intraperitoneal and parenteral routes. Moderately toxic by ingestion and intravenous routes. Experimental teratogenic effects. Human systemic effects by intravenous route: nausea or vomiting, luekopenia (reduced white blood cell count), and changes in dehydrogenase enzymatic activity. Mutation data reported. When heated to decomposition it emits toxic fumes of NO_x.

DAB800   CAS: 1172-18-5   **HR: 3**
DALMANE
mf: $C_{21}H_{23}ClFN_3O \cdot 2ClH$ mw: 460.84

SYNS: BENOZIL * DALMADORM * DALMADORM HYDROCHLORIDE * DALMATE * DORMODOR * FELISON * FLURAZEPAN DIHYDROCHLORIDE * FLURAZEPAM HYDROCHLORIDE * ID 480 DIHYDROCHLORIDE * INSUMIN * LUNIPAK * NSC-78559 * RO 5-6901 * SOMLAN

SAFETY PROFILE: Poison by intravenous and intraperitoneal routes. Moderately toxic by ingestion and subcutaneous routes. Habituating and possibly addictive. An hypnotic and sedative. When heated to decomposition it emits very toxic fumes of F^-, NO_x, and Cl^-.

DAB879   CAS: 469-62-5   **HR: 3**
DARVON
mf: $C_{22}H_{29}NO_2$ mw: 339.52

SYNS: DEXTROPROPOXYPHENE * α-(+)-4-DIMETHYLAMINO-1,2-DIPHENYL-3-METHYL-2-BUTANOL PROPIONATE ESTER * DOLENE * DOLOXENE * (+)-PROPOXYPHENE * d-PROPOXYPHENE * PROXAGESIC * SK 65

SAFETY PROFILE: Poison by ingestion, intraperitoneal, subcutaneous, and intravenous routes. Human systemic effects by ingestion: change in cardiac rate, respiratory depression and coma. When heated to decomposition it emits toxic fumes of NO_x.

DAC000   CAS: 20830-81-3   **HR: 3**
DAUNOMYCIN
mf: $C_{27}H_{29}NO_{10}$ mw: 527.57
PROP: Thin, red needles. Mp: 190° (decomp). Isolated from cultures of a *Streptomyces*.

SYNS: ACETYLADRIAMYCIN * CERUBIDIN * DAUNAMYCIN * DAUNORUBICIN * DAUNORUBICINE * DM * FI6339 * LEUKAEMOMYCIN C * NCI-C04693 * NSC-82151 * RCRA WASTE NUMBER U059 * RP 13057 * 13,057 R.P. * RUBIDOMYCIN * RUBIDOMYCINE * RUBOMYCIN C * RUBOMYCIN C 1 * STREPTOMYCES PEUCETIUS

CONSENSUS REPORTS: IARC Cancer Review: GROUP 2B IMEMDT 7,56,87; Animal Sufficient Evidence IMEMDT 10,145,76. NCI Carcinogenesis Studies (ipr); Clear Evidence: rat CANCAR 40,1935,77; No Evidence: mouse CANCAR 40,1935,77. EPA Genetic Toxicology Program.

SAFETY PROFILE: Suspected carcinogen with experimental carcinogenic, neoplastigenic, and tumorigenic data. Human poison by ingestion. Experimental poison by subcutaneous, intravenous, and intraperitoneal routes. Experimental teratogenic and reproductive effects. Human mutation data reported. When heated to decomposition it emits toxic fumes of NO_x.

DAC800   CAS: 33857-26-0   **HR: 3**
DCDD
mf: $C_{12}H_6Cl_2O_2$ mw: 253.08
PROP: Colorless crystals. Mp: 210°.

SYNS: 2,7-DICHLORODIBENZODIOXIN * 2,7-DICHLORODIBENZO(b,e)(1,4)DIOXIN * 2,7-DICHLORODIBENZO-p-DIOXIN * NCI-C03667

CONSENSUS REPORTS: IARC Cancer Review: Animal Inadequate Evidence IMEMDT 15,41,77. NCI Carcinogenesis Bioassay (feed); Clear Evidence: mouse NCITR* NCI-CG-TR-123,79; No Evidence: rat NCITR* NCI-CG-TR-123,79.

SAFETY PROFILE: An eye irritant. Experimental teratogenic data. Questionable carcinogen with experimental tumorigenic data. When heated to decomposition it emits toxic fumes of Cl^-.

DAD200   CAS: 50-29-3   **HR: 3**
DDT
DOT: NA 2761
mf: $C_{14}H_9Cl_5$ mw: 354.48

PROP: Colorless crystals or white to slightly off-white powder. Odorless or with slight aromatic odor. Mp: 108.5-109°.

SYNS: AGRITAN * ANOFEX * ARKOTINE * AZOTOX * α,α-BIS(p-CHLOROPHENYL)-β,β,β-TRI-CHLORETHANE * 1,1-BIS-(p-CHLOROPHENYL)-2,2,2-TRICHLOROETHANE * 2,2-BIS(p-CHLOROPHENYL)-1,1,1-TRICHLOROETHANE * BOSAN SUPRA * BOVIDERMOL * CHLOROPHENOTHAN * CHLOROPHENOTHANE * CHLOROPHENOTOXUM * CITOX * CLOFENOTANE * p,p′-DDT * DEDELO * DEOVAL * DETOX * DETOXAN * DIBOVAN * DICHLORODIPHENYLTRICHLOROETH-ANE * p,p′-DICHLORODIPHENYLTRICHLOROETHANE * 4,4′-DICHLORODIPHENYLTRICHLOROETHANE * DICHLORODIPHENYLTRICHLOROETHANE (DOT) * DICOPHANE * DIDIGAM * DIDIMAC * DIPHENYLTRICHLOROETHANE * DODAT * DYKOL * ENT 1,506 * ESTONATE * GENI-TOX * GESAFID * GESAPON * GESAREX * GESAROL * GUESAPON * GUESAROL * GYRON * HAVERO-EXTRA * HILDIT * IVORAN * IXODEX * KOPSOL * MICRO DDT 75 * MUTOXIN * NCI-C00464 * NEOCID * PARACHLOROCIDUM * PEB1 * PENTACHLO-RIN * PENTECH * PPZEIDAN * R50 * RCRA WASTE NUMBER U-061 * RUKSEAM * SANTO-BANE * TECH DDT * 1,1,1-TRICHLOOR-2,2-BIS(4-CHLOOR FENYL)-ETHAAN (DUTCH) * 1,1,1-TRI-CHLOR-2,2-BIS(4-CHLOR-PHENYL)-AETHAN (GERMAN) * TRICHLOROBIS (4-CHLOROPHENYL) ETHANE * 1,1,1-TRICHLORO-2,2-DI(4-CHLOROPHENYL)-ETHANE * 1,1′-(2,2,2-TRICHLOROETHYLIDENE)BIS(4-CHLORO-BENZENE) * 1,1,1-TRICLORO-2,2-BIS(4-CLORO-FENI-L)ETANO (ITALIAN) * ZEIDANE * ZERDANE

CONSENSUS REPORTS: IARC Cancer Review: GROUP 2B IMEMDT 7,186,87; Animal Sufficient Evidence IMEMDT 5,83,74; Human Inadequate Evidence IMEMDT 5,83,74. NTP Fourth Annual Report On Carcinogens, 1984. NCI Carcinogenesis Bioassay (feed); No Evidence: mouse, rat NCITR* NCI-CG-TR-131,78. Reported in EPA TSCA Inventory. EPA Genetic Toxicology Program.

OSHA PEL: TWA 1 mg/m³ (skin)
ACGIH TLV: TWA 1 mg/m³
NIOSH REL: (DDT) TWA 0.5 mg/m³; avoid skin contact
DFG MAK: 1 mg/m³
DOT Classification: ORM-A; Label: None.

SAFETY PROFILE: Confirmed carcinogen with experimental carcinogenic, neoplastigenic, and tumorigenic data. Human poison by ingestion and possibly other routes. Experimental poison by ingestion, skin contact, subcutaneous, intravenous, and intraperitoneal routes. Experimental reproductive effects. Human systemic effects by ingestion: anesthetic, convulsions, headache, analgesia, cardiac arrythmias, nausea or vomiting, sweating and unspecified pulmonary changes. Experimental teratogenic effects. Human mutation data reported. An insecticide. When heated to decomposition it emits toxic fumes of Cl^-.

A dose of 20 grams has proved highly dangerous though not fatal to man. This dose was taken by 5 persons who vomited an unknown portion of the material and even so recovered only incompletely after 5 weeks. Smaller doses produced less important symptoms with relatively rapid recovery. Experimental ingestion of 1.5 grams resulted in great discomfort and moderate neurological changes including paraesthesia, tremor, moderate ataxia, exaggeration of part of the reflexes, headache, and fatigue. Vomiting followed only after 11 hours. Recovery was complete on the following day. The fatal dose of DDT for humans is not known. Judging from the literature, no one has ever been killed by DDT in the absence of other insecticides and/or a variety of toxic solvents. However, these common solvent formulations are highly fatal when taken in small doses, partly because of the toxicity of the solvent, and perhaps because of the increased absorbability of the DDT; several fatal cases in humans have been reported. Little is known of the hazard of chronic DDT poisoning. Human volunteers have ingested up to 35 mg/day for 21 months with no ill effects.

DDT and some of its degradation products, particularly DDE, are stored in fat. This storage effect leads to a concentration of DDT at higher levels of the food chain. DDT stored in the fat is at least largely inactive since a greater total dose may be stored in an experimental animal than is sufficient as a lethal dose for that same animal if given at one time. A study based on 75 human cases reported an average of 5.3 ppm of DDT stored in the fat. A higher content of DDT and its derivatives (up to 434 ppm of DDE and 648 ppm of DDT) was found in workers who had very extensive exposure. Without exception, the samples were taken from persons who were either asymptomatic or suffering from some disease completely unrelated to

DDT. Careful hospital examination of workers who had been very extensively exposed and who had volunteered for examination revealed no abnormality which could be attributed to DDT. Much higher levels have been found in humans than have been observed in the fat of experimental animals which were apparently asymptomatic. DDT stored in the fat is eliminated only very gradually when further dosage is discontinued. However, weight loss can speed the release of this stored DDT (and DDE) into the blood. After a single dose, the secretion of DDT in the milk and its excretion in the urine reach their height within a day or two and continue at a lower level thereafter.

DAE400   CAS: 17702-41-9   **HR: 3**
DECABORANE

DOT: UN 1868
mf: $B_{10}H_{14}$ mw: 122.24

PROP: Colorless needles. Mp: 99.7°, d: 0.94. (solid), d: 0.78 (liquid @ 100°), vap press: 19 mm @ 100°.

SYN: DECABORANE(14)

CONSENSUS REPORTS: Reported in EPA TSCA Inventory. EPA Extremely Hazardous Substances List.

OSHA PEL: (Transitional: TWA 0.05 ppm (skin)) TWA 0.05 ppm; STEL 0.15 ppm (skin)
ACGIH TLV: TWA 0.05 ppm; STEL 0.15 ppm (skin)
DFG MAK: 0.05 ppm (0.3 mg/m^3)
DOT Classification: Flammable Solid; Label: Flammable Solid and Poison.

SAFETY PROFILE: Poison by inhalation, ingestion, skin contact, and intraperitoneal routes. Ignites in O_2 at 100°C. Forms impact-sensitive explosive mixtures with ethers (e.g., dioxane) and halocarbons (e.g., carbon tetrachloride). Incompatible with dimethyl sulfoxide. When heated to decomposition it emits toxic fumes of boron oxides.

DAE450   CAS: 25152-84-5   **HR: 1**
trans,trans-2,4-DECADIENAL
mf: $C_{10}H_{16}O$ mw: 152.23

PROP: Yellow liquid; chicken fat odor. D: 0.806-0.876, refr index: 1.514-1.516, flash p: +212°F. Sol in alc, fixed oils; insol in water @ 104°.

SYNS: FEMA No. 3135   *   HEPTENYL ACROLEIN

SAFETY PROFILE: Combustible liquid. When heated to decomposition it emits acrid smoke and irritating fumes.

DAE600   CAS: 15652-38-7   **HR: 3**
DECAFENTIN
mf: $C_{28}H_{36}P \cdot C_{18}H_{15}BrClSn$ mw: 868.99

SYNS: A-36   *   CELA A-36   *   DECYLTRIPHENYL-
PHOSPHONIUM BROMOCHLOROTRIPHENYLSTANNATE
*   (DECYL-TRIPHENYL-PHOSPHONIUM)-TRIPHENYL-
BROM-CHLOR-STANNAT (GERMAN)   *   STANNOPLUS
*   STANNORAM   *   STANNPLOUS

OSHA PEL: TWA 0.1 mg(Sn)/m^3 (skin)
ACGIH TLV: TWA 0.1 mg(Sn)/m^3 (skin) (Proposed: TWA 0.1 mg(Sn)/m^3; STEL 0.2 mg(Sn)/m^3 (skin))
NIOSH REL: (Organotin Compounds) TWA 0.1 mg(Sn)/m^3

SAFETY PROFILE: Poison by skin contact. Moderately toxic by ingestion. When heated to decomposition it emits very toxic fumes of PO_x, Br$^-$, and Cl$^-$. A pesticide.

DAE800   CAS: 91-17-8   **HR: 3**
DECAHYDRONAPHTHALENE

DOT: UN 1147
mf: $C_{10}H_{18}$ mw: 138.28

PROP: Water-white liquid. Mp (cis): −43.3°, mp (trans): −30.7°, bp: (cis): 195.6°, bp: (trans) 187.3°, flash p: 136°F, (CC), autoign temp: 482°F, vap press: (cis) 1 mm @ 22.5°, (trans) 10 mm @ 47.2°, d: (cis) 0.8963 @ 20°/4°, vap d: 4.76, lel: 0.7% @ 212°F, uel: 4.9% @ 212°F.

SYNS: BICYCLO(4.4.0)DECANE   *   DEC   *   DECA-
LIN   *   DECALIN (DOT)   *   DECALIN SOLVENT
*   DE-KALIN   *   DEKALINA (POLISH)   *   NAPHTHA-
LANE   *   NAPHTHANE   *   PERHYDRONAPHTHALENE

CONSENSUS REPORTS: Reported in EPA TSCA Inventory.

DOT Classification: Combustible Liquid; Label: None; Flammable or Combustible Liquid; Label: Flammable Liquid.

SAFETY PROFILE: Questionable carcinogen with experimental carcinogenic and neoplastigenic data. Moderately toxic by inhalation and ingestion. Mildly toxic by skin contact. Human systemic effects by inhalation: conjuctiva irritation, unspecified olfactory and pulmonary system changes. Can cause kidney damage. Mutation data reported. A skin and eye irritant. Flammable when exposed to heat or flame, can react with oxidizing materials. To fight fire, use foam, CO_2, dry chemical. When heated to decomposition it emits acrid smoke and fumes.

DAF200   CAS: 705-86-2   **HR: 1**
Δ-DECALACTONE
mf: $C_{10}H_{18}O_2$ mw: 170.28

PROP: Colorless liquid; coconut, fruity odor, butterlike on dilution. Refr index: 1.456-1.459. Very sol in alc and propylene glycol; insol in water @ 281°.

SYNS: AMYL-Δ-VALEROLACTONE * DECANOLIDE-1,5 * FEMA No. 2361

CONSENSUS REPORTS: Reported in EPA TSCA Inventory.

SAFETY PROFILE: A skin and eye irritant. When heated to decomposition it emits acrid smoke and irritating fumes.

DAG000   CAS: 112-31-2   **HR: 2**
1-DECANAL
mf: $C_{10}H_{20}O$ mw: 156.30

PROP: Found in over 50 sources including citrus oils, citronella and lemongrass. Colorless to light yellow liquid; floral, fatty odor. D: 0.830 @ 15°/4°, bp: 208°, flash p: 185°F. Sol in 80% alc, fixed oils, volatile oils, and mineral oils; insol in water and glycerol.

SYNS: ALDEHYDE C10 * C-10 ALDEHYDE * CAPRALDEHYDE * 1-DECYL ALDEHYDE * FEMA No. 2362

CONSENSUS REPORTS: Reported in EPA TSCA Inventory.

SAFETY PROFILE: Moderately toxic by ingestion. Mildly toxic by skin contact. A severe skin irritant. Combustible liquid. When heated to decomposition it emits acrid smoke and irritating fumes.

DAG200   CAS: 112-31-2   **HR: 2**
1-DECANAL (mixed isomers)

SYN: FEMA No. 2362

CONSENSUS REPORTS: Reported in EPA TSCA Inventory.

SAFETY PROFILE: Moderately toxic by ingestion. A severe skin irritant. When heated to decomposition it emits acrid smoke and fumes.

DAG400   CAS: 124-18-5   **HR: 3**
DECANE
DOT: UN 2247
mf: $C_{10}H_{22}$ mw: 142.29

PROP: Liquid. Mp: −29.7°, bp: 174.1°, lel: 0.8%, uel: 5.4%, flash p: 115°F (CC), d: 0.730 @ 20°/4°, autoign temp: 410°F, vap press: 1 mm @ 16.5°, vap d: 4.90.

SYN: n-DECANE (DOT)

CONSENSUS REPORTS: Reported in EPA TSCA Inventory.

DOT Classification: Flammable or Combustible Liquid; Label: Flammable Liquid.

SAFETY PROFILE: Questionable carcinogen with experimental tumorigenic data. A simple asphyxiant. Narcotic in high concentrations. Flammable when exposed to heat or flame. Can react with oxidizing materials. Moderately explosive in its vapor form. To fight fire, use foam, CO_2, dry chemical.

DAG600   CAS: 2016-57-1   **HR: 3**
1-DECANEAMINE
mf: $C_{10}H_{23}N$ mw: 157.34

PROP: Liquid. Mp: 17°, bp: 95° @ 10 mm, flash p: 210°F, d: 0.79 @ 20°, vap d: 5.5.

SYNS: 1-AMINODECANE * DECYLAMINE

CONSENSUS REPORTS: Reported in EPA TSCA Inventory.

SAFETY PROFILE: Poison by ingestion and skin contact. A skin irritant. Flammable when exposed to heat or flame; can react with oxidizing materials. To fight fire, use alcohol foam, foam, dry chemical. When heated to decomposition it emits toxic fumes of NO_x.

DAH400   CAS: 334-48-5   **HR: 3**
DECANOIC ACID
mf: $C_{10}H_{20}O_2$ mw: 172.30

PROP: White crystals; unpleasant odor. D: 0.8782 @ 50°/4°, bp: 270°, mp: 31.4°. Sol in most organic solvents and in dilute nitric acid; insol in water.

SYNS: CAPRIC ACID * n-CAPRIC ACID * CAPRINIC ACID * CAPRYNIC ACID * n-DECANOIC

ACID * n-DECOIC ACID * DECYLIC ACID * n-DECYLIC ACID * HEXACID 1095 * NEO-FAT 10 * 1-NONANECARBOXYLIC ACID

CONSENSUS REPORTS: Reported in EPA TSCA Inventory.

SAFETY PROFILE: Poison by intravenous route. Mutation data reported. A moderate skin irritant. When heated to decomposition it emits acrid smoke and irritating fumes.

DAI350   CAS: 3913-71-1   **HR: 2**
2-DECENAL
mf: $C_{10}H_{18}O$ mw: 154.28

PROP: Sltly yellow liquid; orange odor. D: 0.836-0.846, refr index: 1.452-1.457. Sol in alc, fixed oils; insol in water.

SYNS: trans-2-DECEN-1-AL * DECENALDEHYDE * FEMA No. 2366

CONSENSUS REPORTS: Reported in EPA TSCA Inventory.

SAFETY PROFILE: Moderately toxic by skin contact. Mildly toxic by ingestion. A severe skin irritant. When heated to decomposition it emits acrid smoke and fumes.

DAI600   CAS: 112-30-1   **HR: 3**
DECYL ALCOHOL
mf: $C_{10}H_{22}O$ mw: 158.32

PROP: Found in sweet orange and a few other essential oils. Colorless, viscous, refractive liquid; floral fruity odor. Mp: 7°, bp: 232.9°, flash p: 180°F (OC), d: 0.8297 @ 20°/4°, refr index: 1.435-1.439, vap press: 1 mm @ 69.5°, vap d: 5.3. Sol in alc, ether, mineral oil, propylene glycol, fixed oils; insol in glycerin water @ 233°.

SYNS: AGENT 504 * ALCOHOL C-10 * ANTAK * C 10 ALCOHOL * CAPRIC ALCOHOL * CAPRINIC ALCOHOL * DECANAL DIMETHYL ACETAL * DECANOL * n-DECANOL * 1-DECANOL (FCC) * n-DECATYL ALCOHOL * n-DECYL ALCOHOL * DECYLIC ALCOHOL * DYTOL S-91 * EPAL 10 * FEMA No. 2365 * LOROL 22 * NONYLCARBINOL * PRIMARY DECYL ALCOHOL * ROYALTAC * SIPOL L10

CONSENSUS REPORTS: Reported in EPA TSCA Inventory.

SAFETY PROFILE: Questionable carcinogen with experimental tumorigenic data. Moderately toxic by skin contact. Mildly toxic by ingestion, inhalation, and possibly other routes. A severe human skin and eye irritant. Combustible when exposed to heat or flame; can react with oxidizing materials. To fight fire, use foam, CO_2, dry chemical. When heated to decomposition it emits acrid smoke and irritating fumes.

DAJ000   CAS: 1322-98-1   **HR: 3**
DECYL BENZENE SODIUM SULFONATE
mf: $C_{16}H_{25}O_3S \cdot Na$ mw: 320.46

SYNS: SODIUM DECYLBENZENESULFONAMIDE * SODIUM DECYLBENZENESULFONATE

CONSENSUS REPORTS: Reported in EPA TSCA Inventory.

SAFETY PROFILE: Poison by intravenous route. Moderately toxic by ingestion. A severe eye irritant. When heated to decomposition it emits toxic fumes of SO_x.

DAJ800   CAS: 8024-14-4   **HR: 2**
DEERTONGUE INCOLORE

PROP: Found in leaves of *Liatris odoratissima*; contains coumarin.

SYNS: DEER'S TONGUE * LIATRIS * LIATRIX OLEORESIN * VANILLA PLANT

SAFETY PROFILE: Moderately toxic by ingestion and skin contact. When heated to decomposition it emits toxic fumes of NO_x.

DAL400   CAS: 23107-12-2   **HR: 3**
DEHYDRORETRONECINE
mf: $C_8H_{11}NO_2$ mw: 153.20

SYNS: 3,8-DIDEHYDRORETRONECINE * (R)-2,3-DIHYDRO-1-HYDROXY-1H-PYRROLIZINE-7-METHANOL

CONSENSUS REPORTS: IARC Cancer Review: Animal Sufficient Evidence IMEMDT 10,333,76.

SAFETY PROFILE: Confirmed carcinogen with experimental carcinogenic and neoplastigenic data. Poison by intraperitoneal route. Human mutation data reported. When heated to decomposition it emits toxic fumes of NO_x.

DAL600   CAS: 84-17-3   **HR: 3**
DEHYDROSTILBESTROL
mf: $C_{18}H_{18}O_2$ mw: 266.36

SYNS: 3,4-BIS(p-HYDROXYPHENYL)-2,4-HEXADIENE
* 3,4-BIS(4-HYDROXYPHENYL)-2,4-HEXADIENE
* CYCLADIENE * DEHYDROSTILBOESTROL
* DIENESTROL * DIENOESTROL * β-DIENOES-
TROL * DIENOL * 4,4'-(1,2-DIETHYLIDENE-1,2-
ETHANEDIYL)BISPHENOL * p,p'-(DIETHYLIDENE-
ETHYLENE)DIPHENOL * 4,4'-(DIETHYLIDENE-
ETHYLENE)DIPHENOL * DINOVEX * DI(p-
OXYPHENYL)-2,4-HEXADIENE * DV * ESTRA-
GARD * ESTRODIENOL * ESTRORAL
* FOLLIDIENE * FOLLORMON * GYNEFOL-
LIN * HORMOFEMIN * 4,4'-HYDROXY-γ,Δ-
DIPHENYL-β,Δ-HEXADIENE * ISODIENES-
TROL * OESTRASID * OESTRODIENE * OES-
TRODIENOL * OESTRORAL * PARA-DIEN
* RESTROL * RETALON * SEXADIEN
* SYNESTROL * TESERENE * WILLNESTROL

CONSENSUS REPORTS: IARC Cancer Review: Animal Inadequate Evidence IMEMDT
21,161,79; Human Limited Evidence IMEMDT
21,161,79

SAFETY PROFILE: Suspected human carcinogen. Human mutation data reported. Experimental reproductive effects. Used as a drug for the
treatment of postmenopausal symptoms. When
heated to decomposition it emits acrid smoke
and irritating fumes.

DAM600   CAS: 57-42-1   **HR: 3**
DEMAROL
mf: $C_{15}H_{21}NO_2$ mw: 247.37

SYNS: DEMEROL * DOLCONTRAL * DOLOSAL
* DOLSIN * ETHYL-1-METHYL-4-PHENYLISONIPE-
COTATE * ETHYL-1-METHYL-4-PHENYLPIPERIDINE-
4-CARBOXYLATE * ISONIPECAINE * LIDOL
* MEPERIDINE * N-METHYL-4-PHENYL-4-
CARBETHOXYPIPERIDINE * 1-METHYL-4-PHENY-
LISONIPECOTIC ACID, ETHYL ESTER * 1-METHYL-
4-PHENYL-PIPERIDIN-4-CARBON-SAEURE-
AETHYLESTER-HYDROCHLORID (GERMAN)
* 1-METHYL-4-PHENYLPIPERIDINE-4-CARBOXYLIC
ACID ETHYL ESTER * NEMEROL * PETHIDINETER
* PETHIDOINE * PHETIDINE * PIPERSAL
* PIRIDOSAL

SAFETY PROFILE: A human poison by an unspecified route. Poison experimentally by ingestion, subcutaneous, intravenous, intradermal,
parenteral, and intraperitoneal routes. Human
systemic effects by unspecified route: changes
in sleep patterns and muscle weakness. Human
reproductive effects by intramuscular route: effects on measurements and viability of newborn.
A pharmaceutical pain killer. When heated to
decomposition it emits toxic fumes of NO_x.

DAM700   CAS: 50-13-5   **HR: 3**
DEMEROL HYDROCHLORIDE
mf: $C_{15}H_{21}NO_2 \cdot ClH$ mw: 283.83

SYNS: ALGIL * ALODAN (GEROT) * ANTIDU-
ROL * CENTRALGIN * CHLORBICYCLENE
(FRENCH) * DEMEROL * DISPADOL * DOLAN-
TAL * DOLANTIN * DOLANTIN HYDROCHLORIDE
* DOLANTOL * DOLAREN * DOLARGAN
* DOLCONTRAL * DOLENAL * DOLENOL
* DOLESTINE * DOLIN * DOLOGAL * DOLO-
NEURINE * DOLOPETHIN * DOLOSAL
* DOLVANOL * ENDOLAT * ETHYL-1-METHYL-
4-PHENYLISONIPECOTATE HYDROCHLORIDE
* ETHYL-1-METHYL-4-PHENYLPIPERIDINE-4-CARBOX-
YLATE HYDROCHLORIDE * ETHYL-1-METHYL-4-
PHENYLPIPERIDYL-4-CARBOXYLATE HYDROCHLORIDE
* ISONIPECAINE HYDROCHLORIDE * LIDOL
* LYDOL * MEFEDINA * MEPADIN * MEPERI-
DINE HYDROCHLORIDE * MEPHEDINE
* 1-METHYL-4-CARBETHOXY-4-PHENYLPIPERIDINE
HYDROCHLORIDE * N-METHYL-4-PHENYL-4-
CARBETHOXYPIPERIDINE HYDROCHLORIDE
* 1-METHYL-4-PHENYL-4-CARBOETHOXYPIPERIDINE
HYDROCHLORIDE * 1-METHYL-4-PHENYLISONIPE-
COTIC ACID ETHYL ESTER HYDROCHLORIDE
* OPERIDINE * PANTALGINE * PENTANTIN
* PETANTIN HYDROCHLORIDE * PETHIDINE
CHLORIDE * PETIDIN * PIRIDOSAL * S 140
* SAUTERALGYL * SPASMEDAL * SPASMO-
DOLIN * SYNELAUDINE * WY 554

SAFETY PROFILE: Poison by ingestion, subcutaneous, intravenous, and intraperitoneal routes.
Moderately toxic by parenteral route. Experimental teratogenic and reproductive effects. Mutation data reported. An analgesic. When heated
to decomposition it emits very toxic fumes of
HCl and NO_x.

DAN000   CAS: 58957-92-9   **HR: 3**
4-DEMETHOXYDAUNOMYCIN
mf: $C_{26}H_{27}NO_9$ mw: 497.54

SYNS: 4-DEMETHOXYDAUNORUBICIN * NSC-
256439

SAFETY PROFILE: Poison by ingestion and
intravenous routes. Mutation data reported.
When heated to decomposition it emits toxic
fumes of NO_x.

DAO600   CAS: 8065-48-3   **HR: 3**
DEMETON-O + DEMETON-S
mf: $C_8H_{19}O_3PS_2 \cdot C_8H_{19}O_3PS_2$ mw: 516.72

PROP: A light brown liquid, sulfur compound odor. .

SYNS: BAY 10756 * BAYER 8169 * DEMETON * DEMOX * DIETHOXY THIOPHOSPHORIC ACID ESTER of 2-ETHYLMERCAPTOETHANOL * O,O-DIETHYL 2-ETHYLMERCAPTOETHYL THIOPHOSPHATE * O,O-DIETHYL O(and S)-2-(ETHYLTHIO)ETHYL PHOSPHORO-THIOATE MIXTURE * E 1059 * ENT 17,295 * MERCAPTOPHOS * PHOSPHOROTHIOIC ACID-O,O-DIETHYL-O-(2-(ETHYLTHIO)ETHYL) ESTER, mixed with O,O-DIETHYL S-(2-(ETHYLTHIO)ETHYL) ESTER (7:3) * SYSTEMOX * SYSTOX * ULV

CONSENSUS REPORTS: EPA Genetic Toxicology Program. EPA Extremely Hazardous Substances List.

OSHA PEL: TWA 0.1 mg/m³ (skin)
ACGIH TLV: TWA 0.01 ppm (skin)
DFG MAK: 0.01 ppm (0.1 mg/m³)

SAFETY PROFILE: A deadly human poison by ingestion. Poison experimentally by ingestion, inhalation, skin contact, intramuscular, intravenous, subcutaneous, and intraperitoneal routes. An experimental teratogen. Human mutation data reported. An insecticide which inhibits cholinesterase in humans and animals and thus causes the buildup of acetylcholine. Doses are cumulative. If illness occurs, it is acute in nature whether caused by a single large dose or by repeated exposure. Persons poisoned with demeton may be expected to show the following symptoms: headache, giddiness, blurred vision, weakness, nausea, diarrhea, and discomfort in the chest. When heated to decomposition it emits very toxic fumes of PO_x and SO_x.

DAO800   CAS: 867-27-6   **HR: 3**
DEMETON-O-METHYL
mf: $C_6H_{15}O_3PS_2$ mw: 230.30

SYNS: BAY 15203 * DEMETON-O-METILE (ITALIAN) * O,O-DIMETHYL-O-(2-AETHYLTHIO-AETHYL MONO-THIOPHOSPHAT (GERMAN) * O,O-DIMETHYL-O-ETHYLMERCAPTOETHYL THIOPHOSPHATE * O,O-DI-METHYL 2-ETHYLMERCAPTOETHYL THIOPHOSPHATE, THIONO ISOMER * O,O-DIMETHYL-O-(2-ETHYL-THIO-ETHYL)-MONOTHIOFOSFAAT (DUTCH) * O,O-DI-METHYL-O-2-(ETHYLTHIO)ETHYL PHOSPHOROTHIOATE * O,O-DIMETIL-O-(2-ETILTIO-ETIL)-MONOTIOFOSFATO (ITALIAN) * ENT 18,862 * β-ETHYLMERCAPTO-ETHYL DIMETHYL THIONOPHOSPHATE * O-(2-(ETHYLTHIO)ETHYL)-O,O-DIMETHYL PHOSPHORO-THIOATE * 2-(ETHYLTHIO)ETHYL DIMETHYL PHOSPHOROTHIONATE * METHYL-DEMETON-O * O-METHYLDEMETON * METHYLLCISTOX * METHYLMERCAPTOPHOS * METHYLSYSTOX * THIOPHOSPHATE de O,O-DIMETHYLE et de O-2-ETHYLTHIO-ETHYLE (FRENCH)

SAFETY PROFILE: Poison by ingestion, skin contact, inhalation, intravenous, and possibly other routes. Experimental teratogenic and reproductive effects. When heated to decomposition it emits very toxic fumes of PO_x and SO_x.

DAP000   CAS: 301-12-2   **HR: 3**
DEMETON-O-METHYL SULFOXIDE
mf: $C_6H_{15}O_3PS_2$ mw: 230.30

SYNS: BAY 21097 * DEMETON-S-METHYL-SUL-FOXID (GERMAN) * DEMETON-S-METHYL SULFOXIDE * DEMETON-METHYL SULPHOXIDE * O,O-DIMETHYL-S-(2-AETHYLSULFINYL-AETHYL)-THIOLPHOS-PHAT (GERMAN) * O,O-DIMETHYL-S-(2-ETHTHIONYL-ETHYL) PHOSPHOROTHIOATE * DIMETHYL-S-(2-ETH-THIONYLETHYL) THIOPHOSPHATE * O,O-DIMETHYL-S-(2-ETHYLSULFINYL-ETHYL)-MONOTHIOFOSFAAT (DUTCH) * O,O-DIMETHYL-S-(2-(ETHYLSULFINYL) ETHYL) PHOSPHOROTHIOATE * O,O-DIMETHYL-S-(2-ETHYLSULFINYL)ETHYL THIOPHOSPHATE * O,O-DIMETHYL-S-ETHYLSULPHINYLETHYL PHOS-PHOROTHIOLATE * O,O-DIMETHYL-S-(3-OXO-3-THIA-PENTYL)-MONOTHIOPHOSPHAT (GERMAN) * O,O-DIMETIL-S-(2-ETIL-SOLFINIL-ETIL)-MONOTIOFOSFATO (ITALIAN) * ENT 24,964 * S-(2-(ETHYLSULFINYL) ETHYL)-O,O-DIMETHYL PHOSPHOROTHIOATE * ISOMETHYLSYSTOX SULFOXIDE * METAISO-SYSTOXSULFOXIDE * METASYSTEMOX * META-SYSTOX-R * METHYL DEMETON-O-SULFOXIDE * METILMERCAPTOFOSOKSID * OXYDEMETON-METHYL * OXYDEMETON-METILE (ITALIAN) * R 2170 * THIOPHOSPHATE de O,O-DIMETHYLE et de S-2-ETHYLSULFINYLETHYLE (FRENCH)

CONSENSUS REPORTS: EPA Genetic Toxicology Program.

SAFETY PROFILE: Poison by ingestion, skin contact, intravenous, intraperitoneal, and possibly other routes. Human mutation data reported. When heated to decomposition it emits very toxic fumes of PO_x and SO_x.

DAP200 CAS: 126-75-0 **HR: 3**
DEMETON-S
mf: $C_8H_{19}O_3PS_2$ mw: 258.36

SYNS: O,O-DIAETHYL-S-(2-AETHYLTHIO-AETHYL) MONOTHIOPHOSPHAT (GERMAN) * DIAETHYLTHIO-PHOSPHORSAEUREESTER des AETHYLTHIOGLYKOL (GERMAN) * DIETHYL-S-(2-ETHIOETHYL)THIOPHOS-PHATE * O,O-DIETHYL-S-(2-ETHTHIOETHYL)PHOS-PHOROTHIOATE * O,O- DIETHYL-S-ETHYL-2-ETHYLMERCAPTOPHOSPHOROTHIOLATE * O,O-DIETHYL-S-2-(ETHYLTHIO)ETHYL PHOSPHOROTHIOATE * O,O-DIETHYL-S-(2-(ETHYLTHIO)ETHYL) PHOSPHORO-THIOLATE (USDA) * O,O-DIETHYL-S-(2-ETHYLTHIO-ETHYL)-MONOTHIOFOSFAAT (DUTCH) * O,O-DIETIL-S-(2-ETILTIO-ETIL)-MONOTIOFOSFATO (ITALIAN) * O,O-DIETYL-S-2-ETYLMERKAPTOETYLTIOFOSFAT (CZECH) * 2-(ETHYLTHIO)-ETHANETHIOL S-ESTER with O,O-DIETHYL PHOSPHOROTHIOATE * ISODEME-TON * IZOSYSTOX (CZECH) * PO-SYSTOX * THIOLDEMETON * THIOL SYSTOX * THIO-PHOSPHATE de O,O-DIETHYLE et de S-(2-ETHYLTHIO-ETHYLE) (FRENCH)

SAFETY PROFILE: Poison by ingestion, intra-peritoneal, and subcutaneous routes. When heated to decomposition it emits very toxic fumes of PO_x and SO_x.

DAP400 CAS: 919-86-8 **HR: 3**
DEMETON-S-METHYL
mf: $C_6H_{15}O_3PS_2$ mw: 230.30

SYNS: BAY 18436 * BAYER 25/154 * O,O-DIMETHYL-S-(2-AETHYLTHIO-AETHYL)-MONOTHIO-PHOSPHAT (GERMAN) * DEMETON-S-METILE (ITAL-IAN) * O,O-DIMETHYL-S-(2-ETHTHIOETHYL)PHOS-PHOROTHIOATE * DIMETHYL-S-(2-ETHTHIOETHYL) THIOPHOSPHATE * O,O-DIMETHYL-S-ETHYLMER-CAPTOETHYL THIOPHOSPHATE * O,O-DIMETHYL-S-ETHYLMERCAPTOETHYL THIOPHOSPHATE, THIOLO ISOMER * O,O-DIMETHYL-S-(2-ETHYLTHIO-ETHYL) MONOTHIOFOSFAAT (DUTCH) * O,O-DIMETHYL-S-(2-(ETHYLTHIO)ETHYL)PHOSPHOROTHIOATE * O,O-DIMETHYL-S-(3-THIA-PENTYL)-MONOTHIO-PHOSPHAT (GERMAN) * O,O-DIMETIL-S-(2-ETILITIO-ETIL)-MONOTIOFOSFATO (ITALIAN) * DURATOX * S-(2-(ETHYLTHIO)ETHYL)-O,O-DIMETHYL PHOSPHO-ROTHIOATE * S-(2-(ETHYLTHIO)ETHYL)DIMETHYL PHOSPHOROTHIOLATE * S-(2-(ETHYLTHIO)ETHYL)-O,O-DIMETHYL THIOPHOSPHATE * ISOMETASYSTOX * ISOMETHYLSYSTOX * METAISOSEPTOX * METAISOSYSTOX * METASYSTOX FORTE * METHYL DEMETON THIOESTER * METHYL ISO-SYSTOX * METHYL-MERCAPTOFOS TEOLOVY

* THIOPHOSPHATE de O,O-DIMETHYLE et de S-2-ETHYLTHIOETHYLE (FRENCH)

CONSENSUS REPORTS: Reported in EPA TSCA Inventory. EPA Extremely Hazardous Substances List.

SAFETY PROFILE: Poison by ingestion, inha-lation, skin contact, intraperitoneal, intrave-nous, and possibly other routes. Mutation data reported. An insecticide. When heated to de-composition it emits very toxic fumes of PO_x and SO_x.

DAP600 CAS: 17040-19-6 **HR: 3**
DEMETON-S-METHYL-SULPHONE
mf: $C_6H_{15}O_5PS_2$ mw: 262.30

SYNS: BAYER 20315 * DEMETON-S-METHYLSUL-FONE * DEMETON-S-METHYLSULFON (GERMAN) * O,O-DIMETHYL-S-(2-AETHYLSULFONYL-AETHYL) THIOLPHOSPHAT (GERMAN) * O,O-DIMETHYL-S-(2-ETHSULFONYLETHYL)PHOSPHOROTHIOATE * DI-METHYL-S-(2-ETHSULFONYLETHYL)THIOPHOSPHATE * O,O-DIMETHYL-S-ETHYL-2-SULFONYLETHYL PHOS-PHOROTHIOLATE * O,O-DIMETHYL-S-ETHYLSULPHO-NYLETHYL PHOSPHOROTHIOLATE * DIOXYDEME-TON-S-METHYL * E 158 * ISOMETASYSTOX SULFONE * ISOMETHYLSYSTOX SULFONE * M 3/158 * METAISOSYSTOX-SOLFON 20 315

SAFETY PROFILE: Poison by ingestion, inha-lation, intraperitoneal, intravenous, and possi-bly other routes. Moderately toxic by skin con-tact. Mutation data reported. An insecticide. When heated to decomposition it emits very toxic fumes of PO_x and SO_x.

DAQ400 CAS: 83-44-3 **HR: 3**
DEOXYCHOLATIC ACID
mf: $C_{24}H_{40}O_4$ mw: 392.64

PROP: A white crystalline powder. Mp: 178°. Sol in alc, acetone; sltly sol in ether and chloro-form; insol in water.

SYNS: CHOLEIC ACID * CHOLEREBIC * CHO-LOREBIC * DEGALOL * DEOXYCHOLIC ACID (FCC) * 7-α-DEOXYCHOLIC ACID * DESOXY-CHOLIC ACID * DESOXYCHOLSAEURE (GERMAN) * 3,12-DIHYDROXYCHOLANIC ACID * 3-α,12-α-DI-HYDROXYCHOLANIC ACID * 3-α,12-α-DIHYDROXY-5-β-CHOLAN-24-OIC ACID * 3-α,12-α-DIHYDROXY-5-β-CHOLANOIC ACID * 3-α,12-α-DIHYDROXYCHOLANSA-EURE (GERMAN) * DROXOLAN * 17-β-(1-METHYL-3-CARBOXYPROPYL)-ETIOCHOLANE-3-α,12-α-DIOL * PYROCHOL * SEPTOCHOL

CONSENSUS REPORTS: Reported in EPA TSCA Inventory.

SAFETY PROFILE: Poison by intraperitoneal route. Moderately toxic by ingestion and intravenous routes. Questionable carcinogen with experimental tumorigenic data. Mutation data reported. When heated to decomposition it emits acrid smoke and irritating fumes.

DAR000   CAS: 55297-96-6   **HR: 3**
14-DEOXY-14-((2-DIETHYLAMINO-ETHYL)MERCAPTOACETOXY)-MUTILIN HYDROGEN FUMARATE
mf: $C_{28}H_{47}NO_4S \cdot C_4H_4O_4$ mw: 609.90

SYNS: SQ 22947   *   TIAMUTIN

SAFETY PROFILE: Poison by intramuscular route. Moderately toxic by ingestion subcutaneous routes. When heated to decomposition it emits very toxic fumes of NO_x and SO_x.

DAR400   CAS: 50-91-9   **HR: 3**
2'-DEOXY-5-FLUOROURIDINE
mf: $C_9H_{11}FN_2O_5$ mw: 246.22

SYNS: DEOXYFLUOROURIDINE   *   1-β-d-2'-DEOXY-RIBOFURANOSYL-5-FLUROURACIL   *   FDUR
*   FLOXURIDIN   *   FLOXURIDINE   *   FLUORODE-OXYURIDINE   *   β-5-FLUORO-2'-DEOXYURIDINE
*   5-FLUORODEOXYURIDINE   *   5-FLUORO-2-DE-OXYURIDINE   *   5-FLUORO-2'-DEOXYURIDINE
*   5-FLUOROURACIL DEOXYRIBOSIDE   *   5-FLUORO-URACIL-2'-DEOXYRIBOSIDE   *   FLUORURIDINE DE-OXYRIBOSE   *   FUDR   *   5-FUDR   *   NSC-27640
*   RO 5-0360

CONSENSUS REPORTS: EPA Genetic Toxicology Program.

SAFETY PROFILE: Poison by ingestion. Moderately toxic by intraperitoneal route. Human systemic effects by intravenous and possibly other routes: hypermotility, diarrhea, nausea, vomiting and other unspecified gastrointestinal and bone marrow changes. Experimental teratogenic and reproductive effects. Human mutation data reported. When heated to decomposition it emits very toxic fumes of F^- and NO_x.

DAR600   CAS: 154-17-6   **HR: 3**
2-DEOXYGLUCOSE
mf: $C_6H_{12}O_5$ mw: 164.18

PROP: Crystals from acetone or butanone. Mp: 142-144°. α-Form: Crystals from isopropanol. Mp: 134-136°.

SYNS: 2-DEOXY-d-ARABINO-HEXOSE   *   2-DEOXY-3-ARABINO-HEXOSE   *   d-2-DEOXYGLUCOSE
*   2-DEOXY-d-GLUCOSE   *   2-DESOXY-d-GLUCOSE (FRENCH)   *   2-DG   *   NSC 15193

CONSENSUS REPORTS: Reported in EPA TSCA Inventory.

SAFETY PROFILE: Poison by subcutaneous route. Moderately toxic by intraperitoneal route. Experimental teratogenic and reproductive effects. When heated to decomposition it emits acrid smoke and fumes.

DAS000   CAS: 54-42-2   **HR: 3**
2'-DEOXY-5-IODOURIDINE
mf: $C_9H_{11}IN_2O_5$ mw: 354.12

SYNS: ALLERGAN 211   *   DENDRID   *   1-(2-DE-OXY-β-d-RIBOFURANOSYL)-5-IODOURACIL   *   1-β-d-2'-DEOXYRIBOFURANOSYL-5-IODOURACIL   *   EMANIL
*   HERPESIL   *   HERPIDU   *   HERPLEX   *   HER-PLEX LIQUIFILM   *   IDEXUR   *   IDOXENE
*   IDOXURIDIN   *   IDOXURIDINE   *   IDU
*   IDUCHER   *   IDULEA   *   IDUOCULOS   *   IDUR
*   IDURIDIN   *   5-IODODEOXYURIDINE   *   5-IODO-2'-DEOXYURIDINE   *   5-IODOURACIL DEOXYRIBOSIDE
*   IUDR   *   5-IUDR   *   JODDEOXIURIDIN
*   KERECID   *   NSC 39661   *   OPHTHALMADINE
*   SK&F 14287   *   STOXIL   *   SYNMIOL

CONSENSUS REPORTS: Reported in EPA TSCA Inventory. EPA Genetic Toxicology Program.

SAFETY PROFILE: Questionable carcinogen with experimental carcinogenic data. Moderately toxic by intraperitoneal route. Experimental teratogenic and reproductive effects. Human mutation data reported. When heated to decomposition it emits very toxic fumes of I^- and NO_x.

DBA000   **HR: 3**
DERRIS ELLIPTICA, root

SYNS: DEGUELIA ROOT   *   DERRIS RESINS
*   DERRIS ROOT   *   TUBA ROOT

SAFETY PROFILE: Poison by ingestion. An insecticide. When heated to decomposition it emits acrid smoke and fumes.

DBA600   CAS: 5626-16-4   **HR: 3**
DESMETHYLDOXEPIN
mf: $C_{17}H_{17}NO$ mw: 251.35

SYNS: DIBENZ(b,e)OXEPIN-$\Delta^{11(6H)}$,γ-PROPYLAMINE * KS 1675

SAFETY PROFILE: Poison by ingestion, intravenous, and intraperitoneal routes. When heated to decomposition it emits toxic fumes of NO_x.

DBA800 CAS: 300-42-5 **HR: 3**
DESOXYEPHEDRINE HYDROCHLORIDE
mf: $C_{10}H_{15}N \cdot ClH$ mw: 185.72

SYNS: A 884 * AMDRAM * AMEDRINE * AMPHEDROXY * AMPHEDROXYN * APAMINE * BOMBITA * C 6379 * CORVITIN * DAROPERVAMIN * DEA OXO-5 * DEOFED * DEOXYEPHEDRINE * DEPOXIN * DESAMINE * DESFEDRIN * DESOSSIEFEDRINA * DES-OXA-D * DESOXEDRINE * DESOXIN * DESOXO-5 * DESOXYFED * DESOXYN * DESOXYPHED * DESTIM * DETREX * DEXOPHRINE * DEXOVAL * N,α-DIMETHYLPHENETHYLAMINE HYDROCHLORIDE * DOPIDRIN * DOXEPHRIN * DOXYFED * DRINALFA * EFFROXINE * ESTIMULEX * EUPHODRIN * FENYPRIN * GEROBIT * GEROVIT * HEROPON * ISOPHEN * KEMODRIN * LANAZINE * MADRINE * METAMFETAMINA * METAMPHETAMIN * METHAMPHETAMINE HYDROCHLORIDE * METANFETAMINA * METHEDRINE * METHEDRINE HYDROCHLORIDE * METHOXYN * METHYLAMPHETAMINE HYDROCHLORIDE * METHYLBENZEDRIN * METHYLISOMIN * N-METHYL-β-PHENYLISOPROPYLAMINHYDROCHLORID (GERMAN) * METHYLPROPAMINE * NEODRINE * NEOPHARMEDRINE * NORODIN * OXYDRENE * OXYFED * PERVITIN * PHILOPON * PREMODRIN * SEMOXYDRINE * SPEED * STIMULEX * TONEDRIN * VONEDRINE

SAFETY PROFILE: Poison by ingestion, intravenous, intraperitoneal, subcutaneous, and intramuscular routes. Human systemic effects by ingestion: altered sleep patterns, anorexia, and change in heart rate. Experimental reproductive effects. An eye irritant. A powerful central nervous system stimulant. When heated to decomposition it emits very toxic fumes of NO_x and HCl.

DBB000 CAS: 7632-10-2 **HR: 3**
DESOXYN
mf: $C_{10}H_{15}N$ mw: 149.26

SYNS: ANADREX * DEOXYEPHEDRINE * DESOXYEPHEDRINE * METHAMPHETAMINE * METHEDRINE * METHYLAMPHETAMINE * N-METHYLAMPHETAMINE * N-METHYL-β-PHENYLISOPROPYLAMIN (GERMAN) * N-METHYL-β-PHENYLISOPROPYLAMINE * PERVERTIN * 1-PHENYL-2-METHYLAMINO-PROPAN (GERMAN) * 1-PHENYL-2-METHYLAMINOPROPANE * α-PHENYL-β-METHYLAMINOPROPANE * STIMULEX

SAFETY PROFILE: Poison by ingestion, intraperitoneal, subcutaneous, and intravenous routes. A powerful central nervous system stimulant. When heated to decomposition it emits toxic fumes of NO_x.

DBB200 CAS: 125-33-7 **HR: 3**
2-DESOXYPHENOBARBITAL
mf: $C_{12}H_{14}N_2O_2$ mw: 218.28

SYNS: 5-AETHYL-5-PHENYL-HEXAHYDROPYRIMIDIN-4,6-DION (GERMAN) * CYRAL * 2-DEOXYPHENOBARBITAL * DESOXYPHENOBARBITONE * 5-ETHYLDIHYDRO-5-PHENYL-4,6(1H,5H)-PYRIMIDINEDIONE * 5-ETHYLHEXAHYDRO-4,6-DIOXO-5-PHENYLPHRIMIDINE * 5-ETHYLHEXAHYDRO-5-PHENYLPYRIMIDINE-4,6-DIONE * 5-ETHYL-5-PHENYLHEXAHYDROPYRIMIDINE-4,6-DIONE * HEXADIONA * HEXAMIDINE * HEXAMIDINE (the antispasmodic) * LEPIMIDIN * LEPSIRAL * MAJSOLIN * MIDONE * MILEPSIN * MISODINE * MISOLYNE * MIZODIN * MIZOLIN * MYLEPSIN * MYLEPSINUM * MYSEDON * MYSOLINE * NCI-C56360 * 5-PHENYL-5-ETHYL-HEXAHYDROPYRIMIDINE-4,6-DIONE * PRILEPSIN * PRIMACIONE * PRIMACLONE * PRIMACONE * PRIMAKTON * PRIMIDON * PRIMIDONE * PRYSOLINE * PYRIMIDONE MEDI-PETS * ROE 101 * SERTAN

SAFETY PROFILE: Poison by ingestion, intraperitoneal, and possibly other routes. Human reproductive effects by ingestion and possibly other routes: effects on newborn including unusual growth statistics, drug dependence, physical and other neonatal changes. Human teratogenic effects include developmental abnormalities of the craniofacial area, skin and skin appendages, and cardiovascular system. Experimental teratogenic and reproductive effects. Human mutation data reported. Addictive. When heated to decomposition it emits toxic fumes of NO_x.

DBB600   CAS: 67293-88-3   **HR: 3**
DEUTERIOMORPHINE
mf: $C_{17}H_{16}D_3NO_3$ mw: 288.37

SAFETY PROFILE: Poison by subcutaneous and intracerebral routes. When heated to decomposition it emits toxic fumes of NO_x.

DBB800   CAS: 7782-39-0   **HR: 3**
DEUTERIUM

DOT: UN 1957
mf: D_2 mw: 4.03

PROP: A gas. Chemically the same as hydrogen. Lel: 5%, uel: 75%

SYN: D₂

DOT Classification: Flammable Gas; Label: Flammable Gas

SAFETY PROFILE: Very dangerous fire and explosion hazard when exposed to heat, flame, sparks, and oxidizers. To fight fire, stop flow of gas.

DBC000   CAS: 14333-26-7   **HR: 3**
DEUTERIUM FLUORIDE
mf: DF   mw: 21.01

PROP: Chemically the same as hydrogen fluoride.
OSHA PEL: TWA 2.5 mg(F)/m³
ACGIH TLV: TWA 2.5 mg(F)/m³
NIOSH REL: TWA (Inorganic Fluorides) 2.5 mg(F)/m³

SAFETY PROFILE: Moderately toxic by inhalation. A dangerously reactive, powerful oxidant. When heated to decomposition it emits toxic fumes of F^-.

DBC800   CAS: 9004-54-0   **HR: 3**
DEXTRAN 1

PROP: A linear water-sol polymer of average molecular weight 200,000.

CONSENSUS REPORTS: Reported in EPA TSCA Inventory.

SAFETY PROFILE: Questionable carcinogen with experimental neoplastigenic, tumorigenic, and teratogenic data. When heated to decomposition it emits acrid smoke and fumes.

DBD000   CAS: 9004-54-0   **HR: 3**
DEXTRAN 2

PROP: A linear, water-sol polymer of average molecular weight 100,000.

CONSENSUS REPORTS: Reported in EPA TSCA Inventory.

SAFETY PROFILE: Questionable carcinogen with experimental neoplastigenic data. When heated to decomposition it emits acrid smoke and fumes.

DBD200   CAS: 9004-54-0   **HR: 3**
DEXTRAN 5

PROP: A highly branched, water-sol polymer.

CONSENSUS REPORTS: Reported in EPA TSCA Inventory.

SAFETY PROFILE: Questionable carcinogen with experimental neoplastigenic data. When heated to decomposition it emits acrid smoke and fumes.

DBD400   CAS: 9004-54-0   **HR: 3**
DEXTRAN 10

PROP: A branched, water-sol polymer of average molecular weight 89,400.

CONSENSUS REPORTS: Reported in EPA TSCA Inventory.

SAFETY PROFILE: Suspected carcinogen with experimental carcinogenic data. When heated to decomposition it emits acrid smoke and fumes.

DBD600   CAS: 9004-54-0   **HR: 3**
DEXTRAN 11

PROP: A highly branched, water-sol polymer of average molecular weight 71,400.

CONSENSUS REPORTS: Reported in EPA TSCA Inventory.

SAFETY PROFILE: Questionable carcinogen with experimental neoplastigenic and tumorigenic data. When heated to decomposition it emits acrid smoke and fumes.

DBD800   CAS: 9004-53-9   **HR: 1**
DEXTRINS
mf: $(C_6H_{10}O_5)_n \cdot xH_2O$

PROP: An intermediate product formed by the hydrolysis of starches. It describes a class of substances. Yellow or white powder or granules. Sol in water; insol in alc and ether; forms colloids.

SYNS: ARTIFICIAL GUM * DEXTRANS * STARCH GUM * TAPIOCA * VEGETABLE GUM

SAFETY PROFILE: Mildly toxic by intravenous route. When heated to decomposition it emits acrid smoke and irritating fumes.

DBF200   CAS: 642-65-9   **HR: 3**
2-DIACETAMIDOFLUORENE
mf: $C_{17}H_{15}NO_2$ mw: 265.33

SYNS: N-ACETYL-N-9H-FLUOREN-2-YL-ACETAMIDE * N-DIACETYL-2-AMINOFLUORENE * 2-DIACETYLAMINOFLUORENE * N,N-DIACETYL-2-AMINOFLUORENE * N,N-DIACETYL-2-FLUORENAMINE * F-diAA * 2-FLUORENYLDIACETAMIDE * N-FLUOREN-2-YLDIACETAMIDE * N-2-FLUORENYLDIACETAMIDE

SAFETY PROFILE: Suspected carcinogen with experimental carcinogenic, neoplastigenic, and tumorigenic data. When heated to decomposition it emits toxic fumes of NO_x.

DBF600   CAS: 102-62-5   **HR: 2**
1,3-DIACETIN
mf: $C_7H_{12}O_5$ mw: 176.19

PROP: Crystals. D: 1.178 @ 15°/15°, bp: 280°, mp: 40°.

SYNS: 1,3-DIACETATE GLYCEROL * 1,2-DIACETATE 1,2,3-PROPANETRIOL * DIACETIN * 1,2-DIACETIN * 2,3-DIACETIN * DIACETYL GLYCERINE * DIGLYCERIDE ACETIC ACID * GLYCEROL DIACETATE * GLYCERYL-1,3-DIACETATE * (HYDROXYMETHYL)ETHYLENE ACETATE

CONSENSUS REPORTS: Reported in EPA TSCA Inventory.

SAFETY PROFILE: Moderately toxic by subcutaneous and intravenous routes. Mildly toxic by ingestion. When heated to decomposition it emits acrid smoke and irritating fumes.

DBF750   CAS: 123-42-2   **HR: 3**
DIACETONE ALCOHOL

DOT: UN 1148
mf: $C_6H_{12}O_2$ mw: 116.18

PROP: Liquid; faint, pleasant odor. Mp: −47 to −54°, bp: 167.9°, flash p: 148°F, d: 0.9306 @ 25°/4°, autoign temp: 1118°F, vap d: 4.00, vap press: 1.1 mm @ 20°, lel: 1.8%, uel: 6.9%, flash p: (acetone free): 136°F.

SYNS: DIACETONALCOHOL (DUTCH) * DIACETONALCOOL (ITALIAN) * DIACETONALKOHOL (GERMAN) * DIACETONE * DIACETONE-ALCOOL (FRENCH) * DIKETONE ALCOHOL * 4-HYDROXY-2-KETO-4-METHYLPENTANE * 4-HYDROXY-4-METHYL-PENTAN-2-ON (GERMAN, DUTCH) * 4-HYDROXY-4-METHYLPENTANONE-2 * 4-HYDROXY-4-METHYL-2-PENTANONE * 4-HYDROXY-4- METHYL PENTAN-2-ONE * 4-IDROSSI-4-METIL-PENTAN-2-ONE (ITALIAN) * 2-METHYL-2-PENTANOL-4-ONE * PYRANTON * TYRANTON

CONSENSUS REPORTS: Reported in EPA TSCA Inventory.

OSHA PEL: TWA 50 ppm
ACGIH TLV: TWA 50 ppm
DFG MAK: 50 ppm (240 mg/m³)
NIOSH REL: (Ketones) TWA 240 mg/m³
DOT Classification: Flammable Liquid; Label: Flammable Liquid; Flammable or Combustible Liquid; Label: Flammable Liquid.

SAFETY PROFILE: Moderately toxic by ingestion and intraperitoneal routes. Mildly toxic by skin contact. Human systemic effects by inhalation: headache, nausea or vomiting, eye and pulmonary changes. A skin, mucous membrane, and severe eye irritant. Can cause anemia and damage to liver and kidneys. Narcotic in high concentration. Flammable when exposed to heat or flame; can react with oxidizing materials. Explosive in the form of vapor when exposed to heat or flame. To fight fire, use alcohol foam, foam, CO_2, dry chemical. When heated to decomposition it emits acrid smoke and irritating fumes.

DBF800   CAS: 1067-33-0   **HR: 3**
DIACETOXYDIBUTYL STANNANE
mf: $C_{12}H_{24}O_4Sn$ mw: 351.05

PROP: Clear, colorless liquid with a slight acetic acid odor. Bp: decomp, fp: 5°-10°, flash p: 290°F (OC), d: 1.31 @ 25°, vap d: 12.1.

SYNS: BA 2726 * BIS(ACETYLOXY)DIBUTYLSTANNANE * DIACETOXYBUTYLTIN * DIACETOXYDIBUTYLTIN * DIBUTYL TIN DIACETATE * FOMREZ SUL-3 * NCI-C02028 * T 1 (CATALYST)

CONSENSUS REPORTS: NCI Carcinogenesis Bioassay (feed); Inadequate Studies: mouse, rat NCITR* NCI-CG-TR-183,79. Reported in EPA TSCA Inventory.

OSHA PEL: TWA 0.1 mg(Sn)/m³ (skin)
ACGIH TLV: TWA 0.1 mg(Sn)/m³ (skin) (Proposed: TWA 0.1 mg(Sn)/m³; STEL 0.2 mg(Sn)/m³ (skin))

NIOSH REL: (Organotin Compounds) TWA 0.1 mg(Sn)/m^3

SAFETY PROFILE: Poison by ingestion and intravenous routes. Combustible when exposed to heat or flame; can react with oxidizing materials. To fight fire, use water, foam, CO_2, dry chemical. When heated to decomposition it emits acrid smoke and irritating fumes.

DBH000   CAS: 95-45-4   **HR: 3**
DIACETYL DIOXIME
mf: $C_4H_8N_2O_2$ mw: 116.14

SYNS: 2,3-DIISONITROSOBUTANE * DIMETHYL-GLYOXIME

CONSENSUS REPORTS: Reported in EPA TSCA Inventory.

SAFETY PROFILE: Poison by ingestion. Mutation data reported. When heated to decomposition it emits toxic fumes of NO_x.

DBH800   CAS: 73622-67-0   **HR: 3**
3,4-DI(ACETYLTHIOMETHYL)-5-HYDROXY-6-METHYLPYRIDINE HYDROBROMIDE
mf: $C_{12}H_{15}NO_3S_2 \cdot BrH$ mw: 366.32

SYNS: 4,5-DI(MERCAPTOMETHYL)-2-METHYL-3-PYRIDINOL DITHIOACETATE HYDROBROMIDE * 4,5-DI-MERCAPTOPYRIDOXINDI-THIOACETAT HYDROBROMID (GERMAN)

SAFETY PROFILE: Poison by subcutaneous and intravenous routes. Moderately toxic by ingestion. When heated to decomposition it emits very toxic fumes of NO_x, SO_x, and HBr.

DBI099   CAS: 10311-84-9   **HR: 3**
DIALIFOR
mf: $C_{14}H_{17}NO_4PS_2$ mw: 393.86

SYNS: S-(2-CHLORO-1-(1,3-DIHYDRO-1,3-DIOXO-2H-ISOINDOL-2-YL)ETHYL)-O,O-DIETHYL PHOSPHORO-DITHIOATE * S-(2-CHLORO-1-PHTHALIMIDOETHYL)-O,O-DIETHYL PHOSPHORODITHIOATE * O,O-DI-ETHYL- S-(2-CHLORO-1-PHTHALIMIDOETHYL)PHOS-PHORODITHIOATE * ENT 27,320 * HERCULES 14503 * PHOSPHORODITHIOIC ACID-S-(2-CHLORO-1-(1,3-DIHYDRO-1,3-DIOXO-2H-ISOINDOL-2-YL)ETHYL-O,O-DIETHYL ESTER * PHOSPHORODITHIOIC ACID-S-(2-CHLORO-1-PHTHALIMIDOETHYL)-O,)-DIETHYL ESTER * TORAK

SAFETY PROFILE: Poison by ingestion and skin contact. Experimental reproductive effects.

When heated to decomposition it emits toxic fumes of SO_x, PO_x, and NO_x.

DBI200   CAS: 2303-16-4   **HR: 3**
DIALLATE
mf: $C_{10}H_{17}Cl_2NOS$ mw: 270.24

PROP: Brown liquid. Bp: 150° @ 9 mm, mp: 25-30°. Sltly sol in water; sol in organic solvents.

SYNS: AVADEX * CP 15,336 * DATC * 2,3-DCDT * DIALLAAT (DUTCH) * DIALLAT (GERMAN) * S-(2,3-DICHLORO-ALLIL)-N,N-DIISOPRO-PIL-MONOTIOCARBAMMATO (ITALIAN) * S-(2,3-DI-CHLOR-ALLYL)-N,N-DIISOPROPYL-MONOTHIOCARBA-MAAT (DUTCH) * 2,3-DICHLORALLYL-N,N-(DIISOPRO-PYL)-THIOCARBAMAT (GERMAN) * DICHLOROALLYL DIISOPROPYLTHIOCARBAMATE * S-2,3-DICHLOROAL-LYL DIISOPROPYLTHIOCARBAMATE * 2,3-DICHLORO-ALLYL-N,N-DIISOPROPYLTHIOLCARBAMATE * 2,3-DICHLORO-2-PROPENE-1-THIOL DIISOPROPYL-CARBAMATE * S-(2,3-DICHLORO-2-PROPENYL)ESTER, BIS(1-METHYLETHYL) CARBAMOTHIOIC ACID * DI-ISOPROPYLTHIOLOCARBAMATE de S-(2,3-DICHLO-ROALLYLE) (FRENCH) * PYRADEX * RCRA WASTE NUMBER U062

CONSENSUS REPORTS: IARC Cancer Review: GROUP 3 IMEMDT 7,56,87; Animal Limited Evidence IMEMDT 30,235,83; Animal Sufficient Evidence IMEMDT 12,69,76. EPA Genetic Toxicology Program. Community Right-To-Know List.

SAFETY PROFILE: Questionable carcinogen with experimental carcinogenic and tumorigenic data. Poison by ingestion and possibly other routes. Moderately toxic by skin contact. Mutation data reported. When heated to decomposition it emits very toxic fumes of Cl$^-$, NO_x, and SO_x.

DBI600   CAS: 124-02-7   **HR: 3**
DIALLYLAMINE
DOT: UN 2359
mf: $C_6H_{11}N$ mw: 97.18

PROP: Liquid, sol in water. D: 0.7889 @ 20°, bp: 112°, fp: −100°, flash p.: 69.8°F.

SYNS: DI-2-PROPENYLAMINE * N-2-PROPENYL-2-PROPEN-1-AMINE

CONSENSUS REPORTS: Reported in EPA TSCA Inventory.

DOT Classification: Flammable Liquid; Label: Flammable Liquid.

SAFETY PROFILE: Poison by skin contact, intraperitoneal, and possibly other routes. Moderately toxic by ingestion and inhalation. Human systemic effects by inhalation route: eye lacrimation, and changes in the trachea or bronchi. A skin and severe eye irritant. A dangerous fire hazard when exposed to heat or flame. When heated to decomposition it emits toxic fumes of NO_x.

DBJ400   CAS: 17381-88-3   **HR: 3**
DIALLYLDIBROMO STANNANE
mf: $C_6H_{10}Br_2Sn$ mw: 360.67

SYN: DIALLYLTIN DIBROMIDE

OSHA PEL: TWA 0.1 mg(Sn)/m^3 (skin)
ACGIH TLV: TWA 0.1 mg(Sn)/m^3 (skin) (Proposed: TWA 0.1 mg(Sn)/m^3; STEL 0.2 mg(Sn)/m^3 (skin))
NIOSH REL: (Organotin Compounds) TWA 0.1 mg(Sn)/m^3

SAFETY PROFILE: Poison by intravenous route. When heated to decomposition it emits toxic fumes of Br$^-$.

DBK000   CAS: 557-40-4   **HR: 3**
DIALLYL ETHER
DOT: UN 2360
mf: $C_6H_{10}O$ mw: 98.16

PROP: Liquid, odor of radishes. Bp: 94.3°, d: 0.805, vap d: 3.38, flash p: 20°F (OC).

SYNS: ALLYLETHER * 3,3'-OXYBIS(1-PROPENE) * PROPENYL ETHER

CONSENSUS REPORTS: Reported in EPA TSCA Inventory.

DOT Classification: Flammable Liquid; Label: Flammable Liquid and Poison.

SAFETY PROFILE: Poison by ingestion. Moderately toxic by skin contact. A skin and eye irritant. A dangerous fire hazard when exposed to heat, flame, or oxidizing materials. To fight fire, use alcohol foam. Reacts with air to form explosive peroxides. Violent explosions have occurred during distillation. When heated to decomposition it emits acrid smoke and fumes.

DBK200   CAS: 999-21-3   **HR: 3**
DIALLYL MALEATE
mf: $C_{10}H_{12}O_4$ mw: 196.22

PROP: Liquid. Vap d: 6.6.

SYNS: MALEIC ACID, DIALLYL ESTER * SIPOMER DAM

CONSENSUS REPORTS: Reported in EPA TSCA Inventory.

SAFETY PROFILE: Poison by ingestion and intraperitoneal routes. Moderately toxic by skin contact. A skin and eye irritant. When heated to decomposition it emits acrid smoke and irritating fumes.

DBL200   CAS: 131-17-9   **HR: 2**
DIALLYL PHTHALATE
mf: $C_{14}H_{14}O_4$ mw: 246.28

PROP: Nearly colorless, oily liquid. Bp: 157°, flash p: 330°F, d: 1.120 @ 20°/20°, vap d: 8.3.

SYNS: DAPON 35 * DAPON R * DI-2-PROPENYL ESTER, 1,2-BENZENEDICARBOXYLIC ACID * NCI-C50657 * PHTHALIC ACID, DIALLYL ESTER * o-PHTHALIC ACID, DIALLYL ESTER

CONSENSUS REPORTS: NTP Carcinogenesis Studies (gavage); Equivocal Evidence: rat NTPTR* NTP-TR-284,85. Reported in EPA TSCA Inventory.

SAFETY PROFILE: Suspected carcinogen with experimental carcinogenic data. Moderately toxic by ingestion, skin contact, intraperitoneal, and subcutaneous routes. An eye irritant. Mutation data reported. Combustible when exposed to heat or flame; can react with oxidizing materials. To fight fire use CO_2 or dry chemical. When heated to decomposition it emits acrid smoke and irritating fumes.

DBL800   CAS: 140-64-7   **HR: 3**
DIAMIDINE
mf: $C_{19}H_{24}N_4O_2 \cdot C_4H_{12}O_8S_2$ mw: 592.75

SYNS: 4,4'-DIAMIDINODIPHENOXYPENTANE DI(β-HYDROXYETHANESULFONATE) * 4,4'-DIAMIDINO-α,omega-DIPHENOXYPENTANE ISETHIONATE * LOMIDIN * LOMIDINE * LOMIDINE ISOETHIONATE * M & B 800 * p,p'-(PENTAMETHYLENEDIOXY)DIBENZAMIDINE BIS(β-HYDROXYETHANESULFONATE) * PENTAMIDINE DIISETHIONATE * PENTAMIDINE ISETHIONATE * 2512 R.P. * R.P. 2512 * USAF XR-10

SAFETY PROFILE: Poison by intraperitoneal, subcutaneous, and intravenous routes. Human systemic effects by intramuscular route: hemorrhage and dermatitis. Mutation data reported. When heated to decomposition it emits very toxic fumes of NO_x and SO_x.

DBM800   CAS: 59-33-6   **HR: 3**
DIAMINIDE MALEATE
mf: $C_{17}H_{23}N_3O \cdot C_4H_4O_4$ mw: 401.51

SYNS: AH * ANISOPYRADAMINE * ANTHISAN MALEATE * ANTIHIST * N-DIMETHYLAMINO-ETHYL-N-p-METHOXY-α-AMINOPYRIDINE MALEATE * 2-((2-(DIMETHYLAMINO)ETHYL)(p-METHOXYBEN-ZYL)AMINO)PYRIDINE BIMALEATE * 2-((2-(DIMETHY-LAMINO)ETHYL)(p-METHOXYBENZYL)AMINO)PYRIDINE MALEATE * N,N-DIMETHYL-N'-(4-METHOXYBENZYL)-N'-(2-PYRIDYL)ETHYLENEDIAMINE MALEATE * HISTATEX * MEPYRAMINE MALEATE * N-p-METHOXYBENZYL-N'-N'-DIMETHYL-N-α-PYRIDYLETHY-LENEDIAMINE MALEATE * MINIHIST * NEOAN-TERGAN MALEATE * PARAMAL * PARAMINYL MALEATE * PYMAFED * PYRA MALEATE * PYRANILAMINE MALEATE * PYRANINYL * PYRANISAMINE MALEATE * PYRILAMINE MA-LEATE * RENSTAMIN * 2786 R.P. MALEATE * STANGEN MALEATE * STATOMIN MALEATE * THYLOGEN MALEATE

CONSENSUS REPORTS: Reported in EPA TSCA Inventory.

SAFETY PROFILE: Questionable carcinogen with experimental tumorigenic data. A human poison by ingestion. An experimental poison by ingestion, subcutaneous, intravenous, and intraperitoneal routes. Experimental reproductive effects. Mutation data reported. An antihistamine. When heated to decomposition it emits toxic fumes of NO_x.

DBN000   CAS: 3407-94-1   **HR: 3**
2,6-DIAMINOACRIDINE
mf: $C_{13}H_{11}N_3$ mw: 209.27

SYNS: ACRAMINE RED * 2,6-ACRIDINEDIAMINE * 3,7-DIAMINOACRIDINE * DIFLAVINE (ACRIDINE)

SAFETY PROFILE: Poison by intraperitoneal and subcutaneous routes. When heated to decomposition it emits toxic fumes of NO_x.

DBN400   CAS: 553-30-0   **HR: D**
3,6-DIAMINOACRIDINE SULPHATE (1:1)
mf: $C_{13}H_{11}N_3 \cdot H_2O_4S$ mw: 307.35

SYNS: 3,6-ACRIDINEDIAMINE SULFATE (2:1) * 3,6-ACRIDINEDIAMINE SULPHATE * 3,6-DIAMI-

NOACRIDINE BISULPHATE * 3,6-DIAMINOACRIDINE SULFATE (1:1) * 3,6-DIAMINOACRIDINIUM MONOHY-DROGEN SULPHATE * 2,8-DIAMINOACRIDINIUM SULPHATE * FLAVINE * FLAVIN SULPHATE * ISOFLAV * NEUTRAL PROFLAVINE SULPHATE * PANCRIDINE * PROFALVINE SULPHATE * PROFLAVINE (SULFATE) * PROFLAVIN SULFATE * SANOFLAVIN

CONSENSUS REPORTS: EPA Genetic Toxicology Program.

SAFETY PROFILE: Mutation data reported. When heated to decomposition it emits very toxic fumes of NO_x and SO_x.

DBN600   CAS: 92-62-6   **HR: 3**
3,6-DIAMINOACRIDINIUM
mf: $C_{13}H_{11}N_3$ mw: 209.27

SYNS: 3,6-ACRIDINEDIAMINE * 3,6-DIAMINOACRI-DINE * 2,8-DIAMINOACRIDINE (EUROPEAN) * 2,8-DIAMINOACRIDINIUM * 3,7-DIAMINO-5-AZAANTHRACENE * ISOFLAV BASE * PROFLAVIN * PROFLAVINE * PROFOLIOL * PROFORMIPHEN * PROFUNDOL * PROFURA * PROGARMED * PRO-GEN * PROGESIC

CONSENSUS REPORTS: IARC Cancer Review: Animal Inadequate Evidence IMEMDT 24,195,80. Reported in EPA TSCA Inventory. EPA Genetic Toxicology Program.

SAFETY PROFILE: Questionable carcinogen. Poison by intravenous and subcutaneous routes. Mutation data reported. When heated to decomposition it emits toxic fumes of NO_x.

DBO000   CAS: 615-05-4   **HR: 3**
2,4-DIAMINOANISOLE
mf: $C_7H_{10}N_2O$ mw: 138.19

SYNS: C.I. 76050 * C.I. OXIDATION BASE 12 * 2,4-DAA * 2,4-DIAMINEANISOLE * 2,4-DIAMI-NOANISOL * 2,4-DIAMINOANISOLE BASE * m-DIAMINOANISOLE 1,3-DIAMINO-4-METHOXYBENZENE * 2,4-DIAMINO-1-METHOXYBENZENE * FURRO L * 4-METHOXY-1,3-BENZENEDIAMINE * p-ME-THOXY-m-PHENYLENEDIAMINE * 4-METHOXY-m-PHENYLENEDIAMINE * 4-MMPD * PELAGOL DA * PELAGOL GREY L * PELAGOL L

CONSENSUS REPORTS: IARC Cancer Review: GROUP 2B IMEMDT 7,56,87; Human Limited Evidence IMEMDT 27,103,82. EPA Genetic Toxicology Program. Reported in EPA TSCA Inventory.

DFG MAK: Animal Carcinogen, Suspected Human Carcinogen.

NIOSH REL: (2,4-diaminoanisole) Reduce to lowest feasible level

SAFETY PROFILE: Confirmed carcinogen. Poison by intraperitoneal route. Moderately toxic by ingestion. Human mutation data reported. A skin irritant. When heated to decomposition it emits toxic fumes of NO_x.

DBO400   CAS: 39156-41-7   **HR: 3**
2,4-DIAMINOANISOLE SULPHATE
mf: $C_7H_{10}N_2O \cdot xH_2O_4S$ mw: 824.75

SYNS: BASF URSOL SLA * C.I. 76051 * C.I. OXIDATION BASE 12A * 2,4-DAA SULFATE * 2,4-DIAMINOANISOLE SULFATE * 2,4-DIAMINO-ANISOL SULPHATE * 2,4-DIAMINO-1-METHOXYBENZENE * 1,3-DIAMINO-4-METHOXYBENZENE SULPHATE * 2,4-DIAMINO-1-METHOXYBENZENE SULPHATE * 2,4-DIAMINOSOLE SULPHATE * DURAFUR BROWN MN * FOURAMINE BA * FOURRINE SLA * FURRO SLA * 4-METHOXY-1,3-BENZENEDIAMINE SULFATE * 4-METHOXY-1,3-BENZENEDIAMINE SULFATE * 4-METHOXY-m-PHENYLENEDIAMINE SULFATE * p-METHOXY-m-PHENYLENEDIAMINE SULPHATE * 4-METHOXY-m-PHENYLENEDIAMINE SULPHATE * 4-MMPD SULPHATE * NAKO TSA * NCI-C01989 * OXIDATION BASE 12A * PELAGOL GREY * RENAL SLA * URSOL SLA * ZOBA SLE

CONSENSUS REPORTS: IARC Cancer Review: Animal Sufficient Evidence IMEMDT 27,103,82; Animal Inadequate Evidence IMEMDT 16,51,78. NTP Fourth Annual Report On Carcinogens, 1984. NCI Carcinogenesis Bioassay (feed); Clear Evidence: mouse, rat NCITR* NCI-CG-TR-84,78. Reported in EPA TSCA Inventory. EPA Genetic Toxicology Program. Community Right-To-Know List.

SAFETY PROFILE: Confirmed carcinogen with experimental carcinogenic, neoplastigenic, and tumorigenic data. Poison by intraperitoneal route. Mutation data reported. When heated to decomposition it emits very toxic fumes of NO_x and SO_x.

DBP909   CAS: 145-49-3   **HR: 3**
1,5-DIAMINOANTHRARUFIN
mf: $C_{14}H_{10}N_2O_4$ mw: 270.26

SYNS: 4,8-DIAMINOANTHRARUFIN * 1,5-DIAMINO-4,8-DIHYDROXY-9,10-ANTHRACENEDIONE

* 4,8-DIAMINO-1,5-DIHYDROXYANTHRAQUINONE * 1,5-DIAMINO-4,8-DIHYDROXYANTHRAQUINONE * leuco-1,5-DIAMINO-4,8-DIHYDROXYANTHRAQUINONE * 1,5-DIHYDROXY-4,8-DIAMINOANTHRACHINON (CZECH) * 1,5-DIHYDROXY-4,8-DIAMINOANTHRAQUINONE

CONSENSUS REPORTS: Reported in EPA TSCA Inventory.

SAFETY PROFILE: Poison by intravenous route. An eye irritant. Mutation data reported. When heated to decomposition it emits toxic fumes of NO_x.

DBT200   CAS: 92-26-2   **HR: 3**
3,6-DIAMINO-2,7-DIMETHYLACRIDINE
mf: $C_{15}H_{15}N_3$ mw: 237.33

SYNS: ACRIDINE YELLOW BASE * 2,8-DIAMINO-3,7-DIMETHYLACRIDINE

CONSENSUS REPORTS: Reported in EPA TSCA Inventory.

SAFETY PROFILE: Poison by subcutaneous route. Mutation data reported. When heated to decomposition it emits toxic fumes of NO_x.

DBU800   CAS: 13426-91-0   **HR: 3**
1,2-DIAMINOETHANE COPPER COMPLEX
DOT: UN 1761
mf: $C_2H_{10}N_2 \cdot xCu$ mw: 506.92

SYNS: CUPRIETHYLENE DIAMINE * CUPRIETHYLENEDIAMINE, solution (DOT)

CONSENSUS REPORTS: Copper and its compounds are on the Community Right-To-Know List.

DOT Classification: Corrosive Material; Label: Corrosive, Poison.

SAFETY PROFILE: A corrosive poison. An irritating and corrosive material to the skin, eyes, and mucous membranes. When heated to decomposition it emits toxic fumes of NO_x.

DBV400   CAS: 1239-45-8   **HR: 3**
2,7-DIAMINO-10-ETHYL-9-PHENYL-PHENANTHRIDINIUM BROMIDE
mf: $C_{21}H_{20}N_3 \cdot Br$ mw: 394.35

SYNS: 3,8-DIAMINO-5-ETHYL-6-PHENYLPHENANTHRIDINIUM BROMIDE * 2,7-DIAMINO-9-PHENYL-10-E-

THYLPHENANTHRIDINIUM BROMIDE * 2,7-DI-
AMINO-9-PHENYLPHENANTHRIDINE ETHOBROMIDE
* DROMILAC * ETHIDIUM BROMIDE * HOMI-
DIUM BROMIDE * RD 1572

CONSENSUS REPORTS: EPA Genetic Toxicology Program.

SAFETY PROFILE: Poison by intraperitoneal
and subcutaneous routes. Human mutation data
reported. When heated to decomposition it emits
very toxic fumes of NO_x and Br^-.

DBX400   CAS: 8048-52-0   **HR: 3**
3,6-DIAMINO-10-METHYLACRIDINIUM
CHLORIDE with 3,6-ACRIDINEDIAMINE
mf: $C_{14}H_{14}N_3 \cdot Cl \cdot C_{13}H_{11}N_3$ mw: 469.03

SYNS: ACRIFLAVIN * ACRIFLAVINE mixture with
PROFLAVINE * ACRIFLAVINIUM CHLORIDE
* ACRIFLAVINIUM CHLORIDUM * ACRIFLAVON
* ANGIFLAN * ASSIFLAVINE * AVLON
* BIALFLAVINA * BIOACRIDIN * BOVOFLAVIN
* BURNOL * BUROFLAVIN * CHOLIFLAVIN
* CHROMOFLAVINE * DIACRID * 3,6-DIAMINO-
ACRIDINE mixture with 3,6-DIAMINO-10-METHYLACRIDI-
NIUM CHLORIDE * 2,8-DIAMINO-10-METHYLACRIDI-
NIUM CHLORIDE mixture with 2,8-DIAMINOACRIDINE
* EUFLAVINE * FLAVACRIDINUM HYDROCHLORI-
CUM * FLAVINE * FLAVIOFORM * FLAVIPIN
* FLAVISEPT * GLYCO-FLAVINE * GONACRINE
* ISRAVIN * MEDIFLAVIN * NEUTRAL ACRIFLA-
VINE * PANFLAVIN * PANTONSILETTEN
* TRACHOSEPT * TRIPLA-ETILO * TRYPAFLA-
VINE * VETAFLAVIN * XANTHACRIDINUM
* ZORIFLAVIN

CONSENSUS REPORTS: IARC Cancer Re-
view: GROUP 3 IMEMDT 7,56,87; Animal
Inadequate Evidence IMEMDT 13,31,77. EPA
Genetic Toxicology Program.

SAFETY PROFILE: Questionable carcinogen
with experimental tumorigenic data. Poison by
subcutaneous route. Human mutation data reported. A topical antiseptic used in the treatment
of gonorrhea. When heated to decomposition
it emits very toxic fumes of NO_x and Cl^-.

DBY800   CAS: 720-69-4   **HR: 3**
4,6-DIAMINO-2-(5-NITRO-2-FURYL)-S-
TRIAZINE
mf: $C_7H_6N_6O_3$ mw: 222.19

CONSENSUS REPORTS: EPA Genetic Toxicology Program.

SAFETY PROFILE: Questionable carcinogen
with experimental carcinogenic data. Mutation

data reported. When heated to decomposition
it emits toxic fumes of NO_x.

DCB000   CAS: 38304-91-5   **HR: 3**
2,4-DIAMINO-6-PIPERIDINOPYRIMI-
DINE-3-OXIDE
mf: $C_9H_{15}N_5O$ mw: 209.29

SYNS: 6-AMINO-1,2-DIHYDRO-1-HYDROXY-2-IMINO-4-
PIPERIDINOPYRIMIDINE * 2,4-DIAMINO-6-PIPERIDI-
NILPIRIMIDINA-3-OSSIDO (ITALIAN) * 2,3-DIHYDRO-
3-HYDROXY-2-IMINO-6-(1-PIPERIDINYL)- 4-PYRIMIDIN-
AMINE * LONITEN * MINOSSIDILE (ITALIAN)
* MINOXIDIL * 6-PIPERIDINO-2,4- DIAMINO-
PYRIMIDINE-3-OXIDE * 6-(1-PIPERIDINYL)- 2,4-
PYRIMIDINEDIAMINE-3-OXIDE * U-10,858

SAFETY PROFILE: Poison by intravenous
route. Moderately toxic by ingestion and intraperitoneal routes. Human systemic effects by
ingestion: thrombocytopenia (reduced numbers
of blood platelets). An antihypertensive agent.
When heated to decomposition it emits toxic
fumes of NO_x.

DCC800   CAS: 141-86-6   **HR: 3**
2,6-DIAMINOPYRIDINE
mf: $C_5H_7N_3$ mw: 109.15

PROP: Crystals. Mp: 120.8°, bp: 285°.

CONSENSUS REPORTS: Reported in EPA
TSCA Inventory.

SAFETY PROFILE: Poison by intravenous and
intraperitoneal routes. Mutation data reported.
When heated to decomposition it emits toxic
fumes of NO_x.

DCE000   CAS: 636-23-7   **HR: 3**
2,4-DIAMINOTOLUENE
DIHYDROCHLORIDE
mf: $C_7H_{10}N_2 \cdot 2ClH$ mw: 195.11

SYN: METATOLYLENEDIAMINE DIHYDROCHLORIDE

CONSENSUS REPORTS: Reported in EPA
TSCA Inventory.

SAFETY PROFILE: Questionable carcinogen
with experimental neoplastigenic data. Poison
by intraperitoneal route. Moderately toxic by
ingestion. When heated to decomposition it emits very toxic fumes of NO_x and HCl.

DCE200   CAS: 615-45-2   **HR: 3**
2,5-DIAMINOTOLUENE DIHYDROCHLORIDE
mf: C₇H₁₀N₂ • 2ClH   mw: 195.11

SYNS: 2-METHYL-1,4-BENZENEDIAMINE DIHYDRO-
CHLORIDE * p-TOLUENEDIAMINE DIHYDROCHLO-
RIDE

CONSENSUS REPORTS: Reported in EPA TSCA Inventory.

SAFETY PROFILE: Poison by ingestion. Experimental teratogenic and reproductive effects. When heated to decomposition it emits very toxic fumes of NO$_x$ and HCl.

DCE400   CAS: 15481-70-6   **HR: 3**
2,6-DIAMINOTOLUENE DIHYDROCHLORIDE
mf: C₇H₁₀N₂ • 2ClH   mw: 195.11

SYN: NCI-C50317

CONSENSUS REPORTS: Carcinogenesis Bioassay Completed; Results Negative NCITR* NCI-CG-TR-200,80. NCI Carcinogenesis Bioassay (feed); No Evidence: mouse, rat NCITR* NCI-CG-TR-200,80.

SAFETY PROFILE: When heated to decomposition it emits very toxic fumes of NO$_x$ and HCl.

DCE600   CAS: 615-50-9   **HR: 3**
2,5-DIAMINOTOLUENE SULFATE
mf: C₇H₁₀N₂ • H₂O₄S   mw: 220.27

SYNS: C.I. 76043 * p-DIAMINOTOLUENE SULFATE
* 2,5-DIAMINOTOLUENE SULPHATE * 2-METHYL-
1,4-BENZENEDIAMINE SULFATE * 2-METHYL-p-
PHENYLENEDIAMINE SULPHATE * NCI-C01832
* p-TOLUENEDIAMINE SULFATE * 2,5-TOLUENEDI-
AMINE SULFATE * TOLUENE-2,5-DIAMINE, SULFATE
(1:1) (8CI) * TOLUENE-2,5-DIAMINE SULPHATE
* p-TOLUENEDIAMINE SULPHATE * TOLUYLENE-
2,5-DIAMINE SULPHATE * p-TOLUYLENEDIAMINE
SULPHATE * p-TOLYLENEDIAMINE SULPHATE

CONSENSUS REPORTS: IARC Cancer Review: Animal Indefinite Evidence IMEMDT 16,97,78. NCI Carcinogenesis Bioassay Completed; Results Indefinite: mouse, rat NCITR* NCI-CG-TR-126,78. Reported in EPA TSCA Inventory.

SAFETY PROFILE: Questionable carcinogen with experimental tumorigenic data. Poison by ingestion and intraperitoneal routes. When heated to decomposition it emits very toxic fumes of NO$_x$ and SO$_x$.

DCF200   CAS: 1455-77-2   **HR: 3**
3,5-DIAMINO-s-TRIAZOLE
mf: C₂H₅N₅   mw: 99.12

SYNS: GUANAZOLE * NCI-C04819 * NSC 1895

CONSENSUS REPORTS: NCI Carcinogenesis Studies (ipr); Equivocal Evidence: rat CANCAR 40,1935,77; No Evidence: mouse CANCAR 40,1935,77. Reported in EPA TSCA Inventory.

SAFETY PROFILE: Human systemic effects by intravenous route: leukopenia (reduced white blood cell count) and thrombocytopenia (reduced blood platelet count). Human mutation data reported. Questionable carcinogen with experimental tumorigenic data. When heated to decomposition it emits toxic fumes of NO$_x$.

DCG000   **HR: 3**
DIAMMINEMALONATO PLATINUM (II)

CONSENSUS REPORTS: Reported in EPA TSCA Inventory.

SAFETY PROFILE: Poison by intraperitoneal route. Mutation data reported. When heated to decomposition it emits toxic fumes of NO$_x$.

DCG800   CAS: 7784-44-3   **HR: 3**
DIAMMONIUM HYDROGEN ARSENATE
DOT: UN 1546
mf: AsH₃O₄ • 2H₃N   mw: 176.03

PROP: White powder or crystals. Mp: decomp to yield NH₃.

SYNS: AMMONIUM ACID ARSENATE * AMMO-
NIUM ARSENATE, solid (DOT) * DIAMMONIUM
ARSENATE * DIAMMONIUM MONOHYDROGEN AR-
SENATE * DIBASIC AMMONIUM ARSENATE
* SECONDARY AMMONIUM ARSENATE

CONSENSUS REPORTS: Arsenic and its compounds are on the Community Right-To-Know List.

OSHA PEL: TWA 0.01 mg(As)/m³
ACGIH TLV: TWA 0.2 mg(As)/m³
NIOSH REL: (Inorganic Arsenic) CL 0.002 mg(As)/m³/15M
DOT Classification: Poison B; Label: Poison.

SAFETY PROFILE: A poison. When heated to decomposition it emits very toxic fumes of As, NO$_x$ and NH$_3$.

DCH000   CAS: 3164-29-2   HR: 3
DIAMMONIUM TARTRATE
DOT: NA 9091
mf: C$_4$H$_6$O$_6$ • 2H$_3$N   mw: 184.18

SYNS: AMMONIUM TARTRATE (DOT) * AMMO-NIUM-d-TARTRATE * 2,3-DIHYDROXYBUTANEDIOIC ACID, DIAMMONIUM SALT * l-TARTARIC ACID, AMMONIUM SALT * TARTARIC ACID, DIAMMONIUM SALT

CONSENSUS REPORTS: Reported in EPA TSCA Inventory.

DOT Classification: ORM-E; Label: None.

SAFETY PROFILE: Poison by intravenous route. Moderately toxic by subcutaneous route. When heated to decomposition it emits very toxic fumes of NH$_3$ and NO$_x$.

DCH200   CAS: 2050-92-2   HR: 3
DIAMYL AMINE
DOT: UN 2841
mf: C$_{10}$H$_{23}$N   mw: 157.34

PROP: Water-white liquid. Bp: 202°, flash p: 124°F, d: 0.777 @ 20°/20°, vap d: 5.42.

SYNS: DI-n-AMYLAMINE (DOT) * DIPENTYLAMINE * PENTYL PENTYLAMINE

CONSENSUS REPORTS: Reported in EPA TSCA Inventory.

DOT Classification: Poison B; Label: St. Andrews Cross, Flammable Liquid.

SAFETY PROFILE: Poison by inhalation, ingestion, and skin contact. A severe skin irritant. Flammable when exposed to heat or flame; can react with oxidizing materials. To fight fire, use alcohol foam, foam, CO$_2$, dry chemical. When heated to decomposition it emits toxic fumes of NO$_x$.

DCH600   CAS: 13256-06-9   HR: 3
DI-n-AMYLNITROSAMINE
mf: C$_{10}$H$_{22}$N$_2$O   mw: 186.34

SYNS: DIAMYLNITROSAMIN (GERMAN) * DIPENTYLNITROSAMINE * DI-n-PENTYLNITROSAMINE * N-NITROSODIPENTYLAMINE * N-NITROSODI-n-PENTYLAMINE

CONSENSUS REPORTS: EPA Genetic Toxicology Program.

SAFETY PROFILE: Questionable carcinogen with experimental carcinogenic and tumorigenic data. Moderately toxic by ingestion and subcutaneous routes. Mutation data reported. When heated to decomposition it emits toxic fumes of NO$_x$.

DCI400   CAS: 35865-33-9   HR: 3
DIANEMYCIN
mf: C$_{47}$H$_{78}$O$_{14}$ mw: 867.25

SAFETY PROFILE: Poison by ingestion, intraperitoneal, and subcutaneous routes. When heated to decomposition it emits acrid smoke and irritating fumes.

DCI600   CAS: 23261-20-3   HR: 3
DIANHYDROGALACTITOL
mf: C$_6$H$_{10}$O$_4$ mw: 146.16

SYNS: DAD * DAG * DIANHYDROCULCITOL * 1,2:5,6-DIANHYDRODULCITOL * 1,2:5,6-DIANHY-DROGALACTITOL * 1,2:5,6-DIEPOXYDULCITOL * DULCITOLDIEPOXIDE * NSC 132313

SAFETY PROFILE: Poison by ingestion, intravenous, subcutaneous, intraperitoneal, and possibly other routes. Mutation data reported. When heated to decomposition it emits acrid smoke and irritating fumes.

DCJ200   CAS: 119-90-4   HR: 3
o-DIANISIDINE
mf: C$_{14}$H$_{16}$N$_2$O$_2$ mw: 244.32

PROP: Colorless crystals. Mp: 137-138°, flash p: 403°F, vap d: 8.5.

SYNS: ACETAMINE DIAZO BLACK RD * AMACEL DEVELOPED NAVY SD * AZOENE FAST BLUE BASE * AZOFIX BLUE B SALT * AZOGNE FAST BLUE B * BLUE BN BALSE * BRENTAMINE FAST BLUE B BASE * CELLITAZOL B * C.I. 24110 * C.I. AZOIC DIAZO COMPONENT 48 * CIBACETE DIAZO NAVY BLUE 2B * C.I. DISPERSE BLACK 6 * DIA-CELLITON FAST GREY G * DIACEL NAVY DC * o-DIANISIDIN (CZECH, GERMAN) * o-DIANISIDINA (ITALIAN) * 3,3'-DIANISIDINE * O,O'DIANISIDINE * DIATO BLUE BASE B * 3,3'-DIMETHOXYBENZIDIN (CZECH) * 3,3'-DIMETHOXYBENZIDINE * 3,3'-DI-METOSSIBENZODINA (ITALIAN) * FAST BLUE B BASE * HILTONIL FAST BLUE B BASE * HILTOSAL FAST BLUE B SALT * HINDASOL BLUE B SALT * KAKO BLUE B SALT * KAYAKU BLUE B BASE * LAKE

BLUE B BASE   *   MEISEI TERYL DIAZO BLUE HR
*   MITSUI BLUE B BASE   *   NAPHTHANIL BLUE B
BASE   *   NEUTROSEL NAVY BN   *   RCRA WASTE
NUMBER U091   *   SANYO FAST BLUE SALT B
*   SETACYL DIAZO NAVY R   *   SPECTROLENE
BLUE B

CONSENSUS REPORTS: IARC Cancer Review: GROUP 2B IMEMDT 7,198,87; Animal Sufficient Evidence IMEMDT 4,41,74. NTP Fourth Annual Report On Carcinogens, 1984. EPA Genetic Toxicology Program. Community Right-To-Know List. Reported in EPA TSCA Inventory.

DFG MAK: Animal Carcinogen, Suspected Human Carcinogen.
NIOSH REL: (Benzidine-based dye) Reduce to lowest feasible level

SAFETY PROFILE: Confirmed carcinogen with experimental tumorigenic data. Moderately toxic by ingestion. Mutation data reported. Combustible when exposed to heat or flame. When heated to decomposition it emits toxic fumes of NO_x.

DCJ400   CAS: 91-93-0   HR: 3
DIANISIDINE DIISOCYANATE
mf: $C_{16}H_{12}N_2O_4$ mw: 296.30

SYNS: 4,4'-DIISOCYANATO-3,3'-DIMETHOXY-1,1'-BI-
PHENYL   *   3,3'-DIMETHOXYBENZIDINE-4,4'-DIISO-
CYANATE   *   3,3'-DIMETHOXY-4,4'-BIPHENYLENE DI-
ISOCYANATE   *   NCI-C02175

CONSENSUS REPORTS: IARC Cancer Review: GROUP 3 IMEMDT 7,56,87; Animal Limited Evidence IMEMDT 39,279,86. NCI Carcinogenesis Bioassay (feed); No Evidence: mouse NCITR* NCI-CG-TR-128,79; Clear Evidence: rat NCITR* NCI-CG-TR-128,79.

NIOSH REL: (Diisocyanates) TWA 0.005 ppm; CL 0.02 ppm/10M

SAFETY PROFILE: Questionable carcinogen with experimental carcinogenic data. Poison by intravenous route. When heated to decomposition it emits toxic fumes of NO_x.

DCJ600   CAS: 13601-02-0   HR: 3
DIAQUODIAMMINEPLATINUM DINITRATE
mf: $H_{10}N_2O_2Pt \cdot N_2O_6$ mw: 389.23

SYN: cis-DIAQUODIAMMINEPLATINUM(II) DINITRATE

SAFETY PROFILE: Poison by intraperitoneal route. Human systemic skin effects by intradermal route. Mutagenic data reported. When heated to decomposition it emits toxic fumes of NO_x.

DCJ800   CAS: 68855-54-9   HR: 1
DIATOMACEOUS EARTH

PROP: Composed of skeletons of small aquatic plants related to algae and contains as much as 88% amorphous silica. White to buff colored solid. Insol in water; sol in hydrofluoric acid.

SYNS: D.E.   *   DIATOMACEOUS SILICA   *   DIATO-
MITE   *   INFUSORIAL EARTH   *   KIESELGUHR

OSHA PEL: (Transitional: TWA 80 mg/m^3/%SiO$_2$) TWA 6 mg/m^3
ACGIH TLV: TWA (nuisance particulate) 10 mg/m^3 of total dust (when toxic impurities are not present, e.g., quartz < 1%).

SAFETY PROFILE: The dust may cause fibrosis of the lungs. Roasting or calcining at high temperatures produces cristobalite and tridymite, thus increasing the fibrogenicity of the material.

DCK700   CAS: 283-66-9   HR: 3
1,6-DIAZA-3,4,8,9,12,13-HEXAOXABI-CYCLO(4.4.4)TETRADECANE
mf: $C_6H_{12}N_2O_6$ mw: 208.17

SYN: HEXAMETHYLENETRIPEROXYDIAMINE

DOT Classification: Forbidden

SAFETY PROFILE: The dry material is a powerful explosive that is heat- and shock-sensitive. Explodes on contact with bromine or sulfuric acid. When heated to decomposition it emits toxic fumes of NO_x.

DCL125   CAS: 2294-47-5   HR: 3
1,4-DIAZIDOBENZENE
mf: $C_6H_4N_6$ mw: 160.14

SYNS: BENZENE, 1,4-DIAZIDO-   *   p-DIAZIDOBEN-
ZENE (DOT)   *   1,4-DIAZIDOBENZENE   *   p-PHENYL-
ENE DIAZIDE

DOT Classification: Forbidden

SAFETY PROFILE: Explodes violently when heated. When heated to decomposition it emits toxic fumes of NO_x.

DCL600   CAS: 629-13-0   **HR: 3**
1,2-DIAZIDOETHANE
mf: $C_2H_4N_6$ mw: 112.09

DOT Classification: Forbidden.

SAFETY PROFILE: Explodes on heating or on contact with sulfuric acid. Upon decomposition it emits toxic fumes of NO_x.

DCM750   CAS: 333-41-5   **HR: 3**
DIAZINON

DOT: NA 2783
mf: $C_{12}H_{21}N_2O_3PS$ mw: 304.38

PROP: Liquid with faint ester-like odor. Bp: 84° @ 0.002 mm, d: 1.116 @ 20°/4°. Miscible in organic solvents.

SYNS: ALFA-TOX * BASUDIN * BASUDIN 10 G * BAZUDEN * DAZZEL * O,O-DIAETHYL-O-(2-ISOPROPYL-4-METHYL-PYRIMIDIN-6-YL)-MONOTHIO-PHOSPHAT (GERMAN) * O,O-DIAETHYL-O-(2-ISOPRO-PYL-4- METHYL)-6-PYRIMIDYL-THIONOPHOSPHAT (GERMAN) * DIANON * DIATERR-FOS * DIAZA-JET * DIAZATOL * DIAZIDE * DIAZINONE * DIAZITOL * DIAZOL * O,O-DIETHYL-O-(2-ISOPROPYL- 4-METHYL-PYRIMIDIN-6-YL)MONOTHIOFOS-FAAT (DUTCH) * O,O-DIETHYL-O-(2-ISOPROPYL-4-METHYL-6-PYRIMIDINYL)PHOSPHOROTHIOATE * O,O-DIETHYL-O-(2-ISOPROPYL-6-METHYL-4-PYRIMI-DINYL) PHOSPHOROTHIOATE * DIETHYL 4-(2-ISOPRO-PYL-6-METHYLPYRIMIDINYL)PHOSPHOROTHIONATE * O,O-DIETHYL-O-(2-ISOPROPYL-4-METHYL-6-PYRIMI-DYL)PHOSPHOROTHIOATE * O,O-DIETHYL-O-(2-ISO-PROPYL-4-METHYL-6-PYRIMIDYL) THIONOPHOSPHATE * O,O-DIETHYL-2-ISOPROPYL-4-METHYLPYRIMIDYL-6-THIOPHOSPHATE * O,O-DIETHYL-O-6-METHYL-2-ISO-PROPYL-4-PYRIMIDINYL PHOSPHOROTHIOATE * O,O-DIETIL-O-(2-ISOPROPIL-4-METIL-PIRIMIDIN-6-IL) MONOTIOFOSFATO (ITALIAN) * DIMPYLATE * DIPOFENE * DIZINON * DYZOL * ENT 19,507 * G 301 * G-24480 * GARDENTOX * GEIGY 24480 * O-2-ISOPROPYL-4-METHYL-PYRIMIDYL-O,O-DIETHYL PHOSPHOROTHIOATE * ISOPROPYLMETHYLPYRIMIDYL DIETHYL THIO-PHOSPHATE * KAYAZINON * KAYAZOL * NCI-C08673 * NEDCIDOL * NEOCIDOL * NIPSAN * NUCIDOL * SAROLEX * SPECTRACIDE * THIOPHOSPHATE de O,O-DI-ETHYLE et de o-2-ISOPROPYL-4- METHYL-6-PYRIMIDYLE (FRENCH)

CONSENSUS REPORTS: NCI Carcinogenesis Bioassay (feed); No Evidence: mouse, rat NCITR* NCI-CG-TR-137,79. Reported in EPA

TSCA Inventory. EPA Genetic Toxicology Program.

OSHA PEL: TWA 0.1 mg/m^3 (skin)
ACGIH TLV: TWA 0.1 mg/m^3
DFG MAK: 1 mg/m^3
DOT Classification: ORM-A; Label: None.

SAFETY PROFILE: Poison by ingestion, skin contact, subcutaneous, intravenous, intraperitoneal, and possibly other routes. Mildly toxic by inhalation. Human systemic effects by ingestion: changes in motor activity, muscle weakness, and sweating. Experimental teratogenic and reproductive effects. A skin and severe eye irritant. Human mutation data reported. When heated to decomposition it emits very toxic fumes of NO_x, PO_x and SO_x.

DCN800   CAS: 623-73-4   **HR: 3**
DIAZOACETIC ESTER
mf: $C_4H_6N_2O_2$ mw: 114.12

SYNS: DAAE * DIAZOACETIC ACID, ETHYL ESTER * DIAZOESSIGSAEURE-AETHYLESTER (GERMAN) * EDA * ETHOXYCARBONYLDIAZOMETHANE * ETHYL DIAZOACETATE

SAFETY PROFILE: Questionable carcinogen with experimental carcinogenic and tumorigenic data. Poison by ingestion and intravenous routes. Can explode. Explodes on contact with tris(dimethylamino) antimony. When heated to decomposition it emits toxic fumes of NO_x.

DCO800   CAS: 820-75-7   **HR: 3**
N-(DIAZOACETYL)GLYCINE
HYDRAZINE
mf: $C_4H_7N_5O_2$ mw: 157.16

SYNS: N-DIAZOACETILGLICINA-IDRAZIDE (ITALIAN) * DIAZOACETYLGLYCINE HYDRAZIDE * N-DIAZO-ACETYL GLYCYLHYDRAZIDE * NSC-58404

CONSENSUS REPORTS: EPA Genetic Toxicology Program.

SAFETY PROFILE: Poison by intravenous route. Moderately toxic by ingestion, intraperitoneal, and subcutaneous routes. Questionable carcinogen with experimental carcinogenic and neoplastigenic data. Mutation data reported. When heated to decomposition it emits toxic fumes including NO_x.

DCP800   CAS: 334-88-3   **HR: 3**
DIAZOMETHANE
mf: CH_2N_2 mw: 42.05

PROP: Yellow gas at ordinary temp. Mp: −145°, bp: −23°, d: 1.45.

SYNS: AZIMETHYLENE * DIAZIRINE

CONSENSUS REPORTS: IARC Cancer Review: GROUP 3 IMEMDT 7,56,87; Animal Sufficient Evidence IMEMDT 7,223,74. EPA Genetic Toxicology Program. Community Right-To-Know List.

OSHA PEL: TWA 0.2 ppm
ACGIH TLV: TWA 0.2 ppm
DFG MAK: Animal Carcinogen, Suspected Human Carcinogen.

SAFETY PROFILE: Confirmed carcinogen with experimental tumorigenic data. A poison irritant by inhalation. A powerful allergen. It can cause pulmonary edema and frequently causes hypersensitivity leading to asthmatic symptoms. Mutation data reported. Highly explosive when shocked, exposed to heat or by chemical reaction. Undiluted liquid or gas may explode on contact with alkali metals, rough surfaces, heat (100°C), high intensity light or shock. When heated to decomposition or on contact with acid or acid fumes it emits highly toxic fumes of NO_x. Incompatible with alkali metals, calcium sulfate.

DCQ600   CAS: 2435-76-9   **HR: 3**
DIAZOURACIL
mf: $C_4H_2N_4O_2$ mw: 138.10

SYNS: 5-DIAZOPYRIMIDINE-2,4(3H)-DIONE * 5-DIAZO-2,4(1H,3H)-PYRIMIDINEDIONE * 5-DIAZOURACIL * 2,4-DIOSSI-5-DIAZOPIRIMIDINA (ITALIAN) * 2,6-DIOXO-5-DIAZOPYRIMIDINE * DU * NSC 23519 * (1,2,3)OXADIAZOLO(5,4-d)PYRIMIDIN-5(4H)-ONE

CONSENSUS REPORTS: Reported in EPA TSCA Inventory.

SAFETY PROFILE: Poison by subcutaneous and intraperitoneal routes. Mutation data reported. When heated to decomposition it emits toxic fumes of NO_x.

DCQ800   CAS: 34493-98-6   **HR: 3**
DIBEKACIN
mf: $C_{18}H_{37}N_5O_8$ mw: 451.60

SYNS: DEBECACIN * DIDEOXYKANAMYCIN B * 3′,4′-DIDEOXYKANAMYCIN B * DKB * ORBICIN

CONSENSUS REPORTS: EPA Genetic Toxicology Program.

SAFETY PROFILE: Poison by intraperitoneal, subcutaneous, intramuscular, and intravenous routes. Moderately toxic by ingestion. Experimental teratogenic and reproductive effects. An antibacterial agent. When heated to decomposition it emits toxic fumes of NO_x.

DCR200   CAS: 55-43-6   **HR: 3**
DIBENAMINE HYDROCHLORIDE
mf: $C_{16}H_{18}ClN \cdot ClH$ mw: 296.26

SYNS: N-(2-CHLOROETHYL)DIBENZYLAMINE HYDROCHLORIDE * DIBENAMINE * N,N-DIBENZYLAMINOETHYL CHLORIDE HYDROCHLORIDE * DIBENZYLCHLORETHAMINE HYDROCHLORIDE * DIBENZYLCHLORETHYLAMINE HYDROCHLORIDE * N,N-DIBENZYL-β-CHLOROETHYLAMINE HYDROCHLORIDE * N,N-DIBENZYL-2-CHLOROETHYLAMINE HYDROCHLORIDE * SYMPATHOLYTIN

SAFETY PROFILE: Poison by subcutaneous, intravenous, and intraperitoneal routes. An adrenergic blocker and diagnostic aid (pheochromocytoma). When heated to decomposition it emits very toxic fumes of Cl^- and NO_x.

DCR400   CAS: 203-20-3   **HR: 3**
DIBENZ(a,j)ACEANTHRYLENE
mf: $C_{24}H_{14}$ mw: 302.38

SYN: 15,16-BENZDEHYDROCHOLANTHRENE

SAFETY PROFILE: Questionable carcinogen with experimental tumorigenic data. Poison by intravenous route. When heated to decomposition it emits acrid smoke and irritating fumes.

DCS200   CAS: 1977-10-2   **HR: 3**
DIBENZACEPIN
mf: $C_{18}H_{18}ClN_3O$ mw: 327.84

SYNS: 2-CHLORO-11-(4-METHYL-1-PIPERAZINYL)-DIBENZO(b,f)(1,4)OXAZEPINE * 2-CHLORO-11-(4-METHYL-1-PIPERAZINYL)-DIBENZO(b,f)(1,4)OXAZEPINE * CL-62362 * CL-71563 * CLOXAZEPINE * DIBENZOAZEPINE * HF3170 * LOXAPINE * LW 3170 * OXILAPINE * S-805 * SUM 3170

SAFETY PROFILE: Poison by ingestion, intraperitoneal, subcutaneous, and intravenous routes. Experimental teratogenic and reproductive effects. A tranquilizer. Many dibenz-azepine compounds have central nervous system effects. When heated to decomposition it emits very toxic fumes of Cl⁻ and NO$_x$.

DCS400   CAS: 226-36-8   HR: 3
DIBENZ(a,h)ACRIDINE
mf: $C_{21}H_{13}N$ mw: 279.35

SYNS: 7-AZADIBENZ(a,h)ANTHRACENE * DB(a,h)AC * DIBENZ(a,d)ACRIDINE * 1,2,5,6-DIBENZACRIDINE * 1,2,5,6-DIBENZOACRIDINE * 1,2,5,6-DINAPHTHACRIDINE

CONSENSUS REPORTS: IARC Cancer Review: GROUP 2B IMEMDT 7,56,87; Animal Sufficient Evidence IMEMDT 32,277,83; IMEMDT 3,247,73. NTP Fourth Annual Report On Carcinogens, 1984. EPA Genetic Toxicology Program.

SAFETY PROFILE: Confirmed carcinogen with experimental carcinogenic and tumorigenic data. When heated to decomposition it emits toxic fumes of NO$_x$.

DCS600   CAS: 224-42-0   HR: 3
DIBENZ(a,j)ACRIDINE
mf: $C_{21}H_{13}N$ mw: 279.35

SYNS: 7-AZADIBENZ(a,j)ANTHRACENE * DB(a,j)AC * DIBENZ(a,f)ACRIDINE * 1,2,7,8-DIBENZACRIDINE * 3,4,5,6-DIBENZACRIDINE * DIBENZO(a,j)ACRIDINE * 3,4,6,7-DINAPHTHACRIDINE

CONSENSUS REPORTS: IARC Cancer Review: GROUP 2B IMEMDT 7,56,87; Animal Sufficient Evidence IMEMDT 32,283,83; IMEMDT 3,254,73. NTP Fourth Annual Report On Carcinogens, 1984. EPA Genetic Toxicology Program.

SAFETY PROFILE: Confirmed carcinogen with experimental carcinogenic and tumorigenic data. Experimental reproductive effects. Mutation data reported. When heated to decomposition it emits toxic fumes of NO$_x$.

DCS800   CAS: 224-53-3   HR: 3
DIBENZ(c,h)ACRIDINE
mf: $C_{21}H_{13}N$ mw: 279.35

SYNS: 14-AZADIBENZ(a,j)ANTHRACENE * 3,4:5,6-DIBENZACRIDINE * 1,2,7,8-DIBENZACRIDINE (FRENCH)

SAFETY PROFILE: Questionable carcinogen with experimental tumorigenic data. Mutation data reported. When heated to decomposition it emits toxic fumes of NO$_x$.

DCT400   CAS: 53-70-3   HR: 3
DIBENZ(a,h)ANTHRACENE
mf: $C_{22}H_{14}$ mw: 278.36

SYNS: 1,2:5,6-BENZANTHRACENE * DBA * DB(a,h)A * 1,2,5,6-DBA * 1,2,5,6-DIBENZAN-THRACEEN (DUTCH) * 1,2:5,6-DIBENZANTHRACENE * 1,2:5,6-DIBENZ(a)ANTHRACENE * DIBENZO(a,h)ANTHRACENE * 1,2:5,6-DIBENZOANTHRACENE * RCRA WASTE NUMBER U063

CONSENSUS REPORTS: IARC Cancer Review: GROUP 2A IMEMDT 7,56,87; Animal Sufficient Evidence IMEMDT 32,299,83; IMEMDT 3,178,73. NTP Fourth Annual Report On Carcinogens, 1984. EPA Genetic Toxicology Program. Reported in EPA TSCA Inventory.

SAFETY PROFILE: Confirmed carcinogen with experimental carcinogenic, tumorigenic, and neoplastigenic data. Poison by intravenous route. Human mutation data reported. When heated to decomposition it emits acrid smoke and irritating fumes.

DCT600   CAS: 224-41-9   HR: 3
DIBENZ(a,j)ANTHRACENE
mf: $C_{22}H_{14}$ mw: 278.36

SYN: 1,2:7,8-DIBENZANTHRACENE

CONSENSUS REPORTS: IARC Cancer Review: GROUP 3 IMEMDT 7,56,87; Animal Limited Evidence IMEMDT 32,309,83.

SAFETY PROFILE: Questionable carcinogen with experimental tumorigenic data. Mutation data reported. When heated to decomposition it emits acrid smoke and irritating fumes.

DCV200   CAS: 298-46-4   HR: 3
5H-DIBENZ(b,f)AZEPINE-5-
CARBOXAMIDE
mf: $C_{15}H_{12}N_2O$ mw: 236.29

SYNS: BISTON * CARBAMAZEPEN * CARBA-MAZEPINE * CARBAMEZEPINE * 5-CARBAMOYL-5H-DIBENZ(b,f)AZEPINE * 5-CARBAMOYL-5H-DIBEN-ZO(b,f)AZEPINE * 5-CARBAMOYLDIBENZO(b,f) AZEPINE * 5-CARBAMYLDIBENZO(b,f)AZEPINE * 5-CARBAMYL-5H-DIBENZO(b,f)AZEPINE * CAR-BAZEPINE * FINLEPSIN * G 32883 * GEIGY 32883 * STAZEPIN * TEGRETAL * TEGRETOL * TELESMIN * TIMONIL

CONSENSUS REPORTS: EPA Genetic Toxicology Program.

SAFETY PROFILE: A human poison by ingestion. Poison experimentally by intraperitoneal route. Human systemic effects by ingestion: aplastic anemia, sleep, hallucinations, distorted perceptions, nausea or vomiting, somnolence, dermatitis, ataxia (loss of muscle coordination), urine volume increase, and agranulo-cytosis. Human reproductive effects. Experimental teratogenic and reproductive effects. An analgesic and anticonvulsant. When heated to decomposition it emits toxic fumes of NO_x.

DCW600   CAS: 4498-32-2   HR: 3
DIBENZEPIN
mf: $C_{18}H_{21}N_3O$ mw: 295.37

SYNS: DIBENZEPINE * 5,10-DIHYDRO-10-(2-(DI-METHYLAMINO)ETHYL)-5-METHYL-11H-DIBENZO(b,e) (1,4)DIAZEPIN-11-ONE * 10-(2-(DIMETHYLAMI-NO)ETHYL)-5,10-DIHYDRO-5-METHYL-11H-DIBEN-ZO(B,E)(1,4)DIAZEPIN-11-ONE * 10-(2-(DIMETHYL-AMINO)ETHYL)-5-METHYL-5H-DIBENZO(b,e)(1,4)DI-AZEPIN-11(10H)-ONE * HF 1927

SAFETY PROFILE: Poison by ingestion, intraperitoneal, intravenous, and subcutaneous routes. Many dibenz-azepine compounds have central nervous system effects. When heated to decomposition it emits toxic fumes of NO_x.

DCW800   CAS: 315-80-0   HR: 3
DIBENZEPINE HYDROCHLORIDE
mf: $C_{18}H_{21}N_3O \cdot ClH$ mw: 331.88

SYNS: DIBENZEPIN HYDROCHLORIDE * HF 1927 * HYDROFLUORIDE-1927 WANDER * 5-METHYL-10-β-DIMETHYLAMINOAETHYL-10,11-DIHYDRO-11-OXO-5-DIBENZO(b,e)(1,4)DIAZEPIN * NEODALIT * NO-VERIL * NOVERYL

SAFETY PROFILE: Poison by ingestion, intravenous, subcutaneous, and intraperitoneal routes. An antidepressant. Many dibenz-azepine compounds have central nervous system effects. When heated to decomposition it emits very toxic fumes of HCl and NO_x.

DCY000   CAS: 194-59-2   HR: 3
7H-DIBENZO(c,g)CARBAZOLE
mf: $C_{20}H_{13}N$ mw: 267.34

PROP: Needles. Mp: 158°.

SYNS: 7-AZA-7H-DIBENZO(c,g)FLUORENE * 3,4,5,6-DIBENZCARBAZOL * 3,4,5,6-DIBENZCARBAZOLE * 3,4,5,6-DIBENZOCARBAZOLE * 3,4,5,6-DINAPH-THACARBAZOLE * 7H-DB(c,g)C

CONSENSUS REPORTS: IARC Cancer Review: GROUP 2B IMEMDT 7,56,87; Animal Sufficient Evidence IMEMDT 32,315,83; IMEMDT 3,260,73. NTP Fourth Annual Report On Carcinogens, 1984.

SAFETY PROFILE: Confirmed carcinogen with experimental carcinogenic, neoplastigenic, and tumorigenic data. Poison by intraperitoneal route. Mutation data reported. When heated to decomposition it emits toxic fumes of NO_x.

DCY200   CAS: 189-64-0   HR: 3
DIBENZO(b,def)CHRYSENE
mf: $C_{24}H_{14}$ mw: 302.38

SYNS: BD(a,h)P * DIBENZO(a,h)PYRENE * 1,2,6,7-DIBENZOPYRENE * 3,4,8,9-DIBENZOPY-RENE * 3,4,8,9-DIBENZPYRENE

CONSENSUS REPORTS: IARC Cancer Review: GROUP 2B IMEMDT 7,56,87; Animal Sufficient Evidence IMEMDT 32,331,83; IMEMDT 3,207,73.

SAFETY PROFILE: Suspected carcinogen with experimental carcinogenic and tumorigenic data. When heated to decomposition it emits acrid smoke and irritating fumes. Mutation data reported.

DCY400   CAS: 191-30-0   HR: 3
DIBENZO(def,p)CHRYSENE
mf: $C_{24}H_{14}$ mw: 302.38

SYNS: BA 51-090462 * DB(a,l)P * DIBEN-ZO(a,d)PYRENE * DIBENZO(a,l)PYRENE * 1,2:3,4-DIBENZOPYRENE * 1,2,9,10-DIBENZOPYRENE * 2,3:4,5-DIBENZOPYRENE * 1,2,3,4-DIBENZPYRENE * 4,5,6,7-DIBENZPYRENE

CONSENSUS REPORTS: IARC Cancer Review: GROUP 2B IMEMDT 7,56,87; Animal Sufficient Evidence IMEMDT 32,343,83; Animal Limited Evidence IMEMDT 3,224,73

SAFETY PROFILE: Suspected carcinogen with experimental tumorigenic data. When heated to decomposition it emits acrid smoke and irritating fumes.

DDA800   CAS: 262-12-4   HR: 3
DIBENZO-p-DIOXIN
mf: $C_{12}H_8O_2$ mw: 184.20

PROP: Crystals. Mp: 123°.

SYNS: DIBENZODIOXIN * DIBENZO(1,4)DIOXIN * DIBENZO(b.e)(1,4)DIOXIN * DIPHENYLENE DIOX-

IDE * NCI-C03656 * OXANTHRENE * PHENODI-OXIN

CONSENSUS REPORTS: IARC Cancer Review: Animal Inadequate Evidence IMEMDT 15,41,77. NCI Carcinogenesis Bioassay Completed; Results Negative NCITR* NCI-CG-TR-122,79.

SAFETY PROFILE: Questionable carcinogen with experimental tumorigenic data. When heated to decomposition it emits acrid smoke and irritating fumes.

DDB600   CAS: 3693-22-9   **HR: 3**
2-DIBENZOFURANAMINE
mf: $C_{12}H_9NO$ mw: 183.22

SYNS: 2-ADO * 3-AMINODIBENZOFURAN * 2-AMINODIPHENYLENE OXIDE

SAFETY PROFILE: Questionable carcinogen with experimental carcinogenic and tumorigenic data. Mutation data reported. When heated to decomposition it emits toxic fumes of NO_x.

DDD000   CAS: 1785-74-6   **HR: 3**
DIBENZOSUBERONE OXIME
mf: $C_{15}H_{13}NO$ mw: 223.29

SYN: 10,11-DIHYDRO-5H-DIBENZO(a,d)CYCLOHEPTEN-5-ONE OXIME

SAFETY PROFILE: Poison by intraperitoneal and intravenous routes. When heated to decomposition it emits toxic fumes of NO_x.

DDE200   CAS: 257-07-8   **HR: 3**
DIBENZ(b,f)(1,4)OXAZEPINE
mf: $C_{13}H_9NO$ mw: 195.23

SYNS: CR * EA 3547

CONSENSUS REPORTS: Reported in EPA TSCA Inventory.

SAFETY PROFILE: Poison by intraperitoneal and intravenous routes. Moderately toxic by ingestion and inhalation. Experimental teratogenic and reproductive effects. A human skin and eye irritant. Questionable carcinogen with experimental carcinogenic, and tumorigenic data. When heated to decomposition it emits toxic fumes of NO_x.

DDG800   CAS: 63-92-3   **HR: 3**
DIBENZYLINE HYDROCHLORIDE
mf: $C_{18}H_{22}ClNO•ClH$ mw: 340.32

SYNS: 688A * BENSYLYT NEN * 2-(N-BENZYL-2-CHLOROETHYLAMINO)-1-PHENOXYPROPANE HYDRO-CHLORIDE * BENZYL(2-CHLOROETHYL)(1-METHYL-2-PHENOXYETHYL)AMINE HYDROCHLORIDE * N-BENZYL-N-PHENOXYISOPROPYL-β-CHLORETHYL-AMINE HYDROCHLORIDE * BENZYLYT * BLOCADREN * N-(2-CHLOROETHYL)-N-(1-ME-THYL-2-PHENOXYETHYL)BENZENEMETHANAMINE HY-DROCHLORIDE * N-(2-CHLOROETHYL)-N-(1-METHYL-2-PHENOXYETHYL)BENZYLAMINE HYDROCHLORIDE * DIBENZYLENE * DIBENZYLIN * DIBENZYRAN * FENOXYBENZAMIN * NCI-C01661 * PHENOXYBENZAMIDE HYDROCHLORIDE * N- PHENOXYISOPROPYL-N-BENZYL-β-CHLORO-ETHYLAMINE HYDROCHLORIDE * N-2-PHENOXY-ISOPROPYL-N- BENZYL-CHLOROETHYLAMINE HYDROCHLORIDE * SKF 688A

CONSENSUS REPORTS: IARC Cancer Review: GROUP 2B IMEMDT 7,56,87; Animal Sufficient Evidence IMEMDT 24,185,80. NCI Carcinogenesis Bioassay Completed; Results Positive: mouse, rat NCITR* NCI-CG-TR-72,78.

SAFETY PROFILE: Suspected carcinogen with experimental carcinogenic data. Poison by intra-peritoneal, intravenous, and subcutaneous routes. Moderately toxic by ingestion. Experimental teratogenic and reproductive effects. A long-acting adrenergic blocker. When heated to decomposition it emits very toxic fumes of NO_x and Cl^-.

DDH000   CAS: 780-24-5   **HR: 3**
DIBENZYLMERCURY
mf: $C_{14}H_{14}Hg$ mw: 382.87

PROP: Colorless crystals, sol in organic solvents.

OSHA PEL: (Transitional: CL 1 mg/10m³) TWA 0.01 mg(Hg)/m³; STEL 0.03 mg/m³ (skin)

ACGIH TLV: TWA 0.01 mg(Hg)/m³; STEL 0.03 mg(Hg)/m³

NIOSH REL: (Inorganic Mercury) TWA 0.05 mg(Hg)/m³

SAFETY PROFILE: Poison by intravenous route. When heated to decomposition it emits toxic fumes of Hg.

DDI450   CAS: 19287-45-7   **HR: 3**
DIBORANE

DOT: UN 1911
mf: B_2H_6 mw: 27.68

PROP: Colorless gas, sickly sweet odor. Mp: $-165.5°$, bp: $-92.5°$, d: 0.447 (liquid @ $-112°$), 0.577 (solid @ $-183°$), vap press: 224 mm @ $-112°$, autoign temp: 38-52°, lel: 0.9%, uel: 98%, flash p: $-90°F$.

SYNS: BOROETHANE * BORON HYDRIDE * DIBORANE(6) * DIBORON HEXAHYDRIDE

CONSENSUS REPORTS: Reported in EPA TSCA Inventory. EPA Extremely Hazardous Substances List.

OSHA PEL: TWA 0.1 ppm
ACGIH TLV: TWA 0.1 ppm
DFG MAK: 0.1 ppm (0.1 mg/m^3)
DOT Classification: Flammable Gas; Label: Flammable Gas and Poison; DOT-IMO: Flammable Gas; Label: Poison Gas and Flammable Gas.

SAFETY PROFILE: Poison by inhalation. An irritant to skin, eyes, and mucous membranes comparable to chlorine, fluorine, arsine, and phosgene. The liquid causes local inflammation, blisters, redness, and swelling. Injuries to central nervous system, liver, and kidneys have also been produced in experimental animals. Similar observations have been reported in humans resulting at times in a reaction resembling metal fume fever. Human exposure to pentaborane has produced signs of severe central nervous system irritation such as drowsiness, dizziness, visual disturbances, muscle twitching and in severe cases, painful muscle spasm. Dangerously flammable when exposed to heat or flame or by chemical reaction. On contact with moisture, hydrogen is usually evolved. Highly explosive when exposed to heat or flame. Explosive reaction with air, tetravinyllead, O_2 above 165°C, octanol oxime + sodium hydroxide, benzene vapor, HNO_3, Cl_2. Violent reaction with halocarbon liquids. Other boron hydrides evolve H_2 upon contact with moisture or can propagate a flame rapidly enough to cause an explosion. Heat can cause these materials to decompose violently or at least to evolve H_2. They also react with water or steam to evolve hydrogen. Reaction with Al or Li forms complex hydrides which may ignite spontaneously in air. Powerful oxidizing agents such as chlorine gas, etc., can react violently with boron hydrides. Pentaborane (stable) is spontaneously flammable in air.

DDJ000 CAS: 10318-26-0 **HR: 3**
DIBROMDULCITOL
mf: $C_6H_{12}Br_2O_4$ mw: 308.00

SYNS: DBD * 1,6-DIBROMODIDEOXYDULCITOL * 1,6-DIBROMO-1,6-DIDEOXYDULCITOL * 1,6-DIBROMO-1,6-DIDEOXYGALACTITOL * 1,6-DIBROMO-1,6-DIDEOXY-d-GALACTITOL * DIBROMODULCITOL * 1,6-DIBROMODULCITOL * ELOBROMOL * GALACTICOL * MITOLAC * MITOLACTOL * NCI-C04795 * NSC-104800

CONSENSUS REPORTS: NCI Carcinogenesis Bioassay Completed; Results Positive: mouse, rat (RRCRBU 52,1,75).

SAFETY PROFILE: Questionable carcinogen with experimental carcinogenic, neoplastigenic, and tumorigenic data. Poison by ingestion. Moderately toxic by intraperitoneal and possibly other routes. Human mutation data reported. An anti-cancer agent taken orally. When heated to decomposition it emits very toxic fumes of Br^-.

DDJ800 CAS: 624-61-3 **HR: 3**
DIBROMOACETYLENE
mf: C_2Br_2 mw: 183.83

PROP: Liquid. Mp: 76° (approx), bp: explodes, d: 2 (approx), vap d: 6.35.

DOT Classification: Forbidden.

SAFETY PROFILE: Ignites spontaneously in air. Explodes when heated. When heated to decomposition it emits toxic fumes of Br^-.

DDK600 CAS: 6305-43-7 **HR: 3**
2,2′-DIBROMOBIACETYL
mf: $C_4H_4Br_2O_2$ mw: 243.90

SYN: α,α′-DIBROMOBIACETYL

CONSENSUS REPORTS: Reported in EPA TSCA Inventory.

SAFETY PROFILE: Poison by intravenous and intraperitoneal routes. When heated to decomposition it emits toxic fumes of Br^-.

DDL400 CAS: 6974-12-5 **HR: 3**
1,4-DIBROMO-2-BUTENE
mf: $C_4H_6Br_2$ mw: 213.92

SYN: TL 80

SAFETY PROFILE: Poison by ingestion and intraperitoneal routes. Moderately toxic by inhalation. A skin and severe eye irritant. When heated to decomposition it emits toxic fumes of Br⁻.

DDL800   CAS: 96-12-8   HR: 3
1,2-DIBROMO-3-CHLOROPROPANE

DOT: UN 2872
mf: $C_3H_5Br_2Cl$ mw: 236.35

PROP: Bp: 196°, flash p: 170°F (TOC).

SYNS: BBC 12 * 1-CHLORO-2,3-DIBROMOPROPANE * 3-CHLORO-1,2-DIBROMOPROPANE * DBCP * DIBROMCHLORPROPAN (GERMAN) * 1,2-DIBROM-3-CHLOR-PROPAN (GERMAN) * DIBROMOCHLORO-PROPANE * 1,2-DIBROMO-3-CLORO-PROPANO (ITALIAN) * 1,2-DIBROOM-3-CHLOORPROPAAN (DUTCH) * FUMAGON * FUMAZONE * NCI-C00500 * NEMABROM * NEMAFUME * NEMAGON * NEMAGONE * NEMAGON SOIL FUMIGANT * NEMANAX * NEMAPAZ * NEMASET * NEMATOCIDE * NEMATOX * NEMAZON * OS 1897 * OXY DBCP * RCRA WASTE NUMBER 7066 * SD 1897

CONSENSUS REPORTS: IARC Cancer Review: GROUP 2B IMEMDT 7,191,87; Animal Sufficient Evidence IMEMDT 15,139,77; Human Limited Evidence IMEMDT 20,83,79; Animal Sufficient Evidence IMEMDT 20,83,79. NTP Fourth Annual Report On Carcinogens, 1984. NCI Carcinogenesis Bioassay Completed; Results Positive: mouse, rat NCITR* NCI-CG-TR-28,78. EPA Genetic Toxicology Program. Community Right-To-Know List. Reported in EPA TSCA Inventory.

OSHA PEL: TWA 0.001 ppm; Cancer Hazard.
DFG MAK: Animal Carcinogen, Suspected Human Carcinogen.
NIOSH REL: CL 0.01 ppm/30M
DOT Classification: Poison B; Label: St. Andrews Cross

SAFETY PROFILE: Confirmed human carcinogen with experimental carcinogenic and teratogenic data. Poison by ingestion, inhalation, and subcutaneous routes. Moderately toxic by skin contact. An eye and severe skin irritant. A suspected human carcinogenic. Narcotic in high concentrations. Has been implicated in causing human male sterility in factory workers. Human mutation data reported. A soil fumigant. Flammable when exposed to heat or flame. When heated to decomposition it emits toxic fumes of Cl⁻ and Br⁻.

DDM000   CAS: 10222-01-2   HR: 3
α,α-DIBROMO-α-CYANOACETAMIDE
mf: $C_3H_2Br_2N_2O$ mw: 241.89

SYNS: DBNPA * DIBROMOCYANOACETAMIDE * 2,2-DIBROMO-3-NITRILOPROPIONAMIDE

CONSENSUS REPORTS: Cyanide and its compounds are on the Community Right-To-Know List. Reported in EPA TSCA Inventory.

SAFETY PROFILE: Poison by ingestion and intravenous routes. A severe skin and eye irritant. When heated to decomposition it emits very toxic fumes of Br⁻ and NO_x.

DDM400   CAS: 996-08-7   HR: 3
DIBROMODIBUTYLSTANNANE
mf: $C_8H_{18}Br_2Sn$ mw: 392.77

PROP: Mp: 20°.

SYNS: DIBROMODIBUTYLTIN * DIBUTYL TIN DIBROMIDE

OSHA PEL: TWA 0.1 mg(Sn)/m³ (skin)
ACGIH TLV: TWA 0.1 mg(Sn)/m³ (skin) (Proposed: TWA 0.1 mg(Sn)/m³; STEL 0.2 mg(Sn)/m³ (skin))
NIOSH REL: (Organotin Compounds) TWA 0.1 mg(Sn)/m³

SAFETY PROFILE: Poison by ingestion. Moderately toxic by skin contact. When heated to decomposition it emits toxic fumes of Br⁻.

DDP000   CAS: 1689-84-5   HR: 3
3,5-DIBROMO-4-HYDROXYBENZONITRILE
mf: $C_7H_3Br_2NO$ mw: 276.93

SYNS: BRITTOX * BROMINAL * BROMINEX * BROMINIL * BROMOXYNIL * BROXYNIL * BUCTRIL * BUCTRIL INDUSTRIAL * BUTIL-CHLOROFOS * CHIPCO BUCTRIL * CHIPCO CRAB-KLEEN * 2,6-DIBROMO-4-CYANOPHENOL * 3,5-DI-BROMO-4-HYDROXYPHENYLCYANIDE * ENT 20,852 * 4-HYDROXY-3,5-DIBROMOBENZONITRILE * MB 10064 * ME4 BROMINAL * NU-LAWN WEEDER * OXYTRIL M

CONSENSUS REPORTS: Cyanide and its compounds are on the Community Right-To-Know List.

SAFETY PROFILE: Poison by ingestion, intravenous, and possibly other routes. An herbicide. When heated to decomposition it emits highly toxic fumes of NO_x, CN⁻ and Br⁻.

DDP600 CAS: 488-41-5 **HR: 3**
1,6-DIBROMOMANNITOL
mf: $C_6H_{12}Br_2O_4$ mw: 308.00

SYNS: DBM * DIBROMANNIT * DIBROMANNI-
TOL * d-DIBROMANNITOL * 1,6-DIBROMO-1,6-
DIDEOXY-d-MANNITOL * 1,6-DIBROMO-1,6-d-
DIDESOXYMANNITOL * MIEOBROMOL * MITO-
BRONITOL * MYEBROL * MYELOBROMOL
* NCI-C04762 * NSC-94100 * R 54

CONSENSUS REPORTS: NCI Carcinogenesis Bioassay Completed; Results Positive:
mouse, rat (RRCRBU 52,1,75).

SAFETY PROFILE: Questionable carcinogen
with experimental carcinogenic, neoplastigenic,
and teratogenic data. Moderately toxic by ingestion, intravenous, intraperitoneal, and subcutaneous routes. Other experimental reproductive
effects. Human mutation data reported. When
heated to decomposition it emits toxic fumes
of Br^-.

DDP800 CAS: 74-95-3 **HR: 3**
DIBROMOMETHANE

DOT: UN 2664
mf: CH_2Br_2 mw: 173.85

PROP: Colorless, heavy liquid. Bp: 95.6°-
97.4°, fp: <50°, d: 2.485 @ 25°/25°, vap d:
6.05.

SYNS: METHYLENE BROMIDE * METHYLENE DI-
BROMIDE * RCRA WASTE NUMBER U068

CONSENSUS REPORTS: Community Right-
To-Know List. Reported in EPA TSCA Inventory.

DOT Classification: Poison B; Label: St. Andrews Cross.

SAFETY PROFILE: A poison. Moderately toxic
by subcutaneous route. Mildy toxic by inhalation. Mutation data reported. Mixtures with potassium explode on light impact. When heated
to decomposition it emits toxic fumes of Br^-.

DDQ800 CAS: 57541-73-8 **HR: 3**
3,4-DIBROMONITROSOPIPERIDINE
mf: $C_5H_8Br_2N_2O$ mw: 271.97

SYN: N-NITROSO-3,4-DIBROMOPIPERIDINE

CONSENSUS REPORTS: EPA Genetic Toxicology Program.

SAFETY PROFILE: Mutation data reported.
Questionable carcinogen with experimental tumorigenic data. Many N-nitroso compounds are
carcinogens. When heated to decomposition it
emits very toxic fumes of Br^- and NO_x.

DDR200 CAS: 696-24-2 **HR: 3**
DIBROMOPHENYLARSINE
mf: $C_6H_5AsBr_2$ mw: 311.85

SYNS: PHENYLARSONOUS DIBROMIDE * PHENYL-
DIBROMOARSINE

CONSENSUS REPORTS: Arsenic and its
compounds are on the Community Right-To-
Know List.

OSHA PEL: TWA 0.5 mg(As)/m^3

SAFETY PROFILE: Poison by skin contact and
intravenous routes. When heated to decomposition it emits very toxic fumes of As and Br^-.

DDS400 CAS: 5221-17-0 **HR: 3**
2,3-DIBROMOPROPIONALDEHYDE
mf: $C_3H_4Br_2O$ mw: 215.89
SYN: DIBROMOPROPANAL

SAFETY PROFILE: Poison by intravenous
route. Mutagenic data reported. When heated
to decomposition it emits toxic fumes of Br^-.

DDT200 CAS: 85-79-0 **HR: 3**
DIBUCAINE
mf: $C_{20}H_{29}N_3O_2$ mw: 343.52

SYNS: 2-BUTOXY-N-(β-DIETHYLAMINOETHYL)
CINCHONINAMIDE * 2-BUTOXY-N-(2-(DIETHYL-
AMINO)ETHYL)CINCHONINAMIDE * 2-BUTOXY-
QUINOLINE-4-CARBOXYLIC ACID DIETHYLAMINO-
ETHYLAMIDE * α-BUTYLOXYCINCHONINIC ACID
DIETHYLETHYLENEDIAMINE * 2-BUTYOXY-
N-(2-(DIETHYLAMINO)ETHYL)-4-QUINOLINECARBOX-
AMIDE * CINCHOCAINE * DERMACAINE
* N-(2-(DIETHYLAMINO)ETHYL)-2-BUTOXYCIN-
CHONINAMIDE * NUPERCAINAL * NUPERCAINE
* SOVCAINE

SAFETY PROFILE: A human poison by ingestion. Poison experimentally by subcutaneous,
intravenous, intraperitoneal, and possibly other
routes. When heated to decomposition it emits
toxic fumes of NO_x.

DDT800 CAS: 111-92-2 **HR: 3**
n-DIBUTYLAMINE
DOT: UN 2248
mf: $C_8H_{19}N$ mw: 129.28

PROP: Liquid. Mp: −59°; bp: 159°, flash p: 125°F (OC), d: 0.76, vap d: 4.46, vap press: 2 mm @ 20°.

SYNS: N-BUTYL-1-BUTANAMINE * DI-n-BUTYL-AMINE * DI(n-BUTYL)AMINE (DOT)

CONSENSUS REPORTS: Reported in EPA TSCA Inventory. EPA Genetic Toxicology Program.

DOT Classification: IMO: Corrosive Material; Label: Corrosive, Flammable Liquid.

SAFETY PROFILE: Poison by ingestion and subcutaneous routes. Moderately toxic by skin contact and inhalation. Corrosive. A severe skin and eye irritant. Mutation data reported. Flammable when exposed to heat or flame; can react with oxidizing materials. To fight fire, use alcohol foam, foam, CO_2, dry chemical. Exothermic reaction with cellulose nitrate does not proceed to ignition. When heated to decomposition it emits toxic fumes of NO_x.

DDU600   CAS: 102-81-8   HR: 3
2-N-DIBUTYLAMINOETHANOL

DOT: UN 2873
mf: $C_{10}H_{23}NO$ mw: 173.34

PROP: Liquid. Bp: 222°, flash p: 220°F (OC), d: 0.85, vap d: 6.0.

SYNS: BU2AE * DIBUTYLAMINOETHANOL * 2-DIBUTYLAMINOETHANOL * 2-N-DIBUTYLAMI-NOETHANOL * 2-DI-n-BUTYLAMINOETHANOL * N,N-DI-n-BUTYLAMINOETHANOL (DOT) * β-N-DI-BUTYLAMINOETHYL ALCOHOL * N,N-DIBUTYL-ETHANOLAMINE * N,N-DIBUTYL-N-(2-HYDROXY-ETHYL)AMINE

CONSENSUS REPORTS: Reported in EPA TSCA Inventory.

OSHA PEL: TWA 2 ppm
ACGIH TLV: TWA 2 ppm (skin)
DOT Classification: Poison B; Label; St. Andrews Cross.

SAFETY PROFILE: Poison by intraperitoneal route. Moderately toxic by ingestion and skin contact. A severe eye and skin irritant. Flammable when exposed to heat or flame; can react with oxidizing materials. To fight fire, use CO_2, dry chemical. When heated to decomposition it emits toxic fumes of NO_x.

DDV600   CAS: 77-58-7   HR: 3
DIBUTYLBIS(LAUROYLOXY)STANNANE
mf: $C_{32}H_{64}O_4Sn$ mw: 631.65

PROP: Pale yellow liquid to colorless solid (when pure). Mp: 23°, bp: non-distillable @ 10 mm, flash p: 455°F (OC), d: 1.066 @ 20°/20°, vap d: 21.8.

SYNS: BIS(DODECANOYLOXY)DI-n-BUTYLSTANNANE * BIS(LAUROYLOXY)DIBUTYLSTANNANE * BIS-(LAUROYLOXY)DI(n-BUTYL)STANNANE * BUTYNO-RATE * DBTL * DIBUTYLBIS(LAUROYLOXY)TIN * DI-n-BUTYLTIN DI(DODECANOATE) * DIBUTYL-TIN DILAURATE (USDA) * DIBUTYLTIN LAURATE * DIBUTYL-ZINN-DILAURAT (GERMAN) * FOMREZ SUL-4 * LAUDRAN DI-n-BUTYLCINICITY (CZECH) * LAURIC ACID, DIBUTYLSTANNYLENE deriv. * LAURIC ACID, DIBUTYLSTANNYLENE SALT * STABILIZER D-22 * THERM CHEK 820 * TIN DI-BUTYL DILAURATE * TINOSTAT

CONSENSUS REPORTS: Reported in EPA TSCA Inventory.

OSHA PEL: TWA 0.1 mg(Sn)/m³ (skin)
ACGIH TLV: TWA 0.1 mg(Sn)/m³ (skin) (Proposed: TWA 0.1 mg(Sn)/m³; STEL 0.2 mg(Sn)/m³ (skin))
NIOSH REL: (Organotin Compounds) TWA 0.1 mg(Sn)/m³

SAFETY PROFILE: Poison by ingestion and intraperitoneal routes. A skin and eye irritant. Avoid the vapor produced by heating. Combustible when exposed to heat or flame; reacts with oxidizers. When heated to decomposition it emits acrid smoke and fumes.

DDV800   CAS: 78-46-6   HR: 3
DIBUTYL BUTANEPHOSPHONATE
mf: $C_{12}H_{27}O_3P$ mw: 250.36

PROP: Colorless liquid, mild odor. Bp: 128° @ 2.5 mm, flash p: 311° (COC), d: 8.62.

SYN: DIBUTYL BUTYLPHOSPHONATE

CONSENSUS REPORTS: Reported in EPA TSCA Inventory.

SAFETY PROFILE: Poison by intraperitoneal and intravenous routes. Combustible when exposed to heat or flame. It can react vigorously with oxidizing materials. To fight fire, use foam, CO_2, or dry chemical. When heated to decomposition it emits toxic fumes of PO_x.

DDW000   CAS: 532-49-0   **HR: 3**
DI-n-BUTYL-CARBAMYLCHOLINE SULPHATE
mf: $C_{30}H_{66}N_4O_4 \cdot O_4S$ mw: 643.06

SYNS: DIBULINESULFAT * DIBULINE SULFATE * DIBUTOLINE * DIBUTOLINE SULFATE * 1-(((DI-BUTYLAMINO)CARBONYL)OXY)-N-ETHYL-N,N-DIMETH-YLETHANAMINIUM SULFATE (2:1) * (2-DIBUTYLCAR-BAMYLOXYETHYL)-DIMETHYLETHYLAMMONIUM SUL-FATE * DIMETHYL-ETHYL-β-HYDROXYETHYL-AM-MONIUM-SULFATE-DI-n-BUTYLCARBAMATE * DIMETHYLETHYL-β-HYDROXYETHYLAMMONIUM SULFATE DIBUTYLURETHAN * ETHYL(2-HYDROXY-ETHYL)DIMETHYL-AMMONIUM SULFATE (SALT), BIS(DI-BUTYLCARBAMATE)

SAFETY PROFILE: Poison by intraperitoneal and subcutaneous routes. When heated to decomposition it emits very toxic fumes of NO_x, NH_3, and SO_x.

DDY000   CAS: 4593-81-1   **HR: 3**
DIBUTYLDICHLOROGERMANE
mf: $C_8H_{18}Cl_2Ge$ mw: 257.75

SYNS: DI-n-BUTYLGERMANEDICHLORIDE * DI-CHLORODIBUTYLGERMANE

CONSENSUS REPORTS: Reported in EPA TSCA Inventory.

SAFETY PROFILE: Poison by intraperitoneal route. Mutation data reported. When heated to decomposition it emits very toxic fumes of Cl^-.

DDY200   CAS: 683-18-1   **HR: 3**
DIBUTYLDICHLOROSTANNANE
mf: $C_8H_{18}Cl_2Sn$ mw: 303.85

PROP: White, crystalline solid. Mp: 43°, bp: 135° @ 10 mm, flash p: 335°F (OC), d: 1.36 @ 50°, vap press: 2 mm @ 100°, vap d: 10.5.

SYNS: CHLORID DI-n-BUTYLCINICITY (CZECH) * D.B.T.C. * DIBUTYLDICHLOROTIN * DIBUTYL-TIN CHLORIDE * DIBUTYLTIN DICHLORIDE * DI-n-BUTYLTIN DICHLORIDE * DI-n-BUTYL-ZINN-DICHLORID (GERMAN) * DICHLORODIBUTYLSTAN-NANE * DICHLORODIBUTYLTIN

CONSENSUS REPORTS: Reported in EPA TSCA Inventory.

OSHA PEL: TWA 0.1 mg(Sn)/m³ (skin)
ACGIH TLV: TWA 0.1 mg(Sn)/m³ (skin) (Proposed: TWA 0.1 mg(Sn)/m³; STEL 0.2 mg(Sn)/m³ (skin))

NIOSH REL: (Organotin Compounds) TWA 0.1 mg(Sn)/m³

SAFETY PROFILE: Poison by ingestion, intravenous, intraperitoneal, and possibly other routes. Moderately toxic by skin contact. A severe skin and eye irritant. Mutation data reported. Combustible when exposed to heat or flame. A dangerous material; emits highly toxic fumes of HCl; will react with water or steam to produce heat and toxic fumes; can react vigorously with oxidizing materials. To fight fire, use water, foam, CO_2, dry chemical.

DDZ000   CAS: 7392-96-3   **HR: 3**
DIBUTYL(DIFORMYLOXY)STANNANE
mf: $C_{10}H_{20}O_4Sn$ mw: 322.99

SYNS: DI-n-BUTYLTIN DIFORMATE * MRAVENCAN DI-n-BUTYLCINICITY (CZECH)

OSHA PEL: TWA 0.1 mg(Sn)/m³ (skin)
ACGIH TLV: TWA 0.1 mg(Sn)/m³ (skin) (Proposed: TWA 0.1 mg(Sn)/m³; STEL 0.2 mg(Sn)/m³ (skin))
NIOSH REL: (Organotin Compounds) TWA 0.1 mg(Sn)/m³

SAFETY PROFILE: Poison by ingestion. A severe skin and eye irritant. When heated to decomposition it emits acrid and irritating fumes.

DEA600   CAS: 3465-74-5   **HR: 3**
DIBUTYLDIPENTANOYLOXYSTANNANE
mf: $C_{18}H_{36}O_4Sn$ mw: 435.23

SYNS: DI-n-BUTYLTIN DIPENTANOATE * DI(PEN-TANOYLOXY)DIBUTYLSTANNANE * VALERAN DI-n-BUTYLCINICITY (CZECH)

OSHA PEL: TWA 0.1 mg(Sn)/m³ (skin)
ACGIH TLV: TWA 0.1 mg(Sn)/m³ (skin) (Proposed: TWA 0.1 mg(Sn)/m³; STEL 0.2 mg(Sn)/m³ (skin))
NIOSH REL: (Organotin Compounds) TWA 0.1 mg(Sn)/m³

SAFETY PROFILE: Poison by ingestion. A severe skin and eye irritant. When heated to decomposition it emits acrid and irritating fumes.

DEB800   CAS: 26818-53-1   **HR: 3**
N,N-DI-sec-BUTYL DITHIOOXAMIDE
mf: $C_{10}H_{20}N_2S_2$ mw: 232.44

SAFETY PROFILE: Poison by intravenous route. When heated to decomposition it emits very toxic fumes of NO_x and SO_x.

DEC000   CAS: 625-22-9   **HR: 3**
DIBUTYL ESTER SULFURIC ACID
mf: $C_8H_{18}O_4S$ mw: 210.32

SYNS: DIBUTYL SULFATE   *   DI-n-BUTYLSULFAT
(GERMAN)

SAFETY PROFILE: Questionable carcinogen with experimental tumorigenic data. Poison by ingestion. Mildly toxic by subcutaneous route. When heated to decomposition it emits toxic fumes of SO_x.

DEC200   CAS: 625-17-2   **HR: 3**
DI-sec-BUTYL FLUOROPHOSPHONATE
mf: $C_8H_{18}FO_3P$ mw: 212.23

SYNS: DI-sec-BUTYL ESTER PHOSPHOROFLUORIDIC
ACID   *   DI-sec-BUTYLFLUOROPHOSPHATE   *   T-1835
*   TL 1266

SAFETY PROFILE: Poison by inhalation. Human systemic effects by inhalation including: miosis (pupillary constriction), somnolence, and respiratory changes. When heated to decomposition it emits very toxic fumes of F^- and PO_x.

DEC600   CAS: 105-75-9   **HR: 3**
DIBUTYL FUMARATE
mf: $C_{12}H_{20}O_4$ mw: 228.32

PROP: Colorless, clear, mobile liquid; typical odor. Bp: 285.1°, fp: −19°, flash p: 300°F (OC), d: 0.986 @ 20°/20°, vap d: 7.88.

SYN: FUMARIC ACID, DIBUTYL ESTER

CONSENSUS REPORTS: Reported in EPA TSCA Inventory.

SAFETY PROFILE: Poison by intraperitoneal route. Mildly toxic by ingestion and skin contact. An eye, skin, and mucous membrane irritant. Combustible when exposed to heat or flame; can react with oxidizing materials. To fight fire, use foam, CO_2, dry chemical. When heated to decomposition it emits acrid smoke and fumes.

DED400   CAS: 2587-84-0   **HR: 3**
DIBUTYL LEAD DIACETATE
mf: $C_{12}H_{24}O_4Pb$ mw: 439.55

SYN: DIACETOXYDIBUTYLPLUMBANE

CONSENSUS REPORTS: Lead and its compounds are on the Community Right-To-Know List.

SAFETY PROFILE: Poison by ingestion, intraperitoneal, and intravenous routes. When heated to decomposition it emits toxic fumes of Pb.

DED600   CAS: 105-76-0   **HR: 3**
DIBUTYL MALEATE
mf: $C_{12}H_{20}O_4$ mw: 228.32

PROP: Liquid. Mp: −85° (sets to a glass), bp: 281°, flash p: 285°F (OC), d: 0.9964 @ 20°/20°, vap d: 7.9.

SYNS: 2-BUTENEDIOIC ACID, DIBUTYL ESTER
*   DBM   *   MALEIC ACID, DIBUTLY ESTER
*   RC COMONOMER DBM   *   STAFLEX DBM

CONSENSUS REPORTS: Reported in EPA TSCA Inventory.

SAFETY PROFILE: Poison by intraperitoneal route. Moderately toxic by ingestion. Mildly toxic by skin contact. An eye and skin irritant. Combustible when exposed to heat or flame; can react with oxidizing materials. To fight fire, use foam, CO_2, dry chemical, alcohol foam. When heated to decomposition it emits acrid smoke and irritating fumes.

DEE000   CAS: 629-35-6   **HR: 3**
DIBUTYLMERCURY
mf: $C_8H_{18}Hg$ mw: 314.85

PROP: Liquid. Bp: 105° @ 10 mm, d: 1.779, vap d: 10.8.

CONSENSUS REPORTS: Mercury and its compounds are on the Community Right-To-Know List. Reported in EPA TSCA Inventory.

OSHA PEL: (Transitional: CL 1 mg/10m³)
   TWA 0.01 mg(Hg)/m³; STEL 0.03 mg/m³
   (skin)
ACGIH TLV: TWA 0.01 mg(Hg)/m³; STEL
   0.03 mg(Hg)/m³
NIOSH REL: (Inorganic Mercury) TWA 0.05
   mg(Hg)/m³

SAFETY PROFILE: Poison by intraperitoneal route. Flammable when exposed to heat or flame. Can react vigorously with oxidizing materials. When heated to decomposition or on contact with acid or acid fumes it emits highly toxic fumes of mercury.

DEE200   CAS: 691-88-3   **HR: 3**
DI-sec-BUTYLMERCURY
mf: $C_8H_{18}Hg$ mw: 314.85

CONSENSUS REPORTS: Mercury and its compounds are on the Community Right-To-Know List.

OSHA PEL: (Transitional: CL 1 mg/10m^3) TWA 0.01 mg(Hg)/m^3; STEL 0.03 mg/m^3 (skin)

ACGIH TLV: TWA 0.01 mg(Hg)/m^3; STEL 0.03 mg(Hg)/m^3

NIOSH REL: (Inorganic Mercury) TWA 0.05 mg(Hg)/m^3

SAFETY PROFILE: Poison by intraperitoneal route. When heated to decomposition it emits toxic fumes of Hg.

DEF400   CAS: 818-08-6   **HR: 3**
DIBUTYLOXOSTANNANE
mf: C$_8$H$_{18}$OSn   mw: 248.95

PROP: White, amorphous powder. Mp: decomp without melting, bulk density: 0.5, vap d: 8.6.

SYNS: DBOT * DIBUTYLOXIDE of TIN * DIBU-TYLOXOTIN * DIBUTYLSTANNANE OXIDE * DIBUTYLTIN OXIDE * DI-n-BUTYLTIN OXIDE * DI-n-BUTYL-ZINN-OXYD (GERMAN) * KYSLICNIK DI-n-BUTYLCINICITY (CZECH)

CONSENSUS REPORTS: Reported in EPA TSCA Inventory.

OSHA PEL: TWA 0.1 mg(Sn)/m^3 (skin)

ACGIH TLV: TWA 0.1 mg(Sn)/m^3 (skin) (Proposed: TWA 0.1 mg(Sn)/m^3; STEL 0.2 mg(Sn)/m^3 (skin))

NIOSH REL: (Organotin Compounds) TWA 0.1 mg(Sn)/m^3

SAFETY PROFILE: Poison by ingestion and intraperitoneal routes. A skin and eye irritant. Flammable when exposed to flame; can react with oxidizing materials. To fight fire use dry chemical, fog, CO$_2$. When heated to decomposition it emits acrid smoke and irritating fumes.

DEF800   CAS: 5510-99-6   **HR: 3**
2,6-DI-sec-BUTYLPHENOL
mf: C$_{14}$H$_{22}$O   mw: 206.36

PROP: Amber liquid. Bp: 152°-165° @ 25 mm, fp: −50°, flash p: 280°F, d: 0.936 @ 25°/4°.

SYN: 2,6-DI-sec-BUTYLFENOL (CZECH)

SAFETY PROFILE: Poison by intravenous route. Moderately toxic by ingestion. A severe skin and eye irritant. Combustible when exposed to heat or flame; can react with oxidizing materials. To fight fire, use foam, CO$_2$, dry chemical. When heated to decomposition it emits acrid and irritating fumes.

DEG200   CAS: 101-96-2   **HR: 3**
N,N′-DI-sec-BUTYL-p-PHENYLENEDIAMINE
mf: C$_{14}$H$_{24}$N$_2$ mw: 220.40

PROP: Liquid. Mp: 17.8°, flash p: 285°F (OC), d: 0.94-0.95 @ 24°/24°.

SYN: TENAMENE 2

CONSENSUS REPORTS: Reported in EPA TSCA Inventory.

SAFETY PROFILE: Poison by ingestion. Moderately toxic by inhalation and skin contact. Corrosive to skin. A mild allergen. Symptoms of exposure are sweating, flushing, shortness of breath and slow pulse. Combustible when exposed to heat or flame; can react with oxidizing materials. To fight fire, use foam, CO$_2$, dry chemical. When heated to decomposition it emits toxic fumes of NO$_x$.

DEG400   CAS: 2655-19-8   **HR: 2**
3,5-DI-tert-BUTYLPHENYLMETHYL-CARBAMATE
mf: C$_{16}$H$_{25}$NO$_2$ mw: 263.42

SYNS: BUTACARB * BUTACARBE (FRENCH)

SAFETY PROFILE: Moderately toxic by ingestion. When heated to decomposition it emits toxic fumes of NO$_x$.

DEG600   CAS: 2528-36-1   **HR: 2**
DIBUTYL PHENYL PHOSPHATE
mf: C$_{14}$H$_{23}$O$_4$P   mw: 286.34

SYN: PHOSPHORIC ACID, DIBUTYL PHENYL ESTER

CONSENSUS REPORTS: Reported in EPA TSCA Inventory.

ACGIH TLV: (Proposed: TWA 0.3 ppm (skin))

SAFETY PROFILE: Moderately toxic by ingestion. When heated to decomposition it emits toxic fumes of PO$_x$.

DEG700   CAS: 107-66-4   **HR: 2**
DIBUTYL PHOSPHATE
mf: C$_8$H$_{19}$PO$_4$ mw: 210.2

PROP: Pale amber liquid. Bp: decomp > 100°.
OSHA PEL: TWA 1 ppm; STEL 2 ppm
ACGIH TLV: TWA 1 ppm; STEL 2 ppm

SAFETY PROFILE: Moderately toxic by ingestion. When heated to decomposition it emits toxic fumes of PO_x.

DEH200   CAS: 84-74-2   HR: 3
DIBUTYL PHTHALATE
DOT: NA 9095
mf: $C_{16}H_{22}O_4$ mw: 278.38

PROP: Oily liquid, mild odor. Bp: 340°, fp: −35°, flash p: 315°F (CC), d: 1.047-1.049 @ 20°/20°, autoign temp: 757°F, vap d: 9.58.

SYNS: o-BENZENEDICARBOXYLIC ACID, DIBUTYL ESTER * BENZENE-o-DICARBOXYLIC ACID DI-n-BUTYL ESTER * n-BUTYL PHTHALATE (DOT) * CELLUFLEX DPB * DBP * DIBUTYL-1,2-BENZENEDICARBOXYLATE * DI-n-BUTYL PHTHALATE * ELAOL * HEXAPLAS M/B * PALATINOL C * POLYCIZER DBP * PX 104 * RCRA WASTE NUMBER U069 * STAFLEX DBP * WITCIZER 300

CONSENSUS REPORTS: Community Right-To-Know List. EPA Genetic Toxicology Program. Reported in EPA TSCA Inventory.

OSHA PEL: TWA 5 mg/m³
ACGIH TLV: TWA 5 mg/m³
DOT Classification: ORM-E; Label: None.

SAFETY PROFILE: Moderately toxic by intraperitoneal and intravenous routes. Mildly toxic by ingestion. Human systemic eye effects by ingestion: hallucinations, distorted perceptions, nausea or vomiting and kidney, ureter or bladder changes. Experimental teratogenic and reproductive effects. Mutation data reported. Combustible when exposed to heat or flame; can react with oxidizing materials. Violent reaction with Cl_2. Incompatible with chlorine. To fight fire, use CO_2, dry chemical. When heated to decomposition it emits acrid smoke and fumes.

DEH600   CAS: 109-43-3   HR: 1
DIBUTYL SEBACATE
mf: $C_{18}H_{34}O_4$ mw: 314.52

PROP: Clear liquid. Bp: 180° @ 3 mm, fp: −11°, flash p: 353°F (COC), d: 0.936 @ 20°/20°, vap d: 10.8.

SYNS: BIS(n-BUTYL)SEBACATE * DECANEDIOIC ACID, DIBUTYL ESTER * DI-n-BUTYL SEBACATE

* KODAFLEX DBS * MONOPLEX DBS * POLYCIZER DBS * PX 404 * SEBACIC ACID, DIBUTYL ESTER * STAFLEX DBS

CONSENSUS REPORTS: Reported in EPA TSCA Inventory.

SAFETY PROFILE: Mildly toxic by ingestion. Experimental reproductive effects. Combustible liquid when exposed to heat or flame; can react with oxidizing materials. To fight fire, use CO_2, dry chemical. When heated to decomposition it emits acrid smoke and fumes.

DEL000   CAS: 79-43-6   HR: 2
DICHLORACETIC ACID
DOT: UN 1764
mf: $C_2H_2Cl_2O_2$ mw: 128.94

PROP: Colorless, corrosive liquid; pungent odor. Mp (a): 10°, (b): −4°, bp: 194°, d: 1.5634 @ 20°/4°, vap press: 1 mm @ 44.0°, vap d: 4.45.

SYNS: BICHLORACETIC ACID * DCA * DICHLORETHANOIC ACID * 2,2-DICHLOROACETIC ACID * DICHLOROETHANOIC ACID * URNER'S LIQUID

CONSENSUS REPORTS: Reported in EPA TSCA Inventory.

DOT Classification: Corrosive Material; Label: Corrosive.

SAFETY PROFILE: Moderately toxic by skin contact and ingestion. It is corrosive to the skin, eyes, and mucous membranes. Questionable carcinogen with experimental tumorigenic data. Will react with water or steam to produce toxic and corrosive fumes. When heated to decomposition it emits toxic fumes of Cl^-.

DEM000   CAS: 5571-97-1   HR: 3
DICHLORMETHAZANONE
mf: $C_{11}H_{11}Cl_2NO_3S$ mw: 308.19

SYNS: DICHLORMEZANONE * 2-(3,4-DICHLOROPHENYL)-3-METHYL-4-METATHIAZANONE-1,1-DIOXIDE * 2-(3,4-DICHLOROPHENYL)TETRAHYDRO-3-METHYL-4H-1,3-THIAZIN-4-ONE-1,1-DIOXIDE * WIN 12267

SAFETY PROFILE: Poison by ingestion and intraperitoneal routes. When heated to decomposition it emits very toxic fumes of Cl^-, SO_x, and NO_x.

DEM800   CAS: 116-54-1   **HR: 3**
DICHLOROACETIC ACID METHYL ESTER

DOT: UN 2299
mf: $C_3H_4Cl_2O_2$ mw: 142.97

PROP: Colorless liquid, ethereal odor. Bp: 143.0°, d: 1.3809 @ 19.2°/19.2°, vap d: 4.93.

SYNS: METHYL DICHLOROACETATE (DOT)
* METHYL DICHLOROETHANOATE

CONSENSUS REPORTS: Reported in EPA TSCA Inventory.

DOT Classification: Corrosive Material; Label: Corrosive; IMO: Poison B; Label: St. Andrews Cross.

SAFETY PROFILE: Poisonous irritant to the skin, eyes, and mucous membranes. Hydrolyzes upon contact with moisture to form a product corrosive to tissue. Dangerous; when heated to decomposition it emits highly toxic fumes of phosgene and Cl^-.

DEN400   CAS: 79-36-7   **HR: 3**
DICHLOROACETYL CHLORIDE

DOT: UN 1765
mf: C_2HCl_3O mw: 147.38

PROP: Fuming liquid, acrid odor, misc in ether. D: 1.5315 @ 16°/4°, bp: 108°, flash p: 151°F, vap d: 5.8.

SYNS: CHLORURE de DICHLORACETYLE (FRENCH)
* DICHLORACETYL CHLORIDE * α,α-DICHLORO-
ACETYL CHLORIDE * 2,2-DICHLOROACETYL CHLO-
RIDE * DICHLOROACETYL CHLORIDE (DOT)
* DICHLOROETHANOYL CHLORIDE

CONSENSUS REPORTS: Reported in EPA TSCA Inventory.

DOT Classification: Corrosive Material; Label: Corrosive.

SAFETY PROFILE: Questionable carcinogen with experimental tumorigenic data. Moderately toxic by ingestion, inhalation, and skin contact. Corrosive to the skin, eyes, and mucous membranes. Flammable when exposed to heat or flame. When heated to decomposition it emits toxic fumes of Cl^-.

DEN600   CAS: 7572-29-4   **HR: 3**
DICHLOROACETYLENE
mf: C_2Cl_2 mw: 94.92

SYN: DICHLOROETHYNE

CONSENSUS REPORTS: IARC Cancer Review: GROUP 3 IMEMDT 7,56,87; Animal Limited Evidence IMEMDT 39,369,86

OSHA PEL: CL 0.1 ppm
ACGIH TLV: CL 0.1 ppm
DFG MAK: Animal Carcinogen, Suspected Human Carcinogen.
DOT Classification: Forbidden.

SAFETY PROFILE: Confirmed carcinogen with experimental carcinogenic data. Poison by inhalation. Central nervous system effects. Can be formed by thermal decomposition (>70°) from trichloroethylene. Symptoms include a disabling nausea and intense jaw pain. Strong explosive when shocked or exposed to heat or air. Can react vigorously with oxidizing materials. When heated to decomposition or on contact with acid or acid fumes it emits highly toxic fumes of Cl^-.

DEP600   CAS: 95-50-1   **HR: 3**
o-DICHLOROBENZENE

DOT: UN 1591
mf: $C_6H_4Cl_2$ mw: 147.00

PROP: Clear liquid. Mp: −17.5°, bp: 180-183°, fp: −22°, flash p: 151°F, d: 1.307 @ 20°/20°, vap d: 5.05, autoign temp: 1198°F, lel: 2.2%, uel: 9.2%.

SYNS: CHLOROBEN * CHLORODEN * CLORO-
BEN * DCB * o-DICHLORBENZENE * o-DI-
CHLOR BENZOL * 1,2-DICHLOROBENZENE (MAK)
* DICHLOROBENZENE, ORTHO, liquid (DOT)
* DILANTIN DB * DILATIN DB * DIZENE
* DOWTHERM E * NCI-C54944 * ODB
* ODCB * ORTHODICHLOROBENZENE * ORTHO-
DICHLOROBENZOL * RCRA WASTE NUMBER U070
* SPECIAL TERMITE FLUID * TERMITKIL

CONSENSUS REPORTS: IARC Cancer Review: GROUP 3 IMEMDT 7,192,87; Animal Inadequate Evidence IMEMDT 7,231,74, IMEMDT 29,213,82; Human Inadequate Evidence IMEMDT 7,231,74, IMEMDT 29,-213,82. Reported in EPA TSCA Inventory. Community Right-To-Know List.

OSHA PEL: CL 50 ppm
ACGIH TLV: CL 50 ppm
DFG MAK: 50 ppm (300 mg/m³)
DOT Classification: ORM-A; Label: None; IMO: Poison B; Label: St. Andrews Cross.

SAFETY PROFILE: Poison by ingestion and intravenous routes. Moderately toxic by inhalation and intraperitoneal routes. An eye, skin, and mucous membrane irritant. Causes liver and kidney injury. Experimental teratogenic and reproductive effects. Questionable carcinogen. Mutation data reported. A pesticide. Flammable when exposed to heat or flame. Can react vigorously with oxidizing materials. To fight fire, use water, foam, CO_2, or dry chemical. Slow reaction with aluminum may lead to explosion during storage in a sealed aluminum container. When heated to decomposition it emits toxic fumes of Cl^-.

DEP699   CAS: 541-73-1   HR: 3
m-DICHLOROBENZENE

DOT: UN 1591
mf: $C_6H_4Cl_2$ mw: 147.00

SYN: 1,3-DICHLOROBENZENE

CONSENSUS REPORTS: Reported in EPA TSCA Inventory. Community Right-To-Know List.

DOT Classification: IMO: Poison B; Label: St. Andrews Cross.

SAFETY PROFILE: A poison. Mutation data reported. When heated to decomposition it emits toxic fumes of Cl^-.

DEP800   CAS: 106-46-7   HR: 3
p-DICHLOROBENZENE

DOT: UN 1592
mf: $C_6H_4Cl_2$ mw: 147.00

PROP: White crystals, penetrating odor. Mp: 53°, bp: 173.4°, flash p: 150°F (CC), d: 1.4581 @ 20.5°/4°, vap press: 10 mm @ 54.8°, vap d: 5.08.

SYNS: p-CHLOROPHENYL CHLORIDE * p-DI-CHLOORBENZEEN (DUTCH) * 1,4-DICHLOORBENZEEN (DUTCH) * p-DICHLORBENZOL (GERMAN) * 1,4-DICHLOR-BENZOL (GERMAN) * DI-CHLORICIDE * 1,4-DICHLOROBENZENE (MAK) * p-DICLOROBENZENE (ITALIAN) * DICHLOROBENZENE, PARA, solid (DOT) * 1,4-DICLOROBENZENE (ITALIAN) * p-DICHLOROBENZOL * EVOLA * NCI-C54955 * PARACIDE * PARA CRYSTALS * PARADI * PARADICHLORBENZOL (GERMAN) * PARADICHLOROBENZENE * PARADICHLOROBENZOL * PARADOW * PARAMOTH * PARANUGGETS * PARAZENE * PDB * PDCB * PERSIA-PERAZOL * RCRA WASTE NUMBER U070 * RCRA

WASTE NUMBER U071 * RCRA WASTE NUMBER U072 * SANTOCHLOR

CONSENSUS REPORTS: IARC Cancer Review: GROUP 2B IMEMDT 7,192,87; Animal Inadequate Evidence IMEMDT 7,231,74; Human Inadequate Evidence IMEMDT 7,231,74; Animal Inadequate Evidence IMEMDT 29,-213,82. Reported in EPA TSCA Inventory. EPA Genetic Toxicology Program. Community Right-To-Know List.

OSHA PEL: (Transitional: TWA 75 ppm) TWA 75 ppm; STEL 110 ppm
ACGIH TLV: TWA 75 ppm; STEL 110 ppm
DFG MAK: 75 ppm (450 mg/m³)
DOT Classification: ORM-A; Label: None; IMO: Poison B; Label: St. Andrews Cross.

SAFETY PROFILE: Suspected carcinogen with experimental carcinogenic data. A human poison by an unspecified route. Moderately toxic to humans by ingestion. Moderately toxic experimentally by ingestion, subcutaneous, and intraperitoneal routes. Mildly toxic by subcutaneous route. Experimental teratogenic and reproductive effects. Human systemic effects by ingestion: unspecified changes in the eyes, lungs, thorax and respiration, and decreased motility or constipation. Can cause liver injury in humans. A human eye irritant. Mutation data reported. A fumigant. Flammable when exposed to heat, flame, or oxidizers. Dangerous; can react vigorously with oxidizing materials. To fight fire, use water, foam, CO_2, dry chemical. When heated to decomposition it emits toxic fumes of Cl^-.

DEQ000   CAS: 5836-73-7   HR: 3
3,4-DICHLOROBENZENE DIAZOTHIOUREA

mf: $C_7H_6Cl_2N_4S$ mw: 249.13

SYNS: CHLOROPROMURITE * (3,4-DICHLOORFENYL-AZO)-THIOUREUM (DUTCH) * 1-(3',4'-DICHLOROBENZENEDIAZOL)-2-THIOUREA * 3,4-DICHLOROBENZENE DIAZOTHIOCARBAMID * 3,4-DICHLOROPHENYLAZOTHIOUREA * 3,4-DICHLOROPHENYL-AZOTHIOUREE (FRENCH) * (3,4-DICHLOR-PHENYL-AZO)-THIOHARNSTOFF (GERMAN) * (3,4-DICLORO-FENIL-AZO)-TIOUREA (ITALIAN) * MURITAN * PROMURIT * PROMURITE

SAFETY PROFILE: A deadly poison by ingestion, intraperitoneal, and possibly other routes. When heated to decomposition it emits very toxic fumes of Cl^-, NO_x, and SO_x.

DEQ600   CAS: 91-94-1   **HR: 3**
3′,3′-DICHLOROBENZIDINE
mf: $C_{12}H_{10}Cl_2N_2$ mw: 253.14

PROP: Crystals. Mp: 133°; insol in water; sol in alc, benzene, and glacial acetic acid.

SYNS: C.I. 23060 * CURITHANE C126 * DCB * 4,4′-DIAMINO-3,3′-DICHLOROBIPHENYL * 4,4′-DIAMINO-3,3′-DICHLORODIPHENYL * 3,3′-DICHLORBENZIDIN (CZECH) * 3,3′-DICHLOROBENZIDINA (SPANISH) * DICHLOROBENZIDINE * 3,3′-DICHLOROBENZIDENE * o,o′-DICHLOROBENZIDINE * DICHLOROBENZIDINE BASE * 3,3′-DICHLORO-4,4′-BIPHENYLDIAMINE * 3,3′-DICHLOROBIPHENYL-4,4′-DIAMINE * 3,3′-DICHLORO-4,4′-DIAMINOBIPHENYL * 3,3′-DICHLORO-4,4′-DIAMINO(1,1-BIPHENYL) * RCRA WASTE NUMBER U073

CONSENSUS REPORTS: IARC Cancer Review: GROUP 2B IMEMDT 7,193,87; Human Inadequate Evidence IMEMDT 29,239,82; Animal Sufficient Evidence IMEMDT 29,239,82; IMEMDT 4,49,74. NTP Fourth Annual Report On Carcinogens, 1984. Reported in EPA TSCA Inventory. Community Right-To-Know List. EPA Genetic Toxicology Program.

OSHA PEL: Carcinogen
ACGIH TLV: Suspected Human Carcinogen.
DFG TRK: 0.1 mg/m^3, Animal Carcinogen, Suspected Human Carcinogen.
NIOSH REL: (Benzidine-based Dye) Reduce to lowest feasible level.

SAFETY PROFILE: Confirmed carcinogen with experimental carcinogenic and tumorigenic data. Mildly toxic by ingestion. Human mutation data reported. When heated to decomposition it emits very toxic fumes of Cl$^-$ and NO$_x$.

DEQ800   CAS: 612-83-9   **HR: 3**
3,3′-DICHLOROBENZIDINE DIHYDROCHLORIDE
mf: $C_{12}H_{10}Cl_2N_2 \cdot 2ClH$ mw: 326.06

SYN: 3,3′-DICHLORO-(1,1′-BIPHENYL)-4,4′-DIAMINE DIHYDROCHLORIDE

CONSENSUS REPORTS: Reported in EPA TSCA Inventory.

OSHA PEL: Carcinogen

SAFETY PROFILE: Confirmed carcinogen. Moderately toxic by ingestion. Mutation data reported. When heated to decomposition it emits very toxic fumes of Cl$^-$ and NO$_x$.

DER000   CAS: 510-15-6   **HR: 3**
4,4′-DICHLOROBENZILIC ACID ETHYL ESTER
mf: $C_{16}H_{14}Cl_2O_3$ mw: 325.20

PROP: Viscous liquid, sometimes yellow, sltly sol in water. Bp: 156-158°, vap press: 2.2 × 10^{-6} mm @ 20°.

SYNS: ACAR * ACARABEN 4E * AKAR * BENZILAN * BENZ-o-CHLOR * CHLORBENZILATE * CHLOROBENZYLATE * COMPOUND 338 * 4,4′-DICHLORBENZILSAEUREAETHYLESTER (GERMAN) * 4,4′-DICHLOROBENZILATE * ENT 18,596 * ETHYL 4-CHLORO-α-(4-CHLOROPHENYL)-α-HYDROXYBENZENEACETATE * ETHYL-p,p′-DICHLOROBENZILATE * ETHYL-4,4′-DICHLOROBENZILATE * ETHYL-4,4′-DICHLORODIPHENYL GLYCOLLATE * ETHYL-4,4′-DICHLOROPHENYL GLYCOLLATE * ETHYL ESTER of 4,4′-DICHLOROBENZILIC ACID * ETHYL-2-HYDROXY-2,2-BIS(4-CHLOROPHENYL)ACETATE * FOLBEX * FOLBEX SMOKE-STRIPS * G 338 * G 23992 * GEIGY 338 * KOP MITE * NCI-C00408 * NCI-C60413 * RCRA WASTE NUMBER U038

CONSENSUS REPORTS: IARC Cancer Review: GROUP 3 IMEMDT 7,56,87; Animal Limited Evidence IMEMDT 30,73,83; Animal Sufficient Evidence IMEMDT 5,75,74. NCI Carcinogenesis Bioassay Completed; Results Positive: mouse NCITR* NCI-CG-TR-75,78. NCI Carcinogenesis Bioassay Completed; Results Indefinite: rat NCITR* NCI-CG-TR-75,78. Community Right-To-Know List. Reported in EPA TSCA Inventory.

SAFETY PROFILE: Suspected carcinogen with experimental carcinogenic, neoplastigenic, and tumorigenic data. Moderately toxic by ingestion and possibly other routes. A skin and eye irritant. A pesticide. When heated to decomposition it emits toxic fumes of Cl$^-$.

DER600   CAS: 51-44-5   **HR: 3**
3,4-DICHLOROBENZOIC ACID
mf: $C_7H_4Cl_2O_2$ mw: 191.01

SYNS: SYNSTIGMINE * SYNTOSTIGMIN * VAGOSTIGMIN

CONSENSUS REPORTS: Reported in EPA TSCA Inventory.

SAFETY PROFILE: Poison by subcutaneous route. When heated to decomposition it emits toxic fumes of Cl$^-$.

DEU200   CAS: 38780-42-6   **HR: 3**
cis-DICHLOROBIS(PYRROLIDINE) PLATINUM(II)
mf: $C_8H_{18}Cl_2N_2Pt$ mw: 408.27

SYN: cis-DIPYRROLIDINEDICHLOROPLATINUM(II)

SAFETY PROFILE: Poison by intraperitoneal route. Questionable carcinogen with experimental tumorigenic data. Mutation data reported. When heated to decomposition it emits very toxic fumes of Cl^- and NO_x.

DEV000   CAS: 764-41-0   **HR: 3**
1,4-DICHLORO-2-BUTENE
mf: $C_4H_6Cl_2$ mw: 125.00

PROP: Colorless liquid. Mp: 1°-3°; bp: 156°; d: 1.183 @ 25°/4°.

SYNS: DCB * 1,4-DCB * 1,4-DICHLOROBUTENE-2 (MAK) * RCRA WASTE NUMBER U074

CONSENSUS REPORTS: Reported in EPA TSCA Inventory. EPA Genetic Toxicology Program.

DFG MAK: Animal Carcinogen, Suspected Human Carcinogen.

SAFETY PROFILE: Confirmed carcinogen with experimental carcinogenic and neoplastigenic data. Poison by ingestion, inhalation, and intravenous routes. Moderately toxic by skin contact. An experimental teratogen. Mutation data reported. A severe skin and eye irritant. When heated to decomposition it emits toxic fumes of Cl^-.

DEV400   CAS: 821-10-3   **HR: 3**
1,4-DICHLORO-2-BUTYNE
mf: $C_4H_4Cl_2$ mw: 122.98

SYN: 1,4-DICHLOROBUTYNE

CONSENSUS REPORTS: Reported in EPA TSCA Inventory.

SAFETY PROFILE: Poison by intravenous route. When heated to decomposition it emits toxic fumes of Cl^-. Probably a dangerous fire and explosion hazard.

DEW000   CAS: 333-25-5   **HR: 3**
DICHLORO(2-CHLOROVINYL)ARSINE OXIDE

SYN: LEWISITE I OXIDE

CONSENSUS REPORTS: Arsenic and its compounds are on the Community Right-To-Know List.

OSHA PEL: TWA 0.5 mg(As)/m^3

SAFETY PROFILE: Poison by ingestion, intravenous, and subcutaneous routes. When heated to decomposition it emits very toxic fumes of Cl^- and As.

DEW400   CAS: 20373-56-2   **HR: 3**
2,6-DICHLORO-N-CYCLOPROPYL-N-ETHYL ISONICOTINAMIDE
mf: $C_{11}H_{12}Cl_2N_2O$ mw: 259.15

SYN: ABBOTT-28440

SAFETY PROFILE: Poison by ingestion and intraperitoneal routes. When heated to decomposition it emits very toxic fumes of Cl^- and NO_x.

DEX000   CAS: 14913-33-8   **HR: 3**
trans-DICHLORODIAMMINE-PLATINUM(II)
mf: $C_{12}H_6N_2Pt$ mw: 300.07

SYNS: trans-DIAMMINEDICHLOROPLATINUM(II) * trans-PLATINUM(II)DIAMMINEDICHLORIDE

CONSENSUS REPORTS: EPA Genetic Toxicology Program.

SAFETY PROFILE: Questionable carcinogen with experimental tumorigenic data. Poison by intraperitoneal route. Human mutation data reported. When heated to decomposition it emits toxic fumes of NO_x and Cl^-.

DEY800   CAS: 1719-53-5   **HR: 3**
DICHLORODIETHYLSILANE

DOT: UN 1767
mf: $C_4H_{10}Cl_2Si$ mw: 157.13

PROP: Liquid. Mp: −96°; bp: 131.0°, d: 1.05, vap d: 5.41; flash p: 75.2°F.

SYN: DIETHYLDICHLOROSILANE (DOT)

CONSENSUS REPORTS: Reported in EPA TSCA Inventory.

DOT Classification: Flammable Liquid; Label: Flammable Liquid; IMO: Corrosive Material; Label: Corrosive, Flammable Liquid.

SAFETY PROFILE: Poison by intraperitoneal route. Moderately toxic by ingestion. Corrosive to tissue. Dangerous fire hazard when exposed to heat, flame, or oxidizers. Can react vigorously with oxidizing materials. To fight fire, use foam, CO_2, dry chemical. When heated to decomposition or in reaction with water or steam it emits toxic and corrosive fumes of Cl^-.

DEZ000   CAS: 866-55-7   **HR: 3**
DICHLORODIETHYLSTANNANE
mf: $C_4H_{10}Cl_2Sn$ mw: 247.73

PROP: Water-white crystals. Mp: 85°, bp: 220°.

SYNS: DIAETHYLZINNDICHLORID (GERMAN) * DICHLORODIETHYLTIN * DIETHYLDICHLOROS-TANNANE * DIETHYLSTANNYL DICHLORIDE * DIETHYLTIN CHLORIDE * DIETHYLTIN DICHLORIDE

OSHA PEL: TWA 0.1 mg(Sn)/m³ (skin)
ACGIH TLV: TWA 0.1 mg(Sn)/m³ (skin) (Proposed: TWA 0.1 mg(Sn)/m³; STEL 0.2 mg(Sn)/m³ (skin))
NIOSH REL: (Organotin Compounds) TWA 0.1 mg(Sn)/m³

SAFETY PROFILE: Poison by ingestion and intravenous routes. When heated to decomposition it emits toxic fumes of Cl^-.

DFA200   CAS: 27156-03-2   **HR: 2**
DICHLORODIFLUOROETHYLENE
DOT: NA 9018
mf: $C_2Cl_2F_2$ mw: 132.92

PROP: Liquid. Vap d: 4.6.

DOT Classification: ORM-A; Label: None.

SAFETY PROFILE: Moderately toxic by inhalation. A skin, eye, and mucous membrane irritant. Will react with water or steam to produce toxic and corrosive fumes. When heated to decomposition it emits toxic fumes of F^- and Cl^-.

DFA400   CAS: 76-38-0   **HR: 3**
2,2-DICHLORO-1,1-DIFLUOROETHYL METHYL ETHER
mf: $C_3H_4Cl_2F_2O$ mw: 164.97

SYNS: ANALGIZER * ANECOTAN * 2,2-DI-CHLORO-1,1-DIFLUORO-1-METHOXYETHANE * INGA-LAN * INGALAN (RUSSIAN) * INHALAN * METHOFLURANE * METHOXANE * METHOXY-FLUORAN * METHOXYFLUORANE * METHOXY-FLURANE * METOFANE * METOXFLURAN * METOXIFLURAN * MOF * NSC-110432 * PENTHRANE * PENTRAN * PENTRANE

CONSENSUS REPORTS: EPA Genetic Toxicology Program.

NIOSH REL: (Waste Anesthetic Gases and Vapors) CL 2 ppm/1H

SAFETY PROFILE: A human poison by ingestion. Mildly toxic by inhalation. Human systemic effects by inhalation: depressed renal function. An experimental teratogen. Human mutation data reported. An eye irritant. When heated to decomposition it emits very toxic fumes of Cl^- and F^-.

DFA600   CAS: 75-71-8   **HR: 1**
DICHLORODIFLUOROMETHANE
DOT: UN 1028
mf: CCl_2F_2 mw: 120.91

PROP: Colorless, almost odorless gas. Mp: −158°, bp: −29°, vap press: 5 atm @ 16.1°.

SYNS: ALGOFRENE TYPE 2 * ARCTON 6 * DIFLUORODICHLOROMETHANE * DWUCHLORO-DWUFLUOROMETAN (POLISH) * ELECTRO-CF 12 * ESKIMON 12 * F 12 * FC 12 * FLUOROCAR-BON-12 * FREON F-12 * FRIGEN 12 * GENE-TRON 12 * HALON * ISCEON 122 * ISOTRON 12 * KAISER CHEMICALS 12 * LEDON 12 * PROPEL-LANT 12 * RCRA WASTE NUMBER U075 * R 12 (DOT) * REFRIGERANT 12 * UCON 12 * UCON 12/HALOCARBON 12

CONSENSUS REPORTS: Reported in EPA TSCA Inventory. EPA Genetic Toxicology Program.

OSHA PEL: TWA 1000 ppm
ACGIH TLV: TWA 1000 ppm
DFG MAK: 1000 ppm (5000 mg/m³)
DOT Classification: Nonflammable Gas; Label: Nonflammable gas.

SAFETY PROFILE: Human systemic effects by inhalation: conjunctiva irritation, fibrosing alveolitis, and liver changes. Narcotic in high concentrations. Nonflammable Gas. Can react violently with Al. When heated to decomposition it emits highly toxic fumes of phosgene, Cl^-, and F^-.

DFB000   CAS: 70281-30-0   **HR: 1**
DICHLORODIFLUOROMETHANE mixed with CHLORODIFLUOROMETHANE
DOT: UN 1078/NA 1954

SYN: DICHLORODIFLUOROMETHANE-CHLORODIFLUO-
ROMETHANE MIXTURE (DOT)

DOT Classification: Nonflammable Gas; Label: Nonflammable Gas (UN 1078); Flammable Gas; Label; Flammable Gas (NA 1954).

SAFETY PROFILE: A simple asphyxiant.
When heated to decomposition it emits very
toxic fumes of Cl⁻ and F⁻.

DFB400   CAS: 56275-41-3   HR: 1
DICHLORODIFLUOROMETHANE with 1,1-DIFLUOROETHANE

DOT: UN 1078/NA 1954
mf: $C_2H_4F_2 \cdot CCl_2F_2$ mw: 186.97

SYNS: DICHLORODIFLUOROMETHANE and DIFLUO-
ROETHANE mixture (constant boiling mixture) (DOT)
* FREON 500 * UCON 500/HALOCARBON 500

DOT Classification: Nonflammable Gas; Label: Nonflammable Gas (UN 1078); Flammable Gas; Label: Flammable Gas (NA 1954).

SAFETY PROFILE: A simple asphyxiant.
When heated to decomposition it emits very
toxic fumes of Cl⁻ and F⁻.

DFC000   HR: 1
DICHLORODIFLUOROMETHANE with TRICHLOROTRIFLUOROETHANE
mf: $CCl_3F \cdot CCl_2F_2$ mw: 258.27

SYN: DICHLORODIFLUOROMETHANE-TRICHLOROTRI-
FLUOROETHANE MIXTURE (DOT)

DOT Classification: Nonflammable Gas; Label: Nonflammable Gas; Flammable Gas; Label: Flammable Gas.

SAFETY PROFILE: Very mildly toxic by inhalation. When heated to decomposition it emits
very toxic fumes of Cl⁻ and F⁻.

DFC200   CAS: 2767-41-1   HR: 3
DICHLORODIHEXYLSTANNANE
mf: $C_{12}H_{26}Cl_2Sn$ mw: 359.97

SYN: DIHEXYLTIN DICHLORIDE

OSHA PEL: TWA 0.1 mg(Sn)/m³ (skin)
ACGIH TLV: TWA 0.1 mg(Sn)/m³ (skin) (Proposed: TWA 0.1 mg(Sn)/m³; STEL 0.2
mg(Sn)/m³ (skin))
NIOSH REL: (Organotin Compounds) TWA
0.1 mg(Sn)/m³

SAFETY PROFILE: Poison by ingestion and
intravenous routes. When heated to decomposition it emits toxic fumes of Cl⁻.

DFC800   CAS: 33770-60-4   HR: 3
(2,5-DICHLORO-3,6-DIHYDROXY-p-BENZOQUINOLATO)MERCURY
mf: $C_6Cl_2HgO_4$ mw: 407.55

SYNS: 2,5-DICHLORO-3,6-DIHYDROXY-p-BENZOQUI-
NONE, MERCURY SALT * (2,5-DICHLORO-3,6-DIHY-
DROXY-p-BENZOQUINONE), MERCURY SALT

CONSENSUS REPORTS: Mercury and its
compounds are on the Community Right-To-
Know List.

OSHA PEL: (Transitional: CL 1 mg/10m³) CL
0.1 mg(Hg)/m³ (skin)
ACGIH TLV: TWA 0.1 mg(Hg)/m³ (skin)
NIOSH REL: (Inorganic Mercury) TWA 0.05
mg(Hg)/m³

SAFETY PROFILE: Poison by intravenous
route. When heated to decomposition it emits
very toxic fumes of Cl⁻ and Hg.

DFD000   CAS: 10331-57-4   HR: 3
5,5'-DICHLORO-2,2'-DIHYDROXY-3,3'-DINITROBIPHENYL
mf: $C_{12}H_6Cl_2N_2O_6$ mw: 345.10

SYNS: BAY 9015 * BAYER 9015 * BILEVON M
* 4,4'-DICHLORO-6,6'-DINITRO-O,O'-BIPHENOL
* 3,3'-DICHLORO-5,5'-DINITRO-O,O'-BIPHENOL
(FRENCH) * 5,5'-DICHLORO-3,3'-DINITRO(1,1'-BIPHE-
NYL)-2,2'-DIOL * ME 3625 * MENICHLOPHOLAN
* NICLOFOLAN

SAFETY PROFILE: A poison by ingestion. An
experimental teratogen. When heated to decomposition it emits very toxic fumes of Cl⁻ and
NO_x.

DFE200   CAS: 118-52-5   HR: 2
1,3-DICHLORO-5,5-DIMETHYL HYDANTOIN
mf: $C_5H_6Cl_2N_2O_2$ mw: 197.03

PROP: Crystals, liberates chlorine on contact
with hot water. Mp: 132°; subl @ 100°, confla-
grates @ 212°; d: 1.5 @ 20°, vap d: 6.8.

SYNS: DACTIN * DAKTIN * DANTOIN * DCA
* DICHLORANTIN * DICHLORODIMETHYLHYDAN-
TOIN * 1,3-DICHLORO-5,5-DIMETHYL-2,4-IMIDAZOLI-

DINEDIONE * 1,3-DICHLORO-5,5'-METHYLHYDAN-
TOIN * HALANE * HYDAN * HYDAN (antiseptic)
* NCI-C03054 * OMCHLOR

CONSENSUS REPORTS: Reported in EPA
TSCA Inventory.

OSHA PEL: TWA 0.2 mg/m^3; STEL 0.4 mg/
m^3
ACGIH TLV: TWA 0.2 mg/m^3; STEL 0.4 mg/
m^3

SAFETY PROFILE: Moderately toxic by ingestion and possibly other routes. Mildly toxic by
inhalation. A severe skin irritant. Mutation data
reported. Avoid excessive contact because of
effects of active chlorine on skin. Some of the
hydantoins are central nervous system depressants. Mixtures with xylene may explode. Will
react with water or steam to produce toxic and
corrosive fumes. When heated to decomposition
it emits toxic fumes of Cl$^-$ and NO$_x$.

DFE259   CAS: 75-78-5   **HR: 3**
DICHLORODIMETHYLSILANE

DOT: UN 1162
mf: C$_2$H$_6$Cl$_2$Si   mw: 129.06

DOT Classification: Flammable Liquid; Label:
Flammable Liquid

SAFETY PROFILE: Probably a skin, eye, and
mucous membrane irritant. Violent reaction on
contact with water. When heated to decomposition it emits toxic fumes of Cl$^-$.

DFE600   CAS: 3883-43-0   **HR: 3**
trans-2,3-DICHLORO-1,4-DIOXANE
mf: C$_4$H$_6$Cl$_2$O$_2$ mw: 157.00

SYN: trans-2,3-DICHLORO-p-DIOXANE

SAFETY PROFILE: Questionable carcinogen
with experimental neoplastigenic and tumorigenic data. Moderately toxic by ingestion and
skin contact. When heated to decomposition it
emits toxic fumes of Cl$^-$.

DFE800   CAS: 28675-08-3   **HR: 2**
DICHLORO DIPHENYL OXIDE
mf: C$_{12}$H$_8$Cl$_2$O   mw: 239.10

PROP: Liquid. Vap d: 8.2.

SYNS: DICHLOROPHENYL ETHER * PHENYL
ETHER DICHLORO

OSHA PEL: TWA 0.5 mg/m^3

SAFETY PROFILE: Moderately toxic by ingestion. When heated to decomposition it emits
toxic fumes of Cl$^-$.

DFF000   CAS: 80-10-4   **HR: 3**
DICHLORODIPHENYLSILANE

DOT: UN 1769
mf: C$_{12}$H$_{10}$Cl$_2$Si   mw: 253.21

PROP: Colorless liquid. Mp: $-22°$, bp: 303°,
d: 1.19 @ 20°, vap d: 8.45.

SYN: DIPHENYL DICHLOROSILANE (DOT)

CONSENSUS REPORTS: Reported in EPA
TSCA Inventory.

DOT Classification: Corrosive Material; Label:
Corrosive.

SAFETY PROFILE: A poison irritant to skin,
eyes, and mucous membranes. Can react vigorously with oxidizing materials. When heated
to decomposition or on contact with acid or
acid fumes it emits it emits toxic fumes of Cl$^-$.

DFF400   CAS: 867-36-7   **HR: 3**
DICHLORODIPROPYLSTANNANE
mf: C$_6$H$_{14}$Cl$_2$Sn   mw: 275.79

PROP: Colorless crystals. Sol in organic solvents. Mp: 81°.

SYNS: DICHLORODIPROPYLTIN * DIPROPYLTIN
CHLORIDE * DIPROPYLTIN DICHLORIDE * DI-n-
PROPYLTIN DICHLORIDE

OSHA PEL: TWA 0.1 mg(Sn)/m^3 (skin)
ACGIH TLV: TWA 0.1 mg(Sn)/m^3 (skin) (Proposed: TWA 0.1 mg(Sn)/m^3; STEL 0.2
mg(Sn)/m^3 (skin))
NIOSH REL: (Organotin Compounds) TWA
0.1 mg(Sn)/m^3

SAFETY PROFILE: Poison by ingestion. When
heated to decomposition it emits toxic fumes
of Cl$^-$.

DFF809   CAS: 75-34-3   **HR: 3**
1,1-DICHLOROETHANE

DOT: UN 2362
mf: C$_2$H$_4$Cl$_2$ mw: 98.96

PROP: Colorless liquid; aromatic, ethereal
odor; hot, saccharine taste. Mp: $-97.7°$, lel:
5.6%, bp: 57.3°, flash p: 22°F (TOC), d: 1.174
@ 20°/4°, vap press: 230 mm @ 25°, vap d:
3.44, autoign temp: 856°F.

SYNS: AETHYLIDENCHLORID (GERMAN) * CHLO-
RINATED HYDROCHLORIC ETHER * CHLORURE
d'ETHYLIDENE (FRENCH) * CLORURO di ETILIDENE
(ITALIAN) * 1,1-DICHLOORETHAAN (DUTCH)
* 1,1-DICHLORAETHAN (GERMAN) * 1,1-DICLORO-
ETANO (ITALIAN) * ETHYLIDENE CHLORIDE
* ETHYLIDENE DICHLORIDE * NCI-C04535
* RCRA WASTE NUMBER U076

CONSENSUS REPORTS: NCI Carcinogenesis Bioassay (gavage); Inadequate Studies:
mouse, rat NCITR* NCI-CG-TR-66,78. Reported in EPA TSCA Inventory.

OSHA PEL: TWA 100 ppm
ACGIH TLV: TWA 200 ppm; STEL 250 ppm
DFG MAK: 100 ppm (400 mg/m^3)
DOT Classification: Flammable Liquid; Label:
   Flammable Liquid.

SAFETY PROFILE: Questionable carcinogen
with experimental tumorigenic data. Moderately
toxic by ingestion. Experimental teratogenic and
reproductive effects. Liver damage reported in
experimental animals. A very dangerous fire
hazard and moderate explosion hazard when
exposed to heat or flame; can react vigorously
with oxidizing materials. To fight fire, use alcohol foam, water, foam, CO_2, dry chemical.
When heated to decomposition it emits highly
toxic fumes of phosgene and Cl^-.

DFH000   CAS: 10072-25-0   HR: 3
9-(2-(DI(2-CHLOROETHYL)AMINO)
ETHYLAMINO)-6-CHLORO-2-METH-
OXYACRIDINE
mf: $C_{20}H_{22}Cl_3N_3O \cdot 2ClH \cdot H_2O$ mw:
517.74

SYNS: 9-(2-(BIS(2-CHLOROETHYL)AMINO)ETHYL-
AMINO)-6-CHLORO-2-METHOXYACRIDINE DIHYDRO-
CHLORIDE * ICR-48b * NSC-34372 * QUINAC-
RINE ETHYL MUSTARD

SAFETY PROFILE: Questionable carcinogen
with experimental carcinogenic data. When
heated to decomposition it emits very toxic
fumes of Cl^- and NO_x.

DFH200   CAS: 598-14-1   HR: 3
DICHLOROETHYLARSINE

DOT: UN 1892
mf: $C_2H_5AsCl_2$ mw: 174.89

PROP: Colorless liquid; fruity, biting, irritating
odor. Mp: −65°, bp: 156° decomp, d: 1.742

@ 14°, vap press: 2.29 mm @ 21.5°, vap d:
6.03.

SYNS: ARSENIC DICHLOROETHANE * DICK (GER-
MAN) * ED * ETHYLARSONOUS DICHLORIDE
* ETHYLIDICHLORARSINE * ETHYLIDICHLOROAR-
SINE (DOT) * TL 214

CONSENSUS REPORTS: Arsenic and its
compounds are on the Community Right-To-
Know List.

OSHA PEL: TWA 0.5 mg(As)/m^3
DOT Classification: Poison B; Label: Poison.

SAFETY PROFILE: A human poison by inhalation. Experimentally, a deadly poison by inhalation and subcutaneous routes, and probably by
ingestion. A severe irritant. A military poison
gas. Can react with oxidizing materials. Will
react with water or steam to produce toxic and
corrosive fumes. Dangerous; on contact with
acid or acid fumes it emits highly toxic fumes
of Cl^-, As and phosgene.

DFH600   CAS: 321-55-1   HR: 2
O,O-DI(2-CHLOROETHYL)-O-(3-
CHLORO-4-METHYLCOUMARIN-7-YL)
PHOSPHATE
mf: $C_{14}H_{14}Cl_3O_6P$ mw: 415.60

SYNS: O,O-BIS(2-CHLOROETHYL)-O-(3-CHLORO-4-
METHYL-7-COUMARINYL) PHOSPHATE * 2-CHLORO-
ETHANOL HYDROGEN PHOSPHATE ESTER with 3-
CHLORO-7-HYDROXY-4-METHYLCOUMARIN
* 2-CHLOROETHANOL PHOSPHATE DIESTER ESTER
with 3-CHLORO-7-HYDROXY-4-METHYLCOUMARIN
* 3-CHLORO-7-HYDROXY-4-METHYLCOUMARIN BIS(2-
CHLOROETHYL)PHOSPHATE * 3-CHLORO-4-METHYL-
UMBELLIFERONE BIS(2-CHLOROETHYL)PHOSPHATE
* DI-(2-CHLOROETHYL)-3-CHLORO-4-METHYL-7-COU-
MARINYL PHOSPHATE * DI-(2-CHLOROETHYL)-3-
CHLORO-4-METHYLCOUMARIN-7-YL PHOSPHATE
* EUSTIDIL * GALLOXON * GALOXANE
* 96H60 * HALOXON * HELMIRANE * HELMI-
RON * HELMIRONE * LOXON * LUXON
* LXON

SAFETY PROFILE: Moderately toxic by ingestion and intraperitoneal routes. Human mutation
data reported. When heated to decomposition
it emits very toxic fumes of PO_x and Cl^-.

DFH800   CAS: 25323-30-2   HR: 2
DICHLOROETHYLENE

DOT: UN 1150
mf: $C_2H_2Cl_2$ mw: 96.94

DOT Classification: Flammable Liquid; Label: Flammable Liquid.

SAFETY PROFILE: Moderately toxic by ingestion. Mildly toxic by inhalation. Flammable when exposed to heat or flame. When heated to decomposition it emits toxic fumes of Cl⁻.

DFI100   CAS: 540-59-0   **HR: 3**
1,2-DICHLOROETHYLENE
mf: $C_2H_2Cl_2$ mw: 96.94

SYNS: ACETYLENE DICHLORIDE * 1,2-DICHLORO-ETHYLENE * DIOFORM * 1,2-DICHLOR-AETHEN (GERMAN) * DICHLORO-1,2-ETHYLENE (FRENCH) * sym-DICHLOROETHYLENE * NCI-C56031

CONSENSUS REPORTS: Reported in EPA TSCA Inventory. Community Right-To-Know List.

OSHA PEL: TWA 200 ppm
ACGIH TLV: TWA 200 ppm
DFG MAK: 200 ppm (790 mg/m³)

SAFETY PROFILE: Poison by inhalation. Moderately toxic by ingestion and other routes. When heated to decomposition it emits highly toxic fumes of Cl⁻.

DFI200   CAS: 156-59-2   **HR: 1**
cis-DICHLOROETHYLENE
mf: $C_2H_2Cl_2$ mw: 96.94

PROP: Colorless liquid, pleasant odor. Mp: −80.5°, bp: 59°, lel: 9.7%, uel: 12.8%, flash p: 39°F, d: 1.2743 @ 25°/4°, vap press: 400 mm @ 41.0°, vap d: 3.34.

SYN: 1,2-DICHLOROETHYLENE

CONSENSUS REPORTS: Reported in EPA TSCA Inventory.

DFG MAK: 200 ppm (790 mg/m³)

SAFETY PROFILE: Mildly toxic by by ingestion and inhalation. In high concentration it is irritating and narcotic. Has produced liver and kidney injury in experimental animals. Mutation data reported. Sometimes thought to be non-flammable, however, it is a dangerous fire hazard when exposed to heat or flame. Reaction with solid caustic alkalies or their concentrated solutions produces chloracetylene gas which ignites spontaneously in air. Reacts violently with N_2O_4; KOH; Na; NaOH. Moderate explosion hazard in the form of vapor when exposed to flame. Can react vigorously with oxidizing materials. To fight fire, use water spray, foam, CO_2, dry chemical. When heated to decomposition it emits toxic fumes of Cl⁻.

DFJ000   CAS: 14096-51-6   **HR: 3**
DICHLORO(ETHYLENEDIAMMINE) PLATINUM(II)
mf: $C_2H_8Cl_2N_2Pt$ mw: 326.11

SYNS: ETHYLENEDIAMINEDICHLORIDE PLATINUM (II) * PLATINUM ETHYLENEDIAMMINE DICHLORIDE

SAFETY PROFILE: Poison by intraperitoneal route. Human mutation data reported. When heated to decomposition it emits very toxic fumes of Cl⁻ and NO_x.

DFJ050   CAS: 111-44-4   **HR: 3**
DICHLOROETHYL ETHER
DOT: UN 1916
mf: $C_4H_8Cl_2O$ mw: 143.02

PROP: Colorless, stable liquid. Bp: 178.5°, fp: −51.9°, flash p: 131°F (CC), d: 1.2220 @ 20°/20°, autoign temp: 696°F, vap press: 0.7 mm @ 20°, vap d: 4.93.

SYNS: BIS(2-CHLOROETHYL) ETHER * BIS(β-CHLOROETHYL) ETHER * CHLOREX * 1-CHLORO-2-(β-CHLOROETHOXY)ETHANE * CHLOROETHYL ETHER * CLOREX * DCEE * 2,2′-DICHLOORETHYL-ETHER (DUTCH) * 2,2′-DICHLOR-DIAETHYLAETHER (GERMAN) * 2,2′-DICHLORETHYL ETHER * β,β-DICHLORODIETHYL ETHER * DICHLOROETHER * DI(β-CHLOROETHYL)ETHER * β,β′-DICHLORO-ETHYL ETHER * sym-DICHLOROETHYL ETHER * 2,2′-DICHLOROETHYL ETHER (MAK) * DICHLORO-ETHYL OXIDE * 2,2′-DICLOROETILETERE (ITALIAN) * DWUCHLORODWUETYLOWY ETER (POLISH) * ENT 4,504 * ETHER DICHLORE (FRENCH) * 1,1′-OXYBIS(2-CHLORO)ETHANE * OXYDE de CHLORETHYLE (FRENCH) * RCRA WASTE NUMBER U025

CONSENSUS REPORTS: IARC Cancer Review: GROUP 3 IMEMDT 7,56,87; Animal Sufficient Evidence IMEMDT 9,117,75. Reported in EPA TSCA Inventory. On Community Right-To-Know List. On EPA Extremely Hazardous Substances List.

OSHA PEL: (Transitional: CL 15 ppm (skin)) TWA 5 ppm; STEL 10 ppm (skin)
ACGIH TLV: TWA 5 ppm; STEL 10 ppm (skin)
DFG MAK: 10 ppm (60 mg/m³)

DOT Classification: IMO: Poison B; Label: Poison.

SAFETY PROFILE: A poison by ingestion, skin contact, and inhalation. A skin, eye, and mucous membrane irritant. Questionable carcinogen with experimental carcinogenic and tumorigenic data. Mutation data reported. Exposure to 1000 ppm for 30 to 60 minutes may result in death within days. The odor is easily detectable at 35 ppm which causes only slight irritation. Flammable when exposed to heat, flame, or oxidants. Dangerous explosion hazard; reacts vigorously with oleum, chlorosulfonic acid. Reacts with water or steam to evolve toxic and corrosive fumes. Can react vigorously with oxidizing materials. To fight fire, use water, foam, mist, fog, spray, dry chemical. When heated to decomposition it emits toxic fumes of Cl^-.

DFJ200   CAS: 63917-06-6   HR: 3
DI-2-CHLOROETHYL MALEATE
mf: $C_8H_{10}Cl_2O_4$ mw: 241.08

SYN: DI(2-CHLOROETHYL) ESTER, MALEIC ACID

SAFETY PROFILE: Poison by ingestion and skin contact. When heated to decomposition it emits toxic fumes of Cl^-.

DFJ400   CAS: 20198-77-0   HR: 3
2,3-DICHLORO-N-ETHYLMALEINIMIDE
mf: $C_6H_5Cl_2NO_2$ mw: 194.02

SYN: N-ETHYL-DICHLOROMALEINIMIDE

SAFETY PROFILE: Poison by intraperitoneal and intravenous routes. Experimental teratogenic and reproductive effects. When heated to decomposition it emits very toxic fumes of Cl^- and NO_x.

DFJ800   CAS: 1125-27-5   HR: 3
DICHLOROETHYLPHENYLSILANE
DOT: UN 2435
mf: $C_8H_{10}Cl_2Si$ mw: 205.17
PROP: Liquid.

SYN: ETHYL PHENYL DICHLOROSILANE (DOT)

CONSENSUS REPORTS: Reported in EPA TSCA Inventory.

DOT Classification: Corrosive Material; Label: Corrosive.

SAFETY PROFILE: Poison by ingestion and inhalation. A poison irritant to skin, eyes, and mucous membranes. Corrosive. Will react with water or steam to produce toxic and corrosive fumes. Can react with oxidizing materials. When heated to decomposition it emits toxic fumes of Cl^- and phenol.

DFK000   CAS: 1789-58-8   HR: 3
DICHLOROETHYLSILANE
DOT: UN 1183
mf: $C_2H_6Cl_2Si$ mw: 129.07

PROP: Liquid. Vap d: 4.45, flash p: < 73.4°F.

SYN: ETHYL DICHLOROSILANE (DOT)

CONSENSUS REPORTS: Reported in EPA TSCA Inventory.

DOT Classification: Flammable Liquid; Label: Flammable Liquid; IMO: Flammable Liquid; Label: Flammable Liquid, Corrosive.

SAFETY PROFILE: Poison by ingestion and inhalation. A severe irritant to skin, eyes, and mucous membranes. Corrosive. Dangerous fire hazard if exposed to heat, open flames or powerful oxidizers. Will react with water or steam to produce heat and toxic and corrosive fumes. To fight fire, use foam, dry chemical, mist, spray. When heated to decomposition it emits toxic fumes of Cl^- and phosgene.

DFK600   CAS: 97-17-6   HR: 3
DICHLOROFENTHION
mf: $C_{10}H_{13}Cl_2O_3PS$ mw: 315.16

PROP: A nonvolatile, residual organic phosphate nematocide and insecticide. Bp: 166° @ 0.1 mm, d: 1.3. Insol in water; sol in most organic solvents.

SYNS: BROMEX * O,O-DIAETHYL-O-2,4-DICHLOR-PHENYL-MONOTHIOPHOSPHAT (GERMAN) * O,O-DIA-ETHYL-O-2,4-DICHLORPHENYL-THIONOPHOSPHAT (GER-MAN) * DICHLOFENTHION * DICHLOFENTION * 2,4-DICHLORO-PHENOL-O-ESTER with O,O-DIETHYL PHOSPHOROTHIOATE * O-2,4-DICHLOROPHENYL-O,O-DIETHYL PHOSPHOROTHIOATE * 2,4-DICHLORO-PHE-NYL DIETHYL PHOSPHOROTHIONATE * O,O-DI-ETHYL-O-(2,4-DICHLOOR-FENYL)-MONOTHIOFOSFAAT (DUTCH) * O,O-DIETHYL-O-(2,4-DICHLOROPHENYL) PHOSPHOROTHIOATE * DIETHYL 2,4-DICHLOROPHE-NYL PHOSPHOROTHIONATE * O,O-DIETHYL-O-2,4-DI-CHLOROPHENYL THIOPHOSPHATE * O,O-DIETIL-O-(2,4-DICLORO-FENIL)-MONOTIOFOSFATO (ITALIAN) * ECP * ENT 17,470 * HEXA-NEMA * MOBI-LAWN * NEMACIDE * THIOPHOSPHATE de O-2,4-

DICHLOROPHENYLE et de O,O-DIETHYLE (FRENCH) * TRI-VC 13 * VC13 NEMACIDE

SAFETY PROFILE: Poison by ingestion. Mildly toxic by skin contact. A very toxic insecticide. When heated to decomposition it emits very toxic fumes of PO_x, SO_x and Cl^-.

DFL000　CAS: 75-43-4　HR: 1
DICHLOROFLUOROMETHANE

DOT: UN 1029
mf: $CHCl_2F$　mw: 102.92

PROP: Heavy, colorless gas. Mp: $-135°$, bp: 8.9°, d: 1.48, vap press: 2 atm @ 28.4°, vap d: 3.82.

SYNS: ALGOFRENE TYPE 5 * ARCTON 7 * DICHLOROMONOFLUOROMETHANE (OSHA, DOT) * DWUCHLOROFLUOROMETAN (POLISH) * FLUORODICHLOROMETHANE * FREON 21 * GENETRON 21

CONSENSUS REPORTS: Reported in EPA TSCA Inventory.

OSHA PEL: (Transitional: TWA 1000 ppm) TWA 10 ppm
ACGIH TLV: TWA 10 ppm
DFG MAK: 10 ppm (45 mg/m^3)
DOT Classification: Nonflammable Gas; Label: Nonflammable Gas.

SAFETY PROFILE: Mildly toxic by inhalation. Experimental reproductive effects. When heated to decomposition it emits very toxic fumes of Cl^- and F^-.

DFM000　CAS: 303-04-8　HR: 3
2,3-DICHLOROHEXAFLUOROBUTENE-2
mf: $C_4Cl_2F_6$　mw: 232.94

SYNS: DCHFB * 2,3-DICHLOROHEXAFLUORO-2-BUTENE * 2,3-DICHLORO-1,1,1,4,4,4-HEXAFLUOROBUTENE-2

SAFETY PROFILE: Poison by inhalation. Moderately toxic by ingestion. When heated to decomposition it emits very toxic fumes of Cl^- and F^-.

DFM099　HR: 3
4,5-DICHLORO-3,3,4,5,6,6-HEXAFLUORO-1,2-DIOXANE

SAFETY PROFILE: Explodes violently when heated. When heated to decomposition it emits toxic fumes of F^- and Cl^-.

DFM800　CAS: 101652-07-7　HR: 3
6,7-DICHLORO-10-(3-(N-(2-HYDROXYETHYL)METHYLAMINO) PROPYL) ISOALLOXAZINE SULFATE
mf: $C_{16}H_{17}Cl_2N_5O_3 \cdot H_2O_4S$　mw: 496.36

SAFETY PROFILE: Poison by intraperitoneal, intravenous, intramuscular, and subcutaneous routes. When heated to decomposition it emits very toxic fumes of SO_x, NO_x and Cl^-.

DFN800　CAS: 1193-54-0　HR: 3
DICHLOROMALEIMIDE
mf: $C_4HCl_2NO_2$　mw: 165.96

SYNS: DICHLOROMALEINIMIDE * 3,4-DICHLORO-2,5-PYRROLIDINEDIONE

SAFETY PROFILE: Poison by intraperitoneal route. Experimental teratogenic and reproductive effects. When heated to decomposition it emits very toxic fumes of Cl^- and NO_x.

DFO000　CAS: 528-74-5　HR: 3
3'5'-DICHLOROMETHOTREXATE
mf: $C_{20}H_{20}Cl_2N_8O_5$　mw: 523.38

SYNS: DCM * DICHLOROAMETHOPTERIN * 3',5'-DICHLOROAMETHOPTERIN * 3',5'-DICHLORO-4-AMINO-4-DEOXY-N$_{10}$-METHYLPTEROGLUTAMIC ACID * N-(3,5-DICHLORO-4-((2,4-DIAMINO-6-PTERIDINYL METHYL)METHYLAMINO)BENZOYL)GLUTAMIC ACID * DICHLOROMETHOTREXATE * NCI-C04875 * NSC-29630

CONSENSUS REPORTS: NCI Carcinogenesis Studies (ipr): Equivocal Evidence: rat; No Evidence: mouse CANCAR 40,1935,77

SAFETY PROFILE: Questionable carcinogen with experimental tumorigenic data. Moderately toxic by intraperitoneal and intravenous routes. When heated to decomposition it emits very toxic fumes of NO_x and Cl^-.

DFP200　CAS: 593-89-5　HR: 3
DICHLOROMETHYLARSINE

DOT: NA 1556
mf: CH_3AsCl_2　mw: 160.86

PROP: Colorless liquid. Bp: 134.5°, fp: $-59°$, flash p: $> 221°F$, d: 1.838 @ 20°/4°, vap press: 10 mm @ 24.3°, vap d: 5.40.

SYNS: METHYLARSINE DICHLORIDE * METHYLARSONOUS DICHLORIDE * METHYLDICHLORARSINE * METHYLDICHLOROARSINE (DOT) * TL 294

CONSENSUS REPORTS: Arsenic and its compounds are on the Community Right-To-Know List.

OSHA PEL: TWA 0.5 mg/(As)/m^3
DOT Classification: Poison A; Label: Poison Gas.

SAFETY PROFILE: Poison irritant to skin, eyes, and mucous membranes and poison by ingestion and inhalation. A blistering type of military poison. It is rapidly detoxified in the body. A moderately persistent gas. Combustible when exposed to heat or flame. To fight fire, use water, foam, CO_2, dry chemical. Explosive reaction with chlorine. Can react vigorously with oxidizing materials. Dangerous; when heated to decomposition or on contact with acid or acid fumes it emits highly toxic fumes of Cl^- and As.

DFP600   CAS: 58-54-8   HR: 3
2,3-DICHLORO-4-(2-METHYLENE-BUTYRL)PHENOXY ACETIC ACID
mf: $C_{13}H_{12}Cl_2O_4$ mw: 303.15

SYNS: CRINURYL * (2,3-DICHLORO-4-(2-METHY-LENEBUTYRYL)PHENOXY)ACETIC ACID * (2,3-DI-CHLORO-4-(2-METHYLENE-1-OXOBUTYL)PHENOXY)ACE-TIC ACID * EDECRIL * EDECRIN * EDECRINA * ENDECRIL * ETACRINIC ACID * ETAKRINIC ACID * ETHACRYNIC ACID * HIDROMEDIN * HYDROMEDIN * (4-(2-METHYLENEBUTYRYL)-2,3-DICHLOROPHENOXY)ACETIC ACID * METHYLENEBU-TYRYL PHENOXYACETIC ACID * MINGIT * MK-595 * OTACRIL * REOMAX * TALADREN * UREGIT

SAFETY PROFILE: Poison by intravenous route. Moderately toxic by ingestion. Human systemic effects by ingestion and intravenous routes: urine volume increase, impaired hearing, and tinnitus (ringing in the ears). A diuretic. When heated to decomposition it emits toxic fumes of Cl^-.

DFP800   CAS: 1123-61-1   HR: 3
DICHLORO-N-METHYLMALEIMIDE
mf: $C_5H_3Cl_2NO_2$ mw: 179.99

SYNS: 2,3-DICHLORO-N-METHYLMALEIMIDE * N-METHYLDICHLOROMALEINIMIDE

SAFETY PROFILE: Poison by intraperitoneal and intravenous routes. An experimental terato-gen. When heated to decomposition it emits very toxic fumes of Cl^- and NO_x.

DFQ800   CAS: 149-74-6   HR: 3
DICHLOROMETHYLPHENYLSILANE
DOT: UN 2437
mf: $C_7H_8Cl_2Si$ mw: 191.14

SYNS: METHYLPHENYLDICHLOROSILANE (DOT) * PHENYLMETHYLDICHLOROSILANE

CONSENSUS REPORTS: Reported in EPA TSCA Inventory. EPA Extremely Hazardous Substances List.

DOT Classification: IMO: Flammable or Combustible Liquid; Label: Flammable Liquid, Corrosive.

SAFETY PROFILE: Poison by inhalation, subcutaneous, and intraperitoneal routes. Corrosive to eyes, skin, and mucous membranes. When heated to decomposition it emits toxic fumes of Cl^-.

DFS000   CAS: 75-54-7   HR: 3
DICHLOROMETHYLSILANE
DOT: UN 1242
mf: CH_4Cl_2Si mw: 115.04

PROP: Colorless liquid, sol in benzene, ether and heptane. Bp: 41°, d: 1.10 @ 27°, flash p: −26°F.

SYNS: METHYL DICHLOROSILANE (DOT) * METHYL-DICHLORSILAN (CZECH)

CONSENSUS REPORTS: Reported in EPA TSCA Inventory.

DOT Classification: Flammable Liquid; Label: Flammable Liquid; IMO: Flammable Liquid; Label: Flammable Liquid, Corrosive.

SAFETY PROFILE: Moderately toxic by inhalation. Corrosive. A severe irritant to skin, eyes, and mucous membranes. Ignites spontaneously in air. A very dangerous fire hazard when exposed to heat or flame. Forms impact-sensitive explosive mixtures with potassium permanganate; lead(II) oxide; lead(IV) oxide; copper oxide; silver oxide. To fight fire, use water, foam, CO_2, mist. When heated to decomposition it emits toxic fumes of Cl^-.

DFT000   CAS: 117-80-6   HR: 3
2,3-DICHLORO-1,4-NAPHTHOQUINONE
DOT: NA 2761
mf: $C_{10}H_4Cl_2O_2$ mw: 227.04

PROP: Golden-yellow crystals. Mp: 193°, vap d: 7.8. Insol in water; moderately sol in organic solvents.

SYNS: ALGISTAT * COMPOUND 604 * DI-CHLONE (DOT) * 2,3-DICHLOR-1,4-NAPHTHOCHINON (GERMAN) * 2,3-DICHLORO-1,4-NAPHTHALENEDIONE * 2,3-DICHLORO-1,4-NAPHTHAQUINONE * DICHLO-RONAPHTHOQUINONE * 2,3-DICHLORONAPHTHOQUI-NONE * 2,3-DICHLORO-α-NAPHTHOQUINONE * 2,3-DICHLORONAPHTHOQUINONE-1,4 * ENT 3,776 * PHYGON * PHYGON PASTE * PHYGON SEED PROTECTANT * PHYGON XL * QUINTAR * QUINTAR 540F * SANQUINON * UNIROYAL * USR 604 * U.S. RUBBER 604

CONSENSUS REPORTS: Reported in EPA TSCA Inventory.

DOT Classification: ORM-E; Label: None.

SAFETY PROFILE: Questionable carcinogen with experimental carcinogenic and neoplasti-genic data. Poison by ingestion and intraperito-neal routes. Mildly toxic by skin contact. A skin, eye, and mucous membrane irritant. Large doses can cause central nervous system depression. A fungicide and algaecide. When heated to decomposition it emits toxic fumes of Cl⁻.

DFT800   CAS: 1836-75-5   **HR: 3**
2,4-DICHLORO-4'-NITRODIPHENYL ETHER
mf: $C_{12}H_7Cl_2NO_3$ mw: 284.10

SYNS: 2,4-DECHLOROPHENYL-p-NITROPHENYL ETHER * 2',4'-DICHLORO-4-NITROBIPHENYL ETHER * 2,4-DICHLORO-1-(4-NITROPHENOXY)BENZENE * 4-(2,4-DICHLOROPHENOXY)NITROBENZENE * 2,4-DICHLOROPHENYL-p-NITROPHENYL ETHER * 2,4-DICHLOROPHENYL-4-NITROPHENYL ETHER * 2,4,-DICHLORPHENYL-4-NITROPHENYLAETHER (GERMAN) * FW 925 * MEZOTOX * NCI-C00420 * NICLOFEN * NIP * NITOFEN * NITRAFEN * NITRAPHEN * NITROCHLOR * 4'-NITRO-2,4-DI-CHLORODIPHENYL ETHER * NITROFEN * NITRO-FENE (FRENCH) * NITROPHEN * NITROPHENE * PREPARATION 125 * TOK * TOK-2 * TOK E * TOK E-25 * TOK E 40 * TOKKORN * TOK WP-50 * TRIZILIN

CONSENSUS REPORTS: IARC Cancer Review: GROUP 2B IMEMDT 7,56,87; Animal Sufficient Evidence IMEMDT 30,271,83. NCI Carcinogenesis Bioassay (feed); No Evidence: rat NCITR* NCI-CG-TR-184,79; Clear Evi-dence: mouse, rat NCITR* NCI-CG-TR-26,78; Clear Evidence: mouse NCITR* NCI-CG-TR-184,79. NTP Fourth Annual Report On Carcinogens, 1984. EPA Genetic Toxicology Program. Community Right-To-Know List. Reported in EPA TSCA Inventory.

SAFETY PROFILE: Confirmed carcinogen with experimental carcinogenic data. Poison by ingestion. Moderately toxic by inhalation and possibly other routes. Mildly toxic by skin contact. Experimental teratogenic and reproductive effects. A skin and severe eye irritant. Mutation data reported. A broad spectrum herbicide. When heated to decomposition it emits very toxic fumes of Cl⁻ and NO_x.

DFU000   CAS: 594-72-9   **HR: 3**
1,1-DICHLORO-1-NITROETHANE
DOT: UN 2650
mf: $C_2H_3Cl_2NO_2$ mw: 143.96

PROP: Liquid. Bp: 124°, flash p: 168°F(OC), d: 1.4153 @ 20°/20°, vap d: 4.97.

SYNS: 1,1-DICHLOOR-1-NITROETHAAN (DUTCH) * 1,1-DICHLOR-1-NITROAETHAN (GERMAN) * DICHLORONITROETHANE * 1,1-DICLORO-1-NI-TROETANO (ITALIAN) * ETHIDE

OSHA PEL: (Transitional: CL 10 ppm) TWA 2 ppm
ACGIH TLV: TWA 2 ppm
DFG MAK: 10 ppm (60 mg/m³)
DOT Classification: Poison B; Label: Poison.

SAFETY PROFILE: Poison by ingestion and intraperitoneal routes. Moderately toxic by inhalation. A strong irritant. Inhalation causes pulmonary edema. A fumigant for produce. Flammable when exposed to heat, flame or oxidizers. Can react vigorously with oxidizing materials. To fight fire, use water, CO_2, dry chemical. When heated to decomposition it emits highly toxic fumes of Cl⁻ and NO_x.

DFU600   CAS: 609-89-2   **HR: 3**
2,4-DICHLORO-6-NITROPHENOL
mf: $C_6H_3Cl_2NO_3$ mw: 208.00

SYN: 2,4-DICHLOR-6-NITROFENOL (CZECH)

CONSENSUS REPORTS: Chlorophenol compounds are on the Community Right-To-Know List.

SAFETY PROFILE: Poison by ingestion. A severe eye irritant. When heated to decomposition it emits very toxic fumes of Cl⁻ and NO$_x$.

DFV400   CAS: 50-65-7   **HR: 3**
2′,5-DICHLORO-4′-NITRO-SALICYLANILIDE
mf: $C_{13}H_8Cl_2N_2O_4$ mw: 327.13

SYNS: BAY 2353 * BAYER 73 * BAYER 2353 * BAYLUSCID * CHEMAGRO 2353 * 5-CHLORO-N-(2-CHLORO-4-NITROPHENYL)-2-HYDROXYBENZAMIDE * 5-CHLORO-2′-CHLORO-4′-NITROSALICYLANILIDE * 2-CHLORO-4-NITROPHENYLAMIDE-6-CHLOROSALICYLIC ACID * N-(2-CHLORO-4-NITROPHENYL)-5-CHLOROSALICYLAMIDE * CLONITRALID * 2′,5-DICHLOR-4′-NITRO-SALIZYLSAEUREANILID (GERMAN) * DICHLOSALE * ENT 25,823 * FENASAL * HL 2447 * 2-HYDROXY-5-CHLORO-N-(2-CHLORO-4-NITROPHENYL)BENZAMIDE * IOMESAN * IOMEZAN * NICLOSAMIDE * PHENASAL * VERMITIN * YOMESAN

SAFETY PROFILE: Poison by intraperitoneal and intravenous routes. Moderately toxic by ingestion. Human mutation data reported. When heated to decomposition it emits very toxic fumes of Cl⁻ and NO$_x$.

DFV600   CAS: 1420-04-8   **HR: 3**
2′,5-DICHLORO-4′-NITROSALI-CYLANILIDE-2-AMINOETHANOL SALT
mf: $C_{13}H_8Cl_2N_2O_4 \cdot C_2H_7NO$ mw: 388.23

SYNS: BAYER 73 * BAYER 25648 * BAYLUSCID * BAYLUSCIDE * 5-CHLORO-N-(2-CHLORO-4-NITRO-PHENYL)-2-HYDROXYBENZAMIDE with 2-AMINOETHANOL (1:1) * CLONITARLID * 5,2′-DICHLORO-4′-NITROSALICYLANILIDE ETHANOLAMINE SALT * 5,2-DICHLORO-4-NITROSALICYLIC ANILIDE-2-AMINOETHANOL SALT * 2′,5-DICHLORO-4′-NITRO-SALICYLOYLANILIDE ETHANOLAMINE SALT * ETHANOLAMINE SALT of 5,2′-DICHLORO-4′-NITROSALICYCLICANILIDE * M 73 * MOLLUSCICIDE BAYER 73 * NCI-C00431 * NICLOSAMIDE * SR 73

CONSENSUS REPORTS: NCI Carcinogenesis Bioassay (feed); Inadequate Studies: mouse, rat NCITR* NCI-CG-TR-91,78.

SAFETY PROFILE: Poison by intraperitoneal route. Moderately toxic by ingestion. Many N-nitroso compounds are carcinogens. A pesticide. When heated to decomposition it emits very toxic fumes of NO$_x$ and Cl⁻.

DFW000   CAS: 69112-96-5   **HR: 3**
2,2′-DICHLORO-N-NITROSODI-PROPYLAMINE
mf: $C_6H_{12}Cl_2N_2O$ mw: 199.10

SYN: NITROSOBIS(2-CHLOROPROPYL)AMINE

SAFETY PROFILE: Questionable carcinogen with experimental tumorigenic data. Many N-nitroso compounds are carcinogens. Mutation data reported. When heated to decomposition it emits very toxic fumes of Cl⁻ and NO$_x$.

DFW200   CAS: 57541-72-7   **HR: 3**
3,4-DICHLORONITROSOPIPERIDINE
mf: $C_5H_8Cl_2N_2O$ mw: 183.05

SYN: N-NITROSO-3,4-DICHLOROPIPERIDINE

CONSENSUS REPORTS: EPA Genetic Toxicology Program.

SAFETY PROFILE: Questionable carcinogen with experimental tumorigenic data. Human mutation data reported. Many N-nitroso compounds are carcinogens. When heated to decomposition it emits very toxic fumes of Cl⁻ and NO$_x$.

DFW600   CAS: 59863-59-1   **HR: 3**
3,4-DICHLORO-N-NITROSO-PYRROLIDINE
mf: $C_4H_6Cl_2N_2O$ mw: 169.02

SAFETY PROFILE: Questionable carcinogen with experimental tumorigenic data. Mutation data reported. Many N-nitroso compounds are carcinogens. When heated to decomposition it emits very toxic fumes of Cl⁻ and NO$_x$.

DFX000   CAS: 30586-10-8   **HR: 1**
DICHLOROPENTANE

DOT: UN 1152
mf: $C_5H_{10}Cl_2$ mw: 141.05

PROP: Clear, light yellow liquid. Bp: 130°, flash p: 106°F (OC), vap d: 4.86, d: 1.06-1.08 @ 20°.

SYN: DICHLOROPENTANES (DOT)

DOT Classification: Flammable or Combustible Liquid; Label: Flammable Liquid.

SAFETY PROFILE: Flammable when exposed to heat or flame. Can react vigorously with oxidizing materials. To fight fire, use water, foam, CO$_2$, dry chemical. When heated to decomposition it emits highly toxic fumes of Cl⁻ and phosgene.

DFX400   CAS: 536-29-8   **HR: 3**
DICHLOROPHENARSINE HYDROCHLORIDE
mf: $C_6H_6AsCl_2NO \cdot ClH$ mw: 290.41

SYNS: 2-AMINO-4-DICHLOROARSINOPHENOL HYDROCHLORIDE * (3-AMINO-4-HYDROXYPHENYL)ARSONOUS DICHLORIDE MONOHYDROCHLORIDE * 3-AMINO-4-HYDROXYPHENYL DICHLORARSINE HYDROCHLORIDE * (3-AMINO-4-HYDROXYPHENYL)DICHLOROARSINE HYDROCHLORIDE * ARSECLOR * CHLORARSOL * CHLORASEN * CLORARSEN * DICHLOROMAPHARSEN * FILARSEN * FONTARSOL * HALARSOL * R.P. 2591

CONSENSUS REPORTS: Arsenic and its compounds, as well as chlorophenol compounds, are on the Community Right-To-Know List.

OSHA PEL: TWA 0.5 mg(As)/m³
ACGIH TLV: TWA 0.2 mg(As)/m³

SAFETY PROFILE: Poison by intravenous, intraperitoneal, and possibly other routes. Moderately toxic by ingestion. Human systemic effects by parenteral route: hypermotility, diarrhea, nausea, vomiting. When heated to decomposition it emits very toxic fumes of As, NO_x, and Cl⁻.

DFX800   CAS: 120-83-2   **HR: 3**
2,4-DICHLOROPHENOL
mf: $C_6H_4Cl_2O$ mw: 163.00

PROP: Colorless crystals. Mp: 45°, bp: 210°, flash p: 237°F, d: 1.383 @ 60°/25°, vap d: 5.62, vap press: 1 mm @ 53.0°.

SYNS: DCP * 2,4-DCP * NCI-C55345 * RCRA WASTE NUMBER U081

CONSENSUS REPORTS: IARC Cancer Review: Human Limited Evidence IMEMDT 41,319,86. Reported in EPA TSCA Inventory. EPA Genetic Toxicology Program. Community Right-To-Know List.

SAFETY PROFILE: Suspected carcinogen with experimental carcinogenic and teratogenic data. Poison by intraperitoneal route. Moderately toxic by ingestion, skin contact, and subcutaneous routes. Combustible when exposed to heat or flame. Can react vigorously with oxidizing

materials. To fight fire, use alcohol foam, foam, CO_2, dry chemical. When heated to decomposition, or on contact with acid or acid fumes it emits highly toxic fumes of Cl⁻.

DFY000   CAS: 87-65-0   **HR: 3**
2,6-DICHLOROPHENOL
mf: $C_6H_4Cl_2O$ mw: 163.00

SYNS: 2,6-DICHLORFENOL (CZECH) * RCRA WASTE NUMBER U082

CONSENSUS REPORTS: Reported in EPA TSCA Inventory. EPA Genetic Toxicology Program. Chlorophenol compounds are on the Community Right-To-Know List.

SAFETY PROFILE: Poison by intraperitoneal route. Moderately toxic by ingestion and subcutaneous routes. A severe skin and eye irritant. When heated to decomposition it emits toxic fumes of Cl⁻.

DFY400   CAS: 97-16-5   **HR: 3**
2,4-DICHLOROPHENOL BENZENESULFONATE
mf: $C_{12}H_8Cl_2O_3S$ mw: 303.16

SYNS: COMPOUND 923 * 2,4-DICHLOROPHENYL BENZENESULFONATE * 2,4-DICHLOROPHENYL BENZENESULPHONATE * 2,4-DICHLOROPHENYL ESTER of BENZENESULFONIC ACID * 2,4-DICHLOROPHENYL ESTER BENZENESULPHONIC ACID * DPBS * EM 923 * GENITE * GENITOL

CONSENSUS REPORTS: Chlorophenol compounds are on the Community Right-To-Know List.

SAFETY PROFILE: Questionable carcinogen with experimental carcinogenic and tumorigenic data. Poison by intravenous route. Moderately toxic by ingestion and possibly other routes. An irritant. A pesticide. When heated to decomposition it emits very toxic fumes of Cl⁻ and SO_x.

DGA200   CAS: 14255-88-0   **HR: 3**
5,6-DICHLORO-1-PHENOXYCARBONYL-2-TRIFLUOROMETHYLBENZIMIDAZOLE
mf: $C_{15}H_7Cl_2F_2N_2O_2$ mw: 375.14

SYNS: 5,6-DICHLORO-2-TRIFLUOROMETHYLBENZIMIDAZOLE-1-CARBOXYLATE * 5,6-DICHLORO-2-(TRIFLUOROMETHYL)-1H-BENZIMIDAZOLE-1-

CARBOXYLIC ACID PHENYL ESTER * ENT 27,438
* FENAZAFLOR * FENOFLURAZOLE * FENOZA-
FLOR * FENZAFLOR * FISONS NC 5016
* LOVOZAL * NC 5016 * NSC 191025 * PHE-
NYL-5,6-DICHLORO-2-TRIFLUOROMETHYL-BENZIMIDA-
ZOLE-1-CARBOXYLATE * TARZOL

SAFETY PROFILE: Poison by ingestion, intraperitoneal, and possibly other routes. Moderately toxic by skin contact. When heated to
decomposition it emits very toxic fumes of F^-,
Cl^-, and NO_x.

DGB000   CAS: 120-36-5   **HR: 3**
2-(2,4-DICHLOROPHENOXY)
PROPIONIC ACID
mf: $C_9H_8Cl_2O_3$ mw: 235.07

SYNS: ACIDE-2-(2,4-DICHLORO-PHENOXY) PROPIONI-
QUE (FRENCH) * ACIDO-2-(2,4-DICLORO-FENOSSI)
PROPIONICO (ITALIAN) * CORNOX RD * CORNOX
RK * DESORMONE * 2-(2,4-DICHLOOR-FENOXY)
PROPIONZUUR (DUTCH) * α-(2,4-DICHLOROPHEN-
OXY) PROPIONIC ACID * DICHLOROPROP * 2-(2,4-
DICHLOR-PHENOXY)-PROPIONSAEURE (GERMAN)
* DICHLORPROP * 2,4-DP * 2-(2,4-DP)
* HEDONAL * HEDONAL DP * HORMATOX
* KILDIP * POLYCLENE * POLYMONE
* POLYTOX * RD 406 * SERITOX 50 * U46
* U46 DP-FLUID * VISKO-RHAP * WEEDONE DP
* WEEDONE 170

CONSENSUS REPORTS: IARC Cancer Review: GROUP 2B IMEMDT 7,156,87; Human
Limited Evidence IMEMDT 41,357,86. Reported in EPA TSCA Inventory.

SAFETY PROFILE: Suspected carcinogen. Poison by ingestion. Moderately toxic by skin contact. An experimental teratogen. Mutation data
reported. A fumigant. When heated to decomposition it emits toxic fumes of Cl^-.

DGB600   CAS: 696-28-6   **HR: 3**
DICHLOROPHENYLARSINE
DOT: NA 1556
mf: $C_6H_5AsCl_2$ mw: 222.93

PROP: Colorless gas or liquid, changes to yellow. Bp: 255-275°, fp: −15.6°, d: 1.654 @
20°, vap press: 0.021 mm @ 20°, vap d: 7.7.

SYNS: FDA * FENILDICLOROARSINA (ITALIAN)
* PHENYLARSINEDICHLORIDE * PHENYLARSON-
OUS DICHLORIDE * PHENYL DICHLOROARSINE (DOT)
* RCRA WASTE NUMBER P036 * TL 69

CONSENSUS REPORTS: Arsenic and its
compounds are on the Community Right-To-
Know List. Reported in EPA TSCA Inventory.
EPA Extremely Hazardous Substances List.

OSHA PEL: TWA 0.5 mg(As)/m³
DOT Classification: Poison B; Label: Poison.

SAFETY PROFILE: Poison by inhalation, ingestion, skin contact, and intravenous routes.
A lachrymator type of military poison gas. When
exposed to heat, water, or steam it reacts to
produce corrosive fumes of Cl^-. When heated
to decomposition it emits highly toxic fumes
of arsenic.

DGC600   CAS: 38780-39-1   **HR: 3**
cis-DICHLORO(o-PHENYLENE-
DIAMINE)PLATINUM(II)
mf: $C_6H_8Cl_2N_2Pt$ mw: 374.15

SYN: DICHLORO(1,2-PHENYLENEDIAMMINE)PLATI-
NUM(II)

SAFETY PROFILE: Poison by intraperitoneal
route. Mutation data reported. When heated to
decomposition it emits very toxic fumes of Cl^-
and NO_x.

DGD600   CAS: 330-55-2   **HR: 3**
3-(3,4-DICHLOROPHENYL)-1-
METHOXYMETHYLUREA
mf: $C_9H_{10}Cl_2N_2O_2$ mw: 249.11

PROP: Solid. Mp: 93-94°. Sltly sol in water;
partially sol in acetone and alc.

SYNS: 3-(3,4-DICHLOOR-FENYL)-1-METHOXY-1-
METHYLUREUM (DUTCH) * 3-(3,4-DICHLORO-FENIL)-
1-METOSSI-1-METIL-UREA (ITALIAN) * 3-(3,4-DICHLO-
ROPHENYL)-1-METHOXY-1-METHYLUREA * N'-(3,4-
DICHLOROPHENYL)-N-METHOXY-N-METHYLUREA
* 1-(3,4-DICHLOROPHENYL)3-METHOXY-3-METH-
YLUREE (FRENCH) * N-(3,4-DICHLOROPHENYL)-N'-
METHYL-N'-METHOXYUREA * 3-(4,5-DICHLORPHE-
NYL)-1-METHOXY-1-METHYLHARNSTOFF (GERMAN)
* 3-(3,4-DICHLOR-PHENYL)-1-METHOXY-1-METHYL-
HARNSTOFF (GERMAN) * DU PONT 326 * DUPONT
HERBICIDE 326 * GARNITAN * HERBICIDE 326
* HOE 2810 * LINEX 4L * LINOROX * LINU-
REX * LINURON * LINURON (HERBICIDE)
* LOREX * LOROX * LOROX LINURON WEED
KILLER * METHOXYDIURON * 1-METHOXY-1-
METHYL-3-(3,4-DICHLOROPHENYL)UREA * PREMALIN
* SARCLEX * SCARCLEX * SINURON

CONSENSUS REPORTS: Reported in EPA TSCA Inventory. EPA Genetic Toxicology Program.

SAFETY PROFILE: Poison by inhalation. Moderately toxic by ingestion. Mutation data reported. A selective herbicide used in farming. When heated to decomposition it emits very toxic fumes of Cl^- and NO_x.

DGD800   CAS: 299-85-4   HR: 3
O-(2,4-DICHLOROPHENYL)-O-METHYL-ISOPROPYLPHOSPHORAMIDOTHIOATE
mf: $C_{10}H_{14}Cl_2NO_2PS$ mw: 314.18

SYNS: O-(2,4-DICHLOROPHENYL)-O-METHYL-N-ISO-PROPYLPHOSPHORAMIDOTHIOATE * DMPA * DOW 1329 * DOWCO 118 * ENT 25,647 * ISOPROPYLPHOSPHORAMIDOTHIOIC ACID-O-2,4-DI-CHLOROPHENYL-O-METHYL ESTER * K 22023 * (1-METHYLETHYL)PHOSPHORAMIDOTHIOIC ACID O-(2,4-DICHLOROPHENYL)-O-METHYL ESTER * OMS 115 * ZYTRON

SAFETY PROFILE: Poison by ingestion. Moderately toxic by skin contact and possibly other routes. An herbicide and plant growth regulator. When heated to decomposition it emits very toxic fumes of Cl^-, NO_x, PO_x, and SO_x.

DGE400   CAS: 644-97-3   HR: 3
DICHLOROPHENYLPHOSPHINE
DOT: UN 2798
mf: $C_6H_5Cl_2P$ mw: 178.98

PROP: Fuming liquid. Bp: 225°, d: 1.319, vap d: 6.17.

SYNS: BENZENE PHOSPHORUS DICHLORIDE (DOT) * PHOSPHENYL CHLORIDE

CONSENSUS REPORTS: Reported in EPA TSCA Inventory.

DOT Classification: Corrosive Material; Label: Corrosive.

SAFETY PROFILE: A poison irritant to skin, eyes, and mucous membranes and poison by ingestion and inhalation. When heated to decomposition it emits very toxic fumes of Cl^- and PO_x.

DGF000   CAS: 24096-53-5   HR: 3
N-(3,5-DICHLOROPHENYL) SUCCINIMIDE
mf: $C_{10}H_7Cl_2NO_2$ mw: 244.08

SYNS: 1-(3,5-DICHLOROPHENYL)-2,5-PYRROLIDINE-DIONE * DIMETHACHLON * OHRIC

SAFETY PROFILE: Questionable carcinogen with experimental tumorigenic data. Moderately toxic by ingestion. When heated to decomposition it emits very toxic fumes of Cl^- and NO_x.

DGF200   CAS: 27137-85-5   HR: 3
(DICHLOROPHENYL)TRICHLOROSI-LANE
DOT: UN 1766
mf: $C_6H_3Cl_5Si$ mw: 280.43

PROP: Straw-colored liquid, sol in benzene and perchloroethylene. (mixture of isomers). D: 1.562, bp: 260°, flash p: 286°F.

SYNS: DICHLOROPHENYLTRICHLOROSILANE (DOT) * TRICHLORO(DICHLOROPHENYL)SILANE

CONSENSUS REPORTS: Reported in EPA TSCA Inventory. EPA Extremely Hazardous Substances List.

DOT Classification: Corrosive Material; Label: Corrosive.

SAFETY PROFILE: Poison by ingestion, inhalation, subcutaneous, and intraperitoneal routes. Corrosive to the eyes, skin, and mucous membranes. On contact with moisture it releases corrosive HCl. Combustible when exposed to heat or flame. When heated to decomposition it emits toxic fumes of Cl^-.

DGF800   CAS: 142-28-9   HR: 2
1,3-DICHLOROPROPANE
mf: $C_3H_6Cl_2$ mw: 112.99

PROP: Colorless liquid. Bp: 125°, d: 1.201 @ 15°, vap d: 3.90, flash p: 69.8°F.

SYN: TRIMETHYLENE DICHLORIDE

CONSENSUS REPORTS: Reported in EPA TSCA Inventory.

SAFETY PROFILE: Moderately toxic by ingestion. Mutation data reported. A very dangerous fire hazard when exposed to heat or flame. When heated to decomposition it emits highly toxic fumes of Cl^- and phosgene.

DGG000   CAS: 8003-19-8   HR: 3
DICHLOROPROPANE-DICHLOROPROPENE MIXTURE
mf: $C_3H_6Cl_2 \cdot C_3H_4Cl_2$ mw: 223.96

PROP: D-D Soil fumigant consists of chlorinated C_3 hydrocarbons (100%), 1,3-dichloropropene, 3,3-dichloropropene, 1,2-dichloropropane, 2,3-dichloropropene, and related C_3 chlorinated hydrocarbons.

SYNS: D-D * DD MIXTURE * DD SOIL FUMIGANT * 1,3-DICHLOROPROPENE and 1,2-DICHLOROPROPANE MIXTURE * DICHLORPROPAN-DICHLORPROPENGEMISCH (GERMAN) * DOWFUME N * ENT 8,420 * NEMAFENE * TELONE * VIDDEN D

SAFETY PROFILE: Poison by ingestion and inhalation. Moderately toxic by skin contact. Severe skin and eye irritant. Mutation data reported. A fumigant. When heated to decomposition it emits toxic fumes of Cl^-.

DGG200   HR: 3
1,2-DICHLOROPROPANE mixed with DICHLOROPROPENE

SYN: DICHLOROPROPENE and PROPYLENE DICHLORIDE MIXTURE (DOT)

DOT Classification: Corrosive Material; Label: Corrosive.

SAFETY PROFILE: Probably a poison. Corrosive to the eyes, skin, and mucous membranes. When heated to decomposition it emits toxic fumes of Cl^-.

DGG400   CAS: 96-23-1   HR: 3
1,3-DICHLORO-2-PROPANOL

DOT: UN 2750
mf: $C_3H_6Cl_2O$ mw: 128.99

PROP: Colorless liquid, ether-like odor. Bp: 174°, d: 1.367 @ 20°/4°, vap press: 1 mm @ 28.0°, vap d: 4.45, flash p: 165°F (OC), mp: −4°.

SYNS: DICHLOROHYDRIN * α-DICHLOROHYDRIN * sym-DICHLOROISOPROPYL ALCOHOL * 1,3-DICHLOROPROPANOL-2 (DOT) * GLYCEROL α,γ-DICHLOROHYDRIN * sym-GLYCEROL DICHLOROHYDRIN * U 25,354

CONSENSUS REPORTS: Reported in EPA TSCA Inventory. EPA Genetic Toxicology Program.

DOT Classification: Poison B; Label: Poison.

SAFETY PROFILE: Poison by ingestion and inhalation. Moderately toxic by skin contact.

Human mutation data reported. A skin irritant. Action may be similar to carbon tetrachloride, but more irritating to mucous membranes. Flammable when exposed to heat, flame, or oxidizers. To fight fire, use alcohol foam, dry chemical, fog, mist, or spray. Dangerous; when heated to decomposition it emits highly toxic fumes of Cl^- and phosgene.

DGG600   CAS: 616-23-9   HR: 3
2,3-DICHLOROPROPANOL
mf: $C_3H_6Cl_2O$ mw: 128.99

SYNS: 1,2-DICHLORO-3-PROPANOL * 1,2-DICHLOROPROPANOL-3 * 2,3-DICHLORO-1-PROPANOL * GLYCEROL-α,β-DICHLOROHYDRIN

CONSENSUS REPORTS: Reported in EPA TSCA Inventory.

SAFETY PROFILE: Poison by ingestion and skin contact. Moderately toxic by inhalation. A skin and severe eye irritant. Mutation data reported. When heated to decomposition it emits toxic fumes of Cl^-.

DGG950   CAS: 542-75-6   HR: 3
1,3-DICHLOROPROPENE
mf: $C_3H_4Cl_2$ mw: 110.97

PROP: Liquid. Bp: 103-110°, flash p: 95°F, d: 1.22, vap d: 3.8.

SYNS: α-CHLOROALLYL CHLORIDE * γ-CHLOROALLYL CHLORIDE * 1,3-DICHLOROPROPENE-1 * α,γ-DICHLOROPROPYLENE * 1,3-DICHLOROPROPYLENE * NCI-C03985 * RCRA WASTE NUMBER U084 * TELONE * TELONE II SOIL FUMIGANT * VIDDEN D

CONSENSUS REPORTS: IARC Cancer Review: GROUP 2B IMEMDT 7,195,87; Human Inadequate Evidence IMEMDT 41,113,86; Animal Sufficient Evidence IMEMDT 41,113,86. NTP Carcinogenesis Studies (gavage); Clear Evidence: mouse, rat NTPTR* NTP-TR-269,86. Reported in EPA TSCA Inventory. EPA Genetic Toxicology Program. Community Right-To-Know List.

OSHA PEL: TWA 1 ppm (skin)
ACGIH TLV: TWA 1 ppm (skin)
DFG MAK: Animal Carcinogen, Suspected Human Carcinogen.

SAFETY PROFILE: Confirmed carcinogen with experimental carcinogenic data. Poison by ingestion. Moderately toxic by skin contact. Mildly toxic by inhalation. A strong irritant.

Mutation data reported. A pesticide. Dangerous fire hazard when exposed to heat, flame, or oxidizers. Reacts vigorously with oxidizing materials. To fight fire, use water, foam, CO_2, dry chemical. When heated to decomposition it emits toxic fumes of Cl^-.

DGH000   CAS: 10061-02-6   **HR: 3**
trans-1,3-DICHLOROPROPENE
mf: $C_3H_4Cl_2$ mw: 110.97

PROP: Flash point 21°C.

SYNS: (E)-1,3-DICHLOROPROPENE * trans-1,3-DI-CHLOROPROPYLENE

CONSENSUS REPORTS: EPA Genetic Toxicology Program.

DFG MAK: Animal Carcinogen, Suspected Human Carcinogen.

SAFETY PROFILE: Human mutation data reported. A dangerous fire hazard when exposed to heat, flame or oxidizers. When heated to decomposition it emits toxic fumes of Cl^-.

DGH200   CAS: 10061-01-5   **HR: 3**
cis-1,3-DICHLOROPROPENE
mf: $C_3H_4Cl_2$ mw: 110.97

PROP: Flash point 21°C.

SYNS: (Z)-1,3-DICHLOROPROPENE * cis-1,3-DI-CHLOROPROPYLENE

CONSENSUS REPORTS: EPA Genetic Toxicology Program.

DFG MAK: Animal Carcinogen, Suspected Human Carcinogen.

SAFETY PROFILE: Confirmed carcinogen with experimental neoplastigenic data. Human mutation data reported. A dangerous fire hazard when exposed to heat, flame or oxidizers. When heated to decomposition it emits toxic fumes of Cl^-.

DGH400   CAS: 78-88-6   **HR: 3**
2,3-DICHLOROPROPENE
mf: $C_3H_4Cl_2$ mw: 110.97

PROP: Flash p: 50°F.

SYNS: 2,3-DICHLORO-1-PROPENE * 2,3-DICHLORO-PROPYLENE

CONSENSUS REPORTS: Reported in EPA TSCA Inventory. EPA Genetic Toxicology Program.

SAFETY PROFILE: Poison by ingestion. Moderately toxic by inhalation and skin contact. Human mutation data reported. A severe skin irritant. A very dangerous fire hazard when exposed to heat, flame, or oxidizers. When heated to decomposition it emits toxic fumes of Cl^-.

DGH800   CAS: 10140-89-3   **HR: 3**
2,3-DICHLORO PROPIONALDEHYDE
mf: $C_3H_4Cl_2O$ mw: 126.97

PROP: Liquid. Vap d: 4.4.

SYNS: 1,2-DICHLORO-3-PROPIONAL * α,β-DICHLO-ROPROPIONALDEHYDE

SAFETY PROFILE: Poison by ingestion and skin contact. Mildly toxic by inhalation. A severe skin and eye irritant. Mutation data reported. When heated to decomposition it emits toxic fumes of Cl^-.

DGI000   CAS: 709-98-8   **HR: 3**
DICHLOROPROPIONANILIDE
mf: $C_9H_9Cl_2NO$ mw: 218.09

PROP: Light brown solid (pure); liquid (technical grade). Mp (pure): 85-89°, bp (technical grade): 91-95°.

SYNS: BAY 30130 * CHEM RICE * CRYSTAL PROPANIL-4 * DCPA * N-(3,4-DICHLOROPHE-NYL)PROPANAMIDE * N-(3,4-DICHLOROPHENYL)PRO-PIONAMIDE * 3,4-DICHLOROPROPIONANILIDE * 3′,4′-DICHLOROPROPIONANILIDE * DIPRAM * DPA * FARMCO PROPANIL * FW 734 * GRASCIDE * HERBAX TECHNICAL * MONT-ROSE PROPANIL * PROPANEX * PROPANID * PROPANIDE * PROPANIL * PROPIONIC ACID-3,4-DICHLOROANILIDE * PROP-JOB * RISELECT * ROGUE * ROSANIL * S 10165 * STAM * STAM F 34 * STAM LV 10 * STAM M-4 * STAMPEDE * STAMPEDE 3E * STAM SUPER-NOX * STREL * SUPERNOX * SURCOPUR * SURPUR * VERTAC

CONSENSUS REPORTS: EPA Genetic Toxicology Program.

SAFETY PROFILE: Poison by ingestion. Moderately toxic by an unspecified route. Mildly toxic by skin contact. Mutation data reported. When heated to decomposition it emits very toxic fumes of Cl^- and NO_x.

DGI400   CAS: 75-99-0   **HR: 2**
2,2-DICHLOROPROPIONIC ACID

DOT: NA 1760
mf: $C_3H_4Cl_2O_2$ mw: 142.97

PROP: White to tan powder.

SYNS: BASFAPON * BASFAPON B * BASFAPON/ BASFAPON N * BASINEX * BH DALAPON * CRISAPON * DALAPON (USDA) * DALAPON 85 * DED-WEED * DEVIPON * α-DICHLOROPRO- PIONIC ACID * α,α-DICHLOROPROPIONIC ACID * DOWPON * DOWPON M * GRAMEVIN * KENAPON * LIROPON * PROPROP * RADA- PON * REVENGE * UNIPON

CONSENSUS REPORTS: EPA Genetic Toxi- cology Program. Reported in EPA TSCA Inven- tory.

OSHA PEL: TWA 1 ppm
ACGIH TLV: TWA 1 ppm
DFG MAK: 1 ppm (6 mg/m^3)
DOT Classification: Corrosive Material; Label: Corrosive.

SAFETY PROFILE: Moderately toxic by inges- tion. Corrosive. A skin irritant. Mutation data reported. When heated to decomposition it emits toxic fumes of Cl$^-$.

DGI600   CAS: 127-20-8   **HR: 2**
α,α-DICHLOROPROPIONIC ACID SODIUM SALT
mf: $C_3H_3Cl_2O_2 \cdot Na$ mw: 164.95

SYNS: BASFAPON B * DALAPON * DALAPON SODIUM * DALAPON SODIUM SALT * 2,2-DICHLO- ROPROPIONIC ACID, SODIUM SALT * DOWPON * 2,2-DPA * GRAMEVIN * NATRIUMSALZ DER 2,2-DICHLORPROPIONSAURE * RADAPON * SO- DIUM DALAPON * SODIUM-α,α-DICHLOROPROPION- ATE * SODIUM-2,2-DICHLOROPROPIONATE * UNIPON

CONSENSUS REPORTS: Reported in EPA TSCA Inventory. EPA Genetic Toxicology Pro- gram.

SAFETY PROFILE: Moderately toxic by inges- tion and possibly other routes. Mutation data reported. When heated to decomposition it emits toxic fumes of Na$_2$O and Cl$^-$.

DGK200   CAS: 320-72-9   **HR: 3**
3,5-DICHLOROSALICYLIC ACID
mf: $C_7H_4Cl_2O_3$ mw: 207.01

SYN: USAF DO-68

CONSENSUS REPORTS: Reported in EPA TSCA Inventory.

SAFETY PROFILE: Poison by intraperitoneal route. When heated to decomposition it emits toxic fumes of Cl$^-$.

DGK300   CAS: 4109-96-0   **HR: 3**
DICHLOROSILANE

DOT: UN 2189
mf: Cl_2H_2Si mw: 101.01

DOT Classification: Poison A; Label:Poison Gas and Flammable Gas

SAFETY PROFILE: Ignites spontaneously in air. Confined mixtures with air are spontane- ously explosive. When heated to decomposition it emits toxic fumes of Cl$^-$.

DGL600   CAS: 1320-37-2   **HR: 1**
DICHLOROTETRAFLUOROETHANE

DOT: UN 1958
mf: $C_2Cl_2F_4$ mw: 170.92

PROP: Colorless gas. Bp: 3.5°.

SYNS: DWUCHLOROCZTEROFLUOROETAN (POLISH) * TETRAFLUORODICHLOROETHANE

OSHA PEL: TWA 1000 ppm
ACGIH TLV: TWA 1000 ppm
DOT Classification: Nonflammable Gas; La- bel: Nonflammable Gas.

SAFETY PROFILE: A mildly toxic irritant; nar- cotic in high concentrations. An asphyxiant. Reacts violently with Al. When heated to de- composition it emits toxic fumes of F$^-$ and Cl$^-$.

DGM600   CAS: 1918-13-4   **HR: 3**
2,6-DICHLOROTHIOBENZAMIDE
mf: $C_7H_5Cl_2NS$ mw: 206.09

SYNS: CHLOROTHIAMIDE * DCBN * 2,6-DI- CHLOROBENZENECARBOTHIOAMIDE * SD 7961 * WL-5792

CONSENSUS REPORTS: EPA Genetic Toxi- cology Program.

SAFETY PROFILE: Poison by ingestion and intraperitoneal route. Moderately toxic by skin contact. Mutation data reported. An herbicide. When heated to decomposition it emits very toxic fumes of Cl$^-$, NO$_x$ and SO$_x$.

DGN200   CAS: 2782-57-2   **HR: 2**
1,3-DICHLORO-s-TRIAZINE-2,4,6(1H,3H,5H)-TRIONE

DOT: UN 2465
mf: $C_3H_2Cl_2N_3O_3$ mw: 198.98

PROP: White crystals, chlorine odor, moderately sol in water. Mp: 225°.

SYNS: ACL 70 * DICHLOROISOCYANURIC ACID * DICHLOROISOCYANURIC ACID, DRY (DOT) * KYSELINA DICHLORISOKYANUROVA (CZECH) * TROCLOSENE

CONSENSUS REPORTS: Reported in EPA TSCA Inventory.

DOT Classification: Oxidizer; Label: Oxidizer.

SAFETY PROFILE: Moderately toxic by ingestion. Human systemic effects by ingestion: ulceration or bleeding from stomach. Autopsy findings include gastrointestinal tract irritation, tissue edema, liver and kidney congestion. A severe eye and skin irritant. When heated to decomposition it emits chlorides and carbon monoxide.

DGP400   CAS: 612-12-4   **HR: 3**
α,α′-DICHLORO-o-XYLENE
mf: $C_8H_8Cl_2$ mw: 175.06

SYNS: 1,2-BIS(CHLOROMETHYL)BENZENE- * o-XYLYLENE DICHLORIDE

CONSENSUS REPORTS: Reported in EPA TSCA Inventory.

SAFETY PROFILE: Poison by intravenous route. When heated to decomposition it emits toxic Cl^-.

DGP900   CAS: 62-73-7   **HR: 3**
DICHLORVOS

DOT: NA 2783
mf: $C_4H_7Cl_2O_4P$ mw: 220.98

PROP: Liquid. Bp: 120° @ 14 mm, bp: 77° @ 1 mm. Sltly sol in water and glycerin; miscible with aromatic and chlorinated hydrocarbon solvents and alcohol.

SYNS: APAVAP * ASTROBOT * ATGARD * BAY 19149 * BENFOS * BIBESOL * BREVINYL * CANOGARD * CEKUSAN * CHLORVINPHOS * CYANOPHOS * CYPONA * DDVF * DDVP * DEDEVAP * DERIBAN * DERRIBANTE * DEVIKOL * (2,2-DICHLOOR-VINYL)-DI-METHYL-FOSFAAT (DUTCH) * DICHLOORVO (DUTCH) * DICHLORFOS (POLISH) * 2,2-DICHLOROETHENOL DIMETHYL PHOSPHATE * 2,2-DICHLOROETHENYL DIMETHYL PHOSPHATE * 2,2-DICHLOROETHENYL PHOSPHORIC ACID DIMETHYL ESTER * DICHLOROPHOS * DICHLOROVAS * (2,2-DICHLORO-VINIL)DIMETILFOSFATO (ITALIAN) * 2,2-DICHLOROVINYL ALCOHOL, DIMETHYL PHOSPHATE * 2,2-DICHLOROVINYL DIMETHYL PHOSPHATE * 2,2-DICHLOROVINYL DIMETHYL PHOSPHORIC ACID ESTER * DICHLOROVOS * DICHLORPHOS * (2,2-DICHLOR-VINYL)-DIMETHYL-PHOSPHAT (GERMAN) * O-(2,2-DICHLORVINYL)-O,O-DIMETHYLPHOSPHAT (GERMAN) * DIMETHYL-2,2-DICHLOROETHENYL PHOSPHATE * DIMETHYL DICHLOROVINYL PHOSPHATE * DIMETHYL-2,2-DICHLOROVINYL PHOSPHATE * O,O-DIMETHYL DICHLOROVINYL PHOSPHATE * O,O-DIMETHYL-O-2,2-DICHLOROVINYL PHOSPHATE * O,O-DIMETHYL-O-(2,2-DICHLOR-VINYL)-PHOSPHAT (GERMAN) * DIVIPAN * DQUIGARD * DUO-KILL * DURAVOS * ENT 20,738 * EQUIGEL * ESTROSEL * ESTROSOL * FECAMA * FLY-DIE * FLY FIGHTER * HERKAL * KRECALVIN * LINDAN * MAFU * MARVEX * MOPARI * NCI-C00113 * NERKOL * NOGOS * NO-PEST * NO-PEST STRIP * NSC-6738 * NUVA * OKO * OMS 14 * PHOSPHATE de DIMETHYLE et de 2,2-DICHLOROVINYLE (FRENCH) * PHOSPHORIC ACID-2,2-DICHLOROETHENYL DIMETHYL ESTER * PHOSVIT * SD-1750 * SZKLARNIAK * TAP 9VP * TASK * TASK TABS * TENAC * TETRAVOS * VAPONA * VAPONITE * VERDICAN * VERDIPOR * VINYLOFOS * VINYLOPHOS

CONSENSUS REPORTS: IARC Cancer Review: GROUP 3 IMEMDT 7,56,87; Animal Inadequate Evidence IMEMDT 20,97,79. NCI Carcinogenesis Bioassay (feed); No Evidence: mouse, rat NCITR* NCI-CG-TR-10,77. EPA Genetic Toxicology Program. Community Right-To-Know List. EPA Extremely Hazardous Substances List.

OSHA PEL: TWA 1 mg/m^3 (skin)
ACGIH TLV: TWA 0.1 ppm (skin)
DFG MAK: 0.1 ppm (1 mg/m^3)
DOT Classification: Poison B; Label: Poison; Poison B; Label: Poison.

SAFETY PROFILE: Poison by ingestion, inhalation, skin contact, subcutaneous, intravenous, intraperitoneal, and possibly other routes. Experimental teratogenic and reproductive effects. Human mutation data reported. A cholinesterase inhibitor, it is used in flea (pest) collars for pets. No neurotoxicity has been observed. It is very rapidly metabolized and excreted. When heated to decomposition it emits very toxic fumes of Cl^- and PO_x.

DGQ875   CAS: 141-66-2   HR: 3
DICROTOPHOS
mf: $C_8H_{16}NO_5P$ mw: 237.22

SYNS: BIDIRL * BIDRIN * C 709 * CARBI-CRON * CIBA 709 * DIAPADRIN * DICROTOFOS (DUTCH) * 3-(DIMETHOXYPHOSPHINYLOXY)-N,N-DI-METHYL-cis-CROTONAMIDE * 3-(DIMETHOXYPHOS-PHINYLOXY)-N,N-DIMETHYLISOCROTONAMIDE * 3-(DIMETHYLAMINO)-1-METHYL-3-OXO-1-PROPENYL DIMETHYL PHOSPHATE * cis-2-DIMETHYLCARBA-MOYL-1-METHYLVINYL DIMETHYLPHOSPHATE * O,O-DIMETHYL-O-(2-DIMETHYL-CARBAMOYL-1-METHYL-VINYL)PHOSPHAT (GERMAN) * O,O-DI-METHYL-O-(N,N-DIMETHYLCARBAMOYL-1-METHYL-VINYL) PHOSPHATE * O,O-DIMETHYL-O-(1,4-DI-METHYL-3-OXO-4-AZA-PENT-1-ENYL)FOSFAAT (DUTCH) * O,O-DIMETHYL-O-(1,4-DIMETHYL-3-OXO-4-AZA-PENT-1-ENYL)PHOSPHATE * DIMETHYLPHOSPHATE ESTER with 3-HYDROXY-N,N-DIMETHYL-cis-CROTONAMIDE * DIMETHYL PHOSPHATE of 3-HYDROXY-N,N-DI-METHYL-cis-CROTONAMIDE * O,O-DIMETIL-O-(1,4-DI-METIL-3-OXO-4-AZA-PENT-1-ENIL)-FOSFATO (ITALIAN) * EKTAFOS * ENT 24,482 * 3-HYDROXYDI-METHYL CROTONAMIDE DIMETHYL PHOSPHATE * 3-HYDROXY-N,N-DIMETHYL-cis-CROTONAMIDE DI-METHYL PHOSPHATE * PHOSPHATE de DIMETHYLE et de 2-DIMETHYLCARBAMOYL-1-METHYL VINYLE (FRENCH) * SD 3562 * SHELL SD-3562

CONSENSUS REPORTS: EPA Farm Worker Reentry (39 FR 16888,74). EPA Genetic Toxicology Program. EPA Extremely Hazardous Substances List.

OSHA PEL: TWA 0.25 mg/m^3 (skin)
ACGIH TLV: TWA 0.25 mg/m^3 (skin)

SAFETY PROFILE: Poison by ingestion, inhalation, skin contact, subcutaneous, intravenous, intraperitoneal, and possibly other routes. Mutation data reported. Used to control the coffee borer and certain economically important pests of cotton. When heated to decomposition it emits very toxic fumes of NO_x and PO_x.

DGR600   CAS: 80-43-3   HR: 1
DI-α-CUMYL PEROXIDE

DOT: UN 2121
mf: $C_{18}H_{22}O_2$ mw: 270.40

SYNS: ACTIVE DICUMYL PEROXIDE * BIS(α,α-DI-METHYLBENZYL)PEROXIDE * CUMENE PEROXIDE * CUMYL PEROXIDE * DICUMYL PEROXIDE (DOT) * DI-CUP * DI-CUP 40 KF * DI-CUPR * DIISO-PROPYLBENZENE PEROXIDE * ISOPROPYLBENZENE PEROXIDE * LUPERCO * LUPEROX * LUPEROX 500R * LUPEROX 500T * VAROX DCP-R * VAROX DCP-T

CONSENSUS REPORTS: Reported in EPA TSCA Inventory.

DOT Classification: Organic Peroxide, Label: Organic Peroxide.

SAFETY PROFILE: Mildly toxic by ingestion and possibly other routes. When heated to decomposition it emits acrid smoke and irritating fumes.

DGT200   CAS: 38780-37-9   HR: 3
cis-DICYCLOBUTYLAMMINEDI-CHLOROPLATINUM(II)
mf: $C_8H_{18}Cl_2N_2Pt$ mw: 408.27

SYNS: cis-BIS(CYCLOBUTYLAMMINE)DICHLORO-PLATINUM(II) * cis-DICHLOROBIS(CYCLOBUTYL-AMMINE)PLATINUM(II)

SAFETY PROFILE: Poison by intraperitoneal route. Mutagenic data reported. When heated to decomposition it emits very toxic fumes of Cl^- and NO_x.

DGT600   CAS: 101-83-7   HR: 3
N,N-DICYCLOHEXYLAMINE

DOT: UN 2565
mf: $C_{12}H_{23}N$ mw: 181.36

PROP: Liquid, fishy odor. Mp: −1°, bp: 256°, flash p: >210°F (OC), d: 0.910, vap d: 6.27.

SYNS: CDHA * N-CYCLOHEXYLCYCLOHEXANA-MINE * DICYCLOHEXYLAMINE (DOT) * DICYKLO-HEXYLAMIN (CZECH) * DODECAHYDRODIPHENYL-AMINE

CONSENSUS REPORTS: IARC Cancer Review: GROUP 3 IMEMDT 7,178,87; Animal Inadequate Evidence IMEMDT 22,55,80. Reported in EPA TSCA Inventory.

DOT Classification: Corrosive Material; Label: Corrosive.

SAFETY PROFILE: Questionable carcinogen with experimental tumorigenic data. Poison by ingestion and subcutaneous routes. Human mutation data reported. Corrosive. A severe skin and eye irritant. Combustible when exposed to heat or flame; can react with oxidizing materials. To fight fire, use alcohol foam, CO_2, dry chemical. When heated to decomposition it emits toxic fumes of NO_x.

DGW000   CAS: 77-73-6   HR: 3
DICYCLOPENTADIENE

DOT: UN 2048
mf: $C_{10}H_{12}$ mw: 132.22

PROP: Colorless crystals. Mp: 32.9°, bp: 166.6°, d: 0.976 @ 35°, vap press: 10 mm @ 47.6°, vap d: 4.55, flash p: 90°F (OC).

SYNS: BICYCLOPENTADIENE * BISCYCLOPENTA-DIENE * 1,3-CYCLOPENTADIENE, DIMER * DICY-CLOPENTADIENE * DICYKLOPENTADIEN (CZECH) * DIMER CYKLOPENTADIENU (CZECH) * 3a,4,7,7a-TETRAHYDRO-4,7-METHANOINDENE

CONSENSUS REPORTS: Reported in EPA TSCA Inventory. EPA Genetic Toxicology Program.

OSHA PEL: TWA 5 ppm
ACGIH TLV: TWA 5 ppm
DOT Classification: IMO: Flammable or Combustible Liquid; Label: Flammable Liquid.

SAFETY PROFILE: Poison by ingestion and intraperitoneal routes. Moderately toxic by inhalation. Mildly toxic by skin contact. A severe skin and moderate eye irritant. Dangerous fire hazard when exposed to heat or flame; can react with oxidizing materials. To fight fire, use alcohol foam. When heated to decomposition it emits acrid smoke and fumes.

DHA400   CAS: 64070-13-9   HR: 3
3′,4′-DIDEOXYKANAMYCIN B SULFATE
mf: $C_{18}H_{37}N_5O_8 \cdot O_4S$ mw: 547.66

SYN: DKB SULFATE

SAFETY PROFILE: Poison by intramuscular and intravenous routes. Moderately toxic by intraperitoneal and subcutaneous routes. When heated to decomposition it emits very toxic fumes of NO_x and SO_x.

DHB400   CAS: 60-57-1   HR: 3
DIELDRIN

DOT: UN 2761
mf: $C_{12}H_8Cl_6O$ mw: 380.90

PROP: White crystals; odorless. Mp: 150°, vap d: 13.2. Insol in water; sol in common organic solvents.

SYNS: ALVIT * COMPOUND 497 * DIELDREX * DIELDRINE (FRENCH) * DIELDRITE * ENT 16,225 * HEOD * HEXACHLOROEPOXYOCTAHY-DRO-ENDO,EXO-DIMETHANONAPHTHALENE * 3,4,5,6,9,9-HEXACHLORO-1a,2,2a,3,6,6a,7,7a-OCTAHY-DRO-2,7:3,6-DIMETHANONAPHTH(2,3-b)OXIRENE * ILLOXOL * INSECTICIDE No. 497 * NCI-C00124 * OCTALOX * PANORAM D-31 * QUINTOX * RCRA WASTE NUMBER P037

CONSENSUS REPORTS: IARC Cancer Review: GROUP 3 IMEMDT 7,196,87; Human Inadequate Evidence IMEMDT 5,125,74; Animal Sufficient Evidence IMEMDT 5,125,74. NCI Carcinogenesis Bioassay (feed); Clear Evidence: mouse NCITR* NCI-CG-TR-21,78; No Evidence: rat NCITR* NCI-CG-TR-22,78; Inadequate Studies: rat NCITR* NCI-CG-TR-21,78.

OSHA PEL: TWA 0.25 mg/m^3 (skin)
ACGIH TLV: TWA 0.25 mg/m^3 (skin)
DFG MAK: 0.25 mg/m^3
NIOSH REL: Lowest reliable detectable level.
DOT Classification: ORM-A; Label: None.

SAFETY PROFILE: A human poison by ingestion and possibly other routes. Poison experimentally by inhalation, ingestion, skin contact, intravenous, intraperitoneal and possibly other routes. Experimental teratogenic and reproductive effects. Absorbed readily through the skin and by other routes. It is a central nervous system stimulant. Questionable carcinogen with experimental carcinogenic, neoplastigenic, and tumorigenic data. Human mutation data reported. An insecticide. Dieldrin is considerably more toxic than DDT by ingestion and skin contact. Dieldrin or its derivatives may accumulate in the body from chronic low dosages. When heated to decomposition it emits toxic fumes of Cl^-.

DHB600   CAS: 298-18-0   HR: 3
dl-DIEPOXYBUTANE
mf: $C_4H_6O_2$ mw: 86.10

PROP: Colorless liquid. Bp: 138°, mp: 4°, d: 1.112 @ 18°/4°.

SYNS: dl-BUTADIENE DIOXIDE * 1,2:3,4-DIANHY-DRO-dl-THREITOL * (±)-1,2:3,4-DIEPOXYBUTANE * dl-1,2:3,4-DIEPOXYBUTANE

CONSENSUS REPORTS: IARC Cancer Review: GROUP 2B IMEMDT 7,56,87; Animal Sufficient Evidence IMEMDT 11,115,76. EPA Genetic Toxicology Program.

SAFETY PROFILE: Suspected carcinogen with experimental carcinogenic, neoplastigenic, and tumorigenic data. Poison by ingestion, inhalation, and skin contact. Mutation data reported. When heated to decomposition it emits acrid smoke and irritating fumes.

DHB800   CAS: 564-00-1   HR: 3
meso-1,2,3,4-DIEPOXYBUTANE
mf: $C_4H_6O_2$ mw: 86.10

SYNS: (R*,S*)-2,2'-BIOXIRANE * 1,2:3,4-DIANHY-DROERYTHRITOL * meso-DIEPOXYBUTANE * (R*,S*)-DIEPOXYBUTANE * ERYTHRITOL ANHYDRIDE

CONSENSUS REPORTS: IARC Cancer Review: GROUP 2B IMEMDT 7,56,87; Animal Sufficient Evidence IMEMDT 11,115,76.

SAFETY PROFILE: Suspected carcinogen with experimental carcinogenic, neoplastigenic, and tumorigenic data. Poison by skin contact and intraperitoneal routes. Mutation data reported. When heated to decomposition it emits acrid smoke and irritating fumes.

DHF000   CAS: 111-42-2   HR: 2
DIETHANOLAMINE
mf: $C_4H_{11}NO_2$ mw: 105.16

PROP: A faintly colored, viscous liquid. Mp: 28°, bp: 269.1° (decomp), flash p: 305°F (OC), d: 1.0919 @ 30°/20°, autoign temp: 1224°F, vap press: 5 mm @ 138°, vap d: 3.65.

SYNS: BIS(2-HYDROXY ETHYL)AMINE * DEA * DIAETHANOLAMIN (GERMAN) * DIETHANOL-AMIN (CZECH) * DIETHYLOLAMINE * 2,2'-DIHY-DROXYDIETHYLAMINE * DI(2-HYDROXYETHYL)-AMINE * DIOLAMINE * 2,2'-IMINOBISETHANOL * 2,2'-IMINODIETHANOL * NCI-C55174

CONSENSUS REPORTS: Community Right-To-Know List. Reported in EPA TSCA Inventory.

OSHA PEL: TWA 3 ppm
ACGIH TLV: TWA 3 ppm

SAFETY PROFILE: Moderately toxic by ingestion, intraperitoneal, and subcutaneous routes. Mildly toxic by skin contact. A severe eye and mild skin irritant. Combustible when exposed to heat or flame; can react with oxidizing materials. To fight fire, use alcohol foam, water, CO_2, dry chemical. When heated to decomposition it emits toxic fumes such as NO_x.

DHF200   CAS: 5716-15-4   HR: 3
DIETHANOLAMMONIUM MALEIC HYDRAZIDE
mf: $C_4H_{11}NO_2 \cdot C_4H_4N_2O_2$ mw: 217.26

SYNS: 6-HYDROXY-3-(2H)-PYRIDAZINONE DIETHA-NOLAMINE * 2,2'-IMINODI-ETHANOL with 1,2-DIHY-DRO-3,6-PYRIDAZINEDIONE (1:1) * MALEIC HYDRA-ZIDE DIETHANOLAMINE SALT * MH-30 * NCI-C54660

CONSENSUS REPORTS: Reported in EPA TSCA Inventory.

SAFETY PROFILE: Questionable carcinogen with experimental tumorigenic data. Moderately toxic by ingestion. Mutation data reported. When heated to decomposition it emits toxic fumes of NO_x and NH_3.

DHG000   CAS: 78-62-6   HR: 1
DIETHOXYDIMETHYLSILANE
DOT: UN 2380
mf: $C_6H_{16}O_2Si$ mw: 148.31

PROP: Liquid. Bp: 113.5°, d: 0.834, vap press: 10 mm @ 13.3°, vap d: 5.1. flash p.: <73.4°F.

SYNS: DIMETHYL-DIETHOXYSILAN (CZECH) * DIMETHYLDIETHOXYSILANE (DOT)

CONSENSUS REPORTS: Reported in EPA TSCA Inventory.

DOT Classification: Flammable Liquid; Label: Flammable Liquid.

SAFETY PROFILE: Mildly toxic by inhalation and ingestion. A skin and eye irritant. A dangerous fire hazard when exposed to heat, flame or oxidizers. When heated to decomposition it emits acrid smoke and irritating fumes.

DHH400   CAS: 950-10-7   HR: 3
2-(DIETHOXYPHOSPHINYLIMINO)-4-METHYL-1,3-DITHIOLANE
mf: $C_8H_{16}NO_3PS_2$ mw: 269.34

SYNS: AC 47470 * AMERICAN CYANAMID CL-47470 * CL-47,470 * CYCLIC PROPYLENE (DIETHOXYPHOSPHINYL)DITHIOIMIDOCARBONATE

* CYTROLANE * p,p-DIETHYL CYCLIC PROPYLENE ESTER of PHOSPHONODITHIOIMIDOCARBONIC ACID * DIETHYL (4-METHYL-1,3-DITHIOLAN-2-YLIDENE) PHOSPHOROAMIDATE * EI-47470 * ENT 25,991 * MEPHOSFOLAN * (4-METHYL-1,3-DITHIOLAN-2-YLIDENE)PHOSPHORAMIDIC ACID, DIETHYL ESTER

CONSENSUS REPORTS: EPA Extremely Hazardous Substances List. Reported in EPA TSCA Inventory. EPA Genetic Toxicology Program.

SAFETY PROFILE: Poison by ingestion and skin contact. When heated to decomposition it emits very toxic fumes of NO_x, PO_x, and SO_x.

DHI000   CAS: 97-96-1   **HR: 2**
DIETHYL ACETALDEHYDE
DOT: UN 1178
mf: $C_6H_{12}O$ mw: 100.18

PROP: Colorless liquid; ungent odor. Bp: 116.8°, flash p: 70°F (OC), fp: −89°, d: 0.808-0.814, vap press: 13.7 mm @ 20°, vap d: 3.45, lel: 1.2%, uel: 7.7%. Misc in alc, ether; sltly sol in water.

SYNS: ALDEHYDE-2-ETHYLBUTYRIQUE (FRENCH) * 2-ETHYLBUTANAL * 2-ETHYLBUTRIC ALDEHYDE * ETHYL BUTYRALDEHYDE * α-ETHYLBUTYRAL-DEHYDE * ETHYL BUTYRALDEHYDE (DOT) * 2-ETHYLBUTYRALDEHYDE (DOT,FCC) * FEMA No. 2426

CONSENSUS REPORTS: Reported in EPA TSCA Inventory.

DOT Classification: Flammable Liquid; Label: Flammable Liquid.

SAFETY PROFILE: Moderately toxic by ingestion. Mildly toxic by inhalation. A skin irritant. Flammable liquid. Can react vigorously with oxidizing materials. To fight fire, use alcohol foam, CO_2, dry chemical. When heated to decomposition it emits acrid smoke and fumes.

DHI200   CAS: 685-91-6   **HR: 3**
N,N-DIETHYLACETAMIDE
mf: $C_6H_{13}NO$ mw: 115.20

PROP: Liquid. Mp: <65°, bp: 180°, flash p: 170°F, d: 0.92, vap d: 4.0.

CONSENSUS REPORTS: Reported in EPA TSCA Inventory.

SAFETY PROFILE: Questionable carcinogen with experimental tumorigenic data. Moderately

toxic by ingestion, intravenous, and intraperitoneal routes. Flammable when exposed to heat or flame. To fight fire, use foam, mist, CO_2, dry chemical. When heated to decomposition it emits toxic fumes of NO_x.

DHI400   CAS: 88-09-5   **HR: 2**
DIETHYLACETIC ACID
mf: $C_6H_{12}O_2$ mw: 116.18

PROP: Colorless, volatile liquid; rancid odor. Mp: −93°, bp: 121.0°, flash p: 78°F (CC), d: 0.917, vap press: 10 mm @ 15.3°, vap d: 4.0, autoign temp: 865°F. Misc in alc, ether, water.

SYNS: 2-ETHYL BUTANOIC ACID * α-ETHYLBU-TYRIC ACID * 2-ETHYLBUTYRIC ACID (FCC) * FEMA No. 2429 * 3-PENTANECARBOXYLIC ACID

CONSENSUS REPORTS: Reported in EPA TSCA Inventory.

SAFETY PROFILE: Moderately toxic by ingestion and skin contact. An irritant to skin and mucous membranes. A severe eye irritant. Narcotic in high concentrations. Flammable liquid. To fight fire, use CO_2, dry chemical, alcohol foam. When heated to decomposition it emits acrid smoke and fumes.

DHI800   CAS: 63019-57-8   **HR: 3**
1-DIETHYLACETYLAZIRIDINE
mf: $C_8H_{15}NO$ mw: 141.24

SYN: DIETHYLACETYLETHYLENEIMINE

SAFETY PROFILE: Questionable carcinogen with experimental neoplastigenic data. Mutation data reported. When heated to decomposition it emits toxic fumes of NO_x.

DHJ200   CAS: 109-89-7   **HR: 2**
DIETHYLAMINE
DOT: UN 1154
mf: $C_4H_{11}N$ mw: 73.16

PROP: Colorless liquid, ammoniacal odor. Mp: −38.9°, bp: 55.5°, flash p: −0.4°F, d: 0.7108 @ 20°/20°, autoign temp: 594°F, vap press: 400 mm @ 38.0°, vap d: 2.53, lel: 1.8%, uel: 10.1%.

SYNS: 2-AMINOPENTANE * DIAETHYLAMIN (GERMAN) * N,N-DIETHYLAMINE * DIETILAMINA (ITALIAN) * DWUETYLOAMINA (POLISH) * N-ETHYL-ETHANAMINE

OSHA PEL: (Transitional: TWA 25 ppm) TWA 10 ppm; STEL 25 ppm
ACGIH TLV: TWA 10 ppm; STEL 25 ppm
DFG MAK: 10 ppm (30 mg/m^3)
DOT Classification: Flammable Liquid; Label: Flammable Liquid.

SAFETY PROFILE: Moderately toxic by ingestion, inhalation, and skin contact. A skin and severe eye irritant. Exposure to strong vapor can cause severe cough and chest pains. Contact with liquid can damage eyes, possibly permanently; contact with skin causes necrosis and vesiculation. A very dangerous fire hazard when exposed to heat, flame or oxidizers. To fight fire, use alcohol foam, CO_2, dry chemical. Explodes on contact with dicyanofurazan. Violent reaction with sulfuric acid. Ignites on contact with cellulose nitrate of sufficiently high surface area. When heated to decomposition it emits toxic fumes of NO_x.

DHJ600   CAS: 3010-02-4   **HR: 3**
N,N-DIETHYLAMINOACETONITRILE
mf: $C_6H_{12}N_2$ mw: 112.20

SYNS: (DIETHYLAMINO)ACETONITRILE * N,N-DI-ETHYLGLYCINONITRILE * NITRIL KISELINY DIETH-YLAMINOOCTOVE (CZECH)

CONSENSUS REPORTS: Cyanide and its compounds are on the Community Right-To-Know List. Reported in EPA TSCA Inventory.

SAFETY PROFILE: Poison by ingestion and skin contact. Moderately toxic by inhalation. A skin and severe eye irritant. When heated to decomposition it emits toxic fumes of NO_x.

DHK400   CAS: 137-58-6   **HR: 3**
2-(DIETHYLAMINO)-2',6'-ACETOXYLIDIDE
mf: $C_{14}H_{22}N_2O$ mw: 234.38

SYNS: ANESTACON * DIETHYLAMINOACETO-2,6-XYLIDIDE * α-DIETHYLAMINOACETO-2,6-XYLIDIDE * α-DIETHYLAMINO-2,6-ACETOXYLIDIDE * DI-ETHYLAMINOACET-2,6-XYLIDIDE * α-DIETHYL-AMINO-2,6-DIMETHYLACETANILIDE * omega-DI-ETHYLAMINO-2,6-DIMETHYLACETANILIDE * α-DIETILAMINO-2,6-DIMETILACETANILIDE (ITALIAN) * DUNCAINE * GRAVOCAIN * ISICAINA * LEOSTESIN * LIDA-MANTLE * LIDOCAINE * LIGNOCAINE * MARICAINE * RUCAINA * SOLCAIN * XILOCAINA (ITAL-

IAN) * XYCIANE * XYLESTESIN * XYLOCAIN * XYLOCITIN * XYLOTOX

CONSENSUS REPORTS: Reported in EPA TSCA Inventory.

SAFETY PROFILE: Poison by ingestion, intravenous, intraperitoneal, and subcutaneous routes. Human systemic effects by ingestion: excitement, hallucinations, distorted perceptions, and changes in heart rate. A local anesthetic. When heated to decomposition it emits toxic fumes of NO_x.

DHK600   CAS: 73-78-9   **HR: 3**
2-DIETHYLAMINO-2',6'-ACETOXYLIDIDE HYDROCHLORIDE
mf: $C_{14}H_{22}N_2O \cdot ClH$ mw: 270.84

SYNS: ANESTACON HYDROCHLORIDE * 2-(DI-ETHYLAMINO)-2',6'-ACETOXYLIDIDE MONOHYDRO-CHLORIDE * α-DIETHYLAMINO-2,5-ACETOXYLIDINE HYDROCHLORIDE * omega-DIETHYLAMINO-2,6-DI-METHYLACETANILIDE HYDROCHLORIDE * 2-(DI-ETHYLAMINO)-N-(2,6-DIMETHYLPHENYL)ACETAMIDE MONOHYDROCHLORIDE * DUNCAINE HYDROCHLO-RIDE * GRAVOCAIN HYDROCHLORIDE * ISICAINE HYDROCHLORIDE * LEOSTESIN HYDROCHLORIDE * LIDOCAINE HYDROCHLORIDE * LIDOTHESIN HY-DROCHLORIDE * LIGNOCAINE HYDROCHLORIDE * RUCAINA HYDROCHLORIDE * S 202 * XYCAINE HYDROCHLORIDE * XYLESTESIN HYDROCHLORIDE * XYLOCAINE HYDROCHLORIDE * XYLOCARD * XYLOCITIN HYDROCHLORIDE * XYLONEURAL * XYLOTOX HYDROCHLORIDE

CONSENSUS REPORTS: Reported in EPA TSCA Inventory. EPA Genetic Toxicology Program.

SAFETY PROFILE: Poison by ingestion, intraperitoneal, intravenous, subcutaneous, intramuscular, and intratracheal routes. Human systemic effects by intravenous route: somnolence, respiratory depression, low blood pressure. A skin and eye irritant. An anesthetic. When heated to decomposition it emits very toxic fumes of NO_x and HCl.

DHL800   CAS: 1027-14-1   **HR: 3**
DIETHYLAMINOACETYL-2,4,6-TRIMETHYLANILINE HYDROCHLORIDE
mf: $C_{15}H_{24}N_2O \cdot ClH$ mw: 284.87

SYNS: 2-DIETHYLAMINO-2',4',6'-TRIMETHYLACETANI-LIDE HYDROCHLORIDE * 2-(DIETHYLAMINO)-2',4',6'-TRIMETHYLACETANILIDE MONOHYDROCHLORIDE

* 2-(DIETHYLAMINO)-N-(2,4,6-TRIMETHYLPHENYL) ACETAMIDE MONOHYDROCHLORIDE * MESIDICAINE HYDROCHLORIDE * MESOCAINE HYDROCHLORIDE * MESOKAIN HYDROCHLORIDE * TRIMECAINE * TRIMECAINE HYDROCHLORIDE * TRIMEKAIN HYDROCHLORIDE * N-sym-TRIMETHYLPHENYLDIETHYL-AMINOACETAMIDE HYDROCHLORIDE

SAFETY PROFILE: Poison by intraperitoneal and subcutaneous routes. When heated to decomposition it emits very toxic fumes of NO_x and HCl.

DHO500   CAS: 100-37-8   HR: 3
2-DIETHYLAMINOETHANOL

DOT: UN 2686
mf: $C_6H_{15}NO$ mw: 117.22

PROP: Colorless, hygroscopic liquid. Bp: 162°, flash p: 140°F (OC), d: 0.8851 @ 20°/20°, vap press: 1.4 mm @ 20°, vap d: 4.03.

SYNS: DEAE * DIAETHYLAMINOAETHANOL (GERMAN) * DIETHYLAMINOETHANOL * β-DIETHYL-AMINOETHANOL * N-DIETHYLAMINOETHANOL * 2-(DIETHYLAMINO)ETHANOL * 2-N-DIETHYLAMI-NOETHANOL * DIETHYLAMINOETHANOL (DOT) * β-DIETHYLAMINOETHYL ALCOHOL * DIETHYL-ETHANOLAMINE * N,N-DIETHYLETHANOLAMINE * N,N-DIETHYL-N-(β-HYDROXYETHYL)AMINE * 2-HYDROXYTRIETHYLAMINE

CONSENSUS REPORTS: Reported in EPA TSCA Inventory.

OSHA PEL: TWA 10 ppm (skin)
ACGIH TLV: TWA 10 ppm (skin)
DFG MAK: 10 ppm (50 mg/m³)
DOT Classification: Flammable or Combustible Liquid; Label: Flammable Liquid.

SAFETY PROFILE: Poison by intraperitoneal and intravenous routes. Moderately toxic by ingestion, skin contact, subcutaneous, intramuscular, and possibly other routes. Human systemic effects by inhalation: nausea or vomiting. A skin and severe eye skin irritant. Combustible liquid. Flammable when exposed to heat or flame; can react with oxidizing materials. To fight fire, use alcohol foam, CO_2, dry chemical. When heated to decomposition it emits toxic fumes of NO_x.

DHO600   CAS: 487-53-6   HR: 3
DIETHYLAMINOETHANOL-p-AMINOSALICYLATE
mf: $C_{13}H_{20}N_2O_3$ mw: 252.35

SYNS: DIETHYLAMINOETHYL-p-AMINOSALICYLATE * 2-DIETHYLAMINOETHYL-p-AMINOSALICYLATE * 2-DIETHYLAMINOETHYL-4-AMINOSALICYLATE * DIETHYLAMINOETHYL-3-HYDROXY-4-AMINOBEN-ZOATE * HYDROXYPROCAINE * m-HYDROXYPRO-CAINE * METAHYDROXYPROCAINE * OXYCAINE * OXYPROCAIN * OXYPROCAINE

SAFETY PROFILE: Poison by intravenous and intraperitoneal routes. When heated to decomposition it emits toxic fumes of NO_x.

DHP200   CAS: 6376-26-7   HR: 3
o-(DIETHYLAMINOETHOXY)BENZANI-LIDE
mf: $C_{19}H_{24}N_2O_2$ mw: 312.45

SYNS: o-DIAETHYLAMINOAETHOXY-BENZANILID (GERMAN) * 2-(2-(DIETHYLAMINO)ETHOXY)BENZANI-LIDE

SAFETY PROFILE: Poison by ingestion, subcutaneous, and intravenous routes. An experimental teratogen. When heated to decomposition it emits toxic fumes of NO_x.

DHQ800   CAS: 17822-74-1   HR: 3
2-(2-(DIETHYLAMINO)ETHOXY)-3-METHYLBENZANILIDE
mf: $C_{20}H_{25}N_2O_2$ mw: 325.47

SAFETY PROFILE: Poison by subcutaneous route. An experimental teratogen. When heated to decomposition it emits toxic fumes of NO_x.

DHS000   CAS: 67-98-1   HR: 3
(p-2-DIETHYLAMINOETHOXYPHENYL)-1-PHENYL-2-p-ANISYLETHANOL
mf: $C_{27}H_{33}NO_3$ mw: 419.61

SYNS: 1-(p-2-DIETHYLAMINOETHOXYPHENYL)-1-PHE-NYL-2-p-ANISYLETHANOL * 1-(4-(2-DIETHYLAMINO-ETHOXY)PHENYL)-1-PHENYL-2-(p-ANISYL)ETHANOL * 1-(p-(2-(DIETHYLAMINO)ETHOXY)PHENYL)-1-PHE-NYL-2-(p-METHOXYPHENYL)ETHANOL * ETHAMOXY-TRIPHETOL * ETHANOXYTRIPHETOL * MER 25

SAFETY PROFILE: Questionable carcinogen with experimental tumorigenic data. Moderately toxic by ingestion. Experimental teratogenic and reproductive effects. When heated to decomposition it emits toxic fumes of NO_x.

DHS200   CAS: 2192-21-4   HR: 3
1-(2-(2-(DIETHYLAMINO)ETHOXY) PHENYL)-3-PHENYL-1-PROPANONE HYDROCHLORIDE
mf: $C_{21}H_{27}NO_2 \cdot ClH$ mw: 361.95

SYNS: ASAMEDOL * BAXACOR * CORODILAN * DIALICOR * (o-β-DIETHYLAMINOETHOXY)-PHE-NYL PROPIOPHENONE HYDROCHLORIDE * ETAFE-NONE HYDROCHLORIDE * HETAPHENONE * L.G. 11,457 HYDROCHLORIDE * PAGANO-COR * β-PHENYL-o-(DIETHYLAMINOETHOXY)PROPIO-PHENONE HYDROCHLORIDE * RELICOR * RELICOR HYDROCHLORIDE

SAFETY PROFILE: Poison by ingestion and intravenous routes. When heated to decomposition it emits very toxic fumes of NO_x and HCl.

DHU000   CAS: 479-50-5   **HR: 3**
1-(2'-DIETHYLAMINO)ETHYLAMINO-4-METHYLTHIOXANTHENONE
mf: $C_{20}H_{24}N_2OS$ mw: 340.52

SYNS: 1-((2-(DIETHYLAMINO)ETHYL)AMINO)-4-METHYL-9H-THIOXANTHEN-9-ONE * LUCANTHON * LUCANTHONE * MIRACIL D * NILODIN

CONSENSUS REPORTS: EPA Genetic Toxicology Program.

SAFETY PROFILE: Poison by intravenous route. Human mutation data reported. When heated to decomposition it emits very toxic fumes of NO_x and SO_x.

DHW200   CAS: 902-83-0   **HR: 3**
2-(DIETHYLAMINO)ETHYLCHLORODI-PHENYLACETATE HYDROCHLORIDE
mf: $C_{20}H_{24}ClNO_2 \cdot ClH$ mw: 382.36

SYNS: 2-CHLORO-2,2-DIPHENYLACETIC ACID-2-(DI-ETHYLAMINO)ETHYL ESTER HYDROCHLORIDE * DIAMINOPHEN * DIAPHEN (NEUROPLEGIC)

SAFETY PROFILE: Poison by intraperitoneal route. When heated to decomposition it emits very toxic fumes of Cl^- and NO_x.

DHX800   CAS: 64-95-9   **HR: 3**
2-DIETHYLAMINOETHYL DIPHENYLACETATE
mf: $C_{20}H_{25}NO_2$ mw: 311.46

SYNS: ADIPHENIN * ADIPHENINE * BENZENE-ACETIC ACID, α-PHENYL-, 2-(DIETHYLAMINO)ETHYL ESTER, (9CI) * 2-DIETHYLAMINOETHYLESTER KY-SELINY DIFENYLOCTOVE * DIFACIL * DIPHACIL * DIPHACYL * DIPHENYLACETIC ACID DIETHYL-AMINOETHYL ESTER * DIPHENYLACETIC ACID, 2-(DIETHYLAMINO)ETHYL ESTER * DIPHENYLACE-TYLDIETHYLAMINOETHANOL * ESTER DWUETYLO-

AMINOETYLOWY KWASU DWUFENYLOOCTOWEGO * PATROVINE * SPASMOLYTIN * TRANSENTINE * TRANZETIL * TRASENTIN * TRASENTINE * TRAZENTYNA * VEGANTINE * WEGANTYNA

SAFETY PROFILE: Poison by intravenous and intraperitoneal routes. Moderately toxic by ingestion and subcutaneous routes. When heated to decomposition it emits toxic fumes of NO_x.

DIB600   CAS: 1227-61-8   **HR: 3**
N-(2-(DIETHYLAMINO)ETHYL)-2-(p-METHOXYPHENOXY)ACETAMIDE
mf: $C_{15}H_{24}N_2O_3$ mw: 280.41

SYNS: MEPHEXAMIDE * 2-(p-METHOXYPHENOXY)-N-(2-(DIETHYLAMINO)ETHYL)ACETAMIDE * MEXE-PHENAMIDE

SAFETY PROFILE: Poison by ingestion and intravenous routes. When heated to decomposition it emits toxic fumes of NO_x.

DIF600   CAS: 14557-50-7   **HR: 3**
2-(2-(DIETHYLAMINO)ETHYL)-2-PHENYL-4-PENTENOIC ACID ETHYL ESTER
mf: $C_{19}H_{29}NO_2$ mw: 303.49

SYN: UCB 6249

SAFETY PROFILE: Poison by ingestion, intra-peritoneal, and intravenous routes. When heated to decomposition it emits toxic fumes of NO_x.

DIG400   CAS: 302-33-0   **HR: D**
2-DIETHYLAMINOETHYLPROPYLDI-PHENYL ACETATE
mf: $C_{23}H_{31}NO_2$ mw: 353.55

SYNS: BCTB * 2-DIETHYLAMINOETHYL-2,2-DIPHE-NYLVALERATE * 2,2-DIPHENYL-VALERIC ACID-2-(DI-ETHYLAMINO)ETHYL)ESTER * HL 8727 * NSC-39690 * α-PHENYL-α-PROPYL-BENZENEACETIC ACID-S-(DIETHYLAMINO)ETHYL ESTER * PROADIFEN * RP5171 * SKF-525-A

SAFETY PROFILE: Experimental reproductive effects. When heated to decomposition it emits toxic fumes of NO_x.

DII400   CAS: 78109-87-2   **HR: 3**
N-(2-(DIETHYLAMINO)ETHYL)-2,4,6-TRIMETHYLBENZAMIDE HYDROCHLORIDE
mf: $C_{16}H_{26}N_2O \cdot ClH$ mw: 298.90

SYNS: C 3235 * N-(2-(DIETHYLAMINO)ETHYL)-β-ISODURYLAMIDE HYDROCHLORIDE

SAFETY PROFILE: Poison by intraperitoneal and subcutaneous routes. An eye irritant. When heated to decomposition it emits very toxic fumes of HCl and NO$_x$.

DIN800   CAS: 29232-93-7   **HR: 2**
2-DIETHYLAMINO-6-METHYLPYRIMIDIN-4-YL DIMETHYL PHOSPHOROTHIONATE
mf: $C_{11}H_{20}N_3O_3PS$ mw: 305.37

SYNS: ACTELIC * ACTELLIC * ACTELLIFOG * BLEX * O-(2-(DIETHYLAMINO)-6-METHYL-4-PYRIMIDINYL)-O,O-DIMETHYL PHOSPHOROTHIOATE * O-(2-DIETHYLAMINO-6-METHYLPYRIMIDIN-4-YL)-O,O-DIMETHYL PHOSPHOROTHIOATE * ENT 27,699GC * METHYL PIRIMIPHOS * PIRIMIFOS-METHYL * PLANT PROTECTION PP511 * PP511 * PYRIMIDINE PHOSPHATE * PYRIMIPHOS METHYL * SILOSAN

SAFETY PROFILE: Moderately toxic by ingestion and possibly other routes. Mutation data reported. When heated to decomposition it emits very toxic fumes of NO$_x$, PO$_x$ and SO$_x$.

DIO200   CAS: 15421-84-8   **HR: 3**
7-DIETHYLAMINO-5-METHYL-s-TRIAZOLO(1,5-a)PYRIMIDINE
mf: $N_5C_{10}H_{15}$ mw: 205.30

SYNS: AR 12008 * N,N-DIETHYL-5-METHYL-(1,2,-4)TRIAZOLO(1,5-a)PYRIMIDINE-7-AMINE * 5-METHYL-7-DIETHYLAMINO-s-TRIAZOLO-(1,5-a)PYRIMIDINE * ROCORNAL * TRAPIDIL * TRAPYMIN

SAFETY PROFILE: Poison by ingestion, intraperitoneal, subcutaneous, and intravenous routes. Experimental teratogenic and reproductive effects. A coronary vasodilator. When heated to decomposition it emits toxic fumes of NO$_x$.

DIR000   CAS: 522-00-9   **HR: 3**
10-(2-DIETHYLAMINOPROPYL)PHENOTHIAZINE
mf: $C_{19}H_{24}N_2S$ mw: 312.51

SYNS: AETHOPROPAZIN * ATHAPROPAZINE * ATHOPROPAZIN * DIBUTIL * 10-(2-DIETHYLAMINO-2-METHYLETHYL)PHENOTHIAZINE * 2-DIETHYLAMINO-1-PROPYL-N-DIBENZOPARATHIAZINE * N,N-DIETHYL-α-METHYL-10H-PHENOTHIAZINE-10-

ETHANAMINE * ETHOPROMAZINE * ETOPROPEZINA * FEMPROPAZINE * FENPROPAZINA * ISOTAZIN * ISOTHAZINE * ISOTHIAZINE * LYSIVANE * PARCIDOL * PARDIDOL * PARDISOL * PARFEZINE * PARKIN * PARKISOL * PARPHEZEIN * PARSIDOL * PARSITAN * PHENOPROPAZINE * PHENOPROZINE * PRODICTAZIN * PRODIERAZINE * PROFENAMINA (ITALIAN) * PROFENAMINUM * ROCHIPEL * ROCIPEL * RODIPAL * RP 3356 * SC 2538 * SKF 2538 * TOMIL * W 483

SAFETY PROFILE: Poison by ingestion, subcutaneous, intraperitoneal, and intravenous routes. An anticholinergic agent used to treat Parkinsons disease. When heated to decomposition it emits very toxic fumes of NO$_x$ and SO$_x$.

DIS700   CAS: 91-66-7   **HR: 2**
N,N-DIETHYLANILINE
DOT: UN 2432
mf: $C_{10}H_{15}N$ mw: 149.26

PROP: Colorless to yellow liquid. D: (25/4) 0.9302, bp: 215-216°, mp: −38°, n (24/D) 1.5394. Volatile with steam. Sltly sol in alc, chloroform, ether. One gram dissolves in 70 mL water at 12°.

SYNS: DIAETHYLANILIN (GERMAN) * N,N-DIETHYLAMINOBENZENE * N,N-DIETHYLANILIN (CZECH) * DIETHYLANILINE * DIETHYLPHENYLAMINE

CONSENSUS REPORTS: Reported in EPA TSCA Inventory.

DOT Classification: Poison B; Label: St. Andrews Cross.

SAFETY PROFILE: Moderately toxic by ingestion, intraperitoneal and possibly other routes. When heated to decomposition it emits toxic fumes of NO$_x$.

DIU000   CAS: 25340-17-4   **HR: 1**
DIETHYL BENZENE
DOT: UN 2049
mf: $C_{10}H_{14}$ mw: 134.24

PROP: Colorless, mobile liquid. Bp: 183.8°, flash p: 134°F, d: 0.868 @ 25°/25°, autoign temp: 743-842°F, vap press: 1 mm @ 20.7°, vap d: 4.62.

CONSENSUS REPORTS: Reported in EPA TSCA Inventory.

DOT Classification: Flammable or Combustible Liquid; Label: Flammable Liquid.

SAFETY PROFILE: Mildly toxic by ingestion. A skin and eye irritant. Flammable when exposed to heat or flame; can react with oxidizing materials. To fight fire, use CO_2, dry chemical. When heated to decomposition it emits acrid smoke and fumes.

DIU200   CAS: 141-93-5   **HR: 2**
m-DIETHYLBENZENE
mf: $C_{10}H_{14}$ mw: 134.24

CONSENSUS REPORTS: Reported in EPA TSCA Inventory.

SAFETY PROFILE: Moderately toxic by ingestion. When heated to decomposition it emits acrid and irritating fumes.

DIU400   CAS: 1709-50-8   **HR: 2**
N,N-DIETHYLBENZENESULFONAMIDE
mf: $C_{10}H_{15}NO_2S$ mw: 213.32

SAFETY PROFILE: Moderately toxic by ingestion. Experimental teratogenic and reproductive effects. When heated to decomposition it emits very toxic fumes of SO_x and NO_x.

DIV000   CAS: 542-63-2   **HR: 3**
DIETHYLBERYLLIUM
mf: $C_4H_{10}Be$ mw: 67.13

PROP: Colorless liquid. Mp: 12°; bp: 110° @ 15 mm; vap d: 2.3.

CONSENSUS REPORTS: Beryllium and its compounds are on the Community Right-To-Know List.

OSHA PEL: (Transitional: TWA 0.002 mg(Be)/m³; CL 0.005; Pk 0.025/30M/8H) TWA 0.002 mg(Be)/m³; STEL 0.005 mg(Be)/m³/30M; CL 0.025 mg(Be)/m³
ACGIH TLV: TWA 0.002 mg/m³, Suspected Human Carcinogen.

SAFETY PROFILE: Confirmed human carcinogen. Very poisonous. Dangerous fire hazard when exposed to heat or flame. Spontaneously flammable in air. Can react vigorously with oxidizing materials. To fight fire, use special extinguishing agents, dry chemical. Explodes on contact with water. Upon decomposition it emits poisonous fumes of BeO.

DIV600   CAS: 2641-56-7   **HR: 3**
DIETHYLBIS(OCTANOYLOXY)STANNANE
mf: $C_{20}H_{40}O_4Sn$ mw: 463.29

SYNS: DIETHYLBIS(1-OXOOCTYL)OXY)STANNANE * DIETHYLTIN DICAPRYLATE * DIETHYLTIN DIOCTANOATE

OSHA PEL: TWA 0.1 mg(Sn)/m³ (skin)
ACGIH TLV: TWA 0.1 mg(Sn)/m³ (skin) (Proposed: TWA 0.1 mg(Sn)/m³; STEL 0.2 mg(Sn)/m³ (skin))
NIOSH REL: (Organotin Compounds) TWA 0.1 mg(Sn)/m³

SAFETY PROFILE: Poison by ingestion. When heated to decomposition it emits acrid and irritating fumes.

DIV800   **HR: 3**
DIETHYLCADMIUM
mf: $C_4H_{10}Cd$ mw: 170.5

PROP: An oil; decomp by moisture. D: 1.6562, mp: −21°, bp: 64°.

CONSENSUS REPORTS: Cadmium and its compounds are on the Community Right-To-Know List.

OSHA PEL: TWA 0.1 mg(Cd)/m³; CL 0.6 mg(Cd)/m³ (fume)
ACGIH TLV: TWA 0.05 mg(Cd)/m³ (Proposed: TWA 0.01 mg(Cd)/m³ (dust), Human Carcinogen); BEI: 10 μg/g creatinine in urine; 10 μg/L in blood.
DFG BAT: Blood 1.5 μg/dL; Urine 15 μg/dL, Suspected Carcinogen.
NIOSH REL: (Cadmium) Reduce to lowest feasible level

SAFETY PROFILE: Confirmed human carcinogen. A poison. A dangerous fire and explosion hazard. Explodes when heated rapidly to 130°C. On exposure to air it forms white fumes which turn brown and explode. The vapor explodes when heated to 180°C. When heated to decomposition it emits highly toxic fumes of cadmium.

DIW200   CAS: 1642-54-2   **HR: 3**
DIETHYLCARBAMAZINE ACID CITRATE
mf: $C_{10}H_{21}N_3O \cdot C_6H_8O_7$ mw: 391.48

SYNS: BANOCIDE * CARICIDE * CARITROL * DICAROCIDE * DIETHYLCARBAMAZANE CITRATE * DIETHYLCARBAMAZINE CITRATE * DIETHYLCARBAMAZINE HYDROGEN CITRATE * 1-DIETHYL-

CARBAMOYL-4-METHYLPIPERAZINE DIHYDROGEN CI-
TRATE * N,N-DIETHYL-4-METHYL-1-PIPERAZINE
CARBOXAMIDE CITRATE * N,N-DIETHYL-4-METHYL-
1-PIPERAZINECARBOXAMIDE DIHYDROGEN CITRATE
* N,N-DIETHYL-4-METHYL-1-PIPERAZINECARBOXAM-
IDE-2-HYDROXY-1,2,3-PROPANETIRCARBOXYLATE
* DITRAZIN * DITRAZIN CITRATE * DITRAZINE
* DITRAZINE CITRATE * ETHODRYL CITRATE
* FRANOCIDE * FRANOZAN * HETRAZAN
* LOXURAN * 1-METHYL-4-DIETHYLCARBAMOYL-
PIPERAZINE CITRATE

CONSENSUS REPORTS: Reported in EPA
TSCA Inventory.

SAFETY PROFILE: Poison by inhalation and
intravenous routes. Moderately toxic by ingestion, subcutaneous, and intraperitoneal routes.
An experimental teratogen. When heated to decomposition it emits toxic fumes of NO_x.

DIW400   CAS: 88-10-8   **HR: 2**
DIETHYLCARBAMOYL CHLORIDE
mf: $C_5H_{10}ClNO$ mw: 135.61

PROP: Liquid. Mp: $-44°$, bp: 190-195°, vap
d: 4.1.

SYNS: DIETHYLCARBAMIC CHLORIDE * DIETHYL-
CARBAMIDOYL CHLORIDE * N,N-DIETHYLCARBA-
MOYL CHLORIDE * DIETHYLCARBAMYL CHLORIDE

CONSENSUS REPORTS: Reported in EPA
TSCA Inventory.

DFG MAK: Suspected Carcinogen.

SAFETY PROFILE: Suspected carcinogen with
experimental carcinogenic data. Moderately
toxic by intraperitoneal route. Mutation data
reported. Reacts with water or steam to produce
toxic and corrosive fumes. When heated to decomposition it emits highly toxic fumes of Cl^-
and NO_x.

DIW600   CAS: 2425-25-4   **HR: 3**
O,O-DIETHYL-S-(CARBETHOXY)
METHYL PHOSPHOROTHIOLATE
mf: $C_8H_{17}O_5PS$ mw: 256.28

SYNS: ACETOPHOS * ACETOXON * O,O-DI-
ETHYL-S-CARBOETHOXYMETHYL PHOSPHOROTHIOATE
* O,O-DIETHYL-S-CARBOETHOXYMETHYL THIOPHOS-
PHATE * PHOSPHOROTHIOIC ACID-O,O-DIETHYL ES-
TER-S-ESTER with ETHYL MERCAPTOACETATE

SAFETY PROFILE: Poison by ingestion, intraperitoneal, and possibly other routes. When

heated to decomposition it emits very toxic
fumes of PO_x and SO_x.

DIX200   CAS: 105-58-8   **HR: 3**
DIETHYL CARBONATE
DOT: UN 2366
mf: $C_5H_{10}O_3$ mw: 118.15

PROP: Colorless liquid, mild odor. Mp: 43°,
bp: 125.8°, flash p: 77°F (OC), d: 0.975 @
20°/4°, vap press: 10 mm @ 23.8°, vap d: 4.07.

SYNS: DEC * DIAETHYLCARBONAT (GERMAN)
* DIETHYL CARBONATE (DOT) * ETHOXYFORMIC
ANHYDRIDE * ETHYL CARBONATE * EUFIN
* NCI-C60899

CONSENSUS REPORTS: Reported in EPA
TSCA Inventory.

DOT Classification: IMO: Flammable or Combustible Liquid; Label: Flammable Liquid.

SAFETY PROFILE: Questionable carcinogen
with experimental tumorigenic and teratogenic
data. Mildly toxic by subcutaneous route. A
dangerous fire hazard when exposed to heat or
flame; can react with oxidizing materials. To
fight fire, use foam, CO_2, dry chemical. When
heated to decomposition it emits acrid smoke
and fumes.

DIX600   CAS: 1757-18-2   **HR: 3**
O,O-DIETHYL-O-(2-CHLORO-1,2,5-
DICHLOROPHENYLVINYL)
PHOSPHOROTHIOATE
mf: $C_{12}H_{14}Cl_3O_3PS$ mw: 375.64

SYNS: AKTON * AXIOM * O-(2-CHLORO-1-(2,5-
DICHLOROPHENYL)-O,O-DIETHYL ESTER PHOSPHORO-
THIOIC ACID * O-(2-CHLORO-1-(2,5-DICHLOROPHE-
NYL)VINYL)-O,O-DIETHYL PHOSPHOROTHIOATE
* ENT 27,102 * SD 9098 * SHELL SD-9098

SAFETY PROFILE: Poison by ingestion and
skin contact. An insecticide. When heated to
decomposition it emits very toxic fumes of Cl^-,
PO_x, and SO_x.

DIX800   CAS: 7173-84-4   **HR: 3**
O,O-DIETHYL-S-p-CHLOROPHENYL
THIOMETHYLPHOSPHOROTHIOATE
mf: $C_{11}H_{16}ClO_3PS_2$ mw: 326.81

SYNS: S-((p-CHLOROPHENYLTHIO)METHYL)-O,O-DI-
ETHYL PHOSPHORODITHIOATE * DANIFOS

SAFETY PROFILE: Poison by ingestion and skin contact. When heated to decomposition it emits very toxic fumes of Cl^-, PO_x, and SO_x.

DIY000   CAS: 814-49-3   **HR: 3**
DIETHYL CHLOROPHOSPHATE
mf: $C_4H_{10}ClO_3P$ mw: 172.56

PROP: Water white liq. Bp: 60° @ 2 mm, d: 1.1915 @ 25°/25°, vap d: 5.94.

SYNS: CHLOROPHOSPHORIC ACID DIETHYL ESTER * DIETHOXYPHOSPHORUS OXYCHLORIDE

CONSENSUS REPORTS: Reported in EPA TSCA Inventory.

SAFETY PROFILE: Deadly poison by skin contact. Poison by ingestion. A cholinesterase inhibitor. Trace HCl catalyzes a hazardous reaction during the preparation of diethyl phosphate from diethyl chlorophosphate. When heated to decomposition it emits very toxic fumes of Cl^- and PO_x.

DIY600   CAS: 100-38-9   **HR: 3**
N,N-DIETHYL CYSTEAMINE
mf: $C_6H_{15}NS$ mw: 133.28

SYNS: N-DIAETHYL CYSTEAMIN (GERMAN) * DIETHYLAMINOETHANETHIOL * 2-(DIETHYLAMINO)ETHANETHIOL * 2-(DIETHYLAMINO)ETHYLMERCAPTAN * β-DIETHYLAMINOETHYL MERCAPTAN * DIETHYLCYSTEAMIN * DIETHYLCYSTEAMINE * N-DIETHYL CYSTEAMINE * DIETHYL(2-MERCAPTOETHYL)AMINE

SAFETY PROFILE: Poison by ingestion, intraperitoneal, and subcutaneous routes. Mildly toxic by inhalation. When heated to decomposition it emits very toxic fumes of NO_x and SO_x.

DIY800   CAS: 104-78-9   **HR: 2**
N,N-DIETHYL-1,3-DIAMINOPROPANE
DOT: UN 2684
mf: $C_7H_{18}N_2$ mw: 130.27

PROP: Liquid. Bp: 165°-170°, flash p: 138°F (OC), d: 0.82, vap d: 4.48.

SYNS: 1-AMINO-3-(DIETHYLAMINO)PROPANE * N-(3-DIETHYLAMINOPROPYL)AMINE * N,N-DIETHYLAMINOPROPYLAMINE * 3-(DIETHYLAMINO) PROPYLAMINE (DOT) * DIETHYLAMINOTRIMETHYLENAMINE

CONSENSUS REPORTS: Reported in EPA TSCA Inventory.

DOT Classification: Corrosive Material; Label: Corrosive, Flammable Liquid.

SAFETY PROFILE: Moderately toxic by ingestion and skin contact. Corrosive to the eyes, skin, and mucous membranes. A sensitizer. Flammable when exposed to heat or flame; can react with oxidizing materials. To fight fire, use foam, CO_2, dry chemical. When heated to decomposition it emits toxic fumes of NO_x.

DIZ100   CAS: 1609-47-8   **HR: 3**
DIETHYL DICARBONATE
mf: $C_6H_{10}O_5$ mw: 162.16

PROP: Viscous liquid; fruity odor. D: 1.12, visc (20°): 1.97 cp. Soluble in alc, esters, ketones, and hydrocarbons.

SYNS: BAYCOVIN * DEPC * DICARBONIC ACID DIETHYL ESTER * DIETHYL DICARBONATE * DIETHYL ESTER of PYROCARBONIC ACID * DIETHYL OXYDIFORMATE * DIETHYL PYROCARBONATE * DIETHYL PYROCARBONIC ACID * DKD * ETHYL PYROCARBONATE * OXYDIFORMIC ACID DIETHYL ESTER * PIREF * PYROCARBONATE d'ETHYLE (FRENCH) * PYROCARBONIC ACID, DIETHYL ESTER * PYROKOHLENSAEURE DIAETHYL ESTER (GERMAN)

CONSENSUS REPORTS: Reported in EPA TSCA Inventory.

SAFETY PROFILE: Poison by ingestion and intraperitoneal routes. Concentrated DEPC is irritating to eyes, mucous membranes, and skin. When heated to decomposition it emits acrid smoke and fumes.

DJA200   CAS: 3152-41-8   **HR: 3**
**O,O-DIETHYL-S-(3,4-
DICHLOROPHENYL-THIO)METHYL
PHOSPHOROTHIOATE**
mf: $C_{11}H_{15}Cl_2O_2PS_3$ mw: 377.31

SYNS: S-((3,4-DICHLOROPHENYLTHIO)METHYL)-O,O-DIETHYL PHOSPHORODITHIOATE * ENT 25,555-X * G 27365 * GEIGY G-27365

SAFETY PROFILE: Poison by ingestion. When heated to decomposition it emits very toxic fumes of Cl^-, PO_x and SO_x.

DJA400   CAS: 78-53-5   **HR: 3**
**O,O-DIETHYL-S-(2-DIETHYLAMINO-
ETHYL) THIOPHOSPHATE**
mf: $C_{10}H_{24}NO_3PS$ mw: 269.38

PROP: Liquid. Bp: 110° @ 0.2 mm, mp: 98°.

SYNS: AMITON * CHIPMAN 6200 * CITRAM * S-(DIETHYLAMINOETHYL)-O,O-DIETHYL PHOSPHO-ROTHIOATE * S-(2-(DIETHYLAMINO)ETHYL)PHOS-PHOROTHIOIC ACID-O,O-DIETHYL ESTER * DI-ETHYL-S-2- DIETHYLAMINOETHYL PHOSPHOROTHIOATE * (2-DIETHYLAMINO)ETHYLPHOSPHOROTHIOIC ACID-O,O-DIETHYL ESTER * O,O-DIETHYL-S-(β-DIETHYL-AMINO)ETHYL PHOSPHOROTHIOLATE * O,O-DI-ETHYL-S-DIETHYLAMINOETHYL PHOSPHOROTHIOLATE * O,O-DIETHYL-S-2-DIETHYLAMINOETHYL PHOSPHOROTHIOATE * O,O-DIETHYL-S-2-DIETHYL-AMINOETHYL PHOSPHOROTHIOLATE * DSDP * ENT 24,980-X * INFERNO * METRAMAC * METRAMAK * R-5,158 * RHODIA-6200 * TETRAM

CONSENSUS REPORTS: EPA Extremely Hazardous Substances List.

SAFETY PROFILE: A deadly poison by ingestion and subcutaneous routes. A cholinesterase inhibitor. An insecticide. When heated to decomposition it emits very toxic fumes of NO_x, PO_x, and SO_x.

DJB000 CAS: 2767-55-7 **HR: 3**
DIETHYLDIIODOSTANNANE
mf: $C_4H_{10}I_2Sn$ mw: 430.63

PROP: Very sltly sol white crystals. Mp: 45°, bp: 240-245° (decomp).

SYN: DIETHYLTIN DIIODIDE

OSHA PEL: TWA 0.1 mg(Sn)/m³ (skin)
ACGIH TLV: TWA 0.1 mg(Sn)/m³ (skin) (Proposed: TWA 0.1 mg(Sn)/m³; STEL 0.2 mg(Sn)/m³ (skin))
NIOSH REL: (Organotin Compounds) TWA 0.1 mg(Sn)/m³

SAFETY PROFILE: Poison by ingestion and intraperitoneal routes. When heated to decomposition it emits toxic fumes of I^-.

DJB200 CAS: 7773-34-4 **HR: 3**
α,α'-DIETHYL-4,4'-DIMETHOXYSTILBENE
mf: $C_{20}H_{24}O_2$ mw: 296.44

SYNS: 3,4-BIS(p-METHOXYPHENYL)-3-HEXENE * DEPOT-OESTROMENINE * DEPOT-OESTROMON * 3,4-DIANISYL-3-HEXENE * trans-α,α'-DIETHYL-4,4'-DIMETHOXYSTILBENE * (E)-1,1'-(1,2-DIETHYL-1,2-ETHENE-DIYL)BIS(4-METHOXYBENZENE) * DIETHYL-

STILBESTROL DIMETHYL ETHER * DIMESTROL * 4,4'-DIMETHOXY-α,β-DIETHYLSTILBENE * STIL-BESTROL DIMETHYL ETHER * SYNTHILA

SAFETY PROFILE: Questionable carcinogen with experimental tumorigenic data. When heated to decomposition it emits acrid and irritating fumes.

DJB600 CAS: 64048-13-1 **HR: 3**
p-DIETHYL-p'-DIMETHYLTHIOPYRO-PHOSPHATE
mf: $C_6H_{16}O_5P_2S_2$ mw: 294.28

SYN: p'-DIETHYL-p-DIMETHYL THIOPYROPHOSPHATE

SAFETY PROFILE: A deadly poison by intramuscular and intraperitoneal routes. When heated to decomposition it emits very toxic fumes of SO_x and PO_x.

DJC000 CAS: 72-56-0 **HR: 3**
DIETHYLDIPHENYL DICHLOROETHANE
mf: $C_{18}H_{20}Cl_2$ mw: 307.28

SYNS: 1,1-BIS(p-ETHYLPHENYL)-2,2-DICHLOROETH-ANE * 2,2-BIS(p-ETHYLPHENYL)-1,1-DICHLOROETH-ANE * 1,1-DICHLORO-2,2-BIS(p-ETHYLPHENYL) ETHANE * 1,1-DICHLORO-2,2-BIS(4-ETHYLPHENYL) ETHANE * 2,2-DICHLORO-1,1-BIS(p-ETHYLPHE-NYL)ETHANE * α,α-DICHLORO-2,2-BIS(p-ETHYL-PHENYL)ETHANE * DI(p-ETHYLPHENYL)DICHLORO-ETHANE * ETHYLAN * p,p-ETHYL DDD * p,p'-ETHYL-DDD * NCI-C02868 * PERTHANE * Q-137

CONSENSUS REPORTS: NCI Carcinogenesis Bioassay (feed); Clear Evidence: mouse NCITR* NCI-CG-TR-156,79; No Evidence: rat NCITR* NCI-CG-TR-156,79.

SAFETY PROFILE: Questionable carcinogen with experimental carcinogenic and tumorigenic data. Poison by intravenous route. Mildly toxic by ingestion. Experimental teratogenic and reproductive effects. Mutation data reported. A pesticide. When heated to decomposition it emits toxic fumes of Cl^-.

DJC400 CAS: 85-98-3 **HR: 3**
1,3-DIETHYL-1,3-DIPHENYLUREA
mf: $C_{17}H_{20}N_2O$ mw: 268.39

PROP: Colorless crystals. Mp: 73°, d: 1.12, bp: 326°, flash p: 302°F (CC), vap d: 9.3.

SYNS: BIS(N-ETHYL-N-PHENYL)UREA * N,N-DI-ETHYLCARBANILIDE * N,N'-DIETHYL-N,N'-DIPHENY-

LUREA   *   sym-DIETHYLDIPHENYLUREA
*   USAF EK-1047

CONSENSUS REPORTS: Reported in EPA TSCA Inventory.

SAFETY PROFILE: Poison by intraperitoneal route. Combustible when exposed to heat or flame. Probably a slight explosion hazard, although it is a component of smokeless explosive mixtures. When heated to decomposition it burns and emits very toxic fumes of NO_x. To fight fire, use dry chemical, CO_2, spray or mist.

DJD400   CAS: 136-92-5   **HR: 3**
DIETHYLDITHIOCARBAMIC ACID
SELENIUM(II) SALT
mf: $C_{10}H_{20}N_2S_4 \cdot Se$ mw: 375.52

SYNS: ETHYL SELENAC   *   SELENIUM DIETHYLDI-
THIOCARBAMATE

CONSENSUS REPORTS: Selenium and its compounds are on the Community Right-To-Know List.

OSHA PEL: TWA 0.2 mg(Se)/m^3
ACGIH TLV: TWA 0.2 mg(Se)/m^3
DFG MAK: 0.1 mg(Se)/m^3

SAFETY PROFILE: Poison by ingestion and intraperitoneal routes. When heated to decomposition it emits very toxic fumes of NO_x, SO_x, and Se.

DJD600   CAS: 111-46-6   **HR: 3**
DIETHYLENE GLYCOL
mf: $C_4H_{10}O_3$ mw: 106.14

PROP: Clear, colorless, practically odorless, syrupy liquid. Bp: 245.8°, fp: −8°, flash p: 255°F, d: 1.1184 @ 20°/20°, autoign temp: 444°F, vap press: 1 mm @ 91.8°, vap d: 3.66.

SYNS: BIS(2-HYDROXYETHYL) ETHER   *   BRECO-
LANE NDG   *   CARBITOL   *   DEACTIVATOR E
*   DEACTIVATOR H   *   DEG   *   DICOL   *   DIGLY-
COL   *   DIHYDROXYDIETHYL ETHER   *   β,β'-DIHY-
DROXYDIETHYL ETHER   *   2,2'-DIHYDROXYETHYL
ETHER   *   DISSOLVANT APV   *   ETHYLENE DIGLY-
COL   *   GLYCOL ETHER   *   GLYCOL ETHYL ETHER
*   3-OXAPENTANE-1,5-DIOL   *   3-OXA-1,5-PENTANE-
DIOL   *   2,2'-OXYBISETHANOL   *   2,2'-OXYDIETHA-
NOL   *   TL4N

CONSENSUS REPORTS: Reported in EPA TSCA Inventory. Glycol ether compounds are on the Community Right-To-Know List.

SAFETY PROFILE: Questionable carcinogen with experimental carcinogenic, tumorigenic, and teratogenic data. Moderately toxic to humans by ingestion. Poison by experimentally inhalation; moderately toxic by ingestion and intravenous routes. An eye and human skin irritant. Combustible when exposed to heat or flame; can react with oxidizing materials. To fight fire, use alcohol foam, water, CO_2, dry chemical. Mixtures with sodium hydroxide decompose exothermically when heated to 230°C and release explosive hydrogen gas. When heated to decomposition it emits acrid smoke and irritating fumes.

DJE400   CAS: 693-21-0   **HR: 3**
DIETHYLENE GLYCOL DINITRATE
DOT: UN 0075
mf: $C_4H_8N_2O_7$ mw: 196.14

PROP: Liquid. Vap d: 6.76.

SYNS: BIS(HYDROXYAETHYL)-AETHER-DINITRAT
(GERMAN)   *   DIETHYLENEGLYCOL DINITRATE, con-
taining at least 25% phlegmatizer (DOT)   *   DIETHYLEN-
GLYKOLDINITRATE (CZECH)   *   DIGLYCOLDINITRAAT
(DUTCH)   *   DIGLYCOL (DINITRATE de) (FRENCH)
*   DIGLYKOLDINITRAT (GERMAN)   *   DI(HYDROXY-
ETHYL) ETHER DINITRATE   *   DINITRATE de DIETHY-
LENE-GLYCOL (FRENCH)   *   DINITRODIGLICOL (ITAL-
IAN)   *   DINITRODIGLYKOL (CZECH)

CONSENSUS REPORTS: Reported in EPA TSCA Inventory. Glycol ether compounds are on the Community Right-To-Know List.

DOT Classification: Forbidden; IMO: Class A Explosive; Label: Explosive A, with phlegmatizer.

SAFETY PROFILE: Moderately toxic by ingestion. Ingestion of this compound can cause a drop in blood pressure and cardiac disturbances. A dangerous fire hazard when exposed to heat or flame; can react vigorously with oxidizing or reducing materials. A dangerous explosive sensitive to heat, shock, and vibration. Used in low freezing dynamites and some permissible explosives. Upon decomposition it emits toxic fumes of NO_x.

DJG600   CAS: 111-40-0   **HR: 3**
DIETHYLENETRIAMINE
DOT: UN 2079
mf: $C_4H_{13}N_3$ mw: 103.20

PROP: Yellow, viscous liquid; mild ammonia-cal odor. Mp: −39°, bp: 207°, flash p: 215°F (OC), d: 0.9586 @ 20°/20°, autoign temp: 750°F, vap press: 0.22 mm @ 20°, vap d: 3.48.

SYNS: AMINOETHYLETHANDIAMINE * N-(2-AMI-NOETHYL)ETHYLENEDIAMINE * 3-AZAPENTANE-1,5-DIAMINE * BIS(2-AMINOETHYL)AMINE * BIS(β-AMINOETHYL)AMINE * D.E.H. 20 * DETA * 2,2′-DIAMINODIETHYLAMINE * 2,2′-IMINOBIS-ETHYLAMINE

CONSENSUS REPORTS: Reported in EPA TSCA Inventory.

OSHA PEL: TWA 1 ppm
ACGIH TLV: TWA 1 ppm (skin)
DOT Classification: Corrosive Material; Label: Corrosive.

SAFETY PROFILE: Poison by skin contact and intraperitoneal routes. Moderately toxic by ingestion. Corrosive. A severe skin and eye irritant. High concentration of vapors causes irritation of respiratory tract, nausea, and vomiting. Repeated exposures can cause asthma and sensitization of skin. Combustible when exposed to heat or flame; can react with oxidizing materials. Mixture with nitromethane is a shock-sensitive explosive. Ignites on contact with cellulose nitrate of high surface area. To fight fire, use alcohol foam. When heated to decomposition it emits toxic fumes of NO_x.

DJG800   CAS: 67-43-6   **HR: 2**
(DIETHYLENETRINITRILO)PENTA-ACETIC ACID
mf: $C_{14}H_{23}N_3O_{10}$ mw: 393.40

SYNS:
((CARBOXYMETHYLIMINO)BIS(ETHYLENENITRILO))TET-RAACETIC ACID * CHEL 330 * CHEL 330 ACID * CHEL DTPA * DIETHYLENETRIAMINEPENTA-ACETIC ACID * 1,1,4,7,7-DIETHYLENETRIAMINE-PENTAACETIC ACID * DTPA * HAMP-EX ACID * MONAQUEST * PENTHAMIL * PERMA KLEER * 3,6,9-TRIS(CARBOXYMETHYL)-3,6,9-TRIAZAUNDECA-NEDIOIC ACID

CONSENSUS REPORTS: Reported in EPA TSCA Inventory.

SAFETY PROFILE: Moderately toxic by intraperitoneal route. Mildly toxic by ingestion and possibly other routes. When heated to decomposition it emits toxic fumes of NO_x.

DJI000   CAS: 2595-54-2   **HR: 3**
O,O-DIETHYL-S-(N-ETHOXYCAR-BONYL-N-METHYLCARBAMOYL-METHYL) PHOSPHORODITHIOATE
mf: $C_{10}H_{20}NO_5PS_2$ mw: 329.40

SYNS: AFOS * O,O-DIAETHYL-S-(3-METHYL-2,4-DI-OXO-5-OXA-3-AZA-HEPTYL)-DITHIOPHOSPHAT (GER-MAN) * O,O-DIETHYL S-(N-ETHOXYCARBONYL-N-METHYLCARBAMOYLMETHYL) PHOSPHOROTHIOLO-THIONATE * O,O-DIETHYL S-(N-METHYL-N-CARBO-ETHOXYCARBAMOYLMETHYL) DITHIOPHOSPHATE * O,O-DIETHYL-S-(3-METHYL-2,4-DIOXO-5-OXA-3-AZA-HEPTYL)-DITHIOFOSFAAT (DUTCH) * O,O-DIETIL-S-(N-ETOSSI-CARBONIL-N-METIL-CARBAMOIL-METIL)-DITIOFOSFATO (ITALIAN) * DITHIOPHOSPHATE de O,O-DIETHYLE et de S-N-METHYL-N-CARBOETHOXY CARBAMOYLMETHYLE (FRENCH) * N-ETHOXYCAR-BONYL-N-METHYLCARBAMOYLMETHYL-O,O-DIETHYL PHOSPHORODITHIOATE * S-((ETHOXYCARBONYL)-METHYLCARBAMOYL)METHYL-O,O-DIETHYL PHOSPHORODITHIOATE * S-(N-ETHOXYCARBONYL-N-METHYLCARBAMOYLMETHYL)-DIETHYL PHOSPHO-RODITHIOATE * MARFOTOKS * MC 474 * MECARBAM * MS 1053 * MS 1143 * MURATOX * MURFOTOX * MUROTOX * MURPHOTOX * MURUTOX * PENNSALT TD-72 * PESTAN

CONSENSUS REPORTS: EPA Genetic Toxicology Program.

SAFETY PROFILE: Poison by ingestion, skin contact, subcutaneous, and possibly other routes. An insecticide. When heated to decomposition it emits very toxic fumes of SO_x, PO_x, and NO_x.

DJJ400   CAS: 358-74-7   **HR: 3**
DIETHYL FLUOROPHOSPHATE
mf: $C_4H_{10}FO_3P$ mw: 156.11

PROP: A liquid with a sweet or fruity odor. Mp: low, bp: 170°, d: 1.15 (approx), vap d: 5.38.

SYNS: FLUOPHOSPHORIC ACID, DIETHYL ESTER * PHOSPHOROFLUORIDIC ACID, DIETHYL ESTER * T-1036 * TL 345

SAFETY PROFILE: Poison by inhalation and skin contact. When heated to decomposition or on contact with acid or acid fumes it emits highly toxic fumes of F^- and PO_x.

DJJ850   CAS: 26645-10-3   **HR: 3**
DIETHYL GOLD BROMIDE
mf: $C_4H_{10}AuBr$ mw: 334.994

DOT Classification: Forbidden

SAFETY PROFILE: Explodes at 70°C. When heated to decomposition it emits toxic fumes of Br⁻.

DJK800   CAS: 16111-62-9   **HR: 2**
DI(2-ETHYLHEXYL)
PEROXYDICARBONATE

DOT: UN 2122/UN 2123/UN 2960
mf: $C_{18}H_{34}O_6$ mw: 346.52

SYNS: BIS(2-ETHYLHEXYL) ESTER, PEROXYDICAR-BONIC ACID * DI-(2-ETHYLHEXYL)PEROXYDICARBO-NATE, technical pure (DOT) * DI-(2-ETHYLHEXYL) ESTER, PEROXYDICARBONIC ACID

CONSENSUS REPORTS: Reported in EPA TSCA Inventory.

DOT Classification: Organic Peroxide; Label: Organic Peroxide.

SAFETY PROFILE: Moderately toxic by ingestion. When heated to decomposition it emits acrid smoke and irritating fumes.

DJL000   CAS: 577-11-7   **HR: 3**
DI-(2-ETHYLHEXYL) SODIUM
SULFOSUCCINATE
mf: $C_{20}H_{38}O_7S \cdot Na$ mw: 445.63

PROP: White, waxlike, plastic solid; octyl alcohol odor. Sol in hexane, glycerin, alc; sltly sol in water.

SYNS: AEROSOL GPG * ALCOPOL O * ALPHA-SOL OT * BEROL 478 * BIS(ETHYLHEXYL) ESTER of SODIUM SULFOSUCCINIC ACID * BIS(2-ETHYL-HEXYL)SODIUM SULFOSUCCINATE * BIS(2-ETHYL-HEXYL)-S-SODIUM SULFOSUCCINATE * 1,4-BIS(2-ETHYLHEXYL) SODIUM SULFOSUCCINATE * 1,4-BIS(2-ETHYLHEXYL)SULFOBUTANEDIOIC ACID ESTER, SODIUM SALT * CELANOL DOS 75 * CLESTOL * COLACE * COMPLEMIX * CONSTONATE * COPROL * DEFILIN * DIOCTLYN * DIOC-TYLAL * DIOCTYL ESTER of SODIUM SULFOSUCCI-NATE * DIOCTYL ESTER of SODIUM SULFOSUCCINIC ACID * DIOCTYL-MEDO FORTE * DIOCTYL SO-DIUM SULFOSUCCINATE (FCC) * DIOCTYL SULFO-SUCCINATE SODIUM SALT * DIOMEDICONE * DIOSUCCIN * DIOTILAN * DIOVAC

* DOCUSATE SODIUM * DOXINATE * DOXOL * DSS * DULSIVAC * DUOSOL * 2-ETHYL-HEXYL SULFOSUCCINATE SODIUM * HUMIFEN WT 27G * KONLAX * KOSATE * LAXINATE * MANOXAL OT * MERVAMINE * MODANE SOFT * MOLATOC * MOLCER * MOLOFAC * MONAWET MD 70E * NEKAL WT-27 * NEVAX * NIKKOL OTP 70 * NORVAL * OBSTON * RAPISOL * REGUTOL * REQUTOL * REVAC * SANMORIN OT 70 * SBO * SOBITAL * SODIUM BIS(2-ETHYLHEXYL) SULFOSUCCINATE * SODIUM DI-(2-ETHYLHEXYL) SULFOSUCCINATE * SODIUM DIOCTYL SULFOSUCCINATE * SODIUM DIOCTYL SULPHOSUCCINATE * SODIUM-2-ETHYL-HEXYLSULFOSUCCINATE * SODIUM SULFODI-(2-ETHYLHEXYL)SULFOSUCCINATE * SOFTIL * SOLIWAX * SOLUSOL-75% * SOLUSOL-100% * SULFIMEL DOS * TEX WET 1001 * TRITON GR-5 * VATSOL OT * VELMOL * WAXSOL * WETAID SR

CONSENSUS REPORTS: Reported in EPA TSCA Inventory.

SAFETY PROFILE: Poison by intravenous route. Moderately toxic by ingestion and intra-peritoneal routes. A skin and severe eye irritant. When heated to decomposition it emits toxic fumes of SO_x and Na_2O.

DJL400   CAS: 1615-80-1   **HR: 3**
1,2-DIETHYLHYDRAZINE
mf: $C_4H_{12}N_2$ mw: 88.18

PROP: Bp: 86°, d: 0.797 @ 26°. Sol in alc and ether.

SYNS: 1,2-DIAETHYLHYDRAZINE (GERMAN) * N-N'-DIETHYLHYDRAZINE * sym-DIETHYLHYDRA-ZINE * HYDRAZOETHANE * HYDROAZOETHANE * RCRA WASTE NUMBER U086 * SDEH

CONSENSUS REPORTS: IARC Cancer Review: GROUP 2B IMEMDT 7,56,87; Animal Sufficient Evidence IMEMDT 4,153,74.

SAFETY PROFILE: Suspected carcinogen with experimental carcinogenic, tumorigenic, and teratogenic data. It is also a transplacental carcinogen. When heated to decomposition it emits toxic fumes of NO_x.

DJL600   CAS: 7699-31-2   **HR: 3**
1,2-DIETHYLHYDRAZINE
DIHYDROCHLORIDE
mf: $C_4H_{12}N_2 \cdot 2ClH$ mw: 161.10

SAFETY PROFILE: Questionable carcinogen with experimental tumorigenic and teratogenic data. When heated to decomposition it emits very toxic fumes of HCl and NO_x.

DJM200   CAS: 60-44-6   **HR: 3**
DIETHYL(2-HYDROXYETHYL)METHYL-AMMONIUM BROMIDE-α-CYCLO-PENTYL-2-THIOPHENEGLYCOLATE
mf: $C_{18}H_{30}NO_3S \cdot Br$ mw: 420.46

SYNS: α-CYCLOPENTYL-2-THIOPHENEGLYCOLATE DIETHYL(2-HYDROXYETHYL)METHYLAMMONIUM BROMIDE * 2-DIETHYLAMINOETHYL-2-CYCLOPEN-TYL-2-(2-THIENYL)HYDROXYACETATE METHOBROMIDE * 2-DIETHYLAMINOETHYL-α-CYCLOPENTYL-2-THIO-PHENEGLYCOLATE METHOBROMIDE * MONODRAL * MONODRAL BROMIDE * PENTHIENATE BROMIDE * WIN 4369

SAFETY PROFILE: Poison by intravenous and subcutaneous routes. Moderately toxic by ingestion. When heated to decomposition it emits very toxic fumes of NO_x, NH_3, SO_x, and Br^-.

DJM600   CAS: 50-10-2   **HR: 3**
DIETHYL(2-HYDROXYETHYL)METHYL-AMMONIUM BROMIDE-α-PHENYLCY-CLOHEXANEGLYCOLATE
mf: $C_{21}H_{34}NO_3 \cdot Br$ mw: 428.47

SYNS: ANTRENIL * ANTRENYL * ANTRENYL BROMIDE * BA-5473 * C 5473 * 2-((CYCLO-HEXYLHYDROXYPHENYLACETYL)OXY)-N,N-DIETHYL-N-METHYLETHANAMINIUM BROMIDE * METACIN * METATSIN * METHACIN * OXIFENON * OXYFENON * OXYPHENON * OXYPHENONIUM * OXYPHENONIUM BROMIDE * SPASMOPHEN

SAFETY PROFILE: Poison by ingestion, intramuscular, intraperitoneal, intravenous, and subcutaneous routes. Human systemic effects by ingestion: change in heart rate. When heated to decomposition it emits very toxic fumes of NO_x, NH_3, and Br^-.

DJN489   **HR: 3**
DIETHYL HYDROXYTIN HYDROPEROXIDE

SAFETY PROFILE: An explosive. When heated to decomposition it emits acrid smoke and fumes.

DJN600   CAS: 78-52-4   **HR: 3**
O,O-DIETHYL-S-2-ISOPROPYLMER-CAPTOMETHYLDITHIOPHOSPHATE
mf: $C_8H_{19}O_2PS_3$ mw: 274.42

SYNS: AMERICAN CYANAMID 12,008 * O,O-DIE-THYL-S-(ISOPROPYLMERCAPTOMETHYL) PHOSPHORO-DITHIOATE * O,O-DIETHYL-S-(ISOPROPYLTHIOME-THYL) PHOSPHORODITHIOATE * ENT 22,865 * EXPERIMENTAL INSECTICIDE 12008 * (ISOPRO-PYLTHIO)-METHANETHIOL-S-ESTER with O,O-DIETHYL PHOSPHORODITHIOATE * TM 12008

SAFETY PROFILE: Poison by ingestion and subcutaneous routes. When heated to decomposition it emits very toxic fumes of PO_x and SO_x.

DJN750   CAS: 96-22-0   **HR: 3**
DIETHYL KETONE
DOT: UN 1156
mf: $C_5H_{10}O$ mw: 86.15

PROP: Colorless, mobile liquid; acetone-like odor. Mp: −42°, bp: 101°, flash p: 55°F, d: 0.8159 @ 19°/4°, vap d: 2.96, autoign temp: 842°F, lel: 1.6%. Sol in water; misc in alc and ether.

SYNS: DEK * DIETHYLCETONE (FRENCH) * DIMETHYLACETONE * METACETONE * METH-ACETONE * PENTANONE-3 * 3-PENTANONE * PROPIONE

CONSENSUS REPORTS: Reported in EPA TSCA Inventory.

OSHA PEL: TWA 200 ppm
ACGIH TLV: TWA 200 ppm
DOT Classification: Label: Flammable Liquid.

SAFETY PROFILE: Moderately toxic by ingestion, intraperitoneal, and intravenous routes. A skin and eye irritant. Mutation data reported. Dangerous fire hazard when exposed to heat or flame; can react vigorously with oxidizing materials. To fight fire, use alcohol foam, foam, CO_2, dry chemical. Reacts with hydrogen peroxide + nitric acid to form a shock- and heat-sensitive explosive peroxide. When heated to decomposition it emits acrid smoke and irritating fumes.

DJN800   CAS: 15773-47-4   **HR: 3**
DIETHYL LEAD DIACETATE
mf: $C_8H_{16}O_4Pb$ mw: 383.43

CONSENSUS REPORTS: Lead and its compounds are on the Community Right-To-Know List.

NIOSH REL: TWA 0.10 mg(Pb)/m^3

SAFETY PROFILE: Poison by ingestion. When heated to decomposition it emits toxic fumes of Pb.

DJO000 CAS: 50-37-3 **HR: 3**
N,N-DIETHYLLYSERGAMIDE
mf: $C_{20}H_{25}N_3O$ mw: 323.48

SYNS: ACID * CUBES * DELYSID * 9,10-DI-DEHYDRO-N,N-DIETHYL-6-METHYL-ERGOLINE-8-β-CAR-BOXAMIDE * HEAVENLY BLUE * LSD * LSD-25 * LYSERGAMID * LYSERGAURE DIETHYLAMID * d-LSD * d-LYSERGIC ACID DIETHYLAMIDE * LYSERGIC ACID DIETHYLAMIDE-25 * LYSERGIDE * LYSERGSAUEREDIAETHYLAMID * PEARLY GATES * ROYAL BLUE * WEDDING BELLS

CONSENSUS REPORTS: EPA Genetic Toxicology Program.

SAFETY PROFILE: Poison by ingestion, subcutaneous, intraperitoneal, and intravenous routes. Mutation data reported. Human systemic effects by ingestion and intramuscular routes: euphoria, hallucinations, distorted perceptions, excitement, anorexia, nausea and vomiting. Experimental teratogenic and reproductive effects. Mutation data reported. A much abused hallucinogen. A federally regulated substance. When heated to decomposition it emits toxic fumes of NO$_x$.

DJO100 CAS: 557-18-6 **HR: 3**
DIETHYL MAGNESIUM

DOT: UN 1367
mf: $C_4H_{10}Mg$ mw: 82.43

DOT Classification: Flammable Solid; Label: Spontaneously Combustible

SAFETY PROFILE: Ignites on contact with moist air, water, or carbon dioxide.

DJO400 CAS: 627-44-1 **HR: 3**
DIETHYL MERCURY
mf: $C_4H_{10}Hg$ mw: 258.73

PROP: Colorless liquid, hazel-like odor. Bp: 159°, d: 2.4660 @ 20°.

CONSENSUS REPORTS: Reported in EPA TSCA Inventory. Mercury and its compounds are on the Community Right-To-Know List.

OSHA PEL: (Transitional: CL 1 mg/10m^3) TWA 0.01 mg(Hg)/m^3; STEL 0.03 mg/m^3 (skin)
ACGIH TLV: TWA 0.01 mg(Hg)/m^3; STEL 0.03 mg(Hg)/m^3
NIOSH REL: TWA 0.05 mg(Hg)/m^3

SAFETY PROFILE: A deadly human poison by inhalation. Poison by intraperitoneal route. An experimental teratogen. Flammable when exposed to heat or flame; can react with oxidizing materials. When heated to decomposition or on contact with acid or acid fumes it emits highly toxic fumes of Hg.

DJP600 CAS: 50285-72-8 **HR: 3**
1,1-DIETHYL-3-METHYL-3-NITROSOUREA
mf: $C_6H_{13}N_3O_2$ mw: 159.22

SYNS: NITROSO-1,1-DIETHYL-3-METHYLUREA * NITROSOMETHYLDIAETHYLHARNSTOFF * NITRO-SOMETHYLDIETHYLUREA * 1-NITROSO-1-METHYL-3,3-DIETHYLUREA

SAFETY PROFILE: Questionable carcinogen with experimental carcinogenic and tumorigenic data. Poison by subcutaneous route. Mutation data reported. When heated to decomposition it emits toxic fumes of NO$_x$.

DJS200 CAS: 59-26-7 **HR: 3**
N,N-DIETHYLNICOTINAMIDE
mf: $C_{10}H_{14}N_2O$ mw: 178.26

SYNS: ANACARDONE * ANACORDONE * AS-TROCAR * BETAPYRIMIDUM * CAMPHOZONE * CARBAMIDAL * CARDAMINE * CARDIAGEN * CARDIAMID * CARDIAMINA * CARDIAMINE * CARDIMON * CITOCOR * CORACON * CORAETHAMIDE * CORAETHAMIDUM * CORALEPT * CORAMINE * CORAVITA * CORAZONE * CORDIAMID * CORDIAMIN * CORDIAMINE * CORDITON * CORDYNIL * COREDIOL * CORESPIN * CORETHAMIDE * CORETONE * CORMED * CORMID * COR-MOTYL * CORNOTONE * COROTONIN * CORO-VIT * CORVITAN * CORVITOL * CORVITONE * CORYWAS * DANAMINE * DIAETHYL-NICOTINAMID (GERMAN) * DIETHYL-NICOTAMIDE * N,N-DIETHYL-3-PYRIDINECARBOXAMIDE * DIETILAMIDE-CARBOPIRIDINA * DINACORYL

* DYNACORYL * DYNAMICARDE * ELITONE
* EUCORAN * HANSACOR * INICARDIO
* KARDIAMID * KARDONYL * KORDIAMIN
* LEPTAMIN * MEDIAMID * NIAMINE
* NICAMIDE * NICETAMIDE * NICETHAMIDE
* NICOR * NICORDAMIN * NICORINE
* NICORYL * NICOTINIC ACID DIETHYLAMIDE
* NIKARDIN * NIKETAMID * NIKETHAROL
* NIKETHYL * NIKETILAMID * NIKORIN
* NIQUETAMIDA * NISETAMIDE * PERCORAL
* PROCARDINE * PROCORMAN * PYRICAROYL
* PYRIDINE-3-CARBOXYDIETHYLAMIDE * PYRI-
DINE-3-CARBOXYLIC ACID DIETHYLAMIDE * REFOR-
MIN * REHORMIN * SALVACARD * SALVACO-
RIN * SOLYACORD * STELLAMINE * STIMINOL
* STIMULIN * TONOCARD * TONOCOR
* VASAZOL * VENTRAMINE

SAFETY PROFILE: Poison by ingestion, intravenous, intraperitoneal, and subcutaneous
routes. When heated to decomposition it emits
toxic fumes of NO_x.

DJT000   CAS: 597-88-6   HR: 3
O,S-DIETHYL-O-(4-NITROPHENYL) THIOPHOSPHATE
mf: $C_{10}H_{14}NO_5PS$ mw: 291.28

SYNS: O,S-DIETHYL-O-(p-NITROPHENYL) PHOSPHORO-
THIOATE * O,S-DIETHYL-O-(4-NITROPHENYL)PHOS-
PHOROTHIOATE * O,S-DIETHYL-O-(4-NITRO-
PHENYL)PHOSPHOROTHIOIC ACID ESTER * O,S-DI-
ETHYL-O-(p-NITROPHENYL)PHOSPHOROTHIOIC ACID
ESTER * S-ETHYL PARATHION * ISOPARATHION

SAFETY PROFILE: Poison by ingestion, skin
contact, intraperitoneal, subcutaneous, and intramuscular routes. When heated to decomposition it emits very toxic fumes of NO_x.

DJT200   CAS: 95-92-1   HR: 3
DIETHYL OXALATE

DOT: UN 2525
mf: $C_6H_{10}O_4$ mw: 146.16

PROP: Colorless, oily, aromatic liquid; decomp
in water. Mp: −40.6°, bp: 185.4°, flash p: 168°F
(OC), d: 1.08426 @ 15°, 1.0785 @ 20°/4°,
vap d: 5.04.

SYNS: DIETHYL ETHANEDIOATE * ETHYL OXA-
LATE * ETHYL OXALATE (DOT) * OXALIC ACID,
DIETHYL ESTER

CONSENSUS REPORTS: Reported in EPA
TSCA Inventory.

DOT Classification: Poison B; Label: St. Andrews Cross.

SAFETY PROFILE: Poison by ingestion. Flammable when exposed to heat or flame; can react
with oxidizing materials. To fight fire, use foam,
CO_2, dry chemical. When heated to decomposition it emits acrid smoke and fumes.

DJT400   CAS: 702-54-5   HR: 3
5,5-DIETHYL-1,3-OXAZIN-2,4-DIONE
mf: $C_8H_{13}NO_3$ mw: 171.22

SYNS: DIETADIONE (ITALIAN) * DIETHADION
* DIETHADIONE * 5,5-DIETHYLDIHYDRO-2H-1,3-
OXAZINE-2,4(3H)-DIONE * 5,5-DIETHYL-1,3-OXAZINE-
2,4-DIONE * 5,5-DIETHYLTETRAHYDRO-2H-1,3-OX-
AZINE-2,4(3H)-DIONE * 5,5-DIETILDIIDRO-1,3-OS-
SAZIN-2,4-DIONE (ITALIAN) * DIETROXINE * DI-
HYDRO-5,5-DIETHYL-2H-1,3-OXAZINE-2,4(3H)-DIONE
* DIIDRO-5,5-DIETIL-2H-1,3-OSSAZIN-2,4(3H)-DIONE
(ITALIAN) * DIOXONE * L 1811 * LEDOSTEN
* LEPTON * PERSISTEN * TOCE * TOCEN

CONSENSUS REPORTS: Reported in EPA
TSCA Inventory.

SAFETY PROFILE: Poison by ingestion, intravenous, intraperitoneal, subcutaneous, and
intramuscular routes. An analeptic (central nervous system stimulant). When heated to decomposition it emits toxic fumes of NO_x.

DJT800   CAS: 514-73-8   HR: 3
3,3′-DIETHYLPENTAMETHINETHIA-CYANINE IODIDE
mf: $C_{23}H_{24}N_2S_2 \cdot I$ mw: 519.51

SYNS: ABMINTHIC * ANELMID * ANGUIFUGAN
* COMPOUND 01748 * DEJO * DELVEX * DI-
ETHYLTHIADICARBOCYANINE IODIDE * 3,3′-DIETH-
YLTHIADICARBOCYANINE IODIDE * DILOMBRIN
* DITHIAZANINE IODIDE * DITHIAZANIN IODIDE
* DITHIAZININE * EASTMAN 7663 * 3-ETHYL-2-
(5-(3-ETHYL-2-BENZOTHIAZOLINYLIDENE)-1,3-PENTA-
DIENYL)BENZOTHIAZOLIUM IODIDE * L-01748
* NETOCYD * NK 136 * OMNI-PASSIN
* PARTEL * TELMICID * TELMID * TELMIDE
* VERCIDON

CONSENSUS REPORTS: Reported in EPA
TSCA Inventory.

SAFETY PROFILE: Poison by ingestion, intraperitoneal, and intravenous routes. When heated
to decomposition it emits very toxic fumes of
I^-, SO_x, and NO_x.

DJU200 CAS: 512-48-1 **HR: 3**
2,2-DIETHYL-4-PENTENAMIDE
mf: $C_9H_{17}NO$ mw: 155.27

SYNS: DIAETHYLALLYLACETAMIDE (GERMAN)
* EPINOVAL * NOVONAL

SAFETY PROFILE: Poison to humans by ingestion. An experimental poison by ingestion, rectal, and intraperitoneal routes. Human systemic effects by ingestion: muscle spasms, cardiac arrythmias, and respiratory depression. When heated to decomposition it emits toxic fumes of NO_x.

DJU600 CAS: 14666-78-5 **HR: 3**
DIETHYL PEROXYDICARBONATE

DOT: UN 2175
mf: $C_6H_{10}O_6$ mw: 178.14

DOT Classification: Organic Peroxide; Label: Organic Peroxide

SAFETY PROFILE: The impure material is a powerful explosive extremely sensitive to heat or impact. When heated to decomposition it emits acrid smoke and fumes.

DJV200 CAS: 93-05-0 **HR: 3**
DIETHYL-p-PHENYLENEDIAMINE
mf: $C_{10}H_{16}N_2$ mw: 164.28

SYN: N,N-DIETHYL-p-PHENYLENEDIAMINE

CONSENSUS REPORTS: Reported in EPA TSCA Inventory.

SAFETY PROFILE: Poison by ingestion, skin contact, subcutaneous, and intravenous routes. Human systemic skin effects by skin contact: hemorrhage, allergic dermatitis, and a primary irritant. When heated to decomposition it emits toxic fumes of NO_x.

DJV800 CAS: 64036-46-0 **HR: 3**
DIETHYL PHENYLTIN ACETATE
mf: $C_{12}H_{18}O_2Sn$ mw: 312.99

SYN: ACETOXYDIETHYLPHENYLSTANNANE

OSHA PEL: TWA 0.1 mg(Sn)/m^3 (skin)
ACGIH TLV: TWA 0.1 mg(Sn)/m^3 (skin) (Proposed: TWA 0.1 mg(Sn)/m^3; STEL 0.2 mg(Sn)/m^3 (skin))
NIOSH REL: (Organotin Compounds) TWA 0.1 mg(Sn)/m^3

SAFETY PROFILE: Poison by ingestion. When heated to decomposition it emits acrid and irritating fumes.

DJW600 CAS: 2524-04-1 **HR: 3**
O,O-DIETHYLPHOSPHOROCHLORIDO-THIOATE

DOT: UN 2751
mf: $C_4H_{10}ClO_2PS$ mw: 188.62

SYNS: CHLORO-PHOSPHONOTHIOIC ACID-O,O-DI-ETHYL ESTER * DIETHYLCHLOROTHIOPHOSPHATE * DIETHYLCHLORTHIOFOSFAT (CZECH) * DIETHYL-THIOPHOSPHORYL CHLORIDE (DOT)

CONSENSUS REPORTS: Reported in EPA TSCA Inventory.

DOT Classification: Corrosive Material; Label: Corrosive, Flammable Liquid.

SAFETY PROFILE: Poison by inhalation and skin contact. Moderately toxic by ingestion. Corrosive. Probably a severe eye and skin irritant. When heated to decomposition it emits very toxic fumes of Cl^-, PO_x, and SO_x.

DJW800 CAS: 2942-58-7 **HR: 3**
DIETHYL PHOSPHOROCYANIDATE
mf: $C_5H_{10}NO_3P$ mw: 163.13

SYNS: DIETHOXYPHOSPHORYL CYANIDE * DIE-THYLCYANOPHOSPHATE * DIETHYL CYANOPHOS-PHONATE

SAFETY PROFILE: Poison by intravenous, intraperitoneal, and subcutaneous routes. When heated to decomposition it emits very toxic fumes of NO_x and PO_x.

DJX000 CAS: 84-66-2 **HR: 3**
DIETHYL PHTHALATE
mf: $C_{12}H_{14}O_4$ mw: 222.26

PROP: Clear, colorless liquid. Mp: $-40.5°$, bp: $302°$, flash p: $325°F$ (OC), d: 1.110, vap d: 7.66.

SYNS: ANOZOL * 1,2-BENZENEDICARBOXYLIC ACID, DIETHYL ESTER * DIETHYL-o-PHTHALATE * ESTOL 1550 * ETHYL PHTHALATE * NCI-C60048 * NEANTINE * PALATINOL A * PHTHALIC ACID, DIETHYL ESTER * PHTHALOL * PHTHALSAEUREDIAETHYLESTER (GERMAN) * PLACIDOL E * RCRA WASTE NUMBER U088 * SOLVANOL

CONSENSUS REPORTS: Reported in EPA TSCA Inventory.

OSHA PEL: TWA 5 mg/m^3
ACGIH TLV: TWA 5 mg/m^3

SAFETY PROFILE: Poison by intravenous route. Moderately toxic by ingestion, subcutaneous, and intraperitoneal routes. Human systemic effects by inhalation: lachrimation, respiratory obstruction, and other unspecified respiratory system effects. An eye irritant and systemic irritant by inhalation. Experimental teratogenic and reproductive effects. Narcotic in high concentrations. Combustible when exposed to heat or flame. To fight fire, use water spray, mist, foam. When heated to decomposition it emits acrid smoke and irritating fumes.

DJY000   CAS: 21600-43-1   HR: 3
3,3-DIETHYL-1-(m-PYRIDYL)TRIAZENE
mf: $C_9H_{14}N_4$ mw: 178.27

SYNS: PYDT * 1-(PYRIDYL-3-)-3,3-DIAETHYL-TRIAZEN (GERMAN) * 1-PYRIDYL-3,3-DIETHYLTRIAZENE * 1-(PYRIDYL-3)-3,3-DIETHYLTRIAZENE * m-PYRIDYL-DIETHYL-TRIAZENE * 1-(3-PYRIDYL)-3,3-DIETHYLTRIAZENE

SAFETY PROFILE: Questionable carcinogen with experimental carcinogenic, neoplastigenic, tumorigenic, and teratogenic data. Poison by ingestion and subcutaneous routes. Human mutation data reported. A transplacental carcinogen. When heated to decomposition it emits toxic fumes of NO_x.

DJY200   CAS: 13593-03-8   HR: 3
O,O-DIETHYL-O-2-QUINOXALYLTHIOPHOSPHATE
mf: $C_{12}H_{15}N_2O_3PS$ mw: 298.32

SYNS: BAY 5821 * BAY 77049 * BAYRUSIL * CHINALPHOS * O,O-DIAETHYL-O-(CHINOXA-LYL-(2))-MONOTHIOPHOSPHAT (GERMAN) * DIETH-QUINALPHION * DIETHQUINALPHIONE * O,O-DI-ETHYL-O-(2-CHINOXALYL)PHOSPHOROTHIOATE * O,O-DIETHYL-O-QUINOXALIN-2-YL PHOSPHOROTH-IOATE * O,O-DIETHYL-O-(2-QUINOXALINYL) PHOS-PHOROTHIOATE * O,O-DIETHYL-O-(2-QUINOXALYL) PHOSPHOROTHIOATE * EKALUX * ENT 27,394 * NSC 190986 * QUINALPHOS * SAN 6538 I * SANDOZ 6538 * SPENCER S-6538 * SRA 7312 * WIE OBEN

SAFETY PROFILE: Poison by ingestion, inhalation, skin contact, parenteral, and intraperitoneal routes. Experimental reproductive effects. An insecticide. When heated to decomposition it emits very toxic fumes of NO_x, PO_x, and SO_x.

DJY600   CAS: 110-40-7   HR: 1
DIETHYL SEBACATE
mf: $C_{14}H_{26}O_4$ mw: 258.40

PROP: Colorless to sltly yellow liquid; faint fruity odor. D: 0.960-0.965, refr index: 1.435. Misc with alc, ether, other organic solvents, fixed oils; insol in water @ 302°.

SYNS: DIETHYL DECANEDIOATE * DIETHYL-1,10-DECANEDIOATE * ETHYL SEBACATE * FEMA No. 2376 * SEBACIC ACID, DIETHYL ESTER

CONSENSUS REPORTS: Reported in EPA TSCA Inventory.

SAFETY PROFILE: Mildly toxic by ingestion. A skin irritant. When heated to decomposition it emits acrid smoke and irritating fumes.

DJY800   CAS: 5117-17-9   HR: 3
N,N-DIETHYLSELENOUREA
mf: $C_5H_{12}N_2Se$ mw: 179.15

SYNS: 1,1-DIETHYL-2-SELENOUREA * USAF B-100

CONSENSUS REPORTS: Reported in EPA TSCA Inventory. Community Right-To-Know List.

OSHA PEL: TWA 0.2 mg(Se)/m^3
ACGIH TLV: TWA 0.2 mg(Se)/m^3
DFG MAK: 0.1 mg(Se)/m^3

SAFETY PROFILE: Poison by intraperitoneal route. When heated to decomposition it emits very toxic fumes of NO_x and Se.

DKA400   CAS: 63528-82-5   HR: 3
α,α'-DIETHYL-4,4'-STILBENEDIOL DISODIUM SALT
mf: $C_{18}H_{18}O_2 \cdot 2Na$ mw: 312.34

SYNS: DES DISODIUM SALT * DIETHYLSTILBES-TROL DISODIUM SALT

SAFETY PROFILE: Experimental teratogenic and reproductive effects. Questionable carcinogen with experimental neoplastigenic data. When heated to decomposition it emits toxic fumes of Na_2O.

DKA600   CAS: 56-53-1   HR: 3
DIETHYLSTILBESTEROL
mf: $C_{18}H_{20}O_2$ mw: 268 data.38

PROP: Small crystals. Mp: 171°.

SYNS: ACNESTROL * AGOSTILBEN * ANTIGES-TIL * BIO-DES * 3,4-BIS(p-HYDROXYPHENYL)-3-

HEXENE * BUFON * CLIMATERINE * COMES-TROL * COMESTROL ESTROBENE * CYREN * DAWE'S DESTROL * DEB * DESMA * DES (synthetic estrogen) * DESTROL * DIASTYL * DIBESTROL * DICORVIN * DI-ESTRYL * trans-4,4'-(1,2-DIETHYL-1,2-ETHENEDIYL)BISPHENOL * 4,4'-(1,2-DIETHYL-1,2-ETHENEDIYL)BIS-PHENOL * α,α'-DIETHYLSTILBENEDIOL * α,α'-DIETHYL-(E)-4,4'-STILBENEDIOL * α,α'-DIETHYL-4,4'-STILBENE-DIOL * trans-α,α'-DIETHYL-4,4'-STILBENEDIOL * 2,2'-DIETHYL-4,4'-STILBENEDIOL * trans-DIETHYL-STILBESTEROL * DIETHYLSTILBESTROL * trans-DI-ETHYLSTILBESTROL * DIETHYLSTILBOESTEROL * trans-DIETHYLSTILBOESTEROL * DIETILESTILBES-TROL (SPANISH) * 4,4'-DIHYDROXYDIETHYLSTIL-BENE * 4,4'-DIHYDROXY-α,β-DIETHYLSTILBENE * 3,4'(4,4'-DIHYDROXYPHENYL)HEX-3-ENE * DIS-TILBENE * DOMESTROL * DYESTROL * ESTIL-BEN * ESTRIL * ESTROBENE * ESTROGEN * ESTROMENIN * ESTROSYN * FOLLIDIENE * FONATOL * GRAFESTROL * GYNOPHARM * HIBESTROL * IDROESTRIL * ISCOVESCO * MAKAROL * MENOSTILBEEN * MICREST * MICROEST * MILESTROL * NEO-OESTRANOL 1 * NSC-3070 * OEKOLP * OESTROGENINE * OESTROL VETAG * OESTROMENIN * OESTRO-MENSIL * OESTROMENSYL * OESTROMIENIN * OESTROMON * PABESTROL * PALESTROL * PERCUTATRINE OESTROGENIQUE ISCOVESCO * PROTECTONA * RCRA WASTE NUMBER U089 * RUMESTROL 1 * RUMESTROL 2 * SEDESTRAN * SERRAL * SEXOCRETIN * SIBOL * SINTES-TROL * STIBILIUM * STIL * STILBESTROL * STILBESTRONE * STILBETIN * STILBOEFRAL * STILBOESTROFORM * STILBOESTROL * STIL-BOFOLLIN * STILBOL * STILKAP * STIL-ROL * SYNESTRIN * SYNTHOESTRIN * SYNTHOFOLIN * SYNTOFOLIN * TAMPOVAGAN STILBOESTROL * TYLOSTERONE * VAGESTROL

CONSENSUS REPORTS: IARC Cancer Review: GROUP 1 IMEMDT 7,273,87; Human Limited Evidence IMEMDT 6,55,74; IMEMDT 21,173,79; Animal Sufficient Evidence IMEMDT 21,173,79; IMEMDT 6,55,74. NTP Fourth Annual Report On Carcinogens, 1984. EPA Genetic Toxicology Program. Reported in EPA TSCA Inventory.

SAFETY PROFILE: Confirmed carcinogen producing skin, liver, and lung tumors in exposed humans as well as uterine and other reproductive system tumors in the female offspring of exposed women. Experimental carcinogenic, neoplastigenic, tumorigenic, and teratogenic data. A transplacental carcinogen. A human teratogen by many routes. Poison by intraperitoneal and subcutaneous routes. Moderately toxic by ingestion and other routes. It causes glandular system effects by skin contact. Human reproductive effects by ingestion: abnormal spermatogenesis; changes in testes, epididymis and sperm duct; menstrual cycle changes or disorders; changes in female fertility; unspecified maternal effects; developmental abnormalities of the fetal urogenital system; germ cell effects in offspring; and delayed effects in newborn. Implicated in male impotence and enlargement of male breasts. Other experimental reproductive effects. Mutation data reported. When heated to decomposition it emits acrid smoke and fumes.

DKA800   CAS: 63019-08-9   **HR: 3**
DIETHYLSTILBESTROL DIPALMITATE
mf: $C_{50}H_{80}O_4$ mw: 745.30

SYNS: α,α'-DIETHYL-4,4'-STILBENEDIOL DIPALMITATE * 4,4'-DIHYDROXY-α,β-DIETHYLSTILBENE PALMITATE

SAFETY PROFILE: Experimental teratogenic and reproductive effects. Questionable carcinogen with experimental tumorigenic data. When heated to decomposition it emits acrid smoke and irritating fumes.

DKB000   CAS: 130-80-3   **HR: 3**
DIETHYLSTILBESTROL DIPROPIONATE
mf: $C_{24}H_{28}O_4$ mw: 380.52

PROP: Crystals. Mp: 104°.

SYNS: CLINESTROL * CYREN B * DESD * DIBESTIL * trans-4,4'-(1,2-DIETHYL-1,2-ETHENE-DIYL)BISPHENOL DIPROPIONATE * α,α'-DIETHYL-4,4'-STILBENEDIOL, DIPROPIONATE * α,α'-DIETHYL-4,4'-STILBENEDIOL trans-DIPROPIONATE * trans-α,α'-DIETHYL-4,4'-STILBENEDIOL DIPROPIONATE * DI-ETHYLSTILBENE DIPROPIONATE * α,α'-DI-ETHYL-4,4'-STILBENEDIOL DIPROPIONYL ESTER * DIETHYLSTILBESTEROL DIPROPIONATE * DI-ETHYLSTILBESTROL PROPIONATE * DIHYDROXYDI-ETHYLSTILBENE DIPROPIONATE * 4,4'-DIHYDROXY-α,β-DIETHYLSTILBENE DIPROPIONATE * DIPROPION-ATO de ESTILBENE (SPANISH) * p,p'-DIPROPIONOXY-trans-α,β-DIETHYLSTILBENE * DISTILBENE * ESTILBEN * ESTILBIN * ESTROBEN * ESTROBENE * ESTROGENIN * ESTROSTILBEN * EUVESTIN * GYNOLETT * HORFEMINE * NEO-OESTRANOL II * OESTROGYNAEDRON

* ORESTOL * PABESTROL * SINCICLAN * STILBESTROL DIETHYL DIPROPIONATE * STIL-BESTROL DIPROPIONATE * STILBESTROL PROPIO-NATE * STILBESTRONATE * STILBOESTROL DI-PROPIONATE * STILBOFAX * STILRONATE * SYNESTRIN * SYNOESTRON * SYNTESTRIN * SYNTESTRINE * WILLESTROL

CONSENSUS REPORTS: IARC Cancer Review: Animal Sufficient Evidence IMEMDT 21,173,79. EPA Genetic Toxicology Program.

SAFETY PROFILE: Confirmed carcinogen with experimental tumorigenic data. Experimental teratogenic and reproductive effects. Human mutagenic data. When heated to decomposition it emits acrid smoke and irritating fumes.

DKB110   CAS: 64-67-5   HR: 3
DIETHYL SULFATE
DOT: UN 1594
mf: $C_4H_{10}O_4S$ mw: 154.20

PROP: Colorless, oily liquid; faint ethereal odor. Mp: $-25°$, Bp: $209.5°$ (decomp to ethyl ether), flash p: $220°F$ (CC), d: 1.172 @ $25°/4°$, autoign temp: $817°F$, vap press: 1 mm @ $47.0°$, vap d: 5.31. Insol in water; decomp by hot water; misc with alc and ether.

SYNS: DIAETHYLSULFAT (GERMAN) * DIETHYL ESTER SULFURIC ACID * ETHYL SULFATE

CONSENSUS REPORTS: IARC Cancer Review: GROUP 2A IMEMDT 7,198,87; Animal Sufficient Evidence IMEMDT 4,277,74. NTP Fourth Annual Report On Carcinogens, 1984. EPA Genetic Toxicology Program. Community Right-To-Know List. Reported in EPA TSCA Inventory.

DFG TRK: 0.03 ppm; Animal Carcinogen, Human Suspected Carcinogen.
DOT Classification: Poison B; Label: Poison

SAFETY PROFILE: Confirmed with experimental carcinogenic and tumorigenic data. Poison by inhalation and subcutaneous route. Moderately toxic by ingestion and skin contact. A severe skin irritant. Mutation data reported. Combustible when exposed to heat or flame; can react with oxidizing materials. Moisture causes liberation of H_2SO_4. Violent reaction with potassium tert-butoxide. Reacts violently with 3,8-dinitro-6-phenylphenanthridine + water. Reaction with iron + water forms the explosive hydrogen gas. To fight fire, use alcohol foam, H_2O foam, CO_2, dry chemicals. When heated to decomposition it emits toxic fumes of SO_x.

DKB600   CAS: 5827-03-2   HR: 3
N,N-DIETHYLTHIOCARBAMYL-O,O-DIISOPROPYLDITHIOPHOSPHATE
mf: $C_{11}H_{24}NO_2PS_3$ mw: 329.51

SYNS: DIETHYLDITHIOCARBAMIC ANHYDRIDE OF O,O-DIISOPROPYL THIONOPHOSPHORIC ACID * DIETHYLDITHIOCARBAMIC ANHYDROSULFIDE * O,O-DIISOPROPYL-S-DIETHYLDITHIOCARBA-MOYLPHOSPHORODITHIOATE * O,O-DIIOSPRO-PYL DITHIOPHOSPHORIC ACID ESTER OF-N,N-S-DIETHYLTHIOCARBAMOYL-O,O-DIISOPROPYL PHOS-PHOROTHIOATE * DIISOPROPYL ESTER OF DI-THIOCARBAMYL PHOSPHOROTHIOIC ACID * ENT 24,725

SAFETY PROFILE: Poison by ingestion. When heated to decomposition it emits very toxic fumes of PO_x, NO_x, and SO_x.

DKC200   CAS: 69226-06-8   HR: 3
2,2-DIETHYL-3-THIOMORPHOLINONE
mf: $C_8H_{15}NOS$ mw: 173.30

SAFETY PROFILE: Poison by intravenous route. Moderately toxic by ingestion and intraperitoneal route. When heated to decomposition it emits very toxic fumes of NO_x and SO_x.

DKC400   CAS: 105-55-5   HR: 3
1,3-DIETHYLTHIOUREA
mf: $C_5H_{12}N_2S$ mw: 132.25

SYNS: N,N'-DIETHYLTHIOCARBAMIDE * N,N'-DIE-THYLTHIOUREA * 1,3-DIETHYL-2-THIOUREA * NCI-C03816 * PENNZONE E * THIATE H * U 15030 * USAF EK-1803

CONSENSUS REPORTS: NCI Carcinogenesis Bioassay (feed); Clear Evidence: rat NCITR* NCI-CG-TR-149,79; No Evidence: mouse NCITR* NCI-CG-TR-149,79. Reported in EPA TSCA Inventory. EPA Genetic Toxicology Program.

SAFETY PROFILE: Questionable carcinogen with experimental carcinogenic data. Poison by ingestion. Moderately toxic by intraperitoneal route. When heated to decomposition it emits very toxic fumes of NO_x and SO_x.

DKC800   CAS: 134-62-3   **HR: 3**
DIETHYL-m-TOLUAMIDE
mf: $C_{12}H_{17}NO$ mw: 191.30

PROP: A liquid, sol in water, alc, and ether. Bp: 160° @ 19 mm, d: 0.996 @ 20°/4°.

SYNS: AI 3-22542 * AUTAN * BAKER'S ANTIFOL * CHEMFORM * DEET * DELPHENE * m-DELPHENE * DET * m-DET * m-DETA * DETAMIDE * DIELTAMID * N,N-DIETHYL-3-METHYLBENZAMIDE * DIETHYLTOLUAMIDE * N,N-DIETHYL-m-TOLUAMIDE * ENT 20,218 * ENT 22,542 * FLYPEL * METADELPHENE * 3-METHYL-N,N-DIETHYLBENZAMIDE * MGK DIETHYLTOLUAMIDE * NAUGATUCK DET * OFF * REPEL * REPPER-DET * REPUDIN-SPECIAL * m-TOLUIC ACID DIETHYLAMIDE

CONSENSUS REPORTS: Reported in EPA TSCA Inventory.

SAFETY PROFILE: Poison by intravenous route. Moderately toxic by ingestion, skin contact, and possibly other routes. Human systemic effects by skin contact: dermatitis. An eye and skin irritant. Experimental reproductive effects by skin contact. Mutation data reported. Can cause central nervous system disturbances. A pesticide. DEET is the active ingredient in most commercial insect repellents. When heated to decomposition it emits toxic fumes of NO_x.

DKD200   CAS: 63980-20-1   **HR: D**
DIETHYL TRIAZENE
mf: $C_4H_{11}N_3$ mw: 101.18

SAFETY PROFILE: Questionable carcinogen with experimental carcinogenic and tumorigenic data. An experimental teratogen. When heated to decomposition it emits toxic fumes of NO_x.

DKE200   CAS: 304-84-7   **HR: 3**
N,N-DIETHYLVANILLAMIDE
mf: $C_{12}H_{17}NO_3$ mw: 223.30

SYNS: DIETHYLAMIDE de VANILLIQUE * 3-METH-OXY-4-HYDROXYBENZOIC ACID DIETHYLAMIDE * VANILLIC ACID DIETHYLAMIDE * VANILLIC ACID-N,N-DIETHYLAMIDE * VANILLINSAEURE-DIA-ETHYLAMID (GERMAN)

SAFETY PROFILE: Poison by ingestion, intravenous, subcutaneous, and intraperitoneal routes. When heated to decomposition it emits toxic fumes of NO_x.18123

DKE400   CAS: 1851-77-0   **HR: 3**
DI(ETHYLXANTHOGEN)TRISULFIDE
mf: $C_6H_{10}O_2S_5$ mw: 274.46

SYNS: BEXT * BIS(ETHOXYTHIOCARBONYL)TRI-SULFIDE * BIS(ETHYLXANTHOGEN) TRISULFIDE * DEFOLIANT 713 * DI-ETHOXYTHIOKARBONYL-TRISULFID * TRISULFIDE, BIS(ETHOXYTHIOCAR-BONYL)

SAFETY PROFILE: Poison by ingestion. Moderately toxic by inhalation. When heated to decomposition it emits toxic fumes of SO_x.

DKE600   CAS: 557-20-0   **HR: 3**
DIETHYLZINC
DOT: UN 1366/UN 2845
mf: $C_4H_{10}Zn$ mw: 123.51

PROP: Liquid. Mp: −28°, bp: 118°, d: 1.2065 @ 20°/4°.

SYNS: ZINC ETHIDE * ZINC ETHYL (DOT)

CONSENSUS REPORTS: Reported in EPA TSCA Inventory. Zinc and its compounds are on the Community Right-To-Know List.

DOT Classification: Flammable Liquid; Label: Flammable Liquid; IMO: Flammable Liquid; Label: Spontaneously Combustible.

SAFETY PROFILE: Presumed to be a poison. Ignites spontaneously in air. Dangerously flammable by spontaneous chemical reaction in air, or with oxidizing materials. A dangerous explosion hazard. Explosive reaction with alkenes + diiodomethane; sulfur dioxide. Reacts violently with bromine; water; nitro compounds. Ignites on contact with air; ozone; methanol; or hydrazine. Reacts violently with non-metal halides (e.g., arsenic trichloride or phosphorus trichloride to produce pyrophoric triethyl arsine or triethyl phosphine. To fight fire, do not use water, foam, or halogenated extinguishing agents. Use dry materials, such as graphite, sand, etc. When heated to decomposition it emits toxic fumes of ZnO.

DKE800   CAS: 63868-62-2   **HR: 3**
DIETROL
mf: $C_{12}H_{17}NO \cdot 2C_4H_6O_6$ mw: 491.50

SYNS: ADPHEN * BACARATE * 3,4-DIMETHYL-2-PHENYLMORPHOLINE BITARTRATE * HOURBESE * LIMIT * MINUS * NEO-NILOREX * OBEPAR * PHENAZINE * PHENDIMETRAZINE BITARTRATE * PLEGINE * REDUCTO * STATOBEX * SYMETRA * TRIMSTAT * TRIMTABS

SAFETY PROFILE: Poison by ingestion, intra-peritoneal, and intravenous routes. When heated to decomposition it emits toxic fumes of NO_x.

DKG400   CAS: 61735-78-2   **HR: 3**
2,10-DIFLUOROBENZO(rst)
PENTAPHENE
mf: $C_{24}H_{12}F_2$ mw: 338.36

SYN: 2,10-DIFLUORODIBENZO(a,i)PYRENE

SAFETY PROFILE: Questionable carcinogen with experimental tumorigenic data. Mutation data reported. When heated to decomposition it emits toxic fumes of F^-.

DKG850   CAS: 75-61-6   **HR: 1**
DIFLUORODIBROMOMETHANE

DOT: UN 1941
mf: CBr_2F_2 mw: 209.83

PROP: Colorless, heavy liquid. Bp: 23.2°, fp: −141°, d: 2.288 @ 15°/4°.

SYNS: DIBROMODIFLUOROMETHANE * FREON 12-B2 * HALON 1202

CONSENSUS REPORTS: Reported in EPA TSCA Inventory.

OSHA PEL: TWA 100 ppm
ACGIH TLV: TWA 100 ppm
DFG MAK: 100 ppm (860 mg/m^3)
DOT Classification: ORM-A; Label: None.

SAFETY PROFILE: Mildly toxic by inhalation. When heated to decomposition it emits very toxic fumes of Br^- and F^-.

DKH200   CAS: 3582-17-0   **HR: 3**
DIFLUORODIMETHYLSTANNANE
mf: $C_2H_6F_2Sn$ mw: 186.77

PROP: White crystals. Water-sol. Bp: decomp < 360°.

SYNS: DIMETHYLTIN DIFLUORIDE * DIMETHYL-TIN FLUORIDE

CONSENSUS REPORTS: Reported in EPA TSCA Inventory.

OSHA PEL: TWA 0.1 mg(Sn)/m^3 (skin)
ACGIH TLV: TWA 0.1 mg(Sn)/m^3 (skin) (Pro-posed: TWA 0.1 mg(Sn)/m^3; STEL 0.2 mg(Sn)/m^3 (skin))
NIOSH REL: (Organotin Compounds) TWA 0.1 mg(Sn)/m^3

SAFETY PROFILE: Poison by intravenous route. When heated to decomposition it emits toxic fumes of F^-.

DKI400   CAS: 368-97-8   **HR: 3**
DIFLUOROPHENYLARSINE
mf: $C_6H_5AsF_2$ mw: 190.03

SYN: PHENYLDIFLUOROARSINE

CONSENSUS REPORTS: Arsenic and its compounds are on the Community Right-To-Know List.

OSHA PEL: TWA 0.5 mg(As)/m^3

SAFETY PROFILE: Poison by skin contact and intravenous routes. When heated to decomposi-tion it emits very toxic fumes of As and F^-.

DKI600   CAS: 22494-42-4   **HR: 3**
5-(2,4-DIFLUOROPHENYL)SALICYLIC
ACID
mf: $C_{13}H_8F_2O_3$ mw: 250.21

SYNS: DIFLUNISAL * 2′,4′-DIFLUORO-4-HY-DROXY-(1,1′-BIPHENYL)-3-CARBOXYLIC ACID * 2′,4′-DIFLUORO-4-HYDROXY-3-BIPHENYLCARBOX-YLIC ACID * 2′,4′-DIFLUORO-4-HYDROXY-(1′,1-DIPHE-NYL)-3-CARBOXYLIC ACID * DOLOBID * DOLOBIL * DOLOBIS * FLOVACIL * FLUNIGET * 2-(HYDROXY)-5-(2,4-DIFLUOROPHENYL)BENZOIC ACID * MK 647

SAFETY PROFILE: Poison by ingestion, subcu-taneous, and intraperitoneal routes. Human sys-temic effects by ingestion: tolerance, and choles-tatic jaundice (due to the stoppage of the flow of bile). An analgesic and anti-inflammatory agent. When heated to decomposition it emits toxic fumes of F^-.

DKK000   **HR: 2**
DIFUMARATE

SYN: 4-(3-(2-CHLOROPHENOTHIAZIN-10-YL)PROPYL-1-PIPERAZINEETHANOL-3,4,5-TRIMETHOXYBENZOATE DI-FUMARATE

SAFETY PROFILE: Moderately toxic by inges-tion. Experimental teratogenic and reproductive effects. When heated to decomposition it emits very toxic fumes of Cl^-, NO_x, and SO_x.

DKK800   CAS: 63906-88-7   **HR: 3**
DIGAMMACAINE
mf: $C_{21}H_{26}N_2O \cdot ClH$ mw: 358.95

SYNS: 1-BENZAMIDO-1-PHENYL-3-PIPERIDINOPRO-PANE HYDROCHLORIDE * N-(3-BENZAMIDO-3-PHE-NYL)PROPYL PIPERIDINE HYDROCHLORIDE

SAFETY PROFILE: Poison by ingestion, intramuscular, subcutaneous, and intravenous routes. When heated to decomposition it emits very toxic fumes of HCl and NO_x.

DKL200   CAS: 8031-42-3   **HR: 3**
DIGITALIS

PROP: Dried whole leaf of *Digitalis purpurea*. Composition: digitoxin (0.2-0.4%), etc.

SYNS: DIGITANNOID * DIGITALIS PURPUREA, LEAF * FOXGLOVE

SAFETY PROFILE: A deadly human poison by intravenous route. An experimental poison by ingestion, intravenous, and intraperitoneal routes. 2.5 grams or 30 cc of the tincture is a toxic dose. An overdose can be fatal. it has been implicated in aplastic anemia. It contains digitalin, digitalein, digitonin, and digitoxin (the most toxic component).

DKL400   CAS: 11024-24-1   **HR: 3**
DIGITONIN
mf: $C_{56}H_{92}O_{29}$ mw: 1229.48

SYN: DIGITIN

CONSENSUS REPORTS: Reported in EPA TSCA Inventory.

SAFETY PROFILE: Poison by ingestion, intravenous, intraperitoneal, and subcutaneous routes. Mutation data reported. When heated to decomposition it emits acrid smoke and irritating fumes.

DKL800   CAS: 71-63-6   **HR: 3**
DIGITOXIN
mf: $C_{41}H_{64}O_{13}$ mw: 765.05

SYNS: ACEDOXIN * ASTHENTHILO * CARDIDI-GIN * CARDIGIN * CARDITOXIN * CRISTAPU-RAT * CRYSTALLINE DIGITALIN * CRYSTODIGIN * DIGILONG * DIGIMED * DIGIMERCK * DIGISIDIN * DIGITALIN * DIGITALINE (FRENCH) * DIGITALINE CRISTALLISEE * DIGITA-LINE NATIVELLE * DIGITALINUM VERUM * DIGITOPHYLLIN * DIGITOXIGENIN-TRIDIGITOXO-SID (GERMAN) * DIGITOXIGENIN TRIDIGITOXOSIDE * DITAVEN * GLUCODIGIN * LANATOXIN * MONO-GLYCOCOARD * MYODIGIN * PURODI-GIN * PURPURID * TARDIGAL * TRI-DIGITOXO-SIDE (GERMAN) * UNIDIGIN

CONSENSUS REPORTS: Reported in EPA TSCA Inventory. EPA Extremely Hazardous Substances List.

SAFETY PROFILE: A deadly poison by most routes. Human systemic effects by ingestion and possibly other routes: cardiac arrythmias, nausea and vomiting. Human reproductive effects by ingestion: reduced viability of newborn. An eye irritant. When heated to decomposition it emits acrid smoke and irritating fumes.

DKM200   CAS: 2238-07-5   **HR: 3**
DIGLYCIDYL ETHER
mf: $C_6H_{10}O_3$ mw: 130.16

PROP: Liquid.

SYNS: BIS(2,3-EPOXYPROPYL)ETHER * DGE * DI(2,3-EPOXYPROPYL) ETHER

CONSENSUS REPORTS: EPA Extremely Hazardous Substances List. Reported in EPA TSCA Inventory. EPA Genetic Toxicology Program.

OSHA PEL: (Transitional: CL 0.5 ppm) TWA 0.1 ppm
ACGIH TLV: TWA 0.1 ppm
DFG MAK: 0.1 ppm (0.6 mg/m^3); Suspected Carcinogen.
NIOSH REL: (Glycidyl Ethers) CL 1 mg/m^3/ 15M

SAFETY PROFILE: Suspected carcinogen with experimental tumorigenic data. Poison by ingestion, inhalation, and intravenous routes. Moderately toxic by skin contact. A severe eye and skin irritant. Mutation data reported. Chronic exposure can cause bone marrow depression. When heated to decomposition it emits acrid smoke and fumes.

DKN400   CAS: 20830-75-5   **HR: 3**
DIGOXIN
mf: $C_{41}H_{64}O_{14}$ mw: 781.05

PROP: White, crystalline powder. Mp: 265°. Glycoside isolated from *Digitalis lanata*.

SYNS: CHLOROFORMIC DIGITALIN * DIGACIN * DIGITALIS GLYCOSIDE * DIGOXIGENIN-TRIDIGI-TOXOSID (GERMAN) * DIGOXINE * HOMOLLE'S DIGITALIN * LANICOR * LANOXIN * ROU-GOXIN * SK-DIGOXIN

CONSENSUS REPORTS: EPA Extremely Hazardous Substances List. Reported in EPA TSCA Inventory. EPA Genetic Toxicology Program.

SAFETY PROFILE: A deadly poison by most routes. Human systemic effects by ingestion: anorexia, cardiac arrythmias, nausea and vomiting. An experimental teratogen. When heated to decomposition it emits acrid and irritating fumes.

DKN600   HR: 3
є-DIGOXIN ACETATE

SYN: є-ACETYLDIGOXIN (GERMAN)

SAFETY PROFILE: Poison by ingestion and intravenous routes. When heated to decomposition it emits acrid and irritating fumes.

DKO000   CAS: 51622-02-7   HR: 3
DIHEPTYLMERCURY
mf: $C_{14}H_{30}Hg$ mw: 399.03

OSHA PEL: (Transitional: CL 1 mg/10m^3) TWA 0.01 mg(Hg)/m^3; STEL 0.03 mg/m^3 (skin)

CONSENSUS REPORTS: Mercury and its compounds are on the Community Right-To-Know List.

ACGIH TLV: TWA 0.01 mg(Hg)/m^3; STEL 0.03 mg(Hg)/m^3
NIOSH REL: (Inorganic Mercury) TWA 0.05 mg(Hg)/m^3

SAFETY PROFILE: Poison by intraperitoneal route. When heated to decomposition it emits toxic fumes of Hg.

DKO600   CAS: 143-16-8   HR: 3
DIHEXYLAMINE
mf: $C_{12}H_{27}N$ mw: 185.40

PROP: Liquid. Bp: 233-243°, flash p: 220°F (OC), d: 0.78, vap d: 6.38.

SYN: DI-N-HEXYLAMINE

CONSENSUS REPORTS: Reported in EPA TSCA Inventory.

SAFETY PROFILE: Poison by ingestion, skin contact, and intravenous routes. A skin irritant. Flammable when exposed to heat or flame; can react with oxidizing materials. To fight fire, use CO_2, dry chemical. When heated to decomposition it emits toxic fumes of NO_x.

DKQ000   CAS: 57-41-0   HR: 3
DIHYDANTOIN
mf: $C_{15}H_{12}N_2O_2$ mw: 252.29

SYNS: ALEVIATIN * ANTISACER * AURANILE * CAUSOIN * CITRULLAMON * COMITAL * CONVUL * DANTEN * DANTINAL * DANTOINAL KLINOS * DANTOINE * DENYL * DIDAN-TDC-250 * DIFENILHIDANTOINA (SPANISH) * DIFENIN * DIFHYDAN * DIHYCON * DI-HYDAN * DILANTIN * DILANTINE * DINTOIN * DIPHANTOIN * DIPHEDAL * DIPHENINE * DIPHENTOIN * DIPHENYLAN * DIPHENYLHYDANTOIN * 5,5-DIPHENYLHYDANTOIN * DIPHENYLHYDANTOINE (FRENCH) * 5,5-DIPHENYLIMIDAZOLIDIN-2,4-DIONE * 5,5-DIPHENYL-2,4-IMIDAZOLIDINEDIONE * DI-PHETINE * DITOINATE * DPH * EKKO CAPSULES * ELEPSINDON * ENKELFEL * EPAMIN * EPANUTIN * EPASMIR '5' * EPDANTOINE SIMPLE * EPELIN * EPIFENYL * EPIHYDAN * EPILAN * EPILANTIN * EPINAT * EPISED * EPTAL * EPTOIN * FENANTOIN * FENIDANTOIN 'S' * FENYLEPSIN * FENYTOINE * GEROT-EPILAN-D * HIDAN * HIDANTILO * HIDANTINA SENOSIAN * HIDANTINA VITORIA * HIDANTOMIN * HYDANTAL * HYDANTOIN * ICTALIS SIMPLE * IDANTOIN * KESSODANTEN * LABOPAL * LEHYDAN * LEPITOIN * LEPSIN * MINETOIN * NCI-C55765 * NEOS-HIDANTOINA * NOVANTOINA * OM-HYDANTOINE * OXYLAN * PHANANTIN * PHENATOINE * RITMENAL * SACERIL * SANEPIL * SILANTIN * SODANTON * SOLANTIN * SYLANTOIC * TACOSAL * THILOPHENYL * TOIN UNICELLES * ZENTRONAL * ZENTROPIL

CONSENSUS REPORTS: IARC Cancer Review: GROUP 2B IMEMDT 7,319,87; Human Limited Evidence IMEMDT 13,201,77; Animal Sufficient Evidence IMEMDT 13,201,77. NTP Fourth Annual Report On Carcinogens, 1984. EPA Genetic Toxicology Program.

SAFETY PROFILE: Confirmed carcinogen producing lymphoma, Hodgkin's disease, tumors of the skin and appendages. Experimental carcinogenic and tumorigenic data. A human poison by ingestion. Poison experimentally by ingestion, subcutaneous, intravenous, and intraperitoneal routes. Moderately toxic by an unspecified route. Experimental teratogenic and reproductive effects. Human systemic effects by ingestion: dermatitis, change in motor activity (specific assay), ataxia (loss of muscle coordination), degenerate brain changes, and jaundice. Human teratogenic effects by ingestion and possibly other routes: developmental abnormalities of the central nervous system, cardiovascular (circulatory) system, musculoskeletal system, craniofacial area, skin and skin appendages, eye, ear, other developmental abnormalities. Effects on newborn include abnormal growth statistics (e.g., reduced weight gain), physical abnormalities, other postnatal measures or effects, and delayed effects. Human mutation data reported. A drug for the treatment of grand mal and psychomotor seizures. When heated to decomposition it emits toxic fumes of NO_x.

DKS800   CAS: 10023-25-3   HR: 3
6,13-DIHYDROBENZO(e)(1)BENZOTHIO-PYRANO(4,3-b)INDOLE
mf: $C_{19}H_{13}NS$ mw: 287.39

SAFETY PROFILE: Questionable carcinogen with experimental neoplastigenic data. Mutation data reported. When heated to decomposition it emits very toxic fumes of SO_x and NO_x.

DKU000   CAS: 17573-23-8   HR: 3
7,8-DIHYDROBENZO(a)PYRENE
mf: $C_{20}H_{14}$ mw: 254.34

CONSENSUS REPORTS: EPA Genetic Toxicology Program.

SAFETY PROFILE: Questionable carcinogen with experimental neoplastigenic data. Mutation data reported. When heated to decomposition it emits acrid smoke and irritating fumes.

DKU400   CAS: 66788-01-0   HR: 3
9,10-DIHYDROBENZO(e)PYRENE
mf: $C_{20}H_{14}$ mw: 254.34

SYN: 9,10-H2 B(e)P

SAFETY PROFILE: Questionable carcinogen with experimental neoplastigenic and tumorigenic data. Mutation data reported. When heated to decomposition it emits acrid smoke and irritating fumes.

DKV150   CAS: 619-01-2   HR: 2
DIHYDROCARVEOL
mf: $C_{10}H_{18}O$ mw: 154.28

PROP: Colorless, oily liquid; spearmint odor. D: 0.921-0.926, refr index: 1.477-1.481, flash p: +153°F. Sol in alc, fixed oils; insol in water.

SYNS: 1,6-DIHYDROCARVEOL * FEMA No. 2379 * 8-p-MENTHEN-2-OL * 6-METHYL-3-ISOPROPYLCYCLOHEXANOL

CONSENSUS REPORTS: Reported in EPA TSCA Inventory.

SAFETY PROFILE: A moderate skin and eye irritant. A combustible liquid. When heated to decomposition it emits acrid smoke and irritating fumes.

DKV175   CAS: 7764-50-3   HR: 2
d-DIHYDROCARVONE
mf: $C_{10}H_{16}O$ mw: 152.26

PROP: Colorless liquid; spearmint-like odor. D: 0.923-0.928, refr index: 1.470-1.474. Sol in alc, fixed oils; insol in water.

SYNS: FEMA No. 3565 * 8-p-MENTHEN-2-ONE * p-MENTH-8-EN-2-ONE * d-2-METHYL-5-(1-METHYLENENYL)-CYCLOHEXANONE

CONSENSUS REPORTS: Reported in EPA TSCA Inventory.

SAFETY PROFILE: Moderately toxic by subcutaneous route. When heated to decomposition it emits acrid smoke and irritating fumes.

DKW000   CAS: 360-68-9   HR: 3
DIHYDROCHOLESTEROL
mf: $C_{27}H_{48}O$ mw: 388.75

SYNS: (3-β,5-β)-CHOLESTAN-3-OL * 3-β-CHOLESTANOL * COPROSTANOL * COPROSTAN-3-β-OL * COPROSTEROL * 3-β-HYDROXYCHOLESTANE * KOPROSTERIN (GERMAN) * STERCORIN * XYMOSTANOL

SAFETY PROFILE: Questionable carcinogen with experimental neoplastigenic data. When heated to decomposition it emits acrid smoke and irritating fumes.

DKW800   CAS: 125-28-0   HR: 3
DIHYDROCODEINE
mf: $C_{18}H_{23}NO_3$ mw: 301.42

SYNS: CODHYDRINE * COHYDRIN * DF 118 * DEHACODIN * DIDRATE * DIHYDRIN * 7,8-DIHYDROCODEINE * DIHYDRONEOPINE * DROCODE * HYDROCODIN * 6-HYDROXY-3-METHOXY-N-METHYL-4,5-EPOXYMORPHINAN * NADEINE * NOVICODIN * PARACODIN * PARACODINE * PARZONE * RAPACODIN

SAFETY PROFILE: Poison by ingestion, intravenous, and subcutaneous routes. Human systemic effects by ingestion and subcutaneous routes: somnolence, miosis (pupillary constriction), and respiratory depression. An analgesic. Can cause drug dependency with repeated doses. When heated to decomposition it emits toxic fumes of NO_x.

DKX000   CAS: 5965-13-9   **HR: 3**
DIHYDROCODEINE BITARTRATE
mf: $C_{18}H_{23}NO_3 \cdot C_4H_6O_6$ mw: 451.52

SYNS: DF 118 * DIHYDROCODEINE ACID TARTRATE * DIHYDROCODEINE TARTRATE * DIHYDROCODEINE TARTRATE (1:1)

SAFETY PROFILE: Poison by ingestion, intraperitoneal, subcutaneous, and parenteral routes. Human systemic effects by intravenous route: irritability. When heated to decomposition it emits toxic fumes of NO_x.

DLC000   CAS: 24909-09-9   **HR: 3**
9,10-DIHYDRO-9,10-DIHYDROXYBENZO (a)PYRENE
mf: $C_{20}H_{14}O_2$ mw: 286.34

SYN: 9,10-DIHYDROBENZO(a)PYRENE-9,10-DIOL

CONSENSUS REPORTS: EPA Genetic Toxicology Program.

SAFETY PROFILE: Questionable carcinogen with experimental neoplastigenic and tumorigenic data by skin contact. Human mutation data reported. When heated to decomposition it emits acrid smoke and irritating fumes.

DLC600   CAS: 37571-88-3   **HR: 3**
trans-4,5-DIHYDRO-4,5-DIHYDROXYBENZO(a)PYRENE
mf: $C_{20}H_{14}O_2$ mw: 286.34

SYNS: (E)-BENZO(a)PYRENE-4,5-DIHYDRODIOL
* trans-4,5-DIHYDROBENZO(a)PYRENE-4,5-DIOL
* trans-4,5-DIHYDROXY-4,5-DIHYDROBENZO(a)PYRENE

CONSENSUS REPORTS: EPA Genetic Toxicology Program.

SAFETY PROFILE: Questionable carcinogen with experimental neoplastigenic data by skin contact. Mutation data reported. When heated to decomposition it emits acrid smoke and irritating fumes.

DLD600   CAS: 68162-13-0   **HR: 3**
trans-3,4-DIHYDRO-3,4-DIHYDROXY-7,12-DIMETHYLBENZ(a)ANTHRACENE
mf: $C_{20}H_{18}O_2$ mw: 290.38

SYN: trans-3,4-DIHYDRO-3,4-DIHYDROXY DMBA

SAFETY PROFILE: Questionable carcinogen with experimental neoplastigenic and tumorigenic data by skin contact. Mutation data reported. When heated to decomposition it emits acrid smoke and irritating fumes.

DLD800   CAS: 65763-32-8   **HR: 3**
trans-8,9-DIHYDRO-8,9-DIHYDROXY-7,12-DIMETHYLBENZ(a)ANTHRACENE
mf: $C_{20}H_{18}O_2$ mw: 290.38

SYNS: (E)-8,9-DIHYDRO-8,9-DIHYDROXY-7,12-DIMETHYLBENZ(a)ANTHRACENE * trans-8,9-DIHYDRO-8,9-DIHYDROXY DMBA

SAFETY PROFILE: Questionable carcinogen with experimental tumorigenic data by skin contact. Mutation data reported. When heated to decomposition it emits acrid smoke and irritating fumes.

DLE000   CAS: 64598-80-7   **HR: 3**
(±)-(1R,2S,3R,4R)-3,4-DIHYDRO-3,4-DIHYDROXY-1,2-EPOXYBENZ(a) ANTHRACENE
mf: $C_{18}H_{14}O_3$ mw: 278.32

SYNS: BA-3,4-DIOL-1,2-EPOXIDE-1 * BA-3,4-DIOL-1,2-EPOXIDE-2 * BENZ(a)ANTHRACENE 3,4-DIOL-1,2-EPOXIDE-2 * (±)-3-α,4-β-DIHYDROXY-1-α,2-α-EPOXY-1,2,3,4-TETRAHYDROBENZ(a)ANTHRACENE * (E)-1,2,3,4-TETRAHYDRO-3-α,4-β-DIHYDROXY-1-α,2-α-EPOXYBENZ(a)ANTHRACENE

CONSENSUS REPORTS: EPA Genetic Toxicology Program.

SAFETY PROFILE: Questionable carcinogen with experimental tumorigenic data by skin contact. Mutation data reported. When heated to decomposition it emits acrid smoke and irritating fumes.

DLF200   CAS: 64521-15-9   **HR: 3**
trans-8,9-DIHYDRO-8,9-DIHYDROXY-7-METHYLBENZ(a)ANTHRACENE
mf: $C_{18}H_{16}O_2$ mw: 264.34

CONSENSUS REPORTS: EPA Genetic Toxicology Program.

SAFETY PROFILE: Questionable carcinogen with experimental tumorigenic data by skin contact. Mutation data reported. When heated to decomposition it emits acrid smoke and irritating fumes.

DLF600   CAS: 67523-22-2   **HR: 3**
7,8-DIHYDRO-7,8-DIHYDROXY-5-METHYLCHRYSENE
mf: $C_{19}H_{16}O_2$ mw: 276.35

SYN: 7,8-DIHYDRO-5-METHYL-7,8-CHRYSENEDIOL

SAFETY PROFILE: Questionable carcinogen with experimental neoplastigenic data by skin contact. Mutation data reported. When heated to decomposition it emits acrid and irritating fumes.

DLH600   CAS: 50-49-7   **HR: 3**
5,6-DIHYDRO-N-(3-(DIMETHYLAMINO)PROPYL)-11H-DIBENZ(b,e)AZEPINE
mf: $C_{19}H_{24}N_2$ mw: 280.45

SYNS: ANTIDEPRIN * BERKOMINE * CENSTIN * 10,11-DIHYDRO-5-(3-(DIMETHYLAMINO)PROPYL)-5H-DIBENZ(b,f)AZEPINE * 2,2'-(3-DIMETHYLAMINOPROPYLAMINO)BIBENZYL * 1-(3-DIMETHYLAMINOPROPYL)-4,5-DIHYDRO-2,3,6,7-DIBENZAZEPINE * 5-(3-DIMETHYLAMINOPROPYL)-10,11-DIHYDRO-5H-DIBENZO(b,f)AZEPINE * 2,2'-(3-DIMETHYLAMINOPROPYLIMINO)DIBENZYL * N-(γ-DIMETHYLAMINOPROPYL)IMINODIBENZYL * DIMIPRESSIN * DPID * DYNAPRIN * DYNA-ZINA * EU-PRAMIN * G-22355 * IM * IMIDOBENZYLE * IMIPRAMINA (ITALIAN) * IMIPRAMINE * IMIPRIN * IMIZIN * IMIZINUM * IMPRA-MINE * INTALPRAM * IRAMIL * IRMIN * MELIPRAMIN * MELIPRAMINE * NELIPRAMIN * PRAZEPINE * PROMIBEN * SURPLIX * TIMOLET * TOFRANIL

CONSENSUS REPORTS: EPA Genetic Toxicology Program.

SAFETY PROFILE: A human poison by ingestion. An experimental poison by ingestion, subcutaneous, intravenous, intraperitoneal, and possibly other routes. Human systemic effects by ingestion: somnolence, hallucinations, distorted perceptions, changes in motor activity, ataxia (loss of muscle coordination), coma, nausea, and vomiting. Experimental teratogenic and reproductive effects. Mutation data reported. When heated to decomposition it emits toxic fumes of NO_x.

DLH630   CAS: 113-52-0   **HR: 3**
10,11-DIHYDRO-5-(3-(DIMETHYLAMINO)PROPYL)-5H-DIBENZ(b,f)AZEPINE HYDROCHLORIDE
mf: $C_{19}H_{24}N_2 \cdot ClH$ mw: 316.91

SYNS: ANTIDEPRIN HYDROCHLORIDE * BERKO-MINE * CENSTIM * CENSTIN * CHIMOREPTIN * CHRYTEMIN * CO CAP IMIPRAMINE 25 * DEPRINOL * 10,11-DIHYDRO-N,N-DIMETHYL-5H-DIBENZ(b,f)AZEPINE-5-PROPANAMINE MONOHYDRO-CHLORIDE * 5-(3-DIMETHYLAMINOPROPYL)-10,11-DIHYDRO-5H-DIBENZ(b,f)AZEPINE HYDROCHLORIDE * N-(3-DIMETHYLAMINOPROPYL)IMINODIBENZYL HY-DROCHLORIDE * N-(γ-DIMETILAMINOPROPIL)-IMINO-DIBENZILE CLORIDRATO (ITALIAN) * DIMIPRESSIN * DYNA-ZINA * EFURANOL * EUPRAMIN * FEINALMIN * G 22150 * G 22355 * IA-PRAM * IMAVATE * IMIDOBENZYLE * IMIDOL * IMILANYLE * IMIPRAMINA (ITALIAN) * IMIPRAMINE * IMIPRAMINE HYDROCHLORIDE * IMIPRAMINE MONOHYDROCHLORIDE * IMIPRIN * IMP HYDROCHLORIDE * INTALPRAM * IPRO-GEN * IRAMIL * JANIMINE * LOFEPRAMINE * MELIPRAMIN * MELIPRAMINE * MELIPRA-MINE HYDROCHLORIDE * MELIPRAMIN HYDRO-CHLORIDE * NSC 114900 * PERSAMINE * PERTOFRAM * PRESAMINE * PROMIBEN * PYRLEUGAN * SK-PRAMINE * SK-PRAMINE HYDROCHLORIDE * SURPLIX * TEPERINE * TIMOLET * TOFRANIL * TOFRANILE

CONSENSUS REPORTS: Reported in EPA TSCA Inventory.

SAFETY PROFILE: Human poison by ingestion. An experimental poison by ingestion, intravenous, subcutaneous, and intraperitoneal routes. Human systemic effects by ingestion: sleep, somnolence, convulsions, muscle contraction or spasticity, coma, blood pressure decrease, dyspnea (difficulty in breathing), paresthesia (abnormal sensations) and kidney changes. Experimental teratogenic and reproductive effects. Mutation data reported. Used in the treatment of depression. When heated to decomposition it emits very toxic fumes of NO_x and HCl.

DLI000   CAS: 52171-93-4   **HR: 3**
3,4-DIHYDRO-1,11-DIMETHYLCHRYSENE
mf: $C_{20}H_{18}$ mw: 258.38

CONSENSUS REPORTS: EPA Genetic Toxicology Program.

SAFETY PROFILE: Questionable carcinogen with experimental neoplastigenic data by skin contact. Mutation data reported. When heated to decomposition it emits acrid smoke and irritating fumes.

DLI200   CAS: 5831-16-3   **HR: 3**
16,17-DIHYDRO-11,17-DIMETHYLCY-CLOPENTA(a)PHENANTHRENE
mf: $C_{19}H_{17}$ mw: 245.36

SYN: 11,17-DIMETHYL-16,17-DIHYDRO-15H-CYCLOPEN-TA(a)PHENANTHRENE

CONSENSUS REPORTS: EPA Genetic Toxicology Program.

SAFETY PROFILE: Questionable carcinogen with experimental tumorigenic data by skin contact. When heated to decomposition it emits acrid smoke and irritating fumes.

DLK200   CAS: 3347-22-6   **HR: 3**
5,10-DIHYDRO-5,10-DIOXONAPHTHO (2,3-b)-p-DITHIIN-2,3-DICARBONITRILE
mf: $C_{14}H_4N_2O_2S_2$ mw: 296.32

SYNS: DELAN * DELAN-COL * 2,3-DICARBONI-TRILO-1,4-DIATHIAANTHRACHINON (GERMAN) * 2,3-DICYANO-1,4-DITHIA-ANTHRAQUINONE * 2,3-DINITRILO-1,4-DITHIA-ANTHRAQUINONE * 2,3-DINITRILO-1,4-DITHIOANTHRACHINON (GERMAN) * 1,4-DITHIAANTHRAQUINONE-2,3-DICARBONITRILE * 1,4-DITHIAANTHRAQUINONE-2,3-DINITRILE * DITHIANON * DITHIANONE * DTA * IT 931 * MV 119A * STAUFFER MV-119A * THYNON

CONSENSUS REPORTS: Cyanide and its compounds are on the Community Right-To-Know List.

SAFETY PROFILE: Poison by ingestion. Moderately toxic by an unspecified route. A fungicide. When heated to decomposition it emits very toxic fumes of NO_x, SO_x, and CN^-.

DLK800   CAS: 511-12-6   **HR: 3**
DIHYDROERGOTAMINE
mf: $C_{33}H_{37}N_5O_5$ mw: 583.65

SYN: DEHYDROERGOTAMINE

SAFETY PROFILE: Poison by intravenous and subcutaneous routes. When heated to decomposition it emits toxic fumes of NO_x.

DLL000   CAS: 5989-77-5   **HR: 3**
DIHYDROERGOTAMINE TARTRATE (2:1)
mf: $C_{66}H_{74}N_{10}O_{10} \cdot C_4H_6O_6$ mw: 1317.60

SAFETY PROFILE: Poison by intravenous, intraperitoneal, and subcutaneous routes. When heated to decomposition it emits toxic fumes of NO_x.

DLL600   CAS: 29734-68-7   **HR: 3**
DIHYDRO-β-ERYTHROIDINE HYDROBROMIDE
mf: $C_{16}H_{21}NO_3 \cdot BrH$ mw: 356.30

SAFETY PROFILE: A deadly poison by intravenous and intraperitoneal routes. When heated to decomposition it emits very toxic fumes of Br^- and NO_x.

DLO200   CAS: 21243-26-5   **HR: 3**
6,11-DIHYDRO-4-FLUORO(1)BENZO-THIOPYRANO(4,3-b)INDOLE
mf: $C_{15}H_{10}FNS$ mw: 255.32

SAFETY PROFILE: Questionable carcinogen with experimental neoplastigenic data. Mutation data reported. When heated to decomposition it emits very toxic fumes of F^-, SO_x, and NO_x.

DLP000   CAS: 63918-74-1   **HR: 3**
6,7-DIHYDRO-6-(2-HYDROXYETHYL)-5H-DIBENZ(c,e)AZEPINE
mf: $C_{16}H_{17}NO$ mw: 239.34

SYN: RO 2-3599

SAFETY PROFILE: Poison by intravenous and intraperitoneal routes. When heated to decomposition it emits toxic fumes of NO_x.

DLQ000   CAS: 6414-38-6   **HR: 3**
DIHYDROISOCODEINE ACID TARTRATE
mf: $C_{18}H_{23}NO_3 \cdot C_4H_6O_6$ mw: 451.52

SYNS: DIHYDROISOCODEINE TARTRATE * DIHYDRO ISOCODEINE TARTRATE (1:1)

SAFETY PROFILE: Poison by subcutaneous route. When heated to decomposition it emits toxic fumes of NO_x.

DLQ400   CAS: 37795-69-0   **HR: 3**
2,3-DIHYDRO-9H-ISOXAZOLO(3,2-b)QUINAZOLIN-9-ONE
mf: $C_{10}H_8N_2O_2$ mw: 188.20

SYN: w-2429

SAFETY PROFILE: Poison by ingestion and intravenous routes. When heated to decomposition it emits toxic fumes of NO_x.

DLQ800   CAS: 21842-58-0   **HR: 3**
12,β,13,α-DIHYDROJERVINE
mf: $C_{26}H_{39}NO_3$ mw: 413.66

SAFETY PROFILE: Poison by ingestion. Experimental teratogenic and reproductive effects. When heated to decomposition it emits toxic fumes of NO_x.

DLS600   CAS: 58-28-6   **HR: 3**
**10,11-DIHYDRO-5-(3-(METHYLAMINO)
PROPYL)-5H-DIBENZ(b,f)AZEPINE
HYDROCHLORIDE**
mf: $C_{18}H_{22}N_2 \cdot ClH$ mw: 302.88

SYNS: DESIPRAMINE HYDROCHLORIDE * DESMETHYLIMIPRAMINE HYDROCHLORIDE * DIMETHYLIMIPRAMINE HYDROCHLORIDE * DMI HYDROCHLORIDE * EX 4355 * G 35020 * GMI * IMIPRAMINEDEMETHYL HYDROCHLORIDE * IRENE * JB 8181 * N-(γ-METHYLAMINOPROPYL)IMINODIBENZYL HYDROCHLORIDE * NORPRAMIN * NORTIMIL * NSC-114901 * PERTOFRAN * PERTOFRANE * RMI9,384A

SAFETY PROFILE: Poison by ingestion, intraperitoneal, subcutaneous, and intravenous routes. An experimental teratogen. Mutation data reported. When heated to decomposition it emits very toxic fumes of NO_x and HCl.

DLS800   CAS: 1563-67-3   **HR: 3**
**2,3-DIHYDRO-2-METHYLBENZO-
PYRANYL-7,N-METHYLCARBAMATE**
mf: $C_{11}H_{13}NO_3$ mw: 207.25

SYNS: A 468 * BAY 48130 * BAY 62863 * BAYER 62863 * C 1120 * DECARBOFURAN * ENT 27,324

SAFETY PROFILE: Poison by ingestion and subcutaneous routes. When heated to decomposition it emits toxic fumes of NO_x.

DLV800   CAS: 3978-86-7   **HR: 3**
**6,11-DIHYDRO-11-(1-METHYL-4-
PIPERIDYLIDENE)-5H-
BENZO(5,6)CYCLOHEPTA (1,2-b)
PYRIDINE DIMALEATE**
mf: $C_{20}H_{22}N_2 \cdot 2C_4H_4O_4$ mw: 522.60

SYNS: AZATADINE DIMALEATE * AZATADINE MELEATE * IDULIAN * OPTIMINE * SCH 10649 * ZADINE

SAFETY PROFILE: Poison by ingestion, subcutaneous, and intraperitoneal routes. An antihistamine. When heated to decomposition it emits toxic fumes of NO_x.

DLW600   CAS: 466-99-9   **HR: 3**
DIHYDROMORPHINONE
mf: $C_{17}H_{19}NO_3$ mw: 285.37

SYNS: DIMO * HYDROMORPHONE * HYMORPHAN * LAUDICON * PARAMORPHAN

SAFETY PROFILE: A deadly human poison by ingestion. An experimental poison by ingestion and subcutaneous route. When heated to decomposition it emits toxic fumes of NO_x.

DLX400   CAS: 124-90-3   **HR: 3**
DIHYDRONE HYDROCHLORIDE
mf: $C_{18}H_{21}NO_4 \cdot ClH$ mw: 351.86

SYNS: DIHYDROOXYCODEINONE HYDROCHLORIDE * DIHYDROXYCODEINONE HYDROCHLORIDE * DINARKON * EUBINE * EUCODAL * EUKODAL * EUTAGEN * 14-HYDROXYDIHYDROCODEINONE HYDROCHLORIDE * OXIKON * OXYCODONE HYDROCHLORIDE * OXYCODON HYDROCHLORIDE * OXYCON * OXYKODAL * OXYKON * PANCODINE * PERCODAN HYDROCHLORIDE * STUPENONE * TECODIN * TECODINE * TEKODIN * THECODIN * THECODINE * THEKODIN

SAFETY PROFILE: Poison by intravenous and subcutaneous routes. When heated to decomposition it emits very toxic fumes of NO_x and HCl.

DLY000   CAS: 146-22-5   **HR: 3**
**1,3-DIHYDRO-7-NITRO-5-PHENYL-2H-
1,4-BENZODIAZEPIN-2-ONE**
mf: $C_{15}H_{11}N_3O_3$ mw: 281.29

SYNS: BENZALIN * CALSMIN * EATAN * EPIBENZALIN * EPINELBON * EUNOCTIN * HIPNAX * HIPSAL * LA 1 * MOGADAN * NELBON * NEOZEPAM * NEUCHLONIC * NITRADOS * NITRAZEPAM * NITRENPAX * 7-NITRO-5-PHENYL-2,3-DIHYDRO-1H-1,4-BENZODIAZEPIN-2-ONE * NSC-58775 * PAXISYN * PELSON * RADEDORM * RELACT * RO 4-5360 * RO 5-3059 * SOMNASED * SOM-

NIBEL * SOMNITE * SONEBON * SONNOLIN * SUREM * UNISOMNIA

CONSENSUS REPORTS: EPA Genetic Toxicology Program.

SAFETY PROFILE: Poison by intraperitoneal and intravenous routes. Moderately toxic by ingestion. Experimental reproductive effects. Mutation data reported. An anticonvulsant and hypnotic agent. When heated to decomposition it emits toxic fumes of NO_x.

DMB000　　CAS: 21820-82-6　　**HR: 3**
5-(2-(3,6-DIHYDRO-4-PHENYL-1(2H)-PYRIDYL)ETHYL)-3-METHYL-2-OXAZOLIDINONE
mf: $C_{17}H_{22}N_2O_2$　　mw: 286.41

SYN: AHR-1680

SAFETY PROFILE: Poison by ingestion, intraperitoneal and intravenous routes. When heated to decomposition it emits toxic fumes of NO_x.

DMC200　　CAS: 110-87-2　　**HR: 3**
DIHYDROPYRAN
mf: C_5H_8O　　mw: 84.13

PROP: Colorless, mobile liquid; ethereal odor. Bp: 85.6°, flash p: 0°F, d: 0.923 @ 20°/4°, vap d: 2.90.

SYNS: Δ²-DIHYDROPYRAN * 3,4-DIHYDROPYRAN * 2H-3,4-DIHYDROPYRAN

CONSENSUS REPORTS: Reported in EPA TSCA Inventory.

DOT Classification: Flammable Liquid; Label: Flammable Liquid.

SAFETY PROFILE: Poison by intraperitoneal route. Very dangerous fire hazard when exposed to heat or flame; can react vigorously with oxidizing materials. Keep away from heat and open flame. To fight fire, use alcohol foam, CO_2, or dry chemical. When heated to decomposition it emits acrid smoke and irritating fumes.

DMC600　　CAS: 123-33-1　　**HR: 3**
1,2-DIHYDROPYRIDAZINE-3,6-DIONE
mf: $C_4H_4N_2O_2$　　mw: 112.10

PROP: Crystals. Mp: > 300°. Sol in water and alc.

SYNS: 1,2-DIHYDRO-3,6-PYRIDAZINEDIONE * ENT 18,870 * 6-HYDROXY-3(2H)-PYRIDAZINONE

* MALEIC ACID HYDRAZIDE * MALEIC HYDRAZIDE * N,N-MALEOYLHYDRAZINE * 1,2,3,6-TETRAHYDRO-3,6-DIOXOPYRIDAZINE

CONSENSUS REPORTS: IARC Cancer Review: GROUP 3 IMEMDT 7,56,87; Animal Inadequate Evidence IMEMDT 4,173,74. Reported in EPA TSCA Inventory.

SAFETY PROFILE: Questionable carcinogen with experimental tumorigenic data. Moderately toxic by ingestion. Mutation data reported. Can cause chronic liver damage and acute central nervous system effects. When heated to decomposition emits highly toxic fumes of NO_x.

DMD600　　CAS: 94-58-6　　**HR: 3**
DIHYDROSAFROLE
mf: $C_{10}H_{12}O_2$　　mw: 164.22

PROP: An oily liquid. Bp: 228°, d: 1.0695 @ 20°.

SYNS: 1,2-(METHYLENEDIOXY)-4-PROPYLBENZENE * 5-PROPYL-1,3-BENZODIOXOLE * 4-PROPYL-1,2-METHYLENEDIOXYBENZENE * RCRA WASTE NUMBER U090

CONSENSUS REPORTS: IARC Cancer Review: GROUP 2B IMEMDT 7,56,87; Animal Sufficient Evidence IMEMDT 10,231,76; Animal Limited Evidence IMEMDT 1,169,72. Reported in EPA TSCA Inventory. EPA Genetic Toxicology Program.

SAFETY PROFILE: Suspected carcinogen with experimental carcinogenic data. Moderately toxic by ingestion and intraperitoneal routes. A skin irritant. When heated to decomposition it emits acrid smoke and irritating fumes.

DME000　　CAS: 128-46-1　　**HR: 3**
DIHYDROSTREPTOMYCIN
mf: $C_{21}H_{41}N_7O_{12}$　　mw: 583.69

SYNS: DHMS * DST

CONSENSUS REPORTS: EPA Genetic Toxicology Program.

SAFETY PROFILE: Poison by intravenous and intramuscular routes. Moderately toxic by subcutaneous and intraperitoneal routes. Human teratogenic effects by unspecified route: developmental abnormalities of the eye and ear. An experimental teratogen. Other experimental reproductive effects. Mutation data reported. A derivative of streptomycin; has anesthetic properties. When heated to decomposition it emits toxic fumes of NO_x.

DMH000   CAS: 81-64-1   HR: 3
1,4-DIHYDROXYANTHRAQUINONE
mf: $C_{14}H_8O_4$ mw: 240.22

PROP: Crystals. Mp: 194°, bp: 450.0°, vap press: 1 mm @ 196.7°, vap d: 8.3.

SYNS: 1,4-DIHYDROXYANTHRACHINON (CZECH) * 1,4-DIHYDROXY-9,10-ANTHRAQUINONE * 1,4-DIOXYANTHRAQUINONE (RUSSIAN) * QUINIZARIN

CONSENSUS REPORTS: Reported in EPA TSCA Inventory.

SAFETY PROFILE: Poison by intravenous route. Moderately toxic by intraperitoneal route. Mutation data reported. An eye irritant. A weak allergen. When heated to decomposition it emits acrid smoke and irritating fumes.

DMI400   CAS: 2373-98-0   HR: 3
3,3'-DIHYDROXYBENZIDINE
mf: $C_{12}H_{12}N_2O_2$ mw: 216.26

SYNS: 6,6'-DIAMINO-m,m'-BIPHENOL * 4,4'-DIAMINO-3,3'-BIPHENYLDIOL * 3,3'-DIOXYBENZIDINE * 3,3'-DWUOKSYBENZYDYNA (POLISH)

SAFETY PROFILE: Suspected carcinogen with experimental carcinogenic, neoplastigenic, and tumorigenic data. Mutation data reported. When heated to decomposition it emits toxic fumes of NO_x.

DMI600   CAS: 131-56-6   HR: 3
2,4-DIHYDROXYBENZOPHENONE
mf: $C_{13}H_{10}O_3$ mw: 214.23

SYNS: 2,4-DIHYDROXYBENZOFENON (CZECH) * EASTMAN INHIBITOR DHPB * QUINSORB 010 * SYNTASE 100 * UF 1 * USAF DO-28 * USAF ND-54 * UVINUL 400

CONSENSUS REPORTS: Reported in EPA TSCA Inventory.

SAFETY PROFILE: Poison by intravenous and intraperitoneal routes. Mildly toxic by ingestion. An eye irritant. When heated to decomposition it emits acrid smoke and irritating fumes.

DMJ400   CAS: 32222-06-3   HR: 3
1a,25-DIHYDROXYCHOLECALCIFEROL

SYNS: CALCITRIOL * 1-α,25-DIHYDROXYCHOLECALCIFEROL * 1,25-DIHYDROXYCHOLECALCIFEROL * DIHYDROXYVITAMIN D3 * 1-α,25-DIHYDROXYVITAMIN D3 * Ro 215535 * ROCALTROL * (5Z,7E)-9,10-SECOCHESTA-5,7,10(19)-TRIENE-1-α,3-β,25-TRIOL * (1-α,3-β,5Z,7E)-9,10-SECOCHOLESTA-5,7,10(19)-TRIENE-1,3,25-TRIOL

SAFETY PROFILE: A deadly poison by ingestion, intraperitoneal, and subcutaneous routes. Experimental teratogenic and reproductive effects. Mutation data reported. Enhances intestinal calcium transport and bone mineral mobilization. When heated to decomposition it emits acrid smoke and irritating fumes.

DML200   CAS: 60864-95-1   HR: 3
(−)-trans-7,8-DIHYDROXY-7,8-DIHYDROBENZO(a)PYRENE
mf: $C_{20}H_{14}O_2$ mw: 286.34

SYN: BP-7,8-DIHYDRODIOL

CONSENSUS REPORTS: EPA Genetic Toxicology Program.

SAFETY PROFILE: Questionable carcinogen with experimental neoplastigenic data. Mutation data reported. When heated to decomposition it emits acrid smoke and fumes.

DMU800   CAS: 551-11-1   HR: 3
7-(3,5-DIHYDROXY-2-(3-HYDROXY-1-OCTENYL)CYCLOPENTYL)-5-HEPTENOIC ACID
mf: $C_{20}H_{34}O_5$ mw: 354.54

SYNS: AMOGLANDIN * DINOPROST * ENZAPROST * ENZAPROST F * PANACELAN * PGF2-α * PROSTAGLANDIN F2-α * PROSTALMON F * PROSTARMON F * PROSTIN F2-α * (5Z,9,α,11,α,13E,15S)-9,11,15-TRIHYDROXYPROSTA-5,13-DIEN-1-OIC ACID * 9,11,15-TRIHYDROXYPROSTA-5,13-DIEN-1-OIC ACID * U-14583

CONSENSUS REPORTS: EPA Genetic Toxicology Program.

SAFETY PROFILE: Poison by subcutaneous, intravenous, and intramuscular routes. Moderately toxic by ingestion. Human and experimental teratogenic and experimental reproductive effects. Human reproductive effects by subcutaneous, intravenous, intramuscular, intraperitoneal, intravaginal, and intraplacental routes: postpartum depression and other maternal effects, abortion and changes in measures of fertility. Human teratogenic effects by intraplacental route: extra embryonic structures. Human systemic effects by intravenous route: hypermotility, diarrhea, nausea or vomiting. Human mutation data reported. When heated to decomposition it emits acrid smoke and fumes.

DMV600   CAS: 7683-59-2   **HR: 3**
3,4-DIHYDROXY-α-((ISOPROPYL-AMINO)METHYL)BENZYL ALCOHOL
mf: $C_{11}H_{17}NO_3$ mw: 211.29

SYNS: A 21 * ALEUDRIN * ALUDRINE * ASIPRENOL * ASMALAR * ASSIPRENOL * BELLASTHMAN * BRONKEPHRINE * DI-HYDROXYPHENYLETHANOLISOPROPYLAMINE * 1-(3,4-DIHYDROXYPHENYL)-2-ISOPROPYLAMINO-ETHANOL * EPINEPHRINE ISOPROPYL HOMOLOG * 4-(1-HYDROXY-2-((1-METHYLETHYL)AMINO)ETHYL)-1,2-BENZENEDIOL * IPA * ISONORENE * ISO-PRENALINE * ISOPROPYDRIN * ISOPROPYLADRE-NALINE * ISOPROPYLAMINOMETHYL-3,4-DI-HYDROXYPHENYL CARBINOL * α-(ISOPROPYL-AMINOMETHYL)PROTOCATECHUYL ALCOHOL * ISOPROPYLARTERENOL * N-ISOPROPYL-β-DIHYDROXYPHENYL-β-HYDROXYETHYLAMINE * ISOPROPYL NORADRENALINE * 1-ISOPROPYL-NORADRENALINE * N-ISOPROPYLNORADRENALINE * ISOPROTERENOL * 1-ISOPROTERENOL * ISORENIN * ISUPREL * ISUPREN * LOMUPREN * NEODRENAL * NEO-EPININE * NORISODRINE * NOVODRIN * PROTERNOL * RESPIFRAL * SAVENTRINE * VAPO-N-ISO * WIN 5162

SAFETY PROFILE: Poison by ingestion, subcutaneous, intravenous, intraperitoneal, and possibly other routes. An experimental teratogen. Human systemic effects by intramuscular route: increased pulse and cardiac rate. A bronchodilator. Mutation data reported. When heated to decomposition it emits toxic fumes of NO_x.

DMW000   CAS: 7361-61-7   **HR: 3**
5,6-DIHYDRO-2-(2,6-XYLIDINO)-4H-1,3-THIAZINE
mf: $C_{12}H_{16}N_2S$ mw: 220.36

SYNS: BAY 1470 * BAY VA 1470 * N-(5,6-DIHY-DRO-4H-1,3-THIAZINYL)-2,6-XYLIDINE * 2-(2,6-DI-METHYLANILINO)-5,6-DIHYDRO-4H-1,3-THIAZINE * 2-(2,6-DIMETHYLPHENYLAMINO)-4H-5,6-DIHYDRO-1,3-THIAZINE * N-(2,6-DIMETHYLPHENYL)-5,6-DIHY-DRO-4H-1,3-THIAZIN-2-AMINE * N-(2,6-DIMETHYL-PHENYL)-5,6-DIHYDRO-4H-1,3-THIAZINE-2-AMINE (9CI)

* ROMPUN * WH 7286 * XYLZIN * XYLAZINE (USDA)

SAFETY PROFILE: Poison by ingestion, subcutaneous, and intravenous routes. When heated to decomposition it emits very toxic fumes of NO_x and SO_x.

DMX200   CAS: 2318-18-5   **HR: 3**
2,12-DIHYDROXY-4-METHYL-11,16-DIOXOSENECIONANIUM
mf: $C_{19}H_{28}NO_6$ mw: 366.48

SYNS: trans-15-ETHYLIDENE-12-β-HYDROXY-4,12-α,13-β-TRIMETHYL 8-OXO-4,8 SECOSENEC-1-ENINE * 12-HYDROXY-4-METHYL-4,8-SECOSENECIO-NAN-8,11,16-TRIONE * NSC-89945 * RENARDIN * RENARDINE * SENKIRKIN * SENKIRKINE

CONSENSUS REPORTS: IARC Cancer Review: GROUP 3 IMEMDT 7,56,87; Animal Limited Evidence IMEMDT 31,231,83; Animal Inadequate Evidence IMEMDT 10,327,76.

SAFETY PROFILE: Questionable carcinogen with experimental neoplastigenic data. Poison by ingestion and intraperitoneal routes. Mutation data reported. When heated to decomposition it emits toxic fumes of NO_x.

DNA200   CAS: 59-92-7   **HR: 3**
l-DIHYDROXYPHENYL-l-ALANINE
mf: $C_9H_{11}NO_4$ mw: 197.21

SYNS: 2-AMINO-3-(3,4-DIHYDROXYPHENYL)PRO-PANOIC ACID * BENDOPA * BIODOPA * BRO-CADOPA * CEREPAP * CIDANDOPA * DA * DEADOPA * DIHYDROXY-l-PHENYLALANINE * (−)-3-(3,4-DIHYDROXYPHENYL)-l-ALANINE * β-(3,4-DIHYDROXYPHENYL)-α-ALANINE * l-α-DI-HYDROXYPHENYLALANINE * l-β-(3,4-DIHYDROXY-PHENYL)ALANINE * l-3,4-DIHYDROXYPHE-NYL-α-ALANINE * β-(3,4-DIHYDROXYPHE-NYL)-l-ALANINE * 3-(3,4-DIHYDROXYPHENYL)-l-ALA-NINE * 3,4-DIHYDROXYPHENYLALANINE * (−)-3,4-DIHYDROXYPHENYLALANINE * 3,4-DI-HYDROXYPHENYL-l-ALANINE * 3,4-DIHYDROXY-l-PHENYLALANINE * l-3,4-DIHYDROXYPHENYL-ALANINE * (−)-DOPA * l-DOPA * DOPAFLEX * DOPAL * DOPARKINE * DOPASOL * DO-PRIN * ELDOPAL * EURODOPA * HELFO DOPA * l-o-HYDROXYTYROSINE * 3-HYDROXY-l-TYRO-SINE * INSULAMINA * LARODOPA * MAIPE-DOPA * PARDA * RO 4-6316 * SOBIODOPA * VELDOPA

CONSENSUS REPORTS: Reported in EPA TSCA Inventory.

SAFETY PROFILE: Questionable human carcinogen producing skin tumors. Poison by ingestion. Moderately toxic by intravenous, intraperitoneal, and possibly other routes. Human systemic effects by ingestion: somnolence, hallucinations and distorted perceptions, toxic psychosis, motor activity changes, ataxia, dyspnea. Experimental teratogenic and reproductive effects. Human mutation data reported. An anticholinergic agent used as an anti-Parkinsonian drug. When heated to decomposition it emits toxic fumes of NO_x.

DNA600 CAS: 13055-82-8 **HR: 3**
7-(3-(2-(3,5-DIHYDROXYPHENYL-2-HYDROXY-ETHYLAMINO)PROPYL) THEOPHYLLINE HYDROCHLORIDE

SYNS: BRONCHODIL * BRONCHOSPASMIN * REPROTEROL HYDROCHLORIDE * 7-(3-((β,3,5-TRIHYDROXYPHENETHYL)AMINO)PROPYL)THEOPHYLLINE MONOHYDROCHLORIDE * W-2946M

SAFETY PROFILE: Poison by intravenous route. Experimental reproductive effects. When heated to decomposition it emits very toxic fumes of HCl and NO_x.

DNA800 CAS: 555-30-6 **HR: 3**
l-(−)-3-(3,4-DIHYDROXYPHENYL)-2-METHYLALANINE
mf: $C_{10}H_{13}NO_4$ mw: 211.24

SYNS: ALDOMET * ALDOMETIL * ALDOMIN * ALPHA MEDOPA * AMD * BAYER 1440 L * BAYPRESOL * l(−)-β-(3,4-DIHYDROXYPHENYL)-α-METHYLALANINE * DOPAMET * DOPEGYT * DOPTAEC * 3-HYDROXY-α-METHYL-l-TYROSINE * HYPERPAX * l-(α-MD) * MEDOMET * MEDOPREN * METHOPLAIN * α-METHYL-l-3,4-DIHYDROXYPHENYLALANINE * l-α-METHYL-3,4-DI-HYDROXYPHENYLALANINE * α-METHYL-β-(3,4-DIHYDROXYPHENYL)-l-ALANINE * l-(−)-α-METHYL-β-(3,4-DIHYDROXYPHENYL)ALANINE * METHYLDOPA *(l-α-METHYLDOPA * α-METHYL-l-DOPA * MK. B51 * MK 351 * NCI-C55721 * NR.C 2294 * PRESINOL * PRESOLISIN * SEDOMETIL * SEMBRINA

CONSENSUS REPORTS: Reported in EPA TSCA Inventory. EPA Genetic Toxicology Program.

SAFETY PROFILE: Poison by intraperitoneal route. Moderately toxic by ingestion and intravenous routes. Experimental teratogenic and reproductive effects. Human reproductive effects by unspecified route: menstrual cycle changes or disorders, effects on newborn including abnormal neonatal measures and growth statistics, biochemical and metabolic changes. Mutation data reported. When heated to decomposition it emits toxic fumes of NO_x.

DNB000 CAS: 2589-47-1 **HR: 3**
17R,21-α-DIHYDROXY-4-PROPYL-AJMALANIUM HYDROGEN TARTRATE
mf: $C_{23}H_{32}N_2O_2 \cdot C_4H_6O_6$ mw: 518.67

SYNS: GT-1012 * NEO-GILURYTMAL * NPA * PRAJMALINE BITARTRATE * PRAJMALINE HYDROGEN TARTRATE * N-PROPYLAJMALINE BITARTRATE * N-PROPYLAJMALINE HYDROGEN TARTRATE * N-PROPYLAJMALINIUM BITARTRATE * N-PROPYLAJMALINIUMHYDROGENTARTRAT (GERMAN) * N⁴-PROPYLAJMALINIUM HYDROGEN TARTRATE

SAFETY PROFILE: Poison by ingestion and intravenous routes. Experimental teratogenic and reproductive effects. Human systemic effects by ingestion: hallucinations and distorted perceptions. An antiarrhythmic agent. When heated to decomposition it emits toxic fumes of NO_x.

DNB200 CAS: 53609-64-6 **HR: 3**
DI(2-HYDROXY-n-PROPYL)AMINE
mf: $C_6H_{14}N_2O_3$ mw: 162.22

SYNS: BHP * N-BIS(2-HYDROXYPROPYL)NITROSAMINE * 2,2′-BISHYDROXYPROPYLNITROSAMINE * DHPN * 2,2′-DIHYDROXY-DI-n-PROPYLNITROSOAMINE * N,N-DI-(2-HYDROXYPROPYL)NITROSAMINE * DIISOPROPANOLNITROSAMINE * DIPN * N-NITROSOBIS(2-HYDROXYPROPYL)AMINE * N-NITROSO-N,N-DI(2-HYDROXYPROPYL)AMINE * N-NITROSO-1,1′-IMINODI-2-PROPANOL * 1,1′-NITROSOIMINODI-2-PROPANOL

SAFETY PROFILE: Suspected carcinogen with experimental carcinogenic, neoplastigenic, and tumorigenic data. Moderately toxic by subcutaneous route. Experimental teratogenic and reproductive effects. Human mutation data reported. When heated to decomposition it emits toxic fumes of NO_x.

DNC200 CAS: 59-00-7 **HR: 3**
4,8-DIHYDROXYQUINALDIC ACID
mf: $C_{10}H_7NO_4$ mw: 205.18

PROP: Sulfur-yellow crystals. Mp: 286°. Insol in water; sol in aqueous alkali, hydroxides, and hot dil HCl.

SYNS: 4,8-DIHYDROXYQUINALDINIC ACID
* 4,8-DIHYDROXYQUINOLINE-2-CARBOXYLIC ACID
* XANTHURENIC ACID

SAFETY PROFILE: Questionable carcinogen with experimental neoplastigenic data. When heated to decomposition it emits toxic fumes of NO_x.

DNE000 CAS: 488-17-5 **HR: 3**
2,3-DIHYDROXYTOLUENE
mf: $C_7H_8O_2$ mw: 124.15

SYNS: 3-METHYL-1,2-BENZENEDIOL * 3-METHYL-CATECHOL * 3-METHYLPYROCATECHOL * 2,3-TO-LUENEDIOL

CONSENSUS REPORTS: Reported in EPA TSCA Inventory.

SAFETY PROFILE: Poison by intravenous route. When heated to decomposition it emits acrid smoke and fumes.

DNE400 CAS: 3468-11-9 **HR: 3**
1,3-DIIMINOISOINDOLINE
mf: $C_8H_7N_3$ mw: 145.18

SYNS: AFASTOGEN BLUE 5040 * 1,3-DIIMINOISOIN-DOLIN (CZECH) * FASTOGEN BLUE FP-3100 * FAS-TOGEN BLUE SH-100 * MODR FRALOSTANOVA 3G (CZECH) * PHTHALIMIDIMIDE * PHTHALO-CYANINE BLUE 01206 * PHTHALOGEN

CONSENSUS REPORTS: Reported in EPA TSCA Inventory.

SAFETY PROFILE: Questionable carcinogen with experimental carcinogenic and tumorigenic data. Poison by ingestion. A severe eye and skin irritant. When heated to decomposition it emits toxic fumes of NO_x.

DNE500 CAS: 624-74-8 **HR: 3**
DIIODOACETYLENE
mf: C_2I_2 mw: 277.83

SYN: DIIODOETHYNE

DOT Classification: Forbidden

SAFETY PROFILE: An explosive sensitive to impact, crushing, or heating to 84°C. When heated to decomposition it emits toxic fumes of I^-.

DNF600 CAS: 83-73-8 **HR: 3**
DIIODOHYDROXYQUIN
mf: $C_9H_5I_2NO$ mw: 396.95

SYNS: DIIODOHYDROXYQUINOLINE * 5,7-DIIODO-8-HYDROXYQUINOLINE * 5,7-DIIODO-OXINE
* 5,7-DIIODO-8-QUINOLINOL * DIODOHYDROXY-QUIN * 8-HYDROXY-5,7-DIIODOQUINOLINE

CONSENSUS REPORTS: Reported in EPA TSCA Inventory.

SAFETY PROFILE: Poison by ingestion, in-travenous, and intraperitoneal routes. Human systemic effects by ingestion: eye effects. When heated to decomposition it emits very toxic fumes of I^- and NO_x.

DNG000 CAS: 305-85-1 **HR: 3**
2,6-DIIODO-4-NITROPHENOL
mf: $C_6H_3I_2NO_3$ mw: 390.90

SYNS: ANCYLOL * DIISOPHENOL * DISOFEN
* DNP * DISOPHENOL

CONSENSUS REPORTS: Reported in EPA TSCA Inventory.

SAFETY PROFILE: Poison by ingestion, intra-peritoneal, subcutaneous, intravenous, and par-enteral routes. An anthelmintic. When heated to decomposition it emits very toxic fumes of I^- and NO_x.

DNG200 CAS: 3861-47-0 **HR: 3**
3,5-DIIODO-4-OCTANOYLOXY-BENZONITRILE
mf: $C_{15}H_{17}I_2NO_2$ mw: 497.13

SYNS: 4-CYANO-2,6-DIJODPHENOL CAPRYSAEUREES-TER (GERMAN) * 3,5-DIIODO-4-HYDROXYBENZONI-TRILE OCTANOATE * 3,5-DIJOD-4-HYDROXY-BENZO-NITRIL CAPRYSAEUREESTER (GERMAN) * IOXYNIL OCTANOATE * M&B 11,461 * RIP-15830
* TOTRIL

CONSENSUS REPORTS: EPA Genetic Toxi-cology Program. Cyanide and its compounds are on the Community Right-To-Know List.

SAFETY PROFILE: Poison by ingestion. Muta-tion data reported. An herbicide. When heated to decomposition it emits very toxic fumes of I^-, NO_x, and CN^-.

DNH125　CAS: 141-04-8　　HR: 2
DIISOBUTYL ADIPATE
mf: $C_{14}H_{26}O_4$　mw: 258.40

SYNS: DIBA * FTAFLEX DIBA * ISOBUTYL ADIPATE

CONSENSUS REPORTS: Reported in EPA TSCA Inventory.

SAFETY PROFILE: Moderately toxic by intraperitoneal route. Mildly toxic by ingestion. Experimental teratogenic and reproductive effects. When heated to decomposition it emits acrid smoke and fumes.

DNH400　CAS: 110-96-3　　HR: 3
DIISOBUTYLAMINE
DOT: UN 2361
mf: $C_8H_{19}N$　mw: 129.28

PROP: Water-white liquid, amine odor. Mp: $-70°$, bp: 139°, flash p: 69.8°F, d: 0.745 @ 20°/4°, vap press: 10 mm @ 30.6°, vap d: 4.46.

SYN: 2-METHYL-N-(2-METHYLPROPYL)-1-PROPANAMINE

CONSENSUS REPORTS: Reported in EPA TSCA Inventory.

DOT Classification: Flammable or Combustible Liquid; Label: Flammable Liquid.

SAFETY PROFILE: Poison by ingestion. A dangerous fire hazard when exposed to heat or flame; can react vigorously with oxidizing materials. To fight fire, use alcohol foam, CO_2, dry chemical. When heated to decomposition it emits toxic fumes of NO_x.

DNH800　CAS: 108-82-7　　HR: 2
DIISOBUTYL CARBINOL
mf: $C_9H_{20}O$　mw: 144.29

PROP: Colorless liquid. Bp: 173.3°, fp: $-65°$, flash p: 165°F, d: 0.8121 @ 20°/20°, vap press: 0.3 mm @ 20°, vap d: 4.98, lel: 0.8% @ 212°F, uel: 6.1% @ 212°F.

SYNS: 2,6-DIMETHYL-4-HEPTANOL * 2,6-DIMETHYL HEPTANOL-4 * sec-NONYL ALCOHOL

CONSENSUS REPORTS: Reported in EPA TSCA Inventory.

SAFETY PROFILE: Moderately toxic by ingestion and intraperitoneal routes. Mildly toxic by skin contact. A powerful systemic irritant by inhalation. A skin and eye irritant. Can cause central nervous system and liver damage when ingested. Flammable when exposed to heat or flame; can react with oxidizing materials. To fight fire, use alcohol foam, foam, CO_2, dry chemical. When heated to decomposition it emits acrid smoke and fumes.

DNI800　CAS: 108-83-8　　HR: 2
DIISOBUTYL KETONE
DOT: UN 1157
mf: $C_9H_{18}O$　mw: 142.27

PROP: Liquid. Bp: 166°, flash p: 140°F, d: 0.81, vap d: 4.9, lel: 0.8% @ 212°F, uel: 6.2% @ 212°F.

SYNS: DIISOBUTILCHETONE (ITALIAN) * DI-ISO-BUTYLCETONE (FRENCH) * DIISOBUTYLKETON (DUTCH, GERMAN) * s-DIISOPROPYLACETONE * 2,6-DIMETHYL-HEPTAN-4-ON (DUTCH, GERMAN) * 2,6-DIMETHYLHEPTAN-4-ONE * 2,6-DIMETHYL-4-HEPTANONE * 2,6-DIMETIL-EPTAN-4-ONE (ITALIAN) * ISOBUTYL KETONE * ISOVALERONE * VALERONE

CONSENSUS REPORTS: Reported in EPA TSCA Inventory.

OSHA PEL: (Transitional: TWA 50 ppm) TWA 25 ppm
ACGIH TLV: TWA 25 ppm
DFG MAK: 50 ppm (290 mg/m^3)
NIOSH REL: (Ketones) TWA 140 mg/m^3
DOT Classification: Flammable or Combustible Liquid; Label: Flammable Liquid; Combustible Liquid; Label: None.

SAFETY PROFILE: Moderately toxic by ingestion and inhalation. Mildly toxic by skin contact. Human systemic effects by inhalation: headache, nausea or vomiting, and unspecified eye effects. An eye and skin irritant. Narcotic in high concentrations. Flammable when exposed to heat or flame; can react with oxidizing materials. To fight fire, use CO_2, dry chemical, water spray, mist or fog. When heated to decomposition it emits acrid smoke and fumes.

DNJ000　CAS: 61947-30-6　　HR: 3
DIISOBUTYLOXOSTANNANE
mf: $C_8H_{18}OSn$　mw: 248.95

SYNS: DIISOBUTYLTIN OXIDE * KYSLICNIK DIISO-BUTYLCINICITY (CZECH)

OSHA PEL: TWA 0.1 mg(Sn)/m^3 (skin)
ACGIH TLV: TWA 0.1 mg(Sn)/m^3 (skin) (Proposed: TWA 0.1 mg(Sn)/m^3; STEL 0.2 mg(Sn)/m^3 (skin))
NIOSH REL: (Organotin Compounds) TWA 0.1 mg(Sn)/m^3

SAFETY PROFILE: Poison by ingestion. An eye and severe skin irritant. When heated to decomposition it emits acrid smoke and fumes.

DNJ400   CAS: 84-69-5   HR: 2
DIISOBUTYL PHTHALATE
mf: $C_{16}H_{22}O_4$ mw: 278.38

PROP: Liquid. Mp: −64°, flash p: 385°F, d: 1.039-1.043, vap d: 9.59.

SYNS: DIBP * HEXAPLAS M/1B * PALATINOL IC

CONSENSUS REPORTS: Reported in EPA TSCA Inventory.

SAFETY PROFILE: Moderately toxic by intraperitoneal route. Mildly toxic by ingestion and skin contact. Experimental teratogenic and reproductive effects. Combustible when exposed to heat or flame. To fight fire, use foam, CO_2, dry chemical. When heated to decomposition it emits acrid smoke and fumes.

DNJ800   CAS: 822-06-0   HR: 3
1,6-DIISOCYANATOHEXANE
DOT: UN 2281
mf: $C_8H_{12}N_2O_2$ mw: 168.22

SYNS: HEXAMETHYLENE DIISOCYANATE * HEX-AMETHYLENE-1,6-DIISOCYANATE * 1,6-HEXAMETH-YLENE DIISOCYANATE (MAK) * HEXAMETHYL-ENEDIISOCYANATE (DOT) * 1,6-HEXANEDIOL DIISO-CYANATE * HMDI * ISOCYANIC ACID, DIESTER with 1,6-HEXANEDIOL * ISOCYANIC ACID, HEXAME-THYLENE ESTER * METHYLENO-BIS-FENYLOIZOCY-JANIAN (POLISH) * SZESCIOMETYLENODWUIZO-CYJANIAN (POLISH) * TL 78

CONSENSUS REPORTS: Reported in EPA TSCA Inventory.

DFG MAK: 0.01 ppm (0.07 mg/m^3)
NIOSH REL: (Diisocyanates) TWA 0.005 ppm; CL 0.02 ppm/10M
DOT Classification: Poison B; Label: Poison.

SAFETY PROFILE: Poison by inhalation and intravenous routes. Moderately toxic by inges-

tion and skin contact. Potentially explosive reaction with alcohols + base. When heated to decomposition it emits toxic fumes of NO_x.

DNK800   CAS: 27215-10-7   HR: 2
DIISOOCTYL ACID PHOSPHATE
DOT: UN 1902
mf: $C_{16}H_{35}O_4P$ mw: 322.48

PROP: A corrosive liquid.

SYN: DIISOOCTYL PHOSPHATE (DOT)

CONSENSUS REPORTS: Reported in EPA TSCA Inventory.

DOT Classification: Corrosive Material; Label: Corrosive.

SAFETY PROFILE: Moderately toxic by irritation to skin, eyes, and mucous membranes. A corrosive compound. When heated to decomposition it emits toxic fumes of PO_x.

DNM200   CAS: 108-18-9   HR: 2
DIISOPROPYLAMINE
DOT: UN 1158
mf: $C_6H_{15}N$ mw: 101.22

PROP: Colorless liquid. Bp: 83-84°, flash p: 19.4°F. D: 0.722 @ 220.0°, vap d: 3.5.

SYNS: DIPA * N-(1-METHYLETHYL)-2-PROPAN-AMINE

CONSENSUS REPORTS: Reported in EPA TSCA Inventory.

OSHA PEL: TWA 5 ppm (skin)
ACGIH TLV: TWA 5 ppm (skin)
DOT Classification: Flammable Liquid; Label: Flammable Liquid.

SAFETY PROFILE: Moderately toxic by ingestion, subcutaneous, and possibly other routes. Mildly toxic by inhalation. Mutation data reported. A severe eye irritant. Inhalation of fumes can cause pulmonary edema. A very dangerous fire hazard when exposed to heat or flame; can react vigorously with oxidizing materials. To fight fire, use alcohol foam, foam, CO_2, dry chemical. When heated to decomposition it emits toxic fumes of NO_x.

DNO200   CAS: 15721-33-2   HR: 3
DIISOPROPYLBERYLLIUM
mf: $C_6H_{14}Be$ mw: 95.19

CONSENSUS REPORTS: Beryllium and its compounds are on the Community Right-To-Know List.

OSHA PEL: (Transitional: TWA 0.002 mg(Be)/m^3; CL 0.005; Pk 0.025/30M/8H) TWA 0.002 mg(Be)/m^3; STEL 0.005 mg(Be)/m^3/30M; CL 0.025 mg(Be)/m^3

ACGIH TLV: TWA 0.002 mg/m^3, Suspected Human Carcinogen.

DFG TRK: 0.002 mg(Be)/m^3. Animal Carcinogen, Suspected Human Carcinogen.

SAFETY PROFILE: Confirmed human carcinogen. Explosive reaction on contact with water. When heated to decomposition it emits toxic fumes of BeO.

DNO900   CAS: 2973-10-6   HR: 3
DIISOPROPYL ESTER SULFURIC ACID
mf: $C_6H_{14}O_4S$ mw: 182.26

SYNS: DI-ISOPROPYLSULFAT (GERMAN) * DI-ISO-PROPYLSULFATE * ISOPROPYL SULFATE

SAFETY PROFILE: Questionable carcinogen with experimental tumorigenic data. Moderately toxic by ingestion and skin contact. When heated to decomposition it emits toxic fumes of SO_x.

DNQ800   CAS: 1071-39-2   HR: 3
DIISOPROPYLMERCURY
mf: $C_6H_{14}Hg$ mw: 286.79

PROP: Liquid. Bp: 63° @ 10 mm, d: 2.0024, vap d: 9.9.

CONSENSUS REPORTS: Mercury and its compounds are on the Community Right-To-Know List.

OSHA PEL: (Transitional: CL 1 mg/10m^3) TWA 0.01 mg(Hg)/m^3; STEL 0.03 mg/m^3 (skin)

ACGIH TLV: TWA 0.01 mg(Hg)/m^3; STEL 0.03 mg(Hg)/m^3

NIOSH REL: (Inorganic Mercury) TWA 0.05 mg(Hg)/m^3

SAFETY PROFILE: Mercury compounds are poisons. When heated to decomposition it emits toxic fumes of Hg.

DNR200   CAS: 23668-76-0   HR: 3
DIISOPROPYLOXOSTANNANE
mf: $C_6H_{14}OSn$ mw: 220.89

PROP: Solid. Insol in water.

SYNS: DIISOPROPYLTIN OXIDE * KYSLICNIK DI-ISOPROPYLCINICITY (CZECH)

OSHA PEL: TWA 0.1 mg(Sn)/m^3 (skin)

ACGIH TLV: TWA 0.1 mg(Sn)/m^3 (skin) (Proposed: TWA 0.1 mg(Sn)/m^3; STEL 0.2 mg(Sn)/m^3 (skin))

NIOSH REL: (Organotin Compounds) TWA 0.1 mg(Sn)/m^3

SAFETY PROFILE: Poison by ingestion. An eye and severe skin irritant. When heated to decomposition it emits acrid smoke and irritating fumes.

DNR309   CAS: 3254-66-8   HR: 3
DIISOPROPYL PARAOXON
mf: $C_{12}H_{18}NO_6P$ mw: 303.28

SYNS: DIISOPROPYL-p-NITROPHENYL PHOSPHATE * O,O-DIISOPROPYL-o,p-NITROPHENYL PHOSPHATE * MIOTICOL * PROPICOL

SAFETY PROFILE: Poison by ingestion and intraperitoneal routes. When heated to decomposition it emits very toxic fumes of NO_x and PO_x.

DNR400   CAS: 105-64-6   HR: 3
DIISOPROPYL PERDICARBONATE
DOT: UN 2133/NA 2134
mf: $C_8H_{14}O_6$ mw: 206.22

PROP: Colorless, crystalline solid. Rapid decomp @ 63°F, mp: 8°-10°, d: 1.080 @ 15.5°/4°. Almost insol in water; miscible with aliphatic and aromatic hydrocarbons, esters, ethers, and chlorinated hydrocarbons.

SYNS: DIISOPROPYL PEROXYDICARBONATE * ISOPROPYL PERCARBONATE * ISOPROPYL PERCARBONATE, stabilized (DOT) * ISOPROPYL PERCARBONATE, unstabilized (DOT) * ISOPROPYL PEROXYDICARBONATE * ISOPROPYL PEROXYDICARBONATE, technically pure (DOT) * PEROXYDICARBONATE D'ISOPROPYLE (FRENCH) * PEROXYDICARBONIC ACID, BIS(1-METHYLETHYL) ESTER

CONSENSUS REPORTS: Reported in EPA TSCA Inventory.

DOT Classification: Organic Peroxide; Label: Organic Peroxide.

SAFETY PROFILE: Moderately toxic by ingestion and skin contact. A severe eye irritant. Very dangerous fire hazard. Dangerously unstable above 10°C. An impact- and heat-sensitive explosive. Solutions may spontaneously explode (the hazard increases with concentration). Storage in sealed containers may be dangerous. Explodes on contact with amines or potassium iodide. May explode on contact with organic matter. When heated to decomposition it emits acrid smoke and fumes.

DNR800   CAS: 2078-54-8   **HR: 3**
2,6-DIISOPROPYLPHENOL
mf: $C_{12}H_{18}O$ mw: 178.30

PROP: A colorless liquid or solid. Bp: 242.4°, fp: 17.9°, flash p: 235°F (CC), d: 0.955 @ 20°/4°.

CONSENSUS REPORTS: Reported in EPA TSCA Inventory.

SAFETY PROFILE: Poison by intravenous and intraperitoneal routes. Combustible when exposed to heat or flame; can react with oxidizing materials. To fight fire, use foam, CO_2, dry chemical. When heated to decomposition it emits acrid smoke and fumes.

DNT000   CAS: 38802-82-3   **HR: 3**
DIISOPROPYLTIN DICHLORIDE
mf: $C_6H_{14}Cl_2Sn$ mw: 275.79

PROP: Colorless crystals. Sol in water. Mp: 84°

SYN: DICHLORODIISOPROPYLSTANNANE

OSHA PEL: TWA 0.1 mg(Sn)/m³ (skin)
ACGIH TLV: TWA 0.1 mg(Sn)/m³ (skin) (Proposed: TWA 0.1 mg(Sn)/m³; STEL 0.2 mg(Sn)/m³ (skin))
NIOSH REL: (Organotin Compounds) TWA 0.1 mg(Sn)/m³

SAFETY PROFILE: Poison by intravenous route. When heated to decomposition it emits toxic fumes of Cl^-.

DNU000   CAS: 630-93-3   **HR: 3**
DILANTIN
mf: $C_{15}H_{11}N_2O_2 \cdot Na$ mw: 274.27

SYNS: ALEPSIN * ANTILEPSIN * ANTISACER * AURANILE * CITRULLAMON * DANTEN * DANTOIN * DENYL * DENYLSODIUM * DERIZENE * DIFENIN * DIFETOIN * DIFHYDAN * DI-HYDAN * DIHYDANTOIN * DILANTIN SODIUM * DI-LEN * DINTOINA * DIPHANTOINE SODIUM * DIPHEDAN * DIPHENATE * DIPHE-

NIN * DIPHENINE SODIUM * DIPHENTOIN * DIPHENYLAN SODIUM * DIPHENYLHYDANTOIN SODIUM * 5,5-DIPHENYLHYDANTOIN SODIUM * 5,5-DIPHENYL-2,4-IMIDAZOLIDINE-DIONE, MONOSODIUM SALT * DI-PHETINE * DITOIN * DIVULSAN * DPH * ENKEFAL * EPAMIN * EPANUTIN * EPELIN * EPIFENYL * EPIHYDAN * EPILAN-D * EPILANTIN * EPINAT * EPTOIN * FENANTOIN * FENITOIN * FENYTOINE * HYDANTIN SODIUM * HYDANTOIN SODIUM * IDANTOIL * IDANTOINAL * LEPITOIN * LEPITOIN SODIUM * MINETOIN * NOVANTOINA * NOVODIPHENYL * OM-HYDANTOINE SODIUM * PHENYTOIN SODIUM * SACERIL * SDPH * SODANTON * SODIUM DIPHENYLHYDANTOIN * SODIUM DIPHENYL HYDANTOINATE * SODIUM-5,5-DIPHENYLHYDANTOINATE * SODIUM-5,5-DIPHENYL-2,4-IMIDAZOLIDINEDIONE * SOLANTOIN * SOLANTYL * SOLUBLE PHENYTOIN * SYLANTOIC * TACOSAL * THILOPHENYT * ZENTROPIL

CONSENSUS REPORTS: IARC Cancer Review: Animal Sufficient Evidence IMEMDT 13,201,77. Reported in EPA TSCA Inventory.

SAFETY PROFILE: Confirmed carcinogen. Poison by ingestion, subcutaneous, intravenous, intraperitoneal, and possibly other routes. Human systemic effects by ingestion: anorexia, respiratory depression, nausea or vomiting, hemorrhage, dermititis, and endocrine effects. Experimental teratogenic and reproductive effects. Mutation data reported. An anticonvulsant and cardiac depressant used for the treatment of grand mal and psychomotor seizures. When heated to decomposition it emits very toxic fumes of NO_x and Na_2O.

DNU200   CAS: 849-55-8   **HR: 3**
DILATOL HYDROCHLORIDE
mf: $C_{19}H_{25}NO_2 \cdot ClH$ mw: 335.91

SYNS: ARLIDIN HYDROCHLORIDE * BUPHENINE HYDROCHLORIDE * DILATYL * p-HYDROXY-α-(1-((1-METHYL-3-PHENYLPROPYL)AMINO)ETHYL)BENZYL ALCOHOL HYDROCHLORIDE * 1-p-HYDROXYPHENYL-2-(1'-METHYL-3'-PHENYLPROPYLAMINO)-1-PROPANOL HYDROCHLORIDE * NYLIDRIN HYDROCHLORIDE * SUPRIFEN PSB HYDROCHLORIDE * VERINA

SAFETY PROFILE: Poison by ingestion, intraperitoneal, and intravenous routes. When heated to decomposition it emits very toxic fumes of Cl^- and NO_x.

DNU300   CAS: 71-68-1   **HR: 3**
DILAUDID
mf: $C_{17}H_{19}NO_3 \cdot ClH$ mw: 321.83

SYNS: DIHYDROMORPHINONE HYDROCHLORIDE
* DILAUDID HYDROCHLORIDE * 4,5-α-EPOXY-3-HY-
DROXY-17-METHYLMORPHINAN-6-ONE HYDROCHLO-
RIDE * HYDROMORPHONE HYDROCHLORIDE
* HYMORPHAN

SAFETY PROFILE: Poison by subcutaneous
and intravenous routes. Experimental teratogenic and reproductive effects. A powerful analgesic. When heated to decomposition it emits
very toxic fumes of NO_x and HCl.

DNU310   CAS: 1421-28-9   **HR: 3**
DILAUDID HYDROCHLORIDE
mf: $C_{17}H_{21}NO_3 \cdot ClH$ mw: 323.85

SYNS: DIHYDROMORPHINE HYDROCHLORIDE
* PARAMORFAN

SAFETY PROFILE: Poison by subcutaneous,
intravenous, and parenteral routes. Moderately
toxic by ingestion. When heated to decomposition it emits very toxic fumes of NO_x and HCl.

DNU400   CAS: 8006-75-5   **HR: 1**
DILL SEED OIL, EUROPEAN TYPE

PROP: From steam distillation of the dried ripe
fruit of *Anethum graveolens* L. (Fam. *Umbelliferae*. Yellowish liquid; caraway odor and taste.
D: 0.890-0.915, refr index: 1.4836 @ 20°. Sol
in fixed oils, mineral oil, and propylene glycol;
insol in glycerin.

SYNS: DILL FRUIT OIL * DILL HERB OIL * DILL
OIL * DILL SEED OIL * DILL WEED OIL

CONSENSUS REPORTS: Reported in EPA
TSCA Inventory.

SAFETY PROFILE: Mildly toxic by ingestion.
A skin irritant. Mutation data reported. When
heated to decomposition it emits acrid smoke
and fumes.

DNV000   CAS: 1165-48-6   **HR: 3**
DIMEFLINE
mf: $C_{20}H_{21}NO_3$ mw: 323.42

SYNS: 8-(DIMETHYLAMINOMETHYL)-7-METHOXY-3-
METHYLFLAVONE * 8-((DIMETHYLAMINO)METHYL)-
7-METHOXY-3-METHYL-2-PHENYLFLAVONE * DW 62
* MALIVAN * N-(7-METHOXY-3-METHYL-4-OXO-2-

PHENYL-4H-CHROMEN-8-YL)METHYL-N,N-DIMETHYL-
AMINE * REANIMIL * REC 7/0267 * REMEFLIN

SAFETY PROFILE: Poison by ingestion, intraperitoneal, intravenous, rectal, and subcutaneous routes. When heated to decomposition it
emits very toxic fumes of NO_x.

DNW000   CAS: 63869-15-8   **HR: 3**
DIMERCUROUS METHANE ARSONATE
mf: $CH_3AsO_3 \cdot 2Hg$ mw: 539.14

SYN: METHANEARSONIC ACID DIMERCURY SALT

CONSENSUS REPORTS: Arsenic and its
compounds, as well as mercury and its compounds, are on the Community Right-To-Know
List.

OSHA PEL: TWA 0.5 mg(As)/m³; (Transi-
    tional: CL 1 mg/10m³) CL 0.1 mg(Hg)/m³
    (skin)
ACGIH TLV: TWA 0.2 mg(As)/m³; 0.1
    mg(Hg)/m³ (skin)
NIOSH REL: (Inorganic Mercury) TWA 0.05
    mg(Hg)/m³

SAFETY PROFILE: Poison by intraperitoneal
route. When heated to decomposition it emits
very toxic fumes of As and Hg.

DNX400   CAS: 2773-92-4   **HR: 3**
DIMETHISOQUIN HYDROCHLORIDE
mf: $C_{17}H_{24}N_2O \cdot ClH$ mw: 308.89

SYNS: 3-BUTYL-1-(2-(DIMETHYLAMINO)ETHOXY)
ISOQUINOLINE HYDROCHLORIDE * 2-((3-BUTYL-1-
ISOQUINOLINYL)OXY)-N,N-DIMETHYLETHANAMINE
MONOHYDROCHLORIDE * 1-(β-DIMETHYLAMINO-
ETHOXY)-3-N-BUTYLISOQUINOLINE HYDROCHLORIDE
* 1-(β-DIMETHYLAMINOETHOXY)-3-N-BUTYL-
ISOQUINOLINE MONOHYDROCHLORIDE * ISO-
CHINOL * PRURALGAN * PRURALGIN * QUO-
TANE * QUOTANE HYDROCHLORIDE

SAFETY PROFILE: Poison by intraperitoneal
and intravenous routes. A topical anesthetic.
When heated to decomposition it emits very
toxic fumes of HCl and NO_x.

DNX500   CAS: 8015-19-8   **HR: 3**
**DIMETHISTERONE and ETHINYL
ESTRADIOL**
mf: $C_{23}H_{32}O_2 \cdot C_{20}H_{24}O_2$ mw: 636.99

SYNS: ETHINYL ESTRADIOL and DIMETHISTERONE
* ORACON * OVIN * SECROVIN

SAFETY PROFILE: Suspected human carcinogen producing uterine tumors. Human reproductive effects by ingestion: abnormalities of the uterus, cervix, and vagina. A steroid. When heated to decomposition it emits acrid smoke and irritating fumes.

DNX600   CAS: 116-01-8   **HR: 3**
DIMETHOATE-ETHYL
mf: $C_6H_{14}NO_3PS_2$ mw: 243.30

SYNS: AMERICAN CYANAMID 18706 * B/77 * O,O-DIMETHYL-S-(N-ETHYLCARBAMOYLMETHYL) DITHIOPHOSPHATE * O,O-DIMETHYL-S-(N-ETHYL-CARBAMOYLMETHYL) PHOSPHORODITHIOATE * EI-18706 * ENT-25,506 * ETHOATE METHYL * S-(2-(ETHYLAMINO-2-OXOETHYL)-O,O-DIMETHYL PHOSPHORODITHIOATE * S-(N-ETHYLCARBA-MOYLMETHYL) DIMETHYL PHOSPHORODITHIOATE * FITIOS * FITIOS B/77 * N-MONOETHYLAMIDE of O,O-DIMETHYLDITHIOPHOSPHORYLACETIC ACID * PHOSHOROTHIOIC ACID-S-(2-(ETHYLAMINO)-2-OXO-ETHYL)-O,O-DIMETHYL ESTER

SAFETY PROFILE: Poison by ingestion, intramuscular, and possibly other routes. Moderately toxic by skin contact. A pesticide. When heated to decomposition it emits very toxic fumes of NO_x, PO_x, and SO_x.

DNX800   CAS: 1113-02-6   **HR: 3**
DIMETHOATE OXYGEN ANALOG
mf: $C_5H_{12}NO_4PS$ mw: 213.21

SYNS: O-ANALOG of DIMETHOATE * BAY 45432 * BAYER 45,432 * DIMETHOATE O-ANALOG * DIMETHOATE PO ISOLOGUE * DIMETHOXON * O,O-DIMETHYL-S-((N-METHYL-CARBAMOYL) METHYL)MONOTHIOFOSFAAT (DUTCH) * O,O-DI-METHYL-S-(N-METHYL-CARBAMOYL)-METHYL-MONO-THIOPHOSPHAT (GERMAN) * O,O-DIMETHYL-S-((METHYLCARBAMOYL)METHYL)PHOSPHOROTHIOATE * O,O-DIMETHYL-S-(N-METHYLCARBAMOYL-METHYL)PHOSPHOROTHIOATE * O,O-DIMETHYL-S-(N-METHYLCARBAMOYLMETHYL) PHOSPHOROTHIO-LATE * DIMETHYL-S-(N-METHYL-CARBAMOYL-METHYL)PHOSPHOROTHIOLATE * O,O-DIMETHYL-S-(N-METHYLCARBAMOYLMETHYL) THIOPHOSPHATE * O,O-DIMETHYL-S-(2-OXO-3-AZABUTYL)-MONO-THIOPHOSPHATE * O,O-DIMETIL-S-(N-METIL-CARBA-MOIL)-METIL-MONOTIOFOSFATO (ITALIAN) * ENT 25,776 * FOLIMAT * OMETHOAT * OMETHOATE * PHOSPHOROTHIOIC ACID, O,O-DI-METHYL S-(2-(METHYLAMINO)-2-OXOETHYL) ESTER * PO-DIMETHOATE * THIOPHOSPHATE de O,O-DI-

ETHYLE et de S-(N-METHYLCARBAMOYL) METHYLE (FRENCH)

SAFETY PROFILE: Poison by ingestion, intraperitoneal, and possibly other routes. Moderately toxic by skin contact. Mutation data reported. An insecticide. When heated to decomposition it emits very toxic fumes of NO_x, PO_x, and SO_x.

DNZ000   CAS: 475-83-2   **HR: 3**
1,2-DIMETHOXY-6a-β-APORPHINE
mf: $C_{19}H_{21}NO_2$ mw: 295.41

PROP: Alkaloid obtained from the lotus *Nelumbo nucifera* and *Nelumbo lutea*.

SYNS: 1-5, 6-DIMETHOXYAPORPHINE * (R)-1,2-DI-METHOXYAPORPHINE * NUCIFERIN * NUCIFE-RINE * (−)-NUCIFERINE * 1-NUCIFERINE * (R)-5,6,6a,7-TETRAHYDRO-1,2-DIMETHOXY-6-METHYL-4H-DIBENZO(de,g)QUINOLINE

SAFETY PROFILE: Poison by ingestion and intraperitoneal routes. When heated to decomposition it emits toxic fumes of NO_x.

DOA800   CAS: 20325-40-0   **HR: 3**
3,3′-DIMETHOXYBENZIDINE DIHYDROCHLORIDE
mf: $C_{14}H_{16}N_2O_2 \cdot 2ClH$ mw: 317.24

SYNS: C.I. DISPERSE BLACK-6-DIHYDROCHLORIDE * o-DIANISIDINE DIHYDROCHLORIDE * 3,3-DI-METHOXY-(1,1′-BIPHENYL)-4,4′-DIAMINE DIHYDRO-CHLORIDE

CONSENSUS REPORTS: Reported in EPA TSCA Inventory.

NIOSH REL: (Benzidine-Based Dye) Reduce to lowest feasible level

SAFETY PROFILE: Questionable carcinogen with experimental carcinogenic and tumorigenic data. Mutation data reported. When heated to decomposition it emits very toxic fumes of NO_x and HCl.

DOE200   CAS: 120-20-7   **HR: 3**
3,4-DIMETHOXYDOPAMINE
mf: $C_{10}H_{15}NO_2$ mw: 181.26

PROP: Colorless to pale yellow liquid. Mp: 15°, bp: 156° @ 10 mm, d: 1.08 @ 28°/4°, vap d: 6.25.

SYNS: DIMETHYOXYDOPAMINE * 3,4-DIMETHOX-YPHENETHYLAMINE * 3,4-DIMETHOXY-β-PHEN-

ETHYLAMINE * DIMETHOXYPHENYLETHYLAMINE * 3,4-DIMETHOXYPHENYLETHYLAMINE * 3,4-DIMETHOXY-β-PHENYLETHYLAMINE * β-(3,4-DIMETHOXYPHENYL)ETHYLAMINE * 2-(3,4-DIMETHOXYPHENYL)ETHYLAMINE * 3,4-DIMETHOXYPHENYLETHYLAMINE (base) * DIMETHYLMESCALINE * DIMPEA * DMPE * DMPEA * HOMOVERATRYLAMINE

CONSENSUS REPORTS: Reported in EPA TSCA Inventory.

SAFETY PROFILE: Poison by intravenous route. Moderately toxic by intraperitoneal route. When heated to decomposition it emits toxic fumes of NO_x.

DOE600   CAS: 110-71-4   HR: 3
1,2-DIMETHOXYETHANE

DOT: UN 2252
mf: $C_4H_{10}O_2$ mw: 90.14

PROP: Liquid; sharp, ethereal odor. D: 0.86877, mp: −58°, bp: 82-83°, n (24/D) 1.3739, flash p: 4.5°C (40°F). Miscible with water and alc; sol in hydrocarbon solvents.

SYNS: DIMETHOXYETHANE * α,β-DIMETHOXYETHANE * 1,2-DIMETHOXYETHANE (DOT) * DIMETHYLCELLOSOLVE * 2,5-DIOXAHEXANE * EGDME * ETHYLENE DIMETHYL ETHER * ETHYLENE GLYCOL DIMETHYL ETHER * GLYCOL DIMETHYL ETHER * GLYME * MONOETHYLENE GLYCOL DIMETHYL ETHER * MONOGLYME

CONSENSUS REPORTS: Reported in EPA TSCA Inventory. Glycol ether compounds are on the Community Right-To-Know List.

DOT Classification: Flammable Liquid; Label: Flammable Liquid.

SAFETY PROFILE: Experimental reproductive effects. Readily forms an explosive peroxide. A very dangerous fire hazard when exposed to heat, flame, or oxidizers. Mixture with lithium tetrahydroaluminate may ignite or explode if heated. When heated to decomposition it emits acrid smoke and fumes.

DOF400   CAS: 117-82-8   HR: 3
DIMETHOXY ETHYL PHTHALATE
mf: $C_{14}H_{18}O_6$ mw: 282.32

PROP: Light-colored, clear liquid; mild aromatic odor. Mp: −40° (forms gel), bp: 190°-

210° @ 4 mm, flash p: 360°F, d: 1.171 @ 20°/20°, vap press: 0.3 mm @ 150°, vap d: 9.75.

SYNS: 1,2-BENZENEDICARBOXYLIC ACID BI(2-METHOXYETHYL)ESTER (9CI) * BIS(METHOXYETHYL) PHTHALATE * BIS(2-METHOXYETHYL) PHTHALATE * DI(2-METHOXYETHYL)PHTHALATE * DMEP * KESSCOFLEX MCP * 2-METHOXYETHYL PHTHALATE * PHTHALIC ACID BIS(2-METHOXYETHYL) ESTER

CONSENSUS REPORTS: EPA Genetic Toxicology Program. Reported in EPA TSCA Inventory.

SAFETY PROFILE: Moderately toxic by ingestion and intraperitoneal routes. Mildly toxic by inhalation. Experimental teratogenic and reproductive effects. Mutation data reported. A pesticide. Combustible when exposed to heat or flame; can react with oxidizing materials. To fight fire use water, foam, CO_2, dry chemical. When heated to decomposition it emits acrid smoke and irritating fumes.

DOG600   CAS: 15589-00-1   HR: 3
2,5-DIMETHOXY-4-METHYLAMPHETAMINE HYDROCHLORIDE
mf: $C_{12}H_{19}NO_2 \cdot ClH$ mw: 245.78

SYNS: 2,5-DIMETHOXY-α,4-DIMETHYLPHENETHYLAMINE HYDROCHLORIDE * 1-(2,5-DIMETHOXY-4-METHYLPHENYL)-2-AMINOPROPANE

SAFETY PROFILE: Poison by ingestion, intraperitoneal and intravenous routes. A central nervous system stimulant. When heated to decomposition it emits very toxic fumes of NO_x and HCl.

DOI400   CAS: 635-85-8   HR: 3
3,4-DIMETHOXYPHENETHYLAMINE HYDROCHLORIDE
mf: $C_{10}H_{15}NO_2 \cdot ClH$ mw: 217.72

SYN: 3,4-DIMETHOXY-β-PHENYLETHYLAMINE HYDROCHLORIDE

SAFETY PROFILE: Poison by intraperitoneal and intravenous routes. When heated to decomposition it emits very toxic fumes of NO_x and HCl.

DOJ200   CAS: 91-10-1   HR: 3
2,6-DIMETHOXYPHENOL
mf: $C_8H_{10}O_3$ mw: 154.18

SYNS: ALDRICH * 1,3-DIMETHYL PYROGALLATE * PYROGALLOL DIMETHYLETHER * PYROGALLOL-1,3-DIMETHYL ETHER * SYRINGOL

CONSENSUS REPORTS: Reported in EPA TSCA Inventory.

SAFETY PROFILE: Poison by intravenous route. Moderately toxic by ingestion. When heated to decomposition it emits acrid smoke and irritating fumes.

DOJ800   CAS: 24973-25-9   **HR: 3**
1-(2,5-DIMETHOXYPHENYL)-2-AMINOPROPANE
mf: $C_{11}H_{17}NO_2 \cdot ClH$ mw: 231.75

SYNS: 2,5-DIMETHOXYAMPHETAMINE HYDROCHLORIDE * 2,5-DIMETHOXY-α-METHYLBENZENEETHANAMINE HYDROCHLORIDE * 2,5-DIMETHOXY-α-METHYLPHENETHYLAMINE HYDROCHLORIDE * 2,5-DIMETHOXY-α-METHYL-β-PHENYLETHYLAMINE HYDROCHLORIDE * β-(2,5-DIMETHOXYPHENYL)ISOPROPYLAMINE HYDROCHLORIDE

SAFETY PROFILE: Poison by intraperitoneal and intravenous routes. Mutation data reported. When heated to decomposition it emits very toxic fumes of NO_x and HCl.

DOK000   CAS: 13078-75-6   **HR: 3**
1-(3,4-DIMETHOXYPHENYL)-2-AMINOPROPANE
mf: $C_{11}H_{17}NO_2 \cdot ClH$ mw: 231.75

SYNS: 3,4-DIMETHOXYAMPHETAMINE HYDROCHLORIDE * 3,4-DIMETHOXY-α-METHYL-β-PHENYLETHYLAMINEHYDROCHLORIDE

SAFETY PROFILE: Poison by intraperitoneal and intravenous routes. When heated to decomposition it emits very toxic fumes of NO_x and HCl.

DOK200   CAS: 6358-53-8   **HR: 3**
1-((2,5-DIMETHOXYPHENYL)AZO)-2-NAPHTHOL
mf: $C_{18}H_{16}N_2O_3$ mw: 308.36

PROP: Mp: 156°. Sltly water-sol; mod sol in alc.

SYNS: C.I. 12156 * C.I. SOLVENT RED 80 * CITRUS RED NO. 2 * 2,5-DIMETHOXYBENZENEAZO-β-NAPHTHOL * 1-((2,5-DIMETHOXYPHENYL)AZO)-2-

NAPHTHALENOL * 2,5-DIMETHOXY-1-(PHENYLAZO)-2-NAPHTHOL * 1-(1-(2,5-DIMETHOXYPHENYL)AZO)-2-NAPHTHOL * 1-(2,5-DIMETHYLOXYPHENYLAZO)-2-NAPHTHOL

CONSENSUS REPORTS: IARC Cancer Review: GROUP 2B IMEMDT 7,56,87; Animal Sufficient Evidence IMEMDT 8,101,75. EPA Genetic Toxicology Program.

SAFETY PROFILE: Suspected carcinogen with experimental carcinogenic data. Mutation data reported. When heated to decomposition it emits toxic fumes of NO_x.

DOL800   CAS: 25601-84-7   **HR: 3**
3-(DIMETHOXYPHOSPHINYLOXY)-N-METHYL-N-METHOXY-cis-CROTONAMIDE
mf: $C_8H_{16}NO_6P$ mw: 253.22

SYNS: CIBA C-2307 * ENT 27,625 * 3-HYDROXY-N-METHOXY-N-METHYL-cis-CROTONAMIDE, DIMETHYL PHOSPHATE * METHOCROTOPHOS * (E)-(3-(METHOXYMETHYLAMINO)-1-METHYL-3-OXO-1-PROPENYL)DIMETHYL PHOSPHATE * NSC 195154

SAFETY PROFILE: A poison by ingestion and skin contact. When heated to decomposition it emits very toxic fumes of NO_x and PO_x.

DON400   CAS: 23435-31-6   **HR: 3**
2′,5′-DIMETHOXYSTILBENAMINE
mf: $C_{16}H_{17}NO_2$ mw: 255.34

SYNS: (trans)-2,5-DIMETHOXY-4′-AMINOSTILBENE * 4-(2,5-DIMETHOXYPHENETHYL)ANILINE * 4-(2-(2,5-DIMETHOXYPHENYL)ETHYL)BENZENAMINE * 4-(2,5-DIMETHOXY)STILBENAMINE * 2,5-DIMETHOXY-4′-STILBENAMINE

SAFETY PROFILE: Suspected carcinogen with experimental carcinogenic data. When heated to decomposition it emits toxic fumes of NO_x.

DOO600   CAS: 534-15-6   **HR: 2**
DIMETHYLACETAL
DOT: UN 2377
mf: $C_4H_{10}O_2$ mw: 90.14

PROP: Colorless liquid; strong aromatic odor. Bp: 61.8°, flash p: 34°F, d: 0.848 @ 25°, vap d: 3.1.

SYNS: ACETALDEHYDE DIMETHYL ACETAL
* 1,1-DIMETHOXYETHANE (DOT) * DIMETHYL AL-
DEHYDE * ETHYLIDENE DIMETHYL ETHER
* METHYL FORMYL

CONSENSUS REPORTS: Glycol ether compounds are on the Community Right-To-Know List. Reported in EPA TSCA Inventory.

DOT Classification: Flammable Liquid; Label: Flammable Liquid.

SAFETY PROFILE: Mildly toxic by inhalation, ingestion, and skin contact. A skin and eye irritant. A very dangerous fire hazard when exposed to heat, flame, or oxidizers. When exposed to heat or flame it can react vigorously with oxidizing materials. To fight fire, use foam, CO_2, dry chemical. When heated to decomposition it emits acrid smoke and irritating fumes.

DOO800   CAS: 127-19-5   **HR: 3**
N,N-DIMETHYLACETAMIDE
mf: C_4H_9NO mw: 87.14

PROP: Liquid. Mp: −20°, bp: 165°, d: 0.9448 @ 15.5°, vap d: 3.01, vap press: 1.3 mm @ 25°, flash p: 171°F (TOC), lel: 2.0%, uel: 11.5% @ 740 mm and 160°.

SYNS: ACETDIMETHYLAMIDE * ACETIC ACID DI-
METHYLAMIDE * DIMETHYLACETAMIDE * DI-
METHYLACETONE AMIDE * DIMETHYLAMIDE
ACETATE * DMA * DMAC * NSC 3138
* U-5954

CONSENSUS REPORTS: Reported in EPA TSCA Inventory.

OSHA PEL: TWA 10 ppm (skin)
ACGIH TLV: TWA 10 ppm (skin)
DFG MAK: 10 ppm (35 mg/m³)

SAFETY PROFILE: Moderately toxic by skin contact, inhalation, intravenous and intraperitoneal routes. Mildly toxic by ingestion. Experimental teratogenic and reproductive effects. A skin irritant. Less toxic than dimethylformamide. Flammable when exposed to heat and flame. A moderate explosion hazard. Violent reaction with halogenated compounds (e.g., carbon tetrachloride, hexachlorocyclohexane) when heated above 90°C. Iron powder catalyzes the reaction so that it initiates at 71°C. When heated to decomposition it emits toxic fumes of NO_x.

DOP200   CAS: 13265-60-6   **HR: 3**
O,O-DIMETHYL-S-(2-(ACETYLAMINO) ETHYL) DITHIOPHOSPHATE
mf: $C_6H_{14}NO_3PS_2$ mw: 243.30

SYNS: S-(2-(ACETYLAMINO)ETHYL)-O,O-DIMETHYL
PHOSPHORODITHIOATE * AMIPHOS * CP 49674
* DAEP * O,O-DIMETHYL-S-(2-ACETAMIDOETHYL)
ESTER PHOSPHORODITHIOIC ACID * O,O-DIMETHYL-
S-(2-ACETYLAMINOETHYL) PHOSPHORODITHIOATE
* N-((O,O-DIMETHYLPHOSPHORODITHIOYL)ETHYL)
ACETAMIDE * ENT 27,346 * MONSANTO CP-49674
* NSC 190945 * PHOSPHORODITHIOIC ACID,
O,O-DIMETHYL ESTER, S-ESTER with N-(2-MERCAPTO-
ETHYL)ACETAMIDE

SAFETY PROFILE: Poison by ingestion, intraperitoneal, subcutaneous, skin contact, and possibly other routes. Experimental teratogenic and reproductive effects. When heated to decomposition it emits very toxic NO_x, PO_x, and SO_x.

DOP600   CAS: 30560-19-1   **HR: 3**
O,S-DIMETHYLACETYLPHOSPHORO-AMIDOTHIOATE
mf: $C_4H_{10}NO_3PS$ mw: 183.18

SYNS: ACEPHAT (GERMAN) * ACEPHATE
* ACETYLPHOSPHORAMIDOTHIOIC ACID-O,S-DI-
METHYL ESTER * CHEVRON RE 12,420 * ENT
27,822 * ORTHENE * ORTHENE-755 * ORTHO
12420 * ORTRAN * ORTRIL * RE 12420
* 75 SP

CONSENSUS REPORTS: EPA Genetic Toxicology Program.

SAFETY PROFILE: Poison by ingestion. Moderately toxic by skin contact and inhalation. Human mutation data reported. When heated to decomposition it emits very toxic fumes of NO_x, PO_x, and SO_x.

DOQ400   CAS: 359-83-1   **HR: 3**
2-(3,3-DIMETHYLALLYL)CYCLAZOCINE
mf: $C_{19}H_{27}NO$ mw: 285.47

SYNS: 2-DIMETHYLALLYL-5,9-DIMETHYL-2'-HYDORX-
YBENZOMORPHAN * 2-(3,3-DIMETHYLALLYL)-2',2'-
HYDROXY-5,9-DIMETHYL-6,7-BENZOMORPHAN
* FORTALGESIC * FORTALIN * FORTRAL
* 1,2,3,4,5,6-HEXAHYDRO-6,11-DIMETHYL-3-(3-
METHYL-2-BUTENYL)-2,6-METHANO-3-BENZAZOCINE
* 2'-HYDROXY-5,9-DIMETHYL-2-(3,3-DIMETHYLAL-
LYL)-6,7-BENZOMORPHAN * dl-2'-HYDROXY-5,9-DI-
METHYL-2-(3,3-DIMETHYLALLYL)-6,7-BENZOMORPHAN
* II-C-2 * KF-1820 * LITICON * 3-(3-METHYL--

2-BUTENYL)-1,2,3,4,5,6-HEXAHYDRO-6,11-DIMETHYL-2,6-
METHANO-3-BENZAZOCIN-8-OL   ✶   NIH 7958
✶   NSC-107430   ✶   PENTAGIN   ✶   PENTAZOCINE
✶   SOSIGON   ✶   TALWAN   ✶   TALWIN   ✶   WIN
20228

SAFETY PROFILE: Poison by ingestion, subcutaneous, intramuscular, intraperitoneal, and intravenous routes. Experimental reproductive effects. Human systemic effects by intramuscular and intravenous routes: wakefulness, euphoria, hallucinations or distorted perceptions, tremors, convulsions, excitement, motor activity changes, muscle weakness, analgesia, withdrawal, parasympathomimetic effects, nausea or vomiting, and dermititis. Can cause drug dependency and other central nervous system effects. An analgesic. When heated to decomposition it emits toxic fumes of NO_x.

DOQ600   CAS: 3639-66-5   **HR: 3**
2-(3,3-DIMETHYLALLYL)-5-ETHYL-2'-
HYDROXY-9-METHYL-6,7-
BENZOMORPHAN
mf: $C_{20}H_{29}NO$ mw: 299.50

SYN: 5-ETHYL-2'-HYDROXY-2(N)-(3-METHYL-2-BUTE-
NYL)-9-METHYL-6,7-BENZOMORPHAN

SAFETY PROFILE: Poison by subcutaneous and intravenous routes. When heated to decomposition it emits toxic fumes of NO_x.

DOQ800   CAS: 124-40-3   **HR: 3**
DIMETHYLAMINE
DOT: UN 1032/UN 1160
mf: C_2H_7N mw: 45.10

SYNS: DIMETHYLAMINE, aqueous solution (DOT)
✶   DIMETHYLAMINE, anhydrous (DOT)   ✶   DIMETHYL-
AMINE, solution (DOT)   ✶   DMA   ✶   N-METHYLMETH-
ANAMINE   ✶   RCRA WASTE NUMBER U092

CONSENSUS REPORTS: EPA Genetic Toxicology Program. Reported in EPA TSCA Inventory.

OSHA PEL: TWA 10 ppm
ACGIH TLV: TWA 10 ppm
DFG MAK: 10 ppm (18 mg/m³)
DOT Classification: Flammable Liquid; Label:
    Flammable Liquid; Flammable Gas; Label:
    Flammable Gas, anhydrous.

SAFETY PROFILE: Poison by ingestion. Moderately toxic by inhalation and intravenous

routes. Mutation data reported. An eye irritant. Corrosive to the eyes, skin, and mucous membranes. A flammable gas. When heated to decomposition it emits toxic fumes of NO_x. Incompatible with acrylaldehyde, fluorine, and maleic anhydride

DOR000   CAS: 124-40-3   **HR: 3**
DIMETHYLAMINE (anhydrous)
mf: C_2H_7N mw: 45.10

PROP: Colorless gas. Bp: 6.88°, flash p: 0°F, fp: −92.19°, d: 0.6804 @ 0°/4°, autoign temp: 752°F, vap d: 1.55, lel: 2.8%, uel: 14.4%. Sol in water, ether, alc.

CONSENSUS REPORTS: Reported in EPA TSCA Inventory.

DOT Classification: Flammable Gas; Label:
    Flammable Gas.

SAFETY PROFILE: Poison by ingestion. An irritant. A very dangerous fire hazard when exposed to heat or flame; can react vigorously with oxidizing materials. Moderately explosive when exposed to flame. To fight fire, stop flow of gas, foam, CO_2, dry chemical. When heated to decomposition it emits toxic fumes of NO_x.

DOR200   CAS: 74-94-2   **HR: 3**
DIMETHYLAMINE BORANE
mf: $C_2H_7N \cdot BH_3$ mw: 58.94

SYNS: BORANE with DIMETHYLAMINE (1:1)
✶   DMAB   ✶   N-METHYLMETHANAMINE with BORANE
(1:1)

CONSENSUS REPORTS: Reported in EPA TSCA Inventory.

SAFETY PROFILE: Poison by ingestion, intraperitoneal, and intravenous routes. A skin and eye irritant. When heated to decomposition it emits toxic fumes of NO_x.

DOR400   CAS: 2032-59-9   **HR: 3**
4-DIMETHYLAMINE m-CRESYL
METHYLCARBAMATE
mf: $C_{11}H_{16}N_2O_2$ mw: 208.29

SYNS: A 363   ✶   AMINOCARB   ✶   AMINOCARBE
(FRENCH)   ✶   BAY 44646   ✶   BAYER 5080   ✶   BAYER
44646   ✶   4-DIMETHYLAMINO-3-CRESYL METHYLCAR-
BAMATE   ✶   4-(DIMETHYLAMINO)-3-METHYLPHENOL
METHYL CARBAMATE (ester)   ✶   (4-DIMETHYLAMINO-3-
METHYL-PHENYL)N-METHYL-CARBAMAAT (DUTCH)
✶   (4-DIMETHYLAMINO-3-METHYL-PHENYL)N-METHYL-

CARBAMATE * (4-DIMETHYLAMINO-3-METHYL-
PHENYL)N-METHYL-CARBAMAT (GERMAN) * 4-(DI-
METHYLAMINO)-m-TOLYL METHYLCARBAMATE
* (4-DIMETILAMINO-3-METIL-FENIL)-N-METIL-CAR-
BAMMATO (ITALIAN) * ENT 25,784 * MATACIL
* N-METHYLCARBAMATE de 4-DIMETHYLAMINO-3-
METHYL PHENYLE (FRENCH) * MITACIL

SAFETY PROFILE: Poison by ingestion, skin
contact, intraperitoneal, subcutaneous, and possibly other routes. Mutation data reported. An
insecticide used for forest insect control. When
heated to decomposition it emits toxic fumes
of NO_x.

DOS000   CAS: 315-18-4   HR: 3
4-(DIMETHYLAMINE)-3,5-XYLYL-N-METHYLCARBAMATE

DOT: NA 2757
mf: $C_{12}H_{18}N_2O_2$ mw: 222.32

PROP: Crystals. Mp: 85°, vap press: <0.1 mm
@ 139°.

SYNS: 4-(DIMETHYLAMINO)-3,5-DIMETHYLPHENOL
METHYLCARBAMATE (ESTER) * 4-(DIMETHYL-
AMINO)-3,5-DIMETHYLPHENYL ESTER, METHYLCAR-
BAMIC ACID * 4-(DIMETHYLAMINO)-3,5-DIMETHYL-
PHENYL-N-METHYLCARBAMATE * 4-(DI-
METHYLAMINO)-3,5-XYLENOL METHYLCARBAMATE
(ESTER) * 4-(DIMETHYLAMINO)-3,5-XYLYL ESTER
METHYLCARBAMIC ACID * 4-DIMETHYLAMINO-3,5-
XYLYL METHYLCARBAMATE * 4-DIMETHYL-
AMINO-3,5-XYLYL-N-METHYLCARBAMATE
* 4-(N,N-DIMETHYLAMINO)-3,5-XYLYL N-METHYL-
CARBAMATE * DOWCO 139 * ENT 25,766
* METHYL-4-DIMETHYLAMINO-3,5-XYLYL CAR-
BAMATE * METHYL-4-DIMETHYLAMINO-3,5-XYLYL
ESTER of CARBAMIC ACID * MEXACARBATE (DOT)
* NCI-C00544 * OMS-47 * ZACTRAN
* ZECTANE * ZECTRAN * ZEXTRAN

CONSENSUS REPORTS: IARC Cancer Re-
view: GROUP 3 IMEMDT 7,56,87; Animal
Inadequate Evidence IMEMDT 12,237,76. NCI
Carcinogenesis Bioassay (feed); No Evidence:
mouse, rat NCITR* NCI-CG-TR-147,78. EPA
Extremely Hazardous Substances List.

DOT Classification: Poison B; Label: Poison.

SAFETY PROFILE: Poison by ingestion, intraperitoneal, and possibly other routes. Moderately toxic by skin contact. Experimental teratogenic and reproductive effects. Questionable
carcinogen with experimental neoplastigenic

data. When heated to decomposition it emits
toxic fumes of NO_x.

DOS200   CAS: 926-64-7   HR: 3
DIMETHYLAMINOACETONITRILE

DOT: UN 2378
mf: $C_4H_8N_2$ mw: 84.14

PROP: Flash p: <73.4°F.

SYNS: N-(CYANOMETHYL)DIMETHYLAMINE
* N,N-DIMETHYLGLYCINONITRILE

CONSENSUS REPORTS: Cyanide and its
compounds are on the Community Right-To-
Know List.

DOT Classification: Flammable or Combustible Liquid; Label: Flammable and Poison.

SAFETY PROFILE: Poison by ingestion, skin
contact, and ocular routes. Moderately toxic
by inhalation. A dangerous fire hazard when
exposed to heat or flame. When heated to decomposition it emits toxic fumes of NO_x and
CN^-.

DOT000   CAS: 58-15-1   HR: 3
DIMETHYLAMINOANTIPYRINE
mf: $C_{13}H_{17}N_3O$ mw: 231.33

PROP: Colorless leaflets, somewhat water-sol.
Mp: 107-109°.

SYNS: AMIDAZOPHEN * AMIDOFEBRIN
* AMIDOPHEN * AMIDOPHENAZONE * AMIDO-
PYRAZOLINE * AMIDOPYRIN * AMINOFENAZONE
(ITALIAN) * AMINOPHENAZONE * AMINOPYRINE
* ANAFEBRINA * BRUFANEUXOL * DAP
* DEREUMA * DIMAPYRIN * DIMETHYLAMINO-
ANALGESINE * 4-(DIMETHYLAMINO)ANTIPYRINE
* DIMETHYLAMINOAZOPHENE * 4-(DIMETHYL-
AMINO)-1,2-DIHYDRO-1,5-DIMETHYL-2-PHENYL-3H-PY-
RAZOL-3-ONE * 4-DIMETHYLAMINO-2,3-DIMETHYL-1-
PHENYL-3-PYRAZOLIN-5-ONE * 4-DIMETHYLAMINO-
2,3-DIMETHYL-1-PHENYL-5-PYRAZOLONE * DIMETH-
YLAMINOPHENAZON (GERMAN) * DIMETHYLAMINO-
PHENAZONE * 4-DIMETHYLAMINOPHENAZONE
* DIMETHYLAMINOPHENYLDIMETHYLPYRAZOLIN
* 4-DIMETHYLAMINO-1-PHENYL-2,3-DIMETHYLPYRA-
ZOLONE * 3-keto-1,5-DIMETHYL-4-DIMETHYLAMINO-
2-PHENYL-2,3-DIHYDROPYRAZOLE * 1,5-DIMETHYL-
4-DIMETHYLAMINO-2-PHENYL-3-PYRAZOLONE
* 2,3-DIMETHYL-4-DIMETHYLAMINO-1-PHENYL-5-PY-
RAZOLONE * DIPIRIN * DIPYRIN * FEBRININA
* FEBRON * ITAMIDONE * MAMALLET-A
* NETSUSARIN * NOVAMIDON * 1-PHENYL-2,3-

DIMETHYL-4-DIMETHYLAMINOPYRAZOL-5-ONE
* 1-PHENYL-2,3-DIMETHYL-4-DIMETHYLAMINOPYRA-
ZOLONE-5 * PIRAMIDON * PIRIDOL * PIROMI-
DINA * POLINALIN * PYRADONE * PYRAMI-
DON * PYRAMIDONE

CONSENSUS REPORTS: Reported in EPA TSCA Inventory. EPA Genetic Toxicology Program.

SAFETY PROFILE: Human poison by unspecified route. Experimental poison by ingestion, subcutaneous, intramuscular, intravenous, intraperitoneal, and possibly other routes. Moderately toxic by parenteral route. Experimental teratogenic and reproductive effects. Questionable carcinogen when mixed with $NaNO_2$ (1:1). Mutation data reported. Can cause bone marrow depression resulting in leucopenia. Has been implicated in development of aplastic anemia. A tranquilizer. When heated to decomposition it emits toxic fumes of NO_x.

DOT300   CAS: 60-11-7   **HR: 3**
4-DIMETHYLAMINOAZOBENZENE
mf: $C_{14}H_{15}N_3$ mw: 225.32

PROP: Yellow, crystalline tablets; insol in water; sol in strong mineral acids and oils.

SYNS: ATUL FAST YELLOW R * BENZENEAZODI-
METHYLANILINE * BRILLIANT FAST YELLOW
* BUTTER YELLOW * CERASINE YELLOW GG
* C.I. 11020 * C.I. SOLVENT YELLOW 2 * DAB
* p-DIMETHYLAMINOAZOBENZEN (CZECH) * DI-
METHYLAMINOAZOBENZENE * N,N-DIMETHYL-4-
AMINOAZOBENZENE * N,N-DIMETHYL-p-AMINO-
AZOBENZENE * p-DIMETHYLAMINOAZOBENZENE
* 4-(N,N-DIMETHYLAMINO)AZOBENZENE * DI-
METHYLAMINOAZOBENZOL * p-DIMETHYLAMINO-
AZOBENZOL (GERMAN) * 4-DIMETHYLAMINOAZO-
BENZOL * 4-DIMETHYLAMINOPHENYLAZOBEN-
ZENE * N,N-DIMETHYL-p-AZOANILINE * N,N-DI-
METHYL-p-PHENYLAZOANILINE * N,N-DIMETHYL-4-
(PHENYLAZO)BENZAMINE * N,N-DIMETHYL-4-(PHE-
NYLAZO)BENZENAMINE * DIMETHYL YELLOW
* DIMETHYL YELLOW-N,N-DIMETHYLANILINE
* DMAB * ENIAL YELLOW 2G * FAST OIL YEL-
LOW B * FAT YELLOW * GRASAL BRILLIANT
YELLOW * JAUNE de BEURRE (FRENCH)
* METHYL YELLOW * OIL YELLOW * OLEAL
YELLOW 2G * ORGANOL YELLOW ADM
* ORIENT OIL YELLOW GG * P.D.A.B. * PETROL
YELLOW WT * RCRA WASTE NUMBER U093
* RESINOL YELLOW GR * RESOFORM YELLOW
GGA * SILOTRAS YELLOW T2G * SOMALIA YEL-

LOW A * STEAR YELLOW JB * SUDAN YELLOW
* TOYO OIL YELLOW G * USAF EK-338
* WAXOLINE YELLOW AD * YELLOW
G SOLUBLE in GREASE * ZLUT MASELNA
(CZECH)

CONSENSUS REPORTS: IARC Cancer Review: GROUP 2B IMEMDT 7,56,87; Animal Sufficient Evidence IMEMDT 8,125,75. NTP Fourth Annual Report On Carcinogens, 1984. EPA Genetic Toxicology Program. Community Right-To-Know List. Reported in EPA TSCA Inventory.

OSHA PEL: Carcinogen

SAFETY PROFILE: Confirmed carcinogen with experimental carcinogenic, neoplastigenic, and tumorigenic data. Poison by ingestion and intraperitoneal routes. Experimental teratogenic and reproductive effects. Human mutation data reported. When heated to decomposition it emits toxic fumes of NO_x.

DOT600   CAS: 443-30-1   **HR: 3**
1-(4-DIMETHYLAMINOBENZAL)INDENE
mf: $C_{18}H_{17}N$ mw: 247.36

SYNS: DABI * (4-DIMETHYLAMINOBENZYL-
IDENE)INDENE * N,N-DIMETHYL-α-INDOLYLIDENE-p-
TOLUIDINE * 4-(1H-INDEN-1-YLIDENEMETHYL)-N,
N-DIMETHYLBENZENAMINE * NSC-80087

SAFETY PROFILE: Questionable carcinogen with experimental tumorigenic and teratogenic data. Moderately toxic by intraperitoneal route. When heated to decomposition it emits toxic fumes of NO_x.

DOT800   CAS: 536-17-4   **HR: 3**
**p-DIMETHYLAMINOBENZALRHODA-
NINE**
mf: $C_{12}H_{12}N_2OS_2$ mw: 264.38

SYNS: p-(DIMETHYLAMINO)BENZAL-5-RHODANINE
* 5-(p-DIMETHYLAMINOBENZAL)RHODANINE
* 5-(p-DIMETHYLAMINOBENZOYLIDENE)RHODANINE
* p-DIMETHYLAMINOBENZYLIDENE RHODAMINE
* USAF PD-20

CONSENSUS REPORTS: Reported in EPA TSCA Inventory.

SAFETY PROFILE: Poison by intraperitoneal route. When heated to decomposition it emits very toxic fumes of NO_x and SO_x.

DOU600 CAS: 140-56-7 **HR: 3**
p-DIMETHYLAMINOBENZENEDIAZOSO-DIUM SULPHONATE
mf: $C_8H_{10}N_3O_3S \cdot Na$ mw: 251.26

PROP: Yellow-brown crystals.

SYNS: BAYER 5072 * DAPA * DAS * DEK-SONAL * DEXON * p-DIMETHYLAMINOBENZENE DIAZO SODIUM SULFONATE * p-(DIMETHYLAMINO) BENZENEDIAZOSULFONATE * p-DIMETHYLAMINO-BENZENEDIAZOSULFONIC ACID, SODIUM SALT * 4-DIMETHYLAMINOBENZENEDIAZOSULFONIC ACID, SODIUM SALT * p-(DIMETHYLAMINO)BENZENEDIAZO-SULPHONATE * p-(DIMETHYLAMINO)BENZENE-DIAZOSULPHONIC ACID, SODIUM SALT * 4-DI-METHYLAMINOBENZENEDIAZOSULPHONIC ACID, SODIUM SALT * p-DIMETHYLAMINOBENZOL-DIAZOSULFONAT (NATRIUMSALZ) (GERMAN) * (4-(DIMETHYLAMINO)PHENYL)DIAZENESULFONIC ACID, SODIUM SALT * 4-((DIMETHYLAMINO) PHENYL)DIAZENESULFONIC ACID, SODIUM SALT * p-(DIMETHYLAMINO)-PHENYLDIAZO-NATRIUMSUL-FONAT (GERMAN) * N,N-DIMETHYL-p-ANILINE-DIAZOSULFONIC ACID SODIUM SALT * FEN-AMINOSULF * GOLD ORANGE MP * LESAN * NCI-C03010 * SODIUM-p-(DIMETHYLAMINO) BENZENEDIAZOSULFONATE * SODIUM-4-(DI-METHYLAMINO)BENZENEDIAZOSULFONATE * SODIUM-p-(DIMETHYLAMINO)BENZENEDIAZOSUL-PHONATE * SODIUM-4-(DIMETHYLAMINO)BENZENE-DIAZOSULPHONATE * SODIUM-(4-(DIMETHYL-AMINO)PHENYL)DIAZENESULFONATE * TRO-PAEOLIN D

CONSENSUS REPORTS: IARC Cancer Review: GROUP 3 IMEMDT 7,56,87; Animal Inadequate Evidence IMEMDT 8,147,75. NCI Carcinogenesis Bioassay (feed); No Evidence: mouse, rat NCITR* NCI-CG-TR-101,78. EPA Genetic Toxicology Program.

SAFETY PROFILE: Poison by ingestion, intravenous, intraperitoneal, and possibly other routes. Experimental teratogenic and reproductive effects. Human mutation data reported. Questionable carcinogen. A fungicide. When heated to decomposition it emits very toxic fumes of NO_x, Na_2O, and SO_x.

DOX000 CAS: 5913-82-6 **HR: 3**
3-β-(DIMETHYLAMINO)CON-5-ENINE-DIHYDROBROMIDE
mf: $C_{24}H_{40}N_2 \cdot 2BrH$ mw: 356.66

SYNS: CONESSINE DIHYDROBROMIDE * KONESSIN DIHYDROBROMIDE * NERIINE DIHYDRBROMIDE * ROQUESSINE DIHYDROBROMIDE * WRIGHTINE DIHYDROBROMIDE

SAFETY PROFILE: Poison by ingestion, intraperitoneal, and intravenous routes. When heated to decomposition it emits toxic fumes of NO_x and HBr.

DOY400 CAS: 60-46-8 **HR: 3**
4-(DIMETHYLAMINO)-2,2-DIPHENYL-VALERAMIDE
mf: $C_{19}H_{24}N_2O$ mw: 296.45

SYNS: AMINOPENTAMIDE * BL 139 * CENTRINE * α-(2-(DIMETHYLAMINO)PROPYL)-α-PHENYLBENZENE-ACETAMIDE * DIMEVAMIDE * α,α-DIPHENYL-γ-DIMETHYLAMINOVALERAMIDE * 3-METHYL-4-DI-METHYLAMINO-2,2-DIPHENYLBUTYRAMIDE * VAL-ERAMIDE-OM

SAFETY PROFILE: Poison by intraperitoneal, intravenous, and possibly other routes. Moderately toxic by ingestion. An anticholinergic. Used in veterinary medicine as an anticonvulsant and anti-emetic. When heated to decomposition it emits toxic fumes of NO_x.

DOY800 CAS: 108-01-0 **HR: 2**
N-DIMETHYLAMINOETHANOL
DOT: UN 2051
mf: $C_4H_{11}NO$ mw: 89.16

PROP: Liquid. Bp: 131°, flash p: 105°F (OC), d: 0.8866 @ 20°/4°, vap d: 3.03.

SYNS: DEANOL * DIMETHYLAETHANOLAMIN (GERMAN) * DIMETHYLAMINOAETHANOL (GERMAN) * DIMETHYLAMINOETHANOL * β-DIMETHYLAMI-NOETHANOL * N,N-DIMETHYLAMINOETHANOL * 2-(DIMETHYLAMINO)ETHANOL * β-DIMETHYL-AMINOETHYL ALCOHOL * DIMETHYLETHANOL-AMINE * N,N-DIMETHYLETHANOLAMINE * DI-METHYLETHANOLAMINE (DOT) * N,N-DIMETHYL-2-HYDROXYETHYLAMINE * N,N-DIMETHYL-N-(2-HY-DROXYETHYL)AMINE * DMAE * β-HYDROXY-ETHYLDIMETHYLAMINE

CONSENSUS REPORTS: Reported in EPA TSCA Inventory.

DOT Classification: Flammable or Combustible Liquid; Label: Flammable Liquid.

SAFETY PROFILE: Moderately toxic by ingestion, inhalation, skin contact, intraperitoneal, and subcutaneous routes. A skin and severe eye irritant. Used medically as a central nervous system stimulant. Flammable when exposed to heat or flame; can react vigorously with oxidizing materials. Ignites spontaneously in contact with cellulose nitrate of high surface area. To fight fire, use alcohol foam, foam, CO_2, dry chemical. When heated to decomposition it emits toxic fumes of NO_x.

DPF600 CAS: 61-50-7 **HR: 3**
3-(2-(DIMETHYLAMINO)ETHYL)INDOLE
mf: $C_{12}H_{16}N_2$ mw: 188.30

SYNS: N,N-DIMETHYLTRYPTAMINE * DMT

SAFETY PROFILE: Poison by intravenous and intraperitoneal routes. Human systemic effects by intramuscular route: pupiliary dilation, hallucinations and distorted perceptions, blood pressure increase. When heated to decomposition it emits toxic fumes of NO_x.

DPG109 CAS: 487-93-4 **HR: 3**
3-(2-DIMETHYLAMINOETHYL)-5-INDOLOL
mf: $C_{12}H_{16}N_2O$ mw: 204.30

SYNS: BUFOTENIN * 3-(β-DIMETHYLAMINO-ETHYL)-5-HYDROXYINDOLE * N,N-DIMETHYL-5-HY-DROXYTRYPTAMINE * N,N-DIMETHYLSEROTONIN * 5-HYDROXY-N,N-DIMETHYLTRYPTAMINE

SAFETY PROFILE: Poison by intraperitoneal route. Human systemic effects with very small amounts taken by intravenous route: psychotropic effects. A modified natural neurotransmitter. When heated to decomposition it emits toxic fumes of NO_x.

DPG600 CAS: 2867-47-2 **HR: 3**
DIMETHYLAMINOETHYL METHACRYLATE
DOT: UN 2522
mf: $C_8H_{15}NO_2$ mw: 157.24

PROP: Liquid, sol in water and organic solvents. D: 0.933 @ 25°, bp: 182-190°, flash p: 165°F (TOC), vap d: 5.4.

SYNS: AGEFLEX FM-1 * 2-(DIMETHYLAMINO)ETH-ANOL METHACRYLATE * 2-(DIMETHYLAMINO) ETHYL ESTER METHACRYLIC ACID * N,N-DIMETH-YLAMINOETHYL METHACRYLATE * β-DIMETH-YLAMINOETHYL METHACRYLATE * 2-(DIMETH-YLAMINO)ETHYL METHACRYLATE * USAF RH-3

CONSENSUS REPORTS: Reported in EPA TSCA Inventory.

DOT Classification: Poison B; Label: Poison.

SAFETY PROFILE: Poison by intraperitoneal route. Moderately toxic by ingestion and inhalation. A skin, eye, and mucous membrane irritant. A powerful lachrymator. Flammable when exposed to sparks, heat, open flame or oxidizers. To fight fire, use alcohol foam, dry chemical, spray. When heated to decomposition it emits toxic fumes of NO_x.

DPH000 CAS: 4724-58-7 **HR: 3**
2-DIMETHYLAMINOETHYL-2-METHYL-BENZHYDRYL ETHER CITRATE
mf: $C_{18}H_{23}NO \cdot C_6H_8O_7$ mw: 461.56

SYNS: N,N-DIMETHYL-2-((o-METHYL-α-PHENYL-BEN-ZYL)OXY)-ETHYLAMINE CITRATE * ORPHENADRINE CITRATE

SAFETY PROFILE: Poison by ingestion, intramuscular, and intravenous routes. Mutation data reported. When heated to decomposition it emits toxic fumes of NO_x.

DPH600 CAS: 4985-15-3 **HR: 3**
5-DIMETHYLAMINOETHYLOXYIMINO-5H-DIBENZO(a,d)CYCLOHEPTA-1,4-DIENE HYDROCHLORIDE
mf: $C_{19}H_{22}N_2O \cdot ClH$ mw: 330.89

SYNS: AGEDAL * BAY 1521 * 5-(DIMETHYLAMI-NOAETHYL-OXYIMINO)-5H-DIBENZO(a,d)CYCLOHEPTA-1,4-DIENHYDROCHLORID (GERMAN) * 5-(DIMETHYL-AMINOOXYIMINO)-5H-DIBENZO(a,b)CYCLOHEPTA-1,4-DIENE HYDROCHLORIDE * 5-(DIMETILAMINOETIL-OXIMINO-5H-DIBENZO(a,d)CICLOEPTA-1,4-DIENE) CLORI-DRATO (ITALIAN) * NOGEDAL * NOXIPTILINE HY-DROCHLORIDE * NOXIPTILIN HYDROCHLORID (GER-MAN) * NOXIPTYLINE HYDROCHLORIDE

SAFETY PROFILE: Poison by ingestion, subcutaneous, intravenous, intramuscular, and intraperitoneal routes. Experimental reproductive effects. When heated to decomposition it emits very toxic fumes of NO_x and HCl.

DPK000 **HR: 3**
anti-8-(N,N-DIMETHYLAMINOMETHYL) DIBENZOBICYCLO(3.2.1)OCTADIENE HYDROCHLORIDE

SYN: 10,11-DIHYDRO-N,N-DIMETHYL-5,10-METHANO-5H-DIBENZO(a,d)CYCLOHEPTENE-12-METHANAMINE HCl

SAFETY PROFILE: Poison by ingestion and intravenous routes. When heated to decomposition it emits very toxic fumes of NO_x and HCl.

DPL000   CAS: 55738-54-0   HR: 3
trans-2-((DIMETHYLAMINO)METHYL-IMINO)-5-(2-(5-NITRO-2-FURYL)VINYL)-1,3,4-OXADIAZOLE
mf: $C_{11}H_{12}N_5O_4$ mw: 277.27

CONSENSUS REPORTS: IARC Cancer Review: GROUP 2B IMEMDT 7,56,87; Animal Limited Evidence IMEMDT 7,147,74. EPA Genetic Toxicology Program.

SAFETY PROFILE: Suspected carcinogen with experimental carcinogenic data. Mutation data reported. When heated to decomposition it emits toxic fumes of NO_x.

DPL200   CAS: 2914-77-4   HR: 3
2-DIMETHYLAMINOMETHYL-1-(m-METHOXYPHENYL)CYCLOHEXANOL
mf: $C_{16}H_{25}NO_2$ mw: 263.42

SAFETY PROFILE: Poison by ingestion route. When heated to decomposition it emits toxic fumes of NO_x.

DPN200   CAS: 605-65-2   HR: 3
5-(DIMETHYLAMINO)-1-NAPHTHALENE-SULFONYL CHLORIDE
mf: $C_{12}H_{12}ClNO_2$ mw: 237.70

SYNS: 1-CHLOROSULFONYL-5-DIMETHYLAMINO-NAPHTHALENE * DANSYL * DANSYL CHLORIDE * DIMETHYLAMINONAPHTHALENESULFONYL CHLORIDE * 1-DIMETHYLAMINONAPHTHALENE-5-SULFONYL CHLORIDE * 1-(DIMETHYLAMINO)-5-NAPHTHALENESULFONYLCHLORIDE * 5-DIMETHYLAMINO-NAPHTHYL-5-SULFONYL CHLORIDE

CONSENSUS REPORTS: Reported in EPA TSCA Inventory.

SAFETY PROFILE: Poison by intravenous route. When heated to decomposition it emits very toxic fumes of Cl^- and NO_x.

DPQ200   CAS: 33804-48-7   HR: 3
4-((p-(DIMETHYLAMINO)PHENYL)AZO)-N-METHYLACETANILIDE
mf: $C_{17}H_{20}N_4O$ mw: 296.41

SYNS: N'-ACETYL-N'-METHYL-4'-AMINO-N,N-DIMETHYL-4-AMINOAZOBENZENE * 4-(N-ACETYL-N-METHYL)AMINO-4'-(N',N'-DIMETHYLAMINO)AZOBENZENE * N',N'-DIMETHYL-4'-AMINO-N-ACETYL-N-MONOMETHYL-4-AMINOAZOBENZENE * N-(4-((4-(DIMETHYLAMINO)PHENYL)AZO)PHENYL)-N-METHYL-ACETAMIDE

SAFETY PROFILE: Poison by intraperitoneal route. Questionable carcinogen with experimental tumorigenic data. Mutation data reported. When heated to decomposition it emits toxic fumes of NO_x.

DPR000   CAS: 30041-69-1   HR: 3
6-((p-(DIMETHYLAMINO)PHENYL)AZO)QUINOLINE
mf: $C_{17}H_{16}N_4$ mw: 276.37

SYNS: N,N-DIMETHYL-4-(6'-QUINOLYLAZO)ANILINE * QUINOLINE-6-AZO-p-DIMETHYLANILINE

CONSENSUS REPORTS: EPA Genetic Toxicology Program.

SAFETY PROFILE: Questionable carcinogen with experimental tumorigenic data. Mutation data reported. When heated to decomposition it emits toxic fumes of NO_x.

DPS200   CAS: 24220-18-6   HR: 3
2-(p-DIMETHYLAMINOPHENYL)-1,6-DIMETHYLQUINOLINIUM CHLORIDE
mf: $C_{19}H_{21}N_2 \cdot Cl$ mw: 312.87

SAFETY PROFILE: Poison by ingestion and intraperitoneal routes. When heated to decomposition it emits very toxic fumes of NO_x and Cl^-.

DPU000   CAS: 1738-25-6   HR: 3
3-(DIMETHYLAMINO)PROPIONITRILE
mf: $C_5H_{10}N_2$ mw: 98.17

PROP: Liquid. Mp: −43°, bp: 170°, d: 0.8617, vap d: 3.35. flash p: <71.6°F.

SYN: β-DIMETHYLAMINOPROPIONITRILE

CONSENSUS REPORTS: Reported in EPA TSCA Inventory. Cyanide and its compounds are on the Community Right-To-Know List.

SAFETY PROFILE: Poison by intravenous route. Moderately toxic by ingestion and skin contact. A dangerous fire hazard when exposed to heat, flame, or oxidizers; can react with oxidizing materials. To fight fire, use foam, CO_2, dry chemical. When heated to decomposition it emits highly toxic fumes of NO_x and CN^-.

DPW600   CAS: 303-54-8   **HR: 3**
5-(3-(DIMETHYLAMINO)PROPYL)-5H-DIBENZ(b,f)AZEPINE
mf: $C_{19}H_{22}N_2$ mw: 278.43

SAFETY PROFILE: Poison by intravenous route. When heated to decomposition it emits toxic fumes of NO_x.

DPY600   CAS: 51003-81-7   **HR: 3**
5-DIMETHYLAMINO-6-PROPYL-5H-INDENO(5,6-d)-1,3-DIOXOLE HYDROCHLORIDE
mf: $C_{15}H_{19}NO_2 \cdot ClH$ mw: 281.81

SYNS: pr-MDI * 2-PROPYL-3-DIMETHYLAMINO-5,6-METHYLENEDIOXYINDENE HYDROCHLORIDE
* 2-N-PROPYL-3-DIMETHYLAMINO-5,6-METHYLENEDI-OXYINDENE HYDROCHLORIDE

SAFETY PROFILE: Poison by intraperitoneal and intravenous routes. When heated to decomposition it emits very toxic fumes of NO_x and HCl.

DQA400   CAS: 60-87-7   **HR: 3**
10-(2-(DIMETHYLAMINO)PROPYL)PHE-NOTHIAZINE
mf: $C_{17}H_{20}N_2S$ mw: 284.45

SYNS: A-91033 * APROBIT * ATOSIL
* AVOMINE * DIMAPP * DIMETHYLAMINO-ISO-PROPYL-PHENTHIAZIN (GERMAN) * (2-DIMETHYL-AMINO-2-METHYL)ETHYL-N-DIBENZOPARATHIAZINE
* N-(2'-DIMETHYLAMINO-2'-METHYL)ETHYLPHENO-THIAZINE * 10-(2-(DIMETHYLAMINO)-2-METHYL-ETHYL)PHENOTHIAZINE * N-DIMETHYLAMINO-2-METHYLETHYL THIODIPHENYLAMINE * (DIMETHYL-AMINO-2-PROPYL-10-PHENOTHIAZINE HYDROCHLO-RIDE (FRENCH) * DIPRAZINE * DIPROZIN
* FARGAN * FENAZIL * FENERGAN * FENE-TAZINA * HIBERNA * HISTARGAN * IERGIGAN
* ISOPHENERGAN * ISOPROMETHAZINE
* LERCIGAN * LERGIGAN * LILLY 1516
* LILLY 01516 * NCI-C60673 * PHARGAN
* PHENERGAN * PHENSEDYL * PILPOPHEN
* PIPOLPHEN * PROAZAIMINE * PROAZAMINE
* PROCIT * PROMAZINAMIDE * PROMETASIN
* PROMETAZIN * PROMETHIAZINE * PROMEZA-THINE * PROREX * PROTAZINE * PROTHAZIN
* PROVIGAN * PYRETHIA * PYRETHIAZINE
* ROMERGAN * 3277 RP * 3389 R.P. * 4182 R.P. * SKF 1498 * SYNALGOS * TANIDIL
* THIERGAN * VALLERGINE * WY 509

SAFETY PROFILE: Poison by ingestion, intravenous, intramuscular, intraperitoneal, and subcutaneous routes. Unspecified human reproductive effects. Human mutation data reported. A severe eye irritant. When heated to decomposition it emits very toxic fumes of NO_x and SO_x.

DQA600   CAS: 58-40-2   **HR: 3**
10-(3-(DIMETHYLAMINO)PROPYL)PHE-NOTHIAZINE
mf: $C_{17}H_{20}N_2S$ mw: 284.45

SYNS: AMPAZINE * N,N-DIMETHYL-10H-PHENO-THIAZINE-10-PROPANAMINE * ESPARIN * LIRA-NOL * NEO-HIBERNEX * PROMAZINE * PRO-TACTYL * SPARINE * VEROPHEN * WY 1094

CONSENSUS REPORTS: EPA Genetic Toxicology Program.

SAFETY PROFILE: Human poison by unspecified route. An experimental poison by ingestion, subcutaneous, intravenous, and intraperitoneal routes. Unspecified human reproductive effects. Experimental reproductive effects. When heated to decomposition it emits very toxic fumes of NO_x and SO_x.

DQB800   CAS: 5585-67-1   **HR: 3**
2-(DIMETHYLAMINO) RESERPILINATE
mf: $C_{26}H_{35}N_3O_5$ mw: 469.64

SYNS: ANTIPRESSINE DIHYDROCHLORIDE
* 2-(DIMETHYLAMINO) RESERPILIN-24-OIC ACID ETHYL ESTER

SAFETY PROFILE: Poison by intraperitoneal route. Moderately toxic by ingestion and subcutaneous routes. When heated to decomposition it emits toxic fumes of NO_x.

DQD400   CAS: 1596-84-5   **HR: 3**
DIMETHYLAMINOSUCCINAMIC ACID
mf: $C_6H_{12}N_2O_3$ mw: 160.20

SYNS: ALAR * ALAR-85 * AMINOZIDE
* B 995 * BERNSTEINSAEURE-2,2-DIMETHYLHY-DRAZID (GERMAN) * B-NINE * BUTANEDIOIC ACID MONO(2,2-DIMETHYLHYDRAZIDE) * DAMINO-ZIDE (USDA) * DIMAS * N-DIMETHYL AMINO-β-CARBAMYL PROPIONIC ACID * N-(DIMETHYL-AMINO)SUCCINAMIC ACID * N-DIMETHYLAMINO-SUCCINAMIDSAEURE (GERMAN) * DMASA
* DMSA * KYLAR * NCI-C03827 * SADH
* SUCCINIC ACID-2,2-DIMETHYLHYDRAZIDE
* SUCCINIC-1,1-DIMETHYL HYDRAZIDE

CONSENSUS REPORTS: EPA Genetic Toxicology Program. NCI Carcinogenesis Bioassay (feed); Clear Evidence: mouse, rat NCITR* NCI-CG-TR-83,78.

SAFETY PROFILE: Suspected carcinogen with experimental carcinogenic, tumorigenic, and teratogenic data. Poison by intraperitoneal route. Moderately toxic by ingestion and possibly other routes. When heated to decomposition it emits toxic fumes of NO_x.

DQE800   CAS: 14144-91-3   **HR: 3**
5-DIMETHYLAMINO-4-TOLYL
METHYLCARBAMATE
mf: $C_{11}H_{16}N_2O_2$ mw: 208.29

SYNS: BAY 42696 * 4-METHYL-3-DIMETHYLAMINO-PHENYL ESTER-N-METHYLCARBAMIC ACID

DOT Classification: Oxidizer; Label: Oxidizer.

SAFETY PROFILE: Poison by ingestion and subcutaneous routes. A strong oxidizing agent. When heated to decomposition it emits toxic fumes of NO_x.

DQF800   CAS: 121-69-7   **HR: 3**
N,N-DIMETHYLANILINE
DOT: UN 2253
mf: $C_8H_{11}N$ mw: 121.20

PROP: Liquid. Mp: 2.5°, bp: 193.1°, flash p: 145°F (CC), d: 0.9557 @ 20°/4°, ULC: 20-25, autoign temp: 700°F, vap press: 1 mm @ 29.5°, vap d: 4.17.

SYNS: (DIMETHYLAMINO)BENZENE * N,N-DI-METHYLBENZENEAMINE * DIMETHYLPHENYLAMINE * N,N-DIMETHYLPHENYLAMINE * DWUMETYLO-ANILINA (POLISH) * NCI-C56428 * VERSNELLER NL 63/10

CONSENSUS REPORTS: Reported in EPA TSCA Inventory. Community Right-To-Know List.

OSHA PEL: (Transitional: TWA 5 ppm (skin))
    TWA 5 ppm; STEL 10 ppm (skin)
ACGIH TLV: TWA 5 ppm; STEL 10 ppm (skin)
DFG MAK: 5 ppm (25 mg/m³)
DOT Classification: Poison B; Label: Poison.

SAFETY PROFILE: Human poison by ingestion. Moderately toxic by inhalation and skin contact. A skin irritant. Physiological action is similar to, but less toxic than aniline. A central nervous system depressant. Flammable when exposed to heat, flame, or oxidizers. Explodes on contact with benzoyl peroxide or diisopropyl peroxydicarbonate. To fight fire, use foam, CO_2, dry chemical. When heated to decomposition it emits highly toxic fumes of aniline and NO_x.

DQG200   CAS: 781-43-1   **HR: 3**
9,10-DIMETHYLANTHRACENE
mf: $C_{16}H_{14}$ mw: 206.30

CONSENSUS REPORTS: EPA Genetic Toxicology Program.

SAFETY PROFILE: Questionable carcinogen with experimental carcinogenic and tumorigenic data. Mutation data reported. When heated to decomposition it emits acrid smoke and irritating fumes.

DQG600   CAS: 593-57-7   **HR: 3**
DIMETHYLARSINE
mf: C_2H_7As mw: 106.07

PROP: Colorless liquid. Bp: 36°, d: 1.213 @ 29°/4°, vap d: 3.65.

SYN: CACODYL HYDRIDE

CONSENSUS REPORTS: Arsenic and its compounds are on the Community Right-To-Know List.

SAFETY PROFILE: Arsenic compounds are generally poisons. Ignites spontaneously in air. It is more toxic than its oxidation products; reacts vigorously with oxidizing agents. To fight fire, exclude O_2, allow fire to burn, or apply water, foam, dry chemical, water spray, or CO_2. When heated to decomposition it emits toxic fumes of As.

DQI200   CAS: 963-89-3   **HR: 3**
7,9-DIMETHYLBENZ(c)ACRIDINE
mf: $C_{19}H_{15}N$ mw: 257.35

SYN: 3,10-DIMETHYL-7,8-BENZACRIDINE (FRENCH)

CONSENSUS REPORTS: EPA Genetic Toxicology Program.

SAFETY PROFILE: Questionable carcinogen with experimental tumorigenic data. Mutation data reported. When heated to decompositions it emits toxic fumes of NO_x.

DQI600   CAS: 53-69-0   **HR: 3**
5,7-DIMETHYL-1,2-BENZACRIDINE
mf: $C_{19}H_{15}N$ mw: 257.35

SYN: 8,10-DIMETHYL-BENZ(a)ACRIDINE

SAFETY PROFILE: Poison by intrarenal route. Mutation data reported. When heated to decomposition it emits toxic fumes of NO_x.

DQI800   CAS: 2381-40-0   **HR: 3**
6,9-DIMETHYL-1,2-BENZACRIDINE
mf: $C_{19}H_{15}N$ mw: 257.35

SYNS: 7,10-DIMETHYLBENZ(c)ACRIDINE * 2,10-DI-METHYL-7,8-BENZACRIDINE (FRENCH)

CONSENSUS REPORTS: EPA Genetic Toxicology Program.

SAFETY PROFILE: Questionable carcinogen with experimental tumorigenic data. Mutation data reported. When heated to decomposition it emits toxic fumes of NO_x.

DQJ200   CAS: 57-97-6   **HR: 3**
DIMETHYLBENZANTHRACENE
mf: $C_{20}H_{16}$ mw: 256.36

SYNS: DBA * DIMETHYLBENZ(a)ANTHRACENE * 7,12-DIMETHYLBENZANTHRACENE * 7,12-DI-METHYLBENZ(a)ANTHRACENE * 9,10-DIMETHYL-BENZANTHRACENE * 9,10-DIMETHYLBENZ(a)AN-THRACENE * 9,10-DIMETHYL-1,2-BENZANTHRA-CENE * 9,10-DIMETHYL-1,2-BENZANTHRAZEN (GERMAN) * DIMETHYLBENZANTHRENE * 7,12-DI-METHYLBENZO(a)ANTHRACENE * 1,4-DIMETHYL-2,3-BENZPHENANTHRENE * DMBA * 7,12-DMBA * NCI-C03918 * RCRA WASTE NUMBER U094

CONSENSUS REPORTS: Reported in EPA TSCA Inventory. EPA Genetic Toxicology Program.

SAFETY PROFILE: Suspected carcinogen with experimental carcinogenic, neoplastigenic, and tumorigenic data. A transplacental carcinogen. Poison by ingestion, intravenous, subcutaneous, intraperitoneal, and intratracheal routes. Experimental teratogenic and reproductive effects. Human mutation data reported. A skin irritant. When heated to decomposition it emits acrid smoke and irritating fumes.

DQL200   CAS: 58429-99-5   **HR: 3**
6,7-DIMETHYL-1,2-BENZANTHRACENE
mf: $C_{20}H_{16}$ mw: 256.36

SYN: 9,10-DIMETHYLBENZ(a)ANTHRACENE

SAFETY PROFILE: Questionable carcinogen with experimental tumorigenic data. Mutation

data reported. When heated to decomposition it emits acrid smoke and irritating fumes.

DQM600   CAS: 22781-23-3   **HR: 3**
2,2-DIMETHYL-1,3-BENZODIOXOL-4-OL METHYLCARBAMATE
mf: $C_{11}H_{13}NO_4$ mw: 223.25

SYNS: BENCARBATE * BENDIOCARB * BICAM ULV * 2,2-DIMETHYL-1,3-BENZDIOXOL-4-YL-N-METHYLCARBAMATE * 2,2-DIMETHYLBENZO-1,3-DI-OXOL-4-YL METHYLCARBAMATE * 2,2-DIMETHYL-4-(N-METHYLAMINOCARBOXYLATO)-1,3-BENZODIOX-OLE * 2,2-DIMETHYL-4-(N-METHYLCARBAMATO)-1,3-BENZODIOXOLE * DYCARB * FICAM * GAR-VOX * 2,3-ISOPROPYLIDENEDIOXYPHENYL METHYL-CARBAMATE * MC6897 * METHYLCARBAMIC ACID-2,3-(ISOPROPYLIDENEDIOXY)PHENYL ESTER * MULTAMAT * NIOMIL * ROTATE * TAT-TOO * TURCAM

CONSENSUS REPORTS: EPA Genetic Toxicology Program.

SAFETY PROFILE: Poison by ingestion. Moderately toxic by skin contact. When heated to decomposition it emits toxic fumes of NO_x.

DQP800   CAS: 103-83-3   **HR: 3**
N,N-DIMETHYLBENZYLAMINE
DOT: UN 2619
mf: $C_9H_{13}N$ mw: 135.23

SYNS: ARALDITE ACCELERATOR 062 * BDMA * BENZYLDIMETHYLAMINE * BENZYL-N,N-DI-METHYLAMINE * N-BENZYLDIMETHYLAMINE * N,N-DIMETHYLBENZENEMETHANAMINE * N-(PHENYLMETHYL)DIMETHYLAMINE * SUMINE 2015

CONSENSUS REPORTS: Reported in EPA TSCA Inventory.

DOT Classification: Corrosive Material; Label: Corrosive, Flammable Liquid.

SAFETY PROFILE: Poison by ingestion. Moderately toxic by inhalation and skin contact. Corrosive. A severe eye and skin irritant. Flammable when exposed to heat of flame. When heated to decomposition it emits toxic fumes of NO_x.

DQQ000   CAS: 1875-92-9   **HR: 3**
DIMETHYLBENZYLAMINE HYDROCHLORIDE
mf: $C_9H_{13}N \cdot ClH$ mw: 171.69

SYNS: DIMETHYLBENZYLAMMONIUM CHLORIDE
* USAF EL-78

SAFETY PROFILE: Poison by intraperitoneal route. When heated to decomposition it emits toxic fumes of NO_x, NH_3 and HCl.

DQQ200   CAS: 100-86-7   HR: 2
DIMETHYL BENZYL CARBINOL
mf: $C_{10}H_{14}O$ mw: 150.24

PROP: White crystalline solid; floral odor. D: 0.972-0.977, flash p: 198°F. Sol in fixed oils mineral oil, propylene glycol; insol in glycerin.

SYNS: BENZYL DIMETHYL CARBINOL * α,α-DI-METHYLPHENETHYL ALCOHOL * 1,1-DIMETHYL-2-PHENYLETHANOL * DMBC * FEMA No. 2393

CONSENSUS REPORTS: Reported in EPA TSCA Inventory.

SAFETY PROFILE: Moderately toxic by ingestion. Combustible liquid. When heated to decomposition it emits acrid smoke and irritating fumes.

DQQ375   HR: 1
DIMETHYL BENZYL CARBINYL ACETATE
mf: $C_{12}H_{16}O_2$ mw: 192.26

PROP: Colorless liquid to solid at room temp; floral, fruity odor. D: 0.995-1.002, refr index: 1.490-1.495, flash p: +212°F. Sol in fixed oils; sltly sol in propylene glycol; insol in water.

SYNS: α,α-DIMETHYLPHENETHYL ACETATE
* FEMA No. 2392

SAFETY PROFILE: Combustible liquid. When heated to decomposition it emits acrid smoke and irritating fumes.

DQQ380   HR: 2
DIMETHYL BENZYL CARBINYL BUTYRATE
mf: $C_{14}H_{20}O_2$ mw: 220.31

PROP: Colorless liquid; prunelike odor. D: 0.960-0.981, refr index: 1.473-1.493 @ 25°, flash p: +151°F. Sol in alc, fixed oils; insol in water, propylene glycol.

SYNS: α,α-DIMETHYLPHENRTHYL BUTYRATE
* FEMA No. 2394

SAFETY PROFILE: Combustible liquid. Use in accordance with good manufacturing practice.

DQR200   CAS: 506-63-8   HR: 3
DIMETHYL BERYLLIUM
mf: C_2H_6Be mw: 39.09

PROP: White needles. Bp: sublimes @ 200°.

CONSENSUS REPORTS: Beryllium and its compounds are on the Community Right-To-Know List.

OSHA PEL: (Transitional: TWA 0.002 mg(Be)/m^3; CL 0.005; Pk 0.025/30M/8H) TWA 0.002 mg(Be)/m^3; STEL 0.005 mg(Be)/m^3/30M; CL 0.025 mg(Be)/m^3
ACGIH TLV: TWA 0.002 mg/m^3, Suspected Human Carcinogen.

SAFETY PROFILE: Confirmed human carcinogen. A poison. Flammable when exposed to heat or flame; can react with oxidizing materials. Explosive reaction on contact with water. Ignites on contact with moist air or carbon dioxide. Upon decomposition it emits highly toxic fumes of BeO.

DQR289   HR: 3
DIMETHYLBERYLLIUM-1,2-DIMETHOXYETHANE

CONSENSUS REPORTS: Beryllium and its compounds are on the Community Right-To-Know List.

SAFETY PROFILE: Confirmed human carcinogen. Ignites spontaneously in air. Upon decomposition it emits highly toxic fumes of BeO.

DQR600   CAS: 657-24-9   HR: 3
1,1-DIMETHYLBIGUANIDE
mf: $C_4H_{11}N_5$ mw: 129.20

SYNS: N,N-DIMETHYLBIGUANIDE * N,N-DIMETH-YLDIGUANIDE * FLUMAMINE * GLUCOPHAGE
* GLUCOPHAGE LA 6023 * GLUEOPHOGE
* LA 6023 * MELBIN * METFORMIN * NNDG

SAFETY PROFILE: Poison by subcutaneous and intraperitoneal routes. Mildly toxic by parenteral route. Experimental teratogenic and reproductive effects. Mutation data reported. When heated to decomposition it emits toxic fumes of NO_x.

DQT200   CAS: 75-83-2   HR: 2
2,2-DIMETHYLBUTANE
DOT: UN 1208
mf: C_6H_{14} mw: 86.20

PROP: Liquid. Bp: 49.7°, mp: −98.2°, flash p: −54°F, fp: −101.9°, d: 0.649, autoign temp: 797°F, vap press: 400 mm @ 31.0°, vap d: 3.00, lel: 1.2%, uel: 7.0%.

SYN: NEOHEXANE (DOT)

CONSENSUS REPORTS: Reported in EPA TSCA Inventory.

OSHA PEL: TWA 500 ppm; STEL 1000 ppm
ACGIH TLV: TWA 500 ppm; STEL 1000 ppm
NIOSH REL: (Alkanes) TWA 350 mg/m^3
DOT Classification: Flammable Liquid; Label: Flammable Liquid.

SAFETY PROFILE: Probably an irritant and narcotic in high concentration. A very dangerous fire and explosion hazard when exposed to heat or flame; can react vigorously with oxidizing materials. Keep away from heat or open flame. To fight fire, use foam, CO_2, dry chemical. When heated to decomposition it emits acrid smoke and irritating fumes.

DQT400 CAS: 79-29-8 **HR: 2**
2,3-DIMETHYLBUTANE

DOT: UN 2457
mf: C_6H_{14} mw: 86.20

PROP: Liquid. Mp: −135°, bp: 58.0°, flash p: −20°F, d: 0.662 @ 20°/4°, autoign temp: 788°F, vap press: 400 mm @ 39.0°, vap d: 3.0, lel: 1.2%, uel: 7.0%.

SYN: ISOHEXANE

CONSENSUS REPORTS: Reported in EPA TSCA Inventory.

OSHA PEL: TWA 500 ppm; STEL 1000 ppm
ACGIH TLV: TWA 500 ppm; STEL 1000 ppm
NIOSH REL: TWA (Alkanes) 350 mg/m^3
DOT Classification: Flammable Liquid; Label: Flammable Liquid.

SAFETY PROFILE: Probably an irritant and narcotic in high concentration. A very dangerous fire and explosion hazard when exposed to heat or flame; can react vigorously with oxidizing materials. Keep away from heat and open flame. To fight fire, use foam, CO_2, dry chemical. When heated to decomposition it emits acrid smoke and irritating fumes.

DQU200 CAS: 3625-18-1 **HR: 3**
5-(1,3-DIMETHYL-2-BUTENYL)-5-ETHYL BARBITURIC ACID

mf: $C_{12}H_{18}N_2O_3$ mw: 238.32

SYNS: 5-(1,3-DIMETHYL-2-BUTENYL)-5-ETHYL-2,4,-6(1H,3H,5H)PYRIMIDINETRIONE * MCNEIL 481

SAFETY PROFILE: Poison by ingestion, intravenous, and intraperitoneal routes. When heated to decomposition it emits toxic fumes of NO_x.

DQU600 CAS: 108-09-8 **HR: 3**
1,3-DIMETHYL BUTYLAMINE

DOT: UN 2379
mf: $C_6H_{15}N$ mw: 101.22

PROP: A liquid. Bp: 106-109°, flash p: 55°F (OC), d: 0.750 @ 20°/20°.

CONSENSUS REPORTS: Reported in EPA TSCA Inventory.

DOT Classification: Flammable Liquid; Label: Flammable Liquid

SAFETY PROFILE: Poison by intravenous route. Moderately toxic by ingestion and skin contact. Mildly toxic by inhalation. A dangerous fire and explosion hazard when exposed to heat or flame; can react vigorously with oxidizing materials. To fight fire, use foam, CO_2, dry chemical. When heated to decomposition it emits toxic fumes of NO_x.

DQW800 CAS: 506-82-1 **HR: 3**
DIMETHYLCADMIUM

mf: C_2H_6Cd mw: 142.47

PROP: Oil, decomp by water, foul odor. D: 1.984; mp: −4.5°; bp: 106°.

CONSENSUS REPORTS: Cadmium and its compounds are on Community Right-To-Know List.

OSHA PEL: TWA 0.1 mg(Cd)/m^3; CL 0.6 mg(Cd)/m^3 (fume)
ACGIH TLV: TWA 0.05 mg(Cd)/m^3 (Proposed: TWA 0.01 mg(Cd)/m^3 (dust), Human Carcinogen); BEI: 10 μg/g creatinine in urine; 10 μg/L in blood.
DFG BAT: Blood 1.5 μg/dL; Urine 15 μg/dL, Suspected Carcinogen.
NIOSH REL: (Cadmium) Reduce to lowest feasible level

SAFETY PROFILE: Confirmed human carcinogen. Contact with air produces the friction-sensitive explosive dimethyl cadmium peroxide. Explodes when heated above 150°C. Ignition may occur on contact with air if the surface area is large.

DQY400　　CAS: 63680-76-2　　HR: 3
DIMETHYLCARBAMIC ESTER of 8-OXYMETHYLQUINOLINIUM METHYLSULFATE
mf: $C_{13}H_{15}N_2O_2 \cdot CH_3O_4S$　　mw: 342.40

SYNS: N,N-DIMETHYLCARBAMIC ACID-8-QUINOLINYL ESTER METHOSULFATE ＊ 8-HYDROXY-1-METHYL-QUINOLINIUM METHYLSULFATE DIMETHYLCARBA-MATE

SAFETY PROFILE: Poison by ingestion and intravenous routes. When heated to decomposition it emits very toxic fumes of NO_x and SO_x.

DQY800　　CAS: 114-80-7　　HR: 3
3-DIMETHYLCARBAMOXYPHENYL TRIMETHYL AMMONIUM BROMIDE
mf: $C_{12}H_{19}N_2O_2 \cdot Br$　　mw: 303.24

SYNS: BENZENAMINIUM, 3-(((DIMETHYLAMINO)CAR-BONYL)OXY)-N,N,N-TRIMETHYL-, BROMIDE (9CI) ＊ CARBAMIC ACID, DIMETHYL-, ester with (m-HY-DROXYPHENYL)TRIMETHYLAMMONIUM BROMIDE ＊ PROSERINE ＊ EUSTIGMIN BROMIDE ＊ (m-HY-DROXYPHENYL)TRIMETHYLAMMONIUM BROMIDE DI-METHYLCARBAMATE ＊ 3-HYDROXYPHENYLTRI-METHYLAMMONIUM BROMIDE DIMETHYLCARBAMIC ESTER ＊ KIRKSTIGMINE BROMIDE ＊ LEOSTIG-MINE BROMIDE ＊ NEOESERINE BROMIDE ＊ NEO-SERINE BROMIDE ＊ NEOSTIGMINE BROMIDE ＊ NEOSTIGMINE METHYL BROMIDE ＊ PHILO-STIGMIN BROMIDE ＊ PROSERINE BROMIDE ＊ PROSTIGMIN BROMIDE ＊ PROSTIGMINE BROMIDE ＊ RCRA WASTE NUMBER U053 ＊ STIGMANOL BROMIDE ＊ STIGMOSAN BROMIDE ＊ SYNSTIGMIN BROMIDE ＊ SYNTHOSTIGMINE BROMIDE ＊ SYNTOSTIGMIN (tablet) ＊ SYNTOSTIG-MIN BROMIDE ＊ SYNTOSTIGMINE BROMIDE ＊ VAGOSTIGMINE BROMIDE

SAFETY PROFILE: Poison by ingestion, subcutaneous, intravenous, and intraperitoneal routes. When heated to decomposition it emits very toxic fumes of Br^-, NH_3, and NO_x.

DQY909　　CAS: 51-60-5　　HR: 3
3-(DIMETHYLCARBAMOXY)PHENYL TRIMETHYLAMMONIUM METHYL SULFATE
mf: $C_{12}H_{19}N_2O_2 \cdot CH_3O_4S$　　mw: 334.43

SYNS: AR-32 ＊ N,N-DIMETHYLCARBAMIC ACID-3-DIMETHYLAMINOPHENYL ESTER METHOSULFATE

＊ DIMETHYLCARBAMIC ACID ESTER with (m-HY-DROXYPHENYL)TRIMETHYLAMMONIUM METHYL SUL-FATE ＊ N,N-DIMETHYLCARBAMIC ACID-3-(TRI-METHYLAMMONIO)PHENYL ESTER METHYLSULFATE ＊ DIMETHYLCARBAMIC ESTER of 3-OXYPHENYLTRI-METHYLAMMONIUM METHYLSULFATE ＊ (3-(DI-METHYLCARBAMOYLOXY)PHENYL)TRIMETHYLAM-MONIUM METHYLSULFATE ＊ EUSTIGMIN METHYL-SULFATE ＊ HODOSTIN ＊ (m-HYDROXYPHENYL)TRIMETHYLAMMONIUM METHYL SULFATE DIMETH-YLCARBAMATE ＊ (3-HYDROXYPHENYL)TRIMETHYL-AMMONIUM METHYL SULFATE DIMETHYLCARBAMIC ESTER ＊ KIRKSTIGMINE METHYL SULFATE ＊ LEOSTIGMINE METHYL SULFATE ＊ NEOESER-INE METHYL SULFATE ＊ NEOSTIGMETH ＊ NEOSTIGMINE METHOSULFATE ＊ NEOSTIGMINE METHYL SULFATE ＊ NEOSTIGMINE MONOMETHYL-SULFATE ＊ NORMASTIGMIN ＊ PHILOSTIGMIN METHYL SULFATE ＊ POLSTIGMINE ＊ PROSERIN ＊ PROSERINE METHYL SULFATE ＊ PROSTIG-MINE METHYLSULFATE ＊ SB-23 ＊ STIGMANOL METHYL SULFATE ＊ STIGMOSAN METHYL SUL-FATE ＊ SYNTHOSTIGMINE METHYL SULFATE ＊ TL-1394 ＊ VAGOSTIGMINE METHYL SUL-FATE

SAFETY PROFILE: A deadly poison by ingestion, intravenous, subcutaneous, intraperitoneal, and intramuscular routes. Experimental reproductive effects. When heated to decomposition it emits very toxic fumes of SO_x, NH_3, and NO_x.

DQY950　　CAS: 79-44-7　　HR: 3
DIMETHYL CARBAMOYL CHLORIDE
DOT: UN 2262
mf: C_3H_6ClNO　　mw: 107.55

PROP: Liquid. Mp: $-33°$, bp: 165-167°, d: 1.678 @ 20°/4°, vap d: 3.73.

SYNS: CHLOROFORMIC ACID DIMETHYLAMIDE ＊ DDC ＊ (DIMETHYLAMINO)CARBONYL CHLORIDE ＊ DIMETHYLCARBAMIC ACID CHLORIDE ＊ DI-METHYLCARBAMIC CHLORIDE ＊ DIMETHYLCAR-BAMIDOYL CHLORIDE ＊ DIMETHYLCARBAMOYL CHLORIDE ＊ N,N-DIMETHYLCARBAMOYL CHLO-RIDE (DOT) ＊ DIMETHYLCARBAMYL CHLORIDE ＊ N,N-DIMETHYLCARBAMYL CHLORIDE ＊ DMCC ＊ RCRA WASTE NUMBER U097 ＊ TL 389

CONSENSUS REPORTS: IARC Cancer Review: GROUP 2A IMEMDT 7,199,87; Animal Sufficient Evidence IMEMDT 12,77,76; Human Inadequate Evidence IMEMDT 12,77,76. NTP

Fourth Annual Report On Carcinogens, 1984. EPA Genetic Toxicology Program. Community Right-To-Know List. Reported in EPA TSCA Inventory.

ACGIH TLV: Suspected Human Carcinogen
DFG MAK: Animal Carcinogen, Suspected Human Carcinogen.
DOT Classification: Corrosive Material; Label: Corrosive.

SAFETY PROFILE: Confirmed carcinogen with experimental carcinogenic, neoplastigenic, and tumorigenic data. Poison by intraperitoneal route. Moderately toxic by inhalation and ingestion. Human mutation data reported. Can cause skin and papillary tumors by skin contact, and squamous cell carcinoma by inhalation. Will react with water or steam to produce toxic and corrosive fumes. A powerful lachrymator. When heated to decomposition it emits very toxic fumes of Cl^- and NO_x.

DRC000   CAS: 4584-46-7   **HR: 3**
DIMETHYL(2-CHLOROETHYL)AMINE HYDROCHLORIDE
mf: $C_4H_{10}ClN \cdot ClH$ mw: 144.06

SYNS: 2-CHLORO-N,N-DIMETHYLETHYLAMINE HYDROCHLORIDE * DIMETHYL-β-CHLOROETHYLAMINE HYDROCHLORIDE

CONSENSUS REPORTS: Reported in EPA TSCA Inventory.

SAFETY PROFILE: Questionable carcinogen with experimental neoplastigenic data. Poison by intraperitoneal and subcutaneous routes. Mutation data reported. When heated to decomposition it emits very toxic fumes of Cl^- and NO_x.

DRE400   CAS: 52171-92-3   **HR: 3**
1,11-DIMETHYLCHRYSENE
mf: $C_{20}H_{16}$ mw: 256.36

SYN: 5,7-DIMETHYLCHRYSENE

CONSENSUS REPORTS: EPA Genetic Toxicology Program.

SAFETY PROFILE: Poison by intraperitoneal and intravenous routes. Questionable carcinogen with experimental neoplastigenic data. Mutation data reported. When heated to decomposition it emits acrid smoke and irritating fumes.

DRF600   CAS: 1467-79-4   **HR: 3**
DIMETHYLCYANAMIDE
mf: $C_3H_6N_2$ mw: 70.11

PROP: Colorless, mobile liquid. Mp: $-41.0°$, bp: $160°$, flash p: $160°F$ (TCC), d: 0.8767 @ $30°$, vap press: 40 mm @ $80°$, vap d: 2.55.

CONSENSUS REPORTS: Reported in EPA TSCA Inventory.

SAFETY PROFILE: Poison by ingestion, skin contact, and intraperitoneal routes. Moderately toxic by inhalation. Flammable when exposed to heat, flame, or oxidizers. Can react with oxidizing materials. To fight fire, use foam, CO_2, dry chemical. When heated to decomposition or in reaction with water or steam it produces toxic fumes of NO_x and CN^- and flammable vapors.

DRF709   CAS: 98-94-2   **HR: 3**
N,N-DIMETHYLCYCLOHEXANAMINE
DOT: UN 2264
mf: $C_8H_{17}N$ mw: 127.26

SYNS: CYCLOHEXYLDIMETHYLAMINE * N-CYCLOHEXYLDIMETHYLAMINE * (DIMETHYLAMINO)CYCLOHEXANE * N,N-DIMETHYLAMINOCYCLOHEXANE * DIMETHYLCYCLOHEXYLAMINE * N,N-DIMETHYLCYCLOHEXYLAMINE (DOT) * POLYCAT 8

CONSENSUS REPORTS: Reported in EPA TSCA Inventory.

DOT Classification: Corrosive Material; Label: Corrosive, Flammable Liquid.

SAFETY PROFILE: Poison by ingestion. Moderately toxic by inhalation. When heated to decomposition it emits toxic fumes of NO_x.

DRF800   CAS: 583-57-3   **HR: 2**
cis-1,2-DIMETHYLCYCLOHEXANE
DOT: UN 2263
mf: C_8H_{16} mw: 112.22

PROP: Flash p: $61.8°F$.

SYNS: o-DIMETHYLCYCLOHEXANE * 1,2-DIMETHYLCYCLOHEXANE (DOT)

DOT Classification: Flammable Liquid; Label: Flammable Liquid.

SAFETY PROFILE: A very dangerous fire hazard when exposed to heat, flame, or oxidizers. When heated to decomposition it emits acrid smoke and fumes.

DRG000   CAS: 591-21-9   **HR: 2**
1,3-DIMETHYLCYCLOHEXANE

DOT: UN 2263
mf: C_8H_{16} mw: 112.22

PROP: Flash p: 42.8°F.

SYN: m-DIMETHYLCYCLOHEXANE

DOT Classification: Flammable Liquid; Label: Flammable Liquid.

SAFETY PROFILE: A very dangerous fire hazard when exposed to heat, flame or oxidizers. When heated to decomposition it emits acrid smoke and fumes.

DRG200   CAS: 589-90-2   **HR: 3**
1,4-DIMETHYLCYCLOHEXANE

DOT: UN 2263
mf: C_8H_{16} mw: 112.24

PROP: Liquid. Mp: 86°, bp: 119.5°, flash p: 50°F (CC), d: 0.77, vap press: 10 mm @ 10.2°, vap d: 3.86.

DOT Classification: Flammable Liquid; Label: Flammable Liquid.

SAFETY PROFILE: Dangerous fire hazard when exposed to heat or flame; can react vigorously with oxidizing materials. Keep away from heat and open flame. To fight fire, use foam, CO_2, dry chemical.

DRJ800   CAS: 78-63-7   **HR: 2**
2,5-DIMETHYL-2,5-DI-(tert-BUTYLPEROXY)HEXANE

DOT: UN 2155
mf: $C_{16}H_{34}O_4$ mw: 290.50

PROP: Colorless to light yellow liquid. D: 0.85, fp: 8°, flash p: >180°F (MOC), bp: 250°. Insol in water; sol in many organic solvents.

SYNS: 2,5-DIMETHYL-2,5-DI-(tert-BUTYLPEROXY)HEX- ANE, technically pure (DOT)   *   TRIGONOX 101-101/45   *   VAROX

CONSENSUS REPORTS: Reported in EPA TSCA Inventory.

DOT Classification: Organic Peroxide; Label: Organic Peroxide.

SAFETY PROFILE: Moderately toxic by intraperitoneal route. Flammable when exposed to heat, flames, or reducing agents. To fight fire, use water spray, foam, dry chemical. When heated to decomposition it emits acrid smoke and irritating fumes. Used in the polymerization of styrene and in cross-linking of various grades of polyethylene.

DRK600   CAS: 19072-57-2   **HR: 3**
2,6-DIMETHYL-1,1-DIETHYLPIPERIDINIUM BROMIDE
mf: $C_{11}H_{24}BrN$ mw: 250.27

SYNS: AGILENE   *   SC-1950

SAFETY PROFILE: Poison by ingestion, intraperitoneal, and intravenous routes. When heated to decomposition it emits very toxic fumes of Br^- and NO_x.

DRL200   CAS: 122-15-6   **HR: 3**
5,5-DIMETHYLDIHYDRORESORCINOL DIMETHYLCARBAMATE
mf: $C_{11}H_{17}NO_3$ mw: 211.29

SYNS: DIMETAN   *   DIMETHYLCARBAMATE de 5,5- DIMETHYL DIHYDRORESORCINOL (FRENCH)   *   DI- METHYLCARBAMIC ACID ester with 3-HYDROXY-5,5-DI- METHYL-2-CYCLOHEXEN-1-ONE   *   5,5-DIMETHYL DIHYDRORESORCINOL-N,N-DIMETHYLCARBAMAT (GER- MAN)   *   5,5-DIMETHYL-4,5-DIHYDRO-3-RESOR- CYL-DIMETHYL-CARBAMAT (GERMAN)   *   (5,5-DI- METHYL-3-OXO-CYCLOHEX-1-EN-YL)-N,N-DIMETHYL- CARBAMAAT (DUTCH)   *   5,5-DIMETHYL-3-OXO-1-CY- CLOHEXEN-1-YL DIMETHYLCARBAMATE   *   5,5-DI- METHYL-3-OXOCYCLOHEX-1-ENYL DIMETHYLCARBA- MATE   *   (5,5-DIMETHYL-3-OXO-CYCLOHEX-1-EN-YL)- N,N-DIMETHYL-CARBAMAT (GERMAN)   *   (5,5-DI- METIL-3-OXO-CICLOES-1-EN-IL)-N,N-DIMETIL-CARBAM- MATO (ITALIAN)   *   ENT 24,738   *   GEIGY 19258   *   3-HYDROXY-5,5-DIMETHYL-2-CYCLOHEXEN-1-ONE DIMETHYLCARBAMATE

SAFETY PROFILE: Poison by ingestion and possibly other routes. When heated to decomposition it emits toxic fumes of NO_x.

DRL600   CAS: 38035-28-8   **HR: 3**
2,3-DIMETHYL-8-(DIMETHYLAMINOMETHYL)-7-METHOXYCHROMONE HYDROCHLORIDE
mf: $C_{15}H_{19}NO_3 \cdot ClH$ mw: 297.81

SAFETY PROFILE: Poison by ingestion, intraperitoneal, and subcutaneous routes. When heated to decomposition it emits very toxic fumes of NO_x and HCl.

DRM000 CAS: 3759-07-7 **HR: 3**
9,9-DIMETHYL-10-DIMETHYLAMINO-PROPYLACRIDAN HYDROGEN TARTRATE
mf: $C_{20}H_{26}N_2 \cdot C_4H_4O_6$ mw: 442.56

SYNS: DIMETACRINE BITARTRATE * DIMETACRIN HYDROGENTARTRATE * DIMETHACRINE TARTRATE * 10-(3-(DIMETHYLAMINO)PROPYL)-9,9-DIMETHYL-ACRIDAN TARTRATE (1:1) * 9,9-DIMETHYL-10-(3-DIMETHYLAMINO)PROPYLACRIDINE TARTRATE * ISOTONIL * ISTONYL * MIROISTONIL * MO 709 * SD 709 * (R-R*,R*))-N,N,9,9-TETRAMETHYL-10(9H)-ACRIDINEPROPANAMINE-2,3-DIHYDROXYBUTANEDIOATE (1:1)

SAFETY PROFILE: Poison by ingestion, intravenous, and intraperitoneal routes. Moderately toxic by subcutaneous route. Experimental teratogenic and reproductive effects. When heated to decomposition it emits toxic fumes of NO_x.

DRN800 CAS: 55556-88-2 **HR: 3**
2,5-DIMETHYLDINITROSOPIPERAZINE
mf: $C_6H_{14}N_4O_2$ mw: 174.24

PROP: Mixture approximately 25% cis and 75% trans conformers.

SYNS: 2,5-DIMETHYL-1,4-DINITROSOPIPERAZINE * 2,5-DIMETHYL-DNPZ * DINITROSO-2,5-DIMETHYL-PIPERAZINE

CONSENSUS REPORTS: EPA Genetic Toxicology Program.

SAFETY PROFILE: Questionable carcinogen with experimental tumorigenic data. Many N-nitroso compounds are carcinogens. Mutation data reported. When heated to decomposition it emits toxic fumes of NO_x.

DRO000 CAS: 55380-34-2 **HR: 3**
2,6-DIMETHYLDINITROSOPIPERAZINE
mf: $C_6H_{14}N_4O_2$ mw: 174.24

SYNS: 2,6-DIMETHYL-DNPZ * DINITROSO-2,6-DIMETHYLPIPERAZINE * N,N'-DINITROSO-2,6-DIMETHYLPIPERAZINE * 1,4-DINITROSO-2,6-DIMETHYLPIPERAZINE * DNDMP

CONSENSUS REPORTS: EPA Genetic Toxicology Program.

SAFETY PROFILE: Questionable carcinogen with experimental carcinogenic, neoplastigenic, and tumorigenic data. Mutation data reported. A model carcinogen and carcinogenic metabolite. When heated to decomposition it emits toxic fumes of NO_x.

DRO200 CAS: 6972-76-5 **HR: 3**
N,N'-DIMETHYL-N,N'-DINITROSO-1,3-PROPANEDIAMINE
mf: $C_5H_{12}N_4O_2$ mw: 160.21

SYNS: DINITROSODIMETHYLPROPANEDIAMINE * N,N'-DINITROSO-N,N'-DIMETHYL-1,3-PROPANEDI-AMINE * NSC 62580

SAFETY PROFILE: Questionable carcinogen with experimental neoplastigenic data. Many N-nitroso compounds are carcinogens. When heated to decomposition it emits toxic fumes of NO_x.

DRP800 CAS: 957-51-7 **HR: 2**
N,N-DIMETHYL-2,2-DIPHENYLACET-AMIDE
mf: $C_{16}H_{17}NO$ mw: 239.34

PROP: White solid; very sltly sol in water; mod sol in acetone, dimethyl formamide, and phenyl cellosolve. Mp: 134.5-135.5°.

SYNS: DIAMIDE * DIF 4 * N,N-DIMETHYLDI-PHENYLACETAMIDE * N,N-DIMETHYL-α,α-DIPHENYLACETAMIDE * N,N-DIMETHYL-α-PHENYLBENZENE-ACETAMIDE * DIMID * DIPHENAMID * DIPHENAMIDE * DIPHENYLAMIDE * 2,2-DIPHENYL-N,N-DIMETHYLACETAMIDE * DYMID * ENIDE * FDN * FENAM * LILLY 34,314 * U 4513

SAFETY PROFILE: Moderately toxic by ingestion, intraperitoneal, subcutaneous, and possibly other routes. Mutation data reported. A pesticide. When heated to decomposition it emits toxic fumes of NO_x.

DRQ200 CAS: 997-95-5 **HR: 3**
2,2'-DIMETHYLDIPROPYLINITROSO-AMINE
mf: $C_8H_{18}N_2O$ mw: 158.28

SYNS: DI-ISO-BUTYLNITROSAMINE * DMDPN * NITROSODIISOBUTYLAMINE * N-NITROSODIISO-BUTYLAMINE * N-NITROSODI-ISO-BUTYLAMINE * N-NITROSO-2,2'-DIMETHYLDI-n-PROPYLAMINE

SAFETY PROFILE: Questionable carcinogen with experimental neoplastigenic and tumorigenic data. Mildly toxic by subcutaneous route. Mutation data reported. Many nitrosamines compounds are carcinogens. When heated to decomposition it emits toxic fumes of NO_x.

DRQ400   CAS: 624-92-0   **HR: 3**
DIMETHYLDISULFIDE

DOT: UN 2381
mf: $C_2H_6S_2$ mw: 94.20

PROP: Liquid. flash p: 44.6°F. Bp: 109.7°, d: 1.0569 @ 25°, vap press: 28.6 mm @ 25°, vap d: 3.24.

CONSENSUS REPORTS: Reported in EPA TSCA Inventory. EPA Extremely Hazardous Substances List.

DOT Classification: Flammable Liquid; Label: Flammable Liquid.

SAFETY PROFILE: Poison by inhalation and possibly other routes. A very dangerous fire hazard when exposed to heat, flame or oxidizers. Can react vigorously with oxidizing materials.

DRR200   CAS: 2540-82-1   **HR: 3**
O,O-DIMETHYL DITHIOPHOSPHORYL-ACETIC ACID-N-METHYL-N-FORMYLAMIDE
mf: $C_6H_{12}NO_4PS_2$ mw: 257.28

SYNS: AFLIX * ANTHIO * ANTIO * CP 53926 * O,O-DIMETHYL-S-(N-FORMYL-N-METHYLCARBA-MOYLMETHYL) PHOSPHORODITHIOATE * O,O-DI-METHYL-S-(3-METHYL-2,4-DIOXO-3-AZA-BUTYL)-DITHIO-FOSFAAT (DUTCH) * O,O-DIMETHYL-S-(3-METHYL-2,4-DIOXO-3-AZA-BUTYL)-DITHIOPHOSPHAT (GERMAN) * O,O-DIMETHYL-S-(N-METHYL-N-FORMYL-CARBA-MOYLMETHYL)-DITHIOPHOSPHAT * O,O-DIMETHYL-S-(N-METHYL-N-FORMYLCARBAMOYLMETHYL)PHOS-PHORODITHIOATE * O,O-DIMETHYL PHOSPHORODI-THIOATE N-FORMYL-2-MERCAPTO-N-METHYLACETAM-IDE-S-ESTER * O,O-DIMETIL-S-(N-FORMIL-N-METIL-CARBAMOIL-METIL)-DITIOFOSFATO (ITALIAN) * ENT 27,257 * FORMOTHION * S-(2-(FORMYL-METHYLAMINO)-2-OXOETHYL)-O,O-DIMETHYLPHOSPHO-RODITHIOATE * N-FORMYL-N-METHYLCARBAMOYL-METHYL-O,O-DIMETHYL PHOSPHORODITHIOATE * S-(N-FORMYL-N-METHYLCARBAMOYLMETHYL)-O,O-DIMETHYL PHOSPHORODITHIOATE * S-(N-FORMYL-N-METHYLCARBAMOYLMETHYL) DIMETHYL PHOSPHO-ROTHIOLOTHIONATE * S 6900 * SAN 244 I * SAN 6913 I * SAN 7107 I * SPENCER S-6900 * VEL 4284

CONSENSUS REPORTS: EPA Extremely Hazardous Substances List.

SAFETY PROFILE: Poison by ingestion, inhalation, skin contact, and intravenous routes. Mutation data reported. When heated to decomposition it emits very toxic fumes of NO_x, PO_x, and SO_x.

DRR400   CAS: 2597-03-7   **HR: 3**
(O,O-DIMETHYLDITHIOPHOSPHORYL-PHENYL)ACETIC ACID ETHYL ESTER
mf: $C_{12}H_{17}O_4PS_2$ mw: 320.38

SYNS: AIMSAN * BAY 33051 * BAYER 18510 * CIDEMUL * CIDIAL * DIMEPHENTHIOATE * DIMEPHENTHOATE * O,O-DIMETHYL-S-(1-CAR-BOETHOXYBENZYL) DITHIOPHOSPHATE * O,O-DI-METHYL-S-α-ETHOXY-CARBONYLBENZYL PHOSPHORO-DITHIOATE * O,O-DIMETHYL-S-(PHENYLACETIC ACID ETHYL ESTER) PHOSPHORODITHIOATE * O,O-DI-METHYL-S-(PHENYL)(CARBOETHOXY)METHYL PHOS-PHORODITHIOATE * (DIMETHYL-S-(PHENYLETHOXY-CARBONYLMETHYL)PHOSPHOROTHIOLOTHIONATE) * ELSAN * ENT 23,438 * ENT 27,386GC * S-α-ETHOXYCARBONYLBENZYL-O,O-DIMETHYL PHOSPHORODITHIOATE * S-α-ETHOXYCARBONYL-BENZYL DIMETHYL PHOSPHOROTHIOLOTHIONATE * ETHYL-α-((DIMETHOXYPHOSPHENOTHIOYL)THIO) BENZENEACETATE * ETHYL-O,O-DIMETHYL PHOS-PHORODITHIOYLPHENYL ACETATE * ETHYL ESTER of O,O-DIMETHYLDITHIOPHOSPHORYL α-PHENYL ACETATE ACID * ETHYL MERCAPTOPHENYL-ACETATE-O,O-DIMETHYL PHOSPHOROCITHIOATE * FENTHOATE * L-561 * MONTECATINI L-561 * NSC 190978 * OMS 1075 * PAP * PAPTHION * PHENDAL * PHENTHOATE * ROGODIAL * S 2940 * TANONE * TH 346-1 * TSIDIAL

SAFETY PROFILE: Poison by ingestion and possibly other routes. Moderately toxic by skin contact. An insecticide used for control of crop pests and mosquitoes. When heated to decomposition it emits very toxic fumes of PO_x and SO_x.

DRS800   CAS: 120-08-1   **HR: 3**
6,7-DIMETHYLESCULETIN
mf: $C_{11}H_{10}O_4$ mw: 206.21

SYNS: AESCULETIN DIMETHYL ETHER * 6,7-DI-METHOXYBENZOPYRAN-2-ONE * 6,7-DIMETHOXY-COUMARIN * ESCOPARONE * ESCULETIN DI-METHYL ETHER * SCOPARON * SCOPARONE

SAFETY PROFILE: Poison by ingestion and intraperitoneal routes. Experimental reproductive effects. An antihypertensive agent. When heated to decomposition it emits acrid smoke and irritating fumes.

DRY600   CAS: 794-00-3   **HR: 3**
7,12-DIMETHYL-5-FLUOROBENZ (a)ANTHRACENE
mf: $C_{20}H_{15}F$ mw: 274.35

SYN: 5-FLUORO-7,12-DIMETHYLBENZ(a)ANTHRACENE

CONSENSUS REPORTS: EPA Genetic Toxicology Program.

SAFETY PROFILE: Questionable carcinogen with experimental tumorigenic data. Human mutation data reported. An initiator. When heated to decomposition it emits toxic fumes of F^-.

DRZ000   CAS: 2023-61-2   **HR: 3**
7,12-DIMETHYL-11-FLUOROBENZ (a)ANTHRACENE
mf: $C_{20}H_{15}F$ mw: 274.35

SYN: 11-FLUORO-7,12-DIMETHYLBENZ(a)ANTHRACENE

CONSENSUS REPORTS: EPA Genetic Toxicology Program.

SAFETY PROFILE: Questionable carcinogen with experimental neoplastigenic and tumorigenic data. An initiator. When heated to decomposition it emits toxic fumes of F^-.

DSA000   CAS: 150-74-3   **HR: 3**
N,N-DIMETHYL-p-((p-FLUOROPHENYL) AZO)ANILINE
mf: $C_{14}H_{14}FN_3$ mw: 243.31

SYNS: 4-(DIMETHYLAMINO)-4'-FLUOROAZOBENZENE
* 4'-FLUORO-N,N-DIMETHYL-4-AMINOAZOBENZENE
* 4'-FLUORO-p-DIMETHYLAMINOAZOBENZENE
* 4'-FLUORO-4-DIMETHYLAMINOAZOBENZENE
* 4'-FLUORO-N,N-DIMETHYL-p-PHENYLAZOANILINE
* p-((p-FLUOROPHENYL)AZO)-N,N-DIMETHYLANILINE
* 4-((4-FLUOROPHENYL)AZO)-N,N-DIMETHYLBENZEN-AMINE

SAFETY PROFILE: Experimental teratogenic and reproductive effects. Questionable carcinogen with experimental tumorigenic data. Mutation data reported. When heated to decomposition it emits very toxic fumes of F^- and NO_x.

DSA800   CAS: 5954-50-7   **HR: 3**
DIMETHYL FLUOROPHOSPHATE
mf: $C_2H_6FO_3P$ mw: 128.05

PROP: Liquid. Mp: low, bp: 149°, d: 1.28, vap d: 4.42.

SYNS: FLUOPHOSPHORIC ACID, DIMETHYL ESTER
* PHOSPHOROFLUORIDIC ACID, DIMETHYL ESTER
* PF-1 * T-1035 * TL 311

SAFETY PROFILE: Poison by inhalation, skin contact, and intravenous routes. When heated to decomposition it emits toxic fumes of F^- and PO_x.

DSB000   CAS: 68-12-2   **HR: 3**
DIMETHYLFORMAMIDE
DOT: UN 2265
mf: C_3H_7NO mw: 73.11

PROP: Colorless, mobile liquid. Bp: 152.8°, lel: 2.2% @ 100°, uel: 15.2% @ 100°, flash p: 136°, fp: −61°, d: 0.9445 @ 25°/4°, autoign temp: 833°F, vap press: 3.7 mm @ 25°, vap d: 2.51.

SYNS: DIMETHYLFORMAMID (GERMAN) * N,N-DI-METHYL FORMAMIDE * N,N-DIMETHYLFORMAMIDE (DOT) * DIMETILFORMAMIDE (ITALIAN) * DI-METYLFORMAMIDU (CZECH) * DMF * DMFA * DWUMETHYLOFORMAMID (POLISH) * N-FORMYL-DIMETHYLAMINE * NCI-C60913 * NSC 5356 * U-4224

CONSENSUS REPORTS: EPA Genetic Toxicology Program. Reported in EPA TSCA Inventory.

OSHA PEL: TWA 10 ppm (skin)
ACGIH TLV: TWA 10 ppm (skin); BEI: 40 mg(N-methylformamide)/g creatinine at end of shift.
DFG MAK: 20 ppm (60 mg/m³)
DOT Classification: Flammable or Combustible Liquid; Label: Flammable Liquid.

SAFETY PROFILE: Moderately toxic by ingestion, intravenous, subcutaneous, intramuscular, and intraperitoneal routes. Mildly toxic by skin contact and inhalation. Experimental teratogenic and reproductive effects. Human mutation data reported. A skin and eye irritant. Flammable when exposed to heat or flame; can react with oxidizing materials. Explosion hazard when exposed to flame. Explosive reaction with bromine, potassium permanganate, triethylaluminum + heat. Forms explosive mixtures with lithium azide (shock sensitive above 200°C), uranium perchlorate. Ignition on contact with chromium trioxide. Violent reaction with chlorine, sodium hydroborate + heat, diisocyanatomethane, carbon tetrachloride + iron, 1,2,3,4,5,6-hexachlorocyclohexane + iron. Vigorous exothermic reaction with magnesium nitrate, sodium + heat, sodium hydride + heat, sulfinyl chloride + traces of iron or zinc, 2,4,6-trichloro-1,3,5-triazine (with gas evolution), and many other materials. Avoid contact with halogenated hydrocarbons, inorganic and organic nitrates, (2,5-dimethyl pyrrole + $P(OCl)_3$), C_6Cl_6, methylene diisocyanates, P_2O_3. To fight fire, use foam, CO_2, dry chemical. When heated to decomposition it emits toxic fumes of NO_x.

DSB200 CAS: 533-74-4 HR: 3
DIMETHYLFORMOCARBOTHIALDINE
mf: $C_5H_{10}N_2S_2$ mw: 162.29

PROP: Crystals, sol in alc. Mp: 107°.

SYNS: BASAMID * BASAMID G * BASAMID-GRANULAR * BASAMID P * BASAMID-PUDER * CARBOTHIALDIN * CARBOTHIALDINE * CRAG 974 * CRAG FUNGICIDE 974 * CRAG NEMACIDE * CRAG 85W * DAZOMET * 3,5-DIMETHYLPERHYDRO-1,3,5-THIADIAZIN-2-THION (CZECH, GERMAN) * 3,5-DIMETHYLTETRAHYDRO-1,3,5-THIADIAZINE-2-THIONE * 3,5-DIMETHYL-1,2,3,5-TETRAHYDRO-1,3,5-THIADIAZINETHIONE-2 * 3,5-DIMETHYL-TETRAHYDRO-1,3,5-2H-THIADIAZINE-2-THIONE * 3,5-DIMETHYL-1,3,5-2H-TETRAHYDROTHIADIAZINE-2-THIONE * 3,5-DIMETHYLTETRAHYDRO-2H-1,3,5-THIADIAZINE-2-THIONE * 3,5-DIMETHYL-2-THIONOTETRAHYDRO-1,3,5-THIADIAZINE * 3,5-DIMETIL-PERIDRO-1,3,5-TIHADIAZIN-2-TIONE (ITALIAN) * DMTT * FENNOSAN B 100 * MICOFUME * MYLON (CZECH) * MYLONE * MYLONE 85 * N 521 * NALCON 243 * NEFUSAN * PREZERVIT * STAUFFER N 521 * TETRAHYDRO-2H-3,5-DIMETHYL-1,3,5-THIADIAZINE-2-THIONE * TETRAHYDRO-3,5-DIMETHYL-2H-1,3,5-THIADIAZINE-2-THIONE * THIAZON * THIAZONE * 2-THIO-3,5-DIMETHYLTETRAHYDRO-1,3,5-THIADIAZINE * TIAZON * TROYSAN 142 * UCC 974

CONSENSUS REPORTS: Reported in EPA TSCA Inventory.

SAFETY PROFILE: Poison by ingestion and intraperitoneal routes. Moderately toxic by subcutaneous and possibly other routes. A skin and severe eye irritant. A mild primary skin irritant and sensitizer. When heated to decomposition it emits very toxic fumes of NO_x and SO_x.

DSD775 CAS: 106-72-9 HR: 1
2,6-DIMETHYL-5-HEPTENAL
mf: $C_9H_{16}O$ mw: 140.23

PROP: Pale yellow liquid; melon odor. D: 0.852-0.858, refr index: 1.443-1.448

SYN: FEMA No. 2497

SAFETY PROFILE: Skin and eye irritant. When heated to decomposition it emits acrid smoke and irritating fumes.

DSF400 CAS: 57-14-7 HR: 3
1,1-DIMETHYLHYDRAZINE
DOT: UN 1163
mf: $C_2H_8N_2$ mw: 60.12

PROP: Colorless liquid, ammonia-like odor. Hygroscopic, water-miscible. Bp: 63.3°, fp: −58°, flash p: 5°F, d: 0.782 @ 25°/4°, vap press: 157 mm @ 25°, vap d: 1.94, autoign temp: 480°F, lel: 2%, uel: 95%.

SYNS: DIMAZINE * DIMETHYLHYDRAZINE * asym-DIMETHYLHYDRAZINE * N,N-DIMETHYLHYDRAZINE * uns-DIMETHYLHYDRAZINE * unsym-DIMETHYLHYDRAZINE * 1,1-DIMETHYLHYDRAZINE (GERMAN) * DIMETHYLHYDRAZINE, unsymmetrical (DOT) * DMH * NIESYMETRYCZNA DWU METYLOHYDRAZYNA (POLISH) * RCRA WASTE NUMBER U098 * UDMH (DOT)

CONSENSUS REPORTS: IARC Cancer Review: GROUP 2B IMEMDT 7,56,87; Animal Sufficient Evidence IMEMDT 4,137,74. NTP Fourth Annual Report On Carcinogens, 1984. EPA Genetic Toxicology Program. Community Right-To-Know List. EPA Extremely Hazardous Substances List. Reported in EPA TSCA Inventory.

OSHA PEL: TWA 0.5 ppm (skin)
ACGIH TLV: TWA 0.5 ppm (skin); Suspected Human Carcinogen; (Proposed: TWA 0.01 ppm (skin); Suspected Human Carcinogen)
DFG MAK: Animal Carcinogen, Suspected Human Carcinogen.
NIOSH REL: (Hydrazines) CL 0.15 mg/m³/2H
DOT Classification: Flammable Liquid; Label: Flammable Liquid and Poison; Flammable Liquid; Label: Flammable Liquid, Corrosive.

SAFETY PROFILE: Confirmed carcinogen with experimental carcinogenic, tumorigenic, and teratogenic data. Poison by ingestion, intraperitoneal, intravenous, and intracerebral routes. Moderately toxic by inhalation and skin contact. Human mutation data reported. A plant growth control agent. Corrosive. A powerful reducing agent. A dangerous fire hazard. It is hypergolic with many oxidants (e.g., dinitrogen tetroxide, hydrogen peroxide, and nitric acid). Dangerous when exposed to heat, flame or oxidizers; can react vigorously with oxidizing materials such as air, fuming HNO_3, $(HNO_3 + N_2O_4)$, NO. A high energy propellant for liquid fueled rockets. To fight fire, use alcohol foam, CO_2, dry chemical. When heated to decomposition it emits highly toxic fumes of NO_x.

DSF600   CAS: 540-73-8   **HR: 3**
1,2-DIMETHYLHYDRAZINE

DOT: UN 2382
mf: $C_2H_8N_2$ mw: 60.12

PROP: Clear, colorless, flammable, hygroscopic liquid; fishy ammonia odor. Flash p: < 73.4°F, bp: 81°, mp: −9°, d: 0.8274 @ 20°/ 4°.

SYNS: 1,2-DIMETHYLHYDRAZIN (GERMAN) * N,N′-DIMETHYLHYDRAZINE * sym-DIMETHYLHYDRAZINE * DIMETHYLHYDRAZINE, symmetrical (DOT) * DMH * symetryczna DWUMETYLOHYDRAZYNA (POLISH) * HYDRAZOMETHANE * RCRA WASTE NUMBER U099 * SDMH

CONSENSUS REPORTS: IARC Cancer Review: GROUP 2B IMEMDT 7,56,87; Animal Sufficient Evidence IMEMDT 4,145,74. EPA Genetic Toxicology Program.

DFG MAK: Animal Carcinogen, Suspected Human Carcinogen.
DOT Classification: Flammable Liquid; Label: Flammable Liquid, Poison.

SAFETY PROFILE: Confirmed carcinogen with experimental carcinogenic, neoplastigenic, and tumorigenic data. Poison by ingestion, intraperitoneal, intravenous, subcutaneous, and intramuscular routes. Moderately toxic by inhalation. Experimental teratogenic and reproductive effects. Human mutation data reported. A very dangerous fire hazard when exposed to heat, flame, or oxidizers. A high energy propellant for liquid fuelled rockets. When heated to decomposition it emits toxic fumes of NO_x.

DSF800   CAS: 306-37-6   **HR: 3**
1,2-DIMETHYLHYDRAZINE DIHYDROCHLORIDE
mf: $C_2H_8N_2 \cdot 2ClH$ mw: 133.04

SYNS: N,N′-DIMETHYLHYDRAZINE DIHYDROCHLORIDE * sym-DIMETHYLHYDRAZINE DIHYDROCHLORIDE * DMH

CONSENSUS REPORTS: Reported in EPA TSCA Inventory. EPA Genetic Toxicology Program.

SAFETY PROFILE: Suspected carcinogen with experimental carcinogenic, neoplastigenic, and tumorigenic data. Poison by ingestion and subcutaneous routes. Experimental reproductive effects. Mutation data reported. A rocket fuel. When heated to decomposition it emits very toxic fumes of HCl and NO_x.

DSG000   CAS: 593-82-8   **HR: 3**
1,1-DIMETHYLHYDRAZINE HYDROCHLORIDE
mf: $C_2H_8N_2 \cdot ClH$ mw: 96.58

NIOSH REL: (Hydrazines) CL 0.15 mg/m³/2H

SAFETY PROFILE: Questionable carcinogen with experimental tumorigenic data. Poison by ingestion, intraperitoneal, and intravenous routes. When heated to decomposition it emits very toxic fumes of HCl and NO_x.

DSG200   CAS: 56400-60-3   **HR: 3**
1,2-DIMETHYLHYDRAZINE HYDROCHLORIDE
mf: $C_2H_8N_2 \cdot ClH$ mw: 96.58

SYNS: sym-DIMETHYLHYDRAZINE HYDROCHLORIDE * DMH

SAFETY PROFILE: Questionable carcinogen with experimental carcinogenic, neoplastigenic, and tumorigenic data. Poison by ingestion, intraperitoneal, subcutaneous, and intravenous routes. Mutation data reported. When heated to decomposition it emits very toxic fumes of HCl and NO_x.

DSG400   CAS: 26049-69-4   **HR: 3**
2-(2,2-DIMETHYLHYDRAZINO)-4-(5-NITRO-2-FURYL)THIAZOLE
mf: $C_9H_{10}N_4O_3S$ mw: 254.29

SYN: DMNT

CONSENSUS REPORTS: EPA Genetic Toxicology Program.

SAFETY PROFILE: Suspected carcinogen with experimental carcinogenic data. Mutation data reported. When heated to decomposition it emits very toxic fumes of NO_x and SO_x.

DSI709   CAS: 50-47-5   HR: 3
DIMETHYLIMIPRAMINE
mf: $C_{18}H_{22}N_2$ mw: 266.42

SYNS: DEMETHYLIMIPRAMINE * DESIMIPRAMINE * DESIPRAMIN * DESIPRAMINE (D4) * DESMETHYLIMIPRAMINE * DMI * DMI 50475 * METHYLAMINOPROPYLIMINODIBENZYL * MONODEMETHYLIMIPRAMINE * NORIMIPRAMINE * PENTOFRAN * PERTOFRAN * PERTOFRANE * SERTOFRAN

SAFETY PROFILE: Human poison by ingestion. Experimental poison by ingestion, intraperitoneal, subcutaneous, intravenous, and possibly other routes. Human systemic effects by ingestion: degenerative brain changes, tremors, coma, and cyanosis. Experimental reproductive effects. Mutation data reported. An antidepressant. Related to diazepam. When heated to decomposition it emits toxic fumes of NO_x.

DSK200   CAS: 119-38-0   HR: 3
DIMETHYL-5-(1-ISOPROPYL-3-METHYLPYRAZOLYL)CARBAMATE
mf: $C_{10}H_{17}N_3O_2$ mw: 211.30

SYNS: DIMETHYLCARBAMATE-d'l-ISOPROPYL-3-METHYL-5-PYRAZOLYLE (FRENCH) * DIMETHYL-CARBAMIC ACID 3-METHYL-1-(1-METHYLETHYL)-1H-PYRAZOL-5-YL ESTER * ENT 19,060 * GEIGY G-23611 * ISOLAN * ISOLANE (FRENCH) * (1-ISOPROPIL-3-METIL-1H-PIRAZOL-5-IL)-N,N-DI-METIL-CARBAMMATO (ITALIAN) * (1-ISOPROPYL-3-METHYL-1H-PYRAZOL-5-YL)-N,N-DIMETHYLCARBA-MAAT (DUTCH) * (1-ISOPROPYL-3-METHYL-1H-PY-RAZOL-5-YL)-N,N-DIMETHYL-CARBAMAT (GERMAN) * ISOPROPYLMETHYLPYRAZOLYL DIMETHYLCARBA-MATE * 1-ISOPROPYL-3-METHYL-5-PYRAZOLYL DIMETHYLCARBAMATE * 1-ISOPROPYL-3-METH-YLPYRAZOLYL-(5)-DIMETHYLCARBAMATE * 5-METHYL-2-ISOPROPYL-3-PYRAZOLYL DIMETHYL-CARBAMATE * PRIMIN * SAOLAN

CONSENSUS REPORTS: EPA Extremely Hazardous Substances List.

SAFETY PROFILE: Poison by ingestion, skin contact, intraperitoneal, and possibly other routes. Questionable carcinogen with experimental tumorigenic data. Mutation data reported. An insecticide. When heated to decomposition it emits toxic fumes of NO_x.

DSK600   CAS: 2674-91-1   HR: 3
O,O-DIMETHYL-S-ISOPROPYL-2-SULFINYLETHYLPHOSPHOROTHIOATE
mf: $C_7H_{17}O_4PS_2$ mw: 260.33

SYNS: S-2-AETHYLSULFINYL-1-METHYL AETHYL-O,O DIMETHYL-MONOTHIOPHOSPHAT * BAY 23655 * BAYER 23655 * ENT 25,674 * ESP * ESTON * ESTOX * S-2-ETHYL-SULFINYL-1-METHYL-ETHYL-O,O-DIMETHYL-MONOTHIOFOSFAAT * S-2-ETHYL-SULPHINYL-1-METHYL-ETHYL-O,O-DIMETHYL PHOSPHO-ROTHIOLATE * S-2-ETIL-SULFINIL-1-METIL-ETIL-O,O-DIMETIL-MONOTIOFOSFATO * METASYSTOX-S * OXYDEPROFOS * OXYPHIONFOS * PHOSPHO-ROTHIOIC ACID, O,O-DIMETHYL S-(ETHYLSULFINYL-(2-ISOPROPYL)) ESTER * S410 * THIOMETAN * THIOPHOSPHATE DE O,O-DIMETHYLE ET DE S-2-(ISO-PROPYLSULFINYL)-ETHYLE

SAFETY PROFILE: Poison by ingestion and intraperitoneal routes. Mutation data reported. When heated to decomposition it emits very toxic fumes of PO_x and SO_x.

DSK800   CAS: 36614-38-7   HR: 3
O,O-DIMETHYL-S-2-(ISOPROPYLTHIO)ETHYLPHOSPHORODITHIOATE
mf: $C_7H_{17}O_2PS_3$ mw: 260.39

SYNS: HODSON * HOSALON * HOSDON GRAN-ULE * S-2-ISOPROPYLTHIOETHYL-O,O-DIMETHYL PHOSPHORODITHIOATE * ISOTHIOATE * PHOS-PHORODITHIOIC ACID-O,O-DIMETHYL-S-(2-((1-METHYL-ETHYL)THIO)ETHYL) ESTER

SAFETY PROFILE: Poison by ingestion and skin contact. When heated to decomposition it emits very toxic fumes of PO_x and SO_x.

DSL600   CAS: 2999-74-8   HR: 3
DIMETHYLMAGNESIUM
DOT: UN 1368
mf: C_2H_6Mg mw: 54.38

DOT Classification: Flammable Solid; Label: Spontaneously Combustible

SAFETY PROFILE: The solid and its solution in ether ignite on contact with water. The powder ignites on contact with moist air. When heated to decomposition it emits irritating fumes of MgO.

DSM000 CAS: 766-39-2 **HR: 3**
α,β-DIMETHYLMALEIC ANHYDRIDE
mf: $C_6H_6O_3$ mw: 126.12

SYN: DIMETHYLMALEIC ANHYDRIDE

CONSENSUS REPORTS: Reported in EPA TSCA Inventory.

SAFETY PROFILE: Questionable carcinogen with experimental tumorigenic data. When heated to decomposition it emits acrid smoke and irritating fumes.

DSN600 CAS: 7203-92-1 **HR: 3**
3,3-DIMETHYL-1-p-METHOXYPHENYL-TRIAZENE
mf: $C_9H_{13}N_3O$ mw: 179.25

SYNS: 1-p-METHOXYFENYL-3,3-DIMETHYLTRIAZEN (CZECH) * 1-(p-METHOXYPHENYL)-3,3-DIMETHYL-TRIAZENE * 1-(4-METHYLOXYPHENYL)-3,3-DIMETH-YLTRIAZINE

CONSENSUS REPORTS: EPA Genetic Toxicology Program.

SAFETY PROFILE: Poison by ingestion. Moderately toxic by subcutaneous route. Question able carcinogen with experimental carcinogenic data. Mutation data reported. When heated to decomposition it emits toxic fumes of NO_x.

DSO000 CAS: 950-37-8 **HR: 3**
O,O-DIMETHYL-S-(5-METHOXY-1,3,4-THIADIAZOLINYL-3-METHYL) DITHIOPHOSPHATE
mf: $C_6H_{11}N_2O_4PS_3$ mw: 302.34

SYNS: CIBA-GEIGY GS 13005 * S-(2,3-DIHYDRO-5-METHOXY-2-OXO-1,3,4-THIADIAZOL-3-METHYL) * (O,O-DIMETHYL)-S-(-2-METHOXY-Δ²-1,3,4-THIADI-AZOLIN-5-ON-4-YLMETHYL)DITHIOPHOSPHATE DI-METHYL PHOSPHOROTHIOLOTHIONATE * O,O-DI-METHYL-S-(2-METHOXY-1,3,4-THIADIAZOL-5-(4H)-ONYL-(4)-METHYL)-DITHIOPHOSPHAT (GERMAN) * O,O-DI-METHYL-S-(2-METHOXY-1,3,4-THIADIAZOL-5(4H)-ONYL-(4)-METHYL) PHOSPHORODITHIOATE * O,O-DI-METHYL-S-((2-METHOXY-1,3,4 (4H)-THIODIAZOL-5-ON-4-YL)-METHYL)DITHIOFOSFAAT (DUTCH) * O,O-DI-METIL-S-((2-METOSSI-1,3,4-(4H)-TIADIAZOL-5-ON-4-IL) METIL)-DITIFOSFATO (ITALIAN) * DMTP (JAPAN) * ENT 27,193 * FISONS NC 2964 * GEIGY 13005 * METHIDATHION * S-((5-METHOXY-2-OXO-1,3,4-THIADIAZOL-3(2H)-YL)METHYL)-O,O-DIMETHYL PHOS-PHORODITHIOATE * SOMONIL * SURPRA-CIDE * ULTRACIDE

CONSENSUS REPORTS: EPA Extremely Hazardous Substances List.

SAFETY PROFILE: Poison by ingestion, skin contact, and possibly other routes. Moderately toxic by inhalation. Human mutation data reported. A severe eye irritant. An insecticide. When heated to decomposition it emits very toxic fumes of NO_x, PO_x, and SO_x.

DSO200 CAS: 23422-53-9 **HR: 3**
N,N-DIMETHYL-N'-(((METHYLAMINO) CARBONYL)OXY)PHENYLMETHANI-MIDAMIDE MONOHYDROCHLORIDE
mf: $C_{11}H_{15}N_3O_2 \cdot ClH$ mw: 257.75

SYNS: CARZOL SP * DICARZOL * m-(((DIMETH-YLAMINO)METHYLENE)AMINO)PHENYLMETHYL CAR-BAMATE,HYDROCHLORIDE * 3-DIMETHYLAMINO-METHYLENEIMINOPHENYL-N-METHYLCARBAMATE, HYDROCHLORIDE * ENT 27,566 * EP-332 * FORMETANATE HYDROCHLORIDE * MORTON EP332 * NOR-AM EP 332 * SCHERING 36056 * SN 36056

CONSENSUS REPORTS: Reported in EPA TSCA Inventory. EPA Extremely Hazardous Substances List.

SAFETY PROFILE: Poison by ingestion. Mildly toxic by skin contact. When heated to decomposition it emits very toxic fumes of NO_x and HCl.

DSP400 CAS: 60-51-5 **HR: 3**
O,O-DIMETHYL METHYLCARBAMOYL-METHYL PHOSPHORODITHIOATE
mf: $C_5H_{12}NO_3PS_2$ mw: 229.27

SYNS: AC-12682 * AMERICAN CYANAMID 12880 * BI-58 * CEKUTHOATE * CL 12880 * CYGON * CYGON INSECTICIDE * DAPHENE * DE-FEND * DEMOS-L40 * DEVIGON * DIMATE 267 * DIMETATE * DIMETHOAAT (DUTCH) * DI-METHOAT (GERMAN) * DIMETHOATE (USDA) * DIMETHOAT TECHNISCH 95% * DIMETHOGEN * O,O-DIMETHYLDITHIOPHOSPHORYLACETIC ACID-N-MONOMETHYLAMIDE SALT * O,O-DIMETHYL-DI-THIOPHOSPHORYLESSIGSAEURE MONOMETHYLAMID (GERMAN) * O,O-DIMETHYL-S-(2-(METHYLAMINO)-2-OXOETHYL) PHOSPHORODITHIOATE * O,O-DI-METHYL-S-(N-METHYL-CARBAMOYL)-METHYL-DITHIO-FOSFAAT (DUTCH) * (O,O-DIMETHYL-S-(N-METHYL-CARBAMOYL-METHYL)-DITHIOPHOSPHAT) (GERMAN) * O,O-DIMETHYL-S-(N-METHYLCARBAMOYLMETHYL)

DITHIOPHOSPHATE * O,O-DIMETHYL-S-(N-METHYL-CARBAMOYLMETHYL) PHOSPHORODITHIOATE * O,O-DIMETHYL-S-(N-METHYLCARBAMYLMETHYL) THIOTHIONOPHOSPHATE * O,O-DIMETHYL-S-(N-MONOMETHYL)-CARBAMYL METHYLDITHIOPHOS-PHATE * O,O-DIMETHYL-S-(2-OXO-3-AZA-BUTYL) DITHIOPHOSPHAT (GERMAN) * O,O-DIMETIL-S-(N-METIL-CARBAMOIL-METIL)-DITIOFOSFATO (ITALIAN) * DIMETON * DIMEVUR * DITHIOPHOSPHATE de O,O-DIMETHYLE et de S(-N-METHYLCARBAMOYL-METHYLE) (FRENCH) * EI-12880 * ENT 24,650 * EXPERIMENTAL INSECTICIDE 12,880 * FERKETH-ION * FORTION NM * FOSFAMID * FOSFOTOX * FOSTION MM * L-395 * LURGO * S-METH-YLCARBAMOYLMETHYL-O,O-DIMETHYL PHOSPHORODI-THIOATE * N-MONOMETHYLAMIDE of O,O-DIMETH-YLDITHIOPHOSPHORYLACETIC ACID * NC-262 * NCI-C00135 * PERFECTHION * PHOSPHA-MID * PHOSPHORODITHIOIC ACID-O,O-DIMETHYL-S-(2-(METHYLAMINO)-2-OXOETHYL) ESTER * RACUSAN * RCRA WASTE NUMBER P044 * REBELATE * ROGODIAL * ROGOR * ROXION U.A. * SINORATOX * TRIMETION

CONSENSUS REPORTS: NCI Carcinogenesis Bioassay (feed); No Evidence: mouse, rat NCITR* NCI-CG-TR-4,77. Reported in EPA TSCA Inventory. EPA Genetic Toxicology Program. EPA Extremely Hazardous Substances List.

SAFETY PROFILE: Questionable carcinogen with experimental carcinogenic data. Poison by ingestion, skin contact, intraperitoneal, subcutaneous, and possibly other routes. Moderately toxic by intravenous route. Experimental teratogenic and reproductive effects. Human mutation data reported. When heated to decomposition it emits very toxic fumes of NO_x, PO_x, and SO_x.

DSP600   CAS: 23135-22-0   **HR: 3**
N′,N′-DIMETHYL-N-((METHYLCARBA-MOYL)OXY)-1-METHYLTHIOOXAMI-MIDIC ACID
mf: $C_7H_{13}N_3O_3S$ mw: 219.29

SYNS: D-1410 * 2-(DIMETHYLAMINO)-N-(((METHYL-AMINO)CARBONYL)OXY)-2-OXOETHANIMIDOTHIOIC ACID METHYL ESTER * 2-DIMETHYLAMINO-1-(METH-YLTHIO)GLYOXAL-o-METHYLCARBAMOYLMONOXIME * N,N-DIMETHYL-α-METHYLCARBAMOYLOXYIMINO-α-(METHYLTHIO)ACETAMIDE * N′,N′-DIMETHYL-N-((METHYLCARBAMOYL)OXY)-1-THIOOXAMIMIDIC ACID METHYL ESTER * DPX 1410 * IN-

SECTICIDE-NEMATICIDE 1410 * METHYL-2-(DIMETH-YLAMINO)-N-(((METHYLAMINO)CARBONYL)OXY)-2-OXO-ETHANIMIDOTHIOATE * METHYL-1-(DIMETHYLCAR-BAMOYL)-N-(METHYLCARBAMOYLOXY)THIO-FORMIMIDATE * S-METHYL-1-(DIMETHYLCARBA-MOYL)-N-((METHYLCARBAMOYL)OXY)THIO-FORMIMIDATE * METHYL-N′,N′-DIMETHYL-N-((METHYLCARBAMOYL)OXY)-1-THIOOXAMIMIDATE * OXAMYL * THIOXAMYL * VYDATE * VYDATE L INSECTICIDE/NEMATICIDE * VYDATE L OXAMYL INSECTICIDE/NEMATOCIDE

CONSENSUS REPORTS: EPA Extremely Hazardous Substances List.

SAFETY PROFILE: Poison by ingestion and inhalation. Moderately toxic by skin contact. When heated to decomposition it emits very toxic fumes of NO_x, and SO_x.

DSQ000   CAS: 122-14-5   **HR: 3**
DIMETHYL-3-METHYL-4-NITRO-PHENYLPHOSPHOROTHIONATE
mf: $C_9H_{12}NO_5PS$ mw: 277.25

SYNS: ACCOTHION * ACEOTHION * AGRIA 1050 * AGRIYA 1050 * AGROTHION * AMERICAN CYANAMID CL-47,300 * ARBOGAL * BAY 41831 * BAYER 41831 * BAYER S 5660 * CEKUTRO-THION * CL 47300 * CP 47114 * CYFEN * CYTEL * CYTEN * O,O-DIMETHYL-O-(3-METHYL-4-NITROFENYL)-MONOTHIOFOSFAAT (DUTCH) * O,O-DIMETHYL-O-(3-METHYL-4-NITRO-PHENYL) MONOTHIOPHOSPHAT (GERMAN) * O,O-DIMETHYL-O-(3-METHYL-4-NITROPHENYL) PHOSPHOROTHIOATE * O,O-DIMETHYL-O-(3-METHYL-4-NITROPHENYL) THIO-PHOSPHATE * O,O-DIMETHYL-O-(3-METHYL) PHOS-PHOROTHIOATE * O,O-DIMETHYL-O-(4-NITRO-3-METHYLPHENYL)THIOPHOSPHATE * O,O-DIMETHYL-O-4-NITRO-m-TOLYL PHOSPHOROTHIOATE * O,O-DI-METIL-O-(3-METIL-4-NITRO-FENIL)-MONOTIOFOSFATO (ITALIAN) * EI 47300 * ENT 25,715 * FALITHION * FENITOX * FENITROTHION * FENITROTION (HUNGARIAN) * FOLETHION * H-35-F 87 (BVM) * 8057HC * KOTION * MEP (Pesticide) * META-THIONE * METATION * METHYLNITROPHOS * MONSANTO CP 47114 * NITROPHOS * NOVA-THION * NUVANOL * OLEOSUMIFENE * OMS 43 * OVADOFOS * PENNWALT C-4852 * PHENI-TROTHION * S 112A * S 5660 * SUMITHIAN * THIOPHOSPHATE de O,O-DIMETHYLE et de O-(3-METHYL-4-NITROPHENYLE) (FRENCH) * VERTHION

CONSENSUS REPORTS: EPA Genetic Toxicology Program. EPA Extremely Hazardous Substances List.

SAFETY PROFILE: Poison by ingestion, inhalation, intravenous, intraperitoneal, and possibly other routes. Moderately toxic by skin contact, intratracheal, and subcutaneous routes. Human systemic effects by ingestion: hypermotility, diarrhea, nausea or vomiting, and dyspnea. Mutation data reported. When heated to decomposition it emits very toxic fumes of NO_x, PO_x and SO_x.

DSR200   CAS: 20241-03-6   **HR: 3**
3,3-DIMETHYL-1-(m-METHYLPHENYL) TRIAZENE
mf: $C_9H_{13}N_3$ mw: 163.25

SYNS: 3,3-DIMETHYL-1-(m-TOLYL)TRIAZENE
* 1-(m-METHYLPHENYL)-3,3-DIMETHYLTRIAZENE
* 1-(3-METHYLPHENYL)-3,3-DIMETHYLTRIAZENE

SAFETY PROFILE: Poison by ingestion and intraperitoneal routes. Moderately toxic by subcutaneous route. Questionable carcinogen with experimental carcinogenic data. Mutation data reported. When heated to decomposition it emits toxic fumes of NO_x.

DSS800   CAS: 3761-42-0   **HR: 3**
O,O-DIMETHYL-o-(4-(METHYLSULFONYL)-m-TOLYL) PHOSPHOROTHIOATE
mf: $C_{10}H_{15}O_5PS_2$ mw: 310.34

SYNS: O,O-DIMETHYL-o-((4-METHYLTHIO)-m-TOLYL) PHOSPHOROTHIOATE SULFONE * FENTHION SULFONE

SAFETY PROFILE: Poison by ingestion and intraperitoneal routes. When heated to decomposition it emits very toxic fumes of PO_x and SO_x.

DST000   CAS: 2032-65-7   **HR: 3**
3,5-DIMETHYL-4-METHYLTHIOPHENYL-N-METHYLCARBAMATE
DOT: NA 2757
mf: $C_{11}H_{15}NO_2S$ mw: 225.33

SYNS: BAY 9026 * BAYER 37344 * 3,5-DIMETHYL-4-(METHYLTHIO)PHENOL METHYLCARBAMATE * 3,5-DIMETHYL-4-METHYL-THIOPHENYL-N-CARBAMAT (GERMAN) * DRAZA * ENT 25,726 * H 321 * MERCAPTODIMETHUR (DOT) * MESUROL * METHIOCARB * METHYL CARBAMIC ACID-4-(METHYLTHIO)-3,5-XYLYL ESTER * 4-METHYLMERCAPTO-3,5-DIMETHYLPHENYL N-METHYLCARBAMATE * 4-METHYLMERCAPTO-3,5-XYLYL METHYLCARBA-

MATE * 4-METHYLTHIO-3,5-DIMETHYLPHENYL METHYLCARBAMATE * 4-(METHYLTHIO)-3,5-XYLENOL METHYLCARBAMATE * 4-(METHYLTHIO)-3,5-XYLYL METHYLCARBAMATE * METMERCAPTURON * OMS-93

CONSENSUS REPORTS: EPA Extemely Hazardous Substances List.

DOT Classification: ORM-E; Label: None.

SAFETY PROFILE: Poison by ingestion, skin contact, intraperitoneal, and possibly other routes. Used as an insecticide, molluscicide and bird repellant. When heated to decomposition it emits very toxic fumes of NO_x and SO_x.

DST200   CAS: 55-37-8   **HR: 3**
O,O-DIMETHYL-O-4-(METHYLTHIO)-3,5-XYLYL PHOSPHOROTHIOATE
mf: $C_{11}H_{17}O_3PS_2$ mw: 292.37

SYNS: BAY 37342 * BAYER 9013 * BAYER 37342 * O,O-DIMETHYL-O-(3,5-DIMETHYL-4-METHYLTHIO-PHENYL) PHOSPHOROTHIOATE * O-(3,5-DIMETHYL-4-(METHYLTHIO)PHENYL)-O,O-DIMETHYL PHOSPHOROTHIOATE * ENT 25,684 * G 347

SAFETY PROFILE: Poison by ingestion. When heated to decomposition it emits very toxic fumes of PO_x and SO_x.

DSU000   CAS: 55-93-6   **HR: 3**
DIMETHYLMYLERAN
mf: $C_8H_{18}O_6S_2$ mw: 272.36

SYNS: DDM * 2,5-DIMETHANESULFOMYLOXYHEXANE * 1,4-DIMETHANESULFONOXY-1,4-DIMETHYL-BUTANE * 2,5-HEXANEDIOL DIMETHYLSULFONATE * NSC-23890

CONSENSUS REPORTS: EPA Genetic Toxicology Program.

SAFETY PROFILE: Poison by intravenous and intraperitoneal routes. Mutation data reported. Used for treatment of chronic granulocytic leukemia. When heated to decomposition it emits very toxic fumes of SO_x.

DSU400   CAS: 86-56-6   **HR: 3**
N,N-DIMETHYL-1-NAPHTHYLAMINE
mf: $C_{12}H_{13}N$ mw: 171.26

SYNS: 1-DIMETHYLAMINONAPHTHALENE * DIMETHYL-α-NAPHTHYLAMINE * α-DIMETHYLNAPHTHYLAMINE * N,N-DIMETHYL-α-NAPHTHYLAMINE

CONSENSUS REPORTS: Reported in EPA TSCA Inventory.

SAFETY PROFILE: Poison by intraperitoneal route. Moderately toxic by ingestion. When heated to decomposition it emits toxic fumes of NO_x.

DSU600   CAS: 607-59-0   **HR: 3**
N,N-DIMETHYL-p-(1-NAPHTHYLAZO) ANILINE
mf: $C_{18}H_{17}N_3$ mw: 275.38

SYNS: DAN * p-DIMETHYLAMINOBENZENEAZO-1-NAPHTHALENE * p-DIMETHYLAMINOBENZENE-1-AZO-1-NAPHTHALENE

SAFETY PROFILE: Questionable carcinogen with experimental carcinogenic and tumorigenic data. Mutation data reported. When heated to decomposition it emits toxic fumes of NO_x.

DSV200   CAS: 4164-28-7   **HR: 3**
DIMETHYLNITRAMINE
mf: $C_2H_6N_2O_2$ mw: 90.10

SYNS: DIMETHYLNITRAMIN (GERMAN) * DIMETH-YLNITROAMINE * DMNM * DMNO * N-NITRO-DIMETHYLAMINE * N-NITRO-DMA

SAFETY PROFILE: Poison by intraperitoneal route. Moderately toxic by ingestion. Questionable carcinogen with experimental tumorigenic data. Mutation data reported. When heated to decomposition it emits toxic fumes of NO_x.

DSV800   CAS: 551-92-8   **HR: 3**
1,2-DIMETHYL-5-NITROIMIDAZOLE
mf: $C_5H_7N_3O_2$ mw: 141.15

SYNS: 1,2-DIMETHYL-5-NITRO-1H-IMIDAZOLE * DIMETRIDAZOLE * EMTRYL * EMTRYLVET * EMTRYMIX * 8595 R.P.

CONSENSUS REPORTS: EPA Genetic Toxicology Program.

SAFETY PROFILE: Questionable carcinogen with experimental neoplastigenic data. Mutation data reported. When heated to decomposition it emits toxic fumes of NO_x.

DSX400   CAS: 7227-92-1   **HR: 3**
3,3-DIMETHYL-1-(p-NITROPHENYL) TRIAZENE
mf: $C_8H_{10}N_4O_2$ mw: 194.22

SYNS: 1-p-NITROFENYL-3,3-DIMETHYLTRIAZEN (CZECH) * 1-(p-NITROPHENYL-3,3-DIMETHYL-

TRIAZEN (GERMAN) * 1-(p-NITROPHENYL)-3,3-DI-METHYL-TRIAZENE * 1-(4-NITROPHENYL)-3,3-DI-METHYLTRIAZENE

SAFETY PROFILE: Poison by subcutaneous route. Moderately toxic by ingestion. Questionable carcinogen with experimental neoplastigenic and tumorigenic data. When heated to decomposition it emits toxic fumes of NO_x.

DSY600   CAS: 138-89-6   **HR: 3**
N,N-DIMETHYL-p-NITROSOANILINE
DOT: UN 1369
mf: $C_8H_{10}N_2O$ mw: 150.20

SYNS: ACCELERINE * p-(DIMETHYLAMINO)NITRO-SOBENZENE * 4-(DIMETHYLAMINO)NITROSOBEN-ZENE * DIMETHYL-p-NITROSOANILINE (DOT) * N,N-DIMETHYL-4-NITROSOBENZENAMINE * DI-METHYL(p-NITROSOPHENYL)AMINE * NCI-C01821 * NDMA * p-NITROSO-N,N-DIMETHYLANILINE * 4-NITROSODIMETHYLANILINE * p-NITROSODI-METHYLANILINE (DOT) * PARANITROSODIMETHYL-ANILIDE * ULTRA BRILLIANT BLUE P

CONSENSUS REPORTS: Reported in EPA TSCA Inventory.

DOT Classification: Flammable Solid; Label: Spontaneously Combustible.

SAFETY PROFILE: Poison by ingestion. Questionable carcinogen with experimental tumorigenic data. Mutation data reported. Flammable when exposed to heat, flame or oxidizers. Violent reaction with acetic anhydride + acetic acid. When heated to decomposition it emits toxic fumes of NO_x.

DTA000   CAS: 1456-28-6   **HR: 3**
2,6-DIMETHYLNITROSOMORPHOLINE
mf: $C_6H_{12}N_2O_2$ mw: 144.20

SYNS: DIMETHYLNITROSOMORPHOLINE * 2,6-DI-METHYL-N-NITROSOMORPHOLINE * DMNM * Me2NMOR * NITROSO-2,6-DIMETHYLMORPHO-LINE * N-NITROSO-2,6-DIMETHYLMORPHOLINE

CONSENSUS REPORTS: EPA Genetic Toxicology Program.

SAFETY PROFILE: Suspected carcinogen with experimental carcinogenic, tumorigenic, and neoplastigenic data. Poison by ingestion and subcutaneous routes. Mutation data reported. Used as a model carcinogenic and carcinogenic metabolite. When heated to decomposition it emits toxic fumes of NO$_x$.

DTB200 CAS: 13256-32-1 **HR: 3**
1,3-DIMETHYLNITROSOUREA
mf: C$_3$H$_7$N$_3$O$_2$ mw: 117.13

SYNS: DIMETHYLNITROSOHARNSTOFF (GERMAN)
* N,N'-DIMETHYLNITROSOUREA * 1,3-DIMETHYL-
N-NITROSOUREA * NITROSODIMETHYLUREA
* N-NITROSODIMETHYLUREA

CONSENSUS REPORTS: EPA Genetic Toxicology Program.

SAFETY PROFILE: Suspected carcinogen with experimental carcinogenic, neoplastigenic, and tumorigenic data. Poison by ingestion and intravenous routes. Experimental teratogenic and reproductive effects. Mutation data reported. When heated to decomposition it emits toxic fumes of NO$_x$.

DTC600 CAS: 122-19-0 **HR: 3**
**DIMETHYLOCTADECYLBENZYL-
AMMONIUM CHLORIDE**
mf: C$_{27}$H$_{50}$N • Cl mw: 424.23

SYNS: AMMONYX 4 * AMMONYX CA SPECIAL
* ARQUAD DM18B-90 * BARQUAT SB-25
* BENZYLDIMETHYLSTEARYLAMMONIUM CHLORIDE
* BENZYLSTEARYLDIMETHYLAMMONIUM CHLORIDE
* CARSOQUAT SDQ-25 * DEHYQUART STC-25
* DIMETHYLBENZYLOCTADECYLAMMONIUM CHLO-
RIDE * INTEXAN SB-85 * J SOFT C 4 * KATA-
MINE AB * NISSAN CATION S2-100 * N-OCTADE-
CYL-N-BENZYL-N,N-DIMETHYLAMMONIUMCHLORIDE
* OCTADECYLDIMETHYLBENZYLAMMONIUM CHLO-
RIDE * ORTHOSAN MB * QUATERNOL 1
* STEARALKONIUM CHLORIDE * STEARYLDI-
METHYLBENZYLAMMONIUM CHLORIDE * STEBAC
* TALLOW BENZYL DIMETHYLAMMONIUM CHLORIDE
* TRITON X-40 * VARISOFT SDC

CONSENSUS REPORTS: Reported in EPA TSCA Inventory.

SAFETY PROFILE: Poison by intraperitoneal route. Moderately toxic by ingestion. A human skin irritant and severe experimental eye irritant. When heated to decomposition it emits very toxic fumes of NO$_x$, NH$_3$, and Cl$^-$.

DTC800 CAS: 5392-40-5 **HR: 2**
3,7-DIMETHYL-2,6-OCTADIENAL
mf: C$_{10}$H$_{16}$O mw: 152.26

PROP: Mobile, pale yellow liquid; strong lemon odor. D: 0.891-0.897 @ 15°, refr index: 1.486-1.490, flash p: 198°F. Sol in 5 volumes of 60% alc; sol in all proportions of benzyl benzoate, diethyl phthalate, glycerin, propylene glycol, mineral oil, fixed oils and 95% alc; insol in water.

SYNS: BUTOBEN * BUTYL p-HYDROXYBENZOATE
* CITRAL (FCC) * FEMA No. 2203 * NCI-C56348
* NERAL

CONSENSUS REPORTS: Reported in EPA TSCA Inventory.

SAFETY PROFILE: Moderately toxic by intraperitoneal route. Mildly toxic by ingestion. Experimental reproductive effects. A human and experimental skin irritant. Combustible liquid. When heated to decomposition it emits acrid smoke and irritating fumes.

DTD000 CAS: 106-24-1 **HR: 3**
3,7-DIMETHYL-(E)-2,6-OCTADIEN-1-OL
mf: C$_{10}$H$_{18}$O mw: 154.28

PROP: Colorless to pale yellow, oily liquid; pleasant geranium odor. D: 0.870-0.890 @ 15°, refr index: 1.469-1.478, mp: 15°, bp: 230°, flash p: 214°F. Sol in fixed oils, propylene glycol; sltly sol in water; insol in glycerin @ 230°.

SYNS: 2,6-DIMETHYL-trans-2,6-OCTADIEN-8-OL
* 3,7-DIMETHYL-trans-2,6-OCTADIEN-1-OL * FEMA
No. 2507 * GERANIOL (FCC) * GERANIOL ALCO-
HOL * GERANIOL EXTRA * GERANYL ALCOHOL
* GUANIOL * LEMONOL

CONSENSUS REPORTS: Reported in EPA TSCA Inventory.

SAFETY PROFILE: Poison by intravenous route. Moderately toxic by ingestion and intramuscular routes. Combustible liquid. When heated to decomposition it emits acrid smoke and irritating fumes.

DTD200 CAS: 106-25-2 **HR: 2**
2-cis-3,7-DIMETHYL-2,6-OCTADIEN-1-OL
mf: C$_{10}$H$_{18}$O mw: 154.28

PROP: Colorless liquid; sweet, rose odor. D: 0.875-0.880, refr index: 1.467-1.478. Sol in alc, chloroform, ether, water @ 227°.

SYNS: 3,7-DIMETHYL-(Z)-2,6-OCTADIEN-1-OL
* FEMA No. 2770 * NEROL (FCC)

CONSENSUS REPORTS: Reported in EPA TSCA Inventory.

SAFETY PROFILE: Moderately toxic by intramuscular route. Mildly toxic by ingestion. A skin irritant. When heated to decomposition it emits acrid smoke and irritating fumes.

DTD800   CAS: 105-87-3   **HR: 1**
trans-3,7-DIMETHYL-2,6-OCTADIEN-1-OL ACETATE
mf: $C_{12}H_{20}O_2$ mw: 196.32

PROP: Colorless, sweet, clear liquid; odor of lavender. D: 0.907-0.918 @ 15°, refr index: 1.458-1.464, bp: 128-129° @ 16 mm, flash p: 219°F. Sol in alc, fixed oils, ether; sltly sol in propylene glycol; insol in water and glycerol.

SYNS: ACETIC ACID GERANIOL ESTER * 3,7-DI-METHYL-2-trans-6-OCTADIENYL ACETATE * trans-3,7-DIMETHYL-2,6-OCTADIEN-1-YL ACETATE * trans-2,6-DIMETHYL-2,6-OCTADIEN-8-YL ETHANOATE * FEMA No. 2509 * GERANIOL ACETATE * GERANYL ACETATE (FCC) * NCI-C54728

CONSENSUS REPORTS: Reported in EPA TSCA Inventory.

SAFETY PROFILE: Mildly toxic by ingestion. Combustible liquid. When heated to decomposition it emits acrid smoke and irritating fumes.

DTE600   CAS: 106-21-8   **HR: 2**
DIMETHYLOCTANOL
mf: $C_{10}H_{22}O$ mw: 158.32

PROP: Colorless liquid; sweet, rose odor. D: 0.26-0.842, refr index: 1.435. Sol in fixed oils, propylene glycol; insol in glycerin.

SYNS: DIHYDROCITRONELLOL * 2,6-DIMETHYL-8-OCTANOL * 3,7-DIMETHYL-1-OCTANOL (FCC) * FEMA No. 2391 * GERANIOL TETRAHYDRIDE * PELARGOL * PERHYDROGERANIOL * TETRA-HYDROGERANIOL

CONSENSUS REPORTS: Reported in EPA TSCA Inventory.

SAFETY PROFILE: Moderately toxic by skin contact. A skin irritant. When heated to decomposition it emits acrid smoke and irritating fumes.

DTF400   CAS: 141-25-3   **HR: 2**
2,6-DIMETHYL-1-OCTEN-8-OL
mf: $C_{10}H_{20}O$ mw: 156.30

PROP: Flash p: +212°F.

SYNS: α-CITRONELLOL * 3,7-DIMETHYL-7-OCTEN-1-OL * FEMA No. 2981 * RHODINOL (FCC)

CONSENSUS REPORTS: Reported in EPA TSCA Inventory.

SAFETY PROFILE: Moderately toxic by intramuscular route. When heated to decomposition it emits acrid smoke and irritating fumes.

DTH000   CAS: 1955-45-9   **HR: 3**
3,3-DIMETHYL-2-OXETHANONE
mf: $C_5H_8O_2$ mw: 100.13

SYNS: 3,3-DIMETHYL-2-OXETANONE * DIMETHYL PROPIOLACTONE * 3,3-DIMETHYL-β-PROPIOLACTONE * NCI-C04126 * PIVALIC ACID LACTONE * PIVALOLACTONE

CONSENSUS REPORTS: NCI Carcinogenesis Bioassay (gavage); No Evidence: mouse NCITR* NCI-CG-TR-140,78; Clear Evidence: rat NCITR* NCI-CG-TR-140,78. Reported in EPA TSCA Inventory.

SAFETY PROFILE: Questionable carcinogen with experimental carcinogenic and tumorigenic data. Poison by ingestion. Mutation data reported. When heated to decomposition it emits acrid smoke and irritating fumes.

DTH400   CAS: 2273-45-2   **HR: 3**
DIMETHYLOXOSTANNANE
mf: C_2H_6OSn mw: 164.77

PROP: White powder. Insol in water.

SYN: DIMETHYLTIN OXIDE

CONSENSUS REPORTS: Reported in EPA TSCA Inventory.

OSHA PEL: TWA 0.1 mg(Sn)/m^3 (skin)
ACGIH TLV: TWA 0.1 mg(Sn)/m^3 (skin) (Proposed: TWA 0.1 mg(Sn)/m^3; STEL 0.2 mg(Sn)/m^3 (skin))
NIOSH REL: (Organotin Compounds) TWA 0.1 mg(Sn)/m^3

SAFETY PROFILE: Poison by intravenous route. When heated to decomposition it emits acrid smoke and irritating fumes.

DTH800   CAS: 3820-53-9   **HR: 3**
DIMETHYL PARANITROPHENYL THIONOPHOSPHATE
mf: $C_8H_{10}NO_5PS$ mw: 263.22

PROP: Crystals. Vap d: 9.1, mp: 38°, d: 1.235 @ 20°/4°.

SYNS: O,O-DIMETHYL-S-p-NITROFENYL ESTER KYSELINY THIOFOSFORECEN (CZECH) * O,O-DIMETHYL-S-(p-NITROPHENYL) PHOSPHOROTHIOATE * O,O-DIMETHYL-S-(4-NITROPHENYL)THIOPHOSPHATE

SAFETY PROFILE: Poison by ingestion and subcutaneous routes. When heated to decomposition it emits very toxic fumes of NO_x, PO_x, and SO_x.

DTJ400   CAS: 122-09-8   HR: 3
α,α-DIMETHYLPHENETHYLAMINE
mf: $C_{10}H_{15}N$ mw: 149.26

SYNS: α,α-DIMETHYLBENZEETHANAMINE * 1,1-DIMETHYL-2-PHENYLETHANAMINE *. α,α-DIMETHYL-β-PHENYLETHYLAMINE * DUROMINE * LIPOPILL * LONAMIN * MG 18370 * MG 18570 * MIRAPRONT * PHENTERMINE * 2-PHENYL-tert-BUTYLAMINE * RCRA WASTE NUMBER P046 * WILPO

CONSENSUS REPORTS: Reported in EPA TSCA Inventory.

SAFETY PROFILE: Poison by ingestion, intravenous, and intraperitoneal routes. Human systemic effects by ingestion: sympathomimetic. Mutation data reported. When heated to decomposition it emits toxic fumes of NO_x.

DTK600   CAS: 2747-31-1   HR: 3
N,N-DIMETHYL-p-PHENYLAZOANILINE-N-OXIDE
mf: $C_{14}H_{15}N_3O$ mw: 241.32

SYNS: DAB-N-OXIDE * 4-DIMETHYLAMINOAZOBENZENE AMINE-N-OXIDE * N,N-DIMETHYLAMINO-AZOBENZENE-N-OXIDE

SAFETY PROFILE: Questionable carcinogen with experimental tumorigenic data. Poison by intraperitoneal route. Moderately toxic by ingestion. When heated to decomposition it emits toxic fumes of NO_x.

DTL200   CAS: 126-27-2   HR: 3
2-DI(N-METHYL-N-PHENYL-tert-BUTYL-CARBAMOYLMETHYL)AMINOETHANOL
mf: $C_{28}H_{41}N_3O_3$ mw: 467.72

SYNS: BETALGIL * N,N-BIS(N-METHYL-N-PHENYL-tert-BUTYLACETAMIDO)-β-HYDROXYETHYLAMINE * EMOREN * FH 099 * H4 099 * 2,2'-((2-HYDROXYETHYL)IMINO BIS(N-(α,α-DIMETHYLPHENETHYL)-N-METHYL-ACETAMIDE * 2,2'-((2-HYDROXYETHYL)IMINO)BIS(N-(1,1-DIMETHYL-2-PHENYLETHYL)-N-METHYLACETAMIDE) * MUCAINE * MUCOXIN * MUTHESA * OXAINE * OXETACAINE * OXETHACAINA (ITALIAN) * OXETHAZINE * STOMACAIN * TEPILTA * TOPICAIN * WY 806

SAFETY PROFILE: Poison by ingestion, intratracheal, intravenous, intraperitoneal, intramuscular, subcutaneous, and implant routes. Mutation data reported. When heated to decomposition it emits toxic fumes of NO_x.

DTL800   CAS: 105-10-2   HR: 3
N,N-DIMETHYL-p-PHENYLENEDIAMINE
mf: $C_8H_{12}N_2$ mw: 136.22

SYNS: p-AMINODIMETHYLANILINE * C.I. 76075 * p-DIMETHYLAMINOPHENYLAMINE * N,N-DIMETHYL-1,4-BENZENEDIAMINE * DIMETHYL-p-PHENYLENEDIAMINE * DMPD

SAFETY PROFILE: Poison by intraperitoneal route. Mutation data reported. When heated to decomposition it emits toxic fumes of NO_x.

DTM600   CAS: 154-99-4   HR: 3
o,p-DIMETHYL-β-PHENYLETHYLHY-DRAZINE DIHYDROGEN SULFATE
SYNS: β-(2,4-DIMETHYLPHENYL)ETHYLHYDRAZINE DIHYDROGEN SULPHATE * LON 41

SAFETY PROFILE: Poison by ingestion and subcutaneous routes. Experimental reproductive effects. When heated to decomposition it emits very toxic fumes of SO_x and NO_x.

DTN200   CAS: 2655-14-3   HR: 3
3,5-DIMETHYLPHENYL-N-METHYLCARBAMATE
mf: $C_{10}H_{13}NO_2$ mw: 179.24

SYNS: DRC 3340 * H-69 * MACBAL * MAQBARL * 3,5-XMC * 3,5-XYLENOL METHYLCARBAMATE * 3,5-XYLENYL-N-METHYLCARBAMATE * 3,5-XYLYL-N-METHYLCARBAMATE

CONSENSUS REPORTS: EPA Genetic Toxicology Program.

SAFETY PROFILE: Poison by ingestion and possibly other routes. When heated to decomposition it emits toxic fumes of NO_x.

DTN800   CAS: 7635-51-0   **HR: 3**
3,4-DIMETHYL-2-PHENYLMORPHO-LINEHYDROCHLORIDE
mf: $C_{12}H_{17}NO \cdot ClH$ mw: 227.76

SYNS: PHENDIMETRAZINE HYDROCHLORIDE
* d-2-PHENYL-3,4-DIMETHYLMORPHOLINE HYDRO-CHLORIDE

SAFETY PROFILE: Poison by ingestion, intraperitoneal, subcutaneous, and intravenous routes. When heated to decomposition it emits very toxic fumes of NO_x and HCl.

DTO000   CAS: 54-77-3   **HR: 3**
1,1-DIMETHYL-4-PHENYLPIPERAZINE IODIDE
mf: $C_{12}H_{19}N_2 \cdot I$ mw: 318.23

SYNS: 1,1-DIMETHYL-4-PHENYLPIPERAZINIUM IODIDE
* DMPP * DMPP IODIDE

SAFETY PROFILE: Poison by intravenous, intraperitoneal, and intramuscular routes. When heated to decomposition it emits very toxic fumes of NO_x and I^-.

DTO200   CAS: 3734-17-6   **HR: 3**
1,2-DIMETHYL-3-PHENYL-3-PYRROLIDYL PROPIONATE
mf: $C_{15}H_{21}NO_2$ mw: 247.37

SYNS: A-1981 * COGESIC * 1,2-DIMETHYL-3-PHENYL-3-PYRROLIDINOL PROPIONATE (ester)
* PRODILIDINE

SAFETY PROFILE: Poison by ingestion, intravenous, intraperitoneal, and subcutaneous routes. When heated to decomposition it emits toxic fumes of NO_x.

DTO800   CAS: 6152-43-8   **HR: 3**
N,N-DIMETHYL-2-(α-PHENYL-o-TOL-OXY)ETHYLAMINE HYDROCHLORIDE
mf: $C_{17}H_{21}NO \cdot ClH$ mw: 291.85

SYNS: BRISTAMIN HYDROCHLORIDE * PHENYLTO-LOXAMINE HYDROCHLORIDE

SAFETY PROFILE: Poison by ingestion, intravenous, and intraperitoneal routes. Moderately toxic by ingestion. When heated to decomposition it emits very toxic fumes of HCl and NO_x.

DTP000   CAS: 7227-91-0   **HR: 3**
3,3-DIMETHYL-1-PHENYLTRIAZENE
mf: $C_8H_{11}N_3$ mw: 149.22

SYNS: 3,3-DIMETHYL-1-PHENYL-1-TRIAZENE
* DMPT * 1-FENYL-3,3-DIMETHYLTRIAZIN
* NSC 3094 * PDMT * PDT * 1-PHENYL-3,3-DI-METHYLTRIAZENE * PHENYLDIMETHYLTRIAZINE
* X 119

CONSENSUS REPORTS: EPA Genetic Toxicology Program.

SAFETY PROFILE: Questionable carcinogen with experimental carcinogenic and tumorigenic data. Poison by ingestion and intraperitoneal routes. Experimental teratogenic and reproductive effects. Human mutation data reported. Decomposes explosively on attempted distillation at atmospheric pressure. When heated to decomposition it emits toxic fumes of NO_x.

DTP800   CAS: 34491-04-8   **HR: 3**
DIMETHYL PHOSPHATE ESTER with 2-CHLORO-N-METHYL-3-HYDROXY-CROTONAMIDE
mf: $C_7H_{13}ClNO_5P$ mw: 257.63

SYNS: CIBA C-768 * ENT 27,357 * NSC 190955

SAFETY PROFILE: Poison by ingestion and subcutaneous routes. When heated to decomposition it emits very toxic Cl^-, NO_x, and PO_x.

DTQ400   CAS: 10265-92-6   **HR: 3**
O,S-DIMETHYL PHOSPHORAMIDO-THIOATE
mf: $C_2H_8NO_2PS$ mw: 141.14

PROP: Crystals. Mp: 40°. Slightly water-sol; sol in alc.

SYNS: ACEPHATE-MET * BAY 71628 * BAYER 71628 * CHEVRON 9006 * CHEVRON ORTHO 9006
* O,S-DIMETHYL ESTER AMIDE of AMIDOTHIOATE
* ENT 27,396 * HAMIDOP * METAMIDOFOS ESTRELLA * METHAMIDOPHOS * MONITOR
* MTD * NSC 190987 * ORTHO 9006 * PIL-LARON * SRA 5172 * TAHMABON * TAMARON
* THIOPHOSPHORSAEURE-O,S-DIMETHYLESTERAMID (GERMAN)

CONSENSUS REPORTS: EPA Extremely Hazardous Substances List.

SAFETY PROFILE: Poison by ingestion, inhalation, skin contact, subcutaneous, and intraperitoneal routes. Human systemic effects by ingestion: fasciculations, pupillary constriction and sweating. A cholinesterase inhibitor type of insecticide. When heated to decomposition it emits very toxic fumes of NO_x, PO_x, and SO_x.

DTQ600   CAS: 2524-03-0   HR: 3
O,O-DIMETHYLPHOSPHOROCHLORI-DOTHIOATE

DOT: NA 2267/UN 2922
mf: $C_2H_6ClO_2PS$ mw: 160.56

SYNS: CHLOROPHOSPHONOTHIOIC ACID-O,O-DI-METHYL ESTER * DIMETHYL CHLOROTHIOPHOS-PHATE (DOT) * DIMETHYLCHLORTHIOFOSAT (CZECH) * O,O-DIMETHYLESTER KYSELINY CHLOR-THIOFOSFORECNE (CZECH) * DIMETHYL PHOS-PHOROCHLORIDOTHIOATE (DOT) * METHYL PCT * PHOSPHOROCHLORIDOTHIOIC ACID-O,O-DI-METHYL ESTER

CONSENSUS REPORTS: Reported in EPA TSCA Inventory. EPA Extremely Hazardous Substances List.

DOT Classification: Corrosive Material; Label: Corrosive.

SAFETY PROFILE: Poison by inhalation. Moderately toxic by ingestion and skin contact. Corrosive. When heated to decomposition it emits very toxic fumes of Cl^-, PO_x and SO_x.

DTR200   CAS: 131-11-3   HR: 2
DIMETHYLPHTHALATE
mf: $C_{10}H_{10}O_4$ mw: 194.20

PROP: Colorless, odorless liquid. Bp: 283.7°, flash p: 295°F (CC), d: 1.189 @ 25°/25°, autoign temp: 1032°F, vap d: 6.69, vap press: 1 mm @ 100.3°.

SYNS: AVOLIN * 1,2-BENZENEDICARBOXYLIC ACID DIMETHYL ESTER * DIMETHYL-1,2-BENZENEDI-CARBOXYLATE * DIMETHYL BENZENEORTHODICAR-BOXYLATE * DMP * ENT 262 * FERMINE * METHYL PHTHALATE * MIPAX * NTM * PALATINOL M * PHTHALIC ACID METHYL ESTER * PHTHALSAEUREDIMETHYLESTER (GERMAN) * RCRA WASTE NUMBER U102 * SOLVANOM * SOLVARONE

CONSENSUS REPORTS: Reported in EPA TSCA Inventory. Community Right-To-Know List. EPA Extremely Hazardous Substances List.

OSHA PEL: TWA 5 mg/m^3
ACGIH TLV: TWA 5 mg/m^3

SAFETY PROFILE: Moderately toxic by ingestion and intraperitoneal routes. Mildly toxic by inhalation. Experimental teratogenic and reproductive effects. Mutation data reported. An eye irritant. A pesticide and insect repellent. Combustible when exposed to heat or flame; can react with oxidizing materials. To fight fire, use CO_2, dry chemical. When heated to decomposition it emits acrid smoke and irritating fumes.

DTR850   HR: 1
DIMETHYLPOLYSILOXANE
mf: $[(CH_3)_2SiO—]$

PROP: Clear, colorless viscous liquid. D: 0.96, refr index: 1.400. Sol in hydrocarbon solvents; insol in water.

SYNS: DIMETHYL SILICONE * POLYDIMETHYLSI-LOXANE

SAFETY PROFILE: Combustible liquid. When heated to decomposition it emits acrid smoke and irritating fumes.

DTS400   CAS: 3282-30-2   HR: 3
2,2-DIMETHYLPROPANOYL CHLORIDE

DOT: UN 2438
mf: C_5H_9ClO mw: 120.59

SYNS: 2,2-DIMETHYLPROPIONYL CHLORIDE * NEOPANTANOYL CHLORIDE * PIVALIC ACID CHLORIDE * PIVALOLYL CHLORIDE * PIVALOYL CHLORIDE * PIVALYL CHLORIDE * TRIMETHYL ACETYL CHLORIDE (DOT)

CONSENSUS REPORTS: Reported in EPA TSCA Inventory.

DOT Classification: Corrosive Material; Label: Corrosive; Flammable Liquid.

SAFETY PROFILE: A corrosive irritant to skin, eyes, and mucous membranes. The liquid is flammable when exposed to heat, flame, or oxidizers. When heated to decomposition it emits toxic fumes of Cl^-.

DTT600   CAS: 23950-58-5   HR: 3
N-(1,1-DIMETHYLPROPYNYL)-3,5-DICHLOROBENZAMIDE
mf: $C_{12}H_{11}Cl_2NO$ mw: 256.14

SYNS: 3,5-DICHLORO-N-(1,1-DIMETHYL-2-PROPYNYL) BENZAMIDE * KERB * PROMAMIDE * PRONAMIDE * PROPYZAMIDE * RCRA WASTE NUMBER U192 * RH 315

SAFETY PROFILE: Questionable carcinogen with experimental carcinogenic and tumorigenic data. Mildly toxic by ingestion. An herbicide. When heated to decomposition it emits very toxic fumes of Cl^- and NO_x.

DTU400   CAS: 5910-89-4   HR: 2
2,3-DIMETHYLPYRAZINE
mf: $C_6H_8N_2$ mw: 108.16

PROP: Colorless liquid; nutty cocoa odor. D: 1.000-1.022 @ 20°, refr index: 1.506-1.509, flash p: 147°F (OC), d: 0.99, vap d: 3.72, bp: 182.2°. Misc with water, organic solvents.

SYNS: 2,3-DIMETHYL-1,4-DIAZINE * FEMA No. 3271

CONSENSUS REPORTS: Reported in EPA TSCA Inventory.

SAFETY PROFILE: Moderately toxic by ingestion and intraperitoneal routes. Combustible liquid. When heated to decomposition it emits toxic fumes of NO_x.

DTU600   CAS: 123-32-0   HR: 2
2,5-DIMETHYLPYRAZINE
mf: $C_6H_8N_2$ mw: 108.16

PROP: Colorless liquid; potato taste. D: 0.980-1.000, refr index: 1.497-1.501, flash p: 147°F (OC), d: 0.99, vap d: 3.72, bp: 182.2°. Misc with water, organic solvents.

SYNS: 2,5-DIMETHYL-1,4-DIAZINE * FEMA No. 3272

CONSENSUS REPORTS: Reported in EPA TSCA Inventory.

SAFETY PROFILE: Moderately toxic by ingestion and intraperitoneal routes. Mutation data reported. Combustible liquid when exposed to heat, open flame, spark, oxidizers. To fight fire use water spray, mist, dry chemical, CO_2, foam. When heated to decomposition it emits toxic fumes of NO_x.

DTU800   CAS: 108-50-9   HR: 2
2,6-DIMETHYLPYRAZINE
mf: $C_6H_8N_2$ mw: 108.16

PROP: White to yellow crystals; nutty, coffee odor. Mp: 48°, d:.965 @ 50°. Sol in water, organic solvents @ 155°.

SYN: FEMA No. 3273

CONSENSUS REPORTS: Reported in EPA TSCA Inventory.

SAFETY PROFILE: Moderately toxic by intraperitoneal route. When heated to decomposition it emits toxic fumes of NO_x.

DTV200   CAS: 21600-42-0   HR: 3
(3,3-DIMETHYL-1-(m-PYRIDYL-N-OXIDE))TRIAZENE
mf: $C_7H_{10}N_4O$ mw: 166.21

SYNS: 3-(3′,3′-DIMETHYLTRIAZENO)-PYRIDIN-N-OXID (GERMAN) * 3-(3′,3′-DIMETHYLTRIAZENO)PYRIDINE-N-OXIDE * PYNDT * 1-(PYRIDYL-3-N-OXID)-3,3-DIMETHYL-TRIAZEN (GERMAN) * 1-(PYRIDYL-3-N-OXIDE)-3,3-DIMETHYLTRIAZENE

CONSENSUS REPORTS: EPA Genetic Toxicology Program.

SAFETY PROFILE: Questionable carcinogen with experimental carcinogenic and tumorigenic data. Poison by intravenous and subcutaneous routes. Human mutation data reported. When heated to decomposition it emits toxic fumes of NO_x.

DTV400   CAS: 333-40-4   HR: 3
S-(4,6-DIMETHYL-2-PYRIMIDINYL)-O,O-DIETHYL PHOSPHORODITHIOATE
mf: $C_{10}H_{17}N_2O_2PS_2$ mw: 292.38

SYNS: ENT 25,737 * STAUFFER R-3413

SAFETY PROFILE: Poison by ingestion. When heated to decomposition it emits very toxic fumes of NO_x, PO_x, and SO_x.

DUB800   CAS: 1145-73-9   HR: 3
N,N-DIMETHYL-4-STILBENAMINE
mf: $C_{16}H_{17}N$ mw: 223.34

SYNS: 4-DIMETHYLAMINOSTILBEN (GERMAN) * N,N-DIMETHYL-4-AMINOSTILBENE * N,N-DIMETHYL-p-STYRYLANILINE * STILBENYL-N,N-DIMETHYLAMINE

SAFETY PROFILE: Poison by ingestion and intraperitoneal routes. Questionable carcinogen with experimental carcinogenic and tumorigenic data. Mutation data reported. When heated to decomposition it emits toxic fumes of NO_x.

DUC000   CAS: 838-95-9   **HR: 3**
(E)-N,N-DIMETHYL-4-STILBENAMINE
mf: $C_{16}H_{17}N$ mw: 223.34

SYNS: trans-p-(DIMETHYLAMINO)STILBENE * trans-4-DIMETHYLAMINOSTILBENE * (E)-N,N,-DIMETHYL-4-(2-PHENYLETHENYL)BENZENAMINE * 4-DIMETHYL-AMINO-trans-STILBENE * trans-N,N-DIMETHYL-4-STIL-BENAMINE

CONSENSUS REPORTS: EPA Genetic Toxicology Program.

SAFETY PROFILE: Poison by ingestion and subcutaneous routes. Questionable carcinogen with experimental carcinogenic and tumorigenic data. Mutation data reported. When heated to decomposition it emits toxic fumes of NO_x.

DUD100   CAS: 77-78-1   **HR: 3**
DIMETHYL SULFATE
DOT: UN 1595
mf: $C_2H_6O_4S$ mw: 126.14

PROP: Colorless, odorless liquid. Mp: $-31.8°$, bp: $188°$, flash p: $182°F$ (OC), d: 1.3322 @ $20°/4°$, vap d: 4.35, autoign temp: $370°F$.

SYNS: DIMETHYLESTER KYSELINY SIROVE (CZECH) * DIMETHYL MONOSULFATE * DIMETHYLSULFAAT (DUTCH) * DIMETHYLSULFAT (CZECH) * DIMETIL-SOLFATO (ITALIAN) * DMS * DMS(METHYL SULFATE) * DWUMETYLOWY SIARCZAN (POLISH) * METHYLE (SULFATE de) (FRENCH) * METHYL SULFATE (DOT) * RCRA WASTE NUMBER U103 * SULFATE de METHYLE (FRENCH) * SULFATE DIMETHYLIQUE (FRENCH) * SULFURIC ACID, DI-METHYL ESTER

CONSENSUS REPORTS: IARC Cancer Review: GROUP 2A IMEMDT 7,200,87; Animal Sufficient Evidence IMEMDT 4,271,74; Human Inadequate Evidence IMEMDT 4,271,74. NTP Fourth Annual Report On Carcinogens, 1984. EPA Genetic Toxicology Program. Community Right To Know List. EPA Extremely Hazardous Substances List. Reported in EPA TSCA Inventory.

OSHA PEL: (Transitional: TWA 1 ppm (skin)) TWA 0.1 ppm (skin)
ACGIH TLV: TWA 0.1 ppm (skin); Suspected Human Carcinogen.
DFG TRK: Production: 0.02 ppm; Use: 0.04 ppm; Animal Carcinogen, Suspected Human Carcinogen.

DOT Classification: Corrosive Material; Label: Corrosive; Poison B; Label: Poison

SAFETY PROFILE: Confirmed carcinogen with experimental carcinogenic, tumorigenic, and teratogenic data. Human poison by inhalation. Experimental poison by ingestion, inhalation, intravenous, and subcutaneous routes. Human mutation data reported. A corrosive irritant to skin, eyes, and mucous membranes. There is no odor or initial irritation to give warning of exposure. On brief, mild exposures, conjunctivitis, catarrhal inflammation of the mucous membranes of the nose, throat, larynx, and trachea and possibly some reddening of the skin develop after the latent period. With longer, heavier exposures, the cornea shows clouding, the irritation changes to the nasopharynx are more marked and after 6 to 8 hours pulmonary edema may develop. Death may occur in 3 or 4 days. The liver and kidneys are frequently damaged. Spilling of the liquid on the skin can cause ulceration and local necrosis. In patients surviving severe exposure, there may be serious injury of the liver and kidneys, with suppression of urine, jaundice, albuminuria and hematuria appearing. Death, resulting from the kidney or liver damage, may be delayed for several weeks. Flammable when exposed to heat, flame, or oxidizers. Can react with oxidizing materials. Violent reaction with NH_4OH and NaN_3. To fight fire, use water, foam, CO_2, dry chemical. When heated to decomposition it emits toxic fumes of SO_x.

DUD400   CAS: 1003-78-7   **HR: 3**
2,4-DIMETHYL SULFOLANE
mf: $C_6H_{12}O_2S$ mw: 148.24

PROP: Solid. Bp: $280°$, flash p: $290°F$ (OC), d: 1.1362 @ $20°/4°$, vap press: 0.006 mm @ $20°$.

SYNS: DMS * TETRAHYDRO-2,4-DIMETHYLTHIO-PHENE-1,1-DIOXIDE

SAFETY PROFILE: Poison by ingestion, intraperitoneal, and intravenous routes. Moderately toxic by skin contact. Combustible when exposed to heat or flame; can react with oxidizing materials. To fight fire, use water, foam, CO_2, dry chemical. When heated to decomposition it emits toxic fumes of SO_x.

DUD800   CAS: 67-68-5   **HR: 3**
DIMETHYL SULFOXIDE
mf: C_2H_6OS mw: 78.14

PROP: Clear, water-white, hygroscopic liquid. Mp: 18.5°, bp: 189°, flash p: 203°F (OC), d: 1.100 @ 20°, vap press: 0.37 mm @ 20°, lel: 2.6%, uel: 28.5%, autoign temp: 419°F.

SYNS: A 10846 * DELTAN * DEMASORB * DEMAVET * DEMESO * DEMSODROX * DERMASORB * DIMETHYL SULPHOXIDE * DIMEXIDE * DIPIRARTRIL-TROPICO * DMS-70 * DMS-90 * DMSO * DOLICUR * DOLIGUR * DOMOSO * DROMISOL * DURASORB * GAMASOL 90 * HYADUR * INFILTRINA * M 176 * METHYLSULFINYLMETHANE * METHYL SULFOXIDE * NSC-763 * RIMSO-50 * SOMIPRONT * SQ 9453 * SULFINYLBIS (METHANE) * SYNTEXAN * TOPSYM

CONSENSUS REPORTS: Reported in EPA TSCA Inventory. EPA Genetic Toxicology Program.

SAFETY PROFILE: Questionable carcinogen with experimental tumorigenic data. Poison by ingestion. Moderately toxic by intravenous and intraperitoneal routes. Mildly toxic by subcutaneous route. Human systemic effects by intravenous route: nausea or vomiting and jaundice. Experimental teratogenic and reproductive effects. A skin and eye irritant. Human mutation data reported. Can cause an anaphylactic reaction, and corneal opacity. It freely penetrates the skin and may carry dissolved chemicals with it into the body. Combustible when exposed to heat or flame; can react with oxidizing materials. To fight fire, use foam, alcohol foam, CO_2, dry chemical. Violent or explosive reaction with many acyl, aryl and non-metal halides (e.g., acetyl chloride, benzenesulfonyl chloride, bromobenzoyl acetanilide, cyanuric chloride, iodine pentafluoride, $Mg(ClO_4)_2$, CH_3Br, NIO_4, oxalyl chloride, P_2O_3, phosphorous trichloride, phosphoryl chloride, silver fluoride, silver difluoride, sodium hydride, sulfur dichloride, disulfur dichloride, sulfuryl chloride, tetrachlorosilane, thionyl chloride. Violent or explosive reaction with boron compounds [e.g., borane, nonahydrononaborate(2-) ion], 4(4'-bromobenzoyl)acetanilide, carbonyl diisothiocyanate, dinitrogen tetraoxide, hexachlorocyclotriphosphazine, copper + trichloroacetic acid, metal alkoxides (e.g., potassium tert-butoxide, sodium isopropoxide), trifluoroacetic acid anhydride. Incompatible with magnesium perchlorate, metal oxosalts, perchloric acid, periodic acid, sulfur trioxide. Forms powerfully explosive mixtures with metal salts of oxoacids (e.g., aluminum perchlorate, sodium perchlorate, iron(III) nitrate). When heated to decomposition it emits toxic fumes of SO_x.

DUE000   CAS: 120-61-6   **HR: 3**
DIMETHYL TEREPHTHALATE
mf: $C_{10}H_{10}O_4$ mw: 194.20

SYNS: 1,4-BENZENE DICARBOXYLIC ACID DIMETHYL ESTER (9CI) * DIMETHYL-1,4-BENZENE DICARBOXYLATE * METHYL-4-CARBOMETHOXY BENZOATE * NCI-C50055 * TEREPHTHALIC ACID METHYL ESTER

CONSENSUS REPORTS: NCI Carcinogenesis Bioassay (feed): Clear Evidence: mouse NCITR* NCI-CG-TR-121,79; No Evidence: rat NCITR* NCI-CG-TR-121,79. Reported in EPA TSCA Inventory.

SAFETY PROFILE: Questionable carcinogen with experimental carcinogenic data. Moderately toxic by intraperitoneal route. Mildly toxic by ingestion. An eye irritant. When heated to decomposition it emits acrid smoke and irritating fumes.

DUG600   CAS: 50847-92-2   **HR: 3**
2,2-DIMETHYL-3-THIOMORPHOLINONE
mf: $C_6H_{11}NOS$ mw: 145.24

SYN: 2,2-DIMETHYL-3-THIOMORPHOLONE

SAFETY PROFILE: Poison by intraperitoneal route. Moderately toxic by ingestion. When heated to decomposition it emits very toxic fumes of NO_x and SO_x.

DUG800   CAS: 2767-47-7   **HR: 3**
DIMETHYLTIN DIBROMIDE
mf: $C_2H_6Br_2Sn$ mw: 308.59

PROP: Colorless crystals. Sol in water and organic solvents. Mp: 76°; bp: 208-213°.

SYN: DIBROMODIMETHYL STANNANE

OSHA PEL: TWA 0.1 mg(Sn)/m³ (skin)
ACGIH TLV: TWA 0.1 mg(Sn)/m³ (skin) (Proposed: TWA 0.1 mg(Sn)/m³; STEL 0.2 mg(Sn)/m³ (skin))
NIOSH REL: (Organotin Compounds) TWA 0.1 mg(Sn)/m³

SAFETY PROFILE: Poison by intravenous route. When heated to decomposition it emits toxic fumes of Br^-.

DUH600   CAS: 55-80-1   HR: 3
N,N-DIMETHYL-p-(m-TOLYLAZO) ANILINE
mf: $C_{15}H_{17}N_3$ mw: 239.35

SYNS: 4-(N,N-DIMETHYLAMINO)-3′-METHYLAZOBEN-ZENE * N,N-DIMETHYL-p-(3′-METHYLPHENYLAZO) ANILINE * N,N-DIMETHYL-4-((3-METHYLPHENYL) AZO)BENZENAMINE * MDAB * 3′-MDAB * 3′-METHYLBUTTERGELB (GERMAN) * 3′-METHYL-DAB * 3′-METHYL-4-DIMETHYLAMINOAZOBENZEN (CZECH) * M′-METHYL-p-DIMETHYLAMINOAZOBEN-ZENE * 3′-METHYL-4-DIMETHYLAMINOAZOBENZENE * 3′-METHYL-N,N-DIMETHYL-4-AMINOAZOBENZENE * 3′-METHYLDIMETHYLAMINOAZOBENZOL (GERMAN)

CONSENSUS REPORTS: Reported in EPA TSCA Inventory. EPA Genetic Toxicology Program.

SAFETY PROFILE: Questionable carcinogen with experimental carcinogenic, neoplastigenic, tumorigenic, and teratogenic data. Moderately toxic by ingestion. Mutation data reported. When heated to decomposition it emits toxic fumes of NO_x.

DUI400   CAS: 64038-56-8   HR: D
5-(3,3-DIMETHYL-1-TRIAZENO)IMIDA-ZOLE-4-CARBOXAMIDE CITRATE
mf: $C_6H_{10}N_6O \cdot C_6H_8O_7$ mw: 374.36

SYN: DTIC CITRATE

SAFETY PROFILE: Experimental reproductive effects. When heated to decomposition it emits toxic fumes of NO_x.

DUK800   CAS: 2164-17-2   HR: 3
1,1-DIMETHYL-3-(α,α,α-TRIFLUORO-m-TOLYL) UREA
mf: $C_{10}H_{11}F_3N_2O$ mw: 232.23

SYNS: C 2059 * CIBA 2059 * COTORAN * COTORAN MULTI 50WP * COTTONEX * 1,1-DI-METHYL-3-(3-TRIFLUOROMETHYLPHENYL)UREA * N,N-DIMETHYL-N′-(3-TRIFLUOROMETHYLPHENYL) UREA * FLUOMETURON * HERBICIDE C-2059 * LANEX * NCI-C08695 * PAKHTARAN * 3-(5-TRIFLUORMETHYLPHENYL)-,1-DIMETHYL-HARNSTOFF (GERMAN) * N-(m-TRIFLUOROMETHYL-

PHENYL)-N′,N′-DIMETHYLUREA * N-(3-TRIFLUORO-METHYLPHENYL)-N′-N′-DIMETHYLUREA * 3-(m-TRI-FLUOROMETHYLPHENYL)-1,1-DIMETHYLUREA

CONSENSUS REPORTS: EPA Genetic Toxicology Program. IARC Cancer Review: GROUP 3 IMEMDT 7,56,87; Animal Inadequate Evidence IMEMDT 30,245,83. NCI Carcinogenesis Bioassay (feed); No Evidence: rat NCITR* NCI-CG-TR-195,80; Equivocal Evidence: mouse NCITR* NCI-CG-TR-195,80. Reported in EPA TSCA Inventory.

SAFETY PROFILE: Questionable carcinogen with experimental carcinogenic data. Moderately toxic by ingestion, intraperitoneal, and possibly other routes. Mutation data reported. When heated to decomposition it emits very toxic fumes of F^- and NO_x.

DUL800   CAS: 5152-30-7   HR: 3
o,o′-DIMETHYLTUBOCURARINE
mf: $C_{40}H_{48}N_2O_6$ mw: 652.90

SYNS: DIMETHYL TUBOCURARINE * o,o-DIMETH-YLTUBOCURARINE * N,N′,o,o-TETRAMETHYL-(+)-TUBOCURINE

SAFETY PROFILE: Poison by intravenous route. When heated to decomposition it emits toxic fumes of NO_x.

DUM200   CAS: 96-31-1   HR: D
1,3-DIMETHYLUREA
mf: $C_3H_8N_2O$ mw: 88.13

PROP: Colorless crystals, water- and alc-sol. D: 1.14, mp: 106°, bp: 270°.

SYNS: N,N′-DIMETHYLHARNSTOFF (GERMAN) * N,N′-DIMETHYLUREA * sym-DIMETHYLUREA * SYMMETRIC DIMETHYLUREA

CONSENSUS REPORTS: Reported in EPA TSCA Inventory.

SAFETY PROFILE: Experimental teratogenic and reproductive effects. Human mutation data reported. When heated to decomposition it emits toxic fumes of NO_x.

DUO400   CAS: 119-48-2   HR: 3
DIMORPHOLAMINE
mf: $C_{20}H_{38}N_4O_4$ mw: 398.62

SYNS: AMIPAN T * N,N′-DIBUTYL-N,N′-DICARBOX-YETHYLENE DIAMINEMORPHOLIDE * N,N′-DIBUTYL-

N,N'-DICARBOXYMORPHOLIDE-ETHYLENEDIAMINE
* N,N'-DI-n-BUTYLETHYLENEDIAMINE-N,N'-DICAR-
BOXYBISMORPHOLIDE * N,N'-1,2-ETHANEDIYLBIS(N-
BUTYL-4-MORPHOLINECARBOXAMIDE) * N,N'-ETHY-
LENEBIS(N-BUTYL-4-MORPHOLINECARBOXAMIDE)
* PRONTODIN * 1064 TH * THERALEPTIQUE
* THERAPTIQUE

SAFETY PROFILE: Poison by ingestion, intra-peritoneal, intravenous, intramuscular, and sub-cutaneous routes. An analeptic agent (stimulant). When heated to decomposition it emits toxic fumes of NO_x.

DUP300   CAS: 148-01-6   HR: 3
DINITOLMIDE
mf: $C_8H_7N_3O_5$ mw: 225.18

PROP: Yellowish solid. Mp: 177°. Very sltly sol in water; sol in acetone, acetonitrile, and dimethyl formamide.

SYNS: COCCIDINE A * COCCIDOT * DINITOL-MID * DINITOLMIDE * 3,5-DINITRO-o-TOLUAMIDE * D.O.T. * 2-METHYL-3,5-DINITROBENZAMIDE * ZOALENE * ZOAMIX

OSHA PEL: TWA 5 mg/m³
ACGIH TLV: TWA 5 mg/m³

SAFETY PROFILE: Poison by intravenous route. Moderately toxic by ingestion. Mutation data reported. A strong exothermic reaction above 248°C has caused industrial explosions. When heated to decomposition it emits toxic fumes of NO_x.

DUP600   CAS: 97-02-9   HR: 3
2,4-DINITROANILINE
mf: $C_6H_5N_3O_4$ mw: 183.14

PROP: Yellow, needle-like crystals; insol in water. Mp: 188°, flash p: 435°F (CC), d: 1.615, vap d: 6.31.

SYNS: 2,4-DINITRANILINE * 2,4-DINITROANILIN (GERMAN) * 2,4-DINITROANILINA (ITALIAN) * 2,4-DINITROBENZENAMIME * DNA * NCI-C60753

CONSENSUS REPORTS: Reported in EPA TSCA Inventory.

SAFETY PROFILE: Poison by ingestion and intraperitoneal routes. Experimental teratogenic and reproductive effects. Mutation data reported. An eye irritant. Combustible and explosive when exposed to heat or flame; can react

with oxidizing materials. To fight fire use CO_2, dry chemical. Mixtures with charcoal ignite at 350°C. Vigorous reaction with chlorine + hydrochloric acid evolves gases. When heated to decomposition emits highly toxic fumes of NO_x.

DUP800   CAS: 119-27-7   HR: 3
2,4-DINITROANISOL
mf: $C_7H_6N_2O_5$ mw: 198.15

PROP: Colorless to yellow crystals. Mp: 89°, bp: sublimes, d: 1.341 @ 20°/4°, vap d: 6.83.

SYNS: 2,4-DINITROANISOLE * α-DINITROANISOLE * 2,4-DINITROPHENYLMETHYL ETHER * 1-METHOXY-2,4-DINITROBENZENE

CONSENSUS REPORTS: Reported in EPA TSCA Inventory.

SAFETY PROFILE: Poison by ingestion. Mutation data reported. When heated to decomposition it emits toxic fumes of NO_x.

DUQ180   CAS: 25154-54-5   HR: 3
DINITROBENZENE
DOT: UN 1597
mf: $C_6H_4N_2O_4$ mw: 168.12

SYNS: DINITROBENZENE, solution (DOT) * DINITROBENZOL solid (DOT)

OSHA PEL: TWA 1 mg/m³ (skin)
ACGIH TLV: TWA 0.15 ppm (skin)
DFG MAK: 0.15 ppm (1 mg/m³); Suspected Carcinogen
DOT Classification: Poison B; Label: Poison, Solid and Solution.

SAFETY PROFILE: Suspected carcinogen. A poison. When heated to decomposition it emits toxic fumes of NO_x.

DUQ200   CAS: 99-65-0   HR: 3
m-DINITROBENZENE
DOT: UN 1597
mf: $C_6H_4N_2O_4$ mw: 168.12

PROP: Yellowish crystals. Mp: 89°, bp: 301°.

SYNS: BINITROBENZENE * 1,3-DINITROBENZENE * 2,4-DINITROBENZENE * 1,3-DINITROBENZOL * DWUNITROBENZEN (POLISH)

CONSENSUS REPORTS: Reported in EPA TSCA Inventory. EPA Genetic Toxicology Program.

OSHA PEL: TWA 1 mg/m^3 (skin)
ACGIH TLV: TWA 0.15 ppm (skin)
DFG MAK: 0.15 ppm (1 mg/m^3); Suspected Carcinogen
DOT Classification: Poison B; Label: Poison.

SAFETY PROFILE: Suspected carcinogen. Human poison by ingestion. Experimental poison by ingestion, intraperitoneal, and intravenous routes. Human systemic effects by skin contact: cyanosis and motor activity changes. Experimental reproductive effects. Mutation data reported. Mixture with nitric acid is a high explosive. Mixture with tetranitromethane is a high explosive very sensitive to sparks. When heated to decomposition it emits toxic fumes of NO$_x$.

DUQ400   CAS: 528-29-0   HR: 3
o-DINITROBENZENE

DOT: UN 1597
mf: C$_6$H$_4$N$_2$O$_4$ mw: 168.12

PROP: Colorless needles or plates. Mp: 118°, bp: 319°, flash p: 302°F (CC), d: 1.571 @ 0°/4°, vap d: 5.79.

SYN: 1,2-DINITROBENZENE

OSHA PEL: TWA 1 mg/m^3 (skin)
ACGIH TLV: TWA 0.15 ppm (skin)
DFG MAK: 0.15 ppm (1 mg/m^3); Suspected Carcinogen
DOT Classification: Poison B; Label: Poison.

SAFETY PROFILE: Suspected carcinogen. Poison by inhalation and ingestion. Moderately toxic by skin contact. Can cause liver, kidney, and central nervous system injury. Combustible when exposed to heat or flame; can react vigorously with oxidizing materials. A severe explosion hazard when shocked or exposed to heat or flame. It is used in bursting charges and to fill artillery shells. Mixtures with nitric acid are highly explosive. To fight fire, use water, CO$_2$, dry chemical. Dangerous; when heated to decomposition it emits highly toxic fumes of NO$_x$ and explodes.

DUQ600   CAS: 100-25-4   HR: 3
p-DINITROBENZENE

DOT: UN 1597
mf: C$_6$H$_4$N$_2$O$_4$ mw: 168.12

PROP: White crystals. Mp: 173°, bp: 299°. Volatile with steam.

SYN: DITHANE A-4

CONSENSUS REPORTS: Reported in EPA TSCA Inventory.

OSHA PEL: TWA 1 mg/m^3 (skin)
ACGIH TLV: TWA 0.15 ppm (skin)
DFG MAK: 0.15 ppm (1 mg/m^3); Suspected Carcinogen
DOT Classification: Poison B; Label: Poison.

SAFETY PROFILE: Suspected carcinogen. Poison by ingestion. Mutation data reported. Mixture with nitric acid is a high explosive. When heated to decomposition it emits toxic fumes of NO$_x$.

DUR800   CAS: 87-31-0   HR: 3
5,7-DINITRO-1,2,3-BENZOXADIAZOLE

DOT: UN 0074
mf: C$_6$H$_2$N$_4$O$_5$ mw: 210.12

SYNS: DDNP * DIAZO * 2-DIAZO-4,6-DINITRO-BENZENE-1-OXIDE * DIAZODINITROPHENOL (DOT) * DIAZODINITROPHENOL, containing, by weight, at least 40% water (DOT) * DIAZODINITROPHENOL, dry (DOT) * INITIATING EXPLOSIVE DIAZODINITROPHENOL (DOT)

DOT Classification: Class A Explosive; Label: Explosive A; Forbidden, Dry.

SAFETY PROFILE: An explosive. When heated to decomposition it emits toxic fumes of NO$_x$.

DUS000   CAS: 1528-74-1   HR: 3
4,4′-DINITROBIPHENYL
mf: C$_{12}$H$_8$N$_2$O$_4$ mw: 244.22

SYN: 4,4′-DINITROBIFENYL (CZECH)

CONSENSUS REPORTS: Reported in EPA TSCA Inventory.

SAFETY PROFILE: Questionable carcinogen with experimental tumorigenic data. Mildly toxic by ingestion. Mutation data reported. An eye irritant. When heated to decomposition it emits toxic fumes of NO$_x$.

DUS600   CAS: 2401-85-6   HR: 3
2,4-DINITRO-1-CHLORO-NAPHTHALENE
mf: C$_{10}$H$_5$ClN$_2$O$_4$ mw: 252.62

SYN: 1-CHLORO-2,4-DINITRONAPHTHALENE

CONSENSUS REPORTS: Reported in EPA TSCA Inventory.

SAFETY PROFILE: Poison by unspecified route. Questionable carcinogen with experimental carcinogenic and neoplastigenic data. When heated to decomposition it emits very toxic fumes of Cl^- and NO_x.

DUS700   CAS: 534-52-1   **HR: 3**
DINITRO-o-CRESOL
mf: $C_7H_6N_2O_5$ mw: 198.15

PROP: Yellow, prismatic crystals. Mp: 85.8°, vap d: 6.82.

SYNS: ANTINONIN * ARBOROL * CAPSINE * CHEMSECT DNOC * DEGRASSAN * DEKRYSIL * DETAL * DINITROCRESOL * 2,4-DINITRO-o-CRESOL * 4,6-DINITRO-o-CRESOL * 4,6-DINITRO-o-CRESOLO (ITALIAN) * DINITRODENDTROXAL * 3,5-DINITRO-2-HYDROXYTOLUENE * 4,6-DINI-TRO-o-KRESOL (CZECH) * 4,6-DINITROKRESOL (DUTCH) * DINITROL * DINITROMETHYL CYCLO-HEXYLTRIENOL * 2,4-DINITRO-6-METHYLPHENOL * DINOC * DINURANIA * DITROSOL * DN-DRY MIX NO.2 * DNOK (CZECH) * DWUNITRO-o-KREZOL (POLISH) * EFFUSAN * ELGETOL * ELIPOL * ENT 154 * EXTRAR * HEDOLIT * K III * KRENITE (OBS.) * KRESAMONE * KREZOTOL 50 * LE DINITROCRESOL-4,6 (FRENCH) * LIPAN * 2-METHYL-4,6-DINITROPHENOL * NITRADOR * NITROFAN * PROKARBOL * RAFEX * RAPHATOX * RCRA WASTE NUMBER P047 * SANDOLIN * SELINON * SINOX * TRIFOCIDE * TRIFRINA * WINTERWASH * ZAHLREICHE BEZEICHNUNGEN (GERMAN)

CONSENSUS REPORTS: Reported in EPA TSCA Inventory. EPA Genetic Toxicology Program. Community Right-To-Know List. EPA Extremely Hazardous Substances List.

OSHA PEL: TWA 0.2 mg/m³ (skin)
ACGIH TLV: TWA 0.2 mg/m³ (skin)
DFG MAK: 0.2 mg/m³)
NIOSH REL: (Dinitro-Ortho-Cresol) TWA 0.2 mg/m³

SAFETY PROFILE: Human poison by unspecified route. Experimental poison by ingestion, inhalation, skin contact, intraperitoneal, intravenous, and possibly other routes. Human systemic effects by ingestion and inhalation: somnolence, headache, brain recordings from specific areas of the central nervous system, cardiac and gastrointestinal changes. Mutation data reported. An eye and skin irritant. Less toxic than the para form, but is still highly toxic. A pesticide.

DUS800   CAS: 1335-85-9   **HR: 3**
DINITRO-o-CRESOL
DOT: UN 1598
mf: $C_7H_6N_2O_5$ mw: 198.15

SYNS: DINITRO-o-CRESOL, liquid (DOT) * DINITRO-o-CRESOL, solid (DOT) * 2-METHYLDINITROPHENOL

NIOSH REL: (Dinitro-Ortho-Cresol) TWA 0.2 mg/m³
DOT Classification: Poison B; Label: Poison; Poison B; Label: St. Andrews Cross.

SAFETY PROFILE: Poison by unspecified route. When heated to decomposition it emits toxic fumes of NO_x.

DUT000   CAS: 497-56-3   **HR: 3**
3,5-DINITRO-o-CRESOL
mf: $C_7H_6N_2O_5$ mw: 198.15

NIOSH REL: (Dinitro-Ortho-Cresol) TWA 0.2 mg/m³

SAFETY PROFILE: Poison by subcutaneous route. When heated to decomposition it emits toxic fumes of NO_x.

DUT200   CAS: 63989-82-2   **HR: 3**
3,5-DINITRO-p-CRESOL
mf: $C_7H_6N_2O_5$ mw: 198.15

PROP: Crystals.

SAFETY PROFILE: Poison by intraperitoneal and possibly other routes. Strong irritant to eyes, skin, and mucous membranes. Can cause brain, liver, and kidney damage by various routes. When heated to decomposition it emits toxic fumes of NO_x.

DUT600   CAS: 609-93-8   **HR: 3**
2,6-DINITRO-p-CRESOL
mf: $C_7H_6N_2O_5$ mw: 198.15

SYNS: DINITRO-p-CRESOL * DNPC * VICTORIA ORANGE * VICTORIA YELLOW

CONSENSUS REPORTS: Reported in EPA TSCA Inventory.

SAFETY PROFILE: Poison by intraperitoneal route. Mutation data reported. When heated to decomposition it emits toxic fumes of NO_x.

DUU600   CAS: 2312-76-7   **HR: 3**
4,6-DINITRO-o-CRESOL SODIUM SALT
mf: $C_7H_5N_2O_5 \cdot Na$ mw: 220.13

PROP: Brilliant, orange-yellow dye.

SYNS: CORODINOC * CRESOTOL * DINITRO-o-
CRESOL SODIUM SALT * 3,5-DINITRO-o-CRESOL SO-
DIUM SALT * 2,4-DINITRO-6-METHYLPHENOL SO-
DIUM SALT * DINOC * DNOC SOLDIUM SALT
* DYNOSOL * EK 54 * ELGETOL * KRENITE
(OBS.) * KREZONITE * 2-METHYL-4,6-DINITRO-
PHENOL SODIUM SALT * SINOX * SODIUM-4,6-DI-
NITRO-o-CRESOXIDE * SODIUM SALT of 4,6-DINITRO-
o-CRESOL

CONSENSUS REPORTS: Reported in EPA
TSCA Inventory.

NIOSH REL: (Dinitro-Ortho-Cresol) TWA 0.2
  mg/m^3

SAFETY PROFILE: Poison by ingestion, skin
contact and subcutaneous routes. Flammable.
A pesticide. When heated to decomposition it
emits toxic fumes of Na_2O.

DUV600   CAS: 1582-09-8   **HR: 3**
2,6-DINITRO-N,N-DIPROPYL-
4-(TRIFLUOROMETHYL)BENZENAMINE
mf: $C_{13}H_{16}F_3N_3O_4$ mw: 335.32

PROP: Technical product contains 84-88 ppm
diproplynitrosoamine NCITR* NCI-CG-TR-
34,78.

SYNS: AGREFLAN * AGRIFLAN 24 * CRISA-
LIN * DIGERMIN * 2,6-DINITRO-N,N-DI-N-PRO-
PYL-α,α,α-TRIFLURO-p-TOLUIDINE * 2,6-DINITRO-4-
TRIFLUORMETHYL-N,N-DIPROPYLANILIN (GERMAN)
* 4-(DI-N-PROPYLAMINO)-3,5-DINITRO-1-TRIFLUORO-
METHYLBENZENE * N,N-DI-N-PROPYL-2,6-DINITRO-4-
TRIFLUOROMETHYLANILINE * N,N-DIPROPYL-4-TRI-
FLUOROMETHYL-2,6-DINITROANILINE * ELANCO-
LAN * L-36352 * LILLY 36,352 * NCI-C00442
* NITRAN * OLITREF * SU SEGURO CARPIDOR
* TREFANOCIDE * TREFICON * TREFLAM
* TREFLAN * TREFLANOCIDE ELANCOLAN
* TRIFLUORALIN (USDA) * TRIFLURALIN
* TRIFLURALINE * α,α,α-TRIFLUORO-2,6-
DINITRO-N,N-DIPROPYL-p-TOLUIDINE * TRIFU-
REX * TRIKEPIN * TRIM

CONSENSUS REPORTS: NCI Carcinogen-
esis Bioassay (feed); Clear Evidence: mouse
NCITR* NCI-CG-TR-34,78; No Evidence: rat
NCITR* NCI-CG-TR-34,78. EPA Genetic Tox-
icology Program. Community Right-To-Know
List.

SAFETY PROFILE: Moderately toxic by ingestion and intraperitoneal routes. Experimental

teratogenic and reproductive effects. Questionable carcinogen with experimental carcinogenic
and tumorigenic data. Human mutation data reported. When heated to decomposition it emits
very toxic fumes of F^- and NO_x.

DUW400   CAS: 70-34-8   **HR: 3**
2,4-DINITRO-1-FLUOROBENZENE
mf: $C_6H_3FN_2O_4$ mw: 186.11

PROP: Crystals; sol in ether, benzene, propylene glycol. Mp: 26°, bp: 137° @ 20 mm.

SYNS: 2,4-DINITROFLUOROBENZENE * 2,4-DNFB
* 1,2,4-FLUORODINITROBENZENE * 1-FLUORO-2,4-
DINITROBENZENE

CONSENSUS REPORTS: Reported in EPA
TSCA Inventory. EPA Genetic Toxicology Program.

SAFETY PROFILE: Poison by ingestion, skin
contact, and subcutaneous routes. A powerful
irritant and vesicant. Mutation data reported.
Solutions in ether may explode when evaporated. When heated to decomposition it emits
highly toxic fumes of NO_x and F^-.

DUX700   CAS: 605-71-0   **HR: 3**
1,5-DINITRONAPHTHALENE
mf: $C_{10}H_6N_2O_4$ mw: 218.17

DFG MAK: Suspected Carcinogen. (all isomers)

SAFETY PROFILE: Mixtures with sulfur or sulfuric acid (used in commercial reactions) may
explode if heated to 120°C. Initiation temperature depends on the quality of the dinitronaphthalene. When heated to decomposition it
emits toxic fumes of NO_x.

DUX800   CAS: 605-69-6   **HR: 3**
2,4-DINITRO-1-NAPHTHOL
mf: $C_{10}H_6N_2O_5$ mw: 234.18

PROP: Yellow needles or leaflets. Mp: 138°,
vap d: 8.08.

SYNS: 2-4 DINITRO-α-NAPHTOL (FRENCH) * GOL-
DEN YELLOW * MANCHESTER YELLOW * MARI-
TUS YELLOW * NAPHTHOL YELLOW * NAPHTHY-
LENE YELLOW * SAFFRON YELLOW

CONSENSUS REPORTS: Reported in EPA
TSCA Inventory.

SAFETY PROFILE: Poison by subcutaneous,
intramuscular, intravenous, and intraperitoneal

routes. Human reproductive effects by skin contact: toxic to the skin.

DUY600   CAS: 25550-58-7   **HR: 3**
DINITROPHENOL

DOT: UN 0076/UN 1320/UN 1599
mf: $C_6H_4N_2O_5$ mw: 184.12

DOT Classification: Poison B; Label: Poison (solution); Class A Explosive; Label: Explosive A and Poison; Poison B; Label: Flammable Liquid and Poison (solution).

SAFETY PROFILE: Poison by ingestion and subcutaneous routes. An explosive. When heated to decomposition it emits toxic fumes of NO_x.

DUY800   CAS: 66-56-8   **HR: 3**
2,3-DINITROPHENOL
mf: $C_6H_4N_2O_5$ mw: 184.12

PROP: Yellow needles. Mp: 144°, d: 1.681 @ 20°, vap d: 6.35.

SAFETY PROFILE: Poison by unspecified route. Inhalation of dust can be fatal. A skin irritant and an allergen. A powerful stimulant of the metabolism by excessive oxidation. Highly explosive when exposed to heat. It is used as a component of some shell and bomb charges.

DUZ000   CAS: 51-28-5   **HR: 3**
2,4-DINITROPHENOL
mf: $C_6H_4N_2O_5$ mw: 184.12

PROP: Yellow crystals. Mp: 112°, d: 1.683 @ 24°, vap d: 6.35.

SYNS: ALDIFEN * CHEMOX PE * 2,4-DINITRO-FENOL (DUTCH) * DINITROFENOLO (ITALIAN) * α-DINITROPHENOL * 2,4-DNP * FENOXYL CARBON N * 1-HYDROXY-2,4-DINITROBENZENE * MAROXOL-50 * NITRO KLEENUP * NSC 1532 * RCRA WASTE NUMBER P048 * SOLFO BLACK B * SOLFO BLACK BB * SOLFO BLACK 2B SUPRA * SOLFO BLACK G * SOLFO BLACK SB * TER-TROSULPHUR BLACK PB * TERTROSULPHUR PBR

CONSENSUS REPORTS: Reported in EPA TSCA Inventory. EPA Genetic Toxicology Program.

SAFETY PROFILE: A deadly human poison by ingestion. An experimental poison by ingestion, inhalation, intravenous, intraperitoneal, subcutaneous, intramuscular, and possibly other routes. Moderately toxic by skin contact. Experimental teratogenic and reproductive effects. Mutation data reported. A skin irritant. Phytotoxic. A pesticide. An explosive. Forms explosive salts with alkalies and ammonia. When heated to decomposition it emits toxic fumes of NO_x.

DVA200   CAS: 573-56-8   **HR: 3**
2,6-DINITROPHENOL
mf: $C_6H_4N_2O_5$ mw: 184.12

PROP: Yellow crystals. Mp: 63°, vap d: 6.35. Sltly sol in cold water, alc. Very sol in chloroform, ether or boiling alc; also sol in fixed alkali solns.

SYN: β-DINITROPHENOL

SAFETY PROFILE: Poison by intramuscular and possibly other routes. Moderately explosive when exposed to heat.

DVA400   CAS: 577-71-9   **HR: 3**
3,4-DINITROPHENOL
mf: $C_6H_4N_2O_5$ mw: 184.12

SAFETY PROFILE: A poison by unspecified routes. When heated to decomposition it emits toxic fumes of NO_x.

DVA600   CAS: 586-11-8   **HR: 3**
3,5-DINITROPHENOL
mf: $C_6H_4N_2O_5$ mw: 184.12

SAFETY PROFILE: A poison by unspecified routes. When heated to decomposition it emits toxic fumes of NO_x.

DVC800   CAS: 63732-56-9   **HR: 3**
2,4-DINITROPHENYLMORPHINE
HYDROCHLORIDE
mf: $C_{23}H_{21}N_3O_7 \cdot ClH$ mw: 487.93

SYN: 2,4-DINITROPHENYL ETHER of MORPHINE

SAFETY PROFILE: Poison by intravenous, intraperitoneal, and subcutaneous routes. Moderately toxic by parenteral route. When heated to decomposition it emits very toxic fumes of HCl and NO_x.

DVD400   CAS: 75321-20-9   **HR: 3**
1,3-DINITROPYRENE
mf: $C_{16}H_8N_2O_4$ mw: 292.26

SYN: DINITROPYRENE

DFG MAK: Suspected Carcinogen.

SAFETY PROFILE: Suspected carcinogen with experimental carcinogenic data. Mutation data reported. When heated to decomposition it emits toxic fumes of NO_x.

DVD600   CAS: 42397-64-8   HR: 3
1,6-DINITROPYRENE
mf: $C_{16}H_8N_2O_4$ mw: 292.26

SYN: DINITROPYRENE

DFG MAK: Suspected Carcinogen.

SAFETY PROFILE: Suspected carcinogen with experimental carcinogenic data. Human mutation data reported. When heated to decomposition it emits toxic fumes of NO_x.

DVD800   CAS: 42397-65-9   HR: 3
1,8-DINITROPYRENE
mf: $C_{16}H_8N_2O_4$ mw: 292.26

SYN: DINITROPYRENE

DFG MAK: Suspected Carcinogen.

SAFETY PROFILE: Suspected carcinogen with experimental carcinogenic data. Human mutation data reported. When heated to decomposition it emits toxic fumes of NO_x.

DVE000   CAS: 1596-52-7   HR: 3
4,6-DINITROQUINOLINE-1-OXIDE
mf: $C_9H_5N_3O_5$ mw: 235.17

SAFETY PROFILE: Questionable carcinogen with experimental carcinogenic data. Human mutation data reported. When heated to decomposition it emits toxic fumes of NO_x.

DVF200   CAS: 140-79-4   HR: 3
DINITROSOPIPERAZINE
mf: $C_4H_8N_4O_2$ mw: 144.16

PROP: White crystals. Mp: 158°, vap d: 4.97.

SYNS: DINITROSOPIPERAZIN (GERMAN) * N,N′-DINITROSOPIPERAZINE * 1,4-DINITROSOPIPERAZINE * DNPZ * NSC 339 * USAF DO-36

CONSENSUS REPORTS: EPA Genetic Toxicology Program. Reported in EPA TSCA Inventory.

SAFETY PROFILE: Suspected carcinogen with experimental carcinogenic, neoplastigenic and tumorigenic data. Poison by ingestion, subcuta-

neous, and intraperitoneal routes. Experimental teratogenic and reproductive effects. Human mutation data reported. When heated to decomposition it emits toxic fumes of NO_x.

DVG600   CAS: 25321-14-6   HR: 3
DINITROTOLUENE
DOT: UN 1600/UN 2038
mf: $C_7H_6N_2O_4$ mw: 182.15

SYNS: DINITROPHENYLMETHANE * ar,ar-DINITRO-TOLUENE * DINITROTOLUENE, liquid (DOT) * DINITROTOLUENE, molten (DOT) * DINITRO-TOLUENE, solid (DOT) * METHYLDINITROBEN-ZENE

CONSENSUS REPORTS: Reported in EPA TSCA Inventory. EPA Genetic Toxicology Program.

OSHA PEL: TWA 1.5 mg/m³ (skin)
DFG MAK: Animal Carcinogen, Suspected Human Carcinogen.
NIOSH REL: (Dinitrotoluene): Reduce to lowest level
DOT Classification: ORM-E; Label: None, Liquid and Solid; Poison B; Label: Flammable Liquid and Poison (Liquid); Poison B; Label: Poison (Solid and Molten).

SAFETY PROFILE: Confirmed carcinogen with experimental tumorigenic data. A poison. Experimental teratogenic and reproductive effects. Mutation data reported. Flammable. When heated to decomposition it emits toxic fumes of NO_x.

DVH000   CAS: 121-14-2   HR: 3
2,4-DINITROTOLUENE
mf: $C_7H_6N_2O_4$ mw: 182.15

PROP: Yellow needles. Mp: 69.5°, bp: 300°, d: 1.521 @ 15°, vap d: 6.27, flash p: 404°F.

SYNS: 2,4-DINITROTOLUOL * DNT * 2,4-DNT * 1-METHYL-2,4-DINITROBENZENE * NCI-C01865 * RCRA WASTE NUMBER U105

CONSENSUS REPORTS: NCI Carcinogenesis Bioassay (feed); No Evidence: mouse NCITR* NCI-CG-TR-54,78; Some Evidence: rat NCITR* NCI-CG-TR-54,78. Reported in EPA TSCA Inventory.

OSHA PEL: TWA 1.5 mg/m³ (skin)
ACGIH TLV: TWA 1.5 mg/m³ (skin)

NIOSH REL: (Dinitrotoluene): Reduce to lowest level

SAFETY PROFILE: Suspected carcinogen with experimental carcinogenic and neoplastigenic data. Poison by ingestion and subcutaneous routes. Experimental reproductive effects. Mutation data reported. An irritant and an allergen. Can cause anemia, methemoglobinemia, cyanosis, and liver damage. Combustible when exposed to heat or flame; can react with oxidizing materials. To fight fire, use water spray or mist, dry chemical. Decomposes when heated to 250°C. There are instances of explosion during manufacture or storage. Mixture with nitric acid is a high explosive. Mixture with sodium carbonate can decompose with significant pressure increase at 210°C. Mixtures with other alkalies may have the same effect. Ignites on contact with sodium oxide. When heated to decomposition it emits toxic fumes of NO_x.

DVI600   CAS: 6379-46-0   HR: 3
4,6-DINITRO-1,2,3-TRICHLORO-BENZENE
mf: $C_6HCl_3N_2O_4$ mw: 271.44

SYNS: 1,2,3-TRICHLORO-4,6-DINITROBENZENE * VANCIDE PB

SAFETY PROFILE: Questionable carcinogen with experimental carcinogenic data. When heated to decomposition it emits very toxic fumes of NO_x and Cl^-.

DVJ200   CAS: 363-24-6   HR: 3
DINOPROSTONE
mf: $C_{20}H_{32}O_5$ mw: 352.52

SYNS: (5Z,11-α,13E,15S)-11,15-DIHYDROXY-9-OXOP-ROSTA-5,13-DIEN-1-OIC ACID * 7-(3-HYDROXY-2-(3-HYDROXY-1-OCTENYL)-5-OXOCYCLOPENTYL)-5-HEP-TENOIC ACID * PGE2 * PROSTAGLANDIN E2 * (−)-PROSTAGLANDIN E2 * (15S)-PROSTAGLANDIN E2 * PROSTIN E2 * U-12062

SAFETY PROFILE: Poison by subcutaneous and intravenous routes. Moderately toxic by ingestion and intraperitoneal routes. Human reproductive effects by intravenous, intraplacental, and intravaginal routes: changes in the uterus, cervix and vagina; termination of pregnancy; and changes in fertility. Experimental teratogenic and reproductive effects. Mutation data reported. When heated to decomposition it emits acrid smoke and irritating fumes.

DVK600   CAS: 141-02-6   HR: 3
DIOCTYL FUMARATE
mf: $C_{20}H_{36}O_4$ mw: 340.56

PROP: Clear, mobile liquid; mild odor. Bp: 211-220°, flash p: 365°F (COC), d: 0.942 @ 20°/20°.

SYNS: BIS(2-ETHYLHEXYL) FUMARATE * 2-BUTEN-EDIOIC ACID BIS(2-ETHYLHEXYL) ESTER * DI(2-ETH-YLHEXYL) FUMARATE * DOF * 2-ETHYLHEXYL FUMARATE * RC COMONOMER DOF

CONSENSUS REPORTS: Reported in EPA TSCA Inventory.

SAFETY PROFILE: Poison by intraperitoneal route. An eye and severe skin irritant. Combustible when exposed to heat or flame; can react with oxidizing materials. To fight fire, use foam, CO_2, dry chemical.

DVL600   CAS: 117-84-0   HR: 2
n-DIOCTYL PHTHALATE
mf: $C_{24}H_{38}O_4$ mw: 390.62

SYNS: o-BENZENEDICARBOXYLIC ACID DIOCTYL ES-TER * 1,2-BENZENEDICARBOXYLIC ACID DI-OCYTL ESTER * CELLUFLEX DOP * DINOPOL NOP * DIOCTYL-o-BENZENEDICARBOXYLATE * DI-OCTYL PHTHALATE * DNOP * OCTYL PHTHAL-ATE * n-OCTYL PHTHALATE * PX-138 * RCRA WASTE NUMBER U107 * VINICIZER 85

CONSENSUS REPORTS: Reported in EPA TSCA Inventory.

SAFETY PROFILE: Mildly toxic by ingestion. Experimental teratogenic and reproductive effects. A skin and severe eye irritant. Used as a plasticizer. When heated to decomposition it emits acrid smoke and irritating fumes.

DVL700   CAS: 117-81-7   HR: 3
DI-sec-OCTYL PHTHALATE
mf: $C_{24}H_{38}O_4$ mw: 390.62

SYNS: BEHP * BIS(2-ETHYLHEXYL)-1,2-BENZENE-DICARBOXYLATE * BIS(2-ETHYLHEXYL)PHTHALATE * BISOFLEX 81 * BISOFLEX DOP * COMPOUND 889 * DAF 68 * DEHP * DI(2-ETHYLHEXYL)OR-THOPHTHALATE * DI(2-ETHYLHEXYL)PHTHALATE * DIOCTYL PHTHALATE * DOP * ETHYLHEXYL PHTHALATE * ERGOPLAST FDO * 2-ETHYLHEXYL PHTHALATE * EVIPLAST 80 * EVIPLAST 81 * FLEXIMEL * FLEXOL DOP * FLEXOL PLASTI-CIZER DOP * GOOD-RITE GP 264 * HATCOL DOP * HERCOFLEX 260 * KODAFLEX DOP * MOL-

LAN O * NCI-C52733 * NUOPLAZ DOP * OC-TOIL * OCTYL PHTHALATE * PALATINOL AH * PHTHALIC ACID DIOCTYL ESTER * PITTSBURGH PX-138 * PLATINOL AH * PLATINOL DOP * RC PLASTICIZER DOP * RCRA WASTE NUMBER U028 * REOMOL DOP * REOMOL D 79P * SICOL 150 * STAFLEX DOP * TRUFLEX DOP * VESTINOL AH * VINICIZER 80 * WITCIZER 312

CONSENSUS REPORTS: IARC Cancer Review: GROUP 2B IMEMDT 7,56,87; Human Inadequate Evidence IMEMDT 29,269,82; Animal Sufficient Evidence IMEMDT 29,269,82. NTP Fourth Annual Report On Carcinogens, 1984. NTP Carcinogenesis Bioassay (feed); Clear Evidence: mouse, rat NTPTR* NTP-TR-217,82. EPA Genetic Toxicology Program. Reported in EPA TSCA Inventory. Community Right-To-Know List.

OSHA PEL: (Transitional: TWA 5 mg/m^3) TWA 5 mg/m^3; STEL 10 mg/m^3
ACGIH TLV: TWA 5 mg/m^3; STEL 10 mg/m^3
DFG MAK: 10 mg/m^3
NIOSH REL: (DEHP) Reduce to lowest feasible level

SAFETY PROFILE: Confirmed carcinogen with experimental carcinogenic and tumorigenic data. Experimental teratogen data. Poison by intravenous route. Human systemic effects by ingestion: gastrointestinal tract effects. A mild skin and eye irritant. When heated to decomposition it emits acrid smoke.

DVL800   CAS: 69226-45-5   HR: 3
DIOCTYL(1,2-PROPYLENEDIOXYBIS (MALEOYLDIOXY))STANNANE
mf: C$_{27}$H$_{42}$O$_8$Sn   mw: 613.38

SYNS: DI-n-OCTYLTIN DI(1,2-PROPYLENEGLYCOLMALEATE) * DI-n-OCTYL-ZINN-DI-(1,2-PROPYLENGLYKOLMALEINAT)(GERMAN)

OSHA PEL: TWA 0.1 mg(Sn)/m^3 (skin)
ACGIH TLV: TWA 0.1 mg(Sn)/m^3 (skin) (Proposed: TWA 0.1 mg(Sn)/m^3; STEL 0.2 mg(Sn)/m^3 (skin))
NIOSH REL: (Organotin Compounds) TWA 0.1 mg(Sn)/m^3

SAFETY PROFILE: Poison by intraperitoneal route. Mildly toxic by ingestion. When heated to decomposition it emits acrid smoke and irritating fumes.

DVN909   CAS: 3594-15-8   HR: 3
DIOCTYLTIN-3,3'-THIODIPROPIONATE
mf: C$_{22}$H$_{42}$O$_4$SSn   mw: 521.39

SYN: 2,2-DIOCTYL-1,3-DIOXA-2-STANNA-7-THIADECAN-4,10-DIONE

OSHA PEL: TWA 0.1 mg(Sn)/m^3 (skin)
ACGIH TLV: TWA 0.1 mg(Sn)/m^3 (skin) (Proposed: TWA 0.1 mg(Sn)/m^3; STEL 0.2 mg(Sn)/m^3 (skin))
NIOSH REL: (Organotin Compounds) TWA 0.1 mg(Sn)/m^3

SAFETY PROFILE: Poison by intravenous route. When heated to decomposition it emits toxic fumes of SO$_x$.

DVQ000   CAS: 123-91-1   HR: 3
DIOXANE
DOT: UN 1165
mf: C$_4$H$_8$O$_2$ mw: 88.11

PROP: Colorless liquid, pleasant odor. Mp: 12°, bp: 101.1°, lel: 2.0%, uel: 22.2%, flash p: 54°F (CC), d: 1.0353 @ 20°/4°, autoign temp: 356°F, vap press: 40 mm @ 25.2°, vap d: 3.03.

SYNS: DIETHYLENE DIOXIDE * 1,4-DIETHYLENE DIOXIDE * DIETHYLENE ETHER * DI(ETHYLENE OXIDE) * DIOKAN * DIOKSAN (POLISH) * DIOSSANO-1,4 (ITALIAN) * DIOXAAN-1,4 (DUTCH) * 1,4-DIOXACYCLOHEXANE * DIOXAN-1,4 (GERMAN) * p-DIOXAN (CZECH) * 1,4-DIOXANE (MAK) * p-DIOXANE * DIOXANNE (FRENCH) * DIOXYETHYLENE ETHER * GLYCOL ETHYLENE ETHER * NCI-C03689 * RCRA WASTE NUMBER U108 * TETRAHYDRO-p-DIOXIN * TETRAHYDRO-1,4-DIOXIN

CONSENSUS REPORTS: IARC Cancer Review: GROUP 2B IMEMDT 7,201,87; Animal Sufficient Evidence IMEMDT 11,247,76. NTP Fourth Annual Report On Carcinogens, 1984. NCI Carcinogenesis Bioassay (oral); Clear Evidence: mouse, rat NCITR* NCI-CG-TR-80, 78. EPA Genetic Toxicology Program. Glycol ether compounds are on the Community Right-To-Know List. Reported in EPA TSCA Inventory.

OSHA PEL: (Transitional: TWA 100 ppm (skin)) TWA 25 ppm (skin)
ACGIH TLV: TWA 25 ppm (skin)
DFG MAK: 50 ppm (180 mg/m^3); Suspected Carcinogen.

NIOSH REL: CL (Dioxane) 1 ppm/30M
DOT Classification: Flammable Liquid; Label: Flammable Liquid.

SAFETY PROFILE: Confirmed carcinogen with experimental carcinogenic, neoplastigenic, and tumorigenic data. Poison by intraperitoneal route. Moderately toxic by ingestion and inhalation. Mildly toxic by skin contact. Human systemic effects by inhalation: lachrimation, conjunctiva irritation, convulsions, high blood pressure, unspecified respiratory and gastrointestinal system effects. Experimental teratogenic and reproductive effects. Mutation data reported. An eye and skin irritant. The irritant effects probably provide sufficient warning, in acute exposures, to enable a worker to leave exposure before being seriously affected. Repeated exposure to low concentrations has resulted in human fatalities, the organs chiefly affected being the liver and kidneys.

A very dangerous fire and explosion hazard when exposed to heat or flame; can react vigorously with oxidizing materials. Violent reaction with (H_2 + Raney Ni), $AgClO_4$. Can form dangerous peroxides when exposed to air. Potentially explosive reaction with nitric acid + perchloric acid, Raney nickel catalyst (above 210°C). Forms explosive mixtures with decaborane (impact-sensitive), triethynylaluminum (sensitive to heating or drying). Violent reaction with sulfur trioxide. Incompatible with sulfur trioxide. To fight fire, use alcohol foam, CO_2, dry chemical. When heated to decomposition it emits acrid smoke and irritating fumes.

DVQ709   CAS: 78-34-2   HR: 3
DIOXATHION
mf: $C_{12}H_{26}O_6P_2S_4$ mw: 456.56

PROP: Nonvolatile, stable solid. Nonflammable. Insol in water.

SYNS: BIS(DITHIOPHOSPHATE de O,O-DIETHYLE) de S,S'-(1,4-DIOXANNE-2,3-DIYLE) (FRENCH) * DELNAV * 1,4-DIOSSAN-2,3-DIYL-BIS(O,O-DIETIL-DITIOFOSFATO) (ITALIAN) * 1,4-DIOXAAN-2,3-DIYL-BIS(O,O-DIETHYL-DITHIOFOSFAAT) (DUTCH) * 2,3-p-DIOXANDITHIOL S,S-BIS(O,O-DIETHYL PHOSPHORODITHIOATE) * 1,4-DIOXAN-2,3-DIYL-BIS(O,O-DIAETHYL-DITHIOPHOSPHAT) (GERMAN) * 1,4-DIOXAN-2,3-DIYL-BIS(O,O-DIETHYLPHOSPHOROTHIOLOTHIONATE) * 1,4-DIOXAN-2,3-DIYL-O,O,O',O'-TETRAETHYL DI-(PHOSPHOROMITHIOATE) * 2,3-p-DIOXANE-S,S-BIS(O,O-DIETHYLPHOSPHOROITHIOATE) * p-DIOX-

ANE-2,3-DITHIOL-S,S-DIESTER with O,O-DIETHYL PHOSPHORODITHIOATE * p-DIOXANE-2,3-DIYL ETHYL PHOSPHORODITHIOATE * ENT 22,897 * NCI-C00395 * PHOSPHORODITHIOIC ACID-S,S'-1,4-DIOXANE-2,3-DIYL O,O,O',O'-TETRAETHYL ESTER

CONSENSUS REPORTS: NCI Carcinogenesis Bioassay (feed); No Evidence: mouse, rat NCITR* NCI-CG-TR-125,78.

OSHA PEL: TWA 0.2 mg/m³ (skin)
ACGIH TLV: TWA 0.2 mg/m³ (skin)

SAFETY PROFILE: Poison by ingestion, inhalation, skin contact, and intraperitoneal routes. A cholinesterase inhibitor. When heated to decomposition it emits very toxic fumes of PO_x and SO_x.

DVR200   CAS: 105-11-3   HR: 3
DIOXIME-p-BENZOQUINONE
mf: $C_6H_6N_2O_2$ mw: 138.14

SYNS: ACTOR Q * 1,4-BENZOQUINONE DIOXINE * 2,5-CYCLOHEXADIENE-1,4-DIONE DIOXIME * DIBENZO PQD * DIOXIME-1,4-CYCLOHEXADIENE-DIONE * DIOXIME-2,5-CYCLOHEXADIENE-1,4-DIONE * G-M-F * NCI-C03850 * PQD * QDO * QUINONE DIOXIME * p-QUINONE DIOXIME * p-QUINONE OXIME

CONSENSUS REPORTS: IARC Cancer Review: GROUP 3 IMEMDT 7,56,87; Animal Limited Evidence IMEMDT 29,185,82. NCI Carcinogenesis Bioassay (feed); Clear Evidence: rat NCITR* NCI-CG-TR-179,79; No Evidence: mouse NCITR* NCI-CG-TR-179,79. Reported in EPA TSCA Inventory.

SAFETY PROFILE: Questionable carcinogen with experimental neoplastigenic and tumorigenic data. Moderately toxic by ingestion. Mutation data reported. When heated to decomposition it emits toxic fumes of NO_x.

DVR600   CAS: 100-79-8   HR: 1
DIOXOLAN
DOT: UN 1166
mf: $C_6H_{12}O_3$ mw: 132.18

PROP: Water-white liquid. Mp: −26.4°, bp: 75°, flash p: 35°F (OC), d: 1.065, vap press: 70 mm @ 20°, vap d: 2.6.

SYNS: CYCLIC (HYDROXYMETHYL)ETHYLENE ACETAL ACETONE * 2,2-DIMETHYL-1,3-DIOXOLANE-4-METHANOL * 2,2-DIMETHYL-5-HYDROXYMETHYL-

1,3-DIOXOLANE * 2,2-DIMETHYL-4-OXYMETHYL-1,3-DIOXOLANE * DIOXOLANE (DOT) * GIE * GLYCEROLACETONE * GLYCEROL DIMETHYL-KETAL * 4-HYDROXYMETHYL-2,2-DIMETHYL-1,3-DIOXOLANE * ISO PROPYLIDENE GLYCEROL * 1,2-o-ISOPROPYLIDENE GLYCEROL * SOLKETAL

CONSENSUS REPORTS: Reported in EPA TSCA Inventory.

DOT Classification: Flammable Liquid; Label: Flammable Liquid.

SAFETY PROFILE: An eye irritant. Mutation data reported. A very dangerous fire hazard when exposed to heat or flame; can react vigorously with oxidizing materials. To fight fire, use alcohol foam, CO_2, dry chemical. When heated to decomposition it emits acrid smoke and fumes.

DVS000   CAS: 6988-21-2   **HR: 3**
o-(1,3-DIOXOLAN-2-YL)PHENYL METHYLCARBAMATE
mf: $C_{11}H_{13}NO_4$ mw: 223.25

SYNS: CIBA 8353 * DIOXACARB * 2-(1,3-DI-OXOLANE-2-YL)PHENYL N-METHYLCARBAMATE * 2-(1,3-DIOXOLAN-2-YL)PHENYL-N-METHYLCARBA-MAT * DU PONT INSECTICIDE 1519 * ELO-CRON * ENT 27,389 * FAMID * NSC 190981

SAFETY PROFILE: Poison by ingestion, intraperitoneal, and possibly other routes. Moderately toxic by skin contact. Mutation data reported. A toxic contact and systemic insecticide. When heated to decomposition it emits toxic fumes of NO_x.

DVT400   CAS: 13754-56-8   **HR: 3**
DIOXOPROMETHAZINE HYDROCHLORIDE
mf: $C_{17}H_{20}N_2O_2S \cdot ClH$ mw: 352.91

SYN: 5,5-DIOXO-10-(2-(DIMETHYLAMINO)PROPYL)PHENOTHIAZINE HYDROCHLORIDE

SAFETY PROFILE: Poison by ingestion, intraperitoneal, intravenous, and subcutaneous routes. When heated to decomposition it emits very toxic fumes of NO_x, SO_x and HCl.

DVV200   CAS: 1118-42-9   **HR: 3**
DIPENTYLTIN DICHLORIDE
mf: $C_{10}H_{22}Cl_2Sn$ mw: 331.91

SYN: DICHLORODIPENTYLSTANNANE

OSHA PEL: TWA 0.1 mg(Sn)/m^3 (skin)
ACGIH TLV: TWA 0.1 mg(Sn)/m^3 (skin) (Proposed: TWA 0.1 mg(Sn)/m^3; STEL 0.2 mg(Sn)/m^3 (skin))
NIOSH REL: (Organotin Compounds) TWA 0.1 mg(Sn)/m^3

SAFETY PROFILE: Poison by intravenous route. When heated to decomposition it emits toxic fumes of Cl$^-$.

DVV600   CAS: 82-66-6   **HR: 3**
DIPHENADIONE
mf: $C_{23}H_{16}O_3$ mw: 340.39

PROP: Pale yellow crystals. Mp: 147°. Sol in acetone and acetic acid.

SYNS: DIDANDIN * DIPAXIN * DIPHACIN * DIPHACINONE * DIPHENACIN * 2-DIPHENYL-ACETYL-1,3-DIKETOHYDRINDENE * 2-DIPHENYLACE-TYL-1,3-INDANDIONE * 2-(DIPHENYLACETYL)INDAN-1,3-DIONE * 2-(DIPHENYLACETYL)-1H-INDENE-1,-3(2H)-DIONE * PID * PROMAR * RAMIK * RATINDAN 1 * U 1363

CONSENSUS REPORTS: EPA Extremely Hazardous Substances List.

SAFETY PROFILE: Poison by ingestion. Inhibits blood clotting, leading to hemorrhages. Action similar to coumadin (warfarin). A pesticide used in rodent control. When heated to decomposition it emits acrid smoke and irritating fumes.

DVX200   CAS: 86-29-3   **HR: 3**
DIPHENYLACETONITRILE
mf: $C_{14}H_{11}N$ mw: 193.26

SYNS: BENZHYDRYLCYANIDE * α-CYANODI-PHENYLMETHANE * DIPAN * DIPHENATRILE * DIPHENYL-α-CYANOMETHANE * DIPHENYL-METHYLCYANIDE * α-PHENYLBENZYLCYANIDE * α-PHENYLPHENYLACETONITRILE * USAF KF-13

CONSENSUS REPORTS: Reported in EPA TSCA Inventory. Cyanide and its compounds are on the Community Right-To-Know List.

SAFETY PROFILE: Questionable carcinogen with experimental carcinogenic and tumorigenic data. Poison by ingestion, intraperitoneal, and intravenous routes. Moderately toxic by subcutaneous and possibly other routes. When heated to decomposition it emits toxic fumes of NO_x and CN$^-$.

DVX600   CAS: 2510-95-4   **HR: 3**
2,3-DIPHENYLACRYLONITRILE
mf: $C_{15}H_{11}N$ mw: 205.27

SYNS: BENZAL-(BENZYL-CYANID) (GERMAN)
* BENZYLIDENEPHENYLACETONITRILE * α-CYA-
NOSTILBENE * α,β-DIPHENYLACRYLONITRILE
* F 2387 * α-PHENYLCINNAMONITRILE * α-(PHE-
NYLMETHYLENE)BENZENEACETONITRILE * α-STIL-
BENECARBONITRILE * USAF A-9789

CONSENSUS REPORTS: Reported in EPA
TSCA Inventory. Cyanide and its compounds
are on the Community Right-To-Know List.

SAFETY PROFILE: Poison by intraperitoneal
and possibly other routes. When heated to decomposition it emits toxic fumes of NO_x and
CN^-.

DVX800   CAS: 122-39-4   **HR: 3**
DIPHENYLAMINE
mf: $C_{12}H_{11}N$ mw: 169.24

PROP: Crystals; floral odor. Mp: 52.9°, bp:
302.0°, flash p: 307°F (CC), d: 1.16, autoign
temp: 1173°F, vap press: 1 mm @ 108.3°, vap
d: 5.82. Sol in benzene, ether, and carbon disulfide.

SYNS: ANILINOBENZENE * BIG DIPPER
* C.I. 10355 * DFA * N,N-DIPHENYLAMINE
* DPA * NO SCALD * N-PHENYLANILINE
* N-PHENYLBENEZENAMINE * SCALDIP

CONSENSUS REPORTS: Reported in EPA
TSCA Inventory. EPA Genetic Toxicology Program.

OSHA PEL: TWA 10 mg/m³
ACGIH TLV: TWA 10 mg/m³

SAFETY PROFILE: Poison by ingestion. Experimental teratogenic and reproductive effects.
Action similar to aniline but less severe. Combustible when exposed to heat or flame. Can
react violently with hexachloromelamine or trichloromelamine. Can react with oxidizing materials. To fight fire, use CO_2, dry chemical. When
heated to decomposition it emits highly toxic
fumes of NO_x.

DVZ000   CAS: 102-09-0   **HR: 3**
DIPHENYL CARBONATE
mf: $C_{13}H_{10}O_3$ mw: 214.23

SYNS: CARBONIC ACID, DIPHENYL ESTER
* PHENYL CARBONATE

CONSENSUS REPORTS: Reported in EPA
TSCA Inventory.

SAFETY PROFILE: Questionable carcinogen
with experimental neoplastigenic and tumorigenic data. When heated to decomposition it
emits acrid smoke and irritating fumes.

DWA600   CAS: 14148-99-3   **HR: 3**
1,2-DIPHENYL-1-(DIMETHYLAMINO)
ETHANE
mf: $C_{16}H_{19}N \cdot ClH$ mw: 261.82

SYNS: (R) (−)-N,N-DIMETHYL-1,2-DIPHENYLETHYL-
AMINE HYDROCHLORIDE * (R)-N,N-DIMETHYL-α-
PHENYLBENZENEETHANAMINE, HYDROCHLORIDE
* SPA

SAFETY PROFILE: Poison by ingestion, intravenous, and subcutaneous routes. When
heated to decomposition it emits very toxic
fumes of HCl and NO_x.

DWB800   CAS: 1241-94-7   **HR: 3**
DIPHENYL-2-ETHYLHEXYL
PHOSPHATE
mf: $C_{20}H_{27}O_4P$ mw: 362.44

SYNS: 2-ETHYL-1-HEXANOL ESTER with DIPHENYL
PHOSPHATE * 2-ETHYLHEXYL DIPHENYL ESTER
PHOSPHORIC ACID * 2-ETHYLHEXYL DIPHENYL-
PHOSPHATE * SANTICIZER 141 (MONSANTO)

CONSENSUS REPORTS: Reported in EPA
TSCA Inventory.

SAFETY PROFILE: Poison by intravenous
route. When heated to decomposition it emits
toxic fumes of PO_x.

DWC600   CAS: 102-06-7   **HR: 3**
DIPHENYLGUANIDINE
mf: $C_{13}H_{13}N_3$ mw: 211.29

PROP: White powder. Mp: 145°, d: 1.115 @
25°.

SYNS: N,N'-DIPHENYLGUANIDINE * 1,3-DIPHE-
NYLGUANIDINE * DPG * DPG ACCELERATOR
* DWUFENYLOGUANIDYNA (POLISH) * MELANI-
LINE * NCI-C60924 * USAF B-19 * USAF EK-1270
* VULCACID D * VULKACIT D/C * VULKAZIT

CONSENSUS REPORTS: Reported in EPA
TSCA Inventory.

SAFETY PROFILE: Poison by ingestion, intraperitoneal and possibly other routes. Experi-

mental teratogenic and reproductive effects. Mutation data reported. When heated to decomposition it emits toxic fumes of NO_x.

DWD800   CAS: 587-85-9   **HR: 3**
DIPHENYLMERCURY
mf: $C_{12}H_{10}Hg$ mw: 354.81

PROP: White crystals, insol in water. D: 2.318, mp: 122° (sublimes), bp: 204° @ 10.5 mm.

CONSENSUS REPORTS: Reported in EPA TSCA Inventory. Mercury and its compounds are on the Community Right-To-Know List.

OSHA PEL: (Transitional: CL 1 mg/10m³) CL 0.1 mg(Hg)/m³ (skin)
ACGIH TLV: TWA 0.1 mg(Hg)/m³ (skin)
NIOSH REL: (Inorganic Mercury) TWA 0.05 mg(Hg)/m³

SAFETY PROFILE: Poison by intraperitoneal route. Moderately toxic by ingestion. Incompatible with non-metal oxides. When heated to decomposition it emits toxic fumes of Hg.

DWF000   CAS: 606-90-6   **HR: 3**
4-(DIPHENYLMETHOXY)-1-METHYL-PIPERIDINE CHLOROTHEOPHYLLINE
mf: $C_{19}H_{23}NO \cdot C_7H_7ClN_4O_2$ mw: 496.06

SYN: P 284

SAFETY PROFILE: Poison by ingestion and intravenous routes. When heated to decomposition it emits very toxic fumes of NO_x and Cl^-.

DWF790   CAS: 60607-34-3   **HR: 3**
1-(3-(4-(DIPHENYLMETHYL)-1-PIPER-AZINYL)PROPYL)-1,3-DIHYDRO-2H-BENZ IMIDAZOL-2-ONE
mf: $C_{27}H_{30}N_4O$ mw: 426.61

SYNS: KW-4354   *   OXATIMIDE   *   OXATOMIDA   *   OXATOMIDE   *   R 35443   *   TINSET

SAFETY PROFILE: Poison by ingestion and intravenous routes. Experimental teratogenic and reproductive effects. When heated to decomposition it emits toxic fumes of NO_x.

DWI000   CAS: 86-30-6   **HR: 3**
DIPHENYLNITROSAMINE
mf: $C_{12}H_{10}N_2O$ mw: 198.24

PROP: Green crystals. Mp: 144°.

SYNS: CURETARD A   *   DELAC J   *   DIPHENYLNI-TROSAMIN (GERMAN)   *   DIPHENYL N-NITROSOAMINE

*   N,N-DIPHENYLNITROSAMINE   *   NAUGARD TJB
*   NCI-C02880   *   NDPA   *   NDPhA   *   N-NITROSO-DIFENYLAMIN (CZECH)   *   NITROSODIPHENYLAMINE
*   N-NITROSODIPHENYLAMINE   *   N-NITROSO-N-PHE-NYLANILINE   *   NITROUS DIPHENYLAMIDE   *   RE-DAX   *   RETARDER J   *   TJB   *   VULCALENT A
*   VULCATARD   *   VULKALENT A (CZECH)
*   VULTROL

CONSENSUS REPORTS: IARC Cancer Review: GROUP 3 IMEMDT 7,56,87; Animal Limited Evidence IMEMDT 27,213,82. NCI Carcinogenesis Bioassay (feed); Clear Evidence: rat NCITR* NCI-CG-TR-164,79; No Evidence: mouse NCITR* NCI-CG-TR-164,79. Reported in EPA TSCA Inventory. EPA Genetic Toxicology Program. Community Right-To-Know List.

SAFETY PROFILE: Questionable carcinogen with experimental carcinogenic and tumorigenic data. Moderately toxic by ingestion and possibly other routes. Human mutation data reported. An eye irritant. Dangerous fire hazard when exposed to heat, flame, or oxidizing materials. Can react vigorously with oxidizing materials. When heated to decomposition it emits highly toxic fumes of NO_x.

DWI400   CAS: 16230-71-0   **HR: 3**
3,3-DIPHENYL-2-OXETANONE
mf: $C_{15}H_{12}O_2$ mw: 224.27

SYNS: 2,2-DIPHENYL-3-HYDROXYPROPIONIC ACID LACTONE   *   α,α-DIPHENYL-β-PROPIOLACTONE

SAFETY PROFILE: Questionable carcinogen with experimental tumorigenic data. When heated to decomposition it emits acrid smoke and irritating fumes.

DWK400   CAS: 467-60-7   **HR: 3**
α,α-DIPHENYL-2-PIPERIDINEMETH-ANOL
mf: $C_{18}H_{21}NO$ mw: 267.40

SYNS: DETARIL   *   GERODYL   *   MERATRAN
*   MRD 108   *   α-(2-PIPERIDYL)BENZHYDROL
*   PIPRADOL   *   α-PIPRADOL   *   PIRIDROL
*   PYRIDROL   *   PYRIDROLE

SAFETY PROFILE: Poison by ingestion, intravenous, and intraperitoneal routes. When heated to decomposition it emits toxic fumes of NO_x.

DWL400   CAS: 10087-89-5   **HR: 3**
1,1-DIPHENYL-2-PROPYNYL-N-
CYCLOHEXYLCARBAMATE
mf: $C_{22}H_{23}NO_2$ mw: 333.46

SYNS: 1,1-DIPHENYL-2-PROPYN-1-OL CYCLOHEXANE-
CARBAMATE * 1,1-DIPHENYL-2-PROPYNYL ESTER
CYCLOHEXANECARBAMIC ACID * ENPROMATE

CONSENSUS REPORTS: EPA Genetic Toxicology Program.

SAFETY PROFILE: Questionable carcinogen
with experimental tumorigenic data. Poison by
intraperitoneal route. Moderately toxic by ingestion. Mutation data reported. When heated to
decomposition it emits toxic fumes of NO_x.

DWM000   CAS: 57-96-5   **HR: 3**
DIPHENYLPYRAZONE
mf: $C_{23}H_{20}N_2O_3S$ mw: 404.51

SYNS: 1,2-DIPHENYL-4-(2′-PHENYLSULFINETHYL)-3,5-
PYRAZOLIDINEDIONE * 4-(PHENYLSULFOXYETHYL)-
1,2-DIPHENYL-3,5-PYRAZOLIDINEDIONE * SULFINPY-
RAZINE * SULFOXYPHENYLPYRAZOLIDINE
* USAF GE-13

SAFETY PROFILE: Poison by ingestion, intravenous, and intraperitoneal routes. When
heated to decomposition it emits very toxic
fumes of NO_x and SO_x.

DWN200   CAS: 60-10-6   **HR: 3**
DIPHENYLTHIOCARBAZONE
mf: $C_{13}H_{12}N_4S$ mw: 256.35

PROP: Bluish-black, crystalline powder. Insol
in water; sparingly sol in alcohol; freely sol in
carbon tetrachloride and chloroform.

SYNS: DITHIZON * DITHIZONE * USAF EK-3092

CONSENSUS REPORTS: Reported in EPA
TSCA Inventory.

SAFETY PROFILE: Poison by intraperitoneal
route. Can cause eye injury and glycosuria.
When heated to decomposition it emits highly
toxic fumes of NO_x and SO_x.

DWN800   CAS: 102-08-9   **HR: 2**
DIPHENYLTHIOUREA
mf: $C_{13}H_{12}N_2S$ mw: 228.33

PROP: White to faint gray powder. Mp: 154°,
bp: decomp, d: 1.32 @ 25°.

SYNS: DFT * N,N′-DIPHENYLTHIOCARBAMIDE
* sym-DIPHENYLTHIOCARBAMIDE * N,N′-DI-

PHENYLTHIOUREA * sym-DIPHENYLTHIOUREA
* 1,3-DIPHENYLTHIOUREA * 1,3-DIPHENYL-2-THIO-
UREA * 2-FENYLOTIOMOCZNIK (POLISH) * RHEN-
OCURE CA * STABILISATOR C * SULFOCARBA-
NILIDE * THIOCARBANILIDE * USAF EK-245
* VALKACIT CA

CONSENSUS REPORTS: Reported in EPA
TSCA Inventory.

SAFETY PROFILE: Moderately toxic by ingestion and intraperitoneal routes. When heated
to decomposition it emits highly toxic fumes
of SO_x and NO_x.

DWO800   CAS: 136-35-6   **HR: 3**
1,3-DIPHENYLTRIAZENE
mf: $C_{12}H_{11}N_3$ mw: 197.26

PROP: Golden yellow crystals. Mp: 98-99°,
bp: explodes, vap d: 6.8.

SYNS: CELLOFOR (CZECH) * DAAB * DIA-
ZOAMINOBENZEN (CZECH) * DIAZOAMINOBENZENE
* p-DIAZOAMINOBENZENE * DIAZOAMINOBENZOL
(GERMAN) * N-(PHENYLAZO)ANILINE

CONSENSUS REPORTS: EPA Genetic Toxicology Program. Reported in EPA TSCA Inventory.

SAFETY PROFILE: Poison by ingestion. Questionable carcinogen with experimental tumorigenic data. Strongly explosive when shocked
or heated to 98°C. Mixture with acetic anhydride
explodes when warmed. When heated to decomposition it emits toxic fumes of NO_x.

DWQ000   CAS: 7727-21-1   **HR: 2**
DIPOTASSIUM PERSULFATE
DOT: UN 1492
mf: $H_2O_8S_2 \cdot 2K$ mw: 272.34

PROP: White, odorless crystals. Mp: decomp
@ 100°, d: 2.477.

SYNS: ANTHION * PEROXYDISULFURIC ACID DI-
POTASSIUM SALT * POTASSIUM PEROXYDISULFATE
* POTASSIUM PEROXYDISULPHATE * POTASSIUM
PERSULFATE (DOT)

CONSENSUS REPORTS: Reported in EPA
TSCA Inventory.

DOT Classification: Oxidizer; Label: Oxidizer.

SAFETY PROFILE: Moderately toxic. An irritant and allergen. A powerful oxidizer. Flamma-

ble when exposed to heat or by chemical reaction. Can react with reducing materials. It liberates oxygen above 100° when dry or @ about 50° when in solution. When heated to decomposition it emits highly toxic fumes of SO_x and K_2O.

DWQ800 CAS: 3248-28-0 **HR: 3**
DIPROPIONYL PEROXIDE
DOT: UN 2132
mf: $C_6H_{10}O_4$ mw: 146.15

SYNS: BIS(1-OXOPROPYL)PEROXIDE * PROPIONYL PEROXIDE (DOT) * PROPIONYL PEROXIDE, not more than 28% in solution

DOT Classification: Organic Peroxide; Label: Organic Peroxide.

SAFETY PROFILE: The pure material explodes at room temperature. When heated to decomposition it emits acrid smoke and fumes.

DWR000 CAS: 142-84-7 **HR: 3**
DIPROPYLAMINE
DOT: UN 2383
mf: $C_6H_{15}N$ mw: 101.22

PROP: Water-white liquid, amine odor. Mp: −40°, bp: 105°, flash p: 63°F (OC), d: 0.741 @ 20°, vap d: 3.5.

SYNS: DI-n-PROPYLAMINE * n-DIPROPYLAMINE * N-PROPYL-1-PROPANAMINE * RCRA WASTE NUMBER U110

CONSENSUS REPORTS: Reported in EPA TSCA Inventory.

DOT Classification: Flammable Liquid; Label: Flammable Liquid.

SAFETY PROFILE: Poision by ingestion. Moderately toxic by skin contact, inhalation, and possibly other routes. A skin irritant. A very dangerous fire hazard, when exposed to heat or flame. Can react with oxidizers. Explosion Hazard is unknown. Keep away from heat and open flame. To fight fire, use foam, CO_2, dry chemical. When heated to decomposition it emits toxic fumes of NO_x.

DWT200 CAS: 34590-94-8 **HR: 1**
DIPROPYLENE GLYCOL METHYL ETHER
mf: $C_7H_{16}O_3$ mw: 148.23

PROP: Liquid. Bp: 190°, d: 0.951, vap d: 5.11, flash p: 185°F.

SYNS: ARCOSOLV * DIPROPYLENE GLYCOL MONOMETHYL ETHER * DOWANOL DPM * DOWANOL-50B * UCAR SOLVENT 2LM

CONSENSUS REPORTS: Reported in EPA TSCA Inventory. Glycol ether compounds are on the Community Right-To-Know List.

OSHA PEL: (Transitional: TWA 100 ppm (skin)) TWA 100 ppm; STEL 150 ppm (skin)
ACGIH TLV: TWA 100 ppm; STEL 150 ppm (skin)
DFG MAK: 50 ppm (300 mg/m³)

SAFETY PROFILE: Mildly toxic by ingestion and skin contact. A skin and eye irritant. A mild allergen. Flammable when exposed to heat or flame; can react with oxidizing materials. To fight fire, use dry chemical, CO_2, mist, foam. When heated to decomposition it emits acrid smoke and irritating fumes.

DWT600 CAS: 123-19-3 **HR: 2**
DIPROPYL KETONE
DOT: UN 2710
mf: $C_7H_{14}O$ mw: 114.21

PROP: Colorless, refractive liquid. Bp: 144°, mp: −32.6°, vap press: 5.2 mm @ 20°, flash p: 120°F (CC), d: 0.815, vap d: 3.93.

SYNS: BUTYRONE (DOT) * GBL * 4-HEPTANONE * HEPTAN-4-ONE * PROPYL KETONE

CONSENSUS REPORTS: Reported in EPA TSCA Inventory.

OSHA PEL: TWA 50 ppm
ACGIH TLV: TWA 50 ppm
DOT Classification: Flammable or Combustible Liquid; Label: Flammable Liquid.

SAFETY PROFILE: Moderately toxic by ingestion, inhalation, and skin contact. Flammable when exposed to heat or flame; can react with oxidizing materials. To fight fire, use CO_2, dry chemical, alcohol foam, fog and mist. When heated to decomposition it emits acrid smoke and fumes.

DWU000 CAS: 628-85-3 **HR: 3**
DIPROPYL MERCURY
mf: $C_6H_{14}Hg$ mw: 286.79

PROP: Colorless liquid, immisc in water. D: 2.0208, bp: 190°.

CONSENSUS REPORTS: Mercury and its compounds are on the Community Right-To-Know List.

OSHA PEL: (Transitional: CL 1 mg/10m^3) TWA 0.01 mg(Hg)/m^3; STEL 0.03 mg/m^3 (skin)
ACGIH TLV: TWA 0.01 mg(Hg)/m^3; STEL 0.03 mg(Hg)/m^3
NIOSH REL: (Inorganic Mercury) TWA 0.05 mg(Hg)/m^3

SAFETY PROFILE: Poison by intraperitoneal route. Violent reaction with iodine. When heated to decomposition it emits toxic fumes of Hg.

DWV000   CAS: 7664-98-4   **HR: 3**
DIPROPYLOXOSTANNANE
mf: C$_6$H$_{14}$OSn   mw: 220.89

SYNS: DIPROPYLTIN OXIDE * KYSLICNIK DI-N-PROPYLCINICITY (CZECH)

OSHA PEL: TWA 0.1 mg(Sn)/m^3 (skin)
ACGIH TLV: TWA 0.1 mg(Sn)/m^3 (skin) (Proposed: TWA 0.1 mg(Sn)/m^3; STEL 0.2 mg(Sn)/m^3 (skin))
NIOSH REL: (Organotin Compounds) TWA 0.1 mg(Sn)/m^3

SAFETY PROFILE: Poison by ingestion. An eye and severe skin irritant. When heated to decomposition it emits acrid smoke and irritating fumes.

DWV400   CAS: 16066-38-9   **HR: 2**
DI-n-PROPYL PEROXYDICARBONATE
DOT: UN 2176
mf: C$_8$H$_{14}$O$_6$ mw: 206.22

SYNS: PEROXYDICARBONIC ACID DIPROPYL ESTER * n-PROPYL PERCARBONATE

CONSENSUS REPORTS: Reported in EPA TSCA Inventory.

DOT Classification: Organic Peroxide; Label: Organic Peroxide.

SAFETY PROFILE: Moderately toxic by ingestion and skin contact. When heated to decomposition it emits acrid smoke and irritating fumes.

DWW200   CAS: 23795-03-1   **HR: 3**
p-(DIPROPYLSULFAMOYL)BENZOIC ACID SODIUM SALT
mf: C$_{13}$H$_{18}$NO$_4$S • Na   mw: 307.37

SYNS: p-(DI-N-PROPYLSULFAMYL)BENZOIC ACID SODIUM SALT * PROBENECID SODIUM SALT

SAFETY PROFILE: Poison by intraperitoneal and intravenous routes. Moderately toxic by ingestion and subcutaneous routes. Human systemic effects by ingestion: proteinurea and damage to the kidney (glomeruli), ureter and bladder. When heated to decomposition it emits very toxic fumes of NO$_x$, Na$_2$O and SO$_x$.

DWW700   CAS: 67730-10-3   **HR: 3**
DIPYRIDO(1,2-a:3′,2′-d)IMIDAZOL-2-AMINE
mf: C$_{10}$H$_8$N$_4$ mw: 184.22

SYNS: 2-AMINODIPYRIDO(1,2-a:3′,2′-d)-IMIDAZOLE * GLU-P-2

CONSENSUS REPORTS: IARC Cancer Review: GROUP 2B IMEMDT 7,56,87; Animal Sufficient Evidence IMEMDT 40,235,86

SAFETY PROFILE: Suspected carcinogen with experimental carcinogenic data. Human mutation data reported. When heated to decomposition it emits toxic fumes of NO$_x$.

DWX000   CAS: 21000-42-0   **HR: 3**
DIPYRIDYL HYDROGEN PHOSPHATE
mf: C$_{12}$H$_{14}$N$_2$ • 2C$_2$H$_6$O$_4$P   mw: 436.2

SYN: DIPYRIDYL PHOSPHATE

SAFETY PROFILE: Poison by ingestion and possibly other routes. Moderately toxic by skin contact. When heated to decomposition it emits very toxic fumes of PO$_x$ and NO$_x$.

DWX200   CAS: 20738-78-7   **HR: 3**
DI-3-PYRIDYLMERCURY
mf: C$_{10}$H$_8$HgN$_2$ mw: 356.79

CONSENSUS REPORTS: Mercury and its compounds are on the Community Right-To-Know List.

OSHA PEL: (Transitional: CL 1 mg/10m^3) CL 0.1 mg(Hg)/m^3 (skin)
ACGIH TLV: TWA 0.1 mg(Hg)/m^3 (skin)
NIOSH REL: (Inorganic Mercury) TWA 0.05 mg(Hg)/m^3

SAFETY PROFILE: Poison by intravenous route. When heated to decomposition it emits very toxic fumes of NO$_x$ and Hg.

DWX800 CAS: 85-00-7 **HR: 3**
DIQUAT

DOT: NA 2781
mf: $C_{12}H_{12}N_2 \cdot 2Br$ mw: 344.08

PROP: Yellow crystals. Mp: 355°. Sol in water.

SYNS: AQUACIDE * DEIQUAT * DEXTRONE * 9,10-DIHYDRO-8a,10,-DIAZONIAPHENANTHRENE DIBROMIDE * 9,10-DIHYDRO-8a,10a-DIAZONIAPHENANTHRENE(1,1′-ETHYLENE-2,2′-BIPYRIDYLIUM)DIBROMIDE * 5,6-DIHYDRO-DIPYRIDO(1,2a;2,1c)PYRAZINIUM DIBROMIDE * 6,7-DIHYDROPYRIDO (1,2a;2′,1′-C)PYRAZINEDIUM DIBROMIDE * DIQUAT DIBROMIDE * 1,1′-ETHYLENE-2,2′-BIPYRIDYLIUM DIBROMIDE * ETHYLENE DIPYRIDYLIUM DIBROMIDE * 1,1-ETHYLENE 2,2-DIPYRIDYLIUM DIBROMIDE * 1,1′-ETHYLENE-2,2′-DIPYRIDYLIUM DIBROMIDE * FB/2 * FEGLOX * PREEGLONE * REGLON * REGLONE * WEEDTRINE-D

CONSENSUS REPORTS: EPA Genetic Toxicology Program.

OSHA PEL: TWA 0.5 mg/m³
ACGIH TLV: TWA 0.5 mg/m³
DOT Classification: ORM-E; Label: None.

SAFETY PROFILE: Poison by ingestion, subcutaneous, intravenous, intraperitoneal, and possibly other routes. Experimental teratogenic and reproductive effects. Human mutation data reported. A skin and eye irritant. When heated to decomposition it emits very toxic fumes of NO_x and Br⁻.

DWY000 CAS: 4032-26-2 **HR: 3**
DIQUAT DICHLORIDE
mf: $C_{12}H_{12}N_2 \cdot 2Cl$ mw: 255.16

SYN: 1,1′-ETHYLENE-2,2′-DIPYRIDINIUM DICHLORIDE

SAFETY PROFILE: Poison by ingestion, intravenous, and subcutaneous routes. When heated to decomposition it emits very toxic fumes of NO_x and Cl⁻.

DWY800 CAS: 1464-43-3 **HR: 3**
3,3′-DISELENODIALANINE
mf: $C_6H_{12}N_2O_4Se_2$ mw: 334.12

SYNS: SELENIUM CYSTINE * SELENOCYSTINE

CONSENSUS REPORTS: Selenium and its compounds are on the Community Right-To-Know List.

OSHA PEL: TWA 0.2 mg(Se)/m³
ACGIH TLV: TWA 0.2 mg(Se)/m³

DFG MAK: 0.1 mg(Se)/m³

SAFETY PROFILE: Poison by intraperitoneal route. Mutation data reported. When heated to decomposition it emits very toxic fumes of NO_x and Se.

DXA000 **HR: 3**
DISILANE
mf: H_6Si_2 mw: 62.22

PROP: Gas, repulsive odor. Mp: −132.5°, bp: −14.5°, d: 0.686 @ −25°/4°.

SYN: SILICOETHANE

SAFETY PROFILE: Poison by inhalation. Dangerous when exposed to heat or flame or by chemical reaction; can react with oxidizing materials. Ignites spontaneously in air. Reacts violently with CCl_4, $CHCl_3$, O_2, and SF_6.

DXB200 CAS: 13464-37-4 **HR: 3**
DISODIUM ARSENITE
mf: $AsO_2 \cdot Na_2$ mw: 152.90

SYNS: ARSENIOUS ACID DISODIUM SALT * SODIUM ARSENITE

CONSENSUS REPORTS: Arsenic and its compounds are on the Community Right-To-Know List.

OSHA PEL: TWA 0.01 mg(As)/m³
ACGIH TLV: TWA 0.2 mg(As)/m³
NIOSH REL: (Inorganic Arsenic) CL 0.002 mg(As)/m³/15M

SAFETY PROFILE: Poison by ingestion, skin contact, and intraperitoneal routes. Experimental reproductive effects. Human mutation data reported. When heated to decomposition it emits toxic fumes of As and Na_2O.

DXC000 CAS: 21259-76-7 **HR: 3**
DISODIUM-N-(3-(CARBOXYMETHYL-THIOMERCURI)-2-METHOXYPROPYL)-α-CAMPHORAMATE
mf: $C_{16}H_{27}HgNO_6S \cdot 2Na$ mw: 608.07

SYNS: N-(γ-CARBOXYMETHYLMERCAPTOMERCURI-β-METHOXY)PROPYLCAMPHORAMIC ACID DISODIUM SALT * DIUCARDYN SODIUM * MERCAPTOMERIN SODIUM * SODIUM MERCAPTOMERIN * THIOMERIN SODIUM

CONSENSUS REPORTS: Mercury and its compounds are on the Community Right-To-Know List.

OSHA PEL: (Transitional: CL 1 mg/10m^3) CL 0.1 mg(Hg)/m^3 (skin)
ACGIH TLV: TWA 0.1 mg(Hg)/m^3 (skin)
NIOSH REL: (Inorganic Mercury) TWA 0.05 mg(Hg)/m^3

SAFETY PROFILE: Poison by subcutaneous, intramuscular and intravenous routes. When heated to decomposition it emits very toxic fumes of Hg, NO$_x$, Na$_2$O, and SO$_x$.

DXC200   CAS: 7775-11-3   HR: 3
DISODIUM CHROMATE

DOT: NA 9145
mf: CrO$_4$ • 2Na   mw: 161.98

SYNS: CHROMATE of SODA * CHROMIUM DISO-DIUM OXIDE * CHROMIUM SODIUM OXIDE * NEUTRAL SODIUM CHROMATE * SODIUM CHRO-MATE (DOT) * SODIUM CHROMATE (VI)

CONSENSUS REPORTS: IARC Cancer Review: Animal Inadequate Evidence IMEMDT 23,205,80; Human Inadequate Evidence IMEMDT 23,205,80. Reported in EPA TSCA Inventory. EPA Genetic Toxicology Program. Chromium and its compounds are on the Community Right-To-Know List.

OSHA PEL: Cl 0.1 mg(CrO$_3$)/m^3
ACGIH TLV: TWA 0.05 mg(CrO)/m^3
NIOSH REL: TWA 25 µg(Cr(VI))/m^3; CL 50 µg/m^3/15M
DOT Classification: ORM-E; Label: None.

SAFETY PROFILE: Poison by skin contact, intraperitoneal, intravenous, subcutaneous, and intradermal routes. A suspected human carcinogenic. Mutation data reported. A powerful oxidizer. When heated to decomposition it emits toxic fumes of Na$_2$O.

DXC400   CAS: 144-33-2   HR: 3
DISODIUM CITRATE
mf: C$_6$H$_6$O$_7$ • 2Na   mw: 236.10

PROP: White crystals or granular powder; odorless. Mp: loses water @ 150°, bp: decomp @ red heat. Sol in water; insol in alc.

SYNS: DISODIUM HYDROGEN CITRATE * NA-TRIUM CITRICUM (GERMAN) * SODIUM CITRATE (FCC)

CONSENSUS REPORTS: Reported in EPA TSCA Inventory.

SAFETY PROFILE: Poison by intravenous route. Moderately toxic by intraperitoneal and subcutaneous routes. When heated to decomposition it emits toxic fumes of Na$_2$O.

DXD000   CAS: 129-67-9   HR: 3
DISODIUM-3,6-ENDOXOHEXAHY-DROPHTHALATE
mf: C$_8$H$_8$O$_5$ • 2Na   mw: 230.14

PROP: Water-sol solid. Mp: 144°.

SYNS: ACCELERATE * AGUATHOL * DES-I-CATE * DINATRIUM-(3,6-EPOXY-CYCLOHEXAAN-1,2-DICARBOXYLAAT) (DUTCH) * DINATRIUM-(3,6-EPOXY-CYCLOHEXAN-1,2-DICARBOXYLAT) (GERMAN) * DISODIUM 3,6-EPOXYCYCLOHEXANE-1,2-DICARBOX-YLATE * DISODIUM 7-OXABICYCLO(2.2.1)HEPTANE-2,3-DICARBOXYLATE * DISODIUM SALT of ENDOT-HALL * DISODIUM SALT of 7-OXABICYCLO(2.2.1)HEP-TANE-2,3-DICARBOXYLIC ACID * ENDOTAL * ENDOTHAL * ENDOTHAL-NATRIUM (DUTCH) * ENDOTHAL-SODIUM * ENDOTHAL WEED KILLER * 3,6-ENDOXOHEXAHYDROPHTHALIC ACID DISODIUM SALT * (3,6-EPOSSI-CICLOESAN-1,2-DICARBOSSILATO) DISODICO (ITALIAN) * 3,6-EPOXY-CYCLOHEXANE 1,2-CARBOXYLATE DISODIQUE (FRENCH) * HERBICIDE 273 * HYDOUT * HY-DROTHOL * NIAGARATHAL * RCRA WASTE NUM-BER P088 * RIPENTHOL * TRI-ENDOTHAL

SAFETY PROFILE: Poison by ingestion, skin contact, and intravenous routes. Very irritating to eyes, skin, and mucous membranes. A defoliant and an herbicide. When heated to decomposition it emits toxic fumes of Na$_2$O.

DXD200   CAS: 142-59-6   HR: 3
DISODIUM ETHYLENE-1,2-BISDITHIOCARBAMATE
mf: C$_4$H$_6$N$_2$S$_4$ • 2Na   mw: 256.34

PROP: Crystals. Sol in water.

SYNS: CARBON D * CHEM BAM * DINATRIUM-AETHYLENBISDITHIOCARBAMAT (GERMAN) * DINA-TRIUM-(N,N'-AETHYLEN-BIS(DITHIOCARBAMAT)) (GER-MAN) * DINATRIUM-(N,N'-ETHYLEEN-BIS(DITHIO-CARBAMAAT)) (DUTCH) * DISODIUM ETHYLENEBIS (DITHIOCARBAMATE) * DITHANE A-40 * DITHANE D-14 * DSE * 1,2-ETHANEDIYLBISCARBAMO-DITHIOIC ACID DISODIUM SALT * N,N'-ETHYLENE BIS(DITHIOCARBAMATE de SODIUM) (FRENCH) * ETHYLENEBIS(DITHIOCARBAMATE) DISODIUM SALT * ETHYLENEBIS(DITHIOCARBAMIC ACID) DISO-DIUM SALT * N,N'-ETILEN-BIS(DITIOCARBAMMATO) di SODIO (ITALIAN) * NABAM * NABAME (FRENCH) * PARZATE * SPRING-BAK

CONSENSUS REPORTS: Reported in EPA TSCA Inventory. EPA Genetic Toxicology Program.

SAFETY PROFILE: Poison by ingestion. Moderately toxic by intraperitoneal route. Experimental teratogenic and reproductive effects. Mutation data reported. When heated to decomposition it emits very toxic fumes of NO_x, Na_2O, and SO_x.

DXD800 CAS: 17013-01-3 **HR: 2**
DISODIUM FUMARATE
mf: $C_4H_2O_4 \cdot 2Na$ mw: 160.64

SYN: SODIUM FUMARATE

SAFETY PROFILE: Moderately toxic by ingestion, intravenous, and intraperitoneal routes. Human systemic effects by ingestion: hypermotility, diarrhea, nausea or vomiting, and other gastrointestinal changes. When heated to decomposition it emits toxic fumes of Na_2O.

DXE000 CAS: 16893-85-9 **HR: 3**
DISODIUM HEXAFLUOROSILICATE

DOT: UN 2674
mf: $F_6Si \cdot 2Na$ mw: 188.07

SYNS: DESTRUXOL APPLEX * (2-)-DISODIUM HEXAFLUOROSILICATE * DISODIUM SILICOFLUORIDE * ENS-ZEM WEEVIL BAIT * ENT 1,501 * FLUOSILICATE de SODIUM * NATRIUMSILICOFLUORID (GERMAN) * ORTHO EARWIG BAIT * ORTHO WEEVIL BAIT * PRODAN * PSC CO-OP WEEVIL BAIT * SAFSAN * SALUFER * SILICON SODIUM FLUORIDE * SODIUM FLUOROSILICATE * SODIUM FLUOSILICATE * SODIUM HEXAFLUOROSILICATE * SODIUM HEXAFLUOSILICATE * SODIUM SILICOFLUORIDE (DOT) * SUPER PRODAN

CONSENSUS REPORTS: Reported in EPA TSCA Inventory.

OSHA PEL: TWA 2.5 mg(F)/m^3
NIOSH REL: (Inorganic Fluorides) TWA 2.5 mg(F)/m^3
DOT Classification: Poison B; Label: St. Andrews Cross.

SAFETY PROFILE: Poison by ingestion and subcutaneous route. A skin and severe eye irritant. An insecticide. When heated to decomposition it emits very toxic fumes of F^- and Na_2O.

DXE500 CAS: 4691-65-0 **HR: 2**
DISODIUM INOSINATE
mf: $C_{10}H_{13}N_4O_8P \cdot 2Na$ mw: 394.22

PROP: Colorless to white crystals; characteristic taste. Sol in water; sltly sol in alc; insol in ether.

SYNS: DISODIUM IMP * DISODIUM-5'-INOSINATE * DISODIUM INOSINE-5'-MONOPHOSPHATE * DISODIUM INOSINE-5'-PHOSPHATE * IMP DISODIUM SALT * 5'-IMP DISODIUM SALT * IMP SODIUM SALT * INOSINE-5'-MONOPHOSPHATE DISODIUM * INOSIN-5'-MONOPHOSPHATE DISODIUM * SODIUM INOSINATE * SODIUM-5'-INOSINATE

CONSENSUS REPORTS: Reported in EPA TSCA Inventory.

SAFETY PROFILE: Moderately toxic by several routes. An experimental teratogen. Mutation data reported. When heated to decomposition it emits toxic fumes of PO_x, NO_x, and Na_2O.

DXE600 CAS: 144-21-8 **HR: 3**
DISODIUM METHANEARSENATE
mf: $CH_3AsO_3 \cdot 2Na$ mw: 183.94

PROP: Crystals, water-sol hydrate. Mp: 132-139°, bp: 165°, fp: -6°, d: 1.15.

SYNS: ANSAR 184 * ANSAR DSMA LIQUID * ARRHENAL * ARSINYL * ARSYNAL * CACODYL NEW * CHIPCO CRAB KLEEN * CLOUT * CRAB-E-RAD * CRALO-E-RAD * DAL-E-RAD 100 * DIARSEN * DIMET * DINATE * DISODIUM METHANEARSONATE * DISODIUM METHYLARSENATE * DISODIUM METHYLARSONATE * DISODIUM MONOMETHYLARSONATE * DISOMAR * DI-TAC * DMA * DREXEL DSMA LIQUID * DSMA LIQUID * JON-TROL * MAA SODIUM SALT * METHAR * METHARSINAT * NAMATE * NEOASYCODILE * SODAR * SODIUM METHANEARSONATE * SODIUM METHARSONATE * SODIUM METHYLARSONATE * SOMAR * STENOSINE * TONARSEN * VERSAR DSMA LQ * WEED BROOM * WEED-E-RAD * WEEDHOE

CONSENSUS REPORTS: EPA Genetic Toxicology Program. Arsenic and its compounds are on the Community Right-To-Know List.

OSHA PEL: TWA 0.5 mg(As)/m^3
ACGIH TLV: TWA 0.2 mg(As)/m^3

SAFETY PROFILE: Confirmed human carcinogen. Moderately toxic by ingestion. Experimental teratogenic and reproductive effects. Dangerous fire hazard by spontaneous chemical reaction. Ignites spontaneously in dry air. Can react vigorously with oxidizing materials, i.e., air, Cl_2. An herbicide. When heated to decomposition it emits toxic fumes of As and Na_2O.

DXE800   CAS: 7631-95-0   **HR: 3**
DISODIUM MOLYBDATE
mf: $MoO_4 \cdot 2Na$ mw: 205.92

SYNS: MOLYBDIC ACID, DISODIUM SALT * NATRIUMMOLYBDAT (GERMAN) * SODIUM MOLYBDATE * SODIUM MOLYBDATE(VI)

CONSENSUS REPORTS: Reported in EPA TSCA Inventory.

OSHA PEL: TWA 5 mg(Mo)/m^3
ACGIH TLV: TWA 5 mg(Mo)/m^3

SAFETY PROFILE: Poison by intraperitoneal route. Moderately toxic by subcutaneous and intravenous routes. Experimental reproductive effects. Mutation data reported. When heated to decomposition it emits toxic fumes of Na_2O.

DXF400   CAS: 53778-51-1   **HR: 3**
DISODIUM-2-(p-(γ-PHENYLPROPYL AMINO)BENZENESULFONAMIDO) PYRIDINE
mf: $C_{20}H_{19}N_3O_8S_3 \cdot 2Na$ mw: 571.58

SYNS: DISODIUM CINNAMYLIDENE BISULFITE derivative of SULFAPYRIDINE * 1-PHENYL-3-(p-2-PYRIDYL-SULFAMOYLANILINO)-1,3-PROPANEDISULFONIC ACID DISODIUM SALT * SOLUPYRIDINE * SULFAPYRIDINE NEUTRAL SOLUBLE

SAFETY PROFILE: Poison by ingestion, intravenous, and subcutaneous routes. When heated to decomposition it emits very toxic fumes of NO_x, Na_2O, and SO_x.

DXF800   CAS: 7758-16-9   **HR: 3**
DISODIUM PYROPHOSPHATE
mf: $H_2O_7P_2 \cdot Na_2$ mw: 221.94

PROP: White, crystalline powder. D: 1.862, mp: 220° (decomp). Sol in water.

SYNS: DINATRIUMPYROPHOSPHAT (GERMAN) * DIPHOSPHORIC ACID, DISODIUM SALT * DISO-

DIUM DIHYDROGEN PYROPHOSPHATE * DISODIUM DIPHOSPHATE * SODIUM ACID PYROPHOSPHATE (FCC) * SODIUM PYROPHOSPHATE

CONSENSUS REPORTS: Reported in EPA TSCA Inventory.

SAFETY PROFILE: Poison by intravenous route. Moderately toxic by ingestion and subcutaneous routes. An irritant to skin, eyes, and mucous membranes. When heated to decomposition it emits toxic fumes of PO_x and Na_2O.

DXG000   CAS: 13410-01-0   **HR: 3**
DISODIUM SELENATE
mf: $O_4Se \cdot 2Na$ mw: 188.94

PROP: Colorless, rhombic crystals. D: 3.098.

SYNS: NATRIUMSELENIAT (GERMAN) * P-40 * SEL-TOX SSO2 and SS-20 * SODIUM SELENATE

CONSENSUS REPORTS: IARC Cancer Review: GROUP 3 IMEMDT 7,56,87. Selenium and its compounds are on the Community Right-To-Know List. EPA Genetic Toxicology Program. Reported in EPA TSCA Inventory.

OSHA PEL: TWA 0.2 mg(Se)/m^3
ACGIH TLV: TWA 0.2 mg(Se)/m^3
DFG MAK: 0.1 mg(Se)/m^3

SAFETY PROFILE: Questionable carcinogen with experimental carcinogenic data. Poison by ingestion, intravenous, subcutaneous, and intraperitoneal routes. Human systemic effects by ingestion: EKG changes, hypermotility, diarrhea and liver impairment. Experimental teratogenic and reproductive effects. Effects similar to arsenic. Mutation data reported. A pesticide. When heated to decomposition it emits toxic fumes of Se and Na_2O.

DXG800   CAS: 15876-67-2   **HR: 3**
DISTIGMINE BROMIDE
mf: $C_{22}H_{32}N_4O_4 \cdot 2Br$ mw: 576.40

SYNS: HEXAMARIUM * 3,3'-(1,6-HEXANEDIYLBIS-((METHYLIMINO)CARBONYL)OXY)BIS(1-METHYLPYRIDINIUMDIBROMIDE) * 3-HYDROXY-1-METHYLPYRIDINIUM BROMIDE HEXAMETHYLENEBIS(METHYLCARBAMATE) * UBRETID * UBRITIL

SAFETY PROFILE: Poison by ingestion, intravenous, and intraperitoneal routes. When heated to decomposition it emits very toxic fumes of NO_x and Br^-.

DXH250 CAS: 97-77-8 HR: 3
DISULFIRAM
mf: $C_{10}H_{20}N_2S_4$ mw: 296.56

PROP: Yellow-white crystals; mp: 72°.

SYNS: ABSTENSIL * ABSTINYL * ALCOPHOBIN * ALK-AUBS * ANTABUS * ANTABUSE * ANTADIX * ANTAENYL * ANTAETHAN * ANTAETHYL * ANTAETIL * ANTALCOL * ANTETAN * ANTETHYL * ANTETIL * ANTEYL * ANTIAETHAN * ANTIETANOL * ANTI-ETHYL * ANTIETIL * ANTIKOL * ANTIVITIUM * AVERSAN * AVERZAN * BIS(DIETHYLAMINO)THIOXOMETHYL)DISULPHIDE * BIS(N,N-DIETHYLTHIOCARBAMOYL) DISULFIDE * BIS(DIETHYLTHIOCARBAMOYL) DISULFIDE * BIS(N,N-DIETHYLTHIOCARBAMOYL)DISULPHIDE * BONIBAL * CONTRALIN * CONTRAPOT * CRONETAL * DICUPRAL * DISETIL * DISULFAN * DISULFURAM * DISULPHURAM * 1,1'-DITHIOBIS(N,N-DIETHYLTHIOFORMAMIDE) * EKAGOM TEDS * EPHORRAN * ESPENAL * ESPERAL * ETABUS * ETHYLDITHIOURAME * ETHYLDITHIURAME * ETHYL THIRAM * ETHYL THIUDAD * ETHYL THIURAD * ETHYL TUADS * ETHYL TUEX * EXHORAN * EXHOR-RAN * HOCA * KROTENAL * NCI-C02959 * NOCBIN * NOXAL * REFUSAL * RO-SULFI-RAM * STOPAETHYL * STOPETHYL * STOP-ETYL * TATD * TENURID * TENUTEX * TETD * TETIDIS * TETRADIN * TETRADINE * TETRAETHYLTHIOPEROXYDICARBONIC DIAMIDE * TETRAETHYLTHIRAM DISULPHIDE * TETRA-ETHYLTHIURAM * TETRAETHYLTHIURAM DISULFIDE * TETRAETHYLTHIURAM DISULPHIDE * N,N,N',N'-TETRAETHYLTHIURAM DISULPHIDE * TETRAETIL * TETURAM * TETURAMIN * THIOSAN * THIOSCABIN * THIRERANIDE * THIURAM E * THIURANIDE * TILLRAM * TIURAM * TTD * TTS * USAF B-33

CONSENSUS REPORTS: IARC Cancer Review: GROUP 3 IMEMDT 7,56,87; Animal Inadequate Evidence IMEMDT 12,85,76. NCI Carcinogenesis Bioassay (feed); No Evidence: mouse, rat NCITR* NCI-CG-TR-16,79. Reported in EPA TSCA Inventory.

OSHA PEL: TWA 2 mg/m^3
ACGIH TLV: TWA 2 mg/m^3
DFG MAK: 2 mg/m^3

SAFETY PROFILE: Questionable carcinogen with experimental neoplastigenic data. A human poison by ingestion. An experimental poison by intraperitoneal route. Toxic symptoms when accompanied by ingestion of alcohol.

DXH325 CAS: 298-04-4 HR: 3
DISULFOTON
DOT: NA 2783
mf: $C_8H_{19}O_2PS_3$ mw: 274.42

SYNS: BAYER 19639 * O,O-DIAETHYL-S-(2-AETHYLTHIO-AETHYL)-DITHIOPHOSPHAT (GERMAN) * O,O-DIAETHYL-S-(3-THIA-PENTYL)-DITHIOPHOSPHAT (GERMAN) * O,O-DIETHYL-S-(2-ETHTHIOETHYL) PHOSPHORODITHIOATE * O,O-DIETHYL-S-(2-ETH-THIOETHYL) THIOTHIONOPHOSPHATE * O,O-DIETH-YL-S-(2-ETHYLMERCAPTOETHYL) DITHIOPHOSPHATE * O,O-DIETHYL-S-(2-ETHYLTHIO-ETHYL)-DITHIOFOS-FAAT (DUTCH) * O,O-DIETHYL-2-ETHYLTHIOETHYL PHOSPHORODITHIOATE * O,O-DIETHYL-S-2-(ETHYL-THIO)ETHYL PHOSPHORODITHIOATE * O,O-DIETIL-S-(2-ETILTIO-ETIL)-DITIOFOSFATO (ITALIAN) * DIMAZ * DISULFATON * DI-SYSTON * DISYSTOX * DITHIODEMETON * DITHIOPHOSPHATE de O,O-DI-ETHYLE et de S-(2-ETHYLTHIO-ETHYLE) (FRENCH) * DITHIOSYSTOX * ENT 23,437 * O,O-ETHYL-S-2(ETHYLTHIO)ETHYL PHOSPHORODITHIOATE * S-2-(ETHYLTHIO)ETHYL O,O-DIETHYL ESTER of PHOSPHORODITHIOIC ACID * FRUMIN AL * M-74 * RCRA WASTE NUMBER P039 * S 276 * SOLVIREX * THIODEMETON * THIODEME-TRON

CONSENSUS REPORTS: EPA Extremely Hazardous Substances List. EPA Genetic Toxicology Program.

OSHA PEL: TWA 0.1 mg/m^3 (skin)
ACGIH TLV: TWA 0.1 mg/m^3
DOT Classification: Poison B; Label: Poison; Poison B; Label: Poison, dry or liquid.

SAFETY PROFILE: Poison by ingestion, inhalation, skin contact, intraperitoneal, intravenous, and possibly other routes. Human mutation data reported. When heated to decomposition it emits very toxic SO_x and PO_x.

DXI600 CAS: 7187-55-5 HR: 3
DITHIAZANINE
mf: $C_{23}H_{23}N_2S_2$ mw: 391.60

SAFETY PROFILE: Poison by ingestion, intravenous, and intraperitoneal routes. When heated to decomposition it emits very toxic fumes of NO_x and SO_x.

DXJ800   CAS: 1141-88-4   **HR: 3**
2,2′-DITHIOBISANILINE
mf: $C_{12}H_{12}N_2S_2$ mw: 248.38

SYNS: BIS(2-AMINOPHENYL)DISULFIDE * BIS(o-AMINOPHENYL)DISULFIDE * 1,1′-BIS(2-AMINOPHE-NYL)DISULFIDE * O,O′-DIAMINO DIPHENYL DISUL-FIDE * O,O-DITHIO-BIS-ANILINE * 2,2′-DITHIODI-ANILINE * USAF AB-315

CONSENSUS REPORTS: Reported in EPA TSCA Inventory.

SAFETY PROFILE: Poison by intravenous and intraperitoneal routes. Moderately toxic by ingestion. A severe eye irritant. When heated to decomposition it emits very toxic fumes of NO_x and SO_x.

DXL400   CAS: 4540-66-3   **HR: D**
2,2′-DITHIOBIS(PYRIDINE-1-OXIDE) MAGNESIUM SULFATE TRIHYDRATE
mf: $C_{10}H_8N_2O_2S_2 \cdot MgOS \cdot 3H_2O$ mw: 378.75

SYN: MDS

SAFETY PROFILE: Experimental teratogenic and reproductive effects. When heated to decomposition it emits very toxic fumes of SO_x and NO_x.

DXL800   CAS: 541-53-7   **HR: 3**
DITHIOBIURET
mf: $C_2H_5N_3S_2$ mw: 135.22

PROP: Crystals. Mp: 181°, bp: decomp, d: 1.522 @ 30°.

SYNS: DTB * RCRA WASTE NUMBER P049 * 2-THIO-1-(THIOCARBAMOYL)UREA * USAF B-44 * USAF EK-P-6281

CONSENSUS REPORTS: Reported in EPA TSCA Inventory. EPA Extremely Hazardous Substances List.

SAFETY PROFILE: Poison by ingestion and intraperitoneal routes. When heated to decomposition it emits highly toxic fumes of SO_x and NO_x.

DXM600   CAS: 1892-29-1   **HR: 3**
DITHIODIGLYCOL
mf: $C_4H_{10}O_2S_2$ mw: 154.26

SYNS: 2,2-DITHIODIETHANOL * USAF TH-9

CONSENSUS REPORTS: Reported in EPA TSCA Inventory.

SAFETY PROFILE: Poison by intraperitoneal route. When heated to decomposition it emits toxic fumes of SO_x.

DXN600   CAS: 333-29-9   **HR: 3**
DITHIOLANE IMINOPHOSPHATE
mf: $C_7H_{14}NO_2PS_3$ mw: 271.37

SYNS: AC-43064 * AMERICAN CYANAMID AC 43,064 * CL-43,064 * CYALANE * CYCLIC ETH-YLENE (DIETHOXYPHOSPHINOTHIOYL)DITHIOIMIDO-CARBONATE * CYCLIC ETHYLENE ESTER of (DI-ETHOXYPHOSPHINOTHIOYL)DITHIOIMIDOCARBONIC ACID * CYLAN * CYOLAN * CYOLANE INSECTICIDE * 2-(DIETHOXYPHOSPHINYLIMINO)-1,3-DITHIOLANE * DIETHYL-N-1,3-DITHIOLANYL-2-IMINO PHOSPHATE * O,O-DIETHYL 1,3-DITHIOLAN-2-YLIDENEPHOSPHORAMIDOTHIOATE * DITHIO-LANE * 1,3-DITHIOLAN-2-YLIDENE-PHOSPHOR-AMIDOTHIOIC ACID DIETHYL ESTER * 1,3-DI-THIOLAN-2-YLIDENE-PHOSPHORAMIDOTHIOIC ACID-O,O-DIETHYL ESTER * ENT 25,809 * IMINO-PHOSPHATE * PHOSFOLAN

SAFETY PROFILE: Poison by ingestion and skin contact. An insecticide. When heated to decomposition it emits very toxic fumes of NO_x and SO_x.

DXO000   CAS: 572-48-5   **HR: 3**
DITHION
mf: $C_{17}H_{21}O_5PS$ mw: 368.41

PROP: Crystals nearly insol in water. Mp: 88°.

SYNS: O,O-DIETHYL-7-HYDROXY-3,4-TETRAMETHY-LENE COUMARINYL PHOSPHOROTHIOATE * O,O-DI-ETHYL-O-(7,8,9,10-TETRAHYDRO-6-OXOBENZO(C)CHRO-MAN-3-YL)PHOSPHOROTHIOATE * O,O-DIETHYL-O-(7,8,9,10-TETRAHYDRO-6-OXO-6H-DIBENZO(b,d)PYRAN-3-YL)PHOSPHOROTHIOATE * O,O-DIETHYL-O-(3,4-TETRAMETHYLENECOUMARINYL-7) THIOPHOSPHATE * DITHIONE * ENT-24,986 * 7-HYDROXY-3,4-TET-RAMETHYLENECOUMARIN-O,O-DIETHYL THIOPHOS-PHATE

SAFETY PROFILE: Poison by ingestion and possibly other routes. When heated to decomposition it emits very toxic fumes of PO_x and SO_x.

DXO200   CAS: 79-40-3   HR: 3
DITHIOOXAMIDE
mf: $C_2H_4N_2S_2$ mw: 120.20

SYNS: DITHIOOXALDIIMIDIC ACID * DITHIOXAM-IDE * ETHANEDITHIOAMIDE * HYDRORUBEANIC ACID * RUBEANE * RUBEANIC ACID * RVK * USAF EK-4394 * USAF MK-6

CONSENSUS REPORTS: Reported in EPA TSCA Inventory.

SAFETY PROFILE: Poison by ingestion, intraperitoneal, and intravenous routes. When heated to decomposition it emits very toxic fumes of NO_x and SO_x.

DXP000   CAS: 27755-15-3   HR: 3
DITOLYLETHANE
mf: $C_{16}H_{18}$ mw: 210.34

SAFETY PROFILE: Moderately toxic by ingestion. Questionable carcinogen with experimental tumorigenic data. When heated to decomposition it emits acrid smoke and irritating fumes.

DXP200   CAS: 97-39-2   HR: 3
DI-o-TOLYLGUANIDINE
mf: $C_{15}H_{17}N_3$ mw: 239.35

PROP: White crystals. Mp: 179°, d: 1.10 @ 20°/4°, vap d: 8.24.

SYNS: DIORTHOTOLYLGUANIDINE * 1,3-DI-o-TO-LYLGUANIDINE * DOTG ACCELERATOR * USAF A-6598 * VULKACIT DOTG/C

CONSENSUS REPORTS: Reported in EPA TSCA Inventory.

SAFETY PROFILE: Poison by ingestion and intraperitoneal routes. When heated to decomposition it emits toxic fumes of NO_x.

DXP600   CAS: 137-97-3   HR: 3
DI-o-TOLYLTHIOUREA
mf: $C_{15}H_{16}N_2S$ mw: 256.39

PROP: Crystals. Mp: 178°, vap d: 8.85.

SYNS: N,N'-BIS(2-METHYLPHENYLTHIOUREA * 1,3-BIS(o-TOLYL)-2-THIOUREA * 2,2'-DIMETHYL-THIOCARBANILIDE * DI-o-TOLUYLTHIOUREA * USAF EK-1651

CONSENSUS REPORTS: Reported in EPA TSCA Inventory.

SAFETY PROFILE: Poison by intraperitoneal route. Moderately toxic by ingestion. When heated to decomposition it emits very toxic fumes of NO_x and SO_x.

DXP800   CAS: 8015-54-1   HR: 3
DITRAN
mf: $C_{20}H_{29}NO_3 \cdot C_{20}H_{29}NO_3 \cdot 2ClH$ mw: 735.92

SYN: JB 329

SAFETY PROFILE: Poison by intravenous, intraperitoneal, and intramuscular routes. Human systemic effects by ingestion and intramuscular routes: visual field changes, hallucinations, distorted perceptions, nausea and vomiting. When heated to decomposition it emits toxic fumes of NO_x and Cl^-.

DXQ500   CAS: 330-54-1   HR: 3
DIURON
DOT: NA 2767
mf: $C_9H_{10}Cl_2N_2O$ mw: 233.11

PROP: Crystals. Mp: 159°. Sltly sol in water and hydrocarbon solvents.

SYNS: AF 101 * CEKIURON * CRISURON * DAILON * DCMU * DIATER * 3-(3,4-DI-CHLOOR-FENYL)-1,1-DIMETHYLUREUM (DUTCH) * DICHLORFENIDIM * 3-(3,4-DICHLOROPHENOL)-1,1-DIMETHYLUREA * N'-(3,4-DICHLOROPHENYL)-N,N-DIMETHYLUREA * 1-(3,4-DICHLOROPHENYL)-3,3-DIMETHYLUREE (FRENCH) * 3-(3,4-DICHLOR-PHE-NYL)-1,1-DIMETHYL-HARNSTOFF (GERMAN) * 3-(3,4-DICLORO-FENYL)-1,1-DIMETIL-UREA (ITALIAN) * 1,1-DIMETHYL-3-(3,4-DICHLOROPHENYL)UREA * DI-ON * DIREX 4L * DIUREX * DIUROL * DIURON 4L * DMU * DREXEL * DREXEL DIURON 4L * DURAN * DYNEX * FARMCO DIURON * HERBATOX * HW 920 * KARMEX * KARMEX DIURON HERBICIDE * KARMEX DW * MARMER * SUP'R FLO * TELVAR * TELVAR DIURON WEED KILLER * UNIDRON * USAF P-7 * USAF XR-42 * VONDURON

CONSENSUS REPORTS: Reported in EPA TSCA Inventory. EPA Genetic Toxicology Program. Chlorophenol compounds are on The Community Right-To-Know List.

OSHA PEL: TWA 10 mg/m^3
ACGIH TLV: TWA 10 mg/m^3
DOT Classification: ORM-E; Label: None.

SAFETY PROFILE: Questionable carcinogen with experimental tumorigenic and teratogenic data. Moderately toxic by ingestion, intraperitoneal, and possibly other routes. Mutation data reported. When heated to decomposition it emits highly toxic fumes of Cl^- and NO_x.

DXQ745   CAS: 1321-74-0   HR: 2
DIVINYLBENZENE
mf: $C_{10}H_{10}$ mw: 130.20

PROP: Pale straw-colored liquid. Bp: 195-200°, mp: −87°, d: 0.918, flash p: 165F°. Not misc in water; sol in eth, methanol.

SYN: VINYLSTYRENE

CONSENSUS REPORTS: Reported in EPA TSCA Inventory.

OSHA PEL: 10 ppm
ACGIH TLV: 10 ppm

SAFETY PROFILE: Moderately toxic by ingestion. A skin and eye irritant. Combustible. When heated to decomposition it emits acrid smoke and irritating fumes.

DXR200   CAS: 77-77-0   HR: 3
DIVINYL SULFONE
mf: $C_4H_6O_2S$ mw: 118.16

SYNS: TL 797 * VINYL SULFONE

SAFETY PROFILE: Poison by ingestion, skin contact, intravenous, subcutaneous, and intraperitoneal routes. A skin and eye irritant.

DXR800   CAS: 60539-20-0   HR: 3
DIXYRAZINE DIHYDROCHLORIDE
mf: $C_{24}H_{33}N_3O_2S \cdot 2ClH$ mw: 500.58

SYN: 2-(2-(4-(2-METHYL-3-PHENOTHIAZIN-10-YLPROPYL)-1-PIPERAZINYL)ETHOXY)ETHANOL DIHYDROCHLORIDE

SAFETY PROFILE: Poison by intravenous and intraperitoneal routes. Moderately toxic by ingestion. When heated to decomposition it emits very toxic fumes of Cl^-, SO_x, and NO_x.

DXS200   CAS: 38222-35-4   HR: 3
DMA
mf: $C_8H_{13}NO_3$ mw: 171.22

SYNS: 2,5-DIHYDROXY-3-DIMETHYLAMINO-5-METHYL-2-CYCLOPENTEN-1-ONE * DIMETHYLAMINO HEXOSE REDUCTIONE

SAFETY PROFILE: Poison by ingestion and intraperitoneal routes. When heated to decomposition it emits toxic fumes of NO_x.

DXS700   HR: 2
Δ-DODECALACTONE
mf: $C_{12}H_{22}O_2$ mw: 198.31

PROP: Colorless to yellow liquid; coconut-fruity odor. Refr index: 1.458-1.461, flash p: +151°F. Very sol in alc, propylene glycol, veg. oil; insol in water.

SYN: FEMA No. 2401

SAFETY PROFILE: Combustible liquid. When heated to decomposition it emits acrid smoke and irritating fumes.

DXT000   CAS: 112-54-9   HR: 1
1-DODECANAL
mf: $C_{12}H_{24}O$ mw: 184.36

PROP: Reported in pine-needle, lime, sweet-orange, and a dozen other essential oils. Colorless to light yellow liquid; fatty odor. D: 0.826-0.836, refr index: 1.433-1.439, flash p: 180°F. Sol in alc, fixed oils, propylene glycol; insol in glycerin, water.

SYNS: C-12 ALDEHYDE, LAURIC * 1-DODECYL ALDEHYDE * DUODECYLIC ALDEHYDE * FEMA No. 2615 * LAURYL ALDEHYDE (FCC)

CONSENSUS REPORTS: Reported in EPA TSCA Inventory.

SAFETY PROFILE: Mildly toxic by ingestion. A human and experimental skin irritant. Combustible liquid. When heated to decomposition it emits acrid smoke and irritating fumes.

DXT800   CAS: 25103-58-6   HR: 3
tert-DODECANETHIOL
mf: $C_{12}H_{26}S$ mw: 202.44

PROP: White to light yellow liquid. Bp: 200-235°, flash p: 205°F (OC), d: 0.85 @ 25°/25°, vap d: 6.98.

SYNS: tert-DODECYLMERCAPTAN * terc.DODECYLMERKAPTAN (CZECH) * tert-DODECYLTHIOL * 2,3,3,4,4,5-HEXAMETHYL-2-HEXANETHIOL

CONSENSUS REPORTS: Reported in EPA TSCA Inventory.

SAFETY PROFILE: Poison by ingestion. Moderately toxic by intraperitoneal route. A skin and eye irritant. Combustible when exposed to heat or flame; can react vigorously with oxidizing materials. To fight fire, use foam, CO_2, dry chemical. When heated to decomposition it emits toxic fumes of SO_x.

DXU200   CAS: 18186-71-5   HR: 3
DODECATRIETHYLAMMONIUM BROMIDE
mf: $C_{18}H_{40}N \cdot Br$ mw: 350.50

SAFETY PROFILE: Poison by ingestion and intraperitoneal routes. When heated to decomposition it emits very toxic fumes of NO_x, NH_3, and Br^-.

DXU400   CAS: 2855-19-8   HR: 3
DODECENE EPOXIDE
mf: $C_{12}H_{24}O$ mw: 184.36

SYN: 1,2-EPOXYDODECANE

CONSENSUS REPORTS: Reported in EPA TSCA Inventory.

SAFETY PROFILE: Questionable carcinogen with experimental tumorigenic data. When heated to decomposition it emits acrid smoke and irritating fumes.

DXV000   CAS: 25377-73-5   HR: 3
DODECENYLSUCCINIC ANHYDRIDE
mf: $C_{16}H_{27}O_3$ mw: 266.38

PROP: Light yellow, clear, viscous oil. Bp: 180-182° @ 5 mm, flash p: 352°F (COC), d: 1.002 @ 25°/4°.

CONSENSUS REPORTS: Reported in EPA TSCA Inventory.

SAFETY PROFILE: Poison by intraperitoneal route. An irritant and sensitizer. Combustible when exposed to heat or flame; can react with oxidizing materials. To fight fire, use foam, CO_2, dry chemical. When heated to decomposition it emits acrid smoke and irritating fumes.

DXV600   CAS: 112-53-8   HR: 3
DODECYL ALCOHOL
mf: $C_{12}H_{26}O$ mw: 186.38

PROP: Colorless solid, liquid above 21°; floral odor. D: 0.830-0.836, refr index: 1.440-1.444, mp: 24°, bp: 259°, flash p: 260°F, autoign temp: 527°F. Sol in 2 parts of 70% alc, fixed oils, propylene glycol; insol in water, glycerin.

SYNS: ALCOHOL C-12 * ALFOL 12 * CACHALOT L-50 * CO 12 * CO-1214 * n-DODECANOL * 1-DODECANOL * n-DODECYL ALCOHOL * DUODECYL ALCOHOL * DYTOL J-68 * EPAL 12 * FEMA No. 2617 * LAURIC ALCOHOL * LAURINIC ALCOHOL * LAURYL 24 * LAURYL ALCO-

HOL (FCC) * n-LAURYL ALCOHOL, PRIMARY * LOROL * MA-1214 * SIPOL L12

CONSENSUS REPORTS: Reported in EPA TSCA Inventory.

SAFETY PROFILE: Questionable carcinogen with experimental tumorigenic data. Moderately toxic by intraperitoneal route. Mildly toxic by ingestion. A severe human skin irritant. Combustible when exposed to heat or flame; can react with oxidizing materials. To fight fire, use dry chemical, CO_2. When heated to decomposition it emits acrid smoke and irritating fumes.

DXW000   CAS: 124-22-1   HR: 3
DODECYLAMINE
mf: $C_{12}H_{27}N$ mw: 185.40

PROP: Oil, amine odor. Fp: 28.3°, vap press: 64 mm @ 170°.

SYN: LAURYLAMINE

CONSENSUS REPORTS: Reported in EPA TSCA Inventory.

SAFETY PROFILE: Poison by intraperitoneal route. Moderately toxic by ingestion. A severe skin and eye irritant. When heated to decomposition it emits toxic fumes of NO_x.

DXW200   CAS: 25155-30-0   HR: 3
DODECYL BENZENE SODIUM SULFONATE
DOT: NA 9146
mf: $C_{18}H_{29}O_3S \cdot Na$ mw: 348.52

PROP: White to light yellow flakes, granules, or powder.

SYNS: AA-9 * ABESON NAM * BIO-SOFT D-40 * CALSOFT F-90 * CONCO AAS-35 * CONOCO C-50 * DETERGENT HD-90 * DODECYLBENZENESULFONIC ACID SODIUM SALT * DODECYLBENZENESULPHONATE, SODIUM SALT * DODECYLBENZENSULFONAN SODNY (CZECH) * MERCOL 25 * NACCANOL NR * NECCANOL SW * PILOT HD-90 * PILOT SF-40 * RICHONATE 1850 * SANTOMERSE 3 * SODIUM DODECYLBENZENESULFONATE (DOT) * SODIUM DODECYLBENZENESULFONATE, DRY * SODIUM LAURYLBENZENESULFONATE * SOLAR 40 * SOL SODOWA KWASU LAURYLOBENZENOSULFONOWEGO (POLISH) * SULFAPOL * SULFAPOLU (POLISH) * SULFRAMIN 85 * SULFRAMIN 40 FLAKES

* SULFRAMIN 40 GRANULAR * SULFRAMIN 1238 SLURRY * p-1′,1′,4′,4′-TETRAMETHYLOKTYLBEN-ZENSULFONAN SODNY (CZECH) * ULTRAWET K

CONSENSUS REPORTS: Reported in EPA TSCA Inventory.

DOT Classification: ORM-E; Label: None.

SAFETY PROFILE: Poison by intravenous route. Moderately toxic by ingestion. A skin and severe eye irritant. When heated to decomposition it emits toxic fumes of Na_2O.

DXX000   CAS: 538-71-6   **HR: 3**
DODECYLDIMETHYL(2-PHENOXY-ETHYL)AMMONIUM BROMIDE
mf: $C_{22}H_{40}NO \cdot Br$ mw: 414.54

SYNS: PHENODODECINIUM BROMIDE * β-PHENOXYETHYLDIMETHYLDODECYLAMMONIUM BROMIDE

CONSENSUS REPORTS: Reported in EPA TSCA Inventory.

SAFETY PROFILE: Poison by intraperitoneal and intravenous routes. When heated to decomposition it emits very toxic fumes of NO_x, NH_3, and Br^-.

DXX200   CAS: 1166-52-5   **HR: 2**
DODECYL GALLATE
mf: $C_{19}H_{30}O_5$ mw: 338.49

SYNS: LAURYL GALLATE * NIPAGALLIN LA * PROGALLIN LA

CONSENSUS REPORTS: Reported in EPA TSCA Inventory.

SAFETY PROFILE: Moderately toxic by ingestion. When heated to decomposition it emits acrid smoke and irritating fumes.

DXX400   CAS: 2439-10-3   **HR: 3**
N-DODECYLGUANIDINE ACETATE
mf: $C_{13}H_{29}N_3 \cdot C_2H_4O_2$ mw: 287.51

PROP: Crystals, sol in hot water and alc. Mp: 136°.

SYNS: AC 5223 * AMERICAN CYANAMID 5223 * APADODINE * CARPENE * CURITAN * CYPREX * CYPREX 65W * N-DODECYLGUANI-DINACETAT (GERMAN) * DODECYLGUANIDINE ACE-TATE * DODGUADINE * DODINE * DODINE ACETATE * DODINE, mixture with GLYODIN

* DOGQUADINE * ENT 16,436 * EXPERIMEN-TAL FUNGICIDE 5223 * LAURYLGUANIDINE ACE-TATE * MELPREX * MILPREX * SYLLIT * TSITREX * VENTUROL * VONDODINE

CONSENSUS REPORTS: Reported in EPA TSCA Inventory. EPA Genetic Toxicology Program.

SAFETY PROFILE: Questionable carcinogen with experimental tumorigenic data. Poison by ingestion. Moderately toxic by skin contact and possibly other routes. A severe eye irritant. A pesticide. When heated to decomposition it emits very toxic fumes of NO_x.

DXY600   CAS: 27193-86-8   **HR: 2**
DODECYLPHENOL
mf: $C_{18}H_{30}O$ mw: 262.48

PROP: Straw-colored liquid, phenolic odor. Bp: 154-168°, flash p: 325°F (OC), d: 0.93 @ 20°/20°, vap d: 9.04.

CONSENSUS REPORTS: Reported in EPA TSCA Inventory.

SAFETY PROFILE: Moderately toxic by ingestion. Mildly toxic by skin contact. Combustible when exposed to heat or flame; can react with oxidizing materials. To fight fire, use CO_2, dry chemical. When heated to decomposition it emits acrid smoke and irritating fumes.

DXZ000   CAS: 7631-98-3   **HR: 3**
N-DODECYLSARCOSINE SODIUM SALT
mf: $C_{15}H_{30}NO_2 \cdot Na$ mw: 279.45

SYN: SODIUM-N-LAURYL SARCOSINE

CONSENSUS REPORTS: Reported in EPA TSCA Inventory.

SAFETY PROFILE: Poison by intravenous route. When heated to decomposition it emits toxic fumes of NO_x and Na_2O.

DYA600   CAS: 1399-80-0   **HR: 3**
DODECYL-p-TOLYL TRIMETHYL AMMONIUM CHLORIDE

PROP: Aqueous preparation containing approximately 40% methyl dodecylbenzyl trimethyl ammonium chloride and 10% methyl dodecylxylene bis(trimethyl ammonium chloride).

SYNS: ALKYL(C9-15)TOLYL METHYLTRIMETHYL AMMONIUM CHLORIDE * HYAMINE 2389 * METHYL DODECYL BENZYL AMMONIUM CHLORIDE

SAFETY PROFILE: Poison by ingestion, intraperitoneal, and intravenous routes. A skin and severe eye irritant. When heated to decomposition it emits very toxic fumes of NO_x, NH_3, and Cl^-.

DYA800   CAS: 4484-72-4   HR: 3
DODECYLTRICHLOROSILANE

DOT: UN 1771
mf: $C_{12}H_{25}Cl_3Si$ mw: 303.81

PROP: Colorless to yellow liquid, readily hydrolyzed by moisture with the production of hydrochloric acid. Bp: 288°, d: 1.026 @ 25°/25°.

SYN: TRICHLORODODECYLSILANE

CONSENSUS REPORTS: Reported in EPA TSCA Inventory.

DOT Classification: Corrosive Material; Label: Corrosive.

SAFETY PROFILE: A poison. A corrosive irritant to the eyes, skin and mucous membranes. When heated to decomposition it emits toxic fumes of Cl^-.

DYC200   CAS: 5796-14-5   HR: 3
l-DOPA HYDROCHLORIDE
mf: $C_9H_{11}NO_4 \cdot ClH$ mw: 233.67

SYNS: l-3-(3,4-DIHYDROXYPHENYL)ALANINE * 3-HYDROXY-l-TYROSINE HYDROCHLORIDE

SAFETY PROFILE: Poison by intravenous route. Moderately toxic by ingestion. Used to treat Parkinson's disease. When heated to decomposition it emits very toxic fumes of NO_x and HCl.

DYC400   CAS: 51-61-6   HR: 3
DOPAMINE
mf: $C_8H_{11}NO_2$ mw: 153.20

SYN: 4-(2-AMINOETHYL)PYROCATECHOL

SAFETY PROFILE: Poison by intravenous, intracervical, and intraperitoneal routes. Experimental teratogenic and reproductive effects. Human mutation data reported. A neurotransmitter. An adrenergic agent. When heated to decomposition it emits toxic fumes of NO_x.

DYC800   CAS: 77-21-4   HR: 3
DORIDEN
mf: $C_{13}H_{15}NO_2$ mw: 217.29

PROP: dl-Form: Crystals from ether or from ethyl acetate + petr ether. Mp: 84°. Freely sol in ethyl acetate, acetone, ether, chloroform; sol in ethanol, methanol. Practically insol in water. d-Form: Crystals. Mp: 102.5-103°, refractive index: (α) (20/D) +176° (methanol). l-Form: Crystals. Mp: 102-103°, refr index: (α) (20/D) -181° (methanol).

SYNS: ALFIMID * CC 11511 * DORIDEN-SED * ELRODORM * 3-ETHYL-3-PHENYL-2,6-DIKETOPIPERIDINE * 3-ETHYL-3-PHENYL-2,6-DIOXOPIPERIDINE * α-ETHYL-α-PHENYLGLUTARIMIDE * 2-ETHYL-2-PHENYLGLUTARIMIDE * 3-ETHYL-3-PHENYL-2,6-PIPERIDINEDIONE * GIMID * GLIMID * GLUTATHIMID * GLUTETHIMID * GLUTETHIMIDE * GLUTETIMIDE * NOXYRON * 3-PHENYL-3-ETHYL-2,6-DIKETOPIPERIDINE * 3-PHENYL-3-ETHYL-2,6-DIOXOPIPERIDINE * α-PHENYL-α-ETHYLGLUTARIC ACID IMIDE * 2-PHENYL-2-ETHYLGLUTARIC ACID IMIDE * α-PHENYL-α-ETHYLGLUTARIMIDE * SARODORMIN

CONSENSUS REPORTS: EPA Genetic Toxicology Program.

SAFETY PROFILE: Poison by ingestion, intraperitoneal, and possibly other routes. Human systemic effects by ingestion: pupillary dilation, ataxia, somnolence, coma, and blood pressure depression. Experimental reproductive effects. When heated to decomposition it emits toxic fumes of NO_x. Caution: May be habit forming. This is a controlled substance (depressant) listed in the U.S. Code of Federal Regulations, Title 21 Part 1308.13 (1985).

DYE400   CAS: 53908-27-3   HR: 3
DOWFUME EB-5

PROP: Contains ethylene dichloride (30%), carbon tetrachloride (63%) and ethylene dibromide (7%).

SAFETY PROFILE: Poison by ingestion. Mildly toxic by skin contact and inhalation. When heated to decomposition it emits very toxic fumes of Cl^- and Br^-.

DYE409   CAS: 1668-19-5   HR: 3
DOXEPIN
mf: $C_{19}H_{21}NO$ mw: 279.41

SYNS: 11-(3-DIMETHYLAMINOPROPYLIDENE)-6,11-DI-HYDRODIBENZ(b,e)OXIPIN * N,N-DIMETHYLDI-BENZ(b,e)OXEPIN-$\Delta^{11(6H)}$-γ-PROPYLAMINE

SAFETY PROFILE: Human poison by ingestion. Experimental poison by ingestion, intravenous, and intraperitoneal routes. A sedative and hypnotic used as an antianxiety agent. When heated to decomposition it emits toxic fumes of NO_x.

DYE600   CAS: 523-87-5   HR: 3
DRAMAMINE
mf: $C_{17}H_{21}NO \cdot C_7H_7ClN_4O_2$ mw: 470.02

SYNS: AMOSYT * ANAUTINE * ANDRAMINE * AVIOMARIN * o-BENZHYDRYLDIMETHYLAMINO-ETHANOL-8-CHLOROTHEOPHYLLINATE * 2-(BENZHYDRYLOXY)-N,N-DIMETHYLETHYLAMINE with 8-CHLOROTHEOPHYLLINE * CHLORANAUTINE * DIAMARIN * DIMENHYDRINATE * DIPHENHYDRINATE * DRAMAMIN * DRAMARIN * DRAMYL * DROMYL * ELDODRAM * ETHYLAMINE-2-(DIPHENYLMETHOXY)-N,N-DIMETHYL, compound with 8-CHLOROTHEOPHYLLINE (1:1) * GRAVINOL * GRAVOL * MENHYDRINATE * NCI-C60639 * NEO-NAVIGAN * NOVAMIN * NOVAMINE * PERMITAL * REISE-ENGLETTEN * SUPREMAL * TEODRAMIN * TRAVELIN * TRAVELMIN * VOMEX A * XAMAMINA

CONSENSUS REPORTS: Reported in EPA TSCA Inventory.

SAFETY PROFILE: Poison by ingestion, intraperitoneal, and intravenous route. A drug much used for motion sickness. Human systemic effects by ingestion: hallucinations and distorted perceptions. When heated to decomposition it emits very toxic fumes of NO_x and Cl^-.

DYF200   CAS: 548-73-2   HR: 3
DROPERIDOL
mf: $C_{22}H_{22}FN_3O_2$ mw: 379.47

SYNS: DEHIDROBENZPERIDOL * DEHYDRO-BENZPERIDOL * DEIDROBENZPERIDOLO * DHBP * DIHIDROBENZPERIDOL * DRIDOL * DROLEPTAN * 1-(1-(3-(p-FLUOROBENZOYL)PROPYL)-1,2,3,6-TETRAHYDRO-4-PYRIDYL)-2 -BENZIMIDAZOLINONE * 1-(1-(4-(p-FLUOROPHENYL-4-OXOBUTYL)-1,2,3,6-TETRAHYDRO-4-PYRIDYL)-2-BENZIMIDAZOLINONE * HALKAN * INAPPIN * INAPSIN * INNOVAN * INNOVAR * INOPSIN * INOVAL * LEPTANAL * LEPTOFEN * MCN-JR-4749 * PROPERI-DOL * R 4749 * SINTOSIAN * THALAMONAL * VETKALM

SAFETY PROFILE: Poison by intravenous, subcutaneous, and intraperitoneal routes. Moderately toxic by ingestion. Human systemic effects by ingestion: wakefulness, tremors, and muscle weakness. An antipsychotic agent. When heated to decomposition it emits very toxic fumes of F^- and NO_x.

DYG000   HR: 3
DYNAMITE
SYN: GELATINE DYNAMITE

DOT Classification: Class A Explosive; Label: Explosive A.

SAFETY PROFILE: A high explosive used industrially in construction and mining. The name generally refers to a mixture containing as its principal explosive ingredient either glyceryl trinitrate (nitroglycerin) or ammonium nitrate, suitably sensitized. It does not apply to black blasting powders, chlorate powders, and other deflagrating mixtures. While this material is a powerful explosive when detonated by shock or heat, it is only moderately hazardous. It can react vigorously with oxidizing materials. Dangerous; shock and heat will explode it; when heated to decomposition it emits highly toxic fumes of NO_x and CO, etc.

An ordinary blasting cap or an electric blasting cap is used for detonating a charge of dynamite. The various classes and grades of dynamite are made from mixtures composed of an explosive compound or a mixture of explosive compounds, a "dope", and an antiacid. If any of the explosive ingredients are in a liquid state they are referred to as the "explosive oil," which is usually composed of glyceryl trinitrate (nitroglycerin) and about 25-30% of ethylene glycol dinitrate. The latter compound depresses the freezing point of the nitroglycerin and renders the dynamite low-freezing. Other compounds may also be used as freezing point depressants. The explosive oil is absorbed by carbonaceous materials that have entirely replaced kieselguhr (diatomaceous earth), formerly used exclusively as the absorbent or "dope" in dynamites. This type of "dope" does not enter into the explosive reaction. Wood pulp is now most commonly used as the absorber, either alone or mixed in suitable proportions with flour, starch, etc.

The absorbents may be mixed with an oxidizer such as sodium nitrate, in which case an active "dope" is formed. For neutralizing any acid that may be present, about 1% of an antiacid (calcium carbonate or zinc oxide) is added to the mixture. The explosive oil is mixed into the "dope". The strength of a kieselguhr dynamite, when detonated, is derived only from the explosive oil, since kieselguhr is inert. A mixture of this kind is known as a straight dynamite. On the other hand, an active "dope", (an admixture of carbonaceous absorbents with an oxidizer), furnishes explosive strength in addition to that derived from the explosive ingredients.

By replacing a part of the explosive oil of a straight dynamite with ammonium nitrate, so that the latter becomes the principal explosive ingredient, a mixture known as an ammonia dynamite is obtained.

When the explosive oil is gelatinized the explosive is known as a gelatin or an ammonia gelatin dynamite.

Blasting gelatin is a gelatinized mass of an elastic nature obtained by incorporating nitrocotton with an explosive oil into which is mixed about 1% of antiacid.

Dynamites may be in bulk form (bag powder) or in cartridge form, the most common size being 1¼ inch in diameter and 8 inches long, although for holes of small diameter, cartridges as small as ⅞ inch in diameter are also used. In large diameter well-drill holes for quarry blasting, cartridge diameters up to 10 inches and lengths up to 30 inches may be used. These upper limits or 50 pounds in weight of each cartridge are imposed by the DOT Regulations, and the maximum length of 30 inches applies to all cartridge diameters between 4 and 10 inches.

An integral part of a stick of dynamite is the paraffined paper wrapper that not only holds the ingredients together but enters into the explosive reaction.

The wrapper also affords some measure of protection from moderate exposure to dampness. For blasting in wet operations, a gelatinized dynamite which resists the absorption of water is used.

The strength of straight dynamite is graded by its explosive oil content (% by weight), while for any other class of dynamite, the strength is determined experimentally in comparison with the various grades of the straight dynamites. For example, a 40% straight dynamite is one which contains 40% of explosive oil; a 40% strength ammonia dynamite, as determined by tests, equals a 40% straight dynamite in strength. In other words a 40% strength ammonia dynamite will release the same energy as an equivalent weight of a 40% straight dynamite.

DYG400 HR: 2
DYSPROSIUM
aw: Dy   aw: 162.5

PROP: Bright, lustrous, silvery metal. Mp: 1409°, bp: 2335°, d: 8.540 @ 25°.

SAFETY PROFILE: It may exhibit an anticoagulant effect. Flammable; an active reducing agent. Reacts violently in air and to halogens.

DYG600   CAS: 10025-74-8   HR: 3
DYSPROSIUM CHLORIDE
mf: Cl_3Dy mw: 268.85

PROP: Shiny, yellow crystals. D: 3.67 @ 0°/4°, mp: 718°, bp: 1500°. A sol salt.

CONSENSUS REPORTS: Reported in EPA TSCA Inventory.

SAFETY PROFILE: Poison by intraperitoneal route. Mildly toxic by ingestion. When heated to decomposition it emits toxic fumes of Cl^-.

DYG800   CAS: 13074-91-4   HR: 3
DYSPROSIUM CITRATE
mf: $C_6H_5O_7 \cdot Dy$ mw: 351.61

SYN: CITRIC ACID, DYSPROSIUM(3+) salt (1:1)

SAFETY PROFILE: Poison by intraperitoneal route. When heated to decomposition it emits acrid smoke and irritating fumes.

DYH000   CAS: 35725-30-5   HR: 3
DYSPROSIUM(III) NITRATE
HEXAHYDRATE (1:3:6)
mf: $N_3O_9 \cdot Dy \cdot 6H_2O$ mw: 456.65

SYN: NITRIC ACID DYSPROSIUM(3+) SALT HEXAHYDRATE

SAFETY PROFILE: Poison by intraperitoneal route. Moderately toxic by ingestion. When heated to decomposition it emits very toxic fumes of NO_x.

E

EAF000   CAS: 506-30-9   **HR: 3**
EICOSANOIC ACID
mf: $C_{20}H_{40}O_2$ mw: 312.60

PROP: White leaflets. Mp: 77°, bp: 328°. Slt decomp in water; very sol in hot absolute alc and ether.

SYNS: ARACHIC ACID * ARACHIDIC ACID

CONSENSUS REPORTS: Reported in EPA TSCA Inventory.

SAFETY PROFILE: Questionable carcinogen with experimental neoplastigenic data by implant route. When heated to decomposition it emits acrid smoke and fumes.

EAG000   CAS: 23315-05-1   **HR: 3**
ELAIOMYCIN
mf: $C_{13}H_{26}N_2O_3$ mw: 258.41

PROP: Metabolite of *Streptomyces hepaticus*.

SYN: d-threo-METHOXY-3-(1-OCTENYL-ONN-AZOXY)-2-BUTANOL

SAFETY PROFILE: Poison by intravenous and subcutaneous routes. Questionable carcinogen with experimental tumorigenic data. Causes tumors of the brain. When heated to decomposition it emits toxic fumes of NO_x.

EAH500   CAS: 50-48-6   **HR: 3**
ELAVIL
mf: $C_{20}H_{23}N$ mw: 277.44

SYNS: AMITRIPTILINE * AMITRIPTYLIN (GERMAN) * AMITRIPTYLINE * DAMILAN * 3,10-DIHYDRO-5H-DIBENZO(a,d)CYCLOHEPTEN-5-YLIDENE-N,N-DIMETHYL-1-PROPANAMINE * 10,11-DIHYDRO-5-(γ-DIMETHYLAMINOPROPYLIDENE)-5H-DIBENZO(a,d)CYCLOHEPTENE * 10,11-DIHYDRO-N,N-DIMETHYL-5H-DIBENZO(a,d)HEPTALENE-Δ⁵-γ-PROPYLAMINE * 5-(3'-DIMETHYLAMINOPROPYLIDENE)-DIBENZO-(a,d)(1,4)-CYCLOHEPTADIENE * 5-(γ-DIMETHYLAMINOPROPYLIDENE)-5H-DIBENZO(a,d)-10,11-DIHYDROCYCLOHEPTENE * 5-(3-DIMETHYLAMINOPROPYLIDENE)-10,11-DIHYDRO-5H-DIBENZO(a,d)CYCLOHEPTENE * 5-(γ-DIMETHYLAMINOPROPYLIDENE)-10,11-DIHYDRO-5H-DIBENZO(A,D)CYCLOHEPTENE * 5-(3-DIMETHYLPROPYLIDENE)DIBENZO(a,d)(1,4)CYCLOHEPTADIENE * ELANIL * LAROXIL * LAROXYL * PROHEPTADIENE * TRYPTIZOL

SAFETY PROFILE: Human poison by an unspecified route. Poison experimentally by ingestion, intraperitoneal, subcutaneous, intramuscular, intravenous, and possibly other routes. Human systemic effects by ingestion, intramuscular and possibly other routes: changes in sleep, headache, paresthesia, convulsions, excitement, somnolence, muscle contractions, change in heart rate, and other cardiac changes. An experimental teratogen. An antidepressant. When heated to decomposition it emits toxic fumes of NO_x.

EAI000   CAS: 549-18-8   **HR: 3**
ELAVIL HYDROCHLORIDE
mf: $C_{20}H_{23}N \cdot ClH$ mw: 313.90

SYNS: AMITID * AMITRIL * AMITRIPTYLINE CHLORIDE * AMITRYPTYLINE HYDROCHLORIDE * DAMILEN HYDROCHLORIDE * DEPREX * 10,11-DIHYDRO-N,N-DIMETHYL-5H-DIBENZO(a,d)-CYCLOHEPTENE-Δ⁵·γ-PROPYLAMINE HCL * 3-(3-DIMETHYLAMINOPROPYLIDENE)-1:2-4:5-DIBENZOCYCLOHEPTA-1:4-DIENE * 5-(3-DIMETHYLAMINOPROPYLIDENE)DIBENZO(a,d)(1,4)CYCLOHEPTADIENE HYDROCHLORIDE * DOMICAL * ELAVIL * ENDEP * LENTIZOL * MIKETORIN * PROHEPTADIEN MONOHYDROCHLORIDE * SAROTEN * SAROTENE * SK-AMITRIPTYLINE * TRYPTIZOL * TRYPTIZOL HYDROCHLORIDE

CONSENSUS REPORTS: Reported in EPA TSCA Inventory.

SAFETY PROFILE: Poison by ingestion, intraperitoneal, intravenous, and subcutaneous routes. Human systemic effects by ingestion: convulsions, respiratory depression, changes in sleep, hallucinations, muscle contractions, somnolence, blood pressure decrease, coma, cyanosis, dyspnea, and ataxia. An experimental teratogen. Mutation data reported. Used in the treatment of depression. When heated to decomposition it emits very toxic fumes of HCl and NO_x.

EAJ000   CAS: 548-43-6   **HR: 3**
ELYMOCLAVINE
mf: $C_{16}H_{18}N_2O$ mw: 254.36

PROP: A close chemical relative of LSD found in ergot fungi and bindweeds of the genus *Ipomoea, fm. convolvulaceae*.

SYNS: ELIMOCLAVIN * ELYMOCLAVIN

SAFETY PROFILE: A poison by intraperitoneal and possibly other routes. Experimental teratogenic and reproductive effects. When heated to decomposition it emits acrid smoke and fumes.

EAJ500   CAS: 19526-81-9   **HR: 3**
EMAZOL RED B
mf: $C_{18}H_{16}N_2O_{10}S_3 \cdot 2Na$ mw: 562.52

CONSENSUS REPORTS: Reported in EPA TSCA Inventory.

SAFETY PROFILE: Questionable carcinogen with experimental tumorigenic data. When heated to decomposition it emits very toxic fumes of NO_x, Na_2O, and SO_x.

EAL100   CAS: 1302-74-5   **HR: 3**
EMERY
mf: Al_2O_3 mw: 101.96

PROP: A varicolored mineral. D: 3.95-4.10.

SYNS: ALUMINUM OXIDE * CORUNDUM
* ELECTROCORUNDUM * EN 237 * KER 710
* KO 7 * KORUND * KU 5-3 * MP 1 (refractory)

OSHA PEL: (Transitional: TWA Total Dust: 15 mg/m^3; Respirable Fraction: 5 mg/m^3) TWA Total Dust: 10 mg/m^3; Respirable Fraction: 5 mg/m^3
ACGIH TLV: TWA (nuisance particulate) 10 mg/m^3 of total dust (when toxic impurities are not present, e.g., quartz < 1%).

SAFETY PROFILE: Questionable carcinogen with experimental carcinogenic data. May cause a pneumoconiosis. It is mainly a nuisance dust.

EAL500   CAS: 483-18-1   **HR: 3**
EMETINE
mf: $C_{29}H_{40}N_2O_4$ mw: 480.71

PROP: White powder or lumps, bitter taste, darkens on exposure. Mp: 74°.

SYNS: CEPHAELINE METHYL ETHER * (−)-EMETINE * NSC-33669 * 6',7',10,11-TETRAMETHOXY-EMETAN

CONSENSUS REPORTS: NCI Carcinogenesis Studies (ipr); Inadequate Studies: mouse, rat NCITR* NCI-CG-TR-43,77

SAFETY PROFILE: A human poison by an unspecified route. A experimental poison by ingestion, intraperitoneal, intravenous, and subcutaneous routes. Human systemic effects by subcutaneous route: muscle weakness, cardiac arrythmias and gastrointestinal effects. Mutation data reported. A severe eye and moderate skin irritant. It is one of the two potent alkaloids obtained from the Brazilian plant ipecac. The therapeutic use of various ipecac preparations has caused many cases of poisoning, in some instances with fatal results. The toxic effects are particularly prominent if the drug is given intravenously. Special care should therefore be exercised when administering it in this manner. The symptoms of intoxication are gastrointestinal irritation and salivation, as well as general edema which follows renal insufficiency, hemoptysis (blood stained sputum), flaccid paralysis, peripheral neuritis (inflammation of the nerve endings), aphonia (loss of voice), difficulties in swallowing, delirium, coma and failure of the heart. The fatal dose is considered to be approximately 2 grams, whether administered over a short or relatively long period. The drug seems to have a cumulative effect. A severe eye irritant. When heated to decomposition it emits highly toxic fumes of NO_x.

EAM000   **HR: 3**
EMETINE ANTIMONY IODIDE

SYN: ANTIMONY EMETINE IODIDE

CONSENSUS REPORTS: Antimony and its compounds are on the Community Right-To-Know List.

OSHA PEL: TWA 0.5 mg(Sb)/m^3
ACGIH TLV: TWA 0.5 mg(Sb)/m^3
NIOSH REL: (Antimony) TWA 0.5 mg/m^3

SAFETY PROFILE: Poison by ingestion. Human systemic effects by ingestion: nausea, vomiting and other gastrointestinal effects. When heated to decomposition it emits very toxic fumes of I^-, NO_x, and Sb.

EAM500   CAS: 8001-15-8   **HR: 3**
EMETINE BISMUTH IODIDE
mf: $C_{29}H_{40}N_2O_4 \cdot BiI_3$ mw: 1070.39

PROP: Composition = 25% emetine and 17% bismuth.

SYNS: BISMUTH EMETINE IODIDE * EMETINE with BISMUTH(III) TRIIODIDE * EMETINE TRIIODOBIS-

MUTH(III) * NSC 44185 * TRIIODO(6′,7′,10,11-TETRAMETHOXYEMETAN)BISMUTH

SAFETY PROFILE: A poison by ingestion and intraperitoneal routes. When heated to decomposition it emits very toxic fumes of I⁻ and NO$_x$.

EAN000 CAS: 316-42-7 **HR: 3**
1-EMETINE DIHYDROCHLORIDE
mf: $C_{29}H_{40}N_2O_4 \cdot 2ClH$ mw: 553.63

SYNS: AMEBICIDE * EMETINE, DIHYDROCHLORIDE * (−)-EMETINE DIHYDROCHLORIDE * EMETINE HYDROCHLORIDE * NSC-33669

CONSENSUS REPORTS: EPA Extremely Hazardous Substances List. Reported in EPA TSCA Inventory.

SAFETY PROFILE: A poison by ingestion, intraperitoneal, subcutaneous, and intravenous routes. A human eye irritant. When heated to decomposition it emits very toxic fumes of Cl⁻ and NO$_x$.

EAP000 CAS: 8015-30-3 **HR: 3**
ENAVID
mf: $C_{21}H_{26}O_2 \cdot C_{20}H_{26}O_2$ mw: 608.93

PROP: Mixture of 98.5% (17-α)-19-norpregn-4-en-20-yn-3-one,17-hydroxy- and 1.5% (17-α)-19-norpregna-1,3,5(10)-trien-20-yn-17-ol,3-methoxy-.

SYNS: CONOVID * CONOVID E * ENIDREL * ENOVID * ENOVID-E * ETHINYLESTRADIOL-3-METHYL ETHER and NORETHYNODRED (1:50) * MESTRANOL mixed with NORETHYNODREL * NORETHANDROL * NORETHYNODREL and ETHINYLESTRADIOL-3-METHYL ETHER (50:1) * NORETHYNODREL mixed with MESTRANOL

CONSENSUS REPORTS: IARC Cancer Review: Animal Sufficient Evidence IMEMDT 6,191,74. EPA Genetic Toxicology Program.

SAFETY PROFILE: Confirmed carcinogen producing liver tumors in women by ingestion. Experimental carcinogenic, neoplastigenic, and tumorigenic data. Human reproductive effects by ingestion: menstrual cycle changes or disorders; abnormalities of the uterus, cervix, and vagina; and changes in fertility. A human teratogen which causes developmental abnormalities of the urogenital system. Experimental reproductive effects. A steroid. When heated to

decomposition it emits acrid smoke and fumes.

EAQ750 CAS: 115-29-7 **HR: 3**
ENDOSULFAN

DOT: NA 2761
mf: $C_9H_6Cl_6O_3S$ mw: 406.91

PROP: A mixture of 2 isomers, brown crystals, nearly insol in water; sol in most organic solvents. Mp (α): 106°, mp (β): 212°, d: 1.745 @ 20°/20°.

SYNS: BENZOEPIN * BEOSIT * BIO 5,462 * CHLORTHIEPIN * CRISULFAN * CYCLODAN * DEVISULPHAN * ENDOCEL * ENDOSOL * ENDOSULPHAN * ENSURE * ENT 23,979 * FMC 5462 * 1,2,3,4,7,7-HEXACHLOROBICYCLO(2.2.1)HEPTEN-5,6-BIOXYMETHYLENE-SULFITE * α,β-1,2,3,4,7,7-HEXACHLOROBICYCLO(2.2.1)-2-HEPTENE-5,6-BISOXYMETHYLENE SULFITE * HEXACHLOROHEXAHYDROMETHANO 2,4,3-BENZODIOXATHIEPIN-3-OXIDE * 6,7,8,9,10,10-HEXACHLORO-1,5,5a,6,9,9a-HEXAHYDRO-6,9-METHANO-2,4,3-BENZODIOXATHIEPIN-3-OXIDE * 1,4,5,6,7,7-HEXACHLORO-5-NORBORNENE-2,3-DIMETHANOL CYCLIC SULFITE * HILDAN * HOE 2,671 * INSECTOPHENE * KOP-THIODAN * MALIX * NCI-C00566 * NIA 5462 * NIAGARA 5,462 * OMS 570 * RCRA WASTE NUMBER P050 * SULFUROUS ACID, cyclic ester with 1,4,5,6,7,7-HEXACHLORO-5-NORBORNENE-2,3-DIMETHANOL * THIFOR * THIMUL * THIODAN * THIOFOR * THIOMUL * THIONEX * THIOSULFAN * THIOSULFAN TIONEL * TIOVEL

CONSENSUS REPORTS: EPA Extremely Hazardous Substances List. NCI Carcinogenesis Bioassay (feed); No Evidence: mouse, rat NCITR* NCI-CG-TR-62,77.

OSHA PEL: TWA 0.1 mg/m³ (skin)
ACGIH TLV: TWA 0.1 mg/m³ (skin)
DOT Classification: Poison B; Label: Poison.

SAFETY PROFILE: Poison by ingestion, inhalation, skin contact, intraperitoneal, subcutaneous, and possibly other routes. Experimental teratogenic and reproductive effects. Questionable carcinogen with experimental tumorigenic and neoplastigenic data. Human mutation data reported. A central nervous system stimulant producing convulsions. A highly toxic organochlorine pesticide which does not accumulate significantly in human tissue. Absorption is normally slow, but is increased by alcohols, oil, and emulsifiers. When heated to decomposition it emits toxic fumes of Cl^- and SO_x.

EAR000   CAS: 145-73-3   HR: 3
ENDOTHAL
mf: $C_8H_{10}O_5$ mw: 186.18

PROP: Solid. Mp: 144°. Sol in water.

SYNS: AQUATHOL * ENDOTHALL * ENDOTHAL TECHNICAL * 3,6-ENDOOXOHEXAHYDROPHTHALIC ACID * 3,6-ENDOXOHEXAHYDROPHTHALIC ACID * 3,6-endo-EPOXY-1,2-CYCLOHEXANEDICARBOXYLIC ACID * HEXAHYDRO-3,6-endo-OXYPHTHALIC ACID * HYDOUT * HYDROTHAL-47 * 7-OXABICYCLO (2.2.1)HEPTANE-2,3-DICARBOXYLIC ACID * RCRA WASTE NUMBER P088 * TRI-ENDOTHAL

SAFETY PROFILE: Poison by ingestion. Very irritating to skin, eyes, and mucus membranes. Causes diarrhea. When heated to decomposition it emits acrid smoke and fumes.

EAS000   CAS: 2778-04-3   HR: 3
ENDOTHION
mf: $C_9H_{13}O_6PS$ mw: 280.25

SYNS: AC-18,737 * O,O-DIMETHYL-S-(5-METHOXY-4-OXO-4H-PYRAN-2-YL)PHOSPHOROTHIOATE * O,O-DIMETHYL-S-(5-METHOXY-PYRON-2-YL)-METHYL) THIOLPHOSPHAT (GERMAN) * O,O-DIMETHYL-S-(5-METHOXYPYRONYL-2-METHYL) THIOPHOSPHATE * ENDOCID * ENDOCIDE * ENT 24,653 * EXOTHION * 5-METHOXY-2-(DIMETHOXYPHOS-PHINYLTHIOMETHYL)PYRONE-4 * S-5-METHOXY-4-OXOPYRAN-2-YLMETHYL DIMETHYL PHOS-PHOROTHIOATE * S-((5-METHOXY-4H-PYRON-2-YL) METHYL)-O,O-DIMETHYL-MONOTHIOFOSFAAT (DUTCH) * S-((5-METHOXY-4H-PYRON-2-YL) METHYL)-O,O-DIMETHYL-MONOTHIOPHOSPHAT(GER-MAN) * S-(5-METHOXY-4-PYRON-2-YLMETHYL) DI-METHYL PHOSPHOROTHIOLATE * S-((5-METOSSI-4H-PIRON-2-IL)-METIL)-O,O-DIMETIL-MONOTIOFOSFATO (ITALIAN) * NIA-5767 * NIAGARA 5767 * PHOS-PHATE 100 * PHOSPHOPYRON * PHOSPHOPYRONE * THIOPHOSPHATE de O,O-DIMETHYLE et de S-((5-ME-THOXY-4-PYRONYL)-METHYLE) (FRENCH)

CONSENSUS REPORTS: EPA Extremely Hazardous Substances List.

SAFETY PROFILE: A poison by ingestion, skin contact, and possibly other routes. A pesticide.

When heated to decomposition it emits very toxic fumes of PO_x and SO_x.

EAS500   CAS: 50-18-0   HR: 3
ENDOXAN
mf: $C_7H_{15}Cl_2N_2O_2P$ mw: 261.11

PROP: Crystals. Mp: 41-45°. Water-sol; sltly sol in organic solvents.

SYNS: ASTA * ASTA B518 * B 518 * N,N-BIS-(β-CHLORAETHYL)-N',O-PROPYLEN-PHOSPHORSA-EURE-ESTER-DIAMID (GERMAN) * 2-(BIS(2-CHLORO-ETHYL)AMINO)-2H-1,3,2-OXAAZAPHOSPHORINE 2-OXIDE * N,N-BIS(2-CHLOROETHYL)-N'-(3-HYDROXYPROPYL)PHOSPHORODIAMIDIC ACID intramol. ESTER * BIS(2-CHLOROETHYL)PHOSPHORAMIDE-CYCLIC PROPANOLAMIDE ESTER * N,N-BIS(2-CHLO-ROETHYL)-N',O-PROPYLENEPHOSPHORIC ACID ESTER DIAMIDE * N,N-BIS(2-CHLOROETHYL)TETRAHYDRO-2H-1,3,2-OXAPHOSPHORIN-2-AMINE-2-OXIDE * N,N-BIS(β-CHLOROETHYL)-N',O-TRIMETHYLENEPHOSPHORIC ACID ESTER DIAMIDE * CB 4564 * CLAFEN * CLAPHENE * CP * CPA * CTX * CY * CYCLOPHOSPHAMIDE * CYCLOPHOSPHAMIDUM * CYCLOPHOSPHAN * CYCLOPHOSPHORAMIDE * CYCLOSTIN * CYTOPHOSPHAN * CYTOXAN * N,N-DI(2-CHLOROETHYL)-N,o-PROPYLENE-PHOS-PHORIC ACID ESTER DIAMIDE * ENDOXANAL * ENDOXAN-ASTA * ENDOXAN R * GENOXAL * HEXADRIN * MITOXAN * NCI-C04900 * NEOSAR * NSC 26271 * 2-H-1,3,2-OXAZAPHOS-PHORINANE * PROCYTOX * RCRA WASTE NUM-BER U058 * SEMDOXAN * SENDUXAN * SK 20501 * ZYKLOPHOSPHAMID (GERMAN)

CONSENSUS REPORTS: IARC Cancer Review: GROUP 1 IMEMDT 7,182,87; Human Sufficient Evidence IMEMDT 26,165,81; Animal Sufficient Evidence IMEMDT 26,165,81; IMEMDT 9,135,75; Human Limited Evidence IMEMDT 9,135,75. NTP Fourth Annual Report On Carcinogens, 1984. NCI Carcinogenesis Studies (ipr); Clear Evidence: mouse, rat RRCRBU 52,1,75. EPA Genetic Toxicology Program.

SAFETY PROFILE: Confirmed human carcinogen producing leukemia, Hodgkin's disease, gastrointestinal, and bladder tumors. Experimental carcinogenic and neoplastigenic data. A human poison by ingestion and many other routes. Human systemic effects by ingestion, intravenous, intraperitoneal, subcutaneous, and possibly other routes: kidney changes (hepatic dysfunction), leukopenia (reduced white blood cell count), nausea and alopecia (loss of hair). Human reproductive and teratogenic effects by multiple routes: spermatogenesis, testical changes, epididymis and sperm duct changes, menstrual cycle changes; fetal developmental abnormalities of the craniofacial area, musculoskeletal and cardiovascular systems. Experimental teratogenic and reproductive effects. Human mutation data reported. A powerful skin irritant. Used as an immunosuppressive agent in nonmalignant diseases. When heated to decomposition it emits highly toxic fumes of PO_x, NO_x and Cl^-.

EAT500   CAS: 72-20-8   **HR: 3**
ENDRIN

DOT: UN 2761
mf: $C_{12}H_8Cl_6O$ mw: 380.90

PROP: White crystals. Mp: decomp @ 200°.

SYNS: COMPOUND 269 * ENDREX * ENDRINE (FRENCH) * ENT 17,251 * HEXACHLOROEPOXYOC-TAHYDRO-endo,endo-DIMETHANONAPHTHALENE * 3,4,5,6,9,9-HEXACHLORO-1a,2,2a,3,6,6a,7,7a-OCTAHY-DRO-2,7:3,6-DIMETHANONAPHTH(2,3-b)OXIRENE * HEXADRIN * MENDRIN * NCI-C00157 * NENDRIN * RCRA WASTE NUMBER P051

CONSENSUS REPORTS: IARC Cancer Review: GROUP 3 IMEMDT 7,56,87; Animal Inadequate Evidence IMEMDT 5,157,74; Human Inadequate Evidence IMEMDT 5,157,74. NCI Carcinogenesis Bioassay (feed); No Evidence: mouse, rat NCITR* NCI-CG-TR-12,79. EPA Genetic Toxicology Program. EPA Extremely Hazardous Substances List.

OSHA PEL: TWA 0.1 mg/m^3 (skin)
ACGIH TLV: TWA 0.1 mg/m^3 (skin)
DFG MAK: 0.1 mg/m^3
DOT Classification: Poison B; Label: Poison; Poison B; Label: Poison, liquid.

SAFETY PROFILE: Poison by ingestion, skin contact, intravenous, and possibly other routes. Experimental teratogenic and reproductive effects. Questionable carcinogen. Mutation data reported. A central nervous system stimulant. Highly toxic to birds, fish and humans. Many cases of fatal poisoning have been attributed to it. Does not accumulate in human tissue. In humans, ingestion of 1 mg/kg has caused symptoms. A dangerous fire hazard. Mixtures with parathion dissolve very exothermically in petro-

leum solvents and may cause an air-vapor explosion.

EAT900   CAS: 13838-16-9   **HR: 2**
ENFLURANE
mf: $C_3H_2ClF_5O$ mw: 184.50

SYNS: ANESTHETIC COMPOUND NO. 347 * 2-CHLORO-1-(DIFLUOROMETHOXY)-1,1,2-TRIFLUORO-ETHANE * 2-CHLORO-1,1,2-TRIFLUOROETHYL DI-FLUOROMETHYL ETHER * COMPOUND 347 * ETHRANE * METHYLFLURETHER * NSC-115944 * OHIO 347

CONSENSUS REPORTS: Reported in EPA TSCA Inventory.

ACGIH TLV: TWA 75 ppm
NIOSH REL: (Waste Anesthetic Gases and Vapors) CL 2 ppm/1H

SAFETY PROFILE: Mildly toxic by inhalation, ingestion, and subcutaneous routes. Experimental reproductive data by inhalation. Human systemic effects by inhalation: decreased urine volume or anuria. Experimental reproductive effects. Human mutation data reported. An eye irritant. Questionable carcinogen with experimental carcinogenic data. An anesthetic. When heated to decomposition it emits very toxic fumes of F^- and Cl^-.

EAV500   CAS: 33419-42-0   **HR: 3**
EPE
mf: $C_{29}H_{32}O_{13}$ mw: 588.61

SYNS: DEMETHYL-EPIODOPHYLLOTOXIN ETHYLI-DENE GLUCOSIDE * 4-DEMETHYLEPIODOPHYLLO-TOXIN-β,d-ETHYLIDENEGLUCOSIDE * 4'-DE-METHYLEPIPODOPHYLLOTOXIN-9-(4,6-O-ETHYLIDENE-β-d-GLUCOPYRANOSIDE * 4-DEMETHYL-EPIPODOPHYLLOTOXIN-β,d-ETHYLIDEN-GLUCOSIDE * 4'-DEMETHYLEPIPODOPHYLLOTOXIN ETHYLIDENE-β,d-GLUCOSIDE * 4'-O-DEMETHYL-1-O-(4,6-O-ETHYLI-DENE-β,d-GLUCOPYRANOSYL)EPIPODOPHYLLOTOXIN * ETOPOSIDE * NK 171 * NSC 141540 * VEPESID * VP 16213

SAFETY PROFILE: Poison by ingestion, intraperitoneal, intravenous, and subcutaneous routes. Human systemic effects by ingestion and inhalation: angranulocytosis, aplastic anemia, and other changes in bone marrow. Experimental teratogenic and reproductive effects. Human mutation data reported. When heated to decomposition it emits acrid smoke and fumes.

EAW000   CAS: 299-42-3   **HR: 3**
EPHEDRINE
mf: $C_{10}H_{15}NO$ mw: 165.26

PROP: White granules. Mp: 79° (dl), bp: 255° (decomp). Sol in ether and chloroform.

SYNS: BIOPHEDRIN * ECIPHIN * EFEDRIN * EPHEDRAL * EPHEDRATE * EPHEDREMAL * EPHEDRIN * 1-EPHEDRINE * l(−)-EPHEDRINE * EPHEDRITAL * EPHEDROL * EPHEDROSAN * EPHEDROTAL * EPHEDSOL * EPHENDRONAL * EPHOXAMIN * FEDRIN * α-HYDROXY-β-METHYL AMINE PROPYLBENZENE * 1-HYDROXY-2-METHYLAMINO-1-PHENYLPROPANE * l-SEDRIN * ISOFEDROL * KRATEDYN * MANADRIN * MANDRIN * (−)-α-(1-METHYLAMINOETHYL)BENZYL ALCOHOL * 1-α-(1-METHYLAMINOETHYL)BENZYL ALCOHOL * 1-2-METHYLAMINO-1-PHENYLPROPANOL * N-METHYLNOREPHEDRINE * NASOL * 1-PHENYL-2-METHYLAMINOPROPANOL * SANEDRINE * VENCIPON * ZEPHROL

CONSENSUS REPORTS: Reported in EPA TSCA Inventory.

SAFETY PROFILE: A human poison by an unspecified route. An experimental poison by intravenous, subcutaneous, intramuscular, intraperitoneal and possibly other routes. Moderately toxic by ingestion and parenteral routes. Causes rapid pulse, rise in blood pressure, and other actions similar to epinephrine. Has been known to cause allergic sensitization. When heated to decomposition it emits toxic fumes of NO_x.

EAW500   CAS: 50-98-6   **HR: 3**
EPHEDRINE HYDROCHLORIDE
mf: $C_{10}H_{15}NO \cdot ClH$ mw: 201.72

CONSENSUS REPORTS: Reported in EPA TSCA Inventory.

SAFETY PROFILE: A poison by subcutaneous and intravenous routes. Human systemic effects by intradermal route: skin effects. When heated to decomposition it emits very toxic fumes of Cl^- and NO_x.

EAX000   CAS: 24221-86-1   **HR: 3**
d-EPHEDRINE HYDROCHLORIDE
mf: $C_{10}H_{15}NO \cdot ClH$ mw: 201.72

CONSENSUS REPORTS: Reported in EPA TSCA Inventory.

SAFETY PROFILE: A poison by intraperitoneal, subcutaneous, and intravenous routes. When heated to decomposition it emits very toxic fumes of HCl and NO_x.

EAX500   CAS: 134-71-4   **HR: 3**
dl-EPHEDRINE HYDROCHLORIDE
mf: $C_{10}H_{15}NO \cdot ClH$ mw: 201.72

SYNS: EPHETONIN * EPHETONINE * dl-α-(1-(METHYLAMINO)ETHYL) BENZYL ALCOHOL HYDROCHLORIDE * 1-PHENYL-2-METHYLAMINOPROPANOL-1 * RACEPHEDRINE HYDROCHLORIDE

CONSENSUS REPORTS: Reported in EPA TSCA Inventory.

SAFETY PROFILE: Poison by subcutaneous, intravenous, and intraperitoneal routes. Moderately toxic by ingestion. When heated to decomposition it emits very toxic fumes of HCl and NO_x.

EAY000   CAS: 50-98-6   **HR: 3**
l-EPHEDRINE HYDROCHLORIDE
mf: $C_{10}H_{15}NO \cdot ClH$ mw: 201.72

SYNS: EPHEDRINE HYDROCHLORIDE * (−)-EPHEDRINE HYDROCHLORIDE * (R-(R*,S*))-α-(1-(METHYLAMINO)ETHYL)BENZENEMETHANOL HYDROCHLORIDE

CONSENSUS REPORTS: Reported in EPA TSCA Inventory.

SAFETY PROFILE: Poison by ingestion, intraperitoneal, intravenous, subcutaneous, and intramuscular routes. Moderately toxic by parenteral route. Human systemic effects by intradermal route: local anesthetic. When heated to decomposition it emits very toxic fumes of HCl and NO_x.

EAY500   CAS: 134-72-5   **HR: 3**
1-EPHEDRINE SULFATE
mf: $C_{20}H_{30}N_2O_2 \cdot H_2O_4S$ mw: 428.60

SYNS: ISOFEDROL * 1-α-(1-(METHYLAMINO)ETHYL)BENZYL ALCOHOL SULFATE * NCI-C55652 * 1-PHENYL-2-METHYLAMINE-PROPANOL-1-SULFATE

CONSENSUS REPORTS: NTP Carcinogenesis Studies (feed); No Evidence: mouse, rat NTPTR* NTP-TR-307,86. Reported in EPA TSCA Inventory.

SAFETY PROFILE: Poison by intravenous, intraperitoneal and subcutaneous routes. Moderately toxic by ingestion. Human systemic effects by intravenous route: increased pulse rate and blood pressure. When heated to decomposition it emits very toxic fumes of NO_x and SO_x.

EAZ000 CAS: 62-32-8 **HR: 3**
EPHININE HYDROCHLORIDE
mf: $C_9H_{13}NO_2 \cdot ClH$ mw: 203.69

SYNS: 3,4-DIHYDROXYPHENYLETHYLMETHYLAMINE HYDROCHLORIDE * 3,4-DIHYDROXYPHENYL-1-METHYLAMINO-2-ETHANE HYDROCHLORIDE * 4-(2-METHYLAMINOETHYL)PYROCATECHOL HYDROCHLORIDE * METHYL-(β-(3,4-DIHYDROXY PHENYL ETHYL) AMINE HYDROCHLORIDE * N-METHYLDOPAMINE HYDROCHLORIDE

SAFETY PROFILE: A poison by subcutaneous and intraperitoneal routes. When heated to decomposition it emits very toxic fumes of NO_x and HCl.

EAZ500 CAS: 106-89-8 **HR: 3**
EPICHLOROHYDRIN
DOT: UN 2023
mf: C_3H_5ClO mw: 92.53

PROP: Colorless, mobile liquid; irritating chloroform-like odor. Bp: 117.9°, fp: − 57.1°, flash p: 105.1°F (OC) (40°C), mp -25.6°C, d: 1.1761 @ 20°/20°, vap press: 10 mm @ 16.6°, vap d: 3.29.

SYNS: 1-CHLOOR-2,3-EPOXY-PROPAAN (DUTCH) * 1-CHLOR-2,3-EPOXY-PROPAN (GERMAN) * 1-CHLORO-2,3-EPOXYPROPANE * 3-CHLORO-1,2-EPOXY-PROPANE * epi-CHLOROHYDRIN * (CHLORO-METHYL)ETHYLENE OXIDE * CHLOROMETHYLOXIRANE * 2-(CHLOROMETHYL)OXIRANE * CHLORO-PROPYLENE OXIDE * γ-CHLOROPROPYLENE OXIDE * 3-CHLORO-1,2-PROPYLENE OXIDE * 1-CLORO-2,3-EPOSSIPROPANO (ITALIAN) * ECH * EPICHLOOR-HYDRINE (DUTCH) * EPICHLORHYDRIN (GERMAN) * EPICHLORHYDRINE (FRENCH) * α-EPICHLORO-HYDRIN * (dl)-α-EPICHLOROHYDRIN * EPICHLOR-OHYDRYNA (POLISH) * EPICHLOROPHYDRIN * EPICLORIDRINA (ITALIAN) * 1,2-EPOXY-3-CHLO-ROPROPANE * 2,3-EPOXYPROPYL CHLORIDE * GLYCEROL EPICHLORHYDRIN * RCRA WASTE NUMBER U041 * SKEKhG

CONSENSUS REPORTS: IARC Cancer Review: GROUP 2A IMEMDT 7,202,87; Animal Sufficient Evidence IMEMDT 11,131,76. NTP Fourth Annual Report On Carcinogens, 1984. EPA Genetic Toxicology Program. Community Right-To-Know List. EPA Extremely Hazardous Substances List. Reported in EPA TSCA Inventory.

OSHA PEL: (Transitional: TWA 5 ppm (skin)) TWA 2 ppm (skin)
ACGIH TLV: TWA 2 ppm (skin)
DFG TRK: 3 ppm; Animal Carcinogen, Suspected Human Carcinogen.
NIOSH REL: Minimize exposure
DOT Classification: Flammable Liquid; Label: Flammable Liquid; IMO: Poison B; Label: Poison, Flammable Liquid.

SAFETY PROFILE: Confirmed carcinogen with experimental carcinogenic, neoplastigenic, tumorigenic, and teratogenic data. Poison by ingestion, intravenous, intraperitoneal, parenteral, and subcutaneous routes. Moderately toxic by skin contact and inhalation. Human mutation data reported. Human systemic effects by inhalation: unspecified effects on the respiratory system, sense of smell, and eyes. A skin and eye irritant. A sensitizer. Flammable when exposed to heat or flame. Explosive reaction with aniline. Reaction with trichloroethylene forms the explosive dichloroacetylene. Ignition on contact with potassium tert-butoxide. Violent reaction with sulfuric acid or isopropylamine. Exothermic polymerization on contact with strong acids, caustic alkalies, aluminum, aluminum chloride, iron(III) chloride, or zinc. When heated to decomposition it emits toxic fumes of Cl^-.

EBB500 CAS: 329-65-7 **HR: 3**
dl-EPINEPHRINE
mf: $C_9H_{13}NO_3$ mw: 183.23

SYN: EPINEPHRINE racemic

SAFETY PROFILE: Very poisonous by intravenous and intraperitoneal routes. When heated to decomposition it emits toxic fumes of NO_x.

EBD700 CAS: 2104-64-5 **HR: 3**
EPN
mf: $C_{14}H_{14}NO_4PS$ mw: 323.32

PROP: Liquid or pale yellow crystals with an aromatic odor. D: 1.268 @ 25°, mp: 36°. Nearly insol in water; sol in organic solvents.

SYNS: O-AETHYL-O-n(4-NITROPHENYL)-PHENYL-MONOTHIOPHOSPHONAT (GERMAN) * ENT 17,798 * O-ESTER-p-NITROPHENOL with O-ETHYL PHENYL PHOSPHONOTHIOATE * ETHOXY-4-NITROPHENOXY-PHENYLPHOSPHINE SULFIDE * O-ETHYL-O-((4-NITRO-

FENYL)-FENYL)-MONOTHIOFOSFONAAT (DUTCH)
* O-ETHYL O-(4-NITROPHENYL)BENZENETHIONOPHO-
SPHONATE * ETHYL-p-NITROPHENYL BENZENE-
THIOPHOSPHATE * ETHYL-p-NITROPHENYL BEN-
ZENETHIONOPHOSPHONATE * ETHYL-p-NITROPHE-
NYL BENZENETHIOPHOSPHONATE * ETHYL-p-NITRO-
PHENYL PHENYLPHOSPHONOTHIOATE * O-ETHYL-
O-p-NITROPHENYL PHENYLPHOSPHONOTHIOLATE
* O-ETHYL-O-(4-NITROPHENYL) PHENYLPHOS-
PHONOTHIOATE * O-ETHYL-O-p-NITROPHENYL
PHENYLPHOSPHOROTHIOATE * ETHYL-p-NITROPHE-
NYL THIONOBENZENEPHOSPHATE * ETHYL-p-NITRO-
PHENYL THIONOBENZENEPHOSPHONATE * O-ETHYL
PHENYL-p-NITROPHENYL THIOPHOSPHONATE
* O-ETIL-O-((4-NTIRO-FENIL)-FENIL)-MONOTIOFOSFO-
NATO (ITALIAN) * PHENYLTHIOPHOSPHONATE de
O-ETHYLE et O-4-NITROPHENYLE (FRENCH) * PIN
* SANTOX * THIONOBENZENEPHOSPHONIC ACID
ETHYL-p-NITROPHENYL ESTER

CONSENSUS REPORTS: EPA Farm Worker
Field Reentry. EPA Extremely Hazardous Substances List.

OSHA PEL: TWA 0.5 mg/m^3 (skin)
ACGIH TLV: TWA 0.5 mg/m^3 (skin)
DFG MAK: 0.5 mg/m^3

SAFETY PROFILE: Poison by ingestion, skin
contact, subcutaneous, and intraperitoneal
routes. A highly toxic insecticide. A cholinesterase inhibitor. This material is extremely hazardous on contact with skin, inhalation or ingestion.
When heated to decomposition it emits highly
toxic fumes of SO_x, PO_x, NO_x, and phosphine.

EBJ500   CAS: 930-22-3   HR: 3
3,4-EPOXY-1-BUTENE
mf: C_4H_6O mw: 70.10

PROP: Liquid. Mp: $-135°$, bp: $67°$, flash p:
$<-58°F$ (CC), d: 0.869, autoign temp: $806°F$,
vap d: 2.41.

SYNS: BUTADIENE MONOXIDE * 1,2-EPOXYBU-
TENE-3

CONSENSUS REPORTS: Reported in EPA
TSCA Inventory.

SAFETY PROFILE: A poison by intraperitoneal
route. Questionable carcinogen with experimental tumorigenic data. Mutation data reported.
A very dangerous fire hazard when exposed to
heat or flame; can react with oxidizing materials.
To fight fire, use CO_2, dry chemical, water

spray. When heated to decomposition it emits
acrid smoke and fumes.

EBM000   CAS: 1250-95-9   HR: 3
EPOXYCHOLESTEROL
mf: $C_{27}H_{46}O_2$ mw: 402.73

SYNS: CHOLESTEROL-α-EPOXIDE * CHOLES-
TEROL-5-α,6-α-EPOXIDE * CHOLESTEROL OXIDE
* CHOLESTEROL-α-OXIDE * 5-α,6-α-EPOXYCHOLES-
TANOL * 5,6-α-EPOXY-5-α-CHOLESTAN-3-β-OL

CONSENSUS REPORTS: EPA Genetic Toxicology Program.

SAFETY PROFILE: Questionable carcinogen
with experimental carcinogenic data. Human
mutation data reported. When heated to decomposition it emits acrid smoke and fumes.

EBR000   CAS: 96-09-3   HR: 3
1,2-EPOXYETHYLBENZENE
mf: C_8H_8O mw: 120.16

PROP: Colorless liquid. Bp: 194.2, flash p:
$165°F$ (OC), fp: $-36.7°$, d: 1.0469 @ $25°/4°$,
vap d: 4.14.

SYNS: EPOXYETHYLBENZENE (8CI) * EPOXYSTY-
RENE * α,β-EPOXYSTYRENE * NCI-C54977
* PHENETHYLENE OXIDE * 1-PHENYL-1,2-EPOXY-
ETHANE * PHENYLETHYLENE OXIDE * PHENYL-
OXIRANE * 1-PHENYLOXIRANE * 2-PHENYLOXI-
RANE * STYRENE EPOXIDE * STYRENE OXIDE
* STYRENE-7,8-OXIDE * STYRYL OXIDE

CONSENSUS REPORTS: IARC Cancer Review: GROUP 2B IMEMDT 7,345,87; Animal Sufficient Evidence IMEMDT 36,245,85;
Animal Inadequate Evidence IMEMDT 19,-
275,79; IMEMDT 11,201,76. Reported in EPA
TSCA Inventory. EPA Genetic Toxicology
Program.

SAFETY PROFILE: Suspected carcinogen with
experimental carcinogenic and tumorigenic
data. Moderately toxic by ingestion, inhalation,
skin contact, and intraperitoneal routes. Experimental teratogenic and reproductive effects. Human mutation data reported. A skin and eye
irritant. Flammable when exposed to heat,
flame, or oxidizers; can react with oxidizing
materials. To fight fire, use foam, CO_2, dry
chemical. When heated to decomposition it
emits acrid smoke and fumes.

EBW500   CAS: 1024-57-3   **HR: 3**
EPOXYHEPTACHLOR
mf: $C_{10}H_5Cl_7O$ mw: 389.30

SYNS: ENT 25,584 * HCE * HEPTACHLOR EPOX-
IDE (USDA) * 1,4,5,6,7,8,8-HEPTACHLORO-2,3-EPOXY-
2,3,3a,4,7,7a-HEXAHYDRO-4,7-METHANOINDENE
* 1,4,5,6,7,8,8-HEPTACHLORO-2,3-EPOXY-3a,4,7,7a-TET-
RAHYDRO-4,7-METHANOINDAN * 2,3,4,5,6,7,7-HEPTA-
CHLORO-1a,1b,5,5a,6,6a-HEXAHYDRO-2,5-METHANO-2H-
INDENO(1,2-b)OXIRENE * VELSICOL 53-CS-17

CONSENSUS REPORTS: IARC Cancer Re-
view: Human Inadequate Evidence IMEMDT
20,129,79; Animal Inadequate Evidence
IMEMDT 5,173,74; Animal Limited Evidence
IMEMDT 20,129,79. EPA Genetic Toxicology
Program.

SAFETY PROFILE: Suspected carcinogen with
experimental carcinogenic data. Poison by ingestion and intravenous routes. Human mutation
data reported. When heated to decomposition
it emits toxic fumes of Cl^-.

EBX500   CAS: 7320-37-8   **HR: 3**
1,2-EPOXYHEXADECANE
mf: $C_{16}H_{32}O$ mw: 240.48

SYNS: HEXADECENE EPOXIDE * NCI-C55538

CONSENSUS REPORTS: Reported in EPA
TSCA Inventory.

SAFETY PROFILE: Questionable carcinogen
with experimental tumorigenic data. When
heated to decomposition it emits acrid smoke
and fumes.

ECD500   CAS: 2443-39-2   **HR: 3**
cis-9,10-EPOXYOCTADECANOIC ACID
mf: $C_{18}H_{34}O_3$ mw: 298.52

SYNS: cis-9,10-EPOXYOCTADECANOATE * EPOX-
YOLEIC ACID * 9,10-EPOXYSTEARIC ACID
* cis-9,10-EPOXYSTEARIC ACID * cis-3-OCTYL-OXI-
RANEOCTANOIC ACID

CONSENSUS REPORTS: IARC Cancer Review: GROUP 3 IMEMDT 7,56,87; Animal
Inadequate Evidence IMEMDT 11,153,76.

SAFETY PROFILE: Questionable carcinogen
with experimental tumorigenic data. When
heated to decomposition it emits acrid smoke
and fumes.

ECH500   CAS: 106-90-1   **HR: 3**
2,3-EPOXYPROPYL ACRYLATE
mf: $C_6H_8O_3$ mw: 128.14

PROP: Insol in water. Bp: 57.2° @ 2 mm,
flash p: 141°F (OC), d: 1.1, vap d: 4.4.

SYNS: 2,3-EPOXY-1-PROPANOL ACRYLATE
* 2,3-EPOXYPROPYL ESTER ACRYLIC ACID
* GLYCIDYL ACRYLATE * GLYCIDYL PROPENATE
* 2-PROPENOIC ACID OXIRANYLMETHYL ESTER

CONSENSUS REPORTS: Reported in EPA
TSCA Inventory.

SAFETY PROFILE: A poison by ingestion, inhalation, and skin contact. Mutation data reported. A skin and severe eye irritant. Flammable when exposed to heat or flame. Can react
vigorously with oxidizers. To fight fire, use
foam, dry chemical, CO_2. When heated to decomposition it emits acrid smoke and fumes.

ECJ000   CAS: 5431-33-4   **HR: 3**
2,3-EPOXYPROPYL OLEATE
mf: $C_{21}H_{38}O_3$ mw: 338.59

SYNS: 2,3-EPOXY-1-PROPANOL OLEATE * 2,3-
EPOXYPROPYL ESTER of OLEIC ACID * GLYCIDOL
OLEATE * GLYCIDYL OCTADECENOATE * GLY-
CIDYL OLEATE * OLEIC ACID GLYCIDYL ESTER
* OXIRANYLMETHYL ESTER of 9-OCTADECENOIC ACID

CONSENSUS REPORTS: IARC Cancer Review: GROUP 3 IMEMDT 7,56,87; Animal
Inadequate Evidence IMEMDT 11,183,76.

SAFETY PROFILE: Poison by intravenous
route. Moderately toxic by ingestion and subcutaneous routes. Mildly toxic by skin contact.
Questionable carcinogen with experimental tumorigenic data. Mutation data reported. When
heated to decomposition it emits acrid smoke
and fumes.

ECK500   **HR: D**
EPOXY RESINS, CURED

SAFETY PROFILE: Most cured resins have little or no toxicity. If curing is incomplete there
may be residues of highly toxic curing agents
such as the organic amines: m-phenylene diamine, diethylene triamine, tetraethylene penta-
mine, and hexamethylene tetramine, as well as
phthalic anhydride and related compounds.
When heated to decomposition they emit highly
toxic fumes.

ECL000 — HR: 3
EPOXY RESINS, UNCURED

SYN: POLYMERS of EPICHLOROHYDRIN and 2,2-BIS(4-HYDROXY PHENYL)PIPERAZINE

SAFETY PROFILE: Animal experiments have shown disturbed blood formation. The degree of toxicity of uncured epoxy resins varies and is partly dependent on the extent of unreacted curing agents. When heated to decomposition it emits acrid smoke and fumes.

ECU750 — CAS: 12126-59-9 — HR: 3
EQUIGYNE

PROP: Conjugated forms of natural mixed estrogens, principally sodium estrone sulfate and sodium equilin sulfate, or synthetic estrogen piperazine estrone sulfate.

SYNS: AMNESTROGEN * CES * CLIMESTRONE * CO-ESTRO * CONEST * CONESTRON * CONJES * CONJUGATED ESTROGENS * CONJU-TABS * EQUIGYNE * ESTRATAB * ESTRIFOL * ESTROATE * ESTROCON * ESTROMED * ESTROPAN * EVEX * FEMACOID * FEMEST * FEM H * FEMOGEN * FORMATRIX * GA-NEAKE * GENISIS * GLYESTRIN * KESTRIN * MENEST * MENOGEN * MENOTAB * MENO-TROL * MILPREM * MSMED * NEO-ESTRONE * NOVOCONESTRON * OESTRILIN * OESTRO-FEMINAL * OESTROPAK MORNING * OVEST * PALOPAUSE * PAR ESTRO * PMB * PREMA-RIN * PRESOMEN * PROMARIT * SK-ESTRO-GENS * SODESTRIN-H * TAG-39 * THEOGEN * TRANSANNON * TROCOSONE * ZESTE

CONSENSUS REPORTS: IARC Cancer Review: Animal Inadequate Evidence IMEMDT 21,147,79; Human Limited Evidence IMEMDT 21,147,79. NTP Fourth Annual Report On Carcinogens, 1984.

SAFETY PROFILE: Confirmed human carcinogen producing tumors of the vascular system and liver. Human reproductive effects by ingestion and possibly other routes: changes in female fertility. When heated to decomposition it emits toxic fumes of Na_2O.

ECV000 — CAS: 517-09-9 — HR: 3
EQUILENIN
mf: $C_{18}H_{18}O_2$ mw: 266.36

PROP: Leaflets from acetone and ethanol; very sltly sol in water.

SYNS: EQUILENINA (SPANISH) * EQUILENINE * 3-HYDROXYESTRA-1,3,5(10),6,8-PENATEN-17-ONE

SAFETY PROFILE: Questionable carcinogen with experimental tumorigenic and teratogenic data. When heated to decomposition it emits acrid smoke and irritating fumes.

ECW000 — CAS: 474-86-2 — HR: 3
EQUILIN
mf: $C_{18}H_{20}O_2$ mw: 268.38

SYNS: 1,3,5,7-ESTRATETRAEN-3-OL-17-ONE * 3-HYDROXYESTRA-1,3,5(10),7-TETRAEN-17-ONE

SAFETY PROFILE: Questionable carcinogen with experimental neoplastigenic and tumorigenic data. When heated to decomposition it emits acrid smoke and irritating fumes.

ECX500 — CAS: 10138-41-7 — HR: 3
ERBIUM CHLORIDE
mf: Cl_3Er mw: 273.61

SYN: ERBIUM TRICHLORIDE

CONSENSUS REPORTS: Reported in EPA TSCA Inventory.

SAFETY PROFILE: Poison by intraperitoneal and subcutaneous routes. Moderately toxic by ingestion. When heated to decomposition it emits toxic fumes of Cl^-.

ECY500 — CAS: 10168-80-6 — HR: 3
ERBIUM(III) NITRATE (1:3)
mf: $N_3O_9 \cdot Er$ mw: 353.29

PROP: Reddish crystals.

SYN: NITRIC ACID, ERBIUM (3+) SALT

CONSENSUS REPORTS: Reported in EPA TSCA Inventory.

SAFETY PROFILE: Poison by intravenous and intraperitoneal routes. When heated to decomposition it emits toxic fumes of NO_x.

ECZ000 — CAS: 13476-05-6 — HR: 3
ERBIUM(III) NITRATE, HEXAHYDRATE (1:3:6)
mf: $N_3O_9 \cdot Er \cdot 6H_2O$ mw: 461.41

PROP: Mp: $-4H_2O$ @ 130°.

SYN: NITRIC ACID, ERBIUM (3+) SALT, HEXAHYDRATE

SAFETY PROFILE: Poison by intraperitoneal and intravenous routes. When heated to decomposition it emits toxic fumes of NO_x.

EDB500   CAS: 129-51-1   HR: 3
ERGOT
mf: $C_{19}H_{23}N_3O_2 \cdot C_4H_4O_4$ mw: 441.53

PROP: Composition: ergot amine, ergosine, ergocristine, ergocryptine, ergocornine, ergosinine, ergocristinine, ergocryptinine, ergotaminine, etc.

SYNS: CORNOCENTIN * CRUDE ERGOT * ERGOMETRINE ACID MALEATE * ERGOMETRINE MALEATE * ERGONOVINE, MALEATE (1:1) (SALT) * ERGOTRATE * ERGOTRATE MALEATE * OXYTOCIC

SAFETY PROFILE: Human poison by unspecified route. Experimental poison by intravenous route. Experimental reproductive effects. Can cause vomiting, diarrhea, thirst, tachycardia, confusion, coma, central nervous system symptoms, gastrointestinal disturbances, gangrene; circulatory changes can follow ingestion. Questionable carcinogen with experimental tumorigenic data. When heated to decomposition it emits toxic fumes of NO_x.

EDC000   CAS: 113-15-5   HR: 3
ERGOTAMINE
mf: $C_{33}H_{35}N_5O_5$ mw: 581.63

PROP: A specific alkaloid present in rye ergot.

SAFETY PROFILE: Poison by intravenous and subcutaneous routes. Human mutation data reported. When heated to decomposition it emits toxic fumes of NO_x.

EDC500   CAS: 379-79-3   HR: 3
ERGOTAMINE TARTRATE
mf: $C_{66}H_{70}N_{10}O_{10} \cdot C_4H_6O_6$ mw: 1313.56

SYNS: ERGAM * ERGATE * ERGOMAR * ERGOSTAT * ERGOTAMINE BITARTRATE * ERGOTARTRATE * ETIN * EXMIGRA * FEMERGIN * GOTAMINE TARTRATE * GYNERGEN * LINGRAINE * LINGRAN * NEO-ERGOTIN * RIGETAMIN * SECAGYN * SECUPAN

CONSENSUS REPORTS: Reported in EPA TSCA Inventory. EPA Genetic Toxicology Program. EPA Extremely Hazardous Substances List.

SAFETY PROFILE: Poison by ingestion, intravenous, and subcutaneous routes. Human

systemic effects by ingestion: hallucinations, distorted perceptions, convulsions, nausea or vomiting. Experimental teratogenic and reproductive effects. When heated to decomposition it emits toxic fumes of NO_x.

EDC650   CAS: 66733-21-9   HR: 3
ERIONITE
mf: $Al_2O_{18}Si_7 \cdot 1/2Ca \cdot 7H_2O \cdot 1/2Na$ mw: 715.68

CONSENSUS REPORTS: IARC Cancer Review: GROUP 1 IMEMDT 7,203,87; Animal Sufficient Evidence, Human Sufficient Evidence IMEMDT 42,225,87.

SAFETY PROFILE: Confirmed carcinogen with experimental carcinogenic and tumorigenic data.

EDH500   CAS: 114-07-8   HR: 3
ERYTHROMYCIN
mf: $C_{37}H_{67}NO_{13}$ mw: 734.05

PROP: White or slightly yellow, crystalline powder; odorless; freely sol in alc, chloroform, and ether; very sltly sol in water. Mp: 133-138°.

SYNS: DOTYCIN * EM * E-MYCIN * ERYCIN * ERYTHROCIN * ERYTHROGRAN * ERYTHROGUENT * ERYTHROMYCIN A * ILOTYCIN * PANTOMICINA * PROPIOCINE * ROBIMYCIN

CONSENSUS REPORTS: EPA Genetic Toxicology Program.

SAFETY PROFILE: Poison by intravenous and intramuscular routes. Moderately toxic by ingestion, intraperitoneal, and subcutaneous routes. An experimental teratogen. Mutation data reported. When heated to decomposition it emits toxic fumes of NO_x.

EDM000   CAS: 20977-05-3   HR: 3
ESCIN, SODIUM SALT
mf: $C_{55}H_{85}O_{24} \cdot Na$ mw: 1153.39

PROP: A mixture of saponins occurring in the seed of the horse chestnut tree.

SYNS: A-4760 * AESCIN SODIUM SALT * AESCUSAN SODIUM SALT * Na-AESCINAT * REPARIL SODIUM SALT * SODIUM AESCINATE

SAFETY PROFILE: Poison by ingestion, intravenous, intraperitoneal, and subcutaneous routes. Experimental teratogenic and reproductive effects. When heated to decomposition it emits toxic fumes of Na$_2$O.

EDN500   HR: D
ESTERS

PROP: A large group of organic compounds which correspond structurally to salts in inorganic chemistry. They are considered to be derived from acids by the replacement of hydrogen by an organic alkyl radical. Esters of acetic acid are called acetates and esters of carbonic acid are called carbonates. The esterification of a fatty acid RCOOH, by an alcohol, R′OH yields the fatty ester RCOOR′. The most common alcohol used is methanol, yielding the methyl ester RCOOCH$_3$. The methyl esters of fatty acids have higher vapor pressures than the corresponding acids.

SAFETY PROFILE: No general statement can be made as to the toxicity of esters. Many are highly volatile and hence can act as asphyxiants or narcotics. Skin contact, as well as inhalation, may be important routes of absorption for those esters which are volatile and have a high solvent action. The degree of toxicity ranges from mildly toxic to poison. Esters generally hydrolyze upon contact with moisture; hence, a rough guide to the toxicity of a given ester may be the sum of the toxicities of the products of hydrolysis. Incompatible with nitrates. When heated to decomposition they emit acrid smoke and fumes.

EDO000   CAS: 50-28-2   HR: 3
ESTRADIOL
mf: C$_{18}$H$_{24}$O$_2$ mw: 272.42

PROP: Needles out of benzene, acetone. Mp: 173-179°, bp: decomp. Sol in dioxone, alc, and ether.

SYNS: ALTRAD * BARDIOL * DIHYDROFOL-LICULAR HORMONE * DIHYDROFOLLICULIN * DIHYDROMENFORMON * DIHYDROTHEELIN * 3,17-β-DIHYDROXYESTRA-1,3,5(10)-TRIENE * 3,17-β-DIHYDROXY-1,3,5(10)-ESTRATRIENE * DIHYDROXYESTRIN * 3,17-β-DIHYDROXYOES-TRA-1,3,5-TRIENE * 3,17-β-DIHYDROXY-1,3,5(10)-OES-TRATRIENE * DIHYDROXYOESTRIN * DIMENFOR-MON * DIMENFORMON PROLONGATUM * DIOGYN * DIOGYNETS * E^2 * 3,17-EPIDIHYDROXYESTRA-TRIENE * 3,17-EPIDIHYDROXYOESTRATRIENE * ESTRADIOL-17-β * α-ESTRADIOL * β-ESTRA-DIOL * 3,17-β-ESTRADIOL * 17-β-ESTRADIOL

* cis-ESTRADIOL * d-ESTRADIOL * d-3,17-β-ESTRADIOL * ESTRALDINE * ESTRA-1,3,5(10)-TRIENE-3,17-β-DIOL * 17-β-ESTRA-1,3,5(10)-TRIENE-3,17-DIOL * 1,3,5-ESTRATRIENE-3,17-β-DIOL * ESTROVITE * FEMESTRAL * FEMOGEN * GYNERGON * GYNESTREL * GYNOESTRYL * LAMDIOL * MACRODIOL * MACROL * MICRODIOL * NORDICOL * NSC-9895 * OESTERGON * OESTRADIOL * α-OESTRADIOL * β-OESTRADIOL * 3,17-β-OESTRADIOL * cis-OESTRADIOL * d-OESTRADIOL * d-3,17-β-OESTRA-DIOL * OESTRADIOL R * OESTRADIOL-17-β * OESTRA-1,3,5(10)-TRIENE-3,17-β-DIOL * 17-β-OES-TRA-1,3,5(10)-TRIENE-3,17-DIOL * OESTROGLANDOL * OESTROGYNAL * 17-β-OH-ESTRADIOL * 17-β-OH-OESTRADIOL * OVAHORMON * OVASTEROL * OVASTEVOL * OVOCICLINA * OVOCYCLIN * OVOCYCLINE * OVOCYLIN * PRIMOFOL * PROFOLIOL * PROGYNON * PROGYNON-DH * SYNDIOL

CONSENSUS REPORTS: IARC Cancer Review: Human Limited Evidence IMEMDT 21,279,79; Animal Sufficient Evidence IMEMDT 21,279,79; IMEMDT 6,99,74. NTP Fourth Annual Report On Carcinogens, 1984. EPA Genetic Toxicology Program.

SAFETY PROFILE: Confirmed carcinogen with experimental carcinogenic, neoplastigenic, tumorigenic, and teratogenic data. A promoter. Human reproductive effects by ingestion: fertility effects. Experimental reproductive effects. Human mutation data reported. A steroid hormone much used in medicine. When heated to decomposition it emits acrid smoke and irritating fumes.

EDP000   CAS: 50-50-0   HR: 3
ESTRADIOL-3-BENZOATE
mf: C$_{25}$H$_{28}$O$_3$ mw: 376.53

PROP: White or slightly yellow to brownish crystalline powder, odorless. Almost insol in water; sol in alc, acetone, and dioxane; sparingly sol in vegetable oils; sltly sol in ether. Mp: 191-196°

SYNS: BENOVOCYLIN * BENZHORMOVARINE * BENZOATE d'OESTRADIOL (FRENCH) * BENZOES-TROFOL * BENZOFOLINE * BENZO-GYNOESTRYL * BENZOIC ACID ESTRADIOL * DIFFOLLISTEROL * DIFOLLICULINE * DIHYDROESTRIN BENZOATE * DIHYDROFOLLICULIN BENZOATE * DIMENFOR-MON BENZOATE * DIMENFORMONE * DIOGYN B * EBZ * ESTON-B * ESTRADIOL BENZOATE

* ESTRADIOL-17-β-BENZOATE * ESTRADIOL-17-β-3-BENZOATE * β-ESTRADIOL BENZOATE * β-ESTRADIOL-3-BENZOATE * 17-β-ESTRADIOL BENZOATE * 17-β-ESTRADIOL-3-BENZOATE * ESTRADIOL MONOBENZOATE * 17-β-ESTRADIOL MONOBENZOATE * ESTRA-1,3,5(10)-TRIENE-3,17-DIOL (17-β)-3-BENZOATE * ESTRA-1,3,5(10)-TRIENE-3,17-β-DIOL, 3-BENZOATE * 1,3,5(10)-ESTRATRIENE-3,17-β-DIOL 3-BENZOATE * FEMESTRONE * FOLLICORMON * FOLLIDRIN * GRAAFINA * de GRAAFINA * GYNECORMONE * GYNFORMONE * HIDROESTRON * HORMOGYNON * HYDROXYESTRIN BENZOATE * MEE * ODB * OESTRADIOL BENZOATE * OESTRADIOL-3-BENZOATE * β-OESTRADIOL BENZOATE * β-OESTRADIOL-3-BENZOATE * 17-β-OESTRADIOL-3-BENZOATE * OESTRADIOL MONOBENZOATE * OESTRAFORM (BDH) * 1,3,5(10)-OESTRATRIENE-3,17-β-DIOL 3-BENZOATE * OVAHORMON BENZOATE * OVASTEROL-B * OVEX * OVOCYCLIN BENZOATE * OVOCYCLIN M * OVOCYCLIN-MB * PRIMOGYN B * PRIMOGYN BOLEOSUM * PRIMOGYN I * PROGYNON B * PROGYNON BENZOATE * RECTHORMONE OESTRADIOL * SOLESTRO * UNISTRADIOL

CONSENSUS REPORTS: IARC Cancer Review: Animal Sufficient Evidence IMEMDT 21,279,79.

SAFETY PROFILE: Confirmed carcinogen with experimental carcinogenic and tumorigenic data. Human reproductive effects by intramuscular route: menstrual cycle changes and disorders. Experimental teratogenic and reproductive effects. Mutation data reported. A steroid. When heated to decomposition it emits acrid smoke and irritating fumes.

EDR000   CAS: 113-38-2   HR: 3
ESTRADIOL DIPROPIONATE
mf: $C_{24}H_{32}O_4$ mw: 384.56

SYNS: AGOFOLLIN * DIMENFORMON DIPROPIONATE * DIOVOCYCLIN * DIOVOCYLIN * DIPROPIONATE d'OESTRADIOL (FRENCH) * DIPROSTRON * ENDOFOLLICOLINA D.P. * ESTRADIOL-3,17-DIPROPIONATE * β-ESTRADIOL DIPROPIONATE * β-ESTRADIOL-3,17-DIPROPIONATE * 3,17-β-ESTRADIOL DIPROPIONATE * 17-β-ESTRADIOL DIPROPIONATE * ESTRA-1,3,5(10)-TRIENE-3,17-DIOL (17-β)-DIPROPIONATE * 1,3,5(10)-ESTRATRIENE-3,17-β-DIOL DIPROPIONATE * ESTROICI * ESTRONEX * FOLLICYCLIN P * NACYCLYL * OESTRADIOL DIPROPIONATE * OESTRADIOL-3,17-DIPROPIONATE

* β-OESTRADIOL DIPROPIONATE * 3,17-β-OESTRADIOL DIPROPIONATE * 17-β-OESTRADIOL DIPROPIONATE * OVOCYCLIN DIPROPIONATE * OVOCYCLIN-P * PROGYNON-DP

CONSENSUS REPORTS: IARC Cancer Review: Animal Sufficient Evidence IMEMDT 21,279,79. EPA Genetic Toxicology Program.

SAFETY PROFILE: Confirmed carcinogen with experimental carcinogenic and tumorigenic data. A poison by intravenous and parenteral routes. Experimental teratogenic and reproductive effects. A drug for the treatment of menopause. When heated to decomposition it emits acrid smoke and irritating fumes.

EDR500   CAS: 22966-79-6   HR: 3
ESTRADIOL MUSTARD
mf: $C_{42}H_{50}Cl_4N_2O_4$ mw: 788.74

PROP: Mp: 40-65° (freeze dried).

SYNS: BIS((4-(BIS(2-CHLOROETHYL)AMINO)BENZENE)ACETATE)ESTRA-1,3,5(10)-TRIENE-3,17-DIOL(17-β) * BIS((4-(BIS(2-CHLOROETHYL)AMINO)BENZENE)ACETATE)OESTRA-1,3,5(10)-TRIENE-3,17-DIOL(17-β) * BIS((p-(BIS(2-CHLOROETHYL)AMINO)PHENYL)ACETATE)ESTRADIOL * BIS((p-(BIS(2-CHLOROETHYL)AMINO)PHENYL)ACETATE)ESTRA-1,3,5(10)-TRIENE-3,17-β-DIOL * BIS((p-(BIS(2-CHLOROETHYL)AMINO)PHENYL)ACETATE)OESTRADIOL * BIS((p-BIS(2-CHLOROETHYL)AMINOPHENYL)ACETATE)OESTRA-1,3,5(10)-TRIENE-3,17-β-DIOL * NCI-C01570 * NSC 112259 * OESTRADIOL MUSTARD

CONSENSUS REPORTS: IARC Cancer Review: GROUP 3 IMEMDT 7,56,87; Animal Limited Evidence IMEMDT 9,217,75. NCI Carcinogenesis Bioassay (gavage); Clear Evidence: mouse NCITR* NCI-CG-TR-59,78; No Evidence: rat NCITR* NCI-CG-TR-59,78.

SAFETY PROFILE: Questionable carcinogen with experimental carcinogenic and neoplastigenic data. When heated to decomposition it emits very toxic fumes of Cl^- and NO_x.

EDS000   CAS: 28014-46-2   HR: 3
ESTRADIOL POLYESTER with PHOSPHORIC ACID
mf: $(C_{18}H_{24}O_2 \cdot H_3O_4P)_x$

SYNS: ESTRADIOL PHOSPHATE POLYMER * ESTRADURIN * (17-β)-ESTRA-1,3,5(10)-TRIENE-3,17-DIOL POLYMER with PHOSPHORIC ACID * OESTRADIOL

PHOSPHATE POLYMER ＊ OESTRADIOL POLYESTER with PHOSPHORIC ACID ＊ PEP ＊ POLY(ESTRADIOL PHOSPHATE) ＊ POLYOESTRADIOL PHOSPHATE

CONSENSUS REPORTS: IARC Cancer Review: Animal Sufficient Evidence IMEMDT 21,279,79.

SAFETY PROFILE: Confirmed carcinogen producing liver tumors. An experimental teratogen. A drug used in cancer treatment. When heated to decomposition it emits toxic fumes of PO_x.

EDS100   CAS: 979-32-8   **HR: 3**
ESTRADIOL-17-VALERATE
mf: $C_{24}H_{32}O_3$ mw: 368.56

SYNS: ALTADIOL ＊ DELADIOL ＊ DELAHORMONE UNIMATIC ＊ DELESTROGEN ＊ DELESTROGEN 4X ＊ DURA-ESTRADIOL ＊ ESTRADIOL VALERATE ＊ ESTRADIOL 17-β-VALERATE ＊ ESTRADIOL VALERIANATE ＊ (17-β)-ESTRA-1,3,5(10)-TRIENE-3,17-DIOL-17-PENTANOATE (9CI) ＊ ESTRAVEL ＊ FEMOGEX ＊ NEOFOLLIN ＊ PHARLON ＊ PROGYNON ＊ PROGYNON-DEPOT ＊ PROGYNOVA

SAFETY PROFILE: Suspected carcinogen with carcinogenic data. Experimental teratogenic and reproductive effects. When heated to decomposition it emits acrid smoke and irritating fumes.

EDU500   CAS: 50-27-1   **HR: 3**
ESTRIOL
mf: $C_{18}H_{24}O_3$ mw: 288.42

PROP: Small, white crystals. D: 0.965, bp: 214.6°

SYNS: AACIFEMINE ＊ COLPOVISTER ＊ DESTRIOL ＊ DEUSLON-A ＊ ESTRA-1,3,5(10)-TRIENE-3,16-α,17-β-TRIOL ＊ 1,3,5-ESTRATRIENE-3-β,16-α,17-β-TRIOL ＊ (16-α,17-β)-ESTRA-1,3,5(10)-TRIENE-3,16,17-TRIOL ＊ ESTRATRIOL ＊ 3,16-α,17-β-ESTRIOL ＊ 16-α,17-β-ESTRIOL ＊ ESTRIOLO (ITALIAN) ＊ FOLLICULAR HORMONE HYDRATE ＊ GYNAESAN ＊ HEMOSTYPTANON ＊ HOLIN ＊ HORMOMED ＊ HORMONIN ＊ 16-α-HYDROXYESTRADIOL ＊ 16-α-HYDROXYOESTRADIOL ＊ KLIMORAL ＊ NSC-12169 ＊ OE3 ＊ OESTRA-1,3,5(10)-TRIENE-3,16-α,17-β-TRIOL ＊ 1,3,5-OESTRATRIENE-3-β-3,16-α,17-β-TRIOL ＊ (16-α,17-β)-OESTRA-1,3,5(10)-TRIENE-3,16,17-TRIOL ＊ OESTRATRIOL ＊ OESTRIOL ＊ 3,16-α,17-β-OESTRIOL ＊ 16-α,17-β-OESTRIOL ＊ ORGASTYPTIN ＊ OVESTERIN ＊ OVESTIN ＊ OVESTINON ＊ OVESTRION ＊ STIPTANON

＊ SYNAPAUSE ＊ THEELOL ＊ THULOL ＊ TRIDESTRIN ＊ 3,16-α,17-β-TRIHYDROXY-Δ-1,3,5-ESTRATRIENE ＊ 3,16-α,17-β-TRIHYDROXY-Δ-1,3,5-OESTRATRIENE ＊ 3,16-α,17-β-TRIHYDROXYESTRA-1,3,5(10)-TRIENE ＊ TRIHYDROXYESTRIN ＊ 3,16-α,17-β-TRIHYDROXYOESTRA-1,3,5(10)-TRIENE ＊ TRIHYDROXYOESTRIN ＊ TRIODURIN ＊ TRIOVEX

CONSENSUS REPORTS: IARC Cancer Review: Animal Limited Evidence IMEMDT 21,327,79; Human Limited Evidence IMEMDT 21,327,79; Animal Inadequate Evidence IMEMDT 6,117,74.

SAFETY PROFILE: Suspected carcinogen with experimental carcinogenic, neoplastigenic, tumorigenic, and teratogenic data. Mutation data reported. A steroid drug for the treatment of menopause. When heated to decomposition it emits acrid smoke and irritating fumes.

EDV000   CAS: 53-16-7   **HR: 3**
ESTRONE
mf: $C_{18}H_{22}O_2$ mw: 270.40

PROP: White crystals. Mp: 254°. Insol in water; sol in alc, benzene, ether, and chloroform.

SYNS: AQUACRINE ＊ CRINOVARYL ＊ CRISTALLOVAR ＊ CRYSTOGEN ＊ DESTRONE ＊ DISYNFORMON ＊ E¹ ＊ ENDOFOLLICULINA ＊ ESTERONE ＊ 1,3,5-ESTRATRIEN-3-OL-17-ONE ＊ 1,3,5(10)-ESTRATRIEN-3-OL-17-ONE ＊ Δ-1,3,5-ESTRATRIEN-3-β-OL-17-ONE ＊ ESTRIN ＊ ESTROL ＊ ESTRON ＊ ESTRONA (SPANISH) ＊ ESTRONE-A ＊ ESTRUGENONE ＊ ESTRUSOL ＊ FEMESTRONE INJECTION ＊ FEMIDYN ＊ FOLIKRIN ＊ FOLIPEX ＊ FOLISAN ＊ FOLLESTRINE ＊ FOLLICULAR HORMONE ＊ FOLLICULIN ＊ FOLLICULINE BENZOATE ＊ FOLLICUNODIS ＊ FOLLIDRIN ＊ GLANDUBOLIN ＊ HIESTRONE ＊ HORMOFOLLIN ＊ HORMOVARINE ＊ 3-HYDROXYESTRA-1,3,5(10)-TRIEN-17-ONE ＊ 3-HYDROXY-17-KETO-ESTRA-1,3,5-TRIENE ＊ 3-HYDROXY-17-KETO-OESTRA-1,3,5-TRIENE ＊ 3-HYDROXY-OESTRA-1,3,5(10)-TRIEN-17-ONE ＊ 3-HYDROXY-1,3,5(10)-OESTRATRIEN-17-ONE ＊ KESTRONE ＊ KETODESTRIN ＊ KETOHYDROXY-ESTRATRIENE ＊ KETOHYDROXYESTRIN ＊ KETOHYDROXYOESTRIN ＊ KOLPON ＊ MENAGEN ＊ MENFORMON ＊ Δ-1,3,5-OESTRATRIEN-3-β-OL-17-ONE ＊ 1,3,5-OESTRATRIEN-3-OL-17-ONE ＊ 1,3,5(10)-OESTRATRIEN-3-OL-17-ONE ＊ OESTRIN ＊ OESTROFORM ＊ OESTRONE ＊ OESTROPEROS ＊ OVEX ＊ OVIFOLLIN ＊ PERLATAN ＊ SOLLICULIN ＊ THEELIN ＊ THELESTRIN ＊ THELYKI-

NIN * THYNESTRON * TOKOKIN * UNDEN * WNYESTRON

CONSENSUS REPORTS: IARC Cancer Review: Human Limited Evidence IMEMDT 21,343,79; Animal Sufficient Evidence IMEMDT 6,123,74; IMEMDT 21,343,79. NTP Fourth Annual Report On Carcinogens, 1984. Reported in EPA TSCA Inventory.

SAFETY PROFILE: Confirmed carcinogen with experimental carcinogenic, neoplastigenic, tumorigenic, and teratogenic data. A poison by intraperitoneal and subcutaneous routes. Human reproductive effects by implantation: spermatogenesis and impotence. Mutation data reported. A steroid drug for the treatment of menopause and ovariectomy symptoms. When heated to decomposition it emits acrid smoke and irritating fumes.

EDV500 CAS: 2393-53-5 HR: 3
ESTRONE BENZOATE
mf: $C_{25}H_{26}O_3$ mw: 374.51

SYNS: BENZOATE d'OESTRONE (FRENCH) * 3-(BENZOYLOXY)ESTRA-1,3,5(10)-TRIEN-17-ONE * 3-HYDROXYESTRA-1,3,5(10)-TRIEN-17-ONE BENZOATE * KETOHYDROXYESTRIN BENZOATE * OESTRONBENZOAT (GERMAN)

CONSENSUS REPORTS: IARC Cancer Review: Animal Limited Evidence IMEMDT 21,343,79.

SAFETY PROFILE: Suspected carcinogen with experimental carcinogenic, neoplastigenic, and tumorigenic data. A steroid. When heated to decomposition it emits acrid smoke and irritating fumes.

EDY000 CAS: 75-04-7 HR: 3
ETHANAMINE (anhydrous)
DOT: UN 1036

SYN: MONOMETHYLAMINE, anhydrous (DOT)

CONSENSUS REPORTS: Reported in EPA TSCA Inventory.

DOT Classification: Flammable Gas; Label: Flammable Gas.

SAFETY PROFILE: A flammable gas. When heated to decomposition it emits toxic fumes such as NO_x.

EDY500 CAS: 75-04-7 HR: 3
ETHANAMINE, aqueous solution
mf: C_2H_7N mw: 45.10

DOT: UN 2270

SYNS: ETHYLAMINE, solution, in water, concentrations up to 70% (DOT) * MONOMETHYLAMINE, aqueous solution (DOT)

CONSENSUS REPORTS: Reported in EPA TSCA Inventory.

DOT Classification: Flammable Liquid; Label: Flammable Liquid; Flammable or Combustible Liquid; Label: Flammable Liquid.

SAFETY PROFILE: A flammable liquid. When heated to decomposition it emits toxic fumes such as NO_x.

EDZ000 CAS: 74-84-0 HR: 2
ETHANE
mf: C_2H_6 mw: 30.08

DOT: UN 1035/UN 1961

PROP: Colorless, odorless gas. Mp: $-172°$, bp: $-88.6°$, lel: 3.0%, uel: 12.5%, fp: $-183.2°$, d: 0.446 @ 0° (liquid), autoign temp: 959°F, vap d: 1.04, flash p: $-202°F$.

SYNS: BIMETHYL * DIMETHYL * ETHANE, compressed (DOT) * ETHANE, refrigerated liquid (DOT) * ETHYL HYDRIDE * METHYLMETHANE

CONSENSUS REPORTS: Reported in EPA TSCA Inventory.

DOT Classification: Flammable Gas; Label: Flammable Gas.

SAFETY PROFILE: A simple asphyxiant. A very dangerous fire hazard when exposed to heat or flame; can react vigorously with oxidizing materials. Moderate explosion hazard when exposed to flame. To fight fire, stop flow of gas. Incompatible with chlorine, dioxygenyl tetrafluoroborate, oxidizing materials, heat or flame. When heated to decomposition it emits acrid smoke and irritating fumes.

EEA500 CAS: 107-15-3 HR: 3
1,2-ETHANEDIAMINE
DOT: UN 1604
mf: $C_2H_8N_2$ mw: 60.12

PROP: Volatile, colorless, hygroscopic liquid; ammonia-like odor. Mp: 8.5°, bp: 117.2°, flash p: 110°F (CC), d: 0.8994 @ 20°/4°, vap press:

10.7 mm @ 20°, vap d: 2.07, autoign temp: 725°F.

SYNS: AETHALDIAMIN (GERMAN) * AETHYLENE-DIAMIN (GERMAN) * 1,2-DIAMINOAETHAN (GERMAN) * 1,2-DIAMINO-ETHAAN (DUTCH) * 1,2-DIAMINO-ETHANE * 1,2-DIAMINO-ETHANO (ITALIAN) * DIMETHYLENEDIAMINE * ETHYLEENDIAMINE (DUTCH) * ETHYLENEDIAMINE (OSHA) * 1,2-ETHYLENEDIAMINE * ETHYLENE-DIAMINE (FRENCH) * NCI-C60402

CONSENSUS REPORTS: Reported in EPA TSCA Inventory. EPA Extremely Hazardous Substances List.

OSHA PEL: TWA 10 ppm
ACGIH TLV: TWA 10 ppm
DOT Classification: Corrosive Material; Label: Corrosive, Flammable Liquid.

SAFETY PROFILE: An irritant poison in humans by inhalation. Experimental poison by inhalation, intraperitoneal, subcutaneous, and intravenous routes. Moderately toxic by ingestion and skin contact. Corrosive. A severe skin and eye irritant. An allergen and sensitizer. Mutation data reported. Flammable when exposed to heat, flame or oxidizers. Can react violently with acetic acid, acetic anhydride, acrolein, acrylic acid, acrylonitrile, allyl chloride, CS_2, chlorosulfonic acid, epichlorohydrin, ethylene chlorohydrin, HCl, mesityl oxide, HNO_3, oleum, $AgClO_4$, H_2SO_4, β-propiolactone, or vinyl acetate. To fight fire, use CO_2, dry chemical, alcohol foam. When heated to decomposition it emits toxic fumes of NO_x and NH_3.

EEB000   CAS: 540-63-6   HR: 3
1,2-ETHANEDITHIOL
mf: $C_2H_6S_2$ mw: 94.20

SYNS: 1,2-DIMERCAPTOETHANE * DITHIOETHYL-ENEGLYCOL * DITHIOGLYCOL * ETHYLENE DIMERCAPTAN * α-ETHYLENE DIMERCAPTAN * ETHYLENE DITHIOGLYCOL * ETHYLENEDITHIOL * ETHYL HYDROPERSULFIDE

CONSENSUS REPORTS: Reported in EPA TSCA Inventory.

SAFETY PROFILE: Poison by ingestion, intraperitoneal, and intravenous routes. When heated to decomposition it emits very toxic fumes of SO_x.

EEC600   CAS: 141-43-5   HR: 3
ETHANOLAMINE
DOT: UN 2491
mf: C_2H_7NO mw: 61.10

PROP: Colorless liquid; ammoniacal odor. Hygroscopic, bp: 170.5°, fp: 10.5°, flash p: 200°F (OC), d: 1.0180 @ 20°/4°, vap press: 6 mm @ 60°, vap d: 2.11. Misc in water and alc; sltly sol in benzene; sol in chloroform.

SYNS: AETHANOLAMIN (GERMAN) * 2-AMINO-AETHANOL (GERMAN) * 2-AMINOETANOLO (ITALIAN) * 2-AMINOETHANOL (MAK) * β-AMINO-ETHYL ALCOHOL * COLAMINE * ETANOLAMINA (ITALIAN) * β-ETHANOLAMINE * ETHANOLA-MINE, solution (DOT) * ETHYLOLAMINE * GLYCI-NOL * 2-HYDROXYETHYLAMINE * β-HYDROXY-ETHYLAMINE * MEA * MONOAETHANOLAMIN (GERMAN) * MONOETHANOLAMINE * OLAMINE * THIOFACO M-50 * USAF EK-1597

CONSENSUS REPORTS: Reported in EPA TSCA Inventory.

OSHA PEL: (Transitional: TWA 3 ppm) TWA 3 ppm; STEL 6 ppm
ACGIH TLV: TWA 3 ppm; STEL 6 ppm
DFG MAK: 3 ppm (8 mg/m^3)
DOT Classification: Corrosive Material; Label: Corrosive.

SAFETY PROFILE: Poison by intraperitoneal route. Moderately toxic by ingestion, skin contact, subcutaneous, intravenous, and intramuscular routes. A corrosive irritant to skin, eyes, and mucous membranes. Flammable when exposed to heat or flame. A powerful base. Reacts violently with acetic acid, acetic anhydride, acrolein, acrylic acid, acrylonitrile, cellulose, chlorosulfonic acid, epichlorohydrin, HCl, HF, mesityl oxide, HNO_3, oleum, H_2SO_4, β-propiolactone, vinyl acetate. To fight fire, use foam, alcohol foam, dry chemical. When heated to decomposition it emits toxic fumes of NO_x.

EEF000   CAS: 17088-21-0   HR: 3
1-ETHENYL PYRENE
mf: $C_{18}H_{12}$ mw: 228.2

SYNS: 1-VINYLPYRENE * 3-VINYLPYRENE

SAFETY PROFILE: Questionable carcinogen with experimental tumorigenic data. Mutation data reported. When heated to decomposition it emits acrid smoke and irritating fumes.

EEG000   CAS: 88-12-0   **HR: 2**
1-ETHENYL-2-PYRROLIDINONE
mf: C_6H_9NO mw: 111.16

PROP: Colorless liquid, water-sol. Bp: 148° @ 100 mm, fp: 13.5°, flash p: 209°F (OC), d: 1.04 @ 25°, autoign temp: 687°F, vap d: 3.8, fire p: 213°F.

SYNS: VINYLBUTYROLACTAM * N-VINYLPYRRO-LIDINONE * N-VINYL-2-PYRROLIDINONE * 1-VI-NYL-2-PYRROLIDINONE * VINYLPYRROLIDONE * N-VINYLPYRROLIDONE * N-VINYL-2-PYRROLI-DONE * 1-VINYL-2-PYRROLIDONE * V-PYROL

CONSENSUS REPORTS: IARC Cancer Review: GROUP 3 IMEMDT 7,56,87. Reported in EPA TSCA Inventory.

SAFETY PROFILE: Moderately toxic by ingestion, inhalation and skin contact. A severe eye irritant. Probably irritating and narcotic in high concentrations. Questionable carcinogen. Combustible when exposed to heat or flame; can react vigorously with oxidizing materials. To fight fire, use alcohol foam, CO_2, dry chemical. When heated to decomposition it emits highly toxic fumes of NO_x.

EEG500   **HR: 2**
ETHERS

PROP: Organic compounds in which an oxygen atom is interposed between two carbon atoms in the structure of the molecule.

SAFETY PROFILE: The simpler ethers such as ethyl ether, isopropyl ether, etc., are powerful narcotics which in large doses can cause death. The danger from ethers is usually acute and seldom chronic. After-effects to ether intoxication are uncommon although continued exposure to small concentrations (not enough to cause an overt symptom) has been known to cause loss of appetite, excessive thirst, and fatigue.

The most common ethers such as ethyl, methyl, and diisopropyl are particularly dangerous fire and explosion hazards when exposed to heat, flame, or sparks. They can react violently with strong oxidizers. Many plant and laboratory fires and explosions have resulted from their high flammability and tendency to form explosive peroxides. The common ethers are easily ignited and have low flash points. The diethyl, ethyl tert-butyl, ethyl tert-pentyl and diisopropyl ethers are very hazardous. Methyl tert-alkyl ethers are relatively safe. Besides the risk of explosion from air mixtures

of ether vapors, ethers tend to form peroxides upon standing. For some ethers peroxide levels do not reach dangerous concentrations (e.g., diethyl ether, ethyl vinyl ether, tetrahydrofuran, p-dioxane, 1,1-diethoxyethane and the dimethyl ethers of ethylene glycol). When ethers containing peroxides are heated they can detonate. It is necessary to control smoking, open flames, or even the use of hot plates in areas where low molecular weight ethers are apt to reach 1% concentration or more in air. Only electrical equipment of explosion-proof type (Group C classification) is permitted to be operated in ether areas. Ethers should not be stored near powerful oxidizers or in areas of high fire hazard. They should be kept cool and the containers electrically grounded to avoid sparks.

Dangerous; shock or heat can cause gaseous ethers to escape from their containers and create flammable or even explosive conditions. Incompatible with oxidizing materials, BI_3.

EEH500   CAS: 57-63-6   **HR: 3**
ETHINYL ESTRADIOL
mf: $C_{29}H_{24}O_2$ mw: 296.44

SYNS: 3,17-β-DIHYDROXY-17-α-ETHYNYL-1,3,5(10)-ES-TRATRIENE * 3,17-β-DIHYDROXY-17-α-ETHYNYL-1,3,5(10)-OESTRATRIENE * ESTROGEN * 17-α-ETHI-NYL-3,17-DIHYDROXY-$\Delta^{1,3,5}$-ESTRATRIENE * 17-α-ETHINYL-3,17-DIHYDROXY-$\Delta^{1,3,5}$-OESTRATRIENE * 17-ETHINYLESTRADIOL * 17-ETHINYL-3,17-ES-TRADIOL * 17-α-ETHINYLESTRADIOL * 17-α-ETHI-NYL-17-β-ESTRADIOL * 17-α-ETHINYLESTRA-1,3,5(10)-TRIENE-3,17-β-DIOL * ETHINYLESTRIOL * ETHI-NYLOESTRADIOL * 17-ETHINYL-3,17-OESTRADIOL * ETHINYL-OESTRANOL * 17-α-ETHINYLOESTRA-1,3,5(10)-TRIENE-3,17-β-DIOL * 17-α-ETHINYL-Δ(SUP 1,3,5(10))OESTRATRIENE-3,17-β-DIOL * ETHINYLOES-TRIOL * 17-ETHYNYL-3,17-DIHYDROXY-1,3,5-OESTRA-TRIENE * ETHYNYLESTRADIOL * 17-α-ETHYNY-LESTRADIOL * 17-α-ETHYNYLESTRADIOL-17-β * 17-α-ETHYNYL-1,3,5(10)-ESTRATRIENE-3,17-β-DIOL * 17-α-ETHYNYLESTRA-1,3,5(10)-TRIENE-3,17-β-DIOL * ETHYNYLOESTRADIOL * 17-ETHYNYLOESTRA-DIOL * 17-α-ETHYNYLOESTRADIOL * 17-α-ETHY-NYL-17-β-OESTRADIOL * 17-α-ETHYNYLOESTRADIOL-17-β * 17-ETHYNYLOESTRA-1,3,5(10)-TRIENE-3,17-β-DIOL * 17-α-ETHYNYL-1,3,5-OESTRATRIENE-3,17-β-DIOL * 17-α-ETHYNYL-1,3,5(10)-OESTRATRIENE-3,17-β-DIOL * 17-α-ETHYNYLOESTRA-1,3,5(10)-TRIENE-3,17-β-DIOL * 19-NOR-17-α-PREGNA-1,3,5(10)-TRIEN-2-YNE-3,17-DIOL * (17-α)-19-NORPREGNA-1,3,5(10)-TRIEN-20-YNE-3,17,DIOL

CONSENSUS REPORTS: IARC Cancer Review: Human Limited Evidence IMEMDT 21,233,79; Animal Sufficient Evidence IMEMDT 6,77,74; IMEMDT 21,233,79. NTP Fourth Annual Report On Carcinogens, 1984. Reported in EPA TSCA Inventory.

SAFETY PROFILE: Confirmed carcinogen with experimental carcinogenic, tumorigenic, and neoplastigenic data. Moderately toxic by ingestion. Human systemic effects by ingestion: glandular effects. Experimental reproductive effects. Human mutation data reported. When heated to decomposition it emits acrid smoke and irritating fumes.

EEH520 CAS: 8015-12-1 **HR: 3**
ETHINYLESTRADIOL and
NORETHINDRONE ACETATE
mf: $C_{22}H_{28}O_3 \cdot C_{20}H_{24}O_2$ mw: 636.94

SYNS: ANOVLAR 21 * CONTROLVAR * ETHINYL OESTRADIOL mixed with NORETHISTERONE ACETATE * GYN-ANOVLAR * GYNONLAR 21 * MINORLAR * MINOVLAR * NORETHINDRONE ACETATE and ETHINYLESTRADIOL * NORETHISTERONE ACETATE mixed with ETHINYL OESTRADIOL * NORLESTRIN * PRIMODOS

SAFETY PROFILE: Suspected human carcinogen producing lung and liver tumors. Experimental neoplastigenic and tumorigenic data. Human and experimental teratogenic and reproductive effects. When heated to decomposition it emits acrid smoke and irritating fumes.

EEH550 CAS: 68-23-5 **HR: 3**
17-α-ETHINYL-5,10-ESTRENOLONE
mf: $C_{20}H_{26}O_2$ mw: 298.46

SYNS: 17-α-ETHINYL-$\Delta^{5,10-19}$-NORTESTOSTERONE * 17-ETHINYL-5(10)-ESTRAENEOLONE * 17-α-ETHINYL-ESTRA(5,10)ENEOLONE * 17-α-ETHYNYL-5(10)-ESTREN-17-OL-3-ONE * 17-α-ETHYNYLESTR-5(10)-EN-17-β-OL-3-ONE * 17-α-ETHYNYL-ESTR-5(10)-EN-3-ON-17-β-OL * 17-α-ETHYNYL-17-HYDROXYESTR-5(10)-EN-3-ONE * 17-α-ETHYNYL-17-HYDROXY-5(10)-ESTREN-3-ONE * 17-α-ETHYNYL-17-β-HYDROXY-5(10)-ESTREN-3-ONE * 17-α-ETHYNYL-17-β-HYDROXYESTR-5(10)-EN-3-ONE * 17-α-ETHYNYL-17-β-HYDROXY-$\Delta^{5(10)}$-ESTREN-3-ONE * 17-α-ETHYNYL-17-β-HYDROXY-$\Delta^{-5(10)}$-ESTREN-3-ONE * 17-α-ETHYNYL-17-β-HYDROXY-3-OXO-$\Delta^{5(10)}$-ESTRENE * 17-α-ETHYNYL-19-NOR-5(10)-ANDROSTEN-17-β-OL-3-ONE * 17-β-HYDROXY-17-α-ETHINYL-5(10)-

ESTREN-3-ONE * 17-HYDROXY-19-NOR-17-α-PREGN-5(10)-EN-20-YN-3-ONE * (17-α)-17-HYDROXY-19-NORPREGN-5(10)-EN-20-YN-3-ONE * 17-HYDROXY(17-α)-19-NORPREGN-5(10)-EN-20-YN-3-ONE * LYNESTROL * NORETHINODREL * 19-NOR-ETHINYL-5,10-TESTOSTERONE * NORETHINYNODREL * NORETHYNODRAL * NORETHYNODREL * 19-NORETHYNODREL * NSC-15432 * SC-4642

CONSENSUS REPORTS: IARC Cancer Review: Animal Limited Evidence IMEMDT 21,461,79; Animal Sufficient Evidence IMEMDT 6,191,74.

SAFETY PROFILE: Suspected carcinogen with experimental tumorigenic data. Human and experimental reproductive effects. Mutation data reported. When heated to decomposition it emits acrid smoke and irritating fumes.

EEH600 CAS: 563-12-2 **HR: 3**
ETHION

DOT: NA 2783
mf: $C_9H_{22}O_4P_2S_4$ mw: 384.49

PROP: Liquid. Mp: −13°, d: 1.220 @ 20°/4°. Sltly sol in water; sol in xylene, chloroform, acetone.

SYNS: AC 3422 * BIS(S-(DIETHOXYPHOSPHINOTHIOYL)MERCAPTO)METHANE * BLADAN * DIETHION * EMBATHION * ENT 24,105 * ETHANOX * ETHIOL * ETHODAN * ETHYL METHYLENE PHOSPHORODITHIOATE * FMC-1240 * FOSFONO 50 * HYLEMOX * ITOPAZ * KWIT * METHANEDITHIOL-S,S-DIESTER with O,O-DIETHYL ESTER PHOSPHORODITHIOIC ACID * METHYLEEN-S,S′-BIS(O,O-DIETHYL-DITHIOFOSFAAT) (DUTCH) * S,S′-METHYLEN-BIS(O,O-DIAETHYL-DITHIOPHOSPHAT) (GERMAN) * METHYLENE-S,S′-BIS(O,O-DIAETHYL-DITHIOPHOSPHAT) (GERMAN) * S,S′-METHYLENE O,O,O′,O′-TETRAETHYL PHOSPHORODITHIOATE * NIAGARA 1240 * NIALATE * PHOSPHOTOX E * RHODIACIDE * RHODOCIDE * RODOCID * RP 8167 * SOPRATHION * O,O,O′,O′-TETRAAETHYL-BIS(DITHIOPHOSPHAT) (GERMAN) * O,O,O′,O′-TETRAETHYL S,S′-METHYLENEBISPHOSPHORDITHIOATE * O,O,O′,O′-TETRAETHYL-S,S′-METHYLENEBISPHOSPHORODITHIOATE * TETRAETHYL S,S′-METHYLENE BIS(PHOSPHORO-THIOLOTHIONATE) * O,O,O′,O′-TETRAETHYL S,S′-METHYLENE DI(PHOSPHORODITHIOATE) * VEGFRU FOSMITE

CONSENSUS REPORTS: EPA: Farm Worker Field Reentry. EPA Genetic Toxicology Program. EPA Extremely Hazardous Substances List.

OSHA PEL: TWA 0.4 mg/m^3 (skin)
ACGIH TLV: TWA 0.4 mg/m^3 (skin)
DOT Classification: Poison B; Label: Poison.

SAFETY PROFILE: Poison by ingestion, skin contact, intraperitoneal and possibly other routes. Human systemic effects by ingestion: flaccid paralysis without anesthesia, motor activity changes, fever and inhibition of cholinesterase. When heated to decomposition it emits highly toxic fumes of SO$_x$ and PO$_x$.

EEI000   CAS: 67-21-0   **HR: 3**
dl-ETHIONINE
mf: C$_6$H$_{13}$NO$_2$S   mw: 163.26

PROP: Crystals. Decomp @ 273°.

SYNS: AETHIONIN * 2-AMINO-4-(ETHYLTHIO)BU-
TYRIC ACID * dl-2-AMINO-4-(ETHYLTHIO)BUTYRIC
ACID * CN 8676 * ETH * ETHIONIN * ETHIO-
NINE * (±)-ETHIONINE * S-ETHYL-HOMOCY-
STEINE * S-ETHYL-dl-HOMOCYSTEINE * NSC 751
* U-1434

CONSENSUS REPORTS: EPA Genetic Toxicology Program. Reported in EPA TSCA Inventory.

SAFETY PROFILE: Suspected carcinogen with experimental carcinogenic and tumorigenic data. Moderately toxic by intraperitoneal route. Mildly toxic by ingestion. Experimental reproductive effects. Mutation data reported. When heated to decomposition it emits toxic fumes of SO$_x$ and NO$_x$.

EEJ000   CAS: 61791-14-8   **HR: 2**
ETHOMEEN C/15

PROP: Polyoxyethylene(5)cocoa amine alkyl links C8-C18 which consists of dodecyl (47%), undecyl (18%), decyl (9%), octyl (8%), hexadecyl (10%) and octadecyl (5%).

CONSENSUS REPORTS: Reported in EPA TSCA Inventory.

SAFETY PROFILE: Moderately toxic by ingestion. An eye irritant. When heated to decomposition it emits acrid smoke and irritating fumes.

EEL000   CAS: 927-80-0   **HR: 3**
ETHOXY ACETYLENE
mf: C$_4$H$_6$O   mw: 70.09

PROP: Insol in water. Flash p: 19.4°F, d: 0.8, vap d: 2.4, bp: 61°.

SYN: ETHOXYETHYNE

SAFETY PROFILE: Potentially explosive decomposition when heated above 100°C. Potentially explosive reaction with ethyl magnesium iodide. A very dangerous fire hazard when exposed to heat, flame or oxidizers. To fight fire, use foam, dry chemical, CO$_2$. When heated to decomposition it emits acrid smoke and fumes.

EEP000   CAS: 73987-52-7   **HR: 3**
ETHOXY CARBONYL DIGOXIN

SYN: CARBAETHOXYDIGOXIN (GERMAN)

SAFETY PROFILE: Poison by ingestion and intravenous routes.

EER500   CAS: 103-75-3   **HR: 2**
2-ETHOXY DIHYDROPYRAN
mf: C$_7$H$_{12}$O$_2$ mw: 128.19

PROP: D: 1.0, bp: 143°, flash p: 111°F (OC).

SYNS: 2-ETHOXY-2,3-DIHYDRO-γ-PYRAN * 2-
ETHOXY-3,4-DIHYDRO-1,2-PYRAN * 2-ETHOXY-3,4-
DIHYDRO-2H-PYRAN

CONSENSUS REPORTS: Reported in EPA TSCA Inventory.

SAFETY PROFILE: Moderately toxic by skin contact. Mildly toxic by ingestion and inhalation. A skin and eye irritant. Flammable when exposed to flame, sparks, and oxidizers. To fight fire, use dry chemical, foam, fog. When heated to decomposition it emits acrid smoke and irritating fumes.

EES350   CAS: 110-80-5   **HR: 2**
2-ETHOXYETHANOL

DOT: UN 1171
mf: C$_4$H$_{10}$O$_2$ mw: 90.14

PROP: Colorless liquid, practically odorless. Bp: 135.1°, lel: 1.8%, uel: 14%, flash p: 202°F(CC), d: 0.9360 @ 15°/15°, autoign temp: 455°F, vap press: 3.8 mm @ 20°, vap d: 3.10.

SYNS: ATHYLENGLYKOL-MONOATHYLATHER (GER-
MAN) * CELLOSOLVE (DOT) * CELLOSOLVE SOL-
VENT * DOWANOL EE * EKTASOLVE EE
* ETHER MONOETHYLIQUE de L'ETHYLENE-GLYCOL
(FRENCH) * ETHYL CELLOSOLVE * ETHYLENE
GLYCOL ETHYL ETHER * ETHYLENE GLYCOL MONO-
ETHYL ETHER * ETHYLENE GLYCOL MONO-

ETHYL ETHER (DOT) * ETOKSYETYLOWY ALKOHOL (POLISH) * GLYCOL ETHER EE * GLYCOL ETHYL ETHER * GLYCOL MONOETHYL ETHER * HYDROXY ETHER * JEFFERSOL EE * NCI-C54853 * OXITOL * POLY-SOLV EE

CONSENSUS REPORTS: Reported in EPA TSCA Inventory. Glycol ether compounds are on the Community Right-To-Know List.

OSHA PEL: TWA 200 ppm (skin)
ACGIH TLV: TWA 5 ppm (skin)
DFG MAK: 20 ppm (75 mg/m^3)
NIOSH REL: (Glycol Ethers): Reduce to lowest level
DOT Classification: Combustible Liquid; Label: None; IMO: Flammable or Combustible Liquid; Label: Flammable Liquid.

SAFETY PROFILE: Moderately toxic by ingestion, skin contact, intravenous, intraperitoneal, and possibly other routes. Mildly toxic by inhalation and subcutaneous routes. Experimental teratogenic and reproductive effects. An eye and skin irritant. Combustible when exposed to heat or flame; can react with oxidizing materials. Moderate explosion hazard in the form of vapor when exposed to heat or flame. Mixture with hydrogen peroxide + polyacrylamide gel + toluene is explosive when dry. To fight fire, use alcohol foam, dry chemical.

EES400   CAS: 111-15-9   HR: 2
2-ETHOXYETHYL ACETATE

DOT: UN 1172
mf: $C_6H_{12}O_3$ mw: 132.18

PROP: Colorless liquid with a mild, pleasant ester-like odor. Bp: 156.4°, flash p: 117°F (COC), lel: 1.7%, fp: −61.7°, d: 0.9748 @ 20°/20°, autoign temp: 715°F, vap press: 1.2 mm @ 20°, vap d: 4.72.

SYNS: ACETATE de CELLOSOLVE (FRENCH) * ACETATE de l'ETHER MONOETHYLIQUE DE L'ETHYLENE-GLYCOL (FRENCH) * ACETATE d'ETHYLGLYCOL (FRENCH) * ACETATO di CELLOSOLVE (ITALIAN) * ACETIC ACID-2-ETHOXYETHYL ESTER * 2-AETHOXY-AETHYLACETAT (GERMAN) * AETHYLENGLYKOLAETHERACETAT (GERMAN) * CELLOSOLVE ACETATE (DOT) * CSAC * EKTASOLVE EE ACETATE SOLVENT * ETHOXY ACETATE * 2-ETHOXYETHANOL ACETATE * 2-ETHOXYETHANOL, ESTER with ACETIC ACID * 2-ETHOXY-ETHYLACETAAT (DUTCH) * ETHOXYETHYL ACETATE * β-ETHOXYETHYL ACETATE * 2-ETHOXYETHYLACETATE

* 2-ETHOXYETHYLE, ACETATE de (FRENCH) * ETHYL CELLOSOLVE ACETAAT (DUTCH) * ETHYLENE GLYCOL ETHYL ETHER ACETATE * ETHYLENE GLYCOL MONOETHYL ETHER ACETATE (MAK, DOT) * ETHYLGLYKOLACETAT (GERMAN) * 2-ETOSSIETIL-ACETATO (ITALIAN) * GLYCOL ETHER EE ACETATE * GLYCOL MONOETHYL ETHER ACETATE * OCTAN ETOKSYETYLU (POLISH) * OXYTOL ACETATE * POLY-SOLV EE ACETATE

CONSENSUS REPORTS: Reported in EPA TSCA Inventory. Glycol ether compounds are on the Community Right-To-Know List.

OSHA PEL: TWA 100 ppm (skin)
ACGIH TLV: TWA 5 ppm (skin)
DFG MAK: 20 ppm (110 mg/m^3)
DOT Classification: Combustible Liquid; Label: None; IMO: Flammable or Combustible Liquid; Label: Flammable Liquid.

SAFETY PROFILE: Moderately toxic by ingestion and intraperitoneal routes. Mildly toxic by skin contact, inhalation, and subcutaneous routes. A skin and eye irritant. Experimental teratogenic and reproductive effects. Flammable when exposed to heat or flame; can react with oxidizing materials. Moderate explosion hazard in the form of vapor when heated. Mild explosions have occurred at the end of distillations. To fight fire, use alcohol foam, CO_2, dry chemical. When heated to decomposition it emits acrid smoke and irritating fumes.

EEX500   CAS: 22960-71-0   HR: 3
N-ETHOXYMORPHOLINO DIAZENIUM FLUOROBORATE
mf: $C_6H_{13}N_2O_2 \cdot BF_4$ mw: 232.02

SAFETY PROFILE: Moderately toxic by subcutaneous route. Questionable carcinogen with experimental carcinogenic data. When heated to decomposition it emits very toxic NO_x and F$^-$.

EFE000   CAS: 150-69-6   HR: 3
4-ETHOXYPHENYLUREA
mf: $C_9H_{12}N_2O_2$ mw: 180.23

PROP: Needle-like crystals. Mp: 174°.

SYNS: p-AETHOXYPHYLHARNSTOFF (GERMAN) * DULCINE * N-(4-ETHOXYPHENYL)UREA * p-ETHOXYPHENYLUREA * NCI-C02073 * PHENETHYLCARBAMID (GERMAN) * p-PHENETOLCARBAMID (GERMAN) * p-PHENETOLCARBAMIDE * p-PHENETOLECARBAMIDE * p-PHENETYLUREA * SUCROL * SUESSTOFF * VALZIN

CONSENSUS REPORTS: IARC Cancer Review: GROUP 3 IMEMDT 7,56,87; Animal Inadequate Evidence IMEMDT 12,97,76.

SAFETY PROFILE: Human poison by ingestion. Moderately toxic experimentally by ingestion. Human systemic effects by ingestion: somnolence, hallucinations, distorted perceptions and changes in motor activity. In adults 20 to 40 grams produces dizziness, nausea, methemoglobinemia, cyanosis, and hypotension. Questionable carcinogen with experimental tumorigenic data. When heated to decomposition it emits toxic fumes of NO_x.

EFE500   CAS: 63815-42-9   HR: 3
4-ETHOXY-β-(1-PIPERIDYL)PROPIO-PHENONE HYDROCHLORIDE
mf: $C_{16}H_{23}NO_2 \cdot ClH$ mw: 297.86

SAFETY PROFILE: Poison by intraperitoneal, subcutaneous, and intravenous routes. When heated to decomposition it emits very toxic fumes of HCl and NO_x.

EFG500   CAS: 63918-98-9   HR: 3
ETHOXY PROPIONALDEHYDE
mf: $C_5H_{10}O_2$ mw: 102.15

PROP: Liquid. Mp: −69.4°, bp: 135.2°, flash p: 100°F (OC), d: 0.918 @ 20°/20°, vap d: 3.63, vap press: 5.5 mm @ 20°.

SAFETY PROFILE: Poison by ingestion. Moderately toxic by inhalation and skin contact. A skin and severe eye irritant. A very dangerous fire hazard when exposed to heat or flame; can react with oxidizing materials. To fight fire, use alcohol foam, CO_2, dry chemical. When heated to decomposition it emits acrid smoke and irritating fumes.

EFH000   CAS: 1331-11-9   HR: 2
ETHOXYPROPIONIC ACID
mf: $C_5H_{10}O_3$ mw: 118.15

PROP: Liquid. Mp: −10.7°, bp: 219°, flash p: 225°F (OC), d: 1.0474, vap d: 4.08.

SAFETY PROFILE: Moderately toxic by skin contact. Mildly toxic by ingestion. A skin and severe eye irritant. Combustible when exposed to heat or flame; can react with oxidizing materials. To fight fire, use alcohol foam, CO_2, dry chemical. When heated to decomposition it emits acrid smoke and irritating fumes.

EFL000   CAS: 112-50-5   HR: 1
ETHOXYTRIGLYCOL
mf: $C_8H_{18}O_4$ mw: 178.26

PROP: Bp: 255.4°, flash p: 275°F (OC), d: 1.0208 @ 20°/20°, vap press: 0.01 mm @ 20°.

SYNS: DOWANOL TE * 2-(2-(2-ETHOXYETHOXY)ETHOXY)ETHANOL * ETHOXYTRIETHYLENE GLYCOL * POLY-SOLV TE * TRIETHYLENE GLYCOL ETHYL ETHER * TRIETHYLENE GLYCOL MONOETHYL ETHER * TRIGLYCOL MONOETHYL ETHER

CONSENSUS REPORTS: Reported in EPA TSCA Inventory. Glycol ether compounds are on the Community Right-To-Know List.

SAFETY PROFILE: Mildly toxic by ingestion and skin contact. An eye irritant. Combustible when exposed to heat or flame; can react with oxidizing materials. To fight fire use foam, alcohol foam, CO_2, dry chemical. When heated to decomposition it emits acrid smoke and irritating fumes.

EFQ500   CAS: 529-65-7   HR: 2
N-ETHYLACETANILIDE
mf: $C_{10}H_{13}NO$ mw: 163.24

PROP: White crystals, faint odor. Mp: 54°, bp: 258°, flash p: 126°F, d: 0.994, vap d: 5.62.

SYNS: ACETETHYLANILIDE * ETHYLACETANILIDE

CONSENSUS REPORTS: Reported in EPA TSCA Inventory.

SAFETY PROFILE: Moderately toxic by ingestion. Flammable when exposed to heat or flame; can react with oxidizing materials. To fight fire, use foam, CO_2, dry chemical. When heated to decomposition it emits toxic fumes of NO_x.

EFR000   CAS: 141-78-6   HR: 3
ETHYL ACETATE
DOT: UN 1173
mf: $C_4H_8O_2$ mw: 88.12

PROP: Colorless liquid; fragrant odor. Mp: −83.6°, bp: 77.15°, ULC: 85-90, lel: 2.2%, uel: 11%, flash p: 24°F, d: 0.8946 @ 25°, autoign temp: 800°F, vap press: 100 mm @ 27.0°, vap d: 3.04. Misc with alc, ether, glycerin, volatile oils, water @ 54°.

SYNS: ACETIC ETHER * ACETIDIN * ACETOXYETHANE * AETHYLACETAT (GERMAN) * ESSIGESTER (GERMAN) * ETHYLACETAAT (DUTCH) * ETHYL ACETIC ESTER * ETHYLE (ACETATE d')

(FRENCH) * ETHYL ETHANOATE * ETILE (ACE-TATO di) (ITALIAN) * FEMA No. 2414 * OCTAN ETYLU (POLISH) * RCRA WASTE NUMBER U112 * VINEGAR NAPHTHA

CONSENSUS REPORTS: Reported in EPA TSCA Inventory. EPA Genetic Toxicology Program.

OSHA PEL: TWA 400 ppm
ACGIH TLV: TWA 400 ppm
DFG MAK: 400 ppm (1400 mg/m^3)
DOT Classification: Flammable Liquid; Label: Flammable Liquid.

SAFETY PROFILE: Poison by inhalation. Moderately toxic by intraperitoneal and subcutaneous routes. Mildly toxic by ingestion. Human systemic effects by inhalation: olfactory changes, conjunctiva irritation, and pulmonary changes. Human eye irritant. Mutation data reported. Irritating to mucous surfaces, particularly the eyes, gums and respiratory passages, and is also mildly narcotic. On repeated or prolonged exposures, it causes conjunctival irritation and corneal clouding. It can cause dermatitis. High concentrations have a narcotic effect and can cause congestion of the liver and kidneys. Chronic poisoning has been described as producing anemia, leucocytosis (transient increase in the white blood cell count), and cloudy swelling, and fatty degeneration of the viscera. A synthetic flavoring substance and adjuvant.

Highly flammable liquid. A very dangerous fire hazard when exposed to heat or flame; can react vigorously with oxidizing materials. Moderate explosion hazard when exposed to flame. Potentially explosive reaction with lithium tetrahydroaluminate. Ignites on contact with potassium tert-butoxide. Violent reaction with chlorosulfonic acid; (LiAlH$_2$ + 2-chloromethyl furan); oleum. To fight fire, use CO$_2$, dry chemical or alcohol foam. When heated to decomposition it emits acrid smoke and irritating fumes.

EFS000 CAS: 141-97-9 **HR: 2**
ETHYL ACETYL ACETATE
mf: C$_6$H$_{10}$O$_3$ mw: 130.16

PROP: Colorless liquid; fruity odor. Bp: 180.8°, fp: −45°, flash p: 185°F (COC), autoign temp: 563°F, d: 1.0261 @ 20°/20°, refr index: 1.418, vap press: 1 mm @ 28.5°, vap d: 4.48. Misc with alc, ether, ethyl acetate, water.

SYNS: ACETOACETIC ACID, ETHYL ESTER * ACETOACETIC ESTER * ACTIVE ACETYL ACE-

TATE * DIACETIC ETHER * EAA * ETHYL ACETOACETATE (FCC) * ETHYL ACETYLACETONATE * ETHYL BENZYL ACETOACETATE * ETHYL-3-OXO-BUTANOATE * ETHYL-3-OXOBUTYRATE * FEMA No. 2415 * 3-OXOBUTANOIC ACID ETHYL ESTER

CONSENSUS REPORTS: Reported in EPA TSCA Inventory.

SAFETY PROFILE: Moderately toxic by ingestion. A skin and eye irritant. Combustible liquid when exposed to heat or flame; can react with oxidizing materials. Explosive reaction when heated with Zn + trimbromoneopentyl alcohol or 2,2,2-tris(bromomethyl)ethanol. To fight fire, use alcohol foam, CO$_2$, dry chemical. When heated to decomposition it emits acrid smoke and irritating fumes.

EFS500 CAS: 107-00-6 **HR: 2**
ETHYL ACETYLENE

DOT: UN 2452
mf: C$_4$H$_6$ mw: 54

PROP: A colorless, highly flammable gas. Bp: 8.3°, d: 0.669 @ 0°/0°, mp: −130°, flash p: <30°F (TOC), <7°C (Gas >8°C).

SYNS: 1-BUTYNE * ETHYL ACETYLENE, INHIBITED (DOT) * ETHYLETHYNE

DOT Classification: Flammable Gas; Label: Flammable Gas.

CONSENSUS REPORTS: Reported in EPA TSCA Inventory.

SAFETY PROFILE: Probably an asphyxiant. A very dangerous fire hazard when exposed to heat, open flame or powerful oxidizers. A dangerous explosion hazard. To fight fire, stop flow of gas.

EFT000 CAS: 140-88-5 **HR: 3**
ETHYL ACRYLATE

DOT: UN 1917
mf: C$_5$H$_8$O$_2$ mw: 100.13

PROP: Colorless liquid; acrid penetrating odor. Bp: 99.8°, fp: <−72°: lel: 1.8%, flash p: 60°F (OC), 48.2°F d: 0.916-0.919, vap press, 29.3 mm @ 20°, vap d: 3.45. Misc with alc, ether; sltly sol in water.

SYNS: ACRYLATE d'ETHYLE (FRENCH) * ACRYLIC ACID ETHYL ESTER * ACRYLSAEUREAETHYLESTER (GERMAN) * AETHYLACRYLAT (GERMAN)

* ETHOXYCARBONYLETHYLENE * ETHYLACRY-
LAAT (DUTCH) * ETHYLAKRYLAT (CZECH)
* ETHYL PROPENOATE * ETHYL-2-PROPENOATE
* ETIL ACRILATO (ITALIAN) * ETILACRILATULUI
(ROMANIAN) * FEMA No. 2418 * NCI-C50384
* 2-PROPENOIC ACID, ETHYL ESTER (MAK)
* RCRA WASTE NUMBER U113

CONSENSUS REPORTS: IARC Cancer Review: GROUP 2B IMEMDT 7,56,87; Animal
Sufficient Evidence IMEMDT 39,81,86; Animal Inadequate Evidence IMEMDT 19,47,79;
Human Inadequate Evidence IMEMDT 19,
47,79. NTP Carcinogenesis Studies (gavage);
Clear Evidence: mouse, rat NTPTR* NTP-TR-
259,86. Reported in EPA TSCA Inventory.
Community Right-To-Know List.

OSHA PEL: (Transitional: TWA 25 mg/m^3
   (skin)) TWA 5 ppm; STEL 25 ppm (skin)
ACGIH TLV: TWA 5 ppm; STEL 25 ppm (Pro-
   posed: 5 ppm; STEL 15 ppm Suspected Hu-
   man Carcinogen)
DFG MAK: 5 ppm (20 mg/m^3)
DOT Classification: Flammable Liquid; Label:
   Flammable Liquid.

SAFETY PROFILE: Suspected carcinogen with
experimental carcinogenic data. Poison by ingestion and inhalation. Moderately toxic by skin
contact and intraperitoneal routes. Human systemic effects by inhalation: eye, olfactory and
pulmonary changes. A skin and eye irritant.
Characterized in its terminal stages by dyspnea,
cyanosis, and convulsive movements. It caused
severe local irritation of the gastro-enteric tract;
and toxic degenerative changes of cardiac, hepatic, renal, and splenic tissues were observed.
It gave no evidence of cumulative effects. When
applied to the intact skin of rabbits, the ethyl
ester caused marked local irritation, erythema,
edema, thickening, and vascular damage. Animals subjected to a fairly high concentrations
of these esters suffered irritation of the mucous
membranes of the eyes, nose, and mouth as
well as lethargy, dyspnea, and convulsive movements. A substance which migrates to food from
packaging materials.
      Flammable liquid. A very dangerous fire
hazard when exposed to heat or flame; can react
vigorously with oxidizing materials. Violent reaction with chlorosulfonic acid. To fight fire,
use CO_2, dry chemical or alcohol foam. When
heated to decomposition it emits acrid smoke
and irritating fumes.

EFT500   CAS: 462-95-3   **HR: 2**
ETHYLAL
DOT: UN 2373
mf: $C_5H_{12}O_2$ mw: 104.17

PROP: Flash p: <69.8°F.

SYN: DIETHOXYMETHANE (DOT)

CONSENSUS REPORTS: Reported in EPA
TSCA Inventory.

DOT Classification: Flammable Liquid; Label:
   Flammable Liquid.

SAFETY PROFILE: Moderately toxic by ingestion. Flammable when exposed to heat or flame;
can react vigorously with oxidizers. When
heated to decomposition it emits acrid smoke
and irritating fumes.

EFU000   CAS: 64-17-5   **HR: 3**
ETHYL ALCOHOL
DOT: UN 1170
mf: C_2H_6O mw: 46.08

PROP: Clear, colorless liquid; fragrant odor
and burning taste. Bp: 78.32°, ULC: 70, lel:
3.3%, uel: 19% @ 60°, fp: <-130°, flash p:
55.6°F, d: 0.7893 @ 20°/4°, autoign temp:
793°F, vap press: 40 mm @ 19°, vap d: 1.59,
refr index: 1.364. Misc in water, alc, chloroform, and ether.

SYNS: ABSOLUTE ETHANOL * AETHANOL (GER-
MAN) * AETHYLALKOHOL (GERMAN) * ALCOHOL
* ALCOHOL, anhydrous * ALCOHOL, dehydrated
* ALCOOL ETHYLIQUE (FRENCH) * ALCOOL ETIL-
ICO (ITALIAN) * ALGRAIN * ALKOHOL (GERMAN)
* ALKOHOLU ETYLOWEGO (POLISH) * ANHYDROL
* COLOGNE SPIRIT * COLOGNE SPIRITS (ALCOHOL)
(DOT) * ETANOLO (ITALIAN) * ETHANOL (MAK)
* ETHANOL 200 PROOF * ETHANOL, solution (DOT)
* ETHYLALCOHOL (DUTCH) * ETHYL ALCOHOL, an-
hydrous * ETHYL HYDRATE * ETHYL HYDROXIDE
* ETYLOWY ALKOHOL (POLISH) * FERMENTATION
ALCOHOL * GRAIN ALCOHOL * JAYSOL
* JAYSOL S * METHYLCARBINOL * MOLASSES
ALCOHOL * NCI-C03134 * POTATO ALCOHOL
* SD ALCOHOL 23-HYDROGEN * SPIRITS of WINE
* SPIRT * TECSOL

CONSENSUS REPORTS: IARC Cancer Review: Human Sufficient Evidence IMEMDT
44,259,88. Reported in EPA TSCA Inventory.
EPA Genetic Toxicology Program.

OSHA PEL: TWA 1000 ppm
ACGIH TLV: TWA 1000 ppm
DFG MAK: 1,000 ppm (1,900 mg/m^3)
DOT Classification: Flammable Liquid; Label: Flammable Liquid; Flammable or Combustible Liquid; Label: Flammable Liquid.

SAFETY PROFILE: Confirmed human carcinogen for ingestion of beverage alcohol. Experimental teratogenic and reproductive effects. Moderately toxic to humans by ingestion. Moderately toxic experimentally by intravenous and intraperitoneal routes. Mildly toxic by inhalation and skin contact. Human systemic effects by ingestion and subcutaneous route: sleep disorders, hallucinations, distorted perceptions, convulsions, motor activity changes, ataxia, coma, antipsychotic, headache, pulmonary changes, alteration in gastric secretion, nausea or vomiting, other gastrointestinal changes, menstrual cycle changes and body temperature decrease. Can also cause glandular effects in humans. Human reproductive effects by ingestion, intravenous, and intrauterine routes: changes in female fertility index. Effects on newborn include: changes in apgar score, neonatal measures or effects and drug dependence. Experimental reproductive effects. Human mutation data reported. An eye and skin irritant.

The systemic effect of ethanol differs from that of methanol. Ethanol is rapidly oxidized in the body to carbon dioxide and water, and in contrast to methanol, no cumulative effect occurs. Though ethanol possesses narcotic properties, concentrations sufficient to produce this effect are not reached in industry. Concentrations below 1,000 ppm usually produce no signs of intoxication. Exposure to concentrations over 1,000 ppm may cause headache, irritation of the eyes, nose and throat, and, if continued for an hour, drowsiness and lassitude, loss of appetite and inability to concentrate. There is no concrete evidence that repeated exposure to ethanol vapor results in cirrhosis of the liver. Ingestion of large doses can cause alcohol poisoning. Repeated ingestions can lead to alcoholism. It is a central nervous system depressant.

Flammable liquid when exposed to heat or flame; can react vigorously with oxidizers. To fight fire, use alcohol foam, CO_2, dry chemical. Explosive reaction with the oxidized coating around potassium metal. Ignites and then explodes on contact with acetic anhydride + sodium hydrogen sulfate. Reacts violently with acetyl bromide (evolves hydrogen bromide), dichloromethane + sulfuric acid + nitrate or nitrite, disulfuryl difluoride, tetrachlorisilane + water, and strong oxidants. Ignites on contact with disulfuric acid + nitric acid, phosphorous(III) oxide, platinum, potassium-tert-butoxide + acids. Forms explosive products in reaction with ammonia + silver nitrate (forms silver nitride and silver fulminate), magnesium perchlorate (forms ethyl perchlorate), nitric acid + silver (forms silver fulminate), silver nitrate (forms ethyl nitrate), silver(I) oxide + ammonia or hydrazine (forms silver nitride and silver fulminate), sodium (evolves hydrogen gas). Incompatible with acetyl chloride, BrF_5, $Ca(OCl)_2$, ClO_3, CrO_3, $Cr(OCl)_2$, (cyanuric acid + H_2O), H_2O_2, HNO_3, (H_2O_2 + H_2SO_4), (I + CH_3OH + HgO), [$Mn(ClO_4)_2$ + 2,2-dimethoxy propane], $Hg(NO_3)_2$, $HClO_4$, perchlorates, (H_2SO_4 + permanganates), $HMnO_4$, KO_2, $KOC(CH_3)_3$, $AgClO_4$, NaH_3N_2, $UO_2(ClO_4)_2$.

EFU400   CAS: 75-04-7   **HR: 3**
ETHYLAMINE

DOT: UN 1036
mf: C_2H_7N mw: 45.10

PROP: Colorless gas or liquid, strong ammoniacal odor. Bp: 16.6°, flammable, lel: 4.95%, uel: 20.75%, fp: −80.6°, flash p: −0.4°F, d: 0.662 @ 20°/4°, autoign temp: 725°F, vap d: 1.56. vap press: 400 mm @ 20°. Miscible with water, alc, ether.

SYNS: AETHYLAMINE (GERMAN) * AMINOETHANE * 1-AMINOETHANE * ETHANAMINE * ETHYLAMINE * ETILAMINA (ITALIAN) * ETYLOAMINA (POLISH) * MONOETHYLAMINE (DOT) * MONOETHYLAMINE, anhydrous (DOT)

CONSENSUS REPORTS: Reported in EPA TSCA Inventory.

OSHA PEL: TWA 10 ppm
ACGIH TLV: TWA 10 ppm
DFG MAK: 10 ppm (18 mg/m^3)
DOT Classification: Flammable Liquid; Label: Flammable Liquid; Flammable Gas; Label: Flammable Gas.

SAFETY PROFILE: A poison by ingestion, skin contact and intravenous routes. Moderately toxic by inhalation. A severe eye irritant. A very dangerous fire hazard when exposed to heat or flame. Moderate explosion hazard when exposed to spark or flame. Keep away from heat and open flame, can react vigorously with oxidizing materials. To fight fire, stop flow of gas, use alcohol foam, dry chemical. Incompatible with cellulose nitrate or oxidizers. When heated to decomposition it emits toxic fumes of NO_x.

EFX000   CAS: 94-09-7   **HR: 3**
ETHYL-4-AMINOBENZOATE
mf: $C_9H_{11}NO_2$ mw: 165.21

PROP: Crystals. Mp: 88-90°.

SYNS: AMERICAINE * p-AMINOBENZOIC ACID ETHYL ESTER * 4-AMINOBENZOIC ACID ETHYL ESTER * ANESTHESIN * ANESTHONE * BENZOCAINE * ETHYL AMINOBENZOATE * ETHYL-p-AMINOBENZOATE * KELOFORM * NORCAIN * ORTHESIN * PARATHESIN * TOPCAINE

CONSENSUS REPORTS: Reported in EPA TSCA Inventory.

SAFETY PROFILE: Poison by ingestion and intraperitoneal routes. Human systemic effects by rectal route: methemoglobinemia/carboxhemoglobinemia in infants. A skin irritant and a mild sensitizer. Local contact may cause contact dermatitis. Used as a topical anesthetic and as a sun-screening agent. When heated to decomposition it emits highly toxic fumes of NO_x.

EGA500   CAS: 110-73-6   **HR: 3**
2-ETHYLAMINOETHANOL
mf: $C_4H_{11}NO$ mw: 89.16

PROP: Flash p: 160°F (OC), d: 0.92, vap d: 3.06, bp: 161°.

SYNS: 2-(ETHYLAMINO)ETHANOL * 2-N-MONOETHYLAMINOETHANOL

CONSENSUS REPORTS: Reported in EPA TSCA Inventory.

SAFETY PROFILE: Poison by skin contact. Moderately toxic by ingestion and intraperitoneal routes. A skin and severe eye irritant. Flammable when exposed to heat or flame; can react vigorously with oxidizers. To fight fire, use alcohol foam, dry chemical, CO_2. When heated to decomposition it emits toxic fumes of NO_x.

EGI000   CAS: 13275-68-8   **HR: 3**
2-ETHYLAMINO-1,3,4-THIADIAZOLE
mf: $C_4H_7N_3S$ mw: 129.20

SYNS: CL 19217 4090L 7-5525 * 2-ETHYLAMINOTHIADIAZOLE * NSC 4730

CONSENSUS REPORTS: Reported in EPA TSCA Inventory.

SAFETY PROFILE: Poison by intraperitoneal and subcutaneous routes. Experimental teratogenic and reproductive effects. When heated to decomposition it emits very toxic fumes of NO_x and SO_x.

EGI750   CAS: 541-85-5   **HR: 2**
ETHYL AMYL KETONE
mf: $C_8H_{16}O$ mw: 128.24

PROP: Liquid; mild, fruity odor. Bp: 157-162°, d: 0.822 @ 20°/20°, flash p: 138°F. Sol in many organic solvents.

SYNS: AMYL ETHYL KETONE * 3-METHYL-5-HEPTANONE * 5-METHYL-3-HEPTANONE

OSHA PEL: TWA 25 ppm
ACGIH TLV: TWA 25 ppm

SAFETY PROFILE: Moderately irritating to skin, eyes, and mucous membranes by inhalation and ingestion. Narcotic in high concentration. When heated to decomposition it emits acrid smoke. Combustible. To fight fire, use foam, CO_2, dry chemical.

EGI755   CAS: 106-68-3   **HR: 2**
ETHYL AMYL KETONE
DOT: UN 2271
mf: $C_8H_{16}O$ mw: 128.24

PROP: Liquid; fruity odor. Bp: 157-162°, d: 0.822 @ 20°/20°, flash p: 138°F.

SYNS: AMYL ETHYL KETONE * EAK * 3-OCTANONE

CONSENSUS REPORTS: Reported in EPA TSCA Inventory.

OSHA PEL: TWA 25 ppm
DOT Classification: Flammable or Combustible Liquid; Label: Flammable Liquid

SAFETY PROFILE: Poison by intraperitoneal route. Moderately irritating to skin, eyes, and mucous membranes by inhalation. Narcotic in high concentration. When heated to decomposition it emits acrid smoke. Dangerously flammable from heat, flame, oxidizers. To fight fire, use foam, CO_2, dry chemical.

EGK000 CAS: 103-69-5 **HR: 3**
N-ETHYLANILINE

DOT: UN 2272
mf: $C_8H_{11}N$ mw: 121.20

PROP: Clear, yellow-brown oil. Mp: $-63.5°$,
bp: 204°, d: 0.958 @ 25°/25°, vap press: 1
mm @ 38.5°, vap d: 4.18, flash p: 185°F (OC).

SYNS: AETHYLANILIN (GERMAN) * ANILINO-
ETHANE * N-ETHYLAMINOBENZENE * ETHYL-
ANILINE * N-ETHYLBENZENAMINE * N-ETHYL-
BENZENAMINO * ETHYLPHENYLAMINE

CONSENSUS REPORTS: Reported in EPA
TSCA Inventory.

DOT Classification: Poison B; Label: St. Andrews Cross.

SAFETY PROFILE: Poison by ingestion and
intraperitoneal routes. Moderately toxic by an
unspecified route. Mildly toxic by skin contact.
An allergen. Flammable when exposed to heat
or flame; can react with oxidizing materials.
To fight fire, use dry chemical, CO_2, foam.
Hypergolic reaction with red fuming nitric acid.
When heated to decomposition or on contact
with acid or acid fumes it emits highly toxic
fumes of aniline and NO_x.

EGK500 CAS: 578-54-1 **HR: 3**
2-ETHYLANILINE

DOT: UN 2273
mf: $C_8H_{11}N$ mw: 121.20

PROP: Yellow liquid, darkens upon standing.
Mp: $-63.5°$, bp: 215°, flash p: 185°F (OC),
d: 0.98 @ 25°/25°, vap d: 4.17.

SYNS: o-AMINOETHYLBENZENE * o-ETHYLANI-
LINE * 2-ETHYLBENZENAMINE

CONSENSUS REPORTS: Reported in EPA
TSCA Inventory.

DOT Classification: Poison B; Label: St. Andrews Cross.

SAFETY PROFILE: A poison. Moderately toxic
by ingestion. Flammable when exposed to heat
or flame; can react with oxidizing materials.
To fight fire, use foam, CO_2, dry chemical.
When heated to decomposition it emits highly
toxic fumes of aniline and NO_x.

EGL000 CAS: 589-16-2 **HR: 3**
4-ETHYLANILINE
mf: $C_8H_{11}N$ mw: 121.20

PROP: D: 0.963, mp: 65.8°, bp: 205.5°. Insol
in water; misc in alc and ether.

SYNS: 1-AMINO-4-ETHYLBENZENE * p-ETHYLANI-
LINE

CONSENSUS REPORTS: Reported in EPA
TSCA Inventory.

SAFETY PROFILE: Poison by ingestion, intravenous, and intraperitoneal routes. Mutation
data reported. When heated to decomposition
it emits toxic fumes of NO_x.

EGM000 CAS: 87-25-2 **HR: 2**
ETHYL ANTHRANILATE
mf: $C_9H_{11}NO_2$ mw: 165.21

PROP: Colorless to amber liquid; floral, orange
blossom odor. D: 1.115-1.120, refr index:
1.5631.566, flash p: +151°F. Sol in alc, fixed
oils, propylene glycol.

SYNS: o-AMINOBENZOIC ACID, ETHYL ESTER
* ETHYL-o-AMINOBENZOATE * FEMA No. 2421

CONSENSUS REPORTS: Reported in EPA
TSCA Inventory.

SAFETY PROFILE: Moderately toxic by ingestion. A skin irritant. Combustible liquid. When
heated to decomposition it emits toxic fumes
of NO_x.

EGP500 CAS: 100-41-4 **HR: 2**
ETHYL BENZENE

DOT: UN 1175
mf: C_8H_{10} mw: 106.18

PROP: Colorless liquid, aromatic odor. Bp:
136.2°, fp: $-94.9°$, flash p: 59°F, d: 0.8669
@ 20°/4°, autoign temp: 810°F, vap press: 10
mm @ 25.9°, vap d: 3.66, lel: 1.2%, uel: 6.8%.
Misc in alc and ether; insol in NH_3; sol in SO_2.

SYNS: AETHYLBENZOL (GERMAN) * EB
* ETHYLBENZEEN (DUTCH) * ETHYLBENZOL
* ETILBENZENE (ITALIAN) * ETYLOBENZEN (POL-
ISH) * NCI-C56393 * PHENYLETHANE

CONSENSUS REPORTS: Reported in EPA
TSCA Inventory. EPA Genetic Toxicology Program. Community Right-To-Know List.

OSHA PEL: (Transitional: TWA 100 ppm
(skin)) TWA 100 ppm; STEL 125 ppm
ACGIH TLV: TWA 100 ppm; STEL 125 ppm;
BEI: 2 g(mandelic acid)/L in urine at end

of shift; 2 ppm ethyl benzene in end-exhaled air prior to next shift.
DFG MAK: 100 ppm (440 mg/m^3)
DOT Classification: Flammable Liquid; Label: Flammable Liquid.

SAFETY PROFILE: Moderately toxic by ingestion and intraperitoneal route. Mildly toxic by inhalation and skin contact. An experimental teratogen. Human systemic effects by inhalation: eye, sleep and pulmonary changes. An eye and skin irritant. Human mutation data reported. The liquid is an irritant to the skin and mucous membranes. A concentration of 0.1% of the vapor in air is an irritant to human eyes, and a concentration of 0.2% is extremely irritating at first, then causes dizziness, irritation of the nose and throat and a sense of constriction in the chest. Exposure of guinea pigs to 1% concentration has been reported as causing ataxia, loss of consciousness, tremor of the extremities and finally death through respiratory failure. The pathological findings were congestion of the brain and lungs with edema.

A very dangerous fire and explosion hazard when exposed to heat or flame; can react vigorously with oxidizing materials. To fight fire, use foam, CO_2, dry chemical. When heated to decomposition it emits acrid smoke and irritating fumes.

EGR000   CAS: 93-89-0   HR: 2
ETHYL BENZOATE
mf: $C_9H_{10}O_2$ mw: 150.19

PROP: Colorless liquid; heavy fruity odor. Mp: −34.6°, bp: 213.4°, flash p: >204°F, d: 1.048 @ 20°/20°, refr index: 1.502-1.506, vap press: 1 mm @ 44.0°, vap d: 5.17, autoign temp: 914°F. Sol in alc, fixed oils, and propylene glycol; insol in glycerin, water @ 212°; misc in petroleum, chloroform, and ether.

SYNS: BENZOIC ETHER * ESSENCE of NIOBE
* FEMA No. 2422

CONSENSUS REPORTS: Reported in EPA TSCA Inventory.

SAFETY PROFILE: Moderately toxic by ingestion. Mildly toxic by skin contact. A skin and eye irritant. Combustible liquid when exposed to heat or flame; can react with oxidizing materials. To fight fire, use foam, CO_2, dry chemical. When heated to decomposition it emits acrid smoke and irritating fumes.

EGV000   CAS: 105-36-2   HR: 3
ETHYL BROMACETATE
DOT: UN 1603
mf: $C_4H_7BrO_2$ mw: 167.02

PROP: Colorless to straw-colored liquid. Bp: 158.8°, fp: <−20°, flash p: 118°F, d: 1.514 @ 13°/4°, vap d: 5.8. Insol in water; misc in alc and ether.

SYNS: BROMOACETIC ACID, ETHYL ESTER
* ETHOXYCARBONYLMETHYL BROMIDE * ETHYL
BROMOACETATE * ETHYL-α-BROMOACETATE
* ETHYL MONOBROMOACETATE

CONSENSUS REPORTS: Reported in EPA TSCA Inventory.

DOT Classification: Poison B; Label: Flammable Liquid and Poison.

SAFETY PROFILE: A poison. An irritant to skin, eyes, and mucous membranes. Questionable carcinogen with experimental neoplastigenic data. Flammable when exposed to heat, flame, and oxidizers. Will react with water or steam to produce toxic and corrosive fumes. To fight fire, use water as a fire blanket. When heated to decomposition or on contact with acid or acid fumes, it emits highly toxic fumes of Br^-.

EGV400   CAS: 74-96-4   HR: 3
ETHYL BROMIDE
DOT: UN 1891
mf: C_2H_5Br mw: 108.98

PROP: Colorless, volatile liquid. Mp: −119°, bp: 38.4°, lel: 6.7%, uel: 11.3%, flash p: < −4°F, d: 1.451 @ 20°/4°, autoign temp: 952°F, vap press: 400 mm @ 21°, vap d 3.76.

SYNS: BROMOETHANE * BROMURE d'ETHYLE
* ETYLU BROMEK (POLISH) * HALON 2001
* MONOBROMOETHANE * NCI-C55481

CONSENSUS REPORTS: EPA Genetic Toxicology Program. Reported in EPA TSCA Inventory.

OSHA PEL: (Transitional: TWA 200 ppm) TWA 200 ppm; STEL 250 ppm
ACGIH TLV: TWA 200 ppm; STEL 250 ppm
DFG MAK: 200 ppm (890 mg/m^3)
DOT Classification: Poison B; Label: Poison.

SAFETY PROFILE: A poison. Moderately toxic by ingestion and intraperitoneal routes. Mildly toxic by inhalation. An eye and skin irritant. Physiologically, it is an anesthetic and narcotic. Its vapors are markedly irritating to the lungs on inhalation for even short periods. It can produce acute congestion and edema. Liver and kidney damage in humans has been reported. It is much less toxic than methyl bromide, but more toxic than ethyl chloride. It is a preparative hazard. Dangerously flammable by heat, open flame (sparks), oxidizers. Moderately explosive when exposed to flame. Reacts with water or steam to produce toxic and corrosive fumes. Vigorous reaction with oxidizing materials. To fight fire, use CO_2, dry chemical. Readily decomposes when heated to emit toxic fumes of Br^-.

EGV500   CAS: 4824-78-6   **HR: 3**
ETHYL BROMOPHOS
mf: $C_{10}H_{12}BrCl_2O_3PS$ mw: 394.06

SYNS: 4-BROMO-2,5-DICHLOROPHENOL-o-ESTER with O,O-DIETHYL PHOSPHOROTHIOATE * O-(4-BROMO-2,5-DICHLOROPHENYL)-O,O-DIETHYL PHOSPHORO-THIOATE * O-(4-BROMO-2,5 DICHLOROPHENYL)-O,O-DIETHYLPHOSPHOROTHIONATE * BROMOFOS-ETHYL * BROMOPHOSETHYL * CELA S-2225 * O,O-DIA-ETHYL-O-(4-BROM-2,5-DICHLOR)-PHENYL-MONOTHIO-PHOSPHAT (GERMAN) * O,O-DIAETHYL-O-(2,5-DICH-LOR-4-BROMPHENYL)-THIONOPHOSPHAT (GERMAN) * O,O-DIETHYL-O-(4-BROOM-2,5-DICHLOOR-FENYL) MONOTHIOFOSFAAT (DUTCH) * O,O-DIETHYL O-2,5-DICHLORO-4-BROMOPHENYL-PHOSPHOROTHIOATE * O,O-DIETHYL O-(2,5-DICHLORO-4-BROMOPHENYL) THIOPHOSPHATE * O,O-DIETIL-O-(4-BROMO-2,5 DI-CLORO-FENIL)-MONOTIOFOSFATO (ITALIAN) * ENT 27,258 * FILARIOL * NEXAGAN * OMS-659 * S 2225 * THIOPHOSPHATE de O,O-DIETHYLE et de O-(2,5-DICHLORO-4-BROMO) PHENYLE (FRENCH)

CONSENSUS REPORTS: Chlorophenol compounds are on the Community Right-To-Know List.

SAFETY PROFILE: Poison by ingestion and possibly other routes. Moderately toxic by skin contact and inhalation. An insecticide. When heated to decomposition it emits very toxic fumes of Br^-, PO_x, SO_x, and Cl^-.

EGW000   CAS: 97-95-0   **HR: 2**
2-ETHYLBUTANOL
DOT: UN 2275/UN 2282
mf: $C_6H_{14}O$ mw: 102.20

PROP: Clear liquid. Bp: 148.9°, flash p: 135°F (COC), d: 0.8328, vap press: 0.9 mm @ 20°, vap d: 3.4.

SYNS: 2-ETHYLBUTANOL-1 * 2-ETHYL-1-BUTANOL * 2-ETHYLBUTYL ALCOHOL * sec-HEXANOL (DOT) * sec-HEXYL ALCOHOL * 3-METHYLOLPENTANE * sec-PENTYLCARBINOL * 3-PENTYLCARBINOL * PSEUDOHEXYL ALCOHOL

CONSENSUS REPORTS: Reported in EPA TSCA Inventory.

DOT Classification: Flammable or Combustible Liquid; Label: Flammable Liquid.

SAFETY PROFILE: Moderately toxic by ingestion and skin contact. A skin and severe eye irritant. Flammable when exposed to heat or flame; can react with oxidizing materials. To fight fire, use dry chemical, CO_2, foam, fog. When heated to decomposition it emits acrid smoke and irritating fumes.

EGW500   CAS: 760-21-4   **HR: 1**
2-ETHYL-1-BUTENE
mf: C_6H_{12} mw: 84.18

PROP: Flash p: $< -4°$, autoign temp: 599°F, d: 0.69, vap d: 2.9, bp: 62°.

CONSENSUS REPORTS: Reported in EPA TSCA Inventory.

SAFETY PROFILE: A human eye irritant. A very dangerous fire hazard when exposed to heat, flames, or oxidizers. To fight fire, use dry chemical, CO_2, foam, spray. When heated to decomposition it emits acrid smoke and irritating fumes.

EGZ000   CAS: 3953-10-4   **HR: 2**
2-ETHYLBUTYLACRYLATE
mf: $C_9H_{16}O_2$ mw: 156.25

PROP: Clear, colorless liquid. Bp: 82° @ 10 mm, fp: −70°, flash p: 125°F (OC), d: 0.8964 @ 20°/20°, vap press: 1.7 mm @ 20°.

SYNS: 2-ETHYLBUTYL ESTER, ACRYLIC ACID * 2-PROPENOIC ACID-2-ETHYLBUTYL ESTER

CONSENSUS REPORTS: Reported in EPA TSCA Inventory.

SAFETY PROFILE: Mildly toxic by ingestion and skin contact. An eye and severe skin irritant. Flammable when exposed to heat or flame; can react with oxidizing materials. To fight fire, use foam, CO_2, dry chemical. When heated to decomposition it emits acrid smoke and irritating fumes.

EHA000   CAS: 617-79-8   **HR: 3**
2-ETHYLBUTYLAMINE
mf: $C_6H_{15}N$ mw: 101.22

PROP: Water-white liquid. Bp: 110-113°, flash p: 64°F (OC), d: 0.739 @ 20°/20°, vap d: 3.5.

SYN: 2-ETHYL-1-BUTANAMINE

SAFETY PROFILE: Poison by ingestion. Moderately toxic by skin contact. A skin and severe eye irritant. A very dangerous fire hazard when exposed to heat or flame; can react vigorously with oxidizing materials. Keep away from heat and open flame. To fight fire, use dry chemical, CO_2, foam. When heated to decomposition it emits toxic fumes of NO_x.

EHA500   CAS: 628-81-9   **HR: 2**
ETHYL BUTYL ETHER
DOT: UN 1179
mf: $C_6H_{14}O$ mw: 102.20

PROP: Colorless liquid. Bp: 92°, mp: −124°, flash p: 40°F, d: 0.7528 @ 20°/20°, vap d: 3.52. Insol in water; misc in alc and ether.

SYN: ETHER ETHYLBUTYLIQUE (FRENCH)

CONSENSUS REPORTS: Reported in EPA TSCA Inventory.

DOT Classification: Flammable Liquid; Label: Flammable Liquid.

SAFETY PROFILE: Moderately toxic by ingestion. A skin and eye irritant. A very dangerous fire hazard when exposed to heat or flame; can react vigorously with oxidizing materials. Keep away from heat and open flame. To fight fire, use alcohol foam, CO_2, dry chemical. When heated to decomposition it emits acrid smoke and irritating fumes.

EHA600   CAS: 106-35-4   **HR: 2**
ETHYL BUTYL KETONE
mf: $C_7H_{14}O$ mw: 114.21

PROP: Clear mobile liquid; fatty odor. Mp: −36.7°, bp: 148°, flash p: 115°F (OC), d: 0.8198 @ 20°/20°, vap d: 3.93. Misc with alc, ether, water @ 149°.

SYNS: AETHYLBUTYLKETON (GERMAN) * n-BU-TYL ETHYL KETONE * EPTAN-3-ONE (ITALIAN)

* ETHYLBUTYLCETONE (FRENCH) * ETHYLBU-TYLKETON (DUTCH) * ETILBUTILCHETONE (ITALIAN) * FEMA No. 2545 * HEPTAN-3-ON (DUTCH, GERMAN) * 3-HEPTANONE * HEPTAN-3-ONE

CONSENSUS REPORTS: Reported in EPA TSCA Inventory.

OSHA PEL: TWA 50 ppm
ACGIH TLV: TWA 50 ppm

SAFETY PROFILE: Moderately toxic by ingestion and inhalation. A skin and eye irritant. Combustible liquid. Can react with oxidizing materials. To fight fire, use foam, CO_2, dry chemical.

EHC000   CAS: 4549-44-4   **HR: 3**
ETHYL-N-BUTYLNITROSAMINE
mf: $C_6H_{14}N_2O$ mw: 130.22

SYNS: AETHYL-N-BUTYL-NITROSOAMIN (GERMAN) * N-ETHYL-N-NITROSOBUTYLAMINE * N-NITRO-SO-N-BUTYLETHYLAMINE * N-NITROSOETHYL-N-BUTYLAMINE

SAFETY PROFILE: Poison by ingestion and intravenous routes. Questionable carcinogen with experimental carcinogenic and tumorigenic data. Mutation data reported. When heated to decomposition it emits toxic fumes of NO_x.

EHE000   CAS: 105-54-4   **HR: 1**
ETHYL n-BUTYRATE
DOT: UN 1180
mf: $C_6H_{12}O_2$ mw: 116.18

PROP: Colorless liquid; banana-pineapple odor. D: 0.874, refr index: 1.391, mp: −100.8°, bp: 121.6°, flash p: 79°F. Sol in water, fixed oils, propylene glycol; misc in alc and ether; insol in glycerin @ 121°.

SYNS: BUTANOIC ACID ETHYL ESTER * BUTYRIC ETHER * ETHYL BUTANOATE * ETHYL BUTY-RATE (DOT,FCC) * FEMA No. 2427

CONSENSUS REPORTS: Reported in EPA TSCA Inventory.

DOT Classification: Flammable or Combustible Liquid; Label: Flammable Liquid.

SAFETY PROFILE: Mildly toxic by ingestion. A skin irritant. Flammable liquid when exposed to heat or flame; can react vigorously with oxidizing materials. When heated to decomposition it emits acrid smoke and irritating fumes.

EHE500 CAS: 110-38-3 **HR: 1**
ETHYL CAPRATE
mf: $C_{12}H_{24}O_2$ mw: 200.36

PROP: Colorless liquid; oily, brandy odor. Bp: 243°, d: 0.863, refr index: 1.424, vap d: 6.9, flash p: +212°F. Sol in fixed oils; insol in glycerin, propylene glycol @ 243°.

SYNS: CAPRIC ACID ETHYL ESTER * DECANOIC ACID, ETHYL ESTER * ETHYL CAPRINATE * ETHYL DECANOATE (FCC) * ETHYL DECYLATE * FEMA No. 2432

CONSENSUS REPORTS: Reported in EPA TSCA Inventory.

SAFETY PROFILE: A skin irritant. Combustible liquid when exposed to heat or flame; can react with oxidizing materials. When heated to decomposition it emits acrid smoke and irritating fumes.

EHF000 CAS: 123-66-0 **HR: 1**
ETHYL CAPROATE
DOT: UN 1177
mf: $C_8H_{16}O_2$ mw: 144.24

PROP: Colorless liquid; mild wine odor. Bp: 163°, flash p: 130°F (OC), d: 0.867-0.871, refr index: 1.406-1.409, vap d: 5.0. Sol in fixed oils; sltly sol in propylene glycol; insol in glycerin @ 166.

SYNS: ETHYL BUTYLACETATE (DOT) * ETHYL HEXANOATE (FCC) * FEMA No. 2439

CONSENSUS REPORTS: Reported in EPA TSCA Inventory.

DOT Classification: Flammable or Combustible Liquid; Label: Flammable Liquid.

SAFETY PROFILE: A skin irritant. Flammable or combustible when exposed to heat or flame; can react with oxidizing materials. When heated to decomposition it emits acrid smoke and irritating fumes. To fight fire, use CO_2, foam, dry chemical.

EHF500 CAS: 63833-90-9 **HR: 3**
2-(N-ETHYL CARBAMOYL HYDROXYMETHYL)FURAN
mf: $C_8H_{11}NO_3$ mw: 169.20

SAFETY PROFILE: Poison by intraperitoneal route. Mutagenic data reported. When heated to decomposition it emits toxic fumes of NO_x.

EHG500 CAS: 105-39-5 **HR: 3**
ETHYL CHLORACETATE
DOT: UN 1181
mf: $C_4H_7ClO_2$ mw: 122.56

PROP: Colorless liquid; fruity, pungent odor. Bp: 143.6°, fp: -26.6° flash p: 100°F, d: 1.159 @ 20°/4°, vap press: 10 mm @ 37.5° vap d: 4.3. Insol in water; misc in alc and ether.

SYNS: CHLOROACETIC ACID, ETHYL ESTER * ETHYL CHLOROACETATE * ETHYL-α-CHLOROACETATE * ETHYL CHLOROETHANOATE * ETHYL MONOCHLORACETATE * ETHYL MONOCHLOROACETATE

CONSENSUS REPORTS: Reported in EPA TSCA Inventory.

DOT Classification: Combustible Liquid; Label: None; Poison B; Label: Flammable Liquid and Poison.

SAFETY PROFILE: Poison by skin contact and subcutaneous routes. A severe eye irritant. Questionable carcinogen with experimental neoplastigenic data. A dangerous fire hazard when exposed to heat or flame; can react vigorously with oxidizing materials. Will react with water or steam to produce toxic and corrosive fumes. Vigorous reaction with sodium cyanide. To fight fire, use water, foam, CO_2, dry chemical. When heated to decomposition it emits highly toxic fumes of Cl^-.

EHH000 CAS: 75-00-3 **HR: 1**
ETHYL CHLORIDE
DOT: UN 1037
mf: C_2H_5Cl mw: 64.52

PROP: Colorless liquid or gas; ether-like odor, burning taste. Sol in water at 0. Bp: 12.3°, lel: 3.8%, uel: 15.4%, fp: -139°, flash p: -58°F (CC), d: 0.9214 @ 0°/4°, autoign temp: 966°F, vap press: 1000 mm @ 20° vap d: 2.22.45; misc in alc and ether.

SYNS: AETHYLCHLORID (GERMAN) * AETHYLIS * AETHYLIS CHLORIDUM * ANODYNON * CHELEN * CHLOORETHAAN (DUTCH) * CHLORETHYL * CHLORIDUM * CHLOROAETHAN (GERMAN) * CHLOROETHANE * CHLORURE D'ETHYLE (FRENCH) * CHLORYL * CHLORYL ANESTHETIC * CLOROETANO (ITALIAN) * CLORURO DI ETILE (ITALIAN) * ETHER CHLORATUS * ETHER HYDRO-

CHLORIC * ETHER MURIATIC * ETYLU CHLOREK (POLISH) * HYDROCHLORIC ETHER * KELENE * MONOCHLORETHANE * MURIATIC ETHER * NARCOTILE * NCI-C06224

CONSENSUS REPORTS: Reported in EPA TSCA Inventory. Community Right-To-Know List.

OSHA PEL: TWA 1000 ppm
ACGIH TLV: TWA 1000 ppm
DFG MAK: 1000 ppm (2600 mg/m^3)
DOT Classification: Flammable Liquid; Label: Flammable Liquid; Flammable Gas; Label; Flammable Gas.

SAFETY PROFILE: Mildly toxic by inhalation. An irritant to skin, eyes, and mucous membranes. The liquid is harmful to the eyes and can cause some irritation. In the case of guinea pigs, the symptoms attending exposure are similar to those caused by methyl chloride, except that the signs of lung irritation are not as pronounced. It gives some warning of its presence because it is irritating, but it is possible to tolerate exposure to it until one becomes unconscious. It is the least toxic of all the chlorinated hydrocarbons. It can cause narcosis, although the effects are usually transient.

A very dangerous fire hazard when exposed to heat or flame; can react vigorously with oxidizing materials. Severe explosion hazard when exposed to flame. Reacts with water or steam to produce toxic and corrosive fumes. Incompatible with potassium. To fight fire, use carbon dioxide. When heated to decomposition it emits toxic fumes of phosgene and Cl$^-$.

EHH500   CAS: 1331-31-3   **HR: 1**
ETHYL CHLORO BENZENE
mf: C$_8$H$_9$Cl   mw: 140.62

PROP: Clear, colorless liquid. Mp: −62.6°, bp: 184.3°, flash p: 147°F, d: 1.05 @ 25°/25°, vap press: 1 mm @ 19.2°, vap d: 4.86.

SYN: CHLOROETHYLBENZENE

SAFETY PROFILE: Mildly toxic by ingestion and skin contact. A skin and eye irritant. Flammable when exposed to heat or flame; can react vigorously with oxidizing materials. To fight fire, use foam, CO$_2$, dry chemical. When heated to decomposition it emits acrid smoke and irritating fumes.

EHJ500   CAS: 38915-14-9   **HR: 3**
9-(3-ETHYL-2-CHLOROETHYL)AMINO-PROPYLAMINO)-4-METHOXYACRIDINE DIHYDROCHLORIDE
mf: C$_{21}$H$_{26}$ClN$_3$O • 2ClH   mw: 444.87

SYNS: ICR 377 * 4-METHOXY-9-(3-(ETHYL-2-CHLOROETHYL)

SAFETY PROFILE: Poison by intravenous route. Questionable carcinogen with experimental neoplastigenic data. Mutation data reported. When heated to decomposition it emits very toxic fumes of Cl$^-$ and NO$_x$.

EHK500   CAS: 541-41-3   **HR: 3**
ETHYL CHLOROFORMATE
DOT: UN 1182
mf: C$_3$H$_5$ClO$_2$ mw: 108.53

PROP: Colorless liquid. Mp: −80.6°, bp: 94°, flash p: 35.6°F, d: 1.138 @ 20°/4°, vap d: 3.74, autoign temp: 932°F. Decomp in water; misc in alc, benzene, ether, and chloroform.

SYNS: CHLOROCARBONATE d'ETHYLE (FRENCH) * CHLOROFORMIC ACID ETHYL ESTER * CLHORA-MEISENSAEUREAETHYLESTER (GERMAN) * ECF * ETHYLCHLOORFORMIAAT (DUTCH) * ETHYL CHLOROCARBONATE (DOT) * ETHYLE, CHLOROFOR-MIAT D' (FRENCH) * ETIL CLOROCARBONATO (ITALIAN) * ETIL CLOROFORMIATO (ITALIAN) * TL 423

CONSENSUS REPORTS: Reported in EPA TSCA Inventory. Community Right-To-Know List.

DOT Classification: Flammable Liquid; Label: Flammable Liquid and Poison; Flammable Liquid; Label: Flammable Liquid, Poison, Corrosive.

SAFETY PROFILE: Poison by ingestion, inhalation, and intraperitoneal routes. Moderately toxic by skin contact. Corrosive. An eye, skin and mucous membrane irritant. A very dangerous fire hazard when exposed to heat or flame; can react vigorously with oxidizing materials. Reacts with water or steam to produce toxic and corrosive fumes. To fight fire, use CO$_2$, dry chemical. When heated to decomposition it emits highly toxic fumes of Cl$^-$.

EHN000   CAS: 103-36-6   **HR: 2**
ETHYL-trans-CINNAMATE
mf: C$_{11}$H$_{12}$O$_2$ mw: 176.23

PROP: Nearly colorless, oily liquid; faint cinnamon odor. D: 1.049 @ 20°/4°, refr index: 1.558-

1.561, bp: 271°, mp: 9°, flash p: +212°F. Misc in alc, ether, fixed oils; insol in glycerin, water @ 272°.

SYNS: ETHYL CINNAMATE (FCC) * ETHYL-β-PHE-NYLACRYLATE * ETHYL-3-PHENYLPROPENOATE * FEMA No. 2430

CONSENSUS REPORTS: Reported in EPA TSCA Inventory.

SAFETY PROFILE: Moderately toxic by ingestion. Combustible liquid. When heated to decomposition it emits acrid smoke and irritating fumes.

EHP000   CAS: 95-04-5   HR: 2
cis-(2-ETHYLCROTONYL) UREA
mf: $C_7H_{12}N_2O_2$ mw: 156.21

SYNS: ACTINE * (Z)-N-(AMINOCARBONYL)-2-ETHYL-2-BUTENAMIDE * ASTYN * CRONIL * CROTURAL * DISTESOL * DISTESSOL * ECTIDA * ECTILUREA * ECTON * ECTYDA * ECTYLCARBAMIDE * ECTYLUREA * ECTYN * EKTYLCARBAMID * (α-ETHYL-cis-CROTONYL) CARBAMIDE * 2-ETHYL-cis-CROTONYLUREA * 2-ETHYLCROTONYLUREA * EUPLACID * LEVANIL * LEVIL * MA-110 * NASTYN * NEOCROSEDIN * NESTYN * NEUROPROCIN * NOSTAL * NOSTIN * PACETYN * SEDAREX * TRANZER * U 8771

SAFETY PROFILE: Moderately toxic by ingestion and intraperitoneal routes. A sedative. When heated to decomposition it emits toxic fumes of NO_x.

EHP500   CAS: 105-56-6   HR: 2
ETHYL CYANOACETATE
DOT: UN 2666
mf: $C_5H_7NO_2$ mw: 113.13

PROP: Colorless to pale straw-colored liquid. Mp: −22.5°, bp: 206°, flash p: 230°F, d: 1.06 @ 25°/25°, vap press: 1 mm @ 67.8°, vap d: 3.9.

SYNS: CYANACETATE ETHYLE (GERMAN) * CYANOACETIC ACID ETHYL ESTER * CYANO-ACETIC ESTER * ETHYL CYANOETHANOATE * MALONIC ACID ETHYL ESTER NITRILE * USAF KF-25

CONSENSUS REPORTS: Reported in EPA TSCA Inventory. Cyanide and its compounds are on the Community Right-To-Know List.

DOT Classification: Poison B; Label: St. Andrews Cross.

SAFETY PROFILE: Moderately toxic by intraperitoneal and subcutaneous routes. Combustible when exposed to heat or flame; can react with oxidizing materials. Will react with water or steam to produce toxic and flammable vapors. To fight fire, use CO_2, dry chemical. When heated to decomposition or on contact with acid or acid fumes it emits highly toxic fumes of CN^-.

EHT000   CAS: 5459-93-8   HR: 2
N-ETHYL(CYCLOHEXYL)AMINE
mf: $C_8H_{17}N$ mw: 127.26

PROP: Sltly sol in water; flash p: 86°F (OC), d: 0.8, vap d: 4.4.

SYN: N-ETHYL-CYCLOHEXYLAMINE

CONSENSUS REPORTS: Reported in EPA TSCA Inventory.

SAFETY PROFILE: Moderately toxic by ingestion, inhalation, and skin contact. A skin irritant. A very dangerous fire hazard when exposed to heat or flame; can react vigorously with oxidizing materials. To fight fire, use alcohol foam, mist, spray, dry chemical.

EHV000   CAS: 26747-87-5   HR: 3
ETHYL DECABORANE
mf: $C_2H_{18}B_{10}$ mw: 150.30

SAFETY PROFILE: Poison by inhalation.

EHV500   CAS: 302-49-8   HR: 3
ETHYL(DI-(1-AZIRIDINYL)PHOSPHINYL) CARBAMATE
mf: $C_7H_{14}N_3O_3P$ mw: 219.21

SYNS: (BIS(1-AZIRIDINYL)PHOSPHINYL)CARBAMIC ACID, ETHYL ESTER * BIS(ETHYLENIMIDO)PHOS-PHORYLURETHAN * ETHYL (BIS(1-AZIRIDINYL)PHOS-PHINYL)CARBAMATE * NSC 37095

SAFETY PROFILE: Poison by intraperitoneal route. When heated to decomposition it emits very toxic fumes of PO_x and NO_x.

EID000   CAS: 389-08-2   HR: 3
1-ETHYL-1,4-DIHYDRO-7-METHYL-4-OXO-1,8-NAPHTHYRIDINE-3-CARBOXYLIC ACID
mf: $C_{12}H_{12}N_2O_3$ mw: 232.26

SYNS: ACIDE 1-ETIL-7-METIL-1,8-NAFTIRIDIN-4-ONE-3-CARBOSSILICO (ITALIAN) * ACIDE NALIDIXICO (ITALIAN) * ACIDE NALIDIXIQUE (FRENCH) * BETAXINA * 3-CARBOXY-1-ETHYL-7-METHYL-1,8-NAPHTHIDIN-4-ONE * CHINOIN * CYBIS * 1,4-DIHYDRO-1-ETHYL-7-METHYL-4-OXO-1,8-NAPH-THYRIDINE-3-CARBOXYLIC ACID * DIXIBEN * 1-ETHYL-7-METHYL-1,4-DIHYDRO-1,8-NAPHTHYRI-DIN-4-ONE-3-CARBOXYLIC ACID * 1-ETHYL-7-METHYL-1,4-DIHYDRO-1,8-NAPHTHYRIDINE-4-ONE-3-CARBOXYLIC ACID * 1-ETHYL-7-METHYL-4-OXO-1,4-DIHYDRO-1,8-NAPHTHYRIDINE-3-CARBOXYLIC ACID * EUCISTEN * INNOXALON * KUSNARIN * NA * NALIDIC ACID * NALIDICRON * NALIDIXIC ACID * NALIDIXIN * NALITUCSAN * NARIGIX * NCI-C56199 * NEGRAM * NEVI-GRAMON * NICELATE * NOGRAM * NSC-82174 * POLEON * SPECIFEN * URALGIN * URIBEN * URODIXIN * UROMAN * URONEG * WIN 18,320 * WINTOMYLON

CONSENSUS REPORTS: EPA Genetic Toxicology Program.

SAFETY PROFILE: Poison by intravenous and intraperitoneal routes. Moderately toxic by ingestion and subcutaneous routes. Human systemic effects by ingestion and possibly other routes: convulsions, hyperglycemia, sweating, and blood changes in children. Experimental teratogenic and reproductive effects. Human mutation data reported. Used as an antibacterial agent and urinary tract antiseptic. When heated to decomposition it emits toxic fumes of NO_x.

EIF000   CAS: 77-81-6   **HR: 3**
ETHYL DIMETHYLAMIDOCYANO-PHOSPHATE
mf: $C_5H_{11}N_2O_2P$ mw: 162.15

PROP: A colorless to brownish liquid. Bp: decomp @ 238°, fp: −49.4°, flash p: 172°F, d: 1.073 @ 25°, vap press: 0.07 mm @ 25°, vap d: 5.63.

SYNS: DIMETHYLAMIDOETHOXYPHOSPHORYL CYAN-IDE * DIMETHYLAMINOCYANPHOSPHORSAEUREA-ETHYLESTER (GERMAN) * DIMETHYLPHOSPHOR-AMIDOCYANIDIC ACID, ETHYL ESTER * ETHYL N,N-DIMETHYLAMINO CYANOPHOSPHATE * ETHYL DIMETHYLPHOSPHORAMIDOCYANIDATE * ETHYL-N,N-DIMETHYLPHOSPHORAMIDOCYANIDATE * GA * Le-100 * MCE * T-2104 * TABUN * TL 1578

CONSENSUS REPORTS: EPA Extremely Hazardous Substances List. Cyanide and its compounds are on the Community Right-To-Know List. Reported in EPA TSCA Inventory.

SAFETY PROFILE: Human poison by inhalation, skin contact, and intravenous routes. Experimental poison by ingestion, inhalation, skin contact, subcutaneous, intravenous, intraperitoneal and intramuscular routes. A nerve gas. Vapor does not penetrate skin; liquid does so rapidly. The primary physiological action is on the sympathetic nervous system, causing a vasoparesis (partial paralysis of the vasomotor nerves which control the diameter of the blood vessels). Vapors when inhaled can cause nausea, vomiting and diarrhea, which can be followed by muscular twitching and convulsions. Flammable when exposed to heat or flame; can react with oxidizing materials. When heated to decomposition it emits very toxic fumes of PO_x, CN^- and NO_x.

EIF500   CAS: 20820-80-8   **HR: 3**
O-ETHYL-S-(2-DIMETHYL AMINO ETHYL)-METHYLPHOSPHONOTHIOATE
mf: $C_7H_{18}NO_2PS$ mw: 211.29

SYN: O-AETHYL-S-(2-DIMETHYLAMINOAETHYL)-METHYLPHOSPHONOTHIOATE (GERMAN)

SAFETY PROFILE: Poison by ingestion, intraperitoneal, intravenous, and intramuscular routes. When heated to decomposition it emits very toxic fumes of NO_x, PO_x, and SO_x.

EIJ000   CAS: 64037-50-9   **HR: 3**
N-ETHYL-N-1-DIMETHYL-3,3-DI-2-THIENYL-2-PROPENAMINE HYDROCHLORIDE
mf: $C_{15}H_{19}NS_2 \cdot ClH$ mw: 313.93

SYNS: 1C50 HYDROCHLORIDE * N,1-DIMETHYL-3,3-DI-2-THIENYL-N-ETHYLALLYLAMINE HYDROCHLO-RIDE * EMETHIBUTIN HYDROCHLORIDE * N-ETHYL-N-1-DIMETHYL-3,3-DI-2-THIENYLALLYLAMINE HYDROCHLORIDE * ETHYLMETHIAMBUTENE HY-DROCHLORIDE * 3-ETHYLMETHYLAMINO-1,1-DI(2'-THIENYL)BUT-1-ENE HYDROCHLORIDE * ETHYL-METHYLTHIAMBUTENE HYDROCHLORIDE * NIH-5145 HYDROCHLORIDE

SAFETY PROFILE: Poison by ingestion, intravenous, and subcutaneous routes. Used as a narcotic analgesic. When heated to decomposition it emits very toxic fumes of NO_x, SO_x, and HCl.33144

EIK000 CAS: 78-78-4 HR: 3
ETHYLDIMETHYLMETHANE

DOT: UN 1265
mf: C_5H_{12} mw: 72.17

PROP: Colorless liquid with pleasant odor. Bp: 27.8°, fp: −160.5°, flash p: < −60°F (CC), vap press: 595 mm @ 21.1°, vap d: 2.48, lel: 1.4%, uel: 7.6%.

SYNS: ISOAMYLHYDRIDE * ISOPENTANE * 2-METHYLBUTANE

CONSENSUS REPORTS: Reported in EPA TSCA Inventory.

NIOSH REL: (Alkanes) TWA 350 mg/m^3
DOT Classification: Flammable Liquid; Label: Flammable Liquid.

SAFETY PROFILE: Probably mildly toxic and narcotic by inhalation. Flammable Liquid. A very dangerous fire and explosion hazard when exposed to heat, flame, or oxidizers. Keep away from sparks, heat, or open flame; can react with oxidizing materials. To fight fire, use foam, CO_2, dry chemical. When heated to decomposition it emits acrid smoke and irritating fumes.

EIL100 HR: 2
2-ETHYL-3,5(6)-DIMETHYLPYRAZINE
mf: $C_8H_{12}N_2$ mw: 136.20

PROP: Colorless to sltly yellow liquid; roasted cocoa odor. D: 0.950-0.980, refr index: 1.500, flash p: 154°F. Sol in water, organic solvents.

SYN: FEMA No. 3149

SAFETY PROFILE: Combustible liquid. When heated to decomposition emits toxic fumes of NO_x.

EIM000 CAS: 17109-49-8 HR: 3
O-ETHYL-S,S-DIPHENYL DITHIOPHOSPHATE
mf: $C_{14}H_{15}O_2PS_2$ mw: 310.38

PROP: A clear yellow to light brown liquid. D: 1.23.

SYNS: O-AETHYL-S,S-DIPHENYL-DITHIOPHOSPHAT (GERMAN) * BAYER 78418 * DITHIOPHOSPHOR-SAEURE-O-AETHYL-S,S-DIPHENYLESTER (GERMAN) * EDDP * EDIFENPHOS * EDIPHENPHOS * O-ETHYL-S,S-DIPHENYL PHOSPHORODITHIOATE * HINOSAN * LUTROL * SRA 7847

SAFETY PROFILE: Poison by ingestion, intra-peritoneal, and possibly other routes. Mutation data reported. A cholinesterase inhibitor. When heated to decomposition it emits very toxic fumes of SO_x and PO_x.

EIN000 CAS: 13194-48-4 HR: 3
O-ETHYL-S,S-DIPROPYLPHOSPHORO-DITHIOATE
mf: $C_8H_{19}O_2PS_2$ mw: 242.36

SYNS: ENT 27,318 * ETHOPROP * ETHOPRO-PHOS * O-ETHYL-S,S-DIPROPYL ESTER, PHOSPHORO-DITHIOIC ACID * JOLT * MOBIL V-C 9-104 * MOCAP * PROPHOS * V-C CHEMICAL V-C 9-104 * VIRGINIA CAROLINA VC 9-104

CONSENSUS REPORTS: EPA Extremely Hazardous Substances List.

SAFETY PROFILE: Poison by ingestion and skin contact. A cholinesterase inhibitor type of insecticide. When heated to decomposition it emits very toxic fumes of PO_x and SO_x.

EIN500 CAS: 759-94-4 HR: 3
S-ETHYL-N,N-DI-N-PROPYLTHIO-CARBAMATE
mf: $C_9H_{19}NOS$ mw: 189.35

SYNS: S-AETHYL-N,N-DIPROPYLTHIOLCARBAMAT (GERMAN) * DIPROPYLCARBAMOTHIOIC ACID-S-ETHYL ESTER * N,N-DIPROPYLTHIOCARBAMIC ACID-S-ETHYL ESTER * EPTAM * EPTC * S-ETHYL-N,N-DIPROPYLTHIOLCARBAMATE * ETHYL DI-N-PRO-PYLTHIOLCARBAMATE * ETHYL-N,N-DIPROPYL-THIOLCARBAMATE * ETHYL-N,N-DI-N-PROPYLTHIOL-CARBAMATE * FDA 1541 * GENEP EPTC * R-1608

CONSENSUS REPORTS: Reported in EPA TSCA Inventory. EPA Genetic Toxicology Program.

SAFETY PROFILE: Poison by ingestion, inha-lation, and intravenous routes. Moderately toxic by skin contact and possibly other routes. An herbicide. When heated to decomposition it emits very toxic fumes of NO_x and SO_x.

EIO000 CAS: 74-85-1 HR: 3
ETHYLENE

DOT: UN 1038/UN 1962
mf: C_2H_4 mw: 28.06

PROP: Soluble in water, alcohol, ether. D: 0.975, colorless gas, sweet odor and taste. Bp: −103.9°, mp: −169.4°, lel: 2.7%, uel: 36%, d: 0.610 @ 0°, autoign temp: 914°F, vap d: 0.98, fp: −181°.

SYNS: ACETENE * ATHYLEN (GERMAN) * BICARBURRETTED HYDROGEN * ELAYL * ETHENE * ETHYLENE, compressed (DOT) * ETHYLENE, refrigerated liquid (DOT) * LIQUID ETHYENE * OLEFIANT GAS

CONSENSUS REPORTS: Reported in EPA TSCA Inventory. Community Right-To-Know List.

DOT Classification: Flammable Gas; Label: Flammable Gas.

SAFETY PROFILE: A simple asphyxiant. High concentrations cause anesthesia. A common air contaminant. It is phytotoxic. A very dangerous fire hazard when exposed to heat or flame. Moderate explosion hazard when exposed to flame. A flammable gas. To fight fire, stop flow of gas, use CO_2, dry chemical or fine water spray. Mixtures with aluminum chloride explode in the presence of nickel catalysts; methyl chloride; or nitromethane. Explosive reaction with bromotrichloromethane (at 120°C/51 bar); carbon tetrachloride (from 25-100°C/30 bar). Explosive reaction with chlorine catalyzed by sunlight or UV light or in the presence of mercury(I) oxide; mercury(II) oxide; or silver oxide. Mixtures with chlorotrifluoroethylene polymerize explosively when exposed to 50 Kv gamma rays at 308 Krad/hr. Has been involved in industrial accidents. Violent polymerization is catalyzed by copper above 400°C/54 bar. Incompatible with $AlCl_3$; (CCl_4 + benzoyl peroxide); (bromtrichloromethane + $AlCl_3$); O_3; CCl_4; Cl_2; NO_x; tetrafluoroethylene; trifluorohypofluorite. When heated to decomposition it emits acrid smoke and irritating fumes.

EIP000   CAS: 2274-11-5   **HR: 3**
ETHYLENE ACRYLATE
mf: $C_8H_{10}O_4$ mw: 170.18

SYNS: ACRYLIC ACID, ETHYLENE ESTER * ACRYLIC ACID, ETHYLENE GLYCOL DIESTER * ETHYLDIOL ACRILATE (RUSSIAN) * ETHYLENE DIACRYLATE * ETHYLENE GLYCOL DIACRYLATE * 2-PROPENOIC ACID-1,2-ETHANEDIYL ESTER

CONSENSUS REPORTS: Reported in EPA TSCA Inventory.

SAFETY PROFILE: Poison by ingestion and inhalation. Moderately toxic by skin contact.

When heated to decomposition it emits acrid smoke and irritating fumes.

EIQ500   **HR: D**
ETHYLENEBIS(DITHIOCARBAMATO) MANGANESE and ZINC ACETATE (50:1)

SYNS: MANEB PLUS ZINC ACETATE (50:1) * ZINC ACETATE PLUS MANEB (1:50)

CONSENSUS REPORTS: Manganese and its compounds, as well as Zinc and its compounds, are on the Community Right-To-Know List.

SAFETY PROFILE: Experimental reproductive effects. When heated to decomposition it emits very toxic fumes of NO_x, ZnO and SO_x.

EIR000   CAS: 12122-67-7   **HR: 3**
ETHYLENE BIS(DITHIOCARBAMATO) ZINC
mf: $C_4H_6N_2S_4 \cdot Zn$ mw: 275.73

PROP: Light-colored powder, insol in water.

SYNS: ASPOR * ASPORUM * BERCEMA * BLIGHTOX * BLITEX * BLIZENE * CARBADINE * CHEM ZINEB * CINEB * CRITTOX * CYNKOTOX * DAISEN * DIPHER * DITHANE Z * DITIAMINA * ENT 14,874 * ((1,2-ETHANEDIYLBIS(CARBAMODITHIOATO))(2-)ZINC * 1,2-ETHANEDIYLBIS(CARBAMODITHIOATO) (2-)-S,S'-ZINC * 1,2-ETHANEDIYLBISCARBAMODITHIOIC ACID, ZINC COMPLEX * 1,2-ETHANEDIYLBISCARBAMODITHIOIC ACID, ZINC SALT * ETHYLENEBIS(DITHIOCARBAMIC ACID), ZINC SALT * ETHYL ZIMATE * HEXATHANE * KUPRATSIN * KYPZIN * LIROTAN * LONACOL * MICIDE * MILTOX * MILTOX SPECIAL * NOVIZIR * NOVOSIR N * PAMOSOL 2 FORTE * PARZATE * PEROSIN * POLYRAM Z * SPERLOX-Z * THIODOW * TIEZENE * TRITOFTOROL * TSINEB (RUSSIAN) * Z-78 * ZEBENIDE * ZEBTOX * ZIDAN * ZIMATE * ZINC ETHYLENEBISDITHIOCARBAMATE * ZINC ETHYLENE-1,2-BISDITHIOCARBAMATE * ZINEB * ZINK-(N,N'-AETHYLEN-BIS(DITHIOCARBAMAT)) (GERMAN) * ZINOSAN

CONSENSUS REPORTS: IARC Cancer Review: GROUP 3 IMEMDT 7,56,87; Animal Inadequate Evidence IMEMDT 12,245,76. Community Right-To-Know List. EPA Genetic Toxicology Program.

SAFETY PROFILE: Moderately toxic by ingestion and possibly other routes. Experimental teratogenic and reproductive effects. Questionable carcinogen with experimental carcinogenic and tumorigenic data. Human mutation data reported. Used as a fungicide. When heated to decomposition it emits very toxic fumes of NO_x, ZnO, and SO_x.

EIS000   CAS: 62207-76-5   **HR: 3**
N,N'-ETHYLENE BIS(3-FLUOROSALI-CYLIDENEIMINATO)COBALT(II)
mf: $C_{16}H_{12}CoF_2N_2O_2$ mw: 361.23

SYNS: BIS(3-FLUOROSALICYLALDEHYDE)-ETHYLENE-DIIMINE-COBALT * FLUOMINE * FLUOMINE DUST

CONSENSUS REPORTS: Cobalt and its compounds are on the Community Right-To-Know List. EPA Extremely Hazardous Substances List. Reported in EPA TSCA Inventory.

NIOSH REL: (Cobalt) TWA 0.1 mg/m³

SAFETY PROFILE: Poison by ingestion and inhalation. A skin and eye irritant. When heated to decomposition it emits very toxic fumes of F^- and NO_x.

EIT000   CAS: 67-42-5   **HR: 3**
(ETHYLENEBIS(OXYETHYLENENI-TRILO))TETRAACETIC ACID
mf: $C_{14}H_{24}N_2O_{10}$ mw: 380.40

SYNS: ETHYLENEDIOXYBIS(ETHYLENEAMINO)TETRA-ACETIC ACID * ETHYLENE GLYCOL BIS(AMINO-ETHYL ETHER)TETRAACETATE * ETHYLENE GLYCOL BIS(β-AMINOETHYL ETHER)TETRAACETATE * ETHYLENE GLYCOL BIS(β-AMINOETHYL ETHER)-N,N'-TET-RAACETIC ACID * ETHYLENE GLYCOL BIS(2-AMINO-ETHYL ETHER)TETRAACETIC ACID * ETHYLENE GLYCOL BIS(2-AMINOETHYL ETHER)-N,N,N',N'-TETRAACETIC ACID * GLYCOL-ETHERDIAMINETET-RAACETIC ACID

CONSENSUS REPORTS: Glycol ether compounds are on the Community Right-To-Know List. Reported in EPA TSCA Inventory.

SAFETY PROFILE: Poison by intraperitoneal route. Moderately toxic by ingestion. When heated to decomposition it emits toxic fumes of NO_x.

EIU800   CAS: 107-07-3   **HR: 3**
ETHYLENE CHLOROHYDRIN

DOT: UN 1135
mf: C_2H_5ClO mw: 80.52

PROP: Colorless liquid; faint, ethereal odor. Mp: −69°, bp: 128.8°, flash p: 140°F (OC), d: 1.197 @ 20°/4°, autoign temp: 797°F, vap press: 10 mm @ 30.3°, vap d: 2.78, lel: 4.9%, uel: 15.9%.

SYNS: AETHYLENECHLORHYDRIN (GERMAN) * 2-CHLOORETHANOL (DUTCH) * 2-CHLORAETHA-NOL (GERMAN) * 2-CHLORETHANOL (GERMAN) * Δ-CHLOROETHANOL * 2-CHLOROETHANOL (MAK) * 2-CHLOROETHYL ALCOHOL * β-CHLOROETHYL ALCOHOL * CHLOROETHYLOWY ALKOHOL (POLISH) * 2-CLOROETANOLO (ITALIAN) * ETHYLEEN-CHLO-ORHYDRINE (DUTCH) * ETHYLENE GLYCOL, CHLO-ROHYDRIN * GLICOL MONOCLORIDRINA (ITALIAN) * GLYCOL CHLOROHYDRIN * GLYCOLMONOCHLO-ORHYDRINE (DUTCH) * GLYCOL MONOCHLOROHY-DRIN * GLYCOMONOCHLORHYDRIN * MONO-CHLORHYDRINE DU GLYCOL (FRENCH) * 2-MONO-CHLOROETHANOL * NCI-C50135

CONSENSUS REPORTS: NTP Carcinogenesis Studies (dermal); No Evidence: mouse, rat NTPTR* NTP-TR-275,85; Reported in EPA TSCA Inventory. EPA Genetic Toxicology Program. EPA Extremely Hazardous Substances List.

OSHA PEL: (Transitional: TWA 5 ppm (skin))
  CL 1 ppm (skin)
ACGIH TLV: CL 1 ppm (skin)
DFG MAK: 1 ppm (3 mg/m³)
DOT Classification: Poison B; Label: Flammable Liquid and Poison.

SAFETY PROFILE: A poison by ingestion, inhalation, skin contact, intraperitoneal, intravenous, subcutaneous, and possibly other routes. Moderately toxic to humans by inhalation. It can affect the nervous system, liver, spleen, and lungs. An experimental teratogen. Mutation data reported. A severe eye and mild skin irritant. Flammable when exposed to heat, flame, or oxidizers. To fight fire, use alcohol foam, CO_2, dry chemical. Violent reaction with chlorosulfonic acid, ethylene diamine, sodium hydroxide. Reacts with water or steam to produce toxic and corrosive fumes. Potentially violent reaction with oxidizing materials. When heated to decomposition it emits highly toxic fumes of Cl^- and phosgene.

EIV000   CAS: 64-02-8   **HR: 3**
N,N'-ETHYLENEDIAMINEDIACETIC ACID TETRASODIUM SALT
mf: $C_{10}H_{12}N_2O_8 \cdot 4Na$ mw: 380.20

SYNS: AQUAMOLLIN * CALSOL * CELON E * CELON H * CELON IS * CHEELOX BF * CHEELOX BR-33 * CHELON 100 * CHEMCOLOX 200 * COMPLEXONE * CONIGON BC * DISTOL 8 * EDATHANIL TETRASODIUM * EDETATE SODIUM * EDETIC ACID TETRASODIUM SALT * EDTA, SODIUM SALT * EDTA TETRASODIUM SALT * ENDRATE TETRASODIUM * N,N'-1,2-ETHANEDIYL-BIS(N-(CARBOXYMETHYL)GLYCINE TETRA-SODIUM SALT * ETHYLENEBIS(IMINODIACETIC ACID) TETRASODIUM SALT * ETHYLENEDIAMINE-TETRAACETIC ACID, TETRASODIUM SALT * HAMP-ENE 100 * HAMP-ENE 215 * HAMP-ENE 220 * HAMP-ENE Na4 * IRGALON * KALEX * KEPMPLEX 100 * KOMPLXON * METAQUEST C * NERVANAID B LIQUID * NERVANID B * NULLAPON B * NULLAPON BF-78 * NULLAPON BFC CONC * PERMA KLEER 50 CRYSTALS * PERMA KLEER TETRA CP * QUESTEX 4 * SEQUESTRENE 30A * SEQUESTRENE Na 4 * SEQUESTRENE ST * SODIUM EDETATE * SODIUM EDTA * SODIUM ETHYLENEDIAMINE-TETRAACETATE * SODIUM ETHYLENEDIAMINETET-RAACETIC ACID * SODIUM SALT of ETHYLENEDIA-MINETETRAACETIC ACID * SYNTES 12A * SYN-TRON B * TETRACEMIN * TETRASODIUM EDTA * TETRASODIUM ETHYLENEDIAMINETETRAACETATE * TETRASODIUM ETHYLENEDIAMINETETRACETATE * TETRASODIUM (ETHYLENEDINITRILO)TETRAACE-TATE * TETRASODIUM SALT of EDTA * TETRASO-DIUM SALT of ETHYLENEDIAMINETETRACETICACID * TETRINE * TRILON B * TST * TYCLAROSOL * VERSENE 100 * VERSENE POWDER * WARKE-ELATE PS-43

CONSENSUS REPORTS: Reported in EPA TSCA Inventory.

SAFETY PROFILE: Poison by intraperitoneal route. A skin and eye irritant. When heated to decomposition it emits toxic fumes of NO_x and Na_2O.

EIX000   CAS: 60-00-4   **HR: 3**
ETHYLENEDIAMINETETRAACETIC ACID

DOT: NA 9117
mf: $C_{10}H_{16}N_2O_8$ mw: 292.28

PROP: Colorless crystals. Mp: decomp @ 240°. Sltly water-sol, insol in common organic solvents.

SYNS: ACIDE ETHYLENEDIAMINETETRACETIQUE (FRENCH) * 3,6-BIS(CARBOXYMETHYL)-3,5-DIAZOOC-

TANEDIOIC ACID * CELON A * CELON ATH * CHEELOX BF ACID * CHEMCOLOX 340 * COMPLEXON II * EDATHAMIL * EDETIC ACID * EDTA (CHELATING AGENT) * EDTA ACID * ENDRATE * N,N'-1,2-ETHANEDIYLBIS(N-(CAR-BOXYMETHYL)GLYCINE * ETHYLENEDIAMINE-TETRAACETATE * ETHYLENEDIAMINE-N,N,N',N'-TETRAACETIC ACID * ETHYLENEDINITRILOTETRA-ACETIC ACID * HAMP-ENE ACID * HAVIDOTE * METAQUEST A * NERVANAID B ACID * NULLAPON BF ACID * PERMA KLEER 50 ACID * QUESTEX 4H * SEQ 100 * SEQUESTRENE AA * SEQUESTRIC ACID * SEQUESTROL * TETRINE ACID * TITRIPLEX * TRICON BW * TRILON BW * VERSENE ACID * VINKEIL 100 * WARKEELATE ACID

CONSENSUS REPORTS: Reported in EPA TSCA Inventory. EPA Genetic Toxicology Program.

DOT Classification: ORM-E; Label: None.

SAFETY PROFILE: Poison by intraperitoneal route. Experimental teratogenic and reproductive effects. Mutation data reported. A general-purpose chelating and complexing agent. When heated to decomposition it emits toxic fumes of NO_x.

EIX500   CAS: 139-33-3   **HR: 3**
ETHYLENEDIAMINETETRAACETIC ACID, DISODIUM SALT
mf: $C_{10}H_{14}N_2O_8 \cdot 2Na$ mw: 336.24

PROP: White crystalline powder. Sol in water.

SYNS: CHELADRATE * CHELAPLEX III * CHE-LATON III * COMPLEXON III * d'E.D.T.A. DISODI-QUE (FRENCH) * DISODIUM DIACID ETHYLENEDI-AMINETETRAACETATE * DISODIUM DIHYDROGEN ETHYLENEDIAMINETETRAACETATE * DISODIUM DIHYDROGEN(ETHYLENEDINITRILO)TETRAACETATE * DISODIUM EDATHAMIL * DISODIUM EDETATE * DISODIUM EDTA (FCC) * DISODIUM ETHYLENE-DIAMINETETRAACETATE * DISODIUM ETHYLENE-DIAMINETETRAACETIC ACID * DISODIUM (ETHYLEN-EDINITRILO)TETRAACETATE * DISODIUM (ETHYLENEDINITRILO)TETRAACETIC ACID * DISO-DIUM SALT of EDTA * DISODIUM SEQUESTRENE * DISODIUM TETRACEMATE * DISODIUM VERSEN-ATE * DISODIUM VERSENE * EDATHAMIL DISO-DIUM * EDETATE DISODIUM * EDTA, DISODIUM SALT * ENDRATE DISODIUM * N,N'-1,2-ETHANE-DIYLBIS(N-(CARBOXYMETHYL)GLYCINE) DISODIUM SALT * ETHYLENEBIS(IMINODIACETIC ACID) DISO-

DIUM SALT * ETHYLENEDIAMINETETRAACETATE DI-
SODIUM SALT * (ETHYLENEDINITRILO)-TETRA-
ACETIC ACID DISODIUM SALT * F 1 (complexon)
* KIRESUTO B * METAQUEST B * PERMA KLEER
50 CRYSTALS DISODIUM SALT * SELEKTON B 2
* SEQUESTRENE SODIUM 2 * SODIUM VERSENATE
* TETRACEMATE DISODIUM * TITRIPLEX III
* TRILON BD * TRIPLEX III * VERESENE DISO-
DIUM SALT * VERSENE SODIUM 2

CONSENSUS REPORTS: Reported in EPA
TSCA Inventory. EPA Genetic Toxicology Program.

SAFETY PROFILE: Poison by intravenous
route. Moderately toxic by ingestion. Experimental teratogenic and reproductive effects. Mutation data reported. The calcium disodium salt
of EDTA is used as a chelating agent in treating
lead poisoning. When heated to decomposition
it emits toxic fumes of NO_x and Na_2O.

EIY500   CAS: 106-93-4   HR: 3
1,2-ETHYLENE DIBROMIDE

DOT: UN 1605
mf: $C_2H_4Br_2$ mw: 187.88

PROP: Colorless, heavy liquid; sweet odor. Bp:
131.4°, fp: 9.3°, flash p: none, d: 2.172 @
25°/25°, 2.1707 @ 25°/4°, vap press: 17.4 mm
@ 30°, vap d: 6.48.

SYNS: AETHYLENBROMID (GERMAN) * BROMO-
FUME * BROMURO di ETILE (ITALIAN) * CELMIDE
* DBE * 1,2-DIBROMAETHAN (GERMAN)
* 1,2-DIBROMOETANO (ITALIAN) * 1,2-DIBROMO-
ETHANE (MAK) * α,β-DIBROMOETHANE * sym-DI-
BROMOETHANE * DIBROMURE d'ETHYLENE
(FRENCH) * 1,2-DIBROOMETHAAN (DUTCH)
* DOWFUME 40 * DOWFUME EDB * DOWFUME
W-8 * DWUBROMOETAN (POLISH) * EDB
* EDB-85 * E-D-BEE * ENT 15,349 * ETHYL-
ENE BROMIDE * FUMO-GAS * GLYCOL BROMIDE
* GLYCOL DIBROMIDE * ISCOBROME D
* KOPFUME * NCI-C00522 * NEPHIS * PEST-
MASTER * PESTMASTER EDB-85 * RCRA WASTE
NUMBER U067 * SOILBROM-40 * SOILBROM-85
* SOILFUME * UNIFUME

CONSENSUS REPORTS: IARC Cancer Review: GROUP 2A IMEMDT 7,204,87; Animal
Sufficient Evidence IMEMDT 15,195,77. NTP
Fourth Annual Report On Carcinogens, 1984.
NCI Carcinogenesis Bioassay (gavage); Clear
Evidence: mouse, rat NCITR* NCI-CG-TR-
86,78; NTP Carcinogenesis Bioassay (inhalation); Clear Evidence: mouse, rat NTPTR*
NTP-TR-210,82. EPA Genetic Toxicology Program. Community Right-To-Know List. Reported in EPA TSCA Inventory.

OSHA PEL: TWA 20 ppm; CL 30 ppm; Pk
50 ppm/5M/8H
ACGIH TLV: Suspected Human Carcinogen
DFG TRK: 0.1 ppm; Animal Carcinogen, Sus-
pected Human Carcinogen.
NIOSH REL: (EDB) 0.045 ppm; CL 1 mg/m^3/
15M
DOT Classification: ORM-A; Label: None;
Poison B; Label: Poison.

SAFETY PROFILE: Confirmed carcinogen with
experimental carcinogenic and neoplastigenic
data. Human poison by ingestion. Experimental
poison by ingestion, skin contact, intraperitoneal and possibly other routes. Moderately toxic
by inhalation and rectal routes. Experimental
teratogenic and reproductive effects. Human
mutation data reported. A severe skin and eye
irritant. An experimental eye irritant. Implicated
in worker sterility. When heated to decomposition it emits toxic fumes of Br^-.

EIY600   CAS: 107-06-2   HR: 3
ETHYLENE DICHLORIDE

DOT: UN 1184
mf: $C_2H_4Cl_2$ mw: 98.96

PROP: Colorless, clear liquid; pleasant odor,
sweet taste. Bp: 83.5°, ULC: 60-70, lel: 6.2%,
uel: 15.9%, fp: −35.7°, flash p: 56°F, d: 1.257
@ 20°/4°, autoign temp: 775°F, vap press: 100
mm @ 29.4°, vap d: 3.35, refr index: 1.445
@ 20°. Sol in alc, ether, acetone, carbon tetrachloride; sltly sol in water.

SYNS: AETHYLENCHLORID (GERMAN) * BICHLO-
RURE d'ETHYLENE (FRENCH) * BORER SOL
* BROCIDE * CHLORURE d'ETHYLENE (FRENCH)
* CLORURO di ETHENE (ITALIAN) * 1,2-DCE
* DESTRUXOL BORER-SOL * 1,2-DICHLOORETHAAN
(DUTCH) * 1,2-DICHLOR-AETHAN (GERMAN)
* DICHLOREMULSION * DI-CHLOR-MULSION
* DICHLORO-1,2-ETHANE (FRENCH) * 1,2-DICHLO-
ROETHANE * α,β-DICHLOROETHANE * sym-DI-
CHLOROETHANE * DICHLOROETHYLENE * 1,2-DI-
CLOROETANO (ITALIAN) * DUTCH LIQUID
* DUTCH OIL * EDC * ENT 1,656 * ETHANE
DICHLORIDE * ETHYLEENDICHLORIDE (DUTCH)
* ETHYLENE CHLORIDE * 1,2-ETHYLENE DICHLO-
RIDE * GLYCOL DICHLORIDE * NCI-C00511
* RCRA WASTE NUMBER U077

CONSENSUS REPORTS: IARC Cancer Review: GROUP 2B IMEMDT 7,56,87; Human Limited Evidence IMEMDT 20,429,79; Animal Sufficient Evidence IMEMDT 20,429,-79. NTP Fourth Annual Report On Carcinogens, 1984. NCI Carcinogenesis Bioassay (gavage); Clear Evidence: mouse-rat NCITR* NCI-CG-TR-55,78. EPA Genetic Toxicology Program. Reported in EPA TSCA Inventory.

OSHA PEL: (Transitional: TWA 50 ppm; CL 100 ppm; PK 200 ppm/5M)) TWA 1 ppm; STEL 2 ppm
ACGIH TLV: TWA 10 ppm
DFG MAK: 20 ppm (80 mg/m^3); Suspected Carcinogen.
NIOSH REL: TWA 1 ppm; CL 2 ppm/15M
DOT Classification: Flammable Liquid; Label: Flammable Liquid; IMO: Flammable Liquid; Label: Flammable Liquid, Poison.

SAFETY PROFILE: Confirmed carcinogen with experimental carcinogenic, neoplastigenic and tumorigenic data. An experimental transplacental carcinogen. A human poison by ingestion. Poison experimentally by intravenous and subcutaneous routes. Moderately toxic by inhalation, skin contact, and intraperitoneal routes. Human systemic effects by ingestion and inhalation: flaccid paralysis without anesthesia (usually neuromuscular blockade), somnolence, cough, jaundice, nausea or vomiting, hypermotility, diarrhea, ulceration or bleeding from the stomach, fatty liver degeneration, change in cardiac rate, cyanosis and coma. It may also cause dermatitis, edema of the lungs, toxic effects on the kidneys, and severe corneal effects. A strong narcotic. Experimental teratogenic and reproductive effects. A skin and severe eye irritant, and strong local irritant. Its smell and irritant effects warn of its presence at relatively safe concentrations. Human mutation data reported.

Flammable liquid. A dangerous fire hazard if exposed to heat, flame or oxidizers. Moderately explosive in the form of vapor when exposed to flame. Violent reaction with Al, N_2O_4, NH_3, dimethylaminopropylamine. Can react vigorously with oxidizing materials and emit vinyl chloride and HCl. To fight fire, use water, foam, CO_2, dry chemicals. When heated to decomposition it emits highly toxic fumes of Cl$^-$ and phosgene.

EJA000   CAS: 6943-65-3   **HR: 3**
ETHYLENE DIISOTHIOURONIUM DIBROMIDE
mf: $C_4H_{10}N_4S_2 \cdot 2BrH$ mw: 340.14

SYNS: 1,2-ETHANEDIYL ESTER CARBAMIMIDOTHIOIC ACID DIHYDROBROMIDE * 2,2'-ETHYLENE-BIS-(2-THIOPSEUDOUREA), DIHYDROBROMIDE * ETHYLENE DIISOTHIOUREA DIHYDROBROMIDE * 2,2-ETHYLENE-DITHIODIPSEUDOUREA DIHYDROBROMIDE

SAFETY PROFILE: Poison by intraperitoneal, intravenous, parenteral and possibly other routes. When heated to decomposition it emits very toxic fumes of NO_x, SO_x and HBr.

EJC500   CAS: 107-21-1   **HR: 3**
ETHYLENE GLYCOL
mf: $C_2H_6O_2$ mw: 62.08

PROP: Colorless, sweet-tasting, hygroscopic liquid. Bp: 197.5°, lel: 3.2%, fp: −13°, flash p: 232°F (CC), d: 1.113 @ 25°/25°, autoign temp: 752°F, vap d: 2.14, vap press: 0.05 mm @ 20°.

SYNS: ATHYLENGLYKOL (GERMAN) * 1,2-DIHY-DROXYETHANE * DOWTHERM SR 1 * 1,2-ETHANE-DIOL * ETHYLENE ALCOHOL * ETHYLENE DIHY-DRATE * GLYCOL * GLYCOL ALCOHOL * LUTROL-9 * MACROGOL 400 BPC * M.E.G. * MONOETHYLENE GLYCOL * NCI-C00920 * NORKOOL * TESCOL * UCAR 17

CONSENSUS REPORTS: EPA Genetic Toxicology Program. Community Right-To-Know List. Reported in EPA TSCA Inventory.

OSHA PEL: CL 50 ppm
ACGIH TLV: CL 50 ppm (vapor)

SAFETY PROFILE: Human poison by ingestion. (Lethal dose for humans reported to be 100 mL.) Moderately toxic to humans by an unspecified route. Moderately toxic experimentally by ingestion, subcutaneous, intravenous and intramuscular routes. Mildly toxic by skin contact. Human systemic effects by ingestion and inhalation: eye lacrimation, general anesthesia, headache, cough, respiratory stimulation, nausea or vomiting, pulmonary, kidney and liver changes. If ingested it causes initial central nervous system stimulation followed by depression. Later, it causes potentially lethal kidney damage. Very toxic in particulate form upon inhalation. An experimental teratogen. Human mutation data reported. A skin, eye and mucous membrane irritant.

Combustible when exposed to heat or flame; can react vigorously with oxidants. Moderate explosion hazard when exposed to flame. Ignites on contact with chromium trioxide, potassium permanganate, and sodium peroxide. Mixtures with ammonium dichromate, silver chlorate, sodium chlorite, and uranyl nitrate ignite when heated to 100°C. Can react violently with chlorosulfonic acid, oleum, H_2SO_4, $HClO_4$, and P_2S_5. Aqueous solutions may ignite silvered copper wires which have an applied D.C. voltage. To fight fire, use alcohol foam, water, foam, CO_2, dry chemical. When heated to decomposition it emits acrid smoke and irritating fumes.

EJD759   CAS: 111-55-7   HR: 2
ETHYLENE GLYCOL DIACETATE
mf: $C_6H_{10}O_4$ mw: 146.16

PROP: Colorless liquid or crystals. Mp: −31°, bp: 191°, flash p: 205°F (OC), d: 1.128 @ 0°/4°, vap press: 1 mm @ 38.3°, vap d: 5.04.

SYNS: 1,2-ETHANEDIOL DIACETATE * ETHYLENE ACETATE * ETHYLENE GLYCOL ACETATE * GLYCOL DIACETATE

CONSENSUS REPORTS: Reported in EPA TSCA Inventory.

SAFETY PROFILE: Moderately toxic by intraperitoneal route. Mildly toxic by ingestion and skin contact. An eye irritant. Combustible when exposed to heat or flame; can react with oxidizing materials. To fight fire, use alcohol foam, CO_2, dry chemical. When heated to decomposition it emits acrid smoke and irritating fumes.

EJE500   CAS: 629-14-1   HR: 2
ETHYLENE GLYCOL DIETHYL ETHER
DOT: UN 1153
mf: $C_6H_{14}O_2$ mw: 118.20

PROP: Colorless liquid, slight ethereal odor. Mp: −74°, bp: 121.4°, flash p: 95°F (OC), d: 0.8417 @ 20°/20°, autoign temp: 406°F, vap d: 6.56, vap press: 9.4 mm.

SYNS: 1,2-DIETHOXYETHANE * DIETHYL CELLOSOLVE (DOT) * ETHYL GLYME

CONSENSUS REPORTS: Reported in EPA TSCA Inventory. Glycol ether compounds are on the Community Right-To-Know List.

DOT Classification: Combustible Liquid; Label: None; Flammable or Combustible Liquid; Label: Flammable Liquid.

SAFETY PROFILE: Moderately toxic by ingestion. Mildly toxic by inhalation. Experimental reproductive effects. An eye irritant. An aprotic solvent. A very dangerous fire hazard when exposed to heat or flame; can react with oxidizing materials. To fight fire, use CO_2, dry chemical.

EJF000   CAS: 629-15-2   HR: 3
ETHYLENE GLYCOL DIFORMATE
mf: $C_4H_6O_4$ mw: 118.10

PROP: Liquid. Mp: −10°, bp: 177°, flash p: 200°F (OC), d: 1.2277 @ 20°/20°, vap d: 4.07.

SYNS: ETHYLENE FORMATE * GLYCOL DIFORMATE

CONSENSUS REPORTS: Reported in EPA TSCA Inventory.

SAFETY PROFILE: Poison by ingestion. A severe eye irritant. Flammable when exposed to heat or flame; can react with oxidizing materials. To fight fire, use CO_2, dry chemical. When heated to decomposition it emits acrid smoke and irritating fumes.

EJG000   CAS: 628-96-6   HR: 3
ETHYLENE GLYCOL DINITRATE
mf: $C_2H_4N_2O_6$ mw: 152.08

PROP: Yellow liquid. Mp: −20°, bp: explodes @ 114°, d: 1.483 @ 8°, vap d: 5.25.

SYNS: DINITROGLICOL (ITALIAN) * DINITROGLYCOL * EGDN * ETHANEDIOL DINITRATE * ETHYLENE DINITRATE * ETHYLENE NITRATE * ETHYLENGLYKOLDINITRAT (CZECH) * GLYCOLDINITRAAT (DUTCH) * GLYCOL DINITRATE * GLYCOL (DINITRATE DE) (FRENCH) * GLYKOLDINITRAT (GERMAN) * NITROGLYCOL * NITROGLYKOL (CZECH)

CONSENSUS REPORTS: Reported in EPA TSCA Inventory.

OSHA PEL: (Transitional: CL 0.2 ppm (skin)) STEL 0.1 mg/m³ (skin)
ACGIH TLV: TWA 0.05 ppm (skin)
DFG MAK: 0.05 ppm (0.3 mg/m³)
NIOSH REL: (Nitroglycerin) CL 0.1 mg/m³/20M
DOT Classification: Forbidden.

SAFETY PROFILE: Poison by subcutaneous route. Moderately toxic by ingestion. Can cause lowered blood pressure leading to headache, dizziness and weakness. Used as an explosive. When heated to decomposition it emits toxic fumes of NO_x.

EJH500   CAS: 109-86-4   HR: 3
ETHYLENE GLYCOL METHYL ETHER

DOT: UN 1188
mf: $C_3H_8O_2$ mw: 76.11

PROP: Colorless liquid; mild, agreeable odor. Misc in water, alc, ether, benzene. Bp: 124.5°, fp: −86.5°, flash p: 115°F (OC), lel: 2.5%, uel: 14%, d: 0.9660 @ 20°/4°, autoign temp: 545°F, vap press: 6.2 mm @ 20°, vap d: 2.62.

SYNS: AETHYLENGLYKOL-MONOMETHYLAETHER (GERMAN) * DOWANOL EM * EGM * EGME * ETHER MONOMETHYLIQUE de l'ETHYLENE-GLYCOL (FRENCH) * ETHYLENE GLYCOL MONOMETHYL ETHER (MAK, DOT) * GLYCOL ETHER EM * GLYCOLMETHYL ETHER * GLYCOL MONO-METHYL ETHER * JEFFERSOL EM * MECS * 2-METHOXY-AETHANOL (GERMAN) * 2-METHOX-YETHANOL (ACGIH) * METHOXYHYDROXYETHANE * METHYL CELLOSOLVE (OSHA, DOT) * METHYL ETHOXOL * METHYL GLYCOL * METHYLGLYKOL (GERMAN) * METHYL OXITOL * METIL CELLO-SOLVE (ITALIAN) * METOKSYETYLOWY ALKOHOL (POLISH) * 2-METOSSIETANOLO (ITALIAN) * MONOMETHYL ETHER of ETHYLENE GLYCOL * POLY-SOLV EM * PRIST

CONSENSUS REPORTS: Reported in EPA TSCA Inventory. Community Right-To-Know List.

OSHA PEL: TWA 25 ppm (skin)
ACGIH TLV: TWA 5 ppm (skin)
DFG MAK: 5 ppm (15 mg/m³)
NIOSH REL: TWA (Glycol Ethers): Reduce to lowest level
DOT Classification: Combustible Liquid; Label: None; Flammable or Combustible Liquid; Label: Flammable Liquid.

SAFETY PROFILE: Moderately toxic to humans by ingestion. Moderately toxic experimentally by ingestion, inhalation, skin contact, intraperitoneal and intravenous routes. Human systemic effects by inhalation: change in motor activity, tremors, and convulsions. Experimental teratogenic and reproductive effects. Mutation data reported. A skin and eye irritant. When used under conditions which do not require the application of heat, this material probably pre-

sents little hazard to health. However, in the manufacture of fused collars which require pressing with a hot iron, cases have been reported showing disturbance of the hemopoietic system with or without neurological signs and symptoms. The blood picture may resemble that produced by exposure to benzene. Two cases reported had severe aplastic anemia with tremors and marked mental dullness. The persons affected had been exposed to vapors of methyl "Cellosolve", ethanol and methanol, ethyl acetate and petroleum naphtha.

Flammable or combustible when exposed to heat or flame. A moderate explosion hazard. Can react with oxidizing materials to form explosive peroxides. To fight fire, use alcohol foam, CO_2, dry chemical. When heated to decomposition it emits acrid smoke and irritating fumes.

EJJ500   CAS: 110-49-6   HR: 3
ETHYLENE GLYCOL MONOMETHYL ETHER ACETATE

DOT: UN 1189
mf: $C_5H_{10}O_3$ mw: 118.15

PROP: Colorless liquid. Bp: 143°, fp: −70°, flash p: 111°F (CC), d: 1.005 @ 20°/20°, vap d: 4.07, lel: 1.7%, uel: 8.2%.

SYNS: ACETATE de L'ETHER MONOMETHYLIQUE de L'ETHYLENE-GLYCOL (FRENCH) * ACETATE de METHYLE GLYCOL (FRENCH) * ACETATO DI ME-TIL CELLOSOLVE (ITALIAN) * AETHYLENGLYKOL-METHYLAETHERACETAT (GERMAN) * ETHYLENE GLYCOL METHYL ETHER ACETATE * GLYCOL ETHER EM ACETATE * GLYCOL MONOMETHYL ETHER ACETATE * MeCsAc * 2-METHOXYA-ETHYLACETAT (GERMAN) * 2-METHOXYETHANOL, ACETATE * 2-METHOXY-ETHYL ACETAAT (DUTCH) * 2-METHOXYETHYL ACETATE (ACGIH) * 2-METH-OXYETHYLE, ACETATE de (FRENCH) * METHYL CELLOSOLYE ACETAAT (DUTCH) * METHYL CELLOSOLVE ACETATE (OSHA, DOT) * METHYL GLYCOL ACETATE * METHYL GLYCOL MONO-ACETATE * METHYLGLYKOLACETAT (GERMAN) * 2-METOSSIETILACETATO (ITALIAN)

CONSENSUS REPORTS: Glycol ether compounds are on the Community Right-To-Know List. Reported in EPA TSCA Inventory.

OSHA PEL: TWA 25 ppm (skin)
ACGIH TLV: TWA 5 ppm (skin)
DFG MAK: 5 ppm (25 mg/m³)

DOT Classification: Flammable or Combustible Liquid; Label: Flammable Liquid.

SAFETY PROFILE: Moderately toxic by ingestion, intraperitoneal and subcutaneous routes. Mildly toxic by inhalation and skin contact. Human systemic effects by inhalation: eye lacrimation, cough and pulmonary changes. Experimental reproductive effects. Mutation data reported. An inhalation irritant in humans. An eye irritant. Flammable when exposed to heat or flame; can react with oxidizing materials. A moderate explosion hazard. To fight fire, use CO_2, dry chemical. When heated to decomposition it emits acrid smoke and irritating fumes.

EJM500   CAS: 111-60-4   **HR: 3**
ETHYLENE GLYCOL STEARATE
mf: $C_{20}H_{40}O_3$ mw: 328.60

SYNS: CLINDROL SEG * EMEREST 2350 * EMPILAN 2848 * ETHYLENE GLYCOL, MONO-STEARATE * GLYCOL MONOSTEARATE * GLYCOL STEARATE * 2-HYDROXYETHYL ESTER STEARIC ACID * IVORIT * LIPO EGMS * MONTHYBASE * MONTHYLE * PARASTARIN * PRODHYBASE ETHYL * S 151 * SEDETOL * STEARIC ACID, MONOESTER with ETHYLENE GLYCOL * TEGO-STEARATE * USAF KE-11

CONSENSUS REPORTS: Reported in EPA TSCA Inventory.

SAFETY PROFILE: Poison by intraperitoneal route. A skin irritant. Used in cosmetics. When heated to decomposition it emits acrid smoke and irritating fumes.

EJM900   CAS: 151-56-4   **HR: 3**
ETHYLENEIMINE

DOT: UN 1185
mf: C_2H_5N mw: 43.08

PROP: Oily, water-white liquid. Pungent ammoniacal odor. Bp: 55-56°, fp: −71.5°, flash p: 12°F, d: 0.832 @ 20°/4°, autoign temp: 608°F, vap press: 160 mm @ 20°, vap d: 1.48, lel: 3.6%, uel: 46%.

SYNS: AETHYLENIMIN (GERMAN) * AMINOETHYLENE * AZACYCLOPROPANE * AZIRANE * AZIRIDIN (GERMAN) * AZIRIDINE * DIHYDROAZIRENE * DIHYDRO-1H-AZIRINE * DIMETHYLENEIMINE * DIMETHYLNIMINE * EI * ENT 50,324 * ETHYLEENIMINE (DUTCH) * ETHYLENE IMINE, INHIBITED (DOT) * ETHYLENI-

MINE * ETHYLIMINE * ETILENIMINA (ITALIAN) * RCRA WASTE NUMBER P054 * TL 337

CONSENSUS REPORTS: IARC Cancer Review: GROUP 3 IMEMDT 7,56,87; Animal Sufficient Evidence IMEMDT 9,37,75. Community Right-To-Know List. EPA Extremely Hazardous Substances List. Reported in EPA TSCA Inventory. EPA Genetic Toxicology Program.

OSHA PEL: TWA 1 mg/m^3 (skin); Carcinogen
ACGIH TLV: TWA 0.5 ppm (skin)
DFG TRK: 0.5 ppm; Animal Carcinogen, Suspected Human Carcinogen.
DOT Classification: Flammable Liquid; Label: Flammable Liquid and Poison.

SAFETY PROFILE: Confirmed carcinogen with experimental carcinogenic, neoplastigenic, tumorigenic, and teratogenic data. Poison by ingestion, skin contact, inhalation, intraperitoneal, and possibly other routes. Human mutation data reported. A skin, mucous membrane, and severe eye irritant. An allergic sensitizer of skin. Causes opaque cornea, keratoconus and necrosis of cornea (experimentally). Has been known to cause severe human eye injury. Drinking of carbonated beverages is recommended as an antidote to this material in stomach.

A very dangerous fire and explosion hazard when exposed to heat, flame or oxidizers. Reacts violently with acids, aluminum chloride + substituted anilines, acetic acid, acetic anhydride, acrolein, acrylic acid, allyl chloride, CS_2, Cl_2, chlorosulfonic acid, epichlorohydrin, glyoxal, HCl, HF, HNO_3, oleum, β-propiolactone, Ag, NaOCl, H_2SO_4; vinyl acetate. Reacts with chlorinating agents (e.g., sodium hypochlorite solution) to form the explosive 1-chloro aziridine. Reacts with silver or its alloys to form explosive silver derivatives. Dangerous; heat and/or the presence of catalytically active metals or chloride ions can cause a violent exothermic reaction. To fight fire, use alcohol foam, CO_2, dry chemical. When heated to decomposition it emits acrid smoke and irritating fumes.

EJN500   CAS: 75-21-8   **HR: 3**
ETHYLENE OXIDE

DOT: UN 1040
mf: C_2H_4O mw: 44.06

PROP: Colorless gas at room temperature. Mp: −111.3°, bp: 10.7°, ULC: 100, lel: 3.0%, uel: 100%, flash p: −4°F, d: 0.8711 @ 20°/20°, autoign temp: 804°F, vap press: 1095 mm @ 20°, vap d: 1.52. Misc in water and alc; very sol in ether.

SYNS: AETHYLENOXID (GERMAN) * AMPROLENE * ANPROLENE * ANPROLINE * DIHYDROOXI-RENE * DIMETHYLENE OXIDE * ENT 26,263 * E.O. * 1,2-EPOXYAETHAN (GERMAN) * EPOXYETHANE * 1,2-EPOXYETHANE * ETHENE OXIDE * ETHYLEENOXIDE (DUTCH) * ETHYLENE (OXYDE d') (FRENCH) * ETILENE (OSSIDO di) (ITALIAN) * ETO * ETYLENU TLENEK (POLISH) * FEMA No. 2433 * MERPOL * NCI-C50088 * OXACYCLO-PROPANE * OXANE * OXIDOETHANE * α,β-OX-IDOETHANE * OXIRAAN (DUTCH) * OXIRANE * OXYFUME * OXYFUME 12 * RCRA WASTE NUMBER U115 * STERILIZING GAS ETHYLENE OXIDE 100% * T-GAS

CONSENSUS REPORTS: IARC Cancer Review: GROUP 2A IMEMDT 7,205,87; Animal Inadequate Evidence IMEMDT 11,157,76; Human Inadequate Evidence IMEMDT 36,189,85; Animal Sufficient Evidence IMEMDT 36, 189,85. NTP Fourth Annual Report On Carcinogens, 1984. Community Right-To-Know List. EPA Extremely Hazardous Substances List. Reported in EPA TSCA Inventory. EPA Genetic Toxicology Program.

OSHA PEL: TWA 1 ppm; Cancer Hazard
ACGIH TLV: TWA 1 ppm; Suspected Human Carcinogen.
DFG TRK: 3 ppm; Animal Carcinogen, Suspected Human Carcinogen.
NIOSH REL: (Oxirane) TWA 0.1 ppm; CL 5 ppm/10M/D
DOT Classification: Flammable Liquid; Label: Flammable Liquid; Flammable Gas; Label: Poison Gas and Flammable Gas.

SAFETY PROFILE: Confirmed human carcinogen with experimental carcinogenic, tumorigenic and neoplastigenic data. Poison by ingestion, intraperitoneal, subcutaneous, intravenous, and possibly other routes. Moderately toxic by inhalation. Human systemic effects by inhalation: convulsions, nausea, vomiting, olfactory and pulmonary changes. Experimental teratogenic and reproductive effects. Mutation data reported. A skin and eye irritant. An irritant to mucous membranes of respiratory tract. High concentrations can cause pulmonary edema.

Highly flammable liquid or gas. Severe explosion hazard when exposed to flame. To fight fire, use alcohol foam, CO_2, dry chemical. Violent polymerization occurs on contact with ammonia, alkali hydroxides, amines, metallic potassium, acids, covalent halides (e.g., aluminum chloride, iron(III) chloride, tin(IV) chloride, aluminum oxide, iron oxide, rust). Explosive reaction with glycerol at 200°. Rapid compression of the vapor with air causes explosions. Incompatible with bases, alcohols, air, m-nitroaniline, trimethyl amine, copper, iron chlorides, iron oxides, magnesium perchlorate, mercaptans, potassium, tin chlorides, contaminants, alkane thiols, bromoethane. When heated to decomposition it emits acrid smoke and irritating fumes.

EJO000   CAS: 8070-50-6   **HR: 3**
ETHYLENE OXIDE, mixed with CARBON DIOXIDE

DOT: UN 1041

PROP: Contains less than 10% carbon dioxide (NTIS** PB225-283).

SYNS: ANHYDRIDE CARBONIQUE et OXYDE d'ETHY-LENE MELANGES (FRENCH) * CARBON DIOXIDE and ETHYLENE OXIDE MIXTURES, with more than 6% ETHYL-ENE OXIDE (DOT) * ETHYLENE OXIDE and CARBON DIOXIDE MIXTURES (DOT) * OXYFUME 20 * OXYFUME 30

DOT Classification: Poison A; Label: Poison Gas and Flammable Gas.

SAFETY PROFILE: A poison. Mildly toxic by inhalation. Used for the sterilization of vacuum chambers.

EJP000   CAS: 1072-53-3   **HR: 3**
ETHYLENE SULFATE
mf: $C_2H_4O_4S$ mw: 124.12

SYNS: 2,2-DIOXIDE-1,3,2-DIOXATHIOLANE * ETHYLENE GLYCOL, CYCLIC SULFATE * GLYCOL SULFATE * SULFURIC ACID, CYCLIC ETHYLENE ESTER

CONSENSUS REPORTS: EPA Genetic Toxicology Program.

SAFETY PROFILE: Questionable carcinogen with experimental tumorigenic and neoplastigenic data. Mutation data reported. When heated to decomposition it emits toxic fumes of SO_x.

EJP500   CAS: 420-12-2   **HR: 3**
ETHYLENE SULFIDE
mf: C_2H_4S mw: 60.12

PROP: Colorless liquid. Bp: 55-56° decomp, d: 1.0368 @ 0°/4°, vap d: 2.07.

SYNS: AETHYLENSULFID (GERMAN) * 2,3-DIHY-DROTHIIRENE * ETHYLENE EPISULFIDE * ETHYL-ENE EPISULPHIDE * ETHYLENE SULPHIDE * THIACYCLOPROPANE * THIIRANE

CONSENSUS REPORTS: IARC Cancer Review: GROUP 3 IMEMDT 7,56,87; Animal Limited Evidence IMEMDT 11,257,76. Reported in EPA TSCA Inventory.

SAFETY PROFILE: Poison by ingestion, intraperitoneal, and subcutaneous routes. Mildly toxic by inhalation. A skin, eye and mucous membrane irritant. Questionable carcinogen with experimental tumorigenic data. Can react with oxidizing materials. When heated to decomposition, or on contact with acid or acid fumes, it emits highly toxic fumes of SO_x.

EJU000   CAS: 60-29-7   **HR: 3**
ETHYL ETHER
DOT: UN 1155
mf: $C_4H_{10}O$ mw: 74.14

PROP: A clear, volatile liquid; sweet, pungent odor. Sol in water; misc in alcohol and ether; sol in chloroform. Mp: −116.2°, bp: 34.6°, ULC: 100, lel: 1.85%, uel: 36%, flash p: −49°F, d: 0.7135 @ 20°/4°, autoign temp: 320°F, vap press: 442 mm @ 20°, vap d: 2.56.

SYNS: AETHER * ANAESTHETIC ETHER * ANESTHESIA ETHER * ANESTHETIC ETHER * DIAETHYLAETHER (GERMAN) * DIETHYL ETHER (DOT) * DIETHYL OXIDE * DWUETYLOWY ETER (POLISH) * ETERE ETILICO (ITALIAN) * ETHER * ETHER ETHYLIQUE (FRENCH) * ETHOXYETHANE * 1,1'-OXYBISETHANE * OXYDE d'ETHYLE (FRENCH) * RCRA WASTE NUMBER U117 * SOLVENT ETHER

CONSENSUS REPORTS: Reported in EPA TSCA Inventory. EPA Genetic Toxicology Program.

OSHA PEL: (Transitional: TWA 400 ppm) TWA 400 ppm; STEL 500 ppm
ACGIH TLV: TWA 400 ppm; STEL 500 ppm
DFG MAK: 400 ppm (1200 mg/m³)
DOT Classification: Flammable Liquid; Label: Flammable Liquid.

SAFETY PROFILE: Moderately toxic to humans by ingestion. Poison experimentally by subcutaneous route. Moderately toxic by intraperitoneal and intravenous routes. Mildly toxic by inhalation. Human systemic effects by inhalation: olfactory changes. Mutation data reported. A severe eye and moderate skin irritant. Ethyl ether is not corrosive or dangerously reactive. It must not be considered safe for individuals to inhale or ingest. It is a depressant of the central nervous system and is capable of producing intoxication, drowsiness, stupor, and unconsciousness. Death due to respiratory failure may result from severe and continued exposure.

A very dangerous fire and explosion hazard when exposed to heat or flame. A storage hazard. It auto-oxidizes to form explosive polymeric 1-oxy-peroxides. Explosive reaction with boron triazide, bromine trifluoride, bromine pentafluoride, perchloric acid, uranyl nitrate + light, wood pulp extracts + heat. Violent reaction or ignition on contact with halogens (e.g., bromine, chlorine), interhalogens (e.g., iodine heptafluoride), oxidants (e.g., silver perchlorate, nitrosyl perchlorate, nitryl perchlorate, chromyl chloride, fluorine nitrate, permanganic acid, nitric acid, hydrogen peroxide, peroxodisulfuric acid, iodine(VII) oxide, sodium peroxide, ozone, and liquid air), sulfur and sulfur compounds (e.g., sulfur when dried with peroxidized ether, sulfunoyl chloride). Can react vigorously with acetyl peroxide, air, bromoazide, ClF_3, CrO_3, $Cr(OCl)_2$, $LiAlH_2$, $NOClO_4$, O_2, $NClO_2$, (H_2SO_4 + permanganates), K_2O_2, [$(C_2H_5)_3Al$ + air], [$(CH_3)_3Al$ + air]. To fight fire, use alcohol foam, CO_2, dry chemical. When heated to decomposition it emits acrid smoke and irritating fumes.

EJY000   CAS: 17013-37-5   **HR: 3**
5-ETHYL-5-(1-ETHYLPROPYL)BARBITURIC ACID
mf: $C_{11}H_{18}N_3O_3$ mw: 240.32

SYNS: 5-ETHYL-5-(1-ETHYLPROPYL)2,4,6(1H,3H,5H)-PYRIMIDINETRIONE * ISOMEBUMAL

SAFETY PROFILE: Poison by intraperitoneal route. When heated to decomposition it emits toxic fumes of NO_x.

EKI000   CAS: 353-03-7   **HR: 3**
ETHYL-10-FLUORODECANOATE
mf: $C_{12}H_{23}FO_2$ mw: 218.35

SYNS: ETHYL-φ-FLUORODECANOATE * ETHYL-9-FLUORONONANECARBOXYLATE

SAFETY PROFILE: Poison by intraperitoneal, parenteral, and possibly other routes. When heated to decomposition it emits toxic fumes of F^-.

EKK500   CAS: 332-97-8   HR: 3
ETHYL-8-FLUORO OCTANOATE
mf: $C_{10}H_{19}FO_2$ mw: 190.29

SYNS: ETHYL-φ-FLUOROOCTANOATE * 8-FLUO-ROOCTANOIC ACID, ETHYL ESTER

SAFETY PROFILE: Poison by intraperitoneal, subcutaneous, parenteral, and possibly other routes. When heated to decomposition it emits toxic fumes of F^-.

EKL000   CAS: 109-94-4   HR: 3
ETHYL FORMATE
DOT: UN 1190
mf: $C_3H_6O_2$ mw: 74.09

PROP: Colorless liquid; sharp, rum-like odor. Mp: $-79°$, bp: $54.3°$, lel: 2.7%, uel: 13.5%, flash p: $-4°F$ (CC), d: 0.9236 @ $20°/20°$, refr index: 1.359, autoign temp: $851°F$, vap press: 100 mm @ $5.4°$, vap d: 2.55. Sol in fixed oils, propylene glycol, water (decomp); sltly sol in mineral oil; insol in glycerin @ $54°$.

SYNS: AETHYLFORMIAT (GERMAN) * AREGINAL * ETHYLE (FORMIATE d') (FRENCH) * ETHYLFOR-MIAAT (DUTCH) * ETHYL FORMIC ESTER * ETHYL METHANOATE * ETILE (FORMIATO di) (ITALIAN) * FEMA No. 2434 * FORMIC ACID, ETHYL ESTER * FORMIC ETHER * MROWCZAN ETYLU (POLISH)

CONSENSUS REPORTS: Reported in EPA TSCA Inventory.

OSHA PEL: TWA 100 ppm
ACGIH TLV: TWA 100 ppm
DFG MAK: 100 ppm (300 mg/m³)
DOT Classification: Flammable Liquid; Label: Flammable Liquid.

SAFETY PROFILE: Moderately toxic by ingestion and subcutaneous routes. Mildly toxic by skin contact and inhalation. A powerful inhalation irritant in humans. A skin and eye irritant. Questionable carcinogen with experimental tumorigenic data. Highly flammable liquid. A very dangerous fire and explosion hazard when exposed to heat, flame, or oxidizers. To fight fire, use alcohol foam, spray, mist, dry chemical. When heated to decomposition it emits acrid smoke and irritating fumes.

EKN000   CAS: 2642-71-9   HR: 3
ETHYL GUTHION
mf: $C_{12}H_{16}N_3O_3PS_2$ mw: 345.40

SYNS: ATHYL-GUSATHION * AZINFOS-ETHYL (DUTCH) * AZINOS * AZINPHOS-AETHYL (GER-MAN) * AZINPHOS ETHYL * AZINPHOS-ETILE (ITALIAN) * BAY 16225 * BAYER 16259 * BENZOTRIAZINE derivative of an ETHYL DITHIOPHOS-PHATE * COTNION-ETHYL * CYRSTHION * O,O-DIAETHYL-S-(4-OXOBENZOTRIAZIN-3-METHYL) DITHIOPHOSPHAT (GERMAN) * O,O-DIAETHYL-S-((4-OXO-3H-1,2,3-BENZOTRIAZIN-3-YL)-METHYL)-DITHIO-PHOSPHAT (GERMAN) * O,O-DIETHYL-S-(4-OXO-3H-1,2,3-BENZOTRIAZINE-3-YL)-METHYL-DITHIOPHOSPHATE * O,O-DIETHYL-S-((4-OXO-3H-1,2,3-BENZOTRIAZIN-3-YL)-METHYL)-DITHIO FOSFAAT (DUTCH) * O,O-DI-ETHYL-S-(4-OXOBENZOTRIAZINO-3-METHYL)PHOSPHO-RODITHIOATE * O,O-DIETHYL PHOSPHORODITH-IOATE S-ester with 3-(MERCAPTOMETHYL)-1,2,3-BENZO-TRIAZIN-4(3H)-ONE * O,O-DIETIL-S-((4-OXO-3H-1,2,3-BENZOTRIAZIN-3-IL)-METIL)-DITIOFOSFATO (ITALIAN) * 3,4-DIHYDRO-4-OXO-3-BENZOTRIAZINYLMETHYL O,O-DIETHYL PHOSPHORODITHIOATE * S-(3,4-DIHY-DRO-4-OXO-1,2,3-BENZOTRIAZIN-3-YLMETHYL) O,O-DI-ETHYL PHOSPHORODITHIOATE * ENT 22,014 * ETHYL GUSATHION * GUSATHION A * GU-THION (ETHYL) * R 1513 * TRIAZOTION (RUSSIAN)

CONSENSUS REPORTS: EPA Extremely Hazardous Substances List.

SAFETY PROFILE: Poison by ingestion, inhalation, skin contact, and intraperitoneal route. A cholinesterase inhibitor type of insecticide. When heated to decomposition it emits toxic fumes of SO_x, PO_x, and NO_x.

EKN050   CAS: 106-30-9   HR: 2
ETHYL HEPTANOATE
mf: $C_9H_{18}O_2$ mw: 158.24

PROP: Colorless liquid; wine-brandy odor. D: 0.867-0.872, refr index: 1.411, flash p: $149°F$. Misc in alc, chloroform, fixed oils; sltly sol in propylene glycol.

SYNS: ETHYL HEPTOATE * FEMA No. 2437

SAFETY PROFILE: Combustible liquid. When heated to decomposition it emits acrid smoke and irritating fumes.

EKR500   CAS: 1632-16-2   HR: 3
2-ETHYL-1-HEXENE
mf: C_8H_{16} mw: 112.24

PROP: Colorless liquid. Bp: 120°, d: 0.7270 @ 20°/20°, vap d: 3.87.

SYNS: 2-ETHYL HEXENE-1 * USAF DO-21

CONSENSUS REPORTS: Reported in EPA TSCA Inventory.

SAFETY PROFILE: Poison by intraperitoneal route. Mildly toxic by inhalation. Combustible when exposed to heat or flame; can react with oxidizing materials. When heated to decomposition it emits acrid smoke and irritating fumes.

EKS500   CAS: 104-75-6   **HR: 3**
2-ETHYL HEXYLAMINE
DOT: UN 2276
mf: $C_8H_{19}N$ mw: 129.28

PROP: A clear, miscible liquid. Bp: 169.2°, flash p: 140°F (OC), d: 0.7894 @ 20°/20°, vap press: 1.2 mm @ 20°, vap d: 4.45.

SYN: 1-AMINO-2-ETHYLHEXAN (CZECH)

CONSENSUS REPORTS: Reported in EPA TSCA Inventory.

DOT Classification: Corrosive Material; Label: Corrosive, Flammable Liquid.

SAFETY PROFILE: Poison by intraperitoneal route. Moderately toxic by ingestion, inhalation, and skin contact. Corrosive. A severe skin and eye irritant. Flammable when exposed to heat or flame; can react with oxidizing materials. To fight fire, use alcohol foam, CO_2, dry chemical. When heated to decomposition it emits toxic fumes of NO_x.

EKV000   CAS: 94-96-2   **HR: 2**
ETHYL HEXYLENE GLYCOL
mf: $C_8H_{18}O_2$ mw: 146.26

PROP: Practically colorless, somewhat viscous, odorless liquid. Bp: 243.1°, flash p: 260°F (OC), fp: −40°, d: 0.9422 @ 20°/20°, vap press: <0.01 mm @ 20°, vap d: 5.03.

SYNS: CARBIDE 6-12 * COMPOUND 6-12 INSECT REPELLENT * ENT 375 * ETHOHEXADIOL * ETHYL HEXANEDIOL * 2-ETHYL-1,3-HEXANEDIOL * 2-ETHYLHEXANE-1,3-DIOL * 2-ETHYLHEXANE-DIOL-1,3 * 2-ETHYL-3-PROPYL-1,3-PROPANEDIOL * 3-HYDROXYMETHYL-n-HEPTAN-4-OL * 6-12-INSECT REPELLENT * OCTYLENE GLYCOL * REPELLENT 612 * RUTGERS 612

CONSENSUS REPORTS: Reported in EPA TSCA Inventory.

SAFETY PROFILE: Moderately toxic by ingestion and skin contact. A skin and severe eye irritant. Used as an insecticide, insect repellent, and in hair care preparations. Combustible when exposed to heat or flame; can react with oxidizing materials. To fight fire, use alcohol foam, foam, dry chemical. When heated to decomposition it emits acrid smoke and irritating fumes.

ELC500   CAS: 3413-58-9   **HR: 3**
ETHYLHYDROCUPREINE HYDROCHLORIDE
mf: $C_{21}H_{28}N_2O_2 \cdot ClH$ mw: 376.97

SYNS: HYDROCUPREINE ETHYL ESTER HYDROCHLORIDE * NEUMOLISINA * NUMOQUIN HYDROCHLORIDE * OPTOCHIN HYDROCHLORIDE * OPTOQUINHYDROCHLORIDE * RHOMBIC

SAFETY PROFILE: Poison by unspecified route. Moderately toxic by subcutaneous route. Human systemic effects by intravenous route: visual field changes and arteriolar constriction. An antiseptic. When heated to decomposition it emits very toxic fumes such as Cl^- and NO_x.

ELD000   CAS: 3031-74-1   **HR: 3**
ETHYL HYDROPEROXIDE
mf: $C_2H_6O_2$ mw: 62.07

SYN: ETHYL HYDROGEN PEROXIDE

DOT Classification: Forbidden.

SAFETY PROFILE: Explodes violently when superheated. The barium salt is heat- and impact-sensitive. Explosive reaction with hydroiodic acid or finely divided silver. When heated to decomposition it emits acrid smoke and irritating fumes.

ELG500   CAS: 13147-25-6   **HR: 3**
ETHYL-2-HYDROXYETHYLNITROS-AMINE
mf: $C_4H_{10}N_2O_2$ mw: 118.16

SYNS: AETHYL-AETHANOL-NITROSOAMIN (GERMAN) * EENA * EHEN * N-ETHYL-N-HYDROXYETHYL-NITROSAMINE * 2-(ETHYLNITROSAMINO)ETHANOL * N-NITROSOAETHYLAETHANOLAMIN (GERMAN) * N-NITROSOETHYLETHANOLAMINE * N-NITROSOETHYL-2-HYDROXYETHYLAMINE * N-NITROSO-N-ETHYL-N-(2-HYDROXYETHYL)AMINE

CONSENSUS REPORTS: IARC Cancer Review: Animal Limited Evidence IMEMDT 17,83,78. EPA Genetic Toxicology Program.

SAFETY PROFILE: Suspected carcinogen with experimental carcinogenic, neoplastigenic, and tumorigenic data. Mutation data reported. Explodes when heated to 170°C. When heated to decomposition it emits toxic fumes of NO_x.

ELL500   CAS: 70-70-2   **HR: 3**
ETHYL-p-HYDROXYPHENYL KETONE
mf: $C_9H_{10}O_2$ mw: 150.19

SYNS: FRENANTOL * FRENOHYPON * H-365 * p-HYDROXYPHENYL-1-PROPANONE * 1-(4-HYDROXYPHENYL)-1-PROPANONE * HYDROXYPROPIOPHENONE * p-HYDROXYPROPIOPHENONE * 4-HYDROXYPROPIOPHENONE * HYPOPHENON * p-OXYPROPIOPHENONE * PAROXON * PAROXYPROPIONE * PHP * POP * PROFENONE * p-PROPIONYLPHENOL * USAF EK-3302

CONSENSUS REPORTS: Reported in EPA TSCA Inventory.

SAFETY PROFILE: Poison by intraperitoneal, subcutaneous, and parenteral routes. When heated to decomposition it emits acrid smoke and irritating fumes.

ELN500   CAS: 75-37-6   **HR: 1**
ETHYLIDENE DIFLUORIDE
DOT: UN 1030
mf: $C_2H_4F_2$ mw: 66.06

PROP: Colorless gas. Mp: −117.0°, bp: −26.5°, d: 1.004 @ 25°, vap d: 2.28.

SYNS: ALGOFRENE TYPE 67 * DIFLUOROETHANE * 1,1-DIFLUOROETHANE (DOT) * ETHYLENE FLUORIDE * ETHYLIDENE FLUORIDE * FC 152a * FREON 152 * GENETRON 100 * HALOCARBON 152A

CONSENSUS REPORTS: Reported in EPA TSCA Inventory. EPA Genetic Toxicology Program.

DOT Classification: Flammable Gas; Label: Flammable Gas.

SAFETY PROFILE: Mildly toxic by inhalation. Mutation data reported. Narcotic in high concentration. A very dangerous fire hazard, when exposed to heat or flame; can react vigorously with oxidizing materials.

ELO500   CAS: 16219-75-3   **HR: 2**
ETHYLIDENE NORBORNENE
mf: C_9H_{12} mw: 120.21

SYNS: 5-ETHYLIDENEBICYCLO(2.2.1)HEPT-2-ENE * 5-ETHYLIDENE-2-NORBORNENE

CONSENSUS REPORTS: Reported in EPA TSCA Inventory.

OSHA PEL: CL 5 ppm
ACGIH TLV: CL 5 ppm

SAFETY PROFILE: Moderately toxic by ingestion. Mildly toxic by inhalation and skin contact. Human systemic effects by inhalation: conjuctiva, olfactory and taste changes. A skin irritant. When heated to decomposition it emits acrid smoke and irritating fumes.

ELS000   CAS: 97-62-1   **HR: 2**
ETHYL ISOBUTYRATE
DOT: UN 2385
mf: $C_6H_{12}O_2$ mw: 116.18

PROP: Colorless, volatile liquid; fruity, aromatic odor. Mp: −88°, bp: 110-111°, d: 0.862, vap press: 40 mm @ 33.8°, vap d: 4.01, refr index: 1.385, flash p: <64.4°F.

SYNS: ETHYL ISOBUTANOATE * ETHYLISOBUTYRATE (DOT) * ETHYL-2-METHYLPROPANOATE * ETHYL-2-METHYLPROPIONATE * FEMA No. 2428 * ISOBUTYRIC ACID, ETHYL ESTER * 2-METHYLPROPIONIC ACID, ETHYL ESTER

CONSENSUS REPORTS: Reported in EPA TSCA Inventory.

DOT Classification: Flammable Liquid; Label: Flammable Liquid.

SAFETY PROFILE: Moderately toxic by intraperitoneal route. A skin irritant. Flammable Liquid. A very dangerous fire hazard when exposed to heat or flame; can react vigorously with oxidizing materials. To fight fire, use foam, CO_2, dry chemical. When heated to decomposition it emits acrid smoke and irritating fumes.

ELS500   CAS: 109-90-0   **HR: 3**
ETHYL ISOCYANATE
DOT: UN 2481
mf: C_3H_5NO mw: 71.09

PROP: Bp: 60°, d: 0.90 @ 20°/4°, vap d: 2.45.

SYNS: ETHYL ISOCYANATE (DOT) * ISOCYANATOETHANE * ISOCYANIC ACID, ETHYL ESTER

CONSENSUS REPORTS: Reported in EPA TSCA Inventory.

DOT Classification: Flammable Liquid; Label: Flammable Liquid and Poison.

SAFETY PROFILE: Poison by intravenous route. Mutation data reported. When heated to decomposition it emits toxic fumes of NO_x.

ELU000   CAS: 1570-45-2   **HR: 3**
ETHYL ISONICOTINATE
mf: $C_8H_9NO_2$ mw: 151.18

SYNS: ISONICOTINIC ACID, ETHYL ESTER * 4-PY-RIDINECARBOXYLIC ACID, ETHYL ESTER

CONSENSUS REPORTS: Reported in EPA TSCA Inventory.

SAFETY PROFILE: Poison by intravenous route. When heated to decomposition it emits toxic fumes of NO_x.

ELX000   CAS: 76-76-6   **HR: 3**
ETHYLISOPROPYLBARBITURIC ACID
mf: $C_9H_{14}N_2O_3$ mw: 198.25

SYNS: 5-ETHYL-5-ISOPROPYLBARBITURIC ACID * 5-ETHYL-5-(1-METHYLETHYL)-2,4,6(1H,3H,5H)-PYRI-MIDINETRIONE * IPRAL * IRENAL * PROBAR-BITAL * PROBARBITONE * VASALGIN

SAFETY PROFILE: Poison by ingestion, intra-peritoneal, subcutaneous, and intravenous routes. When heated to decomposition it emits toxic fumes of NO_x.

ELX500   CAS: 16339-04-1   **HR: 3**
ETHYLISOPROPYLNITROSOAMINE
mf: $C_5H_{12}N_2O$ mw: 116.19

SYNS: AETHYL-ISOPROPYL-NITROSOAMIN (GERMAN) * 1-METHYL-N-NITROSODIETHYLAMINE * N-NITRO-SOETHYLISOPROPYLAMINE

SAFETY PROFILE: Moderately toxic by inges-tion. Questionable carcinogen with experimen-tal carcinogenic and tumorigenic data. When heated to decomposition it emits toxic fumes of NO_x.

ELY700   CAS: 106-33-2   **HR: 1**
ETHYL LAURATE
mf: $C_{14}H_{28}O_2$ mw: 228.37

PROP: Colorless, oily liquid; fruity-floral odor. D: 0.858, refr index: 1.430, flash p: +212°F. Misc in alc, chloroform, ether; insol in water @ 269°.

SYNS: ETHYL DODECANOATE * FEMA No. 2441

SAFETY PROFILE: Combustible liquid. When heated to decomposition it emits acrid smoke and irritating fumes.

EMA500   CAS: 105-53-3   **HR: 1**
ETHYL MALONATE
mf: $C_7H_{12}O_4$ mw: 160.19

PROP: Clear, colorless liquid; fruit-like odor. Bp: 198.9°, fp: −49.8°, flash p: 200°F (OC), d: 1.055 @ 25°/25°, refr index: 1.413-1.416, vap press: 1 mm @ 40.0°, vap d: 5.52. Sol in fixed oils, propylene glycol; sltly sol in alc, water; insol in glycerin, mineral oil @ 200°.

SYNS: CARBETHOXYACETIC ESTER * DICARB-ETHOXYMETHANE * DIETHYL MALONATE (FCC) * DIETHYL PROPANEDIOATE * FEMA No. 2375 * MALONIC ACID, DIETHYL ESTER * MALONIC ES-TER * METHANEDICARBOXYLIC ACID, DIETHYL ES-TER * PROPANEDIOIC ACID, DIETHYL ESTER

CONSENSUS REPORTS: Reported in EPA TSCA Inventory.

SAFETY PROFILE: Mildly toxic by ingestion. A skin irritant. Combustible liquid when ex-posed to heat or flame; can react with oxidizing materials. To fight fire, use water to blanket fire, foam, CO_2, dry chemical. When heated to decomposition it emits acrid smoke and irritat-ing fumes.

EMA600   CAS: 4940-11-8   **HR: 2**
ETHYL MALTOL
mf: $C_7H_8O_3$ mw: 140.15

PROP: White crystalline powder; sweet fruity taste. Mp: 90°. Sol in water, alc, propylene glycol, chloroform.

SYNS: 2-ETHYL-3-HYDROXY-4H-PYRAN-4-ONE * 2-ETHYL PYROMECONIC ACID * 3-HYDROXY-2-ETHYL-4-PYRONE

CONSENSUS REPORTS: Reported in EPA TSCA Inventory.

SAFETY PROFILE: Moderately toxic by inges-tion and subcutaneous routes. Mutation data re-ported. When heated to decomposition it emits acrid smoke and irritating fumes.

EMB100   CAS: 75-08-1   **HR: 2**
ETHYL MERCAPTAN
DOT: UN 2363
mf: C_2H_6S mw: 62.14

PROP: Colorless liquid, penetrating garlic-like odor. Mp: $-147°$, bp: $36.2°$, lel: 2.8%, uel: 18.2%, d: 0.83907 @ $20°/4°$, autoign temp: $570°F$, vap d: 2.14. flash p: $< -0.4°F$.

SYNS: AETHANETHIOL (GERMAN) * AETHYLMER-CAPTAN (GERMAN) * ETANTIOLO (ITALIAN) * ETHAANTHIOL (DUTCH) * ETHANETHIOL * ETHYL HYDROSULFIDE * ETHYLMERCAPTAAN (DUTCH) * ETHYLMERKAPTAN (CZECH) * ETHYL SULFHYDRATE * ETHYL THIOALCOHOL * ETIL-MERCAPTANO (ITALIAN) * LPG ETHYL MERCAPTAN 1010 * THIOETHANOL * THIOETHYL ALCOHOL

CONSENSUS REPORTS: Reported in EPA TSCA Inventory.

OSHA PEL: (Transitional: CL 10 ppm) TWA 0.5 ppm
ACGIH TLV: TWA 0.5 ppm
DFG MAK: 0.5 ppm (1 mg/m^3)
NIOSH REL: (n-Alkane Mono Thiols) CL 0.5 ppm/15M
DOT Classification: Flammable Liquid; Label: Flammable Liquid; Flammable Liquid; Label: Flammable Liquid, Poison.

SAFETY PROFILE: Moderately toxic by ingestion, inhalation, and intraperitoneal routes. A skin and eye irritant. Inhalation causes central nervous system effects in humans. A very dangerous fire hazard when exposed to heat or flame; can react vigorously with oxidizing materials. A moderate explosion hazard when exposed to spark or flame. Violent reaction with $Ca(OCl)_2$. Will react with water or steam to produce toxic and flammable vapors. To fight fire, use CO_2, dry chemical. When heated to decomposition or on contact with acid or acid fumes it emits highly toxic fumes of SO_x.

EME050   CAS: 2597-93-5   **HR: 3**
ETHYLMERCURICHLORENDIMIDE
mf: $C_{11}H_7Cl_6HgNO_2$ mw: 598.48

SYNS: 50-CS-46 * EMMI * N-(ETHYLMERCURI)-1,4,5,6,7,7-HEXACHLOROBICYCLO(2.2.1)HEPT-5-ENE- 2,3-DICARBOXIMIDE * N-ETHYLMERCURI-3,4,5,6,7,7-HEXACHLORO-3,6-ENDOMETHYLENE-1,2,3,6- TETRAHY-DROPHTHALIMIDE * N-ETHYLMERCURI-1,2,3,6-TET-RAHYDRO-3,6-ENDOMETHANO- 3,4,5,6,7,7-HEXACHLO-ROPHTHALIMIDE * 1,4,5,6,77-HEXACHLORO-N-(ETHYLMERCURI)-5-NORBORNENE-2,3-DICARBOX-IMIDE

OSHA PEL: (Transitional: CL 1 mg/10m^3) CL 0.1 mg(Hg)/m^3 (skin)

ACGIH TLV: TWA 0.1 mg(Hg)/m^3 (skin)
NIOSH REL: TWA 0.05 mg(Hg)/m^3

CONSENSUS REPORTS: Mercury and its compounds are on the Community Right-To-Know List.

SAFETY PROFILE: Poison by ingestion and possibly other routes. When heated to decomposition it emits very toxic fumes of Cl^-, Hg, and NO_x.

EME500   CAS: 517-16-8   **HR: 3**
ETHYLMERCURY-p-TOLUENE SULFONAMIDE
mf: $C_{15}H_{17}HgNO_2S$ mw: 475.98

PROP: Crystals; pungent, garlic-like odor; water-insol.

SYNS: CERESAN M * COMPOUND-1452-F * EMTS * N-ETHYLMERCURI-N-PHENYL-p-TOLUEN-ESULFONAMIDE * N-(ETHYLMERCURI)-p-TOLUENE-SULFONANILIDE * N-(ETHYLMERCURI)-p-TOLUENE-SULPHONANILIDE * ETHYLMERCURY p-TO-LUENESULFANILIDE * ETHYLMERCURY-p-TO-LUENESULFONANILIDE * ETHYL(N-PHENYL-p-TO-LUENESULFONAMIDATO)MERCURY * ETHYL(N-PHE-NYL-p-TOLUENESULFONAMIDO)MERCURY * ETHYL(p-TOLUENESULFONANILIDATO)MERCURY * GRANOSAN M * (N-PHENYL-p-TOLUENESULFON-AMIDO)ETHYLMERCURY

CONSENSUS REPORTS: Mercury and its compounds are on the Community Right-To-Know List. EPA Genetic Toxicology Program.

OSHA PEL: (Transitional: CL 1 mg/10m^3) TWA 0.01 mg(Hg)/m^3; STEL 0.03 mg/m^3 (skin)
ACGIH TLV: TWA 0.01 mg(Hg)/m^3; STEL 0.03 mg(Hg)/m^3
NIOSH REL: TWA 0.05 mg(Hg)/m^3

SAFETY PROFILE: Poison by ingestion and possibly other routes. Mutation data reported. A fungicide. When heated to decomposition it emits very toxic fumes of Hg, NO_x, and SO_x.

EMF000   CAS: 97-63-2   **HR: 3**
ETHYL METHACRYLATE
DOT: UN 2277
mf: $C_6H_{10}O_2$ mw: 114.16

PROP: A liquid. Mp: $< -75°$, bp: $119°$, lel: 1.8%, uel: saturation, flash p: $68°F$ (OC), d: 0.911 @ $25°/25°$, vap d: 3.94.

SYNS: ETHYL METHACRYLATE, INHIBITED (DOT) * ETHYL-α-METHYL ACRYLATE * ETHYL-2-METHYLACRYLATE * ETHYL-2-METHYL-2-PROPENOATE * 2-METHYL-2-PROPENOIC ACID, ETHYL ESTER * RCRA WASTE NUMBER U118 * RHOPLEX AC-33 (ROHM and HAAS)

CONSENSUS REPORTS: Reported in EPA TSCA Inventory.

DOT Classification: Flammable Liquid; Label: Flammable Liquid.

SAFETY PROFILE: Moderately toxic by ingestion and intraperitoneal routes. Mildly toxic by inhalation. Experimental teratogenic and reproductive effects. A skin irritant. A very dangerous fire and explosion hazard when exposed to heat, sparks or flame; can react with oxidizing materials. To fight fire, use CO_2, dry chemical. When heated to decomposition it emits acrid smoke and irritating fumes.

EMF500 CAS: 62-50-0 **HR: 3**
ETHYL METHANESULFONATE
mf: $C_3H_8O_3S$ mw: 124.17

SYNS: EMS * ENT 26,396 * ETHYL ESTER of METHANESULFONIC ACID * ETHYL ESTER of METHYLSULFONIC ACID * ETHYL ESTER of METHYLSULPHONIC ACID * ETHYL METHANESULPHONATE * ETHYL METHANSULFONATE * ETHYL METHANSULPHONATE * HALF-MYLERAN * METHANESULPHONIC ACID ETHYL ESTER * METHYLSULFONIC ACID, ETHYL ESTER * NSC 26805 * RCRA WASTE NUMBER U119

CONSENSUS REPORTS: IARC Cancer Review: GROUP 2B IMEMDT 7,56,87; Animal Sufficient Evidence IMEMDT 7,245,74. Reported in EPA TSCA Inventory. EPA Genetic Toxicology Program.

SAFETY PROFILE: Suspected carcinogen with experimental carcinogenic, neoplastigenic, and tumorigenic data. Poison by intraperitoneal route. Experimental teratogenic and reproductive effects. Human mutation data reported. When heated to decomposition it emits toxic fumes of SO_x.

EMO500 CAS: 125-42-8 **HR: 3**
5-ETHYL-5-(1-METHYL-1-BUTENYL) BARBITURATE
mf: $C_{11}H_{16}N_2O_3$ mw: 224.29

SYNS: 5-ETHYL-5-(1-METHYL-1-BUTENYL)BARBITURIC ACID * 5-ETHYL-5-(1-METHYL-1-BUTENYL)-2,4,-6(1H,3H,5H)-PYRIMIDINETRIONE

SAFETY PROFILE: Poison by ingestion and intraperitoneal routes. When heated to decomposition it emits toxic fumes of NO_x.

EMP600 CAS: 7452-79-1 **HR: 2**
ETHYL 2-METHYLBUTYRATE
mf: $C_7H_{14}O_2$ mw: 130.19

PROP: Colorless liquid; strong, apple-like odor. D: 0.861-0.866, refr index: 1.396, flash p: +153°F. Sol in alc, propylene glycol; misc in fixed oils; very sltly sol in water.

SYN: FEMA No. 2443

SAFETY PROFILE: Combustible liquid. When heated to decomposition it emits acrid smoke and irritating fumes.

EMQ500 CAS: 105-40-8 **HR: 3**
ETHYL-N-METHYL CARBAMATE
mf: $C_4H_9NO_2$ mw: 103.14

PROP: Needles. Mp: 54°, bp: 177°.

SYNS: METHYLCARBAMIC ACID, ETHYL ESTER * N-METHYL URETHAN

CONSENSUS REPORTS: Reported in EPA TSCA Inventory.

SAFETY PROFILE: Moderately toxic by subcutaneous route. Experimental teratogenic and reproductive effects. Questionable carcinogen with experimental tumorigenic data. When heated to decomposition it emits toxic fumes of NO_x.

EMT000 CAS: 540-67-0 **HR: 3**
ETHYL METHYL ETHER
DOT: UN 1039
mf: C_3H_8O mw: 60.11

PROP: Colorless liquid or gas. Bp: 11°, lel: 2.0%, uel: 10.1%, flash p: -35°F (CC), d: 0.7260 @ 0°/4°, autoign temp: 374°F, vap d: 2.07.

SYNS: ETHOXYMETHANE * ETHYL METHYL ETHER (DOT) * METHOXYETHANE * METHYL ETHYL ETHER (DOT)

DOT Classification: Flammable Liquid; Label: Flammable Liquid; Flammable Gas; Label: Flammable Gas.

SAFETY PROFILE: Has anesthetic properties. A very dangerous fire and moderate explosion hazard when exposed to heat or flame; can react vigorously with oxidizing materials (e.g., air, O_2). To fight fire use alcohol foam, CO_2, dry chemical.

ENB000   CAS: 2058-66-4   **HR: 3**
N-ETHYL-N-METHYL-p-(PHENYLAZO) ANILINE
mf: $C_{15}H_{17}N_3$ mw: 239.35

SYNS: p-ETHYLMETHYLAMINOAZOBENZENE
* N-ETHYL-N-METHYL-p-AMINOAZOBENZENE
* 4-ETHYLMETHYLAMINOAZOBENZENE
* 4-(METHYLETHYL)AMINOAZOBENZENE
* N-METHYL-N-ETHYL-p-AMINOAZOBENZENE

SAFETY PROFILE: Questionable carcinogen with experimental carcinogenic and tumorigenic data. Mutation data reported. When heated to decomposition it emits toxic fumes of NO_x.

ENB500   CAS: 115-38-8   **HR: 3**
5-ETHYL-N-METHYL-5-PHENYLBAR-BITURIC ACID
mf: $C_{13}H_{14}N_2O_3$ mw: 246.29

SYNS: ENFENEMAL * ENPHENEMAL
* N-ETHYLMETHYLPHENYLBARBITURIC ACID
* 5-ETHYL-1-METHYL-5-PHENYLBARBITURIC ACID
* 5-ETHYL-1-METHYL-5-PHENYL-2,4,6(1H,3H,5H)-PYRI-MIDINETRIONE * 5-ETHYL-5-PHENYL-N-METHYLBAR-BITURIC ACID * ISONAL * ISONAL (ROUSSEL)
* MEBARAL * MEBEREL * MENTA-BAL
* MEPHOBARBITAL * MEPHOBARBITONE
* MEPHYTAL * METHYL-CALMINAL
* 1-METHYL-5-ETHYL-5-PHENYLBARBITURIC ACID
* METHYLPHENOBARBITAL * N-METHYL-PHENOBARBITAL * 1-METHYLPHENOBARBITAL
* METHYLPHENOBARBITONE * N-METHYL-PHENOLBARBITOL * METHYLPHENYLBARBI-TURIC ACID * N-METHYL-5-PHENYL-5-ETHYLBARBI-TAL * 1-METHYL-5-PHENYL-5-ETHYLBARBITURIC ACID * METYLFENEMAL * METYNA * MOR-BUSAN * PHEMETONE * PHEMITON
* PHEMITONE * 5-PHENYL-5-ETHYL-3-METHY-LBARBITURIC ACID * PROMINAL

SAFETY PROFILE: Poison by ingestion and intraperitoneal routes. A human teratogen by an unspecified route with developmental abnormalities of the cardiovascular (circulatory) system. When heated to decomposition it emits toxic NO_x.

ENC000   CAS: 77-83-8   **HR: 1**
ETHYL METHYLPHENYLGLYCIDATE
mf: $C_{12}H_{14}O_3$ mw: 206.26

PROP: Colorless to yellowish liquid; strawberry-like odor. D: 1.086-1.112, refr index: 1.504-1.513, flash p: 273°F. Sol in fixed oils, propylene glycol; insol in glycerin.

SYNS: C-16 ALDEHYDE * EMPG * α-β-EPOXY-β-METHYLHYDROCINNAMIC ACID, ETHYL ESTER
* ETHYL α,β-EPOXY-β-METHYLHYDROCINNAMATE
* ETHYL 2,3-EPOXY-3-METHYL-3-PHENYLPROPIONATE
* ETHYL ESTER of 2,3-EPOXY-3-PHENYLBUTANOIC ACID * FEMA No. 2444 * FRAESEOL
* 3-METHYL-3-PHENYLGLYCIDIC ACID ETHYL ESTER
* STRAWBERRY ALDEHYDE

CONSENSUS REPORTS: Reported in EPA TSCA Inventory.

SAFETY PROFILE: Mildly toxic by ingestion. Combustible liquid. When heated to decomposition it emits acrid smoke and irritating fumes.

ENG500   CAS: 77-67-8   **HR: 2**
2-ETHYL-2-METHYLSUCCINIMIDE
mf: $C_7H_{11}NO_2$ mw: 141.19

SYNS: AETHOSUXIMIDE (GERMAN) * ASAMID
* ATYSMAL * CAPITUS * CI 366 * EMESIDE
* EPILEO PETIT MAL * ETHOSUCCIMIDE
* ETHOSUCCINIMIDE * ETHOSUXIDE * ETHO-SUXIMIDE * 3-ETHYL-3-METHYLPYRROLIDINE-2,5-DIONE * 3-ETHYL-3-METHYL-2,5-PYRROLIDINE-DIONE * α-ETHYL-α-METHYLSUCCINIMIDE
* ETHYMAL * ETOMAL * ETOSUXIMIDE
* H-490 * H 940 * MESENTOL * 3-METHYL-3-ETHYLPYRROLIDINE-2,5-DIONE * γ-METHYL-γ-ETHYL-SUCCINIMIDE * PEMAL * PEMALIN
* PENTINIMID * PETINIMID * PETNIDAN
* PM 671 * PYKNOLEPSINUM * RONTON
* SIMATIN(E) * SUCCIMAL * SUCCIMITIN
* SUXILEP * SUXIMAL * SUXIN * SUXINUTIN
* THETAMID * THILOPEMAL * ZARAONDAN
* ZARODAN * ZARONDAN-SAFT * ZARONTIN
* ZARTALIN

SAFETY PROFILE: Moderately toxic by ingestion, intravenous, subcutaneous and intraperitoneal routes. Experimental teratogenic and reproductive effects. An anticonvulsant. When heated to decomposition it emits toxic fumes of NO_x.

ENJ000   CAS: 458-24-2   **HR: 3**
N-ETHYL-α-METHYL-m-(TRIFLUORO-METHYL)PHENETHYLAMINE
mf: $C_{12}H_{16}F_3N$ mw: 231.29

SYNS: FENFLURAMINE * 3-(TRIFLUOROMETHYL)-N-ETHYL-α-METHYL PHENETHYL AMINE

SAFETY PROFILE: A human poison by ingestion. An experimental poison by ingestion and intraperitoneal routes. Human systemic effects by ingestion: hallucinations, distorted perceptions and autonomic nervous system effects (an adrenergic stimulant.) Experimental reproductive effects. When heated to decomposition it emits very toxic fumes of F^- and NO_x.

ENK000   CAS: 76-58-4   **HR: 3**
ETHYLMORPHINE
mf: $C_{19}H_{23}NO_3$ mw: 313.43

SYNS: CODETHYLINE * (5-α,6-α)-7,8-DIDEHYDRO-4,5-EPOXY-3-ETHOXY-17-METHYLMORPHINAN-6-OL * DIONINE * DIONIN * 3-o-ETHYLMORPHINE

SAFETY PROFILE: Poison by intraperitoneal, intravenous, and subcutaneous routes. Moderately toxic by ingestion. When heated to decomposition it emits toxic fumes of NO_x.

ENK500   CAS: 6746-59-4   **HR: 3**
ETHYL MORPHINE HYDROCHLORIDE DIHYDRATE
mf: $C_{19}H_{23}NO_3 \cdot ClH \cdot 2H_2O$ mw: 385.93

PROP: White, microscopic, crystalline powder. Mp: 125° (decomp), vap d: 13.3.

SYNS: 7,8-DIDEHYDRO-4,5-α-EPOXY-3-ETHOXY-17-METHYLMORPHINAN-6-α-OL HYDROCHLORIDE DIHYDRATE * DIONIN * ETHYLMORPHINE HYDROCHLORIDE

SAFETY PROFILE: Poison by subcutaneous route. Can be habit forming. When heated to decomposition it emits very toxic fumes of NO_x and HCl.

ENL000   CAS: 100-74-3   **HR: 3**
N-ETHYLMORPHOLINE
mf: $C_6H_{13}NO$ mw: 115.20

PROP: Colorless liquid. Bp: 138°, flash p: 89.6°F (OC), d: 0.916 @ 20°/20°, vap d: 4.00.

SYN: 4-ETHYLMORPHOLINE

CONSENSUS REPORTS: Reported in EPA TSCA Inventory.

OSHA PEL: (Transitional: TWA 20 ppm (skin) TWA 5 ppm (skin)

ACGIH TLV: TWA 5 ppm (skin)

SAFETY PROFILE: Poison by intravenous route. Moderately toxic by ingestion. Mildly toxic by inhalation. A skin and severe eye irritant. A very dangerous fire hazard when exposed to heat or flame; can react vigorously with oxidizing materials. To fight fire, use alcohol foam, foam, CO_2, dry chemical. When heated to decomposition it emits toxic fumes of NO_x.

ENL850   CAS: 124-06-1   **HR: 1**
ETHYL MYRISTATE
mf: $C_{16}H_{23}O_2$ mw: 256.42

PROP: Colorless to pale yellow liquid; waxy odor. D: 0.857, refr index: 1.434, flash p: +212°F.

SYN: FEMA No. 2445

SAFETY PROFILE: Combustible liquid. When heated to decomposition it emits acrid smoke and irritating fumes.

ENM500   CAS: 625-58-1   **HR: 3**
ETHYL NITRATE
DOT: UN 1993
mf: $C_2H_5NO_3$ mw: 91.08

PROP: Colorless liquid, pleasant odor, sweet taste. Mp: −112°, bp: 88.7°, lel: 3.8%, flash p: 50°F (CC), d: 1.105 @ 20°/4°, vap d: 3.14.

SYNS: NITRIC ACID, ETHYL ESTER * NITRIC ETHER (DOT)

CONSENSUS REPORTS: Reported in EPA TSCA Inventory.

DOT Classification: Flammable Liquid; Label: Flammable Liquid.

SAFETY PROFILE: Mutation data reported. A very dangerous fire hazard when exposed to heat or flame; can react vigorously with oxidizing materials. A moderate explosion hazard when exposed to heat (explodes @ 185°F). To fight fire, use foam, CO_2, dry chemical, water to blanket fire. Incompatible with Lewis acids. When heated to decomposition it emits toxic fumes of NO_x.

ENN000   CAS: 109-95-5   **HR: 3**
ETHYL NITRITE
DOT: UN 1194
mf: $C_2H_5NO_2$ mw: 75.08

PROP: Colorless or yellowish liquid; highly aromatic, ethereal odor. Decomp on standing. Very sltly sol in water; misc in alc and ether. Bp: 16.4°, lel: 3.0%, uel: 50%, explodes at 194°F, flash p: −31°F (CC), d: 0.900 @ 15.5°, autoign temp: 194°F, vap d: 2.59. Can explode > 90°C.

SYNS: ETHYL NITRITE (DOT) * ETHYL NITRITE, solution (DOT) * NITROSYL ETHOXIDE * NITROUS ACID ETHYL ESTER * NITROUS ETHER (DOT) * NITROUS ETHYL ETHER

CONSENSUS REPORTS: Reported in EPA TSCA Inventory.

DOT Classification: Flammable Liquid; Label: Flammable Liquid.

SAFETY PROFILE: A human poison by an un-specified route. Narcotic in high concentrations. Lowers blood pressure. Methemoglobinemia has been reported. A very dangerous fire and severe explosion hazard when exposed to heat or flame. A powerful oxidizer. May explode when heated above 90°C. Highly dangerous when heated to decomposition or on contact with acid or acid fumes. To fight fire, use foam, CO_2, dry chemical, or water spray. When heated to decomposition it emits toxic fumes of NO_x.

ENR500   CAS: 65986-80-3   HR: 3
(ETHYLNITROSAMINO)METHYL ACETATE
mf: $C_5H_{10}N_2O_3$ mw: 146.17

SYNS: ACETOXYMETHYLETHYLNITROSAMINE * N-(ACETOXY)METHYL-N-ETHYLNITROSAMINE * N-ACETOXYMETHYL-N-NITROSOETHYLAMINE * N-(1-ACETOXYMETHYL)-N-NITROSOETHYL AMINE * AETHYL ACETOXYMETHYLNITROSAMIN (GERMAN) * EAMN * ETHYL ACETOXYMETHYLNITROSAMINE * N-ETHYL-N-(ACETOXYMETHYL)NITROSAMINE

SAFETY PROFILE: Moderately toxic by inges-tion. Questionable carcinogen with experimen-tal carcinogenic and tumorigenic data. Mutation data reported. When heated to decomposition it emits toxic fumes of NO_x.

ENT000   CAS: 32976-88-8   HR: 3
N-ETHYL-N-NITROSOBIURET
mf: $C_4H_8N_4O_3$ mw: 160.16

SYNS: ENBU * ETHYLNITROSOBIURET * N-NITROSO-N-ETHYL BIURET

SAFETY PROFILE: Moderately toxic by inges-tion. Questionable carcinogen with experimen-tal carcinogenic, and tumorigenic data. Experi-mental teratogenic and reproductive effects. When heated to decomposition it emits toxic fumes of NO_x.

ENU000   CAS: 4245-77-6   HR: 3
N-ETHYL-N-NITROSO-N′-NITRO-GUANIDINE
mf: $C_3H_7N_5O_3$ mw: 161.15

SYNS: N-AETHYL-N′-NITRO-N-NITROSOGUANIDIN (GERMAN) * ENNG * N-ETHYL-N′-NITRO-N-NITRO-SOGUANIDINE

CONSENSUS REPORTS: EPA Genetic Toxi-cology Program.

SAFETY PROFILE: Questionable carcinogen with experimental carcinogenic, neoplastigenic, and tumorigenic data. Human mutation data re-ported. When heated to decomposition it emits toxic fumes of NO_x.

ENV000   CAS: 759-73-9   HR: 3
1-ETHYL-1-NITROSOUREA
mf: $C_3H_7N_3O_2$ mw: 117.13

PROP: Pale yellow crystals. Mp: 103° (decomp) 1% water soln @ 20°.

SYNS: AENH (GERMAN) * AETHYLNITROSO-HARNSTOFF (GERMAN) * ENU * N-ETHYL-N-NI-TROSOCARBAMIDE * ETHYLNITROSOUREA * N-ETHYL-N-NITROSO-UREA * NEU * NITROSO-ETHYLUREA * NSC 45403 * RCRA WASTE NUM-BER U176

CONSENSUS REPORTS: IARC Cancer Re-view: GROUP 2A IMEMDT 7,56,87; Human Limited Evidence IMEMDT 17,191,78; Animal Sufficient Evidence IMEMDT 17,191,78; IMEMDT 1,135,72. NTP Fourth Annual Report On Carcinogens, 1984. EPA Genetic Toxicology Program. Community Right-To-Know List. Reported in EPA TSCA Inventory.

SAFETY PROFILE: Confirmed carcinogen with experimental carcinogenic, neoplastigenic, tu-morigenic, and teratogenic data. Poison by in-gestion, subcutaneous, and intravenous routes. Moderately toxic by intraperitoneal route. Hu-man mutation data reported. When heated to decomposition it emits toxic fumes of NO_x.

ENV500   CAS: 139-94-6   **HR: 3**
**1-ETHYL-3-(5-NITRO-2-THIAZOLYL)
UREA**
mf: $C_6H_8N_4O_3S$ mw: 216.24

SYNS: N-ETHYL-N′-(5-NITRO-2-THIAZOLYL)UREA
* HEPZIDE * NCI-C03792 * NITHIAZID
* NITHIAZIDE

CONSENSUS REPORTS: IARC Cancer Review: GROUP 3 IMEMDT 7,56,87; Animal Limited Evidence IMEMDT 31,179,83. NCI Carcinogenesis Bioassay (feed); Clear Evidence: mouse, rat NCITR* NCI-CG-TR-146,79

SAFETY PROFILE: Suspected carcinogen with experimental carcinogenic data. Moderately toxic by ingestion. Mutation data reported. When heated to decomposition it emits very toxic fumes of NO_x and PO_x.

ENW000   CAS: 123-29-5   **HR: 1**
ETHYL NONANOATE
mf: $C_{11}H_{22}O_2$ mw: 186.33

PROP: Colorless liquid; fruity, cognac odor. D: 0.863-0867, refr index: 1.420, flash p: 185°F. Misc with alc, propylene glycol; insol in water.

SYNS: ETHYL NONYLATE * ETHYL PELARGONATE
* FEMA No. 2447 * NONANOIC ACID, ETHYL ESTER
* WINE ETHER

CONSENSUS REPORTS: Reported in EPA TSCA Inventory.

SAFETY PROFILE: Mildly toxic by ingestion. A skin irritant. Combustible liquid. When heated to decomposition it emits acrid smoke and irritating fumes.

ENX500   CAS: 3198-07-0   **HR: 3**
**ETHYLNORADRENALINE
HYDROCHLORIDE**
mf: $C_{10}H_{15}NO_3 \cdot ClH$ mw: 233.72

SYNS: α-(1-AMINOPROPYL)PROTOCATECHUYL ALCO-
HOL HYDROCHLORIDE * BRONKEPHRINE HYDRO-
CHLORIDE * BUTANEFRINE HYDROCHLORIDE
* 1-(3,4-DIHYDROXYPHENYL)-2-AMINO-1-BUTANOL
HYDROCHLORIDE * 1-(3,4-DIHYDROXYPHENYL)-1-
HYDROXY-2-AMINOBUTANE HYDROCHLORIDE
* E.N.E. * E.N.S. * ETHYL NOREPINEPHRINE
HYDROCHLORIDE * α-ETHYLNOREPINEPHRINE HY-
DROCHLORIDE * ETHYLNORSUPRARENIN HYDRO-
CHLORIDE

SAFETY PROFILE: Poison by intravenous and subcutaneous routes. Human systemic effects by subcutaneous, intravenous, and intramuscular routes: heart rate change, blood pressure decrease and pulse pressure increase. When heated to decomposition it emits very toxic fumes of Cl^- and NO_x.

ENY000   CAS: 106-32-1   **HR: 1**
ETHYL OCTANOATE
mf: $C_{10}H_{20}O_2$ mw: 172.30

PROP: Colorless liquid; wine-brandy fruit odor. D: 0.865-0.869, refr index: 1.417, flash p: 185°F. Sol in fixed oils; sltly sol in propylene glycol; insol in glycerin, water @ 209°.

SYNS: ETHYL CAPRYLATE * ETHYL OCTYLATE
* FEMA No. 2449 * OCTANOIC ACID, ETHYL ESTER

CONSENSUS REPORTS: Reported in EPA TSCA Inventory.

SAFETY PROFILE: Mildly toxic by ingestion. A skin irritant. Combustible liquid. When heated to decomposition it emits acrid smoke and irritating fumes.

ENY500   CAS: 122-51-0   **HR: 2**
ETHYL ORTHOFORMATE
DOT: UN 2524
mf: $C_7H_{16}O_3$ mw: 148.23

PROP: Clear liquid, pungent odor. Bp: 145.9°, flash p: 86°F (CC), d: 0.895 @ 20°/20°, vap press: 10 mm @ 40.5°, vap d: 5.11.

SYNS: AETHON * ETHONE * ETHYLESTER KY-
SELINY ORTHOMRAVENCI (CZECH) * 1,1′,1′-(METH-
YLIDYNETRIS(OXY))TRIS(ETHANE) * ORTHOFORMIC
ACID, ETHYL ESTER * ORTHOFORMIC ACID, TRI-
ETHYL ESTER * ORTHOMRAVENCAN ETHYLNATY
(CZECH) * TRIETHOXYMETHANE * TRIETHYL
ORTHOFORMATE

CONSENSUS REPORTS: Reported in EPA TSCA Inventory.

DOT Classification: Flammable or Combustible Liquid; Label: Flammable Liquid.

SAFETY PROFILE: Moderately toxic by ingestion. Mildly toxic by inhalation, skin contact, and subcutaneous routes. A skin and eye irritant. A very dangerous fire hazard when exposed to heat or flame; can react vigorously with oxidizing materials. To fight fire, use foam, CO_2, dry chemical. When heated to decomposition it emits acrid smoke and irritating fumes.

EOD000   CAS: 22750-93-2   **HR: 3**
ETHYL PERCHLORATE
mf: $C_2H_5ClO_4$ mw: 128.52

DOT Classification: Forbidden

SAFETY PROFILE: Possibly the most explosive chemical known. Very sensitive to impact, friction and heat. Upon decomposition it emits toxic fumes of Cl^-.

EOH000   CAS: 101-97-3   **HR: 2**
ETHYL PHENYLACETATE
mf: $C_{10}H_{12}O_2$ mw: 164.22

PROP: Colorless liquid; sweet, honey odor. Bp: 227°, d: 1.033 @ 20°, refr index: 1.496-1.500, vap d: 5.67, flash p: +100 C. Sol in fixed oils; insol in glycerin, propylene glycol, water.

SYNS: BENZENEACETIC ACID, ETHYL ESTER (9CI) * ETHYL BENZENEACETATE * ETHYL PHENACE-TATE * ETHYL-2-PHENYLETHANOATE * ETHYL-α-TOLUATE * FEMA No. 2452 * PHENYLACETIC ACID, ETHYL ESTER * α-TOLUIC ACID, ETHYL ESTER

CONSENSUS REPORTS: Reported in EPA TSCA Inventory.

SAFETY PROFILE: Moderately toxic by ingestion. Combustible liquid. When heated to decomposition it emits acrid smoke and irritating fumes.

EOJ500   CAS: 6368-72-5   **HR: 3**
N-ETHYL-1-((p-(PHENYLAZO)PHENYL) AZO)-2-NAPHTHYLAMINE
mf: $C_{24}H_{21}N_5$ mw: 379.50

SYNS: CERES RED 7B * C.I. 26050 * C.I. SOLVENT RED 19 * N-ETHYL-1-((p-(PHENYLAZO)PHE-NYL)AZO)-2-NAPHTHALENAMINE * N-ETHYL-1-((4-(PHENYLAZO)PHENYL)AZO)-2-NAPHTHALENAMINE * N-ETHYL-1-((4-(PHENYLAZO)PHENYL)AZO)-2-NAPHTHYLAMINE * FAT RED 7B * HEXATYPE CARMINE B * LACQUER RED V3B * OIL VIOLET * ORGANOL BORDEAUX B * (PHENYLAZO-4-PHE-NYLAZO)-1-ETHYLAMINO-2-NAPHTHALENE * 1-(4-PHENYLAZO-PHENYLAZO)-2-ETHYLAMINONAPHTHA-LENE * SOLVENT RED 19 * SPECIAL BLUE X 2137 * SUDAN RED 7B * SUDANROT 7B * TYPOGEN CARMINE

CONSENSUS REPORTS: IARC Cancer Review: GROUP 3 IMEMDT 7,56,87; Animal Inadequate Evidence IMEMDT 8,253,75. Reported in EPA TSCA Inventory.

SAFETY PROFILE: Questionable carcinogen with experimental tumorigenic data. Mutation data reported. When heated to decomposition it emits toxic fumes of NO_x.

EOK000   CAS: 50-06-6   **HR: 3**
5-ETHYL-5-PHENYLBARBITURIC ACID
mf: $C_{12}H_{12}N_2O_3$ mw: 232.26

SYNS: ACIDO-5-FENIL-5-ETILBARBITURICO (ITALIAN) * ADONAL * AEPHENAL * AGRYPNAL * AMYLOFENE * APHENYLBARBIT * APHENY-LETTEN * AUSTROMINAL * BARBAPIL * BARBELLON * BARBENYL * BARBILEHAE (BARBILETTAE) * BARBINAL * BARBIPHENYL * BARBITA * BARBIVIS * BARBONAL * BARBOPHEN * BARDORM * BARTOL * BIALMINAL * BLU-PHEN * CABRONAL * CALMETTEN * CALMINAL * CARDENAL * CODIBARBITA * CORONALETTA * CRATECIL * DAMORAL * DEZIBARBITUR * DORMINA * DORMIRAL * DOSCALUN * DUNERYL * ENSOBARB * ENSODORM * EPANAL * EPIDORM * EPILOL * EPISEDAL * EPSY-LONE * ESKABARB * 5-ETHYL-5-PHENYL-2,4,6-(1H,3H,5H)PYRIMIDINETRIONE * ETILFEN * EUNERYL * FENBITAL * FENEMAL * FENOBARBITAL * FENOSED * FENYLETTAE * GARDENAL * GARDEPANYL * GLYSOLETTEN * HAPLOPAN * HAPLOS * HELIONAL * HENNOLETTEN * HYPNALETTEN * HYPNOGEN * HYPNOLONE * HYPNO-TABLINETTEN * HYS-TEPS * LEFEBAR * LEONAL * LEPHEBAR * LEPINAL * LEPINALETTEN * LINASEN * LIQUITAL * LIXOPHEN * LUBERGAL * LUBROKAL * LUMEN * LUMESETTES * LUMESYN * LUMINAL * LUMOFRIDETTEN * LUPHENIL * LURAMIN * MOLINAL * NEUROBARB * NIRVONAL * NOPTIL * NOVA-PHENO * NUNOL * PARKOTAL * PHARMETTEN * PHENAEMAL * PHENOBAL * PHENOBARBITAL * PHENOBARBITONE * PHENOBARBITURIC ACID * PHENOLURIC * PHENOMET * PHENONYL * PHENOTURIC * PHENYLETHYLBARBITURATE * PHENYL-ETHYL-BARBITURIC ACID * 5-PHENYL-5-ETHYLBARBITURIC ACID * PHENYLETHYLMALONYLUREA * PHENY-LETTEN * PHENYRAL * PHOB * POLCOMINAL * PROMPTONAL * SEDABAR * SEDA-TABLINEN * SEDICAT * SEDIZORIN * SEDLYN * SEDO-FEN * SEDONAL * SEDONETTES * SEDOPHEN * SEVENAL * SK-PHENOBARBITAL * SOLFOTON

* SOMBUTOL * SOMNOLENS * SOMNOLETTEN
* SOMNOSAN * SOMONAL * SPASEPILIN
* STARIFEN * STARILETTAE * STENTAL EXTEN-
TABS * TALPHENO * TEOLAXIN * THENOBAR-
BITAL * THEOLOXIN * THEOMINAL * TRIA-
BARB * TRIDEZIBARBITUR * TRIPHENATOL
* VERSOMNAL * ZADOLETTEN * ZADONAL

CONSENSUS REPORTS: EPA Genetic Toxicology Program. IARC Cancer Review: GROUP 2B IMEMDT 7,313,87; Human Inadequate Evidence IMEMDT 13,157,77.

SAFETY PROFILE: Suspected carcinogen with experimental carcinogenic, neoplastigenic, and tumorigenic data. A human poison by ingestion. An experimental poison by ingestion, intraperitoneal, subcutaneous, intravenous, and rectal routes. Human systemic effects by ingestion: somnolence, motor activity changes, pulmonary changes, allergic dermatitis and fever. A human teratogen. Human reproductive effects by ingestion and possibly other routes: drug dependence and other postnatal measures or effects. Human teratogenic effects include developmental abnormalities of the central nervous system, body wall, musculoskeletal, respiratory, gastrointestinal and urogenital systems. Experimental teratogenic and reproductive effects. Human mutation data reported. Used as a drug in the treatment of epilepsy, and as an hypnotic and sedative. When heated to decomposition it emits toxic fumes of NO_x.

EOK600   CAS: 121-39-1   **HR: 2**
ETHYL PHENYLGLYCIDATE
mf: $C_{11}H_{12}O_3$ mw: 192.23

PROP: Colorless liquid; strong strawberry odor. D: 1.120, refr index: 1.516-1.521. Sol in alc, chloroform, ether; insol in water.

SYNS: ETHYL-α,β-EPOXYHYDROCINNAMATE * ETHYL-α,β-EPOXY-α-PHENYLPROPIONATE * ETHYL-3-PHENYLGLYCIDATE * FEMA No. 2454

CONSENSUS REPORTS: Reported in EPA TSCA Inventory.

SAFETY PROFILE: Moderately toxic by ingestion. Mutation data reported. When heated to decomposition it emits acrid smoke and irritating fumes.

EOL000   CAS: 631-07-2   **HR: 3**
ETHYLPHENYLHYDANTOIN
mf: $C_{11}H_{12}N_2O_2$ mw: 204.25

PROP: Colorless, odorless, crystalline powder. Mp: 199-200°.

SYNS: 5-ETHYL-5-PHENYLHYDANTOIN * 5-ETHYL-5-PHENYL-2,4-IMIDAZOLIDINEDIONE * NIRVANOL

SAFETY PROFILE: Poison by subcutaneous route. Moderately toxic by ingestion. A skin, eye, and mucous membrane irritant. Combustible when exposed to heat or flame; can react with oxidizing materials. When heated to decomposition it emits toxic fumes such as NO_x.

EOQ000   CAS: 1498-40-4   **HR: 2**
ETHYLPHOSPHONOUS DICHLORIDE

DOT: UN 2845
mf: $C_2H_5Cl_2P$ mw: 130.94

SYNS: DICHLOROETHYLPHOSPHINE * DICHLOROMETHYL PHOSPHINE * ETHYL PHOSPHONOUS DICHLORIDE, anhydrous (DOT) * TL 373

DOT Classification: Corrosive Material; Label: Corrosive.

SAFETY PROFILE: Moderately toxic by inhalation. Corrosive. A severe irritant to skin, eyes, and mucous membranes. When heated to decomposition it emits very toxic fumes of PO_x, Cl^-, and phosphine.

EOR000   CAS: 1498-51-7   **HR: 2**
ETHYL PHOSPHORODICHLORIDATE

DOT: NA 1760
mf: $C_2H_5Cl_2O_2P$ mw: 162.94

SYNS: DICHLOROPHOSPHORIC ACID, ETHYL ESTER * PHOSPHORODICHLORIDIC ACID, ETHYL ESTER

CONSENSUS REPORTS: Reported in EPA TSCA Inventory.

DOT Classification: Corrosive Material; Label: Corrosive.

SAFETY PROFILE: A corrosive material which is very toxic to tissue. A severe eye, skin, and mucous membrane irritant. When heated to decomposition it emits very toxic fumes of Cl^- and PO_x.

EOS000   CAS: 104-90-5   **HR: 3**
5-ETHYL-α-PICOLINE

DOT: UN 2300
mf: $C_8H_{11}N$ mw: 121.20

PROP: Liquid. Bp: 174°, d: 0.9184 @ 23°/4°, flash p: 165° (OC).

SYNS: ALDEHYDECOLLIDINE * ALDEHYDINE * COLLIDINE, ALDEHYDECOLLIDINE * 3-ETHYL-6-METHYLPYRIDINE * 5-ETHYL-2-METHYLPYRIDINE * 5-ETHYL-2-PICOLINE * MEP * 2-METHYL-5-ETHYLPYRIDINE * 6-METHYL-3-ETHYLPYRIDINE * METHYL ETHYL PYRIDINE (DOT) * 2-METHYL-5-ETHYLPYRIDINE (DOT)

CONSENSUS REPORTS: Reported in EPA TSCA Inventory. EPA Genetic Toxicology Program.

DOT Classification: Corrosive Material; Label: Corrosive; Poison B; Label: St. Andrews Cross.

SAFETY PROFILE: Poison by ingestion and subcutaneous routes. Moderately toxic by skin contact. Mildly toxic by inhalation. Corrosive. A severe skin and eye irritant. Flammable when exposed to heat or flame; can react vigorously with oxidizers. Potentially explosive reaction with nitric acid at 145°C/14.5 bar. To fight fire, use alcohol foam. When heated to decomposition it emits acrid smoke and irritating fumes.

EOS500   CAS: 766-09-6   **HR: 3**
1-ETHYLPIPERIDINE

DOT: UN 2386
mf: $C_7H_{15}N$ mw: 113.23

PROP: Flash p: 66.2°F.

SYN: N-AETHYLPIPERIDIN (GERMAN)

CONSENSUS REPORTS: Reported in EPA TSCA Inventory.

DOT Classification: Flammable Liquid; Label: Flammable Liquid.

SAFETY PROFILE: Poison by intravenous and subcutaneous routes. An eye irritant. A very dangerous fire hazard when exposed to heat or flame; can react vigorously with oxidizing materials. When heated to decomposition it emits toxic fumes of NO_x.

EPB500   CAS: 105-37-3   **HR: 3**
ETHYL PROPIONATE

DOT: UN 1195
mf: $C_5H_{10}O_2$ mw: 102.15

PROP: Colorless liquid; fruity, rum odor. Bp: 210°F, mp: −72.6°, flash p: 54°F (CC), d: 0.886, refr index: 1.383, autoign temp: 824°F, vap press: 40 mm @ 27.2°, vap d: 3.52, lel:

1.9%, uel: 11%. Misc with alc, ether, propylene glycol; sol in fixed oils; 1 mL in 42 mL water @ 99°.

SYNS: FEMA No. 2456 * PROPIONATE d'ETHYLE (FRENCH) * PROPIONIC ACID, ETHYL ESTER * PROPIONIC ETHER

CONSENSUS REPORTS: Reported in EPA TSCA Inventory.

DOT Classification: Flammable Liquid; Label: Flammable Liquid.

SAFETY PROFILE: Moderately toxic by ingestion and intraperitoneal routes. A skin irritant. Flammable Liquid. A very dangerous fire and explosion hazard when exposed to heat or flame; can react vigorously with oxidizing materials. To fight fire, use foam, CO_2, dry chemical. When heated to decomposition it emits acrid smoke and irritating fumes.

EPC000   CAS: 67050-97-9   **HR: 3**
5-ETHYL-5-(1-PROPYL-1-BUTENYL)
BARBITURIC ACID
mf: $C_{13}H_{20}N_2O_3$ mw: 252.35

SAFETY PROFILE: Poison by ingestion and intraperitoneal routes. When heated to decomposition it emits toxic fumes of NO_x.

EPC125   CAS: 628-32-0   **HR: 3**
ETHYL PROPYL ETHER

DOT: UN 2615
mf: $C_5H_{12}O$ mw: 88.15

PROP: D: 0.8; bp: 147°F; flash p: < −4°F; lel: 1.7%; uel: 9%.

SYNS: 1-ETHOXYPROPANE * PROPYLETHYL ETHER

DOT Classification: Flammable Liquid; Label: Flammable Liquid.

SAFETY PROFILE: Very dangerous fire and explosion hazard when exposed to heat or open flame. To fight fire, use alcohol foam. When heated to decomposition it emits acrid smoke and fumes.

EPC500   CAS: 297-97-2   **HR: 3**
ETHYL PYRAZINYL
PHOSPHOROTHIOATE
mf: $C_8H_{13}N_2O_3PS$ mw: 248.26

PROP: Amber liquid. Mp: −1.7°, bp: 80°, n (25/D) 1.5131, vap press @ 30°: 0.003 mm

Hg. Sltly sol in water; misc with most organic solvents.

SYNS: AC 18133 * AMERCIAN CYANAMID 18133 * CL 18133 * CYNEM * O,O-DIAETHYL-O-(PYRA-ZIN-2YL)-MONOTHIOPHOSPHAT (GERMAN) * O,O-DIAETHYL-O-(2-PYRAZINYL)-THIONOPHOSPHAT (GERMAN) * O,O-DIETHYL-O,2-PYRAZINYL PHOS-PHOROTHIOATE * DIETHYL-O-2-PYRAZINYL PHOSPHOROTHIONATE * O,O-DIETHYL-O-2-PYRAZI-NYL PHOSPHOTHIONATE * O,O-DIETHYL-O-PYRAZI-NYL THIOPHOSPHATE * EN 18133 * ENT 25,580 * EXPERIMENTAL NEMATOCIDE 18,133 * NEMAFOS * NEMAPHOS * NEMATOCIDE * PHOSPHORO-THIOIC ACID-O,O-DIETHYL-O-2-PYRAZINYL ESTER * PYRAZINOL-O-ESTER with O,O-DIETHYL PHOS-PHOROTHIOATE * RCRA WASTE NUMBER P404 * THIONAZIN * ZINOPHOS

CONSENSUS REPORTS: EPA Extremely Hazardous Substances List. Reported in EPA TSCA Inventory. Community Right-To-Know List.

SAFETY PROFILE: Poison by ingestion, skin contact, and ocular routes. A cholinesterase inhibitor type of insecticide. When heated to decomposition it emits highly toxic fumes of NO_x, PO_x, and SO_x.

EPF550   CAS: 78-10-4   **HR: 3**
ETHYL SILICATE
DOT: UN 1292
mf: $C_8H_{20}O_4Si$ mw: 208.37

PROP: Colorless, flammable liquid. Mp: $-77°$, bp: 165-166°. flash p: 125°F (52°C), d: (20/4) 0.933, n (25/D) 1.3818. Viscosity 0.6 cps. Practically insol in water with slow decomp. Miscible with alc.

SYNS: ETHYL ORTHOSILICATE * ETYLU KRZE-MIAN (POLISH) * EXTREMA * SILICATE D'ETHYLE (FRENCH) * SILICIC ACID TETRAETHYL ESTER * TEOS * TETRAETHOXYSILANE * TETRAETHYL ORTHOSILICATE * TETRAETHYL ORTHOSILICATE (DOT) * TETRAETHYL SILICATE * TETRAETHYL SILICATE (DOT)

CONSENSUS REPORTS: Reported in EPA TSCA Inventory.

OSHA PEL: (Transitional: TWA 100 ppm) TWA 10 ppm
ACGIH TLV: TWA 10 ppm
DFG MAK: 100 ppm (850 mg/m³)

DOT Classification: Flammable or Combustible Liquid; Label: Flammable Liquid.

SAFETY PROFILE: Poison by intravenous route. Moderately toxic by other routes. A skin, mucous membrane and severe eye irritant. Narcotic in high concentrations. Flammable when exposed to heat or flame; can react vigorously with oxidizing materials. When heated to decomposition it emits acrid smoke and fumes.

EPH000   CAS: 352-93-2   **HR: 1**
ETHYL SULFIDE
DOT: UN 2375
mf: $C_4H_{10}S$ mw: 90.20

PROP: Liquid, garlic-like odor. Mp: $-102°$, bp: 92-93°, d: 0.837 @ 20°/4°, vap d: 3.11. Flash p: 14°F.

SYNS: DIETHYLSULFID (CZECH) * DIETHYL SUL-FIDE (DOT) * DIETHYLTHIOETHER * ETHYL MONOSULFIDE * ETHYLTHIOETHANE * ETHYL THIOETHER * SULFODOR (CZECH) * 3-THIAPEN-TANE * 1,1'-THIOBISETHANE * THIOETHYL ETHER

CONSENSUS REPORTS: Reported in EPA TSCA Inventory.

DOT Classification: Flammable Liquid; Label: Flammable Liquid and Poison.

SAFETY PROFILE: Mildly toxic by ingestion. A skin and eye irritant. A very dangerous fire hazard when exposed to heat, flame, or sparks; can react vigorously with oxidizers. Reacts with water, steam, acids, or acid fumes to produce toxic and flammable vapors. To fight fire, use water spray or mist, dry chemical, CO_2, foam. When heated to decomposition it yields highly toxic fumes of SO_x.

EPI300   CAS: 842-00-2   **HR: 3**
4-(ETHYLSULFONYL)-1-NAPHTHALENE SULFONAMIDE
mf: $C_{12}H_{13}NO_4S_2$ mw: 299.38

SYNS: ENS * 4-ETHYLSULPHONYLNAPHTHALENE-1-SULFONAMIDE * 4-ETHYLSULPHONYLNAPHTHAL-ENE-1-SULPHONAMIDE * HPA

SAFETY PROFILE: Questionable carcinogen with experimental carcinogenic, neoplastigenic, and tumorigenic data. Mutation data reported. When heated to decomposition it emits very toxic fumes of NO_x and SO_x.

EPJ000 CAS: 20941-65-5 **HR: 3**
ETHYL TELLURAC
mf: $C_{20}H_{40}N_4S_8 \cdot Te$ mw: 720.72

PROP: Orange-yellow powder. D: 1.44, mp: 108-118°.

SYNS: DIETHYLDITHIO CARBAMIC ACID TELLURIUM SALT * NCI-C02857 * TELLURIUM DIETHYLDITHIOCARBAMATE * TETRAKIS(DIETHYLCARBAMODITHIOATO-S,S')TELLURIUM * TETRAKIS(DIETHYLDITHIOCARBAMATO)TELLURIUM

CONSENSUS REPORTS: IARC Cancer Review: GROUP 3 IMEMDT 7,56,87; Animal Inadequate Evidence IMEMDT 12,115,76. NCI Carcinogenesis Bioassay (feed); No Evidence: mouse, rat NCITR* NCI-CG-TR-152,79; Results Indefinite: Mouse, Rat NCITR* NCI-CG-TR-152,79. Reported in EPA TSCA Inventory.

OSHA PEL: TWA 0.1 mg(Te)/m^3
ACGIH TLV: TWA 0.1 mg(Te)/m^3

SAFETY PROFILE: Questionable carcinogen with experimental tumorigenic data. When heated to decomposition it emits very toxic fumes of NO_x, SO_x, and Te.

EPP000 CAS: 542-90-5 **HR: 3**
ETHYL THIOCYANATE
mf: C_3H_5NS mw: 87.15

PROP: D: 0.996, mp: −85.5, bp: 145°. Insol in water; misc in alc and ether.

SYNS: AETHYLRHODANID (GERMAN) * ETHYL RHODANATE * ETHYL SULFOCYANATE * THIOCYANATOETHANE * THIOCYANIC ACID, ETHYL ESTER

CONSENSUS REPORTS: Community Right-To-Know List. Reported in EPA TSCA Inventory.

SAFETY PROFILE: Poison by ingestion, subcutaneous, intraperitoneal, and intravenous routes. When heated to decomposition it emits very toxic fumes of NO_x and SO_x.

EPQ000 CAS: 536-33-4 **HR: 3**
2-ETHYLTHIOISONICOTINAMIDE
mf: $C_8H_{10}N_2S$ mw: 166.26

SYNS: AETINA * AETIVA * AMIDAZIN * BAYER 5312 * ETH * ETHIMIDE * ETHINA * ETHINAMIDE * ETHIONIAMIDE * α-ETHYLISONICOTINIC ACID THIOAMIDE * 2-ETHYLISONICO-TINIC ACID THIOAMIDE * 2-ETHYLISONICOTINIC THIOAMIDE * α-ETHYLISONICOTINOYLTHIOAMIDE * ETHYLISOTHIAMIDE * α-ETHYLISOTHIONICO-TINAMIDE * 2-ETHYLISOTHIONICOTINAMIDE * 2-ETHYL-4-PYRIDINECARBOTHIOAMIDE * 2-ETHYL-4-THIOAMIDYLPYRIDINE * 2-ETHYL-4-THIO-CARBAMOYLPYRIDINE * α-ETHYLTHIOISONICOTI-NAMIDE * ETHYONOMIDE * ETIMID * ETIOCI-DAN * ETIONAMID * ETIONIZINA * ETP * FATOLIAMID * F.I. 58-30 * IRIDOCIN * IRIDOZIN * ISOTHIN * ISOTIAMIDA * ITIOCIDE * NCI-C01694 * NICOTION * NISOTIN * NIZOTIN * RIGENICID * SERTI-NON * TEBERUS * TH 1314 * THIANIDE * THIOAMIDE * THIONIDEN * TRECATOR * TRESCATYL * TRESCAZIDE * TUBERMIN * TUBEROID * TUBEROSON

CONSENSUS REPORTS: IARC Cancer Review: GROUP 3 IMEMDT 7,56,87; Animal Limited Evidence IMEMDT 13,83,77. NCI Carcinogenesis Bioassay (feed); No Evidence: mouse, rat NCITR* NCI-CG-TR-46,78.

SAFETY PROFILE: A human systemic poison. Moderately toxic by ingestion, intraperitoneal, and subcutaneous routes. Human systemic effects by ingestion: jaundice and liver function impairment. It affects the human peripheral nervous system. Experimental teratogenic and reproductive effects. Questionable carcinogen with experimental carcinogenic data. Mutation data reported. Used to treat tuberculosis. When heated to decomposition it emits very toxic fumes of SO_x and NO_x.

EPW500 CAS: 80-40-0 **HR: 3**
ETHYL TOSYLATE
mf: $C_9H_{12}O_3S$ mw: 200.27

PROP: Liquid. Mp: 33°, bp: 221.3°, flash p: 316°F (CC), d: 1.17, vap d: 6.98.

SYNS: ETHYL-p-METHYL BENZENESULFONATE * ETHYL PTS * ETHYL-p-TOLUENESULFONATE * ETHYL-p-TOSYLATE * p-TOLUOLSULFONSAEU-REAETHYL ESTER (GERMAN)

CONSENSUS REPORTS: Reported in EPA TSCA Inventory. EPA Genetic Toxicology Program.

SAFETY PROFILE: Moderately toxic by subcutaneous and intraperitoneal routes. Questionable carcinogen with experimental tumorigenic data. Mutation data reported. Combustible when exposed to heat or flame; can react with oxidizing materials. To fight fire, use CO_2, dry chemical. When heated to decomposition it emits highly toxic fumes of SO_x.

EPY000   CAS: 327-98-0   HR: 3
ETHYL TRICHLOROPHENYLETHYL-PHOSPHONOTHIOATE
mf: $C_{10}H_{12}Cl_3O_2PS$ mw: 333.60

SYNS: O-AETHYL-O-(2,4,5-TRICHLORPHENYL) AETHYLTHIONOPHOSPHONAT (GERMAN) * AGRISIL * AGRITOX * BAYER 37289 * BAYER 5081 * BAYER S 4400 * CHEMAGRO 37289 * ENT 25,712 * O-ETHYL-O-2,4,5-TRICHLOROPHENYL ETHYLPHOSPHONOTHIOATE * FENOPHOSPHON * PHYTOSOL * STAUFFER N-3049 * TRICHLORO-NAT * 2,4,5-TRICHLOROPHENOL-O-ESTER with O-ETHYL ETHYLPHOSPHONOTHIOATE * WIRKSTOFF 37289

CONSENSUS REPORTS: Chlorophenol compounds are on the Community Right-To-Know List. EPA Extremely Hazardous Substances List.

SAFETY PROFILE: Poison by ingestion, skin contact, and possibly other routes. Moderately toxic by inhalation. An insecticide. When heated to decomposition it emits very toxic fumes of Cl^-, PO_x, and SO_x.

EPY500   CAS: 115-21-9   HR: 3
ETHYL TRICHLOROSILANE
DOT: UN 1196
mf: $C_2H_5Cl_3Si$ mw: 163.51

PROP: Liquid. Mp: $-105.6°$, bp: $99.5°$, flash p: $72°F$ (OC), d: 1.24 @ $25°/25°$, vap d: 5.6.

SYNS: ETHYL SILICON TRICHLORIDE * TRICHLO-ROETHYLSILANE * TRICHLOROETHYLSILICANE

CONSENSUS REPORTS: EPA Extremely Hazardous Substances List. Reported in EPA TSCA Inventory.

DOT Classification: Flammable Liquid; Label: Flammable Liquid.

SAFETY PROFILE: Poison by inhalation and intraperitoneal routes. Moderately toxic by ingestion. A very dangerous fire hazard when exposed to heat, flame, or oxidizers; will react with water or steam to produce heat and toxic and corrosive fumes; can react vigorously with oxidizing materials. To fight fire, use CO_2, dry

chemical. When heated to decomposition it emits highly toxic fumes of Cl^- and phosgene.

EQE000   HR: 3
ETHYLUREA and SODIUM NITRITE (2:1)
SYNS: AETHYLHARNSTOFF und NATRIUMNITRIT (GERMAN) * AETHYLHARNSTOFF und NITRIT (GERMAN) * SODIUM NITRITE and ETHYLUREA (1:2)

SAFETY PROFILE: Suspected carcinogen with experimental carcinogenic, neoplastigenic, and tumorigenic data. Experimental teratogenic and reproductive effects. When heated to decomposition it emits toxic fumes of NO_x and Na_2O.

EQF000   CAS: 121-32-4   HR: 2
ETHYL VANILLIN
mf: $C_9H_{10}O_3$ mw: 166.19

PROP: Fine, crystalline needles; vanilla odor. Mp: $76.5°$, flash p: $+212°F$. Sol in alc, chloroform, ether, propylene glycol; sltly sol in water.

SYNS: BOURBONAL * ETHAVAN * ETHOVAN * 3-ETHOXY-4-HYDROXYBENZALDEHYDE * ETHYL-PROTAL * FEMA No. 2464 * 4-HYDROXY-3-ETHOX-YBENZALDEHYDE * PROTOCATECHUIC ALDEHYDE ETHYL ETHER * QUANTROVANIL * VANILLAL * VANIROM

CONSENSUS REPORTS: Reported in EPA TSCA Inventory.

SAFETY PROFILE: Moderately toxic by ingestion, intraperitoneal, subcutaneous, and intravenous routes. A human skin irritant. Mutation data reported. Combustible liquid. When heated to decomposition it emits acrid smoke and irritating fumes.

EQF500   CAS: 109-92-2   HR: 3
ETHYL VINYL ETHER
DOT: UN 1302
mf: C_4H_8O mw: 72.12

PROP: Colorless liquid. Bp: $35.6°$, flash p: $< -50°F$, fp: $-115°$, d: 0.754, autoign temp: $395°F$, vap press: 428 mm @ $20°$, lel: 1.7%, uel: 28%, vap d: 2.5.

SYNS: ETHOXY ETHENE * EVE * VINAMAR * VINYL ETHYL ETHER * VINYL ETHYL ETHER, inhibited (DOT)

CONSENSUS REPORTS: Reported in EPA TSCA Inventory.

DOT Classification: Flammable Liquid; Label: Flammable Liquid.

SAFETY PROFILE: Mildly toxic by ingestion. Mutation data reported. A skin irritant. A very dangerous fire and explosion hazard when exposed to heat or flame; can react vigorously with oxidizing materials. To fight fire, use alcohol foam, foam, CO_2, dry chemical. Explosive polymerization is catalyzed by methane sulfonic acid. When heated to decomposition it emits acrid smoke and irritating fumes.

EQG000　　CAS: 109-92-2　　HR: 2
ETHYL VINYL ETHER (inhibited)

CONSENSUS REPORTS: Reported in EPA TSCA Inventory.

DOT Classification: Flammable Liquid; Label: Flammable Liquid.

SAFETY PROFILE: A very dangerous fire and explosion hazard when exposed to heat or flame.

EQJ500　　CAS: 297-76-7　　HR: 3
ETHYNODIOL ACETATE
mf: $C_{24}H_{32}O_4$　　mw: 384.56

SYNS: CERVICUNDIN * 3-β,17-β-DIACETOXY-17-α-ETHYNYL-4-OESTRENE * 3-β,17-β-DIACETOXY-19-NOR-17-α-PREGN-4-EN-20-YNE * ETHINODIOL DIACE-TATE * ETHYNODIOL DIACETATE * β-ETHYNO-DIOL DIACETATE * 17-α-ETHYNYL-3,17-DIHYDROXY-4-ESTRENE DIACETATE * 17-α-ETHYNYLESTR-4-ENE-3-β,17-β-DIOL ACETATE * 17-α-ETHYNYL-4-ESTRENE-3-β,17-β-DIOL DIACETATE * 17-α-ETHYNYL-4-ES-TRENE-3-β,17-DIOL DIACETATE * 17-α-ETHYNYL-19-NORANDROST-4-ENE-3-β,17-β-DIOL DIACETATE * FEMULEN * LUTO-METRODIOL * METRODIOL * METRODIOL DIACETATE * (3-β,17-α)-19-NOR-PREGN-4-EN-20-YNE-3,17-DIOL DIACETATE * OVULEN 50

CONSENSUS REPORTS: IARC Cancer Review: Animal Limited Evidence IMEMDT 21,387,79; Animal Sufficient Evidence IMEMDT 6,173,74.

SAFETY PROFILE: Suspected carcinogen. Human reproductive effects by ingestion: menstrual cycle changes. Experimental reproductive effects. Mutation data reported. A steroid. When heated to decomposition it emits acrid smoke and irritating fumes.

EQP000　　CAS: 29767-20-2　　HR: 3
ETP
mf: $C_{32}H_{32}O_{13}S$　　mw: 656.70

SYNS: 4'-DEMETHYLEPIPODOPHYLLOTOXIN-9-(4,6-O-2-THENYLIDENE-β-d-GLUCOPYRANOSIDE * 4'-DE-METHYL-EPIPODOPHYLLOTOXIN-β-d-THENYLIDENE-GLUCOSIDE * 4'-DEMETHYL 1-O-(4,6-O,O-(2-THENYLI-DENE)-β-d-GLUCOPYRANOSYL)EPIPODOPHYLLOTOXIN * EPT * NSC-122819 * PTG * TENIPOSIDE * VEHAM-SANDOZ * VEHEM * VM-26 * VUMON

SAFETY PROFILE: Poison by intraperitoneal and subcutaneous routes. Experimental teratogenic and reproductive effects. Human systemic effects by ingestion and intravenous route: anorexia, nausea or vomiting, leukopenia, agranulocytosis and aplastic anemia of the blood, bone marrow changes and hair changes. Human mutation data reported. When heated to decomposition it emits very toxic fumes of SO_x.

EQQ000　　CAS: 8000-48-4　　HR: 3
EUCALYPTUS OIL

PROP: From steam distillation of leaves of *Eucalyptus globulus* Labillardiere. Chief constituent is eucalyptol. Colorless to pale-yellow liquid; spicy odor and taste. Composition: eucalyptol, aldehydes, d-pinene. Mp: −15.4° (approx). D: 0.905-0.925 @ 25°/25°.

SYNS: DINKUM OIL * EUKALYPTUS OEL (GER-MAN) * OIL of EUCALYPTUS

CONSENSUS REPORTS: Reported in EPA TSCA Inventory.

SAFETY PROFILE: A human poison by ingestion. Human systemic effects by ingestion: ciliary eye spasms, somnolence and respiratory depression. A skin irritant. When heated to decomposition it emits acrid smoke and irritating fumes.

EQR500　　CAS: 97-53-0　　HR: 3
EUGENOL
mf: $C_{10}H_{12}O_2$　　mw: 164.22

PROP: Colorless or yellowish liquid; pungent, clove odor. D: 1.064-1.070, refr index: 1.540, bp: 253.5°, flash p: 219°F. Sol in alc, chloroform, ether, volatile oils; very sltly sol in water.

SYNS: 4-ALLYLGUAIACOL * 4-ALLYL-1-HYDROXY-2-METHOXYBENZENE * 4-ALLYL-2-METHOXYPHENOL * CARYOPHYLLIC ACID * EUGENIC ACID

* Fa 100   * FEMA No. 2467   * 1-HYDROXY-2-METHOXY-4-ALLYLBENZENE   * 4-HYDROXY-3-METHOXYALLYLBENZENE   * 1-HYDROXY-2-METHOXY-4-PROP-2-ENYLBENZENE   * 2-METHOXY-4-ALLYLPHENOL   * 2-METHOXY-4-PROP-2-ENYLPHENOL   * 2-METHOXY-4-(2-PROPENYL)PHENOL   * 2-METOKSY-4-ALLILOFENOL (POLISH)   * NCI-C50453   * SYNTHETIC EUGENOL

CONSENSUS REPORTS: IARC Cancer Review: GROUP 3 IMEMDT 7,56,87; Animal Limited Evidence IMEMDT 36,75,85. NTP Carcinogenesis Studies (feed); Equivocal Evidence: mouse NTPTR* NTP-TR-223,83; No Evidence: rat NTPTR* NTP-TR-223,83. Reported in EPA TSCA Inventory. EPA Genetic Toxicology Program.

SAFETY PROFILE: Moderately toxic by ingestion, intraperitoneal, and subcutaneous routes. Human mutation data reported. A human skin irritant. Questionable carcinogen with experimental carcinogenic and tumorigenic data. Combustible liquid. When heated to decomposition it emits acrid smoke and irritating fumes.

EQS000   CAS: 93-28-7   HR: 2
EUGENOL ACETATE
mf: $C_{12}H_{14}O_3$ mw: 206.26

PROP: Solid or pale yellow liquid; mild clove odor. D: 1.87, mp: 29-30°, bp: 281.2°, flash p: +151°F. Insol in water; sol in alc and ether.

SYNS: ACETEUGENOL   * 1-ACETOXY-2-METHOXY-4-ALLYLBENZENE   * ACETYLEUGENOL   * 4-ALLYL-2-METHOXYPHENOL ACETATE   * 1,3,4-EUGENOL ACETATE   * EUGENYL ACETATE   * FEMA No. 2469

CONSENSUS REPORTS: Reported in EPA TSCA Inventory.

SAFETY PROFILE: Moderately toxic by ingestion. A skin irritant. Combustible liquid. When heated to decomposition it emits acrid smoke and irritating fumes.

ERA500   CAS: 10025-76-0   HR: 3
EUROPIUM CHLORIDE
mf: Cl_3Eu mw: 258.31

SYN: EUROPIC CHLORIDE

CONSENSUS REPORTS: Reported in EPA TSCA Inventory.

SAFETY PROFILE: Poison by intraperitoneal route. Moderately toxic by ingestion. When heated to decomposition it emits very toxic fumes of Cl^-.

ERC000   CAS: 10031-53-5   HR: 3
EUROPIUM(III) NITRATE, HEXAHYDRATE (1:3:6)
mf: $N_3O_9 \cdot Eu \cdot 6H_2O$ mw: 446.11

SYN: NITRIC ACID, EUROPIUM(3+) SALT, HEXAHYDRATE

SAFETY PROFILE: Poison by intraperitoneal route. When heated to decomposition it emits very toxic fumes of NO_x.

ERD500   CAS: 56-29-1   HR: 3
EVIPAL
mf: $C_{12}H_{16}N_2O_3$ mw: 236.30

SYNS: BARBIDORM   * CITODON   * CITOPAN   * 5-(1-CYCLOHEXEN-1-YL)-1,5-DIMETHYLBARBITURIC ACID   * 5-(1-CYCLOHEXEN-1-YL)-1,5-DIMETHYL-2,4,-6(1H,3H,5H)-PYRIMIDINETRIONE   * 5-(1-CYCLOHEXENYL-1)-1-METHYL-5-METHYLBARBITURIC ACID   * 5-(Δ-1,2-CYCLOHEXENYL)-5-METHYL-N-METHYL-BARBITURSAEURE (GERMAN)   * CYCLONAL   * CYCLOPAN   * 1,5-DIMETHYL-5-(1-CYCLOHEXENYL)BARBITURIC ACID   * DORICO   * ENHEXYMAL   * ESOBARBITALE (ITALIAN)   * EVIPAN   * HEXABARBITAL   * HEXANASTAB ORAL   * HEXENAL   * HEXENAL (barbiturate)   * HEXOBARBITAL   * HEXOBARBITONE   * METHEXENYL   * N-METHYL-5-CYCLOHEXENYL-5-METHYLBARBITURIC ACID   * METHYLHEXABARBITAL   * METHYLHEXABITAL   * NARCOSAN   * NOCTOVANE   * SOMBUCAPS   * SOMBULEX   * SOMNALERT

SAFETY PROFILE: Poison by subcutaneous, intraperitoneal, intravenous, intrapleural, and rectal routes. Moderately toxic by ingestion. Human mutation data reported. When heated to decomposition it emits toxic fumes of NO_x.

ERF000   HR: 3
EXPLOSIVES, HIGH

SAFETY PROFILE: High explosives (HE) are those which decompose by detonation. This is a very rapid (nearly instantaneous), and hence, violent process. An explosion may be initiated by sudden shock, high temperatures, or a combination of the two. For many explosives the conditions under which they will explode are well known, for example:

An explosion may be initiated by elevated temperature alone:

(1) In the case of mercury fulminate, a 15-second exposure to 200°C or 1 second exposure to 340°C will set it off.

(2) Trinitrotoluene will be set off by exposure to 500°C for 1 second.

(3) Tetryl will detonate in 1000 seconds at 160°C or in 0.1 second at 500°C.

(4) Picric acid will detonate in 9 seconds at 300°C or 1 second at 355°C.

An explosion of HE may also be initiated by severe shock. Sensitivity of explosives to shock may be measured in several ways, such as the impact pendulum method and the drop test. The impact pendulum test operates by allowing a heavy pendulum to swing down over a sample of explosive in a dished, inclined container so arranged that there is very little clearance between the pendulum and the sample. Thus, the effect of contact between the sample and the pendulum bob is one of a combination of shock and rubbing. The height from which the pendulum is allowed to swing to explode the sample is a measure of the sensitivity of the sample to this test. The drop test consists of placing a sample upon an anvil and allowing a 5 pounds weight to drop on it. The height from which the weight must drop to explode the sample is a measure of the sample's sensitivity to shock.

The table below shows the results of a drop test upon several samples. These results must be considered as relative and not by any means absolute. A solid explosive in a tightly fitting container is much more sensitive to shock.

(1) mercury fulminate = 2 in. at 5 lbs.
(2) nitroglycerin = 4 in. at 5 lbs.
(3) tetryl = 8 in. at 5 lbs.
(4) picric acid = 14 in. at 5 lbs.
(5) trinitrotoluene = 20 in. at 5 lbs.
(6) black powder (a low explosive) = 30 in. at 5 lbs.*

*From *Explosions, Their Anatomy and Destructiveness*, by C. S. Robinson (McGraw-Hill).

Another test for explosives is the speed at which a detonation travels. This speed is usually in the range of thousands of m/sec. Speed of detonation is found to be a function of the kind of explosive and state of compaction. There is an optimum state of compaction beyond which the explosive tends to become "dead-pressed," in which state it is difficult to make the whole sample explode. Below the point of optimum compaction the rate of detonation is found to be directly proportional to the density of the sample. Some maximum detonation rates are listed below in meters/sec for some common explosives:

(1) nitroglycerin   8500
(2) PETN   8100
(3) tetryl   7700
(4) picric acid   7400
(5) trinitrotoluene   7400
(6) lead azide   4900
(7) mercury fulminate   4800
(8) ammonlium nitrate   1100
(9) low explosives   1000

It has been found that upon detonation, an explosive can cause a nearby sample of explosive to detonate "sympathetically." The distance over which one charge can detonate another is a function of the amount of energy produced by the first explosion and the medium through which the shock wave is propagated to the second charge of explosive. For instance, the relationship for air (very approximately) would be expected to be: Weight of explosive in lbs/(distance in ft)^3 = 4. Thus, to calculate the maximum distance for a possible sympathetic detonation of 40,000 lbs of explosive, the calculation is:

$$D^3 = (40,000)/4$$
$$D^3 = 10,000$$
$$D = 22 \text{ ft (approximately)}.$$

According to C. S. Robinson, the formula is more nearly:

weight of explosive = 4 × (distance)$^{2.25}$

The power of the shock wave is much more rapidly attenuated in water, wood, etc., than in air, which means that if a shield of water or wood is interposed between piles of explosive the distance between them may be lessened.

Liquid Oxygen: Though not itself explosive, liquid oxygen can be dangerous when blended with highly flammable or carbonaceous materials. In this combination it is used in coal mining, quarrying, strip mining, open-cut ore mining, and in rocket fuels. Its use underground or in confined places is not recommended by the U.S. Bureau of Mines because it evolves a lot of carbon monoxide. This type of explosive has many safety advantages. For instance, it is not itself an explosive until mixed with a flammable absorbent which can be done at the last moment before firing. However, once the explosive has been made up, it is very flammable and when it catches fire it will usually detonate. Liquid oxygen explosives are not stored, as they deteriorate rapidly and lose a great deal of their explosive power in a short time.

A very dangerous fire hazard when exposed to heat or by chemical reaction with powerful oxidizing or reducing agents.

A moderate to dangerous explosion hazard when severely shocked or heated, depending upon the kind of explosive, state of compaction, degree of confinement, etc. Practically all high explosives used commercially require a detonator or cap to set them off, as compared to an igniter needed to set off black blasting powder.

Detonating Devices: To develop the desired disruptive effect of an explosive, some means must be adopted to "set off," "fire," or "detonate" it without killing or maiming the persons doing the blasting. Several devices or methods are being utilized, all with a view to having this work done as safely and efficiently as possible. There are two general types of devices or methods of getting explosives into action, namely, igniters and detonators. The former merely conveys a flame to the explosive mass and ignites it, while the latter transmits (originally through ignition of a small quantity of highly explosive substance by an arc, a flame, or spark) a sharp blow that causes the explosive to disassociate, or detonate, or burn with very great rapidity. Igniters are squibs (plain and electric), fuse, and delay igniters; detonators are blasting caps (plain and electric), delay electric blasting caps, delay electric igniters with caps, and Cordeau-Bickford detonating fuse.

The squib is a small-diameter tube of straw or paper filled with quick-burning powder and having a relatively slow burning "match head" attached to one end; the latter is ignited or lighted by an ordinary match or other flame, and its relatively slow burning allows the person handling the ignition to retire before the fire is communicated to the quick-burning material in the tube. Squibs are by no means either safe or efficient, even though still used to a considerable extent, especially in coal mining. Electric squibs are somewhat similar to ordinary squibs, except that the ignition is accomplished by means of an electric arc; electric squibs are much more satisfactory from a safety viewpoint than ordinary squibs.

A fuse (or, as it is sometimes called, "safety fuse") consists of a fine-grained black powder core covered with cotton hemp or jute to form a ropelike material about 3/16-inch in diameter; one end of the fuse is brought in contact with the powder charge or with a detonating "cap," and the other end (usually several feet away from the explosive) is lighted by a flame from a match or open light. The fine-grained black powder burns gradually and somewhat slowly (about 30 to 40 seconds to the linear foot of fuse) until it reaches the explosive (black powder) or the detonating cap (if some form of dynamite is used), giving the blaster time to get in the clear before the main explosion takes place. Fuses are much safer than squibs, but have their own hazards and must be used with care.

Delay electric igniters usually are a combination of electric igniters and fuses, the latter being ignited by the igniters within the blasting hole, the fuse transmitting the ignition to the explosive. Delay igniters usually are much safer than fuse, particularly for coal-mine use; but they, too, have their hazards. Delay blasting is by no means a safe procedure in coal mining, though it is a standard and relatively safe practice in metal mining and tunneling, if sensible precautions are taken.

Blasting caps or detonators are metallic cylinders (usually copper) closed at one end, about 3/16-inch in diameter, and usually less than 2 inches in length, partly filled with a small amount of relatively easily fired or "detonated" compound, the resultant shock or blow when fired being sufficient, when embedded in dynamite, to fire or detonate the dynamite mass. Ordinary blasting caps usually are fired or detonated by the flame of the fuse, the end of the latter being inserted into the open end of the detonator or cap and placed in contact with the highly explosive material in the interior of the metallic capsule or cap. Caps are extremely hazardous to handle, as they are likely to be detonated by heat, friction, or a relatively moderate blow; however, they are relatively safe if handled carefully. Partial proof of this is the fact that they are manufactured and shipped by the thousands daily and accidents are decidedly rare, primarily because the caps are at all times handled with utmost care.

Electric blasting caps are somewhat similar to ordinary caps or detonators, but the cap is fired by electricity. The electric wires are so placed in the capsule or cap that when attached to an electric current an arc is formed within the cap, which detonates the sensitive explosive material in the cap. A hazard in the use of electric blasting caps is unexpected explosions due to radio or radar induced electric currents which may activate the cap.

Delay electric blasting caps or detonators are somewhat similar to ordinary electric blasting caps, except that several time intervals in blasting are obtained by having the electric arc ignite a short piece of fuse or some slow-burning substance before it reaches the highly sensitive detonating material in the capsule or cap. Numerous time-interval delays are obtained; in general, delay electric blasting caps are relatively safe and effective even in wet holes, though they ought not be used in coal mining if explosions of gas or dust are to be avoided. Delay electric igniters with caps or detonators are a combination of electric igniter and blasting cap, usually with suitable lengths of fuse between to give the desired delay; they have some advantages but are relatively unsafe and should not be used in coal mining.

The Cordeau-Bickford denotating fuse is a combined fuse and detonator in the form of a lead tube about 1/4-inch in diameter filled throughout its length with a high explosive, trinitrotoluene (TNT). It is fired by a fuse and an ordinary detonator or cap or by an electric cap; when fired, it detonates throughout its length (which may be up to or over 100 ft) almost instantaneously, the explosion wave traveling at a rate of about 17,500 ft/sec. Although somewhat expensive, it is relatively safe to handle and is particularly effective in deep-well drill holes in quarry and similar work, as it detonates simultaneously throughout its length, adding effectiveness to the main body of explosive which it detonates. It fires black powder as well as high explosives (dynamite, etc.), and is obtainable in lengths of approximately 500 feet wound on spools.

ERF500   HR: 3
EXPLOSIVES, LOW

PROP: Black powder is composed of saltpeter, charcoal and sulfur in the approximate proportions of 6:1:1. ("A" blasting powder uses KNO_3 and "B" blasting powder uses $NaNO_3$).

SYNS: "A" BLASTING POWDER * "B" BLASTING POWDER * BLACK BLASTING POWDER * GUNPOWDER

SAFETY PROFILE: Low explosives are explosives which deflagrate; this differentiates them both in composition and properties from high explosives, which detonate. A deflagrating explosive is one that burns progressively over a relatively sustained period of time in comparison with a detonating explosive, which decomposes almost instantaneously. A dangerous fire hazard when exposed to heat or flame or by chemical reaction.

Black powder is the most treacherous explosive material used today and it is regarded as one of the worst known explosive hazards. When ignited unconfined it burns with explosive violence and will explode if ignited under even slight confinement. It can be ignited easily by very small sparks, heat, and friction. It is the slowest acting of all explosives. It has a shearing and heaving action tending to blast materials in large, firm fragments. The action derives from a relatively slow development of gas pressure so that it must be carefully loaded and closely confined. It is subject to rapid deterioration in the presence of moisture, but if kept dry it retains its explosive properties for many years. It is used to ignite smokeless powder, propelling charges, airplane flares and bursting charges of hand grenades, as a bursting charge in shrapnel, practice bombs, practice trench-mortar shells, in saluting charges, smoke-puff charges, time and percussion fuses, pellets, primers and primer detonators, and in expelling charges of pyrotechnic signals.

Although most safety men now look upon black blasting powder with disfavor, it is one of the oldest and most generally used explosives in commercial work. It burns with extreme rapidity instead of detonating as high explosives do. It is highly sensitive to flame or sparks or friction and gives off much flame, which is hot and of great length of duration. These properties make it extremely hazardous for use in mines (especially coal mines) and quarries. The gases given off in detonation are not only very hot but frequently contain harmful constituents. Notwithstanding its numerous deficiencies, from a safety standpoint it has action characteristics that make it valuable in both coal mining and quarrying, though it has relatively little utility in metal mining. It is difficult to use effectively in wet places and this is its main disadvantage from an efficiency standpoint.

Most black powder fires start from sparks. Ignition results in an explosion so quickly that no attempt can be made to fight the fire. Every effort should be made to prevent fires from reaching stores of black powder; but if this fails, fire-fighting forces should be withdrawn at least 800 feet from the fire and should protect themselves against an explosion by seeking any cover available or by lying flat on the ground. If an explosion does occur, every effort should be made to prevent flames from spreading to neighboring magazines. Fire-fighting forces should be cautioned against approaching a fire which may involve black powder to avoid being trapped or injured by an explosion.

The following safety rules should be strictly enforced and obeyed. Open no containers in a magazine to which explosives or ammunitions are stored. This should be done only in a building free from all other explosives or ammunitions; or in suitable weather in the open, at least 100 feet from the nearest magazine. The quantity at or near such an operation should be limited to 100 pounds. Safety tools only should be used in opening or closing containers or in other operations involving black powder. Processes should be so laid out as to bring about frequent grounding of all operators handling this material. Safety shoes (non-insulating) should be worn in all rooms where black powder is handled and by all persons engaged in handling black powder. The wearing of all nonconductive shoes, such as rubber is prohibited. If the handling of black powder is carried on over a concrete floor, the floor should be covered with a tarpaulin or other suitable material. Loose black powder is extremely dangerous. Whenever it is necessary to handle loose black powder, not over 50 pounds should be permitted at or near such operations. If black powder is spilled on benches or floors, all work should be stopped until it has been removed and the explosive hazard of any remaining dust or particles has been neutralized with water. Rooms or buildings in which black powder is handled should be inspected frequently for dust, and all such dust should be immediately removed with water. The empty powder containers should be washed out, as explosions are said to have occurred from "empty" containers.

If dry and in good condition, black powder burns rapidly, especially in small grain size, with a yellow or pinkish-blue flame and dense smoke.

Pellet powder is black blasting powder in consolidated (pellet or stick) form rather than in grains or granules, and it has few if any real advantages over black blasting powder, notwithstanding the fairly prevalent idea that it is a "safe" explosive.

Smokeless powders have a somewhat different composition from that of black blasting powder and are used chiefly for sportsmen's ammunition and, more widely, for military purposes. They are decidedly sensitive to flame and impact but ordinarily are so packaged that if reasonable judgment is used they are relatively harmless.

ERG000   HR: 3
EXPLOSIVES, PERMITTED

SAFETY PROFILE: "Permissible" explosives are essentially high explosives (dynamite) modified by the introduction of "dopes." The function of the dopes in general is to decrease flame temperature and to a smaller extent, the length and duration of flame, when the explosive is converted from a solid into a gas; in other words, when it is fired or detonated. The designation "permissible" is given to an explosive of modified dynamite type after it has passed certain tests made by the Federal Bureau of Mines. The permissible character of such explosives depends not only upon the ingredients in the explosive, but also on certain well-defined specifications as to handling and use. As with the dynamites, there are several different types and grades; "permissibles," hydrated "permissibles," organic nitrate "permissibles," nitroglycerin "permissibles," ammonium perchlorate "permissibles," and gelatin "permissibles." Essentially all of those now used to any extent are in either the ammonium nitrate or the gelatin classes.

The ammonium nitrate "permissible" explosives contain relatively little nitroglycerin and relatively large proportions of ammonium nitrate. The latter is an explosive but one less sensitive to impact, sparks, and flames than nitroglycerin. This type of permissible explosive is now used extensively, as it has a rather wide range of strength, rate of detonation, density, size of cartridge, etc., and can be utilized not only in dry but also to some extent in fairly wet holes if charged carefully and fired promptly.

Gelatin "permissible," explosives are more suitable than ammonium nitrate "permissible," for wet holes, and in general are stronger and more violent than the ammonium nitrate types.

All "permissible" explosives are strong, and must be used in relatively small quantities (less than 1.5 pounds) per hole to retain their permissibility, give off considerable quantities of toxic gases on detonation, and, while much safer than black blasting powder or dynamite, must be stored, handled, and used with care.

Classification Upon Basis of Toxic Gases: All "permissible" explosives, when detonated, emit some toxic gases and a much larger volume of nontoxic gases. In order that the toxic products may not become a menace to the life or health of miners, no explosive is now or can become "permissible" if upon detonation it evolves more than 158 liters (5.5 cu ft) of toxic gases per 1.5 pound charge as determined by tests in the Bichel pressure gage. Classification upon the basis of the volume of toxic gases produced by 580 grams (1.5 pounds) of explosive is as follows: *Class A,* not more than 53 liters; *Class B,* between 53 and 106 liters; and *Class C,* between 106 and 158 liters. (These classifications are not to be confused with the I.C.C. Classification of explosives).

Field tests were made with a 1.5 pounds charge of a "permissible" explosive that produced, in the Bichel gage, the maximum allowable quantity of poisonous gases (158 liters per 1.5 pounds); these tests indicated that in a narrow entry, without artificial ventilation. 1800 ppm of carbon monoxide (the only poisonous gas present) was produced, as shown by analysis of an air sample taken 2 minutes after the shot. Another sample of the air taken 2 minutes later contained 800 ppm of carbon monoxide. Under no conditions should miners or shot firers return to the place until the poisonous gases have been removed by adequate ventilation.

It is provided further that, in accordance with the provisions and conditions, explosives enumerated on the "permissible" lists of the Bureau of Mines are "permissible" in use only when they satisfy the following requirements:

1. That the explosive be in all respects similar to the sample submitted by the manufacturer for test, and that the diameters of the cartridges used must be those that have been approved.

2. That electric detonators (not fuse and detonators) be used of not less efficiency than No. 6, the detonation charge of which shall consist of a 1 gram mixture of 80 parts of mercury fulminate and 20 parts of potassium chlorate (or their equivalents), and that the required electric firing must be done by means of a "permissible" type blasting unit.

3. That the explosive be stored in surface magazines under proper conditions, so that it will not undergo change in character, and that after being taken underground it be used in less than 36 hours.

4. That the coal to be blasted be undercut or equivalently relieved; that, to prevent blowthrough, all portions of the borehole must be at least 18 inches from relief in any direction; that, to prevent blowouts, the charge be properly confined with not less than 2 feet of clay (if the length of the hole will not permit the charge desired and 2 feet of stemming, at least half the length of the hole shall be filled with stemming) or other incombustible stemming and not be on the solid; that, to prevent the hole being on the solid; it shall be at least 6 inches shorter than the depth of the undercut or equivalent relief, and, when placed adjacent to the roofs, ribs, or floor, all but 12 inches at the rear of the hole must be at least 6 inches from the adjacent surface as projected into the coal to be blasted, and all parts of the hole shall be free from the adjacent surface as projected into the coal to be blasted; that the shot be not a dependent shot; and that the shot hole be cleaned before charging.

5. That the quantity used for a shot (1) be not in excess of 680 grams (1.5 pounds) when fired in accordance with these requirements and (2) when used under certain additional requirements or restrictions be not in excess of 1,361 grams (3 pounds). The use of charges over 1.5 pounds and not exceeding 3 pounds is approved tentatively pending further investigation. For charges of over 1.5 pounds, the following additional requirements must be observed, (a) Shot holes must be 6 feet or more in length. (b) Explosives must be charged in a continuous train, with no cartridges deliberately deformed or crushed, with all cartridges in contact with each other, and with the end cartridges touching the rear of the hole and the stemming, respectively. (c) Examination for gas must be made in the blasting area before and after a shot is fired. (d) The "permissible" explosive must be one showing toxic gas emission that will place it either in Class A or Class B.

6. That the region in which the blasting is done be kept well protected by rock dust or otherwise in accordance with Bureau of Mines inspection standards.

7. That the shot not be fired in the presence of a dangerous percentage of firedamp. Examination for firedamp to be made at the blasting area before shooting in a gassy mine.

F

FAB000　　　　　　　　　　　**HR: 3**
F III (sugar fraction)

SAFETY PROFILE: Poison by intravenous route. Moderately toxic by intraperitoneal route. Mildly toxic by ingestion and subcutaneous routes. When heated to decomposition it emits acrid smoke and irritating fumes.

FAB600　　　CAS: 52-85-7　　　**HR: 3**
FAMPHUR
mf: $C_{10}H_{16}NO_5PS_2$　　mw: 325.36

PROP: Crystalline powder. Mp: 55°. Very sol in chloroform and carbon tetrachloride; sltly sol in water.

SYNS: AMERICAN CYANAMID-38023 ＊ AMERICAN CYANAMID CL-38,023 ＊ BO-ANA ＊ CL-38023 ＊ CYFLEE ＊ O-(4-((DIMETHYLAMINO)SULFONYL) PHENYL) O,O-DIMETHYL PHOSPHOROTHIOATE ＊ O,O-DIMETHYL-O-(p-(N,N-DIMETHYLSULFAMOYL) PHENYL)PHOSPHOROTHIOATE ＊ DOVIP ＊ ENT 25,644 ＊ FAMFOS ＊ FAMOPHOS ＊ FAMOPHOS WARBEX ＊ FAMPHOS ＊ FANFOS ＊ RCRA WASTE NUMBER P097 ＊ WARBEX

CONSENSUS REPORTS: Reported in EPA TSCA Inventory.

SAFETY PROFILE: Poison by ingestion and intramuscular routes. Moderately toxic by skin contact. A cholinesterase inhibitor. When heated to decomposition it emits toxic fumes of NO_x and SO_x.

FAB800　　　CAS: 4602-84-0　　　**HR: 2**
FARNESOL
mf: $C_{15}H_{26}O$　　mw: 222.41

PROP: trans-Farnesol: Light yellow liquid; mild oily odor. Bp: 111°, n (25/D) 1.4872. Commercial farnesol: Bp: 110-113°, d: (20/4) 0.8871, n (20/D) 1.4870., refr index: 1.487-1.492. Insol in water @ 263°.

SYNS: FARNESYL ALCOHOL ＊ FEMA No. 2478 ＊ 3,7,11-TRIMETHYL-2,6,10-DODECATRIEN-1-OL

CONSENSUS REPORTS: Reported in EPA TSCA Inventory.

SAFETY PROFILE: Moderately toxic by intraperitoneal route. Mildly toxic by ingestion. Mu-

tation data reported. When heated to decomposition it emits acrid smoke and irritating fumes.

FAE000　　　CAS: 3844-45-9　　　**HR: 2**
FD&C BLUE No. 1
mf: $C_{37}H_{36}N_2O_9S_3 \cdot 2Na$　　mw: 794.91

PROP: Dark purple to bronze powder. Sol in water, ether, conc sulfuric acid.

SYNS: ACID SKY BLUE A ＊ AIZEN FOOD BLUE No. 2 ＊ 1206 BLUE ＊ BRILLIANT BLUE FCD No. 1 ＊ BRILLIANT BLUE FCF ＊ CANACERT BRILLIANT BLUE FCF ＊ C.I. 42090 ＊ C.I. ACID BLUE 9, DISODIUM SALT ＊ C.I. FOOD BLUE 2 ＊ COGILOR BLUE 512.12 ＊ COSMETIC BLUE LAKE ＊ D&C BLUE No. 4 ＊ DISPERSED BLUE 12195 ＊ DOLKWAL BRILLIANT BLUE ＊ EDICOL BLUE CL 2 ＊ ERIOGLAUCINE G ＊ FENAZO BLUE XI ＊ FOOD BLUE 2 ＊ FOOD BLUE DYE No. 1 ＊ HEXACOL BRILLIANT BLUE A ＊ INTRACID PURE BLUE L ＊ MERANTINE BLUE EG ＊ USACERT BLUE No. 1

CONSENSUS REPORTS: IARC Cancer Review: GROUP 3 IMEMDT 7,56,87; Animal Sufficient Evidence IMEMDT 16,171,78. Reported in EPA TSCA Inventory.

SAFETY PROFILE: Questionable carcinogen with experimental neoplastigenic and tumorigenic data. Mutation data reported. When heated to decomposition it emits very toxic fumes of NO_x, Na_2O and SO_x.

FAE100　　　CAS: 860-22-0　　　**HR: 3**
FD&C BLUE No. 2
mf: $C_{16}H_{10}N_2O_8S_2 \cdot 2Na$　　mw: 468.38

PROP: Blue-brown to red-brown powder. Sol in water, conc sulfuric acid; sltly sol in alc.

SYNS: ACID BLUE W ＊ ACID LEATHER BLUE IC ＊ A.F. BLUE No. 2 ＊ AIRDALE BLUE IN ＊ AMACID BRILLIANT BLUE ＊ ANILINE CARMINE POWDER ＊ ATUL INDIGO CARMINE ＊ 1311 BLUE ＊ 12070 BLUE ＊ BUCACID INDIGOTINE B ＊ CANACERT INDIGO CARMINE ＊ CARMINE BLUE (BIOLOGICAL STAIN) ＊ C.I. 73015 ＊ C.I. 7581 ＊ C.I. ACID BLUE 74 ＊ C.I. FOOD BLUE 1 ＊ DISODIUM INDIGO-5,5-DISULFONATE ＊ DISODIUM SALT of 1-INDIGOTIN-S,S'-DISULPHONIC ACID ＊ DOLKWAL INDIGO CARMINE ＊ E 132 ＊ GRAPE BLUE A GEIGY ＊ INDIGO CARMINE ＊ INDIGO CARMINE (BIOLOGICAL

STAIN) * INDIGO CARMINE DISODIUM SALT
* INDIGO EXTRACT * INDIGO-KARMIN (GERMAN)
* 5,5'-INDIGOTIN DISULFONIC ACID * INDIGOTINE
* INDIGOTINE DISODIUM SALT * INTENSE BLUE
* L-BLAU 2 (GERMAN) * MAPLE INDIGO CARMINE
* SACHSISCHBLAU * SCHULTZ Nr. 1309 (GERMAN)
* SODIUM 5,5'-INDIGOTIDISULFONATE * SOLUBLE
INDIGO * USACERT BLUE No.2

CONSENSUS REPORTS: Reported in EPA TSCA Inventory. EPA Genetic Toxicology Program.

SAFETY PROFILE: Poison by intravenous route. Moderately toxic by ingestion and subcutaneous routes. Questionable carcinogen with experimental neoplastigenic data. Mutation data reported. When heated to decomposition it emits very toxic fumes of SO_x, NO_x, and Na_2O.

FAF000   CAS: 5141-20-8   HR: 2
FD&C GREEN No. 2
mf: $C_{37}H_{36}N_2O_9S_3 \cdot 2Na$ mw: 794.91

SYNS: ACIDAL LIGHT GREEN SF * ACID BRILLIANT GREEN SF * ACID GREEN A * ACILAN GREEN SFG * A.F. GREEN No. 2 * AMACID GREEN G * C.I. 42095 * C.I. ACID GREEN 5 * C.I. ACID GREEN 5, DISODIUM SALT * C.I. FOOD GREEN 2 * D&C GREEN No. 4 * FAST ACID GREEN N * FD&C GREEN No. 2-ALUMINUM LAKE * FENAZO GREEN 7G * FOOD GREEN 2 * GREEN No. 203 * LEATHER GREEN SF * LICHTGRUEN (GERMAN) * LIGHT GREEN FCF YELLOWISH * LIGHT GREEN LAKE * LIGHT SF YELLOWISH (BIOLOGICAL STAIN) * LISSAMINE LAKE GREEN SF * MERANTINE GREEN SF * MY/68 * PENCIL GREEN SF * SULFO GREEN J * SUMITOMO LIGHT GREEN SF YELLOWISH * WOOL BRILLIANT GREEN SF

CONSENSUS REPORTS: IARC Cancer Review: GROUP 3 IMEMDT 7,56,87; Animal Sufficient Evidence IMEMDT 16,209,78. Reported in EPA TSCA Inventory. EPA Genetic Toxicology Program.

SAFETY PROFILE: Moderately toxic by intravenous route. Questionable carcinogen with experimental carcinogenic and neoplastigenic data. Mutation data reported. When heated to decomposition it emits very toxic fumes of NO_x, Na_2O, and SO_x.

FAG000   CAS: 2353-45-9   HR: 2
FD&C GREEN No. 3
mf: $C_{37}H_{36}N_2O_{10}S_3 \cdot 2Na$ mw: 810.91

PROP: Red to brown-violet powder. Sol in water, conc sulfuric acid.

SYNS: AIZEN FOOD GREEN No. 3 * C.I. 42053 * C.I. FOOD GREEN 3 * FAST GREEN FCF * 1724 GREEN * SOLID GREEN FCF

CONSENSUS REPORTS: IARC Cancer Review: GROUP 3 IMEMDT 7,56,87; Animal Sufficient Evidence IMEMDT 16,187,78. Reported in EPA TSCA Inventory. EPA Genetic Toxicology Program.

SAFETY PROFILE: Questionable carcinogen with experimental neoplastigenic data. Mutation data reported. When heated to decomposition it emits very toxic fumes of NO_x and SO_x.

FAG018   CAS: 3564-09-8   HR: 2
FD&C RED No. 1
mf: $C_{19}H_{16}N_2O_7S_2 \cdot 2Na$ mw: 494.47

SYNS: A.F. RED No. 1 * CERVEN KUMIDINOVA * C.I. 16155 * C.I. FOOD RED 6 * C.I. FOOD RED 6, DISODIUM SALT * DISODIUM 3-HYDROXY-4-((2,4,5-TRIMETHYLPHENYL)AZO)-2,7-NAPHTHALENEDISULFONATE * DISODIUM 3-HYDROXY-4-((2,4,5-TRIMETHYLPHENYL)AZO)-2,7-NAPHTHALENEDISULFONIC ACID * DISODIUM 3-HYDROXY-4-((2,4,5-TRIMETHYLPHENYL)AZO)-2,7-NAPHTHALENEDISULPHONATE * DISODIUM 3-HYDROXY-4-((2,4,5-TRIMETHYLPHENYL)AZO)-2,7-NAPHTHALENEDISULPHONIC ACID * DOLKWAL PONCEAU 3R * EXT. D&C RED No. 15 * 3-HYDROXY-4-((2,4,5-TRIMETHYLPHENYL)AZO)-2,7-NAPHTHALENEDISULPHONIC ACID, DISODIUM SALT * 3-HYDROXY-4-((2,4,5-TRIMETHYLPHENYL)AZO)-2,7-NAPHTHALENEDISULFONIC ACID, DISODIUM SALT * MAPLE PONCEAU 3R * PONCEAU 3R * SODIUM CUMENEAZO-β-NAPHTHOL DISULPHONATE * USACERT RED No. 1

CONSENSUS REPORTS: IARC Cancer Review: GROUP 2B IMEMDT 7,56,87; Animal Sufficient Evidence IMEMDT 8,199,75

SAFETY PROFILE: Suspected carcinogen with experimental carcinogenic and tumorigenic data. Mutation data reported. When heated to decomposition it emits toxic fumes of NO_x and SO_x.

FAG040   CAS: 16423-68-0   HR: 3
FD&C RED No. 3
mf: $C_{20}H_6I_4O_5 \cdot 2Na$ mw: 879.84

PROP: Brown powder. Sol in water, conc sulfuric acid.

SYNS: AIZEN ERYTHROSINE * CALCOCID ERYTH-
ROSINE N * CANACERT ERYTHROSINE BS
* 9-(o-CARBOXYPHENYL)-6-HYDROXY-2,4,5,7-TETRA-
IODO-3-ISOXANTHONE * C.I. 45430 * C.I. ACID
RED 51 * CILEFA PINK B * D&C RED No. 3
* DOLKWAL ERYTHROSINE * DYE FD&C RED No. 3
* E 127 * EBS * EDICOL SUPRA ERYTHROSINE A
* ERYTHROSIN * ERYTHROSINE B-FO (BIOLOGICAL
STAIN) * FOOD RED 14 * HEXACERT RED No. 3
* HEXACOL ERYTHROSINE BS * LB-ROT 1
* MAPLE ERYTHROSINE * NEW PINK BLUISH
GEIGY * 1427 RED * 1671 RED * 2′,4′,5′,7′-TET-
RAIODOFLUORESCEIN, DISODIUM SALT * TETRAIO-
DOFLUORESCEIN SODIUM SALT * USACERT RED
No. 3

CONSENSUS REPORTS: Reported in EPA
TSCA Inventory. EPA Genetic Toxicology Program.

SAFETY PROFILE: Poison by intravenous
route. Moderately toxic by ingestion and possibly other routes. Questionable carcinogen with
experimental tumorigenic data. Human mutation data reported. When heated to decomposition it emits very toxic fumes of Na_2O and I^-.

FAG070   CAS: 81-88-9   **HR: 3**
FD&C RED No. 19
mf: $C_{28}H_{31}N_2O_3 \cdot Cl$ mw: 479.06

SYNS: ACID BRILLIANT PINK B * ADC RHODA-
MINE B * AIZEN RHODAMINE BH * AKIRIKU RHO-
DAMINE B * BASIC VIOLET 10 * BRILLIANT PINK
B * CALCOZINE RED BX * CALCOZINE RHODA-
MINE BX * 9-o-CARBOXYPHENYL-6-DIETHYLAMINO-
3-ETHYLIMINO-3-ISOXANTHENE, 3-ETHOCHLORIDE
* (9-(o-CARBOXYPHENYL)-6-(DIETHYLAMINO)-3H-XAN-
THEN-3-YLIDENE) DIETHYLAMMONIUM CHLORIDE
* CERISE TONER X1127 * CERTIQUAL RHODAMIEN
* C.I. 749 * C.I. BASIC VIOLET 10 * C.I. FOOD
RED 15 * COGILOR RED 321.10 * COSMETIC BRIL-
LIANT PINK BLUISH D CONC * D&C RED No. 19
* DIABASIC RHODAMINE B * DIETHYL-m-AMINO-
PHENOLPHTHALEIN HYDROCHLORIDE * EDICOL SU-
PRA ROSE B * ELCOZINE RHODAMINE B * ERI-
OSIN RHODAMINE B * FOOD RED 15 * GERANIUM
LAKE N * HEXACOL RHODAMINE B EXTRA
* IKADA RHODAMINE B * IRAGEN RED L-U
* MITSUI RHODAMINE BX * 11411 RED * RED
NO 213 * RHEONINE B * RHODAMINE * RHO-
DAMINE S (RUSSIAN) * SICILIAN CERISE TONER A-
7127 * SYMULEX MAGENTA F * SYMULEX PINK F
* TAKAOKA RHODAMINE B * TETRAETHYLDIAM-
INO-o-CARBOXY-PHENYL-XANTHENYL CHLORIDE
* TETRAETHYLRHODAMINE

CONSENSUS REPORTS: IARC Cancer Review: GROUP 3 IMEMDT 7,56,87; Animal
Sufficient Evidence IMEMDT 16,221,78. Reported in EPA TSCA Inventory. EPA Genetic
Toxicology Program.

SAFETY PROFILE: Poison by intraperitoneal
and intravenous routes. Moderately toxic by ingestion. Questionable carcinogen with experimental carcinogenic and tumorigenic data.
Mutation data reported. When heated to
decomposition it emits very toxic fumes of NO_x,
NH_3, and Cl^-.

FAG100   CAS: 25956-17-6   **HR: D**
FD&C RED No. 40
mf: $C_{18}H_{14}N_2O_8S_2Na_2$ mw: 496.42

PROP: Red powder. Sol in water; sltly sol in
abs alc.

SYNS: ALLURA RED AC * C.I. 16035

SAFETY PROFILE: Experimental reproductive
effects. When heated to decomposition it emits
very toxic fumes of NO_x and SO_x.

FAG120   CAS: 1694-09-3   **HR: 3**
FD&C VIOLET No. 1
mf: $C_{39}H_{41}N_3O_6S_2 \cdot Na$ mw: 734.94

SYNS: ACID VIOLET * A.F. VIOLET No 1
* AIZEN FOOD VIOLET No 1 * BENZYL VIOLET
* BENZYL VIOLET 3B * CALCOCID VIOLET 4BNS
* C.I. 42640 * C.I. FOOD VIOLET 2 * COOMASSIE
VIOLET * DISPERSED VIOLET 12197 * FORMYL VI-
OLET S4BN * PERGACID VIOLET 2B * SOLAR VIO-
LET 5BN * WOOL VIOLET

CONSENSUS REPORTS: IARC Cancer Re-
view: GROUP 2B IMEMDT 7,56,87; Animal
Sufficient Evidence IMEMDT 16,153,78. EPA
Genetic Toxicology Program. Reported in EPA
TSCA Inventory.

SAFETY PROFILE: Suspected carcinogen with
experimental carcinogenic and tumorigenic
data. Mutation data reported. When heated to
decomposition it emits very toxic fumes of NO_x,
NH_3, Na_2O, and SO_x.

FAG130   CAS: 85-84-7   **HR: 2**
FD&C YELLOW No. 3
mf: $C_{16}H_{13}N_3$ mw: 247.32

SYNS: A.F YELLOW No. 2 * 1-BENZENE-AZO-β-
NAPHTHYLAMINE * 1-BENZENEAZO-2-NAPHTHYL-

AMINE * CERISOL YELLOW AB * C.I. 11380 * C.I. FOOD YELLOW 10 * C.I. SOLVENT YELLOW 5 * DOLKWAL YELLOW AB * EXT. D&C YELLOW No. 9 * GRASAL YELLOW * JAUNE AB * OIL YELLOW A * 1-(PHENYLAZO)-2-NAPH-THALENAMINE * 1-(PHENYLAZO)-2-NAPH-THYLAMINE * YELLOW AB * YELLOW No. 2

CONSENSUS REPORTS: IARC Cancer Review: GROUP 3 IMEMDT 7,56,87; Animal No Evidence IMEMDT 8,279,75. Reported in EPA TSCA Inventory. EPA Genetic Toxicology Program.

SAFETY PROFILE: Moderately toxic by ingestion and subcutaneous routes. Questionable carcinogen with experimental tumorigenic data. Mutation data reported. When heated to decomposition it emits toxic fumes of NO_x.

FAG135 CAS: 131-79-3 **HR: 2**
FD&C YELLOW No. 4
mf: $C_{17}H_{15}N_3$ mw: 261.35

SYNS: A.F. YELLOW No. 3 * CERISOL YELLOW TB * C.I. 11390 * C.I. FOOD YELLOW 11 * DOLK-WAL YELLOW OB * EXT. D&C YELLOW No. 10 * JAUNE OB * 1-(2-METHYLPHENYL)AZO-2-NAPH-THALENAMINE * 1-((2-METHYLPHENYL)AZO)-2-NAPHTHALENAMINE * 1-(2-METHYLPHENYL)AZO-2-NAPHTHYLAMINE * OIL YELLOW OB * o-TOL-UENE-1-AZO-2-NAPHTHYLAMINE * 1-(o-TOLYLAZO)-2-NAPHTHYLAMINE * YELLOW OB

CONSENSUS REPORTS: IARC Cancer Review: GROUP 3 IMEMDT 7,56,87; Animal Sufficient Evidence IMEMDT 8,287,75. EPA Genetic Toxicology Program.

SAFETY PROFILE: Moderately toxic by ingestion, intraperitoneal, and subcutaneous routes. Questionable carcinogen with experimental tumorigenic data. Mutation data reported. When heated to decomposition it emits toxic fumes of NO_x.

FAG140 CAS: 1934-21-0 **HR: 1**
FD&C YELLOW No. 5
mf: $C_{16}H_9N_4O_9S_2 \cdot 3Na$ mw: 534.38

PROP: Yellow-orange powder. Sol in water, conc sulfuric acid.

SYNS: ACID LEATHER YELLOW T * ACILAN YEL-LOW GG * AIREDALE YELLOW T * AIZEN TAR-TRAZINE * ATUL TARTRAZINE * BUCACID TAR-TRAZINE * CALCOCID YELLOW XX * CANACERT

TARTRAZINE * 3-CARBOXY-5-HYDROXY-1-p-SULFO-PHENYL-4-o-SULFOPHENYLAZOPYRAZOLE TRISODIUM SALT * C.I. 19140 * C.I. FOOD YELLOW 4 * CURON FAST YELLOW 5G * D&C YELLOW No. 5 * DOLKWAL TARTRAZINE * EDICOL SUPRA TAR-TRAZINE N * EGG YELLOW A * EUROCERT TAR-TRAZINE * FENAZO YELLOW T * FOOD YELLOW No. 4 * HEXACOL TARTRAZINE * HYDRAZINE YELLOW * KARO TARTRAZINE * KITON YELLOW T * LAKE YELLOW * MAPLE TARTRAZOL YEL-LOW * NAPHTHOCARD YELLOW O * OXANAL YELLOW T * SHULTZ No. 737 * SUGAI TARTRA-ZINE * TARTAR YELLOW * TARTRAZINE * TARTRAZOL YELLOW * TRISODIUM-3-CARBOXY-5-HYDROXY-1-p-SULFOPHENYL-4-p-SULFOPHENYLAZO-PYRAZOLE * TRISODIUM SALT of 3-CARBOXY-5-HY-DROXY-1-SULFOPHENYLAZOPYRAZOLE * UNITER-TRACID YELLOW TE * USACERT YELLOW No. 5 * VONDACID TARTRAZINE * WOOL YELLOW * XYLENE FAST YELLOW GT * YELLOW LAKE 69

CONSENSUS REPORTS: Reported in EPA TSCA Inventory. EPA Genetic Toxicology Program.

SAFETY PROFILE: Mildly toxic by ingestion. Experimental teratogenic and reproductive effects. Human systemic effects by ingestion: paresthesia and changes in teeth and supporting structures. Human mutation data reported. When heated to decomposition it emits very toxic fumes of NO_x, SO_x, and Na_2O.

FAG150 CAS: 2783-94-0 **HR: 2**
FD&C YELLOW No. 6
mf: $C_{16}H_{10}N_2O_7S_2Na_2$ mw: 452.36

PROP: Orange powder. Sol in water, conc sulfuric acid; sltly sol in abs alc.

SYNS: ACID YELLOW TRA * AIZEN FOOD YELLOW No. 5 * CANACERT SUNSET YELLOW FCF * C.I. 15985 * GELBORANGE-S (GERMAN) * SUNSET YELLOW FCF

CONSENSUS REPORTS: IARC Cancer Review: GROUP 3 IMEMDT 7,56,87; Animal Inadequate Evidence IMEMDT 8,257,75. Reported in EPA TSCA Inventory.

SAFETY PROFILE: Moderately toxic by intraperitoneal route. Questionable carcinogen. When heated to decomposition it emits very toxic fumes of NO_x and SO_x.

FAK000 CAS: 22224-92-6 **HR: 3**
FENAMIPHOS
mf: $C_{13}H_{22}NO_3PS$ mw: 303.39

SYNS: o-AETHYL-o-(3-METHYL-4-METHYLTHIOPHE-
NYL)-ISOPROPYLAMIDO-PHOSPHORSAEURE ESTER
(GERMAN) * BAY 68138 * ENT 27,572 * ETHYL-
3-METHYL-4-(METHYLTHIO)PHENYL(1-METHYLETHYL)
PHOSPHORAMIDATE * ETHYL-4-(METHYLTHIO)-m-
TOLYL ISOPROPYL PHOSPHOR AMIDATE * ISOPRO-
PYLAMINO-o-ETHYL-(4-METHYLMERCAPTO-3-METHYL-
PHENYL)PHOSPHATE * 1-(METHYLETHYL)-ETHYL 3-
METHYL-4-(METHYLTHIO)PHENYL PHOSPHORAMIDATE
* NEMACUR * NSC 195106 * PHANAMIPHOS

CONSENSUS REPORTS: EPA Extremely
Hazardous Substances List.

OSHA PEL: TWA 0.1 mg/m^3 (skin)
ACGIH TLV: TWA 0.1 mg/m^3 (skin)

SAFETY PROFILE: Poison by ingestion, inhalation, and skin contact. When heated to decomposition it emits very toxic fumes of NO_x, PO_x,
and SO_x.

FAK100   CAS: 60168-88-9   **HR: 2**
FENARIMOL
mf: $C_{17}H_{12}Cl_2N_2O$ mw: 331.21

PROP: White, odorless crystals. Mp: 117-119°.
Practically insol in water; sol in most organic
solvents.

SYNS: BLOC * (2-CHLOROPHENYL)-α-(4-CHLORO-
PHENYL)-5-PYRIMIDINEMETHANOL * α-(2-CHLORO-
PHENYL)-α-(4-CHLOROPHENYL)-5-PYRIMIDINEMETHA-
NOL * EL 222 * RIMIDIN * RUBIGAN

CONSENSUS REPORTS: Reported in EPA
TSCA Inventory.

SAFETY PROFILE: Moderately toxic by ingestion. Mutation data reported. When heated to
decomposition it emits toxic fumes of Cl^- and
NO_x.

FAL000   CAS: 13669-70-0   **HR: 3**
FENAZOXINE
mf: $C_{17}H_{19}NO$ mw: 253.37

SYNS: NEFOPAM * 3,4,5,6,7-TETRAHYDRO-5-
METHYL-1-PHENYL-1H-2,5-BENZOXAZOCINE

SAFETY PROFILE: Poison by ingestion, intravenous, intraperitoneal, and intramuscular
routes. Human systemic effects by ingestion:
hallucinations, distorted perceptions, excitement, motor activity changes, increased pulse
rate without blood pressure fall and heart rate
changes. When heated to decomposition it emits
toxic fumes of NO_x.

FAL100   CAS: 43210-67-9   **HR: D**
FENBENDAZOLE
mf: $C_{15}H_{13}N_3O_2S$ mw: 299.37

SYNS: FENBENDAZOL * HOE 881 * PANACUR
* (5-(PHENYLTHIO)-2-BENZIMIDAZOLECARBAMIC
ACID, METHYL ESTER

SAFETY PROFILE: Human mutation data reported. When heated to decomposition it emits
toxic fumes of SO_x and NO_x.

FAP000   CAS: 8006-84-6   **HR: 2**
FENNEL OIL

PROP: From steam distillation of *Foeniculum
vulgare* Miller (Fam. *Umbelliferae*). Colorless
to pale yellow liquid; odor and taste of fennel.

SYNS: BITTER FENNEL OIL * FENCHEL OEL (GER-
MAN) * OIL of FENNEL

CONSENSUS REPORTS: Reported in EPA
TSCA Inventory.

SAFETY PROFILE: Moderately toxic by ingestion. Mutation data reported. A severe skin irritant. When heated to decomposition it emits
acrid smoke and irritating fumes.

FAQ800   CAS: 115-90-2   **HR: 3**
FENSULFOTHION
mf: $C_{11}H_{17}O_4PS_2$ mw: 308.37

SYNS: BAY 25141 * BAYER S767 * CHEMAGRO
25141 * DASANIT * o,o-DIAETHYL-o-4-METHYL-
SULFINYL-PHENYL-MONOTHIOPHOSPHAT (GERMAN)
* o,o-DIETHYL-o-(p-(METHYLSULFINYL)PHENYL)
PHOSPHOROTHIOATE * o,o-DIETHYL-o-p-(METHYL-
SULFINYL)PHENYL THIOPHOSPHATE * DMSP
* ENT 24,945 * S 767 * TERRACUR P

CONSENSUS REPORTS: EPA Genetic Toxicology Program. EPA Extremely Hazardous
Substances List.

OSHA PEL: TWA 0.1 mg/m^3
ACGIH TLV: TWA 0.1 mg/m^3

SAFETY PROFILE: A poison by ingestion, skin
contact, intraperitoneal, and possibly other
routes. A pesticide. When heated to decomposition it emits very toxic fumes of SO_x and PO_x.

FAQ999   CAS: 55-38-9   **HR: 3**
FENTHION
mf: $C_{10}H_{15}O_3PS_2$ mw: 278.34

SYNS: BAY 29493 * BAYCID * BAYER 9007
* BAYTEX * o,o-DIMETHYL-o-4-(METHYLMER-

CAPTO)-3-METHYLPHENYL PHOSPHOROTHIOATE
* O,O-DIMETHYL-p-4-(METHYLMERCAPTO)-3-METHYL-
PHENYL THIOPHOSPHATE * O,O-DIMETHYL-O-(3-
METHYL-4-METHYLMERCAPTOPHENYL)PHOSPHORO-
THIOATE * O,O-DIMETHYL-O-(3-METHYL-4-
METHYLTHIO-FENYL)-MONOTHIOFOSFAAT (DUTCH)
* O,O-DIMETHYL-O-(3-METHYL-4-METHYLTHIO-
PHENYL)-MONOTHIOPHOSPHAT (GERMAN) * O,O-DI-
METHYL-O-3-METHYL-4-METHYLTHIOPHENYL PHOS-
PHOROTHIOATE * O,O-DIMETHYL-O-(3-METHYL-4-
METHYLTHIO-PHENYL)-THIONOPHOSPHAT (GERMAN)
* O,O-DIMETHYL-O-(4-METHYLTHIO-3-METHYLPHE-
NYL) PHOSPHOROTHIOATE * O,O-DIMETHYL-O-
(4-(METHYLTHIO)-m-TOLYL) PHOSPHOROTHIOATE
* O,O-DIMETIL-O-(3-METIL-4-METILTIO-FENIL)-MONO-
TIOFOSFATO (ITALIAN) * DMTP * ENT 25,540
* ENTEX * LEBAYCID * MERCAPTOPHOS
* 4-METHYLMERCAPTO-3-METHYLPHENYL DI-
METHYL THIOPHOSPHATE * MPP * NCI-C08651
* OMS 2 * PHOSPHOROTHIOIC ACID-O,O-DIME-
THYL-O-(3-METHYL-4-METHYLTHIOPHENYLE) (FRENCH)
* QUELETOX * S 1752 * SPOTTON * TALO-
DEX * THIOPHOSPHATE de O,O-DIMETHYLE et de
O-(3-METHYL-4-METHYLTHIOPHENYLE) (FRENCH)
* TIGUVON

CONSENSUS REPORTS: NCI Carcinogenesis Bioassay Completed; Results Negative:
rat NCITR* NCI-CG-TR-103,79; Indefinite:
mouse NCITR* NCI-CG-TR-103,79. EPA Genetic Toxicology Program.

OSHA PEL: TWA 0.2 mg/m^3 (skin)
ACGIH TLV: TWA 0.2 mg/m^3 (skin)
DFG MAK: 0.2 mg/m^3

SAFETY PROFILE: A human poison by an unspecified route. Poison experimentally by ingestion, skin contact, intraperitoneal, intravenous,
and intramuscular routes. Moderately toxic by
inhalation. Human systemic effects by ingestion:
pulse rate increase, hypermotility, diarrhea,
nausea or vomiting. Experimental teratogenic
and reproductive effects. Questionable carcinogen with experimental tumorigenic data. Mutation data reported. When heated to decomposition it emits very toxic fumes of PO$_x$ and
SO$_x$.

FAR100   CAS: 51630-58-1   **HR: 3**
FENVALERATE
mf: C$_{25}$H$_{22}$ClNO$_3$ mw: 419.93

PROP: Clear, yellow, viscous liquid at 23°.
D: 1.17, n (20/D) 1.5533. Solubility at 20° (g/
L): acetone, >450; chloroform, >450; metha-

nol, >450; hexane, 77. Insol in water. Decomp
gradually between 150-300°.

SYNS: BELMARK * α-CYANO-3-PHENOXYBENZYL-
2-(4-CHLOROPHENYL)ISOVALERATE PYDRIN * α-
CYANO-3-PHENOXYBENZYL-2-(4-CHLOROPHENYL)-3-
METHYLBUTYRATE * CYANO(3-PHENOXYPHENYL)
METHYL 4-CHLORO-α-(1-METHYLETHYL)BENZENEACE-
TATE * ECTRIN * PHENVALERATE * PYDRIN
* S 5602 * SANMARTON * SD 43775 * SUMICI-
DIN * SUMIFLY * SUMIPOWER * WL 43775

CONSENSUS REPORTS: Cyanide and its
compounds are on the Community Right-To-
Know List.

SAFETY PROFILE: Poison by ingestion, intravenous, and intracerebral routes. Moderately
toxic by skin contact. Experimental reproductive
effects. Highly toxic to fish and bees. Corrosive,
causes eye damage. A skin irritant. When heated
to decomposition it emits toxic fumes of Cl$^-$,
NO$_x$, and CN$^-$.

FAS000   CAS: 14484-64-1   **HR: 3**
FERBAM
mf: C$_9$H$_{18}$N$_3$S$_6$ • Fe   mw: 416.51

PROP: Black solid, sltly sol in water. Mp: decomp 180°.

SYNS: AAFERTIS * BERCEMA FERTAM 50
* CARBAMATE * DIMETHYLCARBAMODITHIOIC
ACID, IRON COMPLEX * DIMETHYLCARBAMODI-
THIOIC ACID, IRON(3+) SALT * DIMETHYLDITHIO-
CARBAMIC ACID, IRON SALT * DIMETHYLDITHIO-
CARBAMIC ACID, IRON(3+) SALT * EISENDI-
METHYLDITHIOCARBAMAT (GERMAN) * EISEN-
(III)-TRIS(N,N-DIMETHYLDITHIOCARBAMAT)
(GERMAN) * ENT 14,689 * FERBAM 50 * FER-
BAM, IRON SALT * FERBECK * FERMATE FERBAM
FUNGICIDE * FERMOCIDE * FERRADOW
* FERRIC DIMETHYLDITHIOCARBAMATE * FUKLA-
SIN ULTRA * HEXAFERB * HOKMATE * IRON
DIMETHYLDITHIOCARBAMATE * KARBAM BLACK
* KNOCKMATE * NIACIDE * SUP'R FLO FERBAM
FLOWABLE * TRIFUNGOL * TRIS(DIMETHYLCAR-
BAMODITHIOATO-S,S')IRON * TRIS)DIMETHYLDI-
THIOCARBAMATO)IRON * TRIS(N,N-DIMETHYL-
DITHIOCARBAMATO) IRON(111) * VANCIDE FE95

CONSENSUS REPORTS: IARC Cancer Review: GROUP 3 IMEMDT 7,56,87; Animal
Inadequate Evidence IMEMDT 12,121,76. Reported in EPA TSCA Inventory. EPA Genetic
Toxicology Program.

OSHA PEL: (Transitional: TWA Total Dust: 15 mg/m^3; Respirable Fraction: 5 mg/m^3) TWA Total Dust: 10 mg/m^3; Respirable Fraction: 5 mg/m^3
ACGIH TLV: TWA 10 mg/m^3
DFG MAK: 15 mg/m^3

SAFETY PROFILE: Poison by intraperitoneal route. Moderately toxic by ingestion. Experimental teratogenic and reproductive effects. Questionable carcinogen with experimental carcinogenic and tumorigenic data. Mutation data reported. A fungicide. When heated to decomposition it emits very toxic fumes of NO$_x$ and SO$_x$.

FAU000   CAS: 7705-08-0   **HR: 3**
FERRIC CHLORIDE

DOT: UN: 1773/UN 2582
mf: Cl$_3$Fe   mw: 162.20

PROP: Black-brown solid. Mp: 292°, bp: 319.0°, d: 2.90 @ 25°, vap press: 1 mm @ 194.0°.

SYNS: CHLORURE PERRIQUE * FERRIC CHLORIDE, anhydrous (DOT) * FERRIC CHLORIDE, solid (DOT) * FERRIC CHLORIDE, solid, anhydrous (DOT) * FERRIC CHLORIDE, solution (DOT) * FLORES MARTIS * IRON CHLORIDE * IRON(III) CHLORIDE * IRON CHLORIDE, solid (DOT) * IRON SESQUICHLORIDE, solid (DOT) * IRON TRICHLORIDE * PERCHLORURE DE FER

CONSENSUS REPORTS: Reported in EPA TSCA Inventory. EPA Genetic Toxicology Program.

OSHA PEL: TWA 1 mg(Fe)/m^3
ACGIH TLV: TWA 1 mg(Fe)/m^3
DOT Classification: ORM-B; Label: None, anhydrous; Corrosive Material; Label: Corrosive.

SAFETY PROFILE: Poison by intravenous route. Moderately toxic by ingestion. Experimental reproductive effects. Corrosive. Probably an eye, skin, and mucous membrane irritant. Mutation data reported. Reacts with water to produce toxic and corrosive fumes. Catalyzes potentially explosive polymerization of ethylene oxide, chlorine + monomers (e.g., styrene). Forms shock-sensitive explosive mixtures with some metals (e.g., potassium, sodium). Violent reaction with allyl chloride. When heated to decomposition it emits highly toxic fumes of HCl.

FAW000   CAS: 7705-08-0   **HR: 2**
FERRIC CHLORIDE (solution)
mf: FeCl$_3$ mw: 162.2

SYN: IRON(III) CHLORIDE (solution)

CONSENSUS REPORTS: Reported in EPA TSCA Inventory.

DOT Classification: Corrosive Material; Label: Corrosive.

SAFETY PROFILE: Corrosive to the skin, eyes, and mucous membranes. When heated to decomposition it emits very toxic fumes of Cl$^-$.

FAX000   CAS: 7783-50-8   **HR: 3**
FERRIC FLUORIDE

DOT: NA 9120
mf: F$_3$Fe   mw: 112.85

PROP: Green crystals. D: 3.87.

SYNS: IRON FLUORIDE * IRON TRIFLUORIDE

CONSENSUS REPORTS: Reported in EPA TSCA Inventory.

OSHA PEL: TWA 2.5 mg(F)/m^3; TWA 1 mg(Fe)/m^3
ACGIH TLV: TWA 2.5 mg(F)/m^3; 1 mg(Fe)/m^3
NIOSH REL: TWA (Inorganic Fluorides) 2.5 mg(F)/m^3
DOT Classification: ORM-E; Label: None.

SAFETY PROFILE: Poison by intravenous route. When heated to decomposition it emits toxic fumes of F$^-$.

FBA000   CAS: 10028-22-5   **HR: D**
FERRIC SULFATE

DOT: NA 9121
mf: Fe$_2$O$_{12}$S$_3$ mw: 399.88

PROP: Yellow solid.

SYNS: DIIRON TRISULFATE * IRON PERSULFATE * IRON SESQUISULFATE * IRON SULFATE (2:3) * IRON(III) SULFATE * IRON TERSULFATE * SULFURIC ACID, IRON (3$^+$) SALT (3:2)

CONSENSUS REPORTS: Reported in EPA TSCA Inventory.

ACGIH TLV: TWA 1 mg(Fe)/m^3
DOT Classification: ORM-E.

SAFETY PROFILE: When heated to decomposition it emits toxic fumes of SO$_x$ and Fe$^-$.

FBC000 CAS: 102-54-5 **HR: 3**
FERROCENE
mf: $C_{10}H_{10}Fe$ mw: 186.05

PROP: Orange crystals; camphor odor. Mp: 174°, sublimes @ >100°, volatile in steam. Insol in water; sol in alcohol and ether.

SYNS: BISCYCLOPENTADIENYLIRON * DI-2,4-CY-CLOPENTADIEN-1-YL IRON * DICYCLOPENTADIENYL IRON (OSHA, ACGIH) * IRON BIS(CYCLOPENTADIENE) * IRON DICYCLOPENTADIENYL

CONSENSUS REPORTS: Reported in EPA TSCA Inventory.

OSHA PEL: (Transitional: TWA Total Dust: 15 mg/m³; Respirable Fraction: 5 mg/m³) TWA Total Dust: 10 mg/m³; Respirable Fraction: 5 mg/m³
ACGIH TLV: TWA 10 mg/m³

SAFETY PROFILE: Poison by intraperitoneal and intravenous routes. Moderately toxic by ingestion. Questionable carcinogen with experimental tumorigenic data. Mutation data reported. Flammable; reacts violently with NH_4ClO_4. When heated to decomposition it emits acrid smoke and irritating fumes.

FBC100 CAS: 1336-80-7 **HR: 3**
FERROCHOLINATE
mf: $C_6H_{10}FeO_{10} \cdot C_5H_{14}NO$ mw: 402.21

PROP: Greenish-brown, reddish-brown or brown amorphous solid with glistening surface upon fracture. Sol in water, acids, and alkalies.

SYNS: CHELAFER * CHEL-IRON * FERRIC CHO-LINE CITRATE * FERROLIP * IRON CHOLINE CITRATE COMPLEX

SAFETY PROFILE: Poison by intravenous and intraperitoneal routes. Mildly toxic by ingestion. When heated to decomposition it emits toxic fumes of NO_x.

FBD000 CAS: 11114-46-8 **HR: 3**
FERROCHROME (exothermic)

SYNS: CARBON FERROCHROMIUM * CHROME FERROALLOY * CHROMIUM ALLOY, Cr,C,Fe,N,Si * CHROMIUM ALLOY, BASE, Cr,C,Fe,N,Si (FERRO-CHROMIUM) * exothermic FERROCHROME (DOT) * FERROCHROME * FERROCHROME, exothermic (DOT) * FERROCHROMIUM

CONSENSUS REPORTS: IARC Cancer Review: Animal Inadequate Evidence IMEMDT 23,205,80. Reported in EPA TSCA Inventory. Chromium and its compounds are on the Community Right-To-Know List.

OSHA PEL: TWA 1 mg(Cr)/m³
ACGIH TLV: TWA 0.5 mg(Cr)/m³
DOT Classification: ORM-C; Label: None.

SAFETY PROFILE: Poison by inhalation. Questionable carcinogen.

FBE000 CAS: 12604-53-4 **HR: 3**
FERROMANGANESE (exothermic)

SYN: EXOTHERMIC FERROMANGANESE (DOT)

CONSENSUS REPORTS: Reported in EPA TSCA Inventory. Manganese and its compounds are on the Community Right-To-Know List.

SAFETY PROFILE: The dust will burn violently and give off toxic fumes of MnO_2.

FBF000 CAS: 8049-19-2 **HR: D**
FERROPHOSPHORUS

SYN: IRON ALLOY, BASE, Fe,P (FERROPHOSPHORUS)

CONSENSUS REPORTS: Reported in EPA TSCA Inventory.

DOT Classification: ORM-A; Label: None.

SAFETY PROFILE: When heated to decomposition it emits very toxic fumes of PO_x.

FBG000 CAS: 8049-17-0 **HR: 3**
FERROSILICON
mf: FeSi mw: 83.90

DOT: UN 1408

PROP: Crystalline, metallic solid. Fe + Si, d: 5.4. Containing 30% or more but not more than 70% silicon.

SYNS: FERROSILICON, containing more than 30% but less than 90% SILICON (DOT) * IRON-SILICON ALLOY

CONSENSUS REPORTS: Reported in EPA TSCA Inventory.

DOT Classification: ORM-A; Label: None; Flammable Solid; Label: Dangerous When Wet, Poison.

SAFETY PROFILE: Reaction with moisture releases hydrogen and acetylene gases which then ignite; impurities in the alloy may liberate such poisonous and reactive gases as phosphine and arsine. Dry mixtures with sodium hydroxide react incandescently when water is added. Reaction with acid, acid fumes or oxidizing materials can emit toxic fumes. Reaction hazards increase with decreasing particle size.

FBI000   CAS: 7758-94-3   **HR: 3**
FERROUS CHLORIDE

DOT: UN 1759/UN 1760
mf: Cl_2Fe mw: 126.75

PROP: Green to yellow, deliquescent crystals. Mp: 614-670°, bp: 1026°, d: 3.16, vap press: 10 mm @ 700°.

SYNS: IRON(II) CHLORIDE (1:2) * IRON DICHLORIDE * IRON PROTOCHLORIDE

CONSENSUS REPORTS: Reported in EPA TSCA Inventory. EPA Genetic Toxicology Program.

OSHA PEL: TWA 1 mg(Fe)/m^3
ACGIH TLV: TWA 1 mg(Fe)/m^3
DOT Classification: Corrosive Material; Label: Corrosive, solution; ORM-B; Label: None, solid.

SAFETY PROFILE: Poison by intraperitoneal route. Mutation data reported. Corrosive. Probably an irritant to the eyes, skin and mucous membranes. Can react violently with ethylene oxide, K, Na. When heated to decomposition it emits toxic fumes of Cl^-.

FBJ100   CAS: 141-01-5   **HR: 3**
FERROUS FUMARATE
mf: $C_4H_2O_4 \cdot Fe$ mw: 169.91

PROP: Reddish-orange to reddish-brown granular powder; odorless, almost tasteless. D: 2.435. Solubility at 25° in water: 0.14 g/100 mL; in alc <0.01 g/100 mL. Solubility in acid is limited by liberation of fumaric acid.

SYNS: CPIRON * ERCO-FER * ERCOFERRO * FEOSTAT * FEROTON * FERROFUME * FERRONAT * FERRONE * FERROTEMP * FERRUM * FERSAMAL * FIRON * FUMAFER * FUMAR-F * FUMIRON * GALFER * HEMOTON * IRCON * IRON FUMARATE * METERFER * METERFOLIC * ONE-IRON * PALAFER * TOLERON * TOLFERAIN * TOLIFER

CONSENSUS REPORTS: Reported in EPA TSCA Inventory.

OSHA PEL: TWA 1 mg(Fe)/m^3
ACGIH TLV: TWA 1 mg/(Fe)/m^3

SAFETY PROFILE: Poison by intraperitoneal route. Moderately toxic by ingestion and subcutaneous routes. When heated to decomposition it emits acrid smoke and irritating fumes.

FBK000   CAS: 299-29-6   **HR: 3**
FERROUS GLUCONATE
mf: $C_{12}H_{22}O_{14} \cdot Fe \cdot H_2O$ mw: 482.17

PROP: Yellowish-gray or pale greenish-yellow, fine powder or granules with slt odor of burned sugar. Sol in water and glycerin; insol in alc.

SYNS: FERGON * FERGON PREPARATIONS * FERLUCON * FERRONICUM * GLUCO-FERRUM * IROMIN * IRON GLUCONATE * IROX (GADOR) * NIONATE * RAY-GLUCIRON

CONSENSUS REPORTS: Reported in EPA TSCA Inventory.

OSHA PEL: TWA 1 mg(Fe)/m^3
ACGIH TLV: TWA 1 mg(Fe)/m^3

SAFETY PROFILE: Poison by intraperitoneal and intravenous routes. Moderately toxic by ingestion. Human systemic effects by ingestion: hypermotility, diarrhea, nausea, and vomiting. Experimental teratogenic effects. Questionable carcinogen with experimental tumorigenic data. When heated to decomposition it emits acrid smoke and irritating fumes.

FBN100   CAS: 7720-78-7   **HR: 3**
FERROUS SULFATE

DOT: NA 9125
mf: $O_4S \cdot Fe$ mw: 151.91

PROP: Grayish white to buff powder. Slowly sol in water; insol in alc.

SYNS: COPPERAS * DURETTER * DUROFERON * EXSICCATED FERROUS SULFATE * EXSICCATED FERROUS SULPHATE * FEOSOL * FEOSPAN * FER-IN-SOL * FERO-GRADUMET * FERRALYN * FERRO-GRADUMET * FERROSULFAT (GERMAN) * FERROSULFATE * FERRO-THERON * FERSOLATE * GREEN VITRIOL * IRON MONOSULFATE * IRON PROTOSULFATE * IRON(II) SULFATE (1:1) * IRON VITRIOL * IROSPAN * IROSUL * SLOW-FE * SULFERROUS * SULFURIC ACID, IRON(2$^+$) SALT (1:1)

CONSENSUS REPORTS: Reported in EPA TSCA Inventory. EPA Genetic Toxicology Program.

OSHA PEL: TWA 1 mg/(Fe)/m^3
ACGIH TLV: TWA 1 mg/(Fe)/m^3
DOT Classification: ORM-E; Label: None.

SAFETY PROFILE: A human poison by ingestion. Moderately toxic to humans by an unspecified route. An experimental poison by ingestion, intraduodenal, intraperitoneal, intravenous, and subcutaneous routes. Human systemic effects by ingestion: aggression, somnolence, brain recording changes, diarrhea, nausea or vomiting, bleeding from the stomach, coma. Questionable carcinogen with experimental tumorigenic data. Experimental teratogenic and reproductive effects. Mutation data reported. Potentially explosive reaction with methyl isocyanoacetate at 25°. May ignite on contact with arsenic trioxide + sodium nitrate. When heated to decomposition it emits toxic fumes of SO$_x$.

FBO000   CAS: 7782-63-0   **HR: 3**
FERROUS SULFATE HEPTAHYDRATE
mf: O$_4$S • Fe • 7H$_2$O   mw: 278.05

PROP: Pale blue green monoclinic crystals or granules; odorless with a salt taste. D: 2.99-3.08. Sol in water; insol in alc.

SYNS: COPPERAS * FEOSOL * FER-IN-SOL * FERO-GRADUMET * FERROUS SULFATE (FCC) * FESOFOR * FESOTYME * GREEN VITROL * HAEMOFORT * IRONATE * IRON(II) SULFATE (1:1), HEPTAHYDRATE * IRON VITROL * IROSUL * MOL-IRON * PRESFERSUL * SULFERROUS

OSHA PEL: TWA 1 mg(Fe)/m^3
ACGIH TLV: TWA 1 mg(Fe)/m^3

SAFETY PROFILE: Poison by intravenous, intraperitoneal, and subcutaneous routes. Moderately toxic by ingestion and rectal routes. Mutation data reported. When heated to decomposition it emits toxic fumes of SO$_x$.

FBP000   CAS: 12604-58-9   **HR: 3**
FERROVANADIUM DUST

PROP: A gray to black dust.

CONSENSUS REPORTS: Reported in EPA TSCA Inventory.

OSHA PEL: (Transitional: TWA 1 mg/m^3) TWA 1 mg/m^3; STEL 3 mg/m^3
ACGIH TLV: TWA 1 mg/m^3; STEL 3 mg/m^3
DFG MAK: 1 mg/m^3
NIOSH REL: (Vanadium) TWA 1.0 mg(V)/m^3

SAFETY PROFILE: Can cause pulmonary damage. Combustible when exposed to heat or flame.

FBQ000   **HR: 2**
FIBROUS GLASS

SYNS: FIBERGLASS * FIBROUS GLASS DUST (ACGIH) * GLASS * GLASS FIBERS

OSHA PEL: TWA 15 mg/m^3 (total dust); 5 mg/m^3 (nuisance dust)
ACGIH TLV: TWA 10 mg/m^3 (dust)
NIOSH REL: TWA 5 mg/m^3 (total fibrous glass)

SAFETY PROFILE: Questionable carcinogen with experimental carcinogenic, neoplastigenic, and tumorigenic data by inhalation and other routes. Human mutation data reported. Used as thermal and acoustic insulation.

The possibility of lung problems due to inhalation of fine particles or flakes or fibers of fiberglass has often been raised. The extensive medical research so far reported has shown no consistent evidence of chronic health effects in workers who are exposed to manmade vitreous fibers. In some studies where massive doses of fine diameter fibers were implanted into mice, cancer development in the pleura was noted. Also some animal studies involving injection of fibers into the trachae resulted in a minimal fibrosis.

Exposure to glass fibers sometimes causes irritation of the skin and, less frequently, irritation of the eyes, nose, or throat. This is not an allergic reaction, but simply a mechanical irritation. Skin irritation typically is experienced by individuals who are newly exposed to fibrous glass and it usually diminishes after several days of exposure. Good personal and industrial hygiene practices minimize the amount of discomfort experienced.

FBS000   CAS: 9001-33-6   **HR: 3**
FICIN

PROP: A proteolytic enzyme in the crude latex of the fig tree *Ficus*. White powder. Very sol in water.

SYNS: DEBRICIN * FICUS PROTEASE * FICUS PROTEINASE * HIGUEROXYL DELABARRE * TL 367

CONSENSUS REPORTS: Reported in EPA TSCA Inventory.

SAFETY PROFILE: Poison by inhalation and intravenous routes data. Mildly toxic by ingestion. When heated to decomposition it emits toxic fumes.

FBU000 CAS: 59536-65-1 HR: 3
FIREMASTER BP-6

PROP: Consists mainly of penta-, hexa-, and heptabromobiphenyl, with lesser amounts of tetra- and other brominated biphenyls.

SYNS: HEXABROMOBIPHENYL (technical grade) * PBB * POLYBROMINATED BIPHENYLS

CONSENSUS REPORTS: IARC Cancer Review: Animal Inadequate Evidence IMEMDT 18,107,78. Polybrominated biphenyl compounds are on the Community Right-To-Know List.

SAFETY PROFILE: Poison by ingestion. Experimental teratogenic and reproductive effects. Questionable carcinogen with experimental carcinogenic and tumorigenic data. Mutation data reported. When heated to decomposition it emits very toxic Br^-.

FBU509 CAS: 67774-32-7 HR: 3
FIREMASTER FF-1

PROP: 2,4,5,2',4',5'-hexabromobiphenyl is the predominant isomer.

SYNS: 2,4,5,2',4',5'-HEXABROMOBIPHENYL * PBB * POLYBROMINATED BIPHENYL * POLYBROMINATED BIPHENYL (FF-1)

CONSENSUS REPORTS: IARC Cancer Review: GROUP 2B IMEMDT 7,321,87; Human Inadequate Evidence IMEMDT 41,261,86; Animal Sufficient Evidence IMEMDT 41,261,86. NTP Carcinogenesis Studies (gavage); Clear Evidence: mouse, rat NTPTR* NTP-TR-244,83. Polybrominated biphenyl compounds are on the Community Right-To-Know List.

SAFETY PROFILE: Suspected carcinogen with experimental carcinogenic and neoplastigenic data. Experimental teratogenic and reproductive effects. When heated to decomposition it emits very toxic fumes of Br^-.

FBV000 CAS: 8021-29-2 HR: 1
FIR NEEDLE OIL, SIBERIAN

PROP: Found in the needles and twigs of *Abies sibirica* Ledeb. (Fam. *Pinaceae*). Colorless to faintly yellow liquid. Sol in fixed oils, mineral oil; insol in glycerin, propylene glycol.

SYN: PINE NEEDLE OIL

CONSENSUS REPORTS: Reported in EPA TSCA Inventory.

SAFETY PROFILE: Mildly toxic by ingestion. A human and experimental skin irritant. When heated to decomposition it emits acrid smoke and irritating fumes.

FDA880 CAS: 17160-71-3 HR: 3
FLUANISONE HYDROCHLORIDE
mf: $C_{21}H_{25}FN_2O_2 \cdot ClH$ mw: 392.94

SYNS: ANTI-PICA * 4'-FLUORO-4-(4-(o-METHOXY-PHENYL)-1-PIPERAZINYL)BUTYROPHENONE HYDROCHLORIDE * HALOANISONE COMPOSITUM

SAFETY PROFILE: Poison by subcutaneous, intraperitoneal, and intravenous routes. When heated to decomposition it emits very toxic fumes of HCl, NO_x, and F^-.

FDE000 CAS: 30223-48-4 HR: D
FLUORACIZINE
mf: $C_{20}H_{21}F_3N_2OS$ mw: 394.49

SYN: 10-DIETHYLAMINOPROPIONYL-3-TRIFLUORO-METHYL PHENOTHIAZINE HYDROCHLORIDE

SAFETY PROFILE: Experimental teratogenic and reproductive effects. When heated to decomposition it emits very toxic fumes of SO_x, NO_x, and F^-.

FDF000 CAS: 206-44-0 HR: 3
FLUORANTHENE
mf: $C_{16}H_{10}$ mw: 202.26

PROP: A polycyclic hydrocarbon. Colorless solid. Mp: 120°, bp: 367°, vap press: 0.01 mm @ 20°.

SYNS: 1,2-BENZACENAPHTHENE * BENZO-(jk)FLUORENE * IDRYL * 1,2-(1,8-NAPHTHALENE-DIYL)BENZENE * 1,2-(1,8-NAPHTHYLENE)BENZENE * RCRA WASTE NUMBER U120

CONSENSUS REPORTS: IARC Cancer Review: GROUP 3 IMEMDT 7,56,87; Animal No Evidence IMEMDT 32,355,83. Reported in EPA TSCA Inventory. EPA Genetic Toxicology Program.

SAFETY PROFILE: Poison by intravenous route. Moderately toxic by ingestion and skin contact. Questionable carcinogen with experimental tumorigenic data. Human mutation data reported. Combustible when exposed to heat or flame. When heated to decomposition it emits acrid smoke and irritating fumes.

FDI000   CAS: 153-78-6   **HR: 3**
FLUOREN-2-AMINE
mf: $C_{13}H_{11}N$ mw: 181.25

SYNS: AMINOFLUOREN (GERMAN) * 2-AMINO-FLUORENE * 2-FLUORENAMINE * 2-FLUORENE-AMINE

CONSENSUS REPORTS: EPA Genetic Toxicology Program.

SAFETY PROFILE: Suspected carcinogen with experimental carcinogenic, neoplastigenic, and tumorigenic data. Poison by intraperitoneal route. Mutation data reported. When heated to decomposition it emits toxic fumes of NO_x.

FDR000   CAS: 53-96-3   **HR: 3**
N-FLUOREN-2-YL ACETAMIDE
mf: $C_{15}H_{13}NO$ mw: 223.29

SYNS: AAF * 2-AAF * ACETOAMINOFLUORENE * 2-ACETAMIDOFLUORENE * 2-ACETAMINOFLUOR-ENE * 2-ACETYLAMINO-FLUOREN (GERMAN) * N-ACETYL-2-AMINOFLUORENE * 2-ACETYLAMI-NOFLUORENE (OSHA) * AZETYLAMINOFLUOREN (GERMAN) * FAA * 2-FAA * 2-FLUORENYL-ACETAMIDE * N-2-FLUORENYLACETAMIDE * RCRA WASTE NUMBER U005

CONSENSUS REPORTS: NTP Fourth Annual Report On Carcinogens, 1984. Community Right-To-Know List. EPA Genetic Toxicology Program. Reported in EPA TSCA Inventory.

OSHA PEL: Carcinogen

SAFETY PROFILE: Confirmed human carcinogen with experimental carcinogenic, neoplastigenic, and tumorigenic data. Moderately toxic by ingestion and intraperitoneal routes. Experimental teratogenic and reproductive effects. Human mutation data reported. When heated to decomposition it emits toxic fumes of NO_x.

FDZ000   CAS: 3671-71-4   **HR: 3**
N-FLUOREN-2-YL BENZOHYDROXAMIC ACID
mf: $C_{20}H_{15}NO_2$ mw: 301.36

SYNS: N-BENZOYLOXY-ACETYLAMINOFLUORENE * N-(2-FLUORENYL)BENZOHYDROXAMIC ACID

* N-9H-FLUOREN-2-YL-N-HYDROXYBENZAMIDE
* N-HYDROXY-2-BENZOYLAMINOFLUORENE
* N-HYDROXY-N-2-FLUORENYLBENZAMIDE

SAFETY PROFILE: Questionable carcinogen with experimental carcinogenic data. Mutation data reported. When heated to decomposition it emits toxic fumes of NO_x.

FEV000   CAS: 2321-07-5   **HR: 3**
FLUORESCEIN
mf: $C_{20}H_{12}O_5$ mw: 332.32

PROP: Orange-red, crystalline powder. Mp: 314-316° with decomp.

SYNS: 9-(o-CARBOXYPHENYL)-6-HYDROXY-3-ISOXAN-THENONE * 9-(o-CARBOXYPHENYL)-6-HYDROXY-3H-XANTHEN-3-ONE * C.I. 45330 * C.I. 45350 (FREE ACID) * C.I. SOLVENT YELLOW 94 * D&C YEL-LOW NO. 7 * 3′,6′-DIHYDROXYFLUORAN * DIHY-DROXYFLUORANE * 3′,6′-DIHYDROXYSPIRO(ISOBEN-ZOFURAN-1(3H),9′(9H)-XANTHEN)-3-ONE * 3,6-FLUORANDIOL * 3′,6′-FLUORANDIOL * FLUORES-CEINE * HIDACID FLUORESCEIN * RESORCINOL-PHTHALEIN * SOAP YELLOW F * 11712 YELLOW

CONSENSUS REPORTS: Reported in EPA TSCA Inventory. EPA Genetic Toxicology Program.

SAFETY PROFILE: Poison by intravenous and possibly other routes. Moderately toxic by intraperitoneal route. Mutation data reported. When heated to decomposition it emits acrid smoke and irritating fumes.

FEW000   CAS: 518-47-8   **HR: 3**
FLUORESCEIN SODIUM
mf: $C_{20}H_{10}O_5 \cdot 2Na$ mw: 376.28

PROP: Orange-red powder; sol in water; sltly sol in alcohol.

SYNS: AIZEN URANINE * CALCOCID URANINE B4315 * 9-o-CARBOXYPHENYL-6-HYDROXY-3-ISOXAN-THONE, DISODIUM SALT * CERTIQUAL FLUORES-CEINE * C.I. 766 * C.I. ACID YELLOW 73 * C.I. 45350 DISODIUM SALT * D&C YELLOW NO. 8 * DISODIUM-6-HYDROXY-3-OXO-9-XANTHENE-o-BEN-ZOATE * FLUORESCEIN SODIUM B.P * FLUORES-CEIN, SOLUBLE * FLUOR-I-STRIP A.T. * FUL-GLO * FUNDUSCEIN * FURANIUM * HIDACID URA-NINE * NCI-C54706 * RESORCINOL PHTHALEIN SO-DIUM * SODIUM FLUORESCEIN * SODIUM FLUOR-ESCEINATE * SODIUM SALT of HYDROXY-o-CARBOXY-PHENYL-FLUORONE * SOLUBLE

FLUORESCEIN ∗ SPIRO(ISOBENZOFURAN-1(3H),9′-
(9H)XANTHENE-3-ONE, 3′,6′-DIHYDROXY-DISODIUM
SALT ∗ URANIN ∗ URANINE A EXTRA ∗ URA-
NINE USP XII ∗ URANINE YELLOW ∗ 11824 YEL-
LOW ∗ 12417 YELLOW

CONSENSUS REPORTS: Reported in EPA
TSCA Inventory.

SAFETY PROFILE: Moderately toxic by intraperitoneal route. Mildly toxic by ingestion. Experimental reproductive effects. Questionable
carcinogen with experimental tumorigenic data.
Mutation data reported. When heated to decomposition it emits toxic fumes of Na_2O.

FEY000   HR: 2
FLUORIDES

OSHA PEL: TWA 2.5 mg(F)/m^3
ACGIH TLV: TWA 2.5 mg(F)/m^3
DFG MAK: 2.5 mg/m^3; BAT: 7 mg/kg creati-
    nine in urine at end of exposure; 4 mg/kg
    creatinine in urine about 16 hours after end
    of exposure.
NIOSH REL: TWA 2.5 mg(F)/m^3

SAFETY PROFILE: Inorganic fluorides are generally highly irritating and toxic. Acute effects
resulting from exposure to fluorine compounds
are due to HF. Chronic fluorine poisoning, or
"fluorosis," occurs among miners of cryolite,
and consists of a sclerosis of the bones, caused
by fixation of the calcium by the fluorine. There
may also be some calcification of the ligaments.
The teeth are mottled, and there is osteosclerosis
and ostemalacia. The bony and ligamentous
changes are demonstrable by x-ray. The estimated human lethal dose is 2.5 to 5.9 grams
of F$^-$. Large doses can cause very severe nausea, vomiting, diarrhea, abdominal burning, and
cramp-like pains. It is not taken up by the thyroid
and does not interfere with iodine uptake. Can
cause or aggravate attacks of asthma and severe
bone changes, making normal movements painful. Some signs of pulmonary fibrosis are noted.
Some enzyme systems effects are reported. Irritants to the eyes, skin, and mucous membranes.
Loss of weight, anorexia, anemia, wasting and
cachexia, and dental defects are among the common findings in chronic fluorine poisoning.
There may be an eosinophilia and impairment
of growth in young workers. Symptoms of intoxication include gastric, intestinal, circulatory,
respiratory and nervous complaints, and skin
rashes. When heated to decomposition, or on

contact with acid or acid fumes, they emit highly
toxic fumes of F$^-$.

Organic fluorides are generally less toxic
than other halogenated hydrocarbons. Fluoro-
carbons are chemically inert to most materials
but can react violently with barium, sodium,
and potassium. Fluoroamides react violently
with lithium tetrahydroaluminate and with sodium at very high temperatures. Some fluorinated cyclopropenyl methyl ethers react
violently with water or methanol. Some
fluorodinitro compounds of methane and ethane
are sensitive explosives. When heated to decomposition they emit toxic fumes of F$^-$. Common
air contaminants.

FEZ000   CAS: 7782-41-4   HR: 3
FLUORINE

DOT: UN 1045
mf: F_2 mw: 38.00

PROP: Pale yellow gas. Mp: $-218°$, bp: $-187°$,
d: 1.14 @ $-200°$, 1.108 @ $-188°$, vap d:
1.695.

SYNS: BIFLUORIDEN (DUTCH) ∗ FLUOR (DUTCH,
FRENCH, GERMAN, POLISH) ∗ FLUORINE, compressed
(DOT) ∗ FLUORO (ITALIAN) ∗ FLUORURES ACIDE
(FRENCH) ∗ FLUORURI ACIDI (ITALIAN) ∗ RCRA
WASTE NUMBER P056 ∗ SAEURE FLUORIDE (GER-
MAN)

CONSENSUS REPORTS: Reported in EPA
TSCA Inventory.

OSHA PEL: TWA 0.1 ppm
ACGIH TLV: TWA 1 ppm; STEL 2 ppm
DFG MAK: 0.1 ppm (0.2 mg/m^3)
DOT Classification: Nonflammable Gas; La-
    bel: Poison and Oxidizer; Poison A; Label:
    Poison Gas, Oxidizer.

SAFETY PROFILE: A poison gas. A skin, eye,
and mucous membrane irritant. A most powerful
caustic irritant to tissue. Mutation data reported.
A very dangerous fire and explosion hazard.
A powerful oxidizer. Reacts violently with many
materials.

Explosive or potentially explosive reaction
with ammonia, cesium fluoride + fluorocarbox-
ylic acids, cesium heptafluoropropoxide, 1- or
2-fluoriminoperfluoropropane, graphite, halocarbons (e.g., carbon tetrachloride, chloroform,
perfluorocyclobutane, iodoform, 1,2-dichloro-
tetrafluoroethane), liquid hydrocarbons (e.g.,
anthracene, turpentine), hydrogen, hydrogen +

oxygen, hydrogen fluoride + seleninyl fluoride + heat, nitric acid, silver cyanide, sulfur dioxide, carbon monoxide, sodium acetate, sodium bromate, stainless steel, water.

Reacts to form explosive products with alkanes + oxygen (forms peroxides), cyanoguanidine, perchloric acid (forms fluorine perchlorate gas), potassium chlorate (forms fluorine perchlorate gas), potassium hydroxide (forms potassium trioxide). Forms explosive mixtures with acetonitrile + chlorine fluoride, ice.

Ignition or violent reaction on contact with acetylene, ceramic materials, covalent halides (e.g., chromyl chloride, phosphorus pentachloride, phosphorus trichloride, phosphorus trifluoride, boron trichloride, silicon tetrachloride), halogens (e.g., bromine, iodine, chlorine + spark or heating to 100°C), dicyanogen, gaseous hydrocarbons (e.g., town gas, methane, benzene), hydrogen halide gases or concentrated solutions (e.g., hydrogen bromide, hydrogen chloride, hydrogen iodide, hydrogen fluoride), metal acetylides and carbides (e.g., monocesium acetylide, cesium acetylide, lithium acetylide, rubidium acetylide, tungsten carbide, ditungsten carbide, zirconium dicarbide, uranium dicarbide), metal cyano complexes [e.g., potassium hexacyanoferrate(II), lead hexacyanoferrate(III), potassium hexacyanoferrate(III)], metal hydrides (e.g., copper hydride, potassium hydride, sodium hydride), metal iodides (e.g., lead iodide, calcium iodide, mercury iodide, potassium iodide), metals, metal salts, metal silicides (e.g., calcium disilicide, lithium hexasilicide), nickel(IV) oxide, non-metals (e.g., boron, yellow or red phosphorus, selenium, tellurium, silicon, carbon, charcoal, sulfur), oxygenated organic compounds (e.g., methanol, ethanol, 3-methyl butanol, acetaldehyde, trichloroacetaldehyde, acetone, lactic acid, benzoic acid, salicylic acid, ethyl acetate, methyl borate), non-metal oxides (e.g., arsenic trioxide, nitrogen oxide, dinitrogen tetroxide), oxygen + polymers [e.g., phenolformaldehyde resins (bakelite), polyacrylonitrile-butadiene (Buna N), polyamides (nylons), polychloropene (neoprene), polyethylene, polytrifluoopropylmethylsiloxane, polyvinylchloride-vinyl acetate (Tygon), polyvinylidene fluoride-hexafluoropropylene (Viton), polyurethane foam, polymethyl methacrylate (Perspex), polytetrafluooethylene (Teflon)], sulfides (e.g., antimony trisulfide, carbon disulfide vapor, chromium (II) sulfide, hydrogen sulfide, barium sulfide, potassium sulfide, zinc sulfide, molybdenum sulfide), xenon + catalysts (e.g., nickel fluoride, silver difluoride, nickel(III) oxide, silver (I) oxide).

Incandescent reaction with boron nitride, hexalithium disilicide + heat, metal borides, metal oxides (e.g., nickel(II) oxide, alkali metal oxides, alkaline earth oxides), nitrogenous bases (e.g., aniline, dimethylamine, pyridine), gallic acid.

Incompatible with cesium heptafluoro propoxide, cyanoguanidine, halocarbons, hexalithium disilicide, seleninyl fluoride, hydrogen sulfide, oxygen, sodium acetate, sodium bromate, sodium dicyanamides, most organic matter, H-containing molecules, oxides of S, N, P, alkali metals, and alkaline earths. It reacts violently with halogen acids, hydrazine, ClO_2, coke, cyanamide, cyanides, KNO_3, (PbO + glycerol), CCl_4, silicides, silicates, trinitromethane, alkenes, alkyl benzenes, CS_2, $Cr(OCl)_2$, Al, Tl, Sn, Sb, As, natural gas, liquid air, perfluoropropionyl fluoride, polyvinyl chloride acetate. Many reactions go on even at $< -160°$. Reacts with water or steam to produce heat and toxic and corrosive fumes. Used as one component of liquid rocket fuel and in chemical lasers.

FFD000   CAS: 10049-03-3   **HR: 3**
FLUORINE PERCHLORATE
mf: $ClFO_4$ mw: 118.45

PROP: Colorless gas, pungent, acrid odor. Mp: $-167.3°$, bp: $-15.9°$.

SYNS: CHLORINE TETROXYFLUORIDE * PERCHLORYL HYPOFLUORITE

SAFETY PROFILE: A poison by ingestion and inhalation. Corrosive to the skin, eye, and mucous membranes. Very unstable. A powerful oxidizer which can react violently with oxidizable materials. A very dangerous explosion hazard; it explodes on slightest provocation. The liquid explodes on freezing at $-167°C$. The gas explodes when exposed to sparks, flame, or on contact with grease, dust, rubber, or aqueous potassium iodide. Ignites in contact with hydrogen gas. When heated to decomposition it emits highly toxic fumes of F^- and Cl^-.

FFE000   CAS: 1544-46-3   **HR: 3**
FLUOROACETALDEHYDE
mf: C_2H_3FO mw: 62.05

SAFETY PROFILE: Poison by intraperitoneal and subcutaneous routes. When heated to decomposition it emits toxic fumes of F^-.

FFF000   CAS: 640-19-7   **HR: 3**
FLUOROACETAMIDE
mf: C_2H_4FNO mw: 77.07

SYNS: AFL 1081 * COMPOUND 1081 * FAA * FLUORAKIL 100 * 2-FLUOROACETAMIDE * FLUOROACETIC ACID AMIDE * FUSSOL * MEGATOX * MONOFLUOROACETAMIDE * NAVRON * RCRA WASTE NUMBER P057 * RODEX * YANOCK

CONSENSUS REPORTS: EPA Extremely Hazardous Substances List. Reported in EPA TSCA Inventory.

SAFETY PROFILE: A human poison by an unspecified route. Poison experimentally by ingestion, skin contact, intraperitoneal, subcutaneous, intravenous and possibly other routes. Human systemic effects by unspecified route: convulsions, coma, nausea and vomiting. Experimental reproductive effects. Mutation data reported. Used as an insecticide and rodenticide. When heated to decomposition it emits very toxic fumes of F^- and NO_x.

FFG000   CAS: 343-89-5   **HR: 3**
7-FLUORO-2-ACETAMIDO-FLUORENE
mf: $C_{15}H_{12}FNO$ mw: 241.28

SYNS: 7-FLUORO-2-ACETYLAMINOFLUORENE * N-(7-FLUOROFLUORENE-2-YL)ACETAMIDE

SAFETY PROFILE: Questionable carcinogen with experimental carcinogenic and tumorigenic data. Mutation data reported. When heated to decomposition it emits toxic fumes of F^- and NO_x.

FFH000   CAS: 330-68-7   **HR: 3**
FLUOROACETANILIDE
mf: C_8H_8FNO mw: 153.17

SYNS: AFL 1082 * 2-FLUOROACETANILIDE * 2-FLUORO-N-PHENYLACETAMIDE * TL 1312

SAFETY PROFILE: Poison by ingestion, intraperitoneal and possibly other routes. Moderately toxic by inhalation. When heated to decomposition it emits very toxic fumes of F^- and NO_x.

FFJ000   CAS: 503-20-8   **HR: 3**
FLUOROACETONITRILE
mf: C_2H_2FN mw: 59.05

SYN: FLUOROMETHYL CYANIDE

CONSENSUS REPORTS: Cyanide and its compounds are on the Community Right-To-Know List.

SAFETY PROFILE: Poison by intraperitoneal and subcutaneous routes. When heated to decomposition it emits very toxic fumes of F^-, CN^-, and NO_x.

FFR000   CAS: 359-06-8   **HR: 3**
FLUOROACETYL CHLORIDE
mf: C_2H_2ClFO mw: 96.49

CONSENSUS REPORTS: Reported in EPA TSCA Inventory.

SAFETY PROFILE: Poison by inhalation. When heated to decomposition it emits very toxic fumes of Cl^- and F^-.

FFT000   CAS: 364-71-6   **HR: 3**
o-(FLUOROACETYL)SALICYLIC ACID
mf: $C_9H_7FO_4$ mw: 198.16

SYN: SALICYLIC ACID, FLUOROACETATE

SAFETY PROFILE: Poison by ingestion and subcutaneous routes. When heated to decomposition it emits toxic fumes of F^-.

FFV000   CAS: 592-79-0   **HR: 3**
5-FLUORO AMYLAMINE
mf: $C_5H_{12}FN$ mw: 105.18

SYN: 5-FLUOROPENTYLAMINE

SAFETY PROFILE: Poison by intraperitoneal and subcutaneous routes. When heated to decomposition it emits very toxic fumes of F^- and NO_x.

FFX000   CAS: 661-18-7   **HR: 3**
5-FLUOROAMYL THIOCYANATE
mf: $C_6H_{10}FNS$ mw: 147.23

SYN: 5-FLUOROPENTYL THIOCYANATE

SAFETY PROFILE: Poison by intraperitoneal and subcutaneous routes. When heated to decomposition it emits very toxic F^-, NO_x, and SO_x.

FFY000 CAS: 371-40-4 **HR: 3**
4-FLUOROANILINE
DOT: UN 2944
mf: C_6H_6FN mw: 111.13
PROP: D: 1.1724, bp: 187.4°, mp: −1.9°

SYNS: BENZENAMINE, 4-FLUORO-(9CI) * 4-FLUO-
RANILIN * p-FLUOROANILINE * 4-FLUOROBENZE-
NAMINE * p-FLUOROPHENYLAMINE

CONSENSUS REPORTS: Reported in EPA
TSCA Inventory. EPA Genetic Toxicology Program.

DOT Classification: Poison B; Label: St. Andrews Cross.

SAFETY PROFILE: Poison by ingestion. Mutation data reported. A severe skin and eye irritant.
When heated to decomposition it emits very
toxic fumes of NO_x and F^-.

FGA000 CAS: 462-06-6 **HR: 2**
FLUOROBENZENE
DOT: UN 2387
mf: C_6H_5F mw: 96.11
PROP: Colorless liquid. D: 1.024, mp: −41.9°,
flash p: 5°F, d: 1.024, vap d: 3.31, bp: 82.8°
insol in water, misc in alcohol and ether.

SYN: PHENYL FLUORIDE

CONSENSUS REPORTS: Reported in EPA
TSCA Inventory.

DOT Classification: Flammable Liquid; Label:
Flammable Liquid.

SAFETY PROFILE: Mildly toxic by ingestion
and inhalation. A very dangerous fire hazard
when exposed to heat, flame, or oxidizers. To
fight fire, use water spray, mist, foam, dry chemical, CO_2. When heated to decomposition it
emits toxic fumes of F^-.

FGI100 **HR: 3**
6-FLUOROBENZO(a)PYRENE
SAFETY PROFILE: Suspected carcinogen with
experimental carcinogenic and neoplastigenic
data. When heated to decomposition it emits
toxic fumes of F^-.

FGV000 CAS: 20977-50-8 **HR: 3**
**1-(3-(4-FLUOROBENZOYL)PROPYL)-4-
PIPERIDYL-N-ISOPROPYL
CARBAMATE**
mf: $C_{19}H_{27}FN_2O_3$ mw: 350.48

SYN: AL-1021

SAFETY PROFILE: Poison by ingestion and
intravenous routes. Human systemic effects by
ingestion: psychotropic effects. When heated
to decomposition it emits very toxic F^- and
NO_x.

FGW000 CAS: 59921-81-2 **HR: 3**
**1-(4'-FLUOROBENZOYL)-3-PYRRO-
LIDINYLPROPANE MALEATE**
mf: $C_{14}H_{18}FNO \cdot C_4H_4O_4$ mw: 351.41

SYNS: 1-(4'-FLUOROBENZOIL)-3-PIRROLIDINOPRO-
PANO MALEATO (ITALIAN) * 4'-FLUORO-4-(1-PYRRO-
LIDINYL)BUTYROPHENONE MALEATE

SAFETY PROFILE: Poison by ingestion and
intravenous routes. When heated to decomposition it emits very toxic fumes of F^- and NO_x.

FHD000 CAS: 353-17-3 **HR: 3**
4-FLUOROBUTYL THIOCYANATE
mf: C_5H_8FNS mw: 133.20

SAFETY PROFILE: Poison by intraperitoneal
and subcutaneous routes. When heated to decomposition it emits very toxic fumes of F^-,
NO_x, and SO_x.

FHG000 CAS: 1893-33-0 **HR: 3**
FLUOROBUTYROPHENONE
mf: $C_{21}H_{30}FN_3O_2$ mw: 375.54

SYNS: DIPIPERAL * DIPIPERON * DIPIPERONE
* FLOROPIPAMIDE * 1'-(3-(p-FLUOROBENZOYL)
PROPYL)(1,4'-BIPIPERIDINE 1-4'-CARBOXAMIDE
* 1-(3-(p-FLUOROBENZOYL)PROPYL)-4-PIPERIDINO-
ISONIPACOTAMIDE * 4'-FLUORO-4-(4-N-PIPER-
IDINO-4-CARBAMIDOPIPERIDINO)BUTYROPHENONE
* p-FLUORO-γ-(4-PIPERIDINO-4-CARBAMOYLPIPERI-
DINO)BUTYROPHENONE * FPA * MCN-JR-3345
* PIPAMPERONE * PIPANEPERONE
* PIPERONYL * PROPITAN * R 3345

SAFETY PROFILE: Poison by ingestion, subcutaneous, and intravenous routes. An experimental teratogen. When heated to decomposition
it emits very toxic fumes of F^- and NO_x.

FHO000 CAS: 10356-76-0 **HR: 2**
5-FLUORO-2'-DEOXYCYTIDINE
mf: $C_9H_{12}FN_3O_4$ mw: 245.24

SYNS: FCdR * FCDR * 5-FLUOR-DESOXYCYTI-
DIN (GERMAN) * 5-FLUORODEOXYCYTIDINE

SAFETY PROFILE: Moderately toxic by intraperitoneal route. Experimental teratogenic and reproductive effects. Mutation data reported. When heated to decomposition it emits very toxic fumes of F⁻ and NO$_x$.

FIB000   CAS: 353-36-6   HR: 3
FLUOROETHANE
DOT: UN 2453
mf: C_2H_5F mw: 48.06

PROP: Mp: −143.2°, bp: −37.7°, d: 0.8158 @ −37.7°, vap d: 1.66.

SYNS: ETHYL FLUORIDE (DOT) * MONOFLUORO-ETHANE * R161

CONSENSUS REPORTS: Reported in EPA TSCA Inventory.

DOT Classification: Flammable Gas; Label: Flammable Gas.

SAFETY PROFILE: A very dangerous fire hazard when exposed to heat, flames, or oxidizers. To fight fire, stop flow of gas. When heated to decomposition it emits toxic fumes of F⁻.

FIC000   CAS: 144-49-0   HR: 3
FLUOROETHANOIC ACID
DOT: UN 2642
mf: $C_2H_3FO_2$ mw: 78.05

PROP: Colorless solid. Mp: 33°, bp: 165°. Sol in water.

SYNS: ACIDE-MONOFLUORACETIQUE (FRENCH) * ACIDO MONOFLUOROACETIO (ITALIAN) * CYMONIC ACID * FAA * FLUOROACETATE * FLUOROACETIC ACID * 2-FLUOROACETIC ACID * FLUOROACETIC ACID (DOT) * GIFBLAAR POISON * HFA * MFA * MONOFLUORAZIJNZUUR (DUTCH) * MONOFLUORESSIGSAURE (GERMAN) * MONOFLUOROACETATE * MONOFLUOROACETIC ACID

CONSENSUS REPORTS: EPA Extremely Hazardous Substances List. Reported in EPA TSCA Inventory.

DOT Classification: Poison B; Label: Poison.

SAFETY PROFILE: Poison by ingestion, subcutaneous, intraperitoneal, intravenous, and possibly other routes. Affects the human central nervous system, causing convulsions and ventricular fibrillation. When heated to decomposition it emits toxic fumes of F⁻ and Na$_2$O.

FID000   CAS: 63919-01-7   HR: 3
FLUOROETHANOL
mf: C_2H_5FO mw: 64.07

SYNS: MONOFLUORETHANOL * MONOFLUORO-ETHANOL

SAFETY PROFILE: Poison by intraperitoneal and subcutaneous routes. Mildly toxic by inhalation. When heated to decomposition it emits very toxic fumes of F⁻.

FIE000   CAS: 371-62-0   HR: 3
2-FLUOROETHANOL
mf: C_2H_5FO mw: 64.07

SYNS: β-FLUOROETHANOL * TL 741

CONSENSUS REPORTS: EPA Extremely Hazardous Substances List. Reported in EPA TSCA Inventory.

SAFETY PROFILE: Poison by inhalation, intraperitoneal, subcutaneous, and intravenous routes. When heated to decomposition it emits very toxic fumes of F⁻.

FIO000   CAS: 63765-78-6   HR: 3
2-FLUOROETHYL-5-FLUOROHEXOATE
mf: $C_8H_{14}F_2O_2$ mw: 180.22

SAFETY PROFILE: Poison by inhalation, intramuscular, subcutaneous, and intravenous routes. When heated to decomposition it emits very toxic F⁻.

FIM000   CAS: 459-99-4   HR: 3
β-FLUOROETHYL FLUOROACETATE
mf: $C_4H_6F_2O_2$ mw: 124.10

SYNS: 2-FLUOROETHYL FLUOROACETATE * TL 855

SAFETY PROFILE: Poison by inhalation, subcutaneous and parenteral routes. When heated to decomposition it emits toxic fumes of F⁻.

FIN000   CAS: 371-29-9   HR: 3
2-FLUORO ETHYL-γ-FLUORO BUTYRATE
mf: $C_6H_{10}F_2O_2$ mw: 152.16

SYN: β-FLUOROETHYL-γ-FLUOROBUTYRATE

SAFETY PROFILE: Poison by inhalation. When heated to decomposition it emits toxic fumes of F⁻.

FIP999　　　CAS: 4242-33-5　　　**HR: 3**
β-FLUOROETHYLIC ESTER of
XENYLACETIC ACID
mf: $C_{16}H_{15}FO_2$　　mw: 258.31

PROP: A brown solid, sol in organic solvents.
Mp: 60.6°.

SYNS: 2-FLUOROETHYL ESTER DIPHENYLACETIC
ACID　*　LAMBROL　*　M 2060

SAFETY PROFILE: Poison by ingestion and
skin contact. An insecticide. When heated to
decomposition it emits toxic fumes of F⁻.

FIW000　　　CAS: 31540-62-2　　　**HR: 3**
4'-FLUORO-4-(8-FLUORO-2,3,4,5-
TETRAHYDRO-1H-PYRIDO(4,3-b)
INDOL-2-YL)BUTYROPHENONE
HYDROCHLORIDE
mf: $C_{21}H_{20}F_2N_2O \cdot ClH$　　mw: 390.89

SYN: ABBOTT-30360

SAFETY PROFILE: Poison by intramuscular,
ingestion, and intravenous routes. When heated
to decomposition it emits very toxic fumes of
HCl, NO_x, and F⁻.

FJK000　　　CAS: 593-53-3　　　**HR: 2**
FLUOROMETHANE

DOT: UN 2454
mf: CH_3F　　mw: 34.03

PROP: Colorless gas; agreeable, ether-like
odor. D: (liquid) 0.8774 @ −78°, (gas) 1.1951
(air = 1), (gas) 1.0813 (oxygen = 1), mp:
−141.8°, bp: −75.7° @ 872 mm, −78.2° @
760 mm. Freely sol in alc and ether.

SYNS: FREON 41　*　METHYL FLUORIDE (DOT)

CONSENSUS REPORTS: Reported in EPA
TSCA Inventory.

DOT Classification: Flammable Gas; Label:
　Flammable Gas.

SAFETY PROFILE: Narcotic in high concentrations. Acts as a simple asphyxiant. Burns with
evolution of hydrogen fluoride. The flame is
about as colorless as that of alcohol. When
heated to decomposition it emits toxic fumes
of F⁻.

FKI000　　　CAS: 1622-79-3　　　**HR: 3**
4'-FLUORO-4-(4-METHYLPIPERIDINO)
BUTYROPHENONE HYDROCHLORIDE
mf: $C_{16}H_{22}FNO \cdot ClH$　　mw: 299.85

SYNS: BURONIL　*　EUNERPAN　*　FG 5111
*　METHYLPERONE HYDROCHLORIDE　*　γ-(4-
METHYLPIPERIDINE)-p-FLUOROBUTYROPHENONE
HYDROCHLORIDE

SAFETY PROFILE: Poison by ingestion, subcutaneous, and intravenous routes. Experimental
teratogenic and reproductive effects. A neuroleptic drug used to treat anxiety and confusion.
When heated to decomposition it emits very
toxic fumes of F⁻, NO_x, and HCl.

FLJ000　　　CAS: 2804-00-4　　　**HR: 3**
8-(4-p-FLUORO PHENYL-4-OXOBUTYL)
2-METHYL-2,8-DIAZASPIRO(4.5)
DECANE-1,3-DIONE
mf: $C_{19}H_{23}FN_2O_3$　　mw: 346.44

SYNS: F-33　*　FR-33　*　R 7158

SAFETY PROFILE: Poison by ingestion and
intravenous route. When heated to decomposition it emits very toxic fumes of F⁻ and NO_x.

FLQ000　　　CAS: 5675-31-0　　　**HR: 3**
2-FLUORO-2-PROPEN-1-OL
mf: C_3H_5FO　　mw: 76.08

SAFETY PROFILE: Poison by ingestion and
skin contact. Mildly toxic by inhalation. When
heated to decomposition it emits toxic fumes
of F⁻.

FLU000　　　CAS: 1649-18-9　　　**HR: 3**
4'-FLUORO-4-(4-(2-PYRIDYL)-1-
PIPERAZINYL)BUTYROPHENONE
mf: $C_{19}H_{22}FN_3O$　　mw: 327.44

SYNS: AZAPERONE (USDA)　*　AZEPERONE
*　EUCALMYL　*　FLUOPERIDOL　*　1-(3-(4-FLUORO-
BENZOYL)PROPYL)-4-(2-PYRIDYL)PIPERAZINE
*　1-(4-FLUOROPHENYL)-4-(4-(2-PYRIDINYL)-1-PIPERAZI-
NYL)-1-BUTANONE　*　R 1929　*　STRESNIL
*　SUICALM

SAFETY PROFILE: Poison by ingestion, intravenous, intraperitoneal and subcutaneous
routes. When heated to decomposition it emits
very toxic fumes of F⁻ and NO_x.

FLV000　　　CAS: 2266-22-0　　　**HR: 3**
4'-FLUORO-4-(n-(4-PYRROLIDINAMIDO-
4-m-TOLYPIPERIDINO)
BUTYROPHENONE
mf: $C_{27}H_{33}FN_2O_2$　　mw: 436.62

SYNS: MEPERIDIDE　*　METHYLPERIDIDE

SAFETY PROFILE: Poison by subcutaneous and intravenous routes. When heated to decomposition it emits very toxic fumes of F^- and NO_x.

FLZ000   CAS: 7789-21-1   **HR: 3**
FLUOROSULFURIC ACID

DOT: UN 1777
mf: FHO_3S mw: 100.07

PROP: Colorless, fuming, highly corrosive liquid. Mp: $-87.3°$, bp: $163.5°$, d: 1.743 @ $15°$.

SYNS: FLUOROSUFONIC ACID (DOT) * FLUOSULFONIC ACID (DOT)

CONSENSUS REPORTS: Reported in EPA TSCA Inventory.

OSHA PEL: TWA 2.5 mg(F)/m^3
NIOSH REL: (Inorganic Fluorides) TWA 2.5 mg(F)/m^3
DOT Classification: Corrosive Material; Label: Corrosive.

SAFETY PROFILE: Probably a poison by inhalation. A corrosive irritant to the skin, eyes, and mucous membranes.

FMC000   CAS: 352-32-9   **HR: 2**
p-FLUOROTOLUENE

DOT: UN 2388
mf: C_7H_7F mw: 110.14

PROP: Colorless liquid. D: 1.001, bp: $116-117°$. Flash p: 50°F. Insol in water; sol in alc and ether.

CONSENSUS REPORTS: Reported in EPA TSCA Inventory.

DOT Classification: Flammable Liquid; Label: Flammable Liquid; Flammable or Combustible Liquid; Label: Flammable Liquid.

SAFETY PROFILE: Moderately toxic by parenteral route. A very dangerous fire hazard when exposed to heat or flame; can react vigorously with oxidizing materials. When heated to decomposition it emits toxic fumes of F^-.

FMH000   CAS: 19982-87-7   **HR: 3**
4-FLUORO-4'-TRIFLUOROMETHYLBEN-ZOPHENONE GUANYLHYDRAZONE HYDROCHLORIDE
mf: $C_{15}H_{12}F_4N_4 \cdot ClH$ mw: 360.77

SYNS: FTBG * WR 09792

SAFETY PROFILE: Poison by ingestion and intraperitoneal routes. When heated to decomposition it emits very toxic fumes of HCl, F^- and NO_x.

FMJ000   CAS: 139-26-4   **HR: 3**
3-FLUOROTYROSIN
mf: $C_9H_{10}FNO_3$ mw: 199.20

SYNS: m-FLUOROTYROSINE * 3-FLUOROTYROSINE * FLUORTHYRIN * 3-FLUORTYROSIN (GERMAN)

SAFETY PROFILE: Human poison by ingestion. Experimental poison by ingestion, skin contact, and subcutaneous routes. When heated to decomposition it emits very toxic fumes of F^- and NO_x.

FMM000   CAS: 51-21-8   **HR: 3**
FLUOROURACIL
mf: $C_4H_3FN_2O_2$ mw: 130.09

SYNS: ADRUCIL * ARUMEL * CARZONAL * EFFLUDERM (free base) * EFUDEX * EFUDIX * 5-FLUORACIL (GERMAN) * FLUOROBLASTIN * FLUOROPLEX * 5-FLUORPROPYRIMIDINE-2,4-DIONE * 5-FLUORO-2,4-PYRIMIDINEDIONE * 5-FLUORO-2,4(1H,3H)-PYRIMIDINEDIONE * 5-FLUOROURACIL * 5-FLUORURACIL (GERMAN) * FLURACIL * FLURI * FLURIL * 5-FU * NSC-19893 * RO 2-9757 * TIMAZIN * U-8953 * ULUP

CONSENSUS REPORTS: IARC Cancer Review: GROUP 3 IMEMDT 7,210,87; Human Inadequate Evidence IMEMDT 26,217,81; Animal Inadequate Evidence IMEMDT 26,217,81. Reported in EPA TSCA Inventory. EPA Genetic Toxicology Program. EPA Extremely Hazardous Substances List.

SAFETY PROFILE: Poison by ingestion, intraperitoneal, subcutaneous, intravenous, and possibly other routes. Moderately toxic by parenteral and rectal routes. Experimental teratogenic and reproductive effects. Human systemic effects by ingestion, intravenous, and possibly other routes: EKG changes, bone marrow changes, cardiac, pulmonary and gastrointestinal effects. Human mutation data reported. A human skin irritant. Questionable carcinogen. When heated to decomposition it emits very toxic fumes of F^- and NO_x.

FMN000   CAS: 316-46-1   **HR: 3**
5-FLUOROURIDINE
mf: $C_9H_{11}FN_2O_6$ mw: 262.22

SYNS: FUR * 5-FUR

SAFETY PROFILE: Poison by intraperitoneal and subcutaneous routes. Human mutation data reported. When heated to decomposition it emits very toxic fumes of F^- and NO_x.

FMO129 CAS: 2709-56-0 **HR: 3**
cis-(Z)-FLUPENTHIXOL
mf: $C_{23}H_{25}F_3N_2OS$ mw: 434.56

SYNS: EMERGIL * FLUANXOL * FLUPEN-THIXOL * (α,β)-FLUPENTHIXOL * FLUPENTHIX-OLE * FLUPENTIXOL * FLURENTIXOL * FLUX-ANXOL * LC 44 * N 7009 * SIPLARIL * SIPLAROL * 2-TRIFLUOROMETHYL-9-(3-(4-(β-HY-DROXYETHYL-1-PIPERAZINYL)PROPYLIDENE)THIOX-ANTHENE * 4-(3-(2-(TRIFLUOROMETHYL)THIOXAN-THEN-9-YLIDENE)PROPYL)-1-PIPERAZINEETHANOL * 4-(3-(2-(TRIFLUOROMETHYL)-9H-THIOXANTHEN-9-YLIDENE)PROPYL)-1-PIPERAZINEETHANOL

SAFETY PROFILE: Poison by ingestion, intraperitoneal, and intravenous routes. Human systemic effects by ingestion: central nervous system effects. When heated to decomposition it emits very toxic fumes of F^-, NO_x, and SO_x.

FMQ000 CAS: 17617-23-1 **HR: 3**
FLURAZEPAM
mf: $C_{21}H_{23}ClFN_3O$ mw: 387.92

PROP: White rods from ether-petr ether. Mp: 77-82°.

SYNS: 7-CHLORO-1-(2-(DIETHYLAMINO)ETHYL)-5-(2-FLUOROPHENYL)-1H-1,4-BENZODIAZEPIN-2(3H)-ONE * FELMANE * NOCTOSOM * STAURODERM * Ro-5-6901/3

SAFETY PROFILE: Poison by intraperitoneal and intravenous routes. Moderately toxic by ingestion and subcutaneous routes. Experimental reproductive effects. Caution: May be habit forming. This is a controlled substance (depressant) listed in the U.S. Code of Federal Regulations, Title 21 Part 1308.14. When heated to decomposition it emits very toxic fumes of Cl^-, F^- and NO_x.

FMT000 CAS: 59-30-3 **HR: 3**
FOLIC ACID
mf: $C_{19}H_{19}N_7O_6$ mw: 441.45

PROP: A member of the vitamin B complex. Orange-yellow needles or platelets; odorless. Sol in dilute alkali hydroxide and carbonate solns; sltly sol in water; insol in lipid solvents, acetone, alc, chloroform, ether.

SYNS: l-N-(p-(((-2-AMINO-4-HYDROXY-6-PTERIDINYL)METHYL)AMINO)BENZOYL)GLUTAMIC ACID * FOLACIN * FOLATE * FOLCYSTEINE * NSC 3073 * PTEGLU * PTEROYLGLUTAMIC ACID * PTEROYL-l-GLUTAMIC ACID * PTEROYL-MONOGLUTAMIC ACID * PTEROYL-l-MONOGLU-TAMIC ACID * USAF CB-13 * VITAMIN Bc * VITAMIN M

CONSENSUS REPORTS: Reported in EPA TSCA Inventory.

SAFETY PROFILE: Poison by intraperitoneal and intravenous routes. Experimental teratogenic and reproductive effects. Mutation data reported. When heated to decomposition it emits toxic fumes of NO_x.

FMU045 CAS: 944-22-9 **HR: 3**
FONOFOS
mf: $C_{10}H_{15}OPS_2$ mw: 246.34

SYNS: O-AETHYL-S-PHENYL-AETHYL-DITHIOPHOS-PHONAT (GERMAN) * DIFONATE * DYFONATE * DYPHONATE * ENT 25,796 * O-ETHYL-S-PHE-NYL ETHYLDITHIOPHOSPHONATE * O-ETHYL-S-PHE-NYL ETHYLPHOSPHONODITHIOATE * N 2790 * STAUFFER N 2790

CONSENSUS REPORTS: EPA Genetic Toxicology Program. EPA Extremely Hazardous Substances List.

OSHA PEL: TWA 0.1 mg/m^3 (skin)
ACGIH TLV: TWA 0.1 mg/m^3 (skin)

SAFETY PROFILE: Poison by ingestion, skin contact, and possibly other routes. An insecticide. When heated to decomposition it emits very toxic fumes of PO_x and SO_x.

FMU059 CAS: 2650-18-2 **HR: 3**
FOOD BLUE 1
mf: $C_{37}H_{36}N_2O_9S_3 \cdot 2H_3N$ mw: 783.01

SYNS: ACID BLUE 9 * ACILAN TURQUOISE BLUE AE * A.F. BLUE NO. 1 * AIZEN BRILLIANT BLUE FCF * ALPHAZURINE * AMACID BLUE FG CONC * BLEU BRILLIANT FCF * 11388 BLUE * BRIL-LIANT BLUE * BUCACID AZURE BLUE * CALCO-CID BLUE EG * C.I. 671 * C.I. 42090 * C.I. ACID BLUE 9, DIAMMONIUM SALT * C.I. DIRECT BROWN 78, DIAMMONIUM SALT * C.I. FOOD BLUE 2 * D&C BLUE NO. 4 * DISULPHINE LAKE BLUE EG

* EDICOL SUPRA BLUE E6 * ERIOGLAUCINE
* ERIOSKY BLUE * FENAZO BLUE XR * HID-
ACID AZURE BLUE * H.K. FORMULA NO. K. 7117
* KITON PURE BLUE L * MAPLE BRILLIANT BLUE
FCF * NEPTUNE BLUE BRA CONCENTRATION
* PATENT BLUE AE * PEACOCK BLUE X-1756
* SCHULTZ NO. 770 * TRIANTINE LIGHT BROWN
3RN * XYLENE BLUE VSG

CONSENSUS REPORTS: Community Right-To-Know List. Reported in EPA TSCA Inventory. EPA Genetic Toxicology Program.

SAFETY PROFILE: Human poison by intravenous route. Human systemic effects by intravenous route: muscle contractions or spasticity and dyspnea. Questionable carcinogen with experimental neoplastigenic and tumorigenic data. Mutation data reported. When heated to decomposition it emits very toxic fumes of NH_3, NO_x, and SO_x.

FMV000   CAS: 50-00-0   HR: 3
FORMALDEHYDE

DOT: UN 1198/UN 2209
mf: CH_2O mw: 30.03

PROP: Clear, water-white, very sltly acid gas or liquid; pungent odor. Pure formaldehyde is not available commercially because of its tendency to polymerize. It is sold as aqueous solutions containing from 37% to 50% formaldehyde by weight and varying amounts of methanol. Some alcoholic solns are used industrially and the physical properties and hazards may be greatly influenced by the solvent. Lel: 7.0%, uel: 73.0%, autoign temp: 806°F, d: 1.0, bp: −3°F, flash p: (37% methanol-free): 185°F, flash p: (15% methanol-free): 122°F.

SYNS: ALDEHYDE FORMIQUE (FRENCH) * AL-
DEIDE FORMICA (ITALIAN) * BFV * FA
* FANNOFORM * FORMALDEHYD (CZECH, POLISH)
* FORMALDEHYDE, solution (DOT) * FORMALIN
* FORMALIN 40 * FORMALIN (DOT) * FORMA-
LINA (ITALIAN) * FORMALINE (GERMAN)
* FORMALIN-LOESUNGEN (GERMAN) * FORMALITH
* FORMIC ALDEHYDE * FORMOL * FYDE
* HOCH * IVALON * KARSAN * LYSOFORM
* METHANAL * METHYL ALDEHYDE * METHY-
LENE GLYCOL * METHYLENE OXIDE * MORBO-
CID * NCI-C02799 * OPLOSSINGEN (DUTCH)
* OXOMETHANE * OXYMETHYLENE * PARA-
FORM * POLYOXYMETHYLENE GLYCOLS
* RCRA WASTE NUMBER U122 * SUPERLYSOFORM

CONSENSUS REPORTS: IARC Cancer Review: GROUP 2A IMEMDT 7,211,87; Human Inadequate Evidence IMEMDT 29,345,82; Animal Sufficient Evidence IMEMDT 29,345,82. NTP Fourth Annual Report On Carcinogens, 1984. EPA Genetic Toxicology Program. Reported in EPA TSCA Inventory.

OSHA PEL: TWA 1 ppm; STEL 2 ppm (For certain industries: TWA 3 ppm; CL 5 ppm; Pk 10 ppm/30M

ACGIH TLV: TWA 1 ppm; Suspected Human Carcinogen (Proposed: CL 0.3 ppm; Suspected Human Carcinogen)

DFG MAK: 0.5 ppm (0.6 mg/m^3); Suspected Carcinogen.

NIOSH REL: (Formaldehyde) Limit to lowest feasible level.

DOT Classification: Combustible Liquid; Label: None; ORM-A; Label: None; Flammable or Combustible Liquid; Label: Flammable Liquid.

SAFETY PROFILE: Confirmed carcinogen with experimental carcinogenic, tumorigenic, and teratogenic data. Human poison by ingestion. Experimental poison by ingestion, skin contact, inhalation, intravenous, intraperitoneal, and subcutaneous routes. Human systemic effects by inhalation: lacrimation, olfactory changes, aggression and pulmonary changes. Experimental reproductive effects. Human mutation data reported. A human skin and eye irritant. If swallowed it causes violent vomiting and diarrhea which can lead to collapse. Frequent or prolonged exposure can cause hypersensitivity leading to contact dermatitis, possibly of an eczematoid nature. An air concentration of 20 ppm is quickly irritating to eyes. A common air contaminant.

Combustible liquid when exposed to heat or flame; can react vigorously with oxidizers. A moderate explosion hazard when exposed to heat or flame. The gas is a more dangerous fire hazard than the vapor. Should formaldehyde be involved in a fire, irritating gaseous formaldehyde may be evolved. When aqueous formaldehyde solutions are heated above their flash points, a potential for an explosion hazard exists. High formaldehyde concentration or methanol content lowers the flash point. Reacts with NO_x at about 180°, the reaction becomes explosive. Also reacts violently with perchloric acid + aniline, performic acid, nitromethane, magnesium carbonate, H_2O_2. Moderately dangerous, because of irritating vapor which may exist in toxic concentrations locally if storage tank is ruptured. To fight fire, stop flow of gas (for pure form), alcohol foam for 37 percent methanol-free form. When heated to decomposition it emits acrid smoke and fumes.

FMX000   CAS: 541-66-2   **HR: 3**
FORMAL-γ-TRIMETHYLAMMONIUM PROPANEDIOL
mf: $C_7H_{16}NO_2 \cdot I$ mw: 273.14

SYNS: DILVASENE * ((1,3-DIOXOLAN-4-YL) METHYL)TRIMETHYLAMMONIUM IODIDE * 2249F * OXAPROPANIUM IODIDE * N,N,N-TRIMETHYL-1,3-DIOXOLANE-4-METHANAMINIUM IODIDE * VASODI-LATATEUR 2249F

SAFETY PROFILE: Poison by ingestion, subcutaneous, and intravenous routes. When heated to decomposition it emits very toxic fumes of NO_x, NH_3, and I^-.

FMY000   CAS: 75-12-7   **HR: 3**
FORMAMIDE
mf: CH_3NO mw: 45.05

PROP: Colorless, hygroscopic and oily liquid. Mp: 2.5, fp: 2.6°, vap press: 29.7 mm @ 129.4°, flash p: 310°F (COC), bp: 210° decomp, d: 1.134 @ 20°/40°; 1.1292 @ 25°/4°. Misc in water and alc; very sltly sol in ether.

SYNS: CARBAMALDEHYDE * METHANAMIDE

CONSENSUS REPORTS: Reported in EPA TSCA Inventory. EPA Genetic Toxicology Program.

OSHA PEL: TWA 20 ppm; STEL 30 ppm
ACGIH TLV: TWA 10 ppm (skin)

SAFETY PROFILE: Poison by skin contact and subcutaneous routes. Moderately toxic by ingestion, intraperitoneal, and intramuscular routes. An irritant to skin, eyes, and mucous membranes. Experimental teratogenic and reproductive effects. Mutation data reported. An eye irritant. Combustible when exposed to heat or flame; can react vigorously with oxidizing materials. Incompatible with I_2, pyridine, SO_3. When heated to decomposition it emits toxic fumes of NO_x. Has exploded while in storage.

FNA000   CAS: 64-18-6   **HR: 3**
FORMIC ACID
DOT: UN 1779
mf: CH_2O_2 mw: 46.03

PROP: Colorless, fuming liquid; pungent, penetrating odor. Bp: 100.8°, fp: 8.2°, flash p: 156°F (OC), d: 1.2267 @ 15°/4°, 1.220 @ 20°/4°, autoign temp: 1114°F, vap press: 40 mm @ 24.0°, vap d: 1.59, flash p:(90% soln): 122°F, autoign temp (90% soln): 813°F, lel (90% soln) = 18%, uel (90% soln) = 57%. Misc in water, alc, glycerin, ether.

SYNS: ACIDE FORMIQUE (FRENCH) * ACIDO FORMICO (ITALIAN) * AMEISENSAEURE (GERMAN) * AMINIC ACID * FORMYLIC ACID * HYDROGEN CARBOXYLIC ACID * KWAS METANIOWY (POLISH) * METHANOIC ACID * MIERENZUUR (DUTCH) * RCRA WASTE NUMBER U123

CONSENSUS REPORTS: Reported in EPA TSCA Inventory.

OSHA PEL: TWA 5 ppm
ACGIH TLV: TWA 5 ppm (Proposed: 5 ppm; STEL 10 ppm)
DFG MAK: 5 ppm (9 mg/m³)
DOT Classification: Corrosive Material; Label: Corrosive, solution.

SAFETY PROFILE: Poison by intravenous route. Moderately toxic by ingestion and intraperitoneal routes. Mildly toxic by inhalation. Mutation data reported. Corrosive. A skin and severe eye irritant. A substance migrating to food from packaging materials. Flammable liquid when exposed to heat or flame; can react vigorously with oxidizing materials. Explosive reaction with furfuryl alcohol, H_2O_2, $Tl(NO_3)_3 \cdot 3H_2O$, nitromethane, P_2O_5. To fight fire, use CO_2, dry chemical, alcohol foam. When heated to decomposition it emits acrid smoke and irritating fumes.

FNK025   CAS: 100-50-5   **HR: 2**
4-FORMYLCYCLOHEXENE
DOT: UN 2498
mf: $C_7H_{10}O$ mw: 110.17

SYNS: 3-CYCLOHEXENE-1-CARBOXALDEHYDE * 1,2,3,6-TETRAHYDROBENZALDEHYDE (DOT) * 1,2,5,6-TETRAHYDROBENZALDEHYDE

CONSENSUS REPORTS: Reported in EPA TSCA Inventory.

DOT Classification: Corrosive Material: Label: Corrosive; Flammable or Combustible Liquid; Label: Flammable Liquid.

SAFETY PROFILE: Moderately toxic by skin contact and ingestion. Corrosive. An eye, skin, and mucous membrane irritant. When heated to decomposition it emits acrid smoke and irritating fumes.

FNO000   CAS: 689-13-4   **HR: 1**
N-FORMYL-N-HYDROXYGLYCINE
mf: $C_3H_5NO_4$ mw: 119.09

SYNS: ASYMMETRIN * N-FORMYL HYDROXYAMINOACETIC ACID * HADACIDIN * HADACIDINE * HADACIN * NFHAA * NSC 521778

SAFETY PROFILE: Mildly toxic by intraperitoneal route. Experimental teratogenic and reproductive effects. When heated to decomposition it emits toxic fumes of NO_x.

FNW000   CAS: 758-17-8   **HR: 3**
N-FORMYL-N-METHYLHYDRAZINE
mf: $C_2H_6N_2O$ mw: 74.10

SYNS: FORMIC ACID, METHYLHYDRAZIDE * 1-FORMYL-1-METHYLHYDRAZINE * N-METHYL-N-FORMLYHYDRAZINE * MFH

SAFETY PROFILE: Suspected carcinogen with experimental carcinogenic and teratogenic data. Poison by ingestion and possibly other routes. Mutation data reported. When heated to decomposition it emits toxic fumes of NO_x.

FNZ000   CAS: 51-15-0   **HR: 3**
2-FORMYL-1-METHYLPYRIDINIUM CHLORIDE OXIME
mf: $C_7H_9N_2O \cdot Cl$ mw: 172.63

SYNS: 2-FORMYL-N-METHYLPYRIDINIUM OXIME CHLORIDE * 2-(HYDROXYIMINOMETHYL)-1-METHYL-PYRIDINIUM CHLORIDE * 1-METHYL-2-ALDOXIMINO-PYRIDINIUM CHLORIDE * 1-METHYL-2-FORMYLPY-RIDINIUM CHLORIDE OXIME * 1-METHYL-2-PYRIDINIUM ALDOXIME CHLORIDE * N-METHYLPYRIDINIUM CHLORIDE-2-ALDOXIME * 2-PAM CHLORIDE * PRALIDOXIME CHLORIDE * PROTOPAM CHLORIDE * 2-PYRIDINEALDOXIME CHLORIDE * PYRIDINE-2-ALDOXIME METHOCHLORIDE * 2-PYRIDINE ALDOXIME METHYL CHLORIDE * PYRIDINIUM ALDOXIME METHOCHLORIDE

SAFETY PROFILE: Poison by intravenous, intramuscular, and intraperitoneal routes. Moderately toxic by ingestion. Human systemic effects by intravenous route: coma, blood pressure increase, bronchiolar constriction and cyanosis. Used as a cholinesterase reactivator. When heated to decomposition it emits very toxic fumes of Cl^- and NO_x.

FOI000   CAS: 6804-07-5   **HR: D**
2-FORMYLQUINOXALINE-1,4-DIOXIDE CARBOMETHOXYHYDRAZONE
mf: $C_{11}H_{10}N_4O_4$ mw: 262.25

SYNS: CARBADOX (USDA) * FORTIGRO * GS 6244 * MECADOX * (2-QUINOXALINYL-METHYLENE)-HYDRAZINECARBOXYLIC ACID METHYL ESTER-N,N'-DIOXIDE

SAFETY PROFILE: Human mutation data reported. When heated to decomposition it emits toxic fumes of NO_x.

FOM050   CAS: 1332-10-1   **HR: 3**
FOWLER'S SOLUTION

SYNS: ARSENICAL solution * POTASSIUM ARSENITE solution

CONSENSUS REPORTS: IARC Cancer Review: GROUP 1 IMEMDT 7,100,87.

SAFETY PROFILE: Confirmed carcinogen.

FOO000   CAS: 76-13-1   **HR: 1**
FREON 113
mf: $C_2Cl_3F_3$ mw: 187.37

PROP: Colorless gas. Mp: 13.2°, bp: 45.8°, d: 1.5702, autoign temp: 1256°F.

SYNS: ARCTON 63 * ARKLONE P * DAIFLON S 3 * FLUOROCARBON 113 * FREON 113TR-T * FRIGEN 113a * GENETRON 113 * HALOCARBON 113 * ISCEON 113 * KAISER CHEMICALS 11 * KHLADON 113 * R 113 * REFRIGERANT 113 * TRICHLOROTRIFLUOROETHANE * 1,1,2-TRI-CHLORO-1,2,2-TRIFLUOROETHANE (OSHA, ACGIH, MAK) * UCON 113 * UCON FLUOROCARBON 113 * UCON 113/HALOCARBON 113

CONSENSUS REPORTS: Reported in EPA TSCA Inventory.

OSHA PEL: (Transitional: TWA 1000 ppm) TWA 1000 ppm; STEL 1250 ppm
ACGIH TLV: TWA 1000 ppm; STEL 1250 ppm
DFG MAK: 1000 ppm (7600 mg/m^3)

SAFETY PROFILE: Mildly toxic by ingestion and inhalation. Affects the central nervous system in humans. A skin irritant. Combustible when exposed to heat or flame. Incompatible with Al, Ba, Li, Sm, NaK alloy, Ti.

FOO509   CAS: 76-14-2   **HR: 1**
FREON 114
mf: $C_2Cl_2F_4$ mw: 170.92

PROP: Colorless, practically odorless, noncorrosive, nonirritating, nonflammable gas. Faint, ether-like odor in high concentrations. D: 1.5312, mp: −94°, bp: 4.1°, n (0/D) 1.3092. Insol in water; sol in alc and ether.

SYNS: ARCTON 33 * ARCTON 114 * CRYOFLUO-RAN * CRYOFLUORANE * sym-DICHLOROTETRA-FLUOROETHANE * 1,2-DICHLORO-1,1,2,2-TETRA-FLUOROETHANE (MAK) * DICHLOROTETRAFLUORO-ETHANE (OSHA, ACGIH) * F 114 * FC 114 * FLUORANE 114 * FLUOROCARBON 114 * FRIGEN 114 * FRIGIDERM * GENETRON 114 * GENETRON 316 * HALOCARBON 114 * LEDON 114 * PROPELLANT 114 * R 114 * 1,1,2,2-TETRAFLUORO-1,2-DICHLOROETHANE * UCON 114

CONSENSUS REPORTS: Reported in EPA TSCA Inventory.

OSHA PEL: TWA 1000 ppm
ACGIH TLV: TWA 1000 ppm
DFG MAK: 1000 ppm (7000 mg/m^3)

SAFETY PROFILE: An asphyxiant.

FOP000   HR: 1
FUEL OIL

PROP: A petroleum fraction consisting of a complex mixture of aromatic, paraffinic, olefinic, and naphthenic hydrocarbons. Brown, sltly viscous liquid. Flash p: 100°F, d: <1, autoign temp: 494°F.

SYN: DIESEL FUEL (DOT)

DOT Classification: Combustible Liquid; Label: None.

SAFETY PROFILE: Mildly toxic by ingestion. Flammable when exposed to heat or flame; can react vigorously with oxidizing materials. To fight fire, use CO_2, dry chemical. When heated to decomposition it emits acrid smoke and irritating fumes.

FOQ000   CAS: 4368-28-9   HR: 3
FUGU POISON
mf: $C_{11}H_{17}N_3O_8$ mw: 319.31

SYNS: MACULOTOXIN * SPHEROIDINE * TARI-CHATOXIN * TETRODONTOXIN * TETRODOTOXIN * TETRODOXIN * TTX

SAFETY PROFILE: Poison by ingestion, intraperitoneal, subcutaneous, intravenous, and pos-

sibly other routes. When heated to decomposition it emits toxic fumes of NO_x.

FOR000   CAS: 3309-87-3   HR: 3
FUJITHION
mf: $C_8H_{10}ClO_3PS$ mw: 252.66

SYNS: S-(p-CHLOROPHENYL)-O,O-DIMETHYL PHOS-PHOTHIOATE * O,O-DIMETHYL-S-p-CHLOROPHENYL PHOSPHOROTHIOATE

SAFETY PROFILE: Poison by ingestion. Moderately toxic by skin contact. When heated to decomposition it emits very toxic fumes of Cl^-, PO_x, and SO_x.

FOS000   HR: 3
FULMINATES

SAFETY PROFILE: Variable toxicity. A very dangerous fire hazard when exposed to heat or flame. Severe explosion hazard when shocked or exposed to heat or flame.

The fulminates are a group of explosives which are very sensitive to heat, impact, and friction when dry. They should be kept moist until ready for use. If compressed beyond 25,000 psi they become what is known as "dead-pressed," i.e., not capable of being exploded by flame. Fulminates are subject to deterioration when stored in hot climates. They decompose completely and violently when detonated. They can be ignited with a flame or "spit," with a fuse, or with an electrically heated wire. They are widely used as initiators or primers for detonation of high explosives or the ignition of powder. They are commonly used in combination with substances which provide a more prolonged blow and a bigger flame than fulminates alone. In the reinforced type of detonator, fulminates are made more effective by the addition of a more sensitive and powerful high explosive such as tetryl. This material is generally used in the manufacture of caps and detonators for initiating explosions for military, industrial, and sporting purposes.

All precautions required for protection of magazines apply to storage of these materials. They should not be handled when frozen. Wet fulminate of mercury or wet floor coverings containing small quantities of fulminates may be burned on windrows of flammable material. Nonexplosive products are formed by neutralizing fulminates with cold sodium thiosulfate. All floors, tables and walls where the dry fulminates have been used should be washed with this solution. In the manufacture of mercury fulminate, the fumes given off are toxic and flammable. Care is required to prevent fulminate dust from being carried off in the exhaust system: deposits thus made have caused explosions. Careful attention should be given to cleanliness as foreign or gritty materials in the product may cause an unexpected explosion. The floors on which fulminates are used should be covered with 1/16-inch cloth inserted rubber packing or its equal. All cracks and crevices should be covered. The walls of these rooms should be covered with glazed, water-proof material. Frequent washing with neutralizing solution is necessary. In manufacture, the fulminate is dried on muslin squares on a drying table. Drying tables may be heated with hot water or the dry house may be heated with an air blower system to between 50 and 60°. Primer caps and detonators loaded with fulminate of mercury are less sensitive than the dry bulk material but must be handled with great care. Fires involving these assemblies should be treated the same as for the bulk material. They will explode as soon as fire reaches them. Stocks in an assembly or loading room should be kept as small as possible. Examples of fulminates commonly used in the explosive industry are mercury fulminate, copper fulminate, and silver fulminate.

FOS050　　　CAS: 506-85-4　　　HR: 3
FULMINIC ACID
mf: CHNO　　mw: 43.02

SYN: HYDROGEN CYANIDE-N-OXIDE

CONSENSUS REPORTS: Cyanide and its compounds are on the Community Right-To-Know List.

DOT Classification: Forbidden

SAFETY PROFILE: An unstable explosive sensitive to heat, shock, or friction. When heated to decomposition it emits toxic fumes of NO_x.

FOT000　　　CAS: 6029-87-4　　　HR: 3
FULVINE
mf: $C_{16}H_{23}NO_5$　　mw: 309.40

SYN: CRISPATINE

SAFETY PROFILE: Poison by intraperitoneal and possibly other routes. Experimental teratogenic and reproductive effects. Mutation data

reported. When heated to decomposition it emits toxic fumes of NO_x.

FOU000　　　CAS: 110-17-8　　　HR: 3
FUMARIC ACID
DOT: NA 9126
mf: $C_4H_4O_4$　　mw: 116.08

PROP: White crystals; odorless. Mp: 287°, d: 1.635 @ 20°/4°. Bp: 290°. Sol in water, ether; very sltly sol in chloroform.

SYNS: ALLOMALEIC ACID　*　BOLETIC ACID　*　trans-BUTENEDIOIC ACID　*　(E)-BUTENEDIOIC ACID　*　trans-1,2-ETHYLENEDICARBOXYLIC ACID　*　(E)1,2-ETHYLENEDICARBOXYLIC ACID　*　KYSELINA FUMAROVA (CZECH)　*　LICHENIC ACID　*　NSC-2752　*　U-1149　*　USAF EK-P-583

CONSENSUS REPORTS: Reported in EPA TSCA Inventory.

DOT Classification: ORM-E; Label: None.

SAFETY PROFILE: Poison by intraperitoneal route. Mildly toxic by ingestion and skin contact. A skin and eye irritant. Combustible when exposed to heat or flame; can react vigorously with oxidizing materials. When heated to decomposition it emits acrid smoke and irritating fumes.

FOW000　　　CAS: 130-86-9　　　HR: 3
FUMARINE
mf: $C_{20}H_{19}NO_5$　　mw: 353.40

SYNS: BIFLORINE　*　CORYDININE　*　MACLEYINE　*　7-METHYL-2,3:9,10-BIS(METHYLENEDIOXY)-7,13a-SECOBERBIN-13a-ONE　*　PROTOPINE　*　4,6,-7,14-TETRAHYDRO-5-METHYL-BIS(1,3)BENZODIOXOLO(4,5-c:5′,6′-g)AZECIN-13(5H)-ONE

SAFETY PROFILE: Poison by ingestion and intraperitoneal routes. When heated to decomposition it emits toxic fumes of NO_x.

FOY000　　　CAS: 627-63-4　　　HR: 2
FUMARYL CHLORIDE
DOT: UN 1780
mf: $C_4H_2Cl_2O_2$　　mw: 152.96

PROP: Clear, straw-colored liquid. Mp: 160°, d: 1.408 @ 20°/4°.

SYNS: CHLORURE de FUMARYLE (FRENCH)　*　DICHLORID KYSELINY FUMAROVE (CZECH)　*　FUMAROYL CHLORIDE　*　FUMARYLCHLORID (CZECH)　*　TL 189

CONSENSUS REPORTS: Reported in EPA TSCA Inventory.

DOT Classification: Corrosive Material; Label: Corrosive.

SAFETY PROFILE: Moderately toxic by ingestion, inhalation, and skin contact. A skin, eye and mucous membrane irritant. A corrosive agent. Will react with water or steam to produce toxic and corrosive fumes. When heated to decomposition it emits highly toxic fumes of phosgene and HCl.

FPB875 CAS: 35554-44-0 HR: 3
FUNGAFLOR
mf: $C_{14}H_{14}Cl_2N_2O$ mw: 297.20

PROP: Solidified oil. Sltly sol in organic solvents; poorly sol in water.

SYNS: (±)-1-(β-(ALLYLOXY)-2,4-DICHLOROPHEN-ETHYL)IMIDAZOLE * ENILOCONAZOL (SP) * 1-(2-(2,4-DICHLORPHENYL)-2-PROPENYLOXY) AETHYL)-1H-IMIDAZOLE * 1-(2-(2,4-DICHLOROPHE-NYL)-2-(2-PROPENYLOXY)ETHYL)-1H-IMIDAZOLE * IMAVEROL * IMAZALIL * R 23979

SAFETY PROFILE: Poison by ingestion and intraperitoneal routes. Experimental reproductive effects. A skin and eye irritant. When heated to decomposition it emits toxic fumes of Cl⁻ and NO_x.

FPD000 CAS: 11055-06-4 HR: 3
FUNICOLOSIN
mf: $C_{27}H_{41}NO$ mw: 395.69

SAFETY PROFILE: Poison by ingestion and intraperitoneal routes. When heated to decomposition it emits toxic fumes of NO_x.

FPI150 CAS: 3031-51-4 HR: 3
l-FURALTADONE HYDROCHLORIDE
mf: $C_{13}H_{16}N_4O_6 \cdot ClH$ mw: 360.79

PROP: Yellow crystals. Decomposes @ 206°.

SYNS: FURMETHONOL * l-5-(MORPHOLINO-METHYL)-3-((5-NITROFURFURYLIDENE)AMINO)-2-OXAZOLIDINONEHYDROCHLORIDE

CONSENSUS REPORTS: IARC Cancer Review: GROUP 2B IMEMDT 7,56,87; Animal Limited Evidence IMEMDT 7,161,74.

SAFETY PROFILE: Suspected carcinogen with experimental carcinogenic data. Poison by intravenous route. Moderately toxic by ingestion. Mutation data reported. When heated to decomposition it emits very toxic fumes of HCl and NO_x.

FPK000 CAS: 110-00-9 HR: 3
FURAN
DOT: UN 2389
mf: C_4H_4O mw: 68.08

PROP: Water white liquid. Mp: −85.65°, bp: 31.36°, lel: 2.3%, uel: 14.3%, flash p: −32°F, d: 0.937 @ 20°/4°, vap d: 2.35.

SYNS: DIVINYLENE OXIDE * FURFURAN * NCI-C56202 * OXACYCLOPENTADIENE * OXOLE * RCRA WASTE NUMBER U124 * TETROLE

CONSENSUS REPORTS: EPA Extremely Hazardous Substances List. Reported in EPA TSCA Inventory.

DOT Classification: Flammable Liquid; Label: Flammable Liquid.

SAFETY PROFILE: Poison by inhalation and intraperitoneal routes. Probably at least moderately toxic by ingestion and skin contact. A narcotic. Mutation data reported. The exposure concentration limit of 10 ppm together with its low boiling point requires that adequate ventilation be provided in areas where this chemical is handled. Contact with liquid must be avoided since this chemical can be absorbed through the skin. Washing thoroughly with soap and water followed by prolonged rinsing should be done immediately after accidental contact.

A very dangerous fire hazard when exposed to heat or flame; can react with oxidizing materials. Unstabilized, it may form unstable peroxides on exposure to air and should always be tested before distillation. Washing with an aqueous solution of ferrous sulfate slightly acidified with sodium bisulfate will remove these peroxides. Confirm by test. Contact with acids can initiate a violent exothermic reaction. Moderate explosion hazard when exposed to flame. Furan's low boiling point makes it easy to obtain explosive concentrations of the vapor in inadequately ventilated areas. Highly dangerous upon exposure to heat or flame; can react vigorously with oxidizing materials. To fight fire, use CO_2, dry chemical. When heated to decomposition it emits acrid smoke and irritating fumes.

FPQ000   CAS: 9000-21-9   **HR: 2**
FURCELLERAN GUM

PROP: Vegetable gum from *Furcellaria fastigiata* (Fam. *Rodophyceae*) available as an odorless white powder. Sol in warm water.

SYN: BURTONITE 44

CONSENSUS REPORTS: Reported in EPA TSCA Inventory.

SAFETY PROFILE: Moderately toxic by ingestion. When heated to decomposition it emits acrid smoke and fumes.

FPQ875   CAS: 98-01-1   **HR: 3**
FURFURAL

DOT: UN 1199
mf: $C_5H_4O_2$ mw: 96.09

PROP: Colorless-yellowish liquid; almond-like odor. Bp: 161.7° @ 764 mm, lel: 2.1%, uel: 19.3%, flash p: 140°F (CC), d: 1.154-1.158, refr index: 1.522-1.528, autoign temp: 600°F, vap d: 3.31. Sol in water; misc with alc.

SYNS: ARTIFICIAL ANT OIL * FEMA No. 2489 * FURAL * FURALE * 2-FURALDEHYDE * 2-FURANALDEHYDE * 2-FURANCARBONAL * 2-FURANCARBOXALDEHYDE * 2-FURFURAL * FURFURALDEHYDE * FURFURALE (ITALIAN) * FURFUROL * FURFUROLE * 2-FURIL-META-NALE (ITALIAN) * FUROLE * α-FUROLE * 2-FURYL-METHANAL * NCI-C56177 * PYROMUCIC ALDEHYDE * RCRA WASTE NUMBER U125

CONSENSUS REPORTS: EPA Genetic Toxicology Program. Reported in EPA TSCA Inventory.

OSHA PEL: (Transitional: TWA 5 mg/m^3 (skin)) TWA 2 ppm (skin)
ACGIH TLV: TWA 2 ppm (skin)
DFG MAK: 5 ppm (20 mg/m^3)
DOT Classification: Combustible Liquid; Label: None; Flammable or Combustible Liquid; Label: Flammable Liquid.

SAFETY PROFILE: Poison by ingestion, intraperitoneal, subcutaneous, intravenous, and intramuscular routes. Moderately toxic by inhalation and skin contact. Human mutation data reported. A skin and eye irritant. Mutation data reported. The liquid is dangerous to the eyes. The vapor is irritating to mucous membranes and is a central nervous system poison. How-

ever, its low volatility reduces its toxicity effect. Ingestion of furfural has produced cirrhosis of the liver in rats. In industry there is a tendency to minimize the danger of acute effects resulting from exposure to it. This is particularly true because of its low volatility.

Combustible liquid when exposed to heat or flame; can react with oxidizing materials. Moderate explosion hazard when exposed to heat or flame or by chemical reaction. An exothermic polymerization of almost explosive violence can occur upon contact with strong mineral acids or alkalies. Keep away from heat and open flames. Mixture with sodium hydrogen carbonate ignites spontaneously. To fight fire, use alcohol foam, CO_2, dry chemical. When heated to decomposition it emits acrid smoke and irritating fumes.

FPS000   CAS: 494-47-3   **HR: 3**
FURFURAMIDE
mf: $C_{15}H_{12}N_2O_3$ mw: 268.29

PROP: Needles from alcohol. Mp: 117-21°, bp: 250° decomp. Insol in water; decomp in acid; sol in alc and ether.

SYNS: 2-(BIS(FURFURYLIDENAMINO))METHYLFURAN * HYDROFURAMIDE

SAFETY PROFILE: Poison by ingestion. A skin, eye, and mucous membrane irritant. Causes intense pulmonary irritation and reported to cause liver and kidney damage. When heated to decomposition it emits toxic fumes of NO$_x$. A component of fungicides.

FPU000   CAS: 98-00-0   **HR: 3**
FURFURYL ALCOHOL

DOT: UN 2874
mf: $C_5H_6O_2$ mw: 98.11

PROP: Clear, colorless, mobile liquid. Mp: −31°, lel: 1.8%, uel: 16.3%, both between 72-122°, bp: 171° @ 750 mm, flash p: 167°F (OC), d: 1.129 @ 20°/4°, autoign temp: 915°F, vap press: 1 mm @ 31.8°, vap d: 3.37.

SYNS: 2-FURANCARBINOL * 2-FURANMETHANOL * FURFURAL ALCOHOL * 2-FURFURYLALKOHOL (CZECH) * FURYL ALCOHOL * 2-FURYLCARBINOL * α-FURYLCARBINOL * (2-FURYL)METHANOL * 2-HYDROXYMETHYLFURAN * NCI-C56224

CONSENSUS REPORTS: Reported in EPA TSCA Inventory.

OSHA PEL: (Transitional: TWA 50 ppm) TWA 10 ppm; STEL 15 ppm (skin)
ACGIH TLV: TWA 10 ppm; STEL 15 ppm (skin)
DFG MAK: 50 ppm (200 mg/m^3)
NIOSH REL: TWA 200 mg/m^3
DOT Classification: Poison B; Label: St. Andrews Cross.

SAFETY PROFILE: Poison by ingestion, skin contact, and subcutaneous routes. Moderately toxic by inhalation and intraperitoneal routes. Mutation data reported. An eye irritant. Flammable when exposed to heat or flame; can react with oxidizing materials. Moderate explosion hazard when exposed to heat or flame. Reacts violently with acids (e.g., formic acid; cyanoacetic acid + heat). Ignites on contact with 85% hydrogen peroxide. To fight fire, use alcohol foam, CO_2, dry chemical. When heated to decomposition it emits acrid smoke and fumes.

FPW000   CAS: 617-89-0   HR: 3
FURFURYLAMINE

DOT: UN 2526
mf: C_5H_7NO mw: 97.13

PROP: Light straw-colored liquid. Bp: 146°, flash p: 99°F (OC), fp: −70°, d: 1.0502 @ 25°, vap d: 3.35.

SYNS: 2-FURANMETHYLAMINE * 1-(2-FURYL)METHYLAMINE * USAF Q-1

CONSENSUS REPORTS: Reported in EPA TSCA Inventory.

DOT Classification: Flammable or Combustible Liquid; Label: Flammable Liquid.

SAFETY PROFILE: Poison by intraperitoneal route. A skin, eye, and mucous membrane irritant. A dangerous fire hazard when exposed to heat or flame; can react with oxidizing materials. To fight fire, use foam, CO_2, dry chemical. When heated to decomposition it emits toxic fumes of NO_x.

FQJ000   CAS: 2578-75-8   HR: 3
FUROTHIAZOLE
mf: $C_8H_6N_4O_4S$ mw: 254.24

SYN: N-(5-(5-NITRO-2-FURYL)-1,3,4-THIADIAZOL-2-YL)ACETAMIDE

SAFETY PROFILE: Questionable carcinogen with experimental carcinogenic and neoplasti-genic data. Mutation data reported. When heated to decomposition it emits very toxic fumes of SO_x and NO_x.

FQN000   CAS: 3688-53-7   HR: 3
2-(2-FURYL)-3-(5-NITRO-2-FURYL)ACRYLAMIDE
mf: $C_{11}H_8N_2O_5$ mw: 248.21

SYNS: AF-2 (preservative) * FF * FURYLAMIDE * FURYLFURAMIDE * α-2-FURYL-5-NITRO-2-FURA-NACYRLAMIDE * 2-(2-FURYL)-3-(5-NITRO-2-FURYL)ACRYLIC ACID AMIDE * α-(FURYL)-β-(5-NITRO-2-FURYL)ACRYLIC AMIDE * TOFURON

CONSENSUS REPORTS: IARC Cancer Review: GROUP 2B IMEMDT 7,56,87; Human Inadequate Evidence IMEMDT 31,47,83; Animal Sufficient Evidence IMEMDT 31,47,83. EPA Genetic Toxicology Program.

SAFETY PROFILE: Suspected carcinogen with experimental carcinogenic and neoplastigenic data. Poison by ingestion. Experimental teratogenic and reproductive effects. Human mutation data reported. When heated to decomposition it emits toxic fumes of NO_x.

FQR000   CAS: 23255-69-8   HR: 3
FUSARENONE X
mf: $C_{17}H_{22}O_8$ mw: 354.39

PROP: Isolated from the culture filtrate of *Fusarium nivale* (34ZHAD -,163,71).

SYNS: 4-ACETYLOXY-12,13-EPOXY-3,7,15-TRIHY-DROXY-(3-α,4-β,7-β)-TRICHOTHEC-9-EN-8-ONE * NIVALENOL-4-O-ACETATE * 3,7,15-TRIHYDROXY-4-ACETOXY-8-OXO-12,13-EPOXY-Δ^9-TRICHOTHECENE * 3,7,15-TRIHYDROXYSCIRP-4-ACETOXY-9-EN-8-ONE

SAFETY PROFILE: Poison by ingestion, subcutaneous, intravenous, intraperitoneal, and possibly other routes. Experimental reproductive effects. Questionable carcinogen with experimental tumorigenic data. Human mutation data reported. When heated to decomposition it emits acrid smoke and irritating fumes.

FQS000   CAS: 21259-20-1   HR: 3
FUSARIOTOXIN T 2
mf: $C_{24}H_{34}O_9$ mw: 466.58

PROP: A strain of *F. tricinctum* isolated from infected corn.

SYNS: 4,15-DIACETOXY-8-(3-METHYLBUTYRYLOXY)
12,13-EPOXY-Δ-9-TRICHOTHECEN-3-OL * 4-β,15-DI-
ACETOXY-8-α-(3-METHYLBUTYRYLOXY)-3-α-HYDROXY-
12,13-EPOXYTRICHOTHEC-9-ENE * 3-HYDROXY-4,15-
DIACETOXY-8-(3-METHYLBUTYRYLOXY)-12,13-EPOXY-
Δ⁹-TRICHOTHECENE * INSARIOTOXIN * 8-ISOVAL-
ERATE * 8-(3-METHYLBUTYRYLOXY)-DIACETOXY-
SCIRPENOL * NSC 138780 * T-2 MYCOTOXIN
* TOXIN T2 * T (^{2})-TRICHOTHECENE

CONSENSUS REPORTS: IARC Cancer Re-
view: GROUP 3 IMEMDT 7,56,87; Animal
Inadequate Evidence IMEMDT 31,265,83.
EPA Genetic Toxicology Program.

SAFETY PROFILE: Poison by ingestion, intramuscular, subcutaneous, intraperitoneal, intracerebral, and intravenous routes. Moderately
toxic by inhalation. Experimental teratogenic
and reproductive effects. A skin irritant. Questionable carcinogen with experimental neoplastigenic data. Mutation data reported. When
heated to decomposition it emits acrid smoke
and irritating fumes.

FQT000   CAS: 8013-75-0   **HR: 2**
FUSEL OIL

DOT: UN 1201

PROP: Colorless to pale yellow liquid; odorless.
D: 0.807-0.813, refr index: 1.405-1.410. Composition of grain fusel oil is methanol, ethanol,
acetaldehyde, and other alcohols.

SYNS: FEMA No. 2497 * FUSELOEL (GERMAN)
* FUSEL OIL, REFINED (FCC) * HUILE de FUSEL
(FRENCH)

CONSENSUS REPORTS: Reported in EPA
TSCA Inventory.

DOT Classification: Combustible Liquid; Label: None; Flammable Liquid; Label: Flammable Liquid.

SAFETY PROFILE: May contain carcinogens.
Mutation data reported. Flammable liquid when
exposed to heat or flame; can react vigorously
with oxidizing materials. When heated to decomposition it emits acrid smoke and fumes.

G

GAD000   CAS: 3060-41-1   **HR: 2**
p-GABA HYDROCHLORIDE
mf: $C_{10}H_{13}NO_2 \cdot ClH$ mw: 215.70

SYNS: β-(AMINOMETHYL)-BENZENEPROPANOIC ACID HYDROCHLORIDE * β-(AMINOMETHYL)-HYDROCIN-NAMIC ACID HYDROCHLORIDE * FENIBUT HYDRO-CHLORIDE * FENIGAM HYDROCHLORIDE * PHENI-BUT HYDROCHLORIDE * PHENIGAM HYDRO-CHLORIDE * PHENIGAMA HYDROCHLORIDE * PHENYBUT HYDROCHLORIDE * PHENYGAM HY-DROCHLORIDE * PHENYLGAMMA HYDROCHLORIDE * PHGABA HYDROCHLORIDE

SAFETY PROFILE: Moderately toxic by intra-peritoneal route. When heated to decomposition it emits very toxic fumes of Cl^- and NO_x.

GAF000   CAS: 7440-54-2   **HR: 3**
GADOLINIUM
af: Gd   aw: 157.25

PROP: A yellow-white, malleable, and ductile metallic element. A rare earth, stable in dry air; reacts slowly with H_2O. Mp: 1312°, bp: 3233°, d: 7.898 @ 25°.

CONSENSUS REPORTS: Reported in EPA TSCA Inventory.

SAFETY PROFILE: Questionable carcinogen with experimental tumorigenic data. It may act as an anticoagulant. It can react violently with air and halogens.

GAH000   CAS: 10138-52-0   **HR: 3**
GADOLINIUM CHLORIDE
mf: Cl_3Gd mw: 263.60

PROP: White, monoclinic crystals. D: 4.52 @ 0°, mp: approx 609°. Sol in H_2O.

SYN: GADOLINIUM TRICHLORIDE

CONSENSUS REPORTS: Reported in EPA TSCA Inventory.

SAFETY PROFILE: Poison by intraperitoneal route. A skin and eye irritant. When heated to decomposition it emits very toxic fumes of Cl^-.

GAJ000   CAS: 3088-53-7   **HR: 3**
GADOLINIUM CITRATE

SAFETY PROFILE: Poison by intraperitoneal route. When heated to decomposition it emits acrid smoke and irritating fumes.

GAL000   CAS: 10168-81-7   **HR: 3**
GADOLINIUM(III) NITRATE (1:3)
mf: $N_3O_9 \cdot Gd$ mw: 343.28

SYN: NITRIC ACID, GADOLINIUM(3+) SALT

CONSENSUS REPORTS: Reported in EPA TSCA Inventory.

SAFETY PROFILE: Poison by intraperitoneal route. Moderately toxic by ingestion. When heated to decomposition it emits toxic fumes of NO_x.

GAV000   CAS: 59-23-4   **HR: D**
GALACTOSE
mf: $C_6H_{12}O_6$ mw: 180.18

PROP: (α form): Prisms from water or ethanol. Mp: 167°. Freely sol in hot H_2O; sol in pyridine; sltly sol in alc. (β form): Crystals. Mp: 167°.

SYN: d-GALACTOSE

CONSENSUS REPORTS: Reported in EPA TSCA Inventory.

SAFETY PROFILE: Experimental teratogenic and reproductive effects. When heated to de-composition it emits acrid smoke and irritating fumes.

GBE000   CAS: 149-91-7   **HR: 3**
GALLIC ACID
mf: $C_7H_6O_5$ mw: 170.13

PROP: White- to pale fawn-colored, odorless crystals; somewhat water-sol. D: 1.694, mp: 225-250° (decomp), $-H_2O$ @ 100-120°.

SYN: 3,4,5-TRIHYDROXYBENZOIC ACID

CONSENSUS REPORTS: Reported in EPA TSCA Inventory. EPA Genetic Toxicology Pro-gram.

SAFETY PROFILE: Poison by intravenous route. Moderately toxic by intraperitoneal route. Mildly toxic by ingestion. Experimental repro-ductive effects. Mutation data reported. When heated to decomposition it emits acrid smoke and irritating fumes.

GBG000   CAS: 7440-55-3   **HR: 3**
GALLIUM

DOT: UN 2803
af: Ga   aw: 69.72

PROP: A beautiful, lustrous, silvery liquid or metal or a gray solid. Mp: 29.78°, bp: 2403°, d (solid): 5.904 @ 29.6°, d (liquid): 6.905 @ 29.8°.

SYNS: GALLIUM METAL, solid (DOT) * GALLIUM METAL, liquid (DOT)

CONSENSUS REPORTS: Reported in EPA TSCA Inventory.

DOT Classification: ORM-B; Label: None; liquid: ORM-B; Label: None; solid: Corrosive Material; Label: Corrosive.

SAFETY PROFILE: Poison by subcutaneous and intravenous routes. Corrosive; probably an eye, skin, and mucous membrane irritant. It has a metallic taste, causes dermatitis and depression of bone marrow function. Potentially explosive reaction with hydrogen peroxide + hydrochloric acid. Violent or vigorous reaction with halogens. Forms an amalgam with aluminum alloys.

GBM000   CAS: 13450-90-3   **HR: 3**
GALLIUM (3+) CHLORIDE
mf: Cl_3Ga mw: 176.07

PROP: Colorless needles. Mp: 78°.

SYN: GALLIUM CHLORIDE

CONSENSUS REPORTS: Reported in EPA TSCA Inventory. EPA Extremely Hazardous Substances List.

SAFETY PROFILE: Poison by inhalation, subcutaneous, intravenous, and intraperitoneal routes. When heated to decomposition it emits very toxic fumes of Cl^-.

GBO000   CAS: 27905-02-8   **HR: 3**
GALLIUM CITRATE
mf: $C_6H_5O_7 \cdot Ga$ mw: 258.83

SYN: CITRIC ACID, GALLIUM SALT (1:1)

SAFETY PROFILE: Poison by subcutaneous route. When heated to decomposition it emits acrid smoke and irritating fumes.

GBS000   CAS: 13494-90-1   **HR: 3**
GALLIUM(III) NITRATE (1:3)
mf: $N_3O_9 \cdot Ga$ mw: 255.75

PROP: White, deliquescent crystals. Mp: decomp @ 110°, bp: releases Ga_2O_3 @ 200°.

SYNS: GALLIUM NITRATE * NITRIC ACID, GALLIUM(3+) SALT

CONSENSUS REPORTS: Reported in EPA TSCA Inventory.

SAFETY PROFILE: Poison by intraperitoneal and subcutaneous routes. Moderately toxic by ingestion. Human systemic effects by intravenous route: nausea or vomiting, renal function changes, proteinuria, normocytic anemia and thrombocytopenia. A severe skin irritant. Mutation data reported. When heated to decomposition it emits toxic fumes of NO_x.

GBU800   **HR: 1**
GARLIC OIL

PROP: From steam distillation of *Allium sativum* L. (Fam. *Liliaceae*). Clear to yellow liquid; strong odor and taste of garlic. Sol in fixed oils, mineral oil; insol in glycerin, alc, propylene glycol.

SAFETY PROFILE: An eye irritant. When heated to decomposition it emits acrid smoke and irritating fumes.

GBY000   CAS: 8006-61-9   **HR: 3**
GASOLINE

DOT: UN 1203/UN 1257

PROP: Clear, aromatic, volatile liquid; a mixture of aliphatic HC. Flash p: −50°F, d: <1.0, vap d: 3.0-4.0, ULC: 95-100, lel: 1.3%, uel: 6.0%, autoign temp: 536-853°F, bp: Initially 39°, after 10% distilled = 60°, after 50% = 110°, after 90% = 170°, final bp: 204°. Insol in water; freely sol in absolute alc, ether, chloroform, and benzene.

SYNS: CASING HEAD GASOLINE (DOT) * MOTOR FUEL (DOT) * MOTOR SPIRIT (DOT) * NATURAL GASOLINE (DOT) * PETROL (DOT)

CONSENSUS REPORTS: Reported in EPA TSCA Inventory.

OSHA PEL: TWA 300 ppm; STEL 500 ppm
ACGIH TLV: TWA 300 ppm; STEL 500 ppm
DOT Classification: Flammable Liquid; Label: Flammable Liquid.

SAFETY PROFILE: Mildly toxic by inhalation. Human systemic effects by inhalation: cough, conjunctiva irritation, hallucinations or distorted perceptions. Repeated or prolonged dermal exposure causes dermatitis. Can cause blistering of skin. Questionable carcinogen. Inhalation or ingestion can cause central nervous system depression. Pulmonary aspiration can cause severe pneumonitis. Some addiction has been reported from inhalation of fumes. Even brief inhalations of high concentrations can cause a fatal pulmonary edema. The vapors are considered to be moderately poisonous. If its concentration in air is sufficiently high to reduce the oxygen content below that needed to maintain life, it acts as a simple asphyxiant. A human eye irritant. Gasoline is a common air contaminant. A very dangerous fire and explosion hazard when exposed to heat or flame; can react vigorously with oxidizing materials. To fight fire, use foam, CO_2, dry chemical.

GCE100   HR: 3
GASOLINE, UNLEADED

SYNS: UNLEADED GASOLINE * UNLEADED MOTOR GASOLINE

SAFETY PROFILE: Suspected carcinogen with experimental carcinogenic data. Moderately toxic by inhalation. Pulmonary aspiration can cause severe pneumonitis. Skin irritant. Flammable liquid. When heated to decomposition it emits acrid smoke and irritating fumes.

GCK000   CAS: 509-15-9   HR: 3
GELSEMINE
mf: $C_{20}H_{22}N_2O_2$ mw: 322.44

PROP: An alkaloid. Mp: 178°. Sltly sol in water; sol in alc, benzene, chloroform, ether, acetone, and dilute acids.

SYN: GELSEMIN

SAFETY PROFILE: A deadly poison by subcutaneous, intraperitoneal and possibly other routes. A poisonous alkaloid. Can cause muscular weakness and respiratory arrest. Used as a central nervous system stimulant. When heated to decomposition it emits toxic fumes of NO_x.

GCO000   CAS: 1403-66-3   HR: 3
GENTAMICIN

SYNS: GARAMYCIN * GENTAMYCIN * GENTAMYCIN-CREME (GERMAN) * UROMYCINE

CONSENSUS REPORTS: EPA Genetic Toxicology Program.

SAFETY PROFILE: Poison by intravenous, intraperitoneal, intramuscular, and subcutaneous routes. Mildly toxic by ingestion. Experimental

teratogenic and reproductive effects. Mutation data reported. Human systemic effects by intravenous route: vestibular function changes, hallucinations, distorted perceptions, motor activity changes, trigeminal nerve sensory changes, and unspecified kidney changes. Affects the peripheral nervous system by intravenous route. An antibiotic. When heated to decomposition it emits acrid smoke and irritating fumes.

GCS000   CAS: 1405-41-0   HR: 3
GENTAMYCIN SULFATE

SYNS: GARAMYCIN * GENOPTIC * GENOPTIC S.O.P. * GM SULFATE * NSC-82261 * SCH 9724

SAFETY PROFILE: Poison by intravenous, intraperitoneal, and intramuscular routes. Moderately toxic by subcutaneous route. Mutation data reported. When heated to decomposition it emits very toxic fumes of SO_x.

GCY000   CAS: 105-86-2   HR: 1
GERANIOL FORMATE
mf: $C_{11}H_{18}O_2$ mw: 182.29

PROP: Colorless to pale yellow liquid; rose odor. D: 0.906-0.920, refr index: 1.457-1.466, flash p: 205°F. Sol in alc, fixed oils; insol in glycerin, propylene glycol, water @ 216°.

SYNS: trans-3,7-DIMETHYL-2,6-OCTADIEN-1-OL FORMATE * 3,7-DIMETHYL-2,6-OCTADIENYL ESTER FORMIC ACID (E) * trans-3,7-DIMETHYL-2,6-OCTADIEN-1-YL FORMATE * FEMA No. 2514 * FORMIC ACID, GERANIOL ESTER * GERANYL FORMATE (FCC)

CONSENSUS REPORTS: Reported in EPA TSCA Inventory.

SAFETY PROFILE: A human skin irritant and an experimental eye irritant. Combustible liquid. When heated to decomposition it emits acrid smoke and irritating fumes.

GDA000   CAS: 8000-46-2   HR: 1
GERANIUM OIL ALGERIAN TYPE

PROP: From steam distillation of leaves from *Pelargonium graveolens* l'Her (Fam. *Geraniaceae*). Contains geraniol and geranyl tiglate. Yellow liquid; odor of rose and geraniol. D: 0.886-0.898, refr index: 1.454-1.472 @ 20°. Sol in fixed oils, mineral oil; insol in glycerin.

SYNS: GERANIUM OIL * OIL of GERANIUM * OIL of PELARGONIUM * OIL of ROSE GERANIUM

* OIL ROSE GERANIUM ALGERIAN * PELARGO-
NIUM OIL * ROSE GERANIUM OIL ALGERIAN

CONSENSUS REPORTS: Reported in EPA
TSCA Inventory.

SAFETY PROFILE: A skin irritant. When
heated to decomposition it emits acrid smoke
and irritating fumes.

GDE800   HR: 2
GERANYL BENZOATE
mf: $C_{17}H_{22}O_2$ mw: 258.36

PROP: Sltly yellow liquid; floral odor resembling ylang ylang oil. D: 0.978-0.984, refr index: 1.513-1.518, flash p: +212°F. Misc in
alc, chloroform; insol in water @ 305°.

SYNS: 3,7-DIMETHYL-2,6-OCTADIEN-1-YL BENZOATE
* FEMA No. 2511

SAFETY PROFILE: Combustible liquid. When
heated to decomposition it emits acrid smoke
and irritating fumes.

GDE825   HR: 1
GERANYL BUTYRATE
mf: $C_{14}H_{24}O_2$ mw: 224.34

PROP: Colorless to pale yellow liquid; fruity,
roselike odor. D: 0.889-0.904, refr index:
1.455-1.462, flash p: +199°F. Sol in alc, fixed
oils; insol in glycerin, propylene glycol, water
@ 253°.

SYNS: 3,7-DIMETHYL-2,6-OCTADIENE-1-YL BUTYRATE
* FEMA No. 2512

SAFETY PROFILE: Combustible liquid. When
heated to decomposition it emits acrid smoke
and irritating fumes.

GDK000   CAS: 109-20-6   HR: 1
GERANYL ISOVALERATE
mf: $C_{15}H_{26}O_2$ mw: 238.41

SYNS: trans-3,7-DIMETHYL-2,6-OCTADIENYL ISOPEN-
TANOATE * (E)-ISOVALERIC ACID-3,7-DIMETHYL-2,6-
OCTADIENYL ESTER * (E)-3-METHYLBUTYRIC ACID-
3,7-DIMETHYL-2,6-OCTADIENYL ESTER

CONSENSUS REPORTS: Reported in EPA
TSCA Inventory.

SAFETY PROFILE: A skin irritant. When
heated to decomposition it emits acrid smoke
and irritating fumes.

GDM400   HR: 1
GERANYL PHENYLACETATE
mf: $C_{18}H_{24}O_2$ mw: 272.39

PROP: Yellow liquid; honey-rose odor. D:
0.971-0.978, refr index: 1.507-1.511, flash p:
+212°F. Misc in alc, chloroform, ether; insol
in water.

SYNS: 3,7-DIMETHYL-2,6-OCTADIEN-1-YL PHENYL-
ACETATE * FEMA No. 2516

SAFETY PROFILE: Combustible liquid. When
heated to decomposition it emits acrid smoke
and irritating fumes.

GDM450   HR: 1
GERANYL PROPIONATE
mf: $C_{13}H_{22}O_2$ mw: 210.32

PROP: Colorless liquid; rosy, fruity odor. D:
0.896-0.913, refr index: 1.456-1.464, flash p:
+212°F. Sol in alc, fixed oils; insol in glycerin,
propylene glycol, water @ 253°.

SYNS: 3,7-DIMETHYL-2,6-OCTADADIEN-1-YL PROPIO-
NATE * FEMA No. 2517

SAFETY PROFILE: Combustible liquid. When
heated to decomposition it emits acrid smoke
and irritating fumes.

GDS000   CAS: 1310-53-8   HR: 3
GERMANIC OXIDE (crystalline)

CONSENSUS REPORTS: Reported in EPA
TSCA Inventory.

SAFETY PROFILE: Poison by intraperitoneal
route. When heated to decomposition it emits
acrid smoke and irritating fumes.

GDW000   CAS: 13450-92-5   HR: 3
GERMANIUM BROMIDE
mf: Br_4Ge mw: 392.23

PROP: Gray-white crystals. Mp: 26.1°, bp:
186.5°, d: 3.232 @ 29°/29°.

SYN: GERMANIUM TETRABROMIDE

CONSENSUS REPORTS: Reported in EPA
TSCA Inventory.

SAFETY PROFILE: Poison by intravenous
route. When heated to decomposition it emits
very toxic fumes of Br^-.

GDY000   CAS: 10038-98-9   **HR: 3**
GERMANIUM CHLORIDE
mf: Cl$_4$Ge   mw: 214.39

PROP: Colorless, mobile liquid. Fumes in air. Peculiar acidic odor but can be distinguished from that of concentrated HCl. Volatile @ room temp. Sol in benzene, ether, other organic solvents. Mp: −49.5°, bp: 83.1°, d: 1.879 @ 20°/20°.

SYNS: EXTREMA * GERMANIUM TETRACHLORIDE

CONSENSUS REPORTS: Reported in EPA TSCA Inventory. EPA Genetic Toxicology Program.

SAFETY PROFILE: Poison by intravenous route. Mildly toxic by inhalation. A skin, eye, and mucous membrane irritant. Will react violently with water or steam to produce toxic and corrosive fumes. When heated to decomposition emits toxic fumes of Cl⁻.

GEA000   **HR: 2**
GERMANIUM COMPOUNDS

SAFETY PROFILE: Germanium compounds are considered to be of a low order of toxicity, but rare instances of poisoning have been reported in the literature. Experimental LD50 values are typically about 100-1000 mg/kg for parenteral route and 500-5000 mg/kg for ingestion. The animals suffer from hypothermia, diarrhea, and respiratory and cardiac failure. Inhalation of large amounts of GeCl$_4$ by experimental animals causes necrosis of the tracheal epithelium, bronchitis, and interstitial pneumonia. These effects were not apparent with chronic inhalation of 7 mg/m^3. The tetrachloride and tetrafluoride are eye, skin, and mucous membrane irritants. Alkyl germanium compounds are much less toxic than the corresponding tin or lead compounds. Tributyl germanium and germanium tetrachloride are mutagens. Dimethyl germanium is a teratogen. Chronic ingestion of 1000 ppm or 100 ppm of germanium dioxide in water has been shown to inhibit growth in chickens. No effect was seen at 5 ppm. It has been found that the dioxide stimulates the generation of red blood cells, but it is believed to be relatively nontoxic. Buffered germanium dioxides in solution have been found to be nonirritating to the skin. Germanium hydride is a hemolytic gas and has been shown to have toxic properties at a concentration of 100 ppm. It can cause death at a concentration

of 150 ppm. Otherwise, little is known about the toxicity of organic germanium compounds except they may resemble other organometals in having higher toxicity than inorganic forms. When germanium is given in sublethal amounts, it causes a pronounced tolerance to be exhibited. Interest is high in this material because of its close chemical relationship to arsenic.

GEI100   CAS: 7782-65-2   **HR: 3**
GERMANIUM TETRAHYDRIDE
DOT: UN 2192
mf: GeH$_4$ mw: 76.63

PROP: Colorless gas. Mp: −165°, bp: −90°, d: 1.523 @ −142°/4°, sltly sol in hot HCl, decomp in nitric acid.

SYNS: GERMANE (DOT) * GERMANIUM HYDRIDE * MONOGERMANE

CONSENSUS REPORTS: Reported in EPA TSCA Inventory.

ACGIH TLV: TWA 0.2 ppm
DOT Classification: Poison A; Label: Poison Gas and Flammable Gas.

SAFETY PROFILE: Poison by inhalation. A hemolytic gas. Ignites spontaneously in air. Incompatible with Br$_2$.

GEM000   CAS: 77-06-5   **HR: 3**
GIBBERELLIC ACID
mf: C$_{19}$H$_{22}$O$_6$ mw: 346.41

PROP: A plant growth-promoting hormone. White crystals or crystalline powder. Mp: 233-235°. Sltly sol in water, ether; sol in methanol, ethanol, acetone, aqueous solns of sodium bicarbonate and sodium acetate; moderately sol in ethyl acetate.

SYNS: BERELEX * BRELLIN * CEKUGIB * FLORALTONE * GA * GIBBERELLIN * GIBBREL * GIB-SOL * GIB-TABS * GROCEL * NCI-C55823 * PRO-GIBB * 2,4a,7-TRIHYDROXY-1-METHYL-8-METHYLENEGIBB-3-ENE-1,10-CARBOXYLIC ACID 1-4-LACTONE

CONSENSUS REPORTS: EPA Genetic Toxicology Program. Reported in EPA TSCA Inventory.

SAFETY PROFILE: Mildly toxic by ingestion. Questionable carcinogen with experimental tumorigenic data. Mutation data reported. When

heated to decomposition it emits acrid smoke and irritating fumes.

GEO000   CAS: 12002-43-6   HR: 1
GILSONITE

PROP: A black solid hydrocarbon mineral formed from petroleum millions of years ago by geologic processes.

SYN: NCI-C55185

SAFETY PROFILE: A skin, eye, and mucous membrane irritant. An allergen. Has been known to cause photosensitization of skin. Flammable when exposed to heat or open flame. To fight fire use water, foam, dry chemical and CO_2. When heated to decomposition it emits acrid smoke and irritating fumes.

GEQ000   CAS: 8007-08-7   HR: 1
GINGER OIL

PROP: From steam distillation of ground rhizomes of *Zingiber officinale* Roscoe (Fam. *Zingiberaceae*). Yellow liquid; odor of ginger. D: 0.870-0.882, refr index: 1.488 @ 20°. Sol in fixed oils, mineral oil, alc; insol in glycerin, propylene glycol.

CONSENSUS REPORTS: Reported in EPA TSCA Inventory.

SAFETY PROFILE: A skin irritant. Mutation data reported. When heated to decomposition it emits acrid smoke and irritating fumes.

GEU000   CAS: 4562-36-1   HR: 3
GITOXIN
mf: $C_{41}H_{64}O_{14}$ mw: 781.05

PROP: Stout prisms from chloroform and methanol. Decomp @ 285° (rapid heating). Almost insol in chloroform, ethyl acetate and acetone. Dissolves in a mixture of chloroform and alc or pyridine or dil alc.

SYNS: ANHYDROGITALIN * BIGITALIN
* GITOXIGENIN-TRIDIGITOXOSID (GERMAN)
* PSEUDODIGITOXIN

SAFETY PROFILE: Poison by ingestion, intravenous, parenteral, and possibly other routes. When heated to decomposition it emits acrid smoke and irritating fumes.

GEW000   CAS: 7242-04-8   HR: 3
GITOXIN PENTAACETATE
mf: $C_{51}H_{74}O_{19}$ mw: 991.25

PROP: Rhombic crystals. Mp: 151-155°.

SYNS: CARNACID-COR * CORDOVAL * PENGI-TOXIN * PENTAACETYLGITOXIN * PENTA-o-ACE-TYLGITOXIN * PENTAGIT

SAFETY PROFILE: A deadly poison by ingestion, intraperitoneal, and intravenous routes. Used as a cardiotonic agent. When heated to decomposition it emits acrid smoke and irritating fumes.

GFA000   CAS: 15879-93-3   HR: 3
α-d-GLUCOCHLORALOSE
mf: $C_8H_{11}Cl_3O_6$ mw: 309.54

SYNS: AGC * ALFAMAT * ANHYDROGLUCO-CHLORAL * APHOSAL * CHLORALOSANE
* α-CHLORALOSE * CHLOROALOSANE * DULCI-DOR * GLUCOCHLORAL * GLUCOCHLORALOSE
* KALMETTUMSOMNIFERUM * MONOTRICHLOR-AETHYLIDEN-α-GLUCOSE (GERMAN) * MUREX
* SOMIO * 1,2-o-(2,2,2-TRICHLOROETHYLIDENE)-α-d-GLUCOFURANOSE

CONSENSUS REPORTS: Reported in EPA TSCA Inventory.

SAFETY PROFILE: Poison by ingestion, subcutaneous, and intraperitoneal routes. Questionable carcinogen with experimental tumorigenic data. When heated to decomposition it emits toxic fumes of Cl^-.

GFC000   CAS: 124-99-2   HR: 3
GLUCOPROSCILLARIDIN A
mf: $C_{36}H_{52}O_{13}$ mw: 692.88

SYNS: 14-β-HYDROXY-3-β-SCILLOBIOSIDOBUFA-4,-20,22-TRIENOLIDE * SCILLAGLYKOSID A (GERMAN)
* SCILLAREN A * SCILLARENIN-3,6-DEOXY-4-o-β-d-GLUCOPYRANOSYL-α-l-MANNOPYRANOSIDE
* SCILLAREN & RHAMNOSE & GLUCOSE (GERMAN)
* 3-β-SCILLOBIOSIDO-14-β-HYDROXY-Δ-4,20,22-BUFA-TRIENOLID (GERMAN) * TRANSVAALIN

SAFETY PROFILE: Poison by intravenous and possibly other routes. When heated to decomposition it emits acrid smoke and irritating fumes.

GFG000   CAS: 50-99-7   HR: 3
d-GLUCOSE
mf: $C_6H_{12}O_6$ mw: 180.18

PROP: Colorless crystals or white crystalline or granular powder; odorless with sweet taste. D: 1.544, mp: 146°. Sol in water; sltly sol in alc. α Form: (monohydrate) crystals from water. Mp: 83°. α Form: (anhydrous) crystals from hot ethanol or water. Mp: 146°. Very sparingly sol in abs alc, ether, acetone; sol in hot glacial acetic acid, pyridine, aniline. β Form: crystals from hot H_2O + ethanol, from dil acetic acid or from pyridine; mp: 148-155°.

SYNS: CARTOSE * CERELOSE * CORN SUGAR * DEXTROPUR * DEXTROSE (FCC) * DEXTROSE, anhydrous * DEXTROSOL * GLUCOLIN * GLUCOSE * d-GLUCOSE, anhydrous * GLUCOSE LIQUID * GRAPE SUGAR * SIRUP

CONSENSUS REPORTS: Reported in EPA TSCA Inventory. EPA Genetic Toxicology Program.

SAFETY PROFILE: Mildly toxic by ingestion. Experimental reproductive effects. Questionable carcinogen with experimental tumorigenic data. Mutation data reported. Potentially explosive reaction with potassium nitrate + sodium peroxide when heated in a sealed container. Mixtures with alkali release carbon monoxide when heated. When heated to decomposition it emits acrid smoke and irritating fumes.

GFO000 CAS: 56-86-0 **HR: 1**
I-GLUTAMIC ACID
mf: $C_5H_9NO_4$ mw: 147.15

PROP: A nonessential amino acid present in all complete proteins. White crystals or crystalline powder. Mp (dl form): 194°, d (dl form): 1.4601 @ 20°/4°, mp (l form): 224-225°, d (l form): 1.538 @ 20°/4°. Sltly sol in water.

SYNS: α-AMINOGLUTARIC ACID * l-2-AMINOGLUTARIC ACID * 2-AMINOPENTANEDIOIC ACID * 1-AMINOPROPANE-1,3-DICARBOXYLIC ACID * GLUSATE * GLUTACID * GLUTAMIC ACID * α-GLUTAMIC ACID * d-GLUTAMIENSUUR * GLUTAMINIC ACID * l-GLUTAMINIC ACID * GLUTAMINOL * GLUTATON

CONSENSUS REPORTS: Reported in EPA TSCA Inventory.

SAFETY PROFILE: Human systemic effects by ingestion and intravenous route: headache and nausea or vomiting. When heated to decomposition it emits toxic fumes of NO_x.

GFO050 CAS: 56-85-9 **HR: 1**
GLUTAMINE
mf: $C_5H_{10}N_2O_3$ mw: 146.17

PROP: l-Form (natural): Fine opaque needles from water or dil ethanol. Decomp 185-186°. Sol in water; practically insol in methanol, ethanol, ether, benzene, acetone, ethyl acetate, chloroform. dl-Form: prisms from dil acetone. Mp: 185-186°.

SYNS: 2-AMINOGLUTARAMIC ACID * l-2-AMINOGLUTARAMIDIC ACID * CEBROGEN * GLUMIN * GLUTAMIC ACID AMIDE * GLUTAMIC ACID-5-AMIDE * γ-GLUTAMINE * l-GLUTAMINE (9CI, FCC) * LEVOGLUTAMID * LEVOGLUTAMIDE * STIMULINA

CONSENSUS REPORTS: Reported in EPA TSCA Inventory.

SAFETY PROFILE: Mildly toxic by ingestion. Experimental reproductive effects. When heated to decomposition it emits toxic fumes of NO_x.

GFQ000 CAS: 111-30-8 **HR: 3**
GLUTARALDEHYDE
mf: $C_5H_8O_2$ mw: 100.13

SYNS: CIDEX * GLUTARAL * GLUTARALDEHYD (CZECH) * GLUTARDIALDEHYDE * GLUTARIC DIALDEHYDE * NCI-C55425 * 1,5-PENTANEDIAL * 1,5-PENTANEDIONE * POTENTIATED ACID GLUTARALDEHYDE * SONACIDE

CONSENSUS REPORTS: Reported in EPA TSCA Inventory.

OSHA PEL: CL 0.2 ppm
ACGIH TLV: CL 0.2 ppm
DFG MAK: 0.2 ppm (0.8 mg/m^3)

SAFETY PROFILE: Poison by ingestion, intravenous. and intraperitoneal routes. Moderately toxic by inhalation, skin contact, and subcutaneous routes. Experimental teratogenic and reproductive effects. Mutation data reported. A severe eye and human skin irritant. When heated to decomposition it emits acrid smoke and irritating fumes.

GGA000 CAS: 56-81-5 **HR: 2**
GLYCERIN
mf: $C_3H_8O_3$ mw: 92.11

PROP: Colorless or pale yellow liquid; odorless, syrupy, sweet and warm taste. Mp: 17.9 (solidifies at a much lower temp), bp: 290°, ULC: 10-20, flash p: 320°F, d: 1.260 @ 20°/4°, autoign temp: 698°F, vap press: 0.0025 mm @ 50°, vap d: 3.17. Misc with water, alc; insol in chloroform, ether, oils.

SYNS: GLYCERIN, anhydrous * GLYCERINE
* GLYCERIN, SYNTHETIC * GLYCERITOL
* GLYCEROL * GLYCYL ALCOHOL * GROCO-
LENE * MOON * 1,2,3-PROPANETRIOL * STAR
* SUPEROL * SYNTHETIC GLYCERIN * 90 TECH-
NICAL GLYCERINE * TRIHYDROXYPROPANE
* 1,2,3-TRIHYDROXYPROPANE

CONSENSUS REPORTS: Reported in EPA
TSCA Inventory.

OSHA PEL: (Transitional: TWA Total Mist:
   15 mg/m³; Respirable Fraction: 5 mg/m³)
   TWA Total Mist: 10 mg/m³; Respirable Fraction: 5 mg/m³
ACGIH TLV: TWA 10 mg/m³ (mist)

SAFETY PROFILE: Poison by subcutaneous
route. Mildly toxic by ingestion. Human systemic effects by ingestion: headache and nausea
or vomiting. Experimental reproductive effects.
Human mutation data reported. A skin and eye
irritant. In the form of mist it is a nuisance
particulate and inhalation irritant.
   Combustible liquid when exposed to heat,
flame, or powerful oxidizers. Mixtures with hydrogen peroxide are highly explosive. Ignites
on contact with potassium permanganate, calcium hypochlorite. Mixture with nitric acid +
sulfuric acid forms the explosive glyceryl nitrate. Mixture with perchloric acid + lead oxide
forms explosive perchlorate esters. Confined
mixture with chlorine explodes if heated to
70-80°. Can react violently with acetic anhydride, aniline + nitrobenzene, Ca(OCl)₂, CrO₃,
Cr₂O₃, F₂ + PbO, phosphorus triiodide, ethylene oxide + heat, KMnO₄, K₂O₂, AgClO₄,
Na₂O₂, NaH. Energetic reaction with sodium
hydride. Mixture with nitric acid + hydrofluoric
acid is a storage hazard due to gas evolution.
To fight fire, use alcohol foam, CO₂, dry chemical. When heated to decomposition it emits acrid
smoke and fumes.

GGI000   CAS: 38571-73-2   **HR: 3**
**GLYCEROL (TRI(CHLOROMETHYL))
ETHER**
mf: C₆H₁₁Cl₃O₃   mw: 237.52

SYN: TRIS-1,2,3-(CHLOROMETHOXY)PROPANE

CONSENSUS REPORTS: IARC Cancer Review: GROUP 2A IMEMDT 7,56,87; Animal
Sufficient Evidence IMEMDT 15,301,77.

SAFETY PROFILE: Suspected carcinogen with
experimental neoplastigenic and tumorigenic

data. When heated to decomposition it emits
toxic fumes of Cl⁻.

GGS000   CAS: 59-47-2   **HR: 3**
GLYCERYL-o-TOLYL ETHER
mf: C₁₀H₁₄O₃   mw: 182.24

SYNS: A 1141 * AGEFLEX CGE * ANATENSIN
* ANXINE * ATENSIN * AVESYL * AVOXYL
* BDH 312 * CRESODIOL * o-CRESOL GLYCERYL
ETHER * CRESOSSIDIOLO * CRESOSSIPRORAN-
DIOLO * CRESOXYDIOL * CRESOXYPROPANEDIOL
* o-CRESYL-α-GLYCERYL ETHER * CURARIL
* CURARYTHAN * DASERD * DASEROL
* DECONTRACTIL * α,β-DIHYDROXY-γ-(2-METHYL-
PHENOXY)PROPANE * 1,2-DIHYDROXY-3-(2-METHYL-
PHENOXY)PROPANE * DILOXOL * FINDOLAR
* GLUKRESIN * GLYOTOL * KINAVOSYL
* o-KRESOL-GLYCERINAETHER (GERMAN) * KRES-
OXYPROPANDIOL * LISSENPHAN * MC 2303
* MEFENSINA * MEPHATE * MEPHEDAN
* MEPHELOR * MEPHENSIN * MEPHOSAL
* MEPHSON * 3-(2-METHYLPHENOXY)-1,2-PROPAN-
EDIOL * MIANESINA * MOCTYNOL * MYANIL
* MYCOCURAN * MYODETENSINE * MYOLAX
* MYOPAN * MYOSEROL * MYOXANE
* NEMBUSEN * NEPHELOR * ORANIXON
* ORTOL * PROLAX * PROLOXIN * RELAX-
ANT * RELAXAR * RENARCOL * REX REGU-
LANS * RP 3602 * SANSDOLOR * SECONESINZ
* SINAN * SPARTOLOXYN * SQ 1156 * STIL-
ALGIN * THIOXIDIL * TOLANSIN * TOLCIL
* TOLOFREN * 3-o-TOLOXY-1,2-PROPANEDIOL
* TOLSEROL * TOLULOX * TOLYDRIN
* 1-o-TOLYLGLYCEROL ETHER * α-(o-TOLYL)GLY-
CERYL ETHER * 3-(o-TOLYLOXY)PROPANE-1,2-DIOL
* TOLYNOL * TOLYSPAZ * TORULOX
* WALKO-NESIN * XERAL

SAFETY PROFILE: Poison by intraperitoneal,
intravenous, and subcutaneous routes. Moderately toxic by ingestion. When heated to decomposition it emits acrid smoke and irritating
fumes.

GGW000   CAS: 765-34-4   **HR: 3**
GLYCIDALDEHYDE

DOT: UN 2622
mf: C₃H₄O₂   mw: 72.07

PROP: Colorless liquid. Bp: 113°, d: 1.1403
@ 20°/4°.

SYNS: EPIHYDRINALDEHYDE * EPIHYDRINE AL-
DEHYDE * 2,3-EPOXYPROPANAL * 2,3-EPOXY-1-

PROPANAL * 2,3-EPOXYPROPIONALDEHYDE
* GLYCIDAL * GLYCIDYLALDEHDYE * OXI-
RANE-CARBOXALDEHYDE * RCRA WASTE NUMBER
U126

CONSENSUS REPORTS: IARC Cancer Review: GROUP 2B IMEMDT 7,56,87; Animal Sufficient Evidence IMEMDT 11,175,76. EPA Genetic Toxicology Program.

DOT Classification: Flammable or Combustible Liquid; Label: Flammable and Poison.

SAFETY PROFILE: Suspected carcinogen with experimental carcinogenic, neoplastigenic, and tumorigenic data. Poison by ingestion, skin contact, intraperitoneal, and intravenous routes. Moderately toxic by inhalation. Human systemic effects by inhalation: changes in central nervous system electrical activity, olfactory changes, and excitement. Mutation data reported. A human eye irritant. Powerful skin sensitizer and mucous membrane irritant. Flammable when exposed to heat, flame, or oxidizing materials. When heated to decomposition it emits acrid smoke and irritating fumes.

GGW500   CAS: 556-52-5   **HR: 3**
GLYCIDOL
mf: $C_3H_6O_2$ mw: 74.09

PROP: Colorless liquid. D: 1.165 @ 0°/4°, bp: 167° (decomp). Entirely sol in water, alc, and ether.

SYNS: EPIHYDRIN ALCOHOL * 2,3-EPOXYPROPA-
NOL * 2,3-EPOXY-1-PROPANOL * GLYCIDE
* GLYCIDYL ALCOHOL * 3-HYDROXY-1,2-EPOXY-
PROPANE * NCI-C55549 * OXIRANEMETHANOL
* OXIRANYLMETHANOL

CONSENSUS REPORTS: Reported in EPA TSCA Inventory. EPA Genetic Toxicology Program.

OSHA PEL: (Transitional: TWA 50 ppm) TWA 25 ppm
ACGIH TLV: TWA 25 ppm
DFG MAK: 50 ppm (150 mg/m^3)

SAFETY PROFILE: Poison by intraperitoneal route. Moderately toxic by ingestion, inhalation, and skin contact. Experimental teratogenic and reproductive effects. Human mutation data reported. A skin irritant. Animal experiments suggest somewhat lower toxicity than related epoxy compounds. Readily absorbed through the skin. Causes nervous excitation followed by depres-

sion. Explodes when heated or in the presence of strong acids, bases, metals (e.g., copper, zinc) and metal salts (e.g., aluminum chloride, iron(III) chloride, tin(IV) chloride). When heated to decomposition it emits acrid smoke and fumes.

GHA000   CAS: 56-40-6   **HR: 2**
GLYCINE
mf: $C_2H_5NO_2$ mw: 75.08

PROP: The simplest amino acid and the principal amino acid in sugar cane. White crystals; odorless, sweet taste. Mp: 232-236° (decomp), d: 1.1607. Sol in water; insol in alc and ether.

SYNS: AMINOACETIC ACID * GLYCOLIXIR
* HAMPSHIRE GLYCINE

CONSENSUS REPORTS: Reported in EPA TSCA Inventory.

SAFETY PROFILE: Moderately toxic by intravenous route. Mildly toxic by ingestion. When heated to decomposition it emits toxic fumes of NO_x.

GHE000   CAS: 2619-97-8   **HR: 3**
GLYCINE NITROGEN MUSTARD
mf: $C_6H_{11}Cl_2NO_2 \cdot ClH$ mw: 236.54

SYNS: N,N-BIS(β-CHLOROETHYL)GLYCINE HYDRO-
CHLORIDE * N,N-BIS(2-CHLOROETHYL)GLYCINE HY-
DROCHLORIDE * GLYCINE MUSTARD * NSC 17661

SAFETY PROFILE: A deadly poison by intraperitoneal, intravenous, and intracerebral routes. When heated to decomposition it emits very toxic fumes of Cl$^-$ and NO_x.

GIA000   CAS: 36734-19-7   **HR: 2**
GLYCOPHEN
mf: $C_{13}H_{13}Cl_2N_3O_3$ mw: 330.19

SYNS: CHIPCO 26019 * 3-(3,5-DICHLOROPHENYL)-
N-(1-METHYLETHYL)-2,4-DIOXO-1-IMIDAZOLIDINECAR-
BOXAMIDE * GLYCOPHENE * IPRODIONE
* 1-ISOPROPYL CARBAMOYL-3-(3,5-DICHLOROPHE-
NYL)-HYDANTOIN * LFA 2043 * MRC 910
* PROMIDIONE * ROP 500 F * ROVRAL
* RP 26019

SAFETY PROFILE: Moderately toxic by ingestion. When heated to decomposition it emits very toxic fumes of NO_x and Cl$^-$.

GIC000   CAS: 596-51-0   **HR: 3**
GLYCOPYRRONIUM BROMIDE
mf: $C_{19}H_{28}NO_3 \cdot Br$ mw: 398.39

PROP: Crystals from butanone. Mp: 193.2-194.5°.

SYNS: ASECRYL * 1,1-DIMETHYL-3-HYDROXYPYR-ROLIDINIUM BROMIDE-α-CYCLOPENTYLMANDELATE * GASTRODYN * GLYCOPYRROLATE * GLYCO-PYRROLATE BROMIDE * NODAPTON * ROBANUL * ROBINUL * TARODYL * TARODYN

SAFETY PROFILE: Poison by intravenous and intraperitoneal routes. Moderately toxic by ingestion and subcutaneous routes. Experimental reproductive effects. When heated to decomposition it emits very toxic fumes of NO_x and Br^-.

GIK000   CAS: 107-22-2   **HR: 3**
GLYOXAL
mf: $C_2H_2O_2$ mw: 58.04

PROP: Yellow prisms or irregular pieces turning white on cooling. D: 1.14 @ 20°/4°. Opaque @ 10°, mp: 15°, bp: (776) 51°. The vapors are green and burn with a purple flame, n (20.5/D) 1.3826. Sol in anhyd solvents, pH of a 40% aq soln: 2.1-2.7; d: (20/4) 1.27.

SYNS: AEROTEX GLYOXAL 40 * BIFORMAL * BIFORMYL * DIFORMAL * ETHANDIAL * ETHANEDIAL * 1,2-ETHANEDIONE * GLYOXYLALDEHYDE * OXAL * OXALALDE-HYDE

CONSENSUS REPORTS: Reported in EPA TSCA Inventory.

SAFETY PROFILE: Poison by intraperitoneal route. Moderately toxic by ingestion and skin contact. Mutation data reported. A skin and severe eye irritant. A powerful reducing agent. May explode on contact with air. Polymerizes violently on contact with water. During storage it may spontaneously polymerize and ignite. Reacts violently with chlorosulfonic acid, ethylene imine, HNO_3, oleum, NaOH, can cause violent reactions. Can explode during manufacture. When heated to decomposition it emits acrid smoke and irritating fumes.

GIS000   CAS: 7440-57-5   **HR: 1**
GOLD
af: Au   aw: 196.97

PROP: Cubic, yellow, ductile, metallic crystals. Mp: 1064.76°, bp: 2700°, d: 19.3 (liquid)

17.0 @ 1063°, vap press: 1 mm @ 1869°, Hardness (Mohs') 2.5-3.0; (Brinell's) 18.5.

SYNS: BURNISH GOLD * COLLOIDAL GOLD * GOLD FLAKE * GOLD LEAF * GOLD POWDER * MAGNESIUM GOLD PURPLE * SHELL GOLD

CONSENSUS REPORTS: Reported in EPA TSCA Inventory.

SAFETY PROFILE: Poison by intravenous route. Experimental reproductive effects. Questionable carcinogen with experimental tumorigenic data by implantation. Can form explosive compounds with NH_3, NH_4OH + aqua regia, H_2O_2. Incompatible with mixtures containing chlorides, bromides, or iodides (if they can generate nascent halogens), some oxidizing materials (especially those containing halogens), alkali cyanides, thiocyanate solutions, and double cyanides.

GJC000   CAS: 12244-57-4   **HR: 3**
GOLD SODIUM THIOMALATE
mf: $C_4H_3AuO_4S \cdot 2Na$ mw: 390.08

SYNS: AuTM * ((1,2-DICARBOXYETHYL)THIO) GOLD DISODIUM SALT * (DIHYDROGEN MERCAPTO-SUCCINATO)GOLD DISODIUM SALT * DISODIUM AUROTHIOMALATE * (MERCAPTOBUTANEDIOATO (1-))GOLD DISODIUM SALT * MERCAPTOSUCCINIC ACID, GOLD SODIUM SALT * MYOCHRYSINE * MYOCRISIN * SODIUM AUROTHIOMALATE * TAURE(o)DON

SAFETY PROFILE: Poison by subcutaneous and intramuscular routes. Moderately toxic by intravenous route. Human systemic effects by ingestion, intramuscular, parenteral, and possibly other routes: flaccid paralysis without anesthesia, recording from peripheral motor nerve, somnolence, muscle weakness, aggression, renal function tests depressed, dermatitis, proteinuria, hemorrhage, thrombocytopenia, cell count changes, aplastic anemia, and changes in blood, teeth, and supporting structures. Experimental teratogenic and reproductive effects. When heated to decomposition it emits very toxic Na_2O and SO_x.

GJG000   CAS: 10210-36-3   **HR: 3**
GOLD SODIUM THIOSULFATE DIHYDRATE
mf: $O_6S_4 \cdot Au \cdot 3Na \cdot 2H_2O$ mw: 526.22

SYNS: AURICIDINE * AUROCIDIN * AUROLIN
* AUROPEX * AUROPIN * AUROSAN * AURO-
THION * CRISALBINE * NOVACRYSIN * SANO-
CHRYSINE * SODIUM AUROTHIOSULPHATE DIHY-
DRATE * SOLFOCRISOL * THIOCHRYSINE
* THIOSULFURIC ACID, GOLD(1+) SODIUM SALT
(2:1:3), DIHYDRATE

SAFETY PROFILE: Poison by subcutaneous,
intraperitoneal, intravenous, and intramuscular
routes. Human systemic effects by intramuscular
route: dermatitis, granulocytopenia and thrombocytopenia. Used as an antirheumatic agent.
When heated to decomposition it emits very
toxic fumes of SO_x and Na_2O.

GJM000   CAS: 303-45-7   HR: 2
GOSSYPOL
mf: $C_{30}H_{30}O_8$ mw: 518.60

PROP: A polyphenolic yellow pigment isolated
from cottonseed pigment glands. Mp: 180° from
ether; mp: 199° from chloroform, mp: 214° from
ligroin; very sltly sol in methanol, ethanol,
ether, chloroform, DMF; freely sol (with slow
decomp) in dilute solns of ammonia, sodium
carbonate. Insol in water.

SYNS: 2,2'-BIS(1,6,7-TRIHYDROXY-3-METHYL-5-ISO-
PROPYL-8-ALDEHYDONAPHTHALENE * 8-FORMYL-
1,6,7-TRIHYDROXY-5-ISOPROPYL-3-METHYL-2,2'-BIS-
NAPHTHALENE

CONSENSUS REPORTS: EPA Genetic Toxicology Program.

SAFETY PROFILE: Moderately toxic by ingestion. Human reproductive effects by ingestion:
spermatogenesis and male fertility index
changes. Experimental reproductive effects. Human mutation data reported. Can be irritating
to the gastrointestinal tract. In experimental animals, large doses cause edema of lungs, shortness of breath, paralysis. When heated to decomposition it emits acrid smoke and irritating
fumes.

GJO000   CAS: 1405-97-6   HR: 3
GRAMICIDIN
mf: $C_{148}H_{210}N_{30}O_{26}$ mw: 2825.88

PROP: Spear-shaped or lenticular platelets. Mp:
229-230°. Almost insol in water; sol in lower
alc, acetic acid, and pyridine; practically insol
in ether or hydrocarbons.

SAFETY PROFILE: Poison by intravenous, intraperitoneal, and parenteral routes. An antibi-

otic. When heated to decomposition it emits
very toxic fumes of NO_x.

GJU000   CAS: 8016-20-4   HR: 3
GRAPEFRUIT OIL

PROP: From the fresh peel of *Citrus paradisi*
Macfayden (*Citrus decumana* L.). Yellow liquid. Sol in fixed oils, mineral oil; sltly sol in
propylene glycol; insol in glycerin.

SYNS: GRAPEFRUIT OIL, coldpressed * GRAPEFRUIT
OIL, expressed * OIL of GRAPEFRUIT * OIL of SHAD-
DOCK

CONSENSUS REPORTS: Reported in EPA
TSCA Inventory.

SAFETY PROFILE: A skin irritant. Questionable carcinogen with experimental tumorigenic
data. Mutation data reported. When heated to
decomposition it emits acrid smoke and irritating
fumes.

GKC000   CAS: 53216-90-3   HR: 3
GRISEOVIRIDIN
mf: $C_{22}H_{27}N_3O_7S$ mw: 477.58

PROP: Polymorphic crystals. Decomp @ 158-
166°, 194-200° or 230-240°, depending on the
crystal modification. Sol in pyridine; mod sol
in lower alcs; very sltly sol in water and nonpolar
solvents.

SAFETY PROFILE: Poison by intraperitoneal,
subcutaneous, and intravenous routes. When
heated to decomposition it emits very toxic
fumes of SO_x and NO_x.

GKE000   CAS: 126-07-8   HR: 3
GRISOFULVIN
mf: $C_{17}H_{17}ClO_6$ mw: 352.79

SYNS: AMUDANE * BIOGRISIN-FP * ,7-CHLORO-
4,6,2'-TRIMETHOXY-6'-METHYLGRIS-2'-EN-3,4'-DIONE
* DELMOFULVINA * FULCIN * FULCINE
* FULVICAN GRISACTIN * FULVICIN * FULVINA
* FULVISTATIN * FUNGIVIN * GREOSIN
* GRESFEED * GRICIN * GRIFULVIN * GRIS-
ACTIN * GRISCOFULVIN * GRISEFULINE
* GRISEO * (+)-GRISEOFULVIN * GRISEOFUL-
VIN-FORTE * GRISEOFULVINUM * GRISETIN
* GRISOVIN * GRIS-PEG * GRYSIO * GUSER-
VIN * LAMORYL * LIKUDEN * MURFULVIN
* NEO-FULCIN * NSC 34533 * PONCYL
* SPIROFULVIN * SPOROSTATIN * USAF SC-2

CONSENSUS REPORTS: IARC Cancer Review: GROUP 2B IMEMDT 7,56,87; Animal

Sufficient Evidence IMEMDT 10,153,76. EPA Genetic Toxicology Program.

SAFETY PROFILE: Suspected carcinogen with experimental neoplastigenic data. Poison by intravenous and intraperitoneal routes. Moderately toxic by subcutaneous route. Human mutation data reported. Experimental teratogenic and reproductive effects. Used as a antibiotic, pharmaceutical and veterinary drug. When heated to decomposition it emits toxic fumes of Cl⁻.

GKI000 CAS: 90-05-1 **HR: 3**
GUAIACOL
mf: $C_7H_8O_2$ mw: 124.15

PROP: Clear, pale yellow liquid or solid. Characteristic odor, darkens on exposure to air and light. D (crystals): 1.129, d (liquid): about 1.112. Misc with alc, chloroform, ether, oils, glacial acetic acid. Sltly sol in petroleum ether; sol in NaOH soln, mp: 28°, bp: 202-209°, flash p: 180°F (OC), d: 1.097 @ 25°/25°.

SYNS: GUAICOL * o-HYDROXYANISOLE * 2-HYDROXYANISOLE * 1-HYDROXY-2-METHOXY-BENZENE * o-METHOXYPHENOL * 2-METHOXYPHENOL * METHYLCATECHOL * PYROGUAIAC ACID

CONSENSUS REPORTS: Reported in EPA TSCA Inventory. EPA Genetic Toxicology Program.

SAFETY PROFILE: Human poison by ingestion. Experimental poison by intravenous and subcutaneous routes. Mildly toxic by skin contact and inhalation. Human systemic effects by ingestion: tremors and gastrointestinal changes. Human mutation data reported. An eye and severe skin irritant. Ingestion produces burning in the mouth and throat. Flammable when exposed to heat or flame; can react with oxidizing materials. To fight fire, use foam, CO_2, dry chemical. Protect from light. Used as an expectorant. When heated to decomposition it emits acrid smoke and irritating fumes.

GKU000 CAS: 645-43-2 **HR: 3**
GUANETHIDINE MONOSULFATE
mf: $C_{10}H_{22}N_4 \cdot H_2O_4S$ mw: 296.44

SYNS: N-(2-GUANIDINO ETHYL)HEPTAMETHYLENIMINE SULFATE * (2-(HEXAHYDRO-1(2H)-AZOCINYL) ETHYL) GUANIDINE HYDROGEN SULFATE * 2-(OCTAHYDRO-1-AZOCINYL)ETHYL GUANIDINE SULPHATE

SAFETY PROFILE: Poison by intraperitoneal and intravenous routes. Moderately toxic by ingestion. A human eye irritant. When heated to decomposition it emits very toxic fumes of SO_x and NO_x.

GKW000 CAS: 113-00-8 **HR: 3**
GUANIDINE
mf: CH_5N_3 mw: 59.09

PROP: Hygroscopic, colorless crystals. Mp: 50°. Very sol in water; sol in alc.

SYNS: AMINOFORMAMIDINE * AMINOMETHANAMIDINE * CARBAMAMIDINE * CARBAMIDINE * IMINOUREA

SAFETY PROFILE: Poison by intraperitoneal and subcutaneous routes. Moderately toxic by ingestion. On heating to 160° it converts to melamine and NH_3. Keep well closed. When heated to decomposition it emits toxic fumes of NO_x.

GLA000 CAS: 506-93-4 **HR: 3**
GUANIDINE MONONITRATE
DOT: UN 1467
mf: $CH_5N_3 \cdot HNO_3$ mw: 122.11

PROP: White granules. Mp: 214°.

SYN: GUANIDINE NITRATE (DOT)

CONSENSUS REPORTS: Reported in EPA TSCA Inventory.

DOT Classification: Oxidizer; Label: Oxidizer.

SAFETY PROFILE: A powerful oxidizer. Flammable when shocked or exposed to heat or flame. A stable, flashless, non-hygroscopic high explosive used as a blasting explosive in combination with charcoal and inorganic nitrates. Keep away from heat and open flame. When heated to decomposition it emits very toxic fumes of HNO_3 and NO_x.

GLS800 CAS: 5550-12-9 **HR: 2**
GUANYLIC ACID SODIUM SALT
mf: $C_{10}H_{14}N_5O_8P \cdot 2Na$ mw: 409.24

PROP: Colorless to white crystals; characteristic taste. Sol in water; sltly sol in alc; insol in ether.

SYNS: DISODIUM GMP * DISODIUM GUANYLATE (FCC) * DISODIUM-5'-GMP * DISODIUM-5'-

GUANYLATE * GMP DISODIUM SALT * 5'-GMP DISODIUM SALT * GMP SODIUM SALT * SODIUM GMP * SODIUM GUANOSINE-5'-MONOPHOSPHATE * SODIUM GUANYLATE * SODIUM-5'-GUANYLATE

CONSENSUS REPORTS: Reported in EPA TSCA Inventory.

SAFETY PROFILE: Moderately toxic by intraperitoneal, subcutaneous, and intravenous routes. Mildly toxic by ingestion. Mutation data reported. When heated to decomposition it emits toxic fumes of PO_x, NO_x and Na_2O.

GLU000   CAS: 9000-30-0   HR: 1
GUAR GUM

PROP: Yellowish-white powder, dispersible in hot or cold water, obtained from the ground endosperms of *Cyanopsis tetragonoloan* L. Taub (Fam. *Leguminosae*). White powder; odorless. Sol in water; insol in oils, grease, hydrocarbons, ketones, esters.

SYNS: A-20D * BURTONITE V-7-E * CYAMOPSIS GUM * DEALCA TP1 * DECORPA * GALACTASOL * GENDRIV 162 * GUAR * GUAR FLOUR * GUM CYAMOPSIS * GUM GUAR * INDALCA AG * JAGUAR NO. 124 * JAGUAR GUM A-20-D * JAGUAR PLUS * LYCOID DR * NCI-C50395 * REGONOL * REIN GUARIN * SUPER-COL U POWDER * SYNGUM D 46D * UNI-GUAR

CONSENSUS REPORTS: NTP Carcinogenesis Bioassay (feed); No Evidence: mouse, rat NTPTR* NTP-TR-229,82. Reported in EPA TSCA Inventory. EPA Genetic Toxicology Program.

SAFETY PROFILE: Mildly toxic by ingestion. When heated to decomposition it emits acrid smoke and irritating fumes.

GLY000   CAS: 9000-28-6   HR: 1
GUM GHATTI

PROP: The gummy exudation from the stem of *Anogeissus latifolia*. Colorless to pale yellow tears; almost odorless. Sltly sol in water.

SYN: INDIAN GUM

CONSENSUS REPORTS: Reported in EPA TSCA Inventory.

SAFETY PROFILE: Mildly toxic by ingestion. When heated to decomposition it emits acrid smoke and irritating fumes.

GLY100   CAS: 9000-29-7   HR: 2
GUM GUAIAC

PROP: From wood of *guajacum officinale* L. or *Guajacum sanctum* L. (Fam. *Zygophyllaceae*). Brown solid; balsamic odor, sltly acrid taste. Sol in alc, ether, chloroform, solns of alkalies; sltly sol in carbon disulfide, benzene.

SYN: GUAIAC GUM

SAFETY PROFILE: Moderately toxic by ingestion. When heated to decomposition it emits acrid smoke and irritating fumes.

GMG000   CAS: 639-14-5   HR: 3
GYPSOGENIN
mf: $C_{30}H_{46}O_4$ mw: 470.76

PROP: Needles or leaflets from methanol. Mp: 274-276°.

SYNS: ALBSAPOGENIN * ASTRANTIAGENIN D * GITHAGENIN * GYPSOPHILASAPOGENIN * GYPSOPHILASAPONIN * 3-β-HYDROXY-23-OXO-OLEAN-12-EN-28-OIC ACID * SAPONIN-GYPSOPHILA

SAFETY PROFILE: Poison by subcutaneous and intravenous routes. Moderately toxic by ingestion. When heated to decomposition it emits acrid smoke and irritating fumes.

H

HAC000   CAS: 7440-58-6   **HR: 3**
HAFNIUM
af: Hf   aw: 178.49

DOT: UN 1326/UN 2545

PROP: A silvery, ductile, lustrous metal. Mp: 2227°, bp: 4602°, d: 13.31 @ 20°.

SYNS: HAFNIUM METAL, dry (DOT)   *   HAFNIUM METAL, wet (DOT)   *   HAFNIUM, wet with not less than 25% water (DOT)

CONSENSUS REPORTS: Reported in EPA TSCA Inventory.

OSHA PEL: TWA 0.5 mg/m^3
ACGIH TLV: TWA 0.5 mg/m^3
DFG MAK: 0.5 mg/m^3
DOT Classification: Flammable Solid; Label: Flammable Solid, dry and wet; IMO: Flammable Solid; Label: Spontaneously Combustible, dry.

SAFETY PROFILE: A poison by an unspecified route. It is poorly soluble in water and thus is not absorbed efficiently by ingestion. Many hafnium compounds are poisons. Dangerous fire hazard. The powder ignites with friction, heat, sparks, or exposure to air. The damp powder burns explosively. The powder may self-explode. The powder can explode when heated with nitrogen, phosphorus, oxygen, sulfur, nonmetals, or halogens. May explode on contact with hot nitric acid and other oxidants.

HAD500   CAS: 13759-17-6   **HR: 3**
HAFNIUM OXYCHLORIDE
mf: Cl$_2$HfO   mw: 265.39

SYN: HAFNIUM CHLORIDE OXIDE

CONSENSUS REPORTS: Reported in EPA TSCA Inventory.

SAFETY PROFILE: A poison by intraperitoneal route. When heated to decomposition it emits very toxic fumes of Cl$^-$.

HAE500   CAS: 12116-66-4   **HR: 3**
HAFNOCENE DICHLORIDE
mf: C$_{10}$H$_{10}$Cl$_2$Hf   mw: 379.59

SYNS: DICHLOROBIS(eta-CYCLOPENTADIENYL)HAFNIUM   *   DICHLORODICYCLOPENTADIENYLHAFNIUM   *   DICHLORODI-pi-CYCLOPENTADIENYLHAFNIUM

*   DICYCLOPENTADIENYLHAFNIUM DICHLORIDE
*   HAFNIUM DICYCLOPENTADIENE DICHLORIDE

CONSENSUS REPORTS: Reported in EPA TSCA Inventory.

SAFETY PROFILE: A poison by intravenous route. When heated to decomposition it emits very toxic fumes such as Cl$^-$.

HAG500   CAS: 151-67-7   **HR: 3**
HALOTHANE
mf: C$_2$HBrClF$_3$ mw: 197.39

PROP: Nonflammable, highly volatile liquid; characteristic sweetish, not unpleasant odor. D: 1.871 @ 20°/4°, bp: 50.2°, 20° @ 243 mm. Sensitive to light. Miscible with petr. ether, other fat solvents; sltly sol in water.

SYNS: BROMOCHLOROTRIFLUOROETHANE
*   2-BROMO-2-CHLORO-1,1,1-TRIFLUOROETHANE
*   CHALOTHANE   *   FLUOROTANE   *   FLUOTHANE
*   FTOROTAN (RUSSIAN)   *   HALOTAN   *   HALSAN
*   NARCOTANE   *   NARCOTANN NE-SPOFA (RUSSIAN)
*   1,1,1-TRIFLUORO-2-BROMO-2-CHLOROETHANE
*   1,1,1-TRIFLUORO-2-CHLORO-2-BROMOETHANE
*   2,2,2-TRIFLUORO-1-CHLORO-1-BROMOETHANE

CONSENSUS REPORTS: EPA Genetic Toxicology Program.

ACGIH TLV: TWA 50 ppm
DFG MAK: 5 ppm (40 mg/m^3); BAT: 250 μg/dL of trifluoroacetic acid in blood at end of shift.
NIOSH REL: (Waste Anesthetic Gases and Vapors) CL 2 ppm/1H

SAFETY PROFILE: A human poison by ingestion and intravenous routes. Human systemic effects by intravenous route: general anesthetic, heart rate change, cyanosis, by inhalation: hepatitis, nausea, fever. Experimental teratogenic and reproductive effects. A severe eye irritant. Human mutation data reported. Used as a clinical anesthetic. When heated to decomposition it emits very toxic fumes of F$^-$, Cl$^-$, and Br$^-$.

HAH000   CAS: 13382-33-7   **HR: 3**
HALVISOL
mf: C$_{21}$H$_{27}$FN$_2$O$_2$ mw: 358.50

SYNS: ANISOPIROL   *   (+-)-α-(p-FLUOROPHENYL)-4-(o-METHOXYPHENYL)-1-PIPERAZINEBUTANOL

* dl-1-(4-FLUOROPHENYL)-4-(1-(4-(2-METHOXY-PHE-
NYL))-PIPERAZINYL)BUTANOL

SAFETY PROFILE: A poison by subcutaneous and intravenous routes. When heated to decomposition it emits very toxic fumes of F^- and NO_x.

HAI500   CAS: 442-51-3   **HR: 3**
HARMINE
mf: $C_{13}H_{12}N_2O$ mw: 212.27

PROP: An alkaloid isolated from *Banisteria caapi sp.*, a South American narcotic agent.

SYNS: BANISTERINE * LEUCOHARMINE
* 6-METHOXYHARMAN * 7-METHOXY-1-METHYL-
9H-PYRIDO(3,4-b)INDOLE * 1-METHYL-7-METHOXY-β-
CARBOLINE * TELEPATHINE * YAGEINE
* YAJEINE

SAFETY PROFILE: Poison by intravenous and subcutaneous routes. Human systemic effects by intramuscular route: sleep disturbance, tremors, nausea. When heated to decomposition it emits toxic fumes of NO_x.

HAJ500   CAS: 34465-46-8   **HR: 3**
HCDD
mf: $C_{12}H_2Cl_6O_2$ mw: 390.84

PROP: Colorless solid. Mp: 239°.

SYNS: HEXACHLORODIBENZO-p-DIOXIN * 1,2,3,6,
7,8-HEXACHLORODIBENZO-p-DIOXIN

CONSENSUS REPORTS: IARC Cancer Review: Animal Inadequate Evidence IMEMDT 15,41,77.

SAFETY PROFILE: A deadly poison by ingestion. An experimental teratogen. An eye irritant. When heated to decomposition it emits toxic fumes of Cl^-.

HAK000   CAS: 7789-20-0   **HR: D**
HEAVY WATER
mf: D_2O mw: 20.02

PROP: Mp: 3.81°, triple point temp 3.82°, bp: 101.42°. Critical temp 371.5°, d: 1.1044. Heat is evolved on mixing with normal water.

SYNS: DEUTERIUM OXIDE * DIDEUTERIUM OXIDE
* HEAVY WATER-d2 * WATER-d2 (9CI)
* WATER²-H2

CONSENSUS REPORTS: EPA Genetic Toxicology Program. Reported in EPA TSCA Inventory.

SAFETY PROFILE: Experimental reproductive effects.

HAL500   CAS: 303-33-3   **HR: 3**
HELIOTRINE
mf: $C_{16}H_{27}NO_5$ mw: 313.44

SYN: HELIOTRON

CONSENSUS REPORTS: EPA Genetic Toxicology Program.

SAFETY PROFILE: A poison by ingestion, intravenous, intraperitoneal, and possibly other routes. Experimental teratogenic and reproductive effects. Questionable carcinogen with experimental tumorigenic data. Human mutation data reported. When heated to decomposition it emits toxic fumes of NO_x.

HAM500   CAS: 7440-59-7   **HR: 1**
HELIUM
DOT: UN 1046/UN 1963
af: He   aw: 4.00

PROP: Colorless, odorless, tasteless, inert gas. Mp: −272.2° @ 26 atm, bp: −268.9°, d: (gas): 0.1785 g/L @ 0°, d: (liquid): 0.147 @ −270.8°.

SYNS: HELIUM, compressed (DOT) * HELIUM, refrigerated liquid (DOT)

CONSENSUS REPORTS: Reported in EPA TSCA Inventory.

DOT Classification: Nonflammable Gas; Label: Nonflammable Gas.

SAFETY PROFILE: A simple asphyxiant. Nonflammable Gas.

HAN000   CAS: 58933-55-4   **HR: 1**
HELIUM-OXYGEN (mixture)
DOT: NA 1980
mf: HeO_2 mw: 36.00

SYN: HELIUM-OXYGEN mixture (DOT)

DOT Classification: Nonflammable Gas; Label: Nonflammable Gas.

SAFETY PROFILE: A simple asphyxiant.

HAO000   CAS: 13495-01-7   **HR: 3**
HELVETICOSIDE DIHYDRATE
mf: $C_{29}H_{42}O_9 \cdot 2H_2O$ mw: 570.75

PROP: Needles from oil, methanol. Mp: 153-157°.

SYNS: ALLEOSIDE A DIHYDRATE * ERISIMIN DI-
HYDRATE * ERYSIMIN DIHYDRATE

SAFETY PROFILE: A deadly poison by intravenous route. When heated to decomposition it emits acrid smoke and fumes.

HAO875   CAS: 1317-60-8   HR: 3
HEMATITE

PROP: Consists mainly of Fe_2O_3.

SYNS: BLOOD STONE * HAEMATITE * IRON ORE * RED IRON ORE

CONSENSUS REPORTS: NTP Fourth Annual Report on Carcinogens. IARC Cancer Review: Group 3, Indefinite IMSUPP 4,254,82. Reported in EPA TSCA Inventory.

SAFETY PROFILE: Confirmed carcinogen.

HAQ000   CAS: 312-45-8   HR: 3
HEMICHOLINIUM-3-DIBROMIDE
mf: $C_{24}H_{34}N_2O_4 \cdot 2Br$ mw: 574.42

SYNS: 2,2'-(1,1'-BIPHENYL-4,4'-DIYLBIS(2-HYDROXY-4,4-DIMETHYL-MORPHOLINIUM DIBROMIDE * HC-3 * HEMICHOLINE * HEMICHOLINIUM-3 * HEMICHOLINIUM BROMIDE * HEMICHOLINIUM-3-BROMIDE * HEMICHOLINIUM DIBROMIDE

CONSENSUS REPORTS: Reported in EPA TSCA Inventory.

SAFETY PROFILE: A poison by subcutaneous, intraperitoneal, intravenous, and possibly other routes. When heated to decomposition it emits very toxic fumes of NO_x and Br^-.

HAQ500   CAS: 9005-49-6   HR: 2
HEPARIN

SYNS: α-HEPARIN * HEPARINATE * HEPARINIC ACID * HEPARIN SULFATE * LIPO-HEPIN * LIQUAEMIN * LIQUEMIN * NOVOHEPARIN * SUBLINGULA * THROMBOLIQUINE * VETREN * VITAMIN AB * VITRUM AB

SAFETY PROFILE: Moderately toxic by ingestion, intraperitoneal, and intravenous routes. Mutation data reported. When heated to decomposition it emits toxic fumes of NO_x.

HAR000   CAS: 76-44-8   HR: 3
HEPTACHLOR

DOT: NA 2761
mf: $C_{10}H_5Cl_7$ mw: 373.30

PROP: Crystals. Mp: 96°. Nearly insol in water; sol in organic solvents.

SYNS: AGROCERES * 3-CHLOROCHLORDENE * DRINOX * E 3314 * ENT 15,152 * EPTA-CLORO (ITALIAN) * 1,4,5,6,7,8,8-EPTACLORO-3a,4,7,7a-TETRAIDRO-4,7-endo-METANO-INDENE (ITALIAN) * GPKh * H-34 * HEPTACHLOOR (DUTCH) * 1,4,5,6,7,8,8-HEPTACHLOOR-3a,4,7,7a-TETRAHYDRO-4,7-endo-METHANO-INDEEN (DUTCH) * HEPTACHLORE (FRENCH) * 3,4,5,6,7,8,8-HEPTACHLORO-DICYCLOPENTADIENE * 3,4,5,6,7,8,8a-HEPTACHLORODICYCLOPENTADIENE * 1,4,5,6,7,10,10-HEPTACHLORO-4,7,8,9,-TETRAHYDRO-4,7-ENDO-METHYLENEINDENE * 1,4,5,6,7,8,8-HEPTACHLORO-3a,4,7,7a-TETRAHYDRO-4,7-ENDOMETHANOINDENE * 1,4,5,6,7,8,8a-HEPTACHLORO-3a,4,7,7a-TETRAHYDRO-4,7-METHANOINDANE * 1,4,5,6,7,8,8-HEPTACHLORO-3a,4,7,7a-TETRAHYDRO-4,7-METHANOINDENE * 1(3a),4,5,6,7,8,8-HEPTACHLORO-3a(1),4,7,7a-TETRAHYDRO-4,7-METHANOINDENE * 1,4,5,6,7,8,8-HEPTACHLORO-3a,4,7,7a-TETRAHYDRO-4,7-METHANOL-1H-INDENE * 1,4,5,6,7,8,8-HEPTACHLORO-3a,4,7,7,7a-TETRAHYDRO-4,7-METHYLENE INDENE * 1,4,5,6,7,8,8-HEPTACHLOR-3a,4,7,7,7a-TETRAHYDRO-4,7-endo-METHANO-INDEN (GERMAN) * HEPTAGRAN * HEPTAMUL * NCI-C00180 * RCRA WASTE NUMBER P059 * RHODIACHLOR * VELSICOL 104

CONSENSUS REPORTS: IARC Cancer Review: GROUP 3 IMEMDT 7,146,87; Human Inadequate Evidence IMEMDT 20,129,79; Animal Inadequate Evidence IMEMDT 5,173,74; Animal Sufficient Evidence IMEMDT 20,-129,79. NCI Carcinogenesis Bioassay (feed) Clear Evidence: Mouse NCITR* NCI-CG-TR-9,77; Results negative: rat NCITR* NCI-CG-TR-9,77. EPA Genetic Toxicology Program. Community Right-To-Know List.

OSHA PEL: TWA 0.5 mg/m³ (skin)
ACGIH TLV: TWA 0.5 mg/m³ (skin)
DFG MAK: 0.5 mg/m³, Suspected Carcinogen.
DOT Classification: ORM-E; Label; None.

SAFETY PROFILE: Suspected carcinogen with experimental carcinogenic data. A poison by ingestion, skin contact, intraperitoneal, intravenous, and possibly other routes. Human mutation data reported. Acute exposure and chronic doses have caused liver damage. In man, a dose of 1-3 grams can cause serious symptoms, especially where liver impairment is the case. Acute symptoms include tremors, convulsions, kidney damage, respiratory collapse, and death. When heated to decomposition it emits toxic fumes of Cl^-.

NOTE: The EPA has canceled registration of pesticides containing heptachlor with the exception of its use for termite control by subsurface ground insertion external to the dwelling.

HAV450   CAS: 5910-85-0   **HR: 3**
2,4-HEPTADIENAL
mf: $C_7H_{10}O$ mw: 110.17

PROP: Slightly yellow liquid; green odor. Refr index: 1.478-1.480, flash p: 140°F. Sol in alc, fixed oils, water.

SYNS: FEMA No. 3164 * HEPTADIENAL-2,4 * trans,trans-2,4-HEPTDIENAL * 2,4-HEPTDIENAL

SAFETY PROFILE: Poison by skin contact. Moderately toxic by ingestion. A severe skin irritant. Combustible liquid. When heated to decomposition it emits acrid smoke and fumes.

HAX500   CAS: 375-22-4   **HR: 3**
HEPTAFLUOROBUTYRIC ACID
mf: $C_4HF_7O_2$ mw: 214.05

PROP: Colorless liquid; sharp, butyric acid odor. Bp: 210° @ 735 mm.

CONSENSUS REPORTS: Reported in EPA TSCA Inventory.

SAFETY PROFILE: A poison by intraperitoneal route. Probably an eye, skin, and mucous membrane irritant. Will react with water or steam to produce corrosive fumes. When heated to decomposition it emits toxic fumes of F^-.

HBB500   CAS: 111-71-7   **HR: 1**
HEPTANAL
mf: $C_7H_{14}O$ mw: 114.18

PROP: Colorless liquid; penetrating, fruity odor. D: 0.814-0.819, refr index: 1.412-1.420, • mp: −43.3°, bp: 152.8°, flash p: 93°F. Sol in alc, ether, fixed oils; sltly sol in water @ 153°; misc in alc, ether.

SYNS: ENANTHAL * ENANTHALDEHYDE * ENANTHOLE * FEMA No. 2540 * HEPTALDEHYDE * OENANTHALDEHYDE * OENANTHOL

CONSENSUS REPORTS: Reported in EPA TSCA Inventory.

SAFETY PROFILE: Mildly toxic by ingestion. Flammable liquid. When heated to decomposition it emits acrid smoke.

HBC500   CAS: 142-82-5   **HR: 3**
HEPTANE
DOT: UN 1206
mf: C_7H_{16} mw: 100.23

PROP: Colorless liquid. Bp: 98.52, lel: 1.05%, uel: 6.7%, fp: −90.5°, flash p: 25°F (CC), d: 0.684 @ 20°/4°, autoign temp: 433.4° F, vap press: 40 mm @ 22.3°, vap d: 3.45. Sltly sol in alc; misc in ether and chloroform; insol in water.

SYNS: DIPROPYL METHANE * EPTANI (ITALIAN) * GETTYSOLVE-C * HEPTAN (POLISH) * n-HEPTANE * HEPTANEN (DUTCH) * HEPTYL HYDRIDE

CONSENSUS REPORTS: Reported in EPA TSCA Inventory.

OSHA PEL: (Transitional: TWA 500 ppm) TWA 400 ppm; STEL 500 ppm
ACGIH TLV: TWA 400 ppm; STEL 500 ppm
DFG MAK: 500 ppm (2000 mg/m³)
NIOSH REL: TWA (Alkanes) 350 mg/m³
DOT Classification: Flammable Liquid; Label: Flammable Liquid.

SAFETY PROFILE: Poison by intravenous route. Human systemic effects by inhalation: hallucinations. Narcotic in high concentrations. A volatile, flammable liquid when exposed to heat or flame. Can react vigorously with oxidizing materials. Moderately explosive when exposed to heat or flame. Violent reaction with phosphorus + chlorine. To fight fire, use foam, CO_2, dry chemical. When heated to decomposition it emits acrid smoke and fumes.

HBD500   CAS: 1639-09-4   **HR: 3**
1-HEPTANETHIOL
mf: $C_7H_{16}S$ mw: 132.29

SYNS: HEPTYL MERCAPTAN * n-HEPTYLMERCAPTAN * USAF EK-2122

CONSENSUS REPORTS: Reported in EPA TSCA Inventory.

NIOSH REL: (n-Alkane Mono Thiols) CL 0.5 ppm/15M

SAFETY PROFILE: A poison by intraperitoneal route. When heated to decomposition it emits very toxic fumes of SO_x.

HBE500   CAS: 543-49-7   **HR: 2**
2-HEPTANOL
mf: $C_7H_{16}O$ mw: 116.23

PROP: Liquid. Bp: 160.4°, flash p: 160°F (OC), d: 0.8344 @ 0°, vap press: 1 mm @ 14.6°, vap d: 4.01. Insol in water; sol in alc, ether, and benzene.

SYNS: AMYL METHYL CARBINOL * HEPTANOL-2 * 2-HYDROXYHEPTANE * METHYL AMYL CARBINOL

CONSENSUS REPORTS: Reported in EPA TSCA Inventory.

SAFETY PROFILE: Moderately toxic by ingestion and skin contact. A skin and severe eye irritant. Flammable when exposed to heat and flame; can react vigorously with oxidizers. To fight fire, use foam, CO_2, dry chemical.

HBI800   HR: 3
cis-4-HEPTEN-1-AL
mf: $C_7H_{12}O$ mw: 112.17

PROP: Sltly yellow liquid; fatty odor. Refr index: 1.432-1.436, flash p: 68°F. Sol in alc, fixed oils; insol in water.

SYNS: FEMA No. 3289 * 4-HEPTENAL * n-PROPYLIDENE BUTYRALDEHYDE

SAFETY PROFILE: Flammable liquid. When heated to decomposition it emits acrid smoke and irritating fumes.

HBJ000   HR: 2
n-HEPTENE
mf: C_7H_{14} mw: 98.19

2278

PROP: Colorless liquid, insol in water, sol in ether. D: 0.6969 @ 20°, mp: −10°, bp: 93.6, flash p: <30.2°F, autoign temp: 707°F.

SYN: 1-HEPTYLENE

DOT Classification: Flammable Liquid; Label: Flammable Liquid.

SAFETY PROFILE: A simple asphyxiant. Dangerous fire hazard when exposed to heat, flame or oxidizers. Unknown explosion hazard. To fight fire, use foam, dry chemical, CO_2. When heated to decomposition it emits acrid smoke and fumes.

HBL500   CAS: 111-70-6   HR: 2
HEPTYL ALCOHOL
mf: $C_7H_{16}O$ mw: 116.23

PROP: Colorless liquid; citrus odor. Mp: −34.6°, bp: 175.8°, d: 0.824 @ 20°/4°, refr index: 1.423-1.427, flash p: 160°F. Misc in alc, fixed oils, ether; sltly sol in water @ 175°.

SYNS: l'ALCOOL n-HEPTYLIQUE PRIMAIRE (FRENCH) * ENANTHIC ALCOHOL * FEMA No. 2548

* n-HEPTANOL * 1-HEPTANOL * n-HEPTANOL-1 (FRENCH) * 1-HYDROXYHEPTANE

CONSENSUS REPORTS: Reported in EPA TSCA Inventory.

SAFETY PROFILE: Moderately toxic by ingestion and skin contact. Mildly toxic by inhalation. Combustible liquid. Can react with oxidizing materials. When heated to decomposition it emits acrid smoke and fumes.

HBO500   CAS: 112-23-2   HR: 1
HEPTYL FORMATE
mf: $C_8H_{16}O_2$ mw: 144.24

SYNS: FORMIC ACID, HEPTYL ESTER * HEPTANOL, FORMATE * n-HEPTYL METHANOATE

CONSENSUS REPORTS: Reported in EPA TSCA Inventory.

SAFETY PROFILE: A skin irritant. When heated to decomposition it emits acrid smoke and fumes.

HBP000   CAS: 16338-99-1   HR: 3
HEPTYLMETHYLINITROSAMINE
mf: $C_8H_{18}N_2O$ mw: 158.28

SYNS: METHYLHEPTYLNITROSAMIN (GERMAN) * N-METHYL-N-NITROSOHEPTYLAMINE * N-NITROSO-N-METHYLHEPTYLAMINE

SAFETY PROFILE: Moderately toxic by subcutaneous route. Questionable carcinogen with experimental tumorigenic data. Mutation data reported. Many N-nitroso compounds are carcinogens. When heated to decomposition it emits toxic fumes of NO_x.

HBT500   CAS: 561-27-3   HR: 3
HEROIN
mf: $C_{21}H_{23}NO_5$ mw: 369.45

PROP: White, odorless, bitter crystals or crystalline powder. Mp: 173°, bp: 273° @ 12 mm.

SYNS: ACETOMORFINE * ACETOMORPHINE * ASPRON * BOY * DIACEPHIN * DIACETYLMORFIN * DIACETYLMORPHINE * DIAMORFINA * DIAMORPHINE * DIAPHORM * DIASETIELMORFIEN * DIASETILMORFIN * DIASETYLMORFIIMI * DIAZETYLMORPHINE * 7,8-DIHYDRO-4,5-α-EPOXY-17-METHYLMORPHINAN-3,6-α-DIOL DIACETATE * DOOJE * ECLORION * EROINA * "H" * HAIRY * HARRY * HEROIEN * HEROIIN * HEROLAN * HORSE * IEROIN * IROINI * JOY POWDER * MORPHACETIN * MORPHINE DIACETATE * PREZA * SCOT * WHITE STUFF

SAFETY PROFILE: A poison by ingestion,intracerebral and subcutaneous routes. The fatal dose is between 1/6 and 2 grains. Human reproductive effects by subcutaneous and intravenous routes: newborn drug dependence. Experimental reproductive effects. Mutation data reported.

Resembles morphine in its general results, but acts more strongly on the respiration and is therefore more poisonous. Its depressant effects on the cerebrum appear to be greater than that of codeine. Large doses cause excitement and convulsions in animals and humans. The more common symptoms are headache; disturbance of vision; slow, small, regular pulse; restlessness; cramps in the extremities; slight cyanosis; slow and deep respiration, and death from respiratory paralysis. A poisonous, habit-forming drug. When heated to decomposition it emits toxic fumes of NO_x.

HCB000   CAS: 13007-92-6   HR: 3
HEXACARBONYLCHROMIUM
mf: C_6CrO_6 mw: 220.06

SYNS: CHROMIUM CARBONYL (MAK) * CHROMIUM CARBONYL (OC-6-11) (9CI) * CHROMIUM HEXACARBONYL * HEXACARBONYL CHROMIUM

CONSENSUS REPORTS: IARC Cancer Review: Animal Inadequate Evidence IMEMDT 23,205,80; Chromium and its compounds are on the Community Right-To-Know List. Reported in EPA TSCA Inventory.

OSHA PEL: CL 0.1 mg(CrO_3)/m^3
ACGIH TLV: TWA 0.05 mg(Cr)/m^3; Confirmed Human Carcinogen; (Proposed: TWA 0.001; Suspected Human Carcinogen)
DFG TRK: 0.1 mg/m^3 calculated as CrO_3 in that portion of dust that can possibly be inhaled; 0.2 mg/m^3 arc-welding by hand; others 0.1 mg/m^3. Animal Carcinogen, Suspected Human Carcinogen.
NIOSH REL: TWA 0.001 mg(Cr(VI))/m^3

SAFETY PROFILE: Confirmed carcinogen with experimental tumorigenic data. Poison by intravenous route. Explodes at 210°C.

HCC500   CAS: 118-74-1   HR: 3
HEXACHLOROBENZENE
DOT: UN 2729
mf: C_6Cl_6 mw: 284.76

PROP: Monoclinic prisms. Mp: 231°, bp: 323-326°, flash p: 468°F, vap press: 1 mm @ 114.4°,

vap d: 9.8. D: 2.44. Insol in water; sol in benzene; very sltly sol in hot alc; sol in hot ether and chloroform.

SYNS: AMATIN * ANTICARIE * BUNT-CURE * BUNT-NO-MORE * CO-OP HEXA * ESACHLOROBENZENE (ITALIAN) * GRANOX NM * HCB * HEXA C.B. * HEXACHLORBENZOL (GERMAN) * JULIN'S CARBON CHLORIDE * NO BUNT LIQUID * PENTACHLOROPHENYL CHLORIDE * PERCHLOROBENZENE * PHENYL PERCHLORYL * RCRA WASTE NUMBER U127 * SAATBEIZFUNGIZID (GERMAN) * SANOCIDE * SMUT-GO * SNIECIOTOX

CONSENSUS REPORTS: IARC Cancer Review: GROUP 2B IMEMDT 7,219,87; Animal Sufficient Evidence IMEMDT 20,155,79; Human Limited Evidence IMEMDT 20,155,79. NTP Fourth Annual Report On Carcinogens, 1984. Community Right-To-Know List. Reported in EPA TSCA Inventory. EPA Genetic Toxicology Program.

DFG MAK: BAT: 15 μg/dL in plasma/serum.
DOT Classification: IMO: Poison B; Label: St. Andrews Cross.

SAFETY PROFILE: Confirmed carcinogen with experimental carcinogenic and neoplastigenic data. A human poison by an unspecified route. A suspected human carcinogenic. Experimental teratogenic and reproductive effects. Mildly toxic experimentally by inhalation. Mutation data reported. A fungicide. Combustible when exposed to heat or flame. Violent reaction with dimethylformamide. To fight fire, use CO_2, dry chemical. When heated to decomposition it emits highly toxic fumes of Cl^-.

HCD250   CAS: 87-68-3   HR: 3
HEXACHLORBUTADIENE
DOT: UN 2279
mf: C_4Cl_6 mw: 260.74

PROP: Autoign temp: 1130°F, vap d: 8.99.

SYNS: DOLEN-PUR * GP-40-66:120 * HCBD * HEXACHLOR-1,3-BUTADIEN (CZECH) * HEXACHLORO-1,3-BUTADIENE (MAK) * 1,1,2,3,4,4-HEXACHLORO-1,3-BUTADIENE * PERCHLOROBUTADIENE * RCRA WASTE NUMBER U128

CONSENSUS REPORTS: IARC Cancer Review: GROUP 3 IMEMDT 7,56,87; Animal Suspected IMEMDT 20,179,79. Community Right-To-Know List. Reported in EPA TSCA Inventory.

OSHA PEL: TWA 0.02 ppm
ACGIH TLV: TWA 0.02 ppm (skin); Suspected Human Carcinogen.
DFG MAK: Suspected Carcinogen.
DOT Classification: Poison B; Label: St. Andrews Cross.

SAFETY PROFILE: Suspected carcinogen with experimental carcinogenic data. Poison by ingestion, intraperitoneal, and possibly other routes. Moderately toxic by inhalation and skin contact. A skin and eye irritant. Experimental reproductive effects. Mutation data reported. Combustible when exposed to heat or flame; can react vigorously with oxidizing materials. To fight fire, use dry chemical, CO_2, alcohol foam, water spray, fog, mist. Reacts with bromine perchlorate to form an explosive product. When heated to decomposition it emits very toxic fumes of Cl^-. A solvent, heat transfer fluid, transformer, hydraulic fluid, and wash liquor.

HCE000   CAS: 599-52-0   HR: 3
HEXACHLORO-2,5-CYCLOHEXADIEN-1-ONE
mf: C_6Cl_6O mw: 300.76

SYNS: 2,3,4,4,5,6-HEXACHLORCYKLOHEXA-2,5-DIEN-1-ON (CZECH) * HEXACHLORFENOL (CZECH) * HEXACHLORO-2,5-CYCLOHEXADIENONE * USAF DO-65

SAFETY PROFILE: A poison by ingestion and intraperitoneal routes. A eye and skin and eye irritant. When heated to decomposition it emits toxic fumes of Cl^-.

HCE500   CAS: 77-47-4   HR: 3
HEXACHLOROCYCLOPENTADIENE
DOT: UN 2646
mf: C_5Cl_6 mw: 272.75

PROP: Yellow- to amber-colored liquid, pungent odor. Mp: 9.9°, bp: 239°, fp: -2°, flash p: none (OC), d: 1.715 @ 15.5°/15.5°, vap d: 9.42.

SYNS: C-56 * HCCPD * HEXACHLORCYKLOPEN-TADIEN (CZECH) * NCI-C55607 * PCL * RCRA WASTE NUMBER U130

CONSENSUS REPORTS: EPA Extremely Hazardous Substances List. Community Right-To-Know List. Reported in EPA TSCA Inventory.

OSHA PEL: TWA 0.01 ppm

ACGIH TLV: TWA 0.01 ppm
DOT Classification: Corrosive Material; Label: Corrosive; IMO: Poison B; Label: Poison.

SAFETY PROFILE: A deadly poison by inhalation and ingestion. Moderately toxic by skin contact. Experimental teratogenic effects. Corrosive. A severe skin and eye irritant. May explode on contact with sodium. When heated to decomposition it emits toxic fumes of Cl^-.

HCF000   CAS: 57653-85-7   HR: 3
1,2,3,4,7,8-HEXACHLORODIBENZO-p-DIOXIN
mf: $C_{12}H_2Cl_6O_2$ mw: 390.84

CONSENSUS REPORTS: IARC Cancer Review: Animal Inadequate Evidence IMEMDT 15,41,77.

SAFETY PROFILE: A deadly poison by ingestion. Questionable carcinogen. When heated to decomposition it emits toxic fumes of Cl^-.

HCF500   HR: 3
1,2,3,6,7,8-HEXACHLORODIBENZO-p-DIOXIN mixed with 1,2,3,7,8,9-HEXACHLORODIBENZO-p-DIOXIN
SYNS: 1,2,3,7,8,9-HEXACHLORODIBENZO-p-DIOXIN mixed with 1,2,3,6,7,8-HEXACHLORODIBENZO-p-DIOXIN * NCI-C03703

CONSENSUS REPORTS: NCI Carcinogenesis Bioassay (gavage); Clear Evidence: mouse, rat NCITR* NCI-CG-TR-198,80. NCI Carcinogenesis Bioassay (dermal); No Evidence: mouse NCITR* NCI-CG-TR-202,80.

SAFETY PROFILE: Suspected carcinogen with experimental carcinogenic data. A deadly poison by ingestion. When heated to decomposition it emits very toxic fumes of Cl^- and dioxin.

HCI000   CAS: 67-72-1   HR: 3
HEXACHLOROETHANE
DOT: NA 9037
mf: C_2Cl_6 mw: 236.72

PROP: Rhombic, triclinic, or cubic crystals, colorless, camphor-like odor. Mp: 186.6° (sublimes), d: 2.091, vap press: 1 mm @ 32.7°, bp: 186.8° (triple point). Sol in alc, benzene, chloroform, ether, oils; insol in water.

SYNS: AVLOTANE * CARBON HEXACHLORIDE * DISTOKAL * DISTOPAN * DISTOPIN * EGITOL * ETHANE HEXACHLORIDE * ETHYL-

ENE HEXACHLORIDE * FALKITOL * FASCIOLIN * HEXACHLOR-AETHAN (GERMAN) * 1,1,1,2,2,2-HEXACHLOROETHANE * HEXACHLOROETHYLENE * MOTTENHEXE * NCI-C04604 * PERCHLORO-ETHANE * PHENOHEP * RCRA WASTE NUMBER U131

CONSENSUS REPORTS: IARC Cancer Review: GROUP 3 IMEMDT 7,56,87; Animal Limited Evidence IMEMDT 20,467,79. NCI Carcinogenesis Bioassay (gavage); Clear Evidence: mouse NCITR* NCI-CG-TR-68,78. NCI Carcinogenesis Bioassay (gavage); No Evidence: rat NCITR* NCI-CG-TR-68,78. Community Right-To-Know List. Reported in EPA TSCA Inventory. EPA Genetic Toxicology Program.

OSHA PEL: TWA 1 ppm (skin)
ACGIH TLV: TWA 1 ppm
DFG MAK: 1 ppm (10 mg/m^3)
NIOSH REL: (Hexachloroethane) Reduce to lowest level
DOT Classification: ORM-A; Label: None.

SAFETY PROFILE: A poison by intravenous route. Moderately toxic by intraperitoneal route. Mildly toxic by ingestion. Experimental reproductive effects. Questionable carcinogen with experimental carcinogenic data. Liver injury has resulted from exposure to this material. An insecticide. Slightly explosive by spontaneous chemical reaction. Dehalogenation of this material by reaction with alkalies, metals, etc., will produce spontaneous explosive chloroacetylenes. When heated to decomposition it emits highly toxic fumes of Cl$^-$ and phosgene.

HCK500   CAS: 1335-87-1   **HR: 3**
HEXACHLORONAPHTHALENE
mf: C$_{10}$H$_2$Cl$_6$ mw: 334.82

PROP: White solid.

CONSENSUS REPORTS: Community Right-To-Know List. Reported in EPA TSCA Inventory.

OSHA PEL: TWA 0.2 mg/m^3 (skin)
ACGIH TLV: TWA 0.2 mg/m^3 (skin)

SAFETY PROFILE: A poison by ingestion, skin contact, and inhalation. Causes severe acneform eruptions and toxic narcosis of liver. Absorbed by skin. When heated to decomposition it emits toxic fumes of Cl$^-$.

HCL000   CAS: 70-30-4   **HR: 3**
HEXACHLOROPHENE
DOT: UN 2875
mf: C$_{13}$H$_6$Cl$_6$O$_2$ mw: 406.89
PROP: Crystals, water insol. Mp: 165°. Sol in alc, acetone, ether, chloroform, propylene glycol, polyethylene glycols, olive oil, cottonseed oil, dil solns of alkalies.

SYNS: ACIGENA * ALMEDERM * AT 7 * B32 * BILEVON * BIS(2-HYDROXY-3,5,6-TRI-CHLOROPHENYL)METHANE * BIS-2,3,5-TRICHLOR-6-HYDROXYFENYLMETHAN (CZECH) * BIS(3,5,6-TRI-CHLORO-2-HYDROXYPHENYL)METHANE * COMPOUND G-11 * COTOFILM * DERMADEX * 2,2′-DIHYDROXY-3,3′,5,5′,6,6′-HEXACHLORODIPHE-NYLMETHANE * 2,2′-DIHYDROXY-3,5,6,3′,5′,6′-HEXA-CHLORODIPHENYLMETHANE * EXOFENE * FOMAC * FOSTRIL * G-11 * GAMOPHENE * G-ELEVEN * GERMA-MEDICA * HCP * HEXABALM * 2,2′,3,3′,5,5′-HEXACHLORO-6,6′-DIHYDROXYDIPHENYLMETHANE * HEXACHLORO-FEN (CZECH) * HEXACHLOROPHANE * HEXA-CHLOROPHEN * HEXACHLOROPHENE (DOT) * HEXAFEN * HEXIDE * HEXOPHENE * HEXOSAN * ISOBAC 20 * 2,2′-METHYLENEBIS(3,4,6-TRICHLOROPHENOL) * NABAC * NCI-C02653 * NEOSEPT * PHISODANV * PHISOHEX * RCRA WASTE NUMBER U132 * RITOSEPT * SEPTISOL * SEPTOFEN * STERAL * STERASKIN * SURGI-CEN * SUROFENE * TERSASEPTIC * TRICHLORO-PHENE * TURGEX

CONSENSUS REPORTS: IARC Cancer Review: GROUP 3 IMEMDT 7,56,87; Human Inadequate Evidence IMEMDT 20,241,79. NCI Carcinogenesis Bioassay (feed); No Evidence: rat NCITR* NCI-CG-TR-40,78. Reported in EPA TSCA Inventory. Chlorophenols are on the Community Right-To-Know List.

DOT Classification: IMO: Poison B; Label: St. Andrews Cross.

SAFETY PROFILE: A human poison by ingestion. An experimental poison by ingestion, intraperitoneal, and intravenous routes. Moderately toxic by skin contact. Human systemic effects by ingestion: cardiomyopathy (damage to the heart muscle), nausea or vomiting, diarrhea, shock. Unspecified human reproductive effects. Experimental teratogenic and reproductive effects. An eye and human skin irritant. Questionable carcinogen with experimental neoplastigenic and tumorigenic data. Strong concentrations may be irritating, but ordinary use of 1-2% solutions is not.

For many years, the toxicologic hazard of hexachlorophene was unrecognized and the compound had a wide and virtually unrestricted use. However, studies by FDA scientists demonstrated that brain lesions occur from exposure in both rats and monkeys treated at levels only slightly higher than those of persons using soaps, toothpaste, shampoos, and a variety of other household products and cosmetics containing it. The FDA has now restricted sale of hexachlorophene, and most preparations containing higher levels of the compound are available only through prescription. In FDA studies, it was found that 2 weeks after onset of exposure, rats fed 500 ppm (25 mg/kg/day) of hexachlorophene in their diet showed weakness in their hindquarters which progressed to paralysis. Microscopic examination of the brain and spinal cord of these rats revealed a particular edema of the white matter resembling spongy degeneration noted in infants. When the animals were removed from the diet, they recovered gradually over a period of weeks. Similar symptoms were noted in the monkey. Following ingestion of hexachlorophene, early symptoms are primarily gastrointestinal in nature and include anorexia, nausea, vomiting, abdominal cramps, and diarrhea. Dehydration is sometimes severe and may be associated with shock.

Used as a germicidal agent. An additive permitted in the feed and drinking water of animals and/or for the treatment of food-producing animals. Also permitted in food for human consumption. When heated to decomposition it emits its toxic fumes of Cl^-.

HCL500   CAS: 116-16-5   HR: 3
HEXACHLORO-2-PROPANONE

DOT: UN 2661
mf: C_3Cl_6O mw: 264.73

PROP: Liquid. Bp: 203°, fp: −2°, vap d: 9.2.

SYNS: GC-1106 * HCA * HEXACHLOROACE- TONE (DOT) * 1,1,1,3,3,3-HEXACHLORO-2-PROPA- NONE

CONSENSUS REPORTS: Reported in EPA TSCA Inventory.

DOT Classification: IMO: Poison B; Label: Poison.

SAFETY PROFILE: A poison. Moderately toxic by ingestion, inhalation, and skin contact. Muta-

tion data reported. When heated to decomposition it emits toxic fumes of Cl^-.

HCM000   CAS: 1888-71-7   HR: 3
HEXACHLOROPROPENE
mf: C_3Cl_6 mw: 248.73

SYN: HEXACHLOROPROPYLENE

CONSENSUS REPORTS: Reported in EPA TSCA Inventory.

SAFETY PROFILE: A poison by inhalation and intraperitoneal routes. A powerful irritant. When heated to decomposition it emits toxic fumes of Cl^-.

HCO500   CAS: 143-27-1   HR: 3
1-HEXADECANAMINE
mf: $C_{16}H_{35}N$ mw: 241.52

SYNS: ALAMINE 6 * ARMEEN 16D
* CETYLAMIN (GERMAN) * CETYLAMINE
* N-HEXADECYLAMINE * PALMITYLAMINE

CONSENSUS REPORTS: Reported in EPA TSCA Inventory.

SAFETY PROFILE: A poison by intraperitoneal route. Experimental teratogenic and reproductive effects. When heated to decomposition it emits toxic fumes of NO_x.

HCP000   CAS: 36653-82-4   HR: 2
1-HEXADECANOL
mf: $C_{16}H_{34}O$ mw: 242.50

PROP: Solid or leaf-like crystals. Mp: 49.3°, bp: 190° @ 15 mm, d: 0.8176 @ 50°/4°. Insol in H_2O; sol in alc, chloroform, ether.

SYNS: ADOL * ALCOHOL C-16 * ATALCO C
* CACHALOT C-50 * CETAFFINE * CETAL
* CETALOL CA * CETYL ALCOHOL * CETYLIC ALCOHOL * CETYLOL * CO-1670 * CRODACOL- CAS * CYCLAL CETYL ALCOHOL * DYTOL F-11
* EPAL 16NF * ETHAL * ETHOL * HEXA- DECANOL * n-HEXADECANOL * HEXADECAN- 1-OL * HEXADECYL ALCOHOL * n-HEXADECYL ALCOHOL * LOROL 24 * LOXANOL K
* PALMITYL ALCOHOL * PRODUCT 308

CONSENSUS REPORTS: Reported in EPA TSCA Inventory.

SAFETY PROFILE: Moderately toxic by ingestion and intraperitoneal routes. An eye and human skin irritant. Flammable when exposed to heat or flame; can react with oxidizing materials.

To fight fire, use foam, CO_2, dry chemical. When heated to decomposition it emits acrid smoke and fumes.

HCQ000   CAS: 5894-60-0   **HR: 2**
HEXADECYLTRICHLOROSILANE

DOT: UN 1781
mf: $C_{16}H_{33}Cl_3Si$ mw: 359.93

PROP: Colorless to yellow liquid. D: 0.996 @ 25°/25°, bp: 269°, flash p: 295°F (COC).

SYN: TRICHLOROHEXADECYLSILANE

CONSENSUS REPORTS: Reported in EPA TSCA Inventory.

DOT Classification: Corrosive Material; Label: Corrosive.

SAFETY PROFILE: A corrosive irritant to skin, eyes, and mucous membranes. Combustible when exposed to heat or flame. When heated to decomposition or on contact with water it emits toxic fumes of Cl^- and HCl.

HCQ500   CAS: 57-09-0   **HR: 3**
HEXADECYLTRIMETHYLAMMONIUM BROMIDE

mf: $C_{19}H_{42}N \cdot Br$ mw: 364.53

SYNS: ACETOQUAT CTAB * BROMAT * CEE DEE * CENTIMIDE * CETAB * CETAROL * CETAVLON * CETRIMIDE * CETRIMONIUM BROMIDE * CETYLAMINE * CETYLTRIMETHYL-AMMONIUM BROMIDE * N-CETYLTRIMETHYLAM-MONIUM BROMIDE * CIRRASOL-OD * CTAB * CYCLOTON V * N-HEXADECYLTRIMETHYLAM-MONIUM BROMIDE * N-HEXADECYL-N,N,N-TRI-METHYLAMMONIUM BROMIDE * (1-HEXADECYL)TRI-METHYLAMMONIUM BROMIDE * LISSOLAMINE * MICOL * POLLACID * QUAMONIUM * SUTICIDE * TRIMETHYLCETYLAMMONIUM BROMIDE * N,N,N-TRIMETHYL-1-HEXADECANAMI-NIUM BROMIDE * TRIMETHYLHEXADECYLAMMO-NIUM BROMIDE

CONSENSUS REPORTS: Reported in EPA TSCA Inventory.

SAFETY PROFILE: A poison by ingestion, intravenous, intraperitoneal, and subcutaneous routes. Experimental teratogenic and reproductive effects. A skin and severe eye irritant. When heated to decomposition it emits very toxic fumes of NH_3, NO_x, and Br^-.

HCR500   CAS: 592-42-7   **HR: 3**
1,5-HEXADIENE

mf: C_6H_{10} mw: 82.16

PROP: Liquid. D: 0.691, mp: −141°, bp: 59.6°. Flash p: −50.80°F. Insol in water.

SYNS: BIALLYL * DIALLYL * HEXA-1,5-DIENE

CONSENSUS REPORTS: Reported in EPA TSCA Inventory.

DOT Classification: Label: Flammable Liquid.

SAFETY PROFILE: No toxicity data. A very dangerous fire and explosion hazard when exposed to heat, flame, or oxidizers. When heated to decomposition it emits acrid smoke and fumes.

HCV500   CAS: 9011-04-5   **HR: 3**
HEXADIMETHRINE BROMIDE

mf: $C_{10}H_{24}N_2 \cdot xC_3H_6Br_2$ mw: 1585.73

PROP: White hygroscopic, amorphous polymer. Sol in water up to 10%.

SYNS: 1,3-DIBROMOPROPANE polymer with N,N,N′,N′-TETRAMETHYL-1,6-HEXANEDIAMINE * 1,5-DI-METHYL-1,5-d-DIAZAUNDECAMETHYLENE POLYMETHO-BROMIDE * POLYBREME * POLY (N,N,N′,N′-TETRAMETHYL-N-TRIMETHYLENE-HEXAMETHYLENEDIAMMONIUM DIBROMIDE) * N,N,N′,N′-TETRAMETHYLHEXAMETHYLENE-DIAMINE-1,3-DIBROMOPROPANE copolymer

SAFETY PROFILE: A poison by intravenous and intraperitoneal routes. When heated to decomposition it emits very toxic fumes of NO_x and Br^-.

HCY000   CAS: 757-58-4   **HR: 3**
HEXAETHYL TETRAPHOSPHATE

DOT: UN 1611/UN 1612/NA 2783
mf: $C_{12}H_{30}O_{13}P_4$ mw: 506.30

PROP: Liquid. Mp: −40°; bp: decomp above 150°.

SYNS: BLADAN * BLADAN BASE * ETHYL TET-RAPHOSPHATE * HET * HETP * HEXAETHYL TETRAPHOSPHATE, liquid (DOT) * HEXAETHYL TET-RAPHOSPHATE, liquid, containing more than 25% hexaethyl tetraphosphate (DOT) * HTP * RCRA WASTE NUM-BER P062 * TETRAPHOSPHATE HEXAETHYLIQUE (FRENCH)

DOT Classification: Poison B; Label: Poison.

SAFETY PROFILE: A poison by ingestion, skin contact, intraperitoneal, subcutaneous, intravenous, and intramuscular routes. When heated to decomposition it emits toxic fumes of PO_x.

HCZ000   CAS: 684-16-2   **HR: 3**
HEXAFLUOROACETONE

DOT: UN 2420
mf: C_3F_6O mw: 166.03

PROP: A colorless, nonflammable solvent liquid. D: 1.65 @ 25°.

SYNS: 6FK   *   NCI-C56440

CONSENSUS REPORTS: Reported in EPA TSCA Inventory.

OSHA PEL: TWA 0.1 ppm (skin)
ACGIH TLV: TWA 0.1 ppm (skin)
DOT Classification: IMO: Poison A; Label: Poison Gas.

SAFETY PROFILE: A poison by ingestion and possibly by skin contact. Moderately toxic by inhalation. A poisonous irritant to the skin, eyes, and mucous membranes. Experimental teratogenic and reproductive effects. When heated to decomposition it emits toxic fumes of F^-.

HDA000   CAS: 10543-95-0   **HR: 3**
HEXAFLUOROACETONE HYDRATE

DOT: UN 2552
mf: $C_3F_6O \cdot H_2O$ mw: 184.05

SYN: HEXAFLUORO-2-PROPANONE HYDRATE

DOT Classification: IMO: Poison B; Label: Poison.

SAFETY PROFILE: Poison by ingestion and skin contact. When heated to decomposition it emits toxic fumes of F^-.

HDA500   CAS: 34202-69-2   **HR: 3**
HEXAFLUORO ACETONE TRIHYDRATE
mf: $C_3F_6O \cdot 3H_2O$ mw: 220.09

SYN: GC 7787

SAFETY PROFILE: A poison by ingestion and skin contact. Experimental reproductive effects. When heated to decomposition it emits toxic fumes of F^-.

HDB500   CAS: 11111-49-2   **HR: 3**
HEXAFLUORODICHLOROBUTENE

SYN: HFCB

SAFETY PROFILE: A poison by inhalation. When heated to decomposition it emits very toxic fumes of Cl^- and F^-.

HDC000   CAS: 333-36-8   **HR: 3**
HEXAFLUORODIETHYL ETHER
mf: $C_4H_4F_6O$ mw: 182.08

SYNS: BIS(TRIFLUOROETHYL)ETHER   *   BIS(2,2,2-TRIFLUOROETHYL)ETHER   *   FLUOROETHYL   *   FLUOROTHYL   *   FLUROTHYL   *   HFE   *   INDOKLON   *   SF6539   *   SKF 6539

SAFETY PROFILE: A poison by intravenous route. Moderately toxic by intraperitoneal route. When heated to decomposition it emits toxic fumes of F^-.

HDC500   CAS: 920-66-1   **HR: 3**
HEXAFLUOROISOPROPANOL
mf: $C_3H_2F_6O$ mw: 168.05

SYNS: 1,1,1,3,3,3-HEXAFLUORO-2-PROPANOL   *   HFIP

CONSENSUS REPORTS: Reported in EPA TSCA Inventory.

SAFETY PROFILE: A poison by intravenous and intraperitoneal routes. Moderately toxic by ingestion. Mildly toxic by inhalation. A severe eye irritant. When heated to decomposition it emits toxic fumes of F^-.

HDE000   CAS: 16940-81-1   **HR: 3**
HEXAFLUOROPHOSPHORIC ACID

DOT: UN 1782
mf: F_6HP mw: 145.98

PROP: Corrosive, colorless, clear liquid. Mp: 31°, d: 1.65.

SYN: HYDROGEN HEXAFLUOROPHOSPHATE

CONSENSUS REPORTS: Reported in EPA TSCA Inventory.

OSHA PEL: TWA 2.5 $mg(F)/m^3$
NIOSH REL: (Inorganic Fluorides) TWA 2.5 $mg(F)/m^3$
DOT Classification: Corrosive Material; Label: Corrosive.

SAFETY PROFILE: A poison by all routes. A corrosive irritant to skin, eyes, and mucous membranes. When heated to decomposition it emits highly toxic F^- and PO_x.

HDE500   CAS: 13098-39-0   **HR: 3**
HEXAFLUORO-2-PROPANONE SESQUIHYDRATE
mf: $C_3F_6O \cdot 3/2H_2O$ mw: 193.06

SYN: HEXAFLUOROACETONE SESQUIHYDRATE

SAFETY PROFILE: A poison by ingestion, intraperitoneal, and intravenous routes. Experimental reproductive effects. When heated to decomposition it emits toxic fumes of F^-.

HDF000 CAS: 116-15-4 **HR: 3**
HEXAFLUOROPROPENE
DOT: UN 1858
mf: C_3F_6 mw: 150.03

PROP: Gas. Mp: $-156°$, bp: $-29°$, d: 1.583 @ $-40°/4°$.

SYNS: HEXAFLUOROPROPYLENE (DOT) * PERFLUOROPROPENE * PERFLUOROPROPYLENE

CONSENSUS REPORTS: Reported in EPA TSCA Inventory.

DOT Classification: Nonflammable Gas; Label: Nonflammable Gas.

SAFETY PROFILE: Mildly toxic by inhalation. Explosive reaction with Grignard reagents (e.g., phenylmagnesium bromide). Reacts with tetrafluorethylene + air to form explosive peroxides. When heated to decomposition it emits toxic fumes of F^-.

HDG000 CAS: 111-49-9 **HR: 2**
HEXAHYDRO-1H-AZEPINE
DOT: UN 2493

PROP: Flash p: 71.6°F.
mf: $C_6H_{13}N$ mw: 99.20

SYNS: AZACYCLOHEPTANE * 1-AZACYCLOHEPTANE * CYCLOHEXAMETHYLENIMINE * G 0 * HEXAHYDROAZEPINE * HEXAMETHYLENE IMINE (DOT) * HEXAMETHYLENIMINE * HOMOPIPERIDINE * PERHYDROAZEPINE

CONSENSUS REPORTS: Reported in EPA TSCA Inventory.

DOT Classification: Corrosive Material; Label: Corrosive; IMO: Flammable Liquid; Label: Flammable Liquid, Corrosive.

SAFETY PROFILE: Moderately toxic by ingestion and subcutaneous routes. Mildly toxic by inhalation. A corrosive irritant to the eyes, skin, and mucous membranes. A dangerous fire hazard when exposed to heat or flame; can react vigorously with oxidizers. When heated to decomposition it emits toxic fumes of NO_x.

HDQ500 CAS: 15923-42-9 **HR: 3**
1,2,3,4,5,6-HEXAHYDRO-6-METHYLAZEPINO(4,5-b)INDOLE HYDROCHLORIDE
mf: $C_{13}H_{16}N_2 • ClH$ mw: 236.77

SYN: U-22394A

SAFETY PROFILE: A poison by ingestion and intraperitoneal routes. When heated to decomposition it emits very toxic fumes of NO_x and HCl.

HDV500 CAS: 13980-04-6 **HR: 3**
HEXAHYDRO-1,3,5-s-TRIAZINE
mf: $C_3H_6N_6O_3$ mw: 174.15

SYNS: HEXAHYDRO-1,3,5-TRINITROSO-s-TRIAZINE * HEXAHYDRO-1,3,5-TRINITROSO-1,3,5-TRIAZINE * 1,3,5-TRINITROSO-1,3,5-TRIAZACYCLOHEXANE * TRINITROSOTRIMETHYLENETRIAMINE * TRINITROSOTRIMETHYLENTRIAMIN (GERMAN) * TTT

SAFETY PROFILE: A poison by ingestion. Questionable carcinogen with experimental tumorigenic data. Explodes on contact with sulfuric acid. When heated to decomposition it emits toxic fumes of NO_x.

HDY000 CAS: 531-18-0 **HR: 3**
HEXA(HYDROXYMETHYL)MELAMINE
mf: $C_9H_{18}N_6O_6$ mw: 306.33

SYNS: CILAG 61 * HEXAKIS(HYDROXYMETHYL)MELAMINE * HEXAKIS(HYDROXYMETHYL)-1,3,5-TRIAZINE-2,4,6-TRIAMINE * HEXAMETHYLOLMELAMIN (CZECH) * HEXAMETHYLOLMELAMINE * RESLOOM M 75 * (1,3,5-TRIAZINE-2,4,6-TRIYL-TRINITRILO)HEXAKIS METHANOL * (s-TRIAZINE-2,4,6-TRIYLTRINITRILO)HEXAMETHANOL * 2,4,6-TRIS(BIS(HYDROXYMETHYL)AMINO)-s-TRIAZINE * 2,4,6-TRIS(DI(HYDROXYMETHYL)AMINO)-1,3,5-TRIAZINE

CONSENSUS REPORTS: Reported in EPA TSCA Inventory.

SAFETY PROFILE: A poison by intravenous route. A skin and eye irritant. When heated to decomposition it emits toxic fumes of NO_x.

HEA000 CAS: 55-97-0 **HR: 3**
HEXAMETHONIUM BROMIDE
mf: $C_{12}H_{30}N_2 • 2Br$ mw: 362.26

PROP: Crystals. Mp: 274-276°; sol in water, alc, acid to litmus. Aq solns are stable.

SYNS: α,omega-BIS(TRIMETHYL AMMONIUM)HEXANE DIBROMIDE * C 6 * ESAMETINA * GANGLIO-

STAT * HB * HEXAMETHIONIUM BROMIDE * HEXAMETHONIUM DIBROMIDE * HEXAMETHYLENEBIS(TRIMETHYLAMMONIUM) BROMIDE * N,N,N,N′,N′,N′-HEXAMETHYL-1,6-HEXANEDIAMINIUM DIBROMIDE * HEXAMETON * HEXONIUM DIBROMIDE * SIMPATOBLOCK * VEGOLYSEN * VEGOLYSIN

SAFETY PROFILE: A poison by intraperitoneal, subcutaneous, and intravenous routes. Moderately toxic by ingestion. Human reproductive and teratogenic effects by subcutaneous route: abnormal neonatal measurements and developmental abnormalities of the gastrointestinal system. Used to treat hypertension. When heated to decomposition it emits very toxic NH_3, NO_x, and Br^-.

HEB000   CAS: 870-62-2   HR: 3
HEXAMETHONIUM DIIODIDE
mf: $C_{12}H_{30}N_2 \cdot 2I$ mw: 456.24

SYNS: ESAMETONIO IODURO (ITALIAN) * HEXAMETHONIUM IODIDE * HEXAMETHYLENEBIS(TRIMETHYLAMMONIUM IODIDE) * N,N,N,N′,N′,N′-HEXAMETHYL-1,6-HEXANEDIAMINIUM DIIODIDE * HEXATHIDE * HEXONIUM DIIODIDE

SAFETY PROFILE: A poison by subcutaneous and intravenous routes. Moderately toxic by ingestion. When heated to decomposition it emits very toxic fumes of NO_x, NH_3, and I^-.

HEC000   CAS: 87-85-4   HR: 3
HEXAMETHYLBENZENE
mf: $C_{12}H_{18}$ mw: 162.30

PROP: Plates from ethanol. Mp: 165.5°, bp: 265°. Insol in water; very sol in ether.

CONSENSUS REPORTS: Reported in EPA TSCA Inventory.

SAFETY PROFILE: Mildly toxic by ingestion. Questionable carcinogen with experimental neoplastigenic data. Potentially explosive reaction with nitromethane. When heated to decomposition it emits acrid smoke and fumes.

HEE500   CAS: 661-69-8   HR: 3
HEXAMETHYLDITIN
mf: $C_6H_{18}Sn_2$ mw: 327.62

SYNS: HEXAMETHYLDISTANNANE * PENNSALT TD 5032 * TD-5032

OSHA PEL: TWA 0.1 mg(Sn)/m³ (skin)
ACGIH TLV: TWA 0.1 mg(Sn)/m³ (skin) (Proposed: TWA 0.1 mg(Sn)/m³; STEL 0.2 mg(Sn)/m³ (skin))

NIOSH REL: (Organotin Compounds) TWA 0.1 mg(Sn)/m³

SAFETY PROFILE: A poison by ingestion. When heated to decomposition it emits acrid smoke and fumes.

HEG000   CAS: 317-52-2   HR: 3
HEXAMETHYLENE BIS(9-FLUORENYL DIMETHYLAMMONIUM)DIBROMIDE
mf: $C_{36}H_{42}N_2 \cdot Br$ mw: 582.71

SYNS: 1,6-BIS(9 FLUORENYLDIMETHYL-AMMONIUM) HEXANE BROMIDE * HEXAFLUORENIUM DIBROMIDE * HEXAFLURONIUM BROMIDE * HEXAMETHYLENEBIS(DIMETHYL-9-FLUORENYLAMMONIUM BROMIDE) * HEXAMETHYLENEBIS(FLUOREN-9-YLDIMETHYLAMMONIUM BROMIDE) * IN-117 * MILAXEN * MYLAXEN * NSC-19477

SAFETY PROFILE: A poison by ingestion, intraperitoneal, subcutaneous, and intravenous routes. When heated to decomposition it emits very toxic fumes of NH_3, NO_x and Br^-.

HEI000   CAS: 1169-26-2   HR: 3
1-(2-HEXAMETHYLENEIMINOETHYL)-2-OXOCYCLOHEXANECARBOXYLIC ACID BENZYL ESTER HYDROCHLORIDE
mf: $C_{22}H_{31}NO_3 \cdot ClH$ mw: 394.00

SYN: 2-(β-HEXAMETHYLENIMINOAETHYL)CYCLOHEXANON-2-CARBONSAUREBENZYLESTER-HYDROCHLORIDE (GERMAN)

SAFETY PROFILE: A poison by ingestion, intraperitoneal, subcutaneous, and intravenous routes. When heated to decomposition it emits very toxic fumes of HCl and NO_x.

HEI500   CAS: 100-97-0   HR: 3
HEXAMETHYLENETETRAMINE
DOT: UN 1328
mf: $C_6H_{12}N_4$ mw: 140.22

PROP: Odorless, rhombic crystals from alc. Mp: 280° (sublimes), flash p: 482°F, d: 1.33 @ −5°. Very sltly sol in hot ether.

SYNS: ACETO HMT * AMINOFORM * AMMOFORM * AMMONIOFORMALDEHYDE * CYSTAMIN * CYSTOGEN * ESAMETILENTETRAMINA (ITALIAN) * FORMAMINE * FORMIN * HEXAFORM * HEXAMETHYLENAMINE * HEXAMETHYLENEAMINE * HEXAMETHYLENETETRAAMINE

* HEXAMETHYLENTETRAMIN (GERMAN)
* HEXAMINE (DOT) * HEXILMETHYLENAMINE
* HMT * METHAMIN * METHENAMINE
* PREPARATION AF * RESOTROPIN * 1,3,5,7-
TETRAAZAADAMANTANE * URITONE * URO-
TROPIN * UROTROPINE

CONSENSUS REPORTS: EPA Genetic Toxicology Program. Reported in EPA TSCA Inventory.

DOT Classification: IMO: Flammable Solid; Label: Flammable Solid.

SAFETY PROFILE: A poison by subcutaneous route. Moderately toxic by ingestion and intraperitoneal routes. Human mutation data reported. An irritant to skin, eyes, and mucous membranes. Questionable carcinogen with experimental tumorigenic data. Some persons suffer a skin rash if they come in contact with this material or the fumes evolved when it is heated. Pure hexamethylenetetramine may be taken internally in small amounts and has been used in medicine as a urinary antiseptic. Its major industrial use is in the manufacture of phenolic resins.

Combustible when exposed to heat or flame. Can react with oxidizing materials. Explosive reaction with acetic acid + acetic anhydride + ammonium nitrate + nitric acid, 1-bromopenta borane(9) (above 90°C), iodoform (at 178°C), iodine (at 138°C). Reaction with nitric acid + acetic anhydride forms the military explosives RDX and HMX. Reacts violently with Na_2O_2. When heated to decomposition it emits toxic fumes of formaldehyde and NO_x.

HEJ500   CAS: 645-05-6   **HR: 3**
HEXAMETHYLMELAMINE
mf: $C_9H_{18}N_6$ mw: 210.33

PROP: A solid material. Insol in water; sol in acetone.

SYNS: ALTRETAMINE * ENT 50,852 * HEMEL * N,N,N',N',N',N''-HEXAMETHYL-1,3,5-TRIAZINE-2,4,6-TRIAMINE * HEXASTAT * HMM * NCI-C50259 * NSC 13875 * 2,4,6-TRIS(DIMETHYLAMINO)-s-TRIAZINE * 2,4,6-TRIS(DIMETHYLAMINO)-1,3,5-TRIAZINE

CONSENSUS REPORTS: EPA Genetic Toxicology Program.

SAFETY PROFILE: A poison by ingestion, intraperitoneal, and intravenous routes. Experi-

mental teratogenic and reproductive effects. Questionable carcinogen with experimental neoplastigenic data. Human mutation data reported. Human systemic effects by ingestion: nausea or vomiting and leukopenia (reduced white blood cell count). When heated to decomposition it emits toxic fumes of NO_x.

HEK000   CAS: 680-31-9   **HR: 3**
HEXAMETHYL PHOSPHORAMIDE
mf: $C_6H_{18}N_3OP$ mw: 179.24

PROP: Clear, colorless, mobile liquid, spicy odor. Bp: 233°, fp: 6°, d: 1.024 @ 25°/25°, vap d: 6.18.

SYNS: ENT 50,882 * HEXAMETAPOL * HEXAMETHYLPHOSPHORIC ACID TRIAMIDE (MAK) * HEXAMETHYLPHOSPHORIC TRIAMIDE * N,N,N,N,N,N-HEXAMETHYLPHOSPHORIC TRIAMIDE * HEXAMETHYLPHOSPHOROTRIAMIDE * HEXAMETHYLPHOSPHOTRIAMIDE * HEXMETHYLPHOSPHORAMIDE * HMPA * HMPT * HPT * MEMPA * PHOSPHORIC TRIS(DIMETHYLAMIDE) * PHOSPHORYL HEXAMETHYLTRIAMIDE * TRI(DIMETHYLAMINO)PHOSPHINEOXIDE * TRIS-(DIMETHYLAMINO)PHOSPHINE OXIDE * TRIS(DIMETHYLAMINO)PHOSPHORUS OXIDE

CONSENSUS REPORTS: IARC Cancer Review: GROUP 2B IMEMDT 7,56,87; Animal Sufficient Evidence IMEMDT 15,211,77. NTP Fourth Annual Report On Carcinogens, 1984. Community Right-To-Know List. Reported in EPA TSCA Inventory. EPA Genetic Toxicology Program.

ACGIH TLV: Suspected Human Carcinogen. DFG MAK: Animal Carcinogen, Suspected Human Carcinogen.

SAFETY PROFILE: Confirmed carcinogen with experimental carcinogenic and tumorigenic data. Moderately toxic by ingestion, skin contact, intraperitoneal, and intravenous routes. Experimental reproductive effects. Human mutation data reported. When heated to decomposition it emits very toxic fumes of phosphine, PO_x, and NO_x.

HEL500   CAS: 1164-33-6   **HR: 3**
HEXAMID
mf: $C_{18}H_{25}N_3O_3 \cdot ClH$ mw: 367.92

SYNS: 3-(2-(DIETHYLAMINO)ETHYL)-5-ETHYL-5-PHENYLBARBITURIC ACID HYDROCHLORIDE * F 156 * 5,5-PHENYL-AETHYL-3-(β-DIAETHYLAMINO-

AETHYL)-2,4,6-TRIOXO-HEXAHYDROPYRIMIDIN-HCl (GERMAN)

SAFETY PROFILE: A poison by ingestion and intravenous routes. Moderately toxic by subcutaneous route. When heated to decomposition it emits very toxic fumes of HCl and NO_x.

HEM000   CAS: 66-25-1   **HR: 2**
1-HEXANAL

DOT: UN 1207
mf: $C_6H_{12}O$ mw: 100.18

PROP: Colorless liquid; powerful fatty-green odor. Reported in about a dozen essential oils. Mp: $-56.3°$, bp: $128.7°$, flash p: $90°F$ (OC), d: 0.808-0.812, refr index: 1.402-1.407, vap press: 8.6 mm @ $20°$, vap d: 3.45. Sol in alc, fixed oils, propylene glycol; very sltly sol in water.

SYNS: ALDEHYDE C-6 * CAPROALDEHYDE * CAPROIC ALDEHYDE * CAPRONALDEHYDE * n-CAPROYLALDEHYDE * FEMA No. 2557 * HEXALDEHYDE (DOT) * HEXANAL

CONSENSUS REPORTS: Reported in EPA TSCA Inventory.

DOT Classification: Flammable or Combustible Liquid; Label: Flammable Liquid.

SAFETY PROFILE: Mildly toxic by ingestion and inhalation. An irritant to skin and eyes. Flammable liquid. A dangerous fire hazard when exposed to heat or flame; can react vigorously with oxidizing materials. When heated to decomposition it emits acrid smoke and fumes.

HEM500   CAS: 628-02-4   **HR: 3**
HEXANAMIDE
mf: $C_6H_{13}NO$ mw: 115.20

SYNS: CAPROAMIDE * CAPRONAMIDE * NCI-C02142

CONSENSUS REPORTS: Reported in EPA TSCA Inventory.

SAFETY PROFILE: Moderately toxic by ingestion. Questionable carcinogen with experimental carcinogenic data. When heated to decomposition it emits toxic fumes such as NO_x.

HEN000   CAS: 110-54-3   **HR: 3**
n-HEXANE

DOT: UN 1208
mf: C_6H_{14} mw: 86.20

PROP: Colorless clear liquid; faint odor. Bp: $69°$, ULC: 90-95, lel: 1.2%, uel: 7.5%, fp: $-95.6°$, flash p: $-9.4°F$, d: 0.6603 @ $20°/4°$, autoign temp: $437°F$, vap press: 100 mm @ $15.8°$, vap d: 2.97. Insol in water; misc in chloroform, ether, alc. Very volatile liquid.

SYNS: ESANI (ITALIAN) * GETTYSOLVE-B * HEKSAN (POLISH) * HEXANE (DOT) * HEXANEN (DUTCH) * HEXANES (FCC) * NCI-C60571

CONSENSUS REPORTS: Reported in EPA TSCA Inventory.

OSHA PEL: (Transitional: TWA 500 mg/m^3) TWA 50 ppm
ACGIH TLV: TWA 50 ppm; BEI: 5 mg(2,5-hexanedione)/L in urine at end of shift; 40 ppm n-hexane in end-exhaled air during shift.
DFG MAK: 50 ppm (180 mg/m^3)
NIOSH REL: TWA (Alkanes) 350 mg/m^3
DOT Classification: Flammable Liquid; Label: Flammable Liquid.

SAFETY PROFILE: Slightly toxic by ingestion and inhalation. Human systemic effects by inhalation: hallucinations. Experimental teratogenic and reproductive effects. Mutation data reported. An eye irritant. Can cause a motor neuropathy in exposed workers. May be irritating to respiratory tract and narcotic in high concentrations. Inhalation of 5000 ppm for 1/6-hour produces marked vertigo; 2500-1000 ppm for 12 hours produces drowsiness, fatigue, loss of appetite, paresthesia in distal extremities; 2500-500 ppm produces muscle weakness, cold pulsation in extremities, blurred vision, headache, anorexia, and onset of polyneuropathy. 2000 ppm for 1/6-hour produces no symptoms. 1000-500 ppm for 3-6 months produces fatigue, loss of appetite, distal paresthesia. Dangerous if abused.

Flammable liquid. A very dangerous fire and explosion hazard when exposed to heat or flame; can react vigorously with oxidizing materials. Mixtures with dinitrogen tetraoxide may explode at $28°$. To fight fire, use CO_2, dry chemical. When heated to decomposition it emits acrid smoke and fumes.

HEO000   CAS: 124-09-4   **HR: 3**
1,6-HEXANEDIAMINE

DOT: UN 1783/UN 2280
mf: $C_6H_{16}N_2$ mw: 116.24

PROP: Colorless leaflets; odor of piperidine. Mp: 39-42°, bp: 205°. Absorbs water and CO_2

from air; very sol in water; sltly sol in alc, benzene.

SYNS: 1,6-DIAMINOHEXANE * 1,6-HEXAMETH-YLENEDIAMINE * HEXAMETHYLENE DIAMINE, solid (DOT) * HMDA * NCI-C61405

CONSENSUS REPORTS: Reported in EPA TSCA Inventory.

DOT Classification: Corrosive Material; Label: Corrosive, Solid and Solution; IMO: Corrosive Material; Label: Corrosive, Poison, Solution.

SAFETY PROFILE: Poison by intravenous and intraperitoneal routes. Moderately toxic by ingestion, inhalation, and skin contact. An experimental teratogen. A corrosive irritant to skin, eyes and mucous membranes. Combustible when exposed to heat or flame; can react with oxidizing materials.

HER500   CAS: 628-73-9   HR: 3
HEXANENITRILE
mf: $C_6H_{11}N$ mw: 97.18

SYNS: CAPRONITRILE * NC5

CONSENSUS REPORTS: Reported in EPA TSCA Inventory. Cyanide and its compounds are on the Community Right-To-Know List.

SAFETY PROFILE: A poison by intravenous and subcutaneous routes. Moderately toxic by ingestion. When heated to decomposition it emits its toxic fumes of NO_x and CN^-.

HES000   CAS: 111-31-9   HR: 3
1-HEXANETHIOL
mf: $C_6H_{14}S$ mw: 118.26

SYNS: HEXYL MERCAPTAN * USAF EK-4628

CONSENSUS REPORTS: Reported in EPA TSCA Inventory.
NIOSH REL: (n-Alkane Mono Thiols) CL 0.5 ppm/15M

SAFETY PROFILE: A poison by intraperitoneal route. Moderately toxic by inhalation and ingestion. When heated to decomposition it emits very toxic fumes of SO_x.

HET675   CAS: 918-37-6   HR: 3
HEXANITROETHANE
mf: $C_2N_6O_{12}$ mw: 300.06

DOT Classification: Forbidden

SAFETY PROFILE: A powerful oxidant which explodes above 140°C. Explosive reaction with

boron. Hypergolic reaction with dimethyl hydrazine or other strong organic bases. Forms powerfully explosive mixtures with nitrogen containing organic compounds (e.g., 2-nitroaniline). Upon decomposition it emits toxic fumes of NO_x.

HEU000   CAS: 142-62-1   HR: 2
HEXANOIC ACID
DOT: NA 1706
mf: $C_6H_{12}O_2$ mw: 116.18

PROP: Oily, colorless liquid; odor of Limburger cheese. Bp: 205.0°, fp: −3.4°, flash p: 215°F (COC), d: 0.9295 @ 20°/20°, refr index: 1.415-1.418, vap press: 0.18 mm @ 20°, vap d: 4.0, autoign temp: 716°F. Very sol in ether, fixed oils; sltly sol in water.

SYNS: BUTYLACETIC ACID * CAPROIC ACID * n-CAPROIC ACID * CAPRONIC ACID * FEMA No. 2559 * HEXACID 698 * n-HEXANOIC ACID * n-HEXOIC ACID * PENTIFORMIC ACID * PENTYLFORMIC ACID

CONSENSUS REPORTS: Reported in EPA TSCA Inventory.

DOT Classification: Corrosive Material; Label: Corrosive.

SAFETY PROFILE: Moderately toxic by ingestion, skin contact, intraperitoneal, and subcutaneous routes. Mutation data reported. Corrosive. A skin and severe eye irritant. Combustible when exposed to heat or flame; can react with oxidizing materials. To fight fire, use CO_2, dry chemical, fog, mist. When heated to decomposition it emits acrid smoke and fumes.

HEV000   CAS: 591-78-6   HR: 3
2-HEXANONE
mf: $C_6H_{12}O$ mw: 100.18

PROP: Clear liquid. Mp: −56.9°, bp: 127.2°, lel: 1.22%, uel: 8.0%, flash p: 95°F (OC), d: 0.830 @ 0°/4°, vap press: 10 mm @ 38.8°, vap d: 3.45, autoign temp: 991°F. Sltly sol in water; sol in alc and ether.

SYNS: BUTYL METHYL KETONE * n-BUTYL METHYL KETONE * HEXANONE-2 * MBK * METHYL n-BUTYL KETONE (ACGIH) * MNBK

CONSENSUS REPORTS: Reported in EPA TSCA Inventory.

OSHA PEL: (Transitional: TWA 100 ppm) TWA 5 ppm

ACGIH TLV: TWA 5 ppm
DFG MAK: 5 ppm (21 mg/m^3)
NIOSH REL: (Ketones) TWA 4 mg/m^3

SAFETY PROFILE: Moderately toxic by ingestion and intraperitoneal routes. Mildly toxic by inhalation and skin contact. Experimental teratogenic and reproductive effects. Human systemic effects by inhalation: unspecified eye effects, headache, nausea or vomiting. An eye irritant. Dangerous fire and explosion hazard when exposed to heat or flame; can react with oxidizing materials. To fight fire, use alcohol foam, CO_2, dry chemical.

HEW000   CAS: 45776-10-1   HR: 3
1-HEXANOYLAZIRIDINE
mf: $C_8H_{15}NO$ mw: 141.24

SYNS: 1-CAPROYLAZIRIDINE * CAPROYLETHYL-ENEIMINE * HEXANOYLETHYLENEIMINE

SAFETY PROFILE: Questionable carcinogen with experimental neoplastigenic and tumorigenic data. Mutation data reported. When heated to decomposition it emits toxic fumes of NO_x.

HFA300   CAS: 51235-04-2   HR: 2
HEXAZINONE
mf: $C_{12}H_{20}N_4O_2$ mw: 252.36

SYNS: 3-CYCLOHEXYL-6-(DIMETHYLAMINO)-1-METHYL-s-TRIAZINE-2,4(1H,3H)-DIONE * 3-CYCLO-HEXYL-6-(DIMETHYLAMINO)-1-METHYL-1,3,5-TRIAZINE-2,4(1H,3H)-DIONE * DPX 3674 * VELPAR * VELPAR WEED KILLER

CONSENSUS REPORTS: Reported in EPA TSCA Inventory.

SAFETY PROFILE: Moderately toxic by ingestion and intraperitoneal routes. Mildly toxic by skin contact. Experimental reproductive effects. An eye irritant. When heated to decomposition it emits toxic fumes of NO_x.

HFA500   CAS: 505-57-7   HR: 3
2-HEXENAL
mf: $C_6H_{10}O$ mw: 98.16

SYNS: HEX-2-ENAL * HEX-2-EN-1-AL * HEXY-LENIC ALDEHYDE * LEAF ALDEHYDE

CONSENSUS REPORTS: Reported in EPA TSCA Inventory.

SAFETY PROFILE: A poison by intraperitoneal route. Mutation data reported. When heated to decomposition it emits acrid smoke and fumes.

HFA525   HR: 3
trans-2-HEXEN-1-AL
mf: $C_6H_{10}O$ mw: 98.15

PROP: Pale yellow liquid; fruity, vegetable odor. D: 0.841-0.848, refr index: 1.445-1.449, flash p: 100°F. Sol in alc, propylene glycol, fixed oils; very sltly sol in water.

SYN: FEMA No. 2560

SAFETY PROFILE: Flammable liquid. When heated to decomposition it emits acrid smoke and irritating fumes.

HFB000   CAS: 592-41-6   HR: 3
1-HEXENE
DOT: UN 2370
mf: C_6H_{12} mw: 84.158

PROP: Colorless liquid. Bp: 64.5°, mp: −139.9°, flash p: −14.8°F, d: 0.6732 @ 20°/4°, vap press: 310 mm @ 38°, vap d: 3.0, lel: 1.2%, uel: 6.9%

SYNS: BUTYL ETHYLENE * HEXENE * HEXY-LENE

CONSENSUS REPORTS: Reported in EPA TSCA Inventory.

DOT Classification: Flammable Liquid; Label: Flammable Liquid.

SAFETY PROFILE: Moderately toxic irritant to skin, eyes and mucous membranes. A very dangerous fire and explosion hazard when exposed to heat, flame, or oxidizers. Can react vigorously with oxidizing materials. To fight fire, use dry chemical, CO_2, foam. When heated to decomposition it emits acrid smoke and fumes.

HFD500   CAS: 928-95-0   HR: 2
2-HEXEN-1-OL, (E)-
mf: $C_6H_{12}O$ mw: 100.18

PROP: Colorless liquid; fruity-green odor. D: 0.836-0.841, refr index: 0.437-1.442, flash p: 129°F. Sol in alc, propylene glycol, fixed oils; very sltly sol in water.

SYNS: FEMA No. 2562 * trans-2-HEXENOL * 2-HEXENOL * trans-2-HEXEN-1-OL (FCC)

CONSENSUS REPORTS: Reported in EPA TSCA Inventory.

SAFETY PROFILE: Moderately toxic by ingestion. Mildly toxic by skin contact. A skin irritant. Combustible liquid. When heated to decomposition it emits acrid smoke and fumes.

HFE000   CAS: 928-96-1   **HR: 3**
cis-3-HEXENOL
mf: $C_6H_{12}O$ mw: 100.18

PROP: Colorless liquid; powerful grassy-green odor. D: 0.846-0.850, refr index: 1.43-1.441, bp: 137°, flash p: 111°F. Sol in alc, propylene glycol, fixed oils; very sltly sol in water.

SYNS: BLATTERALKOHOL * FEMA No. 2563
* β-γ-HEXENOL * cis-3-HEXEN-1-OL (FCC)
* LEAF ALCOHOL

CONSENSUS REPORTS: Reported in EPA TSCA Inventory.

SAFETY PROFILE: A poison by intraperitoneal route. Mildly toxic by ingestion. Combustible liquid. When heated to decomposition it emits acrid smoke and fumes.

HFE550   **HR: 2**
cis-3-HEXENYL 2-METHYLBUTYRATE
mf: $C_{11}H_{20}O_2$ mw: 184.28

PROP: Colorless liquid; powerful, fruity odor like unripe apples. D: 0.876-0.880, refr index: 1.430, flash p: 153°F. Sol in alc, fixed oils; insol in water.

SYN: FEMA No. 3497

SAFETY PROFILE: Combustible liquid. When heated to decomposition it emits acrid smoke and irritating fumes.

HFF000   CAS: 10138-60-0   **HR: 3**
4-HEXEN-1-YN-3-OL
mf: C_6H_8O mw: 96.14

SYN: 4-HEXENE-1-YNE-3-OL

SAFETY PROFILE: A poison by ingestion, inhalation, and skin contact. A skin irritant. When heated to decomposition it emits acrid smoke and fumes.

HFF300   CAS: 13061-80-8   **HR: 3**
4-HEXEN-1-YN-3-ONE
mf: C_6H_6O mw: 94.12

SYN: 4-HEXENE-1-YNE-3-ONE

SAFETY PROFILE: A poison by ingestion, inhalation, and skin contact. A skin irritant. When heated to decomposition it emits acrid smoke and fumes.

HFG500   CAS: 108-10-1   **HR: 3**
HEXONE
DOT: UN 1245
mf: $C_6H_{12}O$ mw: 100.18

PROP: Colorless mobile liquid; fruity, ethereal odor. Bp: 118°, lel: 1.4%, uel: 7.5%, flash p: 62.6°F, d: 0.796-0.799, fp: −80.2°, autoign temp: 858°F, vap press: 16 mm @ 20°. Misc with alc, ether; sol in alc.

SYNS: FEMA No. 2731 * HEXON (CZECH)
* ISOBUTYL-METHYLKETON (CZECH) * ISOBUTYL METHYL KETONE * ISOPROPYLACETONE
* METHYL-ISOBUTYL-CETONE (FRENCH) * METHYLISOBUTYLKETON (DUTCH, GERMAN) * METHYL ISOBUTYL KETONE (ACGIH, DOT) * METYLOIZOBUTYLOKETON (POLISH) * 4-METHYL-PENTAN-2-ON (DUTCH, GERMAN) * 4-METHYL-2-PENTANON (CZECH) * 2-METHYL-4-PENTANONE * 4-METHYL-2-PENTANONE (FCC) * METILISOBUTILCHETONE (ITALIAN) * 4-METILPENTAN-2-ONE (ITALIAN)
* MIBK * MIK * RCRA WASTE NUMBER U161
* SHELL MIBK

CONSENSUS REPORTS: Community Right-To-Know List.

OSHA PEL: (Transitional: TWA 100 mg/m³) TWA 50 ppm; STEL 75 ppm
ACGIH TLV: TWA 50 ppm; STEL 75 ppm
NIOSH REL: (Ketones) TWA 200 mg/m³
DOT Classification: Flammable Liquid; Label: Flammable Liquid.

SAFETY PROFILE: A poison by intraperitoneal route. Moderately toxic by ingestion. Mildly toxic by inhalation. Very irritating to the skin, eyes, and mucous membranes. A human systemic irritant by inhalation. Narcotic in high concentration. Flammable liquid when exposed to heat, flame, or oxidizers. Ignites on contact with potassium-tert-butoxide. Moderately explosive in the form of vapor when exposed to heat or flame. May form explosive peroxides upon exposure to air. Can react vigorously with reducing materials. To fight fire, use alcohol foam, CO_2, dry chemical. Incompatible with air, potassium-tert-butoxide.

HFI500   CAS: 142-92-7   **HR: 1**
HEXYL ACETATE
mf: $C_8H_{16}O_2$ mw: 144.24

PROP: Colorless liquid; fruity odor. D: 0.878, mp: −60.9°, bp: 171.5°, refr index: 1.407, flash p: 109°F. Insol in water; very sol in alc and ether.

SYNS: ACETIC ACID HEXYL ESTER * FEMA No. 2565 * n-HEXYL ACETATE (FCC) * 1-HEXYL ACETATE * HEXYL ALCOHOL, ACETATE * HEXYL ETHANOATE

CONSENSUS REPORTS: Reported in EPA TSCA Inventory.

SAFETY PROFILE: Mildly toxic by ingestion. Combustible liquid. When heated to decomposition it emits acrid smoke and fumes.

HFJ000   CAS: 108-84-9   **HR: 1**
sec-HEXYL ACETATE

DOT: UN 1233
mf: $C_8H_{16}O_2$ mw: 144.24

PROP: Clear liquid, pleasant odor. Bp: 146.3°, fp: −63.8°, flash p: 113°F (COC), d: 0.8598 @ 20°/20°, vap press: 3.8 mm @ 20°, vap d: 4.97.

SYNS: ACETIC ACID-1,3-DIMETHYLBUTYL ESTER * 1,3-DIMETHYLBUTYL ACETATE * MAAC * METHYLAMYL ACETATE * METHYL AMYL ACETATE (DOT) * METHYLISOAMYL ACETATE * METHYLISOBUTYLCARBINOL ACETATE * METHYLISOBUTYLCARBINYL ACETATE * 4-METHYL-2-PENTANOL, ACETATE * 4-METHYL-2-PENTYL ACETATE

CONSENSUS REPORTS: Reported in EPA TSCA Inventory.

OSHA PEL: TWA 50 ppm
ACGIH TLV: TWA 50 ppm
DFG MAK: 50 ppm (300 mg/m^3)
DOT Classification: Flammable or Combustible Liquid; Label: Flammable Liquid.

SAFETY PROFILE: Mildly toxic by ingestion, skin contact and inhalation. Human systemic effects by inhalation: conjunctiva irritation, unspecified changes in olfactory and respiratory systems. A skin and human eye irritant. Flammable when exposed to heat or flame; can react with oxidizing materials. To fight fire, use alcohol foam, CO_2, dry chemical.

HFJ500   CAS: 111-27-3   **HR: 3**
n-HEXYL ALCOHOL

DOT: UN 2282
mf: $C_6H_{14}O$ mw: 102.20

PROP: Colorless liquid. Bp: 157.2°, fp: −44.6°, flash p: 145°F, d: 0.816-0.821, vap

press: 1 mm @ 24.4°, vap d: 3.52. Misc in alc, ether; sltly sol in water.

SYNS: AMYLCARBINOL * CAPROYL ALCOHOL * EPAL 6 * FEMA No. 2567 * HEXANOL * n-HEXANOL (DOT) * 1-HEXANOL * HEXYL ALCOHOL * 1-HYDROXYHEXANE * PENTYLCARBINOL

CONSENSUS REPORTS: Reported in EPA TSCA Inventory.

DOT Classification: Flammable or Combustible Liquid; Label: Flammable Liquid.

SAFETY PROFILE: Poison by intravenous route. Moderately toxic by ingestion and skin contact. A skin and severe eye irritant. Combustible liquid. Can react with oxidizing materials. To fight fire, use alcohol foam, CO_2, dry chemical.

HFJ600   CAS: 26401-20-7   **HR: 3**
tert-HEXYL ALCOHOL

DOT: UN 2282
mf: $C_6H_{14}O$ mw: 102.20

SYN: tert-HEXANOL (9CI, DOT)

DOT Classification: Flammable or Combustible Liquid; Label: Flammable Liquid.

SAFETY PROFILE: Poison by ingestion and intravenous routes. When heated to decomposition it emits acrid smoke and fumes. Flammable when exposed to heat or flame; can react vigorously with oxidizing materials. A fire hazard.

HFK000   CAS: 111-26-2   **HR: 3**
HEXYLAMINE
mf: $C_6H_{15}N$ mw: 101.22

PROP: Liquid. Mp: −22.9°, bp: 131.4°, flash p: 85°F (OC), d: 0.7675 @ 20°/20°, vap d: 3.49.

SYNS: 1-AMINOHEXANE * 1-HEXANAMINE * N-HEXYLAMINE * MONO-N-HEXYLAMINE

CONSENSUS REPORTS: Reported in EPA TSCA Inventory.

SAFETY PROFILE: A poison by intraperitoneal routes. Moderately toxic by ingestion, inhalation, and skin contact. A severe skin and eye irritant. Dangerous fire hazard when exposed to heat or flame; can react with oxidizing materials. To fight fire, use alcohol foam, CO_2, dry chemical. Upon decomposition it emits toxic fumes of NO_x.

HFO500 CAS: 101-86-0 **HR: 2**
HEXYL CINNAMALDEHYDE
mf: $C_{15}H_{20}O$ mw: 216.35

PROP: Pale yellow liquid; jasmine odor. D 0.953-0.959, refr index: 1.548-1.552. Sol in fixed oils; insol in propylene glycol, glycerin.

SYNS: FEMA No. 2569 * α-HEXYLCINNAMALDE-HYDE (FCC) * HEXYL CINNAMIC ALDEHYDE * α-HEXYLCINNAMIC ALDEHYDE * α-n-HEXYL-β-PHENYLACROLEIN * 2-(PHENYLMETHYLENE)OCTA-NOL

CONSENSUS REPORTS: Reported in EPA TSCA Inventory.

SAFETY PROFILE: Moderately toxic by ingestion. A skin irritant. When heated to decomposition it emits acrid smoke and fumes.

HFP875 CAS: 107-41-5 **HR: 2**
HEXYLENE GLYCOL
mf: $C_6H_{14}O_2$ mw: 118.20

PROP: Mild odor, colorless liquid, water-sol. Bp: 197.1°, fp: −50°, flash p: 205°F (OC), d: 0.9234 @ 20°/20°, vap press: 0.05 mm @ 20°, d: 4.

SYNS: 2,4-DIHYDROXY-2-METHYLPENTANE * DIOLANE * 1,2-HEXANEDIOL * ISOL * 2-METHYL PENTANE-2,4-DIOL * 2-METHYL-2,4-PENTANEDIOL * PINAKON * α,α,α′-TRIMETHYL-TRIMETHYLENE GLYCOL

CONSENSUS REPORTS: Reported in EPA TSCA Inventory.

OSHA PEL: CL 25 ppm
ACGIH TLV: CL 25 ppm

SAFETY PROFILE: Moderately toxic by ingestion and intraperitoneal routes. Mildly toxic by skin contact. Human systemic effects by inhalation: conjunctiva and other eye, olfactory and pulmonary changes. Mutation data reported. Combustible when exposed to heat or flame; can react with oxidizing materials. To fight fire, use foam, CO_2, dry chemicals. When heated to decomposition it emits acrid smoke and fumes.

HFR200 **HR: 2**
HEXYL 2-METHYLBUTYRATE
mf: $C_{11}H_{22}O_2$ mw: 186.30

PROP: Colorless liquid; strong, fresh-green, fruity odor. D: 0.854, refr index: 1.416-1.421,

flash p: 122°F. Sol in alc, fixed oils; insol in water.

SYN: FEMA No. 3499

SAFETY PROFILE: Combustible liquid. When heated to decomposition it emits acrid smoke and irritating fumes.

HFV500 CAS: 136-77-6 **HR: 3**
HEXYLRESORCINOL
mf: $C_{12}H_{18}O_2$ mw: 194.30

PROP: Colorless liquid to pale yellow, heavy liquid becoming solid on standing at room temp; needles from benzene or petr ether. Pungent odor, sharp astringent taste. Bp: 179°, mp: 67.5-69°. Very sol in water; sol in benzene, ether, acetone, chloroform, alc, vegetable oils; sltly sol in petr ether.

SYNS: ASCARYL * CAPROKOL * CRYSTOIDS * CYSTOIDS ANTHELMINTIC * 4-HEXYL-1,3-BEN-ZENEDIOL * 4-HEXYL-1,3-DIHYDROXYBENZENE * HEXYLRESORCIN (GERMAN) * 4-HEXYLRESOR-CINE * p-HEXYLRESORCINOL * 4-HEXYLRESORCI-NOL * 4-n-HEXYLRESORCINOL * NCI-C55787 * S.T. 37 * SUCRETS * WORM-AGEN

CONSENSUS REPORTS: Reported in EPA TSCA Inventory.

SAFETY PROFILE: A poison by ingestion, intraperitoneal, and possibly other routes. Moderately toxic by subcutaneous route. Experimental reproductive effects. Questionable carcinogen with experimental tumorigenic data. Mutation data reported. An eye irritant. Concentrated solutions can cause burns on the skin and mucous membranes in humans. An anthelmintic and topical antiseptic. When heated to decomposition it emits acrid smoke and fumes.

HFX500 CAS: 928-65-4 **HR: 3**
HEXYLTRICHLOROSILANE

DOT: UN 1784
mf: $C_6H_{13}Cl_3Si$ mw: 219.63

CONSENSUS REPORTS: Reported in EPA TSCA Inventory.

DOT Classification: Label: Corrosive.

SAFETY PROFILE: A poison by ingestion and inhalation. Corrosive. A severe irritant to skin, eyes, and mucous membranes. When heated to decomposition or in reaction with water or steam it produces toxic and corrosive fumes of Cl^- and HCl.

HGC000   CAS: 1936-15-8   **HR: D**
HISPACID FAST ORANGE 2G
mf: $C_{16}H_{10}N_2O_7S_2 \cdot 2Na$ mw: 452.38

SYNS: ACIDAL FAST ORANGE * ACID FAST OR-
ANGE EGG * ACID LEATHER ORANGE PGW
* ACID LIGHT ORANGE G * ACID ORANGE 10
* ACILAN ORANGE GX * APOCID ORANGE 2G
* ATUL ACID CRYSTAL ORANGE G * BRASILAN
ORANGE 2G * BUCACID FAST ORANGE G * CAL-
COCID FAST LIGHT ORANGE 2G * CERTICOL OR-
ANGE GS * CETIL LIGHT ORANGE GG * C.I. 27
* C.I. ACID ORANGE 10 * C.I. FOOD ORANGE 4
* CRYSTAL ORANGE 2G * D&C ORANGE No. 3
* ENIACID LIGHT ORANGE G * ERIO FAST ORANGE
AS * FAST LIGHT ORANGE GA * HEXACOL OR-
ANGE GG CRYSTALS * HIDACID FAST ORANGE G
* 7-HYDROXY-8-(PHENYLAZO)-1,3-NAPHTHALENEDI-
SULFONIC ACID, DISODIUM SALT * 7-HYDROXY-8-
(PHENYLAZO)-1,3-NAPHTHALENEDISULPHONIC ACID,
DISODIUM SALT * INK ORANGE JSN * INTRACID
FAST ORANGE G * JAVA ORANGE 2G * KITON
FAST ORANGE G * NAPHTHALENE FAST ORANGE
2GS * NCI-C53838 * NEKLACID FAST ORANGE 2G
* ORANGE #10 * ORANGE G (BIOLOGICAL STAIN)
* ORANGE G DYE * ORANGE G (INDICATOR)
* ORANZ G (POLISH) * 1-PHENYLAZO-2-NAPHTHOL-
6,8-DISULFONIC ACID, DISODIUM SALT * 1-PHENYL-
AZO-2-NAPHTHOL-6,8-DISULPHONIC ACID, DISODIUM
SALT * SCHULTZ NO. 39 * SOLAR LIGHT ORANGE
GX * STANDACOL ORANGE G * SULFACID LIGHT
ORANGE J * TERTRACID LIGHT ORANGE G
* UNITERTRACID LIGHT ORANGE G * VENDACID
LIGHT ORANGE 2G * WOOL ORANGE 2G * XY-
LENE FAST ORANGE G

CONSENSUS REPORTS: IARC Cancer Review: GROUP 3 IMEMDT 7,56,87; Animal
Inadequate Evidence IMEMDT 8,181,75. Reported in EPA TSCA Inventory. EPA Genetic
Toxicology Program.

SAFETY PROFILE: Experimental reproductive
effects. Questionable carcinogen. Mutation data
reported. Used as a drug and cosmetic colorant.
When heated to decomposition it emits very
toxic SO_x, Na_2O, and NO_x.

HGC500   CAS: 569-65-3   **HR: 3**
HISTAMETHIZINE
mf: $C_{25}H_{27}ClN_2$ mw: 390.99

SYNS: ANCOLAN * BONADETTES * BONA-
DOXIN * BONAMINE * CALMONAL
* CHICLIDA * 1-(p-CHLOROBENZHYDRYL)-4-(m-
METHYLBENZYL)DIETHYLENEDIAMINE * 1-p-
CHLOROBENZHYDRYL-4-m-METHYLBENZYLPIPERAZINE
* 1-(p-CHLORO-α-PHENYLBENZYL)-4-(m-METHYLBEN-
ZYL)PIPERAZINE * HISTAMETHINE * HISTAMETI-
ZINE * HISTAMETIZYNE * ITINEROL * LONGI-
FENE * MAREX * MECLIZINE * MECLOZINE
* NAVICALM * NEO-ISTAFENE * NEO-SUPRIMAL
* NEO-SUPRIMEL * PARACHLORAMINE * PERE-
MESIN * POSTAFEN * SABARI * SEA-LEGS
* SIGURAN * SUBARI * SUPRIMAL * TRA-
VELON * UCB 170 * VIBAZINE * VOMISSELS

CONSENSUS REPORTS: Reported in EPA
TSCA Inventory.

SAFETY PROFILE: A poison by intravenous
route. Moderate toxicity by ingestion and intramuscular routes. Human reproductive effects
by an unspecified route: reduced viability of
newborn. Experimental teratogenic and reproductive effects. An antihistamine. When heated
to decomposition it emits very toxic fumes of
Cl^- and NO_x.

HGD000   CAS: 51-45-6   **HR: 3**
HISTAMINE
mf: $C_5H_9N_3$ mw: 111.17

PROP: White crystals. Mp: 83-84°, bp: 210°
@ 18 mm. Very sol in water; sol in hot chloroform; insol in ether.

SYNS: β-AMINOETHYLGLYOXALINE * β-AMINO-
ETHYLIMIDAZOLE * 4-(2-AMINOETHYL)IMIDAZOLE
* ERAMIN * ERGAMINE * ERGOTIDINE
* FREE HISTAMINE * 1H-IMIDAZOLE-4-ETHANA-
MINE * IMIDAZOLE-4-ETHYLAMINE * 4-IMID-
AZOLEETHYLAMINE * 5-IMIDAZOLEETHYLAMINE
* β-IMIDAZOLYL-4-ETHYLAMINE * 2-(4-IMID-
AZOLYL)ETHYLAMINE * 2-IMIDAZOL-4-YL-ETHYL-
AMINE * THERAMINE

SAFETY PROFILE: A poison by intravenous
and subcutaneous routes. Moderately toxic by
intraperitoneal route. Experimental reproductive
effects. Human mutation data reported. A neurotransmitter. The most potent capillary dilator
known. Ingestion or inhalation produces the following effects: flushing followed by pallor, dizziness, fainting, fall in blood pressure, headache, rapid, weak pulse. Allergic effects on skin
(hives) may occur. When heated to decomposition it emits toxic fumes of NO_x.

HGE700   CAS: 71-00-1   **HR: D**
HISTIDINE
mf: $C_6H_9N_3O_2$ mw: 155.18

PROP: l-Histidine, the natural form. White needles, plates, or crystalline powder; sltly bitter

taste. Decomp 287° (softens at 277°). Solubility in water at 25°: 41.9 g/L. Sol in water; very sltly sol in alc; insol in ether.

SYNS: l-α-AMINO-4(OR 5)-IMIDAZOLEPROPIONIC ACID * GLYOXALINE-5-ALANINE * l-HISTIDINE (FCC)

CONSENSUS REPORTS: Reported in EPA TSCA Inventory.

SAFETY PROFILE: Experimental reproductive effects. Human mutation data reported. When heated to decomposition it emits toxic fumes of NO$_x$.

HGF500   **HR: 2**
HOLMIUM
af: Ho   aw: 164.93

PROP: Bright metallic luster; soft, malleable metal; stable in dry air; oxidizes rapidly in moist air. Bp: 2720°, d: 8.78 @ 25°, vap press 2 mm @ 1630°.

SAFETY PROFILE: It may be an anticoagulant like the lanthanides. The toxicity (intravenous administration) of the salts decreases as follows: nitrate > sulfate > 3-sulfoisonicotinate > acetate > propionate > chloride. Can react violently with air or halogens.

HGG000   CAS: 10138-62-2   **HR: 3**
HOLMIUM CHLORIDE
mf: Cl$_3$Ho   mw: 271.28

PROP: Bright yellow, crystalline solid. Mp: 718°.

CONSENSUS REPORTS: Reported in EPA TSCA Inventory.

SAFETY PROFILE: A poison by intraperitoneal route. Mildly toxic by ingestion. When heated to decomposition it emits highly toxic fumes of Cl$^-$.

HGG500   CAS: 13455-50-0   **HR: 3**
HOLMIUM CITRATE
mf: C$_6$H$_5$O$_7$ • Ho   mw: 354.04

SAFETY PROFILE: A poison by intraperitoneal route. When heated to decomposition it emits acrid smoke and fumes.

HGH000   CAS: 35725-31-6   **HR: 3**
HOLMIUM(III) NITRATE, HEXAHYDRATE (1:3:6)
mf: N$_3$O$_9$ • Ho • 6H$_2$O   mw: 459.08

SYN: NITRIC ACID, HOLMIUM(3$^+$) SALT, HEXA-HYDRATE

SAFETY PROFILE: A poison by intraperitoneal route. Moderately toxic by ingestion. When heated to decomposition it emits toxic fumes of NO$_x$.

HGK500   CAS: 117-51-1   **HR: 3**
3-HOMOTETRA HYDRO CANNIBINOL
mf: C$_{22}$H$_{32}$O$_2$ mw: 328.54

SYNS: 3-HEXYL-7,8,9,10-TETRAHYDRO-6,6,9-TRI-METHYL-6H-DIBENZO(B,D)PYRAN-1-OL * 1-HY-DROXY-3-N-HEXYL-6,6,9-TRIMETHYL-7,8,9,10-TETRAHY-DRO-6-DIBENZOPYRAN * PARAHEXYL * PYRA-HEXYL * SYNHEXYL

SAFETY PROFILE: A poison by intravenous route. Moderately toxic by ingestion and intraperitoneal routes. When heated to decomposition it emits acrid smoke and fumes.

HGO500   CAS: 23255-93-8   **HR: 3**
HYCANTHONE METHANESULFONATE
mf: C$_{20}$H$_{24}$N$_2$O$_2$S • CH$_4$O$_3$S; mw: 452.63

SYNS: 1-((2-(DIETHYLAMINO)ETHYL)AMINO)-4-(HY-DROXYMETHYL)-9H-THIOXANTHEN-9-ONE MONOMETHANE-SULFONATE (SALT) * ETRENOL * HCT * HYCANTHONE MESYLATE * HYCAN-THONE METHANESULPHONATE * HYCANTHONE MO-NOMETHANESULPHONATE

CONSENSUS REPORTS: IARC Cancer Review: GROUP 3 IMEMDT 7,56,87; Animal Inadequate Evidence IMEMDT 13,91,77. EPA Genetic Toxicology Program.

SAFETY PROFILE: A poison by intraperitoneal, subcutaneous, intravenous, and intramuscular routes. Moderately toxic by ingestion. Questionable carcinogen with experimental carcinogenic, tumorigenic, and teratogenic data. Human systemic effects by intramuscular route: hallucinations, muscle weakness, nausea or vomiting. Human mutation data reported. When heated to decomposition it emits very toxic fumes of SO$_x$.

HGP000   CAS: 109-78-4   **HR: 3**
HYDRACRYLONITRILE
mf: C$_3$H$_5$NO   mw: 71.09

PROP: Colorless to straw-colored liquid. Bp: 228° decomp, fp: −46°, flash p: 265°F (OC), d: 1.0404 @ 25°, vap press: 0.08 mm @ 25°, vap d: 2.45. Misc with water, acetone, methyl ethyl ketone, and ethanol. Sltly sol in ether; insol in benzene, petr ether, carbon disulfide, and carbon tetrachloride.

SYNS: 2-CYANOETHANOL * 2-CYANOETHYL AL-COHOL * ETHYLENE CYANOHYDRIN * GLYCOL CYANOHYDRIN * β-HPN * 3-HYDROXYPROPANE-NITRILE * β-HYDROXYPROPIONITRILE * 3-HY-DROXYPROPIONITRILE * METHANOLACETONITRILE * USAF RH-7

CONSENSUS REPORTS: Reported in EPA TSCA Inventory. Cyanide compounds are on the Community Right-To-Know List.

SAFETY PROFILE: Poison by inhalation. Moderately toxic by ingestion and intraperitoneal routes. Mildly toxic by skin contact. A skin and eye irritant. Combustible when exposed to heat or flame. Reacts violently with mineral acids (e.g., chlorosulfonic acid; oleum; sulfuric acid); amines or inorganic bases (e.g., NaOH). Reacts with water or steam to produce toxic and flammable vapors. To fight fire, use CO_2, dry chemical, alcohol foam. When heated to decompositon or on contact with acid or acid fumes it emits highly toxic fumes of CN^-.

HGP500   CAS: 304-20-1   HR: 3
HYDRALAZINE HYDROCHLORIDE
mf: $C_8H_8N_4 \cdot ClH$ mw: 196.66

PROP: Yellow crystals. Decomp @ 273°. Very sltly sol in ether.

SYNS: AISELAZINE * APPRESINUM * APREL-AZINE * APRESAZIDE * APRESINE * APRESO-LIN * APRESOLINE-ESIDRIX * APRESOLINE HY-DROCHLORIDE * APREZOLIN * BA 5968 * CIBA 5968 * DRALZINE * HIDRALAZIN * HIPOFTALIN * HYDRALAZINE CHLORIDE * HYDRALAZINE MONOHYDROCHLORIDE * HY-DRALLAZINE HYDROCHLORIDE * HYDRAPRESS * 1-HYDRAZINOPHTHALAZINE HYDROCHLORIDE * 1-HYDRAZINOPHTHLAZINE MONOHYDROCHLORIDE * HYPERAZIN * HYPOPHTHALIN * HYPOS * IPOLINA * LOPRESS * NOR-PRESS 25 * 1(2H)-PHTHALAZINONE HYDRAZONE HYDROCHLO-RIDE * 1(2H)-PHTHLAZINONE, HYDRAZONE, MONOHYDROCHLORIDE * PRAPARAT 5968 * ROLAZINE * SERPASIL APRESOLINE No. 2

CONSENSUS REPORTS: IARC Cancer Review: Animal Limited Evidence IMEMDT 24,85,80. Reported in EPA TSCA Inventory.

SAFETY PROFILE: Suspected carcinogen with experimental neoplastigenic data. A poison by ingestion, subcutaneous, intravenous, and intraperitoneal routes. Human mutation data reported. An antihypertensive agent. When heated to decomposition it emits very toxic NO_x and HCl.

HGS000   CAS: 302-01-2   HR: 3
HYDRAZINE
DOT: UN 2029/UN 2030
mf: H_4N_2 mw: 32.06

PROP: Colorless, oily, fuming liquid or white crystals. Mp: 1.4°, bp: 113.5°, flash p: 100°F (OC), d: 1.1011 @ 15° (liquid), autoign temp: can vary from 74°F in contact with iron rust, 270°F in contact with black iron, 313°F in contact with stainless steel, 518°F in contact with glass. Vap d: 1.1; lel: 4.7%, uel: 100%.

SYNS: DIAMIDE * DIAMINE * HYDRAZINE, anhydrous (DOT) * HYDRAZINE, aqueous solution (DOT) * HYDRAZINE BASE * HYDRAZYNA (POLISH) * RCRA WASTE NUMBER U133

CONSENSUS REPORTS: IARC Cancer Review: GROUP 2B IMEMDT 7,223,87; Animal Sufficient Evidence IMEMDT 4,127,74. NTP Fourth Annual Report On Carcinogens, 1984. EPA Extremely Hazardous Substances List. Community Right-To-Know List. Genetic Toxicology Program. Reported in EPA TSCA Inventory.

OSHA PEL: (Transitional: TWA 1 ppm (skin)) TWA 0.1 ppm (skin)
ACGIH TLV: TWA 0.1 ppm (skin); Suspected Human Carcinogen; (Proposed: 0.01 ppm (skin); Suspected Human Carcinogen)
DFG TRK: 0.1 ppm; Animal Carcinogen, Suspected Human Carcinogen.
NIOSH REL: (Hydrazines) CL 0.04 mg/m^3/2H
DOT Classification: Flammable Liquid; Label: Flammable Liquid and Poison; Corrosive Material; Label: Corrosive, aqueous solution.

SAFETY PROFILE: Confirmed carcinogen with experimental carcinogenic, neoplastigenic, and tumorigenic data. A poison by ingestion, skin contact, intraperitoneal, intravenous, and possibly other routes. Moderately toxic by inhalation. Experimental teratogenic and reproductive effects. Human mutation data reported. A powerful reducing agent which is corrosive to the eyes, skin, and mucous membranes. May cause skin sensitization as well as systemic poisoning. Hydrazine and some of its derivatives may cause damage to the liver and destruction of red blood cells.

Flammable liquid. A very dangerous fire hazard when exposed to heat, flame or oxidizing agents. Severe explosion hazard when exposed to heat or flame or by chemical reaction. Explodes on contact with barium oxide, calcium oxide, chromate salts, chromium dioxide, dicyanofurazan, mercury oxide, trioxygen difluoride, N-haloimides, potassium, silver compounds, sodium hydroxide, titanium compounds (at 130°). Potentially explosive reactions with alkali metals, NH_3, Cl_2, chromates, CuO, Cu++ salts, F_2, metallic oxides, Ni, $Ni(ClO_4)_2$, O_2, liquid O_2, $K_2Cr_2O_7$, $Na_2Cr_2O_7$, tetryl, zinc diamide, $Zn(C_2H_5)_2$. Forms sensitive, explosive mixtures with 2-chloro-5-methylnitrobenzene, metal salts [e.g., cadmium perchlorate, copper chlorate (heat-sensitive), manganese nitrate (heat-sensitive), mercury(I) chloride, mercury-(II) chloride, mercury(I) nitrate, mercury(II) nitrate, tin(II) chloride], methanol + nitromethane, air, lithium perchlorate, sodium perchlorate, sodium. Ignites on contact with cotton waste + heavy metals, dinitrogen oxide, rhenium + alumina, catalysts, nitric acid, hydrogen peroxide, N,2,4,6-tetranitroaniline, rust + heat. Ignites spontaneously in air when absorbed on earth, asbestos, cloth, wood. Violent reaction with 1-chloro-2,4-dinitrobenzene, oxidants (e.g., iron oxide, chlorates, peroxides), thiocarbonyl azide thiocyanate. Vigorous reaction with benzene-seleninic acid or anhydride, carbon dioxide + stainless steel, copper oxide, lead oxide, potassium peroxodisulfate, ruthenium(III) chloride. On contact with metal catalysts (e.g., platinum black, Raney nickel, copper-iron oxide, molybdenum, molybdenum oxides, iridium), it decomposes to ammonia, hydrogen and nitrogen gases which may ignite or explode. A hypergolic reaction with dinitrogen tetraoxide is the basis of a liquid rocket fuel mixture. The vapor will burn without air. It is a powerful explosive. It is very sensitive and must not be used without full and complete instructions from the manufacturer for handling, storage, and disposal. Dangerous; when heated to decomposition it emits highly toxic fumes of NO_x and NH_3.

HGU000   CAS: 57-56-7   HR: 3
HYDRAZINECARBOXAMIDE
mf: CH_5N_3O mw: 75.09

SYNS: AMINOUREA * CARBAMIC ACID HYDRA-ZIDE * CARBAMOYLHYDRAZINE * CARBAMYL-HYDRAZINE * CARBAZAMIDE * SEMICARBAZIDE

CONSENSUS REPORTS: Reported in EPA TSCA Inventory.

SAFETY PROFILE: A poison by ingestion, intraperitoneal, subcutaneous, and intravenous routes. Human systemic effects by intravenous route: convulsions. Questionable carcinogen with experimental tumorigenic data. Mutation data reported. When heated to decomposition it emits toxic fumes of NO_x.

HGU500   CAS: 7803-57-8   HR: 3
HYDRAZINE HYDRATE
mf: $H_4N_2 \cdot H_2O$ mw: 50.08

PROP: Colorless fuming, refractive liquid. Mp: $-51.7°$, bp: $118.5°$ @ 740 mm. D: 1.03 @ 21°. Faint characteristic odor. A strong base, very corrosive; attacks glass, rubber, and cork. Very powerful reducing agent. Misc with water and alc; insol in chloroform and ether.

SYNS: HYDRAZINE MONOHYDRATE * IDRAZINA IDRATA (ITALIAN)

CONSENSUS REPORTS: EPA Genetic Toxicology Program.

NIOSH REL: (Hydrazines) CL 0.04 mg/m^3/2H

SAFETY PROFILE: A poison by ingestion and intravenous routes. Experimental reproductive effects. A corrosive irritant to the eyes, skin, and mucous membranes. Questionable carcinogen with experimental carcinogenic data. Mutation data reported. Incompatible with HgO, Na, $SnCl_2$, 2,4-dinitrochlorobenzene. When heated to decomposition it emits toxic fumes of NO_x.

HGV000   CAS: 2644-70-4   HR: 3
HYDRAZINE HYDROCHLORIDE
mf: $H_4N_2 \cdot ClH$ mw: 68.52

CONSENSUS REPORTS: Reported in EPA TSCA Inventory.

NIOSH REL: (Hydrazines) CL 0.04 mg/m^3/2H

SAFETY PROFILE: A poison by ingestion, intravenous, and intraperitoneal routes. When heated to decomposition it emits very toxic fumes of Cl^- and NO_x.

HGW500   CAS: 10034-93-2   HR: 3
HYDRAZINE SULFATE (1:1)
mf: $H_4N_2 \cdot H_2O_4S$ mw: 130.14

PROP: Colorless crystals. D: 1.378, mp: 85°. Sol in water; insol in alc; very sol in hot water.

SYNS: HYDRAZINE HYDROGEN SULFATE * HYDRAZINE MONOSULFATE * HYDRAZINE SULPHATE * HYDRAZINIUM SULFATE * HYDRAZONIUM SULFATE * HS * IDRAZINA SOLFATO (ITALIAN) * NSC-150014 * SIRAN HYDRAZINU (CZECH)

CONSENSUS REPORTS: IARC Cancer Review: Animal Sufficient Evidence IMEMDT 4,127,74. NTP Fourth Annual Report On Carcinogens, 1984. Community Right-To-Know List. Reported in EPA TSCA Inventory. EPA Genetic Toxicology Program.

NIOSH REL: (Hydrazines) CL 0.04 mg/m^3/2H

SAFETY PROFILE: Confirmed carcinogen with experimental carcinogenic, neoplastigenic, and tumorigenic data. A poison by ingestion and intraperitoneal routes. Human systemic effects by ingestion: paresthesia (abnormal sensations), somnolence, nausea or vomiting. Human mutation data reported. An eye irritant. A reducing agent. When heated to decomposition it emits very toxic fumes of SO$_x$ and NO$_x$.

HHC000   CAS: 109-84-2   **HR: 3**
2-HYDRAZINOETHANOL
mf: C$_2$H$_8$N$_2$O   mw: 76.12

PROP: Colorless, sltly viscous liquid. Mp: −70°, bp: 145-153° @ 25 mm, flash p: 224°F, vap d: 2.63, d: 1.11. Misc with water; sol in lower alcs; sltly sol in ether.

SYNS: BOH * HYDROXYETHYL HYDRAZINE * β-HYDROXYETHYLHYDRAZINE * N-(2-HYDROXYETHYL)HYDRAZINE

CONSENSUS REPORTS: Reported in EPA TSCA Inventory. EPA Genetic Toxicology Program.

SAFETY PROFILE: Poison by ingestion. Questionable carcinogen with experimental carcinogenic data. Mutation data reported. Combustible when exposed to heat or flame; can react with oxidizing materials. To fight fire, use foam, CO$_2$, dry chemical. When heated to decomposition it emits toxic fumes such as NO$_x$.

HHD500   CAS: 26049-68-3   **HR: 3**
2-HYDRAZINO-4-(5-NITRO-2-FURYL)THIAZOLE
mf: C$_7$H$_6$N$_4$O$_3$S   mw: 226.23

SYNS: HNT * 2-HYDRAZINO-4-(5-NITRO-2-FURANYL)THIAZOLE

CONSENSUS REPORTS: EPA Genetic Toxicology Program.

SAFETY PROFILE: Questionable carcinogen with experimental carcinogenic and neoplastigenic data. Mutation data reported. When heated to decomposition it emits very toxic fumes of NO$_x$ and SO$_x$.

HHF500   CAS: 63981-09-9   **HR: 3**
4-HYDRAZINO-2-THIOURACIL
mf: C$_4$H$_6$N$_4$S   mw: 142.20

SYN: 2-THIO-4-HYDRAZINOURACIL

SAFETY PROFILE: A poison by intraperitoneal route. When heated to decomposition it emits very toxic NO$_x$ and SO$_x$.

HHG000   CAS: 122-66-7   **HR: 3**
HYDRAZOBENZENE
mf: C$_{12}$H$_{12}$N$_2$ mw: 184.26

PROP: Light or yellow crystals from ethanol. D: 1.58, mp: 131°, bp: decomp. Very sltly sol in water, insol in acetylene.

SYNS: N,N′-BIANILINE * sym-DIPHENYLHYDRAZINE * 1,2-DIPHENYLHYDRAZINE * HYDRAZOBENZEN (CZECH) * HYDRAZODIBENZENE * NCI-C01854 * RCRA WASTE NUMBER U109

CONSENSUS REPORTS: NTP Fourth Annual Report On Carcinogens, 1984. NCI Carcinogenesis Bioassay (feed); Clear Evidence: mouse, rat NCITR* NCI-CG-TR-92,78. Community Right-To-Know List. Reported in EPA TSCA Inventory.

SAFETY PROFILE: Confirmed carcinogen with experimental carcinogenic and tumorigenic data. Poison by ingestion. Mutation data reported. When heated to decomposition it emits toxic fumes of NO$_x$.

HHG500   CAS: 7782-79-8   **HR: 3**
HYDRAZOIC ACID
mf: HN$_3$ mw: 43.04

PROP: Colorless liquid, very sol in water, intolerable pungent odor. Mp: −80°, bp: 37°, d: 1.09 @ 25°/4°.

SYNS: AZOIMIDE * DIAZOIMIDE * HYDROGEN AZIDE * HYDRONITRIC ACID * STICKSTOFFWASSERSTOFFSAEURE (GERMAN) * TRIAZOIC ACID

CONSENSUS REPORTS: Reported in EPA TSCA Inventory. EPA Genetic Toxicology Program.

DFG MAK: 0.1 ppm (0.27 mg/m^3)

SAFETY PROFILE: Poison by intraperitoneal route. Mildly toxic by inhalation. A severe irritant to skin, eyes, and mucous membranes. Continued inhalation causes central nervous system problems in humans (changes in EEG, somnolence, cough, headache, change in heart rate, fall in blood pressure, collapse, chills, and fever). High concentrations can cause fatal convulsions. Chronic exposure has been reported to cause injury to kidneys and spleen, hypotension, palpitation, ataxia, weakness. A dangerously sensitive explosive hazard when shocked or exposed to heat. Reacts with heavy metals to form very unstable heavy metal azides. Reacts violently with Cd; Cu; Ni; HNO$_3$; F$_2$. When heated to decomposition it emits toxic fumes of NO$_x$.

HHH000   CAS: 13529-51-6   **HR: 3**
2,2'-HYDRAZONODIETHANOL
mf: C$_4$H$_{12}$N$_2$O$_2$ mw: 120.18

SYNS: 1,1-BIS(2-HYDROXYETHYL)HYDRAZINE
* DEH * 1,1-DIETHANOLHYDRAZINE

SAFETY PROFILE: A poison by subcutaneous route. Questionable carcinogen with experimental tumorigenic data. When heated to decomposition it emits toxic fumes such as NO$_x$.

HHH500   **HR: 3**
HYDRIDES

SAFETY PROFILE: Variable toxicity. The highly toxic hydrides of phosphorus, arsenic, sulfur, selenium, tellurium, and boron produce local irritations and destroy red blood cells. They are particularly dangerous because of their volatility and ease of entry into the body. The hydrides of the alkali metals, alkaline earths, aluminum, zirconium, and titanium react with moisture to evolve hydrogen and leave behind the hydroxide of the metallic element which is usually caustic. The primary metallic hydrides include those of calcium, lithium, magnesium, potassium, sodium, and strontium. In the presence of moisture, they are readily converted to hydroxides which are highly irritating to the skin by caustic and thermal action. Similar effects can occur on contact with the eyes and respiratory mucous membranes. The volatile hydrides are flammable, some spontaneously so in air. All hydrides react violently on contact with powerful oxidizing agents. When heated or on contact with moisture or acids, an exothermic reaction evolving hydrogen occurs. Often, enough heat is evolved to cause ignition. Hydrides require special handling instructions which should be obtained from the manufacturers. The volatile hydrides (such as hydrides of boron, arsenic, phosphorus, selenium, tellurium) form explosive mixtures with air. The nonvolatile hydrides (such as sodium, lithium, calcium) readily liberate hydrogen when heated or on contact with moisture or acids. Furthermore, hydrides form dust clouds which can explode upon contact with flames, sparks, heat, or oxidizers. Highly dangerous; when heated, they can ignite at once or liberate explosive hydrogen. They react with moisture or acids to evolve heat and hydrogen. Violent reaction on contact with powerful oxidizers.

HHI500   CAS: 10034-85-2   **HR: 3**
HYDRIODIC ACID

DOT: UN 1787/UN 2197
mf: HI   mw: 127.91

PROP: Colorless when freshly made, but rapidly turns yellowish or brown on exposure to light or air. Keep protected from light and air, preferably not above 3°. Misc with water and alc. Mp: −50.8°, bp: −35.38° @ 5 atm, d: 5.66 g/L @ 0°.

SYNS: HYDRIODIC ACID, solution (DOT) * HYDROGEN IODIDE * HYDROGEN IODIDE, anhydrous (DOT) * HYDROGEN IODIDE solution (DOT)

CONSENSUS REPORTS: Reported in EPA TSCA Inventory.

DOT Classification: Corrosive Material; Label: Corrosive; IMO: Nonflammable Gas; Label: Nonflammable Gas, Corrosive.

SAFETY PROFILE: Poison by ingestion and inhalation. A corrosive and poisonous irritant to skin, eyes and mucous membranes. Explodes on contact with ethyl hydroperoxide. Ignites on contact with magnesium, perchloric acid, potassium + heat, potassium chlorate + heat, oxidants (e.g., fluorine, dinitrogen trioxide, dinitrogen tetraoxide, fuming nitric acid). Violent reaction with HClO$_4$ + Mg, O$_3$, metals. Potentially violent reaction with phosphorus. Reacts with water or steam to produce toxic and corrosive fumes. When heated to decomposition it emits highly toxic fumes of I$^-$.

HHJ000   CAS: 10035-10-6   **HR: 3**
HYDROBROMIC ACID

DOT: UN 1048/UN 1788
mf: BrH   mw: 80.92

PROP: Colorless gas or pale yellow liquid. Mp: −87°, bp: −66.5°, d: 3.50 g/L @ 0°. Misc with water, alc. Keep protected from light.

SYNS: ACIDE BROMHYDRIQUE (FRENCH) * ACIDO BROMIDRICO (ITALIAN) * BROMOWODOR (POLISH) * BROMWASSERSTOFF (GERMAN) * BROOMWATERSTOF (DUTCH) * HYDROBROMIC ACID, anhydrous (DOT) * HYDROGEN BROMIDE (OSHA ACGIH, MAK, DOT)

CONSENSUS REPORTS: Reported in EPA TSCA Inventory.

OSHA PEL: (Transitional: TWA 3 ppm) CL 3 ppm
ACGIH TLV: CL 3 ppm
DFG MAK: 5 ppm (17 mg/m^3)
DOT Classification: Corrosive Material; Label: Corrosive; Nonflammable Gas; Label: Nonflammable Gas, anhydrous; IMO: Poison A; Label: Poison Gas, Corrosive.

SAFETY PROFILE: Mildly toxic by inhalation. A corrosive irritant to the eyes, skin, and mucous membranes. Reacts violently with F_2, Fe_2O_3, NH_3, O_3. When heated to decomposition or in reaction with water or steam it emits toxic and corrosive fumes of Br$^-$ and HBr.

HHJ500   **HR: 3**
HYDROCARBON GAS

PROP: Contains hydrogen, methane, carbon monoxide, lel: 5.3%, uel: 31%, autoign temp: 1200°F.

SYNS: COAL GAS * HYDROCARBON GAS, compressed (DOT) * HYDROCARBON GAS, liquefied (DOT) * HYDROCARBON GAS, nonliquefied (DOT)

DOT Classification: Flammable Gas; Label: Flammable Gas; IMO: Poison A; Label: Poison Gas and Flammable Gas.

SAFETY PROFILE: A poison by inhalation. Very dangerous fire hazard when exposed to heat or flame; can react vigorously with oxidizing materials. Moderately explosive when exposed to heat or flame. To fight fire, stop flow of gas; CO_2, dry chemical, or water spray.

HHL000   CAS: 7647-01-0   **HR: 3**
HYDROCHLORIC ACID

DOT: UN 1050/UN 1789/UN 2186
mf: ClH   mw: 36.46

PROP: Colorless, fuming gas or colorless, fuming liquid; strongly corrosive with pungent odor. Mp: −114.3°, bp: −84.8°, d: 1.639 g/L (gas) @ 0°, 1.194 @ −26° (liquid), vap press: 4.0 atm @ 17.8°. Misc with water, alc.

SYNS: ACIDE CHLORHYDRIQUE (FRENCH) * ACIDO CLORIDRICO (ITALIAN) * CHLOORWATERSTOF (DUTCH) * CHLOROHYDRIC ACID * CHLOROWODOR (POLISH) * CHLORWASSERSTOFF (GERMAN) * HYDROCHLORIC ACID, anhydrous (DOT) * HYDROCHLORIC ACID, solution, inhibited (DOT) * HYDROCHLORIDE * HYDROGEN CHLORIDE (OSHA, ACGIH, MAK, DOT) * HYDROGEN CHLORIDE, anhydrous (DOT) * HYDROGEN CHLORIDE, refrigerated liquid (DOT) * MURIATIC ACID (DOT) * SPIRITS of SALT (DOT)

CONSENSUS REPORTS: EPA Extremely Hazardous Substances List. Community Right-To-Know List. Reported in EPA TSCA Inventory. EPA Genetic Toxicology Program.

OSHA PEL: CL 5 ppm
ACGIH TLV: CL 5 ppm
DFG MAK: 5 ppm (7 mg/m^3)
DOT Classification: Corrosive Material; Label: Corrosive, solution; Nonflammable Gas; Label: Nonflammable Gas; IMO: Nonflammable Gas; Label: Nonflammable Gas, Corrosive.

SAFETY PROFILE: A human poison by an unspecified route. Mildly toxic to humans by inhalation. Moderately toxic experimentally by ingestion. A corrosive irritant to the skin, eyes, and mucous membranes. Mutation data reported. An experimental teratogen. A concentration of 35 ppm causes irritation of the throat after short exposure. In general, hydrochloric acid causes little trouble in industry other than from accidental splashes and burns. It is a common air contaminant and is heavily used in industry.

Nonflammable Gas. Explosive reaction with alcohols + hydrogen cyanide, potassium permanganate, sodium, tetraselenium tetranitride. Ignition on contact with fluorine, hexalithium disilicide, metal acetylides or carbides (e.g., cesium acetylide, rubidium acetylide). Violent reactions with acetic anhydride, 2-amino ethanol, NH_4OH, Ca_3P_2, chlorosulfonic acid,

1,1-difluoroethylene, ethylene diamine, ethylene imine, oleum, $HClO_4$, β-propiolactone, propylene oxide, $(AgClO_4 + CCl_4)$, NaOH, H_2SO_4, U_3P_4, vinyl acetate, CaC_2, CsC_2H, Cs_2C_2, Mg_3B_2, $HgSO_4$, RbC_2H, Rb_2C_2, Na. Vigorous reaction with aluminum, chlorine + dinitroanilines (evolves gas). Potentially dangerous reaction with sulfuric acid releases HCl gas. When heated to decomposition it emits toxic fumes of Cl^-.

HHM000   CAS: 8007-56-5   HR: 3
HYDROCHLORIC ACID, mixed with NITRIC ACID (3:1)
DOT: UN 1798
mf: $ClH \cdot HNO_3$ mw: 99.48

PROP: Yellow, fuming, corrosive, volatile liquid; suffocating odor. Misc with water.

SYNS: AQUA REGIA * NITROHYDROCHLORIC ACID (DOT) * NITROHYDROCHLORIC ACID, DILUTED (DOT) * NITROMURIATIC ACID (DOT)

DOT Classification: Corrosive Material; Label: Corrosive; Label: Corrosive, diluted.

SAFETY PROFILE: A corrosive irritant to the eyes, skin, and mucous membranes. When heated to decomposition it emits very toxic HCl, HNO_3, Cl^-, and NO_x.

HHP000   CAS: 104-53-0   HR: 3
HYDROCINNAMALDEHYDE
mf: $C_9H_{10}O$ mw: 134.19

PROP: Colorless to sltly yellow liquid; strong floral, hyacinth odor. Bp: 221-224°, d: 1.010-1.020, refr index: 1.520-1.532, flash p: 203°F. Misc with alc, ether; insol in water.

SYNS: BENZENEPROPANAL * BENZYLACETALDEHYDE * DIHYDROCINNAMALDEHYDE * FEMA No. 2887 * HYDROCINNAMIC ALDEHYDE * 3-PHENYLPROPANAL * 3-PHENYL-1-PROPANAL * 3-PHENYLPROPIONALDEHYDE (FCC) * β-PHENYLPROPIONALDEHYDE * 3-PHENYLPROPYL ALDEHYDE

CONSENSUS REPORTS: Reported in EPA TSCA Inventory.

SAFETY PROFILE: A poison by intravenous route. A human skin irritant. Combustible liquid. When heated to decomposition it emits acrid smoke and fumes.

HHP050   CAS: 122-97-4   HR: 2
HYDROCINNAMIC ALCOHOL
mf: $C_9H_{12}O$ mw: 136.21

PROP: Colorless sltly viscous liquid; sweet, hyacinth-mignonette odor. D:.998-1.002, refr index: 1.524-1.528, flash p: 228°F. Sol in fixed oils, propylene glycol; insol in glycerin.

SYNS: 3-BENZENEPROPANOL * FEMA No. 2885 * HYDROCINNAMYL ALCOHOL * (3-HYDROXYPROPYL)BENZENE * γ-PHENYLPROPANOL * 3-PHENYLPROPANOL * 3-PHENYL-1-PROPANOL (FCC) * PHENYLPROPYL ALCOHOL * γ-PHENYLPROPYL ALCOHOL * 3-PHENYLPROPYL ALCOHOL

CONSENSUS REPORTS: Reported in EPA TSCA Inventory.

SAFETY PROFILE: Moderately toxic by ingestion. Mildly toxic by skin contact. A skin irritant. Combustible liquid. When heated to decomposition it emits toxic fumes.

HHP500   CAS: 122-72-5   HR: 1
HYDROCINNAMYL ACETATE
mf: $C_{11}H_{14}O_2$ mw: 178.25

PROP: Colorless liquid; spicy, floral odor. D: 1.012, refr index: 1.494, flash p: +212°F. Sol in alc; insol in water.

SYNS: FEMA No. 2890 * 3-PHENYL-1-PROPANOL ACETATE * PHENYLPROPYL ACETATE * 3-PHENYLPROPYL ACETATE (FCC) * 3-PHENYL-1-PROPYL ACETATE

CONSENSUS REPORTS: Reported in EPA TSCA Inventory.

SAFETY PROFILE: Mildly toxic by ingestion. Combustible liquid. When heated to decomposition it emits acrid smoke and fumes.

HHR000   CAS: 125-04-2   HR: 2
HYDROCORTISONE SODIUM SUCCINATE
mf: $C_{25}H_{35}O_9 \cdot Na$ mw: 502.59

PROP: White, odorless, hygroscopic, amorphous solid. Mp: 169-171°. Very sol in water and alc; insol in chloroform; very sltly sol in acetone.

SYNS: A-HYDROCORT * BUCCALSONE * CORLAN * CORTISOL HEMISUCCINATE SODIUM SALT * CORTISOL SODIUM HEMISUCCINATE * CORTISOL SODIUM SUCCINATE * CORTISOL-21-SODIUM SUCCINATE * CORTISOL SUCCINATE, SODIUM SALT * EL-CORTELAN SOLUBLE * EMI-CORLIN

* FLEBOCORTID * HYCORACE * HYDROCORTI-
SONE-21-SODIUM SUCCINATE * 21-(HYDROGEN SUC-
CINATE)CORTISOL, MONOSODIUM SALT * INTRA-
CORT * NORDICORT * ORALSONE * SODIUM
HYDROCORTISONE SUCCINATE * SODIUM HYDRO-
CORTISONE-21-SUCCINATE * SOLU-CORTEF
* SOLU-GLYC * U 4905

SAFETY PROFILE: Moderately toxic by intraperitoneal route. Experimental teratogenic and
reproductive effects. When heated to decomposition it emits toxic fumes of Na_2O.

HHR500   CAS: 119-84-6   HR: 3
HYDROCOUMARIN
mf: $C_9H_8O_2$ mw: 148.17

PROP: Colorless to pale yellow liquid; coconut
odor. D: 1.186, refr index: 1.555, flash p: 266°F.

SYNS: 1,2-BENZODIHYDROPYRONE (FCC) * 2-
CHROMANONE * DIHYDROCOUMARIN * 3,4-DIHY-
DROCOUMARIN * o-HYDROXY-HYDROCINNAMIC
ACID-Δ-LACTONE * FEMA No. 2381 * MELILOTIN
* MELILOTOL * NCI-C55890 * 2-OXOCHROMAN
* USAF DO-12

CONSENSUS REPORTS: Reported in EPA
TSCA Inventory.

SAFETY PROFILE: A poison by intraperitoneal
route. Moderately toxic by ingestion. A skin
irritant. Combustible liquid. When heated to
decomposition it emits acrid smoke and fumes.

HHS000   CAS: 74-90-8   HR: 3
HYDROCYANIC ACID

DOT: UN 1614/UN 1051
mf: CHN   mw: 27.03

PROP: Odor of bitter almonds. Mp: −13.2°,
bp: 25.7°, lel: 5.6%, uel: 40%, flash p: 0°F
(CC), d: 0.6876 @ 20°/4°, autoign temp:
1000°F, vap press: 400 mm @ 9.8°, vap d:
0.932. Misc in water, alc, and ether.

SYNS: ACIDE CYANHYDRIQUE (FRENCH) * ACIDO
CIANIDRICO (ITALIAN) * AERO liquid HCN
* BLAUSAEURE (GERMAN) * BLAUWZUUR (DUTCH)
* CYAANWATERSTOF (DUTCH) * CYANWASSER-
STOFF (GERMAN) * CYCLON * CYCLONE B
* CYJANOWODOR (POLISH) * HCN * HYDRO-
CYANIC ACID, liquefied (DOT) * HYDROCYANIC ACID
(PRUSSIC), unstabilized (DOT) * HYDROGEN CYANIDE
(OSHA, ACGIH) * HYDROGEN CYANIDE, anhydrous, sta-
bilized (DOT) * PRUSSIC ACID (DOT) * PRUSSIC
ACID, unstabilized * RCRA WASTE NUMBER P063
* ZACLON DISCOIDS

CONSENSUS REPORTS: EPA Extremely
Hazardous Substances List. Community Right-
To-Know List. Reported in EPA TSCA Inven-
tory.

OSHA PEL: (Transitional: TWA 10 ppm
(skin)); STEL 4.7 ppm (skin)
ACGIH TLV: CL 10 ppm (skin)
DFG MAK: 10 ppm (11 mg/m^3)
NIOSH REL: (Cyanide) CL 5 mg(CN)/m^3/10M
DOT Classification: Poison A; Label: Poison
    Gas and Flammable Gas; IMO: Poison B;
    Label: Poison (UN 1614); IMO: Poison B;
    Label: Flammable Liquid and Poison; Forbid-
    den, Unstabilized.

SAFETY PROFILE: A deadly human and experimental poison by all routes. Hydrocyanic
acid and the cyanides are true protoplasmic poisons, combining in the tissues with the enzymes
associated with cellular oxidation. They thereby
render the oxygen unavailable to the tissues and
cause death through asphyxia. The suspension
of tissue oxidation lasts only while the cyanide
is present; upon its removal, normal function
is restored provided death has not already occurred. HCN does not combine easily with hemoglobin, but it does combine readily with methemoglobin to form cyanmethemoglobin. This
property is utilized in the treatment of cyanide
poisoning when an attempt is made to induce
methemoglobin formation. The presence of
cherry-red venous blood in cases of cyanide
poisoning is due to the inability of the tissues
to remove the oxygen from the blood. Exposure
to concentrations of 100-200 ppm for periods
of 30-60 minutes can cause death. In cases of
acute cyanide poisoning death is extremely
rapid, although sometimes breathing may continue for a few minutes. In less acute cases,
there is cyanosis, headache, dizziness, unsteadi-
ness of gait, a feeling of suffocation, and nausea.
Where the patient recovers, there is rarely any
disability.

    Very dangerous fire hazard when exposed
to heat, flame or oxidizers. Can polymerize explosively at 50-60C° or in the presence of traces
of alkali. Severe explosion hazard when exposed
to heat or flame or by chemical reaction with
oxidizers. The anhydrous liquid is stabilized
at or below room temperature by the addition
of acid. The gas forms explosive mixtures with
air. Reacts violently with acetaldehyde. To fight
fire, use CO_2, non-alkaline dry chemical, foam.
When heated to decomposition or in reaction

with water, steam, acid or acid fumes it produces highly toxic fumes of CN^-. An insecticide.

HHU500 CAS: 7664-39-3 HR: 3
HYDROFLUORIC ACID

DOT: UN 1790/UN 1052
mf: FH mw: 20.01

PROP: Clear, colorless, fuming, corrosive liquid or gas. Mp: $-83.1°$, bp: $19.54°$, d: 0.901 g/L(gas); 0.699 @ $22°$ (liquid), vap press: 400 mm @ $2.5°$.

SYNS: ACIDE FLUORHYDRIQUE (FRENCH) * ACIDO FLUORIDRICO (ITALIAN) * FLUOROWODOR (POLISH) * FLUORWASSERSTOFF (GERMAN) * FLUORWATERSTOF (DUTCH) * HYDROFLUORIC ACID, anhydrous (DOT) * HYDROFLUORIC ACID, solution (DOT) * HYDROFLUORIDE * HYDROGEN FLUORIDE (OSHA, ACGIH, MAK, DOT) * RCRA WASTE NUMBER U134

CONSENSUS REPORTS: EPA Extremely Hazardous Substances List. Community Right-To-Know List. EPA Genetic Toxicology Program. Reported in EPA TSCA Inventory.

OSHA PEL: (Transitional: TWA 3 ppm (F)) TWA 3 ppm; STEL 6 ppm (F)
ACGIH TLV: CL 3 ppm (F)
DFG MAK: 3 ppm (2 mg/m^3); BAT 7.0 mg/g creatinine in urine at end of shift.
NIOSH REL: (HF) TWA 2.5 mg(F)/m^3; CL 5.0 mg(F)/m^3/15M
DOT Classification: Corrosive Material; Label: Corrosive; IMO: Poison A; Label: Poison Gas, Corrosive; IMO: Corrosive Material; Label: Corrosive, Poison.

SAFETY PROFILE: A human poison by inhalation. A poison experimentally by inhalation, subcutaneous and intraperitoneal routes. A corrosive irritant to skin, eyes (@ 0.05 mg/L), and mucous membranes. Experimental teratogenic and reproductive effects. Mutation data reported. Inhalation of the vapor may cause ulcers of the upper respiratory tract. Concentrations of 50-250 ppm are dangerous, even for brief exposures. Hydrofluoric acid produces severe skin burns which are slow in healing. The subcutaneous tissues may be affected, becoming blanched and bloodless. Gangrene of the affected areas may follow. It is a common air contaminant.

Explosive reaction with cyanogen fluoride, glycerol + nitric acid, sodium (with aqueous acid), methanesulfonic acid (evolves oxygen difluoride which explodes). Violent reaction with As_2O_3, P_2O_5, acetic anhydride, 2-amino ethanol, NH_4OH, $HBiO_3$, bismuthic acid (evolves oxygen), CaO, chlorosulfonic acid, ethylene diamine, ethylene imine, F_2, mercury(II) oxide + organic materials (above $0°C$), n-phenylazopiperidine, potassium permanganate, potassium tetrafluorosilicate(2-)(evolves silicon tetrafluoride gas), (HNO_3 + lactic acid), oleum, β-propiolactone, propylene oxide, Na, NaOH, H_2SO_4, vinyl acetate, HgO, sodium tetrafluoro silicate, n-phenyl azo piperidine. Incandescent reaction of liquid HF with oxides (e.g., arsenic trioxide, calcium oxide). Dangerous storage hazard with nitric acid + lactic acid, nitric acid + propylene glycol (mixtures evolve gas which may burst a sealed container). Reacts with water or steam to produce toxic and corrosive fumes. When heated to decomposition it emits highly corrosive fumes of F^-.

HHW500 CAS: 1333-74-0 HR: 3
HYDROGEN

DOT: UN 1049/UN 1966
mf: H_2 mw: 2.02

PROP: Colorless, odorless, tasteless gas. Mp: $-259.18°$, bp: $-252.8°$, lel: 4.1%, uel: 74.2%, d: 0.0899 g/L, autoign temp: $752°F$, vap d: 0.069.

SYNS: HYDROGEN (DOT) * HYDROGEN, compressed (DOT) * HYDROGEN, refrigerated liquid (DOT)

CONSENSUS REPORTS: Reported in EPA TSCA Inventory.

DOT Classification: Flammable Gas; Label: Flammable Gas.

SAFETY PROFILE: Practically no toxicity except that it may asphyxiate. Highly dangerous fire and severe explosion hazard when exposed to heat, flame or oxidizers. Flammable or explosive when mixed with air, O_2, chlorine. To fight fire, stop flow of gas.

Explodes on contact with bromine trifluoride, chlorine trifluoride, fluorine, hydrogen peroxide + catalysts, acetylene + ethylene. Explodes when heated with calcium carbonate + magnesium, 3,4-dichloronitrobenzene + catalysts, vegetable oils + catalysts, ethylene + nickel catalysts, difluorodiazene (above $90°C$), 2-nitroanisole (above $250°C/34$ bar + 12% catalyst), copper(II) oxide, nitryl fluoride (above

200°C), polycarbon monofluoride (above 500°C).

Forms sensitive explosive mixtures with bromine, chlorine, iodine heptafluoride (heat- or spark-sensitive), chlorine dioxide, dichlorine oxide, iodine heptafluoride (heat- or spark-sensitive), dinitrogen oxide, dinitrogen tetraoxide, oxygen (gas), 1,1,1-trisazidomethylethane + palladium catalyst. Mixtures with liquid nitrogen react with heat to form an explosive product.

Violent reaction or ignition with air + catalysts (platinum and similar metals containing adsorbed oxygen or hydrogen), bromine, iodine, dioxane + nickel, lithium, nitrogen trifluoride, oxygen difluoride, palladium + isopropyl alcohol, 3-methyl-2-penten-4-yn-1-ol, lead trifluoride, bromine fluoride (ignition on contact), nickel + oxygen, fluorine perchlorate (ignition on contact), xenon hexafluoride (violent reaction), nitrogen oxide + oxygen (ignition above 360°C), palladium powder + 2-propanol + air (spontaneous ignition), platinum catalyst, polycarbon monofluoride (ignition above 400°C).

Vigorous exothermic reaction with benzene + Raney nickel catalyst, metals (e.g., lithium, calcium, barium, strontium, sodium, potassium, above 300°C), palladium(II)oxide, palladium trifluoride, 1,1,1-tris(hydroxymethyl) nitromethane + nickel catalyst.

HHW800   CAS: 61788-32-7   **HR: 3**
HYDROGENATED TERPHENYLS

PROP: Complex mixtures of o-, m-, and p-terphenyls in various stages of hydrogenation. Five such stages exist for each of the three above isomers.

CONSENSUS REPORTS: Reported in EPA TSCA Inventory.

OSHA PEL: TWA 0.5 ppm
ACGIH TLV: TWA 0.5 ppm

SAFETY PROFILE: Contact with hot coolant can cause severe damage to lungs, skin, and eyes from burns. May cause chronic damage to liver, kidney, and blood-forming organs; metabolic disorders. Inhalation has caused bronchopneumonia. When heated to decomposition it emits acrid smoke and fumes.

HHX000   CAS: 7647-01-0   **HR: 3**
HYDROGEN CHLORIDE
mf: ClH   mw: 36.46

PROP: Colorless, corrosive, nonflammable gas. Pungent odor, fumes in air. D: 1.639 @ −137.77°, bp: −154.37° @ 1.0 mm.

CONSENSUS REPORTS: EPA Extremely Hazardous Substances List. EPA Genetic Toxicology Program. Reported in EPA TSCA Inventory.

OSHA PEL: CL 5 ppm
ACGIH TLV: CL 5 ppm
DFG MAK: 5 ppm (7 mg/m^3)

SAFETY PROFILE: A highly corrosive irritant to the eyes, skin, and mucous membranes. Mildly toxic by inhalation. Explosive reaction with alcohols + hydrogen cyanide, potassium permanganate, sodium (with aqueous HCl), tetraselenium tetranitride. Ignition on contact with aluminum-titanium alloys (with HCl vapor), fluorine, hexalithium disilicide, metal acetylides or carbides (e.g., cesium acetylide, rubidium acetylide). Violent reaction with 1,1-difluoroethylene. Vigorous reaction with aluminum, chlorine + dinitroanilines (evolves gas). Potentially dangerous reaction with sulfuric acid releases HCl gas. Adsorption of the acid onto silicon dioxide is exothermic.

HHX500   CAS: 7647-01-0   **HR: 3**
HYDROGEN CHLORIDE (AEROSOL)
mf: ClH   mw: 36.46

PROP: Saturated water aerosol mist (NTIS** AD744-829).

CONSENSUS REPORTS: Reported in EPA TSCA Inventory. EPA Genetic Toxicology Program.

SAFETY PROFILE: Mildly toxic by inhalation. A very powerful human skin, eye, and mucous membrane irritant.

HHY500   CAS: 1333-74-0   **HR: 3**
HYDROGEN, CRYOGENIC LIQUID (DOT)

CONSENSUS REPORTS: Reported in EPA TSCA Inventory.

DOT Classification: Label: Flammable Gas.

SAFETY PROFILE: Contact with the liquid can cause frostbite. A very flammable gas. Can explode in air. Forms an explosive mixture with ozone.

HIB000 CAS: 7722-84-1 **HR: 3**
HYDROGEN PEROXIDE

DOT: UN 2014/UN 2015
mf: H_2O_2 mw: 34.02

PROP: Colorless, heavy liquid, or, at low temp, a crystalline solid; bitter taste. D: 1.71 @ $-20°$, 1.46 @ 0°, vap press: 1 mm @ 15.3°, unstable. Mp: $-0.43°$, bp: 152°. Misc with water; sol in ether; insol in petr ether. Decomposed by many organic solvents.

SYNS: ALBONE * DIHYDROGEN DIOXIDE * HIOXYL * HYDROGEN DIOXIDE * HYDROGEN PEROXIDE, solution (over 52% peroxide) (DOT) * HYDROGEN PEROXIDE, stabilized (over 60% peroxide) (DOT) * HYDROPEROXIDE * INHIBINE * OXYDOL * PERHYDROL * PERONE * PEROSSIDO di IDROGENO (ITALIAN) * PEROXAN * PEROXIDE * PEROXYDE d'HYDROGENE (FRENCH) * SUPEROXOL * T-STUFF * WASSERSTOFFPEROXID (GERMAN) * WATERSTOFPEROXYDE (DUTCH)

CONSENSUS REPORTS: IARC Cancer Review: GROUP 3 IMEMDT 7,56,87; Animal Limited Evidence IMEMDT 28,151,82. EPA Extremely Hazardous Substances List. Reported in EPA TSCA Inventory. EPA Genetic Toxicology Program.

OSHA PEL: TWA 1 ppm
ACGIH TLV: TWA 1 ppm
DFG MAK: 1 ppm (1.4 mg/m³)
DOT Classification: Oxidizer; Label: Oxidizer and Corrosive.

SAFETY PROFILE: Moderately toxic by inhalation, ingestion, and skin contact. A corrosive irritant to skin, eyes, and mucous membranes. Questionable carcinogen with experimental tumorigenic data. Human mutation data reported. A very powerful oxidizer.

Pure H_2O_2, its solutions, vapors, and mists are very irritating to body tissue. This irritation can vary from mild to severe depending upon the concentration of H_2O_2. For instance, solutions of H_2O_2 of 35 wt% and over can easily cause blistering of the skin. Irritation caused by H_2O_2 which does not subside upon flushing the affected part with water should be treated by a physician. The eyes are particularly sensitive to this material. It is a common air contaminant.

A dangerous fire hazard by chemical reaction with flammable materials. H_2O_2 is a powerful oxidizer, particularly in the concentrated state. It is important to keep containers covered because uncovered containers are much more prone to react with flammable vapors, gases, etc., and if uncovered, the water from an H_2O_2 solution can evaporate, concentrating the material and thus increasing the fire hazard. For instance, solutions of H_2O_2 in concentration in excess of 65 wt% heat up spontaneously when decomposed to H_2O + 1/2 O_2. Thus, 90 wt% solutions, when caused to decompose rapidly due to the introduction of a catalytic decomposition agent, can get quite hot and perhaps start fires.

A severe explosion hazard when highly concentrated or when pure H_2O_2 is exposed to: heat, mechanical impact, detonation of a blasting cap, or caused to decompose catalytically by metals (in order of decreasing effectiveness: osmium, palladium, platinum, iridium, gold, silver, manganese, cobalt, copper, lead). Explodes on contact with alcohols + H_2SO_4, acetal + acetic acid + heat, acetic acid + n-heterocycles (above 50°), 2-amino-4-methyloxazole + iron(II) catalyst, aromatic hydrocarbons + trifluoroacetic acid, azeliac acid + sulfuric acid (above 45°), benzenesulfonic anhydride, tert-butanol + sulfuric acid, carboxylic acids, 3,5-dimethyl-3-hexanol + sulfuric acid, diphenyl diselenide (above 53°), 2-ethoxyethanol + polyacrylamide gel + toluene + heat, gadolinium hydroxide (above 80°), gallium + hydrochloric acid, hydrogen + palladium catalysts (has caused major industrial explosions), iron(II) sulfate + 2-methylpyridine + sulfuric acid, iron(II) sulfate + nitric acid + sodium carboxymethylcellulose (when evaporated), nitric acid + ketones (e.g., 2-butanone, 3-pentanone, cyclopentanone, cyclohexanone, 3-methylcyclohexanone), trioxane (sensitive to heat, shock, or on contact with lead), methanol + tert-amines + platinum catalysts, nitric acid + soils, nitrogenous bases (e.g., ammonia, hydrazine hydrate, 1,1-dimethylhydrazine), organic compounds (e.g., glycerol, acetic acid, ethanol, aniline, quinoline, 2-phenyl-1,1-dimethylethanol, cellulose, charcoal), organic materials + sulfuric acid (especially if confined), water + oxygenated compounds (e.g., acetaldehyde, acetic acid, acetone, ethanol, formaldehyde, formic acid, methanol, 2-propanol, propionaldehyde), sulfuric acid (during evaporation), tetrahydrothiophene, vinyl acetate, alcohols + tin chloride, P_2O_5, P, H_2O, HNO_3, Sb_2S_3, As_2S_3, Cl_2 + KOH, + chlorosulfonic acid, CuS, FeS, formic acid + organic matter, H_2Se, hydrazine, PbO_2, PbO, PbS, MnO_2, HgO, Hg_2O, MoS_2, organic matter, (2-methyl-1-phenyl-2-propanol + sulfuric acid), $KMnO_4$, $NaIO_3$, thiodiglycol, uns-dimethyl hydrazine, $FeSO_4$ + 2-methylpyridine + H_2SO_4, HgO + HNO_3.

Forms unstable explosive products in reaction with acetaldehyde + desiccants (forms polyethylidine peroxide), acetic acid (forms peracetic acid), acetic + 3-thietanol, acetic anhydride, acetone (forms explosive peroxides), alcohols (products are shock- and heat-sensitive), carboxylic acids (e.g., formic acid, acetic acid, tartaric acid), diethyl ether, ethyl acetate, formic acid + metaboric acid, ketene (forms diacetyl peroxide), mercury(II) oxide + nitric acid (forms mercury(II) peroxide), thiourea + nitric acid, polyacetoxyacrylic acid lactone + poly(2-hydroxyacrylic acid) + sodium hydroxide.

Ignition on contact with furfuryl alcohol, powdered metals (e.g., magnesium, iron), wood. Violent reaction with aluminum isopropoxide + heavy metal salts, charcoal, coal, dimethylphenylphosphine, hydrogen selenide, lithium tetrahydroaluminate, metals (e.g., potassium, sodium, lithium), metal oxides (e.g., cobalt oxide, iron oxide, lead oxide, lead hydroxide, manganese oxide, mercury oxide, nickel oxide), metal salts (e.g., calcium permanganate), methanol + phosphoric acid, 4-methyl-2,4,6-triazatricyclo [5.2.2.0^{2,6}] undeca-8-ene-3,5-dione + potassium hydroxide, α-phenylselenoketones, phosphorus, phosphorus(V) oxide, tin(II) chloride, unsaturated organic compounds.

BEWARE: Although many mixtures of H_2O_2 and organic materials do not explode upon contact, the resultant combination is detonatable either upon catching fire or by impact. The detonation velocity of aqueous solutions of H_2O_2 has been found to be about 6500 m/second for solutions of between 96 wt% and 100 wt% H_2O_2. Another source of H_2O_2 explosions is from sealing the material in strong containers. Under such conditions, even gradual decomposition of H_2O_2 to H_2O + 1/2 O_2 can cause large pressures to build up in the containers which may then burst explosively. Highly dangerous; when heated, shocked, or contaminated, the concentrated material can explode or start fires.

HIB500   CAS: 124-43-6   **HR: 2**
HYDROGEN PEROXIDE with UREA (1:1)

DOT: UN 1511
mf: $CH_4N_2O \cdot H_2O_2$ mw: 94.09

PROP: White crystals. Mp: 75-85° (decomp).

SYNS: CARBAMIDE PEROXIDE * GLY-OXIDE * HYDROGEN PEROXIDE CARBAMIDE * HYDROPERIT * HYPEROL * ORTIZON * PERCARBAMIDE * PERHYDRIT * PERHYDROL-UREA * THENARDOL * UREA DIOXIDE * UREA HYDROGEN PEROXIDE (DOT) * UREA HYDROGEN PEROXIDE SALT * UREA HYDROPEROXIDE * UREA PEROXIDE (DOT)

CONSENSUS REPORTS: Reported in EPA TSCA Inventory.

DOT Classification: Organic Peroxide; Label: Organic Peroxide; IMO: Oxidizer; Label: Oxidizer.

SAFETY PROFILE: An irritant to skin, eyes, and mucous membranes. An FDA over-the-counter drug. When heated to decomposition it emits toxic fumes of NO_x.

HIC000   CAS: 7783-07-5   **HR: 3**
HYDROGEN SELENIDE

DOT: UN 2202
mf: H_2Se mw: 80.98

PROP: Colorless gas. Mp: −64°, bp: −41.4°, d: 3.614 g/L (gas), 2.12 @ −42° (liquid), vap press: 10 atm @ 23.4°. Flammable. Disagreeable odor. Sol in carbonyl chloride and carbon disulfide.

SYNS: ELECTRONIC E-2 * HYDROGEN SELENIDE, anhydrous (DOT) * SELENIUM HYDRIDE

CONSENSUS REPORTS: Selenium and its compounds are on the Community Right-To-Know List. EPA Extremely Hazardous Substances List. Reported in EPA TSCA Inventory.

OSHA PEL: TWA 0.05 ppm (Se)
ACGIH TLV: TWA 0.05 ppm (Se)
DFG MAK: 0.05 ppm (0.2 mg/m^3)
DOT Classification: Flammable Gas; Label: Poison Gas and Flammable Gas; IMO: Poison A; Label: Poison Gas and Flammable Gas.

SAFETY PROFILE: A deadly poison by inhalation. Very poisonous irritant to skin, eyes and mucous membranes. Causes central nervous system effects in humans. An allergen. Can cause damage to the lungs and liver as well as conjunctivitis. It has been found that repeated exposures to concentrations of 0.3 ppm prove fatal to experimental animals by causing a pneumonitis, as well as injury to the liver and spleen. Causes garlic odor of breath, dizziness, nausea. Concentrations of 0.3 ppm are readily detected by odor, but there is no noticeable irritant effect at that level. Concentrations of 1.5 ppm or higher are strongly irritating to the eyes and nasal passages.

As in the case of hydrogen sulfide, the odor of hydrogen selenide in concentrations below 1 ppm disappears rapidly because of olfactory fatigue. The odor and irritating effects do not offer a dependable warning to workmen who may be exposed to gradually increasing amounts and therefore become used to it. Due to its extreme toxicity and irritating effects, it seldom is allowed to reach a concentration in which it is flammable in air. Very little data are available on possible chronic effects of this material, but it is logical to assume that when the concentration of this gas is low enough to avoid the irritant effects, only the systemic effects will be noticeable.

Dangerous fire hazard when exposed to heat or flame; will react vigorously with powerful oxidizing agents, such as H_2O_2, HNO_3. Dangerous; forms explosive mixtures with air; keep away from heat and open flame.

HIC500   CAS: 7783-06-4   **HR: 3**
HYDROGEN SULFIDE

DOT: UN 1053
mf: H_2S mw: 34.08

PROP: Colorless, flammable gas; offensive odor. Mp: −85.5°, bp: −60.4°, lel: 4%, uel: 46%, autoign temp: 500°F, d: 1.539 g/L @ 0°, vap press: 20 atm @ 25.5°, vap d: 1.189.

SYNS: ACIDE SULFHYDRIQUE (FRENCH) * HYDRO-GENE SULFURE (FRENCH) * HYDROGEN SULFURIC ACID * IDROGENO SOLFORATO (ITALIAN) * RCRA WASTE NUMBER U135 * SCHWEFELWASSERSTOFF (GERMAN) * SIARKOWODOR (POLISH) * STINK DAMP * SULFURETED HYDROGEN * SULFUR HYDRIDE * ZWAVELWATERSTOF (DUTCH)

CONSENSUS REPORTS: EPA Extremely Hazardous Substances List. Reported in EPA TSCA Inventory.

OSHA PEL: (Transitional: CL 20 ppm; Pk 50/ 10M) TWA 10 ppm; STEL 15 ppm

ACGIH TLV: TWA 10 ppm; STEL 15 ppm
DFG MAK: 10 ppm (15 mg/m³)
NIOSH REL: (Hydrogen Sulfide) CL 15 mg/ m³/10M
DOT Classification: Flammable Gas; Label: Poison Gas and Flammable Gas.

SAFETY PROFILE: A human poison by inhalation. A severe irritant to eyes and mucous membranes. An asphyxiant. Human systemic effects by inhalation: coma, chronic pulmonary edema. Low concentrations of 20-150 ppm cause irritation of the eyes; slightly higher concentrations may cause irritation of the upper respiratory tract, and if exposure is prolonged, pulmonary edema may result. The irritant action has been explained on the basis that H_2S combines with the alkali present in moist surface tissues to form sodium sulfide, a caustic. With higher concentration the action of the gas on the nervous system becomes more prominent. A 30 minute exposure to 500 ppm results in headache, dizziness, excitement, staggering gait, diarrhea and dysuria, followed sometimes by bronchitis or bronchopneumonia.

The action of small amounts on the nervous system is one of depression; in larger amounts, it stimulates, and with very high amounts the respiratory center is paralyzed. Exposures of 800-1000 ppm may be fatal in 30 minutes, and high concentrations are instantly fatal. Fatal hydrogen sulfide poisoning may occur even more rapidly than that following exposure to a similar concentration of HCN. H_2S does not combine with the hemoglobin of the blood; its asphyxiant action is due to paralysis of the respiratory center. With repeated exposures to low concentrations, conjunctivitis, photophobia, corneal bullae, tearing, pain, and blurred vision are the commonest findings. High concentrations may cause rhinitis, bronchitis, and occasionally pulmonary edema. Exposure to very high concentrations results in immediate death. Chronic poisoning results in headache, inflammation of the conjunctivae and eyelids, digestive disturbances, weight loss and general debility. It is a common air contaminant.

It is an insidious poison since sense of smell may be fatigued. The odor and irritating effects do not offer a dependable warning to workers who may be exposed to gradually increasing amounts and therefore become used to it.

Very dangerous fire hazard when exposed

to heat, flame, or oxidizers. Moderately explosive when exposed to heat or flame. Explodes on contact with oxygen difluoride; nitrogen trichloride; bromine pentafluoride; chlorine trifluoride; dichlorine oxide; silver fulminate. Potentially explosive reaction with copper + oxygen. Explosive reaction when heated with perchloryl fluoride (above 100°C); oxygen (above 280°C). Reacts with 4-bromobenzenediazonium chloride to form an explosive product.

Ignites on contact with metal oxides (e.g., barium peroxide, chromium trioxide, copper oxide, lead dioxide, manganese dioxide, nickel oxide, silver(I) oxide, silver(II) oxide, sodium peroxide, thallium(III) oxide, mercury oxide, calcium oxide, nickel oxide), oxidants (e.g., silver bromate, heptasilver nitrate octaoxide, dibismuth dichromium nonaoxide, mercury(I) bromate, lead(II) hypochlorite, copper chromate, fluorine, nitric acid, sodium peroxide, lead(IV) oxide), rust, soda-lime + air. Reacts violently with NI_3, NF_3, p-bromobenzenediazonium chloride, OF_2, F_2, Cu, ClO, BrF_5, acetaldehyde, (BaO + Hg_2O + air), (BaO + NiO + air), hydrated iron oxide, phenyl diazonium chloride, (NaOH + CaO + air). Incandescent reaction with chromium trioxide. Vigorous reaction with metal powders (e.g., copper, tungsten). When heated to decomposition it emits highly toxic fumes of SO_x. To fight fire, stop flow of gas.

HIG500 CAS: 1435-55-8 **HR: 3**
HYDROQUINIDINE
mf: $C_{20}H_{26}N_2O_2$ mw: 326.48

PROP: Plates from ether, needles from alc. Mp: 169°; very sol in hot alc; sltly sol in water and ether.

SYNS: 10,11-DIHYDRO-6'-METHOXYCINCHONAN-9-OL * DIHYDROQUINIDINE * 10,11-DIHYDROQUINIDINE * HYDROCONQUININE

SAFETY PROFILE: A poison by ingestion and intravenous routes. When heated to decomposition it emits toxic fumes of NO_x.

HIH000 CAS: 123-31-9 **HR: 3**
HYDROQUINONE
DOT: UN 2662
mf: $C_6H_6O_2$ mw: 110.12

PROP: Colorless, hexagonal prisms. Mp: 170.5°, bp: 286.2°, flash p: 329°F (CC), d: 1.358

@ 20°/4°, autoign temp: 960°F (CC), vap press: 1 mm @ 132.4°, vap d: 3.81. Very sol in alc and ether; sltly sol in benzene. Keep well closed and protected from light.

SYNS: ARCTUVIN * p-BENZENEDIOL * 1,4-BENZENEDIOL * BENZOHYDROQUINONE * BENZOQUINOL * BLACK AND WHITE BLEACHING CREAM * 1,4-DIHYDROXY-BENZEEN (DUTCH) * 1,4-DIHYDROXYBENZEN (CZECH) * DIHYDROXYBENZENE * p-DIHYDROXYBENZENE * 1,4-DIHYDROXYBENZENE * 1,4-DIHYDROXY-BENZOL (GERMAN) * 1,4-DIIDROBENZENE (ITALIAN) * p-DIOXOBENZENE * ELDOPAQUE * ELDOQUIN * HYDROCHINON (CZECH, POLISH) * HYDROQUINOL * α-HYDROQUINONE * p-HYDROQUINONE * p-HYDROXYPHENOL * IDROCHINONE (ITALIAN) * NCI-C55834 * β-QUINOL * TECQUINOL * TENOX HQ * USAF EK-356

CONSENSUS REPORTS: IARC Cancer Review: GROUP 3 IMEMDT 7,56,87; Animal Inadequate Evidence IMEMDT 15,155,77. Community Right-To-Know List. EPA Extremely Hazardous Substances List. EPA Genetic Toxicology Program. Reported in EPA TSCA Inventory.

OSHA PEL: TWA 2 mg/m^3
ACGIH TLV: TWA 2 mg/m^3
DFG MAK: 2 mg/m^3
NIOSH REL: (Hydroquinone) CL 2.0 mg/m^3/ 15M
DOT Classification: IMO: Poison B; Label: St. Andrews Cross.

SAFETY PROFILE: A human poison by ingestion. An experimental poison by ingestion, intraperitoneal, intravenous, parenteral, subcutaneous and possibly other routes. Human systemic effects by ingestion: pulse rate increase without fall in blood pressure, cyanosis, coma. An active allergen and a strong skin irritant. Human mutation data reported. A severe human skin irritant. Experimental reproductive data. Questionable carcinogen.

Absorption of this material by tissues can cause symptoms of illness which resemble those induced by its o- or m- isomers. For instance, the ingestion of 1 gram by an adult or a smaller quantity by a child may induce tinnitus, nausea, dizziness, a sensation of suffocation, an increased rate of respiration, vomiting, pallor, muscular twitching, headache, dyspnea, cyanosis, delirium, and collapse. The literature contains reports of fatal cases which have been caused by the ingestion of 5-12 grams. Cases of dermatitis have resulted from skin contact, and have also followed the application of an antiseptic oil which apparently contained traces of hydroquinone added as an antioxidant. The report also contains cases of keratitis and discoloration of the conjunctiva among personnel exposed to this material in concentrations ranging from 10 to 30 mg of the vapor or dust per cubic meter of air. It is considered to be more toxic than phenol. The inhalation of vapors, particularly when liberated at high temperatures, must be avoided. If this material accidentally comes into contact with the skin, it should be removed at once and the affected area washed with plenty of soap and water.

Combustible when exposed to heat or flame; can react with oxidizing materials. Potentially explosive reaction with oxygen at 90°C/100 bar. Violent reaction with NaOH. Slight explosion hazard when exposed to heat. To fight fire, use water, CO_2, dry chemical.

HIM000   CAS: 103-90-2   **HR: 3**
4′-HYDROXYACETANILIDE
mf: $C_8H_9NO_2$ mw: 151.18

SYNS: ABENSANIL * ACAMOL * ACETAGESIC * ACETALGIN * p-ACETAMIDOPHENOL * 4-ACETAMIDOPHENOL * ACETAMINOPHEN * p-ACETAMINOPHENOL * N-ACETYL-p-AMINO-PHENOL * p-ACETYLAMINOPHENOL * ALGO-TROPYL * ALPINYL * ALVEDON * AMADIL * ANAFLON * ANELIX * ANHIBA * APADON * APAMIDE * APAP * BEN-U-RON * BICKIE-MOL * CALPOL * CETADOL * CLIXODYNE * DATRIL * DIAL-A-GESIC * DIROX * DOLI-PRANE * DYMADON * ENELFA * ENERIL * EXDOL * FEBRILIX * FEBRO-GESIC * FEBROLIN * FENDON * FINIMAL * G 1 * GELOCATIL * HEDEX * HOMOOLAN * p-HYDROXYACETANILIDE * 4-HYDROXYACETAN-ILIDE * N-(4-HYDROXYPHENYL)ACETAMIDE * JANUPAP * KORUM * LESTEMP * LIQUA-GESIC * LONARID * LYTECA SYRUP * MO-MENTUM * MULTIN * NAPA * NAPRINOL * NCI-C55801 * NOBEDON * PACEMO * PANADOL * PANETS * PANEX * PANOFEN * PARACETAMOLE * PARACETAMOLO (ITALIAN) * PARACETANOL * PARAPAN * PARASPEN * PARMOL * PEDRIC * PYRINAZINE * SK-Apap * TABALGIN * TAPAR * TEMLO * TEMPANAL * TEMPRA * TRALGON * TUSSAPAP * TYLENOL * VALADOL * VALGESIC

CONSENSUS REPORTS: Reported in EPA TSCA Inventory. EPA Genetic Toxicology Program.

SAFETY PROFILE: Suspected carcinogen with experimental carcinogenic and tumorigenic data. A human poison by ingestion and possibly other routes. An experimental poison by intraperitoneal route. Moderately toxic by subcutaneous, intravenous, and possibly other routes. Human systemic effects by ingestion: changes in exocrine pancreas, diarrhea, nausea, irritability, somnolence, general anesthetic, fever, hepatitis, kidney tubule damage. Experimental teratogenic and reproductive effects. Human mutation data reported. Used as an analgesic and antipyretic. When heated to decomposition it emits toxic fumes of NO_x.

HIM500   CAS: 107-16-4   **HR: 3**
HYDROXYACETONITRILE
mf: C_2H_3NO mw: 57.06

SYNS: CYANOMETHANOL * FORMALDEHYDE CYANOHYDRIN * GLYCOLIC NITRILE * GLY-COLONITRILE * GLYCONITRILE * 2-HYDROXY-ACETONITRILE * HYDROXYMETHYLINITRILE * USAF A-8565

CONSENSUS REPORTS: EPA Extremely Hazardous Substances List. Reported in EPA TSCA Inventory. Cyanide and its compounds are on The Community Right-To-Know List.

NIOSH REL: (Nitriles) CL 5 mg/m³/15M

SAFETY PROFILE: A poison by ingestion, skin contact, inhalation, intraperitoneal, ocular, and subcutaneous routes. A eye irritant. May undergo spontaneous and violent decomposition. Traces of alkali promote violent polymerization. When heated to decomposition it emits toxic fumes of NO_x and CN^-.

HIP000   CAS: 53-95-2   **HR: 3**
N-HYDROXY-N-ACETYL-2-AMINOFLUORENE
mf: $C_{15}H_{13}NO_2$ mw: 239.29

SYNS: FLUORENYL-2-ACETHYDROXAMIC ACID * N-FLUOREN-2-YL ACETOHYDROXAMIC ACID * N-2-FLUORENYL ACETOHYDROXAMIC ACID * N-HYDROXY-AAF * N-HYDROXY-2-ACETAMIDO-FLUORENE * 2-(N-HYDROXYACETAMIDO)FLUORENE * N-HYDROXY-2-ACETYLAMINOFLUORENE

* N-HYDROXY-2-FAA * N-HYDROXY-N-(2-FLUORE-NYL)ACETAMIDE * NOHFAA

CONSENSUS REPORTS: EPA Genetic Toxicology Program.

SAFETY PROFILE: Suspected carcinogen with experimental carcinogenic, neoplastigenic, and tumorigenic data. A poison by intraperitoneal route. Experimental teratogenic and other reproductive effects. Human mutation data reported. When heated to decomposition it emits toxic fumes of NO_x.

HIY500   CAS: 4637-56-3   **HR: 3**
4-(HYDROXYAMINO)QUINOLINE-1-OXIDE
mf: $C_9H_8N_2O_2$ mw: 176.19

SYNS: 4HAQO * N-(4-QUINOLYL)HYDROXYLAMINE-1'-OXIDE

CONSENSUS REPORTS: EPA Genetic Toxicology Program.

SAFETY PROFILE: A poison by intravenous route. Questionable carcinogen with experimental carcinogenic, neoplastigenic, and tumorigenic data. Human mutation data reported. When heated to decomposition it emits toxic fumes of NO_x.

HJF000   CAS: 1689-82-3   **HR: 3**
4-HYDROXYAZOBENZENE
mf: $C_{12}H_{10}N_2O$ mw: 198.24

PROP: Orange, rhombic crystals from ethanol. Mp: 155-156°, bp: 220-230°, very sol in ether.

SYNS: p-BENZENEAZOPHENOL * C.I. SOLVENT YELLOW 7 * p-HYDROXYAZOBENZENE * p-PHENYLAZOPHENOL * 4-PHENYLAZOPHENOL

CONSENSUS REPORTS: IARC Cancer Review: GROUP 3 IMEMDT 7,56,87; Animal Inadequate Evidence IMEMDT 8,157,75. Reported in EPA TSCA Inventory.

SAFETY PROFILE: A poison by intraperitoneal route. Questionable carcinogen. When heated to decomposition it emits toxic fumes of NO_x.

HJF500   CAS: 3567-69-9   **HR: 3**
4-HYDROXY-3,4'-AZODI-1-NAPHTHALENESULFONIC ACID, DISODIUM SALT
mf: $C_{20}H_{12}N_2O_7S_2 \cdot 2Na$ mw: 502.44

SYNS: ACETACID RED B * ACID BRILLIANT RUBINE 2G * ACID CHROME BLUE BA * ACID FAST

RED FB * ACID RUBINE * AIREDALE CARMOISINE * AMACID CHROME BLUE R * ATUL CRYSTAL RED F * AZORUBIN * BRASILAN AZO RUBINE 2NS * BRILLIANT CRIMSON RED * CARMOISIN (GERMAN) * CARMOISINE ALUMINUM LAKE * CARMOISINE SUPRA * CERTICOL CARMOISINE S * CHROME FAST BLUE 2R * C.I. 14720 * C.I. ACID RED 14, DISODIUM SALT * C.I. FOOD RED 3 * CRIMSON EMBL * DIADEM CHROME BLUE R * DISODIUM SALT of 2-(4-SULPHO-1-NAPHTHYLAZO)-1-NAPHTHOL-4-SULPHONIC ACID * DISODIUM 2-(4-SULFO-1-NAPHTHYLAZO)-1-NAPHTHOL-4-SULFONATE * DISODIUM 2-(4-SULPHO-1-NAPHTHYLAZO)-1-NAPHTHOL-4-SULPHONATE * EDICOL SUPRA CARMOISINE WS * ENIACID BRILLIANT RUBINE 3B * EUROCERT AZORUBINE * EXTRACT D&C RED NO. 10 * FENAZO RED C * FOOD RED 5 * FRUIT RED A EXTRA YELLOWISH GEIGY * HEXACOL CARMOISINE * HIDACID AZO RUBINE * 4-HYDROXY-3,4'-AZODI-1-NAPHTHALENESULPHONIC ACID, DISODIUM SALT * 4-HYDROXY-3-((4-SULFO-1-NAPHTHALENYL)AZO)-1-NAPHTHALENESULFONIC ACID, DISODIUM SALT * JAVA RUBINE N * KARMESIN * KENACHROME BLUE 2R * KITON CRIMSON 2R * LIGHTHOUSE CHROME BLUE 2R * NACARAT A EXPORT * NCI-C53849 * NEKLACID RUBINE W * NYLOMINE ACID RED P4B * OMEGA CHROME BLUE FB * POLOXAL RED 2B * PONTACYL RUBINE R * RED #14 * 11959 RED * SCHULTZ Nr. 208 (GERMAN) * SOLAR RUBINE * SOLOCHROME BLUE FB * STANDACOL CARMOISINE * 2-(4-SULFO-1-NAPHTHYLAZO)-1-NAPHTHOL-4-SULFONIC ACID, DISODIUM SALT * TERTRACID RED CA * TERTROCHROME BLUE FB

CONSENSUS REPORTS: IARC Cancer Review: GROUP 3 IMEMDT 7,56,87; Animal Inadequate Evidence IMEMDT 8,83,75. NTP Carcinogenesis Bioassay (feed); No Evidence: mouse, rat NTPTR* NTP-TR-220,82. Reported in EPA TSCA Inventory. EPA Genetic Toxicology Program.

SAFETY PROFILE: Moderately toxic by intraperitoneal and intravenous routes. Questionable carcinogen. Mutation data reported. When heated to decomposition it emits very toxic fumes of SO_x, Na_2O and NO_x.

HJL000   CAS: 120-47-8   **HR: 2**
p-HYDROXYBENZOIC ACID ETHYL ESTER
mf: $C_9H_{10}O_3$ mw: 166.19

SYNS: ASEPTOFORM E * BONOMOLD OE * p-CARBETHOXYPHENOL * EASEPTOL

* ETHYL-p-HYDROXYBENZOATE * ETHYL PARA-BEN * ETHYL PARASEPT * p-HYDROXYBENZOIC ETHYL ESTER * NIPAGIN A * NIPAZIN A * p-OXYBENZOESAEUREAETHYLESTER (GERMAN) * SOLBROL A * TEGOSEPT E

CONSENSUS REPORTS: Reported in EPA TSCA Inventory.

SAFETY PROFILE: Moderately toxic by ingestion and intraperitoneal routes. Experimental teratogenic effects. Mutation data reported. When heated to decomposition it emits acrid smoke and fumes.

HJL500   CAS: 99-76-3   HR: 2
p-HYDROXYBENZOIC ACID METHYL ESTER
mf: $C_8H_8O_3$ mw: 152.16

PROP: Colorless crystals or white crystalline powder; faint odor and burning taste. Sol in alc, ether, and propylene glycol; sltly sol in water, glycerin, fixed oils, benzene, and carbon tetrachloride;

SYNS: ABIOL * ASEPTOFORM * MASEPTOL * METHYLBEN * METHYL CHEMOSEPT * METHYL ESTER of p-HYDROXYBENZOIC ACID * METHYL p-HYDROXYBENZOATE * METHYL p-OXYBENZOATE * METHYLPARABEN (FCC) * METHYL PARAHYDROXYBENZOATE * METHYL PARASEPT * METOXYDE * MOLDEX * NIPAGIN * p-OXYBENZOESAUREMETHYLESTER (GERMAN) * PARABEN * PARASEPT * PARIDOL * PRESERVAL M * SEPTOS * SOLBROL M * TEGOSEPT M

CONSENSUS REPORTS: Reported in EPA TSCA Inventory.

SAFETY PROFILE: Moderately toxic by ingestion, subcutaneous and intraperitoneal routes. Mutation data reported. When heated to decomposition it emits acrid smoke and fumes.

HJO500   CAS: 469-65-8   HR: 3
(p-HYDROXYBENZYL)TARTARIC ACID
mf: $C_{11}H_{12}O_7$ mw: 256.23

PROP: From the bark of the Jamaica dogwood.

SYNS: 2,3-DIHYDROXY-2-((4-HYDROXYPHENYL)METHYL)BUTANEDIOIC ACID * PISCIDEIN * PISCIDIC ACID

SAFETY PROFILE: A poison by intraperitoneal, subcutaneous, and parenteral routes. When heated to decomposition it emits acrid smoke and fumes.

HJQ000   CAS: 5809-59-6   HR: 3
2-HYDROXY-3-BUTENENITRILE
mf: C_4H_5NO mw: 83.10

SYNS: ACROLEIN CYANOHYDRIN * 1-CYANO-2-PROPEN-1-OL * VINYLGLYCOLONITRILE

CONSENSUS REPORTS: Cyanide and its compounds are on the Community Right-To-Know List.

SAFETY PROFILE: A poison by ingestion, skin contact, and inhalation. A skin and severe eye irritant. May polymerize explosively when exposed to light and air above 25°C. A storage hazard. When heated to decomposition it emits toxic fumes of NO_x and CN^-.

HJQ350   CAS: 3817-11-6   HR: 3
4-HYDROXYBUTYLBUTYLNITRO-SAMINE
mf: $C_8H_{18}N_2O_2$ mw: 174.28

SYNS: BBN * BBNOH * BHBN * BUTANOL (4)-BUTYL-NITROSAMINE * BUTYL-BUTANOL(4)-NITROSAMIN * BUTYL-BUTANOL-NITROSAMINE * N-BUTYL-N-(4-HYDROXYBUTYL)NITROSAMINE * n-BUTYL-(4-HYDROXYBUTYL)NITROSAMINE * 4-(BUTYLNITROSAMINO)-1-BUTANOL * 4-(n-BUTYLNITROSAMINO)-1-BUTANOL * DIBUTYLAMINE, 4-HYDROXY-N-NITROSO- * HBBN * NBHA * N-NITROSO-n-BUTYL-(4-HYDROXYBUTYL)AMINE * OH-BBN

CONSENSUS REPORTS: IARC Cancer Review: Animal Sufficient Evidence IMEMDT 17,51,78

SAFETY PROFILE: Confirmed carcinogen with experimental carcinogenic and neoplastigenic data. Moderately toxic by ingestion. Mutation data reported. When heated to decomposition it emits toxic fumes of NO_x.

HJV000   CAS: 57651-82-8   HR: 3
1-HYDROXYCHOLECALCIFEROL
mf: $C_{27}H_{44}O_2$ mw: 400.71

PROP: Mp: 134-136° or 138-139.5°.

SYNS: ALFACALCIDOL * 1-α-DIHYDROXYVITAMIN D3 * α-HCC * HYDROXYCHOLECALCIFEROL * 1-α-HYDROXYCHOLECALCIFEROL * 1-α-HYDROXYVITAMIN D3 * 1-α-OH-CC * 1-α-OH-D³ * 1-α-OH VITAMIN D3 * 9,10-SECOCHOLESTA-5,-7,10(19)-TRIENE-1-α,3-β-DIOL * VITAMIN D³

SAFETY PROFILE: A deadly poison by ingestion, subcutaneous and, intravenous routes. Experimental teratogenic and reproductive effects. When heated to decomposition it emits acrid smoke and fumes.

HJV700   HR: 1
HYDROXYCITRONELLAL DIMETHYL ACETAL
mf: $C_{12}H_{26}O_3$ mw: 218.34

PROP: Colorless liquid; floral odor. D: 0.925, refr index: 1.441, flash p: +212°F. Sol in fixed oils, propylene glycol; insol in glycerin.

SYNS: FEMA No. 2585 * 7-HYDROXY-3,7-DIMETHYL OCTANAL:ACETAL

SAFETY PROFILE: Combustible liquid. When heated to decomposition it emits acrid smoke and irritating fumes.

HKB500   CAS: 1689-83-4   HR: 3
4-HYDROXY-3,5-DIIODOBENZONITRILE
mf: $C_7H_3I_2NO$ mw: 370.91

PROP: Colorless solid. Mp: 213°. Sltly sol in water.

SYNS: ACTRIL * BANTROL * BENTROL * CERTROL * 4-CYANO-2,6-DIIODOPHENOL * 4-CYANO-2,6-DIJODPHENOL (GERMAN) * 3,5-DI-IODO-4-HYDROXYBENZONITRILE * 3,5-DIJOD-4-HY-DROXY-BENZONITRIL (GERMAN) * IOTOX * IOXYNIL * LOXYNIL (GERMAN) * M&B 8873 * OXYTRIL * TOTRIL

CONSENSUS REPORTS: Cyanide and its compounds are on the Community Right-To-Know List.

SAFETY PROFILE: A human poison by ingestion. Very poisonous experimentally by skin contact, ingestion, intravenous, and possibly other routes. An herbicide. When heated to decomposition it emits toxic fumes of I^- and CN^-.

HKC000   CAS: 75-60-5   HR: 3
HYDROXYDIMETHYLARSINE OXIDE
DOT: UN 1572
mf: $C_2H_7AsO_2$ mw: 138.01

PROP: Colorless crystals; odorless. Mp: 192°. Sol in water.

SYNS: ACIDE CACODYLIQUE (FRENCH) * ACIDE DIMETHYLARSINIQUE (FRENCH) * AGENT BLUE * ANSAR * ARSAN * BOLLS-EYE * CACODYLIC ACID (DOT) * CHEXMATE * DILIC * DIMETHYLARSENIC ACID * DIMETHYLARSINIC

ACID * DMAA * ERASE * PHYTAR * RAD-E-CATE 25 * RCRA WASTE NUMBER U136 * SALVO * SILVISAR 510

CONSENSUS REPORTS: IARC Cancer Review: Animal Inadequate Evidence IMEMDT 23,39,80. Arsenic and its compounds are on the Community Right-To-Know List. Reported in EPA TSCA Inventory. EPA Genetic Toxicology Program.

OSHA PEL: TWA 0.5 mg(As)/m³
ACGIH TLV: TWA 0.2 mg(As)/m³
DOT Classification: IMO: Poison B; Label: Poison.

SAFETY PROFILE: Poison by an unspecified route. Moderately toxic by ingestion and intraperitoneal routes. Experimental teratogenic and reproductive effects. A skin and eye irritant. Questionable carcinogen with experimental tumorigenic data. Mutation data reported. Used as an herbicide, defoliant, and silvicide. Hazardous when water solution is in contact with active metals, i.e., Fe, Al, Zn. When heated to decomposition it emits toxic fumes of As.

HKC500   CAS: 124-65-2   HR: 3
HYDROXYDIMETHYLARSINE OXIDE, SODIUM SALT
DOT: UN 1688
mf: $C_2H_6AsO_2 \cdot Na$ mw: 159.99

SYNS: ALKARSODYL * ANSAR 160 * ARSECO-DILE * ARSYCODILE * BOLLS-EYE * CACO-DYLATE de SODIUM (FRENCH) * CACODYLIC ACID SODIUM SALT * CHEMAID * ((DIMETHYLARSINO)OXY)SODIUM-As-OXIDE * DUTCH-TREAT * PHYTAR 560 * RAD-E CATE 16 * SILVISAR * SODIUM CACODYLATE (DOT) * SODIUM DIMETHYLARSINATE * SODIUM DIMETHYLARSONATE * SODIUM SALT of CACODYLIC ACID

CONSENSUS REPORTS: Arsenic and its compounds are on the Community Right-To-Know List. EPA Extremely Hazardous Substances List. Reported in EPA TSCA Inventory.

OSHA PEL: TWA 0.5 mg(As)/m³
ACGIH TLV: TWA 0.2 mg(As)/m³
DOT Classification: IMO: Poison B; Label: Poison.

SAFETY PROFILE: Confirmed human carcinogen. Poison by ingestion. Moderately toxic by subcutaneous route. Experimental teratogenic and other reproductive effects. When heated to decomposition it emits toxic fumes of As and Na$_2$O.

HKE000   CAS: 520-53-6   HR: 3
4-HYDROXY-N,N-DIMETHYLTRYPTAMINE
mf: C$_{12}$H$_{16}$N$_2$O   mw: 204.30

SYNS: CX-59 * 3-(2-(DIMETHYLAMINO)ETHYL)IN-DOL-4-OL * PSILOCINE * PSILOTSIN

SAFETY PROFILE: A poison by intravenous and intraperitoneal routes. When heated to decomposition it emits toxic fumes of NO$_x$.

HKH500   CAS: 365-26-4   HR: 3
p-HYDROXYEPHEDRINE
mf: C$_{10}$H$_{15}$NO$_2$ mw: 181.23

PROP: Crystalline powder. Mp: 152-154°; Sery sltly sol in water, alc, and ether. Very sol in NaOH solns and dil acids.

SYNS: p-HYDROXYPHENYLMETHYLAMINOPROPANOL * 1-(4-HYDROXYPHENYL)-2-METHYLAMINOPROPANOL * α-(1-METHYLAMINOETHYL)-p-HYDROXYBENZYL AL-COHOL

SAFETY PROFILE: A poison by intravenous route. When heated to decomposition it emits toxic fumes of NO$_x$.

HKJ000   CAS: 106-11-6   HR: 3
2-(2-HYDROXYETHOXY)ETHYL ESTER STEARIC ACID
mf: C$_{22}$H$_{44}$O$_4$ mw: 372.66

SYNS: AQUA CERA * ATLAS G 2146 * CERA-SYNT * CLINDROL SDG * DIETHYLENE GLYCOL, MONOESTER with STEARIC ACID * DIETHYLENE GLY-COL MONOSTEARATE * DIETHYLENE GLYCOL STEA-RATE * DIGLYCOL MONOSTEARATE * DIGLYCOL STEARATE * EMCOL DS-50 CAD * GLYCO STEA-RIN * NONEX 411 * PROMUL 5080 * USAF KE-8

CONSENSUS REPORTS: Reported in EPA TSCA Inventory.

SAFETY PROFILE: A poison by intraperitoneal route. When heated to decomposition it emits acrid smoke and fumes.

HKS780   CAS: 2809-21-4   HR: 3
1-HYDROXYETHYLIDENE-1,1-DIPHOSPHONIC ACID
mf: C$_2$H$_8$O$_7$P$_2$ mw: 206.03

SAFETY PROFILE: When heated above 200° it decomposes violently to produce toxic fumes of phosphine, phosphoric acid, and PO$_x$.

HKV000   CAS: 21600-45-3   HR: 3
3-(2-HYDROXYETHYL)-3-METHYL-1-PHENYLTRIAZENE
mf: C$_9$H$_{13}$N$_3$O   mw: 179.25

SYNS: 1-PHENYL-3-METHYL-3-(2-HYDROXYAETHYL)-TRIAZEN (GERMAN) * 1-PHENYL-3-METHYL-3-(2-HY-DROXYETHYL)TRIAZENE

SAFETY PROFILE: A poison by subcutaneous route. Questionable carcinogen with experimental carcinogenic data. When heated to decomposition it emits toxic fumes of NO$_x$.

HKW500   CAS: 13743-07-2   HR: 3
1-(2-HYDROXYETHYL)-1-NITROSOUREA
mf: C$_3$H$_7$N$_3$O$_3$ mw: 133.13

SYNS: HENU * HNU * NITROSO-2-HYDROXY-ETHYLUREA * N-NITROSOHYDROXYETHYLUREA * 1-NITROSO-1-(2-HYDROXYETHYL)UREA

CONSENSUS REPORTS: EPA Genetic Toxicology Program.

SAFETY PROFILE: Suspected carcinogen with experimental carcinogenic and tumorigenic data. A poison by intraperitoneal route. Experimental reproductive effects. Mutation data reported. When heated to decomposition it emits toxic fumes of NO$_x$.

HLC500   CAS: 71-27-2   HR: 3
(2-HYDROXYETHYL)TRIMETHYLAMMO-NIUM CHLORIDE SUCCINATE
mf: C$_{14}$H$_{30}$N$_2$O$_4$ • 2Cl   mw: 361.36

SYNS: ANECTINE * ANECTINE CHLORIDE * BIS (2-DIMETHYLAMINOETHYL)SUCCINATE BIS(METHO-CHLORIDE) * BIS(SUCCINYLDICHLOROCHOLINE) * CHLORSUCCINYLCHOLIN (GERMAN) * CHLO-RURE de SUCCINILCOLINE (FRENCH) * CHOLINE SUCCINATE DICHLORIDE * CLORURO di SUC-CINILCOLINA (ITALIAN) * DIACETYLCHOLINE CHLORIDE * DIACETYLCHOLINE DICHLORIDE * 2-DIMETHYLAMINOETHYL SUCCINATE DIMETHO-CHLORIDE * DITILIN * DITILINE * 2,2′-((1,4-DIOXO-1,4-BUTANEDIYL)BIS(OXY)BIS(N,N,N-TRI-METHYLETHANAMINIUM DICHLORIDE * LISTE-NON * LYSTENON * LYSTHENONE * MIDA-RINE * MYOPLEGINE * PANTOLAX * QUELICIN * QUELICIN CHLORIDE * SCH

CHLORIDE * SCOLINE * SCOLINE CHLORIDE * SKOLIN * SUCCICURAN * SUCCINIC ACID BIS(β-DIMETHYLAMINOETHYL) ESTER, DIHYDRO-CHLORIDE * SUCCINIC ACID BIS(β-DIMETHYLAMINO-ETHYL)ESTER DIMETHOCHLORIDE * SUCCINIC ACID DIESTER with CHOLINE CHLORIDE * SUCCINOYLCHO-LINE CHLORIDE * SUCCINYL-ASTA * SUCCINYL BISCHOLINE CHLORIDE * SUCCINYLBISCHOLINE DI-CHLORIDE * SUCCINYLCHOLINE CHLORIDE * SUCCINYLCHOLINE DICHLORIDE * SUCCINYL-CHOLINE HYDROCHLORIDE * SUCCINYLDICHOLINE CHLORIDE * SUCCINYLFORTE * SUCOSTRIN * SUCOSTRIN CHLORIDE * SURAMETHINIUM * SUXAMETHIONIUM CHLORIDE * SUXAMETHO-NIUM CHLORIDE * SUXAMETHONIUM DICHLORIDE * SUXCERT * SUXETHONIUM CHLORIDE * SUXINYL * ULTRAPAL CHLORIDE

SAFETY PROFILE: A deadly poison by intravenous and intraperitoneal routes. Human systemic effects by intravenous route: metabolic changes in potassium level, cardiac arrythmias. Used as a skeletal muscle relaxant. When heated to decomposition it emits very toxic fumes of NO_x, NH_3, and Cl^-.

HLJ500   CAS: 1689-89-0   **HR: 3**
4-HYDROXY-3-IODO-5-NITROBENZONITRILE
mf: $C_7H_3IN_2O_3$ mw: 290.02

SYNS: DOVENIX * NITROXYNIL * TRODAX

CONSENSUS REPORTS: Cyanide and its compounds are on the Community Right-To-Know List.

SAFETY PROFILE: A poison by ingestion and parenteral routes. When heated to decomposition it emits very toxic I^-, NO_x, and CN^-.

HLM500   CAS: 7803-49-8   **HR: 3**
HYDROXYLAMINE
mf: H_3NO mw: 33.04

PROP: Colorless liquid or white needles. Unstable, hygroscopic, decomp rapidly at room temp. Mp: 34.0°, bp: 110.0°, flash p: explodes at 265°F, d: 1.227, vap press: 10 mm @ 47.2°. Decomp in hot water; very sol in liquid ammonia and methanol; very sltly sol in ether, benzene, carbon disulfide, and chloroform.

SYN: OXAMMONIUM

CONSENSUS REPORTS: Reported in EPA TSCA Inventory. EPA Genetic Toxicology Program.

SAFETY PROFILE: A poison by intraperitoneal, subcutaneous, and possibly other routes. A corrosive irritant to the eye, skin, and mucous membranes. Locally it is irritating, and systemically it can cause methemoglobinemia. Human mutation data reported. Dangerous fire hazard when exposed to heat, flame, and oxidizers. May ignite spontaneously in air if a large surface area is exposed (e.g., precipitate on paper). Explodes in air when heated above 70°C. Explosive reaction with potassium dichromate, chromium trioxide, powdered zinc + heat. Forms the heat-sensitive explosive bis(hydroxylamide) in reaction with zinc or calcium. Ignites on contact with copper(II) sulfate, metals (e.g., sodium), oxidants (e.g., barium peroxide, barium oxide, lead dioxide, potassium permanganate, chlorine), phosphorus chlorides (e.g., phosphorus trichloride, phosphorus pentachloride). Incompatible with carbonyls, pyridine. Vigorous reaction with hypochlorites. When heated to decomposition it emits toxic fumes of NO_x.

HLS500   CAS: 4756-45-0   **HR: 3**
o-(2-HYDROXY-4-METHOXYBENZOYL) BENZOIC ACID
mf: $C_{15}H_{12}O_5$ mw: 272.27

SYN: 2′-CARBOXY-2-HYDROXY-4-METHOXYBENZO-PHENONE(o-(2-HYDROXY-p-ANISOYL)BENZOIC ACID)

CONSENSUS REPORTS: Reported in EPA TSCA Inventory.

SAFETY PROFILE: A poison by intravenous route. When heated to decomposition it emits acrid smoke and fumes.

HLV000   CAS: 1910-36-7   **HR: 3**
N-HYDROXY-N-METHYL-4-AMINOAZOBENZENE
mf: $C_{13}H_{13}N_3O$ mw: 227.29

SYNS: N-HYDROXY-MAB * N-METHYL-N-(p-(PHE-NYLAZO)PHENYL)HYDROXYLAMINE

SAFETY PROFILE: Questionable carcinogen with experimental carcinogenic and tumorigenic data. Mutation data reported. When heated to decomposition it emits toxic fumes of NO_x.

HLX925   CAS: 78246-54-5   **HR: 3**
4-(HYDROXYMETHYL)BENZENEDIAZO-NIUM TETRAFLUOROBORATE
mf: $C_7H_7N_2O \cdot BF_4$ mw: 221.97

SYNS: HMBD * BENZENEDIAZONIUM, 4-(HY-DROXYMETHYL)-, TETRAFLUOROBORATE(1-)

SAFETY PROFILE: Suspected carcinogen with experimental carcinogenic data. Mutation data reported. When heated to decomposition it emits toxic fumes of NO_x, B, and F^-.

HMA500   CAS: 3308-64-3   **HR: 3**
2-HYDROXY-3-METHYL-CHOLANTHRENE
mf: $C_{21}H_{16}O$ mw: 284.37

SYN: 3-METHYLCHOLANTHREN-2-OL

CONSENSUS REPORTS: EPA Genetic Toxicology Program.

SAFETY PROFILE: Questionable carcinogen with experimental carcinogenic and neoplastigenic data. Mutation data reported. When heated to decomposition it emits acrid smoke and fumes.

HMB000   CAS: 5980-33-6   **HR: 3**
7-HYDROXY-4-METHYLCOUMARIN SODIUM
mf: $C_{10}H_7O_3 \cdot Na$ mw: 198.16

SYNS: CANTABILINE SODIUM   *   HYMECROMONE SODIUM   *   METHYL-4-OMBELLIFERONE SODEE (FRENCH)   *   METHYL-4-UMBELLIFERONE SODIUM

SAFETY PROFILE: A poison by intraperitoneal and intravenous routes. When heated to decomposition it emits toxic fumes of Na_2O.

HMB500   CAS: 80-71-7   **HR: 2**
2-HYDROXY-3-METHYL-2-CYCLO-PENTEN-1-ONE
mf: $C_6H_8O_2$ mw: 112.14

PROP: White crystalline power; nutty odor, maple in dilute solutions. Flash p: +212°F. Sol in alc, propylene glycol; sltly sol in fixed oils, water.

SYNS: CORYLON   *   CORYLONE   *   CYCLOTEN   *   FEMA No. 2700   *   MAPLE LACTONE   *   3-METHYLCYCLOPENTANE-1,2-DIONE   *   METHYL CYCLOPENTENOLONE (FCC)

CONSENSUS REPORTS: Reported in EPA TSCA Inventory.

SAFETY PROFILE: Moderately toxic by ingestion and intraperitoneal routes. Human mutation data reported. Combustible liquid. When heated to decomposition it emits acrid smoke and fumes.

HMF000   CAS: 568-75-2   **HR: 3**
7-HYDROXYMETHYL-12-METHYLBENZ (a)ANTHRACENE
mf: $C_{20}H_{16}O$ mw: 272.36

SYNS: 7-HM-12-MBA   *   12-METHYBENZ(a)ANTHRACENE-7-METHANOL   *   7-OHM-MBA   *   7-OHM-12-MBA

CONSENSUS REPORTS: EPA Genetic Toxicology Program.

SAFETY PROFILE: Poison by intravenous route. Experimental teratogenic and reproductive effects. Questionable carcinogen with experimental carcinogenic, neoplastigenic, and tumorigenic data. Mutation data reported. When heated to decomposition it emits acrid smoke and fumes.

HMF500   CAS: 568-70-7   **HR: 3**
12-HYDROXYMETHYL-7-METHYLBENZ (a)ANTHRACENE
mf: $C_{20}H_{16}O$ mw: 272.36

SYNS: 12-HM-7-MBA   *   7-METHYLBENZ(a)ANTHRACENE-12-METHANOL   *   7-METHYL-12-HYDROXYMETHYLBENZ(a)ANTHRACENE

CONSENSUS REPORTS: EPA Genetic Toxicology Program.

SAFETY PROFILE: Questionable carcinogen with experimental carcinogenic and tumorigenic data. Mutation data reported. When heated to decomposition it emits acrid smoke and fumes.

HMG000   CAS: 590-96-5   **HR: 3**
1-HYDROXYMETHYL-2-METHYL-DITMIDE-2-OXIDE
mf: $C_2H_6N_2O_2$ mw: 90.10

SYNS: MAM   *   METHYLAZOXYMETHANOL   *   (METHYL-ONN-AZOXY)METHANOL

CONSENSUS REPORTS: IARC Cancer Review: GROUP 2B IMEMDT 7,56,87; Animal Sufficient Evidence IMEMDT 10,121,76. EPA Genetic Toxicology Program.

SAFETY PROFILE: Suspected carcinogen with experimental tumorigenic data. Experimental teratogenic and reproductive effects. Mutation data reported. When heated to decomposition it emits toxic fumes of NO_x.

HMQ500   CAS: 65229-18-7   **HR: 3**
d-N,N'-(1-HYDROXYMETHYLPROPYL) ETHYLENEDINITROSAMINE
mf: $C_{10}H_{22}N_4O_4$ mw: 262.36

SYNS: DDETA * 2,2'-(ETHYLENEBIS(NITROSO-IMINO))BISBUTANOL * d-N,N'-(1-IDROSSIMETIL PRO-PIL)-ETILENDINITROSAMINA (ITALIAN)

SAFETY PROFILE: Questionable carcinogen with experimental carcinogenic data. When heated to decomposition it emits toxic fumes of NO_x.

HMV000   CAS: 1491-41-4   HR: 3
N-HYDROXYNAPHTHALIMIDE, DIETHYL PHOSPHATE
mf: $C_{16}H_{16}NO_6P$ mw: 349.30

PROP: Tan, crystalline powder. Mp: 177.0°. Sol in methylene chloride; difficultly sol in most organic solvents.

SYNS: B-9002 * BAYER 25820 * CHEMAGRO B-9002 * 2-((DIETHOXYPHOSPHINYL)OXY)-1H-BENZ-(de)ISOQUINOLINE-1,3(2H)-DIONE * O,O-DIETHYL N-HYDROXYNAPHTHALIMIDE PHOSPHATE * ENT 25,567 * N-HYDROXYNAPHTHYLIMIDE DIETHYL PHOSPHATE * MARETIN * NAFTALOFOS * NAPHTHALOPHOS * PHOSPHORIC ACID, DIETHYL ESTER-N-NAPHTHALI-MIDE deriv. * PHOSPHORIC ACID, DIETHYL ESTER, NAPHTHALIMIDO deriv. * PHTALOPHOS * RAME-TIN * RAWETIN * S 940

SAFETY PROFILE: A poison by ingestion and skin contact. A pesticide used in veterinary medicine as a ruminant antihelminitic. A cholinesterase inhibitor. When heated to decomposition it emits toxic fumes of NO_x and PO_x.

HMY000   CAS: 121-19-7   HR: 3
4-HYDROXY-3-NITROBENZENE-ARSONIC ACID
mf: $C_6H_6AsNO_6$ mw: 263.05

SYNS: AKLOMIX-3 * 4-HYDROXY-3-NITROPHENYL-ARSONIC ACID * NCI-C56508 * 3N4HPA * NITRO ACID 100 percent * 2-NITRO-1-HYDROXY-BENZENE-4-ARSONIC ACID * 3-NITRO-4-HYDROXY-BENZENEARSONIC ACID * 3-NITRO-4-HYDROXYPHE-NYLARSONIC ACID * NITROPHENOLARSONIC ACID * NSC-2101 * REN O-SAL * RISTAT * ROXAR-SONE (USDA)

CONSENSUS REPORTS: Arsenic and its compounds are on the Community Right-To-Know List. Reported in EPA TSCA Inventory.

OSHA PEL: TWA 10 $\mu g/m^3$
ACGIH TLV: TWA 0.2 mg(As)/m^3

SAFETY PROFILE: Poison by ingestion and intraperitoneal routes. When heated to decom-position it emits very toxic fumes of NO_x and As.

HNB875   CAS: 54-49-9   HR: 3
HYDROXYNOREPHEDRINE
mf: $C_9H_{13}NO_2$ mw: 167.23

SYNS: 1-α-(1-AMINOETHYL)-m-HYDROXYBENZYL AL-COHOL * ARAMINE * BENZENEMETHANOL, α-(1-AMINOETHYL)-3-HYDROXY-, (R-(R*,S*))-(9CI) * m-HYDROXY NOREPHEDRINE * m-HYDROXY-PROPADRINE * ICORAL B * METARADRINE * METARAMINOL * (-)-METARAMINOL * 1-METARAMINOL * PRESSONEX

SAFETY PROFILE: A poison by ingestion, in-traperitoneal, subcutaneous, and intravenous routes. When heated to decomposition it emits toxic fumes of NO_x.

HNI500   CAS: 129-20-4   HR: 3
p-HYDROXYPHENYLBUTAZONE
mf: $C_{19}H_{20}N_2O_3$ mw: 324.41

SYNS: ARTROFLOG * BM 1 * BUTAFLOGIN * BUTANOVA * BUTAPIRONE * BUTILENE * 4-BUTYL-2-(4-HYDROXYPHENYL)-1-PHENYL-3,5-DI-OXOPYRAZOLIDINE * 4-BUTYL-1-(p-HYDROXYPHE-NYL)-2-PHENYL-3,5-PYRAZOLIDINEDIONE * 4-BUTYL-1-(4-HYDROXYPHENYL)-2-PHENYL-3,5-PYR-AZOLIDINEDIONE * 4-BUTYL-2-(p-HYDROXYPHENYL)-1-PHENYL-3,5-PYRAZOLIDINEDIONE * CROVARIL * DEFLOGIN * 3,5-DIOXO-1-PHENYL-2-(p-HYDROXY-PHENYL)-4-N-BUTYLPYRAZOLIDENE * ETROZOLI-DINA * FLAMARIL * FLANARIL * GLOGAL * FLOGHENE * FLOGISTIN * FLOGITOLO * FLOGODIN * FLOGORIL * FLOGOSTOP * FLOPIRINA * FRABEL * G 27202 * 1-(p-HY-DROXYPHENYL)-2-PHENYL-4-BUTYL-3,5-PYRAZOLIDINE-DIONE * 1-p-HYDROXYPHENYL-2-PHENYL-3,5-DIOXO-4-N-BUTYLPYRAZOLIDINE * IDROBUTAZINA * INFAMIL * IPABUTONA * IRIDIL * ISO-BUTAZINA * ISOBUTIL * METABOLITE I * NEO-FARMADOL * NEOFEN * OFFITRIL * OXALID * OXAZOLIDIN * OXAZOLIDIN-GEIGY * OXIBUTOL * OXI-FENIBUTOL * OXIFENYL-BUTAZON * OXYPHENBUTAZONE * OXYPHENYL-BUTAZONE * 1-PHENYL-2-(p-HYDROXYPHENYL)-3,5-DIOXO-4-BUTYLPYRAZOLIDINE * PIRABUTINA * PIRAFLOGIN * POLIFLOGIL * REMAZIN * REUMOX * RUMAPAX * TANDACOTE * TANDALGESIC * TANDEARIL * TANDERAL * TELIDAL * TENDEARIL * VALIOIL * VISUBUTINA * USAF GE-14

CONSENSUS REPORTS: IARC Cancer Review: GROUP 3 IMEMDT 7,56,87; Human Inadequate Evidence IMEMDT 13,183,77.

SAFETY PROFILE: A poison by ingestion, intraperitoneal, and intravenous routes. Moderately toxic to humans by ingestion. Human systemic effects by ingestion: salivary gland changes, diarrhea, nausea or vomiting, hepatitis, hemorrhage, agranulocytosis, thrombocytopenia, dermatitis, fever, and unspecified endocrine system effects. Experimental reproductive effects. Questionable carcinogen. Mutation data reported. Used as an anti-inflammatory agent. When heated to decomposition it emits toxic fumes of NO_x.

HNL000   CAS: 306-23-0   **HR: 3**
p-HYDROXYPHENYLLACTIC ACID
mf: $C_9H_{10}O_4$ mw: 182.19

SAFETY PROFILE: Questionable carcinogen with experimental carcinogenic and teratogenic data. When heated to decomposition it emits acrid smoke and fumes.

HNO500   CAS: 3983-39-9   **HR: 3**
(m-HYDROXYPHENYL)TRIMETHYLAM-
MONIUM IODIDE, METHYLCARBAMATE
mf: $C_{11}H_{17}N_2O_2 \cdot I$ mw: 336.20

SYNS: CARBAMIC ACID-N-METHYL-3-DIMETHYLAMI-NOPHENYL ESTER METHIODIDE * METHIODIDE of N-METHYLURETHANE of 3-DIMETHYLAMINOPHENOL * METHYLCARBAMIC ACID, (m-(TRIMETHYLAMMO-NIO)PHENYL)ESTER, IODIDE * (3-(METHYLCARB-AMOYLOXY)PHENYL)TRIMETHYLAMMONIUM IODIDE * T-1152 * TL 1178

SAFETY PROFILE: A deadly poison by subcutaneous and intravenous routes. When heated to decomposition it emits very toxic fumes of NH_3, NO_x and I^-.

HNR500   CAS: 64050-79-9   **HR: 3**
(m-HYDROXYPHENYL)TRIMETHYLAM-
MONIUM METHYLSULFATE METHYL-
PHENYLCARBAMATE
mf: $C_{17}H_{21}N_2O_2 \cdot CH_3O_4S$ mw: 396.50

SYN: METHYLPHENYLCARBAMIC ESTER OF 3-OXY-PHENYLTRIMETHYLAMMONIUM METHYLSULFATE

SAFETY PROFILE: A poison by ingestion, subcutaneous, and intravenous routes. When heated to decomposition it emits very toxic fumes of NH_3, NO_x and SO_x.

HNT500   CAS: 630-56-8   **HR: 3**
HYDROXYPROGESTERONE
CAPROATE
mf: $C_{27}H_{40}O_4$ mw: 428.67

PROP: Dense needles. Mp: 119-121°.

SYNS: CAPRON * CORLUTIN L.A. * DELALUTIN * DEPO-PROLUTON * DURALUTON * ESTRALU-TIN * GESTEROL L.A. * 17-α-HEXANOYLOXYP-REGN-4-ENE-3,20-DIONE * HORMOFORT * HPC * 17-HYDROXYPREGN-4-ENE-3,20-DIONE HEXANOATE * 17-α-HYDROXYPROGESTERONE CAPROATE * 17-α-HYDROXY PROGESTERONE-N-CAPROATE * 17-α-HYDROXYPROGESTERONE HEXANOATE * HYDROXON * HYLUTIN * HYPROVAL-PA * IDROGESTENE * LUETOCRIN DEPOT * LUTATE * LUTEOCRIN * LUTOPRON * NEOLUTIN * NSC-17592 * 17-((1-OXOHEXYL)OXY)PREGN-4-ENE-3,20-DIONE * PRIMOLUT DEPOT * PROGESTERONE CAPROATE * PROGESTERONE RETARD PHARLON * PROLUTON DEPOT * RELUTIN * SQUIBB * SYNGYNON * TERALUTIL

CONSENSUS REPORTS: IARC Cancer Review: Animal Inadequate Evidence IMEMDT 21,399,79.

SAFETY PROFILE: Human reproductive effects by an unknown route: behavioral effects on newborn. Experimental teratogenic and reproductive effects. Questionable carcinogen. A steroid. Used to treat menstrual disorders, threatened abortion, and sterility. When heated to decomposition it emits acrid smoke and fumes.

HNT600   CAS: 999-61-1   **HR: 3**
2-HYDROXYPROPYL ACRYLATE
mf: $C_6H_{10}O_3$ mw: 130.16

SYNS: ACRYLIC ACID-2-HYDROXYPROPYL ESTER * β-HYDROXYPROPYL ACRYLATE * 1,2-PROPANE-DIOL-1-ACRYLATE * 2-PROPENOIC ACID-2-HYDROXY-PROPYL ESTER * PROPYLENE GLYCOL MONOACRY-LATE

OSHA PEL: TWA 0.5 ppm (skin)
ACGIH TLV: TWA 0.5 ppm (skin)

SAFETY PROFILE: Poison by ingestion and subcutaneous routes. When heated to decomposition it emits acrid smoke and fumes.

HNU500   CAS: 94-13-3   **HR: 3**
p-HYDROXYPROPYL BENZOATE
mf: $C_{10}H_{12}O_3$ mw: 180.22

PROP: Colorless crystals or white powder. Sltly sol in water; sol in alc, ether.

SYNS: ASEPTOFORM P * BETACIDE P * BONO-MOLD OP * 4-HYDROXYBENZOIC ACID PROPYL ESTER * p-HYDROXYBENZOIC ACID PROPYL ESTER * NIPASOL * p-OXYBENZOESAUREPROPYLESTER (GERMAN) * PARABEN * PARASEPT * PASEPTOL * PRESERVAL P * PROPYL p-HYDROXYBENZOATE * n-PROPYL p-HYDROXYBENZOATE * PROPYLPARABEN (FCC) * PROPYLPARASEPT * PROTABEN P * TEGOSEPT P

CONSENSUS REPORTS: Reported in EPA TSCA Inventory. EPA Genetic Toxicology Program.

SAFETY PROFILE: Poison by intraperitoneal routes. Moderately toxic by subcutaneous route. Mildly toxic by ingestion. An allergen. When heated to decomposition it emits acrid smoke and fumes.

HNV000   CAS: 9004-64-2   HR: 1
HYDROXYPROPYL CELLULOSE

PROP: White powder. Sol in water and organic solvents.

SYNS: HYDROXYPROPYL ETHER of CELLULOSE * KLUCEL

CONSENSUS REPORTS: Reported in EPA TSCA Inventory.

SAFETY PROFILE: Slightly toxic by ingestion. When heated to decomposition it emits acrid smoke and fumes.

HNX000   CAS: 9004-65-3   HR: 1
HYDROXYPROPYL METHYLCELLULOSE

PROP: White fibrous or granular powder. Sol in water, organic solvents; insol in anhydrous alc, ether, and chloroform.

SYN: METHOCEL HG

CONSENSUS REPORTS: Reported in EPA TSCA Inventory.

SAFETY PROFILE: Mildly toxic by intraperitoneal route. When heated to decomposition it emits acrid smoke and fumes.

HNX500   CAS: 61499-28-3   HR: 3
1-((2-HYDROXYPROPYL)NITROSO) AMINO)ACETONE
mf: $C_6H_{12}N_2O_3$ mw: 160.20

SYNS: HPOP * 1-((2-HYDROXYPROPYL)NITROSOAMINO)-2-PROPANONE * N-NITROSO(2-HYDROXYPROPYL)(2-OXOPROPYL)AMINE

SAFETY PROFILE: Suspected carcinogen with experimental carcinogenic, neoplastigenic, and tumorigenic data. A poison by subcutaneous route. Mutation data reported. When heated to decomposition it emits toxic fumes of NO_x.

HNZ000   CAS: 59413-14-8   HR: 2
1-(3-HYDROXYPROPYL)THEOBROMINE
mf: $C_{10}H_{14}N_4O_3$ mw: 238.28

SYNS: 3,7-DIHYDRO-1-(3-HYDROXYPROPYL)-3,7-DIMETHYL-1H-PURINE-2,6-DIONE * 1-(3-HYDROXYPROPYL)-3,7-DIMETHYLXANTHINE * 1-(3-HYDROXYPROPYL)THEOBROMINE * γ-OXYPROPYLTHEOBROMIN (GERMAN) * γ-(γ-OXYPROPYL)-THEOBROMIN (GERMAN)

SAFETY PROFILE: Moderately toxic by intraperitoneal and subcutaneous routes. When heated to decomposition it emits toxic fumes of NO_x.

HOF000   CAS: 26782-43-4   HR: 3
HYDROXYSENKIRKINE
mf: $C_{19}H_{27}NO_7$ mw: 381.47

PROP: Isolated from the plant *Crotalaria laburnifolia*.

SYN: 8,12,18-TRIHYDROXY-4-METHYL-11,16-DIOXOSENECIONANIUM

CONSENSUS REPORTS: IARC Cancer Review: GROUP 3 IMEMDT 7,56,87; Animal Inadequate Evidence IMEMDT 10,265, 76.

SAFETY PROFILE: Poison by ingestion. Questionable carcinogen with experimental tumorigenic data. When heated to decomposition it emits toxic fumes of NO_x.

HOH500   CAS: 79-57-2   HR: 3
5-HYDROXYTETRACYCLINE
mf: $C_{22}H_{24}N_2O_9$ mw: 460.48

SYNS: BIOSTAT * NCI-C56473 * OTC * OXITETRACYCLIN * OXYMYKOIN * OXYTERRACINE * OXYTETRACYCLINE * OXYTETRACYCLINE AMPHOTERIC * RIOMITSIN * RYOMYCIN * TAOMYCIN * TAOMYXIN * TERRAFUNGINE * TERRAMITSIN * TERRAMYCIN * TETRAN

CONSENSUS REPORTS: Reported in EPA TSCA Inventory.

SAFETY PROFILE: A poison by intravenous route. Moderately toxic by ingestion, subcutaneous, and intraperitoneal routes. Human systemic effects by ingestion: hemorrhage, dermatitis, and unspecified effects on teeth and supporting structures. Human reproductive effects by an unspecified route: abnormal postnatal measures or effects. Experimental teratogenic and reproductive effects. Mutation data reported. When heated to decomposition it emits toxic fumes of NO_x.

HOI000   CAS: 2058-46-0   HR: 3
5-HYDROXYTETRACYCLINE HYDROCHLORIDE
mf: $C_{22}H_{24}N_2O_9 \cdot ClH$ mw: 496.94

SYNS: BISOLVOMYCIN * HYDROCYCLIN * LIQUAMYCIN INJECTABLE * NSC 9169 * OTETRYN * OXLOPAR * OXYJECT 100 * OXYTETRACYCLINE HYDROCHLORIDE * TERA-MYCIN HYDROCHLORIDE * TETRAMINE * TET-RAN HYDROCHLORIDE

CONSENSUS REPORTS: NTP Carcinogenesis Studies (feed); Equivocal Evidence: rat NTPTR* NTP-TR-315,87. NTP Carcinogenesis Studies (feed); No Evidence: mouse NTPTR* NTP-TR-315,87. Reported in EPA TSCA Inventory.

SAFETY PROFILE: Poison by intravenous route. Moderately toxic by subcutaneous route. Mildly toxic by ingestion. Experimental teratogenic and reproductive effects. Questionable carcinogen with experimental tumorigenic data. Mutation data reported. When heated to decomposition it emits very toxic fumes of HCl and NO_x.

HOL000   CAS: 64050-03-9   HR: 3
(3-HYDROXY-p-TOLYL)TRIMETHYL-AMMONIUM CHLORIDE,METHYL-CARBAMATE
mf: $C_{12}H_{19}N_2O_2 \cdot Cl$ mw: 258.78

SYN: METHYLCARBAMIC ACID-5-(TRIMETHYLAMMO-NIO)-o-TOLYL ESTER, CHLORIDE

SAFETY PROFILE: A deadly poison by ingestion, intraperitoneal, subcutaneous, intravenous, and implantation routes. When heated to decomposition it emits very toxic fumes of NO_x and Cl^-.

HON000   CAS: 76-87-9   HR: 3
HYDROXYTRIPHENYLSTANNANE
mf: $C_{18}H_{16}OSn$ mw: 367.03

PROP: Mp: 122°.

SYNS: DOWCO 186 * DU-TER * ENT 28,009 * FENOLOVO * FINTINE HYDROXYDE (FRENCH) * FINTIN HYDROXID (GERMAN) * FENTIN HYDROX-IDE * FINTIN HYDROXYDE (DUTCH) * FINTIN IDROSSIDO (ITALIAN) * HAITIN * HYDROXYDE de TRIPHENYL-ETAIN (FRENCH) * HYDROXYTRIPHE-NYLTIN * IDROSSIDO DI STAGNO TRIFENILE (ITAL-IAN) * NCI-C00260 * SUZU H * TPTH * TRI-FENYL-TINHYDROXYDE (DUTCH) * TRIPHENYLTIN HYDROXIDE (USDA) * TRIPHENYLTIN OXIDE * TRIPHENYL-ZINNHYDROXID (GERMAN) * TUBOTIN * VANCIDE KS

CONSENSUS REPORTS: NCI Carcinogenesis Bioassay (feed); No Evidence: mouse, rat NCITR* NCI-CG-TR-139,78. Reported in EPA TSCA Inventory.

OSHA PEL: TWA 0.1 mg(Sn)/m^3 (skin)
ACGIH TLV: TWA 0.1 mg(Sn)/m^3 (skin) (Proposed: TWA 0.1 mg(Sn)/m^3; STEL 0.2 mg(Sn)/m^3 (skin))
NIOSH REL: (Organotin Compounds) TWA 0.1 mg(Sn)/m^3

SAFETY PROFILE: A poison by ingestion and intraperitoneal routes. Moderately toxic by an unspecified route. A severe eye irritant. Experimental teratogenic and reproductive effects. Mutation data reported. When heated to decomposition it emits acrid smoke and fumes.

HOO500   CAS: 127-07-1   HR: 3
HYDROXYUREA
mf: $CH_4N_2O_2$ mw: 76.07

PROP: Needles from ethanol. Mp: 133-136°, bp: decomp. Very sol in water; sol in hot alc.

SYNS: N-(AMINOCARBONYL)HYDROXYLAMINE * BIOSUPRESSIN * CARBAMOHYDROXAMIC ACID * CARBAMOHYDROXIMIC ACID * CARBAMOHY-DROXYAMIC ACID * CARBAMOYL OXIME * N-CARBAMOYLHYDROXYLAMINE * CARBAMYL HYDROXAMATE * HIDRIX * HYDREA * HYDROXYCARBAMINE * HYDROXYLUREA * N-HYDROXYUREA * HYDURA * LITALER * NCI-C04831 * NSC 32065 * ONCO-CARBIDE * OXYUREA * SK 22591 * SQ 1089

CONSENSUS REPORTS: NCI Carcinogenesis Studies (ipr); No Evidence: mouse CAN-

CAR 40,1935,77. NCI Carcinogenesis Studies (ipr); Equivocal Evidence: rat CANCAR 40,1935,77. EPA Genetic Toxicology Program.

SAFETY PROFILE: Human systemic effects by ingestion, intravenous, and possibly other routes: nausea or vomiting, microcytosis (smaller than normal red blood cells), normocytic anemia (reduced red blood cell count), leukopenia (reduced white blood cell count), thrombocytopenia (decrease in the number of blood platelets), and other blood effects. Experimental teratogenic and reproductive effects. Mildly toxic by several routes. Human mutation data reported. Questionable carcinogen with experimental tumorigenic data. When heated to decomposition it emits toxic fumes of NO_x.

HOP000   CAS: 13479-29-3   HR: 3
3-HYDROXYXANTHINE
mf: $C_5H_4N_4O_3$ mw: 168.13

SYNS: 3,7-DIHYDRO-3-HYDROXY-1H-PURINE-2,6-DIONE * XANTHINE-x-N-OXIDE * XANTHINE-3-N-OXIDE

SAFETY PROFILE: A poison by intraperitoneal route. Moderately toxic by subcutaneous route. Experimental teratogenic effects. Questionable carcinogen with experimental carcinogenic and neoplastigenic data. Mutation data reported. When heated to decomposition it emits toxic fumes of NO_x.

HOP259   CAS: 16870-90-9   HR: 3
7-HYDROXYXANTHINE
mf: $C_5H_4N_4O_3$ mw: 168.13

SYN: XANTHINE-7-N-OXIDE

SAFETY PROFILE: Poison by intraperitoneal route. Questionable carcinogen with experimental carcinogenic and neoplastigenic data. When heated to decomposition it emits toxic fumes of NO_x.

HOT500   CAS: 114-49-8   HR: 3
HYOSCINE HYDROBROMIDE
mf: $C_{17}H_{21}NO_4 \cdot BrH$ mw: 384.31

SYNS: BELDAVRIN * EUSCOPOL * HYDRO-SCINE HYDROBROMIDE * HYOSCINE BROMIDE * HYOSCINE F HYDROBROMIDE * (−)-HYOSCINE HYDROBROMIDE * 1-HYOSCINE HYDROBROMIDE * HYOSCYINE HYDROBROMIDE * HYSCO * ISO-SCOPIL * KWELLS * SCOPAMIN * SCOPOL-AMINE BROMIDE * (−)-SCOPOLAMINE BROMIDE

* SCOPOLAMINE HYDROBROMIDE * (−)-SCOPOL-AMINE HYDROBROMIDE * SCOPOLAMINIUM BROMIDE * SCOPOLAMMONIUM BROMIDE * SCOPOS * SEREEN * TRIPTONE

CONSENSUS REPORTS: Reported in EPA TSCA Inventory. EPA Genetic Toxicology Program.

SAFETY PROFILE: A poison by intravenous and intramuscular routes. Moderately toxic by ingestion, subcutaneous, intraduodenal, and intraperitoneal routes. Experimental teratogenic and reproductive effects. Human mutation data reported. When heated to decomposition it emits very toxic fumes of NO_x and HBr.

HOU000   CAS: 101-31-5   HR: 3
(−)-HYOSCYAMINE
mf: $C_{17}H_{23}NO_3$ mw: 289.41

PROP: Mp: 106–108°. Very sol in alc, dil acids. White, crystalline alkaloid.

SYNS: (−)-ATROPINE * DATURINE * HYOSCY-AMINE * 1-HYOSCYAMINE * (−)-TROPIC ACID ES-TER with TROPINE

CONSENSUS REPORTS: Reported in EPA TSCA Inventory.

SAFETY PROFILE: A deadly human poison by an unspecified route. An experimental poison by intravenous route. This is one of the atropine alkaloids and is very toxic, acting very much like atropine. It has the same effect on the central nervous system but twice the effect on the peripheral nerves. The symptoms of poisoning are dryness of the throat and mouth, marked difficulty in swallowing, and a sensation of burning and thirst. The vision becomes impaired through dilation and loss of accommodation, and the eyes present a rather prominent, brilliant, staring appearance. The voice is husky and the tongue is red. When heated to decomposition it emits highly toxic fumes of NO_x.

HOU500   HR: 3
HYPOCHLORITES
PROP: Salts of hypochlorous acid.

SAFETY PROFILE: Toxic by ingestion and inhalation. Powerful irritants to the skin, eyes, and mucous membranes. Flammable by chemical reaction with reducing agents. These are powerful oxidizers particularly at higher temperatures, when chlorine and then oxygen are evolved, or in the presence of moisture or carbon dioxide. With urea, it forms the highly explosive NCl_3. Dangerous; when heated or on contact with acid or acid fumes, they emit highly toxic fumes of Cl^-. React with water or steam to produce toxic and corrosive fumes of Cl^- and HCl.

HOV500   CAS: 7778-54-3   HR: 3
HYPOCHLOROUS ACID, CALCIUM SALT

DOT: UN 1471/ UN 1748/UN 1791/UN 2208/ UN 2880
mf: $Cl_2O_2 \cdot Ca$ mw: 142.98

PROP: White powder. Compound contains 39% or less available chlorine.

SYNS: B-K POWDER   *   BLEACHING POWDER   *   BLEACHING POWDER, CONTAINING 39% OR LESS CHLORINE (DOT)   *   CALCIUM CHLOROHYDROCHLORITE   *   CALCIUM HYPOCHLORIDE   *   CALCIUM HYPOCHLORITE   *   CALCIUM OXYCHLORIDE   *   CAPORIT   *   CCH   *   CHLORIDE of LIME (DOT)   *   CHLORINATED LIME (DOT)   *   HTH   *   HYCHLOR   *   LIME CHLORIDE   *   LO-BAX   *   LOSANTIN   *   PERCHLORON   *   PITTCHLOR   *   PITTCIDE   *   PITTCLOR   *   SENTRY

CONSENSUS REPORTS: Reported in EPA TSCA Inventory.

DOT Classification: ORM-C; Label: None.

SAFETY PROFILE: Moderately toxic by ingestion. Can cause severe irritation of skin and mucous membranes and emit fumes capable of causing pulmonary edema. Mutation data reported. A powerful oxidizer.

The bulk material may ignite or explode in storage. Traces of water may initiate the reaction. A rapid exothermic decomposition above 175°C releases oxygen and chlorine. Moderately explosive in its solid form when heated. Explosive reaction with acetic acid + potassium cyanide, amines, ammonium chloride, carbon or charcoal + heat, carbon tetrachloride + heat, N,N-dichloromethylamine + heat, ethanol, methanol, iron oxide, rust, 1-propanethiol, isobutanethiol, turpentine. Potentially explosive reaction with sodium hydrogen sulfate + starch + sodium carbonate. Reaction with acetylene or nitrogenous bases forms explosive products.

Ignites on contact with algacide, hydroxy compounds (e.g., glycerol, diethylene glycol monomethyl ether, phenol), organic sulfur compounds. Violent reaction with organic matter (above 100°C), sulfur. Vigorous reaction with nitromethane, reducing materials. Flammable by chemical reaction with combustible materials, i.e., anthracene, grease, oil, mercaptans, methyl carbitol, nitromethane, organic matter, propylmercaptan.

Deflagration occurs in contact with combustible substances. Dangerous; when heated to decomposition or on contact with acid or acid fumes, it emits highly toxic fumes of HCl and explodes. Reacts with water or steam to produce toxic and corrosive fumes or Cl^- and HCl.

HOW000   HR: 3
HYPOCHLOROUS ACID, CALCIUM SALT (DRY MIXTURE)
mf: $Cl_2H_2O_2 \cdot Ca$ mw: 145.00

PROP: Compound contains more than 39% available chlorine.

SYN: CALCIUM HYPOCHLORITE MIXTURE, DRY (DOT)

CONSENSUS REPORTS: Reported in EPA TSCA Inventory.

DOT Classification: Label: Oxidizer.

SAFETY PROFILE: A powerful irritant and oxidizer. Potentially explosive. When heated to decomposition or in reaction with water or acids it emits toxic fumes of Cl^-.

HOW500   CAS: 14448-38-5   HR: 3
HYPONITROUS ACID

SYN: N-NITROSOHYDROXYLAMINE
mf: $H_2N_2O_2$ mw: 62.03

PROP: Explosive solid.

DOT Classification: Forbidden.

SAFETY PROFILE: Many N-nitroso compounds are carcinogens. Incompatible with potassium hydroxide. When heated to decomposition it emits toxic fumes of NO_x.

I

IAC000   CAS: 553-68-4   **HR: 3**
IBYLCAINE HYDROCHLORIDE
mf: $C_{13}H_{20}N_2O_2 \cdot ClH$ mw: 272.81

SYNS: BUTETHAMINE HYDROCHLORIDE * 2-ISO-
BUTYLAMINOETHANOL HYDROCHLORIDE ACID SALT,
p-AMINOBENZOIC ACID ESTER * 2-(ISOBUTYLAMINO)
ETHYL-p-AMINOBENZOATE HYDROCHLORIDE
* MONOCAINE HYDROCHLORIDE

SAFETY PROFILE: A poison by intraperitoneal, intravenous, and subcutaneous routes.
Used in veterinary medicine as local anesthetic.
When heated to decomposition it emits very
toxic fumes of Cl^- and NO_x.

IAL000   CAS: 288-32-4   **HR: 3**
IMIDAZOLE
mf: $C_3H_4N_2$ mw: 68.09

PROP: Prisms. Mp: 90-91°, bp: 257°. Sol in
water, ether, chloroform; very sol in alc, pyridine; sltly sol in benzene.

SYNS: 1,3-DIAZA-2,4-CYCLOPENTADIENE * 1,3-
DIAZOLE * GLYOXALIN * GLYOXALINE
* IMIDAZOL * IMINAZOLE * IMUTEX
* MIAZOLE * PYRRO(b)MONAZOLE * USAF EK-
4733 * N,N'-VINYLENEFORMAMIDINE

CONSENSUS REPORTS: Reported in EPA
TSCA Inventory.

SAFETY PROFILE: A poison by subcutaneous
and intraperitoneal routes. Moderately toxic by
ingestion and intravenous routes. Human mutation data reported. When heated to decomposition it emits toxic fumes of NO_x.

IAN000   CAS: 5034-77-5   **HR: 3**
IMIDAZOLE MUSTARD
mf: $C_8H_{12}Cl_2N_6O$ mw: 279.16

SYNS: BIC * 5-(3,3-BIS(2-CHLOROETHYL)-1-TRI-
AZENO)IMIDAZOLE-4-CARBOXAMIDE * NCI-C01616
* NSC-82196 * SRI 2489 * TIC MUSTARD

CONSENSUS REPORTS: NCI Carcinogenesis Studies (ipr); Clear Evidence: mouse, rat
CANCAR 40,1935,77.

SAFETY PROFILE: Suspected carcinogen with
experimental neoplastigenic and tumorigenic
data. Poison by ingestion and intraperitoneal
routes. Experimental teratogenic and reproduc-

tive effects. Human systemic effects by intravenous route: nausea. When heated to decomposition it emits very toxic fumes of Cl^- and NO_x.

IAQ000   CAS: 96-45-7   **HR: 3**
2-IMIDAZOLIDINETHIONE
mf: $C_3H_6N_2S$ mw: 102.17

PROP: White crystals. Water solubility: 9 g/
100 mL @ 30°. Often occurs as a main degradation product of the metal salts of ethylene bis-
dithiocarbamic acid.

SYNS: 4,5-DIHYDROIMIDAZOLE-2(3H)-THIONE
* ETHYLENE THIOUREA * N,N'-ETHYLENETH-
IOUREA * 1,3-ETHYLENE-2-THIOUREA * l'ETHYL-
ENE THIOUREE (FRENCH) * ETU * 2-MERCAPTO-
IMIDAZOLINE * 2-MERKAPTOIMIDAZOLIN (CZECH)
* NA-22 * NCI-C03372 * PENNAC CRA
* RCRA WASTE NUMBER U116 * RODANIN S-62
(CZECH) * SODIUM-22 NEOPRENE ACCELERATOR
* 2-THIOL-DIHYDROGLYOXALINE * USAF EL-62
* VULKACIT NPV/C2 * WARECURE C

CONSENSUS REPORTS: IARC Cancer Re-
view: GROUP 2B IMEMDT 7,207,87; Animal
Sufficient Evidence IMEMDT 7,45,74. NTP
Fourth Annual Report On Carcinogens, 1984.
Community Right-To-Know List. EPA Genetic
Toxicology Program. Reported in EPA TSCA
Inventory.

NIOSH REL: (ETU) Use encapsulated form;
minimize exposure.

SAFETY PROFILE: Confirmed carcinogen with
experimental carcinogenic data. Poison by ingestion and intraperitoneal routes. Experimental
teratogenic and reproductive effects. Mutation
data reported. An eye irritant. When heated to
decomposition it emits very toxic fumes of NO_x
and SO_x.

IAR000   **HR: 3**
**2-IMIDAZOLIDINETHIONE mixed with
SODIUM NITRITE**

SYNS: ETHYLENETHIOUREA mixed with SODIUM NI-
TRITE * SODIUM NITRITE mixed with ETHYLENE-
THIOUREA

SAFETY PROFILE: Suspected carcinogen. 2-
Imidazolidinethione and sodium nitrite are experimental carcinogens. Experimental terato-

genic and reproductive data. Sodium nitrite is a poison. Mutation data reported. When heated to decomposition it emits very toxic fumes of SO_x, Na_2O, and NO_x.

IBA000   CAS: 2465-27-2   **HR: 3**
4,4'-(IMIDOCARBONYL)BIS(N,N-DIMETHYLAMINE) MONOHYDRO CHLORIDE
mf: $C_{17}H_{21}N_3 \cdot ClH \cdot H_2O$ mw: 321.89

SYNS: ADC AURAMINE O * AIZEN AURAMINE * AURAMINE (MAK) * AURAMINE HYDROCHLORIDE * AURAMINE O (BIOLOGICAL STAIN) * AURAMINE YELLOW * 4,4'-BIS(DIMETHYLAMINO) BENZHYDRYLIDENIMINE HYDROCHLORIDE * 4,4'-BIS(DIMETHYLAMINO)BENZOPHENONE-IMINE HYDROCHLORIDE * 1,1-BIS(p-DIMETHYLAMINOPHENYL) METHYLENIMINEHYDROCHLORIDE * CALCOZINE YELLOW OX * 4,4'-CARBONIMIDOYLBIS(N,N-DIMETHYLBENZENAMINE)MONOHYDROCHLORIDE * C.I. 41000 * C.I. BASIC YELLOW 2 * C.I. BASIC YELLOW 2, MONOHYDROCHLORIDE * MITSUI AURAMINE O

CONSENSUS REPORTS: Reported in EPA TSCA Inventory. EPA Genetic Toxicology Program.

DFG MAK: Animal Carcinogen, Suspected Human Carcinogen.

SAFETY PROFILE: Confirmed carcinogen with experimental neoplastigenic and tumorigenic data. Poison by skin contact, ingestion, and intraperitoneal routes. Human mutation data reported. A chelating agent which might disturb trace element metabolism if taken into the body. Used as a biological stain. When heated to decomposition it emits very toxic fumes of NO_x and HCl.

IBB000   CAS: 492-80-8   **HR: 3**
4,4'-(IMIDOCARBONYL)BIS(N,N-DIMETHYLANILINE)
mf: $C_{17}H_{21}N_3$ mw: 267.41

PROP: Yellow needles. Mp: 136°. Insol in water.

SYNS: APYONINE AURAMINE BASE * AURAMINE (MAK) * AURAMINE BASE * BIS(p-DIMETHYLAMINOPHENYL)METHYLENEIMINE * BRILLIANT OIL YELLOW * 4,4'-CARBONIMIDOYLBIS(N,N-DIMETHYLBENZENAMINE) * C.I. 41000B * C.I. BASIC YELLOW 2, FREE BASE * C.I. SOLVENT YELLOW 34 * 4,4'-DIMETHYLAMINOBENZOPHENONIMIDE * GLAURAMINE * RCRA WASTE

NUMBER U014 * TETRAMETHYLDIAMINODIPHENYLACETIMINE * WAXOLINE YELLOW O * YELLOW PYOCTANINE

CONSENSUS REPORTS: IARC Cancer Review: GROUP 2B IMEMDT 7,118,87; Human Sufficient Evidence IMEMDT 1,69,72; Animal Sufficient Evidence IMEMDT 1,69,72. Community Right-To-Know List. Reported in EPA TSCA Inventory.

DFG MAK: Animal Carcinogen, Suspected Human Carcinogen.

SAFETY PROFILE: Confirmed human carcinogen with experimental carcinogenic, neoplastigenic, and tumorigenic data. Poison by intraperitoneal route. Human mutation data reported. Used as an antiseptic. When heated to decomposition it emits toxic fumes of NO_x.

IBM000   CAS: 2152-34-3   **HR: 3**
2-IMINO-5-PHENYL-4-OXAZOLIDINONE
mf: $C_9H_8N_2O_2$ mw: 176.19

SYNS: PHENOXAZOLE * 5-PHENYL-2-IMINO-4-OXAZOLIDINONE * 5-PHENYL-2-IMINO-4-OXOOXAZOLIDINE * PHENYL ISOHYDANTOIN * PHENYLPSEUDOHYDANTOIN

SAFETY PROFILE: A poison by ingestion. Moderately toxic by intraperitoneal route. An experimental teratogen. A central nervous system stimulant. When heated to decomposition it emits toxic fumes of NO_x.

IBP309   CAS: 2207-85-4   **HR: 3**
IMPIRAMINE-N-OXIDE
mf: $C_{19}H_{24}N_2O$ mw: 296.45

PROP: White needle-like crystals. Mp: 120-123° (decomp); sol in methanol, ether, acetone, and benzene. Hygroscopic.

SYNS: 5-(3-(DIMETHYLAMINO)PROPYL)-10,11-DIHYDRO-5H-DIBENZ(b,f)AZEPINE-5-OXIDE * GP 38383 * IPNO

SAFETY PROFILE: A poison by ingestion and intraperitoneal routes. When heated to decomposition it emits toxic fumes of NO_x.

IBS000   CAS: 606-23-5   **HR: 3**
1,3-INDANDIONE
mf: $C_9H_6O_2$ mw: 146.15

PROP: Crystal or liquid. Mp: 129-131° decomp. Very sltly sol in cold water; sol in hot alc, benzene.

SYNS: 1,3-DIKETOHYDRINDENE * 1H-INDENE-1, 3(2H)-DIONE

CONSENSUS REPORTS: Reported in EPA TSCA Inventory.

SAFETY PROFILE: A poison by intraperitoneal route. Experimental teratogenic and reproductive effects. When heated to decomposition it emits acrid smoke and fumes.

IBX000   CAS: 95-13-6   **HR: 3**
INDENE
mf: C_9H_8 mw: 116.17

PROP: Liquid from coal tars. Water-insol, but miscible in organic solvents. D: 0.9968 @ 20°/4°, mp: $-1.8°$, bp: 181.6°.

SYN: INDONAPHTHENE

CONSENSUS REPORTS: Reported in EPA TSCA Inventory.

OSHA PEL: TWA 10 ppm
ACGIH TLV: TWA 10 ppm

SAFETY PROFILE: Moderately toxic by ingestion, inhalation, and subcutaneous routes. Irritating to skin, eyes, and mucous membranes. It has exploded during nitration with (H_2SO_4 + HNO_3). When heated to decomposition it emits acrid smoke and fumes.

IBZ000   CAS: 193-39-5   **HR: 3**
INDENO(1,2,3-cd)PYRENE
mf: $C_{22}H_{12}$ mw: 276.34

SYNS: 2,3-PHENYLENEPYRENE * 2,3-o-PHENYLE-NEPYRENE * 1,10-(o-PHENYLENE)PYRENE * 1,10-(1,2-PHENYLENE)PYRENE * RCRA WASTE NUMBER U137

CONSENSUS REPORTS: IARC Cancer Review: GROUP 2B IMEMDT 7,56,87; Animal Sufficient Evidence IMEMDT 32,373,83; IMEMDT 3,229,73. NTP Fourth Annual Report On Carcinogens, 1984. Reported in EPA TSCA Inventory.

SAFETY PROFILE: Confirmed carcinogen with experimental carcinogenic and tumorigenic data. Mutation data reported. When heated to decomposition it emits acrid smoke and fumes.

ICB000   CAS: 525-66-6   **HR: 3**
INDERAL
mf: $C_{16}H_{21}NO_2$ mw: 259.38

SYNS: AY 64043 * DOCITON * ICI 45520 * 1-ISOPROPYLAMINE-3-(1-NAPHTHYLOXY)-2-PROPA-NOL * 1-ISOPROPYLAMINO-3-(1-NAPHTHYLOXY)-2-PROPANOL * NSC-91523 * PROPANALOL * PROPRANOLOL

SAFETY PROFILE: A deadly human poison by ingestion and intravenous routes. Poison experimentally by ingestion, intraperitoneal, intravenous, and subcutaneous routes. Human systemic effects by ingestion: cardiac arrythmias, hallucinations, hypoglycemia, convulsions; thyroid malfunction. Human reproductive and teratogenic effects by ingestion: extra embryonic structures, abnormal apgar score in newborn, and abnormal growth statistics. Mutation data reported. When heated to decomposition it emits toxic fumes of NO_x.

ICF000   CAS: 7440-74-6   **HR: 3**
INDIUM
af: In   aw: 114.82

PROP: Soft, silvery-white metal. Mp: 156.61°, bp: 2080°, d: 7.31 @ 20°.

CONSENSUS REPORTS: Reported in EPA TSCA Inventory.

OSHA PEL: TWA 0.1 mg(In)/m³
ACGIH TLV: TWA 0.1 mg(In)/m³

SAFETY PROFILE: A poison by subcutaneous route. It affects the liver, heart, kidneys, and the blood. Experimental teratogenic effects. Inhalation of indium compounds may cause damage to the respiratory system. Hydrated indium oxide is a poison by intravenous route. Flammable in the form of dust when exposed to heat or flame. Incandesces. Explosive reaction with dinitrogen tetraoxide + acetonitrile. Violent reaction with mercury(II) bromide at 350°C. Mixtures with sulfur ignite when heated.

ICH000   CAS: 4194-69-8   **HR: 2**
INDIUM CITRATE
mf: $C_{18}H_{15}O_{21} \cdot$ In   mw: 682.15

ACGIH TLV: TWA 0.1 mg/m³

SAFETY PROFILE: Moderately toxic by subcutaneous route. When heated to decomposition it emits acrid smoke and fumes.

ICI000   CAS: 13770-61-1   **HR: 2**
INDIUM NITRATE
mf: InN_3O_9 mw: 300.85

CONSENSUS REPORTS: Reported in EPA TSCA Inventory.

ACGIH TLV: TWA 0.1 mg/m^3

SAFETY PROFILE: Experimental teratogenic and reproductive effects. A severe skin irritant. When heated to decomposition it emits toxic fumes of NO$_x$.

ICJ000   CAS: 13464-82-9   **HR: 3**
INDIUM SULFATE
mf: O$_{12}$S$_3$•In$_2$ mw: 517.82

PROP: Grayish-white, hygroscopic powder. D: 3.44. Sol in water. Keep well-closed.

SYNS: INDISULFAT (GERMAN) * SULFURIC ACID, INDIUM SALT

CONSENSUS REPORTS: Reported in EPA TSCA Inventory.

ACGIH TLV: TWA 0.1 mg/m^3

SAFETY PROFILE: A poison by intravenous and subcutaneous routes. Moderately toxic by ingestion. When heated to decomposition it emits toxic fumes of SO$_x$.

ICK000   CAS: 10025-82-8   **HR: 3**
INDIUM TRICHLORIDE
mf: Cl$_3$In   mw: 221.17

PROP: Yellowish, deliquescent crystals. D: 4.0, mp: 586°, sublimes @ 500°, bp: volatile @ 600°. Very sol in water. Keep tightly closed.

SYN: INDIUM CHLORIDE

CONSENSUS REPORTS: Reported in EPA TSCA Inventory.

ACGIH TLV: TWA 0.1 mg/m^3

SAFETY PROFILE: A poison by subcutaneous, intraperitoneal, and intravenous routes. Mutation data reported. When heated to decomposition it emits toxic fumes of Cl$^-$.

ICM000   CAS: 120-72-9   **HR: 3**
INDOLE
mf: C$_8$H$_7$N   mw: 117.16

PROP: Colorless to yellowish scales; intense fecal odor. Mp: 52°, bp: 253°; volatile with steam. Sol in hot water, alc, ether, petroleum ether; insol in mineral oil, glycerin.

SYNS: 1-AZAINDENE * 1-BENZAZOLE * BENZO-PYRROLE * 2,3-BENZOPYRROLE * FEMA No. 2593 * INDOL (GERMAN) * KETOLE

CONSENSUS REPORTS: Reported in EPA TSCA Inventory.

SAFETY PROFILE: A poison by intraperitoneal and subcutaneous routes. Moderately toxic by ingestion and skin contact. Questionable carcinogen with experimental carcinogenic and tumorigenic data. When heated to decomposition it emits toxic fumes of NO$_x$.

ICN000   CAS: 87-51-4   **HR: 3**
1H-INDOLE-3-ACETIC ACID
mf: C$_{10}$H$_9$NO$_2$ mw: 175.20

PROP: Colorless leaves from benzene. Mp: 165-168°. Very sltly sol in cold water; sol in alc, ether, and acetic acid; insol in chloroform.

SYNS: HETEROAUXIN * IAA * β-INDOLEACETIC ACID * β-INDOLE-3-ACETIC ACID * 3-INDOLE-ACETIC ACID * INDOLYACETIC ACID * α-INDOL-3-YL-ACETIC ACID * β-INDOLYLACETIC ACID * INDOLYL-3-ACETIC ACID * 3-INDOLYLACETIC ACID * RHIZOPIN * φ-SKATOLE CARBOXYLIC ACID

CONSENSUS REPORTS: Reported in EPA TSCA Inventory.

SAFETY PROFILE: A poison by intraperitoneal route. Experimental teratogenic effects. Questionable carcinogen with experimental tumorigenic data. Mutation data reported. When heated to decomposition it emits toxic fumes of NO$_x$.

ICO000   CAS: 1204-06-4   **HR: 3**
INDOLE-3-ACRYLIC ACID
mf: C$_{11}$H$_9$NO$_2$ mw: 187.21

SYNS: 3-INDOLYLACRYLIC ACID * 3-(1-H-INDOL-3-YL)-2-PROPENOIC ACID

SAFETY PROFILE: Questionable carcinogen with experimental carcinogenic, neoplastigenic, and tumorigenic data. Human mutation data reported. When heated to decomposition it emits toxic fumes of NO$_x$.

ICP000   CAS: 133-32-4   **HR: 3**
1H-INDOLE-3-BUTANOIC ACID
mf: C$_{12}$H$_{13}$NO$_2$ mw: 203.26

PROP: White crystals or powder. Mp: 124°. Sol in acetone and ether; insol in water and chloroform.

SYNS: HORMEX ROOTING POWDER * HORMODIN * IBA * INDOLE BUTYRIC * INDOLE BUTYRIC ACID * β-INDOLEBUTYRIC ACID * γ-(INDOLE-3)-

BUTYRIC ACID * 3-INDOLEBUTYRIC ACID
* 3-INDOLYL-γ-BUTYRIC ACID * γ-(3-INDOLYL)BU-
TYRIC ACID * γ-(INDOL-3-YL)BUTYRIC ACID
* INDOLYL-3-BUTYRIC ACID * 4-(INDOL-3-YL)BU-
TYRIC ACID * 4-(INDOLYL)BUTYRIC ACID
* 4-(3-INDOLYL)BUTYRIC ACID * JIFFY GROW
* ROOTONE

CONSENSUS REPORTS: Reported in EPA
TSCA Inventory.

SAFETY PROFILE: A poison by ingestion and
intraperitoneal routes. Mutation data reported.
Used for promoting and accelerating root formation of plant clippings. When heated to decomposition it emits toxic fumes of NO_x.

ICW000 CAS: 771-51-7 HR: 3
3-INDOLYLACETONITRILE
mf: $C_{10}H_8N_2$ mw: 156.20

SYNS: 3-(CYANOMETHYL)INDOLE * 3-INDOLACE-
TONITRILE * INDOLEACETONITRILE * INDOLE-3-
ACETONITRILE * 1H-INDOLE-3-ACETONITRILE
* INDOLYLACETONITRILE * USAF CB-29

CONSENSUS REPORTS: Reported in EPA
TSCA Inventory. Cyanide and its compounds
are on the Community Right-To-Know List.

SAFETY PROFILE: A poison by intraperitoneal
and subcutaneous routes. When heated to decomposition it emits toxic NO_x and CN^-.

IDA000 CAS: 53-86-1 HR: 3
INDOMETHACIN
mf: $C_{19}H_{16}ClNO_4$ mw: 357.81

PROP: Crystals. One form: mp: 155°; another
form: mp: 162°. Sol in ethanol, ether, acetone,
and castor oil; insol in water.

SYNS: AMUNO * ARTRACIN * ARTRINOVO
* ARTRIVIA * N-p-CHLORBENZOYL-5-METHOXY-2-
METHYLINDOLE-3-ACETIC ACID * 1-(p-CHLOROBEN-
ZOYL)-5-METHOXY-2-METHYLINDOLE-3-ACETIC ACID
* 1-(p-CHLOROBENZOYL)-2-METHYL-5-METHOXYIN-
DOLE-3-ACETIC ACID * 1-(p-CHLOROBENZOYL)-2-
METHYL-5-METHOXY-3-INDOLE-ACETIC ACID * α-(1-
(p-CHLOROBENZOYL)-2-METHYL-5-METHOXY-3-INDO-
LYL)ACETIC ACID * 1-p-CLORO-BENZOIL-5-METOXI-
2-METILINDOL-3-ACIDO ACETICO (SPANISH) * CON-
FORTID * DOLOVIN * IDOMETHINE
* IMBRILON * INACID * INDOCID * INDO-
MECOL * INDOMED * INDOMETHAZINE
* INDOMETICINA (SPANISH) * INDOPTIC
* INDO-RECTOLMIN * INDO-TABLINEN * IN-
FLAZON * INTEBAN SP * LAUSIT * METACEN

* METARTRIL * METHAZINE * METINDOL
* MEZOLIN * MIKAMETAN * MOBILAN
* NCI-C56144 * REUMACIDE * SADOREUM
* TANNEX

CONSENSUS REPORTS: Reported in EPA
TSCA Inventory.

SAFETY PROFILE: A poison by ingestion, intravenous, intraperitoneal, and subcutaneous
routes. Human systemic effects by ingestion:
weight loss, diarrhea, necrotic changes to intestines, retinal changes, hemorrhage, hallucinations and distorted perceptions, toxic psychosis,
hepatitis, change in kidney tubules, pulse rate
decrease with fall in blood pressure, urine volume increased or anuria, hematuria (blood in
the urine). Human teratogenic effects by ingestion and intravenous routes: developmental abnormalities of the respiratory system and urogenital system; homeostasis, other neonatal
effects. Experimental teratogenic and reproductive effects. Mutation data reported.
When heated to decomposition it emits very
toxic Cl^- and NO_x.

IDF000 CAS: 909-39-7 HR: 3
INSIDON DIHYDROCHLORIDE
mf: $C_{23}H_{29}N_3O \cdot 2ClH$ mw: 436.47

PROP: Long, rectangular plates from water.
Mp: 90°.

SYNS: 4-(3-(5H-DIBENZ(b,f)AZEPIN-5-YL)PROPYL)-1-PI-
PERAZINEETHANOL DIHYDROCHLORIDE * 5-(γ-(β-
HYDROXYETHYLPIPERAZINO)PROPYL)-5H-DIBEN-
ZO(b,f)AZEPINE DIHYDROCHLORIDE * OPIPRAMOL
DIHYDROCHLORIDE

SAFETY PROFILE: A poison by intraperitoneal, intravenous, and subcutaneous routes.
Moderately toxic by ingestion. Human systemic
effects by ingestion: somnolence. When heated
to decomposition it emits very toxic NO_x and
HCl.

IDJ700 HR: 1
IODATES

SAFETY PROFILE: Salts of iodic acid. Variable
toxicity. Generally eye, skin, and mucous membrane irritants. Powerful oxidizers. Similar to
bromates and chlorates. Contamination of iodates with organic matter may produce explosive
mixtures. Iodate is used in bread as an improving
agent for the dough. When heated to decomposition they emit toxic fumes of I^-.

IDL000
IODIDES
HR: 2

SAFETY PROFILE: Similar in toxicity to bromides. Prolonged absorption of iodides may produce "iodism" which is manifested by skin rash, running nose, headache and irritation of mucous membranes. In severe cases, the skin may show pimples, boils, redness, black and blue spots, hives, and blisters. Weakness, anemia, loss of weight and general depression may occur. Generally very soluble in water and easily absorbed into the body. The iodides of copper(I); lead(II), silver(I) and mercury(II) are poorly soluble in water. When heated to decomposition they can emit highly toxic fumes of I^- and iodine compounds.

IDM000　　CAS: 7553-56-2　　**HR: 3**
IODINE
mf: I_2　　mw: 253.80

PROP: Rhombic, violet-black crystals, metallic luster. Mp: 113.5°, bp: 185.24°, d: 4.93 (Solid @ 25°), vap press: 1 mm @ 38.7°. Characteristic odor, sharp acrid taste, vap press (solid): 0.030 mm @ 0°, very sol in aq solns of HI and iodides.

SYNS: IODE (FRENCH) * IODINE CRYSTALS * IODINE SUBLIMED * IODIO (ITALIAN) * JOD (GERMAN, POLISH) * JOOD (DUTCH)

CONSENSUS REPORTS: Reported in EPA TSCA Inventory.

OSHA PEL: CL 0.1 ppm
ACGIH TLV: CL 0.1 ppm
DFG MAK: 0.1 ppm (1 mg/m^3)

SAFETY PROFILE: A human poison by ingestion and possibly other routes. An experimental poison by intravenous and subcutaneous routes. Moderately toxic by inhalation. Human systemic effects by ingestion: diarrhea. Experimental reproductive effects. Mutation data reported. The effect of iodine vapor upon the body is similar to that of chlorine and bromine, but it is more irritating to the lungs. Serious exposures are seldom encountered in industry due to the low volatility of the solid at ordinary room temperatures. Signs and symptoms are irritation and burning of the eyes, lachrimation, coughing and irritation of the nose and throat. Ingestion of large quantities causes abdominal pain, nausea, vomiting, diarrhea. In severe cases, purging, excessive thirst, and circulatory failure may de-velop. Doses of 2-3 grams have been fatal. Chronic ingestion of large amounts (200 mg/day) results in thyroid disease.

Explosive reaction with acetylene, antimony powder, hafnium powder + heat, tetraamine copper(II) sulfate + ethanol, trioxygen difluoride (possibly ignition), polyacetylene (at 113°C). Forms sensitive, explosive mixtures with potassium (impact- and heat-sensitive), sodium (shock-sensitive), oxygen difluoride (heat-sensitive). Reacts to form explosive products with ammonia, ammonia + lithium 1-heptynide, ammonia + potassium, butadiene + ethanol + mercuric oxide, silver azide.

Ignition on contact with bromine pentafluoride (or violent reaction), chlorine trifluoride, fluorine, metals (powdered) + water, aluminum-titanium alloys + heat, metal acetylides (e.g., cesium acetylide, copper(I) acetylide, lithium acetylide, rubidium acetylide), non-metals (e.g., boron ignites at 700°C), phosphorus, sodium phosphinate. Violent reaction with acetaldehyde, aluminum + diethyl ether, dipropylmercury, titanium (above 113°C). Incandescent reaction with cesium oxide (above 150°C), bromine trifluoride, metal acetylides or carbides [e.g., barium acetylide (above 122°C), calcium acetylide (above 305°C), strontium acetylide (above 182°C), zirconium acetylide (above 400°C)].

Incompatible with ethanol, ethanol + butadiene, ethanol + phosphorus, ethanol + methanol + HgO, formamide + pyridine + sulfur trioxide, formamide, halogens or interhalogens (e.g., chlorine), mercuric oxide, metals (e.g., aluminum, lithium, magnesium), metal carbides (e.g., lithium carbide, zirconium carbide), oxygen, pyridine, sodium hydride, sulfides.

When heated to decomposition it emits toxic fumes of I^- and various iodine compounds. Reacts vigorously with reducing materials.

IDN000　　CAS: 14696-82-3　　**HR: 3**
IODINE AZIDE
mf: IN_3　　mw: 168.93

SYNS: IODINE(I) AZIDE * IODOAZIDE * NITROGEN IODIDE

DOT Classification: Forbidden.

SAFETY PROFILE: A very shock- and friction-sensitive explosive. Incompatible with sulfur-containing alkenes. When heated to decomposition it emits very toxic fumes of I^- and NO_x.

IDS000   CAS: 7790-99-0   **HR: 3**
IODINE MONOCHLORIDE

DOT: UN 1792
mf: ClI   mw: 162.38

PROP: Black crystals or reddish-brown liquid. Exists in α, β forms; crystals α form (stable) black needles; sol in water, alc, ether, CS_2, acetic acid. Red-brown crystals or oily liquid. Mp: (α): 27°, (β): 14°, bp: 97.4 decomp @ 100°; d (α): 3.1822 @ 0°, (β): 3.24 @ 34°.

SYNS: IODINE CHLORIDE * PROTOCHLORURE D'IODE (FRENCH) * WIJS' CHLORIDE

CONSENSUS REPORTS: Reported in EPA TSCA Inventory.

DOT Classification: Corrosive Material; Label: Corrosive.

SAFETY PROFILE: A poison by ingestion. Moderately toxic by skin contact. A corrosive irritant to skin, eyes, and mucous membranes. Moderately explosive when exposed to heat. Reacts with water or steam to produce toxic and corrosive fumes. Dangerous reactions with metals e.g., sodium (mixture explodes on impact), potassium (explodes on contact), aluminum (ignition after a delay period). Reacts violently with Al foil, CdS, PbS, organic matter, P, PCl_3, rubber, Ag_2S, ZnS. When heated to decomposition it emits highly toxic fumes of Cl^- and I^- and may explode.

IDT000   CAS: 7783-66-6   **HR: 3**
IODINE PENTAFLUORIDE

DOT: UN 2495
mf: F_5I mw: 221.90

PROP: Liquid. Mp: 9.43°, bp: 100.5°, d: 3.19 @ 25°. Fumes in air attacks glass, especially when hot.

SYN: PENTAFLUOROIODINE

CONSENSUS REPORTS: Reported in EPA TSCA Inventory.

OSHA PEL: TWA 2.5 mg(F)/m^3
ACGIH TLV: TWA 2.5 mg(F)/m^3
DOT Classification: Oxidizer; Label: Oxidizer and Poison.

SAFETY PROFILE: A poison. Probably an irritant to the eyes, skin and mucous membranes. A powerful oxidizer. Explosive reaction with benzene (above 50°C), diethylaminotrimethyl silane, dimethyl sulfoxide, limonene + tetrafluoroethylene (polymerization), potassium, molten sodium, tetraiodoethylene. Reaction with organic compounds results in charring and then ignition. Violent reaction with water, potassium hydroxide. Incandescent reaction with calcium carbide, potassium hydride, metals and non-metals (e.g., boron, silicon, red phosphorus, sulfur, arsenic, antimony, bismuth, molybdenum, tungsten). When heated to decomposition it emits very toxic fumes of F^- and I^-.

IDW000   CAS: 144-48-9   **HR: 3**
IODOACETAMIDE

mf: C_2H_4INO mw: 184.97

SYNS: α-IODOACETAMIDE * 2-IODOACETAMIDE * MONOIODOACETAMIDE * SURAUTO * USAF D-1

CONSENSUS REPORTS: EPA Genetic Toxicology Program. Reported in EPA TSCA Inventory.

SAFETY PROFILE: A poison by ingestion, intraperitoneal, and intravenous routes. Questionable carcinogen with experimental tumorigenic data. Human mutation data reported. When heated to decomposition it emits very toxic fumes of I^- and NO_x.

IDZ000   CAS: 64-69-7   **HR: 3**
IODOACETIC ACID

mf: $C_2H_3IO_2$ mw: 185.95

PROP: Colorless or white crystals. Mp: 82-83°. Sol in water and alc. Very sltly sol in ether.

SYNS: IA * IODOACETATE * MIA * MONOIODOACETATE * MONOIODOACETIC ACID

CONSENSUS REPORTS: Reported in EPA TSCA Inventory.

SAFETY PROFILE: A poison by ingestion, subcutaneous, and intravenous routes. Experimental teratogenic effects. Questionable carcinogen with experimental neoplastigenic and tumorigenic data. Human mutation data reported. When heated to decomposition it emits toxic fumes of I^-.

IEH000   CAS: 513-48-4   **HR: 3**
2-IODOBUTANE

DOT: UN 2390
mf: C_4H_9I mw: 184.03

PROP: Flash p: 14°F.

SYN: sec-BUTYL IODIDE

CONSENSUS REPORTS: Reported in EPA TSCA Inventory. EPA Genetic Toxicology Program.

DOT Classification: Flammable Liquid; Label: Flammable Liquid.

SAFETY PROFILE: Questionable carcinogen with experimental neoplastigenic data. Mutation data reported. A very dangerous fire hazard when exposed to heat or flame; can react vigorously with oxidizing materials. When heated to decomposition it emits toxic fumes of I⁻.

IEP000   CAS: 75-47-8   **HR: 3**
IODOFORM
mf: CHI_3 mw: 393.72

PROP: Yellow powder or crystals, disagreeable odor. D: 4.1, mp: 120° (approx), bp: subl. Decomp @ high temp evolving iodine, volatile with steam. Very sol in H_2O, benzene, acetone; sltly sol in petr ether.

SYNS: NCI-C04568 * TRIIODOMETHANE

CONSENSUS REPORTS: NCI Carcinogenesis Bioassay (gavage); No Evidence: mouse, rat NCITR* NCI-CG-TR-110,78. Reported in EPA TSCA Inventory.

OSHA PEL: TWA 0.6 ppm (skin)
ACGIH TLV: TWA 0.6 ppm

SAFETY PROFILE: A poison by ingestion. Moderately toxic by inhalation, skin contact, subcutaneous, and possibly other routes. Mutation data reported. Used as an antiseptic, disinfectant on superficial wounds and in female reproductive tract. 1:1 mixtures with hexamethylenetetramine explode at 178°C. Incompatible with mercuric oxide, calomel, silver nitrate, tannin, Balsam Peru directly mixed, Li, acetone. When heated to decomposition it emits toxic fumes of I⁻.

IEY000   CAS: 141-76-4   **HR: 3**
3-IODOPROPIONIC ACID
mf: $C_3H_5IO_2$ mw: 199.98

PROP: (a) Needles from water. D: 1.857, mp: 93-94°, bp: decomp. Sltly sol in water; very sol in alc and ether. (b) Needles. Mp: 44.5-45.5°, bp: 105°. Very sltly sol in water; sol in alc, ether.

CONSENSUS REPORTS: Reported in EPA TSCA Inventory.

SAFETY PROFILE: Moderately toxic by skin contact. Questionable carcinogen with experimental tumorigenic data. Mutation data reported. When heated to decomposition it emits toxic fumes of I⁻.

IFA000   CAS: 777-11-7   **HR: 3**
3-IODO-2-PROPYNYL-2,4,5-TRICHLOROPHENYL ETHER
mf: $C_9H_4Cl_3IO$ mw: 361.38

SYNS: 2,4,5-TRICHLOROPHENYL-γ-IODOPROPARGIL ETHER * 2,4,5-TRICHLOROPHENYL IODOPROPARGYL ETHER

CONSENSUS REPORTS: EPA Genetic Toxicology Program.

SAFETY PROFILE: A poison by intraperitoneal route. Moderately toxic by ingestion. A skin and eye irritant. An FDA over-the-counter drug. An antibacterial agent. When heated to decomposition it emits very toxic Cl⁻ and I⁻.

IFG000   CAS: 17236-22-5   **HR: 3**
3-IODOTETRAHYDROTHIOPHENE-1,1-DIOXIDE
mf: $C_4H_7IO_2S$ mw: 246.07

SYN: TETRAHYDRO-3-IODOTHIOPHENE-1,1-DIOXIDE

SAFETY PROFILE: A poison by ingestion, intraperitoneal, and intravenous routes. When heated to decomposition it emits very toxic I⁻ and SO_x.

IFN000   CAS: 811-73-4   **HR: 3**
IODOTRIMETHYLTIN
mf: C_3H_9ISn mw: 290.71

PROP: White powder, insol in water and organic solvents.

SYNS: IODOTRIMETHYLSTANNANE * TRIMETHYLSTANNYL IODINE * TRIMETHYLTIN IODIDE

OSHA PEL: TWA 0.1 mg(Sn)/m³ (skin)
ACGIH TLV: TWA 0.1 mg(Sn)/m³ (skin) (Proposed: TWA 0.1 mg(Sn)/m³; STEL 0.2 mg(Sn)/m³ (skin))
NIOSH REL: (Organotin Compounds) TWA 0.1 mg(Sn)/m³

SAFETY PROFILE: A poison by intravenous route. When heated to decomposition it emits toxic fumes of I⁻.

IFW000   CAS: 127-41-3   **HR: 1**
α-IONONE
mf: $C_{13}H_{20}O$ mw: 192.33

PROP: Colorless oil; woody, violet odor. D: 0.930, refr index: 1.497-1.502, bp: 136.1. Sol in alc, fixed oils propylene glycol; sltly sol in water; misc in ether; insol in glycerin.

SYNS: α-CYCLOCITRYLIDENEACETONE * FEMA No. 2594 * 4-(2,6,6-TRIMETHYL-2-CYCLOHEXEN-1-YL)-3-BUTEN-2-ONE

CONSENSUS REPORTS: Reported in EPA TSCA Inventory.

SAFETY PROFILE: Mildly toxic by ingestion. When heated to decomposition it emits acrid smoke and fumes.

IFX000   CAS: 14901-07-6   **HR: 1**
β-IONONE
mf: $C_{13}H_{20}O$ mw: 192.33

PROP: Colorless oil; woody odor. D: 0.944, refr index: 1.517-1.522, bp: 140°, flash p: +234°F. Sol in alc, fixed oils, propylene glycol; sltly sol in water; misc in ether; insol in glycerin.

SYNS: β-CYCLOCITRYLIDENEACETONE * FEMA No. 2595 * 4-(2,6,6-TRIMETHYL-1-CYCLOHEXEN-1-YL)-3-BUTEN-2-ONE

CONSENSUS REPORTS: Reported in EPA TSCA Inventory.

SAFETY PROFILE: Mildly toxic by ingestion. Combustible liquid. When heated to decomposition it emits acrid smoke and fumes.

IGF000   CAS: 8012-96-2   **HR: 1**
IPECAC SYRUP

PROP: Dried rhizome and roots of Rio or Brazilian ipecac. Contains emetine, cephaline, emetamine, ipecacuanic acid, psychotrine, methyl psychotaine, resin.

SYNS: DIHYDROTACHY STEROL * IPECACUANHA * SYRUP of IPECAC, U.S.P. * orl-dog LDLo: 5 g/kg

SAFETY PROFILE: Mildly toxic by ingestion. A centrally acting emetic. Human systemic effects by ingestion: nausea, vomiting, blood pressure lowering, change in heart rate, dyspnea. Has caused fatalities after prolonged ingestion. An FDA over-the-counter drug.

IGG000   CAS: 22254-24-6   **HR: 3**
IPRATROPIUM BROMIDE
mf: $C_{20}H_{30}NO_3 \cdot Br$ mw: 412.42

SYNS: ATEM * ATROVENT * 3-α-HYDROXY-8-ISOPROPYL-1-α-H,5-α-H-TROPANIUM BROMIDE (±)-TROPATE * (8r)-3-α-HYDROXY-8-ISOPROPYL-1-α-H,5-α-H-TROPIUMBROMIDE-(±)-TROPATE * IPRATROPIUM-BROMID (GERMAN) * 8-ISOPROPYLNORATROPINE METHOBROMIDE * N-ISOPROPYLNORATROPINIUM BROMOMETHYLATE * ITROP * Sch 1000

SAFETY PROFILE: A poison by intravenous, intraperitoneal and subcutaneous routes. Moderately toxic by ingestion. Human systemic effects by inhalation of very small amounts: gastrointestinal changes. Experimental reproductive effects. Used as a bronchodilator. When heated to decomposition it emits toxic fumes of NO_x and Br^-.

IGH000   CAS: 14885-29-1   **HR: 2**
IPROPRAN
mf: $C_7H_{11}N_3O_2$ mw: 169.21

SYNS: IPRONIDAZOLE (USDA) * 2-ISOPROPYL-1-METHYL-5-NITROIMIDAZOLE * 1-METHYL-2-(1-METHYLETHYL)-5-NITRO-1H-IMIDAZOLE * RO 7-1554

SAFETY PROFILE: Moderately toxic by ingestion. Mutation data reported. When heated to decomposition it emits toxic fumes of NO_x.

IGJ000   CAS: 7439-88-5   **HR: 3**
IRIDIUM
af: Ir   aw: 192.2

PROP: Silver-white very hard metallic element. Mp: 2450°, bp: approx 4500°, d: 22.65 @ 20°/4°. Highest specific gravity of all elements.

SAFETY PROFILE: The pure metal is clinically inert and no toxicity data is available. Most of its compounds are poorly soluble in water and thus are not absorbed efficiently by the body. The chlorides are poison or moderately toxic by ingestion and are eye and skin irritants. There are no reports of acute or chronic health effects to workers handling iridium and its compounds. The ^{190}Ir and ^{192}Ir radioisotopes are used in clinical radiography and most references to the toxicity of iridium relate to these isotopes.

A catalytic metal. The powdered metal may ignite spontaneously in air. Violent reaction or ignition on contact with interhalogens (e.g., bromine pentafluoride; chlorine trifluoride). Alloys with zinc, after extraction with acids, leave heat-sensitive explosive residues. Is attacked by F_2, Cl_2 at red heat; by potassium sulfate or a mixture of potassium hydroxide and nitrate; on fusion; lead; zinc; tin.

IGK800 CAS: 7439-89-6 **HR: 3**
IRON
af: Fe   aw: 55.85

PROP: From decomposition of iron pentacarbonyl: dark grey powder. From electrodeposition: lusterless, gray black powder. From chemical reduction: gray-black powder.

SYNS: ANCOR EN 80/150 * ARMCO IRON
* CARBONYL IRON * IRON, CARBONYL (FCC)
* IRON, ELECTROLYTIC * IRON, ELEMENTAL
* IRON, REDUCED (FCC)

CONSENSUS REPORTS: Reported in EPA TSCA Inventory.

SAFETY PROFILE: Poison by intraperitoneal route. Questionable carcinogen with experimental tumorigenic data. Iron is potentially toxic in all forms and by all routes of exposure. The inhalation of large amounts of iron dust results in iron pneumoconiosis (arc welder's lung). Chronic exposure to excess levels of iron (> 50-100 mg Fe/day) can result in pathological deposition of iron in the body tissues, the symptoms of which are fibrosis of the pancreas, diabetes mellitus, and liver cirrhosis.

As with other metals, it becomes more reactive as it is more finely divided. Ultrafine iron powder is pyrophoric and potentially explosive. Explosive or violent reaction with ammonium nitrate + heat, ammonium peroxodisulfate, chloric acid, chlorine trifluoride, chloroformamidinium nitrate, bromine pentafluoride + heat (with iron powder), air + oil (with iron dust), sodium acetylide. Ignites on contact with chlorine, dinitrogen tetraoxide, liquid fluorine, hydrogen peroxide (with iron powder), nitryl fluoride + heat, peroxyformic acid, potassium perchlorate, potassium dichromate, sodium peroxide (at 240°), polystyrene + friction or spark (iron powder). Mixtures of iron dust with air + water may ignite on drying. Reduced iron reacts with water to produce explosive hydrogen gas. Catalyzes the exothermic polymerization of acetaldehyde.

IGM000 CAS: 10102-50-8 **HR: 3**
IRON(II) ARSENATE (3:2)
DOT: UN 1608
mf: $As_2O_8 \cdot 3Fe$ mw: 445.39

SYNS: ARSENATE of IRON, FERROUS * FERROUS
ARSENATE (DOT) * FERROUS ARSENATE, solid (DOT)
* IRON ARSENATE (DOT)

CONSENSUS REPORTS: Arsenic and its compounds are on the Community Right-To-Know List.

OSHA PEL: TWA 0.01 mg(As)/m³; Cancer Hazard
ACGIH TLV: TWA 0.2 mg(As)/m³; TWA 1 mg/(Fe)/m³
NIOSH REL: (Inorganic Arsenic) CL 0.002 mg(As)/m³/15M
DOT Classification: Poison B; Label: Poison.

SAFETY PROFILE: Confirmed human carcinogen. A deadly poison by various routes. A pesticide. When heated to decomposition it emits toxic fumes of As.

IGN000 CAS: 10102-49-5 **HR: 3**
IRON(III) ARSENATE (1:1)
DOT: UN 1606
mf: $AsO_4 \cdot Fe$ mw: 194.77

SYNS: ARSENATE of IRON, FERRIC * FERRIC AR-
SENATE, solid (DOT)

CONSENSUS REPORTS: Arsenic and its compounds are on the Community Right-To-Know List.

OSHA PEL: TWA 0.01 mg(As)/m³; Cancer Hazard
ACGIH TLV: TWA 0.2 mg(As)/m³; TWA 1 mg/(Fe)/m³
NIOSH REL: (Inorganic Arsenic) CL 0.002 mg(As)/m³/15M
DOT Classification: Poison B; Label: Poison.

SAFETY PROFILE: Confirmed human carcinogen. A deadly poison. A pesticide. When heated to decomposition it emits toxic fumes of As.

IGO000 CAS: 63989-69-5 **HR: 3**
IRON(III)-o-ARSENITE PENTAHYDRATE
DOT: UN 1607
mf: $As_2Fe_2O_6 \cdot Fe_2O_3 \cdot 5H_2O$ mw: 607.34

PROP: Brown-yellow powder.

SYNS: FERRIC ARSENITE, BASIC * FERRIC ARSE-
NITE, solid (DOT)

CONSENSUS REPORTS: Arsenic and its compounds are on the Community Right-To-Know List.

OSHA PEL: TWA 0.01 mg(As)/m³; Cancer Hazard
ACGIH TLV: TWA 0.2 mg(As)/m³; TWA 1 mg/(Fe)/m³

NIOSH REL: (Inorganic Arsenic) CL 0.002 mg(As)/m³/15M

DOT Classification: Poison B; Label: Poison.

SAFETY PROFILE: Confirmed human carcinogen. A deadly poison. When heated to decomposition it emits toxic fumes of As.

IGS000   CAS: 9004-66-4   **HR: 3**
IRON-DEXTRAN COMPLEX

PROP: For human use, it is a sterile dark brown colloidal solvent, water-soluble. Approximate molecular weight is 180,000.

SYNS: A 100 (PHARMACEUTICAL) * CHINOFER * DEXTRAN ION COMPLEX * DEXTROFER 75 * EISENDEXTRAN (GERMAN) * Fe-DEXTRAN * FENATE * FERDEX 100 * FERRIC DEXTRAN * FERRIDEXTRAN * FERRODEXTRAN * FERRO-FLUKIN 75 * FERROGLUCIN * FERROGLUKIN 75 * IMFERON * IMPOSIL * IRO-JEX * IRON DEXTRAN * IRON DEXTRAN INJECTION * IRON HYDROGENATED DEXTRAN * IRONORM INJECTION * MYOFER 100 * POLYFER * PROLONGAL * RCRA WASTE NUMBER U139 * URSOFERRAN

CONSENSUS REPORTS: IARC Cancer Review: GROUP 2B IMEMDT 7,226,87; Human Inadequate Evidence IMEMDT 2,161,73; Animal Sufficient Evidence IMEMDT 2,161,73. NTP Fourth Annual Report On Carcinogens, 1984.

SAFETY PROFILE: Confirmed carcinogen producing tumors at site of application. Experimental carcinogenic, neoplastigenic, and tumorigenic data. Moderately toxic by ingestion and several other routes. Experimental teratogenic and reproductive effects.

IGU000   CAS: 9004-51-7   **HR: 3**
IRON-DEXTRIN COMPLEX

PROP: For human use, it is a clear, brown, colloidal solvent. Approximate molecular weight is 230,000.

SYNS: ASTRAFER * DEXTRIFERRON * DEXTRI-FERRON INJECTION * FERRIGEN * IRON CARBO-HYDRATE COMPLEX * IRON DEXTRIN INJECTION

CONSENSUS REPORTS: IARC Cancer Review: GROUP 3 IMEMDT 7,227,87; Animal Sufficient Evidence IMEMDT 2,161,73.

SAFETY PROFILE: A poison by intravenous route. Questionable carcinogen with experimental neoplastigenic and tumorigenic data. Moderately toxic by intraperitoneal route.

IGV000   CAS: 12068-85-8   **HR: 3**
IRON DISULFIDE
mf: FeS₂   mw: 119.97

SYNS: IRON PYRITES * IRON SULFIDE

CONSENSUS REPORTS: Reported in EPA TSCA Inventory.

SAFETY PROFILE: A poison by inhalation and ingestion. The powdered sulfide ignites spontaneously in air and some air-powder mixtures may be explosive. Trace carbon lowers the ignition temperature in air to 228°C and increases the sensitivity of air-dust mixtures. Heats up spontaneously and ignites with combustibles. Incompatible with water. When heated to decomposition or in reaction with acid or acid fumes it emits very toxic fumes of SO_x.

IGW000   **HR: 3**
IRON DUST

SAFETY PROFILE: Iron dust can cause conjunctivitis, choroiditis, retinitis, and siderosis of tissues if iron contacts and remains in these tissues. Iron ore dust can cause palpebral conjunctivitis, massive pulmonary fibrosis, and an increased incidence of lung cancer. Questionable carcinogen with experimental neoplastigenic data.

Flammable in the form of dust when exposed to heat or flame. Reacts violently with Cl_2; ClF_3; F_2; H_2O_2; NO_2; P; $Na2C_2$; H_2SO_4; air; water; polystyrene. Moderately explosive in the form of dust when exposed to heat or flame. To fight fire, use special mixtures of dry chemical.

IHD000   CAS: 1309-37-1   **HR: 3**
IRON OXIDE
mf: Fe₂O₃   mw: 159.70

SYNS: ANCHRED STANDARD * ANHYDROUS IRON OXIDE * ANHYDROUS OXIDE of IRON * ARMENIAN BOLE * BAUXITE RESIDUE * BLACK OXIDE of IRON * BLENDED RED OXIDES of IRON * BURNTISLAND RED * BURNT SIENNA * BURNT UMBER * CALCOTONE RED * CAPUT MORTUUM * C.I. 77491 * C.I. PIGMENT RED 101 * COLCOTHAR * COLLOIDAL FERRIC OXIDE * CROCUS MARTIS ADSTRINGENS * DEANOX * EISENOXYD * ENGLISH RED * FERRIC OXIDE * FERRUGO * INDIAN RED * IRON(III) OXIDE * IRON OXIDE RED * IRON SESQUIOXIDE * JEWELER'S ROUGE * LEVANOX RED 130A * LIGHT RED * MANUFACTURED IRON OXIDES

* MARS BROWN * MARS RED * NATURAL IRON OXIDES * NATURAL RED OXIDE * OCHRE * PRUSSIAN BROWN * RADDLE * 11554 RED * RED IRON OXIDE * RED OCHRE * ROUGE * RUBIGO * SIENNA * SPECULAR IRON * STONE RED * SUPRA * SYNTHETIC IRON OXIDE * VENETIAN RED * VITRIOL RED * VOGEL'S IRON RED * YELLOW FERRIC OXIDE * YELLOW OXIDE of IRON

CONSENSUS REPORTS: IARC Cancer Review: GROUP 3 IMEMDT 7,216,87; Human Limited Evidence IMEMDT 1,29,72; Animal No Evidence IMEMDT 1,29,72. Reported in EPA TSCA Inventory.

OSHA PEL: Dust and Fume: TWA 10 mg(Fe)/m^3; Rouge (Transitional: Total Dust: 15 mg/m^3; Respirable Fraction: 5 mg/m^3) TWA Total Dust: 10 mg/m^3; Respirable Fraction: 5 mg/m^3
ACGIH TLV: TWA 5 mg(Fe)/m^3 (vapor, dust); Rouge: 10 mg/m^3
DFG MAK: 6 mg/m^3

SAFETY PROFILE: A poison by subcutaneous route. Questionable carcinogen with experimental tumorigenic data. Catalyzes the potentially explosive polymerization of ethylene oxide. Explosive reaction when heated with guanidinium perchlorate. Reaction with carbon monoxide may form an explosive product. Potentially violent reaction with hydrogen peroxide. The wet oxide reacts explosively with molten aluminum-magnesium alloys. Violent reaction when heated with powdered aluminum, calcium disilicide, magnesium, metal acetylides (e.g., calcium acetylide + iron(III) chloride (on ignition), cesium acetylide (incandescent reaction when warmed), rubidium acetylide). Reacts violently with Al, Ca(OCl)$_2$, N$_2$H$_4$, ethylene oxide.

IHE000   **HR: 3**
IRON OXIDE, CHROMIUM OXIDE, and NICKEL OXIDE FUME

SYNS: CHROMIUM OXIDE, NICKEL OXIDE, and IRON OXIDE FUME * NICKEL OXIDE, IRON OXIDE, and CHROMIUM OXIDE FUME

CONSENSUS REPORTS: Nickel and its compounds, as well as chromium and its compounds, are on The Community Right-To-Know List.

OSHA PEL: TWA 1 mg(Ni)/m^3; CL 0.1 mg (CrO$_3$)/m^3

ACGIH TLV: TWA 1 mg(Ni)/m^3; (Proposed: TWA 0.05 mg(Ni)/m^3; Human Carcinogen); 0.05 mg(Cr)/m^3
NIOSH REL: (Chromium (VI)) TWA 0.025 mg(Cr(VI))/m^3; CL 0.05/15M; (Inorganic Nickel) TWA 0.015 mg(Ni)/m^3

SAFETY PROFILE: Confirmed human carcinogen. Human systemic effects by inhalation: cough.

IHF000   CAS: 1309-37-1   **HR: 3**
IRON OXIDE FUME
mf: Fe$_2$O$_3$ mw: 159.70

SYN: ZELAZA TLENKI (POLISH)

OSHA PEL: TWA 10 mg/m^3
ACGIH TLV: TWA 5 mg/m^3, welding fumes.

SAFETY PROFILE: Questionable carcinogen.

IHG000   CAS: 8047-67-4   **HR: 3**
IRON OXIDE, SACCHARATED

PROP: Saccharated oxide of iron.

SYNS: COLLIRON I.V. * FEOJECTIN * FERRIC OXIDE, SACCHARATED * FERRIC SACCHARATE IRON OXIDE (MIX.) * FERRIVENIN * IRON SACCHARATE * IRON SUGAR * IVIRON * NEO-FERRUM * PROFERRIN * SACCHARATED FERRIC OXIDE * SACCHARATED IRON * SUCROFER

CONSENSUS REPORTS: IARC Cancer Review: Animal Sufficient Evidence IMEMDT 2,161,73.

SAFETY PROFILE: Confirmed carcinogen with experimental neoplastigenic and tumorigenic data. A poison by intravenous route.

IHG500   CAS: 13463-40-6   **HR: 3**
IRON PENTACARBONYL
mf: C$_5$FeO$_5$ mw: 195.90

PROP: Yellow to dark red viscous liquid. Mp: −25°, bp: 103.0°, flash p: 5°F, d: 1.453 @ 25°/4°, vap press: 40 mm @ 30.3°.

SYNS: FER PENTACARBONYLE (FRENCH) * IRON CARBONYL * PENTACARBONYLIRON

CONSENSUS REPORTS: EPA Extremely Hazardous Substances List. Reported in EPA TSCA Inventory.

OSHA PEL: TWA 0.1 ppm (Fe); STEL 0.2 ppm
ACGIH TLV: TWA 0.1 ppm (Fe); STEL 0.2 ppm

DFG MAK: 0.1 ppm (0.8 mg/m^3)

SAFETY PROFILE: A poison by inhalation, skin contact, ingestion, subcutaneous, and intravenous routes. Inhalation causes dizziness, nausea and vomiting. If continued, unconsciousness follows. Often there is a delayed reaction of chest pain, cough, and difficult breathing. There may be cyanosis and circulatory collapse. In fatal cases, death occurs from the fourth to eleventh day with pneumonitis and injury to kidneys, liver, and brain. Iron carbonyl is less toxic than nickel carbonyl.

A very dangerous fire and moderate explosion hazard when exposed to heat or flame; can react vigorously with oxidizing materials. Pyrophoric in air! Mixtures with nitrogen oxide explode above 50°C. Violent reaction with zinc + transition metal halides (e.g., cobalt halides, rhodium halides, ruthenium halides). Mixtures with acetic acid + water produce a pyrophoric powder. To fight fire, use water, foam, CO$_2$, dry chemical.

IHH000   HR: 3
IRON-POLYSACCHARIDE COMPLEX

SYN: MUSCULARON

SAFETY PROFILE: A poison by intravenous and intraperitoneal routes. Questionable carcinogen with experimental tumorigenic data. When heated to decomposition it emits acrid smoke and fumes.

IHL000   CAS: 1338-16-5   HR: 3
IRON SORBITOL CITRATE
mf: C$_6$H$_{14}$O$_6$ • C$_6$H$_8$O$_7$ • xFe   mw: 765.29

SYNS: ESZ * IRON SORBITEX * IRON-SORBI-TOL-CITRIC ACID

CONSENSUS REPORTS: IARC Cancer Review: GROUP 3 IMEMDT 7,56,87; Animal No Evidence IMEMDT 2,161,73.

OSHA PEL: TWA 1 mg/(Fe)/m^3
ACGIH TLV: TWA 1 mg/(Fe)/m^3

SAFETY PROFILE: A poison by subcutaneous and intravenous routes. Moderately toxic by ingestion and intramuscular routes. Questionable carcinogen. When heated to decomposition it emits acrid smoke and fumes.

IHO850   CAS: 123-92-2   HR: 3
ISOAMYL ACETATE
mf: C$_7$H$_{14}$O$_2$ mw: 130.21

PROP: Colorless liquid; banana-like odor. Bp: 142.0°, ULC: 55-60, lel: 1% @ 212°F, uel: 7.5%, flash p: 77°F, d: 0.876, refr index: 1.400, autoign temp: 680°F, vap d: 4.49. Misc in alc, ether, ethyl acetate, fixed oils; sltly sol in water; insol in glycerin, propylene glycol.

SYNS: ACETIC ACID, ISOPENTYL ESTER * BANANA OIL * FEMA No. 2055 * ISOAMYL ETHANOATE * ISOPENTYL ACETATE * ISOPENTYL ALCOHOL ACETATE * 3-METHYLBUTYL ACETATE * 3-METHYL-1-BUTYL ACETATE * 3-METHYLBUTYL ETHANOATE * PEAR OIL

CONSENSUS REPORTS: Reported in EPA TSCA Inventory.

OSHA PEL: TWA 100 ppm
ACGIH TLV: TWA 100 ppm

SAFETY PROFILE: Mildly toxic by ingestion, inhalation and subcutaneous routes. Exposure to concentrations of about 1,000 ppm for 1 hour can cause headache, fatigue, pulmonary irritation and serious toxicity effects. Highly flammable liquid when exposed to heat or flame; can react vigorously with reducing materials. Moderately explosive in the form of vapor when exposed to heat or flame. To fight fire, use alcohol foam, CO$_2$, dry chemical. When heated to decomposition it emits acrid smoke and fumes.

IHP000   CAS: 123-51-3   HR: 3
ISOAMYL ALCOHOL

DOT: UN 1105
mf: C$_5$H$_{12}$O   mw: 88.17

PROP: Clear liquid; pungent, repulsive taste. Bp: 132°, ULC: 35-40, lel: 1.2%, uel: 9.0% @ 212°F, flash p: 109°F (CC), d: 0.813, autoign temp: 662°F, vap d: 3.04, mp: −117.2°. Sol in water @ 14°; misc in alc and ether.

SYNS: ALCOOL AMILICO (ITALIAN) * ALCOOL ISOAMYLIQUE (FRENCH) * AMYLOWY ALKOHOL (POLISH) * FERMENTATION AMYL ALCOHOL * ISOAMYL ALKOHOL (CZECH) * ISO-AMYLALKOHOL (GERMAN) * ISOAMYLOL * ISOBUTYLCARBINOL * ISOPENTANOL * ISOPENTYL ALCOHOL * 2-METHYL-4-BUTANOL * 3-METHYL BUTANOL * 3-METHYLBUTAN-1-OL * 3-METHYL-1-BUTANOL (CZECH) * 3-METIL-BUTANOLO (ITALIAN)

CONSENSUS REPORTS: Reported in EPA TSCA Inventory.

OSHA PEL: TWA 100 ppm; STEL 125 ppm
ACGIH TLV: TWA 100 ppm; STEL 125 ppm
DFG MAK: 100 ppm (360 mg/m^3)
DOT Classification: Flammable or Combustible Liquid; Label: Flammable Liquid.

SAFETY PROFILE: A poison by intraperitoneal and intravenous routes. Moderately toxic by ingestion and skin contact. A skin and human eye irritant. Human systemic effects by inhalation: olfactory effects, conjunctiva irritation, respiratory changes. Questionable carcinogen with experimental carcinogenic data. Mutation data reported. Flammable when exposed to heat or flame; can react vigorously with reducing materials. Slight explosion hazard when exposed to flame. Explosive reaction with hydrogen trisulfide. To fight fire, use alcohol foam, CO_2, dry chemical. When heated to decomposition it emits acrid smoke and fumes. Used as a flotation agent, a solvent, and in organic synthesis.

IHP010   CAS: 584-02-1   HR: 3
ISOAMYL ALCOHOL

DOT: UN 2706
mf: $C_5H_{12}O$ mw: 88.17

PROP: Liquid; acetone-like odor. Bp: 115.6°, d: 0.815 @ 25°/4°, flash p: 66°F, lel: 1.2%, uel: 9%. Sol alc, ether; sltly sol in water.

SYNS: DIETHYL CARBINOL * DIETHYLCARBINOL (DOT) * 3-PENTANOL * PENTANOL-3 * PENTAN-3-OL

CONSENSUS REPORTS: Reported in EPA TSCA Inventory.

OSHA PEL: TWA 100 ppm; STEL 125 ppm
ACGIH TLV: TWA 100 ppm; STEL 125 ppm
DFG MAK: 100 ppm (360 mg/m^3)
DOT Classification: Flammable or Combustible Liquid; Label: Flammable Liquid.

SAFETY PROFILE: Moderately toxic by ingestion, skin contact, and intraperitoneal routes. A severe eye and mild skin irritant. Dangerous fire and explosion hazard when exposed to heat, flame, or oxidizing materials. Used as a flotation agent, a solvent, and in organic synthesis. When heated to decomposition it emits acrid smoke and irritating fumes.

IHP100   CAS: 94-46-2   HR: 1
ISOAMYL BENZOATE
mf: $C_{12}H_{16}O$ mw: 176.28

PROP: Colorless to pale yellow liquid; pungent, fruity odor. D: 0.986-0.992, refr index: 1.492, flash p: +212°F.

SYNS: AMYL BENZOATE * BENZOIC ACID, 1-(3-METHYL)BUTYL ESTER * FEMA No. 2058 * ISOPENTYL BENZOATE * 1-(3-METHYL)BUTYL BENZOATE

CONSENSUS REPORTS: Reported in EPA TSCA Inventory.

SAFETY PROFILE: Mildly toxic by ingestion. A skin irritant. Combustible liquid. When heated to decomposition it emits acrid smoke and irritating fumes.

IHP400   CAS: 106-27-4   HR: 2
ISOAMYL BUTYRATE
mf: $C_9H_{18}O_2$ mw: 158.24

PROP: Colorless liquid; fruity odor. D: 0.860, refr index: 1.409-1.414, flash p: 149°F. Sol in alc, fixed oils; insol in glycerin, propylene glycol, water @ 179°.

SYNS: AMYL BUTYRATE * FEMA No. 2060

DOT Classification: Flammable or Combustible Liquid; Label: Flammable Liquid

SAFETY PROFILE: Combustible liquid. When heated to decomposition it emits acrid smoke and irritating fumes.

IHS000   CAS: 110-45-2   HR: 2
ISOAMYL FORMATE

DOT: UN 1109
mf: $C_6H_{12}O_2$ mw: 116.18

PROP: Clear liquid; fruity odor. Bp: 123.3°, d: 0.877 @ 20°, refr index: 1.396, vap press: 10 mm @ 17.1°, flash p: 127°F. Misc with alc, ether, propylene glycol; very sltly sol in water; insol in glycerin.

SYNS: FEMA No. 2069 * FORMIC ACID, ISOPENTYL ESTER * ISOAMYL METHANOATE * ISOPENTYL ALCOHOL, FORMATE * ISOPENTYL FORMATE * 3-METHYLBUTYL FORMATE

CONSENSUS REPORTS: Reported in EPA TSCA Inventory.

DOT Classification: Flammable or Combustible Liquid; Label: Flammable Liquid.

SAFETY PROFILE: Moderately toxic by ingestion. A skin irritant. This material is very irritating and can cause narcosis. The symptoms are usually transient in nature, but it is possible upon severe or prolonged exposure to have serious consequences. Combustible liquid. Can react with oxidizing materials. When heated to decomposition it emits acrid smoke and fumes.

IHU100   HR: 2
ISOAMYL HEXANOATE
mf: $C_{11}H_{22}O_2$ mw: 186.29

PROP: Colorless liquid; fruity odor. D: 0.858-0.863, refr index: 1.418-1.422, flash p: 190°F. Sol in alc, fixed oils; insol in glycerin, propylene glycol, water @ 222°.

SYNS: AMYL HEXANOATE * FEMA No. 2075
* ISOAMYL CAOPROATE * PENTYL HEXANOATE

SAFETY PROFILE: Combustible liquid. When heated to decomposition it emits acrid smoke and irritating fumes.

IHX600   HR: 1
ISOBORNYL ACETATE
mf: $C_{12}H_{20}O_2$ mw: 196.29

PROP: Colorless liquid; camphoraceous, piney, balsamic odor. D: 0.980, refr index: 1.462, flash p: +212°F. Sol in alc, fixed oils; sltly sol in propylene glycol; insol in water @ 227°.

SYN: FEMA No. 2160

SAFETY PROFILE: Combustible liquid. When heated to decomposition it emits acrid smoke and irritating fumes.

IHZ000   CAS: 115-31-1   HR: 2
ISOBORNYL THIOCYANATOACETATE
mf: $C_{13}H_{19}NO_2S$ mw: 253.39

PROP: Yellow, oily liquid; terpene-like odor. D: 1.1465 @ 25°/4°, bp: 95° @ 0.06 mm, flash p: 82°C (180°F). Very sol in alc, benzene, chloroform, and ether; insol in water.

SYNS: BORNATE * CIDALON * ENT 92
* ISOBORNEOL THIOCYANATOACETATE * ISOBORNYL THIOCYANOACETATE * TERPINYL THIOCYANOACETATE * THANISOL * THANITE * THIOCYANATOACETIC ACID ISOBORNYL ESTER

CONSENSUS REPORTS: Reported in EPA TSCA Inventory.

SAFETY PROFILE: Moderately toxic by ingestion. Slightly toxic by skin contact. Very irritating to eyes, mucous membranes, and skin. Flammable when exposed to heat or flame; can

react vigorously with oxidizing materials. When heated to decomposition it emits very toxic fumes of NO_x and SO_x. Used as an FDA over-the-counter drug; an insecticide and fly spray.

IIC000   CAS: 115-11-7   HR: 3
ISOBUTENE
DOT: UN 1055/UN 1075
mf: C_4H_8 mw: 56.12

PROP: Volatile liquid or easily liquefied gas. Bp: −6.9°, fp: −140.3°, flash p: <14°F, d: 0.600, autoign temp: 869°F, lel: 1.8%, uel: 9.6%. Insol in water, very sol in alc, ether, sulfuric acid.

SYNS: γ-BUTYLENE * ISOBUTYLENE (DOT)
* LIQUEFIED PETROLEUM GAS (DOT) * 2-METHYLPROPENE

CONSENSUS REPORTS: Reported in EPA TSCA Inventory.

DOT Classification: Flammable Gas; Label: Flammable Gas.

SAFETY PROFILE: A simple asphyxiant; may have narcotizing action. A very dangerous fire and explosion hazard when exposed to heat or flame. Can react vigorously with oxidizing materials. To fight fire, stop flow of gas. When heated to decomposition it emits acrid smoke and fumes.

IIJ000   CAS: 110-19-0   HR: 3
ISOBUTYL ACETATE
DOT: UN 1213
mf: $C_6H_{12}O_2$ mw: 116.18

PROP: Colorless, neutral liquid; fruit-like odor. Mp: −98.9°, bp: 118°, flash p: 64°F (CC) (18°), d: 0.8685 @ 15°, refr index: 1.389, vap press: 10 mm @ 12.8°, autoign temp: 793°F, vap d: 4.0, lel: 2.4%, uel: 10.5%. Very sol in alc, fixed oils, propylene glycol; sltly sol in water.

SYNS: ACETATE d'ISOBUTYLE (FRENCH) * ACETIC ACID, ISOBUTYL ESTER * ACETIC ACID-2-METHYLPROPYL ESTER * FEMA No. 2175 * 2-METHYLPROPYL ACETATE * 2-METHYL-1-PROPYL ACETATE
* β-METHYLPROPYL ETHANOATE

CONSENSUS REPORTS: Reported in EPA TSCA Inventory.

OSHA PEL: TWA 150 ppm
ACGIH TLV: TWA 150 ppm
DFG MAK: 200 ppm (950 mg/m^3)

DOT Classification: Flammable Liquid; Label: Flammable Liquid.

SAFETY PROFILE: Mildly toxic by ingestion and inhalation. A skin and eye irritant. Upon absorption by the body it can hydrolyze to acetic acid and isobutanol. Highly flammable liquid. A very dangerous fire and moderate explosion hazard when exposed to heat, flame, or oxidizers. To fight fire, use alcohol foam, CO_2, dry chemical. When heated to decomposition it emits acrid smoke and fumes.

IIK000   CAS: 106-63-8   **HR: 2**
ISOBUTYL ACRYLATE

DOT: UN 2527
mf: $C_7H_{12}O_2$ mw: 128.19

SYNS: ACRYLIC ACID ISOBUTYL ESTER * ISOBU-TYL ACRYLATE, INHIBITED (DOT) * ISOBUTYL PROPENOATE * ISOBUTYL-2-PROPENOATE * Z-METHYLPROPYL ACRYLATE * 2-PROPENOIC ACID-2-METHYLPROPYL ESTER

CONSENSUS REPORTS: Reported in EPA TSCA Inventory.

DOT Classification: Flammable or Combustible Liquid; Label: Flammable Liquid.

SAFETY PROFILE: Moderately toxic by skin contact and intraperitoneal routes. Mildly toxic by inhalation and ingestion. A skin irritant. Flammable when exposed to heat or flame; can react vigorously with oxidizing materials. When heated to decomposition it emits acrid smoke and toxic fumes.

IIL000   CAS: 78-83-1   **HR: 3**
ISOBUTYL ALCOHOL

DOT: UN 1212
mf: $C_4H_{10}O$ mw: 74.14

PROP: Clear mobile liquid; sweet odor. Bp: 107.90°, flash p: 82°F, ULC: 40-45, lel: 1.2%, uel: 10.9% @ 212°F, fp: −108°, d: 0.800, autoign temp: 800°F, vap press: 10 mm @ 21.7°, vap d: 2.55. Sltly sol in water; misc with alc and ether.

SYNS: ALCOOL ISOBUTYLIQUE (FRENCH) * FEMA No. 2179 * FERMENTATION BUTYL ALCOHOL * 1-HYDROXYMETHYLPROPANE * ISOBUTANOL (DOT) * ISOBUTYLALKOHOL (CZECH) * ISOPRO-PYLCARBINOL * 2-METHYL PROPANOL * 2-METHYL-1-PROPANOL * 2-METHYLPROPAN-1-OL * 2-METHYLPROPYL ALCOHOL * RCRA WASTE NUMBER U140

CONSENSUS REPORTS: Reported in EPA TSCA Inventory.

OSHA PEL: (Transitional: TWA 100 mg/m^3) TWA 50 ppm
ACGIH TLV: TWA 50 ppm
DFG MAK: 100 ppm (300 mg/m^3)
DOT Classification: Flammable or Combustible Liquid; Label: Flammable Liquid.

SAFETY PROFILE: Poison by intravenous and intraperitoneal routes. Moderately toxic by ingestion and skin contact. Mildly toxic by inhalation. A severe skin and eye irritant. Questionable carcinogen with experimental carcinogenic and tumorigenic data. Mutation data reported. Flammable liquid. Dangerous fire hazard when exposed to heat or flame. Moderately explosive in the form of vapor when exposed to heat, flame or oxidizers. Ignites on contact with chromium trioxide. Reacts with aluminum at 100° to form explosive hydrogen gas. Keep away from heat and open flame. To fight fire, use alcohol foam, CO_2, dry chemical. When heated to decomposition it emits acrid smoke and fumes.

IIM000   CAS: 78-81-9   **HR: 3**
ISOBUTYLAMINE

DOT: UN 1214
mf: $C_4H_{11}N$ mw: 73.16

PROP: Colorless liquid. Mp: −85.5°, bp: 68.6°, flash p: 15°F, d: 0.731 @ 20°/20°, vap press: 100 mm @ 18.8°, autoign temp: 712°F, vap d: 2.5. Misc with water, alc, and ether.

SYNS: 1-AMINO-2-METHYLPROPANE * MONOISO-BUTYLAMINE * VALAMINE

CONSENSUS REPORTS: Reported in EPA TSCA Inventory.

DFG MAK: 5 ppm (15 mg/m^3)
DOT Classification: Flammable Liquid; Label: Flammable Liquid.

SAFETY PROFILE: A poison by ingestion. A powerful irritant to skin, eyes, and mucous membranes. Skin contact can cause blistering. Inhalation can cause headache and dryness of nose and throat. A very dangerous fire hazard when exposed to heat or flame. Can react vigorously with oxidizing materials. To fight fire, use dry chemical, foam, CO_2, alcohol foam. When heated to decomposition it emits toxic fumes of NO_x.

IIN000   CAS: 538-93-2   **HR: 3**
ISOBUTYLBENZENE

DOT: UN 2709
mf: $C_{10}H_{14}$ mw: 134.24

PROP: Liquid. Insol in water; sol in alc and ether. Mp: $-51.5°$, bp: $170.5°$, flash p: $131°F$ (CC), d: 0.867 @ $20°/4°$, autoign temp: $806°F$, vap press: 1 mm @ $14.1°$, vap d: 4.62, lel: 0.8%, uel: 6.0%.

SYN: 2-METHYL-1-PHENYLPROPANE

CONSENSUS REPORTS: Reported in EPA TSCA Inventory.

DOT Classification: Flammable or Combustible; Label: Flammable Liquid.

SAFETY PROFILE: Mildly toxic by ingestion. An irritant and possibly narcotic. Flammable when exposed to heat or flame; can react with oxidizing materials. Moderate explosion hazard when exposed to heat or flame. To fight fire, use foam, CO_2, dry chemical. When heated to decomposition it emits acrid smoke and fumes.

IIP000   CAS: 4439-24-1   **HR: 3**
ISOBUTYL CELLOSOLVE
mf: $C_6H_{14}O_2$ mw: 118.20

PROP: Colorless liquid. D: 0.903 @ $20°/4°$, bp: $171.2°$; misc in water, alc, ether.

SYNS: EKTASOLVE EIB * ETHYLENE GLYCOL MONOISOBUTYL ETHER * 2-ISOBUTOXYETHANOL

CONSENSUS REPORTS: Glycol ether compounds are on the Community Right-To-Know List. Reported in EPA TSCA Inventory.

SAFETY PROFILE: Poison by ingestion. Moderately toxic by skin contact. Mildly toxic by inhalation. A skin irritant. When heated to decomposition it emits acrid smoke and fumes.

IIQ000   CAS: 122-67-8   **HR: 1**
ISOBUTYL CINNAMATE
mf: $C_{13}H_{16}O_2$ mw: 204.29

PROP: Colorless liquid; sweet, fruity odor. D: 1.001, refr index: 1.539-1.541, flash p: $+212°F$. Misc with alc, chloroform, ether, fixed oils; insol in water.

SYNS: CINNAMIC ACID, ISOBUTYL ESTER * FEMA No. 2193 * LABDANOL * 3-PHENYL-2-PROPENOIC ACID, 2-METHYLPROPYL ESTER

CONSENSUS REPORTS: Reported in EPA TSCA Inventory.

SAFETY PROFILE: A skin irritant. Combustible liquid. When heated to decomposition it emits acrid smoke and fumes.

IIR000   CAS: 542-55-2   **HR: 2**
ISOBUTYL FORMATE

DOT: UN 2393
mf: $C_5H_{10}O_2$ mw: 102.15

PROP: Liquid. D: 0.885 @ $20°/4°$, mp: $-95.3°$, bp: $98.2°$, flash p: $<70°F$, autoign temp: $608°F$, lel: 2.0%, uel: 8%. Sol in water @ $22°$; misc in alc and ether.

SYNS: FORMIC ACID, ISOBUTYL ESTER * TETRYL FORMATE

CONSENSUS REPORTS: Reported in EPA TSCA Inventory.

DOT Classification: Flammable Liquid; Label: Flammable Liquid.

SAFETY PROFILE: Moderately toxic by ingestion. A very dangerous fire hazard when exposed to heat, open flame, or oxidizers. A moderate explosion hazard when exposed to heat or flame. To fight fire, use water spray, foam, CO_2, dry chemical. When heated to decomposition it emits acrid smoke and fumes.

IIV000   CAS: 6104-30-9   **HR: 3**
ISOBUTYLIDENEDIUREA
mf: $C_6H_{14}N_4O_2$ mw: 174.24

SYNS: 1,1-DIUREIDISOBUTANE * DIUREIDOISOBU-TANE * IBDU * ISOBUTYLDIUREA * ISOBU-TYLENEDIUREA * 1,1'-ISOBUTYLIDENEBISUREA * ISODUR * N,N''-(2-METHYLPROPYLIDENE)BIS-UREA (9CI)

CONSENSUS REPORTS: Reported in EPA TSCA Inventory.

SAFETY PROFILE: Questionable carcinogen with experimental tumorigenic data. When heated to decomposition it emits toxic fumes of NO_x.

IIW000   CAS: 97-85-8   **HR: 1**
ISOBUTYL ISOBUTYRATE

DOT: UN 2528
mf: $C_8H_{16}O_2$ mw: 144.24

PROP: Liquid with fruity odor. Mp: $-81°$, bp: $147.5°$, d: 0.850-0.860 @ $20°/20°$, vap press: 10 mm @ $39.9°$. Insol in H_2O, misc with alc.

SYNS: ISOBUTYRIC ACID, ISOBUTYL ESTER * ISOBUTYLISOBUTYRATE (DOT) * 2-METHYLPRO-

PYL ISOBUTYRATE * 2-METHYLPROPYLPROPANOIC
ACID-2-METHYLPROPYL ESTER (9CI)

CONSENSUS REPORTS: Reported in EPA TSCA Inventory.

DOT Classification: Flammable or Combustible Liquid; Label: Flammable Liquid.

SAFETY PROFILE: Mildly toxic by ingestion and inhalation. An insect-repellent. Combustible when exposed to heat or flame. Can react with oxidizing materials. When heated to decomposition it emits acrid smoke and fumes.

IIY000 CAS: 97-86-9 HR: 2
ISOBUTYL METHACRYLATE
DOT: UN 2283
mf: $C_8H_{14}O_2$ mw: 142.22

SYNS: ISOBUTYL-α-METHACRYLATE * METHACRYLIC ACID, ISOBUTYL ESTER * 2-METHYL-2-PROPENOIC ACID-2-METHYLPROPYL ESTER * 2-METHYLPROPYL METHACRYLATE

CONSENSUS REPORTS: Reported in EPA TSCA Inventory.

DOT Classification: Flammable or Combustible Liquid; Label: Flammable Liquid.

SAFETY PROFILE: Moderately toxic by intraperitoneal route. Mildly toxic by ingestion. Experimental teratogenic and reproductive effects. Flammable when exposed to heat or flame. When heated to decomposition it emits acrid smoke and fumes.

IJD000 CAS: 542-56-3 HR: 3
ISOBUTYL NITRITE
mf: $C_4H_9NO_2$ mw: 103.14

PROP: Liquid. D: 0.870 @ 22°/4°, bp: 67-68°. Sltly sol in and decomp in water, misc in alc.

SYNS: IBN * NCI-C61052 * NITROUS ACID, ISOBUTYL ESTER * NITROUS ACID, 2-METHYLPROPYL ESTER

CONSENSUS REPORTS: Reported in EPA TSCA Inventory.

SAFETY PROFILE: A poison by ingestion and intraperitoneal routes. Mildly toxic by inhalation. Human systemic effects by ingestion: carboxhemoglobinemia, blood pressure lowering, change in heart rate. Mutation data reported. When heated to decomposition it emits toxic fumes of NO_x.

IJF400 HR: 1
ISOBUTYL PHENYLACETATE
mf: $C_{12}H_{16}O_2$ mw: 192.23

PROP: Colorless liquid; rose, honey-like odor. D: 0.984-0.988, refr index: 1.486, flash p: 241°F. Sol in alc, fixed oils; insol in glycerin, propylene glycol, water.

SYN: FEMA No. 2210

SAFETY PROFILE: Combustible liquid. When heated to decomposition it emits acrid smoke and irritating fumes.

IJN000 CAS: 87-19-4 HR: 2
ISOBUTYL SALICYLATE
mf: $C_{11}H_{14}O_3$ mw: 194.25

PROP: Colorless liquid; orchid odor. D: 1.062-1.066, refr index: 1.507, flash p: 250°F. Sol in fixed oils; insol in glycerin, propylene glycol.

SYNS: FEMA No. 2213 * ISOBUTYL-o-HYDROXYBENZOATE * SALICYLIC ACID, ISOBUTYL ESTER

CONSENSUS REPORTS: Reported in EPA TSCA Inventory.

SAFETY PROFILE: Moderately toxic by ingestion. Combustible liquid. When heated to decomposition it emits acrid smoke and fumes.

IJQ000 CAS: 109-53-5 HR: 3
ISOBUTYL VINYL ETHER
DOT: UN 1304
mf: $C_6H_{12}O$ mw: 100.18

PROP: Liquid. Mp: −112°, bp: 82.9-83.2°, flash p: 16°F, d: 0.76 @ 25°/4°, vap d: 3.45.

SYNS: IVE * VINOFLEX MO 400* * VINYL ISOBUTYL ETHER (DOT) * VINYL ISOBUTYL ETHER, INHIBITED (DOT)

CONSENSUS REPORTS: Reported in EPA TSCA Inventory.

DOT Classification: Flammable Liquid; Label: Flammable Liquid.

SAFETY PROFILE: Very mildly toxic by ingestion, inhalation, and skin contact. A very dangerous fire hazard when exposed to heat, flame, oxidizers. Severe explosion hazard when exposed to sparks or open flame. Can react vigorously with oxidizing materials. When heated to decomposition it emits acrid smoke and fumes. To fight fire, use alcohol foam, CO_2, dry chemical.

IJR000   CAS: 63916-90-5   **HR: 3**
p-ISOBUTYOXYBENZOIC ACID-3-(2′-METHYLPIPERIDINO)PROPYL ESTER
mf: $C_{20}H_{31}NO_3$ mw: 333.52

SAFETY PROFILE: A poison by subcutaneous and intravenous routes. When heated to decomposition it emits toxic fumes of NO_x.

IJS000   CAS: 78-84-2   **HR: 3**
ISOBUTYRALDEHYDE

DOT: UN 2045
mf: C_4H_8O mw: 72.12

PROP: Transparent, colorless, highly refractive liquid; pungent odor. Mp: $-65°$, bp: $64°$, flash p: $-40°F$ (CC), d: 0.783-0.788, autoign temp: $434°F$, lel: 1.6%, uel: 10.6%, vap d: 2.5. Sol in water; misc in alc, ether, benzene, carbon disulfide, acetone, toluene, chloroform.

SYNS: FEMA No. 2220 * ISOBUTANAL * ISOBUTYLALDEHYDE * ISOBUTYL ALDEHYDE (DOT) * ISOBUTYRALDEHYD (CZECH) * ISOBUTYRIC ALDEHYDE * 2-METHYLPROPANAL * 2-METHYL-1-PROPANAL * 2-METHYLPROPIONALDEHYDE * NCI-C60968 * VALINE ALDEHDYE

CONSENSUS REPORTS: Community Right-To-Know List. Reported in EPA TSCA Inventory.

DOT Classification: Flammable Liquid; Label: Flammable Liquid.

SAFETY PROFILE: Moderately toxic by ingestion. Mildly toxic by skin contact and inhalation. A severe skin and eye irritant. Highly flammable liquid. A very dangerous fire hazard when exposed to heat, flame, or oxidizers. Moderately explosive in the form of vapor when exposed to heat or flame. Can react vigorously with reducing materials. When heated to decomposition it emits acrid smoke and fumes. To fight fire, use dry chemical, CO_2, mist, foam.

IJU000   CAS: 79-31-2   **HR: 3**
ISOBUTYRIC ACID

DOT: UN 2529
mf: $C_4H_8O_2$ mw: 88.12

PROP: Colorless liquid; pungent odor of rancid butter. Mp: $-47°$, bp: $154.5°$, flash p: $132°F$ (TOC), d: 0.949 @ $20°/4°$, refr index: 1.392, vap press: 1 mm @ $14.7°$, vap d: 3.04, autoign temp: $935°F$. Misc with alc, chloroform and ether. Misc with alc, fixed oils, glycerin, propylene glycol; insol in water.

SYNS: DIMETHYLACETIC ACID * FEMA No. 2222 * ISOPROPYLFORMIC ACID * α-METHYLPROPIONIC ACID * 2-METHYLPROPANOIC ACID * 2-METHYLPROPIONIC ACID

CONSENSUS REPORTS: Reported in EPA TSCA Inventory.

DOT Classification: Corrosive Material; Label: Corrosive; Flammable or Combustible Liquid; Label: Flammable Liquid.

SAFETY PROFILE: A poison by ingestion. Moderately toxic by skin contact. A corrosive irritant to the eyes, skin and mucous membranes. Flammable liquid when exposed to heat or flame; can react with oxidizing materials. To fight fire, use alcohol foam, CO_2, dry chemical. When heated to decomposition it emits acrid smoke and fumes.

IJV000   CAS: 103-28-6   **HR: 2**
ISOBUTYRIC ACID, BENZYL ESTER
mf: $C_{11}H_{14}O_2$ mw: 178.25

PROP: Colorless liquid; floral, jasmine odor. D: 1.001-1.005, refr index: 1.489, flash p: $212°F$. Sol in alc, fixed oils; sltly sol in propylene glycol; insol in glycerin @ $229°$.

SYNS: BENZYL ISOBUTYRATE (FCC) * BENZYL-2-METHYL PROPIONATE * FEMA No. 2141

CONSENSUS REPORTS: Reported in EPA TSCA Inventory.

SAFETY PROFILE: Moderately toxic by ingestion. Combustible liquid. When heated to decomposition it emits acrid smoke and fumes.

IJW000   CAS: 97-72-3   **HR: 2**
ISOBUTYRIC ANHYDRIDE

DOT: UN 2530
mf: $C_8H_{14}O_3$ mw: 158.22

PROP: Liquid, decomp in water. Bp: $360°F$, d: 0.951-0.956 @ $20°/20°$, vap d: 5.5, flash p: $139°F$, autoign temp: $665°F$.

CONSENSUS REPORTS: Reported in EPA TSCA Inventory.

DOT Classification: Corrosive Material; Label: Corrosive; Flammable or Combustible Liquid; Label: Flammable Liquid.

SAFETY PROFILE: A corrosive irritant to skin, eyes, and mucous membranes. Flammable when exposed to heat, flame, or oxidizers. To fight fire, use alcohol foam, fog, dry chemical, CO_2.

When heated to decomposition it emits acrid smoke and fumes.

IJX000 CAS: 78-82-0 **HR: 3**
ISOBUTYRONITRILE

DOT: UN 2284
mf: C_4H_7N mw: 69.12

PROP: Colorless liquid, sltly sol in water, very sol in alc and ether. D: 0.773 @ 20°/20°, bp: 107°, mp: −75°, flash p: 46.4°F.

SYNS: ISOPROPYL CYANIDE * 2-METHYLPROPIONITRILE

CONSENSUS REPORTS: Cyanide and its compounds are on the Community Right-To-Know List. EPA Extremely Hazardous Substances List. Reported in EPA TSCA Inventory.

NIOSH REL: (Nitriles) TWA 22 mg/m^3
DOT Classification: Flammable Liquid; Label: Flammable Liquid and Poison.

SAFETY PROFILE: A poison by ingestion, skin contact, and subcutaneous routes. Mildly toxic by inhalation. A skin irritant. A very dangerous fire hazard when exposed to heat or flame. When heated to decomposition it emits toxic fumes of NO_x and CN^-.

IJZ000 CAS: 533-28-8 **HR: 3**
ISOCAINE

mf: $C_{16}H_{23}NO_2 \cdot ClH$ mw: 297.86

SYNS: o-AMINOBENZOYL DI(ISOPROPYLAMINO)ETHANOL HYDROCHLORIDE * 3-BENZOXY-1-(2-METHYL-PIPERIDINO)PROPANE HYDROCHLORIDE * dl-3-BENZOXY-1-(2-METHYLPIPERIDINO)PROPANE HYDROCHLORIDE * BENZOYL-γ-(2-METHYLPIPERIDINE) PROPANOL HYDROCHLORIDE * METHCAINE HYDROCHLORIDE * 2-METHYL-1-PIPERIDINEPROPANOL BENZOATE HYDROCHLORIDE * 3-(2-METHYLPIPERIDINO)PROPYL BENZOATE HYDROCHLORIDE * dl-(2-METHYLPIPERIDINO)PROPYL BENZOATE HYDROCHLORIDE * γ-(2-METHYLPIPERIDINO)PROPYL BENZOATE HYDROCHLORIDE * (+−)-γ-(2-METHYLPIPERIDYL) PROPYL BENZOATE HYDROCHLORIDE * NEOTHESIN HYDROCHLORIDE * PIPEROCAINE HYDROCHLORIDE * PIPEROCAINIUM CHLORIDE

SAFETY PROFILE: A poison by intraperitoneal, intravenous, subcutaneous, and intraspinal routes. Used as a local anesthetic. An FDA over-the-counter drug. When heated to decomposition it emits very toxic fumes of NO_x and HCl.

IKC000 CAS: 59-63-2 **HR: 3**
ISOCARBOXAZID

mf: $C_{12}H_{13}N_3O_2$ mw: 231.28

PROP: Crystals from methanol, practically tasteless. Mp: 106°, very sltly sol in hot water; sltly sol in alc, glycerol, and propylene glycol.

SYNS: BENAZIDE * N′-BENZYL N-METHYL-5-ISOXAZOLECARBOXYLHYDRAZIDE-3 * 1-BENZYL-2-(5-METHYL-3-ISOXAZOIYL-CARBONYL)HYDRAZINE * 1-BENZYL-1-(5-METHYL-3-ISOXAZOIYLCARBONYL) HYDRAZINE * BMIH * ENERZER * ISOCARBONAZID * ISOCARBOSSAZIDE * ISOCARBOXAZIDE * ISOCARBOXYZID * MARAPLAN * MARPLAN * MARPLON * 5-METHYL-3-ISOXAZOLECARBOXYLIC ACID-2-BENZYLHYDRAZIDE * RO 5-0831

SAFETY PROFILE: A poison by ingestion, intraperitoneal, and subcutaneous routes. An experimental teratogen. Mutation data reported. A pharmaceutical and veterinary drug. When heated to decomposition it emits toxic fumes of NO_x.

IKF000 CAS: 11071-47-9 **HR: 2**
ISOCTENE

DOT: UN 1216
mf: C_8H_{16} mw: 112.24

SYN: 2,2,4-TRIMETHYL-1-PENTENE

DOT Classification: Label: Flammable Liquid.

SAFETY PROFILE: A dangerous fire hazard when exposed to heat or flame. When heated to decomposition it emits acrid smoke and fumes.

IKG000 CAS: 103-65-1 **HR: 3**
ISOCUMENE

DOT: UN 2364
mf: C_9H_{12} mw: 120.21

PROP: Clear liquid. Insol in water; misc in alc and ether. Mp: −99.5°, bp: 159.2°, flash p: 86°F (CC), d: 0.862, vap press: 10 mm @ 43.4°, vap d: 4.14, autoign temp: 842°F, lel: 0.8%, uel: 6%.

SYNS: 1-PHENYLPROPANE * n-PROPYLBENZENE * PROPYL BENZENE (DOT)

CONSENSUS REPORTS: Reported in EPA TSCA Inventory.

DOT Classification: Flammable or Combustible Liquid; Label: Flammable Liquid.

SAFETY PROFILE: Mildly toxic by ingestion and inhalation. A very dangerous fire hazard when exposed to heat, flame, or oxidizers; can react with oxidizing materials. A moderate explosion hazard in the form of vapor when exposed to heat or flame. To fight fire, use foam, CO_2, dry chemical. When heated to decomposition it emits acrid smoke and fumes.

IKO000   CAS: 465-73-6   HR: 3
ISODRIN
mf: $C_{12}H_8Cl_6$ mw: 364.90

PROP: Crystals. Mp: 240-242°.

SYNS: COMPOUND 711 * ENT 19,244 * EXPERI-MENTAL INSECTICIDE 711 * 1,2,3,4,10,10-HEXA-CHLORO-1,4,4a,5,8,8a-HEXAHYDRO-1,4,5,8-endo,endo-DIMETHANONAPHTHALENE * 1,2,3,4,10,10-HEXA-CHLORO-1,4,4a,5,8,8a-HEXAHYDRO-1,4-endo,endo-5,8-DI-METHANONAPHTHALENE * RCRA WASTE NUMBER P060

CONSENSUS REPORTS: EPA Extremely Hazardous Substances List. Reported in EPA TSCA Inventory. Chlorophenol compounds are on the Community Right-To-Know List.

SAFETY PROFILE: A poison by ingestion, skin contact, and possibly other routes. Causes liver injury, acne, and skin rashes. When heated to decomposition it emits toxic fumes of Cl^-.

IKQ000   CAS: 97-54-1   HR: 2
ISOEUGENOL
mf: $C_{10}H_{12}O_2$ mw: 164.22

PROP: Pale yellow oil; carnation odor. D: 1.079-1.085, refr index: 1.572-1.577, mp: $-10°$, bp: 266°. cis Form: liquid, bp: 133° @ 11 mm, d: 1.088 @ 20°/4°. trans Form: crystals, mp: 33°; bp: 140° @ 12 mm; d: 1.087 @ 20°/4°, flash p: +212°F. Sol in fixed oils, propylene glycol; very sltly sol in water; misc in alc and ether; insol in glycerin.

SYNS: FEMA No. 2468 * 1-HYDROXY-2-METHOXY-4-PROPENYLBENZENE * 4-HYDROXY-3-METHOXY-1-PROPENYLBENZENE * 2-METHOXY-4-PROPENYLPHE-NOL * NCI-C60979 * 4-PROPENYLGUAIACOL

CONSENSUS REPORTS: Reported in EPA TSCA Inventory.

SAFETY PROFILE: Moderately toxic by ingestion. Human mutation data reported. Combustible liquid. When heated to decomposition it emits acrid smoke and fumes.

IKR000   CAS: 93-16-3   HR: 3
1,3,4-ISOEUGENOL METHYL ETHER
mf: $C_{11}H_{14}O_2$ mw: 178.25

PROP: Colorless to pale yellow liquid; clove-carnation odor. D: 1.047, refr index: 1.566, flash p: +212°F. Sol in fixed oils; insol in glycerin, propylene glycol.

SYNS: 1,2-DIMETHOXY-4-PROPENYLBENZENE * FEMA No. 2476 * ISOEUGENYL METHYL ETHER * ISOHOMOGENOL * METHYL ISOEUGENOL (FCC) * 4-PROPENYL VERATROLE

CONSENSUS REPORTS: Reported in EPA TSCA Inventory.

SAFETY PROFILE: Poison by intravenous route. Moderately toxic by intraperitoneal route. A skin irritant. Combustible liquid. When heated to decomposition it emits acrid smoke and fumes.

IKS600   CAS: 107-83-5   HR: 3
ISOHEXANE
DOT: UN 1208/UN 2462
mf: C_6H_{14} mw: 86.20

PROP: Liquid. Bp: 54-60°, lel: 1.0%, uel: 7.0%, flash p: 20°F (CC), d: 0.669, vap d: 3.00, autoign temp: 583°F.

SYN: 2-METHYLPENTANE

CONSENSUS REPORTS: Reported in EPA TSCA Inventory.

OSHA PEL: TWA 500 ppm; STEL 1000 ppm
ACGIH TLV: TWA 500 ppm; STEL 1000 ppm (hexane isomer)
NIOSH REL: (Alkanes) TWA 350 mg/m³
DOT Classification: Flammable Liquid; Label: Flammable Liquid .

SAFETY PROFILE: A human eye irritant. A very dangerous fire hazard when exposed to heat, flame or oxidizers. Severe explosion hazard when exposed to heat or flame. Explosive in the form of vapor when exposed to heat or flame. Keep away from sparks, heat, or open flame; can react vigorously with oxidizing materials. To fight fire, use foam, CO_2, dry chemical. When heated to decomposition it emits acrid smoke and irritating fumes.

IKX000   CAS: 73-32-5   HR: 1
ISOLEUCINE
mf: $C_6H_{13}NO_2$ mw: 131.17

PROP: An essential amino acid; many isomeric forms. White crystalline powder; bitter taste. Mp: (dl): 292° (decomp), (l): 283-284° (decomp). Sltly sol in water; nearly insol in alc; insol in ether.

SYNS: 2-AMINO-3-METHYLPENTANOIC ACID * α-AMINO-β-METHYLVALERIC ACID * l-ISOLEUCINE (FCC)

CONSENSUS REPORTS: Reported in EPA TSCA Inventory.

SAFETY PROFILE: Mildly toxic by intraperitoneal route. When heated to decomposition it emits toxic fumes of NO_x.

IKZ000   CAS: 466-40-0   **HR: 3**
ISOMETHADONE
mf: $C_{21}H_{27}NO$ mw: 309.49

SYNS: 6-DIMETHYLAMINO-5-METHYL-4,4-DIPHENYL-3-HEXANONE * 1,1-DIPHENYL-1-(DIMETHYLAMINO-ISOPROPYL)BUTANONE-2 * ISOAMIDONE II

SAFETY PROFILE: A poison by ingestion and intravenous routes. When heated to decomposition it emits toxic fumes of NO_x.

ILD000   CAS: 54-85-3   **HR: 3**
ISONICOTINIC ACID HYDRAZIDE
mf: $C_6H_7N_3O$ mw: 137.16

PROP: Consists of 12% w/v each of dodecylamine hydrochloride, trimethyl alkyl ammonium chloride, and methyl alkyl dipolyoxypropylene ammonium methyl sulfate.

SYNS: AMIDON * ANDRAZIDE * ANTIMICINA * ANTITUBERKULOSUM * ARMACIDE * ATCOTIBINE * AZUREN * BACILLIN * CEDIN * CEMIDON * CHEMIAZID * CHEMIDON * CORTINAZINE * COTINAZIN * COTINIZIN * DEFONIN * DIBUTIN * DIFORIN * DINACRIN * DITUBIN * EBIDENE * ERALON * ERTUBAN * EUTIZON * EVALON * FIMALENE * HIDRANIZIL * HIDRASONIL * HIDRULTA * HIDRUN * HYCOZID * HYDRAZID * HYDRAZIDE * HYOZID * HYZYD * IDRAZIDE DELL'ACIDO ISONICOTINICO * IDRAZIL * ISCOTIN * ISIDRINA * ISMAZIDE * ISOBICINA * ISOCID * ISOCIDENE * ISOCOTIN * ISOLYN * ISONERIT * ISONEX * ISONIACID * ISONIAZID * ISONIAZIDE * ISONICAZIDE * ISONICID * ISONICO * ISONICOTAN * ISONICOTIL * ISONICOTINHYDRAZID * ISONICOTINOYL HYDRAZIDE * ISONICOTINOYLHYDRAZINE * ISONICOTINSAEUREHYDRAZID * ISONICOTINYL HYDRAZIDE * ISONIDE * ISONIDRIN * ISONIKAZID

* ISONILEX * ISONIN * ISONINDON * ISONIRIT * ISONITON * ISONIZIDE * ISOTEBE * ISOTEBEZID * ISOTINYL * ISOZIDE * ISOZYD * LANIAZID * LANIOZID * MYBASAN * NEOTEBEN * NEOXIN * NEUMANDIN * NEVIN * NIADRIN * NICAZIDE * NICETAL * NICIZINA * NICONYL * NICOTIBINA * NICOTIBINE * NICOZIDE * NIDATON * NIDRAZID * NIKOZID * NIPLEN * NITADON * NITEBAN * NSC 9659 * NYDRAZID * NYSCOZID * PELAZID * PERCIN * PHTHISEN * PYCAZIDE * PYREAZID * PYRICIDIN * PYRIDICIN * 4-PYRIDINECARBOXYLIC ACID, HYDRAZIDE * PYRIZIDIN * RAUMANON * RAZIDE * RETOZIDE * RIFAMATE * RIMICID * RIMIFON * RIMITSID * ROBISELIN * ROBISELLIN * ROXIFEN * SANOHIDRAZINA * SAUTERAZID * SAUTERZID * STANOZIDE * TEBECID * TEBENIC * TEBEXIN * TEBOS * TEEBACONIN * TEKAZIN * TIBAZIDE * TIBEMID * TIBINIDE * TIBISON * TIBIVIS * TIBIZIDE * TIBUSAN * TISIN * TISIODRAZIDA * TIZIDE * TUBAZID * TUBAZIDE * TUBECO * TUBERCID * TUBERIAN * TUBICON * TUBOMEL * TYVID * UNICOCYDE * USAF CB-2 * VAZADRINE * VEDERON * ZINADON * ZONAZIDE

CONSENSUS REPORTS: IARC Cancer Review: GROUP 3 IMEMDT 7,227,87, Animal Sufficient Evidence IMEMDT 4,159,74. EPA Genetic Toxicology Program.

SAFETY PROFILE: A human poison by ingestion. An experimental poison by ingestion, intravenous, subcutaneous, intraperitoneal, and intramuscular routes. Experimental teratogenic and reproductive effects. Human systemic effects by ingestion: peripheral nerve sensory changes, somnolence, respiratory depression, anorexia, sweating, urine changes, toxic psychosis, hepatitis, dermatitis. Human mutation data reported. A skin irritant. Questionable carcinogen with experimental carcinogenic, neoplastigenic, and tumorigenic data. Used as an antitubercular, antibacterial, and anti-actinomycotic agent. When heated to decomposition it emits toxic fumes of NO_x and NH_3.

ILE000   CAS: 54-92-2   **HR: 3**
ISONICOTINIC ACID-2-ISOPROPYLHYDRAZIDE
mf: $C_9H_{13}N_3O$ mw: 179.25

SYNS: EUPHOZID * FOSFAZIDE * IIH * IPN * IPRAZID * IPRONIAZID * IPRONID

* IPRONIN * 1-ISONICOTINOYL-2-ISOPROPYLHY-
DRAZINE * 1-ISONICOTINYL-2-ISOPROPYLHYDRA-
ZINE * N-ISOPROPYL ISONICOTINHYDRAZIDE
* LH * MARSALID * MARSILID * P 887
* RIVIVOL * RO 2-4572 * YATROZIDE

SAFETY PROFILE: A human poison by ingestion. An experimental poison by ingestion, intraperitoneal, intravenous, and possibly other
routes. Moderately toxic by skin contact, intramuscular, and subcutaneous routes. Human systemic effects by ingestion: constipation, anuria,
metabolic changes, change in liver function.
Human reproductive effects by ingestion: impotence. Experimental teratogenic and reproductive effects. Questionable carcinogen with experimental tumorigenic data. Mutation data reported. Used as an antidepressant. When heated
to decomposition it emits toxic fumes of NO_x.

ILK000   CAS: 503-01-5   **HR: 3**
ISONYL
mf: $C_9H_{19}N$ mw: 141.29

PROP: Colorless, oily liquid; characteristic
amine odor, water-insol.

SYNS: ISOMETHEPTENE * 6-METHYLAMINO-2-
METHYLHEPTENE * METHYLISOOCTENYLAMINE
* 2-METHYL-6-METHYLAMINO-2-HEPTENE
* METHYLOCTENYLAMINE * OCTANIL
* OCTIN * OCTINUM * OCTON * N-1,5-TRI-
METHYL-4-HEXENYLAMINE

CONSENSUS REPORTS: Reported in EPA
TSCA Inventory.

SAFETY PROFILE: A poison by ingestion, intravenous, intraperitoneal, and subcutaneous
routes. Can cause headache, nausea, and dizziness in humans. When heated to decomposition
it emits toxic fumes of NO_x.

ILL000   CAS: 26952-21-6   **HR: 2**
ISOOCTYL ALCOHOL
mf: $C_8H_{18}O$ mw: 130.26

SYN: ISOOCTANOL

CONSENSUS REPORTS: Reported in EPA
TSCA Inventory.

OSHA PEL: TWA 50 ppm (skin)
ACGIH TLV: TWA 50 ppm (skin)

SAFETY PROFILE: Moderately toxic by ingestion and skin contact. A skin and severe eye
irritant. When heated to decomposition it emits
acrid smoke and fumes.

ILM000   CAS: 543-82-8   **HR: 3**
2-ISOOCTYL AMINE
mf: $C_8H_{19}N$ mw: 129.28

PROP: dl-Form: Viscous liquid, fishy odor. Bp:
154-156°, n (24/D) 1.4200.

SYNS: AMIDRINE * 2-AMINO-6-METHYLHEPTANE
* 6-AMINO-2-METHYLHEPTANE * α,ε-DIMETHYL-
HEXYLAMINE * 1,5-DIMETHYLHEXYLAMINE
* 2-METHYL-6-AMINOHEPTANE * 2-METHYL-2-
HEPTYLAMINE * 6-METHYL-2-HEPTYLAMINE
* 2-METIL-6-AMINO-EPTANO (ITALIAN) * OCTO-
DRINE * SKF 51 * VAPORPAC

CONSENSUS REPORTS: Reported in EPA
TSCA Inventory.

SAFETY PROFILE: A poison by intraperitoneal
and intramuscular routes. Moderately toxic by
ingestion and subcutaneous routes. When heated
to decomposition it emits toxic fumes of NO_x.

ILO000   CAS: 25168-26-7   **HR: 3**
**ISOOCTYL-2,4-DICHLOROPHENOXY-
ACETATE**
mf: $C_{16}H_{22}Cl_2O_3$ mw: 333.28

SYNS: 2,4-DICHLOROPHENOXYACETIC ACID ISOOC-
TYL ESTER * 2,4-D ISOOCTYL ESTER * ISOOCTYL
ALCOHOL (2,4-DICHLOROPHENOXY)ACETATE
* REED LV 2,4-D * REED LV 400 2,4-D * REED
LV 600 2,4-D * WEEDTRINE-II

CONSENSUS REPORTS: IARC Cancer Review: Animal Inadequate Evidence IMEMDT
15,111,77.

SAFETY PROFILE: A poison by ingestion. Experimental teratogenic and reproductive effects.
Questionable carcinogen with experimental carcinogenic and tumorigenic data. Human mutation data reported. An herbicide. When heated
to decomposition it emits toxic fumes of Cl^-.

ILR100   CAS: 27554-26-3   **HR: 2**
ISOOCTYL PHTHALATE
mf: $C_{24}H_{38}O_4$ mw: 390.62

SYNS: 1,2-BENZENEDICARBOXYLIC ACID, DIISOOC-
TYL ESTER * BIS(6-METHYLHEPTYL)ESTER of
PHTHALIC ACID * CORFLEX 880 * DIISOOCTYL
PHTHALATE * FLEXOL PLASTICIZER DIP * HEXA-
PLAS M/O

CONSENSUS REPORTS: Reported in EPA
TSCA Inventory.

SAFETY PROFILE: Moderately toxic by ingestion. Mildly toxic by skin contact. A skin irri-

tant. When heated to decomposition it emits acrid smoke and irritating fumes.

IMB000 CAS: 110-46-3 **HR: 3**
ISOPENTYL NITRITE
mf: $C_5H_{11}NO_2$ mw: 117.17

PROP: Transparent, flammable liquid; penetrating, fragrant odor. Unstable; decomp on exposure to air and light. D: 0.872 @ 20°/4°, bp: 97-99°, autoign temp: 408°F, vap d: 4.0, flash p: < 73.4°F.

SYNS: ISOAMYL NITRITE * ISOPENTYL ALCO-HOL NITRITE * 3-METHYLBUTANOL NITRITE * 3-METHYLBUTYL NITRITE * NITROUS ACID-3-METHYL BUTYL ESTER * VAPOROLE

CONSENSUS REPORTS: Cyanide and its compounds are on the Community Right-To-Know List. Reported in EPA TSCA Inventory.

SAFETY PROFILE: Poison by intravenous and intraperitoneal routes. Moderately toxic by ingestion. Mildly toxic by inhalation. Mutation data reported. A recreational drug said to enhance sexual enjoyment in humans by inhalation. Dangerous fire hazard when exposed to spark, heat, oxidizers or flame. Forms an explosive mixture in air or O_2. Vapors will explode when heated. When heated to decomposition it emits toxic fumes of NO_x.

IME000 CAS: 87-20-7 **HR: 2**
ISOPENTYL SALICYLATE
mf: $C_{12}H_{16}O_3$ mw: 208.28

PROP: Coorless liquid; pleasant odor. D: 1.047, refr index: 1.503-1.509, flash p: 271°F. Misc with alc, chloroform, ether, fixed oils; insol in glycerin, propylene glycol, water.

SYNS: FEMA No. 2084 * ISOAMYL o-HYDROXYBEN-ZOATE * ISOAMYL SALICYLATE (FCC) * ISOPEN-TYL-2-HYDROXYPHENYL METHANOATE * 3-METH-YLBUTYL 2-HYDROXYBENZOATE * SALICYLIC ACID, ISOPENTYL ESTER

CONSENSUS REPORTS: Reported in EPA TSCA Inventory.

SAFETY PROFILE: Moderately toxic by intravenous route. Experimental reproductive effects. Combustible liquid. When heated to decomposition it emits acrid smoke and fumes.

IMF300 CAS: 25311-71-1 **HR: 3**
ISOPHENPHOS
mf: $C_{15}H_{24}NO_4PS$ mw: 345.43

PROP: Oil. Bp: 120°, d: (20°/4°) 1.13. Vap press at 20°: 0.000004 mm Hg. Solubility in water at 20°: 23.8 mg/kg. Sol in dichloromethane, cyclohexanone, acetone, alc, ether, benzene.

SYNS: 2-(O-AETHYL-N-ISOPROPYLAMINDOTHIO-PHOSPHORYLOXY)-BENZOSAEURE-ISOPROPYLESTER (GERMAN) * AMAZE * BAY-92114 * BAY-SRA-12869 * 2-((ETHOXY((1-METHYLETHYL)AMINO) PHOSPHINOTHIOYL)OXY)BENZOIC ACID 1-METHYL-ETHYL ESTER * O-ETHYL-O-(2-ISOPROPOXY-CAR-BONYL)-PHENYL ISOPROPYLPHOSPHORAMIDO-THIOATE * ISOFENPHOS * ISOPROPYL-PHOS-PHORAMIDOTHIOIC ACID O-ETHYL O-(2-ISOPROPOXY-CARBONYLPHENYL) ESTER * ISOPROPYL SALICY-LATE O-ESTER with O-ETHYLISOPROPYLPHOSPHOR-AMIDOTHIOATE * 1-METHYLETHYL-2-((ETHOXY-((1-METHYLETHYL)AMINO)PHOSPHINOTHIOYL)OXY) BENZOATE * OFTANOL * 40 SD * SALICYLIC ACID ISOPROPYL ESTER O-ESTER with O-ETHYL ISOPROPYLPHOSPHORAMIDOTHIOATE * SRA 12869

SAFETY PROFILE: Poison by ingestion and skin contact. When heated to decomposition it emits toxic fumes of PO_x, SO_x, and NO_x.

IMF400 CAS: 78-59-1 **HR: 3**
ISOPHORONE
mf: $C_9H_{14}O$ mw: 138.23

PROP: Practically water-white liquid. Bp: 215.2°, flash p: 184°F (OC), d: 0.9229, autoign temp: 864°F, vap press: 1 mm @ 38.0°, vap d: 4.77, lel: 0.8%, uel: 3.8%.

SYNS: ISOACETOPHORONE * ISOFORON * ISOFORONE (ITALIAN) * IZOFORON (POLISH) * NCI-C55618 * 1,1,3-TRIMETHYL-3-CYCLOHEXENE-5-ONE * 3,5,5-TRIMETHYL-2-CYCLOHEXENE-1-ONE * 3,5,5-TRIMETHYL-2-CYCLOHEXEN-1-ON (GERMAN, DUTCH) * 3,5,5-TRIMETIL-2-CICLOESEN-1-ONE (ITAL-IAN)

CONSENSUS REPORTS: NTP Carcinogenesis Studies (gavage); Some Evidence: rat NTPTR* NTP-TR-291,86; (gavage); Equivocal Evidence: mouse NTPTR* NTP-TR-291,86. Reported in EPA TSCA Inventory.

OSHA PEL: (Transitional: TWA 25 ppm) TWA 4 ppm
ACGIH TLV: CL 5 ppm
DFG MAK: 5 ppm (28 mg/m³)
NIOSH REL: TWA (Ketones) 23 mg/m³

SAFETY PROFILE: Moderately toxic by ingestion and skin contact. Mildly toxic by inhalation. Human systemic effects by inhalation: olfactory changes, conjunctiva irritation, and respiratory changes. Human systemic irritant by inhalation. A skin and severe eye irritant. Questionable carcinogen with experimental carcinogenic data. Mutation data reported. Considered to be more toxic than mesityl oxide. However, due to its low volatility, it is not a dangerous industrial hazard. The response of guinea pigs and rats to repeated inhalation of the vapors indicates that it is one of the most toxic of the ketones. It is chiefly a kidney poison. It can cause irritation, lachrimation, possible opacity of the cornea, and necrosis of the cornea (experimental). It is irritating at the level of 25 ppm to humans. In animal experiments death during exposure was usually due to narcosis, but occasionally due to irritation of the lungs.

Flammable and explosive when exposed to heat or flame; can react with oxidizing materials. To fight fire, use foam, CO_2, dry chemical.

IMG000   CAS: 4098-71-9   HR: 3
ISOPHORONE DIISOCYANATE

DOT: UN 2290
mf: $C_{12}H_{18}N_2O_2$ mw: 222.32

SYNS: 3-ISOCYANATOMETHYL-3,5,5-TRIMETHYLCY-CLOHEXYLISOCYANATE * ISOPHORONE DIAMINE DI-ISOCYANATE

CONSENSUS REPORTS: EPA Extremely Hazardous Substances List. Reported in EPA TSCA Inventory.

OSHA PEL: TWA 0.005 ppm (skin)
ACGIH TLV: TWA 0.005 ppm (skin)
DFG MAK: 0.01 ppm (0.09 mg/m³)

SAFETY PROFILE: Poison by inhalation. Moderately toxic by skin contact. When heated to decomposition it emits toxic fumes of NO_x.

IMH000   CAS: 3778-73-2   HR: 3
ISOPHOSPHAMIDE
mf: $C_7H_{15}Cl_2N_2O_2P$ mw: 261.11

SYNS: A 4942 * ASTA Z 4942 * N,N-BIS(β-CHLO-ROETHYL)-AMINO-N′-O-PROPYLENE-PHOSPHORIC ACID ESTER DIAMIDE * 2,3-(N,N(₁)-BIS(2-CHLOROETHYL) DIAMIDO-1,3,2-OXAZAPHOSPHORIDINOXY * N,3-BIS(2-CHLOROETHYL)TETRAHYDRO-2H-1,3,2-OXAZA-PHOSPHORIN-2-AMINE 2-OXIDE * N-(2-CHLORA-ETHYL)-N′-(2 CHLOROETHYL)-N′-o-PROPYLEN-

PHOSPHORSAUREESTER-DIAMID (GERMAN) * 3-(2-CHLOROETHYL)-2-((2-CHLOROETHYL)AMINO) PERHYDRO-2H-1,3,2-OXAZAPHOSPHORINE OXIDE * 3-(2-CHLOROETHYL)-2-((2-CHLOROETHYL)AMINO) TETRAHYDRO-2H-1,3,2-OXAZAPHOSPHORINE-2-OXIDE * N-(2-CHLOROETHYL)-N′-(2-CHLOROETHYL)-N′,O-PROPYLENEPHOSPHORIC ACID DIAMIDE * N-(2-CHLOROETHYL)-N′-(2-CHLOROETHYL)-N′,O-PROPYL-ENEPHOSPHORIC ACID ESTER DIAMIDE * CYFOS * HOLOXAN * IFOSFAMID * IFOSFAMIDE * IPHOSPHAMIDE * ISOENDOXAN * ISOFOSFA-MIDE * MITOXANA * MJF 9325 * NAXAMIDE * NCI-C01638 * NSC-109724 * Z 4942

CONSENSUS REPORTS: IARC Cancer Review: GROUP 3 IMEMDT 7,56,87; Animal Limited Evidence IMEMDT 26,237,81. NCI Carcinogenesis Bioassay (ipr); Clear Evidence: mouse, rat NCITR* NCI-CG-TR-32,77. EPA Genetic Toxicology Program.

SAFETY PROFILE: Suspected carcinogen with experimental carcinogenic and neoplastigenic data. A poison by ingestion, intraperitoneal, subcutaneous, intravenous, and possibly other routes. Human systemic effects by ingestion and intravenous routes: nausea or vomiting; proteinuria, hematuria, inflammation, necrosis or scarring of the bladder, and other kidney, ureter, or bladder changes; changes in hair covering the skin; leukopenia (decreased white blood cell count), thrombocytopenia (decrease in the number of blood platelets); hallucinations, distorted perceptions; tumorigenic effects (active as an anti-cancer agent) data. Experimental teratogenic and reproductive effects. Human mutation data reported. When heated to decomposition it emits very toxic Cl^-, NO_x, and PO_x.

IMQ000   CAS: 76-00-6   HR: 3
ISOPRAL
mf: $C_3H_5Cl_3O$ mw: 163.43

PROP: Crystals, camphor-like odor, pungent taste, water-sol. Mp: 50°, bp: 162°.

SYNS: ISOPRAL * TRICHLOROISOPROPANOL * 1,1,1-TRICHLOROISOPROPYL ALCOHOL * 1,1,1-TRICHLORO-2-PROPANOL

SAFETY PROFILE: Poison by unspecified routes. Moderately toxic by ingestion.

IMS000   CAS: 78-79-5   HR: 3
ISOPRENE

DOT: UN 1218
mf: C_5H_8 mw: 68.13

PROP: Colorless, volatile liquid. Mp: −146.7°, bp: 34°, flash p: −65°F, d: 0.6806 @ 20°/4°, autoign temp: 428°F, vap press: 400 mm @ 15.4°, vap d: 2.35; fp: −145.95°. Insol in water; misc in alc and ether.

SYNS: ISOPRENE, INHIBITED (DOT) * β-METHYLBI-VINYL * 2-METHYLBUTADIENE * 2-METHYL-1,3-BUTADIENE (DOT)

CONSENSUS REPORTS: Reported in EPA TSCA Inventory.

DOT Classification: Flammable Liquid; Label: Flammable Liquid.

SAFETY PROFILE: Mildly toxic by inhalation. Irritating to skin, eyes, and mucous membranes. A concentration of 2% in air is not narcotic to mice but produces bronchial irritation. Highly dangerous fire hazard when exposed to heat, flame, or oxidizers. Reacts with air to form dangerously unstable peroxides which can explode after concentration by evaporation. Ignites on contact with oxygen + ozone. Reacts with ozone to form explosive peroxides. Explosive reaction with vinylamine. Violent reaction with chlorosulfonic acid, HNO_3, oleum, H_2SO_4. Can react vigorously with reducing materials. To fight fire, use CO_2, dry chemical. When heated to decomposition it emits acrid smoke and fumes.

IMU000 CAS: 75-33-2 **HR: 3**
ISOPROPANETHIOL
DOT: UN 2402/UN 2703
mf: C_3H_8S mw: 76.17

PROP: Liquid, extremely powerful unpleasant odor. Mp: −130.7°, bp: 58-60°, d: 0.814 @ 60°/60°F, boiling range: 51-55°, flash p: −30°F. Sltly sol in water; misc in alc and ether.

SYNS: ISOPROPYL MERCAPTAN (DOT) * ISOPRO-PYLTHIOL * 2-MERCAPTOPROPANE * 1-METHYL-ETHANETHIOL * 2-PROPANETHIOL * 2-PROPYL MERCAPTAN

CONSENSUS REPORTS: Reported in EPA TSCA Inventory.

DOT Classification: Flammable Liquid; Label: Flammable Liquid.

SAFETY PROFILE: Probably moderately toxic by inhalation. A very dangerous fire hazard when exposed to heat, flame, or oxidizers. When heated to decomposition it emits highly toxic fumes of SO_x.

IMW000 CAS: 513-42-8 **HR: 2**
ISOPROPENYL CARBINOL
DOT: UN 2614
mf: C_4H_8O mw: 72.12

PROP: Liquid. D: 0.852 @ 20°/4°C, bp: 114.5°, sol in water @ 25°.

SYNS: METHALLYL ALCOHOL (DOT) * 2-METHYL-2-PROPEN-1-OL

CONSENSUS REPORTS: Reported in EPA TSCA Inventory.

DOT Classification: Flammable or Combustible Liquid; Label: Flammable Liquid.

SAFETY PROFILE: Moderately toxic by ingestion and skin contact. Mildly toxic by inhalation. A skin irritant. Flammable when exposed to heat or flame. When heated to decomposition it emits acrid smoke and fumes.

INA500 CAS: 109-59-1 **HR: 2**
2-ISOPROPOXYETHANOL
mf: $C_5H_{12}O_2$ mw: 104.17

SYNS: DOWANOL EIPAT * ETHYLENE GLYCOL ISOPROPYL ETHER * ETHYLENE GLYCOL, MONOISO-PROPYL ETHER * β-HYDROXYETHYL ISOPROPYL ETHER * ISOPROPYL CELLOSOLVE * ISOPROPYL GLYCOL * MONOISOPROPYL ETHER of ETHYLENE GLYCOL

CONSENSUS REPORTS: Glycol ether compounds are on the Community Right-To-Know List. Reported in EPA TSCA Inventory.

OSHA PEL: TWA 25 ppm
ACGIH TLV: TWA 25 ppm

SAFETY PROFILE: Moderately toxic by skin contact and intraperitoneal routes. Mildly toxic by inhalation and ingestion. When heated to decomposition it emits acrid smoke and fumes.

INE100 CAS: 108-21-4 **HR: 2**
ISOPROPYL ACETATE
DOT: UN 1220
mf: $C_5H_{10}O_2$ mw: 102.15

PROP: Colorless, aromatic liquid. Mp: −73°, bp: 88.4°, lel: 1.8%, uel: 7.8%, fp: −69.3°, flash p: 40°F, d: 0.874 @ 20°/20°, autoign temp: 860°F, vap press: 40 mm @ 17.0°, d: 3.52. Sltly sol in water; misc in alc, ether, fixed oils.

SYNS: ACETATE d'ISOPROPYLE (FRENCH) * ACE-TIC ACID ISOPROPYL ESTER * ACETIC ACID-1-

METHYLETHYL ESTER (9CI) * 2-ACETOXYPROPANE * FEMA No. 2926 * ISOPROPILE (ACETATO di) (ITALIAN) * ISOPROPYLACETAAT (DUTCH) * ISOPROPYLACETAT (GERMAN) * ISOPROPYL (ACETATE d') (FRENCH) * 2-PROPYL ACETATE

CONSENSUS REPORTS: Reported in EPA TSCA Inventory.

OSHA PEL: (Transitional: TWA 250 ppm) TWA 250 ppm; STEL 310 ppm
ACGIH TLV: TWA 250 ppm; STEL 310 ppm
DFG MAK: 200 ppm (840 mg/m^3)
DOT Classification: Flammable Liquid; Label: Flammable Liquid.

SAFETY PROFILE: Moderately toxic by ingestion. Mildly toxic by inhalation. Human systemic irritant effects by inhalation and systemic eye effects by an unspecified route. Narcotic in high concentration. Chronic exposure can cause liver damage. Highly flammable liquid. Dangerous fire hazard when exposed to heat, flame, or oxidizers. Moderately explosive when exposed to heat or flame. Dangerous; keep away from heat and open flame; can react vigorously with oxidizing materials. To fight fire, use foam, CO$_2$, dry chemical.

INJ000   CAS: 67-63-0   HR: 3
ISOPROPYL ALCOHOL
DOT: UN 1219
mf: C$_3$H$_8$O   mw: 60.11

PROP: Clear, colorless liquid; slt odor, sltly bitter taste. Mp: −88.5 to −89.5°, bp: 82.5°, lel: 2.5%, uel: 12%, flash p: 53°F (CC), d: 0.7854 @ 20°/4°, refr index: 1.377 @ 20°, vap d: 2.07, ULC: 70. fp: −89.5°; autoign temp: 852°F. Misc with water, alc, ether, chloroform; insol in salt solns.

SYNS: ALCOOL ISOPROPILICO (ITALIAN) * ALCOOL ISOPROPYLIQUE (FRENCH) * DIMETHYLCARBINOL * ISOHOL * ISOPROPANOL (DOT) * ISOPROPYLALKOHOL (GERMAN) * LUTOSOL * PETROHOL * PROPAN-2-OL * 2-PROPANOL * i-PROPANOL (GERMAN) * sec-PROPYL ALCOHOL (DOT) * i-PROPYLALKOHOL (GERMAN) * SPECTRAR

CONSENSUS REPORTS: IARC Cancer Review: GROUP 3 IMEMDT 7,229,87. The isopropyl alcohol strong acid manufacturing process is on the Community Right-To-Know List. EPA Genetic Toxicology Program. Reported in EPA TSCA Inventory.

OSHA PEL: (Transitional: TWA 400 ppm) TWA 400 ppm; STEL 500 ppm
ACGIH TLV: TWA 400 ppm; STEL 500 ppm
DFG MAK: 400 ppm (980 mg/m^3)
NIOSH REL: (Isopropyl Alcohol) TWA 400 ppm; CL 800 ppm/15M
DOT Classification: Flammable Liquid; Label: Flammable Liquid.

SAFETY PROFILE: Poison by ingestion and subcutaneous routes. Moderately toxic to humans by an unspecified route. Moderately toxic experimentally by intravenous and intraperitoneal routes. Mildly toxic by skin contact. Human systemic effects by ingestion or inhalation: flushing, pulse rate decrease, blood pressure lowering, anesthesia, narcosis, headache, dizziness, mental depression, hallucinations, distorted perceptions, dyspnea, respiratory depression, nausea or vomiting, coma. Experimental teratogenic and reproductive effects. Questionable carcinogen. Mutation data reported. An eye and skin irritant.

The single lethal dose for a human adult is about 250 mL although as little as 100 mL can be fatal. It can cause corneal burns and eye damage. Acts as a local respiratory irritant and in high concentration as a narcotic. It has good warning properties because it causes a mild irritation of the eyes, nose, and throat at a concentration level of 400 ppm. It may induce a mild narcosis, the effects of which are usually transient, and it is somewhat less toxic than the normal isomer, but twice as volatile.

There is some evidence that humans can acquire a slight tolerance to this material. It is absorbed by the skin, but single or repeated applications on the skin of rats, rabbits, dogs, or human beings induced no untoward effects. It acts very much like ethanol in regard to absorption, metabolism, and elimination but with a stronger narcotic action. Chronic injuries have been detected in animals. Workers producing isopropanol show an excess of sinus and laryngeal cancers. This may all or in part be due to the by-product, isopropyl oil. Humans have ingested up to 20 mL diluted with water and noticed only a sensation of heat and slight lowering of the blood pressure. There are, however, reports of serious illness from as little as 10 mL taken internally. A common air contaminant.

Flammable liquid. A very dangerous fire hazard when exposed to heat, flame, or oxidizers. Moderately explosive when exposed to heat or flame. Reacts with air to form dangerous peroxides. The presence of 2-butanone increases the reaction rate for peroxide formation. Hydrogen peroxide sharply reduces the autoignition temperature. Violent explosive reaction when heated with aluminum isopropoxide + crotonaldehyde + heat. Forms explosive mixtures with trinitromethane, hydrogen peroxide (similar in power and sensitivity to glyceryl nitrate). Reacts with barium perchlorate to form the highly explosive propyl perchlorate. Ignites on contact with dioxgenyl tetrafluoroborate, chromium trioxide, potassium tert-butoxide (after a delay). Reacts with oxygen to form dangerously unstable peroxides. Vigorous reaction with sodium dichromate + sulfuric acid, aluminum (after a delay period). Reacts violently with H_2 + Pd, nitroform, oleum, $COCl_2$, Al triisopropoxide, oxidants. Can react vigorously with oxidizing materials. To fight fire, use CO_2, dry chemical, alcohol foam. When heated to decomposition it emits acrid smoke and fumes.

INK000   CAS: 75-31-0   HR: 3
ISOPROPYLAMINE

DOT: UN 1221
mf: C_3H_9N mw: 59.13

PROP: Colorless liquid, amino odor. Mp: $-101.2°$, flash p: $-35°F$ (OC), d: 0.694 @ $15°/4°$, autoign temp: $756°F$, d: 2.03, bp: 33-34°, lel: 2.3%, uel: 10.4%; misc with H_2O, alc, ether.

SYNS: 2-AMINO-PROPAAN (DUTCH) * 2-AMINO-PROPAN (GERMAN) * 2-AMINOPROPANE * 2-AMINO-PROPANO (ITALIAN) * ISOPROPILAMINA (ITALIAN) * 1-METHYLETHYLAMINE * MONOISO-PROPYLAMINE * 2-PROPANAMINE * sec-PROPYL-AMINE * 2-PROPYLAMINE

CONSENSUS REPORTS: Reported in EPA TSCA Inventory.

OSHA PEL: (Transitional: TWA 5 ppm) TWA 5 ppm; STEL 10 ppm
ACGIH TLV: TWA 5 ppm; STEL 10 ppm
DFG MAK: 5 ppm (12 mg/m³)
DOT Classification: Flammable Liquid; Label: Flammable Liquid.

SAFETY PROFILE: Poison by skin contact. Moderately toxic by ingestion and possibly other routes. Mildly toxic by inhalation. A severe skin and eye irritant. Occasionally contact causes sensitization. Narcotic in high concentra-

tion. Very dangerous fire hazard and moderate explosion hazard when exposed to sparks, heat, flame or oxidizers. Can react vigorously with oxidizing materials. Reacts with perchloryl fluoride to form an explosive liquid. Incompatible with 1-chloro-1,3-epoxypropane. To fight fire, use alcohol foam, foam, CO_2, dry chemical. When heated to decomposition it emits toxic fumes of NO_x.

INT000   CAS: 51-02-5   HR: 3
α-((ISOPROPYLAMINO)METHYL) NAPHTHALENEMETHANOL, HYDROCHLORIDE
mf: $C_{15}H_{19}NO_2 \cdot ClH$ mw: 265.81

SYNS: ALDERLIN HYDROCHLORIDE * ICI 38174 * I.C.I. HYDROCHLORIDE * INETOL * 2-ISOPRO-PYLAMINO-1-(2-NAPHTHYL)ETHANOL HYDROCHLORIDE * α-(((1-METHYLETHYL)AMINO)METHYL)-2-NAPHTHAL-ENEMETHANOL, HYDROCHLORIDE * NAPHTHYLISO-PROTERENOL HYDROCHLORIDE * NETHALIDE HY-DROCHLORIDE * PRONETHALOL * PRONETHALOL HYDROCHLORIDE

CONSENSUS REPORTS: IARC Cancer Review: GROUP 3 IMEMDT 7,56,87; Animal Limited Evidence IMEMDT 13,227,77.

SAFETY PROFILE: A poison by intravenous and intraperitoneal routes. Moderately toxic by ingestion. Questionable carcinogen with experimental tumorigenic data. When heated to decomposition it emits very toxic fumes of NO_x and HCl.

INW000   CAS: 63710-43-0   HR: 3
9-((3-(ISOPROPYLAMINO)PROPYL) AMINO)-1-NITROACRIDINE DIHYDRO-CHLORIDE
mf: $C_{19}H_{22}N_4O_2 \cdot 2ClH$ mw: 411.37

SYNS: N-(1-METHYLETHYL)-N'-(1-NITRO-9-ACRIDI-NYL)-1,3-PROPANEDIAMINE DIHYDROCHLORIDE * 1-NITRO-9-(3-ISOPROPYLAMINOPROPYLAMINE)-ACRI-DINE DIHYDROCHLORIDE * 1-NITRO-9-(3-ISOPROPYL-AMINOPROPYLAMINO)-ACRIDINE DIHYDROCHLORIDE

CONSENSUS REPORTS: EPA Genetic Toxicology Program.

SAFETY PROFILE: A poison by ingestion and intravenous routes. Human mutation data reported. When heated to decomposition it emits very toxic fumes of NO_x and HCl.

INX000　　　CAS: 768-52-5　　**HR: 2**
N-ISOPROPYLANILINE
mf: $C_9H_{13}N$　　mw: 135.23

SYNS: o-AMINOISOPROPYLBENZENE　*　2-AMINO-ISOPROPYLBENZENE　*　o-CUMIDINE　*　o-ISO-PROPYLANILINE　*　2-ISOPROPYL ANILINE　*　2-(1-METHYLETHYL)BENZENAMINE

CONSENSUS REPORTS: Reported in EPA TSCA Inventory.

OSHA PEL: TWA 2 ppm (skin)
ACGIH TLV: TWA 2 ppm (skin)

SAFETY PROFILE: Moderately toxic by ingestion. When heated to decomposition it emits toxic fumes of NO_x.

INY000　　　CAS: 479-92-5　　**HR: 3**
4-ISOPROPYLANTIPYRINE
mf: $C_{14}H_{18}N_2O$　　mw: 230.34

SYNS: 1,2-DIHYDRO-1,5-DIMETHYL-4-((1-METHYL-ETHYL)AMINO)-2-PHENYL-3H-PYRAZOL-3-ONE
*　ISOPROPYLANTIPYRIN　*　ISOPROPYLANTIPYRINE
*　4-ISOPROPYL-2,3-DIMETHYL-1-PHENYL-3-PYRAZO-LIN-5-ONE　*　ISOPROPYLPHENAZONE　*　ISOPYRINE
*　LARODON　*　1-PHENYL-2,3-DIMETHYL-4-ISOPRO-PYL-3-PYRAZOLIN-5-ONE　*　1-PHENYL-2,3-DIMETHYL-4-ISOPROPYLPYRAZOL-5-ONE　*　PROPYPHENAZONE

CONSENSUS REPORTS: Reported in EPA TSCA Inventory.

SAFETY PROFILE: A poison by ingestion and intraperitoneal routes. When heated to decomposition it emits toxic fumes of NO_x.

INZ000　　　CAS: 63020-47-3　　**HR: 3**
5-ISOPROPYL-1:2-BENZANTHRACENE
mf: $C_{21}H_{18}$　　mw: 270.39

SYN: 8-ISOPROPYLBENZ(a)ANTHRACENE

SAFETY PROFILE: Questionable carcinogen with experimental tumorigenic data. When heated to decomposition it emits acrid smoke and fumes.

IOB000　　　CAS: 80-15-9　　**HR: 3**
ISOPROPYLBENZENE HYDROPEROXIDE
DOT: UN 2116
mf: $C_9H_{12}O_2$　　mw: 152.21

PROP: Bp: 153°, flash p: 175°F, d: 1.05. The hydroperoxide of cumene.

SYNS: CUMEENHYDROPEROXYDE (DUTCH)
*　CUMENE HYDROPEROXIDE (DOT)　*　CUMENE HY-DROPEROXIDE, TECHNICALLY PURE (DOT)　*　CU-MENT HYDROPEROXIDE　*　CUMENYL HYDROPEROX-IDE　*　CUMOLHYDROPEROXID (GERMAN)
*　CUMYL HYDROPEROXIDE　*　α-CUMYL HYDRO-PEROXIDE　*　CUMYL HYDROPEROXIDE, TECHNICAL PURE (DOT)　*　α,α-DIMETHYLBENZYL HYDROPEROX-IDE (MAK)　*　HYDROPEROXYDE de CUMENE (FRENCH)　*　HYDROPEROXYDE de CUMYLE (FRENCH)
*　IDROPEROSSIDO di CUMENE (ITALIAN)　*　IDRO-PEROSSIDO di CUMOLO (ITALIAN)　*　RCRA WASTE NUMBER U096

CONSENSUS REPORTS: Community Right-To-Know List. Reported in EPA TSCA Inventory. EPA Genetic Toxicology Program.

DFG MAK: Moderate Skin Effects.
DOT Classification: Organic Peroxide; Label: Organic Peroxide.

SAFETY PROFILE: A poison by ingestion and intraperitoneal routes. Moderately toxic by skin contact, inhalation and subcutaneous routes. Mutation data reported. A skin and eye irritant. Questionable carcinogen with experimental tumorigenic data. A strong oxidizing agent. Flammable when exposed to heat or flame; can react with reducing materials. Its use in industry has resulted in many explosions. Storage above 109°C may cause explosive decomposition. Potentially explosive reactions with acids or reductants. Violent or explosive reaction when heated with solutions of 1,2-dibromo-1,2-diisocyana-toethane polymers in benzene. Violent decomposition on contact with cobalt, copper, copper alloys, lead alloys, mineral acids. Vigorous exothermic reaction on contact with charcoal. When heated to decomposition it emits acrid smoke and fumes. To fight fire, use foam, CO_2, dry chemical.

IOI000　　　CAS: 5419-55-6　　**HR: 3**
ISOPROPYL BORATE
mf: $C_9H_{21}BO_3$　　mw: 188.11

PROP: Colorless liquid. Mp: −59°, bp: 141.0-142.4°, flash p: 82°F (TCC), d: 0.8138 @ 25°.

SYN: TRIISOPROPYL BORATE

CONSENSUS REPORTS: Reported in EPA TSCA Inventory.

DOT Classification: Flammable or Combustible Liquid; Label: Flammable Liquid

SAFETY PROFILE: A poison by intravenous route. Moderately toxic by ingestion. An eye irritant. Dangerous fire hazard when exposed to heat, flame or oxidizers. Can react vigorously with oxidizing materials. To fight fire, use foam, CO_2, dry chemical.

IOJ000 CAS: 1746-77-6 **HR: 3**
ISOPROPYL CARBAMATE
mf: $C_4H_9NO_2$ mw: 103.14

PROP: Prisms. Mp: 60-61°, bp: 200°C. Very sol in water, alc, and ether.

SYNS: CARBAMIC ACID, ISOPROPYL ESTER
* CARBAMIC ACID-1-METHYLETHYL ESTER

CONSENSUS REPORTS: Reported in EPA TSCA Inventory.

SAFETY PROFILE: Moderately toxic by subcutaneous route. Questionable carcinogen with experimental neoplastigenic data. Mutation data reported. When heated to decomposition it emits toxic fumes of NO_x.

IOL000 CAS: 108-23-6 **HR: 3**
ISOPROPYL CHLOROCARBONATE
mf: $C_4H_7ClO_2$ mw: 122.56

PROP: A clear, colorless, volatile liquid with a pungent, irritating odor. D: 1.078 @ 20°/4°, bp: 105°, flash p: 28°C (TOC), 20°C (TCC), fp: −80°, fire p: 40°C (TOC), autoign temp: >500°, vap d: 4.2 @ 20°, refr index: 1.3974 @ 20°, lel: 4%, uel: 15%, vap press: 72 mm @ 70°F, bulk d: 9.0 lbs/gal, % volatile: 100%. Sol in aromatic or aliphatic hydrocarbon solvents, ethyl ether, acetone, and chloroform. Insol in water and alc. Decomp slowly in cold water, faster in hot water. Misc in ether and benzene. A phosgene derivative.

SYNS: CARBONOCHLORIDE ACID-1-METHYL ESTER
* CHLOROFORMIC ACID ISOPROPYL ESTER
* ISOPROPYL CHLOROFORMATE * ISOPROPYL CHLOROMETHANOATE

CONSENSUS REPORTS: EPA Extremely Hazardous Substances List. Reported in EPA TSCA Inventory.

DOT Classification: Flammable Liquid; Label: Flammable Liquid.

SAFETY PROFILE: A poison by skin contact and ingestion. Moderately toxic by inhalation. Ingestion of even small amounts can be fatal. A skin and severe eye irritant. Inhalation of a small amount can cause immediate lachrimation, coughing, choking, and respiratory dis-

tress. Death may result from pulmonary edema which may not appear for several hours after exposure. A skin and severe eye irritant. A dangerous fire and moderate explosion hazard when exposed to heat, spark, or flame. Self-reactive. Iron salts may catalyze a potentially explosive thermal decomposition. Incompatible with water, iron, metal salts, acids, alkalies, amines, alcohols. Stable under refrigeration below 20°, but one reference (1973) reports that it has exploded while stored in a refrigerator. Present day formulations appear to be more stable. Temperatures above 20° can cause decomposition. When heated to decomposition it emits acrid smoke and fumes.

IOT000 CAS: 2275-18-5 **HR: 3**
ISOPROPYL DIETHYLDITHIOPHOS-PHORYLACETAMIDE
mf: $C_9H_{20}NO_3PS_2$ mw: 285.39

SYNS: AC 18682 * AMERICAN CYANAMID 18682
* O,O-DIETHYLDITHIOPHOSPHORYLACETIC ACID-N-MONOISOPROPYLAMIDE * O,O-DIETHYL-S-(N-ISO-PROPYLCARBAMOYLMETHYL) DITHIOPHOSPHATE
* O,O-DIETHYL-S-ISOPROPYLCARBAMOYLMETHYL PHOSPHORODITHIOATE * O,O-DIETHYL-S-(N-ISOPRO-PYLCARBAMOYLMETHYL) PHOSPHORODITHIOATE
* ENT 24,652 * FAC * FAC 20 * FOSTION
* N-ISOPROPYL-2-MERCAPTOACETAMIDE-S-ESTER with O,O-DIETHYL PHOSPHORODITHIOATE * L343
* N-MONOISOPROPYLAMIDE of O,O-DIETHYLDITHIO-PHOSPHORYLACETIC ACID * OLEOFAC * PHOS-PHORODITHIOIC ACID-O,O-DIETHYL ESTER-S-ESTER with N-ISOPROPYL-2-MERCAPTOACETAMIDE * PRO-THOATE * PROTOAT (HUNGARIAN) * TELEFOS
* TRIMETHOATE

CONSENSUS REPORTS: EPA Extremely Hazardous Substances List.

SAFETY PROFILE: A poison by ingestion, inhalation, skin contact, and possibly other routes. An insecticide. When heated to decomposition it emits very toxic NO_x, PO_x, and SO_x.

IOY000 CAS: 94-11-1 **HR: 3**
ISOPROPYL-2,4-D ESTER
mf: $C_{11}H_{12}Cl_2O_3$ mw: 263.13

SYNS: (2,4-DICHLOROPHENOXY)ACETIC ACID, ISO-PROPYL ESTER * (2-4-DICHLOROPHENOXY)ACETIC ACID-1-METHYLETHYL ESTER (9CI) * 2,4-D ISOPRO-PYL ESTER * ESTERON 44 * WEEDONE 128

CONSENSUS REPORTS: IARC Cancer Review: Animal Inadequate Evidence IMEMDT 15,111,77.

SAFETY PROFILE: Moderately toxic by ingestion. Experimental teratogenic and reproductive effects. Questionable carcinogen with experimental tumorigenic data. Used as a pesticide. When heated to decomposition it emits toxic fumes of Cl^-.

IOZ750   CAS: 108-20-3   HR: 2
ISOPROPYL ETHER
DOT: UN 1159
mf: $C_6H_{14}O$ mw: 102.20

PROP: Colorless liquid, ethereal odor, miscible in water. Mp: $-60°$, bp: 68.5°, lel: 1.4%, uel: 7.9%, flash p: $-18°F$ (CC), d: 0.719 @ 25°, autoign temp: 830°F, vap press: 150 mm @ 25°, vap d: 3.52.

SYNS: DIISOPROPYL ETHER * DIISOPROPYL OXIDE * ETHER ISOPROPYLIQUE (FRENCH) * 2-ISOPRO-POXYPROPANE * IZOPROPYLOWY ETER (POLISH)

OSHA PEL: TWA 500 ppm
ACGIH TLV: TWA 250 ppm; STEL 310 ppm
DFG MAK: 500 ppm (2100 mg/m³)
DOT Classification: Flammable Liquid; Label: Flammable Liquid.

SAFETY PROFILE: Moderately toxic by intraperitoneal and possibly other routes. Mildly toxic by ingestion, inhalation, and skin contact. A skin irritant. A very dangerous fire hazard and severe explosion hazard when exposed to heat, flame, sparks, or oxidizers. Under some conditions shock will explode it. Dangerous; on exposure to air it rapidly forms very sensitive, explosive peroxides which precipitate as crystals. Violent reaction with chlorosulfonic acid; HNO_3. Potentially dangerous reaction with propionyl chloride can burst a sealed container. Reacts vigorously with oxidizing materials. To fight fire, use alcohol foam, CO_2, foam, dry chemical. When heated to decomposition it emits acrid smoke and fumes.

IPC000   CAS: 625-55-8   HR: 3
ISOPROPYL FORMATE
DOT: UN 1281
mf: $C_4H_8O_2$ mw: 88.12

PROP: Clear liquid. Bp: 68.3°, flash p: 22°F (CC), d: 0.873, autoign temp: 905°F, vap press: 100 mm @ 17.8°, vap d: 3.03.

SYN: FORMIC ACID, ISOPROPYL ESTER

CONSENSUS REPORTS: EPA Extremely Hazardous Substances List. Reported in EPA TSCA Inventory.

DOT Classification: Flammable Liquid; Label: Flammable Liquid.

SAFETY PROFILE: A poison by ingestion. A toxic fumigant. A very dangerous fire hazard when exposed to heat, spark or flame. Can react vigorously with oxidizing materials. When heated to decomposition it emits acrid smoke and fumes. To fight fire, use alcohol foam, foam, CO_2, dry chemical.

IPD000   CAS: 4016-14-2   HR: 2
ISOPROPYL GLYCIDYL ETHER
mf: $C_6H_{12}O_2$ mw: 116.18

PROP: A liquid.

SYNS: IGE * NCI-C56439

CONSENSUS REPORTS: Glycol ether compounds are on the Community Right-To-Know List. Reported in EPA TSCA Inventory.

OSHA PEL: (Transitional: TWA 50 ppm) TWA 50 ppm; STEL 75 ppm
ACGIH TLV: TWA 50 ppm; STEL 75 ppm
DFG MAK: 50 ppm (240 mg/m³)
NIOSH REL: (Glycidyl Ethers) CL 240 mg/m³/15M

SAFETY PROFILE: Moderately toxic by ingestion. Mildly toxic by inhalation and skin contact. A skin and eye irritant. Mutation data reported. When heated to decomposition it emits acrid smoke and fumes.

IPG000   CAS: 24426-36-6   HR: 3
ISOPROPYL-S HYDROCHLORIDE
mf: $C_7H_{15}Cl_2N \cdot ClH$ mw: 220.59

SYNS: N,N-BIS(2-CHLOROETHYL)ISOPROPYLAMINE HYDROCHLORIDE * 2,2'-DICHLORO-N-ISOPROPYLDI-ETHYLAMINE HYDROCHLORIDE * 2,2'-DICHLORO-1''-METHYLTRIETHYLAMINE HYDROCHLORIDE * ISO-PROPYL BIS(β-CHLOROETHYL)AMINE HYDROCHLORIDE * TL 301 HYDROCHLORIDE

SAFETY PROFILE: A poison by ingestion, subcutaneous, intraperitoneal and intravenous routes. When heated to decomposition it emits very toxic fumes of HCl and NO_x.

IPU000   CAS: 78-44-4   HR: 3
ISOPROPYL MEPROBAMATE
mf: $C_{12}H_{24}N_2O_4$ mw: 260.38

SYNS: APESAN * ARUSAL * BRIANIL * CAPRODAT * CARBAMIC ACID, ESTER with 2-(HDYROXYMETHYL)-1-METHYLPENTYLISOPROPYLCAR-BAMATE * CARBAMIC ACID, ESTER with 2-METHYL-2-PROPYL-1,3-PROPANEDIOL ISOPROPYLCARBAMATE * CARISOL * CARISOMA * CARISOPRODATE * CARISOPRODATUM * CARISOPRODOL * CARLSODAL * CARLSOMA * CARSODOL * CB 8019 * DIOLENE * DOMARAX * FLEXAL * FLEXARTAL * FLEXARTEL * ISOBAMATE * ISOMEPROBAMATE * ISOPROPYLCARBAMIC ACID, ESTER with 2-(HYDROXYMETHYL)-2-METHYL-PENTYL CARBAMATE * N-ISOPROPYL-2-METHYL-2-PROPYL-1,3-PROPANEDIOL DICARBAMATE * MEDI-QUIL * (1-METHYLETHYL)CARBAMIC ACID 2-(((AMINOCARBONYL)OXY)METHYL)-2-METHYL-PENTYL ESTER * 2-METHYL-2-PROPYL-1,3-PRO-PANEDIOL CARBAMATE ISOPROPYLCARBAMATE * MIOARTRINA * MIOLISODAL * MIOLISODOL * MIORATRINA * MIORIL * MIORIODOL * NCI-C56235 * NOSPASM * RELA * RELASOM * RELAX * SANOMA * SCH 7307 * SOMA * SOMADRIL * SOMALGIT * SOMANIL * TONOLYT ISOPROPYL MEPROBA-MATE

SAFETY PROFILE: A poison by intravenous route. Moderately toxic by ingestion and intraperitoneal routes. A skeletal muscle relaxant. When heated to decomposition it emits toxic fumes of NO_x.

IPX000   CAS: 107-44-8   HR: 3
ISOPROPYL METHANEFLUORO-PHOSPHONATE
mf: $C_4H_{10}FO_2P$ mw: 140.11

PROP: Bp: 147°, fp: −58°, d: 1.100 @ 20° vap press: 1.57 mm @ 20°, vap d: 4.86.

SYNS: FLUOROISOPROPOXYMETHYLPHOSPINE OXIDE * GB * IMPF * ISOPROPHYL METHYLPHOSPHO-NOFLUORIDATE * ISOPROPOXYMETHYLPHORYL, FLUORIDE * ISOPROPYL METHYLFLUOROPHOS-PHATE * ISOPROPYL METHYLPHOSPHONOFLUORI-DATE * O-ISOPROPYL METHYLPHOSPHONOFLUORI-DATE * ISOPROPYL-METHYL-PHOSPHORYL FLUO-RIDE * METHYLFLUOROPHOSPHORIC ACID, ISOPROPYL ESTER * METHYLFLUORPHOSPHORSAE-UREISOPROPYLESTER (GERMAN) * METHYLPHOS-PHONOFLUORIDIC ACID ISOPROPYL ESTER * METHYLPHOSPHONOFLUORIDIC ACID-1-METHYL-ETHYL ESTER * MFI * SARIN * SARIN II * T-144 * T-2106 * TL 1618 * TRILONE 46

CONSENSUS REPORTS: EPA Extremely Hazardous Substances List. Reported in EPA TSCA Inventory.

SAFETY PROFILE: A deadly human poison by skin contact and inhalation. (A small drop on the skin can kill a man.) A deadly experimental poison by ingestion, inhalation, skin contact, subcutaneous, intravenous, intramuscular, and intraperitoneal routes. Human systemic effects by inhalation and ingestion: muscle weakness, bronchiolar constriction (including asthma), nausea or vomiting, flaccid paralysis without anesthesia, miosis (pupillary constriction), cholinesterase inhibition. A "nerve gas" used as a chemical warfare agent. To fight fire, use foam, CO_2, dry chemical. When heated to decomposition or reacted with steam, it emits very toxic fumes of F^- and PO_x.

IQE000   CAS: 52061-60-6   HR: 2
1-ISOPROPYL-4-METHYLCYCLOHANE HYDROPEROXIDE
DOT: UN 2125
mf: $C_{10}H_{20}O_2$ mw: 172.30

SYNS: HEXAHYDRO-p-CYMENE HYDROPEROXIDE * p-MENTHANE HYDROPEROXIDE, TECHNICALLY PURE (DOT) * PARAMENTHANE HYDROPEROXIDE (DOT)

DOT Classification: Organic Peroxide; Label: Organic Peroxide.

SAFETY PROFILE: A severe irritant to skin, eyes, and mucous membranes. A very powerful oxidizer. When heated to decomposition it emits acrid smoke and fumes.

IQN000   CAS: 110-27-0   HR: 3
ISOPROPYL MYRISTATE
mf: $C_{17}H_{34}O_2$ mw: 270.44

PROP: Liquid of low viscosity, odorless. Bp: 192.6° @ 20 mm, decomp @ 208°, d: 0.8532 @ 20°. Sol in castor oil, cottonseed oil, acetone, chloroform, ethyl acetate, ethanol, toluene, and mineral oil. Insol in water, glycerol, and propylene glycol. Dissolves many waxes.

SYNS: ISOMYST * KESSCOMIR * TETRADECA-NOIC ACID, ISOPROPYL

CONSENSUS REPORTS: Reported in EPA TSCA Inventory.

SAFETY PROFILE: A human skin irritant. When heated to decomposition it emits toxic smoke and fumes.

IQP000 CAS: 1712-64-7 HR: 3
ISOPROPYL NITRATE
DOT: UN 1222
mf: $C_3H_7NO_3$ mw: 105.11

PROP: Liquid. Bp: 102°, flash p: 51.8°F, uel: 100%.

SYN: NITRIC ACID, ISOPROPYL ESTER

CONSENSUS REPORTS: Reported in EPA TSCA Inventory.

DOT Classification: Flammable Liquid; Label: Flammable Liquid.

SAFETY PROFILE: Mildly toxic by inhalation. A dangerous fire hazard when exposed to heat, spark, or flames. Ignites spontaneously when compressed. The pure vapor ignites spontaneously at very low temperatures and pressures. An explosive of low sensitivity. It can be used as a rocket monopropellant. When heated to decomposition it emits toxic fumes of NO_x. Incompatible with Lewis acids.

IQQ000 CAS: 541-42-4 HR: 3
ISOPROPYL NITRITE
mf: $C_3H_7NO_2$ mw: 89.10

PROP: Flash p: < 50°F.

SYNS: 1-METHYL ETHYL ESTER NITROUS ACID (9CI) * NITROUS ACID, ISOPROPYL ESTER * PROPANOL NITRITE

CONSENSUS REPORTS: Reported in EPA TSCA Inventory.

SAFETY PROFILE: Moderately toxic by inhalation. Can cause vasodilation with fall in blood pressure, tachycardia, headache. Large doses can cause methemoglobinuria and cyanosis. Severe poisoning results in shock which can be fatal. A very dangerous fire hazard when exposed to heat, spark, or flame. When heated to decomposition it emits toxic fumes of NO_x.

IQU000 HR: 3
ISOPROPYL OILS
CONSENSUS REPORTS: IARC Cancer Review: Animal Inadequate Evidence IMEMDT 15,223,77; Human Limited Evidence IMEMDT 15,223,77.

DFG MAK: Suspected Carcinogen.

SAFETY PROFILE: Suspected carcinogen with experimental neoplastigenic data. When heated to decomposition they emit acrid smoke and fumes.

IQZ000 CAS: 99-89-8 HR: 3
p-ISOPROPYLPHENOL
mf: $C_9H_{12}O$ mw: 136.21

PROP: Needles. D: 0.990 @ 20°, mp: 61°, bp: 228.2-229.2°, very sltly sol in water, sol in alc @ 25°, sol in ether @ 25°.

SYNS: AUSTRALOL * p-CUMENOL * 4-ISOPROPYLPHENOL * 4-(1-METHYLETHYL)PHENOL * PRODOX 133

CONSENSUS REPORTS: Reported in EPA TSCA Inventory.

SAFETY PROFILE: A poison by intraperitoneal and intravenous routes. Moderately toxic by ingestion. When heated to decomposition it emits acrid smoke and fumes.

IRF000 CAS: 55-91-4 HR: 3
ISOPROPYL PHOSPHOROFLUORIDATE
mf: $C_6H_{14}FO_3P$ mw: 184.17

PROP: Oily liquid. Mp: −82°, bp: 46° @ 5 mm, d: 1.07 (approx), vap d: 5.24.

SYNS: DFP * DIFLUPYL * DIFLUROPHATE * DIISOPROPOXYPHOSPHORYL FLUORIDE * DIISOPROPYL FLUOROPHOSPHATE * O,O-DIISOPROPYL FLUOROPHOSPHATE * DIISOPROPYL FLUOROPHOSPHONATE * DIISOPROPYLFLUOROPHOSPHORIC ACID ESTER * DIISOPROPYLFLUORPHOSPHORSAEUREESTER (GERMAN) * DIISOPROPYL PHOSPHOFLUORIDATE * DIISOPROPYL PHOSPHOROFLUORIDATE * O,O'-DIISOPROPYL PHOSPHORYL FLUORIDE * DYFLOS * FLOROPRYL * FLUOPHOSPHORIC ACID, DIISOPROPYL ESTER * FLUORODIISOPROPYL PHOSPHATE * FLUOROPRYL * FLUOSTIGMINE * ISOFLUOROPHATE * ISOFLUROPHATE * ISOPROPYL FLUOPHOSPHATE * NEOGLAUCIT * PF-3 * PHOSPHOROFLUORIDIC ACID, DIISOPROPYL ESTER * RCRA WASTE NUMBER P043 * T-1703 * TL 466

CONSENSUS REPORTS: EPA Extremely Hazardous Substances List. Reported in EPA TSCA Inventory.

SAFETY PROFILE: A poison by ingestion, inhalation, skin contact, intraperitoneal, subcutaneous, intramuscular, ocular, and intravenous routes. Moderately toxic by parenteral route. Human systemic effects by inhalation: miosis (pupillary constriction) and headache. Experimental reproductive effects. Used as a basis for "nerve gases." An insecticide. Ingestion can cause damage to eyes, nausea, vomiting, diarrhea, and central nervous system disturbances. An FDA proprietary drug. Used as a miotic agent. When heated to decomposition it emits toxic fumes of F^- and PO_x.

IRU000 CAS: 114-45-4 **HR: 3**
(±)-ISOPROTERENOL SULFATE
mf: $C_{22}H_{34}N_2O_6 \cdot H_2O_4S$ mw: 520.66

PROP: (dl form): Crystals from (acetone + methanol). Mp: 128° (some decomp), sltly sol in alc. Insol in chloroform, ether, benzene. (l Form): Crystals. Mp: 164-165°.

SYNS: dl-α-3,4-DIHYDROXYPHENYL-β-ISOPROPYL-AMINOETHANOL SULFATE * dl-ISOPRENALINE SULFATE * (±)-ISOPRENALINE SULFATE * dl-ISO-PROTERENOL SULFATE

SAFETY PROFILE: A poison by ingestion, intravenous, intraperitoneal, and subcutaneous routes. Mildly toxic by inhalation. Mutation data reported. When heated to decomposition it emits very toxic fumes of NO_x and SO_x.

IRV000 CAS: 52-53-9 **HR: 3**
ISOPTIN
mf: $C_{27}H_{38}N_2O_4$ mw: 454.67

SYNS: CP-16533-1 * D-365 * DILACORAN * 5-((3,4-DIMETHOXYPHENETHYL)METHYLAMINO)-2-(3,4-DIMETHOXYPHENYL)-2-ISOPROPYLVALERONI-TRILE * IPROVERATRIL * α-(N-METHYL-N-HOMOVERATRYL)-γ-AMINOPROPYL)-3,4-DIMETHOXY-PHENYLACETONITRILE * VASOLAN * VERA-PAMIL

CONSENSUS REPORTS: Cyanide and its compounds are on the Community Right-To-Know List.

SAFETY PROFILE: A poison by ingestion, subcutaneous, intraperitoneal, and intravenous routes. Human systemic effects by ingestion: hepatitis, hepatocellular necrosis (diffuse), pulse rate increased, blood pressure lowering, change in cardiac rate, sweating, dyspnea, cyanosis. When heated to decomposition it emits toxic fumes of NO_x and CN^-.

IRX000 CAS: 119-65-3 **HR: 3**
ISOQUINOLINE
mf: C_9H_7N mw: 129.17

PROP: Liquid, pungent odor, almost insol in water, miscible with many organic solvents, sol in dilute acids. Hygroscopic platelets when solid. D: 1.09101 @ 30°/4°, mp: 26.48°, bp: 243°.

SYNS: 2-AZANAPHTHALENE * 2-BENZAZINE * BENZO(c)PYRIDINE * LEUCOLINE

CONSENSUS REPORTS: Reported in EPA TSCA Inventory. EPA Genetic Toxicology Program.

SAFETY PROFILE: A poison by ingestion and intraperitoneal routes. Moderately toxic by skin contact. A severe skin and eye irritant. When heated to decomposition it emits toxic fumes of NO_x.

IRY000 CAS: 94-86-0 **HR: 2**
ISOSAFROEUGENOL
mf: $C_{11}H_{14}O_2$ mw: 178.25

PROP: White crystalline powder; vanilla odor. Flash p: +212°F. Sol fixed oils; insol in water.

SYNS: 6-ETHOXY-m-ANOL * 1-ETHOXY-2-HY-DROXY-4-PROPENYLBENZENE * FEMA No. 2922 * HYDROXY METHYL ANETHOL * PROPENYLGUA-ETHOL (FCC)

CONSENSUS REPORTS: Reported in EPA TSCA Inventory.

SAFETY PROFILE: Moderately toxic by ingestion. Combustible liquid. When heated to decomposition it emits acrid smoke and fumes.

IRZ000 CAS: 120-58-1 **HR: 3**
ISOSAFROLE
mf: $C_{10}H_{10}O_2$ mw: 162.20

PROP: Liquid, odor of anise. Bp: 253°, mp: 8.2°.

SYNS: 1,2-METHYLENEDIOXY-4-PROPENYLBENZENE * 3,4-METHYLENEDIOXY-1-PROPENYL BENZENE * 5-(1-PROPENYL)-1,3-BENZODIOXOLE * 4-PROPE-NYLCATECHOL METHYLENE ETHER * 4-PROPENYL-1,2-METHYLENEDIOXYBENZENE * RCRA WASTE NUMBER U141

CONSENSUS REPORTS: IARC Cancer Review: GROUP 3 IMEMDT 7,56,87; Animal Sufficient Evidence IMEMDT 1,169,72. Reported in EPA TSCA Inventory. EPA Genetic Toxicology Program.

SAFETY PROFILE: Poison by intraperitoneal and intravenous routes. Moderately toxic by ingestion and subcutaneous routes. A skin irritant. Questionable carcinogen with experimental carcinogenic and tumorigenic data. Used as a pesticide. When heated to decomposition it emits acrid smoke and fumes.

ISA000   CAS: 120-62-7   **HR: 3**
ISOSAFROLE-n-OCTYLSULFOXIDE
mf: $C_{18}H_{28}O_3S$ mw: 324.52

PROP: Water-insol; sltly sol in petroleum oils; sol in most organic solvents.

SYNS: ENT 16,634 * ISOSAFROLE, OCTYL SULFOXIDE * 1,2-(METHYLENEDIOXY)-4-(2-(OCTYLSULFINYL)PROPYL)BENZENE * 1-METHYL-2-(3,4-METHYLENEDIOXYPHENYL)ETHYL OCTYL SULFOXIDE * NCI-C02824 * n-OCTYLISOSAFROLE SULFOXIDE * PIPERONYL SULFOXIDE * SULFOX-CIDE * SULFOXIDE * SULFOXYL * SULPHOXIDE

CONSENSUS REPORTS: NCI Carcinogenesis Bioassay (feed); No Evidence: rat NCITR* NCI-CG-TR-124,79. NCI Carcinogenesis Bioassay (feed); Clear Evidence: mouse NCITR* NCI-CG-TR-124,79.

SAFETY PROFILE: Moderately toxic by ingestion. Slightly toxic by skin contact. Questionable carcinogen with experimental carcinogenic, tumorigenic, and teratogenic data. Reacts violently with $HClO_4$. An insecticide. When heated to decomposition it emits highly toxic fumes of SO_x.

ISD000   CAS: 2496-92-6   **HR: 3**
ISO SYSTOX SULFOXIDE
mf: $C_8H_{19}O_4PS_2$ mw: 274.36

SYNS: O,O-DIETHYL-S-(2-ETHTHIONYLETHYL) PHOSPHOROTHIOATE * DIETHYL-S-(2-ETHTHIONYLETHYL) THIOPHOSPHATE * O,O-DIETHYL-S-ETHYL-2-ETHYLMERCAPTO PHOSPHOROTHIOLATE SULFOXIDE

SAFETY PROFILE: A poison by ingestion and intraperitoneal routes. When heated to decomposition it emits very toxic fumes of PO_x and SO_x.

ISE000   CAS: 556-61-6   **HR: 3**
ISOTHIOCYANATOMETHANE

DOT: UN 2477
mf: C_2H_3NS mw: 73.12

PROP: Crystalline. Bp: 119°, d: 1.069. Very sltly sol in water; misc in alc and ether.

SYNS: EP-161E * ISOTHIOCYANATE de METHYLE (FRENCH) * ISOTHIOCYANIC ACID, METHYL ESTER * ISOTIOCIANATO di METILE (ITALIAN) * METHYL-ISOTHIOCYANAAT (DUTCH) * METHYL-ISOTHIOCYANAT (GERMAN) * METHYL ISOTHIOCYANATE (DOT) * METHYL MUSTARD OIL * METHYLSENFOEL (GERMAN) * MIC * MIT * MITC * MORTON WP-161E * TRAPEX * TRAPEXIDE * VORLEX * VORTEX * WN 12

CONSENSUS REPORTS: EPA Extremely Hazardous Substances List. Reported in EPA TSCA Inventory.
DOT Classification: Flammable Liquid; Label: Flammable Liquid and Poison.

SAFETY PROFILE: A poison by ingestion, skin contact, and subcutaneous routes. Very irritating to skin, eyes, and mucous membranes. Human systemic effects by ingestion: convulsions or effects on seizure threshold, change in motor activity, coma. An agricultural chemical and pesticide. Flammable when exposed to heat or flame; can react vigorously with oxidizing materials. When heated to decomposition it emits very toxic fumes of NO_x and SO_x.

ISK000   CAS: 3688-08-2   **HR: 3**
ISOTHIOCYANIC ACID, ETHYLENE ESTER
mf: $C_4H_4N_2S_2$ mw: 144.22

SYNS: AETHYLEN-BIS-THIURAMMONOSULFID (GERMAN) * AETHYLSENFOEL (GERMAN) * AETM (GERMAN) * 1,2-DIISOTHIOCYANATOETHANE * DIMETHYLENE DIISOTHIOCYANATE * EBI * EBIS * ETHYLENEBISISOTHIOCYANATE * ETHYLENE-BIS-THIURAMMONO-SULFIDE * ETHYLENE DIISOTHIOCYANATE * SENFOL (GERMAN)

SAFETY PROFILE: A poison by ingestion and subcutaneous routes. Experimental teratogenic and reproductive effects. When heated to decomposition it emits very toxic fumes of NO_x, NH_3, and SO_x.

ISQ000   CAS: 103-72-0   **HR: 3**
ISOTHIOCYANIC ACID, PHENYL ESTER
mf: C_7H_5NS mw: 135.19

PROP: Pale yellow liquid. Mp: −21°, bp: 221°, d: 1.1282. Insol in water; sol in alc and ether.

SYNS: BENZENE-1-ISOTHIOCYANATE * ISOTHIO-
CYANATOBENZENE * PHENYL ISOTHIOCYANATE
* PHENYL MUSTARD OIL * PHENYLSENFOEL
(GERMAN) * PITC * THIOCARBANIL
* USAF M-4

CONSENSUS REPORTS: Reported in EPA
TSCA Inventory.

SAFETY PROFILE: A poison by ingestion, intraperitoneal, and subcutaneous routes. When
heated to decomposition, or on contact with
acid or acid fumes, it emits highly toxic fumes
of cyanides and SO_x.

ISR000   CAS: 62-56-6   **HR: 3**
ISOTHIOUREA
DOT: UN 2877
mf: CH_4N_2S mw: 76.13

PROP: White powder or crystals. Mp: 177°,
bp: decomp, d: 1.405.

SYNS: PSEUDOTHIOUREA * RCRA WASTE NUM-
BER U219 * SULOUREA * THIOCARBAMATE
* THIOCARBAMIDE * β-THIOPSEUDOUREA
* THIOUREA (DOT) * 2-THIOUREA * THU
* TSIZP 34 * USAF EK-497

CONSENSUS REPORTS: IARC Cancer Re-
view: GROUP 2B IMEMDT 7,56,87; Animal
Sufficient Evidence IMEMDT 7,95,74. NTP
Fourth Annual Report On Carcinogens, 1984.
EPA Genetic Toxicology Program. Reported
in EPA TSCA Inventory.

DOT Classification: Poison B; Label: St. Andrews Cross.

SAFETY PROFILE: Confirmed carcinogen with
experimental carcinogenic, neoplastigenic, and
tumorigenic data. A human poison by an unspecified route. An experimental poison by ingestion
and intraperitoneal routes. Human mutation data
reported. Human systemic effects by ingestion:
hemorrhage, granulocytopenia (reduction in
number of granulocytes), and changes in cell
count (unspecified). May cause depression of
bone marrow with anemia, leukopenia, and
thrombocytopenia. May also cause allergic skin
eruptions. Causes hepatic tumors upon chronic
administration. Experimental teratogenic and reproductive effects. May react violently with
acrolein. Incompatible with acrylaldehyde,
H_2O_2, HNO_3. When heated to decomposition it emits very toxic fumes of NO_x and
SO_x.

ISU000   CAS: 503-74-2   **HR: 3**
ISOVALERIC ACID
DOT: NA 1760
mf: $C_5H_{10}O_2$ mw: 102.15

PROP: Colorless liquid; acid taste, disagreeable
rancid-cheese odor. Solidifies @ −37°, d: 0.931
@ 20°/4°, refr index: 1.403, mp: −34.5°
(−50°), bp: 175-177°. Sol in water @ 16°; misc
in alc, chloroform, ether.

SYNS: DELPHINIC ACID * FEMA No. 3102
* ISOPENTANOIC ACID (DOT) * ISOPROPYLACETIC
ACID * ISOVALERIANIC AICD * 3-METHYLBU-
TANOIC ACID * β-METHYLBUTYRIC ACID
* 3-METHYLBUTYRIC ACID

CONSENSUS REPORTS: Reported in EPA
TSCA Inventory.

DOT Classification: Corrosive Material; Label:
Corrosive.

SAFETY PROFILE: A poison by skin contact.
Moderately toxic by ingestion and intravenous
routes. A corrosive skin and eye irritant. When
heated to decomposition it emits acrid smoke
and fumes.

ISV000   CAS: 2835-39-4   **HR: 3**
ISOVALERIC ACID, ALLYL ESTER
mf: $C_8H_{14}O_2$ mw: 142.22

SYNS: ALLYL ISOVALERATE * ALLYL ISOVALERI-
ANATE * ALLYL 3-METHYLBUTYRATE * FEMA
No. 2045 * 3-METHYLBUTANOIC ACID, 2-PROPENYL
ESTER * 3-METHYLBUTYRIC ACID, ALLYL ESTER
* NCI-C54717 * 2-PROPENYL ISOVALERATE
* 2-PROPENYL 3-METHYLBUTANOATE

CONSENSUS REPORTS: IARC Cancer Re-
view: GROUP 3 IMEMDT 7,56,87; Animal
Limited Evidence IMEMDT 36,69,85. NTP
Carcinogenesis Studies (gavage); Clear Evi-
dence: mouse, rat NTPTR* NTP-TR-253,83.
Reported in EPA TSCA Inventory.

SAFETY PROFILE: Suspected carcinogen with
experimental carcinogenic and tumorigenic
data. A poison by ingestion. Moderately toxic
by skin contact. A skin irritant. When heated
to decomposition it emits acrid smoke and
fumes.

ISW000   CAS: 103-38-8   **HR: 1**
ISOVALERIC ACID, BENZYL ESTER
mf: $C_{12}H_{16}O_2$ mw: 192.28

PROP: Colorless liquid; fruity apple odor. D: 0.985-0.9911, refr index: 1.486, flash p: +212°F. Sol in alc, fixed oils; sltly sol in propylene glycol; insol in glycerin, water @ 246°.

SYNS: BENZYL ISOVALERATE (FCC) * BENZYL-3-METHYLBUTANOATE * BENZYL-3-METHYL BUTY-RATE * FEMA No. 2152 * ISOPENTANOIC ACID, PHENYLMETHYL ESTER * ISOPROPYL ACETIC ACID, BENZYL ESTER * 3-METHYLBUTANOIC ACID, PHENYLETHYL ESTER

CONSENSUS REPORTS: Reported in EPA TSCA Inventory.

SAFETY PROFILE: A skin irritant. Combustible liquid. When heated to decomposition it emits acrid smoke and fumes.

ISX000   CAS: 109-19-3   **HR: 1**
ISOVALERIC ACID, BUTYL ESTER
mf: $C_9H_{18}O_2$ mw: 158.27

PROP: Colorless to pale yellow liquid; fruity odor. Vap d: 5.45, bp: 150°, d: 0.851-0.857, refr index: 1.407. Misc with alc, fixed oils; sltly sol in propylene glycol; insol in water.

SYNS: n-BUTYL ISOPENTANOATE * n-BUTYL ISO-VALERATE * 1-BUTYL ISOVALERATE * BUTYL ISOVALERIANATE * BUTYL 3-METHYLBUTYRATE * FEMA No. 2218 * 3-METHYLBUTANOIC ACID, BUTYL ESTER

CONSENSUS REPORTS: Reported in EPA TSCA Inventory.

SAFETY PROFILE: Mildly toxic by ingestion. A skin irritant. Flammable when exposed to heat, flame, sparks, and oxidizers. To fight fire, use alcohol foam, dry chemical, spray, mist, fog. When heated to decomposition it emits acrid smoke and fumes.

ISY000   CAS: 108-64-5   **HR: 3**
ISOVALERIC ACID, ETHYL ESTER
mf: $C_7H_{14}O_2$ mw: 130.21

PROP: Colorless, oily liquid; apple odor. Flash p: 77°F, d: 0.868 @ 20°/20°, refr index: 1.395-1.399, bp: 135°, mp: −99°. Misc with alc, fixed oils, benzene, ether; sol in propylene glycol; sltly sol in water @ 135°.

SYNS: ETHYL ISOVALERATE (FCC) * FEMA No. 2463 * 3-METHYLBUTANOIC ACID, ETHYL ESTER * 3-METHYLBUTYRIC ACID, ETHYL ESTER

CONSENSUS REPORTS: Reported in EPA TSCA Inventory.

SAFETY PROFILE: Moderately toxic by intraperitoneal route. Mildly toxic by ingestion. A skin irritant. Flammable liquid when exposed to heat, flame or sparks. When heated to decomposition it emits acrid smoke and fumes.

ISZ000   CAS: 35154-45-1   **HR: 1**
(Z)-ISOVALERIC ACID-3-HEXENYL
mf: $C_{11}H_{20}O_2$ mw: 184.31

PROP: Colorless liquid; sweet, apple odor. D: 0.869-0.874, refr index: 1439-1.435. Sol in alc, propylene glycol, fixed oils; insol in water.

SYNS: AI3-35966 * FEMA No. 3498 * cis-3-HEXE-NYL ISOVALERATE (FCC)

CONSENSUS REPORTS: Reported in EPA TSCA Inventory.

SAFETY PROFILE: A skin irritant. When heated to decomposition it emits acrid smoke and fumes.

ITB000   CAS: 659-70-1   **HR: 1**
ISOVALERIC ACID, ISOPENTYL ESTER
mf: $C_{10}H_{20}O_2$ mw: 172.30

PROP: Colorless liquid; apple odor. D: 0.851-0.857, refr index: 1.411, flash p: 162°F. Misc in alc, fixed oils; sltly sol in propylene glycol; insol in water.

SYNS: FEMA No. 2085 * ISOAMYL ISOVALERATE (FCC) * ISOPENTYL ISOVALERATE

CONSENSUS REPORTS: Reported in EPA TSCA Inventory.

SAFETY PROFILE: Mildly toxic by ingestion. A skin irritant. Combustible liquid. When heated to decomposition it emits acrid smoke and fumes.

ITC000   CAS: 556-24-1   **HR: 1**
ISOVALERIC ACID, METHYL ESTER
DOT: UN 2400
mf: $C_6H_{12}O_2$ mw: 116.18

SYNS: 3-METHYLBUTANOIC ACID, METHYL ESTER * METHYL ISOPENTANOATE * METHYL ISOVALER-ATE * METHYLISOVALERATE (DOT) * METHYL-3-METHYLBUTANOATE * METHYL-3-METHYLBUTY-RATE

CONSENSUS REPORTS: Reported in EPA TSCA Inventory.

DOT Classification: Flammable Liquid; Label: Flammable Liquid.

SAFETY PROFILE: Mildly toxic by ingestion and very slightly toxic by inhalation. Flammable when exposed to heat or flame; can react vigorously with oxidizing materials. When heated to decomposition it emits acrid smoke and fumes.

ITD875   CAS: 70288-86-7   **HR: 3**
IVERMECTIN

SYNS: 22,23-DIHYDROAVERMECTIN B1 * HYVERMECTIN * MK 933

SAFETY PROFILE: Poison by subcutaneous route. When heated to decomposition emits toxic fumes of NO_x.

J

JAK000   CAS: 6870-67-3   **HR: 3**
JACOBINE
mf: $C_{18}H_{25}NO_6$ mw: 351.44

PROP: An alkaloid isolated from *S. Jacobaea*.

SYNS: 15,20-EPOXY-15,30-DIHYDRO-12-HYDROXYSE-NECIONAN-11,16-DIONE   *   NSC 89936

CONSENSUS REPORTS: IARC Cancer Review: GROUP 3 IMEMDT 7,56,87; Animal Inadequate Evidence IMEMDT 10,275,76. EPA Genetic Toxicology Program.

SAFETY PROFILE: Poison by intravenous and possibly other routes. Questionable carcinogen. Mutation data reported. When heated to decomposition it emits toxic fumes of NO_x.

JAT000   CAS: 128-58-5   **HR: 1**
JADE GREEN BASE
mf: $C_{36}H_{20}O_4$ mw: 516.56

SYNS: C.I. 59825   *   DIMETHOXYVIOLANTHRONE   *   16,17-DIMETHOXYVIOLANTHRONE   *   ZELEN OSTANTHRENOVA BRILANTNI FFB (CZECH)

CONSENSUS REPORTS: Reported in EPA TSCA Inventory.

SAFETY PROFILE: An eye irritant. When heated to decomposition it emits acrid smoke and fumes.

JCS000   CAS: 469-59-0   **HR: 3**
JERVINE
mf: $C_{27}H_{39}NO_3$ mw: 425.67

PROP: Needles from (methanol + water). Mp: 243.5-244.5°. An alkamine isolated from *Veratrum album*.

SAFETY PROFILE: Poison by ingestion, intravenous, and subcutaneous routes. Experi-mental teratogenic and reproductive effects. A natural toxin found in some plants. When heated to decomposition it emits toxic fumes of NO_x.

JDA100   **HR: 3**
JET FUEL HEF-2

PROP: Mixture of alkylpentaborane derivatives (CRDLR* 3035,60).

SYN: HEF-2

SAFETY PROFILE: Poison by ingestion, inhalation, intravenous, subcutaneous, and intraperitoneal routes. Moderately toxic by skin contact. Flammable when exposed to heat or flame. When heated to decomposition it emits toxic fumes of boron and acrid smoke and fumes.

JEA000   CAS: 8012-91-7   **HR: 2**
JUNIPER BERRY OIL

PROP: A volatile oil. Principal constituents include d-pinene, camphene, 1-terpineol-4, and other oxygenated constituents. From steam distillation of the fruit of *Juniperus communis* L. (Fam. *Cupressaceae*). Colorless to faint green-yellow liquid; aromatic bitter taste. Sol in fixed oils, mineral oil; insol in glycerin, propylene glycol.

SYNS: OIL of JUNIPER BERRY   *   WACHOLDERBEER OEL (GERMAN)

CONSENSUS REPORTS: Reported in EPA TSCA Inventory.

SAFETY PROFILE: Mildly toxic by ingestion. A human skin irritant. An allergen. A systemic irritant. If taken internally, a severe kidney irritation similar to that caused by turpentine may result. When heated to decomposition it emits acrid smoke and fumes.

K

KAJ000   CAS: 40596-69-8   **HR: 2**
KABAT
mf: $C_{19}H_{34}O_3$ mw: 310.53

PROP: Amber liquid. Bp: 100°. Solubility in water: 1.39 ppm. Sol in most organic solvents.

SYNS: ALTOSID * ALTOSID IGR * ALTOSID SR 10 * ENT 70,460 * ISOPROPYL(2E,4E)-11-METHOXY-3,7,11-TRIMETHYL-2,4-DODECADIENOATE * MANTA * METHOPRENE * (E,E)-11-METHOXY-3,7,11-TRI-METHYL-2,4-DODECANDIENOATE * ZR 515

SAFETY PROFILE: Moderately toxic by skin contact. Mildly toxic by ingestion. Mutation data reported. When heated to decomposition it emits acrid smoke and fumes.

KAL000   CAS: 59-01-8   **HR: 3**
KANAMYCIN
mf: $C_{18}H_{36}N_4O_{11}$ mw: 484.58

SYNS: CANTREX * 4,6-DIAMINO-2-HYDROXY-1,3-CYCLOHEXANE-3,6'-DIAMINO-3,6'-DIDEOXYDI-α-d-GLU-COSIDE * 4,6-DIAMINO-2-HYDROXY-1,3-CYCLOHEX-YLENE 3,6'-DIAMINO-3,6'-DIDEOXYDI-d-GLUCOPYRANO-SIDE * KANAMICINA (ITALIAN) * KANAMYCIN A * KANAMYTREX * KANTREX * KM * KM (THE ANTIBIOTIC)

CONSENSUS REPORTS: Reported in EPA TSCA Inventory. EPA Genetic Toxicology Program.

SAFETY PROFILE: Poison by intravenous and intramuscular routes. Moderately toxic by ingestion, intraperitoneal, and subcutaneous routes. Human reproductive effects. Experimental teratogenic and reproductive effects. Mutation data reported. When heated to decomposition it emits toxic fumes of NO_x.

KAM000   CAS: 25389-94-0   **HR: 3**
KANAMYCIN SULFATE
mf: $C_{18}H_{36}N_4O_{11} \cdot O_4S$ mw: 580.64

PROP: Irregular prisms. Decomp over wide range above 250°. Very sol in water; insol in common alc and nonpolar solvents.

SYNS: KANNASYN * KANTREX SULFATE

SAFETY PROFILE: A poison by intravenous route. Moderately toxic by intraperitoneal, subcutaneous, and intramuscular routes. When

heated to decomposition it emits very tox NO_x and SO_x.

KBB600   CAS: 1332-58-7   **HR: 1**
KAOLIN

PROP: Fine white to light yellow powder; earth taste. Insol in eater, alc, dil acids, and alkali solutions.

SYN: CHINA CLAY

OSHA PEL: (Transitional: TWA Respirable Fraction: 15 mg/m³; Respirable Fraction: 5 mg/m³) TWA Total Dust: 10 mg/m³; Respirable Fraction: 5 mg/m³

ACGIH TLV: TWA (nuisance particulate) 10 mg/m³ of total dust (when toxic impurities are not present, e.g., quartz < 1%).

SAFETY PROFILE: A nuisance dust.

KBK000   CAS: 9000-36-6   **HR: 1**
KARAYA GUM

PROP: Dried exudate of the tree, *Sterculia ureus* Roxburgh (Fam. *Sterculiaceae*). Fine, white powder; slt odor of acetic acid. Insol in alc; swells in water to a gel.

CONSENSUS REPORTS: Reported in EPA TSCA Inventory.

SAFETY PROFILE: Very mildly toxic by ingestion. A mild allergen.

KBU000   CAS: 39472-31-6   **HR: 3**
KARMINOMYCIN

SYNS: CARMINOMYCIN * o-DEMETHYLDAUNOMY-CIN

SAFETY PROFILE: Poison by ingestion, intravenous, intraperitoneal, and subcutaneous routes. Human systemic effects by intravenous route: anorexia, hallucinations and distorted perceptions, thrombosis, nausea or vomiting, fatty liver degeneration, impaired liver function, endocrine changes, and leukopenia (reduced white blood cell count). Experimental teratogenic and reproductive effects. Mutation data reported. When heated to decomposition it emits acrid smoke and fumes.

KEA000   CAS: 143-50-0   HR: 3
KEPONE

DOT: NA 2761
mf: $C_{10}Cl_{10}O$ mw: 490.60

PROP: A chlorinated polycyclic ketone, a crystalline material, sltly water-sol, sol in alc, ketones and acetic acid. Mp: decomp @ 350°. Readily hydrates on exposure to room temperature and humidity; normally used as a mono- to trihydrate (NCIPR*).

SYNS: CHLORDECONE * CIBA 8514 * COMPOUND 1189 * 1,2,3,5,6,7,8,9,10,10-DECACHLORO
$(5.2.1.0^{(2,6)}.0^{(3,9)}.0^{(5,8)})$DECANO-4-ONE * DECACHLOROKETONE * DECACHLORO-1,3,4-METHENO-2H-CYCLOBUTA(cd)PENTALEN-2-ONE
* DECACHLOROOCTAHYDROKEPONE-2-ONE
* DECACHLOROOCTAHYDRO-1,3,4-METHENO-2H-CYCLOBUTA(cd)PENTALEN-2-ONE * 1,1a,3,3a,4,5,5,5a,5b,6-DECACHLOROOCTAHYDRO-1,3,4-METHENO-2H-CYCLOBUTA(cd)PENTALEN-2-ONE * DECACHLOROPENTACYCLO$(5.2.1.0^{(2,6)}.0^{(3,9)}.0^{(5,8)})$DECAN-4-ONE
* DECACHLOROPENTACYCLO$(5.3.0.0^{(2,6)}.0^{(4,10)}.0^{(5,9)})$
DECAN-3-ONE * DECACHLOROTETRACYCLODECANONE * DECACHLOROTETRAHYDRO-4,7-METHANOINDENEONE * ENT 16,391 * GENERAL CHEMICALS 1189 * MEREX * NCI-C00191
* RCRA WASTE NUMBER U142

CONSENSUS REPORTS: IARC Cancer Review: GROUP 2B IMEMDT 7,56,87; Human Limited Evidence IMEMDT 20,67,79; Animal Sufficient Evidence IMEMDT 20,67,79. NTP Fourth Annual Report On Carcinogens, 1984. EPA Genetic Toxicology Program.

DFG MAK: Suspected Carcinogen.
NIOSH REL: (Kepone) CL 0.001 mg/m^3/15M
DOT Classification: ORM-E; Label: None.

SAFETY PROFILE: Confirmed carcinogen with experimental carcinogenic data. Poison by ingestion, skin contact, and possibly other routes. Experimental teratogenic and reproductive effects. Mutation data reported. Inhalation, absorption or ingestion by humans can lead to central nervous system, liver and kidney damage, including bizarre symptoms caused by damage to the nervous system. Usually, the symptoms are tremors, ataxia, skin changes, hyperexcitability, hyperactivity, muscle spasms, testicular atrophy, low sperm count, estrogenic effects, sterility, breast enlargement, liver lesions and cancer. An insecticide and fungicide. Registration suspended by the USEPA.

KEK000   CAS: 8008-20-6   HR: 3
KEROSENE

DOT: UN 1223

PROP: A pale yellow to water-white, oily liquid. Bp: 175-325°, ULC: 40, flash p: 150-185°F, d: 0.80 to <1.0, lel: 0.7%, uel: 5.0%, autoign temp: 410°F, vap d: 4.5. Insol in H_2O; misc with other petr solvents. A mixture of petroleum hydrocarbons, chiefly of the methane series having from 10-16 carbon atoms per molecule.

SYNS: COAL OIL * COAL OIL (EXPORT SHIPMENT ONLY) (DOT)

CONSENSUS REPORTS: Reported in EPA TSCA Inventory.

NIOSH REL: (Kerosene) TWA 100 mg/m^3
DOT Classification: Combustible Liquid; Label: None; Flammable or Combustible Liquid; Label: Flammable Liquid.

SAFETY PROFILE: Poison by intravenous and intratracheal routes. Moderately toxic to humans by an unspecified route. Moderately toxic to animals by ingestion. Human systemic effects by ingestion and intravenous routes: somnolence, hallucinations and distorted perceptions, coughing, nausea or vomiting, and fever. Aspiration of vomitus can cause serious pneumonitis, particularly in young children. Combustible when exposed to heat or flame; can react with oxidizing materials. Moderately explosive in the form of vapor when exposed to heat or flame. When heated to decomposition it emits acrid smoke and fumes. To fight fire, use foam, CO_2, dry chemical.

KEU000   CAS: 463-51-4   HR: 3
KETENE

mf: C_2H_2O mw: 42.04

PROP: Colorless gas with disagreeable taste. Decomp in water. Mp: -150°, bp: -56°, vap d: 1.45. Decomp in alc; sol in ether and acetone.

SYNS: CARBOMETHENE * ETHENONE * KETOETHYLENE

CONSENSUS REPORTS: Reported in EPA TSCA Inventory. EPA Genetic Toxicology Program.

OSHA PEL: (Transitional: TWA 0.5 ppm) TWA 0.5 ppm; STEL 1.5 ppm

ACGIH TLV: TWA 0.5 ppm; STEL 1.5 ppm
DFG MAK: 0.5 ppm (0.9 mg/m^3)

SAFETY PROFILE: Poison by inhalation. Moderately toxic by ingestion. Can cause pulmonary edema. Reacts with hydrogen peroxide to form the explosive diacetyl peroxide. When heated to decomposition it emits acrid smoke and fumes.

KFA000   CAS: 674-82-8   **HR: 2**
KETENE DIMER

DOT: UN 2521
mf: $C_4H_4O_2$ mw: 84.08

PROP: Colorless, nonhygroscopic liquid, pungent odor, decomp in H_2O. Mp: −6.5°, bp: 127.4°, d: 1.0897, vap d: 2.9, flash p: 93°F(TOC).

SYNS: 3-BUTENO-β-LACTONE * DIKETENE * DIKETENE, INHIBITED (DOT) * 4-METHYLENE-2-OXETANONE

CONSENSUS REPORTS: Reported in EPA TSCA Inventory.

DOT Classification: Flammable or Combustible Liquid; Label: Flammable Liquid.

SAFETY PROFILE: Moderately toxic by ingestion and skin contact. A skin and severe eye irritant. Flammable when exposed to heat or flame; can react vigorously with oxidizing materials. A violent polymerization reaction is catalyzed by acids, bases, or sodium acetate. A storage hazard. Self-initiated exothermic dimerization is explosive. To fight fire, use alcohol foam. When heated to decomposition it emits acrid smoke and fumes.

KFK000   CAS: 5965-49-1   **HR: 3**
KETOBEMIDONE HYDROCHLORIDE
mf: $C_{15}H_{21}NO_2$ mw: 247.37

SYNS: CLIRADON HYDROCHLORIDE * 1-METHYL-4-(m-HYDROXYPHENYL)PIPERIDINE-4-ETHYLKETONE HYDROCHLORIDE

SAFETY PROFILE: Poison by ingestion and intravenous routes. When heated to decomposition it emits toxic fumes of NO_x.

KGA000   **HR: D**
KETONES

PROP: Liquid organic compounds containing the carbonyl group C=O attached to two alkyl groups. Derived from secondary alcohols by oxidation. Acetone, which is dimethyl ketone, is the most familiar of this group of compounds.

SAFETY PROFILE: No general statement can be made as to the toxicity of ketones. Some are highly volatile and hence may have narcotic or anesthetic effects. Skin absorption, as well as inhalation, may be an important route of entry into the body. None of the ketones has been shown to have a high degree of chronic toxicity. Some are dangerous fire hazards. They react violently with aldehydes; HNO_3; HNO_3 + H_2O_2; $HClO_4$. A variety of peroxides can be formed from the reactions of ketones and hydrogen peroxide. Many of these peroxides are explosives sensitive to heat and shock. Common air contaminants.

KHU000   CAS: 74278-22-1   **HR: 3**
KROMAD

PROP: Contains 5% cadmium sebacate, 5% potassium chromate, 1% malachite green, and 16% thiram (FMCHA2-,D176,80).

CONSENSUS REPORTS: Cadmium and its compounds, as well as chromium and its compounds, are on the Community Right-To-Know List.

OSHA PEL: TWA 0.1 mg(Cd)/m^3; CL 0.6 mg(Cd)/m^3 (fume)
ACGIH TLV: TWA 0.05 mg(Cd)/m^3 (Proposed: TWA 0.01 mg(Cd)/m^3 (dust), Human Carcinogen); BEI: 10 µg/g creatinine in urine; 10 µg/L in blood.
DFG BAT: Blood 1.5 µg/dL; Urine 15 µg/dL, Suspected Carcinogen.
NIOSH REL: (Cadmium) Reduce to lowest feasible level

SAFETY PROFILE: Confirmed human carcinogen. Poison by ingestion. Moderately toxic by skin contact. When heated to decomposition it emits toxic fumes of K_2O, Cd, and Cr.

L

LAC000   CAS: 8016-26-0   **HR: 1**
LABDANUM OIL

PROP: Main constituents are acetophenone, 1,5,5-trimethyl-6-cyclohexanone and ladaniol found in the gum of the shrub *Cistus ladaniferus* L. (Fam. *Cistaceae*). Prepared by steam distillation of the crude gum. Yellow, viscous liquid; powerful balsamic odor. D: 0.905-0.993, refr index: 1.492-1.507 @ 20°, flash p: 187°F. Sol in fixed oils, mineral oil; insol in glycerin, propylene glycol.

SYN: OIL of LABDANUM

SAFETY PROFILE: Mildly toxic by ingestion. A skin irritant. Combustible liquid. When heated to decomposition it emits acrid smoke and fumes.

LAG000   CAS: 50-21-5   **HR: 2**
LACTIC ACID
mf: $C_3H_6O_3$ mw: 90.09

PROP: Yellow to colorless crystals or syrupy 50% liquid. Mp: 16.8°, bp: 122° @ 15 mm, d: 1.249 @ 15°. Volatile with superheated steam. Sol in eater, alc, and furfurol; sltly sol in ether; insol in chloroform, petr ether, carbon disulfide. Misc in water, (alc + ether).

SYNS: ACETONIC ACID * ETHYLIDENELACTIC ACID * 1-HYDROXYETHANECARBOXYLIC ACID * 2-HYDROXYPROPANOIC ACID * 2-HYDROXYPRO-PIONIC ACID * α-HYDROXYPROPIONIC ACID * KYSELINA MLECNA (CZECH) * dl-LACTIC ACID * MILCHSAURE (GERMAN) * MILK ACID * ORDINARY LACTIC ACID * racemic LACTIC ACID

CONSENSUS REPORTS: Reported in EPA TSCA Inventory.

SAFETY PROFILE: Moderately toxic by ingestion and rectal routes. Mutation data reported. A severe skin and eye irritant. Mixtures with nitric acid + hydrofluoric acid may react vigorously and are storage hazards. When heated to decomposition it emits acrid smoke and irritating fumes.

LAH000   CAS: 64059-26-3   **HR: 3**
LACTIC ACID, BERYLLIUM SALT

SYN: BERYLLIUM LACTATE

CONSENSUS REPORTS: Beryllium and its compounds are on The Community Right-To-Know List.

OSHA PEL: (Transitional: TWA 0.002 mg(Be)/m³; CL 0.005; Pk 0.025/30M/8H) TWA 0.002 mg(Be)/m³; STEL 0.005 mg(Be)/m³/30M; CL 0.025 mg(Be)/m³
ACGIH TLV: TWA 0.002 mg(Be)m³, Suspected Carcinogen
NIOSH REL: CL (Beryllium) not to exceed 0.0005 mg(Be)/m³

SAFETY PROFILE: Confirmed carcinogen. Poison by intravenous route. When heated to decomposition it emits very toxic fumes of Be.

LAJ000   CAS: 97-64-3   **HR: 2**
LACTIC ACID, ETHYL ESTER

DOT: UN 1192
mf: $C_5H_{10}O_3$ mw: 118.15

PROP: Colorless liquid; mild odor. Bp: 154°, ULC: 30-35%, lel: 1.55% @ 212°F, flash p: 115°F (CC), flash p (technical): 131°F, d: 1.029-1.032, refr index: 1.410-1.420, autoign temp: 752°F, vap d: 4.07. Very sol in alc, ether, chloroform, water.

SYNS: ACTYLOL * ACYTOL * ETHYL α-HY-DROXYPROPIONATE * ETHYL 2-HYDROXYPROPIO-NATE * ETHYL LACTATE (DOT,FCC) * FEMA No. 2440 * LACTATE d'ETHYLE (FRENCH) * SOLACTOL

CONSENSUS REPORTS: Reported in EPA TSCA Inventory.

DOT Classification: Flammable or Combustible Liquid; Label: Flammable Liquid; Combustible Liquid; Label: None.

SAFETY PROFILE: Moderately toxic by ingestion, intraperitoneal, subcutaneous, and intravenous routes. Flammable or combustible liquid when exposed to heat or flame; can react with oxidizing materials. Slight explosion hazard in the form of vapor when exposed to flame. To fight fire, use foam, CO_2, dry chemical. When heated to decomposition it emits acrid smoke and irritating fumes.

LAL000   CAS: 5905-52-2   **HR: 3**
LACTIC ACID, IRON(2+) SALT (2:1)
mf: $C_6H_{10}O_6 \cdot Fe$ mw: 234.01

703

PROP: Greenish-white crystals; slight peculiar odor. Moderately sol in water; sltly sol in alc.

SYNS: FERROUS LACTATE * IRON(2+) LACTATE

OSHA PEL: TWA 1 mg(Fe)/m^3
ACGIH TLV: TWA 1 mg(Fe)/m^3

SAFETY PROFILE: Poison by ingestion. Questionable carcinogen with experimental tumorigenic data. When heated to decomposition it emits acrid smoke and irritating fumes.

LAM000 CAS: 72-17-3 **HR: 2**
LACTIC ACID, MONOSODIUM SALT
mf: C$_3$H$_5$O$_3$•Na mw: 112.07

PROP: Hygroscopic solid; slt salt taste.

SYNS: 2-HYDROXYPROPANOIC ACID MONOSODIUM SALT * LACOLIN * LACTIC ACID SODIUM SALT * PER-GLYCERIN * SODIUM LACTATE

CONSENSUS REPORTS: Reported in EPA TSCA Inventory.

SAFETY PROFILE: Moderately toxic by intraperitoneal route. An eye irritant. When heated to decomposition it emits toxic fumes of Na$_2$O.

LAQ000 CAS: 78-97-7 **HR: 3**
LACTONITRILE
mf: C$_3$H$_5$NO mw: 71.09

PROP: Straw-colored liquid. Mp: −40°, bp: 103° @ 50 mm, fp: −34°, flash p: 170°F (TCC), d: 0.9834 @ 25°, vap d: 2.45.

SYN: 2-HYDROXYPROPIONITRILE

CONSENSUS REPORTS: EPA Extremely Hazardous Substances List. Reported in EPA TSCA Inventory. Cyanide and its compounds are on the Community Right-To-Know List.

SAFETY PROFILE: Poison by ingestion, skin contact, and subcutaneous routes. Moderately toxic by inhalation. In the presence of alkali, it evolves HCN. Combustible when exposed to heat or flame; can react vigorously with oxidizing materials. To fight fire, use foam, CO$_2$, dry chemical. When heated to decomposition it emits toxic fumes of CN$^-$ and NO$_x$.

LAR000 CAS: 63-42-3 **HR: 2**
LACTOSE
mf: C$_{12}$H$_{22}$O$_{11}$ mw: 342.34

PROP: Colorless, rhombic crystals; faintly sweet taste. D: 1.525 @ 20°, mp: 202° (anhy-

drous), bp: decomp. Sol in water; insol in alc and ether.

SYNS: 4-(β-d-GALACTOSIDO)-d-GLUCOSE * LACTIN * LACTOBIOSE * d-LACTOSE * MILK SUGAR * SACCHARUM LACTIN

CONSENSUS REPORTS: Reported in EPA TSCA Inventory.

SAFETY PROFILE: Moderately toxic by intravenous route. Experimental reproductive effects. Questionable carcinogen with experimental tumorigenic and teratogenic data. Mixtures with oxidants (e.g., potassium chlorate, potassium nitrate, or potassium perchlorate) may be explosion hazards. When heated to decomposition it emits acrid smoke and irritating fumes.

LAS000 CAS: 1332-94-1 **HR: 3**
LAETRILE
mf: C$_{14}$H$_{15}$NO$_7$ mw: 309.30

PROP: Solid. Mp: 214-216°.

SYNS: CYANOPHENYLMETHYL-β-d-GLUCOPYRANOSIDURONIC ACID * 1-MANDELONITRILE-β-GLUCURONIC ACID

CONSENSUS REPORTS: Cyanide and its compounds are on The Community Right-To-Know List.

SAFETY PROFILE: Human poison by ingestion. Experimental poison by intraperitoneal route. Human systemic effects by ingestion and multiple routes: central nervous system and gastrointestinal changes. Mutation data reported. A controversial treatment for cancer. When heated to decomposition it emits toxic fumes of NO$_x$ and CN$^-$.

LAU000 CAS: 17575-22-3 **HR: 3**
LANATOSIDE C
mf: C$_{49}$H$_{76}$O$_{20}$ mw: 985.25

PROP: Long, flat prisms from alc. Decomp @ 248-250° after drying in high vacuum @ 150°. Very sol in pyridine and dioxane, insol in ether, petr ether.

SYNS: CEDILANID * DIGILANID C * ISOLANID * ISOLANIDE * LANATOSID C (GERMAN)

SAFETY PROFILE: Poison by ingestion, intraperitoneal, and intravenous routes. When heated to decomposition it emits acrid smoke and irritating fumes.

LAV000   CAS: 7439-91-0   **HR: 3**
LANTHANUM
af: La   aw: 138.91

PROP: Silvery-white, malleable and ductile metal element soft enough to cut with a knife. Very reactive rare earth metal. Mp: 920°, bp: 3454°, d: 6.166 @ 25°.

SAFETY PROFILE: Poison by intravenous route. Lanthanum and other lanthanons can cause delayed blood clotting leading to hemorrhages. Has caused liver injury in experimental animals. The dust is a dangerous fire hazard when exposed to flame; can react vigorously with oxidizing materials. Violent reaction with nitric acid, phosphorus (above 400°C), air, halogens. Moderately explosive in the form of dust when exposed to flame or by chemical reaction. Incompatible with H_2O, C, N, B, Se, Si, S.

LAW000   CAS: 917-70-4   **HR: 2**
LANTHANUM ACETATE
mf: $C_2H_4O_2 \cdot xLa$ mw: 1032.43

SYNS: LANTHANACETAT (GERMAN) * LANTHANUM TRIACETATE

CONSENSUS REPORTS: Reported in EPA TSCA Inventory.

SAFETY PROFILE: Moderately toxic by intraperitoneal and subcutaneous routes. Mildly toxic by ingestion. Mutation data reported. When heated to decomposition it emits acrid smoke and irritating fumes.

LAX000   CAS: 10099-58-8   **HR: 3**
LANTHANUM CHLORIDE
mf: Cl_3La mw: 245.26

PROP: Heptahydrate: triclinic crystals. Sol in water and alc.

CONSENSUS REPORTS: Reported in EPA TSCA Inventory. EPA Genetic Toxicology Program.

SAFETY PROFILE: Poison by intraperitoneal and intravenous routes. Moderately toxic by subcutaneous route. Mildly toxic by ingestion. Experimental reproductive effects. When heated to decomposition it emits toxic fumes of Cl^-.

LAZ000   CAS: 11138-87-7   **HR: 3**
LANTHANUM EDETATE

SAFETY PROFILE: Poison by intraperitoneal route.

LBA000   CAS: 10099-59-9   **HR: 3**
LANTHANUM NITRATE
mf: $N_3O_9 \cdot La$ mw: 324.94

PROP: Hexahydrate; white, deliquescent crystals. Mp: approx 40°, bp: 126°. Very sol in water, alc. Keep well stoppered.

CONSENSUS REPORTS: Reported in EPA TSCA Inventory.

SAFETY PROFILE: Poison by intraperitoneal route. Mildly toxic by ingestion. Experimental reproductive effects. Mutation data reported. When heated to decomposition it emits toxic fumes of NO_x.

LBD000   CAS: 32854-75-4   **HR: 3**
LAPPACONITINE
mf: $C_{32}H_{44}N_2O_8$ mw: 584.78

PROP: Bitter crystals. Mp: 217-218°; sol in benzene; sltly sol in alc, ether; insol in water. Chief alkaloid in aconitum septentrionale.

SAFETY PROFILE: Poison by ingestion, intraperitoneal, and intravenous routes. When heated to decomposition it emits toxic fumes of NO_x.

LBF500   CAS: 11054-70-9   **HR: 3**
LASALOCID
mf: $C_{35}H_{54}O_8$ mw: 602.89

SYN: ANTIBIOTIC X 537

SAFETY PROFILE: Poison by ingestion and intraperitoneal routes. An eye and skin irritant. When heated to decomposition it emits acrid smoke and irritating fumes.

LBG000   CAS: 303-34-4   **HR: 3**
LASIOCARPINE
mf: $C_{21}H_{33}NO_7$ mw: 411.55

PROP: An alkaloid isolated from *H. Lasiocarpum*.

SYNS: HELIOTRIDINE ESTER with LASIOCARPUM and ANGELIC ACID * NCI-C01478 * RCRA WASTE NUMBER U143

CONSENSUS REPORTS: IARC Cancer Review: GROUP 2B IMEMDT 7,56,87; Animal Limited Evidence IMEMDT 10,281,76. NCI Carcinogenesis Bioassay (feed); No Evidence: mouse, rat NCITR* NCI-CG-TR-39,78. EPA Genetic Toxicology Program.

SAFETY PROFILE: Suspected carcinogen with experimental carcinogenic data. Poison by ingestion, intravenous, intraperitoneal, parenteral, and possibly other routes. Human mutation data reported. When heated to decomposition it emits toxic fumes of NO_x.

LBK000 CAS: 8006-78-8 **HR: 2**
LAUREL LEAF OIL

PROP: Main constituent is cineole. From steam distillation of the leaves of *Laurus nobilis* L. (Fam. *Lauraceae)*. Yellow liquid; aromatic and spicy odor. D: 0.905-0.929, refr index: 1.465 at 20°. Sol in fixed oils, mineral oil, propylene glycol; insol in glycerin.

SYNS: BAY LEAF OIL * OIL of LAUREL LEAF

CONSENSUS REPORTS: Reported in EPA TSCA Inventory.

SAFETY PROFILE: Moderately toxic by ingestion. A skin irritant. When heated to decomposition it emits acrid smoke and irritating fumes.

LBL000 CAS: 143-07-7 **HR: 3**
LAURIC ACID
mf: $C_{12}H_{24}O_2$ mw: 200.36

PROP: Colorless, needle-like crystals; slt odor of bay oil. Mp: 48°, bp: 225° @ 100 mm, d: 0.883, vap press: 1 mm @ 121.0°. Insol in water; sol in chloroform, benzene, alc, ether, and petroleum ether.

SYNS: DODECANOIC ACID * DODECOIC ACID * DUODECYLIC ACID * HYDROFOL ACID 1255 * HYSTRENE 9512 * LAUROSTEARIC ACID * NEO-FAT 12 * NINOL AA62 EXTRA * 1-UNDECANECARBOXYLIC ACID * WECOLINE 1295

CONSENSUS REPORTS: Reported in EPA TSCA Inventory.

SAFETY PROFILE: Poison by intravenous route. Mildly toxic by ingestion. Questionable carcinogen with experimental neoplastigenic data. Mutation data reported. Combustible when exposed to heat or flame; can react with oxidizing materials. When heated to decomposition it emits acrid smoke and irritating fumes.

LBR000 CAS: 105-74-8 **HR: 3**
LAUROYL PEROXIDE

DOT: UN 2124/UN 2893
mf: $C_{24}H_{46}O_4$ mw: 398.70

PROP: White, tasteless, coarse powder; faint odor. Mp: 53-55°.

SYNS: ALPEROX C * BIS(1-OXODODECYL)PEROXIDE * DILAUROYL PEROXIDE * DILAUROYL PER-

OXIDE, TECHNICAL PURE (DOT) * DODECANOYL PEROXIDE * DYP-97 F * LAUROX * LAUROYL PEROXIDE, TECHNICALLY PURE (DOT) * LAURYDOL * LYP 97 * PEROXYDE de LAUROYLE (FRENCH)

CONSENSUS REPORTS: IARC Cancer Review: GROUP 3 IMEMDT 7,56,87; Animal Inadequate Evidence IMEMDT 36,315,85. Reported in EPA TSCA Inventory.

DFG MAK: Mild skin effects.
DOT Classification: Organic Peroxide; Label: Organic Peroxide.

SAFETY PROFILE: Questionable carcinogen with experimental tumorigenic data. A powerful oxidizing agent. It is a corrosive irritant to the eyes and mucous membranes and can cause burns. A dangerous fire hazard. When heated to decomposition it emits acrid smoke and fumes.

LBW000 CAS: 93-23-2 **HR: 3**
LAURYLISOQUINOLINIUM BROMIDE
mf: $C_{21}H_{32}N \cdot Br$ mw: 378.45

PROP: Deep amber, water-sol liquid; pleasant, characteristic odor.

SYNS: 2-DODECYLISOQUINOLINIUM BROMIDE * INTEXSAN LQ75 * ISOTHAN

CONSENSUS REPORTS: Reported in EPA TSCA Inventory.

SAFETY PROFILE: Poison by ingestion. A severe eye irritant. Combustible when exposed to heat or flame. Incompatible with oxidizing materials. An FDA over-the-counter drug. When heated to decomposition emits toxic fumes of Br^- and NO_x.

LBX000 CAS: 112-55-0 **HR: 3**
LAURYL MERCAPTAN
mf: $C_{12}H_{26}S$ mw: 202.44

PROP: Water-white to pale yellow liquid. Mp: $-7°$, bp: 115-177°, flash p: 262°F (OC), d: 0.849 @ 15.5°/15.5°.

SYNS: 1-DODECANETHIOL * DODECYL MERCAPTAN * m-DODECYL MERCAPTAN * 1-DODECYL MERCAPTAN * m-LAURYL MERCAPTAN * 1-MERCAPTODODECANE * NCI-C60935 * PENNFLOAT M * PENNFLOAT S

CONSENSUS REPORTS: Reported in EPA TSCA Inventory.

NIOSH REL: (n-Alkane Mono Thiols) CL 0.5 ppm/15M

SAFETY PROFILE: Mutation data reported. Combustible when exposed to heat or flame. To fight fire, use alcohol foam. When heated to decomposition it emits toxic fumes of SO_x.

LCA000   CAS: 8022-15-9   **HR: 1**
LAVANDIN OIL

PROP: Main constituent is Linalool. Prepared by steam distillation of the flowering stalks of the plants *Lavanoula hybrida reverchon*, Lavandula abrialis (Fam. *Labiatae*), *Lavandula officinalis*, or *Lavandula latifolia*. Yellow liquid; camphoraceous odor of lavender. D: 0.885, refr index: 1.460 @ 20°. Sol in fixed oils, propylene glycol, mineral oil; insol in glycerin.

SYN: OIL of LAVANDIN, ABRIAL TYPE

CONSENSUS REPORTS: Reported in EPA TSCA Inventory.

SAFETY PROFILE: A skin irritant. When heated to decomposition it emits acrid smoke and irritating fumes.

LCD000   CAS: 8000-28-0   **HR: 1**
LAVENDER OIL

PROP: Found in the flowers of *Lavandula officinalis* Chaix et Villars (*Lavabdula vera* De Candolle (Fam. *Labiatae*). The main constituent is linalyl acetate. A colorless to yellow liquid; characteristic odor and taste of lavender flowers. D: 0.875, refr index: 1.459-1.470 @ 20°.

SYNS: LAVENDEL OEL (GERMAN) * OIL of LAVENDER

CONSENSUS REPORTS: Reported in EPA TSCA Inventory.

SAFETY PROFILE: Mildly toxic by ingestion. A skin irritant. When heated to decomposition it emits acrid smoke and irritating fumes.

LCF000   CAS: 7439-92-1   **HR: 3**
LEAD
af: Pb   aw: 207.19

PROP: Bluish-gray, soft metal. Mp: 327.43°, bp: 1740°, d: 11.34 @ 20°/4°. vap press: 1 mm @ 973°.

SYNS: C.I. 77575 * C.I. PIGMENT METAL 4
* GLOVER * LEAD FLAKE * LEAD S2
* OLOW (POLISH) * OMAHA * OMAHA & GRANT
* SI * SO

CONSENSUS REPORTS: IARC Cancer Review: GROUP 2B IMEMDT 7,230,87; Animal Inadequate Evidence IMEMDT 23,325,80. Lead and its compounds are on the Community Right-To-Know List. Reported in EPA TSCA Inventory. EPA Genetic Toxicology Program.

OSHA PEL: TWA 0.05 mg(Pb)/m³
ACGIH TLV: TWA 0.15 mg(Pb)/m³; BEI: 50 µg(lead)/L in blood; 150 µg(lead)/g creatinine in urine.
DFG MAK: 0.1 mg/m³; BAT: 70 µg(lead)/L in blood, 30 µg(lead)/L in blood of women less than 45 years old.
NIOSH REL: TWA (Inorganic Lead) 0.10 mg(Pb)/m³

SAFETY PROFILE: Suspected carcinogen. Poison by ingestion. Moderately toxic by intraperitoneal route. Human systemic effects by ingestion and inhalation: loss of appetite, anemia, malaise, insomnia, headache, irritability, muscle and joint pains, tremors, flaccid paralysis without anesthesia, hallucinations and distorted perceptions, muscle weakness, gastritis and liver changes. The major organ systems affected are the nervous system, blood system, and kidneys. Lead encephalopathy is accompanied by severe cerebral edema, increase in cerebral spinal fluid pressure, proliferation and swelling of endothelial cells in capillaries and arterioles, proliferation of glial cells, neuronal degeneration and areas of focal cortical necrosis in fatal cases. Experimental evidence now suggests that blood levels of lead below 10 µg/dl can have the effect of diminishing the IQ scores of children. Low levels of lead impair neurotransmission and immune system function and may increase systolic blood pressure. Reversible kidney damage can occur from acute exposure. Chronic exposure can lead to irreversible vascular sclerosis, tubular cell atrophy, interstitial fibrosis, and glomerular sclerosis. Severe toxicity can cause sterility, abortion and neonatal mortality and morbidity. Experimental teratogenic and reproductive effects. Human mutation data reported. Very heavy intoxication can sometimes be detected by formation of a dark line on the gum margins, the so-called "lead line."

When lead is ingested, much of it passes through the body unabsorbed, and is eliminated in the feces. The greater portion of the lead that is absorbed is caught by the liver and excreted, in part, in the bile. For this reason, larger amounts of lead are necessary to cause toxic effects by this route, and a longer period of exposure is usually necessary to produce symptoms. On the other hand, upon inhalation, absorption takes place easily from the respiratory tract and symptoms tend to develop more quickly. For industry, inhalation is much more important than is ingestion. For the general population, exposure to lead occurs from inhaled air, dust of various types, and food and water with an approximate 50/50 division between inhalation and ingestion routes. Lead occurs in water in either dissolved or particulate form. At low pH, lead is more easily dissolved. Chemical treatment to soften water increases the solubility of lead. Adults absorb about 5-15% of ingested lead and retain less than 5%. Children absorb about 50% and retain about 30%.

Lead produces a brittleness of the red blood cells so that they hemolyze with but slight trauma; the hemoglobin is not affected. Due to their increased fragility, the red cells are destroyed more rapidly in the body than is normal, producing an anemia which is rarely severe. The loss of circulating red cells stimulates the production of new young cells which, on entering the blood stream, are acted upon by the circulating lead, with resultant coagulation of their basophilic material. These cells after suitable staining, are recognized as "stippled cells." There is no uniformity of opinion regarding the effect of lead on the white blood cells.

In addition to its effect on the red blood cells, lead produces a damaging effect on the organs or tissues with which it comes in contact. No specific or characteristic lesion is produced. Autopsies in deaths attributed to lead poisoning and experimental work on animals have shown pathological lesions of the kidneys, liver, male gonads, nervous system, blood vessels and other tissues. None of these changes, however, has been found consistently. In cases of severe lead poisoning, the amount of lead found in the blood is frequently in excess of 0.07 mg per 100 cc of whole blood. The urinary lead excretion generally exceeds 0.1 mg per liter of urine.

Flammable in the form of dust when exposed to heat or flame. Moderately explosive in the form of dust when exposed to heat or flame. Mixtures of hydrogen peroxide + trioxane explode on contact with lead. Rubber gloves containing lead may ignite in nitric acid. Violent reaction on ignition with chlorine trifluoride; concentrated hydrogen peroxide; ammonium nitrate (below 200° with powdered lead); sodium acetylide (with powdered lead). Incompatible with NaN_3; Zr; disodium acetylide; oxidants. Can react vigorously with oxidizing materials. A common air contaminant. When heated to decomposition it emits highly toxic fumes of Pb.

LCG000   CAS: 301-04-2   **HR: 3**
LEAD ACETATE

DOT: UN 1616
mf: $C_4H_6O_4 \cdot Pb$ mw: 325.29

PROP: Trihydrate: colorless crystals or white granules or powder. Sltly acetic odor, slowly effloresces. D: 2.55, mp: 75° (when rapidly heated), decomp above 200°. Very sol in glycerol.

SYNS: ACETATE de PLOMB (FRENCH) * ACETIC ACID LEAD (2+) SALT * BLEIACETAT (GERMAN) * DIBASIC LEAD ACETATE * LEAD (2+) ACETATE * LEAD(II) ACETATE * LEAD DIACETATE * LEAD DIBASIC ACETATE * NORMAL LEAD ACETATE * PLUMBOUS ACETATE * RCRA WASTE NUMBER U144 * SALT of SATURN * SUGAR of LEAD

CONSENSUS REPORTS: IARC Cancer Review: GROUP 3 IMEMDT 7,230,87; Animal Sufficient Evidence IMEMDT 23,325,80; IMEMDT 1,40,72; Human Limited Evidence IMEMDT 23,325,80. NTP Fourth Annual Report On Carcinogens, 1984. Lead and its compounds are on the Community Right-To-Know List. Reported in EPA TSCA Inventory. EPA Genetic Toxicology Program.

OSHA PEL: TWA 0.05 mg(Pb)/m^3
ACGIH TLV: TWA 0.15 mg(Pb)/m^3
NIOSH REL: (Inorganic Lead) TWA 0.10 mg(Pb)/m^3
DOT Classification: ORM-E; Label: None; Poison B; Label: St. Andrews Cross.

SAFETY PROFILE: Confirmed carcinogen with experimental neoplastigenic and tumorigenic data. Poison by ingestion, intraperitoneal, subcutaneous, and intravenous routes. Experimental teratogenic and reproductive effects. Human mutation data reported. Used as color additive in hair dyes, an insecticide, an astringent, and sedative. Incompatible with $KBrO_3$, acids, soluble sulfates, citrates, tartrates, chlorides, carbonates, alkalies, tannin phosphates, resorcinol, salicylic acid, phenol, chloral hydrate, sulfites, vegetable infusions, tinctures. When heated to decomposition it emits toxic fumes of Pb.

LCH000   CAS: 1335-32-6   **HR: 3**
LEAD ACETATE, BASIC
mf: $C_4H_{10}O_8Pb_3$ mw: 807.71

PROP: White powder.

SYNS: BASIC LEAD ACETATE * BIS(ACETO)DIHY-DROXYTRILEAD * BIS(ACETATO)TETRAHYDROXY-TRILEAD * BLA * LEAD MONOSUBACETATE * LEAD SUBACETATE * MONOBASIC LEAD ACE-TATE * RCRA WASTE NUMBER U146 * SUBACE-TATE LEAD

CONSENSUS REPORTS: IARC Cancer Review: GROUP 3 IMEMDT 7,230,87; Animal Sufficient Evidence IMEMDT 23,325,80; IMEMDT 1,40,72; Human Limited Evidence IMEMDT 23,325,80. Lead and its compounds are on the Community Right-To-Know List. Reported in EPA TSCA Inventory. EPA Genetic Toxicology Program.

SAFETY PROFILE: Experimental reproductive effects. Questionable carcinogen with experimental experimental carcinogenic, neoplastigenic, and tumorigenic data. Mutation data reported. When heated to decomposition it emits toxic fumes of Pb.

LCJ000   CAS: 6080-56-4   **HR: 3**
LEAD ACETATE(II), TRIHYDRATE
mf: $C_4H_6O_4 \cdot Pb \cdot 3H_2O$ mw: 379.35

SYNS: ACETIC ACID, LEAD(+2) SALT TRIHYDRATE * BIS(ACETATO)TRIHYDROXYTRILEAD * BLEIAZE-TAT (GERMAN) * LEAD ACETATE TRIHYDRATE * LEAD DIACETATE TRIHYDRATE * PLUMBOUS ACETATE

CONSENSUS REPORTS: IARC Cancer Review: Animal Sufficient Evidence IMEMDT 1,40,72. EPA Genetic Toxicology Program. Lead and its compounds are on the Community Right-To-Know List.

OSHA PEL: TWA 0.05 mg(Pb)/m^3
NIOSH REL: (Inorganic Lead) TWA 0.10 mg(Pb)/m^3

SAFETY PROFILE: Confirmed carcinogen with experimental carcinogenic data. Poison by intraperitoneal route. Moderately toxic by subcutaneous route. Experimental teratogenic and reproductive effects. Mutation data reported. When heated to decomposition it emits toxic fumes of Pb.

LCK000   CAS: 7784-40-9   **HR: 3**
LEAD ACID ARSENATE
DOT: UN 1617
mf: $AsHO_4 \cdot Pb$ mw: 347.12

PROP: White crystals.

SYNS: ACID LEAD ARSENATE * ACID LEAD ORTHOARSENATE * ARSENATE of LEAD * AR-SINETTE * DIBASIC LEAD ARSENATE * GYPSINE * LEAD ARSENATE * LEAD ARSENATE, solid (DOT) * LEAD ARSENATE (standard) * ORTHO L10 DUST * ORTHO L40 DUST * SCHULTENITE * SECURITY * SOPRABEL * STANDARD LEAD ARSENATE * TALBOT

CONSENSUS REPORTS: IARC Cancer Review: Human Sufficient Evidence IMEMDT 23,39,80; Animal Inadequate Evidence IMEMDT 1,40,72; IMEMDT 1,40,72. Arsenic and its compounds, as well as lead and its compounds, are on the Community Right-To-Know List. Reported in EPA TSCA Inventory.

OSHA PEL: TWA 0.05 mg(Pb)/m^3; 0.01 mg(As)/m^3
ACGIH TLV: TWA 0.15 mg(Pb)/m^3; 0.2 mg(As)/m^3
NIOSH REL: (Inorganic Lead) TWA 0.10 mg(Pb)/m^3; (Inorganic Arsenic) CL 0.002 mg(As)/m^3/15M
DOT Classification: Poison B; Label: Poison.

SAFETY PROFILE: Confirmed human carcinogen. A poison by ingestion. Moderately toxic to humans by an unspecified route. Used as an insecticide and herbicide. When heated to decomposition it emits very toxic fumes of As and Pb.

LCL000   CAS: 10031-13-7   **HR: 3**
LEAD(II) ARSENITE
DOT: UN 1618
mf: $As_2O_4 \cdot Pb$ mw: 421.03

PROP: White powder. D: 5.85. Insol in water; sol in dil HNO_3.

SYN: LEAD ARSENITE, solid (DOT)

CONSENSUS REPORTS: Arsenic and its compounds, as well as lead and its compounds, are on the Community Right-To-Know List.

OSHA PEL: TWA 0.05 mg(Pb)/m^3; 0.01 mg(As)/m^3; Cancer Hazard
ACGIH TLV: TWA 0.15 mg(Pb)/m^3; 0.2 mg(As)/m^3

NIOSH REL: (Inorganic Lead) TWA 0.10 mg(Pb)/m^3; (Inorganic Arsenic) CL 0.002 mg(As)/m^3/15M

DOT Classification: Poison B; Label: Poison.

SAFETY PROFILE: Confirmed human carcinogen. A poison. When heated to decomposition it emits very toxic fumes of Pb and As.

LCM000   CAS: 13424-46-9   **HR: 3**
LEAD(II) AZIDE

DOT: UN 0129
mf: N$_6$Pb   mw: 291.25

PROP: Colorless needles or white powder. Explodes @ 350° or when shocked; very sol in acetic acid; insol in NH$_4$OH.

SYNS: INITIATING EXPLOSIVE LEAD AZIDE, DEXTRINATED TYPE ONLY (DOT) * LEAD AZIDE, DRY (DOT)

CONSENSUS REPORTS: Lead and its compounds are on the Community Right-To-Know List. Reported in EPA TSCA Inventory.

OSHA PEL: TWA 0.05 mg(Pb)/m^3
ACGIH TLV: TWA 0.15 mg(Pb)/m^3
NIOSH REL: (Inorganic Lead) TWA 0.10 mg(Pb)/m^3
DOT Classification: Class A Explosive; Label: Explosive A: Forbidden, Dry.

SAFETY PROFILE: A deadly poison. An explosive sensitive to shock or heating to 250°C. Will explode spontaneously during crystallization. Mixtures with calcium stearate may explode spontaneously. May explode spontaneously after prolonged contact with copper, zinc, or their alloys (e.g., brass). Incompatible with CS$_2$. Used in commercial blasting caps and military ammunition. When heated it emits highly toxic fumes of Pb and NO$_x$.

LCP000   CAS: 598-63-0   **HR: 2**
LEAD CARBONATE
mf: CO$_3$•Pb   mw: 267.20

PROP: White, heavy powder. D: 6.61, decomp @ 400° leaving residue of PbO. Insol in water, alc; sol in acetic acid, dil HNO$_3$ (effervescence).

SYNS: CARBONIC ACID, LEAD(2+) SALT (1:1) * CERUSSETE * DIBASIC LEAD CARBONATE * LEAD(2+) CARBONATE * WHITE LEAD

CONSENSUS REPORTS: IARC Cancer Review: Animal Inadequate Evidence IMEMDT 23,325,80; IMEMDT 1,40,72. Lead and its compounds are on the Community Right-To-Know List. Reported in EPA TSCA Inventory.

OSHA PEL: TWA 0.05 mg(Pb)/m^3
ACGIH TLV: TWA 0.15 mg(Pb)/m^3
NIOSH REL: (Inorganic Lead) TWA 0.10 mg(Pb)/m^3

SAFETY PROFILE: Moderately toxic by ingestion. Human systemic effects by ingestion: gastrointestinal contractions and jaundice. Experimental reproductive effects. Questionable carcinogen. Ignites spontaneously and burns fiercely in fluorine. When heated to decomposition it emits toxic fumes of Pb.

LCQ000   CAS: 7758-95-4   **HR: 3**
LEAD CHLORIDE

DOT: NA 2291
mf: Cl$_2$Pb   mw: 278.09

PROP: White crystals. Mp: 501°, bp: 950°, d: 5.85, vap press: 1 mm @ 547°. Somewhat sol in cold water, more sol in hot water; very sol in ammonium chloride, NH$_4$NO$_3$, alkali hydroxides.

SYNS: LEAD (2+) CHLORIDE * LEAD (II) CHLORIDE * LEAD DICHLORIDE * PLUMBOUS CHLORIDE

CONSENSUS REPORTS: IARC Cancer Review: Animal Inadequate Evidence IMEMDT 23,325,80. Lead and its compounds are on the Community Right-To-Know List. Reported in EPA TSCA Inventory. EPA Genetic Toxicology Program.

OSHA PEL: TWA 0.05 mg(Pb)/m^3
ACGIH TLV: TWA 0.15 mg(Pb)/m^3
NIOSH REL: (Inorganic Lead) TWA 0.10 mg(Pb)/m^3
DOT Classification: ORM-B; Label: None.

SAFETY PROFILE: Moderately toxic by ingestion. Experimental teratogenic and reproductive effects. Questionable carcinogen. Human mutation data reported. Explosive reaction with calcium when heated slightly. When heated to decomposition it emits very toxic fumes of Pb and Cl$^-$.

LCR000   CAS: 7758-97-6   **HR: 3**
LEAD CHROMATE
mf: CrO$_4$•Pb   mw: 323.19

PROP: Yellow or orange-yellow powder. One of the most insol salts. Insol in acetic acid;

sol in solns of fixed alkali hydroxides, dil HNO_3. Mp: 844°. Bp: decomp, d: 6.3.

SYNS: CANARY CHROME YELLOW 40-2250 * CHROMATE de PLOMB (FRENCH) * CHROME GREEN * CHROME LEMON * CHROME YELLOW * CHROMIC ACID, LEAD(2+) SALT (1:1) * CHROMIUM YELLOW * C.I. 77600 * C.I. PIGMENT YELLOW 34 * COLOGNE YELLOW * C.P. CHROME YELLOW LIGHT * CROCOITE * DAINICHI CHROME YELLOW G * GIALLO CROMO (ITALIAN) * KING'S YELLOW * LEAD CHROMATE(VI) * LEIPZIG YELLOW * LEMON YELLOW * PARIS YELLOW * PIGMENT GREEN 15 * PLUMBOUS CHROMATE * PURE LEMON CHROME L3GS

CONSENSUS REPORTS: IARC Cancer Review: GROUP 1 IMEMDT 7,165,87; Animal Inadequate Evidence IMEMDT 2,100,73; Animal Sufficient Evidence IMEMDT 23,205,80; Human Sufficient Evidence IMEMDT 23, 205,80. NTP Fourth Annual Report On Carcinogens, 1984. Lead and its compounds, as well as chromium and its compounds, are on the Community Right-To-Know List. Reported in EPA TSCA Inventory. EPA Genetic Toxicology Program.

OSHA PEL: TWA 0.05 mg(Pb)/m^3; CL 0.1 mg(CrO_3)/m^3
ACGIH TLV: 0.05 mg(Cr)/m^3; Human Carcinogen; (Proposed: TWA 0.05 ppm (Pb), 0.012 ppm (Cr); Suspected Human Carcinogen)
DFG MAK: Suspected Carcinogen.
NIOSH REL: (Chromium(VI)) TWA 0.001 mg(Cr(VI))/m^3; (Inorganic Lead) TWA 0.10 mg(Pb)/m^3

SAFETY PROFILE: Confirmed carcinogen with experimental neoplastigenic and tumorigenic data. Poison by intraperitoneal route. Mildly toxic by ingestion. Human mutation data reported. Potentially explosive reactions with azo-dye stuffs (e.g., dinitroaniline orange, chlorinated para red). Violent reaction with aluminum + dinitronaphthalene + heat. Forms pyrophoric mixtures with sulfur, tantalum, and iron(III) hexacyanoferrate(4^-) (e.g., brunswick green pigment, prussian blue pigment). When heated to decomposition it emits toxic fumes of Pb.

LCS000   CAS: 18454-12-1   HR: 3
LEAD CHROMATE, BASIC
mf: $CrO_4Pb \cdot OPb$ mw: 546.38

PROP: Red, amorphous or crystalline solid. Mp: 920°.

SYNS: ARANCIO CROMO (ITALIAN) * AUSTRIAN CINNABAR * BASIC LEAD CHROMATE * CHINESE RED * CHROME ORANGE * CHROMIUM LEAD OXIDE * C.I. 77601 * C.I. PIGMENT ORANGE 21 * C.I. PIGMENT RED * C.P. CHROME LIGHT 2010 * C.P. CHROME ORANGE DARK 2030 * C.P. CHROME ORANGE MEDIUM 2020 * DAINICHI CHROME ORANGE R * GENUINE ACETATE CHROME ORANGE * GENUINE ORANGE CHROME * INDIAN RED * INTERNATIONAL ORANGE 2221 * IRGACHROME ORANGE OS * LEAD CHROMATE OXIDE (MAK) * LEAD CHROMATE, RED * LIGHT ORANGE CHROME * NO. 156 ORANGE CHROME * ORANGE CHROME * ORANGE NITRATE CHROME * PALE ORANGE CHROME * PERSIAN RED * PURE ORANGE CHROME M * RED LEAD CHROMATE * VYNAMON ORANGE CR

CONSENSUS REPORTS: IARC Cancer Review: Human Sufficient Evidence IMEMDT 23,205,80; Animal Limited Evidence IMEMDT 23,205,80. NTP Fourth Annual Report On Carcinogens, 1984. Lead and its compounds, as well as chromium and its compounds are on the Community Right-To-Know List. Reported in EPA TSCA Inventory.

OSHA PEL: TWA 0.05 mg(Pb)/m^3; CL 0.1 mg(CrO_3)/m^3
ACGIH TLV: TWA 0.05 mg(Cr)/m^3; TWA 0.15 mg(Pb)/m^3
DFG MAK: Suspected Carcinogen.
NIOSH REL: (Chromium(VI)) TWA 0.001 mg(Cr(VI))/m^3; (Inorganic Lead) TWA 0.10 mg(Pb)/m^3

SAFETY PROFILE: Confirmed human carcinogen with experimental carcinogenic, neoplastigenic, and tumorigenic data. Human mutation data reported. When heated to decomposition it emits very toxic fumes of Pb.

LCT000   HR: 3
LEAD COMPOUNDS

CONSENSUS REPORTS: Lead and its compounds are on the Community Right-To-Know List.

SAFETY PROFILE: Some are experimental neoplastigens and tumorigens. Lead poisoning is one of the commonest of occupational diseases. The presence of lead-bearing materials or lead compounds in an industrial plant does not necessarily result in exposure on the part of the worker. The lead must be in such form, and so distributed, as to gain entrance into the body or tissues of the worker in measurable quantity, otherwise no exposure can be said to exist. Some lead compounds are carcinogens of the lungs and kidneys.

Mode of entry into body: 1. By inhalation of the dust, fumes, mists or vapors. (Common air contaminants). 2. By ingestion of lead compounds trapped in the upper respiratory tract or introduced into the mouth on food, tobacco, fingers, or other objects. 3. Through the skin; this route is of special importance in the case of organic compounds of lead, as lead tetraethyl. In the case of the inorganic forms of lead, this route is of no practical importance. Significant quantities of lead can be ingested from water that has been sitting in pipes with lead solder. Some water coolers may also have this type of solder.

Lead is a cumulative poison. Increasing amounts build up in the body and eventually reach a point where symptoms and disability occur.

The toxicity of the various lead compounds appears to depend upon several factors: (1) the solubility of the compound in the body fluids; (2) the fineness of the particles of the compound; solubility is greater in proportion to the fineness of the particles; (3) conditions under which the compound is being used. Where a lead compound is used as a powder, contamination of the atmosphere will be much less if the powder is kept damp. Of the various lead compounds, the carbonate, the monoxide, and the sulfate are considered to be more toxic than metallic lead or other lead compounds. Lead arsenate is very toxic due to the presence of the arsenic radical. Organolead compounds are rapidly absorbed by the respiratory and gastrointestinal systems and through the skin. Tetraethyl lead is converted in the body to triethyl lead which is a more severe neurotoxin than inorganic lead. Diagnostic mobilization of lead with calcium EDTA may be useful in questionable cases. When heated to decomposition they emit toxic fumes of Pb.

LCU000   CAS: 592-05-2   **HR: 3**
LEAD(II) CYANIDE

DOT: UN 1620
mf: C_2N_2Pb mw: 259.23

PROP: White powder.

SYNS: C.I. 77610   *   C.I. PIGMENT YELLOW 48   *   CYANURE de PLOMB (FRENCH)   *   LEAD CYANIDE (DOT)

CONSENSUS REPORTS: Lead and its compounds, as well as cyanide and its compounds, are on the Community Right-To-Know List.

OSHA PEL: TWA 0.05 mg(Pb)/m^3
ACGIH TLV: TWA 0.15 mg(Pb)/m^3
NIOSH REL: (Inorganic Lead) TWA 0.10 mg(Pb)/m^3
DOT Classification: Poison B; Label: Poison.

SAFETY PROFILE: Poison by intraperitoneal route. Violent reaction with Mg. A fire hazard and a powerful oxidizer. When heated to decomposition it emits very toxic fumes of Pb, CN$^-$, and NO$_x$.

LCW000   CAS: 19010-66-3   **HR: 2**
LEAD DIMETHYLDITHOCARBAMATE
mf: $C_6H_{12}N_2S_4 \cdot Pb$ mw: 447.63

PROP: Solid. Mp: 258°, d: 2.5.

SYNS: BIS(DIMETHYLCARBAMODITHIOATO-S,S')LEAD   *   BIS(DIMETHYLDITHIOCARBAMIATO)LEAD   *   DIMETHYLDITHIOCARBAMIC ACID, LEAD SALT   *   METHYL LEDATE   *   NCI-C02891

CONSENSUS REPORTS: IARC Cancer Review: GROUP 3 IMEMDT 7,230,87; Animal Inadequate Evidence IMEMDT 12,131,76. NCI Carcinogenesis Bioassay (feed); No Evidence: mouse, rat NCITR* NCI-CG-TR-151,79. Lead and its compounds are on the Community Right-To-Know List. Reported in EPA TSCA Inventory.

NIOSH REL: (Inorganic Lead) TWA 0.10 mg(Pb)/m^3

SAFETY PROFILE: Questionable carcinogen with experimental tumorigenic data. Mutation data reported. Combustible when exposed to heat or flame. When heated to decomposition it emits very toxic fumes of Pb, NO$_x$, and SO$_x$.

LCX000   CAS: 1309-60-0   **HR: 3**
LEAD DIOXIDE

DOT: UN 1872
mf: O_2Pb mw: 239.19

PROP: Brown, hexagonal crystals or dark brown powder. Mp: decomp @ 290°, d: 9.375. Liberates O_2 when heated. Insol in H_2O; sol in HCl evolving chlorine, sol in alkali iodide solns liberating iodine, sol in hot caustic alkali soln.

SYNS: BIOXYDE de PLOMB (FRENCH) * C.I. 77580 * LEAD BROWN * LEAD(IV) OXIDE * LEAD OXIDE BROWN * LEAD PEROXIDE (DOT) * LEAD SUPEROXIDE * PEROXYDE de PLOMB (FRENCH)

CONSENSUS REPORTS: Reported in EPA TSCA Inventory. Lead and its compounds are on the Community Right-To-Know List.

OSHA PEL: TWA 0.05 mg(Pb)/m^3
ACGIH TLV: TWA 0.15 mg(Pb)/m^3
NIOSH REL: (Inorganic Lead) TWA 0.10 mg(Pb)/m^3
DOT Classification: Oxidizer; Label: Oxidizer.

SAFETY PROFILE: Poison by intraperitoneal route. A powerful oxidizer. Probably a severe eye, skin, and mucous membrane irritant. Explosive reaction with warm potassium or sodium; cesium acetylide at 350°C; boron (when ground); yellow phosphorus (when ground); sulfinyl dichloride. Mixtures with silicon (2:1 silicon/lead dioxide) are used as initiators and heat to 1100°C when exposed to flame. Mixtures with zirconium can deflagrate (burn explosively) and are sensitive to friction, ignition, and static electricity. Violent reaction or ignition with chlorine trifluoride, hydrogen sulfide, nitrogen compounds (e.g., hydroxylamine), red phosphorus, sulfur (when ground), sulfur + sulfuric acid, peroxyformic acid. Violent reactions with powdered aluminum, Al$_4$C$_3$, metal acetylides or carbides H$_2$O$_2$, magnesium, non-metal halides, performic acid, phenyl hydrazine, S(OCl)$_2$. Vigorous reaction with seleninyl chloride, metal sulfides + heat (e.g., calcium sulfide, strontium sulfide, or barium sulfide). Incandescent reaction with powdered molybdenum or tungsten when heated, warm phosphorus trichloride, sulfur dioxide. Metal oxides increase the explosive sensitivity of nitroalkanes (e.g., nitromethane, nitroethane). Can react vigorously with reducing materials. When heated to decomposition it emits toxic fumes of Pb.

LDC000   CAS: 69029-52-3   **HR: 3**
LEAD DROSS (DOT)

DOT: UN 1794

SYNS: LEAD DROSS (CONTAINING 3% OR MORE FREE ACID) (DOT) * LEAD SCRAP (DOT)

CONSENSUS REPORTS: Reported in EPA TSCA Inventory. Lead and its compounds are on the Community Right-To-Know List.

OSHA PEL: TWA 0.05 mg(Pb)/m^3
ACGIH TLV: TWA 0.15 mg(Pb)/m^3
NIOSH REL: (Inorganic Lead) TWA 0.10 mg(Pb)/m^3
DOT Classification: ORM-C; Label: None: Corrosive Material; Label: Corrosive.

SAFETY PROFILE: A corrosive irritant to the eyes, skin and mucous membranes. When heated to decomposition it emits toxic fumes of lead.

LDE000   CAS: 13814-96-5   **HR: 3**
LEAD FLUOBORATE

DOT: NA 2291
mf: B$_2$F$_8$•Pb   mw: 380.81

SYN: TETRAFLUORO BORATE(1-) LEAD (2+)

CONSENSUS REPORTS: Reported in EPA TSCA Inventory. Lead and its compounds are on the Community Right-To-Know List.

OSHA PEL: TWA 0.05 mg(Pb)/m^3; TWA 2.5 mg(Pb)/m^3
ACGIH TLV: TWA 0.15 mg(Pb)/m^3
NIOSH REL: TWA 0.10 mg(Pb)/m^3
DOT Classification: ORM-B; Label: None.

SAFETY PROFILE: Poison by ingestion. When heated to decomposition it emits very toxic fumes of Pb, F$^-$, and BO$_x$.

LDF000   CAS: 7783-46-2   **HR: 2**
LEAD(II) FLUORIDE

DOT: NA 2811
mf: F$_2$Pb   mw: 245.19

PROP: Colorless solid. D: (orthorhombic) 8.445, d: (cubic) 7.750, mp: 824°, bp: 1293°, vap press: 10 mm @ 904°. Low solubility in water. Solubility increases in presence of HNO$_3$ or nitrates.

SYNS: LEAD DIFLUORIDE * LEAD FLUORIDE (DOT) * PLOMB FLUORURE (FRENCH) * PLUMBOUS FLUORIDE

CONSENSUS REPORTS: Lead and its compounds are on the Community Right-To-Know List. Reported in EPA TSCA Inventory.

OSHA PEL: TWA 0.05 mg(Pb)/m^3; TWA 2.5 mg(F)/m^3

ACGIH TLV: TWA 0.15 mg(Pb)/m^3; TWA 2.5 mg(F)/m^3

NIOSH REL: (Inorganic Lead) TWA 0.10 mg(Pb)/m^3

DOT Classification: OMB-B; Label: None.

SAFETY PROFILE: Moderately toxic by subcutaneous route. Vigorous reaction with fluorine. Incompatible with CaC$_2$. When heated to decomposition it emits very toxic fumes of Pb and F$^-$.

LDM000 CAS: 12709-98-7 **HR: 3**
LEAD-MOLYBDENUM CHROMATE

SYNS: CHROMIC ACID, LEAD and MOLYBDENUM SALT * CHROMIC ACID LEAD SALT with LEAD MOLYBDATE * C.I. PIGMENT RED 104 * LEAD CHROMATE, SULPHATE and MOLYBDATE * MOLYBDENUM-LEAD CHROMATE * MOLYBDENUM ORANGE

CONSENSUS REPORTS: Lead and its compounds, as well as chromium and its compounds, are on the Community Right-To-Know List.

OSHA PEL: TWA CL 0.1 mg(CrO$_3$)/m^3; TWA 0.05 mg(Pb)/m^3; TWA 5 mg(Mo)/m^3

ACGIH TLV: TWA 0.05 mg(Cr)/m^3; TWA 5 mg(Mo)/m^3; TWA 0.15 mg(Pb)/m^3

NIOSH REL: (Chromium(VI)) TWA 0.001 mg(Cr(VI))/m^3; (Inorganic Lead) TWA 0.10 mg(BrPb)/m^3

SAFETY PROFILE: Questionable carcinogen with experimental neoplastigenic and tumorigenic data. Human mutation data reported. A powerful oxidizer. Probably a severe eye, skin, and mucous membrane irritant. When heated to decomposition it emits toxic fumes of Pb, chromium trioxide, and Mo.

LDN000 CAS: 1317-36-8 **HR: 2**
LEAD MONOXIDE
mf: OPb   mw: 223.19

PROP: Exists in 2 forms: (1) red to reddish-yellow, tetragonal crystals; stable at ordinary temps. (2) Yellow, orthorhombic crystals; stable > 489°. D: 9.53, mp: 888°. Insol in water, alc; sol in acetic acid, dil HNO$_3$, warm solns of fixed alkali hydroxides.

SYNS: C.I. 77577 * C.I. PIGMENT YELLOW 46 * LEAD OXIDE * LEAD(II) OXIDE * LEAD OXIDE YELLOW * LEAD PROTOXIDE * LITHARGE * LITHARGE YELLOW L-28 * MASSICOT * MASSICOTITE * PLUMBOUS OXIDE * YELLOW LEAD OCHER

CONSENSUS REPORTS: IARC Cancer Review: Animal Inadequate Evidence IMEMDT 23,325,80. Reported in EPA TSCA Inventory. EPA Genetic Toxicology Program. Lead and its compounds are on the Community Right-To-Know List.

OSHA PEL: TWA 0.05 mg(Pb)/m^3
ACGIH TLV: TWA 0.15 mg(Pb)/m^3
NIOSH REL: (Inorganic Lead) TWA 0.10 mg(Pb)/m^3

SAFETY PROFILE: Moderately toxic by ingestion and intraperitoneal routes. Mutation data reported. A skin irritant. Questionable carcinogen. Avoid breathing dust. Wash thoroughly after contact with the material and before eating or smoking. Explosive reaction with rubidium acetylide at 200°C, zirconium + heat, silicon + aluminum + heat, chlorine + ethylene (at 100°C), perchloric acid + glycerol. Violent or explosive thermite reaction when heated with aluminum powder. Violent or explosive reaction with chlorinated rubber (above 200°C), fluoro-elastomers (at 200°C), peroxyformic acid. Violent reaction or ignition with hydrogen trisulfide. May ignite spontaneously with linseed oil, dichloromethylsilane, fluorine + glycerol. Vigorous reaction with silicon + heat. Incandescent reaction with warm aluminum carbide, lithium acetylide, boron, seleninyl chloride. Incompatible with chlorine, perchloric acid, metal acetylides, metals, non-metals. Mixtures of lead oxide with glycerol have been used as a jointing compound and may explode when exposed to powerful oxidizers. When heated to decomposition it emits toxic fumes of Pb. Used in manufacturing of storage batteries, ceramic products, paints, and rubber.

LDO000 CAS: 10099-74-8 **HR: 3**
LEAD(II) NITRATE (1:2)

DOT: UN 1469
mf: N$_2$O$_6$ • Pb   mw: 331.21

PROP: White crystals. Mp: decomp @ 470°, d: 4.53 @ 20°.

SYNS: LEAD DINITRATE * LEAD NITRATE * LEAD (2+) NITRATE * LEAD(II) NITRATE * NITRATE de PLOMB (FRENCH) * NITRIC ACID, LEAD (2+) SALT

CONSENSUS REPORTS: IARC Cancer Review: Animal Inadequate Evidence IMEMDT 23,325,80. Reported in EPA TSCA Inventory.

Lead and its compounds are on the Community Right-To-Know List.

OSHA PEL: TWA 0.05 mg(Pb)/m^3
ACGIH TLV: TWA 0.15 mg(Pb)/m^3
NIOSH REL: (Inorganic Lead) TWA 0.10 mg(Pb)/m^3
DOT Classification: Oxidizer; Label: Oxidizer, Poison.

SAFETY PROFILE: Poison by intraperitoneal route. Moderately toxic by ingestion and possibly other routes. Experimental teratogenic and reproductive effects. Questionable carcinogen. Probably a severe eye, skin, and mucous membrane irritant. Mutation data reported. A powerful oxidizer. Explodes on contact with red hot carbon, cyclopentadienylsodium (at 100-130°C), potassium acetate + heat. Reacts violently with ammonium thiocyanate, carbon, lead hypophosphite. When heated to decomposition it emits very toxic fumes of Pb and NO$_x$. Used as a mordant, a chemical reagent, and in production of matches and pyrotechnics.

LDP000   CAS: 51317-24-9   **HR: 3**
LEAD NITRORESORCINATE
mf: C$_6$H$_5$NO$_4$ • xPb   mw: 1605.45

SYNS: INITIATING EXPLOSIVE LEAD MONONITRO-RESORCINATE (DOT) * LEAD MONONITRORES-ORCINATE (DRY) (DOT)

CONSENSUS REPORTS: Lead and its compounds are on the Community Right-To-Know List.

DOT Classification: Forbidden, Dry; Class A Explosive; Label: Explosive A.

SAFETY PROFILE: Poison by ingestion and inhalation. An explosive. When heated to decomposition it emits very toxic fumes of NO$_x$ and Pb.

LDS000   CAS: 1314-41-6   **HR: 3**
LEAD OXIDE RED
mf: O$_4$Pb$_3$ mw: 685.57

PROP: Bright red powder. Mp: 890° (decomp), bp: 1472°, d: 8.32-9.16, vap press: 1 mm @ 943°.

SYNS: C.I. 77578 * C.I. PIGMENT RED 105
* DILEAD(II) LEAD(IV) OXIDE * GOLD SATINOBRE
* LEAD ORTHOPLUMBATE * LEAD TETRAOXIDE
* MINERAL ORANGE * MINERAL RED * MINIUM
* MINIUM NON-SETTING RL-95 * ORANGE LEAD
* PARIS RED * PLUMBOPLUMBIC OXIDE * RED

LEAD * RED LEAD OXIDE * SANDIX * SATURN RED * TRILEAD TETROXIDE

CONSENSUS REPORTS: Lead and its compounds are on the Community Right-To-Know List. Reported in EPA TSCA Inventory.

OSHA PEL: TWA 0.05 mg(Pb)/m^3
ACGIH TLV: TWA 0.15 mg(Pb)/m^3
NIOSH REL: (Inorganic Lead) TWA 0.10 mg(Pb)/m^3

SAFETY PROFILE: Poison by intraperitoneal route. Moderately toxic by ingestion. Combustible by chemical reaction with reducing agents. An oxidizing agent. Explodes on contact with peroxyformic acid. Ignites on contact with dichloromethylsilane. Incandescent reaction with seleninyl chloride. One percent fresh red lead decreases the explosion temperature of 2,4,6-trinitrotoluene to 192°C. Incompatible with Al; CsHC$_2$; (F$_2$ + glycerol); H$_2$S$_3$; (glycerin + HClO$_4$); RbHC$_2$; (Si + Al); Na; SO$_3$; Ti; Zr. Mixtures of lead oxide with glycerol have been used as a jointing compound and may explode when exposed to powerful oxidizers. When heated to decomposition it emits toxic fumes of Pb.

LDU000   CAS: 7446-27-7   **HR: 3**
LEAD(II) PHOSPHATE (3:2)
mf: O$_8$P$_2$ • 3Pb   mw: 811.51

PROP: Hexagonal, colorless crystals or white powder. Mp: 1014, d: 6.9-7.3. Insol in water, alc; sol in HNO$_3$, fixed alkali hydroxides.

SYNS: BLEIPHOSPHAT (GERMAN) * C.I. 77622
* LEAD ORTHOPHOSPHATE * LEAD PHOSPHATE
* LEAD PHOSPHATE (3:2) * LEAD (2+) PHOSPHATE
* NORMAL LEAD ORTHOPHOSPHATE * PHOS-PHORIC ACID, LEAD (2+) SALT (2:3) * PLUMBOUS PHOSPHATE * TRILEAD PHOSPHATE

CONSENSUS REPORTS: IARC Cancer Review: GROUP 2B IMEMDT 7,230,87; Animal Sufficient Evidence IMEMDT 23,325,80; IMEMDT 1,40,72; Human Limited Evidence IMEMDT 23,325,80. NTP Fourth Annual Report On Carcinogens, 1984. Lead and its compounds are on the Community Right-To-Know List. Reported in EPA TSCA Inventory.

OSHA PEL: TWA 0.05 μg(Pb)/m^3
ACGIH TLV: TWA 0.15 mg(Pb)/m^3
NIOSH REL: TWA 0.10 mg(Pb)/m^3

SAFETY PROFILE: Confirmed carcinogen with experimental carcinogenic and tumorigenic data. A suspected human carcinogenic. When heated to decomposition it emits very toxic fumes of Pb and PO_x.

LDY000   CAS: 7446-14-2   HR: 3
LEAD(II) SULFATE (1:1)

DOT: UN 1794
mf: $O_4S \cdot Pb$ mw: 303.25

PROP: White crystals. Mp: decomp @ 1000°, d: 6.2. Insol in alc; sol in NaOH, ammonium acetate, or tartrate soln, concentrated HI. Practically insol in water; somewhat more sol in dil HCl or HNO_3.

SYNS: ANGLISLITE * BLEISULFAT (GERMAN) * C.I. 77630 * C.I. PIGMENT WHITE 3 * FAST WHITE * FREEMANS WHITE LEAD * LEAD BOTTOMS * LEAD DROSS (DOT) * LEAD SULFATE, solid, containing more than 3% free acid (DOT) * MILK WHITE * MULHOUSE WHITE * SULFATE de PLOMB (FRENCH) * SULFURIC ACID, LEAD (2+) SALT (1:1)

CONSENSUS REPORTS: Lead and its compounds are on the Community Right-To-Know List. Reported in EPA TSCA Inventory.

OSHA PEL: TWA 0.05 mg(Pb)/m^3
ACGIH TLV: TWA 0.15 mg(Pb)/m^3
NIOSH REL: (Inorganic Lead) TWA 0.10 mg(Pb)/m^3
DOT Classification: ORM-E; Label: None; Corrosive Material; Label: Corrosive, Solid.

SAFETY PROFILE: Poison by intraperitoneal route. Moderately toxic by ingestion. Human mutation data reported. A corrosive irritant to skin, eyes and mucous membranes. Violent or explosive reaction with potassium. When heated to decomposition it emits very toxic fumes of Pb and SO_x. Used in batteries, lithography, rapid drying oil varnishes, weighting fabrics.

LDZ000   CAS: 1314-87-0   HR: 2
LEAD SULFIDE

DOT: NA 2991
mf: PbS   mw: 239.25

PROP: Silvery, metallic crystals or black powder. Mp: 1114°, bp: 1281° (subl), d: 7.5, vap press: 1 mm @ 852°. Insol in water; sol in HNO_3, hot dil HCl.

SYNS: C.I. 77640 * GALENA * NATURAL LEAD SULFIDE * PLUMBOUS SULFIDE

CONSENSUS REPORTS: Lead and its compounds are on the Community Right-To-Know List. Reported in EPA TSCA Inventory.

OSHA PEL: TWA 0.05 mg(Pb)/m^3
ACGIH TLV: TWA 0.15 mg(Pb)/m^3
NIOSH REL: TWA 0.10 mg(Pb)/m^3
DOT Classification: ORM-E; Label: None.

SAFETY PROFILE: Moderately toxic by intraperitoneal route. Mildly toxic by ingestion. Violent reaction with ICl, H_2O_2. When heated to decomposition it emits very toxic fumes of Pb and SO_x. Used in glazing earthenware, as a friction additive in clutch facings and disc brakes.

LEE000   CAS: 63918-97-8   HR: 3
LEAD TRINITRORESORCINATE

DOT: UN 0130
mf: $C_6HN_3O_8Pb$ mw: 450.29

PROP: Orange-yellow, monoclinic crystals. Mp: explodes @ 311°, d: 3.1-2.9.

SYNS: INITIATING EXPLOSIVE LEAD STYPHNATE (DOT) * INITIATING EXPLOSIVE LEAD TRINITRORESORCINATE (DOT) * LEAD STYPHNATE (DRY) (DOT) * LEAD TRINITRORESORCINATE (DOT) * LEAD 2,4,6-TRINITRORESORCINOXIDE * STYPHNATE of LEAD (DOT)

CONSENSUS REPORTS: Lead and its compounds are on the Community Right-To-Know List. Reported in EPA TSCA Inventory.

NIOSH REL: (Inorganic Lead) TWA 0.10 mg(Pb)/m^3
DOT Classification: Class A Explosive; Label: Explosive A; Forbidden, Dry.

SAFETY PROFILE: A poisonous material. A very shock-, heat- and friction-sensitive priming explosive. It has detonated spontaneously when dry. Explodes when heated to 311°. Upon decomposition it emits very toxic fumes of NO_x and Pb.

LEG000   HR: 1
LEMONGRASS OIL EAST INDIAN

SYNS: BRITISH EAST INDIAN LEMONGRASS OIL * COCHIN * EAST INDIAN LEMONGRASS OIL * LEMONGRAS OEL (GERMAN) * OIL of LEMONGRASS, EAST INDIAN

SAFETY PROFILE: Mildly toxic by ingestion. A skin irritant. When heated to decomposition it emits acrid smoke and irritating fumes.

LEH000   CAS: 8007-02-1   **HR: 1**
LEMONGRASS OIL WEST INDIAN

PROP: Main constituent is citral. From steam distillation of freshly cut and partially dried grasses of *Cymbopogon citratus* (STAPF) and *Andropogon nardus var. ceriferus* (Hack). Light yellow to brown liquid; light lemon odor. D: 0.869-0.894, refr index: 1.483. Sol in mineral oil, propylene glycol; insol in water,

SYNS: GUATEMALA LEMONGRASS OIL * MADA-GASCAR LEMONGRASS OIL * OIL of LEMONGRASS, WEST INDIAN * WEST INDIAN LEMONGRASS OIL

CONSENSUS REPORTS: Reported in EPA TSCA Inventory.

SAFETY PROFILE: A skin irritant. When heated to decomposition it emits acrid smoke and irritating fumes.

LEI000   CAS: 8008-56-8   **HR: 2**
LEMON OIL

PROP: Expressed from the peel of the fruit of *Citrus limon* L. Burmann filius (Fam. *Rutaceae*). Pale yellow liquid; taste and odor of lemon peel. D: 0.849, refr index: 1.473 @ 20°. Misc with dehydrated alc, glacial acetic acid.

SYNS: CEDRO OIL * LEMON OIL, COLDPRESSED (FCC) * LEMON OIL, EXPRESSED * OIL of LEMON * ZITRONEN OEL (GERMAN)

CONSENSUS REPORTS: Reported in EPA TSCA Inventory.

SAFETY PROFILE: Moderately toxic by ingestion. A skin irritant. Questionable carcinogen with experimental tumorigenic data. When heated to decomposition it emits acrid smoke and irritating fumes.

LEI025   **HR: 1**
LEMON OIL, DESERT TYPE, COLDPRESSED

PROP: Expressed without heat from the peel of the fruit of *Citrus limon* L. Burmann filius (Fam. *Rutaceae*). Pale yellow liquid; taste and odor of lemon peel. D: 0.846, refr index: 1.473 @ 20°. Misc with dehydrated alc, glacial acetic acid.

SYN: OIL of LEMON, DESERT TYPE, COLDPRESSED

SAFETY PROFILE: A skin irritant. When heated to decomposition it emits acrid smoke and irritating fumes.

LEI030   **HR: 1**
LEMON OIL, DISTILLED

PROP: From distillation of fresh peel from *Citrus limon* L. Burmann filius (Fam. *Rutaceae*). Pale yellow liquid; taste and odor of fresh lemon peel. D: 0.842, refr index: 1.470 @ 20°. Misc with dehydrated alc, glacial acetic acid.

SYN: OIL of LEMON, DISTILLED

SAFETY PROFILE: A skin irritant. When heated to decomposition it emits acrid smoke and irritating fumes.

LEN000   CAS: 21609-90-5   **HR: 3**
LEPTOPHOS
mf: $C_{13}H_{10}BrCl_2O_2PS$ mw: 412.07

SYNS: ABAR * O-(4-BROMO-2,5-DICHLOROPHE-NYL)-O-METHYL PHENYLPHOSPHONOTHIOATE * O-(2,5-DICHLORO-4-BROMOPHENYL)-O-METHYL PHE-NYLTHIOPHOSPHONATE * FOSVEL * K62-105 * MBCP * O-METHYL-O-(4-BROMO-2,5-DICHLORO-PHENYL)PHENYL THIOPHOSPHONATE * O-METHYL-O-2,5-DICHLORO-4-BROMOPHENYL PHENYLTHIOPHOS-PHONATE * NK 711 * PHENYLPHOSPHONOTHIOIC ACID O-(4-BROMO-2,5-BROMO-2,5-DICHLOROPHENYL) O-METHYL ESTER * PHOSVEL * PSL * VELSICOL 506 * VELSICOL VCS 506

CONSENSUS REPORTS: EPA Extremely Hazardous Substances List.

SAFETY PROFILE: Poison by ingestion, skin contact, intraperitoneal, and subcutaneous routes. Experimental teratogenic and reproductive effects. Mutation data reported. Used in insecticides. When heated to decomposition it emits very toxic fumes of SO_x, PO_x, Br^-, and Cl^-.

LEO000   CAS: 13093-88-4   **HR: 3**
LEPTRYL
mf: $C_{22}H_{28}N_2O_2S$ mw: 384.58

SYNS: 2-METHOXY-10-(3-(4-HYDROXYPIPERIDINO)-2-METHYLPROPYL)PHENOTHIAZINE * 3-METHOXY-10-(3-(4-HYDROXYPIPERIDYL)-2-METHYLPROPYL)PHE-NOTHIAZINE * 2-METHOXY-10-(2-METHYL-3-(4-HYDROXYPIPERIDINO)PROPYL)PHENOTHIAZINE * 1-(3-(2-METHOXYPHENOTHIAZIN-10-YL)-2-METHYL-PROPYL)-4-PIPERIDINOL * PERIMETAZINE * PERIMETHAZINE * RP 9159 * 9159 RP

SAFETY PROFILE: Poison by ingestion, intraperitoneal, subcutaneous, and intravenous routes. When heated to decomposition it emits very toxic fumes of NO_x and SO_x.

LEP000   CAS: 51473-23-5   **HR: 3**
LERGOTRILE MESYLATE
mf: $C_{17}H_{20}ClN_3 \cdot CH_4O_3S$ mw: 397.96

SYN: d-2-CHLORO-6-METHYLERGOLINE-8-β-ACETONI-TRILE METHANESULFONIC ACID SALT

CONSENSUS REPORTS: Cyanide and its compounds are on the Community Right To Know List.

SAFETY PROFILE: Poison by ingestion and intraperitoneal routes. When heated to decomposition it emits toxic fumes of SO_x, Cl^-, CN^-, and NO_x.

LEQ000   CAS: 63917-01-1   **HR: 3**
LETHANE (SPECIAL)

PROP: A liquid. Bp: 160-190° @ 0.1 mm, bp: 120-125° @ 0.28 mm. Insol in water; misc with hydrocarbons and most organic solvents. A mixture of Lethane 60 (3 parts) and Lethane 384 (1 part).

SAFETY PROFILE: Poison by ingestion, skin contact, and intraperitoneal routes. Moderately toxic by subcutaneous route. Mildly toxic by inhalation. Insecticides with n-butyl carbitol-thiocyanate, etc., in a light petroleum base. Accidental and suicidal poisonings have occurred. Symptoms include drowsiness followed by coma, the limbs becoming flaccid and the appearance of twitching and convulsions. The pupils may dilate and respiration may become labored. Cyanosis and vomiting occur. The lethanes are mild irritants and, in higher doses, narcotic. Can be absorbed by intact skin. When heated to decomposition it emits very toxic fumes of SO_x and NO_x.

LER000   CAS: 328-38-1   **HR: 1**
dl-LEUCINE
mf: $C_6H_{13}NO_2$ mw: 131.17

PROP: dl Form (synthetic form): Leaflets from water; odorless with sweet taste. Mp: 290 (decomp). Sol in water, sltly sol in alc; insol in ether.

SYN: dl-2-AMINO-4-METHYLVALERIC ACID

CONSENSUS REPORTS: Reported in EPA TSCA Inventory.

SAFETY PROFILE: Mildly toxic by intraperitoneal route. When heated to decomposition it emits toxic fumes of NO_x.

LES000   CAS: 61-90-5   **HR: 2**
l-LEUCINE
mf: $C_6H_{13}NO_2$ mw: 131.20

PROP: An essential amino acid; occurs in isomeric forms. White crystals. Mp (dl): 332° with decomp, mp (l): 295°, d: 1.239 @ 18°/4°. l Form (natural): glistening, hexagonal plates from aq alc. D: 1.291 @ 18°, subl @ 145-148°, decomp @ 293-295°. Sol in water, sltly sol in alc; insol in ether.

SYNS: α-AMINOISOCAPROIC ACID * 2-AMINO-4-METHYLPENTANOIC ACID * 2-AMINO-4-METHYLVAL-ERIC ACID * l-2-AMINO-4-METHYLVALERIC ACID * α-AMINO-γ-METHYLVALERIC ACID * LEUCIN (GERMAN) * LEUCINE * 4-METHYLNORVALINE

CONSENSUS REPORTS: Reported in EPA TSCA Inventory.

SAFETY PROFILE: Moderately toxic by subcutaneous route. Experimental teratogenic and reproductive effects. When heated to decomposition it emits toxic fumes of NO_x.

LET000   CAS: 92-23-9   **HR: 3**
LEUCINOCAINE
mf: $C_{17}H_{28}N_2O_2$ mw: 292.47

SYN: 2-(DIETHYLAMINO)-4-METHYL-1-PENTANOL-p-AMINOBENZOATE (ESTER)

SAFETY PROFILE: Poison by subcutaneous and intravenous routes. When heated to decomposition it emits toxic fumes of NO_x.

LEY000   CAS: 57-22-7   **HR: 3**
LEUROCRISTINE
mf: $C_{46}H_{56}N_4O_{10}$ mw: 825.06

SYNS: LCR * NCI-C04864 * NSC-67574 * ON-COVIN * 22-OXOVINCALEUKOBLASTINE * VCR * VINCRISTINE * VINCRYSTINE * VINKRISTIN

CONSENSUS REPORTS: NCI Carcinogenesis Studies (ipr); No Evidence: mouse, rat CANCAR 40,1935,77. EPA Genetic Toxicology Program.

SAFETY PROFILE: Poison by parenteral, intraperitoneal, and intravenous routes. Human systemic effects by parenteral, intravenous, and possibly other routes: sensory change involving peripheral nerves, flaccid paralysis without anesthesia, somnolence, anorexia, convulsions or effect on seizure threshold, nausea or vomiting, changes in blood cell count and bone marrow, pulmonary and gastrointestinal changes. Experimental reproductive effects. Questionable carcinogen with experimental tumorigenic and teratogenic data. Human mutation data reported. A skin irritant. When heated to decomposition it emits toxic fumes of NO_x.

LEZ000　　　　CAS: 2068-78-2　　　**HR: 3**
LEUROCRISTINE SULFATE (1:1)
mf: $C_{46}H_{56}N_4O_{10} \cdot H_2O_4S$　　　mw: 923.14

SYNS: KYOCRISTINE * LILLY 37231 * NSC 67574 * ONCOVIN * VCR SULFATE * VINCRISTINE SULFATE ONCORIN * VINCRISTINSULFAT (GERMAN) * VINCRISUL

CONSENSUS REPORTS: IARC Cancer Review: GROUP 3 IMEMDT 7,372,87; Human Inadequate Evidence IMEMDT 26,365,81; Animal Inadequate Evidence IMEMDT 26,365,81.

SAFETY PROFILE: Poison by intraperitoneal and intravenous routes. Experimental teratogenic and reproductive effects. Questionable carcinogen. Human mutation data reported. When heated to decomposition it emits very toxic fumes of NO_x and SO_x.

LFA000　　　　CAS: 6649-23-6　　　**HR: 3**
LEVAMISOLE
mf: $C_{11}H_{12}N_2S$　　　mw: 204.31

SYNS: 6-PHENYL-2,3,5,6-TETRAHYDROIMIDAZO(2,1-b)THIAZOLE * 2,3,5,6-TETRAHYDRO-6-PHENYLIMIDAZO(2,1-b)THIAZOLE

SAFETY PROFILE: Poison by ingestion, intravenous, intraperitoneal and subcutaneous routes. Human systemic effects by ingestion: coma, skin dermatitis and irritation, and fever. When heated to decomposition it emits very toxic fumes of NO_x and SO_x.

LFA020　　　　CAS: 16595-80-5　　　**HR: 3**
LEVAMISOLE HYDROCHLORIDE
mf: $C_{11}H_{12}N_2S \cdot ClH$　　　mw: 240.77

SYNS: CITARIN L * DECARIS * IMIDAZO(2,1-β)THIAZOLE MONOHYDROCHLORIDE * KW-2-LE-T * LEVAMISOLE * LEV HYDROCHLORIDE * LEVOMYSOL HYDROCHLORIDE * NEMICIDE * NIRATIC HYDROCHLORIDE * NIRATIC-PURON HYDROCHLORIDE * NSC-177023 * R-12,564 * RIPERCOL-L * SOLASKIL * STIMAMIZOL HYDROCHLORIDE * (−)-2,3,5,6-TETRAHYDRO-6-PHENYLIMIDAZO(2,1-b)THIAZOLE HYDROCHLORIDE * 1-(−)-2,3,5,6-TETRAHYDRO-6-PHENYL-IMIDAZO(2,1-B) THIAZOLE HYDROCHLORIDE * 1-TETRAMISOLE HY-

DROCHLORIDE * TRAMISOL * TRAMISOLE * WORM-CHEK

SAFETY PROFILE: Poison by ingestion, intraperitoneal, subcutaneous, intravenous, and intramuscular routes. Human systemic effects by ingestion: thrombocytopenia. Experimental teratogenic and reproductive effects. Mutation data reported. When heated to decomposition it emits very toxic fumes of NO_x, SO_x, and HCl.

LFF000　　　　CAS: 1403-17-4　　　**HR: 3**
LEVORIN

SYNS: CANDEPTIN * CANDIMON * VANOBID

SAFETY PROFILE: Poison by intraperitoneal route. Moderately toxic by ingestion and subcutaneous routes. Experimental teratogenic data reported.

LFG000　　　　CAS: 77-07-6　　　**HR: 3**
LEVORPHANOL
mf: $C_{17}H_{23}NO$　　　mw: 257.41

SYNS: levo-DROMORAN * (−)-3-HYDROXY-N-METHYLMORPHINAN * LEVORPHAN

SAFETY PROFILE: Poison by ingestion, subcutaneous, and intravenous routes. Experimental reproductive effects reported. When heated to decomposition it emits toxic fumes of NO_x.

LFI000　　　　CAS: 7660-25-5　　　**HR: 3**
LEVULOSE
mf: $C_6H_{12}O_6$　　　mw: 180.18

PROP: White, hygroscopic crystals or crystalline powder; odorless with sweet taste. D: 1.6. Sol in methanol, ethanol, water.

SYNS: FRUCTOSE (FCC) * FRUIT SUGAR * FRUTABS * LAEVORAL * LAEVOSAN * LEVUGEN

SAFETY PROFILE: Questionable carcinogen with experimental tumorigenic data. When heated to decomposition it emits acrid smoke and fumes.

LFK000　　　　CAS: 58-25-3　　　**HR: 3**
LIBRIUM
mf: $C_{16}H_{14}ClN_3O$　　　mw: 299.78

SYNS: CD 2 * CDP * CHLORDIAZEPOXIDE * CHLORIDIAZEPIDE * CHLORIDIAZEPOXIDE * CHLORODIAZEPOXIDE * 7-CHLORO-2-METHYL-AMINO-5-PHENYL-3H-1,4-BENZODIAZEPINE 4-OXIDE

∗ 7-CHLORO-2-METHYLAMINO-5-PHENYL-3H-1,4-BEN-ZODIAZEPIN 4-OXIDE ∗ 7-CHLORO-N-METHYL-5-PHE-NYL-3H-1,4-BENZODIAZEPIN-2-AMINE-4-OXIDE ∗ CLOPOXIDE ∗ CLORDIAZEPOSSIDO (ITALIAN) ∗ 7-CLORO-2-METILAMINO-5-FENIL-3H-1,4-BENZOIDI-AZEPINA 4-OSSIDO (ITALIAN) ∗ DECACIL ∗ EDEN ∗ ELENIUM ∗ IFIBRIUM ∗ KALMOCAPS ∗ LIBRAX ∗ LIBRININ ∗ LIBRITABS ∗ ME-SURAL ∗ METHAMINODIAZEPOXIDE ∗ MILDMEN ∗ NAPOTON ∗ PSICOSAN ∗ RADEPUR ∗ VIOPSICOL

CONSENSUS REPORTS: EPA Genetic Toxicology Program.

SAFETY PROFILE: Poison by ingestion, intraperitoneal, intravenous, subcutaneous, and intramuscular routes. Human male reproductive effects by ingestion: impotence. Human systemic effects by ingestion: sleep, euphoria, somnolence, ataxia, and antianxiety. Experimental teratogenic and reproductive effects. Mutation data reported. Has been implicated in development of asplastic anemia. Used as a pharmaceutical and veterinary drug. When heated to decomposition it emits very toxic fumes of NO_x.

LFN300 CAS: 8008-94-4 HR: 2
LICORICE ROOT EXTRACT

SYNS: GLYCYRRHIZA ∗ GLYCYRRHIZAE (LATIN) ∗ GLYCYRRHIZA EXTRACT ∗ GLYCYRRHIZINA ∗ KANZO (JAPANESE) ∗ LICORICE ∗ LICORICE EXTRACT ∗ LICORICE ROOT

SAFETY PROFILE: Moderately toxic by intraperitoneal and subcutaneous routes. Mildly toxic by ingestion. Mutation data reported. When heated to decomposition it emits acrid smoke and irritating fumes.

LFO000 CAS: 23257-56-9 HR: 3
LIDEPRAN HYDROCHLORIDE
mf: $C_{14}H_{19}NO_2 \cdot ClH$ mw: 269.80

SYNS: LEVOPHACETOPERANE HYDROCHLORIDE ∗ LEVOPHACETOPERAN HYDROCHLORIDE ∗ α-PHE-NYL-2-PIPERIDINEMETHANOL ACETATE HYDROCHLO-RIDE ∗ 1-PHENYL-1-(2-PIPERIDYL)-1-ACETOXYMETH-ANE HYDROCHLORIDE ∗ PHENYL-(2-PIPERIDYL)METHYL ACETATE HYDROCHLORIDE ∗ RP 8228

SAFETY PROFILE: Poison by ingestion, intraperitoneal, subcutaneous, and intravenous routes. When heated to decomposition it emits very toxic fumes of NO_x and HCl.

LFU000 CAS: 5989-27-5 HR: 3
d-LIMONENE
mf: $C_{10}H_{16}$ mw: 136.26

PROP: Colorless liquid; citrus odor. Bp: 175.5-176°, d: 0.8402 @ 25°/4°, refr index: 1.471. Misc with alc, fixed oils; sltly sol in glycerin; insol in propylene glycol, water.

SYNS: FEMA No. 2633 ∗ (+)-4-ISOPROPENYL-1-METHYLCYCLOHEXENE ∗ d-(+)-LIMONENE ∗ (+)-R-LIMONENE ∗ d-p-MENTHA-1,8-DIENE ∗ p-MENTHA-1,8-DIENE ∗ (R)-1-METHYL-4-(1-METH-YLETHENYL)-CYCLOHEXENE ∗ NCI-C55572

CONSENSUS REPORTS: Reported in EPA TSCA Inventory.

SAFETY PROFILE: Poison by intravenous route. Moderately toxic by intraperitoneal and intraduodenal routes. Mildly toxic by ingestion. Experimental reproductive effects. Questionable carcinogen with experimental tumorigenic and teratogenic data. Reacts explosively with iodine pentafluoride + tetrafluoroethylene (the pentafluoride reacts exothermically with the inhibitor and initiates explosive polymerization of the TFE). When heated to decomposition it emits acrid smoke and irritating fumes. Used as a food additive, flavor agent, packaging material, as an inhibitor of tetrafluoroethylene polymerization, and as a gallstone solubilizer.

LFV000 CAS: 96-08-2 HR: 3
LIMONENE DIOXIDE
mf: $C_{10}H_{16}O_2$ mw: 168.26

SYNS: 1,2,8,9-DIEPOXYLIMONENE ∗ 1,2:8,9-DI-EPOXYMENTHANE ∗ 1,2:8,9-DIEPOXY-p-MENTHANE ∗ DIPENTENE DIOXIDE ∗ EPOXIDE 269 ∗ 4-(1,2-EPOXY-1-METHYLETHYL)-1-METHYL-7-OXABICYCLO(4.1.0)HEPTANE ∗ UNOXAT EPOXIDE 269

CONSENSUS REPORTS: Reported in EPA TSCA Inventory.

SAFETY PROFILE: Moderately toxic by skin contact and intramuscular routes. Mildly toxic by ingestion. A skin irritant. Questionable carcinogen with experimental tumorigenic data. When heated to decomposition it emits acrid smoke and irritating fumes.

LFW000 CAS: 1317-63-1 HR: D
LIMONITE

PROP: Consists mainly of hydrated sesquioxide of iron.

SYNS: BROWN HEMATITE * BROWN IRON ORE * BROWN IRONSTONE CLAY * IRON SESQUIOXIDE HYDRATED

CONSENSUS REPORTS: IARC Cancer Review: Animal Inadequate Evidence IMEMDT 1,29,72

SAFETY PROFILE: Questionable carcinogen.

LFX000   CAS: 78-70-6   **HR: 2**
LINALOOL
mf: $C_{10}H_{18}O$ mw: 154.28

PROP: Colorless liquid; odor similar to that of bergamot oil and French lavender. D: 0.858-0.868 @ 25°, refr index: 1.461, bp: 195-199°, flash p: 172°F. Sol in alc, ether, fixed oils, propylene glycol; insol in glycerin.

SYNS: ALLO-OCIMENOL * 2,6-DIMETHYL-2,7-OCTADIENE-6-OL * 2,6-DIMETHYLOCTA-2,7-DIEN-6-OL * 3,7-DIMETHYLOCTA-1,6-DIEN-3-OL * 3,7-DIMETHYL-1,6-OCTADIEN-3-OL * FEMA No. 2635 * LINALOL * LINALYL ALCOHOL

CONSENSUS REPORTS: Reported in EPA TSCA Inventory.

SAFETY PROFILE: Moderately toxic by ingestion. Mildly toxic by skin contact. A skin irritant. A synthetic flavoring substance and adjuvant. When heated to decomposition it emits acrid smoke and irritating fumes.

LFY100   CAS: 115-95-7   **HR: 1**
LINALYL ACETATE
mf: $C_{12}H_{20}O_2$ mw: 196.32

PROP: Clear, colorless, oily liquid; odor of bergamot. Bp: 108-110°, d: 0.898-0.914, flash p: 185°F. Sol in alc, ether, diethyl phthalate, benzyl benzoate, mineral oil, fixed oils; sltly sol in propylene glycol; insol in water, glycerin.

SYNS: ACETIC ACID LINALOOL ESTER * BERGAMIOL * 3,7-DIMETHYL-1,6-OCTADIEN-3-OL ACETATE * 3,7-DIMETHYL-1,6-OCTADIEN-3-YL ACETATE * FEMA No. 2636 * LICAREOL ACETATE * LINALOL ACETATE * LINALOOL ACETATE

CONSENSUS REPORTS: Reported in EPA TSCA Inventory.

SAFETY PROFILE: Mildly toxic by ingestion. Combustible liquid. When heated to decomposition it emits acrid smoke and irritating fumes.

LFZ000   CAS: 126-64-7   **HR: 1**
LINALYL BENZOATE
mf: $C_{17}H_{22}O_2$ mw: 258.39

PROP: Found in the essential oils of Ylang-Ylang and Tuberose. Yellow to brown-yellow liquid; tuberose odor. D: 0.980-0.999, refr index: 1.505-1.520, flash p: 208°F. Sol in chloroform, alc, ether; insol in water.

SYNS: 3,7-DIMETHYL-1,6-OCTADIEN-3-OL BENZOATE * 3,7-DIMETHYL-1,6-OCTADIEN-3-YL BENZOATE * 1,5-DIMETHYL-1-VINYL-4-HEXEN-1-OL BENZOATE * 1,5-DIMETHYL-1-VINYL-4-HEXEN-1-YL BENZOATE * FEMA No. 2638

CONSENSUS REPORTS: Reported in EPA TSCA Inventory.

SAFETY PROFILE: A skin irritant. Combustible liquid. When heated to decomposition it emits acrid smoke and irritating fumes.

LGA050   **HR: 2**
LINALYL FORMATE
mf: $C_{11}H_{18}O_2$ mw: 182.26

PROP: Colorless liquid; citrus, herbaceous odor. D: 0.910-0.918, refr index: 1.453-1.458, flash p: 189°F. Sol in alc, fixed oils; sltly sol in propylene glycol, water; insol in glycerin @ 202°.

SYNS: 3,7-DIMETHYL-1,6-OCTADIEN-3-YL FORMATE * FEMA No. 2642

SAFETY PROFILE: Combustible liquid. When heated to decomposition it emits acrid smoke and irritating fumes.

LGB000   CAS: 78-35-3   **HR: 1**
LINALYL ISOBUTYRATE
mf: $C_{14}H_{24}O_2$ mw: 224.38

PROP: Colorless liquid; fresh, rosy odor. D: 0.882-0.888, refr index: 1.446-1.451, flash p: +212°F. Misc with alc, chloroform, ether; insol in water @ 20°.

SYNS: 3,7-DIMETHYL-1,6-OCTADIEN-3-OL ISOBUTYRATE * 3,7-DIMETHYL-1,6-OCTADIEN-3-YL ISOBUTYRATE * 1,5-DIMETHYL-1-VINYL-4-HEXENYL ESTER, ISOBUTYRIC ACID * FEMA No. 2640 * LINALOOL ISOBUTYRATE

CONSENSUS REPORTS: Reported in EPA TSCA Inventory.

SAFETY PROFILE: Combustible liquid. Mildly toxic by ingestion. When heated to decomposition it emits acrid smoke and irritating fumes.

LGC100 HR: 2
LINALYL PROPIONATE
mf: $C_{13}H_{22}O_2$ mw: 210.32

PROP: Colorless liquid; fresh, pear odor. D: 0.893-0.902, refr index: 1.449-1.454, flash p: 189°F. Sol in alc, fixed oils; insol in glycerin @ 226°.

SYN: FEMA No. 2645

SAFETY PROFILE: Combustible liquid. When heated to decomposition it emits acrid smoke and irritating fumes.

LGD000 CAS: 154-21-2 HR: 3
LINCOMYCIN
mf: $C_{18}H_{34}N_2O_6S$ mw: 406.60

PROP: Sol in methanol, ethanol, butanol, isopropanol, ethyl acetate, n-butyl acetate, amylacetate, etc. Moderately sol in water.

SYNS: ALBIOTIC * LINCOCIN * LINCOLCINA * LINCOLNENSIN * LINCOMYCINE (FRENCH) * NSC-70731 * U-10149

CONSENSUS REPORTS: EPA Genetic Toxicology Program.

SAFETY PROFILE: Poison by intramuscular route. Moderately toxic by ingestion and intraperitoneal routes. When heated to decomposition it emits very toxic fumes of SO_x and NO_x.

LGG000 CAS: 60-33-3 HR: 1
LINOLEIC ACID
mf: $C_{18}H_{32}O_2$ mw: 280.50

PROP: Colorless oil, easily oxidized by air. D: 0.9038 @ 18°/4°, mp: −12°, bp: 230° @ 16 mm. Sol in ether and ethanol; misc with dimethyl formamide, fat solvents, oils.

SYNS: LEINOLEIC ACID * 9,12-LINOLEIC ACID * cis,cis-9,12-OCTADECADIENOIC ACID * cis-9,cis-12-OCTADECADIENOIC ACID * 9,12-OCTADECADIENOIC ACID

CONSENSUS REPORTS: Reported in EPA TSCA Inventory.

SAFETY PROFILE: A human skin irritant. Ingestion can cause nausea and vomiting. When heated to decomposition it emits acrid smoke and irritating fumes.

LGK000 CAS: 8001-26-1 HR: 1
LINSEED OIL

PROP: Yellowish liquid, peculiar odor, bland taste. Sltly sol in alc; misc with chloroform, ether, petr ether, carbon disulfide, oil, turpentine. Bp: 343°, mp: −19°, d: 0.93, flash p: (raw oil) 432°F (CC), flash p: (boiled) 403°F (CC), autoign temp: 650°F. From seed of *Linum usitatissimum*.

SYNS: GROCO * L-310

CONSENSUS REPORTS: Reported in EPA TSCA Inventory.

SAFETY PROFILE: An allergen and skin irritant to humans. Combustible liquid when exposed to heat or flame; can react with oxidizing materials. Subject to spontaneous heating. Violent reaction with Cl_2. To fight fire, use CO_2, dry chemical.

LGL000 HR: 2
LIQUEFIED CARBON DIOXIDE
mf: CO_2 mw: 44.0

PROP: Heavy gas or liquid under pressure. Mp: −56.6° @ 3952 mm; bp: −78.5° (subl); d: 1.977 g/L @ 0°; (liquid) 1.101 @ −37°.

SYN: LIQUID CARBONIC GAS

SAFETY PROFILE: Contact with skin or living tissue can cause frostbite-like burns. This material is stable when very cold. Solid CO_2 goes directly (sublimes) to gaseous CO_2 which is mainly an asphyxiant.

LGM000 CAS: 68476-85-7 HR: 2
LIQUEFIED PETROLEUM GAS

DOT: UN 1075

SYNS: LPG * L.P.G. (OSHA, ACGIH) * PETROLEUM GAS, LIQUEFIED

CONSENSUS REPORTS: Reported in EPA TSCA Inventory.

OSHA PEL: TWA 1000 ppm
NIOSH REL: TWA 350 mg/m³; CL 1800 mg/m³/15M
ACGIH TLV: TWA 1000 ppm
DOT Classification: Flammable Gas; Label: Flammable Gas.

SAFETY PROFILE: Olefinic impurities may lend a narcotic effect or it may act as a simple asphyxiant. A very dangerous fire hazard when exposed to heat or flame. Can react with oxidizing materials. To fight fire, use CO_2, dry chemical, water spray. Used as a fuel refrigerant, propellant, and raw material in chemical synthesis.

LGO000   CAS: 7439-93-2   **HR: 3**
LITHIUM

DOT: UN 1415
af: Li   aw: 6.94

PROP: Silver-colored, light metal; mixture of isotopes Li^6 and Li^7. Mp: 179°, bp: 1317°, d: 0.534 @ 25°, vap press: 1 mm @ 723°. Sol in liquid ammonia. Keep under mineral oil or other liquid free from O_2 or water.

SYNS: LITHIUM METAL (DOT) * LITHIUM METAL, IN CARTRIDGES (DOT)

CONSENSUS REPORTS: Reported in EPA TSCA Inventory.

DOT Classification: Flammable Solid; Label: Flammable Solid and Dangerous When Wet.

SAFETY PROFILE: A very dangerous fire hazard when exposed to heat or flame. The powder may ignite spontaneously in air. The solid metal ignites above 180°C. It will burn in oxygen, nitrogen, or carbon dioxide, and will continue to burn in sand or sodium carbonate. The use of most types of fire extinguishers (e.g., water, foam, carbon dioxide, halocarbons, sodium carbonate, sodium chloride, and other dry powders) may cause an explosion. Molten lithium is extremely reactive and attacks such inert materials as sand, concrete, and ceramics.

Explosive reaction with bromobenzene, carbon + lithium tetrachloroaluminate + sulfinyl chloride, diazomethane. Forms very friction- and impact-sensitive explosive mixtures with halogens [e.g., bromine, iodine (above 200°C)], halocarbons (e.g., bromoform, carbon tetrabromide, carbon tetrachloride, carbon tetraiodide, chloroform, dichloromethane, diiodomethane, fluorotrichloromethane, tetrachloroethylene, trichloroethylene, 1,1,2-trichlorotrifluoroethane).

Violent reaction with acetonitrile, sulfur, mercury (potentially explosive), metal oxides [e.g., chromium(III) oxide (at 185°C), molybdenum trioxide (at 180°C), niobium pentoxide (at 320°C), titanium dioxide (at 200-400°C), tungsten trioxide (at 200°C), vanadium pentoxide (at 394°C)], iron(II) sulfide (at 260°C), manganese telluride (at 230°C), hot water, bromine pentafluoride (may ignite with lithium powder), platinum (at about 540°C), trifluoromethyl hypofluorite (at about 170°C), arsenic, beryllium, maleic anhydride, carbides, carbon dioxide, carbon monoxide + water, chlorine, chromium, chromium trichloride, cobalt alloys, iron sulfide, diborane, manganese alloys, nickel alloys, nitric acid, nitrogen, organic matter, oxygen, phosphorus, rubber, silicates, $NaNO_2$, Ta_2O_5, Fe alloys, V, $ZrCl_4$, CHI_3, trifluoromethylhypofluorite.

Ignition on contact with carbon + sulfinyl chloride (when ground), nitric acid (becomes violent), viton (poly(1,1-difluorethylene-hexafluoropropylene), chlorine tri- and penta-fluorides (hypergolic reaction), diborane (forms a complex which is pyrophyoric), hydrogen (above 300°C).

Incandescent reaction with ethylene + heat, nitrogen + metal chlorides [e.g., chromium trichloride, zirconium tetrachloride, nitryl fluoride (at 200°C)]. Incompatible with atmospheric gases, bromine pentafluoride, diazomethane, metal chlorides, metal oxides, non-metal oxides.

When burned it emits toxic fumes of LiO_2 and hydroxide. Reacts vigorously with water or steam to produce heat and hydrogen. Can react vigorously with oxidizing materials. To fight fire, use special mixtures of dry chemical, soda ash, graphite. Note: water, sand, carbon tetrachloride and carbon dioxide are ineffective.

LGQ000   CAS: 50475-76-8   **HR: 3**
LITHIUM ACETYLIDE COMPLEXED with ETHYLENEDIAMINE

DOT: NA 2813

SYN: LITHIUM ACETYLIDE-ETHYLENEDIAMINE COMPLEX

DOT Classification: Flammable Solid; Label: Flammable Solid and Dangerous When Wet.

SAFETY PROFILE: A very flammable, unstable mixture. When heated to decomposition it emits toxic fumes of NO_x.

LGT000   CAS: 7782-89-0   **HR: 3**
LITHIUM AMIDE

DOT: UN 1412
mf: H_2LiN mw: 22.97

PROP: White, crystalline solid or powder. Subl in NH_3 current. Insol in anhydrous ether, benzene, toluene. Mp: 380-400°. D: 1.178 @ 17.50.

SYNS: LITHAMIDE * LITHIUM AMIDE, POWDERED (DOT)

CONSENSUS REPORTS: Reported in EPA TSCA Inventory.

DOT Classification: Flammable Solid; Label: Dangerous When Wet; Flammable Solid; Label: Flammable Solid.

SAFETY PROFILE: A powerful irritant to skin, eyes, and mucous membranes. Flammable when exposed to heat or flame. Ammonia is liberated and lithium hydroxide is formed when this compound is exposed to moisture. Reacts violently with water or steam to produce toxic and flammable vapors. Vigorous reaction with oxidizing materials. Exothermic reaction with acid or acid fumes. When heated to decomposition it emits very toxic fumes of LiO, NH_3, and NO_x. Used in synthesis of drugs, vitamins, steroids, and other organics.

LGU000   CAS: 305-97-5   HR: 3
LITHIUM ANTIMONY THIOMALATE
mf: $C_{12}H_9O_{12}S_3Sb \cdot 6Li$ mw: 604.78

SYNS: ANTHIOLIMINE * ANTHIOMALINE * LITHIUM ANTIMONIOTHIOMALATE * MERCAPTO-SUCCINIC ACID ANTIMONATE(III) HEXALITHIUM SALT * MERCAPTOSUCCINIC ACID-S-ANTIMONY DERIVATIVE LITHIUM SALT * MERCAPTOSUCCINIC ACID, THIOANTHIMONATE(III), DILITHIUM SALT * 2,2',2''-(STIBILIDYNETRIS(THIO)TRIS-BUTANEDIOIC ACID HEXA-LITHIUM SALT

CONSENSUS REPORTS: Antimony and its compounds are on the Community Right-To-Know List.

OSHA PEL: TWA 0.05 mg(Sb)/m^3
ACGIH TLV: TWA 0.05 mg(Sb)/m^3
NIOSH REL: (Antimony) TWA 0.5 mg(Sb)/m^3

SAFETY PROFILE: Poison by intraperitoneal and intravenous routes. Human systemic effects by ingestion and intravenous routes: hallucinations, distorted perceptions, nausea or vomiting, skin dermatitis and fever. An anthelmintic agent. When heated to decomposition it emits very toxic fumes of SO_x, Sb, and Li_2O.

LGZ000   CAS: 554-13-2   HR: 3
LITHIUM CARBONATE (2:1)
mf: $CO_3 \cdot 2Li$ mw: 73.89

PROP: White, light alkaline, crystalline powder. D: 2.11 @ 17.5°; mp: 618°. Insol in alc. @ 17.5°.

SYNS: CAMCOLIT * CANDAMIDE * CARBOLITH * CARBONIC ACID, DILITHIUM SALT * CARBONIC ACID LITHIUM SALT * CEGLUTION * CP-15467-61 * DILITHIUM CARBONATE * ESKALITH * HYP-NOREX * LIMAS * LISKONUM * LITHANE * LITHICARB * LITHINATE * LITHIUM CARBONATE * LITHOBID * LITHONATE * LITHOTABS * NSC-16895 * PLENUR * PRIADEL * QUILO-NUM RETARD

CONSENSUS REPORTS: Reported in EPA TSCA Inventory.

SAFETY PROFILE: Poison by intraperitoneal and intravenous routes. Human carcinogenic and teratogenic data. Moderately toxic by ingestion, subcutaneous, and possibly other routes. Human systemic effects by ingestion and possibly other routes: toxic psychosis, tremors, changes in fluid intake, muscle weakness, increased urine volume, and gastrointestinal changes. Human reproductive effects by ingestion and possibly other routes: effects on newborn including apgar score changes and other neonatal measures or effects. Human teratogenic effects by ingestion and possibly other routes: developmental abnormalities of the cardiovascular system, central nervous system, musculoskeletal and gastrointestinal systems. Experimental reproductive effects. Questionable carcinogen producing leukemia and thyroid tumors. Human mutation data reported. Used in the treatment of manic-depressive psychoses. Incompatible with fluorine.

LHB000   CAS: 7447-41-8   HR: 3
LITHIUM CHLORIDE
mf: ClLi   mw: 42.39

PROP: Cubic, white, deliquescent crystals. Mp: 605°, bp: 1350°, d: 2.068 @ 25°, vap press: 1 mm @ 547°.

SYNS: CHLORKU LITU (POLISH) * CHLORURE de LITHIUM (FRENCH)

CONSENSUS REPORTS: Reported in EPA TSCA Inventory. EPA Genetic Toxicology Program.

SAFETY PROFILE: Human poison by ingestion. Experimental poison by intravenous and intracerebral routes. Moderately toxic by subcutaneous and intraperitoneal routes. Experimental teratogenic and reproductive effects. Questionable carcinogen with experimental neoplastigenic data. Human systemic effects by ingestion:

somnolence, tremors, nausea or vomiting. An eye and severe skin irritant. Human mutation data reported. This material has been recommended and used as a substitute for sodium chloride in "salt-free" diets, but cases have been reported in which the ingestion of lithium chloride has produced dizziness, ringing in the ears, visual disturbances, tremors and mental confusion. In most cases, the symptoms disappeared when use was discontinued. Prolonged absorption may cause disturbed electrolyte balance, impaired renal function. Reaction is violent with BrF_3. When heated to decomposition it emits toxic fumes of Cl^-. Used for dehumidification in the air conditioning industry. Also used to obtain lithium metal.

LHD000　　　　　　　　　　　　　**HR: 3**
LITHIUM CHROMATE
mf: $CrH_2O_4 \cdot 2Li$　　　mw: 131.90

PROP: Yellow, crystalline, deliquescent powder. Mp: $-2H_2O$ @ 150°.

SYNS: CHROMIC ACID, DILITHIUM SALT * CHROMIUM LITHIUM OXIDE * DILITHIUM CHROMATE

CONSENSUS REPORTS: Chromium and its compounds are on the Community Right-To-Know List.

ACGIH TLV: TWA 0.05 mg(Cr)/m^3, Confirmed Human Carcinogen.
DOT Classification: ORM-E; Label: None.

SAFETY PROFILE: A toxic material. Combustible when exposed to heat or flame. An oxidizer. It can react with reducing materials. Potentially explosive reaction with zirconium at 450-600°C. When heated to decomposition it emits toxic fumes of Li_2O.

LHE000　　　　　　　　　　　　　**HR: D**
LITHIUM COMPOUNDS

SAFETY PROFILE: Lithium oxide, hydroxide, carbonate, etc., are strong bases and their solutions in water are very caustic. Otherwise, toxicity of lithium compounds is a function of their solubility in water. The halide salts, except the fluoride, are highly soluble in water. The carbonate, phosphate, oxalate, and fluoride are relatively insoluble in water. Lithium ion has central nervous system toxicity. In industry, the most hazardous lithium compound is the hydride. It produces large amounts of hydrogen gas when exposed to water; this reaction can cause severe damage to exposed tissue. Some lithium compounds, particularly the carbonate, are used in psychiatry. The difference between therapeutic levels of lithium and toxic levels is small. Plasma lithium concentrations of 2 mmol/L are associated with toxic symptoms. Concentrations of 4 mmol/L can be fatal.

The initial effects of lithium exposure are tremors of the hands, nausea, micturition, slurred speech, sluggishness, sleepiness, vertigo, thirst, and increased urine volume. Effects from continued exposure are apathy, anorexia, fatigue, lethargy, muscular weakness, and changes in ECG. Long-term exposure leads to hypothyroidism, leukocytosis, edema, weight gain, polydipsia/polyuria (increased water intake leading to increased urinary output), memory impairment, seizures, kidney damage, shock, hypotension, cardiac arrhythmias, coma, death. Have been implicated in development of aplastic anemia.

LHF000　　　CAS: 7789-24-4　　　**HR: 3**
LITHIUM FLUORIDE
mf: FLi　　mw: 25.94

PROP: Fine, white powder. Mp: 845°, bp: 1681°, d: 2.635 @ 20°, vap press: 1 mm @ 1047°. Sol in acids.

SYNS: LITHIUM FLUORURE (FRENCH) * TLD 100

CONSENSUS REPORTS: Reported in EPA TSCA Inventory.

OSHA PEL: TWA 2.5 mg(F)/m^3
ACGIH TLV: TWA 2.5 mg(F)/m^3
NIOSH REL: (Inorganic Fluorides) TWA 2.5 mg(F)/m^3

SAFETY PROFILE: Poison by ingestion and subcutaneous routes. When heated to decomposition it emits toxic fumes of F^-. Used as a flux in enamils, glasses, glazes, and welding.

LHH000　　　CAS: 7580-67-8　　　**HR: 3**
LITHIUM HYDRIDE
DOT: UN 1414/UN 2805
mf: HLi　　mw: 7.95

PROP: White, translucent, crystals. Mp: 680°, d: 0.76-0.77. Darkens rapidly on exposure to light. Decomp in water liberating LiOH and H_2.

SYN: HYDRURE de LITHIUM (FRENCH)

CONSENSUS REPORTS: Reported in EPA TSCA Inventory. EPA Extremely Hazardous Substances List.

OSHA PEL: TWA 0.025 mg/m^3
ACGIH TLV: TWA 0.025 mg/m^3
DFG MAK: 0.025 mg/m^3
DOT Classification: Flammable Solid; Label: Flammable Solid and Dangerous When Wet.

SAFETY PROFILE: Poison by inhalation. A severe eye, skin, and mucous membrane irritant. Upon contact with moisture, lithium hydroxide is formed. The LiOH formed is very caustic and therefore highly toxic, particularly to lungs and respiratory tract skin and mucous membranes. The powder ignites spontaneously in air. The solid can ignite spontaneously in moist air. Mixtures of the powder with liquid oxygen are explosive. Ignites on contact with dinitrogen oxide, oxygen + moisture. To fight fire, use special mixtures of dry chemical.

LHJ000   CAS: 13840-33-0   **HR: 1**
LITHIUM HYPOCHLORITE
DOT: UN 1471
mf: ClO•Li   mw: 58.39

SYN: LITHIUM HYPOCHLORITE COMPOUND, dry, containing more than 39% available chlorine (DOT)

CONSENSUS REPORTS: Reported in EPA TSCA Inventory.

DOT Classification: Oxidizer; Label: Oxidizer.

SAFETY PROFILE: A powerful oxidizer. An eye, skin, and mucous membrane irritant. When heated to decomposition it emits very toxic fumes of Li$_2$O and Cl$^-$. Used for swimming pool chlorination, and as a laundry bleach.

LHM000   CAS: 26134-62-3   **HR: 3**
LITHIUM NITRIDE
DOT: UN 2806
mf: Li$_3$N   mw: 34.82

PROP: Brownish-red, hexagonal crystals; slowly decomp on contact with moisture. D: 1.3, mp: 845°.

CONSENSUS REPORTS: Reported in EPA TSCA Inventory.

DOT Classification: Flammable Solid; Label: Flammable Solid and Dangerous When Wet.

SAFETY PROFILE: A powerful reducing agent. Upon contact with moisture, it decomposes into lithium hydroxide, lithium compounds and ammonia. The powder may ignite spontaneously in moist air. Flammable at elevated temperatures; ignites and burns intensely in air. Violent reaction with silicon tetrafluoride; copper(I) chloride + heat. To fight fire, use dry chemical, sand, graphite; avoid use of water or carbon tetrachloride. When heated to decomposition it emits very toxic fumes of Li$_2$O and NO$_x$. Used as a strong reducing agent in organic synthesis and a solid electrolyte in lithium batteries.

LHO000   CAS: 12031-80-0   **HR: 2**
LITHIUM PEROXIDE
DOT: UN 1472
mf: Li$_2$O$_2$ mw: 45.88

PROP: Fine, white powder or sandy yellow, granular material. Mp: decomp, d: 2.14 @ 20°.

CONSENSUS REPORTS: Reported in EPA TSCA Inventory.

DOT Classification: Oxidizer; Label: Oxidizer.

SAFETY PROFILE: A powerful oxidizer and irritant to skin, eyes, and mucous membranes. A very dangerous fire hazard because it is an extremely powerful oxidizing agent. Will react with water or steam to produce heat; on contact with reducing materials, can react vigorously.

LHP000   CAS: 68848-64-6   **HR: 3**
LITHIUM SILICON
DOT: UN 1417
PROP: Solid. Composition: Li + Si.

DOT Classification: Flammable Solid; Label: Flammable Solid and Dangerous When Wet.

SAFETY PROFILE: A very dangerous fire hazard in the form of dust when exposed to heat or flame or by chemical reaction with moisture or acids. In contact with water, silane and hydrogen are evolved. Slightly explosive in the form of dust when exposed to flame. Will react with water or steam to produce flammable vapors; on contact with oxidizing materials, can react vigorously; on contact with acid or acid fumes, can emit toxic and flammable fumes. To fight fire, use CO$_2$, dry chemical.

LHS000   CAS: 16853-85-3   **HR: 3**
LITHIUM TETRAHYDROALUMINATE
DOT: UN 1410/UN 1411
mf: AlH$_4$•Li   mw: 37.96

PROP: White, microcrystalline lumps.

SYNS: ALUMINUM LITHIUM HYDRIDE * LITHIUM ALANATE * LITHIUM ALUMINOHYDRIDE * LITHIUM ALUMINUM HYDRIDE (DOT) * LITHIUM ALUMINUM HYDRIDE, ETHEREAL (DOT) * LITHIUM ALUMINUM TETRAHYDRIDE

CONSENSUS REPORTS: Reported in EPA TSCA Inventory.

ACGIH TLV: TWA 2 mg(Al)/m^3

DOT Classification: Flammable Solid; Label: Flammable Solid and Dangerous When Wet; Flammable Liquid; Label: Flammable Liquid, Ethereal; Flammable Liquid; Label: Dangerous When Wet, Flammable Liquid, Ethereal.

SAFETY PROFILE: Stable in dry air at room temperature. It decomposes above 125° forming Al, H$_2$ and lithium hydride. Very powerful reducer. Can ignite if pulverized even in a dry box. Reacts violently with air, acids, alcohols, benzoyl peroxide, boron trifluoride etherate, (2-chloromethyl furan + ethyl acetate), diethylene glycol dimethyl ether, diethyl ether, 1,2-dimethoxyethane, dimethyl ether, methyl ethyl ether, (nitriles + H$_2$O), perfluoro-succinamide, (perfluorosuccinamide + H$_2$O), tetrahydrofuran, water. To fight fire, use dry chemical, including special formulations of dry chemicals as recommended by the supplier of the lithium aluminum hydride. Do not use water, fog, spray or mist. Incompatible with bis(2-methoxyethyl)ether, CO$_2$, BF$_3$, diethyl etherate, dibenzoyl peroxide, 3,5-dibromocyclopentene, 1,2-dimethoxy ethane, ethyl acetate, fluoro amides, pyridine, tetrahydrofuran. Used as a reducing agent in the preparation of pharmaceuticals.

LHT000   CAS: 16949-15-8   **HR: 3**
LITHIUM TETRAHYDROBORATE
DOT: UN 1413
mf: BH$_4$•Li   mw: 21.79

SYN: LITHIUM BOROHYDRIDE (DOT)

CONSENSUS REPORTS: Reported in EPA TSCA Inventory.

DOT Classification: Flammable Solid; Label: Flammable Solid and Dangerous When Wet.

SAFETY PROFILE: Poison by ingestion, inhalation, and skin contact. Flammable; can liberate H$_2$. Incompatible H$_2$O as moisture on fibers of cellulose or as liquid.

LHZ000   CAS: 134-63-4   **HR: 3**
LOBELINE HYDROCHLORIDE
mf: C$_{22}$H$_{27}$NO$_2$•ClH   mw: 373.96

SYNS: (−)-2-(6-(β-HYDROXYPHENETHYL)-1-METHYL-2-PIPERIDYL)-ACETOPHENONE HYDROCHLORIDE * (−)-LOBELINE HYDROCHLORIDE * LOBELIN HYDROCHLORIDE

SAFETY PROFILE: Poison by subcutaneous, intravenous, and intraperitoneal routes. When heated to decomposition it emits toxic fumes of NO$_x$ and HCl.

LIA000   CAS: 9000-40-2   **HR: 1**
LOCUST BEAN GUM
PROP: From the ground endosperms of *Ceratonia ailiqua* (L.) Taub. (Fam. *Leguminosae*). White powder; odorless and tasteless but acquires a leguminous taste when boiled in water. A galactomannan polysaccharide. Mw: 310,000 (approx). Insol in most organic solvents.

SYNS: ALGAROBA * CAROB BEAN GUM * CAROB FLOUR * NCI-C50419 * ST. JOHN'S BREAD * SUPERCOL

CONSENSUS REPORTS: NTP Carcinogenesis Bioassay (feed); No Evidence: mouse, rat NTPTR* NTP-TR-221,82. Reported in EPA TSCA Inventory.

SAFETY PROFILE: Mildly toxic by ingestion. When heated to decomposition it emits acrid smoke and irritating fumes.

LIC000   CAS: 8012-74-6   **HR: 3**
LONDON PURPLE
DOT: UN 1621

SYN: LONDON PURPLE, solid (DOT)

DOT Classification: Poison B; Label: Poison.

SAFETY PROFILE: A poison. When heated to decomposition it emits very toxic fumes of As and NO$_x$.

LII000   CAS: 8016-31-7   **HR: 2**
LOVAGE OIL
PROP: The constituents include d-α-terpineol, butyl dihydrophthalides, butyl tetrahydrophthalides, coumarin, aldehydes and acetic and isovaleric acid. From steam distillation of fresh root of *Levisticum officinale* L. Koch syn. *Angelica levisticum*, Baillon (Fam. *Umbelliferae*). Yellow to green to brown liquid; strong odor and taste. D: 1.034-1.057, refr index: 1.536-1.554

@ 20°. Sol in fixed oils; sltly sol in mineral oil; insol in glycerin, propylene glycol.

CONSENSUS REPORTS: Reported in EPA TSCA Inventory.

SAFETY PROFILE: Moderately toxic by ingestion. A skin irritant. When heated to decomposition it emits acrid smoke and irritating fumes.

LIM000 CAS: 3105-97-3 **HR: 3**
LUCANTHONE METABOLITE
mf: $C_{20}H_{24}N_2O_2S$ mw: 356.52

SYNS: 1-((2-(DIETHYLAMINO)ETHYL)AMINO)-4-(HY-DROXYMETHYL)THIOXANTHEN-9-ONE * 1-((2-(DI-ETHYLAMINO)ETHYL)AMINO)-4-(HYDROXYMETHYL)9H-THIOXANTHEN-9-ONE * HYCANTHON * HYCAN-THONE * NSC-134434 * WIN 24933

CONSENSUS REPORTS: EPA Genetic Toxicology Program.

SAFETY PROFILE: Poison by subcutaneous, intravenous, and intramuscular routes. Moderately toxic by ingestion. Experimental teratogenic and reproductive effects. Human mutation data reported. Questionable carcinogen with experimental carcinogenic data. When heated to decomposition it emits very toxic fumes of NO_x and SO_x.

LIV000 CAS: 21884-44-6 **HR: 3**
LUTEOSKYRIN
mf: $C_{30}H_{22}O_{12}$ mw: 574.52

PROP: Yellow rectangular crystals. Mp: 278° (decomp). Anthraquinoid hepatotoxin of *Penicillium islandicum sopp*.

SYNS: 5H,6H-6,5A,13A,14-(1,2,3,4)BUTANETETRAY-CYCLOOCTA(1,2-B:5,6-B')DINAPHTHALENE * 8,8'-DIHYDROXY-RUGULOSIN * FLAVOMYCELIN * (−)-LUTEOSKYRIN

CONSENSUS REPORTS: IARC Cancer Review: GROUP 3 IMEMDT 7,56,87; Animal Limited Evidence IMEMDT 10,163,76.

SAFETY PROFILE: Poison by ingestion, intraperitoneal, subcutaneous, and intravenous routes. Human mutation data reported. Questionable carcinogen producing leukemia and thyroid tumors. When heated to decomposition it emits acrid smoke and irritating fumes.

LIW000 CAS: 10099-66-8 **HR: 3**
LUTETIUM CHLORIDE
mf: Cl_3Lu mw: 281.32

PROP: Colorless crystals. Sublimes above 750°, mp: 892° ± 2°, sol in H_2O.

CONSENSUS REPORTS: Reported in EPA TSCA Inventory.

SAFETY PROFILE: Poison by intraperitoneal route. Mildly toxic by ingestion. When heated to decomposition it emits toxic fumes of Cl^-.

LIX000 CAS: 63917-04-4 **HR: 3**
LUTETIUM CITRATE

SAFETY PROFILE: Poison by intraperitoneal route. When heated to decomposition it emits acrid smoke and irritating fumes.

LIY000 CAS: 10099-67-9 **HR: 3**
LUTETIUM(III) NITRATE (1:3)
mf: $N_3O_9 \cdot Lu$ mw: 361.00

SYN: NITRIC ACID, LUTETIUM(3+) SALT

CONSENSUS REPORTS: Reported in EPA TSCA Inventory.

SAFETY PROFILE: Poison by intraperitoneal route. When heated to decomposition emits toxic fumes of NO_x.

LJB000 CAS: 583-58-4 **HR: 3**
3,4-LUTIDINE
mf: C_7H_9N mw: 107.17

PROP: Liquid. Bp: 163.5-164.5°.

SYN: 3,4-DIMETHYLPYRIDINE

CONSENSUS REPORTS: Reported in EPA TSCA Inventory.

SAFETY PROFILE: Poison by skin contact. Moderately toxic by ingestion and inhalation. When heated to decomposition it emits toxic fumes of NO_x.

LJE000 CAS: 8015-14-3 **HR: 3**
LYNDIOL
mf: $C_{21}H_{26}O_2 \cdot C_{20}H_{28}O$ mw:594.95

SYNS: LYNESTRENOL mixed with MESTRANOL * LYNESTROL mixed with MESTRANOL * LYNOES-TRENOL mixed with MESTRANOL * MESTRANOL mixed with LYNESTRENOL * MESTRANOL mixed with LYNES-TROL * NORACYCLINE * OVANON * OVARIOS-TAT (FRENCH) * RESTOVAR * SISTOMETRENOL

CONSENSUS REPORTS: EPA Genetic Toxicology Program.

SAFETY PROFILE: Suspected human carcinogen producing liver tumors. Human systemic effects by ingestion: dyspnea, nausea or vomiting, and fever. Experimental teratogenic and reproductive effects. Used as an oral contraceptive. When heated to decomposition it emits acrid smoke and irritating fumes.

LJF000   HR: 3
d-LYSERGIC ACID DIETHYLAMIDE

SYN: LYSERGIC ACID DIETHYLAMIDE, 1-ISOMER

SAFETY PROFILE: Poison by intravenous route. When heated to decomposition it emits toxic fumes of NO_x.

LJG000   HR: 3
d-LYSERGIC ACID DIETHYLAMIDE TARTRATE

SYNS: 9,10-DIDEHYDRO-N,N-DIETHYL-6-METHYL-ERGOLINE-8-β-CARBOXAMIDE-d- TARTRATE with METHANOL (1:2) * LSD TARTRATE * LYSERGIC ACID DIETHYLAMIDE TARTRATE

SAFETY PROFILE: Experimental reproductive effects. Human mutation data reported. When heated to decomposition it emits toxic fumes of NO_x.

LJH000   CAS: 4238-84-0   HR: 3
d-LYSERGIC ACID DIMETHYLAMIDE
mf: $C_{18}H_{21}N_3O$ mw: 295.42

SYNS: DAM-57 * 9,10-DIDEHYDRO-N,N,6-TRIMETHYLERGOLINE-8-β-CARBOXAMIDE

SAFETY PROFILE: Poison by intravenous route. Human systemic effects by ingestion: hallucinations, distorted perceptions, toxic psychosis, arteriolar or venous dilation. When heated to decomposition it emits toxic fumes of NO_x.

LJI000   CAS: 478-99-9   HR: 3
LYSERGIC ACID ETHYLAMIDE
mf: $C_{18}H_{21}N_3O$ mw: 295.42

SYNS: 9,10-DIDEHYDRO-N-ETHYL-6-METHYLERGOLINE-8-β-CARBOXAMIDE, N-ETHYLLYSERGAMIDE * LAE-32 * d-LYSERGIC ACID MONOETHYLAMIDE

SAFETY PROFILE: Poison by intravenous route. Human systemic effects by ingestion: hallucinations, distorted perceptions, toxic psychosis. When heated to decomposition it emits toxic fumes of NO_x.

LJL000   CAS: 60-79-7   HR: 3
d-LYSERGIC ACID-l,2-PROPANOLAMIDE
mf: $C_{19}H_{23}N_3O_2$ mw: 325.45

SYNS: BASERGIN * CORNOCENTIN * 9,10-DIDEHYDRO-N-(α-(HYDROXYMETHYL)ETHYL)-6METHYL-ERGOLINE-8-β-CARBOXAMIDE * ERGOATETRINE * ERGOBASINE * ERGOKLININE * ERGOMETRINE * ERGONOVINE * ERGOTOCINE * ERGOTRATE * ERMETRINE * N-(α-(HYDROXYMETHYL)ETHYL)-d-LYSERGOMIDE * N-(1-(HYDROXYMETHYL)ETHYL)-d-LYSERGOMIDE * d-LYSERGIC ACID-1-HYDROXYMETHYLETHYLAMIDE * LYSERGIC ACID PROPANOLAMIDE * MARGONOVINE * NEOFEMERGEN * SECACORNIN * SECOMETRIN * SYNTOMETRINE

SAFETY PROFILE: Poison by intravenous and possibly other routes. Human systemic effects by intramuscular and possibly other routes: convulsions, excitement, motor activity changes, cyanosis and respiratory depression. When heated to decomposition it emits toxic fumes of NO_x.

LJM000   CAS: 2385-87-7   HR: 3
LYSERGIC ACID PYROLIDATE
mf: $C_{20}H_{23}N_3O$ mw: 321.46

SYNS: LSD-25-PYRROLIDATE * d-LYSERGIC ACID PYRROLIDIDE

SAFETY PROFILE: Poison by intravenous route. Human systemic effects by ingestion: hallucinations, distorted perceptions, toxic psychosis, arteriolar or venous dilation. When heated to decomposition it emits toxic fumes of NO_x.

LJO000   CAS: 657-27-2   HR: 1
l-LYSINE MONOHYDROCHLORIDE
mf: $C_6H_{14}N_2O_2 \cdot ClH$ mw: 182.68

PROP: White powder. Mp: 235-236°. Sol in water; insol in alc and ether. Crystals from dil ethanol. Mp: 263-264° (decomp) when anhydrous.

SYNS: 2,6-DIAMINOHEXANOIC ACID HYDROCHLORIDE * l-LYSINE HYDROCHLORIDE * LYSINE MONOHYDROCHLORIDE

CONSENSUS REPORTS: Reported in EPA TSCA Inventory.

SAFETY PROFILE: Mildly toxic by ingestion. When heated to decomposition it emits very toxic fumes of HCl and NO_x.

LJR000 CAS: 147-20-6 **HR: 3**
LYSSIPOLL
mf: C$_{19}$H$_{23}$NO mw: 281.43

SYNS: ALLERGEN * AN 1041 * BELFENE * 4-(BENZHYDRYLOXY)-1-METHYLPIPERIDINE * DAFEN * DAYFEN * DIAFEN * 4-(DIPHE-NYLMETHOXY)-1-METHYLPIPERIDINE * DIPHENYL-PYRALINE * DIPHENYLPYRILENE * HISPRIL * HISTRYL * HISTYN * MEPIBEN * N-METH-YLPIPERIDYL-(4)-BENZHYDRYLAETHER SALZSAUREN SALZE (GERMAN) * NEARGAL * P 253

SAFETY PROFILE: Poison by ingestion and intravenous routes. An eye irritant. When heated to decomposition it emits toxic fumes of NO$_x$.

M

MAC500   CAS: 3248-93-9   **HR: 3**
MAGENTA BASE
mf: $C_{20}H_{19}N_3$ mw: 301.42

SYNS: C.I. SOLVENT RED 41 * FUCHSINE BASE * ROSANILINE BASE * WAXOLINE RED A

CONSENSUS REPORTS: Reported in EPA TSCA Inventory.

SAFETY PROFILE: Poison by ingestion. A human skin irritant. Mutation data reported. When heated to decomposition it emits toxic fumes of NO_x.

MAC650   CAS: 546-93-0   **HR: 1**
MAGNESITE
mf: $CO_3 \cdot Mg$ mw: 84.32

PROP: Very light, white powder; odorless. D: 3.04; decomp @ 350°. Sol in acids; insol in water and alc.

SYNS: CARBONATE MAGNESIUM * CARBONIC ACID, MAGNESIUM SALT * C.I. 77713 * DCI LIGHT MAGNESIUM CARBONATE * HYDROMAGNESITE * MAGMASTER * MAGNESIA ALBA * MAGNESIUM CARBONATE * MAGNESIUM(II) CARBONATE (1:1) * MAGNESIUM CARBONATE, PRECIPITATED * STAN-MAG MAGNESIUM CARBONATE

CONSENSUS REPORTS: Reported in EPA TSCA Inventory.

OSHA PEL: TWA Total Dust: 15 mg/m³; Respirable Fraction: 5 mg/m³
ACGIH TLV: TWA (nuisance particulate) 10 mg/m³ of total dust (when toxic impurities are not present, e.g., quartz < 1%).

SAFETY PROFILE: Incompatible with formaldehyde. When heated to decomposition it emits acrid smoke and irritating fumes.

MAC750   CAS: 7439-95-4   **HR: 3**
MAGNESIUM

DOT: UN 1418/UN 1869/UN 2950
af: Mg   aw: 24.31

PROP: Hexagonal, silvery-white crystals. Mp: 651°, bp: 1100°, d: 1.74 @ 5°, d: 1.738 @ 20°, vap press: 1 mm @ 621°.

SYNS: MAGNESIO (ITALIAN) * MAGNESIUM BORINGS * MAGNESIUM CLIPPINGS (DOT) * MAGNE-SIUM GRANULES COATED, PARTICLE SIZE NOT LESS THAN 149 MICRONS (DOT) * MAGNESIUM METAL (DOT) * MAGNESIUM PELLETS * MAGNESIUM POWDER (DOT) * MAGNESIUM RIBBONS * MAGNESIUM SCALPINGS (DOT) * MAGNESIUM SCRAP (DOT) * MAGNESIUM SHAVINGS (DOT) * MAGNESIUM SHEET * MAGNESIUM TURNINGS (DOT)

CONSENSUS REPORTS: Reported in EPA TSCA Inventory.

DOT Classification: Label: Flammable Solid and Dangerous When Wet.

SAFETY PROFILE: Poison by ingestion. Inhalation of dust and fumes can cause metal fume fever. The powdered metal ignites readily on the skin causing burns. Particles embedded in the skin can produce gaseous blebs which heal slowly.

A dangerous fire hazard in the form of dust or flakes when exposed to flame or oxidizing agents. In solid form, magnesium is difficult to ignite because heat is conducted rapidly away from the source of ignition; it must be heated above its melting point before it will burn. However, in finely divided form, it may be ignited by a spark or the flame of a match. Magnesium fires do not flare up violently unless there is moisture present. Therefore, it must be kept away from water, moisture, etc. It may be ignited by a spark, match flame, or even spontaneously when the material is finely divided and damp, particularly with water-oil emulsion. Moderately explosive in the form of dust when exposed to flame. Also, magnesium reacts with moisture, acids, etc., to evolve hydrogen, a highly dangerous fire and explosion hazard.

Explosive reaction or ignition with calcium carbonate + hydrogen + heat; gold cyanide + heat; mercury cyanide + heat; silver oxide + heat; fused nitrates; phosphates; or sulfates (e.g., ammonium nitrate, metal nitrates); chloroformamidinium nitrate + water (when ignited with powder). The powder may explode on contact with halocarbons (e.g., chloromethane; chloroform; or carbon tetrachloride); and explodes when sparked in dichlorodifluoromethane. Hypergolic reaction with nitric acid + 2-nitroaniline. Mixtures of powdered magnesium and methanol are more powerful than some military explosives. Mixtures of magnesium powder + water can be detonated. Reacts with acetylenic compounds including traces of acetylene found in ethylene gas to form explosive magnesium acetylide.

Violent reactions with ammonium salts; chlorate salts; beryllium fluoride; boron diiodophosphide; carbon tetrachloride + methanol; 1,1,1-trichloroethane; 1,2-dibromoethane; halogens or interhalogens (e.g., fluorine; chlorine; bromine; iodine vapor; chlorine trifluoride; iodine heptafluoride); hydrogen iodide; metal oxides + heat (e.g., beryllium oxide; cadmium oxide; copper oxide; mercury oxide; molybdenum oxide; tin oxide; zinc oxide); nitrogen (when ignited); silicon dioxide powder + heat; polytetrafluoroethylene powder + heat; sulfur + heat; tellurium + heat; barium peroxide; nitric acid vapor; hydrogen peroxide; ammonium nitrate; sodium iodate + heat; sodium nitrate + heat; dinitrogen tetraoxide (when ignited); lead dioxide. Ignites in carbon dioxide at 780°C; molten barium carbonate + water; fluorocarbon polymers + heat; carbon tetrachloride or trichloroethylene (on impact); dichlorodifluoromethane + heat.

Incompatible with ethylene oxide; metal oxosalts; oxidants; potassium carbonate; Al + $KClO_4$; [$Ba(NO_3)_2$ + BaO_2 + Zn]; bromobenzyl trifluoride; CaC; carbonates; $CHCl_3$; [$CuSO_4$ (anhydrous) + NH_4NO_3 + $KClO_3$ + H_2O]; $CuSO_4$; (H_2 + $CaCO_3$); CH_3Cl; NO_2; liquid oxygen; metal cyanides (e.g., cadmium cyanide; cobalt cyanide; copper cyanide; lead cyanide; nickel cyanide; zinc cyanide); performic acid; phosphates; $KClO_3$; $KClO_4$; $AgNO_3$; $NaClO_4$; (Na_2O_2 + CO_2); sulfates; trichloroethylene; Na_2O_2.

To fight fire, operators and fire fighters can approach a magnesium fire to within a few feet if no moisture is present. Water and ordinary extinguishers, such as CO_2, carbon tetrachloride, etc., should not be used on magnesium fires. G-1 powder or powdered talc should be used on open fires. Dangerous when heated; burns violently in air and emits fumes; will react with water or steam to produce hydrogen.

MAD000   CAS: 142-72-3   **HR: 3**
MAGNESIUM ACETATE
mf: $C_4H_6O_4$•Mg   mw: 142.41

PROP: Tetrahydrate, colorless or white, deliquescent crystals. D: 1.45, mp: approx 80°; Very sol in water and alc. Keep container well closed.

SYNS: ACETIC ACID, MAGNESIUM SALT * MAGNESIUM DIACETATE

CONSENSUS REPORTS: Reported in EPA TSCA Inventory.

SAFETY PROFILE: Poison by intravenous route. When heated to decomposition it emits acrid smoke and irritating fumes.

MAE000   CAS: 10326-21-3   **HR: 3**
MAGNESIUM CHLORATE
DOT: UN 2723
mf: Cl_2O_6•Mg   mw: 191.21

PROP: White, deliquescent crystals or powder; bitter taste. Mp: 35°, bp: decomp @ 120°, d: 1.80 @ 25°. Sltly sol in alc. Keep well closed.

SYNS: CHLORATE SALT of MAGNESIUM * DEFOL-ATE * E-Z-OFF * KRMD 58 * MAGNESIUM DICHLORATE * MAGRON * MC DEFOLIANT * ORTHO MC

CONSENSUS REPORTS: Reported in EPA TSCA Inventory.

DOT Classification: Oxidizer; Label: Oxidizer.

SAFETY PROFILE: Moderately toxic by intraperitoneal route. Mildly toxic by ingestion. Probably an eye, skin, and mucous membrane irritant. Experimental reproductive effects. A defoliant. A powerful oxidizer. Explosive reaction with copper(I) sulfide. Incandescent reaction with antimony(III) sulfide, arsenic(III) sulfide, tin(II) sulfide, tin(IV) sulfide. Incompatible with Al, As, C, charcoal, Cu, MnO_2, metal sulfides, dibasic organic acids, organic matter, P, S. When heated to decomposition it emits toxic fumes of Cl^-.

MAE250   CAS: 7786-30-3   **HR: 3**
MAGNESIUM CHLORIDE
mf: Cl_2Mg mw: 95.21

PROP: Thin, white to opaque, gray granules and/or flakes, deliquescent. Mp: 708° (712° with rapid heating), bp: 1412°, d: 2.325. Sol in water (evolving much heat) and alc.

SYN: DUS-TOP

CONSENSUS REPORTS: Reported in EPA TSCA Inventory. EPA Genetic Toxicology Program.

SAFETY PROFILE: Poison by intraperitoneal and intravenous routes. Moderately toxic by ingestion and subcutaneous routes. Human mutation data reported. In humid environments it causes steel to rust very rapidly. When heated to decomposition it emits toxic fumes of Cl⁻.

MAE750   HR: D
MAGNESIUM COMPOUNDS

SAFETY PROFILE: Variable toxicity. The inhalation of fumes of freshly sublimed magnesium oxide may cause metal fume fever. There is no evidence that magnesium produces true systemic poisoning. Particles of metallic magnesium or magnesium alloy which perforate the skin or gain entry through cuts and scratches may produce a severe local lesion characterized by the evolution of gas and acute inflammatory reaction, frequently with necrosis. The condition has been called a ''chemical gas gangrene.'' Gaseous blebs may develop within 24 hours of the injury. The inflammatory response is marked at the site of injury and there may be signs of lymphangitis. The lesion is very slow to heal. The most serious hazard presented by magnesium is the danger from burns. Protection necessary for personnel handling and processing magnesium is usually no different from that which is necessary for other metals. The toxicity of magnesium compounds is usually that of the anion. When heated to decomposition it emits toxic fumes of MgO.

MAG250   CAS: 18972-56-0   HR: 3
MAGNESIUM HEXAFLUOROSILICATE

DOT: UN 2853
mf: $F_6Si \cdot Mg$ mw: 166.40

SYNS: FLUOSILICATE de MAGNESIUM (FRENCH) * MAGNESIUM FLUOSILICATE * MAGNESIUM SILICOFLUORIDE

OSHA PEL: TWA 2.5 mg(F)/m³
NIOSH REL: TWA 2.5 mg(F)/m³
DOT Classification: Poison B; Label: St. Andrews Cross.

SAFETY PROFILE: Poison by ingestion. Moderately toxic by subcutaneous route. When heated to decomposition it emits toxic fumes of F⁻.

MAG500   CAS: 60616-74-2   HR: 3
MAGNESIUM HYDRIDE

DOT: UN 2010
mf: H_2Mg mw: 26.33

PROP: White crystals. D: 1.419, mp: decomp >200°. Sol in isopropylamine.

DOT Classification: Flammable Solid; Label: Dangerous When Wet.

SAFETY PROFILE: The powder may ignite spontaneously in air and react violently with water. Incompatible with oxygen, water.

MAG750   CAS: 1309-42-8   HR: 1
MAGNESIUM HYDROXIDE
mf: H_2MgO_2 mw: 58.33

PROP: White powder, odorless. D: 2.36, mp: decomp @ 350°. Sol in solns of ammonium salts and dilute acids; almost insol in water and alc.

SYNS: MAGNESIA MAGMA * MAGNESIUM HYDRATE * MILK of MAGNESIA

CONSENSUS REPORTS: Reported in EPA TSCA Inventory.

SAFETY PROFILE: Incompatible with maleic anhydride, P.

MAH000   CAS: 10377-60-3   HR: 3
MAGNESIUM(II) NITRATE (1:2)

DOT: UN 1474
mf: $N_2O_6 \cdot Mg$ mw: 148.33

PROP: The dihydrate, [$Mg(NO_3)_2 \cdot 2H_2O$] mw: 184.37, forms white crystals (prisms). D: 2.0256 @ 25°, mp: 129.0°. The hexahydrate, [$Mg(NO_3)_2 \cdot 6H_2O$] mw: 256.43, forms monoclinic, colorless, deliquescent crystals. D: 1.464, mp: 95°, bp: $-5H_2O$ @ 330°.

SYNS: MAGNESIUM NITRATE (DOT) * NITRIC ACID, MAGNESIUM SALT (2:1)

CONSENSUS REPORTS: Reported in EPA TSCA Inventory.

DOT Classification: Label: Oxidizer.

SAFETY PROFILE: Probably a severe irritant to the eyes, skin, and mucous membranes. A powerful oxidizer. Violent decomposition on contact with dimethylformamide. When heated to decomposition it emits toxic fumes of NO_x.

MAH500   CAS: 1309-48-4   HR: 3
MAGNESIUM OXIDE
mf: MgO   mw: 40.31

PROP: White, bulky, very fine powder; odorless. Mp: 2500-2800°, d: 3.65-3.75. Very sltly sol in water; sol in dil acids; insol in alc.

SYNS: CALCINED BRUCITE * CALCINED MAGNESIA * CALCINED MAGNESITE * MAGNESIA

* MAGNESIA USTA * MAGNEZU TLENEK (POLISH)
* SEAWATER MAGNESIA

CONSENSUS REPORTS: Reported in EPA TSCA Inventory.

OSHA PEL: Fume: (Transitional: TWA Total Dust: 15 mg/m^3; Respirable Fraction: 5 mg/m^3) Total Dust: 10 mg/m^3; Respirable Fraction: 5 mg/m^3

ACGIH TLV: TWA 10 mg/m^3 (fume)

DFG MAK: 6 mg/m^3 (fume)

SAFETY PROFILE: Inhalation of the fumes can produce a febrile reaction and leukocytosis in humans. Questionable carcinogen with experimental tumorigenic data. Violent reaction or ignition in contact with interhalogens (e.g., bromine pentafluoride, chlorine trifluoride). Incandescent reaction with phosphorus pentachloride.

MAH750   CAS: 14452-57-4   HR: 2
MAGNESIUM PEROXIDE

DOT: UN 1476

mf: MgO_2 mw: 56.31

PROP: White powder; tasteless and odorless. Insol in water and slowly decomp evolving O_2. Sol in dil acids. Keep container closed.

SYNS: IXPER 25M * MAGNESIUM PEROXIDE, solid (DOT)

CONSENSUS REPORTS: Reported in EPA TSCA Inventory.

DOT Classification: Oxidizer; Label: Oxidizer.

SAFETY PROFILE: A powerful oxidizer. Probably a severe irritant to the eyes, skin, and mucous membranes. Flammable by chemical reaction with acidic materials and moisture; an oxidizing agent. Dangerous; reacts vigorously with reducing agents; will decompose violently in or near a fire.

MAI000   CAS: 12057-74-8   HR: 3
MAGNESIUM PHOSPHIDE

DOT: UN 2011

mf: Mg_3P_2 mw: 134.87

SYNS: FOSFURI di MAGNESIO (ITALIAN) * MAGNESIUMFOSFIDE (DUTCH) * PHOSPHURE de MAGNESIUM (FRENCH)

CONSENSUS REPORTS: Reported in EPA TSCA Inventory.

DOT Classification: Flammable Solid; Label: Dangerous When Wet, Poison.

SAFETY PROFILE: A poison. Moderately toxic by inhalation. Flammable when exposed to heat, flame, or oxidizing materials. Ignites when heated in chlorine, bromine, or iodine vapors. Incandescent reaction with nitric acid. Reacts with water to evolve flammable phosphine gas. When heated to decomposition it emits toxic fumes of PO_x and phosphine.

MAJ000   CAS: 1343-90-4   HR: 1
MAGNESIUM SILICATE HYDRATE

mf: $Mg_2O_8Si_3 \cdot H_2O$ mw: 278.91

PROP: Fine white powder; odorless and tasteless. Insol in water, alc.

SAFETY PROFILE: A human skin irritant.

MAJ250   CAS: 7487-88-9   HR: 2
MAGNESIUM SULFATE (1:1)

mf: $O_4S \cdot Mg$ mw: 120.37

PROP: Opaque needles or granular crystalline powder; odorless with cooling, bitter, salt taste. Sol in water; slowly sol in glycerin; sltly sol in alc.

SYNS: EPSOM SALTS * MAGNESIUM SULPHATE

CONSENSUS REPORTS: Reported in EPA TSCA Inventory.

SAFETY PROFILE: Moderately toxic by ingestion, intraperitoneal, and subcutaneous routes. An experimental teratogen. Potentially explosive reaction when heated with ethoxyethynyl alcohols (e.g., 1-ethoxy-3-methyl-1-butyn-3-ol). When heated to decomposition it emits toxic fumes of SO_x.

MAJ500   HR: 2
MAGNESIUM SULFATE HEPTAHYDRATE

PROP: Efflorescent crystals or powder, bitter taste. Mp: $-7H_2O$ @ 200°, d: 1.68. Sltly sol in alc.

SYNS: BITTER SALTS * EPSOM SALTS * SULFURIC ACID, MAGNESIUM SALT (1:1) HEPTAHYDRATE

SAFETY PROFILE: Moderately toxic by several routes. Parenteral use or use in presence of renal insufficiency may lead to magnesium intoxication. An anticonvulsant and purgative.

MAK700   CAS: 121-75-5   HR: 3
MALATHION

DOT: NA 2783

mf: $C_{10}H_{19}O_6PS_2$ mw: 330.38

PROP: Brown to yellow liquid; characteristic odor. D: 1.23 @ 25°/4°, mp: 2.9°, bp: 156° @ 0.7 mm. Miscible in organic solvents; sltly water-sol.

SYNS: AMERICAN CYANAMID 4,049 * S-(1,2-BIS(AETHOXY-CARBONYL)-AETHYL)-O,O-DIMETHYL-DITHIOPHASPHAT (GERMAN) * S-(1,2-BIS(CARBETH-OXY)ETHYL)-O,O-DIMETHYL DITHIOPHOSPHATE * S-(1,2-BIS(ETHOXY-CARBONYL)-ETHYL)-O,O-DI-METHYL-DITHIOFOSFAAT (DUTCH) * S-(1,2-BIS(ETHOXYCARBONYL)ETHYL)-O,O-DIMETHYL PHOSPHORODITHIOATE * S-1,2-BIS(ETHOXYCARBO-NYL)ETHYL-O,O-DIMETHYL THIOPHOSPHATE * S-(1,2-BIS(ETOSSI-CARBONIL)-ETIL)-O,O-DIMETIL-DITIOFOSFATO (ITALIAN) * CALMATHION * CAR-BETHOXY MALATHION * CARBETOVUR * CAR-BETOX * CARBOFOS * CARBOPHOS * CELT-HIGN * CHEMATHION * CIMEXAN * COM-POUND 4049 * CYTHION * DETMOL MA * DET-MOL MA 96% * S-(1,2-DICARBETHOXYETHYL)-O,O-DIMETHYLDITHIOPHOSPHATE * DICARBO-ETHOXYETHYL-O,O-DIMETHYL PHOSPHORODITHIOATE * 1,2-DI(ETHOXYCARBONYL)ETHYL-O,O-DI-METHYL PHOSPHORODITHIOATE * S-(1,2-DI(ETHOXYCARBONYL)ETHYL DIMETHYL PHOS-PHOROTHIOLOTHIONATE * DIETHYL (DIMETHOXY-PHOSPHINOTHIOYLTHIO) BUTANEDIOATE * DIETHYL (DIMETHOXYPHOSPHINOTHIOYLTHIO)SUCCINATE * DIETHYL MERCAPTOSUCCINATE-O,O-DIMETHYL DITHIOPHOSPHATE, S-ESTER * DIETHYL MERCAPTO-SUCCINATE-O,O-DIMETHYL PHOSPHORODITHIOATE * DIETHYL MERCAPTOSUCCINATE-O,O-DIMETHYL THIOPHOSPHATE * DIETHYL MERCAPTOSUCCINATE-S-ESTER with O,O-DIMETHYLPHOSPHORODITHIOATE * DIETHYL MERCAPTOSUCCINIC ACID O,O-DIMETHYL PHOSPHORODITHIOATE * (DIMETHOXYPHOSPHINO-THIOYL)THIO)BUTANEDIOIC ACID DIETHYL ESTER * O,O-DIMETHYL-S-(1,2-BIS(ETHOXYCARBONYL) ETHYL)DITHIOPHOSPHATE * O,O-DIMETHYL-S-1,2-(DICARBAETHOXYAETHYL)-DITHIOPHOSPHAT (GER-MAN) * O,O-DIMETHYL-S-(1,2-DICARBETHOXY-ETHYL) DITHIOPHOSPHATE * O,O-DIMETHYL-S-(1,2-DICARBETHOXYETHYL)PHOSPHORODITHIOATE * O,O-DIMETHYL-S-(1,2-DICARBETHOXYETHYL) THIOTHIONOPHOSPHATE * O,O-DIMETHYL-S-1,2-DI(ETHOXYCARBAMYL)ETHYL PHOSPHORODITHIOATE * O,O-DIMETHYL-S-1,2-DIKARBETOXYLETHYLDITIO-FOSFAT (CZECH) * O,O-DIMETHYLDITHIOPHOSPHATE DIETHYLMERCAPTOSUCCINATE * DITHIOPHOSPHATE de O,O-DIMETHYLE et de S-(1,2-DICARBOETHOXY-ETHYLE) (FRENCH) * EL 4049 * EMMATOS * EMMATOS EXTRA * ENT 17,034 * S-ESTER with O,O-DIMETHYL PHOSPHOROTHIOATE * ETHIOLACAR * ETIOL * EXPERIMENTAL INSECTICIDE 4049 * EXTERMATHION * FORMAL * FORTHION * FOSFOTHION * FOSFOTION * FOUR THOU-SAND FORTY-NINE * FYFANON * HILTHION * HILTHION 25WDP * INSECTICIDE No. 4049 * KARBOFOS * KOP-THION * KYPFOS * MALACIDE * MALAFOR * MALAGRAN * MALAKILL * MALAMAR * MALAMAR 50 * MALAPHELE * MALAPHOS * MALASOL * MALASPRAY * MALATHION * MALATHION ULV CONCENTRATE * MALATHIOZOO * MALA-THON * MALATHYL LV CONCENTRATE & ULV CONCENTRATE * MALATION (POLISH) * MALATOL * MALATOX * MALDISON * MALMED * MALPHOS * MALTOX * MALTOX MLT * MERCAPTOSUCCINIC ACID DIETHYL ESTER * MERCAPTOTHION * MER-CAPTOTION (SPANISH) * MLT * MOS-CARDA * NCI-C00215 * OLEOPHOSPHOTHION * ORTHO MALATHION * PHOSPHORODITHIOIC ACID-O,O-DIMETHYL ESTER-S-ESTER with DIETHYL MERCAPTOSUCCINATE * PHOSPHOTHION * PRIODERM * SADOFOS * SADOPHOS * SF 60 * SIPTOX I * SUMITOX * TAK * TM-4049 * VEGFRU MALATOX * VETIOL * ZITHIOL

CONSENSUS REPORTS: IARC Cancer Review: GROUP 3 IMEMDT 7,56,87; Animal No Evidence IMEMDT 30,103,83; NCI Carcinogenesis Bioassay (feed); No Evidence: mouse, rat NCITR* NCI-CG-TR-24,78; No Evidence: rat NCITR* NCI-CG-TR-192,79. EPA Genetic Toxicology Program.

OSHA PEL: (Transitional: TWA Total Dust: 15 mg/m^3; Respirable Fraction: 5 mg/m^3 (skin)) TWA Total Dust:10 mg/m^3' Respirable Fraction: 5 mg/m^3 (skin)
ACGIH TLV: TWA 10 mg/m^3 (skin)
DFG MAK: 15 mg/m^3
NIOSH REL: (Malathion) TWA 15 mg/m^3
DOT Classification: ORM-A; Label: None.

SAFETY PROFILE: A human poison by ingestion. An experimental poison by ingestion, inhalation, intraperitoneal, intravenous, intraarterial, subcutaneous, and possibly other routes. Human systemic effects by ingestion: coma, blood pressure depression, and difficulty in breathing. Questionable carcinogen. Human mutation data reported. Has caused allergic sensitization of the skin. An organic phosphate cholinesterase inhibitor. When heated to decomposition it emits toxic fumes of PO$_x$ and SO$_x$.

MAK900   CAS: 110-16-7   **HR: 2**
MALEIC ACID
DOT: NA 2215
mf: $C_4H_4O_4$ mw: 116.08

PROP: White crystals, faint acidulous odor.
Mp: 130.5°, bp: 135° decomp, d: 1.590 @ 20°/
4°, vap d: 4.0.

SYNS: cis-BUTENEDIOIC ACID * (Z)-BUTENEDIOIC
ACID * cis-1,2-ETHYLENEDICARBOXYLIC ACID
* MALEINIC ACID * MALENIC ACID * TOXILIC
ACID

CONSENSUS REPORTS: Reported in EPA
TSCA Inventory.

DOT Classification: ORM-A; Label:None

SAFETY PROFILE: Moderately toxic by ingestion and skin contact. A skin and severe eye
irritant. Believed to be more toxic than its
isomer, fumeric acid. Combustible when exposed to heat or flame. When heated to decomposition it emits acrid smoke and irritating
fumes.

MAL250   CAS: 128-53-0   **HR: 3**
MALEIC ACID-N-ETHYLIMIDE
mf: $C_6H_7NO_2$ mw: 125.14

PROP: Crystals; irritating odor. Mp: 45°.

SYNS: N-ETHYLMALEIMIDE * USAF B-121

CONSENSUS REPORTS: Reported in EPA
TSCA Inventory.

SAFETY PROFILE: Poison by intraperitoneal
route. Human mutation data reported. Vapors
are highly irritating. When heated to decomposition it emits toxic fumes of NO_x.

MAM000   CAS: 108-31-6   **HR: 3**
MALEIC ANHYDRIDE
DOT: UN 2215
mf: $C_4H_2O_3$ mw: 98.06

PROP: Fused black or white crystals. Mp:
52.8°, bp: 202°, flash p: 215°F (CC), d: 1.48
@ 20°/4°, autoign temp: 890°F, vap press: 1
mm @ 44.0°, vap d: 3.4, lel: 1.4%, uel: 7.1%.
Sol in dioxane, water @ 30° forming maleic
acid; very sltly sol in alc.

SYNS: cis-BUTENEDIOIC ANHYDRIDE * 2,5-DIHY-
DROFURAN-2,5-DIONE * 2,5-FURANDIONE
* MALEIC ACID ANHYDRIDE (MAK) * RCRA WASTE
NUMBER U147 * TOXILIC ANHYDRIDE

CONSENSUS REPORTS: Community Right-
To-Know List. Reported in EPA TSCA Inventory.

OSHA PEL: TWA 0.25 ppm
ACGIH TLV: TWA 0.25 ppm
DFG MAK: 0.2 ppm (0.8 mg/m³)
DOT Classification: ORM-A; Label: None;
   IMO: Corrosive Material; Label: None.

SAFETY PROFILE: Poison by ingestion and
intraperitoneal routes. Moderately toxic by skin
contact. A corrosive irritant to eyes, skin, and
mucous membranes. Can cause pulmonary
edema. Questionable carcinogen with experimental tumorigenic data. A pesticide. Combustible when exposed to heat or flame; can react
vigorously on contact with oxidizing materials.
Explosive in the form of vapor when exposed
to heat or flame. Reacts with water or steam
to produce heat. Violent reaction with bases
(e.g., sodium hydroxide, potassium hydroxide,
calcium hydroxide), alkali metals (e.g., sodium,
potassium), amines (e.g., dimethylamine, triethylamine), lithium, pyridine. To fight fire, use
alcohol foam. Incompatible with cations. When
heated to decomposition (above 150°C) it emits
acrid smoke and irritating fumes.

MAM750   CAS: 541-59-3   **HR: 3**
MALEIMIDE
mf: $C_4H_3NO_2$ mw: 97.08

SYNS: MALEINIMIDE * PYRROLE-2,5-DIONE
* 3-PYRROLINE-2,5-DIONE

CONSENSUS REPORTS: Reported in EPA
TSCA Inventory.

SAFETY PROFILE: Poison by intraperitoneal
and intravenous routes. Experimental teratogenic and reproductive effects. When heated
to decomposition it emits toxic fumes of NO_x.

MAN000   CAS: 6915-15-7   **HR: 2**
MALIC ACID
mf: $C_4H_6O_5$ mw: 134.10

PROP: White or colorless crystals; acid taste.
Exhibits isomeric forms (dl, l and d). D (dl):
1.601, d (d or l): 1.595 @ 20°/40; mp (dl):
128°, mp (d or l): 100°; bp (dl): 150°, bp (d
or l): 140° (decomp). Very sol in water and
alc; sltly sol in ether.

SYNS: HYDROXYSUCCINIC ACID * KYSELINA JA-
BLECNA

CONSENSUS REPORTS: Reported in EPA TSCA Inventory.

SAFETY PROFILE: Moderately toxic by ingestion. A skin and severe eye irritant. When heated to decomposition it emits acrid smoke and irritating fumes.

MAO250   CAS: 109-77-3   HR: 3
MALONONITRILE
DOT: UN 2647
mf: $C_3H_2N_2$ mw: 66.07

PROP: White powder. D: 1.049 @ 34°/4°, mp: 30.5°, bp: 220°, flash p: 266°F (TOC). Sol in water, alc, ether.

SYNS: CYANOACETONITRILE * DICYANOMETH-ANE * DWUMETYLOSULFOTLENKU (POLISH) * MALONIC DINITRILE * METHYLENE CYANIDE * NITRIL KYSELINY MALONOVE (CZECH) * PRO-PANEDINITRILE * RCRA WASTE NUMBER U149 * USAF A-4600

CONSENSUS REPORTS: Cyanide and its compounds are on the Community Right-To-Know List. EPA Extremely Hazardous Substances List. Reported in EPA TSCA Inventory.

NIOSH REL: (Nitriles) TWA 8 mg/m^3
DOT Classification: Poison B; Label: Poison.

SAFETY PROFILE: Poison by ingestion, subcutaneous, intravenous, intraperitoneal, and possibly other routes. A severe eye irritant. Combustible when exposed to heat or flame. Polymerizes violently when heated to 130°C or on contact with strong base. May spontaneously explode when stored at 70-80°C. To fight fire, use water, fog, spray, foam. When heated to decomposition it emits toxic fumes of NO_x and CN^-.

MAO350   CAS: 118-71-8   HR: 2
MALTOL
mf: $C_6H_6O_3$ mw: 126.12

PROP: White crystalline powder; caramel-butterscotch odor. Sol in water, alc, glycerin, propylene glycol.

SYNS: CORPS PRALINE * 3-HYDROXY-2-METHYL-4H-PYRAN-4-ONE * 3-HYDROXY-2-METHYL-γ-PYRONE * 3-HYDROXY-2-METHYL-4-PYRONE * LARIXIC ACID * LARIXINIC ACID * 2-METHYL-3-HYDROXY-4-PYRONE * 2-METHYL-3-OXY-γ-PYRONE * 2-METHYL PYROMECONIC ACID * PALATONE * TALMON * VETOL

CONSENSUS REPORTS: Reported in EPA TSCA Inventory.

SAFETY PROFILE: Moderately toxic by ingestion, intraperitoneal, and subcutaneous routes. A skin irritant. Human mutation data reported. When heated to decomposition it emits acrid smoke and irritating fumes.

MAO500   CAS: 69-79-4   HR: 3
MALTOSE
mf: $C_{12}H_{22}O_{11}$ mw: 342.31

PROP: Colorless needles. D: 1.540 @ 17°, mp: decomp. Very sol in water; very sltly sol in cold alc; insol in ether.

SYNS: 4-(α-d-GLUCOPYRANOSIDO)-α-GLUCOPYRA-NOSE * 4-(α-d-GLUCOSIDO)-d-GLUCOSE * MAL-TOBIOSE * d-MALTOSE * MALT SUGAR * α-MALT SUGAR

CONSENSUS REPORTS: Reported in EPA TSCA Inventory.

SAFETY PROFILE: Experimental teratogenic and reproductive effects. Questionable carcinogen with experimental tumorigenic data. When heated to decomposition it emits acrid smoke and irritating fumes.

MAP000   CAS: 90-64-2   HR: 3
MANDELIC ACID
mf: $C_8H_8O_3$ mw: 152.16

PROP: Large, white crystals or powder; faint odor. Bp: decomp. D: 1.30, mp: 117-119°. Sol in water, alc and ether. Darkens and decomposes on prolonged exposure to light.

SYNS: AMYGDALIC ACID * AMYGDALINIC ACID * α-HYDROXYPHENYLACETIC ACID * α-HYDROXY-α-TOLUIC ACID * racemic MANDELIC ACID * PARAMANDELIC ACID * PHENYLGLYCOLIC ACID * PHENYLHYDROXYACETIC ACID

CONSENSUS REPORTS: Reported in EPA TSCA Inventory.

SAFETY PROFILE: Poison by intramuscular route. Moderately toxic by ingestion. Continued absorption can cause kidney irritation. When heated to decomposition it emits acrid smoke and irritating fumes.

MAP250   CAS: 532-28-5   HR: 3
MANDELIC ACID NITRILE
mf: C_8H_7NO mw: 133.16

PROP: Yellow, viscous liquid. Mp: −10°, bp: 170° decomp, d: 1.124.

SYNS: AMYGDALONITRILE * BENZALDEHYDE CYANOHYDRIN * BENZALDEHYDKYANHYDRIN (CZECH) * HYDROXYPHENYLACETONITRILE * NITRIL KYSELINY MANDLOVE (CZECH) * PHENYLGLYCOLONITRILE

CONSENSUS REPORTS: Cyanide and its compounds are on the Community Right-To-Know List. Reported in EPA TSCA Inventory.

SAFETY PROFILE: Poison by ingestion, intravenous, and subcutaneous routes. Mutation data reported. A severe eye irritant. When heated to decomposition it emits toxic fumes of NO_x and CN^-.

MAP750 CAS: 7439-96-5 **HR: 3**
MANGANESE
af: Mn aw: 54.94

PROP: Reddish-grey or silvery, brittle, metallic element. Mp: 1260°, bp: 1900°, d: 7.20, vap press: 1 mm @ 1292°.

SYNS: COLLOIDAL MANGANESE * MAGNACAT * MANGAN (POLISH) * MANGAN NITRIDOVANY (CZECH) * TRONAMANG

CONSENSUS REPORTS: Manganese and its compounds are on the Community Right-To-Know List. Reported in EPA TSCA Inventory.

OSHA PEL: Fume: (Transitional: CL 5 mg/m^3) TWA 1 mg/m^3; STEL 3 mg/m^3; Compounds: CL 5 mg/m^3
ACGIH TLV: Fume: 1 mg/m^3; STEL 3 mg/m^3; Dust and Compounds: TWA 5 mg/m^3
DFG MAK: 5 mg/m^3

SAFETY PROFILE: Human systemic effects by inhalation: degenerative brain changes, change in motor activity, muscle weakness. A skin and eye irritant. Questionable carcinogen with experimental tumorigenic data. Mutation data reported. Flammable and moderately explosive in the form of dust or powder when exposed to flame. The dust may be pyrophoric in air and may explode when heated in carbon dioxide. Mixtures of aluminum dust and manganese dust may explode in air. Mixtures with ammonium nitrate may explode when heated. The powdered metal ignites on contact with fluorine, chlorine + heat, hydrogen peroxide, bromine pentafluoride, sulfur dioxide + heat. Violent reaction with NO_2 and oxidants. Incandescent reaction with phosphorus, nitryl fluoride, nitric acid. Will react with water or steam to produce hydrogen; can react with oxidizing materials. To fight fire, use special dry chemical.

MAR000 CAS: 7773-01-5 **HR: 3**
MANGANESE(II) CHLORIDE (1:2)
mf: Cl_2Mn mw: 125.84

PROP: Cubic, deliquescent, pink crystals. Mp: 650°, bp: 1190°, d: 2.977 @ 25°. Sol in water.

SYNS: MANGANESE DICHLORIDE * MANGANOUS CHLORIDE

CONSENSUS REPORTS: Manganese and its compounds are on the Community Right-To-Know List. Reported in EPA TSCA Inventory. EPA Genetic Toxicology Program.

OSHA PEL: CL 5 mg(Mn)/m^3
ACGIH TLV: TWA 5 mg(Mn)/m^3

SAFETY PROFILE: Poison by intraperitoneal, subcutaneous, intramuscular, intravenous, parenteral, and possibly other routes. Moderately toxic by ingestion. Experimental teratogenic and reproductive effects. Questionable carcinogen with experimental carcinogenic data. Mutation data reported. Explosive reaction when heated with zinc foil. Reacts violently with potassium or sodium. When heated to decomposition it emits toxic fumes of Cl^-.

MAR500 **HR: 3**
MANGANESE COMPOUNDS

CONSENSUS REPORTS: Manganese and its compounds are on the Community Right-To-Know List.

SAFETY PROFILE: Some are experimental tumorigens. Can cause central nervous system and pulmonary system damage by inhalation of fumes and dust. Very few poisonings have occurred from ingestion. Chronic manganese poisoning is a clearly characterized disease which results from inhalation of fumes or dusts of manganese. Exposure to heavy concentrations of dusts or fumes for as little as three months may produce the condition, but usually cases develop after 1-3 years of exposure. The central nervous system is the chief site of damage. If cases are removed from exposure shortly after appearance of symptoms, some improvement in the patient's condition frequently occurs, though there may be some residual disturbances in gait and speech. When well established, however, the disease results in permanent disability. Exposure to dusts and fumes can possibly increase the incidence of upper respiratory infections and pneumonia. Chronic manganese poisoning usually begins with complaints of languor and sleepiness. This is followed by weakness in the legs and the development of stolid, mask-like faces. The patient speaks with a slow monotonous voice. Then muscular twitching appears, varying from a fine tremor of the hands to coarse, rhythmical movements of the arms, legs, and trunk. Nocturnal cramps of the legs appear about the same time. There is a slight increase in tendon reflexes, ankle and patellar clonus, and a typical Parkinsonian slapping gait. The handwriting may be quite minute. The symptoms may simulate progressive bulbar paralysis, postencephalitic Parkinsonism, multiple sclerosis, amyotrophic lateral sclerosis and progressive lenticular degeneration (Wilson's Disease). Often a history of exposure is the only aid in establishing the diagnosis. Manganese compounds are common air contaminants.

MAS000   CAS: 1313-13-9   HR: 3
MANGANESE DIOXIDE
mf: MnO_2 mw: 86.94

PROP: Tetragonal crystals. Mp: $-O_2$ @ 535°, d: 5.0. Insol in water, nitric or cold sulfuric acid.

SYNS: BLACK MANGANESE OXIDE * BOG MANGANESE * BRAUNSTEIN (GERMAN) * BRUINSTEEN (DUTCH) * CEMENT BLACK * C.I. 77728 * C.I. PIGMENT BLACK 14 * C.I. PIGMENT BROWN 8 * MANGAANBIOXYDE (DUTCH) * MANGAANDIOXYDE (DUTCH) * MANGANDIOXID (GERMAN) * MANGANESE BINOXIDE * MANGANESE (BIOSSIDO di) (ITALIAN) * MANGANESE (BIOXYD de) (FRENCH) * MANGANESE BLACK * MANGANESE (DIOSSIDO di) (ITALIAN) * MANGANESE (DIOXYDE de) (FRENCH) * MANGANESE OXIDE * MANGANESE(IV) OXIDE * MANGANESE PEROXIDE * MANGENESE SUPEROXIDE * PYROLUSITE BROWN

CONSENSUS REPORTS: Manganese and its compounds are on the Community Right-To-Know List. Reported in EPA TSCA Inventory.

OSHA PEL: CL 5 mg(Mn)/m^3
ACGIH TLV: TWA 5 mg(Mn)/m^3

SAFETY PROFILE: Poison by intravenous route. Moderately toxic by subcutaneous route. Experimental reproductive effects. A powerful oxidizer. Flammable by chemical reaction. It must not be heated or rubbed in contact with easily oxidizable matter. Violent thermite reaction when heated with aluminum. Potentially explosive reaction with hydrogen peroxide; peroxomonosulfuric acid; chlorates + heat; anilinium perchlorate. Ignition on contact with hydrogen sulfide. Violent reaction with oxidizers; potassium azide (when warmed); diboron tetrafluoride; Incandescent reaction with calcium hydride; chlorine trifluoride; rubidium acetylide (at 350°C). Vigorous reaction with hydroxylaminium chloride. Incompatible with H_2O_2; H_2SO_5; Na_2O_2. Keep away from heat and flammable materials.

MAS500   CAS: 12427-38-2   HR: 3
MANGANESE(II) ETHYLENEBIS
(DITHIOCARBAMATE)
DOT: UN 2210/UN 2968
mf: $C_4H_7N_2S_4 \cdot Mn$ mw: 266.31

PROP: Yellow powder or crystals; water-sol.

SYNS: AAMANGAN * AKZO CHEMIE MANEB * BASF-MANEB SPRITZPULVER * CHEM NEB * CHLOROBLE M * CR 3029 * DITHANE M 22 SPECIAL * ENT 14,875 * 1,2-ETHANE-DIYLBIS(CARBAMODITHIOATO)(2−)-MANGANESE * 1,2-ETHANEDIYLBISCARBAMODITHIOIC ACID MANGANESE COMPLEX * 1,2-ETHANEDIYLBISCAR-BAMODITHIOIC ACID, MANGANESE(2+) SALT (1:1) * 1,2-ETHANEDIYLBISMANEB, MANGANESE (2+) SALT (1:1) * ETHYLENEBISDITHIOCARBAMATE MANGA-NESE * N,N'-ETHYLENE BIS(DITHIOCARBAMATE MANGANEUX) (FRENCH) * ETHYLENEBIS(DITHIO-CARBAMATO) MANGANESE * ETHYLENEBIS(DITHIO-CARBAMIC ACID) MANGANESE SALT * ETHYLENE-BIS(DITHIOCARBAMIC ACID) MANGANOUS SALT * 1,2-ETHYLENEDIYLBIS(CARBAMODITHIOATO)MAN-GANESE * N,N'-ETILEN-BIS(DITIOCARBAMMATO) di MANGANESE (ITALIAN) * F 10 (PESTICIDE) * GRIFFIN MANEX * KYPMAN 80 * LONOCOL M * MANAM * MANEB * MANEB, stabilized against self-heating (DOT) * MANEB, with not less than 60% maneb (DOT) * MANEBE (FRENCH) * MANGAAN(II)-(N,N'-ETHYLEEN-BIS(DITHIOCARBAMAAT)) (DUTCH) * MANGAN(II)-(N,N'-AETHYLEN-BIS(DITHIOCARBA-MATE)) (GERMAN) * MANGANESE ETHYLENE-1,2-BISDITHIOCARBAMATE * MANGANESE(II) ETHYLENE DI(DITHIOCARBAMATE) * MANZATE * MANZATE MANEB FUNGICIDE * MEB * NESPOR * PLAN-TIFOG 160M * POLYRAM M * REMASAN CHLORO-BLE M * TRIMANGOL * UNICROP MANEB * VANCIDE

CONSENSUS REPORTS: IARC Cancer Review: GROUP 3 IMEMDT 7,56,87; Animal

Inadequate Evidence IMEMDT 12,137,76. Community Right-To-Know List. EPA Genetic Toxicology Program.

OSHA PEL: CL 5 mg(Mn)/m^3
ACGIH TLV: TWA 5 mg(Mn)/m^3
DOT Classification: Flammable Solid; Label: Spontaneously Combustible, Danger When Wet.

SAFETY PROFILE: Moderately toxic by ingestion and possibly other routes. Experimental teratogenic and reproductive effects. Questionable carcinogen with experimental carcinogenic and tumorigenic data. Mutation data reported. A fungicide. May ignite spontaneously in air. Dangerous; when heated to decomposition it emits highly toxic fumes of NO$_x$ and SO$_x$.

MAT250 CAS: 1344-43-0 **HR: 2**
MANGANESE(II) OXIDE
mf: MnO mw: 70.94

PROP: Grass-green powder, sol in acids, insol in water. D: 5.45, mp: 1650°, converted to Mn$_3$O$_4$ if heated in air.

SYNS: CASSEL GREEN * C.I. 77726 * MANGANESE GREEN * MANGANESE MONOXIDE * MANGANOUS OXIDE * NU-MANESE * ROSENSTHIEL

CONSENSUS REPORTS: Manganese and its compounds are on the Community Right-To-Know List. Reported in EPA TSCA Inventory.

OSHA PEL: CL 5 mg(Mn)/m^3
ACGIH TLV: TWA 5 mg(Mn)/m^3

SAFETY PROFILE: Moderately toxic by subcutaneous route. Violent reaction with hydrogen peroxide, Ca(OCl)$_2$, F$_2$, H$_2$O$_2$.

MAT500 CAS: 1317-34-6 **HR: 2**
MANGANESE(III) OXIDE
mf: Mn$_2$O$_3$ mw: 157.88

PROP: Fine, black powder. D: 4.50. Insol in water; sol in HCl evolving chlorine.

SYNS: CASSEL BROWN * C.I. 77727 * C.I. NATURAL BROWN 8 * COLOGNE EARTH * COLOGNE UMBER * CULLEN EARTH * DIMANGANESE TRIOXIDE * MANGANESE MANGANATE * MANGANESE SISQUIOXIDE * MANGANESE TRIOXIDE * MANGANIC OXIDE * RUBENS BROWN * SOLUBLE VANDYKE BROWN * VANDYKE BROWN * WALNUT STAIN

CONSENSUS REPORTS: Manganese and its compounds are on the Community Right-To-Know List. Reported in EPA TSCA Inventory.

OSHA PEL: CL 5 mg(Mn)/m^3
ACGIH TLV: TWA 5 mg(Mn)/m^3

SAFETY PROFILE: Moderately toxic by subcutaneous route.

MAU250 CAS: 7785-87-7 **HR: 3**
MANGANESE(II) SULFATE (1:1)
mf: O$_4$S • Mn mw: 151.00

PROP: Pink granular powder; odorless. Mp: 700°, bp: decomp @ 850°. d: 3.25. Very sol in water, more so in boiling water; insol in alc.

SYNS: MANGANOUS SULFATE * MAN-GRO * NCI-C61143 * SORBA-SPRAY Mn * SULFURIC ACID, MANGANESE(2+) SALT

CONSENSUS REPORTS: Manganese and its compounds are on the Community Right-To-Know List. Reported in EPA TSCA Inventory. EPA Genetic Toxicology Program.

OSHA PEL: CL 5 mg(Mn)/m^3
ACGIH TLV: TWA 5 mg(Mn)/m^3

SAFETY PROFILE: Poison by intraperitoneal route. Questionable carcinogen with experimental neoplastigenic data. Mutation data reported. When heated to decomposition it emits toxic fumes of SO$_x$ and manganese.

MAU800 CAS: 1317-35-7 **HR: 2**
MANGANESE TETROXIDE
mf: Mn$_3$O$_4$ mw: 228.82

PROP: Brownish-black powder. D: 4.7. Insol in water; sol in HCl, liberating chlorine.

SYNS: MANGANESE OXIDE * MANGANOMANGANIC OXIDE * TRIMANGANESE TETRAOXIDE * TRIMANGANESE TETROXIDE

CONSENSUS REPORTS: Manganese and its compounds are on the Community Right-To-Know List. Reported in EPA TSCA Inventory.

OSHA PEL: TWA 1 mg(Mn)/m^3
ACGIH TLV: TWA 1 mg(Mn)/m^3
DFG MAK: 1 mg/m^3

SAFETY PROFILE: Experimental reproductive effects. Reacts violently @ <100°.

MAV750 CAS: 12108-13-3 **HR: 3**
MANGANESE TRICARBONYL
METHYLCYCLOPENTADIENYL
mf: C$_9$H$_7$MnO$_3$ mw: 218.10

SYNS: AK-33X * ANTIKNOCK-33 * CI-2 * COMBUSTION IMPROVER -2 * METHYLCYCLO-PENTADIENYL MANGANESE TRICARBONYL (OSHA) * MMT

CONSENSUS REPORTS: EPA Extremely Hazardous Substances List. Manganese and its compounds are on the Community Right-To-Know List. Reported in EPA TSCA Inventory.

OSHA PEL: TWA 0.2 mg(Mn)/m³ (skin)
ACGIH TLV: TWA 5 mg(Mn)/m³

SAFETY PROFILE: Poison by ingestion, inhalation, skin contact, intravenous, and intraperitoneal routes. A skin irritant. When heated to decomposition it emits toxic fumes of CO.

MAW250   CAS: 15825-70-4   **HR: 3**
MANNITOL HEXANITRATE

DOT: UN 0133
mf: $C_6H_8N_6O_{18}$ mw: 452.17

PROP: Colorless crystals. Bp: explodes @ 120°, d: 1.603 @ O°. Mp: 106-108°. Long needles in regular clusters from alc. Sol in alc, ether; insol in water.

SYNS: DILANGIL * HEXANITROL * HYPERTENAIN * INITIATING EXPLOSIVE NITRO MANNITE (DOT) * MANEXIN * MANICOLE * MANITE * MANNITOL HEXANITRATE, containing, by weight, at least 40% water (DOT) * d-MANNITOL HEXANITRATE * MAXITATE * NITROMANNITE (DOT) * NITROMANNITE (DRY) (DOT) * NITROMANNITOL * SDM No. 5

CONSENSUS REPORTS: Reported in EPA TSCA Inventory.

DOT Classification: Class A Explosive; Label: Explosive A; Forbidden, Dry.

SAFETY PROFILE: Moderately toxic by ingestion and inhalation yielding a fall in blood pressure which may result in weakness, headache, and dizziness. Chronic exposure may produce methemoglobinemia with cyanosis. A powerful explosive sensitive to shock or heat. Upon decomposition it emits toxic fumes of NO_x.

MAW500   CAS: 576-68-1   **HR: 3**
MANNOMUSTINE

mf: $C_{10}H_{22}Cl_2N_2O_4$ mw: 305.24

SYNS: 1,6-BIS(CHLOROETHYLAMINO)-1,6-BIS-DEOXY-d-MANNITOL * 1,6-BIS(CHLOROETHYLAMINO)-1,6-DIDEOXY-d-MANNITE * 1,6-BIS((β-CHLOROETHYL) AMINO)-1,6-DIDEOXY-d-MANNITOL * 1,6-BIS((2-CHLO-ROETHYL)AMINO)-1,6-DIDEOXY-d-MANNITOL * DEGRANOL * MANNIT-LOST (GERMAN) * MANNIT-MUSTARD (GERMAN) * MANNITOL NITROGEN MUSTARD

SAFETY PROFILE: Poison by intraperitoneal and intravenous routes. Human mutation data reported. An antineoplastic agent. When heated to decomposition it emits very toxic fumes of Cl^- and NO_x.

MAW750   CAS: 551-74-6   **HR: 3**
MANNOMUSTINE DIHYDROCHLORIDE
mf: $C_{10}H_{23}Cl_2N_2O_4 \cdot 2ClH$ mw: 378.13

PROP: Crystals from 80% ethanol. Decomp @ 239-241°. Sol in water; sltly sol in ethanol.

SYNS: 1,6-BIS-(CHLOROETHYLAMINO)-1,6-DESOXY-d-MANNITOLDIHYDROCHLORIDE * 1,6-BIS-(CHLOROETHYLAMINO)-1,6-DIDEOXY-d-MANNITEDIHYDROCHLORIDE * 1,6-DIDEOXY-1,6-DI(2-CHLOROETHYLAMINO)-d-MANNITOLDIHYDROCHLORIDE * MANNITOL MUSTARD DIHYDROCHLORIDE * NSC-9698

CONSENSUS REPORTS: IARC Cancer Review: GROUP 3 IMEMDT 7,56,87; Animal Sufficient Evidence IMEMDT 9,157,75.

SAFETY PROFILE: Poison by intravenous, subcutaneous, and parenteral routes. Experimental reproductive effects. Questionable carcinogen with experimental carcinogenic and neoplastigenic data. Human mutation data reported. A drug used for the treatment of malignant neoplasms. When heated to decomposition it emits very toxic fumes of HCl^- and NO_x.

MBU500   CAS: 8015-01-8   **HR: 1**
MARJORAM OIL, SPANISH

PROP: Main constituent is cineole. From steam distillation of the flowering plant material from the shrub *Thymus mastichina* L. (Fam. *Labiatae*). Faintly yellow liquid. D: 0.904-0.920, refr index: 1.463 @ 20°. Sol in fixed oils; insol in glycerin, propylene glycol, mineral oil.

SYNS: OIL of MARJORAM, SPANISH * SPANISH MARJORAM OIL

CONSENSUS REPORTS: Reported in EPA TSCA Inventory.

SAFETY PROFILE: A skin irritant. When heated to decomposition it emits acrid smoke and irritating fumes.

MBU820 CAS: 35846-53-8 **HR: 3**
MAYTANSINE
mf: $C_{34}H_{46}ClN_3O_{10}$ mw: 692.21

PROP: Mp: 171-172°. Active principle found in *Maytenus serrata*.

SYNS: MAITANSINE * MAYSANINE * MAYT * NSC-153858

SAFETY PROFILE: A deadly poison by intraperitoneal and intravenous routes. Moderately toxic by subcutaneous route. Human systemic effects by intravenous route: hallucinations, distorted perceptions, change in motor activity, and nausea or vomiting. An experimental teratogen. Mutation data reported. Used as an antineoplastic agent. When heated to decomposition it emits very toxic fumes of Cl^- and NO_x.

MBX500 CAS: 36236-67-6 **HR: 2**
MECLIZINE HYDROCHLORIDE
mf: $C_{25}H_{27}ClN_2 \cdot ClH$ mw: 427.45

SYNS: BONINE * 1-(p-CHLORO-α-PHENYLBENZYL)-4-(m-METHYLBENZYL)PIPERAZINE HYDROCHLORIDE * MECLOZINE HYDROCHLORIDE

SAFETY PROFILE: Moderately toxic by ingestion and intraperitoneal routes. Experimental teratogenic and reproductive effects. An FDA over-the-counter drug. When heated to decomposition it emits very toxic fumes of NO_x, and Cl^-.

MCA000 CAS: 71-58-9 **HR: 3**
MEDROXYPROGESTERONE ACETATE
mf: $C_{24}H_{34}O_4$ mw: 386.58

PROP: White to off-white, odorless, crystalline powder. Melting range 207-209°. Insol in water; freely sol in chloroform; sparingly sol in alc.

SYNS: 17-α-ACETOXY-6-α-METHYLPREGN-4-ENE-3,20-DIONE * 17-ACETOXY-6-α-METHYLPROGESTERONE * (6-α)-17-(ACETYLOXY)-6-METHYLPREG-4-ENE-3,20-DIONE * DEPO-PROVERA * FARLUTIN * 17-α-HYDROXY-6-α-METHYLPREGN-4-ENE-3,20-DIONE ACETATE * 17-HYDROXY-6-α-METHYLPREGN-4-ENE-3,20-DIONE ACETATE * 17-α-HYDROXY-6-α-METHYL-PROGESTERONE ACETATE * 6-α-METHYL-17-α-ACETOXYPREGN-4-ENE-3,20-DIONE * 6-α-METHYL-17-α-ACETOXYPROGESTERONE * 6-α-METHYL-17-α-HYDROXYPROGESTERONE ACETATE * 6-α-METHYL-4-PREGNENE-3,20-DION-17-α-OL ACETATE * METIPREGNONE * NOGEST * ORAGEST * PERLUTEX * REPROMIX

CONSENSUS REPORTS: IARC Cancer Review: GROUP 2B IMEMDT 7,289,87; Animal

Limited Evidence IMEMDT 21,417,79; IMEMDT 6,157,74; Human Inadequate Evidence IMEMDT 21,417,79. Reported in EPA TSCA Inventory. EPA Genetic Toxicology Program.

SAFETY PROFILE: Suspected carcinogen with experimental carcinogenic, neoplastigenic, and tumorigenic data. Human systemic effects by intravenous route: increased intraocular pressure. Human teratogenic effects by an unspecified route: developmental abnormalities of the urogenital system. Human reproductive effects by multiple routes: spermatogenesis, menstrual cycle changes or disorders, postpartum effects, female fertility effects, abortion, newborn behavioral effects. Experimental teratogenic and reproductive effects. Human mutation data reported. A drug for the treatment of secondary amenorrhoea and dysfunctional uterine bleeding. When heated to decomposition it emits acrid smoke and irritating fumes.

MCA500 CAS: 8064-66-2 **HR: 3**
MEGESTROL ACETATE + ETHINYLOESTRADIOL

SYNS: 17-HYDROXY-6-METHYLPREGNA-4,6-DIENE-3,20-DIONE ACETATE mixed with 19-NOR-17-α-PREGNA-1,3,5(10)-TRIEN-2-YNE-3,17-DIOL * MEGESTROL ACETATE 4 MG., ETHINYLOESTRADIOL 50 μg * MENOQUENS * NEODELPREGNIN * ORACONAL * SERIAL * TRI-ERVONUM * VOLDYS * VOLIDAN

SAFETY PROFILE: Human reproductive effects by ingestion: female fertility effects. Questionable human carcinogen producing normocytic anemia and liver tumors. An oral contraceptive. When heated to decomposition it emits acrid smoke and irritating fumes.

MCB000 CAS: 108-78-1 **HR: 3**
MELAMINE
mf: $C_3H_6N_6$ mw: 126.15

PROP: Monoclinic, colorless prisms. Mp: <250°, bp: sublimes, d: 1.573 @ 250°, vap press: 50 mm @ 315°, vap d: 4.34. Sltly sol in water; very sltly sol in hot alc; insol in ether.

SYNS: AERO * CYANURAMIDE * CYANUROTRIAMIDE * CYANUROTRIAMINE * CYMEL * NCI-C50715 * 2,4,6-TRIAMINO-s-TRIAZINE

CONSENSUS REPORTS: IARC Cancer Review: GROUP 3 IMEMDT 7,56,87; Animal

Inadequate Evidence IMEMDT 39,333,86. NTP Carcinogenesis Bioassay (feed); No Evidence: mouse NTPTR* NTP-TR-245,83; (feed); Clear Evidence: rat NTPTR* NTP-TR-245,83. Community Right-To-Know List. Reported in EPA TSCA Inventory.

SAFETY PROFILE: Moderately toxic by ingestion and intraperitoneal routes. An eye, skin, and mucous membrane irritant. Causes dermatitis in humans. Questionable carcinogen with experimental carcinogenic and tumorigenic data. Mutation data reported. When heated to decomposition it emits toxic fumes of NO_x and CN^-.

MCB380   CAS: 2919-66-6   **HR: D**
MELENGESTROL ACETATE
mf: $C_{25}H_{32}O_4$ mw: 396.57

SYNS: 17-(ACETYLOXY)-6-METHYL-16-METHYLENE-PREGNA-4,6-DIENE-3,20-DIONE (9CI) * 6-DEHYDRO-16-METHYLENE-6-METHYL-17-ACETOXYPROGESTERONE * 17-HYDROXY-6-METHYL-16-METHYLENEPREGNA-4,6-DIENE-3,20-DIONE, ACETATE * MGA * MGA 100 (STEROID)

SAFETY PROFILE: Experimental reproductive effects. When heated to decomposition it yields acrid smoke and irritating fumes.

MCB500   CAS: 3771-19-5   **HR: 3**
MELIPAN
mf: $C_{20}H_{22}O_3$ mw: 310.42

SYNS: 2-METHYL-2-(4-(1,2,3,4-TETRAHYDRO-1-NAPH-THALENYL)PHENOXY)PROPANOIC ACID * α-METHYL-α-(p-1,2,3,4-TETRAHYDRONAPHTH-1-YL-PHENOXY)PROPIONIC ACID * 2-METHYL-2-(4-(1,2,3,4-TETRAHYDRO-1-NAPHTHYL)PHENOXY)PROPANOIC ACID * 2-METHYL-2-(p-(1,2,3,4-TETRAHYDRO-1-NAPHTHYL)PHENOXY)PROPIONIC ACID * NAFENOIC ACID * NAFENOPIN * SU-13437

CONSENSUS REPORTS: IARC Cancer Review: GROUP 2B IMEMDT 7,56,87; Human Limited Evidence IMEMDT 24,125,80; Animal Sufficient Evidence IMEMDT 24,125,80.

SAFETY PROFILE: Suspected carcinogen with experimental carcinogenic and tumorigenic data. Mutation data reported. A drug for the treatment of hypercholesterolemia or hypertriglyceridemia. When heated to decomposition it emits acrid smoke and irritating fumes.

MCB625   **HR: D**
MENTHA ARVENSIS, OIL

SYNS: CORNMINT OIL, PARTIALLY DEMENTHOLIZED * MENTHA ARVENSIS OIL, PARTIALLY DEMENTHOLIZED (FCC)

SAFETY PROFILE: Experimental reproductive effects. When heated to decomposition it emits acrid smoke and irritating fumes.

MCB750   CAS: 99-85-4   **HR: 2**
p-MENTHA-1,4-DIENE
mf: $C_{10}H_{16}$ mw: 136.26

PROP: Colorless liquid; citrus odor. D: 0.841, refr index: 1.4731.477. Sol in alc, fixed oils; insol in water.

SYNS: FEMA No. 3559 * 1-METHYL-4-ISOPROPYL-CYCLOHEXADIENE-1,4 * γ-TERPINENE (FCC)

CONSENSUS REPORTS: Reported in EPA TSCA Inventory.

SAFETY PROFILE: Moderately toxic by ingestion. A skin irritant. When heated to decomposition it emits acrid smoke and irritating fumes.

MCC000   CAS: 99-83-2   **HR: 2**
p-MENTHA-1,5-DIENE
mf: $C_{10}H_{16}$ mw: 136.26

PROP: Colorless to sltly yellow liquid; mint odor. D: 0.835-0.865, refr index: 1.471-1.477, flash p: 120°F. Sol in alc; insol in water.

SYNS: α-FELLANDRENE * FEMA No. 2856 * 4-ISOPROPYL-1-METHYL-1,5-CYCLOHEXADIENE * 5-ISOPROPYL-2-METHYL-1,3-CYCLOHEXADIENE * 2-METHYL-5-ISOPROPYL-1,3-CYCLOHEXADIENE * α-PHELLANDRENE (FCC)

CONSENSUS REPORTS: Reported in EPA TSCA Inventory.

SAFETY PROFILE: Mildly toxic by ingestion. A severe human skin irritant. Incompatible with air. Combustible liquid. When heated to decomposition it emits acrid smoke and irritating fumes.

MCC250   CAS: 138-86-3   **HR: 1**
p-MENTHA-1,8-DIENE
DOT: UN 2052
mf: $C_{10}H_{16}$ mw: 136.26

PROP: Liquid. D: 0.842 @ 20°/4°, mp: −96.9°, bp: 177°. Insol in water; misc in alc and ether.

SYNS: ACINTENE DP * ACINTENE DP DIPENTENE * CAJEPUTENE * CINENE * DIPANOL

* DIPENTENE * INACTIVE LIMONENE * KAUT-
SCHIN * LIMONENE * dl-LIMONENE * 1,8(9)-p-
MENTHADIENE * 1-METHYL-4-ISOPROPENYL-1-CY-
CLOHEXENE * NESOL * Δ-1,8-TERPODIENE
* UNITENE

CONSENSUS REPORTS: Reported in EPA TSCA Inventory.

DOT Classification: Flammable or Combustible Liquid; Label: Flammable Liquid.

SAFETY PROFILE: A skin irritant. Flammable when exposed to heat or flame; can react vigorously with oxidizing materials. When heated to decomposition it emits acrid smoke and irritating fumes.

MCC500   CAS: 5989-54-8   **HR: 1**
(S)-(−)-p-MENTHA-1,8-DIENE
mf: $C_{10}H_{16}$ mw: 136.26

PROP: Colorless liquid; light odor. D: 0.837-0.841, refr index:.469-1.473. Misc in alc, fixed oils; insol in water.

SYNS: 1-LIMONENE * (−)-LIMONENE (FCC)
* 1-METHYL-4-(1-METHYLETHENYL)-(S)-CYCLOHEX-
ENE

CONSENSUS REPORTS: Reported in EPA TSCA Inventory.

SAFETY PROFILE: A skin irritant. When heated to decomposition it emits acrid smoke and irritating fumes.

MCE000   CAS: 80-47-7   **HR: 3**
p-MENTHANE-8-HYDROPEROXIDE
mf: $C_{10}H_{20}O_2$ mw: 172.30

PROP: Clear, pale yellow liquid. D: 0.910-0.925 @ 15.5°/4°.

SYN: p-MENTHANE HYDROPEROXIDE

SAFETY PROFILE: Questionable carcinogen with experimental tumorigenic data. When heated to decomposition it emits acrid smoke and irritating fumes. An irritant and powerful oxidizer.

MCE250   CAS: 1074-95-9   **HR: 2**
p-MENTHAN-3-ONE racemic
mf: $C_{10}H_{18}O$ mw: 154.28

PROP: Several stereoisomers found in nature; 1-menthone found in essential oils of Russian and American peppermint, Geranium, *Andropogon fragrans, Mentha timija, Mentha arvensis*

and others; d-menthone found in essential oils of *Barosma pulchellum, Nepeta japonica maxim* and others; d-isomenthone isolated from *Micromeriabissinica benth., Pelargonium tometosum jacquin,* and others; 1-isomenthone identified in *Reunion geranium, Pelargonium capitatum* and others. Flash p: 156°F.

SYNS: FEMA No. 2667 * 2-ISOPROPYL-5-METHYL-
CYCLOHEXAN-1-ONE, racemic * MENTHONE, racemic

CONSENSUS REPORTS: Reported in EPA TSCA Inventory.

SAFETY PROFILE: Moderately toxic by ingestion. A skin irritant. Combustible liquid. When heated to decomposition it emits acrid smoke and irritating fumes.

MCE750   CAS: 7786-67-6   **HR: 2**
p-MENTH-8-EN-3-OL
mf: $C_{10}H_{18}O$ mw: 154.28

PROP: Colorless liquid; mint odor. D: 0.904-0.913, refr index: 1.470-1.475. Misc in alc, ether, fixed oils; sltly sol in water.

SYNS: FEMA No. 2962 * ISOPULEGOL (FCC)
* 8(9)-p-MENTHEN-3-OL * 1-METHYL-4-ISOPROPE-
NYLCYCLOHEXAN-3-OL

CONSENSUS REPORTS: Reported in EPA TSCA Inventory.

SAFETY PROFILE: Moderately toxic by ingestion and skin contact. When heated to decomposition it emits acrid smoke and irritating fumes.

MCF750   CAS: 89-78-1   **HR: 3**
MENTHOL
mf: $C_{10}H_{20}O$ mw: 156.26

PROP: Hexagonal crystals or granules; peppermint taste and odor. D: 0.890 @ 15°/15°, vap press: 1 mm @ 56.0°, vap d: 5.38, mp: 41-43°, bp: 212°, flash p: +199°F. Very sol in alc, chloroform, ether, petr ether, glacial acetic acid, liquid petrolatum; sltly sol in water.

SYNS: FEMA No. 2665 * HEXAHYDROTHYMOL
* 2-ISOPROPYL-5-METHYL-CYCLOHEXANOL
* p-MENTHAN-3-OL * 1-MENTHOL * 5-METHYL-
2-(1-METHYLETHYL)CYCLOHEXANOL * PEPPERMINT
CAMPHOR

CONSENSUS REPORTS: Reported in EPA TSCA Inventory.

SAFETY PROFILE: Poison by intravenous route. Moderately toxic by ingestion and intraperitoneal routes. A severe eye irritant. Incompatible with phenol, β-naphthol, resorcinol or thymol in trituration, potassium permanganate, chromium trioxide, pyrogallol. Combustible liquid. When heated to decomposition it emits acrid smoke and irritating fumes.

MCG000   CAS: 15356-70-4   **HR: 2**
dl-MENTHOL
mf: $C_{10}H_{20}O$ mw: 156.30

SYNS: FEMA No. 2665 * 4-ISOPROPYL-1-METHYL-CYCLOHEXAN-3-OL * dl-3-p-MENTHANOL * 3-p-MENTHOL * MENTHOL racemic * MENTHOL racemique (FRENCH) * 5-METHYL-2-(1-METHYLETHYL) CYCLOHEXANOL (1-α,2-β,5-α)

CONSENSUS REPORTS: NCI Carcinogenesis Bioassay (feed); No Evidence: mouse, rat NCITR* NCI-GC-TR-98,79. Reported in EPA TSCA Inventory.

SAFETY PROFILE: Moderately toxic by ingestion, intraperitoneal, and subcutaneous routes. A skin irritant. When heated to decomposition it emits acrid smoke and irritating fumes.

MCG250   CAS: 2216-51-5   **HR: 3**
l-MENTHOL
mf: $C_{10}H_{20}O$ mw: 156.30

PROP: Found in high concentrations in oils of Peppermint (*Mentha Piperita*) and Japanese Mint Oil (*Mentha Arvensis*), and in lower concentrations in Reunion Geranium Oil and in a large number of essential oils; prepared by isolation from *Mentha arvensis* Oils.

SYNS: FEMA No. 2665 * (−)-MENTHYL ALCOHOL * (1R-(1-α,2-β,5-α))-5-METHYL-2-(1-METHYLETHYL)CYCLOHEXANOL * U.S.P. MENTHOL

CONSENSUS REPORTS: Reported in EPA TSCA Inventory.

SAFETY PROFILE: Poison by intravenous route. Moderately toxic by ingestion, intraperitoneal, and subcutaneous routes. When heated to decomposition it emits acrid smoke and irritating fumes.

MCG275   CAS: 89-80-5   **HR: 2**
MENTHONE
mf: $C_{10}H_{18}O$ mw: 154.28

PROP: Colorless liquid; mint odor. D: 0.888-0.895, refr index: 1.448-1.453. Sol in alc, fixed oils; very sltly sol in water.

SYNS: FEMA No. 2667 * l-p-MENTHAN-3-ONE * l-MENTHONE (FCC) * p-MENTHONE * trans-MENTHONE * trans-5-METHYL-2-(1-METHYLETHYL)-CYCLOHEXANONE

CONSENSUS REPORTS: Reported in EPA TSCA Inventory.

SAFETY PROFILE: Moderately toxic by ingestion, intravenous and subcutaneous routes. Mutation data reported. When heated to decomposition it emits acrid smoke and irritating fumes.

MCG500   CAS: 16409-45-3   **HR: 1**
dl-MENTHYL ACETATE
mf: $C_{12}H_{22}O_2$ mw: 198.34

PROP: Colorless liquid; characteristic minty odor. D: 0.919 @ 20°/4°, refr index: 0.443-1.450, bp: 227°, flash p: 197°F. Sltly sol in water, glycerin; misc with alc, ether, propylene glycol, fixed oils.

SYNS: FEMA No. 2668 * MENTHOL, ACETATE (8CI) * MENTHYL ACETATE * MENTHYL ACETATE racemic * p-MENTH-3-YL ESTER-dl-ACETIC ACID

SAFETY PROFILE: Mildly toxic by ingestion. A skin irritant. Combustible liquid. When heated to decomposition it emits acrid smoke and irritating fumes.

MCG750   CAS: 2623-23-6   **HR: 1**
l-p-MENTH-3-YL ACETATE
mf: $C_{12}H_{22}O$ mw: 182.34

PROP: Colorless liquid; minty odor. D: 0.919-0.924, refr index: 1.443-1.447. Sol in alc, propylene glycol, fixed oils; sltly sol in water, glycerin.

SYNS: FEMA No. 2668 * l-2-ISOPROPYL-5-METHYL-CYCLOHEXAN-1-OL ACETATE * (−)-MENTHYL ACETATE * l-MENTHYL ACETATE (FCC) * l-p-MENTH-3-YL ACETATE * (R-(1α,2β,5α))-5-METHYL-2-(1-METHYLETHYL)-CYCLOHEXANOL ACETATE (9CI)

CONSENSUS REPORTS: Reported in EPA TSCA Inventory.

SAFETY PROFILE: A skin irritant. When heated to decomposition it emits acrid smoke and irritating fumes.

MCI750   CAS: 33396-37-1   **HR: 3**
MEPROSCILLARIN
mf: $C_{29}H_{44}O_{10}$ mw: 552.73

SYN: CLIFT

SAFETY PROFILE: Poison by ingestion and intravenous routes. Experimental reproductive effects. When heated to decomposition it emits acrid smoke and irritating fumes.

MCJ500 HR: 2
MERCAPTANS

PROP: Compounds containing the -SH group bound to carbon. Also called thiols.

SAFETY PROFILE: Generally they have a very offensive odor which may cause nausea and headache. High concentrations of vapor can produce unconsciousness with cyanosis, cold extremities, and rapid pulse. A common air contaminant. Dangerous; when heated to decomposition they almost always emit highly toxic fumes of SO_x. They may react with water, steam or acids to produce toxic and flammable vapors. Aliphatic mercaptans are flammable. They can react violently with powerful oxidizers such as $Ca(OCl)_2$.

MCK000 CAS: 4822-44-0 HR: 3
α-MERCAPTOACETANILIDE
mf: C_8H_8NOS mw: 166.23

SYNS: 2-MERCAPTOACETANILIDE * THIOGLYCO-LANILIDE * THIOGLYCOLIC ACID ANILIDE * USAF EK-6583

CONSENSUS REPORTS: Reported in EPA TSCA Inventory.

SAFETY PROFILE: Poison by intraperitoneal route. When heated to decomposition it emits very toxic fumes of NO_x and SO_x.

MCM750 CAS: 123-93-3 HR: 3
MERCAPTODIACETIC ACID
mf: $C_4H_6O_4S$ mw: 150.16

PROP: A white powder. Mp: 128°.

SYNS: (CARBOXYMETHYLTHIO)ACETIC ACID * DIMETHYLSULFIDE-α,α′-DICARBOXYLIC ACID * THIODIGLYCOLIC ACID * β,β′-THIODIGLYCOLIC ACID * 2,2′-THIODIGLYCOLIC ACID * THIODIGLY-COLLIC ACID * USAF CB-36 * USAF E-2

CONSENSUS REPORTS: Reported in EPA TSCA Inventory.

SAFETY PROFILE: Poison by intraperitoneal route. When heated to decomposition or on contact with acid or acid fumes it emits toxic fumes of SO_x.

MCN250 CAS: 60-24-2 HR: 3
2-MERCAPTOETHANOL

DOT: UN 2966
mf: C_2H_6OS mw: 78.14

PROP: Water-white, mobile liquid. Bp: 157-158° (decomp) @ 742 mm, flash p: 165°F (COC), d: 1.1168 @ 20°/20°, vap press: 1.0 mm @ 20°, vap d: 2.69. Pure liquid is misc with water, alc, ether, and benzene.

SYNS: EMERY 5791 * 1-ETHANOL-2-THIOL * 2-HYDROXY-1-ETHANETHIOL * 2-HYDROXY-ETHYL MERCAPTAN * 2-ME * MERCAPTOETHA-NOL * β-MERCAPTOETHANOL * MONOTHIO-ETHYLENEGLYCOL * 2-THIOETHANOL * THIOGLYCOL (DOT) * THIOMONOGLYCOL * USAF EK-4196

CONSENSUS REPORTS: Reported in EPA TSCA Inventory.

DOT Classification: Poison B; Label: Poison.

SAFETY PROFILE: Poison by ingestion, skin contact, and intraperitoneal routes. Moderately toxic by intravenous route. A skin and severe eye irritant. Human mutation data reported. Flammable when exposed to heat, flame or oxidizers. To fight fire, use alcohol foam, CO_2, dry chemical. When heated to decomposition it emits highly toxic fumes SO_x.

MCN500 CAS: 51-85-4 HR: 3
β-MERCAPTOETHVLAMINE DISULFIDE
mf: $C_4H_{12}N_2S_2$ mw: 152.30

SYNS: BECAPTAN DISULFURE (FRENCH) * BIS(β-AMINOETHYL)DISULFIDE * CYSTAMINE * CYSTEINAMINE DISULFIDE * CYSTINAMIN (GER-MAN) * CYSTINEAMINE * DECARBOXYCYSTINE * β,β′-DIAMINODIETHYL DISULFIDE * 2,2′-DITHIOBIS(ETHYLAMINE) * MERCAMINE DI-SULFIDE * 2-MERCAPTOETHYLAMINE (OXIDIZED)

SAFETY PROFILE: Poison by intraperitoneal, intramuscular and subcutaneous routes. Mutation data reported. When heated to decomposition it emits very toxic fumes of NO_x and SO_x.

MCO500 CAS: 60-56-0 HR: 3
2-MERCAPTO-1-METHYLIMIDAZOLE
mf: $C_4H_6N_2S$ mw: 114.18

SYNS: BASOLAN * DANANTIZOL * FAVISTAN * FRENTIROX * MERCAPTAZOLE * MERCAZO-LYL * METAZOLO * METHIAMAZOLE * 1-METHYLIMIDAZOLE-2-THIOL * 1-METHYL-2-

MERCAPTOIMIDAZOLE * METIZOL * METO-
THYRINE * 1-METYLO-2-MERKAPTOIMIDAZOLEM
(POLISH) * STRUMAZOLE * TAPAZOLE
* THACAPZOL * THIAMAZOLE * THYCAPSOL
* USAF EL-30

CONSENSUS REPORTS: Reported in EPA
TSCA Inventory.

SAFETY PROFILE: Poison by subcutaneous
route. Moderately toxic by ingestion and intraperitoneal routes. Human teratogenic effects.
Experimental teratogenic and reproductive effects. Questionable carcinogen with experimental neoplastigenic data. Mutation data reported.
An antithyroid drug. When heated to decomposition it emits very toxic fumes of NO_x and
SO_x.

MCQ500   CAS: 4988-64-1   HR: 3
MERCAPTOPURINE RIBONUCLEOSIDE
mf: $C_{10}H_{12}N_4O_4S$ mw: 284.32

SYNS: 6-MERCAPTOPURINE RIBOSIDE * NSC 4911
* RIBOFURANOSIDE, 9H-PURINE-6-THIOL-9 * RIBO-
SYL-6-THIOPURINE * THIONOSINE * 6-THIOPU-
RINE RIBONUCLEOSIDE * 6-THIOPURINE RIBOSIDE
* TIOINOSINE

SAFETY PROFILE: Poison by an unspecified
route. Moderately toxic by ingestion, subcutaneous, and intraperitoneal routes. An experimental
teratogen. Human mutation data reported. When
heated to decomposition it emits very toxic
fumes of SO_x and NO_x.

MCR250   CAS: 67479-03-2   HR: 3
p-MERCAPTO SULFADIAZINE

SYNS: TSD * USAF LO-3

SAFETY PROFILE: Poison by intraperitoneal
route. When heated to decomposition it emits
very toxic fumes of SO_x and NO_x.

MCR750   CAS: 52-67-5   HR: 3
d,3-MERCAPTOVALINE
mf: $C_5H_{11}NO_2S$ mw: 149.23

SYNS: CUPRENIL * CUPRIMINE * DEPEN
* DIMETHYLCYSTEINE * β,β-DIMETHYLCYSTEINE
* d-MERCAPTOVALINE * METALCAPTASE * PCA
* d-PENAMINE * PENICILLAMIN * (S)-PENI-
CILLAMIN * PENICILLAMINE * d-PENICILLAMINE
* REDUCED-d-PENICILLAMINE * TROLOVOL

SAFETY PROFILE: Poison by intraperitoneal
route. Moderately toxic by subcutaneous and
intravenous routes. Mildly toxic by ingestion.

Human systemic effects by ingestion: leukopenia, thrombocytopenia, proteinuria, dermatitis,
and fever. Human teratogenic effects by an unspecified route: developmental abnormalities of
the craniofacial areas, skin, and skin appendages, and body wall. Experimental teratogenic
and reproductive effects. Questionable human
carcinogen producing leukemia. Mutation data
reported. Used in the treatment of rheumatoid
arthritis, metal poisonings, and cystinuria.
When heated to decomposition it emits very
toxic fumes of NO_x and SO_x.

MCS750   CAS: 1600-27-7   HR: 3
MERCURIC ACETATE
DOT: UN 1629
mf: $C_4H_6O_4 \cdot Hg$ mw: 318.69

PROP: White crystals or powder; slt acetic odor.
D: 3.280, mp: 178-180° (overheating causes
decomp). Sol in alc. Keep stoppered and protected from light.

SYNS: ACETIC ACID, MERCURY(2+) SALT * BIS
(ACETYLOXY)MERCURY * DIACETOXYMERCURY
* MERCURIACETATE * MERCURIC DIACETATE
* MERCURY ACETATE * MERCURY(2+) ACETATE
* MERCURY(II) ACETATE * MERCURY DIACETATE
* MERCURYL ACETATE

CONSENSUS REPORTS: EPA Extremely
Hazardous Substances List. Mercury and its
compounds are on the Community Right-To-
Know List. EPA Genetic Toxicology Program.
Reported in EPA TSCA Inventory.

OSHA PEL: (Transitional: CL 1 mg/10m³) CL
0.1 mg(Hg)/m³ (skin)
ACGIH TLV: TWA 0.1 mg(Hg)/m³ (skin)
NIOSH REL: (Inorganic Mercury) TWA 0.05
mg(Hg)/m³
DOT Classification: Poison B; Label: Poison.

SAFETY PROFILE: Poison by ingestion, intravenous, intraperitoneal, and subcutaneous
routes. Moderately toxic by skin contact. Experimental teratogenic and reproductive effects.
When heated to decomposition it emits toxic
fumes of Hg.

MCT500   CAS: 21908-53-2   HR: 3
MERCURIC OXIDE
DOT: UN 1641
mf: HgO   mw: 216.59

PROP: Heavy, bright, orange-red or orange-
yellow powder. Mp: decomp @ 500°, d: 11.14.

Practically insol in water; sol in dil HCl or HNO$_3$. Protect from light.

SYNS: C.I. 77760 * MERCURIC OXIDE, RED * MERCURIC OXIDE, solid (DOT) * MERCURIC OXIDE, YELLOW * MERCURY(II) OXIDE * OXYDE de MERCURE (FRENCH) * QUECKSILBEROXID (GERMAN) * RED OXIDE of MERCURY * RED PRECIPITATE * SANTAR * YELLOW MERCURIC OXIDE * YELLOW OXIDE of MERCURY * YELLOW PRECIPITATE

CONSENSUS REPORTS: EPA Extremely Hazardous Substances List. Mercury and its compounds are on the Community Right-To-Know List. Reported in EPA TSCA Inventory.

OSHA PEL: (Transitional: CL 1 mg/10m^3) CL 0.1 mg(Hg)/m^3 (skin)
ACGIH TLV: TWA 0.1 mg(Hg)/m^3 (skin)
NIOSH REL: (Inorganic Mercury) TWA 0.05 mg(Hg)/m^3
DOT Classification: Poison B; Label: Poison.

SAFETY PROFILE: Poison by ingestion, skin contact, intraperitoneal and intramuscular routes. Experimental teratogenic and reproductive effects. An FDA over-the-counter drug. Used for treating fruit trees. Flammable by chemical reactions. A powerful oxidizer. Explosive reaction with acetyl nitrate, butadiene + ethanol + iodine (at 35°C), chlorine + hydrocarbons (e.g., methane, ethylene), diboron tetrafluoride, hydrogen peroxide + traces of nitric acid, reducing agents (e.g., hydrazine hydrate, phosphinic acid). Forms heat- or impact-sensitive explosive mixtures with non-metals (e.g., phosphorus, sulfur), metals (e.g., magnesium, potassium, sodium-potassium alloy). Reacts violently with hydrogen trisulfide (or ignition), hydrazine hydrate, hydrogen peroxide, hypophosphorus acid, iodine + methanol or ethanol, phospham, acetyl nitrate, S$_2$Cl$_2$, reductants. Incandescent reaction with phospham. When heated to decomposition it emits highly toxic fumes of Hg.

MCU000   CAS: 5970-32-1   **HR: 3**
MERCURIC SALICYLATE
DOT: UN 1644
mf: C$_7$H$_4$HgO$_3$ mw: 336.70

PROP: White-yellow or pinkish, odorless powder. Insol in H$_2$O or alc, sol in warm solns of alkali halides.

SYNS: MERCURIC SALICYLATE, solid (DOT) * MERCURISALICYLIC ACID * MERCURY SALICYLATE * MERCURY SUBSALICYLATE

CONSENSUS REPORTS: Mercury and its compounds are on the Community Right-To-Know List.

OSHA PEL: (Transitional: CL 1 mg/10m^3) CL 0.1 mg(Hg)/m^3 (skin)
ACGIH TLV: TWA 0.1 mg(Hg)/m^3 (skin)
NIOSH REL: (Inorganic Mercury) TWA 0.05 mg(Hg)/m^3
DOT Classification: Poison B; Label: Poison.

SAFETY PROFILE: Poison by subcutaneous route. An FDA over-the-counter drug. Incompatible with alkali iodides. When heated to decomposition it emits toxic fumes of Hg.

MCU250   CAS: 592-85-8   **HR: 3**
MERCURIC SULFOCYANATE
DOT: UN 1646
mf: C$_2$HgN$_2$S$_2$ mw: 316.79

PROP: White, odorless powder; sltly sol in cold water; more sol in boiling water (decomp); sol in dil HCl. Protect from light.

SYNS: BIS(THIOCYANATO)-MERCURY * MERCURIC SULFOCYANTE, solid (DOT) * MERCURIC SULFOCYANIDE * MERCURIC THIOCYANATE * MERCURIC THIOCYANATE, solid (DOT) * MERCURY DITHIOCYANATE * MERCURY THIOCYANATE (DOT) * MERCURY(II) THIOCYANATE * THIOCYANIC ACID, MERCURY(2$^+$) SALT

CONSENSUS REPORTS: Mercury and its compounds are on the Community Right-To-Know List. Reported in EPA TSCA Inventory.

OSHA PEL: (Transitional: CL 1 mg/10m^3) CL 0.1 mg(Hg)/m^3 (skin)
ACGIH TLV: TWA 0.1 mg(Hg)/m^3 (skin)
NIOSH REL: (Inorganic Mercury) TWA 0.05 mg(Hg)/m^3
DOT Classification: Poison B; Label: Poison.

SAFETY PROFILE: A poison by ingestion and intraperitoneal routes. Moderately toxic by skin contact. Thermally unstable and decomposition may be vigorous. When heated to decomposition it emits very toxic fumes of Hg, NO$_x$, SO$_x$, and CN$^-$.

MCU750   CAS: 55-68-5   **HR: 3**
MERCURIPHENYL NITRATE
DOT: UN 1895
mf: C$_6$H$_5$HgNO$_3$ mw: 339.71

PROP: Crystals. Mp: 176-186°. Insol in cold water.

SYNS: MERPHENYL NITRATE * MERSOLITE 7 * NITRIC ACID, PHENYLMERCURY SALT * PHENALCO * PHENITOL * PHENMERZYL NITRATE * PHENYLMERCURIC NITRATE * PHENYLMERCURY NITRATE * PHERMERNITE

CONSENSUS REPORTS: Mercury and its compounds are on the Community Right-To-Know List. Reported in EPA TSCA Inventory. EPA Genetic Toxicology Program.

OSHA PEL: (Transitional: CL 1 mg/10m^3) CL 0.1 mg(Hg)/m^3 (skin)
ACGIH TLV: TWA 0.1 mg(Hg)/m^3 (skin)
NIOSH REL: (Inorganic Mercury) TWA 0.05 mg(Hg)/m^3
DOT Classification: Poison B; Label: Poison.

SAFETY PROFILE: Poison by intravenous and subcutaneous routes. FDA over-the-counter drug. When heated to decomposition it emits very toxic fumes of Hg and NO$_x$.

MCV000   CAS: 129-16-8   HR: 2
MERCUROCHROME
mf: $C_{20}H_{10}Br_2HgO_6 \cdot 2Na$ mw: 752.69

SYNS: ASCEPTICHROME * ASEPTICHROME * CHROMARGYRE * 2,7-DIBROMO-4-HYDROXYMERCURIFLUORESCEINE DISODIUM SALT * DISODIUM 2,7-DIBROM-4-HYDROXY-MERCURI-FLUORESCEIN * DISODIUM 2',7'-DIBROMO-4'-(HYDROXYMERCURY) FLUORESCEIN * DOMF * FLAVUROL * FLUOROCHROME * GALLOCHROME * GYNOCHROME * MERBROMIN * MERCURANINE * MERCUROCHROME-220 SOLUBLE * MERCUROCOL * MERCUROME * MERCUROPHAGE * PLANOCHROME

CONSENSUS REPORTS: Mercury and its compounds are on the Community Right-To-Know List.

NIOSH REL: (Inorganic Mercury) TWA 0.05 mg(Hg)/m^3

SAFETY PROFILE: Poison by intravenous and subcutaneous routes. Mutation data reported. Relatively nonirritating and nontoxic to damaged skin or tissue. A topical antiseptic. An FDA over-the-counter drug. When heated to decomposition it emits very toxic fumes including fumes of Na$_2$O, Br$^-$, and Hg.

MCV250   CAS: 12002-19-6   HR: 3
MERCUROL

DOT: UN 1639

PROP: Colorless to brownish powder. Contains 20% mercury.

SYN: MERCURY NUCLEATE, solid (DOT)

CONSENSUS REPORTS: Mercury and its compounds are on the Community Right-To-Know List.

NIOSH REL: (Inorganic Mercury) TWA 0.05 mg(Hg)/m^3
DOT Classification: Poison B; Label: Poison.

SAFETY PROFILE: A poison. When heated to decomposition it emits toxic fumes of Hg.

MCW000   CAS: 7546-30-7   HR: 3
MERCUROUS CHLORIDE
mf: Cl_2Hg_2 mw: 472.09

PROP: White, odorless, tasteless, heavy powder or crystals. Sublimes @ 400°, d: 7.150. Insol in water, alc, and ether. Protect from light. Sunlight causes it to decomp into mercuric chloride and metallic Hg.

SYNS: CALOGREEN * CALOMEL * CALOMELANO (ITALIAN) * CALOSAN * CHLORURE MERCUREUX (FRENCH) * C.I. 77764 * CLORURO MERCUROSO (ITALIAN) * CYCLOSAN * KALOMEL (GERMAN) * MERCUROCHLORIDE (DUTCH) * MERCURY(I) CHLORIDE * MERCURY MONOCHLORIDE * MERCURY PROTOCHLORIDE * MILD MERCURY CHLORIDE * PRECIPITE BLANC * QUECKSILBER(I)-CHLORID (GERMAN) * QUECKSILBER CHLORUER (GERMAN) * SUBCHLORIDE of MERCURY

CONSENSUS REPORTS: Mercury and its compounds are on the Community Right-To-Know List. EPA Genetic Toxicology Program. Reported in EPA TSCA Inventory.

OSHA PEL: (Transitional: CL 1 mg/10m^3) CL 0.1 mg(Hg)/m^3 (skin)
ACGIH TLV: TWA 0.1 mg(Hg)/m^3 (skin)
NIOSH REL: (Inorganic Mercury) TWA 0.05 mg(Hg)/m^3

SAFETY PROFILE: Poison by ingestion and intraperitoneal routes. Moderately toxic by skin contact. Mutation data reported. A fungicide. An FDA over-the-counter drug. Incompatible with bromides; iodides; alkali chlorides; sulfates; sulfites; carbonates; hydroxides; lime water; ammonia; golden antimony sulfide; cyanides; copper salts; hydrogen peroxide; iodine; iodoform; lead salts; silver salts; sulfides. When heated to decomposition it emits very toxic fumes of Cl$^-$ and Hg.

MCW250   CAS: 7439-97-6   **HR: 3**
MERCURY

DOT: NA 2809
af: Hg   aw: 200.59

PROP: Silvery, heavy, mobile liquid. A liquid metallic element. Mp: −38.89°, bp: 356.9°, d: 13.534 @ 25°, vap press: 2 × 10^{-3} mm @ 25°. Solid: tin-white, ductile, malleable mass which can be cut with a knife.

SYNS: COLLOIDAL MERCURY * KWIK (DUTCH)
* MERCURE (FRENCH) * MERCURIO (ITALIAN)
* MERCURY, METALLIC (DOT) * NCI-C60399
* QUECKSILBER (GERMAN) * QUICK SILVER
* RCRA WASTE NUMBER U151 * RTEC (POLISH)

CONSENSUS REPORTS: Mercury and its compounds are on the Community Right-To-Know List.

OSHA PEL: Vapor: (Transitional: CL 1 mg/$10m^3$) 0.05 mg/m^3 (skin)
ACGIH TLV: TWA 0.05 mg(Hg)/m^3 (vapor, skin)
DFG MAK: 0.1 mg/m^3; BAT: 5 μg/dL in blood.
NIOSH REL: (Inorganic Mercury) TWA 0.05 mg(Hg)/m^3
DOT Classification: Corrosive Material; Label: Corrosive.

SAFETY PROFILE: Poison by inhalation. Corrosive to skin, eyes, and mucous membranes. Human systemic effects by inhalation: wakefulness, muscle weakness, anorexia, headache, tinnitus, hypermotility, diarrhea, liver changes, dermatitis, fever. Experimental teratogenic and reproductive effects. Questionable carcinogen with experimental tumorigenic data. Human mutation data reported. Used in dental applications, electronics, and chemical synthesis.
     May explode on contact with 3-bromopropyne, alkynes + silver perchlorate, ethylene oxide, lithium, methylsilane + oxygen (explodes when shaken), peroxyformic acid, chlorine dioxide, tetracarbonylnickel + oxygen. May react with ammonia to form an explosive product. Mixtures with methyl azide are shock- and spark-sensitive explosives. The vapor ignites on contact with boron diiodophosphide. Reacts violently with acetylenic compounds (e.g., acetylene, sodium acetylide, 2-butyne-1,4-diol + acid), metals (e.g., aluminum, calcium, potassium, sodium, rubidium, exothermic formation of amalgams), Cl_2, ClO_2, CH_3N_3, Na_2C_2, nitromethane. Incompatible with methyl azide, oxidants. When heated to decomposition it emits toxic fumes of Hg.

MCW500   CAS: 10124-48-8   **HR: 3**
MERCURY AMIDE CHLORIDE

DOT: UN 1630
mf: ClH_2HgN mw: 252.07

PROP: White, pulverized lumps or powder.

SYNS: AMINOMERCURIC CHLORIDE * AMMONIATED MERCURY * MERCURIC AMMONIUM CHLORIDE, solid * MERCURIC CHLORIDE, AMMONIATED * MERCURY AMINE CHLORIDE * MERCURY AMMONIATED * WHITE MERCURY PRECIPITATED * WHITE PRECIPITATE

CONSENSUS REPORTS: Mercury and its compounds are on the Community Right-To-Know List. Reported in EPA TSCA Inventory.

OSHA PEL: (Transitional: CL 1 mg/$10m^3$) CL 0.1 mg(Hg)/m^3 (skin)
ACGIH TLV: TWA 0.1 mg(Hg)/m^3 (skin)
NIOSH REL: (Inorganic Mercury) TWA 0.05 mg(Hg)/m^3
DOT Classification: Poison B; Label: Poison.

SAFETY PROFILE: A poison. Explosive reaction with halogens or amine metal salts. When heated to decomposition it emits very toxic fumes of Cl^-, NO_x, and Hg.

MCX000   CAS: 38232-63-2   **HR: 3**
MERCURY(I) AZIDE
mf: Hg_2N_6 mw: 485.22

SYNS: MERCUROUS AZIDE (DOT) * MERCURY AZIDE

CONSENSUS REPORTS: Mercury and its compounds are on the Community Right-To-Know List.

OSHA PEL: (Transitional: CL 1 mg/$10m^3$) CL 0.1 mg(Hg)/m^3 (skin)
ACGIH TLV: TWA 0.1 mg(Hg)/m^3 (skin)
NIOSH REL: TWA 0.05 mg(Hg)/m^3
DOT Classification: Forbidden.

SAFETY PROFILE: Poison. Explodes on heating in air. When heated to decomposition it emits very toxic fumes of NO_x and Hg.

MCX500   CAS: 583-15-3   **HR: 3**
MERCURY(II) BENZOATE

DOT: UN 1631
mf: $C_{14}H_{10}O_4$ • Hg   mw: 442.83

PROP: White, crystalline, odorless powder. Mp: 165°. Very sol in NaCl soln; sltly sol in alc. Protect from light.

SYNS: MERCURIC BENZOATE * MERCURIC BENZOATE, solid (DOT)

CONSENSUS REPORTS: Mercury and its compounds are on the Community Right-To-Know List.

OSHA PEL: (Transitional: CL 1 mg/10m³) CL 0.1 mg(Hg)/m³ (skin)
ACGIH TLV: TWA 0.1 mg(Hg)/m³ (skin)
NIOSH REL: (Inorganic Mercury) TWA 0.05 mg(Hg)/m³
DOT Classification: Poison B; Label: Poison.

SAFETY PROFILE: A poison. When heated to decomposition it emits toxic fumes of Hg.

MCX750   CAS: 10031-18-2   **HR: 3**
MERCURY(I) BROMIDE (1:1)

DOT: UN 1634
mf: BrHg   mw: 280.50

PROP: White-yellow, odorless, tetragonal crystals or powder. Darkens on exposure to light. D: 7.307, vap d: 19.3. Sublimes @ approx 390° (decomp). Insol in water, alc, and ether; decomp by hot HCl or alkali bromides. Protect from light.

SYN: MERCUROUS BROMIDE, solid (DOT)

CONSENSUS REPORTS: Mercury and its compounds are on the Community Right-To-Know List.

OSHA PEL: (Transitional: CL 1 mg/10m³) CL 0.1 mg(Hg)/m³ (skin)
ACGIH TLV: TWA 0.1 mg(Hg)/m³ (skin)
NIOSH REL: (Inorganic Mercury) TWA 0.05 mg(Hg)/m³
DOT Classification: Poison B; Label: Poison.

SAFETY PROFILE: A poison. When heated to decomposition it emits very toxic fumes of Br⁻ and Hg.

MCY000   CAS: 7789-47-1   **HR: 3**
MERCURY(II) BROMIDE (1:2)

DOT: UN 1634
mf: Br₂Hg   mw: 360.41

PROP: White crystals or crystalline powder. Sensitive to light. Mp: 237°, bp: 322° (subl), d: 6.109 @ 25°, vap press: 1 mm @ 136.5°.

Very sol in hot alc, methanol, HCl, HBr, alkali bromide solns; sltly sol in chloroform.

SYNS: MERCURIC BROMIDE * MERCURIC BROMIDE, solid (DOT)

CONSENSUS REPORTS: Mercury and its compounds are on the Community Right-To-Know List. Reported in EPA TSCA Inventory.

OSHA PEL: (Transitional: CL 1 mg/10m³) CL 0.1 mg(Hg)/m³ (skin)
ACGIH TLV: TWA 0.1 mg(Hg)/m³ (skin)
NIOSH REL: (Inorganic Mercury) TWA 0.05 mg(Hg)/m³
DOT Classification: Poison B; Label: Poison.

SAFETY PROFILE: A poison by ingestion, skin contact and intraperitoneal routes. Vigorous reaction with indium at 350°C. Incompatible with sodium and potassium. When heated to decomposition it emits very toxic fumes of Br⁻ and Hg.

MCY475   CAS: 7487-94-7   **HR: 3**
MERCURY(II) CHLORIDE

DOT: UN 1624
mf: Cl₂Hg   mw: 271.50

PROP: White crystals or powder. Mp: 276°, bp: 302°, d: 5.440 @ 25°, vap press: 1 mm @ 136.2°.

SYNS: BICHLORIDE of MERCURY * BICHLORURE de MERCURE (FRENCH) * CALOCHLOR * CHLORID RTUTNATY (CZECH) * CHLORURE MERCURIQUE (FRENCH) * CLORURO di MERCURIO (ITALIAN) * CORROSIVE MERCURY CHLORIDE * CORROSIVE SUBLIMATE * MERCURIC CHLORIDE (DOT) * MERCURY BICHLORIDE * MERCURY PERCHLORIDE * NCI-C60173 * QUECKSILBER CHLORID (GERMAN) * PERCHLORIDE of MERCURY * SULEMA (RUSSIAN) * SUBLIMAT (CZECH) * TL 898

CONSENSUS REPORTS: Mercury and its compounds are on the Community Right-To-Know List. EPA Genetic Toxicology Program. Reported in EPA TSCA Inventory.

OSHA PEL: (Transitional: CL 1 mg/10m³) CL 0.1 mg(Hg)/m³ (skin)
ACGIH TLV: TWA 0.1 mg(Hg)/m³ (skin)
NIOSH REL: (Inorganic Mercury) TWA 0.05 mg(Hg)/m³
DOT Classification: Poison B; Label: Poison.

SAFETY PROFILE: A human poison by ingestion and possibly other routes. Poison experimentally by skin contact and subcutaneous routes. Human systemic effects by ingestion: respiratory obstruction, nausea or vomiting, urine volume decreased or anuria. Human reproductive effects by ingestion: terminates pregnancy. Experimental teratogenic and reproductive effects. Human mutation data reported. A severe eye and skin irritant. Reaction with sodium aci-nitromethanide + acids forms the explosive mercury fulminate. Reacts violently with K, Na. When heated to decomposition it emits toxic fumes of Hg.

MCZ000   HR: 3
MERCURY COMPOUNDS, INORGANIC

CONSENSUS REPORTS: Mercury and its compounds are on the Community Right-To-Know List.

SAFETY PROFILE: Mercury is a general protoplasmic poison; after absorption it circulates in the blood and is stored in the liver, kidneys, spleen, and bone. In industrial poisoning, the principal effect is upon the central nervous system, the mouth, and gums. The cardinal symptoms of industrial mercury poisoning are stomatitis, tremors, and psychic disturbances. Usually the first complaints are of excessive salivation and painful chewing. In severe cases there may be gingivitis with loosening of the teeth, and a dark line on the gum margins resembling the "lead line." The psychic disturbance (so called "erethism") includes loss of memory, insomnia, lack of confidence, irritability, vague fears and depression. The dermatitis produced by fulminate of mercury takes the form of small, discrete ulcers on the exposed parts, and is usually accompanied by conjunctivitis and inflammation of the mucous membranes of the nose and throat. In humans, it is readily absorbed by the respiratory tract (elemental mercury vapor, dusts of mercury compounds), intact skin, and the gastrointestinal tract. Occasional incidental swallowing of metallic mercury man be without harm. Spilled and heated elemental mercury is particularly hazardous. A number of mercury compounds, in addition to the fulminate, can cause skin irritation and be absorbed through the skin. They are strong allergens and common air contaminants. Acute Toxicity: Soluble salts have violent corrosive effects on skin and mucous membranes, cause severe nausea, vomiting, abdominal pain, bloody diarrhea, kidney damage, and death usually within 10 days. Many mercury compounds are explosively unstable or undergo hazardous reactions. When heated to decomposition they emit toxic fumes of Hg.

MDA000   HR: D
MERCURY COMPOUNDS, ORGANIC

CONSENSUS REPORTS: Mercury and its compounds are on the Community Right-To-Know List.

DFG MAK: 0.01 mg/m^3

SAFETY PROFILE: The customary grouping of all organic mercurials in a single category is not fully justified by the toxicity of the compounds. Alkyl mercurials have very high toxicity; aryl compounds, particularly the phenyls, are much less toxic, and the organomercurials used in therapeutics are less toxic. The alkyls and aryls commonly cause skin burns and other forms of irritation, and both can be absorbed through the skin. Fatal poisoning has occurred due to exposure to alkyl mercurials and permanent damage to the brain has been reported. Phenyl mercurials appear to be no more toxic than metallic mercury. Organic mercury compounds, like organic lead compounds, seem to have an affinity for lipoid-containing organs, resulting in central nervous system disturbances such as from tetraethyl lead. These are common air contaminants. Many mercury compounds are explosively unstable or undergo hazardous reactions. When heated to decomposition they emit highly toxic fumes of Hg.

MDA250   CAS: 592-04-1   HR: 3
MERCURY(II) CYANIDE

DOT: UN 1636
mf: C_2HgN_2 mw: 252.63

PROP: Colorless, odorless, transparent prisms; darkened by light. Decomp @ 320°, d: 3.996. Sltly sol in ether.

SYNS: CYANURE de MERCURE (FRENCH) * MERCURIC CYANIDE, solid (DOT)

CONSENSUS REPORTS: Reported in EPA TSCA Inventory. Mercury and its compounds, as well as cyanide and its compounds, are on the Community Right-To-Know List.

OSHA PEL: (Transitional: CL 1 mg/10m^3) CL 0.1 mg(Hg)/m^3 (skin)
ACGIH TLV: TWA 0.1 mg(Hg)/m^3 (skin)

NIOSH REL: (Inorganic Mercury) TWA 0.05 mg(Hg)/m^3
DOT Classification: Poison B; Label: Poison.

SAFETY PROFILE: Poison by ingestion, subcutaneous, intravenous, and intraperitoneal routes. Human systemic effects by ingestion: nausea or vomiting, hypermotility, diarrhea, kidney changes, somnolence. Hydrolyzes to toxic fumes. A friction- and impact-sensitive explosive. It may initiate detonation of liquid hydrogen cyanide. Incompatible with fluorine, magnesium, sodium nitrite. When heated to decomposition it emits very toxic fumes of Hg, NO$_x$, and CN$^-$.

MDA500   CAS: 1335-31-5   HR: 3
MERCURY CYANIDE OXIDE

DOT: UN 1642
mf: C$_2$Hg$_2$N$_2$O   mw: 469.22

PROP: White, orthorhombic crystals or crystalline powder. D: 4.44.

SYNS: MERCURIC OXYCYANIDE * MERCURIC OXYCYANIDE, solid (desensitized) (DOT) * MERCURY OXYCYANIDE

CONSENSUS REPORTS: Mercury and its compounds, as well as cyanide and its compounds, are on the Community Right-To-Know List.

OSHA PEL: (Transitional: CL 1 mg/10m^3) CL 0.1 mg(Hg)/m^3 (skin)
ACGIH TLV: TWA 0.1 mg(Hg)/m^3 (skin)
NIOSH REL: (Inorganic Mercury) TWA 0.05 mg(Hg)/m^3
DOT Classification: Poison B; Label: Poison.

SAFETY PROFILE: Poison by intravenous route. An explosive sensitive to friction, impact, or heat. The commercial product is stabilized by excess mercury(II) cyanide. When heated to decomposition it emits very toxic fumes of Hg, CN$^-$, and NO$_x$.

MDC000   CAS: 628-86-4   HR: 3
MERCURY(II) FULMINATE (dry)

DOT: UN 0135
mf: C$_2$HgN$_2$O$_2$ mw: 284.63

PROP: White solid. Mp: explodes, d: 4.42.

SYNS: FULMINATE of MERCURY, DRY (DOT) * MERCURY FULMINATE (DOT) * RCRA WASTE NUMBER P065

CONSENSUS REPORTS: Mercury and its compounds are on the Community Right-To-Know List. Reported in EPA TSCA Inventory.

OSHA PEL: (Transitional: CL 1 mg/10m^3) CL 0.1 mg(Hg)/m^3 (skin)
ACGIH TLV: TWA 0.1 mg(Hg)/m^3 (skin)
NIOSH REL: (Inorganic Mercury) TWA 0.05 mg(Hg)/m^3
DOT Classification: Forbidden.

SAFETY PROFILE: An explosive sensitive to flame, heat, impact, friction, intense radiation or contact with sulfuric acid. Self-explodes. Dangerously flammable; should be kept moist until used. Incompatible with sulfuric acid. When heated to decomposition it emits very toxic fumes of Hg and NO$_x$.

MDC250   CAS: 628-86-4   HR: 3
MERCURY FULMINATE (wet)

DOT: UN 0135
mf: C$_2$HgN$_2$O$_2$ mw: 284.63

SYNS: FULMINATE of MERCURY, WET (DOT) * INITIATING EXPLOSIVE FULMINATE of MERCURY (DOT) * RCRA WASTE NUMBER P065

CONSENSUS REPORTS: Mercury and its compounds are on the Community Right-To-Know List.

OSHA PEL: (Transitional: CL 1 mg/10m^3) CL 0.1 mg(Hg)/m^3 (skin)
ACGIH TLV: TWA 0.1 mg(Hg)/m^3 (skin)
NIOSH REL: (Inorganic Mercury) TWA 0.05 mg(Hg)/m^3
DOT Classification: Label: Explosive A.

SAFETY PROFILE: An explosive. It can be kept more safely in wet form. When heated to decomposition it emits very toxic fumes of Hg and NO$_x$.

MDC500   CAS: 63937-14-4   HR: 3
MERCURY(I) GLUCONATE

DOT: UN 1637
mf: C$_6$H$_{11}$O$_7$•Hg   mw: 395.76

PROP: White solid.

SYNS: MERCUROUS GLUCONATE * MERCUROUS GLUCONATE, solid (DOT)

CONSENSUS REPORTS: Mercury and its compounds are on the Community Right-To-Know List.

OSHA PEL: (Transitional: CL 1 mg/10m^3) CL 0.1 mg(Hg)/m^3 (skin)

ACGIH TLV: TWA 0.1 mg(Hg)/m^3 (skin)
NIOSH REL: (Inorganic Mercury) TWA 0.05 mg(Hg)/m^3
DOT Classification: Poison B; Label: Poison.

SAFETY PROFILE: A poison. When heated to decomposition it emits toxic fumes of Hg.

MDC750   CAS: 7783-30-4   HR: 3
MERCURY(I) IODIDE

DOT: UN 1638
mf: HgI   mw: 327.49

PROP: Heavy, odorless, yellow, tetragonal crystals or amorphous powder. D: 7.70, mp: 290° when rapidly heated (partial decomp). Insol in water, alc, and ether; sol in solns of mercurous or mercuric nitrates. Protect from light.

SYNS: IODURE de MERCURE (FRENCH) * MERCU-ROUS IODIDE * MERCUROUS IODIDE, solid (DOT) * MERCURY PROTOIODIDE * YELLOW MERCURY IODIDE

CONSENSUS REPORTS: Mercury and its compounds are on the Community Right-To-Know List.

NIOSH REL: (Inorganic Mercury) TWA 0.05 mg(Hg)/m^3
DOT Classification: Poison B; Label: Poison.

SAFETY PROFILE: Poison by ingestion and intraperitoneal routes. When heated to decomposition it emits very toxic fumes of Hg and I$^-$.

MDD000   CAS: 7774-29-0   HR: 3
MERCURY(II) IODIDE

DOT: UN 1638
mf: HgI$_2$ mw: 454.39

PROP: Scarlet red, heavy, odorless, almost tasteless powder. Sensitive to light. D: 6.28, mp: 259°, bp: approx 350° (subl); very sol in alkali iodides, HgCl$_2$, Na$_2$S$_2$O$_3$.

SYNS: HYDRARGYRUM BIJODATUM (GERMAN) * MERCURIC IODIDE * MERCURIC IODIDE, solid (DOT) * MERCURIC IODIDE, RED * MERCURIC IODIDE, solution (DOT) * MERCURY BINIODIDE * RED MERCURIC IODIDE

CONSENSUS REPORTS: Mercury and its compounds are on the Community Right-To-Know List. Reported in EPA TSCA Inventory.

OSHA PEL: (Transitional: CL 1 mg/10m^3) CL 0.1 mg(Hg)/m^3 (skin)

ACGIH TLV: TWA 0.1 mg(Hg)/m^3 (skin)
NIOSH REL: (Inorganic Mercury) TWA 0.05 mg(Hg)/m^3
DOT Classification: Poison B; Label: Poison, solid; Poison B, Label: Poison, solution.

SAFETY PROFILE: A human poison by ingestion. Poison experimentally by ingestion, skin contact, and intraperitoneal routes. An experimental teratogen. Violent reaction with chlorine trifluoride. When heated to decomposition it emits very toxic fumes of Hg and I$^-$.

MDD750   CAS: 115-09-3   HR: 3
MERCURY METHYLCHLORIDE

mf: CH$_3$ClHg   mw: 251.08

PROP: White crystals, characteristic odor. D: 4.063, mp: 170°.

SYNS: CASPAN * CHLOROMETHYLMERCURY * METHYLMERCURIC CHLORIDE * METHYLMER-CURY CHLORIDE * MMC * MONOMETHYL MER-CURY CHLORIDE

CONSENSUS REPORTS: Mercury and its compounds are on the Community Right-To-Know List. EPA Genetic Toxicology Program.

OSHA PEL: (Transitional: CL 1 mg/10m^3) TWA 0.01 mg(Hg)/m^3; STEL 0.03 mg/m^3 (skin)
ACGIH TLV: TWA 0.01 mg(Hg)/m^3; STEL 0.03 mg(Hg)/m^3
NIOSH REL: TWA 0.05 mg(Hg)/m^3

SAFETY PROFILE: Poison by ingestion, intra-muscular, intravenous, and intraperitoneal routes. Experimental teratogenic and reproductive effects. Questionable carcinogen with experimental carcinogenic data. Human mutation data reported. When heated to decomposition it emits very toxic fumes of Cl$^-$ and Hg.

MDE250   CAS: 631-60-7   HR: 3
MERCURY MONOACETATE

DOT: UN 1629
mf: C$_2$H$_3$O$_2$ • Hg   mw: 259.64

PROP: Colorless scales or plates. Mp: decomp. Sol in dil acetic acid; insol in alc, ether.

SYNS: MERCUROUS ACETATE * MERCUROUS ACE-TATE, solid (DOT) * MERCURY ACETATE

CONSENSUS REPORTS: Mercury and its compounds are on the Community Right-To-Know List.

NIOSH REL: (Inorganic Mercury) TWA 0.05 mg(Hg)/m^3
DOT Classification: Poison B; Label: Poison.

SAFETY PROFILE: A poison by ingestion and intraperitoneal routes. Moderately toxic by skin contact. When heated to decomposition it emits toxic fumes of Hg.

MDE750   CAS: 10415-75-5   HR: 3
MERCURY(I) NITRATE (1:1)

DOT: UN 1627
mf: NO$_3$ • Hg   mw: 262.60

SYNS: MERCUROUS NITRATE, solid (DOT) ∗ NITRATE MERCUREUX (FRENCH) ∗ NITRIC ACID, MERCURY(I) SALT

CONSENSUS REPORTS: Mercury and its compounds are on the Community Right-To-Know List. Reported in EPA TSCA Inventory.

OSHA PEL: (Transitional: CL 1 mg/10m^3) CL 0.1 mg(Hg)/m^3 (skin)
ACGIH TLV: TWA 0.1 mg(Hg)/m^3 (skin)
NIOSH REL: (Inorganic Mercury) TWA 0.05 mg(Hg)/m^3
DOT Classification: Oxidizer; Label: Oxidizer; Poison B; Label: Poison.

SAFETY PROFILE: Poison by ingestion and intraperitoneal routes. Moderately toxic by skin contact. A powerful oxidizer. Explodes on contact with red-hot carbon. Mixtures with phosphorus are impact-sensitive explosives. When heated to decomposition it emits very toxic fumes of Hg and NO$_x$.

MDF000   CAS: 10045-94-0   HR: 3
MERCURY(II) NITRATE (1:2)

DOT: UN 1625
mf: N$_2$O$_6$ • Hg   mw: 324.61

PROP: White-yellowish, deliquescent powder. Mp: 79°, bp: decomp, d: 4.39.

SYNS: MERCURIC NITRATE ∗ MERCURY NITRATE ∗ MERCURY PERNITRATE ∗ NITRATE MERCURIQUE (FRENCH) ∗ NITRIC ACID, MERCURY(II) SALT

CONSENSUS REPORTS: Mercury and its compounds are on the Community Right-To-Know List. Reported in EPA TSCA Inventory.

OSHA PEL: (Transitional: CL 1 mg/10m^3) CL 0.1 mg(Hg)/m^3 (skin)
ACGIH TLV: TWA 0.1 mg(Hg)/m^3 (skin)
NIOSH REL: TWA 0.05 mg(Hg)/m^3
DOT Classification: Poison B; Label: Poison.

SAFETY PROFILE: Poison by ingestion, skin contact, intraperitoneal, and subcutaneous routes. A powerful oxidizer. Probably an eye, skin and mucous membrane irritant. Reacts with acetylene to form the explosive mercury acetylide which is sensitive to heat, friction, or contact with sulfuric acid. Reaction with ethanol forms the explosive mercury fulminate. Reaction with isobutene forms an unstable explosive product. Forms explosive mixtures with phosphine (heat- and impact- sensitive), potassium cyanide (heat-sensitive), and sulfur. Violent reaction with phosphinic acid, hypophosphoric acid, unsaturated hydrocarbons, aromatics. Vigorous reaction with petroleum hydrocarbons. When heated to decomposition it emits very toxic fumes of Hg and NO$_x$.

MDF250   CAS: 1191-80-6   HR: 3
MERCURY OLEATE

DOT: UN 1640
mf: C$_{36}$H$_{66}$O$_4$ • Hg   mw: 763.61

PROP: Yellowish-brown, somewhat transparent, ointment-like mass; odor of oleic acid. Practically insol in water; sltly sol in alc and ether; very sol in oils. Protect from light.

SYNS: MERCURIC OLEATE, solid (DOT) ∗ OLEATE of MERCURY

CONSENSUS REPORTS: Mercury and its compounds are on the Community Right-To-Know List. Reported in EPA TSCA Inventory.

OSHA PEL: (Transitional: CL 1 mg/10m^3) CL 0.1 mg(Hg)/m^3 (skin)
ACGIH TLV: TWA 0.1 mg(Hg)/m^3 (skin)
NIOSH REL: TWA 0.05 mg(Hg)/m^3
DOT Classification: Poison B; Label: Poison.

SAFETY PROFILE: A poison. An FDA over-the-counter drug. When heated to decomposition it emits toxic fumes of Hg.

MDF350   CAS: 7784-37-4   HR: 3
MERCURY(II) ORTHOARSENATE

DOT: UN 1623
mf: AsHO$_4$ • Hg   mw: 340.52

PROP: Yellow powder. Mp: decomp. Insol in water; sol in HCl or HNO$_3$.

SYN: MERCURIC ARSENATE

CONSENSUS REPORTS: Mercury and its compounds, as well as arsenic and its compounds, are on the Community Right-To-Know List.

OSHA PEL: 0.01 mg(As)/m^3; Cancer Hazard; (Transitional: CL 1 mg(Hg)/10m^3) CL 0.1 mg(Hg)/m^3 (skin)

ACGIH TLV: TWA 0.1 mg(Hg)/m^3 (skin)

NIOSH REL: TWA 0.05 mg(Hg)/m^3; CL 2 μg/m^3/15M

DOT Classification: Poison B; Label: Poison.

SAFETY PROFILE: Confirmed human carcinogen. A poison. When heated to decomposition it emits very toxic fumes of Hg and As.

MDF750 CAS: 15829-53-5 **HR: 3**
MERCURY(I) OXIDE

DOT: UN 1641
mf: Hg$_2$O   mw: 417.22

PROP: Black to grayish-black powder. Mp: decomp @ 100°; d: 9.8. Insol in water; sol in HNO$_3$. Protect from light.

SYNS: MERCUROUS OXIDE, BLACK, solid (DOT) * QUECKSILBEROXID (GERMAN)

CONSENSUS REPORTS: Mercury and its compounds are on the Community Right-To-Know List. Reported in EPA TSCA Inventory.

OSHA PEL: (Transitional: CL 1 mg/10m^3) CL 0.1 mg(Hg)/m^3 (skin)

ACGIH TLV: TWA 0.1 mg(Hg)/m^3 (skin)

NIOSH REL: TWA 0.05 mg(Hg)/m^3

DOT Classification: Poison B; Label: Poison.

SAFETY PROFILE: A poison. Flammable by chemical reaction, an oxidizer. Explosive reaction with hydrogen peroxide, chlorine + ethylene. Reacts violently with molten potassium, molten sodium, S, (H$_2$S + BaO + air). Forms explosive mixtures with non-metals [e.g., phosphorus (impact-sensitive), sulfur (friction-sensitive)]. Incompatible with alkali metals, reducing materials. Dangerous; when heated to decomposition it emits highly toxic fumes of Hg.

MDG000 CAS: 1312-03-4 **HR: 3**
MERCURY OXIDE SULFATE

DOT: NA 2025
mf: Hg$_3$O$_6$S   mw: 729.83

PROP: Lemon yellow powder; odorless. Bp: volatilizes, d: 6.44, vap d: 25.2. Practically insol in water; sol in acids.

SYNS: BASIC MERCURIC SULFATE * MERCURIC BASIC SULFATE * MERCURIC SUBSULFATE, solid (DOT) * TURPETH MINERAL

CONSENSUS REPORTS: Mercury and its compounds are on the Community Right-To-Know List.

OSHA PEL: (Transitional: CL 1 mg/10m^3) CL 0.1 mg(Hg)/m^3 (skin)

ACGIH TLV: TWA 0.1 mg(Hg)/m^3 (skin)

NIOSH REL: TWA 0.05 mg(Hg)/m^3

DOT Classification: Poison B; Label: Poison.

SAFETY PROFILE: A poison. When heated to decomposition it emits very toxic fumes of Hg and SO$_x$.

MDG250 CAS: 7783-36-0 **HR: 3**
MERCURY(I) SULFATE

DOT: UN 1628
mf: O$_4$S • 2Hg   mw: 497.24

PROP: White, crystalline powder. Mp: decomp, d: 7.56. Sltly sol in water; sol in dil HNO$_3$.

SYN: MERCUROUS SULFATE, solid (DOT)

CONSENSUS REPORTS: Mercury and its compounds are on the Community Right-To-Know List.

OSHA PEL: (Transitional: CL 1 mg/10m^3) CL 0.1 mg(Hg)/m^3 (skin)

ACGIH TLV: TWA 0.1 mg(Hg)/m^3 (skin)

NIOSH REL: TWA 0.05 mg(Hg)/m^3

DOT Classification: Poison B; Label: Poison.

SAFETY PROFILE: A poison by ingestion and intraperitoneal routes. Moderately toxic by skin contact. When heated to decomposition it emits very toxic fumes of Hg and SO$_x$.

MDG500 CAS: 7783-35-9 **HR: 3**
MERCURY(II) SULFATE (1:1)

DOT: UN 1633/UN 1645
mf: O$_4$S • Hg   mw: 296.65

PROP: White, crystalline powder; odorless. Mp: decomp, d: 6.47. Sol in HCl, hot dilute H$_2$SO$_4$, concentrated solns of NaCl. Protect from light.

SYNS: MERCURIC SULFATE, solid (DOT) * MERCURY BISULFATE * MERCURY PERSULFATE * SULFATE MERCURIQUE (FRENCH) * SULFURIC ACID, MERCURY(2+) SALT (1:1)

CONSENSUS REPORTS: Mercury and its compounds are on the Community Right-To-Know List. Reported in EPA TSCA Inventory.

OSHA PEL: (Transitional: CL 1 mg/10m^3) CL 0.1 mg(Hg)/m^3 (skin)
ACGIH TLV: TWA 0.1 mg(Hg)/m^3 (skin)
NIOSH REL: TWA 0.05 mg(Hg)/m^3
DOT Classification: Poison B; Label: Poison.

SAFETY PROFILE: Poison by ingestion and intraperitoneal routes. Moderately toxic by skin contact. When heated to decomposition it emits very toxic fumes of Hg and SO$_x$.

MDI000　　　CAS: 54-64-8　　　**HR: 3**
MERTHIOLATE SODIUM
mf: C$_9$H$_9$HgO$_2$S • Na　　mw: 404.82

SYNS: ((o-CARBOXYPHENYL)THIO)ETHYLMERCURY SODIUM SALT * ELCIDE 75 * ELICIDE * o-(ETHYLMERCURITHIO)BENZOIC ACID SODIUM SALT * ETHYLMERCURITHIOSALICYLIC ACID SODIUM SALT * MERCUROTHIOLATE * MERFAMIN * MERTHIOLATE * MERTHIOLATE SALT * MERTORGAN * MERZONIN SODIUM * SET * SODIUM ETHYLMERCURIC THIOSALICYLATE * SODIUM-o-(ETHYLMERCURITHIO)BENZOATE * SODIUM ETHYLMERCURITHIOSALICYLATE * SODIUM MERTHIOLATE * THIMEROSALATE * THIMEROSOL * THIOMERSALATE

CONSENSUS REPORTS: Mercury and its compounds are on the Community Right-To-Know List. EPA Genetic Toxicology Program.

OSHA PEL: (Transitional: CL 1 mg/10m^3) CL 0.1 mg(Hg)/m^3 (skin)
ACGIH TLV: TWA 0.1 mg(Hg)/m^3 (skin)
NIOSH REL: TWA 0.05 mg(Hg)/m^3

SAFETY PROFILE: Poison by ingestion, subcutaneous, intravenous, and possibly other routes. Experimental teratogenic and reproductive effects. An eye irritant. Questionable carcinogen with experimental neoplastigenic data. An ophthalmic preservative, a topical anti-infective, topical veterinary antibacterial and antifungal agent. An FDA over-the-counter drug. When heated to decomposition it emits very toxic fumes of Hg, Na$_2$O, and SO$_x$.

MDI500　　　CAS: 54-04-6　　　**HR: 3**
MESCALINE
mf: C$_{11}$H$_{17}$NO$_3$　　mw: 211.29

PROP: Crystals. Mp: 35-36°, bp: 180° @ 11 mm. Mod sol in water; sol in alc, chloroform, and benzene; practically insol in ether and petr ether.

SYNS: MEZCALINE * MEZCLINE * 3,4,5-TRIMETHOXYBENZENEETHANAMINE * 3,4,5-TRIMETHOXYPHENETHYLAMINE

SAFETY PROFILE: Poison by intraperitoneal and possibly other routes. Moderately toxic by ingestion and parenteral routes. An experimental teratogen. Human systemic effects by intramuscular route: euphoria and hallucinations, distorted perceptions. A psychotoimetic agent (a drug of abuse). When heated to decomposition it emits toxic fumes of NO$_x$.

MDI750　　　CAS: 832-92-8　　　**HR: 3**
MESCALINE HYDROCHLORIDE
mf: C$_{11}$H$_{17}$NO$_3$ • ClH　　mw: 247.75

PROP: Needles. Mp: 181°. Sol in water and alc.

SYNS: 3,4,5-TRIMETHOXYPHENETHYLAMINE HYDROCHLORIDE * 3,4,5-TRIMETHOXY-β-PHENYLETHYLAMINE HYDROCHLORIDE

SAFETY PROFILE: Poison by intravenous, intraperitoneal, and subcutaneous routes. Moderately toxic by ingestion. When heated to decomposition it emits very toxic fumes of HCl and NO$_x$.

MDJ750　　　CAS: 141-79-7　　　**HR: 3**
MESITYL OXIDE
DOT: UN 1229
mf: C$_6$H$_{10}$O　　mw: 98.16

PROP: Oily, colorless liquid; strong odor. Mp: −59°, bp: 130.0°, flash p: 87°F (CC), d: 0.8539 @ 20°/4°, autoign temp: 652°F, vap press: 10 mm @ 26.0°, vap d: 3.38. Solidifies @ 41.5°; somewhat sol in water @ 20°. Misc in alc and ether and with most organic liquids.

SYNS: ISOBUTENYL METHYL KETONE * ISOPROPYLIDENEACETONE * MESITYLOXID (GERMAN) * MESITYLOXYDE (DUTCH) * METHYL ISOBUTENYL KETONE * 4-METHYL-3-PENTENE-2-ONE * 4-METHYL-3-PENTEN-2-ON (DUTCH, GERMAN) * 2-METHYL-2-PENTEN-4-ONE * 4-METHYL-3-PENTEN-2-ONE * 4-METIL-3-PENTEN-2-ONE (ITALIAN) * OSSIDO di MESITILE (ITALIAN) * OXYDE de MESITYLE (FRENCH)

OSHA PEL: (Transitional: TWA 25 ppm) TWA 15 ppm; STEL 25 ppm

ACGIH TLV: TWA 15 ppm; STEL 25 ppm
DFG MAK: 25 ppm (100 mg/m^3)
NIOSH REL: (Ketones) TWA 40 mg/m^3
DOT Classification: Label: Flammable Liquid.

SAFETY PROFILE: Poison by intraperitoneal route. Moderately toxic by ingestion. Mildly toxic by inhalation and skin contact. Human systemic effects by inhalation: conjunctiva irritation. This compound is highly irritating to all tissues on contact; its vapors also are irritating. High concentrations are narcotic. It is readily absorbed through intact skin. Single exposures tend to indicate that this ketone has greater acute and narcotic action than isophorone. It can have harmful effects upon the kidneys and liver, and may damage the eyes and lungs to a serious degree. Prolonged exposure can injure liver, kidneys, and lungs. It can cause opaque cornea, keratoconus, and extensive necrosis of cornea. Dangerous fire hazard when exposed to heat, sparks, or flame; can react with oxidizing materials. Reacts violently with 2-amino ethanol, chlorosulfonic acid, ethylene diamine, HNO_3, oleum, H_2SO_4. An insect repellent. To fight fire, use alcohol foam, CO_2, dry chemical. When heated to decomposition it emits acrid smoke and irritating fumes.

MDL500   CAS: 2244-11-3   **HR: 3**
MESOXALYLUREA MONOHYDRATE
mf: $C_4H_2N_2O_4 \cdot H_2O$ mw: 160.10

PROP: White crystals, become pink on exposure to air. Colorless, aqueous solution imparts pink color to skin. Mp: 170° (decomp); sol in water and alc.

SYNS: ALLOXAN MONOHYDRATE * MESOXALYL-CARBAMIDE MONOHYDRATE * 2,4,5,6(1H,3H)-PYRIMIDINETETRONE HYDRATE * 2,4,5,6-TETRAOXO-HEXAHYDROPYRIMIDINE HYDRATE

SAFETY PROFILE: Poison by intravenous route. Experimental teratogenic and reproductive effects. Mutation data reported. When heated to decomposition it emits toxic fumes of NO$_x$.

MDN250   CAS: 79-41-4   **HR: 3**
METHACRYLIC ACID
DOT: UN 2531
mf: $C_4H_6O_2$ mw: 86.10

PROP: Corrosive liquid or colorless crystals; repulsive odor. Mp: 16°, bp: 163°, flash p: 171°F

(COC), d: 1.014 @ 25° (glacial), vap press: 1 mm @ 25.5° Sol in warm water; misc with alc, ether.

SYNS: α-METHYLACRYLIC ACID * METHACRYLIC ACID, INHIBITED (DOT) * 2-METHYLPROPENOIC ACID

CONSENSUS REPORTS: Reported in EPA TSCA Inventory.

OSHA PEL: TWA 20 ppm (skin)
ACGIH TLV: TWA 20 ppm
DOT Classification: Corrosive Material; Label: Corrosive.

SAFETY PROFILE: Poison by ingestion and intraperitoneal routes. Moderately toxic by skin contact and possibly other routes. Corrosive to skin, eyes, and mucous membranes. Mutation data reported. Flammable when exposed to heat, flame, or oxidizers. A storage hazard; exothermic polymerization may occur spontaneously. To fight fire, use alcohol foam, spray, mist, dry chemical. When heated to decomposition it emits acrid smoke and irritating fumes.

MDN500   CAS: 79-39-0   **HR: 3**
METHACRYLIC ACID AMIDE
mf: C_4H_7NO mw: 85.12

SYNS: METHACRYLIC AMIDE * 2-METHYLACRYLAMIDE * α-METHYL ACRYLIC AMIDE * 2-METHYLPROPENAMIDE * USAF RH-1

CONSENSUS REPORTS: Reported in EPA TSCA Inventory.

SAFETY PROFILE: Poison by intraperitoneal route. Moderately toxic by ingestion. Human systemic effects by inhalation: degenerative brain changes and liver and kidney changes. When heated to decomposition it emits toxic fumes of NO$_x$.

MDO250   CAS: 3963-95-9   **HR: 3**
METHACYCLINE HYDROCHLORIDE
mf: $C_{22}H_{22}N_2O_8 \cdot ClH$ mw: 478.92

PROP: Yellow, crystalline powder; bitter taste. Decomp @ approx 205°. Sol in water; sltly sol in alc; practically insol in ether, chloroform.

SYNS: ADRIAMICINA * CICLOBIOTIC * GERMICICLIN * GLOBOCICLINA * LONDOMYCIN * MEGAMYCINE * METADOMUS * METHACYCLINE MONOHYDROCHLORIDE * METILENBIOTIC * OPTIMYCIN * PHYSIOMYCINE * RINDEX * RONDOMYCIN

SAFETY PROFILE: Poison by intraperitoneal and intravenous routes. Moderately toxic by ingestion. An antibacterial agent. When heated to decomposition it emits very toxic fumes of NO_x and HCl.

MDO750   CAS: 76-99-3   HR: 3
METHADONE
mf: $C_{21}H_{27}NO$ mw: 309.49

SYNS: ADANON * AMIDONE * DIAMINON * DOLOPHINE * HEPTADONE * HEPTANON * KETALGIN * MECODIN * PHENADONE * PHYSEPTONE * POLAMIDONE

CONSENSUS REPORTS: EPA Genetic Toxicology Program.

SAFETY PROFILE: Poison by ingestion, intraperitoneal, intravenous, subcutaneous, and intraduodenal routes. Human systemic effects by intravenous route: nausea or vomiting, respiratory depression, and coma. *Caution:* Abuse leads to habituation or addiction. When heated to decomposition it emits toxic fumes of NO_x.

MDP000   CAS: 1095-90-5   HR: 3
METHADONE HYDROCHLORIDE
mf: $C_{21}H_{27}NO \cdot ClH$ mw: 345.95

SYNS: ADANON HYDROCHLORIDE * ALTHOSE HYDROCHLORIDE * AMIDONE HYDROCHLORIDE * DIAMINON HYDROCHLORIDE * DIASONE HYDROCHLORIDE * 6-DIMETHYLAMINO-4,4-DIPHENYL-3-HEPTANONE HYDROCHLORIDE * 1,1-DIPHENYL-1-(β-DIMETHYLAMINOPROPYL)BUTANONE-2 HYDROCHLORIDE * DOLOPHINE * DOLOPHINE HYDROCHLORIDE * DOLPHINE * HOECHST 1082

SAFETY PROFILE: A human poison by ingestion. Poison experimentally by ingestion, subcutaneous, intravenous, parenteral, and intraperitoneal routes. Human systemic effects by ingestion: antipsychotic effects and analgesia. Mutation data reported. An analgesic and FDA proprietary drug. A synthetic drug whose action is similar to morphine and heroin, and is almost as addictive. *Caution:* Abuse leads to habituation or addiction. When heated to decomposition it emits very toxic fumes of Cl^- and NO_x.

MDP250   CAS: 5967-73-7   HR: 3
l-METHADONE HYDROCHLORIDE
mf: $C_{21}H_{27}NO \cdot ClH$ mw: 345.95

PROP: dl Form: Crystals. Mp: 241°.

SYNS: 1-6-DIMETHYLAMINO-4,4-DIPHENYL-3-HEPTANONE HYDROCHLORIDE * LEVADONE * LEVOTHYL * POLAMIDON

SAFETY PROFILE: Poison by intravenous, subcutaneous, and intraperitoneal routes. Human systemic effects by ingestion: hallucinations, distorted perceptions, and analgesia. *Caution:* Abuse leads to habituation or addiction. When heated to decomposition it emits very toxic fumes of Cl^- and NO_x.

MDP500   CAS: 15284-15-8   HR: 3
d-METHADONE HYDROCHLORIDE
mf: $C_{21}H_{27}NO \cdot ClH$ mw: 345.95

SYN: d-DOLOPHINE HYDROCHLORIDE

SAFETY PROFILE: Poison by intravenous and intraperitoneal routes. Human systemic effects by ingestion: euphoria, somnolence (general depressed activity), and analgesia. *Caution:* Abuse leads to habituation or addiction. When heated to decomposition it emits toxic fumes of Cl^- and NO_x.

MDQ250   CAS: 438-41-5   HR: 3
METHAMINODIAZEPOXIDE HYDROCHLORIDE
mf: $C_{16}H_{14}ClN_3O \cdot ClH$ mw: 336.24

SYNS: ANSIACAL * A-POXIDE * BENT * BENZODIAPIN * CALMODEN * CEBRUM * CHLORDIAZACHEL * CHLORDIAZEPOXIDE HYDROCHLORIDE * CHLORDIAZEPOXIDE MONOHYDROCHLORIDE * CHLORIDEAZEPOXIDE HYDROCHLORIDE * 7-CHLORO-2-METHYLAMINO-5-PHENYL-3H-1,4-BENZODIAZEPIN, 4-OXIDE, HYDROCHLORIDE * 7-CHLORO-N-METHYL-5-PHENYL-EH-1,4-BENZODIAZEPIN-2-AMINE-4-OXIDE, MONOHYDROCHLORIDE * CORAX * DIAZACHEL (OBS.) * DROXOL * ELENIUM * EQUIBRAL * J-LIBERTY * KALMOCAPS * LABICAN * LENTOTRAN * LIBRIUM * LIBRIUM HYDROCHLORIDE * METHAMINODIAZEPINE HYDROCHLORIDE * MILDMEN * MURCIL * NAPOTON * NOVOSED * PSICHIAL * PSICOSAN * RELIBERAN * RO 5-0690 * SEREN VITA * SK-LYGEN * SOPHIAMIN * TENSINYL * TIMOSIN * TRAKIPEAL * VIANSIN * VIOPSICOL

CONSENSUS REPORTS: EPA Genetic Toxicology Program.

SAFETY PROFILE: Poison by intraperitoneal and intravenous routes. Moderately toxic by ingestion and subcutaneous routes. Mutation data reported. A minor tranquilizer. When heated to decomposition it emits very toxic fumes of HCl and NO_x.

MDQ500   CAS: 826-10-8   HR: 3
l-METHAMPHETAMINE HYDROCHLORIDE
mf: $C_{10}H_{15}N \cdot ClH$ mw: 185.72

PROP: Crystals; bitter taste. Mp: 170-175°. Sol in water, alc, and chloroform; almost insol in ether.

SYNS: ADIPEX * l-DESOXYEPHEDRINE HYDRO-CHLORIDE * (−)-N-α-DIMETHYLPHENETHYLAMINE HYDROCHLORIDE * "METH" * l-N-METHYL-β-PHENYLISOPROPYLAMINE HYDROCHLORIDE * "SPEED" * SYNDROX

SAFETY PROFILE: Poison by ingestion, intravenous, intraperitoneal, and subcutaneous routes. A powerful central nervous system stimulant. *Caution:* Excessive use may lead to tolerance and habituation. When heated to decomposition it emits very toxic fumes of HCl and NO_x.

MDQ750   CAS: 74-82-8   HR: 3
METHANE
DOT: UN 1971/UN 1972
mf: CH_4 mw: 16.05

PROP: Colorless, odorless, tasteless gas. Mp: −182.6°. Bp: −161.5°, lel: 5.3%, uel: 15%, fp: −183.2°. D: 0.554 @ 0°/4° (air = 1) or 0.7168 g/L, autoign temp: 650°, vap d: 0.6, flash p: −368.6°F. Sol in water, alc and ether.

SYNS: FIRE DAMP * MARSH GAS * METHANE, compressed (DOT) * METHANE, refrigerated liquid (DOT) * METHYL HYDRIDE

CONSENSUS REPORTS: Reported in EPA TSCA Inventory.

DOT Classification: Flammable Gas; Label: Flammable Gas.

SAFETY PROFILE: A simple asphyxiant. Very dangerous fire and explosion hazard when exposed to heat or flame. Reacts violently with powerful oxidizers (e.g., bromine pentafluoride, chlorine trifluoride, chlorine, fluorine, iodine heptafluoride, dioxygenyl tetrafluoroborate, dioxygen difluoride, trioxygen difluoride, liquid oxygen, ClO_2, NF_3, OF_2). Incompatible with halogens or interhalogens; air (forms explosive mixtures). Explosive in the form of vapor when exposed to heat or flame. To fight fire, stop flow of gas, CO_2 or dry chemical.

MDR250   CAS: 75-75-2   HR: 3
METHANESULFONIC ACID
mf: CH_4O_3S mw: 96.11

PROP: Solid. D: 1.4812 @ 18°/4°, mp: 20°, bp: 167° @ 10 mm. Sol in water, alc, and ether. Corrosive to iron, steel, brass, copper, and lead.

SYN: WSQ 1

CONSENSUS REPORTS: Reported in EPA TSCA Inventory.

SAFETY PROFILE: Poison by ingestion and intraperitoneal routes. May be corrosive to skin, eyes, and mucous membranes. Explosive reaction with ethyl vinyl ether. Incompatible with hydrogen fluoride. When heated to decomposition it emits toxic fumes of SO_x.

MDT250   CAS: 340-56-7   HR: 3
METHAQUALONE HYDROCHLORIDE
mf: $C_{16}H_{14}N_2O \cdot ClH$ mw: 286.78

PROP: Crystals. Mp: 255-265°. Sol in ether, ethanol; almost insol in water.

SYNS: METHYLQUINAZOLONE HYDROCHLORIDE * 2-METHYL-3-TOLYLCHINAZOLON-4 HYDROCHLORIDE (GERMAN) * 2-METHYL-3-o-TOLYL-4(3H)-QUINAZOLINONE HYDROCHLORIDE * 2-METHYL-3-(o-TOLYL)-4-QUINAZOLONE HYDROCHLORIDE

SAFETY PROFILE: Poison by ingestion, intraperitoneal, intravenous, and possibly other routes. Experimental reproductive effects. When heated to decomposition it emits very toxic fumes of NO_x and HCl.

MDT740   CAS: 59-51-8   HR: 2
dl-METHIONINE
mf: $C_5H_{11}NO_2S$ mw: 149.23

PROP: White crystalline platelets; characteristic odor. Sol in water, dilute acids, and alkalis; very sltly sol in alc; insol in ether.

SYNS: ACIMETION * BANTHIONINE * CYNARON * DYPRIN * LOBAMINE * MEONINE * MERTIONIN * METHILANIN * (±)-METHIONINE * METIONE * NESTON

CONSENSUS REPORTS: EPA Genetic Toxicology Program. Reported in EPA TSCA Inventory.

SAFETY PROFILE: Moderately toxic by ingestion and other routes. Experimental reproductive effects. When heated to decomposition it emits toxic fumes of SO_x and NO_x.

MDT750   CAS: 63-68-3   **HR: 1**
l-METHIONINE
mf: $C_5H_{11}NO_2S$ mw: 149.23

PROP: White, crystalline powder or platelets; faint odor. Mp: 281° (decomp), d: 1.340. Sol in water, dilute acids, and alkalies; insol in abs alc, alc, benzene, acetone, ether.

SYNS: l-α-AMINO-γ-METHYLMERCAPTOBUTYRIC ACID * 2-AMINO-4-(METHYLTHIO)BUTYRIC ACID * l(−)-AMINO-γ-METHYLTHIOBUTYRIC ACID * CYMETHION * LIQUIMETH * METHIONINE * l-(−)-METHIONINE * l-γ-METHYLTHIO-α-AMINO-BUTYRIC ACID

CONSENSUS REPORTS: Reported in EPA TSCA Inventory. EPA Genetic Toxicology Program.

SAFETY PROFILE: Mildly toxic by ingestion and intraperitoneal routes. Human mutation data reported. An essential sulfur-containing amino acid. When heated to decomposition it emits very toxic fumes of NO_x and SO_x.

MDU500   CAS: 309-36-4   **HR: 3**
METHOHEXITAL SODIUM
mf: $C_{14}H_{17}N_2O_3 \cdot Na$ mw: 284.32

PROP: Minute crystals. Sol in water.

SYNS: 5-ALLYL-1-METHYL-5-(1-METHYL-2-PENTYNYL) BARBITURIC ACID SODIUM SALT * BREVIMYTAL * BREVITAL SODIUM * BRIETAL SODIUM * ENALLYNYMAL SODIUM * LILLY 22451 * METHOHEXITONE SODIUM * 1-METHYL-5-ALLYL-5-(1-METHYL-2-PENTYNYL)BARBITURIC ACID SODIUM SALT * SODIUM-dl-5-ALLYL-1-METHYL-5-(1-METHYL-2-PENTYNYL)BARBITURATE * SODIUM METHOHEXITAL * SODIUM METHOHEXITONE * SODIUM A-dl-1-METHYL-5-ALLYL-5-(1-METHYL-2-PENTYNYL)BARBITURATE

SAFETY PROFILE: Poison by intravenous and implant routes. Human systemic effects by intravenous route: blood pressure lowering, gastrointestinal effects, and allergic dermatitis. An FDA proprietary drug. *Caution:* Excessive use may lead to addiction or habituation. Allergenic effects by intravenous route. When heated to decomposition it emits toxic fumes of Na_2O and NO_x.

MDU600   CAS: 16752-77-5   **HR: 3**
METHOMYL
mf: $C_5H_{10}N_2O_2S$ mw: 162.23

PROP: White, crystalline solid; slt sulfurous odor. Mododerately water-sol, mp: 79°.

SYNS: DU PONT INSECTICIDE 1179 * ENT 27,341 * INSECTICIDE 1,179 * LANNATE * MESOMILE * METHYL N-((METHYLAMINO)CARBONYL)OXY)ETHA-NIMIDO)THIOATE * METHYL-N-((METHYLCARBA-MOYL)OXY)THIOACETIMIDATE * S-METHYL N-[(METHYLCARBAMOYL0OXY]THIOACETIMIDATE * 2-METHYLTHIO-ACETALDEHYD-O-(METHYLCARBA-MOYL)-OXIM (GERMAN) * 2-METHYLTHIO-PROPION-ALDEHYD-O-(METHYLCARBAMOYL)-OXIM (GERMAN) * METOMIL (ITALIAN) * NU-BAIT II * NUDRIN * RCRA WASTE NUMBER P066 * 3-THIABUTAN-2-ONE, O-(METHYLCARBAMOYL)OXIME * WL 18236

CONSENSUS REPORTS: EPA Genetic Toxicology Program. EPA Extremely Hazardous Substances List.

OSHA PEL: TWA 2.5 mg/m^3
ACGIH TLV: TWA 2.5 mg/m^3

SAFETY PROFILE: Poison by ingestion, inhalation, and subcutaneous routes. Mildly toxic by skin contact. When heated to decomposition it emits very toxic fumes of NO_x and SO_x.

MDU750   CAS: 522-23-6   **HR: 3**
METHOPHENAZINE DIFUMARATE
mf: $C_{31}H_{36}ClN_3O_5S \cdot C_8H_4O_8$ mw: 826.33

SYNS: FRENOLON DIFUMARATE * METHOPHENA-ZATE ACID FUMARATE * PHRENOLAN * T-82 DI-FUMARATE * 3,4,5-TRIMETHOXY-BENZOIC ACID 2-(4-(3-(2-CHLOROPHENOTHIAZIN-10-YL)PROPYL)-1-PI-PERAZINYL)ETHYL ESTER, DIFUMARATE

SAFETY PROFILE: Poison by intraperitoneal, subcutaneous, and intravenous routes. Moderately toxic by ingestion. When heated to decomposition it emits very toxic fumes of Cl^-, SO_x, and NO_x.

MDV250   CAS: 5985-35-3   **HR: 3**
METHORPHINAN HYDROBROMIDE
mf: $C_{17}H_{23}NO \cdot BrH$ mw: 338.33

SYNS: DROMORAN HYDROBROMIDE * dl-3-HY-DROXY-N-METHYLMORPHINAN HYDROBROMIDE * NU 2206 * RACEMORPHAN HYDROBROMIDE * RO 1-5431

SAFETY PROFILE: Poison by subcutaneous, intraperitoneal, and intravenous routes. When heated to decomposition it emits very toxic fumes of NO_x and HBr.

MDV500   CAS: 59-05-2   HR: 3
METHOTREXATE
mf: $C_{20}H_{22}N_8O_5$ mw: 454.50

SYNS: AMETHOPTERIN * 4-AMINO-4-DEOXY-N[10]-METHYLPTEROYLGLUTAMATE * 4-AMINO-4-DEOXY-N[10]-METHYLPTEROYLGLUTAMIC ACID * 4-AMINO-10-METHYLFOLIC ACID * 4-AMINO-N[10]-METHYL-PTEROYLGLUTAMIC ACID * ANTIFOLAN * N-BISMETHYLPTEROYLGLUTAMIC ACID * CL-14377 * 1-(+)-N-(p-(((2,4-DIAMINO-6-PTERIDINYL)METHYL)METHYLAMINO)BENZOYL)GLUTAMIC ACID * EMT 25,299 * EMTEXATE * HDMTX * METHOPTERIN * METHOTEXTRATE * METHYLAMINOPTERIN * MTX * NCI-C04671 * NSC-740 * R 9985

CONSENSUS REPORTS: IARC Cancer Review: GROUP 3 IMEMDT 7,241,87; Animal Inadequate Evidence IMEMDT 26,267,81; Human Inadequate Evidence IMEMDT 26,267,81. NCI Carcinogenesis Studies (ipr); No Evidence: mouse, rat CANCAR 40,1935,77. Reported in EPA TSCA Inventory.

SAFETY PROFILE: A human poison by intraspinal route. Poison experimentally by ingestion, intravenous, subcutaneous, and intraperitoneal routes. Human teratogenic effects by ingestion: developmental abnormalities of the craniofacial area and the musculoskeletal system. Human systemic effects by multiple routes: thrombocytopenia (decrease in the number of blood platelets), bone marrow changes, other blood changes, cerebral spinal fluid effects, eye effects, blood pressure lowering, cough, dyspnea, fibrosis (pneumoconiosis), cyanosis, gastrointestinal effects, fatty liver degeneration, hepatitis, liver function tests impaired, other liver changes, fever, effects on inflammation or mediation of inflammation. Experimental tumorigenic and teratogenic data. Questionable human carcinogen producing leukemia, Hodgkin's disease, and skin tumors. Human mutation data reported. Experimental reproductive effects. A human eye irritant. An FDA proprietary drug. A chemotherapeutic agent. When heated to decomposition it emits toxic fumes including NO_x.

MDW000   CAS: 61-16-5   HR: 3
METHOXAMINE HYDROCHLORIDE
mf: $C_{11}H_{17}NO_3 \cdot ClH$ mw: 247.71

PROP: Crystals. Mp: 212-216°. Very sol in water; practically insol in ether, benzene, and chloroform.

SYNS: 2-AMINO-1-(2,5-DIMETHOXYPHENYL)-1-PROPANOL HYDROCHLORIDE * α-(1-AMINOETHYL)-2,5-DIMETHOXYBENZYL ALCOHOL HYDROCHLORIDE * β-(2,5-DIMETHOXYPHENYL)-β-HYDROXYISOPROPYLAMINE HYDROCHLORIDE * β-HYDROXY-β-(2,5-DIMETHOXYPHENYL)-ISOPROPYLAMINE HYDROCHLORIDE * PRESSOMIN HYDROCHLORIDE * VASOXINE * VASOXINE HYDROCHLORIDE * VASOXYL HYDROCHLORIDE

SAFETY PROFILE: Poison by ingestion, intravenous, intraperitoneal and possibly other routes. Experimental reproductive effects. An FDA proprietary drug. When heated to decomposition it emits very toxic fumes of HCl and NO_x.

MDW750   CAS: 100-06-1   HR: 2
4'-METHOXYACETOPHENONE
mf: $C_9H_{10}O_2$ mw: 150.19

PROP: Colorless to pale yellow fused solid; hawthorn odor. Flash p: +212°F. Sol in fixed oils, propylene glycol; misc in glycerin.

SYNS: ACETANISOLE (FCC) * p-ACETYLANISOLE * 4-ACETYLANISOLE * BANANOTE * FEMA No. 2005 * LINARODIN * p-METHOXYACETOPHENONE * p-METHOXYPHENYL METHYL KETONE * 4-METHOXYPHENYL METHYL KETONE * NOVATONE

CONSENSUS REPORTS: Reported in EPA TSCA Inventory.

SAFETY PROFILE: Moderately toxic by ingestion. Human systemic effects by inhalation: pulse rate increased without fall in blood pressure and blood pressure elevation. A skin irritant. Combustible liquid. When heated to decomposition it emits acrid smoke and irritating fumes.

MEA500   CAS: 52740-56-4   HR: 3
dl,4-METHOXYAMPHETAMINE HYDROCHLORIDE
mf: $C_{10}H_{15}NO \cdot ClH$ mw: 201.72

SYNS: 4-METHOXYAMPHETAMINE HYDROCHLORIDE * dl-p-METHOXY-α-METHYL-PHENETHYLAMINE HYDROCHLORIDE

SAFETY PROFILE: Poison by ingestion, intravenous, and intraperitoneal routes. When heated to decomposition it emits very toxic fumes of HCl and NO_x.

MEC250   CAS: 3811-49-2   **HR: 3**
2-METHOXY-4H-1,2,3-BENZODIOXA-
PHOSPHORINE-2-SULFIDE
mf: $C_8H_9O_3PS$ mw: 216.20

SYNS: K-9 * PHOSPHOROTHIOIC ACID, CYCLIC
O,O-(METHYLENE-O-PHENYLENE) O-METHYL ESTER
* SALITHION * SALITHION-SUMITOMO

SAFETY PROFILE: Poison by ingestion and subcutaneous routes. Mutation data reported. An insecticide. When heated to decomposition it emits very toxic fumes of SO_x and PO_x.

MEC500   CAS: 52351-96-9   **HR: 3**
6-METHOXYBENZO(a)PYRENE
mf: $C_{21}H_{14}O$ mw: 282.35

SAFETY PROFILE: Questionable carcinogen with experimental tumorigenic data. Mutation data reported. When heated to decomposition it emits acrid smoke and irritating fumes.

MED500   CAS: 105-13-5   **HR: 2**
p-METHOXYBENZYL ALCOHOL
mf: $C_8H_{10}O_2$ mw: 138.18

PROP: Needles or colorless liquid; floral odor. D: 1.113 @ 15°/15°, refr index: 1.543, mp: 25°, bp: 258.8°, flash p: +210°F. Insol in water; sol in alc and ether fixed oils; sltly sol glycerin.

SYNS: ANISE ALCOHOL * ANISIC ALCOHOL
* p-ANISOL ALCOHOL * ANISYL ALCOHOL (FCC)
* FEMA No. 2099 * 4-METHOXYBENZENEMETHA-
NOL * 4-METHOXYBENZYL ALCOHOL

CONSENSUS REPORTS: Reported in EPA TSCA Inventory.

SAFETY PROFILE: Moderately toxic by ingestion. A skin irritant. Combustible liquid. When heated to decomposition it emits acrid smoke and irritating fumes.

MEI450   CAS: 72-43-5   **HR: 3**
METHOXYCHLOR

DOT: NA 2761
mf: $C_{16}H_{15}Cl_3O_2$ mw: 345.66

PROP: Crystals. Mp: 78°, vap d: 12.

SYNS: 2,2-BIS(p-ANISYL)-1,1,1-TRICHLOROETHANE
* 1,1-BIS(p-METHOXYPHENYL)-2,2,2-TRICHLOROETH-
ANE * 2,2-BIS(p-METHOXYPHENYL)-1,1,1-TRICHLORO-
ETHANE * CHEMFORM * DIANISYLTRICHLORETH-
ANE * 2,2-DI-p-ANISYL-1,1,1-TRICHLOROETHANE

* p,p'-DIMETHOXYDIPHENYLTRICHLOROETHANE
* DIMETHOXY-DT * DIMETHOXY-DDT * 2,2-
DI-(p-METHOXYPHENYL)-1,1,1-TRICHLOROETHANE
* DI(p-METHOXYPHENYL)-TRICHLOROMETHYL METH-
ANE * DMDT * p,p'-DMDT * ENT 1,716
* MARALATE * MARLATE * METHOXCIDE
* METHOXO * p,p'-METHOXYCHLOR * ME-
THOXY-DDT * METOKSYCHLOR (POLISH) * ME-
TOX * MOXIE * NCI-C00497 * RCRA WASTE
NUMBER U247 * 1,1,1-TRICHLOR-2,2-BIS(4-METH-
OXY-PHENYL)-AETHAN (GERMAN) * 1,1,1-TRI-
CHLORO-2,2-BIS(p-ANISYL)ETHANE * 1,1,1-TRI-
CHLORO-2,2-BIS(p-METHOXYPHENOL)ETHANOL
* 1,1,1-TRICHLORO-2,2-BIS(p-METHOXYPHENYL)ETH-
ANE * 1,1,1-TRICHLORO-2,2-DI(4-METHOXYPHE-
NYL)ETHANE * 1,1'-(2,2,2-TRICHLOROETHYLIDENE)-
BIS(4-METHOXYBENZENE)

CONSENSUS REPORTS: IARC Cancer Review: GROUP 3 IMEMDT 7,56,87; Animal No Evidence IMEMDT 20,259,79; Animal Inadequate Evidence IMEMDT 5,193,74. NCI Carcinogenesis Bioassay (feed); No Evidence: mouse, rat NCITR* NCI-CG-TR-35,78. EPA Genetic Toxicology Program. Community Right-To-Know List.

OSHA PEL: (Transitional: TWA Total Dust: 10 mg/m^3; Respirable Fraction: 5 mg/m^3 TWA Total Dust: 10 mg/m^3; 5 mg/m^3
ACGIH TLV: TWA 10 mg/m^3
DFG MAK: 15 mg/m^3
DOT Classification: ORM-E; Label: None.

SAFETY PROFILE: Suspected carcinogen with experimental carcinogenic and tumorigenic data. Moderately toxic by ingestion, intraperitoneal, and skin contact. Human systemic effects by skin contact: somnolence. Experimental teratogenic and reproductive effects. Mutation data reported. When heated to decomposition emits highly toxic fumes of Cl^-.

MEJ500   CAS: 61413-39-6   **HR: 3**
5-METHOXYCHRYSENE
mf: $C_{19}H_{14}O$ mw: 258.33

SAFETY PROFILE: Questionable carcinogen with experimental neoplastigenic and tumorigenic data by skin contact. When heated to decomposition it emits acrid smoke and irritating fumes.

MEL500   CAS: 1918-00-9   **HR: 2**
2-METHOXY-3,6-DICHLOROBENZOIC ACID
mf: $C_8H_6Cl_2O_3$ mw: 221.04

SYNS: ACIDO (3,6-DICLORO-2-METOSSI)-BENZOICO (ITALIAN) * BANEX * BANLEN * BANVEL * BANVEL HERBICIDE * BRUSH BUSTER * COMPOUND B DICAMBA * DIANAT (RUSSIAN) * DIANATE * DICAMBA (DOT) * 3,6-DICHLOOR-2-METHOXY-BENZOEIZUUR (DUTCH) * 3,6-DI-CHLOR-3-METHOXY-BENZOESAEURE (GERMAN) * 3,6-DICHLORO-o-ANISIC ACID * 2,5-DICHLORO-6-METHOXYBENZOIC ACID * 3,6-DICHLORO-2-METH-OXYBENZOIC ACID * MDBA * MEDIBEN * VELSICOL COMPOUND "R" * VELSICOL 58-CS-11

CONSENSUS REPORTS: EPA Genetic Toxicology Program.

SAFETY PROFILE: Moderately toxic by ingestion and possibly other routes. Mutation data reported. When heated to decomposition it emits toxic fumes of Cl⁻.

MEO500   CAS: 61738-03-2   HR: 3
1-METHOXY ETHYL ETHYLNITROSAMINE
mf: $C_5H_{12}N_2O_2$ mw: 132.19

SYN: 1-METHOXY-AETHYL-AETHYLNITROSAMIN (GERMAN)

SAFETY PROFILE: Moderately toxic by ingestion. Questionable carcinogen with experimental carcinogenic data. When heated to decomposition it emits toxic fumes of NO_x.

MEO750   CAS: 151-38-2   HR: 3
METHOXYETHYL MERCURIC ACETATE
mf: $C_5H_{10}HgO_3$ mw: 318.74

PROP: Crystals. Water-sol.

SYNS: ACETATO(2-METHOXYETHYL)MERCURY * CEKUSIL UNIVERSAL A * LANDISAN * MEMA * MERCURAN * PANOGEN * RADOSAN

CONSENSUS REPORTS: Mercury and its compounds are on the Community Right-To-Know List.

OSHA PEL: (Transitional: CL 1 mg/10m³) CL 0.1 mg(Hg)/m³ (skin)
ACGIH TLV: TWA 0.1 mg(Hg)/m³; STEL 0.03 mg(Hg)/m³
NIOSH REL: TWA 0.05 mg(Hg)/m³

SAFETY PROFILE: Poison by ingestion. Mutation data reported. A fungicide. When heated to decomposition it emits toxic fumes of Hg.

MEP250   CAS: 123-88-6   HR: 3
2-METHOXYETHYLMERCURY CHLORIDE
mf: C_3H_7ClHgO mw: 295.14

PROP: Crystals.

SYNS: AGALLOL * AGALLOLAT * ARATAN * ATIRAN * BAYTAN * CEKUSIL UNIVERSAL C * CELMER * CERESAN UNIVERSAL NAZBEIZE * CHLORO(2-METHOXYETHYL)MERCURY * EMISAN 6 * FALISAN * GRAMISAN * HIGOSAN * MEMC * MERCHLORATE * METHOXYAETHYL-QUECKSILBERCHLORID (GERMAN) * (β-METHOXY-ETHYL)MERCURIC CHLORIDE * METHOXYETHYL MERCURIC CHLORIDE * 2-METHOXYETHYLMER-CURIC CHLORIDE * β-METHOXYETHYLMERCURY CHLORIDE * METHOXYETHYLMERCURY CHLORIDE * SEDRESAN * TAFASAN * TAYSSATAO * TRIADIMENOL

CONSENSUS REPORTS: Mercury and its compounds are on the Community Right-To-Know List. EPA Genetic Toxicology Program.

OSHA PEL: (Transitional: CL 1 mg/10m³) TWA 0.01 mg(Hg)/m³; STEL 0.03 mg/m³ (skin)
ACGIH TLV: TWA 0.01 mg(Hg)/m³; STEL 0.03 mg(Hg)/m³
NIOSH REL: TWA 0.05 mg(Hg)/m³

SAFETY PROFILE: Poison by ingestion, subcutaneous, and possibly other routes. Human mutation data reported. Used to control pineapple disease of sugarcane. When heated to decomposition it emits very toxic fumes of Hg and Cl⁻.

MEP500   CAS: 61738-05-4   HR: 3
1-METHOXY ETHYL METHYLNITROSAMINE
mf: $C_4H_{10}N_2O_2$ mw: 118.16

SYN: 1-METHOXY-AETHYL-METHYLNITROSAMIN (GERMAN)

SAFETY PROFILE: Poison by ingestion. Questionable carcinogen with experimental carcinogenic data. When heated to decomposition it emits toxic fumes of NO_x.

MEV750   CAS: 61738-04-3   HR: 3
METHOXYMETHYL ETHYL NITROSAMINE
mf: $C_4H_{10}N_2O_2$ mw: 118.16

SYN: METHOXYMETHYL-AETHYLNITROSAMINE (GERMAN)

SAFETY PROFILE: Moderately toxic by ingestion. Questionable carcinogen with experimental carcinogenic data. When heated to decomposition it emits toxic fumes of NO_x.

MEW250   CAS: 39885-14-8   **HR: 3**
**METHOXYMETHYL METHYL-
NITROSAMINE**
mf: $C_3H_8N_2O_2$ mw: 104.13

SYNS: METHOXYMETHYL-METHYLNITROSAMIN (GERMAN) * METHYL(METHOXYMETHYL)NITROS-AMINE * N-NITROSO-N-METHOXYMETHYL-METHYLAMINE

SAFETY PROFILE: Moderately toxic by ingestion and intraperitoneal routes. Questionable carcinogen with experimental carcinogenic and tumorigenic data. When heated to decomposition it emits toxic fumes of NO_x.

MEX250   CAS: 107-70-0   **HR: 2**
4-METHOXY-4-METHYL-2-PENTANONE

DOT: UN 2293
mf: $C_7H_{14}O_2$ mw: 130.21

SYN: 4-METHOXY-4-METHYLPENTAN-2-ONE (DOT)

CONSENSUS REPORTS: Reported in EPA TSCA Inventory.

DOT Classification: Flammable or Combustible Liquid; Label: Flammable Liquid.

SAFETY PROFILE: Moderately toxic by ingestion and skin contact. Mildly toxic by inhalation. A skin irritant. Flammable when exposed to heat or flame, can react vigorously with oxidizing materials. When heated to decomposition it emits acrid smoke and irritating fumes.

MEX350   **HR: 2**
2-METHOXY-3(5)-METHYLPYRAZINE
mf: $C_6H_8N_2O$ mw: 124.14

PROP: Colorless liquid; roasted, hazelnut odor. D: 1.000-1.090 @ 20°, refr index: 1.506, flash p: 131°F. Sol in water, organic solvents.

SYN: FEMA No. 3183

SAFETY PROFILE: Combustible liquid. When heated to decomposition emits toxic fumes of NO_x.

MEY000   CAS: 3131-27-9   **HR: 3**
**4-METHOXYMETHYLPYRIDOXINE
HYDROCHLORIDE**
mf: $C_9H_{13}NO_3 \cdot ClH$ mw: 219.69

SYNS: 4-METHOXYMETHYL-5-HYDROXY-6-METHYL-3-PYRIDINEMETHANOL HYDROCHLORIDE * 4-ME-THOXYMETHYLPYRIDOXOL HYDROCHLORIDE

SAFETY PROFILE: Poison by ingestion, subcutaneous, and intravenous routes. When heated to decomposition it emits very toxic fumes of HCl and NO_x.

MFB400   CAS: 75965-74-1   **HR: 3**
**7-METHOXY-2-NITRONAPHTHO(2,1-b)
FURAN**
mf: $C_{13}H_9NO_4$ mw: 243.23

SYNS: 2-NITRO-7-METHOXYNAPHTHO(2,1-b)FURAN * R7000

SAFETY PROFILE: Suspected carcinogen with experimental experimental carcinogenic and tumorigenic data. Mutation data reported. When heated to decomposition it emits toxic fumes of NO_x.

MFC000   CAS: 140-20-5   **HR: 3**
**METHOXYOXIMERCURIPROPYL-
SUCCINYL UREA**
mf: $C_9H_{16}HgN_2O_6$ mw: 448.86

SYNS: N-((3-(HYDROXYMERCURI)-2-METHOXYPRO-PYL)-CARBAMOYL)SUCCINAMIC ACID * MERALLU-RIDE * METHOXYHYDROXYMERCURIPROPYLSUCCI-NYLUREA

CONSENSUS REPORTS: Mercury and its compounds are on the Community Right-To-Know List.

OSHA PEL: (Transitional: CL 1 mg/10m³) CL 0.1 mg(Hg)/m³ (skin)
ACGIH TLV: TWA 0.1 mg(Hg)/m³ (skin)
NIOSH REL: TWA 0.05 mg(Hg)/m³

SAFETY PROFILE: A human poison by intramuscular route. Poison experimentally by subcutaneous route. When heated to decomposition it emits very toxic fumes of Hg and NO_x.

MFC700   CAS: 150-76-5   **HR: 3**
4-METHOXYPHENOL
mf: $C_7H_8O_2$ mw: 124.15

PROP: White, waxy solid. Mp: 52.5°, bp: 246°, d: 1.55 @ 20°/20°.

SYNS: HYDROQUINONE MONOMETHYL ETHER * MEQUINOL * p-METHOXYPHENOL * MME * MONO METHYL ETHER HYDROQUINONE * USAF AN-7

CONSENSUS REPORTS: Reported in EPA TSCA Inventory. EPA Genetic Toxicology Program.

OSHA PEL: TWA 5 mg/m^3
ACGIH TLV: TWA 5 mg/m^3

SAFETY PROFILE: Poison by intraperitoneal route. Moderately toxic by ingestion. A skin irritant. When heated to decomposition it emits acrid smoke and fumes.

MFE250   CAS: 104-01-8   **HR: 3**
p-METHOXYPHENYLACETIC ACID
mf: C$_9$H$_{10}$O$_3$ mw: 166.19

SYNS: ANISYL FORMATE * 2-(p-ANISYL)ACETIC ACID * HOMOANISIC ACID * 4-METHOXYBEN-ZENEACETIC ACID * p-METHOXYBENZYL FORMATE * 4-METHOXYPHENYLACETIC ACID * MOPA

CONSENSUS REPORTS: Reported in EPA TSCA Inventory.

SAFETY PROFILE: Moderately toxic by ingestion and intraperitoneal routes. Questionable carcinogen with experimental neoplastigenic data. When heated to decomposition it emits acrid smoke and irritating fumes.

MFF250   CAS: 3647-17-4   **HR: 3**
N-(p-METHOXYPHENYL)-1-AZIRIDINECARBOXAMIDE
mf: C$_{10}$H$_{12}$N$_2$O$_2$ mw: 192.24

SYNS: 1-(1-AZIRIDINYL)-N-(p-METHOXYPHENYL) FORMAMIDE * p-METHOXYPHENYL-N-CARBAMOYL-AZIRIDINE

SAFETY PROFILE: Poison by intravenous route. Questionable carcinogen with experimental neoplastigenic data. When heated to decomposition it emits toxic fumes of NO$_x$.

MFF580   **HR: 1**
4-p-METHOXYPHENYL-2-BUTANONE

PROP: Colorless to pale yellow liquid; sweet, floral odor. D: 1.042-1.048, refr index: 1.517-1.521, flash p: +212°F.

SYNS: ANISYLACETONE * FEMA No. 2672

SAFETY PROFILE: Combustible liquid. When heated to decomposition it emits acrid smoke and irritating fumes.

MFK500   CAS: 61-01-8   **HR: 3**
2-METHOXYPROMAZINE
mf: C$_{18}$H$_{22}$N$_2$OS   mw: 314.48

PROP: Crystals. Mp: 44-48°.

SYNS: 10-(3-DIMETHYLAMINOPROPYL)-2-METHOXY-PHENOTHIAZINE * 2-METHOXY-10-(3'-

DIMETHYLAMINOPROPYL)PHENOTHIAZINE * MOPA-ZIN * MOPAZINE * NEOPROMA * RP 4632 * TENTON * TENTONE

SAFETY PROFILE: Poison by intraperitoneal route. Moderately toxic by ingestion. When heated to decomposition it emits very toxic fumes of NO$_x$ and SO$_x$.

MFK750   CAS: 3403-42-7   **HR: 3**
METHOXYPROMAZINE MALEATE
mf: C$_{18}$H$_{22}$N$_2$OS • C$_4$H$_4$O$_4$ mw: 430.56

SYNS: 10-(3-(DIMETHYLAMINO)PROPYL)-2-METHOXY) PHENOTHIAZINE, MALEATE * METHOPROMAZINE MALEATE * TENTONE MALEATE

SAFETY PROFILE: Poison by intravenous and intraperitoneal routes. Moderately toxic by ingestion. When heated to decomposition it emits very toxic fumes of NO$_x$ and SO$_x$.

MFM000   CAS: 5332-73-0   **HR: 3**
3-METHOXYPROPYLAMINE
mf: C$_4$H$_{11}$NO   mw: 89.16

PROP: Colorless liquid. Mp: −75.7°, bp: 116°, flash p: 90°F (TOC), d: 0.8615 @ 30°, vap press: 20 mm @ 30°, vap d: 3.07.

SYN: 3-MPA

CONSENSUS REPORTS: Reported in EPA TSCA Inventory.

SAFETY PROFILE: Poison by intravenous route. Irritating to skin, eyes, and mucous membranes. Dangerous fire hazard when exposed to heat or flame; can react with oxidizing materials. To fight fire, use CO$_2$, dry chemical. When heated to decomposition it emits toxic fumes of NO$_x$.

MFN275   CAS: 484-20-8   **HR: 2**
5-METHOXY PSORALEN
mf: C$_{12}$H$_8$O$_4$ mw: 216.20

PROP: Naturally occurring analog of psoralen and isomer of methoxsalen. Found in a wide variety of plants. Needles from alc. Mp: 188° (subl). Practically insol in boiling water; sltly sol in glacial acetic acid, chloroform, benzene, warm phenol. Sol in abs alc: 1 part in 60.

SYNS: BERGAPTEN * 4-METHOXY-7H-FURO(3,2-g)(1)BENZOPYRAN-7-ONE * PSORADERM

CONSENSUS REPORTS: IARC Cancer Review: GROUP 2A IMEMDT 7,242,87; Animal

Inadequate Evidence IMEMDT 40,327,86; Human Inadequate Evidence IMEMDT 40,327,86.

SAFETY PROFILE: Suspected carcinogen. Mutation data reported. When heated to decomposition it emits acrid smoke and irritating fumes.

MFN285　　　CAS: 3149-28-8　　　**HR: 1**
2-METHOXYPYRAZINE
mf: $C_5H_6N_2O$　　mw: 110.12

PROP: Colorless to yellow liquid; nutty, cocoa-like odor. D: 1.110-1.140 @ 20°, refr index: 1.508. Sol in alc; insol in water @ 61°.

SYN: FEMA No. 3302

SAFETY PROFILE: Skin and eye irritant. When heated to decomposition emits toxic fumes of NO_x.

MFO250　　　CAS: 3949-14-2　　　**HR: 3**
5-METHOXY-3-(2-PYRROLIDINOETHYL) INDOLE
mf: $C_{15}H_{20}N_2O$　　mw: 244.37

SYNS: CT 4436　*　METHOXY-5-PYRROLIDINO-2'-ETHYL-3-INDOLE

SAFETY PROFILE: Poison by intraperitoneal and intravenous routes. When heated to decomposition it emits toxic fumes of NO_x.

MFS500　　　CAS: 2736-21-2　　　**HR: 3**
6-METHOXYTRYPTAMINE
mf: $C_{11}H_{14}N_2O \cdot HCl$　　mw: 226.73

SYN: 3-(2-AMINOETHYL)-6-METHOXYINDOLE HYDROCHLORIDE

SAFETY PROFILE: Poison by intravenous route. Moderately toxic by subcutaneous route. When heated to decomposition it emits very toxic fumes of HCl and NO_x.

MFT500　　　CAS: 127-25-3　　　**HR: 1**
METHYL ABIETATE
mf: $C_{21}H_{32}O_2$　　mw: 316.47

PROP: Colorless to thick yellow liquid; almost odorless. Flash p: 356°F (OC), vap d: 10.9. D: 1.040 @ 20°/20°, bp: 360-365° with decomp. Insol in water, misc in alc and ether, the usual organic solvents, and with aliphatic hydrocarbons. From the esterification of the resinous residue of turpentine.

SYNS: ABIETIC ACID, METHYL ESTER　*　METHYL ESTER of WOOD ROSIN　*　METHYL ESTER of WOOD ROSIN, PARTIALLY HYDROGENATED (FCC)

CONSENSUS REPORTS: Reported in EPA TSCA Inventory.

SAFETY PROFILE: A skin irritant. Probably slightly toxic. Combustible liquid when exposed to heat or flame; can react with oxidizing materials. To fight fire, use CO_2, dry chemical. When heated to decomposition it emits acrid smoke and irritating fumes.

MFT750　　　CAS: 79-16-3　　　**HR: 2**
METHYLACETAMIDE
mf: C_3H_7NO　　mw: 73.11

SYNS: N-METHYLACETAMIDE　*　MONOMETHYLACETAMIDE

CONSENSUS REPORTS: Reported in EPA TSCA Inventory.

SAFETY PROFILE: Moderately toxic by intraperitoneal, subcutaneous and possibly other routes. Mildly toxic by ingestion and intravenous routes. Mutation data reported. When heated to decomposition it emits toxic fumes of NO_x.

MFW100　　　CAS: 79-20-9　　　**HR: 2**
METHYL ACETATE
DOT: UN 1231
mf: $C_3H_6O_2$　　mw: 74.09

PROP: Colorless, volatile liquid. Mp: −98.7°, lel: 3.1%, uel: 16%, bp: 57.8°, ULC: 85-90, flash p: 14°F, d: 0.92438, autoign temp: 935°F, vap press: 100 mm @ 9.4°, vap d: 2.55. Moderately sol in water; misc in alc, ether.

SYNS: ACETATE de METHYLE (FRENCH)　*　ACETIC ACID METHYL ESTER　*　DEVOTON　*　METHYL ETHANOATE　*　METHYLACETAAT (DUTCH)　*　METHYLACETAT (GERMAN)　*　METHYLE (ACETATE de) (FRENCH)　*　METHYLESTER KISELINY OCTOVE (CZECH)　*　METILE (ACETATO di) (ITALIAN)　*　OCTAN METYLU (POLISH)　*　TERETON

CONSENSUS REPORTS: Reported in EPA TSCA Inventory.

OSHA PEL: (Transitional: TWA 200 ppm) TWA 200 ppm; STEL 250 ppm
ACGIH TLV: TWA 200 ppm; STEL 250 ppm
DFG MAK: 200 ppm (610 mg/m^3)
DOT Classification: Flammable Liquid; Label: Flammable Liquid.

SAFETY PROFILE: Moderately toxic by several routes. A human systemic irritant by inhalation. A moderate skin and severe eye irritant. Mutation data reported. Dangerous fire hazard when exposed to heat, flame, or oxidizers. Moderate explosion hazard when exposed to heat or flame. When heated to decomposition it emits acrid smoke and fumes.

MFW250   CAS: 122-00-9   **HR: 2**
4′-METHYL ACETOPHENONE
mf: $C_9H_{10}O$ mw: 134.19

PROP: Colorless liquid; fruity, actophenone odor. D: 0.996-1.004, refr index: 1.530-1.535, flash p: 198°F. Sol in fixed oils, propylene glycol; insol in glycerin.

SYNS: p-ACETYLTOLUENE * FEMA No. 2677 * MELILOTAL * p-METHYL ACETOPHENONE * 1-METHYL-4-ACETYLBENZENE * METHYL-p-TOLYL KETONE

CONSENSUS REPORTS: Reported in EPA TSCA Inventory.

SAFETY PROFILE: Moderately toxic by ingestion. A human skin irritant. Combustible liquid. When heated to decomposition it emits acrid smoke and irritating fumes.

MFW500   CAS: 520-45-6   **HR: 3**
METHYLACETOPYRONONE
mf: $C_8H_8O_4$ mw: 168.16

PROP: White crystals or crystalline powder. Mp: 109°, bp: 269.0°, vap press: 1 mm @ 91.7°, vap d: 5.8. Moderately sol in water and organic solvents.

SYNS: 2-ACETYL-5-HYDROXY-3-OXO-4-HEXENOIC ACID Δ-LACTONE * 3-ACETYL-6-METHYL-2,4-PYRANDIONE * 3-ACETYL-6-METHYLPYRANDIONE-2,4 * 3-ACETYL-6-METHYL-2H-PYRAN-2,4(3H)-DIONE * DEHYDRACETIC ACID * DEHYDROACETIC ACID (FCC) * DHA * DHS

CONSENSUS REPORTS: Reported in EPA TSCA Inventory.

SAFETY PROFILE: Poison by ingestion and intravenous routes. Moderately toxic by intraperitoneal route. Questionable carcinogen with experimental tumorigenic data. Combustible when exposed to heat or flame. When heated to decomposition it emits acrid smoke and irritating fumes.

MFW750   CAS: 36375-30-1   **HR: 3**
METHYL-β-ACETOXYETHYL-β-CHLOROETHYLAMINE
mf: $C_7H_{14}ClNO_2$ mw: 179.67

SYNS: 2-ACETOXY-2′-CHLORO-N-METHYL-DIETHYL-AMINE * N-ACETOXYETHYL-N-CHLOROETHYLMETHYLAMINE * 2-((2-CHLOROETHYL)METHYLAMINO)ETHANOL ACETATE * TL 1428

SAFETY PROFILE: Poison by intravenous and subcutaneous routes. Moderately toxic by inhalation. When heated to decomposition it emits very toxic fumes of Cl^- and NO_x.

MFX000   CAS: 7790-01-4   **HR: 3**
METHYLACETOXYMALONONITRILE
mf: $C_6H_6N_2O_2$ mw: 138.14

SYN: 2-ACETOXYISOSUCCINODINITRILE

CONSENSUS REPORTS: Cyanide and its compounds are on the Community Right-To-Know List.

SAFETY PROFILE: Poison by ingestion and skin contact. Moderately toxic by inhalation. When heated to decomposition it emits toxic fumes of NO_x and CN^-.

MFX250   CAS: 105-45-3   **HR: 2**
METHYL ACETYLACETATE
mf: $C_5H_8O_3$ mw: 116.13

PROP: Colorless liquid. Mp: 27.5°, bp: 170°, flash p: 170°F, autoign temp: 536°F, d: 1.077, vap d: 4.00.

SYNS: ACETOACETIC METHYL ESTER * METHYL-ACETOACETATE * METHYL ACETYLACETONATE * METHYL-3-OXOBUTYRATE * 3-OXOBUTANOIC ACID METHYL ESTER

CONSENSUS REPORTS: Reported in EPA TSCA Inventory.

SAFETY PROFILE: Moderately toxic by ingestion. A skin and severe eye irritant. Flammable when exposed to heat or flame. To fight fire, use foam, CO_2, dry chemical. When heated to decomposition it emits acrid smoke and irritating fumes.

MFX590   CAS: 74-99-7   **HR: 3**
METHYL ACETYLENE
mf: C_3H_4 mw: 40.07

PROP: Gas. Mp: −104°, lel: 1.7%, bp: −23.3°, vap press: 3876 mm @ 20°, d: 1.787 g/L @ 0°, vap d: 1.38.

SYNS: PROPINE * PROPYNE

CONSENSUS REPORTS: Reported in EPA TSCA Inventory.

OSHA PEL: TWA 1000 ppm
ACGIH TLV: TWA 1000 ppm
DFG MAK: 1000 ppm (1650 mg/m^3)

SAFETY PROFILE: This compound is a simple anesthetic and in high concentration is an asphyxiant. Dangerous fire hazard when exposed to heat or flame; can react vigorously with oxidizing materials. Explosive in the form of vapor when exposed to heat or flame. Localized heating of liquid containing cylinders to 95°C may cause an explosion. Product of reaction with silver nitrate ignites at 150°C. A commercial mixture containing 30% propyne in MAPP gas is similar to ethylene in potential hazards and handling requirements. To fight fire, stop flow of gas. When heated to decomposition it emits acrid smoke and irritating fumes.

MFX600   CAS: 59355-75-8   **HR: 2**
METHYL ACETYLENE-PROPADIENE MIXTURE

DOT: UN 1060

SYNS: MAPP * METHYLACETYLENE-PROPADIENE, STABILIZED (DOT) * PROPYNE mixed with PROPADIENE

OSHA PEL: (Transitional: TWA 1000 ppm) TWA 1000 ppm; STEL 1250 ppm
ACGIH TLV: TWA 1000 ppm; STEL 1250 ppm
DFG MAK: 1000 ppm (1650 mg/m^3)
DOT Classification: Flammable Gas; Label: Flammable Gas.

SAFETY PROFILE: A flammable gas mixture. To fight fire, stop flow of gas. When heated to decomposition it emits acrid smoke and irritating fumes.

MGA250   CAS: 78-85-3   **HR: 3**
METHYLACRYLALDEHYDE

DOT: UN 2396
mf: C_4H_6O mw: 70.10

PROP: Colorless liquid. Mp: −81°C, bp: 73.5°, flash p: 35°F (OC), d: 0.830 @ 20°/4°, vap press: 120 mm @ 20°, vap d: 2.42. Sol in water.

SYNS: ISOBUTENAL * METHACRALDEHYDE (DOT) * METHACROLEIN * METHACRYLALDEHYDE (DOT) * METHACRYLIC ALDEHYDE * α-METHYLACROLEIN * 2-METHYLACROLEIN * 2-METHYLPROPENAL (CZECH)

CONSENSUS REPORTS: Reported in EPA TSCA Inventory.

DOT Classification: Flammable Liquid; Label: Flammable Liquid and Poison.

SAFETY PROFILE: Poison by ingestion and skin contact. Moderately toxic by inhalation. Severe eye and skin irritant. Mutation data reported. Dangerously flammable when exposed to heat, flame or oxidizers. Can react vigorously with oxidizing materials. To fight fire, use CO_2, alcohol foam, foam, dry chemical. When heated to decomposition it emits acrid smoke and irritating fumes.

MGA500   CAS: 96-33-3   **HR: 3**
METHYL ACRYLATE

DOT: UN 1919
mf: $C_4H_6O_2$ mw: 86.10

PROP: Colorless liquid; acrid odor. D: 0.9561 @ 20°/4°, mp: −76.5°, bp: 70° @ 608 mm, lel: 2.8%, uel: 25%, fp: −75°, flash p: 27°F (OC), vap press: 100 mm @ 28°, vap d: 2.97. Sol in alc and ether.

SYNS: ACRYLATE de METHYLE (FRENCH) * ACRYLIC ACID METHYL ESTER (MAK) * ACRYLSAEUREMETHYLESTER (GERMAN) * CURITHANE 103 * METHOXYCARBONYLETHYLENE * METHYLACRYLAAT (DUTCH) * METHYL-ACRYLAT (GERMAN) * METHYL ACRYLATE, INHIBITED (DOT) * METHYL PROPENATE * METHYL PROPENOATE * METHYL-2-PROPENOATE * METILACRILATO (ITALIAN) * PROPENOIC ACID METHYL ESTER * 2-PROPENOIC ACID METHYL ESTER

CONSENSUS REPORTS: IARC Cancer Review: GROUP 3 IMEMDT 7,56,87; Animal Inadequate Evidence IMEMDT 39,99,86; Human Inadequate Evidence IMEMDT 19,47,79. Community Right-To-Know List. Reported in EPA TSCA Inventory.

OSHA PEL: TWA 10 ppm (skin)
ACGIH TLV: TWA 10 ppm (skin)
DFG MAK: 5 ppm (18 mg/m^3)
DOT Classification: Flammable Liquid; Label: Flammable Liquid.

SAFETY PROFILE: Poison by ingestion and intraperitoneal routes. Moderately toxic by skin contact. Mildly toxic by inhalation. Human systemic effects by inhalation: olfaction effects, eye effects and respiratory effects. A skin and eye irritant. Questionable carcinogen. Mutation data reported. Chronic exposure has produced injury to lungs, liver, and kidneys in experimental animals. Dangerously flammable when exposed to heat, flame, or oxidizers. Dangerous explosion hazard in the form of vapor when exposed to heat, sparks, or flame. Can react vigorously with oxidizing materials. A storage hazard; it forms peroxides which may initiate exothermic polymerization. To fight fire, use foam, CO_2, dry chemical. When heated to decomposition it emits acrid smoke and irritating fumes.

MGA750   CAS: 126-98-7   HR: 3
METHYLACRYLONITRILE
mf: C_4H_5N mw: 67.10

PROP: Mp: −36°, bp: 90.3°, d: 0.805, vap press: 40 mm @ 12.8°, flash p: 55°F.

SYNS: 2-CYANOPROPENE-1 * ISOPROPENE CYA-NIDE * ISOPROPENYLNITRILE * α-METHYL-ACRYLONITRILE * 2-METHYLPROPENENITRILE * RCRA WASTE NUMBER U152 * USAF ST-40

CONSENSUS REPORTS: EPA Extremely Hazardous Substances List. Cyanide and its compounds are on the Community Right-To-Know List. Reported in EPA TSCA Inventory.

OSHA PEL: TWA 1 ppm (skin)
ACGIH TLV: TWA 1 ppm (skin)

SAFETY PROFILE: Poison by ingestion, inhalation, skin contact and intraperitoneal routes. A skin and eye irritant. A dangerous fire hazard when exposed to heat, flame, or sparks. When heated to decomposition it emits toxic fumes of NO_x and CN^-.

MGA850   CAS: 109-87-5   HR: 2
METHYLAL
DOT: UN 1234
mf: $C_3H_8O_2$ mw: 76.11

PROP: Colorless liquid, pungent odor. Mp: −104.8°, bp: 42.3°, d: 0.864 @ 20°/4°, vap press: 330 mm @ 20°, vap d: 2.63, autoign temp: 459°F. flash p.: −0.4°F.

SYNS: ANESTHENYL * DIMETHOXYMETHANE * DIMETHYL FORMAL * FORMAL * FORMALDE-HYDE DIMETHYLACETAL * METHYLENE DIMETHYL ETHER * METYLAL (POLISH)

OSHA PEL: TWA 1000 ppm
ACGIH TLV: TWA 1000 ppm
DFG MAK: 1000 ppm (3100 mg/m³)
DOT Classification: Flammable Liquid; Label: Flammable Liquid.

SAFETY PROFILE: Moderately toxic by subcutaneous route. Mildly toxic by ingestion and inhalation. Can cause injury to lungs, liver, kidneys, and the heart. A narcotic and anesthetic in high concentrations. A very dangerous fire hazard when exposed to heat, flame, or oxidizers. Moderately explosive when exposed to heat or flame. May ignite or explode when heated with oxygen. To fight fire, use foam, CO_2, dry chemical. When heated to decomposition it emits acrid smoke and irritating fumes.

MGB150   CAS: 67-56-1   HR: 3
METHYL ALCOHOL
DOT: UN 1230
mf: CH_4O mw: 32.05

PROP: Clear, colorless, very mobile liquid; slt alcoholic odor when pure; crude material may have a repulsive pungent odor. Mp: 64.8°, lel: 6.0%, uel: 36.5%, ULC: 70, fp: −97.8°, d: 0.7915 @ 20°/4°, autoign temp: 878°F, vap press: 100 mm @ 21.2°, vap d: 1.11. Misc in water, ethanol, ether, benzene, ketones, and most other organic solvents.

SYNS: ALCOOL METHYLIQUE (FRENCH) * ALCOOL METILICO (ITALIAN) * CARBINOL * COLONIAL SPIRIT * COLUMBIAN SPIRITS (DOT) * METANOLO (ITALIAN) * METHANOL * METHYLALKOHOL (GERMAN) * METHYL HYDROXIDE * METHYLOL * METYLOWY ALKOHOL (POLISH) * MONOHY-DROXYMETHANE * PYROXYLIC SPIRIT * RCRA WASTE NUMBER U154 * WOOD ALCOHOL (DOT) * WOOD NAPHTHA * WOOD SPIRIT

CONSENSUS REPORTS: Community Right-To-Know List. Reported in EPA TSCA Inventory. EPA Genetic Toxicology Program.

OSHA PEL: (Transitional: TWA 200 ppm) TWA 200 ppm; STEL 250 ppm (skin)
ACGIH TLV: TWA 200 ppm; STEL 250 ppm (skin)
DFG MAK: 200 ppm (260 mg/m³); BAT: 30 mg/L in urine at end of shift.
NIOSH REL: TWA 200 ppm; CL 800 ppm/ 15M
DOT Classification: Flammable Liquid; Label: Flammable Liquid, Poison.

SAFETY PROFILE: A human poison by ingestion. Poison experimentally by skin contact. Moderately toxic experimentally by intravenous and intraperitoneal routes. Mildly toxic by inhalation. Human systemic effects by ingestion and inhalation: optic nerve neuropathy, visual field changes, lacrimation, headache, cough, dyspnea, other respiratory effects, nausea or vomiting. Experimental teratogenic and reproductive effects. An eye and skin irritant. Human mutation data reported. A narcotic.

Its main toxic effect is exerted upon the nervous system, particularly the optic nerves and possibly the retinae which can progress to permanent blindness. Once absorbed, methanol is only very slowly eliminated. Coma resulting from massive exposures may last as long as 2-4 days. In the body, the products formed by its oxidation are formaldehyde and formic acid, both of which are toxic. Because of the slow elimination, methanol should be regarded as a cumulative poison. Though single exposures to fumes may cause no harmful effect, daily exposure may result in the accumulation of sufficient methanol in the body to cause illness. Death from ingestion of less than 30 mL has been reported. A common air contaminant.

Flammable liquid. Dangerous fire hazard when exposed to heat, flame, or oxidizers. Explosive in the form of vapor when exposed to heat or flame. Explosive reaction with chloroform + heat, diethyl zinc. Violent reaction with alkyl aluminum salts, acetyle bromide, chloroform + sodium hydroxide, CrO_3, cyanuric chloride, (I + ethanol + HgO), $Pb(ClO_4)_2$, $HClO_4$, P_2O_3, (KOH + $CHCl_3$), nitric acid. Incompatible with beryllium dihydride, metals (e.g., potassium, magnesium), oxidants (e.g., barium perchlorate, bromine, sodium hypochlorite, chlorine, hydrogen peroxide), potassium tertbutoxide, carbon tetrachloride + metals (e.g., aluminum, magnesium, zinc), dichloromethane. Dangerous; can react vigorously with oxidizing materials. To fight fire, use alcohol foam. When heated to decomposition it emits acrid smoke and irritating fumes.

MGC250   CAS: 74-89-5   HR: 3
METHYLAMINE

DOT: UN 1061/UN 1235
mf: CH_5N mw: 31.07

PROP: Colorless gas or liquid; powerful ammoniacal odor. Bp: 6.3°, lel: 4.95%, uel: 20.75%, mp: −93.5°, flash p: 32°F (CC), d: 0.662 @ 20°/4°, autoign temp: 806°F, vap d: 1.07. Fuming liquid when liquefied: d: 0.699 @ −10.8°/4°. Sol in alc; misc with ether.

SYNS: AMINOMETHANE * CARBINAMINE * MERCURIALIN * METHANAMINE (9CI) * METHYLAMINE, anhydrous (DOT) * METHYL-AMINE, aqueous solution (DOT) * METHYLAMINEN (DUTCH) * METILAMINE (ITALIAN) * METYLOAMINA (POLISH) * MONOMETHYLAMINE * MONO-METHYLAMINE, anhydrous (DOT) * MONOMETHYL-AMINE, aqueous solution (DOT)

CONSENSUS REPORTS: Reported in EPA TSCA Inventory.

OSHA PEL: TWA 10 ppm
ACGIH TLV: TWA 10 ppm
DFG MAK: 10 ppm (12 mg/m³)
DOT Classification: Flammable Gas; Label: Flammable Gas, Flammable Liquid.

SAFETY PROFILE: Poison by subcutaneous route. Moderately toxic by inhalation. A severe skin irritant. Mutation data reported. A strong base. Flammable gas at ordinary temperature and pressure. Very dangerous fire hazard when exposed to heat, flame, or sparks. Explosive when exposed to heat or flame. To fight fire, stop flow of gas. Forms an explosive mixture with nitromethane. When heated to decomposition it emits toxic fumes of NO_x.

MGF000   CAS: 63917-71-5   HR: 3
METHYLAMINOCOLCHICIDE
mf: $C_{22}H_{26}N_2O_5$ mw: 398.50

SYN: METHYLCOLCHAMINONE

SAFETY PROFILE: Poison by intravenous and intraperitoneal routes. When heated to decomposition it emits toxic fumes of NO_x.

MGG000   CAS: 109-83-1   HR: 3
2-METHYLAMINOETHANOL
mf: C_3H_9NO mw: 75.11

PROP: Viscous liquid; fishy odor. Corrosive to skin, cork, and metals. A strong base. D: 0.9, vap d: 2.9, bp: 156°, flash p: 165°F (OC). Miscible with water, alc, and ether.

SYNS: β-(METHYLAMINO)ETHANOL * N-METHYL-AMINOETHANOL * N-METHYLETHANOLAMINE * METHYLETHYLOLAMINE * METHYL(β-HYDROXY-ETHYL)AMINE * MONOMETHYL-AMINOAETHANOL (GERMAN) * N-MONOMETHYLAMINOETHANOL * MONOMETHYLAMINOETHANOL * USAF DO-50

CONSENSUS REPORTS: Reported in EPA TSCA Inventory.

SAFETY PROFILE: Poison by intraperitoneal route. Moderately toxic by ingestion and subcutaneous routes. A corrosive irritant to skin, eyes, and mucous membranes. Flammable when exposed to heat, flame, or oxidizers. To fight fire, use alcohol foam. When heated to decomposition it emits toxic fumes such as NO_x.

MGK750 CAS: 52777-39-6 **HR: 3**
1-(3-METHYLAMINOPROPYL)-2-ADAMANTANOL HYDROCHLORIDE
mf: $C_{14}H_{25}NO \cdot ClH$ mw: 259.86

SYNS: 3-(2-HYDROXY-1-ADAMANTYL)-N-METHYL-PROPLYAMINE HYDROCHLORIDE * 2-HYDROXY-N-METHYL-1-ADAMANTANEPROPANAMINE HYDRO-CHLORIDE * 2-HYDROXY-1-(3-METHYL-AMINOPROPYL)ADAMANTANE HYDROCHLORIDE

SAFETY PROFILE: Poison by ingestion and intraperitoneal routes. When heated to decomposition it emits very toxic fumes of NO_x and HCl.

MGN500 CAS: 110-43-0 **HR: 2**
METHYL n-AMYL KETONE
DOT: UN 1110
mf: $C_7H_{14}O$ mw: 114.21

PROP: Colorless, mobile liquid; penetrating, fruity odor. Bp: 151.5°, flash p: 120°F (OC), autoign temp: 991°F, vap d: 3.94, d: 0.8197 @ 15°/4°. Very sltly sol in water; sol in alc and ether.

SYNS: AMYL-METHYL-CETONE (FRENCH) * n-AMYL METHYL KETONE * AMYL METHYL KETONE (DOT) * FEMA No. 2544 * 2-HEPTANONE * METHYL-AMYL-CETONE (FRENCH) * METHYL AMYL KETONE (DOT) * METHYL PENTYL KETONE

CONSENSUS REPORTS: Reported in EPA TSCA Inventory.

OSHA PEL: TWA 100 ppm
ACGIH TLV: TWA 50 ppm
NIOSH REL: (Ketones) TWA 465 mg/m³
DOT Classification: Combustible Liquid; Label: None; IMO: Flammable or Combustible Liquid; Label: Flammable Liquid.

SAFETY PROFILE: Moderately toxic by ingestion. Mildly toxic by inhalation and skin contact. A skin irritant. Flammable or combustible liquid when exposed to heat or flame; can react with oxidizing materials. To fight fire, use foam, CO_2, dry chemical. When heated to decomposition it emits acrid smoke and fumes.

MGN750 CAS: 100-61-8 **HR: 3**
METHYLANILINE
DOT: UN 2294
mf: C_7H_9N mw: 107.17

PROP: Colorless or sltly yellow liquid; becomes brown on exposure to air. Mp: −57°, d: 0.989 @ 20°/4°, bp: 194-197°. Sol in alc, ether; sltly sol in water.

SYNS: ANILINOMETHANE * (METHYLAMINO)BEN-ZENE * N-METHYLAMINOBENZENE * N-METHYL ANILINE (MAK) * N-METHYLBENZENAMINE * METHYLPHENYLAMINE * N-METHYLPHENYL-AMINE * MONOMETHYL ANILINE (OSHA) * N-MONOMETHYLANILINE * N-PHENYLMETHYL-AMINE

CONSENSUS REPORTS: Reported in EPA TSCA Inventory.

OSHA PEL: (Transitional: TWA 2 ppm (skin)) TWA 0.5 ppm (skin)
ACGIH TLV: TWA 0.5 ppm (skin)
DFG MAK: 0.5 ppm (2 mg/m³)
DOT Classification: Poison B; Label: St. Andrews Cross.

SAFETY PROFILE: Poison by ingestion and intravenous routes. Moderately toxic by subcutaneous route. When heated to decomposition it emits toxic fumes of NO_x.

MGO500 CAS: 102-50-1 **HR: 3**
2-METHYL-p-ANISIDINE
mf: $C_8H_{11}NO$ mw: 137.20

SYNS: m-CRESIDINE * 4-METHOXY-2-METHYLANI-LINE * 4-METHOXY-2-METHYLBENZENAMINE * 2-METHYL-4-METHOXYANILINE * NCI-C02993

CONSENSUS REPORTS: IARC Cancer Review: GROUP 3 IMEMDT 7,56,87; Animal Inadequate Evidence IMEMDT 27,91,82. NCI Carcinogenesis Bioassay (gavage); Clear Evidence: rat NCITR* NCI-CG-TR-105,78; (gavage); Inadequate Studies: mouse NCITR* NCI-CG-TR-105,78. Reported in EPA TSCA Inventory.

SAFETY PROFILE: Suspected carcinogen with experimental carcinogenic and tumorigenic data. Mutation data reported. When heated to decomposition it emits toxic fumes of NO_x.

MGO750 CAS: 120-71-8 **HR: 3**
5-METHYL-o-ANISIDINE
mf: $C_8H_{11}NO$ mw: 137.20

SYNS: m-AMINO-p-CRESOL, METHYL ESTER
* 3-AMINO-p-CRESOL METHYL ESTER * 1-AMINO-2-
METHOXY-5-METHYLBENZENE * 3-AMINO-4-METH-
OXYTOLUENE * 2-AMINO-4-METHYLANISOLE
* AZOIC RED 36 * C.I. AZOIC RED 83 * CRESI-
DINE * p-CRESIDINE * KRESIDIN * KREZIDINE
* 2-METHOXY-5-METHYLANILINE * 2-METHOXY-5-
METHYL-BENZENAMINE (9CI) * 4-METHOXY-m-TO-
LUIDINE * 4-METHYL-2-AMINOANISOLE * NCI-
C02982

CONSENSUS REPORTS: IARC Cancer Re-
view: GROUP 2B IMEMDT 7,56,87; Human
Limited Evidence IMEMDT 27,91,82; Animal
Sufficient Evidence IMEMDT 27,91,82. NTP
Fourth Annual Report On Carcinogens, 1984.
NCI Carcinogenesis Bioassay (feed); Clear Evi-
dence: mouse, rat NCITR* NCI-CG-TR-
142,79. Reported in EPA TSCA Inventory.
Community Right-To-Know List.

SAFETY PROFILE: Confirmed carcinogen with
experimental carcinogenic and neoplastigenic
data. Moderately toxic by ingestion. Mutation
data reported. When heated to decomposition
it emits toxic fumes of NO_x.

MGP000　　CAS: 104-93-8　　HR: 2
p-METHYL ANISOLE
mf: $C_8H_{10}O$　　mw: 122.18

PROP: Found in oil of Ylang-Ylang, Cananga,
and others. Colorless liquid; ylang-ylang odor.
D: 0.996-0.970, refr index: 1.510-1.513, flash
p: 144°F. Sol in fixed oils; insol in glycerin,
propylene glycol.

SYNS: p-CRESOL METHYL ETHER * p-CRESYL
METHYL ETHER * FEMA No. 2681 * p-METHOXY-
TOLUENE * 4-METHOXYTOLUENE * 4-METHYL-1-
METHOXYBENZENE * 4-METHYLPHENOL METHYL
ETHER * METHYL-p-TOLYL ETHER * p-TOLYL
METHYL ETHER

CONSENSUS REPORTS: Reported in EPA
TSCA Inventory.

SAFETY PROFILE: Moderately toxic by ingestion. A skin irritant. Combustible liquid. When
heated to decomposition it emits acrid smoke
and irritating fumes.

MGP750　　CAS: 779-02-2　　HR: 3
9-METHYLANTHRACENE
mf: $C_{15}H_{12}$　　mw: 192.27

CONSENSUS REPORTS: Reported in EPA
TSCA Inventory.

SAFETY PROFILE: Questionable carcinogen
with experimental tumorigenic data. Human
mutation data reported. When heated to decomposition it emits acrid smoke and irritating
fumes.

MGQ250　　CAS: 85-91-6　　HR: 3
N-METHYLANTHRANILIC ACID,
METHYL ESTER
mf: $C_9H_{11}NO_2$　　mw: 165.21

PROP: Pale yellow liquid; grape-like odor. D:
1.126-1.132, refr index: 1.578-1.581, flash p:
196°F. Sol in fixed oils; sltly sol in propylene
glycol; insol in water, glycerin.

SYNS: DIMETHYL ANTHRANILATE (FCC) * FEMA
No. 2718 * 2-METHYLAMINO METHYL BENZOATE
* METHYL METHYLAMINOBENZOATE * METHYL-
N-METHYL ANTHRANILATE * MMA

CONSENSUS REPORTS: Reported in EPA
TSCA Inventory.

SAFETY PROFILE: Poison by intravenous
route. Moderately toxic by ingestion. Combustible liquid. When heated to decomposition it
emits toxic fumes of NO_x.

MGQ750　　CAS: 2533-82-6　　HR: 3
METHYLARSENIC SULFIDE
mf: CH_3AsS　　mw: 122.02

SYNS: ASOZIN * BAY 4934 * MAS * METH-
YLARSINE SULFIDE * METHYLARSINIC SULFIDE
* METHYLARSINIC SULPHIDE * METHYLTHIOXO-
ARSINE * MONKIL WP * RHIZOCTOL * (THIO-
ARSENOSO)METHANE * URBASULF

CONSENSUS REPORTS: Arsenic and its
compounds are on the Community Right-To-
Know List.

SAFETY PROFILE: Poison by ingestion. Moderately toxic by skin contact. Mutation data reported. When heated to decomposition it emits
very toxic fumes of As and SO_x.

MGS750　　CAS: 592-62-1　　HR: 3
METHYL AZOXYMETHYL ACETATE
mf: $C_4H_8N_2O_3$　　mw: 132.14

SYNS: MAM AC * MAM ACETATE * METHYL-
AZOXYMETHANOL ACETATE * (METHYL-ONN-
AZOXY)METHANOL, ACETATE (ESTER)

CONSENSUS REPORTS: IARC Cancer Review: GROUP 2B IMEMDT 7,56,87; Animal
Sufficient Evidence IMEMDT 10,121,76.

CONSENSUS REPORTS: EPA Genetic Toxicology Program.

SAFETY PROFILE: Suspected carcinogen with experimental carcinogenic, neoplastigenic and tumorigenic data. Poison by ingestion, intraperitoneal, and intravenous routes. Experimental teratogenic and reproductive effects. Human mutation data reported. When heated to decomposition it emits toxic fumes of NO_x.

MGV500   CAS: 316-49-4   **HR: 3**
4-METHYLBENZ(a)ANTHRACENE
mf: $C_{19}H_{14}$ mw: 242.33

SYN: 4'-METHYL-1:2-BENZANTHRACENE

CONSENSUS REPORTS: EPA Genetic Toxicology Program.

SAFETY PROFILE: Questionable carcinogen with experimental tumorigenic data. Mutation data reported. When heated to decomposition it emits acrid smoke and irritating fumes.

MGV750   CAS: 2319-96-2   **HR: 3**
5-METHYLBENZ(a)ANTHRACENE
mf: $C_{19}H_{14}$ mw: 242.33

SYN: 3-METHYL-1,2-BENZANTHRACENE

CONSENSUS REPORTS: EPA Genetic Toxicology Program.

SAFETY PROFILE: Questionable carcinogen with experimental tumorigenic data. Mutation data reported. When heated to decomposition it emits acrid smoke and irritating fumes.

MGW000   CAS: 316-14-3   **HR: 3**
6-METHYLBENZ(a)ANTHRACENE
mf: $C_{19}H_{14}$ mw: 242.33

SYN: 4-METHYL-1,2-BENZANTHRACENE

CONSENSUS REPORTS: EPA Genetic Toxicology Program.

SAFETY PROFILE: Questionable carcinogen with experimental carcinogenic and tumorigenic data. Mutation data reported. When heated to decomposition it emits acrid smoke and irritating fumes.

MGW250   CAS: 2381-31-9   **HR: 3**
8-METHYLBENZ(a)ANTHRACENE
mf: $C_{19}H_{14}$ mw: 242.33

SYN: 5-METHYL-1,2-BENZANTHRACENE

CONSENSUS REPORTS: EPA Genetic Toxicology Program.

SAFETY PROFILE: Questionable carcinogen with experimental carcinogenic and tumorigenic data. Mutation data reported. When heated to decomposition it emits acrid smoke and irritating fumes.

MGW500   CAS: 2381-16-0   **HR: 3**
9-METHYLBENZ(a)ANTHRACENE
mf: $C_{19}H_{14}$ mw: 242.33

SYN: 6-METHYL-1,2-BENZANTHRACENE

CONSENSUS REPORTS: EPA Genetic Toxicology Program.

SAFETY PROFILE: Questionable carcinogen with experimental tumorigenic data. Mutation data reported. When heated to decomposition it emits acrid smoke and irritating fumes.

MGW750   CAS: 2541-69-7   **HR: 3**
10-METHYL-1,2-BENZANTHRACENE
mf: $C_{19}H_{14}$ mw: 242.33

SYNS: 7-MBA * 10-METHYL-1,2-BENZANTHRACEN (GERMAN) * 7-METHYLBENZ(a)ANTHRACENE

CONSENSUS REPORTS: EPA Genetic Toxicology Program.

SAFETY PROFILE: Questionable carcinogen with experimental carcinogenic, neoplastigenic, and tumorigenic data. Mutation data reported. When heated to decomposition it emits acrid smoke and irritating fumes.

MGX000   CAS: 2381-15-9   **HR: 3**
10-METHYLBENZ(a)ANTHRACENE
mf: $C_{19}H_{14}$ mw: 242.33

SYN: 7-METHYL-1,2-BENZANTHRACENE

CONSENSUS REPORTS: EPA Genetic Toxicology Program.

SAFETY PROFILE: Questionable carcinogen with experimental tumorigenic data. Mutation data reported. When heated to decomposition it emits acrid smoke and irritating fumes.

MGX500   CAS: 2422-79-9   **HR: 3**
12-METHYLBENZ(a)ANTHRACENE
mf: $C_{19}H_{14}$ mw: 242.33

SYN: 9-METHYL-1,2-BENZANTHRACENE

CONSENSUS REPORTS: EPA Genetic Toxicology Program.

SAFETY PROFILE: Questionable carcinogen with experimental carcinogenic and tumorigenic data. Mutation data reported. When heated to decomposition it emits acrid smoke and irritating fumes.

MGZ000 CAS: 1155-38-0 **HR: 3**
7-METHYLBENZ(a)ANTHRACENE-5,6-OXIDE
mf: $C_{19}H_{14}O$ mw: 258.33

SYN: 5,6-EPOXY-5,6-DIHYDRO-7-METHYLBENZ(A) AN-THRACENE

CONSENSUS REPORTS: EPA Genetic Toxicology Program.

SAFETY PROFILE: Questionable carcinogen with experimental neoplastigenic and tumorigenic data. Mutation data reported. When heated to decomposition it emits acrid smoke and irritating fumes.

MHA500 CAS: 101-41-7 **HR: 2**
METHYL BENZENEACETATE
mf: $C_9H_{10}O_2$ mw: 150.19

PROP: Colorless liquid; honey, jasmine odor. D: 1.062, refr index: 1.503-1.509, vap d: 5.18, flash p: 192°F. Sol in alc, fixed oils; insol in glycerin, propylene glycol, water @ 215°.

SYNS: BENZENEACETIC ACID, METHYL ESTER * FEMA No. 2733 * METHYL PHENYLACETATE (FCC) * METHYL-α-TOLUATE * PHENYLACETIC ACID, METHYL ESTER

CONSENSUS REPORTS: Reported in EPA TSCA Inventory.

SAFETY PROFILE: Moderately toxic by ingestion and skin contact. A skin irritant. Combustible liquid. When heated to decomposition it emits acrid smoke and irritating fumes.

MHA750 CAS: 93-58-3 **HR: 2**
METHYL BENZENECARBOXYLATE
DOT: UN 2938
mf: $C_8H_8O_2$ mw: 136.16

PROP: Colorless liquid; fragrant odor. Mp: −12.5°, bp: 199.6°, flash p: 181°F, d: 1.082-1.088, refr index: 1.515, vap press: 1 mm @ 39.0°, vap d: 4.69. Sol in alc, fixed oils, propylene glycol, water @ 30°; misc in alc, ether; insol in glycerin.

SYNS: FEMA No. 2683 * METHYL BENZOATE (FCC) * NIOBE OIL * OIL of NIOBE

CONSENSUS REPORTS: Reported in EPA TSCA Inventory.

DOT Classification: Poison B; Label: St. Andrews Cross

SAFETY PROFILE: Moderately toxic by ingestion. Mildly toxic by skin contact. A skin and eye irritant. Combustible liquid when exposed to heat or flame; can react with oxidizing materials. To fight fire, use foam, CO_2, dry chemical, water to blanket fire. When heated to decomposition it emits acrid smoke and irritating fumes.

MHC250 CAS: 615-15-6 **HR: 3**
METHYL-2-BENZIMIDAZOLE
mf: $C_8H_8N_2$ mw: 132.18

PROP: Needles from water. Mp: 175-176°. Sol in hot water, NaOH; sltly sol in alc and ether.

SYN: 2-METHYLBENZIMIDAZOLE

CONSENSUS REPORTS: Reported in EPA TSCA Inventory.

SAFETY PROFILE: Poison by intravenous route. Moderately toxic by ingestion. Experimental reproductive effects. Mutation data reported. When heated to decomposition it emits toxic fumes of NO_x.

MHF750 CAS: 134-84-9 **HR: 3**
4-METHYL-p-BENZOPHENONE
mf: $C_{14}H_{12}O$ mw: 196.26

SYNS: PHENYL-p-TOLYL KETONE * USAF DO-54

CONSENSUS REPORTS: Reported in EPA TSCA Inventory.

SAFETY PROFILE: Poison by intraperitoneal route. When heated to decomposition it emits acrid smoke and irritating fumes.

MHH000 CAS: 63041-77-0 **HR: 3**
4′-METHYLBENZO(a)PYRENE
mf: $C_{21}H_{14}$ mw: 266.35

SYNS: 7-METHYLBENZO(a)PYRENE * 4′-METHYL-3: 4-BENZPYRENE

SAFETY PROFILE: Questionable carcinogen with experimental neoplastigenic and tumorigenic data. Mutation data reported. When heated to decomposition it emits acrid smoke and irritating fumes.

MHI250 CAS: 553-97-9 **HR: 3**
2-METHYL-p-BENZOQUINONE
mf: $C_7H_6O_2$ mw: 122.13

SYNS: METHYL-p-BENZOQUINONE * METHYL-1,4-BENZOQUINONE * 2-METHYLBENZOQUINONE-1,4

* 2-METHYL-1,4-QUINONE   * p-TOLUQUINONE
* 1,4-TOLUQUINONE

CONSENSUS REPORTS: Reported in EPA TSCA Inventory.

SAFETY PROFILE: Poison by ingestion. When heated to decomposition it emits acrid smoke and irritating fumes.

MHJ500   CAS: 5090-37-9   HR: 3
2-((2-METHYLBENZO(b)THIEN-3-YL)METHYL)-2-IMIDAZOLINE HYDROCHLORIDE
mf: $C_{13}H_{14}N_2S \cdot ClH$ mw: 266.81

SYNS: 4,5-DIHYDRO-2-((2-METHYLBENZO(b)THIEN-3-YL)METHYL)-1H-IMIDAZOLE HYDROCHLORIDE
* ELLSYL   * ELSYL   * EX 10-781   * H 1032
* 2-METHYL-3-(Δ^2)-IMIDAZOLINYLMETHYL)BENZO(b)THIOPHENE HYDROCHLORIDE   * α-METIL-β-(2-METILENE-4,5-DIIDROIMIDAZOLIL)BENZOTIOFANE CLORIDRATO (ITALIAN)   * METIZOLINE HYDROCHLORIDE   * RMI 10,482A

SAFETY PROFILE: Poison by ingestion, intravenous, and intraperitoneal routes. When heated to decomposition it emits very toxic fumes of NO_x, SO_x, and HCl.

MHL000   CAS: 31431-39-7   HR: 3
METHYL-5-BENZOYL BENZIMIDAZOLE-2-CARBAMATE
mf: $C_{16}H_{13}N_3O_3$ mw: 295.32

SYNS: N-2 (5-BENZOYL-BENZIMIDAZOLE) CARBAMATE de METHYLE (FRENCH)   * 5-BENZOYL-2-BENZIMIDAZOLECARBAMIC ACID METHYL ESTER
* N-(BENZOYL-5-BENZIMIDAZOLYL)-2, CARBAMATE de METHYLE (FRENCH)   * MBDZ   * MEBENDAZOLE (USDA)   * OVITELMIN   * PANTELMIN   * R 17635
* TELMIN   * VERMIRAX   * VERMOX

CONSENSUS REPORTS: EPA Genetic Toxicology Program.

SAFETY PROFILE: Poison by ingestion. Moderately toxic by intraperitoneal route. Human mutation data reported. Experimental reproductive effects. When heated to decomposition it emits toxic fumes of NO_x.

MHN750   CAS: 10309-79-2   HR: 3
1-METHYL-2-BENZYLHYDRAZINE
mf: $C_8H_{12}N_2$ mw: 136.22

SYN: 1-BENZYL-2-METHYLHYDRAZINE

SAFETY PROFILE: Poison by subcutaneous route. Questionable carcinogen with experimen-

tal carcinogenic, tumorigenic, and teratogenic data. Mutation data reported. When heated to decomposition it emits toxic fumes of NO_x.

MHP250   CAS: 937-40-6   HR: 3
N-METHYL-N-BENZYLNITROSAMINE
mf: $C_8H_{10}N_2O$ mw: 150.20

SYNS: METHYL-BENZYL-NITROSOAMIN (GERMAN)
* N-METHYL-N-NITROSOBENZYLAMINE   * N-NITROSOBENZYLMETHYLAMINE   * N-NITROSOMETHYLBENZYLAMINE

CONSENSUS REPORTS: EPA Genetic Toxicology Program.

SAFETY PROFILE: Poison by ingestion. Questionable carcinogen with experimental tumorigenic data. Mutation data reported. When heated to decomposition it emits toxic fumes of NO_x.

MHR200   CAS: 74-83-9   HR: 3
METHYL BROMIDE
DOT: UN 1062
mf: CH_3Br mw: 94.95

PROP: Colorless, transparent, volatile liquid or gas; burning taste, chloroform-like odor. Bp: 3.56°, lel: 13.5%, uel: 14.5%, fp: −93°, flash p: none, d: 1.732 @ 0°/0°, autoign temp: 998°F, vap d: 3.27, vap press: 1824 mm @ 25°.

SYNS: BROM-METHAN (GERMAN)   * BROMO METHANE   * BROMO-O-GAS   * BROMOMETANO (ITALIAN)   * BROMURE de METHYLE (FRENCH)   * BROMURO di METILE (ITALIAN)   * BROOMMETHAAN (DUTCH)   * DAWSON 100   * DOWFUME   * DOWFUME MC-2 SOIL FUMIGANT   * EDCO   * EMBAFUME   * FUMIGANT-1 (OBS.)   * HALON 1001
* ISCOBROME   * KAYAFUME   * MB   * MBX
* MEBR   * METAFUME   * METHOGAS   * METHYLBROMID (GERMAN)   * METYLU BROMEK (POLISH)   * MONOBROMOMETHANE   * PESTMASTER (OBS.)   * PROFUME (OBS.)   * R 40B1   * RCRA WASTE NUMBER U029   * ROTOX   * TERABOL
* TERR-O-GAS 100   * ZYTOX

CONSENSUS REPORTS: IARC Cancer Review: GROUP 3 IMEMDT 7,245,87; Human Inadequate Evidence IMEMDT 41,187,86; Animal Limited Evidence IMEMDT 41,187,86. Reported in EPA TSCA Inventory. Community Right-To-Know List. EPA Extremely Hazardous Substances List.

OSHA PEL: (Transitional: CL 20 ppm (skin)) TWA 5 ppm (skin)
ACGIH TLV: TWA 5 ppm (skin)

DFG MAK: 5 ppm (20 mg/m^3), Suspected Carcinogen.
NIOSH REL: Reduce to lowest level
DOT Classification: Poison A; Label: Poison Gas.

SAFETY PROFILE: Suspected carcinogen with experimental carcinogenic data. A human poison by inhalation. Human systemic effects by inhalation: anorexia, nausea or vomiting. Corrosive to skin; can produce severe burns. Human mutation data reported. A powerful fumigant gas which is one of the most toxic of the common organic halides. It is hemotoxic and narcotic with delayed action. The effects are cumulative and damaging to nervous system, kidneys, and lung. Central nervous system effects include blurred vision, mental confusion, numbness, tremors, and speech defects.

Methyl bromide is reported to be eight times more toxic on inhalation than ethyl bromide. Moreover, because of its greater volatility, it is a much more frequent cause of poisoning. Death following acute poisoning is usually caused by its irritant effect on the lungs. In chronic poisoning, death is due to injury to the central nervous system. Fatal poisoning has always resulted from exposure to relatively high concentrations of methyl bromide vapors (from 8,600 to 60,000 ppm). Nonfatal poisoning has resulted from exposure to concentrations as low as 100-500 ppm. In addition to injury to the lung and central nervous system, the kidneys may be damaged with development of albuminuria and, in fatal cases, cloudy swelling and/or tubular degeneration. The liver may be enlarged. There are no characteristic blood changes.

Mixtures of 10-15 percent with air may be ignited with difficulty. Moderately explosive when exposed to sparks or flame. Forms explosive mixtures with air within narrow limits at atmospheric pressure, with wider limits at higher pressure. The explosive sensitivity of mixtures with air may be increased by the presence of aluminum, magnesium, zinc, or their alloys. Incompatible with metals, dimethyl sulfoxide, ethylene oxide. To fight fire, use foam, water, CO$_2$, dry chemical. When heated to decomposition it emits toxic fumes of Br$^-$.

MHR250   CAS: 96-32-2   **HR: 3**
METHYL BROMOACETATE

DOT: UN 2643
mf: C$_3$H$_5$BrO$_2$ mw: 152.99

PROP: Liquid. Bp: 51-52° @ 15 mm.

SYNS: BROMOACETIC ACID METHYL ESTER
* METHYL α-BROMOACETATE * METHYL MONO-BROMOACETATE

CONSENSUS REPORTS: Reported in EPA TSCA Inventory.

DOT Classification: Poison B; Label: Poison.

SAFETY PROFILE: Poison by intravenous route. When heated to decomposition it emits toxic fumes of Br$^-$.

MHS750   CAS: 137-32-6   **HR: 3**
2-METHYL BUTANOL-1
mf: C$_5$H$_{12}$O   mw: 88.15

PROP: Colorless liquid. D: 0.81-0.82 @ 20°, fp: < −70°, bp: 128°, flash p: 122°F (OC), vap d: 3.0, lel: 1.4%, uel: 9.0%. Sltly sol in water; miscible with alc and ether.

SYNS: dl-sec-BUTYLCARBINOL * 2-METHYLBUTANOL

CONSENSUS REPORTS: Reported in EPA TSCA Inventory.

SAFETY PROFILE: Moderately toxic by skin contact and intraperitoneal routes. Mildly toxic by ingestion. An eye, skin, and mucous membrane irritant. Can cause deafness, delerium, headache, nausea, and vomiting. Flammable when exposed to heat, flame, or oxidizers. Explosive in the form of vapor when exposed to heat or flame. Incompatible with H$_2$S$_3$. To fight fire, use alcohol foam, spray, mist, dry chemical. When heated to decomposition it emits acrid smoke and irritating fumes.

MHT000   CAS: 563-46-2   **HR: 2**
2-METHYL-1-BUTENE

DOT: UN 2371/UN 2459
mf: C$_5$H$_{10}$ mw: 70.14

PROP: Colorless; extremely volatile liquid or gas. D: 0.7, vap d: 2.4, bp: 38°, flash p: −4°F. Insol in water.

CONSENSUS REPORTS: Reported in EPA TSCA Inventory.

DOT Classification: Flammable Liquid; Label: Flammable Liquid.

SAFETY PROFILE: A simple asphyxiant. Very dangerous fire hazard when exposed to heat, flame or oxidizers. To fight fire, use dry chemical, CO_2, foam. When heated to decomposition it emits acrid smoke and irritating fumes.

MHT250 CAS: 563-45-1 **HR: 3**
3-METHYL-1-BUTENE

DOT: UN 2371/UN 2561
mf: C_5H_{10} mw: 70.14

PROP: Colorless, very volatile, flammable liquid; disagreeable odor. Bp: 31.11°, d: 0.65 @ 20°/20°, fp: −137.5°, flash p: 19.4°F, vap d: 2.4, lel: 1.5%, uel: 9.1%. Insol in water; sol in alc.

CONSENSUS REPORTS: Reported in EPA TSCA Inventory.

DOT Classification: Flammable liquid; Label: Flammable Liquid.

SAFETY PROFILE: Very dangerous fire hazard when exposed to heat, flame, or oxidizers. Explosive in the form of vapor when exposed to heat or flame. To fight fire, use alcohol foam, mist, spray, dry chemical, CO_2. When heated to decomposition it emits acrid smoke and irritating fumes.

MHU750 CAS: 97-88-1 **HR: 3**
2-METHYL BUTYLACRYLATE

DOT: UN 2227
mf: $C_8H_{14}O_2$ mw: 142.22

PROP: Colorless liquid; ester odor. Bp: 163°, flash p: 126°F (TOC), lel: 2%, uel: 8%, autoign temp: 562°F, vap press: 4.9 mm @ 20°, d: 0.895 @ 20°/4°, vap d: 4.8.

SYNS: BUTIL METACRILATO (ITALIAN) * BUTYL-METHACRYLAAT (DUTCH) * N-BUTYL METHACRY-LATE * BUTYL-2-METHACRYLATE * BUTYL-2-METHYL-2-PROPENOATE * METHACRYLATE de BUTYLE (FRENCH) * METHACRYLSAEUREBUTYL-ESTER (GERMAN) * 2-METHYL-BUTYLACRYLAAT (DUTCH) * 2-METHYL-BUTYLACRYLAT (GERMAN)

CONSENSUS REPORTS: Reported in EPA TSCA Inventory.

DOT Classification: Flammable or Combustible Liquid; Label: Flammable Liquid.

SAFETY PROFILE: Poison by intraperitoneal route. Mildly toxic by ingestion, inhalation, and skin contact. Experimental teratogenic and reproductive effects. A skin irritant. Flammable when exposed to heat or flame. Explosive in the form of vapor when exposed to heat or flame. Violent polymerization can be caused by heat, moisture, oxidizers. To fight fire, use foam, dry chemical, CO_2. When heated to decomposition it emits acrid smoke and irritating fumes.

MHV000 CAS: 110-68-9 **HR: 3**
N-METHYL-n-BUTYLAMINE

DOT: 2945
mf: $C_5H_{13}N$ mw: 87.19

PROP: Liquid. D: 0.7335, bp: 91.1°, vap d: 3.0, flash p: 35.6°. Sol in water.

SYNS: METHYLBUTYLAMINE * N-(METHYL) BU-TYL AMINE

CONSENSUS REPORTS: Reported in EPA TSCA Inventory.

DOT Classification: Flammable Liquid; Label: Flammable Liquid.

SAFETY PROFILE: Poison by intravenous route. Moderately toxic by ingestion, skin contact, and intraperitoneal routes. Mildly toxic by inhalation. A skin and severe eye irritant. A very dangerous fire hazard when exposed to heat, flame, or oxidizers. To fight fire, use alcohol foam. When heated to decomposition it emits toxic fumes of NO_x.

MHV750 CAS: 4435-53-4 **HR: 1**
METHYL-1,3-BUTYLENE GLYCOL ACETATE

DOT: UN 2708
mf: $C_7H_{14}O_3$ mw: 146.21

PROP: Liquid; bitter taste and acrid odor. Bp: 135°, flash p: 170°F, d: 0.952-0.958 @ 20°/20°, vap d: 5.05.

SYNS: ACETIC ACID-3-METHOXYBUTYL ESTER * BUTOXYL * 3-METHOXYBUTYL ACETATE

CONSENSUS REPORTS: Reported in EPA TSCA Inventory.

DOT Classification: Flammable or Combustible Liquid; Label: Flammable Liquid.

SAFETY PROFILE: Mildly toxic by ingestion. A skin and eye irritant. Combustible when exposed to heat or flame; can react with oxidizing materials. To fight fire, use alcohol foam, CO_2, dry chemical. When heated to decomposition it emits acrid smoke and irritating fumes.

MHV859   CAS: 1634-04-4   **HR: 2**
METHYL tert-BUTYL ETHER

DOT: UN 2398
mf: $C_5H_{12}O$ mw: 88.17

SYNS: METHYL 1,1-DIMETHYLETHYL ETHER
* PROPANE, 2-METHOXY-2-METHYL (9CI)

CONSENSUS REPORTS: Community Right-To-Know List. Reported in EPA TSCA Inventory.

DOT Classification: Flammable Liquid; Label: Flammable Liquid.

SAFETY PROFILE: Flammable when exposed to heat or flame. When heated to decomposition it emits acrid smoke and irritating fumes.

MHW350   CAS: 71016-15-4   **HR: 3**
N-3-METHYLBUTYL-N-1-METHYL ACETONYLNITROSAMINE
mf: $C_9H_{18}N_2O_2$ mw: 186.29

SYNS: 3-((ISOPENTYL)NITROSOAMINO)-2-BUTANONE
* MAMBNA

SAFETY PROFILE: Suspected carcinogen with experimental carcinogenic, neoplastigenic, and tumorigenic data. Mutation data reported. When heated to decomposition it emits toxic fumes of NO_x.

MHW500   CAS: 7068-83-9   **HR: 3**
METHYLBUTYLNITROSAMINE
mf: $C_5H_{12}N_2O$ mw: 116.19

SYNS: MBNA * METHYL-BUTYL-NITROSAMIN (GERMAN) * METHYL-N-BUTYLNITROSAMINE
* N-METHYL-N-NITROSOBUTYLAMINE * N-NITROSO-N-BUTYLMETHYLAMINE * N-NITROSO-METHYL-N-BUTYLAMINE * NMBA

CONSENSUS REPORTS: EPA Genetic Toxicology Program.

SAFETY PROFILE: Poison by ingestion, inhalation, intraperitoneal, and subcutaneous routes. Questionable carcinogen with experimental carcinogenic and tumorigenic data. Mutation data reported. When heated to decomposition it emits toxic fumes of NO_x.

MHY000   CAS: 623-42-7   **HR: 2**
METHYL-n-BUTYRATE

DOT: UN 1237
mf: $C_5H_{10}O_2$ mw: 102.13

PROP: Colorless liquid. Mp: $< -97°$, bp: 102.3°, flash p: 57°F (CC), d: 0.898, vap press: 40 mm @ 29.6°, vap d: 3.53. Sltly sol in water; misc in alc and ether.

SYNS: METHYL n-BUTANOATE * METHYL BUTYRATE

CONSENSUS REPORTS: Reported in EPA TSCA Inventory.

DOT Classification: Flammable Liquid; Label: Flammable Liquid.

SAFETY PROFILE: Moderately toxic by ingestion and skin contact. A skin irritant. A very dangerous fire hazard when exposed to heat, flame or oxidizers. Can react vigorously with oxidizing materials. To fight fire, use alcohol foam, CO_2, dry chemical. When heated to decomposition it emits acrid smoke and irritating fumes.

MHY550   CAS: 7568-37-8   **HR: 3**
METHYL CADMIUM AZIDE
mf: CH_3CdN_3 mw: 97.13

CONSENSUS REPORTS: Cadmium and its compounds are on the Community Right-To-Know List.

OSHA PEL: TWA 0.1 mg(Cd)/m³; CL 0.6 mg(Cd)/m³ (fume)
ACGIH TLV: TWA 0.05 mg(Cd)/m³ (Proposed: TWA 0.01 mg(Cd)/m³ (dust), Human Carcinogen); BEI: 10 μg/g creatinine in urine; 10 μg/L in blood.
DFG BAT: Blood 1.5 μg/dL; Urine 15 μg/dL, Suspected Carcinogen.
NIOSH REL: (Cadmium) Reduce to lowest feasible level

SAFETY PROFILE: Confirmed human carcinogen. Hydrolysis reaction in the presence of moisture forms the explosive hydrogen azide gas. When heated to decomposition it emits toxic fumes of Cd and NO_x.

MHZ000   CAS: 598-55-0   **HR: 3**
METHYL CARBAMATE
mf: $C_2H_5NO_2$ mw: 75.07

PROP: Needles. Bp: 177°, mp: 52-54°. Very sol in water, alc.

SYNS: BENDIOCARB * METHYLURETHAN
* METHYLURETHANE * NCI-C55594 * URETHYLANE

CONSENSUS REPORTS: IARC Cancer Review: GROUP 3 IMEMDT 7,56,87; Animal

Inadequate Evidence IMEMDT 12,151,76. EPA Genetic Toxicology Program. Reported in EPA TSCA Inventory.

SAFETY PROFILE: Poison by ingestion and intraperitoneal routes. Questionable carcinogen with experimental carcinogenic and tumorigenic data. Mutation data reported. When heated to decomposition it emits toxic fumes of NO_x.

MIA250   CAS: 2631-40-5   **HR: 3**
METHYLCARBAMIC ACID-o-CUMENYL ESTER
mf: $C_{11}H_{15}NO_2$ mw: 193.27

SYNS: BAY 105807 * ENT 25,670 * ETROFOLAN * HYTOX * ISOPROCARB * ISOPROPYLPHENOL METHYLCARBAMATE * o-ISOPROPYLPHENYL-N-METHYLCARBAMATE * 2-ISOPROPYL-PHENYL-N-METHYLCARBAMATE * KHE 0145 * 2-(1-METHYL-ETHYL)PHENYL METHYLCARBAMATE * MIPC * MIPCIN * MIPSIN

CONSENSUS REPORTS: Reported in EPA TSCA Inventory. EPA Genetic Toxicology Program.

SAFETY PROFILE: Poison by ingestion, intravenous, and intraperitoneal routes. Mildly toxic by skin contact. Used for controlling leafhoppers, planthoppers, and bugs in rice and cacao. When heated to decomposition it emits toxic fumes of NO_x.

MIB750   CAS: 1129-41-5   **HR: 3**
METHYLCARBAMIC ACID-m-TOLYL ESTER
mf: $C_9H_{11}NO_2$ mw: 165.21

SYNS: m-CRESYL ESTER of N-METHYLCARBAMIC ACID * m-CRESYL METHYLCARBAMATE * DICRESYL * DRC 3341 * KUMIAI * METACRATE * 3-METHYLPHENYL-N-METHYLCARBAMATE * MTMC * m-TOLYL-N-METHYLCARBAMATE * 3-TOLYL-N-METHYLCARBAMATE * TSUMACIDE

CONSENSUS REPORTS: EPA Extremely Hazardous Substances List. Reported in EPA TSCA Inventory. EPA Genetic Toxicology Program.

SAFETY PROFILE: Poison by ingestion, skin contact, and possibly other routes. Moderately toxic by inhalation. Mutation data reported. When heated to decomposition it emits toxic fumes of NO_x.

MID250   CAS: 60398-22-3   **HR: 3**
METHYLCARBAMIC ESTER of OXYPHENYLMETHYLDIETHYL-AMMONIUM IODIDE
mf: $C_{13}H_{21}N_2O_2 \cdot I$ mw: 364.26

SYNS: METHIODIDE of N-METHYLURETHANE of 3-DIETHYLAMINOPHENOL * N-METHYLCARBAMIC ACID-3-DIETHYLAMINOPHENYL ESTER, METHIODIDE * N-METHYLCARBAMIC ACID-3-(DIETHYLMETHYLAMMONIO)PHENYL ESTER, IODIDE * (3-(N-METHYLCARBAMOYLOXY)PHENYL)DIETHYLMETHYL-AMMONIUM IODIDE * TL 1217

SAFETY PROFILE: A deadly poison by ingestion, intravenous, and subcutaneous routes. When heated to decomposition it emits very toxic fumes of NO_x, NH_3 and I^-.

MIF000   CAS: 616-38-6   **HR: 3**
METHYL CARBONATE
DOT: UN 1161
mf: $C_3H_6O_3$ mw: 90.09

PROP: Colorless liquid; pleasant odor. Mp: 0.5°, d: 1.065 @ 17°/4°, flash p: 66°F (OC). Bp: 90.91°. Misc with acids and alkalies; sol in most organic solvents; insol in water.

SYN: DIMETHYL CARBONATE

CONSENSUS REPORTS: Reported in EPA TSCA Inventory.

DOT Classification: Flammable Liquid; Label: Flammable Liquid.

SAFETY PROFILE: Moderately toxic by intraperitoneal route. Mildly toxic by ingestion. An irritant. Violent reaction or ignition on contact with potassium-tert-butoxide. A very dangerous fire hazard when exposed to heat, open flames (sparks) or oxidizers. To fight fire, use alcohol foam. When heated to decomposition it emits acrid smoke and irritating fumes.

MIF750   CAS: 3121-61-7   **HR: 3**
METHYL CELLOSOLVE ACRYLATE
mf: $C_6H_{10}O_3$ mw: 130.16

PROP: Liquid. Bp: 61° @ 17 mm, flash p: 180°F (OC), d: 1.0134 @ 20°, vap d: 4.49.

SYNS: ACRYLIC ACID-2-METHOXYETHYL ESTER * ETHYLENE GLYCOL MONOMETHYL ETHER ACRYLATE * GLYCOL MONOMETHYL ETHER ACRYLATE * 2-METHOXYETHANOL, ACRYLATE

CONSENSUS REPORTS: Glycol ether compounds are on the Community Right-To-Know List. Reported in EPA TSCA Inventory.

SAFETY PROFILE: Poison by skin contact. Moderately toxic by ingestion and inhalation. A skin irritant. Flammable when exposed to heat or flame; can react with oxidizing materials. To fight fire, use foam, CO_2, dry chemical. When heated to decomposition it emits acrid smoke and irritating fumes.

MIF765   CAS: 74-87-3   HR: 3
METHYL CHLORIDE

DOT: UN 1063
mf: CH_3Cl mw: 50.49

PROP: Colorless gas; ethereal odor and sweet taste. D: 0.918 @ 20°/4°, mp: −97°, bp: −23.7°, flash p: <32°F, lel: 8.1%, uel: 17%, autoign temp: 1170°F, vap d: 1.78. Sltly sol in water; misc with chloroform, ether, glacial acetic acid; sol in alcohol.

SYNS: ARTIC * CHLOOR-METHAAN (DUTCH) * CHLOR-METHAN (GERMAN) * CHLOROMETHANE * CHLORURE de METHYLE (FRENCH) * CLORO-METANO (ITALIAN) * CLORURO di METILE (ITALIAN) * METHYLCHLORID (GERMAN) * METYLU CHLO-REK (POLISH) * MONOCHLOROMETHANE * RCRA WASTE NUMBER U045

CONSENSUS REPORTS: IARC Cancer Review: GROUP 3 IMEMDT 7,246,87; Human Inadequate Evidence IMEMDT 41,161,86; Animal Inadequate Evidence IMEMDT 41,161,86. Reported in EPA TSCA Inventory. EPA Genetic Toxicology Program.

OSHA PEL: (Transitional: TWA 100; CL 200 ppm; Pk 300 ppm/5M)TWA 50 ppm; STEL 100 ppm
ACGIH TLV: TWA 50 ppm; STEL 100 ppm
DFG MAK: 50 ppm (105 mg/m^3); Suspected Carcinogen.
NIOSH REL: (Monohalomethanes) TWA Reduce to lowest level.
DOT Classification: Flammable Gas; Label: Flammable Gas; IMO: Poison A; Label: Poison Gas and Flammable Gas.

SAFETY PROFILE: Suspected carcinogen. Very mildly toxic by inhalation. Experimental teratogenic and reproductive effects. Human mutation data reported. Human systemic effects by inhalation: convulsions, nausea or vomiting, and unspecified effects on the eye.

Chloromethane has slight irritant properties and may be inhaled without noticeable discomfort. It has some narcotic action, but this effect is weaker than that of chloroform. Acute poisoning, characterized by the narcotic effect, is rare in industry. In exposures to high concentrations, dizziness, drowsiness, incoordination, confusion, nausea and vomiting, abdominal pains, hiccoughs, diplopia, and dimness of vision are followed by delirium, convulsions, and coma. Death may be immediate; however, if the exposure is not fatal, recovery is usually slow. Degenerative changes in the central nervous system are not uncommon. The liver, kidneys, and bone marrow may be affected, with resulting acute nephritis and anemia. Death may occur several days after exposure resulting from degenerative changes in the heart, liver, and especially the kidneys. Repeated exposure to low concentrations causes damage to the central nervous system and, less frequently, to the liver, kidneys, bone marrow and cardiovascular system. Hemorrhages into the lungs, intestinal tract, and dura have been reported. Sprayed on the skin, chloromethane produces anesthesia through freezing of the tissues as it evaporates.

Flammable gas. Very dangerous fire hazard when exposed to heat, flame, or powerful oxidizers. Moderate explosion hazard when exposed to flame and sparks. Explodes on contact with interhalogens (e.g., bromine trifluoride; bromine pentafluoride); magnesium and alloys; potassium and alloys; sodium and alloys; zinc. Potentially explosive reaction with aluminum when heated to 152° in a sealed container. Mixtures with aluminum chloride + ethylene react exothermically and then explode when pressurized to above 30 bar. May ignite on contact with aluminum chloride or powdered aluminum. To fight fire, stop flow of gas and use CO_2, dry chemical, or water spray. When heated to decomposition it emits highly toxic fumes of Cl^-.

MIF775   CAS: 96-34-4   HR: 3
METHYL CHLOROACETATE

DOT: UN 2295
mf: $C_3H_5ClO_2$ mw: 108.53

PROP: Colorless liquid. D: 1.238, mp: −33°, bp: 130-132°. Insol in water; miscible with alc, ether.

SYNS: METHYL CHLOROACETATE (DOT) * METHYL MONOCHLORACETATE * METHYL

MONOCHLOROACETATE * MONOCHLOROACETIC ACID METHYL ESTER

CONSENSUS REPORTS: Reported in EPA TSCA Inventory.

DOT Classification: DOT-IMO: Flammable or Combustible Liquid; Label: Flammable Liquid.

SAFETY PROFILE: Poison by ingestion. Moderately toxic by inhalation and subcutaneous routes. Flammable when exposed to heat or flame; can react vigorously with oxidizing materials. When heated to decomposition it emits toxic fumes of Cl⁻.

MIG000   CAS: 79-22-1   HR: 3
METHYL CHLOROCARBONATE

DOT: UN 1238
mf: $C_2H_3ClO_2$ mw: 94.50

PROP: Colorless liquid. Bp: 71.4°, d: 1.223 @ 20°/4°, vap d: 3.26, flash p: 54°F, autoign temp: 940°F. Sltly sol in water with gradual decomp; misc with alc, benzene, chloroform, and ether.

SYNS: CHLORAMEISENSAEURE METHYLESTER (GERMAN) * CHLOROCARBONATE de METHYLE (FRENCH) * CHLOROCARBONIC ACID METHYL ESTER * CHLOROFORMIC ACID METHYL ESTER * MCF * METHOXYCARBONYL CHLORIDE * METHYLCHLOORFORMIAT (DUTCH) * METHYL CHLOROFORMATE (DOT) * METILCLOROFORMIATO (ITALIAN) * RCRA WASTE NUMBER U156

CONSENSUS REPORTS: EPA Extremely Hazardous Substances List. Reported in EPA TSCA Inventory.

DOT Classification: Flammable Liquid; Label: Flammable Liquid, Poison, Corrosive.

SAFETY PROFILE: Poison by ingestion, inhalation, and intraperitoneal routes. Moderately toxic by skin contact. Human systemic effects by inhalation: conjunctiva irritation and respiratory effects. Corrosive to skin, eyes, and mucous membranes. Very dangerous fire hazard when exposed to heat sources, sparks, flame, or oxidizers. Reacts with water or steam to produce toxic and corrosive fumes. When heated to decomposition it emits toxic fumes of Cl⁻, methyl chloroformate and phosgene.

MIH000   CAS: 63905-05-5   HR: 3
METHYL-β-CHLROETHYL-β-HYDROXYETHLAMINE HYDROCHLORIDE
mf: $C_5H_{12}ClNO \cdot ClH$ mw: 174.09

SAFETY PROFILE: Poison by ingestion, subcutaneous, intraperitoneal, and intravenous routes. When heated to decomposition it emits very toxic fumes of Cl⁻ and NO_x.

MIH275   CAS: 71-55-6   HR: 3
METHYL CHLOROFORM

DOT: UN 2831
mf: $C_2H_3Cl_3$ mw: 133.40

PROP: Colorless liquid. Bp: 74.1°, fp: −32.5°, flash p: none, d: 1.3376 @ 20°/4°, vap press: 100 mm @ 20.0°. Insol in water; sol in acetone, benzene, carbon tetrachloride, methanol, ether.

SYNS: AEROTHENE TT * CHLOROETENE * CHLOROETHENE * CHLOROTHANE NU * CHLOROTHENE * CHLOROTHENE (inhibited) * CHLOROTHENE NU * CHLOROTHENE VG * CHLORTEN * INHIBISOL * METHYLCHLOROFORM * METHYLTRICHLOROMETHANE * NCI-C04626 * RCRA WASTE NUMBER U226 * SOLVENT 111 * STROBANE * α-T * 1,1,1-TCE * 1,1,1-TRICHLOORETHAAN (DUTCH) * 1,1,1-TRICHLORAETHAN (GERMAN) * TRICHLORO-1,1,1-ETHANE (FRENCH) * 1,1,1-TRICHLOROETHANE * α-TRICHLOROETHANE * 1,1,1-TRICLOROETANO (ITALIAN) * TRI-ETHANE

CONSENSUS REPORTS: IARC Cancer Review: GROUP 3 IMEMDT 7,56,87; Animal Inadequate Evidence IMEMDT 20,515,79. NCI Carcinogenesis Bioassay (gavage); Inadequate Studies: mouse, rat NCITR* NCI-CG-TR-3,77. Community Right-To-Know List. Reported in EPA TSCA Inventory. EPA Genetic Toxicology Program.

OSHA PEL: (Transitional: TWA 350 ppm) TWA 350 ppm; STEL 450 ppm
ACGIH TLV: TWA 350 ppm; STEL 450 ppm (Proposed: BEI: 10 mg/L trichloroacetic acid in urine at end of workweek.)
DFG MAK: 200 ppm (1080 mg/m³); BAT: 55 μg/dL in blood after several shifts.
NIOSH REL: (1,1,1-Trichloroethane) CL 350 ppm/15M
DOT Classification: ORM-A; Label: None; Poison B; Label: St. Andrews Cross.

SAFETY PROFILE: Poison by intravenous route. Moderately toxic by ingestion, inhalation, skin contact, subcutaneous, and intraperitoneal routes. Human systemic effects by ingestion and inhalation: conjunctiva irritation, hallucinations or distorted perceptions, motor activity changes, irritability, aggression, hypermotility, diarrhea, nausea or vomiting and other gastrointestinal changes. Experimental teratogenic and reproductive effects. Mutation data reported. A human skin irritant. An experimental skin and severe eye irritant. Narcotic in high concentrations. Questionable carcinogen. Causes a proarrhythmic activity which sensitizes the heart to epinephrine-induced arrhythmias. This sometimes will cause cardiac arrest, particularly when this material is massively inhaled as in drug abuse for euphoria.

Under the proper conditions it can undergo hazardous reactions with aluminum oxide + heavy metals, dinitrogen tetraoxide, inhibitors, metals (e.g., magnesium, aluminum, potassium, potassium-sodium alloy), sodium hydroxide, N_2O_4, oxygen. When heated to decomposition it emits toxic fumes of Cl^-. Used as a cleaning solvent, a chemical intermediate to produce vinylidene chloride, and as a propellant in aerosol cans.

MIJ250　　　CAS: 500-28-7　　　**HR: 2**
METHYLCHLOROTHION
mf: $C_8H_9ClNO_5PS$　　　mw: 297.66

SYNS: O-(3-CHLOOR-4-NITRO-FENYL)-O,O-DIMETHYL-MONOTHIOFOSFAAT (DUTCH) * CHLOORTHION (DUTCH) * O-(3-CHLOR-4-NITRO-PHENYL)-O,O-DIMETHYL-MONOTHIOPHOSPHAT (GERMAN) * O-(3-CHLORO-4-NITROPHENYL) O,O-DIMETHYL PHOSPHOROTHIOATE * CHLORTHION METHYL * CHLORTION (CZECH) * O-(3-CLORO-4-NITRO-FENIL)-O,O-DIMETIL-MONOTIOFOSFATO (ITALIAN) * O,O-DIMETHYL-O-3-CHLOR-4-NITROFENYLTIOFOS-FAT (CZECH) * O,O-DIMETHYL-O-(3-CHLOR-4-NITRO-PHENYL)-MONOTHIOPHOSPHAT (GERMAN) * O,O-DIMETHYL-O-(3-CHLORO-4-NITROPHENYL) PHOSPHOROTHIOATE * DIMETHYL-3-CHLORO-4-NITROPHENYL THIONOPHOSPHATE * O,O-DIMETHYL-O-(3-CHLORO-4-NITROPHENYL) THIOPHOSPHATE * O,O-DIMETHYL-p-NITRO-m-CHLOROPHENYL THIOPHOSPHATE * O,O-DIMETHYL-O-4-NITRO-3-CHLOROPHENYL THIOPHOSPHATE * ENT 18,861 * p-NITRO-m-CHLOROPHENYL DIMETHYL THIONOPHOSPHATE * THIOPHOSPHATE de O,O-DIMETHYLE et de O-3-CHLORO-4-NITROPHENYLE (FRENCH)

SAFETY PROFILE: Moderately toxic by ingestion, skin contact, intraperitoneal, and possibly other routes. An insecticide. Decomposes and then ignites when heated above 270°C. When heated to decomposition it emits very toxic fumes of Cl^-, SO_x, PO_x, and NO_x.

MIJ500　　　CAS: 127-33-3　　　**HR: 3**
METHYLCHLORTETRACYCLINE
mf: $C_{21}H_{21}ClN_2O_8$　　　mw: 464.89

SYNS: 7-CHLORO-6-DEMETHYLTETRACYCLINE DEMETHYLCHLOROTETRACYCLINE * DECLOMYCIN * DEMECLOCYCLINE * DEMETHYLCHLOROTETRA-CYCLIN * DEMETHYLCHLOROTETRACYCLINE * 6-DEMETHYLCHLOROTETRACYCLINE * DEMETHYLCHLORTETRACYCLINE * 6-DEMETHYL-7-CHLO-ROTETRACYCLINE DEMETHYLCHLORTETRACYCLINE * 6-DEMETHYL-7-CHLORTETRACYCLINE * 6-DE-METHYLCHLORTETRACYCLINE * 6-DEMETHYL-7-CHLORTETRACYCLINE * DEMETHYLCHLORTETRA-CYCLINE. BASE * DMCT * LEDERMYCIN * MEXOCINE * RP 10192

SAFETY PROFILE: Poison by intravenous and intraperitoneal routes. Human systemic effects by ingestion: diabetes insipidus, urine volume increase, other changes in urine composition, dermatitis, changes in the nails, allergic rhinitis, serum sickness, effects on cyclic nucleotides. Human reproductive effects by an unspecified route: postnatal measures or effects on newborn. Experimental teratogenic and reproductive effects. Human mutation data reported. When heated to decomposition it emits very toxic fumes of Cl^- and NO_x.

MIJ750　　　CAS: 56-49-5　　　**HR: 3**
3-METHYLCHOLANTHRENE
mf: $C_{21}H_{16}$　　　mw: 268.37

PROP: Pale yellow needles from benzene. Mp: 176.5°, bp: 280° @ 80 mm, d: 1.28 @ 20°. Sol in benzene, xylene, toluene; sltly sol in amyl alc; insol in water.

SYNS: 1,2-DIHYDRO-3-METHYL-BENZ(j)ACEANTHRYL-ENE * 3-MCA * METHYLCHOLANTHRENE * 20-METHYLCHOLANTHRENE * RCRA WASTE NUMBER U157

CONSENSUS REPORTS: Reported in EPA TSCA Inventory. EPA Genetic Toxicology Program.

SAFETY PROFILE: Suspected carcinogen with experimental carcinogenic, neoplastigenic, and tumorigenic data. Poison by intravenous and intraperitoneal routes. Experimental teratogenic and reproductive effects. Human mutation data reported. When heated to decomposition it emits acrid smoke and irritating fumes.

MIL250   CAS: 3343-08-6   **HR: 3**
3-METHYLCHOLANTHRENE-2-ONE
mf: $C_{21}H_{14}O$ mw: 282.35

SYN: 3-METHYLCHOLANTHREN-2-ONE

CONSENSUS REPORTS: EPA Genetic Toxicology Program.

SAFETY PROFILE: Questionable carcinogen with experimental carcinogenic and neoplastigenic data. Mutation data reported. When heated to decomposition it emits acrid smoke and irritating fumes.

MIM250   CAS: 3343-07-5   **HR: 3**
20-METHYLCHOLANTHREN-15-ONE
mf: $C_{21}H_{14}O$ mw: 282.35

SYNS: 15-KETO-20-METHYLCHOLANTHRENE
∗   3-METHYLCHOLANTHREN-1-ONE

CONSENSUS REPORTS: EPA Genetic Toxicology Program.

SAFETY PROFILE: Questionable carcinogen with experimental neoplastigenic and tumorigenic data. Mutation data reported. When heated to decomposition it emits acrid smoke and irritating fumes.

MIN000   CAS: 3351-31-3   **HR: 3**
3-METHYLCHRYSENE
mf: $C_{19}H_{14}$ mw: 242.33

CONSENSUS REPORTS: IARC Cancer Review: GROUP 3 IMEMDT 7,56,87; Animal Limited Evidence IMEMDT 32,379,83. EPA Genetic Toxicology Program.

SAFETY PROFILE: Questionable carcinogen with experimental neoplastigenic and tumorigenic data. Mutation data reported. When heated to decomposition it emits acrid smoke and irritating fumes.

MIN250   CAS: 3351-30-2   **HR: 3**
4-METHYLCHRYSENE
mf: $C_{19}H_{14}$ mw: 242.33

CONSENSUS REPORTS: IARC Cancer Review: GROUP 3 IMEMDT 7,56,87; Animal Limited Evidence IMEMDT 32,379,83. EPA Genetic Toxicology Program.

SAFETY PROFILE: Questionable carcinogen with experimental tumorigenic data. Mutation data reported. When heated to decomposition it emits acrid smoke and irritating fumes.

MIN500   CAS: 3697-24-3   **HR: 3**
5-METHYLCHRYSENE
mf: $C_{19}H_{14}$ mw: 242.33

CONSENSUS REPORTS: IARC Cancer Review: GROUP 3 IMEMDT 7,56,87; Animal Sufficient Evidence IMEMDT 32,379,83. EPA Genetic Toxicology Program.

SAFETY PROFILE: Questionable carcinogen with experimental carcinogenic, neoplastigenic, and tumorigenic data. Mutation data reported. When heated to decomposition it emits acrid smoke and irritating fumes.

MIN750   CAS: 1705-85-7   **HR: 3**
6-METHYLCHRYSENE
mf: $C_{19}H_{14}$ mw: 242.33

CONSENSUS REPORTS: IARC Cancer Review: GROUP 3 IMEMDT 7,56,87; Animal Limited Evidence IMEMDT 32,379,83. EPA Genetic Toxicology Program.

SAFETY PROFILE: Questionable carcinogen with experimental tumorigenic data. Mutation data reported. When heated to decomposition it emits acrid smoke and irritating fumes.

MIO000   CAS: 101-39-3   **HR: 2**
α-METHYLCINNAMALDEHYDE
mf: $C_{10}H_{10}O$ mw: 146.20

PROP: Yellow liquid; cinnamon odor. D: 1.035-1.039, refr index: 1.602-1.607, flash p: 174°F. Sol in fixed oils, propylene glycol; insol in glycerin.

SYNS: FEMA No. 2697   ∗   METHYL CINNAMIC ALDEHYDE   ∗   α-METHYLCINNAMIC ALDEHYDE
∗   α-METHYLCINNIMAL   ∗   2-METHYL-3-PHENYL-2-PROPENAL

CONSENSUS REPORTS: Reported in EPA TSCA Inventory.

SAFETY PROFILE: Moderately toxic by ingestion. A skin irritant. Combustible liquid. When heated to decomposition it emits acrid smoke and irritating fumes.

MIO500   CAS: 103-26-4   **HR: 2**
METHYL CINNAMATE
mf: $C_{10}H_{10}O_2$ mw: 162.20

PROP: White to sltly yellow crystals; fruity odor. D: 1.042 @ 36/0°, mp: 33.4°, bp: 263°, flash p: +212°F. Very sol in alc, ether; sol in fixed oils, glycerin, propylene glycol; insol in water.

SYNS: FEMA No. 2698 * METHYL CINNAMYLATE * METHYL-3-PHENYLPROPENOATE * 3-PHENYL-2-PROPENOIC ACID METHYL ESTER (9CI)

CONSENSUS REPORTS: Reported in EPA TSCA Inventory.

SAFETY PROFILE: Moderately toxic by ingestion. Combustible liquid. When heated to decomposition it emits acrid smoke and irritating fumes.

MIP750   CAS: 92-48-8   HR: 2
6-METHYLCOUMARIN
mf: $C_{10}H_8O_2$ mw: 160.18

PROP: White needles from benzene; coconut odor. Mp: 73-76, flash p: +153°F. Sol in alc and benzene.

SYNS: FEMA No. 2690 * 6-MC * 6-METHYL-2H-1-BENZOPYRAN-2-ONE * 6-METHYLBENZOPYRONE * 6-METHYL-1,2-BENZOPYRONE * 6-METHYLCOU-MARINIC ANHYDRIDE * NCI-C55812 * TONCARINE

CONSENSUS REPORTS: EPA Genetic Toxicology Program. Reported in EPA TSCA Inventory.

SAFETY PROFILE: Poison by subcutaneous route. Moderately toxic by ingestion. A skin irritant. Mutation data reported. Combustible liquid. When heated to decomposition it emits acrid smoke and irritating fumes.

MIQ075   CAS: 137-05-3   HR: 3
METHYL 2-CYANOACRYLATE
mf: $C_9H_{13}NO_2$ mw: 111.11

SYNS: ADHERE * COAPT * α-CYANOACRYLATE ACID METHYL ESTER * 2-CYANOACRYLATE ACID METHYL ESTER * CYANOLYT * EASTMAN 910 * MECRLAT

OSHA PEL: TWA 2 ppm; STEL 4 ppm
ACGIH TLV: TWA 2 ppm; STEL 4 ppm
DFG MAK: 2 ppm (8 mg/m^3)

SAFETY PROFILE: Experimental reproductive effects. A human eye irritant. When heated to decomposition it emits toxic fumes of NO_x.

MIQ740   CAS: 108-87-2   HR: 2
METHYLCYCLOHEXANE
DOT: UN 2296
mf: C_7H_{14} mw: 98.21

PROP: Colorless liquid. Mp: −126.4°, lel: 1.2%, uel: 6.7%, bp: 100.3°, flash p: 25°F (CC), d: 0.7864 @ 0°/4°, 0.769 @ 20°/4°, vap press: 40 mm @ 22.0°, vap d: 3.39, autoign temp: 482°F.

SYNS: CYCLOHEXYLMETHANE * HEXAHYDROTO-LUENE * METYLOCYKLOHEKSAN (POLISH) * SEXTONE B * TOLUENE HEXAHYDRIDE

CONSENSUS REPORTS: Reported in EPA TSCA Inventory.

OSHA PEL: (Transitional: TWA 500 ppm) TWA 400 ppm
ACGIH TLV: TWA 400 ppm
DFG MAK: 500 ppm (2000 mg/m^3)
DOT Classification: Flammable Liquid; Label: Flammable Liquid.

SAFETY PROFILE: Moderately toxic by ingestion. Mildly toxic by inhalation. This material does not cause irritation to the eyes and nose, and even at the level of 500 ppm, exhibits only a very faint odor. Therefore, it cannot be said to have any warning properties. It is believed to be about three times as toxic as hexane, and has caused death by tetanic spasm in animals. In sublethal concentrations, it causes narcosis and anesthesia. Dangerous fire hazard and moderate explosion hazard when exposed to heat, flame or oxidizers. To fight fire, use foam, CO_2, dry chemical. When heated to decomposition it emits acrid smoke and fumes.

MIQ745   CAS: 25639-42-3   HR: 2
METHYLCYCLOHEXANOL
DOT: UN 2617
mf: $C_7H_{14}O$ mw: 114.21

PROP: Colorless, viscous liquid; aromatic, menthol-like odor. Bp: 155-180°, flash p: 154°F (CC), autoign temp: 565°F, d: 0.924 @ 15.5°/15.5°, vap d: 3.93.

SYNS: HEXAHYDROCRESOL * HEXAHYDROMETH-YLPHENOL * METYLOCYKLOHEKSANOL (POLISH)

OSHA PEL: (Transitional: TWA 100 ppm) TWA 50 ppm
ACGIH TLV: TWA 50 ppm
DFG MAK: 50 ppm (235 mg/m^3)

DOT Classification: Flammable or Combustible Liquid; Label: Flammable Liquid.

SAFETY PROFILE: Moderately toxic by ingestion and subcutaneous routes. Mildly toxic by skin contact. Human system effects by inhalation: antipsychotic, unspecified liver and kidney effects. Flammable when exposed to heat, flame or oxidizers. On heating it emits acrid fumes; can react with oxidizing materials. To fight fire, use alcohol foam, CO_2, dry chemical.

MIR250 CAS: 1331-22-2 HR: 2
METHYLCYCLOHEXANONE

DOT: UN 2297
mf: $C_7H_{12}O$ mw: 112.19

PROP: Water-white to pale yellow liquid, acetone-like odor. Mp: -14°C, bp: 160-170°, flash p: 118°F (CC), d: 0.925 @ 15°/5°, vap d: 3.86. Insol in water; sol in ether and alc.

SYN: METYLOCYKLOHEKSANON (POLISH)

CONSENSUS REPORTS: Reported in EPA TSCA Inventory.

DOT Classification: Flammable or Combustible Liquid; Label: Flammable Liquid.

SAFETY PROFILE: Moderately toxic by ingestion. Mildly toxic by skin contact. A toxic compound which can damage the kidneys and the liver. It is similar to cyclohexanol in its toxic action, although it is somewhat less active. Harmful exposure in industry is rare. Experimental animals can withstand prolonged exposures of 0.02-0.05% by volume in air. Flammable when exposed to heat or flame. Can react violently with HNO_3 and other oxidizers. To fight fire, use foam, CO_2, dry chemical. When heated to decomposition it emits acrid smoke and irritating fumes.

MIR500 CAS: 583-60-8 HR: 3
2-METHYLCYCLOHEXANONE
mf: $C_7H_{12}O$ mw: 112.19

PROP: Liquid. D: 0.925 @ 20/4°, mp: −14°, bp: 165.1°. Insol in water; sol in alc and ether.

SYNS: 2-METHYL-CYCLOHEXANON (GERMAN, DUTCH) * 1-METHYLCYCLOHEXAN-2-ONE * o-METHYLCYCLOHEXANONE * 2-METILCICLO-ESANONE (ITALIAN)

CONSENSUS REPORTS: Reported in EPA TSCA Inventory.

OSHA PEL: (Transitional: TWA 100 ppm (skin)) TWA 50 ppm; STEL 75 ppm (skin)
ACGIH TLV: TWA 50 ppm (skin)
DFG MAK: 50 ppm (230 mg/m^3)

SAFETY PROFILE: Poison by intravenous and intraperitoneal routes. Moderately toxic by ingestion and skin contact. When heated to decomposition it emits acrid smoke and irritating fumes.

MIS250 CAS: 2021-21-8 HR: D
N-METHYL-4-CYCLOHEXENE-1,2-DICARBOXIMIDE
mf: $C_9H_{11}NO_2$ mw: 165.21

SYN: N-METHYL-1,2,3,6-TETRAHYDROPHTHALIMIDE

SAFETY PROFILE: An experimental teratogen. When heated to decomposition it emits toxic fumes of NO_x.

MIU500 CAS: 96-37-7 HR: 2
METHYLCYCLOPENTANE

DOT: UN 2298
mf: C_6H_{12} mw: 84.18

PROP: Colorless liquid or solid. Mp: −142.5°, bp: 71.8°, flash p: <20°F, d: 0.750 @ 20°/4°, vap press: 100 mm @ 17.9°, vap d: 2.9. Insol in water; sol in ether.

CONSENSUS REPORTS: Reported in EPA TSCA Inventory.

DOT Classification: Flammable Liquid; Label: Flammable Liquid.

SAFETY PROFILE: Mildly toxic by inhalation. Probably irritating and narcotic in high concentration. Very dangerous fire hazard when exposed to heat, flame, or oxidizers. Can react vigorously with oxidizing materials. To fight fire, use foam, CO_2, dry chemical. When heated to decomposition it emits acrid smoke and irritating fumes.

MIW100 CAS: 8022-00-2 HR: 3
METHYL DEMETON

PROP: An oily liquid. D: 1.20. Sltly sol in water.

SYNS: BAY 15203 * BAYER 21/116 * DEMETON METHYL * DEMETHON-METHYL (MAK) * DURA-TOX * ENT 18,862 * S(and O)-2-(ETHYLTHIO) ETHYL-O,O-DIMETHYL PHOSPHOROTHIOATE * METASYSTOX * METHYL-MERCAPTOPHOS * METHYL SYSTOX

OSHA PEL: TWA 0.5 mg/m^3 (skin)
ACGIH TLV: TWA 0.5 mg/m^3 (skin)
DFG MAK: 0.5 ppm (5 mg/m^3)

SAFETY PROFILE: Poison by ingestion, skin contact, inhalation and possibly other routes. A cholinesterase inhibitor. An insecticide and acaricide.

MIW250   CAS: 2587-90-8   HR: 3
METHYL DEMETON METHYL
mf: C$_5$H$_{13}$O$_3$PS$_2$ mw: 216.27

PROP: Pale yellow oil. Bp: 89° @ 0.15 mm, d: 1.207 @ 20°/4°. Sol in water at room temp, and in organic solvents.

SYNS: CEBETOX * CYMETOX * DEMEPHION * ISONITOX * 2-(METHYLTHIO)-ETHANETHIOL-O,O-DIMETHYL PHOSPHOROTHIOATE * 2-(METHYLTHIO) ETHANETHIOL-S-ESTER with O,O-DIMETHYL PHOSPHO-ROTHIOATE * TINOX

CONSENSUS REPORTS: EPA Extremely Hazardous Substances List.

SAFETY PROFILE: Poison by ingestion and skin contact. Mutation data reported. *Caution:* It is a cholinesterase inhibitor. When heated to decomposition it emits very toxic fumes of PO$_x$ and SO$_x$.

MIW500   CAS: 477-30-5   HR: 3
N-METHYL-N-DESACETYLCOLCHICINE
mf: C$_{21}$H$_{25}$NO$_5$ mw: 371.47

SYNS: ALKALOID H 3, from COLCHICUM ANTUMNALE * CIBA 12669A * COLCEMIDE * COLCHAMINE * COLCHINE, N-DEACETYL-N-METHYL * COLEMID * DEACETYLMETHYLCOLCHICINE * DEACETYL-N-METHYLCOLCHICINE * N-DEACETYL-N-METHYLCOL-CHICINE * DEMECOLCINE * N-DESACETYL-N-METHYLCOLCHICINE * DESMECOLCINE * 6,7-DI-HYDRO-1,2,3,10-TETRAMETHOXY-7-(METHYL-AMINO)-BENZO(α)HEPTALEN-9(5H)-ONE * (S)-6,7-DI-HYDRO-1,2,3,10-TETRAMETHOXY-7-(METHYLAMINO) BENZO(a)HEPTALEN-9(5H)-ONE * KOLCHAMIN * METHYLCOLCHICINE * N-METHYL-N-DEACET-YLCOLCHICINE * NSC 3096 * OMAINE * REICH-STEIN'S F * SANTAVY'S SUBSTANCE F

CONSENSUS REPORTS: EPA Genetic Toxi-cology Program.

SAFETY PROFILE: Poison by ingestion, intra-peritoneal, parenteral, intravenous, and intra-muscular routes. Human systemic effects by in-gestion: (skin and appendages) hair effects.

Human mutation data reported. Experimental teratogenic and reproductive effects. When heated to decomposition it emits toxic fumes of NO$_x$.

MJE500   CAS: 892-17-1   HR: 3
11-METHYL-15,16-DIHYDRO-17-OXOCYCLOPENTA(a)PHENANTHRENE
mf: C$_{18}$H$_{14}$O   mw: 246.32

SYNS: 15,16-DIHYDRO-11-METHYLCYCLOPENTA(a) PHENANTHREN-17-ONE * 15,16-DIHYDRO-11-METHYL-17H-CYCLOPENTA(a)PHENANTHREN-17-ONE * 11-METHYL-15,16-DIHYDRO-17H-CYCLOPENTA(a) PHENANTHREN-17-ONE

CONSENSUS REPORTS: EPA Genetic Toxi-cology Program.

SAFETY PROFILE: Suspected carcinogen with experimental carcinogenic, neoplastigenic, and tumorigenic data. Mutation data reported. When heated to decomposition it emits acrid smoke and irritating fumes.

MJE750   CAS: 27156-32-7   HR: 2
METHYLDIHYDROPYRAN
mf: C$_6$H$_{10}$O   mw: 98.16

SYN: 3,4-DIHYDROMETHYL-2H-PYRAN

SAFETY PROFILE: Moderately toxic by inges-tion and subcutaneous routes. Mildly toxic by inhalation. When heated to decomposition it emits acrid smoke and irritating fumes.

MJG500   CAS: 2275-23-2   HR: 3
N-METHYL-O,O-DIMETHYLTHIOLO-PHOSPHORYL-5-THIA-3-METHYL-2-VALERAMIDE
mf: C$_8$H$_{18}$NO$_4$PS$_2$ mw: 287.36

SYNS: AMERICAN CYANAMID-43073 * O,O-DI-METHYL-S-2-(1-N-METHYLCARBAMOYLETHYLMER-CAPTO)ETHYL THIOPHOSPHATE * O,O-DIMETHYL-S-(2-(1-METHYLCARBAMOYLETHYLTHIO)ETHYL) PHOS-PHOROTHIOATE * DIMETHYL-S-(2-(1-METHYLCAR-BAMOYLETHYLTHIO ETHYL) PHOSPHOROTHIOLATE * ENT 26,613 * KILVAL * N-METHYL-3-THIA-2-METHYL-VALERAMID DER O,O-DIMETHYLTHIOLPHOS-PHORSAEURE (GERMAN) * NPH 83 * TRUCIDOR * VAMIDOATE * VAMIDOTHION

CONSENSUS REPORTS: EPA Genetic Toxi-cology Program.

SAFETY PROFILE: Poison by ingestion, skin contact, and possibly other routes. Mutation data reported. When heated to decomposition it emits very toxic fumes of PO_x, SO_x, and NO_x.

MJH250   CAS: 4386-79-2   HR: 3
((2-METHYL-1,3-DIOXALAN-4-YL) METHYL)TRIMETHYLAMMONIUM IODIDE

mf: $C_8H_{18}NO_2 \cdot I$ mw: 287.17

SYNS: ETHYL-γ-TRIMETHYLAMMONIUM PROPANE-DIOL IODIDE * FOURNEAU 2268 * N,N,N,2-TET-RAMETHYL-1,3-DIOXOLANE-4-METHANAMINIUM IODIDE (9CI) * TRIMETHYL((2-METHYL-1,3-DIOXALAN-4-YL)-METHYL)AMMONIUM IODIDE (8CI)

SAFETY PROFILE: Poison by ingestion, intravenous, and subcutaneous routes. When heated to decomposition it emits very toxic fumes of NH_3, NO_x, and I^-.

MJM200   CAS: 101-14-4   HR: 3
4,4'-METHYLENE BIS(2-CHLOROANILINE)

mf: $C_{13}H_{12}Cl_2N_2$ mw: 267.17

SYNS: BIS AMINE * CURALIN M * CURENE 442 * CYANASET * DI-(4-AMINO-3-CHLOROPHENYL) METHANE * DI-(4-AMINO-3-CLOROFENIL)METANO (ITALIAN) * 4,4'-DIAMINO-3,3'-DICHLORODIPHENYL-METHANE * 3,3'-DICHLOR-4,4'-DIAMINODIPHENYL-METHAN (GERMAN) * 3,3'-DICHLORO-4,4'-DIAMINO-DIPHENYLMETHANE * 3,3'-DICLORO-4,4'-DIAMINO-DIFENILMETANO (ITALIAN) * MBOCA * 4,4'-METHYLENE(BIS)-CHLOROANILINE * METH-YLENE-4,4'-BIS(o-CHLOROANILINE) * p,p'-METHYL-ENEBIS(α-CHLOROANILINE) * 4,4'-METHYLENEBIS (o-CHLOROANILINE) * p,p'-METHYLENEBIS(o-CHLO-ROANILINE) * 4,4'-METHYLENEBIS-2-CHLORO-BENZENAMINE * METHYLENE-BIS-ORTHOCHLO-ROANILINE * 4,4-METILENE-BIS-o-CLOROANILINA (ITALIAN) * MOCA * RCRA WASTE NUMBER U158

CONSENSUS REPORTS: IARC Cancer Review: GROUP 2A IMEMDT 7,246,87; Animal Sufficient Evidence IMEMDT 4,65,74. NTP Fourth Annual Report On Carcinogens, 1984. EPA Genetic Toxicology Program. Community Right-To-Know List. Reported in EPA TSCA Inventory.

OSHA PEL: TWA 0.02 ppm (skin)
ACGIH TLV: TWA 0.02 ppm (skin); Suspected Human Carcinogen.
DFG MAK: Animal Carcinogen, Suspected Human Carcinogen.
NIOSH REL: (MOCA) Lowest detectable limit.

SAFETY PROFILE: Confirmed carcinogen with experimental carcinogenic and tumorigenic data. Poison by intraperitoneal route. Moderately toxic by ingestion. Mutation data reported. When heated to decomposition it emits very toxic fumes of Cl^- and NO_x.

MJM500   CAS: 97-23-4   HR: 3
2,2'-METHYLENEBIS(4-CHLOROPHENOL)

mf: $C_{13}H_{10}Cl_2O_2$ mw: 269.13

PROP: Crystals, nearly insol in water. Mp: 178°, vap press: 10^{-4} mm @ 100°.

SYNS: ANTIPHEN * BIS(5-CHLOR-2-HYDROXYPHE-NYL)-METHAN (GERMAN) * BIS(5-CHLORO-2-HY-DROXYPHENYL)METHANE * BIS-2-HYDROXY-5-CHLORFENYLMETHAN (CZECH) * BIS(2-HYDROXY-5-CHLOROPHENYL)METHANE * DICESTAL * DICHLOORFEEN (DUTCH) * 5,5'-DICHLORO-2,2'-DIHYDROXYDIPHENYLMETHANE * DI-(5-CHLORO-2-HYDROXYPHENYL)METHANE * 2,2'-DIHYDROXY-5,5'-DICHLORODIPHENYLMETHANE * HYOSAN * KORIUM * O,O-METHYLEEN-BIS(4-CHLOORFE-NOL) (DUTCH) * O,O-METILEN-BIS(4-CLOROFENOLO) (ITALIAN) * PANACIDE * PREVENTOL * TENIA-THANE * WESPURIL

CONSENSUS REPORTS: Chlorophenol compounds are on the Community Right-To-Know List. Reported in EPA TSCA Inventory.

SAFETY PROFILE: Poison by intravenous route. Moderately toxic by ingestion. A skin and severe eye irritant. Mutation data reported. Can cause cramps and diarrhea. Possibly similar to DDT. An FDA over-the-counter drug. An anthelmintic. When heated to decomposition it emits toxic fumes of Cl^-.

MJM600   CAS: 5124-30-1   HR: 3
METHYLENE BIS(4-CYCLOHEXYLISOCYANATE

mf: $C_{15}H_{22}NO_2$ mw: 262.39

SYNS: BIS(4-ISOCYANATOCYCLOHEXYL)METHANE * NACCONATE H 12

CONSENSUS REPORTS: Reported in EPA TSCA Inventory.

OSHA PEL: CL 0.01
ACGIH TLV: TWA 0.005 ppm

SAFETY PROFILE: Poison by inhalation. Mildly toxic by ingestion. When heated to decomposition it emits very toxic fumes of NO_x.

MJN000   CAS: 101-61-1   **HR: 3**
4,4′-METHYLENE BIS(N,N′-DIMETHYLANILINE)
mf: $C_{17}H_{22}N_2$ mw: 254.41

SYNS: p,p′-BIS(DIMETHYLAMINO)DIPHENYLMETHANE
* 4,4′-BIS(DIMETHYLAMINO)DIPHENYLMETHANE
* BIS(p-DIMETHYLAMINOPHENYL)METHANE
* BIS(p-(N,N-DIMETHYLAMINO)PHENYL)METHANE
* p,p′-BIS(N,N-DIMETHYLAMINOPHENYL)METHANE
* p,p-DIMETHYLAMINODIPHENYLMETHANE
* METHANE BASE * 4,4′-METHYLENEBIS(N,N-DI-METHYL)BENZENAMINE * MICHLER'S BASE
* MICHLER'S HYDRIDE * MICHLER'S METHANE
* NCI-C01990 * TETRA-BASE * TETRAMETHYL-DIAMINODIPHENYLMETHANE * 4,4′-TETRA-METHYLDIAMINODIPHENYLMETHANE * p,p-TETRA-METHYLDIAMINODIPHENYLMETHANE

CONSENSUS REPORTS: IARC Cancer Review: GROUP 3 IMEMDT 7,56,87; Animal Limited Evidence IMEMDT 27,119,82. NTP Fourth Annual Report On Carcinogens, 1984. NCI Carcinogenesis Bioassay (feed); Clear Evidence: mouse, rat NCITR* NCI-CG-TR-186,79. EPA Genetic Toxicology Program. Reported in EPA TSCA Inventory. Community Right-To-Know List.

DFG MAK: Suspected Carcinogen.

SAFETY PROFILE: Confirmed carcinogen with experimental carcinogenic, neoplastigenic, and tumorigenic data. Moderately toxic by ingestion. Mutation data reported. When heated to decomposition it emits toxic fumes of NO_x.

MJN750   CAS: 139-25-3   **HR: 3**
5,5′-METHYLENEBIS(2-ISOCYANATO)TOLUENE
mf: $C_{17}H_{14}N_2O_2$ mw: 278.33

SYNS: 3,3′-DIMETHYLDIPHENYLMETHANE-4,4′-DIISO-CYANATE * ISOCYANIC ACID, ESTER with DI-o-TO-LUENEMETHANE

CONSENSUS REPORTS: Reported in EPA TSCA Inventory.

NIOSH REL: (Diisocyanates) TWA 0.005 ppm; CL 0.02 ppm/10M

SAFETY PROFILE: Poison by intravenous route. When heated to decomposition it emits toxic fumes of NO_x.

MJO250   CAS: 838-88-0   **HR: 3**
4,4′-METHYLENEBIS(2-METHYLANILINE)
mf: $C_{15}H_{18}N_2$ mw: 226.35

PROP: Mp: 149°.

SYNS: BIS-4-AMINO-3-METHYLFENYLMETHAN (CZECH) * 3,3′-DIMETHYL-4,4′-DIAMINODIPHENYL-METHANE * MBOT * ME-MDA * 4,4′-METHYL-ENEBIS(2-METHYLBENZENAMINE) * 4,4′-METHYLENE DI-o-TOLUIDINE

CONSENSUS REPORTS: IARC Cancer Review: GROUP 2B IMEMDT 7,248,87; Animal Limited Evidence IMEMDT 4,73,74. Reported in EPA TSCA Inventory.

DFG MAK: Animal Carcinogen; Suspected Human Carcinogen.

SAFETY PROFILE: Confirmed carcinogen with experimental carcinogenic data. Moderately toxic by ingestion. An eye irritant. Mutation data reported. When heated to decomposition it emits toxic fumes of NO_x.

MJP400   CAS: 101-68-8   **HR: 3**
METHYLENE BISPHENYL ISOCYANATE
DOT: UN 2489
mf: $C_{15}H_{10}N_2O_2$ mw: 250.27

PROP: Crystals or yellow fused solid. Mp: 37.2°, bp: 194-199° @ 5 mm, d: 1.19 @ 50°, vap press: 0.001 mm @ 40°.

SYNS: BIS(p-ISOCYANATOPHENYL)METHANE
* BIS(1,4-ISOCYANATOPHENYL)METHANE * BIS(4-ISOCYANATOPHENYL)METHANE * CARADATE 30
* DESMODUR 44 * DIFENIL-METAN-DIISOCIANATO (ITALIAN) * DIFENYLMETHAAN-DISSOCYANAAT (DUTCH) * 4-4′-DIISOCYANATE de DIPHENYLMETH-ANE (FRENCH) * 4,4′-DIISOCYANATODIPHENYL-METHANE * DIPHENYLMETHAN-4,4′-DIISOCYANAT (GERMAN) * DIPHENYL METHANE DIISOCYANATE
* p,p′-DIPHENYLMETHANE DIISOCYANATE
* 4,4′-DIPHENYLMETHANE DIISOCYANATE
* DIPHENYLMETHANE 4,4′-DIISOCYANATE (DOT)
* HYLENE M50 * ISONATE * MDI
* METHYLENEBIS(4-ISOCYANATOBENZENE)
* 1,1-METHYLENEBIS(4-ISOCYANATOBENZENE)
* METHYLENEBIS(4-PHENYLENE ISOCYANATE)
* METHYLENEBIS(p-PHENYLENE ISOCYANATE)
* p,p′-METHYLENEBIS(PHENYL ISOCYANATE)
* METHYLENEBIS(p-PHENYL ISOCYANATE)
* METHYLENEBIS(4-PHENYL ISOCYANATE)
* 4,4′-METHYLENEBIS(PHENYL ISOCYANATE)
* 4,4′-METHYLENEDIPHENYL DIISOCYANATE
* METHYLENEDI-p-PHENYLENE DIISOCYANATE
* METHYLENEDI-p-PHENYLENE ISOCYANATE
* 4,4′-METHYLENEDIPHENYL ISOCYANATE
* 4,4′-METHYLENEDIPHENYLENE ISOCYANATE

* METHYLENE DI(PHENYLENE ISOCYANATE) (DOT)
* NACCONATE 300 * NCI-C50668 * RUBI-
NATE 44

CONSENSUS REPORTS: IARC Cancer Review: GROUP 3 IMEMDT 7,56,87. Reported in EPA TSCA Inventory. Community Right-To-Know List.

OSHA PEL: CL 0.02 ppm
ACGIH TLV: 0.005 ppm
DFG MAK: 0.01 ppm (0.1 mg/m^3)
NIOSH REL: (Diisocyanates) TWA 0.005 ppm; CL 0.02 ppm/10M
DOT Classification: Poison B; Label: St. Andrews Cross.

SAFETY PROFILE: Poison by inhalation. Mildly toxic by ingestion. Human systemic effects by inhalation: increased immune response and body temperature. A skin and eye irritant. An allergic sensitizer. Questionable carcinogen. Mutation data reported. When heated to decomposition it emits toxic fumes of NO$_x$ and SO$_x$.

MJP450   CAS: 75-09-2   HR: 3
METHYLENE CHLORIDE

DOT: UN 1593
mf: CH$_2$Cl$_2$ mw: 84.93

PROP: Colorless, volatile liquid; odor of chloroform. Bp: 39.8°, lel: 15.5% in O$_2$, uel: 66.4% in O$_2$, fp: −96.7°, d: 1.326 @ 20°/4°, autoign temp: 1139°F, vap press: 380 mm @ 22°, vap d: 2.93, refr index: 1.424 @ 20L. Sol in water; misc with alc, acetone, chloroform, ether, and carbon tetrachloride.

SYNS: AEROTHENE MM * CHLORURE de METH-
YLENE (FRENCH) * DCM * DICHLOROMETHANE
(MAK, DOT) * FREON 30 * METHANE DICHLORIDE
* METHYLENE BICHLORIDE * METHYLENE DI-
CHLORIDE * METYLENU CHLOREK (POLISH)
* NCI-C50102 * RCRA WASTE NUMBER U080
* SOLMETHINE

CONSENSUS REPORTS: IARC Cancer Review: GROUP 2B IMEMDT 7,194,87; Human Inadequate Evidence IMEMDT 41,43,86; Animal Sufficient Evidence IMEMDT 41,43,86; Animal Inadequate Evidence IMEMDT 20,-449,79. NTP Carcinogenesis Studies (inhalation); Clear Evidence: mouse, rat NTPTR* NTP-TR-306,86. Reported in EPA TSCA Inventory. EPA Genetic Toxicology Program. Community Right-To-Know List.

OSHA PEL: (Transitional: TWA 500 ppm; CL 1000 ppm; Pk 2000/5M/2H)
ACGIH TLV: TWA 50 ppm, Suspected Human Carcinogen
DFG MAK: 100 ppm (360 mg/m^3); BAT: 5% CO-Hb in blood at end of shift; Suspected Carcinogen.
NIOSH REL: (Methylene Chloride) Reduce to lowest feasible level.
DOT Classification: Poison B; Label: St. Andrews Cross.

SAFETY PROFILE: Suspected carcinogen with experimental carcinogenic and tumorigenic data. Poison by intravenous route. Moderately toxic by ingestion, subcutaneous, and intraperitoneal routes. Mildly toxic by inhalation. Human systemic effects by ingestion and inhalation: paresthesia, somnolence, altered sleep time, convulsions, euphoria, and change in cardiac rate. Experimental teratogenic and reproductive effects. An eye and severe skin irritant. Human mutation data reported. It is flammable in the range of 12-19 percent in air but ignition is difficult. It will not form explosive mixtures with air at ordinary temperatures. Mixtures in air with methanol vapor are flammable. It will form explosive mixtures with an atmosphere having a high oxygen content, in liquid O$_2$, N$_2$O$_4$, K, Na, NaK. Explosive in the form of vapor when exposed to heat or flame. Reacts violently with Li, NaK, potassium-tert-butoxide, (KOH + n-methyl-n-nitrosourea). It can be decomposed by contact with hot surfaces and open flame, and then yield toxic fumes which are irritating and give warning of their presence. When heated to decomposition it emits highly toxic fumes of phosgene and Cl$^-$.

MJP750   CAS: 1208-52-2   HR: 2
2,4'-METHYLENEDIANILINE
mf: C$_{13}$H$_{14}$N$_2$ mw: 198.29

SYNS: 2',4-BIS(AMINOPHENYL)METHANE * 2,4'-DI-
AMINODIPHENYLMETHAN (GERMAN) * o,p'-DI-
AMINODIPHENYLMETHANE * 2,4'-DIAMINO-
DIPHENYLMETHANE * 2,4'-DIPHENYLMETHANE-
DIAMINE * 2,4'-METHYLENEBIS(ANILINE)

CONSENSUS REPORTS: Reported in EPA TSCA Inventory.

DFG MAK: Animal Carcinogen, Suspected Human Carcinogen.

SAFETY PROFILE: Moderately toxic by subcutaneous route. When heated to decomposition it emits toxic fumes of NO_x.

MJQ000   CAS: 101-77-9   HR: 3
4,4'-METHYLENEDIANILINE

DOT: UN 2651

mf: $C_{13}H_{14}N_2$ mw: 198.29

PROP: Tan flakes or lumps; faint amine-like odor. Mp: 90°, flash p: 440°F.

SYNS: 4-(4-AMINOBENZYL)ANILINE * BIS-p-AMINOFENYLMETHAN (CZECH) * BIS(p-AMINOPHENYL) METHANE * BIS(4-AMINOPHENYL)METHANE * CURITHANE * DDM * p,p'-DIAMINODIFENYLMETHAN (CZECH) * 4,4'-DIAMINODIPHENYLMETHAN (GERMAN) * DIAMINODIPHENYLMETHANE * p,p'-DIAMINODIPHENYLMETHANE * 4,4'-DIAMINODIPHENYLMETHANE * DI-(4-AMINOPHENYL) METHANE * DIANALINEMETHANE * 4,4'-DIPHENYLMETHANEDIAMINE * EPICURE DDM * MDA * METHYLENEBIS(ANILINE) * 4,4'-METHYLENEBISANILINE * METHYLENEDIANILINE * p,p'-METHYLENEDIANILINE * NCI-C54604 * TONOX

CONSENSUS REPORTS: IARC Cancer Review: GROUP 2B IMEMDT 7,56,87; Animal Sufficient Evidence IMEMDT 39,347,86; Animal Inadequate Evidence IMEMDT 4,79,74. NTP Fourth Annual Report On Carcinogens, 1984. Community Right-To-Know List. Reported in EPA TSCA Inventory.

ACGIH TLV: TWA 0.1 ppm (skin); Suspected Human Carcinogen.
DFG MAK: Animal Carcinogen, Suspected Human Carcinogen.
DOT Classification: Poison B; Label: St. Andrews Cross.

SAFETY PROFILE: Confirmed carcinogen with experimental tumorigenic data. Poison by ingestion, subcutaneous, and intraperitoneal routes. Human systemic effects by ingestion: rigidity, jaundice, other liver changes. An eye irritant. Mutation data reported. It is not rapidly absorbed through the skin. Combustible when exposed to heat or flame. When heated to decomposition it emits highly toxic fumes of aniline and NO_x.

MJQ100   CAS: 13552-44-8   HR: 3
4,4'-METHYLENEDIANILINE
DIHYDROCHLORIDE

mf: $C_{13}H_{14}N_2 \cdot 2ClH$ mw: 271.21

SYN: NCI-C54604

CONSENSUS REPORTS: IARC Cancer Review: Animal Sufficient Evidence IMEMDT 39,347,86. NTP Fourth Annual Report On Carcinogens, 1984. NTP Carcinogenesis Studies (oral); Clear Evidence: mouse, rat NTPTR* NTP-TR-248,83. Reported in EPA TSCA Inventory.

SAFETY PROFILE: Confirmed carcinogen with experimental carcinogenic and neoplastigenic data. When heated to decomposition it emits toxic fumes of NO_x and HCl.

MJQ500   CAS: 156-72-9   HR: 3
METHYLENE DIMETHANESULFONATE

mf: $C_3H_8O_6S_2$ mw: 204.23

SYNS: ENT 51,799 * METHANESULFONIC ACID, METHYLENE ESTER * METHYLENE BIS(METHANESULFONATE)

CONSENSUS REPORTS: EPA Genetic Toxicology Program.

SAFETY PROFILE: Poison by intraperitoneal route. Mutation data reported. When heated to decomposition it emits toxic fumes of SO_x.

MJQ750   CAS: 5625-90-1   HR: 3
4,4'-METHYLENEDIMORPHOLINE

mf: $C_9H_{18}N_2O_2$ mw: 186.29

SYNS: BIS(MORPHOLINO-)METHAN (GERMAN) * BISMORPHOLINO METHANE

SAFETY PROFILE: Moderately toxic by subcutaneous route. Questionable carcinogen with experimental tumorigenic data. When heated to decomposition it emits toxic fumes of NO_x.

MJR750   CAS: 42542-07-4   HR: 3
3,4-METHYLENEDIOXY-α-ETHYL-β-
PHENYLETHYLAMINE

mf: $C_{11}H_{15}NO_2 \cdot ClH$ mw: 229.73

SAFETY PROFILE: Poison by intravenous and intraperitoneal routes. When heated to decomposition it emits very toxic fumes of HCl and NO_x.

MJS750   CAS: 6292-91-7   HR: 3
1-(3,4-METHYLENEDIOXYPHENYL)-2-
AMINOPROPANE

mf: $C_{10}H_{13}NO_2 \cdot ClH$ mw: 215.70

SYNS: 3,4-METHYLENEDIOXY-α-METHYL-β-PHENYLETHYLAMINE HYDROCHLORIDE * α-METHYL-3,4-

METHYLENEDIOXYPHENETHYLAMINE HYDROCHLO-
RIDE

SAFETY PROFILE: Poison by intravenous and intraperitoneal routes. Mutagenic data reported. When heated to decomposition it emits very toxic fumes of HCl and NO_x.

MJT000   CAS: 1653-64-1   **HR: 3**
3,4-METHYLENEDIOXY-β-PHENYL-ETHYLAMINE HYDROCHLORIDE
mf: $C_9H_{11}NO_2 \cdot ClH$ mw: 201.67

SAFETY PROFILE: Poison by intravenous and intraperitoneal routes. When heated to decomposition it emits very toxic fumes of HCl and NO_x.

MJT500   CAS: 6317-18-6   **HR: 3**
METHYLENE DITHIOCYANATE
mf: $C_3H_2N_2S_2$ mw: 130.19

SYN: METHYLENDIRHODANID (CZECH, GERMAN)

CONSENSUS REPORTS: Reported in EPA TSCA Inventory.

SAFETY PROFILE: Poison by ingestion, intravenous, and subcutaneous routes. When heated to decomposition it emits very toxic fumes of NO_x and SO_x.

MJU250   CAS: 2679-01-8   **HR: 3**
METHYLENE GREEN
mf: $C_{17}H_{17}N_4O_2S \cdot Cl$ mw: 364.88

SYN: 3,7-BIS(DIMETHYLAMINO)-4-NITRO-PHENOTHI-AZIN-5-IUM, CHLORIDE

SAFETY PROFILE: Poison by intravenous and intraperitoneal routes. Moderately toxic by ingestion. When heated to decomposition it emits very toxic fumes of Cl^-, SO_x, and NO_x.

MJU750   CAS: 17605-71-9   **HR: 3**
METHYLEPHEDRINE
mf: $C_{11}H_{17}NO$ mw: 179.25

PROP: dl Form: Crystals from petr ether or methanol. Mp: 63.5-64.5°, very sol in usual solvents. d form: Crystals. Mp: 87-87.5°. l Form: Crystals from petr ether. Mp: 87-88°.

SYNS: METHYLEPHEDRIN (GERMAN) * 1-PHENYL-2-DIMETHYLAMINOPROPANOL

SAFETY PROFILE: A poison by intraperitoneal route. Moderately toxic by ingestion. When

heated to decomposition it emits toxic fumes of NO_x.

MJV000   CAS: 554-99-4   **HR: 3**
N-METHYLEPINEPHRINE
mf: $C_{10}H_{15}NO_3$ mw: 197.23

SYNS: 3,4-DIHYDROXY-α-(DIMETHYLAMINOMETHYL) BENZYL ALCOHOL * α-(3,4-DIHYDROXYPHENYL)-β-DIMETHYLAMINOETHANOL * α-(3,4-DIHYDROXYPHE-NYL)-α-HYDROXY-β-DIMETHYLAMINOETHANE * α-(DIMETHYLAMINOMETHYL)PROTOCATECHUYL ALCOHOL * METHADRENE * N-METHYLADRENA-LINE

SAFETY PROFILE: Poison by intravenous, intraperitoneal, and subcutaneous routes. When heated to decomposition it emits toxic fumes of NO_x.

MJW000   CAS: 112-61-8   **HR: 3**
METHYL ESTER STEARIC ACID
mf: $C_{19}H_{38}O_2$ mw: 298.57

PROP: Liquid to semi-solid. Mp: 38°, bp: 215° @ 15 mm, flash p: 307°F (CC), d: 0.860. Sol in water and ether.

SYNS: EMERY 2218 * METHOLENE 2218 * METHYL OCTADECANOATE * METHYL STEA-RATE * OCTADECANOIC ACID, METHYL ESTER

CONSENSUS REPORTS: Reported in EPA TSCA Inventory.

SAFETY PROFILE: Questionable carcinogen with experimental tumorigenic data. Combustible when exposed to heat or flame; can react with oxidizing materials. To fight fire, use CO_2, dry chemical. When heated to decomposition it emits acrid smoke and irritating fumes.

MJW500   CAS: 115-10-6   **HR: 3**
METHYL ETHER
DOT: UN 1033
mf: C_2H_6O mw: 46.08

PROP: Colorless gas, ether odor. Mp: −138.5°, bp: −23.7°, lel: 3.4%, uel: 27%, flash p: −42°F (CC), autoign temp: 662°F, vap d: 1.617, d: 0.661 (air = 1). Sol in alc, water, ether.

SYNS: DIMETHYL ETHER (DOT) * OXYBISMETH-ANE * WOOD ETHER

CONSENSUS REPORTS: Reported in EPA TSCA Inventory.

DOT Classification: Flammable Gas; Label: Flammable Gas.

SAFETY PROFILE: Moderately toxic by inhalation. Very dangerous fire hazard when exposed to heat, flame, or oxidizers. Dangerous explosion hazard when exposed to flame, sparks, etc. Violent reaction with AlH_3, $LiAlH_2$. Keep in closed container away from heat and open flame. To fight fire, stop flow of gas. When heated to decomposition it emits acrid smoke and irritating fumes.

MJY500   CAS: 25057-89-0   **HR: 2**
3-(1-METHYLETHYL)-1H-2,1,3-BENZO-THIAZAIN-4(3H)-ONE-2,2-DIOXIDE
mf: $C_{10}H_{12}N_2O_3S$ mw: 240.30

SYNS: BAS 351-H * BASAGRAN * BENDIOXIDE * BENTAZON * 3-ISOPROPYL-2,1,3-BENZOTHIA-DIAZINON-(4)-2,2-DIOXID (GERMAN) * 3-ISOPROPYL-1H-2,1,3-BENZOTHIADIAZIN-4(3H)-ONE-2,2-DIOXIDE

CONSENSUS REPORTS: Reported in EPA TSCA Inventory. EPA Genetic Toxicology Program.

SAFETY PROFILE: Moderately toxic by ingestion and skin contact. When heated to decomposition it emits very toxic fumes of SO_x and NO_x.

MKA000   CAS: 31218-83-4   **HR: 3**
(E)-1-METHYLETHYL-3-(((ETHYLAMINO) METHOXYPHOSPHINOTHIOYL) OXY-2-BUTENOATE
mf: $C_{10}H_{20}NO_4PS$ mw: 281.34

SYNS: BLOTIC * ENT 27,989 * (3)-O-2-ISOPRO-POXY-CARBONYL-1-METHYLVINYL-O-METHYL ETHYL-PHOSPHORAMIDOTHIOATE * PROPETAMPHOS * SAFROTIN * SAN 52 139 I * SANDOZ 52139 * VEL 4283

SAFETY PROFILE: Poison by ingestion. Moderately toxic by skin contact. When heated to decomposition it emits very toxic fumes of PO_x, SO_x, and NO_x.

MKA250   CAS: 64-65-3   **HR: 3**
3-METHYL-3-ETHYLGLUTARIMIDE
mf: $C_8H_{13}NO_2$ mw: 155.22

SYNS: AHYPNON * BEMEGRIDE * 2,6-DIOXO-4-METHYL-4-ETHYLPIPERIDINE * 4-ETHYL-4-METHYL-2,6-DIOXOPIPERIDINE * β-ETHYL-β-METHYLGLUTAR-IMIDE * 3-ETHYL-3-METHYLGLUTARIMIDE * 4-ETHYL-4-METHYL-2,6-PIPERIDINEDIONE * EUKRATON * MALYSOL * MEGIMIDE

* 4-METHYL-4-ETHYL-2,6-DIOXOPIPERIDINE
* β-METHYL-β-ETHYLGLUTARIMIDE * MIKEDIMIDE

CONSENSUS REPORTS: Reported in EPA TSCA Inventory.

SAFETY PROFILE: Poison by ingestion, intravenous, intraperitoneal, intramuscular, subcutaneous, and parenteral routes. Human systemic effects by ingestion: wakefulness, hallucinations, distorted perceptions, toxic psychosis. An analeptic, central nervous system stimulant; used to counteract barbiturate poisoning. When heated to decomposition it emits toxic fumes of NO_x.

MKA400   CAS: 78-93-3   **HR: 3**
METHYL ETHYL KETONE

DOT: UN 1193/UN 1232
mf: C_4H_8O mw: 72.12

PROP: Colorless liquid; acetone-like odor. Bp: 79.57°, fp: −85.9°, lel: 1.8%, uel: 11.5%, flash p: 22°F (TOC), d: 0.80615 @ 20°/20°, vap press: 71.2 mm @ 20°, autoign temp: 960°F, vap d: 2.42, ULC: 85-90. Misc with alc, ether, fixed oils, and water.

SYNS: AETHYLMETHYLKETON (GERMAN) * 2-BUTANONE (OSHA) * BUTANONE 2 (FRENCH) * ETHYL METHYL CETONE (FRENCH) * ETHYL-METHYLKETON (DUTCH) * ETHYL METHYL KETONE (DOT) * FEMA No. 2170 * MEK * METHYL ACE-TONE (DOT) * METILETILCHETONE (ITALIAN) * METYLOETYLOKETON (POLISH) * RCRA WASTE NUMBER U159

CONSENSUS REPORTS: Community Right-To-Know List. EPA Genetic Toxicology Program. Reported in EPA TSCA Inventory.

OSHA PEL: (Transitional: TWA 200 ppm) TWA 200 ppm; STEL 300 ppm
ACGIH TLV: TWA 200 ppm; STEL 300 ppm; BEI: 2 mg(MEK)/L in urine at end off shift.
DFG MAK: 200 ppm (590 mg/m³)
NIOSH REL: (Ketones) TWA 590 mg/m³
DOT Classification: Flammable Liquid; Label: Flammable Liquid.

SAFETY PROFILE: Moderately toxic by ingestion, skin contact, and intraperitoneal routes. Human systemic effects by inhalation: conjunctiva irritation and unspecified effects on the nose and respiratory system. A strong irritant. Human eye irritation @ 350 ppm. Affects peripheral nervous system and central nervous system.

Highly flammable liquid. Reaction with hydrogen peroxide + nitric acid forms a heat- and shock-sensitive explosive product. Ignition on contact with potassium tert-butoxide. Mixture with 2-propanol will produce explosive peroxides during storage. Vigorous reaction with chloroform + alkali. Incompatible with chlorosulfonic acid, oleum. To fight fire, use alcohol foam, CO_2, dry chemical. When heated to decomposition it emits acrid smoke and fumes.

MKA500   CAS: 1338-23-4   **HR: 3**
METHYL ETHYL KETONE PEROXIDE
DOT: UN 2127
mf: $C_8H_{16}O_4$ mw: 176.24
SYNS: HI-POINT 90 * LUPERSOL * MEKP * MEK PEROXIDE * METHYLETHYLKETONHYDRO-PEROXIDE * NCI-C55447 * QUICKSET EXTRA * RCRA WASTE NUMBER U160 * SPRAYSET MEKP * THERMACURE

CONSENSUS REPORTS: Reported in EPA TSCA Inventory.

OSHA PEL: CL 0.7 ppm
ACGIH TLV: CL 0.2 ppm
DFG MAK: Organic Peroxide, moderate skin irritant.
DOT Classification: Forbidden.

SAFETY PROFILE: Poison by intraperitoneal route. Moderately toxic by ingestion and inhalation. Human systemic effects by ingestion: changes in structure or function of esophagus, nausea or vomiting, other gastrointestinal effects. A moderate skin and eye irritant. Questionable carcinogen with experimental tumorigenic data. A shock-sensitive explosive. When heated to decomposition it emits acrid smoke and irritating fumes.

MKA750   CAS: 624-46-4   **HR: 3**
METHYL ETHYL KETONE SEMICARBAZONE
mf: $C_5H_{11}N_3O$ mw: 129.19
SYN: 2-BUTANONE, SEMICARBAZONE

SAFETY PROFILE: Poison by intravenous and intraperitoneal routes. When heated to decomposition it emits toxic fumes of NO_x.

MKB000   CAS: 10595-95-6   **HR: 3**
N,N-METHYLETHYLNITROSAMINE
mf: $C_3H_8N_2O$ mw: 88.13
SYNS: ETHYLMETHYLNITROSAMINE * METHYLA-ETHYLNITROSAMIN (GERMAN) * METHYLETHYLNI-

TROSAMINE * N-METHYL-N-NITROSO-ETHAMINE * N-METHYL-N-NITROSOETHYLAMINE * NEMA * N-NITROSOETHYLMETHYLAMINE * N-NITROSO-METHYLETHYLAMINE (MAK) * NMEA

CONSENSUS REPORTS: IARC Cancer Review: GROUP 2B IMEMDT 7,56,87; Animal Limited Evidence IMEMDT 17,221,78. EPA Genetic Toxicology Program.

DFG MAK: Animal Carcinogen, Suspected Human Carcinogen.

SAFETY PROFILE: Confirmed carcinogen with experimental carcinogenic and tumorigenic data. Poison by ingestion. Mutation data reported. When heated to decomposition it emits toxic fumes of NO_x.

MKB250   CAS: 50-12-4   **HR: 3**
3-METHYL-5-ETHYL-5-PHENYLHYDANTOIN
mf: $C_{12}H_{14}N_2O_2$ mw: 218.28
SYNS: EPILAN * 5-ETHYL-3-METHYL-5-PHENYLHY-DANTOIN * 5-ETHYL-3-METHYL-5-PHENYL-2,4(3H,5H)-IMIDAZOLEDIONE * 5-ETHYL-3-METHYL-5-PHENYLI-MIDAZOLIDIN-2,4-DIONE * 3-ETHYLNIRVANOL * GEROT-EPILAN * INSULTON * MEPHENYTOIN * MESANTOIN * METHOIN * METHYL HYDAN-TOIN * 3-METHYL-5,5-PHENYLETHYLHYDANTOIN * NSC-34652 * PHENANTOIN * PHENYLETHYL-METHYLHYDANTOIN * SACERNO * SEDANTOI-NAL * TRIANTOIN

SAFETY PROFILE: Poison by ingestion and intraperitoneal routes. Human systemic effects by ingestion: somnolence, hemorrhage, changes in teeth and supporting structures. Human mutation data reported. An experimental teratogen. An FDA proprietary drug used as an anticonvulsant. When heated to decomposition it emits toxic fumes of NO_x.

MKB750   CAS: 72-33-3   **HR: 3**
3-METHYLETHYNYLESTRADIOL
mf: $C_{21}H_{26}O_2$ mw: 310.47
SYNS: COMPOUND 33355 * DELTA-MVE * 17-α-ETHINYL ESTRADIOL 3-METHYL ETHER * ETHINYLESTRADIOL-3-METHYL ETHER * 17-α-ETHINYL OESTRADIOL-3-METHYL ETHER * ETHINYLOESTRADIOL-3-METHYL ETHER * ETHYNYLESTRADIOL-3-METHYL ETHER * 17-ETHYNYLESTRADIOL-3-METHYL ETHER * 17-α-ETHYNYLESTRADIOL-3-METHYL ETHER * (+)-17-α-ETHYNYL-17-β-HYDROXY-3-METHOXY-1,3,5(10)-ESTRATRIENE * (+)-17-α-ETHYNYL-17-β-HY-

DROXY-3-METHOXY-1,3,5(10)-OESTRATRIENE
* 17-α-ETHYNYL-3-METHOXY-1,3,5(10)-ESTRATRIEN-17-β-OL * 17-ETHYNYL-3-METHOXY-1,3,5(10)-ESTRA-TRIEN-17-β-OL * 17-α-ETHYNYL-3-METHOXY-17-β-HY-DROXY-Δ-1,3,5(10)-ESTRATRIENE * 17-α-ETHYNYL-3-METHOXY-17-β-HYDROXY-Δ-1,3,5(10)-OESTRATRIENE
* 17-ETHYNYL-3-METHOXY-1,3,5(10)-OESTRATIEN-17-β-OL * ETHYNYLOESTRADIOL METHYL ETHER
* 17-ETHYNYLOESTRADIOL-3-METHYL ETHER
* 17-α-ETHYNYLOESTRADIOL-3-METHYL ETHER
* 17-α-ETHYNYLOESTRADIOL METHYL ETHER
* MESTRANOL * MESTRENOL * 3-METHOXY-17-α-ETHINYLESTRADIOL * 3-METHOXY-17-α-ETHI-NYLOESTRADIOL * 3-METHOXYETHYNYLESTRADIOL
* 3-METHOXY-17-α-ETHYNOESTRADIOL * 3-METH-OXY-17-α-ETHYNYLESTRADIOL * 3-METHOXYETHY-NYLOESTRADIOL * 3-METHOXY-17-ETHYNYLOES-TRADIOL-17-β * 3-METHOXY-17-α-ETHYNYL-1,3,5(10)-ESTRATRIEN-17-β-OL * 3-METHOXY-17-α-ETHYNYL-1,3,5(10)-OESTRATRIEN-17-β-OL * 3-METHOXY-19-NOR-17-α-PREGNA-1,3,5(10)-TRIEN-10-YN-17-OL
* 3-METHOXY-17-α-19-NORPREGNA-K,3,5(10)-TRIEN-20-YN-17-OL * (17-α)-3-METHOXY-19-NORPREGN-1,3,-5(10)-TRIEN-20-YN-17-OL * 3-METHYLETHY-NYLOESTRADIOL

CONSENSUS REPORTS: IARC Cancer Review: Human Limited Evidence IMEMDT 21,257,79; Animal Sufficient Evidence IMEMDT 6,87,74; IMEMDT 21,257,79. NTP Fourth Annual Report On Carcinogens, 1984.

SAFETY PROFILE: Confirmed carcinogen with experimental neoplastigenic and tumorigenic data. Human reproductive effects by ingestion: changes in ovaries and fallopian tubes, fertility effects. Experimental teratogenic and reproductive effects. Mutation data reported. An FDA proprietary drug. A steroid used in oral contraceptives. When heated to decomposition it emits acrid smoke and irritating fumes.

MKD000   CAS: 453-18-9   **HR: 3**
METHYL FLUOROACETATE
mf: $C_3H_5FO_2$ mw: 92.08

SYNS: FLUOROACETIC ACID METHYL ESTER
* MFA

SAFETY PROFILE: Poison by ingestion, inhalation, skin contact, subcutaneous, intramuscular, intraperitoneal, parenteral, and intravenous routes. Human systemic effects by ingestion of very small amounts: nausea or vomiting. When heated to decomposition it emits toxic fumes of F^-.

MKE000   CAS: 406-20-2   **HR: 3**
METHYL-4-FLUOROBUTYRATE
mf: $C_5H_9FO_2$ mw: 120.14

SYNS: 4-FLUOROBUTYRIC ACID METHYL ESTER
* METHYLESTER KYSELINY 4-FLUORMASELNE
* METHYL-γ-FLUOROBUTYRATE * METHYL-φ-FLUOROBUTYRATE

SAFETY PROFILE: Poison by ingestion, inhalation, intravenous, and subcutaneous routes. When heated to decomposition it emits toxic fumes of F^-.

MKE250   CAS: 2367-25-1   **HR: 3**
METHYL-γ-FLUOROCROTONATE
mf: $C_5H_7FO_2$ mw: 118.12

SYNS: 4-FLUORO-CROTONIC ACID METHYL ESTER
* TL 1183

SAFETY PROFILE: Poison by inhalation. When heated to decomposition it emits toxic fumes of F^-.

MKE750   CAS: 63904-99-4   **HR: 3**
METHYL-γ-FLUORO-β-HYDROXYBUTYRATE
mf: $C_5H_9FO_3$ mw: 136.14

SYNS: γ-FLUORO-β-HYDROXY-BUTYRIC ACID METHYL ESTER * TL 1333

SAFETY PROFILE: Poison by ingestion and inhalation. When heated to decomposition it emits toxic fumes of F^-.

MKF000   CAS: 63732-23-0   **HR: 3**
METHYL-γ-FLUORO-β-HYDROXYTHIOLBUTYRATE
mf: $C_5H_9FO_2S$ mw: 152.20

SYN: 4-FLUORO-3-HYDROXY-BUTANETHIOIC ACID METHYL ESTER

SAFETY PROFILE: Poison by inhalation. When heated to decomposition it emits very toxic fumes of SO_x and F^-.

MKG250   CAS: 421-20-5   **HR: 3**
METHYL FLUOROSULFATE
mf: CH_3FO_3S mw: 114.10

PROP: Liquid; ethereal odor. Bp: 92°, d: 1.427 @ 16°, vap d: 3.94.

SYNS: MAGIC METHYL * METHYL ESTER FLUORO-SULFURIC ACID * METHYL FLUOROSULFONATE * METHYL FLUORSULFONATE

SAFETY PROFILE: Poison by inhalation and ingestion. Moderately toxic by skin contact. A skin, mucous membrane, and severe eye irritant. Mutation data reported. Reacts with water, steam or acids to produce toxic and corrosive fumes. When heated to decomposition it emits toxic fumes of F^- and SO_x.

MKG500 CAS: 123-39-7 **HR: 2**
N-METHYLFORMAMIDE
mf: C_2H_5NO mw: 59.08

PROP: Flash p: <71.6°F.

SYNS: METHYLFORMAMIDE * MONOMETHYL-FORMAMIDE * NSC 3051

CONSENSUS REPORTS: Reported in EPA TSCA Inventory. EPA Genetic Toxicology Program.

SAFETY PROFILE: Moderately toxic by ingestion, intraperitoneal, intravenous, intramuscular, subcutaneous, and possibly other routes. Experimental teratogenic and reproductive effects. An eye irritant. A very dangerous fire hazard when exposed to heat or flame. Violent reaction with benzene sulfonyl chloride. When heated to decomposition it emits toxic fumes of NO_x.

MKG750 CAS: 107-31-3 **HR: 3**
METHYL FORMATE
DOT: UN 1243
mf: $C_2H_4O_2$ mw: 60.06

PROP: Colorless liquid; agreeable odor. Mp: −99.8°, bp: 31.5°, lel: 5.9%, uel: 20%, flash p: −2.2°F, d: 0.98149 @ 15°/4°, 0.975 @ 20°/4°, autoign temp: 869°F, vap press: 400 mm @ 16°/0°, vap d: 2.07. Solidifies at about 100°. Moderately sol in water, methyl alcohol; misc in alc.

SYNS: FORMIATE de METHYLE (FRENCH) * METHYLE (FORMIATE de) (FRENCH) * METHYL-FORMIAAT (DUTCH) * METHYLFORMIAT (GERMAN) * METHYL METHANOATE * METIL (FORMIATO di) (ITALIAN)

CONSENSUS REPORTS: Reported in EPA TSCA Inventory.

OSHA PEL: (Transitional: TWA 100 ppm) TWA 100 ppm; STEL 150 ppm
ACGIH TLV: TWA 100 ppm; STEL 150 ppm
DFG MAK: 100 ppm (250 mg/m³)
DOT Classification: Flammable Liquid; Label: Flammable Liquid.

SAFETY PROFILE: Moderately toxic by ingestion. Inhalation of vapor can cause irritation to nasal passages and conjunctiva, optic neuritis, narcosis, retching, and death from pulmonary irritation. Industrial fatalities have occurred only with exposure to high concentrations. Flammable liquid. Very dangerous fire hazard when exposed to heat or flame; can react vigorously with oxidizing materials. Explosive in the form of vapor when exposed to heat or flame. Reacts with methanol + sodium methoxide to form an explosive product. To fight fire, use alcohol foam, CO_2, dry chemical. When heated to decomposition it emits acrid smoke and irritating fumes.

MKH000 CAS: 534-22-5 **HR: 3**
2-METHYLFURAN
DOT: UN 2301
mf: C_5H_6O mw: 82.11

PROP: Colorless, mobile liquid; ether-like odor. Bp: 63.7°, fp: −88.7°, flash p: −22°F, d: 0.914 @ 20°/4°, vap press: 139 mm @ 20°, vap d: 2.8.

SYNS: METHYLFURAN * SILVAN (CZECH)

CONSENSUS REPORTS: Reported in EPA TSCA Inventory.

DOT Classification: Flammable Liquid; Label: Flammable Liquid.

SAFETY PROFILE: Poison by ingestion. Moderately toxic by inhalation and possibly other routes. An eye irritant. Mutation data reported. Very dangerous fire hazard when exposed to heat or flame; can react vigorously with oxidizing materials. To fight fire, use CO_2, dry chemical. When heated to decomposition it emits acrid smoke and irritating fumes.

MKI000 CAS: 31959-87-2 **HR: 3**
METHYL GAG
SYNS: 1,1'-(METHYLETHANEDILIDENEDINITRILO) BIGUANIDINE DIHYDROCHLORIDE DIHYDRATE * NSC 32946

SAFETY PROFILE: Human systemic effects by an unspecified route: gastrointestinal changes and dermatitis. A skin irritant. Questionable carcinogen with experimental neoplastigenic by ingestion data. When heated to decomposition it emits very toxic fumes of HCl and NO_x.

MKI750   CAS: 471-29-4   **HR: 3**
METHYLGUANIDINE
mf: $C_2H_7N_3$ mw: 73.10

PROP: Colorless, deliquescent, strongly alkaline mass. Mp: decomp. Very sol in water; sol in alc.

SYNS: METHYLGUANIDIN (GERMAN) * MONO-METHYL GUANIDIN (GERMAN) * MONOMETHYL-GUANIDINE

CONSENSUS REPORTS: EPA Genetic Toxicology Program.

SAFETY PROFILE: Poison by subcutaneous, intravenous, and intraperitoneal routes. Mutation data reported. When heated to decomposition it emits toxic fumes of NO_x.

MKK000   CAS: 409-02-9   **HR: 2**
6-METHYL-5-HEPTEN-2-ONE
mf: $C_8H_{14}O$ mw: 126.22

PROP: Sltly yellow liquid; citrus-lemongrass odor. D: 0.846-0.851, refr index:.438-1.442, mp: −67.1, bp: 173-174°, flash p: 122°F. Insol in water; misc in alc, ether, and chloroform.

SYNS: FEMA No. 2707 * METHYL HEPTENONE

CONSENSUS REPORTS: Reported in EPA TSCA Inventory.

SAFETY PROFILE: Moderately toxic by ingestion. A skin irritant. Combustible liquid. When heated to decomposition it emits acrid smoke and irritating fumes.

MKK750   CAS: 360-54-3   **HR: 3**
METHYL HEXAFLUOROISOBUTYRATE
mf: $C_5H_4F_6O_2$ mw: 210.09

SYNS: HEXAFLUOROISOBUTYRIC ACID METHYL ESTER * METHYL-3,3,3-TRIFLUORO-2-(TRIFLUORO-METHYL)PROPIONATE

SAFETY PROFILE: Poison by ingestion, intravenous, and intraperitoneal routes. When heated to decomposition it emits toxic fumes of F^-.

MKN000   CAS: 60-34-4   **HR: 3**
METHYL HYDRAZINE

DOT: UN 1244
mf: CH_6N_2 mw: 46.09

PROP: Colorless, hydroscopic liquid; ammonia-like odor. D: 0.874 @ 25°, mp: −20.9°, bp: 87.8°, vap d: 1.6, flash p: 73.4°F, fp:

−52.4°, autoign temp: 196°, lel: 2.5%, uel: 97 ±2%. Sltly sol in water; sol in alc, hydrocarbons, and ether; misc with water and hydrazine. Strong reducing agent.

SYNS: HYDRAZOMETHANE * 1-METHYL HYDRAZINE * METHYLHYDRAZINE (DOT) * METYLOHY-DRAZYNA (POLISH) * MMH * MONOMETHYL HYDRAZINE * RCRA WASTE NUMBER P068

CONSENSUS REPORTS: Community Right-To-Know List. EPA Extremely Hazardous Substances List. Reported in EPA TSCA Inventory. EPA Genetic Toxicology Program.

OSHA PEL: CL 0.2 ppm (skin))
ACGIH TLV: CL 0.2 ppm; Suspected Human Carcinogen
NIOSH REL: CL 0.08 $mg/m^3/2H$
DOT Classification: Flammable Liquid; Label: Flammable Liquid, Corrosive.

SAFETY PROFILE: Suspected carcinogen with experimental carcinogenic, neoplastigenic, and tumorigenic data. Poison by inhalation, ingestion, skin contact, intraperitoneal, subcutaneous, and intravenous routes. Experimental teratogenic and reproductive effects. Human mutation data reported. Corrosive to skin, eyes, and mucous membranes. May self-ignite in air. Very dangerous fire hazard when exposed to heat or flame. To fight fire, use alcohol foam, CO_2, dry chemical. Explosive in the form of vapor when exposed to heat or flame. A powerful reducing agent. It is hypergolic with many oxidants (e.g., dinitrogen tetraoxide and hydrogen peroxide). When heated to decomposition it emits toxic fumes of NO_x.

MKN250   CAS: 7339-53-9   **HR: 3**
METHYLHYDRAZINE HYDROCHLORIDE
mf: $CH_6N_2 \cdot ClH$ mw: 82.55

NIOSH REL: CL 0.08 $mg/m^3/2H$

SAFETY PROFILE: Poison by ingestion, intraperitoneal, and intravenous routes. When heated to decomposition it emits very toxic fumes of Cl^- and NO_x.

MKR250   CAS: 297-90-5   **HR: 3**
N-METHYL-3-HYDROXYMORPHINAN
mf: $C_{17}H_{23}NO$ mw: 257.41

SYNS: CETARIN * DROMORAN * racemic DRO-MORAN * dl-1,3,4,9,10,10A-HEXAHYDRO-11-METHYL-2H-10,4A-IMINOETHANOPHENANTHREN-6-OL * dl-3-HYDROXY-N-METHYLMORPHINAN * (±)-3-

HYDROXY-N-METHYLMORPHINAN * METHORPHINAN * NU 2206 * RACEMORPHAN * RO 1-5431

SAFETY PROFILE: Poison by ingestion, intravenous, and subcutaneous routes. When heated to decomposition it emits toxic fumes of NO_x.

MKV500   CAS: 876-83-5   **HR: 3**
2-METHYL-1,3-INDANDIONE
mf: $C_{10}H_8O_2$ mw: 160.18

SAFETY PROFILE: Poison by intraperitoneal route. Experimental teratogenic and reproductive effects. When heated to decomposition it emits acrid smoke and irritating fumes.

MKV750   CAS: 83-34-1   **HR: 3**
β-**METHYLINDOLE**
mf: C_9H_9N mw: 131.19

PROP: Leaves from ligroin. Mp: 95°, bp: 265° @ 755 mm. Sol in cold water, alc, chloroform, ether, and benzene.

SYNS: 3-METHYLINDOLE * 3-METHYL-1H-INDOLE * 3-MI * SCATOLE * SKATOL * SKATOLE

CONSENSUS REPORTS: Reported in EPA TSCA Inventory.

SAFETY PROFILE: Poison by ingestion, intravenous, and intraperitoneal routes. Moderately toxic by subcutaneous route. When heated to decomposition it emits toxic fumes of NO_x.

MKW200   CAS: 74-88-4   **HR: 3**
METHYL IODIDE

DOT: UN 2644
mf: CH_3I mw: 141.94

PROP: Colorless liquid, turns brown on exposure to light. Mp: −66.4°, bp: 42.5°, d: 2.279 @ 20°/4°, vap press: 400 mm @ 25.3°, vap d: 4.89. Sol in water @ 15°, misc in alc and ether.

SYNS: IODOMETHANE * IODOMETANO (ITALIAN) * IODURE de METHYLE (FRENCH) * JOD-METHAN (GERMAN) * JOODMETHAAN (DUTCH) * METHYL-JODID (GERMAN) * METHYLJODIDE (DUTCH) * METYLU JODEK (POLISH) * MONOIODURO di METILE (ITALIAN) * RCRA WASTE NUMBER U138

CONSENSUS REPORTS: IARC Cancer Review: GROUP 3 IMEMDT 7,56,87; Animal Limited Evidence IMEMDT 41,213,86. NTP

Fourth Annual Report On Carcinogens, 1984. Community Right-To-Know List. EPA Genetic Toxicology Program. Reported in EPA TSCA Inventory.

OSHA PEL: (Transitional: TWA 5 ppm (skin)) TWA 2 ppm (skin)
ACGIH TLV: TWA 2 ppm (skin); Suspected Human Carcinogen.
DFG MAK: Animal Carcinogen, Suspected Human Carcinogen.
NIOSH REL: (Methylbromide) Reduce to lowest level.
DOT Classification: Poison B; Label: Poison.

SAFETY PROFILE: Confirmed carcinogen with experimental neoplastigenic and tumorigenic data. A poison by ingestion, intraperitoneal, and subcutaneous routes. Moderately toxic by inhalation and skin contact. A human skin irritant. Human mutation data reported. A strong narcotic and anesthetic. Explosive reaction with trialkylphosphines, silver chlorite. Violent reaction with oxygen (at 300-500°C), sodium even in solution. When heated to decomposition it emits toxic fumes of I^-.

MKW450   CAS: 110-12-3   **HR: 2**
METHYL ISOAMYL KETONE

DOT: UN 2302
mf: $C_7H_{14}O$ mw: 114.21

PROP: Colorless, stable liquid; pleasant odor. Bp: 144°, d: 0.8132 @ 20°/20°, fp: −73.9°, flash p: 110°F (OC). Sltly sol in water; misc with most organic solvents.

SYNS: ISOAMYL METHYL KETONE * ISOPENTYL METHYL KETONE * 2-METHYL-5-HEXANONE * 5-METHYL-2-HEXANONE * MIAK

CONSENSUS REPORTS: Reported in EPA TSCA Inventory.

OSHA PEL: TWA 50 ppm
ACGIH TLV: TWA 50 ppm
NIOSH REL: TWA 50 ppm
DOT Classification: Flammable or Combustible Liquid; Label: Flammable Liquid.

SAFETY PROFILE: Moderately toxic by ingestion. Mildly toxic by inhalation and skin contact. Flammable when exposed to heat, flame or oxidizers. To fight fire, use dry chemical, CO_2, foam, fog. When heated to decomposition it emits acrid smoke and irritating fumes.

MKW600   CAS: 108-11-2   **HR: 3**
METHYL ISOBUTYL CARBINOL

DOT: UN 2053
mf: $C_6H_{14}O$ mw: 102.20

PROP: Clear liquid. Bp: 131.8°, fp: $< -90°$ (sets to a glass), flash p: 106°F, d: 0.8079 @ 20°/20°, vap press: 2.8 mm @ 20°, vap d: 3.53, lel: 1.0%, uel: 5.5%.

SYNS: ALCOOL METHYL AMYLIQUE (FRENCH) * ISOBUTYL METHYL CARBINOL * ISOBUTYL-METHYLMETHANOL * MAOH * METHYL AMYL ALCOHOL * METHYLISOBUTYL CARBINOL * 2-METHYL-4-PENTANOL * 4-METHYLPENTANOL-2 * 4-METHYL-2-PENTANOL (MAK) * METILAMIL ALCOHOL (ITALIAN) * 4-METILPENTAN-2-OLO (ITALIAN) * MIBC * MIC * 3-MIC

CONSENSUS REPORTS: Reported in EPA TSCA Inventory.

OSHA PEL: (Transitional: TWA 25 ppm (skin)) TWA 25 ppm; STEL 40 ppm (skin)
ACGIH TLV: TWA 25 ppm; STEL 40 ppm (skin)
DFG MAK: 25 ppm (100 mg/m^3)
DOT Classification: Flammable or Combustible Liquid; Label: Flammable Liquid.

SAFETY PROFILE: Moderately toxic by ingestion, skin contact, and intraperitoneal routes. Mildly toxic by inhalation. A skin and severe eye irritant. Inhalation of high concentrations can cause anesthesia. Flammable when exposed to heat or flame; can react with oxidizing materials. A moderate explosion hazard when exposed to heat or flame. To fight fire, use alcohol foam. When heated to decomposition it emits acrid smoke and fumes.

MKX250   CAS: 624-83-9   **HR: 3**
METHYL ISOCYANATE

DOT: UN 2480
mf: C_2H_3NO mw: 57.06

PROP: Liquid. D: 0.9599 @ 20°/20°, bp: 39.1°, flash p: $<5°F$.

SYNS: ISOCYANATE de METHYLE (FRENCH) * ISO-CYANATOMETHANE * ISOCYANIC ACID, METHYL ESTER * METHYLISOCYANAAT (DUTCH) * METHYL ISOCYANAT (GERMAN) * METHYL ISO-CYANATE, solutions (DOT) * METIL ISOCIANATO (ITALIAN) * RCRA WASTE NUMBER P064 * TL 1450

CONSENSUS REPORTS: Reported in EPA TSCA Inventory.

OSHA PEL: TWA 0.02 ppm (skin)
ACGIH TLV: TWA 0.02 ppm (skin)
DFG MAK: 0.01 ppm (0.025 mg/m^3)
DOT Classification: Flammable Liquid; Label: Flammable Liquid and Poison.

SAFETY PROFILE: Poison by inhalation, ingestion, and skin contact. Human systemic effects by inhalation: conjuctiva irritation, olfactory and pulmonary changes. Mutation data reported. A severe eye, skin, and mucous membrane irritant and a sensitizer. It can be absorbed through the skin. Exposure to high concentrations of the vapor can cause blindness, lung damage, including edema, permanent fibrosis, emphysema, and bronchitis, and gynecological effects. Most deaths are a result of lung tissue damage. Effects of cyanide poisoning have been noted but this may be due to impurities. A very dangerous fire hazard when exposed to heat, flame, or oxidizers. To fight fire, use spray, foam, CO_2, dry chemical. Exothermic reaction with water. When heated to decomposition it emits toxic fumes of NO_x.

MKY500   CAS: 814-78-8   **HR: 3**
METHYL ISOPROPENYL KETONE

DOT: UN 1246
mf: C_5H_8O mw: 84.119

PROP: Flash p: 69.8°F, lel: 1.8%, uel: 9.0%, vap d: 2.9, bp: 98°.

SYNS: 3-METHYL-3-BUTEN-2-ON (GERMAN) * 2-METHYL-1-BUTEN-3-ONE * METHYL ISOPROPENYL KETONE INHIBITED (DOT)

CONSENSUS REPORTS: Reported in EPA TSCA Inventory.

DOT Classification: Flammable Liquid; Label: Flammable Liquid.

SAFETY PROFILE: Poison by ingestion and skin contact. Moderately toxic by inhalation and intraperitoneal route. A skin and severe eye irritant. A dangerous fire hazard when exposed to heat or flame. Explosive in the form of vapor when exposed to heat or flame. When heated to decomposition it emits acrid smoke and irritating fumes.

MLA250   CAS: 99-86-5   **HR: 2**
1-METHYL-4-ISOPROPYLCYCLO-HEXADIENE-1,3
mf: $C_{10}H_{16}$ mw: 136.26

PROP: Colorless liquid; lemon odor. D: 0.834 @ 20°/4°, refr index: 1.475-1.480, bp: 181.5°. Insol in water; misc in alc, ether, fixed oils.

SYNS: FEMA No. 3558 * p-MENTHA-1,3-DIENE * 1-METHYL-4-ISOPROPYL-1,3-CYCLOHEXADIENE * α-TERPINENE (FCC)

CONSENSUS REPORTS: Reported in EPA TSCA Inventory.

SAFETY PROFILE: Moderately toxic by ingestion. When heated to decomposition it emits acrid smoke and irritating fumes.

MLA750   CAS: 563-80-4   HR: 3
METHYL ISOPROPYL KETONE

DOT: UN 2397
mf: $C_5H_{10}O$ mw: 86.15

SYNS: ISOPROPYL METHYL KETONE * 3-METHYL-2-BUTANONE * 3-METHYL BUTAN 2-ONE (DOT) * MIPK

CONSENSUS REPORTS: Reported in EPA TSCA Inventory.

OSHA PEL: TWA 200 ppm
ACGIH TLV: TWA 200 ppm
DOT Classification: Flammable Liquid; Label: Flammable Liquid.

SAFETY PROFILE: Poison by ingestion. Mildly toxic by inhalation and skin contact. Mutation data reported. A skin and eye irritant. Flammable when exposed to heat or flame; can react vigorously with oxidizing materials. When heated to decomposition it emits acrid smoke and irritating fumes.

MLC250   CAS: 5707-69-7   HR: 3
3-METHYL-4,5-ISOXAZOLEDIONE-4-((2-CHLOROPHENYL)HYDRAZONE)
mf: $C_{10}H_8ClN_3O_2$ mw: 237.66

SYNS: 4-(2-CHLOROPHENYLHYDRAZONE)-3-METHYL-5-ISOXAZOLONE * 4-(2-CHLOROPHENYLHYDRA-ZONE)-3-METHYL-5(4H)-ISOXAZOLONE * DRAZOXO-LON * DRAZOXOLONE * GANOCIDE * 3-METHYL-4-((o-CHLOROPHENYL)HYDRAZONE)-4,5-ISOXAZOLEDIONE * 3-METHYL-4-(o-CHLOROPHENYL-HYDRAZONO)-5-ISOXAZOLONE * MIL-COL * PP781 * SAISAN * SOPRACOL * SOPRACOL 781

SAFETY PROFILE: Poison by ingestion and intraperitoneal routes. When heated to decomposition it emits very toxic fumes of Cl⁻ and NO_x.

MLC750   CAS: 75-86-5   HR: 3
2-METHYLLACTONITRILE

DOT: UN 1541
mf: C_4H_7NO mw: 85.12

PROP: Mp: −20°, bp: 82° @ 23 mm, d: 0.932 @ 19°, autoign temp: 1270°F, flash p: 165°F, vap d: 2.93.

SYNS: ACETONCIANHIDRINEI (ROUMANIAN) * ACETONCIANIDRINA (ITALIAN) * ACETON-CYAANHYDRINE (DUTCH) * ACETONCYANHYDRIN (GERMAN) * ACETONECYANHYDRINE (FRENCH) * ACETONE CYANOHYDRIN (DOT) * ACETONKYAN-HYDRIN (CZECH) * CYANHYDRINE d'ACETONE (FRENCH) * α-HYDROXYISOBUTYRONITRILE * 2-HYDROXY-2-METHYLPROPIONITRILE * RCRA WASTE NUMBER P069 * USAF RH-8

CONSENSUS REPORTS: Cyanide and its compounds are on the Community Right-To-Know List. Reported in EPA TSCA Inventory. EPA Extremely Hazardous Substances List.

NIOSH REL: (Nitriles) CL 4 mg/m^3/15M
DOT Classification: Poison B; Label: Poison.

SAFETY PROFILE: Poison by ingestion, skin contact, inhalation, intraperitoneal, and subcutaneous routes. Readily decomposes to HCN and acetone. Keep cool and do not store for long periods. Combustible when exposed to heat or flame. To fight fire, use CO_2, dry chemical, alcohol foam. Vigorous reaction with H_2SO_4. When heated to decomposition it emits toxic fumes of CN⁻.

MLD500   CAS: 7240-57-5   HR: 3
1-METHYLLYSERGIC ACID ETHYLAMIDE
mf: $C_{19}H_{23}N_3O$ mw: 309.45

SYNS: 9,10-DIDEHYDRO-N-ETHYL-1,6-DIMETHYLER-GOLINE-8-β-CARBOXAMIDE * d-1-METHYL LYSERGIC ACID MONOETHYLAMIDE * MLA-74

SAFETY PROFILE: Poison by intravenous route. Human systemic effects by ingestion: psychotropic effects. When heated to decomposition it emits toxic fumes of NO_x.

MLE000   CAS: 75-16-1   HR: 3
METHYLMAGNESIUM BROMIDE (ethyl ether solution)
mf: CH_3BrMg mw: 119.26

DOT: UN 1928

PROP: Concentration of ethyl ether is not over 40%.

SYN: METHYL MAGNESIUM BROMIDE in ETHYL ETHER (DOT)

CONSENSUS REPORTS: Reported in EPA TSCA Inventory.

DOT Classification: Flammable Liquid; Label: Flammable Liquid, Spontaneously Combustible.

SAFETY PROFILE: May ignite spontaneously in air. A very dangerous fire hazard when exposed to heat or flame; can react vigorously with oxidizing materials. When heated to decomposition it emits acrid smoke and irritating fumes.

MLE650   CAS: 74-93-1   **HR: 3**
METHYL MERCAPTAN

DOT: UN 1064
mf: CH_4S mw: 48.11

PROP: Flammable gas; odor of rotten cabbage. Mp: $-123.1°$, vap d: 1.66, lel: 3.9%, uel: 21.8%. Bp: 5.95°, d: 0.8665 @ 20°/4°, solidifies @ $-123°$, flash p: $-0.4°F$.

SYNS: MERCAPTAN METHYLIQUE (FRENCH) * METHAANTHIOL (DUTCH) * METHANETHIOL * METHANTHIOL (GERMAN) * METHVTIOLO (ITALIAN) * METHYLMERCAPTAAN (DUTCH) * METILMERCAPTANO (ITALIAN) * RCRA WASTE NUMBER U153

CONSENSUS REPORTS: EPA Extremely Hazardous Substances List. Reported in EPA TSCA Inventory.

OSHA PEL: (Transitional: CL 10 ppm) TWA 0.5 ppm
ACGIH TLV: TWA 0.5 ppm
DFG MAK: 0.5 ppm (1 mg/m^3)
NIOSH REL: (n-Alkane Monothiols) CL 0.5 ppm/15M
DOT Classification: Flammable Gas; Label: Flammable Gas.

SAFETY PROFILE: Poison by inhalation and possibly other routes. Mutation data reported. A common air contaminant. Very dangerous fire hazard when exposed to heat or flame; can react vigorously with oxidizing materials. Explosive in the form of vapor when exposed to heat or flame. Reacts with water, steam, or acids to produce toxic and flammable vapors. Violent reaction with mercury(II) oxide. To fight fire, use alcohol foam, CO_2, dry chemical. Upon decomposition it emits highly toxic fumes of SO_x.

MLF250   CAS: 502-39-6   **HR: 3**
METHYLMERCURIC DICYANDIAMIDE
mf: $C_3H_6HgN_4$ mw: 298.72

SYNS: AGROSOL * CYANOGUANIDINE METHYLMERCURY DERIV. * CYANO(METHYLMERCURI)GUANIDINE * MEMA * METHYLMERCURIC CYANOGUANIDINE * METHYLMERCURY DICYANDIAMIDE * MMD * MORSODREN * MORTON SOIL DRENCH * PANDRINOX * PANO-DRENCH 4 * PANOGEN * PANOGEN TURF FUNGICIDE * PANOGEN TURF SPRAY * PANOSPRAY 30

CONSENSUS REPORTS: Mercury and its compounds are on the Community Right-To-Know List. EPA Extremely Hazardous Substances List.

OSHA PEL: (Transitional: CL 1 $mg/10m^3$) TWA 0.01 $mg(Hg)/m^3$; STEL 0.03 mg/m^3 (skin)
ACGIH TLV: TWA 0.01 $mg(Hg)/m^3$ (skin); 0.01 $mg(Hg)/m^3$; STEL 0.03 $mg(Hg)/m^3$
NIOSH REL: TWA 0.05 $mg(Hg)/m^3$

SAFETY PROFILE: Poison by ingestion, intraperitoneal, and possibly other routes. Experimental teratogenic and reproductive effects. Mutation data reported. When heated to decomposition it emits very toxic fumes of Hg and NO_x.

MLF500   CAS: 5902-79-4   **HR: 3**
METHYLMERCURICHLORENDIMIDE
mf: $C_{10}H_5Cl_6HgNO_2$ mw: 584.45

SYNS: MEMMI * N-(METHYLMERCURI)-1,4,5,6,7,7-HEXACHLOROBICYCLO(2.2.1)HEPT-5-ENE-2,3-DICARBOXIMIDE * N-METHYLMERCURI-1,2,3,6-TETRAHYDRO-3,6-ENDOMETHANO-3,4,5,6,7,7-HEXACHLOROPHTHALIMIDE * N-METHYLMERCURI-1,2,3,6-TETRAHYDRO-3,6-METHANO-3,4,5,6,7,7-HEXACHLOROPHTHALIMIDE

CONSENSUS REPORTS: Mercury and its compounds are on the Community Right-To-Know List.

OSHA PEL: (Transitional: CL 1 $mg/10m^3$) TWA 0.01 $mg(Hg)/m^3$; STEL 0.03 mg/m^3 (skin)
ACGIH TLV: TWA 0.01 $mg(Hg)/m^3$; STEL 0.03 $mg(Hg)/m^3$

NIOSH REL: TWA 0.05 mg(Hg)/m^3

SAFETY PROFILE: Poison by ingestion and possibly other routes. When heated to decomposition it emits very toxic fumes of Cl$^-$, Hg, and NO$_x$.

MLF550   CAS: 22967-92-6   **HR: 3**
METHYLMERCURY
mf: CH$_3$Hg   mw: 215.63

SYNS: METHYL-MERCURY(1+) (9CI) * METHYL-MERCURY(II) CATION * METHYLMERCURY ION * METHYLMERCURY ION(1+)

CONSENSUS REPORTS: Mercury and its compounds are on the Community Right-To-Know List.

OSHA PEL: (Transitional: CL 1 mg/10m^3) TWA 0.01 mg(Hg)/m^3; STEL 0.03 mg/m^3 (skin)
ACGIH TLV: TWA 0.01 mg(Hg)/m^3; STEL 0.03 mg(Hg)/m^3
DFG MAK: 0.01 mg/m^3
NIOSH REL: TWA 0.05 mg(Hg)/m^3

SAFETY PROFILE: A poison. Experimental reproductive effects. Mutation data reported. Used as a fungicide. When heated to decomposition it emits toxic fumes of Hg.

MLG000   CAS: 1184-57-2   **HR: 3**
METHYLMERCURY HYDROXIDE
mf: CH$_4$HgO   mw: 232.64

SYNS: HYDROXYMETHYLMERCURY * METHYL-MERCURIC HYDROXIDE

CONSENSUS REPORTS: Mercury and its compounds are on the Community Right-To-Know List. EPA Genetic Toxicology Program.

OSHA PEL: (Transitional: CL 1 mg/10m^3) TWA 0.01 mg(Hg)/m^3; STEL 0.03 mg/m^3 (skin)
ACGIH TLV: TWA 0.01 mg(Hg)/m^3; STEL 0.03 mg(Hg)/m^3
NIOSH REL: TWA 0.05 mg(Hg)/m^3

SAFETY PROFILE: Poison by intraperitoneal route. Experimental teratogenic and reproductive effects. Human mutation data reported. When heated to decomposition it emits toxic fumes of Hg.

MLH000   CAS: 86-85-1   **HR: 3**
METHYLMERCURY QUINOLINOLATE
mf: C$_{10}$H$_9$HgNO   mw: 359.79

SYNS: ARTHO LM * LIQUI-SAN * LM SEED PROTECTANT * METASOL * METAZOL * 8-(METHYLMERCURIOXY)QUINOLINE * METHYLMERCURY β-HYDROXYQUINOLATE * METHYLMERCURY 8-HYDROXYQUINOLINATE * METHYLMERCURY OXINATE * METHYLMERCURY OXYQUINOLINATE * ORTHO-LM APPLE SPRAY * ORTHO LM CONCENTRATE * ORTHO LM SEED PROTECTANT * 8-(QUINOLINOLATO)METHYL MERCURY * 8-QUINOLINOL, MERCURY COMPLEX

CONSENSUS REPORTS: Mercury and its compounds are on the Community Right-To-Know List.

OSHA PEL: (Transitional: CL 1 mg/10m^3) TWA 0.01 mg(Hg)/m^3; STEL 0.03 mg/m^3 (skin)
ACGIH TLV: TWA 0.01 mg(Hg)/m^3; STEL 0.03 mg(Hg)/m^3
NIOSH REL: TWA 0.05 mg(Hg)/m^3

SAFETY PROFILE: Poison by ingestion. A pesticide. When heated to decomposition it emits very toxic fumes of Hg and NO$_x$.

MLH250   CAS: 1082-88-8   **HR: 3**
α-METHYLMESCALINE
mf: C$_{12}$H$_{19}$NO$_3$ mw: 225.32

SYNS: 3,4,5-TRIMETHOXYAMPHETAMINE * TRIMETHOXYPHENYL-β-AMINOPROPANE * 3,4,5-TRIMETHOXYPHENYL-β-AMINOPROPANE

SAFETY PROFILE: Poison by intraperitoneal and possibly other routes. Human systemic effects by ingestion: psychotropic effects. When heated to decomposition it emits toxic fumes of NO$_x$.

MLH500   CAS: 66-27-3   **HR: 3**
METHYL MESYLATE
mf: C$_2$H$_6$O$_3$S   mw: 110.14

PROP: Liquid. D: 1.046 @ 16°/4° bp: 126.5° @ 756 mm. Decomp in water; sol in alc and ether.

SYNS: as-DIMETHYL SULPHATE * METHANESULPHONIC ACID METHYL ESTER * METHYL ESTER of METHANESULFONIC ACID * METHYL ESTER of METHANESULPHONIC ACID * METHYLMETHANSULFONAT (GERMAN) * METHYL METHANESULFONATE * METHYL METHANESULPHONATE * METHYL METHANSULFONATE * METHYL METHANSULPHONATE * MMS * NSC-50256

CONSENSUS REPORTS: IARC Cancer Review: GROUP 3 IMEMDT 7,56,87; Animal

Sufficient Evidence IMEMDT 7,253,74. Reported in EPA TSCA Inventory. EPA Genetic Toxicology Program.

SAFETY PROFILE: Poison by ingestion, intraperitoneal, intravenous, and subcutaneous routes. Experimental teratogenic and reproductive effects. Questionable carcinogen with experimental carcinogenic, neoplastigenic, and tumorigenic data. Human mutation data reported. When heated to decomposition it emits toxic fumes of SO_x.

MLH750   CAS: 80-62-6   HR: 3
METHYL METHACRYLATE

DOT: NA 1247
mf: $C_5H_8O_2$ mw: 100.13

PROP: Colorless liquid, very sltly sol in water. Mp: $-50°$, bp: $101.0°$, flash p: $50°F$ (OC), d: 0.936 @ $20°/4°$, vap press: 40 mm @ $25.5°$, vap d: 3.45, lel: 2.1%, uel: 12.5%.

SYNS: DIAKON * METAKRYLAN METYLU (POLISH) * METHACRYLATE de METHYLE (FRENCH) * METHACRYLIC ACID, METHYL ESTER (MAK) * METHACRYLSAEUREMETHYL ESTER (GERMAN) * METHYLMETHACRYLAAT (DUTCH) * METHYL-METHACRYLAT (GERMAN) * METHYL METHACRYLATE MONOMER, INHIBITED (DOT) * METHYL-α-METHYLACRYLATE * METHYL-2-METHYL-2-PROPENOATE * 2-METHYL-2-PROPENOIC ACID METHYL ESTER * METIL METACRILATO (ITALIAN) * MME * "MONOCITE" METHACRYLATE MONOMER * NCI-C50680 * RCRA WASTE NUMBER U162

CONSENSUS REPORTS: IARC Cancer Review: GROUP 3 IMEMDT 7,56,87; Human Inadequate Evidence IMEMDT 19,187,79; Animal Inadequate Evidence IMEMDT 19,187,79. NTP Carcinogenesis Studies (inhalation); No Evidence: mouse, rat NTPTR* NTP-TR-314,86. Reported in EPA TSCA Inventory. Community Right-To-Know List.

OSHA PEL: TWA 100 ppm
ACGIH TLV: TWA 100 ppm
DFG MAK: 50 ppm (210 mg/m^3)
DOT Classification: Flammable Liquid; Label: Flammable Liquid.

SAFETY PROFILE: Moderately toxic by inhalation and intraperitoneal routes. Mildly toxic by ingestion. Human systemic effects by inhalation: sleep effects, excitement, anorexia, and blood pressure decrease. Experimental teratogenic and reproductive effects. Mutation data reported. A skin and eye irritant. Questionable carcinogen with experimental tumorigenic data. A common air contaminant.

A very dangerous fire hazard when exposed to heat or flame; can react with oxidizing materials. Explosive in the form of vapor when exposed to heat or flame. The monomer may undergo spontaneous, explosive polymerization. Reacts in air to form a heat-sensitive explosive product (explodes on evaporation at $60°C$). May ignite on contact with benzoyl peroxide. Potentially violent reaction with the polymerization initiators azoisobutyronitrile, dibenzoyl peroxide, di-tert-butyl peroxide, propionaldehyde. To fight fire, use foam, CO_2, dry chemical. When heated to decomposition it emits acrid smoke and irritating fumes.

MLJ500   CAS: 926-93-2   HR: 3
1-METHYL-6-(1-METHYLALLYL)-2,5-DITHIOBIUREA
mf: $C_7H_{14}N_4S_2$ mw: 218.37

SYNS: AIMAX * AY-61122 * COMPOUND 33,828 * DITHIOCARBAMOYLHYDRAZINE * I.C.I. 33,828 * MATCH * METALLIBURE * METHALLIBURE * 1-α-METHYLALLYLTHIOCARBAMOYL-2-METHYL-THIOCARBAMOYLHYDRAZINE * N-((1-METHYLALLYL)THIOCARBAMOYL)-N'-(METHYLTHIOCARBAMOYL) HYDRAZINE * 1-METHYL-6-(1-METHYLALLYL) DITHIOBIUREA * NSC-69536 * SUISYNCHRON * TURISYNCHRON

SAFETY PROFILE: Poison by intravenous route. Moderately toxic by ingestion. Human reproductive effects by unspecified route: menstrual cycle changes or disorders. Experimental teratogenic and reproductive effects. When heated to decomposition it emits very toxic fumes of SO_x and NO_x.

MLL000   CAS: 13984-07-1   HR: 3
N-METHYL-N-(5-(N'-METHYLANILINO)-2,4-PENTADIENYLIDENE) ANILINIUM CHLORIDE
mf: $C_{19}H_{21}N_2 \cdot Cl$ mw: 312.87

SAFETY PROFILE: Poison by intraperitoneal route. Moderately toxic by ingestion. When heated to decomposition it emits very toxic fumes of NO_x and Cl^-.

MLL250   CAS: 80-48-8   HR: 3
METHYL-p-METHYLBENZENE-SULFONATE
mf: $C_8H_{10}O_3S$ mw: 186.24

PROP: Light brown crystals; crystals of ethyl ligroin. D: 1.230-1.238 @ 25°/25°, vap d: 6.45, mp: 28°. Insol in water; sol in benzene; very sol in alc and ether.

SYNS: METHYLESTER KYSELINY p-TOLUENSULFO-NOVE (CZECH) * METHYL-4-METHYLBENZENESULFO-NATE * METHYL-p-TOLUENESULFONATE * METHYL TOLUENE-4-SULFONATE * METHYL TOSYLATE * METHYL-p-TOSYLATE * p-TOLUOL-SULFONSAEURE METHYL ESTER (GERMAN)

CONSENSUS REPORTS: Reported in EPA TSCA Inventory.

SAFETY PROFILE: Poison by ingestion and subcutaneous route. An eye and severe skin irritant. A vesicant and skin sensitizer. Questionable carcinogen with experimental tumorigenic data. When heated to decomposition it emits toxic fumes of SO_x.

MLL600   CAS: 53955-81-0   **HR: 3**
METHYL 2-METHYLBUTYRATE
mf: $C_6H_{12}O_2$ mw: 116.16

PROP: Colorless liquid; sweet, fruity, apple-like odor. D: 0.879, refr index: 1.393-1.397, flash p: 91°F. Sol in alc, fixed oils; insol in water.

SYNS: FEMA No. 2719 * METHYL 2-METHYLBUTA-NOATE

SAFETY PROFILE: Flammable liquid. When heated to decomposition it emits acrid smoke and irritating fumes.

MLP250   CAS: 36304-84-4   **HR: 3**
d-3-METHYL-N-METHYLMORPHINAN PHOSPHATE
mf: $C_{18}H_{25}N \cdot H_3O_4P$ mw: 353.44

SYNS: ASTOMIN * AT-17 PHOSPHATE * DI-MEMORFAN PHOSPHATE * (9-α,13-α,14-α)-3,17-DI-METHYLMORPHINAN PHOSPHATE * 3,17-DIMETHYL-9-α,13-α,14-α-MORPHINAN PHOSPHATE

SAFETY PROFILE: Poison by intravenous, intraperitoneal and subcutaneous routes. Moderately toxic by ingestion. Experimental teratogenic and reproductive effects. Used as an antitussive agent. When heated to decomposition it emits very toxic fumes of NO_x and PO_x.

MLY000   CAS: 801-52-5   **HR: 3**
N-METHYLMITOMYCIN C
mf: $C_{16}H_{20}N_4O_5$ mw: 348.40

PROP: Antibiotic from *Streptomyces ardus* and *Streptomyces verticillatus*.

SYNS: 8-AZATHIOXANTHINE * ENT 50,825 * METHYLMITOMYCIN * NSC-56410 * PORFIRO-MYCIN * PORFIROMYCINE * PORPHYROMYCIN * PROFIROMYCIN * 5-THIO-1H-θ-TRIAZOLO(4,5-d) PYRIMIDINE-5,7(4H,6H)-DIONE * U-14743

SAFETY PROFILE: Poison by ingestion, intravenous, subcutaneous, and intraperitoneal routes. Human systemic effects by intravenous route: blood effects. Human mutation data reported. When heated to decomposition it emits toxic fumes of NO_x.

MMA250   CAS: 109-02-4   **HR: 3**
N-METHYL MORPHOLINE
DOT: UN 2535

PROP: Liquid. Flash p: 75.2°F, d: 0.9, vap d: 3.5, bp: 115°.
mf: $C_5H_{11}NO$ mw: 101.17

SYNS: METHYLMORPHOLINE (DOT) * 4-METHYL-MORPHOLINE

CONSENSUS REPORTS: Reported in EPA TSCA Inventory.

DOT Classification: Flammable Liquid; Label: Flammable Liquid, Corrosive; Flammable or Combustible Liquid; Label: Flammable Liquid, Corrosive.

SAFETY PROFILE: Moderately toxic by ingestion and skin contact. Mildly toxic by inhalation. A corrosive irritant to skin, eyes and mucous membranes. Flammable when exposed to heat or flame, can react vigorously with oxidizing materials. When heated to decomposition it emits toxic fumes of NO_x.

MMD500   CAS: 58-27-5   **HR: 3**
2-METHYL-1,4-NAPHTHOQUINONE
mf: $C_{11}H_8O_2$ mw: 172.19

SYNS: AQUAKAY * AQUINONE * HEMODAL * KAERGONA * KANONE * KAPPAXAN * KARCON * KAREON * KATIV-G * KAY-KLOT * KAYQUINONE * KIPCA * KLOTTONE * KOAXIN * KOLKLOT * K-THROMBYL * K-VITAN * MENADION * MENADIONE * MENAPHTHON * MENAPHTONE * 2-METHYL-1,4-NAPHTHALENDIONE * 2-METHYL-1,4-NAPHTHAL-ENEDIONE * 2-METHYL-1,4-NAPHTHOCHINON (GER-MAN) * 3-METHYL-1,4-NAPHTHOQUINONE * MITENON * MNQ * NSC 4170 * PANOSINE

* PROKAYVIT * SYNKAY * THYLOQUINONE
·* USAF EK-5185 * VITAMIN K2(O) * VITAMIN K3

CONSENSUS REPORTS: Reported in EPA TSCA Inventory.

SAFETY PROFILE: Poison by ingestion, intraperitoneal, and subcutaneous routes. Moderately toxic by intravenous route. Experimental teratogenic and reproductive effects. Questionable carcinogen with experimental tumorigenic data. Human mutation data reported. When heated to decomposition it emits acrid smoke and irritating fumes.

MME500   CAS: 10546-24-4   **HR: 3**
3-METHYL-2-NAPHTHYLAMINE
mf: $C_{11}H_{11}N$ mw: 157.23

SAFETY PROFILE: Suspected carcinogen with experimental carcinogenic and tumorigenic data. When heated to decomposition it emits toxic fumes of NO_x.

MME809   CAS: 5903-13-9   **HR: 3**
N-METHYL-N-(1-NAPHTHYL) FLUOROACETAMIDE
mf: $C_{13}H_{12}FNO$ mw: 217.26

SYNS: 1-(N-ACETAMIDOFLUOROMETHYL)-NAPHTHALENE * DP X 1410 * FAM * 2-FLUORO-N-METHYL-N-1-NAPHTHALENYLACETAMIDE * 2-FLUORO-N-METHYL-N-1-NAPHTHYLACETAMIDE * N-METHYL-N-(1-NAPHTHYL)MONOFLUOROACETAMIDE * MFNA * MNFA * NISSOL EC

SAFETY PROFILE: Poison by ingestion, skin contact, subcutaneous, intravenous, and intraperitoneal routes. Experimental reproductive effects. When heated to decomposition it emits very toxic fumes of F^- and NO_x.

MMF500   CAS: 598-58-3   **HR: 3**
METHYL NITRATE
mf: CH_3NO_3 mw: 77.05

PROP: Colorless liquid. Bp: 65° (explodes), d: 1.208 @ 20°/4°, vap d: 2.66, mp: −83°. Sltly sol in water; sol in alc, ether.

SYN: NITRIC ACID METHYL ESTER

DOT Classification: Forbidden.

SAFETY PROFILE: Poison by ingestion. Moderately toxic by inhalation. A dangerous fire and explosion hazard by spontaneous chemical reaction. A very shock- and heat-sensitive explosive. Explodes when heated to 65°C. It does not require external O_2 for combustion. A rocket fuel. When heated to decomposition it emits toxic fumes of NO_x.

MMF750   CAS: 624-91-9   **HR: 3**
METHYL NITRITE
mf: CH_3NO_2 mw: 61.05

PROP: Gas above 10.4°F. Mp: −17°, bp: −12°, d: 0.991 @ 15°. Sol in alc, ether.

SYN: NITROUS ACID, METHYL ESTER

CONSENSUS REPORTS: Reported in EPA TSCA Inventory.

DOT Classification: Forbidden.

SAFETY PROFILE: Moderately toxic by inhalation. Mutation data reported. Narcotic in high concentration. A very dangerous fire and explosion hazard when exposed to heat or flame. A heat-sensitive explosive more powerful than ethyl nitrite. When heated to decomposition it emits toxic fumes of NO_x.

MMG000   CAS: 129-15-7   **HR: 3**
2-METHYL-1-NITROANTHRAQUINONE
mf: $C_{15}H_9NO_4$ mw: 267.25

SYNS: 2-METHYL-1-NITRO-9,10-ANTHRACENEDIONE * NCI-C01923 * 1-NITRO-2-METHYLANTHRAQUINONE * 1-N-2-MA (RUSSIAN)

CONSENSUS REPORTS: IARC Cancer Review: GROUP 2B IMEMDT 7,56,87; Animal Sufficient Evidence IMEMDT 27,205,82. NCI Carcinogenesis Bioassay (feed); Clear Evidence: rat NCITR* NCI-CG-TR-29,78; (feed); Clear Evidence: mouse IJCNAW 19,117,77.

SAFETY PROFILE: Suspected carcinogen with experimental carcinogenic and neoplastigenic data. Moderately toxic by intraperitoneal route. Mutation data reported. When heated to decomposition it emits toxic fumes of NO_x.

MMN250   CAS: 443-48-1   **HR: 3**
2-METHYL-5-NITROIMIDAZOLE-1-ETHANOL
mf: $C_6H_9N_3O_3$ mw: 171.18

SYNS: ACROMONA * ANAGIARDIL * ATRIVYL * BAYER 5360 * BEXON * CLONT * CONT * DANIZOL * DEFLAMON-WIRKSTOFF * EFLORAN * ELYZOL * ENTIZOL * 1-(β-ETHYLOL)-2-METHYL-5-NITRO-3-AZAPYRROLE * EUMIN * FLAGEMONA * FLAGESOL * FLAGIL * FLAGYL * GIATRICOL * GINEFLAVIR

* 1-(β-HYDROXYETHYL)-2-METHYL-5-NITROIMID-
AZOLE * 1-(2-HYDROXYETHYL)-2-METHYL-5-NITRO-
IMIDAZOLE * 1-HYDROXYETHYL-2-METHYL-5-NITRO-
IMIDAZOLE * 1-(2-HYDROXY-1-ETHYL)-2-METHYL-5-
NITROIMIDAZOLE * KLION * MERONIDAL
* 2-METHYL-1-(2-HYDROXYETHYL)-5-NITROIMID-
AZOLE * 2-METHYL-3-(2-HYDROXYETHYL)-4-NITRO-
IMIDAZOLE * METRONIDAZ * METRONIDAZOL
* METRONIDAZOLO * MONAGYL * NALOX
* NEO-TRIC * NIDA * NOVONIDAZOL
* NSC-50364 * ORVAGIL * 1-(β-OXYETHYL)-2-
METHYL-5-NITROIMIDAZOLE * RP 8823 * SANA-
TRICHOM * SC 10295 * TRICHAZOL * TRICHO-
CIDE * TRICHOMOL * TRICHOMONACID "PHAR-
MACHIM" * TRICHOPOL * TRICOM
* TRICOWAS B * TRIKOJOL * TRIMEKS
* TRIVAZOL * VAGILEN * VAGIMID * VER-
TISAL

CONSENSUS REPORTS: IARC Cancer Re-
view: GROUP 2B IMEMDT 7,250,87; Animal
Sufficient Evidence IMEMDT 13,113,77. NTP
Fourth Annual Report On Carcinogens, 1984.
EPA Genetic Toxicology Program.

SAFETY PROFILE: Confirmed carcinogen with
experimental carcinogenic, neoplastigenic, and
tumorigenic data. Moderately toxic by ingestion, intraperitoneal, and subcutaneous routes.
Human systemic effects by ingestion: paresthesia, nerve or sheath structural changes, eye
changes, tremors, fever, jaundice and other liver
changes. Experimental teratogenic and reproductive effects. Human mutation data reported.
When heated to decomposition it emits toxic
fumes of NO_x.

MMP000   CAS: 70-25-7   **HR: 3**
N-METHYL-N'-NITRO-N-
NITROSOGUANIDINE

DOT: NA 1325
mf: $C_2H_5N_5O_3$ mw: 147.12

PROP: Crystals.

SYNS: METHYLNITRONITROSOGUANIDINE
* 1-METHYL-3-NITRO-1-NITROSOGUANIDINE
* N-METHYL-N'-NITRO-N-NITROSOGUANIDINE, not ex-
ceeding 25 grams in one outside packaging (DOT)
* N-METHYL-N-NITROSONITROGUANIDIN (GERMAN)
* 1-METHYL-1-NITROSO-3-NITROGUANIDINE
* METHYLNITROSOGUANIDINE * N-METHYL-N-
NITROSO-N'-NITROGUANIDINE * N-METYLO-N'-NI-
TRO-N-NITROZOGOUANIDYNY (POLISH) * MNG
* MNNG * N'-NITRO-N-NITROSO-N-METHYLGUANI-

DINE * N-NITROSO-N-METHYLNITROGUANIDINE
* NSC 9369 * RCRA WASTE NUMBER U163

CONSENSUS REPORTS: IARC Cancer Review: GROUP 2A IMEMDT 7,248,87; Animal
Sufficient Evidence IMEMDT 4,183,74. Reported in EPA TSCA Inventory. EPA Genetic
Toxicology Program.

DOT Classification: Flammable Solid; Label:
Flammable Solid.

SAFETY PROFILE: Suspected carcinogen with
experimental carcinogenic and tumorigenic
data. Poison by ingestion, intraperitoneal, and
intravenous routes. Moderately toxic by subcutaneous route. Experimental teratogenic and reproductive effects. Human mutation data reported. An explosive sensitive to heat or impact.
Flammable when exposed to heat or flame; can
react vigorously with oxidizing materials. When
heated to decomposition it emits very toxic
fumes of NO_x.

MMP750   CAS: 1074-98-2   **HR: 3**
3-METHYL-4-NITROPYRIDINE-1-OXIDE
mf: $C_6H_6N_2O_3$ mw: 154.14

SAFETY PROFILE: Questionable carcinogen
with experimental tumorigenic data. Mutation
data reported. When heated to decomposition
it emits toxic fumes of NO_x.

MMQ250   CAS: 4831-62-3   **HR: 3**
2-METHYL-4-NITROQUINOLINE-1-
OXIDE
mf: $C_{10}H_8N_2O_3$ mw: 204.20

SYN: 4-NITROQUINALDINE-N-OXIDE

SAFETY PROFILE: Questionable carcinogen
with experimental tumorigenic data. Human
mutation data reported. When heated to decomposition it emits toxic fumes of NO_x.

MMS200   CAS: 60153-49-3   **HR: 3**
3-METHYLNITROSAMINO-
PROPIONITRILE
mf: $C_4H_7N_3O$ mw: 113.14

SYNS: MNPN * PROPANENITRILE, 3-(METHYLNI-
TROSOAMINO)

CONSENSUS REPORTS: IARC Cancer Review: GROUP 2B IMEMDT 7,56,87, Animal
Sufficient Evidence IMEMDT 37,263,85

SAFETY PROFILE: Suspected carcinogen with
experimental carcinogenic and tumorigenic

data. When heated to decomposition it emits toxic fumes of NO_x.

MMS500   CAS: 64091-91-4   **HR: 3**
4-(N-METHYL-N-NITROSAMINO)-1-(3-PYRIDYL)-1-BUTANONE
mf: $C_{10}H_{13}N_3O_2$ mw: 207.26

SYNS: 4-(N-METHYL-N-NITROSOAMINO)-4-(3-PYRI-DYL)-1-BUTANONE * N-METHYL-N-NITROSO-4-OXO-4-(3-PYRIDYL)BUTYL AMINE * 4-(NITROSOAMINO-N-METHYL)-1-(3-PYRIDYL)-1-BUTANONE * 4-(N-NITROSO-N-METHYLAMINO)-1-(3-PYRIDYL) 1-BUTANONE * NNK

CONSENSUS REPORTS: IARC Cancer Review: GROUP 2B IMEMDT 7,56,87; Animal Sufficient Evidence IMEMDT 37,209,85. EPA Genetic Toxicology Program.

SAFETY PROFILE: Suspected carcinogen with experimental carcinogenic and neoplastigenic data. Mutation data reported. When heated to decomposition it emits toxic fumes of NO_x.

MMT000   CAS: 7417-67-6   **HR: 3**
METHYLNITROSOACETAMIDE
mf: $C_3H_6N_2O_2$ mw: 102.11

SYNS: METHYLNITROSOACETAMID (GERMAN) * N-METHYL-N-NITROSOACETAMIDE * N-NITROSO-N-METHYLACETAMIDE

CONSENSUS REPORTS: EPA Genetic Toxicology Program.

SAFETY PROFILE: Poison by ingestion. Questionable carcinogen with experimental tumorigenic data. Mutation data reported. When heated to decomposition it emits toxic fumes of NO_x.

MMT500   CAS: 4549-43-3   **HR: 3**
N-METHYL-N-NITROSOALLYLAMINE
mf: $C_4H_8N_2O$ mw: 100.14

SYNS: METHYLALLYLNITROSAMIN (GERMAN) * METHYLALLYLNITROSAMINE * N-METHYL-N-NITROSO-2-PROPEN-1-AMINE * N-NITROSOALLYL-METHYLAMINE * NITROSOMETHYLALLYLAMINE * N-NITROSOMETHYLALLYLAMINE

SAFETY PROFILE: Poison by ingestion and intravenous routes. Questionable carcinogen with experimental tumorigenic data. Mutation data reported. When heated to decomposition it emits toxic fumes of NO_x.

MMT750   CAS: 3684-97-7   **HR: 3**
2-(N-METHYL-N-NITROSO)AMINO-ACETONITRILE
mf: $C_3H_5N_3O$ mw: 99.11

SYNS: N-NITROSOMETHYLAMINACETONITRIL (GERMAN) * N-NITROSOMETHYLAMINOACETONITRILE

CONSENSUS REPORTS: Cyanide and its compounds are on the Community Right-To-Know List.

SAFETY PROFILE: Poison by ingestion. Questionable carcinogen with experimental tumorigenic data. When heated to decomposition it emits toxic fumes of NO_x and CN^-.

MMU250   CAS: 614-00-6   **HR: 3**
N-METHYL-N-NITROSOANILINE
mf: $C_7H_8N_2O$ mw: 136.17

SYNS: N-METHYL-N-NITROSOBENZENAMINE * METHYLPHENYLNITROSAMINE * MNA * NITROSOMETHYLANILINE * N-NITROSO-N-METHYLANILINE * N-NITROSOMETHYLPHENYLAMINE (MAK) * NMA * PHENYLMETHYLNITROSAMINE

CONSENSUS REPORTS: Reported in EPA TSCA Inventory. EPA Genetic Toxicology Program.

DFG MAK: Animal Carcinogen, Suspected Human Carcinogen.

SAFETY PROFILE: Confirmed carcinogen with experimental carcinogenic, neoplastigenic, and tumorigenic data. Poison by ingestion and intraperitoneal routes. Experimental teratogenic and reproductive effects. Mutation data reported. When heated to decomposition it emits toxic fumes of NO_x.

MMV000   CAS: 13860-69-0   **HR: 3**
N-METHYL-N-NITROSOBIURET
mf: $C_3H_6N_4O_3$ mw: 146.13

SYNS: 1-METHYL-1-NITROSOBIURET * N-METHYL-N-NITROSO-N'-CARBAMOYLUREA * N-NITROSO-N-METHYLBIURET

CONSENSUS REPORTS: EPA Genetic Toxicology Program.

SAFETY PROFILE: Moderately toxic by ingestion. Questionable carcinogen with experimental tumorigenic data. Mutation data reported. When heated to decomposition it emits toxic fumes of NO_x.

MMW775   CAS: 25355-61-7   **HR: 3**
1-METHYL-1-NITROSO-3-(p-CHLOROPHENYL)UREA
mf: $C_8H_8ClN_3O_2$ mw: 213.64

SYNS: 1-(p-CHLOROPHENYL)-3-METHYL-3-NITRO-SOUREA ∗ 3-(p-CHLOROPHENYL)-1-METHYL-1-NITRO-SOUREA ∗ N-METHYL-N'-(p-CHLOROPHENYL)-N-NI-TROSOUREA

SAFETY PROFILE: Suspected carcinogen with experimental carcinogenic data. Mutation data reported. When heated to decomposition it emits toxic fumes of Cl^- and NO_x.

MMX000   CAS: 33868-17-6   **HR: 3**
METHYLNITROSOCYANAMIDE
mf: $C_2H_3N_3O$ mw: 85.08

SYN: MNC

CONSENSUS REPORTS: EPA Extremely Hazardous Substances List.

SAFETY PROFILE: Poison by ingestion. Questionable carcinogen with experimental tumorigenic data. Mutation data reported. When heated to decomposition it emits toxic fumes of NO_x.

MMX250   CAS: 615-53-2   **HR: 3**
N-METHYL-N-NITROSOETHYL-CARBAMATE
mf: $C_4H_8N_2O_3$ mw: 132.14

SYNS: ETHYL ESTER of METHYLNITROSO-CARBAMIC ACID ∗ N-METHYL-N-NITROSOCARBAMIC ACID, ETHYL ESTER ∗ METHYLNITROSOURETHAN (GER-MAN) ∗ METHYLNITROSOURETHANE ∗ N-METHYL-N-NITROSO-URETHANE ∗ MNU ∗ NITROSOMETHYLURETHAN (GERMAN) ∗ NITRO-SOMETHYLURETHANE ∗ N-NITROSO-N-METHYLURETHANE ∗ NMUM ∗ NMUT ∗ RCRA WASTE NUMBER U178

CONSENSUS REPORTS: IARC Cancer Review: GROUP 2B IMEMDT 7,56,87; Animal Sufficient Evidence IMEMDT 4,211,74. Reported in EPA TSCA Inventory. EPA Genetic Toxicology Program.

SAFETY PROFILE: Suspected carcinogen with experimental carcinogenic and tumorigenic data. Poison by ingestion, intraperitoneal, subcutaneous, and intravenous routes. Moderately toxic by inhalation. Experimental teratogenic and reproductive effects. Mutation data reported. Has been implicated as a transplacental brain carcinogen. Combustible when exposed to heat, sparks, open flame, and powerful oxidizers. Explodes when heated. A storage hazard. When heated to decomposition it emits toxic fumes of NO_x.

MMY500   CAS: 21561-99-9   **HR: 3**
1-METHYL-1-NITROSO-3-PHENYLUREA
mf: $C_8H_9N_3O_2$ mw: 179.20

SYNS: N-METHYL-N-NITROSO-N'-PHENYLUREA ∗ METHYLPHENYLNITROSOUREA ∗ N-METHYL-N'-PHENYL-N-NITROSOUREA ∗ MPNU ∗ NITROSO-METHYLPHENYLUREA ∗ 3-PHENYL-1-METHYL-1-NI-TROSOHARNSTOFF (GERMAN)

SAFETY PROFILE: Suspected carcinogen with experimental carcinogenic and tumorigenic data. Mutation data reported. When heated to decomposition it emits toxic fumes of NO_x.

MNA000   CAS: 924-46-9   **HR: 3**
N-METHYL-N-NITROSO-1-PROPANAMINE
mf: $C_4H_{10}N_2O$ mw: 102.16

SYNS: METHYL-N-PROPYLNITROSAMINE ∗ METHYLPROPYLNITROSOAMINE ∗ MPN ∗ NITROSOMETHYL-N-PROPYLAMINE ∗ NITROSO-METHYLPROPYLAMINE

CONSENSUS REPORTS: EPA Genetic Toxicology Program.

SAFETY PROFILE: Suspected carcinogen with experimental carcinogenic, neoplastigenic, and tumorigenic data. Poison by subcutaneous route. Experimental reproductive effects. Mutation data reported. When heated to decomposition it emits toxic fumes of NO_x.

MNA250   CAS: 16395-80-5   **HR: 3**
N-METHYL-N-NITROSOPROPIONAMIDE
mf: $C_4H_8N_2O_2$ mw: 116.14

SYNS: METHYLNITROSO-PROPIONAMIDE ∗ METHYL-NITROSOPROPIONSAEUREAMID (GERMAN) ∗ METHYLNITROSOPROPIONYLAMIDE

SAFETY PROFILE: Questionable carcinogen with experimental tumorigenic data. When heated to decomposition it emits toxic fumes of NO_x.

MNA750   CAS: 684-93-5   **HR: 3**
N-METHYL-N-NITROSOUREA
mf: $C_2H_5N_3O_2$ mw: 103.10

SYNS: METHYLNITROSO-HARNSTOFF (GERMAN) ∗ N-METHYL-N-NITROSO-HARNSTOFF (GERMAN)

* METHYLNITROSOUREA * 1-METHYL-1-NITROSO-
UREA * METHYLNITROSOUREE (FRENCH) * MNU
* N-NITROSO-N-METHYLCARBAMIDE * N-NITROSO-
N-METHYL-HARNSTOFF (GERMAN) * NITROSO-
METHYLUREA * N-NITROSO-N-METHYLUREA
* 1-NITROSO-1-METHYLUREA * NMH * NMU
* NSC 23909 * RCRA WASTE NUMBER U177
* SKI 24464 * SRI 859

CONSENSUS REPORTS: IARC Cancer Re-
view: GROUP 2A IMEMDT 7,56,87; Human
Limited Evidence IMEMDT 17,227,78; Animal
Sufficient Evidence IMEMDT 17,227,78,
IMEMDT 1,125,72. NTP Fourth Annual Report
On Carcinogens, 1984. EPA Genetic Toxicol-
ogy Program. Community Right-To-Know List.
Reported in EPA TSCA Inventory.

SAFETY PROFILE: Confirmed carcinogen with
experimental carcinogenic, neoplastigenic, and
tumorigenic data. Poison by ingestion, implant,
intraperitoneal, subcutaneous, and intravenous
routes. Experimental teratogenic and reproductive effects. Human mutation data reported. Explodes at room temperature. Can detonate with
$(KOH + CH_2Cl_2)$. When heated to decomposition it emits toxic fumes of NO_x.

MND275   CAS: 111-12-6   HR: 2
METHYL 2-OCTYNOATE
mf: $C_9H_{14}O_2$ mw: 154.23

PROP: Colorless to sltly yellow liquid; powerful, unpleasant odor; violet odor when diluted.
D: 0.919, refr index: 1.446, flash p: +212°F.
Sol in fixed oils; sltly sol in propylene glycol;
insol in glycerin

SYNS: FEMA No. 2729 * FOLIONE * METHYL
HEPTINE CARBONATE * METHYL 2-OCTINATE

CONSENSUS REPORTS: Reported in EPA
TSCA Inventory.

SAFETY PROFILE: Moderately toxic by ingestion and skin contact. A moderate skin and eye
irritant. A combustible liquid. When heated to
decomposition it emits acrid smoke and irritating
fumes.

MNH000   CAS: 298-00-0   HR: 3
METHYL PARATHION

DOT: NA 2783
mf: $C_8H_{10}NO_5PS$ mw: 263.22

PROP: Crystals. Vap d: 9.1, mp: 37-38°, d:
1.358 @ 20°/4°. Sol in most organic solvents.

SYNS: A-GRO * AZOFOS * AZOPHOS
* BAY E-601 * BAY 11405 * BLADAN-M
* CEKUMETHION * DALF * DEVITHION
* O,O-DIMETHYL-O-p-NITROFENYLESTER KYSELINY
THIOFOSFORECNE (CZECH) * O,O-DIMETHYL-O-(4-NI-
TROFENYL)-MONOTHIOFOSFAAT (DUTCH)
* DIMETHYL p-NITROPHENYL MONOTHIOPHOSPHATE
* O,O-DIMETHYL-O-(4-NITRO-PHENYL)-MONOTHIO-
PHOSPHAT (GERMAN) * O,O-DIMETHYL-O-(p-NITRO-
PHENYL) PHOSPHOROTHIOATE * O,O-DIMETHYL-
O-(4-NITROPHENYL) PHOSPHOROTHIOATE
* DIMETHYL 4-NITROPHENYL PHOSPHOROTHIONATE
* O,O-DIMETHYL-O-(p-NITROPHENYL)-THIONOPHOS-
PHAT (GERMAN) * O,O-DIMETHYL-O-(4-NITROPHE-
NYL)-THIONOPHOSPHAT (GERMAN) * DIMETHYL-p-
NITROPHENYL THIONPHOSPHATE * DIMETHYL
p-NITROPHENYL THIOPHOSPHATE * O,O-DIMETHYL-
O-p-NITROPHENYL THIOPHOSPHATE * DIMETHYL
PARATHION * O,O-DIMETIL-O-(4-NITRO-FENIL)
MONOTIOFOSFATO (ITALIAN) * DREXEL METHYL
PARATHION 4E * ENT 17,292 * FOLIDOL M
* GEARPHOS * ME-PARATHION * MEPATON
* MEPTOX * METACIDE * METAFOS * META-
PHOR * METAPHOS * METHYL-E 605
* METHYL FOSFERNO * METHYL NIRAN
* METHYLTHIOPHOS * METILPARATION (HUN-
GARIAN) * METRON * METYLOPARATION (POL-
ISH) * METYLPARATION (CZECH) * NCI-C02971
* p-NITROPHENYLDIMETHYLTHIONOPHOSPHATE
* NITROX * OLEOVOFOTOX * PARAPEST M-50
* PARATAF * M-PARATHION * PARATHION
METHYL * PARATHION-METILE (ITALIAN)
* PARATOX * PENNCAP-M * RCRA WASTE NUM-
BER P071 * SINAFID M-48 * SIXTY-THREE SPECIAL
E.C. INSECTICIDE * TEKWAISA * THIOPHENIT
* THIOPHOSPHATE de O,O-DIMETHYLE et de O-(4-NI-
TROPHENYLE) (FRENCH) * THYLFAR M-50
* TOLL * VERTAC METHYL PARATHION TECH-
NISCH 80% * VOFATOX * WOFATOS * WOFA-
TOX * WOFOTOX

CONSENSUS REPORTS: IARC Cancer Review: GROUP 3 IMEMDT 7,56,87; Animal
No Evidence IMEMDT 30,131,83. NCI Carcinogenesis Bioassay (feed); No Evidence:
mouse, rat NCITR* NCI-CG-TR-157,79. EPA
Genetic Toxicology Program. EPA Extremely
Hazardous Substances List.

OSHA PEL: TWA 0.2 mg/m³ (skin)
ACGIH TLV: TWA 0.2 mg/m³ (skin)

NIOSH REL: (Methyl Parathion) TWA 0.2
mg/m³

DOT Classification: Poison B; Label: Poison.

SAFETY PROFILE: Poison by inhalation, ingestion, skin contact, subcutaneous, intravenous, intraperitoneal, and possibly other routes. Experimental teratogenic and reproductive effects. Questionable carcinogen. Human mutation data reported. A cholinesterase inhibitor type of insecticide. When heated to decomposition it emits very toxic fumes of NO_x, PO_x, and SO_x.

MNH500   CAS: 54363-49-4   HR: 2
METHYLPENTADIENE

DOT: UN 2461
mf: C_6H_{10} mw: 82.16

PROP: Liquid. Bp: 75-77°, flash p: $<-4°F$, d: 0.7184 @ 20°/4°, vap d: 2.83.

DOT Classification: Flammable Liquid; Label: Flammable liquid.

SAFETY PROFILE: Mildly toxic by inhalation. A skin irritant. A very dangerous fire hazard when exposed to heat, flame, or oxidizers. Keep away from heat and open flame. To fight fire, use foam, CO_2, dry chemical. When heated to decomposition it emits acrid smoke and irritating fumes.

MNI500   CAS: 96-14-0   HR: 3
3-METHYLPENTANE

DOT: UN 1208/UN 2462
mf: C_6H_{14} mw: 86.18

PROP: Flash p: 19.4°F, lel: 1.2%, uel: 7.0%, bp: 63.3°, fp: $-118°$ (sets to a glass), d: 0.664 @ 20°/4°, vap press: 100 mm @ 10.5°, vap d: 2.97.

SYN: DIETHYLMETHYL METHANE

CONSENSUS REPORTS: Reported in EPA TSCA Inventory.

DOT Classification: Flammable Liquid; Label: Flammable Liquid.

SAFETY PROFILE: May have narcotic or anesthetic properties. A very dangerous fire hazard when exposed to heat or flame; can react vigorously with oxidizing materials. Explosive in the form of vapor when exposed to heat or flame. To fight fire, use foam, CO_2, dry chemical. When heated to decomposition it emits acrid smoke and irritating fumes.

MNL775   CAS: 105-29-3   HR: 3
3-METHYL-2-PENTEN-4-YN-1-OL

DOT: UN 2705
mf: C_6H_8O mw: 96.13

DOT Classification: Corrosive Material; Label: Corrosive

SAFETY PROFILE: Decomposes violently when heated above 155°C. May polymerize exothermically when heated above 100°C. Polymerization is catalyzed by traces of acid or base. Reaction with sodium hydroxide forms an explosive salt. The cis- isomer readily cyclizes to form the dangerous 2,3-dimethylfuran. When heated to decomposition it emits acrid smoke and irritating fumes.

MNM500   CAS: 302-66-9   HR: 3
METHYLPENTYNOL CARBAMATE
mf: $C_7H_{11}NO_2$ mw: 141.19

PROP: Sltly water-sol crystals. Mp: 57°, bp: 121° @ 16 mm.

SYNS: ANANSIOL * CALMINOL * CARBAMATE de METHYLPENTINOL (FRENCH) * CARBAMIC ACID-1-ETHYL-1-METHYL-2-PROPYNYL ESTER * CARBAMIC ACID-2-ETHYNYL-2-BUTYL ESTER * 3-CARBAMOYL-OXY-3-METHYL-4-PENTYNE * COMESA * 1-ETHYL-1-METHYL-2-PROPYNYL CARBAMATE * 2-ETHYNYL-2-BUTYL CARBAMATE * FORMARIN * MEPARFYNOL CARBAMATE * MEPENTAMATE * MEPENTAMATO * METHYLPARAFYNOL CARBAMATE * 3-METHYL-PENTIN-(1)-OL-(3) (GERMAN) * 3-METHYL-1-PENTYN-3-OL CARBAMATE * 3-MPC * OBLIVON C * OBLIVON CARBAMATE * OLOSED * OVETTEN * PENTIN C * PLACIDAL * PLACIDAS * PSICOPLEGIL * PSICOSEDINA * TRUSONO * USAF EL-108 * VEREDEN

SAFETY PROFILE: Poison by ingestion and intraperitoneal routes. Moderately toxic by subcutaneous route. A sedative and tranquilizer which can cause central nervous system depression and death by overdose. Toxic effects are enhanced with the use of alcohol and barbiturates. When heated to decomposition it emits toxic fumes of NO_x.

MNN000   CAS: 685-09-6   HR: 3
METHYL PERFLUOROMETHACRYLATE
mf: $C_5H_3F_5O_2$ mw: 190.08

SYNS: 3,3-DIFLUORO-2-(TRIFLUOROMETHYL)ACRYLIC ACID, METHYL ESTER * 3,3-DIFLUORO-2-(TRIFLUOROMETHYL)-2-PROPENOIC ACID, METHYL ESTER

SAFETY PROFILE: Poison by ingestion, intravenous, and intraperitoneal routes. When heated to decomposition it emits toxic fumes of F^-.

MNN250   CAS: 3871-82-7   **HR: 3**
METHYLPERIDOL HYDROCHLORIDE
mf: $C_{22}H_{26}FNO_2 \cdot ClH$ mw: 391.95

SYNS: 4'-FLUORO-4-(4-HYDROXY-4-p-TOLYLPIPERI-DINO)BUTYROPHENONE, HYDROCHLORIDE * 1-(4-FLUOROPHENYL)-4-(4-HYDROXY-4-(4-METHYLPHENYL)-1-PIPERIDINYL)-1-BUTANONE HYDROCHLORIDE * LUVATRENE * METHYLPERIDOL * MOPERONE CHLORHYDRATE * MOPERONE HYDROCHLORIDE * R 1658

SAFETY PROFILE: Poison by ingestion, subcutaneous, and intravenous routes. When heated to decomposition it emits very toxic fumes of F^-, HCl, and NO_x.

MNQ000   CAS: 113-45-1   **HR: 3**
METHYL PHENIDYL ACETATE
mf: $C_{14}H_{19}NO_2$ mw: 233.34

SYNS: CALOCAIN * CENTEDEIN * CENTREDIN * MERIDIL * METHYLPHENIDAN * METHYL PHENIDATE * METHYL α-PHENYL-α-(2-PIPERI-DYL)ACETATE * NCI-C56280 * PHENIDYLATE * α-PHENYL-2-PIPERIDINEACETIC ACID METHYL ESTER * PLIMASINE * RITALIN * RITALINE * RITCHER WORKS

SAFETY PROFILE: Poison experimentally by ingestion, intraperitoneal, intravenous, and subcutaneous routes. Moderately toxic to humans by intravenous route. Human systemic effects by intravenous route: dyspnea. Experimental teratogenic and reproductive effects. When heated to decomposition it emits toxic fumes of NO_x.

MNR250   CAS: 140-39-6   **HR: 2**
4-METHYLPHENYL ACETATE
mf: $C_9H_{10}O_2$ mw: 150.19

PROP: Colorless liquid; strong floral odor. D: 1.044 @ 16°, refr index: 1.499-1.502, bp: decomp @ 360°, mp: 220°, vap d: 5.18, flash p: 203°F. Sol in fixed oils, propylene glycol, misc in alc and ether; insol in water, glycerin.

SYNS: ACETIC ACID-4-METHYLPHENYL ESTER * p-ACETOXYTOLUENE * 4-ACETOXYTOLUENE * p-CRESOL ACETATE * p-CRESYL ACETATE (FCC) * FEMA No. 3073 * 4-METHYLBENZOIC ACID

METHYL ESTER * p-METHYLPHENYL ACETATE * NARCEOL * PARACRESYL ACETATE * p-TOLYL ACETATE * p-TOLYL ETHANOATE

CONSENSUS REPORTS: Reported in EPA TSCA Inventory.

SAFETY PROFILE: Moderately toxic by ingestion and skin contact. Combustible liquid. When heated to decomposition it emits toxic smoke and irritating fumes.

MNR500   CAS: 621-90-9   **HR: 3**
N-METHYL-p-(PHENYLAZO)ANILINE
mf: $C_{13}H_{13}N_3$ mw: 211.29

SYNS: MAB * 4-(METHYLAMINO)AZOBENZENE * N-METHYL-4-AMINOAZOBENZENE * N-METHYL-p-AMINOAZOBENZENE * p-MONOMETHYLAMINOAZO-BENZENE * 4-MONOMETHYLAMINOAZOBENZENE

CONSENSUS REPORTS: EPA Genetic Toxicology Program.

SAFETY PROFILE: Moderately toxic by subcutaneous route. Experimental teratogenic and reproductive effects. Questionable carcinogen with experimental carcinogenic, neoplastigenic, and tumorigenic data. Mutation data reported. When heated to decomposition it emits toxic fumes of NO_x.

MNT075   **HR: 2**
METHYL PHENYLCARBINYL ACETATE
mf: $C_{10}H_{12}O_2$ mw: 164.20

PROP: Colorless liquid; gardenia odor. D: 1.023, refr index: 1.493-1.497, flash p: 176°F. Sol in fixed oils, glycerin; insol in water.

SYNS: FEMA No. 2684 * α-PHENYL ETHYL ACETATE

SAFETY PROFILE: Combustible liquid. When heated to decomposition it emits acrid smoke and irritating fumes.

MNT500   CAS: 20240-98-6   **HR: 3**
1-(2-METHYLPHENYL)-3,3-DIMETHYLTRIAZENE
mf: $C_9H_{13}N_3$ mw: 163.25

SYNS: 3,3-DIMETHYL-1-(o-METHYLPHENYL)TRIAZENE * 3,3-DIMETHYL-1-(o-TOLYL)TRIAZENE * 1-(o-METHYLPHENYL)-3,3-DIMETHYL-TRIAZEN (GERMAN) * 1-(o-METHYLPHENYL)-3,3-DIMETHYL-TRIAZENE

SAFETY PROFILE: Poison by ingestion. Moderately toxic by subcutaneous route. Questionable carcinogen with experimental carcinogenic data. When heated to decomposition it emits toxic fumes of NO_x.

MNU250   CAS: 13256-11-6   **HR: 3**
METHYL-PHENYLETHYL-NITROSAMINE
mf: $C_9H_{12}N_2O$ mw: 164.23

SYNS: N-METHYL-N-NITROSOPHENETHYLAMINE
* METHYL(2-PHENYLAETHYL)NITROSAMIN (GERMAN)
* N-NITROSO-N-METHYL-2-PHENYLETHYLAMINE

SAFETY PROFILE: Poison by ingestion. Questionable carcinogen with experimental carcinogenic and tumorigenic data. Mutation data reported. When heated to decomposition it emits toxic fumes of NO_x.

MNV750   CAS: 1707-14-8   **HR: 3**
3-METHYL-2-PHENYLMORPHOLINE HYDROCHLORIDE
mf: $C_{11}H_{15}NO \cdot ClH$ mw: 213.73

SYNS: A 66 HYDROCHLORIDE   * MARSIN
* 3-METHYL-2-PHENYLTETRAHYDRO-2H-1,4-OXAZINE HYDROCHLORIDE   * NEO-ZINE   * PHENMETRAZINE HYDROCHLORIDE   * 2-PHENYL-3-METHYLTETRAHYDRO-1,4-OXAZINE HYDROCHLORIDE   * PRELUDIN HYDROCHLORIDE   * PROBESE-P HYDROCHLORIDE
* PSYCHAMINE A 66 HYDROCHLORIDE
* USAF GE-1

SAFETY PROFILE: Poison by ingestion, intravenous, intraperitoneal, and subcutaneous routes. Human teratogenic effects by ingestion: developmental abnormalities of the respiratory and gastrointestinal systems, and effects on newborn including neonatal measures or effects. An experimental teratogen. When heated to decomposition it emits very toxic fumes of NO_x and HCl.

MNY750   CAS: 57962-60-4   **HR: 3**
5-METHYL-1-PHENYL-2-(PYRROLIDINYL)IMIDAZOLE
mf: $C_{14}H_{17}N_3$ mw: 227.34

SYN: METHYL-5 PHENYL-1 (PYRROLIDINYL-1)-2 IMIDAZOLE (FRENCH)

SAFETY PROFILE: Poison by ingestion, intravenous, and intraperitoneal routes. When heated to decomposition it emits toxic fumes of NO_x.

MOA250   CAS: 73840-42-3   **HR: 3**
1-METHYL-4-(PHENYLTHIO) PYRIDINIUM IODIDE
mf: $C_{12}H_{12}NS \cdot I$ mw: 329.21

SAFETY PROFILE: Poison by intravenous and intraperitoneal routes. When heated to decomposition it emits very toxic fumes of NO_x, SO_x, and I^-.

MOB250   CAS: 18466-11-0   **HR: 3**
METHYLPHOSPHODITHIOIC ACID-S-(((p-CHLOROPHENYL)THIO)METHYL)-O-METHYL ESTER
mf: $C_9H_{12}ClOPS_2$ mw: 266.75

SYNS: ENT 27,180   * N 4548   * STAUFFER N-4548

SAFETY PROFILE: Poison by ingestion and subcutaneous routes. When heated to decomposition it emits very toxic fumes of Cl^-, PO_x, and SO_x.

MOB399   CAS: 676-97-1   **HR: 3**
METHYL PHOSPHONIC DICHLORIDE

DOT: NA 9206
mf: CH_3Cl_2OP mw: 132.91

CONSENSUS REPORTS: EPA Extremely Hazardous Substances List. Reported in EPA TSCA Inventory.

DOT Classification: Corrosive Material; Label: Corrosive and Poison.

SAFETY PROFILE: Poison by inhalation. A corrosive irritant to the eyes, skin, and mucous membranes. When heated to decomposition it emits toxic fumes of Cl^- and PO_x.

MOC250   CAS: 676-83-5   **HR: 3**
METHYLPHOSPHONOUS DICHLORIDE

DOT: UN 2845

CONSENSUS REPORTS: Reported in EPA TSCA Inventory.

DOT Classification: Flammable Liquid; Label: Flammable Liquid.

SAFETY PROFILE: A poison. A corrosive irritant to the skin, eyes, and mucous membranes. Flammable when exposed to heat or flame; can react vigorously with oxidizing materials. When heated to decomposition it emits very toxic fumes of Cl^- and PO_x.

MOD250   CAS: 109-01-3   **HR: 3**
N-METHYLPIPERAZINE
mf: $C_5H_{12}N_2$ mw: 100.19

PROP: A hygroscopic solid; typical amine-like odor. D: 0.9031 20°/20°, mp: 65.5°, bp: 139°, flash p: 108°F (OC), vap d: 3.5.

SYN: 1-METHYLPIPERAZINE

CONSENSUS REPORTS: Reported in EPA TSCA Inventory.

SAFETY PROFILE: Poison by intraperitoneal route. Moderately toxic by inhalation, ingestion, and skin contact. A skin and severe eye irritant. Flammable when exposed to heat or flame; can react with oxidizing materials. To fight fire, use alcohol foam, CO_2, dry chemical. When heated to decomposition it emits toxic fumes of NO_x.

MOG500   CAS: 626-67-5   HR: 3
N-METHYLPIPERIDINE

DOT: UN 2399
mf: $C_6H_{13}N$ mw: 99.20

PROP: Liquid. D: 0.821 @ 15°, bp: 107°, flash p: <73.4°F.

SYN: 1-METHYLPIPERIDINE (DOT)

CONSENSUS REPORTS: Reported in EPA TSCA Inventory.

DOT Classification: Flammable Liquid; Label: Flammable Liquid.

SAFETY PROFILE: Poison by subcutaneous route. A very dangerous fire hazard when exposed to heat or flame. When heated to decomposition it emits toxic fumes of NO_x.

MOK000   CAS: 73790-27-9   HR: 3
METHYL-4-(3-PIPERIDINOPROPIONYL-AMINO)SALICYLATE, METHIODIDE
mf: $C_{17}H_{25}N_2O_4 \cdot I$ mw: 448.34

SYN: 4-(3-PIPERIDINOPROPIONAMIDO) SALICYCLIC ACID METHYL ESTER, METHIODIDE

SAFETY PROFILE: Poison by intraperitoneal and intravenous routes. When heated to decomposition it emits very toxic fumes of NO_x and I^-.

MON250   CAS: 3321-80-0   HR: 3
N-METHYL-3-PIPERIDYL BENZILATE
mf: $C_{20}H_{23}NO_3$ mw: 325.44

SYNS: BENZILIC ACID-1-METHYL-3-PIPERIDYL ESTER * JB 336

SAFETY PROFILE: Poison by intravenous and intraperitoneal routes. When heated to decomposition it emits toxic fumes of NO_x.

MOO250   CAS: 50-52-2   HR: 3
10-(2-(1-METHYL-2-PIPERIDYL)ETHYL)-2-(METHYLTHIO)PHENOTHIAZINE
mf: $C_{21}H_{26}N_2S_2$ mw: 370.61

SYNS: MALLOROL * MELERIL * MELLARIL * MELLERETTE * MELLERETTEN * MELLERIL * 2-METHYLMERCAPTO-10-(2-N-METHYL-2-PIPERI-DYL)ETHYL)PHENOTHIAZINE * SONAPAX * THIORIDAZIN * THIORIDAZINE * TP-21

CONSENSUS REPORTS: EPA Genetic Toxicology Program.

SAFETY PROFILE: Human poison by ingestion. Experimental poison by ingestion, subcutaneous, and intraperitoneal routes. Human systemic effects by ingestion: visual field and retinal changes, toxic psychosis, parasympatholytic, and heart rate change. Experimental reproductive effects. Mutation data reported. An antipsycotic and sedative. When heated to decomposition it emits very toxic fumes of NO_x and SO_x.

MOO500   CAS: 130-61-0   HR: 3
10-(2-(1-METHYL-2-PIPERIDYL)ETHYL)-2-METHYLTHIOPHENOTHIAZINE HYDROCHLORIDE
mf: $C_{21}H_{26}N_2S_2 \cdot ClH$ mw: 407.07

SYNS: MELLARIL HYDROCHLORIDE * 2-METHYL-MERCAPTO-10-(2-(N-METHYL-2-PIPERIDYL)ETHYLPHE-NOTHIAZINE HYDROCHLORIDE * THIORIDAZINE HY-DROCHLORIDE * THORIDAZINE HYDROCHLORIDE * TIORIDAZIN * TP-21 * USAF SZ-3 * USAF SZ-B

CONSENSUS REPORTS: EPA Genetic Toxicology Program.

SAFETY PROFILE: Poison by ingestion, intravenous and intraperitoneal routes. Mutation data reported. When heated to decomposition it emits very toxic fumes of NO_x, SO_x, and HCl.

MOO750   CAS: 314-03-4   HR: 3
9-(1-METHYL-4-PIPERIDYLIDENE) THIOXANTHENE
mf: $C_{19}H_{19}NS$ mw: 293.45

SYNS: BP 400 * CALMIXENE

SAFETY PROFILE: Poison by ingestion and intravenous routes. When heated to decomposition it emits very toxic fumes of NO_x and SO_x.

MOP500   CAS: 60706-49-2   **HR: 3**
9-(METHYL-2-PIPERIDYL)
METHYLCARBAZOLE
mf: $C_{19}H_{22}N_2$ mw: 278.43

SYN: 9-(1-METHYL-PIPERIDYL-(2)-METHYL)-CARBAZOL
(GERMAN)

SAFETY PROFILE: Poison by intravenous and intraperitoneal routes. Moderately toxic by ingestion and subcutaneous routes. When heated to decomposition it emits toxic fumes of NO_x.

MOR500   CAS: 83-43-2   **HR: 2**
METHYLPREDNISOLONE
mf: $C_{22}H_{30}O_5$ mw: 374.52

PROP: Crystals. Mp: 228-237°.

SYNS: MEDROL * MEDROL DOSEPAK
* MEDRONE * Δ^1-6-α-METHYLHYDROCORTISONE
* 6-α-METHYLPREDNISOLONE * METRISONE
* NSC-19987 * 11-β,17,21-TRIHYDROXY-6-α-METH-
YLPREGNA-1,4-DIENE-3,20-DIONE * 11-β,17-α,21-
TRIHYDROXY-6-α-METHYL-1,4-PREGNADIENE-3,20-
DIONE * URBASON * URBASONE * WYACORT

CONSENSUS REPORTS: Reported in EPA TSCA Inventory.

SAFETY PROFILE: Moderately toxic by intraperitoneal route. A steroid hormone. When heated to decomposition it emits acrid smoke and irritating fumes.

MOR750   CAS: 75-28-5   **HR: 3**
2-METHYLPROPANE
DOT: UN 1075/UN 1969
mf: C_4H_{10} mw: 58.14

PROP: Colorless gas. Bp: −11.7°, lel: 1.9%, uel: 8.5%, fp: −160°, d: 0.5572 @ 20°, autoign temp: 864°F, vap d: 2.01.

SYN: ISOBUTANE

CONSENSUS REPORTS: Reported in EPA TSCA Inventory.

DOT Classification: Label: Flammable Gas.

SAFETY PROFILE: An asphyxiant. A common air contaminant. A very dangerous fire and explosion hazard when exposed to heat, flame, or oxidizers. To fight fire, stop flow of gas. When heated to decomposition it emits acrid smoke and irritating fumes.

MOS000   CAS: 75-66-1   **HR: 3**
2-METHYL-2-PROPANETHIOL
mf: $C_4H_{10}S$ mw: 90.20

PROP: Mobile liquid; heavy skunk odor. Mp: −0.5°, bp: 63.7-64.2°, d: 0.79-0.82 @ 15.5°/15.5°, flash p: <−20°F, vap d: 3.1, n (25/D) 1.41984. Sltly sol in water; very sol in alc, ether, and liquid H_2S.

SYNS: tert-BUTANETHIOL * tert-BUTYL MERCAP-
TAN

CONSENSUS REPORTS: Reported in EPA TSCA Inventory.

SAFETY PROFILE: Moderately toxic by intraperitoneal route. Mildly toxic by ingestion. An eye irritant. A very dangerous fire hazard when exposed to heat or flame. Can react vigorously with oxidizing materials. To fight fire, use alcohol foam, dry chemical, mist, fog. When heated to decomposition or on contact with acid or acid fumes it emits highly toxic fumes of SO_x.

MOT000   CAS: 554-12-1   **HR: 3**
METHYL PROPIONATE
DOT: UN 1248
mf: $C_4H_8O_2$ mw: 88.12

PROP: Colorless liquid. Mp: −87.0°, bp: 79.8°, flash p: 28°F (CC) −2°C, d: 0.937 @ 4°, autoign temp: 876°F, vap press: 40 mm @ 11.0°, vap d: 3.03, lel: 2.50%, uel: 13%, d: 0.915 @ 20°/4°. Sol in water @ 20°; misc in alc and ether.

SYNS: METHYL PROPANOATE * METHYL PROPY-
LATE * PROPANOIC ACID, METHYL ESTER
* PROPIONATE de METHYLE (FRENCH)

CONSENSUS REPORTS: Reported in EPA TSCA Inventory.

DOT Classification: Flammable Liquid; Label: Flammable Liquid.

SAFETY PROFILE: Moderately toxic by ingestion. Mildly toxic by inhalation. A skin irritant. A very dangerous fire hazard when exposed to heat, flame, or oxidizers. Explosive in the form of vapor when exposed to heat or flame. To fight fire, use foam, CO_2, dry chemical. When heated to decomposition it emits acrid smoke and irritating fumes.

MOU750   CAS: 2917-19-3   **HR: 3**
3-(1-METHYLPROPYL)-6-CHLORO-
PHENYL METHYLCARBAMATE
mf: $C_{12}H_{16}ClNO_2$ mw: 241.74

SYNS: CAL CHEM 5655 * CHEVRON RE 5655
* 2-CHLORO-5-(1-METHYLPROPYL)PHENYL METHYL-

CARBAMATE * ENT 27,128 * METHYLCARBAMIC ACID-3-sec-BUTYL-6-CHLOROPHENYL ESTER * METH-YLCARBAMIC ACID-2-CHLORO-5-(1-METHYLPROPYL) PHENYL ESTER * ORTHO-5655 * RE 5655

SAFETY PROFILE: Poison by ingestion. When heated to decomposition it emits very toxic fumes of Cl^- and NO_x.

MOV000   CAS: 3766-81-2   HR: 3
2-(1-METHYLPROPYL)PHENYL METHYLCARBAMATE
mf: $C_{12}H_{17}NO_2$ mw: 207.30

SYNS: BASSA * BAYCARD * BPMC * o-sec-BUTYLPHENYL METHYLCARBAMATE * 2-sec-BUTYL-PHENYL-N-METHYLCARBAMATE * CARVIL * HOPCIN * OSBAC

CONSENSUS REPORTS: Reported in EPA TSCA Inventory. EPA Genetic Toxicology Program.

SAFETY PROFILE: Poison by ingestion, skin contact, intravenous and intraperitoneal routes. Used as an insecticide. When heated to decomposition it emits toxic fumes of NO_x.

MOV500   CAS: 4268-36-4   HR: 3
2-METHYL-2-PROPYLTRIMETHYLENE BUTYLCARBAMATE CARBAMATE
mf: $C_{13}H_{26}N_2O_4$ mw: 274.41

SYNS: 2-(((AMINOCARBONYL)OXY)METHYL)-2-METH-YLPENTYL ESTER BUTYL CARBAMIC ACID * BENVIL * N-N-BUTYL-2-METHYL-2-PROPYL-1,3-PROPANEDIOL DICARBAMATE * N-BUTYL-2-METHYL-2-PROPYL-1,3-PROPANEDIOL DICARBAMATE * CARBAMIC ACID, ESTER with 2-(HYDROXYMETHYL)-2-METHYLPENTYL BUTYLCARBAMATE * CARBAMIC ACID, ESTER with 2-METHYL-2-PROPYL-1,3-PROPANEDIOL BUTYLCAR-BAMATE * EFFISAX * 2-(HYDROXYMETHYL)-2-(METHYLPENTYL) BUTYLCARBAMATE CARBAMATE * 2-(HYDROXYMETHYL)-2-METHYLPENTYL ESTER, CARBAMATE, BUTYL CARBAMIC ACID * IDALENE * 2-METHYL-2-PROPYL-1,3-PROPANEDIOL BUTYLCAR-BAMATE CARBAMATE * NOSPAN * SOLACEN * SOLACIN * TIBAMATO * TYBAMATE * TYBATRAN * W 713

SAFETY PROFILE: Poison by intravenous route. Moderately toxic by ingestion and intraperitoneal routes. Human systemic effects by ingestion: somnolence, hallucinations or distorted perceptions, and nausea or vomiting. When heated to decomposition it emits toxic fumes of NO_x.

MOW750   CAS: 109-08-0   HR: 2
2-METHYLPYRAZINE
mf: $C_5H_6N_2$ mw: 94.13

PROP: Liquid; nutty, cocoa odor. Mp: $-29°$, bp: 133° @ 737 mm, flash p: 122°F (COC), d: 1.0224 @ 25°/25°, refr index: 1.504, vap d: 3.2. Misc with water, alc, acetone, fixed oils.

SYN: FEMA No. 3309

CONSENSUS REPORTS: Reported in EPA TSCA Inventory.

SAFETY PROFILE: Moderately toxic by ingestion and intraperitoneal routes. Mutation data reported. Combustible liquid when exposed to heat, flame or oxidizers. Can react with oxidizing materials. To fight fire, use water spray, foam, dry chemical, CO_2. When heated to decomposition it emits highly toxic fumes of NO_x.

MOX250   CAS: 108-34-9   HR: 3
METHYLPYRAZOLYL DIETHYLPHOSPHATE
mf: $C_8H_{15}N_2O_4P$ mw: 234.22

SYNS: O,O-DIAETHYL-O-(3-METHYL-1H-PYRAZOL-5-YL)-PHOSPHAT (GERMAN) * O,O-DIETHYL-O-(3-METHYL-1H-PYRAZOL-5-YL)-FOSFAAT (DUTCH) * DIETHYL-3-METHYL-5-PYRAZOLYL PHOSPHATE * O,O-DIETHYL-O-(3-METHYL-5-PYRAZOLYL) PHOS-PHATE * O,O-DIETIL-O-(3-METIL-1H-PIRAZOL-5-IL)-FOSFATO (ITALIAN) * ENT 24,723 * 3-METHYL-PYRAZOLYL-5-DIETHYLPHOSPHATE * PHOSPHATE de DIETHYLE et de 3-METHYL-5-PYRAZOLYLE (FRENCH) * PHOSPHORIC ACID-DIETHYL-(3-METHYL-5-PYRAZ-OLYL) ESTER * PIRAZOXON (ITALIAN)

SAFETY PROFILE: Poison by ingestion and subcutaneous routes. When heated to decomposition it emits very toxic fumes of NO_x and PO_x.

MOY000   CAS: 109-06-8   HR: 3
2-METHYLPYRIDINE
DOT: UN 2313
mf: C_6H_7N mw: 93.14

PROP: Colorless liquid; strong unpleasant odor. Mp: $-70°$, bp: 129°, flash p: 102°F (OC), d: 0.95 @ 15°/4°, autoign temp: 1000°F, vap press: 10 mm @ 24.4°, vap d: 3.2. Very sol in water; misc in alc and ether.

SYNS: α-METHYLPYRIDINE * 2-PICOLINE * α-PICOLINE * o-PICOLINE (DOT) * RCRA WASTE NUMBER U191

CONSENSUS REPORTS: Reported in EPA TSCA Inventory.

DOT Classification: Flammable or Combustible Liquid; Label: Flammable Liquid.

SAFETY PROFILE: Poison by intraperitoneal route. Moderately toxic by ingestion and skin contact. Mildly toxic by inhalation. A skin and severe eye irritant. Flammable when exposed to heat and flame. To fight fire, use CO_2, dry chemical. Mixtures with hydrogen peroxide + iron(II) sulfate + sulfuric acid may ignite and then explode. When heated to decomposition it emits toxic fumes of NO_x.

MOY250   CAS: 108-89-4   HR: 3
4-METHYLPYRIDINE

DOT: UN 2313
mf: C_6H_7N mw: 93.14

PROP: Colorless liquid; disagreeable odor. Bp: 145°, fp: 3.7°, d: 0.9571 @ 15°/4°, vap d: 3.21, flash p: 134°F (OC).

SYNS: γ-PICOLINE * 4-PICOLINE * p-PICOLINE (DOT)

CONSENSUS REPORTS: Reported in EPA TSCA Inventory.

DOT Classification: Flammable or Combustible Liquid; Label: Flammable Liquid.

SAFETY PROFILE: Poison by skin contact and intraperitoneal routes. Moderately toxic by ingestion. Mildly toxic by inhalation. A severe skin and eye irritant. Flammable when exposed to heat, flames, oxidizers. To fight fire, use alcohol foam. When heated to decomposition it emits toxic fumes of NO_x.

MPB250   CAS: 120-94-5   HR: 3
1-METHYLPYRROLIDINE
mf: $C_5H_{11}N$ mw: 85.15

PROP: Colorless to yellow liquid; penetrating amine-like odor. Bp: 80.5°, fp: −90°, d: 0.8054 @ 20°/20°, flash p: 37.4°F, vap d: 2.9.

SYN: N-METHYLTETRAHYDROPYRROLE

CONSENSUS REPORTS: Reported in EPA TSCA Inventory.

SAFETY PROFILE: Poison by intraperitoneal and intravenous routes. This material is strongly

alkaline. Liquid and vapors are corrosive to the skin, eyes, or mucous membranes. A very dangerous fire hazard; keep away from sparks, heat sources, and powerful oxidizers. Keep in closed containers. To fight fire, use alcohol foam. When heated to decomposition it emits highly toxic fumes of NO_x.

MPD000   CAS: 7236-83-1   HR: 3
3-(1-METHYL-2-PYRROLIDINYL)INDOLE
mf: $C_{13}H_{16}N_2$ mw: 200.31

SAFETY PROFILE: Poison by intraperitoneal route. When heated to decomposition it emits toxic fumes of NO_x.

MPD250   CAS: 3671-00-9   HR: 3
3-(1-METHYL-3-PYRROLIDINYL)INDOLE
mf: $C_{13}H_{16}N_2$ mw: 200.31

SAFETY PROFILE: Poison by ingestion and intraperitoneal routes. When heated to decomposition it emits toxic fumes of NO_x.

MPE250   CAS: 1982-37-2   HR: 3
10-((1-METHYL-3-PYRROLIDINYL)METHYL)PHENOTHIAZINE
mf: $C_{18}H_{20}N_2S$ mw: 296.46

SYNS: DILOSYN * DISYNCRAM * DISYNCRAN * METHDILAZINE * MJ 5022 * NCI-C60720 * PRODUCT 5022 * TACARYL * TACAZYL * TACRYL

SAFETY PROFILE: Poison by ingestion and intraperitoneal routes. Human systemic effects by ingestion: somnolence, dyspnea and gastrointestinal changes. When heated to decomposition it emits very toxic fumes of SO_x and NO_x.

MPF200   CAS: 872-50-4   HR: 2
N-METHYLPYRROLIDONE
mf: C_5H_9NO mw: 99.15

PROP: Colorless liquid; mild odor. Bp: 202°, fp: −24°, flash p: 204°F (OC), d: 1.027 @ 25°/4°, vap d: 3.4.

SYNS: N-METHYL-2-PYRROLIDINONE * 1-METHYL-2-PYRROLIDINONE * 1-METHYL-5-PYRROLIDINONE * N-METHYLPYRROLIDINONE * METHYLPYRROLIDONE * N-METHYL-2-PYRROLIDONE * 1-METHYL-2-PYRROLIDONE * M-PYROL * NMP

CONSENSUS REPORTS: Reported in EPA TSCA Inventory.

DFG MAK: 100 ppm (400 mg/m³)

SAFETY PROFILE: Moderately toxic by intraperitoneal and intravenous routes. Mildly toxic by ingestion, skin contact, and possibly other routes. Experimental teratogenic and reproductive effects. Combustible when exposed to heat, open flame or powerful oxidizers. To fight fire, use foam, CO_2, dry chemical. When heated to decomposition it emits toxic fumes of NO_x.

MPG250   CAS: 14628-06-9   **HR: 3**
8-(METHYLQUINOLYL)-N-METHYL CARBAMATE
mf: $C_{12}H_{12}N_2O_2$ mw: 216.26

SYNS: CIBA C-7824 * ENT 27,407 * GIEGY GS-13798 * GS-13,798 * NSC 190997

SAFETY PROFILE: Poison by ingestion, intraperitoneal, and subcutaneous routes. When heated to decomposition it emits toxic fumes of NO_x.

MPH500   CAS: 504-15-4   **HR: 3**
5-METHYLRESORCINOL
mf: $C_7H_8O_2$ mw: 124.15

SYNS: 1,3-DIHYDROXY-5-METHYLBENZENE * 3,5-DIHYDROXYTOLUENE * 5-METHYL-1,3-BENZENDIOL * 5-METHYLRESORCINOL ORCINOL * ORCIN * ORCINOL

CONSENSUS REPORTS: EPA Genetic Toxicology Program. Reported in EPA TSCA Inventory.

SAFETY PROFILE: Poison by subcutaneous and intravenous routes. Moderately toxic by ingestion, intraperitoneal, and possibly other routes. Mildly toxic by skin contact. When heated to decomposition it emits acrid smoke and irritating fumes.

MPI000   CAS: 119-36-8   **HR: 3**
METHYL SALICYLATE
mf: $C_8H_8O_3$ mw: 152.16

PROP: From steam distillation of leaves from *Gaultheria procumbens* L. (Fam. *Ericacaae*) or from the bark of *Betula lenta* L. (Fam. *Betulaceae*). Colorless, yellowish or reddish oily liquid; odor and taste of wintergreen. Mp: $-8.6°$, bp: 223.3°, ULC: 20-25, flash p: 214°F (CC), fp: $-1.2°$, d: 1.1840 @ 25°/25°, refr index: 1.535, autoign temp: 850°F, vap press: 1 mm @ 54.0°, vap d: 5.24. Sltly sol in water @ 222° (decomp); sol in chloroform, ether, alc, glacial acetic acid.

SYNS: ACIDE ANISIQUE (FRENCH) * ACIDE METHYL-o-BENZOIQUE (FRENCH) * o-ANISIC ACID * BETULA OIL * FEMA No. 2745 * GAULTHERIA OIL, ARTIFICIAL * o-HYDROXYBENZOIC ACID, METHYL ESTER * 2-HYDROXYBENZOIC ACID METHYL ESTER * o-METHOXYBENZOIC ACID * 2-METHOXYBENZOIC ACID * METHYL-o-HYDROXYBENZOATE * METYLESTER KYSELINY SALICYLOVE (CZECH) * NATURAL WINTERGREEN OIL * OIL of WINTERGREEN * SALICYLIC ACID, METHYL ESTER * SWEET BIRCH OIL * SYNTHETIC WINTERGREEN OIL * TEABERRY OIL * WINTERGREEN OIL (FCC) * WINTERGREEN OIL, SYNTHETIC

CONSENSUS REPORTS: Reported in EPA TSCA Inventory.

SAFETY PROFILE: Human poison by ingestion. Moderately toxic to humans by an unspecified route. Moderately toxic experimentally by ingestion, intraperitoneal, intravenous, and subcutaneous routes. Human systemic effects by ingestion: flaccid paralysis without anesthesia, general anesthesia, dyspnea, and nausea or vomiting. Experimental teratogenic and reproductive effects. A severe skin and eye irritant. Ingestion of relatively small amounts has caused severe poisoning and death. Combustible liquid when exposed to heat or flame; can react with oxidizing materials. To fight fire, use CO_2, dry chemical. When heated to decomposition it emits acrid smoke and irritating fumes.

MPI750   CAS: 681-84-5   **HR: 3**
METHYL SILICATE
DOT: UN 2606
mf: $C_4H_{12}O_4Si$ mw: 152.25

PROP: Clear liquid. Vap d: 5.25.

SYNS: METHYL ESTER of o-SILICIC ACID * METHYL ORTHOSILICATE * TETRAMETHOXYSILANE * TETRAMETHYLSILICATE * TL 199

CONSENSUS REPORTS: Reported in EPA TSCA Inventory.

OSHA PEL: TWA 1 ppm
ACGIH TLV: TWA 1 ppm
DOT Classification: Flammable Liquid; Label: Flammable Liquid and Poison.

SAFETY PROFILE: Poison by intravenous and intraperitoneal routes. Moderately toxic by ingestion and inhalation. Mildly toxic by skin contact. A severe eye irritant. This material can cause extensive necrosis (experimentally), keratoconus, and opaque cornea. It also causes severe human eye injuries, as well as necrosis of corneal cells, which progresses long after exposure has ceased. It is destructive and its effects resist treatment. Permanent blindness is possible from exposure to it. The kidney seems to be most subject to injury regardless of the mode of exposure. Pulmonary edema has also occurred. This material is more toxic than either ethyl silicate or silicic acid, although it has been thought that the injury caused is largely due to the action of the silicic acid. Flammable when exposed to heat or flame; can react vigorously with oxidizing materials. Potentially violent reaction with metal hexafluorides (e.g., rhenium; molybdenum; tungsten). When heated to decomposition it emits acrid smoke and irritating fumes.

MPK250   CAS: 98-83-9   HR: 1
α-METHYL STYRENE

DOT: UN 2303
mf: C_9H_{10} mw: 118.19

PROP: Colorless liquid. D: 0.862 @ 20°/4°, mp: −96.0°, bp: 152.4°. Insol in water; misc in alc and ether.

SYNS: ISOPROPENIL-BENZOLO (ITALIAN) * ISOPROPENYL-BENZEEN (DUTCH) * ISOPROPENYLBENZENE * ISOPROPENYL-BENZOL (GERMAN) * α-METIL-STIROLO (ITALIAN) * α-METHYL-STYREEN (DUTCH) * α-METHYL-STYROL (GERMAN) * as-METHYLPHENYLETHYLENE * 2-PHENYLPROPENE * β-PHENYLPROPENE * 2-PHENYLPROPYLENE * β-PHENYLPROPYLENE

CONSENSUS REPORTS: Reported in EPA TSCA Inventory.

OSHA PEL: (Transitional: TWA CL 100 ppm) TWA 50 ppm; STEL 100 ppm
ACGIH TLV: TWA 50 ppm; STEL 100 ppm
DFG MAK: 100 ppm (480 mg/m^3)
DOT Classification: Flammable Liquid.

SAFETY PROFILE: Mildly toxic by inhalation. Human systemic effects by inhalation: irritant effects. A skin and eye irritant. Flammable when exposed to heat or flame; can react vigorously with oxidizing materials. When heated to decomposition it emits acrid smoke and irritating fumes.

MPK500   CAS: 25013-15-4   HR: 2
METHYL STYRENE (mixed isomers)

DOT: UN 2618
mf: C_9H_{10} mw: 118.19

PROP: A mixture containing 55-70% m-vinyltoluene and 30-45% p-vinyltoluene.

SYN: VINYLTOLUENE (mixed isomers) (OSHA)

CONSENSUS REPORTS: Reported in EPA TSCA Inventory.

OSHA PEL: TWA 100 ppm
ACGIH TLV: TWA 50 ppm; STEL 100 ppm
DFG MAK: 100 ppm (480 mg/m^3)
DOT Classification: Flammable or Combustible Liquid; Label:Flammable Liquid

SAFETY PROFILE: Moderately toxic by ingestion. Human systemic effects by inhalation: irritant effects. When heated to decomposition it emits acrid smoke and irritating fumes.

MPN500   CAS: 58-18-4   HR: 3
17-METHYLTESTOSTERONE
mf: $C_{20}H_{30}O_2$ mw: 302.50

SYNS: ANDROMETH * ANDROSAN * ANDROSAN (tablets) * ANDROSTEN * 4-ANDROSTENE-17-α-METHYL-17-β-OL-3-ONE * ANERTAN * ANERTAN (tablets) * DELATESTRYL * DIANABOL * DUMOGRAN * GLOSSO STERANDRYL * HOMANDREN * HORMALE * 17-β-HYDROXY-17-METHYLANDROST-4-EN-3-ONE * MALESTRONE * MALOGEN * MASENONE * MASTESTONA * MESTERONE * METANDREN * 17-METHYLTESTOSTERON * METHYLTESTOSTERONE * 17-α-METHYLTESTOSTERONE * METRONE * M.T. MUCORETTES * NABOLIN * NEO-HOMBREOL-M * NSC-9701 * NU MAN * ORAVIRON * ORETON-M * ORETON METHYL * STENOLON * STERONYL * SYNANDRETS * SYNANDROTABS * TESTHORMONE * TESTORA * TESTOVIRON * TESTRED

CONSENSUS REPORTS: Reported in EPA TSCA Inventory. EPA Genetic Toxicology Program.

SAFETY PROFILE: Poison by intraperitoneal route. Moderately toxic by ingestion. Human teratogenic effects by ingestion: developmental abnormalities of the urogenital system. Experimental teratogenic and reproductive effects. Questionable human carcinogen producing liver tumors. A synthetic androgenic steroid. When heated to decomposition it emits acrid smoke and irritating fumes.

MPT000 CAS: 556-64-9 **HR: 3**
METHYL THIOCYANATE
mf: C_2H_3NS mw: 73.12

PROP: Liquid. D: 1.068 @ 20°, mp: −51°, bp: 130-133°. Very sltly sol in water; misc in alc and ether.

SYNS: METHYLRHODANID (GERMAN) * METHYL SULFOCYANATE

CONSENSUS REPORTS: Reported in EPA TSCA Inventory. EPA Extremely Hazardous Substances List.

SAFETY PROFILE: Poison by ingestion, intravenous, and subcutaneous routes. When heated to decomposition it emits very toxic fumes of NO_x and SO_x.

MPU000 CAS: 342-69-8 **HR: 3**
METHYLTHIOINOSINE
mf: $C_{11}H_{14}N_4O_4S$ mw: 298.35

SYNS: 6-METHYLMERCAPTOPURINE RIBONUCLEOSIDE * 6-METHYLMERCAPTOPURINE RIBOSIDE * 6-METHYL-MP-RIBOSIDE * 6-METHYL-9-RIBOFURANOSYLPURINE-6-THIOL * 6-METHYLTHIOINOSINE * 6-(METHYLTHIO)PURINE RIBONUCLEOSIDE * 6-METHYLTHIOPURINE RIBOSIDE * NCI-C04784 * NSC 40774 * β-d-RIBOSYL-6-METHYLTHIOPURINE * SQ 21977

CONSENSUS REPORTS: NCI Carcinogenesis Studies (ipr); Equivocal Evidence: mouse CANCAR 40,1935,77; (ipr); No Evidence: rat CANCAR 40,1935,77.

SAFETY PROFILE: Poison by intraperitoneal and possibly other routes. Experimental teratogenic and reproductive effects. Questionable carcinogen with experimental tumorigenic data. Mutation data reported. When heated to decomposition it emits very toxic fumes of SO_x and NO_x.

MPW500 CAS: 56-04-2 **HR: 3**
6-METHYLTHIOURACIL
mf: $C_5H_6N_2OS$ mw: 142.19

PROP: Bitter crystals or colorless liquid; odor of onions. Decomp @ 326-331°, sublimes readily. Very sltly sol in ether, cold water, alkaline hydroxides, NH_3; sltly sol in alc, acetone. Almost insol in benzene, chloroform.

SYNS: ALKIRON * ANTIBASON * BASECIL * BASETHYRIN * 2,3-DIHYDRO-6-METHYL-2-THIOXO-4(1H)-PYRIMIDINONE * 2-MERCAPTO-4-HY-DROXY-6-METHYLPYRIMIDINE * 2-MERCAPTO-6-METHYLPYRIMID-4-ONE * 2-MERCAPTO-6-METHYL-4-PYRIMIDONE * METACIL * METHIACIL * METHIOCIL * 6-METHYL-2-THIO-2,4-(1H3H)PYRIMIDINEDIONE * METHYLTHIOURACIL * 4-METHYL-2-THIOURACIL * 6-METHYL-2-THIOURACIL * 4-METHYLURACIL * 6-METIL-TIOURACILE (ITALIAN) * MTU * MURACIL * ORCANON * PROSTRUMYL * RCRA WASTE NUMBER U164 * STRUMACIL * THIMECIL * THIOMECIL * 2-THIO-6-METHYL-1,3-PYRIMIDIN-4-ONE * 6-THIO-4-METHYLURACIL * THIOMIDIL * 2-THIO-4-OXO-6-METHYL-1,3-PYRIMIDINE * THIORYL * THIOTHYMIN * THIOTHYRON * THIURYL * THYREONORM * THYREOSTAT * THYRIL * TIOMERACIL * TIORALE M * TIOTIRON * USAF EK-6454

CONSENSUS REPORTS: IARC Cancer Review: GROUP 2B IMEMDT 7,56,87; Animal Sufficient Evidence IMEMDT 7,53,74. Reported in EPA TSCA Inventory.

SAFETY PROFILE: Suspected carcinogen with experimental carcinogenic, neoplastigenic, and tumorigenic data. Poison by intraperitoneal route. Moderately toxic by ingestion. Human teratogenic and reproductive effects by an unspecified route: developmental abnormalities of the endocrine system and effects on newborn including neonatal measures or effects. Experimental teratogenic and reproductive effects. Used to treat hyperthyroidism. When heated to decomposition it emits very toxic fumes of NO_x and SO_x.

MPY000 CAS: 2058-62-0 **HR: 3**
N-METHYL-p-(m-TOLYLAZO)ANILINE
mf: $C_{14}H_{15}N_3$ mw: 225.32

SYNS: N-METHYL-3′-METHYL-p-AMINOAZOBENZENE * N-METHYL-3′-METHYL-4-AMINOAZOBENZENE * 3′-METHYL-4-MONOMETHYLAMINOAZOBENZENE

SAFETY PROFILE: Questionable carcinogen with experimental tumorigenic data. Mutation data reported. When heated to decomposition it emits toxic fumes of NO_x.

MQC500 CAS: 75-79-6 **HR: 3**
METHYLTRICHLOROSILANE
DOT: UN 1250
mf: CH_3Cl_3Si mw: 149.48

SYNS: METHYL-TRICHLORSILAN (CZECH) * TRICHLOROMETHYLSILANE

CONSENSUS REPORTS: Reported in EPA TSCA Inventory. EPA Extremely Hazardous Substances List.

DOT Classification: Flammable Liquid; Label: Flammable Liquid, Corrosive.

SAFETY PROFILE: Poison by inhalation and intraperitoneal routes. Moderately toxic by ingestion. A corrosive irritant to skin, eyes, and mucous membranes. Flammable when exposed to heat or flame; can react vigorously with oxidizing materials. When heated to decomposition it emits toxic fumes of Cl^-.

MQE000 CAS: 31185-56-5 **HR: 2**
5-METHYL-2-TRIFLUOROMETHYL-OXAZOLIDINE
mf: $C_5H_8F_3NO$ mw: 155.14

SAFETY PROFILE: Moderately toxic by intraperitoneal route. When heated to decomposition it emits very toxic fumes of F^- and NO_x.

MQG500 CAS: 4426-51-1 **HR: 3**
4-METHYL TRIMETHYLENE SULFITE
mf: $C_4H_8O_3S$ mw: 136.18

SYNS: 1,3-BUTANEDIOL, CYCLIC SULFITE * NSC-60195

SAFETY PROFILE: Poison by intravenous and intraperitoneal routes. When heated to decomposition it emits toxic fumes of SO_x.

MQH000 CAS: 5137-55-3 **HR: 3**
METHYLTRIOCTYLAMMONIUM CHLORIDE
mf: $C_{25}H_{54}N \cdot Cl$ mw: 404.25

SYNS: ALIQUAT 336 * N-METHYL-N,N-DIOCTYL-1-OCTANAMINIUM CHLORIDE * METHYLTRICAPRY-LYLAMMONIUMCHLORIDE * TRICAPRYLMETHYL-AMMONIUM CHLORIDE * TRICAPRYLYLMETHYL-AMMONIUM CHLORIDE * TRIOCTYLMETHYL-AMMONIUM CHLORIDE

CONSENSUS REPORTS: Reported in EPA TSCA Inventory.

SAFETY PROFILE: Poison by ingestion and intraperitoneal routes. When heated to decomposition it emits very toxic fumes of NO_x, NH_3, and Cl^-.

MQH750 CAS: 953-17-3 **HR: 3**
METHYL TRITHION
mf: $C_9H_{12}ClO_2PS_3$ mw: 314.81

SYNS: ((p-CHLOROPHENYL)THIO)METHANETHIOL-S-ESTER with O,O-DIMETHYL PHOSPHORODITHIOATE * S-(((p-CHLOROPHENYL)THIO)METHYL) O,O-DI-METHYL PHOSPHORODITHIOATE * S-(((4-CHLORO-PHENYL)THIO)METHYL) O,O-DIMETHYLPHOSPHORO-DITHIOATE * DIMETHYL-p-CHLOROPHENYLTHIO-METHYL DITHIOPHOSPHATE * O,O-DIMETHYL-S-(p-CHLOROPHENYLTHIOMETHYL)PHOSPHORODITHIOATE * O,O-DIMETHYLTHIOPHOSPHORIC ACID, p-CHLORO-PHENYL ESTER * ENT 25,599 * G-29288 * GEIGY G-29288 * METHYLCARBOPHENOTHION * R-1492 * STAUFFER R-1492 * TRI-ME

SAFETY PROFILE: Poison by ingestion, skin contact and possibly other routes. A cholinesterase inhibitor type of insecticide. When heated to decomposition it emits very toxic fumes of Cl^-, PO_x, and SO_x.

MQL750 CAS: 107-25-5 **HR: 3**
METHYL VINYL ETHER
DOT: UN 1087
mf: C_3H_6O mw: 58.09

PROP: Colorless, easily liquefied gas or colorless liquid. Bp: 6.0°, d: 0.7500, vap d: 2.0, fp: −121.6°, vap press: 1052 mm @ 20°, flash p: −68.8°F, lel: 2.6%, uel: 39.0%.

SYNS: METHOXYETHENE * VINYL METHYL ETHER (DOT)

CONSENSUS REPORTS: Reported in EPA TSCA Inventory.

DOT Classification: Flammable Gas; Label: Flammable Gas.

SAFETY PROFILE: Mildly toxic by ingestion. A very dangerous fire hazard when exposed to heat, flame, or oxidizers. Explosive in the form of vapor when exposed to heat or flame. Can react vigorously with oxidizing materials. To fight fire, stop flow of gas. Potentially explosive reaction with halogens (e.g., bromine; chlorine) or hydrogen halides (e.g., hydrogen bromide; hydrogen chloride). Reaction with acids forms acetaldehyde. Weak acids catalyze the exothermic polymerization of the ether. The unstabilized ether can form dangerous peroxides. When heated to decomposition it emits acrid smoke and irritating fumes.

MQN000 CAS: 5974-19-6 **HR: 3**
METHYL VIOLET 6B
mf: $C_{31}H_{34}N_3 \cdot Cl$ mw: 484.13

SYN: PENTAMETHYLBENZYL-p-ROSANILINE CHLO-RIDE

SAFETY PROFILE: Poison by ingestion. When heated to decomposition it emits very toxic fumes of NO_x and Cl^-.

MQP500   CAS: 29605-96-7   **HR: 3**
METHYSERGIDE DIMALEATE
mf: $C_{21}H_{27}N_3O_2 \cdot 2C_4H_4O_4$ mw: 585.67

PROP: Decomp above 165°. Sol in methanol, less sol in water; insol in abs ethanol.

SYNS: 1-(HYDROXYMETHYL)PROPYLAMIDE of 1-METHYL-(+)-LYSERGIC ACID HYDROGEN MALEATE * METHYLSERGIDE BIMALEATE * SANSERT

SAFETY PROFILE: Poison by ingestion and intravenous routes. Experimental reproductive effects. When heated to decomposition it emits toxic fumes of NO_x.

MQQ000   CAS: 5800-19-1   **HR: 3**
METIAPINE
mf: $C_{19}H_{21}N_3S$ mw: 323.49

SYN: 2-METHYL-11-(4-METHYL-1-PIPERAZINYL)-DIBEN-ZO(b,f)(1,4)THIAZEPINE

SAFETY PROFILE: Poison by intraperitoneal route. Moderately toxic by ingestion. Experimental reproductive effects. When heated to decomposition it emits very toxic fumes of NO_x and SO_x.

MQQ500   CAS: 5377-20-8   **HR: 3**
METOMIDATE
mf: $C_{13}H_{14}N_2O_2$ mw: 230.29

SYN: METHYL 1-(α-METHYLBENZYL)IMIDAZOLE-5-CARBOXYLATE

SAFETY PROFILE: Poison by ingestion. When heated to decomposition it emits toxic fumes of NO_x.

MQR000   CAS: 14008-44-7   **HR: 3**
METOPIMAZINE
mf: $C_{22}H_{27}N_3O_3S_2$ mw: 445.64

PROP: Solid. Mp: 170-171°.

SYNS: 10-(3-(4-CARBAMOYLPIPERIDINE)PROPYL)-2-(METHANESULFONYL)PHENOTHIAZINE * EXP 999 * 2-METHYLSULFONYL-10-(3-(4-CARBAMOYLPIPER-IDINO)PROPYL)PHENOTHIAZINE * 1-(3-(2-(METHYL-SULFONYL)PHENOTHIAZIN-10-YL)PROPYL)ISONIPECOT-

AMIDE * 1-(3-(2-METHYLSULFONYL)PHENOTHIAZIN-10-YL)PROPYL)-4-PIPERIDINE CARBOXAMIDE * 1-(3-(2-(METHYLSULFONYL)-10H-PHENOTHIAZIN-10-YL)PROPYL)-4-PIPERIDINE CARBOXAMIDE * RP 9965 * VOGALENE

SAFETY PROFILE: Poison by intravenous, intraperitoneal, and subcutaneous routes. Moderately toxic by ingestion. An anti-emetic agent. When heated to decomposition it emits very toxic fumes of SO_x and NO_x.

MQR200   CAS: 1178-29-6   **HR: 3**
METOSERPATE HYDROCHLORIDE
mf: $C_{24}H_{32}N_2O_5 \cdot ClH$ mw: 465.04

SYNS: METHYL-18-EPIRESERPATE METHYL ETHER HYDROCHLORIDE * PACITRAN * SU-9064 * SU 8842 HYDROCHLORIDE

SAFETY PROFILE: Poison by ingestion and intravenous routes. When heated to decomposition it emits toxic fumes of NO_x and HCl.

MQR275   CAS: 21087-64-9   **HR: 3**
METRIBUZIN
mf: $C_8H_{14}N_4OS$ mw: 214.32

SYNS: 4-AMINO-6-tert-BUTYL-3-(METHYLTHIO)-1,2,4-TRIAZIN-5-ONE * 4-AMINO-6-tert-BUTYL-3-METHYL-THIO-as-TRIAZIN-5-ONE * 4-AMINO-6-(1,1-DIMETH-YLETHYL)-3-(METHYLTHIO)-1,2,4-TRIAZIN-5(4H)-ONE * BAY 61597 * BAY DIC 1468 * BAYER 94337 * BAYER 6159H * BAYER 6443H * DIC 1468 * LEXONE * SENCOR * SENCORAL * SEN-CORER * SENCOREX

CONSENSUS REPORTS: EPA Genetic Toxicology Program.

OSHA PEL: TWA 5 mg/m^3
ACGIH TLV: TWA 5 mg/m^3

SAFETY PROFILE: Poison by ingestion. When heated to decomposition it emits very toxic fumes of NO_x and SO_x.

MQR500   CAS: 826-39-1   **HR: 3**
MEVASIN HYDROCHLORIDE
mf: $C_{11}H_{21}N \cdot ClH$ mw: 203.79

SYNS: INVERSINE HYDROCHLORIDE * MECAMINE HYDROCHLORIDE * MECAMYLAMINE HYDROCHLO-RIDE * MEKAMIN HYDROCHLORIDE * 3-METH-YLAMINOISOCAMPHANE HYDROCHLORIDE * N-METHYL-dl-ISOBORNYLAMINE HYDROCHLORIDE * N,2,3,3-TETRAMETHYL-2-NORBORNANAMINE HY-DROCHLORIDE

SAFETY PROFILE: Poison by ingestion, intraperitoneal, intravenous, and subcutaneous routes. When heated to decomposition it emits very toxic fumes of HCl and NO_x.

MQR750   CAS: 7786-34-7   **HR: 3**
MEVINPHOS
mf: $C_7H_{13}O_6P$ mw: 224.17

SYNS: APAVINPHOS * α-2-CARBOMETHOXY-1-METHYLVINYL DIMETHYL PHOSPHATE * 2-CARBO-METHOXY-1-PROPEN-2-YL DIMETHYL PHOSPHATE * CMDP * COMPOUND 2046 * 3-((DIMETHOXY-PHOSPHINYL)OXY)-2-BUTENOIC ACID METHYL ESTER * O,O-DIMETHYL-O-(2-CARBOMETHOXY-1-METHYLVI-NYL) PHOSPHATE * DIMETHYL-1-CARBOMETHOXY-1-PROPEN-2-YL PHOSPHATE * O,O-DIMETHYL-O-2-METHOXYCARBONYL-1-METHYL-VINYL-PHOSPHAT (GERMAN) * DIMETHYL ESTER PHOSPHORIC ACID ESTER with METHYL 3-HYDROXYCROTONATE * DIMETHYL 2-METHOXYCARBONYL-1-METHYLVINYL PHOSPHATE * DIMETHYL METHOXYCARBONYLPRO-PENYL PHOSPHATE * DIMETHYL (1-METHOXY-CARBOXYPROPEN-2-YL)PHOSPHATE * O,O-DI-METHYL O-(1-METHYL-2-CARBOXYVINYL) PHOSPHATE * DURAPHOS * ENT 22,374 * FOSDRIN * GESFID * GESTID * 3-HYDROXYCROTONIC ACID METHYL ESTER DIMETHYL PHOSPHATE * MENIPHOS * MENITE * (2-METHOXY-CARBONYL-1-METHYL-VINYL)-DIMETHYL-PHOSPHAT (GERMAN) * (2-METHOXYCARBONYL-1-METHYL-VINYL)-DIMETHYL-FOSFAAT (DUTCH) * 2-METHOXY-CARBONYL-1-METHYLVINYL DIMETHYLPHOSPHATE * 1-METHOXYCARBONYL-1-PROPEN-2-YL DIMETHYL PHOSPHATE * (1-METHOXYCARBOXYPROPEN-2-YL) PHOSPHORIC ACID, DIMETHYL ESTER * METHYL-3-(DIMETHOXYPHOSPHINYLOXY)CROTONATE * (2-METOSSICARBONIL-1-METIL-VINIL)-DIMETIL-FOS-FATO (ITALIAN) * MEVINFOS (DUTCH) * OS 2046 * PHOSDRIN (OSHA) * PHOSFENE * PHOSPHATE de DIMETHYLE et de 2-METHOXYCARBONYL-1 METHYL-VINYLE (FRENCH) * PHOSPHENE (FRENCH)

CONSENSUS REPORTS: EPA Genetic Toxicology Program. EPA Extremely Hazardous Substances List.

OSHA PEL: (Transitional: TWA 0.1 mg/m³ (skin)) TWA 0.01 ppm; STEL 0.03 ppm (skin)
ACGIH TLV: TWA 0.01 ppm; STEL 0.03 ppm (skin)
DFG MAK: 0.01 ppm (0.1 mg/m³)
DOT Classification: Poison B; Label: Poison.

SAFETY PROFILE: Poison by ingestion, inhalation, skin contact, subcutaneous, intravenous, and intraperitoneal routes. Human systemic effects by ingestion: peripheral motor nerve recording changes. An insecticide. When heated to decomposition it emits toxic fumes of PO_x.

MQS225   CAS: 3704-09-4   **HR: 3**
MIBOLERONE
mf: $C_{20}H_{30}O_2$ mw: 302.50

PROP: Crystalline solid. Solubility in deionized water: 0.0454 mg/mL @ 37°.

SYNS: CHEQUE * (7-α,17-β)-17-HYDROXY-7,17-DI-METHYL-ESTR-4-EN-3-ONE (9CI) * 17-β-HYDROXY-7-α,17-DIMETHYLESTR-4-EN-3-ONE * MATENON * MIBOLERON * U 10997

SAFETY PROFILE: Experimental teratogenic and reproductive effects. Questionable carcinogen with experimental neoplastigenic data. When heated to decomposition it emits acrid smoke and irritating fumes.

MQS250   CAS: 12001-26-2   **HR: 2**
MICA

PROP: Containing less than 1% crystalline silica.

SYNS: MICA SILICATE * SUZORITE MICA

OSHA PEL: (Transitional: TWA 20 mppcf) TWA Respirable Fraction: 3 mg/m³
ACGIH TLV: TWA Respirable Fraction: 3 mg/m³

SAFETY PROFILE: The dust is injurious to lungs.

MQS500   CAS: 90-94-8   **HR: 3**
MICHLER'S KETONE
mf: $C_{17}H_{20}N_2O$ mw: 268.39

PROP: Leaves from ethanol. Mp: 172°, bp: >360° decomp. Insol in water; very sol in benzene; sol in alc; very sltly sol in ether.

SYNS: p,p'-BIS(N,N-DIMETHYLAMINO)BENZOPHE-NONE * 4,4'-BIS(DIMETHYLAMINO)BENZOPHENONE * BIS(p-(N,N-DIMETHYLAMINO)PHENYL)KETONE * BIS(4-(DIMETHYLAMINO)PHENYL)METHANONE * p,p'-MICHLER'S KETONE * NCI-C02006 * TETRAMETHYLDIAMINOBENZOPHENONE

CONSENSUS REPORTS: NCI Carcinogenesis Bioassay (feed); Clear Evidence: mouse, rat NCITR* NCI-CG-TR-181,79. NTP Fourth Annual Report On Carcinogens, 1984. Reported in EPA TSCA Inventory. EPA Genetic Toxicology Program.

DFG MAK: Suspected Carcinogen.

SAFETY PROFILE: Confirmed human carcinogen with experimental carcinogenic and neoplastigenic data. Poison by ingestion. Mutation data reported. When heated to decomposition it emits toxic fumes of NO_x.

MQU750   CAS: 57-53-4   **HR: 3**
MILTOWN
mf: $C_9H_{18}N_2O_4$ mw: 218.29

SYNS: AMEPROMAT * AMOSENE * ANASTRESS * ANATHYLMON * ANDAKSIN * ANDAXIN * ANEURAL * ANEUXRAL * ANSIATAN * ANSIL * ANSIOWAS * ANURAL * ANXIETIL * APASCIL * ARCOBAN * ARTOLON * ATRAXINE * AYERMATE * BAMD 400 * BIOBAMAT * BROBAMATE * CALMADIN * CALMAX * CALMIREN * CANQUIL-400 * CAP-O-TRAN * CIRPONYL * CRESTANIL * CYPRON * DAPAZ * DICANDIOL * 2,2-DI (CARBAMOYLOXYMETHYL)PENTANE * DIVERON * DORMABROL * ECUANIL * EDENAL * ENORDEN * EPICUR * EQUANIL SUSPENSION * EQUILIUM * EQUINIL * ERINA * ESTASIL * FAS-CILE * GADEXYL * HARMONIN * HARTOL * HOLBAMATE * IPSOTIAN * KESSOBAMATE * KLORT * LARTEN * LEPENIL * LEPETOWN * LETYL * LI-BIOLAN * MADIOL * MAR BATE * MARGONIL * MENDEL * MEPAMTIN * MEPAVLON * MEPIOSINE * MEPOSED * MEPRANIL * MEPROBAM * MEPROBAMAT (GERMAN) * MEPROBAMATE * MEPROBAMATO (ITALIAN) * MEPROCOMPREN * MEPROCON CMC * MEPRODIL * MEPROLEAF * MEPROSAN * MEPROTABS * MEPROZINE * MEPTRAN * 2-METHYL-2-N-PROPYL-1,3-PROPANEDIOL DI-CARBAMATE * 2-METHYL-2-PROPYLTRIMETHYLENE CARBAMATE * METRACTYL * MILPREM * MILTANN * NEO-TRAN * NEPHENTINE * OROLEVOL * PANCALMA * PAN-TRANQUIL * PEREQUIL * PLACIDON * PROCALMIDOL * PROQUANIL * QUIETIDON * RESTENIL * ROBAMATE * SEDABAMATE * SERIL * SPANTRAN * TRANQUILAN * TRELMAR * URBIL * VISTABAMATE * WARDAMATE * WYSEALS * ZIRPON

CONSENSUS REPORTS: Reported in EPA TSCA Inventory. EPA Genetic Toxicology Program.

SAFETY PROFILE: Human poison by unspecified routes. Moderately toxic to humans and experimentally by ingestion. Experimental poison by intravenous, intraperitoneal, and subcutaneous routes. Human systemic effects by ingestion: coma, blood pressure decrease, regional or general arteriolar constriction, dyspnea, cyanosis, respiratory depression, nausea or vomiting and allergic skin dermatitis. Experimental teratogenic and reproductive effects. Mutation data reported. Implicated in aplastic anemia. Used as a tranquilizer. When heated to decomposition it emits toxic fumes of NO_x.

MQV250   CAS: 1401-55-4   **HR: 3**
MIMOSA TANNIN

SYNS: ACACIA MOLLISSIMA TANNIN * TANNIN from MIMOSA

SAFETY PROFILE: Poison by intravenous and intraperitoneal routes. Questionable carcinogen with experimental tumorigenic data. When heated to decomposition it emits acrid smoke and irritating fumes.

MQV500   **HR: D**
MINERAL DUSTS

SAFETY PROFILE: Variable toxicity. From the economic and toxicity standpoints, the most important are those containing free silica which can cause silicosis upon inhalation of sufficient quantity. These include sand, sandstone, quartz, and flint. They consist mainly of silica in the form of quartz; diatomaceous earth, which is essentially amorphous silica; and granite, which contains 20-40% quartz. Minerals that contain combined silica in the form of silicates but no free silica are generally less capable of causing silicosis. Asbestos, however, can cause a fibrotic lung condition of its own, known as asbestosis, and lung cancer. (See also various asbestos entries.) Mica and talc dust are also considered somewhat hazardous. Non-siliceous minerals, like limestone, marble, dolomite, etc., which do not contain toxic elements, do not ordinarily present any significant dust hazard. Minerals containing toxic elements, such as cryolite which contains fluorine, and pyrolusite, which contains manganese, may cause systemic poisoning upon inhalation or ingestion of sufficient quantity. In any event, the minerals are usually less reactive than synthetic compounds of the same elements and, in fact, may be relatively inert by comparison. These are common air contaminants.

MQV750   CAS: 8012-95-1   HR: 3
MINERAL OIL

PROP: Colorless, oily liquid; practically tasteless and odorless. D: 0.83-0.86 (light), 0.875-0.905 (heavy), flash p: 444°F (OC), ULC: 10-20. Insol in water and alc; sol in benzene, chloroform, and ether. A mixture of liquid hydrocarbons from petroleum.

SYNS: ADEPSINE OIL * ALBOLINE * BAYOL F * BLANDLUBE * CRYSTOSOL * DRAKEOL * FONOLINE * GLYMOL * KAYDOL * KONDREMUL * MINERAL OIL, WHITE (FCC) * MOLOL * NEO-CULTOL * NUJOL * OIL MIST, MINERAL (OSHA, ACGIH) * PAROL * PAROLEINE * PARRAFIN OIL * PENETECK * PENRECO * PERFECTA * PETROGALAR * PETROLATUM, liquid * PRIMOL 335 * PROTOPET * SAXOL * TECH PET F * WHITE MINERAL OIL

CONSENSUS REPORTS: Reported in EPA TSCA Inventory.

OSHA PEL: Oil Mist: TWA 5 mg/m^3
ACGIH TLV: TWA 5 mg/m^3; STEL 10 mg/m^3

SAFETY PROFILE: A human teratogen by inhalation which causes testicular tumors in the fetus. Inhalation of vapor or particulates can cause aspiration pneumonia. An eye irritant. Questionable human carcinogen producing gastrointestinal tumors. Combustible liquid when exposed to heat or flame. To fight fire, use dry chemical, CO_2, foam. When heated to decomposition it emits acrid smoke and fumes.

MQV755   CAS: 64741-49-7   HR: 3
MINERAL OIL, PETROLEUM CONDENSATES, VACUUM TOWER

SYNS: CONDENSATES (PETROLEUM), VACUUM TOWER (9CI) * VACUUM RESIDUUM

CONSENSUS REPORTS: IARC Cancer Review: GROUP 1 IMEMDT 7,252,87; Animal Sufficient Evidence IMEMDT 33,87,84. Reported in EPA TSCA Inventory.

SAFETY PROFILE: Confirmed carcinogen. When heated to decomposition it emits acrid smoke and irritating fumes.

MQV760   CAS: 64742-18-3   HR: 3
MINERAL OIL, PETROLEUM DISTILLATES, ACID-TREATED HEAVY NAPHTHENIC

SYNS: ACID-TREATED HEAVY NAPHTHENIC DISTILLATE * DISTILLATES (PETROLEUM), ACID-TREATED HEAVY NAPHTHENIC (9CI)

CONSENSUS REPORTS: IARC Cancer Review: GROUP 1 IMEMDT 7,252,87; Animal Sufficient Evidence IMEMDT 33,87,84. Reported in EPA TSCA Inventory.

SAFETY PROFILE: Confirmed carcinogen. When heated to decomposition it emits acrid smoke and irritating fumes.

MQV765   CAS: 64742-20-7   HR: 3
MINERAL OIL, PETROLEUM DISTILLATES, ACID-TREATED HEAVY PARAFFINIC

SYNS: ACID-TREATED HEAVY PARAFFINIC DISTILLATE * DISTILLATES (PETROLEUM), ACID-TREATED HEAVY PARAFFINIC (9CI)

CONSENSUS REPORTS: IARC Cancer Review: GROUP 1 IMEMDT 7,252,87; Animal Sufficient Evidence IMEMDT 33,87,84. Reported in EPA TSCA Inventory.

SAFETY PROFILE: Confirmed carcinogen. When heated to decomposition it emits acrid smoke and irritating fumes.

MQV770   CAS: 64742-19-4   HR: 3
MINERAL OIL, PETROLEUM DISTILLATES, ACID-TREATED LIGHT NAPHTHENIC

SYNS: ACID-TREATED LIGHT NAPHTHENIC DISTILLATE * DISTILLATES (PETROLEUM), ACID-TREATED LIGHT NAPHTHENIC (9CI)

CONSENSUS REPORTS: IARC Cancer Review: GROUP 1 IMEMDT 7,252,87; Animal Sufficient Evidence IMEMDT 33,87,84. Reported in EPA TSCA Inventory.

SAFETY PROFILE: Confirmed carcinogen. When heated to decomposition it emits acrid smoke and irritating fumes.

MQV775   CAS: 64742-21-8   HR: 3
MINERAL OIL, PETROLEUM DISTILLATES, ACID-TREATED LIGHT PARAFFINIC

SYNS: ACID-TREATED LIGHT PARAFFINIC DISTILLATE * DISTILLATES (PETROLEUM), ACID-TREATED LIGHT PARAFFINIC (9CI)

CONSENSUS REPORTS: IARC Cancer Review: GROUP 1 IMEMDT 7,252,87; Animal Sufficient Evidence IMEMDT 33,87,84. Reported in EPA TSCA Inventory.

SAFETY PROFILE: Confirmed carcinogen. When heated to decomposition it emits acrid smoke and irritating fumes.

MQV780   CAS: 64741-53-3   **HR: 3**
MINERAL OIL, PETROLEUM DISTILLATES, HEAVY NAPHTHENIC

SYNS: DISTILLATES (PETROLEUM), HEAVY NAPHTHENIC (9CI) * HEAVY NAPHTHENIC DISTILLATE * HEAVY NAPHTHENIC DISTILLATES (PETROLEUM)

CONSENSUS REPORTS: IARC Cancer Review: GROUP 1 IMEMDT 7,252,87; Animal Sufficient Evidence IMEMDT 33,87,84. Reported in EPA TSCA Inventory.

SAFETY PROFILE: Confirmed carcinogen with experimental neoplastigenic data. Mutation data reported. When heated to decomposition it emits acrid smoke and irritating fumes.

MQV785   CAS: 64741-51-1   **HR: 3**
MINERAL OIL, PETROLEUM DISTILLATES, HEAVY PARAFFINIC

SYNS: DISTILLATES (PETROLEUM), HEAVY PARAFFINIC (9CI) * HEAVY PARAFFINIC DISTILLATE

CONSENSUS REPORTS: IARC Cancer Review: GROUP 1 IMEMDT 7,252,87; Animal Sufficient Evidence IMEMDT 33,87,84. Reported in EPA TSCA Inventory.

SAFETY PROFILE: Confirmed carcinogen. Mutation data reported.

MQV790   CAS: 64742-52-5   **HR: 3**
MINERAL OIL, PETROLEUM DISTILLATES, HYDROTREATED HEAVY NAPHTHENIC

SYNS: DISTILLATES (PETROLEUM), HYDROTREATED HEAVY NAPHTHENIC (9CI) * HYDROTREATED HEAVY NAPHTHENIC DISTILLATE * HYDROTREATED HEAVY NAPHTHENIC DISTILLATES (PETROLEUM) * PETROLEUM DISTILLATES, HYDROTREATED HEAVY NAPHTHENIC

CONSENSUS REPORTS: IARC Cancer Review: GROUP 1 IMEMDT 7,252,87; Animal Inadequate Evidence IMEMDT 33,87,84. Reported in EPA TSCA Inventory.

SAFETY PROFILE: Confirmed carcinogen with experimental tumorigenic data. Mutation data reported. When heated to decomposition it emits acrid smoke and irritating fumes.

MQV795   CAS: 64742-54-7   **HR: 3**
MINERAL OIL, PETROLEUM DISTILLATES, HYDROTREATED HEAVY PARAFFINIC

SYNS: DISTILLATES (PETROLEUM), HYDROTREATED HEAVY PARAFFINIC (9CI) * HYDROTREATED HEAVY PARAFFINIC DISTILLATE

CONSENSUS REPORTS: IARC Cancer Review: GROUP 1 IMEMDT 7,252,87; Animal Inadequate Evidence IMEMDT 33,87,84. Reported in EPA TSCA Inventory.

SAFETY PROFILE: Confirmed carcinogen. When heated to decomposition it emits acrid smoke and irritating fumes.

MQV800   CAS: 64742-53-6   **HR: 3**
MINERAL OIL, PETROLEUM DISTILLATES, HYDROTREATED LIGHT NAPHTHENIC

SYNS: DISTILLATES (PETROLEUM), HYDROTREATED LIGHT NAPHTHENIC (9CI) * HYDROTREATED LIGHT NAPHTHENIC DISTILLATE * HYDROTREATED LIGHT NAPHTHENIC DISTILLATES (PETROLEUM)

CONSENSUS REPORTS: IARC Cancer Review: GROUP 1 IMEMDT 7,252,87; Animal Sufficient Evidence IMEMDT 33,87,84. Reported in EPA TSCA Inventory.

SAFETY PROFILE: Confirmed carcinogen with experimental neoplastigenic data. When heated to decomposition it emits acrid smoke and irritating fumes.

MQV805   CAS: 64742-55-8   **HR: 3**
MINERAL OIL, PETROLEUM DISTILLATES, HYDROTREATED LIGHT PARAFFINIC

SYNS: DISTILLATES (PETROLEUM), HYDROTREATED LIGHT PARAFFINIC (9CI) * HYDROTREATED LIGHT PARAFFINIC DISTILLATE

CONSENSUS REPORTS: IARC Cancer Review: GROUP 1 IMEMDT 7,252,87; Animal Sufficient Evidence IMEMDT 33,87,84 Reported in EPA TSCA Inventory.

SAFETY PROFILE: Confirmed carcinogen. When heated to decomposition it emits acrid smoke and irritating fumes.

MQV810   CAS: 64741-52-2   **HR: 3**
MINERAL OIL, PETROLEUM DISTILLATES, LIGHT NAPHTHENIC

SYNS: DISTILLATES (PETROLEUM), LIGHT NAPHTHE-NIC (9CI)   *   LIGHT NAPHTHENIC DISTILLATE   *   LIGHT NAPHTHENIC DISTILLATES (PETROLEUM)

CONSENSUS REPORTS: IARC Cancer Review: GROUP 1 IMEMDT 7,252,87; Animal Sufficient Evidence IMEMDT 33,87,84. Reported in EPA TSCA Inventory.

SAFETY PROFILE: Confirmed carcinogen with experimental neoplastigenic data. When heated to decomposition it emits acrid smoke and irritating fumes.

MQV815   CAS: 64741-50-0   **HR: 3**
MINERAL OIL, PETROLEUM DISTILLATES, LIGHT PARAFFINIC

SYNS: DISTILLATES (PETROLEUM), LIGHT PARAF-FINIC (9CI)   *   LIGHT PARAFFINIC DISTILLATE

CONSENSUS REPORTS: IARC Cancer Review: GROUP 1 IMEMDT 7,252,87; Animal Sufficient Evidence IMEMDT 33,87,84. Reported in EPA TSCA Inventory.

SAFETY PROFILE: Confirmed carcinogen. Mutation data reported. When heated to decomposition it emits acrid smoke and irritating fumes.

MQV835   CAS: 64742-64-9   **HR: 3**
MINERAL OIL, PETROLEUM DISTILLATES, SOLVENT-DEWAXED LIGHT NAPHTHENIC

SYNS: DISTILLATES (PETROLEUM), SOLVENT-DE-WAXED LIGHT NAPHTHENIC (9CI)   *   SOLVENT-DE-WAXED LIGHT NAPHTHENIC DISTILLATE

CONSENSUS REPORTS: IARC Cancer Review: GROUP 1 IMEMDT 7,252,87; Animal Sufficient Evidence IMEMDT 33,87,84. Reported in EPA TSCA Inventory.

SAFETY PROFILE: Confirmed carcinogen. When heated to decomposition it emits acrid smoke and irritating fumes.

MQV840   CAS: 64742-56-9   **HR: 3**
MINERAL OIL, PETROLEUM DISTILLATES, SOLVENT-DEWAXED LIGHT PARAFFINIC

SYNS: DISTILLATES (PETROLEUM), SOLVENT-DE-WAXED LIGHT PARAFFINIC (9CI)   *   SOLVENT-DE-WAXED LIGHT PARAFFINIC DISTILLATE

CONSENSUS REPORTS: IARC Cancer Review: GROUP 1 IMEMDT 7,252,87; Animal Sufficient Evidence IMEMDT 33,87,84. Reported in EPA TSCA Inventory.

SAFETY PROFILE: Confirmed carcinogen. When heated to decomposition it emits acrid smoke and irritating fumes.

MQV852   CAS: 64741-97-5   **HR: 3**
MINERAL OIL, PETROLEUM DISTILLATES, SOLVENT-REFINED LIGHT NAPHTHENIC

SYNS: DISTILLATES (PETROLEUM), SOLVENT-RE-FINED LIGHT NAPHTHENIC (9CI)   *   SOLVENT-REFINED LIGHT NAPHTHENIC DISTILLATE

CONSENSUS REPORTS: IARC Cancer Review: GROUP 1 IMEMDT 7,252,87; Animal Sufficient Evidence IMEMDT 33,87,84. Reported in EPA TSCA Inventory.

SAFETY PROFILE: Confirmed carcinogen. When heated to decomposition it emits acrid smoke and irritating fumes.

MQV855   CAS: 64741-89-5   **HR: 3**
MINERAL OIL, PETROLEUM DISTILLATES, SOLVENT-REFINED LIGHT PARAFFINIC

SYNS: DISTILLATES (PETROLEUM), SOLVENT-RE-FINED LIGHT PARAFFINIC (9CI)   *   SOLVENT-REFINED LIGHT PARAFFINIC DISTILLATE

CONSENSUS REPORTS: IARC Cancer Review: GROUP 1 IMEMDT 7,252,87; Animal Sufficient Evidence IMEMDT 33,87,84. Reported in EPA TSCA Inventory.

SAFETY PROFILE: Confirmed carcinogen. When heated to decomposition it emits acrid smoke and irritating fumes.

MQV857   CAS: 64742-11-6   **HR: 3**
MINERAL OIL, PETROLEUM EXTRACTS, HEAVY NAPHTHENIC DISTILLATE SOLVENT

SYNS: EXTRACTS (PETROLEUM), HEAVY NA-PHTHENIC DISTILLATE SOLVENT (9CI)   *   HEAVY NAPHTHENIC DISTILLATE SOLVENT EXTRACT

CONSENSUS REPORTS: IARC Cancer Review: GROUP 1 IMEMDT 7,252,87; Animal Sufficient Evidence IMEMDT 33,87,84. Reported in EPA TSCA Inventory.

SAFETY PROFILE: Confirmed carcinogen. Experimental reproductive data. When heated to decomposition it emits acrid smoke and irritating fumes.

MQV859   CAS: 64742-04-7   HR: 3
MINERAL OIL, PETROLEUM EXTRACTS, HEAVY PARAFFINIC DISTILLATE SOLVENT

SYNS: EXTRACTS (PETROLEUM), HEAVY PARAFFINIC DISTILLATE SOLVENT (9CI) * HEAVY PARAFFINIC DISTILLATE, SOLVENT EXTRACT

CONSENSUS REPORTS: IARC Cancer Review: GROUP 1 IMEMDT 7,252,87; Animal Sufficient Evidence IMEMDT 33,87,84. Reported in EPA TSCA Inventory.

SAFETY PROFILE: Confirmed carcinogen. When heated to decomposition it emits acrid smoke and irritating fumes.

MQV860   CAS: 64742-03-6   HR: 3
MINERAL OIL, PETROLEUM EXTRACTS, LIGHT NAPHTHENIC DISTILLATE SOLVENT

SYNS: EXTRACTS (PETROLEUM), LIGHT NAPHTHENIC DISTILLATE SOLVENT (9CI) * LIGHT NAPHTHENIC DISTILLATE, SOLVENT EXTRACT

CONSENSUS REPORTS: IARC Cancer Review: GROUP 1 IMEMDT 7,252,87; Animal Sufficient Evidence 33,87,84. Reported in EPA TSCA Inventory.

SAFETY PROFILE: Confirmed carcinogen. When heated to decomposition it emits acrid smoke and irritating fumes.

MQV862   CAS: 64742-05-8   HR: 3
MINERAL OIL, PETROLEUM EXTRACTS, LIGHT PARAFFINIC DISTILLATE SOLVENT

SYNS: EXTRACTS (PETROLEUM), LIGHT PARAFFINIC DISTILLATE SOLVENT (9CI) * LIGHT PARAFFINIC DISTILLATE, SOLVENT EXTRACT

CONSENSUS REPORTS: IARC Cancer Review: GROUP 1 IMEMDT 7,252,87; Animal Sufficient Evidence IMEMDT 33,87,84. Reported in EPA TSCA Inventory.

SAFETY PROFILE: Confirmed carcinogen. When heated to decomposition it emits acrid smoke and irritating fumes.

MQV863   CAS: 64742-10-5   HR: 3
MINERAL OIL, PETROLEUM EXTRACTS, RESIDUAL OIL SOLVENT

SYNS: EXTRACTS (PETROLEUM), RESIDUAL OIL SOLVENT (9CI) * RESIDUAL OIL SOLVENT EXTRACT

CONSENSUS REPORTS: IARC Cancer Review: GROUP 1 IMEMDT 7,252,87; Animal Sufficient Evidence IMEMDT 33,87,84. Reported in EPA TSCA Inventory.

SAFETY PROFILE: Confirmed carcinogen. When heated to decomposition it emits acrid smoke and irritating fumes.

MQV865   CAS: 64742-68-3   HR: 3
MINERAL OIL, PETROLEUM NAPHTHENIC OILS, CATALYTIC DEWAXED HEAVY

SYNS: CATALYTIC-DEWAXED HEAVY NAPHTHENIC DISTILLATE * NAPHTHENIC OILS (PETROLEUM), CATALYTIC DEWAXED HEAVY(9CI)

CONSENSUS REPORTS: IARC Cancer Review: GROUP 1 IMEMDT 7,252,87; Animal Sufficient Evidence IMEMDT 33,87,84. Reported in EPA TSCA Inventory.

SAFETY PROFILE: Confirmed carcinogen. When heated to decomposition it emits acrid smoke and irritating fumes.

MQV867   CAS: 64742-69-4   HR: 3
MINERAL OIL, PETROLEUM NAPHTHENIC OILS, CATALYTIC DEWAXED LIGHT

SYNS: CATALYTIC-DEWAXED LIGHT NAPHTHENIC DISTILLATE * NAPHTHENIC OILS (PETROLEUM), CATALYTIC DEWAXED LIGHT (9CI)

CONSENSUS REPORTS: IARC Cancer Review: GROUP 1 IMEMDT 7,252,87; Animal Sufficient Evidence IMEMDT 33,87,84. Reported in EPA TSCA Inventory.

SAFETY PROFILE: Confirmed carcinogen. When heated to decomposition it emits acrid smoke and irritating fumes.

MQV868   CAS: 64742-70-7   HR: 3
MINERAL OIL, PETROLEUM PARAFFIN OILS, CATALYTIC DEWAXED HEAVY

SYNS: CATALYTIC-DEWAXED HEAVY PARAFFINIC DISTILLATE * PARAFFIN OILS (PETROLEUM), CATALYTIC DEWAXED HEAVY (9CI)

CONSENSUS REPORTS: IARC Cancer Review: GROUP 1 IMEMDT 7,252,87; Animal

Sufficient Evidence IMEMDT 33,87,84. Reported in EPA TSCA Inventory.

SAFETY PROFILE: Confirmed carcinogen. When heated to decomposition it emits acrid smoke and irritating fumes.

MQV870   CAS: 64742-71-8   **HR: 3**
MINERAL OIL, PETROLEUM PARAFFIN OILS, CATALYTIC DEWAXED LIGHT

SYNS: CATALYTIC-DEWAXED LIGHT PARAFFINIC DISTILLATE * PARAFFIN OILS (PETROLEUM), CATALYTIC DEWAXED LIGHT (9CI)

CONSENSUS REPORTS: IARC Cancer Review: GROUP 1 IMEMDT 7,252,87; Animal Sufficient Evidence IMEMDT 33,87,84. Reported in EPA TSCA Inventory.

SAFETY PROFILE: Confirmed carcinogen. When heated to decomposition it emits acrid smoke and irritating fumes.

MQV872   CAS: 64742-17-2   **HR: 3**
MINERAL OIL, PETROLEUM RESIDUAL OILS, ACID-TREATED

SYNS: ACID-TREATED RESIDUAL OIL * RESIDUAL OILS (PETROLEUM), ACID-TREATED (9CI)

CONSENSUS REPORTS: IARC Cancer Review: GROUP 1 IMEMDT 7,252,87; Animal Sufficient Evidence IMEMDT 33,87,84.

SAFETY PROFILE: Confirmed carcinogen. When heated to decomposition it emits acrid smoke and irritating fumes.

MQV875   CAS: 8042-47-5   **HR: 3**
MINERAL OIL, SLAB OIL

SYNS: SLAB OIL (9CI) * WHITE MINERAL OIL

CONSENSUS REPORTS: IARC Cancer Review: GROUP 1 IMEMDT 7,252,87; Animal Inadequate Evidence IMEMDT 33,87,84. Reported in EPA TSCA Inventory.

SAFETY PROFILE: Confirmed carcinogen. When heated to decomposition it emits acrid smoke and irritating fumes.

MQW500   CAS: 2385-85-5   **HR: 3**
MIREX
mf: $C_{10}Cl_{12}$ mw: 545.50

PROP: Very white, odorless crystals. Decomp @ 485°. Water-insol; sol in dioxane and benzene.

SYNS: BICHLORENDO * CG-1283 * DECHLORANE 4070 * DODECACHLOROOCTAHYDRO-1,3,4-METHENO-2H-CYCLOBUTA(c,d)PENTALENE * 1,1a,2,2,3,3a,4,5,5,5a,5b,6-DODECACHLOROOCTAHYDRO-1,3,4-METHENO-1H-CYCLOBUTA(c,d)PENTALENE * DODECACHLOROPENTACYCLODECANE * DODECACHLOROPENTACYCLO(3,2,2,0^{2,6},0^{3,9},0^{5,10})DECANE * ENT 25,719 * FERRIAMICIDE * HEXACHLOROCYCLOPENTADIENEDIMER * 1,2,3,4,5,5-HEXACHLORO-1,3-CYCLOPENTADIENE DIMER * HRS 1276 * NCI-C06428 * PERCHLORODIHOMOCUBANE * PERCHLOROPENTACYCLODECANE * PERCHLOROPENTACYCLO(5.2.1.0^{2,6}.0^{3,9}.0^{5,8})DECANE

CONSENSUS REPORTS: IARC Cancer Review: GROUP 2B IMEMDT 7,56,87; Human Limited Evidence IMEMDT 20,283,79; Animal Sufficient Evidence IMEMDT 20,283,79; IMEMDT 5,203,74. NTP Fourth Annual Report On Carcinogens, 1984. EPA Genetic Toxicology Program.

SAFETY PROFILE: Confirmed carcinogen with experimental carcinogenic and tumorigenic data. Poison by ingestion. Moderately toxic by inhalation and skin contact. Experimental teratogenic and reproductive effects. Mutation data reported. A persistent insecticide which is toxic to non-target species. It can bioaccumulate.

MQW750   CAS: 18378-89-7   **HR: 3**
MITHRAMYCIN
mf: $C_{52}H_{76}O_{24}$ mw: 1085.28

PROP: Antibiotic substance isolated from the fermentation broth of three strains of an unidentified *Streptomyces* species.

SYNS: A-2371 * ANTIBIOTIC LA 7017 * AUREOLIC ACID * AURLELIC ACID * MITHRACIN * MITHRAMYCIN A * MITRAMYCIN * NSC 24559 * PA 144

SAFETY PROFILE: A deadly poison by intravenous, intraperitoneal, subcutaneous, and possibly other routes. Moderately toxic by ingestion. Human systemic effects by ingestion: blood thrombocytopenia. Experimental teratogenic and reproductive effects. Human mutation data reported. When heated to decomposition it emits acrid smoke and irritating fumes.

MQY325   CAS: 63642-17-1   **HR: 3**
MNCO
mf: $C_7H_{14}N_4O_4$ mw: 218.25

SYNS: Nᐃ-(N-METHYL-N-NITROSOCARBAMOYL)-1-OR-
NITHINE * N⁵-(METHYLNITROSOCARBAMOYL)-1-OR-
NITHINE * N⁵-(N-METHYL-N-NITROSOCARBAMOYL)-1-
ORNITHINE

SAFETY PROFILE: Suspected carcinogen with
experimental carcinogenic and tumorigenic
data. Mutation data reported. When heated to
decomposition it emits toxic fumes of NO_x.

MRA250   CAS: 11015-37-5   HR: 3
MOENOMYCIN

PROP: Produced by *Streptomyces roseoflavus*.

SYNS: BAMBERMYCIN * FLAVOMYCIN
* FLAVOPHOSPHOLIPOL * MENOMYCIN
* MOENOMYCIN A

SAFETY PROFILE: Poison by intravenous
route. Moderately toxic by subcutaneous route.

MRC250   CAS: 7439-98-7   HR: 3
MOLYBDENUM
af: Mo   aw: 95.94

PROP: Cubic, silver-white metallic crystals or
gray-black powder. Mp: 2622°, bp: approx
4825°, d: 10.2, vap press: 1 mm @ 3102°.

SYN: MOLYBDATE

CONSENSUS REPORTS: Reported in EPA
TSCA Inventory.

OSHA PEL: Soluble Compounds: TWA 5
    mg(Mo)/m³; Insoluble Compounds: (Transi-
    tional: TWA Total Dust: 15 mg/m³; Respira-
    ble Fraction: 5 mg/m³) TWA Total Dust: 10
    mg/m³; Respirable Fraction: 5 mg/m³
ACGIH TLV: Soluble Compounds: TWA 5
    mg(Mo)/m³; Insoluble Compounds: TWA 10
    mg(Mo)/m³
DFG MAK: (insoluble compounds) 15 mg/m³;
    (soluble compounds) 5 mg/m³

SAFETY PROFILE: Poison by intraperitoneal
and intratracheal routes. Mutation data reported.
Experimental teratogenic and reproductive effects. Flammable or explosive in the form of
dust; when exposed to heat or flame. Violent
reaction with oxidants (e.g., bromine trifluoride,
bromine pentafluoride, chlorine trifluoride, potassium perchlorate, nitryl fluoride, fluorine,
iodine pentafluoride, sodium peroxide, lead dioxide). When heated to decomposition it emits
toxic fumes of Mo.

MRC750   HR: 3
MOLYBDENUM COMPOUNDS

SAFETY PROFILE: Poison by subcutaneous
and intraperitoneal routes. Molybdenum and its
compounds are highly toxic based upon animal
experiments. Symptoms of acute poisoning include severe gastrointestinal irritation with diarrhea, coma, and deaths from heart failure. Experimental animals exposed to high levels (57 mg
Mo/m³)of molybdenum dust for 120 days (4
hours/day) accumulated Mo in the lungs, spleen,
and heart, and showed a decrease of DNA and
RNA in the liver, kidneys, and spleen. Workers
exposed to Mo or MoO_3 (concentrations of 1-
19 mg Mo/m³) over a period of 3-7 years have
suffered from pneumoconiosis. Inhalation of
molybdenum dust from alloys or carbides can
cause "hard-metal lung disease". It is suggested
that suitable precautions should be taken against
human inhalation of significant amounts of the
more soluble molybdenum compounds. MoO_3
and Na_2MoO_4 are soluble. $CaMoO_4$, MoO, and
MoS_2 are insoluble. Hexavalent molybdenum
compounds are readily absorbed through the
gastrointestinal tract. Coal-fired electrical power
plants can be significant sources of molybdenum. Application of some fertilizers may raise
molybdenum concentrations in ground water.
Molybdenum is rapidly excreted by the body.
Molybdenum is an important trace element in
the normal growth and development of plants.
It is found also in animal tissue, although its
precise function is unknown. It is a common
air contaminant.

MRD500   CAS: 10241-05-1   HR: 3
MOLYBDENUM PENTACHLORIDE
DOT: UN 2508
mf: Cl_5Mo mw: 273.19

PROP: Green-black solid, dark-red as liquid
or vapor. Hygroscopic, reacting with water and
air. Mp: 194°, bp: 268°, d: 2.9. Sol in dry
ether, dry alc, and other anhydrous organic solvents.

CONSENSUS REPORTS: Reported in EPA
TSCA Inventory. EPA Genetic Toxicology Program.

OSHA PEL: TWA 5 mg(Mo)/m³
ACGIH TLV: TWA 5 mg(Mo)/m³
DOT Classification: ORM-B; Label: None;
    Corrosive Material; Label: Corrosive.

SAFETY PROFILE: A poison. A corrosive irritant to skin, eyes, and mucous membranes. Reacts with moisture to form hydrochloric acid. When heated to decomposition it emits toxic fumes of Mo and Cl^-.

MRE000   CAS: 1313-27-5   **HR: 3**
MOLYBDENUM TRIOXIDE
mf: MoO_3 mw: 143.94

PROP: White or yellow to sltly bluish powder or granules. Mp: 795°; bp: 1155°; d: 4.696 @ 26°/4°. Sol in 1000 parts water, in concentrated mineral acids, solutions of alkali hydroxides. Sol in ammonia or potassium bitartrate, solidifying to a yellowish-white mass.

SYNS: MOLYBDENUM(VI) OXIDE * MOLYBDIC ANHYDRIDE * MOLYBDIC TRIOXIDE

CONSENSUS REPORTS: Reported in EPA TSCA Inventory. EPA Extremely Hazardous Substances List.

OSHA PEL: TWA 5 mg(Mo)/m^3
ACGIH TLV: TWA 5 mg(Mo)/m^3

SAFETY PROFILE: Poison by ingestion, subcutaneous, and intraperitoneal routes. Human systemic effects by inhalation: pulmonary fibrosis and cough. A powerful irritant. Questionable carcinogen with experimental neoplastigenic data. Explodes on contact with molten magnesium. Violent reaction with interhalogens (e.g., bromine pentafluoride, chlorine trifluoride). Incandescent reaction with hot sodium, potassium, or lithium. When heated to decomposition it emits toxic fumes of Mo.

MRE225   CAS: 17090-79-8   **HR: 3**
MONENSIC ACID
mf: $C_{36}H_{62}O_{11}$ mw: 670.98

PROP: Crystals. Mp: 103-105° (monohydrate). Very stable under alkaline conditions. Sltly sol in water; more sol in hydrocarbons; very sol in other organic solvents.

SYNS: A 3823A * ELANCOBAN * MONELAN * MONENSIN (USDA) * MONENSIN A

SAFETY PROFILE: Poison by ingestion and intraperitoneal routes. An eye and skin irritant. When heated to decomposition it emits acrid smoke and irritating fumes.

MRF000   CAS: 7558-63-6   **HR: 2**
MONOAMMONIUM GLUTAMATE
mf: $C_5H_9NO_4 \cdot H_3N$ mw: 164.19

PROP: White crystalline powder; odorless. Sol in water; insol in common organic solvents.

SYNS: AMMONIUMGLUTAMINAT (GERMAN) * MAG * MONOAMMONIUM l-GLUTAMATE

CONSENSUS REPORTS: Reported in EPA TSCA Inventory.

SAFETY PROFILE: Moderately toxic by intraperitoneal route. When heated to decomposition it emits toxic fumes including NO_x and NH_3.

MRG000   CAS: 55398-86-2   **HR: 2**
MONOCHLORO DIPHENYL OXIDE
mf: $C_{12}H_9ClO$ mw: 204.66

SYNS: MONOCHLOROPHENYLETHER * PHENYL ETHER MONO-CHLORO

OSHA PEL: TWA 500 μg/m^3

SAFETY PROFILE: Moderately toxic by ingestion. When heated to decomposition it emits toxic fumes of Cl^-.

MRH000   CAS: 315-22-0   **HR: 3**
MONOCROTALINE
mf: $C_{16}H_{23}NO_6$ mw: 325.40

PROP: Prisms from absolute ethanol. Decomp @ 197-198°.

SYNS: CROTALINE * 14,19-DIHYDRO-12,13-DIHYDROXY(13-α,14-α)-20-NORCROTALANAN-11,15-DIONE * MONOCRATILIN * NCI-C56462 * NSC 28693

CONSENSUS REPORTS: IARC Cancer Review: GROUP 2B IMEMDT 7,56,87; Animal Limited Evidence IMEMDT 10,291,76. EPA Genetic Toxicology Program.

SAFETY PROFILE: Suspected carcinogen with experimental carcinogenic data. Poison by ingestion, intravenous, intraperitoneal, subcutaneous, and possibly other routes. Human mutation data reported. When heated to decomposition it emits toxic fumes of NO_x.

MRH209   CAS: 6923-22-4   **HR: 3**
MONOCROTOPHOS
mf: $C_7H_{14}NO_5P$ mw: 223.19

PROP: A reddish-brown solid; mild ester odor. Bp: 125°.

SYNS: APADRIN * AZODRIN * BILOBRAN * CRISODRIN * CRISODIN * 3-(DIMETHOXYPHOSPHINYLOXY)N-METHYL-cis-CROTONAMIDE * O,O-DIMETHYL-O-(2-N-METHYLCARBAMOYL-1-METHYL-VINYL)-FOSFAAT (DUTCH) * O,O-DIMETHYL-O-(2-N-METHYLCARBAMOYL-1-METHYL)-

VINYL-PHOSPHAT (GERMAN) * O,O-DIMETHYL-O-(2-N-METHYLCARBAMOYL-1-METHYL-VINYL) PHOSPHATE * DIMETHYL-1-METHYL-2-(METHYLCARBAMOYL) VINYLPHOSPHATE, cis * (E)-DIMETHYL 1-METHYL-3-(METHYLAMINO)-3-OXO-1-PROPENYL PHOSPHATE * DIMETHYL PHOSPHATE ESTER OF 3-HY-DROXY-N-METHYL-cis-CROTONAMIDE * DIMETHYL PHOSPHATE OF 3-HYDROXY-N-METHYL-cis-CROTONAMINE * O,O-DIMETIL-O-(2-N-METIL-CARBAMOIL-1-METIL-VINIL)-FOSFATO (ITALIAN) * ENT 27,129 * HAZODRIN * 3-HYDROXY-N-METHYL-cis-CROTONAMIDE DIMETHYL PHOSPHATE * cis-1-METHYL-2-METHYL CARBAMOYL VINYL PHOSPHATE * MONOCIL 40 * MONOCRON * NUVACRON * PHOSPHATE de DIMETHYLE et de 2-METHYLCARBAMOYL-1-METHYL VINYLE (FRENCH) * PHOSPHORIC ACID, DIMETHYL ESTER, ESTER with cis-3-HYDROXY-N-METHYLCROTONAMIDE * PIL-LARDRIN * PLANTDRIN * SHELL SD 9129 * SUSVIN * ULVAIR

CONSENSUS REPORTS: EPA Genetic Toxicology Program. EPA Extremely Hazardous Substances List.

OSHA PEL: TWA 0.25 mg/m^3
ACGIH TLV: TWA 0.25 mg/m^3

SAFETY PROFILE: Poison by ingestion, inhalation, skin contact, intraperitoneal, subcutaneous, and intravenous routes. Mutation data reported. Use may be restricted. When heated to decomposition it emits very toxic NO_x and PO_x.

MRI750   CAS: 39801-14-4   **HR: 3**
8-MONOHYDRO MIREX
mf: $C_{10}HCl_{11}$ mw: 511.06

SYNS: HYDROMIREX * PHOTOMIREX * 1,2,-3,4,5,5,6,7,9,10,10-UNDECACHLORO-PENTACYCLO(5.3.0.0^{2,6}.0^{3,9}.0^{4,8})DECANE

SAFETY PROFILE: Poison by ingestion. Experimental teratogenic and reproductive effects. Questionable carcinogen with experimental tumorigenic data. Mutation data reported. When heated to decomposition it emits toxic fumes of Cl$^-$.

MRJ250   CAS: 29674-96-2   **HR: 3**
MONOMETHYLHYDRAZINE NITRATE
mf: $CH_6N_2 \cdot HNO_3$ mw: 109.11

SYN: METHYLHYDRAZINIUM NITRATE

SAFETY PROFILE: Poison by ingestion, skin contact, intravenous, and intraperitoneal routes. An impact-sensitive explosive. When heated to decomposition it emits toxic fumes of NO_x.

MRJ750   CAS: 5632-47-3   **HR: 3**
MONONITROSOPIPERAZINE
mf: $C_4H_9N_3O$ mw: 115.16

SYNS: N-NITROSOPIPERAZINE * 1-NITROSOPIPERA-ZINE

SAFETY PROFILE: Moderately toxic by ingestion. Questionable carcinogen with experimental carcinogenic, neoplastigenic, and tumorigenic data. Mutation data reported. When heated to decomposition it emits toxic fumes of NO_x.

MRK500   CAS: 19473-49-5   **HR: 1**
MONOPOTASSIUM GLUTAMATE
mf: $C_5H_8NO_4 \cdot K$ mw: 185.24

PROP: White, free-flowing, hygroscopic crystalline powder; practically odorless. Freely sol in water; sltly sol in alc.

SYNS: l-GLUTAMIC ACID, MONOPOTASSIUM SALT * MONOPOTASSIUM l-GLUTAMATE (FCC) * MPG * POTASSIUM GLUTAMATE * POTASSIUM GLUTA-MINATE

CONSENSUS REPORTS: Reported in EPA TSCA Inventory. EPA Genetic Toxicology Program.

SAFETY PROFILE: Mildly toxic by ingestion and possibly other routes. Human systemic effects by ingestion: headache. When heated to decomposition it emits toxic fumes of K_2O and NO_x.

MRL500   CAS: 142-47-2   **HR: 2**
MONOSODIUM GLUTAMATE
mf: $C_5H_9NO_4 \cdot Na$ mw: 170.14

PROP: White or almost white crystals or powder; slt peptone-like odor, meal-like taste. Very sol in water; sltly sol in alc.

SYNS: ACCENT * AJINOMOTO * CHINESE SEASONING * GLUTACYL * GLUTAMIC ACID, SODIUM SALT * GLUTAMMATO MONOSODICO (ITALIAN) * GLUTAVENE * MONOSODIOGLUTAMMATO (ITALIAN) * MONOSODIUM-l-GLUTAMATE (FCC) * α-MONOSODIUM GLUTAMATE * MSG * NATRIUMGLUTAMINAT (GERMAN) * RL-50 * SODIUM GLUTAMATE * SODIUM l-GLUTAMATE * l(+) SODIUM GLUTAMATE * VETSIN * ZEST

CONSENSUS REPORTS: Reported in EPA TSCA Inventory. EPA Genetic Toxicology Program.

SAFETY PROFILE: Moderately toxic by intravenous route. Mildly toxic by ingestion and other routes. Human systemic effects by ingestion and intravenous routes: somnolence, hallucinations and distorted perceptions, headache, dyspnea, nausea or vomiting. Experimental teratogenic and reproductive effects. The cause of "Chinese restaurant syndrome." When heated to decomposition it emits toxic fumes of NO_x and Na_2O.

MRL750   CAS: 2163-80-6   HR: 3
MONOSODIUM METHYLARSONATE
mf: $CH_4AsO_3 \bullet Na$ mw: 161.96

SYNS: ANSAR 170 * ARSONATE liquid * ASAZOL * BUENO * DACONATE 6 * DAL-E-RAD * HERB-ALL * HERBAN M * MERGE * MESA-MATE * MESAMATE CONCENTRATE * METHYL-ARSENIC ACID, SODIUM SALT * MONATE * MONOSODIUM ACID METHANEARSONATE * MONOSODIUM ACID METHARSONATE * MONO-SODIUM METHANEARSONATE * MONOSODIUM METHANEARSONIC ACID * MSMA * NCI-C60071 * PHYBAN * SILVISAR 550 * SODIUM ACID METHANEARSONATE * SODIUM METHANEARSO-NATE * TARGET MSMA * TRANS-VERT * WEED 108 * WEED-E-RAD * WEED-HOE

CONSENSUS REPORTS: Arsenic and its compounds are on the Community Right-To-Know List. EPA Genetic Toxicology Program.

OSHA PEL: TWA 0.5 mg(As)/m^3
ACGIH TLV: TWA 0.2 mg(As)/m^3

SAFETY PROFILE: Poison by unspecified route. Moderately toxic by ingestion. A skin and eye irritant. When heated to decomposition it emits toxic fumes of As and Na_2O.

MRN500   CAS: 480-16-0   HR: 2
MORIN
mf: $C_{15}H_{10}O_7$ mw: 302.25

PROP: Anhydrous needles from absolute alc. Decomp @ 285-290°. Crystalized with 1 or 2 mols water; sltly sol in water, ether, and acetic acid; very sol in alc.

SYNS: A1-MORIN * AURANTICA * BOIS D'ARC (FRENCH) * CALICO YELLOW * C.I. NATURAL YELLOW 8 * 2-2-(2,4-DIHYDROXYPHENYL)-3,5,7-TRI-HYDROXY-4H-1-BENZOPYRAN-4-ONE * 2'-HYDROXY-PELARGIDENOLON 1522 * OSAGE ORANGE * OSAGE ORANGE CRYSTALS * OSAGE ORANGE EXTRACT * 2',3,4',5,7-PENTAHYDROXYFLAVONE * TOXYLON POMIFERUM

CONSENSUS REPORTS: Reported in EPA TSCA Inventory.

SAFETY PROFILE: Moderately toxic by intraperitoneal route. Mutation data reported. When heated to decomposition it emits acrid smoke and irritating fumes.

MRO500   CAS: 57-27-2   HR: 3
(−)-MORPHINE
mf: $C_{17}H_{19}NO_3$ mw: 285.37

PROP: White, crystalline alkaloid. Short, orthorhombic, columnar prisms from anisole. Decomp @ 254°, mp: 197°, sublimes @ 190-200°.

SYNS: MORFINA (ITALIAN) * MORPHIA * MORPHINA * MORPHINE * MORPHINISM * MORPHINUM * MORPHIUM * 4a,5,7a,8-TET-RAHYDRO-12-METHYL-9H-9,9c-IMINOETHANO-PHENANTHRO(4,5-bcd)FURAN-3,5-DIOL

SAFETY PROFILE: A human poison by an unspecified route. Poison experimentally by ingestion, intracerebral, intraperitoneal, subcutaneous, intravenous, and possibly other routes. Human reproductive effects by an unspecified route: effects on newborn including drug dependence. Experimental reproductive effects. Morphine is the constituent of opium most responsible for its toxic effects. When taken orally, the effects of morphine poisoning begin to appear in 20-40 minutes; if taken hypodermically, the symptoms appear much earlier and narcotism is more likely to follow the early symptoms. Abuse leads to habituation or addiction. Individual susceptibility varies greatly and children are more susceptible than adults. When heated to decomposition it emits toxic fumes of NO_x.

MRO750   CAS: 52-26-6   HR: 3
MORPHINE HYDROCHLORIDE
mf: $C_{17}H_{19}NO_3 \bullet ClH$ mw: 321.83

PROP: Trihydrate: White flakes or crystalline powder; bitter taste. Mp: approx 200° (decomp). Dissolves slowly in glycerol; insol in chloroform, ether.

SYNS: 7,8-DIDEHYDRO-4,5-α-EPOXY-17-METHYLMOR-PHINAN-3,6-α-DIOL HYDROCHLORIDE * 7,8-DIDEHY-DRO-4,5-α-EPOXY-17-METHYLMORPHINE HYDROCHLO-RIDE * MORPHINE CHLORHYDRATE * MORPHINE CHLORIDE

CONSENSUS REPORTS: EPA Genetic Toxicology Program.

SAFETY PROFILE: Poison by ingestion, intraperitoneal, intravenous, parenteral, and subcutaneous routes. Experimental teratogenic and reproductive effects. Abuse leads to habituation or addiction. When heated to decomposition it emits very toxic fumes of NO_x and HCl.

MRP250   CAS: 64-31-3   HR: 3
MORPHINE SULFATE
mf: $C_{34}H_{38}N_2O_6 \cdot H_2O_4S$ mw: 668.82

SYN: MORPHINE SULPHATE

SAFETY PROFILE: Poison by subcutaneous, intravenous, intraperitoneal, and intramuscular routes. Moderately toxic by ingestion and parenteral routes. Experimental teratogenic and reproductive effects. Mutation data reported. Used as a narcotic. Abuse leads to habituation or addiction. When heated to decomposition it emits very toxic fumes of NO_x and SO_x.

MRP750   CAS: 110-91-8   HR: 3
MORPHOLINE

DOT: UN 1760/UN 2054
mf: C_4H_9NO mw: 87.14

PROP: Colorless, hygroscopic oil; amine odor. Bp: 128.9°, fp: $-7.5°$, flash p: 100°F (OC), autoign temp: 590°F, vap press: 10 mm @ 23°, vap d: 3.00, mp: $-4.9°$, d: 1.007 @ 20°/4°. Volatile with steam; misc with water evolving some heat; misc with acetone, benzene, ether, castor oil, methanol, ethanol, ethylene, glycol, linseed oil, turpentine, pine oil. Immiscible with concentrated NaOH solns.

SYNS: DIETHYLENEIMIDE OXIDE   *   DIETHYLENE IMIDOXIDE   *   DIETHYLENE OXIMIDE   *   DIETHYL-ENIMIDE OXIDE   *   MORPHOLINE, AQUEOUS MIXTURE (DOT)   *   1-OXA-4-AZACYCLOHEXANE   *   TETRAHYDRO-p-ISOXAZINE   *   TETRAHYDRO-1,4-ISOXAZINE   *   TETRAHYDRO-1,4-OXAZINE   *   TETRAHYDRO-2H-1,4-OXAZINE

CONSENSUS REPORTS: Reported in EPA TSCA Inventory. EPA Genetic Toxicology Program.

OSHA PEL: (Transitional: TWA 20 ppm (skin) TWA 20 ppm (skin); STEL 30 ppm (skin)
ACGIH TLV: TWA 20 ppm; STEL 30 ppm (skin); (Proposed: TWA 20 ppm (skin))
DFG MAK: 20 ppm (70 mg/m³)
DOT Classification: Flammable Liquid; Label: Flammable Liquid; Corrosive Material; Label: Corrosive, aqueous solution.

SAFETY PROFILE: Moderately toxic by ingestion, inhalation, skin contact, intraperitoneal, and possibly other routes. Questionable carcinogen with experimental neoplastigenic data. Mutation data reported. A corrosive irritant to skin, eyes, and mucous membranes. Can cause kidney damage. Flammable liquid. A very dangerous fire hazard when exposed to flame, heat, or oxidizers; can react with oxidizing materials. To fight fire, use alcohol foam, CO_2, dry chemical. Mixtures with nitromethane are explosive. May ignite spontaneously in contact with cellulose nitrate of high surface area. When heated to decomposition it emits highly toxic fumes of NO_x.

MRQ750   CAS: 5299-64-9   HR: 3
4-MORPHOLINENONYLIC ACID
mf: $C_{13}H_{25}NO_3$ mw: 227.39

SYNS: AI 318284   *   N-MORPHOLINO NONANAMIDE   *   4-NONANOYLMORPHOLINE   *   PELARGONIC MORPHOLIDE

SAFETY PROFILE: Poison by inhalation and intravenous routes. Human systemic effects by inhalation: lacrimation, deviated nasal septum, and cough. An eye and intense mucous membrane irritant. When heated to decomposition it emits toxic fumes of NO_x.

MRU250   CAS: 144-41-2   HR: 3
MORPHOTHION
mf: $C_8H_{16}NO_4PS_2$ mw: 285.34

PROP: Colorless solid. Mp: 65°. Sol in acetone, dioxane, acetonitrile.

SYNS: O,O-DIMETHYL-S-((MORFOLINO-CARBONYL) METHYL)-DITHIOFOSFAAT (DUTCH)   *   O,O-DI-METHYL-S-(MORPHOLINOCARBAMOYLMETHYL) DITHIO-PHOSPHATE   *   O,O-DIMETHYL-S-((MORPHOLINO-CAR-BONYL)-METHYL)-DITHIOPHOSPHAT (GERMAN)   *   O,O-DIMETHYL MORPHOLINOCARBONYLMETHYL PHOSPHORODITHIOATE   *   O,O-DIMETHYL-S-(MOR-PHOLINOCARBONYLMETHYL) PHOSPHORODITHIOATE   *   DIMETHYL S-(MORPHOLINOCARBONYLMETHYL) PHOSPHOROTHIOLOTHIONATE   *   O,O-DIMETIL-S-((MORFOLINO-CARBONIL)-METIL)-DITIOFOSFATO (ITALIAN)   *   DITHIOPHOSPHATE de O,O-DIMETHYLE et de S-((MORPHOLINOCARBONYLE)-METHYLE) (FRENCH)   *   4-(MERCAPTOACETYL)MORPHOLINE O,O-DIMETHYL PHOSPHORODITHIOATE   *   MORFOTHION (DUTCH)   *   PHOSPHORODITHIOIC ACID, O,O-DIMETHYL S-(MOR-PHOLINOCARBONYLMETHYL) ESTER

SAFETY PROFILE: Poison by ingestion, skin contact, and possibly other routes. A cholinesterase inhibitor. When heated to decomposition it emits very toxic fumes of PO_x, SO_x, and NO_x.

MRV250 CAS: 992-21-2 **HR: 3**
MUCOMYCIN
mf: $C_{29}H_{33}N_4O_{10}$ mw: 597.66

SYNS: N^2-(((+)-5-AMINO-5-CARBOXYPENTYLAMINO) METHYL)TETRACYCLINE * ARMYL * LYME-CYCLINE * N-LYSINOMETHYLTETRACYCLINE * TETRACICLINA-l-METILENLISINA (ITALIAN) * TETRACYCLINE-l-METHYLENE LYSINE * TETRA-LISAL * TETRALYSAL

SAFETY PROFILE: Poison by intravenous and possibly other routes. When heated to decomposition it emits toxic fumes of NO_x.

MRW000 CAS: 58-34-4 **HR: 3**
MULTERGAN METHYL SULFATE
mf: $C_{18}H_{23}N_2S \cdot CH_3O_4S$ mw: 410.59

SYNS: METHYLPHENAZONIUM METHOSULFATE * MULTERGAN * MULTEZIN * PADISAL * N-(β-(10-PHENOTHIAZINYL)PROPYL)TRIMETHYL-AMMONIUM METHYL SULFATE * PMS * PRO-THAZIN METHOSULFATE * RP 3554 * N,N,N-α-TETRAMETHYL-10H-PHENOTHIAZINE-10-ETHANAMI-NIUM METHYL SULFATE * THIAZINAMIUM METHYL SULFATE * TRIMETHYL (1-METHYL-2-PHENO-THIAZIN-10-YLETHYL)AMMONIUM METHYL SULFATE * TRIMETHYL(1-METHYL-2-(10-PHENOTHIAZINYL) ETHYL)AMMONIUM METHYL SULFATE * VALAN

SAFETY PROFILE: Poison by ingestion and intraperitoneal routes. An antihistamine and anticholinergic agent. When heated to decomposition it emits very toxic fumes of SO_x, NH_3, and NO_x.

MRW250 CAS: 300-54-9 **HR: 3**
MUSCARINE
mf: $C_9H_{20}NO_2$ mw: 174.30

SYNS: MUSCARIN * dl-MUSCARINE * MUSK * MUSKARIN * TRIMETHYL(TETRAHYDRO-4-HY-DROXY-5-METHYLFURFURYL)AMMONIUM

SAFETY PROFILE: Human poison by an unspecified route. Experimental poison by ingestion, intravenous, intraperitoneal, and subcuta-

neous routes. When heated to decomposition it emits toxic fumes of NO_x and NH_3.

MRZ150 CAS: 123-35-3 **HR: 1**
MYRCENE
mf: $C_{10}H_{16}$ mw: 136.26

PROP: Colorless to pale yellow liquid; sweet, balsamic odor. D: 0.789, refr index: 1.466-1.471, flash p: 99°F. Sol in alc, fixed oils; insol in water.

SYNS: FEMA No. 2762 * 3-METHYLENE-7-METHYL-1,6-OCTADIENE * 7-METHYL-3-METHYLENE-1,6-OC-TADIENE

CONSENSUS REPORTS: Reported in EPA TSCA Inventory.

SAFETY PROFILE: A moderate skin and eye irritant. Flammable liquid. When heated to decomposition it emits acrid smoke and irritating fumes.

MSA250 CAS: 544-63-8 **HR: 3**
MYRISTIC ACID
mf: $C_{14}H_{28}O_2$ mw: 228.36

PROP: White or faintly yellow crystals from methanol. Mp: 58.5°, bp: 250.5° @ 100 mm, d: 0.8622 @ 54°/4°. Sol in abs alc, methanol, ether, petroleum ether, benzene, chloroform; insol in water.

SYNS: CRODACID * EMERY 655 * HYDROFOL ACID 1495 * HYSTRENE 9014 * TETRADECANOIC ACID * n-TETRADECOIC ACID * 1-TRIDECANE-CARBOXYLIC ACID * UNIVOL U 316S

CONSENSUS REPORTS: Reported in EPA TSCA Inventory.

SAFETY PROFILE: Poison by intravenous route. Mutation data reported. A human skin irritant. When heated to decomposition it emits acrid smoke and irritating fumes.

MSB500 CAS: 2748-88-1 **HR: 3**
MYRISTYL-γ-PICOLINIUM CHLORIDE
mf: $C_{20}H_{36}N \cdot Cl$ mw: 326.02

SYNS: QUATRESIN * WET-TONE B

CONSENSUS REPORTS: Reported in EPA TSCA Inventory.

SAFETY PROFILE: Poison by ingestion, intraperitoneal, intravenous, and subcutaneous routes. When heated to decomposition it emits very toxic fumes of NO_x and Cl^-.

N

NAG400   CAS: 300-76-5   HR: 3
NALED

DOT: NA 2783
mf: $C_4H_7Br_2Cl_2O_4P$ mw: 380.80

PROP: Slightly sol in aliphatic hydrocarbons, very sol in aromatic hydrocarbons. Mp: 27.0°.

SYNS: ARTHODIBROM * BROMCHLOPHOS
* BROMEX * DIBROM * O-(1,2-DIBROM-2,2-DI-
CHLORAETHYL)-O,O-DIMETHYL-PHOSPHAT (GERMAN)
* 1,2-DIBROMO-2,2-DICHLOROETHYL DIMETHYL PHOS-
PHATE * O-(1,2-DIBROMO-2,2-DICLORO-ETIL)-O,O-DI-
METIL-FOSTATO (ITALIAN) * O-(1,2-DIBROOM-2,2-
DICHLOOR-ETHYL)-O,O-DIMETHYL-FOSFAAT (DUTCH)
* O,O-DIMETHYL-O-(1,2-DIBROMO-2,2-DICHLORO-
ETHYL)PHOSPHATE * DIMETHYL-1,2-DIBROMO-2,2-
DICHLOROETHYL PHOSPHATE (OSHA) * O,O-DI-
METHYL-O-2,2-DICHLORO-1,2-DIBROMOETHYL PHOS-
PHATE * ENT 24,988 * HIBROM * ORTHO 4355
* ORTHODIBROM * ORTHODIBROMO
* PHOSPHATE de O,O-DIMETHLE et de O-(1,2-DI-
BROMO-2,2-DICHLORETHYLE) (FRENCH)
* RE-4355

OSHA PEL: (Transitional: TWA 3 mg/m^3)
   TWA 3 mg/m^3 (skin)
ACGIH TLV: TWA 3 mg/m^3 (skin)

SAFETY PROFILE: Poison by ingestion and inhalation. Moderately toxic by skin contact. A human skin irritant and a severe experimental skin irritant. Mutation data reported. A cholinesterase inhibitor. When heated to decomposition it emits very toxic fumes of Br$^-$, Cl$^-$, and PO$_x$.

NAH500   CAS: 835-31-4   HR: 3
NAPHAZOLINE
mf: $C_{14}H_{14}N_2$ mw: 210.30

SYNS: ANTAN * IMIDIN * NAPHTHIZINE
* α-NAPHTHYLMETHYL IMIDAZOLINE * 2-(α-NAPH-
THYLMETHYL)-IMIDAZOLINE * 2-(1-NAPHTHYL-
METHYL)-2-IMIDAZOLINE * PRIVINE * RHINAZINE
* SANORIN

SAFETY PROFILE: Poison by subcutaneous, intraperitoneal, and intravenous routes. When heated to decomposition it emits toxic fumes of NO$_x$.

NAI500   CAS: 8030-30-6   HR: 3
NAPHTHA

DOT: UN 1255/UN 1256/UN 1271/UN 2553

PROP: Dark straw-colored to colorless liquid. Bp: 149-216°, flash p: 107°F (CC), d: 0.862-0.892, autoign temp: 531°F. Sol in benzene, toluene, xylene, etc. Made from American coal oil and consists chiefly of pentane, hexane, and heptane.

SYNS: AROMATIC SOLVENT * BENZIN * COAL
TAR NAPHTA * HI-FLASH NAPHTHAETHYLEN
* NAPHTA (DOT) * NAPHTHA DISTILLATE (DOT)
* NAPHTHA PETROLEUM (DOT) * NAPHTHA, SOL-
VENT (DOT) * PETROLEUM BENZIN * PETROLEUM
DISTILLATES (NAPHTHA) * PETROLEUM ETHER
(DOT) * PETROLEUM NAPHTHA (DOT) * PETRO-
LEUM SPIRIT (DOT) * SKELLY-SOLVE-F * VM & P
NAPHTHA (ACGIH)

CONSENSUS REPORTS: Reported in EPA TSCA Inventory.

OSHA PEL: TWA 100 ppm
ACGIH TLV: TWA 300 ppm
NIOSH REL: TWA 350 mg/m^3; CL 1800 mg/m^3/15M
DOT Classification: Flammable Liquid; Label: Flammable Liquid.

SAFETY PROFILE: A human poison via intravenous route. Human systemic effects by intravenous route: dyspnea, respiratory stimulation, and other unspecified respiratory effects. Mildly toxic by inhalation. Can cause unconsciousness which may go into coma, stentorious breathing, and bluish tint to the skin. Recovery follows removal from exposure. In mild form, intoxication resembles drunkenness. On a chronic basis, no true poisoning; sometimes headache, lack of appetite, dizziness, sleeplessness, indigestion, and nausea. A common air contaminant. Flammable when exposed to heat or flame; can react with oxidizing materials. Keep containers tightly closed. Slight explosion hazard. To fight fire, use foam, CO$_2$, dry chemical.

NAJ500   CAS: 91-20-3   HR: 3
NAPHTHALENE

DOT: UN 1334/UN 2304
mf: $C_{10}H_8$ mw: 128.18

PROP: Aromatic odor; white, crystalline, volatile flakes. Mp: 80.1°, bp: 217.9°, flash p: 174°F (OC), d: 1.162, lel: 0.9%, uel: 5.9%, vap press: 1 mm @ 52.6°, vap d: 4.42, autoign temp: 1053°F (567°C). Sol in alc, benzene; insol in water; very sol in ether, CCl_4, CS_2, hydronaphthalenes, in fixed and volatile oils.

SYNS: CAMPHOR TAR * MIGHTY 150 * MOTH BALLS (DOT) * MOTH FLAKES * NAFTALEN (POLISH) * NAPHTHALENE, crude or refined (DOT) * NAPHTHALENE, molten (DOT) * NAPHTHALIN (DOT) * NAPHTHALINE * NAPHTHENE * NCI-C52904 * RCRA WASTE NUMBER U165 * TAR CAMPHOR * WHITE TAR

CONSENSUS REPORTS: Reported in EPA TSCA Inventory. EPA Genetic Toxicology Program. Community Right-To-Know List.

OSHA PEL: (Transitional: TWA 10 ppm) TWA 10 ppm; STEL 15 ppm
ACGIH TLV: TWA 10 ppm; STEL 15 ppm
DFG MAK: 10 ppm (50 mg/m^3)
DOT Classification: ORM-A; Label: None; Flammable Solid; Label: Flammable Solid.

SAFETY PROFILE: Human poison by ingestion and possibly other routes. Experimental poison by ingestion, intravenous, and intraperitoneal routes. Moderately toxic by subcutaneous route. Experimental reproductive effects. An eye and skin irritant. Can cause nausea, headache, diaphoresis, hematuria, fever, anemia, liver damage, vomiting, convulsions, and coma. Poisoning may occur by ingestion of large doses, inhalation, or skin absorption. Questionable carcinogen with experimental tumorigenic data. Flammable when exposed to heat or flame; reacts with oxidizing materials. Explosive reaction with dinitrogen pentaoxide. Reacts violently with CrO_3, aluminum chloride + benzoyl chloride. Fires in the benzene scrubbers of coke oven gas plants have been attributed to oxidation of naphthalene. Explosive in the form of vapor or dust when exposed to heat or flame. To fight fire, use water, CO_2, dry chemical. When heated to decomposition it emits acrid smoke and irritating fumes.

NAM000   CAS: 2243-62-1   **HR: 3**
1,5-NAPHTHALENEDIAMINE
mf: $C_{10}H_{10}N_2$ mw: 158.22

SYNS: 1,5-DIAMINONAPHTHALENE * 1,5-NAPH-THYLENEDIAMINE * NCI-C03021

CONSENSUS REPORTS: IARC Cancer Review: GROUP 3 IMEMDT 7,56,87; Animal Limited Evidence IMEMDT 27,127,82. NCI Carcinogenesis Bioassay (feed); Clear Evidence: mouse, rat NCITR* NCI-CG-TR-143,78. EPA Genetic Toxicology Program.

SAFETY PROFILE: Suspected carcinogen with experimental carcinogenic, neoplastigenic, tumorigenic, and teratogenic data. Mutation data reported. When heated to decomposition it emits toxic fumes of NO_x.

NAM500   CAS: 3173-72-6   **HR: 2**
1,5-NAPHTHALENE DIISOCYANATE
mf: $C_{12}H_6N_2O_2$ mw: 210.20

PROP: White to light yellow crystals.

SYNS: 1,5-DIISOCYANATONAPHTHALENE * ISO-CYANIC ACID-1,5-NAPHTHYLENE ESTER

CONSENSUS REPORTS: IARC Cancer Review: GROUP 3 IMEMDT 7,56,87. Reported in EPA TSCA Inventory.

DFG MAK: 0.01 ppm (0.09 mg/m^3)

SAFETY PROFILE: A powerful allergen. An irritant. Questionable carcinogen. When heated to decomposition it emits toxic fumes of NO_x.

NAP500   CAS: 91-60-1   **HR: 3**
2-NAPHTHALENETHIOL
mf: $C_{10}H_8S$ mw: 160.24

PROP: Crystals from ethanol, disagreeable odor. Mp: 81°, bp: 286°. Very sol in ethanol, ether, petr ether; sltly sol in water; sltly volatile with steam.

SYNS: 2-MERCAPTONAPHTHALENE * NAPHTHA-LENE-2-THIOL * β-NAPHTHYL MERCAPTAN * 2-NAPHTHYL MERCAPTAN * 2-NAPHTHYL THIOL * THIO-β-NAPHTHOL * β-THIONAPHTHOL * USAF CY-4

CONSENSUS REPORTS: Reported in EPA TSCA Inventory.

SAFETY PROFILE: Poison by ingestion and intraperitoneal routes. A mosquito larvicide. When heated to decomposition it emits highly toxic fumes of SO_x.

NAQ500   CAS: 2668-92-0   **HR: 3**
NAPHTHALOXIMIDODIETHYL THIOPHOSPHATE
mf: $C_{16}H_{16}NO_5PS$ mw: 365.36

SYNS: O,O-DIETHYL-o-NAPHTHALIMIDE PHOSPHO-ROTHIOATE * O,O-DIETHYL-o-NAPHTHALOXIMIDO PHOSPHOROTHIOATE * O,O-DIETHYL-o-NAPH-THALOXIMIDOPHOSPHOROTHIONATE * O,O-DI-ETHYL-o-NAPHTHYLAMIDOPHOSPHOROTHIOATE * ENT 24,970 * N-HYDROXYNAPHTHALIMIDE-O,O-DIETHYL PHOSPHOROTHIOATE * NAPHTHALOXI-MIDE-O,O-DIETHYL PHOSPHOROTHIOATE * PHOS-PHOROTHIOIC ACID-O,O-DIETHYL ESTER, -o-NAPHTHA-LIMIDO DERIVATIVE * PHOSPHOROTHIOIC ACID-O,O-DIETHYL-o-NAPHTHYLAMIDOESTER

SAFETY PROFILE: Poison by ingestion. When heated to decomposition it emits very toxic SO_x, PO_x, and NO_x.

NAR000   CAS: 1338-24-5   **HR: 2**
NAPHTHENIC ACID

DOT: UN 9137

PROP: Odorless crystals. D: 1.034, mp: 31°, bp: 233°. Sltly water-sol.

SYNS: AGENAP * NAPHID * SUNAPTIC ACID B * SUNAPTIC ACID C

CONSENSUS REPORTS: Reported in EPA TSCA Inventory.

DOT Classification: ORM-E; Label: None.

SAFETY PROFILE: Moderately toxic by ingestion and intraperitoneal routes. When heated to decomposition it emits acrid smoke and irritating fumes.

NAR500   CAS: 61789-51-3   **HR: 2**
NAPHTHENIC ACID, COBALT SALT

DOT: UN 2001

PROP: Brown, amorphous powder or bluish-red solid. Flash p: 120°F, d: 0.9, autoign temp: 529°F. Water-insol; sol in oil, alc, ether. Contains 6% cobalt.

SYNS: COBALT NAPHTHENATE, POWDER (DOT) * NAPHTHENATE de COBALT (FRENCH)

CONSENSUS REPORTS: Cobalt and its compounds are on the Community Right-To-Know List. Reported in EPA TSCA Inventory.

DOT Classification: Flammable Solid; Label: Flammable Solid.

SAFETY PROFILE: Moderately toxic by ingestion. Flammable when exposed to heat or flame. When heated to decomposition it emits acrid smoke and irritating fumes.

NAS000   CAS: 1338-02-9   **HR: 3**
NAPHTHENIC ACID, COPPER SALT

PROP: A solid. Flash p: 100°F, d: 1.055. Contains 8% copper.

SYNS: COPPER NAPHTHENATE * COPPER UVER-SOL * CUPRINOL * WITTOX C

CONSENSUS REPORTS: Copper and its compounds are on the Community Right-To-Know List. Reported in EPA TSCA Inventory.

SAFETY PROFILE: Poison by ingestion. A pesticide. A dangerous fire hazard when exposed to heat or flame; can react with oxidizing materials. To fight fire, use foam, CO_2, dry chemical.

NAS500   CAS: 61790-14-5   **HR: 3**
NAPHTHENIC ACID, LEAD SALT
mf: $C_7H_{12}O_2 \cdot xPb$ mw: 1578.52

PROP: Contains 24% lead.

SYNS: CYCLOHEXANECARBOXYLIC ACID, LEAD SALT * LEAD NAPHTHENATE

CONSENSUS REPORTS: IARC Cancer Review: Animal Inadequate Evidence IMEMDT 23,325,80. Lead and its compounds are on the Community Right-To-Know List. Reported in EPA TSCA Inventory.

NIOSH REL: (Inorganic Lead) TWA 0.10 $mg(Pb)/m^3$

SAFETY PROFILE: A poison. Moderately toxic by intraperitoneal route. Mildly toxic by ingestion. Questionable carcinogen with experimental tumorigenic data. When heated to decomposition it emits toxic fumes of lead.

NAT500   CAS: 192-65-4   **HR: 3**
NAPHTHO(1,2,3,4-def)CHRYSENE
mf: $C_{24}H_{14}$ mw: 302.38
SYNS: DB(a,e)P * DIBENZO(a,e)PYRENE * 1,2,4,5-DIBENZOPYRENE

CONSENSUS REPORTS: IARC Cancer Review: GROUP 2B IMEMDT 7,56,87; Animal Sufficient Evidence IMEMDT 32,327,83; IMEMDT 3,201,73.

SAFETY PROFILE: Suspected carcinogen with experimental neoplastigenic and tumorigenic data. When heated to decomposition it emits acrid smoke and irritating fumes.

NAW000   CAS: 1321-67-1   **HR: 3**
NAPHTHOL
mf: $C_{10}H_8O$ mw: 144.18

PROP: Monoclinic. D: 1.217 @ 4°, mp: 122-123°, bp: 285-286°. Sol in cold and hot water and in chloroform; very sol in alc and benzene.

SAFETY PROFILE: Poison by subcutaneous route. When heated to decomposition it emits acrid smoke and irritating fumes.

NAW500   CAS: 90-15-3   HR: 2
1-NAPHTHOL
mf: $C_{10}H_8O$ mw: 144.18

PROP: Colorless crystals; odor of phenol, disagreeable taste. Mp: 96°, bp: 282.5°, d: 1.0954 @ 98.7°/4°, vap press: 1 mm @ 94.0°. Very sltly sol in water; sol in alc and ether.

SYNS: BASF URSOL ERN * C.I. 76605 * C.I. OXIDATION BASE 33 * DURAFUR DEVELOPER D * FOURAMINE ERN * FOURRINE 99 * FOURRINE ERN * FURRO ER * α-HYDROXYNAPHTHALENE * 1-HYDROXYNAPHTHALENE * NAKO TRB * 1-NAPHTHALENOL * α-NAPHTHOL * TERTRAL ERN * URSOL ERN * ZOBA ERN

CONSENSUS REPORTS: Reported in EPA TSCA Inventory. EPA Genetic Toxicology Program.

SAFETY PROFILE: Moderately toxic by ingestion and skin contact. Experimental teratogenic and reproductive effects. A severe eye and skin irritant. Mutation data reported. Ingestion of large amounts can cause nephritis, vomiting, diarrhea, circulatory collapse, anemia, convulsions, and death. Can cause kidney irritation and injury to cornea and lens of the eye. Combustible when exposed to heat or flame. When heated to decomposition it emits acrid smoke and irritating fumes.

NAX000   CAS: 135-19-3   HR: 3
2-NAPHTHOL
mf: $C_{10}H_8O$ mw: 144.18

PROP: White to yellowish-white crystals; slt phenolic odor. Flash p: 307°F, vap press: 10 mm @ 145.5°, vap d: 4.97 mp: 121-123°, bp: 285-286°, d: 1.22. Darkens with age or exposure to light. Sublimes when heated; distills in vacuum. Sltly sol in water; more sol in boiling water, glycerol, olive oil, solns of alkali hydroxides. Very sol in alc, ether; sol in chloroform. Protect from light.

SYNS: AZOGEN DEVELOPER A * C.I. 37500 * C.I. AZOIC COUPLING COMPONENT 1 * C.I. DEVELOPER 5 * DEVELOPER A * DEVELOPER AMS

* DEVELOPER BN * DEVELOPER SODIUM * β-HYDROXYNAPHTHALENE * 2-HYDROXYNAPHTHALENE * ISONAPHTHOL * β-MONOXYNAPHTHALENE * β-NAFTOL (DUTCH) * 2-NAFTOL (DUTCH) * β-NAFTOLO (ITALIAN) * 2-NAFTOLO (ITALIAN) * 2-NAPHTHALENOL * NAPHTHOL B * β-NAPHTHOL * β-NAPHTHYL ALCOHOL * β-NAPHTHYL HYDROXIDE * 2-NAPHTOL (FRENCH) * β-NAPHTOL (GERMAN)

CONSENSUS REPORTS: Reported in EPA TSCA Inventory. EPA Genetic Toxicology Program.

SAFETY PROFILE: Poison by ingestion and subcutaneous routes. Moderately toxic by unspecified route. Mutation data reported. A skin and eye irritant. Combustible when exposed to heat or flame. To fight fire, use CO_2 dry chemical. Incompatible with antipyrine, camphor, phenol, ferric salts, menthol, potassium permanganate and other oxidizing materials, urethane. When heated to decomposition it emits toxic fumes of Na_2O.

NBA500   CAS: 130-15-4   HR: 3
1,4-NAPHTHOQUINONE
mf: $C_{10}H_6O_2$ mw: 158.16

PROP: Yellow triclinic; odor of benzoquinone. Mp: 125-126°, D: 1.422. Very sltly sol in cold water; very sol in hot alc; sol in ether, benzene, chloroform, carbon bisulfide, acetic acid, alkali hydroxide solns. Volatile with steam.

SYNS: 1,4-DIHYDRO-1,4-DIKETONAPHTHALENE * 1,4-NAPHTHALENEDIONE * α-NAPHTHOQUINONE * RCRA WASTE NUMBER U166 * USAF CY-10

CONSENSUS REPORTS: Reported in EPA TSCA Inventory.

SAFETY PROFILE: Poison by ingestion, intravenous, subcutaneous, and intraperitoneal routes. Experimental reproductive effects. Questionable carcinogen with experimental tumorigenic data. When heated to decomposition it emits acrid smoke and irritating fumes.

NBE000   CAS: 134-32-7   HR: 3
1-NAPHTHYLAMINE
DOT: UN 2077
mf: $C_{10}H_9N$ mw: 143.20

PROP: White crystals, reddening on exposure to air; unpleasant odor. Mp: 50°, bp: 300.8°, flash p: 315°F, d: 1.131, vap press: 1 mm @

104.3°, vap d: 4.93. Sublimes, volatile with steam. Sol in 590 parts water; very sol in alc, ether. Keep well closed and away from light.

SYNS: ALFANAFTILAMINA (ITALIAN) * ALFA-NAF-TYLOAMINA (POLISH) * 1-AMINONAFTALEN (CZECH) * 1-AMINONAPHTHALENE * C.I. AZOIC DIAZO COMPONENT 114 * 1-NAFTILAMINA (SPANISH) * α-NAFTYLAMIN (CZECH) * 1-NAFTYLAMINE (DUTCH) * NAPHTHALIDINE * 1-NAPHTHYLAMIN (GERMAN) * α-NAPHTHYLAMINE * RCRA WASTE NUMBER U167

CONSENSUS REPORTS: IARC Cancer Review: GROUP 3 IMEMDT 7,260,87; Animal Inadequate Evidence IMEMDT 4,87,74; Human Limited Evidence IMEMDT 4,87,74. EPA Genetic Toxicology Program. Community Right-To-Know List. Reported in EPA TSCA Inventory.

OSHA PEL: Carcinogen
DOT Classification: Poison B; Label: St. Andrews Cross.

SAFETY PROFILE: Confirmed carcinogen with experimental tumorigenic data. Along with β-naphthylamine and benzidine, it has been incriminated as a cause of urinary bladder cancer. Poison by subcutaneous and intraperitoneal routes. Moderately toxic by ingestion. Mutation data reported. Combustible when exposed to heat or flame. Incompatible with nitrous acid. To fight fire, use dry chemical, CO_2, mist, spray. When heated to decomposition it emits toxic fumes of NO_x.

NBE500   CAS: 91-59-8   **HR: 3**
β-NAPHTHYLAMINE

DOT: UN 1650
mf: $C_{10}H_9N$ mw: 143.20

PROP: White to faint pink, lustrous leaflets; faint aromatic odor. Mp: 111.5°, d: 1.061 @ 98°/4°, vap press: 1 mm @ 108.0°. Bp: 306° (also listed as 294°). Sol in hot water, alc, and ether.

SYNS: 2-AMINONAFTALEN (CZECH) * 2-AMINO-NAPHTHALENE * BETA-NAFTYLOAMINA (POLISH) * C.I. 37270 * FAST SCARLET BASE B * NA * β-NAFTYLAMIN (CZECH) * β-NAFTILAMINA (ITALIAN) * 2-NAFTYLAMINE (DUTCH) * β-NAFT-YLOAMINA (POLISH) * 2-NAPHTHALAMINE * 2-NAPHTHALENAMINE * β-NAPHTHYLAMIN (GERMAN) * 2-NAPHTHYLAMIN (GERMAN) * 6-NAPH-THYLAMINE * 2-NAPHTHYLAMINE * 2-NAPH-

THYLAMINE MUSTARD   *   RCRA WASTE NUMBER U168 * USAF CB-22

CONSENSUS REPORTS: IARC Cancer Review: GROUP 1 IMEMDT 7,261,87; Animal Sufficient Evidence IMEMDT 4,97,74; Human Sufficient Evidence IMEMDT 4,97,74. NTP Fourth Annual Report On Carcinogens, 1984. Community Right-To-Know List. EPA Genetic Toxicology Program. Reported in EPA TSCA Inventory.

OSHA PEL: Carcinogen
ACGIH TLV: Confirmed Human Carcinogen.
DFG MAK: Human Carcinogen.
DOT Classification: Poison B; Label: Poison.

SAFETY PROFILE: Confirmed carcinogen with experimental neoplastigenic, tumorigenic, and teratogenic data. Long and continued exposure to even small amounts may produce tumors and cancers of the bladder. Poison by intraperitoneal route. Moderately toxic by ingestion and possibly other routes. Human mutation data reported. A very toxic chemical in any of its physical forms, such as flake, lump, dust, liquid, or vapor. It can be absorbed into the body through the lungs, the gastrointestinal tract, or the skin. Combustible when exposed to heat or flame. At elevated temperatures it evolves a vapor which is flammable and explosive. Incompatible with nitrous acid. When heated to decomposition it emits toxic fumes of NO_x.

NBI500   CAS: 613-47-8   **HR: 3**
2-NAPHTHYLHYDROXYLAMINE
mf: $C_{10}H_9NO$ mw: 159.20

SYNS: N-HYDROXY-2-AMINONAPHTHALENE * N-HYDROXY-2-NAPHTHYLAMINE * NHA

CONSENSUS REPORTS: EPA Genetic Toxicology Program.

SAFETY PROFILE: Questionable carcinogen with experimental carcinogenic, neoplastigenic, and tumorigenic data. Human mutation data reported. When heated to decomposition it emits toxic fumes of NO_x.

NBJ500   CAS: 7090-25-7   **HR: 3**
1-NAPHTHYL METHYLNITROSOCARBAMATE
mf: $C_{12}H_{10}N_2O_3$ mw: 230.24

SYNS: DENAPON, NITROSATED (JAPANESE) * METHYL-NITROSOCARBAMIC ACID-1-NAPHTHYL ES-

TER * 1-NAPHTHYL-N-METHYL-N-NITROSOCARBA-MATE * N-NITROSOCARBARYL * NITROSO-NAC

CONSENSUS REPORTS: IARC Cancer Review: Animal Sufficient Evidence IMEMDT 12,37,76. EPA Genetic Toxicology Program.

SAFETY PROFILE: Confirmed carcinogen with experimental carcinogenic, neoplastigenic, and tumorigenic data. Human mutation data reported. When heated to decomposition it emits toxic fumes of NO_x.

NBL000 CAS: 93-46-9 **HR: 3**
2-NAPHTHYL-p-PHENYLENEDIAMINE
mf: $C_{26}H_{20}N_2$ mw: 360.48

SYNS: ACETO DIPP * AGERITE WHITE * DI-β-NAPHTHYL-p-PHENYLDIAMINE * DI-β-NAPHTHYL-p-PHENYLENEDIAMINE * N,N'-DI-β-NAPHTHYL-p-PHE-NYLENEDIAMINE * sym-DI-β-NAPHTHYL-p-PHENYL-ENEDIAMINE * DNPD * DWU-β-NAFTYLO-p-FENY-LODWUAMINA (POLISH) * NONOX CL * TISPERSE MB-2X

CONSENSUS REPORTS: Reported in EPA TSCA Inventory.

SAFETY PROFILE: Mutation data reported. A human skin irritant. An experimental skin and eye irritant. Questionable carcinogen with experimental tumorigenic data. When heated to decomposition it emits toxic fumes of NO_x.

NBR000 CAS: 500-38-9 **HR: 1**
NDGA
mf: $C_{18}H_{22}O_4$ mw: 302.40

PROP: Crystals from acetic acid. Mp: 184-185°. Sol in methanol, ethanol, and ether; sltly sol in hot water and chloroform; nearly insol in benzene and petroleum ether.

SYNS: 1,4-BIS(3,4-DIHYDROXYPHENYL)-2,3-DIMETH-YLBUTANE * DIHYDRONORGUAIARETIC ACID * β,γ-DIMETHYL-α,Δ-BIS(3,4-DIHYDROXYPHENYL)BU-TANE * 4,4'-(2,3-DIMETHYLTETRAMETHYLENE) DIPYROCATECHOL * NORDIHYDROGUAIARETIC ACID * NORDIHYDROGUAIRARETIC ACID

SAFETY PROFILE: Moderately toxic by intraperitoneal route. An antioxidant food additive. When heated to decomposition it emits acrid smoke and irritating fumes.

NBS500 CAS: 23327-57-3 **HR: 3**
NEFOPAM HYDROCHLORIDE
mf: $C_{17}H_{19}NO \cdot ClH$ mw: 289.83

SYNS: ACUPAN * AJAN * FENAZOXINE HYDRO-CHLORIDE * 5-METHYL-1-PHENYL-3,4,5,6-TETRAHY-DRO-1H-2,5-BENZOXAZOCINE HYDROCHLORIDE * SINALGICO

SAFETY PROFILE: Poison by ingestion, intravenous, intraperitoneal, and intramuscular routes. When heated to decomposition it emits very toxic fumes of HCl and NO_x.

NBT500 CAS: 76-74-4 **HR: 3**
NEMBUTAL
mf: $C_{11}H_{18}N_2O_3$ mw: 226.31

SYNS: DORSITAL * ETHAMINAL * 5-ETHYL-5-(1-METHYLBUTYL)BARBITURIC ACID * 5-ETHYL-5-(1-METHYLBUTYL)MALONYLUREA * 5-ETHYL-5-(1-METHYLBUTYL)-2,4,6(1H,3H,5H)-PYRIMIDINETRIONE (9CI) * MEBUBARBITAL * NEODORM (NEW) * PENTABARBITONE * PENTOBARBITAL * PENTOBARBITURATE * PENTOBARBITURIC ACID * RIVADORM

SAFETY PROFILE: Human poison by ingestion. Experimental poison by ingestion, intraperitoneal, intravenous, intramuscular, and subcutaneous routes. Experimental reproductive effects. A sedative, hypnotic and anticonvulsant. When heated to decomposition it emits toxic fumes of NO_x.

NBU000 CAS: 57-33-0 **HR: 3**
NEMBUTAL SODIUM
mf: $C_{11}H_{17}N_2O_3 \cdot Na$ mw: 248.29

PROP: White, crystalline powder. Sol in water and alc; insol in ether.

SYNS: AUROPAN * BARPENTAL * BIOSEDAN * BUTYLONE * CARBRITAL * CONTINAL * DIABUTAL * EMBUTAL * ETAMINAL SODIUM * ETHAMINAL SODIUM * 5-ETHYL-5-(1-METHYLBU-TYL)BARBITURIC ACID SODIUM SALT * 5-ETHYL-5-(1-METHYLBUTYL)-2,4,6(1H,3H,5H)-PYRIMIDINETRIONE MONOSODIUM SALT (9CI) * EUTHATAL * IPRAL SODIUM * ISOBARB * MEBUBARBITAL SODIUM * MEBUMAL NATRIUM * MEBUMAL SODIUM * MINTAL * NAPENTAL * PACIFAN * PALA-PENT * PENBAR * PENTABARBITAL SODIUM * PENTAL * PENTOBARBITONE SODIUM * PENTONAL * PENTYL * PROPYLMETHYL-CARBINYLETHYL BARBITURIC ACID SODIUM SALT * RIVADORN * SAGATAL * SODITAL * SO-DIUM ETHAMINAL * SODIUM 5-ETHYL-5-(1-METHYL-BUTYL)BARBITURATE * SODIUM NEMBUTAL * SODIUM-PENT * SODIUM PENTABARBITAL * SODIUM PENTABARBITONE * SODIUM PENTO-BARBITAL * SODIUM PENTOBARBITONE * SO-

DIUM PENTOBARBITURATE · * SOLUBLE PENTO-BARBITAL * SOMNOPENTYL * SONISTAN * SONTOBARBITAL NABITONE * SOPENTAL * SOTYL * VETBUTAL

SAFETY PROFILE: Poison by ingestion, intraperitoneal, subcutaneous, intravenous, intraduodenal, intramuscular, intracerebral, parenteral, and rectal routes. Human systemic effects by ingestion: wakefulness, change in motor activity, ataxia and antipsychotic effects. Experimental teratogenic and reproductive effects. Mutation data reported. When heated to decomposition it emits toxic fumes of NO_x and Na_2O.

NBV000   CAS: 63681-05-0   HR: 3
NEOANTERGAN PHOSPHATE
mf: $C_{17}H_{23}N_3O \cdot H_3O_4P$ mw: 383.43

SYN: N-α-PYRIDYL-N-p-METHOXYBENZYL-N′,N′-DI-METHYLETHYLENEDIAMINE PHOSPHATE

SAFETY PROFILE: Poison by intravenous, intraperitoneal, and subcutaneous routes. When heated to decomposition it emits very toxic fumes of PO_x and NO_x.

NBV500   CAS: 9014-02-2   HR: 3
NEOCARZINOSTATIN

PROP: An acidic, single-chain polypeptide with a molecular weight of approximately 10,700 isolated from the culture filtrate of *Streptomyces carzinostaticus* Var F41.

SYNS: NCS * NEOCARCINOSTATIN * NEOCAR-ZINOSTATIN K * NSC 69856 * ZINOSTATIN

CONSENSUS REPORTS: EPA Genetic Toxicology Program.

SAFETY PROFILE: A deadly poison by ingestion, intraperitoneal, subcutaneous and intravenous routes. Experimental teratogenic and reproductive effects. Human mutation data reported. When heated to decomposition it emits acrid smoke and irritating fumes.

NBW000   CAS: 64093-79-4   HR: 3
NEOCHROMIUM
mf: $CrHO_5S$ mw: 165.07

SYNS: BASIC CHROMIC SULFATE * BASIC CHROMIC SULPHATE * BASIC CHROMIUM SULFATE * BASIC CHROMIUM SULPHATE * CHROMIUM HYDROXIDE SULFATE * CHROMIUM SULFATE * CHROMIUM SULFATE, BASIC * CHROMIUM SULPHATE * KOREON * MONOBASIC CHROMIUM SULFATE * MONOBASIC CHROMIUM SULPHATE * SULFURIC ACID, CHROMIUM SALT, BASIC

CONSENSUS REPORTS: Chromium and its compounds are on the Community Right-To-Know List.

OSHA PEL: TWA 0.5 mg(Cr)/m^3
ACGIH TLV: TWA 0.5 mg(Cr)/m^3

SAFETY PROFILE: Questionable carcinogen with experimental neoplastigenic and tumorigenic data. Human mutation data reported. When heated to decomposition it emits toxic fumes of SO_x.

NBX000   CAS: 7440-00-8   HR: 3
NEODYMIUM
af: Nd   aw: 144.24

PROP: It is a bright, silvery, lustrous, very reactive rare earth metal. Bp: 3127°, d: 7.003, mp: approx 1024°.

CONSENSUS REPORTS: Reported in EPA TSCA Inventory.

SAFETY PROFILE: Poison by intravenous route. Human systemic effects by intracerebral route: blood changes. It may be an anticoagulant lanthanon. Care in handling is advised. Flammable in the form of dust when exposed to heat or flame. Slight explosion hazard in the form of dust when exposed to flame. Can react violently with air; halogens; N_2. Violent reaction with phosphorus above 400°C. Many of its compounds are poisons.

NBY000   CAS: 10024-93-8   HR: 3
NEODYMIUM CHLORIDE
mf: Cl_3Nd mw: 250.59

PROP: Large, purple prisms. Sol in water and alc.

CONSENSUS REPORTS: Reported in EPA TSCA Inventory.

SAFETY PROFILE: Poison by intravenous, intraperitoneal, subcutaneous, and possibly other routes. Moderately toxic by ingestion. A skin and eye irritant. When heated to decomposition it emits very toxic fumes of Cl^-.

NCB000   CAS: 10045-95-1   HR: 3
NEODYMIUM(III) NITRATE (1:3)
mf: $N_3O_9 \cdot Nd$ mw: 330.27

PROP: Triclinic crystals.

SYN: NITRIC ACID, NEODYMIUM SALT

CONSENSUS REPORTS: Reported in EPA TSCA Inventory.

SAFETY PROFILE: Poison by intraperitoneal and intravenous routes. Moderately toxic by ingestion. When heated to decomposition it emits very toxic fumes of NO_x.

NCB500   CAS: 16454-60-7   **HR: 3**
NEODYMIUM(III) NITRATE, HEXAHYDRATE (1:3:6)
mf: $N_3O_9 \cdot Nd \cdot 6H_2O$ mw: 438.39

SYN: NITRIC ACID, NEODYMIUM (3+) SALT, HEXAHYDRATE

SAFETY PROFILE: Poison by intraperitoneal and intravenous routes. Moderately toxic by ingestion. When heated to decomposition it emits toxic fumes of NO_x.

NCC000   CAS: 1313-97-9   **HR: 2**
NEODYMIUM OXIDE
mf: Nd_2O_3 mw: 336.48

PROP: Blue powder, hexagonal structure. Very stable; sol in dil acids.

CONSENSUS REPORTS: Reported in EPA TSCA Inventory.

SAFETY PROFILE: Moderately toxic by intraperitoneal route.

NCE000   CAS: 1404-04-2   **HR: 3**
NEOMYCIN

PROP: An antibiotic.

SYNS: MYACYNE * MYCIFRADIN * NEOMCIN * NIVEMYCIN * VONAMYCIN POWDER V

SAFETY PROFILE: Poison by intraperitoneal, intravenous, and subcutaneous routes. Moderately toxic by ingestion. Human systemic effects by ingestion and possibly other routes: changes in acuity, liver tubule changes and decreased urine volume or anuria. Mutation data reported. When heated to decomposition it emits acrid smoke and irritating fumes.

NCG000   CAS: 1405-10-3   **HR: 3**
NEOMYCIN SULFATE

SYNS: BIOSOL VETERINARY * FRADIOMYCIN SULFATE * LIDAMYCIN CREME * MYCAIFRADIN SULFATE * MYCIFRADIN-N * MYCIGIENT * NEOBIOTIC * NEO-MANTLE CREME * NEOMIX * NEOMYCINE SULFATE * NEOMYCIN SULPHATE * OTOBIOTIC * QUINTESS-N * USAF CB-19

CONSENSUS REPORTS: Reported in EPA TSCA Inventory. EPA Genetic Toxicology Program.

SAFETY PROFILE: Poison by intraperitoneal, intramuscular, intravenous, and subcutaneous routes. Human systemic effects by ingestion: somnolence, hallucinations and distorted perceptions and anorexia. A human skin irritant. When heated to decomposition it emits very toxic fumes of SO_x.

NCG500   CAS: 7440-01-9   **HR: 1**
NEON

DOT: UN 1065/UN 1913
af: Ne   aw: 20.18

PROP: Colorless, inert, gaseous element; odorless. Mp: $-248.67°$, bp: $-245.9°$. D: (liquid): 1.204 @ $-245.9°$; (gas) 0.89994 g/L @ $0°$.

SYNS: NEON, compressed (DOT) * NEON, refrigerated liquid (DOT)

CONSENSUS REPORTS: Reported in EPA TSCA Inventory.

ACGIH TLV: Simple Asphyxiant
DOT Classification: Nonflammable Gas; Label: Nonflammable Gas.

SAFETY PROFILE: An inert asphyxiant gas.

NCH000   CAS: 463-82-1   **HR: 3**
NEOPENTANE

DOT: UN 1265/UN 2044
mf: C_5H_{12} mw: 72.17

PROP: Liquid or gas. Solidifies @ $-19.8°$, bp: $9.5°$, d: 0.613° @ $0°/0°$ (liquid), flash p: $<19.4°F$, lel: 1.4%, uel: 7.5%. Insol in water.

SYNS: 2,2-DIMETHYLPROPANE (DOT) * tert-PENTANE (DOT)

CONSENSUS REPORTS: Reported in EPA TSCA Inventory.

NIOSH REL: TWA 120 ppm; CL 610 ppm/ 15M
DOT Classification: Flammable Liquid; Label: Flammable Liquid; Flammable Gas; Label: Flammable Gas.

SAFETY PROFILE: Poison by intraperitoneal route. Both the gas and the liquid are flammable when exposed to heat or flame; can react vigorously with oxidizing materials. When heated to decomposition it emits acrid smoke and irritating fumes.

NCI500   CAS: 126-99-8   **HR: 3**
NEOPRENE
DOT: UN 1991
mf: C_4H_5Cl mw: 88.54

PROP: Colorless liquid. An oil-resistant synthetic rubber made by the polymerization of chloroprene. D: 0.958 @ 20°/20°, bp: 59.4°, flash p: −4°F, lel: 4.0%, uel: 20%, vap d: 3.0, brittle point: −35°, softens @ approx 80°. Sltly sol in water; misc in alc and ether.

SYNS: 2-CHLOOR-1,3-BUTADIEEN (DUTCH) * 2-CHLOR-1,3-BUTADIEN (GERMAN) * CHLORO-BUTADIENE * 2-CHLOROBUTA-1,3-DIENE * 2-CHLORO-1,3-BUTADIENE * CHLOROPREEN (DUTCH) * CHLOROPREN (GERMAN, POLISH) * CHLORO-PRENE * β-CHLOROPRENE (OSHA, MAK) * CHLO-ROPRENE, inhibited (DOT) * CHLOROPRENE, uninhibited (DOT) * 2-CLORO-1,3-BUTADIENE (ITALIAN) * CLOROPRENE (ITALIAN)

CONSENSUS REPORTS: IARC Cancer Review: GROUP 3 IMEMDT 7,160,87; Animal Inadequate Evidence IMEMDT 19,131,79; Human Inadequate Evidence IMEMDT 19,131,79. Reported in EPA TSCA Inventory. Community Right-To-Know List.

OSHA PEL: (Transitional: TWA 25 ppm (skin)) TWA 10 ppm (skin)
ACGIH TLV: TWA 10 ppm (skin)
DFG MAK: 10 ppm (36 mg/m³)
NIOSH REL: CL (Chloroprene) 1 ppm/15M
DOT Classification: Flammable Liquid; Label: Flammable Liquid, inhibited; Forbidden, un-inhibited; Flammable Liquid; Label: Flammable Liquid, Poison, inhibited.

SAFETY PROFILE: Poison by ingestion, intravenous, and subcutaneous routes. Moderately toxic by inhalation. Experimental teratogenic and reproductive effects. Human mutation data reported. Human exposure has caused dermatitis, conjunctivitis, corneal necrosis, anemia, temporary loss of hair, nervousness, and irritability. Exposure to the vapor can cause respiratory tract irritation leading to asphyxia. Other effects are central nervous system depression, drop in blood pressure, severe degenerative changes in the liver, kidneys, lungs, and other vital organs. Questionable carcinogen. A very dangerous fire hazard when exposed to heat or flame. Explosive in the form of vapor when exposed to heat or flame. To fight fire, use alcohol foam. Auto-oxidizes in air to form an unstable peroxide which catalyzes exothermic polymerization of the monomer. Incompatible with liquid or gaseous fluorine. When heated to decomposition it emits toxic fumes of Cl^-

NCJ000   CAS: 69343-45-9   **HR: 3**
NEOPSICAINE HYDROCHLORIDE
mf: $C_{19}H_{25}NO_4 \cdot ClH$ mw: 367.91

SYNS: 3-(BENZOYLOXY)-8-METHYL-8-AZABICYCLO (3.2.1)OCTANE-2-CARBOXYLIC ACID PROPYL ESTER, HYDROCHLORIDE (1R-(2-ENDO,3-EXO)) * PSI-CAINE-NEU HYDROCHLORIDE * PSICAIN-NEW HY-DROCHLORIDE

SAFETY PROFILE: Poison by intraperitoneal, intravenous, and subcutaneous routes. An eye irritant. When heated to decomposition it emits very toxic fumes of NO_x and HCl.

NCJ500   CAS: 457-60-3   **HR: 3**
NEOSALVARSAN
mf: $C_{13}H_{13}As_2N_2H_2O_4S \cdot Na$ mw: 466.17
PROP: Yellow powder.

SYNS: ((5-(3-AMINO-4-HYDROXYPHENYL)ARSENO)-2-HYDROXYANILINO)METHANOL SULFOXYLATE SODIUM * ARSEVAN * ARSPHENAMINE METHYLENESUL-FOXYLIC ACID SODIUM SALT * COLLUNOVAR * COLLUNOVER * 3,3'-DIAMINO-4,4'-DIHYDROXY ARSENOBENZENE METHYLENESULFOXYLATE SODIUM * MIARSENOL * NEOARSOLUIN * NEOARSPHEN-AMINE * NOVARSAN * NOVARSENOBENZOL * NOVARSENOBILLON * VETARSENOBILLON

CONSENSUS REPORTS: Arsenic and its compounds are on the Community Right-To-Know List.

OSHA PEL: TWA 0.5 mg(As)m³
ACGIH TLV: TWA 0.2 mg(As)/m³

SAFETY PROFILE: Poison by intravenous and subcutaneous routes. When heated to decomposition it emits very toxic fumes of As, NO_x, Na_2O, and SO_x.

NCL500   CAS: 59-42-7   **HR: 3**
NEOSYNEPHRINE
mf: $C_9H_{13}NO_2$ mw: 167.23

SYNS: 1-α-HYDROXY-β-METHYLAMINO-3-HYDROXY-1-ETHYLBENZENE * (R)-3-HYDROXY-α-((METHYL-AMINO)METHYL)BENZENEMETHANOOL * 1-m-HY-DROXY-α-((METHYLAMINO)METHYL)-BENZYL ALCOHOL * (−)-m-HYDROXY-α-(METHYLAMINOMETHYL) BENZYL ALCOHOL * 1-1-(m-HYDROXYPHENYL)-2-METHYLAMINOETHANOL * 1-(3-HYDROXY-

PHENYL)-N-METHYLETHANOLAMINE * ISOPHRIN * MESATON * METAOXEDRIN * METASYMPATOL * METASYNEPHRINE * m-METHYLAMINO-ETHANOLPHENOL * MEZATON * R(−)-MEZATON * m-OXEDRINE * (−)-m-OXEDRINE * PHENYLEPHRINE * (−)-PHENYLEPHRINE * R(−)-PHENYLEPHRINE * m-SYMPATHOL * m-SYMPATOL * m-SYNEPHRINE * VISADRON

SAFETY PROFILE: Poison by ingestion, subcutaneous, intravenous, intraperitoneal, and intraduodenal routes. Human systemic effects by ocular route: blood pressure increase. Experimental reproductive effects. A nasal decongestant. When heated to decomposition it emits toxic fumes of NO_x.

NCN500 HR: 3
NEPTUNIUM
af: Np   aw: 237.048

PROP: Exists in α, β, and γ forms. Mp: 640°, bp: 3902°, d: 20.45 @ 20°. The first synthetic transuranium element discovered. It is a silvery, radioactive, chemically active metal. Emits α particles and has a half-life of 2.2×10^6 years.

SAFETY PROFILE: Strong radiotoxicity. The soluble compounds, including the nitrates and organic complexes (e.g., tributyl phosphate), are the most important industrially.

NCP500 CAS: 7783-33-7 HR: 3
NESSLER REAGENT
DOT: UN 1643
mf: HgI_42K mw: 786.39

SYNS: CHANNING'S SOLUTION * MERCURIC POTASSIUM IODIDE * MERCURIC POTASSIUM IODIDE, solid (DOT) * MERCURY(II) POTASSIUM IODIDE * POTASSIUM IODOHYDRAGYRATE * POTASSIUM MERCURIC IODIDE * POTASSIUM TETRAIODOMERCURATE (II) * TETRAIODOMERCURATE(2-), DIPOTASSIUM

CONSENSUS REPORTS: Mercury and its compounds are on the Community Right-To-Know List. Reported in EPA TSCA Inventory.
OSHA PEL: (Transitional: CL 1 mg/10m³) CL 0.1 mg(Hg)/m³ (skin)
ACGIH TLV: TWA 0.1 mg(Hg)/m³ (skin)
NIOSH REL: (Inorganic Mercury) TWA 0.05 mg(Hg)/m³
DOT Classification: Poison B; Label: Poison.
SAFETY PROFILE: A poison. Moderately toxic by skin contact and intraperitoneal routes. When

heated to decomposition it emits very toxic fumes of Hg, K_2O, and I^-.

NCQ820 CAS: 464-45-9 HR: 1
NGAI CAMPHOR
mf: $C_{10}H_{18}O$ mw: 154.28

SYNS: 1-2-BORNANOL * (−)-BORNEOL * (1S,2R,4S)-(−)-1-BORNEOL * 1-BORNYL ALCOHOL * 1-2-CAMPHANOL * LINDEROL

CONSENSUS REPORTS: Reported in EPA TSCA Inventory. EPA Genetic Toxicology Program.

SAFETY PROFILE: Mildly toxic by ingestion. A skin irritant. When heated to decomposition it emits acrid smoke and irritating fumes.

NCQ900 CAS: 59-67-6 HR: 3
NIACIN
mf: $C_6H_5NO_2$ mw: 123.12

PROP: The anti-pellagra vitamin. Colorless needles or white crystalline powder; slt odor. Mp: 236°, subl above mp, d: 1.473. Sol in water and boiling alc; insol in most lipid solvents. Nonhygroscopic and stable in air.

SYNS: ACIDE NICOTINIQUE (FRENCH) * ACIDUM NICOTINICUM * AKOTIN * ANTI-PELLAGRA VITAMIN * APELAGRIN * BIONIC * 3-CARBOXYPYRIDINE * DASKIL * DAVITAMON PP * DIREKTAN * EFACIN * NAH * NAOTIN * NICACID * NICAMIN * NICANGIN * NICO * NICO-400 * NICOBID * NICOCAP * NICOCIDIN * NICOCRISINA * NICODAN * NICODELMINE * NICOLAR * NICONACID * NICONAT * NICONAZID * NICOROL * NICOSIDE * NICO-SPAN * NICOSYL * NICOTAMIN * NICOTENE * NICOTIL * NICOTINE ACID * NICOTINIC ACID * NICOTINIPCA * NICOTINOYL HYDRAZINE * NICOTINSAURE (GERMAN) * NICOVASAN * NICOVASEN * NICOVEL * NICYL * NIPELLEN * PELLAGRAMIN * PELLAGRA PREVENTIVE FACTOR * PELLAGRIN * PELONIN * PEVITON * PP FACTOR * P.P. FACTOR-PELLAGRA PREVENTIVE FACTOR * PYRIDINE-3-CARBONIC ACID * PYRIDINE-β-CARBOXYLIC ACID * PYRIDINE-3-CARBOXYLIC ACID * 3-PYRIDINECARBOXYLIC ACID * PYRIDINE-CARBOXYLIQUE-3 (FRENCH) * S115 * SK-NIACIN * TINIC * VITAPLEX N * WAMPOCAP

CONSENSUS REPORTS: Reported in EPA TSCA Inventory.

SAFETY PROFILE: Poison by intraperitoneal route. Moderately toxic by ingestion, intravenous, and subcutaneous routes. Questionable carcinogen with experimental carcinogenic data. When heated to decomposition it emits toxic fumes of NO_x.

NCR000   CAS: 98-92-0   HR: 2
NIACINAMIDE

mf: $C_6H_6N_2O$ mw: 122.14

PROP: Colorless needles or white crystalline powder; odorless with a bitter taste. Mp: 129°, d: 1.40. Very sol in water, ether, glycerin.

SYNS: ACID AMIDE * AMIDE PP * AMINICOTIN * AMIXICOTYN * AMNICOTIN * AUSTROVIT PP * BENICOT * DELONIN AMIDE * DIPEGYL * DIPIGYL * ENDOBION * FACTOR PP * HANSAMID * INOVITAN PP * NAM * NANDERVIT-N * NIACEVIT * NIAMIDE * NICAMIDE * NICAMINA * NICAMINDON * NICASIR * NICOBION * NICOFORT * NICOGEN * NICOMIDOL * NICOSAN 2 * NICOTA * NICOTAMIDE * NICOTILAMIDE * NICOTILILAMIDO * LO..VE ACID AMIDE * NICOTINIC ACID AMIDE * NICOTINIC AMIDE * NICOTINSAUREAMID (GERMAN) * NICOTOL * NICOTYLAMIDE * NICOVEL * NICOVIT * NICOVITOL * NICOZYMIN * NIKO-TAMIN * NIKOTINSAEUREAMID (GERMAN) * NIOCINAMIDE * NIOZYMIN * PELMIN * PELMINE * PELONIN AMIDE * PP-FACTOR * PYRIDINE-3-CARBOXYLIC ACID AMIDE * 3-PYRIDINECARBOXYLIC ACID AMIDE * VI-NICOTYL * VI-NICTYL * VITAMIN B3 * VITAMIN PP * WITAMINA PP

CONSENSUS REPORTS: Reported in EPA TSCA Inventory.

SAFETY PROFILE: Moderately toxic by ingestion, intravenous, intraperitoneal, and subcutaneous routes. Mutation data reported. When heated to decomposition it emits toxic fumes of NO_x.

NCS000   CAS: 62765-93-9   HR: 2
NIAX CATALYST ESN

PROP: Mixture of 95% dimethylaminopropionitrile and 5% bis-dimethylaminoethyl ether.

CONSENSUS REPORTS: Cyanide and its compounds are on the Community Right-To-Know List.

NIOSH REL: (Niax ESN): Exposure to be minimized

SAFETY PROFILE: Moderately toxic by ingestion, skin contact, and intraperitoneal routes.

When heated to decomposition it emits toxic fumes of NO_x and CN^-.

NCW500   CAS: 7440-02-0   HR: 3
NICKEL

af: Ni   aw: 58.71

PROP: A silvery-white, hard, malleable and ductile metal. D: 8.90 @ 25°, vap press: 1 mm @ 1810°. Crystallizes as metallic cubes. Mp: 1455°, bp: 2730°. Stable in air at room temp.

SYNS: C.I. 77775 * Ni 270 * NICKEL 270 * NICKEL (DUST) * NICKEL (ITALIAN) * NICKEL PARTICLES * NICKEL SPONGE * Ni 0901-S * Ni 4303T * NP 2 * RANEY ALLOY * RANEY NICKEL

CONSENSUS REPORTS: IARC Cancer Review: GROUP 1 IMEMDT 7,264,87; Animal Sufficient Evidence IMEMDT 11,75,76; Animal Inadequate Evidence IMEMDT 2,126,73. NTP Fourth Annual Report On Carcinogens, 1984. Community Right-To-Know List. Reported in EPA TSCA Inventory.

OSHA PEL: TWA Soluble: Compounds: 0.1 mg(Ni)/m^3; Insoluble Compounds: 1 mg(Ni)/m^3

ACGIH TLV: TWA 1 mg(Ni)/m^3; (Proposed: TWA 0.05 mg(Ni)/m^3; Human Carcinogen)

DFG TRK: 0.5 mg/m^3; Human Carcinogen.

NIOSH REL: (Inorganic Nickel) TWA 0.015 mg(Ni)/m^3

SAFETY PROFILE: Confirmed carcinogen with experimental carcinogenic, neoplastigenic, and tumorigenic data. Poison by ingestion, intratracheal, intraperitoneal, subcutaneous, and intravenous routes. Experimental teratogenic and reproductive effects. Ingestion of soluble salts causes nausea, vomiting, diarrhea. Mutation data reported. Hypersensitivity to nickel is common and can cause allergic contact dermatitis, pulmonary asthma, conjunctivitis, and inflammatory reactions around nickel-containing medical implants and prostheses. Powders may ignite spontaneously in air. Reacts violently with F_2, NH_4NO_3, hydrazine, NH_3, (H_2 + dioxane), performic acid, P, Se, S, (Ti + $KClO_3$). Incompatible with oxidants (e.g., bromine pentafluoride, peroxyformic acid, potassium perchlorate, chlorine, nitryl fluoride, ammonium nitrate). Raney-nickel catalysts may initiate hazardous reactions with ethylene + aluminum chloride, p-dioxane, hydrogen, hydrogen + oxygen, magnesium silicate, methanol, organic solvents + heat, sulfur compounds. Nickel catalysts have caused many industrial accidents.

NCX000   CAS: 373-02-4   **HR: 3**
NICKEL(II) ACETATE (1:2)
mf: $C_4H_6O_4 \cdot Ni$ mw: 176.81

PROP: Green prisms. Mp: decomp, d: 1.798.

SYNS: ACETIC ACID, NICKEL(2+) SALT * NICKEL-OUS ACETATE

CONSENSUS REPORTS: IARC Cancer Review: GROUP 1 IMEMDT 7,264,87. Nickel and its compounds are on the Community Right-To-Know List. Reported in EPA TSCA Inventory.

OSHA PEL: (Transitional: TWA 1 mg/m^3) TWA 0.1 mg (Ni)/m^3
ACGIH TLV: TWA 0.1 mg(Ni)/m^3; (Proposed: TWA 0.05 mg(Ni)/m^3; Human Carcinogen)
NIOSH REL: (Inorganic Nickel) TWA 0.015 mg(Ni)/m^3

SAFETY PROFILE: Confirmed carcinogen with experimental neoplastigenic and tumorigenic data. Poison by ingestion, intraperitoneal, and subcutaneous routes. Experimental reproductive effects. Mutation data reported. When heated to decomposition it emits irritating fumes.

NCX500   CAS: 6018-89-9   **HR: 3**
NICKEL ACETATE TETRAHYDRATE
mf: $C_4H_6O_4 \cdot Ni \cdot 4H_2O$ mw: 248.89

CONSENSUS REPORTS: Nickel and its compounds are on the Community Right-To-Know List.

OSHA PEL: (Transitional: TWA 1 mg/m^3) TWA 0.1 mg (Ni)/m^3
ACGIH TLV: TWA 0.1 mg(Ni)/m^3; (Proposed: TWA 0.05 mg(Ni)/m^3; Human Carcinogen)
NIOSH REL: (Inorganic Nickel) TWA 0.015 mg(Ni)/m^3

SAFETY PROFILE: Suspected carcinogen. Poison by intraperitoneal route. Mutation data reported. When heated to decomposition it emits acrid smoke and irritating fumes.

NCY125   CAS: 12255-10-6   **HR: 3**
NICKEL ARSENIDE SULFIDE
mf: AsNiS   mw: 165.69

SYN: NICKEL SULFARSENIDE

OSHA: Cancer Hazard

SAFETY PROFILE: Confirmed human carcinogen with experimental carcinogenic and neoplastigenic data. When heated to decomposition it emits toxic fumes of Ni, As, and SO$_x$.

NCY500   CAS: 3333-67-3   **HR: 3**
NICKEL(II) CARBONATE (1:1)
mf: $CNiO_3$ mw: 118.72

PROP: Rhombic, light green crystals. Mp: decomp.

SYNS: BASIC NICKEL CARBONATE * CARBONIC ACID, NICKEL SALT (1:1) * C.I. 77779 * NICKEL-OUS CARBONATE

CONSENSUS REPORTS: IARC Cancer Review: GROUP 1 IMEMDT 7,264,87; Animal Sufficient Evidence IMEMDT 11,75,76. NTP Fourth Annual Report On Carcinogens, 1984. Nickel and its compounds are on the Community Right-To-Know List. Reported in EPA TSCA Inventory.

OSHA PEL: TWA 1 mg(Ni)/m^3
ACGIH TLV: TWA 1 mg(Ni)/m^3; (Proposed: TWA 0.05 mg(Ni)/m^3; Human Carcinogen)
DFG TRK: 0.5 mg/m^3; Human Carcinogen.
NIOSH REL: (Inorganic Nickel) TWA 0.015 mg(Ni)/m^3

SAFETY PROFILE: Confirmed carcinogen with experimental carcinogenic and tumorigenic data. Poison by subcutaneous route. Mutation data reported.

NCZ000   CAS: 13463-39-3   **HR: 3**
NICKEL CARBONYL

DOT: UN 1259
mf: C_4NiO_4 mw: 170.75

PROP: Colorless, volatile liquid or needles. Bp: 43°, mp: −19.3°, lel: 2% @ 20°, d: 1.3185 @ 17°, vap press: 400 mm @ 25.8°, flash p: <−4°. Oxidizes in air. Sol in alc, benzene, chloroform, acetone, carbon tetrachloride.

SYNS: NICKEL CARBONYLE (FRENCH) * NICHEL TETRACARBONILE (ITALIAN) * NICKEL TETRACARBONYL * NICKEL TETRACARBONYLE (FRENCH) * NIKKELTETRACARBONYL (DUTCH) * RCRA WASTE NUMBER P073

CONSENSUS REPORTS: IARC Cancer Review: GROUP 1 IMEMDT 7,264,87; Animal Limited Evidence IMEMDT 2,126,73; IMEMDT 11,75,76. NTP Fourth Annual Report On Carcinogens, 1984. EPA Extremely Hazardous Substances List. Nickel and its compounds are on the Community Right-To-Know List. Reported in EPA TSCA Inventory.

OSHA PEL: TWA 0.001 ppm (Ni)
ACGIH TLV: TWA 0.05 ppm (Ni); (Proposed: TWA 0.05 mg(Ni)/m^3; Human Carcinogen)
DFG TRK: 0.1 ppm; Animal Carcinogen, Suspected Human Carcinogen.
NIOSH REL: (Nickel Carbonyl) TWA 0.001 ppm
DOT Classification: Poison B; Label: Poison, Flammable Liquid; Flammable Liquid; Label: Flammable Liquid and Poison.

SAFETY PROFILE: Confirmed carcinogen with experimental carcinogenic and tumorigenic data. A human poison by inhalation. Poison experimentally by inhalation, intravenous, subcutaneous, and intraperitoneal routes. Experimental teratogenic and reproductive effects. Human systemic effects by inhalation: somnolence, fever and other pulmonary changes. Vapors may cause coughing, dyspnea (difficult breathing), irritation, congestion and edema of the lungs, tachycardia (rapid pulse), cyanosis, headache, dizziness, weakness. Toxicity by inhalation is believed to be caused by both the nickel and carbon monoxide liberated in the lungs. Chronic exposure may cause cancer of lungs, nasal sinuses. Sensitization dermatitis is fairly common. Probably the most hazardous compound of nickel in the workplace. A common air contaminant. It is lipid soluble and can cross biological membranes (e.g., lung alveolus, blood-brain barrier, placental barrier).

A very dangerous fire hazard when exposed to heat, flame, or oxidizers. Moderate explosion hazard when exposed to heat or flame. Explodes when heated to about 60°. Explosive reaction with liquid bromine, mercury + oxygen, oxygen + butane. Violent reaction with dinitrogen tetraoxide, air, oxygen. Reacts with tetrachloropropadiene to form the extremely sensitive explosive dicarbonyl trichloropropenyl dinickel chloride dimer. Can react with oxidizing materials. To fight fire, use water, foam, CO_2, dry chemical. When heated to decomposition or on contact with acid or acid fumes it emits highly toxic fumes of carbon monoxide.

NDA000   CAS: 7791-20-0   **HR: 3**
NICKEL(II) CHLORIDE HEXAHYDRATE (1:2:6)
mf: $Cl_2Ni \cdot 6H_2O$ mw: 237.73

PROP: A: Yellow, deliquescent scales; b: monoclinic, green crystals. A: $NiCl_2$, b: $NiCl_2 \cdot 6H_2O$; mw (a): 129.60, mw (b): 237.70, mp (a): sublimes, bp (a): 987°, d (a): 3.55, vap press: 1 mm @ 671°.

CONSENSUS REPORTS: Nickel and its compounds are on the Community Right-To-Know List.

OSHA PEL: (Transitional: TWA 1 mg/m^3) TWA 0.1 mg (Ni)/m^3
ACGIH TLV: TWA 0.1 mg(Ni)/m^3; (Proposed: TWA 0.05 mg(Ni)/m^3; Human Carcinogen)
NIOSH REL: (Inorganic Nickel) TWA 0.015 mg(Ni)/m^3

SAFETY PROFILE: Suspected carcinogen. Poison by intraperitoneal and intravenous routes. Mutation data reported. Violent reaction with K. When heated to decomposition it emits very toxic fumes of Cl^-.

NDA500   CAS: 1271-28-9   **HR: 3**
NICKEL, COMPOUND with pi-CYCLOPENTADIENYL (1:2)
mf: $C_{10}H_{10} \cdot Ni$ mw: 188.91
SYNS: pi-CYCLOPENTADIENYL COMPOUND with NICKEL * DI-pi-CYCLOPENTADIENYLNICKEL * NICKEL BISCYCLOPENTADIENE * NICKELOCENE

CONSENSUS REPORTS: IARC Cancer Review: GROUP 1 IMEMDT 7,264,87; Animal Sufficient Evidence IMEMDT 11,75,76; Animal Inadequate Evidence IMEMDT 2,126,73. NTP Fourth Annual Report On Carcinogens, 1984. Nickel and its compounds are on the Community Right-To-Know List. Reported in EPA TSCA Inventory.

NIOSH REL: TWA 15 µg(Ni)/m^3

SAFETY PROFILE: Confirmed carcinogen with experimental carcinogenic, neoplastigenic, and tumorigenic data. Poison by intraperitoneal and intramuscular routes. Moderately toxic by ingestion. When heated to decomposition it emits acrid smoke and irritating fumes.

NDB000   **HR: 3**
NICKEL COMPOUNDS
CONSENSUS REPORTS: Nickel and its compounds are on the Community Right-To-Know List.

OSHA PEL: TWA Soluble: Compounds: 0.1 mg(Ni)/m^3; Insoluble Compounds: 1 mg(Ni)/m^3

ACGIH TLV: TWA 1 mg/m^3; (Proposed: TWA 0.05 mg(Ni)/m^3; Human Carcinogen)

DFG MAK: Human Carcinogen.

NIOSH REL: (Inorganic Nickel) TWA 0.015 mg(Ni)/m^3

SAFETY PROFILE: Many are human carcinogens by inhalation. Nickel and many of its compounds are poisons and carcinogens. All airborne nickel contaminating dusts are regarded as carcinogenic by inhalation. Nickel carbonyl is probably the most hazardous compound of nickel in the workplace. It is carcinogenic and highly irritating to the lungs and can produce asphyxia by decomposing to form carbon monoxide. Nickel chloride (NiCl$_2$), sulfate (NiSO$_4$ • 6H$_2$O), nitrate [Ni(NO$_3$)$_2$ • 6H$_2$O], carbonate (NiCO$_3$), hydroxide [Ni(OH)$_2$] and acetate [Ni(COOCH$_3$)$_2$] are the salts of greatest commercial importance.

Ingestion of large doses of nickel compounds (1-3 mg/kg) has been shown to cause intestinal disorders, convulsions, and asphyxia. Hypersensitivity to nickel is common and can cause allergic contact dermatitis, pulmonary asthma, conjunctivitis, and inflammatory reactions around nickel-containing medical implants and prostheses. The most common effect resulting from exposure to nickel compounds is the development of "nickel itch". It occurs primarily in persons doing nickel-plating and is most frequent under conditions of high temperature and humidity, when the skin is moist, and mainly affects the hands and arms. There is marked variation in individual susceptibility to the dermatitis.

Nickel refinery workers experience increased mortality rates from cancer of the lungs and nasal cavities attributed to inhalation of airborne nickel compounds. Cancer develops in rodents after administration of Ni$_3$S$_2$, NiO, and Ni(CO)$_4$. Nickel chloride, sulfate, carbonate, and carbonyl are experimental teratogens.

Pulmonary damage develops in rodents chronically exposed to aerosols of nickel dust, NiCl$_2$, or NiO. Divalent nickel salts cause hyperglycemia, immune system effects, kidney damage, liver damage, and heart effects in experimental animals by parenteral administration. These compounds are common air contaminants.

NDB500   CAS: 557-19-7   **HR: 3**
NICKEL CYANIDE (solid)

DOT: UN 1653
mf: C$_2$N$_2$Ni   mw: 110.75

PROP: Apple-green plates or powder.

SYNS: NICKEL CYANIDE (DOT) * RCRA WASTE NUMBER P074

CONSENSUS REPORTS: Cyanide and its compounds, as well as nickel and its compounds, are on the Community Right-To-Know List. Reported in EPA TSCA Inventory.

OSHA PEL: (Transitional: TWA 1 mg/m^3) TWA 0.1 mg (Ni)/m^3

ACGIH TLV: TWA 0.1 mg(Ni)/m^3; (Proposed: TWA 0.05 mg(Ni)/m^3; Human Carcinogen)

DOT Classification: Poison B; Label: Poison.

SAFETY PROFILE: Suspected carcinogen. A poison. Incandescent reaction when heated with magnesium. When heated to decomposition it emits very toxic fumes of CN$^-$.

NDC000   CAS: 14708-14-6   **HR: 2**
NICKEL(II) FLUOBORATE
mf: B$_2$F$_8$ • Ni   mw: 232.33

SYNS: NICKEL BOROFLUORIDE * NICKEL FLUOROBORATE * NICKELOUS TETRAFLUOROBORATE * NICKEL(II) TETRAFLUOROBORATE * TL 1091

CONSENSUS REPORTS: Nickel and its compounds are on the Community Right-To-Know List. Reported in EPA TSCA Inventory.

OSHA PEL: (Transitional: TWA 1 mg/m^3) TWA 0.1 mg (Ni)/m^3; 2.5 mg(F)/m^3

ACGIH TLV: TWA 0.1 mg(Ni)/m^3; (Proposed: TWA 0.05 mg(Ni)/m^3; Human Carcinogen)

NIOSH REL: (Inorganic Nickel) TWA 0.015 mg(Ni)/m^3

SAFETY PROFILE: Suspected carcinogen. Moderately toxic by ingestion and inhalation. When heated to decomposition it emits very toxic fumes of F$^-$.

NDC500   CAS: 10028-18-9   **HR: 3**
NICKEL(II) FLUORIDE (1:2)
mf: F$_2$Ni   mw: 96.71

PROP: Green crystals. D: 4.63. Sltly watersol; decomp by boiling water; insol in alc, ether.

SYNS: NICKEL DIFLUORIDE * NICKELOUS FLUORIDE

CONSENSUS REPORTS: Nickel and its compounds are on the Community Right-To-Know List. Reported in EPA TSCA Inventory.

OSHA PEL: (Transitional: TWA 1 mg/m^3) TWA 0.1 mg (Ni)/m^3; 2.5 mg(F)/m^3
ACGIH TLV: TWA 0.1 mg(Ni)/m^3; (Proposed: TWA 0.05 mg(Ni)/m^3; Human Carcinogen)
NIOSH REL: (Inorganic Nickel) TWA 0.015 mg(Ni)/m^3

SAFETY PROFILE: Suspected carcinogen. Poison by intravenous route. Reacts violently with potassium. Chronic exposure may cause mottling of teeth, changes in bones. When heated to decomposition it emits toxic fumes of F$^-$.

NDD000   CAS: 26043-11-8   **HR: 3**
NICKEL(II) FLUOSILICATE (1:1)
mf: F$_6$Si • Ni   mw: 200.80

SYN: HEXAFLUOROSILICATE (2−), NICKEL

CONSENSUS REPORTS: Nickel and its compounds are on the Community Right-To-Know List. Reported in EPA TSCA Inventory.

OSHA PEL: (Transitional: TWA 1 mg/m^3) TWA 0.1 mg (Ni)/m^3
ACGIH TLV: TWA 1 mg(Ni)/m^3; (Proposed: TWA 0.05 mg(Ni)/m^3; Human Carcinogen)
NIOSH REL: (Inorganic Nickel) TWA 0.015 mg(Ni)/m^3

SAFETY PROFILE: Suspected carcinogen. Poison by ingestion. When heated to decomposition it emits toxic fumes of F$^-$.

NDD500   CAS: 56668-59-8   **HR: 3**
NICKEL-GALLIUM ALLOY

PROP: Nickel (60%) - gallium (40%) alloy.

SYN: GALLIUM-NICKEL ALLOY

CONSENSUS REPORTS: Nickel and its compounds are on the Community Right-To-Know List.

OSHA PEL: TWA 1 mg(Ni)/m^3
ACGIH TLV: TWA 1 mg(Ni)/m^3; (Proposed: TWA 0.05 mg(Ni)/m^3; Human Carcinogen)
NIOSH REL: (Inorganic Nickel) TWA 0.015 mg(Ni)/m^3

SAFETY PROFILE: Suspected carcinogen with experimental tumorigenic data.

NDE000   CAS: 12054-48-7   **HR: 3**
NICKEL(II) HYDROXIDE

DOT: NA 9140
mf: H$_2$NiO$_2$ mw: 92.73

PROP: Light green crystals or amorphous.

SYNS: NICKEL HYDROXIDE (DOT) * NICKELOUS HYDROXIDE

CONSENSUS REPORTS: IARC Cancer Review: GROUP 1 IMEMDT 7,264,87; Animal Sufficient Evidence IMEMDT 11,75,76. Nickel and its compounds are on the Community Right-To-Know List. Reported in EPA TSCA Inventory.

OSHA PEL: (Transitional: TWA 1 mg/m^3) TWA 0.1 mg (Ni)/m^3
ACGIH TLV: TWA 0.1 mg(Ni)/m^3; (Proposed: TWA 0.05 mg(Ni)/m^3; Human Carcinogen)
NIOSH REL: (Inorganic Nickel) TWA 0.015 mg(Ni)/m^3
DOT Classification: ORM-E; Label: None.

SAFETY PROFILE: Confirmed carcinogen with experimental carcinogenic and tumorigenic data. Poison by subcutaneous route.

NDE500   CAS: 59978-65-3   **HR: 3**
NICKEL IRON SULFIDE
mf: FeNi$_4$S$_4$ mw: 418.93

SYNS: IRON NICKEL SULFIDE * NICKEL-IRON SULFIDE MATTE

CONSENSUS REPORTS: Nickel and its compounds are on the Community Right-To-Know List.

OSHA PEL: TWA 1 mg(Ni)/m^3
ACGIH TLV: TWA 1 mg(Ni)/m^3; (Proposed: TWA 0.05 mg(Ni)/m^3; Human Carcinogen)
NIOSH REL: (Inorganic Nickel) TWA 0.015 mg(Ni)/m^3

SAFETY PROFILE: Suspected carcinogen with experimental carcinogenic and neoplastigenic data. When heated to decomposition it emits toxic fumes of SO$_x$.

NDF500   CAS: 1313-99-1   **HR: 3**
NICKEL MONOXIDE
mf: NiO   mw: 74.71

PROP: Cubic, green-black crystals; yellow when hot. Mp: 1900°, d: 7.45. Insol in water; sol in acids.

SYNS: BUNSENITE * C.I. 77777 * GREEN NICKEL OXIDE * NICKELOUS OXIDE * NICKEL OXIDE (MAK) * NICKEL(II) OXIDE (1:1) * NICKEL PROTOXIDE

CONSENSUS REPORTS: IARC Cancer Review: GROUP 1 IMEMDT 7,264,87; Animal Inadequate Evidence IMEMDT 2,126,73; Animal Sufficient Evidence IMEMDT 11,75,76. NTP Fourth Annual Report On Carcinogens, 1984. Nickel and its compounds are on the Community Right-To-Know List. Reported in EPA TSCA Inventory.

OSHA PEL: TWA 1 mg(Ni)/m^3
ACGIH TLV: TWA 1 mg(Ni)/m^3; (Proposed: TWA 0.05 mg(Ni)/m^3; Human Carcinogen)
DFG TRK: 0.5 mg/m^3; Human Carcinogen.
NIOSH REL: (Inorganic Nickel) TWA 0.015 mg(Ni)/m^3

SAFETY PROFILE: Confirmed carcinogen with experimental carcinogenic and tumorigenic data. Poison by intratracheal, intravenous, and subcutaneous routes. Mutation data reported. Can react violently with fluorine, hydrogen peroxide, hydrogen sulfide, iodine, barium oxide + air.

NDG000   CAS: 13138-45-9   **HR: 3**
NICKEL(II) NITRATE (1:2)
DOT: UN 2725
mf: N$_2$O$_6$ • Ni   mw: 182.73

PROP: Green, deliquescent crystals. Mp: 56.7°, bp: 136.7°, d: 2.05.

SYNS: NICKEL NITRATE (DOT) * NITRIC ACID, NICKEL(II) SALT

CONSENSUS REPORTS: Nickel and its compounds are on the Community Right-To-Know List. Reported in EPA TSCA Inventory.

OSHA PEL: (Transitional: TWA 1 mg/m^3) TWA 0.1 mg (Ni)/m^3
ACGIH TLV: TWA 0.1 mg(Ni)/m^3; (Proposed: TWA 0.05 mg(Ni)/m^3; Human Carcinogen)
NIOSH REL: (Inorganic Nickel) TWA 0.015 mg(Ni)/m^3
DOT Classification: Oxidizer; Label:Oxidizer.

SAFETY PROFILE: Suspected carcinogen. Poison by intravenous route. Experimental reproductive effects. Mutation data reported. A powerful oxidizer. When heated to decomposition it emits very toxic fumes of NO$_x$.

NDG500   CAS: 13478-00-7   **HR: 2**
NICKEL(II) NITRATE, HEXAHYDRATE (1:2:6)
mf: N$_2$O$_6$ • Ni • 6H$_2$O   mw: 290.85

PROP: Green, deliquescent crystals. Mp: 56.7°, bp: 136.7°, d: 2.05. Sol in 0.4 parts water or alc. Keep well closed.

SYNS: NICKEL(2+) NITRATE, HEXAHYDRATE * NITRIC ACID, NICKEL(2+) SALT, HEXAHYDRATE

CONSENSUS REPORTS: Nickel and its compounds are on the Community Right-To-Know List.

OSHA PEL: (Transitional: TWA 1 mg/m^3) TWA 0.1 mg (Ni)/m^3
ACGIH TLV: TWA 0.1 mg(Ni)/m^3; (Proposed: TWA 0.05 mg(Ni)/m^3; Human Carcinogen)
NIOSH REL: (Inorganic Nickel) TWA 0.015 mg(Ni)/m^3

SAFETY PROFILE: Suspected carcinogen. Moderately toxic by ingestion. When heated to decomposition it emits toxic fumes of NO$_x$.

NDH000   CAS: 7718-54-9   **HR: 3**
NICKELOUS CHLORIDE
DOT: NA 9139
mf: Cl$_2$Ni   mw: 129.61

SYNS: NICKEL CHLORIDE (DOT) * NICKEL(II) CHLORIDE (1:2)

CONSENSUS REPORTS: Nickel and its compounds are on the Community Right-To-Know List. Reported in EPA TSCA Inventory. EPA Genetic Toxicology Program.

OSHA PEL: (Transitional: TWA 1 mg/m^3) TWA 0.1 mg (Ni)/m^3
ACGIH TLV: TWA 0.1 mg(Ni)/m^3; (Proposed: TWA 0.05 mg(Ni)/m^3; Human Carcinogen)
NIOSH REL: (Inorganic Nickel) TWA 0.015 mg(Ni)/m^3
DOT Classification: ORM-E; Label: None.

SAFETY PROFILE: Suspected carcinogen. Poison by ingestion, intravenous, intramuscular, and intraperitoneal routes. Experimental teratogenic and reproductive effects. Mutation data reported. When heated to decomposition it emits very toxic fumes of Cl$^-$.

NDH500   CAS: 1314-06-3   **HR: 3**
NICKEL PEROXIDE
mf: Ni$_2$O$_3$ mw: 165.42

PROP: Gray-black powder. Mp: $-O_2$ @ 600°, d: 4.83. Decomp about 600° into NiO and O_2. Insol in water; very sltly sol in cold acid, dissolved by hot HCl evolving Cl_2; dissolved by hot H_2SO_4 or HNO_3 evolving O_2.

SYNS: DINICKEL TRIOXIDE * NICKELIC OXIDE * NICKEL OXIDE * NICKEL OXIDE PEROXIDE * NICKEL SISQUIOXIDE * NICKEL TRIOXIDE

CONSENSUS REPORTS: Nickel and its compounds are on the Community Right-To-Know List. Reported in EPA TSCA Inventory.

OSHA PEL: (Transitional: TWA 1 mg/m^3) TWA 0.1 mg (Ni)/m^3
ACGIH TLV: TWA 1 mg(Ni)/m^3; (Proposed: TWA 0.05 mg(Ni)/m^3; Human Carcinogen)
NIOSH REL: (Inorganic Nickel) TWA 0.015 mg(Ni)/m^3

SAFETY PROFILE: Suspected carcinogen. Poison by subcutaneous route. Mutation data reported. Hazardous reaction with hydrogen peroxide. Presence of the oxide increases the sensitivity of nitroalkanes (e.g., nitromethane, nitroethane, 1-nitropropane) to heat.

NDI000   CAS: 14220-17-8   **HR: 3**
NICKEL POTASSIUM CYANIDE
mf: $C_4N_4Ni \cdot 2K$ mw: 240.99

SYNS: DIPOTASSIUM NICKEL TETRACYANIDE * DIPOTASSIUM TETRACYANONICKELATE * POTASSIUM TETRACYANONICKELATE * POTASSIUM TETRACYANONICKELATE(II)

CONSENSUS REPORTS: Nickel and its compounds, as well as cyanide and its compounds, are on the Community Right-To-Know List. Reported in EPA TSCA Inventory.

OSHA PEL: (Transitional: TWA 1 mg/m^3) TWA 0.1 mg (Ni)/m^3
ACGIH TLV: TWA 0.1 mg(Ni)/m^3; (Proposed: TWA 0.05 mg(Ni)/m^3; Human Carcinogen)
NIOSH REL: (Inorganic Nickel) TWA 0.015 mg(Ni)/m^3

SAFETY PROFILE: Suspected carcinogen. Poison by ingestion. Mutation data reported. When heated to decomposition it emits very toxic fumes of NO_x, K_2O and CN$^-$.

NDI500   **HR: 3**
NICKEL REFINERY DUST

CONSENSUS REPORTS: IARC Cancer Review: Human Sufficient Evidence IMEMDT 2,126,73. Nickel and its compounds are on the Community Right-To-Know List.

OSHA PEL: TWA 1 mg(Ni)/m^3
ACGIH TLV: TWA 1 mg(Ni)/m^3; (Proposed: TWA 0.05 mg(Ni)/m^3; Human Carcinogen)
NIOSH REL: (Inorganic Nickel) TWA 0.015 mg(Ni)/m^3

SAFETY PROFILE: Confirmed carcinogen with experimental carcinogenic and neoplastigenic data. A human carcinogenic. Moderately toxic by intramuscular route. When heated to decomposition it emits toxic fumes of SO_x.

NDJ000   CAS: 13520-61-1   **HR: 3**
NICKEL(2+) SALT PERCHLORIC ACID HEXAHYDRATE
mf: $Cl_2O_8 \cdot Ni \cdot 6H_2O$ mw: 365.73

SYN: NICKEL(2+) PERCHLORATE, HEXAHYDRATE

CONSENSUS REPORTS: Nickel and its compounds are on the Community Right-To-Know List.

OSHA PEL: (Transitional: TWA 1 mg/m^3) TWA 0.1 mg (Ni)/m^3
ACGIH TLV: TWA 0.1 mg(Ni)/m^3; (Proposed: TWA 0.05 mg(Ni)/m^3; Human Carcinogen)
NIOSH REL: TWA 15 µg(Ni)/m^3

SAFETY PROFILE: Suspected carcinogen. Poison by intraperitoneal route. A powerful oxidizer. Mixtures with 2,2-dimethoxypropane explode when heated above 65°C. When heated to decomposition it emits toxic fumes of Cl$^-$.

NDJ399   CAS: 12255-80-0   **HR: 3**
NICKEL SUBARSENIDE
mf: As_2Ni_5 mw: 443.39

SYN: NICKEL ARSENIDE (As$_2$-Ni$_5$)

OSHA: Cancer Hazard

SAFETY PROFILE: Confirmed human carcinogen with experimental carcinogenic data. Moderately toxic by ingestion. When heated to decomposition it emits toxic fumes of As.

NDJ400   CAS: 12256-33-6   **HR: 3**
NICKEL SUBARSENIDE
mf: As_8Ni_{11} mw: 1245.17

SYN: NICKEL ARSENIDE (As$_8$-Ni$_{11}$)

OSHA: Cancer Hazard

SAFETY PROFILE: Confirmed human carcinogen with experimental carcinogenic data. When heated to decomposition it emits toxic fumes of As.

NDJ475 CAS: 12137-13-2 **HR: 3**
NICKEL SUBSELENIDE
mf: Ni_3Se_2 mw: 334.05

SYNS: NICKEL SELENIDE * NICKEL SELENIDE (3:2) CRYSTALLINE

OSHA PEL: TWA Soluble: Compounds: 0.1 mg(Ni)/m^3; Insoluble Compounds: 1 mg(Ni)/m^3; TWA 0.2 mg(Se)/m^3
ACGIH TLV: TWA 1 mg(Ni)/m^3; (Proposed: TWA 0.05 mg(Ni)/m^3; Human Carcinogen); TWA 0.2 mg(Se)/m^3
DFG TRK: 0.5 mg/m^3; Human Carcinogen.
NIOSH REL: (Inorganic Nickel) TWA 0.015 mg(Ni)/m^3

SAFETY PROFILE: Confirmed carcinogen with experimental carcinogenic data. Mutation data reported.

NDJ500 CAS: 12035-72-2 **HR: 3**
NICKEL SUBSULFIDE
mf: Ni_3S_2 mw: 240.25

SYNS: HEAZLEWOODITE * NICKEL SUBSULPHIDE * NICKEL SULFIDE * α-NICKEL SULFIDE (3:2) CRYSTALLINE * NICKEL SULPHIDE * NICKEL TRI-TADISULPHIDE

CONSENSUS REPORTS: IARC Cancer Review: GROUP 1 IMEMDT 7,264,87; Animal Sufficient Evidence IMEMDT 2,126,73, IMEMDT 11,75,76. NTP Fourth Annual Report On Carcinogens, 1984. Nickel and its compounds are on the Community Right-To-Know List. Reported in EPA TSCA Inventory.

OSHA PEL: TWA 1 mg(Ni)/m^3
ACGIH TLV: TWA 1 mg(Ni)/m^3; (Proposed: TWA 0.05 mg(Ni)/m^3; Human Carcinogen)
NIOSH REL: TWA 0.015 mg(Ni)/m^3

SAFETY PROFILE: Confirmed carcinogen with experimental carcinogenic, neoplastigenic, and tumorigenic data. Poison by intraperitoneal route. Experimental teratogenic and reproductive effects. Human mutation data reported. When heated to decomposition it emits toxic fumes of SO$_x$.

NDK000 CAS: 13770-89-3 **HR: 3**
NICKEL (II) SULFAMATE
mf: $H_4N_2NiO_6S_2$ mw: 250.89

CONSENSUS REPORTS: Nickel and its compounds are on the Community Right-To-Know List. Reported in EPA TSCA Inventory.

OSHA PEL: (Transitional: TWA 1 mg/m^3) TWA 0.1 mg (Ni)/m^3
ACGIH TLV: TWA 0.1 mg(Ni)/m^3; (Proposed: TWA 0.05 mg(Ni)/m^3; Human Carcinogen)
NIOSH REL: (Inorganic Nickel) TWA 0.015 mg(Ni)/m^3

SAFETY PROFILE: Suspected carcinogen. Poison by intraperitoneal route. When heated to decomposition it emits very toxic fumes of SO$_x$ and NO$_x$.

NDK500 CAS: 7786-81-4 **HR: 3**
NICKEL SULFATE
DOT: NA 9141
mf: $O_4S \cdot Ni$ mw: 154.77

PROP: Cubic yellow crystals. Mp: $-SO_3$ @ 840°, d: 3.68.

SYNS: NCI-C60344 * NICKELOUS SULFATE * NICKEL(II) SULFATE (1:1) * NICKEL(II) SULFATE * NICKEL SULFATE(1:1) * NICKEL(2+)SULFATE(1:1) * SULFURIC ACID, NICKEL(2+)SALT * SULFURIC ACID, NICKEL(2+) SALT (1:1)

CONSENSUS REPORTS: Nickel and its compounds are on the Community Right-To-Know List. Reported in EPA TSCA Inventory. EPA Genetic Toxicology Program.

OSHA PEL: (Transitional: TWA 1 mg/m^3) TWA 0.1 mg (Ni)/m^3
ACGIH TLV: TWA 0.1 mg(Ni)/m^3; (Proposed: TWA 0.05 mg(Ni)/m^3; Human Carcinogen)
NIOSH REL: (Inorganic Nickel) TWA 0.015 mg(Ni)/m^3
DOT Classification: ORM-E; Label: None.

SAFETY PROFILE: Suspected carcinogen with experimental tumorigenic data. Poison by intravenous, intraperitoneal, and subcutaneous routes. Human mutation data reported. A human skin irritant. When heated to decomposition it emits very toxic fumes of SO$_x$.

NDL000 CAS: 10101-97-0 **HR: 3**
NICKEL(II) SULFATE HEXAHYDRATE (1:1:6)
mf: $NiO_4S \cdot 6H_2O$ mw: 262.89

SYNS: NICKEL MONOSULFATE HEXAHYDRATE * NICKEL SULFATE HEXAHYDRATE * NICKEL (II) SULFATE HEXAHYDRATE * NICKEL SULPHATE HEXAHYDRATE * SULFURIC ACID, NICKEL(2+) SALT, HEXAHYDRATE

CONSENSUS REPORTS: Nickel and its compounds are on the Community Right-To-Know List. EPA Genetic Toxicology Program.

OSHA PEL: (Transitional: TWA 1 mg/m^3) TWA 0.1 mg (Ni)/m^3
ACGIH TLV: TWA 0.1 mg(Ni)/m^3; (Proposed: TWA 0.05 mg(Ni)/m^3; Human Carcinogen)
NIOSH REL: (Inorganic Nickel) TWA 0.015 mg(Ni)/m^3

SAFETY PROFILE: Suspected carcinogen. Poison by intravenous and subcutaneous routes. Experimental reproductive effects. Human mutation data reported. When heated to decomposition it emits toxic fumes of SO$_x$.

NDL100   CAS: 11113-75-0   **HR: 3**
NICKEL SULFIDE
mf: NiS   mw: 90.77

SYNS: NICKEL MONOSULFIDE * NICKELOUS SULFIDE * NICKEL(II) SULFIDE * α-NICKEL SULFIDE (1:1) CRYSTALLINE

CONSENSUS REPORTS: Nickel and its compounds are on the Community Right-To-Know List.

OSHA PEL: TWA 1 mg(Ni)/m^3
ACGIH TLV: TWA 1 mg(Ni)/m^3; (Proposed: TWA 0.05 mg(Ni)/m^3; Human Carcinogen)
DFG TRK: 0.5 mg/m^3; Human Carcinogen.
NIOSH REL: TWA 0.015 mg(Ni)/m^3

SAFETY PROFILE: Confirmed carcinogen with experimental carcinogenic and neoplastigenic data. Mutation data reported. When heated to decomposition it emits toxic fumes of SO$_x$.

NDL500   CAS: 12035-39-1   **HR: 3**
NICKEL TITANIUM OXIDE
mf: NiO$_3$Ti   mw:154.61

SYNS: NICKEL-TITANATE * TITANIUM NICKEL OXIDE

CONSENSUS REPORTS: Nickel and its compounds are on the Community Right-To-Know List. Reported in EPA TSCA Inventory.

OSHA PEL: TWA 1 mg(Ni)/m^3
ACGIH TLV: TWA 0.1 mg(Ni)/m^3; (Proposed: TWA 0.05 mg(Ni)/m^3; Human Carcinogen)
NIOSH REL: (Inorganic Nickel) TWA 0.015 mg(Ni)/m^3

SAFETY PROFILE: Suspected carcinogen with experimental carcinogenic and tumorigenic data.

NDM000   CAS: 27848-84-6   **HR: 3**
NICOTERGOLINE
mf: C$_{24}$H$_{26}$BrN$_3$O$_3$ mw: 484.44

SYNS: 8-β-((5-BROMONICOTINOYLOXY)METHYL)-1,6-DIMETHYL-10-α-METHOXYERGOLINE * FI 6714 * 10-METHOXY-1,6-DIMETHYLERGOLINE-8-METHANOL-5-BROMO-3-PYRIDINECARBOXYLATE (ESTER) * 10-METHOXY-1,6-DIMETHYL-ERGOLIN-8-β-METHANOL-(5-BROMNICOTINAT) (GERMAN) * 1-METHYL-LUMILYSERGOL-8-(5-BROMONICOTINATE)-10-METHYL ETHER * MNE * NARGOLINE * NICERGOLIN (GERMAN) * NICERGOLINE * NIMERGOLINE * SERMION

SAFETY PROFILE: Poison by intravenous route. Moderately toxic by ingestion and subcutaneous routes. A vasodilator. When heated to decomposition it emits very toxic fumes of Br$^-$ and NO$_x$.

NDN000   CAS: 54-11-5   **HR: 3**
NICOTINE
DOT: UN 1654
mf: C$_{10}$H$_{14}$N$_2$ mw: 162.26

PROP: An alkaloid from tobacco. In its pure state, a colorless and almost odorless oil; sharp burning taste. Mp: $< -80°$, bp: 247.3° (partial decomp), lel: 0.75%, uel: 4.0%, d: 1.0092 @ 20°, autoign temp: 471°F, vap press: 1 mm @ 61.8°, vap d: 5.61. Volatile with steam; misc with water below 60°; very sol in alc, chloroform ether, petr ether, and kerosene oils.

SYNS: BLACK LEAF * DESTRUXOL ORCHID SPRAY * EMO-NIK * ENT 3,424 * FLUX MAAG * FUMETOBAC * MACH-NIC * 1-METHYL-2-(3-PYRIDYL)PYRROLIDINE * 3-(N-METHYLPYRROLIDINO)PYRIDINE * (S)-3-(1-METHYL-2-PYRROLIDINYL) PYRIDINE (9CI) * 3-(1-METHYL-2-PYRROLIDINYL) PYRIDINE * l-3-(1-METHYL-2-PYRROLIDYL)PYRIDINE * (−)-3-(1-METHYL-2-PYRROLIDYL)PYRIDINE * NIAGARA P.A. DUST * NICOCIDE * NICO-DUST * NICO-FUME * NICOTINA (ITALIAN) * (−)-NICOTINE * l-NICOTINE * NICOTINE ALKALOID * NICOTINE, liquid (DOT) * NICOTINE, solid (DOT) * NIKOTIN (GERMAN) * NIKOTYNA (POLISH) * ORTHO N-4 DUST * ORTHO N-5 DUST * PYRIDINE, 3-(TETRAHYDRO-1-METHYLPYRROL-2-YL) * β-PYRIDYL-α-N-METHYLPYRROLIDINE * RCRA WASTE NUMBER P075 * TENDUST * dl-TETRAHYDRONICOTYRINE * XL ALL INSECTICIDE

CONSENSUS REPORTS: EPA Extremely Hazardous Substances List. Reported in EPA TSCA Inventory. EPA Genetic Toxicology Program.

OSHA PEL: TWA 0.5 mg/m^3 (skin)
ACGIH TLV: TWA 0.5 mg/m^3 (skin)
DFG MAK: 0.07 ppm; 0.5 mg/m^3
DOT Classification: Poison B; Label: St. Andrews Cross; Poison B; Label: Poison.

SAFETY PROFILE: A deadly human poison by unspecified route. Experimental poison by ingestion, skin contact, intraperitoneal, subcutaneous, intravenous, intramuscular, parenteral, intratracheal, and intraduodenal routes. Human systemic effects by rectal route: hallucinations, distorted perceptions, nausea or vomiting. Human teratogenic effects by ingestion: developmental abnormalities of the cardiovascular system. Human blood pressure effects. Can be absorbed by intact skin. Experimental teratogenic and reproductive effects. "Nicotinism", poisoning by nicotine, is characterized by stimulation and subsequent depression of the central and autonomic nervous systems. Death can result from respiratory paralysis. Mutation data reported. Used as a pesticide and in veterinary medicine as an external parasiticide. Combustible when exposed to heat or flame. Moderately explosive in the form of vapor when exposed to heat or flame. Can react with oxidizing materials. To fight fire, use alcohol foam, dry chemical, CO_2. When heated to decomposition it emits toxic fumes of NO_x, CO, and other highly toxic fumes.

NDP400   CAS: 2820-51-1   HR: 3
NICOTINE HYDROCHLORIDE

DOT: UN 1656
mf: $C_{10}H_{14}N_2 \cdot xClH$ mw: 417.48

SYNS: CHLORHYDRATE de NICOTINE (FRENCH) * NICOTINE HYDROCHLORIDE (d,l) * NICOTINE HYDROCHLORIDE, solution (DOT)

DOT Classification: Poison B; Label: Poison.

SAFETY PROFILE: Poison by intravenous, subcutaneous, and intraperitoneal routes. When heated to decomposition it emits very toxic fumes of Cl$^-$, NO_x, and CO.

NDR000   CAS: 29790-52-1   HR: 3
NICOTINE MONOSALICYLATE

DOT: UN 1657
mf: $C_{10}H_{14}N_2 \cdot C_7H_6O_3$ mw: 300.39

PROP: Mp: 118°. Very sol in water or alc.

SYN: NICOTINE SALICYLATE (DOT)

DOT Classification: Poison B; Label: Poison.

SAFETY PROFILE: A poison. Symptoms of exposure: Extreme nausea, vomiting, evacuation of bowel and bladder, mental confusion, twitching, convulsions. Base is readily absorbed through mucous membranes and intact skin. Institute treatment immediately. When heated to decomposition it emits toxic fumes of NO_x and CO.

NDR500   CAS: 65-30-5   HR: 3
NICOTINE SULFATE

DOT: UN 1658
mf: $C_{20}H_{26}N_4 \cdot O_4S$ mw: 418.56

SYNS: 1-1-METHYL-2-(3-PYRIDYL)-PYRROLIDINE SULFATE * (S)-3-(1-METHYL-2-PYRROLIDINYL)PYRIDINE SULFATE (2:1) * 1-3-(1-METHYL-2-PYRROLIDYL)PYRIDINE SULFATE * NICOTINE SULFATE (2:1) * NICOTINE SULFATE, liquid (DOT) * NICOTINE SULFATE, solid (DOT) * SULFATE de NICOTINE (FRENCH)

CONSENSUS REPORTS: Reported in EPA TSCA Inventory.

DOT Classification: Poison B; Label: Poison.

SAFETY PROFILE: Poison by ingestion and skin contact. When heated to decomposition it emits very toxic fumes of SO_x and organic fumes.

NDS500   CAS: 65-31-6   HR: 3
NICOTINE TARTRATE (1:2)

DOT: UN 1659
mf: $C_{10}H_{14}N_2 \cdot 2C_4H_6O_6$ mw:462.46

SYNS: (S)-3-(1-METHYL-2-PYRROLIDINYL-PYRIDINE (R-(R,R))-2,3-DIHYDROXYBUTANEDIOATE (1:2) * NICOTINE ACID TARTRATE * NICOTINE BITARTRATE * NICOTINE HYDROGEN TARTRATE * (−)-NICOTINE HYDROGEN TARTRATE * NICOTINE TARTRATE * NICOTINE TARTRATE (DOT) * TARTRATE de NICOTINE (FRENCH)

DOT Classification: Label: Poison.

SAFETY PROFILE: Poison by ingestion, intravenous, intraperitoneal, subcutaneous, and possibly other routes. Experimental teratogenic and reproductive effects. When heated to decomposition it emits toxic fumes of NO_x and CO.

NDY000   CAS: 555-84-0   **HR: 3**
NIFURADENE
mf: $C_8H_8N_4O_4$ mw: 224.20

PROP: Lemon-yellow solid from nitromethane.
Mp: 261.5-263° (decomp).

SYNS: NF 246 * 1-(((5-NITRO-2-FURANYL)METH-
YLENE)AMINO)-2-IMIDAZOLIDINONE * N-(5-NI-
TRO-2-FURFURYLIDENE)-1-AMINO-2-IMIDAZOLIDONE
* 1-((5-NITROFURFURYLIDENE)AMINO)-2-IMIDAZOLIDI-
NONE * N-(5-NITRO-2-FURFURYLIDENEAMINO)-2-IMI-
DAZOAIDINONE * NSC-6470 * OXAFURADENE
* OXYFURADENE * RENAFUR

CONSENSUS REPORTS: IARC Cancer Review: GROUP 2B IMEMDT 7,56,87; Animal
Limited Evidence IMEMDT 7,181,74.

SAFETY PROFILE: Suspected carcinogen with
experimental carcinogenic data. Moderately
toxic by ingestion and intraperitoneal routes.
When heated to decomposition it emits toxic
fumes of NO_x.

NDY500   CAS: 3570-75-0   **HR: 3**
NIFURTHIAZOLE
mf: $C_8H_6N_4O_4S$ mw: 254.24

PROP: Bright yellow plates. Mp: 215.5°.

SYNS: AS-17665 * FNT * FORMIC 2-(4-(5-NITRO-
FURYL)-2-THIAZOLYL)HYDRAZIDE * 2-(2-FORMYLHY-
DRAZINO)-4-(5-NITRO-2-FURYL)THIAZOLE * NE-
FURTHIAZOLE * 2-(4-(5-NITRO-2-FURANYL)-2-THIAZO-
LYL)-HYDRAZINECARBOXALDEHYDE * NSC-525334

CONSENSUS REPORTS: IARC Cancer Re-
view: GROUP 2B IMEMDT 7,56,87; Animal
Sufficient Evidence IMEMDT 7,151,74. EPA
Genetic Toxicology Program.

SAFETY PROFILE: Suspected carcinogen with
experimental carcinogenic and neoplastigenic
data. Mutation data reported. When heated to
decomposition it emits very toxic fumes of NO_x
and SO_x.

NDZ000   **HR: 1**
NIOBIUM
af: Nb   aw: 92.906

PROP: Steel gray, cubic crystals. Mp: 2468
± 10°; bp: 4742°; d: 8.57. Sol in aqua regia,
fused alkali. An element which occurs throughout nature.

SAFETY PROFILE: An eye and severe skin
irritant. Can cause kidney damage. No reports

of human intoxication. Experimentally, there
is a moderate fibrogenic effect on the lungs after
intratracheal administration. Some niobium is
found in all parts of the body. Flammable in
the form of dust when exposed to flame or by
chemical reaction. Moderately explosive in the
form of dust when exposed to flame. Ignites
in fluorine, chlorine (at 205°C). Incandescent
reaction with bromine trifluoride.

NEA000   CAS: 10026-12-7   **HR: 3**
NIOBIUM CHLORIDE
mf: Cl_5Nb mw: 270.16

PROP: Yellow, very deliquescent, monoclinic
crystals. Decomp in moist air evolving HCl.
D: 2.75, mp: 204.7-209.5°, bp: approx 250°,
subl @ 125°. Sol in HCl, carbon tetrachloride.

SYNS: COLUMBIUM PENTACHLORIDE * NIOBIUM
PENTACHLORIDE

CONSENSUS REPORTS: Reported in EPA
TSCA Inventory. EPA Genetic Toxicology Program.

SAFETY PROFILE: Poison by intraperitoneal
route. Moderately toxic by ingestion. May cause
kidney injury. When heated to decomposition
it emits very toxic fumes of Cl^-.

NEC000   CAS: 1432-75-3   **HR: 3**
NITRALAMINE HYDROCHLORIDE
mf: $C_{10}H_{13}ClN_2O_2S \cdot ClH$ mw: 297.22

SYNS: 2-((o-CHLORO-α-(NITROMETHYL)BENZYL)
THIO)ETHYLAMINE HYDROCHLORIDE * ((α-NITRO-
METHYL)-o-CHLOROBENZYLTHIO)ETHYLAMINE HY-
DROCHLORIDE

SAFETY PROFILE: Poison by intravenous and
intraperitoneal routes. When heated to decomposition it emits very toxic fumes of SO_x, NO_x,
and Cl^-.

NED000   **HR: 3**
NITRATES

PROP: Organic nitrates are usually termed nitro
compounds. These compounds are a combination of the nitro (—NO_2) group and an organic
radical. However, this term is often used to
denote nitric acid esters of an organic material.
Inorganic nitrates are compounds of metals
which are combined with the mono-valent NO_3
radical.

SAFETY PROFILE: Large amounts taken by
mouth may have serious or even fatal effects.

The symptoms are dizziness, abdominal cramps, vomiting, bloody diarrhea, weakness, convulsions and collapse. Small, repeated doses may lead to weakness, general depression, headache and mental impairment. Also, there is some implication of increased cancer incidence among those exposed.

Flammable by spontaneous chemical reaction; practically all nitrates are powerful oxidizing agents. Some nitrates may explode when shocked, exposed to heat or flame or by spontaneous chemical reaction (see also EXPLOSIVES, HIGH). All the inorganic nitrates act as oxygen carriers; under proper conditions, these can give up their oxygen to other materials which may in turn detonate. Ammonium nitrate has all the properties of the other nitrates, but it is also able to detonate by itself under certain conditions. It is therefore a high explosive, although very insensitive to impact and difficult to detonate. In the pure state, it requires a combination of an initiator and a high explosive. It is a relatively safe high explosive which, however, must be stored in a cool, ventilated place away from acute fire hazards and easily oxidized materials. Ammonium nitrate must not be confined because, if a fire should start, confinement can cause detonation with extremely violent results.

Violent reaction with Al; BP; cyanide; esters; PN_2H; P; NaCN; $SnCl_2$; sodium hypophosphite; thiocyanates. Dangerous disaster hazard due to fire and explosion hazard. When heated to decomposition it emits toxic fumes of NO_x. They are powerful oxidizing agents which may cause violent reaction with reducing materials. Nitrates should be protected carefully in storage.

NED500 CAS: 7697-37-2 **HR: 3**
NITRIC ACID

DOT: UN 2031/UN 2032
mf: HNO_3 mw: 63.02

PROP: Transparent, colorless or yellowish, fuming, suffocating, caustic and corrosive liquid. Mp: $-42°$, bp: $86°$, d: 1.50269 @ $25°/4°$.

SYNS: ACIDE NITRIQUE (FRENCH) * ACIDO NITRICO (ITALIAN) * AQUA FORTIS * AZOTIC ACID * AZOTOWY KWAS (POLISH) * HYDROGEN NITRATE * NITRIC ACID, over 40% (DOT) * SALPETERSAURE (GERMAN) * SALPETERZUUROPLOSSINGEN (DUTCH)

CONSENSUS REPORTS: Reported in EPA TSCA Inventory. EPA Genetic Toxicology Program. Community Right-To-Know List.

OSHA PEL: (Transitional: TWA 2 ppm) TWA 2 ppm; STEL 4 ppm
ACGIH TLV: TWA 2 ppm; STEL 4 ppm
DFG MAK: 10 ppm (25 mg/m^3)
NIOSH REL: (Nitric Acid) TWA 2 ppm
DOT Classification: Corrosive Material; Label: Corrosive; Oxidizer; Label: Oxidizer and Corrosive.

SAFETY PROFILE: Human poison by an unspecified route. Experimental teratogenic and reproductive effects. Corrosive to eyes, skin, mucous membranes, and teeth. Causes upper respiratory irritation which may seem to clear up only to return in a few hours and more severely. Depending on environmental factors the vapor will consist of a mixture of the various oxides of nitrogen and nitric acid. Flammable by chemical reaction with reducing agents. It is a powerful oxidizing agent.

Will react with water or steam to produce heat and toxic and corrosive fumes. To fight fire, use water. When heated to decomposition emits highly toxic fumes of NO_x and hydrogen nitrate.

NEE500 CAS: 7697-37-2 **HR: 3**
NITRIC ACID (RED FUMING)

DOT: UN 2032
mf: HNO_3 mw:63.02

PROP: Colorless to yellow to red corrosive liquid. D: > 1.480. Contains from 8 to 17% NO_2.

SYNS: NITRIC ACID, FUMING (DOT) * NITRIC ACID, RED FUMING (DOT) * NITROUS FUMES * RED FUMING NITRIC ACID * RFNA

CONSENSUS REPORTS: EPA Genetic Toxicology Program.

DOT Classification: Corrosive Material; Label: Corrosive, Oxidizer and Poison.

SAFETY PROFILE: Poison by inhalation. A corrosive irritant to skin, eyes, and mucous membranes. A very dangerous fire hazard and very powerful oxidizing agent. Can react explosively with many reducing agents. Will react with water or steam to produce heat and toxic, corrosive, and flammable vapors. When heated to decomposition it emits highly toxic fumes of NO_x.

NEF000   HR: 3
NITRIC ACID (WHITE FUMING)

SYNS: WFNA   *   WHITE FUMING NITRIC ACID

SAFETY PROFILE: Moderately toxic by inhalation. A corrosive irritant to skin, eyes, and mucous membranes. A very dangerous fire hazard and a very powerful oxidizing agent. Can react explosively with many reducing agents. Will react with water or steam to produce heat and toxic, corrosive, and flammable vapors. When heated to decomposition it emits highly toxic fumes of NO_x.

NEG100   CAS: 10102-43-9   HR: 3
NITRIC OXIDE

DOT: UN 1660
mf: NO   mw: 30.01

PROP: Colorless gas, blue liquid and solid. Mp: $-161°$, bp: -151.18, d: 1.3402 g/L; liquid, 1.269 @ $-150°$; gas, 1.04.

SYNS: BIOXYDE d'AZOTE (FRENCH)   *   NITROGEN MONOXIDE   *   OXYDE NITRIQUE (FRENCH)   *   RCRA WASTE NUMBER P076   *   STICKMONOXYD (GERMAN)

Reported in EPA TSCA Inventory. EPA Extremely Hazardous Substances List.

OSHA PEL: TWA 25 ppm
ACGIH TLV: TWA 25 ppm
NIOSH REL: (Oxides of Nitrogen) TWA 25 ppm
DOT Classification: Poison A; Label: Poison Gas.

SAFETY PROFILE: A poison gas. A severe eye, skin, and mucous membrane irritant. A systemic irritant by inhalation. Mutation data reported. Exposure may occur whenever nitric acid acts upon organic material, such as wood, sawdust, and refuse; it occurs when nitric acid is heated, and when organic nitro compounds are burned, for example, celluloid, cellulose nitrate (guncotton), and dynamite. The action of nitric acid upon metals, as in metal etching and pickling, also liberates the fumes. In high temperature welding, as with the oxyacetylene or electric torch, the nitrogen and oxygen of the air unite to form oxides of nitrogen. Automobile exhaust and power plant emissions are also sources of NO_x. Exposure occurs in many manufacturing processes when nitric acid is made or used. Oxides of nitrogen have been implicated as a cause of acid rain.

The oxides of nitrogen are somewhat soluble in water reacting with it in the presence of oxygen to form nitric and nitrous acids. This is the action that takes place deep in the respiratory system. The acids formed are irritating and can cause congestion in the throat and bronchi and edema of the lungs. The acids are neutralized by the alkalies present in the tissues with the formation of nitrates and nitrites. The latter may cause some arterial dilation, fall in blood pressure, headache and dizziness, and there may be some formation of methemoglobin. However, the nitrite effect is of secondary importance.

Because of their relatively low solubility in water, the nitrogen oxides are initially only slightly irritating to the mucous membranes of the upper respiratory tract. Their warning power is therefore low, and dangerous amounts of the fumes may be breathed before the worker notices any real discomfort. Higher concentrations (60-150 ppm) cause immediate irritation of the nose and throat with coughing and burning in the throat and chest. These symptoms often clear upon breathing fresh air, and the worker may feel well for several hours. Some 6-24 hours after exposure, a sensation of tightness and burning in the chest develops, followed by shortness of breath, sleeplessness, and restlessness. Dyspnea and air hunger may increase rapidly with development of cyanosis and loss of consciousness followed by death. In cases which recover from the pulmonary edema, there is usually no permanent disability, but pneumonia may develop later. Concentrations of 100-150 ppm are dangerous for short exposures of 30-60 minutes. Concentrations of 200-700 ppm may be fatal after even very short exposures.

Continued exposure to low concentrations of the fumes, insufficient to cause pulmonary edema, is said to result in chronic irritation of the respiratory tract with cough, headache, loss of appetite, dyspepsia, corrosion of the teeth, and gradual loss of strength.

Exposure to NO_x is always potentially serious, and persons so exposed should be kept under close observation for at least 48 hours.

An oxidizer. The liquid is a sensitive explosive. Explosive reaction with carbon disulfide (when ignited); methanol (when ignited); pentacarbonyl iron (at 50°C); phosphine + oxygen; sodium diphenylketyl; dichlorine oxide; fluorine; nitrogen trichloride; ozone; perchloryl fluoride (at 100-300°C); vinyl chloride. Reacts to form explosive products with dienes (e.g., 1,3-butadiene, cyclopentadiene, propadiene).

Can react violently with acetic anhydride, Al, amorphous boron, BaO, BCl_3, $CsHC_2$, calcium, carbon + potassium hydrogentartrate, charcoal, ClO, pyrophoric chromium, 1,2-dichloroethane, dichloroethylene, ethylene, fuels, hydrocarbons, hydrogen + oxygen, Na_2O, unsdimethyl hydrazine, NH_3, $CHCl_3$, Fe, Mg, Mn, CH_2Cl_2, olefins, phosphorus, PNH_2, PH_3, potassium, potassium sulfide, propylene, rubidium acetylide, Na, S, WC, trichloroethylene, 1,1,1-trichloroethane, uns-tetrachloroethane, uranium, uranium dicarbide. Will react with water or steam to produce heat and corrosive fumes; can react vigorously with reducing materials.

NEH000   HR: 2
NITRIDES

PROP: Compounds of $N(3-)$ as the anion, such as Li_3N, Ca_3N_2, etc.

SAFETY PROFILE: The details of the toxicity of nitrides as a group are unknown. However, many nitrides react with moisture to evolve ammonia. This gas is an irritant to mucous membranes. To the extent that many nitrides evolve flammable ammonia gas upon contact with moisture, nitrides can be fire hazards. A moderate explosion hazard. When heated to decomposition they emit toxic fumes of NH_3.

NEH500   HR: 3
NITRILES

PROP: Nitriles are organic compounds containing the ($-C\equiv N$) grouping, e.g., acrylonitrile ($CH_2:CHC\equiv N$).

CONSENSUS REPORTS: Cyanides and its compounds are on the Community Right-To-Know List.

SAFETY PROFILE: Nitriles are organic cyanides; acrylonitrile, propionitrile, and some others resemble cyanides in toxicity. Other nitriles, such as cyanamides and cyanates, have no cyanide effect. Can react violently with ($LiAlH_4$ + H_2O). The nitriles may be used as insecticides. Many are flammable. When heated to decomposition they emit highly toxic fumes of CN^-.

NEI000   CAS: 18662-53-8   HR: 3
NITRILOTRIACETIC ACID TRISODIUM SALT MONOHYDRATE
mf: $C_6H_6NO_6 \cdot 3Na \cdot H_2O$ mw: 275.12

SYNS: N,N-BIS(CARBOXYMETHYL)GLYCINE TRISODIUM SALT MONOHYDRATE * NCI-C01445 * NITRILOACETIC ACID TRISODIUM SALT MONOHYDRATE * NTA SODIUM HYDRATE * TRISODIUM NITRILOTRIACETATE MONOHYDRATE

CONSENSUS REPORTS: NCI Carcinogenesis Bioassay (feed); Clear Evidence: mouse, rat NCITR* NCI-CG-TR-6,77. Cyanides and its compounds are on the Community Right-To-Know List.

SAFETY PROFILE: Suspected carcinogen with experimental carcinogenic, tumorigenic, and teratogenic data. Moderately toxic by intraperitoneal route. Human mutation data reported. When heated to decomposition it emits toxic fumes of NO_x, CN^-, and Na_2O.

NEJ000   HR: 3
NITRITES

PROP: Salts of nitrous acid.

SAFETY PROFILE: Large amounts taken by mouth may produce nausea, vomiting, cyanosis (due to methemoglobin formation), collapse, and coma. Repeated small doses cause a fall in blood pressure, rapid pulse, headache, and visual disturbances. They have been implicated in an increased incidence of cancer. They may react with organic amines in the body to form carcinogenic nitrosamines. Organic nitrites are used to treat angina pectoris. Fire hazards are variable. They are generally powerful oxidizers. On contact with readily oxidized materials, a violent reaction such as a fire or explosion may ensue. Explosion hazards are also variable. Organic nitrites may decompose violently in contact with NH_4; salts; cyanide; KCN. Dangerous; shock may explode them; can react vigorously with reducing materials. When heated to decomposition they emit highly toxic fumes of NO_x.

NEJ500   CAS: 602-87-9   HR: 3
5-NITROACENAPHTHENE
mf: $C_{12}H_9NO_2$ mw: 199.22

SYNS: 1,2-DIHYDRO-5-NITRO-ACENAPHTHYLENE * 5-NAN * NCI-C01967 * 5-NITROACENAPHTHYLENE * 5-NITROACENAPTHENE * 5-NITRONAPHTHALENE ETHYLENE

CONSENSUS REPORTS: IARC Cancer Review: GROUP 2B IMEMDT 7,56,87; Animal Sufficient Evidence IMEMDT 16,319,78. NCI Carcinogenesis Bioassay (feed); Clear Evidence: mouse, rat NCITR* NCI-CG-TR-118,78. Reported in EPA TSCA Inventory. EPA Genetic Toxicology Program.

DFG MAK: Animal Carcinogen, Suspected Human Carcinogen.

SAFETY PROFILE: Confirmed carcinogen with experimental carcinogenic, neoplastigenic, and teratogenic data. Mutation data reported. When heated to decomposition it emits toxic fumes of NO_x.

NEL000   CAS: 1777-84-0   HR: 3
3-NITRO-p-ACETOPHENETIDIDE
mf: $C_{10}H_{12}N_2O_4$ mw: 224.24

PROP: Yellow needles in water. M: 103-104°. Sol in abs alc, ether, and chloroform.

SYNS: 4-ACETAMINO-2-NITROPHENETOLE * N-(4-ETHOXY-3-NITRO)PHENYLACETAMIDE * N-(4-ETHOXYPHENYL)-3'-NITROACETAMIDE * NCI C01978 * 2-NITRO-4-ACETAMINOFENETOL (CZECH) * 3-NITRO-p-ACETOPHENETIDE * 5-NITRO-p-ACETOPHENETIDIDE * 3'-NITRO-p-ACETOPHENETIDIN

CONSENSUS REPORTS: NCI Carcinogenesis Bioassay (feed); Clear Evidence: mouse NCITR* NCI-CG-TR-133,79; (feed); No Evidence: rat NCITR* NCI-CG-TR-133,79. Reported in EPA TSCA Inventory.

SAFETY PROFILE: Moderately toxic by ingestion. An eye irritant. Mutation data reported. Questionable carcinogen with experimental carcinogenic, neoplastigenic, and tumorigenic data. When heated to decomposition it emits toxic fumes of NO_x.

NEM480   CAS: 119-34-6   HR: 3
2-NITRO-4-AMINOPHENOL
mf: $C_6H_6N_2O_3$ mw: 154.14

SYNS: 4-AMINO-2-NITROPHENOL * C.I. 76555 * FOURRINE 57 * FOURRINE BROWN PR * FOURRINE BROWN PROPYL * 4-HYDROXY-3-NITROANILINE * NCI-C03963 * o-NITRO-p-AMINOPHENOL * OXIDATION BASE 25

CONSENSUS REPORTS: IARC Cancer Review: GROUP 3 IMEMDT 7,56,87; Animal Inadequate Evidence IMEMDT 16,43,78. NCI Carcinogenesis Bioassay (feed); No Evidence: mouse NCITR* NCI-CG-TR-94,78; Clear Evidence: rat NCITR* NCI-CG-TR-94,78. Reported in EPA TSCA Inventory. EPA Genetic Toxicology Program.

DFG MAK: Suspected Carcinogen.

SAFETY PROFILE: Suspected carcinogen with experimental carcinogenic data. Very poisonous by intraperitoneal route. Moderately toxic by ingestion. A severe eye irritant. Mutation data reported. When heated to decomposition it emits toxic fumes of NO_x.

NEN500   CAS: 99-09-2   HR: 3
m-NITROANILINE
DOT: UN 1661
mf: $C_6H_6N_2O_2$ mw: 138.14

PROP: Yellow, rhombic crystals. D: 0.9011 @ 25°/4°, mp: 114°, bp: 306.4°. Sol in water, alc, and ether.

SYNS: m-AMINONITROBENZENE * 1-AMINO-3-NITROBENZENE * AZOBASE MNA * C.I. 37030 * C.I. AZOIC DIAZO COMPONENT 7 * DAITO ORANGE BASE R * DEVOL ORANGE R * DIAZO FAST ORANGE R * FAST ORANGE R SALT * HILTONIL FAST ORANGE R BASE * MNA * NAPHTOELAN ORANGE R BASE * m-NITROAMINOBENZENE * m-NITRANILINE * 3-NITROANILINE * 3-NITROBENZENAMINE * m-NITROPHENYLAMINE * ORANGE BASE IRGA I

CONSENSUS REPORTS: Reported in EPA TSCA Inventory. EPA Genetic Toxicology Program.

DOT Classification: Poison B; Label: Poison.

SAFETY PROFILE: Poison by ingestion and intraperitoneal routes. Mutation data reported. Absorbed through the skin and by inhalation of the dust. Acute exposure may cause methemoglobinemia cyanosis. Chronic exposure may cause liver damage. Decomposes exothermically at 247°C. Possibly explosive reaction with ethylene oxide at 130°C. When heated to decomposition it emits toxic fumes of NO_x.

NEO000   CAS: 88-74-4   HR: 3
o-NITROANILINE
DOT: UN 1661
mf: $C_6H_6N_2O_2$ mw: 138.14

PROP: Orange-yellow crystals. Mp: 69-71°, bp: 284.5°, vap press: 1 mm @ 104°, d: 0.9015 @ 25°/4°. Sltly sol in cold water; sol in hot water, alc, and ether.

SYNS: 1-AMINO-2-NITROBENZENE * AZOENE FAST ORANGE GR SALT * BRENTAMINE FAST ORANGE GR

BASE * C.I. 37025 * C.I. AZOIC DIAZO COMPO-
NENT 6 * DEVOL ORANGE B * FAST ORANGE
BASE GR * HILTONIL FAST ORANGE GR BASE
* NATASOL FAST ORANGE GR SALT * o-NITRANI-
LINE * ORANGE BASE CIBA II * ORTHONITROANI-
LINE (DOT)

CONSENSUS REPORTS: Reported in EPA
TSCA Inventory.

DOT Classification: Poison B; Label: Poison.

SAFETY PROFILE: A poison. Moderately toxic
by ingestion. Mildly toxic by skin contact. Mutation data reported. Mixtures with magnesium
are hypergolic on contact with nitric acid. Forms
extremely explosive addition compounds with
hexanitroethane. Vigorous reaction with sulfuric
acid above 200°C. When heated to decomposition it emits toxic fumes of NO_x.

NEO500   CAS: 100-01-6   **HR: 3**
p-NITROANILINE

DOT: UN 1661
mf: $C_6H_6N_2O_2$ mw: 138.14

PROP: Bright yellow powder. Mp: 148.5°, bp:
332°, flash p: 390°F (CC), d: 1.424, vap press:
1 mm @ 142.4°. Sol in water, alc, ether, benzene, methanol.

SYNS: p-AMINONITROBENZENE * 1-AMINO-4-NI-
TROBENZENE * AZOFIX RED GG SALT * AZOIC
DIAZO COMPONENT 37 * C.I. 37035 * C.I. AZOIC
DIAZO COMPONENT 37 * C.I. DEVELOPER 17
* DEVELOPER P * DIAZO FAST RED GG * FAST
RED BASE GG * FAST RED 2G SALT * NAPHTO-
ELAN RED GG BASE * NCI-C60786 * p-NITRANI-
LINE * 4-NITRANILINE * NITRAZOL CF EXTRA
* p-NITROANILINA (POLISH) * 4-NITROANILINE
(MAK) * 4-NITROBENZENAMINE * p-NITROPHE-
NYLAMINE * PARANITROANILINE, solid (DOT)
* PNA * RCRA WASTE NUMBER P077 * RED 2G
BASE * SHINNIPPON FAST RED GG BASE

CONSENSUS REPORTS: Reported in EPA
TSCA Inventory. EPA Genetic Toxicology Program.

OSHA PEL: (Transitional: TWA 6 mg/m³
(skin)) TWA 3 mg/m³ (skin)
ACGIH TLV: TWA 3 mg/m³ (skin)
DFG MAK: 1 ppm (6 mg/m³)
DOT Classification: Poison B; Label: Poison.

SAFETY PROFILE: Poison by ingestion, intravenous and intraperitoneal routes. Moderately toxic by intramuscular route. Mutation data

reported. Acute symptoms of exposure are headache, nausea, vomiting, weakness and stupor,
cyanosis and methemoglobinemia. Chronic exposure can cause liver damage. Combustible
when exposed to heat or flame. To fight fire,
use water spray or mist, foam, dry chemical,
CO_2. Vigorous reaction with sulfuric acid above
200°C. Reaction with sodium hydroxide at
130°C under pressure may produce the explosive
sodium 4-nitrophenoxide. When heated to decomposition it emits toxic fumes of NO_x.

NEQ500   CAS: 99-59-2   **HR: 3**
5-NITRO-o-ANISIDINE
mf: $C_7H_8N_2O_3$ mw: 168.17

PROP: Red needles from alc. D: 1.207 @ 156°,
mp: 118°. Sol in hot benzene, alc, and acetic
acid; very sltly sol in ligroin.

SYNS: 2-AMINO-1-METHOXY-4-NITROBENZENE
* 3-AMINO-4-METHOXYNITROBENZENE * 2-AMINO-
4-NITROANISOLE * o-ANISIDINE NITRATE
* AZOAMINE SCARLET * AZOGENE ECARLATE R
* AZOIC DIAZO COMPONENT 13 BASE * C.I. AZOIC
DIAZO COMPONENT 13 * C.I. 37130 * FAST SCAR-
LET R * 2-METHOXY-5-NITROANILINE * 2-METH-
OXY-5-NITROBENZENAMINE * NCI-C01934
* 3-NITRO-6-METHOXYANILINE * 5-NITRO-2-ME-
THOXYANILINE

CONSENSUS REPORTS: IARC Cancer Re-
view: GROUP 3 IMEMDT 7,56,87; Animal
Limited Evidence IMEMDT 27,133,82. NCI
Carcinogenesis Bioassay (feed); Clear Evi-
dence: mouse, rat NCITR* NCI-CG-TR-
127,78. Reported in EPA TSCA Inventory. EPA
Genetic Toxicology Program. Community
Right-To-Know List.

SAFETY PROFILE: Suspected carcinogen with
experimental carcinogenic and tumorigenic
data. Moderately toxic by ingestion. Mutation
data reported. When heated to decomposition
it emits toxic fumes of NO_x.

NER000   CAS: 91-23-6   **HR: 3**
o-NITROANISOLE

DOT: UN 2730
mf: $C_7H_7NO_3$ mw: 153.15

PROP: Colorless crystals. D: 1.254 @ 20°/4°,
mp: 9.5-10.5°, bp: 277°. Insol in water; sol in
alc, ether.

SYNS: 2-METHOXYNITROBENZENE * 1-METHOXY-
2-NITROBENZENE * NCI-C60388 * 2-NITROANISOLE
* o-NITROPHENYL METHYL ETHER

CONSENSUS REPORTS: Reported in EPA TSCA Inventory. EPA Genetic Toxicology Program.

DOT Classification: Poison B; Label: St. Andrews Cross.

SAFETY PROFILE: A poison. Moderately toxic by ingestion. Mutation data reported. Explosive reaction with sodium hydroxide + zinc. Vigorous reaction with hydrogen + catalyst (at 250°C/34bar). When heated to decomposition it emits toxic fumes of NO_x.

NER500   CAS: 100-17-4   **HR: 3**
p-NITROANISOLE

DOT: UN 2730
mf: $C_7H_7NO_3$ mw: 153.15

PROP: Prisms from alc. D: 1.233 @ 20°, mp: 54°, bp: 260°. Sltly sol in cold petr ether; sol in water; very sol in alc, boiling petr ether, and ether.

SYNS: 1-METHOXY-4-NITROBENZENE * p-METHOXYNITROBENZENE * 4-METHOXYNITROBENZENE * p-NITROANISOL * 4-NITROANISOLE

CONSENSUS REPORTS: Reported in EPA TSCA Inventory. EPA Genetic Toxicology Program.

DOT Classification: Poison B; Label: St. Andrews Cross.

SAFETY PROFILE: A poison. Moderately toxic by ingestion and intraperitoneal routes. Mutation data reported. Can explode in presence of Ni. When heated to decomposition it emits toxic fumes of NO_x.

NEX000   CAS: 98-95-3   **HR: 3**
NITROBENZENE

DOT: UN 1662
mf: $C_6H_5NO_2$ mw: 123.12

PROP: Bright yellow crystals or yellow, oily liquid; odor of volatile almond oil. Mp: 6°, bp: 210-211°, ULC: 20-30%, lel: 1.8% @ 200°F, flash p: 190°F (CC), d: 1.205 @ 15°/4°, autoign temp: 900°F, vap press: 1 mm @ 44.4°, vap d: 4.25. Volatile with steam; sol in about 500 parts water; very sol in alc, benzene, ether, oils.

SYNS: ESSENCE OF MIRBANE * ESSENCE OF MYRBANE * MIRBANE OIL * NCI-C60082 * NITRO-

BENZEEN (DUTCH) * NITROBENZENE, liquid (DOT) * NITROBENZEN (POLISH) * NITROBENZOL (DOT) * NITROBENZOL, liquid (DOT) * OIL of MIRBANE (DOT) * OIL of MYRBANE * RCRA WASTE NUMBER U169

CONSENSUS REPORTS: Community Right-To-Know List. EPA Extremely Hazardous Substances List. Reported in EPA TSCA Inventory.

OSHA PEL: TWA 1 ppm (skin)
ACGIH TLV: TWA 1 ppm (skin)
DFG MAK: 1 ppm (5 mg/m³); BAT: 100 μg/L of aniline in blood after several shifts.
DOT Classification: Poison B; Label: Poison.

SAFETY PROFILE: Human poison by an unspecified route. Poison experimentally by subcutaneous and intravenous routes. Moderately toxic by ingestion, skin contact, and intraperitoneal routes. Human systemic effects by ingestion: general anesthetic, respiratory stimulation, and vascular changes. Experimental reproductive effects. Mutation data reported. An eye and skin irritant. Can cause cyanosis due to formation of methemoglobin. It is absorbed rapidly through the skin. The vapors are hazardous.

An oxidant. Combustible when exposed to heat and flame. Moderate explosion hazard when exposed to heat or flame. Explosive reaction with solid or concentrated alkali + heat (e.g., sodium hydroxide or potassium hydroxide), aluminum chloride + phenol (at 120°C), aniline + glycerol + sulfuric acid, nitric + sulfuric acid + heat. Forms explosive mixtures with aluminum chloride, oxidants (e.g., fluorodinitromethane, uronium perchlorate, tetranitromethane, sodium chlorate, nitric acid, nitric acid + water, peroxodisulfuric acid, dinitrogen tetraoxide), phosphorous pentachloride, potassium, sulfuric acid. Reacts violently with aniline + glycerin, N_2O, $AgClO_4$. To fight fire, use water, foam, CO_2, dry chemical. Incompatible with potassium hydroxide. When heated to decomposition it emits toxic fumes of NO_x.

NFA500   CAS: 22751-24-2   **HR: 3**
3-NITROBENZENEDIAZONIUM PERCHLORATE
mf: $C_6H_4ClN_3O_6$ mw: 249.57

DOT Classification: Forbidden.

SAFETY PROFILE: An extremely shock- and heat-sensitive explosive. Upon decomposition it emits toxic fumes of Cl^- and NO_x.

NFB500 CAS: 98-47-5 **HR: 3**
3-NITROBENZENESULFONIC ACID
mf: $C_6H_5NO_5S$ mw: 203.18

PROP: Hygroscopic leaflets. Mp: 70°, bp: decomp. Very sol in water, sol in alc and alkali; insol in ether.

SYNS: KYSELINA NITROBENZEN-m-SULFONOVA (CZECH) * m-NITROBENZENESULFONIC ACID

CONSENSUS REPORTS: Reported in EPA TSCA Inventory.

SAFETY PROFILE: A skin and severe eye irritant. Decomposes violently at about 200°. Mixture with sulfuric acid + sulfur trioxide may explode above 150°C. When heated to decomposition it emits toxic fumes of SO_x and NO_x.

NFJ000 CAS: 2338-12-7 **HR: 3**
5-NITROBENZOTRIAZOLE
DOT: UN 0385
mf: $C_6H_4N_4O_2$ mw: 164.14

SYNS: 5-NITROBENZOTRIAZOL (DOT) * 5-NITRO-1H-BENZOTRIAZOLE * 6-NITRO-1H-BENZOTRIAZOLE

CONSENSUS REPORTS: Reported in EPA TSCA Inventory.

DOT Classification: Class A Explosive; Label: Explosive A.

SAFETY PROFILE: Poison by intraperitoneal and intravenous routes. An explosive. When heated to decomposition it emits toxic fumes of NO_x.

NFJ500 CAS: 98-46-4 **HR: 3**
3-NITROBENZOTRIFLUORIDE
DOT: UN 2306
mf: $C_7H_4F_3NO_2$ mw: 191.12

PROP: Thin, pale, straw-colored, oily liquid with aromatic odor. Mp: −5°, bp: 202.8°, flash p: 217°F (OC), d: 1.437 @ 15.5°/15.5°, vap press: 0.3 mm @ 25°.

SYNS: m-NITROBENZOTRIFLUORIDE (DOT) * m-NITROTRIFLUOROTOLUENE * m-NITROTRI-FLUORTOLUOL(GERMAN) * m-(TRIFLUOROMETHYL) NITROBENZENE * 3-TRIFLUOROMETHYLNITROBEN-ZENE * α,α,α-TRIFLUORO-m-NITROTOLUENE * USAF MA-5

CONSENSUS REPORTS: Reported in EPA TSCA Inventory.

DOT Classification: Poison B; Label: Poison.

SAFETY PROFILE: Poison by intraperitoneal route. Moderately toxic by ingestion, inhalation, and subcutaneous routes. Combustible when exposed to heat or flame. When heated to decomposition it emits very toxic fumes of F^- and NO_x.

NFP500 CAS: 86-00-0 **HR: 2**
o-NITROBIPHENYL
mf: $C_{12}H_9NO_2$ mw: 199.22

PROP: Light yellow- to reddish-colored liquid or crystalline solid. Mp: 35°, bp: 330°, flash p: 354°F, d: 1.189 @ 40°/15.5°, autoign temp: 356°F, vap press: 2 mm @ 140°, vap d: 5.9. Insol in water; sol in methanol, ethanol, tetrahydrofurfuryl alc, acetone, dimethyl-formamide.

SYNS: 2-NITRODIPHENYL * o-NITRODIPHENYL

SAFETY PROFILE: Moderately toxic by ingestion. Mutation data reported. An irritant. Combustible when exposed to heat or flame. To fight fire, use CO_2, dry chemical. When heated to decomposition it emits toxic fumes of NO_x.

NFQ000 CAS: 92-93-3 **HR: 3**
4-NITROBIPHENYL
mf: $C_{12}H_9NO_2$ mw: 199.22

PROP: Needles from alc. Mp: 113-114°, bp: 340°. Insol in water; sltly sol in cold alc; very sol in ether.

SYNS: p-NITRODIPHENYL * p-PHENYL-NITROBEN-ZENE * 4-PHENYL-NITROBENZENE * PNB

CONSENSUS REPORTS: IARC Cancer Review: GROUP 3 IMEMDT 7,56,87; Animal Limited Evidence IMEMDT 4,113,74. Reported in EPA TSCA Inventory.

OSHA: Carcinogen.
ACGIH TLV: Confirmed Human Carcinogen.
DFG MAK: Animal Carcinogen, Suspected Human Carcinogen.

SAFETY PROFILE: Confirmed carcinogen with experimental carcinogenic, neoplastigenic, and tumorigenic data. Poison by intraperitoneal route. Moderately toxic by ingestion. Mutation data reported. When heated to decomposition it emits toxic fumes of NO_x.

NFS500 **HR: 3**
NITRO CARBO NITRATE

DOT Classification: Blasting Agent; Label: Blasting Agent.

SAFETY PROFILE: An explosive and oxidizer. When heated to decomposition it emits toxic fumes of NO_x and NH_3.

NFS525   CAS: 100-00-5   HR: 3
p-NITROCHLOROBENZENE

DOT: UN 1578
mf: $C_6H_4ClNO_2$ mw: 157.56

PROP: D: 1.520, mp: 83°, bp: 242°, flash p: 110°. Insol in water; sltly sol in alc; very sol in CS_2 and ether.

SYNS: 1-CHLOOR-4-NITROBENZEEN (DUTCH) * 1-CHLOR-4-NITROBENZOL (GERMAN) * p-CHLORONITROBENZENE * 4-CHLORONITROBENZENE * 1-CHLORO-4-NITROBENZENE * 4-CHLORO-1-NITROBENZENE * 1-CLORO-4-NITROBENZENE (ITALIAN) * p-NITROCHLOORBENZEEN (DUTCH) * p-NITROCHLOROBENZENE solid (DOT) * p-NITROCHLOROBENZOL (GERMAN) * p-NITROCLOROBENZENE (ITALIAN) * PNCB

CONSENSUS REPORTS: Reported in EPA TSCA Inventory.

OSHA PEL: TWA 1 mg/m³ (skin)
ACGIH TLV: TWA 0.1 ppm (skin)
DFG MAK: 1 mg/m³
DOT Classification: Poison B; Label: Poison.

SAFETY PROFILE: A poison by ingestion. Questionable carcinogen with experimental carcinogenic data. Mutation data reported. May explode on heating. Potentially violent reaction with sodium methoxide. When heated to decomposition it emits very toxic fumes of NO_x and Cl^-.

NFS700   CAS: 121-17-5   HR: 3
3-NITRO-4-CHLOROBENZOTRI-FLUORIDE

DOT: UN 2307
mf: $C_7H_3ClF_3NO_2$ mw: 225.56

SYNS: 4-CHLORO-3-NITRO-α,α,α-TRIFLUOROTOLUENE * 3-NITRO-4-CHLORO-α,α,α-TRIFLUOROTOLUENE

DOT Classification: Poison B; Label: Poison.

SAFETY PROFILE: Poison by ingestion and intravenous routes. Human mutation data reported. When heated to decomposition it emits toxic fumes of F^-, Cl^-, and NO_x.

NFT000   CAS: 2463-84-5   HR: 3
p-NITRO-o-CHLOROPHENYL DIMETHYL THIONOPHOSPHATE
mf: $C_8H_9ClNO_5PS$ mw: 297.66

PROP: White solid. Mp: 52°. Insol in water; sol in acetone, cyclohexane, xylene, toluene, and ethylacetate.

SYNS: O-(4-CHLOOR-3-NITRO-FENYL)-O,O-DIMETHYL-MONOTHIOFOSFAAT (DUTCH) * O-(4-CHLOR-3-NITRO-PHENYL)-O,O-DIMETHYL-MONOTHIOPHOSPHAT (GERMAN) * O-(2-CHLORO-4-NITROPHENYL) O,O-DIMETHYL PHOSPHOROTHIOATE * O-(4-CLORO-3-NITRO-FENIL)-O,O-DIMETIL-MONOIIOFOSFATO (ITALIAN) * O,O-DIMETHYL O-2-CHLORO-4-NITROPHENYL PHOSPHOROTHIOATE * DIMETHYL-2-CHLORONITROPHENYL THIOPHOSPHATE * ENT 17,035 * EXPERIMENTAL INSECTICIDE 4124 * ISOCHLOORTHION (DUTCH) * ISOMERIC CHLORTHION * THIOPHOSPHATE de O,O-DIMETHYLE et de O-4-CHLORO-3-NITROPHENYLE (FRENCH)

SAFETY PROFILE: Poison by ingestion. Moderately toxic by skin contact. When heated to decomposition it emits very toxic fumes of PO_x, SO_x, NO_x, and Cl^-.

NFT459   HR: 3
NITRO COMPOUNDS

PROP: Compounds of the form $C-NO_2$ or $N-NO_2$.

SAFETY PROFILE: The presence of a C— or N— linked nitro group in an organic compound can significantly decrease its reactivity and stability. Nitrate esters, with the nitro group linked to O, are very unstable. Nitro alkanes are mild oxidants but high temperatures and pressures may cause them to react violently. Polynitroalkanes may be explosive. Alkalies react with nitro alkanes to form explosive metal salts. The presence of metal oxides increases the thermal sensitivity of the lower nitro alkanes (e.g., nitromethane, nitroethane, and 1-nitropropane. Many nitro alkenes are highly reactive and may be explosive. Compounds with more than one nitro group (polynitroalkyls, such as trinitromethane and dinitroacetonitrile) are generally explosive.

NFT500   HR: 3
NITRO COMPOUNDS OF AROMATIC HYDROCARBONS

SAFETY PROFILE: The mono-, di-, and trinitrobenzenes are absorbed chiefly through the skin and through inhalation of the dust or vapor when these materials are heated. The dinitrobenzenes are believed to be somewhat more toxic than the mononitrobenzene and more toxic than aniline. The effect of di- and trinitrobenzene on the body is similar to that of aniline and mononitrobenzene with reduction of the oxygen-carrying power of the blood and depression of the nervous system being responsible for most of the symptoms following acute exposure. Poisoning with the solid nitro compounds is usually slower and less severe than is the case with the liquid nitro and amino benzenes since absorption is less rapid. Thus, chronic poisoning occurs more frequently than acute, the picture observed in the chronic form being one of anemia, moderate cyanosis, fatigue, slight dizziness, headache, insomnia, and loss of weight. Prolonged chronic exposure may result in damage to the liver and kidneys, with production of acute yellow atrophy, toxic hepatitis, and fatty degeneration of the kidneys. The introduction of one or more Cl atoms into the nitrobenzene ring results in the formation of chloronitrobenzene compounds or nitrochlors. The chloromono-nitrobenzenes have essentially the same toxic effect as nitrobenzene. The Cl derivatives of dinitrobenzene, on the other hand, while resembling dinitrobenzene in their systemic effects are much more irritating to the skin. They act as direct irritants and, in addition, may cause sensitization.

Dangerous; many of these compounds are highly flammable and some are explosive, especially those with more than one nitro group on the ring (polynitroaryls, such as trinitrobenzene, trinitrotoluene, tetranitro-N-methylaniline, trinitrophenol). The presence of alkali increases the thermal sensitivity of the explosive materials. Industrial explosions have occurred in this manner. When heated to decomposition they evolve highly toxic fumes of NO_x.

NFY500   CAS: 79-24-3   **HR: 3**
NITROETHANE

DOT: UN 2842
mf: $C_2H_5NO_2$ mw: 75.08

PROP: Oily, colorless liquid. Agreeable odor. Mp: $-90°$, bp: $114.0°$, fp: $-50°$, d: 1.052 @ $20°/20°$, autoign temp: $778°F$, flash p: $106°F$, decomp @ $335-382°$, lel: 4.0%, vap press: 15.6 mm @ $20°$, vap d: 2.58. Sol in water, acid, and alkali; misc in alc, chloroform, and ether.

SYN: NITROETAN (POLISH)

CONSENSUS REPORTS: Reported in EPA TSCA Inventory. EPA Genetic Toxicology Program.

OSHA PEL: TWA 100 ppm

ACGIH TLV: TWA 100 ppm
DFG MAK: 100 ppm (310 mg/m³)
DOT Classification: Flammable or Combustible Liquid; Label: Flammable Liquid.

SAFETY PROFILE: Poison by intraperitoneal route. Moderately toxic by ingestion. Causes injury to liver and kidneys. An eye and mucous membrane irritant. A dangerous fire hazard when exposed to heat, flame, or oxidizers. To fight fire, use alcohol foam, CO_2, dry chemical; water can blanket fire. Incompatible with $Ca(OH)_2$, hydrocarbons, hydroxides, inorganic bases, KOH, NaOH, metal oxides. Explodes when heated. When heated to decomposition it emits toxic fumes of NO_x.

NGB000   CAS: 607-57-8   **HR: 3**
2-NITROFLUORENE
mf: $C_{13}H_9NO_2$ mw: 211.23

PROP: Needles from 50% acetic acid. Mp: 157-158°.

CONSENSUS REPORTS: Reported in EPA TSCA Inventory. EPA Genetic Toxicology Program.

SAFETY PROFILE: Moderately toxic by intraperitoneal route. Questionable carcinogen with experimental neoplastigenic and tumorigenic data. Human mutation data reported. When heated to decomposition it emits toxic fumes of NO_x.

NGC000   CAS: 555-15-7   **HR: 3**
5-NITRO-2-FURALDEHYDE OXIME
mf: $C_5H_4N_2O_4$ mw: 156.11

SYNS: MICOFUR * NIFUROXIME * 5-NITRO-2-FURALDOXIME * NITROFUROXIME * USAF EA-5

SAFETY PROFILE: Poison by intravenous and intraperitoneal routes. Mutation data reported. When heated to decomposition it emits toxic fumes of NO_x.

NGE000   CAS: 67-20-9   **HR: 3**
NITROFURANTOIN
mf: $C_8H_6N_4O_5$ mw: 238.18

SYNS: BENKFURAN * BERKFURIN * CHEMIO-FURAN * CYANTIN * DANTAFUR * FURADAN-TIN * FURADONIN * FURANTOIN * FUROBAC-TINA * ITURAN * MACRODANTIN * NCI-C55196 * N-(5-NITROFURFURYLIDENE)-1-AMINOHYDANTOIN * N-(5-NITRO-2-FURFURYLIDENE)-1-AMINOHYDANTOIN * 1-((5-NITROFURFURYLIDENE)AMINO)HYDANTOIN

* NSC 2107 * N-TOIN * ORAFURAN * PARFU-RAN * URIZEPT * USAF EA-2 * WELFURIN * ZOOFURIN

CONSENSUS REPORTS: Reported in EPA TSCA Inventory. EPA Genetic Toxicology Program.

SAFETY PROFILE: Poison by ingestion and intraperitoneal routes. Human systemic effects by ingestion and possibly other routes: peripheral motor nerve recording changes, ataxia, changes in urine composition and hemolysis with or without anemia. Human reproductive effects by ingestion: spermatogenesis. Experimental reproductive effects. Questionable carcinogen with experimental neoplastigenic and teratogen data. Human mutation data reported. When heated to decomposition it emits toxic fumes of NO_x.

NGE500 CAS: 59-87-0 **HR: 3**
NITROFURAZONE
mf: $C_6H_6N_4O_4$ mw: 198.16

PROP: Odorless, lemon-yellow crystals; bitter aftertaste. Darkens upon prolonged exposure to light. Decomp @ 236-240°. Very sltly sol in water; sltly sol in alc, propylene glycol; sol in alkaline solns; insol in ether.

SYNS: ALDOMYCIN * ALFUCIN * AMIFUR * BABROCID * BIOFUREA * CHEMOFURAN * COCAFURIN * COXISTAT * DERMOFURAL * DYNAZONE * ELDEZOL * FEDACIN * FLAVAZONE * FRACINE * FURACILLIN * FURACINETTEN * FURACOCCID * FURACORT * FURACYCLINE * FURALDON * FURAN-OF-TENO * FURAPLAST * FURASEPTYL * FURA-ZONE * FURESOL * FURFURIN * FUVACILLIN * HEMOFURAN * IBIOFURAL * MAMMEX * MONOFURACIN * NCI-C56064 * NEFCO * NF * NIFUZON * 5-NITROFURALDEHYDE SEMI-CARBAZIDE * 6-NITROFURALDEHYDE SEMICARBA-ZIDE * 5-NITRO-2-FURALDEHYDE SEMICARBAZONE * 5-NITROFURAN-2-ALDEHYDE SEMICARBAZONE * 5-NITRO-2-FURANCARBOXALDEHYDE SEMICARBA-ZONE * 2((5-NITRO-2-FURANYL)METHYLENE)HY-DRAZINECARBOXAMIDE * 5-NITROFURFURAL SEMICARBAZONE * (5-NITRO-2-FURFURYLIDENE-AMINO)UREA * NITROZONE * NSC-2100 * OTOFURAN * SANFURAN * SPRAY-DERMIS * SPRAY-FORAL * U-6421 * USAF EA-4 * VABROCID * VADROCID * VET-ERINARY NITROFURAZONE * YATROCIN

CONSENSUS REPORTS: IARC Cancer Review: GROUP 2B IMEMDT 7,56,87; Animal Inadequate Evidence IMEMDT 7,171,74. Reported in EPA TSCA Inventory. EPA Genetic Toxicology Program.

SAFETY PROFILE: Suspected carcinogen with experimental carcinogenic, neoplastigenic, and tumorigenic data. Poison by ingestion and intraperitoneal routes. Moderately toxic by subcutaneous route. Experimental teratogenic and reproductive effects. A human sensitizer. Human mutation data reported. When heated to decomposition it emits toxic fumes of NO_x.

NGG500 CAS: 67-45-8 **HR: 3**
3-((5-NITROFURFURYLIDENE)AMINO)-2-OXAZOLIDONE
mf: $C_8H_7N_3O_5$ mw: 225.18

SYNS: BIFURON * CORIZIUM * DIAFURON * ENTEROTOXON * FURAXONE * FURAZOL * FURAZOLIDON * FURAZOLIDONE (USDA) * FURAZON * FURIDON * FUROVAG * FU-ROX * FUROXAL * FUROXANE * FUROXONE SWINE MIX * FUROZOLIDINE * GIARDIL * GIARLAM * MEDARON * NEFTIN * NG-180 * NICOLEN * NIFULIDONE * NIFURAN * 3-(((5-NITRO-2-FURANYL)METHYLENE)AMINO)-2-OX-AZOLIDINONE * NITROFURAZOLIDONE * NITRO-FURAZOLIDONUM * 3-(5′-NITROFURFURALAMINO)-2-OXAZOLIDONE * NITROFUROXON * N-(5-NITRO-2-FURFURYLIDENE)-3-AMINOOXAZOLIDINE-2-ONE * N-(5-NITRO-2-FURFURYLIDENE)-3-AMINO-2-OXAZOLI-DONE * 3-((5-NITROFURYLIDENE)AMINO)-2-OXAZOLI-DONE * 5-NITRO-N-(2-OXO-3-OXAZOLIDINYL)-2-FU-RANMETHANIMINE * PURADIN * ROPTAZOL * SCLAVENTEROL * TIKOFURAN * TOPAZONE * TRICHOFURON * TRICOFURON * USAF EA-1 * VIOFURAGYN

CONSENSUS REPORTS: IARC Cancer Review: GROUP 3 IMEMDT 7,56,87; Animal Inadequate Evidence IMEMDT 31,141,83. Reported in EPA TSCA Inventory. EPA Genetic Toxicology Program.

SAFETY PROFILE: Poison by ingestion and intraperitoneal routes. Human systemic effects by ingestion: dyspnea, respiratory depression and rosinophillis. Experimental reproductive effects. Human mutation data reported. Questionable carcinogen. When heated to decomposition it emits toxic fumes of NO_x.

NGI500 CAS: 712-68-5 **HR: 3**
2-(5-NITRO-2-FURYL)-5-AMINO-1,3,4-THIADIAZOLE
mf: $C_6H_4N_4O_3S$ mw: 212.20

SYNS: 2-AMINO-5-(5-NITRO-2-FURYL)-1,3,4-THIADI-AZOLE * 5-AMINO-2-(5-NITRO-2-FURYL)-1,3,4-THIADI-AZOLE * FURIDIAZINE * 5-(5-NITRO-2-FURANYL)-1,3,4-THIADIAZOL-2-AMINE * 5-(5-NITRO-2-FURYL)-2-AMINO-1,3,4-THIADIAZOLE

CONSENSUS REPORTS: IARC Cancer Review: GROUP 2B IMEMDT 7,56,87; Animal Limited Evidence IMEMDT 7,143,74.

SAFETY PROFILE: Suspected carcinogen with experimental carcinogenic and neoplastigenic data. When heated to decomposition it emits very toxic fumes of NO_x and SO_x.

NGI800   CAS: 75198-31-1   **HR: 3**
3-(5-NITRO-2-FURYL)-IMIDAZO(1,2-a) PYRIDINE
mf: $C_{11}H_7N_3O_3$ mw: 229.21

SYN: NFIP

SAFETY PROFILE: Suspected carcinogen with experimental carcinogenic data. When heated to decomposition it emits toxic fumes of NO_x.

NGM500   CAS: 24554-26-5   **HR: 3**
N-(4-(5-NITRO-2-FURYL)-2-THIAZOLYL)FORMAMIDE
mf: $C_8H_5N_3O_4S$ mw: 239.22

SYNS: FANFT * 2-FORMYLAMINO-4-(5-NITRO-2-FU-RYL)THIAZOLE * N-(4-(5-NITRO-2-FURYL)-2-THIAZO-LYL)FORMAMID (GERMAN)

CONSENSUS REPORTS: EPA Genetic Toxicology Program.

SAFETY PROFILE: Suspected carcinogen with experimental carcinogenic, tumorigenic, and teratogenic data. Mutation data reported. When heated to decomposition it emits very toxic fumes of SO_x and NO_x.

NGN500   CAS: 42011-48-3   **HR: 3**
N-(4-(5-NITRO-2-FURYL)-2-THIAZOLYL)-2,2,2-TRIFLUOROACETAMIDE
mf: $C_9H_4F_3N_3O_4S$ mw: 307.22

SYNS: 2-(2,2,2-TRIFLUOROACETAMIDO)-4-(5-NITRO-2-FURYL)THIAZOLE * 2,2,2-TRIFLUORO-N-(4-(5-NITRO-2-FURYL)-2-THIAZOLYL)ACETAMIDE

SAFETY PROFILE: Suspected carcinogen with experimental carcinogenic data. Mutation data reported. When heated to decomposition it emits very toxic fumes of F^-, NO_x, and SO_x.

NGP500   CAS: 7727-37-9   **HR: 1**
NITROGEN
DOT: UN 1066/UN 1977
mf: N_2 mw: 28.02

PROP: Colorless gas, colorless liquid or cubic crystals at low temp. Mp: $-210.0°$, d: 1.2506 g/L @ 0°, d (liquid): 0.808 @ $-195.8°$. Condenses to a liquid; sltly sol in water; sol in liquid ammonia, alc.

SYNS: NITROGEN, compressed (DOT) * NITROGEN GAS * NITROGEN, refrigerated liquid (DOT)

CONSENSUS REPORTS: Reported in EPA TSCA Inventory.

DOT Classification: Nonflammable Gas; Label: Nonflammable Gas.

SAFETY PROFILE: Low toxicity. In high concentrations it is a simple asphyxiant. The release of nitrogen from solution in the blood, with formation of small bubbles, is the cause of most of the symptoms and changes found in compressed air illness (caisson disease). It is a narcotic at high concentration and high pressure. Both the narcotic effects and the bends are hazards of compressed air atmospheres such as found in underwater diving. Nonflammable Gas. Can react violently with lithium, neodymium, titanium under the proper conditions.

NGP510   CAS: 7727-37-9   **HR: 3**
NITROGEN (cryogenic liquid)

CONSENSUS REPORTS: Reported in EPA TSCA Inventory.

DOT Classification: Nonflammable Gas; Label: Nonflammable Gas.

SAFETY PROFILE: Nitrogen (liquid) can explode during use.

NGQ500   CAS: 10025-85-1   **HR: 3**
NITROGEN CHLORIDE
mf: Cl_3N mw: 120.37

PROP: Volatile, yellowish oil or rhombic crystals; pungent odor. Mp: $< -40°$, explodes @ 93°, bp: $< 71°$, d: 1.653, vap press: 150 mm @ 20°.

SYNS: CHLORINE NITRIDE (NITROGEN) TRICHLORIDE * TRICHLORAMINE * TRICHLORINE NITRIDE

DOT Classification: Forbidden.

SAFETY PROFILE: An irritant to the eyes, skin, mucous membranes, and a systemic central nervous system irritant. An explosive sensitive to impact, light, and ultrasound. The solid explodes on melting. The liquid explodes above 60°C. Concentrated solutions are also explosive. Explosive decomposition is initiated by contact with: concentrated ammonia, arsenic, dinitrogen tetraoxide, hydrogen sulfide, hydrogen trisulfide, nitrogen oxide, organic matter, ozone, phosphine, phosphorus, potassium cyanide, potassium hydroxide solutions, selenium, hydrogen chloride, hydrogen fluoride, hydrogen bromide, hydrogen iodide. Mixtures with chlorine + hydrogen are potentially explosive. Upon decomposition it emits toxic fumes of Cl^- and NO_x.

NGR500   CAS: 10102-44-0   HR: 3
NITROGEN DIOXIDE

DOT: NA 1067
mf: NO_2 mw: 46.01

PROP: Colorless solid to yellow liquid; irritating odor. Mp: −9.3° (yellow liquid), bp: 21° (red-brown gas with decomp), d: 1.491 @ 0°, vap press: 400 mm @ 80°. Liquid below 21.15°. Sol in concentrated sulfuric acid, nitric acid. Corrosive to steel when wet.

SYNS: AZOTE (FRENCH) * AZOTO (ITALIAN) * NITRITO * NITROGEN PEROXIDE, liquid (DOT) * RCRA WASTE NUMBER P078 * STICKSTOFFDI-OXID (GERMAN) * STIKSTOFDIOXYDE (DUTCH)

CONSENSUS REPORTS: Reported in EPA TSCA Inventory. EPA Genetic Toxicology Program.

OSHA PEL: (Transitional: CL 5 ppm) STEL 1 ppm
ACGIH TLV: TWA 3 ppm; STEL 5 ppm
DFG MAK: 5 ppm (9 mg/m^3)
NIOSH REL: CL (Oxides of Nitrogen) 1 ppm/ 15M
DOT Classification: Poison A; Label: Poison Gas and Oxidizer.

SAFETY PROFILE: Experimental poison by inhalation. Moderately toxic to humans by inhalation. Human systemic effects by inhalation: pulmonary vascular resistance changes, cough, dyspnea and other pulmonary changes. Experimental teratogenic and reproductive effects. Mutation data reported. Violent reaction with cyclohexane, F_2, formaldehyde and alcohols, nitrobenzene, petroleum, toluene. When heated to decomposition it emits toxic fumes of NO_x.

NGS500   CAS: 13847-65-9   HR: 3
NITROGEN FLUORIDE OXIDE

mf: F_3NO mw: 87.01

SYNS: AMOX * TRIFLUOROAMINE OXIDE

OSHA PEL: TWA 2.5 mg(F)/m^3
ACGIH TLV: TWA 2.5 mg(F)/m^3
NIOSH REL: (Inorganic Fluorides) TWA 2.5 mg(F)/m^3

SAFETY PROFILE: Poison by inhalation and intraperitoneal routes. A skin, eye, and mucous membrane irritant. When heated to decomposition it emits very toxic fumes of F^- and NO_x.

NGU000   CAS: 10024-97-2   HR: 2
NITROGEN OXIDE

DOT: UN 1070/UN 2201
mf: N_2O mw: 44.02

PROP: Colorless gas, liquid or cubic crystals; slt sweet odor. Mp: −90.8°, bp: −88.49°, d: 1.977 g/L (liquid 1.226 @ −89°).

SYNS: DINITROGEN MONOXIDE * FACTITIOUS AIR * HYPONITROUS ACID ANHYDRIDE * LAUGHING GAS * NITROUS OXIDE (DOT) * NITROUS OXIDE, compressed (DOT) * NITROUS OXIDE, refrigerated liquid (DOT)

CONSENSUS REPORTS: Reported in EPA TSCA Inventory. EPA Genetic Toxicology Program.

ACGIH TLV: 50 ppm
NIOSH REL: (Waste Anesthetic Gases and Vapors) TWA 25 ppm
DOT Classification: Nonflammable Gas; Label: Nonflammable Gas, Oxidizer, compressed.

SAFETY PROFILE: Moderately toxic by inhalation. Human systemic effects by inhalation: general anesthetic, decreased pulse rate without blood pressure fall and body temperature decrease. Experimental teratogenic and reproductive effects. Mutation data reported. An asphyxiant. Does not burn but is flammable by chemical reaction and supports combustion. Moderate explosion hazard, it can form an explosive mixture with air. Violent reaction with Al, B, hydrazine, LiH, LiC_6H_5, PH_3, Na, WC. Also self-explodes at high temperatures.

NGU500   CAS: 10544-72-6   HR: 3
NITROGEN TETROXIDE

mf: N_2O_4 mw: 92.02

DOT: NA 1067

PROP: Nitrogen tetroxide is a dimer of nitrogen dioxide.

SYNS: DINITROGEN DIOXIDE * DINITROGEN TETROXIDE (DOT) * NITROGEN TETROXIDE, liquid (DOT)

CONSENSUS REPORTS: Reported in EPA TSCA Inventory.

DOT Classification: Poison A; Label: Poison Gas and Oxidizer.

SAFETY PROFILE: A poison. Moderately toxic by inhalation. When heated to decomposition it emits toxic fumes of NO_x.

NGW000   CAS: 7783-54-2   HR: 3
NITROGEN TRIFLUORIDE
DOT: UN 2451
mf: F_3N mw: 71.01

PROP: Colorless gas; odor of mold. Mp: $-208.5°$, bp: $-129°$, d (liquid): 1.537 @ $-129°$; d: (liquid @ bp:) 1.885; insol in water.

SYN: NITROGEN FLUORIDE

CONSENSUS REPORTS: Reported in EPA TSCA Inventory.

OSHA PEL: TWA 10 ppm
ACGIH TLV: TWA 10 ppm
NIOSH REL: (Inorganic Fluorides) TWA 2.5 mg(F)/m^3
DOT Classification: Poison A; Label: Poison Gas; Nonflammable Gas; Label: Nonflammable Gas.

SAFETY PROFILE: A poison. Mildly toxic by inhalation. Prolonged absorption may cause mottling of teeth, skeletal changes. Severe explosion hazard by chemical reaction with reducing agents, particularly when under pressure. A very dangerous fire hazard; a very powerful oxidizer; otherwise inert at normal temperatures and pressures. Reacts violently when ignited with H_2. When pure (dry) it does not attack glass or mercury at normal temperatures. Can react violently with NH_3, CO, diborane, H_2, H_2S, CH_4, tetrafluorohydrazine. Can react vigorously with reducing materials. Particularly hazardous under pressure. Incompatible with charcoal, hydrogen-containing compounds, tetrafluorohydrazine. When heated to decomposition it emits highly toxic fumes of F^-.

NGW500   CAS: 13444-85-4   HR: 3
NITROGEN TRIIODIDE
mf: NI_3 mw: 394.7

PROP: Black crystals. Mp: explodes, bp: sublimes in vacuum.

SYN: NITROGEN IODIDE

DOT Classification: Forbidden.

SAFETY PROFILE: A severe explosion hazard when shocked, exposed to heat or flame, or by spontaneous chemical reaction. It has no known uses as an explosive because it is far too sensitive in the dry state to store or handle safely. If this material must be worked with, it should be kept wet. A convenient way of keeping it wet is with ether; when it is needed in the dry state, it simply has to be taken out into the open and the ether will evaporate, leaving it perfectly dry. When dry, it will explode when given the slightest touch, vibration, or rise in temperature. Even a puff of air directed into it can cause it to detonate. It is a high explosive and is very violent. Incompatible with O_3, H_2S, Cl_2, Br_2, and acids.

NGY000   CAS: 55-63-0   HR: 3
NITROGLYCERIN
DOT: UN 0143/UN 1204
mf: $C_3H_5N_3O_9$ mw: 227.11

PROP: Colorless to yellow liquid; sweet taste. Mp: 13°, bp: explodes @ 218°, d: 1.599 @ 15°/15°, vap press: 1 mm @ 127°, vap d: 7.84, autoign temp: 518°F, decomp @ 50-60°, volatile @ 100°. Misc with ether, acetone, glacial acetic acid, ethyl acetate, benzene, nitrobenzene, pyridine, chloroform, ethylene bromide, dichloroethylene; sltly sol in petr ether, glycerol.

SYNS: ANGININE * BLASTING GELATIN (DOT) * BLASTING OIL * GLONOIN * GLYCERINTRINITRATE (CZECH) * GLYCEROL, NITRIC ACID TRIESTER * GLYCEROL TRINITRATE * GLYCEROL(TRINITRATE de) (FRENCH) * GLYCEROLTRINTRAAT (DUTCH) * GLYCERYL NITRATE * GLYCERYL TRINITRATE * GLYCERYL TRINITRATE, solution up to 1% in alcohol (DOT) * GTN * KLAVI KORDAL * MYOCON * NG * NIGLYCON * NIONG * NITRIC ACID TRIESTER OF GLYCEROL * NITRINE-TDC * NITROGLICERINA (ITALIAN) * NITROGLICERYNA (POLISH) * NITROGLYCERIN, liquid, desensitized (DOT) * NITROGLYCERIN, liquid, not desensitized (DOT) * NITROGLYCERINE * NITROGLYCEROL * NITROGLYN * NITROL * NITROLINGUAL

* NITROLOWE * NITRONET * NITRONG * NITRO-SPAN * NITROSTAT * NK-843 * NTG * PERGLOTTAL * 1,2,3-PROPANET-RIOL, TRINITRATE * 1,2,3-PROPANETRIYL NI-TRATE * RCRA WASTE NUMBER P081 * SK-106N * SOUP * TNG * TRINITRIN * TRI-NITROGLYCERIN * TRINITROGLYCEROL

CONSENSUS REPORTS: Reported in EPA TSCA Inventory. Community Right-To-Know List.

OSHA PEL: (Transitional: TWA CL 0.2 ppm (skin)) STEL: 0.1 mg/m^3 (skin)
ACGIH TLV: TWA 0.05 ppm (skin)
DFG MAK: 0.05 ppm (0.5 mg/m^3) (skin)
NIOSH REL: CL (Nitroglycerin or EGDN) 0.1 mg/m^3/20M
DOT Classification: Flammable Liquid; Label: Explosive A, desensitized; Forbidden, not de-sensitized; Flammable Liquid; Label: Flam-mable Liquid.

SAFETY PROFILE: Human poison by an un-specified route. Poison experimentally by inges-tion, intraperitoneal, subcutaneous, and intrave-nous routes. Experimental reproductive effects. A skin irritant. Questionable carcinogen with experimental tumorigenic and teratogenic data. Mutation data reported. It can cause respiratory difficulties and death due to respiratory paralysis by ingestion. The acute symptoms of nitroglyce-rin poisoning are headaches, nausea, vomiting, abdominal cramps, convulsions, methemoglo-binemia, circulatory collapse and reduced blood pressure, excitement, vertigo, fainting, respira-tory rales and cyanosis. Toxic effects may occur by ingestion, inhalation of dust, or absorption through intact skin. Used as a vasodilator and as an explosive.

A very dangerous fire hazard when ex-posed to heat, flame, or by spontaneous chemi-cal reaction. A severe explosion hazard when shocked or exposed to O$_3$, heat or flame. Nitro-glycerin is a powerful explosive, very sensitive to mechanical shock, heat, or UV radiation. Small quantities of it can readily be detonated by a hammer blow on a hard surface, particularly when it has been absorbed in filter paper. It explodes when heated to 215°C. Frozen nitro-glycerin is somewhat less sensitive than the liq-uid. However, a half-thawed or partially thawed mixture is more sensitive than either one. When heated to decomposition it emits toxic fumes of NO$_x$.

NGY500   CAS: 53569-64-5   **HR: 3**
NITROGLYCERIN mixed with ETHYLENE GLYCOL DINITRATE (1:1)
mf: C$_3$H$_5$N$_3$O$_9$ • C$_2$H$_4$N$_2$O$_6$ mw: 379.19

SYN: ETHYLENE GLYCOL DINITRATE mixed with NI-TROGLYCERIN (1:1)

OSHA PEL: (Transitional: TWA CL 0.2 ppm (skin)) STEL: 0.1 mg/m^3 (skin)
ACGIH TLV: TWA 0.05 ppm (skin)
NIOSH REL: CL (Nitroglycerin or EGDN) 0.1 mg/m^3/20M

SAFETY PROFILE: A human poison. Poison by subcutaneous route. A high explosive. When heated to decomposition it emits very toxic fumes of NO$_x$.

NHA500   CAS: 556-88-7   **HR: 3**
α-NITROGUANIDINE
DOT: UN 0282/UN 1336
mf: CH$_4$N$_4$O$_2$ mw: 104.09

PROP: Yellow solid, high explosive. Usually stable needles. Mp: 246°, decomp: 225-250°. Sltly sol in alc, concentration acids, cold solns of alkalies; sol in water; very sltly sol in ether.

SYNS: NITROGUANIDINE * 1-NITROGUANIDINE * 2-NITROGUANIDINE * NITROGUANIDINE, contain-ing less than 20% water (DOT) * NITROGUANIDINE DRY (DOT) * PICRITE (THE EXPLOSIVE)

CONSENSUS REPORTS: Reported in EPA TSCA Inventory.

DOT Classification: Class A Explosive; Label: Explosive A.

SAFETY PROFILE: Poison by intraperitoneal route. Mutation data reported. A very dangerous fire hazard when exposed to heat, flame or by chemical reaction with oxidizers. A severe ex-plosion hazard when shocked or exposed to heat or flame. It is about as powerful as TNT. It is normally mixed with colloided nitrocellulose or ammonium nitrate and paraffin wax. Can react vigorously with oxidizing materials and the derivatives can be explosive. The mercury and silver salts and other derivatives are much more impact-sensitive. When heated to decom-position it emits highly toxic fumes of NO$_x$.

NHE000   CAS: 4812-22-0   **HR: 3**
3-NITRO-3-HEXENE
mf: C$_6$H$_{11}$NO$_2$ mw: 129.18

SAFETY PROFILE: Poison by intraperitoneal route. Moderately toxic by ingestion and skin contact. A severe eye and skin irritant. Questionable carcinogen with experimental neoplastigenic data. When heated to decomposition it emits toxic fumes of NO_x.

NHI500   CAS: 7046-61-9   HR: 3
NITROIMINODIETHYLENEDIISOCYANIC ACID

mf: $C_6H_8N_4O_4$ mw: 200.18

SYN: 3-NITRO-3-AZAPENTANE-1,5-DIISOCYANATE

CONSENSUS REPORTS: Reported in EPA TSCA Inventory.
NIOSH REL: (Diisocyanates) TWA 0.005 ppm; CL 0.02 ppm/10M

SAFETY PROFILE: Poison by intravenous route. When heated to decomposition it emits toxic fumes of NO_x.

NHM500   CAS: 75-52-5   HR: 3
NITROMETHANE

DOT: UN 1261
mf: CH_3NO_2 mw: 61.05

PROP: An oily liquid; moderate to strong, disagreeable odor. Bp: 101°, lel: 7.3%, fp: −29°, flash p: 95°F (CC), d: 1.1322 @ 25°/4°, autoign temp: 785°F, vap press: 27.8 mm @ 20°, vap d: 2.11. Sltly sol in water; sol in alc, ether.

SYNS: NITROCARBOL   *   NITROMETAN (POLISH)

CONSENSUS REPORTS: Reported in EPA TSCA Inventory.

OSHA PEL: TWA 100 ppm
ACGIH TLV: TWA 100 ppm
DFG MAK: 100 ppm ($250 \, mg/m^3$)
DOT Classification: Flammable Liquid; Label: Flammable Liquid; Flammable or Combustible Liquid; Label: Flammable Liquid.

SAFETY PROFILE: Poison by ingestion and intraperitoneal routes. Moderately toxic by intravenous route. Mildly toxic by inhalation. In humans it may cause anorexia, nausea, vomiting, diarrhea, and kidney injury and liver damage.

A very dangerous fire hazard when exposed to heat, oxidizers, or flame. May explode by detonation, heat, or shock. Its sensitivity is increased when mixed with acids, bases, acetone, aluminum powder, ammonium salts + organic solvents, bis(2-aminoethyl)amine, 1,2-diaminoethane + N,2,4,6-tetranitro-N-methyl aniline, haloforms (e.g., chloroform, bromoform), hydrazine + methanol. Ignites when mixed with alkyl metal halides (e.g., diethylaluminum bromide, dimethylaluminum bromide, ethylaluminum bromide iodide, methyl zinc iodide, methylaluminum diiodide). Can react violently with $AlCl_3$ + organic matter, $Ca(OH)_2$, m-methyl aniline, $Ca(OCl)_2$, hexamethylbenzene, hydrocarbons, inorganic bases, hydroxides, organic amines, KOH, formaldehyde, nitric acid, metal oxides, 1,2-diaminomethane, lithium perchlorate, sodium hydride. Reacts with aqueous silver nitrate to form the explosive silver fulminate. When heated to decomposition it emits toxic fumes of NO_x.

NHP500   CAS: 5863-35-4   HR: 2
NITROMIFENE CITRATE

mf: $C_{27}H_{28}N_2O_4 \cdot C_6H_8O_7$ mw: 636.71

SYNS: CI-628   *   CI-628 CITRATE   *   CN-55945-27   *   1-(2-(p-(α-(p-METHOXYPHENYL)-β-NITROSTYRYL)PHENOXY)ETHYL)PYRROLIDINE CITRATE (1:1)   *   1-(2-(p-(α-(p-METHOXYPHENYL)-β-NITROSTYRYL)PHENOXY)ETHYL)PYRROLIDINE MONOCITRATE   *   PARKE DAVIS CI-628

SAFETY PROFILE: Moderately toxic by ingestion. Experimental teratogenic and reproductive effects. When heated to decomposition it emits toxic fumes of NO_x.

NHQ000   CAS: 86-57-7   HR: 3
1-NITRONAPHTHALENE

mf: $C_{10}H_7NO_2$ mw: 173.18

PROP: Yellow crystals. Bp: 304°, flash p: 327°F (CC), d: 1.331 @ 4°/4°, vap d: 5.96. Mp: 59-61°. Insol in water; sol in CS_2, alc, chloroform, and ether.

SYNS: NCI-C01956   *   α-NITRONAPHTHALENE

CONSENSUS REPORTS: NCI Carcinogenesis Bioassay (feed); No Evidence: mouse, rat NCITR* NCI-CG-TR-64,78. Reported in EPA TSCA Inventory.

DFG MAK: Suspected Carcinogen.

SAFETY PROFILE: Poison by ingestion and intraperitoneal routes. Mutation data reported. A skin, eye and mucous membrane irritant. Combustible when exposed to heat or flame. To fight fire, use CO_2, dry chemical or water spray. Explosive reaction with nitric acid + sulfuric acid above 60°C. Forms a sensitive explosive mixture with tetranitromethane. When heated to decomposition it emits toxic fumes of NO_x.

NHQ500   CAS: 581-89-5   **HR: 3**
2-NITRONAPHTHALENE
mf: $C_{10}H_7NO_2$ mw: 173.18

PROP: Colorless in ethanol. Mp: 79°, bp: 165° @ 15 mm. Insol in water; very sol in alc and ether.

SYN: β-NITRONAPHTHALENE

CONSENSUS REPORTS: EPA Genetic Toxicology Program.

DFG TRK: 0.035 ppm; Animal Carcinogen, Suspected Human Carcinogen.

SAFETY PROFILE: Confirmed carcinogen with experimental tumorigenic data. Moderately toxic by ingestion and intraperitoneal routes. Mutation data reported. A skin and lung irritant. Combustible when exposed to heat or flame. When heated to decomposition it emits toxic fumes of NO_x.

NIE000   CAS: 554-84-7   **HR: 3**
3-NITROPHENOL
DOT: UN 1663
mf: $C_6H_5NO_3$ mw: 139.12

PROP: Monoclinic crystals. Mp: 97°, bp: 194° @ 70 mm, d: 1.485 @ 20°/4°. Decomposes when distilled at ordinary pressure. Sol in hot water and dil acids, caustic solns; insol in petr ether.

SYN: m-NITROPHENOL (DOT)

CONSENSUS REPORTS: EPA Genetic Toxicology Program. Reported in EPA TSCA Inventory.

SAFETY PROFILE: Poison by ingestion, subcutaneous, and intraperitoneal routes. Moderately toxic by skin contact. A skin and severe eye irritant. When heated to decomposition it emits toxic fumes of NO_x.

NIE500   CAS: 88-75-5   **HR: 3**
o-NITROPHENOL
DOT: UN 1663
mf: $C_6H_5NO_3$ mw: 139.12

PROP: Light yellow crystals; aromatic odor. Mp: 45°, bp: 214.5°, d: 1.495 @ 20°, vap press: 1 mm @ 49.3°. Sol in water; very sol in alc, ether, benzene, CS; volatile with steam.

SYNS: 2-HYDROXYNITROBENZENE * 2-NITROPHE-NOL

CONSENSUS REPORTS: EPA Genetic Toxicology Program. Reported in EPA TSCA Inventory. Community Right-To-Know List.

DOT Classification: Poison B; Label: St. Andrews Cross.

SAFETY PROFILE: Poison by ingestion, subcutaneous, and intraperitoneal routes. Moderately toxic by intramuscular route. Can cause liver and kidney damage. The liquid phenol reacts violently with KOH. Product of reaction with chlorosulfuric acid decomposes violently at 24°C. When heated to decomposition it emits toxic fumes of NO_x.

NIF000   CAS: 100-02-7   **HR: 3**
4-NITROPHENOL
DOT: UN 1663
mf: $C_6H_5NO_3$ mw: 139.12

PROP: Colorless to sltly yellow; odorless crystals, sweet then burning taste. D: 1.270 @ 120°/4°, mp: 113-114° (subl). Sltly sol in cold water; very sol in alc, chloroform, ether; sol in alkali solns, hydroxides, and carbonates.

SYNS: 4-HYDROXYNITROBENZENE * NCI-C55992 * 4-NITROFENOL (DUTCH) * p-NITROPHENOL (DOT) * PARANITROFENOL (DUTCH) * PARANITROFEN-OLO (ITALIAN) * PARANITROPHENOL (FRENCH, GERMAN) * RCRA WASTE NUMBER U170

CONSENSUS REPORTS: EPA Genetic Toxicology Program. Community Right-To-Know List. Reported in EPA TSCA Inventory.

SAFETY PROFILE: Poison by ingestion, subcutaneous, intraperitoneal, intravenous, intramuscular and possibly other routes. Moderately toxic by skin contact. Human mutation data reported. Its exothermic decomposition causes a dangerous fast pressure increase. Mixtures with diethyl phosphite may explode when heated. When heated to decomposition it emits toxic fumes of NO_x.

NIJ500   CAS: 98-72-6   **HR: 3**
4-NITROPHENYLARSONIC ACID
mf: $C_6H_6AsNO_5$ mw: 247.05

SYNS: NITARSONE * 4-NITROBENZENEARSONIC ACID * p-NITROPHENYLARSONIC ACID * RAS-26

CONSENSUS REPORTS: Arsenic and its compounds are on the Community Right-To-Know List.

OSHA PEL: TWA 0.5 mg(As)/m³

ACGIH TLV: TWA 0.2 mg(As)m^3

SAFETY PROFILE: Poison by ingestion and intravenous routes. When heated to decomposition it emits very toxic fumes of NO_x and As.

NIM000   CAS: 1224-64-2   **HR: 3**
p-NITROPHENYLDI-N-BUTYLPHOSPHINATE
mf: $C_{14}H_{22}NO_4P$ mw: 299.34

SYNS: DIBUTYL-PHOSPHINIC ACID, 4-NITROPHENYL ESTER * NIBUFIN * p-NITROPHENYLDIBUTYL-PHOSPHINATE

SAFETY PROFILE: A deadly poison by intraperitoneal, intramuscular, intravenous, and subcutaneous routes. When heated to decomposition it emits very toxic fumes of PO_x and NO_x.

NIM500   CAS: 311-45-5   **HR: 3**
p-NITROPHENYL DIETHYLPHOSPHATE
mf: $C_{10}H_{14}NO_6P$ mw: 275.22

PROP: Oily liquid; slt odor. Bp: 170° @ 1 mm, d: 1.2736 @ 20°/4°. Sltly water-sol; freely sol in organic solvents.

SYNS: CHINORTA * O,O'-DIAETHYL-p-NITROPHE-NYLPHOSPHAT (GERMAN) * DIAETHYL-p-NITROPHENYLPHOSPHORSAEUREESTER (GERMAN) * DIETHYL-p-NITROFENYL ESTER KYSELINY FOSFO-RECNE (CZECH) * DIETHYL p-NITROPHENYL PHOS-PHATE * O,O-DIETHYL O-p-NITROPHENYL PHOS-PHATE * DIETHYL PARAOXON * O,O-DIETHYL PHOSPHORIC ACID O-p-NITROPHENYL ESTER * O,O-DIETYL-o-p-NITROFENYLFOSFAT (CZECH) * E 600 * ENT 16,087 * ESTER 25 * ETHYL p-NITRO-PHENYL ETHYLPHOSPHATE * ETHYL PARAOXON * ETICOL * FOSFAKOL * HC 2072 * MINT-ACO * MINTACOL * MIOTISAL * MIOTISAL A * P-NITROPHENOL, ESTER with DIETHYL PHOS-PHATE * OXYPARATHION * PARAOXON * PARAOXONE * PAROXAN * PESTOX 101 * PHOSPHACOL * PHOSPHORIC ACID DIETHYL 4-NITROPHENYL ESTER * RCRA WASTE NUMBER P401 * SOLUGLACIT * TS 219

SAFETY PROFILE: A deadly poison by ingestion, intraperitoneal, intravenous, subcutaneous, intramuscular, parenteral and possibly other routes. Mutation data reported. A cholinesterase inhibitor. An insecticide. When heated to decomposition it emits toxic fumes of PO_x and NO_x.

NIN000   CAS: 3015-74-5   **HR: 3**
p-NITROPHENYL ETHYLBUTYLPHOSPHONATE
mf: $C_{12}H_{18}NO_5P$ mw: 287.28

SAFETY PROFILE: A deadly poison by intravenous and subcutaneous routes. When heated to decomposition it emits very toxic fumes of PO_x and NO_x.

NIX500   CAS: 108-03-2   **HR: 3**
1-NITROPROPANE

DOT: UN 2608
mf: $C_3H_7NO_2$ mw: 89.11

PROP: Colorless liquid. Bp: 132°, fp: −108°, flash p: 93°F (TCC), d: 1.003 @ 20°/20°, autoign temp: 789°F, vap press: 7.5 mm @ 20°, vap d: 3.06, lel: 2.2%. Sltly sol in water; misc with alc, ether, and many organic solvents.

SYN: 1-NP

CONSENSUS REPORTS: Reported in EPA TSCA Inventory.

OSHA PEL: TWA 25 ppm
ACGIH TLV: TWA 25 ppm
DFG MAK: 25 ppm (90 mg/m^3)
DOT Classification: Flammable or Combustible Liquid; Label: Flammable Liquid.

SAFETY PROFILE: Poison by ingestion and intraperitoneal routes. Mildly toxic by inhalation. A human eye irritant. Human systemic effects by inhalation: conjunctiva irritation. Very dangerous fire hazard when exposed to heat, open flame, or oxidizers. Reacts violently with $Ca(OH)_2$, hydrocarbons, hydroxides, inorganic bases. May explode on heating. Metal oxides increase its sensitivity to thermal ignition. To fight fire, use alcohol foam, CO_2, dry chemical, water spray. When heated to decomposition it emits toxic fumes of NO_x.

NIY000   CAS: 79-46-9   **HR: 3**
2-NITROPROPANE

DOT: UN 2608
mf: $C_3H_7NO_2$ mw: 89.11

PROP: Colorless liquid. Bp: 120°, fp: −93°, flash p: 82°F (TCC), d: 0.992 @ 20°/20°, autoign temp: 802°F, vap press: 10 mm @ 15.8°, vap d: 3.06, lel: 2.6%. Misc with organic solvents; sol in water, alc, and ether.

SYNS: DIMETHYLNITROMETHANE * ISONITRO-PROPANE * NIPAR S-20 * NIPAR S-20 SOLVENT

* NIPAR S-30 SOLVENT * NITROISOPROPANE
* β-NITROPROPANE * 2-NP * RCRA WASTE NUMBER U171

CONSENSUS REPORTS: IARC Cancer Review: GROUP 2B IMEMDT 7,56,87; Human Inadequate Evidence IMEMDT 29,331,82; Animal Sufficient Evidence IMEMDT 29,331,82. NTP Fourth Annual Report On Carcinogens, 1984. Community Right-To-Know List. EPA Genetic Toxicology Program. Reported in EPA TSCA Inventory.

OSHA PEL: (Transitional: TWA 25 ppm) TWA 10 ppm
ACGIH TLV: TWA 10 ppm; Suspected Human Carcinogen.
DFG TRK: 5 ppm; Animal Carcinogen, Suspected Human Carcinogen.
DOT Classification: Flammable or Combustible Liquid; Label: Flammable Liquid.

SAFETY PROFILE: Confirmed carcinogen with experimental carcinogenic and tumorigenic data. Poison by intraperitoneal route. Moderately toxic by ingestion and inhalation. Human systemic effects by inhalation: anorexia, hypermotility, diarrhea, nausea or vomiting. Experimental teratogenic and reproductive effects. Mutation data reported. Can cause liver and kidney injury, methemoglobinemia, and cyanosis. Very dangerous fire hazard when exposed to heat, open flame, or oxidizers. May explode on heating. Violent reactions with chlorosulfonic acid, oleum. May react with amines + heavy metal oxides (e.g., mercury oxide or silver oxide) to form explosive salts. May ignite on contact with mixtures of carbon + hopcalite which are used in some respirators. Hopcalite is a catalyst consisting of coprecipitated copper(II) oxide and manganese(IV) oxide. To fight fire, use alcohol foam, CO_2, dry chemical, water spray. When heated to decomposition it emits toxic fumes of NO_x.

NJA000   CAS: 5522-43-0   HR: 3
3-NITROPYRENE
mf: $C_{16}H_9NO_2$ mw: 247.26

SYN: 1-NITROPYRENE

CONSENSUS REPORTS: IARC Cancer Review: GROUP 3 IMEMDT 7,56,87; Animal Limited Evidence IMEMDT 33,209,84. Reported in EPA TSCA Inventory.

DFG MAK: Suspected Carcinogen.

SAFETY PROFILE: Suspected carcinogen with experimental carcinogenic, neoplastigenic, and tumorigenic data. Human mutation data reported. When heated to decomposition it emits toxic fumes of NO_x.

NJA500   CAS: 1124-33-0   HR: 3
4-NITROPYRIDINE-N-OXIDE
mf: $C_5H_4N_2O_3$ mw: 140.11

SYN: 4-NITROPYRIDINE-1-OXIDE

CONSENSUS REPORTS: EPA Extremely Hazardous Substances List. EPA Genetic Toxicology Program. Reported in EPA TSCA Inventory.

SAFETY PROFILE: Poison by ingestion and skin contact. Questionable carcinogen with experimental tumorigenic data. Mutation data reported. Mixtures with diethyl-1,4-dihydro-2,6-dimethylpyridine-3,5-dicarboxylate explode when heated above 130°C. When heated to decomposition it emits toxic fumes of NO_x.

NJD500   CAS: 607-35-2   HR: 3
8-NITROQUINOLINE
mf: $C_9H_6N_2O_2$ mw: 174.17

PROP: Monoclinic crystals from alcohol. Mp: 88-89°. Sol in hot water, alc, ether, and benzene.

CONSENSUS REPORTS: EPA Genetic Toxicology Program. Reported in EPA TSCA Inventory.

SAFETY PROFILE: Poison by intraperitoneal route. Questionable carcinogen with experimental carcinogenic data. Mutation data reported. When heated to decomposition it emits toxic fumes of NO_x.

NJF000   CAS: 56-57-5   HR: 3
4-NITROQUINOLINE-N-OXIDE
mf: $C_9H_6N_2O_3$ mw: 190.17

SYNS: 4-NITROCHINOLIN N-OXID (SWEDISH)
* 4-NITROQUINOLINE-1-OXIDE * 4-NQO

CONSENSUS REPORTS: EPA Genetic Toxicology Program.

SAFETY PROFILE: Suspected carcinogen with experimental carcinogenic, neoplastigenic, and tumorigenic data. Poison by intraperitoneal and subcutaneous routes. Experimental reproductive effects. Human mutation data reported. When heated to decomposition it emits toxic fumes of NO_x.

NJH000   HR: 3
NITROSAMINES

PROP: Compounds which have the chemical group $=N—N=O$ attached to an alkyl or aryl group. They are formed by reaction between an amine and NO_x or nitrites.

SAFETY PROFILE: Confirmed carcinogen of the lung, nasal sinus, brain, esophagus, stomach, liver, bladder, and kidney. They are often produced in food as by-products from processing and preparation. They are often produced in food as by-products from processing and preparation. They are found in whisky, herbicides, and cosmetics as well as in tanneries, rubber factories, and iron foundries. They can be formed within the body by reaction of amine-containing foods or drugs with the nitrites resulting from bacterial conversion of nitrates.

NJK500   CAS: 40580-89-0   HR: 3
1-NITROSOAZACYCLOTRIDECANE
mf: $C_{12}H_{24}N_2O$ mw: 212.38

SYNS: NDMI * N-NITROSODODECAMETHYLENE-IMINE * NITROSODODECAMETHYLENIMINE * N-NITROSODODECAMETHYLENIMINE

SAFETY PROFILE: Questionable carcinogen with experimental carcinogenic and tumorigenic data. Mutation data reported. When heated to decomposition it emits toxic fumes of NO_x.

NJL000   CAS: 15216-10-1   HR: 3
1-NITROSOAZETIDINE
mf: $C_3H_6N_2O$ mw: 86.11

SYNS: N-NITROSAZETIDINE * NITROSO-AZETIDIN (GERMAN) * NITROSOAZETIDINE * N-NITROSOAZ-ETIDINE * NITROSOTRIMETHYLENEIMINE * N-NITROSOTRIMETHYLENEIMINE

SAFETY PROFILE: Moderately toxic by ingestion. Questionable carcinogen with experimental tumorigenic data. Mutation data reported. Many N-nitroso compounds are carcinogens. When heated to decomposition it emits toxic fumes of NO_x.

NJM500   CAS: 60414-81-5   HR: 3
N-NITROSOBIS(2-ACETOXYPROPYL) AMINE
mf: $C_{10}H_{18}N_2O_5$ mw: 246.30

SYNS: BAP * 1,1-(N-NITROSOIMINO)DI-2-PROPA-NOL, DIACETATE

SAFETY PROFILE: Questionable carcinogen with experimental carcinogenic, neoplastigenic, and tumorigenic data. Mutation data reported. When heated to decomposition it emits toxic fumes of NO_x.

NJN000   CAS: 60599-38-4   HR: 3
N-NITROSOBIS(2-OXOPROPYL)AMINE
mf: $C_6H_{10}N_2O_3$ mw: 158.18

SYNS: BIS-(2-OXOPROPYL)-N-NITROSAMINE * BOP * DI-OXO-DI-N-PROPYLNITROSAMINE * 2,2'-DIOXO-DI-N-PROPYLNITROSAMINE * 2,2'-DIOXO-N-NITROSODIPROPYLAMINE * N,N-DI(2-OXOPROPYL)NITROSAMINE * 2,2'-DIOXOPROPYL-N-PROPYLNITROSAMINE * DOPN * N-NITROSO-N,N-DI(2-OXYPROPYL)AMINE * (NITROSOIMINO)DIACETONE

SAFETY PROFILE: Suspected carcinogen with experimental carcinogenic, neoplastigenic, tumorigenic, and teratogenic data. Poison by ingestion and subcutaneous routes. Human mutation data reported. When heated to decomposition it emits toxic fumes of NO_x.

NJT500   HR: 3
NITROSO COMPOUNDS

PROP: Compounds of the form $C—N=O$ or $N—N=O$. Organic nitrogen compounds.

SAFETY PROFILE: Usually highly toxic carcinogens, teratogens, and mutagens by almost all routes of exposure. Some of these compounds may have hazardous instabilities under the appropriate conditions. When heated to decomposition they emit very toxic fumes of NO_x.

NJT550   HR: 3
N-NITROSO COMPOUNDS

PROP: A class of organic compounds of the form $R_2—N—N=O$ or $R=N—N=O$.

SAFETY PROFILE: Many members of this class are toxins, carcinogens, teratogens, and mutagens. Sources of exposure to N-nitroso compounds are: formation in the environment and absorption from food, water, air, or industrial and consumer products; formation in the body from precursors in food, water, or air; from tobacco; and from naturally occurring compounds. Some are used in the production of rubber and they may be formed as by-products in industrial processes. Nitrosamines have been found in food and cosmetics. N-nitroso compounds can be formed from the reaction of nitrates with nitrite under acidic conditions. These conditions can occur in the environment, mouth, and stomach. Nitrites are formed in the mouth by the action of bacteria on nitrates. Nitrosatable substances in the environment include secondary and tertiary amines, quaternary ammonium compounds, ureas, carbamates, and guanidines. Many of the resulting N-nitroso compounds are experimental carcinogens and mutagens.

NJW500   CAS: 55-18-5   HR: 3
N-NITROSODIETHYLAMINE
mf: $C_4H_{10}N_2O$ mw: 102.16

PROP: Yellow oil. D: 0.9422 @ 20°/4°, bp: 47° @ 5 mm, bp: 176.9°. Sol in water, alc, and ether.

SYNS: DANA * DEN * DENA * DIAETHYLNI-TROSAMIN (GERMAN) * DIETHYLNITROSAMINE * N,N-DIETHYLNITROSAMINE * DIETHYLNITRO-SOAMINE * N-ETHYL-N-NITROSO-ETHANAMINE * NDEA * N-NITROSODIAETHYLAMINE (GERMAN) * NITROSODIETHYLAMINE * RCRA WASTE NUM-BER U174

CONSENSUS REPORTS: IARC Cancer Review: GROUP 2A IMEMDT 7,56,87; Animal Sufficient Evidence IMEMDT 1,107,72, IMEMDT 17,83,78, IMEMDT 28,151,82; Human Limited Evidence IMEMDT 17,83,78. NTP Fourth Annual Report On Carcinogens, 1984. NCI Carcinogenesis Studies (ipr); Clear Evidence: mouse, rat RRCRBU 52,1,75. Reported in EPA TSCA Inventory. Community Right-To-Know List.

DFG MAK: Animal Carcinogen, Suspected Human Carcinogen.

SAFETY PROFILE: Confirmed carcinogen with experimental carcinogenic, neoplastigenic, and tumorigenic data. Poison by ingestion, intravenous, intraperitoneal, subcutaneous, and possibly other routes. Experimental teratogenic and reproductive effects. Human mutation data reported. A transplacental carcinogen. When heated to decomposition it emits toxic fumes of NO_x.

NJY000   CAS: 16813-36-8   HR: 3
1-NITROSO-5,6-DIHYDROURACIL
mf: $C_4H_5N_3O_3$ mw: 143.12

SYNS: DIHYDRO-1-NITROSO-2,4(1H,3H)-PYRIMIDINE-DIONE * 5,6-DIHYDRO-1-NITROSOURACIL * NDHU * NO-DHU

CONSENSUS REPORTS: EPA Genetic Toxicology Program.

SAFETY PROFILE: Moderately toxic by intraperitoneal route. Questionable carcinogen with experimental carcinogenic and tumorigenic data. Mutation data reported. When heated to decomposition it emits toxic fumes of NO_x.

NKA000   CAS: 601-77-4   HR: 3
N-NITROSODIISOPROPYLAMINE
mf: $C_6H_{14}N_2O$ mw: 130.22

SYNS: DIISOPROPYLNITROSAMIN (GERMAN) * N-NITROSODI-i-PROPYLAMINE (MAK)

DFG MAK: Animal Carcinogen, Suspected Human Carcinogen.

SAFETY PROFILE: Confirmed carcinogen with experimental carcinogenic and tumorigenic data. Moderately toxic by ingestion. When heated to decomposition it emits toxic fumes of NO_x.

NKA600   CAS: 62-75-9   HR: 3
N-NITROSODIMETHYLAMINE
mf: $C_2H_6N_2O$ mw: 74.10

PROP: Yellow liquid; sol in water, alc, and ether. Bp: 152°, d: 1.005 @ 20°/4°.

SYNS: DIMETHYLNITROSAMIN (GERMAN) * DI-METHYLNITROSAMINE * N,N-DIMETHYLNITROSA-MINE * DIMETHYLNITROSOAMINE * DMN * DMNA * N-METHYL-N-NITROSOMETHANAMINE * NDMA * NITROSODIMETHYLAMINE * RCRA WASTE NUMBER P082

CONSENSUS REPORTS: IARC Cancer Review: GROUP 2A IMEMDT 7,56,87; Animal Sufficient Evidence IMEMDT 17,125,78, IMEMDT 1,95,72; Human Limited Evidence IMEMDT 17,125,78; Human Inadequate Evidence IMEMDT 1,95,72. Reported in EPA TSCA Inventory. EPA Genetic Toxicology Program. Community Right-To-Know List. EPA Extremely Hazardous Substances List.

OSHA PEL: Carcinogen
ACGIH TLV: Suspected Human Carcinogen
DFG MAK: Animal Carcinogen, Suspected Human Carcinogen.

SAFETY PROFILE: Confirmed carcinogen with experimental carcinogenic, neoplastigenic, and tumorigenic data. A transplacental carcinogen. Human poison by ingestion. Experimental poison by ingestion, inhalation, intraperitoneal, subcutaneous, intravenous, and possibly other routes. Human systemic effects by ingestion: ulceration or bleeding from small intestine, nausea or vomiting, and fever. Experimental teratogenic and reproductive effects. Human mutation data reported. Has caused fatal liver disease in humans. When heated to decomposition it emits toxic fumes of NO_x.

NKB500   CAS: 156-10-5   **HR: 3**
p-NITROSODIPHENYLAMINE
mf: $C_{12}H_{10}N_2O$ mw: 198.24

PROP: Green plates with bluish luster (from benzene) or steel blue prisms or plates (from ether + H_2O). Mp: 144-145°. Sltly sol in water or petr ether; very sol in alc, ether, benzene, chloroform.

SYNS: NAUGARD TKB * NCI-C02244 * p-NITRO-SODIFENYLAMIN (CZECH) * 4-NITROSODIPHENYL-AMINE * p-NITROSO-N-PHENYLANILINE * 4-NI-TROSO-N-PHENYLANILINE * 4-NITROSO-N-PHENYL-BENZENAMINE * N-PHENYL-p-NITROSOANILINE * TKB

CONSENSUS REPORTS: IARC Cancer Review: GROUP 3 IMEMDT 7,56,87; Animal Inadequate Evidence IMEMDT 27,227,82. NTP Fourth Annual Report On Carcinogens, 1984. NCI Carcinogenesis Bioassay (feed); Clear Evidence: mouse, rat NCITR* NCI-CG-TR-190,79. Community Right-To-Know List. Reported in EPA TSCA Inventory.

SAFETY PROFILE: Confirmed carcinogen with experimental carcinogenic and neoplastigenic data. Poison by intravenous route. Moderately toxic by ingestion. Mutation data reported. An eye irritant. When heated to decomposition it emits toxic fumes of NO_x.

NKB700   CAS: 621-64-7   **HR: 3**
N-NITROSODI-N-PROPYLAMINE
mf: $C_6H_{14}N_2O$ mw: 130.22

SYNS: DI-n-PROPYLNITROSAMINE * DIPROPYL-NITROSOAMINE * DPN * DPNA * NDPA * N-NITROSODIPROPYLAMINE * N-NITROSO-N-DI-PROPYLAMINE * N-NITROSO-N-PROPYL-1-PROPAN-AMINE * N-NITROSO-N-PROPYLPROPANAMINE * RCRA WASTE NUMBER U111

CONSENSUS REPORTS: IARC Cancer Review: GROUP 2B IMEMDT 7,56,87; Animal Sufficient Evidence IMEMDT 17,177,78; Human Limited Evidence IMEMDT 17,177,78.

NTP Fourth Annual Report On Carcinogens, 1984. EPA Genetic Toxicology Program. Community Right-To-Know List. Reported in EPA TSCA Inventory.

DFG MAK: Animal Carcinogen, Suspected Human Carcinogen.

SAFETY PROFILE: Confirmed carcinogen with experimental carcinogenic, neoplastigenic, and tumorigenic data. Moderately toxic by ingestion and subcutaneous routes. Experimental teratogenic and reproductive effects. Human mutation data reported. When heated to decomposition it emits toxic fumes of NO_x.

NKC000   CAS: 17608-59-2   **HR: 3**
N-NITROSOEPHEDRINE
mf: $C_{10}H_{14}N_2O_2$ mw: 194.26

SYNS: α-(1-(N-METHYL-N-NITROSOAMINO)ETHYL) BENZYL ALCOHOL * 2-(N-METHYL-N-NITROSO-AMINO)-1-PHENYL-1-PROPANOL

SAFETY PROFILE: Suspected carcinogen with experimental carcinogenic data. Poison by intraperitoneal route. Mutation data reported. When heated to decomposition it emits toxic fumes of NO_x.

NKD000   CAS: 612-64-6   **HR: 3**
N-NITROSO-N-ETHYL ANILINE
mf: $C_8H_{10}N_2O$ mw: 150.20

PROP: Yellow oil. D: 1.087 @ 20°/4°, bp: 119-120° @ 15 mm. Insol in water.

SYNS: ETHYLNITROSOANILINE * N-ETHYL-N-NI-TROSOBENZENAMINE * NEA * NITROSOETHYL-ANILINE * N-NITROSOETHYLPHENYLAMINE (MAK)

CONSENSUS REPORTS: EPA Genetic Toxicology Program.

DFG MAK: Animal Carcinogen, Suspected Human Carcinogen.

SAFETY PROFILE: Confirmed carcinogen. Poison by ingestion and intraperitoneal routes data. Experimental teratogenic and reproductive effects. Mutation data reported. Many N-nitroso compounds are carcinogens. When heated to decomposition it emits toxic fumes of NO_x.

NKE500   CAS: 614-95-9   **HR: 3**
N-NITROSO-N-ETHYLURETHAN
mf: $C_5H_{10}N_2O_3$ mw: 146.17

SYNS: AETHYLNITROSOURETHAN (GERMAN) * ENU * ETHYLNITROSOCARBAMIC ACID, ETHYL

ESTER * N-ETHYL-N-NITROSOCARBAMIC ACID ETHYL ESTER * N-ETHYL-N-NITROSOURETHANE * NEU * NITROSOETHYLURETHAN

CONSENSUS REPORTS: EPA Genetic Toxicology Program.

SAFETY PROFILE: Poison by intravenous route. Experimental reproductive effects. Questionable carcinogen with experimental carcinogenic and tumorigenic data. Mutation data reported. When heated to decomposition it emits toxic fumes of NO_x.

NKF000   CAS: 13256-13-8   **HR: 3**
N-NITROSO-N-ETHYLVINYLAMINE
mf: $C_4H_8N_2O$ mw: 100.14

SYNS: AETHYL-VINYL-NITROSOAMIN (GERMAN) * N-ETHYL-N-NITROSOETHENAMINE * N-ETHYL-N-NITROSOETHENYLAMINE * N-ETHYL-N-NITROSOVINYLAMINE * ETHYLVINYLNITROSAMINE * N-NITROSOETHYLVINYLAMINE * VINYLETHYLNITROSAMIN (GERMAN) * VINYLETHYLNITROSAMINE

CONSENSUS REPORTS: EPA Genetic Toxicology Program.

SAFETY PROFILE: Poison by ingestion, intravenous, and subcutaneous routes. Experimental teratogenic and reproductive effects. Questionable carcinogen with experimental carcinogenic and tumorigenic data. When heated to decomposition it emits toxic fumes of NO_x.

NKF500   CAS: 2508-20-5   **HR: 3**
2-NITROSOFLUORENE
mf: $C_{13}H_9NO$ mw: 195.23

CONSENSUS REPORTS: EPA Genetic Toxicology Program.

SAFETY PROFILE: Questionable carcinogen with experimental carcinogenic and neoplastigenic data. Mutation data reported. When heated to decomposition it emits toxic fumes of NO_x.

NKH000   CAS: 674-81-7   **HR: 3**
NITROSOGUANIDINE
mf: CH_4N_4O mw: 88.09

PROP: A solid.

SYNS: INITIATING EXPLOSIVE NITROSOGUANIDINE (DOT) * NITROSOGUANIDIN (GERMAN) * N-NITROSOGUANIDINE

DOT Classification: Class A Explosive; Label: Explosive A.

SAFETY PROFILE: Poison by intraperitoneal route. Many N-nitroso compounds are carcinogens. An explosive. Dangerous when stored in sealed containers as it decomposes to release nitrogen. When heated to decomposition it emits toxic fumes of NO_x.

NKI000   CAS: 932-83-2   **HR: 3**
N-NITROSOHEXAHYDROAZEPINE
mf: $C_6H_{12}N_2O$ mw: 128.20

SYNS: HEXAHYDRO-1-NITROSO-1H-AZEPINE * N-6-MI * N-NITROSOAZACYCLOHEPTANE * NITROSOHEXAMETHYLENIMINE * N-NITROSO-HEXAMETHYLENEIMINE * N-NITROSOPERHY-DROAZEPINE

SAFETY PROFILE: Suspected carcinogen with experimental carcinogenic and tumorigenic data. Poison by ingestion, intraperitoneal, and subcutaneous routes. Experimental teratogenic and reproductive effects. Mutation data reported. When heated to decomposition it emits toxic fumes of NO_x.

NKK500   CAS: 3715-92-2   **HR: 3**
N-NITROSOIMIDAZOLIDINETHIONE
mf: $C_3H_5ON_3S$ mw: 131.17

SYNS: N-NITROSOETHYLENETHIOUREA * NO-ETU

CONSENSUS REPORTS: EPA Genetic Toxicology Program.

SAFETY PROFILE: Poison by ingestion. Experimental teratogenic and reproductive effects. Questionable carcinogen with experimental neoplastigenic data. Mutation data reported. Many N-nitroso compounds are carcinogens. When heated to decomposition it emits very toxic fumes of NO_x and SO_x.

NKL000   CAS: 3844-63-1   **HR: 3**
1-NITROSOIMIDAZOLIDINONE
mf: $C_3H_5N_3O_2$ mw: 115.11

SYNS: ETHYLENENITROSOUREA * N-NITRO-2-IMIDAZOLIDONE * 1-NITRO-2-IMIDAZOLIDONE * N-NITROSO-IMIDAZOLIDON (GERMAN) * 1-NITROSO-2-IMIDAZOLIDINONE * N-NITROSOIMIDAZOLIDONE * NSC 73438 * SRI 1869

CONSENSUS REPORTS: EPA Genetic Toxicology Program.

SAFETY PROFILE: Poison by intraperitoneal and subcutaneous routes. Mutation data reported. Questionable carcinogen with experimental tumorigenic data. Many N-nitroso compounds are carcinogens. When heated to decomposition it emits toxic fumes of NO_x.

NKM000   CAS: 1116-54-7   **HR: 3**
NITROSOIMINO DIETHANOL
mf: $C_4H_{10}N_2O_3$ mw: 134.16

SYNS: BIS(β-HYDROXYAETHYL)NITROSAMIN (GERMAN) * BIS(β-HYDROXYETHYL)NITROSAMINE * DIAETHANOLNITROSAMIN (GERMAN) * DIETHANOLNITROSOAMINE * 2,2'-DIHYDROXY-N-NITROSODIETHYLAMINE * 2,2'-IMINODI-N-NITROSOETHANOL * NCI-C55583 * NDELA * N-NITROSOAMINODIETHANOL * N-NITROSOBIS(2-HYDROXYETHYL)AMINE * N-NITROSODIETHANOLAMIN (GERMAN) * N-NITROSODIETHANOLAMINE (MAK) * 2,2'-(NITROSOIMINO)BISETHANOL * RCRA WASTE NUMBER U173

CONSENSUS REPORTS: IARC Cancer Review: GROUP 2B IMEMDT 7,56,87; Animal Sufficient Evidence IMEMDT 17,77,78; Human Limited Evidence IMEMDT 17,77,78. NTP Fourth Annual Report On Carcinogens, 1984. Reported in EPA TSCA Inventory.

DFG MAK: Animal Carcinogen, Suspected Human Carcinogen.

SAFETY PROFILE: Confirmed carcinogen with experimental carcinogenic, neoplastigenic, and tumorigenic data. Mildly toxic by ingestion. Mutation data reported. When heated to decomposition it emits toxic fumes of NO_x.

NKO400   CAS: 71752-69-7   **HR: 3**
NITROSOISOPROPANOLUREA
mf: $C_4H_9N_3O_3$ mw: 147.16

SYNS: 1-(2-HYDROXYPROPYL)-1-NITROSOUREA * NITROSO-2-HYDROXY-N-PROPYLUREA * N-NITROSO-2-HYDROXY-N-PROPYLUREA

SAFETY PROFILE: Suspected carcinogen with experimental carcinogenic and tumorigenic data. Mutation data reported. When heated to decomposition it emits toxic fumes of NO_x.

NKT500   CAS: 5432-28-0   **HR: 3**
N-NITROSO-N-METHYLCYCLOHEXYLAMINE
mf: $C_7H_{14}N_2O$ mw: 142.23

SYNS: METHYLCYCLOHEXYLNITROSAMIN (GERMAN) * METHYLCYCLOHEXYLNITROSAMINE * N-METHYL-N-NITROSOCYCLOHEXYLAMINE * N-NITROSOMETHYLCYCLOHEXYLAMINE

SAFETY PROFILE: Poison by ingestion, intravenous, and intraperitoneal routes. Questionable carcinogen with experimental tumorigenic data. Mutation data reported. Many N-nitroso compounds are carcinogens. When heated to decomposition it emits toxic fumes of NO_x.

NKU000   CAS: 55090-44-3   **HR: 3**
NITROSOMETHYL-n-DODECYLAMINE
mf: $C_{13}H_{28}N_2O$ mw: 228.43

SYNS: N-METHYL-N-NITROSOLAURYLAMINE * N-NITROSO-N-METHYL-N-DODECYLAMIN (GERMAN) * N-NITROSO-N-METHYL-N-DODECYLAMINE * NMDDA

CONSENSUS REPORTS: EPA Genetic Toxicology Program.

SAFETY PROFILE: Suspected carcinogen with experimental carcinogenic and tumorigenic data. Moderately toxic by ingestion. Mutation data reported. When heated to decomposition it emits toxic fumes of NO_x.

NKU875   CAS: 35631-27-7   **HR: 3**
NITROSO-5-METHYLOXAZOLIDONE
mf: $C_4H_8N_2O_2$ mw: 116.14

SYNS: 5-METHYL-3-NITROSO-1,3-OXAZOLIDINE * NITROSO-5-METHYL-1,3-OXAZOLIDINE * N-NITROSO-5-METHYL-1,3-OXAZOLIDINE

SAFETY PROFILE: Suspected carcinogen with experimental carcinogenic and tumorigenic data. Mutation data reported. When heated to decomposition it emits toxic fumes of NO_x.

NKV000   CAS: 55984-51-5   **HR: 3**
N-NITROSOMETHYL-2-OXOPROPYLAMINE
mf: $C_4H_8N_2O_2$ mw: 116.14

SYNS: 1-(METHYLNITROSOAMINO)2-PROPANONE * MOP * NMOP

SAFETY PROFILE: Suspected carcinogen with experimental carcinogenic and tumorigenic data. Poison by subcutaneous route. Mutation data reported. When heated to decomposition it emits toxic fumes of NO_x.

NKW500   CAS: 16339-07-4   **HR: 3**
1-NITROSO-4-METHYLPIPERAZINE
mf: $C_5H_{11}N_3O$ mw: 129.19

SYNS: N'-METHYL-N-NITROSOPIPERAZINE * 1-METHYL-4-NITROSOPIPERAZINE * N-NITROSO-

N'-METHYLPIPERAZIN (GERMAN) * N-NITROSO-N'-
METHYLPIPERAZINE

SAFETY PROFILE: Poison by ingestion. Questionable carcinogen with experimental tumorigenic data. Mutation data reported. Many N-nitroso compounds are carcinogens. When heated to decomposition it emits toxic fumes of NO_x.

NKY000   CAS: 4549-40-0   HR: 3
N-NITROSOMETHYLVINYLAMINE
mf: $C_3H_6N_2O$ mw: 86.11

SYNS: N-METHYL-N-NITROSO-ETHENYLAMINE
* N-METHYL-N-NITROSOVINYLAMINE * METHYL-
VINYLNITROSAMIN (GERMAN) * METHYLVINYLNI-
TROSAMINE * MVNA * NMVA * RCRA WASTE
NUMBER P084

CONSENSUS REPORTS: IARC Cancer Review: GROUP 2B IMEMDT 7,56,87; Animal Sufficient Evidence IMEMDT 17,257,78; Human Limited Evidence IMEMDT 17,257,78. Community Right-To-Know List. EPA Genetic Toxicology Program.

SAFETY PROFILE: Suspected carcinogen with experimental tumorigenic data. Poison by ingestion and inhalation. When heated to decomposition it emits toxic fumes of NO_x.

NKZ000   CAS: 59-89-2   HR: 3
4-NITROSOMORPHOLINE
mf: $C_4H_8N_2O_2$ mw: 116.14

SYNS: N-NITROSOMORPHOLIN (GERMAN) * NITRO-
SOMORPHOLINE * N-NITROSOMORPHOLINE (MAK)
* NMOR

CONSENSUS REPORTS: IARC Cancer Review: GROUP 2B IMEMDT 7,56,87; Animal Sufficient Evidence IMEMDT 17,263,78; Human Limited Evidence IMEMDT 17,263,78. Community Right-To-Know List. EPA Genetic Toxicology Program.

DFG MAK: Animal Carcinogen, Suspected Human Carcinogen.

SAFETY PROFILE: Confirmed carcinogen with experimental carcinogenic, neoplastigenic, and tumorigenic data. Poison by ingestion, intraperitoneal, subcutaneous, and intravenous routes. Moderately toxic by inhalation. Human mutation data reported. When heated to decomposition it emits toxic fumes of NO_x.

NLD500   CAS: 16543-55-8   HR: 3
N'-NITROSONORNICOTINE
mf: $C_9H_{11}N_3O$ mw: 177.23

SYNS: 1'-NITROSO-1'-DEMETHYLNICOTINE
* 1-NITROSO-2-(3-PYRIDYL)PYRROLIDINE * 3-(1-NI-
TROSO-2-PYRROLIDINYL)PYRIDINE * (s)-3-(1-NI-
TROSO-2-PYRROLIDINYL)PYRIDINE * NNN

CONSENSUS REPORTS: IARC Cancer Review: GROUP 2B IMEMDT 7,56,87; Animal Sufficient Evidence IMEMDT 17,281,78; Human Limited Evidence IMEMDT 17,281,78. NTP Fourth Annual Report On Carcinogens, 1984. Community Right-To-Know List. EPA Genetic Toxicology Program.

SAFETY PROFILE: Confirmed carcinogen with experimental carcinogenic, neoplastigenic, and tumorigenic data. Mutation data reported. When heated to decomposition it emits toxic fumes of NO_x.

NLE000   CAS: 39884-52-1   HR: 3
N-NITROSOOXAZOLIDINE
mf: $C_3H_6N_2O_2$ mw: 102.11

SYNS: N-NITROSOOXAZOLIDIN (GERMAN) * NI-
TROSOOXAZOLIDONE * N-NITROSO-1,3-OXAZOLI-
DINE * 3-NITROSOOXAZOLIDINE

SAFETY PROFILE: Suspected carcinogen with experimental carcinogenic and tumorigenic data. Moderately toxic by ingestion. Mutation data reported. When heated to decomposition it emits toxic fumes of NO_x.

NLF200   CAS: 104-91-6   HR: 3
NITROSOPHENOL
mf: $C_6H_5NO_2$ mw: 123.12

PROP: Pale yellow, orthorhombic needles. Mp: 144° (decomp). Sltly sol in water; sol in dilute alkalies, alc, ether, and acetone.

SYNS: p-NITROSOPHENOL * 4-NITROSOPHENOL
* QUINONE MONOXIME * QUINONE OXIME

CONSENSUS REPORTS: Reported in EPA TSCA Inventory.

SAFETY PROFILE: Poison by parenteral and intraperitoneal routes. Mutation data reported. An irritant and sensitizer. Many nitroso compounds are carcinogens. A very dangerous fire and explosion hazard. When exposed to heat or flame, it burns explosively. Contamination by acid or alkali may cause ignition. Can heat spontaneously and cause fire. When heated to decomposition it emits toxic fumes of NO_x.

NLH000   CAS: 13256-23-0   **HR: 3**
N-NITROSO-4-PICOLYLETHYLAMINE
mf: $C_8H_{11}N_3O$ mw: 165.22

SYNS: AETHYL-4-PICOLYLNITROSAMIN (GERMAN)
* 4-((ETHYLNITROSAMINO)METHYL)PYRIDINE

SAFETY PROFILE: Poison by ingestion and intravenous routes. Questionable carcinogen with experimental tumorigenic data. Many N-nitroso compounds are carcinogens. When heated to decomposition it emits toxic fumes of NO_x.

NLJ500   CAS: 100-75-4   **HR: 3**
N-NITROSOPIPERIDINE
mf: $C_5H_{10}N_2O$ mw: 114.17

PROP: Light yellow oil. D: 1.063 @ 18.5°/4°, bp: 217-218°. Sol in water; very sol in acid solns.

SYNS: HEXAHYDRO-N-NITROSOPYRIDINE * NITROSOPIPERIDIN (GERMAN) * N-NITROSO-PIPERIDIN (GERMAN) * 1-NITROSOPIPERIDINE * N-N-PIP * NPIP * NO-PIP * RCRA WASTE NUMBER U179

CONSENSUS REPORTS: IARC Cancer Review: GROUP 2B IMEMDT 7,56,87; Human Limited Evidence IMEMDT 17,287,78; Animal Sufficient Evidence IMEMDT 17,287,78; IMEMDT 28,151,82. NTP Fourth Annual Report On Carcinogens, 1984. Community Right-To-Know List. EPA Genetic Toxicology Program. Reported in EPA TSCA Inventory.

DFG MAK: Animal Carcinogen, Suspected Human Carcinogen.

SAFETY PROFILE: Confirmed carcinogen with experimental carcinogenic, neoplastigenic, and tumorigenic data. Poison by ingestion, intravenous, and subcutaneous routes. Experimental reproductive effects. Human mutation data reported. When heated to decomposition it emits toxic fumes of NO_x.

NLM500   CAS: 39603-53-7   **HR: 3**
1-(NITROSOPROPYLAMINO)-2-PROPANOL
mf: $C_6H_{14}N_2O_2$ mw: 146.22

SYNS: β-HYDROXYPROPYLPROPYLNITROSAMINE
* (2-HYDROXYPROPYL)PROPYLNITROSOAMINE
* N-NITROSO-2-HYDROXY-N-PROPYL-N-PROPYLAMINE

CONSENSUS REPORTS: IARC Cancer Review: Animal Sufficient Evidence IMEMDT 17,177,78.

SAFETY PROFILE: Confirmed carcinogen with experimental carcinogenic and tumorigenic data. Moderately toxic by subcutaneous route. Experimental reproductive effects. Mutation data reported. When heated to decomposition it emits toxic fumes of NO_x.

NLO500   CAS: 816-57-9   **HR: 3**
N-NITROSO-N-PROPYLUREA
mf: $C_4H_9N_3O_2$ mw: 131.16

SYNS: NITROSOPROPYLUREA * NITROSO-N-PROPYLUREA * NPU * PNU * N-PROPYLNITROSOHARNSTOFF (GERMAN) * N-PROPYLNITROSUREA * 1-PROPYL-1-NITROSOUREA

CONSENSUS REPORTS: EPA Genetic Toxicology Program.

SAFETY PROFILE: Suspected carcinogen with experimental carcinogenic, neoplastigenic, and tumorigenic data. Moderately toxic by ingestion. Experimental teratogenic and reproductive effects. Mutation data reported. When heated to decomposition it emits toxic fumes of NO_x.

NLP500   CAS: 930-55-2   **HR: 3**
N-NITROSOPYRROLIDINE
mf: $C_4H_8N_2O$ mw: 100.14

SYNS: N-NITROSOPYRROLIDIN (GERMAN) * 1-NITROSOPYRROLIDINE * NO-PYR * N-N-PYR * NPYR * RCRA WASTE NUMBER U180 * TETRAHYDRO-N-NITROSOPYRROLE

CONSENSUS REPORTS: IARC Cancer Review: GROUP 2B IMEMDT 7,56,87; Animal Sufficient Evidence IMEMDT 17,313,78; Human Limited Evidence IMEMDT 17,313,78. NTP Fourth Annual Report On Carcinogens, 1984. EPA Genetic Toxicology Program. Reported in EPA TSCA Inventory.

DFG MAK: Animal Carcinogen, Suspected Human Carcinogen.

SAFETY PROFILE: Confirmed carcinogen with experimental carcinogenic, neoplastigenic, and tumorigenic data. Poison by ingestion and subcutaneous routes. Human mutation data reported. When heated to decomposition it emits toxic fumes of NO_x.

NLR500   CAS: 13256-22-9   **HR: 3**
N-NITROSOSARCOSINE
mf: $C_3H_6N_2O_3$ mw: 118.11

SYNS: N-METHYL-N-NITROSOGLYCINE * N-NITROSOMETHYLGLYCINE * NITROSO SARKOSIN (GERMAN)

CONSENSUS REPORTS: IARC Cancer Review: GROUP 2B IMEMDT 7,56,87; Animal Sufficient Evidence IMEMDT 17,327,78; Human Limited Evidence IMEMDT 17,327,78. NTP Fourth Annual Report On Carcinogens, 1984.

SAFETY PROFILE: Confirmed carcinogen with experimental tumorigenic data. Mildly toxic by ingestion. When heated to decomposition it emits toxic fumes of NO_x.

NLY750   CAS: 88208-15-5   **HR: 3**
NITROSO-3,4,5-TRIMETHYLPIPERAZINE
mf: $C_7H_{15}N_3O$ mw: 157.25

SYNS: N-NITROSO-3,4,5-TRIMETHYLPIPERAZINE
* 1-NITROSO-3,4,5-TRIMETHYLPIPERAZINE

SAFETY PROFILE: Suspected carcinogen with experimental carcinogenic and tumorigenic data. Mutation data reported. When heated to decomposition it emits toxic fumes of NO_x.

NMB000   CAS: 9056-38-6   **HR: 3**
NITROSTARCH

DOT: UN 0146

PROP: Solid.

SYNS: NITROSTARCH, containing less than 20% water
(DOT) * NITROSTARCH, DRY (DOT)

CONSENSUS REPORTS: Reported in EPA TSCA Inventory.

DOT Classification: Class A Explosive; Label: Explosive A.

SAFETY PROFILE: A very dangerous fire and explosion hazard when exposed to heat, flame, shock, or oxidizers. It is a powerful high explosive. Nitrostarch is not a definite compound, but a mixture of various nitric acid esters of starch with different degrees of nitration. When heated to decomposition it emits toxic fumes of NO_x.

NMH000   CAS: 2696-92-6   **HR: 3**
NITROSYL CHLORIDE

DOT: UN 1069
mf: ClNO   mw: 65.46

PROP: Yellow gas, irritating odor. Mp: $-64.5°$, bp: $-5.8°$, d (liquid): 1.250 @ 30°, vap d: 2.3, vap press: 76 mm @ 50°. Nonexplosive, very corrosive. Liquid @ $-5.5°$, solid @ $-61.5°$. Sol in fuming H_2SO_4.

SYN: NITROGEN OXYCHLORIDE

CONSENSUS REPORTS: Reported in EPA TSCA Inventory.

DOT Classification: Nonflammable Gas; Label: Nonflammable Gas; Poison A; Label: Poison Gas, Corrosive.

SAFETY PROFILE: Poison by inhalation and ingestion. A corrosive irritant to skin, eyes, and mucous membranes. Inhalation may cause pulmonary edema and hemorrhage. Potentially explosive reaction with acetone + platinum. Mixtures with hydrogen + oxygen ignite spontaneously. When heated to decomposition it emits very toxic fumes of Cl^- and NO_x.

NMH500   CAS: 7789-25-5   **HR: 3**
NITROSYL FLUORIDE
mf: FNO   mw: 49.01

PROP: Colorless gas, often bluish because of impurities. Mp: $-132.5°$, bp: $-59.9°$; d: (liquid @ bp) 1.326, (solid) 1.719.

SYN: NITROGEN OXYFLUORIDE

SAFETY PROFILE: A poison. A severe irritant to skin, eyes, and mucous membranes. Reacts vigorously with glass, corrodes quartz. Explosive reaction with alkenes, oxygen difluoride. Incandescent reaction with metals (e.g., antimony, bismuth, tin, sodium), non-metals (e.g., arsenic, boron, red phosphorus, silicon). When heated to decomposition it emits highly toxic fumes of F^- and NO_x.

NMI000   CAS: 15605-28-4   **HR: 3**
NITROSYL PERCHLORATE
mf: $ClNO_5$ mw: 129.46

SAFETY PROFILE: Below 100° it slowly decomposes; at about 115-120°C it speeds up and explodes. A powerful oxidizer. Explosive reaction with pentaammineazidocobalt(III) perchlorate + phenyl isocyanate; organic materials (e.g., pinene; acetone; ethanol; ether); primary aromatic amines (e.g., aniline; toluidines; xylidines; mesidine). Ignition on mixing with urea. Incompatible with metal salts. When heated to decomposition it emits very toxic fumes of NO_x and Cl^-.

NMJ000   CAS: 7782-78-7   **HR: 3**
NITROSYLSULFURIC ACID

DOT: UN 2308
mf: HNO_5S mw: 127.07

PROP: Prisms. Decomp @ 73.5°. Sol in sulfuric acid, decomp in water.

SYNS: CHAMBER CRYSTALS * NITRO ACID SULFITE * NITROSONIUM BISULFITE * NITROSYL HYDROGEN SULFATE * NITROSYL SULFATE * SULFURIC ACID, MONOANHYDRIDE with NITROUS ACID

CONSENSUS REPORTS: Reported in EPA TSCA Inventory.

DOT Classification: Corrosive Material; Label: Corrosive.

SAFETY PROFILE: A poison. A corrosive irritant to skin, eyes, and mucous membranes. Explosive reaction above 50°C with 2-chloro-4,6-dinitroaniline and 4-chloro-2,6-dinitroaniline. Potentially explosive reaction with dinitroaniline. When heated to decomposition it emits toxic fumes of SO_x and NO_x.

NML000   CAS: 61-57-4   **HR: 3**
NITROTHIAZOLE
mf: $C_6H_6N_4O_3S$ mw: 214.22

SYNS: AMBILHAR * BA 32644 * BA 32644 CIBA * CIBA 32644 * CIBA 32644-BA * NIRIDAZOLE * NITRIDAZOLE * NITROTHIAMIDAZOL * NITROTHIAMIDAZOLE * 1-(5-NITRO-2-THIAZOLYL)IMIDAZOLIDIN-2-ONE * 1-(5-NITRO-2-THIAZOLYL)-2-IMIDAZOLIDINONE * 1-(5-NITRO-2-THIAZOLYL)-2-IMIDAZOLINONE * 1-(5-NITRO-2-THIAZOLYL)-2-OXOTETRAHYDROIMIDAZOL * 1-(5-NITRO-2-THIAZOLYL)-2-OXOTETRAHYDROIMIDAZOLE * NTOI

CONSENSUS REPORTS: IARC Cancer Review: GROUP 2B IMEMDT 7,56,87; Animal Sufficient Evidence IMEMDT 13,123,77. EPA Genetic Toxicology Program.

SAFETY PROFILE: Suspected carcinogen with experimental carcinogenic and neoplastigenic data. Poison by intraperitoneal route. Moderately toxic by ingestion. Experimental reproductive effects. Human mutation data reported. Used as an amoebicide and schistosomicidal agent. When heated to decomposition it emits very toxic fumes of SO_x and NO_x.

NMO500   CAS: 99-08-1   **HR: 3**
m-NITROTOLUENE

DOT: UN 1664
mf: $C_7H_7NO_2$ mw: 137.15

PROP: Liquid. Mp: 15.1°, flash p: 233°F (CC), d: 1.1630 @ 15°/4°, vap press: 1 mm @ 50.2°, vap d: 4.72, bp: 231.9°. Misc with alc, ether; sol in benzene; sol in water @ 30°.

SYNS: 3-METHYLNITROBENZENE * m-METHYLNITROBENZENE * MNT * 3-NITROTOLUENE * 3-NITROTOLUOL

CONSENSUS REPORTS: Reported in EPA TSCA Inventory.

OSHA PEL: (Transitional: TWA 5 ppm (skin)) TWA 2 ppm (skin)
ACGIH TLV: TWA 2 ppm (skin)
DFG MAK: 5 ppm (30 mg/m³)
DOT Classification: Poison B; Label: Poison.

SAFETY PROFILE: Poison by ingestion. Combustible when exposed to heat, flame, or oxidizers. To fight fire, use water, CO_2, dry chemical. Probably an explosive. When heated to decomposition it emits toxic fumes of NO_x.

NMO525   CAS: 88-72-2   **HR: 3**
o-NITROTOLUENE

DOT: UN 1664
mf: $C_7H_7NO_2$ mw: 137.15

PROP: Yellowish liquid. Mp: −10°, bp: 222.3°, flash p: 223°F (CC), d: 1.1622 @ 19°/15°, vap press: 1 mm @ 50°, vap d: 4.72. Insol in water; sol in SO_2 and petr ether; misc in alc, benzene, and ether. Sltly sol in NH_3.

SYNS: 2-METHYLNITROBENZENE * o-METHYLNITROBENZENE * 2-NITROTOLUENE * ONT

CONSENSUS REPORTS: Reported in EPA TSCA Inventory.

OSHA PEL: (Transitional: TWA 5 ppm (skin)) TWA 2 ppm (skin)
ACGIH TLV: TWA 2 ppm (skin)
DFG MAK: 5 ppm (30 mg/m³)
DOT Classification: Poison B; Label: Poison.

SAFETY PROFILE: A poison. Moderately toxic by ingestion. Mucous membrane effects by inhalation. Mutation data reported. Combustible when exposed to heat or open flame. To fight fire, use water spray, fog, foam, CO_2. Potentially explosive reaction with alkali (e.g., sodium hydroxide). When heated to decomposition it emits toxic fumes of NO_x.

NMO550   CAS: 99-99-0   **HR: 3**
p-NITROTOLUENE

DOT: UN 1664
mf: $C_7H_7NO_2$ mw: 137.15

PROP: Yellowish crystals. Bp: 238.3°, flash p: 223°F (CC), d: 1.286, vap press: 1 mm @

53.7°, vap d: 4.72. Mp: 53-54°. Insol in water; sol in alc, benzene, ether, chloroform, and acetone.

SYNS: 4-METHYLNITROBENZENE * p-METHYL NITROBENZENE * NCI-C60537 * 4-NITROTOLUENE * 4-NITROTOLUOL * PNT

CONSENSUS REPORTS: Reported in EPA TSCA Inventory.

OSHA PEL: (Transitional: TWA 5 ppm (skin)) TWA 2 ppm (skin)
ACGIH TLV: TWA 2 ppm (skin)
DFG MAK: 5 ppm (30 mg/m^3)
DOT Classification: Poison B; Label: Poison.

SAFETY PROFILE: A poison. Moderately toxic by ingestion and intraperitoneal routes. Mildly toxic by skin contact. Mutation data reported. Combustible when exposed to heat or flame. To fight fire, use CO_2, dry chemical, foam. The residue from vacuum distillation may explode spontaneously. Reacts with sodium to form an ignitable product. Violent reaction with concentrated sulfuric acid (above 160°C), sulfuric acid + sulfur trioxide (above 52°C). Mixtures with tetranitromethane are sensitive high explosives. May explode on standing. It has been involved in plant scale explosions. When heated to decomposition it emits toxic fumes of NO_x.

NMO600   CAS: 1321-12-6   **HR: 3**
mixo-NITROTOLUENE
mf: $C_7H_7NO_2$ mw: 137.14

DFG MAK: 5 ppm (30 mg/m^3)

SAFETY PROFILE: A poison. May decompose explosively if heated above 190°C. When heated to decomposition it emits toxic fumes of NO_x.

NMP500   CAS: 99-55-8   **HR: 3**
5-NITRO-o-TOLUIDINE
mf: $C_7H_8N_2O_2$ mw: 152.17

SYNS: 2-AMINO-4-NITROTOLUENE * AZOFIX SCARLET G SALT * AZOGENE FAST SCARLET G * C.I. 37105 * C.I. AZOIC DIAZO COMPONENT 12 * DAINICHI FAST SCARLET G BASE * DAITO SCARLET BASE G * DEVOL SCARLET B * DIABASE SCARLET G * DIAZO FAST SCARLET G * FAST RED SG BASE * FAST SCARLET G * HILTONIL FAST SCARLET G BASE * KAYAKU SCARLET G BASE * LAKE SCARLET G BASE * LITHOSOL ORANGE R BASE * 6-METHYL-3-NITROANILINE * 2-METHYL-5-NITRO-BENZENEAMINE * MITSUI SCARLET G BASE

* NAPHTHANIL SCARLET G BASE * NAPHTOELAN FAST SCARLET G SALT * NCI-C01843 * 4-NITRO-2-AMINOTOLUENE (MAK) * PNOT * RCRA WASTE NUMBER U181 * SCARLET BASE CIBA II * SUGAI FAST SCARLET G BASE * SYMULON SCARLET G BASE

CONSENSUS REPORTS: NCI Carcinogenesis Bioassay (feed); Clear Evidence: mouse NCITR* NCI-CG-TR-107,78. Reported in EPA TSCA Inventory.

DFG MAK: Animal Carcinogen, Suspected Human Carcinogen.

SAFETY PROFILE: Confirmed carcinogen with experimental carcinogenic data. Moderately toxic by ingestion. Mutation data reported. Decomposes exothermically when heated to 150°C. When heated to decomposition it emits toxic fumes of NO_x.

NMQ000   CAS: 464-10-8   **HR: 3**
NITROTRIBROMOMETHANE
mf: CBr_3NO_2 mw: 297.75

PROP: Crystals. Bp: 127°, mp: 103°, d: 2.79 @ 18°.

SYNS: BROMOPICRIN * NITROBROMOFORM * TRIBROMONITROMETHANE

CONSENSUS REPORTS: Reported in EPA TSCA Inventory.

SAFETY PROFILE: Poison by intraperitoneal route. In vapor form it is highly toxic by ingestion and inhalation and on contact with skin, eyes, and mucous membranes. An explosive. When heated to decomposition it emits very toxic fumes of Br^- and NO_x.

NMQ500   CAS: 556-89-8   **HR: 3**
NITROUREA
DOT: UN 0147
mf: $CH_3N_3O_3$ mw: 105.06

PROP: Crystals.

SYN: m-NITROCARBAMIDE

CONSENSUS REPORTS: Reported in EPA TSCA Inventory.

DOT Classification: Class A Explosive; Label: Explosive A.

SAFETY PROFILE: A very dangerous fire hazard when exposed to heat or flame. A severe explosion hazard when shocked or exposed to heat. Can react vigorously with oxidizing materials. It is a high explosive. Incompatible with mercuric and silver salts. When heated to decomposition it emits highly toxic fumes of NO_x.

NMR000 CAS: 7782-77-6 **HR: 2**
NITROUS ACID
mf: HNO_2 mw: 47.02

PROP: Pale blue solution.

SYN: NITROSYL HYDROXIDE

CONSENSUS REPORTS: EPA Genetic Toxicology Program. Reported in EPA TSCA Inventory.

SAFETY PROFILE: Mutation data reported. Flammable by chemical reaction; a powerful oxidizer. Explodes on contact with phosphorus trichloride. Reacts violently with PH_3 and PCl_3. Reactions with 1-amino-5-nitrophenol, ammonium decahydroborate($2-$), hydrazine (product is hydrogen azide) may give explosive products. Incompatible with anilines (e.g., 4-bromoaniline, 2-chloroaniline, 3-chloroaniline, 2-nitroaniline, 3-nitroaniline, 4-nitroaniline, aniline), semicarbazone, silver nitrate. When heated to decomposition it emits highly toxic fumes of NO_x.

NMS000 CAS: 25168-04-1 **HR: 3**
NITROXYLENE
DOT: NA 1665
mf: $C_8H_9NO_2$ mw: 151.18

PROP: Light yellow liquid. Mp: 2°, bp: 244°, d: 1.135 @ 15°/4°.

SYNS: NITRODIMETHYLBENZENE * NITROXYLOL (DOT)

DOT Classification: Poison B; Label: Poison.

SAFETY PROFILE: A poison. Moderately toxic by ingestion. When heated to decomposition it emits toxic fumes of NO_x.

NMT000 CAS: 13444-90-1 **HR: 3**
NITRYL CHLORIDE
mf: $ClNO_2$ mw: 81.46

PROP: Corrosive, toxic, colorless gas. Decomp > 120°, vap d: 2.81 g/L @ 100°, bp: $-14.3°$, mp: $-145°$, d: (liquid): 1.37 @ 0°.

SYN: NITROXYL CHLORIDE

SAFETY PROFILE: A poison by inhalation. A corrosive irritant to skin, eyes, and mucous membranes. A powerful oxidizer. The gas or liquid may attack organic matter with explosive violence. Violent reaction with ammonia or sulfur dioxide. Incompatible with tin(II) bromide or tin(II) iodide. Reacts with water or steam to produce corrosive fumes. When heated to decomposition it emits toxic fumes of Cl^- and NO_x.

NMT500 CAS: 10022-50-1 **HR: 3**
NITRYL FLUORIDE
mf: FNO_2 mw: 65.01

PROP: Colorless gas; pungent odor. Mp: $-166.0°$, bp: $-72.4°$, d (liquid): 1.796 @ bp, d (solid): 1.924.

SAFETY PROFILE: Poison by inhalation. A severe irritant to skin, eyes, and mucous membranes. A powerful oxidizing agent. This gas is intensely reactive. Explosive reaction with hydrogen at 200-300°C. Ignites on contact with antimony, arsenic, boron, iodine, phosphorus, selenium. Ignites when warmed with bismuth, carbon, chromium, lead, sulfur. Incandescent reaction with aluminum, cadmium, cobalt, iron, molybdenum, nickel, potassium, sodium, thorium, titanium, tungsten, uranium, vanadium, zinc, zirconium, lithium (at 200-300°C), manganese (at 200-300°C). Incompatible with metals, non-metals. When heated to decomposition it emits toxic fumes of F^- and NO_x.

NMV760 CAS: 557-48-2 **HR: 2**
trans,cis-2,6-NONADIENAL
mf: $C_9H_{14}O$ mw: 138.23

PROP: Slightly yellow liquid; powerful, violet, cucumber odor. D: 0.850-0.870, refr index: 1.470. Sol in alc, fixed oils; insol in water.

SYNS: CUCUMBER ALDEHYDE * FEMA No. 3317 * trans,cis-2,6-NONADIENAL * trans-2,cis-6-NONADIENAL * 2,6-NONADIENAL * VIOLET LEAF ALDEHYDE

CONSENSUS REPORTS: Reported in EPA TSCA Inventory.

SAFETY PROFILE: A moderate skin irritant. When heated to decomposition it emits acrid smoke and irritating fumes.

NMW500 CAS: 124-19-6 **HR: 2**
1-NONANAL
mf: $C_9H_{18}O$ mw: 142.27

PROP: Found in at least 20 essential oils, including rose and citrus oils and several species of pine oil (FCTXAV 11, 95,73). Colorless to light yellow liquid; citrus-rose odor. D: 0.820-0.830, refr index: 1.422-1.429, flash p: 162°F. Sol in alc; fixed oils, propylene glycol; insol in glycerin.

SYNS: ALDEHYDE C-9 * C-9 ALDEHYDE * FEMA No. 2782 * NCI-C61018 * 1-NONALDE-HYDE * 1-NONYL ALDEHYDE * PELARGONIC AL-DEHYDE

CONSENSUS REPORTS: Reported in EPA TSCA Inventory.

SAFETY PROFILE: A severe skin irritant. Combustible liquid. When heated to decomposition it emits acrid smoke and irritating fumes.

NMX000   CAS: 111-84-2   **HR: 3**
NONANE
DOT: UN 1920
mf: C_9H_{20} mw: 128.29

PROP: Colorless liquid. Mp: −53.7°, bp: 150.7°, lel: 0.8%, uel: 2.9%, flash p: 88°F (CC), d: 0.718 @ 20°/4°, autoign temp: 374°F, vap press: 10 mm @ 38.0°, vap d: 4.41. Insol in water; sol in abs alc and ether.

SYN: SHELLSOL 140

CONSENSUS REPORTS: Reported in EPA TSCA Inventory.

OSHA PEL: TWA 200 ppm
ACGIH TLV: TWA 200 ppm
DOT Classification: Flammable or Combustible Liquid; Label: Flammable Liquid.

SAFETY PROFILE: Poison by intravenous route. Mildly toxic by inhalation. Irritating to respiratory tract. Narcotic in high concentrations. A very dangerous fire hazard when exposed to heat or flame; can react with oxidizing materials. Explosive in the form of vapor when exposed to heat or flame. To fight fire, use CO_2, dry chemical. When heated to decomposition it emits acrid smoke and irritating fumes.

NMY000   CAS: 112-05-0   **HR: 3**
NONANOIC ACID
mf: $C_9H_{18}O_2$ mw: 158.27

PROP: Oily, colorless liquid. Bp: 254°, mp: 12°, d: 0.9055 @ 20°/4°. Very sltly sol in water.

SYNS: CIRRASOL 185A * EMFAC 1202 * HEXA-CID C-9 * n-NONOIC ACID * n-NONYLIC ACID * 1-OCTANECARBOXYLIC ACID * PELARGIC ACID * PELARGON (RUSSIAN) * PELARGONIC ACID

CONSENSUS REPORTS: Reported in EPA TSCA Inventory.

SAFETY PROFILE: Poison by intravenous route. Moderately toxic by ingestion. A severe skin and eye irritant. When heated to decomposition it emits acrid smoke and irritating fumes.

NNA300   CAS: 2463-53-8   **HR: 2**
2-NONENAL
mf: $C_9H_{16}O$ mw: 140.25

PROP: White to sltly yellow liquid; fatty, violet odor. D: 0.850-0.870, refr index: 1.457. Sol in alc, fixed oils; insol in water.

SYNS: FEMA No. 3213 * HEPTYLIDENE ALDEHYDE * β-HEXYLACROLEIN * 2-NONEN-1-AL * α-NO-NENYL ALDEHYDE * trans-2-NONENAL (FCC)

CONSENSUS REPORTS: Reported in EPA TSCA Inventory.

SAFETY PROFILE: Moderately toxic by skin contact. Mildly toxic by ingestion. A severe skin irritant. When heated to decomposition it emits acrid smoke and irritating fumes.

NNB400   CAS: 143-13-5   **HR: 2**
NONYL ACETATE
mf: $C_{11}H_{22}O_2$ mw: 186.29

PROP: Colorless liquid; fruity odor. D: 0.864, refr index: 1.422, flash p: +153°F. Sol in alc, ether; insol in water.

SYN: FEMA No. 2788

SAFETY PROFILE: Combustible liquid. When heated to decomposition it emits acrid smoke and irritating fumes.

NNB500   CAS: 143-08-8   **HR: 2**
n-NONYL ALCOHOL
mf: $C_9H_{20}O$ mw: 144.29

PROP: Colorless liquid; rose-citrus odor. D: 0.827 @ 20°/4°, refr index: 1.43-1.435, mp: −5°, bp: 213.5°, flash p: 169°F. Insol in water; misc in alc, ether, chloroform.

SYNS: ALCOHOL C-9 * FEMA No. 2789 * NON-ALOL * 1-NONANOL * NONAN-1-OL * NONYL ALCOHOL * OCTYL CARBINOL * PELARGONIC ALCOHOL

CONSENSUS REPORTS: Reported in EPA TSCA Inventory.

SAFETY PROFILE: Moderately toxic by ingestion. Mildly toxic by skin contact and inhalation.

Combustible liquid. When heated to decomposition it emits acrid smoke and irritating fumes.

NNE000 CAS: 5283-67-0 HR: 2
NONYLTRICHLOROSILANE

DOT: UN 1799
mf: $C_9H_{19}Cl_3Si$ mw: 261.72

CONSENSUS REPORTS: Reported in EPA TSCA Inventory.

DOT Classification: Corrosive Material; Label: Corrosive.

SAFETY PROFILE: A corrosive irritant to skin, eyes, and mucous membranes. When heated to decomposition it emits toxic fumes of Cl^-.

NNM000 CAS: 492-41-1 HR: 3
(−)-NOREPHEDRINE
mf: $C_9H_{13}NO$ mw: 151.23

SYNS: α-(1-AMINOETHYL)-BENZYL ALCOHOL * 2-AMINO-1-PHENYL-1-PROPANOL * FENILPROPANOLAMINA (ITALIAN) * MYDRIATIN * PHENYLPROPANOLAMINE * PPA * PROPADRINE * USAF CS-6

CONSENSUS REPORTS: Reported in EPA TSCA Inventory.

SAFETY PROFILE: A human poison by ingestion. Poison experimentally by intravenous, subcutaneous, and intraperitoneal routes. Moderately toxic by an unspecified route. Human systemic effects by ingestion: sleep, increased pulse rate without blood pressure decrease and chronic pulmonary edema or congestion. When heated to decomposition it emits toxic fumes of NO_x.

NNO500 CAS: 51-41-2 HR: 3
I-NOREPINEPHRINE
mf: $C_8H_{11}NO_3$ mw: 169.20

PROP: Microcrystals. Decomp @ 216.5-218°.

SYNS: ADRENOR * AKTAMIN * 1-2-AMINO-1-(3,4-DIHYDROXYPHENYL)ETHANOL * (R)-4-(2-AMINO-1-HYDROXYETHYL)-1,2-BENZENEDIOL * 1-α-(AMINO-METHYL)-3,4-DIHYDROXYBENZYL ALCOHOL * (−)-α-(AMINOMETHYL)PROTOCATECHUYL ALCOHOL * ARTERENOL * 1-ARTERENOL * 1-1-(3,4-DIHYDROXYPHENYL)-2-AMINOETHANOL * 1-3,4-DIHYDROXYPHENYLETHANOLAMINE * LEVARTERENOL * LEVOARTERENOL * LEVONORADRENALINE * LEVONOREPINEPHRINE * LEVOPHED * (−)NORADREC * NORADRENALIN * NORADRENA-LINA (ITALIAN) * NORADRENALINE * (−)-NORADRENALINE * d-(−)-NORADRENALINE * 1-NORADRENALINE * NORADRENLINE * NORARTRINAL * NOREPINEPHRINE * (−)-NOREPINEPHRINE * NOREPIRENAMINE * SYMPATHIN E

SAFETY PROFILE: Poison by ingestion, intraperitoneal, subcutaneous, and intravenous routes. Experimental teratogenic and reproductive effects. Human mutation data reported. A sympathomimetic vasopressor. When heated to decomposition it emits toxic fumes of NO_x.

NNP500 CAS: 68-22-4 HR: 3
19-NORETHISTERONE
mf: $C_{20}H_{26}O_2$ mw: 298.46

SYNS: 17-α-ETHINYLESTRA-4-EN-17-β-OL-3-ONE * 17-α-ETHINYL-17-β-HYDROXY-Δ:4-ESTREN-3-ONE * 17-α-ETHINYL-19-NORTESTOSTERONE * 17-α-ETHYNYL-4-ESTREN-17-OL-3-ONE * 17-α-ETHYNYL-17-HYDROXY-4-ESTREN-3-ONE * 17-α-ETHYNYL-17-β-HYDROXY-19-NORANDROST-4-EN-3-ONE * 17-α-ETHYNYL-19-NORANDROST-4-EN-17-β-OL-3-ONE * 17-α-ETHYNYL-19-NOR-4-ANDROSTEN-17-β-OL-3-ONE * 17-α-ETHYNYL-19-NORTESTOSTERONE * (17-α)-17-HYDROXY-19-NORPREGN-4-EN-20-YN-3-ONE * 17-β-HYDROXY-19-NORPREGN-4-EN-20-YN-3-ONE * 17-HYDROXY-19-NOR-17-α-PREGN-4-EN-20-YN-3-ONE * 19-NOR-ETHINYL-4,5-TESTOSTERONE * 19-NOR-17-α-ETHYNYLANDROSTEN-17-β-OL-3-ONE * 19-NOR-17-α-ETHYNYL-17-β-HYDROXY-4-ANDROSTEN-3-ONE * 19-NOR-17-α-ETHYNYLTESTOSTERONE * NORLUTIN

CONSENSUS REPORTS: IARC Cancer Review: Animal Limited Evidence IMEMDT 21,441,79; Animal Sufficient Evidence IMEMDT 6,179,74. NTP Fourth Annual Report On Carcinogens, 1984. EPA Genetic Toxicology Program.

SAFETY PROFILE: Confirmed carcinogen with experimental carcinogenic and tumorigenic data. Mildly toxic by ingestion. Human systemic effects by ingestion: dermatitis and androgenic effects. Human reproductive effects by ingestion, implant, and possibly other routes: spermatogenesis; testes, epididymis, sperm duct changes; impotence; male breast development; other male effects; ovaries, fallopian tube changes; menstrual cycle changes or disorders uterus, cervix, vagina effects; postpartum effects; changes in female fertility. Human teratogenic effects by ingestion and possibly other routes: developmental abnormalities of the musculoskeletal system and urogenital system; and behavioral effects in the newborn. Experimental teratogenic and reproductive effects. Human mutation data reported. When heated to decomposition it emits acrid smoke and irritating fumes.

NNQ500   CAS: 6533-00-2   **HR: 3**
NORGESTREL
mf: $C_{21}H_{28}O_2$ mw: 312.49

PROP: Crystals from diethyl ether-hexane. Mp: 142-143°.

SYNS: 13-ETHYL-17-α-ETHYNYLGON-4-EN-17-β-OL-3-ONE * 13-ETHYL-17-α-ETHYNYL-17-β-HYDROXY-4-GONEN-3-ONE * dl-13-β-ETHYL-17-α-ETHYNYL-17-β-HYDROXYGON-4-EN-3-ONE * (±)-13-ETHYL-17-α-ETH-YNYL-17-HYDROXYGON-4-EN-3-ONE * dl-13-β-ETHYL-17-α-ETHYNYL-19-NORTESTOSTERONE * (±)-13-ETHYL-17-HYDROXY-18,19-DINOR-17-α-PREGN-4-EN-20-YN-3-ONE * 17-ETHYNYL-18-METHYL-19-NORTESTOS-TERONE * FH 122-A * 17-β-HYDROXY-18-METHYL-19-NOR-17-α-PREGN-4-EN-20-YN-3-ONE * LD NORGES-TREL (FRENCH) * 18-METHYL-17-α-ETHYNYL-19-NORTESTOSTERONE * MONOVAR * d(−)-NORGES-TREL * (±)-NORGESTREL * α-NORGESTREL * d-NORGESTREL * dl-NORGESTREL * POSTINOR * SH 850 * SH 70850 * WY 3707

CONSENSUS REPORTS: IARC Cancer Review: Animal Inadequate Evidence IMEMDT 6,201,74, IMEMDT 21,479,79; Human Inadequate Evidence IMEMDT 21,479,79.

SAFETY PROFILE: Human reproductive effects by ingestion and implant: menstrual cycle changes or disorders and female fertility index changes. Experimental reproductive effects. Questionable carcinogen with experimental neoplastigenic data. An oral contraceptive. When heated to decomposition it emits acrid smoke and irritating fumes.

NNY000   CAS: 72-69-5   **HR: 3**
NORTRIPTYLINE
mf: $C_{19}H_{21}N$ mw: 263.41

SYNS: DEMETHYLAMITRIPTYLENE * DESMETH-YLAMITRIPTYLINE * 10,11-DIHYDRO-N-METHYL-5H-DIBENZO(a,d)CYCLOHEPTANE-Δ,γ-PROPYLAMINE * 5-(3-(METHYLAMINO)PROPYLIDENE)DIBENZO(a,e)CYCLOHEPTA(1,5)DIENE * 5-(3-METHYLAMINOPRO-PYLIDENE)-10,11-DIHYDRO-5H-DIBENZO(a,d)CYCLOHEP-TENE * NORAMITRIPTYLINE

SAFETY PROFILE: Poison by ingestion, intraperitoneal, and intravenous routes. When heated to decomposition it emits toxic fumes of NO_x.

NOB000   CAS: 1476-53-5   **HR: 3**
NOVOBIOCIN, MONOSODIUM SALT
mf: $C_{31}H_{35}N_2O_{11}$ • Na   mw: 634.67

SYNS: ALBAMYCIN * ALBAMYCIN SODIUM * CATHOMYCIN SODIUM * CATHOMYCIN SODIUM LYOVAC * INAMYCIN * MONOSODIUM NOVO-BIOCIN * NOVOBIOCIN MONOSODIUM * NOVO-BIOCIN, SODIUM derivative * SODIUM ALBAMYCIN * SODIUM NOVOBIOCIN * U-6591

SAFETY PROFILE: Poison by intraperitoneal and subcutaneous routes. Moderately toxic by ingestion and subcutaneous routes. An antibiotic with serious side effects which include liver and blood disease. It may also promote the development of resistant strains of staphylococcus. When heated to decomposition it emits toxic fumes of NO_x and Na_2O.

NOD500   CAS: 63039-90-7   **HR: 3**
NU-1932
mf: $C_{17}H_{25}NO_2$ • ClH   mw: 311.89

SYNS: 3-β-AETHYL-1-METHYL-4-PHENYL-4-α-PIPERI-DYLPROPIONAT HYDROCHLORID (GERMAN) * 3-β-AETHYL-1-METHYL-4-PHENYL-4-α-PROPIONYLOXYPIPER-IDIN HYDROCHLORID (GERMAN) * MEPRODINE (GER-MAN)

SAFETY PROFILE: Poison by ingestion, subcutaneous, intravenous, and intraperitoneal routes. When heated to decomposition it emits very toxic fumes of HCl and NO_x.

NOE500   CAS: 475-83-2   **HR: 3**
1-NUCIFERINE
mf: $C_{19}H_{21}NO_2$ mw: 295.41

PROP: Alkaloid obtained from the lotus *Nelumbo nucifera* and *Nelumbo lutea*.

SYNS: 1,2-DIMETHOXY-6a-β-APORPHINE * 1-5, 6-DIMETHOXYAPORPHINE * (R)-1,2-DIMETHOXYAPOR-PHINE * (−)-NUCIFERINE

SAFETY PROFILE: Poison by ingestion. When heated to decomposition it emits toxic fumes of NO_x.

NOF000   **HR: 1**
NUISANCE DUSTS and AEROSOLS

SAFETY PROFILE: Variable toxicity depending upon composition. Causes local irritation of eyes, nose, throat, and lungs. Some may lead to chronic bronchitis, emphysema, and bronchial asthma. Dermatitis may result from short contact. Asthma, angioneurotic edema, hives, etc., may result from short periods of inhalation. A topic eczema, angioneurotic edema, hives, etc., may also result from prolonged contact. A common air contaminant. Nuisance aerosols do evoke some tissue response in the lung upon inhalation of sufficient amounts. However, this reaction is potentially reversible and leaves no scar tissue.

NOF500   CAS: 61-12-1   **HR: 3**
NUPERCAINE HYDROCHLORIDE
mf: $C_{20}H_{29}N_3O_2 \cdot ClH$ mw: 379.98

PROP: Crystals. Mp: decomp @ 90-98°, vap d: 13.1.

SYNS: BENZOLIN * BENZOLIN HYDROCHLORIDE * BUTOXYCINCHONINIC ACID DIETHYLETHYLENE-DIAMIDE HYDROCHLORIDE * 2-N-BUTOXY-N-(2-DI-ETHYLAMINOETHYL)CINCHONINAMIDE HYDROCHLO-RIDE * 2-BUTOXY-N-(2-DIETHYLAMINOETHYL) CINCHONINAMIDE HYDROCHLORIDE * 2-BUTOXY-N-(2-DIETHYLAMINOETHYL)CINCHONINIC ACID AMIDE HYDROCHLORIDE * CINCAINE HYDROCHLORIDE * CINCHOCAINE HYDROCHLORIDE * CINCHO-CAINIUM CHLORIDE * DIBUCAIN * DIBUCAINE HYDROCHLORIDE * PERCAIN * PERCAINE HYDROCHLORIDE * SOVCAIN * SOVCAINE HYDROCHLORIDE * SOVKAIN * SOVOCAINE HYDROCHLORIDE

SAFETY PROFILE: Poison by ingestion, subcutaneous, intravenous, intraspinal, parenteral, intraperitoneal, and possibly other routes. Mutation data reported. A severe eye irritant. A local anesthetic. When heated to decomposition it emits very toxic fumes of HCl and NO_x.

NOG500   CAS: 8008-45-5   **HR: 2**
NUTMEG OIL, EAST INDIAN

PROP: Major components are α- and β-pinene, camphene, myristicin, dipentene and sabanene. Found in fruit of *Myristica fragrans Houttuyn* (Fam. *Myristicaceae*). Prepared by steam distillation of dried nutmeg. Colorless to pale yellow liquid; odor and taste of nutmeg. East Indian: d: 0.880-0.910, refr index: 1.474-1.488; West Indian: d: 0.854-0.880, refr index: 1.469-1.476 @20°. Sol in fixed oils, mineral oil; sltly sol in cold alc; very sol in hot alc, chloroform, ether; insol in glycerin, propylene glycol.

SYNS: MYRISTICA OIL * NUTMEG OIL * OIL of MYRISTICA * OIL of NUTMEG

CONSENSUS REPORTS: Reported in EPA TSCA Inventory.

SAFETY PROFILE: Moderately toxic by ingestion. Experimental reproductive effects. Mutation data reported. A skin irritant. When heated to decomposition it emits acrid smoke and irritating fumes.

NOH000   CAS: 63428-83-1   **HR: 1**
NYLON
mf: $(C_6H_{11}NO)_n$

PROP: Crystalline solid. Sol in phenol, cresols, xylene, and formic acid. Insol in alc, esters, ketones, hydrocarbons. Film used for implant study.

SYNS: AMILAN * ASHLENE * CAPROLON * ENKALON * GRILON * KAPRON * MIRLON * PERLON * PHRILON * POLYAMID (GERMAN) * SILON * TROGAMID T * VYDYNE

SAFETY PROFILE: Questionable carcinogen with experimental tumorigenic data by implant. Reacts violently with F_2. When heated to decomposition it emits toxic fumes of NO_x.

NOH500   CAS: 1400-61-9   **HR: 3**
NYSTATIN
mf: $C_{46}H_{83}NO_{18}$ mw: 938.30

PROP: Yellow to light-tan powder; odor suggestive of cereals. Mp: decomp > 160°. Sparingly sol in methanol and ethanol; very sltly sol in water; insol in chloroform, ether and benzene.

SYNS: BIOFANAL * CANDEX * CANDIO-HER-MAL * DIASTATIN * MORONAL * MYCOSTA-TIN * MYCOSTATIN 20 * NILSTAT * NYSTAN * NYSTATINE * NYSTAVESCENT * O-V STATIN

CONSENSUS REPORTS: EPA Genetic Toxicology Program.

SAFETY PROFILE: Poison by intraperitoneal and intravenous routes. Moderately toxic by subcutaneous route. Mildly toxic by ingestion. Experimental teratogenic and reproductive effects. Mutation data reported. An antibiotic. When heated to decomposition it emits toxic fumes of NO_x.

O

OAH000 CAS: 27858-07-7 **HR: 2**
OCTABROMODIPHENYL
mf: $C_{12}H_2Br_8$ mw: 785.42

SYNS: BB-8 * BROMKAL 80 * OBB * OCTA-
BROMOBIPHENYL * ar,ar,ar,ar,ar',ar',ar',ar' OCTA-
BROMO-1,1'-BIPHENYL

CONSENSUS REPORTS: Reported in EPA
TSCA Inventory.

SAFETY PROFILE: Moderately toxic by ingestion. Experimental teratogenic and reproductive
effects. Irritating to skin, eyes, and mucous
membranes. Can cause kidney and liver enlargement and thyroid hyperplasia; it is stored in
fatty tissue. When heated to decomposition it
emits toxic fumes of Br^-.

OAJ000 CAS: 3268-87-9 **HR: 3**
OCTACHLORODIBENZODIOXIN
mf: $C_{12}Cl_8O_2$ mw: 459.72

PROP: Colorless crystals. Mp: 239°.

SYNS: NCI-C03678 * OCDD * OCTACHLORODI-
BENZO(b,e)(1,4)DIOXIN * OCTACHLORODIBENZO-p-DI-
OXIN * 1,2,3,4,6,7,8,9-OCTACHLORODIBENZODIOXIN

CONSENSUS REPORTS: IARC Cancer Review: Animal Inadequate Evidence IMEMDT
15,41,77.

SAFETY PROFILE: Poison by ingestion. Experimental reproductive effects. An eye irritant.
Questionable carcinogen with experimental tumorigenic data. When heated to decomposition
it emits toxic fumes of Cl^-.

OAN000 CAS: 297-78-9 **HR: 3**
**1,3,4,5,6,8,8-OCTACHLORO-
1,3,3a,4,7,7a-HEXAHYDRO-4,7-
METHANO ISOBENZOFURAN**
mf: $C_9H_4Cl_8O$ mw: 411.73

SYNS: CP 14,957 * ENT 25,545 * ENT 25,545-X
* ISOBENZAN * OCTACHLORO-HEXAHYDRO-METH-
ANOISOBENZOFURAN * 1,3,4,5,6,7,10,10-OCTA-
CHLORO-4,7-endo-METHYLENE-4,7,8,9-TETRAHYDRO-
PHTHALAN * 1,3,4,5,6,7,8,8-OCTACHLORO-2-OXA-
3a,4,7,7a-TETRAHYDRO-4,7-METHANOINDENE
* OMTAN * R 6700 * SD 4402 * SHELL 4402
* SHELL WL 1650 * TELODRIN * WL 1650

SAFETY PROFILE: Poison by ingestion, skin
contact, intraperitoneal, intravenous, and possibly other routes. Questionable carcinogen with
experimental tumorigenic data. Used as an insecticide. When heated to decomposition it
emits toxic fumes of Cl^-.

OAP000 CAS: 2234-13-1 **HR: 3**
OCTACHLORONAPHTHALENE
mf: $C_{10}Cl_8$ mw: 403.70

CONSENSUS REPORTS: Community Right-
To-Know List. Reported in EPA TSCA Inventory.

OSHA PEL: (Transitional: TWA 0.1 mg/m^3
(skin)) TWA 0.1 mg/m^3 (skin); STEL 0.3
mg/m^3
ACGIH TLV: TWA 0.1 mg/m^3 (skin); STEL
0.3 mg/m^3

SAFETY PROFILE: Poison by inhalation, ingestion and skin contact. When heated to decomposition it emits highly toxic fumes of Cl^-.

OAT000 CAS: 2223-93-0 **HR: 3**
**OCTADECANOIC ACID, CADMIUM
SALT**
mf: $C_{36}H_{72}O_4 \cdot Cd$ mw: 681.48

SYNS: CADMIUM STEARATE * KADMIUMSTEARAT
(GERMAN) * STEARIC ACID, CADMIUM SALT

CONSENSUS REPORTS: EPA Extremely
Hazardous Substances List. Cadmium and its
compounds are on the Community Right-To-
Know List. Reported in EPA TSCA Inventory.

OSHA PEL: TWA 0.1 mg(Cd)/m^3; CL 0.6
mg(Cd)/m^3 (fume)
ACGIH TLV: TWA 0.05 mg(Cd)/m^3 (Proposed:
TWA 0.01 mg(Cd)/m^3 (dust), Human Carcinogen); BEI: 10 μg/g creatinine in urine;
10 μg/L in blood.
DFG BAT: Blood 1.5 μg/dL; Urine 15 μg/
dL, Suspected Carcinogen.
NIOSH REL: (Cadmium) Reduce to lowest feasible level

SAFETY PROFILE: Confirmed carcinogen.
Poison by inhalation. Moderately toxic by ingestion. Human systemic effects by inhalation: hallucinations or distorted perceptions; nausea or
vomiting, other gastrointestinal effects; weight
loss or decreased weight gain; cardiac effects.
When heated to decomposition it emits toxic
fumes of Cd.

OAV000   CAS: 31566-31-1   **HR: 3**
OCTADECANOIC ACID, MONOESTER
with 1,2,3-PROPANETRIOL
mf: $C_{21}H_{42}O_4$ mw: 358.63

PROP: Pure white- or cream-colored, wax-like solid; faint odor. Mp: 58-59°, d: 0.97. Sol in (hot) alc, oils, and hydrocarbons.

SYNS: ABRACOL S.L.G * ADMUL * ADVAWAX 140 * ALDO HMS * ALDO MS * ALDO MSA * ALDO MSLG * ALDO-28 * ALDO-72 * ARLACEL 161 * ARLACEL 169 * ARMOSTAT 801 * ATMOS 150 * ATMUL 67 * ATMUL 84 * ATMUL 124 * CEFATIN * CELINHOL -A * CERASYNT 1000-D * CERASYNT S * CERA-SYNT SD * CERASYNT SE * CERASYNT WM * CITOMULGAN M * CYCLOCHEM GMS * DER-MAGINE * DISTEARIN * DREWMULSE TP * DREWMULSE V * DRUMULSE AA * EMCOL CA * EMEREST 2400 * EMEREST 2401 * EMCOL MSK * EMUL P.7 * ESTOL 603 * GLYCERIN MONO-STEARATE * GLYCEROL MONOSTEARATE * GLYCERYL MONOSTEARATE * GROCOR 5500 * GROCOR 6000 * HODAG GMS * IMWITOR 191 * IMWITOR 900K * KESSCO 40 * LIPO GMS 410 * LIPO GMS 450 * LIPO GMS 600 * MONELGIN * MONOSTEARIN * OGEEN 515 * OGEEN GRB * OGEEN M * ORBON * PROTACHEM GMS * SEDETINE * STARFOL GMS 450 * STARFOL GMS 600 * STARFOL GMS 900 * STEARIC ACID, MONOESTER with GLYCEROL * STEARIC MONOGLYC-ERIDE * TEGIN * TEGIN 503 * TEGIN 515 * UNIMATE GMS * USAF KE-7 * WITCONOL MS * WITCONOL MST

CONSENSUS REPORTS: Reported in EPA TSCA Inventory.

SAFETY PROFILE: Poison by intraperitoneal route. When heated to decomposition it emits acrid smoke and irritating fumes.

OAX000   CAS: 112-92-5   **HR: 3**
1-OCTADECANOL
mf: $C_{18}H_{38}O$ mw: 270.56

PROP: Colorless solid or flakes. Mp: 58°, bp: 202° @ 10 mm, d: 0.8124 @ 59°/4°.

SYNS: ADOL * ADOL 68 * ATALCO S * CO-1895 * CO-1897 * CRODACOL-S * DECYL OCTYL ALCOHOL * DYTOL E-46 * LOROL 28 * OCTADECANOL * n-OCTADECANOL * OCTA DECYL ALCOHOL * n-OCTADECYL ALCOHOL * POLAAX * SIPOL S * SIPONOL S

* STEAROL * STEARYL ALCOHOL * STERAFFINE * USP XIII STEARYL ALCOHOL

CONSENSUS REPORTS: Reported in EPA TSCA Inventory.

SAFETY PROFILE: Mildly toxic by ingestion. Questionable carcinogen with experimental neo-plastigenic data. Flammable when exposed to heat or flame; can react with oxidizing materials. To fight fire, use foam, CO_2, dry chemical. When heated to decomposition it emits acrid smoke and irritating fumes.

OBC000   CAS: 124-30-1   **HR: 3**
OCTADECYLAMINE
mf: $C_{18}H_{39}N$ mw: 269.58

SYNS: N-OCTADECYLAMINE * OKTADECYLAMIN (CZECH) * STEARYLAMINE

CONSENSUS REPORTS: Reported in EPA TSCA Inventory.

SAFETY PROFILE: Poison by intraperitoneal route. A skin irritant. When heated to decompo-sition it emits toxic fumes of NO_x.

OBI000   CAS: 112-04-9   **HR: 2**
OCTADECYLTRICHLOROSILANE
DOT: UN 1800
mf: $C_{18}H_{37}Cl_3Si$ mw: 387.99

PROP: Bp: 223° @ 10 mm, d: 0.984 @ 25°.

SYN: TRICHLOROOCTADECYLSILANE

CONSENSUS REPORTS: Reported in EPA TSCA Inventory.

DOT Classification: Corrosive Material; Label: Corrosive.

SAFETY PROFILE: A corrosive irritant to skin, eyes and mucous membranes. Reacts with water or steam to produce toxic and corrosive fumes. When heated to decomposition it emits toxic fumes of Cl^-.

OBO000   CAS: 360-89-4   **HR: 1**
1,1,1,2,3,4,4,4-OCTAFLUORO-2-BUTENE
DOT: UN 2422
mf: C_4F_8 mw: 200.04

PROP: Bp: 1.2°, fp: −135°.

SYNS: FC-1318 * OCTAFLUOROBUTENE-2 * OCTAFLUOROBUT-2-ENE (DOT) * PERFLUORO-BUT-2-ENE * PERFLUORO-2-BUTENE (DOT)

CONSENSUS REPORTS: Reported in EPA TSCA Inventory. EPA Genetic Toxicology Program.

DOT Classification: Nonflammable Gas; Label: Nonflammable Gas.

SAFETY PROFILE: Mildly toxic by inhalation. Mutation data reported. When heated to decomposition it emits toxic fumes of F^-.

OBY000   CAS: 20917-49-1   **HR: 3**
OCTAHYDRO-1-NITROSOAZOCINE
mf: $C_7H_{14}N_2O$ mw: 142.23

SYNS: NHMI * N-NITROSOAZACYCLOOCTANE * NITROSOHEPTAMETHYLENEIMINE * N-NITROSO-HEPTAMETHYLENEIMINE * NITROSO-HEPTAMETH-YLENIMIN (GERMAN)

SAFETY PROFILE: Poison by ingestion and subcutaneous routes. Questionable carcinogen with experimental carcinogenic and tumorigenic data. Mutation data reported. When heated to decomposition it emits toxic fumes of NO_x.

OCE000   CAS: 104-50-7   **HR: 1**
γ-OCTALACTONE
mf: $C_8H_{14}O_2$ mw: 142.22

PROP: Colorless to pale yellow liquid; coconut odor. D: 0.970-0.980, refr index: 1.443-1.447. Sol in alc; sltly sol in water.

SYNS: γ-n-BUTYL-γ-BUTYROLACTONE * FEMA No. 2798 * 5-HYDROXYOCTANOIC ACID LACTONE * OCTANOLIDE-1,4 * TETRAHYDRO-6-PROPYL-2H-PYRAN-2-ONE

CONSENSUS REPORTS: Reported in EPA TSCA Inventory.

SAFETY PROFILE: Mildly toxic by ingestion. A skin irritant. When heated to decomposition it emits acrid smoke and irritating fumes.

OCM000   CAS: 152-16-9   **HR: 3**
OCTAMETHYLPYROPHOSPHORAMIDE
mf: $C_8H_{24}N_4O_3P_2$ mw: 286.30

PROP: Viscous liquid. Mp: 20-21°, bp: 154° @ 2.0 mm, d: 1.09 @ 25°/4°. Miscible with water; sol in most organic solvents; almost insol in higher aliphatic hydrocarbons.

SYNS: BIS(BISDIMETHYLAMINOPHOSPHONOUS)ANHY-DRIDE * BIS(DIMETHYLAMINO)PHOSPHONOUS ANY-HYDRIDE * BIS(DIMETHYLAMINO)PHOSPHORIC ANHYDRIDE * BIS-N,N,N',N'-TETRAMETHYLPHOS-PHORODIAMIDIC ANHYDRIDE * ENT 17,291 * LETHA LAIRE G-59 * OCTAMETHYL-DIFOS-FORZUUR-TETRAMIDE (DUTCH) * OCTAMETH-YLDIPHOSPHORAMIDE * OCTAMETHYL-DIPHOS-PHORSAEURE-TETRAMID (GERMAN) * OCTA-METHYL PYROPHOSPHORTETRAMIDE * OCTA-METHYL TETRAMIDO PYROPHOSPHATE * OMPA * OMPACIDE * OMPATOX * OMPAX * OTTO-METIL-PIROFOSFORAMMIDE (ITALIAN) * PESTOX * PESTOX 3 * PESTOX III * PYROPHOS-PHORIC ACID OCTAMETHYLTETRAAMIDE * PYROPHOSPHORYLTETRAKISDIMETHYLAMIDE * RCRA WASTE NUMBER P085 * SCHRADAN * SCHRADANE (FRENCH) * SYSTAM * SYS-TOPHOS * SYTAM * TETRAKISDIMETHYLAMINO-PHOSPHONOUS ANHYDRIDE

CONSENSUS REPORTS: EPA Extremely Hazardous Substances List. Reported in EPA TSCA Inventory.

SAFETY PROFILE: Poison by ingestion, inhalation, skin contact, intraperitoneal, intravenous, subcutaneous, ocular and possibly other routes. Human systemic effects by ingestion: a cholinesterase inhibitor. Has been found to inhibit peripheral cholinesterase without pronounced effects on the central nervous system. An insecticide. When heated to decomposition it emits toxic fumes of NO_x and PO_x.

OCO000   CAS: 124-13-0   **HR: 1**
1-OCTANAL
mf: $C_8H_{16}O$ mw: 128.24

PROP: Found in about 20 essential oils, including a number of citrus oils. Colorless to light yellow liquid; fatty-orange odor. Bp: 163.4, flash p: 125°F (CC), d: 0.821 @ 20°/4°, refr index: 1.417-1.425, vap d: 4.41. Sol in alc, fixed oils, propylene glycol; insol in glycerin.

SYNS: ALDEHYDE C-8 * C-8 ALDEHYDE * FEMA No. 2797 * OCTANALDEHYDE * n-OCTYL ALDEHYDE

CONSENSUS REPORTS: Reported in EPA TSCA Inventory.

SAFETY PROFILE: Mildly toxic by ingestion and skin contact. A skin and eye irritant. Combustible when exposed to heat or flame; can react with oxidizing materials. To fight fire, use foam, CO_2, dry chemical.

OCU000   CAS: 111-65-9   **HR: 3**
OCTANE
DOT: UN 1262
mf: C_8H_{18} mw: 114.26

PROP: Clear liquid. Bp: 125.8°, lel: 1.0%, uel: 4.7%; fp: −56.5°, flash p: 56°F, d: 0.7036 @ 20°/4°, autoign temp: 428°F, vap press: 10 mm @ 19.2°, vap d: 3.86. Insol in water, sltly sol in alc, ether; misc with benzene.

SYNS: OKTAN (POLISH) * OKTANEN (DUTCH) * OTTANI (ITALIAN)

CONSENSUS REPORTS: Reported in EPA TSCA Inventory.

OSHA PEL: (Transitional: TWA 500 ppm) TWA 300 ppm; STEL 375 ppm
ACGIH TLV: TWA 300 ppm; STEL 375 ppm
DFG MAK: 500 ppm (2350 mg/m³)
NIOSH REL: (Alkanes) TWA 350 mg/m³
DOT Classification: Flammable Liquid; Label: Flammable Liquid.

SAFETY PROFILE: May act as a simple asphyxiant. A narcotic in high concentration. Human dermal exposure to undiluted octane for five hours resulted in blister formation but no anesthesia; one hour caused diffuse burning sensation. A very dangerous fire hazard and severe explosion hazard when exposed to heat, flame, or oxidizers. When heated to decomposition it emits acrid smoke and irritating fumes.

OCY000   CAS: 124-07-2   **HR: 2**
OCTANOIC ACID
mf: $C_8H_{16}O_2$ mw: 144.24

PROP: Colorless, oily liquid; unpleasant odor, burning rancid taste. D: 0.91 @ 20°, bp: 240°, mp: 17°. Sltly sol in water; sol in most organic solvents.

SYNS: C-8 ACID * CAPRYLIC ACID * n-CAPRYLIC ACID * 1-HEPTANECARBOXYLIC ACID * HEXACID 898 * NEO-FAT 8 * OCTIC ACID * n-OCTOIC ACID * n-OCTYLIC ACID

CONSENSUS REPORTS: Reported in EPA TSCA Inventory.

SAFETY PROFILE: Moderately toxic by intravenous route. Mildly toxic by ingestion. Mutation data reported. A skin irritant. Yields irritating vapors which can cause coughing. When heated to decomposition it emits acrid smoke and irritating fumes.

OCY100   CAS: 20296-29-1   **HR: 1**
3-OCTANOL
mf: $C_8H_{18}O$ mw: 130.28

PROP: Colorless liquid; strong, nutty odor. D: 0.816-0.821, refr index: 1.425. Sol in alc, fixed oils; insol in water.

SYNS: AMYLETHYLCARBINOL * ETHYLAMYLCARBINOL * ETHYL-n-AMYLCARBINOL * FEMA No. 3581 * OCTANOL-3

SAFETY PROFILE: A moderate skin and eye irritant. When heated to decomposition it emits acrid smoke and irritating fumes.

ODG000   CAS: 111-13-7   **HR: 2**
2-OCTANONE
mf: $C_8H_{16}O$ mw: 128.24

PROP: Colorless liquid; pleasant apple odor. D: 0.813-0.818, refr index: 1.414-1.418, mp: −20.9°, bp: 173.5°, vap d: 4.4, flash p: 160°F. Sltly sol in water; sol in alc, hydrocarbons, ether, esters.

SYNS: FEMA No. 2802 * METHYL HEXYL KETONE (FCC)

CONSENSUS REPORTS: Reported in EPA TSCA Inventory.

SAFETY PROFILE: Moderately toxic by an unspecified route. A skin irritant. Combustible liquid when exposed to heat, flame, or oxidizers. To fight fire, use foam, alcohol foam. When heated to decomposition it emits acrid smoke and irritating fumes.

ODW000   CAS: 3391-86-4   **HR: 3**
1-OCTEN-3-OL
mf: $C_8H_{16}O$ mw: 128.24

SYNS: AMYLVINYLCARBINOL * MATSUTAKE ALCOHOL (JAPANESE)

CONSENSUS REPORTS: Reported in EPA TSCA Inventory.

SAFETY PROFILE: Poison by ingestion and intravenous routes. Moderately toxic by skin contact. When heated to decomposition it emits acrid smoke and irritating fumes.

ODY000   CAS: 6168-86-1   **HR: 3**
OCTIN HYDROCHLORIDE
mf: $C_9H_{19}N \cdot ClH$ mw: 177.75

SYNS: ISOMETHEPTENE HYDROCHLORIDE * 2-METHYLAMINOISOOCTANE HYDROCHLORIDE * METHYLAMINO-METHYLHEPTENE HYDROCHLORIDE

SAFETY PROFILE: Poison by intraperitoneal, intravenous, parenteral, and subcutaneous routes. An adrenergic agent. When heated to decomposition it emits very toxic fumes of Cl^- and NO_x.

OEG000   CAS: 112-14-1   **HR: 2**
1-OCTYL ACETATE
mf: $C_{10}H_{20}O_2$ mw: 172.30

PROP: Colorless liquid; orange-jasmine odor. D: 0.865, refr index: 1.418-1.421, mp: $-38.5°$, bp: 210°, flash p: 190°F. Insol in water; misc with alc, ether, fixed oils.

SYNS: ACETATE C-8 * ACETIC ACID, OCTYL ESTER * CAPRYLYL ACETATE * FEMA No. 2806 * 1-OCTANOL ACETATE * n-OCTANYL ACETATE * OCTYL ACETATE * n-OCTYL ACETATE * OCTYL ALCOHOL ACETATE

CONSENSUS REPORTS: Reported in EPA TSCA Inventory.

SAFETY PROFILE: Moderately toxic by ingestion. A skin irritant. Combustible liquid. When heated to decomposition it emits acrid smoke and irritating fumes.

OEG100   **HR: 1**
3-OCTYL ACETATE
mf: $C_{10}H_{20}O_2$ mw: 172.27

PROP: Colorless liquid; rosy, minty odor. D: 0.856-0.860, refr index: 1.414, fp: 190°. Sol in alc, propylene glycol, fixed oils; sltly sol in water.

SYN: FEMA No. 3583

SAFETY PROFILE: Combustible liquid. When heated to decomposition it emits acrid smoke and irritating fumes.

OEI000   CAS: 111-87-5   **HR: 3**
OCTYL ALCOHOL
mf: $C_8H_{18}O$ mw: 130.26

PROP: Colorless liquid. D: 0.827 @ 20° 16/4°, mp: $-16.7°$, bp: 194.5°, flash p: 178°F. Sol in water; misc in alc, ether, and chloroform. Found in several citrus oils and at least 10 other natural sources.

SYNS: ALCOHOL C-8 * ALFOL 8 * CAPRYL ALCOHOL * CAPRYLIC ALCOHOL * DYTOL M-83 * EPAL 8 * FEMA No. 2800 * HEPTYL CARBINOL * 1-HYDROXYOCTANE * LOROL 20 * OCTANOL

* n-OCTANOL * 1-OCTANOL (FCC) * OCTILIN * OCTYL ALCOHOL, NORMAL-PRIMARY * PRIMARY OCTYL ALCOHOL * SIPOL L8

CONSENSUS REPORTS: Reported in EPA TSCA Inventory.

SAFETY PROFILE: Poison by intravenous route. Moderately toxic by ingestion. Mutation data reported. A skin irritant. Combustible liquid when exposed to heat or flame; can react with oxidizing materials. To fight fire, use water foam, fog, alcohol foam, dry chemical, CO_2.

OES000   CAS: 113-48-4   **HR: 2**
N-OCTYL BICYCLOHEPTENE DICARBOXIMIDE
mf: $C_{17}H_{25}NO_2$ mw: 275.43

SYNS: BICYCLO(2.2.1)HEPTENE-2-DICARBOXYLIC ACID, 2-ETHYLHEXYLIMIDE * ENDOMETHYLENE-TETRAHYDROPHTHALIC ACID, N-2-ETHYLHEXYL IMIDE * ENT 8,184 * N-(2-ETHYLHEXYL)BICYCLO-(2,2,1)-HEPT-5-ENE-2,3-DICARBOXIMIDE * N-2-ETHYLHEXYL-IMIDEENDOMETHYLENETETRAHYDROPHTHALIC ACID * N-(2-ETHYLHEXYL)-5-NORBORNENE-2,3-DICARBOXIMIDE * 2-(2-ETHYLHEXYL)-3a,4,7,7a-TETRAHYDRO-4,7-METHANO-1H-ISOINDOLE-1,3(2H)-DIONE * MGK-264 * OCTACIDE 264 * N-OCTYLBICYCLO-(2.2.1)-5-HEPTENE-2,3-DICARBOXIMIDE * PYRODONE * SYNERGIST 264 * VAN DYK 264

SAFETY PROFILE: Moderately toxic by ingestion, skin contact, and intraperitoneal routes. Experimental reproductive effects. Large doses can cause central nervous system stimulation followed by depression. When heated to decomposition it emits toxic fumes of NO_x.

OFA000   CAS: 1034-01-1   **HR: 1**
OCTYL GALLATE
mf: $C_{15}H_{22}O_5$ mw: 282.37

CONSENSUS REPORTS: Reported in EPA TSCA Inventory.

SAFETY PROFILE: Mildly toxic by ingestion. When heated to decomposition it emits acrid smoke and irritating fumes.

OFI000   CAS: 7530-07-6   **HR: 2**
OCTYLPEROXIDE

DOT: NA 2129
mf: $C_{16}H_{34}O_2$ mw: 258.50

SYN: CAPRYLYL PEROXIDE, solution (DOT)

DOT Classification: Label: Organic Peroxide.

SAFETY PROFILE: A powerful oxidizer. Probably a severe eye, skin, and mucous membrane irritant. When heated to decomposition it emits acrid smoke and irritating fumes.

OGE000   CAS: 5283-66-9   **HR: 2**
OCTYLTRICHLOROSILANE

DOT: UN 1801
mf: $C_8H_{17}Cl_3Si$ mw: 247.69

PROP: Fuming liquid.

CONSENSUS REPORTS: Reported in EPA TSCA Inventory.

DOT Classification: Corrosive Material; Label: Corrosive.

SAFETY PROFILE: A corrosive irritant to skin, eyes, and mucous membranes. Will react with water or steam to produce toxic and corrosive fumes. When heated to decomposition it emits toxic fumes of Cl^-.

OGK000   CAS: 8015-79-0   **HR: 3**
OIL of CALAMUS, GERMAN

PROP: Extract of *Acorus calamus L., araceae.* Containing: asarone, eugenol; esters of acetic and heptylic acids. Volatile oil. Yellow to yellowish-brown liquid (viscid); aromatic odor, bitter taste. D: 0.960-0.9707 @ 20°/20°. Very sltly sol in water, misc with alc. Keep well closed, cool, and protected from light.

SYNS: CALAMUS OIL   *   KALMUS OEL (GERMAN)   *   OIL of SWEET FLAG

CONSENSUS REPORTS: Reported in EPA TSCA Inventory.

SAFETY PROFILE: Poison by intraperitoneal route. Moderately toxic by ingestion. Questionable carcinogen with experimental tumorigenic data. When heated to decomposition it emits acrid smoke and irritating fumes.

OGM000   **HR: 3**
OIL GAS

PROP: A gas derived from petroleum. Composition: illuminants 4.2%, carbon monoxide 10.4%, hydrogen 47.6%, methane 27.0%, carbon dioxide 4.6%, nitrogen 5.8%, oxygen 0.4%. Lel: 4.8%; uel: 32.5%; autoign temp: 637°F.

SAFETY PROFILE: A poison. A very dangerous fire hazard when exposed to heat or flame; can react vigorously with oxidizing materials. Explosive in the form of vapor when exposed to heat or flame. To fight fire, use CO_2, dry chemical, water spray.

OGO000   CAS: 8008-26-2   **HR: 2**
OIL of LIME, DISTILLED

PROP: From distillation of juice or crushed fruit of *Citrus aurantofolia* Swingle. Colorless to green-yellow liquid. Sol in fixed oils, mineral oil; insol glycerin, propylene glycol.

SYNS: DISTILLED LIME OIL   *   LIME OIL   *   LIME OIL, DISTILLED (FCC)   *   OILS, LIME

CONSENSUS REPORTS: Reported in EPA TSCA Inventory.

SAFETY PROFILE: A skin irritant. Questionable carcinogen with experimental tumorigenic data. Mutation data reported. When heated to decomposition it emits acrid smoke and irritating fumes.

OGQ100   CAS: 8007-12-3   **HR: 2**
OIL of MACE

PROP: From steam distillation of dried arillode of the ripe seed of *Myristica fragrans* Houtt. (Fam. *Myristicaceae*). Colorless to pale yellow liquid; odor and taste of nutmeg. East Indian: d: 0.880-0.930, refr index: 1.474-1.488; West Indian: d: 0.854-0.880, refr index: 1.469-1.480 @20°. Sol in fixed oils, mineral oil; sltly sol in cold alc; very sol in hot alc, chloroform, ether; insol in glycerin, propylene glycol.

SYNS: NCI-C56484   *   MACE OIL   *   OIL of NUTMEG, EXPRESSED

CONSENSUS REPORTS: Reported in EPA TSCA Inventory. EPA Genetic Toxicology Program.

SAFETY PROFILE: Moderately toxic by ingestion. A skin irritant. Human ingestion causes symptoms similar to volatile oil of nutmeg. When heated to decomposition it emits acrid smoke and irritating fumes.

OGY000   CAS: 8008-57-9   **HR: 2**
OIL of ORANGE

PROP: Yellow to deep-orange liquid; characteristic orange taste and odor. D: 0.842-0.846 @ 25°/25°, refr index: 1.472 @ 20°. Sol in 2 vols 90% alc, in 1 vol glacial acetic acid; sltly sol in water; misc with abs alc, carbon disulfide.

Keep well closed, cool, and protected from light. Oil expressed from the peel of *Citrus sinensis* L. Osbeck (Fam. *Rutaceae*).

SYNS: NEAT OIL of SWEET ORANGE * OIL of SWEET ORANGE * ORANGE OIL * ORANGE OIL, COLDPRESSED (FCC) * SWEET ORANGE OIL

CONSENSUS REPORTS: Reported in EPA TSCA Inventory.

SAFETY PROFILE: A skin irritant. Questionable carcinogen with experimental neoplastigenic data. When heated to decomposition it emits acrid smoke and irritating fumes.

OHA000   CAS: 85-86-9   **HR: 3**
OIL RED
mf: $C_{22}H_{16}N_4O$ mw: 352.42

SYNS: BENZENEAZOBENZENEAZO-β-NAPHTHOL * CERASINROT * C.I. SOLVENT RED 23 * D&C RED NO. 17 * FETTSCHARLACH * OIL SCARLET * ORGANOL SCARLET * 1-((4-(PHENYLAZO)PHEN-YL)AZO)-2-NAPHTHALENOL * 1-(p-PHENYLAZOPHEN-YLAZO)-2-NAPHTHOL * ROUGE CERASINE * SOMALIA RED III * SUDAN III * TETRAZO-BENZENE-β-NAPHTHOL * TONY RED

CONSENSUS REPORTS: IARC Cancer Review: GROUP 3 IMEMDT 7,56,87; Animal Inadequate Evidence IMEMDT 8,241,75. Reported in EPA TSCA Inventory.

SAFETY PROFILE: Poison by intraperitoneal route. Moderately toxic by subcutaneous and intrapleural routes. Questionable carcinogen. Mutation data reported. When heated to decomposition it emits toxic fumes of NO_x.

OHG000   CAS: 115-71-9   **HR: 2**
OIL of SANDALWOOD, EAST INDIAN
mf: $C_{15}H_{24}O$ mw: 220.39

PROP: From steam distillation of the ground dried wood of *Santalus album L.*. Colorless to sltly yellow viscous liquid; sandalwood odor. D: 0.965-0.973, refr index: 1.505. Very sol in alc, fixed oils, propylene glycol; in in water, glycerin.

SYNS: 5-(2,3-DIMETHYLTRICYCLO(2.2.1.0²·⁶)HEPT-3-YL)-2-METHYL-2-PENTEN-1-OL * FEMA No. 3006 * SANDALWOOD OIL, EAST INDIAN * α-SANTALOL (FCC)

CONSENSUS REPORTS: Reported in EPA TSCA Inventory.

SAFETY PROFILE: Moderately toxic by ingestion. A skin irritant. When heated to decomposition it emits acrid smoke and irritating fumes.

OHI000   CAS: 8006-80-2   **HR: 3**
OIL of SASSAFRAS

PROP: Yellow to reddish-yellow liquid; characteristic odor and taste of sassafras. D: 1.065-1.077 @ 25°/25°. Very sltly sol in water; sol in 2 vols 90% alc. Keep well closed, cool, and protected from light. 80% Safrol.

CONSENSUS REPORTS: Reported in EPA TSCA Inventory.

SAFETY PROFILE: Human poison by ingestion. When heated to decomposition it emits acrid smoke and irritating fumes.

OHO000   CAS: 6696-47-5   **HR: 3**
OLEANDOMYCIN HYDROCHLORIDE
mf: $C_{35}H_{61}NO_{12} \cdot ClH$ mw: 724.43

PROP: Long needles from ethyl acetate. Mp: 134-135°. Very sol in water.

SYN: OLEANDOMYCIN MONOHYDROCHLORIDE

SAFETY PROFILE: Poison by intravenous route. Moderately toxic by ingestion and subcutaneous routes. When heated to decomposition it emits very toxic fumes of NO_x and HCl.

OHO200   CAS: 7060-74-4   **HR: 3**
OLEANDOMYCIN PHOSPHATE
mf: $C_{35}H_{61}NO_{12} \cdot H_3O_4P$ mw: 785.97

SYN: MATROMYCIN

SAFETY PROFILE: Poison by intravenous route. Moderately toxic by ingestion and subcutaneous routes. When heated to decomposition it emits toxic fumes of PO_x and NO_x.

OHQ000   CAS: 465-16-7   **HR: 3**
OLEANDRIN
mf: $C_{32}H_{48}O_9$ mw: 576.80

PROP: From the leaves of *Nerium oleander L., Apocynaceae (Laurier rose)*. Crystals from dil methanol. Mp: 250°. Practically insol in water; sol in alc, chloroform.

SYNS: CORRIGEN * FOLIANDRIN * FOLINERIN * FOLINEVIN * NERIOL * NERIOLIN * NERIO-STENE * OLEANDRINE

SAFETY PROFILE: Poison by subcutaneous, intravenous, and possibly other routes. When

heated to decomposition it emits acrid smoke and irritating fumes.

OHS000 HR: 2
OLEFINS

PROP: Unsaturated aliphatic hydrocarbons having one or more double bonds.

SAFETY PROFILE: Unsaturated aliphatic hydrocarbons do not differ greatly from paraffins, particularly insofar as their toxic effect on working personnel is concerned. Ethylene and some of its homologs occur in manufactured and natural gases. Ethylene can be used as an anesthetic, and on inhalation in sufficient quantity it can be an asphyxiant. However, the greatest hazard from its use is the danger of fire and explosion. Prolonged or repeated exposures to high concentrations of various olefins have caused certain toxic effects in animals, such as liver damage and hyperplasia of the bone marrow (due to butene-2), but no corresponding effects have been discovered in human beings due to industrial exposures. The diolefins, butadiene and isoprene, are more irritating than paraffins or mono-olefins of the same volatility. The α-olefins (e.g., 1-octene, 1-octadecene) are particularly reactive because the double bond is on the first carbon. In general the olefins have comparatively low toxicity, but are fire and explosion hazards.

OHU000 CAS: 112-80-1 HR: 3
OLEIC ACID
mf: $C_{18}H_{34}O_2$ mw: 282.52

PROP: Colorless liquid; odorless when pure. Mp: 6°, bp: 360.0°, flash p: 372°F (CC), d: 0.895 @ 25°/25°, autoign temp: 685°F, vap press: 1 mm @ 176.5°, bp: 286° @ 100 mm. Insol in water; misc in alc and ether.

SYNS: CENTURY CD FATTY ACID * EMERSOL 210 * EMERSOL 213 * EMERSOL 6321 * EMERSOL 233LL * EMERSOL 221 LOW TITER WHITE OLEIC ACID * EMERSOL 220 WHITE OLEIC ACID * GLYCON RO * GLYCON WO * GROCO 2 * GROCO 4 * GROCO 5L * HY-PHI 1055 * HY-PHI 1088 * HY-PHI 2066 * HY-PHI 2088 * HY-PHI 2102 * INDUSTRENE 105 * INDUSTRENE 205 * INDUSTRENE 206 * K 52 * l'ACIDE OLEIQUE (FRENCH) * METAUPON * NEO-FAT 90-04 * NEO-FAT 92-04 * cis-Δ^9-OCTADECENOIC ACID * cis-OCTADEC-9-ENOIC ACID * cis-9-OCTADECENOIC ACID * 9,10-OCTADECENOIC ACID * PAMOLYN * RED OIL

* TEGO-OLEIC 130 * VOPCOLENE 27 * WECO-LINE OO * WOCHEM NO. 320

CONSENSUS REPORTS: Reported in EPA TSCA Inventory.

SAFETY PROFILE: Poison by intravenous route. Mildly toxic by ingestion. Mutation data reported. A human and experimental skin irritant. Questionable carcinogen with experimental tumorigenic data. Combustible when exposed to heat or flame. To fight fire, use CO_2, dry chemical. The peroxidized acid explodes on contact with aluminum. Potentially dangerous reaction with perchloric acid + heat. When heated to decomposition it emits acrid smoke and irritating fumes.

OHW000 CAS: 112-62-9 HR: 2
cis-OLEIC ACID, METHYL ESTER
mf: $C_{19}H_{36}O_2$ mw: 296.55

PROP: Oil. D: 0.874 @ 20°/4°, bp: 168-170°. Insol in water; misc in alc and ether.

SYNS: EMEREST 2301 * EMEREST 2801 * EMERY 2219 * EMERY 2310 * EMERY OLEIC ACID ESTER 2301 * KEMESTER 105 * KEMESTER 115 * KEMESTER 205 * KEMESTER 213 * METHYL-9-OCTADECENOATE * METHYL cis-9-OCTADECENOATE * METHYL (Z)-9-OCTADECENOATE * METHYL OLEATE * (Z)-9-OCTADECENOIC ACID METHYL ESTER

CONSENSUS REPORTS: Reported in EPA TSCA Inventory.

SAFETY PROFILE: Questionable carcinogen with experimental tumorigenic data by skin contact. When heated to decomposition it emits acrid smoke and irritating fumes.

OHY000 CAS: 143-18-0 HR: 1
OLEIC ACID, POTASSIUM SALT
mf: $C_{18}H_{34}O_2 \cdot K$ mw: 321.62

SYNS: POTASSIUM cis-9-OCTADECENOIC ACID * POTASSIUM OLEATE

CONSENSUS REPORTS: Reported in EPA TSCA Inventory.

SAFETY PROFILE: An eye irritant. When heated to decomposition it emits toxic fumes of K_2O.

OIA000 CAS: 143-19-1 HR: 3
OLEIC ACID, SODIUM SALT
mf: $C_{18}H_{33}O_2 \cdot Na$ mw: 304.50

PROP: White powder; slt tallow odor. Mp: 232-235°.

SYNS: EUNATROL * SODIUM OLEATE

CONSENSUS REPORTS: Reported in EPA TSCA Inventory.

SAFETY PROFILE: Poison by intravenous route. Migrates to food from packaging materials. Combustible when exposed to heat or flame. When heated to decomposition it emits toxic fumes of Na_2O.

OIM000   CAS: 8050-07-5   **HR: 1**
OLIBANUM GUM

PROP: Contains 3-8% volatile oil (pinene, dipentene, etc.), 60% resins, 20% gum (polysaccharide fraction) and 6-8% bassorin. A gum from the trees *Boswellia carterii* Birdw. and other *Boswellia* species (Fam. *Burseraceae*).

SYN: FRANKINCENSE GUM

CONSENSUS REPORTS: Reported in EPA TSCA Inventory.

SAFETY PROFILE: A skin irritant. When heated to decomposition it emits acrid smoke and irritating fumes.

OIQ000   CAS: 8001-25-0   **HR: 2**
OLIVE OIL

PROP: Yellow oil; pleasing, delicate flavor. Mp: $-6°$, flash p: 437°F (CC), autoign temp: 650°F, d: 0.909-0.915 @ 25°/25°. Becomes rancid on exposure to air. Sltly sol in alc; misc with ether, chloroform, carbon disulfide. From fruit of *Olea europaea* .

CONSENSUS REPORTS: Reported in EPA TSCA Inventory. EPA Genetic Toxicology Program.

SAFETY PROFILE: Moderately toxic by intraperitoneal route. A human skin irritant. Combustible when exposed to heat or flame; can react with oxidizing materials. Some spontaneous heating. To fight fire, use CO_2, dry chemical. When heated to decomposition it emits acrid smoke and irritating fumes.

OIY000   CAS: 26354-18-7   **HR: 3**
OMP 2
mf: $(C_{16}H_{32}O_2Sn \cdot C_5H_8O_2)_x$

PROP: Trialkyltin methacrylate polymer (NTIS** AD-A062-138).

SYNS: 2-METHYL-2-PROPENOIC ACID METHYL ESTER, POLYMER with TRIBUTYL(92-METHYL-1-OXO-2-PROPENYL)OXY)STANNANE * TRIBUTYL(METHACRYLOYLOXY)-STANNANE POLYMER with METHYL METHACRYLATE (8CI)

CONSENSUS REPORTS: Reported in EPA TSCA Inventory.

SAFETY PROFILE: Poison by ingestion and inhalation. A skin and eye irritant. When heated to decomposition it emits acrid smoke and irritating fumes.

OJD200   **HR: 1**
ONION OIL

PROP: From steam distillation of bulbs of *Allium ceoa* L. (Fam. *Lillaceae*). Clear amber liquid; strong pungent odor and taste of onion. Sol in fixed oils, mineral oil, alc; insol in glycerin, propylene glycol.

SYN: OIL of ONION

SAFETY PROFILE: Skin irritant. When heated to decomposition it emits acrid smoke and irritating fumes.

OJG000   **HR: 3**
OPIUM

SYN: GUM OPIUM

SAFETY PROFILE: Poison by ingestion. Mutation data reported. Use may lead to habituation and addiction. A narcotic, sedative, analgesic, and hypnotic. Source of morphine, codeine, papaverine, thebaine, etc. Can cause nausea, vomiting, constipation and respiratory problems. Combustible when exposed to heat or flame.

OJM000   **HR: 3**
ORGANOMETALS

PROP: Compounds containing carbon and a metal. Ordinarily metallic carbonates (calcium carbonate, etc.) are excluded and also metallic salts of common organic acids. Examples of organic metal compounds are Grignard compounds, such as methyl magnesium iodide (CH_3MgI), and metallic alkyls, such as butyllithium (C_4H_9Li), tetraethyllead, triethyl aluminum, tetrabutyl titanate, sodium methylate, copper phthalocyanine, and metallocenes. Also, there are many organotin compounds, such as monoalkyltins, monoaryltins, dialkyltins, diaryltins, trialkyltins, triaryltins, tetraalkyltins and tetraaryltins.

SAFETY PROFILE: Many are highly toxic or flammable. As an example, organotin compounds are poisons by ingestion and intravenous routes. Irritating to skin, eyes, and mucous membranes. Can damage lung tissue and the liver. Trialkyltins are most toxic as a group. Next are the dialkyltins and the monoalkyltins. In each major organotin group the ethyltin derivative is the most toxic, followed by the methyltins. This group of compounds is constantly growing in importance, but there is relatively little toxicity information on most of them. Alkyl compounds of lead, tin, mercury, and aluminum are known to be highly toxic. Less is known about other organometals, but for the most part they are highly reactive chemically and therefore dangerous, if only on direct contact. Until specific toxicity data become available, it is prudent to exercise great caution in handling organometals, particularly the alkyl forms. Many organolithium compounds are explosive.

OJO000   CAS: 8007-11-2   HR: 3
ORIGANUM OIL

PROP: Main constituent is carvacrol. From steam distillation of the herb *Thymus capitatus* Hoffm. et Link. Yellow to dark red brown liquid; pungent spicy odor of thyme oil. D: 0.935-0.960, refr index: 1.502 @ 20°. Sol in fixed oil, propylene glycol, mineral oil; insol in glycerin.

SYN: OIL of ORIGANUM

CONSENSUS REPORTS: Reported in EPA TSCA Inventory.

SAFETY PROFILE: Poison by skin contact. Moderately toxic by ingestion. A severe skin irritant. When heated to decomposition it emits acrid smoke and irritating fumes.

OJW000   CAS: 341-69-5   HR: 3
ORPHENADRINE HYDROCHLORIDE
mf: $C_{18}H_{23}NO \cdot ClH$ mw: 305.88

PROP: Crystals. Mp: 156-157°. Sol in water, alc, chloroform; sltly sol in acetone, benzene; almost insol in ether.

SYNS: BF 5930 * BG 5930 * BROCADISIPAL * BROCASIPAL * BS 5930 * 2-DIMETHYLAMINO-ETHYL-2-METHYLBENZHYDRYL ETHERHYDROCHLO-RIDE * N,N-DIMETHYL-2-(o-METHYL-α-PHENYLBEN-ZYLOXY)ETHYLAMINE HYDROCHLORIDE * DISIPAL HYDROCHLORIDE * MEPHENAMIN HYDROCHLORIDE * MEPHENAMINE HYDROCHLORIDE

SAFETY PROFILE: Poison by ingestion, intravenous, intraperitoneal, and subcutaneous routes. Experimental teratogenic and reproductive effects. When heated to decomposition it emits toxic fumes of NO_x and HCl.

OKE000   CAS: 7440-04-2   HR: 3
OSMIUM
af: Os   aw: 190.20

PROP: A lustrous, bluish-white, extremely hard and dense, brittle metal. Bp: 5027°, d: 22.57, mp: approx 2700°.

SYN: METALLIC OSMIUM

CONSENSUS REPORTS: Reported in EPA TSCA Inventory.

SAFETY PROFILE: Poison by intravenous route. An irritant to eyes and mucous membranes. The principal effects of exposure are ocular disturbances and an asthmatic condition caused by inhalation. Furthermore, it causes dermatitis and ulceration of the skin upon contact. When osmium is heated, it gives off a pungent, poisonous fume of osmium tetraoxide. One case of osmium poisoning reported in the literature resulted from the inhalation of osmium tetraoxide, which gave rise to a capillary bronchitis and dermatitis. The vapor has a pronounced and nauseating odor which should be taken as a warning of a possibly toxic concentration in the atmosphere, and personnel should immediately move to an area of fresh air. The metal itself is not highly toxic. Flammable in the form of dust when exposed to heat or flame. Slight explosion hazard in the form of dust when exposed to heat or flame. Violent reaction or ignition with chlorine trichloride or oxygen difluoride. Ignites when heated to 100°C with fluorine. Incandescent reaction in phosphorus vapor. When heated to decomposition it emits toxic fumes of OsO_4.

OKK000   CAS: 20816-12-0   HR: 3
OSMIUM TETROXIDE

DOT: UN 2471
mf: O_4Os mw: 254.20

PROP: (A) Monoclinic, colorless crystals; (B) yellow mass; pungent, chlorine-like odor. Mp (A): 39.5°, mp: (B): 41°, bp: 130° (sublimes), d: 4.906 @ 22°, vap press (A): 10 mm @ 26.0°, vap press (B): 10 mm @ 31.3°. Sol in benzene.

SYNS: OSMIC ACID * OSMIUM(VIII) OXIDE * RCRA WASTE NUMBER P087

CONSENSUS REPORTS: Community Right-To-Know List. EPA Genetic Toxicology Program. Reported in EPA TSCA Inventory.

OSHA PEL: (Transitional: TWA 0.002 mg/m^3 (Os)) TWA 0.0002 ppm; STEL 0.0006 ppm (Os)
ACGIH TLV: TWA 0.0002 ppm; STEL 0.0006 ppm (Os)
DFG MAK: 0.0002 ppm (0.002 mg/m^3)
DOT Classification: Poison B; Label: Poison.

SAFETY PROFILE: Poison by ingestion, inhalation, and intraperitoneal routes. Human systemic effects by inhalation: lacrimation and other eye effects and structural or functional changes in trachea or bronchi. Experimental reproductive effects. Mutation data reported. Explodes on contact with 1-methylimidazole. Catalytic decomposition of hydrogen peroxide can be hazardous.

OKS000   CAS: 630-60-4   HR: 3
OUABAIN
mf: C$_{29}$H$_{44}$O$_{12}$ mw: 584.73

PROP: A natural plant product.

SYNS: ACOCANTHERIN * ASTROBAIN * GRATIBAIN * GRATUS STROPHANTHIN * G-STROPHANTHIN * OUABAGENIN-l-RHAMNOSID (GERMAN) * OUABAGENIN-l-RHAMNOSIDE * OUBAIN * OUABAINE * PUROSTROPHAN * STROPHANTHIN G * STROPHOPERM

CONSENSUS REPORTS: EPA Extremely Hazardous Substances List. Reported in EPA TSCA Inventory.

SAFETY PROFILE: Poison by ingestion, intramuscular, intraperitoneal, intravenous, subcutaneous, parenteral and possibly other routes. Moderately toxic by intraduodenal route. A cardiac stimulant. When heated to decomposition it emits acrid smoke and irritating fumes.

OKY000   HR: 3
OXALATES
PROP: Salts of oxalic acid.

SAFETY PROFILE: Poisons by ingestion and inhalation. Powerful irritants. Oxalates are corrosive to tissue and produce local irritation. When ingested they have a caustic effect on the mouth, esophagus, and stomach. The soluble oxalates are readily absorbed from the gastrointestinal tract and can cause severe damage to the kidneys. Oxalates are common components of poisonous plants. When heated to decomposition they emit toxic and irritating fumes.

OLA000   CAS: 144-62-7   HR: 3
OXALIC ACID
mf: C$_2$H$_2$O$_4$ mw: 90.04

PROP: Orthorhombic colorless crystals. Mp: 101°, sublimes @ 150°, d: 1.653. Sol in water, abs alc, and ether

SYNS: ACIDE OXALIQUE (FRENCH) * ACIDO OSSALICO (ITALIAN) * ETHANEDIOIC ACID * ETHANEDIONIC ACID * KYSELINA STAVELOVA (CZECH) * NCI-C55209 * OXAALZUUR (DUTCH) * OXALSAEURE (GERMAN)

CONSENSUS REPORTS: Reported in EPA TSCA Inventory.

OSHA PEL: (Transitional: TWA 1 mg/m^3) TWA 1 mg/m^3; STEL 2 mg/m^3
ACGIH TLV: TWA 1 mg/m^3; STEL 2 mg/m^3

SAFETY PROFILE: Poison by ingestion, skin contact, and subcutaneous routes. Moderately toxic by an unspecified route. A skin and severe eye irritant. Acute oxalic poisoning results from ingestion of a solution of the acid. There is marked corrosion of the mouth, esophagus and stomach with symptoms of vomiting, burning and abdominal pain, collapse and sometimes convulsions. Death may follow quickly. The systemic effects are attributed to the removal by the oxalic acid of the calcium in the blood. The renal tubules become obstructed by the insoluble calcium oxalate, and there is profound kidney disturbance. The chief effects of inhalation of the dusts or vapor are severe irritation of the eyes and upper respiratory tract, gastrointestinal disturbances, albuminuria, gradual loss of weight, increasing weakness and nervous system complaints, ulceration of the mucous membranes of the nose and throat, epistaxis, headache, irritation and nervousness. Oxalic acid has a caustic action on the skin and may cause dermatitis; a case of early gangrene of the fingers resembling that caused by phenol has been described. More severe cases may show albuminuria, chronic cough, vomiting, pain in the back and gradual emaciation and weakness. The skin lesions are characterized by cracking and fissuring of the skin and the development of slow-healing ulcers. The skin may be bluish in color, and the nails brittle and yellow. Violent reaction with furfuryl alcohol, Ag, NaClO$_3$, NaOCl. When heated to decomposition it emits acrid smoke and irritating fumes.

OLO000   CAS: 471-46-5   **HR: 3**
OXAMIDE
mf: C$_2$H$_4$N$_2$O$_2$ mw: 88.08

PROP: Triclinic needles. Decomp @ 350°, d: 1.667 @ 20°/4°. Sltly sol in hot water, alc.

SYNS: AMID KYSELINY STAVELOVE (CZECH) * 1-CARBAMOYLFORMIMIDIC ACID * ETHANE-DIAMIDE * OXALAMIDE * OXALIC ACID DIAMIDE * OXAMID (CZECH) * OXAMIMIDIC ACID

CONSENSUS REPORTS: Reported in EPA TSCA Inventory.

SAFETY PROFILE: Poison by ingestion and intraperitoneal routes. An eye irritant. When heated to decomposition it emits toxic fumes of NO$_x$.

OLS000   CAS: 10039-54-0   **HR: 3**
OXAMMONIUM SULFATE
DOT: UN 2865
mf: H$_6$N$_2$O$_2$ • H$_2$O$_4$S   mw: 164.16

PROP: A crystalline material. Mp: 177°.

SYNS: BIS(HYDROXYLAMINE) SULFATE * HYDROXYLAMINE NEUTRAL SULFATE * HYDROXYLAMINE SULFATE * HYDROXYLAMINE SULFATE (2:1) * HYDROXYLAMMONIUM SULFATE

CONSENSUS REPORTS: Reported in EPA TSCA Inventory.

DOT Classification: Corrosive Material; Label: Corrosive.

SAFETY PROFILE: Poison by intraperitoneal route. Mutation data reported. A corrosive irritant to skin, eyes, and mucous membranes. Moderately explosive when exposed to heat or by chemical reaction. In the presence of alkalies at elevated temperatures, free hydroxylamine is liberated and may decompose explosively. When heated to decomposition it emits toxic fumes of SO$_x$ and NO$_x$.

OMG000   CAS: 60607-34-3   **HR: 3**
OXATIMIDE
mf: C$_{27}$H$_{30}$N$_4$O   mw: 426.61

SYNS: 1-(3-(4-(DIPHENYLMETHYL)-1-PIPERAZINYL) PROPYL)-2-BENZIMIDAZOLINONE * 1-(3-(4-(DIPHE-NYLMETHYL)-1-PIPERAZINYL)PROPYL)-1,3-DIHYDRO-2H-

BENZ IMIDAZOL-2-ONE * KW-4354 * OXATOMIDA * OXATOMIDE * R 35443 * TINSET

SAFETY PROFILE: Poison by ingestion, intra-peritoneal, and intravenous routes. Experimental reproductive effects. Used to treat allergies and asthma. When heated to decomposition it emits toxic fumes of NO$_x$.

OMW000   CAS: 503-30-0   **HR: 2**
OXETANE
mf: C$_3$H$_6$O   mw: 58.09

PROP: Oil; agreeable odor. D: 0.8930 @ 25°/4°, bp: 480 @ 750 mm.

SYNS: 1,3-PROPYLENE OXIDE * TRIMETHYLENE OXIDE * TRIMETHYLENOXID (GERMAN)

CONSENSUS REPORTS: Reported in EPA TSCA Inventory.

SAFETY PROFILE: Moderately toxic by subcutaneous route. May be narcotic in high concentrations. Questionable carcinogen with experimental tumorigenic data. When heated to decomposition it emits acrid smoke and irritating fumes.

ONY000   CAS: 1707-95-5   **HR: 3**
2-(3-OXO-1-INDANYLIDENE)-1,3-INDANDIONE
mf: C$_{18}$H$_{10}$O$_3$ mw: 274.28

SYN: BINDON

SAFETY PROFILE: Poison by intraperitoneal route. Experimental teratogenic and reproductive effects. When heated to decomposition it emits acrid smoke and irritating fumes.

OOE000   CAS: 1949-20-8   **HR: 3**
OXOLAMINE CITRATE
mf: C$_{14}$H$_{19}$N$_3$O • C$_6$H$_8$O$_7$ mw:437.50

PROP: Crystals. Sltly sol in water and alc.

SYNS: 5-β-DIETHYLAMINOETHYL-3-PHENYL-1,2,4-OXADIAZOLE CITRATE * 3-PHENYL-5-(β-(DIETHYLAMI-NO)ETHYL)-1,2,4-OXADIAZOLE CITRATE

SAFETY PROFILE: Poison by intraperitoneal route. Moderately toxic by ingestion. Experimental teratogenic and reproductive effects. Questionable carcinogen with experimental carcinogenic data. When heated to decomposition it emits toxic fumes of NO$_x$.

OPM000   CAS: 101-80-4   **HR: 3**
4,4′-OXYDIANILINE
mf: $C_{12}H_{12}N_2O$ mw: 200.26

PROP: Colorless crystals. Mp: 187°, bp: >300°.

SYNS: p-AMINOPHENYL ETHER * 4-AMINOPHENYL ETHER * BIS(4-AMINOPHENYL)ETHER * BIS(p-AMINOPHENYL)ETHER * DADPE * 4,4′-DIAMINOBIPHENYLOXIDE * DIAMINODIPHENYL ETHER * 4,4-DIAMINODIPHENYL ETHER * p,p′-DIAMINODIPHENYL ETHER * 4,4′-DIAMINODIPHENYL OXIDE * 4,4′-DIAMINOPHENYL ETHER * NCI-C50146 * OXYBIS(4-AMINOBENZENE) * 4,4′-OXYBISANILINE * p,p′-OXYBIS(ANILINE) * 4,4′-OXYBISBENZENAMINE * OXYDIANILINE * p,p′-OXYDIANILINE * 4,4′-OXYDIPHENYLAMINE * OXYDI-p-PHENYLENEDIAMINE

CONSENSUS REPORTS: IARC Cancer Review: GROUP 2B IMEMDT 7,56,87; Animal Sufficient Evidence IMEMDT 29,203,82; Animal Inadequate Evidence IMEMDT 16,301,78. NCI Carcinogenesis Bioassay (feed); Clear Evidence: mouse, rat NCITR* NCI-CG-TR-205,80. Reported in EPA TSCA Inventory.

DFG MAK: Animal Carcinogen, Suspected Human Carcinogen.

SAFETY PROFILE: Confirmed carcinogen with experimental carcinogenic, neoplastigenic, and tumorigenic data. Poison by intraperitoneal route. Moderately toxic by ingestion. Mutation data reported. Experimental reproductive effects. Mutation data reported. When heated to decomposition it emits toxic fumes of NO_x.

OQS000   CAS: 2497-07-6   **HR: 3**
OXYDISULFOTON
mf: $C_8H_{19}O_3PS_3$ mw: 290.42

SYNS: BAY 23323 * O,O-DIETHYL-S-((ETHYLSULFINYL)ETHYL)PHOSPHORODITHIOATE * O,O-DIETHYL S-(2-(ETHYLSULFINYL)ETHYL) PHOSPHORODITHIOATE * DISULFOTON DISULIDE * DISULFOTON SULFOXIDE * DISYSTON SULFOXIDE * ETHYLTHIOMETON SULFOXIDE

SAFETY PROFILE: Poison by ingestion and skin contact. When heated to decomposition it emits very toxic fumes of SO_x and PO_x.

OQW000   CAS: 7782-44-7   **HR: 3**
OXYGEN

DOT: UN 1072/UN 1073
mf: O_2 mw: 32.00

PROP: Colorless, odorless, tasteless gas, liquid, or hexagonal crystals. Supports combustion. D (liquid): 1.14 @ −183.0°, d (solid): 1.426 @ −252.5°, vap d: 1.429 @ 0°. D: (gas) 1.429 g/L @ 0°, mp: −218.4°, bp: −182.96°. One vol gas dissolves in 32 vols water @ 20°, dissolves in 7 vols alc @ 20°. Sol in other organic liquids to a greater extent than water.

SYNS: OXYGEN, compressed (DOT) * OXYGEN, refrigerated liquid (DOT)

CONSENSUS REPORTS: Reported in EPA TSCA Inventory. EPA Genetic Toxicology Program.

DOT Classification: Nonflammable Gas; Label: Nonflammable Gas, Oxidizer; Nonflammable Gas; Label: Oxidizer.

SAFETY PROFILE: Human systemic effects by inhalation: cough and other pulmonary changes. Human teratogenic effects by inhalation: developmental abnormalities of the fetal cardiovascular system. Mutation data reported. Not toxic as gas. In liquid form it can cause severe ''burns'' and tissue damage on contact with the skin due to extreme cold.

An oxidant. Though itself nonflammable, it is essential to combustion. Even a slight increase in the oxygen content of the air above the normal 21% greatly increases the oxidation or burning rate (and the hazard) of many materials. Exclusion of O_2 from the neighborhood of a fire is one of the principal methods of extinguishment. Avoid smoking, flames, electric sparks. Liquid O_2 can explode on contact with readily oxidizable materials, especially at high temperatures. Under the proper conditions of temperature, pressure, and reagent concentration it can react violently with acetaldehyde, acetylene, acetone, secondary-alcohols (e.g., 2-propanol, 2-butanol) aluminum, $Al(BH_4)_3$, AlH_3, aluminum-titanium alloys, alkali metals (lithium, cesium, potassium, rubidium, sodium, potassium), ammonia, ammonia + platinum, asphalt, CCl_4, chlorinated hydrocarbons, cyanogen, barium, benzene, 1,4-benzenediol + 1-propanol, benzoic acid, $Be(BH_4)_2$, biological materials + ether, BAs_2Br_3, B_2H_{10}, B_2H_6, boron tribromide, boron trichloride, bromine + chlorotrifluoroethylene, butane + $Ni(CO)_4$, carbon disulfide, carbon disulfide + mercury + anthracene, carbon monoxide, CsH, calcium, calcium phosphide, copper + hydrogen sulfide, $C_{10}H_{14}$, cyclohexane-1,2-dione bis(phenylhydrazone), cyclooctatetraene, diborane, diboron tetrafluoride, dimethoxymethane, dimethylketene, dimethyl sulfide, diphenyl ethylene, disilane, ethers (e.g., diethyl ether, diisopropyl ether, tetrahydrofuran, dioxane, ethyl ether), fibrous fabrics, fluorine + hydrogen, fuels, germanium, glycerol, halocarbons (e.g., 1,1,1-trichloroethane, trichloroethylene, chlorotrifluoroethylene, bromotrifluoroethylene), hydrazine, hydrocarbons (e.g., 1,1-diphenylethylene, gasoline, cyclohexane, ethylene, cumene, p-xylene, buten-3-yne), hydrocarbons + promoters (e.g., methyl nitrate, nitromethane, ethyl nitrate, tetrafluorohydrazine), hydrogen, hydrogen sulfide, lithiated dialkylnitrosoamines, magnesium, metals, metal hydrides (e.g., sodium hydride, uranium hydride, lithium hydride, potassium hydride, rubidium hydride, cesium hydride, magnesium hydride), methane, methoxycyclooctatetraene, 4-methoxytoluene, $Ni(CO)_4$ + butane, non-metal hydrides (e.g., diborane, tetraborane(10), phosphine, pentaborane(11), pentaborane(9), decaborane(14), aluminum tetrahydroborate), oil films, organic matter, $(OF_2 + H_2O)$, phosphorus, phosphorus tribromide, phosphorus trifluoride, phosphorus(III) oxide, polymers [e.g., foam rubber, neoprene, polytetrafluoroethylene (teflon)], polytetrafluoroethylene + stainless steel, polyurethane, polyvinyl chloride, propylene oxide, K_2O_2, rhenium, trirhenium nonachloride, rubber + ozone, rubberized fabric, selenium, NaH, sodium hydroxide + tetramethyldisiloxane, strontium, tetracarbonylnickel, tetracarbonylnickel + mercury, tetrafluoroethylene, tetrafluorohydrazine, tetrasilane, titanium and alloys, trisilane, CH_2Cl_2, oil, paraformaldehyde, wood, charcoal. Compressed O_2 is shipped in steel cylinders under high pressure. If these containers are broken due to shock or exposed to high temperature, an explosion and fire may result.

ORA000 CAS: 7783-41-7 **HR: 3**
OXYGEN DIFLUORIDE
DOT: UN 2190
mf: F_2O mw: 54.00

PROP: Colorless gas or yellowish-brown liquid. Reacts slowly with water. D: (liquid) 1.90 @ −224°, mp: −223.8°, bp: −144.8°.

SYNS: FLUORINE MONOXIDE * FLUORINE OXIDE
* OXYGEN FLUORIDE

OSHA PEL: (Transitional: TWA 0.1 mg/m^3) CL 0.05 ppm
ACGIH TLV: CL 0.05 ppm
DOT Classification: Poison A; Label: Poison Gas.

SAFETY PROFILE: Poison by inhalation. Human systemic effects by inhalation: chronic pulmonary edema or congestion. A corrosive skin, eye and mucous membrane irritant. Attacks lungs with delayed appearance of symptoms. A very powerful oxidizer. Must be kept away from contact with reducing agents. Explosive reaction with adsorbents (e.g., silica gel, alumina, molecular sieve), diborane, halogens + heat, metal halides, aluminum chloride, antimony pentachloride (at 150°C), tungsten + heat, hydrogen sulfide, liquid nitrogen oxide, nitrosyl fluoride, charcoal, sulfur tetrafluoride. Forms spark-sensitive explosive mixtures with water or combustible gases (e.g., carbon monoxide, hydrogen, methane). Ignites on contact with diborane tetrafluoride, non-metals (e.g., red phosphorus, boron powder, silicon), phosphorus(V) oxide, nitrogen oxide gas. Incandescent reaction with metals (e.g., aluminum, barium, cadmium, magnesium, strontium, zinc, zirconium, lithium (above 400°C), potassium (above 400°C), sodium. Incompatible with NH_3, As_2O_3, Cl_2 + Cu, CrO_3, Ir, O_3, O_2 + H_2O, Pd, Pt, Rh, Ru, SiO_2. When heated to decomposition it emits highly toxic fumes of F$^-$.

ORQ000 CAS: 50-10-2 **HR: 3**
OXYPHENONIUM BROMIDE
mf: $C_{21}H_{34}NO_3 \cdot Br$ mw: 428.47

SYNS: ANTRENIL * ANTRENYL * ANTRENYL BROMIDE * DIETHYL(2-HYDROXYETHYL)METHYL-AMMONIUM BROMIDE α-PHENYLCYCLOHEXANEGLYCOLATE * ETHANAMINIUM, 2-((CYCLOHEXYLHYDROXYPHENYLACETYL)OXY)-N,N-DIETHYL-N-METHYL-, BROMIDE (9CI) * METACIN * METATSIN * METHACIN * OXIFENON * OXYFENON * OXYPHENON * OXYPHENONIUM * SPASMOPHEN

SAFETY PROFILE: Poison by ingestion, intravenous, and subcutaneous routes. Human toxic effects by ingestion. When heated to decomposition it emits very toxic fumes of NO_x, NH_3, and Br$^-$.

ORS000   CAS: 39603-54-8   **HR: 2**
β-OXYPROPYLPROPYLNITROSA-
MINE
mf: $C_6H_{12}N_2O_2$ mw: 144.20

SYNS: N-NITROSO-2-OXO-N-PROPYL-N-PROPYLAMINE
* 1-(NITROSOPROPYLAMINO)-2-PROPANONE
* 2-OXI-PROPYL-PROPYLNITROSAMIN (GERMAN)
* 2-OXO-PROPYL-PROPYLNITROSAMINE * (2-OXO-
PROPYL)PROPYLNITROSOAMINE

SAFETY PROFILE: Suspected carcinogen with
experimental carcinogenic, neoplastigenic, and
tumorigenic data. Moderately toxic by subcutaneous route. Mutation data reported. When
heated to decomposition it emits toxic fumes
of NO_x.

ORW000   CAS: 10028-15-6   **HR: 3**
OZONE
mf: O_3 mw: 48.00

PROP: Unstable colorless gas or dark blue liquid; characteristic odor. Mp: $-193°$, bp:
$-111.9°$, d (gas): 2.144 g/L, 1.71 @ $-183°$.
D: (liquid) 1.614 g/mL @ $-195.4°$.

SYNS: OZON (POLISH) * TRIATOMIC OXYGEN

CONSENSUS REPORTS: Reported in EPA
TSCA Inventory. EPA Genetic Toxicology Program.

OSHA PEL: (Transitional: TWA 0.1 ppm)
  TWA 0.1 ppm; STEL 0.3 ppm
ACGIH TLV: TWA CL 0.1 ppm
DFG MAK: 0.1 ppm (0.2 mg/m³)

SAFETY PROFILE: A human poison by inhalation. Human systemic effects by inhalation: visual field changes, lacrimation, headache, decreased pulse rate with fall in blood pressure,
blood pressure decrease, dermatitis, cough,
dyspnea, respiratory stimulation and other pulmonary changes. Experimental teratogenic and
reproductive effects. Human mutation data reported. A skin, eye, upper respiratory system
and mucous membrane irritant. Questionable
carcinogen with experimental neoplastigenic
and tumorigenic data. Can be a safe water disinfectant in low concentration. Concentration of
0.015 ppm of ozone in air produces a barely
detectable odor. Concentrations of 1 ppm produce a disagreeable sulfur-like odor and may
cause headache and irritation of eyes and the
upper respiratory tract; symptoms disappear after leaving the exposure.

A powerful oxidizing agent. Dangerous
chemical reaction with acetylene, alkenes, al-
kylmetals (e.g., dimethylzinc, diethylzinc), antimony, aromatic compounds (e.g., benzene,
aniline), benzene + oxygen + rubber, bromine,
charcoal + potassium iodide, citronellic acid,
combustible gases (e.g., carbon monoxide,
ethylene, nitrogen oxide, ammonia, phosphine),
(diallyl methyl carbinol + acetic acid), trans-
2,3-dichloro-2-butene, dicyanogen, dienes +
oxygen, diethyl ether, 1,1-difluoroethylene,
N_2O_5, ethylene, ethylene + formyl fluoride,
fluoroethylene, liquid hydrogen, hydrogen +
oxygen difluoride, hydrogen bromide, hydrogen
iodide, 4-hydroxy-4-methyl-1,6-heptadiene,
23-hydroxy-2,2,4-trimethyl-3-pentenoic acid
lactone, isopropylidene compounds, nitrogen,
NO_2, NO, nitrogen trichloride, nitrogen triiodide, nitroglycerin, organic liquids, organic
matter, oxygen + rubber powder, oxygen fluorides (e.g., dioxygen difluoride, dioxygen
trifluoride), silica gel, stibine, tetrafluorohydrazine, tetramethylammonium hydroxide,
trifluoroethylene, unsaturated acetals. A severe
explosion hazard in liquid form when shocked,
exposed to heat or flame, or in concentrated
form by chemical reaction with powerful reducing agents. Incompatible with rubber, dinitrogen
tetraoxide.

ORY000   **HR: 3**
OZONE mixed with NITROGEN OXIDES
(53%:47%)

SYN: NITROGEN OXIDES mixed with OZONE (47%:53%)

SAFETY PROFILE: Poison by inhalation. Human systemic effects by inhalation: central nervous system effects.

ORY499   **HR: 3**
OZONIDES

SYN: TRIOXOLANES

SAFETY PROFILE: Many are unstable explosives. The presence of peroxides is thought to
be the cause of instability. Polymeric alkene
ozonides (e.g., trans-2-butene ozonide) are
shock-sensitive explosives. They are decomposed by the catalytic action of powdered palladium, platinum, silver, or iron(II) salts.

P

PAD250 CAS: 7440-05-3 **HR: 1**
PALLADIUM
af: Pd   aw: 106.4

PROP: A steely white, stable metal; can be annealed to be soft and ductile. Mp: 1555°, bp: 3167°, d: 12.02 @ 20°/4. Volatile at high temps.

SAFETY PROFILE: May be a skin sensitizer. This metal in the form of palladium chloride has been administered orally in dosage of about 1 grain daily in the treatment of tuberculosis without apparent ill effects. In the laboratory, palladium appears to bind to many cell components; blocks the action of a number of enzymes and interferes with the use of energy by nerves and muscles; induces lung malfunction and produces abnormal fetuses. Lethal intravenous doses cause appetite loss, hemolysis, renal deposition and bone marrow damage. Poorly absorbed by the body when ingested. Palladium dust can be a fire and explosion hazard. Combustible in the form of dust when exposed to heat, or flame. Explosive reaction with hydrogen + hydrogen peroxide. Reaction with formic acid or sodium tetrahydroborate releases explosive hydrogen gas. Violent reaction with isopropyl alcohol, OF_2S. Under the proper conditions it undergoes hazardous reactions with aluminum, arsenic, carbon, methanol, ozonides, sulfur.

PAD500 CAS: 7647-10-1 **HR: 3**
PALLADIUM(2^+) CHLORIDE
mf: Cl_2Pd mw: 177.30

PROP: Dark brown, deliquescent crystals. D: 4.0 @ 18°, mp: 678-680° (decomp). Sol in water, alc, acetone, and hydrochloric acid.

SYNS: NCI-C60184 * PALLADIUM CHLORIDE * PALLADOUS CHLORIDE

CONSENSUS REPORTS: Reported in EPA TSCA Inventory. EPA Genetic Toxicology Program.

SAFETY PROFILE: Poison by ingestion, intraperitoneal, intravenous, and intratracheal routes. Experimental reproductive effects. A skin irritant. Questionable carcinogen with experimental carcinogenic data. Human mutation data reported. When heated to decomposition it emits highly toxic fumes of Cl^-.

PAE000 CAS: 8014-19-5 **HR: 1**
PALMAROSA OIL

PROP: From steam distillation of the grass *Cymbopogon Martini* Stapf. Var. Motia, mainly *Geraniol*. Yellow oily liquid. D: 0.879-0.892, refr index: 1.473 @ 20°. Sol in fixed oils, propylene glycol, mineral oil; insol in glycerin.

SYNS: GERANIUM OIL, EAST INDIAN TYPE * GERANIUM OIL, TURKISH TYPE * OIL of PALMAROSA

CONSENSUS REPORTS: Reported in EPA TSCA Inventory.

SAFETY PROFILE: A skin irritant. When heated to decomposition it emits acrid smoke and irritating fumes.

PAE250 CAS: 57-10-3 **HR: 3**
PALMITIC ACID
mf: $C_{17}H_{32}O_2$ mw: 256.48

PROP: Colorless plates or white crystalline powder; slt characteristic odor and taste. D: 0.849 @ 70°/4°, mp: 63-64°, bp: 271.5° @ 100 mm. Insol in water; very sltly sol in petr ether; sol in absolute ether, chloroform.

SYNS: CETYLIC ACID * EMERSOL 140 * EMERSOL 143 * HEXADECANOIC ACID * n-HEXADECOIC ACID * HEXADECYLIC ACID * HYDROFOL * HYSTRENE 8016 * INDUSTRENE 4516 * 1-PENTADECANECARBOXYLIC ACID

CONSENSUS REPORTS: Reported in EPA TSCA Inventory.

SAFETY PROFILE: A poison by intravenous route. A human skin irritant. Questionable carcinogen with experimental neoplastigenic data. When heated to decomposition it emits acrid smoke and irritating fumes.

PAG200 CAS: 81-13-0 **HR: 2**
d-PANTHENOL
mf: $C_9H_{19}NO_4$ mw: 205.29

PROP: Viscous, somewhat hygroscopic liquid; sltly bitter taste. D: (20/20) 1.2, bp: 118-120°, easily decomp on distillation. Freely sol in water, alc, methanol, ether; sltly sol in glycerin. Natural pH about 9.5.

SYNS: ALCOPAN-250 * BEPANTHEN * BEPANTHENE * BEPANTOL * COZYME * DEXPAN-

THENOL (FCC) * d-(+)-2,4-DIHYDROXY-N-(3-HY-
DROXYPROPYL)-3,3-DIMETHYLBUTYRAMIDE
* D-P-A INJECTION * ILOPAN * MOTILYN
* PANADON * PANTHENOL * d(+)-PANTHENOL
(FCC) * PANTHODERM * PANTOL
* PANTOTHENOL * d-PANTOTHENOL * PAN-
TOTHENYL ALCOHOL * d-PANTOTHENYL AL-
COHOL * d(+)-PANTOTHENYL ALCOHOL
* THENALTON * ZENTINIC

CONSENSUS REPORTS: Reported in EPA
TSCA Inventory.

SAFETY PROFILE: Moderately toxic by intravenous route. When heated to decomposition
it emits toxic fumes of NO_x.

PAG500   CAS: 9001-73-4   **HR: D**
PAPAIN

PROP: White to gray, sltly hygroscopic powder. Sol in water and glycerin; insol in other
common organic solvents. The most thermo-
static enzyme known, digests protein. Isolated
from the latex of the green fruit and leaves of
Carcia papaya L..

SYNS: ARBUZ * CAROID * NEMATOLYT
* PAPAYOTIN * SUMMETRIN * TROMASIN
* VEGETABLE PEPSIN * VELARDON * VER-
MIZYM

CONSENSUS REPORTS: Reported in EPA
TSCA Inventory.

SAFETY PROFILE: Experimental teratogenic
and reproductive effects. An allergen. When
heated to decomposition it emits toxic fumes
of NO_x.

PAG750   CAS: 63905-64-6   **HR: 3**
**PAPAVERIN CARBOXYLIC ACID,
SODIUM SALT**
mf: $C_{21}H_{20}NO_6 \cdot Na$ mw: 405.41

SYNS: 1-(3,4)-DIMETHOXYBENZYL-6,7-DIMETHOXY-
ISOQUINOLINE-3-CARBOXYLIC ACID, SODIUM SALT
* 6,7-DIMETHOXY-1-VERATRYLISOQUINOLINE-3-CAR-
BOXYLIC ACID SODIUM SALT

SAFETY PROFILE: Poison by ingestion, intravenous, and subcutaneous routes. Human
systemic effects by intravenous route: sensory
changes in peripheral nerves, spasticity. When
heated to decomposition it emits toxic fumes
of NO_x and Na_2O.

PAH000   CAS: 58-74-2   **HR: 3**
PAPAVERINE
mf: $C_{20}H_{21}NO_4$ mw: 339.42

PROP: Colorless, rhombic needles. Mp: 147°,
bp: decomp, d: 1.337 @ 20°/4°. Insol in water;
sol in hot benzene, glacial acetic acid, acetone;
sltly sol in chloroform, carbon tetrachloride,
petr ether.

SYNS: 1-((3,4-DIMETHOXYPHENYL)METHYL)-6,7-DIME-
THOXYISOQUINOLINE * 6,7-DIMETHOXY-1-VERA-
TRYLISOQUINOLINE * PAPANERINE * PAPAVE-
RINA (ITALIAN)

SAFETY PROFILE: Poison by ingestion, intramuscular, subcutaneous, intradermal, intraperitoneal, and intravenous routes. Its central nervous system action is about midway between
morphine and codeine, and large doses do not
produce the amount of excitement caused by
codeine or the soporific action of morphine.
Mutation data reported. A cerebral vasodilator
and smooth muscle relaxant. Combustible when
exposed to heat or flame. When heated to decomposition it emits toxic fumes of NO_x.

PAH750   CAS: 8002-74-2   **HR: 3**
PARAFFIN

PROP: Colorless or white, translucent wax;
odorless. D: approx 0.90, mp: 50-57°. Insol
in water, alc; sol in benzene, chloroform, ether,
carbon disulfide, oils; misc with fats.

SYNS: PARAFFIN WAX * PARAFFIN WAX FUME
(ACGIH)

CONSENSUS REPORTS: Reported in EPA
TSCA Inventory.

OSHA PEL: Fume: TWA 2 mg/m³ (fume)
ACGIH TLV: Fume: TWA 2 mg/m³ (fume)

SAFETY PROFILE: Questionable carcinogen
with experimental tumorigenic data by implant
route. Many paraffin waxes contain carcinogens.

PAH770   **HR: D**
PARAFFIN HYDROCARBONS

SAFETY PROFILE: The effects of the paraffin
hydrocarbons vary with the volatility. The gaseous hydrocarbons, such as methane, ethane,
etc., have but slight anesthetic effects and are
hazardous only when present in sufficient concentration to dilute the oxygen to a point below
that which is necessary to sustain life. With
the volatile liquid hydrocarbons, or with the
next higher fraction, the anesthetic action pre-

dominates, and with the higher molecular weights or with the less volatile compounds, the anesthetic increases, but at the same time an irritant action becomes more pronounced. For information concerning toxic and hazardous properties of these materials, see the individual compounds. Paraffins are common air contaminants. Can be a dangerous fire hazard depending on volatility.

PAH800   CAS: 63449-39-8   **HR: 3**
PARAFFIN WAXES and HYDROCARBON WAXES, CHLORINATED (C12, 60% CHLORINE)

SYN: CHLORINATED PARAFFINS (C12, 60% CHLORINE)

CONSENSUS REPORTS: NTP Carcinogenesis Studies (gavage): Clear Evidence: mouse,rat NTPTR* NTP-TR-308,86. Reported in EPA TSCA Inventory.

SAFETY PROFILE: Suspected carcinogen with experimental carcinogenic and neoplastigenic data. When heated to decomposition it emits acrid smoke and irritating fumes.

PAI000   CAS: 30525-89-4   **HR: 3**
PARAFORMALDEHYDE

DOT: UN 2213
mf: $(CH_2O)_n$

PROP: White crystals; odor of formaldehyde. Flash p: 158°F, autoign temp: 572°F. Sltly sol in cold water; moderately sol in hot water yielding formaldehyde.

SYNS: FLO-MOR * FORMAGENE * PARAFORSN * TRIFORMOL * TRIOXYMETHYLENE

CONSENSUS REPORTS: Reported in EPA TSCA Inventory.

DOT Classification: ORM-A; Label: None; DOT-IMO: Flammable Solid; Label: None.

SAFETY PROFILE: Moderately toxic by ingestion. A severe eye and skin irritant. Mutation data reported. Flammable when exposed to heat or flame; can react with oxidizing materials. To fight fire, use alcohol foam, CO_2, dry chemical. Incompatible with liquid oxygen. Dangerous; when heated to decomposition it emits toxic formaldehyde gas.

PAI250   CAS: 123-63-7   **HR: 3**
PARALDEHYDE

DOT: UN 1264
mf: $C_6H_{12}O_3$ mw: 132.18

PROP: Colorless liquid; disagreeable taste, aromatic odor. Mp: 12.6°, lel: 1.3%, bp: 124.4° @ 752 mm, flash p: 62.6°F, d: 0.9943 @ 20°/4°, autoign temp: 460°F, vap d: 4.55. Sol in water; misc with alc, ether, oils, chloroform.

SYNS: ACETALDEHYDE, TRIMER * ELALDEHYDE * PARACETALDEHYDE * PARAL * PARALDEHYD (GERMAN) * PARALDEIDE (ITALIAN) * PCHO * RCRA WASTE NUMBER U182 * TRIACETALDEHYDE (FRENCH) * 2,4,6-TRIMETHYL-1,3,5-TRIOXAAN (DUTCH) * 2,4,6-TRIMETHYL-s-TRIOXANE * 2,4,6-TRIMETHYL-1,3,5-TRIOXANE * s-TRIMETHYLTRIOXY-METHYLENE * 2,4,6-TRIMETIL-1,3,5-TRIOSSANO (ITALIAN)

CONSENSUS REPORTS: Reported in EPA TSCA Inventory.

DOT Classification: Flammable Liquid; Label: Flammable Liquid.

SAFETY PROFILE: A human poison by rectal route. Moderately toxic to humans by an unspecified route. Moderately toxic experimentally by inhalation, ingestion, intraperitoneal, and subcutaneous routes. Human systemic effects by rectal route: necrotic changes. A skin and severe eye irritant. Low doses produce hypnotic and analgesic effects. Larger doses depress the nervous system with loss of reflexes, coma, and respiratory depression leading to respiratory paralysis and death. Chronic effects include weight loss, muscular weakness, and mental fatigue. However, poisoning is rare. A hypnotic agent. Dangerous fire hazard when exposed to heat, flame, or oxidizers. Slight explosion hazard when exposed to heat or flame. Dangerous; keep away from heat and open flame. To fight fire, use alcohol foam, CO_2, dry chemical. Potentially violent reaction with nitric acid. Incompatible with alkalies, hydrocyanic acid, iodides, oxidizers. When heated to decomposition it emits acrid smoke and irritating fumes.

PAI990   CAS: 4685-14-7   **HR: 3**
PARAQUAT
mf: $C_{12}H_{14}N_2$ mw: 186.28

SYNS: DIMETHYL VIOLOGEN * GRAMOXONE S * METHYL VIOLOGEN (2+) * PARAQUAT DICATION

CONSENSUS REPORTS: EPA Genetic Toxicology Program.

OSHA PEL: Respirable Dust: (Transitional: TWA 0.5 mg/m³ (skin)) TWA 0.1 mg/m³ (skin)

ACGIH TLV: TWA 0.1 mg/m^3

SAFETY PROFILE: Poison by ingestion and intraperitoneal routes. Mutation data reported. Causes ulceration of digestive tract, diarrhea, vomiting, renal damage, jaundice, edema, hemorrhage, fibrosis of lung, and death from anoxia may result. When heated to decomposition it emits toxic fumes of NO$_x$.

PAJ000   CAS: 1910-42-5   **HR: 3**
PARAQUAT DICHLORIDE
mf: C$_{12}$H$_{14}$N$_2$ • 2Cl   mw: 257.18

PROP: Yellow solid. Sol in water.

SYNS: CEKUQUAT * CRISQUAT * DEXTRONE * N,N'-DIMETHYL-4,4'-BIPYRIDINIUM DICHLORIDE * 1,1'-DIMETHYL-4,4'-DIPYRIDYLIUM CHLORIDE * 1,1'-DIMETHYL-4,4'-DIPYRIDINIUM-DICHLORID (GERMAN) * DIMETHYL VICLOGEN CHLORIDE * ESGRAM * GRAMOZONE * METHYLVIOLOGEN * OK 622 * PARAQUAT CHLORIDE

CONSENSUS REPORTS: EPA Extremely Hazardous Substances List.

SAFETY PROFILE: A human poison by ingestion. Poison experimentally by ingestion, skin contact, intraperitoneal, intravenous, and subcutaneous routes. Human systemic effects by ingestion: headache; cough, dyspnea, other pulmonary effects; hypermotility, diarrhea, nausea or vomiting, and other gastrointestinal effects. Experimental reproductive effects. Has a delayed damaging effect on the lung alveoli. Has caused fatal poisoning in humans with severe injury to lungs. Has been implicated in aplastic anemia. An eye irritant. Mutation data reported. The National Institute of Drug Abuse (NIDA), USA, has concluded that contamination of marihuana with the herbicide paraquat may pose a serious threat to marihuana (cannabis) smokers, and issued a warning that marihuana contaminated with the herbicide paraquat could lead to permanent lung damage for regular and heavy users of marihuana. The maximum level of contamination permitted for domestic uses is 0.05 ppm; but paraquat has been found in marihuana samples at levels ranging from 3 to 2,204 ppm, averaging 452 ppm. It tends to concentrate in lung tissue whether it is ingested or inhaled, and produces a condition called fibrosis which reduces the capacity of the lung to absorb oxygen. *Inhalation* of paraquat creates a greater risk of lung damage than *ingestion* of an identical amount. An herbicide. When heated to decomposition it emits toxic fumes of Cl$^-$ and NO$_x$.

PAJ500   CAS: 10048-32-5   **HR: 3**
PARASCORBIC ACID
mf: C$_6$H$_8$O$_2$ mw: 112.14

PROP: Oily liquid; sweet, aromatic odor. Bp: 104-105° @ 14 mm, 119-123° @ 22 mm; d: 1.079 @ 18°/4°. Sol in water; very sol in alc, ether.

SYNS: (S)-(+)-5,6-DIHYDRO-6-METHYL-2H-PYRAN-2-ONE * γ-HEXENOLACTONE * 2-HEXEN-5,1-OLIDE * D''-HEXENOLLACTONE * 5-HYDROXY-2-HEXENOIC ACID LACTONE * PARASORBIC ACID * (+)-PARASORBINSAEURE (GERMAN) * SORBIC OIL

CONSENSUS REPORTS: IARC Cancer Review: GROUP 3 IMEMDT 7,56,87; Animal Limited Evidence IMEMDT 10,199,76.

SAFETY PROFILE: Poison by intraperitoneal and intravenous routes. Mildly toxic by skin contact. Questionable carcinogen with experimental neoplastigenic data. When heated to decomposition it emits acrid smoke and irritating fumes.

PAK000   CAS: 56-38-2   **HR: 3**
PARATHION
DOT: NA 2783
mf: C$_{10}$H$_{14}$NO$_5$PS   mw: 291.28

PROP: Pale-yellow liquid. Bp: 375°, mp: 6°. Very sol in alcs, esters, ethers, ketones, aromatic hydrocarbons; insol in water, petr ether, kerosene.

SYNS: AAT * AATP * AC 3422 * ACC 3422 * ALLERON * APHAMITE * ARALO * B 404 * BAY E-605 * BAYER E-605 * BLADAN * BLADAN F * COMPOUND 3422 * COROTHION * CORTHION * CORTHIONE * DANTHION * O,O-DIAETHYL-O-(4-NITROPHENYL)-MONOTHIOPHOSPHAT (GERMAN) * O,O-DIETHYL-O-(4-NITRO-FENIL)-MONOTHIOFOSFAAT (DUTCH) * O,O-DIETHYL-O-p-NITROFENYLESTER KYSELINYTHIOFOSFORECNE (CZECH) * O,O-DIETHYL-O-p-NITROFENYLTIOFOSFAT (CZECH) * O,O-DIETHYL-O-4-NITROPHENYLPHOSPHO-ROTHIOATE * O,O-DIETHYL-O-(p-NITROPHENYL) PHOSPHOROTHIOATE * O,O-DIETHYL-O-(4-NITROPHE-NYL) PHOSPHOROTHIOATE * DIETHYL-4-NITROPHE-NYL PHOSPHOROTHIONATE * DIETHYL-p-NITROPHE-NYLTHIONOPHOSPHATE * DIETHYL-p-NITROPHENYL-THIOPHOSPHATE * O,O-DIETHYL-O-(p-NITROPHENYL)

THIONOPHOSPHATE * O,O-DIETHYL-O-p-NITRO-PHENYL THIOPHOSPHATE * O,O-DIETHYL-O-4-NITRO-PHENYL THIOPHOSPHATE * DIETHYLPARATHION * O,O-DIETIL-O-(4-NITRO-FENIL)-MONOTIOFOS-FATO (ITALIAN) * DNTP * DPP * DREXEL PARATHION 8E * E 605 * ECATOX * EKATIN WF & WF ULV * EKATOX * ENT 15,108 * ETHLON * ETHYL PARATHION * FOLIDOL * FOLIDOL E605 * FOLIDOL E & E 605 * FOS-FERMO * FOSFERNO * FOSFEX * FOSFIVE * FOSOVA * FOSTERN * FOSTOX * GEAR-PHOS * GENITHION * KOLPHOS * KYPTHION * LETHALAIRE G-54 * LIROTHION * MURFOS * NCI-C00226 * NIRAN * NIRAN E-4 * p-NI-TROPHENOL, O-ESTER with O,O-DIETHYLPHOSPHORO-THIOATE * NITROSTIGMIN (GERMAN) * NITRO-STIGMINE * NIUIF-100 * NOURITHION * OLEO-FOS 20 * OLEOPARAPHENE * OLEOPARATHION * ORTHOPHOS * PAC * PANTHION * PARA-DUST * PARAMAR * PARAMAR 50 * PARA-PHOS * PARATHENE * PARATHION, liquid (DOT) * PARATHION-ETHYL * PARAWET * PESTOX PLUS * PETHION * PHOSKIL * PHOSPHEMOL * PHOSPHENOL * PHOSPHOROTHIOIC ACID, O,O-DI-ETHYL-O-(4-NITROPHENYL) ESTER * PHOSPHOSTIG-MINE * RB * RCRA WASTE NUMBER P089 * RHODIASOL * RHODIATOX * RHODIATROX * SELEPHOS * SIXTY-THREE SPECIAL E.C. INSECTI-CIDE * SNP * SOPRATHION * STABILIZED ETHYL PARATHION * STATHION * STRATHION * SULPHOS * SUPER RODIATOX * T-47 * THIOPHOS * THIOPHOS 3422 * THIOPHOS-PHATE de O,O-DIETHYLE et de O-(4-NITROPHENYLE) (FRENCH) * TIOFOS * TOX 47 * VAPOPHOS * VITREX

CONSENSUS REPORTS: IARC Cancer Review: GROUP 3 IMEMDT 7,56,87; Human Inadequate Evidence IMEMDT 30,153,83; Animal Inadequate Evidence IMEMDT 30,153,83. NCI Carcinogenesis Bioassay (feed); Clear Evidence: rat NCITR* NCI-CG-TR-70,79; (feed); No Evidence: mouse NCITR* NCI-CG-TR-70,79. EPA Farm Worker Field Reentry. EPA Extremely Hazardous Substances List. Community Right-To-Know List. EPA Genetic Toxicology Program.

OSHA PEL: TWA 0.1 mg/m^3 (skin)
ACGIH TLV: TWA 0.1 mg/m^3 (skin) (Proposed: BEI: 0.5 mg/L total p-nitrophenol in urine at end of shift.)
DFG MAK: 0.1 mg/m^3; BAT: 500 μg/L p-nitrophenol in urine after several shifts.

NIOSH REL: TWA 0.05 mg/m^3
DOT Classification: Poison B; Label: Poison.

SAFETY PROFILE: A deadly poison by all routes. Human systemic effects by ingestion: general anesthetic; pulmonary effects; and kidney, ureter, bladder effects. Experimental teratogenic and reproductive effects. Questionable carcinogen with experimental carcinogenic and tumorigenic data. Human mutation data reported. A cholinesterase inhibitor. Parathion, like the other organic phosphorus poisons, acts as an irreversible inhibitor of the enzyme cholinesterase and thus allows the accumulation of large amounts of acetylcholine. When a critical level of cholinesterase depletion is reached, grave symptoms appear. Whether death is actually caused entirely by cholinesterase depletion or by the disturbance of a number of enzymes is not yet known. Recovery is apparently complete if a poisoned animal or man has time to reform a critical amount of cholinesterase. The organism exposed remains susceptible to relatively low dosages of parathion until the cholinesterase level has regenerated. Small doses at frequent intervals are, therefore, more or less additive. There is not, however, at the present time, any indication that, when recovery from a given exposure is entirely complete, the exposed organism is prejudiced in any way. Combustible when exposed to heat or flame. Violent reaction with endrin. Highly dangerous; shock can shatter the container releasing the contents. A broad spectrum insecticide in agricultural applications. When heated to decomposition it emits highly toxic fumes of NO$_x$, PO$_x$, SO$_x$.

PAK250   CAS: 56-38-2   **HR: 3**
PARATHION (mixture, dry)

DOT: NA 2783

SYN: O,O-DIETHYL-O-(p-NITROPHENYL) ESTER (DRY MIXTURE) PHOSPHOROTHIOIC ACID

DOT Classification: Poison B; Label: Poison.

SAFETY PROFILE: A powerful poison and experimental carcinogen. Questionable carcinogen. When heated to decomposition it emits toxic fumes of PO$_x$, SO$_x$, and NO$_x$.

PAL750   CAS: 8000-68-8   **HR: 2**
PARSLEY OIL

PROP: From steam distillation of above ground parts (herb oil) or ripe seed (seed oil) of *Petro-*

selinium sativum Hoffm. (Fam. *Umbelligerae*). Yellow to light brown liquid; odor of parsley. D (herb oil): 0.908-0.940, (seed oil): 1.040; refr index (herb oil): 1.503-1.530 @ 20°, (seed oil): 1.513-1.522 @ 20°. Sol in fixed oils, mineral oil; sltly sol in propylene glycol; insol in glycerin.

SYNS: OIL of PARSLEY * PARSLEY HERB OIL (FCC) * PARSLEY SEED OIL (FCC) * PETERSILIENSAMEN OEL (GERMAN)

CONSENSUS REPORTS: Reported in EPA TSCA Inventory.

SAFETY PROFILE: Moderately toxic by ingestion. A human skin irritant. When heated to decomposition it emits acrid smoke and irritating fumes.

PAM000   CAS: 113-42-8   **HR: 3**
PARTERGIN
mf: $C_{20}H_{25}N_3O_2$ mw: 339.48

SYNS: BASOFORTINA * 9,10-DIDEHYDRO-N-(α-(HYDROXYMETHYL)PROPYL)-6-METHYL-ERGOLINE-8-β-CARBOXAMIDE * ME 277 * METHERGINE * METHYLERGOBASINE * METHYLERGOBREVIN * METHYLERGOMETRIN * METHYLERGOMETRINE * METHYLERGONOVIN * METHYLERGONOVINE

SAFETY PROFILE: Poison by ingestion and intravenous routes. When heated to decomposition it emits toxic fumes of NO_x.

PAN100   CAS: 434-07-1   **HR: 3**
PAVISOID
mf: $C_{21}H_{32}O_3$ mw: 332.53

PROP: Crystals from ethyl acetate. Mp: 178-180°.

SYNS: ADROIDIN * ADROYD * ANADROL * ANADROYD * ANAPOLON * ANASTERON * ANASTERONAL * ANASTERONE * BECOREL * CI-406 * 4,5-DIHYDRO-2-HYDROXYMETHYLENE-17-α-METHYLTESTOSTERONE * DYNASTEN * HMD * 17-β-HYDROXY-2-HYDROXYMETHYLENE-17-α-METHYL-3-ANDROSTANONE * 17-β-HYDROXY-2-(HYDROXYMETHYLENE)-17-α-METHYL-5-α-ANDROSTAN-3-ONE * 17-β-HYDROXY-2-(HYDROXYMETHYLENE)-17-METHYL-5-α-ANDROSTAN-3-ONE * 17-HYDROXY-2-(HYDROXYMETHYLENE)-17-METHYL-5-α-17-β-ANDROST-3-ONE * 2-HYDROXYMETHYLENE-17-α-METHYL-5-α-ANDROSTAN-17-β-OL-3-ONE * 2-HYDROXYMETHYLENE-17-α-METHYL-DIHYDROTESTOSTERONE * 2-(HYDROXYMETHYLENE)-17-α-METHYLDIHYDROTESTOSTERONE * 2-HYDROXYMETHYLENE-17-α-

METHYL-17-β-HYDROXY-3-ANDROSTANONE * METHABOL * 17-α-METHYL-2-HYDROXYMETHYLENE-17-HYDROXY-5-α-ANDROSTAN-3-ONE * NASTENON * NSC-26198 * OXIMETHOLONUM * OXIMETOLONA * OXITOSONA-50 * OXYMETHALONE * OXYMETHENOLONE * OXYMETHOLONE * PARDROYD * PLENASTRIL * PROTANABOL * ROBORAL * SYNASTERON * ZENALOSYN

CONSENSUS REPORTS: IARC Cancer Review: Human Inadequate Evidence IMEMDT 13,131,77. NTP Fourth Annual Report On Carcinogens, 1984.

SAFETY PROFILE: Confirmed human carcinogen producing liver tumors. Human systemic effects by ingestion: impaired liver function. Experimental teratogenic and reproductive effects. When heated to decomposition it emits acrid smoke and irritating fumes.

PAO000   CAS: 8002-03-7   **HR: 1**
PEANUT OIL

PROP: Straw-yellow to greenish-yellow or nearly colorless oil; nutty odor and bland taste. Mp: 2.7°, flash p: 540°F, d: 0.92, autoign temp: 833°F. Misc with ether, petr ether, chloroform, carbon disulfide; sol in benzene, carbon tetrachloride, oils; very sltly sol in alc. From seed of *Arachis hypogaea*.

SYNS: ARACHIS OIL * EARTHNUT OIL * GROUNDNUT OIL * INDIGENOUS PEANUT OIL * KATCHUNG OIL * PECAN SHELL POWDER

CONSENSUS REPORTS: Reported in EPA TSCA Inventory.

SAFETY PROFILE: A human skin irritant and mild allergen. Questionable carcinogen with experimental tumorigenic data. Mutation data reported. Combustible when exposed to heat or flame; can react with oxidizing materials. Slight spontaneous heating. To fight fire, use CO_2, dry chemical. When heated to decomposition it emits acrid smoke and irritating fumes.

PAP250   CAS: 26864-56-2   **HR: 3**
PENFLURIDOL
mf: $C_{28}H_{27}ClF_5NO$ mw: 524.01

PROP: White microcrystals. Mp: 105-107°. Sltly sol in water.

SYNS: 4-(4-CHLORO-α,α,α-TRIFLUORO-m-TOLYL)-1-(4,4-BIS(p-FLUOROPHENYL)BUTYL)-4-PIPERIDINOL * McN-JR-16,341 * R 16341 * SEMAP * TLP-607

SAFETY PROFILE: Poison by ingestion and intravenous routes. Experimental teratogenic and reproductive effects. A neuroleptic agent. When heated to decomposition it emits very toxic fumes of Cl^-, F^-, and NO_x.

PAP550 CAS: 2219-30-9 **HR: 3**
PENICILLAMINE HYDROCHLORIDE
mf: $C_5H_{11}NO_2S \cdot ClH$ mw: 185.69

SYNS: DISTAMINE * METALCAPTASE * d-PENI-CILLAMINE HYDROCHLORIDE * USAF EL-23

SAFETY PROFILE: A poison by intraperitoneal route. Moderately toxic by ingestion and intravenous route. Human systemic effects by ingestion: dermatitis. When heated to decomposition it emits very toxic fumes of NO_x, SO_x, and HCl.

PAP750 CAS: 90-65-3 **HR: 3**
PENICILLIC ACID
mf: $C_8H_{10}O_4$ mw: 170.18

PROP: Needles from petr ether. Mp: 83-84°. Sltly sol in cold water, hot petr ether; very sol in hot water, alc, ether, benzene chloroform; insol in pentane-hexane.

SYNS: γ-KETO-β-METHOXY-Δ-METHYLENE-Δα-HEXE-NOIC ACID * 3-METHOXY-5-METHYL-4-OXO-2,5-HEX-ADIENOIC ACID * PA * PENCILLIC ACID

CONSENSUS REPORTS: IARC Cancer Review: GROUP 3 IMEMDT 7,56,87; Animal Sufficient Evidence IMEMDT 10,211,76. EPA Genetic Toxicology Program.

SAFETY PROFILE: Poison by intravenous, subcutaneous, intraperitoneal, and possibly other routes. Moderately toxic by ingestion. Experimental reproductive effects. Questionable carcinogen with experimental neoplastigenic data. Human mutation data reported. When heated to decomposition it emits acrid smoke and irritating fumes.

PAQ000 CAS: 1406-05-9 **HR: 3**
PENICILLIN
mf: $(CH_3)_2C_5H_3NSO(COOH)NHCOOR$ (bicyclic)

PROP: A group of isomeric and closely related antibiotic compounds with outstanding bacterial activity. An extract from *Penicillium notatum*. Different varieties of penicillin are produced by adding the proper precursors to the nutrient solution.

SYN: PENIZILLIN (GERMAN)

CONSENSUS REPORTS: EPA Genetic Toxicology Program.

SAFETY PROFILE: Poison by intraperitoneal and subcutaneous routes. Moderately toxic by intravenous route. Human reproductive effects by ingestion: abortion. Human systemic effects by intramuscular route: dermatitis. Experimental reproductive effects. Has been implicated in aplastic anemia. When heated to decomposition it emits very toxic fumes of NO_x and SO_x.

PAR500 **HR: 3**
PENNYROYAL OIL

SYN: AMERICAN PENNYROYAL OIL

SAFETY PROFILE: Experimental poison by ingestion. A skin irritant. When heated to decomposition it emits acrid smoke and irritating fumes.

PAT750 CAS: 19624-22-7 **HR: 3**
PENTABORANE(9)

DOT: UN 1380
mf: B_5H_9 mw: 63.14

PROP: Colorless gas or liquid; bad odor. Mp: −46.6°, d: 0.61 @ 0°, vap d: 2.2, vap press: 66 mm @ 0°, lel: 0.42%, bp: 60.°

CONSENSUS REPORTS: EPA Extremely Hazardous Substances List. Reported in EPA TSCA Inventory.

OSHA PEL: (Transitional: 0.005 ppm) TWA 0.005 ppm; STEL 0.015 ppm
ACGIH TLV: TWA 0.005 ppm; STEL 0.013 ppm
DFG MAK: 0.005 ppm (0.01 mg/m³)
DOT Classification: Flammable Liquid; Label: Spontaneously Combustible, Poison; Label: Flammable Liquid and Poison.

SAFETY PROFILE: Poison by inhalation and intraperitoneal routes. Dangerous fire hazard by chemical reaction; spontaneously flammable in air. Dangerous explosion hazard. To fight fire, use special fire-fighting materials; water is not effective; reacts violently with halogenated extinguishing agents. Get instructions from supplier. Explosive reaction with oxygen. Forms shock-sensitive solutions in solvents containing carbonyl, ether, or ester functions; or halogens. Incompatible with dimethyl sulfoxide. Upon decomposition it emits toxic fumes of B.

PAT799   CAS: 18433-84-6   **HR: 3**
PENTABORANE(11)
mf: B_5H_{11} mw: 65.16

SAFETY PROFILE: Ignites spontaneously in air. When heated to decomposition it emits toxic fumes of B.

PAV250   CAS: 25201-35-8   **HR: 3**
PENTACHLOROACETOPHENONE
mf: $C_8Cl_5H_3O$ mw: 292.36

SYN: 2',3',4',5',6'-PENTACHLOROACETOPHENONE

SAFETY PROFILE: Poison by intraperitoneal route. Moderately toxic by ingestion and skin contact. When heated to decomposition it emits toxic fumes of Cl^-.

PAW250   CAS: 42279-29-8   **HR: 3**
PENTACHLORO DIPHENYL OXIDE
mf: $C_{12}H_5Cl_5O$ mw: 342.42

SYNS: ETHER, PENTACHLOROPHENYL * PHENYL ETHER PENTACHLORO

OSHA PEL: TWA 0.5 mg/m^3

SAFETY PROFILE: Poison by ingestion and possibly other routes. When heated to decomposition it emits toxic fumes of Cl^-.

PAW500   CAS: 76-01-7   **HR: 3**
PENTACHLOROETHANE

DOT: UN 1669
mf: C_2HCl_5 mw: 202.28

PROP: Colorless liquid; chloroform-like odor. Mp: $-29°$, bp: 161-162°, d: 1.6728 @ 25°/4°. Insol in water; misc in alc and ether.

SYNS: ETHANE PENTACHLORIDE * NCI-C53894 * PENTACHLOORETHAAN (DUTCH) * PENTACHLO-RAETHAN (GERMAN) * PENTACHLORETHANE (FRENCH) * PENTACLOROETANO (ITALIAN) * PENTALIN * RCRA WASTE NUMBER U184

CONSENSUS REPORTS: IARC Cancer Review: GROUP 3 IMEMDT 7,56,87; Animal Limited Evidence IMEMDT 41,99,86. NTP Carcinogenesis Bioassay (gavage); Clear Evidence: mouse NTPTR* NTP-TR-232,82; (gavage); No Evidence: rat NTPTR* NTP-TR-232,82. Reported in EPA TSCA Inventory.

DFG MAK: 5 ppm (40 mg/m^3)
DOT Classification: Poison B; Label: Poison.

SAFETY PROFILE: Poison by inhalation and intravenous routes. Moderately toxic by inges-tion and subcutaneous routes. An irritant. Questionable carcinogen with experimental carcinogenic data. Flammable when exposed to heat or flame. Moderately explosive by spontaneous chemical reaction. To fight fire, use water, CO_2, dry chemical. Dehalogenation by reaction with alkalies; metals, etc. will produce spontaneously explosive chloroacetylenes. Violent reaction with NaK alloy + bromoform. Mixtures with potassium are very shock-sensitive explosives. When heated to decomposition it emits highly toxic fumes of Cl^-.

PAW750   CAS: 1321-64-8   **HR: 3**
PENTACHLORONAPHTHALENE
mf: $C_{10}H_3Cl_5$ mw: 300.38

PROP: White solid.

CONSENSUS REPORTS: Reported in EPA TSCA Inventory.

OSHA PEL: TWA 0.5 mg/m^3 (skin)
ACGIH TLV: TWA 0.5 mg/m^3
DFG MAK: 0.5 mg/m^3

SAFETY PROFILE: Poison by ingestion, inhalation, and skin contact. An irritant. Action similar to chlorinated naphthalenes and chlorinated diphenyls. Dangerous; when heated to decomposition it emits highly toxic fumes of Cl^-.

PAX000   CAS: 82-68-8   **HR: 3**
PENTACHLORONITROBENZENE
mf: $C_6Cl_5NO_2$ mw: 295.32

PROP: Colorless crystals. Mp: 146°, bp: 328°, vap press: 0.013 mm @ 25°.

SYNS: AVICOL * BATRILEX * BRASSICOL * EARTHCIDE * FARTOX * FOLOSAN * FOMAC 2 * FUNGICLOR * GC 3944-3-4 * KOBU * KOBUTOL * KP 2 * NCI-C00419 * OLPISAN * PCNB * PENTACHLORNITROBEN-ZOL (GERMAN) * PENTAGEN * PKhNB * QUIN-TOCENE * QUINTOZEN * QUINTOZENE * RCRA WASTE NUMBER U185 * SANICLOR 30 * TERRACHLOR * TERRAFUN * TILCAREX * TRI-PCNB * TRITISAN

CONSENSUS REPORTS: IARC Cancer Review: GROUP 3 IMEMDT 7,56,87; Animal Sufficient Evidence IMEMDT 5,211,74. NCI Carcinogenesis Bioassay (feed); No Evidence: mouse, rat NCITR* NCI-CG-TR-61,78. EPA Extremely Hazardous Substances List. Reported in EPA TSCA Inventory. EPA Genetic Toxicology Program.

ACGIG TLV: (Proposed: 0.5 mg/m^3)

SAFETY PROFILE: Moderately toxic by ingestion and possibly other routes. Experimental reproductive effects. Questionable carcinogen with experimental carcinogenic data. Mutation data reported. Used as a fungicide. Dangerous; when heated to decomposition it emits highly toxic fumes of NO$_x$ and Cl$^-$.

PAX250   CAS: 87-86-5   **HR: 3**
PENTACHLOROPHENOL

DOT: UN 2020
mf: C$_6$HCl$_5$O   mw: 266.32

PROP: Dark-colored flakes and sublimed needle crystals; characteristic odor. Mp: 191°, bp: 310° (decomp), d: 1.978, vap press: 40 mm @ 211.2°. Sol in ether, benzene; very sol in alc; insol in water; sltly sol in cold petr ether.

SYNS: CHEM-TOL * CHLOROPHEN * CRYPTO-GIL OL * DOWCIDE 7 * DOWICIDE 7 * DOWICIDE EC-7 * DOWICIDE G * DOW PENTACHLORO-PHENOL DP-2 ANTIMICROBIAL * DUROTOX * EP 30 * FUNGIFEN * GLAZD PENTA * GRUNDIER ARBEZOL * LAUXTOL * LAUXTOL A * LIROPREM * NCI-C54933 * NCI-C55378 * NCI-C56655 * PCP * PENCHLOROL * PENTA * PENTACHLOORFENOL (DUTCH) * PENTACHLORO-FENOL * PENTACHLOROPHENATE * 2,3,4,5,6-PEN-TACHLOROPHENOL * PENTACHLOROPHENOL, DOWI-CIDE EC-7 * PENTACHLOROPHENOL, DP-2 * PENTACHLOROPHENOL (GERMAN) * PENTA-CHLOROPHENOL, TECHNICAL * PENTACLORO-FENOLO (ITALIAN) * PENTACON * PENTA-KIL * PENTASOL * PENWAR * PERATOX * PER-MACIDE * PERMAGARD * PERMASAN * PER-MATOX DP-2 * PERMATOX PENTA * PERMITE * PRILTOX * RCRA WASTE NUMBER U242 * SANTOBRITE * SANTOPHEN * SANTOPHEN 20 * SINITUHO * TERM-I-TROL * THOMPSON'S WOOD FIX * WEEDONE

CONSENSUS REPORTS: IARC Cancer Review: Human Limited Evidence IMEMDT 41,319,86; Animal Inadequate Evidence IMEMDT 20,303,79. Chlorophenol compounds are on The Community Right-To-Know List. Reported in EPA TSCA Inventory. EPA Genetic Toxicology Program.

OSHA PEL: TWA 0.5 mg/m^3 (skin)
ACGIH TLV: TWA 0.5 mg/m^3 (skin); BEI: 2 mg(total PCP)/L in urine prior to last shift of workweek; 5 mg(free PCP)/L in plasma at end of shift.

DFG MAK: 0.05 ppm (0.5 mg/m^3); BAT: 1000 μg/L in plasma/serum.
DOT Classification: ORM-E; Label: None.

SAFETY PROFILE: Suspected human carcinogen with experimental tumorigenic data. Human poison by ingestion. Poison experimentally by ingestion, skin contact, intraperitoneal, and subcutaneous routes. An experimental teratogen. A skin irritant. Mutation data reported. Acute poisoning is marked by weakness with changes in respiration, blood pressure, and urinary output. Also causes dermatitis, convulsions, and collapse. Chronic exposure can cause liver and kidney injury. Dangerous; when heated to decomposition it emits highly toxic fumes of Cl$^-$.

PBB750   CAS: 115-77-5   **HR: 1**
PENTAERYTHRITOL
mf: C$_5$H$_{12}$O$_4$ mw: 136.17

PROP: Crystals. Mp: 262°, d: 1.38 @ 25°/4°.

SYNS: AUXINUTRIL * 2,2-BIS(HYDROXYMETHYL)-1,3-PROPANEDIOL * HERCULES P6 * METHANE TETRAMETHYLOL * MONOPENTEK * PE * PENTAERYTHRITE * PENTEK * TETRAHY-DROXYMETHYLMETANE * TETRAKIS(HYDROXY-METHYL)METHANE * TETRAMETHYLOLMETHANE

CONSENSUS REPORTS: Reported in EPA TSCA Inventory.

OSHA PEL: (Transitional: TWA Total Dust: 15 mg/m^3; Respirable Fraction: 5 mg/m^3) TWA Total Dust: 10 mg/m^3; Respirable Fraction: 5 mg/m^3
ACGIH TLV: TWA (nuisance particulate) 10 mg/m^3 of total dust (when toxic impurities are not present, e.g., quartz < 1%).

SAFETY PROFILE: Mildly toxic by ingestion. A nuisance dust. Flammable from heat or flame or oxidizers. Mixtures with thiophosphoryl chloride react when heated to form a product which ignites and then explodes on contact with air. Used in coatings, stabilizers, explosives, P.E.T.N resins, drugs, insecticides, and lubricants. When heated to decomposition it emits acrid smoke and irritating fumes.

PBC250   CAS: 78-11-5   **HR: 3**
PENTAERYTHRITOL TETRANITRATE

DOT: UN 0150/UN 0411
mf: C$_5$H$_8$N$_4$O$_{12}$ mw: 316.17

PROP: Crystals. Mp: 138-140°, bp: explodes @ 205-215°, d: 1.773 @ 20°/4°. Sol in acetone; insol in water; sltly sol in alc, ether.

SYNS: ANGICAP * ANGITET * ANTORA * ARCOTRATE * BARITRATE * 2,2-BISDIHY-DROXYMETHYL-1,3-PROPANEDIOL TETRANITRATE * 2,2-BIS(HYDROXYMETHYL)-1,3-PROPANEDIOL TET-RANITRATE * CHOT * DELTRATE-20 * 1,3-DINI-TRATO-2,2-BIS(NITRATOMETHYL)PROPANE * DUO-TRATE * EL PETN * ERINIT * HASETHROL * INITIATING EXPLOSIVE PENTAERYTHRITE TETRANI-TRATE (DOT) * KAYTRATE * LOWETRATE * MARTRATE-45 * METRANIL * MYCARDOL * MYOTRATE "10" * NCI-C55743 * NEO-CORO-VAS * NEOPENTANETETRAYL NITRATE * NI-PERYT * NIPERYTH * NITROPENTA * NITRO-PENTAERYTHRITE * NITROPENTAERYTHRITOL * PENCARD * PENTAERYTHRITE TETRANITRATE * PENTAERYTHRITE TETRANITRATE (DOT) * PENTAERYTHRITE TETRANITRATE, desensitized, wet (DOT) * PENTAERYTHRITE TETRANITRATE, dry (DOT) * PENTAERYTHRITE THERANITRATE, with not less than 7% wax (DOT) * PENTAERYTHRITOL TETRANITRATE, diluted * PENTAFIN * PENTESTAN-80 * PENTE-TRATE UNICELLES * PENTRATE * PENTRIOL * PENTRYATE 80 * PERGITRAL * PERIDEX-LA * PERITRATE * PERITYL * PREVANGOR * QUINTRATE * RYTHRITOL * SDM NO. 23 * SUBICARD * TENTRATE-20 * TETRANITROPEN-TAERYTHRITE * TETRASULE * TRANITE D-LAY * VASITOL * VASODIATOL * VASO-80 UNICE-LIES

CONSENSUS REPORTS: Reported in EPA TSCA Inventory.

DOT Classification: Class A Explosive; Label: Explosive A, desensitized, wet: Forbidden, dry.

SAFETY PROFILE: Human systemic effects by ingestion: dermatitis. Effects are similar to nitroglycerin, i.e., headache, weakness, and fall in blood pressure. Severe explosion hazard when shocked or exposed to heat. It explodes at 215°C. On decomposition it emits highly toxic fumes of NO_x; can react vigorously with oxidizing materials. Used in detonators and explosive specialities.

PBI500   CAS: 54-95-5   **HR: 3**
1,5-PENTAMETHYLENETETRAZOLE
mf: $C_6H_{10}N_4$ mw: 138.20

PROP: White, crystalline powder. Mp: 57-58°. Sol in water, alc and ether.

SYNS: α,β-CYCLOPENTAMETHYLENETETRAZOLE * PENTAMETHYLENETETRAZOL * PENTAMETHY-LENE-1,5-TETRAZOLE * PENTYLENETETRAZOL * 6,7,8,9-TETRAHYDRO-5-AZEPOTETRAZOLE * 6,7,8,9-TETRAHYDRO-5H-TETRAZOLOAZEPINE * 7,8,9,10-TETRAZABICYCLO(5.3.0)-8,10-DECADIENE * 1,2,3,3A-TETRAZACYCLOHEPTA-8A,2-CYCLOPENTA-DIENE

CONSENSUS REPORTS: Reported in EPA TSCA Inventory.

SAFETY PROFILE: A human poison by ingestion and intravenous routes. Poison experimentally by ingestion, intravenous, intraperitoneal, subcutaneous, rectal, and parenteral routes. When heated to decomposition it emits toxic fumes of NO_x.

PBK250   CAS: 109-66-0   **HR: 3**
PENTANE

DOT: UN 1265
mf: C_5H_{12} mw: 72.17

PROP: Colorless liquid. Bp: 36.1°, flash p: <−40°F, fp: −129.8°, d: 0.626 @ 20°/4°, autoign temp: 588°F, vap press: 400 mm @ 18.5°, vap d: 2.48, lel: 1.5%, uel: 7.8%. Sol in water; misc in alc, ether, organic solvents.

SYNS: AMYL HYDRIDE (DOT) * PENTAN (POLISH) * PENTANEN (DUTCH) * PENTANI (ITALIAN)

CONSENSUS REPORTS: Reported in EPA TSCA Inventory.

OSHA PEL: (Transitional: TWA 1000 ppm; STEL 750 ppm) TWA 600 ppm; STEL 750 ppm
ACGIH TLV: TWA 600 ppm; STEL 750 ppm
DFG MAK: 1000 ppm (2950 mg/m³)
NIOSH REL: TWA 350 mg/m³
DOT Classification: Flammable Liquid; Label: Flammable Liquid.

SAFETY PROFILE: Moderately toxic by intravenous route. Narcotic in high concentration. The liquid can cause blisters on contact. Flammable liquid. Highly dangerous fire hazard when exposed to heat, flame, or oxidizers. Severe explosion hazard when exposed to heat or flame. Shock can shatter metal containers and release contents. To fight fire, use foam, CO_2, dry chemical. When heated to decomposition it emits acrid smoke and irritating fumes.

PBK500   CAS: 462-94-2   **HR: 3**
1,5-PENTANEDIAMINE
mf: $C_5H_{14}N_2$ mw: 102.21

PROP: Colorless, thick liquid; characteristic odor. Mp: 9°, bp: 178-180°, d: 0.873 @ 25°/4°. Very sol in water and alc; sltly sol in ether.

SYNS: ANIMAL CONIINE * CADAVERIN * CADAVERINE * 1,5-DIAMINOPENTANE * PENTAMETHYLENEDIAMINE * 1,5-PENTAMETH-YLENEDIAMINE

SAFETY PROFILE: Poison by intravenous, rectal, and subcutaneous routes. Moderately toxic by skin contact. An irritant, sensitizer, and allergen. Mutgenic data. When heated to decomposition it emits highly toxic fumes of NO_x.

PBM000   CAS: 110-66-7   **HR: 2**
1-PENTANETHIOL
DOT: UN 1111
mf: $C_5H_{12}S$ mw: 104.23

PROP: Water-white to yellow liquid. D: 0.857 @ 20°, bp: 123.64°, flash p: 65°F, vap press: 13.8 mm @ 25°, vap d: 3.59. Insol in water; misc in alc and ether.

SYNS: AMYL HYDROSULFIDE * n-AMYL MERCAPTAN * AMYL MERCAPTAN (DOT) * AMYL SULFHYDRATE * AMYL THIOALCOHOL * MERCAPTAN AMYLIQUE (FRENCH) * PENTYL MERCAPTAN

CONSENSUS REPORTS: Reported in EPA TSCA Inventory.

NIOSH REL: CL 0.5 ppm/15M
DOT Classification: Label: Flammable Liquid.

SAFETY PROFILE: Moderately toxic by inhalation. A weak sensitizer and allergen. Local contact may cause contact dermatitis. Dangerous fire hazard when exposed to heat or flame; can react vigorously with oxidizing materials. Hypergolic reaction with concentrated nitric acid. To fight fire, use foam, CO_2, dry chemical.

PBM750   CAS: 6032-29-7   **HR: 2**
2-PENTANOL
DOT: UN 1105
mf: $C_5H_{12}O$ mw: 88.17

PROP: Colorless liquid. Bp: 119.3°, flash p: 105°F (OC), ULC:40-45, uel: 9.0%, lel: 1.2%, fp: −50°, d: 0.8169 @ 20°/20°, autoign temp: 650-725°F, vap d: 3.04. Slightly sol in water; misc in alc, and ether.

SYNS: sec-AMYL ALCOHOL (DOT) * METHYL PROPYL CARBINOL * PENTANOL-2 * sec-PENTYL ALCOHOL

CONSENSUS REPORTS: Reported in EPA TSCA Inventory.

DOT Classification: Flammable Liquid; Label: Flammable Liquid: Flammable or Combustible Liquid; Label: Flammable Liquid.

SAFETY PROFILE: Moderately toxic by ingestion and intraperitoneal routes. A narcotic. A skin and severe eye irritant. Flammable when exposed to heat or flame; can react with oxidizing materials. A severe explosion hazard when exposed to heat or flame. To fight fire, use alcohol foam, dry chemical. When heated to decomposition it emits acrid smoke and irritating fumes.

PBN250   CAS: 107-87-9   **HR: 3**
2-PENTANONE
DOT: UN 1249
mf: $C_5H_{10}O$ mw: 86.15

PROP: Water-white liquid; fruity, ethereal odor. D: 0.801-0.806, vap d: 3.0, bp: 216°F, flash p: 45°F, autoign temp: 941°F, lel: 1.5%, uel: 8.2%. Sltly sol in water; misc with alc, ether.

SYNS: ETHYL ACETONE * FEMA No. 2842 * METHYL-PROPYL-CETONE (FRENCH) * METHYL-n-PROPYL KETONE * METHYL PROPYL KETONE (ACGIH, DOT) * METYLOPROPYLOKETON (POLISH) * MPK

OSHA PEL: (Transitional: TWA 200 ppm; STEL 250 ppm) TWA 200 ppm; STEL 250 ppm
ACGIH TLV: TWA 200 ppm; STEL 250 ppm
DFG MAK: 200 ppm (700 mg/m^3)
NIOSH REL: TWA 530 mg/m^3
DOT Classification: Label: Flammable Liquid.

SAFETY PROFILE: Moderately toxic by ingestion and intraperitoneal routes. Mildly toxic by skin contact and inhalation. Human systemic effects by inhalation: headache, nausea, irritation of the respiratory passages, eyes, and skin. A skin irritant. Mutation data reported. A highly flammable liquid. A very dangerous fire hazard when exposed to heat or flame; can react vigorously with oxidizing materials. An explosion hazard in the form of vapor when exposed to heat or flame. To fight fire, use alcohol foam.

Mixtures with bromine trifluoride may explode during evaporation. When heated to decomposition it emits acrid smoke and irritating fumes.

PBS000   CAS: 60-44-6   **HR: 3**
PENTHIENATE BROMIDE
mf: $C_{18}H_{30}NO_3S \cdot Br$ mw: 420.46

SYNS: α-CYCLOPENTYL-2-THIOPHENEGLYCOLATE DIETHYL(2-HYDROXYETHYL)METHYLAMMONIUM BROMIDE * 2-DIETHYLAMINOETHYL 2-CYCLOPENTYL-2-(2-THIENYL)HYDROXYACETATE METHOBROMIDE * 2-DIETHYLAMINOETHYL α-CYCLOPENTYL-2-THIOPHENEGLYCOLATE METHOBROMIDE * DIETHYL(2-HYDROXYETHYL)METHYLAMMONIUM BROMIDE, α-CYCLOPENTYL-2-THIOPHENEGLYCOLATE

SAFETY PROFILE: Poison by intravenous and subcutaneous routes. Moderately toxic by ingestion. When heated to decomposition it emits very toxic fumes of NH_3, NO_x, SO_x, and Br^-.

PBS250   CAS: 115-58-2   **HR: 3**
PENTOBARBITAL
mf: $C_{11}H_{18}N_2O_3$ mw: 226.31

SYNS: 5-AETHYL-5-PENTYL-(2′)-BARBITURSAEURE (GERMAN) * 5-ETHYL-5-PENTYLBARBITURIC ACID

SAFETY PROFILE: Poison by intraperitoneal route. When heated to decomposition it emits toxic fumes of NO_x.

PBT500   CAS: 71-73-8   **HR: 3**
PENTOTHAL SODIUM
mf: $C_{11}H_{17}N_2O_2S \cdot Na$ mw: 264.35

SYNS: 5-ETHYLDIHYDRO-5-(1-METHYLBUTYL)-2-THI-OXO-4,6(1H,5H)-PYRIMIDINEDIONE, MONOSODIUM SALT * 5-ETHYL-5-(1-METHYLBUTYL)-2-THIOBARBITURIC ACID MONOSODIUM * FARMOTAL * HYPNOSTAN * INTRAVAL SODIUM * LEOPENTAL * MONOSODIUM-5-ETHYL-5-(1-METHYLBUTYL) THIOBARBITURATE * NESDONAL SODIUM * PENTHIOBARBITAL SODIUM * RAVONAL * SODIUM-5-ETHYL-5-(1-METHYLBUTYL)-2-THIOBARBITURATE * SODIUM PENTHIOBARBITAL * SODIUM PENTOTHAL * SODIUM PENTOTHIOBARBITAL * SODIUM THIOPENTAL * SODIUM THIOPENTOBARBITAL * SODIUM THIOPENTONE * SOLUBLE THIOPENTONE * THIOMEBUMAL SODIUM * THIONEMBUTAL * THIOPENTAL SODIUM * THIOPENTAL SODIUM SALT * THIOPENTONE SODIUM * THIOTHAL SODIUM * THIPENTAL SODIUM * TIOPENTAL SODIUM * TRAPANAL * TRAPANAL SODIUM

SAFETY PROFILE: Poison by ingestion, intraperitoneal, rectal, subcutaneous, and intravenous routes. Human systemic effects by intraarterial route: acute arterial occlusion; by rectal route: respiratory depression, body temperature decrease, general anesthetic. Experimental reproductive effects. An intravenous anesthetic. When heated to decomposition it emits toxic fumes of NO_x and Na_2O.

PBV000   CAS: 75-85-4   **HR: 3**
tert-PENTYL ALCOHOL
DOT: UN 1105
mf: $C_5H_{12}O$ mw: 88.17

PROP: Colorless liquid. Mp: −11.9°, bp: 101.8°, flash p: 105°F (CC), d: 0.809, autoign temp: 819°F, vap press: 10 mm @ 17.2°, lel: 1.2%, uel: 9%, vap d: 3.03. Sltly sol in water; sol in alc and ether.

SYNS: tert-AMYL ALCOHOL (DOT) * AMYLENE HYDRATE * DIMETHYLETHYLCARBINOL * 2-METHYL BUTANOL-2 * 2-METHYL-2-BUTANOL * 3-METHYLBUTAN-3-OL * tert-PENTANOL

CONSENSUS REPORTS: Reported in EPA TSCA Inventory.

DOT Classification: Flammable or Combustible Liquid; Label: Flammable Liquid.

SAFETY PROFILE: Moderately toxic to humans by an unspecified route. Moderately toxic experimentally by ingestion, intraperitoneal, subcutaneous and rectal routes. Narcotic in high concentration. Flammable when exposed to heat, flame, or oxidizing materials. Moderately explosive in the form of vapor when exposed to heat or flame. A hypnotic agent. When heated to decomposition it emits acrid smoke and irritating fumes.

PBV500   **HR: 2**
PENTYLAMINE (mixed isomers)
mf: $C_5H_{13}N$ mw: 87.17

PROP: Water-white liquid. Mp: −55°, bp: 104°, flash p: 45°F (OC), d: 0.7614 @ 20°/4°, vap d: 3.01, lel: 2.2%, uel: 22%.

SYN: AMYLAMINE (mixed isomers) (DOT)

DOT Classification: Label: Flammable Liquid.

SAFETY PROFILE: Moderately toxic by ingestion and skin contact. A severe skin irritant. Dangerous fire hazard when exposed to heat or flame; can react with oxidizing materials. Moderately explosive in the form of vapor when exposed to heat or flame. To fight fire, use alcohol foam, dry chemical. When heated to decomposition it emits toxic fumes of NO_x.

PBV750   CAS: 2188-67-2   **HR: 3**
2,N-PENTYLAMINOETHYL-p-AMINOBENZOATE
mf: $C_{14}H_{22}N_2O_2$ mw: 250.38

SYNS: p-AMINOBENZOIC ACID-2-N-AMYLAMINO-ETHYL ESTER * 2-N-AMYLAMINOETHYL-p-AMINO-BENZOATE * AMYLCAINE * AMYLSINE * NAEPAINE

SAFETY PROFILE: Poison by intravenous and subcutaneous routes. When heated to decomposition it emits toxic fumes of NO_x.

PBW500   CAS: 543-59-9   **HR: 3**
PENTYL CHLORIDE
DOT: UN 1107
mf: $C_5H_{11}Cl$ mw: 106.61

PROP: Water-white liquid; sweet odor. Mp: −99°, bp: 108.2°, flash p: 54°F (OC), d: 0.883 @ 20°/4°, autoign temp: 500°F, vap d: 3.67, lel: 1.4%, uel: 8.6%.

SYNS: n-AMYL CHLORIDE * AMYL CHLORIDE (DOT) * 1-CHLOROPENTANE

CONSENSUS REPORTS: Reported in EPA TSCA Inventory.

DOT Classification: Label: Flammable Liquid.

SAFETY PROFILE: Dangerous fire hazard when exposed to heat or flame; can react with oxidizing materials. Moderately explosive in the form of vapor when exposed to heat or flame. To fight fire, use foam, CO_2, dry chemical. Dangerous; when heated to decomposition it emits highly toxic fumes of phosgene and Cl^-.

PBW750   CAS: 12789-46-7   **HR: 2**
PENTYL ESTER PHOSPHORIC ACID
DOT: UN 2819
mf: $C_5H_{13}O_4P$ mw: 168.15

SYN: AMYL ACID PHOSPHATE (DOT)

DOT Classification: Corrosive Material; Label: Corrosive.

SAFETY PROFILE: Corrosive and irritating to the skin, eyes, and mucous membranes. When heated to decomposition it emits toxic fumes of PO_x.

PBX000   CAS: 693-65-2   **HR: 3**
PENTYL ETHER
mf: $C_{10}H_{22}O$ mw: 158.32

PROP: Liquid. Mp: −69.3°, bp: 187°, flash p: 135°F (OC), d: 0.783 @ 20°/4°, vap d: 5.46, autoign temp: 340°F.

SYNS: N-AMYL ETHER * AMYL ETHER * DI-PENTYL ETHER * 1,1-OXYBIS PENTANE

CONSENSUS REPORTS: Reported in EPA TSCA Inventory.

SAFETY PROFILE: Poison by intravenous routes. Flammable when exposed to heat or flame; reacts with oxidizing materials. To fight fire, use alcohol foam, dry chemical. When heated to decomposition it emits acrid smoke and irritating fumes.

PBX500   CAS: 10589-74-9   **HR: 3**
n-PENTYLNITROSOUREA
mf: $C_6H_{13}N_3O_2$ mw: 159.22

SYNS: 1-AMYL-1-NITROSOUREA * n-AMYLNITRO-SOUREA * ANU * 1-NITROSO-1-PENTYLUREA

SAFETY PROFILE: Suspected carcinogen with experimental carcinogenic and tumorigenic data. Moderately toxic by ingestion. Mutation data reported. When heated to decomposition it emits toxic fumes of NO_x.

PBY750   CAS: 107-72-2   **HR: 2**
PENTYLTRICHLOROSILANE
DOT: UN 1728
mf: $C_5H_{11}Cl_3Si$ mw: 205.60

SYN: AMYL TRICHLOROSILANE

CONSENSUS REPORTS: Reported in EPA TSCA Inventory.

DOT Classification: Corrosive Material; Label: Corrosive.

SAFETY PROFILE: Moderately toxic by ingestion and skin contact. Mildly toxic by inhalation. A corrosive irritant to the eyes, skin, and mucous membranes. When heated to decomposition it emits toxic fumes of Cl^-.

PCB250   CAS: 8006-90-4   **HR: 2**
PEPPERMINT OIL

PROP: From steam distillation of *Mentha piperita* L. (Fam. *Labiatae*). Colorless to pale yellow liquid; strong odor and taste of peppermint. D:

0.896-0.908 @ 25°/25°, refr index: 1.459 @ 20°.

SYN: PFEFFERMINZ OEL (GERMAN)

CONSENSUS REPORTS: Reported in EPA TSCA Inventory.

SAFETY PROFILE: Moderately toxic by ingestion and intraperitoneal routes. An allergen. Mutation data reported. When heated to decomposition it emits acrid smoke and irritating fumes.

PCC000   CAS: 9076-25-9   **HR: 3**
PEPTICHEMIO

PROP: Made up of 6 peptides of m-(di-(2-chloroethyl)amino-1-phenylalanine.

SYNS: NSC 247516 * PEP * PTC

SAFETY PROFILE: Poison by intravenous, intraperitoneal, and subcutaneous routes. Human mutation data reported. When heated to decomposition it emits very toxic fumes of Cl^- and NO_x.

PCD000   **HR: 3**
PERCHLORATES

PROP: Composition: combinations with the monovalent $-ClO_4$ radical.

SAFETY PROFILE: Perchlorates are unstable materials, and are irritant to the body wherever they come in contact with it. Avoid skin contact. Flammable by chemical reaction; powerful oxidizers. All perchlorates are potentially hazardous when in contact with reducing materials. Moderate explosion hazard when shocked or exposed to heat or by chemical reaction. Perchlorates, when mixed with carbonaceous material, form explosive mixtures. Many perchlorates of nitrogenous bases (e.g., hydroxylamine, urea, methylamaine, ethylamine, isopropylamine, 4-ethylpyridine, diaminoethane) and organic perchlorates are explosives. Diazonium perchlorates are very dangerous. All perchlorates are considered to be fire and explosive hazards when associated with carbonaceous materials or finely divided metals. This is also true of the presence of calcium hydride;; sulfur; powdered magnesium; aluminum; zinc. They react violently with benzene; CaH_2; charcoal; olefins; ethanol; SrH_2; S; H_2SO_4; and reducing materials. To fight fire, use water or foam. When heated to decomposition it emits toxic fumes of Cl^-.

PCD250   CAS: 7601-90-3   **HR: 3**
PERCHLORIC ACID

DOT: UN 1802
mf: $ClHO_4$ mw: 100.46

PROP: Colorless, fuming, unstable liquid. Mp: −112°, bp: 19° @ 11 mm, d: 1.768 @ 22°.

CONSENSUS REPORTS: Reported in EPA TSCA Inventory.

DOT Classification: Label: Oxidizer.

SAFETY PROFILE: Poison by ingestion and subcutaneous routes. A severe irritant to the eyes, skin, and mucous membranes. A powerful oxidizer. A severe explosion hazard; the anhydrous form can explode spontaneously. Potentially explosive reaction with acetic anhydride + acetic acid + organic materials, acetic anhydride + organic materials + transition metals (e.g., chromium, iron, nickel), acetonitrile, alcohols, azo dyes + orthoperiodic acid, bis(2-hydroxyethyl)terephthalate + ethanol + ethylene glycol, bismuth (above 110°C), antimony (above 110°C), carbon, charcoal + chromium trioxide + heat, cellulose and derivatives + heat, combustible materials, dehydrating agents, dichloromethane + dimethylsulfoxide, diethyl ether, dimethyl ether, dioxane + nitric acid + heat, fecal material + nitric acid, graphitic carbon + nitric acid, hydrofluoric acid + structural materials, iron(II) sulfate, nitric acid + organic matter + heat, nitric acid + pyridine + sulfuric acid, nitrogenous epoxides, organic materials + sodium hydrogen carbonate (above 200°C), phenyl acetylene (at −78°C), sodium phosphinate + heat, sulfuric acid + organic materials, sulfur trioxide. Reacts to form explosive products with aniline + formaldehyde, ethylbenzene + thallium triacetate (at 65°C), fluorine (forms fluorine perchlorate), glycerol + lead oxide, hydrogen + heat, hydrogen halides, phosphine, pyridine, sulfoxides. Violent reaction or ignition with acetic acid, acetic acid + acetic anhydride, acetic anhydride, acetic anhydride + carbon tetrachloride + 2-methyl cyclohexanone, antimony compounds, azo pigments, bis-1,2-diaminopropane-cis-dichlorochromium(III) perchlorate, carbon, 1,3-bis-(di-n-cyclopentadienyl iron)-2-propen-1-one, CH_3OH, CCl_4, copper dichromium tetraoxide (at 120°C), DNA, dibutyl sulfoxide, dimethyl sulfoxide, ethylbenzene, glycol ethers, glycols, HNO_3, HCl, H_2SO_4, hypophosphites, iron sulfate, iodides, ketones, PbO + glycerin, methanol + triglycerides, 2-methylpropene + metal oxides, 2-methyl cyclohexanone, NI_3, nitrogenous epoxides, nitrosophenol, o-periodic acid, oleic acid, organophosphorus compounds, paper, P_2O_5 + $CHCl_3$, P_2O_5, P_2Zn_3, sodium iodide + hydroiodic acid, sodium phosphinate, steel, sulfinyl chloride, SO_3, trichloroethylene, vegetable matter, wood, zinc phosphide. When heated to decomposition it emits toxic fumes of Cl^-.

PCD500   CAS: 7790-98-9   HR: 3
PERCHLORIC ACID, AMMONIUM SALT
DOT: UN 0402/UN 1442
mf: $ClO_4 \cdot H_4N$ mw: 117.50

PROP: White crystals. Mp: decomp, d: 1.95.

SYN: AMMONIUM PERCHLORATE (DOT)

CONSENSUS REPORTS: Reported in EPA TSCA Inventory.

DOT Classification: Label: Oxidizer; Class A Explosive; Label: EXPLOSIVE A

SAFETY PROFILE: Moderately toxic by parenteral route. Flammable when exposed to heat or flame or by spontaneous chemical reaction with reducing materials. A very powerful oxidizer which has caused explosions in industry. Ignites violently with combustibles. Severe explosion hazard; decomposes at 130° and explodes at 380°. When contaminated by powdered carbon, ferrocene, S, organic matter, powdered metals, nitryl perchlorate, potassium periodate, potassium permanganate it becomes impact sensitive. Potentially explosive reactions with carbon (above 240°C), dichromium trioxide (at 270°C), cadmium oxide (at 260°C), zinc oxide (at 200°C), copper chromite, copper oxide, iron oxide, potassium permanganate, potassium dichromate, mono-, di-, tri-, or tetra-methylammonium perchlorates, metal perchlorates (e.g., lithium perchlorate, zinc perchlorate), nitrophenol-formaldehyde polymer. Mixtures with aluminum or copper burn violently when ignited. Mixtures with ethylene dinitrate ignite when stored at 60°C. When heated to decomposition it emits toxic fumes of NH_3 and Cl^-.

PCD750   CAS: 13465-95-7   HR: 3
PERCHLORIC ACID, BARIUM SALT • 3H₂O
DOT: UN 1447
mf: $Cl_2O_8 \cdot Ba \cdot 3H_2O$ mw: 336.24

PROP: Colorless crystals. Mp: decomp @ 400°, d: 2.74.

SYN: BARIUM PERCHLORATE (DOT)

CONSENSUS REPORTS: Barium and its compounds are on the Community Right-To-Know List. Reported in EPA TSCA Inventory.

OSHA PEL: TWA 0.5 mg(Ba)/m^3
DOT Classification: Oxidizer; Label: Oxidizer and Poison.

SAFETY PROFILE: A poison. An unstable material. An oxidizer. When refluxed with an alcohol highly explosive alkyl perchlorates are formed. When heated to decomposition it emits toxic fumes of Cl^-. A dessicant.

PCE000   CAS: 10034-81-8   HR: 2
PERCHLORIC ACID, MAGNESIUM SALT
DOT: UN 1475
mf: $Cl_2O_8 \cdot Mg$ mw: 223.21

PROP: White, hygroscopic crystals. Mp: decomp @ 251°, d: 2.60 @ 25°.

SYNS: ANHYDRONE * DEHYDRITE * MAGNESIUM PERCHLORATE * PERCHLORATE de MAGNESIUM (FRENCH)

CONSENSUS REPORTS: Reported in EPA TSCA Inventory.

DOT Classification: Label: Oxidizer.

SAFETY PROFILE: Moderately toxic by intraperitoneal route. A powerful oxidizer which has caused many explosions in industry. Potentially explosive reactions with alkenes (above 220°C), ammonia; aryl hydrazine + ether; dimethyl sulfoxide + heat; ethylene oxide; fluorobutane + water; organic materials; phosphorus; trimethyl phosphate. Reacts to form explosive products with ethanol (forms ethyl perchlorate); cellulose + dinitrogen tetraoxide + oxygen (forms cellulose nitrate). Avoid contact with mineral acids; butyl fluorides; hydrocarbons. A drying agent. When heated to decomposition it emits toxic fumes of MgO and Cl^-.

PCE750   CAS: 7601-89-0   HR: 3
PERCHLORIC ACID, SODIUM SALT
DOT: UN 1502
mf: $ClO_4 \cdot Na$ mw: 122.44

PROP: Colorless, deliquescent crystals. Mp: 482° (decomp).

SYNS: NATRIUMPERCHLORAAT (DUTCH) * NATRI-
UMPERCHLORAT (GERMAN) * PERCHLORATE de SO-
DIUM (FRENCH) * SODIO (PERCLORATO DI) (ITAL-
IAN) * SODIUM PERCHLORATE * SODIUM
PERCHLORATE (DOT)

CONSENSUS REPORTS: Reported in EPA
TSCA Inventory.

DOT Classification: Label: Oxidizer.

SAFETY PROFILE: Moderately toxic by ingestion and intraperitoneal routes. A powerful oxidizer. Forms explosive mixture with acetone,
1,3-butylene glycol, 2,3-butylene glycol, CaH_2,
charcoal, diaminoethane, dimethyl formamide,
ethanolamine, ethylene glycol, formamide, galactose, glycerol, hydrazine, water, NH_4NO_3,
Mg, reducing agents, SrH_2, urea. When heated
to decomposition it emits toxic fumes of Cl^-
and Na_2O.

PCF275   CAS: 127-18-4   HR: 3
PERCHLOROETHYLENE

DOT: UN 1897
mf: C_2Cl_4 mw: 165.82

PROP: Colorless liquid; chloroform-like odor.
Mp: $-23.35°$, bp: $121.20°$, d: 1.6311 @ 15°/
4°, vap press: 15.8 mm @ 22°, vap d: 5.83.

SYNS: ANKILOSTIN * ANTISOL 1 * CARBON BI-
CHLORIDE * CARBON DICHLORIDE * CZTERO-
CHLOROETYLEN (POLISH) * DIDAKENE * DOW-
PER * ENT 1,860 * ETHYLENE TETRACHLORIDE
* FEDAL-UN * NCI-C04580 * NEMA * PER-
AWIN * PERCHLOORETHYLEEN, PER (DUTCH)
* PERCHLOR * PERCHLORAETHYLEN, PER (GER-
MAN) * PERCHLORETHYLENE * PERCHLORETH-
YLENE, PER (FRENCH) * PERCLENE * PERCLORO-
ETILENE (ITALIAN) * PERCOSOLVE * PERK
* PERKLONE * PERSEC * RCRA WASTE NUMBER
U210 * TETLEN * TETRACAP * TETRACHLOOR-
ETHEEN (DUTCH) * TETRACHLORAETHEN (GERMAN)
* TETRACHLOROETHENE * TETRACHLOROETH-
YLENE (DOT) * 1,1,2,2-TETRACHLOROETHYLENE
* TETRACLOROETENE (ITALIAN) * TETRALENO
* TETRALEX * TETRAVEC * TETROGUER
* TETROPIL

CONSENSUS REPORTS: IARC Cancer Review: GROUP 2B IMEMDT 7,355,87; Animal
Limited Evidence IMEMDT 20,491,79. NCI
Carcinogenesis Bioassay (gavage); Clear Evidence: mouse NCITR* NCI-CG-TR-13,77; (inhalation); Clear Evidence: mouse, rat NTPTR*
NTP-TR-311,86; (gavage); Inadequate Studies:

rat NCITR* NCI-CG-TR-13,77. Reported in
EPA TSCA Inventory. EPA Genetic Toxicology
Program. Community Right-To-Know List.

OSHA PEL: (Transitional: TWA 100 ppm; CL
  200 ppm; Pk 600 ppm/5M)TWA 25 ppm
ACGIH TLV: TWA 50 ppm; STEL 200 ppm
  (Proposed: BEI: 7 mg/L trichloroacetic acid
  in urine at end of workweek.)
DFG MAK: 50 ppm (345 mg/m³); BAT: blood
  100 µg/dl
NIOSH REL: (Tetrachloroethylene) Minimize
  workplace exposure.
DOT Classification: Poison B; Label: St. Andrews Cross; ORM-A; Label: None.

SAFETY PROFILE: Suspected carcinogen with
experimental carcinogenic and neoplastigenic
data. Experimental poison by intravenous route.
Moderately toxic to humans by inhalation with
the following effects: local anesthetic, conjunctiva irritation, general anesthesia, hallucinations, distorted perceptions, coma, and pulmonary changes. Moderately experimentally toxic
by ingestion, inhalation, intraperitoneal, and
subcutaneous routes. Experimental teratogenic
and reproductive effects. Human mutation data
reported. An eye and severe skin irritant. The
liquid can cause injuries to the eyes; however,
with proper precautions it can be handled safely.
The symptoms of acute intoxication from this
material are the result of its effects upon the
nervous system. Can cause dermatitis, particularly after repeated or prolonged contact with
the skin. Irritates the gastrointestinal tract upon
ingestion. It may be handled in the presence
or absence of air, water, and light with any of
the common construction materials at temperatures up to 140°. This material is extremely
stable and resists hydrolysis. A common air
contaminant. Reacts violently under the proper
conditions with Ba; Be; Li; N_2O_4; metals;
NaOH. When heated to decomposition it emits
highly toxic fumes of Cl^-.

PCF300   CAS: 594-42-3   HR: 3
PERCHLOROMETHYL MERCAPTAN

DOT: UN 1670
mf: CCl_4S mw: 185.87

PROP: Yellow, oily liquid. Bp: slt decomp @
149°, d: 1.700 @ 20°, vap d: 6.414.

SYNS: CLAIRSIT * MERCAPTAN METHYLIQUE
PERCHLORE (FRENCH) * PCM * PERCHLOR-
METHYLMERKAPTAN (CZECH) * RCRA WASTE

NUMBER P118   *   TRICHLOROMETHANE SULFENYL CHLORIDE   *   TRICHLOROMETHYLSULFENYL CHLORIDE   *   TRICHLOROMETHYLSULPHENYL CHLORIDE

CONSENSUS REPORTS: EPA Extremely Hazardous Substances List. Reported in EPA TSCA Inventory.

OSHA PEL: TWA 0.1 ppm
ACGIH TLV: TWA 0.1 ppm
DOT Classification: Poison B; Label: Poison.

SAFETY PROFILE: Poison by ingestion, inhalation and intravenous routes. A severe skin, eye and mucous membrane irritant. When heated to decomposition it emits very toxic fumes of Cl^- and SO_x.

PCF750   CAS: 7616-94-6   HR: 2
PERCHLORYL FLUORIDE
mf: $ClFO_3$ mw: 102.45

PROP: Colorless, noncorrosive gas; characteristic sweet odor. Mp: $-146°$, bp: $-46.8°$, d: (liquid): 1.434. d: (gas): 0.637.

SYNS: CHLORINE FLUORIDE OXIDE   *   CHLORINE OXYFLUORIDE

CONSENSUS REPORTS: Reported in EPA TSCA Inventory.

OSHA PEL: (Transitional: TWA 3 ppm) TWA 3 ppm; STEL 6 ppm
ACGIH TLV: TWA 3 ppm; STEL 6 ppm

SAFETY PROFILE: Moderately toxic by inhalation route. Forms methemoglobin in the body and destroys red cells causing anemia, anorexia and cyanosis. Recovery is said to be rapid, leaving no permanent physiological damage. Can be absorbed through the skin. Its odor can be detected as low as 10 ppm although this cannot be relied upon as an indication of toxic concentration in air. While nonflammable, it supports combustion. It is a powerful oxidizer. Moderately explosive. Potentially explosive reactions with combustible gases or vapors, benzene + aluminum trichloride, benzocyclobutene + butyllithium + potassium tert-butoxide, calcium acetylide, potassium cyanide, potassium thiocyanate, sodium iodide, charcoal, ethyl-4-fluorobenzoylacetate, hydrocarbons, hydrogen sulfide, nitrogen oxide, sulfur dichloride, vinylidene chloride, 3α-hydroxy-5β-androstane-11,17-dione-17-hydrazone, lithiated compounds, 2-lithio(dimethylaminomethyl)ferroxene, methyl-2-bromo-5,5-ethylene dioxy

(2.2.1)bicycloheptane-7-carboxylate, aliphatic heterocyclic amines, sodium methoxide + methanol, vinylidene chloride. Reacts to form explosive products with nitrogenous bases (e.g., isopropylamine, isobutylamine, aniline, phenyl hydrazine, 1,2-diphenyl hydrazine), sawdust, lampblack. Violent reaction with finely divided organic materials. A fluorinating agent in chemical synthesis, and as an oxidant in rocket fuel. When heated to decomposition it emits toxic fumes of F^- and Cl^-.

PCG500   CAS: 76-42-6   HR: 3
PERCODAN
mf: $C_{18}H_{21}NO_4$ mw: 315.40

SYNS: DIHYDROHYDROXYCODEINONE   *   DIHYDRO-14-HYDROXYCODEINONE   *   14-HYDROXYDIHYDROCODEINONE   *   OXYCODEINONE

SAFETY PROFILE: Poison by intravenous route. Moderately toxic by subcutaneous route. When heated to decomposition it emits toxic fumes of NO_x.

PCI750   CAS: 553-84-4   HR: 3
PERILLA KETONE
mf: $C_{10}H_{14}O_2$ mw: 166.24

PROP: A potent lung toxin from the mint plant, *Perilla frutescens* .

SYNS: 1-(3-FURANYL)-4-METHYL-1-PENTANONE   *   β-FURYL ISOAMYL KETONE   *   1-(3-FURYL)-4-METHYL-1-PENTANONE   *   PURPLE MINT PLANT EXTRACT

SAFETY PROFILE: Poison by intravenous and intraperitoneal routes. A potent pulmonary edemagenic agent (experimental). May also be hazardous to humans. When heated to decomposition it emits acrid smoke and irritating fumes.

PCJ250   CAS: 13444-71-8   HR: 3
O-PERIODIC ACID
mf: HIO_4 mw: 191.91

SAFETY PROFILE: Powerful oxidizer. Potentially explosive reaction with dimethyl sulfoxide. Reaction with tetraethylammonium hydroxide forms an explosive product. Incompatible with azo-pigments. Used as a sweetening agent in Japan. When heated to decomposition it emits toxic fumes of I^-.

PCJ400   HR: 1
PERLITE

PROP: Average density of 0.13. Expands when finely ground and heated. Natural glass, amorphous mineral consisting of fused sodium potassium aluminum silicate, containing $<1\%$ quartz.

ACGIH TLV: TWA (nuisance particulate) 10 mg/m^3 of total dust (when toxic impurities are not present, e.g., quartz $< 1\%$).

SAFETY PROFILE: A nuisance dust.

PCJ500   HR: 3
PERMANGANATES

PROP: Compounds containing an MnO$_4^-$ radical.

CONSENSUS REPORTS: Manganese and its compounds are on The Community Right-To-Know List.

SAFETY PROFILE: Poisons. Many are strong oxidizing agents, hence irritating. Flammable by chemical reaction with reducing agents. Moderately explosive when shocked or exposed to heat. Silver permanganate and other metallic permanganates may detonate when exposed to high temperatures or when they are involved in fires or severely shocked. Store in a cool, ventilated area, away from acute fire hazards and easily oxidized materials. They may be disposed of by dissolving in water since practically all permanganates are soluble in water. They can react vigorously on contact with reducing materials. Incompatible with acetic acid, acetic anhydride, H$_2$SO$_4$ + C$_6$H$_6$.

PCJ750   CAS: 13446-10-1   HR: 2
PERMANGANIC ACID AMMONIUM SALT

DOT: NA 9190
mf: MnO$_4$ • H$_4$N   mw: 136.99

SYN: AMMONIUM PERMANGANATE

CONSENSUS REPORTS: Manganese and its compounds are on the Community Right-To-Know List.

OSHA PEL: CL 5 mg(Mn)/m^3
ACGIH TLV: TWA 5 mg(Mn)/m^3
DOT Classification: Label: Oxidizer.

SAFETY PROFILE: Probably an irritant. A powerful oxidizer and therefore a fire hazard. Heat can cause this material to self-react violently. When heated to decomposition it emits toxic fumes of NO$_x$ and NH$_3$.

PCK000   CAS: 7787-36-2   HR: 2
PERMANGANIC ACID, BARIUM SALT

DOT: UN 1448
mf: Mn$_2$O$_8$ • Ba   mw: 375.22

SYN: BARIUM PERMANGANATE

CONSENSUS REPORTS: Barium and its compounds and manganese and its compounds are on the Community Right-To-Know List.

ACGIH TLV: TWA 0.5 mg(Ba)/m^3
DOT Classification: Label: Oxidizer.

SAFETY PROFILE: Probably an irritant. A powerful oxidizer. When heated to decomposition it emits acrid smoke and irritating fumes.

PCK500   CAS: 84-97-9   HR: 3
PERNAZINE
mf: C$_{20}$H$_{25}$N$_3$S   mw: 339.54

SYNS: N-METHYL-PIPERAZINYL-N'-PROPYL-PHENO-THIAZIN (GERMAN) * 10-(3-(4-METHYL-1-PIPERAZIN-YL)PROPYL)-10H-PHENOTHIAZINE (9CI) * N-(3-(4-METHYL-1-PIPERAZINYL)PROPYL)PHENOTHIAZINE * PERAZINE * PSYTOMIN * TAXILAN

SAFETY PROFILE: Poison by intravenous and intraperitoneal routes. Moderately toxic by ingestion and subcutaneous routes. Human toxic effects by unspecified route. When heated to decomposition it emits very toxic fumes of NO$_x$ and SO$_x$.

PCL000   HR: 3
PEROXIDES, INORGANIC

SAFETY PROFILE: Variable toxicity. They may cause injury on contact with skin or mucous membranes. Moderate to dangerous fire hazard by chemical reaction with reducing agents and contaminants; strong oxidizing agents; contact with moisture may produce much heat. Moderate explosion hazard; heat, shock, or catalysts can cause violent decomposition. Contact with reducing agents may give rise to explosively violent reactions.

PCL250   HR: 3
PEROXIDES, ORGANIC

PROP: Organic compounds containing the —OO— group.

SAFETY PROFILE: Often highly toxic and irritating to the skin, eyes, and mucous membranes. Dangerous fire hazard by chemical reaction with reducing agents or exposure to heat. They readily release oxygen and thus are powerful oxidizers. Severe explosion hazard when shocked, exposed to heat, or by spontaneous chemical reaction. Many peroxides are very unstable. Upon contact with reducing materials, such as organic matter, thiocyanates, an explosion can occur.

Many solvents form dangerous levels of peroxides during storage: e.g., dipropyl ether, divinylacetylene, vinylidene chloride, potassium amide, sodium amide. Other compounds form peroxides in storage but concentration is required to reach dangerous levels: e.g., diethyl ether, ethyl vinyl ether, tetrahydrofuran, p-dioxane, 1,1-diethoxyethane, ethylene glycol dimethyl ether, propyne, butadiyne, dicyclopentadiene, cyclohexene, tetrahydronaphthalenes, deca-hydronaphthalenes. Some monomeric materials can form peroxides which catalyze hazardous polymerization reactions: e.g., acrylic acid, acrylonitrile, butadiene, 2-chlorobutadiene, chlorotrifluoroethylene, methyl methacrylate, styrene, tetrafluoroethylene, vinyl acetate, vinylacetylene, vinyl chloride, vinylidine chloride, vinylpyridine. Compounds which contain one or two ether functions are especially susceptible to peroxide formation.

Peroxyacids (RCO • OOH) are some of the most powerful oxidants of the organic peroxides. Some of the simple peroxyacids are peroxy-formic acid, peroxy-acetic acid, peroxypivalic acid, peroxytrifluoroacetic acid. Traces of transition metals (e.g., cobalt, iron, manganese, nickel, vanadium) can catalyze explosive decomposition of these acids.

Handle all peroxides or peroxide-containing materials with great care.

PCL500   CAS: 79-21-0   HR: 3
PEROXYACETIC ACID
DOT: NA 2131
mf: $C_2H_4O_3$ mw: 76.06

PROP: Not over 40% peracetic acid and not over 6% hydrogen peroxide. Colorless liquid; strong odor. Bp: 105°, explodes @ 110°, flash p: 105°F (OC), d: 1.15 @ 20°. Water-sol. Powerful oxidizer.

SYNS: ACETYL HYDROPEROXIDE * ACIDE PERACETIQUE (FRENCH) * ETHANEPEROXOIC ACID * HYDROPEROXIDE, ACETYL * PERACETIC ACID (MAK) * PERACETIC ACID, solution (DOT) * PEROXYACETIC ACID, maximum concentration 43% in acetic acid (DOT)

CONSENSUS REPORTS: EPA Extremely Hazardous Substances List. Community Right-To-Know List. Reported in EPA TSCA Inventory.

DFG MAK: Very strong skin effects.
DOT Classification: Organic Peroxide; Label: Organic Peroxide, Corrosive.

SAFETY PROFILE: Poison by ingestion. Moderately toxic by inhalation and skin contact. A corrosive eye, skin, and mucous membrane irritant. Questionable carcinogen with experimental tumorigenic data by skin contact. Flammable when exposed to heat or flames. Severe explosion hazard when exposed to heat or by spontaneous chemical reaction. Explodes violently at 110°C. A powerful oxidizing agent. Explosive reaction with acetic anhydride, 5-p-chlorophenyl-2,2-dimethyl-3-hexanone. Violent reaction with ether solvents (e.g., tetrahydrofuran, diethyl ether), metal chloride solutions (e.g., calcium chloride, potassium chloride, sodium chloride), olefins, organic matter. Dangerous, keep away from combustible materials. When heated to decomposition it emits acrid smoke and irritating fumes. To fight fire, use water, foam, CO_2. Used as a polymerization initiator, curing agent, and cross-linking agent.

PCM000   CAS: 93-59-4   HR: 3
PEROXYBENZOIC ACID
mf: $C_7H_6O_3$ mw: 138.13

PROP: Leaflets. Mp: 42°, bp: explodes @ 80-100°. Insol in water, sol in alc and ether.

SYNS: BENZENECARBOPEROXOIC ACID (9CI) * BENZOYLHYDROGEN PEROXIDE * BENZOYL HYDROPEROXIDE * PERBENZOIC ACID

SAFETY PROFILE: Moderately irritating to skin, eyes, and mucous membranes by ingestion and inhalation. Questionable carcinogen with experimental tumorigenic data by skin contact. A dangerous fire hazard when exposed to heat, flame, or reducing materials. A powerful oxidizing agent. Severe explosion hazard when exposed to heat or flame. Violent reaction with olefins. Can react vigorously with reducing materials. Avoid evaporation. When heated to decomposition it emits acrid smoke and irritating fumes.

PCM250   CAS: 13709-32-5   HR: 3
PEROXYDISULFURYL DIFLUORIDE
mf: $F_2O_6S_2$ mw: 198.13

SAFETY PROFILE: A powerful irritant and corrosive to skin, eyes, and mucous membranes. A very powerful oxidizer. A dangerous fire hazard. Ignites organic materials immediately on contact. Explosive reaction with carbon monoxide, dichloromethane. Violent reaction with boron nitride. When heated to decomposition it emits very toxic fumes of F^- and SO_x.

PCM500   CAS: 107-32-4   HR: 3
PEROXYFORMIC ACID
mf: CH_2O_3 mw: 62.03

SYN: METHANEPEROXOIC ACID

SAFETY PROFILE: A powerful irritant and an oxidizer. A dangerous fire hazard when exposed to heat, flame, or reducing materials. Unstable and shock-sensitive. 80% solution is explosive. Extremely dangerous when moved. Violent reaction with carbon, red phosphorus, silicon, formaldehyde, benzaldehyde, aniline, alkenes. Self reactive. Incompatible with metals, nonmetals, organic materials. When heated to decomposition it emits acrid smoke and irritating fumes.

PCN500   HR: 3
PEROXYMONOPHOSPHORIC ACID
mf: H_3O_5P mw: 87.0

SYN: PEROXOMONOPHOSPHORIC ACID

SAFETY PROFILE: Very irritating and corrosive to tissues of skin, eyes, mucous membranes. Powerful oxidizer. A dangerous fire hazard when exposed to heat, flame, or reducing materials. Causes ignition on contact with organic materials. Hazardous reaction with potassium permanganate + coal. When heated to decomposition it emits toxic fumes of PO_x.

PCN750   CAS: 7722-86-3   HR: 3
PEROXYMONOSULFURIC ACID
mf: H_2O_5S mw: 114.08

SYN: PEROXOMONOSULFURIC ACID

SAFETY PROFILE: Strong irritant. Powerful oxidizer. An explosive. Explosive reaction acetone; alcohols; aromatics (e.g., aniline; benzene; phenol); platinum; manganese dioxide; silver. Incompatible with acetone; catalysts; fibers. When heated to decomposition it emits toxic fumes of SO_x.

PCO000   CAS: 26604-66-0   HR: 3
PEROXYNITRIC ACID
mf: HNO_4 mw: 79.02

SAFETY PROFILE: A poison. Very irritating and corrosive to tissue. Decomposes explosively. A dangerous fire hazard when exposed to heat, flame, or reducing materials. Upon decomposition it emits toxic fumes of NO_x.

PCO250   CAS: 359-48-8   HR: 3
PEROXYTRIFLUOROACETIC ACID
mf: $C_2HF_3O_3$ mw: 130.03

SAFETY PROFILE: A poison. A powerful oxidizer. Corrosive and irritating to skin, eyes, mucous membranes. Reaction with 4-iodo-3,5-dimethylisoxazole forms an explosive by-product. When heated to decomposition it emits toxic fumes of F^-.

PCR000   CAS: 91845-41-9   HR: 3
PETASITES JAPONICUS MAXIM

PROP: Dried flower stalk of *Petasites Japonicus Maxim*.

SYNS: COLTS FOOT * FUKI-NO-TOH (JAPANESE)

SAFETY PROFILE: Suspected carcinogen with experimental carcinogenic and tumorigenic data. When heated to decomposition it emits acrid smoke and irritating fumes.

PCR250   CAS: 8002-05-9   HR: 3
PETROLEUM

DOT: NA 1267/UN 1268

PROP: A thick flammable, dark yellow to brown or green-black liquid. D: 0.780-0.970, flash p: 20-90°F. Insol in water; sol in benzene, chloroform, ether. Consists of a mixture of hydrocarbons from C_2H_6 and up, chiefly of the paraffins, cycloparaffins, or of cyclic aromatic hydrocarbons, with small amounts of benzene hydrocarbons, sulfur, and oxygenated compounds

SYNS: BASE OIL * COAL LIQUID * COAL OIL * CRUDE OIL * PETROLEUM CRUDE * ROCK OIL * SENECA OIL

CONSENSUS REPORTS: Reported in EPA TSCA Inventory.

DOT Classification: Combustible Liquid; Label: None: Flammable Liquid; Label: Flammable Liquid.

SAFETY PROFILE: Questionable carcinogen with experimental carcinogenic, neoplastigenic, and tumorigenic data by skin contact. A dangerous fire hazard when exposed to heat, flame, or powerful oxidizers. To fight fire, use foam, CO_2, dry chemical. When heated to decomposition it emits acrid smoke and irritating fumes.

PCR500   CAS: 8052-42-4   HR: 3
PETROLEUM ASPHALT

PROP: Steam refined asphalt.

SYNS: ASPHALT, PETROLEUM * PETROLEUM ROOFING TAR * ROAD ASPHALT

CONSENSUS REPORTS: Reported in EPA TSCA Inventory.

SAFETY PROFILE: Questionable carcinogen with experimental neoplastigenic data skin contact. When heated to decomposition it emits acrid smoke and irritating fumes.

PCS250   CAS: 8002-05-9   HR: 1
PETROLEUM DISTILLATE

DOT: UN 1268

SYN: NAPHTHA

CONSENSUS REPORTS: Reported in EPA TSCA Inventory.

OSHA PEL: (Transitional: TWA 500 ppm) TWA 400 ppm
DOT Classification: Combustible Liquid; Label: None; Flammable Liquid; Label: Flammable Liquid.

SAFETY PROFILE: Human systemic effects by parenteral route: cough, dyspnea, nausea or vomiting. Flammable or combustible liquid when exposed to heat or flame. When heated to decomposition it emits acrid smoke and irritating fumes. Used as a vehicle for pesticides.

PCT250   CAS: 64475-85-0   HR: 3
PETROLEUM SPIRITS

DOT: UN 1271

PROP: Volatile, clear, colorless and non-fluorescent liquid. Mp: $< -73°$, bp: 40-80°, ULC: 95-100, lel: 1.1%, uel: 5.9%, flash p: $<0°F$, d: 0.635-0.660, autoign temp: 550°F, vap d: 2.50.

SYNS: BENZINE * BENZOLINE * CANADOL * HERBITOX * LIGROIN * MINERAL SPIRITS * MINERAL THINNER * MINERAL TURPENTINE * PAINTERS' NAPHTHA * REFINED SOLVENT NAPHTHA * SKELLY-SOLVE S * SOLVENT NAPHTHA

* STODDARD SOLVENT * VARNISH MAKERS' NAPHTHA * VARNISH MAKERS' AND PAINTERS' NAPHTHA * VARSOL * VM&P NAPHTHA * WHITE SPIRITS

CONSENSUS REPORTS: Reported in EPA TSCA Inventory.

OSHA PEL: TWA 300 ppm; STEL 400 PPM
ACGIH TLV: TWA 300 ppm
NIOSH REL: TWA 350 mg/m³; CL 1800 mg/m³/15M

SAFETY PROFILE: Moderately toxic to humans by an unspecified route. Mildly toxic by inhalation and intraperitoneal routes. Ingestion can cause a burning sensation, vomiting, diarrhea, drowsiness, and, in severe cases, pulmonary edema. Inhalation of concentrated vapors can cause intoxication resembling that from alcohol, headache, nausea, coma, and hemorrhage to various vital organs. Highly dangerous fire hazard when exposed to heat, flame sparks, or oxidizing materials. Explosive in the form of vapor when exposed to heat or flame. Highly dangerous; keep away from heat or flame.! To fight fire, use foam, CO_2, dry chemical. When heated to decomposition it emits acrid smoke and irritating fumes.

PCW250   CAS: 85-01-8   HR: 3
PHENANTHRENE
mf: $C_{14}H_{10}$ mw: 178.24

PROP: Solid or monoclinic crystals. Mp: 100°, bp: 339°, d: 1.179 @ 25°, vap press: 1 mm @ 118.3°, vap d: 6.14. Insol in water; sol in CS_2, benzene, and hot alc; very sol in ether.

SYNS: PHENANTHREN (GERMAN) * PHENANTRIN

CONSENSUS REPORTS: IARC Cancer Review: GROUP 3 IMEMDT 7,56,87; Animal Inadequate Evidence IMEMDT 32,419,83. Reported in EPA TSCA Inventory. EPA Genetic Toxicology Program.

OSHA PEL: TWA 0.2 mg/m³

SAFETY PROFILE: Poison by intravenous route. Moderately toxic by ingestion. Mutation data reported. A human skin photosensitizer. Questionable carcinogen with experimental neoplastigenic and tumorigenic data by skin contact. Combustible when exposed to heat or flame; can react vigorously with oxidizing materials. To fight fire, use water, foam, CO_2, dry chemical. When heated to decomposition it emits acrid smoke and irritating fumes.

PDB000   CAS: 578-94-9   **HR: 3**
PHENARSAZINE CHLORIDE

DOT: UN 1698
mf: $C_{12}H_9AsClN$ mw: 277.59

PROP: Light yellow to green granules; irr odor. Mp: 195°, bp: 410° (decomp), d: 1.65, vap press: very low @ 20°, vap d: 9.6. Very sltly sol in water; sltly sol in benzene, brass. Corrodes iron, bronze.

SYNS: ADAMSITE * 5-AZA-10-ARSENAANTHRA-CENE CHLORIDE * 10-CHLORO-5,10-DIHYDROARSA-CRIDINE * 10-CHLORO-5,10-DIHYDROPHENARSAZINE * DIPHENYLAMINECHLORARSINE * DIPHENYL-AMINECHLOROARSINE (DOT) * DM

CONSENSUS REPORTS: Arsenic and its compounds are on the Community Right-To-Know List.

DOT Classification: Irritating Material; Label: Irritant; Poison B; Label: Poison.

SAFETY PROFILE: Human poison by inhalation. Poison experimentally by intravenous route. Human systemic effects by inhalation: changes in function or structure of salivary glands, nausea or vomiting, cough. May be irritating to skin, eyes, and mucous membranes. A vomiting type of poison gas (non-persistent). When heated to decomposition it emits very toxic fumes of As and Cl⁻.

PDB500   CAS: 92-82-0   **HR: 3**
PHENAZINE
mf: $C_{12}H_8N_2$ mw: 180.22

PROP: Pale yellow crystals. Mp: 171°, bp: > 360° (subl). Very sltly sol in water; sol in cold and hot alc, ether.

SYNS: AZOPHENYLENE * DIBENZOPARADIAZINE * DIBENZOPYRAZINE

CONSENSUS REPORTS: Reported in EPA TSCA Inventory.

SAFETY PROFILE: Poison by intraperitoneal and intravenous routes. Questionable carcinogen with experimental tumorigenic data. When heated to decomposition it emits toxic fumes of NO_x.

PDC000   CAS: 91-75-8   **HR: 3**
PHENAZOLINE
mf: $C_{17}H_{19}N_3$ mw: 265.39

SYNS: ANTASTEN * ANTAZOLINE * ANTIHIS-TAL * ANTISTINE * AZALONE * BEN-A-HIST

* 2-(N-BENZYLANILINOMETHYL)-2-IMIDAZOLINE
* 4,5-DIHYDRO-N-PHENYL-N-PHENYLMETHYL-1H-IMIDAZOLE-2-METHANAMINE * HISTOSTAB
* IMIDAMINE * 5512-M * 2-(N-PHENYL-N-BEN-ZYLAMINOMETHYL)IMIDAZOLINE * 2-PHENYL-BEN-ZYL-AMINO-METHYLIMIDAZOLIN (GERMAN)

SAFETY PROFILE: Poison by ingestion, subcutaneous, and intraperitoneal routes. An eye irritant. When heated to decomposition it emits toxic fumes of NO_x.

PDC250   CAS: 136-40-3   **HR: 3**
PHENAZOPYRIDINIUM CHLORIDE
mf: $C_{11}H_{11}N_5 \cdot ClH$ mw: 249.73

PROP: Red crystals; sltly bitter taste. Sltly sol in cold water, alc; sol in acetic acid; insol in acetone, benzene, chloroform, ether.

SYNS: AZODINE * AZODIUM * AZODYNE
* AZO GANTRISIN * AZO GASTANOL * AZO-MANDELAMINE * AZOMINE * AZO-STANDARD
* AZO-STAT * AZOTREX * BARIDIUM
* BISTERIL * CYSTAMINE "MCCLUNG"
* CYSTOPYRIN * CYSTURAL * 2,6-DIAMINO-3-PHENYLAZOPYRIDINE HYDROCHLORIDE * 2,6-DI-AMINO-3-(PHENYLAZO)PYRIDINE MONOHYDROCHLO-RIDE * DI-AZO * DIRIDONE * DOLONIL
* EUCISTIN * GIRACID * MALLOFEEN
* MALLOPHENE * NC 150 * NCI-C01672
* NEFRECIL * PAP * PDP * PHENAZO
* PHENAZODINE * PHENAZOPYRIDINE HYDRO-CHLORIDE * PHENYLAZODIAMINOPYRIDINE HYDROCHLORIDE * β-PHENYLAZO-α,α'-DIAMINO-PYRIDINE HYDROCHLORIDE * 3-PHENYLAZO-2,6-DIAMINOPYRIDINE HYDROCHLORIDE * PHE-NYLAZO-α,α'-DIAMINOPYRIDINE MONOHY-DROCHLORIDE * PHENYLAZOPYRIDINE HY-DROCHLORIDE * 3-(PHENYLAZO)-2,6-PYRI-DINEDIAMINE, HYDROCHLORIDE * PHENYL-AZO TABLETS * PHENYL-IDIUM * PHENYL-IDIUM 200 * PIRID * PIRIDACIL * PYRAZO-DINE * PYRAZOFEN * PYREDAL * PYRIDACIL
* PYRIDENAL * PYRIDENE * PYRIDIATE
* PYRIDIUM * PYRIDIVITE * PYRIPYRIDIUM
* PYRIZIN * SEDURAL * SULADYNE
* SULODYNE * THIOSULFIL-A FORTE
* URAZIUM * URIDINAL * URIPLEX * UROBI-OTIC-250 * URODINE * UROFEEN * UROMIDE
* UROPHENYL * UROPYRIDIN * UROPYRINE
* UTOSTAN * VESTIN * W 1655

CONSENSUS REPORTS: IARC Cancer Review: GROUP 2B IMEMDT 7,312,87; Animal Sufficient Evidence IMEMDT 24,163,80; Human Limited Evidence IMEMDT 24,163,80;

Animal Inadequate Evidence IMEMDT 8, 117,75. NTP Fourth Annual Report On Carcinogens, 1984. NCI Carcinogenesis Bioassay (feed); Clear Evidence: mouse, rat NCITR* NCI-CG-TR-99,78.

SAFETY PROFILE: Confirmed carcinogen with experimental carcinogenic and tumorigenic data. A poison by intraperitoneal and intravenous routes. Moderately toxic by ingestion. Human systemic effects by ingestion: somnolence, cyanosis, diarrhea, nausea or vomiting, anuria or decreased urine volume, normocytic anemia, methemoglobinemia-carboxhemoglobinemia, dehydration, changes in blood sodium levels. When heated to decomposition it emits very toxic fumes of NO_x and HCl.

PDC750   CAS: 2275-14-1   HR: 3
PHENCAPTON

DOT: NA 2783
mf: $C_{11}H_{13}Cl_2O_2PS_3$ mw: 375.29

SYNS: O,O-DIAETHYL-S((2,5-DICHLOR-PHENYL-THIO)-METHYL)-DITHIOPHOSPHAT (GERMAN) * S-(2,5-DI-CHLOROPHENYLTHIOMETHYL) O,O-DIETHYL PHOSPHO-RODITHIOATE * 2,5-DICHLOROPHENYLTHIOMETHYL O,O-DIETHYL PHOSPHORODITHIOATE * S-(2,5-DI-CHLOROPHENYLTHIOMETHYL) DIETHYL PHOS-PHOROTHIOLOTHIONATE * O,O-DIETHYL-S-(2,5-DICHLOROPHENYLTHIOMETHYL) DITHIOPHOSPHATE * O,O-DIETHYL-S-(2,5-DICHLOROPHENYLTHIOMETHYL) DITHIOPHOSPHORAN * O,O-DIETHYL-S-(2,5-DICHLO-ROPHENYLTHIOMETHYL) PHOSPHORODITHIOATE * O,O-DIETHYL-S-(2,5-DICHLOROPHENYLTHIOMETHYL) PHOSPHOROTHIOLOTHIONATE * DITHIOPHOSPHATE de-O,O-DIETHYLE et de S(2,5-DICHLOROPHENYL) THIO-METHYLE (FRENCH) * EENKAPTON (DUTCH) * ENT 25,585 * GEIGY G-28029 * PRZEDZIORKO-FOS (POLISH)

DOT Classification: ORM-A; Label: None.

SAFETY PROFILE: Poison by ingestion and possibly other routes. Moderately toxic by skin contact. A cholinesterase inhibitor. When heated to decomposition it emits very toxic fumes of Cl^-, PO_x, and SO_x.

PDD300   CAS: 65-29-2   HR: 3
(v-PHENENYLTRIS(OXYETHYLENE)) TRIS(TRIETHYLAMMONIUM IODIDE)
mf: $C_{30}H_{60}N_3O_3 \cdot 3I$ mw: 891.63

SYNS: BENZCURINE IODIDE * F 2559 * FLAXE-DIL * GALLAMINE * RELAXAN * RETENSIN

* RP 3697 * SYNCURARINE * TRICURAN * 1,2,3-TRI(β-DIETHYLAMINOETHOXY)BENZENE TRIETHIODIDE * TRI(β-DIETHYLAMINOETHOXY)-1,2,3-BENZENE TRI-IODOETHYLATE * TRIIODO-ETHYLATE de GALLAMINE (FRENCH) * TRIIODO-ETHYLATE OF TRI(DIETHYLAMINOETHYLOXY)-1,2,3-BENZENE * 1,2,3-TRIS(2-DIETHYLAMINOETHOXY) BENZENE TRIETHIODIDE * 1,2,3-TRIS(2-DIETHYL-AMINOETHOXY)BENZENE TRIS(ETHYLIODIDE) * 1,2,3-TRIS(2-TRIETHYLAMMONIUM ETHOXY)BEN-ZENE TRIIODIDE

SAFETY PROFILE: Poison by ingestion, subcutaneous, intravenous, parenteral, intraduodenal, intraperitoneal, and intramuscular routes. When heated to decomposition it emits very toxic fumes of NH_3, NO_x, and I^-.

PDD500   CAS: 156-43-4   HR: 3
PHENETHIDINE

DOT: UN 2311
mf: $C_8H_{11}NO$ mw: 137.20

SYNS: p-AMINOPHENETOLE * 4-AMINOPHENETOLE * p-ETHOXYANILINE * 4-ETHOXYANILINE * p-PHENETIDINE (DOT)

CONSENSUS REPORTS: Reported in EPA TSCA Inventory.

DOT Classification: Poison B; Label: St. Andrews Cross.

SAFETY PROFILE: Poison by inhalation. Moderately toxic by ingestion and intraperitoneal routes. Caution: It can be absorbed through the skin. A skin and eye irritant. Mutation data reported. When heated to decomposition it emits toxic fumes of NO_x.

PDD750   CAS: 60-12-8   HR: 3
PHENETHYL ALCOHOL
mf: $C_8H_{10}O$ mw: 122.18

PROP: Colorless liquid; floral odor of roses. Mp: $-27°$, bp: $220°$, flash p: $216°F$, d: 1.0245 @ $15°$, vap d: 4.21. Misc with alc, ether; sol in fixed oils, glycerin, propylene glycol.

SYNS: BENZYL CARBINOL * FEMA No. 2858 * PHENETHANOL * β-PHENETHYL ALCOHOL * 2-PHENETHYL ALCOHOL * β-PHENYLETHANOL * 2-PHENYLETHANOL * β-PHENYLETHYL ALCOHOL * 2-PHENYLETHYL ALCOHOL

CONSENSUS REPORTS: Reported in EPA TSCA Inventory.

SAFETY PROFILE: Poison by ingestion and intraperitoneal routes. Moderately toxic by skin contact. A skin and eye irritant. Experimental teratogenic effects. Causes severe central nervous system injury to experimental animals. Combustible when exposed to heat or flame; can react with oxidizing materials. To fight fire, use CO_2, dry chemical. When heated to decomposition it emits acrid smoke and irritating fumes.

PDE000   CAS: 98-85-1   HR: 3
α-PHENETHYL ALCOHOL

DOT: UN 2937
mf: $C_8H_{10}O$ mw: 122.18

PROP: Colorless liquid; hyacinth odor. Bp: 204°, fp: 21.4°, d: 1.015 @ 20°/20°, refr index: 1.525, vap press: 0.1 mm @ 20°, vap d: 4.21, flash p: 205°F (OC). Sol in fixed oils, propylene glycol; very sol in glycerin.

SYNS: BENZENEMETHANOL, α-METHYL- * ETHANOL, 1-PHENYL- * FEMA No. 2685 * 1-FENYLETHANOL * FENYL-METHYLKARBINOL * α-METHYLBENZYL ALCOHOL (FCC) * METHYLPHENYLCARBINOL * METHYPHENYLMETHANOL * NCI-C55685 * 1-PHENYLETHANOL * PHENYLMETHYLCARBINOL * STYRALLYL ALCOHOL * STYRALYL ALCOHOL

CONSENSUS REPORTS: Reported in EPA TSCA Inventory. EPA Genetic Toxicology Program.

DOT Classification: Poison B; Label: St. Andrews Cross.

SAFETY PROFILE: Poison by ingestion and subcutaneous routes. Moderately toxic by skin contact. A skin and severe eye irritant. Combustible when exposed to heat or flame; can react with oxidizing materials. To fight fire, use alcohol foam, foam, CO_2, dry chemical.

PDE250   CAS: 64-04-0   HR: 3
β-PHENETHYLAMINE

mf: $C_8H_{11}N$ mw: 121.20

PROP: Colorless to sltly yellow liquid; fishy odor. Bp: 194.5-195°, d: 0.96 @ 15.5°/15.5°, vap d: 4.18. Sol in water; very sol in alc, ether.

SYNS: β-AMINOETHYLBENZENE * 1-AMINO-2-PHENYLETHANE * β-PHENYLAETHYLAMIN (GERMAN) * 1-PHENYL-2-AMINO-ATHAN (GERMAN) * 1-PHENYL-2-AMINOETHANE * PHENYLETHYLAMINE * φ-PHENYLETHYLAMINE * 2-PHENYLETHYLAMINE

CONSENSUS REPORTS: Reported in EPA TSCA Inventory.

SAFETY PROFILE: Poison by ingestion, intraperitoneal, subcutaneous, intracervical, and intravenous routes. A strong base. A skin irritant and possible sensitizer. When heated to decomposition it emits toxic fumes of NO_x.

PDF750   CAS: 103-48-0   HR: 1
PHENETHYL ISOBUTYRATE

mf: $C_{12}H_{16}O_2$ mw: 192.28

PROP: Colorless to light yellow liquid; fruity, rosy odor. D: 0.9871.486-1.490, flash p: +212°F. Sol in alc, fixed oils; insol in water @ 230°.

SYNS: BENZYLCARBINOL ISOBUTYRATE * BENZYLCARBINYL ISOBUTYRATE * FEMA No. 2862 * PHENYLETHYL ISOBUTYRATE * β-PHENYLETHYL ISOBUTYRATE * 2-PHENYLETHYL ISOBUTYRATE * 2-PHENYLETHYL-2-METHYLPROPIONATE

CONSENSUS REPORTS: Reported in EPA TSCA Inventory.

SAFETY PROFILE: Mildly toxic by ingestion. Combustible liquid. When heated to decomposition it emits acrid smoke and irritating fumes.

PDF775   CAS: 140-26-1   HR: 1
PHENETHYL ISOVALERATE

mf: $C_{13}H_{18}O_2$ mw: 206.31

PROP: Colorless to sltly yellow liquid; fruity, rosy odor. D: 0.973, refr index: 1.484, flash p: +212°F. Sol in alc, fixed oils; insol in water @ 263°.

SYNS: FEMA No. 2871 * 3-METHYL-BUTANOIC ACID 2-PHENYLETHYL ESTER * PHENETHYL ESTER ISOVALERIC ACID * PHENYLETHYL ISOVALERATE * β-PHENYLETHYL ISOVALERATE * 2-PHENYLETHYL-3-METHYLBUTIRATE

CONSENSUS REPORTS: Reported in EPA TSCA Inventory.

SAFETY PROFILE: Mildly toxic by ingestion. Combustible liquid. When heated to decomposition it emits acrid smoke and irritating fumes.

PDF790   HR: 1
2-PHENETHYL 2-METHYLBUTYRATE

mf: $C_{13}H_{18}O_2$ mw: 206.28

PROP: Colorless liquid; floral, fruity odor. D: 0.973, refr index: 1.484, flash p: +212°F. Sol in alc, fixed oils; insol in water.

SYN: FEMA No. 3632

SAFETY PROFILE: Combustible liquid. When heated to decomposition it emits acrid smoke and irritating fumes.

PDI000   CAS: 102-20-5   **HR: 2**
PHENETHYL PHENYLACETATE
mf: $C_{16}H_{16}O_2$ mw: 240.32

PROP: Colorless to sltly yellow liquid above 26°; rosy, hyacinth odor. D: 1.079-1.082, flash p: +212°F. Sol in alc; insol in water.

SYNS: BENZENEACETIC ACID, 2-PHENYLETHYL ESTER * BENZYLCARBINYL-α-TOLUATE * FEMA No. 2866 * PHENYLACETIC ACID, PHENETHYL ESTER * β-PHENYLETHYL PHENYLACETATE * 2-PHENYLETHYL PHENYLACETATE * 2-PHENYLETHYL-α-TOLUATE

CONSENSUS REPORTS: Reported in EPA TSCA Inventory.

SAFETY PROFILE: Moderately toxic by ingestion. Combustible liquid. When heated to decomposition it emits acrid smoke and irritating fumes.

PDK200   **HR: 1**
PHENETHYL SALICYLATE
mf: $C_{15}H_{14}O_3$ mw: 242.27

PROP: White crystals; balsamic odor. Solidification point: 41°, flash p: +212°F. Sol in alc; insol in water.

SYN: FEMA No. 2868

SAFETY PROFILE: Combustible liquid. When heated to decomposition it emits acrid smoke and irritating fumes.

PDM250   CAS: 404-82-0   **HR: 3**
PHENFLUORAMINE HYDROCHLORIDE
mf: $C_{12}H_{16}F_3N \cdot ClH$ mw: 267.75

SYNS: N-ETHYL-α-METHYL-m-TRIFLUOROMETHYL-PHENETHYLAMINE * N-ETHYL-α-METHYL-m-(TRIFLUOROMETHYL)PHENETHYLAMINE HYDROCHLORIDE * FENFLURAMINE HYDROCHLORIDE * PONDERAL * PONDERAX * PONDIMIN * 1-(3-TRIFLUOROMETHYLPHENYL)-2-ETHYLAMINO-PROPANE HYDROCHLORIDE

SAFETY PROFILE: Poison by ingestion, intravenous, and intraperitoneal routes. Human systemic effects by ingestion: mydriasis, change in motor activity, nausea. When heated to decomposition it emits very toxic fumes of F^-, NO_x, and HCl.

PDN000   CAS: 55-52-7   **HR: 3**
PHENIPRAZINE
mf: $C_9H_{14}N_2$ mw: 150.25

SYNS: CASTRON * CATRAL * CATRAN * CATRONIAZIDE * CAVODIL * DICATRON * FENILISOPROPILIDRAZINA * 2-HYDRAZINO-1-PHENYLPROPANE * JB 516 * KATRON * KATRONIAZID * (α-METHYLPHENETHYL)HYDRAZINE * MIRAL * P 1142 * PHENIZINE * 1-PHENYL-2-HYDRAZINOPROPANE * β-PHENYLISOPROPYLHYDRAZINE * PHENYLISOPROPYLHYDRAZINE * PIH * PSICOSTEN * RUN

SAFETY PROFILE: Poison by ingestion, intraperitoneal, subcutaneous, and intravenous routes. Human systemic effects by an unspecified route: visual field effects. Mutation data reported. When heated to decomposition it emits toxic fumes of NO_x.

PDN500   CAS: 537-05-3   **HR: 3**
PHENODIANISYL HYDROCHLORIDE
mf: $C_{23}H_{25}N_3O_3 \cdot ClH$ mw: 427.97

PROP: Crystals, odorless. Mp: 176°. Very sol in alc; insol in water, oils.

SYNS: ACOINE * AKOIN HYDROCHLORID (GERMAN) * N,N'-BIS(4-METHOXYPHENYL)-N''-(4-ETHOXYPHENYL)GUANIDINE HYDROCHLORIDE * α,γ-DI-p-ANISYL-β-(ETHOXYPHENYL)GUANIDINE HYDROCHLORIDE * DIANISYL-MONOPHENETHYLGUANIDINE HYDROCHLORIDE * DIPARAANISYL-MONOPHENETHYL-GUANIDIN-HYDROCHLORID (GERMAN) * 2-(4-ETHOXYPHENYL)-1,3-BIS(4-METHOXYPHENYL) GUANIDINE HYDROCHLORIDE * GUANICAINE * PHENODIANISYL

SAFETY PROFILE: Poison by ingestion, intravenous, and subcutaneous routes. Solutions are decomposed by light. When heated to decomposition it emits very toxic fumes of HCl and NO_x.

PDN750   CAS: 108-95-2   **HR: 3**
PHENOL

DOT: UN 1671/UN 2312/NA 2821
mf: C_6H_6O mw: 94.12

PROP: White, crystalline mass which turns pink or red if not perfectly pure; burning taste, distinctive odor. Mp: 40.6°, bp: 181.9°, flash p: 175°F (CC), d: 1.072, autoign temp: 1319°F, vap press: 1 mm @ 40.1°, vap d: 3.24. Sol in water; misc in alc, ether.

SYNS: ACIDE CARBOLIQUE (FRENCH) * BAKER'S P AND S LIQUID and OINTMENT * BENZENOL * CARBOLIC ACID * CARBOLSAURE (GERMAN) * FENOL (DUTCH, POLISH) * FENOLO (ITALIAN) * HYDROXYBENZENE * MONOHYDROXYBENZENE * MONOPHENOL * NCI-C50124 * OXYBENZENE * PHENIC ACID * PHENOL ALCOHOL * PHENOL, molten (DOT) * PHENOLE (GERMAN) * PHENYL HYDRATE * PHENYL HYDROXIDE * PHENYLIC ACID * PHENYLIC ALCOHOL * RCRA WASTE NUMBER U188

CONSENSUS REPORTS: NCI Carcinogenesis Bioassay (oral); No Evidence: mouse, rat NCITR* NCI-CG-TR-203,80. EPA Extremely Hazardous Substances List. Community Right-To-Know List. Reported in EPA TSCA Inventory. EPA Genetic Toxicology Program.

OSHA PEL: TWA 5 ppm (skin)
ACGIH TLV: TWA 5 ppm (skin); BEI: 250 mg(total phenol)/g creatinine in urine at end of shift.
DFG MAK: 5 ppm (19 mg/m^3); BAT: 300 mg/L at end of shift.
NIOSH REL: TWA 20 mg/m^3; CL 60 mg/m^3/15M
DOT Classification: Poison B; Label: Poison.

SAFETY PROFILE: Human poison by ingestion. An experimental poison by ingestion, subcutaneous, intravenous, parenteral, and intraperitoneal routes. Moderately toxic by skin contact. A severe eye and skin irritant. Questionable carcinogen with experimental carcinogenic and neoplastigenic data. Human mutation data reported. Absorption of phenolic solutions through the skin may be very rapid, and can cause death within 30 minutes to several hours by exposure of as little as 64 square inches of skin. Lesser exposures can cause damage to the kidneys, liver, pancreas and spleen, and edema of the lungs. Ingestion can cause corrosion of the lips, mouth, throat, esophagus and stomach, and gangrene. Ingestion of 15 grams has killed. Chronic exposures can cause death from liver and kidney damage. Dermatitis resulting from contact with phenol or phenol-containing products is fairly common in industry. A common air contaminant.

Combustible when exposed to heat, flame, or oxidizers. Potentially explosive reaction with aluminum chloride + nitromethane (at 110°C/100 bar), formaldehyde, peroxydisulfuric acid, peroxymonosulfuric acid, sodium nitrite + heat. Violent reaction with aluminum chloride + nitrobenzene (at 120°C), sodium nitrate + trifluoroacetic acid, butadiene. Can react with oxidizing materials. To fight fire, use alcohol foam, CO_2, dry chemical. When heated to decomposition it emits acrid smoke and irritating fumes.

PDP250   CAS: 92-84-2   **HR: 3**
PHENOTHIAZINE
mf: $C_{12}H_9NS$ mw: 199.28

PROP: Yellow, rhombic leaflets or diamond-shaped plates from toluene or butanol. Mp: 185.1°, sublimes at 130° at 1 mm, bp: 371°. Freely sol in benzene; sol in ether, hot acetic acid; sltly sol in alc and in mineral oils; practically insol in petr ether, chloroform, water.

SYNS: AFI-TIAZIN * AGRAZINE * ANTIVERM * BIVERM * CONTAVERM * DIBENZOPARATHIAZINE * DIBENZOTHIAZINE * DIBENZO-1,4-THIAZINE * ENT 38 * FEENO * FENOTHIAZINE (DUTCH) * FENOTIAZINA (ITALIAN) * FENOVERM * FENTIAZIN * HELMETINA * LETHELMIN * NEMAZENE * NEMAZINE * ORIMON * PADOPHENE * PENTHAZINE * PHENEGIC * PHENOSAN * PHENOVERM * PHENOVIS * PHENOXUR * PHENTHIAZINE * RECONOX * SOUFRAMINE * THIODIFENYLAMINE (DUTCH) * THIODIPHENYLAMIN (GERMAN) * THIODIPHENYLAMINE * TIODIFENILAMINA (ITALIAN) * VERMITIN * WURM-THIONAL * XL-50

CONSENSUS REPORTS: EPA Genetic Toxicology Program. Reported in EPA TSCA Inventory.

OSHA PEL: TWA 5 mg/m^3 (skin)
ACGIH TLV: TWA 5 mg/m^3 (skin)

SAFETY PROFILE: Poison by intravenous route. Moderately toxic to humans by ingestion. Experimental reproductive effects. An insecticide. Large doses, i.e., heavy exposure, may cause hemolytic anemia and toxic degeneration of the liver. Can cause skin irritation and photosensitization. Dangerous; when heated to decomposition or on contact with acid or acid fumes it emits highly toxic fumes of SO_x and NO_x.

PDS900   **HR: 1**
PHENOXYETHYL ISOBUTYRATE
mf: $C_{12}H_{16}O_3$ mw: 208.26

PROP: Colorless liquid; honey, roselike odor. D: 1.044, refr index: 1.492, flash p: +212°F. Misc in alc, chloroform, ether; insol in water.

SYN: FEMA No. 2873

SAFETY PROFILE: Combustible liquid. When heated to decomposition it emits acrid smoke and irritating fumes.

PDT250 CAS: 59-96-1 HR: 3
N-PHENOXYISOPROPYL-N-BENZYL-β-CHLOROETHYLAMINE
mf: $C_{18}H_{22}ClNO$ mw: 303.86

SYNS: A 688 * BENSYLYTE * 2-(N-BENZYL-2-CHLOROETHYLAMINO)-1-PHENOXYPROPANE * BENZYL(2-CHLOROETHYL)-(1-METHYL-2-PHENOXY-ETHYL)AMINE * BENZYLT * N-(2-CHLORO-ETHYL)-N-(1-METHYL-2-PHENOXYETHYL)BENZENE-METHANAMINE * N-(2-CHLOROETHYL)-N-(1-METHYL-2-PHENOXYETHYL)BENZYLAMINE * DIBENYLIN * DIBENYLINE * DIBEN-ZYLINE * NSC 37448 * PHENOXYBENZAMINE

CONSENSUS REPORTS: IARC Cancer Review: Animal Sufficient Evidence IMEMDT 24,185,80; Animal Limited Evidence IMEMDT 9,223,75

SAFETY PROFILE: Confirmed carcinogen with experimental carcinogenic and neoplastigenic data. Poison by intravenous and intracerebral routes. Moderately toxic by ingestion. Human reproductive effects by ingestion: spermatogenesis. Experimental reproductive effects. When heated to decomposition it emits very toxic fumes of Cl^- and NO_x.

PDW500 CAS: 437-38-7 HR: 3
PHENTANYL
mf: $C_{22}H_{28}N_2O$ mw: 336.52

SYNS: FENTANEST * FENTANIL * FENTANYL * N-PHENETHYL-4-(N-PROPIONYLANILINO)PIPERIDINE * 1-PHENETHYL-4-N-PROPIONYLANILINOPIPERIDINE * N-PHENYL-N-(1-(2-PHENYLETHYL)-4-PIPERIDINYL)PROPANAMIDE (9CI) * R 4263 * SENTONIL

SAFETY PROFILE: Poison by intravenous and intraperitoneal routes. Human systemic effects by intravenous route: somnolence, respiratory depression. When heated to decomposition it emits toxic fumes of NO_x.

PDW750 CAS: 990-73-8 HR: 3
PHENTANYL CITRATE
mf: $C_{22}H_{28}N_2O \cdot C_6H_8O_7$ mw: 528.66

SYNS: FENTANEST * FENTANYL CITRATE * LEPTANAL * McN-JR 4263 * MCN-JR-4263-49 * PENTANYL * N-(1-PHENETHYL-4-PIPERIDI-NYL)

PROPIONANILIDE DIHYDROGEN CITRATE * N-(1-PHENETHYL-4-PIPERIDYL)PROPIONANILIDE CITRATE * N-(1-PHENETHYL-4-PIPERIDYL)PROPIONANILIDE DI-HYDROGEN CITRATE * R 4263 * R 5240 * SUBLIMAZE * SUBLIMAZE CITRATE

SAFETY PROFILE: Poison by ingestion, subcutaneous, and intravenous routes. When heated to decomposition it emits toxic fumes of NO_x.

PDX000 CAS: 101-48-4 HR: 2
PHENYLACETALDEHYDE DIMETHYL ACETAL
mf: $C_{10}H_{14}O_2$ mw: 166.24

PROP: Colorless liquid; strong odor. D: 1.000-1.006, refr index: 1.493, flash p: 194°F. Sol in fixed oils, propylene glycol; insol in glycerin.

SYNS: (2,2-DIMETHOXYETHYL)-BENZENE (9CI) * 1,1-DIMETHOXY-2-PHENYLETHANE * FEMA No. 2876 * HYSCYLENE P * PHENACETALDEHYDE DI-METHYL ACETAL * α-TOLYL ALDEHYDE DIMETHYL ACETAL * VIRIDINE

CONSENSUS REPORTS: Reported in EPA TSCA Inventory.

SAFETY PROFILE: Moderately toxic by ingestion. Combustible liquid. When heated to decomposition it emits acrid smoke and irritating fumes.

PDY500 CAS: 4075-79-0 HR: 3
4′-PHENYLACETANILIDE
mf: $C_{14}H_{13}NO$ mw: 211.28

SYNS: 4-ACETAMIDOBIPHENYL * 4-ACETYLAMI-NOBIPHENYL * 4-BIPHENYLACETAMIDE * N-4-BI-PHENYLACETAMIDE * N-(4-BIPHENYLYL)ACETAMIDE * p-PHENYLACETANILIDE

CONSENSUS REPORTS: Reported in EPA TSCA Inventory.

SAFETY PROFILE: Questionable carcinogen with experimental carcinogenic and tumorigenic data. Mutation data reported. When heated to decomposition it emits toxic fumes of NO_x. Used in the manufacture of plastics, resins, rubber, synthetics, dyes, and pigments.

PDY850 CAS: 103-82-2 HR: 2
PHENYLACETIC ACID
mf: $C_8H_8O_2$ mw: 136.16

PROP: Leaflets on distillation in vac; plates, tablets from petr ether; disagreeable odor of geranium. Mp: 76.5°, bp: 265.5°, d (77/4) 1.091, flash p: +212°F. Sltly sol in cold water; freely sol in hot water; sol in alc and ether. Solubility @ 25° in chloroform (moles/L): 4.422; in carbon tetrachloride: 1.842; in acetylene tetrachloride: 4.513; in trichlorethylene: 3.299; in tetrachlorethylene: 1.558; in pentachloroethane: 3.252.

SYNS: BENZENACETIC ACID * BENZENEACETIC ACID * FEMA No. 2878 * omega-PHENYLACETIC ACID * α-TOLUIC ACID

CONSENSUS REPORTS: Reported in EPA TSCA Inventory.

SAFETY PROFILE: Moderately toxic by ingestion, subcutaneous, and intraperitoneal routes. An experimental teratogen. Combustible liquid. When heated to decomposition it emits acrid smoke and irritating fumes.

PEA750   CAS: 140-29-4   **HR: 3**
PHENYLACETONITRILE

DOT: UN 2470
mf: C_8H_7N mw: 117.16

PROP: Oily liquid; aromatic odor. Mp: −23.8°, bp: 233.5°, d: 1.0214 @ 15°/15°, vap press: 1 mm @ 60.0°. Insol in water; misc in alc, ether.

SYNS: BENZENEACETONITRILE * BENZYL CYANIDE * BENZYL NITRILE * (CYANOMETHYL)BENZENE * α-CYANOTOLUENE * φ-CYANOTOLUENE * 2-PHENYLACETONITRILE * PHENYLACETONITRILE, liquid (DOT) * α-TOLUNITRILE * USAF KF-21

CONSENSUS REPORTS: EPA Extremely Hazardous Substances List. Community Right-To-Know List. Reported in EPA TSCA Inventory.

DOT Classification: Poison B; Label: St. Andrews Cross.

SAFETY PROFILE: Poison by ingestion, inhalation, skin contact, subcutaneous, and intraperitoneal routes. A skin irritant. Explosive reaction with sodium hypochlorite. When heated to decomposition it emits very toxic fumes of CN^- and NO_x.

PEC500   CAS: 673-06-3   **HR: 1**
d-PHENYLALANINE
mf: $C_9H_{11}NO_2$ mw: 165.21

PROP: Needles from alc, white crystalline platlets. Mp: 104-105°; Sol in hot water; very sltly sol in alc; sltly sol petr ether.

SYNS: dl-α-AMINO-β-PHENYLPROPIONIC ACID * NCI-C60195 * d-β-PHENYLALANINE * dl-PHENYLALANINE (FCC)

CONSENSUS REPORTS: Reported in EPA TSCA Inventory.

SAFETY PROFILE: Mildly toxic by intraperitoneal route. Human systemic effects by ingestion: nausea, hypermotility, diarrhea. When heated to decomposition it emits toxic fumes of NO_x.

PEC750   CAS: 63-91-2   **HR: 1**
l-PHENYLALANINE
mf: $C_9H_{11}NO_2$ mw: 165.21

PROP: White crystals or crystalline powder; slt odor and bitter taste. Mp: decomp @ 275-283°. Sol in water; very sltly sol in alc, ether.

SYNS: (S)-α-AMINOBENZENEPROPANOIC ACID * α-AMINOHYDROCINNAMIC ACID * α-AMINO-β-PHENYLPROPIONIC ACID * ANTIBIOTIC FN 1636 * PAL * PHENYLALANINE * PHENYL-α-ALANINE * (S)-PHENYLALANINE * β-PHENYLALANINE * l-β-PHENYLALANINE * 3-PHENYLALANINE

CONSENSUS REPORTS: Reported in EPA TSCA Inventory.

SAFETY PROFILE: Mildly toxic by intraperitoneal route. Experimental reproductive effects. When heated to decomposition it emits toxic fumes of NO_x.

PED750   CAS: 148-82-3   **HR: 3**
l-PHENYLALANINE MUSTARD
mf: $C_{13}H_{18}Cl_2N_2O_2$ mw: 305.23

SYNS: ALANINE NITROGEN MUSTARD * ALKERAN * AT-290 * l-3-(p-(BIS(2-CHLOROETHYL)AMINO)PHENYL)ALANINE * p-N-BIS(2-CHLOROETHYL)AMINO-l-PHENYLALANINE * 3-(p-(p-(BIS(2-CHLOROETHYL)AMINO)PHENYL)-l-ALANINE * 4-(BIS(2-CHLOROETHYL)AMINO)-l-PHENYLALANINE * CB 3025 * p-N-DI(CHLOROETHYL)AMINOPHENYLALANINE * p-DI-(2-CHLOROETHYL)AMINO-l-PHENYLALANINE * 3-p-(DI(2-CHLOROETHYL)AMINO)-PHENYL-l-ALANINE * MELPHALAN * NCI-C04853 * NSC-8806 * l-PAM * PHENYLALANINE NITROGEN MUSTARD * RCRA WASTE NUMBER U150 * l-SARCOLYSIN * p-l-SARCOLYSIN * SK-15673

CONSENSUS REPORTS: IARC Cancer Review: GROUP 1 IMEMDT 7,239,87; Animal Sufficient Evidence; Human Limited Evidence IMEMDT 9,167,75. NTP Fourth Annual Report On Carcinogens, 1984. NCI Carcinogenesis

Studies (ipr); Clear Evidence: mouse, rat RRCRBU 52,1,75. EPA Genetic Toxicology Program.

SAFETY PROFILE: Confirmed human carcinogen producing leukemia and Hodgkin's disease. Poison by ingestion, intravenous, and intracerebral routes. An experimental carcinogen. Human systemic effects by ingestion: nausea. Human reproductive effects by ingestion: menstrual changes. Mutation data reported. A skin irritant. Used as a poison gas. When heated to decomposition, it emits toxic fumes of Cl^- and NO_x.

PEE750   CAS: 1698-60-8   **HR: 2**
1-PHENYL-4-AMINO-5-CHLORPYRIDAZ-6-ONE
mf: $C_{10}H_8ClN_3O$ mw: 221.66

SYNS: 5-AMINO-4-CHLORO-2,3-DIHYDRO-3-OXO-2-PHE-NYLPYRIDAZINE * 5-AMINO-4-CHLORO-2-PHENYL-3(2H)-PYRIDAZINONE * BUREX (CZECH) * CHLO-RIDAZON * 1-FENYL-4-AMINO-5-CHLOR-6-PYRIDAZI-NON (CZECH) * HS-119-1 * PCA * PHENOSANE * 1-PHENYL-4-AMINO-5-CHLOROPYRIDAZON-(6) (GER-MAN) * 1-PHENYL-4-AMINO-5-CHLORO-6-PYRIDA-ZONE * 1-PHENYL-4-AMINO-5-CHLOROPYRIDAZONE-6 * PYRAMINE * PYRAMIN RB * PYRAZON * PYRAZONE * PYRAZONL

CONSENSUS REPORTS: Reported in EPA TSCA Inventory. EPA Genetic Toxicology Program.

SAFETY PROFILE: Moderately toxic by ingestion and intraperitoneal routes. A severe eye irritant. Used as a preemergence and early post-emergence herbicide. When heated to decomposition it emits very toxic fumes of Cl^- and NO_x.

PEG250   CAS: 613-37-6   **HR: 3**
p-PHENYLANISOLE
mf: $C_{13}H_{12}O$ mw: 184.25

PROP: Leaves from alc. Mp: 90°. Sol in hot alc.

SYNS: p-METHOXYBIPHENYL * 4-METHOXYBIPHE-NYL

CONSENSUS REPORTS: Reported in EPA TSCA Inventory.

SAFETY PROFILE: Questionable carcinogen with experimental tumorigenic data. When heated to decomposition it emits acrid smoke and irritating fumes.

PEG500   CAS: 91-40-7   **HR: 3**
N-PHENYLANTHRANILIC ACID
mf: $C_{13}H_{11}NO_2$ mw: 213.25

PROP: Needles from alc. Mp: 185-187°, decomp 183-184°. Very sltly sol in hot water; sol in hot alc; very sltly sol in ether.

SYNS: o-ANILINOBENZOIC ACID * 2-ANILINOBEN-ZOIC ACID * 2-CARBOXYDIPHENYLAMINE * DIPHENYLAMINE-2-CARBOXYLIC ACID * FEN-AMIC ACID * PA * 2-(PHENYLAMINO)BENZOIC ACID * PHENYLANTHRANILIC ACID

CONSENSUS REPORTS: Reported in EPA TSCA Inventory.

SAFETY PROFILE: Poison by intravenous and intraperitoneal routes. When heated to decomposition it emits toxic fumes of NO_x.

PEI000   CAS: 60-09-3   **HR: 3**
p-(PHENYLAZO)ANILINE
mf: $C_{12}H_{11}N_3$ mw: 197.26

PROP: Yellow crystals. Mp: 128°, bp: 360°. Sltly sol in hot water; sol in hot alc and ether.

SYNS: AAB * AMINOAZOBENZENE * p-AMINO-AZOBENZENE * 4-AMINOAZOBENZENE * 4-AMINO-1,1'-AZOBENZENE * p-AMINOAZOBENZOL * 4-AMINOAZOBENZOL * p-AMINODIPHENYLIMIDE * ANILINE YELLOW * 4-BENZENEAZOANILINE * BRASILAZINA OIL YELLOW G * CERES YELLOW R * C.I. 11000 * C.I. SOLVENT BLUE 7 * C.I. SOLVENT YELLOW 1 * FAST SPIRIT YELLOW AAB * OIL SOLUBLE ANILINE YELLOW * OIL YELLOW AAB * ORGANOL YELLOW * PARAPHENOLAZO ANILINE * 4-(PHENYLAZO)ANILINE * 4-(PHENYL-AZO)BENZENAMINE * p-PHENYLAZOPHENYLAMINE * SOLVENT YELLOW 1 * SUDAN YELLOW R * USAF EK-1375

CONSENSUS REPORTS: IARC Cancer Review: GROUP 2B IMEMDT 7,56,87; Animal Sufficient Evidence IMEMDT 8,53,75. Community Right-To-Know List. Reported in EPA TSCA Inventory.

SAFETY PROFILE: Suspected carcinogen with experimental neoplastigenic and tumorigenic data. Poison by intraperitoneal route. Experimental reproductive effects. Mutation data reported. Used as a dye for lacquer, varnish, wax products, oil stains, and styrene resins. When heated to decomposition it emits toxic fumes of NO_x.

PEJ250   CAS: 22670-79-7   **HR: 3**
N-PHENYLAZO-N-METHYLTAURINE SODIUM SALT
mf: $C_9H_{12}N_3O_3S \cdot Na$ mw: 265.29

SYNS: 3-METHYL-1-PHENYL-3-(2-SULFOETHYL)TRIAZENE SODIUM SALT * 1-PHENYL-3-METHYL-3-(2-SULFOAETHYL) NATRIUM SALZ (GERMAN) * 1-PHENYL-3-METHYL-3-(2-SULFOETHYL)TRIAZENE, SODIUM SALT

SAFETY PROFILE: Poison by subcutaneous route. Questionable carcinogen with experimental neoplastigenic data. When heated to decomposition it emits very toxic fumes of NO_x, Na_2O, and SO_x.

PEJ500   CAS: 842-07-9   **HR: 3**
1-(PHENYLAZO)-2-NAPHTHOL
mf: $C_{16}H_{12}N_2O$ mw: 248.30

SYNS: ATUL ORANGE R * BENZENEAZO-β-NAPHTHOL * BENZENE-1-AZO-2-NAPHTHOL * 1-BENZOAZO-2-NAPHTHOL * BRILLIANT OIL ORANGE R * CALCOGAS ORANGE NC * CALCO OIL ORANGE 7078 * CAMPBELLINE OIL ORANGE * CARMINAPH * CERES ORANGE R * CEROTINORANGE G * C.I. 12055 * C.I. SOLVENT YELLOW 14 * DISPERSOL YELLOW PP * DUNKELGELB * ENIAL ORANGE I * FAST OIL ORANGE * FAST ORANGE * FETTORANGE R * GRASAN ORANGE R * HIDACO OIL ORANGE * LACQUER ORANGE VG * MOTIORANGE R * NCI-C53929 * OIL ORANGE * OLEAL ORANGE R * ORANGE A l'HUILE * ORANGE INSOLUBLE OLG * ORANGE PEL * ORANGE RESENOLE NO. 3 * ORANGE SOLUBLE A l'HUILE * ORGANOL ORANGE * ORIENT OIL ORANGE PS * PETROL ORANGE Y * 1-(PHENYLAZO)-2-NAPHTHALENOL * 1-PHENYLAZO-β-NAPHTHOL * PLASTORESIN ORANGE F4A * PYRONALORANGE * RESINOL ORANGE R * RESOFORM ORANGE G * SANSEL ORANGE G * SCHARLACH B * SILOTRAS ORANGE TR * SOLVENT YELLOW 14 * SOMALIA ORANGE I * SOUDAN I * SPIRIT ORANGE * SPIRIT YELLOW I * STEARIX ORANGE * SUDAN ORANGE R * TERTROGRAS ORANGE SV * TOYO OIL ORANGE * WAXAKOL ORANGE GL * WAXOLINE YELLOW I

CONSENSUS REPORTS: IARC Cancer Review: GROUP 3 IMEMDT 7,56,87; Animal Sufficient Evidence IMEMDT 8,225,75. NTP Carcinogenesis Bioassay (feed); Clear Evidence: rat NTPTR* NTP-TR-226,82. Community Right-To-Know List. Reported in EPA TSCA Inventory. EPA Genetic Toxicology Program.

SAFETY PROFILE: Questionable carcinogen with experimental carcinogenic, neoplastigenic, and tumorigenic data. When heated to decomposition it emits toxic fumes of NO_x. Used for coloring hydrocarbon solvents, oils, fats, waxes, shoe and floor polishes, and gasoline.

PEK000   CAS: 532-82-1   **HR: 3**
4-PHENYLAZO-m-PHENYLENEDIAMINE
mf: $C_{12}H_{12}N_4 \cdot ClH$ mw: 248.74

SYNS: ASTRA CHRYSOIDINE R * BRASILAZINA ORANGE Y * BRILLIANT OIL ORANGE Y BASE * CALCOZINE CHRYSOIDINE Y * CALCOZINE ORANGE YS * CHRYSOIDIN * CHRYSOIDINE * CHRYSOIDINE A * CHRYSOIDINE B * CHRYSOIDINE C CRYSTALS * CHRYSOIDINE G * CHRYSOIDINE GN * CHRYSOIDINE HR * CHRYSOIDINE(II) * CHRYSOIDINE J * CHRYSOIDINE M * CHRYSOIDINE ORANGE * CHRYSOIDINE PRL * CHRYSOIDINE PRR * CHRYSOIDINE SL * CHRYSOIDINE SPECIAL (biological stain and indicator) * CHRYSOIDINE SS * CHRYSOIDINE Y * CHRYSOIDINE Y BASE NEW * CHRYSOIDINE Y CRYSTALS * CHRYSOIDINE Y EX * CHRYSOIDINE YGH * CHRYSOIDINE YL * CHRYSOIDINE YN * CHRYSOIDINE Y SPECIAL * CHRYSOIDIN FB * CHRYSOIDIN Y * CHRYSOIDIN YN * CHRYZOIDYNA F.B. (POLISH) * C.I. 11270 * C.I. BASIC ORANGE 2 * C.I. BASIC ORANGE 3 * C.I. BASIC ORANGE 2, MONOHYDROCHLORIDE * C.I. SOLVENT ORANGE 3 * 2,4-DIAMINOAZOBENZENE HYDROCHLORIDE * DIAZOCARD CHRYSOIDINE G * ELCOZINE CHRYSOIDINE Y * LEATHER ORANGE HR * 4-(PHENYLAZO)-1,3-BENZENEDIAMINE MONOHYDROCHLORIDE * 4-(PHENYLAZO)-m-PHENYLENEDIAMINE MONOHYDROCHLORIDE * PURE CHRYSOIDINE YBH * PURE CHRYSOIDINE YD * PYRACRYL ORANGE Y * SUGAI CHRYSOIDINE * TERTROPHENE BROWN CG

CONSENSUS REPORTS: IARC Cancer Review: GROUP 3 IMEMDT 7,169,87; Animal Sufficient Evidence IMEMDT 8,91,75. Reported in EPA TSCA Inventory. EPA Genetic Toxicology Program.

SAFETY PROFILE: Moderately toxic by ingestion and subcutaneous routes. Questionable carcinogen with experimental tumorigenic data. Mutation data reported. When heated to decomposition it emits very toxic fumes of NO_x and HCl. Used as a colorant in textiles, paper, leather, inks, wood, and biological stains.

PEK250   CAS: 94-78-0   **HR: 3**
3-(PHENYLAZO)-2,6-PYRIDINEDIAMINE
mf: $C_{11}H_{11}N_5$ mw: 213.27

SYNS: AP * 2,6-DIAMINO-3-PHENYLAZOPYRIDINE * DIRIDONE * DPP * GASTRACID * GASTROTEST * MALLOPHENE * NC 150 * PHENAZODINE * PHENAZOPYRIDINE * PHENYLAZO TABLET * PIRID * PYRAZOFEN * PYRIDACIL * PYRIDIUM * PYRIPYRIDIUM * SEDURAL * URIDINAL * URODINE * W 1655

CONSENSUS REPORTS: IARC Cancer Review: Animal Inadequate Evidence IMEMDT 8,117,75.

SAFETY PROFILE: Moderately toxic by intraperitoneal route. Questionable carcinogen with experimental neoplastigenic data. Used as a local anesthetic. When heated to decomposition it emits toxic fumes of NO_x.

PEM750   CAS: 511-55-7   **HR: 3**
8-(p-PHENYLBENZYL)ATROPINIUM BROMIDE
mf: $C_{30}H_{34}NO_3 \cdot Br$ mw: 536.56

PROP: Crystals. Decomp @ 220-222°.

SYNS: N-(p-BIPHENYLMETHYL)-ATROPINIUM BROMIDE * N,4-BIPHENYL-METHYL-dl-TROPEYL-α-TROPINIUMBROMIDS (GERMAN) * p-BIPHENYLMETHYL-(dl-TROPYL-α-TROPINIUM)BROMIDE * DENDREPAR * 4-DIPHENYLMETHYL-dl-TROPYLTROPINIUM BROMIDE * 4-DIPHENYLMETHYLTROPYLTROPINIUM BROMIDE * GASTRIPON * 3-α-HYDROXY-8-(p-PHENYLBENZYL)-1-α-H,5-α-H-TROPANIUM BROMIDE, (±)-TROPATE * N-399 * XENYTROPIUM BROMIDE

SAFETY PROFILE: Poison by intravenous, subcutaneous, and intraperitoneal routes. Moderately toxic by ingestion. Experimental reproductive effects. When heated to decomposition it emits very toxic fumes of NO_x and Br^-.

PEO500   CAS: 108-86-1   **HR: 2**
PHENYL BROMIDE
DOT: UN 2514
mf: C_6H_5Br mw: 157.02

PROP: Colorless, clear, mobile liquid. Mp: −30.7, bp: 156.2°, flash p: 124°F, d: 1.497, vap press: 10 mm @ 40°, vap d: 5.41, autoign temp: 1051°F.

SYNS: BROMOBENZENE (DOT) * MONOBROMOBENZENE * NCI-C55492

CONSENSUS REPORTS: Reported in EPA TSCA Inventory. EPA Genetic Toxicology Program.

DOT Classification: Flammable or Combustible Liquid; Label: Flammable Liquid.

SAFETY PROFILE: Moderately toxic by ingestion, subcutaneous, and intraperitoneal routes. Mildly toxic by inhalation. An eye and mucous membrane irritant. Mutation data reported. Flammable when exposed to heat or flame. Can react with oxidizing materials. To fight fire, use water to blanket fire, foam, CO_2, water spray or mist, dry chemical. Violent reaction with bromobutane + sodium when heated above 30°C. When heated to decomposition it emits toxic fumes of Br^-.

PEQ750   CAS: 104-68-7   **HR: 2**
PHENYL CARBITOL
mf: $C_{10}H_{14}O_3$ mw: 182.24

PROP: Liquid. Bp: 207° @ 55 mm, fp: −50°, d: 1.1158 @ 20°/20°, vap press: <0.01 mm @ 20°, vap d: 6.28.

SYNS: DIETHYLENE GLYCOL MONOPHENYL ETHER * DIETHYLENE GLYCOL PHENYL ETHER * 2-(2-PHENOXYETHOXY)ETHANOL

CONSENSUS REPORTS: Glycol ether compounds are on the Community Right-To-Know List. Reported in EPA TSCA Inventory.

SAFETY PROFILE: Moderately toxic by ingestion and skin contact. A skin and severe eye irritant. Some glycol ethers have dangerous human reproductive effects. When heated to decomposition it emits acrid smoke and irritating fumes.

PER000   CAS: 122-99-6   **HR: 2**
PHENYL CELLOSOLVE
mf: $C_8H_{10}O_2$ mw: 138.18

PROP: Clear liquid. Mp: 14°, bp: 242°, flash p: 250°F.

SYNS: AROSOL * DOWANOL EP * DOWANOL EPH * EMERESSENCE 1160 * EMERY 6705 * ETHYLENE GLYCOL MONOPHENYL ETHER * ETHYLENE GLYCOL PHENYL ETHER * 2-FENOXYETHANOL (CZECH) * FENYL-CELLOSOLVE (CZECH) * GLYCOL MONOPHENYL ETHER * β-HYDROXYETHYL PHENYL ETHER * 1-HYDROXY-2-PHENOXYETHANE * PHENOXETHOL * PHENOXETOL * PHENOXYETHANOL

* 2-PHENOXYETHANOL * PHENOXYETHYL AL-
COHOL * PHENOXYTOL * PHENYLMONOGLY-
COL ETHER * ROSE ETHER

CONSENSUS REPORTS: Glycol ether compounds are on the Community Right-To-Know List. Reported in EPA TSCA Inventory.

SAFETY PROFILE: Moderately toxic by ingestion and skin contact. A skin and severe eye irritant. Mutation data reported. Some glycol ethers have dangerous human reproductive effects. Combustible when exposed to heat or flame; can react vigorously with oxidizing materials. When heated to decomposition it emits acrid smoke and irritating fumes. To fight fire, use CO_2, dry chemical. Used as a solvent for ester type resins.

PER250   CAS: 48145-04-6   **HR: 2**
PHENYL CELLOSOLVE ACRYLATE
mf: $C_{11}H_{12}O_3$ mw: 192.23

SYN: 2-PHENOXY-ETHANOL, ACRYLATE

CONSENSUS REPORTS: Reported in EPA TSCA Inventory.

SAFETY PROFILE: Moderately toxic by skin contact. Mildly toxic by ingestion. A skin irritant. When heated to decomposition it emits acrid smoke and irritating fumes.

PET500   CAS: 13492-01-8   **HR: 3**
PHENYLCYCLOPROMINE SULFATE
mf: $C_{18}H_{20}N_2 \cdot O_4S$ mw: 360.46

SYNS: 1-AMINO-2-PHENYLCYCLOPROPANE SULFATE
* CYCLOPROPANAMINE, 2-PHENYL-, trans-(+-)-, SUL-
FATE (2:1) * PARNATE * trans,D,L-2-PHENYLCY-
CLOPROPYLAMINE SULFATE * TRANCYLPROMINE
SULFATE * TRANSAMINE SULFATE * TRANYLCY-
PRAMINE SULFATE * TRANYLCYPROMINE SULFATE
* TRANYLCYPROMINE SULPHATE

SAFETY PROFILE: Poison by ingestion, intravenous, and intraperitoneal routes. Human toxic effects by ingestion. When heated to decomposition it emits very toxic fumes of SO_x and NO_x.

PET750   CAS: 3721-28-6   **HR: 3**
trans-2-PHENYLCYCLOPROPYLAMINE
mf: $C_9H_{11}N$ mw: 133.21

SYNS: PARNATE * trans-2-PHENYL-1-AMINOCYCLO-
PROPANE * SKF 385 * TRANILCYPROMINE
* TRANSAMINE * TRANYLCYPRAMINE * TRA-
NYLCYPROMINE

SAFETY PROFILE: Poison by ingestion, intraperitoneal and subcutaneous routes. Experimental reproductive effects. Mutation data reported. When heated to decomposition it emits toxic fumes of NO_x.

PEU500   CAS: 13056-98-9   **HR: 3**
1-PHENYL-3,3-DIETHYLTRIAZENE
mf: $C_{10}H_{15}N_3$ mw: 177.28

SYNS: 3,3-DIETHYL-1-PHENYLTRIAZENE * 1-FE-
NYL-3,3-DIETHYLTRIAZEN (CZECH) * 1-PHENYL-3,3-
DIAETHYLTRIAZEN (GERMAN)

CONSENSUS REPORTS: EPA Genetic Toxicology Program.

SAFETY PROFILE: Moderately toxic by ingestion and subcutaneous routes. Experimental teratogenic and reproductive effects. Questionable carcinogen with experimental carcinogenic data. Mutation data reported. When heated to decomposition it emits toxic fumes of NO_x.

PEV500   CAS: 1754-58-1   **HR: 3**
O-PHENYL-N,N'-DIMETHYL
PHOSPHORODIAMIDATE
mf: $C_8H_{13}N_2O_2P$ mw: 200.20

SYNS: DIAMIDAFOS * DIAMIDFOS * DOWCO
169 * NELLITE

SAFETY PROFILE: Poison by ingestion and skin contact. When heated to decomposition it emits very toxic fumes of PO_x and NO_x. A pesticide used on tobacco to control rootknot nematodes.

PEY000   CAS: 108-45-2   **HR: 3**
m-PHENYLENEDIAMINE

DOT: UN 1673
mf: $C_6H_8N_2$ mw: 108.16

PROP: White crystals. Mp: 63°, bp: 286°, d: 1.139, vap press: 1 mm @ 99.8°. Sol in water, methanol, ethanol, chloroform, acetone; sltly sol in ether, carbon tetrachloride; very sltly sol in benzene, toluene.

SYNS: 3-AMINOANILINE * m-AMINOANILINE
* APCO 2330 * m-BENZENEDIAMINE * 1,3-BEN-
ZENEDIAMINE * C.I. 76025 * DEVELOPER 11
* m-DIAMINOBENZENE * 1,3-DIAMINOBENZENE
* DIRECT BROWN BR * m-FENYLENDIAMIN
(CZECH) * METAPHENYLENEDIAMINE * 1,3-PHE-
NYLENEDIAMINE * m-PHENYLENEDIAMINE (DOT)
* PHENYLENEDIAMINE, META, solid (DOT)

CONSENSUS REPORTS: IARC Cancer Review: GROUP 3 IMEMDT 7,56,87; Animal Inadequate Evidence IMEMDT 16,111,78. EPA Genetic Toxicology Program. Reported in EPA TSCA Inventory.

ACGIH TLV: (Proposed: 0.1 mg/m^3)
DOT Classification: Poison B; Label: St. Andrews Cross.

SAFETY PROFILE: Poison by ingestion, intravenous, subcutaneous, intraperitoneal, and possibly other routes. Mildly toxic by skin contact. Questionable carcinogen with experimental tumorigenic and teratogenic data. Mutation data reported. Combustible when exposed to heat or flame. A hair dye ingredient. When heated to decomposition it emits toxic fumes of NO$_x$.

PEY250   CAS: 95-54-5   HR: 2
o-PHENYLENEDIAMINE

DOT: UN 1673
mf: C$_6$H$_8$N$_2$ mw: 108.16

PROP: Tan crystals. Mp: 104°, bp: 257°. Sltly sol in water; very sol in alc, chloroform, ether.

SYNS: 2-AMINOANILINE * o-BENZENEDIAMINE * 1,2-BENZENEDIAMINE * C.I. 76010 * C.I. OXIDATION BASE 16 * o-DIAMINOBENZENE * 1,2-DIAMINOBENZENE * EK 1700 * NSC 5354 * ORTHAMINE * 1,2-PHENYLENEDIAMINE (DOT)

CONSENSUS REPORTS: Reported in EPA TSCA Inventory. EPA Genetic Toxicology Program.

ACGIH TLV: (Proposed TWA 0.1 mg/m^3; Suspected Human Carcinogen)
DOT Classification: Poison B; Label: St. Andrews Cross.

SAFETY PROFILE: Poison by ingestion and intraperitoneal routes. Moderately toxic by subcutaneous route. Mildly toxic by skin contact. Mutation data reported. A pesticide and pharmaceutical. When heated to decomposition it emits toxic fumes of NO$_x$.

PEY500   CAS: 106-50-3   HR: 3
p-PHENYLENEDIAMINE

DOT: UN 1673
mf: C$_6$H$_8$N$_2$ mw: 108.16

PROP: White-sltly red crystals. Mp: 146°, flash p: 312°F, vap d: 3.72, bp: 267°. Sol in alc, chloroform, ether.

SYNS: p-AMINOANILINE * 4-AMINOANILINE * BASF URSOL D * p-BENZENEDIAMINE * 1,4-BENZENEDIAMINE * BENZOFUR D * C.I. 76060 * C.I. DEVELOPER 13 * C.I. OXIDATION BASE 10 * DEVELOPER 13 * DEVELOPER PF * p-DIAMINOBENZENE * 1,4-DIAMINOBENZENE * DURAFUR BLACK R * FENYLENODWUAMINA (POLISH) * FOURAMINE D * FOURRINE D * FOURRINE 1 * FUR BLACK 41867 * FUR BROWN 41866 * FURRO D * FUR YELLOW * FUTRAMINE D * NAKO H * ORSIN * PARA * PARAPHENYLEN-DIAMINE * PELAGOL D * PELAGOL DR * PELAGOL GREY D * PELTOL D * 1,4-PHENYLENEDIAMINE * PHENYLENEDIAMINE, PARA, solid (DOT) * PPD * RENAL PF * SANTOFLEX IC * TERTRAL D * URSOL D * USAF EK-394 * VULKANOX 4020 * ZOBA BLACK D

CONSENSUS REPORTS: IARC Cancer Review: GROUP 3 IMEMDT 7,56,87; Animal Inadequate Evidence IMEMDT 16,125,78. Community Right-To-Know List. Reported in EPA TSCA Inventory. EPA Genetic Toxicology Program.

OSHA PEL: TWA 0.1 mg/m^3 (skin)
ACGIH TLV: TWA 0.1 mg/m^3 (skin); (Proposed TWA 0.1 mg/m^3)
DFG MAK: 0.1 mg/m^3
DOT Classification: ORM-A; Label: None: Poison B; Label: St. Andrews Cross.

SAFETY PROFILE: Poison by ingestion, subcutaneous, intravenous, intraperitoneal, and possibly other routes. Mildly toxic by skin contact. A human skin irritant. Questionable carcinogen with experimental tumorigenic data. Mutation data reported. Implicated in aplastic anemia. Can cause fatal liver damage. The p-form is more toxic and a stronger irritant than the o- and m- isomers. When used as a hair dye it caused vertigo, anemia, gastritis, exfoliative dermatitis, and death. Has caused asthma and other respiratory symptoms in the fur dying industry. Combustible when exposed to heat or flame; can react vigorously with oxidizing materials. To fight fire, use water, CO$_2$, dry chemical. When heated to decomposition it emits acrid smoke and irritating fumes.

PEY750   CAS: 541-69-5   HR: 3
m-PHENYLENEDIAMINE
HYDROCHLORIDE

mf: C$_6$H$_8$N$_2$ • 2ClH   mw: 181.08

PROP: Colorless needles. Very sol in water; sltly sol in alc, ether.

SYNS: m-AMINOANILINE DIHYDROCHLORIDE
* 3-AMINOANILINE DIHYDROCHLORIDE * m-BEN-
ZENEDIAMINE DIHYDROCHLORIDE * 1,3-BENZENEDI-
AMINE HYDROCHLORIDE * m-DIAMINOBENZENE
DIHYDROCHLORIDE * 1,3-DIAMINOBENZENE DIHY-
DROCHLORIDE * 1,3-PHENYLENEDIAMINE DIHYDRO-
CHLORIDE * USAF EK-206

CONSENSUS REPORTS: IARC Cancer Review: Animal Inadequate Evidence IMEMDT
16,111,78. Reported in EPA TSCA Inventory.

SAFETY PROFILE: Poison by intraperitoneal
and possibly other routes. Questionable carcinogen with experimental tumorigenic data. When
heated to decomposition it emits very toxic
fumes of HCl and NO_x.

PFA500   CAS: 4044-65-9   HR: 3
1,4-PHENYLENEDIISOTHIOCYANIC ACID
mf: $C_8H_4N_2S_2$ mw: 192.26

PROP: Tasteless, odorless, colorless crystals.
Mp: 132°.

SYNS: BISCOMATE * BITOSCANATE * 1,4-DI-
ISOTHIOCYANATOBENZENE * ISOTHIOCYANIC ACID-
p-PHENYLENE ESTER * JONIT * PHENYLENE-1,4-
DIISOTHIOCYANATE * PHENYLENE THIOCYANATE

CONSENSUS REPORTS: Cyanide and its
compounds are on the Community Right-To-
Know List. EPA Extremely Hazardous Substances List. Reported in EPA TSCA Inventory.

SAFETY PROFILE: Poison by ingestion and
intraperitoneal routes. Human systemic effects
by ingestion: hallucinations, nausea. When
heated to decomposition it emits very toxic
fumes of NO_x, CN^-, and SO_x.

PFA850   CAS: 101-84-8   HR: 2
PHENYL ETHER
mf: $C_{12}H_{10}O$ mw: 170.22

PROP: Colorless crystals, geranium odor. Mp:
28°, bp: 257°, flash p: 239°F, d: 1.0728 @
20°, vap d: 5.86, autoign temp: 1148°F, lel:
0.8%, uel: 1.5%.

SYNS: BIPHENYL OXIDE * DIPHENYL ETHER
* DIPHENYL OXIDE * GERANIUM CRYSTALS
* PHENOXYBENZENE

CONSENSUS REPORTS: Reported in EPA
TSCA Inventory.

OSHA PEL: Vapor: TWA 1 ppm
ACGIH TLV: TWA 1 ppm; STEL 2 ppm (vapor)
DFG MAK: 1 ppm (7 mg/m^3)

SAFETY PROFILE: Moderately toxic by ingestion. Prolonged exposure damages liver, spleen,
kidneys, and thyroids and upsets gastrointestinal
tract. A skin and eye irritant. Combustible when
exposed to heat or flame; can react with oxidizing materials. To fight fire, use water, foam,
CO_2, dry chemical. When heated to decomposition it emits acrid smoke and irritating fumes.

PFA860   CAS: 8004-13-5   HR: 2
PHENYL ETHER-BIPHENYL MIXTURE
mf: $C_{12}H_{10} \cdot C_{12}H_{10}O$ mw: 324.44

PROP: Eutectic mixture 73.5% phenylether and
26.5% biphenyl by weight.

SYNS: BIPHENYL, mixed with BIPHENYL OXIDE (3:7)
* 1,1'-BIPHENYL, mixed with 1,1'-OXYBIS(BENZENE)
* BIPHENYL-DIPHENYL ETHER mixture * DINIL
* DINYL * DIPHENYL mixed with DIPHENYL OXIDE
* DIPHYL * DOWTHERM * DOWTHERM A

OSHA PEL: Vapor: TWA 1 ppm

SAFETY PROFILE: Poison by inhalation. Moderately toxic by ingestion. Human systemic effects by inhalation: unspecified effects on the
sense of smell, conjunctiva irritation, and unspecified respiratory effects. A mild skin and
eye irritant. When heated to decomposition it
emits acrid smoke and irritating fumes.

PFB250   CAS: 103-45-7   HR: 2
2-PHENYLETHYL ACETATE
mf: $C_{10}H_{12}O_2$ mw: 164.22

PROP: Colorless liquid; sweet, rosy, honey
odor. Mp: 164.2°, bp: 223.6°, fp: $<-20°$, flash
p: 230°F, d: 1.032 @ 25°/25°, refr index: 1.497-
1.501. Sol in alc, fixed oils, propylene glycol;
insol in glycerin, water @ 232°.

SYNS: ACETIC ACID-2-PHENYLETHYL ESTER
* BENZYLCARBINYL ACETATE * FEMA No. 2857
* β-PHENETHYL ACETATE * 2-PHENETHYL ACE-
TATE * β-PHENYLETHYL ACETATE

CONSENSUS REPORTS: Reported in EPA
TSCA Inventory.

SAFETY PROFILE: Moderately toxic by ingestion. Mildly toxic by skin contact. Combustible
when exposed to heat or flame; can react vigorously with oxidizing materials. To fight fire,
use alcohol foam, CO_2 and dry chemical. When

heated to decomposition it emits acrid smoke and irritating fumes.

PFC500   CAS: 51-71-8   **HR: 3**
2-PHENYLETHYLHYDRAZINE
mf: $C_8H_{12}N_2$ mw: 136.22

SYNS: 1-HYDRAZINO-2-PHENYLETHANE * NARDIL * PHENELZINE * PHENETHYLHYDRAZINE * β-PHENYLETHYLHYDRAZINE * STINERVAL * W 1544

SAFETY PROFILE: Poison by ingestion, intraperitoneal, and subcutaneous routes. Human systemic effects by ingestion: ataxia, somnolence. Experimental reproductive effects. Mutation data reported. Used as an antidepressant. When heated to decomposition it emits toxic fumes of NO_x.

PFC750   CAS: 156-51-4   **HR: 3**
β-PHENYLETHYLHYDRAZINE SULFATE
mf: $C_8H_{12}N_2 \cdot H_2O_4S$ mw: 234.30

SYNS: ALACINE * ALAZIN * ALAZINE * EP-411 * ESTINERVAL * FELAZINE * FENELZIN * 1-HYDRAZINO-2-PHENYLETHANE HYDROGEN SULPHATE * KALGAN * MAO-REM * MONOPHEN * MONOTEN * N-1544A * NARDELZINE * NARDIL * P 1531 * PHENALZINE * PHENALZINE DIHYDROGEN SULFATE * PHENALZINE HYDROGEN SULPHATE * PHENELZIN * PHENELZINE ACID SULFATE * PHENELZINE BISULPHATE * PHENELZINE SULFATE * PHENETHYLHYDRAZINE SULFATE (1:1) * PHENLINE * PHENODYNE * PHENYLAETHYL-HYDRAZIN * β-PHENYLETHYLHYDRAZINE DIHYDROGEN SULFATE * 2-PHENYLETHYLHYDRAZINE DIHYDROGEN SULPHATE * β-PHENYLETHYLHYDRAZINE HYDROGEN SULPHATE * PHENYLETHYLHYDRAZINE SULPHATE * S 1544 * STINERVAL

CONSENSUS REPORTS: IARC Cancer Review: GROUP 3 IMEMDT 7,312,87; Human Inadequate Evidence IMEMDT 24,175,80; Animal Limited Evidence IMEMDT 24,175,80. EPA Genetic Toxicology Program.

SAFETY PROFILE: Poison by ingestion, intraperitoneal, intravenous, and subcutaneous routes. Human systemic effects by ingestion: wakefulness, blood pressure lowering, constipation. Questionable carcinogen with experimental neoplastigenic data. Used as a drug for the treatment of depression. When heated to decomposition it emits very toxic fumes of SO_x and NO_x.

PFH000   CAS: 122-60-1   **HR: 3**
PHENYL GLYCYDYL ETHER
mf: $C_9H_{10}O_2$ mw: 150.19

SYNS: 1,2-EPOXY-3-PHENOXYPROPANE * 2,3-EPOXYPROPYLPHENYL ETHER * FENYL-GLYCIDY-LETHER (CZECH) * GLYCIDYL PHENYL ETHER * PGE * PHENOL-GLYCIDAETHER (GERMAN) * PHENOL GLYCIDYL ETHER (MAK) * 3-PHENOXY-1,2-EPOXYPROPANE * PHENOXYPROPENE OXIDE * PHENOXYPROPYLENE OXIDE * PHENYL-2,3-EPOXYPROPYL ETHER

CONSENSUS REPORTS: Reported in EPA TSCA Inventory. EPA Genetic Toxicology Program.

OSHA PEL: (Transitional: TWA 10 ppm) TWA 1 ppm
ACGIH TLV: TWA 1 ppm
DFG MAK: 1 ppm (6 mg/m³), Suspected Carcinogen.
NIOSH REL: (Glycidyl Ethers) CL 5 mg/m³/15M

SAFETY PROFILE: Suspected carcinogen with experimental carcinogenic data. Moderately toxic by ingestion, skin contact, and subcutaneous routes. A severe eye and skin irritant. Experimental reproductive effects. Mutation data reported. When heated to decomposition it emits acrid smoke and irritating fumes. Used as a chemical intermediate.

PFI000   CAS: 100-63-0   **HR: 3**
PHENYLHYDRAZINE

DOT: UN 2572
mf: $C_6H_8N_2$ mw: 108.16

PROP: Yellow, monoclinic crystals or oil. Mp: 19.6°, bp: 243.5° (decomp), flash p: 192°F (CC), d: 1.0978 @ 20°/4°, vap press: 1 mm @ 71.8°, vap d: 3.7. Sltly sol in hot water; misc in alc, chloroform, ether, benzene.

SYNS: FENILIDRAZINA (ITALIAN) * FENYLHYDRA-ZINE (DUTCH) * HYDRAZINE-BENZENE * HYDRA-ZINOBENEZENE * PHENYLHYDRAZIN (GERMAN)

CONSENSUS REPORTS: Reported in EPA TSCA Inventory.

OSHA PEL: (Transitional: TWA 5 ppm (skin)) TWA 5 ppm (skin); STEL 10 ppm
ACGIH TLV: TWA 5 ppm (skin); STEL 10 ppm; Suspected Human Carcinogen; (Proposed: TWA 0.1 ppm (skin); Suspected Human Carcinogen)

DFG MAK: 5 ppm (22 mg/m^3), Suspected Carcinogen.
NIOSH REL: CL 0.6 mg/m^3/2H
DOT Classification: Poison B; Label: Poison.

SAFETY PROFILE: Suspected carcinogen with experimental carcinogenic data. Poison by ingestion, subcutaneous, intravenous, and possibly other routes. Experimental reproductive effects. Mutation data reported. Ingestion or subcutaneous injection can cause hemolysis of red blood cells. Other effects are damage to the spleen, liver, kidneys, and bone marrow. The most common effect of occupational exposure is the development of dermatitis which, in sensitized persons, may be quite severe. Systemic effects include anemia and general weakness, gastrointestinal disturbances and injury to the kidneys. Flammable when exposed to heat, flame, or oxidizers. To fight fire, use alcohol foam. Violent reaction with 2-phenylamino-3-phenyloxazirane. Reacts with perchloryl fluoride to form an explosive product. Vigorous reaction with lead(IV) oxide. Used as a chemical reagent, in organic synthesis, and in the manufacture of dyes and drugs. Dangerous; when heated to decomposition it emits highly toxic fumes of NO$_x$; can react with oxidizing materials.

PFI250   CAS: 59-88-1   **HR: 3**
PHENYLHYDRAZINE HYDROCHLORIDE
mf: $C_6H_8N_2 \cdot ClH$ mw: 144.62

PROP: Leaflet crystals from alc. Mp: 245°. Very sol in water; sol in alc; insol in ether.

SYNS: PHENYLHYDRAZINE MONOHYDROCHLORIDE
* PHENYLHYDRAZIN HYDROCHLORID (GERMAN)
* PHENYLHYDRAZINIUM CHLORIDE

CONSENSUS REPORTS: Reported in EPA TSCA Inventory. EPA Extremely Hazardous Substances List.

NIOSH REL: CL 0.6 mg/m^3/2H

SAFETY PROFILE: Poison by ingestion and subcutaneous routes. Experimental reproductive effects. Questionable carcinogen with experimental neoplastigenic and tumorigenic data. Mutation data reported. When heated to decomposition it emits very toxic fumes of NO$_x$ and HCl.

PFJ250   CAS: 100-65-2   **HR: 3**
β-PHENYLHYDROXYLAMINE
mf: C_6H_7NO mw: 109.14

PROP: Colorless needles. Mp: 81-82°. Sol in hot and cold water; very sol in alc and ether; very sltly sol in ligroin.

SYNS: NCI-C60093 * N-PHENYLHYDROXYLAMINE

CONSENSUS REPORTS: Reported in EPA TSCA Inventory.

SAFETY PROFILE: Poison by ingestion and subcutaneous routes. Human systemic effects by skin contact: primary irritation. Preparative hazard. When heated to decomposition it emits toxic fumes of NO$_x$.

PFK250   CAS: 103-71-9   **HR: 3**
PHENYL ISOCYANATE
DOT: UN 2487
mf: C_7H_5NO mw: 119.13

PROP: Liquid, acrid odor. Mp: −30° approx, bp: 158-168°, d: 1.1 @ 20°, vap press: 1 mm @ 10.6°, flash p: 132°. Decomp in water, alc; very sol in ether.

SYNS: CARBANIL * ISOCYANIC ACID, PHENYL ESTER * MONDUR P * PHENYLCARBIMIDE
* PHENYL CARBONIMIDE

CONSENSUS REPORTS: Reported in EPA TSCA Inventory.

DOT Classification: Poison B; Label: Flammable Liquid and Poison.

SAFETY PROFILE: A poison. Moderately toxic by ingestion. Mutation data reported. An irritant. Flammable when exposed to heat or flame; can react vigorously with oxidizing materials. Has exploded when stirred with (cobalt pentammine triazoperchlorate + nitrosyl perchlorate). When heated to decomposition it emits toxic fumes of CN$^-$ and NO$_x$.

PFL850   CAS: 108-98-5   **HR: 3**
PHENYL MERCAPTAN
DOT: UN 2337
mf: C_6H_6S mw: 110.18

PROP: Liquid, repulsive odor. Bp: 168.3°, d: 1.0728 @ 25°/4°.

SYNS: BENZENETHIOL (DOT) * RCRA WASTE NUMBER P014 * THIOPHENOL (DOT) * USAF XR-19

CONSENSUS REPORTS: Reported in EPA TSCA Inventory. EPA Extremely Hazardous Substances List.

OSHA PEL: TWA 0.5 ppm
ACGIH TLV: TWA 0.5 ppm
NIOSH REL: CL 0.5 mg/m^3/15M
DOT Classification: Poison B; Label: Flammable Liquid and Poison.

SAFETY PROFILE: Poison by ingestion, inhalation, skin contact and intraperitoneal routes. A severe eye irritant. Can cause severe dermatitis. Exposure may cause headache and dizziness. When heated to decomposition or on contact with acids it emits toxic fumes of SO$_x$.

PFM500   CAS: 100-56-1   **HR: 3**
PHENYL MERCURIC CHLORIDE
mf: C$_6$H$_5$ClHg   mw: 313.15

PROP: Colorless leaves from benzene. Mp: 251°, bp: sublimes. Insol in water; sltly sol in hot alc; sol in pyridine, ether, benzene.

SYNS: CHLORID FENYLRTUTNATY (CZECH) * (CHLOROMERCURI)BENZENE * FENYLMERCURI-CHLORID (CZECH) * MERCURIPHENYL CHLORIDE * MERFAZIN * MERSOLITE 2 * PHENYL CHLOROMERCURY * PHENYLMERCURY CHLORIDE * PHENYLQUECKSILBERCHLORID (GERMAN) * PMC * STOPSPOT

CONSENSUS REPORTS: Mercury and its compounds are on the Community Right-To-Know List.

OSHA PEL: (Transitional: CL 1 mg/10m^3) CL 0.1 mg(Hg)/m^3 (skin)
ACGIH TLV: TWA 0.1 mg(Hg)/m^3 (skin)
NIOSH REL: TWA 0.05 mg(Hg)/m^3

SAFETY PROFILE: Poison by ingestion, intraperitoneal, subcutaneous, and possibly other routes. Human mutation data reported. When heated to decomposition it emits very toxic fumes of Cl$^-$ and Hg.

PFM750   CAS: 12040-56-1   **HR: 3**
PHENYLMERCURIC DINAPHTHYLMETHANEDISULFONATE
mf: C$_{33}$H$_{26}$Hg$_2$O$_6$S$_2$ mw: 983.89

SYNS: DIPHENYLMERCURIDINAPHTHYLMETHANE-DISULFONATE * 2-NAPHTHALENESULFONIC ACID-3,3'-METHYLENEDI-PHENYL-MERCURY

CONSENSUS REPORTS: Mercury and its compounds are on the Community Right-To-Know List.

OSHA PEL: (Transitional: CL 1 mg/10m^3) CL 0.1 mg(Hg)/m^3 (skin)

ACGIH TLV: TWA 0.1 mg(Hg)/m^3 (skin)
NIOSH REL: TWA 0.05 mg(Hg)/m^3

SAFETY PROFILE: Poison by ingestion and intraperitoneal routes. When heated to decomposition it emits very toxic fumes of SO$_x$ and Hg. Used as a bactericide and fungicide to treat wool, leather, and textiles.

PFN000   CAS: 14235-86-0   **HR: 3**
PHENYLMERCURIC DINAPHTHYLMETHANEDISULFONATE
mf: C$_{33}$H$_{24}$Hg$_2$O$_6$S$_2$ mw: 981.87

SYNS: BIS(PHENYLMERCURI)METHYLENEDINAPH-THALENESULFONATE * CONOTRANE * FIBROTAN * HYDRAPHEN * HYDRARGAPHEN * METHYLENEDINAPHTHALENESULFONIC ACID BIS-PHENYLMERCURI SALT * PENOTRANE * PHENYL MERCURIC FIXTAN * PHENYLMERCURIC 3,3'-METHYLENEBIS(2-NAPHTHALENESULFONATE) * PHENYLMERCURY METHYLENEDINAPHTHALENE-SULFONATE * P.M.F. * SEPTOTAN * VERSO-TRANE

CONSENSUS REPORTS: Mercury and its compounds are on the Community Right-To-Know List.

OSHA PEL: (Transitional: CL 1 mg/10m^3) CL 0.1 mg(Hg)/m^3 (skin)
ACGIH TLV: TWA 0.1 mg(Hg)/m^3 (skin)
NIOSH REL: TWA 0.05 mg(Hg)/m^3

SAFETY PROFILE: Poison by ingestion. A severe eye irritant. When heated to decomposition it emits very toxic fumes of Hg and SO$_x$.

PFP500   CAS: 2279-64-3   **HR: 3**
PHENYL MERCURY UREA
mf: C$_7$H$_8$HgN$_2$O   mw: 336.76

SYNS: ABAVIT * LEYTOSAN * PHENYLMER-CURIC UREA * PHENYLMERCURIUREA

CONSENSUS REPORTS: Mercury and its compounds are on the Community Right-To-Know List.

OSHA PEL: (Transitional: CL 1 mg/10m^3) CL 0.1 mg(Hg)/m^3 (skin)
ACGIH TLV: TWA 0.1 mg(Hg)/m^3 (skin)
NIOSH REL: TWA 0.05 mg(Hg)/m^3

SAFETY PROFILE: Poison by an unspecified route. When heated to decomposition it emits very toxic fumes of Hg and NO$_x$.

PFS500   CAS: 16033-21-9   **HR: 3**
1-PHENYL-3-MONOMETHYLTRIAZENE
mf: $C_7H_9N_3$ mw: 135.19

SYN: PMT

SAFETY PROFILE: Poison by subcutaneous route. Questionable carcinogen with experimental neoplastigenic data. Mutation data reported. When heated to decomposition it emits toxic fumes of NO_x.

PFS750   CAS: 92-53-5   **HR: 3**
PHENYL MORPHOLINE
mf: $C_{10}H_{13}NO$ mw: 163.24

PROP: Crystals from ethanol, ether. D: 1.058 @ 270°, mp: 57°, bp: 259.9°. Sol in water, alc, ether.

CONSENSUS REPORTS: Reported in EPA TSCA Inventory.

SAFETY PROFILE: Poison by skin contact. Moderately toxic by ingestion. An eye irritant. When heated to decomposition it emits toxic fumes of NO_x.

PFT500   CAS: 135-88-6   **HR: 3**
N-PHENYL-β-NAPHTHYLAMINE
mf: $C_{16}H_{13}N$ mw: 219.30

PROP: Rhombic crystals from methanol. Mp: 107-108°, bp: 395.5°. Insol in water; sol in hot benzene; very sol in hot alc, ether.

SYNS: ACETO PBN * AGERITE POWDER * ANILINONAPHTHALENE * 2-ANILINONAPHTHA-LENE * ANTIOXIDANT 116 * ANTIOXIDANT PBN * N-(2-NAPHTHYL)ANILINE * 2-NAPHTHYLPHE-NYLAMINE * β-NAPHTHYLPHENYLAMINE * NCI-C02915 * NEOZONE D * NILOX PBNA * NONOX D * PBNA * PHENYL-β-NAPHTHYLA-MINE * PHENYL-2-NAPHTHYLAMINE * N-PHENYL-2-NAPHTHYLAMINE * STABILIZATOR AR

CONSENSUS REPORTS: IARC Cancer Review: GROUP 3 IMEMDT 7,318,87; Human Inadequate Evidence IMEMDT 16,325,78; Animal Limited Evidence IMEMDT 16,325,78. Reported in EPA TSCA Inventory.

ACGIH TLV: Suspected Human Carcinogen. DFG MAK: Suspected Carcinogen.

SAFETY PROFILE: Suspected carcinogen with experimental carcinogenic, neoplastigenic, and tumorigenic data. Moderately toxic by ingestion. Human mutation data reported. When heated to decomposition it emits toxic fumes of NO_x.

PFV250   CAS: 638-21-1   **HR: 3**
PHENYLPHOSPHINE
mf: C_6H_7P mw: 110.10

PROP: Needles from aq alc. Mp: 164-165°, bp: 305-308°. Insol in water; sol in alkali; very sol in alc and ether.
OSHA PEL: CL 0.05 ppm
ACGIH TLV: CL 0.05 ppm

SAFETY PROFILE: Poison by inhalation. Ignites spontaneously in air. When heated to decomposition it emits toxic fumes of PO_x.

PFW100   CAS: 824-72-6   **HR: 3**
PHENYL PHOSPHONYL DICHLORIDE
DOT: UN 2799
mf: $C_6H_5Cl_2OP$ mw: 178.99

DOT Classification: Corrosive Material; Label: Corrosive

SAFETY PROFILE: A storage hazard. May explode in a sealed bottle due to the release of hydrogen chloride gas. When heated to decomposition it emits toxic fumes of Cl^- and PO_x.

PFW500   CAS: 14684-25-4   **HR: 2**
PHENYLPHOSPHORODICHLORIDO-THIOUS ACID
DOT: UN 2799
mf: $C_6H_5Cl_2PS$ mw: 211.04

SYN: BENZENE PHOSPHORUS THIODICHLORIDE

DOT Classification: Corrosive Material; Label: Corrosive.

SAFETY PROFILE: A corrosive irritant to skin, eyes, and mucous membranes. When heated to decomposition it emits very toxic fumes of Cl^-, PO_x, and SO_x.

PFX000   CAS: 92-54-6   **HR: 3**
1-PHENYLPIPERAZINE
mf: $C_{10}H_{14}N_2$ mw: 162.26

PROP: Pale yellow oil. D: 1.0621 @ 20°/4°, bp: 286.5°, mp: 18.8°, flash p: 285°F. Insol in water; sol in alc, ether.

SYN: N-PHENYLPIPERAZINE

CONSENSUS REPORTS: Reported in EPA TSCA Inventory.

SAFETY PROFILE: Poison by ingestion and skin contact. A skin and severe eye irritant. Combustible when exposed to heat or flame. It supports combustion and decomposes to yield toxic fumes of NO_x. To fight fire, use water, foam, dry chemical.

PGA750 CAS: 673-31-4 **HR: 3**
3-PHENYL-1-PROPANOL CARBAMATE
mf: $C_{10}H_{13}NO_2$ mw: 179.24

SYNS: ACTOZINE * ANSEPRON * BENZENEPROPANOL CARBAMATE * CARBAMIC ACID-3-PHENYLPROPYL ESTER * 1-CARBAMOYLOXY-3-PHENYLPROPANE * EIRENAL * EXTACOL * FENPROBAMATO * GAMAQUIL * Hg 532 * MH-532 * PALMITA * γ-PHENYLPROPYLCARBAMAT (GERMAN) * γ-PHENYLPROPYL CARBAMATE * QUAMAQUIL * SPANTOL * TRANQUIL

SAFETY PROFILE: Poison by intravenous and intraperitoneal routes. Moderately toxic by ingestion. Used as a tranquilizer and muscle relaxant. When heated to decomposition it emits toxic fumes of NO_x.

PGE000 CAS: 3567-38-2 **HR: 3**
1-PHENYL-2-PROPYNYL CARBAMATE
mf: $C_{10}H_9NO_2$ mw: 175.20

SYNS: CARFIMAT * CFC * α-ETHYNYLBENZYL CARBAMATE * EQUILIUM * NIRVOTIN * PHENYLETHYNLCARBINOL CARBAMATE

SAFETY PROFILE: Poison by ingestion, intraperitoneal, and possibly other routes. When heated to decomposition it emits toxic fumes of NO_x.

PGF000 CAS: 25332-09-6 **HR: 3**
PROPHENPYRIDAMINE HYDROCHLORIDE
mf: $C_{16}H_{20}N_2 \cdot ClH$ mw: 276.84

SYNS: 1-PHENYL-1-(2-PYRIDYL)-3-DIMETHYLAMINOPROPANE HYDROCHLORIDE * TRIMETON

SAFETY PROFILE: Poison by ingestion, intravenous, intraperitoneal, and subcutaneous routes. When heated to decomposition it emits very toxic fumes of NO_x and HCl.

PGG750 CAS: 118-55-8 **HR: 2**
PHENYL SALICYLATE
mf: $C_{13}H_{10}O_3$ mw: 214.23

PROP: White, small crystals; pleasant odor and taste. D: 1.250 @ 20°/4°, mp: 41.4°, bp: 172-

173° @ 12 mm. Sol in water, ether, benzene; very sol in hot alc.

SYN: SALOL

CONSENSUS REPORTS: Reported in EPA TSCA Inventory.

SAFETY PROFILE: Moderately toxic by ingestion. Experimental teratogenic and reproductive effects. When heated to decomposition it emits acrid smoke and irritating fumes.

PGH000 CAS: 304-06-3 **HR: 3**
3-PHENYLSALICYLIC ACID
mf: $C_{13}H_{10}O_3$ mw: 214.23

PROP: Rhombic crystals from alc. D: 1.250 @ 20°/4°, mp: 41.4°, bp: 172-173° @ 12 mm. Sol in water, ether, benzene; very sol in hot alc.

SYN: USAF DO-59

CONSENSUS REPORTS: Reported in EPA TSCA Inventory.

SAFETY PROFILE: Poison by intraperitoneal route. When heated to decomposition it emits acrid smoke and irritating fumes.

PGM750 CAS: 645-48-7 **HR: 3**
1-PHENYLTHIOSEMICARBAZIDE
mf: $C_7H_9N_3S$ mw: 167.25

PROP: Prisms from alc. Mp: 200-201°. Sltly sol in water, ether; sol in hot alc.

SYNS: USAF EK-5426 * USAF EL-45

CONSENSUS REPORTS: Reported in EPA TSCA Inventory.

SAFETY PROFILE: Poison by ingestion and intraperitoneal routes. Human mutation data reported. When heated to decomposition it emits very toxic fumes of NO_x and SO_x.

PGN250 CAS: 103-85-5 **HR: 3**
1-PHENYL-2-THIOUREA
mf: $C_7H_8N_2S$ mw: 152.23

PROP: Needle-like crystals; bitter taste. Mp: 154°, d: 1.3. Sol in water, alc, aq ether.

SYNS: NCI-C02017 * PHENYLTHIOCARBAMIDE * 1-PHENYLTHIOUREA * N-PHENYLTHIOUREA * α-PHNEYLTHIOUREA * PTC * PTU * RCRA WASTE NUMBER P093 * U 6324 * USAF EK-1569

CONSENSUS REPORTS: NCI Carcinogenesis Bioassay (feed); No Evidence: mouse, rat

NCITR* NCI-CG-TR-148,78. EPA Extremely Hazardous Substances List. Reported in EPA TSCA Inventory.

SAFETY PROFILE: Poison by ingestion and intraperitoneal routes. Experimental teratogenic effects. When heated to decomposition or on contact with acid or acid fumes it emits highly toxic fumes of SO_x and NO_x. Used in medical genetics and production of rodenticide.

PGP500   CAS: 132-45-6   **HR: 3**
2-PHENYLVALERIC ACID-
2-(DIETHYLAMINO)ETHYL ESTER
HYDROCHLORIDE
mf: $C_{17}H_{27}NO_2 \cdot ClH$ mw: 313.91

SYNS: 2-DIETHYLAMINOETHYL-α-PHENYLVALERATE HYDROCHLORIDE * 2-DIETHYLAMINOETHYL-α-PRO-PYLTOLUATE HYDROCHLORIDE * α-PHENYL-VALE-RATE du DIETHYLAMINO-ETHANOL CHLORHYDRATE (FRENCH) * PROPIVANE * PROSPASMIN * PROSPASMINE * PROSPASMINE HYDROCHLORIDE * TROPISTON

SAFETY PROFILE: Poison by intraperitoneal and intravenous routes. Moderately toxic by ingestion and intradermal routes. When heated to decomposition it emits very toxic fumes of NO_x and HCl.

PGP750   CAS: 1322-78-7   **HR: 1**
PHENYL XYLYL KETONE
mf: $C_{15}H_{14}O$ mw: 210.29

SYN: AR,AR-DIMETHYLBENZOPHENONE

SAFETY PROFILE: Mildly toxic by ingestion and skin contact. An eye irritant. Mild skin and eye irritant. When heated to decomposition it emits acrid smoke and irritating fumes.

PGR000   CAS: 108-73-6   **HR: 2**
PHLOROGLUCINOL
mf: $C_6H_6O_3$ mw: 126.12

PROP: White crystals; sweet taste. Mp: 218°, sublimes with decomp. Sol in ether; sltly water sol.

SYNS: BENZENE-s-TRIOL * BENZENE-1,3,5-TRIOL * 1,3,5-BENZENETRIOL * 3,5-DIHYDROXYPHENOL * DILOSPAN S * 5-OXYRESORCINOL * PHLORO-GLUCIN * s-TRIHYDROXYBENZENE * sym-TRIHY-DROXYBENZENE * 1,3,5-TRIHYDROXYBENZENE * 1,3,5-TRIHYDROXYCYCLOHEXATRIENE

CONSENSUS REPORTS: Reported in EPA TSCA Inventory. EPA Genetic Toxicology Program.

SAFETY PROFILE: Moderately toxic by subcutaneous and intraperitoneal routes. Mildly toxic by ingestion. Mutation data reported. When heated to decomposition it emits acrid smoke and irritating fumes. Used in diazo-type printing and textile dyeing, in microscopy as a bone specimen decalcifier.

PGS000   CAS: 298-02-2   **HR: 3**
PHORATE
mf: $C_7H_{17}O_2PS_3$ mw: 260.39

PROP: Liquid. Bp: 118-120° @ 0.8 mm, d: 1.156. @ 25°/4°. Insol in water; misc with carbon tetrachloride, dioxane, xylene.

SYNS: O,O-DIAETHYL-S-(AETHYLTHIO-METHYL)-DI-THIOPHOSPHAT (GERMAN) * O,O-DIETHYL-S-ETHYL-MERCAPTOMETHYL DITHIOPHOSPHONATE * O,O-DI-ETHYL-S-ETHYLTHIOMETHYL DITHIOPHOSPHONATE * O,O-DIETHYL-ETHYLTHIOMETHYL PHOSPHORODI-THIOATE * O,O-DIETHYL-S-ETHYLTHIOMETHYL THIOTHIONOPHOSPHATE * O,O-DIETIL-S-(ETILTIO-METIL)-DITIOFOSFATO (ITALIAN) * DITHIOPHOS-PHATE de O,O-DIETHYLE et d'ETHYLTHIOMETHYLE (FRENCH) * ENT 24,042 * FORAAT (DUTCH) * GRANUTOX * PHORAT (GERMAN) * PHORATE-10G * RAMPART * RCRA WASTE NUMBER P094 * THIMET * TIMET * VEGFRU * VERGFRU FORATOX

CONSENSUS REPORTS: EPA Extremely Hazardous Substances List. EPA Genetic Toxicology Program.

OSHA PEL: TWA 0.05 mg/m^3 (skin)
ACGIH TLV: TWA 0.05 mg/m^3 (skin)

SAFETY PROFILE: Poison by ingestion, skin contact, and intravenous routes. Experimental reproductive effects. Mutation data reported. A cholinesterase inhibitor. When heated to decomposition it emits toxic fumes of PO_x and SO_x.

PGS250   CAS: 17673-25-5   **HR: 3**
PHORBOL
mf: $C_{20}H_{27}O_6$ mw: 363.47

PROP: Anhydrous crystals. Two forms: mp: 162-163° and 233-234°. decomp @ 250-251°.

CONSENSUS REPORTS: EPA Genetic Toxicology Program.

SAFETY PROFILE: Experimental reproductive effects. A skin irritant. Questionable carcinogen with experimental carcinogenic and tumorigenic data. When heated to decomposition it emits acrid smoke and irritating fumes.

PGT250   CAS: 24928-17-4   **HR: 2**
PHORBOL-12,13-DIDECANOATE
mf: $C_{40}H_{55}O_8$ mw: 663.95

SYN: PDD

SAFETY PROFILE: A skin irritant. Questionable carcinogen with experimental tumorigenic data. Mutation data reported. When heated to decomposition it emits acrid smoke and irritating fumes.

PGV000   CAS: 16561-29-8   **HR: 3**
PHORBOL MYRISTATE ACETATE
mf: $C_{36}H_{56}O_8$ mw: 616.92

SYNS: PENTAHYDROXY-TIGLIADIENONE-MONOACE-TATE(C)MONOMYRISTATE(B) * PHORBOL ACETATE, MYRISTATE * PHORBOL MONOACETATE MONOMY-RISTATE * PMA * 12-TETRADECANOYLPHORBOL-13-ACETATE * 12-o-TETRADEKANOYLPHORBOL-13-ACETAT (GERMAN) * TPA

CONSENSUS REPORTS: EPA Genetic Toxicology Program.

SAFETY PROFILE: Deadly poison by intravenous route. Experimental reproductive effects. Human mutation data reported. A skin irritant. Questionable carcinogen with experimental carcinogenic, neoplastigenic, and tumorigenic data. When heated to decomposition it emits acrid smoke and irritating fumes.

PGV500   CAS: 56937-68-9   **HR: 1**
PHORBOLOL MYRISTATE ACETATE
mf: $C_{36}H_{58}O_8$ mw: 618.94

SYNS: PHORBOLOL ACETATE MYRISTATE
* TPA-3-β-OL

SAFETY PROFILE: A skin irritant. When heated to decomposition it emits acrid smoke and irritating fumes.

PGV750   CAS: 37415-55-7   **HR: 3**
PHORBOL-12-o-TIGLYL-13-BUTYRATE
mf: $C_{28}H_{40}O_8$ mw: 504.68

SYN: 12-o-TIGLYL-PHORBOL-13-BUTYRATE

SAFETY PROFILE: A skin irritant. Questionable carcinogen with experimental tumorigenic data. When heated to decomposition it emits acrid smoke and irritating fumes.

PGW250   CAS: 504-20-1   **HR: 2**
PHORONE
mf: $C_9H_{14}O$ mw: 138.23

PROP: Solid or greenish liquid. Mp: 28°, flash p: 185°F (OC), d: 0.879, vap press: 1 mm @ 42.0°, vap d: 4.8, bp: 198-199°. Sol in water, alc, and ether.

SYNS: DIISOPROPYLIDENE ACETONE * sym-DIISO-PROPYLIDENE ACETONE * 2,6-DIMETHYL-2,5-HEPTA-DIEN-4-ONE * PHORON (GERMAN)

CONSENSUS REPORTS: Reported in EPA TSCA Inventory.

SAFETY PROFILE: Moderately toxic by subcutaneous route. Flammable when exposed to heat or flame; can react with oxidizing materials. To fight fire, use foam, CO_2, dry chemical. When heated to decomposition it emits acrid smoke and irritating fumes.

PGW750   CAS: 947-02-4   **HR: 3**
PHOSFOLAN
mf: $C_7H_{14}NO_3PS_2$ mw: 255.31

SYNS: AC 47031 * AMERICAN CYANAMID 47031 * C.I. 47031 * CYCLIC ETHYLENE(DIETHOXYPHOS-PHINOTHIOYL)DITHIOIMIDOCARBONATE * CYCLIC ETHYLENE P,P-DIETHYL PHOSPHONODITHIOIMIDO-CARBONATE * CYLAN * CYOLANE * CYO-LANE INSECTICIDE * (DIETHOXYPHOSPHINYL)DI-THIOIMIDOCARBONIC ACID CYCLIC ETHYLENE ESTER * 2-(DIETHOXYPHOSPHINYLIMINO)-1,3-DITHIOLANE * P,P-DIETHYL CYCLIC ETHYLENE ESTER OF PHOS-PHONODITHIOIMIDOCARBONIC ACID * EI 47031 * ENT 25,830 * 1,2-ETHANEDITHIOL, CYCLIC ESTER with P,P-DIETHYL PHOSPHONODITHIOIMIDO-CARBONATE * 1,2-ETHANEDITHIOL, CYCLIC S,S-ESTER with PHOSPHONODITHIOIMIDOCARBONIC ACID P,P-DIETHYL ESTER

CONSENSUS REPORTS: EPA Extremely Hazardous Substances List. Reported in EPA TSCA Inventory.

SAFETY PROFILE: Poison by ingestion and skin contact. An insecticide used against leaf-feeding larvae of cotton insect pests. When heated to decomposition it emits very toxic fumes of PO_x, SO_x, and NO_x.

PGX000   CAS: 75-44-5   **HR: 3**
PHOSGENE
DOT: UN 1076
mf: CCl_2O mw: 98.91

PROP: Colorless, poison gas or volatile liquid; odor of new mown hay or green corn. Mp: −118°, bp: 8.3°, d: 1.37 @ 20°, vap press:

1180 mm @ 20°, vap d: 3.4. Very sltly sol in water; very sol in benzene and acetic acid; decomp sltly in water.

SYNS: CARBONE (OXYCHLORURE de) (FRENCH) * CARBONIO (OSSICLORURO di) (ITALIAN) * CARBON OXYCHLORIDE * CARBONYLCHLORID (GERMAN) * CARBONYL CHLORIDE * CHLOROFORMYL CHLORIDE * DIPHOSGENE * FOSGEEN (DUTCH) * FOSGEN (POLISH) * FOSGENE (ITALIAN) * KOOLSTOFOXYCHLORIDE (DUTCH) * NCI-C60219 * PHOSGEN (GERMAN) * RCRA WASTE NUMBER P095

CONSENSUS REPORTS: EPA Extremely Hazardous Substances List. Community Right-To-Know List. Reported in EPA TSCA Inventory.

OSHA PEL: TWA 0.1 ppm
ACGIH TLV: TWA 0.1 ppm
DFG MAK: 0.1 ppm (0.4 mg/m^3)
NIOSH REL: TWA 0.1 ppm; CL 0.2 ppm/15M
DOT Classification: Poison A; Label: Poison Gas.

SAFETY PROFILE: A human poison by inhalation. A severe eye, skin, and mucous membrane irritant. In the presence of moisture, phosgene decomposes to form hydrochloric acid and carbon monoxide. This occurs in the bronchioles and alveoli of the lungs resulting in pulmonary edema followed by bronchopneumonia and occasionally lung abscess. There is little immediate irritating effect upon the respiratory tract, and the warning properties of the gas are therefore very slight. There may be no immediate warning that dangerous concentrations are being inhaled. After a latent period of 2 to 24 hours, the patient complains of burning in the throat and chest, shortness of breath and increasing dyspnea. Where the exposure has been severe, the development of pulmonary edema may be so rapid that the patient dies within 36 hours after exposure. In cases where the exposure has been less, pneumonia may develop several days after the occurrence of the accident. In patients who recover, no permanent residual disability is thought to occur. A common air contaminant.

Under the appropriate conditions it undergoes hazardous reactions with Al, tert-butyl azido formate, 2,4-hexadiyn-1,6-diol, isopropyl alcohol, K, Na, sodium azide, hexafluoroisopropylideneamino lithium, lithium. When heated to decomposition or on contact with water or steam it will react to produce toxic and corrosive fumes of Cl$^-$. Caution: Arrangements should be made for monitoring its use.

PGX500   HR: 2
PHOSPHATES

SAFETY PROFILE: Alkali metal phosphates are strong caustics and therefore powerful irritants. Superphosphate is $Ca(H_2PO_4)_2/CaSO_4$. Triple superphosphate contains P_2O_5. Both are used as fertilizers. Organophosphates are often highly toxic pesticides. For an example of organic phosphates, see PARATHION.

PGX750   HR: 3
PHOSPHIDES

PROP: A combination of a cation + elemental phosphorus.

SAFETY PROFILE: Phosphides are particularly dangerous because they tend to decompose to the very toxic phosphine upon contact with moisture or acids. Dangerous fire hazard by chemical reaction, particularly with moisture. Moderate explosion hazard. They react with water, steam, acid, or acid fumes to produce toxic and flammable phosphine gas. Can react vigorously with oxidizing materials. Dangerous; when heated to decomposition they may emit highly toxic fumes of PO_x.

PGY000   CAS: 7803-51-2   HR: 3
PHOSPHINE

DOT: UN 2199
mf: H$_3$P   mw: 34.00

PROP: Colorless gas; foul odor of decaying fish. Mp: −132.5°, bp: −87.5°, d: 1.529 g/L @ 0°, autoign temp: 212°F, lel: 1%. Sltly sol in water.

SYNS: CELPHOS * DELICIA * DETIA GAS EX-B * FOSFOROWODOR (POLISH) * HYDROGEN PHOSPHIDE * PHOSPHORUS TRIHYDRIDE * PHOSPHORWASSERSTOFF (GERMAN) * RCRA WASTE NUMBER P096

CONSENSUS REPORTS: EPA Extremely Hazardous Substances List. Reported in EPA TSCA Inventory.

OSHA PEL: (Transitional: TWA 0.3 ppm) TWA 0.3 ppm; STEL 1 ppm
ACGIH TLV: TWA 0.3 ppm; STEL 1 ppm
DFG MAK: 0.1 ppm (0.15 mg/m^3)
DOT Classification: Poison A; Label: Flammable Gas and Poison Gas.

SAFETY PROFILE: A poison by inhalation. A very toxic gas whose effects are not completely understood. The chief effects are central nervous system depression and lung irritation. There may be pulmonary edema, dilation of the heart, and hyperemia of the visceral organs. Inhalation can cause coma and convulsions leading to death within 48 hours. However, most cases recover without after-effects. Chronic poisoning, characterized by anemia, bronchitis, gastrointestinal disturbances and visual, speech and motor disturbances, may result from continued exposure to very low concentrations.

Very dangerous fire hazard by spontaneous chemical reaction. Moderately explosive when exposed to flame. Explosive reaction with dichlorine oxide, silver nitrate, concentrated nitric acid, nitrogen trichloride, oxygen. Reacts with mercury(II) nitrate to form an explosive product. Ignition or violent reaction with air, boron trichloride, Br_2, Cl_2, aqueous halogen solutions, iodine, metal nitrates, NO, NCl_3, NO_3, N_2O, HNO_2, K + NH_3, oxidants. The organic derivatives of phosphine (phosphines) react vigorously with halogens. To fight fire, use CO_2, dry chemical, or water spray. Dangerous; when heated to decomposition it emits highly toxic fumes of PO_x. Used as a fumigant, doping agent for electronic components, and in chemical synthesis.

PHA500   CAS: 1071-83-6   HR: 3
N-(PHOSPHONOMETHYL)GLYCINE
mf: $C_3H_8NO_5P$ mw: 169.09

SYNS: GLYPHOSATE * MON 0573

SAFETY PROFILE: Poison by intraperitoneal route. Moderately toxic by ingestion. Used as an herbicide. When heated to decomposition it emits very toxic fumes of NO_x and PO_x.

PHA750   CAS: 5776-49-8   HR: 2
PHOSPHORAMIDE MUSTARD
CYCLOHEXYLAMINE SALT
mf: $C_4H_{11}Cl_2N_2O_2P \cdot C_6H_{13}N$ mw: 320.24

SYNS: N,N-BIS(2-CHLOROETHYL)PHOSPHORODI-
AMIDIC ACID, CYCLOHEXYL AMMONIUM SALT
* N-LOST-PHOSPHORSAUREDIAMID (GERMAN)
* NLPD * NSC-69945 * OMF 59

SAFETY PROFILE: Moderately toxic by ingestion. An experimental teratogen. Human mutation data reported. When heated to decomposition it emits very toxic fumes of NH_3, Cl^-, PO_x, and NO_x.

PHB250   CAS: 7664-38-2   HR: 3
PHOSPHORIC ACID
DOT: UN 1805
mf: H_3O_4P mw: 98.00

PROP: Colorless liquid or rhombic crystals. Mp: 42.35°, loses $1/2H_2O$ @ 213°, fp: 42.4°, d: 1.864 @ 25°, vap press: 0.0285 mm @ 20°. Misc with water, alc.

SYNS: ACIDE PHOSPHORIQUE (FRENCH) * ACIDO
FOSFORICO (ITALIAN) * FOSFORZUUROPLOSSINGEN
(DUTCH) * ORTHOPHOSPHORIC ACID * PHOS-
PHORSAEURELOESUNGEN (GERMAN)

CONSENSUS REPORTS: Community Right-To-Know List. Reported in EPA TSCA Inventory. EPA Genetic Toxicology Program.

OSHA PEL: (Transitional: TWA 1 mg/m^3) TWA 1 mg/m^3; STEL 3 mg/m^3
ACGIH TLV: TWA 1 mg/m^3; STEL 3 mg/m^3
DOT Classification: Corrosive Material; Label: Corrosive.

SAFETY PROFILE: Human poison by an unspecified route. Moderately toxic by ingestion and skin contact. A corrosive irritant to eyes, skin, and mucous membranes, and a systemic irritant by inhalation. A common air contaminant. A strong acid. Mixtures with nitromethane are explosive. Reacts with chlorides + stainless steel to form explosive hydrogen gas. Potentially violent reaction with sodium tetrahydroborate. Dangerous; when heated to decomposition it emits toxic fumes of PO_x.

PHD250   CAS: 3254-63-5   HR: 3
PHOSPHORIC ACID DIMETHYL-
p-(METHYLTHIO)PHENYL ESTER
mf: $C_9H_{13}O_4PS$ mw: 248.25

SYNS: O,O-DIMETHYL O-(4-METHYLMERCAPTOPHE-
NYL)PHOSPHATE * DIMETHYL-p-(METHYLTHIO)PHE-
NYL PHOSPHATE * ENT 25,734 * 4-METHYLTHIO-
PHENYLDIMETHYL PHOSPHATE

CONSENSUS REPORTS: EPA Extremely Hazardous Substances List.

SAFETY PROFILE: Poison by ingestion, skin contact, and subcutaneous routes. When heated to decomposition it emits very toxic fumes of SO_x and PO_x.

PHD750   CAS: 2255-17-6   HR: 3
PHOSPHORIC ACID DIMETHYL-4-
NITRO-m-TOLYL ESTER
mf: $C_9H_{12}NO_6P$ mw: 261.19

SYNS: O,O-DIMETHYL-O-(3-METHYL-4-NITROPHENYL) PHOSPHORATE * FENITROXON * 4-NITRO-m-CRESOL DIMETHYL PHOSPHATE * OXOSUMITHION * SUMIOXON

SAFETY PROFILE: Poison by ingestion, intravenous, and intraperitoneal routes. When heated to decomposition it emits very toxic fumes of NO_x and PO_x.

PHE500   CAS: 1623-24-1   HR: 2
PHOSPHORIC ACID, ISOPROPYL ESTER

DOT: UN 1793
mf: $C_3H_9O_4P$ mw: 140.09

SYNS: ISOPROPYL ACID PHOSPHATE solid * ISOPROPYL PHOSPHORIC ACID

CONSENSUS REPORTS: Reported in EPA TSCA Inventory.

DOT Classification: Corrosive Material; Label: Corrosive.

SAFETY PROFILE: A highly corrosive material. Very irritating to the skin, eyes, and mucous membranes. When heated to decomposition it emits toxic fumes of PO_x.

PHF250   CAS: 13779-41-4   HR: 3
PHOSPHORODIFLUORIDIC ACID

DOT: UN 1768
mf: F_2HO_2P mw: 101.98

CONSENSUS REPORTS: Reported in EPA TSCA Inventory.

NIOSH REL: TWA 2.5 mg(F)/m³
DOT Classification: Corrosive Material; Label: Corrosive.

SAFETY PROFILE: Toxic and corrosive material. Very irritating to the skin, eyes, and mucous membranes. When heated to decomposition it emits very toxic fumes of PO_x and F^-. Used as a catalyst.

PHF500   CAS: 13779-41-4   HR: 2
PHOSPHORODIFLUORIDIC ACID (anhydrous)

mf: HPO_2F_2 mw: 102

PROP: Mp: −75°. Bp: 116°, d: 1.583 @ 25°/4°, vap d: 3.52.

SYN: DIFLUOROPHOSPHORIC ACID, anhydrous (DOT)

CONSENSUS REPORTS: Reported in EPA TSCA Inventory.

DOT Classification: Corrosive Material; Label: Corrosive.

SAFETY PROFILE: A corrosive irritant to the eyes, skin, and mucous membranes. When heated to decomposition it emits very toxic fumes of F^- and PO_x.

PHF750   CAS: 371-86-8   HR: 3
PHOSPHORODI(ISOPROPYLAMIDIC) FLUORIDE

DOT: UN 2783
mf: $C_6H_{15}FN_2OP$ mw: 182.21

SYNS: BIS(ISOPROPYLAMIDO) FLUOROPHOSPHATE * BIS(MONOISOPROPYLAMINO)FLUOROPHOSPHATE * BIS(MONOISOPROPYLAMINO)FLUOROPHOSPHINE OXIDE * N,N'-DIISOPROPIL-FOSFORODIAMMIDO-FLUORURO (ITALIAN) * DI(ISOPROPYLAMIDO)PHOSPHORYLFLUORIDE * N,N'-DIISOPROPYL-DIAMIDO-FOSFORZUUR-FLUORIDE (DUTCH) * N,N'-DIISOPROPYL-DIAMIDO-PHOSPHORSAEURE-FLUORID (GERMAN) * N,N'-DIISOPROPYLDIAMIDOPHOSPHORYL FLUORIDE * N,N'-DIISOPROPYLPHOSPHORODIAMIDIC FLUORIDE * FLUOROBISISOPROPYLAMINO- PHOSPHINE OXIDE * FLUORURE de N,N'-DIISOPROPYLE PHOSPHORODIAMIDE (FRENCH) * ISOPESTOX * MIPAFOX (DOT) * PESTON XV * PESTOX 15 * PESTOX XV

DOT Classification: ORM-A; Label: None.

SAFETY PROFILE: Poison by ingestion, subcutaneous, intraperitoneal, and possibly other routes. When heated to decomposition it emits very toxic fumes of F^-, NO_x, and PO_x.

PHI500   CAS: 640-15-3   HR: 3
PHOSPHORODITHIOIC ACID, O,O-DIMETHYL-S-(2-ETHYLTHIO)ETHYL ESTER

mf: $C_6H_{15}O_2PS_3$ mw: 246.36

PROP: Colorless liquid. Sol in acetone, dioxane, acetonitrile.

SYNS: BAY 23129 * COMPOUND M-81 * O,O-DIMETHYL-S-(2-AETHYLTHIO-AETHYL)-DITHIO PHOSPHAT (GERMAN) * O,O-DIMETHYL-S-(CARBONYLMETHYL-MORPHOLINO) PHOSPHORODITHIOATE * O,O-DIMETHYL-S-(2-ETHYLMERCAPTOETHYL) DITHIOPHOSPHATE * O,O-DIMETHYL-S-2-ETHYLMERKAPTO-ETHYLESTER KYSELINY DITHIOFOSFORECNE (CZECH) * O,O-DIMETHYL-S-(2-ETHYLTHIO-ETHYL)-DITHIO-FOSFAAT (DUTCH) * O,O-DIMETHYL S-(2-(ETHYL-THIO)ETHYL) PHOSPHORODITHIOATE * O,O-DI-

METIL-S-(ETILTIO-ETIL)-DITIOFOSFATO (ITALIAN)
* DITHIOMETON (FRENCH) * DITHIOPHOS-
PHATE de O,O-DIMETHYLE et de S-(2-ETHYLTHIO-
ETHYLE) (FRENCH) * EKATIN * EKATIN AERO-
SOL * EKATINE-25 * EKATIN ULV * 2-ETH-
YLTHIOETHYL O,O-DIMETHYL PHOSPHORODI-
THIOATE * S-(2-(ETHYLTHIO)ETHYL) O,O-DI-
METHYLPHOSPHORODITHIONATE * S-(2-(ETHYL-
THIO)ETHYL)DIMETHYL PHOSPHOROTHIOLOTHIO-
NATE * INTRATHION * INTRATION * LUXIS-
TELM * M 81 * SAN 230 * THIAMETON
* THIOMETON

SAFETY PROFILE: Poison by ingestion, skin
contact, intravenous, and possibly other routes.
Mutation data reported. A skin and severe eye
irritant. A cholinesterase inhibitor. When heated
to decomposition it emits very toxic fumes of
PO_x and SO_x.

PHJ250   CAS: 13537-32-1   HR: 2
PHOSPHOROFLUORIDIC ACID

DOT: UN 1776
mf: FH_2O_3P mw: 99.99

SYNS: FLUOROPHOSPHORIC ACID, anhydrous
* MONOFLUOROPHOSPHORIC ACID, anhydrous

NIOSH REL: TWA 2.5 mg(F)/m^3
DOT Classification: Corrosive Material; Label:
Corrosive.

SAFETY PROFILE: A corrosive and irritating
material to skin, eyes, and mucous membranes.
When heated to decomposition it emits very
toxic fumes of F^- and PO_x.

PHK250   CAS: 3734-95-0   HR: 3
PHOSPHOROTHIOIC ACID-S-(((1-CYA-
NO-1-METHYL-ETHYL)CARBAMOYL)
METHYL)-O,O-DIETHYL ESTER
mf: $C_{10}H_{19}N_2O_4PS$ mw: 294.34

SYNS: α-CYANOISOPROPYLAMIDE OF THE O,O-DI-
ETHYLTHIOPHOSPHORYL ACETIC ACID * S-(((1-
CYANO-1-METHYL-ETHYL)CARBAMOYL)METHYL) O,O-
DIETHYL PHOSPHOROTHIOATE * S-N-(1-CYANO-1-
METHYLETHYL)CARBAMOYLMETHYL DIETHYL PHOS-
PHOROTHIOLATE * O,O-DIAETHYL-S-1-METHYL)
AETHYL)-CARBAMOYL-METHYL-MONOTHIOPHOSPHAT
(GERMAN) * O,O-DIETHYL-S-((2-CYAAN-2-METHYL-
ETHYL)-CARBAMOYL)-METHYL-MONOTHIOFOSFAAT
(DUTCH) * O,O-DIETHYL-S-N-(A-CYANOISOPROPYL)
CARBOMOYLMETHYL PHOSPHOROTHIOATE * O,O-
DIETIL-S-((2-CIAN-2-METIL-ETIL)-CARBAMOIL)-METIL-
MONOTIOFOSFATO (ITALIAN) * TARTAN * THIO-

PHOSPHATE de S-N-(1-CYANO-1-METHYLETHYL)
CARBAMOYLMETHYLE et de O,O-DIETHYLE (FRENCH)

CONSENSUS REPORTS: Cyanide and its
compounds are on the Community Right To
Know List.

SAFETY PROFILE: Poison by ingestion and
skin contact. When heated to decomposition it
emits very toxic fumes of CN^-, PO_x, SO_x, and
NO_x.

PHO250   CAS: 63980-61-0   HR: 3
PHOSPHOROUS ACID TRIS(2-
FLUOROETHYLESTER)
mf: $C_6H_{12}F_3O_3P$ mw: 220.15

SYNS: 2-FLUOROETHANOL, PHOSPHITE (3:1)
* TL 833

SAFETY PROFILE: Poison by inhalation.
When heated to decomposition it emits very
toxic fumes of F^- and PO_x.

PHO500   CAS: 7723-14-0   HR: 3
PHOSPHORUS (red)

DOT: UN 1338/UN 1381/UN 2447
af: P   aw: 30.97

PROP: Reddish-brown powder. Bp: 280° (with
ignition), mp: 590° @ 43 atm, d: 2.34, autoign
temp: 500°F in air, vap d: 4.77.

SYN: PHOSPHORUS, AMORPHOUS, RED (DOT)

CONSENSUS REPORTS: EPA Extremely
Hazardous Substances List.

DFG MAK: 0.1 mg/m^3
DOT Classification: Flammable Solid and Poison; Label: Flammable Solid and Poison.

SAFETY PROFILE: A human poison by an unspecified route. May have white phosphorus as
an impurity. Generally less reactive than white
phosphorus. Dangerous fire hazard when exposed to heat or by chemical reaction with oxidizers. Can also react with reducing materials.
Moderate explosion hazard by chemical reaction
or on contact with organic materials. May explode on impact. To fight fire, use water. Explosive reaction with chlorosulfuric acid, hydroiodic acid, magnesium perchlorate, chromyl
chloride. Forms sensitive explosive mixtures
with metal halogenates (e.g., chlorates, bromates, or iodates of barium, calcium, magnesium, potassium, sodium, zinc), ammonium nitrate, mercury(I) nitrate, silver nitrate, sodium

nitrate, potassium permanganate. Violent reaction or ignition with alkalies + heat, fluorine, chlorine, liquid bromine, antimony pentachloride. Reacts with hot alkalies or hydroiodic acid to form phosphine gas which then ignites. Incompatible with cyanogen iodide, halogen azides, halogen oxides (e.g., chlorine dioxide, dichlorine oxide, oxygen difluoride, trioxygen difluoride), interhalogens (e.g., bromine trifluoride, bromine pentafluoride, chlorine trifluoride, iodine trichloride, iodine pentafluoride), hexalithium disilicide, hydrogen peroxide, metal acetylides (e.g., rubidium acetylide, cesium acetylide, lithium acetylide, sodium acetylide, potassium acetylide), antimony pentachloride, metal oxides (e.g., copper oxide, manganese dioxide, lead oxide, mercury oxide, silver oxide, chromium trioxide), metal peroxides (e.g., lead peroxide, potassium peroxide, sodium peroxide), metals (e.g., beryllium, copper, manganese, thorium, zirconium, cerium, lanthanum, neodymium, praseodymium, osmium, platinum), metal sulfates (e.g., barium sulfate, calcium sulfate), nitric acid, nitrogen halides, nitrosyl fluoride, nitryl fluoride, non-metal halides (e.g., boron triiodide, seleninyl chloride, sulfuryl chloride, disulfuryl chloride, disulfur dibromide), non-metal oxides (e.g., nitrogen oxide, dinitogen tetraoxide, dinitrogen pentaoxide, sulfur trioxide, oxygen, peroxyformic acid, potassium nitride, selenium, sodium chlorite, sulfur, sulfuric acid, peroxides, oxidizing materials. When heated to decomposition it emits toxic fumes of PO_x.

PHO740   CAS: 7723-14-0   **HR: 3**
PHOSPHORUS (white in water)

CONSENSUS REPORTS: EPA Extremely Hazardous Substances List. Community Right-To-Know List. Reported in EPA TSCA Inventory.

DOT Classification: Flammable Solid and Poison; Label: Flammable Solid and Poison.

SAFETY PROFILE: A poison. Very flammable solid which must be kept under water. When heated to decomposition it emits toxic fumes of PO_x.

PHO750   CAS: 7723-14-0   **HR: 3**
PHOSPHORUS (yellow)

DOT: UN 1381/UN 2447
mf: P_4 mw: 123.88

PROP: Cubic crystals; colorless to yellow, waxlike solid. Mp: 44.1°, bp: 280°, flash p: spontaneously flammable in air, d: 1.82, autoign temp: 86°F, vap press: 1 mm @ 76.6°, vap d: 4.42.

SYNS: BONIDE BLUE DEATH RAT KILLER * FOSFORO BIANCO (ITALIAN) * GELBER PHOSPHOR (GERMAN) * PHOSPHORE BLANC (FRENCH) * PHOSPHORUS (white) * RAT-NIP * TETRAFOSFOR (DUTCH) * TETRAPHOSPHOR (GERMAN) * WEISS PHOSPHOR (GERMAN) * WHITE PHOSPHORUS

CONSENSUS REPORTS: EPA Extremely Hazardous Substances List. Reported in EPA TSCA Inventory.

OSHA PEL: TWA 0.1 mg/m^3
ACGIH TLV: TWA 0.1 mg/m^3
DOT Classification: Flammable Solid and Poison; Label: Flammable Solid and Poison.

SAFETY PROFILE: Human poison by ingestion. Experimental poison by ingestion and subcutaneous routes. Experimental reproductive effects. Human systemic effects by ingestion: fluid intake, sweating, nausea, diarrhea, cyanosis, cardiomyopathy. Toxic quantities have an acute effect on the liver and can cause severe eye damage. Inhalation can cause photophobia with myosis, dilation of the pupils, retinal hemorrhage, congestion of the blood vessels, and rarely an optic neuritis. Chronic exposure by inhalation or ingestion can cause anemia, gastrointestinal effects, and brittleness of the long bones leading to spontaneous fractures. The most common symptom, however, of chronic phosphorous poisoning is necrosis of the jaw (phossy-jaw).

    More reactive than red phosphorus. Dangerous fire hazard when exposed to heat, flame or by chemical reaction with oxidizers. Ignites spontaneously in air. Very reactive. If combustion occurs in a confined space, it will remove the oxygen and cause asphyxiation. Dangerous explosion hazard by chemical reaction with: alkaline hydroxides, NH_4NO_3, SbF_5, $Ba(BrO_3)_2$, Be, BI_3, $Ca(BrO_3)_2$, $Mg(BrO_3)_2$, $K(BrO_3)$, $NaBrO_3$, $Zn(BrO_3)_2$, Br_2, halogens, BrF_3, BrN_3, (chlorates of Ba, Ca, Mg, K, Na, Zn), (iodates of Ba, Ca, Mg, K, Na, Zn), Ce, Cs, $CsHC_2$, Cs_3N, (charcoal + air), ClO_2, (Cl_2 + heptane), ClO, ClF_3, ClO_3, chlorosulfonic acid, CrO_3, $Cr(OCl)_2$, Cu, NCl, IBr, ICl, IF_5, Fe, La, PbO_2, Li, Li_2C_2, Li_6CS, $Mg(ClO_4)_2$, Mn, HgO, $HgNO_3$, Nd, Ni, nitrates, NBr, NO_2, NBr_3, NCl_3, NOF, FNO_2, O_2, performic acid,

Pt, K, KOH, K_3N, $KMnO_4$, K_2O_2, Rb, $RbHC_2$, Se_2Cl_2, $SeOCl_2$, $SeOF_2$, SeF_4, $AgNO_3$, Ag_2O, Na, Na_2C_2, $NaClO_2$, NaOH, Na_2O_2, S, SO_3, H_2SO_4, Th, $VOCl_2$, Zr, peroxyformic acid, chloro sulfuric acid, halogen azides, hexalithium disilicide. Can react vigorously with oxidizing materials. To fight fire, use water. Used in fertilizers, tracer bullets, incendiaries manufacturing, rat poison, and gas analysis. When heated to decomposition it emits highly toxic fumes of PO_x.

PHQ000   HR: D
PHOSPHORUS COMPOUNDS, INORGANIC

SAFETY PROFILE: Variable toxicity. Most inorganic phosphates (except phosphine) have low toxicity, but in large doses they may cause serious disturbances, particularly in calcium metabolism. Red phosphorus and phosphates are relatively harmless. White (yellow) phosphorus is highly toxic by several routes. The phosphorus halides decompose violently with water to form the halide acid and are thus severe irritants. Phosphorus sulfides behave similarly. Metaphosphates may be highly toxic, causing irritation and hemorrhages in the stomach, as well as liver and kidney damage. Phosphorus trichloride is the most used of the phosphorus halide compounds. Phosphoryl chloride is used to synthesize phosphate esters. Common air contaminants. When heated to decomposition it emits highly toxic fumes of PO_x.

PHQ500   CAS: 7783-55-3   HR: 3
PHOSPHORUS FLUORIDE
mf: F_3P mw: 87.97

PROP: Colorless gas. Mp: $-152°$, bp: $-102°$, d: 3.907 g/L.

SYN: PHOSPHOROUS TRIFLUORIDE

CONSENSUS REPORTS: Reported in EPA TSCA Inventory.

NIOSH REL: TWA 2.5 mg(F)/m^3

SAFETY PROFILE: Moderately toxic by inhalation. A severe eye, skin, and mucous membrane irritant. Explodes on contact with dioxygen difluoride. Violent reaction or ignition with borane; diborane; F_2; hexafluoroisopropylideneamino lithium; O_2. Will react with water or steam to produce toxic and corrosive fumes. Dangerous; when heated to decomposition it emits highly toxic fumes of F^- and PO_x.

PHQ750   CAS: 12037-82-0   HR: 3
PHOSPHORUS HEPTASULFIDE
DOT: UN 1339

PROP: Light yellow crystals; light gray powder or fused solid. Mp: $310°$, bp: $523°$, d: 2.19 @ $17°$.

DOT Classification: Flammable Solid; Label: Flammable Solid.

SAFETY PROFILE: A poison by ingestion. Flammable when exposed to heat or flame; can react vigorously with oxidizing materials. When heated to decomposition it emits very toxic fumes of PO_x and SO_x.

PHQ800   CAS: 10025-87-3   HR: 3
PHOSPHORUS OXYCHLORIDE
DOT: UN 1810
mf: Cl_3OP mw: 153.32

PROP: Colorless to sltly yellow, fuming liquid. Mp: $1.2°$, bp: $105.1°$, d: 1.685 @ $15.5°$, vap press: 40 mm @ $27.3°$, vap d: 5.3.

SYNS: PHOSPHORUS OXYTRICHLORIDE * PHOSPHORYL CHLORIDE

CONSENSUS REPORTS: Reported in EPA TSCA Inventory.

OSHA PEL: TWA 0.1 ppm
ACGIH TLV: TWA 0.1 ppm
DFG MAK: 0.2 ppm (1 mg/m^3)
DOT Classification: Corrosive; Label: Corrosive.

SAFETY PROFILE: Poison by inhalation and ingestion. A corrosive eye, skin, and mucous membrane irritant. Potentially explosive reaction with water evolves hydrogen chloride and phosphine which then ignites. Explosive reaction with 2,6-dimethylpyridine N-oxide, dimethyl sulfoxide, ferrocene-1,1'-dicarboxylic acid, pyridine N-oxide (above 60°C), sodium + heat. Violent reaction or ignition with BI_3, carbon disulfide, 2,5-dimethyl pyrrole + dimethyl formamide, organic matter, zinc powder. Reacts with water or steam to produce heat and toxic and corrosive fumes. Incompatible with carbon disulfide, N,N-dimethyl-formamide, 2,5-dimethylpyrrole, 2,6-dimethylpyridine N-oxide, dimethylsulfoxide, ferrocene-1,1-dicarboxylic acid, water, zinc. When heated to decomposition it emits highly toxic fumes of Cl^- and PO_x.

PHR250   CAS: 7789-69-7   **HR: 3**
PHOSPHORUS PENTABROMIDE

DOT: UN 2691
mf: PBr_5 mw: 430.56

PROP: Yellow, crystalline mass. Mp: decomp, bp: decomp @ 106°. Sol in carbon disulfide.

SYNS: PENTABROMO PHOSPHORANE * PENTA-BROMO PHOSPHORUS * PHOSPHORIC BROMIDE

CONSENSUS REPORTS: Reported in EPA TSCA Inventory.

DOT Classification: Corrosive Material; Label: Corrosive.

SAFETY PROFILE: A poison. Corrosive to the eyes, skin and mucous membranes. Flammable by chemical reaction. Contact with moisture can cause a violent reaction and evolution of heat. Incompatible with water or steam to produce heat and toxic and corrosive fumes. When heated to decomposition it emits highly toxic fumes of Br^- and PO_x.

PHR500   CAS: 10026-13-8   **HR: 3**
PHOSPHORUS PENTACHLORIDE

DOT: UN 1806
mf: Cl_5P mw: 208.22

PROP: Yellowish-white, fuming, crystalline mass; pungent odor. Mp: (under press) 148° decomp, bp: subl @ 160°, d: 4.65 g/L @ 296°, vap press: 1 mm @ 55.5°.

SYNS: FOSFORO(PENTACHLORURO di) (ITALIAN) * FOSFORPENTACHLORIDE (DUTCH) * PHOSPHORE-(PENTACHLORURE de) (FRENCH) * PHOSPHORIC CHLORIDE * PHOSPHORPENTACHLORID (GERMAN) * PHOSPHORUS PERCHLORIDE * PIECIOCHLOREK FOSFORU (POLISH)

CONSENSUS REPORTS: EPA Extremely Hazardous Substances List. Reported in EPA TSCA Inventory.

OSHA PEL: TWA 1 mg/m^3
ACGIH TLV: TWA 0.85 mg/m^3
DFG MAK: 1 mg/m^3
DOT Classification: Corrosive; Label: Corrosive.

SAFETY PROFILE: Poison by inhalation. Moderately toxic by ingestion. A severe eye, skin, and mucous membrane irritant. Corrosive to body tissues. Flammable by chemical reaction. Explosive reaction with chlorine dioxide + chlo-rine, sodium, urea + heat. Reacts to form explosive products with carbamates, 3'-methyl-2-nitrobenzanilide (product explodes on contact with air). Ignites on contact with fluorine. Reacts violently with moisture, ClO_3, hydroxylamine, magnesium oxide, nitrobenzene, phosphorus (III) oxide, K. To fight fire, use CO_2, dry chemical. Incompatible with aluminum, chlorine dioxide, chlorine, diphosphorus trioxide, fluorine, hydroxylamine, magnesium oxide, 3'-methyl-2-nitrobenzanilide, nitrobenzene, sodium, urea, water. Will react with water or steam to produce heat and toxic and corrosive fumes. Used as a catalyst, chlorinating and dehydrating agent. When heated to decomposition it emits highly toxic fumes of Cl^- and PO_x.

PHR750   CAS: 7647-19-0   **HR: 3**
PHOSPHORUS PENTAFLUORIDE

DOT: UN 2198
mf: PF_5 mw: 125.98

PROP: Colorless gas, fumes strongly in air. Mp: −93.8°, bp: −84.6°, d: (gas) 5.805 g/L.

CONSENSUS REPORTS: EPA Extremely Hazardous Substances List. Reported in EPA TSCA Inventory.

OSHA PEL: TWA 2.5 mg(F)/m^3
ACGIH TLV: TWA 2.5 mg(F)/m^3
DOT Classification: Poison A; Label: Poison Gas.

SAFETY PROFILE: A poisonous gas. Violently irritating to skin, eyes, and mucous membranes. Inhalation may cause pulmonary edema. Reacts with water or steam to produce toxic and corrosive fumes. When heated to decomposition it emits highly toxic fumes of F^- and PO_x.

PHS000   CAS: 1314-80-3   **HR: 3**
PHOSPHORUS PENTASULFIDE

DOT: UN 1340
mf: P_2S_5 mw: 222.24

PROP: Gray to yellow-green, crystalline, deliquescent mass. Bp: 514°, d: 2.09, autoign temp: 287°F. Mp: 286-290°.

SYNS: PENTASULFURE de PHOSPHORE (FRENCH) * PHOSPHORIC SULFIDE * PHOSPHORUS PERSULFIDE * RCRA WASTE NUMBER U189 * SIRNIK FOSFORECNY (CZECH) * SULFUR PHOSPHIDE * THIOPHOSPHORIC ANHYDRIDE

CONSENSUS REPORTS: Reported in EPA TSCA Inventory.

OSHA PEL: (Transitional: TWA 1 mg/m^3) TWA 1 mg/m^3; STEL 3 mg/m^3
ACGIH TLV: TWA 1 mg/m^3; STEL 3 mg/m^3
DFG MAK: 1 mg/m^3
DOT Classification: Flammable Solid; Label: Flammable Solid and Dangerous When Wet.

SAFETY PROFILE: A poison by ingestion. A severe eye and skin irritant. Readily liberates toxic hydrogen sulfide and phosphorus pentoxide and evolves heat on contact with moisture. Dangerous fire hazard in the form of dust when exposed to heat or flame. Spontaneous heating in the presence of moisture. Moderate explosion hazard in solid form by spontaneous chemical reaction. Reacts with water, steam, or acids to produce toxic and flammable vapors; can react vigorously with oxidizing materials. Incompatible with air, alcohols, water. To fight fire use CO$_2$ snow, dry chemical or sand. Used as an intermediate in manufacturing lubricant additives, insecticides and fertilizer agents. When heated to decomposition it emits highly toxic fumes of SO$_x$ and PO$_x$.

PHS250 CAS: 1314-56-3 **HR: 3**
PHOSPHORUS PENTOXIDE

DOT: NA 1807
mf: O$_5$P$_2$ mw: 141.94

PROP: Deliquescent crystals. D: 2.30; mp: 340°, sublimes @ 360°.

SYNS: DIPHOSPHORUS PENTOXIDE * PHOSPHORIC ANHYDRIDE * PHOSPHORUS(V) OXIDE * POX * PO$_x$

CONSENSUS REPORTS: EPA Extremely Hazardous Substances List. Reported in EPA TSCA Inventory.

DFG MAK: 1 mg/m^3
DOT Classification: Corrosive Material; Label: Corrosive.

SAFETY PROFILE: Poison by inhalation. A corrosive irritant to the eyes, skin, and mucous membranes. With the appropriate conditions it undergoes hazardous reactions with formic acid, hydrogen fluoride, inorganic bases, iodides, metals, methyl hydroperoxide, oxidants (e.g., bromine, pentafluoride, chlorine trifluoride, perchloric acid, oxygen difluoride, hydrogen peroxide), 3-propynol, water. When heated to decomposition it emits toxic fumes of PO$_x$.

PHS500 CAS: 1314-85-8 **HR: 3**
PHOSPHORUS SESQUISULFIDE

DOT: UN 1341
mf: P$_4$S$_3$ mw: 220.06

PROP: Yellow, crystalline mass. Mp: 172.5°, bp: 407°, d: 2.03, autoign temp: 212°F.

SYNS: PHOSPHORUS(III) SULFIDE(IV) * SESQUISULFURE de PHOSPHORE (FRENCH) * TETRAPHOSPHORUS TRISULFIDE * TRISULFURATED PHOSPHORUS

DOT Classification: Flammable Solid; Label: Flammable Solid and Dangerous When Wet.

SAFETY PROFILE: Poison by ingestion. Flammable by spontaneous ignition. When heated to decomposition it emits very toxic fumes of PO$_x$ and SO$_x$.

PHT000 CAS: 56280-76-3 **HR: 3**
PHOSPHORUS TRIAZIDE
mf: N$_9$P mw: 157.04

SAFETY PROFILE: Highly explosive. May explode at room temperature. When heated to decomposition it emits very toxic fumes of NO$_x$, PO$_x$, and PH$_3$.

PHT250 CAS: 7789-60-8 **HR: 3**
PHOSPHORUS TRIBROMIDE

DOT: UN 1808

SYNS: PHOSPHOROUS BROMIDE (DOT) * TRIBROMOPHOSPHINE

mf: Br$_3$P mw: 270.70

PROP: Mp: −40°, bp: 175.3°, d: 2.852 @ 15°, vap press: 10 mm @ 47.8°.

CONSENSUS REPORTS: Reported in EPA TSCA Inventory.

DOT Classification: Corrosive Material; Label: Corrosive.

SAFETY PROFILE: Probably highly toxic. A corrosive irritant to the eyes, skin, and mucous membranes. Will react with water, steam or acids to produce heat, toxic, and corrosive fumes. Violent reaction or ignition with calcium hydroxide + sodium carbonate, phenylpropanol, sulfuric acid, oleum, fluorosulfuric acid, chlorosulfuric acid, 1,1,1-tris(hydroxymethyl) methane, water, potassium, sodium, RuO$_4$. When heated to decomposition it emits very toxic fumes of Br$^-$ and PO$_x$.

PHT275 CAS: 7719-12-2 **HR: 3**
PHOSPHORUS TRICHLORIDE

DOT: UN 1809
mf: Cl_3P mw: 137.32

PROP: Clear, colorless, fuming liquid. Mp: $-111.8°$, bp: $76°$, d: 1.574 @ $21°$, vap press: 100 mm @ $21°$, vap d: 4.75. Decomp by water and alc; sol in benzene, chloroform, and ether.

SYNS: CHLORIDE of PHOSPHORUS * FOSFORO(TRI-CLORURO di) (ITALIAN) * FOSFORTRICHLORIDE (DUTCH) * PHOSPHORE(TRICHLORURE de) (FRENCH) * PHOSPHORTRICHLORID (GERMAN) * PHOSPHO-RUS CHLORIDE * TROJCHLOREK FOSFORU (POLISH)

CONSENSUS REPORTS: EPA Extremely Hazardous Substances List. Reported in EPA TSCA Inventory.

OSHA PEL: (Transitional: TWA 0.5 ppm) TWA 0.2 ppm; STEL 0.5 ppm
ACGIH TLV: TWA 0.2 ppm; STEL 0.5 ppm
DFG MAK: 0.5 ppm (3 mg/m^3)
DOT Classification: Corrosive Material; Label: Corrosive.

SAFETY PROFILE: Poison by inhalation. Moderately toxic by ingestion. A corrosive irritant to skin, eyes (at 2 ppm), and mucous membranes. Potentially explosive reaction with chlorobenzene + sodium, dimethyl sulfoxide, molten sodium, chromyl chloride, nitric acid, sodium peroxide, oxygen (above 100°C), tetravinyl lead. Reacts with carboxylic acids (e.g., acetic acid) to form violently unstable products. Violent reaction or ignition with Al, chromium pentafluoride, diallyl phosphite + allyl alcohol, F_2, hexafluoroisopropylideneaminolithium, hydroxylamine, iodine chloride, PbO_2, HNO_2, organic matter, potassium, selenium dioxide, sulfur acids (e.g., sulfuric acid, fluorosulfuric acid, oleum). Violent reaction with water evolves hydrogen chloride and diphosphane gas which then ignite. Incompatible with metals or oxidants. Will react with water, steam, or acids to produce heat and toxic and corrosive fumes; can react with oxidizing materials. To fight fire, use CO_2, dry chemical. Used as a chlorinating agent, catalyst, and chemical intermediate. Dangerous; when heated to decomposition it emits highly toxic fumes of Cl^- and PO_x.

PHT500 CAS: 1314-24-5 **HR: 3**
PHOSPHORUS TRIOXIDE

DOT: UN 2578
mf: O_3P_2 mw: 110.0

PROP: Transparent, monoclinic crystals or colorless liquid. D: 2.135 @ $21°/4°$, mp: $23.8°$, bp: $173.1°$. Sol in benzene.

SYN: DIPHOSPHORUS TRIOXIDE

DOT Classification: Corrosive Material; Label: Corrosive.

SAFETY PROFILE: A poison. A corrosive irritant to the eyes, skin, and mucous membranes. Melted material readily ignites in air. Incompatible with ammonia, disulfur dichloride, halogens, oxygen, phosphorus pentachloride, sulfur, sulfuric acid, water. When heated to decomposition it emits toxic fumes of PO_x.

PHT750 CAS: 12165-69-4 **HR: 2**
PHOSPHORUS TRISULFIDE

DOT: UN 1343
mf: P_2S_3 mw: 158.12

PROP: Gray-yellow crystals. Mp: $290°$, bp: $490°$.

DOT Classification: Flammable Solid; Label: Flammable Solid.

SAFETY PROFILE: Can react with oxidizers, water, or steam to emit toxic fumes of H_2S. When heated to decomposition it emits very toxic fumes of PO_x and SO_x.

PHU000 CAS: 7789-59-5 **HR: 3**
PHOSPHORYL BROMIDE

DOT: UN 1939/UN 2576
mf: Br_3OP mw: 286.70

PROP: Colorless plates. Mp: $56°$, bp: $190°$, d: 2.882.

SYNS: PHOSPHOROUS OXYBROMIDE * PHOSPHORYL TRIBROMIDE

CONSENSUS REPORTS: Reported in EPA TSCA Inventory.

DOT Classification: Corrosive Material; Label: Corrosive.

SAFETY PROFILE: Poison by ingestion, inhalation, and skin contact. A corrosive irritant to skin, eyes, and mucous membranes. A corrosive material. Reacts with steam, water to produce much heat with toxic fumes. When heated to decomposition it emits very toxic fumes of Br^- and PO_x.

PHU500   CAS: 520-52-5   **HR: 3**
O-PHOSPHORYL-4-HYDROXY-N,N-DIMETHYLTRYPTAMINE
mf: $C_{12}H_{17}N_2O_4P$ mw: 284.28

SYNS: CY-39 * 3-(2-(DIMETHYLAMINO)ETHYL)-1H-INDOL-4-OL DIHYDROGEN PHOSPHATE ESTER * 3-2'-DIMETHYLAMINOETHYLINDOL-4-PHOSPHATE * 3-(2-DIMETHYLAMINOETHYL)INDOL-4-YL DIHYDROGEN PHOSPHATE * INDOCYBIN * PSILOCIN PHOSPHATE ESTER * PSILOCIPIN * PSILOTSIBIN * TEONANACATL

CONSENSUS REPORTS: EPA Genetic Toxicology Program.

SAFETY PROFILE: Poison by intravenous route. Moderately toxic by intraperitoneal route. Human systemic effects by ingestion and intraperitoneal routes: euphoria, hallucinations, toxic psychosis, muscle weakness; nausea or vomiting; visual field changes. When heated to decomposition it emits very toxic fumes of NO_x and PO_x.

PHW000   CAS: 86-54-4   **HR: 3**
1(2H)-PHTHALAZINONE HYDRAZONE
mf: $C_8H_8N_4$ mw: 160.20

SYNS: APRESOLIN * APPRESSIN * APREZOLIN * BA5968 * C-5068 * C 5968 * CIBA 5968 * HIDRALAZIN * HIPOFTALIN * HYDRALAZINE * HYDRALLAZINE * HYDRAZINOPHTHALAZINE * 1-HYDRAZINOPHTHALAZINE * HYPOPHTHALIN * IDRALAZINA (ITALIAN)

CONSENSUS REPORTS: IARC Cancer Review: GROUP 3 IMEMDT 7,222,87; Human Inadequate Evidence IMEMDT 24,85,80.

SAFETY PROFILE: Poison by ingestion, intravenous, intraperitoneal, and subcutaneous routes. Human systemic effects by ingestion: allergic dermatitis. Human teratogenic effects by an unspecified route: developmental abnormalities of the blood and lymphatic system. Questionable carcinogen. Mutation data reported. When heated to decomposition it emits toxic fumes of NO_x.

PHW250   CAS: 88-99-3   **HR: 2**
PHTHALIC ACID
mf: $C_8H_6O_4$ mw: 166.14

PROP: Crystals. Mp: >230°, d: 1.59, bp: 155° decomp. Sol in water, alc; sltly sol in ether; insol in chloroform.

SYNS: ACIDE PHTALIQUE (FRENCH) * BENZENE-1,2-DICARBOXYLIC ACID * o-BENZENEDICARBOX-YLIC ACID * 1,2-BENZENEDICARBOXYLIC ACID * o-DICARBOXYBENZENE

CONSENSUS REPORTS: Reported in EPA TSCA Inventory.

SAFETY PROFILE: Moderately toxic by intraperitoneal route. Slightly toxic by ingestion. Skin and mucous membrane irritant. Combustible when heated. In the form of dust (anhydride) it can explode. Mixtures with sodium nitrite explode when heated. Violent reaction with HNO_3. When heated to decomposition it emits acrid smoke and irritating fumes. Used in synthesis of dyes and dyestuffs, in medicines and perfumes.

PHW750   CAS: 85-44-9   **HR: 3**
PHTHALIC ANHYDRIDE
DOT: UN 2214
mf: $C_8H_4O_3$ mw: 148.12

PROP: White, crystalline needles. Mp: 131.2°, lel: 1.7%, uel: 10.4%, bp: 295° (sublimes), flash p: 305°F (CC), d: 1.527 @ 4°, autoign temp: 1058°F, vap press: 1 mm @ 96.5°, vap d: 5.10. Very sltly sol in water; sol in alc; sltly sol in ether.

SYNS: ANHYDRIDE PHTALIQUE (FRENCH) * ANIDRIDE FTALICA (ITALIAN) * 1,2-BENZENEDICARBOXYLIC ACID ANHYDRIDE * 1,3-DIOXOPHTHALAN * ESEN * FTAALZUURANHYDRIDE (DUTCH) * FTALOWY BEZWODNIK (POLISH) * 1,3-ISOBENZO-FURANDIONE * NCI-C03601 * 1,3-PHTHALAN-DIONE * PHTHALIC ACID ANHYDRIDE * PHTHAL-SAEUREANHYDRID (GERMAN) * RCRA WASTE NUMBER U190 * RETARDER AK * RETARDER ESEN * RETARDER PD

CONSENSUS REPORTS: NCI Carcinogenesis Bioassay (feed); No Evidence: mouse, rat NCITR* NCI-CG-TR-159,79. Community Right-To-Know List. Reported in EPA TSCA Inventory.

OSHA PEL: (Transitional: TWA 2 ppm) TWA 1 ppm
ACGIH TLV: TWA 1 ppm
DFG MAK: 5 mg/m³
DOT Classification: Corrosive; Label: Corrosive.

SAFETY PROFILE: Poison by ingestion. Experimental teratogenic effects. A corrosive eye, skin, and mucous membrane irritant. A common air contaminant. Combustible when exposed to heat or flame; can react with oxidizing materials. Moderate explosion hazard in the form of dust when exposed to flame. The production of this material has caused many industrial explosions. Mixtures with copper oxide or sodium nitrite explode when heated. Violent reaction with nitric acid + sulfuric acid above 80°C. To fight fire, use CO_2, dry chemical. Used in plasticizers, polyester resins, and alkyd resins, dyes and drugs.

PHX250 CAS: 732-11-6 **HR: 3**
PHTHALIMIDOMETHYL-O,O-DIMETHYL PHOSPHORODITHIOATE
mf: $C_{11}H_{12}NO_4PS_2$ mw: 317.33

SYNS: APPA * DECEMTHION P-6 * (O,O-DI-METHYL-PHTHALIMIDIOMETHYL-DITHIOPHOSPHATE) * O,O-DIMETHYL S-(N-PHTHALIMIDOMETHYL) DITHIO-PHOSPHATE * O,O-DIMETHYL S-PHTHALIMIDO-METHYL PHOSPHORODITHIOATE * ENT 25,705 * FTALOPHOS * IMIDAN * KEMOLATE * N-(MERCAPTOMETHYL)PHTHALIMIDE S-(O,O-DIMETHYL PHOSPHORODITHIOATE) * PERCOLATE * PHOS-MET * PHOSPHORODITHIOIC ACID, S-((1,3-DIHYDRO-1,3-DIOXO-ISOINDOL-2-YL)METHYL) O,O-DIMETHYL ESTER * PHTHALIMIDO-O,O-DIMETHYL PHOSPHO-RODITHIOATE * PHTHALOPHOS * PMP * PRO-LATE * R 1504 * SMIDAN * STAUFFER R 1504

CONSENSUS REPORTS: EPA Extremely Hazardous Substances List.

SAFETY PROFILE: A human poison by ingestion. Poison experimentally by inhalation, ingestion, and possibly other routes. Moderately toxic by skin contact. Human systemic effects by inhalation: lacrimation, somnolence, and olfaction effects. Experimental teratogenic and reproductive effects. Mutation data reported. When heated to decomposition it emits very toxic fumes of NO_x, PO_x, and SO_x.

PHX550 CAS: 626-17-5 **HR: 3**
m-PHTHALODINITRILE
mf: $C_8H_4N_2$ mw: 128.14

PROP: Colorless crystals; water insol; sol in benzene, acetone; vap d: 4.42; mp: 138°; bp: subl.

SYNS: 1,3-BENZENEDICARBONITRILE * m-DICYA-NOBENZENE * 1,3-DICYANOBENZENE * DINITRILE of ISOPHTHALIC ACID * IPN * ISOFTALODINITRIL (CZECH) * ISOPHTHALODINITRILE * ISOPHTHAL-ONITRILE * NITRIL KYSELINY ISOFTALOVE (CZECH) * m-PDN

CONSENSUS REPORTS: Reported in EPA TSCA Inventory. Cyanide and its compounds are on the Community Right-To-Know List.

OSHA PEL: TWA 5 mg/m^3
ACGIH TLV: TWA 5 mg/m^3

SAFETY PROFILE: Poison by ingestion and possibly other routes. An eye irritant. When heated to decomposition it emits toxic fumes of NO_x and CN^-.

PHY000 CAS: 91-15-6 **HR: 3**
PHTHALONITRILE
mf: $C_8H_4N_2$ mw: 128.14

SYNS: o-DICYANOBENZENE * 1,2-DICYANOBEN-ZENE * PHTHALIC ACID DINITRILE * PHTHALODI-NITRILE * o-PHTHALODINITRILE * USAF ND-09

CONSENSUS REPORTS: Cyanide and its compounds are on the Community Right-To-Know List. Reported in EPA TSCA Inventory.

SAFETY PROFILE: Poison by ingestion, subcutaneous, and intraperitoneal routes. Questionable carcinogen with experimental tumorigenic data. When heated to decomposition it emits toxic fumes of CN^- and NO_x.

PIA500 CAS: 57-47-6 **HR: 3**
PHYSOSTIGMINE
mf: $C_{15}H_{21}N_3O_2$ mw: 275.39

SYNS: ERSERINE * ESERINE * ESEROLEIN, METHYLCARBAMATE (ESTER) * METHYL-CARBAMIC ACID, ESTER with ESEROLINE * PHYSOSTOL

CONSENSUS REPORTS: EPA Extremely Hazardous Substances List. Reported in EPA TSCA Inventory.

SAFETY PROFILE: A human poison by an unspecified route. Poison experimentally by ingestion, subcutaneous, intramuscular, intravenous, and intraperitoneal routes. Human systemic effects by ingestion: nausea, dyspnea, coma. Normally administered by injection. Poisoning can occur as a result of a mistake in dosage or due to hypersensitivity of the patient within 5 to 25 minutes after administration. Death usually results from respiratory paralysis. Combustible when exposed to heat or flame. When heated to decomposition it emits toxic fumes of NO_x.

PIA750 CAS: 57-64-7 **HR: 3**
PHYSOSTIGMINE SALICYLATE (1:1)
mf: $C_{15}H_{21}N_3O_2 \cdot C_7H_6O_3$ mw: 413.52

SYNS: ESERINE SALICYLATE * PHYSOSTOL SALI-
CYLATE * SALICYLIC ACID with PHYSOSTIGMINE
(1:1) * TL-1380

CONSENSUS REPORTS: EPA Extremely
Hazardous Substances List. Reported in EPA
TSCA Inventory.

SAFETY PROFILE: Poison by ingestion, subcutaneous, intramuscular, intravenous, and intraperitoneal routes. Human central and peripheral
nervous system effects by ingestion. Experimental reproductive effects. When heated to decomposition it emits toxic fumes of NO_x.

PIB250   CAS: 83-86-3   HR: 3
PHYTIC ACID
mf: $C_6H_{18}O_{24}P_6$ mw: 660.06

SYNS: ALKOVERT * FYTIC ACID * HEXAKIS-
(DIHYDROGEN PHOSPHATE) MYO-INOSITOL * INOSI-
THEXAPHOSPHORSAURE (GERMAN) * INOSITOL
HEXAPHOSPHATE * MYO-INOSISTOL HEXAKISPHOS-
PHATE * MYO-INOSITOL HEXAPHOSPHATE
* SAURE DES PHYTINS (GERMAN)

CONSENSUS REPORTS: Reported in EPA
TSCA Inventory.

SAFETY PROFILE: Poison by intravenous
route. When heated to decomposition it emits
toxic fumes of PO_x.

PIB900   CAS: 1918-02-1   HR: 3
PICLORAM
mf: $C_6H_3Cl_3N_2O_2$ mw: 241.46

PROP: Crystals. Mp: 218°.

SYNS: AMDON GRAZON * 4-AMINO-3,5,6-TRICHLO-
ROPICOLINIC ACID * 4-AMINO-3,5,6-TRICHLORO-2-
PICOLINIC ACID * 4-AMINO-3,5,6-TRICHLORPICO-
LINSAEURE (GERMAN) * ATCP * BOROLIN
* CHLORAMP (RUSSIAN) * K-PIN * NCI-C00237
* TORDON * TORDON 10K * TORDON 22K
* TORDON 101 MIXTURE * 3,5,6-TRICHLORO-4-
AMINOPICOLINIC ACID

CONSENSUS REPORTS: NCI Carcinogenesis Bioassay (feed); No Evidence: mouse
NCITR* NCI-CG-TR-23,78; Clear Evidence:
rat NCITR* NCI-CG-TR-23,78

OSHA PEL: (Transitional: Total Dust: 15 mg/
    m³; Respirable Fraction: 5 mg/m³) TWA To-
    tal Dust: 10 mg/m³; Respirable Fraction: 5
    mg/m³
ACGIH TLV: TWA 10 mg/m³

SAFETY PROFILE: Moderately toxic by ingestion. Questionable carcinogen with experimental carcinogenic, neoplastigenic, tumorigenic,
and teratogenic data. Mutation data reported.
When heated to decomposition it emits very
toxic fumes of Cl^- and NO_x.

PIB920   CAS: 108-99-6   HR: 3
3-PICOLINE
DOT: UN 2313
mf: C_6H_7N mw: 93.14

PROP: Colorless liquid; sweetish, not unpleasant odor. D: (15°/4°) 0.9613, bp: 143-144°, n
(24/D) 1.5043. Misc with water, alc, ether.

SYNS: 3-METHYLPYRIDINE * β-PICOLINE
* m-PICOLINE (DOT)

CONSENSUS REPORTS: Reported in EPA
TSCA Inventory.

DOT Classification: Flammable or Combusti-
    ble Liquid; Label: Flammable Liquid.

SAFETY PROFILE: Poison by intravenous and
intraperitoneal routes. Moderately toxic by ingestion. Flammable when exposed to heat or
flame; can react vigorously with oxidizing materials. When heated to decomposition it emits
toxic fumes of NO_x.

PIC250   CAS: 466-24-0   HR: 3
PICRACONITINE
mf: $C_{32}H_{45}NO_{10}$ mw: 603.78

SYNS: BENZACONINE * BENZOYLACONINE
* ISACONITINE

SAFETY PROFILE: Poison by intravenous and
intraperitoneal routes. When heated to decomposition it emits toxic fumes of NO_x.

PID000   CAS: 88-89-1   HR: 3
PICRIC ACID
DOT: UN 0154/UN 1344
mf: $C_6H_3N_3O_7$ mw: 229.12

PROP: Yellow crystals or yellow liquid; very
bitter. Mp: 121.8°, bp: explodes > 300°, flash
p: 302°F, d: 1.763, autoign temp: 572°F, vap
d: 7.90.

SYNS: ACIDE PICRIQUE (FRENCH) * ACIDO PI-
CRICO (ITALIAN) * CARBAZOTIC ACID * C.I. 10305
* 2-HYDROXY-1,3,5-TRINITROBENZENE * MELINITE
* NITROXANTHIC ACID * PHENOL TRINITRATE
* PICRONITRIC ACID * PIKRINEZUUR (DUTCH)
* PIKRINSAEURE (GERMAN) * PIKRYNOWY KWAS
(POLISH) * 2,4,6-TRINITROFENOL (DUTCH)

* 2,4,6-TRINITROFENOLO (ITALIAN) * 1,3,5-TRINI-TROPHENOL * 2,4,6-TRINITROPHENOL

CONSENSUS REPORTS: Community Right-To-Know List. Reported in EPA TSCA Inventory. EPA Genetic Toxicology Program.

OSHA PEL: TWA 0.1 mg/m^3 (skin)
ACGIH TLV: TWA 0.1 mg/m^3; STEL 0.3 mg/m^3 (skin) (Proposed: TWA 0.1 mg/m^3)
DFG MAK: 0.1 mg/m^3
DOT Classification: Class A Explosive.

SAFETY PROFILE: Poison by ingestion, subcutaneous, and possibly other routes. Mutation data reported. An irritant and an allergen. Skin contact can cause local and systemic allergic reactions. Combustible when exposed to heat or flame; can react vigorously with oxidizing materials. Very unstable. A severe explosion hazard when shocked or exposed to heat. It forms salts easily, and many of its salts, known as picrates, are more sensitive explosives than picric acid. It forms unstable salts with concrete; NH_3; bases; and metals (e.g., copper; lead; mercury; and zinc). Many of these are heat-, friction-, or impact-sensitive. Mixtures with uronium perchlorate are extremely powerful explosives. Mixtures with aluminum and water ignite after a delay period. Can react vigorously with reducing materials. Used in synthesis of dyes, as a drug, to manufacture explosives and matches, to etch copper and make colored glass.

PIE500   CAS: 124-87-8   **HR: 3**
PICROTOXIN

DOT: UN 1584
mf: $C_{13}H_{18}O_7 \cdot C_{15}H_{16}O_6$ mw: 578.62

PROP: Dried fruit of *Anamerta cocculus* (*L.*) containing meni-spermine, paramenispermine, 1% picrotoxin, pictrotoxic acid, cocculine alkaloid, and 5% fat.

SYNS: COCCULIN * COCCULUS * COCCULUS solid (DOT) * COQUES DU LEVANT (FRENCH) * FISH BERRY * INDIAN BERRY * ORIENTAL BERRY * PICROTIN, compounded with PICROTOXININ (1:1) * PICROTOXINE

CONSENSUS REPORTS: EPA Extremely Hazardous Substances List.

DOT Classification: Poison B; Label: Poison.

SAFETY PROFILE: A human poison by ingestion and possibly other routes. Poison experimentally by ingestion, intraperitoneal, subcutaneous, intravenous, intramuscular, and intracerebral routes. Human systemic effects by ingestion: somnolence, gastrointestinal effects. An alkaloid convulsant poison. When heated to decomposition it emits acrid smoke and irritating fumes.

PIF000   CAS: 92-13-7   **HR: 3**
PILOCARPINE
mf: $C_{11}H_{16}N_2O_2$ mw: 208.29

PROP: Colorless or yellow, hygroscopic, needle-like crystals. Mp: 34°, bp: 260° @ 5 mm.

SYNS: ALMOCARPINE * (3S-cis)-3-ETHYLDIHYDRO-4-((1-METHYL-1H-IMIDAZOL-5-YL)METHYL)-2(3H)-FURANONE * α-ETHYL-β-(HYDROXYMETHYL)-1-METHYL-IMIDAZOLE-5-BUTYRIC ACID, γ-LACTONE * PILOCARPOL

CONSENSUS REPORTS: Reported in EPA TSCA Inventory.

SAFETY PROFILE: A human poison by subcutaneous and possibly other routes. Poison experimentally by ingestion, intravenous, intraperitoneal, and subcutaneous routes. A very poisonous alkaloid which is used to remove excess fluid accumulations from the body. Its action on the sweat glands makes it a powerful sudorific. It very rarely causes death, but when it does, it is by paralysis of the heart or edema of the lungs. Dangerous; on heating to decomposition it emits toxic fumes of NO_x.

PIF750   CAS: 7681-93-8   **HR: 3**
PIMARICIN
mf: $C_{33}H_{47}NO_{13}$ mw: 665.81

PROP: An antibiotic produced by a strain of *Streptomyces chattanoogensis*.

SYNS: ANTIBIOTIC A-5283 * CL 12,625 * MYCOPHYT * MYPROZINE * NATACYN * NATAMYCIN * PIMAFUCIN * TENNECETIN

SAFETY PROFILE: Poison by intravenous, intramuscular, subcutaneous, and intraperitoneal routes. Moderately toxic by ingestion. When heated to decomposition it emits toxic fumes of NO_x. Used as an antibacterial agent.

PIG730   CAS: 8016-45-3   **HR: 2**
PIMENTA LEAF OIL

PROP: Main constituent is eugenol. From steam distillation of the shrub *Pimenta officinalis* Lindl. (Fam. *Myrtaceae)*. Pale yellow to brown liquid; spicy odor. D: 1.037-1.050, refr index: 1.531 @ 20°. Sol in propylene glycol, fixed oils; insol in glycerin, mineral oil.

SYN: OIL of PIMENTA LEAF

CONSENSUS REPORTS: Reported in EPA TSCA Inventory.

SAFETY PROFILE: Moderately toxic by ingestion. A severe skin irritant. When heated to decomposition it emits acrid smoke and irritating fumes.

PIG740 HR: 2
PIMENTA OIL

PROP: Contains eugenol. Distilled from the fruit of *Pimenta officinalis* Lindley (Fam. *Myrtaceae*). Yellow to red-yellow liquid; odor and taste of allspice. D: 1.018-1.048, refr index: 1.527-1.540 @ 20°.

SYNS: ALLSPICE * PIMENTA BERRIES OIL * PIMENTO OIL

SAFETY PROFILE: A weak sensitizer which may cause dermatitis on local contact. Eugenol is moderately toxic. Combustible.

PIH175 CAS: 83-26-1 HR: 3
PINDONE

DOT: UN 2472
mf: $C_{14}H_{14}O_3$ mw: 230.28

PROP: Yellow crystals. Mp: 108°.

SYNS: CHEMRAT * 2-(2,2-DIMETHYL-1-OXOPROPYL)-1H-INDENE-1,3(2H)-DIONE * PINDON (DUTCH) * PIVACIN * PIVAL * PIVALDION (ITALIAN) * PIVALDIONE (FRENCH) * 2-PIVALOYL-INDAAN-1,3-DION (DUTCH) * 2-PIVALOYL-INDAN-1,3-DION (GERMAN) * 2-PIVALOYL-1,3-INDANDIONE * 2-PIVALOYLINDANE-1,3-DIONE * 2-PIVALYL-1,3-INDANDIONE * PIVALYL VALONE * PIVALYN * TRI-BAN * 2-(TRIMETIL-ACETIL)-INDAN-1,3-DIONE (ITALIAN)

CONSENSUS REPORTS: Reported in EPA TSCA Inventory.

OSHA PEL: TWA 0.1 mg/m^3
ACGIH TLV: TWA 0.1 mg/m^3
DOT Classification: Poison B; Label: Poison; Poison B; Label: St. Andrews Cross.

SAFETY PROFILE: Poison by ingestion, intravenous, and parenteral routes. Causes reduced blood clotting which leads to hemorrhaging. Used as an anticoagulant and rodenticide.

When heated to decomposition it emits acrid smoke and irritating fumes.

PIH250 CAS: 80-56-8 HR: 3
2-PINENE

DOT: UN 2368
mf: $C_{10}H_{16}$ mw: 136.26

PROP: Liquid; odor of turpentine. Mp: −55°, bp: 155°, flash p: 91°F, d: 0.8592 @ 20°/4°, refr index: 1.464-1.468, vap press: 10 mm @ 37.3°, vap d: 4.7, autoign temp: 491°F. Insol in water; sol in alc, chloroform, ether, glacial acetic acid, fixed oils.

SYNS: ACINTENE A * FEMA No. 2902 * α-PINENE (FCC) * 2,6,6-TRIMETHYLBICYCLO(3.1.1)-2-HEPT-2-ENE

CONSENSUS REPORTS: Reported in EPA TSCA Inventory.

DOT Classification: Flammable or Combustible Liquid; Label: Flammable Liquid.

SAFETY PROFILE: A deadly poison by inhalation. Moderately toxic by ingestion. An eye, mucous membrane, and severe human skin irritant. Flammable liquid. A dangerous fire hazard when exposed to heat, flame, or oxidizing materials. To fight fire, use foam, CO$_2$, dry chemical. Explodes on contact with nitrosyl perchlorate.

PIH400 CAS: 8000-26-8 HR: 1
PINE NEEDLE OIL, DWARF

PROP: From steam distillation of needles of *Pinus mugo* turra var. *pumilio* (Haenke) Zenari (Fam. *Pinaceae*). Colorless to yellow liquid; pleasant odor and a bitter, pungent taste. D: 0.853-0.871, refr index: 1.475 @ 20°.

SYNS: DWARF PINE NEEDLE OIL * KNEE PINE OIL * LATSCHENKIEFEROL * OIL of MOUNTAIN PINE * PINUS MONTANA OIL * PINUS PUMILIO OIL

CONSENSUS REPORTS: Reported in EPA TSCA Inventory.

SAFETY PROFILE: Mildly toxic by ingestion. A human skin irritant. When heated to decomposition it emits acrid smoke and irritating fumes.

PIH500 CAS: 8000-26-8 HR: 1
PINE NEEDLE OIL, SCOTCH

PROP: Volatile oil from steam distillation of *Pinus sylvestris* L. (Fam. *Pinaceae*) constituted of dipentene, pinene, sylvestrene, cadinene and bornyl acetate. Yellow liquid; penetrating odor. Bp: 200-220°, flash p: 172°F (CC), d: 0.86, refr index: 1.473 @ 20°. Sol in fixed oils, mineral oil; sltly sol in propylene glycol; insol in glycerin.

SYNS: KIEFERNADEL OEL (GERMAN) * SCOTCH PINE NEEDLE OIL

CONSENSUS REPORTS: Reported in EPA TSCA Inventory.

SAFETY PROFILE: Mildly toxic by ingestion. A weak allergen and a mild irritant. Flammable when exposed to heat or flame; can react vigorously with oxidizing materials. To fight fire, use foam, CO_2, dry chemical. When heated to decomposition it emits acrid smoke and irritating fumes.

PIH750 CAS: 8002-09-3 **HR: 2**
PINE OIL

DOT: UN 1272

PROP: Pale yellow liquid; penetrating odor. Bp: 200-220°, flash p: 172°F (CC), d: 0.86, flash p: (steam distilled): 138°F. Insol in water; sol in organic solvents.

SYNS: OIL of PINE * OLEUM ABIETIS * TERPENTIN OEL (GERMAN) * UNIPINE * YARMOR * YARMOR PINE OIL

CONSENSUS REPORTS: Reported in EPA TSCA Inventory.

DOT Classification: Combustible Liquid; Label: None.

SAFETY PROFILE: Moderately toxic by ingestion. Mildly toxic by skin contact. A weak allergen and a severe irritant to skin and mucous membranes. Human systemic effects by ingestion: excitement, ataxia, headache. Flammable when exposed to heat or flame; can react with oxidizing materials. Moderate spontaneous heating. To fight fire, use foam, CO_2, dry chemical. Used as an odorant, disinfectant, solvent, wetting agent, and frothing agent.

PII250 CAS: 52212-02-9 **HR: 3**
PIPECURIUM BROMIDE
mf: $C_{35}H_{62}N_4O_4 \cdot 2Br$ mw: 762.83

SYNS: ARDUAN * 2-β,16-β-(4'-DIMETHYL-1'-PIPERAZINO)-3-α,17-β-DIACETOXY-5-α-ANDROSTANE 2BR * PIPECURONIUM BROMIDE * RGH-1106

SAFETY PROFILE: A deadly poison by ingestion, intramuscular, intravenous, subcutaneous,

and intraperitoneal routes. Used as a skeletal muscle relaxant. When heated to decomposition it emits very toxic fumes of Br^- and NO_x.

PII750 CAS: 71-78-3 **HR: 3**
PIPERADROL HYDROCHLORIDE
mf: $C_{18}H_{21}NO \cdot ClH$ mw: 303.86

SYNS: α,α-DIPHENYL-2-PIPERIDINEMETHANOL HYDROCHLORIDE * α-(2-PIPERIDYL)BENZHYDROL HYDROCHLORIDE * PIPRADOL HYDROCHLORIDE * PIPRADROL HYDROCHLORIDE * PIRIDROL HYDROCHLORIDE * PYRIDROL

SAFETY PROFILE: Poison by ingestion, subcutaneous, intravenous, and intraperitoneal routes. When heated to decomposition it emits very toxic fumes of HCl and NO_x.

PIJ000 CAS: 110-85-0 **HR: 2**
PIPERAZINE

DOT: UN 2579/UN 2685
mf: $C_4H_{10}N_2$ mw: 86.16

PROP: Colorless, rhombic crystals. Mp: 106°, bp: 146°, flash p: 190°F (OC), d: 1.1, vap d: 3.0. Very sol in water, glycerol, glycols; insol in ether.

SYNS: ANTIREN * 1,4-DIETHYLENEDIAMINE * N,N-DIETHYLENE DIAMINE (DOT) * DISPERMINE * HEXAHYDRO-1,4-DIAZINE * HEXAHYDROPYRAZINE * LUMBRICAL * PIPERAZIDINE * PIPERAZIN (GERMAN) * PIPERAZINE, anhydrous * PYRAZINE HEXAHYDRIDE

CONSENSUS REPORTS: Reported in EPA TSCA Inventory.

SAFETY PROFILE: Moderately toxic by ingestion, skin contact, intravenous and subcutaneous routes. Mildly toxic by inhalation. A skin and severe eye irritant. Excessive absorption can cause urticaria, vomiting, diarrhea, blurred vision, and weakness. Flammable when exposed to heat or flame; can react vigorously with oxidizing materials. Explodes on contact with dicyanofurazan. To fight fire, use alcohol foam, mist, dry chemical, water spray. When heated to decomposition it emits highly toxic fumes of NO_x.

PIK000 CAS: 142-64-3 **HR: 2**
PIPERAZINE DIHYDROCHLORIDE
mf: $C_4H_{10}N_2 \cdot 2ClH$ mw: 159.08

SYNS: DIHYDROCHLORIDE SALT OF DIETHYLENEDI-AMINE * DOWZENE DHC * PIPERAZINE HYDRO-CHLORIDE

CONSENSUS REPORTS: Reported in EPA TSCA Inventory. EPA Genetic Toxicology Program.

OSHA PEL: TWA 5 mg/m^3
ACGIH TLV: TWA 5 mg/m^3

SAFETY PROFILE: Moderately toxic by intraperitoneal route. Mildly toxic by ingestion. When heated to decomposition it emits very toxic fumes of NO$_x$ and HCl. Used in making fiber, pharmaceuticals, and insecticides.

PIK250   CAS: 21416-87-5   **HR: 3**
2,6-PIPERAZINEDIONE-4,4′-
PROPYLENE DIOXOPIPERAZINE
mf: C$_{11}$H$_{16}$N$_4$O$_4$ mw: 268.31

SYNS: (±)-1,2-BIS(3,5-DIOXOPIPERAZINE-1-YL)PRO-PANE * (±)-1,2-BIS(3,5-DIOXOPIPERAZINYL)PROPANE * ICRF-159 * 4,4′-(1-METHYL-1,2-ETHANEDIYL)BIS-2,6-PIPERAZINEDIONE * NCI-C01627 * NSC-129943 * RAZOXIN * (±)-(3,5,3′,5′-TETRAOXO)-1,2-DIPIPER-AZINOPROPANE

CONSENSUS REPORTS: NCI Carcinogenesis Bioassay (ipr); Clear Evidence: mouse, rat NCITR* NCI-CG-TR-78,78. EPA Genetic Toxicology Program.

SAFETY PROFILE: Suspected carcinogen with experimental carcinogenic and tumorigenic data. Moderately toxic by intraperitoneal route. Experimental teratogenic and reproductive effects. Human systemic effects by ingestion: nausea, thrombocytopenia, leukopenia. Mutation data reported. When heated to decomposition it emits toxic fumes of NO$_x$.

PIL500   CAS: 110-89-4   **HR: 3**
PIPERIDINE

DOT: UN 2401
mf: C$_5$H$_{11}$N   mw: 85.17

PROP: Clear, colorless liquid; amine-like odor. Mp: −7°, bp: 106°, flash p: 37.4°F, d: 0.8622 @ 20°/4°, vap press: 40 mm @ 29.2°, vap d: 3.0. Misc with water; sol in alc, benzene, chloroform.

SYNS: AZACYCLOHEXANE * CYCLOPENTIMINE * CYPENTIL * HEXAHYDROPYRIDINE * HEXA-ZANE * PENTAMETHYLENEIMINE * PIPERIDIN (GERMAN)

CONSENSUS REPORTS: EPA Extremely Hazardous Substances List. Reported in EPA TSCA Inventory. EPA Genetic Toxicology Program.

DOT Classification: Flammable Liquid; Label-:Flammable Liquid

SAFETY PROFILE: Poison by ingestion, skin contact, and intraperitoneal routes. Moderately toxic by subcutaneous route. Mildly toxic by inhalation. Experimental reproductive effects by inhalation. A skin irritant. Mutation data reported. A very dangerous fire hazard when exposed to heat, flame, or oxidizers. Can react vigorously with oxidizing materials. To fight fire, use alcohol foam, CO$_2$, dry chemical. Explodes on contact with 1-perchloryl-piperidine, dicyanofurazan, N-nitrosoacetanilide. When heated to decomposition it emits highly toxic fumes of NO$_x$. Used in agriculture and pharmaceuticals, and as an intermediate for rubber accelerators.

PIO750   CAS: 55792-21-7   **HR: 3**
PIPERIDINOETHYL-2-HEPTOXYPHE-NYLCARBAMOATE HYDROCHLORIDE
mf: C$_{21}$H$_{34}$N$_2$O$_3$ • ClH   mw: 399.03

SYNS: HEPTACAINE * 2-HEPTYLOXYCARBANILIC ACID-2-(1-PIPERIDINYL)ETHYL ESTER HYDROCHLORIDE * (2-(HEPTYLOXY)PHENYL)CARBAMIC ACID-2-(1-PIPER-IDINYL)ETHYL ESTER HYDROCHLORIDE * N-(2-(HEP-TYLOXYPHENYLCARBAMOYLOXY)ETHYL)PIPERIDI-NIUM CHLORIDE

SAFETY PROFILE: Poison by intravenous and intraperitoneal routes. Moderately toxic by subcutaneous route. When heated to decomposition it emits very toxic fumes of NO$_x$ and HCl.

PIT250   CAS: 63918-29-6   **HR: 3**
2-(1-PIPERIDINO)-2-(2-THENYL)ETHYL-AMINE MALEATE
mf: C$_{11}$H$_{18}$N$_2$S • C$_4$H$_4$O$_4$ mw: 326.45

SYN: CIBA CO. 2825

SAFETY PROFILE: Poison by intravenous, intraperitoneal, and subcutaneous routes. Moderately toxic by ingestion. When heated to decomposition it emits very toxic fumes of NO$_x$ and SO$_x$.

PIV750   CAS: 32248-37-6   **HR: 3**
PIPEROCAINE
mf: C$_{16}$H$_{23}$NO$_2$ mw: 261.40

SYNS: 3-BENZOXY-1-(2-METHYLPIPERIDINO)PROPANE * BENZOYL-γ-(2-METHYLPIPERIDINO)PROPANOL * ISOCAINE BASE * 2-METHYL-1-PIPERIDINOPROPANOL, BENZOATE * (2-METHYLPIPERIDINO)PROPYL BENZOATE * γ-(2-METHYLPIPERIDYL)PROPYL BENZOATE * METYCAINE * NEOTHESIN

SAFETY PROFILE: Poison by intravenous and intraperitoneal routes. Moderately toxic by subcutaneous route. When heated to decomposition it emits toxic fumes of NO_x.

PIW000   CAS: 2622-26-6   **HR: 3**
PIPEROCYANOMAZINE
mf: $C_{21}H_{23}N_3OS$ mw: 365.53

SYNS: 2-CYANO-10-(3-(4-HYDROXYPIPERIDINO)PROPYL)PHENOTHIAZINE * 2-CYANO-10-(3-(4-HYDROXY-1-PIPERIDYL)PROPYL)PHENOTHIAZINE * CYANO-3-(HYDROXY-4 PIPERIDYL-1)-3 PROPYL)-10-PHENOTHIAZINE (FRENCH) * F.I. 6145 * 10-(3-(4-HYDROXY-PIPERIDINO)PROPYL)PHENOTHIAZINE-2-CARBONITRILE * IC 6002 * NEMACTIL * NEULACTIL * NEULEPTIL * PERICIAZINE * PERICYAZINE * PROPERICIAZINE * 6909 RP * RP 8908 * SKF 20,716 * WH 7508

CONSENSUS REPORTS: Cyanide and its compounds are on the Community Right-To-Know List.

SAFETY PROFILE: Poison by ingestion, intraperitoneal, intravenous, and subcutaneous routes. Used as an antipsychotic agent. When heated to decomposition it emits very toxic fumes of CN^-, NO_x, and SO_x.

PIW250   CAS: 120-57-0   **HR: 2**
PIPERONAL
mf: $C_8H_6O_3$ mw: 150.14

PROP: Colorless, lustrous crystals; floral odor. Mp: 37°, bp: 263°, vap press: 1 mm @ 87.0°. Very sol in alc, ether; sol in propylene glycol, fixed oils; insol water, glycerin.

SYNS: 3,4-BENZODIOXOLE-5-CARBOXALDEHYDE * 3,4-DIHYDROXYBENZALDEHYDE METHYLENE KETAL * DIOXYMETHYLENE-PROTOCATECHUIC ALDEHYDE * FEMA No. 2911 * HELIOTROPIN * 3,4-METHYLENE-DIHYDROXYBENZALDEHYDE * 3,4-METHYLENEDIOXYBENZALDEHYDE * PIPERONALDEHYDE * PIPERONYL ALDEHYDE * PROTOCATECHUIC ALDEHYDE METHYLENE ETHER

CONSENSUS REPORTS: Reported in EPA TSCA Inventory.

SAFETY PROFILE: Moderately toxic by ingestion and intraperitoneal routes. Can cause central nervous system depression. A skin irritant. Combustible when exposed to heat or flame; can react with oxidizing materials.

PIX000   CAS: 326-61-4   **HR: 2**
PIPERONYL ACETATE
mf: $C_{10}H_{10}O_4$ mw: 194.20

PROP: Colorless to light yellow liquid; heliotrope odor.

SYNS: HELIOTROPYL ACETATE * 3,4-METHYLENEDIOXYBENZYL ACETATE

CONSENSUS REPORTS: Reported in EPA TSCA Inventory.

SAFETY PROFILE: Moderately toxic by ingestion. A skin irritant. When heated to decomposition it emits acrid smoke and irritating fumes.

PIX250   CAS: 51-03-6   **HR: 3**
PIPERONYL BUTOXIDE
mf: $C_{19}H_{30}O_5$ mw: 338.49

PROP: Light brown liquid; mild odor. Bp: 180° @ 1 mm, flash p: 340°F, d: 1.04-1.07 @ 20°/20°. Misc with methanol, ethanol, benzene.

SYNS: BUTACIDE * BUTOCIDE * BUTOXIDE * α-(2-(2-BUTOXYETHOXY)ETHOXY)-4,5-METHYLENEDIOXY-2-PROPYLTOLUENE * α-(2-(2-n-BUTOXYETHOXY)-ETHOXY)-4,5-METHYLENEDIOXY-2-PROPYLTOLUENE * 5-((2-(2-BUTOXYETHOXY)ETHOXY)METHYL)-6-PROPYL-1,3-BENZODIOXOLE * BUTYL CARBITOL 6-PROPYLPIPERONYL ETHER * BUTYLCARBITYL (6-PROPYLPIPERONYL) ETHER * ENT 14,250 * FAC 5273 * FMC 5273 * 3,4-METHYLENDIOXY-6-PROPYLBENZYL-n-BUTYL-DIAETHYLENGLYKOLAETHER (GERMAN) * (3,4-METHYLENEDIOXY-6-PROPYLBENZYL)(BUTYL)DIETHYLENE GLICOL ETHER * 3,4-METHYLENEDIOXY-6-PROPYLBENZYL-n-BUTYL DIETHYLENEGLYCOL ETHER * NCI-C02813 * NIA 5273 * NUSYN-NOXFISH * PB * PRENTOX * 6-(PROPYLPIPERONYL)-BUTYL CARBITYL ETHER * 6-PROPYLPIPERONYL BUTYL DIETHYLENE GLYCOL ETHER * 5-PROPYL-4-(2,5,8-TRIOXA-DODECYL)-1,3-BENZODIOXOL (GERMAN) * PYBUTHRIN * PYRENONE 606 * SYNPREN-FISH

CONSENSUS REPORTS: IARC Cancer Review: GROUP 3 IMEMDT 7,56,87; Animal No Evidence IMEMDT 30,183,83. NCI Carcinogenesis Bioassay (feed); No Evidence:

mouse, rat NCITR* NCI-CG-TR-120,79. Glycol ether compounds are on the Community Right-To-Know List. Reported in EPA TSCA Inventory.

SAFETY PROFILE: Poison by skin contact. Moderately toxic by ingestion and intraperitoneal routes. Experimental reproductive effects. Many glycol ether compounds have dangerous human reproductive effects. Questionable carcinogen with experimental tumorigenic data. Mutation data reported. Combustible when exposed to heat or flame; can react with oxidizing materials. To fight fire, use foam, CO_2, dry chemical. When heated to decomposition it emits acrid smoke and irritating fumes.

PIY500   CAS: 98-77-1   **HR: 3**
PIP-PIP
mf: $C_{11}H_{22}N_2S_2$ mw: 246.47

SYNS: PENTAMETHYLENEDITHIOCARBAMATE * PIPERIDINIUM * "522" RUBBER ACCELERATOR

CONSENSUS REPORTS: Reported in EPA TSCA Inventory.

SAFETY PROFILE: Poison by ingestion. A human skin irritant. An allergen. When heated to decomposition it emits very toxic fumes of NO_x and SO_x.

PJA000   CAS: 125-51-9   **HR: 3**
PIPTAL
mf: $C_{22}H_{28}NO_3 \cdot Br$ mw: 434.42

SYNS: BENZILIC ACID ESTER with 1-ETHYL-3-HYDROXY-1-METHYLPIPERIDINIUM BROMIDE * 1-ETHYL-3-HYDROXY-1-METHYL-PIPERIDINIUM BROMIDE BENZILATE * N-ETHYL-3-PIPERIDYLBENZILATE METHOBROMIDE * 1-ETHYL-3-PIPERIDYL BENZILATE METHYLBROMIDE * JB-323 * PIPENZOLATE BROMIDE * PIPENZOLATE METHYLBROMIDE * QPB

SAFETY PROFILE: Poison by intravenous route. Moderately toxic by ingestion and subcutaneous routes. When heated to decomposition it emits very toxic fumes of Br^- and NO_x.

PJA500   CAS: 75-98-9   **HR: 3**
PIVALIC ACID
mf: $C_5H_{10}O_2$ mw: 102.15

PROP: Crystals. Mp: 35.5°, bp: 164°, d: 0.91. Very sol in alc, ether; somewhat sol in water.

SYNS: 2,2-DIMETHYLPROPANOIC ACID * α,α-DIMETHYLPROPIONIC ACID * 2,2-DIMETHYLPROPIONIC ACID * NEOPENTANOIC ACID * tert-PENTANOIC ACID * PROPANOIC ACID * TRIMETHYLACETIC ACID

CONSENSUS REPORTS: Reported in EPA TSCA Inventory.

SAFETY PROFILE: Moderately toxic by ingestion and skin contact. Questionable carcinogen with experimental tumorigenic data. When heated to decomposition it emits acrid smoke and irritating fumes.

PJD000   CAS: 15663-27-1   **HR: 3**
cis-PLATINOUS DIAMMINE DICHLORIDE
mf: $Cl_2H_6N_2Pt$ mw: 300.07

SYNS: CACP * CDDP * CISPLATINO (SPANISH) * CISPLATYL * CPDC * CPDD * DDP * cis-DDP * cis-DIAMINEDICHLOROPLATINUM * cis-DICHLORODIAMMINE PLATINUM(II) * NCI-C55776 * NEOPLATIN * NSC-119875 * PEYRONE'S CHLORIDE * PLATIBLASTIN * cis-PLATIN * PLATINEX * PLATINOL * cis-PLATINUM(II) DIAMINEDICHLORIDE

CONSENSUS REPORTS: IARC Cancer Review: GROUP 2A IMEMDT 7,170,87; Animal Limited Evidence IMEMDT 26,151,81. Reported in EPA TSCA Inventory. EPA Genetic Toxicology Program.

OSHA PEL: TWA 0.002 mg(Pt)/m^3
ACGIH TLV: TWA 0.002 mg(Pt)/m^3

SAFETY PROFILE: Suspected carcinogen with experimental carcinogenic and tumorigenic data. Poison by ingestion, intramuscular, subcutaneous, intravenous, and intraperitoneal routes. Human systemic effects by intravenous, intradermal, and possibly other routes: nausea or vomiting, change in auditory acuity, depressed renal function tests, changes in bone marrow, change in kidney tubules, hallucinations, corrosive to skin. Experimental teratogenic and reproductive effects. Human mutation data reported. When heated to decomposition it emits very toxic fumes of Cl^- and NO_x.

PJD250   CAS: 10025-99-7   **HR: 3**
PLATINOUS POTASSIUM CHLORIDE
mf: $Cl_4Pt \cdot K_2$ mw: 415.09

PROP: Ruby red. Mp: decomp @ 250°, d: 3.499 @ 24°. Sol in water.

SYNS: POTASSIUM CHLOROPLATINITE * POTASSIUM PLATINOCHLORIDE * POTASSIUM TETRACHLOROPLATINATE(II)

CONSENSUS REPORTS: Reported in EPA TSCA Inventory.

OSHA PEL: TWA 0.002 mg(Pt)/m^3
ACGIH TLV: TWA 0.002 mg(Pt)/m^3

SAFETY PROFILE: Human poison by ingestion. Poison experimentally by intraperitoneal route. Corrosive to human skin by intradermal route. Mutation data reported. When heated to decomposition it emits toxic fumes of Cl$^-$ and K$_2$O. Used as a catalyst for hydroformulations, photocatalysts, and dissociation of water.

PJD500   CAS: 7440-06-4   **HR: 3**
PLATINUM
af: Pt   aw: 195.09

PROP: Silvery-white, malleable, ductile metal; stable in air. Mp: 1772°, bp: 3827°, d: 21.45 @ 20°.

SYNS: C.I. 77795 * LIQUID BRIGHT PLATINUM * PLATIN (GERMAN) * PLATINUM BLACK * PLATINUM SPONGE

CONSENSUS REPORTS: Reported in EPA TSCA Inventory.

OSHA PEL: TWA (metal) 1 mg/m^3; (soluble salts as Pt) 0.002 mg/m^3
ACGIH TLV: TWA (metal) 1 mg/m^3; (soluble salts as Pt) 0.002 mg/m^3
DFG MAK: 0.002 mg/m^3

SAFETY PROFILE: Questionable carcinogen with experimental tumorigenic data by implant route. Finely divided platinum is a powerful catalyst and can be dangerous to handle. Used catalysts are especially dangerous and may be explosive. May undergo hazardous reactions with aluminum, acetone, arsenic, carbon + methanol, nitrosyl chloride, dioxygen difluoride, ethanol, hydrazine, hydrogen + air, hydrogen peroxide, lithium, methyl hydroperoxide, ozonides, peroxymonosulphuric acid, phosphorus, selenium, tellurium, vanadium dichloride + water.

PJE000   CAS: 10025-65-7   **HR: 3**
PLATINUM CHLORIDE
mf: Cl$_2$Pt   mw: 265.99

PROP: Grayish-green powder. D: 5.87. Insol in water, alc, ether, benzene, chloroform.

SYNS: MURIATE of PLATINUM * PLATINOUS CHLORIDE

CONSENSUS REPORTS: Reported in EPA TSCA Inventory.

SAFETY PROFILE: Poison by ingestion and possibly other routes. A skin irritant. Human mutation data reported. When heated to decomposition it emits toxic fumes of Cl$^-$.

PJE250   CAS: 13454-96-1   **HR: 3**
PLATINUM(IV) CHLORIDE
mf: Cl$_4$Pt   mw: 336.89

SYN: PLATINUM TETRACHLORIDE

CONSENSUS REPORTS: Reported in EPA TSCA Inventory. EPA Genetic Toxicology Program.

OSHA PEL: TWA 0.002 mg(Pt)/m^3
ACGIH TLV: TWA 0.002 mg(Pt)/m^3

SAFETY PROFILE: Poison by ingestion and intravenous routes. Experimental reproductive effects. Mutation data reported. A severe skin irritant. When heated to decomposition it emits toxic fumes of Cl$^-$.

PJE500   **HR: 2**
PLATINUM COMPOUNDS

SAFETY PROFILE: cis-[Pt(NH$_3$)$_2$Cl$_2$] is an experimental carcinogen. Exposure to complex platinum salts has been shown to cause symptoms of intoxication such as wheezing, coughing, running of the nose, chest tightness, shortness of breath and cyanosis. Furthermore, many people working with platinum salts are troubled with dermatitis. They may become sensitized after years of exposure. Symptoms of platinum allergy include rhinitis, conjunctivitis, asthma, urticaria and contact dermatitis. Mainly the ionic platinum chloro compounds [e.g., (NHN$_4$)$_2$(PtCl$_6$), (NH$_4$)$_2$(PtCl$_4$), H$_2$(PtCl$_6$)] are responsible for this sensitivity. The bromide and iodide compounds are less effective. These platinum compounds form a platinum-protein conjugate which is the true allergen. Tetrachloroplatinates are mutagens. This seems only to be true of complex platinum salts. It does not include the complex salts of the other precious metals. Platinum amine nitrates and perchlorates either detonate when heated or are impact-sensitive.

PJI000   **HR: 3**
PLUTONIUM COMPOUNDS

SAFETY PROFILE: The toxicity of plutonium compounds is based first upon the very high radiotoxicity of the plutonium atom and secondly upon whatever atoms or combinations of atoms they might contain. Very dangerous! Any disaster which could cause quantities of plutonium or plutonium compounds to be scattered about the environment can cause great ecological stress and render areas of the land unfit for public occupancy. Long-term storage in plastic containers is not recommended as the alpha particles can cause stress cracks and the potential for leakage.

PJI500 CAS: 14913-29-2 **HR: 3**
PLUTONIUM NITRATE (solution)

DOT: NA 9185

DOT Classification: Label: Radioactive.

SAFETY PROFILE: All plutonium compounds are extremely dangerous. When heated to decomposition it emits toxic fumes of NO_x and radioactive fumes of Pu.

PJJ000 CAS: 9000-55-9 **HR: 3**
PODOPHYLLIN

PROP: Light yellow powder or small yellow fragile lumps; bitter, acrid taste.

SYNS: PODOPHYLLUM * PODOPHYLLUM RESIN

SAFETY PROFILE: Poison by ingestion, subcutaneous, intraperitoneal, and possibly other routes. An irritant to skin, eyes, and mucous membranes. Questionable carcinogen with experimental neoplastigenic data. Combustible when exposed to heat or flames. When heated to decomposition it emits acrid smoke and irritating fumes.

PJJ325 CAS: 2438-32-6 **HR: 3**
POLARAMINE MALEATE
mf: $C_{16}H_{19}ClN_2 \cdot C_4H_4O_4$ mw: 390.90

SYNS: (+)-2-(p-CHLORO-α-(2-(DIMETHYLAMINO) ETHYL)BENZYL)PYRIDINE MALEATE * (+)-CHLORPHENIRAMINE MALEATE * d-CHLORPHENIRAMINE MALEATE * S-(+)-CHLORPHENIRAMINE MALEATE * DEXCHLOROPHENIRAMINE MALEATE * DEXCHLORPHENIRAMINE MALEATE * DEXTROCHLORPHENIRAMINE MALEATE * POLARAMIN * PORAMINE MALEATE

SAFETY PROFILE: Poison by ingestion, intravenous, and intraperitoneal routes. When heated to decomposition it emits toxic fumes of Cl^- and NO_x.

PJJ750 **HR: 3**
POLONIUM
af: Po aw: 210

PROP: A low melting, volatile, radioactive, naturally occurring metallic element. Mp: 254°, bp: 962°, d: 9.4.

SYN: RADIUM F

SAFETY PROFILE: Suspected carcinogen. Severe radiotoxicity. Very dangerous to handle. Radiation Hazard: Natural isotope ^{210}Po (radium-F, Uranium Series), $T_{0.5} = 138$ D. Decays to stable ^{206}Pb by alphas of 5.3 MeV. When heated to decomposition it emits toxic and radioactive fumes of Po.

PJK000 **HR: 3**
POLONIUM CARBONYL
mf: PoCO mw: 237.01

SAFETY PROFILE: Suspected carcinogen. Poison by ingestion, inhalation, intravenous, and subcutaneous routes. When heated to decomposition it emits toxic and radioactive fumes of Po.

PJL750 CAS: 1336-36-3 **HR: 3**
POLYCHLORINATED BIPHENYL

DOT: UN 2315

PROP: Bp: 340-375°, flash p: 383°F (COC), d: 1.44 @ 30°. A series of technical mixtures consisting of many isomers and compounds that vary from mobile oily liquids to white crystalline solids and hard noncrystalline resins. Technical products vary in composition, in the degree of chlorination and possibly according to batch.

SYNS: AROCLOR * CHLOPHEN * CHLOREXTOL * CHLORINATED BIPHENYL * CHLORINATED DIPHENYL * CHLORINATED DIPHENYLENE * CHLORO BIPHENYL * CHLORO-1,1-BIPHENYL * CLOPHEN * DYKANOL * FENCLOR * INERTEEN * KANECHLOR * MONTAR * NOFLAMOL * PCB (DOT, USDA) * PHENOCHLOR * POLYCHLORINATED BIPHENYL * POLYCHLOROBIPHENYL * PYRALENE * PYRANOL * SANTOTHERM * SOVOL * THERMINOL FR-1

CONSENSUS REPORTS: IARC Cancer Review: GROUP 2A IMEMDT 7,322,87; Human Limited Evidence IMEMDT 18,43,78. NTP Fourth Annual Report On Carcinogens, 1984. EPA Extremely Hazardous Substances List. Reported in EPA TSCA Inventory.

DFG MAK: Suspected Carcinogen.
NIOSH REL: TWA (Polychlorinated Biphenyls) 0.001 mg/m^3
DOT Classification: ORM-E; Label: None.

SAFETY PROFILE: Confirmed carcinogen with carcinogenic and tumorigenic data. Moderately toxic by ingestion. Some are poisons by other routes. Experimental reproductive effects.

Like the chlorinated naphthalenes, the chlorinated diphenyls have two distinct actions on the body, namely, a skin effect and a toxic action on the liver. This hepato-toxic action of the chlorinated diphenyls appears to be increased if there is exposure to carbon tetrachloride at the same time. The higher the chlorine content of the diphenyl compound, the more toxic it is liable to be. Oxides of chlorinated diphenyls are more toxic than the unoxidized materials. In persons who have suffered systemic intoxication, the usual signs and symptoms are nausea, vomiting, loss of weight, jaundice, edema and abdominal pain. Where the liver damage has been severe the patient may pass into a coma and die.

Combustible when exposed to heat or flame. When heated to decomposition they emit highly toxic fumes of Cl$^-$.

PJM000   CAS: 11104-28-2   **HR: 2**
**POLYCHLORINATED BIPHENYL
(AROCLOR 1221)**

SYNS: AROCHLOR 1221   *   CHLORODIPHENYL (21% Cl)

CONSENSUS REPORTS: IARC Cancer Review: Human Limited Evidence IMEMDT 18,43,78.

NIOSH REL: TWA (Polychlorinated Biphenyls) 0.001 mg/m^3

SAFETY PROFILE: Suspected human carcinogen. Moderately toxic by ingestion and skin contact. Experimental reproductive effects. When heated to decomposition it emits toxic fumes of Cl$^-$. Used in heat transfer, hydraulic fluids, lubricants, and insecticides.

PJM250   CAS: 11141-16-5   **HR: 2**
**POLYCHLORINATED BIPHENYL
(AROCLOR 1232)**

SYNS: AROCLOR 1232   *   CHLORODIPHENYL (32% Cl)

CONSENSUS REPORTS: IARC Cancer Review: Human Limited Evidence IMEMDT 18,43,78.

NIOSH REL: TWA (Polychlorinated Biphenyls) 0.001 mg/m^3

SAFETY PROFILE: Suspected human carcinogen. Moderately toxic by skin contact. Mildly toxic by ingestion. When heated to decomposition it emits toxic fumes of Cl$^-$. Used in heat transfer, hydraulic fluids, lubricants, and insecticides.

PJM500   CAS: 53469-21-9   **HR: 3**
**POLYCHLORINATED BIPHENYL
(AROCLOR 1242)**

SYNS: AROCHLOR 1242   *   AROCLOR 1242
*   CHLORIERTE BIPHENYLE, CHLORGEHALT 42% (GERMAN)   *   CHLORODIPHENYL (42% Cl) (OSHA)
*   CLORODIFENILI, CLORO 42% (ITALIAN)   *   DIPHENYLE CHLORE, 42% de CHLORE (FRENCH)
*   GECHLOREERDEDIFENYL (DUTCH)   *   PCB's

CONSENSUS REPORTS: IARC Cancer Review: Human Limited Evidence IMEMDT 18,43,78. EPA Genetic Toxicology Program.

OSHA PEL: TWA 1 mg/m^3 (skin)
ACGIH TLV: TWA 1 mg/m^3 (skin)
DFG MAK: 0.1 ppm (1 mg/m^3)
NIOSH REL: TWA (Polychlorinated Biphenyls) 0.001 mg/m^3

SAFETY PROFILE: Suspected human carcinogen. Poison by subcutaneous route. Mildly toxic by ingestion. Human systemic effects by inhalation: pulmonary and liver effects. Moderately toxic by ingestion. Experimental reproductive effects. Mutation data reported. When heated to decomposition it emits toxic fumes of Cl$^-$. Used in heat transfer, hydraulic fluids, lubricants, and insecticides.

PJM750   CAS: 12672-29-6   **HR: 3**
**POLYCHLORINATED BIPHENYL
(AROCLOR 1248)**

SYNS: AROCLOR 1248   *   CHLORODIPHENYL (48% Cl)

CONSENSUS REPORTS: IARC Cancer Review: Human Limited Evidence IMEMDT 18,43,78.

NIOSH REL: TWA (Polychlorinated Biphenyls) 0.001 mg/m^3

SAFETY PROFILE: Suspected human carcinogen. Moderately toxic by skin contact. Experimental teratogenic and reproductive effects. When heated to decomposition it emits toxic fumes of Cl⁻. Used in heat transfer, hydraulic fluids, lubricants, and insecticides.

PJN000   CAS: 11097-69-1   **HR: 3**
POLYCHLORINATED BIPHENYL (AROCLOR 1254)

PROP: Composed of 11% tetra-, 49% penta-, 34% hexa- and 6% heptachlorobiphenyls.

SYNS: AROCHLOR 1254 * AROCLOR 1254 * CHLORIERTE BIPHENYLE, CHLORGEHALT 54% (GERMAN) * CHLORODIPHENYL (54% Cl) (OSHA) * CLORODIFENILI, CLORO 54% (ITALIAN) * DIPHENYLE CHLORE, 54% de CHLORE (FRENCH) * NCI-C02664 * PCB's

CONSENSUS REPORTS: IARC Cancer Review: GROUP 2A IMEMDT 7,322,87; Animal Sufficient Evidence IMEMDT 7,261,74; Animal Limited Evidence IMEMDT 18,43,78; Human Limited Evidence IMEMDT 18,43,78. NCI Carcinogenesis Bioassay (feed); Some Evidence: rat NCITR* NCI-CG-TR-38,78. EPA Genetic Toxicology Program.

OSHA PEL: TWA 0.5 mg/m³ (skin)
ACGIH TLV: TWA 0.5 mg/m³ (skin)
NIOSH REL: TWA (Polychlorinated Biphenyls) 0.001 mg/m³

SAFETY PROFILE: Suspected carcinogen with experimental carcinogenic and neoplastigenic data. Poison by intravenous route. Moderately toxic by ingestion and intraperitoneal routes. Experimental teratogenic and reproductive effects. Mutation data reported. When heated to decomposition it emits toxic fumes of Cl⁻. Used in heat transfer, hydraulic fluids, lubricants, and insecticides.

PJN250   CAS: 11096-82-5   **HR: 3**
POLYCHLORINATED BIPHENYL (AROCLOR 1260)

PROP: Composed of 12% penta-, 38% hexa-, 41% hepta-, 8% octa- and 1% nonachlorobiphenyls.

SYNS: AROCHLOR 1260 * AROCLOR 1260 * CHLORODIPHENYL (60% Cl) * CLOPHEN A60 * PHENOCLOR DP6

CONSENSUS REPORTS: IARC Cancer Review: Animal Limited Evidence IMEMDT 18,43,78; Human Limited Evidence IMEMDT 18,43,78.

NIOSH REL: TWA (Polychlorinated Biphenyls) 0.001 mg/m³

SAFETY PROFILE: Suspected carcinogen with carcinogenic and neoplastigenic data. Moderately toxic by ingestion and skin contact. Experimental reproductive effects. Mutation data reported. When heated to decomposition it emits highly toxic fumes of Cl⁻. Used in heat transfer, hydraulic fluids, lubricants, and insecticides.

PJN500   CAS: 37324-23-5   **HR: 3**
POLYCHLORINATED BIPHENYL (AROCLOR 1262)

SYNS: AROCLOR 1262 * CHLORODIPHENYL (62% Cl)

CONSENSUS REPORTS: IARC Cancer Review: Human Limited Evidence IMEMDT 18,43,78.

DFG MAK: 0.1 ppm (1 mg/m³)
NIOSH REL: TWA 0.001 mg/m³

SAFETY PROFILE: Suspected human carcinogen. Moderately toxic by skin contact. When heated to decomposition it emits toxic fumes of Cl⁻. Used in heat transfer, hydraulic fluids, lubricants, and insecticides.

PJN750   CAS: 11100-14-4   **HR: 3**
POLYCHLORINATED BIPHENYL (AROCLOR 1268)

SYNS: AROCLOR 1268 * CHLORODIPHENYL (68% Cl)

CONSENSUS REPORTS: IARC Cancer Review: Human Limited Evidence IMEMDT 18,43,78.

NIOSH REL: TWA 0.001 mg/m³

SAFETY PROFILE: Suspected human carcinogen. Moderately toxic by skin contact. Used in heat transfer, hydraulic fluids, lubricants, and insecticides. When heated to decomposition it emits toxic fumes of Cl⁻.

PJO000   CAS: 37324-24-6   **HR: 3**
POLYCHLORINATED BIPHENYL (AROCLOR 2565)

SYN: AROCLOR 2565

CONSENSUS REPORTS: IARC Cancer Review: Human Limited Evidence IMEMDT 18,43,78.

NIOSH REL: TWA 0.001 mg/m^3

SAFETY PROFILE: Suspected human carcinogen. Moderately toxic by skin contact. Mildly toxic by ingestion. When heated to decomposition it emits toxic fumes of Cl$^-$. Used in heat transfer, hydraulic fluids, lubricants, and insecticides.

PJO250   CAS: 11120-29-9   **HR: 3**
POLYCHLORINATED BIPHENYL (AROCLOR 4465)

SYN: AROCLOR 4465

CONSENSUS REPORTS: IARC Cancer Review: Human Limited Evidence IMEMDT 18,43,78.

NIOSH REL: TWA (Polychlorinated Biphenyls) 0.001 mg/m^3

SAFETY PROFILE: Suspected human carcinogen. Moderately toxic by skin contact. Mildly toxic by ingestion. When heated to decomposition it emits toxic fumes of Cl$^-$. Used in heat transfer, hydraulic fluids, lubricants, and insecticides.

PJO500   CAS: 37353-63-2   **HR: 3**
POLYCHLORINATED BIPHENYL (KANECHLOR 300)

PROP: Average content: 60% trichlorobiphenyl, 23% tetrachlorobiphenyl, 17% dichlorobiphenyl, 1% pentachlorobiphenyl.

SYN: KANECHLOR 300

CONSENSUS REPORTS: IARC Cancer Review: Animal Limited Evidence IMEMDT 7,261,74, IMEMDT 18,43,78; Human Limited Evidence IMEMDT 18,43,78.

NIOSH REL: TWA (Polychlorinated Biphenyls) 0.001 mg/m^3

SAFETY PROFILE: Suspected human carcinogen. Used in heat transfer, hydraulic fluids, lubricants, and insecticides. When heated to decomposition it emits toxic fumes of Cl$^-$.

PJO750   CAS: 12737-87-0   **HR: 3**
POLYCHLORINATED BIPHENYL (KANECHLOR 400)

PROP: Average content: 44% tetrachlorbiphenyl, 33% trichlorobiphenyl, 16% pentachlorobiphenyl, 5% hexachlorobiphenyl, 3% dichlorobiphenyl.

SYNS: KANECHLOR 400   *   KC-400

CONSENSUS REPORTS: IARC Cancer Review: Animal Limited Evidence IMEMDT 7,261,74, IMEMDT 18,43,78; Human Limited Evidence IMEMDT 18,43,78.

NIOSH REL: TWA (Polychlorinated Biphenyls) 0.001 mg/m^3

SAFETY PROFILE: Suspected carcinogen with experimental neoplastigenic data. Experimental teratogenic and reproductive effects. Human systemic effects by ingestion: dermatitis, sweating. When heated to decomposition it emits toxic fumes of Cl$^-$.

PJP000   CAS: 37317-41-2   **HR: 3**
POLYCHLORINATED BIPHENYL (KANECHLOR 500)

PROP: Average content, 55% pentachlorobiphenyl, 26.5% tetrachlorobiphenyl, 12.8% hexachloro biphenyl and 5% trichlorobiphenyl.

SYNS: KANECHLOR 500   *   KC-500

CONSENSUS REPORTS: IARC Cancer Review: Human Limited Evidence IMEMDT 18,43,78; Animal Limited Evidence IMEMDT 18,43,78; Animal Sufficient Evidence IMEMDT 7,261,74.

NIOSH REL: TWA (Polychlorinated Biphenyls) 0.001 mg/m^3.

SAFETY PROFILE: Suspected carcinogen with experimental carcinogenic and tumorigenic data. Experimental teratogenic and reproductive effects. When heated to decomposition it emits toxic fumes of Cl$^-$. Used in heat transfer, hydraulic fluids, lubricants, and insecticides.

PJR000   CAS: 9016-00-6   **HR: 3**
POLYDIMETHYL SILOXANE

PROP: A water-insoluble polymer of high viscosity.

SYNS: DIMETHICONE 350   *   DOW CORNING 346 * GEON * GOOD-RITE * GUM * HYCAR * LATEX * METHYL SILICONE * POLY(OXY(DIMETHYLSILYLENE))

SAFETY PROFILE: Experimental reproductive effects. Questionable carcinogen with experimental neoplastigenic data. When heated to decomposition it emits acrid smoke and irritating fumes. Used as a release material, foam preventative, and surface active agent.

PJR250   CAS: 63394-02-5   HR: 3
POLYDIMETHYLSILOXANE RUBBER

SYNS: POLYSILICONE * SILASTIC * SILICONE RUBBER

SAFETY PROFILE: Questionable carcinogen with experimental carcinogenic and tumorigenic data. When heated to decomposition it emits acrid smoke and irritating fumes.

PJS750   CAS: 9002-88-4   HR: 3
POLYETHYLENE
mf: $(C_2H_4)_n$

PROP: Odorless. The high molecular weight compounds are tough, white leathery, resinous. D: 0.92 @ 20°/4°, mp: 85-110°. Sol in hot benzene; insol in water.

SYNS: AGILENE * ALKATHENE * BAKELITE DYNH * DIOTHENE * ETHENE POLYMER * ETHYLENE HOMOPOLYMER * ETHYLENE POLYMERS * HOECHST PA 190 * MICROTHENE * POLYETHYLENE AS * POLYWAX 1000 * TENITE 800

CONSENSUS REPORTS: IARC Cancer Review: GROUP 3 IMEMDT 7,56,87; Animal Sufficient Evidence IMEMDT 19,157,79; Human Inadequate Evidence IMEMDT 19,157,79. Reported in EPA TSCA Inventory.

SAFETY PROFILE: Questionable carcinogen with experimental tumorigenic data by implant. Reacts violently with F_2. When heated to decomposition it emits acrid smoke and irritating fumes.

PJT000   CAS: 25322-68-3   HR: 2
POLYETHYLENE GLYCOL
mf: $H(OC_2H_4)_nOH$

PROP: Clear liquid or white solid. D: 1.110-1.140 @ 20°, mp: 4-10°, flash p: 471°F. Sol in organic solvents, aromatic hydrocarbons.

SYNS: ALKAPOL PEG-200 * CARBOWAX * α-HYDROXY-omega-HYDROXY-POLY(OXY-1,2-ETHANEDIYL) * JEFFOX * JORCHEM 400 ML * LUTROL * PEG * PLURACOL P-410 * POLY(ETHYLENE OXIDE) * POLY-G SERIES * POLYOX

CONSENSUS REPORTS: Reported in EPA TSCA Inventory. EPA Genetic Toxicology Program.

SAFETY PROFILE: Moderately toxic by intraperitoneal and intravenous routes. Slightly toxic by ingestion. An eye irritant. Questionable carcinogen with experimental tumorigenic data. Combustible liquid when exposed to heat or flame. To fight fire, use water, foam, dry chemical. When heated to decomposition it emits acrid smoke and irritating fumes.

PJT200   CAS: 25322-68-3   HR: 1
POLYETHYLENE GLYCOL 200
mf: $H(OC_2H_4)_nOH$

PROP: Viscous, hydroscopic liquid with n about 4; slt characteristic odor. D (25°/25°) 1.127.

SYNS: CARBOWAX * JEFFOX * NYCOLINE * PEG 200 * PLURACOL E * POLYAETHYLENGLYCOLE 200 (GERMAN) * POLY-G * POLYGLYCOL E * SOLBASE

CONSENSUS REPORTS: EPA Genetic Toxicology Program. Reported in EPA TSCA Inventory.

SAFETY PROFILE: Mildly toxic by ingestion. Caution: Solvent action on some plastics. When heated to decomposition it emits acrid smoke and irritating fumes.

PJT225   CAS: 25322-68-3   HR: 1
POLYETHYLENE GLYCOL 300
mf: $(C_6H_{11}NO)_n$

SYNS: POLYAETHYLENGLYKOLE 300 (GERMAN) * PEG 300

CONSENSUS REPORTS: EPA Genetic Toxicology Program. Reported in EPA TSCA Inventory.

SAFETY PROFILE: Mildly toxic by ingestion. When heated to decomposition it emits acrid smoke and irritating fumes.

PJT230   CAS: 25322-68-3   HR: 1
POLYETHYLENE GLYCOL 400
mf: $H(OC_2H_4)_nOH$

PROP: Liquid with n about 8.2 to 9.1. Mw: 380-420, d: 1.128, mp: 4-8°.

SYNS: PEG 400 * POLYAETHYLENGLYKOLE 400 (GERMAN) * POLY G 400

CONSENSUS REPORTS: EPA Genetic Toxicology Program. Reported in EPA TSCA Inventory.

SAFETY PROFILE: Low toxicity by ingestion, intravenous, and intraperitoneal routes. When

heated to decomposition it emits acrid smoke and irritating fumes.

PJT240   CAS: 25322-68-3   HR: 1
POLYETHYLENE GLYCOL 600
mf: $H(OC_2H_4)_nOH$

PROP: Liquid with n about 12.5 to 13.9.mw: 570-630, d: 1.128, mp: 20-25°.

SYNS: PEG 600   *   POLYAETHYLENGLYKOLE 600 (GERMAN)

CONSENSUS REPORTS: EPA Genetic Toxicology Program. Reported in EPA TSCA Inventory.

SAFETY PROFILE: Low toxicity by ingestion. An eye irritant. When heated to decomposition it emits acrid smoke and irritating fumes.

PJT250   CAS: 25322-68-3   HR: 3
POLYETHYLENE GLYCOL 1000
mf: $H(OC_2H_4)_nOH$

SYNS: CARBOWAX 1000   *   MACROGOL 1000   *   PEG 1000   *   POLYAETHYLENGLYKOLE 1000 (GERMAN)   *   POLYGLYCOL 1000   *   POLYGLYCOL E1000

CONSENSUS REPORTS: Reported in EPA TSCA Inventory. EPA Genetic Toxicology Program.

SAFETY PROFILE: Moderately toxic by intraperitoneal and intravenous routes. Mildly toxic by ingestion. Questionable carcinogen with experimental tumorigenic data. When heated to decomposition it emits acrid smoke and irritating fumes.

PJT500   CAS: 25322-68-3   HR: 1
POLYETHYLENE GLYCOL 1500
mf: $H(OC_2H_4)_nOH$

PROP: White, free-flowing powder. D: 1.15-1.21 @ 25°/25°, fp: 44-48°.

SYNS: CARBOWAX 1500   *   α-HYDRO-omega-HYDROXY-POLY(OXY-1,2-ETHANEDIYL)   *   PEG 1500   *   POLYAETHYLENGLYKOLE 1500 (GERMAN)   *   POLYOXYETHYLENE 1500

CONSENSUS REPORTS: Reported in EPA TSCA Inventory. EPA Genetic Toxicology Program.

SAFETY PROFILE: Mildly toxic by ingestion. A human skin irritant. When heated to decomposition it emits acrid smoke and irritating fumes.

PJT750   CAS: 25322-68-3   HR: 1
POLYETHYLENE GLYCOL 4000
mf: $H(OC_2H_4)_nOH$

PROP: White, free-flowing powder or white flakes. D: 1.20-1.21 @ 25°/25°Fp: 54-58°.

SYNS: CARBOWAX 4000   *   CARSONON PEG-4000   *   MACROGOL 4000   *   PEG 4000   *   POLYAETHYLENGLYKOLE 4000 (GERMAN)   *   POLYGLYCOL 4000   *   POLYGLYCOL E-4000   *   POLYGLYCOL E-4000 USP   *   POLYOXYETHYLENE (75)

CONSENSUS REPORTS: Reported in EPA TSCA Inventory. EPA Genetic Toxicology Program.

SAFETY PROFILE: Mildly toxic by ingestion. A skin irritant. When heated to decomposition it emits acrid smoke and irritating fumes.

PJU000   CAS: 25322-68-3   HR: 1
POLYETHYLENE GLYCOL 6000
mf: $H(OC_2H_4)_nOH$

PROP: White, waxy solid. Mp: 58-62°, flash p: >887°F. Water-sol.

SYNS: CARBOWAX 6000   *   PEG 6000   *   POLYAETHYLENGLYKOLE 6000 (GERMAN)

CONSENSUS REPORTS: Reported in EPA TSCA Inventory. EPA Genetic Toxicology Program.

SAFETY PROFILE: Mildly toxic by ingestion. Mutation data reported. A skin irritant. Combustible when exposed to heat or flame. When heated to decomposition it emits acrid smoke and irritating fumes.

PJU500   CAS: 9005-08-7   HR: 3
POLYETHYLENE GLYCOL DISTEARATE

PROP: Polyethylene glycol distearate, low molecular weight.

SYNS: POLYETHYLENE GLYCOL 300 DISTEARATE   *   POLYETHYLENE GLYCOL 400 (DI) STEARATE   *   POLYETHYLENE GLYCOL 600 (DI) STEARATE   *   POLYGLYCOL DISTEARATE

CONSENSUS REPORTS: Reported in EPA TSCA Inventory.

SAFETY PROFILE: Poison by intravenous route. When heated to decomposition it emits acrid smoke and irritating fumes.

PJV250   CAS: 9004-99-3   **HR: 3**
POLYETHYLENE GLYCOL MONOSTEARATE

SYNS: POLYOXYETHYLENE-8-MONOSTEARATE
* POLYOXYETHYLENE(8)STEARATE

CONSENSUS REPORTS: Reported in EPA TSCA Inventory.

SAFETY PROFILE: Very slightly toxic by ingestion. Questionable carcinogen with experimental tumorigenic data. When heated to decomposition it emits acrid smoke and irritating fumes.

PJY500   CAS: 25038-54-4   **HR: 3**
POLY(IMINOCARBONYLPENTA-METHYLENE)
mf: $(C_6H_{11}NO)_n$

SYNS: AKULON * ALKAMID * AMILAN CM 1001
* 6-AMINOHEXANOIC ACID HOMOPOLYMER * BON-
AMID * CAPRAN 80 * CAPROAMIDE POLYMER
* CAPROLACTAM OLIGOMER * epsilon-CAPROLAC-
TAM POLYMERE (GERMAN) * CAPRON * CHEM-
LON * DANAMID * DULL 704 * DURETHAN BK
* ERTALON 6SA * GRILON * HEXAHYDRO-2H-
AZEPIN-2-ONE HOMOPOLYMER * ITAMID
* KAPROLIT * KAPROLON * KAPROMIN
* KAPRON * MARANYL F 114 * METAMID
* MIRAMID WM 55 * NYLON-6 * ORGAMIDE
* PA 6 (polymer) * PLASKON 201 * POLICAPRAN
* POLYAMIDE 6 * POLY(epsilon-AMINOCAPROIC
ACID) * POLYCAPROAMIDE * POLY(epsilon-CAPRO-
AMIDE) * POLYCAPROLACTAM * POLY(epsilon-
CAPROLACTAM) * POLY(IMINO(1-OXO-1,6-HEXA-
NEDIYL)) * RELON P * SPENCER 401 * STILON
* TARLON XB * TARNAMID T * ULTRA-
MID BMK * VIDLON * WIDLON * ZYTEL 211

CONSENSUS REPORTS: IARC Cancer Review: GROUP 3 IMEMDT 7,56,87; Animal Inadequate Evidence IMEMDT 19,115,75. Reported in EPA TSCA Inventory.

SAFETY PROFILE: Moderately toxic by ingestion. Mildly toxic by inhalation. Questionable carcinogen with experimental neoplastigenic data by implant route. When heated to decomposition it emits toxic fumes of NO_x.

PKA850   **HR: 2**
POLYMERS, WATER INSOLUBLE

SAFETY PROFILE: Many produce local tumors of the soft tissues surrounding the site of implantation.

PKA860   **HR: 2**
POLYMERS, WATER SOLUBLE

SAFETY PROFILE: Many produce local tumors of the soft tissues surrounding the site of implantation and in the lungs, mucosal contact areas, organs, and tissues of retention and deposition.

PKB500   CAS: 9011-14-7   **HR: 3**
POLYMETHYLMETHACRYLATE
mf: $(C_5H_8O_2)_n$

SYNS: ACRYLITE * ACRYPET * ALUTOR M 70
* CMW BONE CEMENT * CRINOTHENE * DEGA-
LAN S 85 * DELPET 50M * DIAKON * DISPA-
SOL M * DV 400 * ELVACITE * KALLOCRYL K
* KALLODENT CLEAR * KORAD * LPT
* LUCITE * METAPLEX NO * METHACRYLIC
ACID METHYL ESTER POLYMERS * METHYL METH-
ACRYLATE HOMOPOLYMER * METHYL METHACRY-
LATE POLYMER * METHYL METHACRYLATE RESIN
* 2-METHYL-2-PROPENOIC ACID METHYL ESTER HO-
MOPOLYMER * ORGANIC GLASS E 2 * OSTEO-
BOND SURGICAL BONE CEMENT * PALACOS
* PARAGLAS * PARAPLEX P 543 * PERSPEX
* PLEXIGLAS * PLEXIGUM M 920 * PMMA
* PONTALITE * REPAIRSIN * RESARIT 4000
* RHOPLEX B 85 * ROMACRYL * SHINKOLITE
* SOL * STELLON PINK * SUMIPLEX LG
* SUPERACRYL AE * SURGICAL SIMPLEX
* TENSOL 7 * VEDRIL

CONSENSUS REPORTS: IARC Cancer Review: GROUP 3 IMEMDT 7,56,87; Human Inadequate Evidence IMEMDT 19,187,79; Animal Sufficient Evidence IMEMDT 19,187,79. Reported in EPA TSCA Inventory.

SAFETY PROFILE: Questionable carcinogen with experimental tumorigenic data by implant route. When heated to decomposition it emits acrid smoke and irritating fumes. Used as the main constituent of acrylic sheet, molding, and extrusion powers.

PKC000   CAS: 1406-11-7   **HR: 3**
POLYMYXIN

PROP: A series of antibiotic substances, polypeptide (basic), sol in water. Colorless powder. Decomp @ 228-230°.

SYN: B-71

SAFETY PROFILE: Poison by intraperitoneal, subcutaneous, and intravenous routes. An additive permitted in food for human consumption.

PKC250 CAS: 1404-24-6 **HR: 3**
POLYMYXIN A

SAFETY PROFILE: Poison by intraperitoneal, subcutaneous, intravenous, and intracerebral routes. When heated to decomposition it emits acrid smoke and irritating fumes.

PKF000 CAS: 9016-45-9 **HR: 2**
POLYOXYETHYLENE (9) NONYL PHENYL ETHER

SYNS: ARKOPAL N-090 * CARSONON N-9 * CONCO NI-90 * IGEPAL CO-630 * NEUTRONYX 600 * PEG-9 NONYL PHENYL ETHER * POLYETHYLENE GLYCOL 450 NONYL PHENYL ETHER * PROTACHEM 630 * REWOPOL HV-9 * TERGITOL TP-9 (NONIONIC)

CONSENSUS REPORTS: Reported in EPA TSCA Inventory. Glycol ethers are on the Community Right-To-Know List.

SAFETY PROFILE: Moderately toxic by ingestion and skin contact. A severe eye and mild skin irritant in humans. Many glycol ethers cause dangerous human reproductive effects. When heated to decomposition it emits acrid smoke and irritating fumes.

PKF500 CAS: 9002-93-1 **HR: 2**
POLY(OXYETHYLENE)-p-tert-OCTYLPHENYL ETHER
mf: $(C_2H_4O)_n$ $C_{14}H_{22}O$

PROP: Mixture in which *n* varies from 5 to 15. Pale yellow, viscous liquid. D: 1.0595. Miscible with water, alc, acetone; sol in benzene, toluene; insol in petr ether.

SYNS: ALFENOL 3 * ALFENOL 9 * ANTAROX A-200 * CONCO NIX-100 * HYDROL SW * HYONIC PE-250 * IGEPAL CA-63 * MARLOPHEN 820 * NEUTRONYX 605 * OCTOXINOL * OCTOXYNOL * OCTOXYNOL 3 * OCTOXYNOL 9 * OCTYL PHENOL CONDENSED with 12-13 MOLES ETHYLENE OXIDE * p-tert-OCTYLPHENOXYPOLYETHOXYETHANOL * OPE 30 * PEG-9 OCTYL PHENYL ETHER * POLYETHYLENE GLYCOL MONOETHER with p-tert-OCTYLPHENYL * POLYETHYLENE GLYCOL MONO(4-OCTYLPHENYL) ETHER * POLYETHYLENE GLYCOL MONO(4-tert-OCTYLPHENYL) ETHER * POLYETHYLENE GLYCOL MONO(p-tert-OCTYLPHENYL) ETHER * POLYETHYLENE GLYCOL MONO(p-(1,1,3,3-TETRAMETHYLBUTYL)PHENYL) ETHER * POLYETHYLENE GLYCOL OCTYLPHENOL ETHER * POLYETHYLENE GLYCOL 450 OCTYL PHENYL ETHER * POLYETHYLENE GLYCOL p-OCTYLPHENYL ETHER * POLYETHYLENE GLYCOL p-tert-OCTYLPHENYL ETHER * POLYETHYLENE GLYCOL p-1,3,3,-TETRAMETHYLBUTYLPHENYL ETHER * POLYOXYETHYLENE MONO(OCTYLPHENYL) ETHER * POLYOXYETHYLENE (9) OCTYLPHENYL ETHER * POLYOXYETHYLENE (13) OCTYLPHENYL ETHER * PRECEPTIN * TRITON X 35 * TRITON X 45 * TRITON X 100 * TRITON X 102 * TRITON X 165 * TRITON X 305 * TRITON X 405 * TRITON X 705 * TX 100

CONSENSUS REPORTS: Glycol ether compounds are on the Community Right-To-Know List. Reported in EPA TSCA Inventory.

SAFETY PROFILE: Moderately toxic by ingestion and intravenous routes. Experimental reproductive effects. Human mutation data reported. An eye and human skin irritant. Many glycol ethers cause dangerous human reproductive effects. When heated to decomposition it emits toxic fumes of NO_x. A surfactant.

PKI500 CAS: 25322-69-4 **HR: 1**
POLYPROPYLENE GLYCOL
mf: $(C_3H_8O_2)_n$

PROP: Clear, colorless liquid. Mw: 400-2000, mp: does not crystallize, flash p: +390°F, d: 1.002-1.007. Sol in water, aliphatic ketones and alcs; insol in ether, aliphatic hydrocarbons.

SYNS: ALKAPOL PPG-1200 * JEFFOX * POLYPROPYLENGLYKOL (CZECH)

CONSENSUS REPORTS: Reported in EPA TSCA Inventory.

SAFETY PROFILE: Mildly toxic by ingestion. A skin and eye irritant. Combustible liquid when exposed to heat or flame; can react with oxidizing materials. To fight fire, use foam, CO_2, dry chemical. When heated to decomposition it emits acrid smoke and irritating fumes.

PKI750 CAS: 25322-69-4 **HR: 3**
POLYPROPYLENE GLYCOL 750

CONSENSUS REPORTS: Reported in EPA TSCA Inventory.

SAFETY PROFILE: Poison by ingestion, intraperitoneal, and intravenous routes. When heated to decomposition it emits acrid smoke and irritating fumes.

PKL000 CAS: 9005-64-5 **HR: 2**
POLYSORBATE 20

PROP: Lemon to amber colored liquid; characteristic odor, bitter taste. Sol in water, alc, ethyl acetate, methanol, dioxane; insol in mineral oil, mineral spirits.

SYNS: GLYCOSPERSE L-20X * POLYOXYETHYLENE (20) SORBITAN MONOLAURATE

CONSENSUS REPORTS: Reported in EPA TSCA Inventory.

SAFETY PROFILE: Moderately toxic by intravenous route. Mildly toxic by ingestion. A human skin irritant. When heated to decomposition it emits acrid smoke and irritating fumes.

PKL030 CAS: 9005-67-8 **HR: 3**
POLYSORBATE 60
mf: $C_{64}H_{126}O_{26}$ mw: 1311.90

PROP: Lemon to orange colored oily liquid; faint odor and bitter taste. Sol in water, aniline, ethyl acetate, toluene; insol in mineral oil, vegetable oil.

SYNS: CAPMUL * LGYCOSPERSE S-20 * LIPOSORB S-20 * POLYOXYETHYLENE SORBITAN MONOSTEARATE * POLYOXYETHYLENE 20 SORBITAN MONOSTEARATE * SORBITAN, MONOOCTADECANOATE, POLY(OXY-1,2-ETHANEDIYL) DERIVATIVES * TWEEN 60

CONSENSUS REPORTS: Reported in EPA TSCA Inventory.

SAFETY PROFILE: Moderately toxic by intravenous route. Experimental reproductive effects. Questionable carcinogen with experimental tumorigenic data. When heated to decomposition it emits acrid smoke and irritating fumes.

PKL100 CAS: 9005-65-6 **HR: 3**
POLYSORBATE 80

PROP: Yellow to orange oily liquid; faint odor, bitter taste. Sol in water, alc, fixed oils, ethyl acetate, toluene; insol in mineral oil.

SYNS: ARMOTAN PMO-20 * ATLOX 1087 * CAPMUL POE-O * CRILL 10 * DREWMULSE POE-SMO * DURFAX 80 * EMSORB 6900 * ETHOXYLATED SORBITAN MONOOLEATE * GLYCOSPERSE O-20 * HODAG SVO 9 * LIPOSORB O-20 * MONITAN * MONTANOX 80 * NCI-C60286 * NIKKOL TO * OLOTHORB * POLYOXYETHYL-ENE SORBITAN MONOOLEATE * POLYOXYETHYL-ENE SORBITAN OLEATE * POLYSORBAN 80 * POLYSORBATE 80, U.S.P. * PROTASORB O-20 * ROMULGIN O * SORBIMACROGOL OLEATE * SORBITAL O 20 * SORETHYTAN (20) MONO-OLEATE * SORLATE * SVO 9 * TWEEN 80

CONSENSUS REPORTS: Reported in EPA TSCA Inventory.

SAFETY PROFILE: Moderately toxic by intravenous route. Mildly toxic by ingestion. Experimental reproductive effects. Questionable carcinogen with experimental tumorigenic data. Human mutation data reported. An eye irritant. When heated to decomposition it emits acrid smoke and irritating fumes.

PKL500 CAS: 9009-54-5 **HR: 3**
POLYURETHANE FOAM

SYNS: ETHERON SPONGE * NCI-C56451 * POLYFOAM PLASTIC SPONGE * POLYFOAM SPONGE * POLYURETHANE ESTER FOAM * POLYURETHANE ETHER FOAM * POLYURETHANE SPONGE

CONSENSUS REPORTS: IARC Cancer Review: GROUP 3 IMEMDT 7,56,87; Animal Sufficient Evidence IMEMDT 19,303,79.

SAFETY PROFILE: Questionable carcinogen with experimental tumorigenic data. When heated to decomposition it emits acrid toxic fumes of CN^- and NO_x.

PKL750 CAS: 25931-01-5 **HR: 3**
POLYURETHANE Y-195
mf: $(C_{15}H_{10}N_2O_2 \cdot C_6H_{10}O_4 \cdot C_2H_6O_2)_x$

SYNS: ADIPIC ACID, POLYMER with ETHYLENE GLYCOL and METHYLENEDI-p-PHENYLENE ISOCYANATE * AMCHEM R 14 * HEXANEDIOIC ACID, POLYMER with 1,3-ETHANEDIOL and 1,1'-METHYLENEBIS(4-ISOCY-ANATOBENZENE) * MUL F 66 * R 14 * Y 195

CONSENSUS REPORTS: IARC Cancer Review: Animal Sufficient Evidence IMEMDT 19,303,79. Reported in EPA TSCA Inventory.

SAFETY PROFILE: Confirmed carcinogen with experimental tumorigenic data. When heated to decomposition it emits toxic fumes of NO_x.

PKM000 **HR: 3**
POLYURETHANE Y-217

SAFETY PROFILE: Confirmed carcinogen with experimental tumorigenic data. When heated to decomposition it emits very toxic fumes of NO_x and CN^-.

PKM250   CAS: 26375-23-5   HR: 3
POLYURETHANE Y-218
mf: $(C_{15}H_{10}N_2O_2 \cdot C_6H_{10}O_4 \cdot C_4H_{10}O_2)_x$

SYNS: ADIPIC ACID, POLYMER with 1,4-BUTANEDIOL and METHYLENEDI-p-PHENYLENE ISOCYANATE
* HEXANEDIOIC ACID, POLYMER with 1,4-BUTANEDIOL and 1,1'-METHYLENEBIS(4-ISOCYANATOBENZENE)
* PANDEX * TEXIN 445D * TPU 10M * Y 218

CONSENSUS REPORTS: IARC Cancer Review: Animal Sufficient Evidence IMEMDT 19,303,79.

SAFETY PROFILE: Confirmed carcinogen with experimental tumorigenic data. When heated to decomposition it emits very toxic fumes of CN^- and NO_x.

PKM500   CAS: 32238-28-1   HR: 3
POLYURETHANE Y-221
mf: $(C_{15}H_{10}N_2O_2 \cdot C_{10}H_{14}O_4 \cdot C_6H_{10}O_4 \cdot C_4H_{10}O_2)_x$

SYNS: ADIPIC ACID, POLYMER with 1,4-BUTANEDIOL, METHYLENEDI-p-PHENYLENE ISOCYANATE and 2,2'-(p-PHENYLENEDIOXY)DIETHANOL * Y 221

CONSENSUS REPORTS: IARC Cancer Review: Animal Sufficient Evidence IMEMDT 19,303,79. Reported in EPA TSCA Inventory.

SAFETY PROFILE: Confirmed carcinogen with experimental tumorigenic data. When heated to decomposition it emits very toxic fumes of CN^- and NO_x.

PKM750   HR: 3
POLYURETHANE Y-222

SAFETY PROFILE: Confirmed carcinogen with experimental tumorigenic data. When heated to decomposition it emits very toxic fumes of CN^- and NO_x.

PKN000   CAS: 52292-20-3   HR: 3
POLYURETHANE Y-223

SYNS: TECOFLEX HR * Y-223

CONSENSUS REPORTS: IARC Cancer Review: Animal Sufficient Evidence IMEMDT 19,303,79. Reported in EPA TSCA Inventory.

SAFETY PROFILE: Confirmed carcinogen with experimental tumorigenic data. When heated

to decomposition it emits very toxic fumes of CN^- and NO_x.

PKN250   HR: 3
POLYURETHANE Y-224

SAFETY PROFILE: Confirmed carcinogen with experimental tumorigenic data. When heated to decomposition it emits very toxic fumes of CN^- and NO_x.

PKN500   CAS: 56779-19-2   HR: 3
POLYURETHANE Y-225

SYN: 1,4-BUTANEDIAMINE, 2-METHYL-, POLYMER with α-HYDRO-omega-HYDROXYPOLY(OXY-1,4-BUTANEDIYL) and 1,1'-METHYLENEBIS(4-ISOCYANATOCYCLOHEXANE)

CONSENSUS REPORTS: IARC Cancer Review: Animal Sufficient Evidence IMEMDT 19,303,79.

SAFETY PROFILE: Confirmed carcinogen with experimental tumorigenic data. When heated to decomposition it emits very toxic fumes of CN^- and NO_x.

PKN750   CAS: 56386-98-2   HR: 3
POLYURETHANE Y-226

CONSENSUS REPORTS: IARC Cancer Review: Animal Sufficient Evidence IMEMDT 19,303,79.

SAFETY PROFILE: Confirmed carcinogen with experimental tumorigenic data. When heated to decomposition it emits very toxic fumes of CN^- and NO_x.

PKO000   CAS: 56631-46-0   HR: 3
POLYURETHANE Y-227

CONSENSUS REPORTS: IARC Cancer Review: Animal Sufficient Evidence IMEMDT 19,303,79.

SAFETY PROFILE: Confirmed carcinogen with experimental tumorigenic data. When heated to decomposition it emits very toxic fumes of CN^- and NO_x.

PKO500   CAS: 27083-55-2   HR: 3
POLYURETHANE Y-290
mf: $(C_{15}H_{10}N_2O_2 \cdot C_6H_{10}O_4 \cdot C_4H_{10}O_2 \cdot C_2H_6O_2)_x$

SYNS: E6 * PPE201 * P07 * TEXIN 192A * TPU 2T

CONSENSUS REPORTS: IARC Cancer Review: Animal Sufficient Evidence IMEMDT 19,303,79.

SAFETY PROFILE: Confirmed carcinogen with experimental tumorigenic data. When heated to decomposition it emits very toxic fumes of CN^- and NO_x.

PKP000   CAS: 25805-16-7   **HR: 3**
POLYURETHANE Y-302
mf: $(C_{15}H_{10}N_2O_2 \cdot C_4H_{10}O_2)_x$

SYNS: 1,4-BUTANEDIOL POLYMER with 1,1'-METHY-LENEBIS(4-ISOCYANATOBENZENE) * ISOCYANIC ACID, METHYLENEDI-p-PHENYLENE ESTER, POLYMER with 1,4-BUTANEDIOL * SANPRENE LQX 31 * Y 302

CONSENSUS REPORTS: IARC Cancer Review: Animal Sufficient Evidence IMEMDT 19,303,79.

SAFETY PROFILE: Confirmed carcinogen with experimental tumorigenic data by implant route. When heated to decomposition it emits very toxic fumes of CN^- and NO_x.

PKP250   CAS: 25036-33-3   **HR: 3**
POLYURETHANE Y-304

CONSENSUS REPORTS: IARC Cancer Review: Animal Sufficient Evidence IMEMDT 19,303,79.

SAFETY PROFILE: Confirmed carcinogen with experimental tumorigenic data. When heated to decomposition it emits very toxic fumes of CN^- and NO_x.

PKP750   CAS: 9002-89-5   **HR: 3**
POLYVINYL ALCOHOL

PROP: Colorless, amorphous powder. Mp: decomp over 200°, flash p: 175°F (OC), d: 1.329. Polymer of average molecular weight 120,000.

SYNS: ELVANOL * ETHENOL HOMOPOLYMER (9CI) * GELVATOLS * GOHSENOLS * POLY(VINYL ALCOHOL) * VINYL ALCOHOL POLYMER

CONSENSUS REPORTS: IARC Cancer Review: GROUP 3 IMEMDT 7,56,87; Animal Limited Evidence IMEMDT 19,341,79; Human Inadequate Evidence IMEMDT 19,341,79.

SAFETY PROFILE: Questionable carcinogen with experimental carcinogenic and tumorigenic data by implant route. Flammable when exposed to heat or flame; can react with oxidizing materials. Slight explosion hazard in the form of dust when exposed to flame. To fight fire, use alcohol foam, CO_2, dry chemical. When heated to decomposition it emits acrid smoke and irritating fumes.

PKQ000   CAS: 25951-54-6   **HR: 3**
POLYVINYLBROMIDE
mf: $(C_2H_3Br)_x$

PROP: Commercial PVBR is a 40% aqueous suspension in which PVBR constitutes about 90% of the solids.

SYNS: BROMOETHYLENE POLYMER * POLYBRO-MOETHYLENE * PVBR

CONSENSUS REPORTS: IARC Cancer Review: Animal Inadequate Evidence IMEMDT 19,367,79.

SAFETY PROFILE: Questionable carcinogen with experimental carcinogenic data. When heated to decomposition it emits toxic fumes of Br^-.

PKQ059   CAS: 9002-86-2   **HR: 2**
POLYVINYL CHLORIDE
mf: $(C_2H_3Cl)_n$

PROP: Polymers with molecular weights ranging from 60,000-150,000 (CNREA8 15,333,55). White powder, d: 1.406.

SYNS: ARMODOUR * ARON COMPOUND HW * ASTRALON * ATACTIC POLY(VINYL CHLORIDE) * BLACAR 1716 * BOLATRON * BONLOID * BREON * CARINA * CHLOROETHENE HOMO-POLYMER * CHLOROETHYLENE POLYMER * CHLOROSTOP * COBEX (polymer) * CONTIZELL * CORVIC 55/9 * DACOVIN * DANUVIL 70 * DARVIC 110 * DARVIS CLEAR 025 * DECELITH H * DENKA VINYL SS 80 * DIAMOND SHAMROCK 40 * DORLYL * DUROFOL P * DYNADUR * E 62 * E 66P * EKAVYL SD 2 * E-PVC * ESCAMBIA 2160 * EUROPHAN * EXON 605 * FC 4648 * FLOCOR * GAFCOTE * GENO-THERM * GEON * GEON LATEX 151 * GUT-TAGENA * HALVIC 223 * HISHIREX 502 * HISPAVIC 229 * HOSTALIT * IGELITE F * IMPROVED WILT PRUF * KAYLITE * KLEGE-CELL * KOROSEAL * LONZA G * LUCOFLEX * LUCOVYL PE * LUTOFAN * MARVINAL * MIRREX MCFD 1025 * MOVINYL 100 * MYRA-FORM * NCI-C60797 * NIKA-TEMP * NIKAVI-NYL SG 700 * NIPEON A 21' * NIPOL 576

* NORVINYL * NOVON 712 * ONGROVIL S 165 * OPALON * ORTUDUR * PANTASOTE R 873 * PARCLOID * PATTINA V 82 * PEVIKON D 61 * PLIOVIC * POLIVINIT * POLY(CHLOROETHYL-ENE) * POLYTHERM * POLYVINYLCHLORID (GER-MAN) * PROTOTYPE III SOFT * PVC (MAK) * QSAH 7 * QUIRVIL * QYSA * RAVINYL * RUCON B 20 * S 65 (polymer) * SCON 5300 * SICRON * S-LON * SOLVIC * SP 60 (CHLO-ROCARBON) * SUMILIT EXA 13 * SUMITOMO PX 11 * TAKILON * TECHNOPOR * TENNECO 1742 * TK 1000 * TROVIDUR * TROVITHERN HTL * U 1 (polymer) * ULTRON * UNICHEM * VERON P 130/1 * VESTOLIT B 7021 * VINIKA KR 600 * VINIKULON * VINIPLAST * VINIPLEN P 73 * VINNOL E 75 * VINOFLEX * VINYL-CHLON 4000LL * VINYL CHLORIDE HOMOPOLYMER * VINYL CHLORIDE POLYMER * VYGEN 85 * WELVIC G 2/5 * WILT PRUF * WINIDUR * X-AB * YUGOVINYL

CONSENSUS REPORTS: IARC Cancer Review: GROUP 3 IMEMDT 7,56,87; Human Inadequate Evidence IMEMDT 19,377,79; IARC Cancer Review: Animal Inadequate Evidence IMEMDT 19,377,79. Reported in EPA TSCA Inventory.

DFG MAK: 6 mg/m^3 (dust)

SAFETY PROFILE: Chronic inhalation of dusts can cause pulmonary damage, blood effects, abnormal liver function. ''Meat wrappers asthma'' has resulted from the cutting of PVC films with a hot knife. Can cause allergic dermatitis. Questionable carcinogen with experimental tumorigenic data. Reacts violently with F_2. When heated to decomposition it emits toxic fumes of Cl$^-$ and phosgene.

PKQ250   CAS: 9003-39-8   **HR: 1**
POLY(1-VINYL-2-PYRROLIDINONE) HOMOPOLYMER
mf: $(C_6H_9ON)_n$

PROP: A free-flowing, white, amorphous powder. D: 1.23-1.29. Sol in water, chlorinated hydrocarbons, alc, amines, nitroparaffins, and lower molecular weight fatty acids.

SYNS: AGENT AT 717 * ALBIGEN A * ALDACOL Q * AT 717 * BOLINAN * 1-ETHENYL-2-PYRRO-LIDINONE HOMOPOLYMER * 1-ETHENYL-2-PYRROLI-DINONE POLYMERS * GANEX P 804 * HEMODESIS * HEMODEZ * K25 (polymer) * KOLLIDON * LUVISKOL * MPK 90 * NCI C60582 * NEO-COMPENSAN * PERAGAL ST * PERISTON

* PLASDONE * POLYCLAR L * POLY(1-(2-OXO-1-PYRROLIDINYL)ETHYLENE) * POLYVIDONE * POLY(n-VINYLBUTYROLACTAM) * POLYVINYL-PYRROLIDONE * POVIDONE (USP XIX) * PROTA-GENT * PVP (FCC) * SUBTOSAN * VINISIL * N-VINYLBUTYROLACTAM POLYMER * N-VINYL-PYRROLIDONE POLYMER

CONSENSUS REPORTS: IARC Cancer Review: GROUP 3 IMEMDT 7,56,87. Reported in EPA TSCA Inventory.

SAFETY PROFILE: Mildly toxic by intraperitoneal and intravenous routes. Questionable carcinogen. When heated to decomposition it emits toxic fumes of NO$_x$.

PKQ500   CAS: 9003-39-8   **HR: 3**
POLY(1-VINYL-2-PYRROLIDINONE)
Hueper's polymer No. 1

PROP: Polymer of average molecular weight 20,000.

SYNS: NCI-C60582 * PVP 1

CONSENSUS REPORTS: IARC Cancer Review: Animal Limited Evidence IMEMDT 19,461,79. Reported in EPA TSCA Inventory.

SAFETY PROFILE: Suspected carcinogen with experimental carcinogenic data. When heated to decomposition it emits toxic fumes of NO$_x$.

PKQ750   CAS: 9003-39-8   **HR: 3**
POLY(1-VINYL-2-PYRROLIDINONE)
Hueper's polymer No. 2

PROP: Polymer of average molecular weight 20,000.

SYNS: NCI-C60582 * PVP 2

CONSENSUS REPORTS: IARC Cancer Review: Animal Limited Evidence IMEMDT 19,461,79. Reported in EPA TSCA Inventory.

SAFETY PROFILE: Suspected carcinogen with experimental neoplastigenic and tumorigenic data. When heated to decomposition it emits toxic fumes of NO$_x$.

PKR000   CAS: 9003-39-8   **HR: 3**
POLY(1-VINYL-2-PYRROLIDINONE)
Hueper's polymer No. 3

PROP: Polymer of average molecular weight 50,000.

SYNS: NCI-C60582 * PVP 3

CONSENSUS REPORTS: IARC Cancer Review: Animal Limited Evidence IMEMDT 19,461,79. Reported in EPA TSCA Inventory.

SAFETY PROFILE: Suspected carcinogen with experimental carcinogenic and tumorigenic data. When heated to decomposition it emits toxic fumes of NO_x.

PKR250   CAS: 9003-39-8   **HR: 3**
POLY(1-VINYL-2-PYRROLIDINONE)
Hueper's polymer No. 4

PROP: Polymer of average molecular weight 300,000.

SYNS: NCI-C60582   *   PVP 4

CONSENSUS REPORTS: IARC Cancer Review: Animal Limited Evidence IMEMDT 19,461,79. Reported in EPA TSCA Inventory.

SAFETY PROFILE: Suspected carcinogen with experimental carcinogenic data. When heated to decomposition it emits toxic fumes of NO_x.

PKR500   CAS: 9003-39-8   **HR: 3**
POLY(1-VINYL-2-PYRROLIDINONE)
Hueper's polymer No. 5

PROP: Polymer of average molecular weight 10,000.

SYN: PVP 5

CONSENSUS REPORTS: IARC Cancer Review: Animal Limited Evidence IMEMDT 19,461,79. Reported in EPA TSCA Inventory.

SAFETY PROFILE: Suspected carcinogen with experimental carcinogenic and tumorigenic data. When heated to decomposition it emits toxic fumes of NO_x.

PKR750   CAS: 9003-39-8   **HR: 3**
POLY(1-VINYL-2-PYRROLIDINONE)
Hueper's polymer No. 6

PROP: Polymer of average molecular weight 50,000.

SYNS: NCI-C60582   *   PVP 6

CONSENSUS REPORTS: IARC Cancer Review: Animal Limited Evidence IMEMDT 19,461,79. Reported in EPA TSCA Inventory.

SAFETY PROFILE: Suspected carcinogen with experimental carcinogenic data. When heated to decomposition it emits toxic fumes of NO_x.

PKS000   CAS: 9003-39-8   **HR: 3**
POLY(1-VINYL-2-PYRROLIDINONE)
Hueper's polymer No. 7

SYNS: NCI-C60582   *   PVP 7

CONSENSUS REPORTS: IARC Cancer Review: Animal Limited Evidence IMEMDT 19,461,79.

SAFETY PROFILE: Suspected carcinogen with experimental neoplastigenic data. When heated to decomposition it emits toxic fumes of NO_x.

PKS250   CAS: 26837-42-3   **HR: 3**
POLYVINYL SULFATE, POTASSIUM SALT

SYNS: POTASSIUM SALT OF POLYVINYL SULFATE   *   PVSK

CONSENSUS REPORTS: Reported in EPA TSCA Inventory.

SAFETY PROFILE: Poison by intraperitoneal and subcutaneous routes. When heated to decomposition it emits toxic fumes of SO_x and K_2O.

PKS750   CAS: 65997-15-1   **HR: 1**
PORTLAND CEMENT

PROP: Fine gray powder composed of compounds of lime, aluminum, silica and iron oxide as $(4CaO \cdot Al_2O_3 \cdot Fe_2)_3$, $(3CaOAl_2O_3)$, $(3CaO \cdot SiO_2)$, and $(2CaOSiO_2)$. Small amounts of magnesia, sodium, potassium, chromium and sulfur are also present in combined form. Containing less than 1% crystalline silica.

SYNS: CEMENT, PORTLAND   *   PORTLAND CEMENT SILICATE

CONSENSUS REPORTS: Reported in EPA TSCA Inventory.

OSHA PEL: (Transitional: TWA 50 mppcf) TWA Total Dust: 10 mg/m^3; Respirable Fraction: 5 mg/m^3

ACGIH TLV: TWA (nuisance particulate) 10 mg/m^3 of total dust (when toxic impurities are not present, e.g., quartz $< 1\%$).

SAFETY PROFILE: A nuisance dust. A skin irritant.

PKT250   CAS: 7440-09-7   **HR: 3**
POTASSIUM

DOT: UN 1420/UN 2257
af: K   aw: 39.10

PROP: Soft ductile, silvery-white, very reactive metal. Mp: 63.65°, bp: 774°, d: 0.862 @ 20°.

SYN: POTASSIUM, METAL (DOT)

CONSENSUS REPORTS: Reported in EPA TSCA Inventory.

DOT Classification: Label: Flammable Solid and Dangerous When Wet.

SAFETY PROFILE: The toxicity of potassium compounds is almost always that of the anion, not of potassium. A dangerous fire hazard. Metallic potassium reacts with moisture to form potassium hydroxide and hydrogen. The reaction evolves much heat, causing the potassium to melt and spatter. The reaction also ignites the hydrogen, which burns, or if there is any confinement, may explode. It can ignite spontaneously in moist air. Store under mineral oil. Potassium metal will form the peroxide (K_2O_2) and the superoxide (KO_3 or K_2O_4) at room temperature even when stored under mineral oil. These oxides can explode on contact with organic materials. Metal which has oxidized on storage under oil may explode violently when handled or cut. Oxide-coated potassium should be destroyed by burning.

A violent explosion hazard with many materials under required conditions of temperature, pressure, and state of division. When heated to decomposition it emits toxic fumes of K_2O.

Danger: burning potassium is difficult to extinguish; dry powdered soda ash or graphite or special mixtures of dry chemical are recommended.

PKT500   CAS: 7440-09-7   **HR: 3**
POTASSIUM (liquid alloy)

DOT: UN 1420

SYN: POTASSIUM, metal liquid alloy (DOT)

CONSENSUS REPORTS: Reported in EPA TSCA Inventory.

DOT Classification: Label: Flammable Solid and Dangerous When Wet.

SAFETY PROFILE: A very dangerous fire hazard. When heated to decomposition in air it emits toxic fumes of K_2O.

PKU250   CAS: 7789-29-9   **HR: 3**
POTASSIUM ACID FLUORIDE

DOT: NA 1811
mf: FK • FH   mw: 78.11

PROP: Colorless crystals. Mp: decomp.

SYNS: BIFLUORURE de POTASSIUM (FRENCH) * POTASSIUM BIFLUORIDE * POTASSIUM HYDROGEN FLUORIDE

CONSENSUS REPORTS: Reported in EPA TSCA Inventory.

NIOSH REL: TWA 2.5 mg(F)/m³

SAFETY PROFILE: A poison by all routes. Corrosive to the eyes, skin, and mucous membranes. A very reactive, dangerous material. When heated to decomposition it emits toxic fumes of F^- and K_2O.

PKV500   CAS: 10124-50-2   **HR: 3**
POTASSIUM ARSENITE

DOT: UN 1678
mf: AsH_3O_3 • xK   mw: 399.65

PROP: White, hygroscopic powder. Sol in water.

SYNS: ARSENENOUS ACID, POTASSIUM SALT * ARSENITE de POTASSIUM (FRENCH) * ARSONIC ACID, POTASSIUM SALT * KALIUMARSENIT (GERMAN) * NSC 3060 * POTASSIUM METAARSENITE

CONSENSUS REPORTS: IARC Cancer Review: Human Sufficient Evidence IMEMDT 23,39,80; Animal Inadequate Evidence IMEMDT 23,39,80, IMEMDT 2,48,73. EPA Extremely Hazardous Substances List. Arsenic and its compounds are on the Community Right-To-Know List.

OSHA PEL: TWA 0.01 mg(As)/m³; Cancer Hazard
ACGIH TLV: TWA 0.2 mg(As)/m³
NIOSH REL: CL (Inorganic Arsenic) 0.002 mg(As)/m³/15M
DOT Classification: Poison B; Label: Poison.

SAFETY PROFILE: Confirmed human carcinogen producing skin and liver tumors. Poison by ingestion, skin contact, subcutaneous, and intravenous routes. Human mutation data reported. Human systemic effects: dermatitis, liver changes. When heated to decomposition it emits toxic fumes of As and K_2O. Used in veterinary medicine and for chronic dermatitis in humans.

PKW760   CAS: 582-25-2   **HR: 2**
POTASSIUM BENZOATE
mf: $C_7H_5O_2$ • K   mw: 160.22

SAFETY PROFILE: Combustible when exposed to heat or flame. When heated to decomposition it emits acrid smoke and irritating fumes.

PKX250   CAS: 7778-50-9   **HR: 3**
POTASSIUM BICHROMATE

DOT: UN 1479
mf: $Cr_2K_2O_7$ mw: 294.20

PROP: Bright, yellowish-red, transparent crystals; bitter, metallic taste. Mp: 398°, bp: decomp @ 500°, d: 2.69.

SYNS: BICHROMATE OF POTASH * CHROMIC ACID, DIPOTASSIUM SALT * DIPOTASSIUM DICHRO-MATE * IOPEZITE * KALIUMDICHROMAT (GER-MAN) * POTASSIUM DICHROMATE(VI)

CONSENSUS REPORTS: IARC Cancer Review: Human Inadequate Evidence IMEMDT 23,205,80; Animal Inadequate Evidence IMEMDT 23,205,80. Chromium and its compounds are on the Community Right-To-Know List. Reported in EPA TSCA Inventory. EPA Genetic Toxicology Program.

OSHA PEL: CL 0.1 mg(CrO_3)/m^3
ACGIH TLV: TWA 0.05 mg(CrO_3)/m^3
NIOSH REL: TWA (Chromium(VI)) 0.025 mg(Cr(VI))/m^3; CL 0.05/15M
DOT Classification: ORM-A; Label: None.

SAFETY PROFILE: Human poison by ingestion. An experimental poison by ingestion, intraperitoneal, intravenous, and subcutaneous routes. Questionable carcinogen. Human mutation data reported. Flammable by chemical reaction. A powerful oxidizer. Explosive reaction with hydrazine. Reacts violently or ignites with H_2SO_4 + acetone, hydroxylamine, ethylene glycol (above 100°C). Forms pyrotechnic mixtures with boron + silicon, iron (ignites 1090°C), tungsten (ignites at 1700°C). Reacts with sulfuric acid to form the strong oxidant chromic acid. Used in photomechanical processing, chrome pigment production and wool preservation methods. When heated to decomposition it emits toxic fumes of K_2O.

PKX500   CAS: 23746-34-1   **HR: 3**
POTASSIUM BIS(2-HYDROXYETHYL) DITHIOCARBAMATE
mf: $C_5H_{10}NO_2S_2$ • K   mw: 219.38

SYNS: BIS(2-HYDROXYETHYL)CARBAMODITHIOIC ACID, MONOPOTASSIUM SALT * BIS(2-HYDROXY-ETHYL)DITHIOCARBAMIC ACID, MONOPOTASSIUM

SALT * BIS(2-HYDROXYETHYL)DITHOCARBAMIC ACID, POTASSIUM SALT

CONSENSUS REPORTS: IARC Cancer Review: GROUP 3 IMEMDT 7,56,87; Animal Sufficient Evidence IMEMDT 12,183,76. Reported in EPA TSCA Inventory.

SAFETY PROFILE: Questionable carcinogen with experimental carcinogenic and tumorigenic data. When heated to decomposition it emits very toxic fumes of K_2O, SO_x, and NO_x. Used as an analytical reagent for quantitative determination of mercury, gold, and copper.

PKX750   CAS: 7646-93-7   **HR: 2**
POTASSIUM BISULFATE

DOT: UN 2509
mf: HO_4S • K   mw: 136.17

PROP: White, deliquescent crystals. D: 2.24; mp: 197°. Sol in water.

SYNS: ACID POTASSIUM SULFATE * MONOPOTAS-SIUM SULFATE * POTASSIUM ACID SULFATE * POTASSIUM BISULPHATE * POTASSIUM HYDRO-GEN SULFATE, solid (DOT) * SAL ENIXUM * SULFURIC ACID, MONOPOTASSIUM SALT

CONSENSUS REPORTS: Reported in EPA TSCA Inventory.

DOT Classification: ORM-B; Label: None; Corrosive Material; Label: Corrosive.

SAFETY PROFILE: Moderately toxic by ingestion. A corrosive irritant to the skin, eyes and mucous membranes. When heated to decomposition it emits toxic fumes of SO_x and K_2O. Can form an explosive mixture.

PKY250   CAS: 13762-51-1   **HR: 3**
POTASSIUM BOROHYDRATE

DOT: UN 1870
mf: BH_3 • K   mw: 52.94

PROP: White, water-sol crystals. D: 1.177, mp: >400° (decomp).

SYNS: BOROHYDRURE de POTASSIUM (FRENCH) * POTASSIUM BOROHYDRIDE (DOT) * TETRAHYDROBORATE(1−) POTASSIUM

CONSENSUS REPORTS: Reported in EPA TSCA Inventory.

DOT Classification: Flammable Solid; Label: Dangerous When Wet.

SAFETY PROFILE: Poison by ingestion. Burns quietly in air. When heated to decomposition it emits toxic fumes of K_2O. .PT1

PKY300   CAS: 7758-01-2   **HR: 3**
POTASSIUM BROMATE

DOT: UN 1484
mf: $BrO_3 \cdot K$ mw: 167.01

PROP: White crystals or crystalline powder. Mp: 350° (approx), decomp @ 370°, d: 3.27 @ 17.5°. Sol in water; sltly sol in alc.

SYN: BROMIC ACID, POTASSIUM SALT

CONSENSUS REPORTS: IARC Cancer Review: GROUP 2B IMEMDT 7,56,87; Animal Sufficient Evidence IMEMDT 40,207,86. Reported in EPA TSCA Inventory.

DOT Classification: Oxidizer; Label: Oxidizer.

SAFETY PROFILE: Suspected carcinogen with experimental carcinogenic data. A poison by ingestion. A powerful oxidizer. An irritant to skin, eyes, and mucous membranes. Mutation data reported. Mixtures with sulfur may ignite. Violent reaction with Al, Al + dinitrotoluene @ 290°, As, C, Cu, $Pb(C_2H_3O_2)_2$, metal sulfides, organic matter, P, S. Aqueous solutions react violently with selenium. When heated to decomposition it emits very toxic fumes of Br^- and K_2O.

PKY500   CAS: 7758-02-3   **HR: 1**
POTASSIUM BROMIDE
mf: BrK   mw: 119.01

PROP: Colorless, cubic, sltly hygroscopic crystals. Mp: 730°, bp: 1380°, d: 2.75 @ 25°, vap press: 1 mm @ 795°.

SYN: BROMIDE SALT OF POTASSIUM

CONSENSUS REPORTS: Reported in EPA TSCA Inventory.

SAFETY PROFILE: Large doses can cause central nervous system depression. Prolonged inhalation may cause skin eruptions. Mutation data reported. Violent reaction with BrF_3. When heated to decomposition it emits toxic fumes of K_2O and Br^-.

PKY750   CAS: 865-47-4   **HR: 3**
POTASSIUM-tert-BUTOXIDE
mf: C_4H_9KO mw: 112.20

SAFETY PROFILE: Probably very toxic and irritating to skin, eyes, and mucous membranes.

A powerful very reactive base. Ignites on contact with acids or reactive solvents (e.g., acetone; butanone; butyl acetate; acetic acid; ethanol; propanol; isopropanol; methanol; $CHCl_3$; carbon tetrachloride; 1-chloro-2,3-epoxypropane; chloroform; 1,2-dichloromethane; diethyl sulfate; dimethyl carbonate; epichlorohydrin; ethyl acetate; ethyl methyl ketone; sulfuric acid; isopropanol; 4-methyl-2-butanone; methyl isobutyl ketone; n-butyl acetate; n-propyl formate; propanol; propyl formate). Ignites when heated in air. When heated to decomposition it emits toxic fumes of K_2O.

PLA000   CAS: 584-08-7   **HR: 3**
POTASSIUM CARBONATE (2:1)
mf: $CO_3 \cdot 2K$ mw: 138.21

PROP: White, deliquescent, granular, translucent powder; odorless with alkaline taste. D: 2.428 @ 19°, mp: 891°, bp: decomposes. Sol in water; insol in alc.

SYNS: CARBONIC ACID, DIPOTASSIUM SALT
* KALIUMCARBONAT (GERMAN) * K-GRAN
* PEARL ASH * POTASH

CONSENSUS REPORTS: Reported in EPA TSCA Inventory.

SAFETY PROFILE: Poison by ingestion. A strong caustic. Incompatible with KCO, chlorine trifluoride, magnesium. When heated to decomposition it emits toxic fumes of K_2O.

PLA250   CAS: 3811-04-9   **HR: 3**
POTASSIUM CHLORATE

DOT: UN 1485/UN 2427
mf: $ClO_3 \cdot K$ mw: 122.55

PROP: Transparent, colorless crystals or white powder; cooling, saline taste. Mp: 368.4°, bp: decomp @ 400°, d: 2.32.

SYNS: BERTHOLLET SALT * CHLORATE de POTASSIUM (FRENCH) * CHLORATE OF POTASH (DOT)
* FEKABIT * KALIUMCHLORAAT (DUTCH)
* KALIUMCHLORAT (GERMAN) * OXYMURIATE OF POTASH * PEARL ASH * POTASH CHLORATE (DOT) * POTASSIO (CHLORATO di) (ITALIAN)
* POTASSIUM CHLORATE (DOT) * POTASSIUM (CHLORATE de) (FRENCH) * POTASSIUM OXYMURIATE * POTCRATE * SALT OF TARTER

CONSENSUS REPORTS: Reported in EPA TSCA Inventory.

DOT Classification: Oxidizer; Label: Oxidizer.

SAFETY PROFILE: Moderately toxic to humans by an unspecified route. Moderately toxic experimentally by ingestion, and intraperitoneal routes. A gastrointestinal tract and kidney irritant. Can cause hemolysis of red blood cells and methemoglobinemia. Toxic dose to a human is about 5 grams.

A powerful oxidizer and very reactive material. It has been the cause of many industrial explosions. May explode on heating. Explosive reactions with ammonium chloride, aqua regia + rutheniun, sulfur dioxide solutions in ether or ethanol. Reacts with fluorine to form the explosive gas fluorine perchlorate.

Forms sensitive explosive mixtures with agricultural materials (e.g., peat, powdered sulfur, sawdust, thiuram), aluminum + antimony trisulfide powders, arsenic trisulfide, carbon, charcoal + potassium nitrate + sulfur, charcoal + sulfur, cyanides, cyanoguanidine, hydrocarbons, manganese dioxide + traces of organic matter, manganese dioxide + potassium hydroxide, metal + wood, metal phosphides (e.g., tricopper diphosphide, trimercury tetraphosphide), metal phosphinates (e.g., barium phosphinate), finely divided metals (e.g., aluminum, copper, magnesium, zinc, germanium, titanium, zirconium, steel, chromium), metal phosphides (e.g., tricopper diphosphide, trimercury tetraphosphide), metal sulfides (e.g., antimony trisulfide, silver sulfide), metal thiocyanates (e.g., ammonium thiocyanate, barium thiocyanate), nitric acid + organic materials, powdered non-metals (e.g., arsenic, carbon, phosphorus, sulfur, boron), reducing agents (e.g., calcium hydride, strontium hydride, sodium phosphinate, calcium phosphinate, barium phosphinate), sugars (e.g., glucose), sulfur, sulfur + metal derivatives (e.g., cobalt, cobalt oxide, copper nitride, copper sulfate, copper chlorate), sulfuric acid, sodium amide, tannic acid.

Violent reaction or ignition with NH_3, NH_4Cl, NH_4^+ salts, ammonium sulfate, Sb_2S_3, As, barium hypophosphite, BaS, calcium hypophosphite, CaS, charcoal, Cu_3P_2, fabrics, gallic acid, hydrogen iodide, lactose, (Mg + $CuSO_4$ (anhydrous) + NH_4NO_3 + H_2O), MnO_2, dinickel trioxide, dibasic organic acids, organic matter, $NaNH_2$, sugar + sulfuric acid, sucrose, SO_2, H_2SO_4, thiocyanates, thorium dicarbide, sodium amide, fabrics, KOH, metal hypophosphites.

When heated to decomposition it emits very toxic fumes of Cl^- and K_2O. Used in the manufacture of soap, glass, and pottery.

PLA500   CAS: 7447-40-7   **HR: 3**
POTASSIUM CHLORIDE
mf: ClK   mw: 74.55

PROP: Colorless or white crystals or powder; odorless with salty taste. D: 1.987, mp: 773° (sublimes @ 1500°). Sol in water; sltly sol in alc; insol in abs alc.

SYNS: CHLORID DRASELNY (CZECH) * CHLORO-POTASSURIL * DIPOTASSIUM DICHLORIDE * EMPLETS POTASSIUM CHLORIDE * ENSEAL * KALITABS * KAOCHLOR * KAON-Cl * KAY CIEL * K-LOR * KLOTRIX * K-PRENDE-DOME * PFIKLOR * POTASSIUM MONOCHLORIDE * POTAVESCENT * REKAWAN * SLOW-K * TRIPOTASSIUM TRICHLORIDE

CONSENSUS REPORTS: Reported in EPA TSCA Inventory

SAFETY PROFILE: A human poison by ingestion. Poison experimentally by ingestion, intravenous, and intraperitoneal routes. Moderately toxic by subcutaneous route. Human systemic effects by ingestion: nausea, blood clotting changes, cardiac arrhythmias. An eye irritant. Mutation data reported. Explosive reaction with BrF_3, sulfuric acid + potassium permanganate. When heated to decomposition it emits toxic fumes of K_2O and Cl^-.

PLB250   CAS: 7789-00-6   **HR: 3**
POTASSIUM CHROMATE(VI)
mf: $CrO_4 \cdot 2K$ mw: 194.20

PROP: Rhombic, yellow crystals. Mp: 975°, d: 2.73 @ 18°. Sol in water; insol in alc.

SYNS: BIPOTASSIUM CHROMATE * CHROMATE OF POTASSIUM * DIPOTASSIUM CHROMATE * DIPO-TASSIUM MONOCHROMATE * NEUTRAL POTASSIUM CHROMATE * TARAPACAITE

CONSENSUS REPORTS: IARC Cancer Review: Human Inadequate Evidence IMEMDT 23,205,80; Animal Inadequate Evidence IMEMDT 23,205,80. Reported in EPA TSCA Inventory. EPA Genetic Toxicology Program. Chromium and its compounds are on the Community Right-To-Know List.

OSHA PEL: CL 0.1 mg(CrO_3)/m³
ACGIH TLV: TWA 0.05 mg(Cr)/m³

NIOSH REL: TWA 0.025 mg(Cr(VI))/m^3; CL 0.05/15M

DOT Classification: ORM-E; Label: None.

SAFETY PROFILE: Poison by intravenous, subcutaneous, and intramuscular routes. Questionable carcinogen with experimental tumorigenic data. Human mutation data reported. A powerful oxidizer. When heated to decomposition it emits toxic fumes of K_2O. Used as a mordant for wool, in the oxidizing and treatment of dyes on materials.

PLB500   CAS: 10141-00-1   **HR: 3**
POTASSIUM CHROMIC SULFATE
mf: Cr • 2H$_2$O$_4$S • K   mw: 287.26

SYNS: CHROME ALUM * CHROME POTASH ALUM * CHROMIC POTASSIUM SULFATE * CHROMIC POTASSIUM SULPHATE * CHROMIUM POTASSIUM SULFATE (1:1:2) * CHROMIUM POTASSIUM SULPHATE * CRYSTAL CHROME ALUM * POTASSIUM CHROMIC SULPHATE * POTASSIUM CHROMIUM ALUM * POTASSIUM DISULPHATOCHROMATE(III) * SULFURIC ACID, CHROMIUM (3+) POTASSIUM SALT (2:1:1)

CONSENSUS REPORTS: Chromium and its compounds are on the Community Right-To-Know List. Reported in EPA TSCA Inventory. EPA Genetic Toxicology Program.

OSHA PEL: TWA 0.5 mg(Cr)/m^3
ACGIH TLV: TWA 0.5 mg(Cr)/m^3

SAFETY PROFILE: Many chromates are carcinogens. Mutation data reported. When heated to decomposition it emits toxic fumes of K_2O.

PLB750   CAS: 866-84-2   **HR: 3**
POTASSIUM CITRATE
mf: C$_6$H$_5$O$_7$ • 3K   mw: 306.41

PROP: Colorless transparent crystals or white powder; odorless with salty taste. D: 1.98, decomp when heated to 230°. Deliquescent, sol in water and glycerol; almost insol in alc.

SYNS: CITRIC ACID, TRIPOTASSIUM SALT * TRIPOTASSIUM CITRATE MONOHYDRATE

CONSENSUS REPORTS: Reported in EPA TSCA Inventory.

SAFETY PROFILE: Poison by intravenous route. When heated to decomposition it emits toxic fumes of K_2O.

PLC250   CAS: 590-28-3   **HR: 3**
POTASSIUM CYANATE
mf: CNO • K   mw: 81.12

PROP: Colorless crystals. Mp: 700-900° (decomp), d: 2.056 @ 20°. Sol in water; very sltly sol in alc.

SYNS: AERO CYANATE * ALICYANATE * BONIDE KRAB CRABGRASS KILLER * BULPUR * CYANIC ACID, POTASSIUM SALT * DED-WEED CRABGRASS KILLER * D & P DOUBLE O CRABGRASS KILLER * DUPONT PC CRABGRASS KILLER * GREEN CROSS CRABGRASS KILLER * KALIUM-CYANAT (GERMAN) * MILLER P.C. WEEDKILLER * P.C. 80 CRABGRASS KILLER * POTASSIUM ISO-CYANATE * WEEDANOL CYANOL * WEEDONE CRAB GRASS KILLER

CONSENSUS REPORTS: Reported in EPA TSCA Inventory. EPA Genetic Toxicology Program. Cyanide and its compounds are on the Community Right-To-Know List.

SAFETY PROFILE: Poison by intraperitoneal route. Moderately toxic by ingestion. Causes irritation of the gastrointestinal tract. An herbicide. It is said to be slowly metabolized in the body to cyanide but does not have high toxicity of cyanides. When heated to decomposition it emits very toxic fumes of CN$^-$ and K_2O.

PLC500   CAS: 151-50-8   **HR: 3**
POTASSIUM CYANIDE

DOT: UN 1680
mf: CN • K   mw: 65.12

PROP: Colorless water soln. Slt odor of bitter almonds.

SYNS: CYANIDE of POTASSIUM * CYANURE de POTASSIUM (FRENCH) * HYDROCYANIC ACID, POTASSIUM SALT * KALIUM-CYANID (GERMAN) * POTASSIUM CYANIDE, solution (DOT) * RCRA WASTE NUMBER P098

CONSENSUS REPORTS: Cyanide and its compounds are on the Community Right-To-Know List. Reported in EPA TSCA Inventory. EPA Genetic Toxicology Program.

OSHA PEL: TWA 5 mg(CN)/m^3
ACGIH TLV: TWA 5 mg(CN)/m^3 (skin)
DFG MAK: 5 mg(cn)/m^3
NIOSH REL: CL (Cyanide) 5 mg(CN)/m^3/10M
DOT Classification: Poison B; Label: Poison, solid and solution.

SAFETY PROFILE: A deadly human poison by ingestion. A experimental poison by ocular, subcutaneous, intravenous, intramuscular, and intraperitoneal routes. Experimental teratogenic and reproductive effects. Mutation data reported. Reacts with acids or acid fumes to liberate deadly HCN. When heated to decomposition it emits very toxic fumes of K_2O, CN^-, and NO_x.

PLC750   CAS: 151-50-8   **HR: 3**
POTASSIUM CYANIDE (solid)
mf: CN • K   mw: 65.12

PROP: White, deliquescent crystals; faint odor of bitter almonds. Mp: 622.5°, bp: 1625°, d: 1.52 @ 16°. Sol in water, glycerol; sltly sol in alc.

CONSENSUS REPORTS: Cyanide and its compounds are on the Community Right-To-Know List. EPA Extremely Hazardous Substances List. Reported in EPA TSCA Inventory.

DOT Classification: Poison B; Label: Poison.

SAFETY PROFILE: A deadly human poison by an unspecified route. Ingestion, inhalation or absorption through injured skin may cause poisoning. Strong solutions are corrosive to skin, eyes, and mucous membranes. Reacts with acids and acid fumes to liberate deadly HCN. Explosive reaction with nitrogen trichloride, sodium nitrite + heat, perchoryl fluoride (at 100-300°C), mercury(II) nitrate (if heated in a closed container). Incompatible with iodine. When heated to decomposition it emits very toxic fumes of K_2O, CN^-, and NO_x.

PLD000   CAS: 2244-21-5   **HR: 2**
POTASSIUM DICHLOROISOCYANU-RATE
DOT: NA 2465
mf: $C_3HCl_2N_3O_3$ • K   mw: 237.07

PROP: White, sltly hygroscopic, crystalline powder or granules; chlorine odor. Mp: 250° (decomp).

SYNS: DICHLOROISOCYANURIC ACID POTASSIUM SALT * DICHLORO-s-TRIAZINE-2,4,6(1H,3H,5H)-TRIONE POTASSIUM DERIV * DICHLOR-s-TRIAZIN-2,4,6(1H,3H,5H)TRIONE POTASSIUM * POTASSIUM DI-CHLORO-s-TRIAZINETRIONE

CONSENSUS REPORTS: Reported in EPA TSCA Inventory.

DOT Classification: Label: Oxidizer.

SAFETY PROFILE: Moderately toxic by ingestion. A skin and severe eye irritant. Causes

emaciation, weakness, lethargy, diarrhea, weight loss. Autopsy indicates gastrointestinal tract irritation, tissue edema, liver and kidney congestion. A powerful oxidizer. When heated to decomposition it emits very toxic fumes of K_2O, Cl^- and NO_x.

PLE260   CAS: 12030-88-5   **HR: 3**
POTASSIUM DIOXIDE
DOT: UN 2466
mf: KO_2 mw: 71.10

DOT Classification: Oxidizer; Label:Oxidizer

SAFETY PROFILE: Explosive reaction when heated with carbon, 2-aminophenol + tetrahydrofuran (at 65°C). Forms a friction-sensitive explosive mixture with hydrocarbons. Violent reaction with diselenium dichloride, ethanol, potassium-sodium alloy. May ignite on contact with organic compounds. Incandescent reaction with metals (e.g., arsenic, antimony, copper, potassium, tin, and zinc). When heated to decomposition it emits toxic fumes of K_2O.

PLF500   CAS: 7789-23-3   **HR: 3**
POTASSIUM FLUORIDE
DOT: UN 1812
mf: FK   mw: 58.10

PROP: White, crystalline, deliquescent powder; sharp saline taste. Bp: 1500°, d: 2.48, vap press: 1 mm @ 885°, mp: 859.9°. Very sol in boiling water.

SYNS: FLUORURE de POTASSIUM (FRENCH)
* POTASSIUM FLUORIDE, solution (DOT) * POTAS-SIUM FLUORURE (FRENCH)

CONSENSUS REPORTS: Reported in EPA TSCA Inventory. EPA Genetic Toxicology Program.

OSHA PEL: TWA 2.5 mg(F)/m^3
ACGIH TLV: TWA 2.5 mg(F)/m^3
NIOSH REL: TWA (Inorganic Fluorides) 2.5 mg(F)/m^3
DOT Classification: Corrosive Material; Label: Corrosive, solution; ORM-B; Label: None; Poison B; Label: St. Andrews Cross.

SAFETY PROFILE: Poison by ingestion and intraperitoneal routes. Moderately toxic by subcutaneous route. Experimental teratogenic effects. A corrosive irritant to the eyes, skin, and mucous membranes. A very reactive material. When heated to decomposition it emits toxic fumes of K_2O and F^-. Used in etching glass, a preservative, insecticide, and in organic synthesis.

PLG000   CAS: 23745-86-0   **HR: 3**
POTASSIUM FLUOROACETATE
DOT: UN 2628
mf: $C_2H_3FO_2 \cdot K$ mw:117.15

PROP: The potassium salt of monofluoroacetic acid was once designated as potassium cymonate.

SYNS: DICHAPETULUM CYMOSUM (HOOK) ENGL * GIFBLAAR

DOT Classification: Poison B; Label: Poison.

SAFETY PROFILE: Poison by ingestion, intravenous, parenteral, and subcutaneous routes. When heated to decomposition it emits toxic fumes of F^- and K_2O.

PLG800   CAS: 299-27-4   **HR: 2**
POTASSIUM GLUCONATE
mf: $C_6H_{12}O_7 \cdot K$ mw: 235.28

PROP: Yellowish-white crystals or powder; mild, sltly salty taste. Decomp at 180°. Freely sol in water, glycerin; practically insol in abs alc, ether, benzene, chloroform.

SYNS: d-GLUCONIC ACID, MONOPOTASSIUM SALT (9CI) * GLUCONIC ACID POTASSIUM SALT * GLUCONSAN K * KALIUM-BETA * KAON * KAON ELIXIR * KATORIN * K-IAO * POTALIUM * POTASORAL * POTASSIUM d-GLUCONATE * POTASSURIL * SIROKAL

CONSENSUS REPORTS: Reported in EPA TSCA Inventory.

SAFETY PROFILE: Moderately toxic by intraperitoneal route. Mildly toxic by ingestion. When heated to decomposition it emits toxic fumes of K_2O.

PLH750   CAS: 16871-90-2   **HR: 3**
POTASSIUM HEXAFLUOROSILICATE
DOT: UN 2655
mf: $F_6Si \cdot 2K$ mw: 220.29

PROP: White, fine powder or crystals. D: 2.27, mp: decomp. Sltly sol in cold water; insol in alc.

SYNS: POTASSIUM FLUOSILICATE * POTASSIUM SILICOFLUORIDE (DOT)

CONSENSUS REPORTS: Reported in EPA TSCA Inventory.

OSHA PEL: TWA 2.5 mg(F)/m^3
NIOSH REL: TWA (Inorganic Fluorides) 2.5 mg(F)/m^3
DOT Classification: Poison B; Label: St. Andrews Cross.

SAFETY PROFILE: A poison. Moderately toxic by ingestion and subcutaneous routes. Ingestion can cause vomiting and diarrhea. A strong irritant. Incompatible with hydrofluoric acid. When heated to decomposition it emits toxic fumes of K_2O and F^-. Used as a porcelain enamel frit, a ceramic and glass ingredient, flux, and sand inhibitor.

PLI000   CAS: 16919-27-0   **HR: 3**
POTASSIUM HEXAFLUOROTITANATE
mf: $F_6Ti \cdot 2K$ mw: 240.10

SYNS: FLUOTITANATE de POTASSIUM (FRENCH) * TITANIUM POTASSIUM FLUORIDE

CONSENSUS REPORTS: Reported in EPA TSCA Inventory.

NIOSH REL: TWA 2.5 mg(F)/m^3

SAFETY PROFILE: Poison by subcutaneous route. When heated to decomposition it emits toxic fumes of K_2O and F^-

PLJ250   CAS: 7693-26-7   **HR: 3**
POTASSIUM HYDRIDE
mf: HK   mw: 40.11

PROP: White needles. Mp: decomp, d: 1.43-1.47.

SAFETY PROFILE: Dangerous fire hazard by chemical reaction. Ignites spontaneously in air. Moderate explosion hazard when exposed to heat or by chemical reaction. Will react with water, steam or acids to produce H_2 which then ignites. Can react vigorously with oxidizing materials. To fight fire, use CO_2, dry chemical. Potentially explosive reactions with o-2,4-dinitrophenylhydroxylamine; fluoroalkenes. Ignites on contact with air; oxygen + moisture; fluorine. Incompatible with Cl_2; acetic acid; acrolein; acrylonitrile; (CaC + Cl_2); ClO_2; (H_2O_2 + Cl_2); ($CHFl_3$ + CH_3OH); 1,2-dichloroethylene; maleic anhydride; (n-methyl-n-nitrosourea + CH_2Cl_2); nitroethane; NCl_3; nitromethane; nitroparaffins; o-nitrophenol; nitropropane; n-nitrosomethylurea; (nitrosomethylurea + CH_2Cl_2);

H_2O; trichloroethylene; tetrahydrofuran; tetrachlorethane. When heated to decomposition it emits highly toxic fumes of K_2O.

PLJ500   CAS: 1310-58-3   **HR: 3**
POTASSIUM HYDROXIDE

DOT: UN 1813/UN 1814
mf: HKO   mw: 56.11

PROP: White, deliquescent pieces, lumps or sticks having crystalline fracture. Mp: $360° \pm 7°$, bp: 1320°, d: 2.044. Sol in water, alc.

SYNS: CAUSTIC POTASH * CAUSTIC POTASH, dry, solid, flake, bead, or granular (DOT) * CAUSTIC POTASH, liquid or solution (DOT) * HYDROXYDE de POTASSIUM (FRENCH) * KALIUMHYDROXID (GERMAN) * KALIUMHYDROXYDE (DUTCH) * LYE * POTASSA * POTASSE CAUSTIQUE (FRENCH) * POTASSIO (IDROSSIDO di) (ITALIAN) * POTASSIUM HYDRATE (DOT) * POTASSIUM HYDROXIDE, dry, solid, flake, bead, or granular (DOT) * POTASSIUM HYDROXIDE, liquid or solution (DOT) * POTASSIUM (HYDROXYDE de) (FRENCH)

CONSENSUS REPORTS: Reported in EPA TSCA Inventory.

OSHA PEL: CL 2 mg/m³
ACGIH TLV: CL 2 mg/m³
DOT Classification: Corrosive Material; Label: Corrosive; Label: Corrosive, solution.

SAFETY PROFILE: Poison by ingestion. An eye irritant and severe human skin irritant. Very corrosive to the eyes, skin, and mucous membranes. Mutation data reported. Ingestion may cause violent pain in throat and epigastrium, hematemesis, collapse. Stricture of esophagus may result if not immediately fatal. Above 84° it reacts with reducing sugars to form the poisonous carbon monoxide gas. Violent, exothermic reaction with water. Potentially explosive reaction with bromoform + crown ethers, chlorine dioxide, nitrobenzene, nitromethane, nitrogen trichloride, peroxidized tetrahydrofuran, 2,4,6-trinitrotoluene. Reaction with ammonium hexachloroplatinate(2−) + heat forms a heat-sensitive explosive product. Violent reaction or ignition under the appropriate conditions with acids, alcohols, p-bis(1,3-dibromoethyl)benzene, cyclopentadiene, germanium, hyponitrous acid, maleic anhydride, nitroalkanes, 2-nitrophenol, potassium peroxodisulphate, sugars, 2,2,3,3-tetrafluoropropanol, thorium dicarbide. When heated to decomposition it emits toxic fumes of K_2O.

PLJ750   CAS: 1310-58-3   **HR: 3**
POTASSIUM HYDROXIDE (solution)

DOT: UN 1813/UN 1814
mf: HKO   mw: 56.11

PROP: Clear liquid.

SYN: POTASSIUM HYDRATE (solution)

CONSENSUS REPORTS: Reported in EPA TSCA Inventory.

DOT Classification: Corrosive Material; Label: Corrosive.

SAFETY PROFILE: Very corrosive to the eyes, skin, and mucous membranes. When heated to decomposition it emits toxic fumes of K_2O.

PLK250   CAS: 7758-05-6   **HR: 3**
POTASSIUM IODATE
mf: $IO_3 \cdot K$ mw: 214.00

PROP: Colorless crystals or white crystalline powder. Mp: 560°, d: 3.89. Sol in water; insol in alc.

SYN: IODIC ACIODIC ACID, POTASSIUM SALT

CONSENSUS REPORTS: Reported in EPA TSCA Inventory.

SAFETY PROFILE: Poison by ingestion and intraperitoneal routes. A trace mineral added to animal feeds. Potentially explosive reaction with charcoal + ozone, metals (e.g., powdered aluminum, copper), arsenic carbon, phosphorus, sulfur, alkali metal hydrides, alkaline earth metal hydrides, antimony sulfide, arsenic sulfide, copper sulfide, tin sulfide, metal cyanides, metal thiocyanates, manganese dioxide, phosphorus. Violent reaction with organic matter. When heated to decomposition it emits very toxic fumes of I^- and K_2O.

PLK500   CAS: 7681-11-0   **HR: 2**
POTASSIUM IODIDE
mf: IK   mw: 166.00

PROP: Colorless or white granules. Mp: 723°, bp: 1420°, d: 3.13, vap press: 1 mm @ 745°. Sltly hygroscopic. Sol in water, glycerin, alc.

CONSENSUS REPORTS: Reported in EPA TSCA Inventory.

SAFETY PROFILE: Poison by intravenous route. Moderately toxic by ingestion and intraperitoneal routes. Human teratogenic effects by ingestion: developmental abnormalities of the endocrine system. Experimental teratogenic and reproductive effects. Mutation data reported. Explosive reaction with charcoal + ozone, trifluoroacetyl hypofluorite, fluorine perchlorate. Violent reaction or ignition on contact with diazonium salts, diisopropyl peroxydicarbonate, bromine pentafluoride, chlorine trifluoride. Incompatible with oxidants, BrF_3, FClO, metallic salts, calomel. When heated to decomposition it emits very toxic fumes of K_2O and I^-.

PLL500   CAS: 7757-79-1   HR: 3
POTASSIUM NITRATE

DOT: UN 1486
mf: KNO_3 mw: 101.11

PROP: Transparent, colorless or white crystalline powder or crystals; odorless with a cooling, pungent, salty taste. Mp: 334°, bp: decomp @ 400°, d: 2.109 @ 16°. Sol in glycerol, water; moderately sol in alc.

SYNS: KALIUMNITRAT (GERMAN) * NITER * NITRE * NITRIC ACID, POTASSIUM SALT * SALTPETER * VICKNITE

CONSENSUS REPORTS: Reported in EPA TSCA Inventory.

DOT Classification: Oxidizer; Label: Oxidizer.

SAFETY PROFILE: Poison by intravenous route. Moderately toxic by ingestion. Experimental teratogenic and reproductive effects. Mutation data reported. Ingestion of large quantities may cause gastroenteritis. Chronic exposure can cause anemia, nephritis, and methemoglobinemia. Heated reaction with calcium hydroxide + polychlorinated phenols forms extremely toxic chlorinated benzodioxins.

A powerful oxidizer. Gunpowder is a mixture of potassium nitrate + sulfur + charcoal. Explosive reaction with aluminum + barium nitrate + potassium perchlorate + water (in storage), boron + laminac + trichloroethylene. Forms explosive mixtures with lactose, powdered metals (e.g., titanium, antimony, germanium), metal sulfides (e.g., antimony trisulfide, barium sulfide, calcium sulfide, germanium monosulfide, titanium disulfide, arsenic disulfide, molybdenum disulfide), non-metals (e.g., boron, carbon, white phosphorus, arsenic), organic materials, phosphides (e.g., copper(II) phosphide, copper monophosphide), reducing agents (e.g., sodium phosphinate, sodium thiosulfate), sodium acetate. Can react violently un-

der the appropriate conditions with 1,3-bis(trichloromethyl)benzene, boron phosphide, F_2, calcium silicide, charcoal, chromium nitride, Na hypophosphite, (Na_2O_2 + dextrose), red phosphorus, (S + As_2S_3), thorium dicarbide, trichloroethylene, zinc, zirconium. When heated to decomposition it emits very toxic fumes of NO_x and K_2O.

PLM500   CAS: 7758-09-0   HR: 3
POTASSIUM NITRITE (1:1)

DOT: UN 1488
mf: $NO_2 \cdot K$ mw: 85.11

PROP: White or sltly yellowish, deliquescent prisms or sticks. Mp: 387°, bp: decomp, d: 1.915. Very sol in water; sltly sol in alc.

SYNS: NITROUS ACID, POTASSIUM SALT * POTASSIUM NITRITE (DOT)

CONSENSUS REPORTS: Reported in EPA TSCA Inventory.

DOT Classification: Oxidizer; Label: Oxidizer.

SAFETY PROFILE: Poison by ingestion. Experimental teratogenic and reproductive effects. Nitrites have been implicated in an increased incidence of cancer. Mutation data reported. Flammable when exposed to heat or flame. A powerful oxidizing material. Slight explosion hazard when exposed to heat. It will explode at 1000°F. Explosive reaction with potassium amide + heat, potassium cyanide or other cyanide salts + heat. Violent reaction or ignition with ammonium salts (e.g., ammonium sulfate), boron. Upon decomposition it emits toxic fumes of K_2O.

PLN100   CAS: 23705-25-1   HR: 3
POTASSIUM OCTACYANODICOBALTATE
mf: $C_8Co_2K_8N_8$ mw: 638.79

CONSENSUS REPORTS: Cyanide and its compounds, as well as cobalt and its compounds, are on the Community Right-To-Know List.

OSHA PEL: TWA 5 mg(CN)/m^3
ACGIH TLV: TWA 5 mg(CN)/m^3 (skin)
DFG MAK: 5 mg/m^3
NIOSH REL: (Cyanide) CL 5 mg(CN)/m^3/10M

SAFETY PROFILE: Many cyanide compounds are poisons. Ignites spontaneously in air. A very unstable material. When heated to decomposition it emits toxic fumes of CN^-, NO_x, and K_2O.

PLO500　　CAS: 7778-74-7　　HR: 3
POTASSIUM PERCHLORATE
DOT: UN 1489
mf: $ClO_4 \cdot K$　　mw: 138.55

PROP: Colorless crystals or white powder. Decomp @ 400° and with organic matter. D: 2.52, mp: 610° ± 10°. Insol in alc.

SYNS: PERIODIN　*　POTASSIUM HYPERCHLORIDE

CONSENSUS REPORTS: Reported in EPA TSCA Inventory.

DOT Classification: Oxidizer; Label: Oxidizer.

SAFETY PROFILE: Experimental reproductive effects. A powerful oxidizer. Severe irritant to skin, eyes, and mucous membranes. Has been implicated in aplastic anemia. Absorption can cause methemoglobinemia and kidney injury.

It has been involved in many industrial explosions. Explodes on contact with aluminum + barium nitrate + potassium nitrate + water. Forms explosive mixtures with aluminum powder + titanium dioxide, ethylene glycol (240°C), cotton lint (245°C), furfural (270°C), lactose, metal powders (e.g., aluminum, iron, magnesium, molybdenum, nickel, tantalum, titanium), sulfur, titanium hydride. Reaction with ethanol + heat forms the explosive ethyl perchlorate. Violent reaction or ignition under the proper conditions with aluminum + aluminum fluoride, barium chromate + tungsten or titanium, boron + magnesium + silicone rubber, ferrocenium diamminetetrakis(thiocyanato-N) chromate(1−), potassium hexacyanocobaltate(3−), Al + Mg, charcoal, F_2, Ni + Ti, reducing agents. When heated to decomposition it emits very toxic fumes of K_2O and Cl^-

PLP000　　CAS: 7722-64-7　　HR: 3
POTASSIUM PERMANGANATE
DOT: UN 1490
mf: $MnO_4 \cdot K$　　mw: 158.04

PROP: Dark purple crystals with a blue metallic sheen; sweetish astringent taste. Mp: decomp @ <240°, d: 2.703.

SYNS: CAIROX　*　CHAMELEON MINERAL　*　C.I. 77755　*　CONDY'S CRYSTALS　*　KALIUMPERMANGANAAT (DUTCH)　*　KALIUMPERMANGANAT (GERMAN)　*　PERMANGANATE de POTASSIUM (FRENCH)　*　PERMANGANATE of POTASH (DOT)　*　POTASSIO (PERMANGANATO di) (ITALIAN)　*　POTASSIUM (PERMANGANATE de) (FRENCH)

CONSENSUS REPORTS: Manganese and its compounds are on the Community Right-To-Know List. Reported in EPA TSCA Inventory. EPA Genetic Toxicology Program.

OSHA PEL: CL 5 mg(Mn)/m^3
ACGIH TLV: TWA 5 mg(Mn)/m^3
DOT Classification: Oxidizer; Label: Oxidizer.

SAFETY PROFILE: A human poison by ingestion. Poison experimentally by ingestion and intravenous routes. Moderately toxic by subcutaneous route. Human systemic effects by ingestion: dyspnea, nausea, other gastrointestinal effects. Experimental reproductive effects. Mutation data reported. A strong irritant due to its oxidizing properties. Used as a topical antibacterial agent, a chemical reagent.

Flammable by chemical reaction. A powerful oxidizer. A dangerous explosion hazard; handle with care. Explosions may occur in contact with organic or readily oxidizable materials, either when dry or in solution. Dangerous; keep away from combustible materials.

Explodes on contact with acetic acid, acetic anhydride, ammonium nitrate, dimethylformamide, formaldehyde, concentrated hydrochloric acid, potassium chloride + sulfuric acid, sulfuric acid + water. Forms sensitive explosive mixtures with aluminum powder + ammonium nitrate + glyceryl nitrate + nitrocellulose, ammonium perchlorate, arsenic, phosphorus, sulfur, slag wool, titanium.

Ignites on contact with Al_4C_3, dimethyl sulfoxide, ethylene glycol, H_2S_3, HCl, H_2SO_4, (H_2SO_4 + organic matter), (H_2SO_4 + KCl), NH_4ClO_4, NH_3, NH_4, NO_3, NH_2OH, organic matter, wood, oxygenated organic compounds (e.g., ethylene glycol, propane-1,2-diol, erythritol, mannitol, triethanolamine, 3-chloropropane-1,2-diol, acetaldehyde, isobutyraldehyde, benzaldehyde, acetylacetone, esters of ethylene glycol, lactic acid, acetic acid, oxalic acid).

Violent reaction or ignition under the proper conditions with acetone + tert-butylamine, alcohols + nitric acid, aluminum carbide, ammonia + sulfuric acid, antimony, coal + peroxomonosulfuric acid, dichloromethylsilane, dimethyl sulfoxide, ethanol + sulfuric acid, glycerol, concentrated hydrofluoric acid, hydrogen peroxide, hydrogen trisulfide, hydroxylamine, carbon, organic nitro compounds, polypropylene, 3,4,4′-trimethyldiphenyl sulfone.

When heated to decomposition it emits toxic fumes of K_2O.

PLP250   CAS: 17014-71-0   HR: 3
POTASSIUM PEROXIDE

DOT: UN 1491
mf: KO_2 mw: 71.1

PROP: Yellow, amorphous mass (white crystals). Mp: 490°.

CONSENSUS REPORTS: Reported in EPA TSCA Inventory.

DOT Classification: Oxidizer; Label: Oxidizer.

SAFETY PROFILE: Dangerous fire hazard by spontaneous chemical reaction. It is a very powerful oxidizer. Fires of this material should be handled like sodium peroxide fires. Moderate explosion hazard by spontaneous chemical reaction. Explodes on contact with water. Violent reactions with air, Sb, As, O_2, K. Vigorous reaction on contact with reducing materials. On contact with acid or acid fumes, it can emit toxic fumes. Incompatible with carbon, diselenium dichloride, ethanol, hydrocarbons, metals. When heated to decomposition it emits toxic fumes of K_2O.

PLQ500   CAS: 20770-41-6   HR: 2
POTASSIUM PHOSPHIDE

DOT: UN 2012
mf: K_3P mw: 148.27

PROP: A solid.

SYN: PHOSPHURE de POTASSIUM (FRENCH)

CONSENSUS REPORTS: Reported in EPA TSCA Inventory.

DOT Classification: Flammable Solid; Label: Dangerous When Wet, Poison

SAFETY PROFILE: Moderately toxic by inhalation. When heated to decomposition it emits very toxic fumes of PO_x, K_2O, and PH_3.

PLR250   CAS: 16731-55-8   HR: 3
POTASSIUM PYROSULFITE

DOT: NA 2693
mf: $O_5S_2 \cdot K$ mw: 183.22

PROP: Monoclinic plates or white crystalline powder; sulfur dioxide odor. Mp: decomp; d: 2.3. Sol in water; insol in alc.

SYNS: POTASSIUM METABISULFITE (DOT, FCC)
* PYROSULFUROUS ACID, DIPOTASSIUM SALT

CONSENSUS REPORTS: Reported in EPA TSCA Inventory. EPA Genetic Toxicology Program.

DOT Classification: ORM-B; Label: None.

SAFETY PROFILE: Experimental reproductive effects. A very irritating material. Questionable carcinogen with experimental tumorigenic data. When heated to decomposition it emits toxic fumes of SO_x and K_2O.

PLR750   CAS: 7790-59-2   HR: 3
POTASSIUM SELENATE
mf: $O_4Se \cdot 2K$ mw: 221.16

PROP: Colorless crystals. D: 3.07. Sol in water.

SYN: SELENIC ACID, DIPOTASSIUM SALT

CONSENSUS REPORTS: Selenium and its compounds are on the Community Right-To-Know List. Reported in EPA TSCA Inventory. EPA Genetic Toxicology Program.

OSHA PEL: TWA 0.2 mg(Se)/m^3
ACGIH TLV: TWA 0.2 mg(Se)/m^3
DFG MAK: 0.1 mg(Se)/m^3

SAFETY PROFILE: Poison by intravenous route. Moderately toxic by ingestion. Experimental reproductive effects. When heated to decomposition it emits toxic fumes of Se and K_2O.

PLS500   CAS: 11135-81-2   HR: 3
POTASSIUM SODIUM ALLOY

DOT: UN 1422

PROP: Low-melting alloy of sodium and potassium metals.

SYN: SODIUM POTASSIUM ALLOY, liquid and solid (DOT)

DOT Classification: Label: Flammable Solid; Label: Flammable Solid and Dangerous When Wet.

SAFETY PROFILE: A low melting alloy of Na and K. Its toxicity is due to either Na or K alone. Corrosive to the eyes, skin, and mucous membranes. Upon contact with moisture it reacts violently to evolve H_2, much heat, and a highly caustic residue of NaOH or KOH. Oxidation forms Na_2O and K_2O which are powerful caustics.

A dangerous fire and explosion hazard. Violent or explosive reaction with O_2, water, moisture, steam, halogens, oxidizers, acids or acid fumes, giving off much heat, hydrogen, toxic and corrosive fumes, often spattering either red-hot particles or actually flaming particles. A severe explosion hazard, will react explosively under the appropriate conditions with moisture, acids, acid fumes, solid CO_2, carbon disulfide, halocarbons (e.g., CH_3Cl, carbon tetrachloride, chloroform, bromoform, 1,1,1-trichloroethane, 1,1,2-trichlorotrifluoroethane, tetrachloroethane, CH_2Cl_2, CH_2I_2), ammonium sulfate + NH_4 + NO_3, HgO, metal halides (e.g., silver halides, zinc chloride, iron(III) chloride), metal oxides (e.g., silver oxide, mercury oxide), nitrogen containing explosives (e.g., ammonium nitrate, ammonium sulfate, picric acid, nitrobenzene), oxalyl bromide, oxalyl chloride, pentachloroethane, K oxides, KO_2, Si, $NaHCO_3$, polytetrafluoroethylene. Reacts vigorously with oxidizing materials.

To fight fire use G-1 powder, dry sodium chloride, dry sodium carbonate, dry calcium carbonate, dry sand, resin-coated sodium chloride, or dry soda ash. Never use water, graphite, carbon dioxide, halocarbons or foam.

Dangerous; when heated it emits highly toxic fumes of Na_2O and K_2O. Used as a liquid coolant for nuclear reactor cores.

PLS750 CAS: 590-00-1 **HR: 2**
POTASSIUM SORBATE
mf: $C_6H_7O_2 \cdot K$ mw: 150.23

PROP: White crystals, crystalline powder, or pellets. Mp: 270° (decomp): d: 1.363 @ 25°/20°. Sol in alc, water.

SYNS: 2,4-HEXADIENOIC ACID POTASSIUM SALT * SORBIC ACID, POTASSIUM SALT * SORBISTAT-K * SORBISTAT-POTASSIUM

CONSENSUS REPORTS: Reported in EPA TSCA Inventory. EPA Genetic Toxicology Program.

SAFETY PROFILE: Moderately toxic by intraperitoneal route. Mildly toxic by ingestion. Mutation data reported. When heated to decomposition it emits toxic fumes of K_2O.

PLT000 CAS: 7778-80-5 **HR: 2**
POTASSIUM SULFATE (2:1)
mf: $O_4S \cdot 2K$ mw: 174.26

PROP: Colorless to white, odorless crystals; bitter salty taste. D: 2.66, mp: 1067°. Sol in water; insol in alc.

SYN: SULFURIC ACID, DIPOTASSIUM SALT

CONSENSUS REPORTS: Reported in EPA TSCA Inventory.

SAFETY PROFILE: Moderately toxic to humans by ingestion. Moderately toxic experimentally by subcutaneous route. Swallowing large doses causes severe gastrointestinal tract effects. When heated to decomposition it emits toxic fumes of K_2O and SO_x.

PLT250 CAS: 1312-73-8 **HR: 3**
POTASSIUM SULFIDE (2:1)

DOT: UN 1382/UN 1847
mf: K_2S mw: 110.26

PROP: Red, crystalline mass; deliquescent in air. Mp: 912°, d: 1.805 @ 14°.

SYNS: HEPAR SULFUROUS * POTASSIUM MONOSULFIDE

CONSENSUS REPORTS: Reported in EPA TSCA Inventory.

DOT Classification: Flammable Solid; Label: Flammable Solid.

SAFETY PROFILE: Poison by ingestion and inhalation. Emits H_2S in contact with acids, steam. A flammable solid. Unstable; may explode on percussion or rapid heating. Ignites on contact with nitrogen oxide. When heated to decomposition it emits very toxic fumes of K_2O and SO_x.

PLT500 CAS: 10117-38-1 **HR: D**
POTASSIUM SULFITE
mf: $O_3S \cdot 2K$ mw: 158.26

PROP: White crystals or granular powder; odorless. Sol in water; sltly sol in alc.

SYN: SULFUROUS ACID, DIPOTASSIUM SALT

CONSENSUS REPORTS: Reported in EPA TSCA Inventory.

SAFETY PROFILE: When heated to decomposition it emits toxic fumes of SO_x and K_2O.

PLU500 CAS: 591-89-9 **HR: 3**
POTASSIUM TETRACYANOMERCURATE(II)

DOT: UN 1626
mf: $C_2HgN_2 \cdot 2CKN$ mw: 382.87

SYNS: MERCURIC POTASSIUM CYANIDE (DOT) * MERCURIC POTASSIUM CYANIDE, solid (DOT)

CONSENSUS REPORTS: Cyanide and its compounds, as well as mercury and its compounds, are on the Community Right-To-Know List.

OSHA PEL: (Transitional: CL 1 mg/10m^3) CL 0.1 mg(Hg)/m^3 (skin)
ACGIH TLV: TWA 0.1 mg(Hg)/m^3 (skin)
NIOSH REL: TWA (Inorganic Mercury) 0.05 mg(Hg)/m^3
DOT Classification: Poison B; Label: Poison.

SAFETY PROFILE: A poison. May explode on contact with ammonia. When heated to decomposition it emits very toxic fumes of NO$_x$, Hg, K$_2$O, and CN$^-$.

PLV750   CAS: 333-20-0   **HR: 3**
POTASSIUM THIOCYANATE
mf: CNS • K   mw: 97.18

PROP: Colorless, deliquescent crystals. D: 1.89, mp: about 173°.

SYNS: POTASSIUM RHODANATE * POTASSIUM RHODANIDE * POTASSIUM SULFOCYANATE * POTASSIUM THIOCYANIDE

CONSENSUS REPORTS: Reported in EPA TSCA Inventory.

SAFETY PROFILE: A human poison by ingestion. Poison experimentally by ingestion, intramuscular, subcutaneous, and intravenous routes. Large doses can cause skin eruptions, psychoses and collapse. Incompatible with calcium chlorite and perchloryl fluoride. When heated to decomposition it emits very toxic fumes of CN$^-$, K$_2$O, SO$_x$, and NO$_x$.

PLW500   CAS: 11103-86-9   **HR: 3**
POTASSIUM ZINC CHROMATE HYDROXIDE
mf: Cr$_2$HO$_9$Zn$_2$ • K   mw: 418.85

SYNS: BUTTERCUP YELLOW * CHROMIC ACID, POTASSIUM ZINC SALT (2:2:1) * CITRON YELLOW * POTASSIUM ZINC CHROMATE * ZINC CHROME * ZINC YELLOW

CONSENSUS REPORTS: IARC Cancer Review: Animal Inadequate Evidence IMEMDT 23,205,80. Chromium and its compounds, as well as zinc and its compounds, are on the Community Right-To-Know List. Reported in EPA TSCA Inventory.

OSHA PEL: (Transitional: 1 mg(CrO$_3$)/10m^3) CL 0.1 mg(CrO$_3$)/m^3
ACGIH TLV: TWA 0.01 mg(Cr)/M^3; Confirmed Human Carcinogen
DFG MAK: Human Carcinogen.
NIOSH REL: (Chromium (VI)) TWA 0.001 mg(Cr(VI))/m^3

SAFETY PROFILE: Confirmed carcinogen. When heated to decomposition it emits toxic fumes of ZnO and K$_2$O. Used as a corrosion inhibiting pigment and in steel priming.

PLX500   CAS: 7440-10-0   **HR: 2**
PRASEODYMIUM
af: Pr   aw: 140.9077

PROP: Yellowish metal. Mp: 935°, bp: 3290°, d: (a) 6.772, (b) 6.64.

SAFETY PROFILE: As a lanthanon, it may depress coagulation of the blood. Limited data suggest moderate toxicity. Flammable in the form of dust when exposed to heat or flame or by chemical reaction. Fine dust ignites readily. Incompatible with air or halogens.

PLX750   CAS: 10361-79-2   **HR: 3**
PRASEODYMIUM CHLORIDE
mf: Cl$_3$Pr   mw: 247.26

CONSENSUS REPORTS: Reported in EPA TSCA Inventory.

SAFETY PROFILE: Poison by intraperitoneal, subcutaneous, and intravenous routes. Moderately toxic by ingestion. A skin and eye irritant. When heated to decomposition it emits toxic fumes of Cl$^-$.

PLY250   CAS: 10361-80-5   **HR: 3**
PRASEODYMIUM(III) NITRATE (1:3)
mf: N$_3$O$_9$ • Pr   mw: 326.94

SYN: NITRIC ACID, PRASEODYMIUM(3+) SALT

CONSENSUS REPORTS: Reported in EPA TSCA Inventory.

SAFETY PROFILE: Poison by intraperitoneal and intravenous routes. Moderately toxic by ingestion. When heated to decomposition it emits toxic fumes of NO$_x$.

PLZ000   CAS: 53-03-2   **HR: 3**
PREDNISONE
mf: C$_{21}$H$_{26}$O$_5$ mw: 358.47

PROP: White, odorless, crystalline powder. Mp: 235° (with some decomp). Very sltly sol in water; sltly sol in alc, chloroform, methanol, and dioxane.

SYNS: ANCORTONE * BICORTONE * COLISONE * CORTAN * CORTANCYL * Δ-CORTELAN * CORTIDELT * Δ-CORTISONE * Δ¹-CORTISONE * Δ-CORTONE * COTONE * DACORTIN * DECORTANCYL * DECORTIN * DECORTISYL * Δ-1-DEHYDROCORTISONE * 1-DEHYDROCORTISONE * DEKORTIN * DELTACORTELAN * DELTACORTISONE * DELTACORTONE * DELTA-DOME * DELTISONE * 17,21-DIHYDROXYPREGNA-1,4-DIENE-3,11,20-TRIONE * ENCORTON * HOSTACORTIN * IN-SONE * JUVASON * LISACORT * METACORTANDRACIN * NCI-C04897 * NSC 10023 * ORASONE * PARACORT * PRECORT * PREDNICEN-M * PREDNILONGA * PREDNISON * PREDNIZON * 1,4-PREGNADIENE-17-α,21-DIOL-3,11,20-TRIONE * RECTODELT * SERVISONE * SK-PREDNISONE * SUPERCORTIL * U 6020 * ULTRACORTEN * WOJTAB * ZENADRID (VETERINARY)

CONSENSUS REPORTS: IARC Cancer Review: GROUP 3 IMEMDT 7,326,87; Human Inadequate Evidence IMEMDT 26,293,81; Animal Inadequate Evidence IMEMDT 26,293,81. NCI Carcinogenesis Studies (ipr); No Evidence: mouse CANCAR 40,1935,77; (ipr); Equivocal Evidence: rat CANCAR 40,1935,77. Reported in EPA TSCA Inventory.

SAFETY PROFILE: Poison by intraperitoneal and subcutaneous routes. Moderately toxic by intramuscular route. Human systemic effects by ingestion and possibly other routes: sensory change involving peripheral nerves. Experimental reproductive effects. Questionable carcinogen with experimental tumorigenic data. Mutation data reported. Has been implicated in aplastic anemia.

PMA000   CAS: 50-24-8   **HR: 3**
PREDONIN
mf: $C_{21}H_{28}O_5$ mw: 360.49

SYNS: CODELCORTONE * CO-HYDELTRA * Δ¹-CORTISOL * DECORTIN H * Δ¹-DEHYDROCORTISOL * Δ¹-DEHYDROHYDROCORTISONE * 1-DEHYDROHYDROCORTISONE * DELCORTOL * DELTA-CORTEF * DELTACORTENOL * DELTA-CORTRIL * DELTA F * DELTA-STAB * DEXACORTIDELT HOSTACORTIN H * DI-ADRESON F * DICORTOL * DYDELTRONE * FERNISOLONE * HOSTACORTIN * HYDELTRA * HYDELTRONE * Δ¹-HYDROCORTISONE * HYDRODELTALONE

* HYDRODELTISONE * HYDRORETROCORTIN * METACORTANDRALONE * METICORTELONE * METI-DERM * PARACORTOL * PARACOTOL * PRECORTANCYL * PRECORTISYL * PREDNEDOME * PREDNELAN * PREDNIS * PREDNISOLONE * PREDONINE * 1,4-PREGNADIENE-3,20-DIONE-11-β,17-α,21-TRIOL * 1,4-PREGNADIENE-11-β,17-α,21-TRIOL-3,20-DIONE * SCHERISOLON * STERANE * STEROLONE * 11-β,17,21-TRIHYDROXYPREGNA-1,4-DIENE-3,20-DIONE * 11-β,17-α,21-TRIHYDROXYPREGNA-1,4-DIENE-3,20-DIONE * 11-β,17-α,21-TRIHYDROXY-1,4-PREGNADIENE-3,20-DIONE * ULACORT * ULTRACORTENE-H

CONSENSUS REPORTS: Reported in EPA TSCA Inventory. EPA Genetic Toxicology Program.

SAFETY PROFILE: A poison by intravenous and subcutaneous routes. Moderately toxic by ingestion and intraperitoneal routes. Human teratogenic effects by an unspecified route: developmental abnormalities of the central nervous system; effects on embryo or fetus: fetal death, extra embryonic structures. Human reproductive effects by an unspecified route: stillbirth. Experimental teratogenic and reproductive effects. Human mutation data reported. When heated to decomposition it emits acrid smoke and irritating fumes.

PMB000   CAS: 12126-59-9   **HR: 3**
PREMARIN

SYNS: CEE * CONJUGATED EQUINE ESTROGEN * ESTROGENS, CONJUGATES

CONSENSUS REPORTS: IARC Cancer Review: Human Limited Evidence IMEMDT 21,147,79; Animal Inadequate Evidence IMEMDT 21,147,79. NTP Fourth Annual Report On Carcinogens, 1984.

SAFETY PROFILE: Confirmed carcinogen with experimental tumorigenic data. Poison by intraperitoneal route. Human reproductive effects by ingestion: changes in female fertility. Experimental teratogenic effects. A steroid. When heated to decomposition it emits acrid smoke and irritating fumes.

PMC250   CAS: 968-58-1   **HR: 3**
PRIDINOL HYDROCHLORIDE
mf: $C_{20}H_{25}NO \cdot ClH$ mw: 331.92

PROP: Crystals. Decomp @ 238°. Sol in alc.

SYNS: α,α-DIPHENYL-1-PIPERIDINEPROPANOL HYDROCHLORIDE * 1,1-DIPHENYL-3-PIPERIDINO-1-PRO-

PANOL HYDROCHLORIDE * 1,1-DIPHENYL-3-(1-PIPER-IDYL)-1-PROPANOL HYDROCHLORIDE * MITANOLINE * PAR KS-12 * α-(2-PIPERIDYLETHYL)BENZHYDROL HYDROCHLORIDE * PRIDINOL

SAFETY PROFILE: Poison by intraperitoneal and intravenous routes. When heated to decomposition it emits very toxic fumes of NO_x and HCl.

PME250   CAS: 671-16-9   **HR: 3**
PROCARBAZINE
mf: $C_{12}H_{19}N_3O$ mw: 221.34

SYNS: IBENZMETHYZINE * 2-(p-ISOPROPYL CAR-BAMOYL BENZYL)-1-METHYLHYDRAZINE * N-ISO-PROPYL-α-(2-METHYLHYDRAZINO)-p-TOLUAMIDE * MATULANE * 4-((2-METHYLHYDRAZINO)METHYL)-N-ISOPROPYLBENZAMIDE * 1-METHYL-2-(-ISOPROPYLCARBAMOYL)BENZYL)HYDRAZINE * MIH * NATULAN * NSC-77213 * PCB * RO 4-6467

CONSENSUS REPORTS: NTP Fourth Annual Report On Carcinogens, 1984.

SAFETY PROFILE: Confirmed carcinogen with experimental carcinogenic, neoplastigenic, and tumorigenic data. Poison by intravenous route. Moderately toxic by intraperitoneal route. Has been implicated as a brain carcinogen. Experimental teratogenic and reproductive effects. Mutation data reported. When heated to decomposition it emits toxic fumes of NO_x.

PME500   CAS: 366-70-1   **HR: 3**
PROCARBAZINE HYDROCHLORIDE
mf: $C_{12}H_{19}N_3O \cdot ClH$ mw: 257.80

PROP: Crystals. Mp: 223-236°.

SYNS: IBENZMETHYZINE HYDROCHLORIDE * IBENZMETHYZIN HYDROCHLORIDE * IBZ * 1-(p-ISOPROPYLCARBAMOYLBENZYL)-2-METHYLHYDRAZINE HYDROCHLORIDE * 2-(p-(ISO-PROPYLCARBAMOYL)BENZYL)-1-METHYLHYDRAZINE HYDROCHLORIDE * N-ISOPROPYL-p-(2-METH-YLHYDRAZINOMETHYL)BENZAMIDEHYDROCHLORIDE * N-ISOPROPYL-α-(2-METHYLHYDRAZINO)-p-TOLUAM-IDE HYDROCHLORIDE * MATULANE * MBH * N-(1-METHYLETHYL)-4-((2-METHYLHYDRAZINO)-METHYL)BENZAMIDE MONOHYDROCHLORIDE * p-(N'-METHYLHYDRAZINOMETHYL)-N-ISOPROPYL)-BENZAMIDE * p-(N'-METHYLHYDRAZINOMETHYL)-N-ISOPROPYLBENZAMIDE HYDROCHLORIDE * 1-METHYL-2-p-(ISOPROPYLCARBAMOYL)BENZOHY-DRAZINE HYDROCHLORIDE * 1-METHYL-2-(p-ISOPRO-

PYLCARBAMOYLBENZYL)HYDRAZINE HYDROCHLORIDE * MIH HYDROCHLORIDE * NATHULANE * NATULAN * NATULANAR * NATULAN HY-DROCHLORIDE * NCI-C01810 * NSC-77213 * PCB HYDROCHLORIDE * PROCARBAZIN (GERMAN) * RO 4-6467

CONSENSUS REPORTS: IARC Cancer Review: GROUP 2A IMEMDT 7,327,87; Human Limited Evidence IMEMDT 26,311,81; Animal Sufficient Evidence IMEMDT 26,311,81. NTP Fourth Annual Report On Carcinogens, 1984. NCI Carcinogenesis Bioassay (ipr); Clear Evidence: mouse, rat NCITR* NCI-CG-TR-19,79; (ipr); Clear Evidence: mouse, rat RRCRBU 52,1,75. EPA Genetic Toxicology Program.

SAFETY PROFILE: Confirmed carcinogen with experimental carcinogenic, neoplastigenic, and tumorigenic data. Poison by an unspecified route. Moderately toxic by ingestion, subcutaneous, intravenous, intraperitoneal, and possibly other routes. Experimental teratogenic and reproductive effects. Mutation data reported. When heated to decomposition it emits very toxic fumes of NO_x and HCl. Used as a chemotherapeutic agent.

PMF250   CAS: 84-02-6   **HR: 3**
PROCHLORPERAZINE HYDROGEN MALEATE
mf: $C_{20}H_{24}ClN_3S \cdot 2C_4H_4O_4$ mw: 606.14

SYNS: 2-CHLORO-10-(3-(4-METHYL-1-PIPERAZINYL)-PROPYL-10H-PHENOTHIAZINE-(Z)-2-BUTENEDIOATE (1:2) * 2-CHLORO-10-(3-(1-METHYL-4-PIPERAZINYL)PROPYL)PHENOTHIAZINE, DIMALEATE * 2-CHLORO-10-(3-(4-METHYL-1-PIPERAZINYL)PROPYL)PHENO-THIAZINE DIMALEATE * 2-CHLORO-10-(3-(4-METHYL-1-PIPERAZINYL)PROPYL)PHENOTHIAZINE MALEATE * COMPAZINE * EMETIRAL * METERAZIN MALEATE * PASOTOMIN * PROCHLOROPRO-AZINE HYDROGEN MALEATE * PROCHLORPERAZINE BIMALEATE * PROCHLORPERAZINE DIMALEATE * PROCHLORPERAZINE MALEATE * PROCHLOR-PERIZINE MALEATE * STEMETIL DIMALEATE

SAFETY PROFILE: Poison by ingestion, intravenous, and subcutaneous routes. When heated to decomposition it emits very toxic fumes of Cl^-, NO_x, and SO_x.

PMF500   CAS: 58-38-8   **HR: 3**
PROCHLORPROMAZINE
mf: $C_{20}H_{24}ClN_3S$ mw: 373.98

SYNS: CHLORO-3 (N-METHYLPIPERAZINYL-3 PRO-
PYL)-10 PHENOTHIAZINE (FRENCH) * 2-CHLORO-10-(3-
(1-METHYL-4-PIPERAZINYL)-PROPYL)-PHENOTHIAZINE
* 2-CHLORO-10-(3-(4-METHYL-1-PIPERAZINYL)PROPYL)
PHENOTHIAZINE * 3-CHLORO-10-(3-(1-METHYL-4-
PIPERAZINYL)PROPYL)PHENOTHIAZINE * CHLORPER-
AZINE * COMPAZINE * N-(γ-(4'-METHYLPIPERAZI-
NYL-1')PROPYL)-3-CHLOROPHENOTHIAZINE * NIPO-
DAL * NOVAMIN * PROCHLOROPERAZINE
* PROCHLORPEMAZINE * PROCHLORPERAZINE
* 6140 RP * STEMETIL * TEMENTIL

SAFETY PROFILE: Poison by ingestion, subcutaneous, intravenous, and intraperitoneal routes.
Experimental teratogenic and reproductive effects. Human systemic effects by ingestion:
headache, blood pressure elevation. Implicated
in aplastic anemia. When heated to decomposition it emits very toxic fumes of SO_x, NO_x,
and Cl^-.

PMG750 HR: 3
PRODUCER GAS

PROP: Composed of carbon monoxide, hydrogen, air and steam. Lel: 20-30%, fuel:
70-80%.

SAFETY PROFILE: Poison. Dangerous fire
hazard when exposed to flame. Explosive in
the form of vapor by spark or flame when mixed
with air in the range of 20.7-73.7%. Dangerous;
can react vigorously with oxidizing materials.
To fight fire, use CO_2, dry chemical, water
spray.

PMH250 CAS: 952-23-8 HR: 3
PROFLAVINE MONOHYDROCHLORIDE
mf: $C_{13}H_{11}N_3 \cdot ClH$ mf: 245.73

SYNS: 3,6-ACRIDINEDIAMINE, MONOHYDROCHLO-
RIDE (9CI) * 3,6-DIAMINOACRIDINE MONOHYDRO-
CHLORIDE * 3,6-DIAMINOACRIDINIUM CHLORIDE
* 3,6-DIAMINOACRIDINIUM CHLORIDE HYDROCHLO-
RIDE * 2,8-DIAMINOACRIDINIUM CHLORIDE MONO-
HYDROCHLORIDE * PROFLAVINE HYDROCHLORIDE

CONSENSUS REPORTS: IARC Cancer Review: Animal Indefinite Evidence IMEMDT
24,195,80.

SAFETY PROFILE: Poison by subcutaneous
route. Questionable carcinogen. Mutation data
reported. When heated to decomposition it emits
very toxic fumes of NO_x and HCl. Used as a
drug, as a disinfectant, and as a topical antiseptic.

PMH500 CAS: 57-83-0 HR: 3
PROGESTERONE
mf: $C_{31}H_{30}O_2$ mw: 314.51

PROP: A female sex hormone. White, crystalline powder; odorless. D: 1.166 @ 23°, mp:
127-131°. Practically insol in water; sol in alc,
acetone, and dioxane; sparingly sol in oils.

SYNS: CORLUTIN * CORLUVITE * CORPORIN
* CORPUS LUTEUM HORMONE * CYCLOGEST
* Δ⁴-PREGNENE-3,20-DIONE * GLANDUCORPIN
* HORMOFLAVEINE * HORMOLUTON
* LINGUSORBS * LIPO-LUTIN * LUCORTEUM
SOL * LUTEAL HORMONE * LUTEOHORMONE
* LUTEOSAN * LUTEX * LUTOCYCLIN
* LUTROMONE * NALUTRON * NSC-9704
* PERCUTACRINE * PIAPONON * 3,20-PREG-
NENE-4 * PREGNENEDIONE * PREGNENE-3,20-
DIONE * PREGN-4-ENE-3,20-DIONE * 4-PREGNENE-
3,20-DIONE * PROGEKAN * PROGESTEROL
* β-PROGESTERONE * PROGESTERONUM
* PROGESTIN * PROGESTONE * PROLIDON
* SYNGESTERONE * SYNOVEX S * SYN-
TOLUTAN

CONSENSUS REPORTS: IARC Cancer Review: Animal Limited Evidence IMEMDT
21,491,79; Animal Sufficient Evidence
IMEMDT 6,135,74. NTP Fourth Annual Report
On Carcinogens, 1984. EPA Genetic Toxicology Program. Reported in EPA TSCA Inventory.

SAFETY PROFILE: Confirmed carcinogen with
experimental carcinogenic, neoplastigenic, and
tumorigenic data. Poison by intravenous and
intraperitoneal routes. Human teratogenic effects by ingestion, parenteral, and possibly other
routes: developmental abnormalities of the urogenital system. Human male reproductive effects by intramuscular route: changes in spermatogenesis, the prostate, seminal vesicle,
Cowper's gland, and accessory glands; impotence, and breast development. Human female
reproductive effects by ingestion, parenteral,
and intravaginal routes: fertility changes; menstrual cycle changes and disorders; uterus, cervix, and vagina changes. Experimental teratogenic and reproductive effects. Human mutation
data reported. When heated to decomposition
it emits acrid smoke and irritating fumes.

PMI500 CAS: 53-60-1 HR: 3
PROMAZINE HYDROCHLORIDE
mf: $C_{17}H_{20}N_2S \cdot ClH$ mw: 320.91

PROP: White to sltly yellow, practically odorless, crystalline powder. Decomp @ 181°. Sol in water, methanol, ethanol, chloroform; insol in ether, benzene.

SYNS: 10-(γ-DIMETHYLAMINO-N-PROPYL)PHENOTHIAZINE HYDROCHLORIDE * 10-(3-(DIMETHYLAMINO) PROPYL)PHENOTHIAZINE HYDROCHLORIDE * SPARINE HYDROCHLORIDE

CONSENSUS REPORTS: Reported in EPA TSCA Inventory.

SAFETY PROFILE: Poison by ingestion, subcutaneous, intravenous, intraperitoneal, and intramuscular routes. Human systemic effects by ingestion: general anesthesia, tremors, antipsychotic effects. An additive permitted in food for human consumption; also permitted in the feed and drinking water of animals and/or for the treatment of food-producing animals. When heated to decomposition it emits toxic fumes of NO_x, SO_x, and HCl.

PMJ000   HR: 3
PROMETHIUM
af: Pm   aw: 147.

PROP: [147]Pm: Metallic solid. Mp: 1080°, bp: 2460°, d: 7.22. A rare earth. The 145 isotope has a half life of 18 years. The 147 isotope has a half life of 2.64 years. The 147 isotope is the only one available.

SAFETY PROFILE: A poison. Radiotoxic metal.

PMJ750   CAS: 74-98-6   HR: 3
PROPANE

DOT: UN 1075/UN 1978
mf: C_3H_8 mw: 44.11

PROP: Colorless gas. Bp: −42.1°, lel: 2.3%, uel: 9.5%, fp: −187.1°, flash p: −156°F, d: 0.5852 @ −44.5°/4°, autoign temp: 842°F, vap d: 1.56. Sol in water, alc, ether.

SYNS: DIMETHYLMETHANE * PROPYL HYDRIDE

CONSENSUS REPORTS: Reported in EPA TSCA Inventory.

OSHA PEL: TWA 1000 ppm
ACGIH TLV: Asphyxiant
DFG MAK: 1000 ppm (1800 mg/m³)
DOT Classification: Flammable Gas; Label: Flammable Gas.

SAFETY PROFILE: Central nervous system effects at high concentrations. An asphyxiant. Flammable gas. Highly dangerous fire hazard when exposed to heat or flame; can react vigorously with oxidizers. Explosive in the form of vapor when exposed to heat or flame. Explosive reaction with ClO_2. Violent exothermic reaction with barium peroxide + heat. To fight fire, stop flow of gas. When heated to decomposition it emits acrid smoke and irritating fumes.

PMK000   CAS: 542-78-9   HR: 3
PROPANEDIAL
mf: $C_3H_4O_2$ mw: 72.07

SYNS: MALONALDEHYDE * MALONDIALDEHYDE * MALONIC ALDEHYDE * MALONIC DIALDEHYDE * MALONODIALDEHYDE * MALONYLDIALDEHYDE * NCI-C54842 * 1,3-PROPANEDIAL * 1,3-PROPANEDIALDEHYDE * 1,3-PROPANEDIONE

CONSENSUS REPORTS: IARC Cancer Review: GROUP 3 IMEMDT 7,56,87; Animal Inadequate Evidence IMEMDT 36,163,85. EPA Genetic Toxicology Program.

SAFETY PROFILE: Moderately toxic by ingestion. Questionable carcinogen with experimental carcinogenic data. Human mutation data reported. When heated to decomposition it emits acrid smoke and irritating fumes.

PMK250   CAS: 78-90-0   HR: 2
1,2-PROPANEDIAMINE

DOT: UN 2258
mf: $C_3H_{10}N_2$ mw: 74.15

PROP: Flash p: 92°F (OC), d: 0.9, vap d: 2.6, bp: 118.9°.

SYNS: 1,2-DIAMINOPROPANE * PROPYLENEDIAMINE * PROPYLENE DIAMINE (DOT)

CONSENSUS REPORTS: Reported in EPA TSCA Inventory.

DOT Classification: Flammable Liquid; Label: Flammable Liquid, Corrosive; Flammable or Combustible Liquid; Label: Flammable Liquid, Corrosive.

SAFETY PROFILE: Moderately toxic by ingestion, skin contact, and subcutaneous routes. A corrosive irritant to eyes, skin, and mucous membranes. Dangerous fire hazard when exposed to heat, flames, oxidizers. To fight fire, use alcohol foam. When heated to decomposition it emits toxic fumes of NO_x. Used as an intermediate in production of petroleum and polymer additives, and surfactants.

PML000   CAS: 57-55-6   **HR: 3**
1,2-PROPANEDIOL
mf: $C_3H_8O_2$ mw: 76.11

PROP: Colorless viscous liquid; practically odorless. Bp: 188.2°, flash p: 210°F (OC), lel: 2.6%, uel: 12.6%, d: 1.0362 @ 25°/25°, autoign temp: 700°F, vap press: 0.08 mm @ 20°, vap d: 2.62, fp: −59°. Hygroscopic; misc with water, acetone, chloroform; sol in essential oils; immisc with fixed oils.

SYNS: 1,2-DIHYDROXYPROPANE * DOWFROST * METHYLETHYLENE GLYCOL * METHYL GLYCOL * MONOPROPYLENE GLYCOL * PG 12 * PROPANE-1,2-DIOL * PROPYLENE GLYCOL (FCC) * PROPYLENE GLYCOL USP * α-PROPYLENEGLYCOL * 1,2-PROPYLENE GLYCOL * SIRLENE * SOLAR WINTER BAN * TRIMETHYL GLYCOL

CONSENSUS REPORTS: Reported in EPA TSCA Inventory. EPA Genetic Toxicology Program.

SAFETY PROFILE: Slightly toxic by ingestion, skin contact, intraperitoneal, intravenous, subcutaneous, and intramuscular routes. Human systemic effects by ingestion: general anesthesia, convulsions, changes in surface EEG. Experimental teratogenic and reproductive effects. An eye and human skin irritant. Mutation data reported. Combustible liquid when exposed to heat or flame; can react with oxidizing materials. Explosive in the form of vapor when exposed to heat or flame. May react with hydrofluoric acid + nitric acid + silver nitrate to form the explosive silver fulminate. To fight fire, use alcohol foam. When heated to decomposition it emits acrid smoke and irritating fumes.

PML400   CAS: 1120-71-4   **HR: 3**
PROPANE SULTONE
mf: $C_3H_6O_3S$ mw: 122.15

SYNS: 3-HYDROXY-1-PROPANESULFONIC ACID γ-SULTONE * 3-HYDROXY-1-PROPANESULPHONIC ACID SULFONE * 3-HYDROXY-1-PROPANESULPHONIC ACID SULTONE * 1,2-OXATHIOLANE-2,2-DIOXIDE * 1-PROPANESULFONIC ACID-3-HYDROXY-γ-SULTONE * 1,3-PROPANE SULTONE (MAK) * RCRA WASTE NUMBER U193

CONSENSUS REPORTS: IARC Cancer Review: GROUP 2B IMEMDT 7,56,87; Animal Sufficient Evidence IMEMDT 4,253,74. NTP Fourth Annual Report On Carcinogens, 1984. Community Right-To-Know List. Reported in EPA TSCA Inventory. EPA Genetic Toxicology Program.

ACGIH TLV: Suspected Human Carcinogen.
DFG MAK: Animal Carcinogen, Suspected Human Carcinogen.

SAFETY PROFILE: Confirmed carcinogen with experimental carcinogenic, neoplastigenic, and tumorigenic data. Poison by subcutaneous route. Moderately toxic by skin contact and intraperitoneal routes. Experimental teratogenic and reproductive effects. Human mutation data reported. Implicated as a human brain carcinogen. A skin irritant. When heated to decomposition it emits toxic fumes of SO_x.

PML500   CAS: 107-03-9   **HR: 2**
PROPANETHIOL
DOT: UN 2402/UN 2704
mf: C_3H_8S mw: 76.17

PROP: Flash p: −4°F.

SYNS: 3-MERCAPTOPROPANOL * PROPANE-1-THIOL * PROPYL MERCAPTAN * N-PROPYL MERCAPTAN

CONSENSUS REPORTS: Reported in EPA TSCA Inventory.

NIOSH REL: CL 0.5 ppm/15M
DOT Classification: Label: Flammable Liquid.

SAFETY PROFILE: Moderately toxic by ingestion and intraperitoneal routes. Mildly toxic by inhalation. A severe eye irritant. Explodes on contact with calcium hypochlorite. Very dangerous fire hazard when exposed to heat or flame. When heated to decomposition it emits toxic fumes of SO_x.

PMN450   CAS: 107-19-7   **HR: 3**
PROPARGYL ALCOHOL
DOT: NA 1986
mf: C_3H_4O mw: 56.07

PROP: Moderately volatile liquid; geranium odor. D: 0.9715 @ 20°/4°, mp: −48° to −52°, bp: 114-115°, flash p: 33°C (97°F) (OC), vap press: 11.6 mm @ 20°, vap d: 1.93.

SYNS: ETHYNYLCARBINOL * ETHYNYLMETHANOL * 1-PROPYNE-3-OL * 2-PROPYN-1-OL

* 3-PROPYNOL * 2-PROPYNYL ALCOHOL
* RCRA WASTE NUMBER P102

CONSENSUS REPORTS: Reported in EPA TSCA Inventory.

OSHA PEL: TWA 1 ppm (skin)
ACGIH TLV: TWA 1 ppm (skin)
DFG MAK: 2 ppm (5 mg/m^3)
DOT Classification: Flammable Liquid; Label: Flammable Liquid and Poison.

SAFETY PROFILE: Poison by ingestion, skin contact, and subcutaneous routes. Moderately toxic by inhalation. A central nervous system depressant. A skin and mucous membrane irritant. Dangerous fire hazard when exposed to heat or flame; can ignite. To fight fire, use foam, CO_2, dry chemical. Potentially explosive reactions with alkalies (when dried); sulfuric acid. Ignites on contact with phosphorus pentaoxide. Violent reaction with mercury(II) sulfate + sulfuric acid + water (at 70°C). Incompatible with oxidizing materials. When heated to decomposition it emits acrid smoke and irritating fumes. Used as a corrosion inhibitor, solvent stabilizer, soil fumigant, and chemical intermediate.

PMN500　　　CAS: 106-96-7　　　HR: 3
PROPARGYL BROMIDE

DOT: UN 2345
mf: C_3H_3Br　　mw: 118.97

PROP: An almost colorless liquid; sharp odor. Bp: 88-90°, fp: −61.07°, flash p: 65°F (COC), d: 1.564-1.570, vap d: 6.87.

SYNS: γ-BROMOALLYLENE * 3-BROMOPROPYNE (DOT) * 3-BROMO-1-PROPYNE

CONSENSUS REPORTS: EPA Extremely Hazardous Substances List. Reported in EPA TSCA Inventory.

DOT Classification: Flammable Liquid; Label: Flammable Liquid.

SAFETY PROFILE: A deadly poison by ingestion. A dangerous fire hazard when exposed to heat or flame. The aerated liquid may be ignited by pressure. A dangerous, extremely shock-sensitive explosive. It can detonate when heated to 220°C, by impact (especially when mixed with chloropicrin) or when heated while confined. May explode on contact with copper; high copper alloys; mercury; or silver. Mixtures with trichloronitromethane are shock- and heat-sensitive explosives. Can react vigorously with

oxidizing materials. To fight fire, use water, foam, CO_2, dry chemical. When heated to decomposition it emits highly toxic fumes of Br^-.

PMN850　　　CAS: 139-40-2　　　HR: 2
PROPAZINE
mf: $C_9H_{16}ClN_7O_2$　　mw: 229.75

SYNS: 2,4-BIS(ISOPROPYLAMINO)-6-CHLORO-s-TRI-AZINE * 2,4-BIS(PROPYLAMINO)-6-CHLOR-1,3,5-TRI-AZIN (GERMAN) * GESAMIL * MILOGARD * PLANTULIN * PRIMATOL P * PROPASIN * PROZINEX

SAFETY PROFILE: Moderately toxic by ingestion. Moderate eye irritation. Questionable carcinogen with experimental tumorigenic data. When heated to decomposition it emits toxic fumes of NO_x and Cl^-.

PMO500　　　CAS: 115-07-1　　　HR: 3
PROPENE

DOT: UN 1075/UN 1077
mf: C_3H_6　　mw: 42.09

PROP: A gas. D: (gas) 1.49 (air = 1.0), d: (liquid) 0.581 @ 0°. Mp: −185°, bp: −47.7°, autoign temp: 860°F, vap press: 10 atm @ 19.8°, lel: 2.4%, uel: 10.1%, vap d: 1.5, flash p: −162°F.

SYNS: METHYLETHENE * METHYLETHYLENE * NCI-C50077 * 1-PROPENE * PROPYLENE (DOT)

CONSENSUS REPORTS: IARC Cancer Review: GROUP 3 IMEMDT 7,56,87. NTP Carcinogenesis Studies (inhalation); No Evidence: mouse, rat NTPTR* NTP-TR-272,85. EPA Extremely Hazardous Substances List. Reported in EPA TSCA Inventory.

DOT Classification: Flammable Gas; Label: Flammable Gas.

SAFETY PROFILE: A simple asphyxiant. No irritant effects from high concentrations in gaseous form. When compressed to liquid form, can cause skin burns from freezing effects on tissue of rapid evaporation. Questionable carcinogen. Very dangerous fire hazard when exposed to heat, flame, or oxidizers. Explosive in the form of vapor when exposed to heat or flame. Under unusual conditions, i.e., 955 atm. pressure and 327°C, it has been known to explode. Explodes on contact with trifluoromethyl hypofluorite. Explosive polymerization is initiated by lithium nitrate + sulfur dioxide. Reacts with oxides of nitrogen to form an explosive product. Dangerous; can react vigorously with oxidizing materials. To fight fire, stop flow of gas. Used in production of fabricated polymers, fibers, and solvents, in production of plastic products and resins.

PMP500   CAS: 9003-07-0   **HR: 2**
PROPENE POLYMERS
mf: $(C_3H_6)_n$

PROP: Solid material. Mp: about 165°, d: 0.90-0.92. Insol in organic materials.

SYNS: ADMER PB 02 * AMCO * AMERFIL * AMOCO 1010 * ATACTIC POLYPROPYLENE * AVISUN * AZDEL * BEAMETTE * BICOLENE P * CARLONA P * CELGARD 2500 * CHISSO 507B * CLYSAR * COATHYLENE PF 0548 * DAPLEN AD * DEXON E 117 * EASTBOND M 5 * ELPON * ENJAY CD 460 * EPOLENE M 5K * GERFIL * HERCOFLAT 135 * HERCULON * HOSTALEN PP * HULS P 6500 * ICI 543 * ISOTACTIC POLYPROPYLENE * J 400 * LAMBETH * LUPAREEN * MARLEX 9400 * MAURYLENE * MERAKLON * MOPLEN * MOSTEN * NOBLEN * NOVAMONT 2030 * NOVOLEN * OLETAC 100 * PAISLEY POLYMER * PELLON 2506 * POLYPRO 1014 * POLYPROPENE * POLYPROPYLENE * POLYTAC * POPROLIN * PROFAX * PROPATHENE * 1-PROPENE HOMOPOLYMER (9CI) * PROPOLIN * PROPOPHANE * PROPYLENE POLYMER * REXALL 413S * REXENE * SHELL 5520 * SHOALLOMER * SYNDIOTACTIC POLYPROPYLENE * TENITE 423 * TRESPAPHAN * TUFFLITE * ULSTRON * VISCOL 350P * W 101 * WEX 1242

CONSENSUS REPORTS: IARC Cancer Review: GROUP 3 IMEMDT 7,56,87; Animal Limited Evidence IMEMDT 19,213,79; Human Inadequate Evidence IMEMDT 19,213,79. Reported in EPA TSCA Inventory.

SAFETY PROFILE: Questionable carcinogen. When heated to decomposition it emits acrid smoke and irritating fumes. Used in injection molding for auto parts, in bottle caps, and container closures.

PMP750   CAS: 6842-15-5   **HR: 1**
PROPENE TETRAMER

DOT: UN 2850
mf: $C_{12}H_{24}$ mw: 168.36

PROP: Colorless liquid. Mp: −31.5°, bp: 213°, d: 0.76 @ 20°/4°, vap press: 1 mm @ 47.2°, vap d: 5.81, flash p: <212°F, autoign temp: 491°F.

SYNS: AMSCO TETRAMER * DODECENE * DODECYLENE * PROPYLENE TETRAMER (DOT) * TETRAPROPYLENE

CONSENSUS REPORTS: Reported in EPA TSCA Inventory.

DOT Classification: Flammable or Combustible Liquid; Label: Flammable Liquid.

SAFETY PROFILE: Probably irritating and narcotic in high concentration. Combustible when exposed to heat or flame; can react with oxidizing materials. To fight fire, use foam, CO_2, dry chemical. When heated to decomposition it emits acrid smoke and irritating fumes.

PMQ750   CAS: 104-46-1   **HR: 3**
p-PROPENYLANISOLE
mf: $C_{10}H_{12}O$ mw: 148.22

PROP: Leaves from alc or light yellow liquid above 23°; sweet taste with anise odor. D: 0.991 @ 20°/20°, refr index: 1.557-1.561, mp: 22.5°, bp: 235.3°, flash p: 198°F. Very sltly sol in water; misc in abs alc, ether, chloroform.

SYNS: ACINTENE O * ANETHOLE (FCC) * ANISE CAMPHOR * ARIZOLE * FEMA No. 2086 * ISOESTRAGOLE * p-METHOXY-β-METHYLSTYRENE * 1-(p-METHOXYPHENYL)PROPENE * 1-METHOXY-4-PROPENYLBENZENE * 4-METHOXYPROPENYLBENZENE * MONASIRUP * NAULI ''GUM'' * OIL of ANISEED * p-1-PROPENYLANISOLE * 4-PROPENYLANISOLE * p-PROPENYLPHENYL METHYL ETHER

CONSENSUS REPORTS: Reported in EPA TSCA Inventory.

SAFETY PROFILE: Poison by ingestion. Questionable carcinogen with experimental tumorigenic data. Combustible liquid. When heated to decomposition it emits acrid smoke and irritating fumes.

PMR750   CAS: 590-21-6   **HR: 3**
PROPENYL CHLORIDE
mf: C_3H_5Cl mw: 76.53

PROP: Liquid. Mp: −137.4°, bp: 22.65°, flash p: <21°F, d: 0.9189°, lel: 4.5%, uel: 16%. Insol in water.

SYNS: 1-CHLOROPROPENE * 1-CHLORO-1-PROPENE

CONSENSUS REPORTS: Reported in EPA TSCA Inventory.

SAFETY PROFILE: Moderately toxic by ingestion. Very mildly toxic by skin contact and inhalation. An eye irritant. Mutation data reported. Questionable carcinogen with experimental neoplastigenic data. Very dangerous fire hazard when exposed to heat, flames (sparks) or oxidizers. Explosive in the form of vapor when exposed to heat or flame. To fight fire, use alcohol foam, dry chemical, mist spray, fog. When heated to decomposition it emits toxic fumes of Cl^-.

PMS500   CAS: 1797-74-6   **HR: 2**
2-PROPENYL PHENYLACETATE
mf: $C_{11}H_{12}O_2$ mw: 176.23

PROP: Colorless to light yellow liquid; fruity odor of banana and honey.

SYNS: ALLYL PHENYLACETATE * BENZENEACETIC ACID, 2-PROPENYL ESTER * PHENYLACETIC ACID ALLYL ESTER

CONSENSUS REPORTS: Reported in EPA TSCA Inventory.

SAFETY PROFILE: Moderately toxic by ingestion. A human skin irritant. When heated to decomposition it emits acrid smoke and irritating fumes.

PMT100   CAS: 57-57-8   **HR: 3**
β-PROPIOLACTONE
mf: $C_3H_4O_2$ mw: 72.07

SYNS: BETAPRONE * BPL * HYDRACRYLIC ACID β-LACTONE * 3-HYDROXYPROPIONIC ACID LACTONE * PROPANOLIDE * PROPIOLACTONE * 1,3-PROPIOLACTONE * 3-PROPIOLACTONE * β-PROPIONOLACTONE * β-PROPRIOLACTONE (OSHA) * β-PROPROLACTONE

CONSENSUS REPORTS: IARC Cancer Review: GROUP 2B IMEMDT 7,56,87; Animal Sufficient Evidence IMEMDT 4,259,74. NTP Fourth Annual Report On Carcinogens, 1984. EPA Genetic Toxicology Program. Community Right-To-Know List. EPA Extremely Hazardous Substances List. Reported in EPA TSCA Inventory.

OSHA: Carcinogen.
ACGIH TLV: TWA 0.5 ppm; Suspected Human Carcinogen.

DFG MAK: Animal Carcinogen, Suspected Human Carcinogen.

SAFETY PROFILE: Confirmed carcinogen with experimental carcinogenic, neoplastigenic, and tumorigenic data. Poison by inhalation. Moderately toxic by intraperitoneal route. An initiator. Human mutation data reported. When heated to decomposition it emits acrid smoke and irritating fumes.

PMT750   CAS: 123-38-6   **HR: 2**
PROPIONALDEHYDE
DOT: UN 1275
mf: C_3H_6O mw: 58.09

PROP: Colorless, mobile liquid; suffocating odor. Mp: $-81°$, bp: $48°$, flash p: 15-19°F (OC), d: 0.807 @ 20°/4°, lel: 2.9%, uel: 17%, vap d: 2.0, autoign temp: 405°F. Misc with alc, ether, water @ 49°.

SYNS: ALDEHYDE PROPIONIQUE (FRENCH) * FEMA No. 2923 * METHYLACETALDEHYDE * NCI-C61029 * PROPALDEHYDE * PROPANAL * PROPIONIC ALDEHYDE * PROPYL ALDEHYDE * PROPYLIC ALDEHYDE

CONSENSUS REPORTS: Community Right-To-Know List. Reported in EPA TSCA Inventory.

DOT Classification: Flammable Liquid; Label: Flammable Liquid.

SAFETY PROFILE: Moderately toxic by skin contact, ingestion and subcutaneous routes. Mildly toxic by inhalation. A skin and severe eye irritant. Flammable liquid. Dangerous fire hazard when exposed to heat or flame; reacts vigorously with oxidizers. Explosive in the form of vapor when exposed to heat or flame. Vigorous polymerization reaction with methyl methacrylate. To fight fire, use alcohol foam, CO_2, dry chemical. When heated to decomposition it emits acrid smoke and irritating fumes.

PMU750   CAS: 79-09-4   **HR: 3**
PROPIONIC ACID
DOT: UN 1848
mf: $C_3H_6O_2$ mw: 74.09

PROP: Oily liquid; pungent, disagreeable, rancid odor. D: 0.998 @ 15°/4°, mp: $-21.5°$, bp: 141.1°, vap press: 10 mm @ 39.7°, vap d: 2.56, autoign temp: 955°F. Misc in water, alc, ether, chloroform.

SYNS: ACIDE PROPIONIQUE (FRENCH) * CAR-
BOXYETHANE * ETHANECARBOXYLIC ACID
* ETHYLFORMIC ACID * METACETONIC ACID
* METHYL ACETIC ACID * PROPANOIC ACID
* PROPIONIC ACID, solution containing not less than 80%
acid (DOT) * PROPIONIC ACID GRAIN PRESERVER
* PROZOIN * PSEUDOACETIC ACID * SENTRY
GRAIN PRESERVER * TENOX P GRAIN PRESERVA-
TIVE

CONSENSUS REPORTS: Reported in EPA
TSCA Inventory.

OSHA PEL: TWA 10 ppm
ACGIH TLV: TWA 10 ppm
DFG MAK: 10 ppm (30 mg/m^3)
DOT Classification: Corrosive Material; Label:
  Corrosive; Label: Corrosive, solution; Label:
  Corrosive, Flammable Liquid.

SAFETY PROFILE: Poison by intraperitoneal
route. Moderately toxic by ingestion, skin contact, and intravenous routes. A corrosive irritant
to eyes, skin, and mucous membranes. Flammable liquid. Highly flammable when exposed to
heat, flame, or oxidizers. To fight fire, use alcohol foam. When heated to decomposition it
emits acrid smoke and irritating fumes.

PMV250   CAS: 540-42-1   **HR: 1**
PROPIONIC ACID, ISOBUTYL ESTER
DOT: UN 2394
mf: C$_7$H$_{14}$O$_2$ mw: 130.21

SYNS: ISOBUTY PROPIONATE (DOT) * 2-METHYL-
PROPYL PROPIONATE * PROPANOIC ACID, 2-METH-
YLPROPYL ESTER

CONSENSUS REPORTS: Reported in EPA
TSCA Inventory.

DOT Classification: Flammable Liquid; Label:
  Flammable Liquid.

SAFETY PROFILE: Mildly toxic by ingestion.
Flammable when exposed to heat or flame, can
react vigorously with oxidizing materials. When
heated to decomposition it emits acrid smoke
and irritating fumes.

PMV500   CAS: 123-62-6   **HR: 2**
PROPIONIC ANHYDRIDE
DOT: UN 2496
mf: C$_6$H$_{10}$O$_3$ mw: 130.16

PROP: Liquid; very rancid odor. Mp: −45°,
bp: 167.0°, flash p: 165°F (OC), d: 1.012, vap

press: 1 mm @ 20.6°, vap d: 4.49. Decomp
in water, alc; sol in methanol, ethanol, ether,
chloroform.

SYNS: METHYLACETIC ANHYDRIDE * PROPANOIC
ANHYDRIDE * PROPIONIC ACID ANHYDRIDE
* PROPIONYL OXIDE

CONSENSUS REPORTS: Reported in EPA
TSCA Inventory.

DOT Classification: Corrosive Material; Label:
  Corrosive.

SAFETY PROFILE: Moderately toxic by ingestion. Mildly toxic by skin contact. A corrosive
irritant to skin, eyes, and mucous membranes.
Flammable when exposed to heat or flame; can
react with oxidizing materials. To fight fire,
use CO$_2$, dry chemical. When heated to decomposition it emits acrid smoke and irritating
fumes. Used as an esterifying agent and dehydrating agent.

PMV750   CAS: 107-12-0   **HR: 3**
PROPIONONITRILE
DOT: UN 2404
mf: C$_3$H$_5$N   mw: 55.09

PROP: Colorless liquid; ethereal odor. Bp:
97.1°, d: 0.783 @ 21°/4°, vap d: 1.9, flash p:
36°F, lel: 3.1%; mp: −91.8°. Misc with alc,
ether.

SYNS: CYANOETHANE * ETHER CYANATUS
* ETHYL CYANIDE * HYDROCYANIC ETHER
* PROPANENITRILE * PROPIONIC NITRILE
* RCRA WASTE NUMBER P101

CONSENSUS REPORTS: Cyanide and its
compounds are on the Community Right-To-
Know List. EPA Extremely Hazardous Substances List. Reported in EPA TSCA Inventory.

NIOSH REL: TWA (Nitriles) 14 mg/m^3
DOT Classification: Flammable Liquid; Label:
  Flammable Liquid and Poison.

SAFETY PROFILE: Poison by ingestion, skin
contact, intravenous, intraperitoneal, and possibly other routes. Moderately toxic by inhalation.
Experimental teratogenic effects. An eye irritant. Dangerous fire hazard when exposed to
heat, flame (sparks), oxidizers. Mixture with
N-bromosuccinimide may explode when heated.
To fight fire, use water spray, foam, mist, CO$_2$,
dry chemical. When heated to decomposition
it emits toxic fumes of NO$_x$ and CN$^-$. Used

as a solvent in petroleum refining, and as a raw material for drug manufacture.

PMW500   CAS: 79-03-8   **HR: 2**
PROPIONYL CHLORIDE
DOT: UN 1815
mf: C_3H_5ClO mw: 92.53

PROP: Mp: $-94°$, bp: $80°$, flash p: $53.6°$, d: 1.065, vap d: 3.2.

SYNS: PROPANOYL CHLORIDE * PROPIONIC ACID CHLORIDE * PROPIONIC CHLORIDE

CONSENSUS REPORTS: Reported in EPA TSCA Inventory.

DOT Classification: Flammable liquid; Label: Corrosive, Flammable Liquid.

SAFETY PROFILE: A corrosive irritant to skin, eyes, and mucous membranes. Dangerous fire hazard when exposed to heat or flame; can react vigorously with oxidizing materials. Reacts with water or steam to produce toxic and corrosive fumes. Exothermic reaction with diisopropyl ether produces much gas. The reaction may be dangerous if confined. To fight fire, use CO_2, dry chemical; do not use water. When heated to decomposition it emits highly toxic fumes of Cl^-.

PMY300   CAS: 114-26-1   **HR: 3**
PROPOXUR
mf: $C_{11}H_{15}NO_3$ mw: 209.27

PROP: A white to tan, crystalline solid; sltly sol in water; sol in all polar organic solvents.

SYNS: APROCARB * BAY 9010 * BAYER 39007 * BAYGON * BIFEX * BLATTANEX * BOY-GON * ENT 25,671 * o-IMPC * INVISI-GARD * ISOCARB * o-ISOPROPOXYPHENYL METHYL-CARBAMATE * o-ISOPROPOXYPHENYL-N-METHYL-CARBAMATE * 2-ISOPROPOXYPHENYL-N-METHYL-CARBAMATE * 2-(1-METHYLETHOXY)PHENOL METHYLCARBAMATE * N-METHYL-2-ISOPRO-POXYPHENYLCARBAMATE * OMS-33 * PHC * PROPOKSURU (POLISH) * PROPYON * SEN-DRAN * SUNCIDE * TUGON FLIEGENKUGEL * UNDEN

CONSENSUS REPORTS: EPA Genetic Toxicology Program. Community Right-To-Know List.

OSHA PEL: TWA 0.5 mg/m^3
ACGIH TLV: TWA 0.5 mg/m^3
DFG MAK: 2 mg/m^3

SAFETY PROFILE: A poison via ingestion, subcutaneous, intraperitoneal, intravenous, intramuscular, and possibly other routes. Moderately toxic by inhalation and skin contact. Experimental reproductive effects. Mutation data reported. Moderately irritating to skin. When heated to decomposition it emits toxic fumes of NO_x.

PNA500   CAS: 1639-60-7   **HR: 3**
d-PROPOXYPHENE HYDROCHLORIDE
mf: $C_{22}H_{29}NO_2 \cdot ClH$ mw: 375.98

PROP: Bitter crystals. Mp: $163-168.5°$. Sol in water, alc, chloroform, acetone; insol in benzene, ether.

SYNS: ALGAFAN * ANTALVIC * DARVON HY-DROCHLORIDE * DEPRANCOL * DEPROMIC * DEVELIN * DEXTROPROPOXYPHENE HYDRO-CHLORIDE * DEXTROPROXYPHEN HYDROCHLORIDE * d-4-DIMETHYLAMINO-3-METHYL-1,2-DIPHENYL-2-BU-TANOL PROPIONATE HYDROCHLORIDE * s-α-(2-(DI-METHYLAMINO)-1-METHYLETHYL)-α-PHENYLBEN-ZENEETHANOL PROPIOATE HYDROCHLORIDE * (+)-1,2-DIPHENYL-2-PROPIONOXY-3-METHYL-4-DI-METHYLAMINOBUTANE HYDROCHLORIDE * DOLENE * DOLOCAP * DOLOXENE * DORAPHEN * ERANTIN * FEMADOL * HARMAR * PROPOX * PROPOXYCHEL * PROPOXYPHENE HYDROCHLORIDE * (+)-PROPOXYPHENE HYDRO-CHLORIDE * α-PROPOXYPHENE HYDROCHLORIDE * α-d-PROPOXYPHENE HYDROCHLORIDE * d-PROPOXYPHENE MONOHYDROCHLORIDE * PROXAGESIC

CONSENSUS REPORTS: Reported in EPA TSCA Inventory.

SAFETY PROFILE: A human poison by ingestion. An experimental poison by ingestion, intraperitoneal, subcutaneous, intravenous, and intramuscular routes. Human systemic effects by ingestion: ataxia, cyanosis, anorexia, chronic pulmonary edema, sleep disturbance. Experimental teratogenic and reproductive effects. When heated to decomposition it emits very toxic fumes of HCl and NO_x.

PNC250   CAS: 109-60-4   **HR: 3**
n-PROPYL ACETATE
DOT: UN 1276
mf: $C_5H_{10}O_2$ mw: 102.15

PROP: Clear, colorless liquid; pleasant odor. Mp: $-92.5°$, bp: $101.6°$, flash p: $58°F$, lel:

2.0%, uel: 8.0%, d: 0.887, autoign temp: 842°F, vap press: 40 mm @ 28.8°, vap d: 3.52. Misc with alc, ether; sol in water.

SYNS: ACETATE de PROPYLE NORMAL (FRENCH) * ACETIC ACID, n-PROPYL ESTER * 1-ACETOXY-PROPANE * OCTAN PROPYLU (POLISH) * PROPYL ACETATE * 1-PROPYL ACETATE

CONSENSUS REPORTS: Reported in EPA TSCA Inventory.

OSHA PEL: (Transitional: TWA 200 ppm) TWA 200 ppm; STEL 250 ppm
ACGIH TLV: TWA 200 ppm; STEL 250 ppm
DFG MAK: 200 ppm (840 mg/m^3)
DOT Classification: Flammable Liquid; Label: Flammable Liquid.

SAFETY PROFILE: Moderately toxic by intraperitoneal and subcutaneous routes. Mildly toxic by ingestion and inhalation. Human systemic effects by inhalation: lachrimation, cough. A skin irritant. A narcotic at high concentrations. Isopropyl acetate is slightly less narcotic than normal propyl acetate. Dangerous fire hazard when exposed to heat, flame, or oxidizers. Explosive in the form of vapor when exposed to heat or flame. Can react vigorously with oxidizing materials. To fight fire, use alcohol foam, CO_2, dry chemical. When heated to decomposition it emits acrid smoke and irritating fumes.

PND000   CAS: 71-23-8   **HR: 3**
n-PROPYL ALCOHOL

DOT: UN 1274
mf: C_3H_8O mw: 60.11

PROP: Clear liquid; alc-like odor. Mp: −127°, bp: 97.19°, flash p: 59°F (CC), ULC: 55-60, d: 0.8044 @ 20°/4°, lel: 2.1%, uel: 13.5%, autoign temp: 824°F, vap press: 10 mm @ 14.7°, vap d: 2.07. Misc in water, alc, and ether.

SYNS: ALCOOL PROPILICO (ITALIAN) * ALCOOL PROPYLIQUE (FRENCH) * ETHYL CARBINOL * 1-HYDROXYPROPANE * OPTAL * OSMOSOL EXTRA * n-PROPANOL * PROPANOL-1 * 1-PROPANOL * PROPANOLE (GERMAN) * PROPANOLEN (DUTCH) * PROPANOLI (ITALIAN) * PROPYL ALCOHOL * 1-PROPYL ALCOHOL * n-PROPYL ALKOHOL (GERMAN) * PROPYLIC ALCOHOL * PROPYLOWY ALKOHOL (POLISH)

CONSENSUS REPORTS: Reported in EPA TSCA Inventory. EPA Genetic Toxicology Program.

OSHA PEL: (Transitional: TWA 200 ppm) TWA 200 ppm; STEL 250 ppm
ACGIH TLV: TWA 200 ppm; STEL 250 ppm (skin)
DOT Classification: Flammable Liquid; Label: Flammable Liquid.

SAFETY PROFILE: Poison by subcutaneous route. Moderately toxic by inhalation, ingestion, intraperitoneal, and intravenous routes. Mildly toxic by skin contact. A skin and severe eye irritant. Questionable carcinogen with experimental carcinogenic data. Mutation data reported. Dangerous fire hazard when exposed to heat, flame, or oxidizers. Explosive in the form of vapor when exposed to heat or flame. Ignites on contact with potassium-tert-butoxide. Dangerous upon exposure to heat or flame; can react vigorously with oxidizing materials. To fight fire, use alcohol foam, CO_2, dry chemical. When heated to decomposition it emits acrid smoke and irritating fumes.

PND250   CAS: 107-10-8   **HR: 3**
PROPYLAMINE

DOT: UN 1277
mf: C_3H_9N mw: 59.13

PROP: Colorless, alkaline liquid; strong ammonia odor. D: 0.7191 @ 20°/20°, mp: −83°, bp: 48-49°, vap press: 248 mm @ 20°, flash p: −35°F, autoign temp: 604°F, lel: 2.0%, uel: 10.4%. Misc water, alc, ether.

SYNS: 1-AMINOPROPANE * MONO-N-PROPYL-AMINE * PROPANAMINE * N-PROPYLAMINE * RCRA WASTE NUMBER U194

CONSENSUS REPORTS: Reported in EPA TSCA Inventory.

DOT Classification: Flammable Liquid; Label: Flammable Liquid.

SAFETY PROFILE: Moderately toxic by inhalation, ingestion, skin contact and possibly other routes. A skin and severe eye irritant. Possibly a skin sensitizer. Very dangerous fire hazard when exposed to heat, flame, or oxidizers. Explosive in the form of vapor when exposed to heat or flame. To fight fire, use alcohol foam. When heated to decomposition it emits toxic fumes of NO_x. Incompatible with triethynyl aluminum.

PNE250   CAS: 104-45-0   **HR: 1**
p-n-PROPYL ANISOLE
mf: $C_{10}H_{14}O$ mw: 150.24

PROP: Colorless to pale yellow liquid; anise odor. D: 0.940, refr index: 1.502-1.506, flash p: 185°F. Sol in fixed oils; insol in glycerin, propylene glycol.

SYNS: DIHYDROANETHOLE * FEMA No. 2930 * 1-METHOXY-4-PROPYLBENZENE * 4-PROPYLANISOLE * 4-n-PROPYLANISOLE

CONSENSUS REPORTS: Reported in EPA TSCA Inventory.

SAFETY PROFILE: Mildly toxic by ingestion. Mutation data reported. Combustible liquid. When heated to decomposition it emits acrid smoke and irritating fumes.

PNF500 CAS: 1114-71-2 **HR: 3**
S-PROPYL
BUTYLETHYLTHIOCARBAMATE
mf: $C_{10}H_{21}NOS$ mw: 203.38

PROP: Liquid. Bp: 142° @ 20 mm.

SYNS: BUTYLETHYLTHIOCARBAMIC ACID S-PROPYL ESTER * PEBC * PEBULATE * S-PROPYL-N-AETHYL-N-BUTYL-THIOCARBAMAT (GERMAN) * PROPYL-ETHYLBUTYLTHIOCARBAMATE * N-PROPYL-N-ETHYL-N-(N-BUTYL)THIOCARBAMATE * PROPYLETHYL-N-BUTYLTHIOCARBAMATE * PROPYL N-ETHYL-N-BUTYLTHIOCARBAMATE * S-(N-PROPYL)-N-ETHYL-N-N-BUTYLTHIOCARBAMATE * N-PROPYL-N-ETHYL-N-(N-BUTYL)THIOLCARBAMATE * PROPYL ETHYLBUTYLTHIOLCARBAMATE * R-2061 * STAUFFER R-2061 * TILLAM (RUSSIAN) * TILLAM-6-E

SAFETY PROFILE: Moderately toxic by ingestion. Causes violent vomiting when accompanied by alcohol ingestion. Questionable carcinogen with experimental tumorigenic data. When heated to decomposition it emits highly toxic fumes of SO_x and NO_x.

PNG250 CAS: 627-12-3 **HR: 3**
PROPYL CARBAMATE
mf: $C_4H_9NO_2$ mw: 103.14

PROP: Crystals. Bp: 196°, mp: 60°, vap press: 1 mm @ 52.4°. Very sol in water, alc, ether.

SYNS: CARBAMIC ACID, PROPYL ESTER * N-PROPYL CARBAMATE * PROPYL URETHANE

CONSENSUS REPORTS: IARC Cancer Review: GROUP 3 IMEMDT 7,56,87; Animal Sufficient Evidence IMEMDT 12,201,76. Reported in EPA TSCA Inventory.

SAFETY PROFILE: Moderately toxic by subcutaneous route. Experimental teratogenic and reproductive effects. Questionable carcinogen with experimental neoplastigenic and tumorigenic data. Mutation data reported. When heated to decomposition it emits toxic fumes of NO_x.

PNH000 CAS: 109-61-5 **HR: 3**
PROPYL CHLOROCARBONATE
DOT: UN 2740
mf: $C_4H_7ClO_2$ mw: 122.56

PROP: Colorless liquid. D: 1.090 @ 20°/4°, bp: 114-115° @ 768 mm. Insol and sltly decomp in water, alc; misc in ether, benzene.

SYNS: CARBONOCHLORIDIC ACID, PROPYL ESTER * CHLOROFORMIC ACID PROPYL ESTER * PROPYL CHLOROFORMATE * n-PROPYL CHLOROFORMATE (DOT)

CONSENSUS REPORTS: EPA Extremely Hazardous Substances List. Reported in EPA TSCA Inventory.

DOT Classification: Flammable or Combustible Liquid; Label: Flammable, Poison, Corrosive.

SAFETY PROFILE: Poison by skin contact. Moderately toxic by ingestion and inhalation. A corrosive irritant to skin, eyes, and mucous membranes. Flammable when exposed to heat or flame, can react vigorously with oxidizing materials. When heated to decomposition it emits toxic fumes of Cl^-. Used as a reactive intermediate to polymerization initiators.

PNJ400 CAS: 78-87-5 **HR: 2**
PROPYLENE DICHLORIDE
mf: $C_3H_6Cl_2$ mw: 112.99

PROP: Colorless liquid. Bp: 96.8°, flash p: 60°F, d: 1.1593 @ 20°/20°, vap press: 40 mm @ 19.4°, vap d: 3.9, autoign temp: 1035°F, lel: 3.4%, uel: 14.5%.

SYNS: BICHLORURE de PROPYLENE (FRENCH) * 1,2-DICHLOROPROPANE * α,β-DICHLOROPROPANE * DWUCHLOROPROPAN (POLISH) * ENT 15,406 * NCI-C55141 * PROPYLENE CHLORIDE * α,β-PROPYLENE DICHLORIDE * RCRA WASTE NUMBER U083

CONSENSUS REPORTS: IARC Cancer Review: GROUP 3 IMEMDT 7,56,87; Animal Limited Evidence IMEMDT 41,131,86. NTP Carcinogenesis Studies (gavage); Equivocal Evidence: rat NTPTR* NTP-TR-263,86; Some Evidence: mouse NTPTR* NTP-TR-263,86. Reported in EPA TSCA Inventory. EPA Genetic Toxicology Program. Community Right-To-Know List.

OSHA PEL: (Transitional: TWA 75 ppm) TWA 75 ppm; STEL 110 ppm
ACGIH TLV: TWA 75 ppm; STEL 110 ppm
DFG MAK: 75 ppm (350 mg/m^3)
DOT Classification: Flammable Liquid; Label: Flammable Liquid.

SAFETY PROFILE: Moderately toxic by inhalation and ingestion. Mildly toxic by skin contact. An eye irritant. Questionable carcinogen with experimental carcinogenic data. Mutation data reported. Can cause liver, kidney, and heart damage. Can cause dermatitis. One of the more toxic chlorinated hydrocarbons. A suggested order of increasing toxicity is dichloromethane, trichloroethylene, carbon tetrachloride, dichloropropane, dichloroethane. Animals exposed to high concentrations often showed marked visceral congestion, fatty degeneration of the liver, kidney, and, less frequently, of the heart. They also showed areas of coagulation and necrosis of the liver. There was found to be a heavy mortality among mice exposed to 400 ppm concentrations. A very dangerous fire hazard when exposed to heat or flame. Reacts with aluminum to form aluminum chloride. This reaction, when confined, can lead to explosion. Can react vigorously with oxidizing materials. To fight fire, use water, foam, CO_2, dry chemical. When heated to decomposition it emits toxic fumes of Cl$^-$.

PNJ750   CAS: 9005-37-2   **HR: 1**
PROPYLENE GLYCOL ALGINATE
mf: (C$_9$H$_{14}$O$_7$)$_8$ mw: 1873.6

PROP: White fibrous or granular powder; odorless and tasteless. Sol in water and dilute organic acids.

SYNS: HYDROXY PROPYL ALGINATE   *   KELCO-LOID

CONSENSUS REPORTS: Reported in EPA TSCA Inventory.

SAFETY PROFILE: Mildly toxic by ingestion. When heated to decomposition it emits acrid smoke and irritating fumes.

PNL000   CAS: 6423-43-4   **HR: 3**
PROPYLENE GLYCOL DINITRATE
mf: C$_3$H$_6$N$_2$O$_6$ mw: 166.11

SYNS: PGDN   *   1,2-PROPYLENE GLYCOL DINITRATE   *   PROPYLENE GLYCOL-1,2-DINITRATE

CONSENSUS REPORTS: Reported in EPA TSCA Inventory.

OSHA PEL: TWA 0.05 ppm
ACGIH TLV: TWA 0.05 ppm (skin)
DFG MAK: 0.05 ppm (0.3 mg/m^3)

SAFETY PROFILE: Poison by ingestion and subcutaneous routes. Moderately toxic by intraperitoneal and intravenous routes. Human systemic effects by inhalation: conjunctiva irritation, headache. An eye irritant. When heated to decomposition it emits toxic fumes of NO$_x$.

PNL250   CAS: 107-98-2   **HR: 2**
PROPYLENE GLYCOL MONOMETHYL ETHER
mf: C$_4$H$_{10}$O$_2$ mw: 90.14

PROP: Colorless liquid. Mp: −96.7°, bp: 120°, flash p: 100°F, d: 0.919 @ 25°/25°.

SYNS: DOWANOL 33B   *   DOWTHERM 209   *   METHOXY ETHER of PROPYLENE GLYCOL   *   1-METHOXY-2-PROPANOL   *   POLY-SOLVE MPM   *   PROPYLENE GLYCOL METHYL ETHER   *   α-PROPYLENE GLYCOL MONOMETHYL ETHER   *   PROPYLENGLYKOL-MONOMETHYLAETHER (GERMAN)

CONSENSUS REPORTS: Glycol ether compounds are on the Community Right-To-Know List. Reported in EPA TSCA Inventory.

OSHA PEL: TWA 100 ppm; STEL: 150 ppm
ACGIH TLV: TWA 100 ppm; STEL: 150 ppm
DFG MAK: 100 ppm (375 mg/m^3)

SAFETY PROFILE: Moderately toxic by intravenous route. Mildly toxic by ingestion, inhalation, and skin contact. Human systemic effects by inhalation: general anesthesia, nausea. A skin and eye irritant. An experimental teratogen. Many glycol ethers have dangerous human reproductive effects. Very dangerous fire hazard when exposed to heat or flame; can react with oxidizing materials. To fight fire, use foam, CO_2, dry chemical. When heated to decomposition it emits acrid smoke and irritating fumes. Used as a solvent and in solvent-sealing of cellophane.

PNL400   CAS: 75-55-8   **HR: 3**
PROPYLENE IMINE

DOT: UN 1921
mf: C$_3$H$_7$N   mw: 57.11

PROP: Liquid. Vap d: 2.0, flash p: 14°F.

SYNS: 2-METHYLAZACYCLOPROPANE * 2-METH-YLAZIRIDINE * METHYLETHYLENIMINE * 2-METHYLETHYLENIMINE * 1,2-PROPYLENEIMINE * PROPYLENE IMINE, INHIBITED (DOT) * RCRA WASTE NUMBER P067

CONSENSUS REPORTS: IARC Cancer Review: GROUP 2B IMEMDT 7,56,87; Animal Limited Evidence IMEMDT 9,61,75. NTP Fourth Annual Report On Carcinogens, 1984. EPA Genetic Toxicology Program. Reported in EPA TSCA Inventory. EPA Extremely Hazardous Substances List. Community Right-To-Know List.

OSHA PEL: TWA 2 ppm (skin)
ACGIH TLV: TWA 2 ppm (skin), Suspected Human Carcinogen.
DFG MAK: Animal Carcinogen, Suspected Human Carcinogen.
DOT Classification: Flammable Liquid; Label: Flammable Liquid.

SAFETY PROFILE: Confirmed carcinogen with experimental carcinogenic data. Poison by ingestion and skin contact. Moderately toxic by inhalation. Mutation data reported. Severe eye irritant. Implicated as a brain carcinogen. A very dangerous fire hazard when exposed to heat or flame; can react vigorously with oxidizing materials. Polymerizes explosively on exposure to acids or acid fumes. A storage hazard. When heated to decomposition it emits toxic fumes of NO_x.

PNL600   CAS: 75-56-9   **HR: 3**
PROPYLENE OXIDE

DOT: UN 1280
mf: C_3H_6O mw: 58.09

PROP: Colorless liquid; ethereal odor. Bp: 33.9°, lel: 2.8%, uel: 37%, fp: −104.4°, flash p: −35°F (TOC), d: 0.8304 @ 20°/20°, vap press: 400 mm @ 17.8°, vap d: 2.0. Sol in water, alc, and ether.

SYNS: EPOXYPROPANE * 1,2-EPOXYPROPANE * 2,3-EPOXYPROPANE * METHYL ETHYLENE OXIDE * METHYL OXIRANE * NCI-C50099 * OXYDE de PROPYLENE (FRENCH) * PROPENE OXIDE * PROPYLENE EPOXIDE * 1,2-PROPYLENE OXIDE

CONSENSUS REPORTS: IARC Cancer Review: GROUP 2A IMEMDT 7,328,87; Human Inadequate Evidence IMEMDT 36,227,85; Animal Sufficient Evidence IMEMDT 36,227,85; Animal Limited Evidence IMEMDT 11,191,76. Carcinogenesis Studies (inhalation); Some Evidence: rat NTPTR* NTP-TR-267,85; Clear Evidence: mouse NTPTR* NTP-TR-267,85. Reported in EPA TSCA Inventory. EPA Genetic Toxicology Program. Community Right-To-Know List. EPA Extremely Hazardous Substances List.

OSHA PEL: (Transitional: TWA 100 ppm) TWA 20 ppm
ACGIH TLV: TWA 20 ppm
DFG MAK: Animal Carcinogen, Suspected Human Carcinogen.
DOT Classification: Flammable Liquid; Label: Flammable Liquid.

SAFETY PROFILE: Confirmed carcinogen with experimental carcinogenic, neoplastigenic, and tumorigenic data. Poison by intraperitoneal route. Moderately toxic by ingestion, inhalation, and skin contact. Experimental reproductive effects. Human mutation data reported. A severe skin and eye irritant. Flammable liquid. A very dangerous fire and explosion hazard when exposed to heat or flame. Explosive reaction with epoxy resin and sodium hydroxide. Forms explosive mixtures with oxygen. Reacts with ethylene oxide + polyhydric alcohol to form the thermally unstable polyether alcohol. Incompatible with NH_4OH, chlorosulfonic acid, HCl, HF, HNO_3, oleum, H_2SO_4. Dangerous; can react vigorously with oxidizing materials. Keep away from heat and open flame. To fight fire, use alcohol foam, CO_2, dry chemical. When heated to decomposition it emits acrid smoke and fumes.

PNM000   CAS: 111-43-3   **HR: 3**
PROPYL ETHER

DOT: UN 2384
mf: $C_6H_{14}O$ mw: 102.20

PROP: Colorless liquid. Mp: −122°, bp: 90°, d: 0.736 @ 20°/4°, flash p: 70°F, autoign temp: 419°F. Sltly sol in water; sol in alc, ether; very volatile.

SYNS: DIPROPYL ETHER * DI-n-PROPYLETHER * DIPROPYL OXIDE * 1,1'-OXYBISPROPANE

CONSENSUS REPORTS: Reported in EPA TSCA Inventory.

DOT Classification: Flammable Liquid; Label: Flammable Liquid.

SAFETY PROFILE: Poison by intravenous route. Possibly narcotic. Dangerous fire hazard when exposed to heat, flame, or oxidizers. Forms explosive peroxides. Dangerous upon exposure to heat or flame; can react vigorously with oxidizing materials. When heated to decomposition it emits acrid smoke and irritating fumes.

PNM500   CAS: 110-74-7   **HR: 3**
n-PROPYL FORMATE

DOT: UN 1281
mf: $C_4H_8O_2$ mw: 88.12

PROP: Colorless liquid, pleasant odor. Mp: −93°, bp: 82°, flash p: 27°F (CC), d: 0.901 @ 20°, vap press: 100 mm @ 29.5°, vap d: 3.03, autoign temp: 851°F, lel: 2.3%; misc alc, ether; sltly sol in water.

SYNS: FORMIATE de PROPYLE (FRENCH) * PROPYL FORMATE (DOT) * PROPYL METHANOATE

CONSENSUS REPORTS: Reported in EPA TSCA Inventory.

DOT Classification: Flammable Liquid; Label: Flammable Liquid.

SAFETY PROFILE: Moderately toxic by ingestion. An irritant to skin, eyes, and mucous membranes. Narcotic in high concentration. Dangerous fire hazard when exposed to heat, flame, or oxidizers. Ignites on contact with potassium-tert-butoxide. Explosive in the form of vapor when exposed to heat or flame. To fight fire, use alcohol foam. When heated to decomposition it emits acrid smoke and irritating fumes.

PNM750   CAS: 121-79-9   **HR: 3**
n-PROPYL GALLATE
mf: $C_{10}H_{12}O_5$ mw: 212.22

PROP: Odorless, fine, ivory powder or crystals; sltly bitter taste. Mp: 147-149°. Sltly sol in water; sol in alc and ether.

SYNS: GALLIC ACID, PROPYL ESTER * NIPA 49 * NIPAGALLIN P * PROGALLIN P * n-PROPYL ESTER of 3,4,5-TRIHYDROXYBENZOIC ACID * PROPYL GALLATE * n-PROPYL-3,4,5-TRIHYDROXYBENZOATE * TENOX PG * 3,4,5-TRIHYDROXYBENZENE-1-PROPYLCARBOXYLATE * 3,4,5-TRIHYDROXYBENZOIC ACID, n-PROPYL ESTER

CONSENSUS REPORTS: NTP Carcinogenesis Bioassay (feed); No Evidence: mouse, rat NTPTR* NTP-TR-240,82. Reported in EPA TSCA Inventory.

SAFETY PROFILE: Poison by ingestion and intraperitoneal routes. Experimental teratogenic and reproductive effects. Questionable carcinogen with experimental tumorigenic data. Mutation data reported. Combustible when exposed to heat or flame; can react with oxidizing materials. When heated to decomposition it emits acrid smoke and irritating fumes.

PNP000   CAS: 110-78-1   **HR: 3**
PROPYL ISOCYANATE

DOT: UN 2428
mf: C_4H_7NO mw: 85.12

SYNS: 1-ISOCYANATOPROPANE * ISOCYANIC ACID, PROPYL ESTER * m-PROPYL ISOCYANATE * 1-PROPYL ISOCYANATE

CONSENSUS REPORTS: Reported in EPA TSCA Inventory.

DOT Classification: Flammable Liquid; Label: Flammable Liquid and Poison.

SAFETY PROFILE: Poison by intravenous route. Flammable when exposed to heat or flame, can react vigorously with oxidizing materials. When heated to decomposition it emits toxic fumes of NO_x.

PNQ500   CAS: 627-13-4   **HR: 3**
n-PROPYL NITRATE

DOT: UN 1865
mf: $C_3H_7NO_3$ mw: 105.11

PROP: Pale yellow liquid; sickly odor. Bp: 110.5°, d: 1.054 @ 20°/4°, flash p: 68°F, autoign temp: 347°F (in air), lel: 2%, uel: 100%. Very sltly sol in water; sol in alc, ether.

SYNS: NITRATE de PROPYLE NORMAL (FRENCH) * NITRIC ACID, PROPYL ESTER * PROPYL NITRATE

CONSENSUS REPORTS: Reported in EPA TSCA Inventory.

OSHA PEL: (Transitional: TWA 25 ppm) TWA 25 ppm; STEL 40 ppm
ACGIH TLV: TWA 25 ppm; STEL 40 ppm
DFG MAK: 25 ppm (110 mg/m³)
DOT Classification: Flammable Liquid; Label: Flammable Liquid.

SAFETY PROFILE: Poison by intravenous route. Inhalation can cause a hypotension and methemoglobinemia. Dangerous fire hazard when exposed to heat, flame, or oxidizers. Explosive in the form of vapor when exposed to heat or flame. A shock-sensitive explosive. It can be desensitized by the addition of 1-2% propane, butane, chloroform, dimethyl ether, or diethyl ether. When heated to decomposition it emits toxic fumes of NO_x. Used as a fuel ignition promoter, chemical intermediate, and in the manufacture of rocket fuels.

PNR250   CAS: 66017-91-2   **HR: 3**
PROPYLNITROSAMINOMETHYL ACETATE
mf: $C_6H_{12}N_2O_3$ mw: 160.20

SYNS: ACETOXYMETHYLPROPYLNITROSAMINE
* N-(ACETOXY)METHYL-N-n-PROPYLNITROSAMINE
* N-NITROSO-N-(1-ACETOXYMETHYL)PROPYL AMINE
* PAMN * PROPYL ACETOXYMETHYLNITROS-
AMINE * N-PROPYL-N-(ACETOXYMETHYL)NITROS-
AMINE

SAFETY PROFILE: Moderately toxic by ingestion. Questionable carcinogen with experimental carcinogenic and tumorigenic data. Mutation data reported. When heated to decomposition it emits toxic fumes of NO_x.

PNX000   CAS: 51-52-5   **HR: 3**
6-PROPYL-2-THIOURACIL
mf: $C_7H_{10}N_2OS$ mw: 170.25

PROP: White, bitter, crystalline powder. Mp: 219-221°. Insol in ether, chloroform, benzene; very sol in aq solns of ammonia; very sltly sol in water.

SYNS: 2,3-DIHYDRO-6-PROPYL-2-THIOXO-4(1H)-PYRI-
MIDINONE * 2-MERCAPTO-4-HYDROXY-6-N-PROPYL-
PYRIMIDINE * 2-MERCAPTO-6-PROPYL-4-PYRIMIDONE
* 2-MERCAPTO-6-PROPYLPYRIMID-4-ONE * PROCA-
SIL * PROPACIL * PROPILTHIOURACIL
* 6-PROPIL-TIOURACILE (ITALIAN) * PROPYCIL
* 6-PROPYL-2-THIO-2,4(1H,3H)PYRIMIDINEDIONE
* PROPYL-THIORIST * PROPYLTHIOURACIL
* 4-PROPYL-2-THIOURACIL * 6-N-PROPYLTHIOURA-
CIL * 6-N-PROPYL-2-THIOURACIL * PROPYL-THY-
RACIL * PROPYTHIOURACIL * PROTHIUCIL
* PROTHIURONE * PROTHYCIL * PROTHYRAN
* PROTIURAL * PTU (THYREOSTATIC) * 2-THIO-
4-OXO-6-PROPYL-1,3-PYRIMIDINE * 2-THIO-6-PROPYL-
1,3-PYRIMIDIN-4-ONE * 6-THIO-4-PROPYLURACIL
* THIURAGYL * THYREOSTAT II * T 72

CONSENSUS REPORTS: IARC Cancer Review: GROUP 2B IMEMDT 7,329,87; Animal Sufficient Evidence IMEMDT 7,67,74. NTP Fourth Annual Report On Carcinogens, 1984. Reported in EPA TSCA Inventory.

SAFETY PROFILE: Confirmed carcinogen with experimental carcinogenic, neoplastigenic, and tumorigenic data. Poison by an unspecified route. Moderately toxic by ingestion. Human teratogenic effects by ingestion: developmental abnormalities of the endocrine system and changes in newborn viability. Experimental teratogenic and reproductive effects. Human mutation data reported. When heated to decomposition it emits very toxic fumes of SO_x and NO_x.

PNX250   CAS: 141-57-1   **HR: 2**
n-PROPYLTRICHLOROSILANE

DOT: UN 1816
mf: $C_3H_7Cl_3Si$ mw: 177.54

PROP: Vap d: 6.15, flash p: 100°F.

SYNS: PROPYLTRICHLOROSILANE (DOT) * TRI-
CHLOROPROPYLSILANE

CONSENSUS REPORTS: Reported in EPA TSCA Inventory.

DOT Classification: Corrosive Material; Label: Corrosive; Label: Corrosive, Flammable Liquid.

SAFETY PROFILE: A corrosive and irritating material to skin, eyes, and mucous membranes. A dangerous fire hazard when exposed to heat or flame. Will react with water or steam to produce toxic and corrosive fumes; can react with oxidizing materials. To fight fire, use foam, CO_2, dry chemical. When heated to decomposition it emits toxic fumes of Cl^-.

POC750   CAS: 38562-01-5   **HR: 3**
PROSTAGLANDIN F2-α-THAM
mf: $C_{20}H_{34}NO_5 \cdot C_4H_{11}NO_3$ mw: 475.70

SYNS: 7-(3,5-DIHYDROXY-2-(3-HYDROXY-1-OCTE-
NYL)CYCLOPENTYL)-5-HEPTENOIC ACID, THAM
* 7-(3,5-DIHYDROXY-2-(3-HYDROXY-1-OCTENYL)CY-
CLOPENTYL)-5-HEPTENOIC ACID, TRIMETHAMINE
SALT * DINOPROST TROMETHAMINE (USDA)
* 583E * LUTALYSE * PGF2-α THAM
* PGF2-α TRIS SALT * PGF2-α TROMETHAMINE
* PROSTAGLANDIN F2-α THAM SALT
* PROSTAGLANDIN F2a TROMETHAMINE
* THAM * TROMETHAMINE PROSTAGLANDIN F2-α
* U-14

SAFETY PROFILE: Poison by intraperitoneal, subcutaneous, intravenous, and intramuscular routes. Moderately toxic by ingestion. Human reproductive effects by intervaginal route: terminates pregnancy, effects on fertility. Experimental teratogenic and reproductive effects. When heated to decomposition it emits toxic fumes of NO_x.

POD000 CAS: 114-80-7 HR: 3
PROSTIGMINE BROMIDE
mf: $C_{12}H_{19}N_2O_2 \cdot Br$ mw: 303.24

SYNS: 3-DIMETHYLCARBAMOXYPHENYLTRIMETHYL-AMMONIUM BROMIDE * (m-HYDROXYPHENYL)TRI-METHYLAMMONIUM BROMIDE DIMETHYLCARBAMATE * 3-HYDROXYPHENYLTRIMETHYLAMMONIUM BROMIDE DIMETHYLCARBAMIC ESTER * SYNTHOSTIG-MINE BROMIDE

SAFETY PROFILE: Poison by ingestion, subcutaneous, intravenous, and intraperitoneal routes. When heated to decomposition it emits very toxic fumes of Br^-, NH_3, and NO_x.

POD500 HR: 3
PROTACTINIUM
af: Pa aw: 231.036

PROP: A bright, lustrous metal. Mp: 1600°, d: 15.37, vap press: 5×10^{-5} mm @ 1927°. Natural isotope ^{231}Pa (Actinium series), $T_{0.5}$ = 3×10^4 Y., decays to radioactive ^{227}Ac by alphas of 5.0 MeV. Artificial isotope ^{233}Pa (Neptunium Series), $T_{0.5}$ = 27D, decays to radioactive ^{233}U by betas of 0.15 (37%), 0.26 (58%), 0.57 (5%) MeV; emits gammas of 0.02-0.42 MeV. Natural isotope ^{234}Pa (Uranium Series), $T_{0.5}$ = 6.7H, decays to radioactive 234U by betas of 0.23-1.36 MeV, emits gammas of 0.04-0.8 MeV.

SAFETY PROFILE: Confirmed carcinogen. A highly radiotoxic metallic element. An alpha emitter. It is a general hazard if absorbed systemically. The dust and fumes are hazardous if inhaled. A severe radiation hazard.

POF250 CAS: 1225-55-4 HR: 3
PROTRIPTYLINE HYDROCHLORIDE
mf: $C_{19}H_{21}N \cdot ClH$ mw: 299.87

SYNS: 5-(3-METHYLAMINOPROPYL)-5H-DIBENZO(a,d) CYCLOHEPTENE HYDROCHLORIDE * N-METHYL-5H-DIBENZO(A,D)CYCLOHEPTENE-5-PROPYLAMINE HYDROCHLORIDE * NORMETHYL EX4442 * TRIP-TIL HYDROCHLORIDE * VIVACTIL

SAFETY PROFILE: Poison by ingestion, intraperitoneal, intravenous, and subcutaneous routes. When heated to decomposition it emits very toxic fumes of NO_x and HCl.

POG250 HR: 3
PSEUDOACONITINE
mf: $C_{36}H_{51}NO_{12}$ mw: 689.78

PROP: White crystals or syrupy mass. Mp: 214° (decomp). Insol in water; sol in alc, ether.

SAFETY PROFILE: Poison by ingestion, inhalation, and skin contact. When heated to decomposition it emits highly toxic fumes of NO_x.

POH750 CAS: 127-91-3 HR: 1
PSEUDOPINENE
mf: $C_{10}H_{16}$ mw: 136.26

PROP: Colorless liquid; pine odor. D: 0.864, refr index: 1.477, flash p: 88°F. Sol in fixed oils; insol in water, propylene glycol, glycerin

SYNS: 6,6-DIMETHYL-2-METHYLENEBICY-CLO(3.1.1)HEPTANE * FEMA No. 2903 * NOPINEN * NOPINENE * β-PINENE (FCC) * 2(10)-PINENE * PSEUDOPINEN

CONSENSUS REPORTS: Reported in EPA TSCA Inventory.

SAFETY PROFILE: Mildly toxic by ingestion. A skin irritant. Flammable liquid. When heated to decomposition it emits acrid smoke and irritating fumes.

POJ000 CAS: 126-17-0 HR: 3
PURAPURIDINE
mf: $C_{27}H_{43}NO_2$ mw: 413.71

SYNS: SOLANCARPIDINE * SOLANIDINE-S * SOLASOD-5-EN-3-β-OL * SOLASODINE

SAFETY PROFILE: Poison by intraperitoneal route. Moderately toxic by ingestion. Experimental teratogenic and reproductive effects. When heated to decomposition it emits toxic fumes of NO_x.

POK000 CAS: 50-44-2 HR: 3
PURINE-6-THIOL
mf: $C_5H_4N_4S$ mw: 152.19

SYNS: 1,7-DIHYDRO-6H-PURINE-6-THIONE * ISMI-PUR * LEUKERAN * LEUPURIN * MERCALEU-KIN * MERCAPTOPURIN (GERMAN) * 6-MERCAP-TOPURIN * 6-MERCAPTOPURINE * 7-MERCAPTO-1,3,4,6-TETRAZAINDENE * MERCAPURIN * MERN * MP * NCI-C04886 * NSC 755 * PURIMETHOL * 3H-PURINE-6-THIOL * 6-PURINETHIOL * PU-RINETHOL * THIOHYPOXANTHINE * 6-THIOXOPU-RINE * U-4748

CONSENSUS REPORTS: IARC Cancer Review: GROUP 3 IMEMDT 7,240,87; Animal Inadequate Evidence IMEMDT 26,249,81; Human Inadequate Evidence IMEMDT 26,249,81. NCI Carcinogenesis Studies (ipr); Equivocal Evidence: rat CANCAR 40,1935,77; (ipr); Clear Evidence: mouse CANCAR 40,1935,77. EPA Genetic Toxicology Program.

SAFETY PROFILE: Poison by ingestion, intraperitoneal, subcutaneous, parenteral, and intravenous routes. Human systemic effects by ingestion: dermatitis. Experimental teratogenic and reproductive effects. Questionable human carcinogen producing Hodgkin's disease and leukemia. Human mutation data reported. When heated to decomposition it emits very toxic fumes of SO_x and NO_x.

POL500   CAS: 98-96-4   HR: 3
PYRAZINECARBOXAMIDE
mf: $C_5H_5N_3O$ mw: 123.13

SYNS: ALDINAMID * 2-CARBAMYL PYRAZINE * D-50 * EPRAZIN * MK 56 * NCI-C01785 * PYRAZINAMIDE * PYRAZINEAMIDE * PYRAZINE CARBOXYLAMIDE * PYRAZINOIC ACID AMIDE * TEBRAZID

CONSENSUS REPORTS: NCI Carcinogenesis Bioassay (feed); No Evidence: rat NCITR* NCI-CG-TR-48,78; (feed); Inadequate Studies: mouse NCITR* NCI-CG-TR-48,78. Reported in EPA TSCA Inventory.

SAFETY PROFILE: Moderately toxic by ingestion, subcutaneous, and intraperitoneal routes. Questionable carcinogen with experimental tumorigenic data. Human mutation data reported. When heated to decomposition it emits toxic fumes of NO_x.

PON250   CAS: 129-00-0   HR: 3
PYRENE
mf: $C_{16}H_{10}$ mw: 202.26

PROP: Colorless solid, solutions have a slight blue color. Mp: 156°, d: 1.271 @ 23°, bp: 404°. Insol in water; fairly sol in organic solvents. (A condensed ring hydrocarbon).

SYNS: BENZO(def)PHENANTHRENE * PYREN (GERMAN) * β-PYRINE

CONSENSUS REPORTS: IARC Cancer Review: GROUP 3 IMEMDT 7,56,87; Animal No Evidence IMEMDT 32,431,83. EPA Extremely Hazardous Substances List. Reported in EPA TSCA Inventory. EPA Genetic Toxicology Program.

OSHA PEL: TWA 0.2 mg/m^3

SAFETY PROFILE: Poison by inhalation. Moderately toxic by ingestion and intraperitoneal routes. A skin irritant. Questionable carcinogen with experimental tumorigenic data. Human mutation data reported. When heated to decomposition it emits acrid smoke and irritating fumes.

POO000   CAS: 97-11-0   HR: 2
PYRETHRIN
mf: $C_{21}H_{28}O_3$ mw: 328.49

SYNS: 2-CYCLOPENTENYL-4-HYDROXY-3-METHYL-2-CYCLOPENTEN-1-ONE CHRYSANTHEMATE * 3-(2-CYCLOPENTEN-1-YL)-2-METHYL-4-OXO-2-CYCLOPENTEN-1-YL CHRYSANTHEMUMATE * 3-(2-CYCLOPENTENYL)-2-METHYL-4-OXO-2-CYCLOPENTENYL CHRYSANTHEMUMMONOCARBOXYLATE * CYCLOPENTENYLRETHONYL CHRYSANTHEMATE * ENT 22,952

SAFETY PROFILE: Moderately toxic by ingestion and possibly other routes. When heated to decomposition it emits acrid smoke and irritating fumes.

POO100   CAS: 121-29-9   HR: 3
PYRETHRIN II
mf: $C_{22}H_{28}O_5$ mw: 372.50

PROP: Viscous liquid. Bp: 200° @ 0.1 mm (decomp).

SYNS: CHRYSANTHEMUMDICARBOXYLIC ACID MONOMETHYL ESTER PYRETHROLONE ESTER * ENT 7,543 * PYRETHRIN * PYRETHROLONE CHRYSANTHEMUM DICARBOXLIC ACIDMETHYL ESTER ESTER * PYRETHROLONE ESTER of CHRYSANTHEMUMDICARBOXYLIC ACID MONOMETHYL ESTER * (+)-PYRETHRONYL (+)-PYRETHRATE * PYRETRIN II

CONSENSUS REPORTS: Reported in EPA TSCA Inventory.

SAFETY PROFILE: Poison experimentally by ingestion and intravenous routes. Moderately toxic to humans by unspecified route. An allergen. When heated to decomposition it emits acrid smoke and irritating fumes. An insecticide.

POO250   CAS: 8003-34-7   HR: 3
PYRETHRINS
DOT: NA 9184

PROP: Viscous liquid. Bp: 170° @ 0.1 mm (decomp).

SYNS: BUHACH * CHRYSANTHEMUM CINERAREA-EFOLIUM * CINERIN I or II * DALMATION INSECT FLOWERS * FIRMOTOX * INSECT POWDER * JASMOLIN I or II * PYRETHRIN I or II * PYRETHRUM (ACGIH) * PYRETHRUM (INSECTICIDE) * TRIESTE FLOWERS

OSHA PEL: TWA 5 mg/m^3
ACGIH TLV: TWA 5 mg/m^3
DFG MAK: 5 mg/m^3
DOT Classification: ORM-E; Label: None.

SAFETY PROFILE: Moderately toxic to humans by ingestion. Poison experimentally by ingestion, intraperitoneal, and intravenous route. Experimental reproductive effects. An allergen. It is rapidly detoxified in the gastrointestinal tract, but can cause gastrointestinal, respiratory, and central nervous system effects. A dose of 15 grams has caused the death of a child. Chronic exposures can cause liver damage. Combustible when exposed to heat or flame. When heated to decomposition it emits acrid smoke and irritating fumes. An insecticide.

POP000   CAS: 119-12-0   HR: 3
PYRIDAPHENTHION
mf: C$_{14}$H$_{17}$N$_2$O$_4$PS   mw: 340.36

SYNS: AMERICAN CYANAMID 12,503 * CL 12503 * O,O-DIETHYL O-(2,3-DIHYDRO-3-OXO-2-PHENYL-6-PYRIDAZINYL)PHOSPHOROTHIOATE * O,O-DIETHYL-PHOSPHOROTHIOATE, O-ESTER with 6-HYDROXY-2-PHE-NYL-3(2H)-PYRIDAZINONE * O-(1,6)-DIHYDRO-6-OXO-1-PHENYLPYRIDAZIN-3-LY), O,O-DIETHYL PHOSPHO-ROTHIOATE * ENT 23,968 * OFNACK * OFUNACK * PYRIDAFENTHION

SAFETY PROFILE: Poison by intraperitoneal route. Moderately toxic by ingestion and skin contact. When heated to decomposition it emits very toxic fumes of SO$_x$, PO$_x$, and NO$_x$. Used to control chewing and sucking insects on rice, fruits, vegetables, and cereals.

POP250   CAS: 110-86-1   HR: 3
PYRIDINE

DOT: UN 1282
mf: C$_5$H$_5$N   mw: 79.11

PROP: Colorless liquid; sharp, penetrating, empyreumatic odor; burning taste. Bp: 115.3°, lel: 1.8%, uel: 12.4%, fp: −42°, flash p: 68°F (CC), d: 0.982, autoign temp: 900°F, vap press: 10

mm @ 13.2°, vap d: 2.73. Volatile with steam; misc with water, alc, ether.

SYNS: AZABENZENE * AZINE * NCI-C55301 * PIRIDINA (ITALIAN) * PIRYDYNA (POLISH) * PYRIDIN (GERMAN) * RCRA WASTE NUMBER U196

CONSENSUS REPORTS: Community Right-To-Know List. Reported in EPA TSCA Inventory. EPA Genetic Toxicology Program.

OSHA PEL: TWA 5 ppm
ACGIH TLV: TWA 5 ppm
DFG MAK: 5 ppm (15 mg/m^3)
DOT Classification: Flammable Liquid; Label: Flammable Liquid; Flammable Liquid; Label: Flammable Liquid, Poison.

SAFETY PROFILE: Poison by intraperitoneal route. Moderately toxic by ingestion, skin contact, intravenous, and subcutaneous routes. Mildly toxic by inhalation. A skin and severe eye irritant. Mutation data reported. Can cause central nervous system depression, gastrointestinal upset, and liver and kidney damage. Dangerous fire hazard when exposed to heat, flame, or oxidizers. Severe explosion hazard in the form of vapor when exposed to flame or spark. Reacts violently with chlorosulfonic acid, chromium trioxide, dinitrogen tetraoxide, HNO$_3$, oleum, perchromates, β-propiolactone, AgClO$_4$, H$_2$SO$_4$. Incandescent reaction with fluorine. Reacts to form pyrophoric or explosive products with bromine trifluoride, trifluoromethyl hypofluorite. Mixtures with formamide + iodine + sulfur trioxide are storage hazards which release carbon dioxide and sulfuric acid. Incompatible with oxidizing materials. Reacts with maleic anhydride (above 150°C) evolving carbon dioxide. To fight fire, use alcohol foam. When heated to decomposition it emits highly toxic fumes of NO$_x$. Used as an intermediate for pesticides production, in pharmaceuticals, as a solvent, and to denature alcohol.

POR500   CAS: 5344-27-4   HR: 3
4-PYRIDINEETHANOL
mf: C$_7$H$_9$NO   mw: 123.17

SYN: 4-ETHANOLPYRIDINE

CONSENSUS REPORTS: Reported in EPA TSCA Inventory.

SAFETY PROFILE: Poison by intravenous route. When heated to decomposition it emits toxic fumes of NO$_x$.

POS750   CAS: 94-63-3   **HR: 3**
PYRIDINIUM-2-ALDOXIME-N-METHYLIODIDE
mf: $C_7H_9N_2O \cdot I$ mw: 264.08

PROP: Water-sol crystals. Mp: 214°.

SYNS: 2-FORMYL-1-METHYLPYRIDINIUM IODIDE OXIME * 2-FORMYL-N-METHYLPYRIDINIUM OXIME IODIDE * 2-HYDROXYIMINOMETHYL-1-METHYL-PYRIDINIUM IODIDE * 1-METHYL-2-ALDOXIMINO-PYRIDINIUM IODIDE * 1-METHYL-2-HYDROXYIMINO-METHYLPYRIDINIUM IODIDE * N-METHYLPYRIDINE-2-ALDOXIME IODIDE * N-METHYLPYRIDINIUM-2-ALDOXIME IODIDE * NSC-7760 * PAM (CZECH) * 2-PAM IODIDE * PRALIDOXIME IODIDE * PRALIDOXIME METHIODIDE * PROTOPAM IODIDE * 2-PYRIDINALDOXIM METHOJODID (GERMAN) * PYRIDIN-2-ALDOXIN (CZECH) * 2-PYRIDINE AL-DOXIME IODOMETHYLATE * PYRIDINE-2-AL-DOXIME METHIODIDE * PYRIDINE-2-ALDOXIME METHYL IODIDE

SAFETY PROFILE: Poison by subcutaneous, intravenous, intramuscular, and intraperitoneal routes. Moderately toxic by ingestion. Used as an antidote to the cholinesterase inhibitors of the parathion group. When heated to decomposition it emits highly toxic fumes of NO_x and I^-.

PPC100   CAS: 15598-34-2   **HR: 3**
PYRIDINIUM PERCHLORATE
mf: $C_5H_6ClNO_4$ mw: 179.56

DOT Classification: Forbidden

SAFETY PROFILE: An explosive sensitive to impact or heating above 335°C. Addition of ammonium perchlorate decreases the temperature required for initiation. When heated to decomposition it emits toxic fumes of Cl^- and NO_x.

PPK250   CAS: 65-23-6   **HR: 2**
PYRIDOXOL
mf: $C_8H_{11}NO_3$ mw: 169.20

SYNS: ADERMINE * BEESIX * GRAVIDOX * HYDOXIN * 3-HYDROXY-4,5-DIMETHYLOL-α-PICOLINE * 5-HYDROXY-6-METHYL-3,4-PYRIDINEDI-METHANOL * 3-HYDROXY-2-PICOLINE-4,5-DIMETH-ANOL * 2-METHYL-4,5-BIS(HYDROXYMETHYL)-3-HYDROXYPYRIDINE * 2-METHYL-3-HYDROXY-4,5-BIS(HYDROXYMETHYL)PYRIDINE * 2-METHYL-3-HYDROXY-4,5-DIHYDROXYMETHYL-PYRIDIN

(GERMAN) * 2-METHYL-3-HYDROXY-4,5-DI(HY-DROXYMETHYL)PYRIDINE * PYRODOXIN * PYRIDOXINE * VITAMIN B6

CONSENSUS REPORTS: Reported in EPA TSCA Inventory.

SAFETY PROFILE: Moderately toxic by ingestion, subcutaneous, intravenous, and intraperitoneal routes. When heated to decomposition it emits toxic fumes of NO_x.

PPK500   CAS: 58-56-0   **HR: 3**
PYRIDOXOL HYDROCHLORIDE
mf: $C_8H_{11}NO_3 \cdot ClH$ mw: 205.66

PROP: Commercial form of pyridoxine (Vitamin B_6). Colorless to white platelets or crystalline powder; odorless. Mp: 204-206° (decomp). Sol in water, alc, acetone; sltly sol in other organic solvents; insol in ether.

SYNS: ADERMINE HYDROCHLORIDE * BECILAN * BENADON * CAMPOVITON 6 * HEXABETALIN * HEBABIONE HYDROCHLORIDE * HEXAVIBEX * HEXERMIN * HEXOBION * 3-HYDROXY-4,5-DIMETHYLOL-α-PICOLINE HYDROCHLORIDE * 5-HYDROXY-6-METHYL-3,4-PYRIDINEDICARBINOL HYDROCHLORIDE * 5-HYDROXY-6-METHYL-3,4-PYRI-DINEDIMETHANOL HYDROCHLORIDE * 2-METHYL-3-HYDROXY-4,5-BIS(HYDROXYMETHYL)PYRIDINE HYDROCHLORIDE * PYRIDIPCA * PYRIDOXINE HYDROCHLORIDE (FCC) * PYRIDOXINIUM CHLORIDE * PYRIDOXINUM HYDROCHLORICUM (HUNGARIAN) * VITAMIN B6-HYDROCHLORIDE

CONSENSUS REPORTS: Reported in EPA TSCA Inventory.

SAFETY PROFILE: Poison by intravenous route. Moderately toxic by ingestion, intramuscular, and subcutaneous routes. Human reproductive effects by ingestion and intramuscular routes: postpartum changes. Experimental teratogenic effects. Human mutation data reported. When heated to decomposition it emits very toxic fumes of NO_x and HCl.

PPL500   CAS: 19992-69-9   **HR: 3**
1-(PYRIDYL-3)-3,3-DIMETHYL TRIAZENE
mf: $C_7H_{10}N_4$ mw: 150.21

SYNS: 1-(PYRIDYL-3)-3,3-DIMETHYL-TRIAZEN (GERMAN) * 1-(m-PYRIDYL)-3,3-DIMETHYL-TRIAZENE

SAFETY PROFILE: Poison by ingestion and subcutaneous routes. Experimental reproductive effects. Questionable carcinogen with experimental neoplastigenic data. Mutation data reported. When heated to decomposition it emits toxic fumes of NO_x.

PPP750   CAS: 53558-25-1   HR: 3
PYRIMINYL
mf: $C_{13}H_{12}N_4O_3$ mw: 272.29

SYNS: N-(4-NITROPHENYL)-N'-(3-PYRIDINYLMETHYL) UREA * N-3-PYRIDYLMETHYL-N'-p-NITROPHENYL-UREA

CONSENSUS REPORTS: EPA Extremely Hazardous Substances List. Reported in EPA TSCA Inventory.

SAFETY PROFILE: Human poison by ingestion. Human systemic effects by ingestion: hallucinations, distorted perceptions, muscle weakness, nausea. When heated to decomposition it emits toxic fumes of NO_x.

PPQ500   CAS: 87-66-1   HR: 3
PYROGALLOL
mf: $C_6H_6O_3$ mw: 126.12

PROP: White, lustrous crystals. Bp: 309°, d: 1.453 @ 4°/4°, vap press: 10 mm @ 167.7°, mp: 131-133°. Sltly sol in benzene, chloroform.

SYNS: 1,2,3-BENZENETRIOL * C.I. 76515 * C.I. OXIDATION BASE 32 * FOURAMINE BROWN AP * FOURRINE PG * PYROGALLIC ACID * 1,2,3-TRIHYDROXYBENZEN (CZECH) * 1,2,3-TRI-HYDROXYBENZENE

CONSENSUS REPORTS: Reported in EPA TSCA Inventory.

SAFETY PROFILE: Human poison by ingestion and subcutaneous routes. An experimental poison by ingestion, subcutaneous, intravenous, and intraperitoneal routes. Experimental teratogenic and reproductive effects. Questionable carcinogen with experimental tumorigenic data. Mutation data reported. Readily absorbed through the skin. Human systemic effects by ingestion: convulsions, dyspnea, gastrointestinal effects. A severe skin and eye irritant. Incompatible with alkalies; NH_3; antipyrine; phenol; iron and lead salts; iodine; $KMnO_4$. When heated to decomposition it emits acrid smoke and irritating fumes. Used as a topical antibacterial agent, as an intermediate, hair dye component, and analytical reagent.

PPR500   CAS: 7791-27-7   HR: 3
PYROSULFURYL CHLORIDE
DOT: UN 1817
mf: $Cl_2O_5S_2$ mw: 215.02

PROP: Colorless, mobile, fuming liquid. Mp: −37°, bp: 151°, d: 1.83, (gas): 9.6 g/L.

SYNS: CHLOROSULFONIC ANHYDRIDE * DISULFUR PENTOXYDICHLORIDE * DISULFURYL CHLORIDE * DISULFURYL DICHLORIDE * PYRO SULFURYL CHLORIDE (DOT) * PYROSULPHURYL CHLORIDE (DOT)

DOT Classification: Corrosive Material; Label: Corrosive.

SAFETY PROFILE: A very poisonous material which is also corrosive to the eyes, skin, and mucous membranes. Violent reaction with water. Vigorous reaction with phosphorus. When heated to decomposition it emits very toxic fumes of Cl^- and SO_x.

PPS250   CAS: 109-97-7   HR: 3
PYRROLE
mf: C_4H_5N mw: 67.10

PROP: Colorless liquid, darkens on standing; mild nutty odor. Fp: −24°, flash p: 102°F (TCC), d: 0.968 @ 20°/4°, refr index: 1.507, vap d: 2.31, bp: 130-131° @ 761 mm. Sltly sol in water; very sol in alc, fixed oils, benzene, ether; insol in alkali.

SYNS: 1-AZA-2,4-CYCLOPENTADIENE * AZOLE * DIVINYLENIMINE * FEMA No. 3386 * IMIDOLE * MONOPYRROLE

CONSENSUS REPORTS: Reported in EPA TSCA Inventory.

SAFETY PROFILE: Poison by subcutaneous, intraperitoneal, and possibly other routes. Flammable when exposed to heat or flame; can react with oxidizing materials. To fight fire, use foam, CO_2, dry chemical. Violent reaction with 2-nitrobenzaldehyde. When heated to decomposition it emits highly toxic fumes of NO_x.

PPS500   CAS: 123-75-1   HR: 3
PYRROLIDINE
DOT: UN 1922
mf: C_4H_9N mw: 71.14

PROP: Colorless, mobile liquid; penetrating, amine-like odor. Fp: −63°, flash p: 37°F (TCC), d: 0.8618 @ 20°/4°, vap press: 128 mm @ 39°, vap d: 2.45; bp: 88.5-89°. Fumes in air. Misc with water; sol in alc, ether, chloroform.

SYNS: AZACYCLOPENTANE * TETRAHYDROPYRROLE * TETRAMETHYLENIMINE

CONSENSUS REPORTS: Reported in EPA TSCA Inventory.

DOT Classification: Flammable Liquid; Label: Flammable Liquid.

SAFETY PROFILE: Poison by ingestion and intravenous routes. Moderately toxic by inhalation. Dangerous fire hazard when exposed to heat or flame; can react vigorously with oxidizing materials. To fight fire, use alcohol foam, CO_2, dry chemical. When heated to decomposition it emits highly toxic fumes of NO_x.

Q

QAK000   CAS: 72-44-6   **HR: 3**
QUAALUDE
mf: $C_{16}H_{14}N_2O$ mw: 250.32

SYNS: CATEUDYL * CITEXAL * CI-705
* CN 38703 * 3,4-DIHYDRO-2-METHYL-4-OXO-3-o-
TOLYLQUINAZOLINE * DORMIGOA * DORMO-
GEN * DORMUTIL * DORSEDIN * FADORMIR
* HOLODORM * HYMINAL * HYPCOL
* HYPTOR BASE * IPNOFIL * MAOA
* MEQUIN * MELSEDIN BASE * MELSOMIN
* METAQUALON * METHAQUALONE
* METHAQUALONEINONE * 2-METHYL-3-
(2-METHYLPHENYL)-4-QUINAZOLINONE * 2-METH-
YL-3-(2-METHYLPHENYL)-4(3H)-QUINAZOLINONE
* 2-METHYL-3-o-TOLYL-4(3H)-CHINAZOLINON (GER-
MAN) * 2-METHYL-3-o-TOLYL-4(3H)-CHINAZO-
LONE * (2-METHYL-3-(o-TOLYL)-3,4-DIHYDRO-
4-(QUINAZOLINONE) * 2-METHYL-3-(o-TOLYL)-
3,4-DIHYDRO-4-QUINAZOLINONE * 2-METHYL-3-
TOLYL-4-OXYBENSDIAZINE * 2-METHYL-3-o-TOLYL-
4(3H)-QUINAZOLINONE * 2-METHYL-3-o-TOLYL-4-
QUINAZOLONE * 2-METHYL-3-(2-TOLYL)QUIN-
AZOL-4-ONE * METOLQUIZOLONE * MOLLINOX
* MOTOLON * MOZAMBIN * MTQ * NOBE-
DORM * NOCTILENE * NORMI-NOX * OMNYL
* OPTINOXAN * ORTHONAL * ORTONAL
* PAREST * PARMINAL * PRO-DORM * QZ 2
* REVONAL * RORER 148 * ROUQUALONE
* SINDESVEL * SOMBEROL * SOMNAFAC
* SOMNOMED * SONAL * SOVERIN * TORI-
NAL * TUAZOLE * TUAZOLONE

CONSENSUS REPORTS: Reported in EPA
TSCA Inventory.

SAFETY PROFILE: Human poison by ingestion. Experimental poison by ingestion, intravenous, and intraperitoneal routes. Moderately
toxic by parenteral route. Human systemic effects by ingestion: convulsions or effect on seizure threshold, nausea or vomiting, and pulmonary changes. Experimental teratogenic and
reproductive effects. A controlled drug which
is often abused. When heated to decomposition
it emits toxic fumes of NO_x.

QBJ000   CAS: 1401-55-4   **HR: 3**
QUEBRACHO TANNIN

SYNS: SCHINOPSIS LORENTZII TANNIN * TANNIN
from QUEBRACHO

SAFETY PROFILE: Poison by intraperitoneal
and intravenous routes. Questionable carcinogen with experimental tumorigenic data. When
heated to decomposition it emits acrid smoke
and irritating fumes.

QBS000   CAS: 64719-39-7   **HR: 3**
QUELAMYCIN
mf: $C_{27}H_{27}O_{11} \cdot 2Fe(2+) \cdot Fe(3+)$ mw:
709.07

SYNS: NSC-267703 * TRIFERRIC ADRIAMYCIN
* TRIFERRIC DOXORUBICIN

SAFETY PROFILE: Poison by intravenous and
intraperitoneal routes. Human systemic effects
by intravenous route: blood effects. Mutation
data reported. When heated to decomposition
it emits acrid smoke and irritating fumes.

QCA000   CAS: 117-39-5   **HR: 3**
QUERCETIN
mf: $C_{15}H_{10}O_7$ mw: 302.25

SYNS: C.I. 75670 * C.I. NATURAL RED 1
* C.I. NATURAL YELLOW 10 * CYANIDELONON
1522 * 2-(3,4-DIHYDROXYPHENYL)-3,5,7-TRIHY-
DROXY-4H-1-BENZOPYRAN-4-ONE * MELETIN
* NCI-C60106 * 3,5,7,3',4'-PENTAHYDROXYFLA-
VONE * QUERCETINE * QUERCETOL * QUER-
CITIN * QUERTINE * SOPHORETIN * 3',4',5,7-
TETRAHYDROXYFLAVAN-3-OL * T-GELB BZW,
GRUN 1 * XANTHAURINE

CONSENSUS REPORTS: IARC Cancer Review: GROUP 3 IMEMDT 7,56,87; Animal
Limited Evidence IMEMDT 31,213,83. Reported in EPA TSCA Inventory. EPA Genetic
Toxicology Program.

SAFETY PROFILE: Poison by ingestion, subcutaneous, and intravenous routes. Experimental
teratogenic and reproductive effects. Questionable carcinogen with experimental carcinogenic,
neoplastigenic, and tumorigenic data. Human
mutation data reported. Used as a pharmaceutical and veterinary drug. When heated to decomposition it emits acrid smoke and irritating
fumes.

QDJ000   CAS: 64046-79-3   **HR: 3**
QUINACRINE MUSTARD
mf: $C_{23}H_{28}Cl_3N_3O$ mw: 468.89

SYNS: 9-(4-(BIS-β-CHLOROETHYLAMINO)-1-METHYL-BUTYLAMINO)-6-CHLORO-2-METHOXYACRIDINE * NSC-3424

SAFETY PROFILE: A deadly poison by intravenous and intraperitoneal routes. Human mutagenic data reported. When heated to decomposition it emits very toxic fumes of Cl^- and NO_x.

QDS000   CAS: 4213-45-0   **HR: 3**
QUINACRINE MUSTARD DIHYDROCHLORIDE
mf: $C_{23}H_{28}Cl_3N_3O \cdot 2ClH$ mw: 541.81

SYNS: 9-(4-BIS(2-CHLOROETHYL)AMINO-1-METHYL-BUTYLAMINO)-6-CHLORO-2-METHOXYACRIDINE DI-HYDROCHLORIDE * ICR 10 * 2-METHOXY-6-CHLORO-9-(4-BIS(2-CHLOROETHYL)AMINO-1-METH-YLBUTYLAMINO)ACRIDINE DIHYDROCHLORIDE * 2-METHOXY-6-CHLORO-9-(3-(ETHYL-2-CHLO-ROETHYL)AMINOPROPYLAMINO)ACRIDINE DIHYDRO-CHLORIDE * QUINACRINE MUSTARD

CONSENSUS REPORTS: EPA Genetic Toxicology Program. Reported in EPA TSCA Inventory.

SAFETY PROFILE: Questionable carcinogen with experimental neoplastigenic data. Human mutation data reported. When heated to decomposition it emits very toxic fumes of Cl^- and NO_x.

QFS000   CAS: 56-54-2   **HR: 3**
QUINIDINE
mf: $C_{20}H_{24}N_2O_2$ mw: 324.46

SYNS: CHINIDIN (GERMAN) * CIN-QUIN * CONCHININ * CONQUININE * 6'-METH-OXYCINCHONAN-9-OL * α-(6-METHOXY-4-QUINOLYL)-5-VINYL-2-QUINUCLIDINEMETHANOL * 6-METHOXY-α-(5-VINYL-2-QUINUCLIDINYL)-4-QUINOLINEMETHANOL * NCI-C56246 * PITAYINE * QUINICARDINE * QUINIDEX * (+)-QUINIDINE * β-QUININE

CONSENSUS REPORTS: Reported in EPA TSCA Inventory.

SAFETY PROFILE: Poison by ingestion, subcutaneous, intravenous, intramuscular, and intraperitoneal routes. An eye irritant. Implicated in aplastic anemia. When heated to decomposition it emits toxic fumes of NO_x.

QHJ000   CAS: 130-95-0   **HR: 3**
QUININE
mf: $C_{20}H_{24}N_2O_2$ mw: 324.46

PROP: Bulky, white, amorphous powder or crystals; bitter taste. Mp: 174.9°.

SYNS: CHININ (GERMAN) * (8-α,9R)-6'-METHOXY-CINCHONAN-9-OL * 6-METHOXYCINCHONINE * α-(6-METHOXY-4-QUINOYL)-5-VINYL-2-QUINCLIDI-NEMETHANOL * (−)-QUININE

CONSENSUS REPORTS: Reported in EPA TSCA Inventory.

SAFETY PROFILE: Human poison by unspecified route. Experimental poison by subcutaneous, intravenous, intramuscular, and intraperitoneal routes. Moderately toxic experimentally by ingestion. Human systemic effects by ingestion: visual field changes, tinnitus and nausea or vomiting. Human teratogenic effects by ingestion: developmental abnormalities of the central nervous system, body wall, musculoskeletal, cardiovascular and hepatobiliary systems. Experimental teratogenic and reproductive effects. Mutation data reported. Can cause temporary loss of vision. Quinine dermatitis is an occupational hazard to barbers particularly, and generally to people who work with quinine tonics, medicaments, or cosmetics. An irritant to mucous membranes. Combustible when exposed to heat or flame. Decomposes on exposure to light. When heated to decomposition it emits toxic fumes of NO_x. Used to treat malaria.

QIJ000   CAS: 60-93-5   **HR: 3**
QUININE DIHYDROCHLORIDE
mf: $C_{20}H_{24}N_2O_2 \cdot 2ClH$ mw: 397.38

PROP: White needles or crystalline powder; odorless with very bitter taste. Sol in water, alc, glycerin; sltly sol in chloroform; very sltly sol in ether.

SYNS: ACID QUININE HYDROCHLORIDE * CHININ-DIHYDROCHLORID (GERMAN) * 6'-METHOXYCIN-CHONAN-9-OL DIHYDROCHLORIDE * QUININE BIMU-RIATE * (−)-QUININE DIHYDROCHLORIDE

CONSENSUS REPORTS: Reported in EPA TSCA Inventory.

SAFETY PROFILE: Poison by intravenous and subcutaneous routes. Moderately toxic by ingestion. Mutation data reported. When heated to decomposition it emits very toxic fumes of NO_x and HCl.

QIS000 CAS: 73771-81-0 **HR: 3**
QUININE ETHIODIDE
mf: $C_{22}H_{29}N_2O_2 \cdot I$ mw: 480.43

SYNS: (8-α,9R)-1-ETHYL-9-HYDROXY-6'-METHOXYCIN-CHONAN-1-IUM IODIDE * 6-(1-HYDROXY-1-(6-METH-OXY-4-QUINOLINYL)METHYL-1-ETHYL-3-VINYLQUINU-CLIDINIUM, IODIDE * 6-(HYDROXY(6-METHOXY-4-QUINOLINYL)METHYL)-1-ETHYL-3-VINYL-QUINUCLI-DINIUM, IODIDE

SAFETY PROFILE: Poison by intravenous route. When heated to decomposition it emits very toxic fumes of I^- and NO_x.

QJS000 CAS: 130-89-2 **HR: 3**
QUININE HYDROCHLORIDE
mf: $C_{20}H_{24}N_2O_2 \cdot ClH$ mw: 360.92

SYNS: QUININE CHLORIDE * QUININE MONOHY-DROCHLORIDE * QUININE MURIATE

CONSENSUS REPORTS: Reported in EPA TSCA Inventory.

SAFETY PROFILE: Poison by ingestion, subcutaneous, intravenous, intramuscular, intraperitoneal, and possibly other routes. Human systemic effects by intravenous route: convulsions or effect on seizure threshold, muscle contraction or spasticity, and nausea or vomiting. Mutation data reported. Used as a local anesthetic. When heated to decomposition it emits very toxic fumes of NO_x and HCl.

QMA000 CAS: 804-63-7 **HR: 3**
QUININE SULFATE
mf: $C_{20}H_{24}N_2O_2 \cdot O_4S$ mw: 420.52

PROP: Fine white needlelike crystals; odorless with a very bitter taste. Sol in water, alc; sltly sol in chloroform.

SYNS: QUININE BISULFATE * QUININE HYDROGEN SULFATE

CONSENSUS REPORTS: Reported in EPA TSCA Inventory.

SAFETY PROFILE: Human poison by ingestion. Human systemic effects by ingestion: flaccid paralysis without anesthesia, visual field changes, tinnitus, motor activity changes and blood angranulocytosis. Experimental reproductive effects. Mutation data reported. When heated to decomposition it emits very toxic fumes of SO_x and NO_x.

QMJ000 CAS: 91-22-5 **HR: 3**
QUINOLINE
DOT: UN 2656
mf: C_9H_7N mw: 129.17

PROP: Refractive, colorless liquid; peculiar odor. Mp: $-14.5°$, bp: 237.7°, d: 1.0900 @ 25°/4°, autoign temp: 896°F, vap press: 1 mm @ 59.7°, vap d: 4.45. Sol in water, CS_2; misc in alc, ether.

SYNS: 1-AZANAPHTHALENE * 1-BENZAZINE * 1-BENZINE * BENZO(b)PYRIDINE * CHINO-LEINE * CHINOLIN (CZECH) * CHINOLINE * LEUCOL * LEUCOLINE * LEUKOL * USAF EK-218

CONSENSUS REPORTS: Reported in EPA TSCA Inventory. EPA Genetic Toxicology Program. Community Right-To-Know List.

DOT Classification: Poison B; Label: St. Andrews Cross, Flammable Liquid; ORM-E; Label: None.

SAFETY PROFILE: Poison by ingestion, subcutaneous, and intraperitoneal routes. Moderately toxic by skin contact. A skin and severe eye irritant. Mutation data reported. Questionable carcinogen with experimental neoplastigenic and tumorigenic data. It can cause retinitis similar to that caused by naphthalene but without causing opacity of the lens. Combustible when exposed to heat or flame. Its preparation has caused many industrial explosions. Potentially explosive reaction with hydrogen peroxide. Violent reaction with dinitrogen tetraoxide, perchromates. Incompatible with linseed oil + thionyl chloride, maleic anhydride. Unpredictably violent. When heated to decomposition it emits toxic fumes of NO_x.

QPA000 CAS: 148-24-3 **HR: 3**
8-QUINOLINOL
mf: C_9H_7NO mw: 145.17

PROP: White crystals or powder. Mp: 76°, bp: 267°. Very sltly sol in cold water; sltly sol in ether; sol in alc, dilute alkali.

SYNS: BIOQUIN * FENNOSAN * HYDROXYBEN-ZOPYRIDINE * 8-HYDROXY-CHINOLIN (GERMAN) * 8-HYDROXYQUINOLINE * NCI-C55298 * 8-OQ * OXINE * OXYBENZOPYRIDINE * OXYCHINO-LIN * o-OXYCHINOLIN (GERMAN) * OXYQUINO-LINE * 8-OXYQUINOLINE * PHENOPYRIDINE * 8-QUINOL * QUINOPHENOL * TUMEX * USAF EK-794

CONSENSUS REPORTS: IARC Cancer Review: GROUP 3 IMEMDT 7,56,87; Animal Inadequate Evidence IMEMDT 13,101,77. NTP Carcinogenesis Studies (feed); No Evidence: mouse, rat NTPTR* NTP-TR-276,85. Reported in EPA TSCA Inventory. EPA Genetic Toxicology Program.

SAFETY PROFILE: Poison by intraperitoneal and subcutaneous routes. Moderately toxic by ingestion and possibly other routes. Questionable carcinogen with experimental carcinogenic, neoplastigenic, and tumorigenic data. Experimental teratogenic effects. A central nervous system stimulant. Human mutation data reported. Combustible when exposed to heat or flame. When heated to decomposition it emits highly toxic fumes of NO_x.

QQS200   CAS: 106-51-4   **HR: 3**
QUINONE
mf: $C_6H_4O_2$ mw: 108.10

PROP: Yellow crystals; characteristic irritating odor. Mp: 115.7°, bp: sublimes, d: 1.318 @ 20°/4°.

SYNS: BENZO-CHINON (GERMAN) * 1,4-BENZO-QUINE * 1,4-BENZOQUINONE * BENZOQUINONE (DOT) * p-BENZOQUINONE * CHINON (DUTCH, GERMAN) * p-CHINON (GERMAN) * CHINONE * CYCLOHEXADEINEDIONE * 1,4-CYCLOHEXA-DIENEDIONE * 2,5-CYCLOHEXADIENE-1,4-DIONE * 1,4-CYCLOHEXADIENE DIOXIDE * 1,4-DIOSSIBEN-ZENE (ITALIAN) * 1,4-DIOXYBENZENE * 1,4-DI-OXY-BENZOL (GERMAN) * NCI-C55845 * p-QUI-NONE * RCRA WASTE NUMBER U197 * USAF P-220

CONSENSUS REPORTS: IARC Cancer Review: GROUP 3 IMEMDT 7,56,87; Animal Inadequate Evidence IMEMDT 15,255,77. Reported in EPA TSCA Inventory. Community Right-To-Know List. EPA Genetic Toxicology Program.

OSHA PEL: TWA 0.01 ppm
ACGIH TLV: TWA 0.1 ppm
DFG MAK: 0.1 ppm (0.4 mg/m³)
DOT Classification: Poison B; Label: Poison.

SAFETY PROFILE: Poison by ingestion, subcutaneous, intraperitoneal, and intravenous routes. Questionable carcinogen with experimental tumorigenic data by skin contact. Human mutation data reported. Quinone has a characteristic, irritating odor. Causes severe damage to the skin and mucous membranes by contact with it in the solid state, in solution, or in the form of condensed vapors. Locally, it causes discoloration, severe irritation, erythema, swelling, and the formation of papules and vesicles, whereas prolonged contact may lead to necrosis. When the eyes become involved, it causes dangerous disturbances of vision. The moist material self heats and decomposes exothermically above 60°C. When heated to decomposition it emits acrid smoke and fumes.

QTS000   CAS: 59-40-5   **HR: 2**
N-(2-QUINOXALINYL)SULFANILAMIDE
mf: $C_{14}H_{12}N_4O_2S$ mw: 300.36

SYNS: 2-p-AMINOBENZENESULFONAMIDOQUINOXA-LINE * 2-p-AMINOBENZENESULPHONAMIDO-QUINOXALINE * N^1-2-QUINOXALINYLSULFANILAM-IDE * N'-2-QUINOXALYLSULFANILAMIDE * SULFABENZPYRAZINE * 2-SULFANILAMIDO-QUINOXALINE * SULFAQUINOXALINE

CONSENSUS REPORTS: Reported in EPA TSCA Inventory.

SAFETY PROFILE: Moderately toxic by ingestion. When heated to decomposition it emits very toxic fumes of NO_x and SO_x.

R

RAQ000
HR: D
RADIATION

PROP: Electromagnetic radiation (also called *radiant energy*) is emitted from matter in the form of photons (quanta), each having an associated electromagnetic wave having frequency (v) and wavelength (λ). The various forms of radiant energy are characterized by their wavelength, and together they comprise the electromagnetic spectrum, the components of which are as follows: (1) cosmic gamma rays, (2) gamma rays from radioactive disintegration of atomic nuclei, (3) x-rays, (4) ultraviolet rays, (5) visible light rays, (6) infrared, (7) microwave, and (8) radio (Hertzian) and electric rays. Radiation having the shortest wavelength is the most penetrating. Quanta are not electrically charged and have no mass, their velocity of propagation is the same, and all display the properties characteristic of light having a dual nature (wave-like and corpuscular). Infrared radiation is that part of the electromagnetic spectrum between visible light and the microwave region, i.e., $7000\text{ÅI} - 2.2 \times 10^6$ ÅI. All objects at a temperature greater than 0°K emit IR radiation to cooler surfaces, and the hotter the emitter the shorter the emitted IR wavelength. When the emitter is hot enough, visible ($4000\text{ÅI} - 7000\text{ÅI}$), and even UV ($100\text{ÅI} - 4000\text{ÅI}$), radiation is also emitted.

SAFETY PROFILE: The main physical effect of exposure to infrared radiation is heating. This is also true for biological tissue. In the case of the eye, there is very sensitive tissue available for exposure to IR radiation. "Near IR," ($7800\text{ÅI} - 14000\text{ÅI}$) is blamed for many eye cataracts. The eyes may be easily protected by wearing goggles. Ultraviolet (UV) radiation is that part of the EM spectrum between 100ÅI and 4000ÅI. The UV-A band of UV extends from $3150\text{ÅI} - 4000\text{ÅI}$ and is called "Black light" or "near UV." This band can cause thermal skin burns, skin pigmentation and photoreactions. It does not, in general, cause eye injury. From 2800ÅI to 3150ÅI is "mid-UV," or erythemal region. This band produces photokeratitis and possibly skin cancer. The UV band from $1000\text{ÅI} - 2800\text{ÅI}$ is the UV-C band. It is known as "far UV" or "short UV." This band

of UV is germicidal and viricidal, and destroys molds and yeasts as well. There is a sub-region of UV-C from $1700\text{ÅI} - 2200\text{ÅI}$ which produces ozone. The whole UV region can damage human skin and eyes. In eyes, it can cause blepharitis, conjunctivitis, keratitis, and keratoconjunctivitis. Skin exposure to solar UV can cause erythema, tanning, chronic skin exposure to solar UV leads to tanning, elastosis (dry, leathery, deeply wrinkled skin) and an incidence of non-melanoma skin cancer.

Type of radiation	Wavelength ÅI	
cosmic	0.0005	-0.005
gamma	0.005	-1.4
X	0.1	-100
UV	100	-4000
visible	4000	-7000
infrared	7000	-2,000,000

RAQ010
HR: D
RADIATION, IONIZING

Extremely short-wavelength, highly energetic, penetrating rays of the following types: (a) gamma rays emitted by radioactive elements and radioisotopes (decay of atomic nucleus); (b) x-rays generated by sudden stoppage of fast-moving electrons; (c) subatomic charged particles (electrons, protons, deuterons) when accelerated in a cyclotron or betatron. The term is restricted to electromagnetic radiation at least as energetic as x-rays, and to charged particles of similar energies. Neutrons also may induce ionization. Such radiation is strong enough to remove electrons from any atoms in its path, leading to the formation of free radicals.

SAFETY PROFILE: These short-lived but highly reactive particles initiate decomposition of many organic compounds. Thus, ionizing radiation can cause mutations in DNA and in cell nuclei, adversely affect protein and amino acid mechanisms, impair or destroy body tissue, and attack bone marrow, the source of red blood cells. Exposure to ionizing radiation for even a short period is highly dangerous, and for an extended period may be lethal. The study of the chemical effects of such radiation is called radiation chemistry or (in the case of body reactions) radiation biochemistry.

RAV000   HR: 3
RADIUM
af: Ra   aw: 226.025

PROP: A radioactive earth metal. Brilliant white, tarnishes in air. Decomp in water. Mp: 700°, bp: 1737°, d: 5.5.

SAFETY PROFILE: A highly radiotoxic element. 1 grams = 3.7×10^{10} disintegrations per second. Inhalation, ingestion, or bodily exposure can lead to lung cancer, bone cancer, osteitis, skin damage, and blood dyscrasias. A common air contaminant. Radium replaces calcium in the bone structure and is a source of irradiation to the blood-forming organs. The ingestion of luminous dial paint prepared from radium caused death in many of the early dial painters before the hazard was fully understood. The data on these workers have been the source of many of the radiation precautions and the maximum permissible levels for internal emitters which are now accepted. ^{226}Ra is the parent of radon and the precautions described under ^{222}Rn should be followed. ^{228}Ra is a member of the thorium series. It was a common constituent of luminous paints, and while its low beta energy was not a hazard, its daughters in the series may have been a causative agent in the deaths of the radium dial painters following World War I. It is metabolized the same as any other radium isotope and it is a source of thoron. The precautions recommended under ^{220}Rn should be followed. Highly dangerous; must be kept heavily shielded and stored away from possible dissemination by explosion, flood, etc.

Radiation Hazard: Natural isotope ^{223}Ra (Actinium-X, Actinium Series), $T_{\frac{1}{2}}$ = 11.4 D, decays to radioactive ^{219}Rn by alphas of 5.5-5.7 MeV. Natural isotope ^{224}Ra (Thorium-X, Thorium Series), $T_{\frac{1}{2}}$ = 3.6 D, decays to radioactive ^{220}Rn by alphas of 5.7 MeV. Natural isotope ^{226}Ra (Uranium Series), $T_{\frac{1}{2}}$ = 1600 Y, decays to radioactive ^{222}Rn by alphas of 4.8 MeV. Natural isotope ^{228}Ra (Mesothorium = 1, Thorium Series), $T_{\frac{1}{2}}$ = 6.7 Y, decays to radioactive ^{228}Ac by betas of 0.05 MeV.

RBA000   HR: 3
RADON
af: Rn   aw: 222

PROP: Colorless, odorless, inert gas; very dense. Bp: $-62°$; d (gas @ 1 atm and 0°): 9.73 g/L, (liquid @ bp): 4.4.

SAFETY PROFILE: A common air contaminant. Radon is a noble gas and thus is relatively unreactive. Radiation Hazard: Natural isotope ^{220}Rn (Thoron, Thorium Series), $T_{\frac{1}{2}}$ = 55s, decays to radioactive ^{216}Po by alphas of 6.3 MeV. Natural isotope ^{222}Rn (Uranium Series), $T_{\frac{1}{2}}$ = 3.8 D, decays to radioactive ^{218}Po by alphas of 5.5 MeV. The permissible levels are given for ^{222}Rn in equilibrium with its daughters. The chief hazard from this isotope is inhalation of the gaseous element and its solid daughters, which are collected on the normal dust of the air. This material is deposited in the lungs and has been considered to be a major causative agent in the high incidence of lung cancer found in uranium miners. Radon and its daughters build up to an equilibrium value in about a month from radium compounds, while the build-up from uranium compounds is negligible. Good ventilation of areas where radium is handled or stored is recommended to prevent accumulation of hazardous concentrations of Rn and its daughters. Accumulation of radon in homes has been implicated in increased incidence of lung cancers. This accumulation is found in well insulated buildings located over land which has concentrations of uranium.

RBF100   CAS: 26538-44-3   HR: D
RALGRO
mf: $C_{18}H_{26}O_5$ mw: 322.44

SYNS: 6-(6,10-DIHYDROXYUNDECYL)-β-RESORCYLIC ACID-mu-LACTONE * FRIDERON * MK-188 * P1496 * RALABOL * RALONE * ZEARALANOL * ZEARANOL * ZERANOL (USDA)

CONSENSUS REPORTS: Reported in EPA TSCA Inventory.

SAFETY PROFILE: Experimental reproductive effects. When heated to decomposition it emits acrid smoke and irritating fumes.

RBP000   HR: 3
RARE EARTHS

Modern ion exchange techniques have eased the separation of rare earths from their ores and from one another.

SAFETY PROFILE: The rare earths are moderately to highly toxic. The rare earth elements exhibit low toxicity by ingestion exposure. However, the intraperitoneal route is highly toxic while the subcutaneous route is poison to moderately toxic. The production of skin and lung granulomas after exposure to them requires extensive protection to prevent such exposure. Toxicity from exposure to rare earth radionuclides is related to absorbed radiation dose. The rare earth radionuclides have proven useful clinically in radiohypophysectomy, treatment of mammary and prostatic carcinoma and Cushing's syndrome, diabetic retinopathy and carcinoma in other body tissues and organs. Rare earth chelates have proven useful diagnostic agents in brain, lung and renal scanning, and in determining regional blood flow and renal function. (Haley T. J., 1965 *J. Pharm Sci* 54, 663.) They were first used for cigarette lighter flints, in Welsbach mantles for increasing the brightness of gas lights, and in Coleman Lanterns. Additional uses include control rods for atomic reactors utilizing their large cross-section capture values for neutrons, the addition of cerium to increase the life of nickel-chrome resistant wire, radiothulium in portable roentgenographic equipment, new types of alloys, lasers, masers, microwave devices, phosphors, insulators, capacitors, semiconductors, ferroelectrics, and color television.

RBU000   CAS: 5471-51-2   HR: 3
RASPBERRY KETONE
mf: $C_{10}H_{12}O_2$ mw: 164.22

PROP: White solid; raspberry odor. Mp: 81-86°, flash p: +212°F.

SYNS: FEMA No. 2588 * FRAMBINONE * 4-(4-HYDROXPHENYL)-2-BUTANONE * p-HYDROXYBENZYL ACETONE * 1-(p-HYDROXYPHENYL)-3-BUTANONE * 4-(p-HYDROXYPHENYL)-2-BUTANONE (FCC) * OXYPHENALON * RHEOSMIN

CONSENSUS REPORTS: Reported in EPA TSCA Inventory.

SAFETY PROFILE: Poison by intraperitoneal route. Moderately toxic by ingestion. Combustible liquid. When heated to decomposition it emits acrid smoke and irritating fumes.

RCA375   CAS: 21416-67-1   HR: 3
RAZOXANE
mf: $C_{11}H_{16}N_4O_4$ mw: 268.31

SYNS: ICI 59118 * ICRF 159 * 4,4'-PROPYLENEDI-2,6-PIPERAZINEDIONE * RAZOXIN

SAFETY PROFILE: Suspected human carcinogen producing leukemia and skin tumors. Moderately toxic by intraperitoneal route. Human effects: normocytic anemia and thrombocytopenia. Human mutation data reported. When heated to decomposition it emits toxic fumes of NO_x.

RCF000   HR: 3
RED SQUILL

SYNS: BONIDE TOPZOL RAT BAITS and KILLING SYRUP * RAT-O-CIDE RAT BAIT * RAT'S END * RODINE * ROUGH & READY RAT BAIT & RAT PASTE * SCILLIROSIDE GLYCOSIDE * SILMURIN * SQUILL * TOPZOL * URGENEA MARITIMA

SAFETY PROFILE: Poison by ingestion and intraperitoneal routes. Human systemic effects by ingestion: nausea or vomiting, decreased pulse rate and fall in blood pressure. When heated to decomposition it emits acrid smoke and irritating fumes.

RDK000   CAS: 50-55-5   HR: 3
RESERPINE
mf: $C_{33}H_{40}N_2O_9$ mw: 608.75

PROP: White or pale buff to sltly yellow powder, odorless. Mp: 264-265° (decomp). Insol in water; very sltly sol in alc; sol in chloroform and acetic acid.

SYNS: ENT 50,146 * METHYLRESERPATE 3,4,5-TRIMETHOXYBENZOIC ACID * METHYL RESERPATE 3,4,5-TRIMETHOXYBENZOIC ACID ESTER * NCI-C50157 * RAUSERPIN * RAUWOLEAF * SERPASIL * SERPASIL APRESOLINE * 3,4,5-TRIMETHOXYBENZOYL METHYL RESERPATE * USAF CB-27 * YOHIMBAN-16-CARBOXYLIC ACID DERIVATIVE of BENZ(G)INDOLO(2,3-A)QUINOLIZINE

CONSENSUS REPORTS: IARC Cancer Review: GROUP 3 IMEMDT 7,330,87; Animal Inadequate Evidence IMEMDT 10,217,76; Human Limited Evidence IMEMDT 24,211,80; Animal Limited Evidence IMEMDT 24,211,80. NCI Carcinogenesis Bioassay (feed); Clear Evidence: mouse, rat NCITR* NCI-CG-TR-193,80. Reported in EPA TSCA Inventory.

SAFETY PROFILE: Suspected human carcinogen producing tumors of the skin and brain. Poison by ingestion, intravenous, subcutaneous and intraperitoneal routes. Mutation data reported. Human reproductive and teratogenic effects by ingestion and possibly other routes:

stillbirth, reduced viability, and other neonatal measures or effects. In humans, 0.014 mg/kg causes psychotropic effects. Experimental teratogenic and reproductive effects. A medicine with side effects. Used as an additive permitted in the feed and drinking water of animals and/or for the treatment of food-producing animals. Also permitted in food for human consumption. A sedative. When heated to decomposition it emits toxic fumes of NO_x.

RDP000   **HR: 2**
RESIN (solution)

SYNS: RESIN, solution, in flammable liquid (DOT)
* RESIN, solution (resin compound, liquid) (DOT)
* SOLUBOND 0-869 * SOLUBOND 3520

DOT Classification: Flammable Liquid; Label: Flammable Liquid; Flammable or Combustible Liquid; Label: Flammable Liquid.

SAFETY PROFILE: Flammable when exposed to heat or flame; can react vigorously with oxidizing materials. When heated to decomposition it emits acrid smoke and irritating fumes.

REA000   CAS: 108-46-3   **HR: 3**
RESORCINOL

DOT: UN 2876
mf: $C_6H_6O_2$ mw: 110.12

PROP: Very white crystals, become pink on exposure to light when not perfectly pure; unpleasant sweet taste. Mp: 110°, bp: 280.5°, flash p: 261°F (CC), d: 1.285 @ 15°, autoign temp: 1126°F, vap press: 1 mm @ 108.4°, vap d: 3.79. Very sol in alc, ether, glycerol; sltly sol in chloroform; sol in water.

SYNS: m-BENZENEDIOL * 1,3-BENZENEDIOL
* C.I. 76505 * C.I. DEVELOPER 4 * C.I. OXIDATION BASE 31 * DEVELOPER R * m-DIHYDROXY-
BENZENE * 1,3-DIHYDROXYBENZENE * m-DIOXY-
BENZENE * DURAFUR DEVELOPER G * FOUR-
AMINE RS * FOURRINE 79 * m-HYDROQUINONE
* 3-HYDROXYCYCLOHEXADIEN-1-ONE * m-HY-
DROXYPHENOL * 3-HYDROXYPHENOL * NAKO
TGG * NCI-C05970 * PELAGOL GREY RS
* RCRA WASTE NUMBER U201 * RESORCIN
* RESORCINE

CONSENSUS REPORTS: IARC Cancer Review: GROUP 3 IMEMDT 7,56,87; Animal Inadequate Evidence IMEMDT 15,155,77. Re-

ported in EPA TSCA Inventory. EPA Genetic Toxicology Program.

OSHA PEL: TWA 10 ppm; STEL 20 ppm
ACGIH TLV: TWA 10 ppm; STEL 20 ppm
DOT Classification: ORM-E; Label: None; Poison B; Label: St. Andrews Cross.

SAFETY PROFILE: Human poison by ingestion. Experimental poison by ingestion, intraperitoneal, parenteral, and subcutaneous routes. Moderately toxic experimentally by skin contact and intravenous routes. Questionable carcinogen with experimental tumorigenic data. A skin and severe eye irritant. It can cause systemic poisoning by acting both as a blood and nerve poison. In a suitable solvent, this material can readily be absorbed through human skin and can cause local hyperemia, itching, dermatitis, edema, and corrosion associated with enlargement of regional lymph glands as well as serious systemic disorders such as restlessness, methemoglobinemia, cyanosis, convulsions, tachycardia, dyspnea and death. These same symptoms can be induced by ingestion of the material. Human mutation data reported. For poisoning, treat symptomatically. Get medical advice. Used as a topical antiseptic and keratolytic agent.

Combustible when exposed to heat or flame; can react with oxidizing materials. To fight fire, use water, CO_2, dry chemical. Potentially explosive reaction with concentrated nitric acid. Incompatible with acetanilide, alkalies, ferric salts, spirit nitrous ether, urethan. When heated to decomposition it emits acrid smoke and irritating fumes.

REF000   CAS: 101-90-6   **HR: 3**
RESORCINOL DIGLYCIDYL ETHER
mf: $C_{12}H_{14}O_4$ mw: 222.26

SYNS: ARALDITE ERE 1359 * m-BIS(2,3-EPOXYPRO-
POXY)BENZENE * 1,3-BIS(2,3-EPOXYPROPOXY)BEN-
ZENE * m-BIS(GLYCIDYLOXY)BENZENE * 1,3-DI-
GLYCIDYLOXYBENZENE * DIGLYCIDYL RESORCINOL
ETHER * ERE 1359 * NCI-C54966 * 2,2′-(1,3-PHE-
NYLENEBIS(OXYMETHYLENE))BISOXIRANE * RDGE
* RESORCINOL BIS(2,3-EPOXYPROPYL)ETHER
* RESORCINYL DIGLYCIDYL ETHER

CONSENSUS REPORTS: IARC Cancer Review: GROUP 2B IMEMDT 7,56,87; Animal Sufficient Evidence IMEMDT 36,181,85; Animal Inadequate Evidence IMEMDT 11,125,76. NTP Carcinogenesis Studies (gavage); Clear Evidence: mouse, rat NTPTR* NTP-TR-257,86. Reported in EPA TSCA Inventory.

SAFETY PROFILE: Suspected carcinogen with experimental carcinogenic and tumorigenic data. Poison by intraperitoneal route. Moderately toxic by ingestion. Mutation data reported. A skin irritant. When heated to decomposition it emits acrid smoke and irritating fumes.

RFP000   CAS: 480-54-6   **HR: 3**
RETRORSINE
mf: $C_{18}H_{25}NO_6$ mw: 351.44

SYNS: 12,18-DIHYDROXY-SENECIONAN-11,16-DIONE
* β-LONGILOBINE * cis-RETRONECIC ACID ESTER of
RETRONECINE

CONSENSUS REPORTS: IARC Cancer Review: GROUP 3 IMEMDT 7,56,87; Animal Limited Evidence IMEMDT 10,303,76.

SAFETY PROFILE: Poison by ingestion, intraperitoneal, intravenous, and possibly other routes. Questionable carcinogen with experimental neoplastigenic and tumorigenic data. Mutation data reported. When heated to decomposition it emits toxic fumes of NO_x.

RFU000   CAS: 15503-86-3   **HR: 3**
RETRORSINE-N-OXIDE
mf: $C_{18}H_{25}NO_7$ mw: 367.44

SYNS: ISATIDINE * cis-RETRONECIC ACID ESTER of
RETRONECINE-N-OXIDE

CONSENSUS REPORTS: IARC Cancer Review: GROUP 3 IMEMDT 7,56,87; Animal Sufficient Evidence IMEMDT 10,269,76.

SAFETY PROFILE: Poison by ingestion and intraperitoneal routes. Moderately toxic by intravenous route. Questionable carcinogen with experimental neoplastigenic and tumorigenic data. Mutation data reported. When heated to decomposition it emits toxic fumes of NO_x.

RGF000   CAS: 7440-15-5   **HR: 3**
RHENIUM
af: Re   aw: 186.20

PROP: Hexagonal, close-packed crystals; black to silver gray. Mp: 3180°, bp: approx 5900°, d: 21.02.

SAFETY PROFILE: No reported cases of human toxicity. Experimentally, the Re^{3+} cation is more toxic than the ReO_4^- anion. Symptoms of Re^{3+} poisoning in rats are sedation, ab-

dominal irritation, and death from cardiovascular collapse. Symptoms of ReO_4^- toxicity in rats include severe sedation and ataxia, tonic convulsions, and cardiovascular collapse. In cats, rhenium causes transient hypertension with tachycardia and transient auricular and ventricular fibrillations. In experimental animals, inhalation of rhenium dust causes pulmonary fibrosis.

Radiation Hazard: Natural (63%) isotope ^{187}Re, $T_{\frac{1}{2}} = 4 \times 10^{10}$ Y, decays to stable ^{187}Os by betas of less than 0.10 MeV. Flammable in the form of dust when exposed to heat or flame. Violent reaction with F_2 @ 125°. Ignites in oxygen at 300°C.

RGP000   CAS: 13569-63-6   **HR: 3**
RHENIUM TRICHLORIDE
mf: Cl_3Re mw: 292.55

CONSENSUS REPORTS: Reported in EPA TSCA Inventory.

SAFETY PROFILE: Poison by intraperitoneal and possibly other routes. When heated to decomposition it emits toxic fumes of Cl^-.

RGW000   CAS: 989-38-8   **HR: 3**
RHODAMINE 6G EXTRA BASE
mf: $C_{28}H_{30}N_2O_3 \cdot ClH$ mw: 479.06

SYNS: C.I. 45160 * C.I. BASIC RED 1, MONOHY-
DROCHLORIDE * NCI-C56122 * RHODAMINE 6G
(biological stain) * RHODAMINE 6GEX ETHYL ESTER

CONSENSUS REPORTS: IARC Cancer Review: GROUP 3 IMEMDT 7,56,87; Animal Sufficient Evidence IMEMDT 16,233,78. Reported in EPA TSCA Inventory. Community Right-To-Know List.

SAFETY PROFILE: Poison by intraperitoneal route. Questionable carcinogen with experimental tumorigenic data. Mutation data reported. When heated to decomposition it emits very toxic fumes of Cl^- and NO_x.

RHA000   CAS: 141-11-7   **HR: 1**
RHODINYL ACETATE
mf: $C_{12}H_{22}O_2$ mw: 198.34

PROP: Mixture of acetates of geraniol and l-citronellol, found in geranium oil. Colorless to sltly yellow liquid; fresh rose odor. D: 0.895-0.908, refr index: 1.450-1.458. Sol in alc and fixed oils; insol in glycerin, propylene glycol, and water @ 237°.

SYNS: α-CITRONELLYL ACETATE * 3,7-DIMETHYL-7-OCTEN-1-OL ACETATE * FEMA No. 2981 * RHODINOL ACETATE

CONSENSUS REPORTS: Reported in EPA TSCA Inventory.

SAFETY PROFILE: A skin irritant. When heated to decomposition it emits acrid smoke and irritating fumes.

RHF000   CAS: 7440-16-6   **HR: 2**
RHODIUM
af: Rh   aw: 102.91

PROP: A silvery-white, metallic element. Mp: 1966°, bp: 3727°, d: 2.41 @ 20°.

CONSENSUS REPORTS: Reported in EPA TSCA Inventory.

OSHA PEL: TWA Metal, Fume, Insoluble Compounds: 0.1 mg(Rh)/m^3; Soluble Compounds: 0.001 mg(Rh)/m^3
ACGIH TLV: TWA (Metal) 1 mg/m^3, (insoluble compounds as Rh) 1 mg/m^3, (soluble compounds as Rh) 0.01 mg/m^3

SAFETY PROFILE: Handle carefully. It may be a sensitizer but not to the same extent as platinum. Most rhodium compounds have only moderate toxicity by ingestion. Flammable when exposed to heat or flame. Violent reaction with chlorine bromine pentafluoride, trifluoride, OF$_2$. A catalytic metal.

RHK000   CAS: 10049-07-7   **HR: 3**
RHODIUM(III) CHLORIDE (1:3)
mf: Cl$_3$Rh   mw: 209.26

SYNS: RHODIUM CHLORIDE * RHODIUM TRICHLORIDE

CONSENSUS REPORTS: Reported in EPA TSCA Inventory. EPA Genetic Toxicology Program.

OSHA PEL: TWA 0.1 mg(Rh)/m^3
ACGIH TLV: TWA 1 mg(Rh)/m^3

SAFETY PROFILE: Poison by ingestion, intraperitoneal, and intravenous routes. Experimental reproductive effects. Questionable carcinogen with experimental carcinogenic data. Mutation data reported. Incompatible with penta carbonyl iron + zinc. When heated to decomposition it emits toxic fumes of Cl$^-$.

RIK000   CAS: 83-88-5   **HR: 3**
RIBOFLAVINE
mf: C$_{17}$H$_{20}$N$_4$O$_6$ mw: 376.37

PROP: Orange to yellow crystals; slt odor. Mp: 282° (decomp). Sltly sol in water, alc; insol in ether, chloroform.

SYNS: BEFLAVINE * 6,7-DIMETHYL-9-d-RIBITYL-ISOALLOXAZINE * 7,8-DIMETHYL-10-d-RIBITYLISO-ALLOXAZINE * 7,8-DIMETHYL-10-(d-RIBO-2,3,4,5-TET-RAHYDROXYPENTYL)ISOALLOXAZINE * FLAVAXIN * HYFLAVIN * HYRE * LACTOFLAVIN * LACTOFLAVINE * RIBIPCA * RIBODERM * RIBOFLAVIN * RIBOFLAVINEQUINONE * VITAMIN B2 * VITAMIN G

CONSENSUS REPORTS: Reported in EPA TSCA Inventory.

SAFETY PROFILE: Poison by intravenous route. Moderately toxic by intraperitoneal and subcutaneous routes. Mutation data reported. When heated to decomposition it emits toxic fumes of NO$_x$.

RKP000   CAS: 13292-46-1   **HR: 3**
RIFAMYCIN AMP
mf: C$_{43}$H$_{58}$N$_4$O$_{12}$ mw: 823.05

SYNS: ARCHIDYN * ARFICIN * DIONE 21-ACETATE * L-5103 * 3-(4-METHYLPIPERAZINYLIMINO-METHYL)-RIFAMYCIN SV * 8-(4-METHYLPIPERAZI-NYLIMINOMETHYL) RIFAMYCIN SV * 8-(((4-METHYL-1-PIPERAZINYL)IMINO)METHYL)RIFAMYCIN SV * NSC 113926 * R/AMP * RIFA * RIFADINE * RIFAGEN * RIFALDAZINE * RIFALDIN * RIFAMATE * RIFAMPICIN * RIFAMPICINE (FRENCH) * RIFAMPICINUM * RIFAMPIN * RIFAPRODIN * RIFINAH * RIFOBAC * RIFOLDIN * RIFORAL * RIMACTAN * RIMACTAZID * TUBOCIN

CONSENSUS REPORTS: IARC Cancer Review: Animal Limited Evidence IMEMDT 24,243,80; Human Inadequate Evidence IMEMDT 24,243,80.

SAFETY PROFILE: Suspected carcinogen with experimental neoplastigenic data. Poison by intraperitoneal and intravenous routes. Moderately toxic to humans by ingestion. Moderately experimentally toxic by ingestion and subcutaneous routes. Human systemic effects by ingestion: conjunctiva irritation, iritis (inflammation of the iris), other eye effects, and skin dermatitis. Experimental teratogenic and reproductive effects. Human mutation data reported. When heated to decomposition it emits toxic fumes of NO_x.

RLK890   CAS: 25875-51-8   **HR: 2**
ROBENIDINE
mf: $C_{15}H_{13}Cl_2N_5$ mw: 334.23

PROP: Crystals from ethanol. Mp: 289-290°.

SYNS: 1,3-BIS((p-CHLOROBENZYLIDENE)AMINO) GUANIDINE * CARBONIMIDIC DIHYDRAZIDE, BIS((4-CHLOROPHENYL)METHYLENE)- * CHEM-COCCIDE * CHEMOCCIDE * CHIMCOCCIDE * KHIMCOCCID * KHIMCOECID * KHIMKOKTSID * KHIMKOKTSIDE

SAFETY PROFILE: Moderately toxic by ingestion. When heated to decomposition it emits toxic fumes of Cl^- and NO_x.

RMA000   CAS: 50471-44-8   **HR: 1**
RONILAN
mf: $C_{12}H_9Cl_2NO_3$ mw: 286.12

SYNS: BAS 352 F * 3-(3,5-DICHLOROPHENYL)-5-ETHENYL-5-METHYL-2,4-OXAZOLIDINEDIONE * 3-(3,5-DICHLOROPHENYL)-5-METHYL-5-VINYL-2,4-OXAZOLIDINEDIONE * VINCLOZOLIN (GER-MAN)

SAFETY PROFILE: Mildly toxic by ingestion. Mutation data reported. When heated to decomposition it emits very toxic fumes of Cl^- and NO_x.

RMA500   CAS: 299-84-3   **HR: 3**
RONNEL
mf: $C_8H_8Cl_3O_3PS$ mw: 321.54

PROP: White powder. Mp: 41°, vap press: 8 × 10^{-4} mm.

SYNS: DERMAFOSU (POLISH) * DERMAPHOS * O,O-DIMETHYL-O-2,4,5-TRICHLOROPHENYL PHOS-PHOROTHIOATE * DIMETHYL TRICHLOROPHENYL THIOPHOSPHATE * O,O-DIMETHYL-O-(2,4,5-TRICHLO-ROPHENYL)THIOPHOSPHATE * O,O-DIMETHYL-O (2,4,5-TRICHLORPHENYL)-THIONOPHOSPHAT(GERMAN) * DOW ET 14 * DOW ET 57 * ECTORAL * ENT 23,284 * ET 14 * ET 57 * ETROLENE * FENCHLOORFOS (DUTCH) * FENCHLORFOS * FENCHLORFOSU (POLISH) * FENCHLOROPHOS * FENCHLORPHOS * KARLAN * KORLAN * KORLANE * NANCHOR * NANKER * NAN-KOR * THIOPHOSPHATE de O,O-DIMETHYLE et de O-(2,4,5-TRICHLOROPHENYLE) (FRENCH) * O-(2,4,5-TRICHLOOR-FENYL)-O,O-DIMETHYL-MONOTHIOFOS-

FAAT (DUTCH) * TRICHLOROMETAFOS * 2,4,5-TRICHLOROPHENOL, O-ESTER with O,O-DIMETHYL PHOSPHOROTHIOATE * O-(2,4,5-TRICHLOR-PHENYL)-O,O-DIMETHYL-MONOTHIOPHOSPHAT (GERMAN) * O-(2,4,5-TRICLORO-FENIL)-O,O-DIMETIL-MONOTIO-FOSFATO (ITALIAN) * TROLEN * TROLENE * VIOZENE

CONSENSUS REPORTS: Chlorophenol compounds are on the Community Right-To-Know List.

OSHA PEL: (Transitional: TWA 15 mg/m^3) TWA 10 mg/m^3
ACGIH TLV: TWA 10 mg/m^3

SAFETY PROFILE: Poison by ingestion, intra-peritoneal, and possibly other routes. Moderately toxic by skin contact. A cholinesterase inhibitor. Experimental teratogenic and reproductive effects. When heated to decomposition it emits very toxic fumes of Cl^-, PO_x, and SO_x.

RMK020   CAS: 569-61-9   **HR: 3**
p-ROSANILINE HYDROCHLORIDE
mf: $C_{19}H_{17}N_3 \cdot ClH$ mw: 323.85

SYNS: 4-((4-AMINOPHENYL)(4-IMINO-2,5-CYCLOHEXA-DIEN-1-YLIDENE)METHYL), MONOCHLORIDE * BASIC PARAFUCHSINE * CALCOZINE MAGENTA N * C.I. 42500 * C.I. BASIC RED 9, MONOHYDROCHLORIDE * p-FUCHSIN * FUCHSINE DR-001 * FUCH-SINE SPC * 4,4′-((4-IMINO-2,5-CYCLOHEXADIEN-1-YLIDENE)METHYLENE)DIANILINE MONOHYDROCHLO-RIDE-o-TOLUIDINE * NCI-C54739 * PARAFUCHSIN (GERMAN) * PARA-MAGENTA * PARARO-SANILINE * PARAROSANILINE CHLORIDE * PARAROSANILINE HYDROCHLORIDE * p-ROSANILINE HCL * SCHULTZ-TAB NO. 779 (GERMAN) * 4,4′4″-TRIAMINOTRIPHENYL-METHAN-HYDROCHLORID (GERMAN)

CONSENSUS REPORTS: IARC Cancer Review: GROUP 3 IMEMDT 7,238,87; Animal Limited Evidence IMEMDT 4,57,74; Human Inadequate Evidence IMEMDT 4,57,74. EPA Genetic Toxicology Program. Reported in EPA TSCA Inventory.

SAFETY PROFILE: Suspected carcinogen with experimental carcinogenic and tumorigenic data. Mildly toxic by ingestion. Mutation data reported. When heated to decomposition it emits very toxic fumes of HCl and NO_x.

RMU000 CAS: 8000-25-7 **HR: 1**
ROSEMARY OIL

PROP: Constituents are α-pinene, camphene, and cineole. From steam distillation of flowering tops of *Rosmarinus officinalis* L. (Fam. *Labiatae*). Colorless to pale yellow liquid; odor of rosemary. D: 0.894-0.912, refr index: 1.464 @ 20°.

SYNS: ROSEMARIE OIL * ROSMARIN OIL (GERMAN)

CONSENSUS REPORTS: Reported in EPA TSCA Inventory.

SAFETY PROFILE: Mildly toxic by ingestion. A skin irritant. When heated to decomposition it emits acrid smoke and irritating fumes.

RNA000 CAS: 8007-01-0 **HR: 1**
ROSE OIL

PROP: Volatile oil from steam distillation of fresh flowers of *Rosa gallica* L. and *Rosa Damascena* Mill. and varieties of these species (Fam. *Rosaceae*). Colorless to yellow liquid; odor and taste of rose. D: 0.848-0.863 @ 30°/15°, refr index: 1.457 @ 30°.

SYN: ROSEN OEL (GERMAN)

CONSENSUS REPORTS: Reported in EPA TSCA Inventory.

SAFETY PROFILE: Mildly toxic by ingestion. When heated to decomposition it emits acrid smoke and irritating fumes.

RNZ000 CAS: 83-79-4 **HR: 3**
ROTENONE
mf: $C_{23}H_{22}O_6$ mw: 394.45

PROP: Orthorhombic plates. Mp: 165-166° (dimorphic form mp: 185-186°). D: 1.27 @ 20°. Almost insol in water; sol in alc, acetone, carbon tetrachloride, chloroform, ether, and other organic solvents. Decomp on exposure to light and air.

SYNS: BARBASCO * CENOL GARDEN DUST * CHEM FISH * CHEM-MITE * CUBE * CUBE EXTRACT * CUBE-PULVER * CUBE ROOT * CUBOR * CUREX FLEA DUSTER * DACTINOL * DERIL * DERRIN * DERRIS * DRI-KIL * ENT 133 * EXTRAX * FISH-TOX * GREEN CROSS WARBLE POWDER * HAIARI * LIQUID DERRIS * MEXIDE * NCI-C55210 * NICOULINE * NOXFISH * PARADERIL * POWDER and ROOT * PRENTOX * RO-KO * RONONE * ROTEFIVE

* ROTEFOUR * ROTENONA (SPANISH) * ROTESSENOL * ROTOCIDE * TUBATOXIN

OSHA PEL: TWA 5 mg/m^3
ACGIH TLV: TWA 5 mg/m^3
DFG MAK: 5 mg/m^3

SAFETY PROFILE: Human poison by ingestion and possibly other routes. Experimental poison by ingestion, intraperitoneal, and possibly other routes. Experimental teratogenic and reproductive effects. Mutation data reported. A skin and eye irritant. Questionable carcinogen with experimental neoplastigenic and tumorigenic data. Acute poisoning causes numbness, nausea, vomiting, and tremors. Chronic exposure injures liver and kidneys. It is toxic to animals and very toxic to fish, but leaves no harmful residue on vegetable crops. When heated to decomposition it emits acrid smoke and irritating fumes. Used as an insecticide and as a fish poison.

ROU000 **HR: 3**
RUBBER SOLVENT

SYNS: LACQUER DILUENT * NAPHTHA * SKELLY-SOLVE-L

ACGIH TLV: TWA 400 ppm
NIOSH REL: TWA (Petroleum Solvent) 350 mg/m^3; CL 1800 mg/m^3/15M

SAFETY PROFILE: Mildly toxic by inhalation. A very dangerous fire hazard when exposed to heat or flame. Explosive in the form of vapor when exposed to heat or flame. To fight fire, use foam, alcohol foam. When heated to decomposition it emits acrid smoke and irritating fumes.

RPA000 CAS: 7440-17-7 **HR: 3**
RUBIDIUM

DOT: UN 1423
af: Rb aw: 85.47

PROP: Soft, silvery-white metal. Mp: 38.89°, bp: 688°, d (solid): 1.532 @ 20°, d (liquid): 1.475 @ 39°.

SYNS: RUBIDIUM METAL (DOT) * RUBIDIUM METAL, IN CARTRIDGES (DOT)

CONSENSUS REPORTS: Reported in EPA TSCA Inventory.

DOT Classification: Flammable Solid; Label: Flammable Solid & Dangerous When Wet.

SAFETY PROFILE: A very reactive alkali metal (more reactive than potassium or cesium). In the body, rubidium substitutes for potassium as an intracellular ion. The ratio of Rb/K intake is important in the toxicology of rubidium. A ratio above 40% is dangerous. In rats, a failure to gain weight is the first symptom, followed by ataxia and hyperirritability. Symptoms include: skin ulcers, poor hair coat, sensitivity, and extreme nervousness leading to convulsions and death.

A very dangerous fire and explosion hazard when exposed to heat or flame or by chemical reaction with oxidizers. Ignites on contact with air, oxygen, and halogens. Reaction with water, moisture or steam forms explosive hydrogen gas which then ignites. Explodes in contact with liquid bromine. Can react explosively with air; halogens; mercury; non-metals; vanadium chloride oxide; moisture; acids; oxidizers. Violent reaction with vanadium trichloride oxide (at 60°C); Cl_2O_2; P. Molten rubidium ignites in sulfur vapor and reacts vigorously with carbon. RbOH is more basic than KOH. Storage and handling: Keep under benzene, petroleum, or other liquids not containing O_2. When heated to decomposition it emits toxic fumes of Rb_2O.

RPF000 CAS: 7791-11-9 **HR: 2**
RUBIDIUM CHLORIDE
mf: ClRb mw: 120.92

PROP: White, crystalline powder. Mp: 715°, bp: 1390°, d: 2.76.

CONSENSUS REPORTS: Reported in EPA TSCA Inventory.

SAFETY PROFILE: Moderately toxic by ingestion and intraperitoneal routes. Mutation data reported. Reacts violently with BrF_3. When heated to decomposition it emits toxic fumes of Cl^-, RbCl, and Rb_2O.

RPK000 CAS: 13446-73-6 **HR: 3**
RUBIDIUM DICHROMATE
mf: $Cr_2O_7Rb_2$ mw: 386.94

PROP: Crystals. D: 3.02-3.13

CONSENSUS REPORTS: Chromium and its compounds are on the Community Right-To-Know List.

OSHA PEL: CL 0.1 mg(CrO_3)/m^3
ACGIH TLV: TWA 0.05 mg(Cr)/m^3
NIOSH REL: TWA 0.025 mg(Cr(VI))/m^3; CL 0.05/15M

SAFETY PROFILE: Suspected carcinogen. A poison. A powerful oxidizer. When heated to decomposition it emits toxic fumes of Rb_2O.

RPP000 CAS: 13446-74-7 **HR: 3**
RUBIDIUM FLUORIDE
mf: FRb mw: 104.47

PROP: Colorless crystals. Mp: 775°, bp: 1410°, d: 3.557, vap press: 1 mm @ 921°.

NIOSH REL: TWA 2.5 mg(F)/m^3

SAFETY PROFILE: Poison as a soluble fluoride. When heated to decomposition it emits toxic fumes of Rb_2O and F^-.

RPZ000 CAS: 1310-82-3 **HR: 2**
RUBIDIUM HYDROXIDE
DOT: UN 2677/UN 2678
mf: HORb mw: 102.48

PROP: Grayish-white, deliquescent mass; strong base. Mp: 300°, d: 3.203 @ 11°.

SYN: RUBIDIUM HYDROXIDE, solid and solution (DOT)

CONSENSUS REPORTS: Reported in EPA TSCA Inventory.

OSHA PEL: TWA 2.5 mg(F)/m^3
DOT Classification: Corrosive Material; Label: Corrosive.

SAFETY PROFILE: Moderately toxic by ingestion. A powerful, corrosive irritant to skin, eyes and mucous membranes. When heated to decomposition it emits toxic fumes of Rb_2O.

RRP000 **HR: 3**
RUSSIAN COMFREY ROOTS

SYNS: COMFREY, RUSSIAN * SYMPHYTUM OFFICINALE L

CONSENSUS REPORTS: IARC Cancer Review: Animal Limited Evidence IMEMDT 31,239,83

SAFETY PROFILE: Questionable carcinogen with experimental carcinogenic data. When heated to decomposition it emits acrid smoke and irritating fumes.

RRU000   CAS: 7440-18-8   **HR: 3**
RUTHENIUM
af: Ru   aw: 101.07

PROP: Lustrous, hard metal, hexagonal crystals. D: 12.45 @ 20°/4°, mp: approx 2450°, bp: approx 4150°. Stable in air.

SAFETY PROFILE: Most ruthenium compounds are poisons. Ruthenium is retained in the bones for a long time. Flammable in the form of dust when exposed to heat or flame. Violent reaction with ruthenium oxide. Explosive reaction with aqua regia + potassium chlorate. When heated to decomposition it emits very toxic fumes of RuO_x and Ru which are highly injurious to the eyes and lung and can produce nasal ulcerations.

RRZ000   CAS: 10049-08-8   **HR: 3**
RUTHENIUM CHLORIDE
mf: Cl_3Ru mw: 207.42

PROP: α Form: Black lustrous crystals. Insol in alc, water. β Form: Dark brown, fluffy, hexagonal crystals. Sol in alc.

SYN: RUTHENIUM TRICHLORIDE

CONSENSUS REPORTS: EPA Genetic Toxicology Program. Reported in EPA TSCA Inventory.

SAFETY PROFILE: Poison by intraperitoneal route. Incompatible with penta carbonyl iron; zinc. When heated to decomposition it emits toxic fumes of RuO_x and Cl^-.

RSF000   **HR: 3**
RUTHENIUM COMPOUNDS

SAFETY PROFILE: Most ruthenium compounds are poisons or moderately toxic. Ruthenium red is an antagonist of Ca^{2+}, inhibits Ca^{2+} transport and binding in mitochondrial membranes, and inhibits Ca^{2+}-ATPase activity. They resemble osmium compounds in that when heated in air, they evolve fumes which are injurious to the eyes and lungs and can produce nasal ulcerations. When heated to decomposition they emit toxic fumes of RuO_x and Ru.

RSZ000   CAS: 15662-33-6   **HR: 3**
RYANIA
mf: $C_{25}H_{35}NO_9$ mw: 493.61

PROP: The powdered stem of *Ryania speciosa*, of proven insecticidal activity.

SYNS: BONIDE RYATOX * GROUND RYANIA SPECISA(VAHL) STEMWOOD (ALKOLOID RYANODINE) * RYANEXEL * RYANIA POWDER * RYANIA SPECIOSA * RYANICIDE * RYANODINE

SAFETY PROFILE: Human poison by ingestion. Experimental poison by ingestion. Moderately toxic experimentally by skin contact. Human systemic effects by ingestion: weakness, respiratory changes, diarrhea, gastrointestinal disturbances, tremors, convulsions, coma and death. Used as an insecticide. Flammable when exposed to heat or flame. To fight fire, use CO_2, mist, spray, foam. When heated to decomposition it emits toxic fumes of NO_x.

S

SAC000   CAS: 8001-23-8   **HR: 1**
SAFFLOWER OIL

PROP: From *Carthanus tinctorius*, consists of triglycerides of linoleic acid. Light yellow oil. D: 0.9211 @ 25°/25°. Sol in oil and fat solvents.

SYN: SAFFLOWER OIL (UNHYDROGENATED) (FCC)

CONSENSUS REPORTS: Reported in EPA TSCA Inventory.

SAFETY PROFILE: A human skin irritant. Ingestion of large doses can cause vomiting. When heated to decomposition it emits acrid smoke and irritating fumes.

SAD000   CAS: 94-59-7   **HR: 3**
SAFROL
mf: $C_{10}H_{10}O_2$ mw: 162.20

PROP: Colorless liquid or crystals; sassafras odor. Mp: 11°, bp: 234.5°, d: 1.0960 @ 20°, vap press: 1 mm @ 63.8°. Insol in water; very sol in alc; misc with chloroform, ether.

SYNS: 5-ALLYL-1,3-BENZODIOXOLE * ALLYL-CATECHOL METHYLENE ETHER * ALLYLDIOXY-BENZENE METHYLENE ETHER * 1-ALLYL-3,4-METH-YLENEDIOXYBENZENE * 4-ALLYL-1,2-METHYLENE-DIOXYBENZENE * m-ALLYLPYROCATECHIN METHYLENE ETHER * 4-ALLYLPYROCATECHOL FORMALDEHYDE ACETAL * ALLYLPYROCATECHOL METHYLENE ETHER * 1,2-METHYLENEDIOXY-4-ALLYLBENZENE * 3,4-METHYLENEDIOXY-ALLY-BENZENE * 5-(2-PROPENYL)-1,3-BENZODIOXOLE * RCA WASTE NUMBER U203 * RHYUNO OIL * SAFROLE * SAFROLE MF * SHIKIMOLE * SHIKOMOL

CONSENSUS REPORTS: IARC Cancer Review: GROUP 3 IMEMDT 7,56,87; Animal Sufficient Evidence IMEMDT 10,231,76, IMEMDT 1,169,72. NTP Fourth Annual Report On Carcinogens, 1984. Community Right-To-Know List. EPA Genetic Toxicology Program. Reported in EPA TSCA Inventory.

SAFETY PROFILE: Confirmed carcinogen with experimental carcinogenic and neoplastigenic data. Poison by intraperitoneal and intravenous routes. Moderately toxic by ingestion and subcutaneous routes. Experimental reproductive effects. Human mutation data reported. A skin irritant. Combustible when exposed to heat or flame. When heated to decomposition it emits acrid smoke and irritating fumes.

SAE500   CAS: 8022-56-8   **HR: 2**
SAGE OIL, DALMATIAN TYPE

PROP: Main constituent is thujone. From steam distillation of leaves from *Salvia officinalis* l. Yellow liquid; thujone odor and taste. D: 0.903-0.925, refr index: 1.457 @ 20°. Sol in fixed oils, mineral oil; sltly sol in propylene glycol; insol in glycerin.

SYNS: DALMATIAN SAGE OIL * SAGE OIL * SALBEI OEL (GERMAN)

CONSENSUS REPORTS: Reported in EPA TSCA Inventory.

SAFETY PROFILE: Moderately toxic by ingestion. Mutation data reported. A human skin irritant. When heated to decomposition it emits acrid smoke and irritating fumes.

SAE550   CAS: 8022-56-8   **HR: 2**
SAGE OIL, SPANISH TYPE

PROP: From steam distillation of plants from *Salvia lavandulaefolia* Vahl. or *Salvia hispanorium* Lag. (Fam. *Labiatae*). Colorless to yellow oil. D: 0.909-0.932, refr index: 1.468 @ 20°. Sol in fixed oils, glycerin, mineral oil, propylene glycol.

SYNS: SAGE OIL * SALBEI OEL (GERMAN)

CONSENSUS REPORTS: Reported in EPA TSCA Inventory.

SAFETY PROFILE: Moderately toxic by ingestion. When heated to decomposition it emits acrid smoke and irritating fumes.

SAH000   CAS: 65-45-2   **HR: 3**
SALICYLAMIDE
mf: $C_7H_7NO_2$ mw: 137.15

PROP: White to sltly pink crystals or powder; somewhat bitter taste. Mp: 140°. Sol in hot water, alc, chloroform, ether.

SYNS: ACKET * AFKO-SAL * ALGAMON * ALGIAMIDA * AMIDOSAL * AMID-SAL * ANAMID * BENESAL * CIDAL * DOLOMIDE * DROPSPRIN * H.P. 34 * o-HYDROXYBENZ-AMIDE * 2-HYDROXYBENZAMIDE * LIQUIPRIN

* NOVECYL * OHB * ORAMID * PANITHAL
* RASPBERIN * SALAMID * SALAMIDE
* SALICILAMIDE (ITALIAN) * SALICIM * SALI-
CYLAMID * SALIPUR * SALIZELL * SALRIN
* SALYMID * SAM * SAMID * URTOSAL

CONSENSUS REPORTS: Reported in EPA
TSCA Inventory.

SAFETY PROFILE: Poison by intravenous and
intraperitoneal routes. Moderately toxic by ingestion. Experimental teratogenic and reproductive effects. Can cause dizziness, drowsiness,
nausea, vomiting, epigastric distress, allergic
reactions and blood dyscrasias in average to
large doses. Used as an analgesic, antipyretic
and anti-inflammatory agent. When heated to
decomposition it emits toxic fumes of NO_x.

SAI000   CAS: 69-72-7   **HR: 3**
SALICYLIC ACID
mf: $C_7H_6O_3$ mw: 138.13

PROP: D: 1.443 @ 20°/4°, mp: 158.3°, bp:
211° @ 20 mm ±. Sol in water, alc, ether.

SYNS: ACIDO SALICILICO (ITALIAN) * o-HY-
DROXYBENZOIC ACID * 2-HYDROXYBENZOIC ACID
* KERALYT * ORTHOHYDROXYBENZOIC ACID
* RETARDER W * SA * SAX

CONSENSUS REPORTS: Reported in EPA
TSCA Inventory. EPA Genetic Toxicology Program.

SAFETY PROFILE: Poison by ingestion, intravenous, and intraperitoneal routes. Moderately toxic by subcutaneous route. Human systemic effects by skin contact: ear tinnitus.
Mutation data reported. A skin and severe eye
irritant. Experimental teratogenic and reproductive effects. Incompatible with iron salts; spirit
nitrous ether; lead acetate; iodine. Used in the
manufacture of aspirin. When heated to decomposition it emits acrid smoke and irritating
fumes.

SAL000   CAS: 118-61-6   **HR: 2**
SALICYLIC ETHYL ESTER
mf: $HO \cdot C_6H_4 \cdot CO_2 \cdot C_2H_5$ mw: 166.18

PROP: Colorless liquid; wintergreen odor. D:
1.127, refr index: 1.520, mp: 1.3°, bp: 233-
234°. Sol in alc, ether, acetic acid, fixed oils;
sltly sol in water, glycerin.

SYNS: ETHYL-o-HYDROXYBENZOATE * ETHYL
SALICYLATE (FCC) * FEMA No. 2458 * SALICYLIC
ETHER

CONSENSUS REPORTS: Reported in EPA
TSCA Inventory.

SAFETY PROFILE: Moderately toxic by ingestion and subcutaneous routes. A skin irritant.
When heated to decomposition it emits acrid
smoke and irritating fumes.

SAP500   CAS: 139-93-5   **HR: 3**
SALVARSAN
mf: $C_{12}H_{12}As_2N_2O_2 \cdot 2ClH$ mw: 439.02

SYNS: ARSENPHENOLAMINE HYDROCHLORIDE
* ARSPHENAMINE * 3,3'-DIAMINO-4,4'-DIHY-
DROXYARSENOBENZENE DIHYDROCHLORIDE
* EHRLICH 606 * PHENARSENAMINE

CONSENSUS REPORTS: Arsenic and its
compounds are on the Community Right-To-
Know List.

SAFETY PROFILE: Poison by intravenous
route. Implicated in aplastic anemia. When
heated to decomposition it emits very toxic
fumes of As, NO_x, and HCl.

SAQ500   **HR: 3**
SAMARIUM
af: Sm   aw: 150.36

PROP: Bright, yellow, lustrous, stable metal.
Mp: 1072°, bp: 1778°, d (α): 7.536, d (β):
7.40.

SAFETY PROFILE: As a lanthanon, it may
cause impairment of blood clotting. Flammable
in the form of dust when exposed to flame or
by spontaneous chemical reaction with oxidizers. Ignites at 150° in air. Reacts with water
to form explosive hydrogen gas. Can react violently with halogens. Potentially explosive reaction with 1,1,2-trichlorotrifluoroethane.

SAR000   CAS: 10465-27-7   **HR: 3**
SAMARIUM ACETATE
mf: $C_6H_9O_6 \cdot Sm$ mw: 327.50

SYNS: ACETIC ACID, SAMARIUM SALT * SAMA-
RIUMACETAT (GERMAN)

CONSENSUS REPORTS: Reported in EPA
TSCA Inventory.

SAFETY PROFILE: Poison by intravenous and
subcutaneous routes. When heated to decomposition it emits acrid smoke and irritating fumes.

SAR500   CAS: 10361-82-7   **HR: 3**
SAMARIUM(III) CHLORIDE
mf: Cl_3Sm mw: 256.70

PROP: White-yellowish powder. D: 4.465, mp: 686°.

CONSENSUS REPORTS: Reported in EPA TSCA Inventory.

SAFETY PROFILE: Poison by intraperitoneal and subcutaneous routes. A skin and eye irritant. When heated to decomposition it emits toxic fumes of Cl⁻.

SAT000 CAS: 13759-83-6 **HR: 3**
SAMARIUM(III) NITRATE, HEXAHYDRATE (1:3:6)
mf: $N_3O_9 \cdot Sm \cdot 6H_2O$ mw: 444.50

PROP: Pale yellow crystals. Mp: 78-79°, d: 2.375.

SYNS: NITRIC ACID, SAMARIUM(3+) SALT, HEXAHY-DRATE * SAMARIUM NITRAT (GERMAN)

SAFETY PROFILE: Poison by intraperitoneal and intravenous routes. Moderately toxic by ingestion and subcutaneous routes. When heated to decomposition it emits toxic fumes of NO_x.

SAU400 **HR: 1**
SANTALYL ACETATE

PROP: Mixture of α- and β-isomers from acetylation of santalol. Colorless to sltly yellow liquid; sandalwood odor. D: 0.980, refr index: 1.488-1.491, flash p: +212°F. Sol in alc; insol in water.

SYN: FEMA No. 3007

SAFETY PROFILE: Combustible liquid. When heated to decomposition it emits acrid smoke and irritating fumes.

SAV000 CAS: 91-53-2 **HR: 3**
SANTOQUINE
mf: $C_{14}H_{19}NO$ mw: 217.34

PROP: Clear, light yellow liquid. Mp: <0°, bp: 125° @ 2 mm, vap d: 7.48, d: 1.030 @ 25°, refr index: 1.57.

SYNS: 1,2-DIHYDRO-6-ETHOXY-2,2,4-TRIMETHYL-QUINOLINE * 1,2-DIHYDRO-2,2,4-TRIMETHYL-6-ETH-OXYQUINOLINE * EMQ * EQ * 6-ETHOXY-1,2-DIHYDRO-2,2,4-TRIMETHYLQUINOLINE * ETHOXY-QUIN (FCC) * ETHOXYQUINE * 6-ETHOXY-2,2,4-TRIMETHYL-1,2-DIHYDROQUINOLINE * NIFLEX * NIX-SCALD * SANTOFLEX A * SANTOFLEX AW * SANTOQUIN * STOP-SCALD * 2,2,4-TRIMETHYL-6-ETHOXY-1,2-DIHYDROQUINOLINE * USAF B-24

CONSENSUS REPORTS: EPA Genetic Toxicology Program. Reported in EPA TSCA Inventory.

SAFETY PROFILE: Poison by intraperitoneal route. Moderately toxic by ingestion. Mutation data reported. Combustible when exposed to heat or flame; can react with oxidizing materials. When heated to decomposition it emits toxic fumes of NO_x.

SAX500 CAS: 11031-48-4 **HR: 3**
SARKOMYCIN
mf: $C_7H_8O_3$ mw: 140.15

PROP: Oily liquid. Sol in water, methanol, ethanol, butanol, ethyl acetate; sltly sol in ether. Isolated from *Streptomyces sp.*.

SYNS: 2-METHYLENE-3-OXO-CYCLOPENTANECAR-BOXYLIC ACID * SARCOMYCIN

SAFETY PROFILE: Moderately toxic by intravenous, subcutaneous, and intraperitoneal routes. Mildly toxic by ingestion. Experimental teratogenic and reproductive effects. Questionable carcinogen with experimental tumorigenic data. Mutation data reported. Used as an antibiotic. When heated to decomposition it emits acrid smoke and irritating fumes.

SAY900 **HR: 3**
SASSAFRAS

SYN: SASSAFRAS ALBIDUM

SAFETY PROFILE: A skin irritant. Questionable carcinogen with experimental neoplastigenic data. When heated to decomposition it emits acrid smoke and irritating fumes.

SBA000 CAS: 8016-68-0 **HR: 3**
SAVORY OIL (SUMMER VARIETY)

PROP: From steam distillation of *Saturiea hortensis* L. (Fam. *Labiatae*). Light yellow to dark brown liquid; spicy odor. D: 0.875-0.954, refr index: 1.486-1.505 @ 20°. Sol in fixed oils, mineral oil; insol in glycerin, propylene glycol.

CONSENSUS REPORTS: Reported in EPA TSCA Inventory.

SAFETY PROFILE: Poison by skin contact. Moderately toxic by ingestion. A severe skin irritant. When heated to decomposition it emits acrid smoke and irritating fumes.

SBB500   **HR: 2**
SCANDIUM
af: Sc   aw: 44.9559

PROP: A naturally occurring isotope.

SAFETY PROFILE: Should be handled carefully. Flammable in the form of dust when exposed to heat or flame or by chemical reaction with oxidizers. Can react violently with halogens; air.

SBC000   CAS: 10361-84-9   **HR: 3**
SCANDIUM CHLORIDE
mf: Cl_3Sc mw: 151.31

PROP: White, deliquescent solid. Mp: 960°. Sol in water; insol in alc.

SYN: SCANDIUM (3+) CHLORIDE

CONSENSUS REPORTS: Reported in EPA TSCA Inventory.

SAFETY PROFILE: Poison by intraperitoneal route. Moderately toxic by ingestion. When heated to decomposition it emits toxic fumes of Cl^-.

SBC500   CAS: 85-83-6   **HR: 3**
SCARLET RED
mf: $C_{24}H_{20}N_4O$ mw: 380.48

SYNS: BRASILAZINA OIL RED B * CALCO OIL RED D * C.I. 258 * C.I. SOLVENT RED 24 * 2′,3-DI-METHYL-4-(2-HYDROXYNAPHTHYLAZO)AZOBENZENE * FAST OIL RED B * FAT RED B * 1-((2-METHYL-4-((2-METHYLPHENYL)AZO)PHENYL)AZO)-2-NAPHTHA-LENOL * PHENOPLASTE ORGANOL RED B * RUBRUM SCARLATINUM * o-TOLUENEAZO-o-TOLUENEAZO-β-NAPHTHOL * o-TOLUENEAZO-o-TOLUENE-β-NAPHTHOL * o-TOLYLAZO-o-TOLYLAZO-β-NAPHTHOL * o-TOLYLAZO-o-TOLYLAZO-2-NAPH-THOL * 1-((4-(o-TOLYLAZO)-o-TOLYL)AZO)-2-NAPH-THOL)

CONSENSUS REPORTS: IARC Cancer Review: GROUP 3 IMEMDT 7,56,87. Reported in EPA TSCA Inventory.

SAFETY PROFILE: Questionable carcinogen with experimental tumorigenic data. Mutation data reported. When heated to decomposition it emits toxic fumes of NO_x.

SBG000   CAS: 51-34-3   **HR: 3**
SCOPOLAMINE
mf: $C_{17}H_{21}NO_4$ mw: 303.39

PROP: Thick, colorless, syrupy liquid alkaloid. Mp: 55°. Very sol in hot water, alc, ether, chloroform, acetone; sltly sol in benzene, petr ether. Decomp on standing.

SYNS: ATROCHIN * ATROQUIN * 6-β,7-β-EPOXY-3-α-TROPANYL S-(−)-TROPATE * EPOXYTRO-PINE TROPATE * HYOSCINE * (−)-HYOSCINE * HYOSOL * ISOPTO HYOSCINE * 9-METHYL-3-OXA-9-AZATRICYCLO(3.3.1.0^{2,4})NONAN-7-OL,TROPATE (ESTER) * OSCINE * SCOPINE TROPATE * (−)-SCOPOLAMINE * TROPIC ACID, ESTER with SCOPINE * TROPIC ACID, 9-METHYL-3-OXA-9-AZATRICYCLO(3.3.1.0^{2,4})NON-7-YL ESTER

CONSENSUS REPORTS: EPA Genetic Toxicology Program.

SAFETY PROFILE: Poison by intravenous, intraperitoneal, and subcutaneous routes. Moderately toxic by ingestion. Human systemic effects from very small amounts by subcutaneous and intramuscular routes: changes in surface EEG, hallucinations, distorted perceptions, and excitement. It can cause the individual who is affected to lose a certain amount of his normal inhibitory control. It is for that reason that it has been called "truth serum". Experimental reproductive effects. Human mutation data reported. In many cases of poisoning from this material, and even to a certain extent following its medical application, there is retention of the urine caused by paralysis of the bladder, and catheterization is necessary. The fatal dose is variable. Death has occurred from as little as 0.6 mg, while recovery has occurred from doses of 7-15 mg. An anticholinergic drug. When heated to decomposition it emits highly toxic fumes of NO_x.

SBM500   CAS: 76-73-3   **HR: 3**
SECONAL
mf: $C_{12}H_{18}N_2O_3$ mw: 238.32

SYNS: 5-ALLYL-5-(1-METHYLBUTYL)BARBITURIC ACID * 5-ALLYL-5-(1-METHYLBUTYL)MALONYLUREA * BARBOSEC * EVRONAL * HYPOTROL * IMESONAL * IMMENOCTAL * IMMENOX * MEBALLYMAL * 5-(1-METHYLBUTYL)-5-(2-PROPE-NYL)-2,4,6(1H,3H,5H)-PYRIMIDINITRIONE * QUINAL-BARBITAL * QUINALBARBITONE * SECOBARBI-TAL * SECOBARBITONE * TRISOMNIN

SAFETY PROFILE: Human poison by ingestion. Experimental poison by ingestion, intraperitoneal, subcutaneous and intravenous routes. Human systemic effects by ingestion:

changes in motor activity, coma and nausea or vomiting. When heated to decomposition it emits toxic fumes of NO_x.

SBN500   CAS: 7783-08-6   HR: 3
SELENIC ACID
DOT: UN 1905
mf: H_2O_4Se mw: 144.98

PROP: Colorless liquid or colorless, hexagonal prisms. Mp: 58°, bp: 260° (decomp), d (solid): 2.951 @ 15°, (liquid): 2.609 @ 15°. Very sol in water; sol in sulfuric acid; insol in ammonia; decomp in alc. Very deliquescent.

SYN: SELENIC ACID, liquid (DOT)

CONSENSUS REPORTS: Selenium and its compounds are on the Community Right-To-Know List. EPA Genetic Toxicology Program. Reported in EPA TSCA Inventory.

OSHA PEL: TWA 0.2 mg(Se)/m³
ACGIH TLV: TWA 0.2 mg(Se)/m³
DFG MAK: 0.1 mg(Se)/m³
DOT Classification: Corrosive Material; Label: Corrosive.

SAFETY PROFILE: Selenium compounds are poisons. A corrosive irritant to skin, eyes, and mucous membranes. When heated to decomposition it emits toxic fumes of Se.

SBO500   CAS: 7782-49-2   HR: 3
SELENIUM
DOT: UN 2658
af: Se   aw: 78.96

PROP: Steel gray, nonmetallic element. Mp: 170-217°, bp: 690°, d: 4.81-4.26, vap press: 1 mm @ 356°. Insol in water and alc; very sltly sol in ether.

SYNS: C.I. 77805 * COLLOIDAL SELENIUM * ELEMENTAL SELENIUM * SELEN (POLISH) * SELENIUM ALLOY * SELENIUM BASE * SELENIUM DUST * SELENIUM ELEMENTAL * SELENIUM HOMOPOLYMER * SELENIUM METAL POWDER, NON-PYROPHORIC (DOT) * VANDEX

CONSENSUS REPORTS: IARC Cancer Review: GROUP 3 IMEMDT 7,56,87. Selenium and its compounds are on the Community Right-To-Know List. Reported in EPA TSCA Inventory.

OSHA PEL: TWA 0.2 mg(Se)/m³
ACGIH TLV: TWA 0.2 mg(Se)/m³

DFG MAK: 0.1 mg(Se)/m³
DOT Classification: Poison B; Label: St. Andrews Cross.

SAFETY PROFILE: Poison by inhalation, intravenous, and possibly other routes. Experimental reproductive effects. Questionable carcinogen with experimental tumorigenic and teratogenic data. Occupational exposure has caused pallor, nervousness, depression, garlic odor of breath and sweat, gastrointestinal disturbances, and dermatitis. Liver damage in experimental animals. Chronic ingestion of 5 mg of selenium per day resulted in 49% morbidity in 5 Chinese villages. The main symptoms were brittle hair with intact follicles, new hair with no pigment, brittle nails with spots and streaks, skin lesions, peripheral anesthesia, acroparaesthesia, pain, and hyperreflexia. Similar effects have been seen in populations with selenium blood levels of 800 μg/L. In cattle, "alkali disease" is associated with consumption of grain or plants containing 5-25 mg/kg of selenium. The symptoms are lack of vitality, loss of appetite, emaciation, deformation and shedding of hoofs, loss of hair, and erosion of joints. Consumption of plants grown in seleniferous areas can cause effects in humans and animals. Selenosis in humans has occurred from ingestion of 3.2 mg selenium per day. Selenium is an essential trace element for many species.

Reacts to form explosive products with metal amides. Can react violently with barium carbide, bromine pentafluoride, calcium carbide, chlorates, chlorine trifluoride, chromic oxide (CrO_3), fluorine, lithium carbide, lithium silicon ($Li_6 Si_2$), metals, nickel, nitric acid, sodium, nitrogen trichloride, oxygen, potassium, potassium bromate, rubidium carbide, zinc, silver bromate, strontium carbide, thorium carbide, uranium. When heated to decomposition it emits toxic fumes of Se.

SBP500   HR: 3
SELENIUM COMPOUNDS

CONSENSUS REPORTS: Selenium and its compounds are on the Community Right-To-Know List.

OSHA PEL: TWA 0.2 mg(Se)/m³
ACGIH TLV: TWA 0.2 mg(Se)/m³
DFG MAK: 0.1 mg(Se)/m³

SAFETY PROFILE: Poison by inhalation and intravenous routes. Some selenium compounds are experimental carcinogens. Selenium in small amounts is essential for normal growth of some animals. Deficiency or excess is associated with serious disease in livestock. Long-term exposure may be a cause of amyotrophic lateral sclerosis in humans, just as it may cause "blind staggers" in cattle. Elemental selenium has low acute systemic toxicity, but dust or fumes can cause serious irritation of the respiratory tract. Hydrogen selenide resembles other hydrides in being highly toxic, and selenium oxychloride is a vesicant. Some organoselenium compounds have the high toxicity of other organometals. Inorganic selenium compounds can cause dermatitis. Garlic odor of breath is a common symptom. Pallor, nervousness, depression, digestive disturbances, and death have been reported in cases of chronic exposure. Selenium compounds are common air contaminants. When heated to decomposition they emit toxic fumes of Se.

SBQ000   CAS: 144-34-3   HR: 3
SELENIUM DIMETHYLDITHIO-CARBAMATE
mf: $C_{12}H_{24}N_4S_8 \cdot Se$ mw: 559.84

PROP: Yellow powder, crystals. D: 1.58, melting range: 140-172°.

SYNS: METHYL SELENAC * TETRAKIS(DIMETHYL-CARBAMODITHIOATO-S,S′)SELENIUM

CONSENSUS REPORTS: IARC Cancer Review: GROUP 2B IMEMDT 7,56,87; Animal Inadequate Evidence IMEMDT 12,161,76. Selenium and its compounds are on the Community Right-To-Know List. Reported in EPA TSCA Inventory.

OSHA PEL: TWA 0.2 mg(Se)/m^3
ACGIH TLV: TWA 0.2 mg(Se)/m^3
DFG MAK: 0.1 mg(Se)/m^3

SAFETY PROFILE: Suspected carcinogen. Selenium compounds are poisons. When heated to decomposition it emits very toxic fumes of Se, SO$_x$, and NO$_x$.

SBR000   CAS: 7488-56-4   HR: 3
SELENIUM(IV) DISULFIDE (1:2)
DOT: UN 2657
mf: S_2Se mw: 143.08

PROP: Red-yellow crystals. Mp: <100°, bp: decomp.

SYNS: EXSEL * RCA WASTE NUMBER U205 * SELENIUM DISULPHIDE (DOT) * SELENIUM SULFIDE * SELSUN BLUE

CONSENSUS REPORTS: Selenium and its compounds are on the Community Right-To-Know List. Reported in EPA TSCA Inventory.

OSHA PEL: TWA 0.2 mg(Se)/m^3
ACGIH TLV: TWA 0.2 mg(Se)/m^3
DFG MAK: 0.1 mg(Se)/m^3
DOT Classification: Poison B; Label: Poison.

SAFETY PROFILE: Poison by ingestion. Used in shampoos. When heated to decomposition it emits very toxic fumes of SO$_x$ and Se.

SBR500   HR: 2
SELENIUM(IV) DISULFIDE SHAMPOO (2.5%)
CONSENSUS REPORTS: Selenium and its compounds are on the Community Right-To-Know List.

OSHA PEL: TWA 0.2 mg(Se)/m^3
ACGIH TLV: TWA 0.2 mg(Se)/m^3
DFG MAK: 0.1 mg(Se)/m^3

SAFETY PROFILE: A severe eye irritant. When heated to decomposition it emits very toxic fumes of Se and SO$_x$.

SBS000   CAS: 7783-79-1   HR: 3
SELENIUM HEXAFLUORIDE
DOT: UN 2194
mf: F_6Se mw: 192.96

PROP: Colorless gas. Mp: −39° (subl @ −40.6°), bp: −34.5°, d: 3.25 @ −25°.

SYN: SELENIUM FLUORIDE

CONSENSUS REPORTS: Selenium and its compounds are on the Community Right-To-Know List.

OSHA PEL: TWA 0.05 ppm (Se)
ACGIH TLV: TWA 0.05 ppm (Se)
DFG MAK: 0.1 mg(Se)/m^3
DOT Classification: Poison A; Label: Poison Gas.

SAFETY PROFILE: Poison by inhalation. When heated to decomposition it emits very toxic fumes of F$^-$ and Se.

SBT000   CAS: 7446-34-6   HR: 3
SELENIUM MONOSULFIDE
mf: SSe   mw: 111.02

PROP: Orange-yellow tablets or powder. Mp: 111.03°, bp: decomp @ 118-119°, d: 3.056 @ 0°.

SYNS: NCI-C50033 * SELENIUM SULFIDE * SELENIUM SULPHIDE * SELENSULFID (GERMAN) * SULFUR SELENIDE

CONSENSUS REPORTS: NTP Fourth Annual Report On Carcinogens, 1984. NCI Carcinogenesis Bioassay (dermal); Inadequate Studies: mouse NCITR* NCI-CG-TR-197,80; (gavage); Clear Evidence: mouse, rat NCITR* NCI-CG-TR-194,80. Selenium and its compounds are on the Community Right-To-Know List.

OSHA PEL: TWA 0.2 mg(Se)/m^3
ACGIH TLV: TWA 0.2 mg(Se)/m^3
DFG MAK: 0.1 mg(Se)/m^3

SAFETY PROFILE: Confirmed carcinogen with experimental carcinogenic data. Poison by ingestion. When heated to decomposition it emits very toxic fumes of SO$_x$ and Se.

SBT500   CAS: 7791-23-3   HR: 3
SELENIUM OXYCHLORIDE

DOT: UN 2879
mf: Cl$_2$OSe   mw: 165.86

PROP: Colorless-yellowish liquid. Mp: 8.5°, bp: 176.4°, d: 2.42 @ 22°, vap press: 1 mm @ 34.8°.

SYNS: SELENINYL CHLORIDE * SELENIUM CHLO-RIDE OXIDE

CONSENSUS REPORTS: Selenium and its compounds are on the Community Right-To-Know List. EPA Extremely Hazardous Substances List. Reported in EPA TSCA Inventory.

OSHA PEL: TWA 0.2 mg(Se)/m^3
ACGIH TLV: TWA 0.2 mg(Se)/m^3
DFG MAK: 0.1 mg(Se)/m^3
DOT Classification: Corrosive Material; Label: Corrosive and Poison

SAFETY PROFILE: Poison by skin contact and subcutaneous routes. Human systemic effects by skin contact with very small amounts: primary irritant, corrosive. Explodes on contact with potassium, white phosphorus. Ignites on contact with antimony. Vigorous reaction with metal oxides (e.g., silver oxide, lead(II) oxide, lead(IV) oxide, Lead(II)(IV) oxide). When heated to decomposition it emits very toxic fumes of Cl$^-$ and Se.

SBW000   HR: 2
SELSUN

SYN: NCI-C54546

CONSENSUS REPORTS: NCI Carcinogenesis Bioassay (dermal); No Evidence: mouse NCITR* NCI-CG-TR-199,80. Selenium and its compounds are on the Community Right-To-Know List.

OSHA PEL: TWA 0.2 mg(Se)/m^3
ACGIH TLV: TWA 0.2 mg(Se)/m^3
DFG MAK: 0.1 mg(Se)/m^3

SAFETY PROFILE: A severe eye irritant. Used as a pharmaceutical and veterinary drug. When heated to decomposition it emits very toxic fumes of Se, SO$_x$, PO$_x$, Na$_2$O, and NO$_x$.

SBW500   CAS: 563-41-7   HR: 3
SEMICARBAZIDE HYDROCHLORIDE
mf: CH$_5$N$_3$O • ClH   mw: 111.55

PROP: Prisms from dilute alc. Decomp @ 175-185°, mp: 176° (decomp). Very sol in water; very sltly sol in hot alc; insol in anhydrous ether.

SYNS: AMIDOUREA HYDROCHLORIDE * AMINO-UREA HYDROCHLORIDE * CARBAMYLHYDRAZINE HYDROCHLORIDE * CH * HYDRAZINECARBOX-AMIDE MONOHYDROCHLORIDE

CONSENSUS REPORTS: IARC Cancer Review: GROUP 3 IMEMDT 7,56,87; Animal Sufficient Evidence IMEMDT 12,209,76. Reported in EPA TSCA Inventory. EPA Extremely Hazardous Substances List.

SAFETY PROFILE: Poison by ingestion. Experimental reproductive effects. Questionable carcinogen with experimental neoplastigenic and teratogenic data. When heated to decomposition it emits very toxic fumes of NO$_x$ and HCl.

SBX500   CAS: 480-81-9   HR: D
SENECIPHYLLINE
mf: C$_{18}$H$_{23}$NO$_5$ mw: 333.42

PROP: Small, rhombic platelets from hot alcohol or acetone. Mp: 217-218°. Easily sol in chloroform, ethylene chloride; less sol in alc, acetone; difficultly sol in ether, ligroin. An alkaloid isolated from S. stenocephalus.

SYNS: 13,19-DIDEHYDRO-12-HYDROXY-SENECIONAN-11,16-DIONE * JACOBINE * SENECIPHYLLIN

CONSENSUS REPORTS: IARC Cancer Review: GROUP 3 IMEMDT 7,56,87; Animal Inadequate Evidence IMEMDT 10,319,76

SAFETY PROFILE: Questionable carcinogen. Mutation data reported. When heated to decomposition it emits toxic fumes of NO_x.

SCC000　　　CAS: 3792-59-4　　　**HR: 3**
S-SEVEN
mf: $C_{14}H_{13}Cl_2O_2PS$　　mw: 347.20

SYNS: O-ETHYL-O-2,4-DICHLOROPHENYL THIONO-BENZENEPHOSPHONATE * PHENYLPHOSPHONO-THIOIC ACID, O-(2,4-DICHLOROPHENYL), O-ETHYL ESTER

SAFETY PROFILE: Poison by ingestion. Moderately toxic by subcutaneous route. When heated to decomposition it emits very toxic fumes of Cl^-, SO_x, and PO_x.

SCE000　　　CAS: 138-59-0　　　**HR: 3**
SHIKIMIC ACID
mf: $C_7H_{10}O_5$　　mw: 174.17

PROP: Isolated from Bracken.

SYNS: BRACKEN FERN TOXIC COMPONENT * SHIKIMATE * 3,4,5-TRIHYDROXY-1-CYCLOHEX-ENE-1-CARBOXYLIC ACID

CONSENSUS REPORTS: IARC Cancer Review: GROUP 3 IMEMDT 7,56,87; Animal Inadequate Evidence IMEMDT 40,47,86. EPA Genetic Toxicology Program.

SAFETY PROFILE: Moderately toxic by intraperitoneal route. Questionable carcinogen with experimental tumorigenic data. Mutation data reported. When heated to decomposition it emits acrid smoke and irritating fumes.

SCH000　　　CAS: 7631-86-9　　　**HR: 1**
SILICA, AMORPHOUS FUMED
mf: O_2Si　　mw: 60.09

PROP: A finely powdered microcellular silica foam with minimum SiO_2 content of 89.5%. Insol in water; sol in hydrofluoric acid.

SYNS: ACTICEL * AEROSIL * AMORPHOUS SILICA DUST * AQUAFIL * CAB-O-GRIP II * CAB-O-SIL * CAB-O-SPERSE * CATALOID * COLLOI-DAL SILICA * COLLOIDAL SILICON DIOXIDE * DAVISON SG-67 * DICALITE * DRI-DIE INSEC-TICIDE 67 * ENT 25,550 * FLO-GARD * FOSSIL FLOUR * FUMED SILICA * FUMED SILICON DIOX-

IDE * HI-SEL * LO-VEL * LUDOX * NAL-COAG * NYACOL * NYACOL 830 * NYACOL 1430 * SANTOCEL * SG-67 * SILICA AEROGEL * SILICA, AMORPHOUS * SILICIC ANHYDRIDE * SILICON DIOXIDE (FCC) * SILIKILL * SYN-THETIC AMORPHOUS SILICA * VULKASIL

CONSENSUS REPORTS: IARC Cancer Review: GROUP 3 IMEMDT 7,341,87; Animal Inadequate Evidence IMEMDT 42,209,88; Human Inadequate Evidence IMEMDT 42,209,88. Reported in EPA TSCA Inventory.

OSHA PEL: (Transitional: TWA 80 mg/m³/%SiO_2) TWA 6 mg/m³
ACGIH TLV: (Proposed TWA 2 mg/m³ (Respirable Dust))

SAFETY PROFILE: Poison by intraperitoneal, intravenous, and intratracheal routes. Moderately toxic by ingestion. Much less toxic than crystalline forms. Questionable carcinogen with experimental carcinogenic data. Mutation data reported. Does not cause silicosis.

SCI000　　　CAS: 7631-86-9　　　**HR: 1**
SILICA, AMORPHOUS HYDRATED
mf: O_2Si　　mw: 60.09

SYNS: SILICA AEROGEL * SILICA GEL * SILICA XEROGEL * SILICIC ACID

CONSENSUS REPORTS: IARC Cancer Review: Animal Inadequate Evidence IMEMDT 42,209,88; Human Inadequate Evidence IMEMDT 42,209,88.

OSHA PEL: (Transitional: TWA 80 mg/m³/%SiO_2) TWA 6 mg/m³
ACGIH TLV: TWA (nuisance particulate) 10 mg/m³ of total dust (when toxic impurities are not present, e.g., quartz < 1%).

SAFETY PROFILE: The pure unaltered form is considered nontoxic. Some deposits contain small amounts of crystalline quartz which is therefore fibrogenic. When diatomaceous earth is calcined (with or without fluxing agents) some silica is converted to cristobalite and is therefore fibrogenic. Tridymite has never been detected in calcined diatomaceous earth.

SCI500　　　　　　　　　　　　　**HR: 3**
SILICA (CRYSTALLINE)
mf: SiO_2　　mw: 60.09

PROP: Transparent, tasteless crystals or amorphous powder. Mp: 1710°, bp: 2230°, d (amorphous): 2.2, d (crystalline): 2.6, vap press: 10 mm @ 1732°. Practically insoluble in water or acids. Dissolves readily in HF, forming silicon tetrafluoride.

SYNS: AGATE * AMETHYST * CHALCEDONY * CHERTS * CRISTOBALITE * FLINT * ONYX * PURE QUARTZ * ROSE QUARTZ * SAND * SILICA FLOUR * SILICON DIOXIDE * TRIDYMITE * TRIPOLI

CONSENSUS REPORTS: IARC Cancer Review: Animal Sufficient Evidence IMEMDT 42,209,88; Human Limited Evidence IMEMDT 42,209,88.

OSHA PEL: (Transitional: TWA Respirable Fraction: $10 \text{ mg/m}^3/2(\%SiO_2+2)$; Total Dust: $30 \text{ mg/m}^3/2(\%SiO_2+2)$) Respirable Fraction: TWA 0.05 mg/m^3
ACGIH TLV: TWA Respirable Fraction: 0.05 mg/m^3
DFG MAK: 0.15 mg/m^3
NIOSH REL: TWA 50 μg/m^3

SAFETY PROFILE: Moderately toxic as an acute irritating dust. From the point of view of numbers of workers exposed and cases of disability produced, silica is the chief cause of pulmonary dust disease. The prolonged inhalation of dusts containing free silica may result in the development of a disabling pulmonary fibrosis known as silicosis. The Committee on Pneumoconiosis of the American Public Health Association defines silicosis as "a disease due to the breathing of air containing silica (SiO_2) characterized by generalized fibrotic changes and the development of miliary nodules in both lungs, and clinically by shortness of breath, decreased chest expansion, lessened capacity for work, absence of fever, increased susceptibility to tuberculosis (some or all of which symptoms may be present), and characteristic x-ray findings."

Silica occurs in the pure state in nature as highly fibrogenic quartz. It is the main constituent of relatively much less toxic sand, sandstone, tripoli and diatomaceous earth. It is present in crystalline form in high amounts (up to 35%) in granite. Exposure to silica occurs in hard rock mining, in foundries, in manufacture of porcelain and pottery, in the spraying of vitreous enamels, in sandblasting, in granite-cutting and tombstone-making, in the manufacture of silica firebrick and other refractories, in grinding

and polishing operations where natural abrasive wheels are used, and other occupations.

The duration of exposure which is associated with the development of silicosis varies widely for different occupations. Thus, the average duration of exposure required for the development of silicosis in sand-blasters is 2-10 years, in moulders and granite cutters, about 30 years, and in hard rock miners, 10-15 years. There is also much variation in individual susceptibility; certain workers show radiological evidence of the disease years before their fellow workmen who are similarly exposed. Such susceptible individuals are, fortunately, rather rare.

The action of crystalline silica on the lungs results in the production of a diffuse, nodular fibrosis in which the parenchyma and the lymphatic systems are involved. This fibrosis is, to a certain extent, progressive, and may continue to increase for several years after exposure is terminated. Where the pulmonary reserve is sufficiently reduced, the worker complains of shortness of breath on exertion. This is the first and most common symptom in cases of uncomplicated silicosis. If severe, it may incapacitate the worker for heavy, or even light, physical exertion, and in extreme cases there may be shortness of breath even while at rest. The most common physical sign of silicosis is a limitation of expansion of the chest. There may be a dry cough, sometimes very troublesome. The characteristic radiographic appearance is one of diffuse, discrete nodulation, scattered throughout both lung fields. Where the disease advances, the shortness of breath becomes worse, and the cough more productive and troublesome. There is no fever or other evidence of systemic reaction. Further progress of the disease results in marked fatigue, extreme dyspnea and cyanosis, loss of appetite, pleuritic pain and total incapacity to work. If tuberculosis does not supervene, the condition may eventually cause death either from cardiac failure or from destruction of lung tissue, with resultant anoxemia. In the later stages, the x-ray may show large conglomerate shadows, due to the coalescence of the silicotic nodules, with areas of emphysema between them.

Silica in some forms is used as an additive permitted in the feed and drinking water of animals and/or for the treatment of food-producing animals. It is also permitted in food for human consumption. It is a common air contaminant. Reacts violently with ClF_3, MnF_3, OF_2.

SCJ000   CAS: 14464-46-1   **HR: 3**
SILICA, CRYSTALLINE-CRISTOBALITE
mf: O_2Si mw: 60.09

PROP: White, cubic-system crystals formed from quartz at temperatures above 1000°C (NTIS** PB246-697).

SYNS: CALCINED DIATOMITE   *   CRISTOBALITE

CONSENSUS REPORTS: IARC Cancer Review: Animal Sufficient Evidence IMEMDT 42,209,88; Human Limited Evidence IMEMDT 42,209,88; GROUP 2A IMEMDT 7,341,87. Reported in EPA TSCA Inventory.

OSHA PEL: (Transitional: TWA Respirable Fraction: (10 mg/m³/2(%SiO_2+2); Total Dust: 30 mg/m³/2(%SiO_2+2)) TWA Respirable Fraction: 0.05 mg/m³
ACGIH TLV: TWA Respirable Fraction: 0.05 mg/m³
DFG MAK: 0.15 mg/m³
NIOSH REL: TWA 50 µg/m³

SAFETY PROFILE: Confirmed carcinogen with experimental carcinogenic and tumorigenic data. Poison by intratracheal route. Human systemic effects by inhalation: cough, dyspnea, fibrosis. About twice as toxic as silica in causing silicosis.

SCJ500   CAS: 14808-60-7   **HR: 3**
SILICA, CRYSTALLINE-QUARTZ
mf: O_2Si mw: 60.09

PROP: Mp: 1710°, bp: 2230°, d: 2.6.

SYNS: AGATE   *   AMETHYST   *   CHALCEDONY   *   CHERTS   *   FLINT   *   ONYX   *   PURE QUARTZ   *   QUARTZ   *   QUAZO PURO (ITALIAN)   *   ROSE QUARTZ   *   SAND   *   SILICA FLOUR (powdered crystalline silica)   *   SILICIC ANHYDRIDE

CONSENSUS REPORTS: IARC Cancer Review: Animal Sufficient Evidence IMEMDT 42,209,88; Human Limited Evidence IMEMDT 42,209,88; GROUP 2A IMEMDT 7,341,87. Reported in EPA TSCA Inventory.

OSHA PEL: (Transitional: TWA Respirable Fraction: 10 mg/m³/2(%SiO_2+2); Total Dust: 30 mg/m³/2(%SiO_2+2)) TWA Respirable Fraction: 0.1 mg/m³
ACGIH TLV: TWA Respirable Fraction: 0.1 mg/m³
DFG MAK: 0.15 mg/m³
NIOSH REL: TWA 50 µg/m³; 3000000 fibers/m³

SAFETY PROFILE: Confirmed carcinogen with experimental carcinogenic, tumorigenic, and neoplastigenic data. Experimental poison by intratracheal and intravenous routes. Human systemic effects by inhalation: cough, dyspnea, liver effects. Incompatible with OF_2, vinylacetate.

SCK000   CAS: 15468-32-3   **HR: 3**
SILICA, CRYSTALLINE-TRIDYMITE
mf: O_2Si mw: 60.09

PROP: White or colorless platelets or orthorhombic (crystals) formed from quartz @ temperatures >870° (NTIS** PB246-697).

SYNS: TRIDIMITE (FRENCH)   *   TRIDYMITE

CONSENSUS REPORTS: IARC Cancer Review: Animal Sufficient Evidence IMEMDT 42,209,88; Human Limited Evidence IMEMDT 42,209,88; GROUP 2A IMEMDT 7,341,87.

OSHA PEL: (Transitional: TWA Respirable: 10 mg/m³/2(%SiO_2+2); Total Dust: TWA 30 mg/m³/2(%SiO_2+2)) TWA 0.05 mg/m³
ACGIH TLV: TWA Respirable Fraction: 0.05 mg/m³
DFG MAK: 0.15 mg/m³
NIOSH REL: TWA 50 µg/m³

SAFETY PROFILE: Confirmed carcinogen with experimental tumorigenic data. Poison by intratracheal route. Human systemic effects by inhalation: cough, dyspnea. About twice as toxic as silica in causing silicosis.

SCK500   **HR: 3**
SILICA FLOUR

PROP: A finely ground *crystalline* silica sometimes marketed as "Amorphous." It is *not* amorphous.

SAFETY PROFILE: Toxic by inhalation. It has shown a very high incidence of silicosis among "silica flour" workers.

SCK600   CAS: 60676-86-0   **HR: 3**
SILICA, FUSED
mf: O_2Si mw: 60.09

PROP: Made up of spherical submicroscopic particles under 0.1 micron in size.

SYNS: AMORPHOUS FUSED SILICA   *   FUSED QUARTZ   *   FUSED SILICA (ACGIH)   *   QUARTZ GLASS   *   SILICA, AMORPHOUS FUSED   *   SILICA, VITREOUS   *   SILICON DIOXIDE   *   VITREOUS QUARTZ

CONSENSUS REPORTS: IARC Cancer Review: Animal Sufficient Evidence IMEMDT 42,209,88; GROUP 2A IMEMDT 7,341,87; Human Limited Evidence IMEMDT 42,209,88. Reported in EPA TSCA Inventory.

OSHA PEL: (Transitional: TWA Respirable: 10 mg/m^3/2(%SiO$_2$+2); Total Dust: TWA 30 mg/m^3/2(%SiO$_2$+2)) TWA 0.1 mg/m^3
ACGIH TLV: TWA Respirable Fraction: 0.1 mg/m^3

SAFETY PROFILE: Confirmed carcinogen. Poison by intraperitoneal, intravenous, and intratracheal routes.

SCL000   CAS: 7699-41-4   **HR: 1**
SILICA, GEL and AMORPHOUS PRECIPITATED
mf: H$_2$O$_3$Si   mw: 78.11

SYNS: KIESELSAURE (GERMAN) * METASILICIC ACID * PRECIPITATED SILICA * SILICA GEL * SILICIC ACID

CONSENSUS REPORTS: IARC Cancer Review: Animal Inadequate Evidence IMEMDT 42,39,87; Human Inadequate Evidence IMEMDT 42,39,87. Reported in EPA TSCA Inventory.

OSHA PEL: (Transitional: TWA 80 mg/m^3/%SiO$_2$) TWA 6 mg/m^3
ACGIH TLV: TWA (nuisance particulate) 10 mg/m^3 of total dust (when toxic impurities are not present, e.g., quartz < 1%).

SAFETY PROFILE: Poison by intravenous route. An eye irritant and nuisance dust. Questionable carcinogen.

SCM500   **HR: D**
SILICATES

PROP: Widely occurring compounds containing silicon, oxygen, and one or more metals with or without hydrogen.

SAFETY PROFILE: Soluble alkaline silicates act locally like mild alkalies. The dust of certain silicates, such as asbestos (hydrated magnesium silicate) and talc, can produce fibrotic changes in the lungs and are implicated as experimental carcinogens. React violently with Li.

SCN000   **HR: 2**
SILICATE SOAPSTONE

PROP: Containing less than 1% crystalline silica.

SYN: SOAPSTONE

OSHA PEL: (Transitional: TWA 20 mppcf) TWA Total Dust: 6 mg/m^3; Respirable Fraction: 3 mg/m^3
ACGIH TLV: TWA Respirable Fraction: 3 mg/m^3; 6 mg/m^3 of total dust (when toxic impurities are not present, e.g., quartz < 1%).

SAFETY PROFILE: Less toxic than quartz.

SCN500   CAS: 15191-85-2   **HR: 3**
SILICIC ACID, BERYLLIUM SALT
mf: O$_4$Si • 2Be   mw: 110.11

PROP: Colorless crystals. D: 3.0.

SYNS: BERYLLIUM ORTHOSILICATE * BERYLLIUM SILICATE * BERYLLIUM SILICIC ACID * ORTHO-SILICATE * PHENACITE * PHENAKITE * PHEN-AZITE

CONSENSUS REPORTS: Beryllium and its compounds are on the Community Right-To-Know List. IARC Cancer Review: GROUP 2A IMEMDT 7,127,87; Animal Sufficient Evidence IMEMDT 23,143,80.

OSHA PEL: (Transitional: TWA 0.002 mg(Be)/m^3; CL 0.005; Pk 0.025/30M/8H) TWA 0.002 mg(Be)/m^3; STEL 0.005 mg(Be)/m^3/30M; CL 0.025 mg(Be)/m^3
ACGIH TLV: TWA 0.002 mg(Be)/m^3; Suspected Carcinogen
NIOSH REL: CL not to exceed 0.0005 mg/(Be)/m^3

SAFETY PROFILE: Suspected carcinogen with experimental carcinogenic and tumorigenic data. When heated to decomposition it emits toxic fumes of BeO.

SCO500   CAS: 16961-83-4   **HR: 3**
SILICOFLUORIC ACID

DOT: NA 1778
mf: F$_6$Si • 2H   mw: 144.11

PROP: Transparent, colorless, fuming liquid. Bp: decomp.

SYNS: ACIDE FLUOROSILICIQUE (FRENCH) * ACIDE FLUOSILICIQUE (FRENCH) * ACIDO FLUO-SILICICO (ITALIAN) * FLUOROSILICIC ACID * FLUOSILICIC ACID * HEXAFLUOROKIESELSAIURE (GERMAN) * HEXAFLUOROKIEZELZUUR (DUTCH) * HEXAFLUOROSILICATE(2-) DIHYDROGEN * HEXAFLUOSILICIC ACID * HYDROFLUOSILICIC ACID * HYDROGEN HEXAFLUOROSILICATE

* HYDROSILICOFLUORIC ACID * KIEZELFLUOR-
WATERSTOFZUUR (DUTCH) * SAND ACID

CONSENSUS REPORTS: Reported in EPA
TSCA Inventory.

NIOSH REL: TWA 2.5 mg(F)/m^3
DOT Classification: Corrosive Material; Label:
  Corrosive.

SAFETY PROFILE: Poison by subcutaneous
route. A corrosive irritant to skin, eyes and
mucous membranes. Will react with water or
steam to produce toxic and corrosive fumes.
When heated to decomposition it emits toxic
fumes of F$^-$.

SCP000   CAS: 7440-21-3   HR: 3
SILICON

DOT: UN 1346
af: Si   aw: 28.09

PROP: Cubic, steel-gray crystals or dark brown
powder. Mp: 1420°, bp: 2600°, d: 2.42 or 2.3
@ 20°, vap press: 1 mm @ 1724°. Almost
insol in water; sol in molten alkali oxides.

CONSENSUS REPORTS: Reported in EPA
TSCA Inventory.

OSHA PEL: (Transitional: TWA Total Dust:
  15 mg/m^3; Respirable Fraction: 5 mg/m^3)
  TWA Total Dust: 10 mg/m^3 of total; Respirable Fraction: 5 mg/m^3
ACGIH TLV: TWA (nuisance particulate) 10
  mg/m^3 of total dust (when toxic impurities
  are not present, e.g., quartz < 1%).
DOT Classification: Flammable Solid.

SAFETY PROFILE: Does not occur freely in
nature, but is found as silicon dioxide (silica)
and as various silicates. Elemental Si is flammable when exposed to flame or by chemical reaction with oxidizers. Violent reactions with alkali
carbonates, oxidants, (Al + PbO), Ca, Cs$_2$C$_2$,
Cl$_2$, CoF$_2$, F$_2$, IF$_5$, MnF$_3$, Rb$_2$C$_2$, FNO, AgF,
NaK alloy. When heated it will react with water
or steam to produce H$_2$; can react with oxidizing
materials.

SCQ000   CAS: 409-21-2   HR: 3
SILICON CARBIDE
mf: CSi   mw: 40.10

PROP: Bluish-black, iridescent crystals. Mp:
2600°, bp: subl > 2000°, decomp @ 2210°,
d: 3.17.

SYNS: CARBOLON * CARBON SILICIDE
* CARBORUNDEUM * CARBORUNDUM * KZ 3M
* KZ 5M * KZ 7M * SILICON MONOCARBIDE
* SILUNDUM

CONSENSUS REPORTS: Reported in EPA
TSCA Inventory.

OSHA PEL: (Transitional: TWA Total Dust:
  15 mg/m^3; Respirable Fraction: 5 mg/m^3)
  TWA Total Dust: 10 mg/m^3; Respirable Fraction: 5 mg/m^3
ACGIH TLV: TWA (nuisance particulate) 10
  mg/m^3 of total dust (when toxic impurities
  are not present, e.g., quartz < 1%).
DFG MAK: 4 mg/m^{3}0

SAFETY PROFILE: Questionable carcinogen
with experimental neoplastigenic data.

SCQ500   CAS: 10026-04-7   HR: 3
SILICON CHLORIDE

DOT: UN 1818
mf: Cl$_4$Si   mw: 169.89

PROP: Colorless, fuming liquid; suffocating
odor. Mp: −70°, bp: 57.57°, d: 1.482. Misc
with benzene, ether, chloroform, petr ether.

SYNS: CHLORID KREMICITY (CZECH) * EXTREMA
* SILICIO(TETRACLORURO di) * SILICIUMTETRA-
CHLORID (GERMAN) * SILICIUMTETRACHLORIDE
(DUTCH) * SILICIUM(TETRACHLORURE de) (FRENCH)
* SILICON TETRACHLORIDE (DOT) * TETRACHLO-
ROSILANE * TETRACHLORURE de SILICIUM (FRENCH)

CONSENSUS REPORTS: Reported in EPA
TSCA Inventory.

DOT Classification: Corrosive Material; Label:
  Corrosive.

SAFETY PROFILE: Mildly toxic by inhalation.
A corrosive irritant to eyes, skin, and mucous
membranes. Reacts with water to form HCl.
Violent reaction with Na, K. When heated to
decomposition it emits toxic fumes of Cl$^-$.

SDC000   HR: D
SILICONES

SYN: SILOXANES

CONSENSUS REPORTS: Organosilicon oxide polymers such as −R$_2$Si−O, where R is a
monovalent organic radical.

SAFETY PROFILE: Most of the silicones that
have been studied are only slightly toxic and
mildly irritating, however some may be severe

irritants. May be spontaneously flammable in air. There can be toxicity due to contamination of silicones by components of manufacture.

SDF650   CAS: 7783-61-1   HR: 3
SILICON FLUORIDE

DOT: UN 1859
mf: F_4Si mw: 104.09

PROP: Colorless gas, very pungent odor. Mp: $-77°$, bp: $-65°$ @ 181 mm, d: 4.67.

SYNS: SILICON TETRAFLUORIDE (DOT) * TETRA-FLUOROSILANE

CONSENSUS REPORTS: Reported in EPA TSCA Inventory.

OSHA PEL: TWA 2.5 mg(F)/m^3
ACGIH TLV: TWA 2.5 mg(F)/m^3
NIOSH REL: (Inorganic Fluorides) TWA 2.5 mg(F)/m^3
DOT Classification: Nonflammable Gas; Label: Nonflammable Gas; Poison A; Label: Poison Gas, Corrosive.

SAFETY PROFILE: A poison. A corrosive irritant to skin, eyes, and mucous membranes. When heated to decomposition it emits toxic fumes of F^-.

SDH575   CAS: 7803-62-5   HR: 3
SILICON TETRAHYDRIDE

DOT: UN 2203
mf: H_4Si mw: 32.13

PROP: Gas with repulsive odor; slowly decomp by water. D: 0.68 @ $-185°$, mp: $-185°$, bp: 112°, fp: $-200°$.

SYNS: MONOSILANE * SILANE * SILICANE

CONSENSUS REPORTS: Reported in EPA TSCA Inventory.

OSHA PEL: TWA 5 ppm
ACGIH TLV: TWA 5 ppm
DOT Classification: Flammable Gas; Label: Flammable Gas.

SAFETY PROFILE: Mildly toxic by inhalation. Silanes are irritating to skin, eyes, and mucous membranes. Easily ignited in air. Explosive reaction or ignition on contact with halogens or covalent halides (e.g., bromine, chlorine, carbonyl chloride, antimony pentachloride, tin(IV) chloride). Ignites in oxygen. Can react with oxidizers. It may self-explode. When heated to decomposition it burns or explodes.

SDI500   CAS: 7440-22-4   HR: 2
SILVER

af: Ag   aw: 107.868

PROP: Soft, ductile, malleable, lustrous, white metal. Mp: 961.93°, bp: 2212°, d: 10.50 @ 20°.

SYNS: ARGENTUM * C.I. 77820 * SHELL SILVER * SILBER (GERMAN) * SILVER ATOM

CONSENSUS REPORTS: Silver and its compounds are on the Community Right-To-Know List. Reported in EPA TSCA Inventory.

OSHA PEL: Metal, Dust, and Fume: TWA 0.01 mg/m^3
ACGIH TLV: TWA (metal) 0.1 mg/m^3, (soluble compounds as Ag) 0.01 mg/m^3
DFG MAK: 0.01 mg/m^3

SAFETY PROFILE: Human systemic effects by inhalation: skin effects. Inhalation of dusts can cause argyrosis. Questionable carcinogen with experimental tumorigenic data. Flammable in the form of dust when exposed to flame or by chemical reaction with C_2H_2, NH_3, bromoazide, ClF_3, ethylene imine, H_2O_2, oxalic acid, H_2SO_4, tartaric acid. Incompatible with acetylene, acetylene compounds, aziridine, bromine azide, 3-bromopropyne, carboxylic acids, copper + ethylene glycol, electrolytes + zinc, ethanol + nitric acid, ethylene oxide, ethyl hydroperoxide, ethyleneimine, iodoform, nitric acid, ozonides, peroxomonosulfuric acid, peroxyformic acid.

SDJ000   CAS: 7659-31-6   HR: 3
SILVER ACETYLIDE

mf: C_2Ag_2 mw: 239.76

CONSENSUS REPORTS: Silver and its compounds are on the Community Right-To-Know List.

DOT Classification: Forbidden.

SAFETY PROFILE: Severe explosion hazard. A more powerful detonator than copper acetylide. Explodes when heated to 120-140°C. Formed when silver-containing solutions contact acetylene. Upon decomposition it emits acrid smoke and irritating fumes.

SDM500   CAS: 13863-88-2   HR: 3
SILVER AZIDE

mf: AgN_3 mw: 149.87

CONSENSUS REPORTS: Silver and its compounds are on the Community Right-To-Know List.

OSHA PEL: TWA 0.01 mg(Ag)/m^3
ACGIH TLV: TWA 0.01 mg(Ag)/m^3
DOT Classification: Forbidden.

SAFETY PROFILE: Explodes when heated above 270°C or on impact. Pure silver azide, explodes @ 340°. An electric field or irradiation by electron pulses can explode the crystals. Shock sensitive when dry and has detonated @ 250 C. Solutions in aqueous ammonia explode above 100°C. Reacts to form more explosive products with iodine (forms iodine azide); bromine and other halogens. The presence of metal oxides or metal sulfides increases the azides sensitivity to explosion. Mixtures with sulfur dioxide are explosive. When heated to decomposition it emits toxic fumes of NO_x.

SDO500   HR: 3
SILVER COMPOUNDS

CONSENSUS REPORTS: Silver and its compounds are on the Commnuity Right-To-Know List.

SAFETY PROFILE: The water-soluble silver compounds are irritating to the skin and mucous membranes and may cause death if ingested. 50 mg of silver collargol is lethal after intravenous injection. Autopsy shows pulmonary edema, hemorrhage, and necrosis of the bone marrow, liver and kidney. The absorption of silver compounds into the circulation and the subsequent deposition of the reduced silver in various tissues of the body may result in the production of a generalized greyish pigmentation of the skin and mucous membranes, a condition known as argyria. Ingestion of 1-30 grams of soluble silver salts or long-term inhalation of a total 1-8 grams of silver can cause argyrosis. The introduction of fine particles of silver through breaks in the skin produces a local pigmentation at the site of the injury. 1 mg/m^3 of silver dust causes skin effects. The condition develops slowly, usually after 2-25 years of exposure. Pigmentation is noticeable first in conjunctivae, and later in the mucous membranes of the mouth and gums and in the skin. There are no constitutional symptoms or physical disability. Persons exhibiting the condition, and who subsequently died from unrelated disease, showed, on autopsy, a deposition of silver in the blood vessel walls, kidneys, testes, pituitary, choroid plexus, and mucous membranes of the nose, maxillary antra, trachea and bronchi. Once deposited, there is no known method by which the silver can be eliminated; the pigmentation is permanent.

SDP000   CAS: 506-64-9   HR: 3
SILVER CYANIDE

DOT: UN 1684
mf: CAgN   mw: 133.89

PROP: White, odorless, tasteless powder which darkens upon exposure to light. Mp: 320° (decomp), d: 3.95.

SYNS: CYANURE d'ARGENT (FRENCH) * KYANID STRIBRNY (CZECH) * RCRA WASTE NUMBER P104

CONSENSUS REPORTS: Silver and its compounds, as well as cyanide and its compounds, are on the Community Right-To-Know List. Reported in EPA TSCA Inventory.

DOT Classification: Poison B; Label: Poison.

SAFETY PROFILE: Deadly poison by ingestion. A skin and severe eye irritant. When heated to decomposition it emits very toxic fumes of CN^- and NO_x. Incompatible with phosphorus tricyanide, fluorine. Used in silver plating.

SDQ500   CAS: 7783-95-1   HR: 3
SILVER(II) FLUORIDE
mf: AgF_2 mw: 145.87

PROP: White when pure; usually a grey-black or brownish solid. D: 4.7, mp: 690°.

SYNS: ARGENT FLUORURE (FRENCH) * ARGENTIC FLUORIDE * SILVER DIFLUORIDE

CONSENSUS REPORTS: Silver and its compounds are on the Community Right-To-Know List. Reported in EPA TSCA Inventory.

OSHA PEL: TWA 0.01 mg(Ag)/m^3; 2.5 mg(F)/m^3
ACGIH TLV: TWA 0.01 mg(Ag)/m^3; 2.5 mg(F)/m^3
NIOSH REL: (Inorganic Fluorides) TWA 2.5 mg(F)/m^3

SAFETY PROFILE: Poison by subcutaneous route. Powerful oxidizing agent. Mixtures with boron + water are explosive. When heated to decomposition it emits toxic fumes of F^-.

SDR759 CAS: 13092-75-6 **HR: 3**
SILVER MONOACETYLIDE
mf: C_2HAg mw: 132.90

CONSENSUS REPORTS: Silver and its compounds are on the Community Right-To-Know List.

DOT Classification: Forbidden.

SAFETY PROFILE: Severe explosion hazard. When heated to decomposition it emits acrid smoke and irritating fumes.

SDS000 CAS: 7761-88-8 **HR: 3**
SILVER(I) NITRATE (1:1)

DOT: UN 1493
mf: $NO_3 \cdot Ag$ mw: 169.88

PROP: Mp: 212°, bp: 444° (decomp), d: 4.352 @ 19°. Very sol in ammonia, water; sltly sol in ether.

SYNS: LUNAR CAUSTIC * NITRATE d'ARGENT (FRENCH) * NITRIC ACID, SILVER(1+) SALT * SILBERNITRAT * SILVER(1+) NITRATE * SILVER NITRATE (DOT)

CONSENSUS REPORTS: Silver and its compounds are on the Community Right-To-Know List. EPA Genetic Toxicology Program. Reported in EPA TSCA Inventory.

OSHA PEL: TWA 0.01 mg(Ag)/m^3
ACGIH TLV: TWA 0.01 mg(Ag)/m^3
DOT Classification: Oxidizer; Label: Oxidizer.

SAFETY PROFILE: Human poison by an unspecified route. Experiemental poison by ingestion, intravenous, subcutaneous, and intraperitoneal routes. Experimental reproductive effects. Human mutation data reported. A severe eye irritant. A powerful caustic and irritant to skin, eyes, and mucous membranes. Swallowing can cause severe gastroenteritis that may be fatal. Questionable carcinogen with experimental tumorigenic data. A powerful oxidizer. Incompatible with acetylene, acetylides, alkalies, aluminum, antimony salts, arsenic, arsenites, bromides, carbon, carbonates, chlorides, ClF_3, chlorosulfuric acid, copper, creosote, ethanol, ferrous salts, hypophosphites, iodides, Mg powder with H_2O, morphine salts, NH_3 with KOH to yield black Ag_3N, oils, PH_3, phosphates, phosphonium iodide, phosphorous, plastics, sulfur, tannic acid, tartrates, thiocyanates, vegetable decoctions and extracts, zinc

with NH_3 with KOH. When heated to decomposition it emits toxic fumes of NO_x.

SDW000 **HR: 3**
SILVER PEROXYCHROMATE
mf: $AgCrO_5$ mw: 239.87

CONSENSUS REPORTS: Silver and its compounds, as well as chromium and its compounds, are on the Community Right-To-Know List.

SAFETY PROFILE: Confirmed carcinogen. An oxidant. When mixed with H_2SO_4 @ $-80°$ it explodes on slow warming to $-30°$.

SEA000 **HR: 2**
SLUDGE ACID

SYNS: ACID, SLUDGE (DOT)

DOT Classification: Corrosive Material; Label: Corrosive.

SAFETY PROFILE: A corrosive irritant to skin, eyes, and mucous membranes. When heated to decomposition it emits very toxic fumes of SO_x and NO_x.

SEB000 **HR: 2**
SMOG

PROP: An atmospheric combination of smoke, fog, and industrial gases. Composition: Contents vary, but sulfur dioxide, oxides of nitrogen, and ozone are common components; others are sulfides, fluorides, chlorides, carbon particles and various hydrocarbons.

SAFETY PROFILE: Moderately irritating to eyes and mucous membranes. Numerous chronic effects have been reported in susceptible populations. A common air contaminant. Possibly carcinogenic.

SEC000 **HR: 3**
SMOKE CONDENSATE, CIGARETTE

SYNS: CIGARETTE SMOKE CONDENSATE * CSC * TOBACCO SMOKE CONDENSATE * TOBACCO TAR

CONSENSUS REPORTS: IARC Cancer Review: Human Sufficient Evidence IMEMDT 38,309,86, Animal Sufficient Evidence IMEMDT 38,309,86.

SAFETY PROFILE: Confirmed carcinogen with experimental carcinogenic, neoplastigenic, and tumorigenic data. Experimental reproductive effects. Human mutation data reported.

SED400   HR: 3
SMOKELESS TOBACCO

PROP: A variety of habituating substances containing tobacco as the major ingredient and used without burning. Tobacco is a product of the leaves and stems of two species of Nicotiana, *N. Tabacum* (grown in North America and Western Europe) and *N. Rustica* (grown in the USSR and India). There is considerable evidence that many if not all of the forms of smokeless tobacco are human carcinogens.

The smokeless tobaccos are introduced into the body through the mouth (chewing tobacco, snuff, misshri, gudakhu, shammah, khaini, nass, naswar or in combination with betel quid) or nose (snuff).

The various smokeless tobacco products are:

Chewing Tobacco is placed between the cheek and gum and chewed slowly. There are three main types: plug, twist/roll, and loose-leaf.

Fine-Cut tobacco was formerly classified in the United States as chewing tobacco and is now placed in the category of moist fine-cut snuff.

Gudakhu is a paste of powdered tobacco, molasses and other ingredients used in parts of India to clean teeth.

Khaini is a mixture of tobacco and lime formed into a ball and placed in the mouth.

Kiwam is made from processed tobacco leaves. After the stalks and stems are removed, the leaves are soaked and boiled in water with flavorings and spices, crushed, then strained, leaving a paste which is chewed.

Loose-Leaf tobacco is prepared from fermented cigar leaves, sweetened with sugars, syrups, liquors, and other flavoring materials. It is packaged as batches of loose pieces or cut strips.

Mainpuri tobacco is a chewed mixture of tobacco with slaked lime, areca nut, camphor, and cloves. It is used in India.

Mishri is prepared from roasted or half-burnt tobacco which has been baked till black on a hot metal plate and then powdered. It is used primarily to clean teeth but is also used as chewing tobacco. Synonyms are masheri and misheri.

Nass is a mixture of tobacco, lime, wood-ash, and cottonseed oil, chewed in Iran and the central Asian region of the USSR.

Naswar is a mixture of powdered tobacco, slaked lime, and indigo placed on the bottom of the mouth or behind the lower lip. It is used in Afghanistan and Pakistan.

Pattiwala tobacco is a sun-cured tobacco leaf chewed with or without lime. It is used in India.

Pill is dried and pelleted Kiwam paste.

Plug tobacco is made from enriched tobacco leaves or leaf fragments wrapped in fine tobacco and pressed into flat bars or rolls. It is chewed.

Shammah is a mixture of powdered tobacco leaves with calcium or sodium carbonate and other materials, including ash, placed in the cheek or behind the lower lip. It is used in southern Saudi Arabia.

Snuff is taken through the mouth or the nose. Moist snuff is finely cut tobacco plus flavorings with a moisture content of up to 50 percent. It is placed in the cheek. Dry snuff has a moisture content of less than 10 percent and may have flavorings. It may be sniffed through the nose, placed behind the lower lip or in the cheek. Oriental snuff is about 50 percent heated calcium carbonate and calcium phosphate with some powdered cuttle-fish bone. In southern Africa, snuff is made from powdered tobacco leaves, plant ash, and sometimes oils, lemon juice, and herbs. In the United States, ''dipping'', refers to the ingestion use of snuff.

Twist/Roll tobacco is stripped tobacco leaves rolled or twisted like a length of rope.

Zarda is tobacco leaf broken into small pieces and boiled in water with lime and spices to dryness and then colored with vegetable dyes. It is usually chewed mixed with areca nut and spices.

SYNS: CHEWING TOBACCO * GUDAKHU (INDIA) * KHAINI (INDIA) * KIWAM (INDIA) * MASHERI (INDIA) * MISHERI (INDIA) * MISHRI (INDIA) * NASS (IRAN) * NASWAR (PAKISTAN and AFGHANISTAN) * PILLS (INDIA) * SHAMMAH (SAUDI ARABIA) * SNUFF * ZARDA (INDIA)

SAFETY PROFILE: Tobaccos contain from 0.5-5% alkaloids predominantly as l-nicotine (>85%). Nicotine is strongly addictive and is the chief cause of tobacco dependence. It is a mild stimulant. It readily forms salts with most acids. These salts are poorly absorbed through the mucous membranes whereas the base is easily absorbed. This explains the practice of combining lime or other alkali in conjunction with ingestion tobacco use. Nicotine and some of the other tobacco alkaloids are experimental teratogens and mutagens.

There are several known classes of carcinogens present in the smokeless tobaccos: N-nitrosamines, polynuclear aromatic hydrocarbons (PAH's), heavy metals (arsenic trioxide, lead, cadmium and nickel compounds), and radionuclides (^{226}Ra, ^{210}Pb and ^{210}Po). Of these, nitrosamines are present in the highest concentration (in the range of mg/kg). The concentrations of the nitrosamines are 100 times higher in tobacco than in other consumer products. Nitrosamine concentrations are higher in chewing tobacco than in cigarette smoke. The major nitrosamines in tobacco are N'-nitrososornornicotine (NNN), 4-(methylnitrosamine)-1-(3-pyridyl)-1-butanone (NNK) and N'-nitrosoanatabine (NAT). They are probably generated during curing, fermentation, and aging of the tobacco leaf from the tobacco alkaloids: nicotine, nornicotine, anatabine, anabasine, continine, myosmine, 2,3'-dipyridyl and N'-formylnornicotine. They may also form in the mouth.

There is sufficient evidence that the ingestion use of snuff, chewing tobacco, and tobacco mixed with lime is carcinogenic to humans. Evidence suggests that the ingestion use of other smokeless tobacco preparations and the nasal use of snuff is carcinogenic to humans. Oral precancerous lesions are commonly observed in smokeless tobacco users.

SEE500   CAS: 7440-23-5   HR: 3
SODIUM

DOT: UN 1428/UN 1429
af: Na   aw: 22.9898

PROP: Light, soft, ductile, malleable, silver-white metal. Mp: 97.81°, bp: 881.4°, d: 0.9710 @ 20°, autoign temp: > 115° in dry air, vap press: 1.2 mm @ 400°.

SYNS: NATRIUM * SODIUM METAL (DOT)

CONSENSUS REPORTS: Reported in EPA TSCA Inventory.

DOT Classification: Label: Flammable Solid and Dangerous When Wet

SAFETY PROFILE: Metallic sodium reacts exothermally with the moisture of body or tissue surfaces, causing thermal and chemical burns. Sodium in elemental form is highly reactive.

Sodium reacts violently with water to form sodium hydroxide. A very dangerous fire hazard when exposed to heat and moisture. Under the appropriate conditions, it can react violently with moisture, air, $AlBr_3$, $AlCl_3$, AlF_3, NH_4 chlorocuprate, NH_4NO_3, $SbBr_3$, $SbCl_3$, SbI_3, $AsCl_3$, AsI_3, $BiBr_3$, $BiCl_3$, BiI_3, Bi_2O_3, BBr_3, bromoazide, CO_2, $CO + NH_3$, CCl_4, Cl_2, ClF_3, $CrCl_4$, CrO_3, $CoBr$, $CoCl$, $CuCl_2$, CuO, $FeBr_3$, $FeCl_3$, $FeBr_2$, $FeCl_2$, FeI_2, hydrazine hydrate, H_2O_2, H_2S, HCl, HF, F_2, 1,2-dichloroethylene, dichloromethane, Br_2, hydroxylamine, iodine, iodine monochloride, iodine pentafluoride, lead oxide, maleic anhydride, manganous chloride, mercuric bromide, mercuric chloride, mercuric fluoride, mercuric iodide, mercurous chloride, mercurous oxide, methyl chloride, molybdenum trioxide, monoammonium phosphate, nitric acid, nitrogen peroxide, nitrosyl fluoride, nitrous oxide, phosgene, phosphorus, phosphorous pentafluoride, phosphorus pentoxide, phosphorus tribromide, phosphorus trichloride, phosphoryl chloride, potassium oxides, potassium ozonide, potassium superoxide, selenium, silicon tetrachloride, silver bromide, silver chloride, silver fluoride, silver iodide, sodium peroxide, stannic chloride, stannic iodide with sulfur, stannic oxide, stannous chloride, sulfur, sulfur dibromide, sulfur dichloride, sulfur dioxide, sulfuric acid, tellurium, tetrachloroethane, thallous bromide, thiophosphoryl bromide, trichloroethylene, vanadium pentachloride, vanadyl chloride, zinc bromide, any oxidizing material. Decomposes moisture to evolve hydrogen and heat. Reacts exothermally with halogens, acids, and halogenated hydrocarbons.

Heated sodium is spontaneously flammable in air. Can be safely stored under liquid hydrocarbons. Dangerous explosion hazard when exposed to moisture in any form! Keep away from water at all times! When heated in air it emits toxic fumes of sodium oxide. Reacts with water or steam to produce heat, hydrogen, and flammable vapors. Can react vigorously to explosively with oxidizing materials. To fight fire, use soda ash, dry sodium chloride or graphite, in order of preference. When heated to decomposition it emits toxic fumes of Na_2O.

SEF500   CAS: 7440-23-5   HR: 3
SODIUM (dispersions)

PROP: Finely divided metallic sodium suspended in toluene, xylene, naphtha, kerosene, etc.

SYN: SODIUM, METAL DISPERSION IN ORGANIC SOLVENT

CONSENSUS REPORTS: Reported in EPA TSCA Inventory.

DOT Classification: Label: Flammable solid and Dangerous When Wet

SAFETY PROFILE: A very dangerous fire hazard when exposed to heat or flame or by chemical reaction. These are very reactive forms of sodium which, if carelessly handled, may catch fire. After sodium has been extinguished, the burning organic vapor can be dealt with by very cautious use of a carbon dioxide extinguisher. To extinguish, see SODIUM. Do not use carbon tetrachloride. Moderate explosion hazard by chemical reaction; will react with water or steam to produce heat and hydrogen; on contact with oxidizing materials it can react vigorously, and on contact with acid or acid fumes it can emit toxic fumes. When heated it loses the solvent and emits highly toxic fumes of Na_2O.

SEF600　　　CAS: 7440-23-5　　　HR: 3
SODIUM (liquid alloy)

DOT: NA 1421
mf: Na　　mw: 22.99

SYN: SODIUM, metal liquid alloy (DOT)

DOT Classification: Label: Flammable Solid and Dangerous When Wet

SAFETY PROFILE: Flammable when exposed to heat or flame; can react vigorously with oxidizing materials. When heated to decomposition it emits toxic fumes of Na_2O.

SEG500　　　CAS: 127-09-3　　　HR: 3
SODIUM ACETATE
mf: $C_2H_3O_2 \cdot Na$　　mw: 82.04

PROP: White granular powder. Autoign temp: 1125°F, d: 1.45, mp: 58°. Decomp @ higher temp. Sol in water, alc.

SYNS: ACETIC ACID, SODIUM SALT * NATRIUM-ACETAT (GERMAN) * SODIUM ACETATE, anhydrous (FCC)

CONSENSUS REPORTS: Reported in EPA TSCA Inventory. EPA Genetic Toxicology Program.

SAFETY PROFILE: Poison by intravenous route. Moderately toxic by ingestion. A skin and eye irritant. Migrates to food from packaging materials. Violent reaction with F_2, KNO_3, diketene. When heated to decomposition it emits toxic fumes of Na_2O.

SEG800　　　CAS: 7681-38-1　　　HR: 2
SODIUM ACID SULFATE (solid)

DOT: UN 1821/UN 2837
mf: $HO_4S \cdot Na$　　mw: 120.06

PROP: White crystals or granules. Mp: >315° (decomp), d: 2.435 @ 13°. Sol in water.

SYNS: GBS * NITRE CAKE * SODIUM ACID SULFATE * SODIUM ACID SULFATE, solution (DOT) * SODIUM BISULFATE, fused * SODIUM BISULFATE, solid (DOT, FCC) * SODIUM BISULFATE, solution (DOT) * SODIUM HYDROGEN SULFATE, solid (DOT) * SODIUM HYDROGEN SULFATE, solution (DOT) * SODIUM PYROSULFATE * SULFURIC ACID, MONOSODIUM SALT

CONSENSUS REPORTS: Reported in EPA TSCA Inventory.

DOT Classification: ORM-B; Label: None; Corrosive Material; Label: Corrosive, solid and solution

SAFETY PROFILE: A corrosive irritant to skin, eyes, and mucous membranes. Mutation data reported. Reacts with moisture to form sulfuric acid. Mixtures with calcium hypochlorite + starch + sodium carbonate explode when compressed. Violent reaction with acetic anhydride + ethanol may lead to ignition and a vapor explosion. Incompatible with calcium hypochlorite. When heated to decomposition it emits toxic fumes of SO_x and Na_2O.

SEH000　　　CAS: 9005-38-3　　　HR: 3
SODIUM ALGINATE
mf: $(C_6H_7O_6Na)_n$　　mw: 198.11

PROP: Colorless to slight yellow filamentous or granular solid or powder; odorless and tasteless. In water it forms a viscous colloidal soln; insol in ether, alc, chloroform.

SYNS: ALGIN * ALGINATE KMF * ALGIN (POLYSACCHARIDE) * ALGIPON L-1168 * AMNUCOL * ANTIMIGRANT C 45 * CECALGINE TBV * COHASAL-1H * DARID QH * DARILOID QH * DUCKALGIN * HALLTEX * K'-ALGILINE * KELCO GEL LV * KELCOSOL * KELGIN * KELGUM * KELSET * KELSIZE * KELTEX * KELTONE * LAMITEX * MANUCOL

* MANUCOL DM * MANUTEX * MEYPRALGIN R/LV * MINUS * MOSANON * NOURALGINE * OG 1 * PECTALGINE * PROCTIN * PROTA-CELL 8 * PROTANAL * PROTATEK * SNOW ALGIN H * SODIUM POLYMANNURONATE * STIPINE * TAGAT * TRAGAYA

CONSENSUS REPORTS: Reported in EPA TSCA Inventory.

SAFETY PROFILE: Poison by intravenous and intraperitoneal routes. When heated to decomposition it emits toxic fumes of Na_2O.

SEM000   CAS: 1344-00-9   HR: 1
SODIUM ALUMINOSILICATE

PROP: Fine, white, amorphous powder or beads; odorless and tasteless. Insol in water, alc, and other organic solvents.

SYNS: NCI-C55505 * SODIUM SILICOALUMINATE

CONSENSUS REPORTS: Reported in EPA TSCA Inventory.

SAFETY PROFILE: An irritant to skin, eyes, and mucous membranes. When heated to decomposition it emits toxic fumes of Na_2O.

SEM500   CAS: 13770-96-2   HR: 3
SODIUM ALUMINUM TETRAHYDRIDE

DOT: UN 2835
mf: $AlH_4 \cdot Na$ mw: 54.01

PROP: White, crystalline material; stable in dry air but sensitive to moisture. Mp: 183°; d: 1.24. Sol in tetrahydrofuran.

SYNS: ALUMINUM SODIUM HYDRIDE * SAH 22 * SODIUM ALUMINUM HYDRIDE (DOT) * SODIUM TETRAHYDROALUMINATE(1−) * (T-4) SODIUM, TETRAHYDROALUMINATE(1−) (9CI)

ACGIH TLV: TWA 2 mg(Al)/m³
DOT Classification: Flammable Solid; Label: Flammable Solid and Dangerous When Wet

SAFETY PROFILE: Flammable when exposed to heat or flame. May ignite and explode on contact with water. Reacts violently with tetrahydrofuran when heated. When heated to decomposition it emits toxic fumes of Na_2O.

SEN000   CAS: 7782-92-5   HR: 3
SODIUM AMIDE

DOT: UN 1425
mf: H_2NNa mw: 39.02

PROP: White, crystalline powder. Mp: 210°, bp: 400°.

SYN: SODAMIDE

CONSENSUS REPORTS: Reported in EPA TSCA Inventory.

DOT Classification: Label: Flammable Solid and Dangerous When Wet

SAFETY PROFILE: An intense irritant to tissue, skin, and eyes. Flammable by chemical reaction. Ignites or explodes with heat or grinding. Explosive reaction with moisture; chromium trioxide; potassium chlorate; halocarbons (e.g., 1,1-diethoxy-2-chloroethane); oxidants; sodium nitrite; air. Can become explosive in storage. Violent reaction with dinitrogen tetraoxide. Will react with water or steam to produce heat and toxic and corrosive fumes of sodium hydroxide and ammonia. When heated to decomposition it emits highly toxic fumes of NH_3 and Na_2O.

SEY500   CAS: 7784-46-5   HR: 3
SODIUM ARSENITE
mf: $AsO_2 \cdot Na$ mw: 129.91

DOT: UN 1686/UN 2027

PROP: White or grayish white powder. Commercially: 95%-98% pure. Very sol in water; sltly sol in alc.

SYNS: ARSENENOUS ACID, SODIUM SALT (9CI) * ARSENIOUS ACID, SODIUM SALT * ARSENITE de SODIUM (FRENCH) * ATLAS "A" * CHEM PELS C * CHEM-SEN 56 * KILL-ALL * PENITE * PRODALUMNOL * PRODALUMNOL DOUBLE * SODANIT * SODIUM ARSENITE, liquid (solution) (DOT) * SODIUM ARSENITE, solid (DOT) * SODIUM METAARSENITE

CONSENSUS REPORTS: IARC Cancer Review: GROUP 1 IMEMDT 7,100,87; Animal Inadequate Evidence IMEMDT 23,39,80; Human Sufficient Evidence IMEMDT 23,39,80; Animal No Evidence IMEMDT 2,48,73. Arsenic and its compounds are on the Community Right To Know List. Reported in EPA TSCA Inventory. EPA Genetic Toxicology Program. EPA Extremely Hazardous Substances List.

OSHA PEL: TWA 0.01 mg(As)/m³
ACGIH TLV: TWA 0.2 mg(As)/m³
NIOSH REL: CL (Inorganic Arsenic) 0.002 mg(As)/m³/15M
DOT Classification: Poison B; Label: Poison

SAFETY PROFILE: Confirmed human carcinogen. Human poison by ingestion. Experimental poison by ingestion, skin contact, intravenous, intramuscular, and intraperitoneal routes. Experimental teratogenic and reproductive effects. Human mutation data reported. Used as a herbicide and pesticide. When heated to decomposition it emits toxic fumes of As and Na_2O.

SEZ000   CAS: 7784-46-5   **HR: 3**
SODIUM ARSENITE (liquid)

CONSENSUS REPORTS: Arsenic and its compounds are on the Community Right To Know List. Reported in EPA TSCA Inventory.

NIOSH REL: CL 2 $\mu g/m^3$/15M
DOT Classification: Poison B; Label: Poison.

SAFETY PROFILE: Confirmed human carcinogen. A deadly poison. When heated to decomposition it emits toxic fumes of As and Na_2O.

SFA000   CAS: 26628-22-8   **HR: 3**
SODIUM AZIDE

DOT: UN 1687
mf: N_3Na mw: 65.02

PROP: Colorless, hexagonal crystals. Mp: decomp, d: 1.846. Insol in ether; sol in liquid ammonia.

SYNS: AZIDE * AZIUM * AZOTURE de SODIUM (FRENCH) * KAZOE * NATRIUMAZID (GERMAN) * NATRIUMMAZIDE (DUTCH) * NCI-C06462 * NSC 3072 * RCA WASTE NUMBER P105 * SODIUM, AZOTURE de (FRENCH) * SODIUM, AZOTURO di (ITALIAN) * U-3886

CONSENSUS REPORTS: Reported in EPA TSCA Inventory. EPA Genetic Toxicology Program. EPA Extremely Hazardous Substances List.

OSHA PEL: As NH_3: CL 0.1 ppm; As NaN_3: Cl 0.3 mg/m^3 (skin)
ACGIH TLV: CL 0.3 mg/m^3
DFG MAK: 0.07 ppm (0.2 mg/m^3)
DOT Classification: Poison B; Label: Poison

SAFETY PROFILE: Poison by ingestion, skin contact, intraperitoneal, intravenous, subcutaneous and possibly other routes. Human systemic effects by ingestion: general anesthesia, somnolence, and kidney changes. Questionable carcinogen with experimental tumorigenic data. Human mutation data reported.

An unstable explosive sensitive to impact. Violent reaction with benzoyl chloride combined with KOH, Br_2, barium carbonate, CS_2, $Cr(OCl)_2$, Cu, Pb, HNO_3, $BaCO_3$, H_2SO_4, hot water, $(CH_3)_2SO_4$, dibromomalononitrile, sulfuric acid. Incompatible with acids, ammonium chloride + trichloroacetonitrile, phosgene, cyanuric chloride, 2,5-dinitro-3-methylbenzoic acid + oleum, trifluroracryloyl chloride. Reacts with heavy metals (e.g., brass, copper, lead) to form dangerously explosive heavy metal azides. When heated to decomposition it emits very toxic fumes of NO_x and Na_2O.

SFB000   CAS: 532-32-1   **HR: 2**
SODIUM BENZOATE
mf: $C_7H_5O_2 \cdot Na$ mw: 144.11

PROP: White crystalline solid; odorless. Sol in water and alc.

SYNS: ANTIMOL * BENZOATE of SODA * BENZOATE SODIUM * BENZOESAEURE (NA-SALZ) (GERMAN) * BENZOIC ACID, SODIUM SALT * SOBENATE * SODIUM BENZOIC ACID

CONSENSUS REPORTS: Reported in EPA TSCA Inventory. EPA Genetic Toxicology Program.

SAFETY PROFILE: Poison by subcutaneous, and intravenous routes. Moderately toxic by ingestion, intramuscular and intraperitoneal routes. Experimental teratogenic and reproductive effects. Mutation data reported. Larger doses of 8-10 grams by mouth may cause nausea and vomiting. Small doses have little or no effect. Combustible when exposed to heat or flame. When heated to decomposition it emits toxic fumes of Na_2O.

SFB500   CAS: 63915-76-4   **HR: 3**
SODIUM BERYLLIUM MALATE
mf: $C_8H_6Be_4Na_2O_{12} \cdot 7H_2O$ mw: 502.30

CONSENSUS REPORTS: Beryllium and its compounds are on the Community Right To Know List.

OSHA PEL: (Transitional: TWA 0.002 mg(Be)/m^3; CL 0.005; Pk 0.025/30M/8H) TWA 0.002 mg(Be)/m^3; STEL 0.005 mg(Be)/m^3/30M; CL 0.025 mg(Be)/m^3
ACGIH TLV: TWA 0.002 mg(Be)/m^3, Suspected Carcinogen
NIOSH REL: CL (Beryllium) not to exceed 0.0005 mg(Be)/m^3

SAFETY PROFILE: Confirmed carcinogen. Poison by intravenous route. When heated to decomposition it emits toxic fumes of BeO and Na_2O.

SFC000   CAS: 63915-77-5   **HR: 3**
SODIUM BERYLLIUM TARTRATE
mf: $C_8H_4Be_4Na_2O_{13} \cdot 10H_2O$ mw: 570.34

CONSENSUS REPORTS: Beryllium and its compounds are on the Community Right To Know List.

OSHA PEL: (Transitional: TWA 0.002 mg(Be)/m^3; CL 0.005; Pk 0.025/30M/8H) TWA 0.002 mg(Be)/m^3; STEL 0.005 mg(Be)/m^3/30M; CL 0.025 mg(Be)/m^3
ACGIH TLV: TWA 0.002 mg(Be)/m^3, Suspected Carcinogen
NIOSH REL: CL (Beryllium) not to exceed 0.0005 mg(Be)/m^3

SAFETY PROFILE: Confirmed carcinogen. Poison by subcutaneous and intravenous routes. When heated to decomposition it emits toxic fumes of BeO and Na_2O.

SFE000   CAS: 7631-90-5   **HR: 3**
SODIUM BISULFITE

DOT: NA 2693
mf: $HO_3S \cdot Na$ mw: 104.06

PROP: White, crystalline powder; odor of sulfur dioxide, disagreeable taste. D: 1.48, mp: decomp. Very sol in hot or cold water; sltly sol in alc.

SYNS: BISULFITE de SODIUM (FRENCH) * HYDROGEN SULFITE SODIUM * SODIUM ACID SULFITE * SODIUM BISULFITE * SODIUM BISULFITE (1:1) * SODIUM BISULFITE, solid (DOT) * SODIUM BISULFITE, solution (DOT) * SODIUM HYDROGEN SULFITE * SODIUM HYDROGEN SULFITE, solid (DOT) * SODIUM HYDROGEN SULFITE, solution (DOT) * SODIUM SULHYDRATE * SULFUROUS ACID, MONOSODIUM SALT

CONSENSUS REPORTS: Reported in EPA TSCA Inventory. EPA Genetic Toxicology Program.

OSHA PEL: TWA 5 mg/m^3
ACGIH TLV: TWA 5 mg/m^3
DOT Classification: ORM-B; Label: None; Corrosive Material; Label: Corrosive

SAFETY PROFILE: Poison by intravenous and intraperitoneal routes. Moderately toxic by ingestion. A corrosive irritant to skin, eyes, and mucous membranes. Mutation data reported.

An allergen. When heated to decomposition it emits toxic fumes of SO_x and Na_2O.

SFE500   CAS: 1303-96-4   **HR: D**
SODIUM BORATE
mf: $B_4O_7 \cdot 2Na$ mw: 201.22

PROP: White crystals. Mp: 741°, bp: 1575° (decomp), d: 2.367. Slowly soluble in water.

SYNS: BORATES, TETRA, SODIUM SALT, anhydrous (OSHA, ACGIH) * SODIUM BORATE anhydrous

OSHA PEL: 10 mg/m^3 (anhydrous, decahydrate, pentahydrate)
ACGIH TLV: TWA 1 mg/m^3

SAFETY PROFILE: Experimental reproductive effects. When heated to decomposition it emits toxic fumes of Na_2O, boron.

SFF000   CAS: 1303-96-4   **HR: 3**
SODIUM BORATE DECAHYDRATE
mf: $B_4O_7 \cdot 2Na \cdot 10H_2O$ mw: 381.42

PROP: Hard, odorless crystals, granules or crystalline powder. D: 1.73, mp: 75° (when rapidly heated).

SYNS: ANTIPYONIN * BORACSU * BORATES, TETRA, SODIUM SALT, anhydrous (OSHA, ACGIH) * BORAX (8CI) * BORAX DECAHYDRATE * BORICIN * GERTLEY BORATE * JAIKIN * NEOBOR * POLYBOR * SODIUM BIBORATE * SODIUM BIBORATE DECAHYDRATE * SODIUM PYROBORATE * SODIUM PYROBORATE DECAHYDRATE * SODIUM TETRABORATE * SODIUM TETRABORATE DECAHYDRATE

CONSENSUS REPORTS: Reported in EPA TSCA Inventory.

OSHA PEL: TWA 10 mg/m^3
ACGIH TLV: TWA 5 mg/m^3

SAFETY PROFILE: Experimental poison by subcutaneous route. Moderately toxic to humans by ingestion. Moderately toxic experimentally by ingestion, intravenous and intraperitoneal routes. Experimental reproductive effects. Mutation data reported. Ingestion of 5-10 grams of borax by children can cause severe vomiting, diarrhea, shock, death. Incompatible with acids; metallic salts. When heated to decomposition it emits toxic fumes of Na_2O, boron. Used in ant poisons, for fly control around refuse and manure piles, as a larvicide, in manufacture of glazes, enamels, cleaning compounds, and in soldering metals.

SFF500   CAS: 16940-66-2   **HR: 3**
SODIUM BOROHYDRIDE
mf: $BH_4 \cdot Na$ mw: 37.84

DOT: UN 1426

PROP: White to gray-white, microcrystalline powder or lumps. Hygroscopic. Mp: >400°C (vacuum) (decomp), d: 1.07. Reacts with hot water; sol in liquid ammonia and "Cellosolve" ether.

SYNS: BOROHYDRURE de SODIUM (FRENCH) * SODIUM TETRAHYDROBORATE(1-)

CONSENSUS REPORTS: Reported in EPA TSCA Inventory.

DOT Classification: Flammable Solid; Label: Dangerous When Wet

SAFETY PROFILE: Poison by ingestion and intraperitoneal routes. A strong alkali. A severe eye, skin, and mucous membrane irritant.

Ignites in air above 288°C when exposed to spark. Potentially explosive reaction with aluminum chloride + bis(2-methoxyethyl) ether. Reacts with ruthenium salts to form a solid product which explodes when touched or on contact with water. Reacts to form dangerously explosive hydrogen gas on contact with alkali, water and other protic solvents (e.g., methanol, ethanol, ethylene glycol, phenol), aluminum chloride + bis(2-methoxyethyl)ether. Reacts violently with anhydrous acids (e.g., sulfuric, phosphoric, fluorophosphoric) to form diborane. Violent exothermic reaction with dimethyl formamide has caused industrial explosions. Mixtures with sulfuric acid may ignite. Incompatible with palladium, diborane + bis(2-methoxyethyl) ether, polyglycols, dimethylacetamide,-oxidizers, metal salts, finely divided metallic precipitates of cobalt, nickel, copper, iron and possibly other metals. Emits flammable vapors on contact with acid fumes. Materials sensitive to polymerization under alkaline conditions, such as acrylonitrile, may polymerize upon contact with sodium borohydride. Avoid storage in glass containers. When heated to decomposition it emits toxic fumes of Na_2O.

SFG000   CAS: 7789-38-0   **HR: 3**
SODIUM BROMATE

DOT: UN 1494
mf: $BrO_3 \cdot Na$ mw: 150.90

PROP: White crystals or crystalline powder. Odorless. Mp: 381°, d: 3.339 @ 17.5°.

SYNS: BROMATE de SODIUM (FRENCH) * BROMIC ACID, SODIUM SALT * DYETONE

CONSENSUS REPORTS: Reported in EPA TSCA Inventory.

DOT Classification: Oxidizer; Label: Oxidizer

SAFETY PROFILE: Poison by ingestion, intravenous, subcutaneous, and intraperitoneal routes. A powerful oxidizer. Violent reactions with Al, As, C, Cu, oil, F_2, metal sulfides, organic matter, P, S. Mixtures with grease are shock-sensitive explosives at 120°C. When heated to decomposition it emits toxic fumes of Na_2O and Br^-.

SFO000   CAS: 497-19-8   **HR: 3**
SODIUM CARBONATE (2:1)
mf: $CO_3 \cdot 2Na$ mw: 105.99

PROP: White, odorless, small crystals or crystalline powder; alkali taste. Mp: 851°, bp: decomp, d: 2.509 @ 0°. Hygroscopic; sol in water.

SYNS: CARBONIC ACID, DISODIUM SALT * CRYSTOL CARBONATE * DISODIUM CARBONATE * SODA ASH * TRONA

CONSENSUS REPORTS: Reported in EPA TSCA Inventory. EPA Genetic Toxicology Program.

SAFETY PROFILE: Poison by intraperitoneal route. Moderately toxic by inhalation and subcutaneous routes. Mildly toxic by ingestion. Experimental reproductive effects. A skin and eye irritant. It migrates to food from packaging materials. Can react violently with Al, P_2O_5, H_2SO_4, F_2, Li, 2,4,6-trinitro-toluene. When heated to decomposition it emits toxic fumes of Na_2O.

SFO500   CAS: 9004-32-4   **HR: 3**
SODIUM CARBOXYMETHYL CELLULOSE

PROP: A synthetic cellulose gum (the sodium salt of carboxy methyl cellulose not less than 99.5% on a dry weight basis, with maximum substitution of 0.95 carboxymethyl groups per anhydroglucose unit, and with a minimum viscosity of 25 centipoises for 2% weight aqueous solutions at 25°). Colorless, odorless, hygroscopic powder or granules. Insol in most organic solvents.

SYNS: AC-DI-SOL NF * AQUAPLAST * B10 * BLANOSE BWM * B 10 (polysaccharide) * CAR-

BOXYMETHYL CELLULOSE * CARBOXYMETHYL CELLULOSE, SODIUM * CARBOXYMETHYL CELLULOSE, SODIUM SALT * CARMETHOSE * CELLOFAS * CELLOGEL C * CELLPRO * CELLUFIX FF 100 * CELLUGEL * CELLULOSE GLYCOLIC ACID, SODIUM SALT * CELLULOSE GUM * CELLULOSE SODIUM GLYCOLATE * CMC * CM-CELLULOSE Na SALT * CMC 7H * CMC SODIUM SALT * COLLOWELL * COPAGEL PB 25 * COURLOSE A 590 * DAICEL 1150 * FINE GUM HES * GLIKOCEL TA * KMTS 212 * LOVOSA * LUCEL (polysaccharide) * MAJOL PLX * MODOCOLL 1200 * NACM-CELLULOSE SALT * NYMCEL S * POLYFIBRON 120 * SANLOSE SN 20A * SARCELL TEL * S 75M * SODIUM CELLULOSE GLYCOLATE * SODIUM CMC * SODIUM CM-CELLULOSE * SODIUM SALT of CARBOXYMETHYLCELLULOSE * TYLOSE 666 * UNISOL RH

CONSENSUS REPORTS: Reported in EPA TSCA Inventory.

SAFETY PROFILE: Mildly toxic by ingestion. Experimental reproductive effects. Questionable carcinogen with experimental neoplastigenic data. It migrates to food from packaging materials. When heated to decomposition it emits toxic fumes of Na_2O.

SFQ000   CAS: 9005-46-3   **HR: 3**
SODIUM CASEINATE

PROP: Coarse, white powder; odorless. Insol in water, alc.

SYNS: CASEIN and CASEINATE SALTS (FCC) * CASEIN-SODIUM * CASEIN, SODIUM COMPLEX * CASEINS, SODIUM COMPLEXES * NUTROSE

CONSENSUS REPORTS: Reported in EPA TSCA Inventory.

SAFETY PROFILE: Questionable carcinogen with experimental tumorigenic data. When heated to decomposition it emits toxic fumes of Na_2O.

SFS000   CAS: 7775-09-9   **HR: 3**
SODIUM CHLORATE

DOT: UN 1495/UN 2428
mf: $ClO_3 \cdot Na$ mw: 106.44

PROP: Colorless, odorless crystals; cooling, saline taste. Mp: 248-261°, bp: decomp, d: 2.490 @ 15°.

SYNS: ASEX * ATLACIDE * ATRATOL * B-HERBATOX * CHLORATE of SODA (DOT)

* CHLORATE SALT of SODIUM * CHLORAX * CHLORSAURE (GERMAN) * DE-FOL-ATE * DESOLET * DREXEL DEFOL * DROP LEAF * EVAU-SUPER * FALL * GRAIN SORGHUM HARVEST-AID * GRANEX O * HARVEST-AID * KLOREX * KUSA-TOHRU * KUSATOL * NATRIUMCHLORAAT (DUTCH) * NATRIUMCHLORAT (GERMAN) * ORTHO C-1 DEFOLIANT & WEED KILLER * OXYCIL * RASIKAL * SHED-A-LEAF * SHED-A-LEAF "L" * SODA CHLORATE (DOT) * SODIO (CLORATO di) (ITALIAN) * SODIUM (CHLORATE de) (FRENCH) * SODIUM CHLORATE, aqueous solution (DOT) * TRAVEX * TUMBLEAF * UNITED CHEMICAL DEFOLIANT NO. 1 * VAL-DROP

CONSENSUS REPORTS: Reported in EPA TSCA Inventory.

DOT Classification: Oxidizer; Label: Oxidizer

SAFETY PROFILE: Human poison by unspecified routes. Moderately toxic experimentally by ingestion and intraperitoneal routes. Human systemic effects by ingestion: blood hemolysis with or without anemia, methemoglobinemia-carboxhemoglobinemia and pulmonary changes. Mutation data reported. A skin, mucous membrane, and eye irritant. Damages the red blood cells of humans when ingested.

A powerful oxidizer. It can explode on contact with flame or sparks (static discharge) and has caused many industrial explosions. May react explosively with agricultural materials (e.g., peat, powdered sulfur, sawdust, urotropine, thiuram), alkenes + potassium osmate, aluminum + rubber, ammonium salts, grease, leather, powdered metals, non-metals, sulfides, cyanides, cyanoborane oligomer, nitrobenzene, organic matter, paint + polyethylene, phosphorus, sodium phosphinate. Violent reaction or ignition with aluminum, ammonium sulfate, Sb_2S_3, arsenic, arsenic trioxide, 1,3-bis(trichloromethylbenzene) + heat, carbon, charcoal, MnO_2, phosphorus, potassium cyanide, osmium + heat, paper, sulfuric acid, thiocyanates, triethylene glycol + wood, wood, zinc. Can also react violently with nitrobenzene, paper, metal sulfides, dibasic organic acids, organic matter. When heated to decomposition it emits toxic fumes of Cl^- and Na_2O.

SFT000   CAS: 7647-14-5   **HR: 2**
SODIUM CHLORIDE
mf: ClNa   mw: 58.44

PROP: Colorless, transparent crystals or white, crystalline powder. Mp: 801°, bp: 1413°, d:

2.165, vap press: 1 mm @ 865°. Sol in water, glycerin.

SYNS: COMMON SALT * DENDRITIS * EXTRA FINE 200 SALT * EXTRA FINE 325 SALT * HALITE * H.G. BLENDING * NATRIUMCHLORID (GERMAN) * PUREX * ROCK SALT * SALINE * SALT * SEA SALT * STERLING * TABLE SALT * TOP FLAKE * USP SODIUM CHLORIDE * WHITE CRYSTAL

CONSENSUS REPORTS: Reported in EPA TSCA Inventory. EPA Genetic Toxicology Program.

SAFETY PROFILE: Poison by intraperitoneal and intracervical routes. Moderately toxic by ingestion, intravenous, and subcutaneous routes. Human systemic effects by ingestion: blood pressure increase. Human reproductive effects by intraplacental route: terminates pregnancy. Experimental teratogenic and reproductive effects. Human mutation data reported. A skin and eye irritant. When bulk sodium chloride is heated to high temperature, a vapor is emitted which is irritating, particularly to the eyes. Ingestion of large amounts of sodium chloride can cause irritation of the stomach. Improper use of salt tablets may produce this effect. Potentially explosive reaction with dichloromaleic anhydride + urea. Electrolysis of mixtures with nitrogen compounds may form the explosive nitrogen trichloride. Reaction with burning lithium forms the dangerously reactive sodium. The molten salt at 1100° reacts explosively with water. Violent reaction with BrF_3. When heated to decomposition it emits toxic fumes of Cl^- and Na_2O.

SFT500   CAS: 7758-19-2   **HR: 3**
SODIUM CHLORITE

DOT: UN 1496
mf: $ClNaO_2$ mw: 90.44

PROP: White crystals or crystalline powder. Bp: decomp @ 180-200°.

SYN: TEXTILE

CONSENSUS REPORTS: Reported in EPA TSCA Inventory.

DOT Classification: Oxidizer; Label: Oxidizer

SAFETY PROFILE: Poison by ingestion. Experimental teratogenic and reproductive effects. Questionable carcinogen with experimental carcinogenic data. Mutation data reported. May

act as an irritant due to its oxidizing power. A powerful oxidizing agent; ignited by friction, heat, or shock. An explosive sensitive to impact or heating to 200°. Potentially explosive reaction with acids, oils, organic matter, oxalic acid + water, zinc. Violent reaction or ignition with carbon (above 60°); ethylene glycol (at 100°); phosphorus (above 50°); sodium dithionate; sulfur containing materials. Can react vigorously on contact with reducing materials. When heated to decomposition it emits highly toxic fumes of Cl^- and Na_2O. Used as a bleaching agent.

SFU000   CAS: 7758-19-2   **HR: 3**
SODIUM CHLORITE (solution)

DOT: UN 1908
mf: $ClNaO_2$ mw: 90.44

PROP: Solution contains 42% or less sodium chlorite.

CONSENSUS REPORTS: Reported in EPA TSCA Inventory.

DOT Classification: Corrosive Material; Label: Corrosive

SAFETY PROFILE: Poison by ingestion. A corrosive irritant to skin, eyes, and mucous membranes. When heated to decomposition it emits toxic fumes of Cl^- and Na_2O.

SFU500   CAS: 3926-62-3   **HR: 3**
SODIUM CHLOROACETATE

DOT: UN 2659
mf: $C_2H_2ClO_2 \cdot Na$ mw: 116.48

PROP: White, free-flowing, odorless powder. Mp: decomp @ 200°.

SYNS: CHLOROACETIC ACID SODIUM SALT * CHLOROCTAN SODNY (CZECH) * DOW DEFOLIANT * MONOXONE * SMA * SMCA * SODIUM MONOCHLORACETATE

CONSENSUS REPORTS: Reported in EPA TSCA Inventory.

DOT Classification: Poison B; Label: St. Andrews Cross

SAFETY PROFILE: Poison by ingestion and intravenous routes. When heated to decomposition it emits toxic fumes of Cl^- and Na_2O. Used as an herbicide.

SFW000 CAS: 361-09-1 **HR: 3**
SODIUM CHOLATE
mf: $C_{24}H_{39}O_5 \cdot Na$ mw: 430.62

SYNS: CHOLIC ACID, MONOSODIUM SALT
* CHOLIC ACID, SODIUM SALT * DS-Na * SO-
DIUM CHOLIC ACID * OX BILE EXTRACT * PURI-
FIED OXGALL * TRIHYDROXY-3,7,12-CHOLANATE de
Na (FRENCH) * (3-α,5-β,7-α,12-α)3,7,12-TRIHYDROXY-
CHOLAN-24-OIC ACID, MONOSODIUM SALT

CONSENSUS REPORTS: Reported in EPA
TSCA Inventory.

SAFETY PROFILE: Poison by intravenous
route. When heated to decomposition it emits
toxic fumes of Na_2O.

SFW500 CAS: 13517-17-4 **HR: 3**
SODIUM CHROMATE DECAHYDRATE
mf: $CrO_42Na \cdot 10H_2O$ mw: 342.18

SYN: CHROMIC ACID, DISODIUM SALT, DECAHY-
DRATE

CONSENSUS REPORTS: Chromium and its
compounds are on the Community Right To
Know List.

OSHA PEL: CL 0.1 mg(CrO_3)/m^3
ACGIH TLV: TWA 0.05 mg(Cr)/m^3
NIOSH REL: TWA 0.025 mg(Cr(VI))/m^3; CL
    0.05 mg/m^3/15M

SAFETY PROFILE: Confirmed human carcinogen. When heated to decomposition it emits
toxic fumes of Na_2O.

SFZ000 **HR: D**
SODIUM COMPOUNDS

SAFETY PROFILE: Variable toxicity. Sodium
ion as such is practically nontoxic. The toxicity
of sodium compounds is frequently, though not
always, due to the anion involved. The hydroxide is very corrosive, being strongly basic. Even
here it is the concentration of hydroxyl ion which
is responsible for the caustic action of this material. When heated to decomposition it emits
toxic fumes of Na_2O.

SGA500 CAS: 143-33-9 **HR: 3**
SODIUM CYANIDE

DOT: UN 1689
mf: CNNa   mw: 49.01

PROP: White, deliquescent, crystalline powder. Mp: 563.7°, bp: 1496°, vap press: 1 mm
@ 817°.

SYNS: CIANURO di SODIO (ITALIAN) * CYANIDE of
SODIUM * CYANOBRIK * CYANOGRAN
* CYANURE de SODIUM (FRENCH) * CYMAG
* HYDROCYANIC ACID, SODIUM SALT * KYANID
SODNY (CZECH) * RCRA WASTE NUMBER P106
* SODIUM CYANIDE, solid and solution (DOT)

CONSENSUS REPORTS: Cyanide and its
compounds are on the Community Right To
Know List. Reported in EPA TSCA Inventory.

OSHA PEL: TWA 5 mg(CN)/m^3 (skin)
ACGIH TLV: TWA 5 mg(CN)/m^3 (skin)
DFG MAK: 5 mg(CN)/m^3
NIOSH REL: CL 5 mg(CN)/m^3/10M
DOT Classification: Poison B; Label: Poison,
    solid and solution

SAFETY PROFILE: A deadly human poison
by ingestion and possibly other routes. A deadly
experimental poison by ingestion, intraperitoneal, subcutaneous, intravenous, parenteral, intramuscular, and ocular routes. Human systemic
effects by ingestion: hallucinations, distorted
perceptions, muscle weakness and gastritis. Experimental teratogenic and reproductive effects.

The volatile cyanides resemble hydrocyanic acid physiologically, inhibiting tissue oxidation and causing death through asphyxia.
Cyanogen is probably as toxic as hydrocyanic
acid; the nitriles are generally considered somewhat less toxic, probably because of their lower
volatility. The nonvolatile cyanide salts appear
to be relatively nonhazardous systemically, so
long as they are not ingested and care is taken
to prevent the formation of hydrocyanic acid.
Workers, such as electroplaters and picklers,
who are daily exposed to cyanide solutions may
develop a "cyanide" rash, characterized by
itching and by macular, papular, and vesicular
eruptions. Frequently there is secondary infection. Exposure to small amounts of cyanide compounds over long periods of time is reported
to cause loss of appetite, headache, weakness,
nausea, dizziness and symptoms of irritation
of the upper respiratory tract and eyes.

Flammable by chemical reaction with heat,
moisture, acid. Many cyanides evolve hydrocyanic acid rather easily. This is a flammable
gas and is highly toxic. Carbon dioxide from
the air is sufficiently acidic to liberate hydrocyanic acid from cyanide solutions. Explodes
if melted with nitrite or chlorate @ about 450°.
Violent reaction with F_2, Mg, nitrates, HNO_3,
nitrites. Upon contact with acid, acid fumes,
water or steam, they will produce toxic and

flammable vapors of CN^- and Na_2O. Used in the extraction of gold and silver ores, in electroplating and in insecticides.

SGC000   CAS: 139-05-9   HR: 3
SODIUM CYCLAMATE
mf: $C_6H_{12}NO_3S \cdot Na$ mw: 201.24

PROP: White, crystalline powder; practically odorless. Sol in water; almost insol in alc, benzene, chloroform, and ether.

SYNS: ASSUGRIN * ASSUGRIN FEINUSS * ASSUGRIN VOLLSUSS * ASUGRYN * CYCLAMATE * CYCLAMATE SODIUM * CYCLAMIC ACID SODIUM SALT * CYCLOHEXANESULFAMIC ACID, MONOSODIUM SALT * CYCLOHEXANESULPHAMIC ACID, MONOSODIUM SALT * CYCLOHEXYL SULPHAMATE SODIUM * DULZOR-ETAS * HACHI-SUGAR * IBIOSUC * NATREEN * NATRIUMZYKLAMATE (GERMAN) * SODIUM CYCLOHEXANESULFAMATE * SODIUM CYCLOHEXANESULPHAMATE * SODIUM CYCLOHEXYL AMIDOSULPHATE * SODIUM CYCLOHEXYL SULFAMATE * SODIUM CYCLOHEXYL SULFAMIDATE * SODIUM CYCLOHEXYL SULPHAMATE * SODIUM SUCARYL * SUCARYL SODIUM * SUCCARIL * SUCROSA * SUESSETTE * SUESTAMIN * SUGARIN * SUGARON

CONSENSUS REPORTS: IARC Cancer Review: GROUP 3 IMEMDT 7,178,87; Animal Limited Evidence IMEMDT 22,55,80. Reported in EPA TSCA Inventory. EPA Genetic Toxicology Program.

SAFETY PROFILE: Moderately toxic by intravenous and intraperitoneal routes. Mildly toxic by ingestion. Experimental teratogenic and reproductive effects. Questionable carcinogen with experimental neoplastigenic and tumorigenic data. Human mutation data reported. When heated to decomposition it emits very toxic fumes of Na_2O, SO_x, and NO_x.

SGD000   CAS: 4418-26-2   HR: 3
SODIUM DEHYDROACETIC ACID
mf: $C_8H_7O_4 \cdot Na$ mw: 190.14

PROP: White powder; odorless with slt characteristic taste. Mp: 109-111°. Sol in water, propylene glycol, glycerin.

SYNS: DEHYDROACETIC ACID, SODIUM SALT * DHA-SODIUM * HARVEN * 4-HEXENOIC ACID, 2-ACETYL-5-HYDROXY-3-OXO, Δ-LACTONE, SODIUM derivative * 3-(1-HYDROXYETHYLIDENE)-6-METHYL-2H-PYRAN-2,4(3H)-DIONE, SODIUM SALT * SODIUM DEHYDROACETATE (FCC)

CONSENSUS REPORTS: Reported in EPA TSCA Inventory. EPA Genetic Toxicology Program.

SAFETY PROFILE: Poison by intravenous route. Moderately toxic by ingestion. Experimental teratogenic and reproductive effects. Mutation data reported. When heated to decomposition it emits toxic fumes of Na_2O.

SGF500   CAS: 136-30-1   HR: 3
SODIUM DIBUTYLDITHIOCARBAMATE
mf: $C_9H_{18}NS_2 \cdot Na$ mw: 227.39

SYNS: BUTYL NAMATE * DIBUTYLDITHIOCARBAMIC ACID SODIUM SALT * PENNAC * SODIUM DBDT * TEPIDONE * TEPIDONE RUBBER ACCELERATOR * USAF B-35 * VULCACURE

CONSENSUS REPORTS: Reported in EPA TSCA Inventory.

SAFETY PROFILE: Poison by intraperitoneal route. When heated to decomposition it emits very toxic fumes of NO_x, SO_x, and Na_2O.

SGG000   CAS: 2156-56-1   HR: 2
SODIUM DICHLOROACETATE
mf: $C_2HCl_2O_2 \cdot Na$ mw: 150.92

SYNS: DICHLOROACETATE SODIUM SALT * DICHLOROACETIC ACID SODIUM SALT * DICHLOROCTAN SODNY (CZECH)

SAFETY PROFILE: Moderately toxic by intravenous route. Mildly toxic by ingestion. Experimental reproductive effects. Mutation data reported. When heated to decomposition it emits toxic fumes of Cl^- and Na_2O.

SGG500   CAS: 2893-78-9   HR: 2
SODIUM DICHLOROCYANURATE
DOT: UN 2465
mf: $C_3HCl_2N_3O_3 \cdot Na$ mw: 220.96

PROP: White crystals; chlorine odor. Mp: 230-250°, water-sol.

SYNS: ACL 60 * CDB 63 * DICHLOROISOCYANURIC ACID SODIUM SALT (DOT) * DIKONIT * DIMANIN C * FI CLOR 60S * OCI 56 * SDIC * SIMPLA * SODIUM DICHLORISOCYANURATE * SODIUM DICHLOROISOCYANURATE * SODIUM 1,3-DICHLORO-1,3,5-TRIAZINE-2,4-DIONE-6-OXIDE * 1-SODIUM 3,5-DICHLORO-s-TRIAZINE-2,4,6-TRIONE * 1-SODIUM 3,5-DICHLORO-1,3,5-TRIAZINE-2,4,6-TRIONE * SODIUM DICHLORO-s-TRIAZINETRIONE, dry, containing more than 39% available chlorine (DOT)

* SODIUM SALT of DICHLORO-s-TRIAZINETRIONE

CONSENSUS REPORTS: Reported in EPA TSCA Inventory.

DOT Classification: Oxidizer; Label: Oxidizer

SAFETY PROFILE: Moderately toxic to humans and animals by ingestion. Experimental teratogenic and reproductive effects. A severe skin and eye irritant. Human systemic effects by ingestion: ulceration or bleeding from stomach. The other main toxic effects were gastrointestinal irritation, salivation, lacrimation, dyspnea, weakness, emaciation, lethargy, diarrhea, coma and (following very high dosage) deaths after 1-8 days, with autopsy showing irritation of stomach and gastrointestinal tract, liver dysfunction, and lung congestion. The concentrated material may be a little more toxic, due to greater gastrointestinal irritation. In the dry form, it is not appreciably irritating to dry skin. However, when moist, the concentrated material is irritating to skin, and also may cause severe eye irritation.

A powerful oxidizer. Incompatible with combustible materials, ammonium salts, nitrogenous materials. Used to chlorinate swimming pools and in cleaning, bleaching, disinfecting, sanitizing. When heated to decomposition it emits very toxic fumes of Cl^-, NO_x, and Na_2O.

SGH500   CAS: 2702-72-9   **HR: 3**
SODIUM 2,4-DICHLOROPHENOXY-ACETATE
mf: $C_8H_5Cl_2O_3$ • Na   mw: 243.02

SYNS: AGRION * 2,4-DICHLOROPHENOXYACETIC ACID, SODIUM SALT * DICONIRT D * 2,4-D SODIUM SALT * FERNOXENE * HORMIT * PIELIK E * SODIUM 2,4-D * SPRAY-HORMITE * SPRITZ-HORMIT

CONSENSUS REPORTS: Reported in EPA TSCA Inventory. EPA Genetic Toxicology Program.

SAFETY PROFILE: Poison by ingestion, intraperitoneal, subcutaneous, and intravenous routes. Human systemic effects by inhalation: anorexia, gastrointestinal and liver changes. Experimental teratogenic and reproductive effects. Mutation data reported. When heated to decomposition it emits toxic fumes of Cl^- and Na_2O.

SGI000   CAS: 10588-01-9   **HR: 3**
SODIUM DICHROMATE
DOT: UN 1479
mf: Cr_2O_7 • 2Na   mw: 261.98

PROP: Anhydrous. Mp: 356.7°, decomp @ about 400°, d: 2.35 @ 13°. Very sol in water.

SYNS: BICHROMATE de SODIUM (FRENCH) * BICHROMATE of SODA * CHROMIC ACID, DISODIUM SALT * CHROMIUM SODIUM OXIDE * DISODIUM DICHROMATE * NATRIUMBICHROMAAT (DUTCH) * NATRIUMDICHROMAAT (DUTCH) * NATRIUMDICHROMAT (GERMAN) * SODIO (DICROMATO di) (ITALIAN) * SODIUM BICHROMATE * SODIUM CHROMATE * SODIUM DICHROMATE(VI) * SODIUM DICHROMATE de (FRENCH)

CONSENSUS REPORTS: IARC Cancer Review: GROUP 1 IMEMDT 7,165,87; Animal Inadequate Evidence IMEMDT 2,100,73; IMEMDT 23,205,80; Human Inadequate Evidence IMEMDT 23,205,80. Chromium and its compounds are on the Community Right To Know List. Reported in EPA TSCA Inventory. EPA Genetic Toxicology Program.

OSHA PEL: CL 0.1 mg/(CrO_3)/m^3
ACGIH TLV: TWA 0.05 mg(Cr)/m^3
NIOSH REL: TWA 0.025 mg(Cr(VI))/M^3; CL 0.05 mg/M^3/15M
DOT Classification: ORM-A; Label: None

SAFETY PROFILE: Confirmed carcinogen with experimental tumorigenic data. Poison by ingestion, skin contact, intravenous, intraperitoneal, and subcutaneous routes. Human systemic effects by ingestion: cough, nausea or vomiting, and sweating. Human mutation data reported. A caustic and irritant. A powerful oxidizer. Potentially explosive reaction with acetic anhydride, ethanol + sulfuric acid + heat, hydrazine. Violent reaction or ignition with boron + silicon (pyrotechnic), organic residues + sulfuric acid, 2-propanol + sulfuric acid, sulfuric acid + trinitrotoluene. Incompatible with hydroxylamine. When heated to decomposition it emits toxic fumes of Na_2O.

SGJ000   CAS: 148-18-5   **HR: 3**
SODIUM DIETHYLDITHIOCARBAMATE
mf: $C_5H_{10}NS_2$ • Na   mw: 171.27

PROP: Crystals. Mp: 95°, d: 1.1 @ 20°/20°, vap d: 5.9.

SYNS: CUPRAL * DDC * DEDC * DEDK * DIETHYLCARBAMODITHIOIC ACID, SODIUM SALT

* DIETHYLDITHIOCARBAMATE SODIUM * DI-ETHYLDITHIOCARBAMIC ACID SODIUM * DIETHYL-DITHIOCARBAMIC ACID, SODIUM SALT * DIETHYL SODIUM DITHIOCARBAMATE * DITHIOCARB * DITHIOCARBAMATE * NCI-C02835 * SODIUM DEDT * SODIUM N,N-DIETHYLDITHIOCARBAMATE * SODIUM SALT of N,N-DIETHYLDITHIOCARBAMIC ACID * THIOCARB * USAF EK-2596

CONSENSUS REPORTS: IARC Cancer Review: GROUP 3 IMEMDT 7,56,87; Animal Inadequate Evidence IMEMDT 12,217,76. NCI Carcinogenesis Bioassay (feed); No Evidence: mouse, rat NCITR* NCI-CG-TR-172,79. Reported in EPA TSCA Inventory.

SAFETY PROFILE: Moderately toxic by ingestion, intraperitoneal, and subcutaneous routes. Experimental teratogenic and reproductive effects. Questionable carcinogen with experimental neoplastigenic data. Human mutation data reported. When heated to decomposition it emits very toxic fumes of NO_x, SO_x, and Na_2O. Used as a pesticide.

SGM500   CAS: 128-04-1   **HR: 2**
SODIUM N,N-DIMETHYLDITHIO-CARBAMATE
mf: $C_3H_6NS_2 \cdot Na$ mw: 143.21

PROP: Crystals.

SYNS: ACETO SDD 40 * ALCOBAM NM * BROG-DEX 555 * CARBON S * DIBAM * DIMETHYL-DITHIOCARBAMIC ACID, SODIUM SALT * DMDK * METHYL NAMATE * SDDC * SHARSTOP 204 * STA-FRESH 615 * STERISEAL LIQUID #40 * THIOSTOP N * VINSTOP * VULNOPOL NM * WING STOP B

CONSENSUS REPORTS: Reported in EPA TSCA Inventory.

SAFETY PROFILE: Moderately toxic by ingestion and intraperitoneal routes. Mutation data reported. When heated to decomposition it emits very toxic fumes of NO_x, SO_x, and Na_2O.

SHF500   CAS: 7681-49-4   **HR: 3**
SODIUM FLUORIDE
DOT: UN 1690
mf: FNa   mw: 41.99

PROP: Clear, lustrous crystals or white powder or balls. Mp: 993°, bp: 1700°, d: 2 @ 41°, vap press: 1 mm @ 1077°.

SYNS: ALCOA SODIUM FLUORIDE * ANTIBULIT * CAVI-TROL * CHEMIFLUOR * CREDO

* DISODIUM DIFLUORIDE * FDA 0101 * F1-TABS * FLORIDINE * FLOROCID * FLOZENGES * FLUORAL * FLUORIDENT * FLUORID SODNY (CZECH) * FLUORIGARD * FLUORINEED * FLUORINSE * FLUORITAB * FLUOR-O-KOTE * FLUORURE de SODIUM (FRENCH) * FLURA-GEL * FLURCARE * FUNGOL B * GEL II * GELU-TION * GLEEM * IRADICAV * KARIDIUM * KARIGEL * KARI-RINSE * LEA-COV * LEMOFLUR * LURIDE * NAFEEN * NaFPAK * Na FRINSE * NATRIUM FLUORIDE * NCI-C55221 * NUFLUOR * OSSALIN * OSSIN * PEDIAFLOR * PEDIDENT * PENNWHITE * PERGANTENE * PHOS-FLUR * POINT TWO * PREDENT * RAFLUOR * RESCUE SQUAD * ROACH SALT * SODIUM FLUORIDE, solid and solution (DOT) * SODIUM FLUORURE (FRENCH) * SO-DIUM HYDROFLUORIDE * SODIUM MONOFLUORIDE * SO-FLO * STAY-FLO * STUDAFLUOR * SUPER-DENT * T-FLUORIDE * THERA-FLUR-N * TRISODIUM TRIFLUORIDE * VILLIAUMITE

CONSENSUS REPORTS: Reported in EPA TSCA Inventory. EPA Genetic Toxicology Program.

OSHA PEL: TWA 2.5 mg(F)/m³
ACGIH TLV: TWA 2.5 mg(F)/m³
NIOSH REL: TWA (Inorganic Fluorides) 2.5 mg(F)/m³
DOT Classification: ORM-B; Label: None; Corrosive Material; Label: Corrosive, solution; Poison B; Label: St. Andrews Cross.

SAFETY PROFILE: Human poison by ingestion and possibly other routes. Experimental poison by ingestion, skin contact, intravenous, intraperitoneal, subcutaneous, and intramuscular routes. Human systemic effects by ingestion and intradermal routes: paresthesia, ptosis (drooping of the eyelid from sympathetic innervation), tremors, fluid intake, muscle weakness, headache, EKG changes, cyanosis, respiratory depression, hypermotility, diarrhea, nausea or vomiting, salivary gland changes, changes in teeth and supporting structures and other musculo-skeletal changes, and increased immune response. Experimental teratogenic and reproductive effects. Human mutation data reported. A corrosive irritant to skin, eyes, and mucous membranes. Experimental reproductive effects. Questionable carcinogen with experimental tumorigenic data. It is very phytotoxic. When heated to decomposition it emits toxic fumes of F^- and Na_2O. Used in chemical cleaning, for fluoridation of drinking water, as a fungicide and insecticide.

SHG500 CAS: 62-74-8 **HR: 3**
SODIUM FLUOROACETATE
DOT: UN 2629
mf: $C_2H_2FO_2 \cdot Na$ mw: 100.03

PROP: Fine, white powder. Sol in water.

SYNS: 1080 * COMPOUND NO. 1080 * FLUOR-ACETATO di (ITALIAN) * FLUOROACETIC ACID, SO-DIUM SALT * FLUORESSIGAEURE (GERMAN) * FRATOL * FURATOL * MONOFLUORESSIG-SAURES NATRIUM (GERMAN) * NATRIUMFLUORACE-TAAT (DUTCH) * NATRIUMFLUORACETAT (GERMAN) * RATBANE 1080 * RCRA WASTE NUMBER P058 * SODIO, FLUORACETATO di (ITALIAN) * SODIUM FLUOACETATE * SODIUM FLUOACETIC ACID * SODIUM FLUORACETATE de (FRENCH) * SODIUM MONOFLUOROACETATE * TL 869 * YASOKNOCK

CONSENSUS REPORTS: Reported in EPA TSCA Inventory. EPA Extremely Hazardous Substances List.

OSHA PEL: (Transitional: TWA 0.05 mg/m^3 (skin)) TWA 0.05 mg/m^3 (skin); STEL 0.15 mg/m^3 (skin)
ACGIH TLV: TWA 0.05 mg/m^3 (skin); STEL 0.15 mg/m^3 (skin)
DFG MAK: 0.05 mg/m^3
DOT Classification: Poison B: Label: Poison.

SAFETY PROFILE: A deadly human poison by ingestion and possibly other routes. Experimental poison by ingestion, intraperitoneal, subcutaneous, intravenous, intramuscular and possibly other routes. A very highly toxic water-soluble salt used mainly as an immediate action rodenticide. It is rapidly absorbed by the gastrointestinal tract but slowly by the skin unless the skin is abraided or cut. It operates by blocking the Krebs cycle by formation of fluorocitric acid, which inhibits aconitase. It has an effect on either or both the cardiovascular and nervous systems in all species and, in some species, the skeletal muscles. Humans have mixed responses with the cardiac feature predominating. By a direct action on the heart, contractile power is lost which leads to declining blood pressure. Ventricular premature contractions and arrhythmias are seen in all species, including humans. The central nervous system is directly attacked by sodium fluoroacetate. In humans, the action on the central nervous system produces epileptiform convulsive seizures followed by severe depression. The dangerous dose for humans is 0.5-2 mg/kg. Other species vary considerably in their response to this material with primates and birds being the most resistant and carnivora and rodents being the most susceptible. Most domestic animals show a susceptibility falling between the two extremes indicated above. When heated to decomposition it emits highly toxic fumes of Na_2O and F^-.

SHJ000 CAS: 141-53-7 **HR: 2**
SODIUM FORMATE
mf: $CHO_2 \cdot Na$ mw: 68.01

PROP: White, deliquescent crystals. Mp: 253°, d: 1.92 @ 20°.

SYN: SALACHLOR

CONSENSUS REPORTS: Reported in EPA TSCA Inventory.

SAFETY PROFILE: Moderately toxic by ingestion, intravenous, subcutaneous, and possibly other routes. Combustible when exposed to heat or flame. When heated to decomposition it emits toxic fumes of Na_2O.

SHK800 CAS: 527-07-1 **HR: 1**
SODIUM GLUCONATE
mf: $C_6H_{12}O_7 \cdot Na$ mw: 219.17

PROP: White to tan granular or crystalline powder. Very sol in water; sltly sol in alc; insol in ether.

SYNS: GLONSEN * GLUCONATO di SODIO (ITAL-IAN) * GLUCONIC ACID SODIUM SALT * MONOSO-DIUM GLUCONATE * PASEXON 100T * PMP SO-DIUM GLUCONATE * SODIUM d-GLUCONATE

CONSENSUS REPORTS: Reported in EPA TSCA Inventory.

SAFETY PROFILE: Low toxicity by intravenous route. When heated to decomposition it emits acrid smoke and irritating fumes.

SHL500 CAS: 7009-49-6 **HR: 3**
SODIUM HEXACYCLONATE
mf: $C_9H_{15}O_3 \cdot Na$ mw: 194.23

SYNS: CYCLOHEXANEACETIC ACID, 1-(HYDROXY-METHYL)-, MONOSODIUM SALT (9CI) * ESACICLO-NATO * GEVILON * GO 186 * HEXACYCLONAS * HEXACYCLONATE SODIUM * 1-(HYDROXY-METHYL)CYCLOHEXANEACETIC ACID, SODIUM SALT * NEURYL * REPRISCAL * SODIUM 1-(HY-DROXYMETHYL)CYCLOHEXANEACETATE * SODIUM β,β-PENTAMETHYLENE-γ-HYDROXYBUTYRATE

SAFETY PROFILE: Poison by ingestion, intraperitoneal, and intravenous routes. When heated to decomposition it emits toxic fumes of Na_2O.

SHM500   CAS: 10124-56-8   HR: 3
SODIUM HEXAMETAPHOSPHATE
mf: $O_{18}P_6 \cdot 6Na$ mw: 611.76

PROP: White powder or flakes. Sol in water.

SYNS: CALGON * CHEMI-CHARL * HEXAMETA-PHOSPHATE, SODIUM SALT * HMP * MEDI-CAL-GON * PHOSPHATE, SODIUM HEXAMETA * POLYPHOS * SHMP

CONSENSUS REPORTS: Reported in EPA TSCA Inventory.

SAFETY PROFILE: Poison by intravenous route. Moderately toxic by intraperitoneal and subcutaneous routes. Mildly toxic by ingestion. When heated to decomposition it emits toxic fumes of PO_x and Na_2O.

SHO500   CAS: 7646-69-7   HR: 3
SODIUM HYDRIDE
DOT: UN 1427
mf: HNa   mw: 24.00

PROP: Microcrystalline, white to brownish-gray powder; reacts with water. Mp: 800° (decomp), d: 0.9.

SYN: NAH 80

CONSENSUS REPORTS: Reported in EPA TSCA Inventory.

DOT Classification: Flammable Solid; Label: Flammable Solid and Dangerous When Wet

SAFETY PROFILE: The powder ignites spontaneously in air. Flammable when exposed to heat or flame. Potentially explosive reaction with water, diethyl succinate + ethyltrifluoroacetate (above 60°C), dimethyl sulfoxide + heat, sulfur dioxide. Ignition or violent reaction with dimethylformamide (above 50°C), ethyl 2,2,3-trifluoropropionate, oxygen (at 230°C). Incompatible with acetylene + moisture, glycerol, halogens, sulphur. Normal fire extinguishers are unsuitable; use sand, ashes, sodium chloride. The commercial material may contain traces of sodium. When heated to decomposition it emits toxic fumes of Na_2O.

SHQ500   CAS: 1333-83-1   HR: 3
SODIUM HYDROGEN FLUORIDE
DOT: UN 2439
mf: F_2HNa mw: 62.00

PROP: White powder. D: 2.08. Sol in water to 42,000 ppm @ 20°C.

SYNS: HYDROFLUORIC ACID, SODIUM SALT (2:1) * SODIUM ACID FLUORIDE * SODIUM BIFLUORIDE (VAN) * SODIUM FLUORIDE(Na(HF_2)) * SODIUM HYDROGEN DIFLUORIDE

CONSENSUS REPORTS: Reported in EPA TSCA Inventory.

NIOSH REL: TWA 2.5 mg(F)/m^3
DOT Classification: Corrosive Material; Label: Corrosive

SAFETY PROFILE: This material is very toxic to humans by ingestion; between 1 teaspoonful and 1 ounce may be fatal. Inhalation of dust may cause irritation to respiratory tract. Skin contact may result in irritation and ulceration; eye contact may cause burns. To fight fire use water, foam, CO2, dry chemicals. When heated to decomposition it emits toxic fumes of F^- and Na_2O.

SHR000   CAS: 16721-80-5   HR: 3
SODIUM HYDROSULFIDE
DOT: UN 2318/NA 2922
mf: HNaS   mw: 56.06

SYNS: SODIUM BISULFIDE * SODIUM HYDROGEN SULFIDE * SODIUM HYDROSULFIDE, solution (DOT) * SODIUM HYDROSULPHIDE, with less than 25% water of crystallization (DOT) * SODIUM HYDROSULPHIDE, solid (DOT) * SODIUM MERCAPTAN * SODIUM MERCAPTIDE * SODIUM SULFHYDRATE

CONSENSUS REPORTS: Reported in EPA TSCA Inventory.

DOT Classification: Corrosive Material; Label: Corrosive; Flammable Solid; Label: Spontaneously Combustible; Flammable Solid; Label: Flammable Solid

SAFETY PROFILE: Poison by intraperitoneal and subcutaneous routes. Mutation data reported. A corrosive irritant to skin, eyes, and mucous membranes. Flammable when exposed to heat or flame. Spontaneous combustion. Reacts violently with diazonium salts. Readily yields H_2S. When heated to decomposition it emits toxic fumes of SO_x and Na_2O.

SHR500   CAS: 7775-14-6   **HR: 3**
SODIUM HYDROSULPHITE

DOT: UN 1384
mf: $O_4S_2 \cdot 2Na$ mw: 174.10

PROP: White or yellow-white crystals. Mp: decomp @ 52°. Decomp in water (hot); sltly sol in cold water; insol in acids.

SYNS: D-OX * HYDROLIN * K-BRITE * REDUCTONE * SODIUM DITHIONITE (DOT) * SODIUM HYDROSULFITE (DOT) * SODIUM SULF-OXYLATE * VATROLITE * V-BRITE * VIR-CHEM * VIRTEX CC * VIRTEX D * VIRTEX L * VIRTEX RD

CONSENSUS REPORTS: Reported in EPA TSCA Inventory.

DOT Classification: Flammable Solid; Label: Spontaneously Combustible; Flammable Solid; Label: Flammable Solid

SAFETY PROFILE: Toxic and an irritant. An allergen. Flammable when exposed to heat or flame. Ignites on contact with water or sodium chlorite. Decomposes violently when heated to 190°C and emits toxic fumes of SO_x and Na_2O.

SHS000   CAS: 1310-73-2   **HR: 3**
SODIUM HYDROXIDE

DOT: UN 1823/UN 1824
mf: HNaO   mw: 40.00

PROP: White, pieces, lumps or sticks. Mp: 318.4°, bp: 1390°, d: 2.120 @ 20°/4°, vap press: 1 mm @ 739°. Deliquescent; sol in water and alc.

SYNS: CAUSTIC SODA * CAUSTIC SODA, bead (DOT) * CAUSTIC SODA, dry (DOT) * CAUSTIC SODA, flake (DOT) * CAUSTIC SODA, granular (DOT) * CAUSTIC SODA, liquid (DOT) * CAUSTIC SODA, solid (DOT) * CAUSTIC SODA, solution (DOT) * HYDROXYDE de SODIUM (FRENCH) * LEWIS-RED DEVIL LYE * LYE (DOT) * NATRIUMHYDROXID (GERMAN) * NATRIUMHYDROXYDE (DUTCH) * SODA LYE * SODIO(IDROSSIDO di) (ITALIAN) * SODIUM HYDRATE (DOT) * SODIUM HYDROXIDE, bead (DOT) * SODIUM HYDROXIDE, dry (DOT) * SODIUM HYDROXIDE, flake (DOT) * SODIUM HYDROXIDE, granular (DOT) * SODIUM HYDROXIDE, solid (DOT) * SODIUM(HYDROXYDE de) (FRENCH) * WHITE CAUSTIC

CONSENSUS REPORTS: Reported in EPA TSCA Inventory. EPA Genetic Toxicology Program.

OSHA PEL: (Transitional: TLV 2 mg/m³) CL 2 mg/m³
ACGIH TLV: Cl 2 mg/m³
DFG MAK: 2 mg/m³
NIOSH REL: (Sodium Hydroxide) CL 2 mg/m³/15M
DOT Classification: Corrosive Material; Label: Corrosive

SAFETY PROFILE: Poison by intraperitoneal route. Moderately toxic by ingestion. Mutation data reported. A corrosive irritant to skin, eyes, and mucous membranes. This material, both solid and in solution, has a markedly corrosive action upon all body tissue causing burns and frequently deep ulceration, with ultimate scarring. Mists, vapors, and dusts of this compound cause small burns, and contact with the eyes rapidly causes severe damage to the delicate tissue. Ingestion causes very serious damage to the mucous membranes or other tissues with which contact is made. It can cause perforation and scarring. Inhalation of the dust or concentrated mist can cause damage to the upper respiratory tract and to lung tissue, depending upon the severity of the exposure. Thus, effects of inhalation may vary from mild irritation of the mucous membranes to a severe pneumonitis.

A strong base. Vigorous reaction with 1,2,4,5-tetrachlorobenzene has caused many industrial explosions and forms the extremely toxic 2,3,7,8-tetrachlorodibenzodioxin. Mixtures with aluminum + arsenic compounds form the poisonous gas arsine. Potentially explosive reaction with bromine, 4-chlorobutyronitrile, 4-chloro-2-methylphenol (in storage), nitrobenzene + heat, sodium tetrahydroborate, 2,2,2-trichloroethanol, zirconium + heat. Reacts to form explosive products with ammonia + silver nitrate (forms silver nitride), N,N'-bis(trinitroethyl)urea (in storage), cyanogen azide, glycols above 230° (e.g., ethylene glycol, diethylene glycol), 3-methyl-2-penten-4-yn-1-ol, trichloroethylene (forms dichloroacetylene). Caution: Under the proper conditions of temperature, pressure, and state of division, it can ignite or react violently with acetic acid, acetaldehyde, acetic anhydride, acrolein, acrylonitrile, allyl alcohol, allyl chloride, Al, benzene-1,4-diol, chlorine trifluoride, chloroform + methanol, chlorohydrin, chloronitro-toluenes, chlorosulfonic acid, 1,2-dichloroethylene, ethylene cyanhydrin, glyoxal, HCl, HF, hydroquinone, maleic anhydride, HNO_3, nitroethane, nitromethane, nitroparaffins, nitropropane, pentol, oleum, P, P_2O_5, β-propiolactone, H_2SO_4, (CH_3OH + tetrachloro-benzene), tetrahydrofuran, water, cinnamaldehyde, diborane + octanol oxime, 2,2-dichloro-3,3-dimethylbutane, 4-methyl-2-nitrophenol, 1,1,1-trichloroethanol, trichloronitromethane, zinc.

Dangerous material to handle. When heated to decomposition it emits toxic fumes of Na_2O.

SHS500　　　CAS: 1310-73-2　　　**HR: 3**
SODIUM HYDROXIDE (liquid)
DOT: UN 1823/UN 1824
mf: HNaO　　mw: 40.00

PROP: Clear to slightly turbid, colorless liquid.

SYNS: CAUSTIC SODA, solution ＊ LYE, solution ＊ SODA LYE ＊ SODIUM HYDRATE, solution ＊ SODIUM HYDROXIDE, solution (FCC) ＊ WHITE CAUSTIC, solution

CONSENSUS REPORTS: Reported in EPA TSCA Inventory. Community Right-To-Know List.

DOT Classification: Corrosive Material; Label: Corrosive

SAFETY PROFILE: Poison by intraperitoneal route. Moderately toxic by ingestion. Mutation data reported. A corrosive irritant to skin, eyes, and mucous membranes. When heated to decomposition it emits toxic fumes of Na_2O.

SHU500　　　CAS: 7681-52-9　　　**HR: 3**
SODIUM HYPOCHLORITE
DOT: UN 1791
mf: ClHO • Na　　mw: 75.45

PROP: Mp: decomp.

SYNS: ANTIFORMIN ＊ B-K LIQUID ＊ CARREL-DAKIN SOLUTION ＊ CHLOROS ＊ CHLOROX ＊ CLOROX ＊ DAKINS SOLUTION ＊ HYCLORITE ＊ MILTON ＊ SURCHLOR

CONSENSUS REPORTS: Reported in EPA TSCA Inventory. EPA Genetic Toxicology Program.

DOT Classification: ORM-B; Label: None

SAFETY PROFILE: Human mutation data reported. An eye irritant. Corrosive and irritating by ingestion and inhalation. The anhydrous salt is highly explosive and sensitive to heat or fric-

tion. Explosive reaction with formic acid (at 55°), phenylacetonitrile. Reacts to form explosive products with amines, ammonium salts [e.g., ammonium acetate, $(NH_4)_2CO_3$, ammonium nitrate, ammonium oxalate, $(NH_4)_3PO_4$], aziridine, methanol. Violent reaction with phenyl acetonitrile, cellulose, ethylene imine. Solutions in water are storage hazards due to oxygen evolution. When heated to decomposition it emits toxic fumes of Na_2O and Cl^-. Used as a bleach.

SHV000　　　CAS: 7681-53-0　　　**HR: 3**
SODIUM HYPOPHOSPHITE
mf: H_2O_2P • Na　　mw: 87.98

PROP: Colorless, pearly, crystalline plates or white granular powder; bittersweet, saline taste. Deliquescent; sol in water; sltly sol in alc.

SYNS: NATRIUMHYPOPHOSPHIT (GERMAN) ＊ SODIUM PHOSPHINATE

CONSENSUS REPORTS: Reported in EPA TSCA Inventory.

SAFETY PROFILE: Poison by subcutaneous route. Moderately toxic by intraperitoneal route. Flammable when exposed to heat or flame. Aqueous solutions may explode on evaporation. Potentially explosive reaction with oxidants (e.g., chlorates, nitrates). Heat causes it to evolve phosphine. It can explode. When heated to decomposition it emits toxic fumes of PO_x and Na_2O.

SHW000　　　CAS: 7681-82-5　　　**HR: 2**
SODIUM IODIDE
mf: INa　　mw: 149.89

PROP: Cubic, colorless crystals. Mp: 651°, bp: 1300°, d: 3.667, vap press: 1 mm @ 767°.

SYNS: ANAYODIN ＊ IODURIL ＊ JODID SODNY ＊ NATRIUMJODID (GERMAN) ＊ SODIUM IODINE ＊ SODIUM MONOIODIDE

CONSENSUS REPORTS: Reported in EPA TSCA Inventory.

SAFETY PROFILE: Moderately toxic by ingestion, intravenous and intraperitoneal routes. Human teratogenic effects by ingestion: developmental abnormalities of the endocrine system. Human reproductive effects by ingestion: effects on newborn including postnatal measurements. A skin and eye irritant. Reacts violently with BrF_3, $HClO_4$, oxidants. When heated to decomposition it emits toxic fumes of I^- and Na_2O.

SIA500 CAS: 540-72-7 **HR: 3**
SODIUM ISOTHIOCYANATE
mf: CHNS • Na mw: 82.08

PROP: Colorless, deliquescent crystals or white powder. Mp: 287°.

SYNS: HAIMASED * NATRIUMRHODANID (GERMAN) * SCYAN * SODIUM RHODANATE * SODIUM RHODANIDE * SODIUM SULFOCYANATE * SODIUM SULFOCYANIDE * SODIUM THIOCYANATE * SODIUM THIOCYANIDE * THIOCYANATE SODIUM * USAF EK-T-434

CONSENSUS REPORTS: Reported in EPA TSCA Inventory.

SAFETY PROFILE: Poison by ingestion, intravenous, intratracheal, and subcutaneous routes. Moderately toxic by intraperitoneal route. Large doses taken internally cause vomiting, convulsions. Chronic poisoning is manifested by weakness, confusion, diarrhea, and skin rashes. When heated to decomposition it emits very toxic fumes of NO_x, SO_x, and Na_2O.

SIB600 CAS: 151-21-3 **HR: 3**
SODIUM LAURYL SULFATE
mf: $C_{12}H_{26}O_4S$ • Na mw: 289.43

PROP: White to cream-colored crystals, flakes or powder; slt odor. Sol in water.

SYNS: AQUAREX METHYL * AVIROL 118 CONC * CARSONOL SLS * CONCO SULFATE WA * CYCLORYL 21 * DETERGENT 66 * DODECYL ALCOHOL, HYDROGEN SULFATE, SODIUM SALT * DODECYL SODIUM SULFATE * DODECYL SULFATE, SODIUM SALT * DREFT * DUPONOL * EMERSAL 6400 * EMULSIFIER NO. 104 * HEXAMOL SLS * IRIUM * LANETTE WAX-S * LAURYL SODIUM SULFATE * LAURYL SULFATE, SODIUM SALT * MAPROFIX 563 * MAPROFIX WAC-LA * NCI-C50191 * NEUTRAZYME * ORVUS WA PASTE * PRODUCT NO. 161 * QUOLAC EX-UB * REWOPOL NLS 30 * RICHONOL C * SIPEX OP * SIPON WD * SLS * SODIUM DODECYL SULFATE * SODIUM MONODODECYL SULFATE * SOLSOL NEEDLES * STANDAPOL 112 CONC * STEPANOL WAQ * STERLING WAQ-COSMETIC * SULFOPON WA 1 * SULFOTEX WALA * SULFURIC ACID, MONODODECYL ESTER, SODIUM SALT * TARAPON K 12 * TEXAPON ZHC * TREPENOL WA * ULTRA SULFATE SL-1

CONSENSUS REPORTS: Reported in EPA TSCA Inventory.

SAFETY PROFILE: Poison by intravenous and intraperitoneal routes. Moderately toxic by ingestion. Experimental teratogenic and reproductive effects. A human skin irritant. An experimental eye and severe skin irritant. A mild allergen. Mutation data reported. When heated to decomposition it emits toxic fumes of SO_x and Na_2O.

SID000 CAS: 57-30-7 **HR: 3**
SODIUM LUMINAL
mf: $C_{12}H_{11}N_2O_3$ • Na mw: 254.24

PROP: White crystals.

SYNS: 5-ETHYL-5-PHENYLBARBITURIC ACID SODIUM * 5-ETHYL-5-PHENYLBARBITURIC ACID SODIUM SALT * 5-ETHYL-5-PHENYL-2,4,6-(1H,3H,5H)PYRIMIDINE-TRIONE MONOSODIUM SALT * GARDENAL SODIUM * LUMINAL SODIUM * PBS * PHENEMALUM * PHENOBAL SODIUM * PHENOBARBITAL ELIXIR * PHENOBARBITAL Na * PHENOBARBITAL SODIUM * PHENOBARBITAL SODIUM SALT * PHENOBARBITONE SODIUM * PHENOBARBITONE SODIUM SALT * PHENYLETHYLBARBITURIC ACID, SODIUM SALT * SODIUM 5-ETHYL-5-PHENYLBARBITURATE * SODIUM PHENOBARBITAL * SODIUM PHENOBARBITONE * SODIUM PHENYLETHYLBARBITURATE * SODIUM PHENYLETHYLMALONYLUREA * SOL PHENOBARBITAL * SOL PHENOBARBITONE * SOLUBLE PHENOBARBITAL * SOLUBLE PHENOBARBITONE

CONSENSUS REPORTS: IARC Cancer Review: Animal Sufficient Evidence IMEMDT 13,157,77. EPA Genetic Toxicology Program.

SAFETY PROFILE: Confirmed carcinogen with experimental carcinogenic, neoplastigenic, and tumorigenic data. Poison by ingestion, intravenous, intraperitoneal, intraduodenal, and subcutaneous routes. Human systemic effects by ingestion: nausea or vomiting and coma. Experimental teratogenic and reproductive effects. Mutation data reported. Used to treat epilepsy, as an hypnotic and sedative. When heated to decomposition it emits toxic fumes of NO_x and Na_2O.

SIG500 CAS: 2492-26-4 **HR: 2**
SODIUM 2-MERCAPTOBENZO-THIAZOLE
mf: $C_7H_4NS_2$ • Na mw: 189.23

SYNS: 2-MERCAPTOBENZOTHIAZOLE SODIUM DERIVATIVE * 2-MERCAPTOBENZOTHIAZOLE SODIUM SALT

CONSENSUS REPORTS: Reported in EPA TSCA Inventory.

SAFETY PROFILE: Moderately toxic by ingestion. When heated to decomposition it emits very toxic fumes of NO_x, SO_x, and Na_2O.

SIH500   CAS: 492-18-2   **HR: 3**
SODIUM MERSALYL
mf: $C_{13}H_{16}HgNO_6 \cdot Na$ mw: 505.88

SYNS: 3-(α-CARBOXY-o-ANISAMIDO)-2-METHOXYPRO-PYL HYDROXYMERCURY, MONOSODIUM SALT * o-((3-HYDROXYMERCURI-2-METHOXYPROPYL)CAR-BAMOYL)PHENOXYACETIC ACID MONOSODIUM SALT * N-(γ-HYDROXYMERCURI-β-METHOXYPROPYL)SALI-CYLAMIDE-o-ACETIC ACID SODIUM SALT * IGROSIN * MERCURAMIDE * MERCURITAL * MERCUSAL * MERSALIN * MERSALYL * SALURIN * SALYRGAN * SODIUM o-((3-(HYDROXYMERCURI)-2-METHOXYPROPYL)CARBAMOYL)PHENOXY ACETATE * SODIUM SALICYL-(γ-HYDROXYMERCURI-β-METH-OXYPROPYL)AMIDE-o-ACETATE * URAGAN

CONSENSUS REPORTS: Mercury and its compounds are on the Community Right To Know List.

OSHA PEL: (Transitional: CL 1 mg/10m³) CL 0.1 mg(Hg)/m³ (skin)
ACGIH TLV: TWA 0.1 mg(Hg)m³ (skin)
NIOSH REL: (Inorganic Mercury) TWA 0.05 mg(Hg)/m³

SAFETY PROFILE: Poison by intravenous, intramuscular, and intraperitoneal routes. Used as a diuretic agent. When heated to decomposition it emits very toxic fumes of Hg, NO_x, and Na_2O.

SII000   CAS: 7681-57-4   **HR: 3**
SODIUM METABISULFITE
DOT: NA 2693
mf: $O_5S_2 \cdot 2Na$ mw: 190.10

PROP: Colorless crystals or white to yellowish powder; odor of sulfur dioxide. Sol in water; sltly sol in alc.

SYNS: DISODIUM PYROSULFITE * SODIUM META-BOSULPHITE * SODIUM PYROSULFITE

CONSENSUS REPORTS: Reported in EPA TSCA Inventory. EPA Genetic Toxicology Program.

OSHA PEL: TWA 5 mg/m³
ACGIH TLV: TWA 5 mg/m³

DOT Classification: ORM-B; Label: None

SAFETY PROFILE: Poison by intravenous route. Moderately toxic by parenteral route. Experimental reproductive effects. Mutation data reported. When heated to decomposition it emits toxic fumes of SO_x and Na_2O.

SII500   CAS: 10361-03-2   **HR: 2**
SODIUM METAPHOSPHATE
mf: $O_3P \cdot Na$ mw: 101.96

PROP: Sodium metaphosphate exists as a number of different molecular species, some of which exhibit various crystalline forms. The vitreous sodium phosphates having a Na_2O/P_2O_3 mole ratio near unity are classified as sodium metaphosphates. The term also extends to short-chain vitreous compositions, the compounds of which exhibit the polyphosphate formula $Na_{n+2}P_nO_{3n+1}$ with n as low as 4-5. In such as $(NaPO_3)$, n may be a small integer <3 (cyclic molecules) or a large number (polymers). Amorphous white solids. Very sol in water.

SYNS: GRAHAM'S SALT * METAFOS * SODIUM HEXAMETAPHOSPHATE * SODIUM POLYPHOS-PHATES, GLASSY * SODIUM TETRAPOLYPHOSPHATE

CONSENSUS REPORTS: Reported in EPA TSCA Inventory.

SAFETY PROFILE: Moderately toxic by intra-peritoneal route. When heated to decomposition it emits toxic fumes of Na_2O and PO_x.

SIK450   CAS: 124-41-4   **HR: 3**
SODIUM METHYLATE
DOT: UN 1431
mf: $CH_3O \cdot Na$ mw: 54.03

PROP: White, amorphous, free-flowing powder. Decomp in air above 127°; decomp by water. Sol in methyl and ethyl alc, fats, esters.

SYNS: METHANOL, SODIUM SALT * SODIUM METHOXIDE * SODIUM METHYLATE, DRY (DOT)

CONSENSUS REPORTS: Reported in EPA TSCA Inventory.

DOT Classification: Flammable Solid; Label: Flammable Solid and Dangerous When Wet

SAFETY PROFILE: A corrosive and irritating material. It hydrolyzes into methanol and sodium hydroxide. May ignite spontaneously in moist air. Flammable when exposed to heat or flame. Ignites on contact with water. Violent reaction with (CHCl$_3$ + CH$_3$OH), (methyl azide + dimethylmalonate), FClO$_3$. When heated to decomposition it emits toxic fumes of Na$_2$O.

SIN500 CAS: 12401-86-4 **HR: 3**
SODIUM MONOXIDE

DOT: UN 1825
mf: Na$_2$O   mw: 61.98

PROP: White-gray, deliquescent crystals. Bp: 1275° (sublimes), d. 2.27.

SYNS: CALCINED SODA * DISODIUM MONOXIDE * DISODIUM OXIDE * SODIUM MONOXIDE, solid (DOT) * SODIUM OXIDE

DOT Classification: Corrosive Material; Label: Corrosive

SAFETY PROFILE: Very corrosive and irritating to skin, eyes, and mucous membranes. Can react violently with water; nitric oxide (above 100°C). Ignites when mixed with 2,4-dinitrotoluene. Mixtures with phosphorus(V) oxide react violently when warmed or on contact with moisture. When heated to decomposition it emits toxic fumes of Na$_2$O.

SIO900 CAS: 7631-99-4 **HR: 3**
SODIUM(I) NITRATE (1:1)

DOT: UN 1498
mf: NO$_3$•Na   mw: 85.00

PROP: Colorless, transparent, odorless crystals; saline, sltly bitter taste. Mp: 306.8°, bp: decomp @ 380°, d: 2.261. Deliquescent in moist air; sol in water, sltly sol in alc.

SYNS: CHILE SALTPETER * CUBIC NITER * NITRATE de SODIUM (FRENCH) * NITRATINE * NITRIC ACID, SODIUM SALT * SODA NITER * SODIUM NITRATE (DOT)

CONSENSUS REPORTS: Reported in EPA TSCA Inventory. EPA Genetic Toxicology Program.

DOT Classification: Oxidizer; Label: Oxidizer

SAFETY PROFILE: Poison by intravenous route. Moderately toxic by ingestion. Questionable carcinogen with experimental tumorigenic data. Human mutation data reported. A powerful oxidizer. It will ignite with heat or friction. Explodes when heated to over 1000°F, or when mixed with cyanides, sodium hypophosphite, boron phosphide. Forms explosive mixtures

with aluminum powder, antimony powder, barium thiocyanate, metal amidosulfates, sodium, sodium phosphinate, sodium thiosulfate, sulfur + charcoal (gunpowder). Potentially violent reaction or ignition when mixed with bitumen, organic matter, calcium-silicon alloy, jute + magnesium chloride, magnesium, metal cyanides, non-metals, perosyformic acid, phenol + trifluoroacetic acid. Incompatible with acetic anhydride, barium thiocyanate, wood. A dangerous disaster hazard. When heated to decomposition it emits toxic fumes of NO$_x$ and Na$_2$O.

SIP500 CAS: 5064-31-3 **HR: 3**
SODIUM NITRILOTRIACETATE
mf: C$_6$H$_6$NO$_6$•3Na   mw: 257.10

SYNS: HAMPSHIRE NTA * NITRILOTRIACETIC ACID, TRISODIUM SALT * NTA * TRISODIUM NITRILOTRIACETATE * TRISODIUM NITRILOTRIACETIC ACID

CONSENSUS REPORTS: Reported in EPA TSCA Inventory.

SAFETY PROFILE: Poison by intraperitoneal route. Moderately toxic by ingestion. Experimental reproductive effects. Questionable carcinogen with experimental neoplastigenic data. Mutation data reported. When heated to decomposition it emits toxic fumes of NO$_x$ and Na$_2$O.

SIQ500 CAS: 7632-00-0 **HR: 3**
SODIUM NITRITE

DOT: UN 1500
mf: NO$_2$•Na   mw: 69.00

PROP: Sltly yellowish or white crystals, sticks or powder; slt salty taste. Mp: 271°, bp: decomp @ 320°, d: 2.168. Deliquescent in air; sol in water, sltly sol in alc.

SYNS: ANTI-RUST * DIAZOTIZING SALTS * DUSITAN SODNY (CZECH) * ERINITRIT * FILMERINE * NATRIUM NITRIT (GERMAN) * NCI-C02084 * NITRITE de SODIUM (FRENCH) * NITROUS ACID, SODIUM SALT

CONSENSUS REPORTS: Reported in EPA TSCA Inventory. EPA Genetic Toxicology Program.

DOT Classification: Oxidizer; Label: Oxidizer

SAFETY PROFILE: They may react with organic amines in the body to form carcinogenic nitrosamines. Human poison by ingestion. Experimental poison by ingestion, subcutaneous, intravenous, and intraperitoneal routes. Human systemic effects by ingestion: motor activity changes, coma, decreased blood pressure with possible pulse rate increase without fall in blood pressure, arteriolar or venous dilation, nausea or vomiting, and blood methemoglobinemia-carboxhemoglobinemia. Experimental teratogenic and reproductive effects. An eye irritant. Questionable carcinogen with experimental neoplastigenic and tumorigenic data. Human mutation data reported.

Flammable; a strong oxidizing agent. In contact with organic matter, will ignite by friction. Explodes when heated to over 1000°F or on contact with cyanides, NH_4^+ salts, cellulose, Li, (K + NH_3), $Na_2S_2O_3$. Incompatible with aminoguanidine salts, butadiene, phthalic acid, phthalic anhydride, reducants, sodium amide, sodium disulphite, sodium thiocyanate, urea, wood. When heated to decomposition it emits toxic fumes of NO_x and Na_2O.

SIT500   CAS: 63915-74-2   **HR: 3**
SODIUM NITRITE, mixed with POTASSIUM NITRITE

DOT Classification: Oxidizer; Label: Oxidizer

SAFETY PROFILE: Both components are poisons. A powerful oxidizer. When heated to decomposition it emits toxic fumes of NO_x, K_2O, and Na_2O.

SIU500   CAS: 14402-89-2   **HR: 3**
SODIUM NITROFERRICYANIDE
mf: $C_5FeN_6O \cdot 2Na$ mw: 261.94

SYNS: DISODIUM NITROSYLPENTACYANOFERRATE * NIPRIDE * NITROPRUSSIDNATRIUM (GERMAN) * SODIUM NITROPRUSSATE * SODIUM NITROPRUSSIDE * SODIUM NITROSYLPENTACYANOFERRATE * SODIUM NITROSYLPENTACYANOFERRATE(III)

CONSENSUS REPORTS: Cyanide and its compounds are on the Community Right To Know List. Reported in EPA TSCA Inventory.

SAFETY PROFILE: Human poison by inhalation and intravenous routes. Experimental poison by ingestion, intraperitoneal, and intravenous routes. Human systemic effects by intravenous and possibly other routes: increased intracranial pressure, general anesthesia, change in heart rate and metabolic acidosis. Experimental teratogenic and reproductive effects. Used as a vasodilator for short-term treatment of se-

vere hypertension. Mixtures with sodium nitrite explode when heated. When heated to decomposition it emits toxic fumes of NO_x, CN^-, and Na_2O.

SIY250   CAS: 13721-39-6   **HR: 3**
SODIUM ORTHOVANADATE
mf: $O_4V \cdot 3Na$ mw: 183.91

PROP: Colorless, hexagonal prisms. Mp: 850-866°.

SYNS: SODIUM VANADATE * SODIUM VANADIUM OXIDE * TRISODIUM ORTHOVANADATE * VANADIC(II) ACID, TRISODIUM SALT

CONSENSUS REPORTS: Reported in EPA TSCA Inventory.

NIOSH REL: (Vanadium Compounds) CL 0.05 mg(V)/m^3/15M

SAFETY PROFILE: Poison by ingestion, intraperitoneal, intravenous, and subcutaneous routes. Mutation data reported. When heated to decomposition it emits toxic fumes of VO_x and Na_2O.

SJA000   CAS: 131-52-2   **HR: 3**
SODIUM PENTACHLOROPHENATE

DOT: UN 2567
mf: $C_6Cl_5O \cdot Na$ mw: 288.30

PROP: Tan powder.

SYNS: DOW DORMANT FUNGICIDE * DOWICIDE G-ST * NAPCLOR-G * PENTACHLOROPHENATE SODIUM * PENTACHLOROPHENOL, SODIUM SALT * PENTACHLOROPHENOXY SODIUM * PENTAPHENATE * SANTOBRITE * SODIUM PCP * SODIUM PENTACHLOROPHENATE (DOT) * SODIUM PENTACHLOROPHENOL * SODIUM PENTACHLOROPHENOLATE * SODIUM PENTACHLOROPHENOXIDE * WEEDBEADS

CONSENSUS REPORTS: EPA Extremely Hazardous Substances List. Chlorophenol compounds are on the Community Right-To-Know List. Reported in EPA TSCA Inventory. EPA Genetic Toxicology Program.

DOT Classification: ORM-A; Label: None: Poison B; Label: Poison

SAFETY PROFILE: Poison by ingestion, inhalation, skin contact, intravenous, intraperitoneal, subcutaneous, and intratracheal routes. Experimental reproductive effects. Mutation data reported. When heated to decomposition it emits toxic fumes of Cl^- and Na_2O.

SJC000 CAS: 10101-50-5 **HR: 3**
SODIUM PERMANGANATE

DOT: UN 1503
mf: $MnO_4 \cdot Na$ mw: 141.93

PROP: Purple to red-black crystals. Mp: decomp.

SYNS: PERMANGANATE de SODIUM (FRENCH)
* PERMANGANIC ACID, SODIUM SALT

CONSENSUS REPORTS: Manganese and its compounds are on the Community Right To Know List. Reported in EPA TSCA Inventory.

OSHA PEL: CL 5 mg(Mn)/m³
ACGIH TLV: TWA 5 mg(Mn)/m³
DOT Classification: Oxidizer; Label: Oxidizer

SAFETY PROFILE: Probably a severe irritant to the skin, eyes, and mucous membranes. A powerful oxidizer and fire hazard. Explosive reaction with acetic acid, acetic anhydride. Reacts vigorously with combustibles. When heated to decomposition it emits toxic fumes of Na_2O.

SJC500 CAS: 1313-60-6 **HR: 3**
SODIUM PEROXIDE

DOT: UN 1504
mf: Na_2O_2 mw: 77.98

PROP: White powder turning yellow when heated. Mp: decomp @ 460°, bp: decomp, d: 2.805.

SYNS: DISODIUM DIOXIDE * DISODIUM PEROXIDE
* FLOCOOL 180 * SODIUM DIOXIDE * SODIUM
OXIDE (Na2-O2) * SOLOZONE

CONSENSUS REPORTS: Reported in EPA TSCA Inventory.

DOT Classification: Oxidizer; Label: Oxidizer

SAFETY PROFILE: A severe irritant to skin, eyes, and mucous membranes. Dangerous fire hazard by chemical reaction; a powerful oxidizing agent. Reacts explosively or violently under the appropriate conditions with water, acids, powdered metals, acetic acid, acetic anhydride, Al, (Al + CO_2), aluminum + aluminum chloride, almond oil, $(NH_4)_2S_2O_8$, aniline, Sb, As, benzene, boron nitride, calcium acetylide, charcoal, Cu, cotton wool, (KNO_3 + dextrose), diethyl ether, fibrous materials + water, glucose

+ potassium nitrate, hexamethylene-tetramine, hydrogen sulfide, hydroxy compounds (e.g., ethanol, ethylene glycol, glycerol, sugar), magnesium, (Mg + CO_2), MnO_2, metals, metals + carbon dioxide + water, non-metals (e.g., carbon, phosphorus, antimony, arsenic, boron, sulfur, selenium), non-metal halides (e.g., diselenium dichloride, disulfur dichloride, phosphorus trichloride), organic matter, paraffin, K, silver chloride + charcoal, soap, Na, sodium dioxide, SCl, Sn, Zn, wood, peroxyformic acid, reducing materials. Will react with water or steam to produce heat and toxic fumes. To fight fire, use carbon dioxide or dry chemical. Combustible materials ignited by contact with sodium peroxide should be smothered with soda ash, salt or dolomite mixtures. Chemical fire extinguishers should not be used. If the fire cannot be smothered, it should be flooded with large quantities of water from a hose. When heated to decomposition it emits toxic fumes of Na_2O.

SJE000 CAS: 7775-27-1 **HR: 3**
SODIUM PERSULFATE

DOT: UN 1505
mf: $O_8S_2 \cdot 2Na$ mw: 238.10

PROP: White, crystalline powder. Sol in water; decomp by alc.

SYNS: PERSULFATE de SODIUM (FRENCH) * SODIUM PEROXYDISULFATE

CONSENSUS REPORTS: Reported in EPA TSCA Inventory.

DOT Classification: Oxidizer; Label: Oxidizer

SAFETY PROFILE: Poison by intraperitoneal and intravenous routes. A powerful oxidizer; can cause fires. When heated to decomposition it emits toxic fumes of SO_x and Na_2O.

SJF000 CAS: 139-02-6 **HR: 3**
SODIUM PHENOXIDE

DOT: UN 2497
mf: $C_6H_5O \cdot Na$ mw: 116.10

PROP: White, deliquescent crystals.

SYNS: PHENOL SODIUM SALT * SODIUM CARBOLATE * SODIUM PHENATE * SODIUM PHENOLATE, solid (DOT)

CONSENSUS REPORTS: Reported in EPA TSCA Inventory.

DOT Classification: Corrosive Material; Label: Corrosive

SAFETY PROFILE: Poison by subcutaneous route. A corrosive irritant to skin, eyes, and mucous membranes. When heated to decomposition it emits toxic fumes of Na_2O.

SJH090   CAS: 7558-79-4   HR: 3
SODIUM PHOSPHATE, DIBASIC

DOT: NA 9147
mf: $HO_4P \cdot 2Na$ mw: 141.96

PROP: Colorless, translucent crystals or white powder. Sol in water; very sltly sol in alc.

SYNS: DIBASIC SODIUM PHOSPHATE * DISODIUM HYDROGEN PHOSPHATE * DISODIUM MONOHYDROGEN PHOSPHATE * DISODIUM ORTHOPHOSPHATE * DISODIUM PHOSPHATE * DISODIUM PHOSPHORIC ACID * DSP * EXSICCATED SODIUM PHOSPHATE * NATRIUMPHOSPHAT (GERMAN) * PHOSPHORIC ACID, DISODIUM SALT * SODA PHOSPHATE * SODIUM HYDROGEN PHOSPHATE * SODIUM MONOHYDROGEN PHOSPHATE (2:1:1)

CONSENSUS REPORTS: Reported in EPA TSCA Inventory.

DOT Classification: ORM-E; Label: None

SAFETY PROFILE: Poison by intravenous route. Moderately toxic by intraperitoneal, subcutaneous, and intramuscular routes. Mildly toxic by ingestion. A skin and eye irritant. When heated to decomposition it emits toxic fumes of PO_x and Na_2O.

SJH100   CAS: 7558-80-7   HR: 3
SODIUM PHOSPHATE, MONOBASIC
mf: $H_2O_4P \cdot Na$ mw: 119.98

PROP: White crystalline powder or granules; odorless. Hygroscopic; sol in water; insol in alc.

SYNS: MONOSODIUM DIHYDROGEN PHOSPHATE * MONOSODIUM PHOSPHATE * MONOSORB XP-4 * PRIMARY SODIUM PHOSPHATE * SODIUM ACID PHOSPHATE * SODIUM BIPHOSPHATE * SODIUM BIPHOSPHATE anhydrous * SODIUM DIHYDROGEN PHOSPHATE (1:2:1)

CONSENSUS REPORTS: Reported in EPA TSCA Inventory.

SAFETY PROFILE: Poison by intramuscular route. Mildly toxic by ingestion. A human and experimental eye irritant. When heated to de-

composition it emits toxic fumes of PO_x and Na_2O.

SJH200   CAS: 7601-54-9   HR: 2
SODIUM PHOSPHATE, TRIBASIC
mf: $O_4P \cdot 3Na$ mw: 163.94

PROP: White crystals or crystalline powder; odorless. Sol in water; insol in alc.

SYNS: DRI-TRI * EMULSIPHOS 440/660 * NUTRIFOS STP * PHOSPHORIC ACID, TRISODIUM SALT * SODIUM PHOSPHATE * SODIUM PHOSPHATE, anhydrous * TRIBASIC SODIUM PHOSPHATE * TRINATRIUMPHOSPHAT (GERMAN) * TRISODIUM ORTHOPHOSPHATE * TRISODIUM PHOSPHATE * TROMETE * TSP

CONSENSUS REPORTS: Reported in EPA TSCA Inventory.

DOT Classification: ORM-E; Label: None.

SAFETY PROFILE: Moderately toxic by intravenous route. Mutation data reported. A strong, caustic material. When heated to decomposition it emits toxic fumes of Na_2O and PO_x.

SJI500   CAS: 12058-85-4   HR: 2
SODIUM PHOSPHIDE

DOT: UN 1432
mf: PNa_3 mw: 99.94

PROP: Red crystals. Mp: decomp.

SYN: PHOSPHURE de SODIUM (FRENCH)

CONSENSUS REPORTS: Reported in EPA TSCA Inventory.

DOT Classification: Flammable Solid; Label: Flammable Solid and Dangerous When Wet

SAFETY PROFILE: Moderately toxic by inhalation. Flammable when exposed to heat or flame. Reacts violently with water to yield phosphine. When heated to decomposition it emits toxic fumes of PO_x and Na_2O.

SJK000   CAS: 9003-04-7   HR: 1
SODIUM POLYACRYLATE

CONSENSUS REPORTS: Reported in EPA TSCA Inventory.

SAFETY PROFILE: An eye irritant. When heated to decomposition it emits toxic fumes of Na_2O.

SJL500   CAS: 137-40-6   **HR: 2**
SODIUM PROPIONATE
mf: $C_3H_5O_2 \cdot Na$ mw: 96.07

PROP: Transparent crystals or granules; nearly odorless. Very sol in water; sltly sol in alc.

SYNS: NATRIUMPROPIONAT (GERMAN) * PROPANOIC ACID, SODIUM SALT

CONSENSUS REPORTS: Reported in EPA TSCA Inventory.

SAFETY PROFILE: Moderately toxic by skin contact and subcutaneous routes. Mildly toxic by unspecified routes. An allergen. When heated to decomposition it emits toxic fumes of Na_2O.

SJN700   CAS: 128-44-9   **HR: 3**
SODIUM SACCHARIN
mf: $C_7H_4NO_3S \cdot Na$ mw: 205.17

PROP: White crystals or crystalline powder; odorless, very sweet taste. Sol in water, alc.

SYNS: ARTIFICIAL SWEETENING SUBSTANZ GENDORF 450 * CRISTALLOSE * CRYSTALLOSE * DAGUTAN * KRISTALLOSE * MADHURIN * ODA * SACCHARIN * SACCHARIN SOLUBLE * SACCHARIN, SODIUM * SACCHARIN, SODIUM SALT * SACCHARINE SOLUBLE * SACCHARINNATRIUM * SACCHAROIDUM NATRICUM * SAXIN * SODIUM 1,2 BENZISOTHIAZOLIN-3-ONE-1,1-DIOXIDE * SODIUM o-BENZOSULFIMIDE * SODIUM BENZOSULPHIMIDE * SODIUM o-BENZOSULPHIMIDE * SODIUM 2-BENZOSULPHIMIDE * SODIUM SACCHARIDE * SODIUM SACCHARINATE * SODIUM SACCHARINE * SOLUBLE GLUSIDE * SOLUBLE SACCHARIN * SUCCARIL * SUCRA * o-SULFONBENZOIC ACID IMIDE SODIUM SALT * SULPHOBENZOIC IMIDE, SODIUM SALT * SWEETA * SYKOSE * WILLOSETTEN

CONSENSUS REPORTS: IARC Cancer Review: GROUP 2B IMEMDT 7,334,87; Animal Sufficient Evidence IMEMDT 22,111,80. EPA Genetic Toxicology Program. Reported in EPA TSCA Inventory.

SAFETY PROFILE: Suspected carcinogen with experimental carcinogenic, neoplastigenic, and tumorigenic data. Moderately toxic by ingestion and intraperitoneal routes. A promoter. Experimental teratogenic and reproductive effects. Human mutation data reported. When heated to decomposition it emits very toxic fumes of SO_x, Na_2O, and NO_x.

SJO000   CAS: 54-21-7   **HR: 3**
SODIUM SALICYLATE
mf: $C_7H_5O_3 \cdot Na$ mw: 160.11

PROP: White, odorless crystals, scales or powder.

SYNS: ALYSINE * ARDALL * AROALL * CLIN * DIURETIN * ENTEROSALICYL * ENTEROSALIL * 2-HYDROXYBENZOIC ACID MONOSODIUM SALT * o-HYDROXYBENZOIC SODIUM SALT * IDOCYL NOVUM * KERASALICYL * KEROSAL * MAGSALYL * NADISAL * NEO-SALICYL * PARBOCYL-REV * SALICYLIC ACID, SODIUM SALT * SALISOD * SALSONIN * SODIUM o-HYDROXYBENZOATE * SODIUM SALICYLIC ACID

CONSENSUS REPORTS: Reported in EPA TSCA Inventory.

SAFETY PROFILE: Experimental poison by subcutaneous route. Moderately toxic to humans by ingestion. Moderately toxic experimentally by ingestion, intraperitoneal and intravenous routes. Human systemic effects by multiple and unspecified routes: toxic psychosis, excitement, respiratory stimulation, nausea or vomiting and sweating. Experimental teratogenic and reproductive effects. Mutation data reported. A powerful irritant which affects the central nervous system. Incompatible with ferric salts, mineral acids, iodine, lead acetate, silver nitrate, sodium phosphate powder. When heated to decomposition it emits toxic fumes of Na_2O.

SJT500   CAS: 10102-18-8   **HR: 3**
SODIUM SELENITE
DOT: UN 2630
mf: $O_3Se \cdot 2Na$ mw: 172.94

PROP: White crystals.

SYNS: DISODIUM SELENITE * NATRIUMSELENIT (GERMAN) * SELENIOUS ACID, DISODIUM SALT

CONSENSUS REPORTS: IARC Cancer Review: GROUP 3 IMEMDT 7,56,87; Animal Inadequate Evidence IMEMDT 9,245,75. Reported in EPA TSCA Inventory. EPA Genetic Toxicology Program. EPA Extremely Hazardous Substances List. Selenium and its compounds are on the Community Right To Know List.

OSHA PEL: TWA 0.2 mg(Se)/m^3
ACGIH TLV: TWA 0.2 mg(Se)/m^3
DFG MAK: 0.1 mg(Se)/m^3

DOT Classification: Poison B; Label: Poison

SAFETY PROFILE: Poison by ingestion, intraperitoneal, intravenous, subcutaneous, intracervical, parenteral, and intramuscular routes. Experimental teratogenic and reproductive effects. Questionable carcinogen. Human mutation data reported. When heated to decomposition it emits toxic fumes of Se and Na_2O.

SJU000   CAS: 6834-92-0   **HR: 3**
SODIUM SILICATE
mf: $O_3Si \cdot 2Na$ mw: 122.07

SYNS: B-W * CRYSTAMET * DISODIUM META-SILICATE * DISODIUM MONOSILICATE * METSO 20 * METSO BEADS 2048 * METSO BEADS, DRYMET * METSO PENTABEAD 20 * ORTHOSIL * SODIUM METASILICATE * SODIUM METASILI-CATE, anhydrous * WATER GLASS

CONSENSUS REPORTS: Reported in EPA TSCA Inventory.

SAFETY PROFILE: Poison by ingestion and intraperitoneal routes. A caustic material which is a severe eye, skin, and mucous membrane irritant. Experimental reproductive effects. Ingestion causes gastrointestinal tract upset. Violent reaction with F_2. When heated to decomposition it emits toxic fumes of Na_2O. Used in cosmetics.

SJV000   CAS: 7757-81-5   **HR: 2**
SODIUM SORBATE
mf: $C_6H_7O_2 \cdot Na$ mw: 134.12

SYN: SORBIC ACID, SODIUM SALT

SAFETY PROFILE: Moderately toxic by intraperitoneal route. Mildly toxic by ingestion. Mutation data reported. Migrates to food from packaging material. When heated to decomposition it emits toxic fumes of Na_2O.

SJV500   CAS: 822-16-2   **HR: 3**
SODIUM STEARATE
mf: $C_{18}H_{36}O_2 \cdot Na$ mw: 306.52

SYNS: OCTADECANOIC ACID, SODIUM SALT * SODIUM OCTADECANOATE * STEARIC ACID, SODIUM SALT

CONSENSUS REPORTS: Reported in EPA TSCA Inventory.

SAFETY PROFILE: Poison by intravenous and possibly other routes. When heated to decomposition it emits toxic fumes of Na_2O.

SJW475   CAS: 127-58-2   **HR: 2**
SODIUM SULFAMERAZINE
mf: $C_{11}H_{12}N_4O_2S \cdot Na$ mw: 287.32

PROP: Crystals; bitter, caustic taste. Hygroscopic. On prolonged exposure to humid air, it absorbs CO_2 with the liberation of sulfamerazine and becomes incompletely sol in water. Its solns are alkaline to phenolphthalein (pH 10 or more). One gram dissolves in 3.6 mL water. Sltly sol in alc; insol in ether, chloroform.

SYNS: 4-AMINO-N-(4-METHYL-2-PYRIMIDINYL)-BEN-ZENESULFONAMIDE MONOSODIUM SALT * N^1-(4-METHYL-2-PYRIMIDINYL)SULFANILAMIDE SODIUM SALT * SODIUM SULPHAMERAZINE * SOLUBLE SULFAMERAZINE * SOLUMEDINE * SUL-FAMERAZINE SODIUM

CONSENSUS REPORTS: Reported in EPA TSCA Inventory.

SAFETY PROFILE: Moderately toxic by ingestion, subcutaneous, intraperitoneal, and intravenous routes. When heated to decomposition it emits toxic fumes of SO_x, NO_x, and Na_2O.

SJY000   CAS: 7757-82-6   **HR: 2**
SODIUM SULFATE (2:1)
mf: $O_4S \cdot 2Na$ mw: 142.04

PROP: White crystals or powder; odorless. Mp: 888°, d: 2.671. Sol in water, glycerin; insol alc.

SYNS: DISODIUM SULFATE * NATRIUMSUFAT (GERMAN) * SALT CAKE * SODIUM SULFATE anhydrous * SODIUM SULPHATE * SULFURIC ACID, DISODIUM SALT * THENARDITE * TRONA

CONSENSUS REPORTS: Reported in EPA TSCA Inventory. EPA Genetic Toxicology Program.

SAFETY PROFILE: Moderately toxic by intravenous route. Mildly toxic by ingestion. Experimental teratogenic and reproductive effects. Violent reaction with Al. When heated to decomposition it emits toxic fumes of SO_x and Na_2O.

SJY500   CAS: 1313-82-2   **HR: 3**
SODIUM SULFIDE (anhydrous)
DOT: UN 1385
mf: Na_2S mw: 78.04

PROP: Amorphous, yellow-pink or white, deliquescent crystals. Mp: 1180°, d: 1.856 @ 14°.

SYNS: SODIUM MONOSULFIDE * SODIUM SUL-PHIDE

CONSENSUS REPORTS: Reported in EPA TSCA Inventory.

DOT Classification: Label: Flammable Solid

SAFETY PROFILE: Flammable when exposed to heat or flame. Unstable and can explode on rapid heating or percussion. Reacts violently with carbon; diazonium salts; n,n-dichloromethylamine; o-nitroaniline diazonium salt; water. When heated to decomposition it emits toxic fumes of SO_x and Na_2O.

SJZ000   CAS: 7757-83-7   **HR: 3**
SODIUM SULFITE (2:1)
mf: $O_3S \cdot 2Na$ mw: 126.04

PROP: Hexagonal prisms or white powder; odorless with salty, sulfurous taste. Bp: decomp, d: 2.633 @ 15.4°. Sol in water; sltly sol in alc.

SYNS: DISODIUM SULFITE * EXSICATED SODIUM SULFITE * NATRIUMSULFID (GERMAN) * SODIUM SULFITE, anhydrous * SODIUM SULPHITE * SULFTECH * SULFUROUS ACID, SODIUM SALT (1:2)

CONSENSUS REPORTS: Reported in EPA TSCA Inventory. EPA Genetic Toxicology Program.

SAFETY PROFILE: Poison by intravenous and subcutaneous routes. Moderately toxic by ingestion and intraperitoneal routes. Human mutation data reported. When heated to decomposition it emits very toxic fumes of Na_2O and SO_x. A reducing agent.

SKC500   CAS: 10102-20-2   **HR: 3**
SODIUM TELLURITE
mf: $O_3Te \cdot 2Na$ mw: 221.58

SYNS: SODIUM TELLURATE(IV) * TELLUROUS ACID, DISODIUM SALT

CONSENSUS REPORTS: EPA Extremely Hazardous Substances List. Reported in EPA TSCA Inventory.

OSHA PEL: TWA 0.1 mg(Te)/m^3
ACGIH TLV: TWA 0.1 mg(Te)/m^3

SAFETY PROFILE: Human poison by ingestion and parenteral routes. Experimental poison by ingestion, intravenous, and intraperitoneal routes. Human mutation data reported. When heated to decomposition it emits toxic fumes of Te and Na_2O.

SKF000   CAS: 12206-14-3   **HR: 3**
SODIUM TETRAPEROXYCHROMATE
mf: $CrNa_3O_8$ mw: 248.97

CONSENSUS REPORTS: Chromium and its compounds are on the Community Right To Know List.

OSHA PEL: CL 0.1 mg(CrO_3)/m^3
ACGIH TLV: TWA 0.05 mg(Cr)/m^3
NIOSH REL: TWA 0.025 mg(Cr(VI))/m^3; CL 0.05 mg/m^3/15M

SAFETY PROFILE: Confirmed human carcinogen. Explodes when heated to 115°C. When heated to decomposition it emits toxic fumes of Na_2O.

SKG500   CAS: 12058-74-1   **HR: 3**
SODIUM TETRAVANADATE
mf: $O_{11}V_4 \cdot 2Na$ mw: 425.74

NIOSH REL: CL 0.05 mg(V)/m^3/15M

SAFETY PROFILE: Human poison by intravenous route. Experimental poison by intravenous, subcutaneous, intramuscular, and parenteral routes. When heated to decomposition it emits toxic fumes of Na_2O and VO_x.

SKH500   CAS: 367-51-1   **HR: 3**
SODIUM THIOGLYCOLATE
mf: $C_2H_3O_2S \cdot Na$ mw: 114.10

PROP: Hygroscopic crystals.

SYNS: MERCAPTOACETIC ACID SODIUM SALT * SODIUM MERCAPTOACETATE * SODIUM THIOGLYCOLLATE * THIOGLYCOLATESODIUM * THIOGLYCOLLIC ACID, SODIUM SALT * USAF EK-5199

CONSENSUS REPORTS: Reported in EPA TSCA Inventory.

SAFETY PROFILE: Poison by intravenous and intraperitoneal routes. Moderately toxic by ingestion. A human skin irritant. This material yields hydrogen sulfide on decomposition. A death has been attributed to the absorption of toxic decomposition products from the use of this material in a hair permanent waving solution. When heated to decomposition it emits toxic fumes of SO_x and Na_2O.

SKI000   CAS: 7772-98-7   **HR: 2**
SODIUM THIOSULFATE
mf: $O_3S_2 \cdot 2Na$ mw: 158.10

PROP: Colorless crystals or crystalline powder. Sol in water; insol in alc.

SYNS: HYPO * SODIUM HYPOSULFITE * SO-
DIUM THIOSULFATE, anhydrous

CONSENSUS REPORTS: Reported in EPA
TSCA Inventory.

SAFETY PROFILE: Moderately toxic by subcutaneous route. Incompatible with metal nitrates,
sodium nitrite. When heated to decomposition
it emits very toxic fumes of Na_2O and SO_x.

SKI500   CAS: 10102-17-7   **HR: 2**
SODIUM THIOSULFATE,
PENTAHYDRATE
mf: $O_3S_2 \cdot 2Na \cdot 5H_2O$ mw: 248.20

PROP: Monoclinic, colorless, odorless crystals.
Mp: 48° (rapid heating), d: 1.69.

SYNS: AMETOX * ANTICHLOR * HYPO
* NSC-45624 * SODIUM HYPOSULFITE * SO-
DOTHIOL * SULFOTHIORINE * THIOSULFURIC
ACID, DISODIUM SALT, PENTAHYDRATE

SAFETY PROFILE: Moderately toxic by intravenous route. Human systemic effects by ingestion: cyanosis. Large doses internally have
a cathartic action. Violent reaction with $NaNO_2$.
When heated to decomposition it emits toxic
fumes of SO_x and Na_2O.

SKM500   CAS: 7785-84-4   **HR: 3**
SODIUM TRIMETAPHOSPHATE
mf: $O_9P_3 \cdot 3Na$ mw: 305.88

PROP: White crystals or white crystalline powder. Sol in water.

SYN: TRIMETAPHOSPHATE SODIUM

CONSENSUS REPORTS: Reported in EPA
TSCA Inventory.

SAFETY PROFILE: Poison by intravenous
route. Moderately toxic by intraperitoneal route.
When heated to decomposition it emits toxic
fumes of PO_x and Na_2O.

SKN000   CAS: 13573-18-7   **HR: 3**
SODIUM TRIPOLYPHOSPHATE
mf: $O_{10}P_3 \cdot 5Na$ mw: 367.86

PROP: White granules or powder. Sltly hygroscopic; sol in water.

SYNS: ARMOFOS * NATRIUMTRIPOLYPHOSPHAT
(GERMAN) * PENTASODIUM TRIPHOSPHATE
* POLY * POLYGON * SODIUM TRIPHOSPHATE
* STPP * TRIPHOSPHORIC ACID, SODIUM SALT
* TRIPOLY * TRIPOLYPHOSPHATE

SAFETY PROFILE: Poison by intravenous
route. Moderately toxic by ingestion, subcutaneous, and intraperitoneal routes. Ingestion of
large doses of sodium phosphates causes catharsis. Sodium meta and pyrophosphates can cause
hemorrhages from the intestine if taken internally in large doses. When heated to decomposition it emits toxic fumes of PO_x and Na_2O.

SKN500   CAS: 13472-45-2   **HR: 3**
SODIUM TUNGSTATE
mf: $O_4W \cdot 2Na$ mw: 293.83

PROP: White, rhombic crystals. Mp: 698°, d:
4.179.

SYN: TUNGSTIC ACID, DISODIUM SALT

CONSENSUS REPORTS: Reported in EPA
TSCA Inventory.

ACGIH TLV: TWA 5 mg(W)/m^3
NIOSH REL: (Tungsten) TWA 1 mg(W)/m^3

SAFETY PROFILE: Poison by ingestion, intravenous, intramuscular, and subcutaneous
routes. Mutation data reported. When heated
to decomposition it emits toxic fumes of Na_2O.

SKP000   CAS: 13718-26-8   **HR: 3**
SODIUM VANADATE
mf: $O_3V \cdot Na$ mw: 121.93

SYNS: SODIUM METAVANADATE * VANADIC
ACID, MONOSODIUM SALT

CONSENSUS REPORTS: Reported in EPA
TSCA Inventory.

NIOSH REL: (Vanadium Compounds) CL 0.05
mg(V)/m^3/15M

SAFETY PROFILE: Poison by ingestion, intraperitoneal, subcutaneous, intravenous, and possibly other routes. When heated to decomposition it emits toxic fumes of Na_2O and VO_x.

SKS500   CAS: 96-64-0   **HR: 3**
SOMAN
mf: $C_7H_{16}FO_2P$ mw: 182.20

SYNS: 3,3-DIMETHYL-2-BUTANOL METHYLPHOS-
PHONOFLUORIDATE * 3,3-DIMETHYL-n-BUT-2-YL
METHYLPHOSPHONOFLUORIDATE * 3,3-DIMETHYL-2-
BUTYL METHYLPHOSPHONOFLUORIDATE * FLUORO-
METHYL(1,2,2-TRIMETHYLPROPOXY)PHOSPHINE OXIDE
* GD * METHYLFLUORPHOSPHORSAEUREPINA-
KOLYLESTER (GERMAN) * METHYLPHOSPHONO-
FLUORIDIC ACID, 3,3-DIMETHYL-2-BUTYL ESTER

* METHYLPHOSPHONOFLUORIDIC ACID 1,2,2-TRI-METHYLPROPYL ESTER * METHYL PINACOLYLOXY PHOSPHORYLFLUORIDE * METHYL PINACOLYL PHOSPHONOFLUORIDATE * PINACOLOXYMETHYL-PHOSPHORYL FLUORIDE * PINACOLYL METHYL-FLUOROPHOSPHONATE * PINACOLYL METHYL-PHOSPHONOFLUORIDATE * PINACOLYL METHYLPHOSPHONOFLUORIDE * PINACOLYLOXY METHYLPHOSPHORYL FLUORIDE * PMFP * PYNACOLYL METHYLFLUOROPHOSPHONATE * 1,2,2-TRIMETHYLPROPYL METHYLPHOSPHONO-FLUORIDATE

SAFETY PROFILE: A deadly human poison by inhalation and skin contact. A deadly experimental poison by inhalation, skin contact, subcutaneous, intravenous, intramuscular, and intraperitoneal routes. An extremely toxic military nerve gas. When heated to decomposition it emits very toxic fumes of F^- and PO_x.

SKS750   HR: 3
SOOT

PROP: Soot is defined as a brown-to-black substance incidentally produced during the incomplete and uncontrolled combustion of any carbonaceous material. It is a mixture of colloidal carbon, organic tars and refractory inorganics whose composition depends on combustion conditions. It is not unusual for the tarry component to account for more than 50 wt% of the soot, particularly, when produced by inefficient combustion of coal or wood. Can be distinguished from carbon black on the basis of differences in physical and chemical properties.

SAFETY PROFILE: Confirmed human carcinogen producing skin, scrotum, or lung tumors. The tarry component and, to a lesser extent, trace inorganic impurities, are believed responsible for the known health hazards attributed to soot, i.e., chronic contact or long-term inhalation can lead to cancer.

SKU000   CAS: 110-44-1   HR: 3
SORBIC ACID
mf: $C_6H_8O_2$ mw: 112.14

PROP: Colorless needles or white powder; characteristic odor. Bp: 228° (decomp), mp: 134.5°, flash p: 260°F (COC), vap press: 0.01 mm @ 20°, vap d: 3.87. Sol in hot water; very sol in alc, ether.

SYNS: (2-BUTENYLIDENE)ACETIC ACID * CROTYL-IDENE ACETIC ACID * HEXADIENIC ACID

* HEXADIENOIC ACID * 2,4-HEXADIENOIC ACID * trans-trans-2,4-HEXADIENOIC ACID * 1,3-PENTA-DIENE-1-CARBOXYLIC ACID * 2-PROPENYLACRYLIC ACID * SORBISTAT

CONSENSUS REPORTS: Reported in EPA TSCA Inventory.

SAFETY PROFILE: Moderately toxic by intraperitoneal and subcutaneous routes. Mildly toxic by ingestion. Experimental reproductive effects. A severe human and experimental skin irritant. Questionable carcinogen with experimental tumorigenic data. Mutation data reported. Combustible when exposed to heat or flame; can react with oxidizing materials. To fight fire, use water. When heated to decomposition it emits acrid smoke and irritating fumes.

SKV150   CAS: 1338-41-6   HR: 1
SORBITAN MONOSTEARATE
mf: $C_{24}H_{46}O_6$ mw: 430.70

PROP: Cream to tan-colored waxy solid; bland odor and taste. Insol in cold water, mineral spirits, acetone; dispersible in warm water; sol above 50° in mineral oil, ethyl acetate.

SYNS: ANHYDRO-d-GLUCITOL MONOOCTADECA-NOATE * ANHYDROSORBITOL STEARATE * ARLACEL 60 * ARMOTAN MS * CRILL 3 * CRILL K 3 * DREWSORB 60 * DURTAN 60 * EMSORB 2505 * GLYCOMUL S * HODAG SMS * IONET S 60 * LIPOSORB S * LIPOSORB S-20 * MONTANE 60 * MS 33 * MS 33F * NEWCOL 60 * NIKKOL SS 30 * NISSAN NONION SP 60 * NONION SP 60 * NONION SP 60R * RIKEMAL S 250 * SORBITAN C * SORBITAN MONOOCTADECA-NOATE * SORBITAN STEARATE * SORBON S 60 * SORGEN 50 * SPAN 55 * SPAN 60

CONSENSUS REPORTS: EPA Genetic Toxicology Program.

SAFETY PROFILE: Very mildly toxic by ingestion. Experimental reproductive effects. When heated to decomposition it emits acrid smoke and irritating fumes.

SKV200   CAS: 50-70-4   HR: 1
SORBITOL
mf: $C_6H_{14}O_6$ mw: 182.20

PROP: White crystalline powder; odorless with sweet taste. D: 1.47 @ −5°, mp: 93° (metastable form); 97.5°, (stable form), bp: 105°. Sol in water; sltly sol in methanol, ethanol, acetic acid, phenol, and acetamide; almost insol in other organic solvents.

SYNS: CHOLAXINE * DIAKARMON * GLUCITOL * d-GLUCITOL * GULITOL * l-GULITOL * KARION * NIVITIN * SIONIT * SIONON * SORBICOLAN * SORBITE * d-SORBITOL * SORBO * SORBOL * SORBOSTYL * SORVI-LANDE

CONSENSUS REPORTS: Reported in EPA TSCA Inventory. EPA Genetic Toxicology Program.

SAFETY PROFILE: Mildly toxic by ingestion. Human systemic effects by ingestion: hypermotility and diarrhea. When heated to decomposition it emits acrid smoke and irritating fumes.

SKY000   CAS: 8008-79-5   **HR: 1**
SPEARMINT OIL

PROP: From steam distillation of the plant *Mentha spicata* L. (Common Spearmint), or of *Mentha cardiaca* Gerard ex Baker (Scotch Spearmint) (Fam. *Labiatae*). Contains principally carvone, phellandrene, limonene, and either dihydrocarveol acetate or dihydrocuminic acetate (FCTXAV 16, 637,78). Colorless or greenish-yellow liquid; odor and taste of spearmint.

SYN: OIL of SPEARMINT

CONSENSUS REPORTS: Reported in EPA TSCA Inventory.

SAFETY PROFILE: Mildly toxic by ingestion. Mutation data reported. A skin irritant and an allergen. When heated to decomposition it emits acrid smoke and irritating fumes. Used as a flavoring agent.

SLB500   CAS: 84837-04-7   **HR: 2**
SPIKE LAVENDER OIL

PROP: From steam distillation of the plant *Lavandula latifolia* Vill. (*Lavandula spica*, D.C.)(Fam. *Labiatae*). The main constituents are linalool and cineole. Yellow liquid; lavender odor. D: 0.893-0.909, refr index: 1.463 @ 20°. Sol in fixed oils, propylene glycol; sltly sol in glycerin, mineral oil.

SYNS: LAVENDER OIL, SPIKE * OIL of SPIKE LAVENDER

SAFETY PROFILE: Moderately toxic by ingestion. A skin irritant. When heated to decomposition it emits acrid smoke and irritating fumes.

SLD000   CAS: 55-63-0   **HR: 3**
SPIRIT of GLYCERYL TRINITRATE

DOT: UN 0143/UN 1204

PROP: Clear, colorless liquid. Composition: 1.0-1.1% glycerol trinitrate in alcoholic solution. D: 0.814-0.820 @ 25°. Miscible with alc, chloroform, ether; very sltly sol in water.

SYNS: GLYCERYL TRINITRATE, solution * RCRA WASTE NUMBER P081 * SPIRIT of GLONOIN * SPIRITS of NITROGLYCERIN (DOT) * SPIRIT of TRINITROGLYCERIN

CONSENSUS REPORTS: Reported in EPA TSCA Inventory.

DOT Classification: Flammable Liquid; Label: Flammable Liquid; Class A Explosive; Label: Explosive A

SAFETY PROFILE: Likely to produce violent headache when tasted or applied to the skin. Flammable when exposed to heat or flame. If the alcohol evaporates, the residue is nitroglycerin; a shock-sensitive explosive. Upon contact with oxidizing materials the mixture can react vigorously. If spilled, immediately pour NaOH solution over the dry residue; friction or shock will explode it. Incompatible with alkalies; carbonates; HCl; HI. When heated to decomposition it emits highly toxic fumes.

SLG500   **HR: 3**
SPRENGEL EXPLOSIVES

SAFETY PROFILE: This type of explosive is a mixture of nitrobenzene and fuming nitric acid. It is a powerful and cheap explosive and would have many uses except that it is limited by practical disadvantages. The components have to be mixed in glass shortly before the explosive is used. This requires preparation and equipment not always available at the site of the explosion. This material can be destroyed by burning small quantities at a time.

SLI325   CAS: 21736-83-4   **HR: 2**
STANILO
mf: $C_{14}H_{24}N_2O_7 \cdot 2ClH$ mw: 405.32

SYNS: DECAHYDRO-4a,7,9-TRIHYDROXY-2-METHYL-6,8-BIS(METHYLAMINO)-4H-PYRANO(2,3-b)(1,4)BENZODI-OXIN-4-ONE DIHYDROCHLORIDE, (2R-(2-α,4a-β,5a-β,6-β,7-β,8-β,9-α,9a-α,10a-β))- * SPECTINOMYCIN DIHYDROCHLORIDE * SPECTINOMYCIN HYDROCHLORIDE

SAFETY PROFILE: Moderately toxic by intraperitoneal route. When heated to decomposition it emits toxic fumes of NO_x and HCl.

SLJ000 CAS: 7081-44-9 **HR: 2**
STAPHYBIOTIC
mf: $C_{19}H_{17}ClN_3O_5S \cdot Na \cdot H_2O$ mw: 475.91

SYNS: BACTOPEN * BRL-1621 * 6-(3-(o-CHLORO-PHENYL)-5-METHYL-4-ISOXAZOLECARBOXAMIDEO)-3,3-DIMETHYL-7-OXO-4-THIA-1-AZABICYCLO(3.2.0)HEP-TANE-2-CARBOXYLIC ACID, SODIUM SALT, MONOHY-DRATE * CLOXACILLIN SODIUM MONOHYDRATE * CLOXAPEN * CLOXYPEN * EKVACILLIN * GELSTAPH * METHOCILLIN-S * ORBENIN SODIUM HYDRATE * P-25 * PROSTAPHLIN-A * SODIUM CLOXACILLIN MONOHYDRATE * STAPHOBRISTOL-250 * TEGOPEN * TEPOGEN

SAFETY PROFILE: Moderately toxic by intraperitoneal, intramuscular, subcutaneous and intravenous routes. Mildly toxic by ingestion. When heated to decomposition it emits very toxic fumes of Cl^-, NO_x, Na_2O, and SO_x.

SLJ500 CAS: 9005-25-8 **HR: 1**
STARCH DUST

OSHA PEL: Total Dust: 15 mg/m³; Respirable Fraction: 5 mg/m³
ACGIH TLV: TWA (nuisance particulate) 10 mg/m³ of total dust (when toxic impurities are not present, e.g., quartz < 1%).

SAFETY PROFILE: A nuisance dust. An allergen. Flammable when exposed to flame, can react with oxidizing materials. Moderately explosive when exposed to flame.

SLK000 CAS: 57-11-4 **HR: 3**
STEARIC ACID
mf: $C_{18}H_{36}O_2$ mw: 284.54

PROP: White, amorphous solid; slt odor and taste of tallow. Mp: 69.3°, bp: 383°, flash p: 385°F (CC), d: 0.847, autoign temp: 743°F, vap press: 1 mm @ 173.7°, vap d: 9.80. Sol in alc, ether, acetone, chloroform; insol in water.

SYNS: CENTURY 1240 * DAR-CHEM 14 * EMER-SOL 120 * GLYCON DP * GLYCON S-70 * GLY-CON TP * GROCO 54 * 1-HEPTADECANECARBOX-YLIC ACID * HYDROFOL ACID 1655 * HY-PHI 1199 * HYSTRENE 80 * INDUSTRENE 5016 * KAM 1000

* KAM 2000 * KAM 3000 * NEO-FAT 18-61 * NEO-FAT 18-S * OCTADECANOIC ACID * PEARL STEARIC * STEAREX BEADS * STEARO-PHANIC ACID * TEGOSTEARIC 254

CONSENSUS REPORTS: Reported in EPA TSCA Inventory. EPA Genetic Toxicology Program.

SAFETY PROFILE: Poison by intravenous route. A human skin irritant. Questionable carcinogen with experimental tumorigenic data by implantation route. Combustible when exposed to heat or flame. Heats spontaneously. To fight fire, use CO_2, dry chemical. When heated to decomposition it emits acrid smoke and irritating fumes.

SLK500 CAS: 7460-84-6 **HR: 3**
STEARIC ACID-2,3-EPOXYPROPYL ESTER
mf: $C_{21}H_{40}O_3$ mw: 340.61

SYNS: 2,3-EPOXY-1-PROPANOL STEARATE * 2,3-EPOXYPROPYL ESTER of STEARIC ACID * 2,3-EPOXYPROPYL STEARATE * GLYCIDOL STEA-RATE * GLYCIDYL OCTADECANOATE * GLYCI-DYL STEARATE * OXIRANYLMETHYL ESTER of OC-TADECANOIC ACID

CONSENSUS REPORTS: IARC Cancer Review: GROUP 3 IMEMDT 7,56,87; Animal Inadequate Evidence IMEMDT 11,187,76.

SAFETY PROFILE: Questionable carcinogen with experimental tumorigenic data. Mutation data reported. When heated to decomposition it emits acrid smoke and irritating fumes.

SLP000 CAS: 10048-13-2 **HR: 3**
STERIGMATOCYSTIN
mf: $C_{18}H_{12}O_6$ mw: 328.34

PROP: A metabolite of *Aspergillus versicolor*.

SYN: 3a,12c-DIHYDRO-8-HYDROXY-6-METHOXY-7H-FURO(3',2':4,5)FURO(2,3-C)XANTHEN-7-ONE

CONSENSUS REPORTS: IARC Cancer Review: GROUP 2B IMEMDT 7,56,87; Animal Sufficient Evidence IMEMDT 10,245,76; Animal Limited Evidence IMEMDT 1,175,72. EPA Genetic Toxicology Program.

SAFETY PROFILE: Suspected carcinogen with experimental carcinogenic and tumorigenic data. Poison by ingestion and intraperitoneal routes. Human mutation data reported. When

heated to decomposition it emits acrid smoke and irritating fumes.

SLQ000   CAS: 7803-52-3   **HR: 3**
STIBINE

DOT: UN 2676
mf: H_3Sb mw: 124.78

PROP: Colorless gas, disagreeable odor. Mp: $-88°$, bp: $-18.4°$, d: 2.204 g/mL @ bp. Gas is sltly sol in water; very sol in alc, carbon disulfide, and organic solvents.

SYNS: ANTIMONWASSERSTOFFES (GERMAN) * ANTIMONY HYDRIDE * ANTIMONY TRIHYDRIDE * ANTYMONOWODOR (POLISH) * HYDROGEN ANTI-MONIDE

CONSENSUS REPORTS: Antimony and its compounds are on the Community Right To Know List.

OSHA PEL: TWA 0.1 ppm
ACGIH TLV: TWA 0.1 ppm
DFG MAK: 0.1 ppm (0.5 mg/m^3)
DOT Classification: Poison A; Label: Poison Gas and Flammable Gas

SAFETY PROFILE: Poison by inhalation. Potentially explosive decomposition at 200°C. Flammable when exposed to heat or flame. Explosive reaction with ammonia + heat, chlorine, concentrated nitric acid, ozone. Incompatible with oxidants. The decomposition products are hydrogen and metallic antimony. When heated to decomposition it emits toxic fumes of Sb. Used as a fumigating agent.

SLR000   CAS: 588-59-0   **HR: 3**
STILBENE
mf: $C_{14}H_{12}$ mw: 180.26

PROP: Colorless or sltly yellow crystals. Mp: 124-125°, bp: 306-307°, d: 0.9707. Insol in water; sol in 90 parts cold alc and 13 parts boiling alc; freely sol in benzene and ether.

SYNS: DIPHENYLETHYLENE * STILBEN (GERMAN)

CONSENSUS REPORTS: Reported in EPA TSCA Inventory.

SAFETY PROFILE: Poison by intravenous route. Moderately toxic by intraperitoneal route. Violent reaction with O_2. When heated to decomposition it emits acrid smoke and irritating fumes.

SLU500   CAS: 8052-41-3   **HR: 3**
STODDARD SOLVENT

PROP: Clear, colorless liquid. Composed of 85% nonane and 15% trimethyl benzene. Bp: 220-300°, flash p: 100-110°F, lel: 1.1%, uel: 6%, autoign temp: 450°F, d: 1.0. Insol in water; misc with abs alc, benzene, ether, chloroform, carbon tetrachloride, carbon disulfide, and some oils (not castor oil). Stoddard solvent to a first approximation contains 85% nonane and 15% trimethylbenzene.

SYNS: NAPHTHA SAFETY SOLVENT * VARNOLINE * WHITE SPIRITS

CONSENSUS REPORTS: Reported in EPA TSCA Inventory.

OSHA PEL: (Transitional: TWA 500 ppm) TWA 100 ppm
ACGIH TLV: TWA 100 ppm
NIOSH REL: TWA 350 mg/m^3; CL 1800 mg/m^3/15M

SAFETY PROFILE: Mildly toxic by inhalation. A human eye irritant. Flammable when exposed to heat or flame. Explosive in the form of vapor when exposed to heat or flame. When heated to decomposition it emits acrid fumes and may explode; can react with oxidizing materials. To fight fire, use foam, CO_2, dry chemical.

SLV500   CAS: 8063-18-1   **HR: 3**
STRAMONIUM

PROP: *Datura stramonium* have 0.25-0.45% alkaloids consisting of atropine, hyoscyamine, and scopolamine.

SYNS: ANGEL TULIP * DATURA STRAMONIUM * DEVIL'S APPLE * DHUTRA * JAMESTOWN WEED * JIMSON WEED * POMME EPINEUSE (FRENCH) * STECKAPFUL (GERMAN) * STRA-MONA (ITALIAN) * THORN APPLE

SAFETY PROFILE: Human and experimental poison by ingestion. When heated to decomposition it emits acrid smoke and irritating fumes.

SLW500   CAS: 57-92-1   **HR: 3**
STREPTOMYCIN
mf: $C_{21}H_{39}N_7O_{12}$ mw: 581.67

PROP: An antibiotic. It is a base and readily forms salts with anions.

SYNS: AGRIMYCIN 17 * CHEMFORM * GEROX * HOKKO-MYCIN * NSC 14083 * STREPCEN * STREPTOMICINA (ITALIAN) * STREPTOMYCIN A

* STREPTOMYCINE * STREPTOMYCINUM
* STREPTOMYZIN (GERMAN)

CONSENSUS REPORTS: EPA Genetic Toxicology Program.

SAFETY PROFILE: Poison by intravenous and subcutaneous routes. Moderately toxic by ingestion and intraperitoneal routes. Human systemic effects by ingestion and intraperitoneal routes: change in vestibular functions, blood pressure decrease, eosinophiis, respiratory depression, and other pulmonary changes. Human reproductive and teratogenic effects by unspecified routes: developmental abnormalities of the eye and ear and effects on newborn including postnatal measures or effects. Toxic to kidneys and central nervous system. Has been implicated in aplastic anemia. Experimental teratogenic and reproductive effects. Human mutation data reported. When heated to decomposition it emits toxic fumes of NO_x.

SMD000 CAS: 18883-66-4 **HR: 3**
STREPTOZOTICIN
mf: $C_8H_{15}N_3O_7$ mw: 265.26

PROP: Plateletes. Mp: 115° (decomp).

SYNS: 2-DEOXY-2-(((METHYLNITROSOAMINO)CAR-
BONYL)AMINO)-d-GLUCOPYRANOSE * 2-DEOXY-2-
(3-METHYL-3-NITROSOUREIDO)-d-GLUCOPYRANOSE
* 2-DEOXY-2-(3-METHYL-3-NITROSOUREIDO)-α(and β)-
d-GLUCOPYRANOSE * N-d-GLUCOSYL(2)-N'-NITROSO-
METHYLHARNSTOFF (GERMAN) * N-d-GLUCOSYL-(2)-
N'-NITROSOMETHYLUREA * NCI-C03167 * NSC
85598 * NSC-85998 * RCRA WASTE NUMBER U206
* STR * STREPTOZOCIN * STRZ * STZ
* U-9889 * ZANOSAR

CONSENSUS REPORTS: IARC Cancer Review: GROUP 2B IMEMDT 7,56,87; Human Limited Evidence IMEMDT 17,337,78; Animal Sufficient Evidence IMEMDT 17,337,78. NTP Fourth Annual Report On Carcinogens, 1984. NCI Carcinogenesis Studies (ipr); Clear Evidence: mouse, rat RRCRBU 52,1,75. EPA Genetic Toxicology Program.

SAFETY PROFILE: Confirmed carcinogen with experimental carcinogenic, neoplastigenic, and tumorigenic data. Experimental poison by ingestion, intravenous, parenteral, subcutaneous, and intraperitoneal routes. Moderately toxic to humans by intravenous route. Human systemic effects by ingestion and intravenous routes: nausea or vomiting, impaired liver function and

kidney changes. Human mutation data reported. Experimental teratogenic and reproductive effects. When heated to decomposition it emits toxic fumes of NO_x.

SMD500 **HR: 3**
STRONTIUM
af: Sr aw: 87.62

PROP: Silvery-white metal. D: 2.6, mp: 757° ± 1°, bp: 1366°; vap press: 10 mm @ 898°.

SAFETY PROFILE: It resembles calcium in its metabolism and behavior. The stable form has low toxicity. Ignites spontaneously in air. Moderately explosive in the form of dust when exposed to flame or by spontaneous chemical reaction. Reacts vigorously with water or steam to evolve hydrogen. Reaction with halogens may lead to ignition. Can be stored under liquid hydrocarbons. Vigorous reaction on contact with oxidizing materials. Highly dangerous in the form of the radioactive isotopes ^{90}Sr.

SME500 CAS: 91724-16-2 **HR: 3**
STRONTIUM ARSENITE
DOT: UN 1691
mf: As_2O_4Sr mw: 301.46

PROP: White powder.

SYNS: ARSENIOUS ACID, STRONTIUM SALT
* STRONTIUM ARSENITE, solid (DOT)

CONSENSUS REPORTS: Arsenic and its compounds are on the Community Right To Know List.

OSHA PEL: TWA 0.01 mg(As)/m³; Cancer
Hazard
NIOSH REL: CL 0.002 mg(As)/m³/15M
DOT Classification: Poison B; Label: Poison

SAFETY PROFILE: Confirmed human carcinogen. A deadly poison. When heated to decomposition it emits toxic fumes of As.

SMH000 CAS: 7789-06-2 **HR: 3**
STRONTIUM CHROMATE (1:1)
DOT: NA 9149
mf: $CrO_4 • Sr$ mw: 203.62

PROP: Monoclinic, yellow crystals. D: 3.895 @ 15°.

SYNS: CHROMIC ACID, STRONTIUM SALT (1:1)
* C.I. PIGMENT YELLOW 32 * DEEP LEMON YEL-
LOW * STRONTIUM CHROMATE (VI) * STRON-
TIUM CHROMATE 12170 * STRONTIUM YELLOW

CONSENSUS REPORTS: IARC Cancer Review: GROUP 1 IMEMDT 7,165,87; Animal Sufficient Evidence IMEMDT 2,100,73; IMEMDT 23,205,80; Human Sufficient Evidence IMEMDT 23,205,80. NTP Fourth Annual Report On Carcinogens, 1984. Chromium and its compounds are on the Community Right To Know List. Reported in EPA TSCA Inventory.

OSHA PEL: CL 0.1 mg(CrO_3)/m^3
ACGIH TLV: TWA 0.05 mg(Cr)/m^3; Confirmed Human Carcinogen; (Proposed: TWA 0.001 mg(Cr)/m^3; Suspected Human Carcinogen)
DFG TRK: 0.1 mg/m^3; Animal Carcinogen, Suspected Human Carcinogen.
NIOSH REL: TWA 0.0001 mg(Cr(VI))/m^3
DOT Classification: ORM-E; Label: None

SAFETY PROFILE: Confirmed human carcinogen with experimental carcinogenic and tumorigenic data. Moderately toxic by ingestion. Mutation data reported.

SMH500   HR: 1
STRONTIUM COMPOUNDS

SAFETY PROFILE: The strontium ion has a low order of toxicity. It is chemically and biologically similar to calcium. Strontium salicylate is the most toxic compound. The oxides and hydroxides are moderately caustic materials. Symptoms of acute toxicity are excessive salivation, vomiting, colic and diarrhea, and possibly respiratory failure. The gastrointestinal absorption of soluble strontium ranges from 5 to 25%. Workers in strontium salt plants have reduced activity of choline esterase and acetylcholine. Drinking water with 13 mg Sr/L caused impaired tooth development in 1 year old children. As with other compounds, the toxicity of a given compound may be a function of the anion. Compounds are highly dangerous if they contain the radioactive isotope ^{90}Sr.

SMI500   CAS: 7783-48-4   HR: 2
STRONTIUM FLUORIDE
mf: F_2Sr mw: 125.62

PROP: Cubic, odorless; crystals or white powder. Mp: 1190°, d: 4.24, bp: 2460. Decomposes by strong acids.

CONSENSUS REPORTS: Reported in EPA TSCA Inventory.

NIOSH REL: TWA 2.5 mg(F)/m^3

SAFETY PROFILE: Moderately toxic by intravenous route. Mildly toxic by ingestion. When heated to decomposition it emits toxic fumes of F^-.

SMK000   CAS: 10042-76-9   HR: 2
STRONTIUM(II) NITRATE (1:2)

DOT: UN 1507
mf: $N_2O_6 \cdot Sr$ mw: 211.64

PROP: White powder. Mp: 570°, d: 2.986.

SYNS: NITRATE de STRONTIUM (FRENCH) * NITRIC ACID, STRONTIUM SALT

CONSENSUS REPORTS: Reported in EPA TSCA Inventory.

DOT Classification: Oxidizer; Label: Oxidizer

SAFETY PROFILE: Moderately toxic by ingestion and intraperitoneal routes. A powerful oxidizer. When heated to decomposition it emits toxic fumes of NO_x.

SMK500   CAS: 1314-18-7   HR: 2
STRONTIUM PEROXIDE

DOT: UN 1509
mf: O_2Sr mw: 119.62

PROP: White powder. Mp: decomp, d: 4.56.

CONSENSUS REPORTS: Reported in EPA TSCA Inventory.

DOT Classification: Oxidizer; Label: Oxidizer

SAFETY PROFILE: A powerful oxidizer. A skin, eye, and mucous membrane irritant. Mixtures with organic materials readily ignite with friction or on contact with moisture.

SMM000   CAS: 1314-96-1   HR: 3
STRONTIUM SULFIDE
mf: SSr   mw: 119.68

PROP: Cubic, light gray crystals. D: 3.70 @ 15°.

SYNS: C.I. 77847 * STRONTIUM MONOSULFIDE * STRONTIUM SULPHIDE

CONSENSUS REPORTS: Reported in EPA TSCA Inventory.

SAFETY PROFILE: Poison by inhalation and ingestion. Readily decomposes to yield H_2S. Incompatible with lead(IV) oxide. When heated to decomposition it emits toxic fumes of SO_x.

SMN500 CAS: 57-24-9 **HR: 3**
STRYCHNINE

DOT: UN 1692
mf: $C_{21}H_{22}N_2O_2$ mw: 334.45

PROP: Hard, white, crystalline alkaloid; very bitter taste. Mp: 268°, bp: 270°, d: 1.359 @ 18°.

SYNS: CERTOX * DOLCO MOUSE CEREAL * KWIK-KIL * MOLE DEATH * MOUSE-NOTS * MOUSE-RID * MOUSE-TOX * PIED PIPER MOUSE SEED * RCRA WASTE NUMBER P108 * RO-DEX * SANASEED * STRICNINA (ITALIAN) * STRYCHNIDIN-10-ONE * STRYCHNIN (GERMAN) * STRYCHNINE, solid and liquid (DOT) * STRYCHNOS

CONSENSUS REPORTS: Reported in EPA TSCA Inventory. EPA Extremely Hazardous Substances List.

OSHA PEL: TWA 0.15 mg/m^3
ACGIH TLV: TWA 0.15 mg/m^3
DFG MAK: 0.15 mg/m^3
DOT Classification: Poison B; Label: Poison; Poison B; Label: St. Andrews Cross

SAFETY PROFILE: Human poison by ingestion and possibly other routes. Experimental poison by ingestion, intravenous, subcutaneous, intraperitoneal, and possibly other routes. Experimental reproductive effects. An allergen. Lethal dose to man: 30-60 mg/kg. If ingested, the time of action depends upon the condition of the stomach, that is, whether empty or full, and the nature of the food present. If taken by subcutaneous injection, the place of administration of the injection will affect the time of action. The first symptoms are a feeling of uneasiness with a heightened reflex of irritability, followed by muscular twitching in some parts of the body. With larger doses, this is followed by a sense of impending suffocation. Convulsive movements begin which have the effect of mechanically causing the patient to cry out or to shriek; then follow the characteristic spasms which set in with violence. These are at first clonic and then tonic. There are successive attacks of spasms. With each successive attack, the symptoms become more violent, eventually resulting in death. A rodenticide. When heated to decomposition it emits toxic fumes of NO$_x$.

SMO000 CAS: 66-32-0 **HR: 3**
STRYCHNINE MONONITRATE
mf: $C_{21}H_{22}N_2O_2 \cdot HNO_3$ mw: 397.47

PROP: Colorless, odorless needles or white, crystalline powder. Insol in ether.

SYN: STRYCHNINE NITRATE

SAFETY PROFILE: Poison by ingestion, intravenous, and subcutaneous routes. When heated to decomposition it emits very toxic fumes of NO$_x$.

SMP000 CAS: 60-41-3 **HR: 3**
STRYCHNINE SULFATE (2:1)
mf: $C_{21}H_{22}N_2O_2 \cdot 1/2H_2O_4S$ mw: 383.49

SYNS: STRYCHININE SULFATE * STRYCHNIDIN-10-ONE, SULFATE (2:1)

CONSENSUS REPORTS: EPA Extremely Hazardous Substances List.

SAFETY PROFILE: Poison by ingestion, intraperitoneal, intravenous, and subcutaneous routes. When heated to decomposition it emits very toxic fumes of SO$_x$ and NO$_x$.

SMP500 CAS: 82-71-3 **HR: 3**
STYPHNIC ACID

DOT: UN 0219
mf: $C_6H_3N_3O_8$ mw: 245.12

PROP: Hexagonal, yellow crystals; astringent taste. Mp: (dry) 175.5°. Very sol in alc, ether.

SYNS: 2,4-DIHYDROXY-1,3,5-TRINITROBENZENE * 1,3-DIHYDROXY-2,4,6-TRINITROBENZENE * 3-HYDROXY-2,4,6-TRINITROPHENOL * 2,4,6-TRINITROBENZENE-1,3-DIOL * 2,4,6-TRINITRO-1,3-BENZENEDIOL * 2,4,6-TRINITRORESORCINOL * TRINITRORESORCINOL (DOT) * TRINITRORESORCINOL, DRY (DOT) * TRINITRORESORCINOL, wetted with less than 20% water (DOT)

CONSENSUS REPORTS: Reported in EPA TSCA Inventory.

DOT Classification: Class A Explosive; Label: Explosive A

SAFETY PROFILE: Very explosive. Upon decomposition it emits toxic fumes of NO$_x$.

SMQ000 CAS: 100-42-5 **HR: 3**
STYRENE

DOT: UN 2055
mf: C_8H_8 mw: 104.16

PROP: Colorless, refractive, oily liquid. Mp: −31°, bp: 146°, lel: 1.1%, uel: 6.1%, flash p: 88°F, d: 0.9074 @ 20°/4°, autoign temp: 914°F, vap d: 3.6, fp: −33°, ULC: 40-50. Very sltly sol in water; misc in alc, and ether.

SYNS: CINNAMENE * CINNAMENOL * DIAREX HF 77 * ETHENYLBENZENE * NCI-C02200 * PHENETHYLENE * PHENYLETHENE * PHENYLETHYLENE * STIROLO (ITALIAN) * STYREEN (DUTCH) * STYREN (CZECH) * STYRENE MONOMER (ACGIH) * STYRENE MONOMER, inhibited (DOT) * STYROL (GERMAN) * STYROLE * STYROLENE * STYRON * STYROPOR * VINYLBENZEN (CZECH) * VINYLBENZENE * VINYLBENZOL

CONSENSUS REPORTS: IARC Cancer Review: GROUP 2B IMEMDT 7,345,87; Animal Sufficient Evidence IMEMDT 19,231,79; Human Inadequate Evidence IMEMDT 19,231,79. NCI Carcinogenesis Bioassay (gavage); Inadequate Studies: mouse, rat NCITR* NCI-CG-TR-170,79; (gavage). Reported in EPA TSCA Inventory. EPA Genetic Toxicology Program. Community Right To Know List.

OSHA PEL: (Transitional: TWA 100 ppm; CL 200; Pk 600/5M/3H) TWA 50 ppm; STEL: 100 ppm
ACGIH TLV: TWA 50 ppm; STEL: 100 ppm (skin); BEI: 1 g(mandelic acid)/L in urine at end of shift; 40 ppb styrene in mixed-exhaled air prior to shift; 18 ppm styrene in mixed-exhaled air during shift; 0.55 mg/L styrene in blood end of shift; 0.02 mg/L styrene in blood prior to shift.
DFG MAK: 20 ppm (85 mg/m^3); BAT: 2g/L of mandelic acid in urine at end of shift.
NIOSH REL: (Styrene) TWA 50 ppm; CL 100 ppm
DOT Classification: Flammable Liquid; Label: Flammable Liquid; Flammable or Combustible Liquid; Label: Flammable Liquid

SAFETY PROFILE: Suspected carcinogen. Experimental poison by ingestion, inhalation, and intravenous routes. Moderately toxic experimentally by intraperitoneal route. Mildly toxic to humans by inhalation. Human systemic effects by inhalation: eye and olfactory changes. It can cause irritation and violent itching of the eyes @ 200 ppm, lacrimation, and severe human eye injuries. Its toxic effects are usually transient and result in irritation and possible narcosis. Experimental teratogenic and reproductive effects. A human skin irritant. An experimental skin and eye irritant. Human mutation data reported.

The monomer has been involved in several industrial explosions. It is a storage hazard above 32°C. A very dangerous fire hazard when exposed to flame, heat or oxidants. Explosive in the form of vapor when exposed to heat or flame. Reacts with oxygen above 40°C to form a heat-sensitive explosive peroxide. Violent or explosive polymerization may be initiated by alkali metal-graphite composites, butyllithium, dibenzoyl peroxide, other initiators (e.g., azoisobutyronitrile, di-tert-butyl peroxide). Reacts violently with chlorosulfonic acid, oleum, sulfuric acid, chlorine + iron(III) chloride (above 50°C). May ignite when heated with air + polymerizing polystyrene. Can react vigorously with oxidizing materials. To fight fire, use foam, CO_2, dry chemical. When heated to decomposition it emits acrid smoke and irritating fumes.

SMQ500   CAS: 9003-53-6   **HR: 2**
STYRENE POLYMER
mf: $(C_8H_8)_n$

DOT: UN 2211

SYNS: A 3-80 * AFCOLENE * ATACTIC POLYSTYRENE * BACTOLATEX * BAKELITE SMD 3500 * BASF III * BEXTRENE XL 750 * BICOLASTIC A 75 * BUSTREN * CADCO 0115 * CARINEX GP * COPAL Z * COSDEN 550 * DENKA QP3 * DIAREX 43G * DORVON * DOW 860 * DYLENE * DYLITE F 40 * ESBRITE * ESCOREZ 7404 * ESTYRENE G 20 * ETHENYLBENZENE HOMOPOLYMER * FOSTER GRANT 834 * GEDEX * HI-STYROL * HOSTYREN S * HT-F 76 * IT 40 * KB (POLYMER) * KRASTEN 1.4 * LACQREN 550 * LUSTREX * MX 5517-02 * NBS 706 * OWISPOL GF * PICCOLASTIC * POLIGOSTYRENE * POLYSTROL D * POLYSTYRENE * POLYSTYRENE BEADS (DOT) * POLYSTYRENE LATEX * POLYSTYROL * PRINTEL'S * REXOLITE 1422 * RHODOLNE * SHELL 300 * STYRAFOIL * STYRAGEL * STYRENE POLYMERS * STYROFOAM * STYROLUX * STYRON * TOPOREX 855-51 * TROLITUL * UBATOL U 2001 * VESTYRON * VINYLBENZENE POLYMER * VINYL PRODUCTS R 3612

CONSENSUS REPORTS: IARC Cancer Review: GROUP 3 IMEMDT 7,56,87; Animal Limited Evidence IMEMDT 19,231,79. Reported in EPA TSCA Inventory.

DOT Classification: ORM; Label: None.

SAFETY PROFILE: Questionable carcinogen with experimental tumorigenic data by implant. When heated to decomposition it emits acrid smoke and irritating fumes.

SMR000   CAS: 9003-55-8   **HR: 1**
STYRENE POLYMER with 1,3-BUTADIENE

SYNS: AFCOLAC B 101 * ANDREZ * BASE 661 * 1,3-BUTADIENE-STYRENE COPOLYMER * BUTADIENE-STYRENE POLYMER * 1,3-BUTADIENE-STYRENE POLYMER * BUTADIENE-STYRENE RESIN * BUTADIENE-STYRENE RUBBER (FCC) * BUTAKON 85-71 * DIAREX 600 * DIENOL S * DOW 209 * DOW LATEX 612 * DST 50 * DURANIT * EDISTIR RB 268 * ETHENYLBENZENE POLYMER with 1,3-BUTADIENE * GOODRITE 1800X73 * HISTYRENE S 6F * HYCAR LX 407 * K 55E * KOPOLYMER BUTADIEN STYRENOVY (CZECH) * KRO 1 * LITEX CA * LYTRON 5202 * MARBON 9200 * NIPOL 407 * PHAROS 100.1 * PLIOFLEX * PLIOLITE S5 * POLY-BUTADIENE-POLYSTYRENE COPOLYMER * POLYCO 2410 * RICON 100 * SBS * SD 354 * S6F HISTYRENE RESIN * SKS 85 * SOIL STABILIZER 661 * SOLPRENE 300 * STYRENE-BUTADIENE COPOLYMER * STYRENE-1,3-BUTADIENE COPOLYMER * STYRENE-BUTADIENE POLYMER * SYNPOL 1500 * THERMOPLASTIC 125 * TR 201 * UP 1E * VESTYRON HI

CONSENSUS REPORTS: IARC Cancer Review: GROUP 3 IMEMDT 7,56,87; Human Inadequate Evidence IMEMDT 19,231,79. Reported in EPA TSCA Inventory.

SAFETY PROFILE: An eye irritant. Questionable carcinogen. When heated to decomposition it emits acrid smoke and irritating fumes.

SMY000   CAS: 110-15-6   **HR: 2**
SUCCINIC ACID
mf: $C_4H_6O_4$ mw: 118.10

PROP: Colorless or white crystals; odorless with acid taste. Mp: 185°, bp: 235° (decomp), d: 1.564 @ 15°/4°. Sol in water; very sol in alc, ether, acetone, glycerin.

SYNS: AMBER ACID * BERNSTEINSAURE (GERMAN) * BUTANEDIOIC ACID * 1,2-ETHANEDICARBOXYLIC ACID * ETHYLENESUCCINIC ACID

CONSENSUS REPORTS: Reported in EPA TSCA Inventory.

SAFETY PROFILE: Moderately toxic by subcutaneous route. A severe eye irritant. When heated to decomposition it emits acrid smoke and irritating fumes.

SNB000   CAS: 123-25-1   **HR: 1**
SUCCINIC ACID, DIETHYL ESTER
mf: $C_8H_{14}O_4$ mw: 174.22

PROP: Colorless, mobile liquid; pleasant odor. Flash p: 230°F. Sol in alc, ether, fixed oils, water.

SYNS: BUTANEDIOIC ACID, DIETHYL ESTER * DIETHYL SUCCINATE (FCC) * ETHYL SUCCINATE * FEMA No. 2377

CONSENSUS REPORTS: Reported in EPA TSCA Inventory.

SAFETY PROFILE: Mildly toxic by ingestion. A skin and eye irritant. Combustible liquid. Reaction with ethyl trifluoroacetate + sodium hydride may cause a fire or explosion. When heated to decomposition it emits acrid smoke and irritating fumes.

SNC000   CAS: 108-30-5   **HR: 3**
SUCCINIC ANHYDRIDE
mf: $C_4H_4O_3$ mw: 100.08

PROP: Colorless needles. Mp: 119.6°, bp: 261°, d: 1.104, vap press: 1 mm @ 92.0°. Very sltly sol in water, petr ether; sltly sol in ether.

SYNS: BERNSTEINSAURE-ANHYDRID (GERMAN) * BUTANEDIOIC ANHYDRIDE * DIHYDRO-2,5-FURANDIONE * 2,5-DIKETOTETRAHYDROFURAN * NCI-C55696 * SUCCINIC ACID ANHYDRIDE * SUCCINYL OXIDE * TETRAHYDRO-2,5-DIOXOFURAN

CONSENSUS REPORTS: IARC Cancer Review: GROUP 3 IMEMDT 7,56,87; Animal Inadequate Evidence IMEMDT 15,265,77. Reported in EPA TSCA Inventory. EPA Genetic Toxicology Program.

SAFETY PROFILE: Experimental teratogenic effects. A severe eye irritant. Mutation data reported. Questionable carcinogen with experimental neoplastigenic data. When heated to decomposition it emits acrid smoke and irritating fumes.

SNC500   CAS: 123-23-9   **HR: 3**
SUCCINIC PEROXIDE
DOT: UN 2135/UN 2962
mf: $C_8H_{10}O_8$ mw: 234.18

PROP: Fine white powder, odorless with tart taste. Mp: 125° (decomp). Mod sol in water.

SYNS: BIS(3-CARBOXYPROPIONYL) PEROXIDE * 3,3′-(DIOXYDICARBONYL)DIPROPIONIC ACID

* SUCCINIC ACID PEROXIDE (DOT) * SUCCINYL PEROXIDE

CONSENSUS REPORTS: Reported in EPA TSCA Inventory.

DOT Classification: Organic Peroxide; Label: Organic Peroxide

SAFETY PROFILE: Poison by intraperitoneal route. An irritant. Explodes on contact with flame. When heated to decomposition it emits acrid smoke and irritating fumes.

SNE000   CAS: 110-61-2   **HR: 3**
SUCCINONITRILE
mf: $C_4H_4N_2$ mw: 80.10

PROP: Colorless, odorless, waxy material. Mp: 58.1°, bp: 267°, flash p: 270°F, d: 1.022 @ 25°, vap press: 2 mm @ 100°, vap d: 2.1. Sltly sol in ether, water, alc; sol in acetone.

SYNS: 1,4-BUTANEDINITRILE * DEPRELIN * s-DICYANOETHANE * 1,2-DICYANOETHANE * DINILE * ETHYLENE CYANIDE * ETHYLENE DICYANIDE * SUCCINIC ACID DINITRILE * SUCCINIC DINITRILE * SUCCINODINITRILE * SUXIL * USAF A-9442

CONSENSUS REPORTS: Cyanide and its compounds are on the Community Right To Know List. Reported in EPA TSCA Inventory.

NIOSH REL: (Nitriles) TWA 20 mg/m³

SAFETY PROFILE: Poison by ingestion, intraperitoneal, and subcutaneous routes. Experimental teratogenic and reproductive effects. Combustible when exposed to heat or flame. Decomposes exothermically above 195°C. Can react with oxidizing materials. To fight fire, use alcohol foam, CO_2, dry chemical. When heated to decomposition, or on contact with acid or acid fumes, it emits highly toxic fumes of NO_x and CN^-.

SNH000   CAS: 57-50-1   **HR: 1**
SUCROSE
mf: $C_{12}H_{22}O_{11}$ mw: 342.34

PROP: White crystals; sweet taste. D: 1.587 @ 25°/4°, mp: 170-186° (decomp). Sol in water, alc; insol in ether.

SYNS: BEET SUGAR * CANE SUGAR * CONFECTIONER'S SUGAR * α-d-GLUCOPYRANOSYL β-d-FRUCTOFURANOSIDE * (α-d-GLUCOSIDO)-β-d-FRUCTOFURANOSIDE * GRANULATED SUGAR * NCI-C56597 * ROCK CANDY * SACCHAROSE * SACCHARUM * SUGAR

CONSENSUS REPORTS: Reported in EPA TSCA Inventory. EPA Genetic Toxicology Program.

OSHA PEL: TWA Total Dust: 15 mg/m³; Respirable Fraction: 5 mg/m³
ACGIH TLV: TWA (nuisance particulate) 10 mg/m³ of total dust (when toxic impurities are not present, e.g., quartz < 1%).

SAFETY PROFILE: Mildly toxic by ingestion. Experimental teratogenic and reproductive effects. Mutation data reported. Vigorous reaction with nitric acid or sulfuric acid (forms carbon monoxide and carbon dioxide). When heated to decomposition it emits acrid smoke and irritating fumes.

SNJ000   CAS: 57-68-1   **HR: 3**
SULFADIMETHYLDIAZINE
mf: $C_{12}H_{14}N_4O_2S$ mw: 278.36

PROP: Crystals; odorless. Mp: 176° (also a range reported of from 178-179°, 198-199°, and 205-207°). Sol in acetone, water, ether; sltly sol in alc.

SYNS: A-502 * 2-(p-AMINOBENZENESULFONAMIDO)-4,6-DIMETHYLPYRIMIDINE * 6-(4'-AMINOBENZOL-SULFONAMIDO)-2,4-DIMETHYLPYRIMIDIN (GERMAN) * (p-AMINOBENZOLSULFONYL)-2-AMINO-4,6-DIMETHYLPYRIMIDIN (GERMAN) * AZOLMETAZIN * CREMOMETHAZINE * DIAZYL * N¹-(4,6-DIMETHYL-2-PYRIMIDINYL)SULFANILAMIDE * N-(4,6-DIMETHYL-2-PYRIMIDYL)SULFANILAMIDE * 4,6-DIMETHYL-2-SULFANILAMIDOPYRIMIDINE * DIMEZATHINE * MERMETH * METAZIN * NCI-C56600 * NEASINA * PIRMAZIN * PRIMAZIN * SA 111 * SEAZINA * SPANBOLET * SULFADIMERAZINE * SULFADIMETHYLPYRIMIDINE * SULFADIMETINE * SULFADIMEZINE * SULFADIMIDINE * SULFADINE * SULFADSIMESINE * SULFA-ISODIMERAZINE * SULFAISODIMIDINE * SULFAMETHIAZINE * SULFAMETHIN * SULFAMEZATHINE * 2-SULFANILAMIDO-4,6-DIMETHYLPYRIMIDINE * SULFISOMIDIN * SULFISOMIDINE * SULFODIMESIN * SULFODIMEZINE * SULMET * SULPHADIMETHYLPYRIMIDINE * SULPHADIMIDINE * SUPERSEPTIL * VERTOLAN

CONSENSUS REPORTS: Reported in EPA TSCA Inventory.

SAFETY PROFILE: Moderately toxic by intravenous and intraperitoneal routes. Mildly toxic by ingestion. Experimental teratogenic and reproductive effects. Questionable carcinogen with experimental tumorigenic data. When heated to decomposition it emits very toxic fumes of SO_x and NO_x.

SNK000   CAS: 723-46-6   **HR: 3**
SULFAMETHOXAZOL
mf: $C_{10}H_{11}N_3O_3S$ mw: 253.30

SYNS: 4-AMINO-N-(5-METHYL-3-ISOXAZOLYL)BEN-ZENESULFONAMIDE * 3-(p-AMINOPHENYLSULPHON-AMIDO)-5-METHYLISOXAZOLE * AZO-GANTANOL * BACTRIM * CO-TRIMOXAZOLE * EUSAPRIM * FECTRIM * GANTANOL * N'-(5-METHYL-3-ISOXAZOLE)SULFANILAMIDE * N'-(5-METHYL-3-ISOXAZOLYL)SULFANILAMIDE * N'-(5-METHYLISOXAZOL-3-YL)SULPHANILAMIDE * N^1-(5-METHYL-3-ISOXAZOLYL)SULPHANILAMIDE * 5-METHYL-3-SULFANILAMIDOISOXAZOLE * 5-METHYL-3-SULPHANIL-AMIDOISOXAZOLE * METOXAL * MS 53 * RADONIL * RO 4-2130 * SIM * SINOMIN * SEPTRA * SEPTRAN * SULFAMETHALAZOLE * SULFAMETHOXAZOLE * SULFAMETHYLISOXAZOLE * 3-SULFANILAMIDO-5-METHYLISOXAZOLE * SULFISOMEZOLE * SUL-PHAMETHALAZOLE * SULPHAMETHOXAZOL * SULPHAMETHOXAZOLE * SULPHAMETHYLISOX-AZOLE * 3-SULPHANILAMIDO-5-METHYLISOXAZOLE * SULPHISOMEZOLE * TRIB * TRIMETOPRIM-SULFA

CONSENSUS REPORTS: IARC Cancer Review: GROUP 3 IMEMDT 7,348,87; Human Inadequate Evidence IMEMDT 24,285,80; Animal Limited Evidence IMEMDT 24,285,80. Reported in EPA TSCA Inventory.

SAFETY PROFILE: Moderately toxic by ingestion and intraperitoneal routes. Questionable carcinogen with experimental tumorigenic data. When heated to decomposition it emits very toxic fumes of NO_x and SO_x.

SNK500   CAS: 5329-14-6   **HR: 3**
SULFAMIC ACID
mf: H_3NO_3S mw: 97.10

DOT: UN 2967

PROP: White crystals. Mp: 200° (decomp), bp: decomp, d: 203 @ 12°.

SYNS: AMIDOSULFONIC ACID * AMIDOSULFURIC ACID * AMINOSULFONIC ACID * KYSELINA AMI-DOSULFONOVA (CZECH) * KYSELINA SULFAMINOVA (CZECH) * SULFAMIDIC ACID * SULPHAMIC ACID (DOT)

CONSENSUS REPORTS: Reported in EPA TSCA Inventory.

DOT Classification: Corrosive Material; Label: Corrosive

SAFETY PROFILE: Poison by intraperitoneal route. Moderately toxic by ingestion. A human skin irritant. A corrosive irritant to skin, eyes, and mucous membranes. A substance which migrates to food from packaging materials. Violent or explosive reactions with chlorine, metal nitrates + heat, metal nitrites + heat, fuming HNO_3. When heated to decomposition it emits very toxic fumes of SO_x and NO_x.

SNM500   CAS: 63-74-1   **HR: 3**
SULFANILAMIDE
mf: $C_6H_8N_2O_2S$ mw: 172.22

PROP: Crystals. Mp: 164.5-166.5°. Sol in glycerol, propylene glycol, HCl; almost insol in chloroform, ether, benzene, petr ether.

SYNS: ALBEXAN * ALBOSAL * AMBESIDE * p-AMINOBENZENESULFAMIDE * p-AMINOBEN-ZENESULFONAMIDE * 4-AMINOBENZENESULFONAM-IDE * p-AMINOPHENYLSULFONAMIDE * 4-AMINO-PHENYLSULFONAMIDE * p-ANILINESULFONAMIDE * ANILINE-p-SULFONIC AMIDE * ANTISTREPT * BACTERAMID * COLLOMIDE * COLSULANYDE * COPTICIDE * DIPRON * ESTREPTOCIDA * F 1162 * FOURNEAU 1162 * GERISON * GOMBARDOL * LUSIL * LYSOCOCCINE * NEOCOCCYL * ORGASEPTINE * PABS * PRONTALBIN * PRONTOSIL I * PROSEPTINE * PROSEPTOL * PYSOCOCCINE * RUBIAZOL A * SEPTAMIDE ALBUM * SEPTINAL * SEPTOPLEX * STOPTON ALBUM * STREPAMIDE * STREPTA-GOL * STREPTOCLASE * STREPTOL * STREP-TOSIL * STREPTOZONE * STREPTROCIDE * p-SULFAMIDOANILINE * SULFAMIDYL * SULFANA * SULFANALONE * SULFANIL * SULFOCIDINE * SULFONAMIDE * SULFONAM-IDE P * SULPHANILAMIDE * THERAPOL * WHITE STREPTOCIDE

CONSENSUS REPORTS: Reported in EPA TSCA Inventory. EPA Genetic Toxicology Program.

SAFETY PROFILE: Poison by ingestion and intraperitoneal routes. Moderately toxic by subcutaneous and intravenous routes. Human teratogenic effects by unspecified route: developmental abnormalities of the blood and lymphatic systems (including the spleen and bone marrow). Experimental teratogenic and reproductive effects. Questionable carcinogen with experimental carcinogenic data. Mutation data reported. Implicated in aplastic anemia. When heated to decomposition it emits very toxic fumes of NO_x and SO_x.

SNN300   CAS: 122-11-2   HR: 2
6-SULFANILAMIDO-2,4-DIMETHOXYPYRIMIDINE
mf: $C_{12}H_{14}N_4O_4S$ mw: 310.36

PROP: Crystals from dil alc. Mp: 201-203°. Sol in dil HCl and in aq solns of sodium carbonate. Solubility in water at 37° (mg/100 mL): 4.6 at pH 4.10; 29.5 at pH 6.7; 58.0 at pH 7.06.

SYNS: ABCID * AGRIBON * ALBON * 4-AMINO-N-(2,6-DIMETHOXY-4-PYRIMIDINYL)BENZENE-SULFONAMIDE * ARNOSULFAN * BACTROVET * DEPOSUL * DIASULFA * DIASULFYL * DI-METAZINA * 2,6-DIMETHOXY-4-(p-AMINOBENZENE-SULFONAMIDO)PYRIMIDINE * N^1-(2,6-DIMETHOXY-4-PYRIMIDINYL)SULFANILAMIDE * DIMETHOXYSUL-FADIAZINE * 2,4-DIMETHOXY-6-SULFANILAMIDO-1,3-DIAZINE * 2,6-DIMETHOXY-4-SULFANILAMIDO-PYRIMIDINE * DINOSOL * DORISUL * FUXAL * MADRIBON * MADRIGID * MADRIQID * MADROXIN * MADROXINE * MAXULVET * MEMCOZINE * METOXIDON * NEOSTREPAL * OMNIBON * PERSULFEN * RADONIN * REDIFAL * ROSCOSULF * SCANDISIL * SDM * SDMO * SUDINE * SULDIXINE * SULFADIMETHOXIN * SULFADIMETHOXINE * SULFADIMETHOXYDIAZINE * SULFAD-IMETOSSINA (ITALIAN) * SULFADIMETOXIN * SULFASOL * SULFASTOP * SULFOPLAN * SULPHADIMETHOXINE * SULXIN * SYMBIO * THERACANZAN

SAFETY PROFILE: Moderately toxic by intraperitoneal, intravenous, and subcutaneous routes. Experimental teratogenic and reproductive effects. When heated to decomposition it emits toxic fumes of SO_x and NO_x.

SNS000   HR: D
SULFATES

SAFETY PROFILE: Variable toxicity. In general the toxic properties of substances containing the sulfate radical is that of the material (cation) with which the sulfate (anion) is combined. Violent reaction with Al; Mg. When heated to decomposition they emit toxic fumes of SO_x.

SNT000   HR: D
SULFIDES

SAFETY PROFILE: Variable toxicity. The alkaline sulfides (potassium, calcium, ammonium, and sodium) are similar in action to alkalies. They cause softening and irritation of the skin. If ingested, they are corrosive and irritating through the liberation of hydrogen sulfide and free alkali. Hydrogen sulfide is especially toxic. Sulfides of the heavy metals are generally insoluble and hence have little toxic action except through the liberation of hydrogen sulfide. Sulfides are used as fungicides. Flammable when exposed to flame or by spontaneous chemical reaction. Many sulfides ignite easily in air at room temperature. Others require a higher temperature or the presence of an oxidizer. Upon contact with moisture or acids, hydrogen sulfide is evolved. Many powerful oxidizers on contact with sulfides ignite violently. Many sulfides react violently and explosively on contact with powerful oxidizers. Hydrogen sulfide evolved can form explosive mixtures with air. They react with water, steam or acids to produce toxic and flammable vapors of hydrogen sulfide. When heated to decomposition they emit highly toxic fumes of SO_x.

SNT500   HR: 2
SULFITES

SAFETY PROFILE: Fairly large doses of sulfites can be tolerated since they are rapidly oxidized to sulfates, although if swallowed they may cause irritation of the stomach by liberating sulfurous acid. Experimentally, large doses of sodium sulfite have been shown to cause retarded growth, nerve irritation, atrophy of bone marrow, depression, and paralysis. They will react with water, steam or acids to produce a toxic and corrosive material. When heated to decomposition they emit highly toxic fumes of SO_x.

SNY000   HR: 1
SULFONATES

SAFETY PROFILE: Variable toxicity. Usually irritating. When heated to decomposition or on contact with acid or acid fumes, they emit highly toxic fumes of SO_x.

SOA500   CAS: 80-08-0   HR: 3
4,4′-SULFONYLDIANILINE
mf: $C_{12}H_{12}N_2O_2S$ mw: 248.32

PROP: Crystals. Mp: 176°, vap d: 8.3. Nearly insol in water; sol in acetone, alc.

SYNS: AVLOSULPHONE * BIS(p-AMINOPHENYL) SULFONE * BIS(4-AMINOPHENYL) SULFONE * BIS(p-AMINOPHENYL)SULPHONE * BIS(4-AMINO-PHENYL)SULPHONE * CROYSULFONE * DADPS * DAPSONE * DDS * DIAMINODIFENILSULFONA (SPANISH) * DIAMINO-4,4'-DIPHENYL SULFONE * p,p'-DIAMINODIPHENYL SULFONE * 4,4'-DIAMI-NODIPHENYL SULFONE * DIAMINO-4,4'-DIPHENYL SULPHONE * p,p-DIAMINODIPHENYL SULPHONE * DI(p-AMINOPHENYL) SULFONE * DI(4-AMINOPHE-NYL)SULFONE * DI(p-AMINOPHENYL)SULPHONE * DI(4-AMINOPHENYL)SULPHONE * DIAPHENYL-SULFONE * DIAPHENYLSULPHON * DIAPHENYL-SULPHONE * DIPHONE * DISULONE * DSS * DUBRONAX * DUMITONE * EPORAL * 1358F * F 1358 * MALOPRIM * METABOLITE C * NCI-C01718 * NOVOPHONE * NSC-6091 * SULFONA * 1,1'-SULFONYLBIS(4-AMINOBENZENE) * 4,4'-SULFONYLBISANILINE * p,p-SULFONYLBIS-BENZAMINE * 4,4'-SULFONYLBISBENZAMINE * p,p-SULFONYLBISBENZENAMINE * p,p'-SULFO-NYLDIANILINE * SULPHADIONE * SULPHON-MERE * 1,1'-SULPHONYLBIS(4-AMINOBENZENE) * p,p-SULPHONYLBISBENZAMINE * 4,4'-SULPHO-NYLBISBENZAMINE * p,p-SULPHONYLBISBENZEN-AMINE * 4,4'-SULPHONYLBISBENZENAMINE * SULPHONYLDIANILINE * p,p-SULPHONYLDIANI-LINE * TARIMYL * UDOLAC * WR 448

CONSENSUS REPORTS: IARC Cancer Review: GROUP 3 IMEMDT 7,185,87; Animal Limited Evidence IMEMDT 24,59,80; Human Inadequate Evidence IMEMDT 24,59,80. NCI Carcinogenesis Bioassay (feed); No Evidence: mouse NCITR* NCI-CG-TR-20,77; (feed); Clear Evidence: rat NCITR* NCI-CG-TR-20,77. Reported in EPA TSCA Inventory.

SAFETY PROFILE: Poison by ingestion, intraperitoneal, and subcutaneous routes. Moderately toxic by unspecified route. Human systemic effects by ingestion: retinal changes, somnolence, cyanosis, jaundice, change in tubules and other kidney changes, hemolysis with or without anemia and effect on joints. Experimental reproductive effects. Can cause hepatitis, dermatitis, and neuritis. Questionable carcinogen with experimental carcinogenic and neoplastigenic data. Human mutation data reported. Used in leprosy treatment and veterinary medicine. When heated to decomposition it emits very toxic fumes of NO_x and SO_x.

SOD100   CAS: 3689-24-5   **HR: 3**
SULFOTEP

DOT: UN 1704
mf: $C_8H_{20}O_5P_2S_2$ mw: 322.34

PROP: A liquid almost insol in water.

SYNS: BAYER-E 393 * BIS-O,O-DIETHYLPHOSPHO-ROTHIONIC ANHYDRIDE * BLADAFUME * BLADA-FUN * DITHIO * DITHIODIPHOSPHORIC ACID, TET-RAETHYL ESTER * DITHIOFOS * DITHIONE * DI(THIOPHOSPHORIC) ACID, TETRAETHYL ESTER * DITHIOPYROPHOSPHATE de TETRAETHYLE (FRENCH) * DITHIOTEP * E393 * ENT 16,273 * ETHYL THIOPYROPHOSPHATE * LETHALAIRE G-57 * PIROFOS * PLANT DITHIO AEROSOL * PLANT-FUME 103 SMOKE GENERATOR * PYROPHOSPHORO-DITHIOIC ACID, TETRAETHYL ESTER * PYROPHOS-PHORODITHIOIC ACID-O,O,O,O-TETRAETHYL ESTER * RCRA WASTE NUMBER P109 * SULFOTEPP * TEDP (OSHA, MAK) * O,O,O,O-TETRAAETHYL-DITHIONOPYROPHOSPHAT (GERMAN) * O,O,O,O-TET-RAETHYL-DITHIO-DIFOSFAAT (DUTCH) * TETRA-ETHYL DITHIONOPYROPHOSPHATE * TETRAETHYL DITHIOPYROPHOSPHATE * O,O,O,O-TETRAETHYL DITHIOPYROPHOSPHATE * TETRAETHYL DITHIO PYROPHOSPHATE, liquid (DOT) * O,O,O,O-TETRAETIL-DITIO-PIROFOSFATO (ITALIAN) * THIOTEPP

OSHA PEL: TWA 0.2 mg/m³ (skin)
ACGIH TLV: TWA 0.2 mg/m³ (skin)
DFG MAK: 0.015 ppm (0.2 mg/m³)
DOT Classification: Poison B; Label: Poison; Poison B; Label: St. Andrews Cross.

SAFETY PROFILE: Poison by ingestion, skin contact, inhalation, intramuscular, intraperitoneal, subcutaneous and intravenous routes. A cholinesterase inhibitor type of insecticide. When heated to decomposition it emits toxic fumes of PO_x and SO_x.

SOD500   CAS: 7704-34-9   **HR: 3**
SULFUR

DOT: UN 1350/UN 2448
af: S   aw: 32.06

PROP: Rhombic yellow crystals or yellow powder. Mp: 119°, bp: 444.6°, flash p: 405°F (CC), d: 2.07, d (liquid): 1.803, autoign temp: 450°F, vap press: 1 mm @ 183.8°. Insol in water; sltly sol in alc, ether; sol in carbon disulfide, benzene, toluene.

SYNS: BENSULFOID * BRIMSTONE * COLLOI-DAL SULFUR * COLLOKIT * COLSUL * CORO-

SUL D AND S * COSAN * CRYSTEX * FLOW-
ERS of SULPHUR (DOT) * GROUND VOCLE SULPHUR
* HEXASUL * KOCIDE * KOLOFOG * KOLO-
SPRAY * KUMULUS * MAGNETIC 70,90 and 95
* MICROFLOTOX * PRECIPITATED SULFUR
* SOFRIL * SPERLOX-S * SPERSUL * SPER-
SUL THIOVIT * SUBLIMED SULFUR * SULFIDAL
* SULFORON * SULFUR FLOWER (DOT) * SUL-
KOL * SUPER COSAN * SULPHUR (DOT) * SUL-
PHUR, lump or power (DOT) * SULPHUR, molten (DOT)
* SULSOL * TECHNETIUM TC 99M SULFUR COL-
LOID * TESULOID * THIOLUX * THIOVIT

CONSENSUS REPORTS: Reported in EPA
TSCA Inventory.

DOT Classification: ORM-C; Label: None;
Flammable Solid; Label: Flammable Solid

SAFETY PROFILE: Poison by ingestion, intravenous and intraperitoneal routes. A human
eye irritant. A fungicide. Chronic inhalation can
cause irritation of mucous membranes. Combustible when exposed to heat or flame or by
chemical reaction with oxidizers. Explosive in
the form of dust when exposed to flame. Can
react violently with halogens, carbides, halogenates, halogenites, zinc, uranium, tin, sodium,
lithium, nickel, palladium, phosphorus, potassium, indium, calcium, boron, aluminum, (aluminum + niobium pentoxide), ammonia, ammonium nitrate, ammonium perchlorate, BrF_5,
BrF_3, (Ca + VO + H_2O), $Ca(OCl)_2$, Ca_3P_2,
Cs_3N, charcoal, (Cu + chlorates), ClO_2, ClO,
ClF_3, CrO_3, $Cr(OCl)_2$, hydrocarbons, IF_5, IO_5,
PbO_2, $Hg(NO_3)_2$, HgO, Hg_2O, NO_2, P_2O_3,
(KNO_3 + As_2S_3), K_3N, $KMnO_4$, $AgNO_3$,
Ag_2O, NaH, ($NaNO_3$ + charcoal), (Na + SnI_4),
SCl_2, Tl_2O_3, F_2. Can react with oxidizing materials. To fight fire, use water or special mixtures
of dry chemical. When heated it burns and emits
highly toxic fumes of SO_x.

SOG500   CAS: 10545-99-0   HR: 3
SULFUR DICHLORIDE

DOT: UN 1828
mf: Cl_2S mw: 102.96

PROP: Reddish-brown liquid; pungent odor.
Mp: $-78°$, bp: $59°$, d: 1.621 @ $15°/15°$, vap
d: 3.55.

SYNS: CHLORIDE of SULFUR (DOT) * CHLORINE
SULFIDE * DICHLOROSULFANE * MONOSULFUR
DICHLORIDE * SULFUR CHLORIDE * SULFUR
CHLORIDE (MONO) (DOT)

CONSENSUS REPORTS: Reported in EPA
TSCA Inventory.

DOT Classification: Corrosive Material; Label:
Corrosive

SAFETY PROFILE: Poison irritant and corrosive to skin, eyes, and mucous membranes.
Flammable when exposed to heat or flame. Reacts violently with Al, NH_3, K, Na, acetone,
dimethyl sulfoxide, water, oxidants, metals,
hexafluoro isopropylidene amino lithium. Reactive with water, steam. When heated to decomposition it emits very toxic fumes of SO_x
and Cl^-.

SOH500   CAS: 7446-09-5   HR: 3
SULFUR DIOXIDE

DOT: UN 1079
mf: O_2S mw: 64.06

PROP: Colorless gas or liquid under pressure;
pungent odor. Mp: $-75.5°$, bp: $-10.0°$, d (liquid): 1.434 @ $0°$, vap d: 2.264 @ $0°$, vap
press: 2538 mm @ $21.1°$. Sol in water.

SYNS: BISULFITE * FERMENICIDE LIQUID
* FERMENICIDE POWDER * SCHWEFELDIOXYD
(GERMAN) * SIARKI DWUTLENEK (POLISH)
* SULFUROUS ACID ANHYDRIDE * SULFUROUS AN-
HYDRIDE * SULFUROUS OXIDE * SULFUR OXIDE
* SULPHUR DIOXIDE, LIQUEFIED (DOT)

CONSENSUS REPORTS: EPA Extremely
Hazardous Substances List. Reported in EPA
TSCA Inventory. EPA Genetic Toxicology Pro-
gram.

OSHA PEL: (Transitional: TWA 5 ppm) TWA
2 ppm; STEL 5 ppm
ACGIH TLV: TWA 2 ppm; STEL 5 ppm
DFG MAK: 2 ppm (5 mg/m^3)
NIOSH REL: (Sulfur Dioxide) TWA 0.5 ppm
DOT Classification: Nonflammable Gas; La-
bel: Nonflammable Gas; Poison A; Label:
Poison Gas

SAFETY PROFILE: A poison gas. Experimental reproductive effects. Human mutation data
reported. Human systemic effects by inhalation:
pulmonary vascular resistance, respiratory depression and other pulmonary changes. Questionable carcinogen with experimental tumorigenic and teratogenic data. It chiefly affects the
upper respiratory tract and the bronchi. It may
cause edema of the lungs or glottis, and can

produce respiratory paralysis. A corrosive irritant to eyes, skin, and mucous membranes. This material is so irritating that it provides its own warning of toxic concentration. Levels of 400-500 ppm are immediately dangerous to life. Its toxicity is comparable to that of hydrogen chloride. However, less than fatal concentration can be borne for fair periods of time with no apparent permanent damage. It is a common air contaminant.

A nonflammable gas. It reacts violently with acrolein, Al, $CsHC_2$, Cs_2O, chlorates, ClF_3, Cr, FeO, F_2, Mn, KHC_2, $KClO_3$, Rb_2C_2, Na, Na_2C_2, SnO, lithium acetylene carbide diammino. Will react with water or steam to produce toxic and corrosive fumes. Incompatible with halogens, or interhalogens, lithium nitrate, metal acetylides, metal oxides, metals, polymeric tubing, potassium chlorate, sodium hydride. When heated to decomposition it emits toxic fumes of SO_x.

SOl000   CAS: 2551-62-4   HR: 1
SULFUR HEXAFLUORIDE

DOT: UN 1080
mf: F_6S mw: 146.06

PROP: Colorless gas. Mp: $-51°$ (subl @ $-64°$), vap d: 6.602, d (liquid): 1.67 @ $-100°$.

SYNS: HEXAFLUORURE de SOUFRE (FRENCH) * SULFUR FLUORIDE

CONSENSUS REPORTS: Reported in EPA TSCA Inventory.

OSHA PEL: TWA 1000 ppm
ACGIH TLV: TWA 1000 ppm
DFG MAK: 1000 ppm (6000 mg/m³)
DOT Classification: Nonflammable Gas; Label: Nonflammable Gas

SAFETY PROFILE: This material is chemically inert in the pure state and is considered to be physiologically inert as well. However, as it is ordinarily obtainable, it can contain variable quantities of the low sulfur fluorides. Some of these are toxic, very reactive chemically and corrosive in nature. These materials can hydrolyze on contact with water to yield hydrogen fluoride, which is highly toxic and very corrosive. In high concentrations and when pure it may act as a simple asphyxiant. Incompatible with disilane. Vigorous reaction with disilane. May explode. When heated to decomposition emits highly toxic fumes of F^- and SO_x.

SOl500   CAS: 7664-93-9   HR: 3
SULFURIC ACID

DOT: UN 1830/UN 1832
mf: H_2O_4S mw: 98.08

PROP: Colorless oily liquid; odorless. Mp: 10.49°, d: 1.834, vap press: 1 mm @ 145.8°, bp: 290°, decomp @ 340°. Misc with water and alc (liberating great heat).

SYNS: ACIDE SULFURIQUE (FRENCH) * ACIDO SOLFORICO (ITALIAN) * BOV * DIPPING ACID * HYDROOT * MATTING ACID (DOT) * NORDHAUSEN ACID (DOT) * OIL of VITRIOL (DOT) * SCHWEFELSAEURELOESUNGEN (GERMAN) * SPENT SULFURIC ACID (DOT) * SULPHURIC ACID * VITRIOL BROWN OIL * VITRIOL, OIL of (DOT) * ZWAVELZUUROPLOSSINGEN (DUTCH)

CONSENSUS REPORTS: Reported in EPA TSCA Inventory.

OSHA PEL: TWA 1 mg/m³
ACGIH TLV: TWA 1 mg/m³; STEL 3 ppm
DFG MAK: 1 mg/m³
NIOSH REL: (Sulfuric Acid) TWA 1 mg/m³
DOT Classification: Corrosive Material; Label: Corrosive

SAFETY PROFILE: Human poison by unspecified route. Experimental poison by inhalation. Moderately toxic by ingestion. A severe eye irritant. Extremely irritating, corrosive, and toxic to tissue resulting in rapid destruction of tissue, causing severe burns. If much of the skin is involved, it is accompanied by shock, collapse and symptoms similar to those seen in severe burns. Repeated contact with dilute solutions can cause a dermatitis, and repeated or prolonged inhalation of a mist of sulfuric acid can cause inflammation of the upper respiratory tract leading to chronic bronchitis. Sensitivity to sulfuric acid or mists or vapors varies with individuals. Normally 0.125-0.50 ppm may be mildly annoying and 1.5-2.5 ppm can be definitely unpleasant, 10-20 ppm is unbearable. Workers exposed to low concentrations of the vapor gradually lose their sensitivity to its irritating action. Inhalation of concentrated vapor or mists from hot acid or oleum can cause rapid loss of consciousness with serious damage to lung tissue. Severe exposure may cause a chemical pneumonitis, erosion of the teeth due to exposure to strong acid fumes has been recognized in industry.

This is a very powerful, acidic oxidizer

which can ignite or explode on contact with many materials, i.e., acetic acid, acetone cyanhydrin, (acetone + HNO_3), (acetone + $K_2Cr_2O_7$), acetonitrile, acrolein, acrylonitrile, (acrylonitrile + H_2O), (alcohols + H_2O_2), allyl alcohol, allyl chloride, NH_4OH, 2-amino ethanol, NH_4, triperchromate, aniline, (bromates + metals), BrF_5, n-butyraldehyde, carbides, $CoHC_2$, chlorates, (metals + chlorates), ClF_3, chlorosulfonic acid, Cu_3N, diisobutylene, (dimethyl benzylcarbinol + H_2O_2), epichlorohydrin, ethylene cyanhydrin, ethylene diamine, ethylene glycol, ethylene imine, fulminates, HCl, H_2, IF_7, (indene + HNO_3), Fe, isoprene, Li_6Si_2, Hg_3N_2, mesityl oxide, metals, (HNO_3 + glycerides), p-nitrotoluene, perchlorates, $HClO_4$, (C_6H_6 + permanganates), pentasilver trihydroxydiamino phosphate, (1-phenyl-2-methyl propyl alcohol + H_2O_2), P, $P(OCN)_3$, picrates, potassium-tert-butoxide, $KClO_3$, $KMnO_4$, ($KMnO_4$ + KCl), ($KMnO_4$ + H_2O), β-propiolactone, $RbHC_2$, propylene oxide, pyridine, Na, Na_2CO_3, NaOH, steel, styrene monomer, water, vinyl acetate, (HNO_3 + toluene). When heated it emits highly toxic fumes; will react with water or steam to produce heat; can react with oxidizing or reducing materials. When heated to decomposition it emits toxic fumes of SO_x.

SOI520　　CAS: 8014-95-7　　HR: 3
SULFURIC ACID, fuming

DOT: NA 1831
mf: $H_2O_4S \cdot O_3S$　　mw: 178.14

PROP: Heavy, fuming, yellow liquid. H_2SO_4 + up to 80% SO_3. A solution of sulfuric anhydride (sulfur trioxide) in anhydrous sulfuric acid (NTIS** PB233-098).

SYNS: DISULPHURIC ACID * DITHIONIC ACID * FUMING SULFURIC ACID (DOT) * OLEUM (DOT) * PYROSULPHURIC ACID

NIOSH REL: TWA 1 mg/m³
DOT Classification: Corrosive Material; Label: Corrosive and Corrosive, poison

SAFETY PROFILE: A poison. Moderately toxic by inhalation. A corrosive irritant to skin, eyes, and mucous membranes. A very dangerous fire hazard by chemical reaction with reducing agents and carbohydrates. A severe explosion hazard by chemical reaction with acetic acid, acetic anhydride, acetonitrile, acrolein, acrylic

acid, acrylonitrile, allyl alcohol, allyl chloride, 2-amino ethanol, NH_4OH, aniline, cresol, n-butyraldehyde, cumene, dichloroethyl ether, diethylene glycol monomethyl ether, diisobutylene, epichlorohydrin, ethyl acetate, ethylene cyanohydrin, ethylene diamine, ethylene glycol, ethylene glycol monoethyl ether acetate, ethylene imine, glyoxal, HCl, HF, isoprene, isopropyl alcohol, mesityl oxide, methyl ethyl ketone, HNO_3, 2-nitropropane, β-propiolacetone, propylene oxide, pyridine, NaOH, styrene monomer, vinylidene chloride, sulfolane, vinyl acetate. Will react with water or steam to produce heat and toxic and corrosive fumes. Can react vigorously with reducing materials. When heated to decomposition it emits highly toxic fumes of SO_x.

SOI530　　CAS: 7664-93-9　　HR: 3
SULFURIC ACID (mist)
mf: H_2O_4S　　mw: 98.08

PROP: The airborne form of sulfuric acid is an aerosol of droplets of varying diameter of aq sulfuric acid solution.

CONSENSUS REPORTS: EPA Extremely Hazardous Substances List. Reported in EPA TSCA Inventory.

ACGIH TLV: TWA 1 mg/m³
NIOSH REL: TWA 1 mg/m³

SAFETY PROFILE: Poison by inhalation. Human systemic effects by inhalation: mouth effects. When heated to decomposition it emits toxic fumes of SO_x.

SON510　　CAS: 10025-67-9　　HR: 3
SULFUR MONOCHLORIDE

DOT: UN 1828
mf: Cl_2S_2　　mw: 135.02

PROP: Amber to yellowish-red, oily, fuming liquid; penetrating odor. mp: $-80°$, bp: $138.0°$, flash p: 245°F (CC), d: 1.6885 @ 15.5°/15.5°, autoign temp: 453°F, vap press: 10 mm @ 27.5°, vap d: 4.66. Decomp in water.

SYNS: CHLORIDE of SULFUR (DOT) * DISULFUR DICHLORIDE * SIARKI CHLOREK (POLISH) * SULFUR CHLORIDE * SULFUR CHLORIDE(DI) (DOT) * SULFUR SUBCHLORIDE * THIOSULFUROUS DICHLORIDE

CONSENSUS REPORTS: Reported in EPA TSCA Inventory.

OSHA PEL: (Transitional: TWA 1 ppm) CL 1 ppm

ACGIH TLV: CL 1 ppm

DFG MAK: 1 ppm (6 mg/m^3)

DOT Classification: Corrosive Material; Label: Corrosive

SAFETY PROFILE: Poison by ingestion and inhalation. A fuming, corrosive liquid very irritating to skin, eyes, and mucous membranes. It decomposes on contact with water to form the highly irritating hydrogen chloride, thiosulfuric acid, and sulfur. Its toxic effects are irritating to the upper respiratory tract, although the results of intoxication are usually transitory in nature. However, if hydrolysis is not complete in the upper respiratory tract, injury to the bronchioles and alveoli can result. A fire hazard when in contact with organic matter, P_2O_3, Na_2O_2, water, $Cr(OCl)_2$. Combustible when exposed to heat or flame. Will react with water or steam to produce heat and toxic and corrosive fumes. Can react with oxidizing materials. To fight fire, use CO_2, dry chemical. When heated to decomposition it emits highly toxic fumes of Cl^- and SO_x.

SOO500   CAS: 7782-99-2   HR: 3
SULFUROUS ACID

DOT: UN 1833

mf: H_2SO_3 mw: 82.08

PROP: Colorless liquid; suffocating sulfur odor (in solution only). D: approx 1.03.

SYNS: SULFUR DIOXIDE, solution   *   SCHWEFLIGE SAURE (GERMAN)

CONSENSUS REPORTS: Reported in EPA TSCA Inventory.

DOT Classification: Corrosive Material; Label: Corrosive

SAFETY PROFILE: A poison by ingestion and inhalation. A corrosive irritant to skin, eyes, and mucous membranes. Human systemic effects by ingestion: nausea or vomiting, hypermotility, diarrhea, and other gastrointestinal effects. When heated to decomposition it emits highly toxic fumes of SO_x.

SOP000   CAS: 2312-35-8   HR: 3
SULFUROUS ACID, 2-(p-tert-BUTYLPHENOXY)CYCLOHEXYL-2-PROPYNYL ESTER

DOT: NA 2765

mf: $C_{19}H_{26}O_4S$ mw: 350.51

SYNS: BPPS   *   2-(p-tert-BUTYLPHENOXY)CYCLOHEXYL PROPARGYL SULFITE   *   2-(p-tert-BUTYLPHENOXY)CYCLOHEXYL 2-PROPYNYL SULFITE   *   COMITE   *   2-(4-(1,1-DIMETHYLETHYL)PHENOXY)CYCLOHEXYL 2-PROPYNYL ESTER, SULFUROUS ACID   *   2-(4-(1,1-DIMETHYLETHYL)PHENOXY)CYCLOHEXYL 2-PROPYNYL SULFITE   *   DO 14   *   ENT 27,226   *   NAUGATUCK D-014   *   OMAIT   *   OMITE   *   PROPARGITE (DOT)   *   UNIROYAL D014   *   U.S. RUBBER D-014

DOT Classification: ORM-E; Label: None

SAFETY PROFILE: Poison by skin contact. Moderately toxic by ingestion and possibly other routes. When heated to decomposition it emits toxic fumes of SO_x.

SOP500   CAS: 140-57-8   HR: 3
SULFUROUS ACID, 2-(p-tert-BUTYLPHENOXY)-1-METHYLETHYL-2-CHLOROETHYL ESTER

mf: $C_{15}H_{23}ClO_4S$ mw: 334.89

PROP: Liquid. D: 1.145-1.1620, mp: −31.7°, bp: 175° @ 0.1 mm, vap press: <10 mm @ 25°. Misc with many organic solvents; insol in water.

SYNS: ACARACIDE   *   ARACIDE   *   ARAMITE   *   ARAMITEARARAMITE-15W   *   ARATRON   *   BUTYLPHENOXYISOPROPYL CHLOROETHYL SULFITE   *   2-(p-BUTYLPHENOXY)ISOPROPYL 2-CHLOROETHYL SULFITE   *   2-(4-tert-BUTYLPHENOXY)ISOPROPYL-2-CHLOROETHYL SULFITE   *   2-(p-tert-BUTYLPHENOXY)ISOPROPYL 2'-CHLOROETHYL SULPHITE   *   2-(p-tert-BUTYLPHENOXY)-1-METHYLETHYL 2-CHLOROETHYL ESTER of SULPHUROUS ACID   *   2-(p-BUTYLPHENOXY)-1-METHYLETHYL 2-CHLOROETHYL SULFITE   *   2-(p-tert-BUTYLPHENOXY)-1-METHYLETHYL-2-CHLOROETHYL SULFITE ESTER   *   2-(p-tert-BUTYLPHENOXY)-1-METHYLETHYL 2'-CHLOROETHYL SULPHITE   *   2-(p-tert-BUTYLPHENOXY)-1-METHYLETHYL SULPHITE of 2-CHLOROETHANOL   *   1-(p-tert-BUTYLPHENOXY)-2-PROPANOL-2-CHLOROETHYL SULFITE   *   CES   *   2-CHLOROETHANOL-2-(p-tert-BUTYLPHENOXY)-1-METHYLETHYL SULFITE   *   2-CHLOROETHANOL ESTER with 2-(p-tert-BUTYLPHENOXY)-1-METHYLETHYL SULFITE   *   β-CHLOROETHYL-β'-(p-tert-BUTYLPHENOXY)-α'- METHYLETHYL SULFITE   *   β-CHLOROETHYL-β-(p-tert-BUTYLPHENOXY)-α-METHYLETHYL SULPHITE   *   2-CHLOROETHYL 1-METHYL-2-(p-tert-BUTYLPHENOXY)ETHYL SULPHATE   *   2-CHLOROETHYL SULFUROUS ACID-2-(4-(1,1-DIMETHYLETHYL)PHENOXY)-1-METHYLETHYL ESTER   *   2-CHLOROETHYL SULPHITE of 1-(p-tert-BUTYLPHENOXY)-2-PROPANOL

* COMPOUND 88R * ENT 16,519 * NIAGARAMITE
* ORTHO-MITE * 88-R

CONSENSUS REPORTS: IARC Cancer Review: GROUP 2B IMEMDT 7,56,87; Animal Sufficient Evidence IMEMDT 5,39,74. NTP Fourth Annual Report On Carcinogens, 1984.

SAFETY PROFILE: Confirmed carcinogen with experimental carcinogenic, neoplastigenic, and tumorigenic data. Experimental poison by intraperitoneal route. Moderately toxic to humans by ingestion. Moderately toxic experimentally by ingestion. Experimental reproductive effects. A pesticide. When heated to decomposition it emits toxic fumes of Cl^- and SO_x.

SOQ450   CAS: 5714-22-7   HR: 3
SULFUR PENTAFLUORIDE
mf: $F_{10}S_2$ mw: 254.12

SYNS: SULFUR DECAFLUORIDE * TL 70

OSHA PEL: (Transitional: TWA 0.025 ppm) CL 0.01 ppm
ACGIH TLV: CL 0.01 ppm
DFG MAK: 0.025 ppm (0.25 mg/m^3)

SAFETY PROFILE: Poison by intravenous route. Moderately toxic by inhalation. When heated to decomposition it emits very toxic fumes of F^- and SO_x.

SOR000   CAS: 7783-60-0   HR: 3
SULFUR TETRAFLUORIDE
DOT: UN 2418
mf: F_4S mw: 108.06

PROP: Gas. Bp: $-40°$, mp: $-124°$.

SYN: TETRAFLUOROSULFURANE

CONSENSUS REPORTS: Reported in EPA TSCA Inventory. EPA Extremely Hazardous Substances List.

OSHA PEL: CL 0.1 ppm
ACGIH TLV: CL 0.1 ppm
NIOSH REL: (Inorganic Fluorides) TWA 2.5 mg(F)/m^3
DOT Classification: Poison A; Label: Poison Gas

SAFETY PROFILE: Poison by inhalation. A powerful irritant. Will react with water, steam or acids to yield toxic and corrosive fumes. Incompatible with dioxygen difluoride. When heated to decomposition it emits very toxic fumes of F^- and SO_x.

SOR500   CAS: 7446-11-9   HR: 3
SULFUR TRIOXIDE
DOT: UN 1829
mf: O_3S mw: 80.06

PROP: It exists in three forms; the most valuable commercially is the γ form (mp: 16.8°, bp: 44.8°) which has a strong tendency to polymerize to the straight chain β form (mp β: 32.5°) and subsequently to the cross-linked α form (mp α: 62°). When the β or α forms are melted they tend to revert to the γ form liquid or ice-like crystals. SO_3 (β) asbestos-like crystals; vap press: (β) 433 mm @ 250°, vap press (α): 344 mm, vap d: 2.76.

SYNS: SULFAN * SULFURIC ANHYDRIDE (DOT) * SULFURIC OXIDE * SULFUR TRIOXIDE, STABILIZED (DOT)

CONSENSUS REPORTS: EPA Extremely Hazardous Substances List. Reported in EPA TSCA Inventory.

DOT Classification: Corrosive Material; Label: Corrosive

SAFETY PROFILE: Poison by inhalation. Human systemic effects by inhalation: cough and other pulmonary and olfactory changes. A corrosive irritant to skin, eyes, and mucous membranes. Violent reaction with O_2F_2, PbO, $NClO_2$, $HClO_4$, P, tetrafluorethylene, acetonitrile, sulfuric acid, dimethyl sulfoxide, dioxan, water, diphenylmercury, formamide, iodine, pyridine, metal oxides. Reacts with steam to form corrosive, toxic fumes of sulfuric acid. When heated to decomposition it emits toxic fumes of SO_x.

SOT000   CAS: 7791-25-5   HR: 3
SULFURYL CHLORIDE
DOT: UN 1834
mf: Cl_2O_2S mw: 134.96

PROP: Colorless liquid; pungent odor. Mp: $-54.1°$, bp: 69.1°, d: 1.6674, vap press: 100 mm @ 17.8°, vap d: 4.65.

SYNS: SULFONYL CHLORIDE * SULFURIC OXYCHLORIDE

CONSENSUS REPORTS: Reported in EPA TSCA Inventory.

DOT Classification: Corrosive Material; Label: Corrosive

SAFETY PROFILE: A corrosive irritant to skin, eyes, and mucous membranes. Questionable carcinogen with experimental tumorigenic data. Can explode with PbO_2. Will react with water or steam to produce heat and toxic and corrosive fumes. Incompatible with alkalies, diethyl ether, dimethyl sulfoxide, dinitrogen pentaoxide, lead dioxide, phosphorus. When heated to decomposition it emits highly toxic fumes of Cl^- and SO_x.

SOU500   CAS: 2699-79-8   **HR: 3**
SULFURYL FLUORIDE

DOT: UN 2191
mf: F_2O_2S mw: 102.06

PROP: Colorless gas. Mp: $-137°$, bp: $-55°$, d: 3.72 g/L.

SYNS: FLUORURE de SULFURYLE (FRENCH) * SULFURIC OXYFLUORIDE * VIKANE * VIKANE FUMIGANT

CONSENSUS REPORTS: Reported in EPA TSCA Inventory.

OSHA PEL: (Transitional: TWA 5 ppm) TWA 5 ppm; STEL 10 ppm
ACGIH TLV: TWA 5 ppm; STEL 10 ppm
DOT Classification: Nonflammable Gas; Label: Nonflammable Gas

SAFETY PROFILE: Poison by ingestion. Mildly toxic by inhalation. Accidental human exposure caused nausea, vomiting, cramps, itching. May be narcotic in high concentration. Can react with water, steam. When heated to decomposition it emits very toxic fumes of F^- and SO_x.

SOU625   CAS: 35400-43-2   **HR: 3**
SULPROFOS
mf: $C_{12}H_{19}O_2PS_3$ mw: 322.46

SYNS: BAY-NTN-9306 * BOLSTAR * O-ETHYL-O-(4-METHYLMERCAPTO)PHENYL)-S-N-PROPYLPHOS-PHOROTHIONOTHIOLATE * O-ETHYL-O-(4-(METHYL-THIO)PHENYL)PHOSPHORODITHIOIC ACID-S-PROPYL ESTER * O-ETHYL-O-(4-(METHYLTHIO)PHENYL) S-PROPYL PHOSPHORODITHIOATE * HELOTHION

OSHA PEL: TWA 1 mg/m^3
ACGIH TLV: TWA 1 mg/m^3

SAFETY PROFILE: Poison by ingestion. Moderately toxic by skin contact. When heated to decomposition it emits very toxic fumes of PO_x and SO_x.

SOW000   CAS: 50-02-2   **HR: 3**
SUPERPREDNOL
mf: $C_{22}H_{29}FO_5$ mw: 392.51

SYNS: AEROSEB-DEX * AZIUM * CORSONE * DECADERM * DECADRON * DECASONE * DECASPRAY * DECTANCYL * 1-DEHYDRO-16-α-METHYL-9-α-FLUOROHYDROCORTISONE * DELTA-FLUORENE * DERGRAMIN * DERONIL * DESA-DRENE * DESAMETASONE * DEXA * DEXA-CORT * DEXA-CORTIDELT * DEXADELTONE * DEXAMETH * DEXAMETHASONE ALCOHOL * DEXONE * DEXTELAN * DEZONE * DXMS * Δ^1-9-α-FLUORO-16-α-METHYLCORTISOL * 9-α-FLUORO-16-α-METHYLPREDNISOLONE * 9-α-FLUORO-16-α-METHYL-1,4-PREGNADIENE-11-β,17-α,21-TRIOL-3,20-DIONE * 4-α-FLUORO-16-α-METHYL-11-β,17,21-TRIHYDROXYPREGNA-1,4-DIENE-3,20-DIONE * 9-FLUORO-11-β,17,21-TRIHYDROXY-16-α-METHYL-PREGNA-1,4-DIENE-3,20-DIONE * 9-α-FLUORO-11-β,17-α,21-TRIHYDROXY-16-α-METHYLPREGNA-1,4-DIENE-3,20-DIONE * FORTECORTIN * GAMMACORTEN * HEXADECADROL * HEXADROL * MAXIDEX * 16-α-METHYL-9-α-FLUORO-1-DEHYDROCORTISOL * 16-α-METHYL-9-α-FLUORO-Δ^1-HYDROCORTISONE * 16-α-METHYL-9-α-FLUOROPREDNISOLONE * 16-α-METHYL-9-α-FLUORO-1,4-PREGNADIENE-11-β,17-α,21-TRIOL-3,20-DIONE * 16-α-METHYL-9-α-FLUORO-11-β,17-α,21-TRIHYDROXYPREGNA-1,4-DIENE-3,20-DIONE * MEXIDEX * MILLICORTEN * MK 125 * ORADEXON * SK-DEXAMETHASONE

CONSENSUS REPORTS: Reported in EPA TSCA Inventory.

SAFETY PROFILE: Poison by intraperitoneal and subcutaneous routes. Experimental teratogenic and reproductive effects. Mutation data reported. When heated to decomposition it emits toxic fumes of F^-.

SOX500   CAS: 337-47-3   **HR: 3**
SURITAL SODIUM
mf: $C_{12}H_{17}N_2O_2S \cdot Na$ mw: 276.36

SYNS: 5-ALLYL-5-(1-METHYLBUTYL)-2-THIOBARBITU-RATE SODIUM * 5-ALLYL-5-(1-METHYLBUTYL)-2-THIO-BARBITURIC ACID SODIUM SALT * BURITAL SODIUM * SODIUM 5-ALLYL-5-(1-METHYLBUTYL)-2-THIOBARBITURATE * SODIUM THIAMYLAL * SURITAL * SURITAL SODIUM derivative * SURITAL SODIUM SALT * THIAMYLAL SODIUM * THIOMYLAL SODIUM

SAFETY PROFILE: Poison by ingestion, subcutaneous, intravenous, and intraperitoneal routes. Experimental reproductive effects. When heated to decomposition it emits very toxic fumes of NO_x, SO_x, and Na_2O.

SOY000 CAS: 122-10-1 **HR: 3**
SWAT
mf: $C_9H_{15}O_8P$ mw: 282.21

SYNS: BOMYL * DIMETHYL 1,3-BIS(CARBOME-THOXY)-1-PROPEN-2-YL PHOSPHATE * DIMETHYL-1,3-DI(CARBOMETHOXY)-1-PROPEN-2-YL PHOSPHATE * DIMETHYL 3-(DIMETHOXYPHOSPHINYLOXY) GLUTACONATE * DIMETHYL 3-HYDROXYGLUTA-CONATE DIMETHYL PHOSPHATE * ENT 24,833 * FLY BAIT GRITS * GC 3707 * GENERAL CHEMICALS 3707 * 3-HYDROXYGLUTACONIC ACID, DIMETHYL ESTER, DIMETHYL PHOSPHATE * 3-HYDROXY-2-PENTANEDIOIC ACID, DIMETHYL ESTER, DIMETHYL PHOSPHATE * PHOSPHORIC ACID, DIMETHYL ESTER, ESTER with DIMETHYL 3-HYDROXYGLUTACONATE

SAFETY PROFILE: Poison by ingestion and skin contact. Used as an insecticide. When heated to decomposition it emits very toxic fumes of PO_x.

SPA000 CAS: 3441-64-3 **HR: 3**
SYDNOPHEN HYDROCHLORIDE
mf: $C_{11}H_{13}N_3O \cdot ClH$ mw: 239.73

SYNS: 3-(α-METHYLPHENETHYL)SYDONE IMINE MONOHYDROCHLORIDE * 3-(β-PHENYLISOPROPYL)-SIDNONIMINE HYDROCHLORIDE * SIDNOFEN * SYDNONE IMINE, 3-(1-METHYL-2-PHENYLETHYL)-, MONOHYDROCHLORIDE * SYDNOPHENE

SAFETY PROFILE: Poison by intraperitoneal, intravenous, and subcutaneous routes. When heated to decomposition it emits very toxic fumes of NO_x and HCl.

SPC500 CAS: 61-76-7 **HR: 3**
m-SYNEPHRINE HYDROCHLORIDE
mf: $C_9H_{13}NO_2 \cdot ClH$ mw: 203.6

SYNS: (−)-m-HYDROXY-α-((METHYLAMINO)METHYL) BENZYLALCOHOL HYDROCHLORIDE * 1-m-HY-DROXY-α-(METHYLAMINOMETHYL)BENZYL ALCOHOL HYDROCHLORIDE * 1-1-(m-HYDROXYPHENYL)-2-METHYL-AMINOETHANOLHYDROCHLORIDE * m-METHYLAMINOETHANOLPHENOL HYDROCHLO-RIDE * d-(−)-PHENYLEPHRINE HYDROCHLORIDE

CONSENSUS REPORTS: Reported in EPA TSCA Inventory.

SAFETY PROFILE: Poison by ingestion, intra-peritoneal, subcutaneous, intravenous, and in-tramuscular routes. Mutation data reported. When heated to decomposition it emits very toxic fumes of HCl and NO_x.

SPF000 CAS: 4891-54-7 **HR: 3**
SYSTOX SULFONE
mf: $C_8H_{19}O_5PS_2$ mw: 290.36

SYNS: O,O-DIETHYL-2-ETHYLMERCAPTOETHYL THIO-PHOSPHATE, THIONO ISOMER * O,O-DIETHYL-O-(2-ETHYLSULFONYLETHYL)PHOSPHOROTHIOATE * DIETHYL-2-ETHYLSULFONYLETHYL THIONOPHOS-PHATE * THIONODEMETON SULFONE

SAFETY PROFILE: Poison by ingestion and possibly other routes. When heated to decompo-sition it emits very toxic fumes of PO_x and SO_x.

T

TAA100   CAS: 93-76-5   **HR: 3**
2,4,5-T

DOT: UN 2765
mf: $C_8H_5Cl_3O_3$ mw: 255.48

PROP: Crystals; light tan solid. Mp: 151-153°. Usually has 2,3,7,8-TCDD as a minor component.

SYNS: ACIDE 2,4,5-TRICHLORO PHENOXYACETIQUE (FRENCH) * ACIDO (2,4,5-TRICLORO-FENOSSI)-ACET-ICO (ITALIAN) * AMINE 2,4,5-T FOR RICE * BCF-BUSHKILLER * BRUSH-OFF 445 MLD VOLATILE BRUSH KILLER * BRUSH RHAP * BRUSHTOX * DACAMINE * DEBROUSSAILLANT CONCENTRE * DECAMINE 4T * DED-WEED BRUSH KILLER * DED-WEED LV-6 BRUSH KIL and T-5 BRUSH KIL * DINOXOL * ENVERT-T * ESTERCIDE T-2 and T-245 * ESTERON 245 BE * ESTERON BRUSH KILLER * FARMCO FENCE RIDER * FORRON * FORST U 46 * FORTEX * FRUITONE A * INVERTON 245 * LINE RIDER * PHORTOX * RCRA WASTE NUMBER U232 * REDDON * REDDOX * SPONTOX * SUPER D WEEDONE * TIPPON * TORMONA * TRANSAMINE * TRIBUTON * (2,4,5-TRI-CHLOOR-FENOXY)-AZIJNZUUR (DUTCH) * 2,4,5-TRI-CHLOROPHENOXYACETIC ACID * (2,4,5-TRICHLOR-PHENOXY)-ESSIGSAEURE (GERMAN) * TRINOXOL * TRIOXON * TRIOXONE * U 46 * VEON 245 * VERTON 2T * VISKO RHAP LOW VOLATILE ESTER * WEEDAR * WEEDONE

CONSENSUS REPORTS: IARC Cancer Review: GROUP 2B IMEMDT 7,156,87; Animal Inadequate Evidence IMEMDT 15,273,77; Human Inadequate Evidence IMEMDT 15,273,77; Human Limited Evidence IMEMDT 41,357,86. Reported in EPA TSCA Inventory. EPA Genetic Toxicology Program.

OSHA PEL: TWA 10 mg/m^3
ACGIH TLV: TWA 10 mg/m^3
DFG MAK: 10 mg/m^3
DOT Classification: ORM-A; Label: None.

SAFETY PROFILE: Suspected carcinogen with experimental neoplastigenic and tumorigenic data. Poison by ingestion. Moderately toxic by unspecified route. Experimental teratogenic and reproductive effects. Mutation data reported. A highly toxic chlorinated phenoxy acid herbicide which is rapidly excreted after ingestion. Readily absorbed by inhalation and ingestion routes, slowly by skin contact. Signs of intoxication include weakness, lethargy, anorexia, diarrhea, ventricular fibrillation and/or cardiac arrest and death. The teratogenicity is due in part to 2,3,7,8-TCDD, which is present as a contaminant. When heated to decomposition it emits toxic fumes of Cl⁻.

TAB750   CAS: 14807-96-6   **HR: 3**
TALC
mf: $H_2O_3Si \cdot 3/4Mg$ mw: 96.33

PROP: White to grayish-white, fine powder; odorless and tasteless. Powdered native hydrous magnesium silicate. Insol in water, cold acids, or alkalies. Containing less than 1% crystalline silica.

SYNS: AGALITE * AGI TALC, BC 1615 * ALPINE TALC USP, BC 127 * ALPINE TALC USP, BC 141 * ALPINE TALC USP, BC 662 * ASBESTINE * C.I. 77718 * DESERTALC 57 * EMTAL 596 * FIBRENE C 400 * LO MICRON TALC 1 * LO MICRON TALC, BC 1621 * LO MICRON TALC USP, BC 2755 * METRO TALC 4604 * METRO TALC 4608 * METRO TALC 4609 * MISTRON FROST P * MISTRON RCS * MISTRON 2SC * MISTRON STAR * MISTRON SUPER FROST * MISTRON VAPOR * MP 12-50 * MP 25-38 * MP 45-26 * NCI-C06008 * NO. 907 METRO TALC * NYTAL * OOS * OXO * PURTALC USP * SIERRA C-400 * SNOWGOOSE * STEAWHITE * SUPREME DENSE * TALCUM

CONSENSUS REPORTS: IARC Cancer Review: Animal Inadequate Evidence IMEMDT 42,185,87; Human Inadequate Evidence IMEMDT 42,185,87; GROUP 3 IMEMDT 7,349,87. Reported in EPA TSCA Inventory.

OSHA PEL: (Transitional: TWA 20 mppcf (containing no asbestos fibers)) TWA 2 mg/m^3
ACGIH TLV: TWA 2 mg/m^3, respirable dust (use asbestos TLV if asbestos fibers are present)
DFG MAK: 2 mg/m^3

SAFETY PROFILE: The talc with less than 1 percent asbestos is regarded as a nuisance dust. A human skin irritant. Prolonged or repeated exposure can produce a form of pulmonary fibrosis (talc pneumoconiosis) which may be due to asbestos content. Questionable carcinogen with experimental tumorigenic data. A common air contaminant.

TAB775 CAS: 14807-96-6 **HR: 3**
TALC, containing asbestos fibers
mf: $H_2O_2Si \cdot 3/4Mg$ mw: 96.33

CONSENSUS REPORTS: IARC Cancer Review: GROUP 1 IMEMDT 7,349,87; Human Sufficient Evidence IMEMDT 42,185,87; Reported in EPA TSCA Inventory.

ACGIH TLV: Human Carcinogen; TWA > 2 mg/m^3, Respirable Dust

SAFETY PROFILE: Confirmed human carcinogen with experimental tumorigenic data.

TAC000 CAS: 8002-26-4 **HR: 1**
TALL OIL

PROP: Composition: Rosin acids, oleic and linoleic acids. Dark brown liquid; acrid odor. D: 0.95, flash p: 360°F.

SYNS: LIQUID ROSIN * TALLOL

SAFETY PROFILE: A mild allergen. A substance which migrates to food from packaging materials. Combustible when exposed to heat or flame; can react with oxidizing materials. To fight fire, use dry chemical, CO_2. When heated to decomposition it emits acrid smoke and irritating fumes.

TAD500 CAS: 8008-31-9 **HR: 1**
TANGERINE OIL

PROP: Expressed from the peels of Dancy and related varieties of *Citrus reticulata Blanco*. The components include d-limonene, n-octylaldehyde, n-decylaldehyde, citral, linalool, citronella, cadinene, terpenes, aldehydes, alcohols, and esters (FCTXAV 16, 637,78). Red orange to brown orange liquid; orange-like odor. Sol in fixed oils, mineral oil; sltly sol in propylene glycol; insol in glycerin.

SYNS: TANGERINE OIL, COLDPRESSED (FCC) * TANGERINE OIL, EXPRESSESED (FCC)

CONSENSUS REPORTS: Reported in EPA TSCA Inventory.

SAFETY PROFILE: A skin irritant. When heated to decomposition it emits acrid smoke and irritating fumes.

TAD750 CAS: 1401-55-4 **HR: 3**
TANNIC ACID
mf: $C_{76}H_{52}O_{46}$ mw: 1701.28

PROP: From the nutgalls of *Quercus infectoria Oliver* or seed pods of *Caesalpinia spinosa* or the nutgalls of various sumac species. Yellowish-white or brown, bulky powder or flakes; odorless with astringent taste. Mp: 200°, flash p: 390°F (OC), autoign temp: 980°F. Very sol in water, alc, acetone; almost insol in benzene, chloroform, ether, petr ether, carbon disulfide.

SYNS: D'ACIDE TANNIQUE (FRENCH) * GALLO-TANNIC ACID * GALLOTANNIN * GLYCERITE * TANNIN

CONSENSUS REPORTS: IARC Cancer Review: GROUP 3 IMEMDT 7,56,87; Animal Sufficient Evidence IMEMDT 10,253,76. Reported in EPA TSCA Inventory. EPA Genetic Toxicology Program.

SAFETY PROFILE: Poison by ingestion, intramuscular, intravenous, and subcutaneous routes. Moderately toxic by parenteral route. Experimental reproductive effects. Questionable carcinogen with experimental carcinogenic and tumorigenic data. Mutation data reported. Combustible when exposed to heat or flame. To fight fire, use water. Incompatible with salts of heavy metals; oxidizing materials. When heated to decomposition it emits acrid smoke and irritating fumes.

TAE750 CAS: 7440-25-7 **HR: 3**
TANTALUM
af: Ta aw: 180.948

PROP: Gray, very hard, malleable, ductile metal. Mp: 2996°; bp: 5429°; d: 16.69. Insol in water.

SYN: TANTALUM-181

CONSENSUS REPORTS: Reported in EPA TSCA Inventory.

OSHA PEL: TWA 5 mg/m^3
ACGIH TLV: TWA 5 mg/m^3
DFG MAK: 5 mg/m^3

SAFETY PROFILE: Some industrial skin injuries from tantalum have been reported. Systemic industrial poisoning however, is apparently unknown. Questionable carcinogen with experimental tumorigenic data. The dry powder ignites spontaneously in air. Incompatible with bromine trifluoride, fluorine, lead chromate.

TAF000 CAS: 7721-01-9 **HR: 3**
TANTALUM CHLORIDE
mf: Cl₅Ta mw: 358.20

SYN: TANTALUM PENTACHLORIDE

CONSENSUS REPORTS: Reported in EPA TSCA Inventory.

SAFETY PROFILE: Poison by intraperitoneal route. Moderately toxic by ingestion. When heated to decomposition it emits toxic fumes of Cl⁻.

TAF250 CAS: 7783-71-3 **HR: 3**
TANTALUM FLUORIDE
mf: F₅Ta mw: 275.95

PROP: Deliquescent, refractive prisms. D: 4.74 @ 20°; mp: 96.8°; bp: 229.5°; vap press: 100 mm @ 130°. Sol in water, concentrated nitric acid; very sol in fuming nitric acid; sltly sol in hot carbon disulfide, hot carbon tetrachloride.

SYN: TANTALIUM PENTAFLUORIDE

CONSENSUS REPORTS: Reported in EPA TSCA Inventory.

NIOSH REL: TWA 2.5 mg(F)/m³

SAFETY PROFILE: Poison by intravenous route. When heated to decomposition it emits toxic fumes of F⁻.

TAF700 CAS: 8016-88-4 **HR: 2**
TARRAGON OIL

PROP: From steam distillation of leaves, stems, and flowers from *Artemesia dracunculus* L. Pale yellow to amber liquid; spicy licorice and sweet basil odor. Sol in fixed oils, mineral oil; insol in propylene glycol, glycerin.

SYN: ESTRAGON OIL

CONSENSUS REPORTS: Reported in EPA TSCA Inventory.

SAFETY PROFILE: Moderately toxic by ingestion. A skin irritant. When heated to decomposition it emits acrid smoke and irritating fumes.

TAF750 CAS: 87-69-4 **HR: 2**
TARTARIC ACID
mf: C₄H₆O₆ mw: 150.10

PROP: Colorless to translucent crystals or white powder; odorless with an acid taste. Sol in water, alc.

SYNS: 2,3-DIHYDROSUCCINIC ACID * 2,3-DIHY-DROXYBUTANEDIOC ACID

CONSENSUS REPORTS: Reported in EPA TSCA Inventory.

SAFETY PROFILE: Moderately toxic by intravenous route. Mildly toxic by ingestion. Reaction with silver produces the unstable silver tartrate. When heated to decomposition it emits acrid smoke and irritating fumes.

TAI000 CAS: 1746-01-6 **HR: 3**
TCDD
mf: C₁₂H₄Cl₄O₂ mw: 321.96

PROP: Colorless needles. Mp: 305°.

SYNS: 2,3,7,8-CZTEROCHLORODWUBENZO-p-DWUOK-SYNY (POLISH) * DIOKSYNY (POLISH) * DIOXINE * DIOXIN (herbicide contaminant) * NCI-C03714 * TCDBD * 2,3,7,8-TCDD * 2,3,7,8-TETRACHLO-RODIBENZO(b,e)(1,4)DIOXAN * 2,3,6,7-TETRACHLO-RODIBENZO-p-DIOXIN * 2,3,7,8-TETRACHLORODI-BENZO-p-DIOXIN * 2,3,7,8-TETRACHLORODIBENZO-1,4-DIOXIN * TETRADIOXIN

CONSENSUS REPORTS: IARC Cancer Review: GROUP 2B IMEMDT 7,350,87; Human Inadequate Evidence IMEMDT 15,41,77; Animal Inadequate Evidence IMEMDT 15,41,77. NTP Fourth Annual Report On Carcinogens, 1984. NTP Carcinogenesis Bioassay (gavage); Clear Evidence: mouse, rat NTPTR* NTP-TR-209,82; (dermal). EPA Genetic Toxicology Program.

DFG MAK: Animal Carcinogen, Suspected Human Carcinogen.
NIOSH REL: (Dioxin) Reduce to lowest feasible level.

SAFETY PROFILE: Confirmed carcinogen with experimental carcinogenic, neoplastigenic, and tumorigenic data. One of the most toxic synthetic chemicals. A deadly experimental poison by ingestion, skin contact, intraperitoneal, and possibly other routes. Human systemic effects by skin contact: allergic dermatitis. Experimental teratogenic and reproductive effects. Human mutation data reported. An eye irritant.

TCDD is the most toxic member of the 75 dioxins. It causes death in rats by hepatic cell necrosis. Death can follow a lethal dose by weeks. Acute and subacute exposure result in wasting, hepatic necrosis, thymic atrophy, hemorrhage, lymphoid depletion, chloracne. A by-product of the manufacture of polychlorinated phenols. It is found at low levels in 2,4, 5-T, 2,4,5-trichlorophenol and hexachlorophene.

It is also formed during various combustion processes. Incineration of chemical wastes, including chlorophenols, chlorinated benzenes, and biphenyl ethers, may result in the presence of TCDD in flue gases, fly ash, and soot particles. It is immobile in contaminated soil and may be retained for years. TCDD has the potential for bio-accumulation in animals. An accident in Seveso, Italy and inadvertent soil contamination in Missouri have resulted in abandonment of the contaminated areas. When heated to decomposition it emits toxic fumes of Cl^-.

TAI250   CAS: 9002-84-0   HR: 1
TEFLON
mf: $(C_2F_4)_n$

PROP: Grayish-white, tough plastic. Chemically very inert.

SYNS: AFLON * ALGLOFLON * ALGOFLON SV * ALKATHENE RXDG33 * AMIP 15m * BALFON 7000 * BDH 29-801 * CHROMOSORB T * DIXON 164 * DLX-6000 * DUROID 5870 * EK 1108GY-A * ETHICON PTFE * FLUO-KEM * FLUON * FLUOROFLEX * FLUOROLON 4 * FLUOROPAK 80 * FLUORPLAST 4 * FTORLON 4 * FTORO-PLAST 4 * GORE-TEX * HALON TFEG 180 * HEYDEFLON * HOSTAFLON * MOLYKOTE 522 * POLIFEN * POLITEF * POLY(ETHYLENE TETRAFLUORIDE) * POLYFENE * POLYFLON * POLYTEF * POLYTETRAFLUOROETHENE * POLYTETRAFLUOROETHYLENE * PTFE * SOREFLON 604 * TARFLEN * TEFLON (various) * TETRAFLUOROETHENE HOMOPOLYMER * TETRAFLUOROETHENE POLYMER * TETRAFLUOROETHYLENE HOMOPOLYMER * TETRAFLUOROETHYLENE POLYMERS * TETRAN PTFE * UNON P * VALFLON * VELFLON * ZITEX H 662-124

CONSENSUS REPORTS: IARC Cancer Review: GROUP 3 IMEMDT 7,56,87; Animal Sufficient Evidence IMEMDT 19,285,79; Human Inadequate Evidence IMEMDT 19,285,79. Reported in EPA TSCA Inventory.

SAFETY PROFILE: The finished polymerized compound is inert under ordinary conditions. Questionable carcinogen with experimental tumorigenic data by implant. There have been reports of "polymer fume fever" in humans exposed to the unfinished product dust or to pyrolysis products which also are irritants. Smoking should be prohibited in areas where this material is being fabricated or, in general, where there may be dust from it. Exposure to pyrolysis or decomposition products appear to be the chief health-related problem. Incompatible with fluorine, sodium potassium alloy. Under the proper conditions it undergoes hazardous reactions with boron, magnesium or titanium. When heated to above 750°F it decomposes to yield highly toxic fumes of F^-.

TAI500   CAS: 113-92-8   HR: 3
TELDRIN
mf: $C_{16}H_{19}ClN_2 \cdot C_4H_4O_4$ mw: 390.90

SYNS: ALLERCLOR * ALLERGIN * ALLERGISAN * ALUNEX * ANTAGONATE * CARBINOXAMIDE MALEATE * CHLORMENE * dl-2(-p-CHLORO-α-2-(DIMETHYLAMINO)ETHYLBENZYL)PYRIDINE BIMALEATE * 1-p-CHLOROPHENYL-1-(2-PYRIDYL)-3-DIMETHYLAMINOPROPANE MALEATE * CHLOROPROPHENYPYRIDAMINE MALEATE * CHLORPHENIRAMINE MALEATE * CHLOR-TRIMETON * CHLOR-TRIMETON MALEATE * CHLOR-TRIPOLON * CLOROPIRIL * C-METON * 1-(N,N-DIMETHYL-AMINO)-3-(p-CHLOROPHENYL-3-α-PYRIDYL)PROPANE MALEATE * HISTADUR * HISTADUR DURA-TABS * HISTALEN * HISTAPAN * IBIOTON * LORPHEN * M.P. CHLORCAPS T.D. * NCI-C55265 * NEORESTAMIN * PIRIEX * PIRITON * POLARONIL (GERMAN) * PYRIDAMAL-100 * SYNISTAMIN

CONSENSUS REPORTS: NTP Carcinogenesis Studies (gavage); No Evidence: mouse, rat NTPTR* NTP-TR-317,86. Reported in EPA TSCA Inventory.

SAFETY PROFILE: Poison by ingestion, intravenous, and subcutaneous routes. Experimental reproductive effects. Used as an antihistamine. When heated to decomposition it emits very toxic fumes of Cl^- and NO_x.

TAJ000   CAS: 13494-80-9   HR: 3
TELLURIUM
af: Te   aw: 127.60

PROP: Silvery-white, metallic, lustrous element; quite brittle. Mp: 449.5°; bp: 989.8°; d: 6.24 @ 20°; vap press: 1 mm @ 520°. Insol in water, benzene, carbon disulfide.

SYNS: NCI-C60117 * TELLOY * TELLUR (POLISH)

CONSENSUS REPORTS: Reported in EPA TSCA Inventory.

OSHA PEL: TWA 0.1 mg(Te)/m^3
ACGIH TLV: TWA 0.1 mg(Te)/m^3
DFG MAK: 0.1 mg/m^3

SAFETY PROFILE: Poison by ingestion and intratracheal routes. Exposure causes nausea, vomiting, tremors, convulsions, respiratory arrest, central nervous system depression, and garlic odor to breath. Aerosols of tellurium, tellurium dioxide, and hydrogen telluride cause irritation of the respiratory system and may lead to the development of bronchitis and pneumonia. Experimental teratogenic and reproductive effects. Under the proper conditions it undergoes hazardous reactions with halogens (e.g., chlorine, fluorine), interhalogens (e.g., bromine pentafluoride, chlorine fluorine, chlorine trifluoride), metals (e.g., cadmium, potassium, sodium, platinum, tin, zinc), hexalithium disilicide, silver bromate, silver iodate. When heated to decomposition it emits toxic fumes of Te.

TAJ010 CAS: 13494-80-9 **HR: 1**
TELLURIUM (dust or fume)

CONSENSUS REPORTS: Reported in EPA TSCA Inventory. EPA Extremely Hazardous Substances List.

ACGIH TLV: TWA 0.1 mg/m^3

SAFETY PROFILE: May cause irritation of the respiratory system and lead to bronchitis and pneumonia. When heated to decomposition it emits toxic fumes of Te.

TAJ500 **HR: 2**
TELLURIUM COMPOUNDS

SAFETY PROFILE: Elemental tellurium has relatively low toxicity. It is converted in the body to dimethyl telluride which imparts a garlic-like odor to the breath and sweat. Heavy exposures may, in addition, result in headache, drowsiness, metallic taste, loss of appetite, nausea, tremors, convulsions, and respiratory arrest. Various tellurium salts may also produce similar symptoms. Large doses can be fatal, as was the case following accidental administration of 2 grams of sodium tellurite. Workers in an iron foundry exposed to less than 0.1 mg Te/m^3 developed a garlic-like odor in breath, sweat, and urine, as well as anorexia, nausea, depression, somnolence, itchy skin, and metallic taste. When heated or on contact with acid or acid fumes they emit highly toxic fumes.

TAK250 CAS: 7783-80-4 **HR: 3**
TELLURIUM HEXAFLUORIDE

DOT: UN 2195
mf: F$_6$Te   mw: 241.60

PROP: Colorless gas; repulsive odor. Mp: $-37.6°$, bp: $-38.9°$ (subl); d: (solid) 4.006 @ $-191°$; (liquid) 2.499 @ $-10°$.

CONSENSUS REPORTS: EPA Extremely Hazardous Substances List. Reported in EPA TSCA Inventory.

OSHA PEL: TWA 0.02 ppm
ACGIH TLV: TWA 0.02 ppm
DOT Classification: Poison A; Label: Poison Gas.

SAFETY PROFILE: Poison by inhalation. Human skin (systemic) effects. When heated to decomposition it emits very toxic fumes of F$^-$ and Te.

TAL250 CAS: 3383-96-8 **HR: 3**
TEMEPHOS
mf: C$_{16}$H$_{20}$O$_6$P$_2$S$_3$ mw: 466.48

PROP: White crystals. Mp: 30°.

SYNS: ABATE * ABATHION * AC 52160 * AMERICAN CYANAMID AC 52,160 * BIOTHION * BITHION * CL 52160 * DIFENTHOS * O,O-DIMETHYL PHOSPHOROTHIOATE-O,O-DIESTER with 4,4'-THIODIPHENOL * ECOPRO * EI 52160 * ENT 27,165 * EXPERIMENTAL INSECTICIDE 52160 * NIMITEX * NIMITOX * SWEBATE * TEMEFOS * TEMOPHOS * TETRAMETHYL-O,O'-THIODI-p-PHENYLENE PHOSPHOROTHIOATE * O,O,O'O',-TETRAMETHYL-O,O'-THIODI-p-PHENYLENE PHOSPHOROTHIOATE * O,O'-(THIODI-4,1-PHENYLENE)BIS(O,O-DIMETHYL PHOSPHOROTHIOATE) * O,O'-(THIODI-p-PHENYLENE)-O,O,O',O'-TETRAMETHYL BIS(PHOSPHOROTHIOATE)

CONSENSUS REPORTS: Reported in EPA TSCA Inventory.

OSHA PEL: (Transitional: Total Dust: 15 mg/m^3; Respirable Fraction: 5 mg/m^3) TWA Total Dust: 10 mg/m^3; Respirable Fraction: 5 mg/m^3
ACGIH TLV: TWA 10 mg/m^3

SAFETY PROFILE: Poison by ingestion. Moderately toxic by skin contact and possibly other routes. An experimental teratogen. A skin irritant. A cholinesterase inhibitor type of insecticide. When heated to decomposition it emits toxic fumes of PO$_x$ and SO$_x$.

TAL750   **HR: 2**
TERBIUM
af: Tb   aw: 158.9254

PROP: A silvery-gray, soft ductile, malleable metallic element. Easily oxidized in air. Mp: 1356°, bp: 3041°, d: 8.234.

SAFETY PROFILE: As a lanthanon it may impair blood coagulation. Fire hazard in the form of dust in air or on contact with halogens. Incompatible with air; halogens.

TAM000   CAS: 10042-88-3   **HR: 3**
TERBIUM CHLORIDE
mf: Cl₃Tb   mw: 265.27

CONSENSUS REPORTS: Reported in EPA TSCA Inventory.

SAFETY PROFILE: Poison by intraperitoneal route. Moderately toxic by ingestion. An eye and severe skin irritant. When heated to decomposition it emits very toxic fumes of Cl⁻.

TAM500   CAS: 13482-49-0   **HR: 3**
TERBIUM CITRATE
mf: C₆H₈O₇ • Tb   mw: 351.06

SYN: 2-HYDROXY-1,2,3-PROPANECARBOXYLIC ACID TERBIUM (3+) SALT (1:1)

SAFETY PROFILE: Poison by intraperitoneal route. When heated to decomposition it emits acrid smoke and irritating fumes.

TAV750   CAS: 126-92-1   **HR: 3**
TERGITOL 08
mf: C₈H₁₈O₄S • Na   mw: 233.31

SYNS: EMERSAL 6465 * 2-ETHYL-1-HEXANOL HYDROGEN SULFATE, SODIUM SALT * 2-ETHYL-1-HEXANOL SULFATE SODIUM SALT * 2-ETHYLHEXYL SODIUM SULFATE * MONO(2-ETHYLHEXYL)SULFATE SODIUM SALT * NCI-C50204 * NIA PROOF 08 * PROPASTE 6708 * SIPEX BOS * SODIUM ETASULFATE * SODIUM ETHASULFATE * SODIUM(2-ETHYLHEXYL)ALCOHOL SULFATE * SODIUM 2-ETHYLHEXYL SULFATE * SULFURIC ACID, MONO(2-ETHYLHEXYL)ESTER, SODIUM SALT (8CI) * TERGEMIST * TERGIMIST * TERGITOL ANIONIC 08

CONSENSUS REPORTS: Reported in EPA TSCA Inventory.

SAFETY PROFILE: Poison by intraperitoneal route. Moderately toxic by ingestion and skin contact. A skin and eye irritant. When heated to decomposition it emits very toxic fumes of SOₓ and Na₂O.

TBC500   CAS: 8001-50-1   **HR: 3**
TERPENE POLYCHLORINATES

PROP: Chlorinated mixed terpenes.

SYNS: DICHLORICIDE MOTHPROOFER * ENT 19,442 * STROBANE

CONSENSUS REPORTS: IARC Cancer Review: GROUP 3 IMEMDT 7,56,87; Animal Sufficient Evidence IMEMDT 5,219,74.

SAFETY PROFILE: Poison by ingestion and possibly other routes. Questionable carcinogen with experimental carcinogenic data. When heated to decomposition it emits toxic fumes of Cl⁻.

TBC620   CAS: 92-06-8   **HR: 2**
m-TERPHENYL
mf: C₁₈H₁₄   mw: 230.32

SYNS: m-DIPHENYLBENZENE * ISODIPHENYLBENZENE * SANTOWAX M * 1,3-TERPHENYL * m-TRIPHENYL

CONSENSUS REPORTS: Reported in EPA TSCA Inventory.

OSHA PEL: (Transitional: TWA CL 1 ppm) CL 0.5 ppm
ACGIH TLV: TWA CL 0.5 ppm

SAFETY PROFILE: Moderately toxic by ingestion. Combustible when exposed to heat or flame. To fight fire, use water, CO₂, dry chemical. When heated to decomposition it emits acrid smoke and irritating fumes.

TBC640   CAS: 84-15-1   **HR: 2**
o-TERPHENYL
mf: C₁₈H₁₄   mw: 230.32

SYN: 1,2-DIPHENYLBENZENE

CONSENSUS REPORTS: Reported in EPA TSCA Inventory.

OSHA PEL: (Transitional: TWA CL 1 ppm) CL 0.5 ppm
ACGIH TLV: TWA CL 0.5 ppm

SAFETY PROFILE: Moderately toxic by ingestion. Combustible when exposed to heat or flame. To fight fire, use water, CO₂, dry chemical. When heated to decomposition it emits acrid smoke and irritating fumes.

TBC750   CAS: 92-94-4   **HR: 2**
p-TERPHENYL
mf: C₁₈H₁₄   mw: 230.32

PROP: Leaves or needles. D: 1.234 @ 0/4°, mp: 212-213°, bp: 276°, flash p: 405°F (OC), vap d: 7.95. Sol in hot benzene; very sol in hot alc, sltly sol in ether.

SYNS: p-DIPHENYLBENZENE * 1,4-DIPHENYLBEN-ZENE * 4-PHENYLBIPHENYL * 4-PHENYLDIPHE-NYL * SANTOWAX * p-TRIPHENYL

CONSENSUS REPORTS: Reported in EPA TSCA Inventory.

OSHA PEL: (Transitional: TWA CL 1 ppm) CL 0.5 ppm
ACGIH TLV: TWA CL 0.5 ppm

SAFETY PROFILE: Moderately toxic by ingestion. Combustible when exposed to heat or flame. To fight fire, use water, CO_2, dry chemical. When heated to decomposition it emits acrid smoke and irritating fumes.

TBD000   CAS: 26140-60-3   **HR: 2**
TERPHENYLS
mf: $C_{18}H_{14}$ mw: 230.32

SYNS: DELOWAS S * DELOWAX OM * DIPHE-NYLBENZENE * GILOTHERM OM 2 * TERBENZENE * TRIPHENYL

CONSENSUS REPORTS: Reported in EPA TSCA Inventory.

OSHA PEL: (Transitional: TWA CL 1 ppm) CL 0.5 ppm
ACGIH TLV: TWA CL 0.5 ppm

SAFETY PROFILE: Moderately toxic by ingestion. Combustible when exposed to heat or flame. To fight fire, use water, CO_2, dry chemical. When heated to decomposition it emits acrid smoke and irritating fumes.

TBD500   CAS: 8006-39-1   **HR: 1**
TERPINEOL
mf: $C_{10}H_{18}O$ mw: 154.28

PROP: A mixture of α, β, and γ isomers. Colorless, viscous liquid; lilac odor. D: 0.930-0.936, refr index: 1.482, flash p: 196°F. Sltly sol in water, glycerin.

SYNS: FEMA No. 3045 * p-MENTH-1-EN-8-OL * MIXTURE of p-METHENOLS * α-TERPINEOL (FCC) * TERPINEOLS

SAFETY PROFILE: Mildly toxic by ingestion. A skin irritant. Combustible liquid. When heated to decomposition it emits acrid smoke and irritating fumes.

TBE000   CAS: 586-62-9   **HR: 1**
TERPINOLENE
DOT: UN 2541
mf: $C_{10}H_{16}$ mw: 136.26

PROP: Colorless liquid. Bp: 185°, d: 0.855, flash p: 100°F (CC). Insol in water; misc in alc, ether. Mixture of p-mentha-1,4(8)-diene and p-mentha-2,4(8)-diene.

SYN: 1-METHYL-4-(1-METHYLETHYLIDENE)CYCLO-HEXENE

CONSENSUS REPORTS: Reported in EPA TSCA Inventory.

DOT Classification: Flammable or Combustible Liquid; Label: Flammable Liquid.

SAFETY PROFILE: Mildly toxic by ingestion. A very dangerous fire hazard when exposed to heat or flame. To fight fire, use foam, CO_2, dry chemical. Can react with oxidizing materials. When heated to decomposition it emits acrid smoke and irritating fumes.

TBE250   CAS: 80-26-2   **HR: 1**
TERPINYL ACETATE
mf: $C_{12}H_{20}O_2$ mw: 196.32

PROP: Colorless liquid; sweet, herbaceous odor. D: 0.966 @ 20/4°, refr index: 1.464, mp: <−50°, bp: 220° decomp, flash p: 212°F. Insol in water; sol in alc, fixed oils, mineral oil, propylene glycol.

SYNS: FEMA No. 3047 * α-TERPINEOL ACETATE

CONSENSUS REPORTS: Reported in EPA TSCA Inventory.

SAFETY PROFILE: Mildly toxic by ingestion. Combustible liquid. When heated to decomposition it emits acrid smoke and irritating fumes.

TBE600   **HR: 1**
TERPINYL PROPIONATE
mf: $C_{13}H_{22}O_2$ mw: 210.32

PROP: Colorless to sltly yellow liquid; sweet, floral, lavender-like odor. D: 0.944, refr index: 1.461, flash p: +212°F. Sol in glycerin; misc in alc, chloroform, ether, fixed oils; sltly sol in propylene glycol; insol in water @ 240°.

SYNS: FEMA No. 3053 * MENTHEN-1-YL-8 PROPIO-NATE

SAFETY PROFILE: Combustible liquid. When heated to decomposition it emits acrid smoke and irritating fumes.

TBF500   CAS: 58-22-0   **HR: 3**
TESTOSTERONE
mf: $C_{19}H_{28}O_2$ mw: 288.47

PROP: Crystals. Mp: 155°. Insol in water; sol in alc and ether.

SYNS: ANDROLIN * ANDRONAQ * ANDROST-4-EN-17β-OL-3-ONE * Δ⁴-ANDROSTEN-17(β)-OL-3-ONE * ANDRUSOL * CRISTERONE T * GENO-CRISTAUZ GREMY * HOMOSTERONE * 17-β-HYDROXY-Δ⁴-ANDROSTEN-3-ONE * 17-β-HYDROXYANDROST-4-EN-3-ONE * 17-β-HYDROXY-4-ANDROSTEN-3-ONE * 17-HYDROXY-(17-β)-ANDROST-4-EN-3-ONE * 7-β-HYDROXYANDROST-4-EN-3-ONE * MALESTRONE (AMPS) * MERTESTATE * NEO-TESTIS * ORETON-F * ORQUISTERONE * PERANDREN * PERCUTACRINE ANDROGENIQUE * PRIMOTEST * PROMOTESTON * SUSTANONE * SYNANDROL F * TESLEN * TESTANDRONE * TESTICULOSTERONE * TESTOBASE * TESTOPROPON * TESTOSTEROID * trans-TESTOSTERONE * TESTOSTERONE HYDRATE * TESTOSTOSTERONE * TESTOVIRON SCHERING * TESTOVIRON T * TESTRONE * TESTRYL * VIRORMONE * VIROSTERONE

CONSENSUS REPORTS: IARC Cancer Review: Animal Sufficient Evidence IMEMDT 6,209,74, IMEMDT 21,519,79; Human Limited Evidence IMEMDT 21,519,79. Reported in EPA TSCA Inventory. EPA Genetic Toxicology Program.

SAFETY PROFILE: Confirmed carcinogen with experimental neoplastigenic data. Poison by intraperitoneal route. Human teratogenic effects by unspecified route: developmental abnormalities of the urogenital system. Experimental teratogenic and reproductive effects. Human mutation data reported. Workers engaged in manufacture and packaging have shown effects from this hormone, i.e., enlargement of the breasts in male workers. A promoter. When heated to decomposition it emits acrid smoke and irritating fumes. Used as a drug for the treatment of hypogonadism and metastatic breast cancer.

TBG000   CAS: 57-85-2   **HR: 3**
TESTOSTERONE PROPIONATE
mf: $C_{22}H_{32}O_3$ mw: 344.54

SYNS: AGOVIRIN * ANDROGEN * ANDROSAN * Δ⁴-ANDROSTENE-17-β-PROPIONATE-3-ONE * ANDROTESTON * ANDROTEST P * ANDRUSOL-P * ANERTAN * AQUAVIRON * BIO-TESTICU-

LINA * ENARMON * HOMANDREN (amps) * HORMOTESTON * MASENATE * NASDOL * NEO-HOMBREOL * NSC 9166 * OKASA-MASCUL * ORCHIOL * ORCHISTIN * ORETON * ORETON PROPIONATE * 17-(1-OXOPROPOXY)-(17-β)-ANDROST-4-EN-3-ONE * PANESTIN * PERANDREN * PROPIOKAN * RECTHORMONE TESTOSTERONE * STERANDRYL * SYNANDROL * SYNERONE * TELIPEX * TESTAFORM * TESTEX * TESTODET * TESTODRIN * TESTOGEN * TESTONIQUE * TESTORMOL * TESTOSTERON PROPIONATE * TESTOSTERONE-17-PROPIONATE * TESTOSTERONE-17-β-PROPIONATE * TESTOVIRON * TESTOXYL * TESTREX * TOSTRIN * TP * UNITESTON * VULVAN

CONSENSUS REPORTS: IARC Cancer Review: Animal Sufficient Evidence IMEMDT 21,519,79. Reported in EPA TSCA Inventory.

SAFETY PROFILE: Confirmed carcinogen with experimental neoplastigenic and tumorigenic data. Moderately toxic by ingestion and intraperitoneal routes. Human male reproductive effects by intramuscular and parenteral routes: changes in spermatogenesis, testes, epididymis, and sperm duct. Human female reproductive effects by intramuscular and parenteral routes: menstrual cycle changes or disorders and effects on fertility. Experimental teratogenic and reproductive effects. Mutation data reported. When heated to decomposition it emits acrid smoke and irritating fumes.

TBM250   CAS: 1461-25-2   **HR: 3**
TETRABUTYLSTANNANE
mf: $C_{16}H_{36}Sn$ mw: 347.21

SYNS: TETRA-n-BUTYLCIN (CZECH) * TETRABUTYLTIN

CONSENSUS REPORTS: Reported in EPA TSCA Inventory.

OSHA PEL: TWA 0.1 mg(Sn)/m³ (skin)
ACGIH TLV: TWA 0.1 mg(Sn)/m³ (skin) (Proposed: TWA 0.1 mg(Sn)/m³; STEL 0.2 mg(Sn)/m³ (skin))
NIOSH REL: (Organotin Compounds) TWA 0.1 mg(Sn)/m³

SAFETY PROFILE: Poison by intravenous, intraperitoneal and parenteral routes. Moderately toxic by ingestion and skin contact. An eye irritant. When heated to decomposition it emits acrid smoke and irritating fumes.

TBN250 **HR: D**
TETRACHLOROACETONE

PROP: Liquid. Bp: 180-182°. Sltly decomp; very sol in benzene, alc, ether.

SAFETY PROFILE: Experimental reproductive effects. Mutation data reported. When heated to decomposition it emits toxic fumes of Cl⁻.

TBO000 CAS: 15721-02-5 **HR: 3**
TETRACHLOROBENZIDINE
mf: $C_{12}H_8Cl_4N_2$ mw: 322.02

SYNS: 2,2′,5,5′-TETRACHLOROBENZIDINE * 3, 3′,6,6′-TETRACHLOROBENZIDINE * 2,2′,5,5′-TETRA-CHLORO-(1,1′-BIPHENYL)-4,4′-DIAMINE, (9CI) * 2,2′,5,5′-TETRACHLORO-4,4′-DIAMINODIPHENYL

CONSENSUS REPORTS: IARC Cancer Review: GROUP 3 IMEMDT 7,56,87; Animal Inadequate Evidence IMEMDT 27,141,82. Reported in EPA TSCA Inventory.

SAFETY PROFILE: Suspected carcinogen with experimental carcinogenic and tumorigenic data. When heated to decomposition it emits very toxic fumes of NO_x and Cl⁻.

TBP000 CAS: 76-11-9 **HR: 1**
1,1,1,2-TETRACHLORO-2,2-DIFLUOROETHANE
mf: $C_2Cl_4F_2$ mw: 203.82

SYNS: HALOCARBON 112a * REFRIGERANT 112a

CONSENSUS REPORTS: Reported in EPA TSCA Inventory.

OSHA PEL: TWA 500 ppm
ACGIH TLV: TWA 500 ppm
DFG MAK: 1000 ppm (8340 mg/m³)

SAFETY PROFILE: Mildly toxic by inhalation. When heated to decomposition it emits very toxic fumes of Cl⁻ and F⁻. Used as a refrigerant.

TBP050 CAS: 76-12-0 **HR: 2**
1,1,2,2-TETRACHLORO-1,2-DIFLUOROETHANE
mf: $C_2Cl_4F_2$ mw: 203.82

PROP: Liquid. Bp: 92.8°, d: 1.6447 @ 25°, vap d: 7.03.

SYNS: 1,2-DIFLUORO-1,1,2,2-TETRACHLOROETHANE * F-112 * FREON 112 * GENETRON 112 * HALOCARBON 112 * REFRIGERANT 112

CONSENSUS REPORTS: Reported in EPA TSCA Inventory.

OSHA PEL: TWA 500 ppm
ACGIH TLV: TWA 500 ppm
DFG MAK: 500 ppm (4170 mg/m³)

SAFETY PROFILE: Moderately toxic by ingestion. Mildly toxic by inhalation. A skin and eye irritant. When heated to decomposition it emits toxic fumes of F⁻ and Cl⁻.

TBP250 CAS: 31242-94-1 **HR: 3**
TETRACHLORODIPHENYL OXIDE
mf: $C_{12}H_6Cl_4O$ mw: 307.98

SYNS: PHENYL ETHER TETRACHLORO * TETRA-CHLOROPHENYL ETHER

OSHA PEL: TWA 0.5 mg/m³

SAFETY PROFILE: Poison by ingestion. When heated to decomposition it emits toxic fumes of Cl⁻.

TBP750 CAS: 25322-20-7 **HR: 3**
TETRACHLOROETHANE

DOT: UN 1702
mf: $C_2H_2Cl_4$ mw: 167.84

CONSENSUS REPORTS: Reported in EPA TSCA Inventory.

NIOSH REL: (Tetrachloroethane) Reduce to lowest feasible level
DOT Classification: ORM-A; Label: None.

SAFETY PROFILE: Poison by ingestion and inhalation. Moderately toxic by intraperitoneal route. Mildly toxic by skin contact. Experimental reproductive effects. Mutation data reported. When heated to decomposition it emits toxic fumes of Cl⁻.

TBQ000 CAS: 630-20-6 **HR: 3**
1,1,1,2-TETRACHLOROETHANE
mf: $C_2H_2Cl_4$ mw: 167.84

PROP: Liquid. D: 1.588 @ 20/4°, bp: 129-130°. Sol in water; misc in alc, ether.

SYNS: NCI-C52459 * RCRA WASTE NUMBER U208

CONSENSUS REPORTS: IARC Cancer Review: GROUP 3 IMEMDT 7,56,87; Animal Limited Evidence IMEMDT 41,87,86. NTP Carcinogenesis Bioassay (gavage); Clear Evidence: mouse NTPTR* NTP-TR-237,82; (gavage); No Evidence: rat NTPTR* NTP-TR-237,82. Reported in EPA TSCA Inventory.

SAFETY PROFILE: A skin and severe eye irritant. Questionable carcinogen with experimental carcinogenic data. Incompatible with dinitrogen tetraoxide, 2,4-dinitrophenyl disulfide, potassium, potassium hydroxide, nitrogen tetroxide, sodium, sodium potassium alloy. When heated to decomposition it emits very toxic fumes of Cl^-.

TBQ100   CAS: 79-34-5   HR: 3
1,1,2,2-TETRACHLOROETHANE

DOT: UN 1702
mf: $C_2H_2Cl_4$ mw: 167.84

PROP: Heavy, colorless, mobile liquid; chloroform-like odor. Mp: −43.8°, bp: 146.4°, d: 1.600 @ 20°/4°.

SYNS: ACETYLENE TETRACHLORIDE * BONO-FORM * CELLON * 1,1,2,2-CZTEROCHLOROETAN (POLISH) * 1,1-DICHLORO-2,2-DICHLOROETHANE * NCI-C03554 * RCRA WASTE NUMBER U209 * TCE * TETRACHLORETHANE * 1,1,2,2-TETRA-CHLOORETHAAN (DUTCH) * 1,1,2,2-TETRACHLORA-ETHAN (GERMAN) * 1,1,2,2-TETRACHLORETHANE (FRENCH) * sym-TETRACHLOROETHANE * TETRA-CHLORURE d'ACETYLENE (FRENCH) * 1,1,2,2-TETRA-CLOROETANO (ITALIAN) * WESTRON

CONSENSUS REPORTS: IARC Cancer Review: GROUP 3 IMEMDT 7,354,87; Animal Limited Evidence IMEMDT 20,477,79; NCI Carcinogenesis Bioassay (gavage); Clear Evidence: mouse NCITR* NCI-CG-TR-27,78; Some Evidence: rat NCITR* NCI-CG-TR-27,78. Reported in EPA TSCA Inventory. EPA Genetic Toxicology Program. Community Right-To-Know List.

OSHA PEL: (Transitional: TWA 5 ppm (skin)) TWA 1 ppm (skin)
ACGIH TLV: TWA 1 ppm (skin)
DFG MAK: 1 ppm (7 mg/m^3); Suspected Carcinogen.
NIOSH REL: (1,1,2,2-Tetrachlorethane) Reduce to lowest level.
DOT Classification: IMO: Poison B; Label: Poison.

SAFETY PROFILE: Suspected carcinogen with experimental carcinogenic and tumorigenic data. Poison by inhalation, ingestion, and intraperitoneal routes. Moderately toxic by several other routes. Mutation data reported. Human central nervous system effects by ingestion and inhalation: general anesthesia, somnolence, hallucinations, and distorted perceptions. Considered the most toxic of the common chlorinated hydrocarbons. Considered to be a very severe industrial hazard and its use has been restricted or even forbidden in certain countries. It is not an inert solvent. Reacts violently with N_2O_4, 2,4-dinitrophenyl disulfide and on contact with sodium or potassium. When heated in contact with solid potassium hydroxide, spontaneously flammable chloro- or dichloro-acetylene gas is evolved. Any water can cause appreciable hydrolysis, even at room temperature, and both hydrolysis and oxidation become comparatively rapid above 110°. When heated to decomposition it emits toxic fumes of Cl^-.

A strong irritant of eyes and mucous membranes. A concentration of 3 ppm produces a detectable odor, thus an initial warning effect. Its narcotic action is stronger than that of chloroform, but because of its low volatility, narcosis is less severe and much less common in industrial poisoning than in the case of other chlorinated hydrocarbons. The toxic action of this material is chiefly on the liver where it produces acute yellow atrophy and cirrhosis. Fatty degeneration of the kidneys and heart, hemorrhage into the lungs and serous membranes, and edema of the brain have also been found in fatal cases. Some reports indicate a toxic action on the central nervous system with changes in the brain and in the peripheral nerves. The effect on the blood is one of hemolysis with appearance of young cells in the circulation and a monocytosis. Due to its solvent action on the natural skin oils, dermatitis is not uncommon.

The initial symptoms resulting from exposure to the vapor are lacrimation, salivation and irritation of the nose and throat. Continued exposure to high concentrations results in restlessness, dizziness, nausea, vomiting, and narcosis. The latter, however, is rare in industry. More commonly, exposure is less severe and most complaints are vague and related to the digestive and nervous systems. The patient's symptoms gradually progress to a more serious illness with development of toxic jaundice, liver tenderness, etc., and possibly albuminuria and edema. With serious liver damage, the jaundice increases and toxic symptoms appear with somnolence, delirium, convulsions and coma usually preceding death.

TBQ750   CAS: 1897-45-6   HR: 3
TETRACHLOROISOPHTHALONITRILE
mf: $C_8Cl_4N_2$ mw: 265.90

SYNS: BRAVO * BRAVO 6F * BRAVO-W-75 * CHLOROALONIL * CHLOROTHALONIL

* CHLORTHALONIL (GERMAN) * DAC 2797
* DACONIL * DACONIL 2787 FLOWABLE FUNGICIDE
* DACOSOIL * 1,3-DICYANOTETRACHLOROBEN-
ZENE * EXOTHERM * EXOTHERM TERMIL
* FORTURF * NCI-C00102 * NOPCOCIDE
* SWEEP * TCIN * m-TCPN * TERMIL
* 2,4,5,6-TETRACHLORO-3-CYANOBENZONITRILE
* m-TETRACHLOROPHTHALONITRILE * TPN (pesti-
cide)

CONSENSUS REPORTS: IARC Cancer Re-
view: GROUP 3 IMEMDT 7,56,87; Animal
Limited Evidence IMEMDT 30,319,83. NCI
Carcinogenesis Bioassay (feed); Clear Evi-
dence: rat NCITR* NCI-CG-TR-41,78. Cya-
nide and its compounds are on the Community
Right-To-Know List. Reported in EPA TSCA
Inventory. EPA Genetic Toxicology Program.

SAFETY PROFILE: Moderately toxic by intraperitoneal route. Mildly toxic by ingestion.
Questionable carcinogen with experimental carcinogenic data. When heated to decomposition
it emits very toxic fumes of Cl^-, NO_x, and
CN^-.

TBR000   CAS: 1335-88-2   HR: 3
TETRACHLORONAPHTHALENE
mf: $C_{10}H_4Cl_4$ mw: 265.94

PROP: Crystals. Mp: 182°.

SYN: HALOWAX

CONSENSUS REPORTS: Reported in EPA
TSCA Inventory.

OSHA PEL: TWA 2 mg/m^3
ACGIH TLV: TWA 2 mg/m^3

SAFETY PROFILE: Probably a poison. When
heated to decomposition it emits highly toxic
fumes of Cl^-.

TBR250   CAS: 2438-88-2   HR: 3
TETRACHLORONITROANISOLE
mf: $C_7H_3Cl_4NO_3$ mw: 290.91

SYNS: ENT 22,335 * NCI-C03032 * 4-NITRO-2,
3,5,6-TETRACHLORANISOLE * TCNA * 1,2,4,5-TET-
RACHLORO-3-METHOXY-6-NITROBENZENE (9CI)
* 2,3,5,6-TETRACHLORO-4-NITROANISOLE

CONSENSUS REPORTS: NCI Carcinogenesis Bioassay (feed); No Evidence: mouse, rat
NCITR* NCI-CG-TR-114,78.

SAFETY PROFILE: Poison by ingestion. When
heated to decomposition it emits very toxic
fumes of Cl^- and NO_x.

TBT000   CAS: 58-90-2   HR: 3
2,4,5,6-TETRACHLOROPHENOL
mf: $C_6H_2Cl_4O$ mw: 231.88

SYNS: DOWICIDE 6 * RCRA WASTE NUMBER U212
* 2,3,4,6-TETRACHLOROPHENOL

CONSENSUS REPORTS: IARC Cancer Review: Human Limited Evidence IMEMDT
41,319,86. Chlorophenol compounds are on the
Community Right-To-Know List. Reported in
EPA TSCA Inventory.

SAFETY PROFILE: Suspected carcinogen with
experimental carcinogenic data. Poison by ingestion, skin contact, intraperitoneal, and subcutaneous routes. May be a human carcinogenic.
Experimental teratogenic and reproductive effects. When heated to decomposition it emits
toxic fumes of Cl^-. Used as a disinfectant and
a preservative for wood, latex and leather.

TBV750   CAS: 6012-97-1   HR: 3
2,3,4,5-TETRACHLOROTHIOPHENE
mf: C_4Cl_4S mw: 221.90

SYNS: 2,3,4,5-CHLOROTHIOPHENE * ENT 25,764
* IF (fumigant) * PENN SALT TD-183 * PENPHENE
* PERCHLOROTHIOPHENE * TCTP * TD-183
* TETRACHLOROTHIOFENE * TETRACHLOROTHIO-
PHENE

SAFETY PROFILE: Poison by ingestion, skin
contact, intravenous, and intraperitoneal routes.
Moderately toxic by inhalation. When heated
to decomposition it emits very toxic fumes of
Cl^- and SO_x.

TBW100   CAS: 961-11-5   HR: 3
TETRACHLORVINPHOS
mf: $C_{10}H_9Cl_4O_4P$ mw: 365.96

SYNS: 2-CHLORO-1-(2,4,5-TRICHLOROPHENYL)VINYL
DIMETHYL PHOSPHATE * 2-CHLORO-1-(2,4,5-TRI-
CHLOROPHENYL(VINYL PHOSPHORIC ACID DIMETHYL
ESTER * O,O-DIMETHYL-O-2-CHLOR-1-(2,4,5-TRI-
CHLORPHENYL)-VINYL-PHOSPHAT (GERMAN)
* IPO 8 * NCI C00168 * PHOSPHORIC ACID,
2-CHLORO-1-(2,4,5-TRICHLOROPHENYL)ETHENYL DI-
METHYL ESTER * 2,4,5-TRICHLORO-α-(CHLORO-
METHYLENE)BENZYL PHOSPHATE

CONSENSUS REPORTS: NCI Carcinogen-
esis Bioassay (feed); Results Positive: Mouse,
Rat NCITR* NCI-CG-TR-33,78. Community
Right-To-Know List.

SAFETY PROFILE: Suspected carcinogen with
experimental carcinogenic, neoplastigenic, and

tumorigenic data. Poison by ingestion. Experimental reproductive effects. When heated to decomposition it emits toxic fumes of Cl^- and PO_x.

TBW250   CAS: 14323-41-2   **HR: 2**
TETRACYANONICKELATE(2−)
DIPOTASSIUM, HYDRATE
mf: $C_4N_4Ni \cdot K \cdot H_2O$ mw: 219.91

SYN: POTASSIUM CYANONICKELATE HYDRATE

CONSENSUS REPORTS: Nickel and its compounds, as well as cyanide and its compounds, are on the Community Right-To-Know List.

OSHA PEL: (Transitional: TWA 1 mg/m^3)
   TWA 0.1 mg (Ni)/m^3
ACGIH TLV: TWA 0.1 mg(Ni)/m^3; (Proposed: TWA 0.05 mg(Ni)/m^3; Human Carcinogen)

SAFETY PROFILE: Suspected human carcinogen. Mutation data reported. Many nickel compounds are poisons. When heated to decomposition it emits very toxic fumes of CN^-, K_2O, and NO_x.

TBX000   CAS: 60-54-8   **HR: 3**
TETRACYCLINE
mf: $C_{22}H_{24}N_2O_8$ mw: 444.48

PROP: Produced by *Streptomyces albo-niger*. Trihydrate: crystals. Decomp @ 170-175°.

SYNS: ABRAMYCIN * ABRICYCLINE * ACHRO-MYCIN * AGROMICINA * AMBRAMICINA * AMBRAMYCIN * BIO-TETRA * BRISTACICLIN α * BRISTACYCLINE * CEFRACYCLINE SUSPEN-SION * CRISEOCICLINE * CYCLOMYCIN * DEMOCRACIN * DESCHLOROBIOMYCIN * HOSTACYCLIN * LIQUAMYCIN * 6-METHYL-1,11-DIOXY-2-NAPHTHACENECARBOXAMIDE * NEOCYCLINE * OLETETRIN * φMYCIN * PANMYCIN * POLYCYCLINE * PUROCYCLINA * ROBITET * SANCLOMYCINE * SIGMAMYCIN * SK-TETRACYCLINE * STECLIN * T-125 * TETRABON * TETRACYCLINE I * TETRACYN * TETRADECIN * TETRAVERINE * TSIKLOMIT-SIN

CONSENSUS REPORTS: EPA Genetic Toxicology Program.

SAFETY PROFILE: Human poison by multiple routes. Experimental poison by intraperitoneal, intravenous, and subcutaneous routes. Moderately toxic by ingestion. Human systemic effects by ingestion: somnolence, decreased motility

or constipation and urine volume decrease or anuria. Human reproductive effects by unspecified routes: effects on newborn including postnatal measures or effects and delayed effects on newborn. Experimental teratogenic and reproductive effects. Mutation data reported. When heated to decomposition it emits toxic fumes of NO_x.

TBX250   CAS: 64-75-5   **HR: 3**
TETRACYCLINE HYDROCHLORIDE
mf: $C_{22}H_{24}N_2O_8 \cdot ClH$ mw: 480.94

PROP: Very sol in water; sol in methanol, ethanol; insol in ether, hydrocarbon solvents.

SYNS: ACHROMYCIN * ACHROMYCIN HYDRO-CHLORIDE * AMBRACYN * ARTOMYCIN * BRISTACYCLINE * CEFRACYCLINE TABLETS * CHLORHYDRATE de TETRACYCLINE (FRENCH) * CYCLOPAR * DIACYCINE * DUMOCYCIN * MEDAMYCIN * MEPHACYCLIN * NCI-C55561 * PALTET * PANMYCIN HYDROCHLORIDE * PARTREX * PIRACAPS * POLYCYCLINE HY-DROCHLORIDE * QIDTET * QUADRACYCLINE * QUATREX * REMICYCLIN * RICYCLINE * RO-CYCLINE * SK-TETRACYCLINE * STECLIN HYDROCHLORIDE * SUBAMYCIN * SUPRAMYCIN * T-250 CAPSULES * TC HYDROCHLORIDE * TEFILIN * TELINE * TELOTREX * TETRA-BAKAT * TETRABLET * TETRACAPS * TETRA-CICLINA CLORIDRATO (ITALIAN) * TETRACOMPREN * TETRACYCLINE CHLORIDE * TETRA-D * TETRALUTION * TETRA-WEDEL * TETROSOL * TOPICYCLINE * TOTOMYCIN * TRIPHACYLIN * U-5965 * UNICIN * UNIMYCIN * VETQUA-MYCIN-324

CONSENSUS REPORTS: Reported in EPA TSCA Inventory.

SAFETY PROFILE: Poison by intraperitoneal, intravenous, and possibly other routes. Moderately toxic by ingestion and subcutaneous routes. Experimental teratogenic and reproductive effects. Mutation data reported. When heated to decomposition it emits very toxic fumes of HCl and NO_x.

TBX750   CAS: 629-59-4   **HR: 3**
TETRADECANE
mf: $C_{14}H_{30}$ mw: 198.44

PROP: Colorless liquid. D: 0.765 @ 20/4°, mp: 5.5°, bp: 252-255°, lel: 0.5%, flash p: 212°F, vap press: 1 mm @ 76.4°, vap d: 6.83, autoign temp: 396°F. Insol in water; very sol in alc and ether.

SAFETY PROFILE: Probably irritating and narcotic in high concentrations. Questionable carcinogen with experimental tumorigenic data. Combustible when exposed to heat or flame. Moderate explosion hazard in the form of vapor when exposed to heat or flame. Can react with oxidizing materials. To fight fire, use foam, CO_2, dry chemical. When heated to decomposition it emits acrid smoke and irritating fumes.

TBY250   CAS: 27196-00-5   **HR: 1**
TETRADECANOL, mixed isomers
mf: $C_{14}H_{30}O$ mw: 214.44

SYNS: MYRISTYL ALCOHOL (mixed isomers) * TETRADECYL ALCOHOL

CONSENSUS REPORTS: Reported in EPA TSCA Inventory.

SAFETY PROFILE: Mildly toxic by ingestion and skin contact. Combustible when exposed to heat or flame; can react with oxidizing materials. To fight fire, use CO_2, dry chemical. When heated to decomposition it emits acrid smoke and irritating fumes.

TCA500   CAS: 139-08-2   **HR: 1**
TETRADECYL DIMETHYL BENZYLAMMONIUM CHLORIDE
mf: $C_{23}H_{42}N \cdot Cl$ mw: 368.11

SYNS: ARQUAD DM14B-90 * N,N-DIMETHYL-N-TETRADECYLBENZENEMETHANAMINIUM, CHLORIDE (9CI) * NISSAN CATION M2-100

CONSENSUS REPORTS: Reported in EPA TSCA Inventory.

SAFETY PROFILE: A skin and eye irritant. When heated to decomposition it emits very toxic fumes of NO_x, NH_3, and Cl^-.

TCE250   CAS: 112-60-7   **HR: 1**
TETRAETHYLENE GLYCOL
mf: $C_8H_{18}O_5$ mw: 194.26

PROP: Colorless to pale straw-colored liquid. Bp: 327.3°, fp: −6°, flash p: 360°F (OC), d: 1.1248 @ 20/20°, vap press: 1 mm @ 153.9°. Misc in water.

SYNS: HI-DRY * 2,2'-(OXYBIS(ETHYLENEOXY)) DIETHANOL

CONSENSUS REPORTS: Reported in EPA TSCA Inventory.

SAFETY PROFILE: Mildly toxic by ingestion. A skin and eye irritant. Combustible when exposed to heat or flame; can react with oxidizing materials. To fight fire, use alcohol foam, water, CO_2, dry chemical. When heated to decomposition it emits acrid smoke and irritating fumes.

TCE500   CAS: 112-57-2   **HR: 3**
TETRAETHYLENEPENTAMINE
DOT: UN 2320
mf: $C_8H_{23}N_5$ mw: 189.36

PROP: Viscous, hygroscopic liquid. Bp: 333°, flash p: 325°F (OC), d: 0.9980 @ 20/20°, vap press: <0.01 mm @ 20°.

SYNS: D.E.H. 26 * 1,4,7,10,13-PENTAAZATRIDECANE

CONSENSUS REPORTS: Reported in EPA TSCA Inventory.

DOT Classification: Corrosive Material; Label: Corrosive.

SAFETY PROFILE: Poison by ingestion and intravenous routes. Moderately toxic by skin contact. Mutation data reported. A corrosive irritant to skin, eyes, and mucous membranes. Combustible when exposed to heat or flame. Can react with oxidizing materials. To fight fire, use CO_2, dry chemical. When heated to decomposition it emits toxic fumes of NO_x.

TCF000   CAS: 78-00-2   **HR: 3**
TETRAETHYL LEAD
DOT: NA 1649
mf: $C_8H_{20}Pb$ mw: 323.47

PROP: Colorless, oily liquid; pleasant characteristic odor. Mp: 125-150°, bp: 198-202° with decomp, d: 1.659 @ 18°, vap press: 1 mm @ 38.4°, flash p: 200°F.

SYNS: CZTEROETHLEK OLOWIU (POLISH) * NCI-C54988 * RCRA WASTE NUMBER P110 * TEL * TETRAETHYLPLUMBANE

CONSENSUS REPORTS: IARC Cancer Review: GROUP 3 IMEMDT 7,230,87; Animal Inadequate Evidence IMEMDT 23,325,80; IMEMDT 2,150,73. EPA Extremely Hazardous Substances List. Reported in EPA TSCA Inventory. EPA Genetic Toxicology Program.

OSHA PEL: TWA 0.075 mg(Pb)/m^3 (skin)
ACGIH TLV: TWA 0.1 mg(Pb)/m^3 (skin)
DFG MAK: 0.01 ppm (0.075 mg/m^3)

DOT Classification: Poison B; Label: Poison and Poison, Flammable Liquid.

SAFETY PROFILE: Human poison by an unspecified route. Experimental poison by ingestion, intraperitoneal, intravenous, subcutaneous, and parenteral routes. Moderately toxic by inhalation and skin contact. Experimental teratogenic and reproductive effects. Questionable carcinogen with experimental carcinogenic data. Lead compounds are particularly toxic to the central nervous system. It is a solvent for fatty materials and has some solvent action on rubber as well. The fact that it is a lipoid solvent makes it an industrial hazard because it can cause intoxication not only by inhalation but also by absorption through the skin. Decomposes when exposed to sunlight or allowed to evaporate, forms triethyl lead, which is also a poisonous compound, as one of its decomposition products. May cause lead exposure intoxication by coming in contact with the skin. A common air contaminant.

Flammable when exposed to heat, flame, or oxidizers. Can react vigorously with oxidizing materials. Exposure to air for several days may cause explosive decomposition. To fight fire, use dry chemical, CO_2, mist, foam. When heated to decomposition it emits toxic fumes of Pb.

TCF250   CAS: 107-49-3   HR: 3
TETRAETHYL PYROPHOSPHATE

DOT: NA 2783
mf: $C_8H_{20}O_7P_2$ mw: 290.22

PROP: Water-white to amber hygroscopic liquid. D: 1.20.

SYNS: BIS-O,O-DIETHYLPHOSPHORIC ANHYDRIDE ∗ BLADAN ∗ DIPHOSPHORIC ACID THETRAETHYL ESTER ∗ ENT 18,771 ∗ FOSVEX ∗ GRISOL ∗ HEPT ∗ HEXAMITE ∗ KILLAX ∗ KILMITE 40 ∗ LETHALAIRE G-52 ∗ LIROHEX ∗ MORTOPAL ∗ NIFOS T ∗ PYROPHOSPHATE de TETRAETHYLE (FRENCH) ∗ RCRA WASTE NUMBER P111 ∗ TEPP ∗ O,O,O,O-TETRAAETHYL-DIPHOSPHAT, BIS(O,O-DIA-ETHYLPHOSPHORSAEURE-ANHYDRID (GERMAN) ∗ O,O,O,O-TETRAETHYL-DIFOSFAAT (DUTCH) ∗ TETRAETHYL PYROFOSFAAT (BELGIAN) ∗ TETRAETHYL PYROPHOSPHATE, liquid (DOT) ∗ O,O,O,O-TETRAETIL-PIROFOSFATO (ITALIAN) ∗ TETRASTIGMINE ∗ TETRON ∗ TETRON-100 ∗ VAPOTONE

CONSENSUS REPORTS: EPA Extremely Hazardous Substances List.

OSHA PEL: TWA 0.05 mg/m^3 (skin)
ACGIH TLV: TWA 0.004 mg/m^3 (skin)
DFG MAK: 0.0005 ppm (0.05 mg/m^3)
DOT Classification: Poison B; Label: Poison, liquid.

SAFETY PROFILE: Human poison by ingestion and intramuscular routes. Experimental poison by ingestion, skin contact, intraperitoneal, intramuscular, subcutaneous, parenteral, intravenous, and possibly other routes. Human systemic effects by ingestion, intramuscular, and parenteral routes: paresthesia, wakefulness, excitement, muscle contraction or spasticity, nausea or vomiting and other gastrointestinal changes. The action is similar to that of parathion; causing an irreversible inhibition of the cholinesterase molecules and the consequent accumulation of large amounts of acetylcholine. Small doses at frequent intervals are largely additive. When heated to decomposition it emits toxic fumes of PO$_x$.

TCF280   CAS: 107-49-3   HR: 3
TETRAETHYL PYROPHOSPHATE MIXTURE (liquid)

DOT: NA 2783
mf: $C_8H_{20}O_7P_2$ mw: 290.22

PROP: Water white to amber, hygroscopic liquid. D: 1.20, decomp @ 170-213°, bp: 138° @ 2.3 mm. Misc with water, acetone, methanol, ethanol, benzene, chloroform.

SYNS: FOSVEX ∗ PYROPHOSPHORIC ACID, TETRA-ETHYL ESTER (liquid mixture) ∗ TEP ∗ TETRON ∗ VAPTONE

DOT Classification: Poison B; Label: Poison.

SAFETY PROFILE: Poison by ingestion and skin contact. A cholinesterase inhibitor type of insecticide. The effects of chronic exposure to small doses is additive. When heated to decomposition it emits toxic fumes of PO$_x$.

TCH000   HR: 3
TETRAFLUOROBORATE(1−)
compound with p-AMINOBENZOIC ACID 2-(DIETHYLAMINO)ETHYL ESTER

PROP: Colorless liquid. Bp: decomp @ 130°.

SYN: PROCAINE FLUOBORATE

NIOSH REL: TWA 2.5 mg(F)/m^3
DOT Classification: Label: Corrosive.

SAFETY PROFILE: Poison by intravenous route. A corrosive irritant to skin, eyes, and mucous membranes. When heated to decomposition it emits toxic fumes of F^- and NO_x.

TCH500   CAS: 116-14-3   **HR: 2**
TETRAFLUOROETHYLENE
DOT: UN 1081
mf: C_2F_4 mw: 100.02

PROP: Colorless gas. Mp: $-142.5°$, bp: $-78.4°$. lel: 11%; uel: 60%.

SYNS: FLUOROPLAST 4 * PERFLUOROETHENE * PERFLUOROETHYLENE * TETRAFLUORETHYLENE * TETRAFLUOROETHENE * TETRAFLUOROETHY-LENE, inhibited (DOT)

CONSENSUS REPORTS: IARC Cancer Review: GROUP 3 IMEMDT 7,56,87. Reported in EPA TSCA Inventory.

DOT Classification: Flammable Gas; Label: Flammable Gas.

SAFETY PROFILE: Mildly toxic by inhalation. Can act as an asphyxiant and may have other toxic properties. Questionable carcinogen. The gas is flammable when exposed to heat or flame. The inhibited monomer will explode if ignited. Explosive in the form of vapor when exposed to heat or flame. Will explode at pressures above 2.7 bar if terpene inhibitor is not added. Iodine pentafluoride depletes the limonene inhibitor and then causes explosive polymerization of the monomer. Mixtures with hexafluoropropene and air form an explosive peroxide. Reacts violently with SO_3, air, difluoromethylene dihypofluorite, dioxygen difluoride, iodine pentafluoride, oxygen. When heated to decomposition it emits highly toxic fumes of F^-.

TCI000   CAS: 10036-47-2   **HR: 3**
TETRAFLUORO HYDRAZINE
DOT: UN 1955
mf: N_2F_4 mw: 104.0

PROP: Colorless gas or liquid; white solid when pure. Mp: $-163°$, bp: $-73°$, d (liquid): 1.5 @ $-100°$.

SYNS: DINITROGEN TETRAFLUORIDE * PER-FLUORO HYDRAZINE

CONSENSUS REPORTS: Reported in EPA TSCA Inventory.

OSHA PEL: TWA 2.5 mg(F)/m^3
ACGIH TLV: TWA 2.5 mg(F)/m^3
NIOSH REL: TWA (Inorganic Fluorides) 2.5 mg(F)/m^3
DOT Classification: Poison A; Label: Poison Gas.

SAFETY PROFILE: A poison. An unstable explosive gas sensitive to light, heat, or contact with air or steel. At high pressures it can explode due to shock or blast. Flammable when exposed to heat or flame. Potentially explosive reaction with hydrocarbons, hydrogen, organic materials, reducing agents, oxygen. Forms explosive mixtures with alkenyl nitrates, nitrogen trifluoride. When heated to decomposition it emits highly toxic fumes of F^- and NO_x.

TCI250   CAS: 63886-77-1   **HR: 3**
TETRAFLUORO-m-PHENYLENE DIAMINE DIHYDROCHLORIDE
mf: $C_6H_4F_4N_2 \cdot 2ClH$ mw: 253.04

SAFETY PROFILE: Poison by intraperitoneal route. Questionable carcinogen with experimental carcinogenic data. When heated to decomposition it emits very toxic fumes of F^-, NO_x, and HCl.

TCJ100   CAS: 1321-16-0   **HR: 2**
TETRAHYDROBENZALDEHYDE
mf: C_7H_9N mw: 110.17

SYN: CYCLOHEXENECARBOXALDEHYDE

DOT Classification: Flammable or Combustible Liquid; Label: Flammable Liquid.

SAFETY PROFILE: Moderately toxic by ingestion, skin contact, and inhalation. A skin and severe eye irritant. The liquid is flammable when exposed to heat or flame. When heated to decomposition it emits toxic fumes of NO_x.

TCM000   CAS: 5957-75-5   **HR: 3**
1-trans-Δ^8-TETRAHYDROCANNABINOL
mf: $C_{21}H_{30}O_2$ mw: 314.51

SYNS: $(-)-\Delta^6$-3,4-trans-TETRAHYDROCANNABINOL * $(-)-\Delta^8$-trans-TETRAHYDROCANNABINOL * Δ^6-THC * Δ^8-THC

CONSENSUS REPORTS: EPA Genetic Toxicology Program.

SAFETY PROFILE: Poison by intravenous and intraperitoneal routes. Moderately toxic by ingestion. Human mutation data reported. An hallucinatory drug. When heated to decomposition it emits acrid smoke and irritating fumes.

TCM250   CAS: 1972-08-3   HR: 3
1-trans-Δ^9-TETRAHYDROCANNABINOL
mf: $C_{21}H_{30}O_2$ mw: 314.51

SYNS: ABBOTT 40566 * 3-PENTYL-6,6,9-TRIME-THYL-6a,7,8,10a-TETRAHYDRO-6H-DIBENZO(b,d)PYRAN-1-OL * SP 104 * (1)-Δ^1-TETRAHYDROCANNABINOL * Δ^1-TETRAHYDROCANNABINOL * (−)-Δ^1-3,4-trans-TETRAHYDROCANNABINOL * (−)-Δ^9-trans-TETRAHYDROCANNABINOL * trans-Δ^9-TETRAHYDROCANNABINOL * Δ^9-TETRAHYDROCANNABINON * THC * Δ^1-THC * Δ^9-THC * 6,6,9-TRIMETHYL-3-PENTYL-7,8,9,10-TETRAHYDRO-6H-DIBENZO(B,D)PYRAN-1-OL

CONSENSUS REPORTS: EPA Genetic Toxicology Program.

SAFETY PROFILE: Poison by intraperitoneal and intravenous routes. Moderately toxic by ingestion. Experimental teratogenic and reproductive effects. Questionable carcinogen with experimental tumorigenic data. Human mutation data reported. An hallucinatory drug. When heated to decomposition it emits acrid smoke and irritating fumes.

TCR750   CAS: 109-99-9   HR: 3
TETRAHYDROFURAN
DOT: UN 2056
mf: C_4H_8O mw: 72.12

PROP: Colorless, mobile liquid; ether-like odor. Bp: 65.4°, flash p: 1.4°F (TCC), lel: 1.8%, uel: 11.8%, fp: −108.5°, d: 0.888 @ 20/4°, vap press: 114 mm @ 15°, vap d: 2.5, autoign temp: 610°F. Misc with water, alc, ketones, esters, ethers, and hydrocarbons.

SYNS: BUTYLENE OXIDE * CYCLOTETRAMETHYLENE OXIDE * DIETHYLENE OXIDE * 1,4-EPOXYBUTANE * FURANIDINE * HYDROFURAN * NCI-C60560 * OXACYCLOPENTANE * OXOLANE * RCRA WASTE NUMBER U213 * TETRAHYDROFURAAN (DUTCH) * TETRAHYDROFURANNE (FRENCH) * TETRAIDROFURANO (ITALIAN) * TETRAMETHYLENE OXIDE * THF

CONSENSUS REPORTS: Reported in EPA TSCA Inventory.

OSHA PEL: (Transitional: TWA 200 ppm) TWA 200 ppm; STEL 250 ppm
ACGIH TLV: TWA 200 ppm; STEL 250 ppm

DFG MAK: 200 ppm (590 mg/m³)
DOT Classification: Flammable Liquid; Label: Flammable Liquid.

SAFETY PROFILE: Moderately toxic by ingestion and intraperitoneal routes. Mildly toxic by inhalation. Human systemic effects by inhalation: general anesthesia. Mutation data reported. Irritant to eyes and mucous membranes. Narcotic in high concentrations. Reported as causing injury to liver and kidneys.

Flammable liquid. A very dangerous fire hazard when exposed to heat, flames, oxidizers. Explosive in the form of vapor when exposed to heat or flame. In common with ethers, unstabilized tetrahydrofuran forms thermally explosive peroxides on exposure to air. Stored THF must always be tested for peroxide prior to distillation. Peroxides can be removed by treatment with strong ferrous sulfate solution made slightly acidic with sodium bisulfate. Caustic alkalies deplete the inhibitor in THF and may subsequently cause an explosive reaction. Explosive reaction with KOH, NaAlH₂, NaOH, sodium tetrahydroaluminate. Reacts with 2-aminophenol + potassium dioxide to form an explosive product. Reacts with lithium tetrahydroaluminate or borane to form explosive hydrogen gas. Violent reaction with metal halides (e.g., hafnium tetrachloride, titanium tetrachloride, zirconium tetrachloride). Vigorous reaction with bromine, calcium hydride + heat. Can react with oxidizing materials. To fight fire, use foam, dry chemical, CO_2. When heated to decomposition it emits acrid smoke and irritating fumes.

TCS500   CAS: 4795-29-3   HR: 3
TETRAHYDROFURFURYLAMINE
DOT: UN 2943
mf: $C_5H_{11}NO$ mw: 101.17

SYN: USAF Q-2

CONSENSUS REPORTS: Reported in EPA TSCA Inventory.

DOT Classification: Flammable or Combustible Liquid; Label: Flammable Liquid.

SAFETY PROFILE: Poison by intraperitoneal route. Flammable liquid when exposed to heat or flame. When heated to decomposition it emits toxic fumes of NO_x.

TCU600   HR: 2
TETRAHYDROLINALOOL
mf: $C_{10}H_{22}O_2$ mw: 158.29

PROP: Colorless liquid; floral odor. D: 0.923, refr index: 1.431, flash p: 183°F. Sol in alc, fixed oils; insol in water.

SYNS: 3,7-DIMETHYL-3-OCTANOL * FEMA No. 3060

SAFETY PROFILE: Combustible liquid. When heated to decomposition it emits acrid smoke and irritating fumes.

TCW750   CAS: 33401-94-4   **HR: 3**
(E)-4,5,6-TETRAHYDRO-1-METHYL-2-(2-(2-THIENYL)ETHENYL)PYRIMIDINE
mf: $C_{11}H_{14}N_2S \cdot C_4H_6O_6$ mw: 356.43

SYNS: BANMINTH * CP 10423-18 * PYRANTEL TARTRATE * PYREQUAN TARTRATE * (E)-1,4,5,6-TETRAHYDRO-1-METHYL-2-(2-(2-THIENYL)VINYL)PYRI-MIDINE TARTARATE (1:1)

SAFETY PROFILE: Poison by ingestion and intravenous routes. When heated to decomposition it emits very toxic fumes of NO_x and SO_x.

TDA500   CAS: 52-31-3   **HR: 3**
TETRAHYDROPHENOBARBITAL
mf: $C_{12}H_{16}N_2O_3$ mw: 236.30

SYNS: ADORM * AMNOSED * CAVONLY * CYCLOBARBITAL * CYCLOBARBITOL * CY-CLOBARBITONE * CYCLODORM * CYCLOHEXE-NYL-ETHYL BARBITURIC ACID * 5-(1-CYCLOHEXE-NYL)-5-ETHYLBARBITURIC ACID * 5-(1-CYCLO-HEXEN-1-YL)-5-ETHYLBARBITURIC ACID * 5-(1-CY-CLOHEXEN-1-YL)-5-ETHYL-2,4,6(1H,3H,5H)-PYRIMI-DINETRIONE * 5-ETHYL-5-CYCLOHEXENYLBAR-BITURIC ACID * ETHYLHEXABITAL * FANO-DORMO * HEXEMAL * IRIFAN * NAMURON * PALINUM * PHANODORM * PHANODORN * PHILODORM * PRALUMIN * PRO-SONIL * SONAFORM

SAFETY PROFILE: Poison by ingestion, subcutaneous, intravenous, and intraperitoneal routes. Human systemic effects by ingestion: pulmonary consolidation. Used as a central nervous system depressant, hypnotic and sedative. When heated to decomposition it emits toxic fumes of NO_x.

TDB000   CAS: 85-43-8   **HR: 2**
TETRAHYDROPHTHALIC ACID ANHYDRIDE
DOT: UN 2698
mf: $C_8H_8O_3$ mw: 152.16

PROP: White powder. Mp: 101.9°, bp: 195° @ 50 mm, flash p: 315°F (OC), d: 1.375 @ 25/20°, vap press: 0.01 mm @ 20°, vap d: 5.25.

SYNS: ANHYDRID KYSELINY TETRAHYDROFTALOVE (CZECH) * MALEIC ANHYDRIDE adduct of BUTADIENE * TETRAHYDROFTALANHYDRID (CZECH) * 3a,4,7,7a-TETRAHYDRO-1,3-ISOBENZOFURANDIOINE * TETRAHYDROPHTHALIC ANHYDRIDE * Δ^4-TETRAHYDROPHTHALIC ANHYDRIDE * 1,2,3,6-TETRAHYDRO PHTHALIC ANHYDRIDE * THPA

CONSENSUS REPORTS: Reported in EPA TSCA Inventory.

DOT Classification: Corrosive Material; Label: None.

SAFETY PROFILE: Moderately toxic by intraperitoneal route. Mildly toxic by ingestion. A corrosive irritant to skin, eyes, and mucous membranes. Combustible when exposed to heat or flame. Will react with water or steam to produce heat; can react with oxidizing materials. To fight fire, use water, foam, CO_2, dry chemical. When heated to decomposition it emits acrid smoke and irritating fumes.

TDC730   CAS: 110-01-0   **HR: 2**
TETRAHYDROTHIOPHENE
DOT: UN 2412
mf: C_4H_8S mw: 88.17

DOT Classification: Flammable Liquid; Label: Flammable Liquid

SAFETY PROFILE: Potentially explosive reaction with hydrogen peroxide. When heated to decomposition it emits toxic fumes of SO_x.

TDK000   CAS: 75-57-0   **HR: 3**
TETRAMETHYLAMMONIUM CHLORIDE
mf: $C_4H_{12}N \cdot Cl$ mw: 109.62

SYN: USAF AN-8

CONSENSUS REPORTS: Reported in EPA TSCA Inventory.

SAFETY PROFILE: Poison by intraperitoneal, subcutaneous, and possibly other routes. When heated to decomposition it emits very toxic fumes of NO_x, NH_3, and Cl^-.

TDK500   CAS: 75-59-2   **HR: 3**
TETRAMETHYLAMMONIUM HYDROXIDE
DOT: UN 1835
mf: $C_4H_{12}N \cdot HO$ mw: 91.18

PROP: Liquid. D: 1.

SYNS: HYDROXYDE de TETRAMETHYLAMMONIUM (FRENCH) * TM * TETRAMETHYL AMMONIUM HYDROXIDE, liquid (DOT)

CONSENSUS REPORTS: Reported in EPA TSCA Inventory.

DOT Classification: Corrosive Material; Label: Corrosive.

SAFETY PROFILE: Poison by subcutaneous route. A powerful caustic. A corrosive irritant to skin, eyes, and mucous membranes. When heated to decomposition it emits toxic fumes of NO_x and NH_3.

TDQ750   CAS: 110-18-9   HR: 2
TETRAMETHYL ETHYLENE DIAMINE
DOT: UN 2372
mf: $C_6H_{16}N_2$ mw: 116.24

SYNS: 1,2-BIS-(DIMETHYLAMINO-ETHANE (DOT) * 1,2-DI-(DIMETHYLAMINO)ETHANE (DOT) * PROPAMINE D * TEMED * TETRAMEEN * N,N,N',N'-TETRAMETHYL-1,2-DIAMINOETHANE * N,N,N',N'-TETRAMETHYLETHYLENEDIAMINE * TMEDA

CONSENSUS REPORTS: Reported in EPA TSCA Inventory.

DOT Classification: Flammable Liquid; Label: Flammable Liquid.

SAFETY PROFILE: Moderately toxic by ingestion. Mildly toxic by skin contact. A skin and severe eye irritant. Flammable when exposed to heat or flame; can react with oxidizing materials. When heated to decomposition it emits toxic fumes of NO_x.

TDR500   CAS: 75-74-1   HR: 3
TETRAMETHYL LEAD
mf: $C_4H_{12}Pb$ mw: 267.35

PROP: Colorless liquid. Mp: −18°F, lel: 1.8%, bp: 110°, d: 1.99, vap d: 9.2, flash p: 100°F.

SYNS: TETRAMETHYLPLUMBANE * TML

CONSENSUS REPORTS: EPA Extremely Hazardous Substances List. Lead and its compounds are on the Community Right-To-Know List. Reported in EPA TSCA Inventory.

OSHA PEL: TWA 0.075 mg(Pb)/m³ (skin)
ACGIH TLV: TWA 0.15 mg(Pb)/m³ (skin)
DFG MAK: 0.01 ppm (0.075 mg/m³)

SAFETY PROFILE: Poison by ingestion, intraperitoneal, parenteral, and intravenous routes.

Moderately toxic by skin contact. Experimental teratogenic and reproductive effects. Lead and its compounds have dangerous central nervous system effects. A very dangerous fire hazard when exposed to heat, flame, or oxidizers. Moderate explosion hazard in the form of vapor when exposed to flame. May explode when heated above 90°C. Explosive reaction with tetrachlorotrifluoromethyl phosphorane. Can react vigorously with oxidizing materials. To fight fire, use water, foam, CO_2, dry chemical. When heated to decomposition it emits toxic fumes of Pb. Used as an octane enhancer for gasoline.

TDR750   CAS: 51-80-9   HR: 3
N,N,N′N′-TETRAMETHYLMETHANE-
DIAMINE
DOT: NA 9069
mf: $C_5H_{14}N_2$ mw: 102.21

PROP: Liquid.

SYNS: N,N,N′,N′-TETRAMETHYLDIAMINOMETHAN (GERMAN) * TETRAMETHYL METHYLENE DIAMINE (DOT)

CONSENSUS REPORTS: Reported in EPA TSCA Inventory.

DOT Classification: ORM-A; Label: None.

SAFETY PROFILE: Poison by intraperitoneal route. A very dangerous fire hazard when exposed to powerful oxidizers, heat or open flame. When heated to decomposition it emits toxic fumes of NO_x.

TDV725   CAS: 1124-11-4   HR: 3
TETRAMETHYLPYRAZINE
mf: $C_8H_{12}N_2$ mw: 136.22

PROP: White crystals or powder; fermented soybean odor. Mp: 85-90°, sol in alc, propylene glycol, fixed oils; sltly sol in water.

SYNS: FEMA No. 3237 * 2,3,5,6-TETRAMETHYL PYRAZINE (FCC)

CONSENSUS REPORTS: Reported in EPA TSCA Inventory.

SAFETY PROFILE: Poison by intravenous and intraperitoneal routes. Moderately toxic by ingestion. When heated to decomposition it emits toxic fumes of NO_x.

TDW250   CAS: 3333-52-6   HR: 3
TETRAMETHYLSUCCINONITRILE
mf: $C_8H_{12}N_2$ mw: 136.22

PROP: Crystallizes in plates; almost no odor. Mp: 169° (sublimes).

SYN: TMSN

CONSENSUS REPORTS: Cyanide and its compounds are on the Community Right-To-Know List.

OSHA PEL: TWA 0.5 ppm (skin)
ACGIH TLV: TWA 0.5 ppm (skin)
DFG MAK: 0.5 ppm (3 mg/m^3)
NIOSH REL: (Nitriles) CL 6 mg/m^3/15M

SAFETY PROFILE: Poison by ingestion, intraperitoneal, and intravenous routes. Experimental teratogenic and reproductive effects. A human skin irritant and allergen. In the preparation of sponge rubber, an azo compound is used which decomposes to form tetramethylsuccinonitrile or TSN. Rats exposed to a concentration of 90 ppm exhibit their first convulsion after 1.5-2 hours or less. Rats exposed to concentration of 5.5 ppm exhibited their first convulsions in 27-31 hours and were dead in from 31-46 hours. Absorbed by skin. The fatal dose in humans is thought to be about 25 mg/kg of body weight. TSN is slowly detoxified by the body. This nitrile is different from other nitriles in that thiosulfate is a poor antidote for intoxication. When heated to decomposition it emits toxic fumes of CN$^-$ and NO$_x$.

TDW500   CAS: 108-62-3   **HR: 3**
2,4,6,8-TETRAMETHYL-1,3,5,7-TETROXOCANE

DOT: UN 1332
mf: C$_8$H$_{16}$O$_4$ mw: 176.24

SYNS: ACETALDEHYDE, TETRAMER * ANTIMILACE * ARIOTOX * CEKUMETA * CORRY'S SLUG DEATH * HALIZAN * META * METACETALDEHYDE * METALDEHYD (GERMAN) * METALDEHYDE (DOT) * METALDEIDE (ITALIAN) * METASON * NAMEKIL * SLUGTOX

DOT Classification: Flammable Solid; Label: Flammable Solid.

SAFETY PROFILE: Human poison by ingestion. Moderately toxic experimentally by ingestion and possibly other routes. Human systemic effects by ingestion: convulsions or effect on seizure threshold. Experimental reproductive effects. Mutation data reported. A flammable solid. When heated to decomposition it emits acrid smoke and irritating fumes.

TDX250   CAS: 632-22-4   **HR: 2**
1,1,3,3-TETRAMETHYLUREA
mf: C$_5$H$_{12}$N$_2$O   mw: 116.19

PROP: Liquid, fat odor. Bp: 177°, mp: −1.2°, d: 0.969, flash p: 167°F. Very sol in alc, ether; misc with water.

SYNS: TEMUR * TETRAMETHYLUREA * TETRAMETHYLUREE (FRENCH) * TMU

CONSENSUS REPORTS: Reported in EPA TSCA Inventory.

SAFETY PROFILE: Moderately toxic by ingestion and intravenous routes. Experimental teratogenic and reproductive effects. Human mutation data reported. Flammable when exposed to heat, flame, and oxidizers. To fight fire, use foam, mist, spray, dry chemicals. When heated to decomposition it emits toxic fumes of NO$_x$.

TDY000   CAS: 3698-54-2   **HR: 3**
TETRANITROANILINE
mf: C$_6$H$_3$N$_5$O$_8$ mw: 273.14

PROP: Solid. Mp: 170°, bp: explodes @ 237°.

SYNS: TETRANITRANILINE (FRENCH) * 2,3,4,6-TETRANITROANILINE * TNA

SAFETY PROFILE: Moderately toxic by subcutaneous route. For fire hazard, see NITRATES. A severe explosion hazard when shocked or exposed to heat (at 237°C). Tetranitroaniline is a powerful and sensitive high explosive, similar to tetryl. It deteriorates in the presence of moisture. Incompatible with reducing materials. When heated to decomposition it emits toxic fumes of NO$_x$.

TDY075   CAS: 4591-46-2   **HR: 3**
N,2,4,6-TETRANITROANILINE
mf: C$_6$H$_3$N$_5$O$_8$ mw: 273.12

DOT Classification: Forbidden

SAFETY PROFILE: The impure material deflagrates (burns explosively) when heated to 50°C. When heated to decomposition it emits toxic fumes of NO$_x$.

TDY250   CAS: 509-14-8   **HR: 3**
TETRANITROMETHANE

DOT: UN 1510
mf: CN$_4$O$_8$ mw: 196.05

PROP: Colorless or yellow liquid. Mp: 13°, bp: 125.7°, d: 1.650 @ 13°, vap press: 10 mm @ 22.7°. Insol in water; very sol in alc, ether.

SYNS: NCI-C55947   *   RCRA WASTE NUMBER P112   *   TNM

CONSENSUS REPORTS: EPA Extremely Hazardous Substances List. Reported in EPA TSCA Inventory.

OSHA PEL: TWA 1 ppm
ACGIH TLV: TWA 1 ppm
DFG MAK: 1 ppm (8 mg/m^3)
DOT Classification: Oxidizer; Label: Oxidizer.

SAFETY PROFILE: Poison by ingestion, inhalation, intravenous and intraperitoneal routes. Irritating to the skin, eyes, mucous membranes, and respiratory passages, and does serious damage to the liver. It occurs as an impurity in crude TNT, and is thought to be mainly responsible for the irritating properties of that material. It can cause pulmonary edema, mild methemoglobinemia and fatty degeneration of the liver and kidneys.

A powerful oxidizer. A very dangerous fire hazard. A severe explosion hazard when shocked or exposed to heat. May explode during distillation. Potentially explosive reaction with ferrocene, pyridine, sodium ethoxide. Mixtures with amines (e.g., aniline) ignite spontaneously and may explode. Mixtures with cotton or toluene may explode when ignited. Forms sensitive and powerful explosive mixtures with nitrobenzene, 1-nitrotoluene, 4-nitrotoluene, 1,3-dinitrobenzene, 1-nitronaphtahlene, other oxygen-deficient explosives, hydrocarbons. Can react vigorously with oxidizing materials. Incompatible with aluminum. When heated to decomposition it emits highly toxic fumes of NO$_x$. Used as an oxidizer in rocket propellants and as an explosive.

TDY600   CAS: 641-16-7   **HR: 3**
2,3,4,6-TETRANITROPHENOL
mf: C$_6$H$_4$N$_2$O$_9$ mw: 248.11

DOT Classification: Forbidden

SAFETY PROFILE: A powerful explosive. When heated to decomposition it emits toxic fumes of NO$_x$.

TED750   CAS: 3440-75-3   **HR: 3**
TETRAPROPYL LEAD
mf: C$_{12}$H$_{28}$Pb   mw: 379.59

PROP: Colorless liquid. D: 1.44, bp: 126° @ 13 mm. Sol in benzene.

CONSENSUS REPORTS: Lead and its compounds are on the Community Right-To-Know List. Reported in EPA TSCA Inventory.

SAFETY PROFILE: Poison by ingestion and parenteral routes. When heated to decomposition it emits toxic fumes of Pb.

TEE500   CAS: 7722-88-5   **HR: 3**
TETRASODIUM PYROPHOSPHATE
mf: O$_7$P$_2$•4Na   mw: 265.90

PROP: White crystalline powder. Mp: 988°, d: 2.534. Sol in water; insol in alc.

SYNS: NATRIUMPYROPHOSPHAT   *   PHOSPHOTEX *   PYROPHOSPHATE   *   SODIUM PYROPHOSPHATE (FCC)   *   TETRANATRIUMPYROPHOSPHAT (GERMAN) *   TETRASODIUM DIPHOSPHATE   *   TETRASODIUM PYROPHOSPHATE, ANHYDROUS   *   TSPP   *   VICTOR TSPP

CONSENSUS REPORTS: Reported in EPA TSCA Inventory.

OSHA PEL: TWA 5 mg/m^3
ACGIH TLV: TWA 5 mg/m^3

SAFETY PROFILE: Poison by ingestion, intraperitoneal, intravenous, and subcutaneous routes. It is not a cholinesterase inhibitor. When heated to decomposition it emits toxic fumes of PO$_x$ and Na$_2$O.

TEF500   CAS: 109-27-3   **HR: 3**
TETRAZENE
DOT: UN 0114
mf: C$_2$H$_8$N$_{10}$O   mw: 188.20

PROP: Crystals.

SYNS: 4-AMIDINO-1-(NITROSAMINOAMIDINO)-1-TETRAZENE   *   GUANYL NITROSAMINO GUANYL TETRAZENE, containing, by weight, at least 30% water (DOT)   *   GUANYL NITROSAMINO GUANYL TETRAZENE (DOT)   *   1-GUANYL-4-NITROSAMINOGUANYLTETRAZENE   *   INITIATING EXPLOSIVE-TETRAZENE (DOT)   *   TETRACENE   *   TETRACENE EXPLOSIVE

CONSENSUS REPORTS: Reported in EPA TSCA Inventory.

DOT Classification: Class A Explosive; Label: Explosive.

SAFETY PROFILE: Many nitrosamines are carcinogens. A very dangerous fire hazard. A shock- and heat-sensitive high explosive which evolves much flame. Highly dangerous. Upon decomposition it emits highly toxic fumes of NO_x.

TEG250   CAS: 479-45-8   **HR: 3**
TETRYL

DOT: UN 0208
mf: $C_7H_5N_5O_8$ mw: 287.17

PROP: Yellow, monoclinic crystals. Mp: 130°, bp: explodes @ 187°, d: 1.57 @ 19°.

SYNS: N-METHYL-N,2,4,6-TETRANITROANILINE * PICRYLMETHYLNITRAMINE * PICRYLNITRO-METHYLAMINE * TETRALITE * N,2,4,5-TETRANITRO-N-METHYLANILINE * 2,4,6-TETRYL * TRINITROPHENYLMETHYLNITRAMINE * 2,4,6-TRINITROPHENYLMETHYLNITRAMINE * 2,4,6-TRINITROPHENYL-N-METHYLNITRAMINE

OSHA PEL: (Transitional: TWA 1.5 mg/m³ (skin)) TWA 0.1 mg/m³ (skin)
ACGIH TLV: TWA 1.5 mg/m³
DOT Classification: Class A Explosive; Label: Explosive A.

SAFETY PROFILE: Mutation data reported. An irritant, sensitizer, and allergen. The chief effect from exposure is dermatitis. Conjunctivitis is followed by iridocyclitis, and keratitis can occur. Sensitization produced by exposure may play a part in these symptoms. Gastrointestinal effects and anemia have also been reported.

A powerful oxidant. A dangerous fire and explosion hazard. A high explosive sensitive to shock, friction, or heat. More sensitive to shock and friction than TNT. Explodes on contact with trioxygen difluoride. Ignites on contact with hydrazine. When heated to decomposition it emits toxic fumes of NO_x.

TEH500   CAS: 50-35-1   **HR: 3**
THALIDOMIDE
mf: $C_{13}H_{10}N_2O_4$ mw: 258.25

PROP: Needles. Mp: 269-271°. Sltly sol in water, methanol, ethanol, acetone, ethylacrylate; very sol in dioxane; sol in ether.

SYNS: ALGOSEDIV * ASIDON 3 * ASMADION * ASMAVAL * BONBRAIN * CALMORE * CALMOREX * CONTERGAN * CORRONAROBETIN * 2,6-DIOXO-3-PHTHALIMIDOPIPERIDINE * 2-(2-6-DIOXO-3-PIPERIDINYL)1H-ISOINDOLE-1,3(2H)-DIONE * N-(2,6-DIOXO-3-PIPERIDYL)PHTHALIMIDE * DISTAVAL * DISTAXAL * DISTOVAL * ECTILURAN * ENTEROSEDIV * GASTRINIDE * GLUPAN * GLUTANON * GRIPPEX * HIPPU-ZON * IMIDA-LAB * IMIDAN (PEYTA) * IMIDENE * ISOMIN * K 17 * KEDAVON * KEVADON * LULAMIN * NEAUFATIN * NEO * NEOSEDYN * NEOSYDYN * NEURODYN * NEUROSEDIN * NEVRODYN * NIBROL * NOCTOSEDIV * NOXODYN * NSC-66847 * PANGUL * PANTOSEDIV * α-PHTHALIMIDO-GLUTARIMIDE * 2-PHTHALIMIDOGLUTARIMIDE * α-(N-PHTHALIMIDO)GLUTARIMIDE * 3-PHTHALIMIDOGLUTARIMIDE * N-PHTHALOYLGLUTAMIMIDE * N-PHTHALYLGLUTAMIC ACID IMIDE * N-PHTHALYL-GLUTAMINSAEURE-IMID (GERMAN) * α-N-PHTHALYLGLUTARAMIDE * POLY-GIRON * POLYGRIPAN * PREDNI-SEDIV * PRO-BAN M * PROFARMIL * PSYCHOLIQUID * PSYCHOTABLETS * QUETIMID * QUIETOPLEX * SANDORMIN * SEDALIS SEDI-LAB * SEDIMIDE * SEDIN * SEDISPERIL * SEDOVAL * SHIN-NAITO S * SHINNIBROL * SLEEPAN * SLIPRO * SOFTENIL * SOFTENON * TALARGAN * TALIMOL * TELAGAN * TELARGAN * TELARGEAN * TENSIVAL * THALIN * THALINETTE * THEOPHILCHOLINE * ULCERFEN * VALGIS * VALGRAINE * YODOMIN

CONSENSUS REPORTS: EPA Genetic Toxicology Program.

SAFETY PROFILE: Poison by ingestion. Moderately toxic by skin contact and intraperitoneal routes. Human teratogenic effects by ingestion: developmental abnormalities of the musculoskeletal, cardiovascular and possibly other systems. Experimental teratogenic and reproductive effects. Questionable carcinogen with experimental tumorigenic data. Human mutation data reported. It was commonly used as a prescription drug in Europe in the late 1950s and early 1960s. Its use was discontinued because it was discovered to cause serious congenital abnormalities in the fetus, notably amelia and phocomelia (absence or deformity of the limbs including hands and feet) when taken by a woman during early pregnancy. When heated to decomposition it emits toxic fumes of NO_x. Used as a sedative and hypnotic.

TEI000   CAS: 7440-28-0   **HR: 3**
THALLIUM
af: Tl   aw: 204.37

PROP: Bluish-white, soft, malleable metal. Mp: 303.5°, bp: 1457°, d: 11.85 @ 20°, vap press: 1 mm @ 825°.

SYN: RAMOR

CONSENSUS REPORTS: Thallium and its compounds are on the Community Right-To-Know List. Reported in EPA TSCA Inventory.

OSHA PEL: TWA 0.1 mg(Tl)/m^3 (skin)
ACGIH TLV: TWA 0.1 mg(Tl)/m^3 (skin)
DFG MAK: 0.1 mg/m^3

SAFETY PROFILE: Human poison by unspecified route. Human systemic effects by ingestion: nerve or sheath structural changes, extra-ocular muscle changes, sweating, and other effects. Flammable in the form of dust when exposed to heat or flame. Violent reaction with F_2. When heated to decomposition it emits toxic fumes of Tl. Used as a rodenticide and fungicide, and in lenses and prisms, in high-density liquids.

TEI250   CAS: 563-68-8   HR: 3
THALLIUM ACETATE
mf: $C_2H_3O_2 \cdot Tl$ mw: 263.42

PROP: Silk-white crystals. Mp: 110°, d: 3.68. Sol in water, alc.

SYNS: RCRA WASTE NUMBER U214 * THALLIUM (1+) ACETATE * THALLIUM(I) ACETATE * THALLIUM MONOACETATE * THALLOUS ACETATE

CONSENSUS REPORTS: Thallium and its compounds are on the Community Right-To-Know List. EPA Genetic Toxicology Program. Reported in EPA TSCA Inventory.

OSHA PEL: TWA 0.1 mg(Tl)/m^3 (skin)
ACGIH TLV: TWA 0.1 mg(Tl)/m^3 (skin)

SAFETY PROFILE: Human poison by unspecified routes. Experimental poison by ingestion, intravenous, intraperitoneal, and subcutaneous routes. Experimental reproductive effects. Mutation data reported. When heated to decomposition it emits toxic fumes of Tl.

TEJ250   CAS: 7791-12-0   HR: 3
THALLIUM CHLORIDE
mf: ClTl   mw: 239.82

PROP: Colorless or white powder. Mp: 430°, bp: 720°, d: 7.00, vap press: 10 mm @ 517°.

SYNS: RCRA WASTE NUMBER U216 * THALLIUM (1+) CHLORIDE * THALLIUM MONOCHLORIDE * THALLOUS CHLORIDE

CONSENSUS REPORTS: EPA Extremely Hazardous Substances List. Thallium and its compounds are on the Community Right-To-Know List. EPA Genetic Toxicology Program. Reported in EPA TSCA Inventory.

OSHA PEL: TWA 0.1 mg(Tl)/m^3 (skin)
ACGIH TLV: TWA 0.1 mg(Tl)/m^3 (skin)

SAFETY PROFILE: Poison by ingestion and intraperitoneal routes. Experimental teratogenic and reproductive effects. Mutation data reported. Incompatible with F_2. When heated to decomposition it emits very toxic fumes of Cl^- and Tl.

TEJ500   HR: 3
THALLIUM COMPOUNDS

CONSENSUS REPORTS: Thallium and its compounds are on the Community Right-To-Know List.

SAFETY PROFILE: Extremely toxic. The lethal dose for a man by ingestion is 0.5-1.0 gram. Effects are cumulative and with continuous exposure toxicity occurs at much lower levels. Major effects are on the nervous system, skin, and cardiovascular tract. The peripheral nervous system can be severely affected with dying-back of the longest sensory and motor fibers. Reproductive organs and the fetus are highly susceptible. Acute poisoning has followed the ingestion of toxic quantities of a thallium-bearing depilatory and accidental or suicidal ingestion of rat poison. Acute poisoning results in swelling of the feet and legs, arthralgia, vomiting, insomnia, hyperesthesia and paresthesia of the hands and feet, mental confusion, polyneuritis with severe pains in the legs and loins, partial paralysis of the legs with reaction of degeneration, angina-like pains, nephritis, wasting and weakness, and lymphocytosis and eosinophilia. About the 18th day, complete loss of the hair on the body and head may occur. Fatal poisoning has been known to occur. Recovery requires months and may be incomplete. Industrial poisoning is reported to have caused discoloration of the hair (which later falls out), joint pain, loss of appetite, fatigue, severe pain in the calves of the legs, albuminuria, eosinophilia, lymphocytosis, and optic neuritis followed by atrophy. Cases of industrial poisoning are rare, however. Thallium is an experimental teratogen. When heated to decomposition they emit highly toxic fumes of Tl.

TEK750   CAS: 10102-45-1   **HR: 3**
THALLIUM NITRATE

DOT: UN 2727
mf: NO₃ • Tl   mw: 266.38

PROP: Cubic crystals. Mp: 206°, bp: 430°, d: 5.55. Decomp @ 450°.

SYNS: NITRIC ACID, THALLIUM(1+) SALT * RCRA WASTE NUMBER U217 * THALLIUM MONONITRATE * THALLOUS NITRATE

CONSENSUS REPORTS: Thallium and its compounds are on the Community Right-To-Know List. Reported in EPA TSCA Inventory.

OSHA PEL: TWA 0.1 mg(Tl)/m³ (skin)
ACGIH TLV: TWA 0.1 mg(Tl)/m³ (skin)
DOT Classification: Poison B; Label:Poison

SAFETY PROFILE: Poison by ingestion, intravenous, intraperitoneal, and subcutaneous routes. Human systemic effects by ingestion: hypermotility, diarrhea, nausea or vomiting, and dehydration. Mutation data reported. When heated to decomposition it emits very toxic fumes of Tl and NO$_x$.

TEL050   CAS: 1314-32-5   **HR: 3**
THALLIUM(III) OXIDE
mf: O₃Tl₂   mw: 456.74

PROP: Hexagonal black crystals, amorphous prisms. Mp: 717° ± 5°, bp: −O₂ @ 875°, d(amorphous): 9.65 @ 21°, d(hexagonal): 10.19 @ 22°.

SYNS: DITHALLIUM TRIOXIDE * RCRA WASTE NUMBER P113 * THALLIC OXIDE * THALLIUM OXIDE * THALLIUM (3+) OXIDE * THALLIUM PEROXIDE * THALLIUM SESQUIOXIDE

CONSENSUS REPORTS: Thallium and its compounds are on the Community Right-To-Know List. Reported in EPA TSCA Inventory.

OSHA PEL: TWA 0.1 mg(Tl)/m³ (skin)
ACGIH TLV: TWA 0.1 mg(Tl)/m³ (skin)

SAFETY PROFILE: Poison by ingestion, intraperitoneal, and intravenous routes. Combustible by chemical reaction. Evolves O₂ @ 875°. Mixtures with sulfur or antimony trisulfide explode when ground. Hydrogen sulfide ignites and may explode weakly on contact with the oxide. When heated to decomposition it emits toxic fumes of Tl.

TEL750   CAS: 10031-59-1   **HR: 3**
THALLIUM SULFATE

DOT: NA 1707
mf: O₄S • xTl   mw: 1526.65

PROP: Colorless crystals. Mp: 632°, bp: decomp, d: 6.77.

SYNS: RATOX * SULFURIC ACID, THALLIUM SALT * THALLIUM SULFATE, solid (DOT) * ZELIO

CONSENSUS REPORTS: Thallium and its compounds are on the Community Right-To-Know List. EPA Extremely Hazardous Substances List. Reported in EPA TSCA Inventory.

OSHA PEL: TWA 0.1 mg(Tl)/m³ (skin)
ACGIH TLV: TWA 0.1 mg(Tl)/m³ (skin)
DOT Classification: Poison B; Label: Poison.

SAFETY PROFILE: Poison to humans by ingestion. Experimental poison by ingestion, intravenous, and subcutaneous routes. Its main hazard is due to its cumulation, especially in liver, brain, and skeletal muscle; readily absorbed by gastrointestinal tract and skin. A cellular toxicant like arsenic. Fatal human dose is about 500 mg of thallium. Intake of thallium causes depilation. Many reported fatalities. When heated to decomposition it emits very toxic fumes of SO$_x$ and Tl. Pesticide for control of rats, moles, and house mice.

TEM000   CAS: 7446-18-6   **HR: 3**
THALLIUM(I) SULFATE (2:1)
mf: O₄S • 2Tl   mw: 504.80

SYNS: C.F.S. * CSF-GIFTWEIZEN * DITHALLIUM SULFATE * DITHALLIUM(1+) SULFATE * ECCOTHAL * M7-GIFTKOERNER * RATTENGIFTKONSERVE * RCRA WASTE NUMBER P115 * SULFURIC ACID, DITHALLIUM(1+) SALT (8CI, 9CI) * SULFURIC ACID, THALLIUM(1+) SALT (1:2) * THALLOUS SULFATE

CONSENSUS REPORTS: Thallium and its compounds are on the Community Right-To-Know List. Reported in EPA TSCA Inventory.

OSHA PEL: TWA 0.1 mg(Tl)/m³ (skin)
ACGIH TLV: TWA 0.1 mg(Tl)/m³ (skin)

SAFETY PROFILE: Human poison by ingestion. Experimental poison by ingestion and subcutaneous routes. Human systemic effects by ingestion: nerve or sheath structural changes, wakefulness, somnolence, excitement, change in heart rate, and nausea or vomiting. When heated to decomposition it emits very toxic fumes of Tl and SO_x. Used as a rat poison, ant bait, and a reagent in analytical chemistry.

TEM250   CAS: 63906-56-9   HR: 3
THALLIUM(II) SULFATE (1:1)
mf: $O_4S \cdot Tl$ mw: 300.43

SYN: SULFURIC ACID, THALLIUM(2+) SALT

CONSENSUS REPORTS: Thallium and its compounds are on the Community Right-To-Know List.

OSHA PEL: TWA 0.1 mg(Tl)m^3 (skin)
ACGIH TLV: TWA 0.1 mg(Tl)/m^3 (skin)

SAFETY PROFILE: Poison by ingestion. When heated to decomposition it emits very toxic fumes of Tl and SO_x.

TEO250   CAS: 91-80-5   HR: 3
THENYLPYRAMINE
mf: $C_{14}H_{19}N_3S$ mw: 261.42

SYNS: A 3322 * AH-42 * 2-((2-(DIMETHYL-AMINO)ETHYL)-2-THENYLAMINO)PYRIDINE * N,N-DIMETHYL-N'-2-PYRIDINYL-N'-(2-THIENYLMETHYL)-1,2-ETHANEDIAMINE * N,N-DIMETHYL-N'-PYRID-2-YL-N'-2-THENYLETHYLENEDIAMINE * DORMIN * HISTADYL * LULAMIN * LULLAMIN * METHAPYRILENE * NCI-C55550 * PARADOR-MALENE * PYRATHYN * N-(α-PYRIDYL)-N-(α-THENYL)-N',N'-DIMETHYLETHYLENEDIAMINE * PYRINISTAB * PYRINISTOL * RCRA WASTE NUMBER U155 * REST-ON * RESTRYL * SEMI-KON * SLEEPWELL * TENALIN * THENYLENE * THIONYLAN

SAFETY PROFILE: Poison by ingestion, subcutaneous, intraperitoneal, and intravenous routes. Mutation data reported. When heated to decomposition it emits very toxic fumes of SO_x and NO_x. Used as an antihistamine.

TEO500   CAS: 83-67-0   HR: 3
THEOBROMINE
mf: $C_7H_8N_4O_2$ mw: 180.19

PROP: White powder, bitter tasting alkaloid. Mp: 357°, sublimes @ 290°-295°. Moderately sol in ammonia; almost insol in benzene, ether, chloroform, carbon tetrachloride.

SYNS: 3,7-DIHYDRO-3,7-DIMETHYL-1H-PURINE-2,6-DIONE * 3,7-DIMETHYLXANTHINE * DIUROBRO-MINE * SANTHEOSE * SC 15090 * TEOBROMIN

* THEOSALVOSE * THEOSTENE * THESAL * THESODATE

CONSENSUS REPORTS: Reported in EPA TSCA Inventory. EPA Genetic Toxicology Program.

SAFETY PROFILE: Poison by ingestion. Moderately toxic by subcutaneous route. Human systemic effects by ingestion: central nervous system and gastrointestinal changes. Experimental teratogenic and reproductive effects. Human mutation data reported. When heated to decomposition it emits toxic fumes of NO_x. Used as a diuretic, smooth muscle relaxant, cardiac stimulant, and vasodilator.

TEP000   CAS: 58-55-9   HR: 3
THEOPHYLLINE
mf: $C_7H_8N_4O_2$ mw: 180.19

PROP: Monoclinic, odorless needles; bitter taste. Mp: 270-274°. Sol in hot water, alkali hydroxides, ammonia, dil HCl, HNO_3; sltly sol in ether.

SYNS: ACET-THEOCIN * AMINOPHYLLINE * 3,7-DIHYDRO-1,3-DIMETHYL-1H-PURINE-2,6-DIONE * 1,3-DIMETHYLXANTHINE * ELIXICON * ELIX-OPHYLLIN * ELIXOPHYLLINE * LANOPHYLLIN * LIQUOPHYLLINE * NSC 2066 * OPTIPHYLLIN * PARKOPHYLLIN * PSEUDOTHEOPHYLLINE * SLO-PHYLLIN * SOLOSIN * TEFAMIN * TEOFYLLAMIN * THEAL TABL. * THEOCIN * THEOFOL * THEOGRAD * THEOLAIR * THEOLIX * THEOPHYL-225 * THEOPHYLLIN * THEOPHYLLINE, anhydrous

CONSENSUS REPORTS: Reported in EPA TSCA Inventory. EPA Genetic Toxicology Program.

SAFETY PROFILE: Human poison by ingestion, parenteral, intravenous, and rectal routes. Experimental poison by multiple routes. Human systemic effects by ingestion, subcutaneous, intravenous, intramuscular, and rectal routes: somnolence, tremor, convulsions or effect on seizure threshold, coma, heart arrythmias, heart rate change, cyanosis, respiratory stimulation, salivary gland changes, nausea or vomiting, fever and other metabolic effects. Experimental teratogenic and reproductive effects. Human mutation data reported. Used as a diuretic, cardiac stimulant, smooth muscle relaxant, and to treat asthma. When heated to decomposition it emits toxic fumes of NO_x.

TER000 HR: 3
"THERMIT"

PROP: Composition: Fe_2O_3 + Al.

SAFETY PROFILE: Dangerous when exposed to heat or flame. The violent reaction of Fe_2O_3 + Al is typical of a series of the oxide-metal "thermite" reactions. They are very difficult to stop, as they supply their own oxygen. They may attain a temperature of about 2500°. The presence of manganese dioxide may cause explosions. Keep away from combustible materials.

TER500 CAS: 19525-20-3 HR: 3
THIABENDAZOLE HYDROCHLORIDE
mf: $C_{10}H_7N_3S \cdot ClH$ mw: 237.72

SYN: 2-(4-THIAZOLYL)-BENZIMIDAZOLE, HYDROCHLORIDE

SAFETY PROFILE: Poison by intravenous route. Moderately toxic by ingestion and intraperitoneal routes. When heated to decomposition it emits very toxic fumes of HCl, SO_x, and NO_x.

TET300 CAS: 67-03-8 HR: 3
THIAMINE HYDROCHLORIDE
mf: $C_{12}H_{17}N_4OS \cdot ClH \cdot Cl$ mw: 337.30

PROP: Small white hygroscopic crystals or crystalline powder; nut-like odor. Mp: 248° (decomp). Sol in water, glycerin; sltly sol in alc; insol in ether, benzene.

SYNS: THIAMINE CHLORIDE HYDROCHLORIDE * THIAMINE DICHLORIDE * THIAMIN HYDROCHLORIDE * THIAMINIUM CHLORIDE HYDROCHLORIDE * USAF CB-20 * VITAMIN B¹ * VITAMIN B HYDROCHLORIDE

CONSENSUS REPORTS: Reported in EPA TSCA Inventory.

SAFETY PROFILE: Poison by intravenous and intraperitoneal routes. Mildly toxic by ingestion. The vitamin is destroyed by alkalies and alkaline drugs such as phenobarbital sodium and by oxidizing and reducing agents. When heated to decomposition it emits very toxic fumes of HCl, Cl^-, SO_x, and NO_x.

TET500 CAS: 532-43-4 HR: 3
THIAMINE MONONITRATE
mf: $C_{12}H_{17}N_4OS \cdot NO_3$ mw: 327.40

PROP: White crystals or crystalline powder; slt characteristic odor. Mp: 196-200° (decomp).

Non-hygroscopic; sltly sol in water, alc, and chloroform.

SYNS: 3-(4-AMINO-2-METHYLPYRIMIDYL-5-METHYL)-4-METHYL-5,β-HYDROXYETHYLTHIAZOLIUM NITRATE * THIAMINE NITRATE * VITAMIN B1 MONONITRATE * VITAMIN B1 NITRATE

CONSENSUS REPORTS: Reported in EPA TSCA Inventory.

SAFETY PROFILE: Poison by intravenous and intraperitoneal routes. A powerful oxidizer. When heated to decomposition it emits very toxic fumes of NO_x and SO_x.

TET800 CAS: 55297-95-5 HR: 3
THIAMUTILIN
mf: $C_{28}H_{47}NO_4S$ mw: 493.82

PROP: Crystals from acetone. Mp: 147-148° (after stirring in ethyl acetate and drying at 60° and 80° overnight).

SYNS: 14-DEOXY-14-((2-DIETHYLAMINOETHYL-THIO)-ACETOXY)MUTILINE * 14-DESOSSI-14-((2-DIETIL-AMINOETIL)MERCAPTO-ACETOSSI)MUTILIN IDROGENO FUMARATO (ITALIAN) * 14-DESOXY-14-((DIETHYL-AMINOETHYL)-MERCAPTO ACETOXYL)-MUTILIN HYDROGEN FUMARATE * DYNALIN INJECTABLE * DYNAMUTILIN * 81723 HFU * SQ 14055 * SQ 22947 * TIAMULIN * TIAMULINA (ITALIAN)

SAFETY PROFILE: Poison by intramuscular and intravenous routes. Moderately toxic by ingestion and subcutaneous routes. When heated to decomposition it emits toxic fumes of SO_x and NO_x.

TEX000 CAS: 148-79-8 HR: 3
2-(THIAZOL-4-YL)BENZIMIDAZOLE
mf: $C_{10}H_7N_3S$ mw: 201.26

PROP: White-to-tan; odorless. Mp: 304°. Insol in water; sltly sol in alc, acetone; very sltly sol in ether, chloroform.

SYNS: APL-LUSTER * ARBOTECT * 4-(2-BENZIMIDAZOLYL)THIAZOLE * BOVIZOLE * EPROFIL * EQUIZOLE * LOMBRISTOP * MERTEC * METASOL TK-100 * MINTEZOL * MINZOLUM * MK 360 * MYCOZOL * NEMAPAN * OMNIZOLE * POLIVAL * TBDZ * TECTO * THIABEN * THIABENDAZOLE (USDA) * THIABENZOLE * 2-(4-THIAZOLYL)BENZIMIDAZOLE * 2-(4'-THIAZOLYL)BENZIMIDAZOLE * 2-(4-THIAZOLYL)-1H-BENZIMIDAZOLE * THIBENZOLE * TOP FORM WORMER

CONSENSUS REPORTS: EPA Genetic Toxicology Program. Reported in EPA TSCA Inventory.

SAFETY PROFILE: Poison by ingestion. Experimental teratogenic and reproductive effects. Mutation data reported. When heated to decomposition it emits toxic fumes of SO_x and NO_x.

TEX250   CAS: 72-14-0   **HR: 3**
N¹-2-THIAZOLYLSULFANILAMIDE
mf: $C_9H_9N_3O_2S_2$ mw: 255.33

SYNS: 2-(p-AMINOBENZENESULFONAMIDO)THIAZOLE * 2-(p-AMINOBENZENESULPHONAMIDO)THIAZOLE * 4-AMINO-N-2-THIAZOLYLBENZENESULFONAMIDE * AZOSEPTALE * CERAZOL (suspension) * CHEMOSEPT * DUATOK * ELEUDRON * FORMOSULFATHIAZOLE * M+B 760 * NEOSTREPSAN * NORSULFASOL * NORSULFAZOLE * PLANOMIDE * POLISEPTIL * RP 2990 * STREPTOSILTHIAZOLE * SULFAMUL * 2-SULFANILAMIDOTHIAZOLE * 2-(SULFANILYLAMINO)THIAZOLE * SULFATHIAZOL * SULFATHIAZOLE (USDA) * 2-SULFONAMIDOTHIAZOLE * SULPHATHIAZOLE * SULZOL * THIACOCCINE * THIAZAMIDE * THIOZAMIDE * USAF SN-9

CONSENSUS REPORTS: Reported in EPA TSCA Inventory. EPA Genetic Toxicology Program.

SAFETY PROFILE: Human poison by unspecified route. Experimental poison by intraperitoneal route. Moderately toxic by intravenous, subcutaneous, and parenteral routes. Mildly toxic by ingestion. Human systemic effects by unspecified route: conjuctiva irritation, tubule changes and allergic skin dermatitis. Questionable carcinogen with experimental tumorigenic data. Mutation data reported. When heated to decomposition it emits very toxic fumes of NO_x and SO_x.

TEX600   CAS: 51707-55-2   **HR: 2**
THIDIAZURON
mf: $C_9H_8N_4OS$ mw: 220.27

SYNS: DEFOLIT * DROPP * N-PHENYL-N'-1,2,3-THIADIAZOL-5-YL-UREA * SN 49537 * (N-1,2,3-THIADIAZOLYL-5)-N'-PHENYLUREA

SAFETY PROFILE: Moderately toxic by ingestion. Experimental reproductive effects. When heated to decomposition it emits toxic fumes of SO_x and NO_x.

TFA000   CAS: 62-55-5   **HR: 3**
THIOACETAMIDE
mf: C_2H_5NS mw: 75.14

PROP: Colorless leaflets; mercaptan odor. Mp: 113°. Very sol in water; sltly sol in alc and ether.

SYNS: ACETOTHIOAMIDE * ETHANETHIOAMIDE * RCRA WASTE NUMBER U218 * TAA * THIACETAMIDE * USAF CB-21 * USAF EK-1719

CONSENSUS REPORTS: IARC Cancer Review: GROUP 2B IMEMDT 7,56,87; Animal Sufficient Evidence IMEMDT 7,77,74. NTP Fourth Annual Report On Carcinogens, 1984. EPA Genetic Toxicology Program. Community Right-To-Know List. Reported in EPA TSCA Inventory.

SAFETY PROFILE: Confirmed carcinogen with experimental carcinogenic, neoplastigenic, and tumorigenic data. Poison by ingestion and intraperitoneal routes. Moderately toxic by subcutaneous route. Experimental teratogenic and reproductive effects. Human mutation data reported. Exposure has caused liver damage. When heated to decomposition it emits very toxic fumes of NO_x and SO_x.

TFA500   CAS: 507-09-5   **HR: 3**
THIOACETIC ACID
mf: C_2H_4OS mw: 76.12

DOT: UN 2436

PROP: Colorless liquid; pungent, disagreeable odor. Mp: <−17°, bp: 93°, d: 1.074 @ 10/4°. flash p: <73.4°. Sol in water; misc in alc, ether.

SYNS: ACETYL MERCAPTAN * ETHANETHIOIC ACID * ETHANETHIOLIC ACID * METHANECARBOTHIOLIC ACID * THIACETIC ACID * THIOLACETIC ACID * THIONOACETIC ACID * USAF EK-P-737

CONSENSUS REPORTS: Reported in EPA TSCA Inventory.

DOT Classification: Flammable Liquid; Label: Flammable Liquid.

SAFETY PROFILE: Poison by intraperitoneal route. A very dangerous fire hazard when exposed to heat or flame. When heated to decomposition it emits toxic fumes of SO_x.

TFC600   CAS: 96-69-5   **HR: 3**
4,4′-THIOBIS(6-tert-BUTYL-m-CRESOL)
mf: $C_{22}H_{30}O_2S$ mw: 358.58

PROP: Light gray to tan powder. Mp: 150°, d: 1.10.

SYNS: BIS(3-tert-BUTYL-4-HYDROXY-6-METHYLPHE-NYL) SULFIDE * BIS(4-HYDROXY-5-tert-BUTYL-2-METHYLPHENYL) SULFIDE * DISPERSE MB-61 * SANTONOX * SNATOWHITE CRYSTALS * THIOALKOFEN BM 4 * 4,4'-THIOBIS(2-tert-BUTYL-5-METHYLPHENOL) * 4,4'-THIOBIS(6-tert-BUTYL-3-METHYLPHENOL) * 4,4'-THIOBIS(3-METHYL-6-tert-BU-TYLPHENOL) * 1,1'-THIOBIS(2-METHYL-4-HYDROXY-5-tert-BUTYLBENZENE) * USAF B-15 * YOSHINOX S

CONSENSUS REPORTS: Reported in EPA TSCA Inventory.

OSHA PEL: (Transitional: TWA Total Dust: 15 mg/m³; Respirable Fraction: 5 mg/m³) TWA Total Dust: 10 mg/m³; Respirable Fraction: 5 mg/m³
ACGIH TLV: TWA 10 mg/m³

SAFETY PROFILE: Poison by intraperitoneal route and probably by ingestion and inhalation. Mutation data reported. When heated to decomposition it emits highly toxic fumes of SO_x.

TFD500 CAS: 123-28-4 **HR: 1**
THIOBIS(DODECYL PROPIONATE)
mf: $C_{30}H_{58}O_4S$ mw: 514.94

PROP: White crystalline flakes; characteristic sweetish odor. Sol in organic solvents; insol in water.

SYNS: BIS(DODECYLOXYCARBONYLETHYL) SULFIDE * DIDODECYL-3,3'-THIODIPROPIONATE * DILA-URYLESTER KYSELINY β',β'-THIODIPROPIONOVE (CZECH) * DILAURYL THIODIPROPIONATE * DILAURYL-β-THIODIPROPIONATE * DILAURYL-β',β'-THIODIPROPIONATE * DILAURYL-3,3'-THIODIPROPIONATE

CONSENSUS REPORTS: Reported in EPA TSCA Inventory.

SAFETY PROFILE: An eye irritant. When heated to decomposition it emits toxic fumes of SO_x.

TFD750 CAS: 91-71-4 **HR: 3**
THIOCARBAMIZINE
mf: $C_{21}H_{17}AsN_2O_5S_2$ mw: 516.44

SYNS: 2,2'-(((4-((AMINOCARBONYL)AMINO)PHENYL) ARSINIDENEBIS(THIO))BIS) BENZOIC ACID * p-(BIS (o-CARBOXYPHENYLMERCAPTO)-ARSINO)-PHENYL-UREA * p-CARBAMIDOPHENYL-BIS(2-CARBOXYPHE-NYLMERCAPTO)ARSINE * 4-CARBAMIDOPHENYL

BIS(o-CARBOXYPHENYLTHIO)ARSENITE * p-CAR BAMIDOPHENYL-DI(1'-CARBOXYPHENYL-2') THIOAR-SENITE * S,S-DIESTER with DITHIO-p-UREIDOBEN-ZENEARSONOUS ACID o-MERCAPTOBENZOIC ACID * DIESTER with o-MERCAPTOBENZOIC ACID DITHIO-p-UREIDOBENZENEARSONOUS ACID * o-MERCAPTO-BENZOIC ACID, DIESTER with DITHIO-p-UREIDOBEN-ZENEARSONOUS ACID * THIOCARBAMISIN * (p-UREIDOBENZENEARSYLENEDITHIO)DI-o-BENZOIC ACID * (p-UREIDOPHENYLARSYLENEDITHIO)DI-o-BENZOIC ACID

CONSENSUS REPORTS: Arsenic and its compounds are on the Community Right-To-Know List.

OSHA PEL: TWA 0.5 mg(As)/m³
ACGIH TLV: TWA 0.2 mg(As)/m³

SAFETY PROFILE: Poison by intraperitoneal and intravenous routes. Moderately toxic by ingestion. When heated to decomposition it emits very toxic fumes of As, SO_x, and NO_x.

TFE500 **HR: D**
THIOCYANATES

SAFETY PROFILE: Variable toxicity. Thiocyanates are not normally dissociated into cyanide; they have a low acute toxicity. Prolonged absorption may produce various skin eruptions, running nose, and occasionally dizziness, cramps, nausea, vomiting, and mild or severe disturbances of the nervous system. Violent reactions have occurred when mixed with chlorates; nitrates; HNO_3; organic peroxides; peroxides; $KClO_3$; $NaClO_3$. Metal thiocyanates are oxidized explosively by chlorates, nitrates at 400° in intimate mixture, HNO_3, or spark or flame ignition. When heated to decomposition or on contact with acid or acid fumes they emit highly toxic fumes of CN^-.

TFI000 CAS: 139-65-1 **HR: 3**
4,4'-THIODIANILINE
mf: $C_{12}H_{12}N_2S$ mw: 216.32

PROP: Needles. Mp: 108°.

SYNS: BIS(p-AMINOPHENYL)SULFIDE * BIS(4-AMI-NOPHENYL) SULFIDE * BIS(p-AMINOPHENYL)SUL-PHIDE * BIS(4-AMINOPHENYL) SULPHIDE * p,p'-DIAMINODIPHENYL SULFIDE * 4,4'-DIAMINODIPHE-NYL SULFIDE * p,p'-DIAMINODIPHENYL SULPHIDE * DI(p-AMINOPHENYL) SULFIDE * DI(p-AMINOPHE-NYL)SULPHIDE * NCI-C01707 * THIOANILINE * 4,4'-THIOANILINE * 4,4'-THIOBIS(ANILINE)

* 4,4′-THIOBISBENZENAMINE   *   p,p-THIODIANILINE
* THIODI-p-PHENYLENEDIAMINE

CONSENSUS REPORTS: IARC Cancer Review: GROUP 2B IMEMDT 7,56,87; Human Limited Evidence IMEMDT 27,147,82; Animal Sufficient Evidence IMEMDT 27,147,82; Animal Limited Evidence IMEMDT 16,343,78. NCI Carcinogenesis Bioassay (feed); Clear Evidence: mouse, rat NCITR* NCI-CG-TR-47,78. Reported in EPA TSCA Inventory. Community Right-To-Know List.

DFG MAK: Animal Carcinogen, Suspected Human Carcinogen.

SAFETY PROFILE: Confirmed carcinogen with experimental carcinogenic and tumorigenic data. Poison by intravenous route. Moderately toxic by ingestion. May be a human carcinogenic. Experimental reproductive effects. Mutation data reported. When heated to decomposition it emits very toxic fumes of NO_x and SO_x.

TFJ100   CAS: 68-11-1   **HR: 3**
THIOGLYCOLIC ACID
DOT: UN 1940
mf: $C_2H_4O_2S$ mw: 92.12

PROP: Liquid, strong odor. Mp: $-16.5°$, bp: $108°$ @ 15 mm. Misc with water, alc, ether, chloroform, and benzene.

SYNS: ACIDE THIOGLYCOLIQUE (FRENCH)
* MERCAPTOACETATE * MERCAPTOACETIC ACID
* 2-MERCAPTOACETIC ACID * α-MERCAPTOACETIC ACID * 2-THIOGLYCOLIC ACID * THIOVANIC ACID
* USAF CB-35

CONSENSUS REPORTS: Reported in EPA TSCA Inventory.

OSHA PEL: TWA 1 ppm (skin)
ACGIH TLV: TWA 1 ppm (skin)
DOT Classification: Corrosive Material; Label: Corrosive.

SAFETY PROFILE: Poison by ingestion, skin contact, intraperitoneal, and intravenous routes. Moderately toxic by subcutaneous route. A corrosive irritant to skin, eyes, and mucous membranes. When heated to decomposition it emits toxic fumes of SO_x.

TFJ250   CAS: 64039-27-6   **HR: 3**
β-THIOGUANINE DEOXYRIBOSIDE
mf: $C_{10}H_{13}N_5O_3S \cdot H_2O$ mw: 301.36

SYNS: 2-AMINO-9-(2-DEOXY-β-d-RIBOFURANOSYL)-9H-PURINE-6-THIOL HYDRATE * β-DEOXYTHIOGUANOSINE * β-2′-DEOXY-6-THIOGUANOSINE MONOHYDRATE * NSC-71261 * β-TGDR

SAFETY PROFILE: Poison by intravenous and intraperitoneal routes. Moderately toxic by ingestion. Questionable carcinogen with experimental carcinogenic and neoplastigenic data. When heated to decomposition it emits very toxic fumes of NO_x and SO_x.

TFL000   CAS: 7719-09-7   **HR: 3**
THIONYL CHLORIDE
DOT: UN 1836
mf: Cl_2OS mw: 118.96

PROP: Colorless to yellow to red liquid; suffocating odor. Mp: $-105°$, bp: $78.8°$ @ 746 mm, d: 1.640 @ 15.5/15.5°, vap press: 100 mm @ 21.4°. Misc with benzene, chloroform, carbon tetrachloride.

SYNS: SULFINYL CHLORIDE * SULFUR CHLORIDE OXIDE * SULFUROUS DICHLORIDE * SULFUROUS OXYCHLORIDE * THIONYL DICHLORIDE

CONSENSUS REPORTS: Reported in EPA TSCA Inventory.

OSHA PEL: CL 1 ppm
ACGIH TLV: CL 1 ppm
DOT Classification: Corrosive Material; Label: Corrosive.

SAFETY PROFILE: Moderately toxic by inhalation. The material itself is more toxic than sulfur dioxide. Has a pungent odor similar to that of sulfur dioxide; it fumes upon exposure to air. Violent reaction with water releases hydrogen chloride and sulfur dioxide. Both these decomposition products constitute serious toxicity hazards. A corrosive irritant which causes burns to the skin and eyes. A powerful chlorinating agent. Potentially explosive reaction with ammonia, bis(dimethylamino)sulfoxide (above 80°C), chloryl perchlorate, 1,2,3-cyclo hexanetrione trioxime + sulfur dioxide, dimethyl sulfoxide, hexafluoroisopropylideneaminolithium. Violent reaction or ignition with 2,4-hexadiyn-1-6-diol, o-nitrobenzoyl acetic acid, o-nitrophenyl-acetic acid, sodium (ignites at 300°C). Incompatible with ammonia, dimethyl formamide + trace iron or zinc, linseed oil + quinoline, toluene + ethanol + water. When heated to decomposition it emits toxic fumes of SO_x and Cl^-.

TFL250   CAS: 7783-42-8   **HR: 2**
THIONYL FLUORIDE
mf: F$_2$OS   mw: 86.06

PROP: Colorless gas; suffocating odor. Bp: $-44°$, mp: $-130°$. d (liq): 1.780 @ $-100°$, (solid): 2.095 @ $-183°$. Sol in ether, benzene.

SYNS: FLUORURE de THIONYLE (FRENCH) * SULFUR DIFLUORIDE MONOXIDE * SULFUR DIFLUORIDE OXIDE * SULFUROUS OXYFLUORIDE * THIONYL DIFLUORIDE

CONSENSUS REPORTS: Reported in EPA TSCA Inventory.

OSHA PEL: TWA 2.5 mg(F)/m^3
ACGIH TLV: TWA 2.5 mg(F)/m^3
NIOSH REL: (Inorganic Fluorides) TWA 2.5 mg(F)/m^3

SAFETY PROFILE: Moderately toxic by inhalation. A severe irritant to skin, eyes, and mucous membranes. When heated to decomposition or on contact with water or steam it emits highly toxic and corrosive fumes of SO$_x$ and F$^-$.

TFM250   CAS: 110-02-1   **HR: 3**
THIOPHENE
mf: C$_4$H$_4$S   mw: 84.14

DOT: UN 2414

PROP: Clear, colorless liquid; slt aromatic odor similar to benzene. D: 1.0573 @ 25/4°, mp: $-38.3°$, bp: 84.4°, flash p: 21.2°F, vap press: 40 mm @ 12.5°, vap d: 2.9. Insol in water; misc with most organic solvents. May be heated to 850° without decomposition.

SYNS: CP 34 * DIVINYLENE SULFIDE * HUILE H50 * THIACYCLOPENTADIENE * THIAPHENE * THIOFURAM * THIOFURAN * THIOFURFURAN * THIOLE * THIOPHEN * THIOTETROLE * USAF EK-1860

CONSENSUS REPORTS: Reported in EPA TSCA Inventory.

DOT Classification: Flammable Liquid; Label: Flammable Liquid.

SAFETY PROFILE: Poison by intraperitoneal route. Mildly toxic by inhalation. A very dangerous fire hazard when exposed to heat or flame. Explosive reaction with N-nitrosoacetanilide. Violent or explosive reaction with nitric acid. Incompatible with oxidizing materials. To fight fire, use foam, CO$_2$, dry chemical. When heated to decomposition it emits highly toxic fumes of SO$_x$.

TFN500   CAS: 463-71-8   **HR: 3**
THIOPHOSGENE

DOT: UN 2474
mf: CCl$_2$S   mw: 114.97

PROP: Reddish liquid. Bp: 73.5°, d: 1.5085 @ 15°. Decomp in water, alc; sol in ether.

SYNS: CARBON CHLOROSULFIDE * CARBONOTHIOIC DICHLORIDE * DICHLOROTHIOCARBONYL * THIOCARBONIC DICHLORIDE * THIOCARBONYL CHLORIDE (DOT) * THIOCARBONYL DICHLORIDE * THIOFOSGEN (CZECH) * THIOKARBONYLCHLORID (CZECH)

CONSENSUS REPORTS: Reported in EPA TSCA Inventory.

DOT Classification: Poison B; Label: Poison.

SAFETY PROFILE: Poison by intravenous route. Moderately toxic by ingestion. A skin, mucous membrane, and severe eye irritant. When heated to decomposition it emits very toxic fumes of Cl$^-$ and SO$_x$.

TFO000   CAS: 3982-91-0   **HR: 2**
THIOPHOSPHORYL CHLORIDE
mf: Cl$_3$PS   mw: 169.38

DOT: UN 1837

PROP: Colorless, mobile liquid; pungent odor. Bp: 125°, fp: $-35°$, flash p: none, d: 1.63 @ 25/4°, vap press: 22 mm @ 25°, vap d: 5.86.

SYNS: PHOSPHOROTHIOIC TRICHLORIDE * PHOSPHOROTHIONIC TRICHLORIDE * PHOSPHOROUS SULFOCHLORIDE * PHOSPHOROUS THIOCHLORIDE * PHOSPHOROUS TRICHLORIDE SULFIDE * THIOPHOSPHORYL TRICHLORIDE * TRICHLOROPHOSPHINE SULFIDE * TL 262

CONSENSUS REPORTS: Reported in EPA TSCA Inventory.

DOT Classification: Corrosive Material; Label: Corrosive.

SAFETY PROFILE: Moderately toxic by ingestion and inhalation. A corrosive irritant to skin, eyes, and mucous membranes. Explosive reaction with methylmagnesium iodide. Explosive reaction with pentaerythritol + heat. Reacts with water or steam to produce toxic and corrosive fumes. When heated to decomposition it emits highly toxic fumes of PO$_x$, SO$_x$, and Cl$^-$.

TFP000   CAS: 75-18-3   **HR: 3**
2-THIOPROPANE
mf: C₂H₆S   mw: 62.14

DOT: UN 1164

PROP: Colorless liquid; disagreeable odor. Mp: −83.2°, lel: 2.2%, uel: 19.7%, flash p: <0°F, bp: 37.5-38°, d: 0.8458 @ 21/4°, vap d: 2.14, autoign temp: 403°F. Insol in water; sol in alc, ether.

SYNS: DIMETHYLSULFID (CZECH) * DIMETHYL SULFIDE (DOT) * DIMETHYL SULPHIDE * DMS * EXACT-S * METHYL SULFIDE (DOT) * METHYL SULPHIDE * METHYLTHIOMETHANE * SULFURE de METHYLE (FRENCH) * 2-THIAPROPANE

CONSENSUS REPORTS: EPA Extremely Hazardous Substances List. Reported in EPA TSCA Inventory.

DOT Classification: Flammable Liquid; Label: Flammable Liquid.

SAFETY PROFILE: Poison by inhalation. Moderately toxic by ingestion and possibly other routes. A skin and severe eye irritant. A very dangerous fire hazard when exposed to heat or flame. Explosive in the form of vapor when exposed to heat or flame. Can react vigorously with oxidizing materials. To fight fire, use CO₂, dry chemical. When heated to decomposition it emits highly toxic fumes of SO_x and may explode.

TFQ000   CAS: 79-19-6   **HR: 3**
THIOSEMICARBAZIDE
mf: CH₅N₃S   mw: 91.15

PROP: Needles from water. Mp: 182-184°. Sol in water, alc.

SYNS: N-AMINOTHIOUREA * HYDRAZINECAR-BOTHIOAMIDE * RCRA WASTE NUMBER P116 * THIOCARBAMYLHYDRAZINE * 3-THIOSEMICAR-BAZIDE * TSC * USAF EK-1275

CONSENSUS REPORTS: EPA Extremely Hazardous Substances List. Reported in EPA TSCA Inventory.

SAFETY PROFILE: Poison by ingestion, intraperitoneal, and intravenous routes. Questionable carcinogen with experimental tumorigenic data. Human mutation data reported. When heated to decomposition it emits very toxic fumes of NO_x and SO_x.

TFQ500   **HR: 1**
THIOSULFATES

SAFETY PROFILE: Up to 12 grams of sodium thiosulfate can be taken daily by mouth with no ill effects except catharsis. Most of the thiosulfates are low in acute toxicity. When heated to decomposition they emit highly toxic fumes of SO_x.

TFQ750   CAS: 52-24-4   **HR: 3**
THIOTRIETHYLENEPHOSPHORAMIDE
mf: C₆H₁₂N₃PS   mw: 189.24

SYNS: CBC 806495 * GIROSTAN * NCI-C01649 * NSC-6396 * ONCOTEPA * ONCOTIOTEPA * 1,1′,1″-PHOSPHINOTHIOYLIDYNETRISAZIRIDINE * PHOSPHOROTHIOIC ACID TRIETHYLENETRIAMIDE * SK 6882 * TESPAMINE * THIOFOZIL * THIOPHOSPHAMIDE * THIO-TEP * TIOFOS-FAMID * TIOFOZIL * TRIAZIRIDINYLPHOSPHINE SULFIDE * N,N′,N″-TRI-1,2-ETHANEDIYLPHOSPHORO-THIOIC TRIAMIDE * N,N′,N″-TRI-1,2-ETHANEDIYL-THIOPHOSPHORAMIDE * TRI(ETHYLENEIMINO)THIOPHOSPHORAMIDE * N,N′,N″-TRIETHYL-ENEPHOSPHOROTHIOIC TRIAMIDE * N,N′,N″-TRIETHYLENETHIOPHOSPHAMIDE * N,N′, N″-TRIETHYLENETHIOPHOSPHORAMIDE * TRIETHYLENETHIOPHOSPHOROTRIAMIDE * TRIS(1-AZIRIDINYL)PHOSPHINE SULFIDE * TRIS (ETHYLENIMINO)THIOPHOSPHATE * TSPA

CONSENSUS REPORTS: IARC Cancer Review: GROUP 2A IMEMDT 7,368,87; Human Limited Evidence IMEMDT 9,85,75; Animal Sufficient Evidence IMEMDT 9,85,75. NTP Fourth Annual Report On Carcinogens, 1984. NCI Carcinogenesis Bioassay (ipr); Clear Evidence: mouse, rat NCITR* NCI-CG-TR-58,78. EPA Genetic Toxicology Program.

SAFETY PROFILE: Confirmed human carcinogen producing leukemia. Poison by ingestion, intraperitoneal, intravenous, and subcutaneous routes. Human systemic effects by parenteral and possibly other routes: paresthesia, bone marrow changes and leukemia. Experimental teratogenic and reproductive effects. Human mutation data reported. When heated to decomposition it emits very toxic fumes of PO_x, SO_x, and NO_x.

TFR250   CAS: 141-90-2   **HR: 3**
2-THIOURACIL
mf: C₄H₄N₂OS   mw: 128.16

PROP: Small crystals; bitter taste. Practically insol in water, alc, ether, acids; sol in alkalies.

SYNS: ANTAGOTHYROID * ANTAGOTHYROIL * DERACIL * 2,3-DIHYDRO-2-THIOXO-4(1H)-PYRIMI-DINONE * 6-HYDROXY-2-MERCAPTOPYRIMIDINE * 4-HYDROXY-2(1H)-PYRIMIDINETHIONE * 2-MER-CAPTO-4-HYDROXYPYRIMIDINE * 2-MERCAPTO-4-PYRIMIDINOL * 2-MERCAPTO-4-PYRIMIDONE * 2-MERCAPTOPYRIMID-4-ONE * NOBILEN * 2-THIO-6-OXYPYRIMIDINE * 2-THIO-1,3-PYRIMIDIN-4-ONE * THIOURACIL * 6-THIOURA-CIL * TIOURACYL (POLISH) * TU * 2-TU

CONSENSUS REPORTS: IARC Cancer Review: GROUP 3 IMEMDT 7,56,87; Animal Sufficient Evidence IMEMDT 7,85,74. Reported in EPA TSCA Inventory. EPA Genetic Toxicology Program.

SAFETY PROFILE: Moderately toxic by ingestion. Human teratogenic effects by unspecified routes: developmental abnormalities of the central nervous system, craniofacial area, and endocrine system. Human reproductive effects by unspecified route: effects on newborn including viability index changes. Experimental teratogenic effects. Questionable carcinogen with experimental neoplastigenic and tumorigenic data. Mutation data reported. When heated to decomposition it emits very toxic fumes of NO_x and SO_x. Used in the treatment of hyperthyroidism, angina pectoris, and congestive heart failure.

TFS350   CAS: 137-26-8   HR: 3
THIRAM

DOT: UN 2771
mf: $C_6H_{12}N_2S_4$ mw: 240.44

PROP: Crystals. Mp: 156°, d: 1.30, bp: 129° @ 20 mm. Insol in water; sol in alc, ether, acetone, and chloroform.

SYNS: AATACK * ACCELERATOR THIURAM * ACETO TETD * ARASAN * AULES * BIS((DI-METHYLAMINO)CARBONOTHIOYL) DISULPHIDE * BIS(DIMETHYL-THIOCARBAMOYL)-DISULFID (GER-MAN) * BIS(DIMETHYLTHIOCARBAMOYL) DISULFIDE * CHIPCO THIRAM 75 * CYURAM DS * DISOL-FURO DI TETRAMETILTIOURAME (ITALIAN) * DISUL-FURE de TETRAMETHYLTHIOURAME (FRENCH) * α,α'-DITHIOBIS(DIMETHYLTHIO)FORMAMIDE * 1,1'-DITHIOBIS(N,N-DIMETHYLTHIO)FORMAMIDE * N,N'-(DITHIODICARBONOTHIOYL)BIS(N-METHYL-METHANAMINE) * EKAGOM TB * FALITIRAM * FERMIDE * FERNACOL * FERNASAN * FERNIDE * FLO PRO T SEED PROTECTANT * HERMAL * HERMAT TMT * HERYL *

HEXATHIR * KREGASAN * MERCURAM * METHYL THIRAM * METHYL THIURAMDISULFIDE * METHYL TUADS * NOBECUTAN * NOMERSAN * NORMERSAN * PANORAM 75 * POLYRAM UL-TRA * POMARSOL * POMASOL * PURALIN * RCRA WASTE NUMBER U244 * REZIFILM * ROYAL TMTD * SADOPLON * SPOTRETE * SQ 1489 * TERSAN * TERAMETHYL THIURAM DISULFIDE * TETRAMETHYLDIURANE SULPHITE * TETRAMETHYLENETHIURAM DISULPHIDE * TETRAMETHYLTHIOCARBAMOYLDISULPHIDE * TETRAMETHYLTHIORAMDISULFIDE (DUTCH) * TETRAMETHYL-THIRAM DISULFID (GERMAN) * TETRAMETHYLTHIURAM BISULFIDE * TETRA-METHYLTHIURAM DISULFIDE * TETRAMETHYL THIURANE DISULFIDE * N,N,N',N'-TETRAMETHYL-THIURAM DISULFIDE * N,N-TETRAMETHYLTHIURAM DISULPHIDE * TETRAMETHYLTHIURUM DISUL-FIDE * TETRAPOM * TETRASIPTON * TETRA-THIURAM DISULFIDE * TETRATHIURAM DISULPHIDE * THILLATE * THIMER * THIOSAN * THIO-TEX * THIOTOX * THIRAMAD * THIRAME (FRENCH) * THIRASAN * THIULIX * THIURAD * THIURAM * THIURAMIN * THIURAMYL * THYLATE * TIRAMPA * TIURAM (POLISH) * TIURAMYL * TMTD * TMTDS * TRAMETAN * TRIDIPAM * TRIPOMOL * TTD * TUADS * TUEX * TULISAN * USAF B-30 * USAF EK-2089 * USAF P-5 * VANCIDA TM-95 * VANCIDE TM * VUAGT-I-4 * VULCAFOR TMTD * VULKA-CIT MTIC * VULKACIT THIURAM * VULKACIT THIURAM/C

CONSENSUS REPORTS: IARC Cancer Review: GROUP 3 IMEMDT 7,56,87; Human Inadequate Evidence IMEMDT 12,225,76; Animal Inadequate Evidence IMEMDT 12,225,76. EPA Genetic Toxicology Program. Reported in EPA TSCA Inventory.

OSHA PEL: TWA 5 mg/m³
ACGIH TLV: TWA 5 mg/m³ (Proposed: TWA 1 mg/m³)
DFG MAK: 5 mg/m³
DOT Classification: ORM-A; Label: None.

SAFETY PROFILE: Poison by ingestion and intraperitoneal routes. Questionable carcinogen with experimental tumorigenic and teratogenic data. Mutation data reported. Affects human pulmonary system. A mild allergen and irritant. Acute poisoning in experimental animals produced liver, kidney, and brain damage. Dangerous in a fire; see NITROGEN MONOXIDE and SULFUR DIOXIDE.

TFS750   CAS: 7440-29-1   **HR: 3**
THORIUM
af: Th   aw: 232.00

DOT: UN 2975

PROP: Silvery-white, air stable, soft, ductile metal. D: 11.72; mp: 1842 ± 30°. A radioactive material.

SYNS: THORIUM-232 * THORIUM METAL, PYROPHORIC (DOT)

CONSENSUS REPORTS: Reported in EPA TSCA Inventory.

DOT Classification: Radioactive Material; Label: Radioactive and Flammable Solid.

SAFETY PROFILE: Suspected carcinogen. Taken internally as ThO_2, it has proven to be a carcinogenic due to its radioactivity. On an acute basis it has caused dermatitis. Flammable in the form of dust when exposed to heat or flame, or by chemical reaction with oxidizers. The powder may ignite spontaneously in air. Potentially hazardous reactions with chlorine, fluorine, bromine, oxygen, phosphorus, silver, sulfur, air, nitryl fluoride, peroxyformic acid.

TFT000   CAS: 10026-08-1   **HR: 3**
THORIUM CHLORIDE
mf: Cl_4Th mw: 373.80

PROP: White, odorless crystals. D: 4.59, mp: 770°, bp: 921°. Sol in water, alc.

SYNS: TETRACHLOROTHORIUM * THORIUM TETRACHLORIDE

CONSENSUS REPORTS: Reported in EPA TSCA Inventory. EPA Genetic Toxicology Program.

SAFETY PROFILE: Poison by intravenous route. Moderately toxic by intraperitoneal and subcutaneous routes. When heated to decomposition it emits toxic fumes of Cl^-.

TFT250   CAS: 15457-87-1   **HR: 3**
THORIUM HYDRIDE
mf: H_4Th mw: 236.07

PROP: Black, metallic crystals. D: 8.24, mp: decomp explosively @ red heat. Reacts with water.

SAFETY PROFILE: Suspected carcinogen. Explodes on heating in air. The powder ignites spontaneously in air.

TFT500   CAS: 13823-29-5   **HR: 3**
THORIUM(IV) NITRATE
mf: $H_4N_4O_{12}$ • Th   mw: 484.08

DOT: UN 2976

PROP: White, crystalline mass. Sol in water, alc.

SYNS: NITRIC ACID, THORIUM(4+) SALT * THORIUM (4+) NITRATE * THORIUM TETRANITRATE

CONSENSUS REPORTS: Reported in EPA TSCA Inventory.

DOT Classification: Radioactive Material; Label: Radioactive and Oxidizer.

SAFETY PROFILE: Poison by intraperitoneal, intravenous, and intratracheal routes. Moderately toxic by ingestion. Experimental reproductive effects. Radioactive. An oxidizing material; when in contact with readily combustible substances, will cause violent combustion or ignition. When heated to decomposition it emits toxic fumes of NO_x.

TFT750   CAS: 1314-20-1   **HR: 3**
THORIUM OXIDE
mf: O_2Th mw: 264.00

PROP: Heavy, white crystalline powder. D: 9.7, mp: 3390°. Insol in water, alkalies; slowly sol in acids.

SYNS: THORIA * THORIUM DIOXIDE * THOROTRAST * THORTRAST * UMBRATHOR

CONSENSUS REPORTS: NTP Fourth Annual Report On Carcinogens, 1984. Community Right-To-Know List. Reported in EPA TSCA Inventory.

SAFETY PROFILE: Confirmed human carcinogen producing angiosarcoma, liver and kidney tumors, lymphoma and other tumors of the blood system, and tumors at the application site.

TFU500   CAS: 299-75-2   **HR: 3**
I-THREITOL-1,4-BISMETHANE-SULFONATE
mf: $C_6H_{14}O_6S_2$ mw: 246.32

SYNS: CB 2562 * 1,4-DIMETHANESULFONATE THREITOL * (2s,3s)-1,4-DIMETHANESULFONATE TREITOL * NSC-39069 * TREOSULFAN * TRESULFAN

CONSENSUS REPORTS: IARC Cancer Review: GROUP 1 IMEMDT 7,363,87; Human Sufficient Evidence IMEMDT 26,341,81. EPA Genetic Toxicology Program.

SAFETY PROFILE: Confirmed carcinogen. Poison by intravenous route. Human mutation data reported. When heated to decomposition it emits toxic fumes of SO_x.

TFU750 CAS: 72-19-5 **HR: 2**
l-THREONINE
mf: $C_4H_9NO_3$ mw: 119.14

PROP: An essential amino acid. Colorless crystals or white crystalline powder; slt sweet taste. Mp: 255-257° with decomp. Sol in water; very sol in hot water; insol in alc, chloroform, ether.

SYNS: l-2-AMINO-3-HYDROXYBUTYRIC ACID * THREONINE

CONSENSUS REPORTS: Reported in EPA TSCA Inventory.

SAFETY PROFILE: Moderately toxic by intraperitoneal route. When heated to decomposition it emits toxic fumes of NO_x.

TFW000 CAS: 546-80-5 **HR: 3**
THUJONE
mf: $C_{10}H_{16}O$ mw: 152.26

PROP: A flavor constituent. A major component of Wormwood Oil, (*Artemisia absinthium, L*) which is the principal ingredient of absinthe, a liquor. Occurs as α, l, (−) or β, d, (+) called isothujone.

SYNS: (1S-1-α,4-α,5-α)-4-METHYL-1-(1-METHYLETHYL)-BICYCLO(3.1.0)HEXAN-3-ONE * (1S,4R,5R)-(−)-3-THU-JANONE * THUJON * (−)-THUJONE * α-THU-JONE * l-THUJONE

CONSENSUS REPORTS: Reported in EPA TSCA Inventory.

SAFETY PROFILE: Poison by intravenous, intraperitoneal, and subcutaneous routes. Moderately toxic by ingestion. Serious physiological consequences from abuse of absinthe (mainly in France), led to its abolition in 1915. Wormwood is still used in concentrations of less than 10 ppm in flavored wines. Thujon at 30 mg/kg causes convulsions associated with lesions of the cerebral cortex. Little is known of Thujone metabolism. Both forms occur in Wormwood oil, Oak Moss. The α form is major constituent of Cedar Leaf Oil or Oil of Thuja, Sage. The β form occurs in Tansy, Yarrow. When heated to decomposition it emits acrid smoke and irritating fumes.

TFW250 **HR: 3**
THULIUM
af: Tm aw: 168.934

PROP: A bright, silvery-gray, lustrous, soft, malleable, ductile metallic element. Mp: 1545°, bp: 1727°, d: 9.333.

SAFETY PROFILE: As a lanthanide it has probably at least a moderate degree of toxicity. Flammable in the form of dust when exposed to flame. Explosive in the form of dust when exposed to heat or flame. Violent reaction with air; halogens.

TFW500 CAS: 13537-18-3 **HR: 3**
THULIUM CHLORIDE
mf: Cl_3Tm mw: 275.28

CONSENSUS REPORTS: Reported in EPA TSCA Inventory.

SAFETY PROFILE: Poison by intraperitoneal route. Mildly toxic by ingestion. A skin and eye irritant. When heated to decomposition it emits toxic fumes of Cl^-.

TFX250 CAS: 35725-33-8 **HR: 3**
THULIUM(III) NITRATE, HEXAHYDRATE (1:3:6)
mf: $N_3O_9 \cdot Tm \cdot 6H_2O$ mw: 463.08

SYN: NITRIC ACID, THULIUM(3+) SALT, HEXAHY-DRATE

SAFETY PROFILE: Poison by intraperitoneal route. When heated to decomposition it emits toxic fumes of NO_x.

TFX500 CAS: 8007-46-3 **HR: 2**
THYME OIL

PROP: From distillation of flowering plant *Thymus vulgaris* L. (Fam. *Labiatae*). Colorless to reddish-brown liquid; pleasant odor, sharp taste. D: 0.930 @ 25/25°, refr index: 1.495 @ 20°.

SYNS: OIL of THYME * THYMIAN OEL (GERMAN) * THYM OIL

CONSENSUS REPORTS: Reported in EPA TSCA Inventory.

SAFETY PROFILE: Moderately toxic by ingestion. Mutation data reported. An allergen and an irritant. Combustible when exposed to heat or flame. When heated to decomposition it emits acrid smoke and irritating fumes.

TFX750 CAS: 8007-46-3 **HR: 2**
THYME OIL RED

PROP: Main constituents are thymol, carvacrol. Found in plants *Thymus vulgaris L.* and *Thymus zygis L.*.

SYN: SPANISH THYME OIL

SAFETY PROFILE: Mildly toxic by ingestion. A severe skin irritant. When heated to decomposition it emits acrid smoke and irritating fumes.

TFX810 CAS: 89-83-8 **HR: 3**
THYMOL
mf: $C_{10}H_{14}O$ mw: 150.24

PROP: Colorless, translucent crystals; pungent, caustic taste. Mp: 51°, bp: 233°, d: 0.972, vap press: 1 mm @ 64°. Sol in water, alkali; very sol in alc, ether, chloroform.

SYNS: p-CYMEN-3-OL * 3-p-CYMENOL * 3-HY-DROXY-p-CYMENE * 3-HYDROXY-1-METHYL-4-ISOPROPYLBENZENE * ISOPROPYL CRESOL * 6-ISOPROPYL-m-CRESOL * 2-ISOPROPYL-5-METHYLPHENOL * 1-METHYL-3-HYDROXY-4-ISOPRO-PYLBENZENE * 5-METHYL-2-ISOPROPYL-1-PHENOL * 5-METHYL-2-(1-METHYLETHYL)PHENOL * THYME CAMPHOR * THYMIC ACID * m-THYMOL

CONSENSUS REPORTS: Reported in EPA TSCA Inventory. EPA Genetic Toxicology Program.

SAFETY PROFILE: Poison by ingestion, intravenous, intraperitoneal, and subcutaneous routes. Experimental reproductive effects. Mutation data reported. An allergen. Incompatible with acetanilide. When heated to decomposition it emits acrid smoke and irritating fumes. An FDA over the counter drug used as an antibacterial and antifungal agent.

TGB250 CAS: 7440-31-5 **HR: 3**
TIN
af: Sn   aw: 118.71

PROP: Cubic, gray, crystalline metallic element. Mp: 231.9°, stabilizes <18°, d: 7.31, vap press: 1 mm @ 1492°, bp: 2507°.

SYNS: SILVER MATT POWDER * TIN (α) * TIN FLAKE * TIN POWDER * ZINN (GERMAN)

CONSENSUS REPORTS: Reported in EPA TSCA Inventory.

OSHA PEL: Organic Compounds: TWA 0.1 mg(Sn)/m³ (skin); Inorganic Compounds (except oxides): TWA 2 mg/m³
ACGIH TLV: TWA metal, oxide and inorganic compounds (except SnH_4) as Sn 2 mg/m³; organic compounds 0.1 mg/m³ (skin) (Proposed: TWA 0.1 mg(Sn)/m³; STEL 0.2 mg(Sn)/m³ (skin))
DFG MAK: Inorganic 2 mg/m³, organic 0.1 mg/m³
NIOSH REL: (Organotin Compounds) TWA 0.1 mg(Sn)/m³

SAFETY PROFILE: Questionable carcinogen with experimental tumorigenic data by implant route. Combustible in the form of dust when exposed to heat or by spontaneous chemical reaction with Br_2, BrF_3, Cl_2, ClF_3, $Cu(NO_3)$, K_2O_2, S.

TGB750 CAS: 7789-67-5 **HR: 3**
TIN(IV) BROMIDE (1:4)
mf: Br_4Sn mw: 438.33

PROP: White crystalline mass. Mp: 31°, bp: 202°, d (liquid): 3.340 @ 35°, vap press: 10 mm @ 72.7°.

SYNS: STANNIC BROMIDE * TIN PERBROMIDE * TIN TETRABROMIDE

CONSENSUS REPORTS: Reported in EPA TSCA Inventory.

OSHA PEL: TWA 2 mg(Sn)/m³
ACGIH TLV: TWA 2 mg(Sn)/m³

SAFETY PROFILE: Poison by intravenous route. Violent reaction with NO_2Cl. When heated to decomposition it emits toxic fumes of Br^-.

TGC000 CAS: 7772-99-8 **HR: 3**
TIN(II) CHLORIDE (1:2)
mf: Cl_2Sn mw: 189.59

DOT: NA 1759

PROP: Colorless crystals. D: 2.71, mp: 37-38°. Sol in less than its own weight of water; very sol in hydrochloric acid (dilute or conc); sol in alc, ethyl acetate, glacial acetic acid, sodium hydroxide solution.

SYNS: C.I. 77864 * NCI-C02722 * STANNOUS CHLORIDE (FCC) * STANNOUS CHLORIDE, solid (DOT) * TIN DICHLORIDE * TIN PROTOCHLORIDE

CONSENSUS REPORTS: NTP Carcinogenesis Bioassay (feed); No Evidence: mouse, rat

NTPTR* NTP-TR-231,82. Reported in EPA TSCA Inventory. EPA Genetic Toxicology Program.

OSHA PEL: TWA 2 mg(Sn)/m^3
ACGIH TLV: TWA 2 mg(Sn)/m^3
DOT Classification: ORM-B; Label: None.

SAFETY PROFILE: Poison by ingestion, intraperitoneal, intravenous, and subcutaneous routes. Experimental reproductive effects. Human mutation data reported. Potentially explosive reaction with metal nitrates. Violent reactions with hydrogen peroxide, ethylene oxide, hydrazine hydrate, nitrates, K, Na. Ignition on contact with bromine trifluoride. A vigorous reaction with calcium acetylide is initiated by flame. When heated to decomposition it emits toxic fumes of Cl$^-$.

TGC250   CAS: 7646-78-8   HR: 3
TIN(IV) CHLORIDE (1:4)
mf: Cl$_4$Sn   mw: 260.49

DOT: UN 1827

PROP: Colorless, fuming caustic liquid or crystals. Mp: −33°, bp: 114.1°, d: 2.232, vap press: 10 mm @ 10°.

SYNS: ETAIN (TETRACHLORURE d') (FRENCH) * LIBAVIUS FUMING SPIRIT * STAGNO (TETRACLO-RURO di) (ITALIAN) * STANNIC CHLORIDE, anhydrous (DOT) * TIN CHLORIDE, fuming (DOT) * TIN PER-CHLORIDE (DOT) * TIN TETRACHLORIDE, anhydrous (DOT) * TINTETRACHLORIDE (DUTCH) * ZINNTET-RACHLORID (GERMAN)

CONSENSUS REPORTS: Reported in EPA TSCA Inventory. EPA Genetic Toxicology Program.

OSHA PEL: TWA 2 mg(Sn)/m^3
ACGIH TLV: TWA 2 mg(Sn)/m^3
DOT Classification: Corrosive Material; Label: Corrosive.

SAFETY PROFILE: Poison by intraperitoneal route. Moderately toxic by inhalation. A corrosive irritant to skin, eyes, and mucous membranes. Combustible by chemical reaction. Upon contact with moisture, considerable heat is generated. Violent reaction with K, Na, turpentine, ethylene oxide, alkyl nitrates. Dangerous; hydrochloric acid is liberated on contact with moisture or heat. When heated to decomposition it emits toxic fumes of Cl$^-$.

TGC500   HR: D
TIN COMPOUNDS

OSHA PEL: Organic Compounds: TWA 0.1 mg(Sn)/m^3 (skin); Inorganic Compounds (except oxides): TWA 2 mg/m^3
ACGIH TLV: TWA metal, oxide and inorganic compounds (except SnH$_4$) as Sn 2 mg/m^3; organic compounds 0.1 mg/m^3 (skin) (Proposed: TWA 0.1 mg(Sn)/m^3; STEL 0.2 mg(Sn)/m^3 (skin))
DFG MAK: Inorganic 2 mg/m^3, organic 0.1 mg/m^3
NIOSH REL: (Organotin Compounds) TWA 0.1 mg(Sn)/m^3

SAFETY PROFILE: Variable toxicity. Elemental tin and inorganic tin compounds have low toxicity and are poorly absorbed when ingested. Some inorganic tin salts are irritating or can liberate toxic fumes on decomposition. The latter is particularly true of tin halogens. Tin hydride is highly toxic with effects similar to arsenic hydride. Inhalation of tin dusts over a period of years may cause pneumoconiosis. Some of the organic tin compounds are strong poisons. Short chain alkyl tin compounds (e.g., ethyl and methyl compounds) are particularly toxic. Generally alkyl tin compounds are more toxic than aryl compounds and short chain compounds are more toxic than long chain compounds. The toxicity increases with the number of alkyl groups. Tetramethyl tin chloride and triethyl tin chloride are very toxic to the nervous system. They are lipid soluble and can be absorbed through the skin. Symptoms recede slowly. The concentration of tin in condensed milk from cans may reach 160 ppm and could become hazardous for babies. Some alkyl tin compounds have high ecotoxicity. They have been used in marine paints to prevent growth on boat hulls, but may have too much environmental effect for this purpose.

TGD000   CAS: 58-14-0   HR: 3
TINDURIN
mf: C$_{12}$H$_{13}$ClN$_4$ mw: 248.74

SYNS: CD * CHLORIDIN * CHLORIDINE * 5-(4'-CHLOROPHENYL)-2,4-DIAMINO-6-ETHYLPYRIMI-DINE * 5-(4-CHLOROPHENYL)-6-ETHYL-2,4-PYRIMI-DINEDIAMINE * DARACLOR * DARAPRAM * DARAPRIM * DARAPRIME * 2,4-DIAMINO-5-p-CHLOROPHENYL-6-ETHYLPYRIMIDINE * 2,4-DI-AMINO-5-(4-CHLOROPHENYL)-6-ETHYLPYRIMIDINE * DIAMINOPYRITAMIN * ERBAPRELINA

* KHLORIDIN * MALACID * MALOCID
* MALOCIDE * MALOPRIM * NCI-C01683
* NSC 3061 * PIRIMECIDAN * PIRIMETAMINA
(SPANISH) * 4753 R.P. * WR 2978

CONSENSUS REPORTS: IARC Cancer Review: GROUP 3 IMEMDT 7,56,87; Animal Limited Evidence IMEMDT 13,233,77. NCI Carcinogenesis Bioassay (feed); Inadequate Studies: mouse NCITR* NCI-CG-TR-77,78; (feed); No Evidence: rat NCITR* NCI-CG-TR-77,78. EPA Genetic Toxicology Program.

SAFETY PROFILE: Poison by ingestion, subcutaneous, and intraperitoneal routes. Experimental teratogenic and reproductive effects. Questionable carcinogen. Human mutation data reported. When heated to decomposition it emits very toxic fumes of Cl^- and NO_x. Used as an antimalarial drug for humans and to treat toxoplasmosis in hogs.

TGD500   CAS: 10294-70-9   HR: 3
TIN(II) IODIDE
mf: I_2Sn mw: 372.49

SYN: STANNOUS IODIDE

CONSENSUS REPORTS: Reported in EPA TSCA Inventory.

OSHA PEL: TWA 2 mg(Sn)/m^3
ACGIH TLV: TWA 2 mg(Sn)/m^3

SAFETY PROFILE: Poison by intravenous route. When heated to decomposition it emits toxic fumes of I^-.

TGD750   CAS: 7790-47-8   HR: 3
TIN(IV) IODIDE (1:4)
mf: I_4Sn mw: 626.29

PROP: Red cubic crystals. Mp: 144.5°, bp: 364°, d: 4.473 @ 0°.

SYNS: STANNIC IODIDE * TIN TETRAIODIDE

CONSENSUS REPORTS: Reported in EPA TSCA Inventory.

OSHA PEL: TWA 2 mg(Sn)/m^3
ACGIH TLV: TWA 2 mg(Sn)/m^3

SAFETY PROFILE: Poison by intravenous route. Strong reaction with NO_2Cl; (K + S); (Na + S). When heated to decomposition it emits toxic fumes of I^-.

TGE500   CAS: 25324-56-5   HR: 3
TIN (IV) PHOSPHIDE
mw: PSn   mw: 149.66

DOT: UN 1433

PROP: Silver-white crystals. D: 6.56.

SYN: STANNIC PHOSPHIDE (DOT)

CONSENSUS REPORTS: Reported in EPA TSCA Inventory.

OSHA PEL: TWA 2 mg(Sn)/m^3
ACGIH TLV: TWA 2 mg(Sn)/m^3
DOT Classification: Flammable Solid; Label: Flammable Solid and Dangerous When Wet.

SAFETY PROFILE: A flammable solid. Reacts with moisture or acid fumes to liberate highly toxic phosphine gas. When heated to decomposition it emits toxic fumes of PO_x.

TGF250   CAS: 7440-32-6   HR: 3
TITANIUM
DOT: UN 1352/UN 2546/UN 2878
af: Ti   aw: 47.90

PROP: Dark gray amorphous powder or lustrous white metal. D: 4.5 @ 20°, autoign temp: 1200° for solid metal in air, 250° for powder, mp: 1677°, bp: 3277°.

SYNS: CONTIMET 30 * C.P. TITANIUM * IMI 115
* NCI-C04251 * OREMET * TITANIUM ALLOY
* TITANIUM METAL POWDER, DRY (DOT) * TITANIUM SPONGE GRANULES (DOT) * TITANIUM SPONGE POWDERS (DOT)

CONSENSUS REPORTS: Reported in EPA TSCA Inventory.

DOT Classification: Flammable Solid; Label: Flammable Solid and Spontaneously Combustible.

SAFETY PROFILE: Questionable carcinogen with experimental tumorigenic data. The dust may ignite spontaneously in air. Flammable when exposed to heat or flame or by chemical reaction. Titanium can burn in an atmosphere of carbon dioxide, nitrogen or air. Also reacts violently with BrF_3, CuO, PbO, (Ni + $KClO_3$), metaloxy salts, halocarbons, halogens, CO_2, metal carbonates, Al, water, AgF, O_2, nitryl fluoride, HNO_3, O_2, $KClO_3$, KNO_3, $KMnO_4$, steam @ 704°, trichloroethylene, trichlorotrifluoroethane. Ordinary extinguishers are often ineffective against titanium fires. Such fires require special extinguishers designed for metal fires. In airtight enclosures, titanium fires can be controlled by the use of argon or helium. titanium, in the absence of moisture, burns slowly, but evolves much heat. The application of water to burning titanium can cause an explosion. Finely divided titanium dust and powders, like most metal powders, are potential explosion hazards when exposed to sparks, open flame or high heat sources.

TGG250   CAS: 7705-07-9   **HR: 2**
TITANIUM CHLORIDE
mf: Cl_3Ti mw: 154.25

DOT: UN 2441

PROP: Colorless to light yellow liquid; fumes in moist air. Mp: $-30°$, bp: $136.4°$, d: 1.772 @ 25/25°, vap press: 10 mm @ 21.3°.

SYNS: TAC 121 * TAC 131 * TITANIUM (III) CHLORIDE * TITANIUM TRICHLORIDE * TITANIUM TRICHLORIDE, PYROPHORIC (DOT) * TITANOUS CHLORIDE * TRICHLORO TRITANIUM

CONSENSUS REPORTS: Reported in EPA TSCA Inventory.

DOT Classification: Flammable Solid; Label: Spontaneously Combustible, Corrosive.

SAFETY PROFILE: A corrosive irritant to skin, eyes, and mucous membranes. A severe corrosive because it liberates heat and hydrochloric acid upon contact with moisture. If spilled on skin, wipe off with dry cloth before applying water. May ignite spontaneously in air. Flammable when exposed to heat or flame. Reacts violently with K, HF. When heated to decomposition it emits toxic fumes of Cl^-.

TGG500   **HR: D**
TITANIUM COMPOUNDS

SAFETY PROFILE: This material is generally considered to be physiologically inert. There are no reported cases in the literature where titanium as such has caused human intoxication. The dusts of titanium or most titanium compounds such as titanium oxide may be placed in the nuisance category. Titanium tetrachloride, however, is an irritant and corrosive material, because when exposed to moisture, it hydrolyzes to hydrogen chloride.

TGG760   CAS: 13463-67-7   **HR: 1**
TITANIUM DIOXIDE
mf: O_2Ti mw: 79.90

PROP: White amorphous powder. Mp: 1860° (decomp), d: 4.26. Insol in water, hydrochloric acid, dil sulfuric acid, alc.

SYNS: 1700 WHITE * A-FIL CREAM * ATLAS WHITE TITANIUM DIOXIDE * AUSTIOX * BAYERITIAN * BAYERTITAN * BAYTITAN * CALCOTONE WHITE T * C.I. 77891 * C.I. PIGMENT WHITE 6 * COSMETIC WHITE C47-5175 * C-WEISS 7 (GERMAN) * FLAMENCO * HOMBITAN * HORSE HEAD A-410 * KH 360 * KRONOS TITANIUM DIOXIDE * LEVANOX WHITE RKB * NCI-C04240 * RAYOX * RUNA RH20 * RUTILE * TIOFINE * TIOXIDE * TITANDIOXID (SWEDEN) * TITANIUM OXIDE * TRIOXIDE(S) * TRONOX * UNITANE O-110 * ZOPAQUE

CONSENSUS REPORTS: NCI Carcinogenesis Bioassay (feed); No Evidence: mouse, rat NCITR* NCI-CG-TR-97,79. Reported in EPA TSCA Inventory. EPA Genetic Toxicology Program. Community Right-To-Know List.

OSHA PEL: (Transitional: TWA Total Dust: 15 mg/m^3; Respirable Fraction: 5 mg/m^3) TWA Total Dust: 10 mg/m^3; Respirable Fraction: 5 mg/m^3

ACGIH TLV: TWA (nuisance particulate) 10 mg/m^3 of total dust (when toxic impurities are not present, e.g., quartz < 1%).
DFG MAK: 6 mg/m^3

SAFETY PROFILE: A human skin irritant. Questionable carcinogen with experimental carcinogenic, neoplastigenic, and tumorigenic data. A common air contaminant and nuisance dust. Violent or incandescent reaction with metals at high temperatures (e.g., aluminum, calcium, magnesium, potassium, sodium, zinc, lithium).

TGH350   CAS: 7550-45-0   **HR: 3**
TITANIUM TETRACHLORIDE

DOT: UN 1838
mf: Cl_4Ti mw: 189.70

SYNS: TETRACHLORURE de TITANE (FRENCH) * TITAANTETRACHLORED (DUTCH) * TITANE (TETRACHLORURE de) (FRENCH) * TITANIO TETRACHLORURO di (ITALIAN) * TITANIUM CHLORIDE * TITANTETRACHLORID (GERMAN)

CONSENSUS REPORTS: EPA Extremely Hazardous Substances List. Community Right-To-Know List. Reported in EPA TSCA Inventory.

DOT Classification: Corrosive Material; Label: Corrosive.

SAFETY PROFILE: Poison by inhalation. A corrosive irritant to skin, eyes, and mucous membranes. When heated to decomposition it emits toxic fumes of Cl⁻.

TGI000   HR: D
TOBACCO LEAF, NICOTIANA GLAUCA

SYN: NICOTIANA GLAUCA

SAFETY PROFILE: The smoke produced by burning tobacco contains the highly toxic alkaloid, nicotine, tars and phenols, carbon monoxide, cyanides, nitrates, nitrites, carcinogenic, co-carcinogenic and perhaps 100 other chemicals, alpha-emitters, etc. Experimental teratogenic and reproductive effects. A nicotine-containing dried leaf of the tobacco plant. Habitual inhalation of tobacco smoke is considered a leading cause of lung cancer and circulatory problems, cardiac problems, etc. Combustible when exposed to heat or flame.

TGI100   HR: 3
TOBACCO PLANT

PROP: Large annual or perennial shrubs with leaves that are often broad, hairy, and sticky. The trumpet-shaped flowers are white, yellow, green-yellow or red. The seed capsule holds many small seeds. *N. tabacum* is the principal commercial tobacco in the western countries. *N. rustica*, native to South America, is found sporadically across the United States and is the most widely cultivated tobacco in the Orient. *N. longiflora* is commonly cultivated as a garden ornamental. *N. attenuata* grows in the region bounded by Idaho, Baja California, and Texas. *N. glauca* is native to South America and now grows in the southwestern United States, Hawaii, Mexico, and the West Indies.

SYNS: NICOTIANA ATTENUATA * NICOTIANA GLAUCA * NICOTIANA LONGIFLORA * NICOTIANA RUSTICA * NICOTIANA TABACUM * PAKA (HAWAII) * TABAC (FRENCH) * TABACO (SPANISH)

SAFETY PROFILE: Confirmed human carcinogen by several routes. The whole plant contains poisonous nicotine and other chemically related alkaloids. The primary alkaloid in *N. tabacum* is nicotine. The primary alkaloid in *N. glauca* is anabasine. Ingestion of any part of the plant can cause salivation, nausea, vomiting, distorted perceptions, convulsions vasomotor collapse and respiratory failure. Most serious poisonings result from ingestion of the leaves in salad, use of infusions as enemas, or skin absorption of alkaloids during commercial harvesting.

TGJ750   CAS: 119-93-7   HR: 3
o-TOLIDINE
mf: $C_{14}H_{16}N_2$ mw: 212.32

PROP: White to reddish crystals. Mp: 129-131°C. Very sltly sol in water; sol in alc, ether, acetic acid.

SYNS: BIANISIDINE * 4,4′-BI-o-TOLUIDINE * C.I. 37230 * C.I. AZOIC DIAZO COMPONENT 113 * (4,4′-DIAMINE-3,3′-DIMETHYL(1,1′-BIPHENYL) * 4,4′-DIAMINO-3,3′-DIMETHYLBIPHENYL * 4,4′-DIAMINO-3,3′-DIMETHYLDIPHENYL * DIAMINODITOLYL * 3,3′-DIMETHYLBENZIDIN * 3,3′-DIMETHYLBENZIDINE * 3,3′-DIMETHYL-4,4′-BIPHENYLDIAMINE * 3,3′-DIMETHYLBIPHENYL-4,4′-DIAMINE * 3,3′-DIMETHYL-(1,1′-BIPHENYL)-4,4′-DIAMINE * 3,3′-DIMETHYL-4,4′-DIPHENYLDIAMINE * 3,3′-DIMETHYL-DIPHENYL-4,4′-DIAMINE * 4,4′-DI-o-TOLUIDINE * FAST DARK BLUE BASE R * RCRA WASTE NUMBER U095 * o-TOLIDIN * 2-TOLIDIN (GERMAN) * 2-TOLIDINA (ITALIAN) * TOLIDINE * 3,3′-TOLIDINE * o,o′-TOLIDINE * 2-TOLIDINE

CONSENSUS REPORTS: IARC Cancer Review: GROUP 2B IMEMDT 7,56,87; Animal Limited Evidence IMEMDT 1,87,72. NTP Fourth Annual Report On Carcinogens, 1984. EPA Genetic Toxicology Program. Community Right-To-Know List. Reported in EPA TSCA Inventory.

ACGIH TLV: Suspected Human Carcinogen
DFG MAK: Animal Carcinogen, Suspected Human Carcinogen.
NIOSH REL: (o-Toluidine) CL 0.02 mg/m³/60M; avoid skin contact.

SAFETY PROFILE: Confirmed carcinogen with experimental carcinogenic and tumorigenic data. Poison by intraperitoneal route. Moderately toxic by ingestion. Human mutation data reported. When heated to decomposition it emits toxic fumes of NO_x.

TGK750   CAS: 108-88-3   HR: 3
TOLUENE

DOT: UN 1294
mf: C_7H_8 mw: 92.15

PROP: Colorless liquid; benzol-like odor. Mp: −95 to −94.5°, bp: 110.4°, flash p: 40°F (CC), ULC: 75-80, lel: 1.27%, uel: 7%, d: 0.866 @ 20°/4°, autoign temp: 996°F, vap press: 36.7

mm @ 30°, vap d: 3.14. Insol in water; sol in acetone; misc in absolute alc, ether, chloroform.

SYNS: ANTISAL 1a * BENZENE, METHYL- * METHACIDE * METHANE, PHENYL- * METHYLBENZENE * METHYLBENZOL * NCI-C07272 * PHENYLMETHANE * RCRA WASTE NUMBER U220 * TOLUEEN (DUTCH) * TOLUEN (CZECH) * TOLUOL * TOLUOL (DOT) * TOLUOLO (ITALIAN) * TOLU-SOL

CONSENSUS REPORTS: Community Right-To-Know List. Reported in EPA TSCA Inventory. EPA Genetic Toxicology Program.

OSHA PEL: (Transitional: TWA 200 ppm; CL 300 ppm; Pk 500 ppm/10M/8H) TWA 100 ppm; STEL 150 ppm
ACGIH TLV: TWA 100 ppm; STEL 150 ppm; BEI: 1 mg(toluene)/L in venous blood end of shift; 20 ppm toluene in end-exhaled air during shift.
DFG MAK: 100 ppm (380 mg/m^3); BAT: 340 μg/dl in blood at end of shift.
NIOSH REL: (Toluene) TWA 100 ppm; CL 200 ppm/10M
DOT Classification: Flammable Liquid; Label: Flammable Liquid.

SAFETY PROFILE: Poison by intraperitoneal route. Moderately toxic by intravenous, subcutaneous, and possibly other routes. Mildly toxic by inhalation. Human systemic effects by inhalation: CNS recording changes, hallucinations or distorted perceptions, motor activity changes, antipsychotic, psychophysiological test changes and bone marrow changes. Experimental teratogenic and reproductive effects. Mutation data reported. A human eye irritant. An experimental skin and severe eye irritant.

Toluene is derived from coal tar, and commercial grades usually contain small amounts of benzene as an impurity. Inhalation of 200 ppm of toluene for 8 hours may cause impairment of coordination and reaction time; with higher concentrations (up to 800 ppm) these effects are increased and are observed in a shorter time. In the few cases of acute toluene poisoning reported, the effect has been that of a narcotic, the workman passing through a stage of intoxication into one of coma. Recovery following removal from exposure has been the rule. An occasional report of chronic poisoning describes an anemia and leucopenia, with biopsy showing a bone marrow hypoplasia. These effects, however, are less common in people working with toluene, and they are not as severe. At 200-500 ppm, headache, nausea, eye irritation, loss of appetite, a bad taste, lassitude, impairment of coordination and reaction time are reported, but are not usually accompanied by any laboratory or physical findings of significance. With higher concentrations, the above complaints are increased and in addition, anemia, leukopenia and enlarged liver may be found in rare cases. A common air contaminant.

Flammable liquid. A very dangerous fire hazard when exposed to heat, flame or oxidizers. Explosive in the form of vapor when exposed to heat or flame. Explosive reaction with 1,3-dichloro-5,5-dimethyl-2,4-imidazolididione; dinitrogen tetraoxide; concentrated nitric acid; H_2SO_4 + HNO_3; N_2O_4; $AgClO_4$; BrF_3; UF_6. Forms an explosive mixture with tetranitromethane. Can react vigorously with oxidizing materials. To fight fire, use foam, CO_2, dry chemical. When heated to decomposition it emits acrid smoke and irritating fumes.

TGL500   CAS: 25376-45-8   **HR: 3**
TOLUENEDIAMINE
mf: $C_7H_{10}N_2$ mw: 122.19

DOT: NA 1709

SYNS: DIAMINOTOLUENE * ar-METHYLBENZENE-DIAMINE * METHYLPHENYLENEDIAMINE * TOLYLENEDIAMINE

CONSENSUS REPORTS: Community Right-To-Know List. Reported in EPA TSCA Inventory.

DOT Classification: ORM-A; Label: None.

SAFETY PROFILE: Probably a poison. When heated to decomposition it emits toxic fumes of NO_x.

TGL750   CAS: 95-80-7   **HR: 3**
TOLUENE-2,4-DIAMINE
mf: $C_7H_{10}N_2$ mw: 122.19

DOT: UN 1709

PROP: Prisms. Mp: 99°, bp: 280°, vap press: 1 mm @ 106.5°.

SYNS: 3-AMINO-p-TOLUIDINE * 5-AMINO-o-TOLU-IDINE * AZOGEN DEVELOPER H * BENZOFUR MT * C.I. 76035 * C.I. OXIDATION BASE * DEVELOPER H * 1,3-DIAMINO-4-METHYLBENZENE * 2,4-DIAMINO-1-METHYLBENZENE * 2,4-DIAMINO-TOLUEN (CZECH) * DIAMINOTOLUENE * 2,4-

DIAMINOTOLUENE   *   2,4-DIAMINO-1-TOLUENE
*   2,4-DIAMINOTOLUOL   *   EUCANINE GB
*   FOURAMINE   *   FOURRINE M   *   META TOLUY-
LENE DIAMINE   *   4-METHYL-1,3-BENZENEDIAMINE
*   4-METHYL-m-PHENYLENEDIAMINE   *   MTD
*   NAKO TMT   *   NCI-C02302   *   PELAGOL GREY J
*   PONTAMINE DEVELOPER TN   *   RCRA WASTE
NUMBER U221   *   RENAL MD   *   TDA   *   2,4-TOLA-
MINE   *   m-TOLUENEDIAMINE   *   2,4-TOLUENEDIA-
MINE   *   m-TOLUYLENDIAMIN (CZECH)   *   m-TOLUY-
LENEDIAMINE   *   2,4-TOLUYLENEDIAMINE (DOT)
*   m-TOLYENEDIAMINE   *   m-TOLYLENEDIAMINE
*   TOLYLENE-2,4-DIAMINE   *   2,4-TOLYLENEDIAMINE
*   4-m-TOLYLENEDIAMINE   *   ZOBA GKE   *   ZOGEN
DEVELOPHER H

CONSENSUS REPORTS: IARC Cancer Re-
view: GROUP 2B IMEMDT 7,56,87; Animal
Sufficient Evidence IMEMDT 16,83,78. NTP
Fourth Annual Report On Carcinogens, 1984.
NCI Carcinogenesis Bioassay (feed); Clear
Evidence: mouse, rat NCITR* NCI-CG-TR-
162,79. Community Right-To-Know List. Re-
ported in EPA TSCA Inventory. EPA Genetic
Toxicology Program.

DFG MAK: Animal Carcinogen, Suspected Hu-
    man Carcinogen.
DOT Classification: Poison B; Label: St. An-
    drews Cross.

SAFETY PROFILE: Confirmed carcinogen with
experimental carcinogenic data. Poison by ingestion, intraperitoneal, and subcutaneous routes.
Experimental reproductive effects. Human mutation data reported. A skin and eye irritant.
This material has a marked toxic action upon
the liver and can cause fatty degeneration of
that organ. When heated to decomposition it
emits toxic fumes of NO_x.

TGM000   CAS: 95-70-5   **HR: 3**
TOLUENE-2,5-DIAMINE
mf: $C_7H_{10}N_2$ mw: 122.19

PROP: Colorless, crystalline tablets. Mp: 64°,
bp: 274°.

SYNS: 4-AMINO-2-METHYLANILINE   *   C.I. 76042
*   2,5-DIAMINOTOLUENE   *   2-METHYL-1,4-BENZENE-
DIAMINE   *   2-METHYL-p-PHENYLENEDIAMINE
*   p-TOLUENEDIAMINE   *   p-TOLUYLENDIAMINE
*   TOLUYLENE-2,5-DIAMINE   *   p,m-TOLYLENEDI-
AMINE

CONSENSUS REPORTS: IARC Cancer Review: GROUP 3 IMEMDT 7,56,87; Animal

Inadequate Evidence IMEMDT 16,97,78. Reported in EPA TSCA Inventory. EPA Genetic
Toxicology Program.

SAFETY PROFILE: Poison by ingestion and
subcutaneous routes. A skin irritant. Mutation
data reported. Questionable carcinogen. Has a
toxic action upon the liver and can cause fatty
degeneration of that organ. Its total effect upon
the body seems to take place three different
ways. It is toxic to the central nervous system.
It produces jaundice by action on the liver and
spleen, and it produces anemia by destruction
of the red blood cells. In this action it is quite
similar to aniline, although by no means identical with it. Its high boiling point and the fact
that the material is solid at room temperature
makes it somewhat less hazardous than aniline,
particularly at ordinary working temperatures.
The literature contains a reference to a permanent injury to an eye due to the use of this
material as an eyelash dye. It is considered to
be an irritating dye material. When heated to
decomposition it emits toxic fumes of NO_x.

TGM740   CAS: 26471-62-5   **HR: 3**
TOLUENE-1,3-DIISOCYANATE
DOT: UN 2078
mf: $C_9H_6N_2O_2$ mw: 174.17

SYN: BENZENE-, 1,3-DIISOCYANATOMETHYL-
*   DESMODUR T100   *   DIISOCYANATOMETHYLBEN-
ZENE   *   DIISOCYANATOTOLUENE   *   HYLENE-T
*   ISOCYANIC ACID, METHYLPHENYLENE ESTER
*   METHYL-meta-PHENYLENE DIISOCYANATE
*   METHYLPHENYLENE ISOCYANATE   *   MONDUR-
TD   *   MONDUR-TD-80   *   NACCONATE-100
*   NIAX ISOCYANATE TDI   *   RCRA WASTE NUMBER
U223   *   RUBINATE TDI   *   RUBINATE TDI 80/20
*   T 100   *   TDI   *   TDI-80   *   TDI 80-20
*   TOLUENE DIISOCYANATE   *   TOLYLENE DIISO-
CYANATE   *   TOLYLENE ISOCYANATE

CONSENSUS REPORTS: IARC Cancer Review: GROUP 2B IMEMDT 7,56,87, Animal
Sufficient Evidence IMEMDT 39,287,86; Human Inadequate Evidence IMEMDT 39,287,86.
NTP Fourth Annual Report On Carcinogens,
1984. NTP Carcinogenesis Studies (gavage):
Clear Evidence: mouse, rat NTPTR* NTP-TR-
251,86. Reported in EPA TSCA Inventory.

NIOSH REL: TDI-air:10H TWA 0.005 ppm;CL
    0.02 ppm/10M
DOT Classification: Poison B; LABEL: Poison

SAFETY PROFILE: Confirmed carcinogen with experimental carcinogenic and neoplastigenic data. Poison by inhalation. Moderately toxic by ingestion. Severe skin irritant. Human mutation data reported. Capable of producing severe dermatitis and bronchial spasm. A common air contaminant. Combustible when exposed to heat or flame. Explosive in the form of vapor when exposed to heat or flame. To fight fire, use dry chemical, CO_2. Potentially violent polymerization reaction with bases or acyl chlorides. Reaction with water releases carbon dioxide. Storage in polyethylene containers is hazardous due to absorption of water through the plastic. When heated to decomposition it emits highly toxic fumes of NO_x.

TGM750   CAS: 584-84-9   HR: 3
TOLUENE-2,4-DIISOCYANATE
mf: $C_9H_6N_2O_2$ mw: 174.17

PROP: Clear, faintly yellow liquid; sharp, pungent odor. Mp: 19.5-21.5°, d (liquid): 1.2244 @ 20/4°, bp: 251°, flash p: 270°F (OC), vap d: 6.0, lel: 0.9%, uel: 9.5%. Misc with alc (decomp), ether, acetone, carbon tetrachloride, benzene, chlorobenzene, kerosene, olive oil.

SYNS: CRESORCINOL DIISOCYANATE * DESMO-DUR T80 * DI-ISOCYANATE de TOLUYLENE * DI-ISO-CYANATOLUENE * 2,4-DIISOCYANATO-1-METHYLBENZENE (9CI) * 2,4-DIISOCYANATO-TOLUENE * DIISOCYANAT-TOLUOL * ISOCYANIC ACID, METHYLPHENYLENE ESTER * ISOCYANIC ACID, 4-METHYL-m-PHENYLENE ESTER * HYLENE T * HYLENE TCPA * HYLENE TLC * HYLENE TM * HYLENE TM-65 * HYLENE TRF * 4-METHYL-PHENYLENE DIISOCYANATE * 4-METHYL-PHENYL-ENE ISOCYANATE * MONDUR TD * MONDUR TD-80 * MONDUR TDS * NACCONATE 1OO * NCI-C50533 * NIAX TDI * NIAX TDI-P * RCRA WASTE NUMBER U223 * RUBINATE TDI 80/20 * TDI * 2,4-TDI * TDI-80 * TDI (OSHA) * TOLUEEN-DIISOCYANAAT * TOLUEN-DISOCI-ANATO * TOLUENE DIISOCYANATE * 2,4-TOL-UENEDIISOCYANATE * TOLUILENODWUIZOCYJA-NIAN * TULUYLENDIISOCYANAT * TOLUYLENE-2,4-DIISOCYANATE * m-TOLYLENE DIISOCYANATE * TOLYLENE-2,4-DIISOCYANATE * 2,4-TOLYLENE-DIISOCYANATE

CONSENSUS REPORTS: IARC Cancer Review: GROUP 2B IMEMDT 7,56,87; Human Inadequate Evidence IMEMDT 39,287,86; Animal Sufficient Evidence IMEMDT 39,287,86.

NTP Fourth Annual Report On Carcinogens, 1984. Community Right-To-Know List. EPA Extremely Hazardous Substances List. Reported in EPA TSCA Inventory.

OSHA PEL: (Transitional: CL 0.02 ppm) TWA 0.005 ppm; STEL 0.02 ppm
ACGIH TLV: TWA 0.005 ppm; STEL 0.02 ppm
DFG MAK: 0.01 ppm (0.07 mg/m^3
NIOSH REL: (Diisocyanates) TWA 0.005 ppm; CL 0.02 ppm/10M
DOT Classification: Poison B; Label: Poison.

SAFETY PROFILE: Confirmed carcinogen. Poison by ingestion, inhalation, and intravenous routes. Human systemic effects by inhalation: unspecified changes to the eyes and sense of smell, respiratory obstruction, cough, sputum, and other pulmonary and gastrointestinal changes. Mutation data reported. A severe skin and eye irritant. Capable of producing severe dermatitis and bronchial spasm. A common air contaminant. Combustible when exposed to heat or flame. Explosive in the form of vapor when exposed to heat or flame. To fight fire, use dry chemical, CO_2. Potentially violent polymerization reaction with bases or acyl chlorides. Reaction with water releases carbon dioxide. Storage in polyethylene containers is hazardous due to absorption of water through the plastic. When heated to decomposition it emits highly toxic fumes of NO_x.

TGM800   CAS: 91-08-7   HR: 3
TOLUENE-2,6-DIISOCYANATE
mf: $C_9H_6N_2O_2$ mw: 174.17

SYNS: 2,6-DIISOCYANATO-1-METHYLBENZENE * 2,6-DIISOCYANATOTOLUENE * HYLENE TM * 2-METHYL-m-PHENYLENE ESTER, ISOCYANIC ACID * 2-METHYL-m-PHENYLENE ISOCYANATE * NIAX TDI * 2,6-TDI * 2,6-TOLUENE DIISOCYANATE * TOLYLENE-2,6-DIISOCYANATE * m-TOLYLENE DI-ISOCYANATE

CONSENSUS REPORTS: IARC Cancer Review: GROUP 2B IMEMDT 7,56,87; Human Inadequate Evidence IMEMDT 39,287,86; Animal Sufficient Evidence IMEMDT 39,287,86. Reported in EPA TSCA Inventory. Community Right-To-Know List. EPA Hazardous Substances List.

DFG MAK: 0.01 ppm (0.07 mg/m^3)
NIOSH REL: (Diisocyanates) TWA 0.005 ppm; CL 0.02 ppm/10M

SAFETY PROFILE: Suspected carcinogen. Poison by ingestion and inhalation. Human systemic effects by inhalation: olfactory, eye and pulmonary changes. When heated to decomposition it emits toxic fumes of NO_x.

TGN250   CAS: 88-19-7   **HR: 3**
o-TOLUENESULFONAMIDE
mf: $C_7H_9NO_2S$ mw: 171.23

PROP: Tetragonal prisms. Mp: 156°. Sol in water, alc.

SYNS: o-METHYLBENZENESULFONAMIDE * 2-METHYLBENZENESULFONAMIDE * ONCO-CARBIDE * ORTHO-TOLUOL-SULFONAMID (GERMAN) * OTS * OXYUREA * TOLUENE-2-SULFONAMIDE

CONSENSUS REPORTS: IARC Cancer Review: GROUP 2B IMEMDT 7,334,87; Animal Limited Evidence IMEMDT 22,111,80. Reported in EPA TSCA Inventory. EPA Genetic Toxicology Program.

SAFETY PROFILE: Suspected carcinogen with experimental tumorigenic data. Mildly toxic by ingestion. Experimental reproductive effects. Mutation data reported. An eye irritant. When heated to decomposition it emits very toxic fumes of NO_x and SO_x. Used as a chemical intermediate in the production of saccharin.

TGO750   CAS: 100-53-8   **HR: 3**
α-TOLUENETHIOL
mf: C_7H_8S mw: 124.21

PROP: A water-white, mobile liquid; strong odor. Bp: 194.8°, flash p: 158°F (CC), d: 1.058 @ 20°, vap d: 4.28.

SYNS: BENZYL MERCAPTAN * BENZYLTHIOL * (MERCAPTOMETHYL)BENZENE * α-MERCAPTO-TOLUENE * PHENYLMETHANETHIOL * PHENYL-METHYL MERCAPTAN * THIOBENZYL ALCOHOL * α-TOLUOLTHIOL * α-TOLYL MERCAPTAN * USAF EK-1509

CONSENSUS REPORTS: Reported in EPA TSCA Inventory.

SAFETY PROFILE: Poison by intraperitoneal route. Moderately toxic by ingestion. An eye irritant. Questionable carcinogen with experimental tumorigenic data. Flammable when exposed to heat or flame. Can react vigorously with oxidizing materials. To fight fire, use foam, CO_2, dry chemical, water spray, mist, fog. When heated to decomposition and on contact

with acid or acid fumes it emits highly toxic fumes of SO_x.

TGQ500   CAS: 108-44-1   **HR: 3**
m-TOLUIDINE
DOT: UN 1708
mf: C_7H_9N mw: 107.17

PROP: Colorless liquid. Mp: −50.5°, bp: 203.3°, d: 0.989 @ 20/4°, vap press: 1 mm @ 41°, vap d: 3.90. Sltly sol in water; sol in alc, ether.

SYNS: 3-AMINO-1-METHYLBENZENE * 3-AMINO-PHENYLMETHANE * 3-AMINOTOLUEN (CZECH) * m-AMINOTOLUENE * 3-AMINOTOLUENE * m-METHYLANILINE * 3-METHYLANILINE * m-METHYLBENZENAMINE * 3-METHYLBENZEN-AMINE * m-TOLUIDIN (CZECH) * 3-TOLUIDINE * m-TOLYLAMINE

CONSENSUS REPORTS: Reported in EPA TSCA Inventory.

OSHA PEL: TWA 2 ppm (skin)
ACGIH TLV: TWA 2 ppm (skin)

SAFETY PROFILE: Poison by ingestion and intraperitoneal routes. A skin and eye irritant. Flammable when exposed to heat or flame. Can react vigorously on contact with oxidizing materials. To fight fire, use foam, CO_2, dry chemical. When heated to decomposition it emits highly toxic fumes of NO_x.

TGQ750   CAS: 95-53-4   **HR: 3**
o-TOLUIDINE
mf: C_7H_9N mw: 107.17

DOT: UN 1708

PROP: Colorless liquid. Mp: −16.3°, bp: 200-202°, ULC: 20-25, flash p: 185° (CC), d: 1.004 @ 20/4°, autoign temp: 900°F, vap press: 1 mm @ 44°, vap d: 3.69. Sltly sol in water, dilute acid; sol in alc and ether.

SYNS: 1-AMINO-2-METHYLBENZENE * 2-AMINO-1-METHYLBENZENE * o-AMINOTOLUENE * 2-AMI-NOTOLUENE * C.I. 37077 * 1-METHYL-2-AMINO-BENZENE * 2-METHYL-1-AMINOBENZENE * o-METHYLANILINE * 2-METHYLANILINE * o-METHYLBENZENAMINE * 2-METHYLBENZEN-AMINE * o-TOLUIDIN (CZECH) * 2-TOLUIDINE * o-TOLUIDYNA (POLISH) * o-TOLYLAMINE

CONSENSUS REPORTS: IARC Cancer Review: GROUP 2B IMEMDT 7,362,87; Human Inadequate Evidence IMEMDT 16,349,78; Human Limited Evidence IMEMDT 27,155,82; Animal Inadequate Evidence IMEMDT 16, 349,78. NTP Fourth Annual Report On Carcinogens, 1984. EPA Genetic Toxicology Program. Community Right-To-Know List. Reported in EPA TSCA Inventory.

OSHA PEL: TWA 5 ppm (skin)
ACGIH TLV: TWA 2 ppm (skin); Suspected Human Carcinogen.
DFG MAK: Animal Carcinogen, Suspected Human Carcinogen.
DOT Classification: Poison B; Label: Poison.

SAFETY PROFILE: Confirmed carcinogen with experimental neoplastigenic and tumorigenic data. Poison by ingestion and intraperitoneal routes. Moderately toxic by skin contact. Human systemic effects by inhalation: urine volume increase, hematuria and blood methemoglobinemia-carboxhemoglobinemia. Human mutation data reported. A skin and eye irritant. Human mucous membrane effects. Can produce severe systemic disturbances. The main portal of entry into the body is the respiratory tract, particularly in cases of industrial exposure. The symptoms produced are headache, weakness, difficulty in breathing, air hunger, psychic disturbances, and marked irritation of the kidneys and bladder. The literature does not yield any good data for comparing the toxicity of the o-, m- and p-isomers. Their behavior is generally comparable to that of aniline. It has been determined experimentally that a concentration of about 100 ppm is the maximum endurable for 1 hour without serious consequences and that from 6-23 ppm is endurable for several hours without serious disturbances.

Flammable when exposed to heat or flame. Hypergolic reaction with red fuming nitric acid. Can react with oxidizing materials. To fight fire, use foam, CO_2, dry chemical. When heated to decomposition it emits highly toxic fumes of NO_x.

TGR000   CAS: 106-49-0   **HR: 3**
p-TOLUIDINE
DOT: UN 1708
mf: C_7H_9N mw: 107.17

PROP: Colorless leaflets. Mp: 44.5°, bp: 200.4°, flash p: 188°F (CC), d: 1.046 @ 20/4°, autoign temp: 900°F, vap press: 1 mm @ 42°, vap d: 3.90. Sol in water, dilute acid, CS_2; very sol in alc, ether.

SYNS: 4-AMINO-1-METHYLBENZENE * 4-AMINO-TOLUEN (CZECH) * p-AMINOTOLUENE * 4-AMINO-TOLUENE * C.I. 37107 * C.I. AZOIC COUPLING COMPONENT 107 * p-METHYLANILINE * 4-METH-YLANILINE * p-METHYLBENZENAMINE * 4-ME-THYLBENZENAMINE * NAPHTOL AS-KG * NAPHTOL AS-KGLL * p-TOLUIDIN (CZECH) * 4-TOLUIDINE * TOLYLAMINE * p-TOLYLA-MINE

CONSENSUS REPORTS: Reported in EPA TSCA Inventory. EPA Genetic Toxicology Program.

OSHA PEL: TWA 2 ppm (skin)
ACGIH TLV: TWA 2 ppm (skin); Suspected Human Carcinogen.

SAFETY PROFILE: Poison by ingestion and intraperitoneal routes. Mutation data reported. A severe skin and eye irritant. Flammable when exposed to heat, flame, or oxidizers. Can react vigorously on contact with oxidizing materials. To fight fire, use foam, CO_2, dry chemical. When heated to decomposition it emits highly toxic fumes of NO_x.

TGS000   CAS: 3209-30-1   **HR: 3**
TOLUIDINE BLUE
mf: $C_{28}H_{22}N_2O_{10}S_2 \cdot 2Na$ mw: 656.62

SYNS: C.I. 63340 * 6,6'-((4,8-DIHYDROXY-1,5-AN-THRAQUINONYLENE)DIIMINO) DI-m-TOLUENE SULFONIC ACID DISODIUM SALT

SAFETY PROFILE: Poison by intravenous route. Mutation data reported. When heated to decomposition it emits very toxic fumes of NO_x, Na_2O and SO_x.

TGS500   CAS: 636-21-5   **HR: 3**
o-TOLUIDINE HYDROCHLORIDE
mf: $C_7H_9N \cdot ClH$ mw: 143.63

PROP: Monoclinic prisms. Mp: 218-220°, bp: 242°. Sol in water; sltly sol in alc.

SYNS: 1-AMINO-2-METHYLBENZENE HYDROCHLO-RIDE * 2-AMINO-1-METHYLBENZENE HYDRO-CHLORIDE * 2-AMINOTOLUENE HYDROCHLORIDE * o-AMINOTOLUENE HYDROCHLORIDE * 1-METHYL-2-AMINOBENZENE HYDROCHLORIDE * 2-METHYL-1-AMINOBENZENE HYDROCHLORIDE * o-METHYLANILINE HYDROCHLORIDE * 2-METH-YLANILINE HYDROCHLORIDE * o-METHYL-BENZENAMINE HYDROCHLORIDE * 2-METHYL-BENZENAMINE HYDROCHLORIDE * NCI-C02335

* RCRA WASTE NUMBER U222 * 2-TOLUIDINE HYDROCHLORIDE * o-TOLYLAMINE HYDROCHLORIDE

CONSENSUS REPORTS: IARC Cancer Review: Animal Sufficient Evidence IMEMDT 27,155,82. NTP Fourth Annual Report On Carcinogens, 1984. NCI Carcinogenesis Bioassay (feed); Clear Evidence: mouse, rat NCITR* NCI-CG-TR-153,79. EPA Genetic Toxicology Program. Community Right-To-Know List. Reported in EPA TSCA Inventory.

SAFETY PROFILE: Confirmed carcinogen with experimental carcinogenic data. Poison by intraperitoneal route. Moderately toxic by ingestion. Mutation data reported. When heated to decomposition it emits very toxic fumes of HCl and NO_x.

TGS750   CAS: 540-23-8   HR: 3
p-TOLUIDINE HYDROCHLORIDE
mf: $C_7H_9N \cdot ClH$ mw: 143.63

PROP: Needles from acetic ether. Mp: 243°, bp: 257.5°. Sol in water, alc; insol in ether, benzene.

SYNS: 4-AMINOTOLUENE HYDROCHLORIDE * 4-METHYLANILINE HYDROCHLORIDE * 4-METHYLBENZENAMINE HYDROCHLORIDE * p-TOLUIDINIUM CHLORIDE

CONSENSUS REPORTS: Reported in EPA TSCA Inventory. EPA Genetic Toxicology Program.

SAFETY PROFILE: Poison by intraperitoneal route. Moderately toxic by ingestion. Questionable carcinogen with experimental carcinogenic and tumorigenic data. When heated to decomposition it emits very toxic fumes of NO_x and HCl.

TGW000   CAS: 2646-17-5   HR: 3
1-(o-TOLYLAZO)-2-NAPHTHOL
mf: $C_{17}H_{14}N_2O$ mw: 262.33

SYNS: A.F.ORANGE No. 2 * AIZEN FOOD ORANGE No. 2 * ATUL OIL ORANGE T * C.I. 12100 * C.I. SOLVENT ORANGE 2 * D&C ORANGE No. 2 * DOLKWAL ORANGE SS * EXTRACT D&C ORANGE No. 4 * FAT ORANGE II * HEXACOL OIL ORANGE SS * LACQUER ORANGE V * 1-((2-METHYLPHENYL)AZO)-2-NAPHTHALENOL * OIL ORANGE O'PEL * OIL ORANGE SS * OLEAL ORANGE SS * ORANGE 3R SOLUBLE IN GREASE * ORGANOL ORANGE 2R * TOLUENE-2-AZONAPHTHOL-2 * o-TOLUENO-AZO-β-NAPHTHOL * 1-(o-TOLYLAZO)-β-NAPHTHOL

CONSENSUS REPORTS: IARC Cancer Review: GROUP 2B IMEMDT 7,56,87; Animal Sufficient Evidence IMEMDT 8,165,75. Reported in EPA TSCA Inventory. EPA Genetic Toxicology Program.

SAFETY PROFILE: Suspected carcinogen with experimental carcinogenic and neoplastigenic data. Poison by intravenous route. Mildly toxic by ingestion. When heated to decomposition it emits toxic fumes of NO_x. Used to color cosmetics, varnishes, oils, fats and waxes, petroleum products.

TGY075   CAS: 106-43-4   HR: 2
p-TOLYL CHLORIDE
DOT: UN 2238
mf: C_7H_7Cl mw: 126.59

PROP: Liquid. Bp: 162.4°, d: (20/4) 1.0697, mp: 7.5°. Sltly sol in water; sol in alc, benzene, chloroform, ether.

SYNS: 4-CHLORO-1-METHYLBENZENE * 4-CHLOROTOLUENE * p-CHLOROTOLUENE (DOT)

CONSENSUS REPORTS: EPA Genetic Toxicology Program. Reported in EPA TSCA Inventory.

DOT Classification: Flammable or Combustible Liquid; Label: Flammable Liquid.

SAFETY PROFILE: Moderately toxic by ingestion and possibly other routes. Mildly toxic by inhalation. Flammable when exposed to heat or flame. When heated to decomposition it emits toxic fumes of Cl^-.

THA250   CAS: 103-93-5   HR: 2
p-TOLYL ISOBUTYRATE
mf: $C_{11}H_{14}O_2$ mw: 178.25

PROP: Colorless liquid; characteristic odor. D: 0.990-0.996, refr index: 1.485, flash p: +212°F. Sol in alc; insol in water.

SYNS: p-CRESYL ISOBUTYRATE * FEMA No. 3075 * ISOBUTYRIC ACID, p-TOLYL ESTER * PARACRESYL ISOBUTYRATE

CONSENSUS REPORTS: Reported in EPA TSCA Inventory.

SAFETY PROFILE: Moderately toxic by ingestion and skin contact. Combustible liquid. When heated to decomposition it emits acrid smoke and irritating fumes.

THG000   CAS: 622-51-5   **HR: 3**
p-TOLYLUREA
mf: $C_8H_{10}N_2O$ mw: 150.20

PROP: Plates from alc. Mp: 188°. Very sltly sol in cold water; sol in hot alc.

SYNS: 4-METHYLPHENYLUREA * NCI-C02153 * p-TOLYCARBAMIDE * p-TOLYUREA

SAFETY PROFILE: Moderately toxic by ingestion. Questionable carcinogen with experimental carcinogenic data. When heated to decomposition it emits toxic fumes of NO_x.

THG250   CAS: 17406-45-0   **HR: 3**
TOMATINE
mf: $C_{50}H_{83}NO_{21}$ mw: 1034.34

PROP: Antifungal substance in wilt-resistant tomato plants. Needles. Mp: 263-268°. Sol in ethanol, methanol, dioxane, propylene alc; almost insol in water, ether, petr ether.

SYNS: LYCOPERSICIN * A''-TOMATIDINE * TOMATIDINE GLYCOSIDE * TOMATIN * α-TOMATINE

SAFETY PROFILE: Poison by intravenous and intraperitoneal routes. Moderately toxic by ingestion and subcutaneous routes. When heated to decomposition it emits toxic fumes of NO_x.

THJ250   CAS: 9000-65-1   **HR: 1**
TRAGACANTH GUM

PROP: from the shrub *Astragalus gummifier* Labillardiere. Powder is white, pieces are white to pale yellow, translucent, and horny; odorless with mucilaginous taste.

SYNS: GUM TRAGACANTH * TRAGACANTH

CONSENSUS REPORTS: Reported in EPA TSCA Inventory.

SAFETY PROFILE: Mildly toxic by ingestion. A mild allergen. Combustible when exposed to heat or flame. When heated to decomposition it emits acrid smoke and irritating fumes.

THJ500   CAS: 27203-92-5   **HR: 3**
TRAMADOL
mf: $C_{16}H_{25}NO_2$ mw: 263.42

SYNS: CG 315 * (±)-trans-2-((DIMETHYLAMINO)METHYL-1-(m-METHOXYPHENYL)CYCLOHEXANOL * TRAMAL

SAFETY PROFILE: Poison by ingestion, subcutaneous, intravenous, and intramuscular routes.

When heated to decomposition it emits toxic fumes of NO_x.

THL750   CAS: 3736-86-5   **HR: 3**
TRENTADIL HYDROCHLORIDE
mf: $C_{20}H_{27}N_5O_3 \cdot ClH$ mw: 421.98

SYNS: BAMIFYLLINE HYDROCHLORIDE * BAMIPHYLLINE HYDROCHLORIDE * BAX 2793Z * BENZETAMOPHYLLINE HYDROCHLORIDE * 8-BENZYL-7-(2-(ETHYL(2-HYDROXYETHYL)AMINO)ETHYL)THEOPHYLLINE, HYDROCHLORIDE * TRENTADIL

SAFETY PROFILE: Poison by ingestion, intraperitoneal, and intravenous routes. When heated to decomposition it emits very toxic fumes of NO_x and HCl.

THM500   CAS: 102-76-1   **HR: 3**
TRIACETYL GLYCERIN
mf: $C_9H_{14}O_6$ mw: 218.23

PROP: Colorless oily liquid; slt fatty odor and taste. Mp: −78°, bp: 258°, flash p: 280°F (COC), d: 1.161, autoign temp: 812°F, vap d: 7.52. Sol in water; misc with alc, ether, chloroform.

SYNS: ENZACTIN * FEMA No. 2007 * FUNGACETIN * GLYCERINE TRIACETATE * GLYCEROL TRIACETATE * GLYCERYL TRIACETATE * GLYPED * KESSCOFLEX TRA * KODAFLEX TRIACETIN * 1,2,3-PROPANETRIOL TRIACETATE * TRIACETIN (FCC) * VANAY

CONSENSUS REPORTS: Reported in EPA TSCA Inventory.

SAFETY PROFILE: Poison by ingestion. Moderately toxic by intraperitoneal, subcutaneous, and intravenous routes. An eye irritant. Combustible when exposed to heat, flame, or powerful oxidizers. To fight fire, use alcohol foam, water, CO_2, dry chemical. When heated to decomposition it emits acrid smoke and irritating fumes.

THN000   CAS: 102-70-5   **HR: 3**
TRIALLYLAMINE

DOT: UN 2610
mf: $C_9H_{15}N$ mw: 137.25

PROP: Liquid. D: 0.800 @ 20°/4°, mp: < −70°, bp: 150-151°, flash p: 103°F (TOC).

SYN: N-N-DI-2-PROPENYL-2-PROPEN-1-AMINE

CONSENSUS REPORTS: Reported in EPA TSCA Inventory.

DOT Classification: Flammable or Combustible Liquid; Label: Flammable Liquid.

SAFETY PROFILE: Poison by skin contact and intraperitoneal routes. Moderately toxic by ingestion, inhalation, and possibly other routes. An eye and severe skin irritant. Human systemic effects by inhalation: structural or functional changes in trachea or bronchi. Flammable when exposed to heat, flame or oxidizers. To fight fire, use foam, alcohol foam, fog. When heated to decomposition it emits toxic fumes of NO_x.

THN500   CAS: 101-37-1   **HR: 3**
TRIALLYL CYANAURATE
mf: $C_{12}H_{15}N_3O_3$ mw: 243.24

PROP: Bp: 120° @ 5 mm, fp: 27.3°, flash p: >176°F (TOC), d: 1.1133 @ 30°, vap press: 1 mm @ 100°.

SYNS: TRIPROPARGYL CYANURATE * 2,4,6-TRIPROP-2-YNYLOXY-s-TRIAZINE * 2,4,6-TRIS(ALLYLOXY)TRIAZINE

CONSENSUS REPORTS: Reported in EPA TSCA Inventory.

SAFETY PROFILE: Poison by intravenous route. Flammable when exposed to heat, flame, or oxidizers. To fight fire, use spray, foam, dry chemical. When heated to decomposition or on contact with acid or acid fumes it emits highly toxic fumes of CN^- and NO_x.

THP000   CAS: 548-61-8   **HR: 3**
TRIAMINOTRIPHENYLMETHANE
mf: $C_{19}H_{19}N_3$ mw: 289.41

PROP: Leaves from water. Mp: 148°. Sltly sol in cold water; sol in abs alc and benzene.

SYNS: LEUCOPARAFUCHSIN * LEUCOPARAFUCHSINE * 4,4′,4′′-METHYLIDYNETRIANILINE * 4,4′,4′′-METHYLIDYNETRISBENZENEAMINE * p,p′,p′′-TRIAMINOTRIPHENYLMETHANE * 4,4′,4′′-TRIAMINOTRIPHENYLMETHANE * TRIS-4-AMINOFENYLMETHAN (CZECH)

SAFETY PROFILE: Moderately toxic by ingestion. An eye irritant. Questionable carcinogen with experimental tumorigenic data. When heated to decomposition it emits toxic fumes of NO_x.

THP250   CAS: 17168-85-3   **HR: 3**
TRIAMMINEDIPEROXOCHROMIUM(IV)
mf: $CrH_9N_3O_4$ mw: 167.09

SAFETY PROFILE: Suspected carcinogen. Chromium compounds are generally poisons. May explode with heat or shock. May explode at 120°C. An oxidizer. When heated to decomposition it emits toxic fumes of NO_x.

THV000   CAS: 1329-86-8   **HR: 3**
TRIBROMOETHANOL
mf: $C_2H_3Br_4O$ mw: 282.78

PROP: Crystals; ethereal odor, aromatic taste. Bp: 92° @ 10 mm, mp: 79-82°, decomp @ 70°. Sltly water sol; sol in alc, organic solvents.

SYNS: AVERTIN * BROMETHOL * ETHOBROM * NARCOLAN * NARKOLAN * TRIBROMETHANOL * TRIBROMOETHYL ALCOHOL

SAFETY PROFILE: Poison by ingestion, intravenous, intraperitoneal, and rectal routes. Experimental reproductive effects. When heated to decomposition it emits toxic fumes of Br^-. An anesthetic drug.

THV500   CAS: 73941-35-2   **HR: 3**
2,4,5-TRIBROMOIMIDAZOLE CADMIUM SALT (2:1)
mf: $C_6Br_6N_4 \cdot Cd$ mw: 719.96

SYN: CADMIUM salt of 2,4-5-TRIBROMOIMIDAZOLE

CONSENSUS REPORTS: Cadmium and its compounds are on the Community Right-To-Know List.

OSHA PEL: TWA 0.2 mg(Cd)/m³; CL 0.6 mg(Cd)/m³ (dust)
ACGIH TLV: TWA 0.05 mg(Cd)/m³ (Proposed: TWA 0.01 mg(Cd)/m³ (dust), Human Carcinogen); BEI: 10 μg/g creatinine in urine; 10 μg/L in blood.
NIOSH REL: (Cadmium) Reduce to lowest feasible level.

SAFETY PROFILE: Confirmed human carcinogen. Poison by intravenous route. When heated to decomposition it emits very toxic fumes of Br^-, Cd, and NO_x.

THX250   CAS: 102-82-9   **HR: 3**
TRIBUTYLAMINE
DOT: UN 2542
mf: $C_{12}H_{27}N$ mw: 185.40

PROP: A colorless liquid. Mp: $-70°$, bp: $213°$, flash p: 187°F (OC) d: 0.78-0.79, vap d: 6.38. Insol in water; sol in alc, ether.

SYNS: TRI-n-BUTYLAMINE * TRIS-N-BUTYLAMINE

CONSENSUS REPORTS: Reported in EPA TSCA Inventory.

DOT Classification: Corrosive Material; Label: Corrosive.

SAFETY PROFILE: Poison by ingestion, inhalation, skin contact, and subcutaneous routes. A central nervous system stimulant, irritant, and sensitizer. A corrosive irritant to skin, eyes, and mucous membranes. Flammable when exposed to heat, flame, or oxidizers. Can react with oxidizing materials. To fight fire, use foam, CO_2, dry chemical. When heated to decomposition it emits toxic fumes of NO_x.

THX500   CAS: 122-56-5   HR: 3
TRI-n-BUTYL BORANE
mf: $C_{12}H_{27}B$ mw: 182.20

PROP: Colorless pyroforic liquid. Mp: $34°$, bp: $170° @ 222$ mm, d: $0.747 @ 25°$, vap press: 1 mm @ 20°, flash p: $-32°F$. Insol in water; sol in most organic solvents.

SYNS: BORIC ACID, TRIBUTYL ESTER * TBB * TRIBUTYLBORINE

CONSENSUS REPORTS: Reported in EPA TSCA Inventory.

SAFETY PROFILE: Poison by intravenous route. Moderately toxic by ingestion. A very dangerous fire hazard when exposed to heat or flame; can ignite spontaneously. When heated to decomposition it emits acrid smoke and irritating fumes.

THX750   CAS: 688-74-4   HR: 2
TRI-n-BUTYL BORATE
mf: $C_{12}H_{27}BO_3$ mw: 230.20

PROP: Colorless, mobile liquid; odor like n-butanol. Bp: $230°$, fp: $<-70°$, flash p: 200°F (COC), d: $0.847 @ 28°$, vap d: 7.95.

SYNS: BORESTER 2 * BORIC ACID, TRI-sec-BUTYL ESTER * BUTYL BORATE * n-BUTYL BORATE * TRIBUTOXYBORANE * TRI-n-BUTOXYBORANE * TRIBUTYL BORATE

CONSENSUS REPORTS: Reported in EPA TSCA Inventory.

SAFETY PROFILE: Moderately toxic by ingestion and intraperitoneal routes. An eye irritant.

Flammable when exposed to heat, flame or oxidizers. To fight fire use foam, CO_2, dry chemical. When heated to decomposition or on contact with acid or acid fumes it can emit toxic fumes; on contact with oxidizing materials it can react vigorously.

TIA250   CAS: 126-73-8   HR: 3
TRIBUTYL PHOSPHATE
mf: $C_{12}H_{27}O_4P$ mw: 266.36

PROP: Colorless odorless liquid. Bp: $289°$ (decomp), mp: $<-80°$, flash p: 295°F (COC), d: $0.982 @ 20°$, vap d: 9.20. Sol in water; misc in alc, ether.

SYNS: CELLUPHOS 4 * TBP * TRIBUTILFOS-FATO (ITALIAN) * TRIBUTYLE (PHOSPHATE DE) (FRENCH) * TRIBUTYLFOSFAAT (DUTCH) * TRI-BUTYLPHOSPHAT (GERMAN) * TRI-n-BUTYL PHOSPHATE

CONSENSUS REPORTS: Reported in EPA TSCA Inventory.

OSHA PEL: (Transitional: TWA 5 mg/m^3) TWA 0.2 ppm
ACGIH TLV: TWA 0.2 ppm

SAFETY PROFILE: Poison by intraperitoneal and intravenous routes. Moderately toxic by ingestion, inhalation, and subcutaneous routes. A skin, eye, and mucous membrane irritant. Combustible when exposed to heat or flame. To fight fire, use CO_2, dry chemical, fog, mist. When heated to decomposition it emits toxic fumes of PO_x.

TIB000   CAS: 5488-45-9   HR: 3
TRIBUTYL(8-QUINOLINOLATO)TIN
mf: $C_{21}H_{33}NOSn$ mw: 434.24

SYN: (8-QUINOLINOLATO)TRIBUTYLSTANNANE

OSHA PEL: TWA 0.1 mg(Sn)/m^3 (skin)
ACGIH TLV: TWA 0.1 mg(Sn)/m^3 (skin) (Proposed: TWA 0.1 mg(Sn)/m^3; STEL 0.2 mg(Sn)/m^3 (skin))
NIOSH REL: (Organotin Compounds) TWA 0.1 mg(Sn)/m^3

SAFETY PROFILE: Poison by intravenous route. When heated to decomposition it emits toxic fumes of NO_x.

TIF250   CAS: 28801-69-6   HR: 3
TRIBUTYLTIN NEODECANOATE
mf: $C_{22}H_{46}O_2Sn$ mw: 461.37

SYNS: 4,4-DIMETHYLOCTANOIC ACID, TRIBUTYL-STANNYL ESTER * (4,4-DIMETHYLOCTANOYLOXY) TRIBUTYLSTANNANE * HYDROXYTRIBUTYL-STANNANE-4,4-DIMETHYLOCTANOATE * TRIBUTYL (NEODECANOYLOXY)STANNANE

OSHA PEL: TWA 0.1 mg(Sn)/m^3

ACGIH TLV: TWA 0.1 mg(Sn)/m^3 (skin) (Proposed: TWA 0.1 mg(Sn)/m^3; STEL 0.2 mg(Sn)/m^3 (skin))

NIOSH REL: (Organotin Compounds) TWA 0.1 mg(Sn)/m^3

SAFETY PROFILE: Poison by intravenous route. Moderately toxic by ingestion. When heated to decomposition it emits acrid smoke and irritating fumes.

TIG250   CAS: 150-50-5   **HR: 3**
S,S,S-TRIBUTYL TRITHIOPHOSPHITE
mf: $C_{12}H_{27}PS_3$ mw: 298.54

PROP: Colorless liquid; mild characteristic odor. Bp: 142-145° @ 4.5 mm, flash p: 295°F (COC), d: 0.987 @ 20°/4°.

SYNS: CHEMAGRO B-1776 * DELEAF DEFOLIANT * EASY OFF-D * FOLEX * MERPHOS * PHOS-PHOROTRITHIOUS ACID, S,S,S-TRIBUTYL ESTER * TRIBUTYL PHOSPHOROTRITHIOITE * S,S,S-TRIBUTYL PHOSPHOROTRITHIOITE

SAFETY PROFILE: Poison by intraperitoneal and possibly other routes. Moderately toxic by ingestion and skin contact. A cholinesterase inhibitor. Combustible when exposed to heat or flame. Can react vigorously with oxidizing materials. When heated to decomposition it emits highly toxic fumes of PO$_x$ and SO$_x$. Used as a defoliant.

TIG750   CAS: 60-01-5   **HR: 3**
TRIBUTYRIN
mf: $C_{15}H_{26}O_6$ mw: 302.41

PROP: Colorless, oily liquid; bitter taste. Mp: −75°, d: 1.0356 @ 20/20°, bp: 305-310°, flash p: +212°F. Insol in water; very sol in alc, ether, chloroform.

SYNS: BUTANOIC ACID, 1,2,3-PROPANETRIYL ESTER * BUTYRIC ACID TRIESTER with GLYCERIN * BUTYRYL TRIGLYCERIDE * FEMA No. 2223 * GLYCEROL TRIBUTYRATE * KODAFLEX * TRIBUTYROIN

CONSENSUS REPORTS: Reported in EPA TSCA Inventory.

SAFETY PROFILE: Poison by intravenous route. Moderately toxic by ingestion. Questionable carcinogen with experimental tumorigenic data. Combustible liquid. When heated to decomposition it emits acrid smoke and irritating fumes.

TIH000   CAS: 12380-95-9   **HR: 3**
TRICADMIUM DINITRIDE
mf: Cd_3N_2 mw: 365.21

SYN: CADMIUM NITRIDE

CONSENSUS REPORTS: Cadmium compounds are on the Community Right-To-Know List.

OSHA PEL: TWA 0.2 mg(Cd)/m^3; CL 0.6 mg(Cd)/m^3 (dust)

ACGIH TLV: TWA 0.05 mg(Cd)/m^3 (Proposed: TWA 0.01 mg(Cd)/m^3 (dust), Human Carcinogen); BEI: 10 μg/g creatinine in urine; 10 μg/L in blood.

NIOSH REL: (Cadium) Reduce to lowest feasible level.

SAFETY PROFILE: Confirmed human carcinogen. Many cadmium compounds are poisons. Explodes violently on shock or heating. Explodes on contact with water, acids, or bases. When heated to decomposition it emits very toxic fumes of NO$_x$ and Cd.

TII250   CAS: 76-03-9   **HR: 3**
TRICHLOROACETIC ACID
DOT: UN 1839/UN 2564
mf: $C_2HCl_3O_2$ mw: 163.38

PROP: Colorless, rhombic, deliquescent crystals. Bp: 197.5°, fp: 57.7°, flash p: none, d: 1.6298 @ 61°/4°, vap press: 1 mm @ 51.0°.

SYNS: ACETO-CAUSTIN * ACIDE TRICHLOR-ACETIQUE (FRENCH) * ACIDO TRICLOROACETICO (ITALIAN) * AMCHEM GRASS KILLER * DOW SODIUM TCA INHIBITED * KONESTA * SODIUM TCA, solution * TCA * TRICHLOORAZIJNZUUR (DUTCH) * TRICHLORESSIGSAEURE (GERMAN) * TRICHLOROACETIC ACID, solid (DOT) * TRICHLOROACETIC ACID, solution (DOT) * TRICHLOROETHANOIC ACID * VARITOX

CONSENSUS REPORTS: Reported in EPA TSCA Inventory. EPA Genetic Toxicology Program.

OSHA PEL: TWA 1 ppm
ACGIH TLV: TWA 1 ppm

DOT Classification: Corrosive Material; Label: Corrosive, solid; Corrosive Material; Label: Corrosive, solution.

SAFETY PROFILE: Poison by ingestion and subcutaneous routes. Moderately toxic by intraperitoneal route. Questionable carcinogen with experimental carcinogenic data. Mutation data reported. A corrosive irritant to skin, eyes, and mucous membranes. When heated to decomposition it emits toxic fumes of Cl⁻ and Na₂O. Used as an herbicide.

TIJ150   CAS: 76-02-8   HR: 2
TRICHLOROACETYL CHLORIDE

DOT: UN 2442
mf: C₂Cl₄O   mw: 181.82

SYNS: TRICHLOROACETIC ACID CHLORIDE
* TRICHLOROACETOCHLORIDE

CONSENSUS REPORTS: EPA Extremely Hazardous Substances List. Reported in EPA TSCA Inventory.

DOT Classification: Corrosive Material; Label: Corrosive.

SAFETY PROFILE: Moderately toxic by inhalation and ingestion. A corrosive irritant to skin, eyes, and mucous membranes. When heated to decomposition it emits toxic fumes of Cl⁻.

TIK250   CAS: 120-82-1   HR: 3
1,2,4-TRICHLOROBENZENE

DOT: UN 2321
mf: C₆H₃Cl₃   mw: 181.44

PROP: Colorless liquid. Mp: 17°, bp: 213°, flash p: 230°F (CC), d: 1.454 @ 25°/25°, vap press: 1 mm @ 38.4°, vap d: 6.26. Sol in water.

SYNS: 1,2,4-TRICHLOROBENZENE, liquid (DOT)
* unsym-TRICHLOROBENZENE * TROJCHLOROBEN-
ZEN (POLISH)

CONSENSUS REPORTS: Community Right-To-Know List. Reported in EPA TSCA Inventory.

OSHA PEL: CL 5 ppm
ACGIH TLV: CL 5 ppm
DFG MAK: 5 ppm (40 mg/m³)
DOT Classification: Poison B; Label: St. Andrews Cross.

SAFETY PROFILE: Poison by ingestion. Moderately toxic by intraperitoneal route. Experi-

mental teratogenic and reproductive effects. A skin irritant. Combustible when exposed to heat or flame. Can react vigorously with oxidizing materials. To fight fire, use water, foam, CO₂, dry chemical. When heated to decomposition it emits toxic fumes of Cl⁻.

TIL360   CAS: 2431-50-7   HR: 3
2,3,4-TRICHLOROBUTENE-1
mf: C₄H₅C₁₃   mw: 159.44

SYN: 1-BUTENE, 2,3,4-TRICHLORO-

CONSENSUS REPORTS: Reported in EPA TSCA Inventory.

DFG MAK: Animal Carcinogen; Suspected Human Carcinogen.

SAFETY PROFILE: Confirmed carcinogen. Poison by ingestion. When heated to decomposition it emits toxic fumes of Cl⁻.

TIN000   CAS: 79-00-5   HR: 3
1,1,2-TRICHLOROETHANE
mf: C₂H₃Cl₃   mw: 133.40

PROP: Liquid; pleasant odor. Bp: 114°, fp: −35°, d: 1.4416 @ 20°/4°, vap press: 40 mm @ 35.2°.

SYNS: ETHANE TRICHLORIDE * NCI-C04579
* RCRA WASTE NUMBER U227 * β-T * 1,1,2-TRI-
CHLORETHANE * β-TRICHLOROETHANE * 1,2,2-
TRICHLOROETHANE * TROJCHLOROETAN(1,1,2) (POL-
ISH) * VINYL TRICHLORIDE

CONSENSUS REPORTS: IARC Cancer Review: GROUP 3 IMEMDT 7,56,87; Animal Limited Evidence IMEMDT 20,533,79. NCI Carcinogenesis Bioassay (gavage); No Evidence: rat NCITR* NCI-CG-TR-74,78; (gavage); Clear Evidence: mouse NCITR* NCI-CG-TR-74,78. Community Right-To-Know List. Reported in EPA TSCA Inventory.

OSHA PEL: TWA 10 ppm (skin)
ACGIH TLV: TWA 10 ppm (skin)
DFG MAK: 10 ppm (55 mg/m³); Suspected Carcinogen.

SAFETY PROFILE: Suspected carcinogen with experimental carcinogenic data. Poison by ingestion, intravenous, and subcutaneous routes. Moderately toxic by inhalation, skin contact, and intraperitoneal routes. Experimental reproductive effects. Mutation data reported. An eye and severe skin irritant. Has narcotic properties and acts as a local irritant to the eyes, nose, and lungs. It may also be injurious to the liver and kidneys. Incompatible with potassium. When heated to decomposition it emits toxic fumes of Cl⁻.

TIN750   CAS: 75-94-5   **HR: 3**
TRICHLOROETHENYLSILANE

DOT: UN 1305
mf: C₂H₃Cl₃Si   mw: 161.49

PROP: Fuming liquid. Bp: 90.6°, d: 1.265 @ 25/25°, flash p: 16°F.

SYNS: SILANE, VINYL TRICHLORO 1-150 * TRICHLORO(VINYL)SILANE * TRICHLOROVINYL SILICANE * UNION CARBIDE A-150 * VINYLSILICON TRICHLORIDE * VINYL TRICHLOROSILANE (DOT) * VINYL TRICHLOROSILANE, INHIBITED (DOT)

CONSENSUS REPORTS: Reported in EPA TSCA Inventory.

DOT Classification: Flammable Liquid; Label: Flammable Liquid, Corrosive.

SAFETY PROFILE: Moderately toxic by ingestion, inhalation, and skin contact. A severe eye, and skin irritant. A corrosive irritant to skin, eyes and mucous membranes. A very dangerous fire hazard when exposed to heat or flame. Reacts violently with water; moist air or steam to produce toxic and corrosive fumes. When heated to decomposition it emits toxic fumes of Cl⁻.

TIO750   CAS: 79-01-6   **HR: 3**
TRICHLOROETHYLENE

DOT: UN 1710
mf: C₂HCl₃   mw: 131.38

PROP: Clear, colorless, mobile liquid; characteristic sweet odor of chloroform. D: 1.4649 @ 20°/4°, bp: 86.7°, flash p: 89.6°F (but practically nonflammable), lel: 12.5%, uel: 90% @ > 30°, mp: −73°, fp: −86.8°, autoign temp: 788°F, vap press: 100 mm @ 32°, vap d: 4.53, refr index: 1.477 @ 20°. Immiscible with water; misc with alc, ether, acetone, carbon tetrachloride.

SYNS: ACETYLENE TRICHLORIDE * ALGYLEN * ANAMENTH * BENZINOL * BLACOSOLV * CECOLENE * 1-CHLORO-2,2-DICHLOROETHYLENE * CHLORYLEA * CHORYLEN * CIRCOSOLV * CRAWHASPOL * DENSINFLUAT * 1,1-DICHLORO-2-CHLOROETHYLENE * DOW-TRI * DUKERON * ETHINYL TRICHLORIDE * ETHYLENE TRICHLORIDE * FLECK-FLIP * FLUATE * GERMALGENE * LANADIN * LETHURIN * NARCOGEN * NARKOSOID * NCI-C04546 * NIALK * PERM-A-CHLOR * PETZINOL * RCRA WASTE NUMBER U228 * THRETHYLENE * TRIAD * TRIASOL * TRICHLOORETHEEN (DUTCH) * TRICHLOORETHYLEEN, TRI (DUTCH) * TRICHLORAETHEN (GERMAN) * TRICHLORAETHYLEN, TRI (GERMAN) * TRICHLORAN * TRICHLORETHENE (FRENCH) * TRICHLORETHYLENE, TRI (FRENCH) * TRICHLOROETHENE * 1,2,2-TRICHLOROETHYLENE * TRI-CLENE * TRICLORETENE (ITALIAN) * TRICLOROETILENE (ITALIAN) * TRIELINA (ITALIAN) * TRILENE * TRIMAR * TRI-PLUS * VESTROL * VITRAN * WESTROSOL

CONSENSUS REPORTS: IARC Cancer Review: GROUP 3 IMEMDT 7,364,87; Animal Limited Evidence IMEMDT 20,545,79; Human Inadequate Evidence IMEMDT 20,545,79; Animal Sufficient Evidence IMEMDT 11,263,76. NCI Carcinogenesis Bioassay (gavage); No Evidence: rat NCITR* NCI-CG-TR-2,76; (gavage); Clear Evidence: mouse NCITR* NCI-CG-TR-2,76. Community Right-To-Know List. Reported in EPA TSCA Inventory. EPA Genetic Toxicology Program.

OSHA PEL: (Transitional: TWA 100 ppm; CL 200 ppm; Pk 300 ppm/5M/2H)TWA 50 ppm; STEL 200 ppm
ACGIH TLV: TWA 50 ppm; STEL 200 ppm; BEI: 320 mg(trichloroethanol)/g creatinine in urine at end of shift; 0.5 ppm trichloroethylene in end-exhaled air prior to shift and end of work week.
DFG MAK: Suspected Carcinogen; 50 ppm (270 mg/m³); BAT: 500 μg/dL in blood at end of shift or work week.
NIOSH REL: (Trichloroethylene) TWA 250 ppm; (Waste Anesthetic Gases) CL 2 ppm/1H
DOT Classification: ORM-A; Label: None; Poison B; Label: St. Andrews Cross.

SAFETY PROFILE: Suspected carcinogen with experimental carcinogenic and tumorigenic data. Experimental poison by intravenous and subcutaneous routes. Moderately toxic experimentally by ingestion and intraperitoneal routes. Mildly toxic to humans by ingestion and inhalation. Mildly toxic experimentally by inhalation. Human systemic effects by ingestion and inhalation: eye effects, somnolence, hallucinations or distorted perceptions, gastrointestinal changes and jaundice. Experimental teratogenic and reproductive effects. Human mutation data reported. An eye and severe skin irritant. Inhalation of high concentrations causes narcosis and anesthesia. A form of addiction has been observed in exposed workers. Prolonged inhalation of moderate concentrations causes headache and drowsiness. Fatalities following severe, acute exposure have been attributed to ventricular fibrillation resulting in cardiac failure. There is damage to liver and other organs from chronic exposure. A common air contaminant.

High concentrations of trichloroethylene vapor in high-temperature air can be made to burn mildly if plied with a strong flame. Though such a condition is difficult to produce, flames or arcs should not be used in closed equipment which contains any solvent residue or vapor. Reacts with alkali, epoxides [e.g., 1-chloro-2,3-epoxypropane, 1,4-butanediol mono-2,3-epoxypropylether, 1,4-butanediol di-2,3-epoxypropylether, 2,2-bis((4(2′,3′-epoxypropoxy)phenyl)propane] to form the spontaneously flammable gas dichloroacetylene. Can react violently with Al, Ba, N_2O_4, Li, Mg, liquid O_2, O_3, KOH, KNO_3, Na, NaOH, Ti. Reacts with water under heat and pressure to form HCl gas. When heated to decomposition it emits toxic fumes of Cl^-.

TIP500   CAS: 75-69-4   HR: 2
TRICHLOROFLUOROMETHANE
mf: CCl_3F mw: 137.36

PROP: Colorless liquid. Mp: −111°, bp: 24.1°, d: 1.484 @ 17.2°.

SYNS: ALGOFRENE TYPE 1 * ARCTON 9 * ELECTRO-CF 11 * ESKIMON 11 * FLUOROCARBON NO. 11 * FLUOROTRICHLOROMETHANE (OSHA) * FLUOROTROJCHLOROMETAN (POLISH) * FREON 11 * FREON MF * FRIGEN 11 * GENETRON 11 * HALOCARBON 11 * ISCEON 131 * ISOTRON 11 * LEDON 11 * MONOFLUOROTRICHLOROMETHANE * NCI-C04637 * RCRA WASTE NUMBER U121 * TRICHLOROMONOFLUOROMETHANE * UCON REFRIGERANT 11

CONSENSUS REPORTS: NCI Carcinogenesis Bioassay (gavage); No Evidence: mouse NCITR* NCI-CG-TR-106,78; (gavage); Inadequate Studies: rat NCITR* NCI-CG-TR-106,78. Reported in EPA TSCA Inventory.

OSHA PEL: (Transitional: TWA 1000 ppm) CL 1000 ppm

ACGIH TLV: CL 1000 ppm
DFG MAK: 1000 ppm (5600 mg/m³)

SAFETY PROFILE: High concentrations cause narcosis and anesthesia in humans. Human systemic effects by inhalation: conjunctiva irritation, fibrosing alveolitis and liver changes. Experimental poison by inhalation. Moderately toxic by intraperitoneal route. Reacts violently with aluminum, barium, or lithium. When heated to decomposition it emits highly toxic fumes of F^- and Cl^-. Used as an aerosol propellant, refrigerant, and blowing agent for polymeric foams.

TIQ250   CAS: 52-68-6   HR: 3
((2,2,2-TRICHLORO-1-HYDROXYETHYL) DIMETHYLPHOSPHONATE)
DOT: NA 2783
mf: $C_4H_8Cl_3O_4P$ mw: 257.44

SYNS: AEROL 1 (pesticide) * AGROFOROTOX * ANTHON * BAY 15922 * BAYER 15922 * BAYER L 13/59 * BILARCIL * BOVINOX * BRITON * BRITTEN * CEKUFON * CHLORAK * CHLORFOS * CHLOROFOS * CHLOROFTALM * CHLOROPHOS * CHLOROPHTHALM * CHLOROXYPHOS * CICLOSOM * CLOROFOS (RUSSIAN) * COMBOT EQUINE * DANEX * DEP (pesticide) * DEPTHON * DETF * DIMETHOXY-2,2,2-TRICHLORO-1-HYDROXY-ETHYL-PHOSPHINE OXIDE * O,O-DIMETHYL-(1-HYDROXY-2,2,2-TRICHLORAETHYL)PHOSPHONSAEURE ESTER (GERMAN) * O,O-DIMETHYL-(1-HYDROXY-2,2,2-TRICHLORATHYL)-PHOSPHAT (GERMAN) * O,O-DIMETHYL-(1-HYDROXY-2,2,2-TRICHLORO)ETHYL PHOSPHATE * DIMETHYL-1-HYDROXY-2,2,2-TRICHLOROETHYL PHOSPHONATE * O,O-DIMETHYL-(1-HYDROXY-2,2,2-TRICHLOROETHYL)PHOSPHONATE * O,O-DIMETHYL-1-OXY-2,2,2-TRICHLOROETHYL PHOSPHONATE * O,O-DIMETHYL-(2,2,2-TRICHLOOR-1-HYDROXY-ETHYL)-FOSFONAAT (DUTCH) * O,O-DIMETHYL-(2,2,2-TRICHLOR-1-HYDROXY-AETHYL)PHOSPHONAT (GERMAN) * DIMETHYLTRICHLOROHYDROXYETHYL PHOSPHONATE * DIMETHYL-2,2,2-TRICHLORO-1-HYDROXYETHYLPHOSPHONATE * O,O-DIMETHYL-2,2,2-TRICHLORO-1-HYDROXYETHYL PHOSPHONATE * O,O-DIMETIL-(2,2,2-TRICLORO-1-IDROSSI-ETIL)-FOSFONATO (ITALIAN) * DIMETOX * DIPTERAX * DIPTEREX * DIPTEREX 50 * DIPTEVUR * DITRIFON * DYLOX * DYLOX-METASYSTOX-R * DYREX * DYVON * ENT 19,763 * EQUINOACID * EQUINO-AID * FLIBOL E * FLIEGENTELLER * FOROTOX * FOSCHLOR * FOSCHLOREM (POLISH) * FOSCHLOR R-50 * 1-HYDROXY-

2,2,2-TRICHLOROETHYLPHOSPHONIC ACID DIMETHYL ESTER * HYPODERMACID * LEIVASOM * LOISOL * MASOTEN * MAZOTEN * METHYL CHLOROPHOS * METIFONATE * METRIFONATE * METRIPHONATE * NCI-C54831 * NEGUVON * NEGUVON A * PHOSCHLOR R50 * POLFOSCHLOR * PROXOL * RICIFON * RITSIFON * SATOX 20WSC * SOLDEP * SOTIPOX * TRICHLOORFON (DUTCH) * TRICHLORFON (USDA) * 2,2,2-TRICHLORO-1-HYDROXY-ETHYL-PHOSPHONATE, DIMETHYL ESTER * (2,2,2-TRICHLORO-1-HYDROXYETHYL)PHOSPHONIC ACID DIMETHYL ESTER * TRICHLOROPHON * TRICHLORPHENE * TRICHLORPHON * TRICHLORPHON FN * TRINEX * TUGON * TUGON FLY BAIT * TUGON STABLE SPRAY * VERMICIDE BAYER 2349 * VOLFARTOL * VOTEXIT * WEC 50 * WOTEXIT

CONSENSUS REPORTS: IARC Cancer Review: GROUP 3 IMEMDT 7,56,87; Animal Inadequate Evidence IMEMDT 30,207,83. Community Right-To-Know List. EPA Genetic Toxicology Program.

DOT Classification: ORM-A; Label: None.

SAFETY PROFILE: Poison by ingestion, inhalation, intraperitoneal, subcutaneous, intravenous, and intramuscular routes. Moderately toxic by skin contact and possibly other routes. Experimental teratogenic and reproductive effects. Questionable carcinogen with experimental carcinogenic and tumorigenic data. Human mutation data reported. An eye irritant. When heated to decomposition it emits very toxic fumes of Cl^- and PO_x.

TIQ750　　　　CAS: 87-90-1　　　**HR: 2**
N,N′,N″-TRICHLOROISOCYANURIC ACID

DOT: NA 2468
mf: $C_3Cl_3N_3O_3$　　　mw: 232.41

PROP: White crystals; chlorine odor. Mp: 225-230° (decomp). Moderately sol in water.

SYNS: ACL 85 * CBD 90 * FICHLOR 91 * FI CLOR 91 * ISOCYANURIC CHLORIDE * KYSELINA TRICHLOISOKYANUROVA (CZECH) * NSC-405124 * SYMCLOSEN * SYMCLOSENE * TRICHLORINATED ISOCYANURIC ACID * TRICHLOROCYANURIC ACID * TRICHLOROISOCYANIC ACID * TRICHLOROISOCYANURIC ACID * 1,3,5-TRICHLOROISOCYANURIC ACID * TRICHLORO-s-TRIAZINETRIONE * 1,3,5-TRICHLORO-1,3,5-TRIAZINE-

TRIONE * TRICHLORO-s-TRIAZINE-2,4,6(1H,3H,5H)-TRIONE * 1,3,5-TRICHLORO-2,4,6-TRIOXOHEXAHYDRO-s-TRIAZINE

CONSENSUS REPORTS: Reported in EPA TSCA Inventory.

DOT Classification: Oxidizer; Label: Oxidizer.

SAFETY PROFILE: Moderately toxic to humans and experimentally by ingestion. Mildly toxic experimentally by skin contact. Human systemic effects by ingestion: ulceration or bleeding from stomach. A severe skin and eye irritant. Toxicity symptoms include emaciation, lethargy, weakness and delayed death. Autopsy shows inflammation of gastrointestinal tract, liver discoloration and kidney hyperemia.

A powerful oxidizer. Forms an explosive product with cyanuric acid + sodium hydroxide. Potentially violent reaction with combustible materials. When heated to decomposition it emits very toxic fumes of Cl^- and NO_x. Used to chlorinate swimming pools.

TIT500　　　　CAS: 1321-65-9　　　**HR: 3**
TRICHLORONAPHTHALENE
mf: $C_{10}H_5Cl_3$　　　mw: 231.50

PROP: A white solid.

SYNS: HALOWAX * NIBREN WAX * SEEKAY WAX

CONSENSUS REPORTS: Reported in EPA TSCA Inventory.

OSHA PEL: TWA 5 mg/m³ (skin)
ACGIH TLV: TWA 5 mg/m³ (skin)
DFG MAK: 5 mg/m³

SAFETY PROFILE: A poison. The chlorinated naphthalenes have toxic effects on the skin and liver.

TIV750　　　　CAS: 95-95-4　　　**HR: 3**
2,4,5-TRICHLOROPHENOL
mf: $C_6H_3Cl_3O$　　　mw: 197.44

PROP: Colorless needles or gray flakes. Bp: 252°, fp: 57.0°, d: 1.678 @ 25/4°, vap press: 1 mm @ 72.0°, mp: 61-63°. Insol in water; sol in CCl_4, alc, benzene, ether.

SYNS: COLLUNOSOL * DOWICIDE 2 * DOWICIDE B * NCI-C61187 * NURELLE * PREVENTOL I * RCRA WASTE NUMBER U230

CONSENSUS REPORTS: IARC Cancer Review: Human Limited Evidence IMEMDT

41,319,86; Animal Inadequate Evidence IMEMDT 20,349,79. Chlorophenol compounds are on the Community Right-To-Know List. Reported in EPA TSCA Inventory.

SAFETY PROFILE: Suspected carcinogen with experimental neoplastigenic data. Poison by intraperitoneal, intravenous, and possibly other routes. Moderately toxic by ingestion and subcutaneous routes. Experimental reproductive effects. When heated to decomposition it emits toxic fumes of Cl⁻ and explodes.

TIW000   CAS: 88-06-2   **HR: 3**
2,4,6-TRICHLOROPHENOL
mf: $C_6H_3Cl_3O$ mw: 197.44

PROP: Colorless needles or yellow solid; strong phenolic odor. Mp: 68°, bp: 244.5°, fp: 62°, d: 1.490 @ 75/4°, vap press: 1 mm @ 76.5°. Sol in water; very sol in alc, ether.

SYNS: DOWICIDE 2S * NCI-C02904 * OMAL * PHENACHLOR * RCRA WASTE NUMBER U231 * 2,4,6-TRICHLORFENOL (CZECH)

CONSENSUS REPORTS: IARC Cancer Review: Animal Inadequate Evidence IMEMDT 20,349,79; Human Limited Evidence IMEMDT 41,319,86. NTP Fourth Annual Report On Carcinogens, 1984. NCI Carcinogenesis Bioassay (feed); Clear Evidence: mouse, rat NCITR* NCI-CG-TR-155,79. Chlorophenol compounds are on the Community Right-To-Know List. Reported in EPA TSCA Inventory. EPA Genetic Toxicology Program.

SAFETY PROFILE: Confirmed carcinogen with experimental carcinogenic data. Poison by intraperitoneal route. Moderately toxic by ingestion and skin contact. A skin and severe eye irritant. When heated to decomposition it emits toxic fumes of Cl⁻. Used as a germicide and preservative.

TIX500   CAS: 93-72-1   **HR: 3**
α-(2,4,5-TRICHLOROPHENOXY) PROPIONIC ACID
mf: $C_9H_7Cl_3O_3$ mw: 269.51

PROP: Crystals. Mp: 182°. Sltly water-sol.

SYNS: ACIDE 2-(2,4,5-TRICHLORO-PHENOXY) PROPIONIQUE (FRENCH) * ACIDO 2-(2,4,5-TRICLORO-FENOSSI)-PROPIONICO (ITALIAN) * AMCHEM 2,4,5-TP * AQUA-VEX * COLOR-SET * DED-WEED * DOUBLE STRENGTH * FENOPROP * FENORMONE * FRUITONE T * HERBICIDES, SILVEX

* KURAN * KURON * KUROSAL * MILLER NU SET * PROPON * RCRA WASTE NUMBER U233 * SILVEX (USDA) * SILVI-RHAP * STA-FAST * 2,4,5-TC * 2,4,5-TCPPA * 2,4,5-TP * 2-(2,4,5-TRICHLOOR-FENOXY)-PROPIONZUUR (DUTCH) * 2-(2,4,5-TRICHLOROPHENOXY)PROPIONIC ACID * 2,4,5-TRICHLOROPHENOXY-α-PROPIONIC ACID * 2-(2,4,5-TRICHLOR-PHENOXY)-PROPIONSAEURE (GERMAN) * WEED-B-GON

CONSENSUS REPORTS: IARC Cancer Review: Human Limited Evidence IMEMDT 41,357,86.

SAFETY PROFILE: Suspected carcinogen. Moderately toxic by ingestion and possibly other routes. Experimental teratogenic and reproductive effects. When heated to decomposition it emits toxic fumes of Cl⁻.

TJA750   CAS: 98-13-5   **HR: 3**
TRICHLOROPHENYLSILANE
DOT: UN 1804
mf: $C_6H_5Cl_3Si$ mw: 211.55

SYNS: PHENYLSILICON TRICHLORIDE * PHENYL TRICHLOROSILANE (DOT) * SILICON PHENYL TRICHLORIDE

CONSENSUS REPORTS: EPA Extremely Hazardous Substances List. Reported in EPA TSCA Inventory.

DOT Classification: Corrosive Material; Label: Corrosive.

SAFETY PROFILE: Poison by inhalation and intravenous routes. Moderately toxic by ingestion and skin contact. A corrosive irritant to skin, eyes, and mucous membranes. When heated to decomposition it emits toxic fumes of Cl⁻.

TJB600   CAS: 96-18-4   **HR: 3**
1,2,3-TRICHLOROPROPANE
mf: $C_3H_5Cl_3$ mw: 147.43

PROP: Bp: 142°, d: 1.414 @ 20°/20°, flash p: 180°F (OC).

SYNS: ALLYL TRICHLORIDE * GLYCEROL TRICHLOROHYDRIN * GLYCERYL TRICHLOROHYDRIN * NCI-C60220 * TRICHLOROHYDRIN

CONSENSUS REPORTS: Reported in EPA TSCA Inventory.

OSHA PEL: (Transitional: TWA 50 ppm) TWA 10 ppm

ACGIH TLV: TWA 10 ppm (skin)
DFG MAK: 50 ppm (300 mg/m^3)

SAFETY PROFILE: Poison by ingestion and possibly other routes. Moderately toxic by inhalation and skin contact. Experimental reproductive effects. A skin and severe eye irritant. Mutation data reported. Moderately flammable by heat, flames (sparks), or powerful oxidizers. When heated to decomposition it yields highly toxic Cl$^-$. To fight fire, use water (as a blanket), spray, mist, dry chemical.

TJD500   CAS: 10025-78-2   **HR: 3**
TRICHLOROSILANE

DOT: UN 1295
mf: Cl$_3$HSi   mw: 135.45

PROP: Colorless, very volatile liquid. Mp: $-126.5°$, bp: 31.8°, flash p: $-18.4°$F (OC), d: 1.35 @ 0°, vap press: 400 mm @ 14.5°, vap d: 4.7, autoign temp: 219°F. Sol in benzene, carbon disulfide, chloroform, carbon tetrachloride. Fumes in air. Decomp in water.

SYNS: SILICI-CHLOROFORME (FRENCH) * SILICI-UMCHLOROFORM (GERMAN) * SILICOCHLOROFORM * TRICHLOORSILAAN (DUTCH) * TRICHLOROMO-NOSILANE * TRICHLORSILAN (GERMAN) * TRICLOROSILANO (ITALIAN)

CONSENSUS REPORTS: Reported in EPA TSCA Inventory.

DOT Classification: Flammable Liquid; Label: Flammable Liquid; Flammable Solid; Label: Dangerous When Wet, Flammable Liquid, Corrosive.

SAFETY PROFILE: Moderately toxic by ingestion and inhalation. A corrosive irritant to skin, eyes and mucous membranes. A very dangerous fire hazard when exposed to heat, flame, or by chemical reaction. May be ignited by spark or impact. Spontaneously flammable in air. Explosive reaction with acetonitrile + diphenyl sulfoxide. Will react with water or steam to produce heat and toxic and corrosive fumes. Can react vigorously with oxidizing materials. To fight fire, use CO$_2$, dry chemical. When heated to decomposition it emits toxic fumes of Cl$^-$.

TJE750   CAS: 6379-69-7   **HR: 3**
TRICHOTHECIN
mf: C$_{19}$H$_{24}$O$_5$ mw: 332.43

PROP: Needles. Mp: 118°. Sltly sol in water; very sol in organic solvents.

SYNS: 12,13-EPOXY-4-HYDROXYTRICHOTHEC-9-EN-8-ONE CROTONATE * 12,13-EPOXY-4-((1-OXO-2-BUTE-NYL)OXY)TRICHOTHEC-9-EN-8-ONE

SAFETY PROFILE: Poison by intravenous and subcutaneous routes. A skin irritant. When heated to decomposition it emits acrid smoke and irritating fumes.

TJN750   CAS: 97-93-8   **HR: 3**
TRIETHYLALUMINUM

DOT: UN 1102
mf: C$_6$H$_{15}$Al   mw: 114.19

PROP: Fp: $-52.5°$, d: 0.837 @ 20°, vap press: 4 mm @ 83°, flash p: $< -63°$F, bp: 194°.

SYN: TEA

CONSENSUS REPORTS: Reported in EPA TSCA Inventory.

ACGIH TLV: TWA 2 mg(Al)/m^3
DOT Classification: Flammable Solid; Label: Spontaneously Combustible.

SAFETY PROFILE: Extremely destructive to living tissue. A very dangerous fire hazard when exposed to heat or flame. Ignites spontaneously in air. Explodes violently in water. To fight fire, use CO$_2$, dry sand, dry chemical. Do not use water, foam or halogenated fire-fighting agents. Explosive reaction with alcohols (e.g., methanol, ethanol, propynol), carbon tetrachloride, N,N-dimethylformamide + heat. Incompatible with halogenated hydrocarbons, triethyl borine. When heated to decomposition it emits acrid smoke and irritating fumes.

TJO000   CAS: 121-44-8   **HR: 2**
TRIETHYLAMINE

DOT: UN 1296
mf: C$_6$H$_{15}$N   mw: 101.22

PROP: Colorless liquid; ammonia odor. Mp: $-114.8°$, bp: 89.5°, flash p: 20°F (OC), d: 0.7255 @ 25/4°, vap d: 3.48, lel: 1.2%, uel: 8.0%. Misc in water, alc, ether.

SYNS: (DIETHYLAMINO)ETHANE * N,N-DIETHYLE-THANAMINE * TEN * TRIAETHYLAMIN (GERMAN) * TRIETILAMINA (ITALIAN)

CONSENSUS REPORTS: Reported in EPA TSCA Inventory.

OSHA PEL: (Transitional: TWA 25 ppm) TWA 10 ppm; STEL 15 ppm
ACGIH TLV: TWA 10 ppm; STEL 15 ppm
DFG MAK: 10 ppm (40 mg/m^3)
DOT Classification: Flammable Liquid; Label: Flammable Liquid.

SAFETY PROFILE: Moderately toxic by ingestion and skin contact. Mildly toxic by inhalation. Experimental reproductive effects. Mutation data reported. A skin and severe eye irritant. Can cause kidney and liver damage. A very dangerous fire hazard when exposed to heat, flame, or oxidizers. Explosive in the form of vapor when exposed to heat or flame. Complex with dinitrogen tetraoxide explodes below 0°C when undiluted with solvent. Exothermic reaction with maleic anhydride above 150°C. Can react with oxidizing materials. Incompatible with N_2O_4. To fight fire, use CO_2, dry chemical, alcohol foam. When heated to decomposition it emits toxic fumes of NO_x.

TJP250 CAS: 97-94-9 **HR: 3**
TRIETHYLBORANE
mf: $C_6H_{15}B$ mw: 98.02

PROP: Colorless liquid. Mp: −93°, d: 0.6961 @ 23°.

SYN: TRIETHYLBORINE

CONSENSUS REPORTS: Reported in EPA TSCA Inventory.

SAFETY PROFILE: Poison by ingestion and intraperitoneal routes. Mildly toxic by inhalation. Animal experiments show that the vapor is a poison which causes pulmonary irritation and convulsions. A very dangerous fire hazard by spontaneous chemical reaction with oxidizers. Spontaneously flammable in air. Explodes in oxygen atmospheres. Hypergolic reaction with triethylaluminum. Ignites on contact with chlorine; bromine; or other halogens. Will react with water or steam to produce toxic and flammable vapors. To fight fire, do NOT use halogenated extinguishing agents. When heated to decomposition or upon contact with air it emits toxic acrid smoke and irritating fumes.

TJP750 CAS: 77-93-0 **HR: 2**
TRIETHYL CITRATE
mf: $C_{12}H_{20}O_7$ mw: 276.32

PROP: Colorless oily liquid; odorless. Bp: 294°, flash p: 303°F (COC), d: 1.136 @ 25°, vap press: 1 mm @ 107.0°. Sltly sol in water; misc in alc, ether.

SYNS: CITROFLEX 2 * ETHYL CITRATE
* 2-HYDROXY,1,2,3-PROPANETRICARBOXYLIC ACID, TRIETHYL ESTER * TEC

CONSENSUS REPORTS: Reported in EPA TSCA Inventory.

SAFETY PROFILE: Moderately toxic by intraperitoneal route. Mildly toxic by ingestion and inhalation. Combustible liquid when exposed to heat or flame. To fight fire, use dry chemical, CO_2. When heated to decomposition it emits acrid smoke and irritating fumes.

TJP775 CAS: 12075-68-2 **HR: 3**
TRIETHYL DIALUMINUM TRICHLORIDE

DOT: UN 1925
mf: $C_6H_{15}Al_2Cl_3$ mw: 247.51

DOT Classification: Flammable Solid; Label: Spontaneously Combustible

SAFETY PROFILE: Mixtures with carbon tetrachloride explode at room temperature. When heated to decomposition it emits toxic fumes of Cl^-.

TJQ000 CAS: 112-27-6 **HR: 3**
TRIETHYLENE GLYCOL
mf: $C_6H_{14}O_4$ mw: 150.20

PROP: Odorless, colorless liquid; hygroscopic. Fp: −7.3°, flash p: 350°F, d: 1.122 @ 25/25°, lel: 0.9%, uel: 9.2%, autoign temp: 700°F, vap press: 1 mm @ 114°, vap d: 5.17, bp: 285°. Misc in water, alc, benzene; insol in petroleum ether; very sltly sol in ether.

SYNS: DI-β-HYDROXYETHOXYETHANE * 3,6-DIOXAOCTANE-1,8-DIOL * 2,2′-(1,2-ETHANEDIYLBIS(OXY)) BISETHANOL * 2,2′-ETHYLENEDIOXYDIETHANOL * 2,2′-ETHYLENEDIOXYETHANOL * ETHYLENE GLYCOL-BIS-(2-HYDROXYETHYL ETHER) * ETHYLENE GLYCOL DIHYDROXYDIETHYL ETHER * GLYCOL BIS(HYDROXYETHYL) ETHER * TEG * TRIGEN * TRIGLYCOL

CONSENSUS REPORTS: Glycol ether compounds are on the Community Right-To-Know List. Reported in EPA TSCA Inventory.

SAFETY PROFILE: Poison by intravenous route. Mildly toxic to humans and experimentally by ingestion. Experimental reproductive effects. A skin irritant. Many glycol ether compounds have dangerous human reproductive effects. Combustible when exposed to heat or flame. Can react with oxidizing materials. Explosive in the form of vapor when exposed to heat, flame or spark. To fight fire, use alcohol foam, dry chemical. When heated to decomposition it emits acrid smoke and irritating fumes.

TJR000   CAS: 112-24-3   HR: 3
TRIETHYLENETETRAMINE

DOT: UN 2259
mf: $C_6H_{18}N_4$ mw: 146.28

PROP: Mod viscous, yellowish liquid. Bp: 278°, mp: 12°; flash p: 275°F, d: 0.982, vap press: <0.01 mm @ 20°, autoign temp: 640°F. Very sol in water, ether.

SYNS: ARALDITE HARDENER HY 951 * ARALDITE HY 951 * N,N'-BIS(2-AMINOETHYL)-1,2-DIAMINO-ETHANE * N,N'-BIS(2-AMINOETHYL)ETHYLENEDI-AMINE * N,N'-BIS(2-AMINOETHYL)-1,2-ETHYLENEDI-AMINE * DEH 24 * 3,6-DIAZAOCTANE-1,8-DIAMINE * HY 951 * TECZA * TETA * 1,4,7,10-TETRA-AZADECANE * TRIEN * TRIENTINE

CONSENSUS REPORTS: Reported in EPA TSCA Inventory.

DOT Classification: Corrosive Material; Label: Corrosive.

SAFETY PROFILE: Poison by intravenous route. Moderately toxic by ingestion, and skin contact. Experimental teratogenic and reproductive effects. Mutation data reported. A corrosive irritant to skin, eyes and mucous membranes. Causes skin sensitization. Combustible when exposed to heat or flame. Ignites on contact with cellulose nitrate of high surface area. Can react with oxidizing materials. To fight fire, use CO_2, dry chemical, alcohol foam. When heated to decomposition it emits toxic fumes of NO_x.

TJS500   CAS: 562-95-8   HR: 3
TRIETHYL LEAD FLUOROACETATE
mf: $C_2H_2FO_2 \cdot C_6H_{15}Pb$ mw: 371.44

SYN: FLUOROACETIC ACID, TRIETHYLLEAD SALT

CONSENSUS REPORTS: Lead and its compounds are on the Community Right-To-Know List.

NIOSH REL: (Inorganic Lead) TWA 0.10 mg(Pb)/m^3

SAFETY PROFILE: Poison by subcutaneous route. Human systemic effects by inhalation: pulmonary system effects. When heated to decomposition it emits very toxic fumes of F$^-$ and Pb.

TJT800   CAS: 122-52-1   HR: 2
TRIETHYL PHOSPHITE

DOT: UN 2323
mf: $C_6H_{15}O_3P$ mw: 166.18

SYN: FOSFORYN TROJETYLOWY (CZECH)

CONSENSUS REPORTS: Reported in EPA TSCA Inventory.

DOT Classification: Flammable or Combustible Liquid; Label: Flammable Liquid.

SAFETY PROFILE: Moderately toxic by ingestion. A skin and eye irritant. Flammable when exposed to heat or flame. When heated to decomposition it emits toxic fumes of PO_x.

TJY100   CAS: 75-63-8   HR: 1
TRIFLUOROBROMOMETHANE

DOT: UN 1009
mf: $CBrF_3$ mw: 148.92

SYNS: BROMOFLUOROFORM * BROMOTRIFLUO-ROMETHANE * F-13B1 * FREON 13B1 * HALON 1301 * TRIFLUOROMONOBROMOMETHANE

CONSENSUS REPORTS: Reported in EPA TSCA Inventory.

OSHA PEL: TWA 1000 ppm
ACGIH TLV: TWA 1000 ppm
DFG MAK: 1000 ppm (6100 mg/m^3)
DOT Classification: Nonflammable Gas; Label: Nonflammable Gas.

SAFETY PROFILE: Mildly toxic by inhalation. Incompatible with aluminum. When heated to decomposition it emits toxic fumes of F$^-$ and Br$^-$.

TKA250   CAS: 76-05-1   HR: 3
TRIFLUOROETHANOIC ACID

DOT: UN 2699
mf: $C_2HF_3O_2$ mw: 114.03

PROP: Colorless liquid; strong pungent odor. Mp: −15.25°, bp: 71.1° @ 734 mm, d: 1.535 @ 0°.

SYNS: PERFLUOROACETIC ACID * TRIFLUOR-ACETIC ACID * TRIFLUOROACETIC ACID (DOT)

CONSENSUS REPORTS: Reported in EPA TSCA Inventory. EPA Genetic Toxicology Program.

DOT Classification: Corrosive Material; Label: Corrosive.

SAFETY PROFILE: Poison by ingestion and intraperitoneal routes. Moderately toxic by intravenous route. Mildly toxic by inhalation. A corrosive irritant to skin, eyes, and mucous membranes. When heated to decomposition it emits toxic fumes of F^-. Used as a strong organic acid catalyst.

TKB250   CAS: 406-90-6   **HR: 3**
2,2,2-TRIFLUOROETHYL VINYL ETHER
mf: $C_4H_5F_3O$ mw: 126.09

SYNS: FLOROXENE * FLUOOXENE * FLUORO-MAR * FLUOROXENE * FLUORXENE * FLU-ROXENE * (2,2,2-TRIFLUOROETHOXY)ETHENE

CONSENSUS REPORTS: Reported in EPA TSCA Inventory. EPA Genetic Toxicology Program.

NIOSH REL: (Waste Anesthetic Gases) CL 2 ppm/1H

SAFETY PROFILE: Human poison by ingestion. Experimental poison by inhalation. Moderately toxic experimentally by intraperitoneal route. Human systemic effects by inhalation: jaundice and liver function tests impaired. Experimental teratogenic and reproductive effects. Mutation data reported. When heated to decomposition it emits toxic fumes of F^-. Used as an anesthetic.

TKB310   CAS: 1493-13-6   **HR: 3**
TRIFLUOROMETHANE SULFONIC ACID
mf: CHF_3O_3S mw: 150.08

SYN: TRIFLIC ACID

SAFETY PROFILE: A corrosive irritant to the skin, eyes, and mucous membranes. A strong acid. Violent reaction with acyl chlorides or aromatic hydrocarbons evolves toxic hydrogen chloride gas. When heated to decomposition it emits toxic fumes of F^- and SO_x.

TKK500   CAS: 749-13-3   **HR: 3**
TRIFLUPERIDOL
mf: $C_{22}H_{23}F_4NO_2$ mw: 409.46

SYNS: 4'-FLUORO-4-(4-HYDROXY-4-(α,α,α-TRIFLUORO-m-TOLYL)PIPERIDINO)BUTYROPHENONE * 4-FLUORO-4,4-IDROSSI-4-(m-TRIFLUOROMETIL-FENIL)-PIPERIDINO-BUTIRROFENONE (ITALIAN) * 1-(4-FLUOROPHENYL)-4-(4-HYDROXY-4-3-(TRIFLUOROMETHYL)PHENYL)-1-PI-

PERIDINYL)-1-BUTANONE * MCN-JR-2498 * PSICO-PERIDOL-R * PSYCHOPERIDOL * R-2498 * TRIFLUPERIDOLO (ITALIAN) * TRIPERIDOL

SAFETY PROFILE: Poison by ingestion, subcutaneous, and intraperitoneal routes. Experimental teratogenic and reproductive effects. When heated to decomposition it emits very toxic fumes of F^- and NO_x.

TKL000   CAS: 146-54-3   **HR: 3**
TRIFLUPROMAZINE
mf: $C_{18}H_{19}F_3N_2S$ mw: 352.45

SYNS: 10-(3-(DIMETHYLAMINO)PROPYL-2-(TRIFLUO-ROMETHYL) PHENOTHIAZINE * N,N-DIMETHYL-2-(TRIFLUOROMETHYL)-10H-PHENOTHIAZINE-10-PROPA-NAMINE * VESPRIN

CONSENSUS REPORTS: EPA Genetic Toxicology Program.

SAFETY PROFILE: Poison by ingestion, intravenous, and intraperitoneal routes. Mutation data reported. When heated to decomposition it emits very toxic fumes of F^-, NO_x, and SO_x.

TKO250   CAS: 1421-63-2   **HR: 3**
2',4',5'-TRIHYDROXY-BUTYROPHENONE
mf: $C_{10}H_{12}O_4$ mw: 196.22

PROP: Yellow-tan crystals. Mp: 149-153°, d: 6.0 lb/gal @ 20°. Very sltly sol in water; sol in alc, propylene glycol.

SYNS: THBP * 2,4,5-TRIHYDROXYBUTYROPHE-NONE * USAF EK

CONSENSUS REPORTS: Reported in EPA TSCA Inventory.

SAFETY PROFILE: Poison by intraperitoneal route. Mutation data reported. When heated to decomposition it emits acrid smoke and irritating fumes.

TKP500   CAS: 102-71-6   **HR: 3**
TRIHYDROXYTRIETHYLAMINE
mf: $C_6H_{15}NO_3$ mw: 149.22

PROP: Pale yellow viscous liquid. Mp: 21.2°, bp: 360°, flash p: 355°F (CC), d: 1.1258 @ 20/20°, vap press: 10 mm @ 205°, vap d: 5.14.

SYNS: DALTOGEN * NITRILO-2,2',2''-TRIETHANOL * 2,2',2''-NITRILOTRIETHANOL * STEROLAMIDE * THIOFACO T-35 * TRIAETHANOLAMIN-NG * TRIETHANOLAMIN * TRIETHANOLAMINE * TRIETHYLOLAMINE * TRI(HYDROXYETHYL)

AMINE   *   2,2',2''-TRIHYDROXYTRIETHYLAMINE   *   TRIS(2-HYDROXYETHYL)AMINE   *   TROLAMINE

CONSENSUS REPORTS: Cyanide and its compounds are on the Community Right-To-Know List. Reported in EPA TSCA Inventory. EPA Genetic Toxicology Program.

SAFETY PROFILE: Moderately toxic by intraperitoneal route. Mildly toxic by ingestion. Liver and kidney damage has been demonstrated in animals from chronic exposure. A human and experimental skin irritant. An eye irritant. Questionable carcinogen with experimental carcinogenic data. Combustible liquid when exposed to heat or flame; can react vigorously with oxidizing materials. To fight fire, use alcohol foam, CO_2, dry chemical. When heated to decomposition it emits toxic fumes of NO_x and CN^-.

TKR500   CAS: 100-99-2   **HR: 3**
TRIISOBUTYL ALUMINUM

DOT: UN 1930
mf: $(C_4H_9)_3Al$ mw: 198.3

PROP: Clear, colorless liquid. D: 0.7859 @ 20°, vap press: 1 mm @ 47°, flash p: <4°, fp: 4.3, bp: decomp.

SYNS: TRIISOBUTYLALANE   *   TRIS(2-METHYLPROPYL)ALUMINIUM

CONSENSUS REPORTS: Reported in EPA TSCA Inventory.

ACGIH TLV: TWA 2 mg(Al)/m^3
DOT Classification: Flammable Solid; Label: Spontaneously Combustible.

SAFETY PROFILE: A poison. Extremely destructive to living tissue. A very dangerous fire hazard; ignites on exposure to air. Incompatible with moisture, acids, air, alcohols, amines, halogens. To fight fire, use CO_2, dry sand, dry chemical. Do not use water, foam or halogenated extinguishing agents. When heated to decomposition it emits acrid smoke and irritating fumes.

TKT750   CAS: 19464-55-2   **HR: 3**
TRIISOPROPYLTIN ACETATE
mf: $C_{11}H_{24}O_2Sn$ mw: 307.04

SYN: ACETOXYTRIISOPROPYLSTANNANE

OSHA PEL: TWA 0.1 mg(Sn)/m^3 (skin)
ACGIH TLV: TWA 0.1 mg(Sn)/m^3 (skin) (Proposed: TWA 0.1 mg(Sn)/m^3; STEL 0.2 mg(Sn)/m^3 (skin))

NIOSH REL: (Organotin Compounds) TWA 0.1 mg(Sn)/m^3

SAFETY PROFILE: Poison by ingestion and intravenous routes. When heated to decomposition it emits acrid smoke and irritating fumes.

TKV000   CAS: 552-30-7   **HR: 1**
TRIMELLITIC ANHYDRIDE
mf: $C_9H_4O_5$ mw: 192.13

PROP: Crystals. Mp: 162°, bp: 240-245° @ 14 mm. Sol in acetone, ethyl acetate, dimethylformamide.

SYNS: ANHYDROTRIMELLIC ACID   *   1,2,4-BENZENETRICARBOXYLIC ACID ANHYDRIDE   *   1,2,4-BENZENETRICARBOXYLIC ACID, CYCLIC 1,2-ANHYDRIDE   *   1,2,4-BENZENETRICARBOXYLIC ANHYDRIDE   *   4-CARBOXYPHTHALIC ANHYDRIDE   *   1,3-DIHYDRO-1,3-DIOXO-5-ISOBENZOFURANCARBOXYLIC ACID   *   1,3-DIOXO-5-PHTHALANCARBOXYLIC ACID   *   DIPHENYLMETHANE-4,4'-DIISOCYANATE-TRIMELLIC ANHYDRIDE-ETHOMID HT POLYMER   *   NCI-C56633   *   TMA   *   TMAN   *   TRIMELLIC ACID ANHYDRIDE   *   TRIMELLIC ACID-1,2-ANHYDRIDE   *   TRIMELLITIC ACID CYCLIC-1,2-ANHYDRIDE

CONSENSUS REPORTS: Reported in EPA TSCA Inventory.

OSHA PEL: TWA 0.005 ppm
ACGIH TLV: TWA 0.005 ppm
DFG MAK: 0.005 ppm (0.04 mg/m^3)

SAFETY PROFILE: Mildly toxic by ingestion. Has caused pulmonary edema from inhalation. Irritant to lungs and air passages. May be a powerful allergen. Typical attack consists of breathlessness, wheezing, cough, running nose, immunological sensitization and asthma symptoms. When heated to decomposition it emits acrid smoke and irritating fumes.

TKW000   CAS: 12136-15-1   **HR: 3**
TRIMERCURY DINITRIDE
mf: Hg_3N_2 mw: 629.78

SYN: MERCURY NITRIDE

CONSENSUS REPORTS: Mercury and its compounds are on the Community Right-To-Know List.

DOT Classification: Forbidden

SAFETY PROFILE: Mercury compounds are poisons. An explosive sensitive to friction, impact, heating or contact with sulfuric acid. Incompatible with sulfuric acid. When heated to decomposition it emits very toxic fumes of Hg and NO$_x$.

TKX500   CAS: 5688-80-2   HR: 3
3,4,5-TRIMETHOXYAMPHETAMINE HYDROCHLORIDE

mf: $C_{12}H_{19}NO_3 \cdot ClH$ mw: 261.78

SYNS: α-METHYL-3,4,5-TRIMETHOXYPHENETHYL-AMINE HYDROCHLORIDE * 3,4,5-TRIMETHOXY-α-METHYL-β-PHENYLETHYLAMINE HYDROCHLORIDE * 1-(3,4,5-TRIMETHOXYPHENYL)-2-AMINOPROPANE

SAFETY PROFILE: Poison by intraperitoneal and intravenous routes. Human systemic effects by ingestion: central nervous system effects. When heated to decomposition it emits very toxic fumes of NO$_x$ and HCl.

TLD272   CAS: 75-24-1   HR: 3
TRIMETHYLALUMINUM

DOT: UN 1103
mf: C_3H_9Al mw: 72.09

SYNS: TRIMETHYLALANE * TRIMETHYLALUMI-NIUM (DOT)

ACGIH TLV: TWA 2 mg(Al)/m^3
DOT Classification: Flammable Solid; Label: Spontaneously Combustible

SAFETY PROFILE: Extremely pyrophoric flammable solid. Mixtures with dichlorodi-μ-chlorobis (pentamethylcyclopentadienyl) dir-hodium + air ignite and burn violently.

TLD500   CAS: 75-50-3   HR: 3
TRIMETHYLAMINE

DOT: UN 1083/UN 1297
mf: C_3H_9N mw: 59.13

PROP: Colorless gas. Pungent, fishy, ammonia-cal odor; saline taste. Bp: 2.87°, lel: 2%, uel: 11.6%, fp: −117.1°, d: 0.662 @ −5°, autoign temp: 374°F, vap d: 2.0, flash p: 20°F (CC). Misc with alc; sol in ether, benzene, toluene, xylene, chloroform.

SYNS: N,N-DIMETHYLMETHANAMINE * TMA * TRIMETHYLAMINE, anhydrous (DOT) * TRI-METHYLAMINE, aqueous solution (DOT) * TRI-METHYLAMINE, aqueous solutions containing not more than 30% of trimethylamine (DOT)

CONSENSUS REPORTS: Reported in EPA TSCA Inventory.

OSHA PEL: TWA 10 ppm; STEL 15 ppm
ACGIH TLV: TWA 10 ppm; STEL 15 ppm
DOT Classification: Flammable Gas; Label: Flammable Gas, Anhydrous; Flammable Liquid; Label: Flammable Liquid.

SAFETY PROFILE: Poison by intravenous route. Moderately toxic by subcutaneous and rectal routes. Mildly toxic by inhalation. A very dangerous fire hazard when exposed to heat or flame. Self-reactive. Moderately explosive in the form of vapor when exposed to heat or flame. Can react with oxidizing materials. To fight fire, stop flow of gas. Potentially explosive reaction with bromine + heat, ethylene oxide, tri-ethynylaluminum. When heated to decomposition it emits toxic fumes of NO$_x$.

TLG250   CAS: 137-17-7   HR: 3
2,4,5-TRIMETHYLANILINE

mf: $C_9H_{13}N$ mw: 135.23

SYNS: 1-AMINO-2,4,5-TRIMETHYLBENZENE * psi-CUMIDINE * NCI-C02299 * PSEUDOCUMI-DINE * 1,2,4-TRIMETHYL-5-AMINOBENZENE * 2,4,5-TRIMETHYLANILIN (CZECH) * 2,4,5-TRI-METHYLBENZENAMINE

CONSENSUS REPORTS: IARC Cancer Review: GROUP 3 IMEMDT 7,56,87; Animal Limited Evidence IMEMDT 27,177,82. NCI Carcinogenesis Bioassay (feed); Clear Evidence: mouse, rat NCITR* NCI-CG-TR-160,79.

DFG MAK: Animal Carcinogen, Suspected Human Carcinogen.

SAFETY PROFILE: Confirmed carcinogen with experimental carcinogenic and tumorigenic data. Moderately toxic by ingestion. Mutation data reported. When heated to decomposition it emits toxic fumes of NO$_x$. Used as a dye, pigment, and printing ink.

TLG500   CAS: 88-05-1   HR: 3
2,4,6-TRIMETHYLANILINE

mf: $C_9H_{13}N$ mw: 135.23

SYNS: AMINOMESITYLENE * 2-AMINOMESITYL-ENE * 1-AMINO-2,4,6-TRIMETHYLBENZEN (CZECH) * 2-AMINO-1,3,5-TRIMETHYLBENZENE * MESIDIN (CZECH) * MESIDINE * MESITYLAMINE * MEZIDINE * 2,4,6-TRIMETHYLBENZENAMINE

CONSENSUS REPORTS: IARC Cancer Review: GROUP 3 IMEMDT 7,56,87; Animal Inadequate Evidence IMEMDT 27,177,82.

EPA Extremely Hazardous Substances List. Reported in EPA TSCA Inventory.

SAFETY PROFILE: Poison by inhalation and possibly other routes. Moderately toxic by ingestion and possibly other routes. A skin and severe eye irritant. Questionable carcinogen with experimental carcinogenic data. Mutation data reported. When heated to decomposition it emits toxic fumes of NO_x.

TLG750   CAS: 21436-97-5   **HR: 3**
2,4,5-TRIMETHYLANILINE HYDROCHLORIDE
mf: $C_9H_{13}N \cdot ClH$ mw: 171.69

SYNS: 1-AMINO-2,4,5-TRIMETHYLBENZENE HYDROCHLORIDE * psi-CUMIDINE HYDROCHLORIDE * PSEUDOCUMIDINE HYDROCHLORIDE * 1,2,4-TRIMETHYL-5-AMINOBENZENE HYDROCHLORIDE * 2,4,5-TRIMETHYLBENZENAMINE HYDROCHLORIDE

CONSENSUS REPORTS: IARC Cancer Review: Animal Inadequate Evidence IMEMDT 27,177,82.

SAFETY PROFILE: Poison by intraperitoneal route. Moderately toxic by ingestion. Questionable carcinogen with experimental carcinogenic, neoplastigenic, and tumorigenic data. When heated to decomposition it emits very toxic fumes of NO_x and HCl.

TLH000   CAS: 6334-11-8   **HR: 3**
2,4,6-TRIMETHYLANILINE HYDROCHLORIDE
mf: $C_9H_{13}N \cdot ClH$ mw: 171.69

SYNS: AMINOMESITYLENE HYDROCHLORIDE * 2-AMINOMESITYLENE HYDROCHLORIDE * 2-AMINO-1,3,5-TRIMETHYLBENZENE HYDROCHLORIDE * MESIDINE HYDROCHLORIDE * MESITYLAMINE HYDROCHLORIDE * 2,4,6-TRIMETHYLBENZENAMINE HYDROCHLORIDE

CONSENSUS REPORTS: IARC Cancer Review: Animal Inadequate Evidence IMEMDT 27,177,82.

SAFETY PROFILE: Suspected carcinogen with experimental carcinogenic and neoplastigenic data. Poison by intraperitoneal route. Moderately toxic by ingestion. When heated to decomposition it emits very toxic fumes of NO_x and HCl.

TLI500   CAS: 51787-42-9   **HR: 3**
7,9,11-TRIMETHYLBENZ(c)ACRIDINE
mf: $C_{20}H_{17}N$ mw: 271.38

SYN: 1,3,10-TRIMETHYL-7,8-BENZACRIDINE (FRENCH)

SAFETY PROFILE: Questionable carcinogen with experimental tumorigenic data. Mutation data reported. When heated to decomposition it emits toxic fumes of NO_x.

TLJ750   CAS: 20627-33-2   **HR: 3**
4,9,10-TRIMETHYL-1,2-BENZANTHRACENE
mf: $C_{21}H_{18}$ mw: 270.39

SYN: 6,7,12-TRIMETHYLBENZ(a)ANTHRACENE

SAFETY PROFILE: Questionable carcinogen with experimental tumorigenic data. Mutation data reported. When heated to decomposition it emits acrid smoke and irritating fumes.

TLK750   CAS: 13345-64-7   **HR: 3**
7,8,12-TRIMETHYLBENZ(a) ANTHRACENE
mf: $C_{21}H_{18}$ mw: 270.39

SYNS: 7,8,12-TMBA * 5:9:10-TRIMETHYL-1:2-BENZANTHRACENE

SAFETY PROFILE: Poison by intravenous route. Questionable carcinogen with experimental carcinogenic, tumorigenic, and teratogenic data. Mutation data reported. When heated to decomposition it emits acrid smoke and irritating fumes.

TLL250   CAS: 25551-13-7   **HR: 1**
TRIMETHYL BENZENE
mf: C_9H_{12} mw: 120.21

SYN: TRIMETHYL BENZENE (mixed isomers)

CONSENSUS REPORTS: Reported in EPA TSCA Inventory.

OSHA PEL: TWA 25 ppm
ACGIH TLV: TWA 25 ppm

SAFETY PROFILE: Mildly toxic by ingestion. A skin and eye irritant. Flammable when exposed to heat, flame, and oxidizers. When heated to decomposition it emits acrid smoke and irritating fumes.

TLL750   CAS: 95-63-6   **HR: 2**
1,2,4-TRIMETHYL BENZENE
mf: C_9H_{12} mw: 120.21

PROP: Liquid. Mp: 120.19°, d: 0.888 @ 4/4°, fp: −61°, bp: 168.89°, flash p: 130°F, autoign temp: 959°F. Insol in water; sol in alc, benzene, and ether.

SYNS: ASYMMETRICAL TRIMETHYL BENZENE * psi-CUMENE * PSEUDOCUMENE * PSEUDO-CUMOL * 1,2,5-TRIMETHYL BENZENE * as-TRIMETHYL BENZENE

CONSENSUS REPORTS: Reported in EPA TSCA Inventory. Community Right-To-Know List.

SAFETY PROFILE: Moderately toxic by intraperitoneal route. Mildly toxic by inhalation. Can cause central nervous system depression, anemia, bronchitis. Flammable when exposed to heat, flame, or oxidizers. To fight fire, use foam, alcohol foam, mist. When heated to decomposition it emits acrid smoke and irritating fumes.

TLM050   CAS: 108-67-8   **HR: 3**
1,3,5-TRIMETHYL BENZENE

DOT: UN 2325
mf: C_9H_{12} mw: 120.21

PROP: A liquid; peculiar odor. Mp: $-44.8°$, d: 0.8637 @ $20°/4°$, bp: $164.7°$, autoign temp: $1022°F$. Insol in water; misc in alc, benzene, and ether.

SYNS: FLEET-X * MESITYLENE * sym-TRIMETHYL BENZENE * TRIMETHYL BENZENE * TRIMETHYL BENZOL

CONSENSUS REPORTS: Reported in EPA TSCA Inventory.

OSHA PEL: TWA 25 ppm
ACGIH TLV: TWA 25 ppm
DOT Classification: Flammable or Combustible Liquid; label: Flammable Liquid.

SAFETY PROFILE: Poison by inhalation. Moderately toxic by intraperitoneal route. Human systemic effects by inhalation: sensory changes involving peripheral nerves, somnolence (general depressed activity), and structural or functional change in trachea or bronchi. Reports of leukopenia and thrombocytopenia in experimental animals. Mutation data reported. Flammable when exposed to heat or flame; can react vigorously with oxidizing materials. Violent reaction with HNO_3. To fight fire, use water spray, fog, foam, CO_2. When heated to decomposition it emits acrid smoke and irritating fumes.

TLN000   CAS: 121-43-7   **HR: 2**
TRIMETHYL BORATE

DOT: UN 2416
mf: $C_3H_9BO_3$ mw: 103.93

PROP: Colorless liquid. Decomp in water; misc in alc, ether. Mp: $-29°$, bp: $68°$, flash p: $<73°F$, d: 0.92 @ $20°$, vap d: 3.59.

SYNS: BORESTER O * METHYL BORATE * TRIMETHOXYBORINE

CONSENSUS REPORTS: Reported in EPA TSCA Inventory.

DOT Classification: Flammable or Combustible Liquid; Label: Flammable Liquid.

SAFETY PROFILE: Moderately toxic by ingestion, skin contact, and intraperitoneal routes. An eye irritant. A very dangerous fire hazard when exposed to heat, flame, or oxidizers. Moderately explosive when exposed to flame. Will react with water or steam to produce toxic and flammable vapors. To fight fire, use dry chemical, CO_2, spray, foam. When heated to decomposition it emits acrid smoke and irritating fumes.

TLN250   CAS: 75-77-4   **HR: 3**
TRIMETHYL CHLOROSILANE

DOT: UN 1298
mf: C_3H_9ClSi mw: 108.66

PROP: Colorless liquid. Bp: $57°$, d: 0.854 @ $25/25°$, flash p: $-18°F$. Sol in benzene, ether, perchloroethylene.

SYNS: CHLOROTRIMETHYLSILICANE * TL 1163

CONSENSUS REPORTS: EPA Extremely Hazardous Substances List. Reported in EPA TSCA Inventory.

DOT Classification: Flammable Liquid; Label: Flammable Liquid and Flammable Liquid, Corrosive.

SAFETY PROFILE: Moderately toxic by inhalation and intraperitoneal routes. A corrosive irritant to skin, eyes, and mucous membranes. Questionable carcinogen with experimental neoplastigenic data. Mutation data reported. A very dangerous fire hazard when exposed to heat or flame. Violent reaction with water or hexafluoroisopropylideneamino lithium. A preparative hazard. To fight fire, use foam, alcohol foam, and fog. When heated to decomposition it emits toxic fumes of Cl^-. An intermediate in the production of silicones.

TLN750   CAS: 3482-37-9   **HR: 3**
TRIMETHYLCOLCHICINIC ACID
mf: $C_{19}H_{21}NO_5$ mw: 343.41

SYNS: (s)-7-AMINO-6,7-DIHYDRO-10-HYDROXY-1,2,3-TRIMETHOXYBENZO(a)HEPTALEN-9(5H)-ONE * DEACETYLCHOLCHICEINE * N-DEACETYLCHOLCHICEINE * DESACETYLCHOLCHICEINE * TMCA

SAFETY PROFILE: Poison by intraperitoneal and possibly other routes. Mutation data reported. When heated to decomposition it emits toxic fumes of NO_x.

TLP500   CAS: 147-47-7   **HR: 2**
2,2,4-TRIMETHYL-1,2-DIHYDROQUINOLINE
mf: $C_{12}H_{15}N$ mw: 173.28

SYNS: ACETONE ANIL * 1,2-DIHYDRO-2,2,4-TRIMETHYLQUINOLINE * FLECTOL H * NCI-C60902

CONSENSUS REPORTS: Reported in EPA TSCA Inventory.

SAFETY PROFILE: Moderately toxic by ingestion. When heated to decomposition it emits toxic fumes of NO_x.

TLP750   CAS: 127-48-0   **HR: 3**
3,3,5-TRIMETHYL-2,4-DIKETOOXAZOLIDINE
mf: $C_6H_9NO_3$ mw: 143.16

SYNS: A 2297 * ABSENTOL * ABSETIL * CONVENIXA * EDION * EPIDIONE * EPIDONE * ETYDION * MINOALEUIATIN * PETIDION * PETIDON * PETILEP * PITMAL * TIOXANONA * TREDIONE * TRICIONE * TRIDILONA * TRIDIONE * TRIDONE * TRILIDONA * TRIMEDAL * TRIMEDONE * TRIMETADIONE * TRIMETHADIONE * TRIMETHIN * 3,5,5-TRIMETHYL-2,4-OXAZOLIDINEDIONE * TRIMETIN * TRIOZANONA * TROXIDONE * TROMEDONE

SAFETY PROFILE: Human poison by unspecified routes. Moderately toxic experimentally by ingestion, subcutaneous, intraperitoneal, and intravenous routes. Human reproductive effects by ingestion and possibly other routes: effects on newborn including physical and other postnatal measures or effects. Human teratogenic effects by ingestion and possibly other routes: developmental abnormalities of the craniofacial, musculoskeletal, cardiovascular, and urogenital systems. Experimental teratogenic and reproductive effects. When heated to decomposition it emits toxic fumes of NO_x.

TLR500   CAS: 544-13-8   **HR: 3**
1,3-TRIMETHYLENEDINITRILE
mf: $C_5H_6N_2$ mw: 94.13

PROP: Colorless liquid. D: 0.989 @ 15/4°, mp: −29°, bp: 286.4°. Sol in water; insol in ether.

SYNS: 1,3-DICYANOPROPANE * GLUTARIC ACID DINITRILE * GLUTARODINITRILE * GLUTARONITRILE * PENTANEDINITRILE * PYROTARTARIC ACID NITRILE

CONSENSUS REPORTS: Cyanide and its compounds are on the Community Right-To-Know List. Reported in EPA TSCA Inventory.

SAFETY PROFILE: Poison by subcutaneous and possibly other routes. Moderately toxic by ingestion and possibly other routes. When heated to decomposition it emits very toxic fumes of NO_x and CN^-.

TLT750   CAS: 60597-20-8   **HR: 3**
TRIMETHYLHYDRAZINE HYDROCHLORIDE
mf: $C_3H_{10}N_2 \cdot ClH$ mw: 110.61

SAFETY PROFILE: Suspected carcinogen with experimental carcinogenic data. When heated to decomposition it emits very toxic fumes of HCl and NO_x.

TLU750   CAS: 3475-63-6   **HR: 3**
1,1,3-TRIMETHYL-3-NITROSOUREA
mf: $C_4H_9N_3O_2$ mw: 131.16

SYNS: N-NITROSO-TRIMETHYLHARNSTOFF (GERMAN) * NITROSOTRIMETHYLUREA * N-NITROSOTRIMETHYLUREA * TRIMETHYLNITROSOHARNSTOFF (GERMAN) * N-TRIMETHYL-N-NITROSOUREA

CONSENSUS REPORTS: EPA Genetic Toxicology Program.

SAFETY PROFILE: Poison by ingestion and intravenous routes. Experimental teratogenic and reproductive effects. Questionable carcinogen with experimental tumorigenic data. Mutation data reported. Many N-nitroso compounds are carcinogens. When heated to decomposition it emits toxic fumes of NO_x.

TLX600   CAS: 149-73-5   **HR: 2**
TRIMETHYL ORTHOFORMATE
mf: $C_4H_{10}O_3$ mw: 106.14

PROP: Colorless liquid; pungent odor. Vap d: 3.67, flash p: 59°F.

SYNS: METHYLESTER KYSELINY ORTHOMRAVENCI (CZECH) * METHYL ORTHOFORMATE * ORTHO-FORMIC ACID, TRIMETHYL ESTER * ORTHOMRAVEN-CAN METHYLNATY (CZECH) * TRIMETHOXYMETH-ANE

CONSENSUS REPORTS: Reported in EPA TSCA Inventory.

SAFETY PROFILE: Moderately toxic by ingestion. Mildly toxic by inhalation. A skin and eye irritant. A very dangerous fire hazard when exposed to heat or flame; can react with oxidizing materials. Hazardous to prepare. To fight fire, use CO_2, fog, haze. When heated to decomposition it emits acrid smoke and irritating fumes.

TLY000   CAS: 64047-30-9   HR: 3
TRIMETHYL-2-OXEPANONE (mixed isomers)
mf: $C_9H_{16}O_2$ mw: 156.25

SYN: TRIMETHYL-ε-LACTONE (mixed isomers)

CONSENSUS REPORTS: IARC Cancer Review: Animal Sufficient Evidence IMEMDT 19,303,79.

SAFETY PROFILE: Confirmed carcinogen. Mildly toxic by ingestion and skin contact. When heated to decomposition it emits acrid smoke and irritating fumes.

TLY500   CAS: 540-84-1   HR: 3
2,2,4-TRIMETHYLPENTANE

DOT: UN 1262
mf: C_8H_{18} mw: 114.26

PROP: Clear liquid; odor of gasoline. Bp: 99.2°, fp: −116°, flash p: 10°F, d: 0.692 @ 20/4°, autoign temp: 779°F, vap press: 40.6 mm @ 21°, vap d: 3.93, lel: 1.1%, uel: 6.0%.

SYNS: ISOBUTYLTRIMETHYLETHANE * ISOOC-TANE (DOT)

CONSENSUS REPORTS: Reported in EPA TSCA Inventory.

NIOSH REL: TWA (Alkanes) 350 mg/m³
DOT Classification: Flammable Liquid; Label: Flammable Liquid.

SAFETY PROFILE: High concentrations can cause narcosis. A very dangerous fire hazard when exposed to heat, flame, oxidizers. Can react vigorously with reducing materials. Explosive in the form of vapor when exposed to heat or flame. To fight fire, use CO_2, dry chemical.

When heated to decomposition it emits acrid smoke and irritating fumes.

TMD000   CAS: 2686-99-9   HR: 3
3,4,5-TRIMETHYLPHENYL METHYLCARBAMATE
mf: $C_{11}H_{15}NO_2$ mw: 193.27

SYNS: ENT 25,843 * LANDRIN * OMS-597 * SD 8530 * SHELL SD-8530

SAFETY PROFILE: Poison by ingestion, intraperitoneal, intravenous, and intramuscular routes. When heated to decomposition it emits toxic fumes of NO_x.

TMD250   CAS: 512-56-1   HR: 3
TRIMETHYL PHOSPHATE
mf: $C_3H_9O_4P$ mw: 140.09

PROP: Liquid. D: 1.97 @ 19.5/0°, bp: 197.2°. Sol in alc, water, ether.

SYNS: METHYL PHOSPHATE * NCI-C03781 * PHOSPHORIC ACID, TRIMETHYL ESTER * TMP * O,O,O-TRIMETHYL PHOSPHATE

CONSENSUS REPORTS: NCI Carcinogenesis Bioassay (gavage); Clear Evidence: mouse, rat NCITR* NCI-CG-TR-81,78. Reported in EPA TSCA Inventory. EPA Genetic Toxicology Program.

DFG MAK: Suspected Carcinogen.

SAFETY PROFILE: Suspected carcinogen with experimental carcinogenic, neoplastigenic, and tumorigenic data. Moderately toxic by ingestion, skin contact, intraperitoneal, intravenous, and possibly other routes. Experimental teratogenic and reproductive effects. Human mutation data reported. Explodes when heat distilled. When heated to decomposition it emits toxic fumes of PO_x.

TMD500   CAS: 121-45-9   HR: 2
TRIMETHYL PHOSPHITE

DOT: UN 2329
mf: $C_3H_9O_3P$ mw: 124.09

PROP: Colorless liquid. D: 1.046 @ 20/4°, vap d: 4.3, bp: 232-234°F, flash p: 130°F (OC). Insol in water; sol in hexane, benzene, acetone, alc, ether, carbon tetrachloride, kerosene.

SYNS: FOSFORYN TROJMETYLOWY (CZECH) * METHYL PHOSPHITE * PHOSPHOROUS ACID, TRI-METHYL ESTER * TRIMETHOXYPHOSPHINE

CONSENSUS REPORTS: Reported in EPA TSCA Inventory.

OSHA PEL: TWA 2 ppm
ACGIH TLV: TWA 2 ppm
DOT Classification: Flammable or Combustible Liquid; Label: Flammable Liquid.

SAFETY PROFILE: Moderately toxic by ingestion and skin contact. A severe skin and eye irritant. Flammable when exposed to heat, flame, or oxidizers. To fight fire, use water, foam, fog, CO_2. Violent explosive reaction on contact with magnesium perchlorate or trimethyl platinum(IV) azide tetramer. When heated to decomposition it emits toxic fumes of PO_x. An intermediate in the production of pesticides, fire retardants, and organic phosphorous additives.

TME270   CAS: 14667-55-1   **HR: 2**
2,3,5-TRIMETHYLPYRAZINE
mf: $C_7H_{10}N_2$ mw: 122.19

PROP: Colorless to sltly yellow liquid; baked potato, peanut odor. D: 0.960-0.990 @ 20°, refr index: 1.503, flash p: +153°F. Sol in water and organic solvents.

SYNS: FEMA No. 3244   *   TRIMETHYLPYRAZINE

CONSENSUS REPORTS: Reported in EPA TSCA Inventory.

SAFETY PROFILE: Moderately toxic by ingestion. Combustible liquid. When heated to decomposition emits toxic fumes of NO_x.

TMH750   CAS: 2489-77-2   **HR: 3**
1,1,3-TRIMETHYL-2-THIOUREA
mf: $C_4H_{10}N_2S$ mw: 118.22

PROP: Trimethylthiourea tested in NCITR* NCI-CG-TR-129 contained 15% 1,3-dimethyl-2-thiourea and 5% Zeolex 80 NCITR* NCI-CG-TR-129,79.

SYNS: NCI-C02186 * TRIMETHYLTHIOUREA
 * N,N,N′-TRIMETHYLTHIOUREA

CONSENSUS REPORTS: NCI Carcinogenesis Bioassay (feed); No Evidence: mouse NCITR* NCI-CG-TR-129,79. Reported in EPA TSCA Inventory.

SAFETY PROFILE: Poison by ingestion. Questionable carcinogen with experimental carcinogenic data. When heated to decomposition it emits very toxic fumes of NO_x and SO_x.

TMI000   CAS: 1118-14-5   **HR: 3**
TRIMETHYLTIN ACETATE
mf: $C_5H_{12}O_2Sn$ mw: 222.86

SYN: ACETOXYTRIMETHYLSTANNANE

OSHA PEL: TWA 0.1 mg(Sn)/m^3 (skin)
ACGIH TLV: TWA 0.1 mg(Sn)/m^3 (skin) (Proposed: TWA 0.1 mg(Sn)/m^3; STEL 0.2 mg(Sn)/m^3 (skin))
NIOSH REL: (Organotin Compounds) TWA 0.1 mg(Sn)/m^3

SAFETY PROFILE: Poison by ingestion. When heated to decomposition it emits acrid smoke and irritating fumes.

TMI500   CAS: 63869-87-4   **HR: 3**
TRIMETHYLTIN SULPHATE
mf: $C_3H_{10}O_4SSn$ mw: 260.88

SYN: TRIMETHYLSTANNANE SULPHATE

OSHA PEL: TWA 0.1 mg(Sn)/m^3 (skin)
ACGIH TLV: TWA 0.1 mg(Sn)/m^3 (skin) (Proposed: TWA 0.1 mg(Sn)/m^3; STEL 0.2 mg(Sn)/m^3 (skin))
NIOSH REL: (Organotin Compounds) TWA 0.1 mg(Sn)/m^3

SAFETY PROFILE: Poison by ingestion and intraperitoneal routes. When heated to decomposition it emits toxic fumes of SO_x.

TMJ750   CAS: 86-21-5   **HR: 3**
TRIMETON
mf: $C_{16}H_{20}N_2$ mw: 240.38

SYNS: p-AMINOSALICYLSAURES SALZ (GERMAN)
 * 2-(α-(2-DIMETHYLAMINOETHYL)BENZYL)PYRIDINE
 * 2-(3-DIMETHYLAMINO-1-PHENYLPROPYL)PYRIDINE
 * N,N-DIMETHYL-3-PHENYL-3-(2-PYRIDYL)PROPYL-AMINE * NCI-C60695 * 1-PHENYL-1-(2-PYRIDYL)-3-DIMETHYLAMINOPROPANE * 3-PHENYL-3-(2-PYRIDYL)-N,N-DIMETHYLPROPYLAMINE

SAFETY PROFILE: Poison by ingestion, intraperitoneal, and intravenous routes. Human systemic effects by ingestion: central nervous system effects. When heated to decomposition it emits toxic fumes of NO_x.

TMK250   CAS: 630-72-8   **HR: 3**
TRINITROACETONITRILE
mf: $C_2N_4O_6$ mw: 176.05

CONSENSUS REPORTS: Cyanide and its compounds are on the Community Right-To-Know List.

DOT Classification: Forbidden.

SAFETY PROFILE: An explosive sensitive to friction, impact or rapid heating to 220°C. When heated to decomposition it emits toxic fumes of CN^- and NO_x.

TMK500 CAS: 99-35-4 **HR: 3**
1,3,5-TRINITROBENZENE
mf: $C_6H_3N_3O_6$ mw: 213.12

DOT: UN 0214

PROP: Yellow crystals. Mp: 122°, bp: decomp, d: 1.760 @ 20/4°.

SYNS: RCRA WASTE NUMBER U234 * TNB * TRINITROBENZEEN (DUTCH) * TRINITROBEN-ZENE * TRINITROBENZENE, dry (DOT) * TRINITRO-BENZOL (GERMAN)

CONSENSUS REPORTS: Reported in EPA TSCA Inventory.

DOT Classification: Class A Explosive; Label: Explosive A.

SAFETY PROFILE: Poison by intravenous route. Moderately toxic by ingestion. Mutation data reported. A severe explosion hazard when shocked or exposed to heat. Trinitrobenzene is considered a powerful high explosive and has more shattering power than TNT. Although it is less sensitive to impact than TNT, it is not used much because it is difficult to produce. The complex with potassium trimethyl stannate explodes at room temperature. Forms heat-sensitive explosive complexes with alkyl or aryl metallates (e.g., lithium or potassium salts of trimethyl-, triethyl-, or triphenyl-germanate, -silanate, or -stannate. Can react vigorously with reducing materials. When heated to decomposition it emits highly toxic fumes of NO_x and explodes.

TML000 CAS: 129-66-8 **HR: 3**
TRINITROBENZOIC ACID (dry)
mf: $C_7H_3N_3O_8$ mw: 257.13

DOT: UN 0215/UN 1355

PROP: Orthorhombic crystals. Mp: 228.7°. Sol @ 25° (2.05% in water, 26.6% in alc, 14.7% in ether), sol in methanol; sltly sol in benzene.

SYN: 2,4,6-TRINITROBENZOIC ACID

DOT Classification: Class A Explosive; Label: Explosive A.

SAFETY PROFILE: An explosive. A hazard in preparation. Reacts with heavy metals to form heat- or impact-sensitive explosive salts. When heated to decomposition it emits toxic fumes of NO_x.

TML325 CAS: 28260-61-9 **HR: 3**
TRINITROCHLOROBENZENE

DOT: UN 0155
mf: $C_6H_2ClN_3O_6$ mw: 247.56

SYN: TRINITROCHLOROBENZENE (DOT)

DOT Classification: Class A Explosive; Label: Explosive A.

SAFETY PROFILE: Mutation data reported. An explosive. When heated to decomposition it emits toxic fumes of Cl^- and NO_x.

TML500 CAS: 602-99-3 **HR: 3**
2,4,6-TRINITRO-m-CRESOL

DOT: UN 0216
mf: $C_7H_5N_3O_7$ mw: 243.15

PROP: Yellow crystals. Mp: 106°, bp: explodes @ 150°.

SYNS: CRESYLITE * 3-METHYL-2,4,6-TRINITRO-PHENOL * TRINITRO-m-CRESOL * TRINITRO-m-CRESOLIC ACID * TRINITROMETACRESOL (DOT)

DOT Classification: Class A Explosive; Label: Explosive A.

SAFETY PROFILE: Poison by intraperitoneal route. A severe explosion hazard when shocked or exposed to heat. Explodes when heated above 150°C. Trinitrocresol is not as powerful a high explosive as TNT or picric acid. Can react vigorously with oxidizing materials. When heated to decomposition it emits highly toxic fumes of NO_x and explodes.

TMM000 CAS: 918-54-7 **HR: 3**
2,2,2-TRINITROETHANOL
mf: $C_2H_3N_3O_7$ mw: 181.08

SYN: TRINITROETHANOL (DOT)

DOT Classification: Forbidden.

SAFETY PROFILE: Poison by intraperitoneal route. A shock-sensitive explosive. When heated to decomposition it emits toxic fumes of NO_x.

TMM250 CAS: 129-79-3 **HR: 3**
2,4,7-TRINITROFLUOREN-9-ONE
mf: $C_{13}H_5N_3O_7$ mw: 315.21

SYNS: 2,4,7-TRINITRO-9-FLUORENONE * 2,4,7-TRINITROFLUORENONE (MAK)

CONSENSUS REPORTS: Reported in EPA TSCA Inventory.

DFG MAK: Suspected Carcinogen.

SAFETY PROFILE: Suspected carcinogen with experimental tumorigenic data. Mildly toxic by ingestion. Human mutation data reported. A skin and eye irritant. When heated to decomposition it emits highly toxic fumes of NO_x.

TMM500   CAS: 517-25-9   **HR: 3**
TRINITROMETHANE
mf: CHN_3O_6 mw: 151.05

PROP: Mp: 15°, d: 1.469, bp: decomp > 25°. Sol in water.

SYN: NITROFORM

CONSENSUS REPORTS: Reported in EPA TSCA Inventory.

DOT Classification: Forbidden.

SAFETY PROFILE: Poison by ingestion and intraperitoneal routes. Moderately toxic by inhalation. Irritating to skin, eyes, and mucous membranes. Inhalation can cause headache and nausea. Causes mild narcosis. A very dangerous explosion hazard; explodes when heated rapidly. Dissolution is exothermic and solutions of more than 50% can explode. Mixtures of 90% trinitromethane + 10% isopropyl alcohol in polyethylene bottles have exploded. Frozen mixtures with 2-propanol (10%) explode when thawed. Can explode during distillation. Mixtures with divinyl ketone can explode at 4°C. When heated to decomposition it emits toxic fumes of NO_x.

TMN000   CAS: 75321-19-6   **HR: 2**
1,3,6-TRINITROPYRENE
mf: $C_{16}H_7N_3O_6$ mw: 337.26

SYN: TRINITROPYRENE

DFG MAK: Suspected Carcinogen.

SAFETY PROFILE: Suspected carcinogen. Mutation data reported. When heated to decomposition it emits toxic fumes of NO_x.

TMN490   CAS: 118-96-7   **HR: 3**
2,4,6-TRINITROTOLUENE

DOT: UN 0209/UN 1356
mf: $C_7H_5N_3O_6$ mw: 227.15

PROP: Colorless, monoclinic crystals. Mp: 80.7°, bp: 240° (explodes), flash p: explodes, d: 1.654. Sol in hot water, alc, ether.

SYNS: ENTSUFON * NCI-C56155 * TNT
* α-TNT * TNT-TOLITE (FRENCH) * TOLIT

* TOLITE   * 2,4,6-TRINITROLUEEN (DUTCH)
* TRINITROTOLUENE * s-TRINITROTOLUENE
* TRINITROTOLUENE, dry (DOT) * sym-TRINITRO-
TOLUENE * s-TRINITROTOLUOL * sym-TRINITRO-
TOLUOL * 2,4,6-TRINITROTOLUOL (GERMAN)
* TRITOL * TROJNITROTOLUEN (POLISH)
* TROTYL * TROTYL OIL

CONSENSUS REPORTS: Reported in EPA TSCA Inventory. EPA Genetic Toxicology Program.

OSHA PEL: (Transitional: TWA 1.5 mg/m^3 (skin)) TWA 0.5 mg/m^3 (skin)
ACGIH TLV: TWA 0.5 mg/m^3 (skin)
DFG MAK: 0.01 ppm (0.1 mg/m^3)
DOT Classification: Class A Explosive; Label: Explosive A.

SAFETY PROFILE: Poison by subcutaneous route. Moderately toxic by ingestion. Human systemic effects by ingestion: hallucinations or distorted perceptions, cyanosis and gastrointestinal changes. Experimental reproductive effects. Mutation data reported. A skin irritant. Has been implicated in aplastic anemia. Can cause headache, weakness, anemia, liver injury. May be absorbed through skin.

Flammable or explosive when exposed to heat or flame. Moderate explosion hazard; will detonate under strong shock. It detonates at around 240°C but can be distilled safely under reduced pressure. It is a comparatively insensitive explosive. In small quantities it will burn quietly if not confined. However, sudden heating of any quantity will cause it to detonate; the accumulation of heat when large quantities are burning will cause detonation. In other respects it is one of the most stable of all high explosives and there are but a few restrictions for its handling. It is for this reason, from the military standpoint, that TNT is quantitatively the most used. It requires a fall of 130 cm for a 2 kg weight to detonate it. It is one of the most powerful high explosives. It can be detonated by the usual detonators and blasting caps (at least a No. 6). For full efficiency, the use of a high velocity initiator, such as tetryl, is required. TNT is one of those explosives containing an oxygen deficiency. In other words, the addition of products which are oxygen-rich can enhance its explosive power. Also mono- and dinitrotoluene may be added for reduction of the temperature of the explosion and to make the explosion flashless. Various materials are added to TNT to make what is known as permissible explosives. TNT may be regarded as the equivalent of 40% dynamite and can be used underwater. It is also used in the manufacture of a detonator fuse known as Cordeau Detonant. For the military, TNT finds use in all types of bursting charges, including armor-piercing types, although it is somewhat too sensitive to be ideal for this purpose, and has since been replaced to a great extent by ammonium picrate. It is a relatively expensive explosive and does not compete seriously with dynamite for general commercial use.

Highly dangerous; explodes with shock or heating to 297°C. Various materials can reduce the explosive temperature: red lead (to 192°C); sodium carbonate (to 218°C); potassium hydroxide (to 192°C). Mixtures with sodium dichromate + sulfuric acid may ignite spontaneously. Reacts with nitric acid + metals (e.g., lead or iron) to form explosive products more sensitive to shock, friction or contact with nitric or sulfuric acids. Reacts with potassium hydroxide dissolved in methanol to form explosive aci-nitro salts. Bases (e.g., sodium hydroxide; potassium iodide; tetramethyl ammonium octahydrotriborate) induce deflagration in molten TNT. Can react vigorously with reducing materials. When heated to decomposition it emits highly toxic fumes of NO_x.

TMO250   CAS: 2467-12-1   **HR: 2**
TRI-n-OCTYL BORATE
mf: $C_{24}H_{51}BO_3$ mw: 398.56

PROP: Colorless liquid; odor of octyl alc. Bp: 192-194° @ 2 mm, flash p: 370°F (COC), d: 0.846 @ 23°, vap d: 13.7.

SYN: BORIC ACID, TRI-n-OCTYL ESTER

CONSENSUS REPORTS: Reported in EPA TSCA Inventory.

SAFETY PROFILE: Moderately toxic by ingestion. An eye irritant. Combustible when exposed to heat or flame; can react with oxidizing materials. To fight fire, use foam, CO_2, dry chemical. When heated to decomposition it emits acrid smoke and irritating fumes.

TMO600   CAS: 78-30-8   **HR: 3**
TRIORTHOCRESYL PHOSPHATE
mf: $C_{21}H_{21}O_4P$ mw: 368.39

PROP: Colorless liquid. Mp: −25 to −30°, bp: 410° (slt decomp), flash p: 437°F, d: 1.17,

autoign temp: 725°F, vap d: 12.7. Insol in water; sol in alc and ether.

SYNS: o-CRESYL PHOSPHATE * PHOSFLEX 179-C * PHOSPHORIC ACID, TRI-o-CRESYL ESTER * TOCP * TOFK * PHOSPHORIC ACID, TRIS (2-METHYLPHENYL) ESTER * o-TOLYL PHOSPHATE * TOTP * TRICRESYL PHOSPHATE * TRI-o-CRESYL PHOSPHATE * o-TRIKESYLPHOSPHATE (GERMAN) * TRI 2-METHYLPHENYL PHOSPHATE * TRI-2-TOLYL PHOSPHATE * TRIS(o-CRESYL) PHOSPHATE * TRIS(o-METHYLPHENYL)PHOSPHATE * TRIS(o-TOLYL)-PHOSPHATE * TRI-o-TOLYL PHOSPHATE * TROJKREZYLU FOSFORAN (POLISH)

CONSENSUS REPORTS: Reported in EPA TSCA Inventory.

OSHA PEL: (Transitional: TWA 0.1 mg/m^3) TWA 0.1 mg/m^3 (skin)
ACGIH TLV: TWA 0.1 mg/m^3 (skin)

SAFETY PROFILE: Poison by subcutaneous, intramuscular, intravenous and intraperitoneal routes. Moderately toxic by ingestion and possibly other routes. Most of the cases of tri-o-cresyl phosphate poisoning have followed its ingestion. In 1930, some 15,000 persons were affected in the United States, and of these, 10 died. The responsible material was found to be an alcoholic drink known as Jamaica ginger, or "jake." This beverage had been adulterated with about 2% of tri-o-cresyl phosphate. The affected persons developed a polyneuritis, which progressed, in many cases, with degeneration of the peripheral motor nerves, the anterior horn cells and the pyramidal tracts. Sensory changes were absent. Since 1930 there have been several other outbreaks of poisoning following ingestion of the material. Tri-o-cresyl phosphate is more toxic than the m-form, and much more so than tri-p-cresyl phosphate or triphenyl phosphate.

Combustible when exposed to heat or flame. Can react with oxidizing materials. To fight fire, use CO_2, dry chemical. When heated to decomposition it emits highly toxic fumes of PO_x.

TMP000   CAS: 110-88-3   **HR: 3**
s-TRIOXANE
mf: $C_3H_6O_3$ mw: 90.09

PROP: Stable, cyclic trimer of formaldehyde, having characteristic ethanol and chloroform-like odors. Crystalline solid. Mp: 64°, bp:

114.5°, sublimes readily, lel: 3.6%, uel: 28.7%, flash p: 113°F (OC), d: 1.17, @ 65°, autoign temp: 777°F, vap press: 13 mm @ 25°, vap d: 3.1. Very sol in water, alc, ketones, ether, acetone, chlorinated and aromatic hydrocarbons, organic solvents; sltly sol in pentane, petr ether.

SYNS: POLYOXYMETHYLENE * TRIOSSIMETHLENE (ITALIAN) * TRIOXANE * 1,3,5-TRIOXANE * TRIOXYMETHYLEEN (DUTCH) * TRIOXYMETHYLEN (GERMAN) * TRIOXYMETHYLENE

CONSENSUS REPORTS: Reported in EPA TSCA Inventory.

SAFETY PROFILE: Moderately toxic by ingestion. Mildly toxic by skin contact. A severe eye and skin irritant. Can evolve toxic formaldehyde fumes when heated strongly or in contact with strong acids or acid fumes. Flammable when exposed to heat, flame or oxidizers. May explode when heated. Explosive in the form of vapor when exposed to heat or flame. Explodes on impact, possibly due to peroxide contamination. Mixtures with hydrogen peroxide are explosives sensitive to heat, shock or contact with lead. Mixtures with liquid oxygen are highly explosive. Incompatible with oxidizing materials. To fight fire, use foam, CO_2, or dry chemical. When heated to decomposition it emits acrid smoke and irritating fumes.

TMP750   CAS: 91-81-6   **HR: 3**
TRIPELENNAMINE
mf: $C_{16}H_{21}N_3$ mw: 255.40

PROP: Oily liquid; amine odor. Bp: 167-172° @ 0.1 mm. Misc with water.

SYNS: BENZOXALE * 2-(BENZYL(2-DIMETHYL AMINOETHYL)AMINO)PYRIDINE * N-BENZYL-N',N'-DIMETHYL-N-2-PYRIDYLETHYLENE DIAMINE * BENZYL-(α-PYRIDYL)-DIMETHYLAETHYLENDIAMIN (GERMAN) * CIZARON * DEHISTIN * β-DIMETHYLAMINO ETHYL-2-PYRIDYLAMINOTOLUENE * β-DIMETHYLAMINOETHYL-2-PYRIDYLBENZYL-AMINE * N,N-DIMETHYL-N'-BENZYL-N'-(α-PYRIDYL) ETHYLENEDIAMINE * NCI-C60662 * PBZ * PIRIBENZIL * PYRIBENZAMINE * PYRINAMINE BASE * RESISTAMINE * TONARIL * TRIPELENAMINE * TRIPELENNAMINA (ITALIAN)

SAFETY PROFILE: Poison by ingestion and intraperitoneal routes. Human mutation data reported. Has been implicated in aplastic anemia. Used as an antihistamine. Addicts have added it to paregoric to make "blue velvet," which can cause a euphoria by injection. When heated to decomposition it emits toxic fumes of NO_x.

TMQ250   CAS: 6304-33-2   **HR: 3**
2,3,3-TRIPHENYLACRYLONITRILE
mf: $C_{21}H_{15}N$ mw: 281.37

SYNS: α,β-DIPHENYLCINNAMONITRILE * α-(DIPHENYLMETHYLENE)BENZENEACETIC ACID * TRIPHENYLACRYLONITRILE * α,β,β-TRIPHENYLACRYLONITRILE * TRIPHENYLCYANOETHYLENE

CONSENSUS REPORTS: Cyanide and its compounds are on the Community Right-To-Know List.

SAFETY PROFILE: Poison by ingestion and intravenous routes. Questionable carcinogen with experimental carcinogenic data. When heated to decomposition it emits toxic fumes of NO_x and CN^-.

TMQ500   CAS: 603-34-9   **HR: 2**
TRIPHENYLAMINE
mf: $C_{18}H_{15}N$ mw: 245.34

PROP: Crystals. D: 0.774 @ 0/0°, mp: 127°, bp: 365°.

SYN: N,N-DIPHENYLANILINE

CONSENSUS REPORTS: Reported in EPA TSCA Inventory.

OSHA PEL: TWA 5 mg/m³
ACGIH TLV: TWA 5 mg/m³

SAFETY PROFILE: Moderately toxic by ingestion. When heated to decomposition it emits toxic fumes of NO_x.

TMS250   CAS: 58-72-0   **HR: 3**
TRIPHENYLETHYLENE
mf: $C_{20}H_{16}$ mw: 256.36

SYN: 1,1,2-TRIPHENYLETHYLENE

CONSENSUS REPORTS: Reported in EPA TSCA Inventory.

SAFETY PROFILE: Experimental teratogenic and reproductive effects. Questionable carcinogen with experimental tumorigenic data. Human mutation data reported. When heated to decomposition it emits acrid smoke and irritating fumes.

TMT750 CAS: 115-86-6 **HR: 3**
TRIPHENYL PHOSPHATE
mf: $C_{18}H_{15}O_4P$ mw: 326.30

PROP: Colorless, odorless, crystalline solid. Mp: 49-50°, bp: 245° @ 11 mm, flash p: 428°F (CC), d: 1.268 @ 60°, vap press: 1 mm @ 193.5°. Insol in water; sol in alc, benzene, ether, chloroform and acetone.

SYNS: CELLUFLEX TPP * PHOSPHORIC ACID, TRIPHENYL ESTER * TPP

CONSENSUS REPORTS: Reported in EPA TSCA Inventory.

OSHA PEL: TWA 3 mg/m³
ACGIH TLV: TWA 3 mg/m³

SAFETY PROFILE: Poison by subcutaneous route. Moderately toxic by ingestion and possibly other routes. Absorbed slowly, particularly by skin contact. Not a potent cholinesterase inhibitor. Combustible when exposed to heat or flame. To fight fire, use CO_2, dry chemical. When heated to decomposition it emits toxic fumes of PO_x.

TMU000 CAS: 603-35-0 **HR: 2**
TRIPHENYLPHOSPHINE
mf: $C_{18}H_{15}P$ mw: 262.30

PROP: Odorless crystals. Mp: 79°, bp: >360°, d: 1.194, flash p: 356°F (OC), vap d: 9.0. Insol in water; sol in HCl, benzene; sltly sol in alc; very sol in ether.

CONSENSUS REPORTS: Reported in EPA TSCA Inventory.

SAFETY PROFILE: Moderately toxic by ingestion. Mildly toxic by inhalation. Combustible when exposed to heat or flame. Slight explosion hazard in the form of vapor when exposed to flame. Can react vigorously with oxidizing materials. To fight fire, use dry chemical, fog, CO_2. When heated to decomposition it emits highly toxic fumes of phosphine and PO_x.

TMU250 CAS: 101-02-0 **HR: 3**
TRIPHENYL PHOSPHITE
mf: $C_{18}H_{15}O_3P$ mw: 310.30

PROP: Water white to pale yellow solid or oily liquid; clean and pleasant odor. D: 1.184 @ 25/25°, mp: 22-25°, bp: 155-160° @ 0.1 mm, flash p: 425°F (OC). Insol in water.

SYNS: EFED * PHOSPHOROUS ACID, TRIPHENYL ESTER * TRIFENOXYFOSFIN (CZECH) * TRIFENYLFOSFIT (CZECH)

CONSENSUS REPORTS: Reported in EPA TSCA Inventory.

SAFETY PROFILE: Poison by intraperitoneal and subcutaneous routes. Moderately toxic by ingestion and possibly other routes. An experimental eye and severe human skin irritant. Combustible when exposed to heat or flame. To fight fire, use CO_2, mist, dry chemical. When heated to decomposition it emits toxic fumes of PO_x.

TMX500 CAS: 1317-95-9 **HR: 3**
TRIPOLI

PROP: Finely granulated white or gray siliceous rock. A form of crystalline silica.

OSHA PEL: (Transitional: TWA Respirable: 10 mg/m³/2(%SiO₂+2); Total Dust: TWA 30 mg/m³/2(%SiO₂+2)) TWA 0.1 mg/m³
ACGIH TLV: TWA 0.1 mg/m³ (of contained respirable quartz dust)

SAFETY PROFILE: The prolonged inhalation of dusts containing free silica may result in the development of a disabling pulmonary fibrosis known as silicosis.

TMY250 CAS: 102-69-2 **HR: 3**
TRI-N-PROPYLAMINE
mf: $C_9H_{21}N$ mw: 143.31

DOT: UN 2260

PROP: Liquid. Mp: −93°, bp: 156°, flash p: 105°F (OC), d: 0.75, vap d: 4.9. Very sltly sol in water.

SYNS: N,N-DIPROPYL-1-PROPANAMINE * TRIPROPYLAMINE (DOT)

CONSENSUS REPORTS: Reported in EPA TSCA Inventory.

DOT Classification: Flammable Liquid; Label: Flammable Liquid, Corrosive; Flammable or Combustible Liquid; Label: Flammable Liquid, Corrosive.

SAFETY PROFILE: Poison by ingestion. Moderately toxic by skin contact, inhalation, and possibly other routes. A corrosive irritant to skin, eyes, and mucous membranes. Flammable when exposed to heat, flame, or oxidizers. Can react with oxidizing materials. To fight fire, use foam, CO_2, dry chemical. When heated to decomposition it emits toxic fumes of NO_x.

TNC500   CAS: 126-72-7   **HR: 3**
TRIS
mf: $C_9H_{15}Br_6O_4P$ mw: 697.67

PROP: Crystals. D: 2.24, flash p: > 112°.

SYNS: ANFRAM 3PB * APEX 462-5 * BROMKAL P 67-6HP * 2,3-DIBROMO-1-PROPANOL, PHOSPHATE (3:1) * 2,3-DIBROMO-1-PROPANOL PHOSPHATE * (2,3-DIBROMOPROPYL) PHOSPHATE * FIREMASTER T23P-LV * FLACAVON R * FLAMMEX AP * FYROL HB32 * NCI-C03270 * PHOSPHORIC ACID, TRIS(2,3-DIBROMOPROPYL) ESTER * RCRA WASTE NUMBER U235 * TDBP (CZECH) * TRIS (flame retardant) * TRIS(DIBROMOPROPYL)PHOSPHATE * TRIS(2,3-DIBROMOPROPYL) PHOSPHATE * TRIS (2,3-DIBROMOPROPYL) PHOSPHORIC ACID ESTER * TRIS-2,3-DIBROMPROPYL ESTER KYSELINY FOSFORECNE (CZECH) * USAF DO-41 * ZETIFEX ZN

CONSENSUS REPORTS: IARC Cancer Review: GROUP 2A IMEMDT 7,341,87; Animal Sufficient Evidence IMEMDT 20,575,79; Human Limited Evidence IMEMDT 20,575,79. NTP Fourth Annual Report On Carcinogens, 1984. NCI Carcinogenesis Bioassay (feed); Clear Evidence: mouse, rat NCITR* NCI-CG-TR-76,78. Community Right-To-Know List. Reported in EPA TSCA Inventory. EPA Genetic Toxicology Program.

SAFETY PROFILE: Confirmed carcinogen with experimental carcinogenic, neoplastigenic, and tumorigenic data. Poison by intraperitoneal route. Moderately toxic by ingestion and possibly other routes. Experimental teratogenic and reproductive effects. Human mutation data reported. An eye and severe skin irritant. Can cause testicular atrophy and sterility. Once used as a flame retardant additive to synthetic textiles and plastics, particularly in children's sleepwear. Use discontinued because it can be absorbed by human skin, or chewed or sucked off of sleepwear by infants. May be flammable when exposed to heat or flame. When heated to decomposition it emits very toxic fumes of Br^- and PO_x.

TND000   CAS: 68-76-8   **HR: 3**
TRIS(1-AZIRIDINYL)-p-BENZOQUINONE
mf: $C_{12}H_{13}N_3O_2$ mw: 231.28

SYNS: BAYER 3231 * 1,1′,1″ -(3,6-DIOXO-1,4-CYCLOHEXADIENE-1,2,4-TRIYL)TRISAZIRIDINE * NSC-29215 * ONCOVEDEX * PRENIMON * RIKER 601 * 10257 R.P. * TEIB * TRENIMON * TRIAZICHON (GERMAN) * TRIAZIQUINONE * TRIAZI-

QUONE * 2,3,5-TRI-(1-AZIRIDINYL)-p-BENZOQUINONE * 2,3,5-TRIETHYLENEIMINO-1,4-BENZOQUINONE * TRIETHYLENIMINOBENZOQUINONE * TRISAETHYLENIMINOBENZOCHIN (GERMAN) * 2,3,5-TRIS(AZIRIDINO)-1,4-BENZOQUINONE * 2,3,5-TRIS(1-AZIRIDINO)-p-BENZOQUINONE * TRIS(AZIRIDINYL)-p-BENZOQUINONE * 2,3,5-TRIS(1-AZIRIDINYL)-p-BENZOQUINONE * 2,3,5-TRIS(AZIRIDINYL)-1,4-BENZOQUINONE * 2,3,5-TRIS(1-AZIRIDINYL)-2,5-CYLOHEXADIENE-1,4-DIONE * 2,3,5-TRISETHYLENEIMINOBENZOQUINONE * TRISETHYLENEIMINOQUINONE * 2,3,5-TRIS(ETHYLENIMINO)BENZOQUINONE * 2,3,5-TRIS(ETHYLENIMINO)-p-BENZOQUINONE * 2,3,5-TRIS(ETHYLENIMINO)-1,4-BENZOQUINONE

CONSENSUS REPORTS: IARC Cancer Review: GROUP 3 IMEMDT 7,367,87; Animal Sufficient Evidence IMEMDT 9,67,75; Human Inadequate Evidence IMEMDT 9,67,75. Community Right-To-Know List. EPA Genetic Toxicology Program.

SAFETY PROFILE: Poison by intraperitoneal, intravenous, and parenteral routes. Experimental teratogenic and reproductive effects. Questionable carcinogen with experimental experimental carcinogenic data. Human mutation data reported. When heated to decomposition it emits toxic fumes of NO_x. Used as a drug for the treatment of neoplastic diseases.

TND250   CAS: 545-55-1   **HR: 3**
TRIS-(1-AZIRIDINYL)PHOSPHINE OXIDE
DOT: UN 2501
mf: $C_6H_{12}N_3OP$ mw: 173.18

PROP: Colorless crystals. Mp: 41°, bp: 90° @ 23 mm. Sol in water, alc, ether.

SYNS: APHOXIDE * APO * 1-AZIRIDINYL PHOSPHINE OXIDE (TRIS) (DOT) * CBC 906288 * ENT 24,915 * IMPERON FIXER T * NSC 9717 * 1,1′,1″-PHOSPHINYLIDYNETRISAZIRIDINE * PHOSPHORIC ACID TRIETHYLENE IMIDE * PHOSPHORIC ACID TRIETHYLENEIMINE (DOT) * SK-3818 * TEF * TEPA * TRIAETHYLENPHOSPHORSAEUREAMID (GERMAN) * TRIAZIRIDINOPHOSPHINE OXIDE * TRI(AZIRIDINYL)PHOSPHINE OXIDE * TRI-1-AZIRIDINYL)PHOSPHINE OXIDE * N,N′,N″-TRI-1,2-ETHANEDIYL PHOSPHORIC TRIMIDE * TRIETHYLENEPHOSPHOROTRIAMIDE * TRIS(1-AZIRIDINE)PHOSPHINE OXIDE * TRIS(N-ETHYLENE)PHOSPHOROTRIAMIDATE

CONSENSUS REPORTS: IARC Cancer Review: GROUP 3 IMEMDT 7,56,87; Animal

Inadequate Evidence IMEMDT 9,75,75. EPA Genetic Toxicology Program.

DOT Classification: Label: Corrosive; Poison B; Label: Poison.

SAFETY PROFILE: Poison by ingestion, skin contact, intravenous, intraperitoneal, and possibly other routes. Experimental teratogenic and reproductive effects. Questionable carcinogen with experimental carcinogenic and neoplastigenic data. Human mutation data reported. A corrosive irritant to the skin, eyes, and mucous membranes. When heated to decomposition it emits very toxic fumes of PO_x and NO_x. Used as an acaricide and in the permanent press treatment of cotton.

TND500   CAS: 51-18-3   **HR: 3**
TRISAZIRIDINYLTRIAZINE
mf: $C_9H_{12}N_6$ mw: 204.27

PROP: Small crystals. Water-sol. Decomp @ 139°.

SYNS: DRP 859025 * ENT 25,296 * M-9500 * NSC 9706 * PERSISTOL * R-246 * SEM (CYTOSTATIC) * SK1133 * TRETAMINE * TRIAETHYLENMELAMIN (GERMAN) * TRIAMELIN * 1,1′,1″-s-TRIAZINE-2,4,6-TRIYLTRISAZIRIDINE * TRIAZIRIDINYL TRIAZINE * TRIETHANOMELAMINE * 2,4,6-TRI(ETHYLENEIMINO)-1,3,5-TRIAZINE * 2,4,6-TRIETHYLENEIMINO-s-TRIAZINE * TRIETHYLENEMELAMINE * 2,4,6-TRIETHYLENIMINO-s-TRIAZINE * 2,4,6-TRIETHYLENIMINO-1,3,5-TRIAZINE * 2,4,6-TRIS(1-AZIRIDINYL)-s-TRIAZINE * 2,4,6-TRIS(1′-AZIRIDINYL)-1,3,5-TRIAZINE * TRIS(ETHYLENEIMINO)TRIAZINE * 2,4,6-TRIS(ETHYLENEIMINO)-s-TRIAZINE * TRISETHYLENEIMINO-1,3,5-TRIAZINE * 2,4,6-TRIS(ETHYLENIMINO)-s-TRIAZINE

CONSENSUS REPORTS: IARC Cancer Review: GROUP 3 IMEMDT 7,56,87; Animal Sufficient Evidence IMEMDT 9,95,75. EPA Genetic Toxicology Program.

SAFETY PROFILE: Poison by ingestion, intraperitoneal, intramuscular, intravenous, subcutaneous, and possibly other routes. Experimental teratogenic and reproductive effects. Questionable carcinogen with experimental neoplastigenic and tumorigenic data. Human mutation data reported. Can cause gastrointestinal tract disturbances and bone marrow depression. When heated to decomposition it emits highly toxic fumes of NO_x. Used as an antineoplastic agent and as an insect sterilant.

TNF500   CAS: 817-09-4   **HR: 3**
TRIS(2-CHLOROETHYL)AMMONIUM CHLORIDE
mf: $C_6H_{12}Cl_3N \cdot ClH$ mw: 241.00

SYNS: LEKAMIN * NSC-30211 * R-47 * SINALOST * SK-100 * TRI(β-CHLOROETHYL)AMINE HYDROCHLORIDE * TRI-(2-CHLOROETHYL)AMINE HYDROCHLORIDE * TRICHLORMETHINE * TRICHLORMETHINIUM CHLORIDE * TRICHLORTRIAETHYLAMIN-HYDROCHLORID (GERMAN) * 2,2′,2″-TRICHLOROTRIETHYLAMINE HYDROCHLORIDE * TRILLEKAMIN * TRIMITAN * TRIMUSTINE * TRIMUSTINE HYDROCHLORIDE * TRIS-(β-CHLOROETHYL)AMINE HYDROCHLORIDE * TRIS-(2-CHLOROETHYL)AMINE HYDROCHLORIDE * TRIS-(2-CHLOROETHYL)AMINE MONOHYDROCHLORIDE * TRIS-N-LOST * TS-160

CONSENSUS REPORTS: IARC Cancer Review: GROUP 3 IMEMDT 7,56,87; Animal Inadequate Evidence IMEMDT 9,229,75. EPA Genetic Toxicology Program.

SAFETY PROFILE: Poison by ingestion, subcutaneous, intravenous, and intraperitoneal routes. Human systemic effects by ingestion and intravenous routes: somnolence, anorexia, headache, thrombosis distant from injection site, nausea or vomiting, and leukopenia. Experimental reproductive effects. Mutation data reported. Questionable carcinogen with experimental carcinogenic data. When heated to decomposition it emits very toxic fumes of Cl^-, NH_3, and NO_x. Used as an antineoplastic agent.

TNK250   CAS: 57-39-6   **HR: 3**
TRIS(1-METHYLETHYLENE) PHOSPHORIC TRIAMIDE
mf: $C_9H_{18}N_3OP$ mw: 215.27

PROP: Amber-colored liquid; amine odor. Bp: 118-125° @ 1 mm, d: 1.079 @ 25/25°. Misc with water and all organic solvents.

SYNS: C 3172 * ENT 50,003 * MAPO * METEPA * METHAPHOXIDE * METHYL APHOXIDE * 1,1′,1″-PHOSPHINYLIDYNETRIS(2-METHYL)AZIRIDINE * TRIS(2-METHYL-1-AZIRIDINYL)PHOSPHINE OXIDE * TRIS(2-METHYLAZIRIDIN-1-YL)PHOSPHINE OXIDE * N,N′,N″-TRIS(1-METHYLETHYLENE)PHOSPHORAMIDE

CONSENSUS REPORTS: IARC Cancer Review: GROUP 3 IMEMDT 7,56,87; Animal Inadequate Evidence IMEMDT 9,107,75. Reported in EPA TSCA Inventory. EPA Genetic Toxicology Program.

SAFETY PROFILE: Poison by ingestion, skin contact, intraperitoneal, and subcutaneous routes. Experimental teratogenic and reproductive effects. Questionable carcinogen with experimental carcinogenic data. Animal experiments suggest cholinesterase inhibition, possibly due to metabolic products of this material in the body. When heated to decomposition it emits very toxic fumes of NO_x and PO_x.

TNL250   CAS: 150-38-9   **HR: 3**
TRISODIUM EDETATE
mf: $C_{10}H_{13}N_2O_8 \cdot 3Na$ mw: 358.22

SYNS: EDETATE TRISODIUM * EDTA TRISODIUM SALT * N,N′-1,2-ETHANEDIYLBIS(N-CARBOXY-METHYL)GLYCINE, TRISODIUM SALT * ETHYLENE-DIAMINEACETIC ACID TRISODIUM SALT * ETHYL-ENEDIAMINETETRAACETICACID, TRISODIUM SALT * NCI-C03974 * NEVANAID-B POWDER * PERMA KLEER 50, TRISODIUM SALT * SEQUESTRENE Na3 * SEQUESTRENE TRISODIUM * SEQUESTRENE TRISODIUM SALT * TRILON AO * TRISODIUM EDTA * TRISODIUM ETHYLENEDIAMINETETRA-ACETATE * TRISODIUM HYDROGEN ETHYLENEDI-AMINETETRAACETATE * TRISODIUM HYDROGEN (ETHYLENEDINITRILO)TETRAACETATE * TRI-SODIUM VERSENATE * VERSENE 9

CONSENSUS REPORTS: Reported in EPA TSCA Inventory.

SAFETY PROFILE: Poison by intraperitoneal route. Moderately toxic by ingestion. When heated to decomposition it emits toxic fumes of NO_x and Na_2O.

TNP250   CAS: 786-19-6   **HR: 3**
TRITHION
mf: $C_{11}H_{16}ClO_2PS_3$ mw: 342.87

PROP: Amber liquid. Bp: 82° @ 0.1 mm, d: 1.29 @ 20°. Essentially insol in water; misc in common solvents.

SYNS: ACARITHION * AKARITHION * CARBO-FENOTHION (DUTCH) * S-((p-CHLOROPHENYLTHIO)-METHYL)-O,O-DIETHYL PHOSPHORODITHIOATE * S-(4-CHLOROPHENYLTHIOMETHYL)DIETHYL PHOS-PHOROTHIOLOTHIONATE * DAGADIP * O,O-DIA-ETHYL-S-((4-CHLOR-PHENYL-THIO)-METHYL)DITHIO-PHOSPHAT (GERMAN) * O,O-DIETHYL-S-(4-CHLOOR-FENYL-THIO)-METHYL)-DITHIOFOSFAAT (DUTCH) * O,O-DIETHYL-S-p-CHLORFENYLTHIOMETHYLESTER KYSELINY DITHIOFOSFORECNE (CZECH) * O,O-DI-ETHYL-S-p-CHLORLPHENYLTHIOMETHYL DITHIOPHOS-PHATE * O,O-DIETHYL-P-CHLOROPHENYLMERCAP-TOMETHYL DITHIOPHOSPHATE * O,O-DIETHYL-S-(4-CHLOROPHENYLTHIOMETHYL) DITHIOPHOSPHATE * O,O-DIETHYL-S-(p-CHLOROPHENYLTHIOMETHYL) PHOSPHORODITHIOATE * O,O-DIETHYL-DITHIOPHOS-PHORIC ACID, p-CHLOROPHENYLTHIOMETHYL ESTER * O,O-DIETIL-S-((4-CLORO-FENIL-TIO)-METILE)-DITIO-FOSFATO (ITALIAN) * DITHIOPHOSPHATE de O,O-DI-ETHYLE et de (4-CHLORO-PHENYL) THIOMETHYLE (FRENCH) * ENDYL * ENT 23,708 * GARRA-THION * LETHOX * NEPHOCARP * OLEO-AKARITHION * R-1303 * STAUFFER R-1,303 * TRITHION MITICIDE

CONSENSUS REPORTS: EPA Farm Worker Field Reentry. EPA Extremely Hazardous Substances List.

SAFETY PROFILE: Poison by ingestion, skin contact, and intraperitoneal routes. Moderately toxic by subcutaneous route. A cholinesterase inhibitor. When heated to decomposition it emits very toxic fumes of SO_x, PO_x, and Cl^-.

TNP500   CAS: 1330-78-5   **HR: 3**
TRITOLYL PHOSPHATE
DOT: UN 2574
mf: $C_{21}H_{21}O_4P$ mw: 368.39

SYNS: CELLUFLEX 179C * CRESYL PHOSPHATE * DISFLAMOLL TKP * DURAD * FLEXOL PLASTI-CIZER TCP * FYRQUEL 150 * IMOL S 140 * KRONITEX * LINDOL * NCI-C61041 * PHOSPHATE de TRICRESYLE (FRENCH) * PHOS-PHORIC ACID, TRITOLYL ESTER * TRICRESILFOSFATI (ITALIAN) * TRICRESYLFOSFATEN (DUTCH) * TRICRESYL PHOSPHATE * TRICRESYLPHOS-PHATE, with more than 3% ortho isomer (DOT) * TRI-KRESYLPHOSPHATE (GERMAN) * TRIS(TOLYLOXY) PHOSPHINE OXIDE

CONSENSUS REPORTS: Reported in EPA TSCA Inventory.

DOT Classification: Poison B; Label: Poison.

SAFETY PROFILE: Poison by ingestion. Moderately toxic by skin contact. Human systemic effects by ingestion: flaccid paralysis without anesthesia, motor activity changes and muscle weakness. Experimental reproductive effects. An eye and skin irritant. When heated to decomposition it emits toxic fumes of PO_x.

TNT500   CAS: 22089-22-1   **HR: 3**
TROPHOSPHAMIDE
mf: $C_9H_{18}Cl_3N_2O_2P$ mw: 323.61

SYNS: A-4828 * ASTA Z 4828 * 2-(BIS(2-CHLORO-ETHYL)AMINO)-3-(2-CHLOROETHYL)TETRAHYDRO-2H-

1,3,2-OXAPHOSPHORINE-2-OXIDE * 3-(2-CHLORO-ETHYL)-2-(BIS(2-CHLOROETHYL)AMINO)PERHYDRO-2H-1,3,2-OXAZAPHOSPHORINE-2-OXIDE * CYCLOPHOS-PHAMIDE-N-MONOCHLOROETHYL derivative * IXOTEN * NSC 109723 * TFF * N,N,N'-TRIS(2-CHLORA-ETHYL)-N',O-PROPYLEN-PHOSPHORSAUREESTER-DIAMID (GERMAN) * N,N,N'-TRIS(2-CHLOROETHYL)-N',O-PROPYLENE PHOSPHORIC ACID ESTER DIAMIDE * N,N,3-TRIS(2-CHLOROETHYL)TETRAHYDRO-2H-1,3,2-OXAPHOSPHORIN-2-AMINE-2-OXIDE * TRI-FOSFAMIDE * TRILOFOSFAMIDA * TRILOPHOSPHAMIDE * TRISFOSFAMIDE * TRISPHOSPHAMIDE * TROFOSFAMID * TROPHOSPHAMID * Z 4828

CONSENSUS REPORTS: EPA Genetic Toxicology Program.

SAFETY PROFILE: Poison by intraperitoneal, subcutaneous, and intravenous routes. Moderately toxic by ingestion. Human mutation data reported. Human systemic effects by unspecified routes: hematuria, luekopenia and thrombocytopenia. When heated to decomposition it emits very toxic fumes of Cl^-, NO_x, and PO_x.

TNU000   CAS: 132-17-2   HR: 3
TROPINE BENZOHYDRYL ETHER METHANESULFONATE
mf: $C_{21}H_{25}NO \cdot CH_4O_3S$ mw: 403.58

SYNS: BENZATROPINE METHANESULFONATE * BENZOTROPINE MESYLATE * BENZOTROPINE METHANESULFONATE * BENZTROPINE MESYLATE * BENZTROPINE METHANESULFONATE * 3-DIPHE-NYLMETHOXYTROPANE MESYLATE * 3-DIPHENYL-METHOXYTROPANE METHANESULFONATE

SAFETY PROFILE: Poison by ingestion, intravenous, subcutaneous, and intraperitoneal routes. Human systemic effects by ingestion: psychotropic effects. When heated to decomposition it emits very toxic fumes of NO_x and SO_x.

TNW500   CAS: 54-12-6   HR: 3
dl-TRYPTOPHAN
mf: $C_{11}H_{12}N_2O_2$ mw: 204.25

PROP: White crystals or crystalline powder; odorless. Sol in water, dil acids, alkalies; sltly sol in alc. Optically inactive.

CONSENSUS REPORTS: Reported in EPA TSCA Inventory. EPA Genetic Toxicology Program.

SAFETY PROFILE: Questionable carcinogen with experimental carcinogenic data. When heated to decomposition it emits toxic fumes of NO_x.

TNX000   CAS: 73-22-3   HR: 3
l-TRYPTOPHANE
mf: $C_{11}H_{12}N_2O_2$ mw: 204.25

PROP: An essential amino acid; occurs in isomeric forms. Mp: decomp 289°. The l and dl forms are: White crystals or crystalline powder; slt bitter taste; (dl) sltly sol in water; (l) Sol in water, hot alc, alkali hydroxides; insol in chloroform.

SYNS: l-α-AMINO-3-INDOLEPROPRIONIC ACID * α'-AMINO-3-INDOLEPROPRIONIC ACID * α-AMINO-INDOLE-3-PROPRIONIC ACID * 2-AMINO-3-INDOL-3-YL-PROPRIONIC ACID * EH 121 * INDOLE-3-ALANINE * 1-β-3-INDOLYL-ALANINE * NCI-C01729 * (−)-TRYPTOPHAN * l-TRYPTOPHAN (FCC) * TRYPTOPHANE

CONSENSUS REPORTS: NCI Carcinogenesis Bioassay (feed); No Evidence: mouse, rat NCITR* NCI-CG-TR-71,78. Reported in EPA TSCA Inventory.

SAFETY PROFILE: Moderately toxic by intraperitoneal route. Experimental teratogenic and reproductive effects. Questionable carcinogen with experimental tumorigenic data. Human mutation data reported. When heated to decomposition it emits toxic fumes of NO_x.

TNX275   CAS: 62450-06-0   HR: 3
TRYPTOPHAN P1
mf: $C_{13}H_{13}N_3$ mw: 211.29

SYNS: 3-AMINO-1,4-DIMETHYL-γ-CARBOLINE * 3-AMINO-1,4-DIMETHYL-5H-PYRIDO(4,3-b)INDOLE * 1,4-DIMETHYL-5H-PYRIDO(4,3-b)INDOL-3-AMINE * TRP-P-1 * dl-TRYPTOPHAN, pyrolyzate 1

CONSENSUS REPORTS: IARC Cancer Review: GROUP 2B IMEMDT 7,56,87; Animal Sufficient Evidence IMEMDT 31,247,83. EPA Genetic Toxicology Program.

SAFETY PROFILE: Suspected carcinogen with experimental carcinogenic and neoplastigenic data. Poison by ingestion. Human mutation data reported. When heated to decomposition it emits toxic fumes of NO_x.

TOA000   CAS: 57-94-3   HR: 3
TUBOCURARINE HYDROCHLORIDE
mf: $C_{38}H_{44}N_2O_6 \cdot 2Cl$ mw: 694.74

SYNS: AMERIZOL * CURARIN-HAF * DELACU-
RARINE * DEXTROTUBOCURARINE CHLORIDE
* d-7',12'-DIHYDROXY-6,6'-DIMETHOXY-2,2',2'-TRI-
METHYLTUBOCURARANIUM CHLORIDE * INTOCOS-
TRIN * d-PARACURARINE CHLORIDE * TUBADIL
* TUBARINE * TUBOCURARINE CHLORIDE
* (+)-TUBOCURARINE CHLORIDE * d-TUBOCURA-
RINE CHLORIDE * TUBOCURARINE, CHLORIDE, HY-
DROCHLORIDE, (+)- (8CI) * d-TUBOCURARINE DI-
CHLORIDE * d-TUBOCURARINE HYDROCHLORIDE
* (+)-TUBOCURARINE HYDROCHLORIDE

CONSENSUS REPORTS: EPA Genetic Toxicology Program.

SAFETY PROFILE: Poison by ingestion, intravenous, intraperitoneal, and subcutaneous
routes. Human toxicity: Large doses and overdoses may cause respiratory paralysis and hypotension. When heated to decomposition it emits
very toxic fumes of NO_x and Cl^-. Used as a
muscle relaxant.

TOA500   HR: 2
TUNG NUT MEALS

SAFETY PROFILE: Toxic by ingestion. Contact causes dermatitis. Ingestion causes nausea,
vomiting, cramps, diarrhea and tenesmus, thirst,
dizziness, lethargy and disorientation. Large
doses can cause fever, tachycardia and respiratory effects. Flammable in the form of dust when
exposed to heat or flame. Processed material
must be cooled thoroughly before storage so
as not to over dry; can react with oxidizing
materials.

TOA510   HR: 2
TUNG NUT OIL

SYN: CHINAWOOD OIL

SAFETY PROFILE: Toxic by ingestion. Contact causes dermatitis. Ingestion causes nausea,
vomiting, cramps, diarrhea and tenesmus, thirst,
dizziness, lethargy and disorientation. Large
doses can cause fever, tachycardia and respiratory effects. Combustible when exposed to heat
or flame. Can react with oxidizing materials.

TOA750   CAS: 7440-33-7   HR: 2
TUNGSTEN
af: W   aw: 183.85

PROP: A steely-gray to white, cuttable, forge-
able and spinnable metallic element. Mp: 3410°,
d: 19.3 @ 20°, bp: 5900°.

SYN: WOLFRAM

CONSENSUS REPORTS: Reported in EPA
TSCA Inventory.

OSHA PEL: TWA (insoluble compounds)
5 mg(W)/m^3; STEL 10 mg(W)/m^3; (soluble compounds) 1 mg(W)/m^3; STEL 3
mg(W)/m^3

ACGIH TLV: TWA (insoluble compounds)
5 mg(W)/m^3; STEL 10 mg(W)/m^3; (soluble compounds) 1 mg(W)/m^3; STEL 3
mg(W)/m^3

NIOSH REL: (Tungsten, Insoluble) TWA 5
mg(W)/m^3

SAFETY PROFILE: Mildly toxic by an unspecified route. A skin and eye irritant. Flammable
in the form of dust when exposed to flame.
The powdered metal may ignite on contact with
air or oxidants (e.g., bromine pentafluoride, bromine, chlorine trifluoride, potassium perchlorate, potassium dichromate, nitryl fluoride, fluorine, oxygen difluoride, iodine pentafluoride,
hydrogen sulfide, sodium peroxide, lead(IV) oxide.

TOC500   HR: 2
TUNGSTEN COMPOUNDS

OSHA PEL: TWA (insoluble compounds)
5 mg(W)/m^3; STEL 10 mg(W)/m^3; (soluble compounds) 1 mg(W)/m^3; STEL 3
mg(W)/m^3

ACGIH TLV: TWA (insoluble compounds)
5 mg(W)/m^3; STEL 10 mg(W)/m^3; (soluble compounds) 1 mg(W)/m^3; STEL 3
mg(W)/m^3

SAFETY PROFILE: Tungsten compounds are
considered somewhat more toxic than those of
molybdenum. However, industrially, this element does not constitute an important health
hazard. Exposure is related chiefly to the dust
arising from the crushing and milling of the
two chief ores of tungsten, namely, scheelite
and wolframite. The feeding of 2, 5, and 10%
of diet as tungsten metal over a period of 70
days has shown no marked effect upon the
growth of rats, as measured in terms of gain
in weight. Sodium tungstate (Na_2WO_4), the
most soluble salt, is moderately toxic by ingestion. Large overdoses cause central nervous system disturbances, diarrhea, respiratory failure
and death in experimental animals. Ammonium-
p-tungstate has been found to be much less toxic
to rats upon ingestion than either tungstic oxide
or sodium tungstate. Tungsten carbide (WC)
is chronically toxic to humans by inhalation

although the effect may be due to cobalt content. Heavy exposure to the dust or the ingestion of large amounts of the soluble compounds produces changes in body weight, behavior, blood cells, choline esterase activity and sperm in experimental animals.

TOD625   **HR: D**
TURMERIC

SYN: OLEORESIN TUMERIC

SAFETY PROFILE: Human mutation data reported. When heated to decomposition it emits acrid smoke and irritating fumes.

TOD750   CAS: 8006-64-2   **HR: 3**
TURPENTINE

DOT: UN 1299

PROP: Colorless liquid, characteristic odor. Bp: 154-170°, lel: 0.8%, flash p: 95°F (CC), d: 0.854-0.868 @ 25/25°, autoign temp: 488°F, vap d: 4.84, ULC: 40-50.

SYNS: OIL of TURPENTINE * OIL of TURPENTINE, RECTIFIED * SPIRIT of TURPENTINE * SPIRITS of TURPENTINE * TEREBENTHINE (FRENCH) * TERPENTIN OEL (GERMAN) * TURPENTINE OIL, RECTIFIER * TURPENTINE STEAM DISTILLED

CONSENSUS REPORTS: Reported in EPA TSCA Inventory.

OSHA PEL: TWA 100 ppm
ACGIH TLV: TWA 100 ppm
DFG MAK: 100 ppm (560 mg/m^3)
DOT Classification: Flammable Liquid; Label: Flammable Liquid; Combustible Liquid; Label: None; Flammable or Combustible Liquid; Label: Flammable Liquid.

SAFETY PROFILE: An experimental poison by intravenous route. Moderately toxic to humans by ingestion and possibly other routes. Mildly toxic experimentally by ingestion and inhalation. Human systemic effects by ingestion and inhalation: conjunctiva irritation, other olfactory and eye effects, hallucinations or distorted perceptions, antipsychotic, headache, pulmonary and kidney changes. A human eye irritant. Irritating to skin and mucous membranes. Can cause serious irritation of kidneys. Questionable carcinogen with experimental tumorigenic data. A common air contaminant.

A very dangerous fire hazard when exposed to heat or flame; can react vigorously with oxidizing materials. Avoid impregnation of combustibles with turpentine. Keep cool and ventilated. Spontaneous heating is possible. Moderate explosion hazard in the form of vapor when exposed to flame; can react violently with $Ca(OCl)_2$; Cl_2; CrO_3; $Cr(OCl)_2$; $SnCl_4$; hexachloromelamine; trichloromelamine. To fight fire, use foam, CO_2, dry chemical. When heated to decomposition it emits acrid smoke and irritating fumes.

TOE600   CAS: 1401-69-0   **HR: 3**
TYLOSIN
mf: $C_{45}H_{77}NO_{17}$ mw: 904.23

PROP: Crystals from water. Mp: 128-132°. Sol in water at 25°: 5 mg/mL. Sol in lower alc, esters, and ketones, in chlorinated hydrocarbons, benzene, ether.

SYNS: TYLAN * TYLON

SAFETY PROFILE: Poison by intravenous route. Moderately toxic by ingestion and intraperitoneal routes. When heated to decomposition it emits toxic fumes of NO_x.

TOE750   CAS: 11032-12-5   **HR: 3**
TYLOSIN HYDROCHLORIDE
mf: $C_{45}H_{77}NO_{17} \cdot ClH$ mw: 940.69

PROP: Crystals. Mp: 141-145°.

SAFETY PROFILE: Poison by intravenous route. When heated to decomposition it emits very toxic fumes of NO_x and HCl.

TOG300   CAS: 60-18-4   **HR: D**
l-TYROSINE
mf: $C_9H_{11}NO_3$ mw: 181.21

PROP: Colorless, silky needles or white crystalline powder. Sol in water, dil mineral acids, alkaline solutions; sltly sol in alc.

SYNS: l-β-(p-HYDROXYPHENYL)ALANINE * TYROSINE * l-p-TYROSINE * p-TYROSINE

CONSENSUS REPORTS: Reported in EPA TSCA Inventory.

SAFETY PROFILE: Experimental reproductive effects. When heated to decomposition it emits acrid smoke and irritating fumes.

U

UJA800   HR: 1
γ-UNDECALACTONE
mf: $C_{11}H_{20}O_2$ mw: 184.28

PROP: Colorless to slightly yellow liquid; peach odor. D: 0.825, refr index: 1.430, flash p: 279°F. Sol in fixed oils, propylene glycol; insol in glycerine, water @ 223°.

SYNS: ALDEHYDE C-14 PURE * FEMA No. 3091 * PEACH ALDEHYDE

SAFETY PROFILE: Combustible liquid. When heated to decomposition it emits acrid smoke and irritating fumes.

UJJ000   CAS: 112-44-7   HR: 1
1-UNDECANAL
mf: $C_{11}H_{22}O$ mw: 170.33

PROP: Colorless to sltly yellow liquid; sweet, fatty, floral odor. Mp: −4°, bp: 117° @ 18 mm, flash p: 235°F (COC), d: 0.830 @ 20/4°, refr index: 1.430, vap press: 0.04 mm @ 20°, vap d: 5.94. Sol in fixed oils, propylene glycol; glycerin, water @ 223°. Reported in lemon and mandarin oils.

SYNS: ALDEHYDE-14 * 1-DECYL ALDEHYDE * FEMA No. 3092 * HENDECANAL * HENDECAN-ALDEHYDE * UNDECANAL * n-UNDECANAL * UNDECANALDEHYDE * UNDECYL ALDEHYDE * N-UNDECYL ALDEHYDE * UNDECYLIC ALDE-HYDE

CONSENSUS REPORTS: Reported in EPA TSCA Inventory.

SAFETY PROFILE: A skin irritant. Combustible liquid when exposed to heat or flame. To fight fire, use CO_2, dry chemical. When heated to decomposition it emits acrid smoke and irritating fumes.

UJS000   CAS: 1120-21-4   HR: 2
UNDECANE
DOT: UN 2330
mf: $C_{11}H_{24}$ mw: 156.35

PROP: Colorless liquid. D: 0.7402 @ 20/4°, fp: −25.75°, bp: 195.6°, flash p: 149°F (OC), vap d: 5.4. Insol in water.

SYNS: HENDECANE * n-UNDECANE

CONSENSUS REPORTS: Reported in EPA TSCA Inventory.

DOT Classification: Flammable or Combustible Liquid; Label: Flammable Liquid.

SAFETY PROFILE: Moderately toxic by intravenous route. Flammable when exposed to heat, flame or oxidizers. To fight fire, use foam, mist, dry chemical. When heated to decomposition it emits acrid smoke and irritating fumes.

UKS000   CAS: 112-12-9   HR: 2
2-UNDECANONE
mf: $C_{11}H_{22}O$ mw: 170.33

PROP: Colorless liquid. Mp: 12°, bp: 223°, flash p: 192°F (CC), d: 0.829 @ 30°, vap d: 5.9. Insol in water.

SYNS: 2-HENDECANONE * METHYL NONYL KE-TONE * METHYL-n-NONYL KETONE * MGK DOG AND CAT REPELLENT * NONYL METHYL KETONE

CONSENSUS REPORTS: Reported in EPA TSCA Inventory.

SAFETY PROFILE: Moderately toxic by ingestion. Flammable when exposed to heat or flame; can react with oxidizing materials. To fight fire, use CO_2, dry chemical. When heated to decomposition it emits acrid smoke and irritating fumes.

ULJ000   CAS: 112-45-8   HR: 1
10-UNDECENAL
mf: $C_{11}H_{20}O$ mw: 168.31

PROP: Colorless to light yellow liquid; rose odor. D: 0.840-0.850, refr index: 1.441-1.447, flash p: 212°F. Sol in fixed oils, propylene glycol; insol in water @ 235°, glycerin.

SYNS: ALDEHYDE C-11, UNDECYLENIC * FEMA No. 3095 * HENDECENAL * 1-UNDECEN-10-AL * UNDECYLENALDEHYDE * 10-UNDECYLENEAL-DEHYDE * UNDECYLENIC ALDEHYDE

CONSENSUS REPORTS: Reported in EPA TSCA Inventory.

SAFETY PROFILE: A skin irritant. Combustible liquid. When heated to decomposition it emits acrid smoke and irritating fumes.

UNA000 CAS: 112-42-5 **HR: 2**
UNDECYL ALCOHOL
mf: $C_{11}H_{24}O$ mw: 172.35

PROP: Colorless liquid; fatty-floral odor. D: 0.820-0.840, refr index: 1.437-1.443, mp: 19°, bp: 131° @ 15 mm, flash p: 234°F. Sol in fixed oils; insol in water.

SYNS: ALCOHOL C-11 * FEMA No. 3097 * HENDECANOIC ALCOHOL * 1-HENDECANOL * HENDECYL ALCOHOL * n-HENDECYLENIC ALCO-HOL * n-UNDECANOL

CONSENSUS REPORTS: Reported in EPA TSCA Inventory.

SAFETY PROFILE: Moderately toxic by ingestion. A skin irritant. Combustible liquid. When heated to decomposition it emits acrid smoke and irritating fumes.

UNS000 CAS: 7440-61-1 **HR: 3**
URANIUM

DOT: UN 2979
af: U aw: 238.00

PROP: A heavy, silvery-white, malleable, ductile, softer-than-steel, metallic element. Mp: 1132°, bp: 3818°, d: 18.95 (ca). Radioactive material.

SYN: URANIUM METAL, PYROPHORIC (DOT)

CONSENSUS REPORTS: Reported in EPA TSCA Inventory.

OSHA PEL: (Transitional: TWA Soluble Compounds: 0.05 mg(U)/m³; Insoluble Compounds 0.25 mg(U)/m³) TWA Soluble Compounds: 0.05 mg(U)/m³; Insoluble Compounds 0.2 mg(U)/m³; STEL 0.6 mg(U)/m³
ACGIH TLV: TWA 0.2 mg(U)/m³; STEL 0.6 mg(U)/m³
DFG MAK: 0.25 mg/m³
DOT Classification: Radioactive Material; Label: Radioactive and Flammable.

SAFETY PROFILE: A highly toxic element on an acute basis. The permissible levels for soluble compounds are based on chemical toxicity, while the permissible body level for insoluble compounds is based on radiotoxicity. The high chemical toxicity of uranium and its salts is largely shown in kidney damage which may not be reversible. Acute arterial lesions may occur after acute exposures. The most soluble uranium compounds are UF_6, $UO_2(NO_3)_2$, UO_2Cl_2, UO_2F_2, and uranyl acetates, sulfates, and carbonates. Some moderately soluble compounds are UF_4, UO_2, UO_4, $(NH_4)_2U_2O_7$, UO_3, and uranyl nitrates. The rapid passage of soluble uranium compounds through the body tends to allow relatively large amounts of radiation to be absorbed. Soluble uranium compounds may be absorbed through the skin. The least soluble compounds are high-fired UO_2, U_3O_8, and uranium hydrides and carbides. The high toxicity effect of insoluble compounds is largely due to lung irradiation by inhaled particles. This material is transferred from the lungs of animals quite slowly.

A very dangerous fire hazard in the form of a solid or dust when exposed to heat or flame. It can react violently with air, Cl_2, F_2, HNO_3, NO, Se, S, water, NH_3, BrF_3, trichloroethylene, nitryl fluoride. During storage it may form a pyrophoric surface due to effects of air and moisture. Depleted uranium (the ^{238}U-by-product of the uranium enrichment process, with relatively low radioactivity) is used in armor-piercing shells, ship or aircraft ballast, and counterbalances. Uranium is also used in making colored ceramic glazes.

UOS000 CAS: 7783-81-5 **HR: 3**
URANIUM FLUORIDE (low specific activity)

DOT: UN 2978
mf: F_6U mw: 352.00

PROP: Containing 0.7% or less U-235.

SYN: URANIUM HEXAFLUORIDE, LOW SPECIFIC ACTIVITY (containing 0.7% or less U-235) (DOT)

OSHA PEL: (Transitional: TWA 0.05 mg(U)/m³) TWA Soluble Compounds: 0.05 mg(U)/m³
ACGIH TLV: TWA 0.2 mg(U)/m³; STEL 0.6 mg(U)/m³; 2.5 mg(F)/m³
DOT Classification: Radioactive Material; Label: Radioactive and Corrosive.

SAFETY PROFILE: Radioactive toxicity. A corrosive irritant to skin, eyes and mucous membranes. When heated to decomposition it emits toxic fumes of F^-.

UPA000 CAS: 13598-56-6 **HR: 3**
URANIUM(III) HYDRIDE
mf: H_3U mw: 241.06

SAFETY PROFILE: A radioactive material. The powder ignites spontaneously in air or on contact with water. Potentially explosive reaction with halocarbons.

UPS000 CAS: 541-09-3 **HR: 3**
URANIUM OXYACETATE
DOT: NA 9180
mf: $C_4H_6O_6U \cdot 2H_2O$ mw: 424.19

PROP: Mp: loses $2H_2O$ @ 110°, bp: 275° (decomp), d: 2.893 @ 15°.

SYNS: URANIUM ACETATE * URANYL ACETATE

CONSENSUS REPORTS: Reported in EPA TSCA Inventory.

OSHA PEL: TWA 0.05 mg(U)/m^3
ACGIH TLV: TWA 0.2 mg(U)/m^3
DOT Classification: Radioactive Material; Label: Radioactive.

SAFETY PROFILE: Poison by intraperitoneal route. A radioactive material.

UQA000 CAS: 13536-84-0 **HR: 3**
URANIUM OXYFLUORIDE
mf: F_2O_2U mw: 308.00

SYNS: URANIUM FLUORIDE OXIDE * URANYL FLUORIDE

CONSENSUS REPORTS: Reported in EPA TSCA Inventory.

OSHA PEL: TWA 0.05 mg(U)/m^3; 2.5 mg(F)/m^3
ACGIH TLV: TWA 0.2 mg(U); 2.5 mg(F)/m^3
NIOSH REL: TWA 2.5 mg(F)/m^3

SAFETY PROFILE: Poison by intravenous route. When heated to decomposition it emits toxic fumes of F$^-$.

UQJ000 CAS: 10026-10-5 **HR: 3**
URANIUM TETRACHLORIDE
mf: Cl_4U mw: 379.80

PROP: Cubic, dark green-gray deliquescent crystals. Mp: 590°, bp: 791°, d: 4.725 @ 25/4°. Freely sol in water (decomp); insol in hydrocarbons, ethyl, ether. Should be stored in sealed ampules.

SYN: URANIUM(IV) CHLORIDE

CONSENSUS REPORTS: Reported in EPA TSCA Inventory.

OSHA PEL: TWA 0.05 mg(U)/m^3
ACGIH TLV: TWA 0.2 mg(U)/m^3; STEL 0.6 mg(U)/m^3

SAFETY PROFILE: Poison by intraperitoneal route. When heated to decomposition it emits toxic fumes of Cl$^-$.

URS000 CAS: 13520-83-7 **HR: 3**
URANYL NITRATE HEXAHYDRATE
DOT: UN 2980
mf: $N_2O_8U \cdot 6H_2O$ mw: 502.14

PROP: Rhombic, deliquescent, yellow crystals. Mp: 60.2°, bp: 118°, decomp @ 100°, d: 2.807 @ 13°.

SYNS: BIS(NITRATO)DIOXOURANIUM HEXAHYDRATE
* DINITRATODIOXOURANIUM, HEXAHYDRATE
* URANYL NITRATE HEXAHYDRATE, solution (DOT)

OSHA PEL: TWA 0.05 mg(U)/m^3
ACGIH TLV: TWA 0.2 mg(U)/m^3; STEL 0.6 mg(U)/m^3
DOT Classification: Radioactive Material; Label: Radioactive and Corrosive.

SAFETY PROFILE: Poison by ingestion, subcutaneous, intravenous, and intraperitoneal routes. Mutation data reported. A corrosive irritant to skin, eyes, and mucous membranes. A radioactive material. When heated to decomposition it emits toxic fumes of NO$_x$.

USA000 CAS: 10102-06-4 **HR: 3**
URANYL NITRATE (solid)
DOT: UN 2981
mf: N_2O_8U mw: 394.02

SYN: BIS(NITRATO-O.O')DIOXO URANIUM (solid)

CONSENSUS REPORTS: Reported in EPA TSCA Inventory.

OSHA PEL: (Transitional: TWA0.25 mg(U)/m^3) TWA 0.2 mg(U)/m^3; STEL 0.6 mg(U)/m^3
ACGIH TLV: TWA 0.2 mg(U)/m^3; STEL 0.6 mg(U)/m^3
DOT Classification: Label: Radioactive and Oxidizer.

SAFETY PROFILE: Poison by ingestion and inhalation. Human mutation data reported. A corrosive irritant to skin, eyes, and mucous membranes. A radioactive material. A powerful explosive and oxidizer. Incompatible with cellulose. Ether solutions in sunlight may explode. When heated to decomposition it emits toxic fumes of NO$_x$.

USS000   CAS: 57-13-6   **HR: 3**
UREA

mf: CH_4N_2O mw: 60.07

PROP: White crystals. Mp: 132.7°, bp: decomp, d: (solid) 1.335. Sol in water, alc; sltly sol in ether.

SYNS: CARBAMIDE * CARBAMIDE RESIN * CARBAMIMIDIC ACID * CARBONYL DIAMIDE * CARBONYLDIAMINE * ISOUREA * NCI-C02119 * PRESPERSION, 75 UREA * PSEUDOUREA * SUPERCEL 3000 * UREAPHIL * UREOPHIL * UREVERT * VARIOFORM II

CONSENSUS REPORTS: Reported in EPA TSCA Inventory. EPA Genetic Toxicology Program.

SAFETY PROFILE: Moderately toxic by ingestion, intravenous, and subcutaneous routes. Human reproductive effects by intraplacental route: fertility effects. Experimental reproductive effects. Human mutation data reported. A human skin irritant. Questionable carcinogen with experimental carcinogenic and neoplastigenic data. Reacts with sodium hypochlorite or calcium hypochlorite to form the explosive nitrogen trichloride. Incompatible with $NaNO_2$, P_2Cl_5, nitrosyl perchlorate. Preparation of the ^{15}N-labeled urea is hazardous. When heated to decomposition it emits toxic fumes of NO_x.

UVA000   CAS: 51-79-6   **HR: 3**
URETHANE

mf: $C_3H_7NO_2$ mw: 89.11

PROP: Colorless, odorless crystals. Mp: 49°, bp: 184°, d: 0.9862, vap press: 10 mm @ 77.8°, vap d: 3.07. Very sol in water, alc, ether.

SYNS: A 11032 * AETHYLCARBAMAT (GERMAN) * AETHYLURETHAN (GERMAN) * CARBAMIC ACID, ETHYL ESTER * CARBAMIDSAEURE-AETHYLESTER (GERMAN) * ESTANE 5703 * ETHYL CARBAMATE * ETHYLURETHAN * ETHYL URETHANE * o-ETHYLURETHANE * LEUCETHANE * LEUCOTHANE * NSC 746 * PRACARBAMIN * PRACARBAMINE * RCRA WASTE NUMBER U238 * U-COMPOUND * URETAN ETYLOWY (POLISH) * URETHAN

CONSENSUS REPORTS: IARC Cancer Review: GROUP 2B IMEMDT 7,56,87; Animal Sufficient Evidence IMEMDT 7,111,74. NTP Fourth Annual Report On Carcinogens, 1984. Community Right-To-Know List. Reported in EPA TSCA Inventory. EPA Genetic Toxicology Program.

DFG MAK: Animal Carcinogen, Suspected Human Carcinogen.

SAFETY PROFILE: Confirmed carcinogen with experimental carcinogenic, neoplastigenic, and tumorigenic data. A transplacental carcinogen. Moderately toxic by ingestion, intraperitoneal, subcutaneous, intramuscular, parenteral, intravenous, and possibly other routes. Experimental teratogenic and reproductive effects. Human mutation data reported. Causes depression of bone marrow and occasionally focal degeneration in the brain. Can also produce central nervous system depression, nausea and vomiting. Has been found in over 1000 beverages sold in the United States. The most heavily contaminated liquors are bourbons, sherries, and fruit brandies (some had 1,000 to 12,000 ppb urethane). Many whiskeys, table and dessert wines, brandies and liqueurs contain potentially hazardous amounts of urethane. The allowable limit for urethane in alcoholic beverages is 125 ppb. It is formed as a side product during processing.

Hot aqueous acids or alkalies decompose urethane to ethanol, carbon dioxide and ammonia. Reacts with phosphorus pentachloride to form an explosive product. When heated it emits toxic fumes of NO_x. Used as an intermediate in the manufacture of pharmaceuticals, pesticides, and fungicides.

V

VAD000   CAS: 54965-21-8   **HR: 2**
VALBAZEN
mf: $C_{12}H_{15}N_3O_2S$ mw: 265.36

PROP: Colorless crystals. Mp: 208-210°.

SYNS: ALBENDAZOLE (USDA) * METHYL 5-(PRO-PYLTHIO)-2-BENZIMIDAZOLECARBAMATE * ((PRO-PYLTHIO)-5-1H-BENZIMIDAZOLYL-2) CARBAMATE de METHYLE (FRENCH) * (5-(PROPYLTHIO)-1H-BENZIMI-DAZOL-2-YL)CARBAMIC ACID METHYL ESTER * 5-(PROPYLTHIO)-2-CARBOMETHOXYAMINOBENZI-MIDAZOLE * SKF 62979 * ZENTAL

SAFETY PROFILE: Moderately toxic by ingestion. Experimental teratogenic and reproductive effects. When heated to decomposition it emits toxic fumes of SO_x and NO_x.

VAG000   CAS: 110-62-3   **HR: 2**
n-VALERALDEHYDE

DOT: UN 2058
mf: $C_5H_{10}O$ mw: 86.15

PROP: Liquid. Flash p: 53.6°F, bp: 102-103°, d: 0.8095 @ 20/4°. Very sltly sol in water; misc with organic solvents.

SYNS: AMYL ALDEHYDE * BUTYL FORMAL * PENTANAL * n-PENTANAL * VALERAL * VALERIANIC ALDEHYDE * VALERIC ACID AL-DEHYDE * VALERIC ALDEHYDE * VALERYLAL-DEHYDE

CONSENSUS REPORTS: Reported in EPA TSCA Inventory.

OSHA PEL: TWA 50 ppm
ACGIH TLV: TWA 50 ppm
DOT Classification: Flammable Liquid; Label: Flammable Liquid.

SAFETY PROFILE: Moderately toxic by ingestion. Mildly toxic by inhalation and skin contact. A severe eye and skin irritant. A very dangerous fire hazard when exposed to heat or flame. When heated to decomposition it emits acrid smoke and irritating fumes.

VAQ000   CAS: 109-52-4   **HR: 2**
VALERIC ACID

DOT: NA 1760
mf: $C_5H_{10}O_2$ mw: 102.15

PROP: Colorless, mobile liquid; penetrating, rancid odor. D: 0.940 @ 20/4°, refr index:

1.405-1.14 @ 25°, mp: −34.5°, bp: 186.4°, flash p: 203°F. Sol in water; misc in alc, ether.

SYNS: BUTANECARBOXYLIC ACID * 1-BUTANE-CARBOXYLIC ACID * FEMA No. 3101 * PENTA-NOIC ACID * n-PENTANOIC ACID * PROPYL-ACETIC ACID * VALERIANIC ACID * n-VALERIC ACID

CONSENSUS REPORTS: Reported in EPA TSCA Inventory.

DOT Classification: Corrosive Material; Label: Corrosive.

SAFETY PROFILE: Moderately toxic by ingestion, intravenous, and subcutaneous routes. Mildly toxic by inhalation. A corrosive irritant to skin, eyes, and mucous membranes. Combustible liquid. When heated to decomposition it emits acrid smoke and irritating fumes. Used in perfumes.

VAV000   CAS: 108-29-2   **HR: 2**
4-VALEROLACTONE
mf: $C_5H_8O_2$ mw: 100.13

PROP: Colorless, mobile liquid; sweet, herbaceous odor. Mp: −31°, bp: 205-206.5°, flash p: 205°F (COC), d: 1.047-1.054, refr index: 1.43, vap d: 3.45. Misc in alc, fixed oils, water.

SYNS: FEMA No. 3103 * 4-HYDROXYPENTANOIC ACID LACTONE * 4-HYDROXYVALERIC ACID LACTONE * γ-METHYL-γ-BUTYROLACTONE * 4-METHYL-γ-BUTYROLACTONE * γ-PENTAL-ACTONE * 4-PENTANOLIDE * γ-VALEROL-ACTONE (FCC)

CONSENSUS REPORTS: Reported in EPA TSCA Inventory.

SAFETY PROFILE: Moderately toxic by ingestion. A skin irritant. Combustible liquid when exposed to heat or flame; can react with oxidizing materials. To fight fire, use water, foam, CO_2, dry chemical. When heated to decomposition it emits acrid smoke and irritating fumes.

VBA000   CAS: 638-29-9   **HR: 2**
VALERYL CHLORIDE

DOT: UN 2502
mf: C_5H_9ClO mw: 120.59

CONSENSUS REPORTS: Reported in EPA TSCA Inventory.

DOT Classification: Corrosive Material; Label: Corrosive.

SAFETY PROFILE: A corrosive irritant to skin, eyes and mucous membranes. When heated to decomposition it emits toxic fumes of Cl^-.

VBK000 CAS: 90-22-2 HR: 3
VALETHAMATE BROMIDE
mf: $C_{19}H_{32}NO_2 \cdot Br$ mw: 386.43

PROP: Crystals from ethanol and ether or acetone. Mp. 100-101°. Freely sol in water and alc; practically insol in ether.

SYNS: 2-DIETHYLAMINOETHYL 3-METHYL-2-PHENYL-VALERATE METHYLBROMIDE * 2-DIETHYLAMINO-ETHYL 2-PHENYL-3-METHYLVALERATEMETHYL BROMIDE * DIETHYL(2-HYDROXYETHYL) METHYLAMMONIUMBROMIDE ESTER * DIETHYL(2-HYDROXYETHYL)METHYL-AMMONIUM BROMIDE 3-METHYL-2-PHENYLVALERATE * DIETHYL(2-HY-DROXYETHYL)METHYLAMMONIUM 3-METHYL-2-PHENYLVALERATE BROMIDE * 3-METHYL-2-PHENYL-VALERIC ACID 2-DIETHYLAMINOETHYL ESTER METHYL BROMIDE * 3-METHYL-2-PHENYLVALERIC ACID * PHENYLMETHYLVALERIANSAEURE-β-DIAETHYL-AMINOAETHYLESTER-BROMMET HYLAT (GERMAN) * VALETHAMATE

SAFETY PROFILE: Poison by ingestion, subcutaneous, and intravenous routes. When heated to decomposition it emits very toxic fumes of NO_x, NH_3 and Br^-.

VBP000 CAS: 72-18-4 HR: 1
VALINE
mf: $C_5H_{11}NO_2$ mw: 117.17

PROP: White, crystalline solid; characteristic taste. Mp (dl): 298° (decomp), mp (l): 315°, d (l): 1.230. Sol in water; very sltly sol in alc; insol in ether. An essential amino acid.

SYNS: l-(+)-α-AMINOISOVALERIC ACID * l-VALINE (FCC)

CONSENSUS REPORTS: Reported in EPA TSCA Inventory.

SAFETY PROFILE: Mutation data reported. When heated to decomposition it emits toxic fumes of NO_x.

VCP000 CAS: 7440-62-2 HR: 3
VANADIUM
af: V   aw: 50.94

PROP: A bright, white, soft, ductile metal; sltly radioactive. Bp: 3000°, d: 6.11 @ 18.7°, mp: 1917°. Insol in water.

CONSENSUS REPORTS: Reported in EPA TSCA Inventory.

OSHA PEL: (Transitional: Respirable Dust: Cl 0.5 $mg(V_2O_5)/m^3$; Fume: Cl 0.1 $mg(V_2O_5)/m^3$) Respirable Dust and Fume: TWA 0.05 $mg(V_2O_5)/m^3$
NIOSH REL: TWA 1.0 $mg(V)/m^3$

SAFETY PROFILE: Poison by subcutaneous route. Questionable carcinogen with experimental tumorigenic data. Flammable in dust form from heat, flame or sparks. Violent reaction with BrF_3, Cl_2, lithium, nitryl fluoride, oxidants. When heated to decomposition it emits toxic fumes of VO_x.

VCZ000 HR: D
VANADIUM COMPOUNDS
NIOSH REL: (Vanadium Compounds) CL 0.05 $mg(V)/m^3/15M$

SAFETY PROFILE: Variable toxicity. Vanadium compounds act chiefly as an irritant to the conjunctivae and respiratory tract. Acute and chronic exposure can give rise to conjunctivitis, rhinitis, reversible irritation of the respiratory tract, and to bronchitis, bronchospasms, and asthma-like diseases in more severe cases. There is still some controversy as to the effects of industrial exposure on other systems of the body. Responses are mostly acute, seldom chronic. The first report of human vanadium poisoning described rather widespread systemic effects, consisting of polycythemia, followed by red blood cell destruction and anemia, loss of appetite, pallor and emaciation, albuminuria and hematuria, gastrointestinal disorders, nervous complaints and cough, sometimes severe enough to cause hemoptysis. More recent reports describe symptoms which, for the most part, are restricted to the conjunctivae and respiratory system, no evidence being found of disturbances of the gastrointestinal tract, kidneys, blood or central nervous system. Vanadate (VO_3^-) is a potent inhibitor of the sodium pump, an enzyme universally present in eukaryotic organisms. The absorption of V_2O_5 by inhalation is nearly 100%. Though certain workers believe that it is only the pentoxide which is harmful, other investigators have found that patronite dust (chiefly vanadium sulfide) is quite toxic to animals, causing acute pulmonary edema. Acute poisoning in animals by ingestion of vanadium compounds causes nervous disturbances, paralysis of legs, respiratory failure, convulsions, bloody diarrhea, and death. Poisoning by inhalation causes bleeding of the nose and acute bronchitis. Some compounds have reported mutation effects. VF_5 and the oxyhalogenides of pentavalent vanadium (VOF_3, $VOCl_3$, $VOBr_3$) are volatile. Vanadium compounds are common air contaminants. The fumes are highly toxic. The major use of vanadium and its alloys is in the steel industry. When heated to decomposition it emits toxic fumes of VO_x.

VDP000 CAS: 7727-18-6 **HR: 3**
VANADIUM OXYTRICHLORIDE

DOT: UN 2443
mf: Cl_3OV mw: 173.29

PROP: Yellow, deliquescent liquid. Mp: $-77°$ $\pm 2°$, bp: 126.7°, d: 1.811 @ 32°.

SYNS: TRICHLOROOXOVANADIUM * VANADIUM TRICHLORIDE OXIDE * VANADYL TRICHLORIDE

CONSENSUS REPORTS: Reported in EPA TSCA Inventory.

NIOSH REL: (Vanadium Compounds) CL 0.05 mg(V)/m³/15M
DOT Classification: Corrosive Material; Label: Corrosive.

SAFETY PROFILE: Poison by ingestion. A corrosive irritant to skin, eyes, and mucous membranes. Explosive reaction with sodium. Violently hygroscopic. Violent reaction with rubidium (at 60°C), potassium. When heated to decomposition it emits toxic fumes of VO_x and Cl^-.

VDU000 CAS: 1314-62-1 **HR: 3**
VANADIUM PENTOXIDE (dust)

DOT: UN 2862
mf: O_5V_2 mw: 181.88

PROP: Yellow to red, crystalline powder. Mp: 690°, bp: decomp @ 1750°, d: 3.357 @ 18°.

SYNS: ANHYDRIDE VANADIQUE (FRENCH) * C.I. 77938 * RCRA WASTE NUMBER P120 * VANADIC ANHYDRIDE * VANADIO, PENTOSSIDO di (ITALIAN) * VANADIUM DUST and FUME (ACGIH) * VANADIUM(V) OXIDE * VANADIUM PENTAOXIDE * VANADIUMPENTOXID (GERMAN) * VANADIUM PENTOXIDE, non-fused form (DOT) * VANADIUM, PENTOXYDE de (FRENCH) * VANADIUMPENTOXYDE (DUTCH) * WANADU PIECIOTLENEK (POLISH)

CONSENSUS REPORTS: Reported in EPA TSCA Inventory. EPA Genetic Toxicology Program.

OSHA PEL: (Transitional: Respirable Dust: Cl 0.5 mg(V_2O_5)/m³; Fume: Cl 0.1 mg(V_2O_5)/m³) Respirable Dust and Fume: TWA 0.05 mg(V_2O_5)/m³
ACGIH TLV: TWA 0.05 mg(V_2O_5)/m³
DFG MAK: (fine dust) 0.05 mg/m³
NIOSH REL: (Vanadium Compounds) CL 0.05 mg(V)/m³/15M
DOT Classification: ORM-E; Label: None; Poison B; Label: Poison.

SAFETY PROFILE: Poison by ingestion, inhalation, intraperitoneal, subcutaneous, intratracheal and intravenous routes. Human systemic effects by inhalation: bronchiolar constriction, including asthma, cough, dyspnea, sputum, and conjunctiva irritation. Experimental teratogenic and reproductive effects. Mutation data reported. A respiratory irritant, causes skin pallor, greenish-black tongue, chest pain, cough, dyspnea, palpitation, lung changes. When ingested it causes gastrointestinal tract disturbances. May also cause a papular skin rash. Mixtures with calcium + sulfur + water may ignite spontaneously. The absorption of V_2O_5 by inhalation is nearly 100%. Incompatible with ClF_3, Li, peroxyformic acid. When heated to decomposition it emits acrid smoke and irritating fumes of VO_x.

VDZ000 CAS: 1314-62-1 **HR: 3**
VANADIUM PENTOXIDE (fume)
mf: O_5V_2 mw: 181.88

SYN: VANADIUM DUST and FUME (ACGIH)

CONSENSUS REPORTS: EPA Extremely Hazardous Substances List. Reported in EPA TSCA Inventory. EPA Genetic Toxicology Program.

OSHA PEL: (Transitional: Fume: Cl 0.1 mg(V_2O_5)/m³) Respirable Dust and Fume: TWA 0.05 mg(V_2O_5)/m³
ACGIH TLV: TWA 0.05 mg(V_2O_5)/m³
NIOSH REL: (Vanadium Compound) CL 0.05 mg(V)/m³/15M

SAFETY PROFILE: A poison by several routes. Can react violently with (Ca + S + H_2O), ClF_3, Li. When heated to decomposition it emits toxic fumes of VO_x.

VEA000 CAS: 1314-34-7 **HR: 3**
VANADIUM SESQUIOXIDE
DOT: UN 2860
mf: O_3V_2 mw: 149.88

PROP: Black crystals. Mp: 1970°, d: 4.87 @ 18°.

SYNS: VANADIC OXIDE * VANADIUM OXIDE * VANADIUM TRIOXIDE

CONSENSUS REPORTS: Reported in EPA TSCA Inventory.

NIOSH REL: (Vanadium Compound) CL 0.05 mg(V)/m^3/15M

SAFETY PROFILE: Poison by ingestion, subcutaneous, and intratracheal routes. Ignites when heated in air. When heated to decomposition it emits toxic fumes of VO$_x$.

VEF000 CAS: 7632-51-1 **HR: 3**
VANADIUM TETRACHLORIDE
DOT: UN 2444
mf: Cl_4V mw: 192.74

PROP: Reddish-brown liquid. Mp: $-28 \pm 2°$, bp: 148.5°, d: 1.816 @ 30°.

SYN: VANADIUM CHLORIDE

CONSENSUS REPORTS: Reported in EPA TSCA Inventory.

NIOSH REL: (Vanadium Compounds) CL 0.05 mg(V)/m^3/15M
DOT Classification: Corrosive Material; Label: Corrosive.

SAFETY PROFILE: Poison by ingestion. A corrosive irritant to skin, eyes, and mucous membranes. When heated to decomposition it emits toxic fumes of VO$_x$ and Cl$^-$.

VEP000 CAS: 7718-98-1 **HR: 3**
VANADIUM TRICHLORIDE
DOT: UN 2475
mf: Cl_3V mw: 157.29

PROP: Pink crystals. Mp: decomp, d: 3.00 @ 18°.

SYN: VANADIUM(III) CHLORIDE

CONSENSUS REPORTS: Reported in EPA TSCA Inventory.

NIOSH REL: (Vanadium Compounds) CL 0.05 mg(V)/m^3/15M
DOT Classification: Corrosive Material; Label: Corrosive.

SAFETY PROFILE: Poison by ingestion and subcutaneous routes. A corrosive irritant to skin, eyes, and mucous membranes. Extremely violent reaction with methyl magnesium iodide and other Grignard reagents. When heated to decomposition it emits toxic fumes of VO$_x$ and Cl$^-$.

VEZ000 CAS: 27774-13-6 **HR: 3**
VANADYL SULFATE
DOT: UN 2931/NA 9152
mf: O_5SV mw: 163.00

PROP: Blue crystals.

SYNS: C.I. 77940 * OXYSULFATOVANADIUM

CONSENSUS REPORTS: Reported in EPA TSCA Inventory.

NIOSH REL: (Vanadium Compounds) CL 0.05 mg(V)/m^3/15M
DOT Classification: ORM-E; Label: None; Poison B; Label: Poison.

SAFETY PROFILE: Poison by intravenous, intraperitoneal, and subcutaneous routes. Mutation data reported. When heated to decomposition it emits toxic fumes of VO$_x$ and SO$_x$.

VFK000 CAS: 121-33-5 **HR: 2**
VANILLIN
mf: $C_8H_8O_3$ mw: 152.16

PROP: White, crystalline needles; vanilla odor. D: 1.056, bp: 285°, mp: 80-81°. Sol in 125 parts water, 20 parts glycerin, 2 parts 95% alc, chloroform, ether.

SYNS: FEMA No. 3107 * 4-HYDROXY-m-ANISALDEHYDE * 4-HYDROXY-3-METHOXYBENZALDEHYDE * LIOXIN * 3-METHOXY-4-HYDROXYBENZALDEHYDE * METHYLPROTOCATECHUALDEHYDE * VANILLA * VANILLALDEHYDE * VANILLIC ALDEHYDE * p-VANILLIN * ZIMCO

CONSENSUS REPORTS: Reported in EPA TSCA Inventory.

SAFETY PROFILE: Moderately toxic by ingestion, intraperitoneal, subcutaneous, and intravenous routes. Experimental reproductive effects. Human mutation data reported. Can react violently with Br$_2$, HClO$_4$, potassium-tert-butoxide, tert-chlorobenzene + NaOH, formic acid + thallium nitrate. When heated to decomposition it emits acrid smoke and irritating fumes.

VGP000   CAS: 51-43-4   **HR: 3**
VASOTONIN
mf: $C_9H_{13}NO_3$ mw: 183.23

SYNS: ADNEPHRINE * ADRENAL * 1-ADRENA-
LIN * ADRENALIN-MEDIHALER * ADRENAMINE
* ADRENAN * ADRENAPAX * ADRENASOL
* ADRENATRATE * ADRENODIS * ADREN-
OHORMA * ADRENUTOL * ADRINE * ASMA-
TANE MIST * ASTHMA METER MIST * ASTMAHA-
LIN * BALMADREN * BERNARENIN * BIO-
RENINE * BOSMIN * BREVIRENIN * BRONK-
AID MIST * CHELAFRIN * CORISOL * 3,4-DI-
HYDROXY-α-((METHYLAMINO)METHYL)BENZYL
ALCOHOL * 1-1-(3,4-DIHYDROXYPHENYL)-2-METHYL-
AMINOETHANOL * DRENAMIST * DYLEPHRIN
* DYSPNE-INHAL * EPIFRIN * EPINEPHRAN
* EPINEPHRINE * (−)-EPINEPHRINE * (R)-EPI-
NEPHRINE * 1-EPINEPHRINE * 1-EPINEPHRINE
(synthetic) * EPIRENAMINE * EPIRENAN
* EPITRATE * ESPHYGMOGENINA * EXADRIN
* GLYCIRENAN * HAEMOSTASIN * HEKTALIN
* HEMISINE * HEMOSTASIN * (R)-4-(1-HY-
DROXY-2-(METHYLAMINO)ETHYL)-1,2-BENZENEDIOL
(9CI) * HYPERNEPHRIN * HYPORENIN
* INTRANEFRIN * KIDOLINE * LEVORENIN
* LYOPHRIN * MEDIHALER-EPI * METANEPHRIN
* METHYLARTERENOL * MUCIDRINA * MYO-
STHENINE * MYTRATE * NEPHRIDINE * NIER-
ALINE * PARANEPHRIN * PRIMATENE MIST
* RCRA WASTE NUMBER P042 * RENAGLADIN
* RENALEPTINE * RENALINA * RENOFORM
* RENOSTYPRICIN * RENOSTYPTIN * SCUREN-
ALINE * SINDRENINA * SOLADREN * SPHYG-
MOGENIN * STRYPTIRENAL * SUPRACAPSULIN
* SUPRADIN * SUPRANEPHRANE * SUPRA-
NEPHRINE * SUPRANOL * SUPRARENIN
* SUPREL * SURENINE * SUSPHRINE
* SYMPATHIN I * TAKAMINA * TOKAMINA
* TONOGEN * VAPONEFRIN * VASOCONSTRIC-
TINE * VASOCONSTRICTOR * VASODRINE
* VASOTON

CONSENSUS REPORTS: Reported in EPA
TSCA Inventory. EPA Genetic Toxicology Program.

SAFETY PROFILE: Human poison by subcutaneous route. Experimental poison by ingestion,
skin contact, subcutaneous, intraperitoneal, intravenous, and intramuscular routes. Experimental teratogenic and reproductive effects.
Mutation data reported. When heated to
decomposition it emits toxic fumes of NO_x.
Used as an adrenergic, sympathomimetic, vasoconstrictor, bronchodilator, and cardiac stimulant.

VHU000   CAS: 71-62-5   **HR: 3**
VERATRIDINE
mf: $C_{36}H_{51}NO_{11}$ mw: 673.88

PROP: Yellow-white powder. Mp: 180°. Sol
in water; sltly sol in ether.

SYNS: 4,9-EPOXYCEVANE-3,4,12,14,16,17,20-HEPTOL
3-(3,4-DIMETHOXYBENZOATE) * VERATRINE (AMOR-
PHOUS) * 3-VERATROYLVERACEVINE

SAFETY PROFILE: Poison by intraperitoneal,
subcutaneous, and intravenous routes. Combustible when exposed to heat or flame. When
heated to decomposition it emits toxic fumes
of NO_x.

VHZ000   CAS: 8051-02-3   **HR: 3**
VERATRINE

PROP: A powder from the plant *Schoenocaulon
officinale*. A botanical insecticide. The active
ingredients are a group of alkaloids known as
veratrin, i.e., cevadine and veratridine.

SYNS: ASAGRAEA OFFICINALIS * CAUSTIC BAR-
LEY * CEVADILLA * CEVADINE * ENT 123
* SABACIDE * SABADILLA * SABANE DUST
* VERATRIDINE * VERATRIN (GERMAN)

SAFETY PROFILE: Human poison by ingestion. Experimental poison by ingestion, intraperitoneal, subcutaneous, and possibly other
routes. Experimental teratogenic and reproductive effects. Ingestion causes severe gastrointestinal tract disturbances, burning in the mouth,
vomiting, diarrhea, and cramps. Also produces
headache, dizziness, slow pulse and weakness.
Large doses cause death by circulatory and respiratory failure. It is a powerful irritant to skin
and mucous membranes. Less toxic than rotenone. Inhalation causes violent sneezing. When
heated to decomposition it emits toxic fumes
of NO_x. Used to kill lice.

VIZ000   CAS: 65072-04-0   **HR: 3**
VERILOID

SYNS: ALKALOIDS, VERATRUM * ALKAVERVIR
* AMERICAN HELLEBORE * AMERICAN VERATRUM
* GREEN HELLEBORE * INDIAN POKE * VERA-
TRUM VIRIDE * VERATRUM VIRIDE ALKALOIDS EX-
TRACT * VERTAVIS

SAFETY PROFILE: Poison by ingestion, intravenous, subcutaneous, and intraperitoneal routes.

VKP000 CAS: 125-44-0 **HR: 3**
VINBARBITAL SODIUM
mf: $C_{11}H_{15}N_2O_3 \cdot Na$ mw: 246.27

SYNS: DELVINAL SODIUM * 5-ETHYL-5-(1-METHYL-1-BUTENYL)BARBITURIC ACID SODIUM SALT * 5-ETHYL-5-(1-METHYL-1-BUTENYL)-2,4,6(1H,3H,5H)-PYRIMIDINETRIONE SODIUM SALT * SODIUM DELVINAL * SODIUM-5-ETHYL-5-(1-METHYL-1-BUTENYL) BARBITURATE * SODIUM VINBARBITAL

SAFETY PROFILE: Poison by ingestion and intraperitoneal routes. Used as a sedative. When heated to decomposition it emits toxic fumes of Na_2O and NO_x.

VKZ000 CAS: 865-21-4 **HR: 3**
VINCALEUKOBLASTINE
mf: $C_{46}H_{58}N_4O_9$ mw: 811.08

SYNS: NCI-C04842 * NDC 002-1452-01 * NINCALUICOLFLASTINE * NSC 47842 * VINBLASTIN * VINBLASTINE * VINCALEUCOBLASTIN * VINCOBLASTINE * VLB

CONSENSUS REPORTS: NCI Carcinogenesis Studies (ipr); No Evidence: mouse CANCAR 40,1935,77; (ipr); Clear Evidence: rat CANCAR 40,1935,77. EPA Genetic Toxicology Program.

SAFETY PROFILE: Human poison by intravenous route. Experimental poison by intravenous, subcutaneous, and intraperitoneal routes. Human systemic effects by intravenous, ocular, and possibly other routes: visual field changes, conjunctiva irritation and other eye effects, cardiomyopathy including infarction, and changes in bone marrow. Experimental teratogenic and reproductive effects. Questionable carcinogen with experimental tumorigenic data. Human mutation data reported. When heated to decomposition it emits toxic fumes of NO_x. Used as an antineoplastic agent.

VLA000 CAS: 143-67-9 **HR: 3**
VINCALEUKOBLASTINE SULFATE (1:1) (SALT)
mf: $C_{46}H_{58}N_4O_9 \cdot H_2O_4S$ mw: 909.16

SYNS: EXAL * 29060 LE * NSC 49842 * VELBAN * VELBE * VINBLASTINE SULFATE * VINCALEUKOBLASTINE SULFATE * VLB MONOSULFATE

CONSENSUS REPORTS: IARC Cancer Review: GROUP 3 IMEMDT 7,371,87; Animal Inadequate Evidence IMEMDT 26,349,81; Human Inadequate Evidence IMEMDT 26,349,81. EPA Genetic Toxicology Program.

SAFETY PROFILE: Poison by intraperitoneal and intravenous routes. Human systemic effects by intravenous route: blood luekopenia and hair changes. Experimental teratogenic and reproductive effects. Questionable carcinogen. Human mutation data reported. When heated to decomposition it emits very toxic fumes of NO_x and SO_x.

VLU200 CAS: 83768-87-0 **HR: 3**
VINTHIONINE
mf: $C_6H_{11}NO_2S$ mw: 161.24

SYNS: S-ETHENYL-dl-HOMOCYSTEINE * S-VINYL-dl-HOMOCYSTEINE

SAFETY PROFILE: Suspected carcinogen with experimental carcinogenic data. Mutation data reported. When heated to decomposition it emits toxic fumes of SO_x and NO_x.

VLU250 CAS: 108-05-4 **HR: 3**
VINYL ACETATE
DOT: UN 1301
mf: $C_4H_6O_2$ mw: 86.10

PROP: Colorless, mobile liquid; polymerizes to solid on exposure to light. Mp: $-92.8°$, bp: 73°, flash p: 18°F, d: 0.9335 @ 20°, autoign temp: 800°F, vap press: 100 mm @ 21.5°, lel: 2.6%, uel: 13.4%, vap d: 3.0. Misc in alc, ether. Somewhat sol in water.

SYNS: ACETIC ACID ETHENYL ESTER * ACETIC ACID VINYL ESTER * 1-ACETOXYETHYLENE * ETHENYL ACETATE * OCTAN WINYLU (POLISH) * VAC * VINILE (ACETATO di) (ITALIAN) * VINYL A MONOMER * VINYLACETAT (GERMAN) * VINYLACETAAT (DUTCH) * VINYLE (ACETATE de) (FRENCH) * VYAC * ZESET T

CONSENSUS REPORTS: IARC Cancer Review: GROUP 3 IMEMDT 7,56,87; Animal Inadequate Evidence IMEMDT 19,341,79; IMEMDT 39,113,86; Human Inadequate Evidence IMEMDT 39,113,86. Reported in EPA TSCA Inventory. Community Right-To-Know List. EPA Extremely Hazardous Substances List.

OSHA PEL: TWA 10 ppm; STEL 20 ppm
ACGIH TLV: TWA 10 ppm; STEL 20 ppm
DFG MAK: 10 ppm (35 mg/m^3)
NIOSH REL: (Vinyl Acetate) CL 15 mg/m^3/
15M
DOT Classification: Label: Flammable Liquid.

SAFETY PROFILE: Moderately toxic by ingestion, inhalation, and intraperitoneal routes. A skin and eye irritant. Questionable carcinogen with experimental carcinogenic data. Human mutation data reported. Highly dangerous fire hazard when exposed to heat, flame, or oxidizers. A storage hazard, it may undergo spontaneous exothermic polymerization. Reaction with air or water to form peroxides which catalyze an exothermic polymerization reaction has caused several large industrial explosions. Reaction with hydrogen peroxide forms the explosive peracetic acid. Reacts with oxygen above 50°C to form an unstable explosive peroxide. Reacts with ozone to form the explosive vinyl acetate ozonide. Solution polymerization of the acetate dissolved in toluene has resulted in large industrial explosions. Polymerization reaction with dibenzoyl peroxide + ethyl acetate may release ignitable and explosive vapors. The vapor may react vigorously with dessicants (e.g., silica gel or alumina). Incompatible (explosive) with 2-amino ethanol, chlorosulfonic acid, ethylenediamine, ethyleneimine, HCl, HF, HNO$_3$, oleum, peroxides, H$_2$SO$_4$.

VMP000   CAS: 593-60-2   **HR: 3**
VINYL BROMIDE

DOT: UN 1085
mf: C$_2$H$_3$Br   mw: 106.96

PROP: A gas. Mp: −138°, bp: 15.6°, d: 1.51. Insol in water; misc in alc, ether.

SYNS: BROMOETHENE   *   BROMOETHYLENE
*   BROMURE de VINYLE (FRENCH)   *   VINILE (BROMURO di) (ITALIAN)   *   VINYLBROMID (GERMAN)
*   VINYL BROMIDE, inhibited (DOT)   *   VINYLE (BROMURE de) (FRENCH)

CONSENSUS REPORTS: IARC Cancer Review: GROUP 2A IMEMDT 7,56,87; Animal Sufficient Evidence IMEMDT 39,133,86; Animal Inadequate Evidence IMEMDT 19,367,79. Community Right-To-Know List. Reported in EPA TSCA Inventory. EPA Genetic Toxicology Program.

OSHA PEL: TWA 5 ppm
ACGIH TLV: TWA 5 ppm; Suspected Human Carcinogen.
DFG MAK: Human Carcinogen.
NIOSH REL: (Vinyl Bromide) Lowest Detectable Level
DOT Classification: Flammable Gas; Label: Flammable Gas.

SAFETY PROFILE: Confirmed carcinogen with experimental carcinogenic, neoplastigenic, and tumorigenic data. Moderately toxic by ingestion. Mutation data reported. A very dangerous fire hazard when exposed to heat or flame. Can react violently with oxidizing materials. May polymerize in sunlight. To fight fire, use CO$_2$, dry chemical or water spray. When heated to decomposition it emits toxic fumes of Br$^-$.

VMZ000   CAS: 111-34-2   **HR: 3**
VINYL BUTYL ETHER

DOT: UN 2352
mf: C$_6$H$_{12}$O   mw: 100.18

PROP: Liquid. Mp: −112.7°, bp: 94.2°, flash p: −9°, d: 0.7803 @ 20°/20°, vap d: 3.45.

SYNS: BUTOXYETHENE   *   BUTYL VINYL ETHER
*   BUTYL VINYL ETHER (inhibited)   *   1-(ETHENYLOXY) BUTANE   *   VINYL-n-BUTYL ETHER

CONSENSUS REPORTS: Reported in EPA TSCA Inventory.

DOT Classification: Flammable Liquid; Label: Flammable Liquid.

SAFETY PROFILE: Mildly toxic by ingestion, skin contact, and inhalation. A skin and eye irritant. A very dangerous fire hazard when exposed to heat or flame. To fight fire, use foam, CO$_2$, dry chemical, alcohol foam. Moderately explosive by spontaneous chemical reaction. Can react with oxidizing materials. When heated to decomposition it emits acrid smoke and irritating fumes.

VNF000   CAS: 123-20-6   **HR: 3**
VINYL BUTYRATE

DOT: UN 2838
mf: C$_6$H$_{10}$O$_2$ mw: 114.16

PROP: D: 0.9, vap d: 4.0, bp: 116°, flash p: 68°F (OC), lel: 1.4%, uel: 8.8%.

SYNS: BUTYRIC ACID, VINYL ESTER   *   VINYL BUTYRATE, INHIBITED (DOT)

DOT Classification: Flammable Liquid; Label: Flammable Liquid.

SAFETY PROFILE: Mildly toxic by inhalation and ingestion. A skin and eye irritant. A very dangerous fire hazard when exposed to heat, flame, or oxidizers. Explosive in the form of vapor when exposed to heat or flame. To fight fire, use alcohol foam, fog, mist, CO_2. When heated to decomposition it emits acrid smoke and irritating fumes.

VNK000 CAS: 15805-73-9 **HR: 3**
VINYL CARBAMATE
mf: $C_3H_5NO_2$ mw: 87.09

SYN: CARBAMIC ACID, VINYL ESTER

SAFETY PROFILE: Poison by intraperitoneal route. Questionable carcinogen with experimental neoplastigenic data. Human mutation data reported. When heated to decomposition it emits toxic fumes of NO_x.

VNP000 CAS: 75-01-4 **HR: 3**
VINYL CHLORIDE
DOT: UN 1086
mf: C_2H_3Cl mw: 62.50

PROP: Colorless liquid or gas (when inhibited); faintly sweet odor. Mp: $-160°$; bp: $-13.9°$, lel: 4%, uel: 22%; flash p: 17.6°F (COC), fp: $-159.7°$, d (liquid): 0.9195 @ 15/4°, vap press: 2600 mm @ 25°, vap d: 2.15, autoign temp: 882°F. Sltly sol in water; sol in alc; very sol in ether.

SYNS: CHLORETHENE * CHLORETHYLENE * CHLOROETHENE * CHLOROETHYLENE * CHLORURE de VINYLE (FRENCH) * CLORURO di VINILE (ITALIAN) * ETHYLENE MONOCHLORIDE * MONOCHLOROETHENE * MONOCHLOROETHYL-ENE (DOT) * RCRA WASTE NUMBER U043 * TROVIDUR * VC * VCM * VINILE (CLO-RURO di) (ITALIAN) * VINYLCHLORID (GERMAN) * VINYL CHLORIDE MONOMER * VINYL C MONO-MER * VINYLE(CHLORURE de) (FRENCH) * WI-NYLU CHLOREK (POLISH)

CONSENSUS REPORTS: IARC Cancer Review: GROUP 1 IMEMDT 7,373,87; Animal Sufficient Evidence IMEMDT 19,377,79; IMEMDT 7,291,74; Human Limited Evidence IMEMDT 7,291,74; Human Sufficient Evidence IMEMDT 19,377,79. NTP Fourth Annual Report On Carcinogens, 1984. Community Right-To-Know List. Reported in EPA TSCA Inventory. EPA Genetic Toxicology Program.

OSHA PEL: TWA 1 ppm; CL 5 ppm/15M; Cancer Suspect Agent
ACGIH TLV: TWA 5 ppm; Human Carcinogen.
DFG TRK: Existing installations: 3 ppm, Human Carcinogen; Others: 2 ppm.
NIOSH REL: (Vinyl Chloride) Lowest Detectable Level
DOT Classification: Flammable Gas; Label: Flammable Gas.

SAFETY PROFILE: Confirmed human carcinogen producing liver and blood tumors. Poison by inhalation. Moderately toxic by ingestion. Human reproductive effects by inhalation: changes in spermatogenesis. Experimental teratogenic data. Human mutation data reported. A severe irritant to skin, eyes, and mucous membranes. Causes skin burns by rapid evaporation and consequent freezing. In high concentration it acts as an anesthetic. Chronic exposure has shown liver injury. Circulatory and bone changes in the fingertips have been reported in workers handling unpolymerized materials.

A very dangerous fire hazard when exposed to heat, flame, or oxidizers. Large fires of this material are practically inextinguishable. A severe explosion hazard in the form of vapor when exposed to heat or flame. Long-term exposure to air may result in formation of peroxides which can initiate explosive polymerization of the chloride. Can react vigorously with oxidizing materials. Can explode on contact with oxides of nitrogen. Obtain instructions for its use from the supplier storing or handling this material. To fight fire, stop flow of gas. When heated to decomposition it emits highly toxic fumes of Cl^-.

VOA000 CAS: 106-87-6 **HR: 3**
VINYL CYCLOHEXENE DIOXIDE
mf: $C_8H_{12}O_2$ mw: 140.20

PROP: Colorless liquid. D: 1.098 @ 20/20°, bp: 227°, flash p: 230°F.

SYNS: CHISSONOX 206 * EP-206 * 1,2-EPOXY-4-(EPOXYETHYL)CYCLOHEXANE * 1-EPOXYETHYL-3,4-EPOXYCYCLOHEXANE * 3-(EPOXYETHYL)-7-OXABICYCLO(4.1.0)HEPTANE * 3-(1,2-EPOXYETHYL)-7-OXABICYCLO(4.1.0)HEPTANE * 4-(1,2-EPOXY-ETHYL)-7-OXABICYCLO(4.1.0)HEPTANE * 4-(EPOXY-ETHYL)-7-OXABICYCLO(4.1.0)HEPTANE * ERLA-2270 * ERLA-2271 * 1-ETHYLENEOXY-3,4-EPOXYCYCLO-

HEXANE * NCI-C60139 * 3-OXIRANYL-7-OXA-BICYCLO(4.1.0)HEPTENE * UCET TEXTILE FINISH 11-74 (OBS.) * UNOX EPOXIDE 206 * VINYL CYCLO-HEXENE DIEPOXIDE * 4-VINYLCYCLOHEXENE DI-EPOXIDE * 4-VINYL-1-CYCLOHEXENE DIEPOXIDE * 4-VINYL-1,2-CYCLOHEXENE DIEPOXIDE * 1-VI-NYL-3-CYCLOHEXENE DIOXIDE * 4-VINLYCYCLO-HEXENE DIOXIDE * 4-VINYL-1-CYCLOHEXENE DIOX-IDE (MAK)

CONSENSUS REPORTS: IARC Cancer Review: GROUP 3 IMEMDT 7,56,87; Animal Sufficient Evidence IMEMDT 11,141,76. Reported in EPA TSCA Inventory.

OSHA PEL: TWA 10 ppm (skin)
ACGIH TLV: TWA 10 ppm (skin); Suspected Human Carcinogen.
DFG MAK: Animal Carcinogen, Suspected Human Carcinogen.

SAFETY PROFILE: Confirmed carcinogen with experimental carcinogenic and tumorigenic data. Poison by unspecified route. Moderately toxic by ingestion and skin contact. Mildly toxic by inhalation. Mutation data reported. A severe skin irritant. Combustible when exposed to heat or flame. To fight fire, use water, foam, dry chemical. When heated to decomposition it emits acrid smoke and irritating fumes.

VOP000   CAS: 109-93-3   **HR: 3**
VINYL ETHER

DOT: UN 1167
mf: C_4H_6O mw: 70.10

PROP: Colorless liquid; very volatile. Bp: 29°, ULC: 100, lel: 1.7%, uel: 27%, flash p: $< -22°F$ (CC), d: 0.774 @ 20/20°, autoign temp: 680°F, vap d: 2.41. Very sltly sol in water; misc in alc, ether.

SYNS: DIVINYL ETHER (DOT) * DIVINYL ETHER, INHIBITED (DOT) * DIVYNYL OXIDE * ETHENYL-OXYETHENE * 1,1'-OXYBISETHENE * VINES-THENE * VINESTHESIN * VINETHEN * VINE-THENE * VINETHER * VINIDYL * VINYDAN

DOT Classification: Flammable Liquid; Label: Flammable Liquid.

SAFETY PROFILE: Mildly toxic by inhalation. Mutation data reported. Prolonged exposure causes liver injury. A very dangerous fire hazard when exposed to heat or flame; can react vigorously with oxidizing materials. A severe explosion hazard in the form of vapor when exposed to heat or flame. Forms peroxides when exposed

to air or oxygen. Hypergolic reaction with concentrated nitric acid. To fight fire, use CO_2, dry chemical. When heated to decomposition it emits acrid smoke and irritating fumes. Used as an inhalation anesthestic.

VPA000   CAS: 75-02-5   **HR: 3**
VINYL FLUORIDE

DOT: UN 1860
mf: $CH_2:CHF$ mw: 46

PROP: Colorless gas. Mp: $-160.5°$, bp: $-72°$, lel: 2.6%, uel: 21.7%. Insol in water; sol in alc, ether.

SYNS: FLUOROETHENE * FLUOROETHYLENE * MONOFLUOROETHYLENE

CONSENSUS REPORTS: Reported in EPA TSCA Inventory.

NIOSH REL: (Vinyl Chloride) TWA 1 ppm; CL 5 ppm/15M
DOT Classification: Label: Flammable Gas.

SAFETY PROFILE: A poison. Mutation data reported. A very dangerous fire hazard. To fight fire, stop flow of gas. When heated to decomposition it emits toxic fumes of F^-.

VPK000   CAS: 75-35-4   **HR: 3**
VINYLIDENE CHLORIDE

DOT: UN 1303
mf: $C_2H_2Cl_2$ mw: 96.94

PROP: Colorless, volatile liquid. Bp: 31.6°, lel: 7.3%, uel: 16.0%, fp: $-122°$, flash p: 0°F (OC), d: 1.213 @ 20°/4°, autoign temp: 1058°F.

SYNS: CHLORURE de VINYLIDENE (FRENCH) * 1-1-DCE * 1,1-DICHLOROETHENE * 1,1-DI-CHLOROETHYLENE * NCI-C54262 * RCRA WASTE NUMBER U078 * SCONATEX * VDC * VINYLI-DENE CHLORIDE (II) * VINYLIDENE DICHLORIDE * VINYLIDINE CHLORIDE

CONSENSUS REPORTS: IARC Cancer Review: GROUP 3 IMEMDT 7,376,87; Human Inadequate Evidence IMEMDT 39,195,86, IMEMDT 19,439,79; Animal Limited Evidence IMEMDT 39,195,86; Animal Sufficient Evidence IMEMDT 19,439,79. EPA Genetic Toxicology Program. Reported in EPA TSCA Inventory. Community Right-To-Know List.

OSHA PEL: TWA 1 ppm
ACGIH TLV: TWA 5 ppm; STEL 20 ppm

DFG MAK: Suspected Carcinogen.
DOT Classification: Flammable Liquid; Label:
Flammable Liquid

SAFETY PROFILE: Suspected carcinogen with experimental carcinogenic, neoplastigenic, and tumorigenic data. Poison by inhalation, ingestion, and intravenous routes. Moderately toxic by subcutaneous route. Human systemic effects by inhalation: general anesthesia, liver and kidney changes. Experimental teratogenic and reproductive effects. Mutation data reported. A very dangerous fire hazard when exposed to heat or flame. Moderately explosive in the form of gas when exposed to heat or flame. It forms explosive peroxides upon exposure to air. Potentially explosive reaction with chlorotrifluoroethylene at 180°C. Reaction with ozone forms dangerous products. Explosive reaction with perchloryl fluoride when heated above 100°C. Also can explode spontaneously. Reacts violently with chlorosulfonic acid, HNO_3, oleum. Can react vigorously with oxidizing materials. To fight fire, use alcohol foam, CO_2, dry chemical. When heated to decomposition it emits toxic fumes of Cl^-.

VPP000 CAS: 75-38-7 **HR: 3**
VINYLIDENE FLUORIDE
DOT: UN 1959
mf: $C_2H_2F_2$ mw: 64.04

PROP: Colorless gas. Bp: $< -70°$, lel: 5.5%, uel: 21.3%.

SYNS: 1,1-DIFLUOROETHYLENE (DOT, MAK)
* HALOCARBON 1132A * NCI-C60208 * VDF

CONSENSUS REPORTS: IARC Cancer Review: GROUP 3 IMEMDT 7,56,87; Animal Inadequate Evidence IMEMDT 39,227,86. Reported in EPA TSCA Inventory.

DFG MAK: Suspected Carcinogen.
DOT Classification: Flammable Gas; Label:
Flammable Gas.

SAFETY PROFILE: Suspected carcinogen with experimental neoplastigenic data. Mildly toxic by inhalation. Mutation data reported. A very dangerous fire hazard when exposed to heat, flames, or oxidizers. Explosive in the form of vapor when exposed to heat or flame. Violent reaction with hydrogen chloride when heated under pressure. To fight fire, stop flow of gas. When heated to decomposition it emits toxic fumes of F^-.

VQK000 CAS: 105-38-4 **HR: 2**
VINYL PROPIONATE
mf: $C_5H_8O_2$ mw: 100.13

PROP: Liquid. D: 0.9173 @ 20/20°, bp: 95°, fp: $-81.1°$, flash p: 34°F (OC), vap d: 3.3. Almost insol in water.

SYN: PROPANOIC ACID, ETHENYL ESTER

CONSENSUS REPORTS: Reported in EPA TSCA Inventory.

SAFETY PROFILE: Mildly toxic by ingestion and inhalation. A skin and eye irritant. A very dangerous fire hazard when exposed to heat or flame. To fight fire, use alcohol foam, mist, fog. When heated to decomposition it emits acrid smoke and irritating fumes.

VQK650 CAS: 25013-15-4 **HR: 2**
VINYL TOLUENE
mf: C_9H_{10} mw: 118.19

SYNS: METHYL STYRENE * NCI-C56406
* VINYLTOLUENE * VINYL TOLUENES (mixed isomers), inhibited (DOT)

CONSENSUS REPORTS: Reported in EPA TSCA Inventory.

OSHA PEL: TWA 100 ppm
ACGIH TLV: TWA 50 ppm; STEL 100 ppm
DOT Classification: Flammable or Combustible Liquid; Label: Flammable Liquid.

SAFETY PROFILE: Moderately toxic by ingestion and inhalation. Human systemic effects by inhalation: eye and olfactory effects. Experimental teratogenic and reproductive effects. Mutation data reported. A skin and eye irritant. Flammable when exposed to heat or flame; can react vigorously with oxidizing materials. When heated to decomposition it emits acrid smoke and irritating fumes.

VRF000 CAS: 11006-76-1 **HR: 2**
VIRGINIAMYCIN

PROP: White powder. Decomp @ 138-140°. Sltly sol in water and dil acid; sol in methanol, ethanol, acetone, benzene; almost insol in ligroin.

SYNS: ANTIBIOTIC NO. 899 * ESKALIN V
* MIKAMYCIN * OSTREOGRYCIN * PATRICIN
* PRISTINAMYCIN * PYOSTACINE * RP7293
* SKF 7988 * STAFAC * STAPHYLOMYCIN
* STAPYOCINE * STREPTOGRAMIN * VERNA-
MYCIN * VIRGIMYCIN

SAFETY PROFILE: Moderately toxic by ingestion, intraperitoneal, and subcutaneous routes.
Used as an antibiotic.

VSK600   CAS: 68-26-8   **HR: 3**
VITAMIN A
mf: $C_{20}H_{30}O$ mw: 286.50

PROP: Light yellow to red oil; mild fishy odor.
Very sol in chloroform, ether; sol in abs alc,
vegetable oil; insol in glycerin, water.

SYNS: ACON * AFAXIN * AGIOLAN * AL-
PHALIN * ALPHASTEROL * ANATOLA * ANTI-
INFECTIVE VITAMIN * ANTIXEROPHTHALMIC VITA-
MIN * AORAL * APEXOL * AQUASYNTH
* AVIBON * AVITA * AVITOL * BIOSTEROL
* CHOCOLA A * 3,7-DIMETHYL-9-(2,6,6-TRI-
METHYL-1-CYCLOHEXEN-1-YL)-2,4,6,8-NONATETRAEN-
1-OL * DISATABS TABS * DOFSOL * EPITELIOL
* HI-A-VITA * LARD FACTOR * MYVPACK
* OLEOVITAMIN A * OPHTHALAMIN * PREPA-
LIN * RETINOL * all-trans RETINOL * RETROVI-
TAMIN A * TESTAVOL * VAFLOL * VI-ALPHA
* VITAMIN A1 * VITAMIN A1 ALCOHOL
* all-trans-VITAMIN A ALCOHOL * VITAVEL-A
* VITPEX * VOGAN * VOGAN-NEU

CONSENSUS REPORTS: Reported in EPA
TSCA Inventory. EPA Genetic Toxicology Program.

SAFETY PROFILE: Moderately toxic by ingestion. Human teratogenic effects by ingestion:
developmental abnormalities of the craniofacial
area and urogenital system. Experimental teratogenic and reproductive effects. Human mutation
data reported. When heated to decomposition
it emits acrid smoke and irritating fumes.

VSK900   CAS: 127-47-9   **HR: 3**
VITAMIN A ACETATE
mf: $C_{22}H_{32}O_2$ mw: 328.54

SYNS: CRYSTALETS * MYVAK * MYVAX
* RETINOL ACETATE * RETINYL ACETATE
* all-trans-RETINYL ACETATE * trans-VITAMIN A
ACETATE * VITAMIN A ALCOHOL ACETATE

CONSENSUS REPORTS: Reported in EPA
TSCA Inventory.

SAFETY PROFILE: Moderately toxic by ingestion. Experimental teratogenic and reproductive
effects. Questionable carcinogen with experimental neoplastigenic data. Mutation data reported. When heated to decomposition it emits
acrid smoke and irritating fumes.

VSP000   CAS: 79-81-2   **HR: 1**
VITAMIN A PALMITATE
mf: $C_{36}H_{60}O_2$ mw: 524.96

SYNS: AQUASOL * AROVIT * RETINOL PALMI-
TATE * RETINYL PALMITATE

CONSENSUS REPORTS: Reported in EPA
TSCA Inventory. EPA Genetic Toxicology Program.

SAFETY PROFILE: Mildly toxic by ingestion.
Experimental teratogenic and reproductive effects. Human mutation data reported. When
heated to decomposition it emits acrid smoke
and irritating fumes.

VSU000   CAS: 65-22-5   **HR: 3**
VITAMIN B₆ HYDROCHLORIDE
mf: $C_8H_9NO_3 \cdot ClH$ mw: 203.64

SYNS: 3-HYDROXY-5-(HYDROXYMETHYL)-2-METH-
YLISONICOTINALDEHYDE, HYDROCHLORIDE
* 2-METHYL-3-HYDROXY-4-FORMYL-5-HYDROXY-
METHYLPYRIDINE HYDROCHLORIDE * PYRIDOXAL
HYDROCHLORIDE

CONSENSUS REPORTS: Reported in EPA
TSCA Inventory.

SAFETY PROFILE: Poison by intramuscular,
intravenous, and intraperitoneal routes. Moderately toxic by ingestion and subcutaneous
routes. When heated to decomposition it emits
very toxic fumes of NO_x and HCl.

VSZ000   CAS: 68-19-9   **HR: 3**
VITAMIN B₁₂ COMPLEX
mf: $C_{63}H_{88}CoN_{14}O_{14}P$ mw: 1355.55

PROP: The anti-pernicious anemia vitamin. All
vitamin B₁₂ compounds contain the cobalt atom
in its trivalent state. There are at least three
active forms: cyanocobalamin, hydroxycobala-
min, and nitrocobalamin. Dark red crystals or
crystalline powder. Very hygroscopic; sltly sol
in water; sol in alc; insol in acetone, chloroform,
ether.

SYNS: ANACOBIN * B-12 * BERUBIGEN
* BETALIN 12 CRYSTALLINE * BEVATINE-12
* BEVIDOX * BYLADOCE * CABADON M
* COBADOCE FORTE * COBALIN * COBAMIN
* COBIONE * COTEL * COVIT * CRYSTAMIN
* CRYSTWEL * CYANO-B12 * CYANOCOBALA-
MIN * CYCOLAMIN * CYKOBEMINET * CYRE-
DIN * CYTACON * CYTAMEN * CYTOBION
* DEPINAR * DIMETHYLBENZIMIDAZOLYCOBA-
MIDE * 5,6-DIMETHYLBENZIMIDAZOLYCOBAMIDE

CYANIDE * DISTIVIT (B12 PEPTIDE) * DOBETIN * DOCEMINE * DOCIBIN * DOCIGRAM * DODECABEE * DODECAVITE * DODEX * DUCOBEE * DUODECIBIN * EMBIOL * EMOCICLINA * ERITRONE * ERYCYTOL * ERYTHROTIN * EUHAEMON * EXTRINSIC FACTOR * FACTOR II (VITAMIN) * FRESMIN * HEMO-B-DOZE * HEMOMIN * HEPAGON * HEPAVIS * HEPCOVITE * LACTOBACILLUS LACTIS DORNER FACTOR * LLD FACTOR * MAC-RABIN * MEGABION * MEGALOVEL * MILBE-DOCE * NAGRAVON * NORMOCYTIN * PER-NAEMON * PERNAEVIT * PERNIPURON * PLECYAMIN * POYAMIN * REBRAMIN * REDAMINA * REDISOL * RHODACRYST * RUBESOL * RUBRAMIN * RUBRIPCA * RUBROCITOL * SYTOBEX * VIBALT * VIBISONE * VIRUBRA * VITAMIN B12 (FCC) * VITARUBIN * VITA-RUBRA * VITRAL * VI-TWEL

CONSENSUS REPORTS: Cobalt and its compounds are on the Community Right-To-Know List. Reported in EPA TSCA Inventory. EPA Genetic Toxicology Program.

SAFETY PROFILE: Poison by subcutaneous route. Moderately toxic by intraperitoneal route. Experimental teratogenic and reproductive effects. When heated to decomposition it emits very toxic fumes of PO_x and NO_x.

VSZ100   CAS: 50-14-6   HR: 3
VITAMIN D2
mf: $C_{28}H_{44}O$ mw: 396.72

PROP: White crystals; odorless. Mp: 115-118°. Insol in water; sol in alc, chloroform, ether, and fatty oils.

SYNS: d-ARTHIN * CALCIFEROL * CALCIFERON 2 * CONDACAPS * CONDOCAPS * CONDOL * CRTRON * CRYSTALLINA * DARAL * DAVITAMON D * DAVITIN * DECAPS * DEE-OSTEROL * DEE-RON * DEE-RONAL * DEE-ROUAL * DELTALIN * DERATOL * DETALUP * DIACTOL * DIVIT URTO * DORAL * DRISDOL * ERGOCALCIFEROL * ERGORONE * ERGOSTEROL, ACTIVATED * ERGOSTEROL, IRRADIATED * ERTRON * FORTODYL * GELTABS * HI-DERATOL * INFRON * IRRADIATED ERGOSTA-5,7,22-TRIEN-3-β-OL * METADEE * MULSIFEROL * MYKOSTIN * OLEOVITAMIN D * OSTELIN * RADIOSTOL * RADSTERIN * 9,10,SECOERGOSTA-5,7,10(19),22-TETRAEN-3-β-OL * SHOCK-FEROL * STEROGYL * VIGANTOL * VIOSTEROL * VITAVEL-D

CONSENSUS REPORTS: EPA Extremely Hazardous Substances List.

SAFETY PROFILE: Poison by ingestion, intraperitoneal, intravenous, and intramuscular routes. Human systemic effects by ingestion: anorexia, nausea or vomiting, and weight loss. Experimental teratogenic and reproductive effects. When heated to decomposition it emits acrid smoke and irritating fumes.

VSZ450   CAS: 59-02-9   HR: D
VITAMIN E
mf: $C_{29}H_{50}O_2$ mw: 430.79

PROP: dl-Form: Sltly viscous, pale yellow oil; d-form: red liquid; odorless. Natural α-tocopherol has been crystallized. Mp: 2.5-3.5°, d: (25°/4°) 0.950, bp: (0.1) 200-220°. Practically insol in water; freely sol in oils, fats, acetone, alc, chloroform, ether, other fat solvents. Gradually darkens on exposure to light.

SYNS: ALMEFROL * ANTISTERILITY VITAMIN * COVI-OX * DENAMONE * EMIPHEROL * ENDO E * EPHYNAL * EPROLIN * EPSILAN * ESORB * ETAMICAN * ETAVIT * EVION * EVITAMINUM * ILITIA * PHYTOGERMINE * PROFECUNDIN * SPAVIT * SYNTOPHEROL * d-α-TOCOPHEROL (FCC) * dl-α-TOCOPHEROL (FCC) * (R,R,R)-α-TOCOPHEROL * α-TOCOPHEROL * (2R,4'R,8'R)-α-TOCOPHEROL * TOKOPHARM * 5,7,8-TRIMETHYLTOCOL * VASCUALS * VERROL * VITAPLEX E * VITAYONON * VITEOLIN

CONSENSUS REPORTS: Reported in EPA TSCA Inventory.

SAFETY PROFILE: Experimental reproductive effects. Mutation data reported. When heated to decomposition it emits acrid smoke and irritating fumes.

VTF000   CAS: 595-33-5   HR: 3
VOLIDAN
mf: $C_{24}H_{32}O_4$ mw: 384.56

SYNS: 17-α-ACETOXY-6-DEHYDRO-6-METHYLPRO-GESTERONE * 17-ACETOXY-6-METHYLPREGNA-4,6-DIENE-3,20-DIONE * 17-α-ACETOXY-6-METHYL-PREGNA-4,6-DIENE-3,20-DIONE * 17-α-ACETOXY-6-METHYL-4,6-PREGNADIENE-3,20-DIONE * BDH 1298 * 6-DEHYDRO-6-METHYL-17-α-ACETOXYPROGESTER-ONE * DMAP * 17-HYDROXY-6-METHYLPREGNA-4,6-DIENE-3,20-DIONE ACETATE * MEGACE * MEGESTROL ACETATE * MEGESTRYL ACETATE

* 6-METHYL-17-α-ACETOXYPREGNA-4,6-DIENE-3,20-
DIONE * 6-METHYL-6-DEHYDRO-17-α-ACETOXYPRO-
GESTERONE * 6-METHYL-6-DEHYDRO-17-α-ACETYL-
PROGESTERONE * 6-METHYL-17-α-HYDROXY-Δ^6-PRO-
GESTERONE ACETATE * 6-METHYL-$\Delta^{4,6}$-PREG-
NADIEN-17-α-OL-3,20-DIONE ACETATE * NSC-71423
* OVABAN * SC10363

CONSENSUS REPORTS: IARC Cancer Review: Animal Limited Evidence IMEMDT
21,431,79.

SAFETY PROFILE: Suspected carcinogen with
experimental carcinogenic data. Poison by intravenous route. Human reproductive effects by
ingestion and implant routes: effects on ovaries
and fallopian tubes, menstrual cycle changes
and female fertility index changes. Mutation
data reported. Experimental teratogenic and reproductive effects. When heated to decomposition it emits acrid smoke and irritating fumes.
An FDA proprietary drug used to treat endome-
triosis and breast cancer. A steroid.

W

WAK000 CAS: 91-84-9 **HR: 3**
**WAIT'S GREEN MOUNTAIN
ANTIHISTAMINE**
mf: $C_{17}H_{23}N_3O$ mw: 285.43

SYNS: AFKO-HIST * ANHISTABS * ANHISTOL
* ANTALERGAN * ANTAMINE * ANTHISAN
* COPSAMINE * CORADON * N-DIMETHYL-
AMINO-AETHYL-N-p-METHOXY-BENZYL-α-AMINO-PYRI-
DIN-MALEAT (GERMAN) * 2-((2-(DIMETHYLAMINO)
ETHYL)-(p-METHOXYBENZYL)AMINO)PYRIDINE
* DIPANE * DORANTAMIN * ENRUMAY
* HARVAMINE * HISTACAP * HISTALON
* HISTAN * HISTAPYRAN * HISTASAN
* ISAMIN * KRIPTIN * MARANHIST * ME-
PYRAMIN (GERMAN) * MEPYREN * MINIHIST
* N-p-METHOXYBENZYL-N',N'-DIMETHYL-N-α-PYRI-
DYLETHYLENEDIAMINE * N-(p-METHOXYBENZYL)-
N',N'-DIMETHYL-N-2-PYRIDYLETHYLENEDIAMINE
* NCI-C60651 * NEOANTERGAN * NEOBRIDAL
* NYSCAPS * PARAMINYL * PARMAL
* PYMAFED * PYRA * PYRAMAL * PYRANI-
SAMINE * PYRILAMINE * R.D. 2786 * RP 2786
* STAMINE * STANGEN * STATOMIN
* THYLOGEN

SAFETY PROFILE: Human poison by an unspecified route. An experimental poison by ingestion, intraperitoneal, subcutaneous, and intravenous routes. Human systemic effects by
unspecified route: sleep effects, somnolence and
muscle contraction or spasticity. Experimental
teratogenic and reproductive effects. An eye irritant. When heated to decomposition it emits
toxic fumes of NO_x. Used as an antihistamine.

WAT000 CAS: 481-39-0 **HR: 3**
WALNUT EXTRACT
mf: $C_{10}H_6O_3$ mw: 174.16

SYNS: C.I. 75500 * C.I. NATURAL BROWN 7
* 5-HYDROXY-1,4-NAPHTHALENEDIONE * 5-HY-
DROXY-1,4-NAPHTHOQUINONE

SAFETY PROFILE: Poison by ingestion. Questionable carcinogen with experimental neoplastigenic data. Mutation data reported. When
heated to decomposition it emits acrid smoke
and irritating fumes.

WAT200 CAS: 81-81-2 **HR: 3**
WARFARIN
mf: $C_{19}H_{16}O_4$ mw: 308.35

PROP: Colorless, odorless, tasteless crystals.
Mp: 161°. Sol in acetone, dioxane; sltly sol in
methanol, ethanol; very sol in alkaline aqusol,
insol in water, benzene.

SYNS: 3-(α-ACETONYLBENZYL)-4-HYDROXYCOUMA-
RIN * ARAB RAT DETH * ATHROMBINE-K
* BRUMIN * COMPOUND 42 * d-CON * CO-
RAX * COUMADIN * COUMAFENE * DETH-
MORE * EASTERN STATES DUOCIDE * 4-HY-
DROXY-3-(3-OXO-1-FENYL-BUTYL) CUMARINE (DUTCH)
* 4-HYDROXY-3-(3-OXO-1-PHENYL-BUTYL)-CUMARIN
(GERMAN) * 4-IDROSSI-3-(3-OXO-)-FENIL-BUTIL)-CU-
MARINE (ITALIAN) * KUMADER * LIQUA-TOX
* MOUSE PAK * 3-(α-PHENYL-β-ACETYLETHYL)-4-
HYDROXYCOUMARIN * 3-(1'-PHENYL-2'-ACETYL-
ETHYL)-4-HYDROXYCOUMARIN * (PHENYL-1
ACETYL-2 ETHYL)-3-HYDROXY-4 COUMARINE (FRENCH)
* PROTHROMADIN * RAT-A-WAY * RAT-B-GON
* RAT-GARD * RAT & MICE BAIT * RATS-NO-
MORE * RCRA WASTE NUMBER P001 * RO-DETH
* ROUGH & READY MOUSE MIX * SOLFARIN
* SPRAY-TROL BRANCH RODEN-TROL * TWIN
LIGHT RAT AWAY * WARFARINE (FRENCH)
* ZOOCOUMARIN (RUSSIAN)

CONSENSUS REPORTS: Reported in EPA
TSCA Inventory. EPA Extremely Hazardous
Substances List.

OSHA PEL: TWA 0.1 mg/m^3
ACGIH TLV: TWA 0.1 mg/m^3
DFG MAK: 0.5 mg/m^3

SAFETY PROFILE: A human poison by ingestion. Poison by ingestion, inhalation, and intravenous routes. Moderately toxic by skin contact, subcutaneous, and intraperitoneal routes.
Human systemic effects by ingestion: hemorrhage. Human reproductive effects by ingestion,
intramuscular, and possibly other routes: fetal
death and physical abnormalities at birth. Human teratogenic effects include developmental
abnormalities of the craniofacial area, musculoskeletal system, and respiratory system. An
experimental teratogen. Used as an oral anticoagulant and as a rodenticide. When heated
to decomposition it emits acrid smoke and
fumes.

WBJ000 **HR: 3**
WELDING FUMES
ACGIH TLV: TWA 5 mg/m^3

SAFETY PROFILE: When welding is done on a surface coated with cadmium, toxic fumes of cadmium are evolved. When zinc-coated surfaces are welded, toxic quantities of zinc oxide may be liberated. When painted surfaces are welded, lead or other pigment fumes may be liberated. And when fluoride fluxes are used in welding, very toxic fluoride fumes are evolved. When oily surfaces are welded, offensive and toxic fumes can be liberated, and when the welding torch is improperly ignited, carbon monoxide, which is very toxic, may be evolved. Also, NO_x is formed. It is therefore considered hazardous to inhale excessive amounts of welding fumes. It is also possible to inhale sufficient quantities of iron oxide from welding to cause siderosis. Metal fume fever is a common reaction. It is characterized by chills, fever, sweating, and leukocytosis coming on several hours after exposure. Recovery is usually complete in 24-48 hours and there are no significant after effects. Safety goggles are required to protect against spatter. Light-filtering goggles are required to shield the eyes against the intense UV light from the arc.

WBS000   **HR: 2**
WHISKEY

PROP: Light yellow-amber liquid. Pleasant to fruity odor. D: 0.923-0.935 @ 15.56°; 47%-53% of ethanol, by volume, flash p: 80.0°F (CC). Made by distillation of fermented malted grains, i.e., corn, rye, or barley. After distillation, whiskey is aged in wooden containers for up to several years. The aging extracts such components as acids and esters from the wood and promotes oxidation of components of raw whiskey and some reactions between organic components to form new flavors.

THR: The carcinogen urethane is sometimes found in whiskey. The whiskey or wine equivalent of 1 ounce of pure ethanol per capita per day is often cited as healthful to adults to relieve stress and promote relaxation. However, it is often abused which can lead to habituation with consequent liver damage, malnutrition, and a wide variety of other physical and mental problems. A fire hazard when exposed to heat or flame. To fight fire, use water, water spray, alcohol foam, CO_2, dry chemical.

WCA000   **HR: 2**
WINE

PROP: An alcoholic beverage made from the fermented juice of grapes, other fruits or plants. Contains from 7-20% ethanol by volume. Concentrations of alcohol higher than those produced naturally are obtained by fortifying with pure ethanol. The distinctive colors, tastes, bouquets of wines are usually produced by adding coloring matter, sugar, acetic acid, salts, and higher fatty acids.

SAFETY PROFILE: Some wines contain the carcinogen urethane. Moderately toxic. Some of the additives to wines have been known to cause allergic reactions in humans.

X

XCA000 CAS: 69-89-6 **HR: 3**
XANTHINE
mf: $C_5H_4N_4O_2$ mw: 152.13

PROP: Scales or plates. Decomp on heating without melting, partial sublimation. Sol in water and mineral acids; less sol in alc; very sol in NH_4OH and NaOH solns.

SYNS: 3,7-DIHYDRO-1H-PURINE-2,6-DIONE * 2,6-DIOXOPURINE * ISOXANTHINE * PSEUDOXANTHINE * PURINE-2,6-DIOL * 9H-PURINE-2,6-DIOL * 2,6(1,3)-PURINEDION * PURINE-2,6-(1H,3H)-DIONE * USAF CB-17 * XAN * XANTHIC OXIDE

CONSENSUS REPORTS: Reported in EPA TSCA Inventory. EPA Genetic Toxicology Program.

SAFETY PROFILE: Moderately toxic by intraperitoneal route. Questionable carcinogen with experimental neoplastigenic data. When heated to decomposition it emits toxic fumes of NO_x.

XCJ000 CAS: 53-46-3 **HR: 3**
XANTHINE BROMIDE
mf: $C_{21}H_{26}NO_3 \cdot Br$ mw: 420.39

SYNS: ASABAINE * AVAGAL * BANTHIN * BANTHINE * BANTHINE BROMIDE * β-DIETHYLAMINOETHYL XANTHENE-9-CARBOXYLATE METHOBROMIDE * β-DIETHYLAMINOETHYL 9-XANTHENECARBOXYLATE METHOBROMIDE * DIETHYL (2-HYDROXYETHYL)METHYLAMMONIUMBROMIDE XANTHENE-9-CARBOXYLATE * DOLADENE * ETHANAMINIUM, N,N-DIETHYL-N-METHYL-2-((9H-XANTHEN-9-YLCARBONYL)OXY)-, BROMIDE (9CI) * FRENOGASTRICO * GASTRON * GASTROSEDAN * MANTHELINE * METANTYL * METAXAN * METHANIDE * METHANTHELINE BROMIDE * METHANTHELINIUM BROMIDE * METHANTHINE BROMIDE * METHELINA * MTB 51 * RESOBANTIN * SC 2910 * ULCINE * ULCUDEXTER * VAGAMIN * VAGANTIN * XANTELINE * XANTHENE-9-CARBOXYLIC ACID, ESTER with DIETHYL(2-HYDROXYETHYL)METHYLAMMONIUM BROMIDE

SAFETY PROFILE: Poison by ingestion, intraperitoneal, and intravenous routes. Moderately toxic by subcutaneous routes. When heated to decomposition it emits very toxic fumes of NO_x, NH_3, and Br^-.

XDJ000 CAS: 298-81-7 **HR: 3**
XANTHOTOXIN
mf: $C_{12}H_8O_4$ mw: 216.20

SYNS: AMMOIDIN * 6-HYDROXY-7-METHOXY-5-BENZOFURANACRYLIC ACID Δ-LACTONE * MELADININ * MELADININE * MELOXINE * METHOXADOME * METHOXSALEN * 8-METHOXY-(FURANO-3'.2':6.7-COUMARIN) * 9-METHOXY-7H-FURO(3,2-g) BENZOPYRAN-7-ONE * 8-METHOXY-2',3',6,7-FUROCOUMARIN * 8-METHOXY-4',5',6,7-FUROCOUMARIN * 8-METHOXYPSORALEN * 9-METHOXYPSORALEN * 8-MOP * 8-MP * NCI-C55903 * OXSORALEN * OXYPSORALEN * PRORALONE-MOP

CONSENSUS REPORTS: IARC Cancer Review: GROUP 1 IMEMDT 7,243,87; Human Inadequate Evidence IMEMDT 24,101,80; Animal Inadequate Evidence IMEMDT 24,101,80. Reported in EPA TSCA Inventory. EPA Genetic Toxicology Program.

SAFETY PROFILE: Confirmed carcinogen. Poison by intraperitoneal route. Moderately toxic by ingestion and subcutaneous routes. Human mutation data reported. When heated to decomposition it emits acrid smoke and irritating fumes. A drug used to treat skin diseases.

XDS000 CAS: 7440-63-3 **HR: 1**
XENON

DOT: UN 2036/UN 2591
af: Xe   aw: 131.29

PROP: Colorless, gaseous element. D (gas): 5.8878 g/L, d (liq): 3.52 @ −109°, mp: −112°, bp: −107°.

SYN: XENON, refrigerated liquid (DOT)

CONSENSUS REPORTS: Reported in EPA TSCA Inventory.

DOT Classification: Nonflammable Gas; Label: Nonflammable Gas.

SAFETY PROFILE: An inert gas which acts as a simple asphyxiant. For a discussion of toxicity effects, see ARGON. A common air contaminant.

XGS000 CAS: 1330-20-7 **HR: 2**
XYLENE

DOT: UN 1307
mf: C_8H_{10} mw: 106.18

PROP: A clear liquid. Bp: 138.5°, flash p: 100°F (TOC), d: 0.864 @ 20°/4°, vap press: 6.72 mm @ 21°. Composition: as nonaromatics 0.07%, toluene 14%, ethyl benzene 19.27%, p-xylene 7.84%, m-xylene 65.01%, o-xylene 7.63%, C9 and aromatics 0.04%.

SYNS: DIMETHYLBENZENE * KSYLEN (POLISH) * METHYL TOLUENE * NCI-C55232 * RCRA WASTE NUMBER U239 * VIOLET 3 * XILOLI (ITALIAN) * XYLENEN (DUTCH) * XYLOL (DOT) * XYLOLE (GERMAN)

CONSENSUS REPORTS: Reported in EPA TSCA Inventory. EPA Genetic Toxicology Program. Community Right-To-Know List.

OSHA PEL: (Transitional: TWA 100 ppm) TWA 100 ppm; STEL 150 ppm
ACGIH TLV: TWA 100 ppm; STEL 150 ppm; BEI: 1.5 g(methyl hippuric acids)/g creatinine in urine end of shift.
DFG MAK: (all isomers) 100 ppm (440 mg/m^3); BAT: 150 µg/dL in blood at end of shift.
NIOSH REL: (Xylene) TWA 100 ppm; CL 200 ppm/10M
DOT Classification: Flammable Liquid; Label: Flammable Liquid; Flammable or Combustible Liquid; Label: Flammable Liquid.

SAFETY PROFILE: Moderately toxic by intraperitoneal and subcutaneous routes. Mildly toxic by ingestion and inhalation. Human systemic effects by inhalation: olfactory changes, conjunctiva irritation and pulmonary changes. Experimental teratogenic and reproductive effects. Mutation data reported. A human eye irritant. An experimental skin and severe eye irritant. Some temporary corneal effects are noted, as well as some conjunctival irritation by instillation (adding drops to the eyes one at a time). Irritation can start @ 200 ppm. A very dangerous fire hazard when exposed to heat or flame; can react with oxidizing materials. To fight fire, use foam, CO$_2$, dry chemical. When heated to decomposition it emits acrid smoke and irritating fumes.

XHA000   CAS: 108-38-3   HR: 3
m-XYLENE
DOT: UN 1307
mf: C$_8$H$_{10}$ mw: 106.18

PROP: Colorless liquid. Mp: −47.9°, bp: 139°, lel: 1.1%, uel: 7.0%, flash p: 77°F, d: 0.864

@ 20/4°, vap press: 10 mm @ 28.3°, vap d: 3.66, autoign temp: 986°F. Insol in water; misc with alc, ether, and some organic solvents.

SYNS: m-DIMETHYLBENZENE * 1,3-DIMETHYLBENZENE * 1,3-XYLENE * m-XYLOL (DOT)

CONSENSUS REPORTS: Community Right-To-Know List. Reported in EPA TSCA Inventory.

OSHA PEL: (Transitional: TWA 100 ppm) TWA 100 ppm; STEL 150 ppm
ACGIH TLV: TWA 100 ppm; STEL 150 ppm; BEI: methyl hippuric acids in urine end of shift 1.5 g/g creatinine
NIOSH REL: (Xylene) TWA 100 ppm; CL 200 ppm/10M
DOT Classification: Flammable or Combustible Liquid; Label: Flammable Liquid; Flammable Liquid; Label: Flammable Liquid.

SAFETY PROFILE: Moderately toxic by intraperitoneal route. Mildly toxic by ingestion, skin contact, and inhalation. Human systemic effects by inhalation: motor activity changes, ataxia and irritability. Experimental teratogenic and reproductive effects. A severe skin irritant. A common air contaminant. A very dangerous fire hazard when exposed to heat or flame; can react with oxidizing materials. Explosive in the form of vapor when exposed to heat or flame. To fight fire, use foam, CO$_2$, dry chemical. When heated to decomposition it emits acrid smoke and irritating fumes.

XHJ000   CAS: 95-47-6   HR: 3
o-XYLENE
DOT: UN 1307
mf: C$_8$H$_{10}$ mw: 106.18

PROP: Colorless liquid. D: 0.880 @ 20/4°, mp: −25.2°, bp: 144.4°, flash p: 62.6°F. lel: 1.0%, uel: 6.0%. Insol in water; misc in abs alc, ether.

SYNS: o-DIMETHYLBENZENE * 1,2-DIMETHYLBENZENE * o-METHYLTOLUENE * 1,2-XYLENE * o-XYLOL (DOT)

CONSENSUS REPORTS: Community Right-To-Know List. Reported in EPA TSCA Inventory.

OSHA PEL: (Transitional: TWA 100 ppm) TWA 100 ppm; STEL 150 ppm
ACGIH TLV: TWA 100 ppm; STEL 150 ppm; BEI: methyl hippuric acids in urine end of shift 1.5 g/g creatinine

NIOSH REL: (Xylene) TWA 100 ppm; CL 200 ppm/10M

DOT Classification: Flammable or Combustible Liquid; Label: Flammable Liquid; Flammable Liquid; Label: Flammable Liquid.

SAFETY PROFILE: Moderately toxic by intraperitoneal route. Mildly toxic by ingestion and inhalation. An experimental teratogen. A common air contaminant. A very dangerous fire hazard when exposed to heat or flame. Explosive in the form of vapor when exposed to heat or flame. To fight fire, use foam, CO_2, dry chemical. Incompatible with oxidizing materials. When heated to decomposition it emits acrid smoke and irritating fumes.

XHS000   CAS: 106-42-3   HR: 3
p-XYLENE

DOT: UN 1307
mf: C_8H_{10} mw: 106.18

PROP: Clear plates. Bp: 138.3°, lel: 1.1%, uel: 7.0%, flash p: 77°F (CC), d: 0.8611 @ 20/4°, vap press: 10 mm @ 27.3°, vap d: 3.66, autoign temp: 986°F, mp: 13-14°. Insol in water; sol in alc, ether, organic solvents.

SYNS: CHROMAR * p-DIMETHYLBENZENE * 1,4-DIMETHYLBENZENE * p-METHYLTOLUENE * SCINTILLAR * 1,4-XYLENE * p-XYLOL (DOT)

CONSENSUS REPORTS: Community Right-To-Know List. Reported in EPA TSCA Inventory.

OSHA PEL: (Transitional: TWA 100 ppm) TWA 100 ppm; STEL 150 ppm
ACGIH TLV: TWA 100 ppm; STEL 150 ppm; BEI: methyl hippuric acids in urine end of shift 1.5 g/g creatinine
NIOSH REL: (Xylene) TWA 100 ppm; CL 200 ppm/10M
DOT Classification: Flammable Liquid; Label: Flammable Liquid; Flammable or Combustible Liquid; Label: Flammable Liquid.

SAFETY PROFILE: Moderately toxic by intraperitoneal route. Mildly toxic by ingestion and inhalation. Experimental teratogenic and reproductive effects. May be narcotic in high concentrations. Chronic toxicity not established, but is less toxic than benzene. A very dangerous fire hazard when exposed to heat or flame; can react with oxidizing materials. Explosive in the form of vapor when exposed to heat or flame. To fight fire, use foam, CO_2, dry chemical.

Potentially explosive reaction with acetic acid + air, 1,3-dichloro-5,5-dimethyl-2,4-imidazolidindione, nitric acid + pressure. When heated to decomposition it emits acrid smoke and irritating fumes.

XHS800   CAS: 1477-55-0   HR: 2
m-XYLENE-α,α′-DIAMINE
mf: $C_8H_{12}N_2$ mw: 136.22

SYNS: 1,3-BIS-AMINOMETHYLBENZEN (CZECH) * MXDA * m-PHENYLENEBIS(METHYLAMINE) * m-XYLYLENDIAMIN (CZECH)

CONSENSUS REPORTS: Reported in EPA TSCA Inventory.

OSHA PEL: TWA CL 0.1 mg/m³ (skin)
ACGIH TLV: TWA CL 0.1 mg/m³ (skin)

SAFETY PROFILE: Moderately toxic by skin contact and ingestion. Mildly toxic by inhalation. A severe skin and eye irritant. When heated to decomposition it emits toxic fumes of NO_x. Used to make polyamide fibers and resins and as a curing agent.

XIJ000   CAS: 3634-83-1   HR: 2
m-XYLENE-α,α′-DIISOCYANATE
mf: $C_{10}H_8N_2O_2$ mw: 188.20

SYN: XYLYLENDIISOKYANAT (CZECH)

NIOSH REL: (Diisocyanates) TWA 0.005 ppm; CL 0.02 ppm/10M

SAFETY PROFILE: Mildly toxic by ingestion. A severe skin and eye irritant. When heated to decomposition it emits very toxic fumes of NO_x.

XKA000   CAS: 1300-71-6   HR: 3
XYLENOL

DOT: UN 2261
mf: $C_8H_{10}O$ mw: 122.18

PROP: The six isomers of xylenol are sltly sol in water; very sol in alc, chloroform, ether, benzene; sol in NaOH soln.

SYNS: DIMETHYLPHENOL * XILENOLI (ITALIAN) * XYLENOLEN (DUTCH)

CONSENSUS REPORTS: Reported in EPA TSCA Inventory.

DOT Classification: Poison B; Label: Poison.

SAFETY PROFILE: A poison. When heated to decomposition it emits acrid smoke and irritating fumes.

XKJ000   CAS: 526-75-0   **HR: 3**
2,3-XYLENOL

DOT: UN 2261
mf: $C_8H_{10}O$ mw: 122.18

PROP: Needles in water. Mp: 75°, bp: 218°. Sol in water, alc.

SYNS: 2,3-DIMETHYLPHENOL   *   o-XYLENOL (DOT)

CONSENSUS REPORTS: Reported in EPA TSCA Inventory.

DOT Classification: Poison B; Label: Poison.

SAFETY PROFILE: Poison by intravenous route. When heated to decomposition it emits acrid smoke and irritating fumes.

XKJ500   CAS: 105-67-9   **HR: 3**
2,4-XYLENOL

DOT: UN 2261
mf: $C_8H_{10}O$ mw: 122.18

SYNS: 2,4-DIMETHYLPHENOL   *   4,6-DIMETHYL-PHENOL   *   1-HYDROXY-2,4-DIMETHYLBENZENE   *   RCRA WASTE NUMBER U101   *   m-XYLENOL   *   m-XYLENOL (DOT)

CONSENSUS REPORTS: Reported in EPA TSCA Inventory.

DOT Classification: DOT-IMO: Poison B; Label: Poison.

SAFETY PROFILE: Poison by intravenous and intraperitoneal routes. Moderately toxic by ingestion and skin contact. Questionable carcinogen with experimental carcinogenic data. When heated to decomposition it emits acrid smoke and irritating fumes.

XKS000   CAS: 95-87-4   **HR: 3**
2,5-XYLENOL

DOT: UN 2261
mf: $C_8H_{10}O$ mw: 122.18

PROP: Crystals. Mp: 74.5°, bp: 211.5-213.5°.

SYNS: 2,5-DIMETHYLPHENOL   *   3,6-DIMETHYL-PHENOL   *   2,5-DMP   *   6-METHYL-m-CRESOL   *   p-XYLENOL (DOT)   *   1,2,5-XYLENOL

CONSENSUS REPORTS: Reported in EPA TSCA Inventory.

DOT Classification: Poison B; Label: Poison.

SAFETY PROFILE: Poison by ingestion. Moderately toxic by an unspecified route. When heated to decomposition it emits acrid smoke and irritating fumes. Questionable carcinogen with experimental tumorigenic data. Used in disinfectants, solvents, pharmaceuticals, plasticizers, and wetting agents.

XLS000   CAS: 108-68-9   **HR: 3**
3,5-XYLENOL
mf: $C_8H_{10}O$ mw: 122.18

PROP: White crystals. Mp: 64°, bp: 219.5°, d: 1.0362, vap press: 1 mm @ 62°. Sltly sol in water; sol in alc.

SYNS: 3,5-DIMETHYLPHENOL   *   3,5-DMP   *   1,3,5-XYLENOL

CONSENSUS REPORTS: Reported in EPA TSCA Inventory. EPA Genetic Toxicology Program.

SAFETY PROFILE: Poison by intraperitoneal route. Moderately toxic by ingestion. A severe eye irritant. Questionable carcinogen with experimental tumorigenic data. When heated to decomposition it emits acrid smoke and irritating fumes.

XMA000   CAS: 1300-73-8   **HR: 3**
XYLIDINE

DOT: UN 1711
mf: $C_8H_{11}N$ mw: 121.20

PROP: Usually liquid (except for o-4-xylidine). Bp: 213-226°, flash p: 206° (CC), d: 0.97-0.99, vap d: 4.17. Sltly sol in water; sol in alc.

SYNS: ACID LEATHER BROWN 2G   *   ACID ORANGE 24   *   AMINODIMETHYLBENZENE   *   11460 BROWN   *   DIMETHYLANILINE   *   DIMETHYLPHENYLAMINE   *   RESORCINE BROWN J   *   RESORCINE BROWN R   *   XILIDINE (ITALIAN)   *   XYLIDINEN (DUTCH)

CONSENSUS REPORTS: Reported in EPA TSCA Inventory.

OSHA PEL: (Transitional: TWA 5 ppm (skin)) TWA 0.2 ppm (skin)

ACGIH TLV: TWA 2 ppm (Proposed: TWA 0.5 ppm (skin); Suspected Human Carcinogen.)

DFG MAK: (all isomers except 2,4-xylidene) 5 ppm (25 mg/m³)

DOT Classification: Poison B; Label: Poison.

SAFETY PROFILE: Poison by intravenous route. Moderately toxic by ingestion. This material, which so closely resembles aniline in its toxic effects, is actually twice as toxic as aniline. It can cause injury to the blood and the liver. It does not necessarily give any alarm or warning, such as cyanosis, headache, and dizziness which characterizes aniline poisoning. Thus, it may be considered a more insidious poison than aniline, and severe and possibly fatal intoxication may come about through skin absorption. Combustible when exposed to heat or flame. Can react vigorously with oxidizing materials. To fight fire, use foam, CO_2, dry chemical. When heated to decomposition emits toxic fumes of NO_x.

XMJ000 CAS: 87-59-2 **HR: 3**
2,3-XYLIDINE
DOT: UN 1711
mf: $C_8H_{11}N$ mw: 121.20

PROP: Liquid. D: 0.991 @ 15°, mp: $<-15°$, bp: 220°. Very sltly sol in water; sol in alc, ether.

SYNS: 2,3-DIMETHYLANILINE * 2,3-DIMETHYL-BENZENAMINE * 2,3-DIMETHYLPHENYLAMINE * o-XYLIDINE (DOT) * 2,3-XYLYLAMINE

CONSENSUS REPORTS: Reported in EPA TSCA Inventory.

DFG MAK: (all isomers except 2,4-xylidene) 5 ppm (25 mg/m³)
DOT Classification: Poison B; Label: Poison.

SAFETY PROFILE: A poison. Moderately toxic by ingestion. Mutation data reported. When heated to decomposition it emits toxic fumes of NO_x.

XMS000 CAS: 95-68-1 **HR: 3**
2,4-XYLIDINE
DOT: UN 1711
mf: $C_8H_{11}N$ mw: 121.20

PROP: Liquid. Bp: 214°, mp: 16°, d: 0.978 @ 19.6/4°. Very sltly sol in water.

SYNS: 1-AMINO-2,4-DIMETHYLBENZENE * 4-AMINO-1,3-DIMETHYLBENZENE * 4-AMINO-3-METHYLTOLUENE * 4-AMINO-1,3-XYLENE * 2,4-DIMETHYLANILINE * 2,4-DIMETHYLBENZENAMINE * 2,4-DIMETHYLPHENYLAMINE * 2-METHYL-p-TOLUIDINE * 4-METHYL-o-TOLUIDINE * 2,4-XYLIDENE (MAK) * m-XYLIDINE (DOT) * m-4-XYLIDINE

CONSENSUS REPORTS: IARC Cancer Review: GROUP 3 IMEMDT 7,56,87; Animal Inadequate Evidence IMEMDT 16,367,78. Reported in EPA TSCA Inventory.

DFG MAK: 5 ppm (25 mg/m³); Suspected Carcinogen.
DOT Classification: Poison B; Label: Poison.

SAFETY PROFILE: Suspected carcinogen. Poison by ingestion. Mutation data reported. When heated to decomposition it emits toxic fumes of NO_x.

XNA000 CAS: 95-78-3 **HR: 3**
2,5-XYLIDINE
DOT: UN 1711
mf: $C_8H_{11}N$ mw: 121.20

PROP: Colorless oil. Bp: 214°, d: 0.979 @ 21/4°, mp: 155°. Very sltly sol in water.

SYNS: 1-AMINO-2,5-DIMETHYLBENZENE * 3-AMINO-1,4-DIMETHYLBENZENE * 2-AMINO-1,4-XYLENE * 2,5-DIMETHYLANILINE * 2,5-DIMETHYLBENZENAMINE * 2,5-DIMETHYLPHENYLAMINE * 5-METHYL-o-TOLUIDINE * 6-METHYL-m-TOLUIDINE * p-XYLIDINE (DOT)

CONSENSUS REPORTS: IARC Cancer Review: GROUP 3 IMEMDT 7,56,87; Animal Inadequate Evidence IMEMDT 16,377,78. Reported in EPA TSCA Inventory.

DFG MAK: (all isomers except 2,4-xylidene) 5 ppm (25 mg/m³)
DOT Classification: Poison B; Label: Poison.

SAFETY PROFILE: A poison. Moderately toxic by ingestion. Questionable carcinogen. Mutation data reported. When heated to decomposition it emits toxic fumes of NO_x.

XNJ000 CAS: 87-62-7 **HR: 2**
2,6-XYLIDINE
mf: $C_8H_{11}N$ mw: 121.20

PROP: Liquid. D: 0.980 @ 15°, mp: 10-12°, bp: 216-217°.

SYNS: 2,6-DIMETHYLANILINE * 2,6-DIMETHYL-BENZENAMINE * NCI-C56188 * o-XYLIDINE * 2,6-XYLYLAMINE

CONSENSUS REPORTS: Community Right-To-Know List. Reported in EPA TSCA Inventory.

DFG MAK: (all isomers except 2,4-xylidene) 5 ppm (25 mg/m^3)

SAFETY PROFILE: Moderately toxic by ingestion. When heated to decomposition it emits toxic fumes of NO$_x$.

XOA000   CAS: 108-69-0   **HR: 2**
3,5-XYLIDINE
mf: C$_8$H$_{11}$N   mw: 121.20

PROP: An oil. D: 0.972 @ 20/4°, bp: 221-222°.

SYNS: 3,5-DIMETHYLANILINE * 3,5-DIMETHYL-BENZENAMINE * 3,5-DIMETHYLPHENYLAMINE * 3,5-XYLYLAMINE

CONSENSUS REPORTS: Reported in EPA TSCA Inventory.

DFG MAK: (all isomers except 2,4-xylidene) 5 ppm (25 mg/m^3)

SAFETY PROFILE: Moderately toxic by ingestion. When heated to decomposition it emits toxic fumes of NO$_x$.

XOJ000   CAS: 21436-96-4   **HR: 3**
2,4-XYLIDINE HYDROCHLORIDE
mf: C$_8$H$_{11}$N • ClH   mw: 157.66

SYNS: 1-AMINO-2,4-DIMETHYLBENZENE HYDRO-CHLORIDE * 4-AMINO-1,3-DIMETHYLBENZENE HY-DROCHLORIDE * 4-AMINO-3-METHYLTOLUENE HYDROCHLORIDE * 4-AMINO-1,3-XYLENE HYDROCHLORIDE * 2,4-DIMETHYLANILINE HYDRO-CHLORIDE * 2,4-DIMETHYLBENZENAMINE HYDRO-CHLORIDE * 4-METHYL-o-TOLUIDINE HYDROCHLO-RIDE * 2-METHYL-p-TOLUIDINE HYDROCHLORIDE * m-XYLIDINE HYDROCHLORIDE

SAFETY PROFILE: Moderately toxic by ingestion and intraperitoneal routes. Questionable carcinogen with experimental neoplastigenic data. When heated to decomposition it emits very toxic fumes of NO$_x$ and HCl.

XOS000   CAS: 51786-53-9   **HR: 3**
2,5-XYLIDINE HYDROCHLORIDE
mf: C$_8$H$_{11}$N • ClH   mw: 157.66

SYNS: 1-AMINO-2,5-DIMETHYLBENZENE HYDRO-CHLORIDE * 3-AMINO-1,4-DIMETHYLBENZENE HY-DROCHLORIDE * 5-AMINO-1,4-DIMETHYLBENZENE HYDROCHLORIDE * 2-AMINO-4-METHYLTOLUENE HYDROCHLORIDE * 2-AMINO-1,4-XYLENE HYDRO-CHLORIDE * 2,5-DIMETHYLANILINE HYDROCHLO-

RIDE * 2,5-DIMETHYLBENZENAMINE HYDROCHLO-RIDE * 5-METHYL-o-TOLUIDINE HYDROCHLORIDE * 6-METHYL-m-TOLUIDINE HYDROCHLORIDE * p-XYLIDINE HYDROCHLORIDE

SAFETY PROFILE: Moderately toxic by intraperitoneal route. Questionable carcinogen with experimental carcinogenic and tumorigenic data. When heated to decomposition it emits very toxic fumes of NO$_x$ and HCl.

XPJ000   CAS: 87-99-0   **HR: 2**
XYLITOL
mf: C$_5$H$_{12}$O$_5$ mw: 152.17

PROP: White crystals or crystalline powder; sweet taste with cooling sensation. Mp: 92-96°. Sol in water; sltly sol in alc.

SYNS: KLINIT * XYLITE (SUGAR)

CONSENSUS REPORTS: Reported in EPA TSCA Inventory.

SAFETY PROFILE: Moderately toxic by intravenous route. Mildly toxic by ingestion. When heated to decomposition it emits acrid smoke and irritating fumes. A sugar.

XRA000   CAS: 3118-97-6   **HR: 3**
1-(2,4-XYLYLAZO)-2-NAPHTHOL
mf: C$_{18}$H$_{16}$N$_2$O   mw: 276.36

SYNS: A.F. RED NO. 5 * AIZEN FOOD RED NO. 5 * BRASILAZINA OIL SCARLET 6G * BRILLIANT OIL SCARLET B * CALCO OIL SCARLET BL * CERES ORANGES RR * CERISOL SCARLET G * CEROTIN-SCHARLACH G * C.I. 12140 * C.I. SOLVENT OR-ANGE 7 * 1-((2,4-DIMETHYLPHENYL)AZO)-2-NAPH-THALENOL * EXTRACT D&C RED NO. 14 * FAST OIL ORANGE II * FAT RED (YELLOWISH) * FAT SCARLET 2G * FETTORANGE B * GRASAN ORANGE 3R * LACQUER ORANGE VR * MOTIROT G * OIL ORANGE KB * OIL ORANGE N EXTRA * OIL ORANGE R * OIL ORANGE 2R * OIL OR-ANGE X * OIL ORANGE XO * OIL RED GRO * OIL RED O * OIL RED RO * OIL RED XO * OIL SCARLET * OIL SCARLET 371 * OIL SCAR-LET APYO * OIL SCARLET BL * OIL SCARLET 6G * OIL SCARLET L * OIL SCARLET YS * ORANGE INSOLUBLE OLG * ORANGE INSOLUBLE RR * ORANGE OIL KB * PONCEAU INSOLUBLE OLG * PYRONALROT R * RED B * RED NO. 5 * RESIN SCARLET 2R * RESOFORM ORANGE R * ROT B * ROT GG FETTLOESLICH * SOMALIA ORANGE A2R * SOMALIA ORANGE 2R * SOUDAN II * SUDAN AX * SUDAN ORANGE * SUDAN

ORANGE RPA * SUDAN ORANGE RRA * SUDAN RED * SUDAN SCARLET 6G * SUDAN X * WAXAKOL VERMILION L * 1-XYLYLAZO-2-NAPHTHOL * 1-(o-XYLYLAZO)-2-NAPHTHOL

CONSENSUS REPORTS: IARC Cancer Review: GROUP 3 IMEMDT 7,56,87; Animal Sufficient Evidence IMEMDT 8,233,75. Reported in EPA TSCA Inventory. Community Right-To-Know List. EPA Genetic Toxicology Program.

SAFETY PROFILE: Questionable carcinogen with experimental carcinogenic data. Mutation data reported. When heated to decomposition it emits toxic fumes of NO_x.

XRS000 CAS: 28258-59-5 **HR: 3**
XYLYL BROMIDE
DOT: UN 1701
mf: C_8H_9Br mw: 185.08

PROP: Colorless liquid. Bp: 212-215° (slt decomp), d: 1.371 @ 23°. Almost insol in water; sol in alc, ether.

SYN: BROMURE de XYLYLE (FRENCH)

DOT Classification: Irritating Material; Label: Irritant; Poison B; Label: Poison.

SAFETY PROFILE: A human poison by inhalation. A powerful irritant. When heated to decomposition it emits toxic fumes of Br^-.

Y

YBJ000 CAS: 146-48-5 **HR: 3**
YOHIMBINE
mf: $C_{21}H_{26}N_2O_3$ mw: 354.49

PROP: Colorless needles from water and alc. Mp: 234°. Sltly sol in water, ether; sol in alc, chloroform, hot benzene.

SYNS: APHRODINE * APHROSOL * CORYNINE * 17-HYDROXYYOHIMBAN-16-CARBOXYLIC ACID METHYL ESTER * QUEBRACHIN * QUEBRACHINE * YOHIMBIC ACID METHYL ESTER

SAFETY PROFILE: Poison by ingestion, subcutaneous, intravenous, and intraperitoneal routes. Cases of poisoning have occurred from its use as an aphrodisiac. Upon local application it produces anesthesia. However, absorption of it can give rise to toxic symptoms, such as salivation, increased respiration, and repeated defecation. With reference to the circulatory system, there may be a fall in blood pressure and sometimes myocardial damage, involving particularly the conduction system of the heart with a resultant decrease in the efficiency of the heart. An adrenergic blocker used to treat arteriosclerosis and angina pectoris. Formerly used as a local anesthetic and mydriatic (pupillary dilator). When heated to decomposition it emits toxic fumes of NO_x.

YCJ000 CAS: 3458-22-8 **HR: 3**
YOSHI 864
mf: $C_8H_{19}NO_6S_2 \cdot ClH$ mw: 325.86

SYNS: N,N-BIS(METHYLSULFONEPROPOXY)AMINE HYDROCHLORIDE * COMPOUND 864 * 3,3'-IMIDODI-1-PROPANOL, DIMETHANESULFONATE (ester), HYDROCHLORIDE * IPD * NCI-C01547 * NSC 102627 * SAKURAI NO. 864

CONSENSUS REPORTS: NCI Carcinogenesis Bioassay (ipr); Clear Evidence: mouse, rat NCITR* NCI-CG-TR-18,78

SAFETY PROFILE: Poison by intraperitoneal and intravenous routes. Human systemic effects by intravenous route: somnolence, hypermotility, diarrhea, nausea or vomiting. Questionable carcinogen with experimental neoplastigenic data. Human mutation data reported. When heated to decomposition it emits very toxic fumes of NO_x, SO_x, and HCl.

YDA000 CAS: 7440-64-4 **HR: 3**
YTTERBIUM
af: Yb aw: 173.04

PROP: A bright, silvery, lustrous soft, malleable, ductile, and fairly stable element. Mp: 824°, bp: 1193°, d: 6.977. A rare earth.

CONSENSUS REPORTS: Reported in EPA TSCA Inventory.

SAFETY PROFILE: As a lanthanon it may have an anticoagulant action on blood. Questionable carcinogen with experimental tumorigenic data. Flammable in the form of dust when reacted with air; halogens.

YDJ000 CAS: 10361-91-8 **HR: 3**
YTTERBIUM CHLORIDE
mf: Cl_3Yb mw: 279.39

PROP: Hexahydrate, deliquescent needles or crystals. D: 2.575; mp: 150-155°.

SYN: YTTERBIUM TRICHLORIDE

CONSENSUS REPORTS: Reported in EPA TSCA Inventory.

SAFETY PROFILE: Poison by intraperitoneal route. Mildly toxic by ingestion. An experimental teratogen. A skin and eye irritant. When heated to decomposition it emits toxic fumes of Cl^-.

YEA000 CAS: 13839-85-5 **HR: 3**
YTTERBIUM(III) NITRATE, HEXAHYDRATE (1:3:6)
mf: $N_3O_9 \cdot Yb \cdot 6H_2O$ mw: 467.19

SYN: NITRIC ACID, YTTERBIUM(3+) SALT, HEXAHYDRATE

SAFETY PROFILE: Poison by intraperitoneal route. Moderately toxic by ingestion. When heated to decomposition it emits toxic fumes of NO_x.

YEJ000 CAS: 7440-65-5 **HR: 2**
YTTRIUM
af: Y aw: 88.9059

PROP: Hexagonal, gray-black, metallic, rare earth element. Mp: 1509°, bp: 3200°, d: 4.472.

SYN: YTTRIUM-89

CONSENSUS REPORTS: Reported in EPA TSCA Inventory.

OSHA PEL: TWA 1 mg(Y)/m^3
ACGIH TLV: TWA 1 mg(Y)/m^3
DFG MAK: 5 mg(Y)/m^3

SAFETY PROFILE: As a lanthanon, it may have an anticoagulant effect on the blood. Flammable in the form of dust when reacted with air; halogens.

YES000   CAS: 10361-92-9   **HR: 3**
YTTRIUM CHLORIDE
mf: Cl$_3$Y   mw: 195.26

PROP: Hexahydrate, colorless, deliquescent crystals. Sol in water, alc.

SYN: YTTRIUM TRICHLORIDE

CONSENSUS REPORTS: Reported in EPA TSCA Inventory.

ACGIH TLV: TWA 1 mg(Y)/m^3

SAFETY PROFILE: Poison by intraperitoneal route. When heated to decomposition it emits toxic fumes of Cl$^-$.

YFJ000   CAS: 10361-93-0   **HR: 3**
YTTRIUM(III) NITRATE (1:3)
mf: N$_3$O$_9$•Y   mw: 274.94

PROP: Hexahydrate, deliquescent crystals. Sol in water.

SYN: NITRIC ACID, YTTRIUM(3+) SALT

CONSENSUS REPORTS: Reported in EPA TSCA Inventory.

ACGIH TLV: TWA 1 mg(Y)/m^3

SAFETY PROFILE: Poison by intraperitoneal route. Moderately toxic by intravenous route. Experimental reproductive effects. Questionable carcinogen with experimental tumorigenic data. When heated to decomposition it emits toxic fumes of NO$_x$.

YGA000   CAS: 1314-36-9   **HR: 2**
YTTRIUM OXIDE
mf: O$_3$Y$_2$ mw: 225.82

PROP: White powder. D: 4.84.

SYN: YTTRIA

CONSENSUS REPORTS: Reported in EPA TSCA Inventory.

ACGIH TLV: TWA 1 mg(Y)/m^3

SAFETY PROFILE: Moderately toxic by intraperitoneal route.

Z

ZBJ000   CAS: 7440-66-6   **HR: 3**
ZINC

DOT: UN 1383/UN 1436
af: Zn   aw: 65.37

PROP: Bluish-white, lustrous, metallic element. Mp: 419.8°, bp: 908°, d: 7.14 @ 25°, vap press: 1 mm @ 487°. Stable in dry air.

SYNS: BLUE POWDER * C.I. 77945 * C.I. PIGMENT BLACK 16 * C.I. PIGMENT METAL 6 * EMANAY ZINC DUST * GRANULAR ZINC * JASAD * MERRILLITE * PASCO * ZINC DUST * ZINC POWDER * ZINC, POWDER OR DUST, NON-PYROPHORIC (DOT) * ZINC, POWDER OR DUST, PYROPHORIC (DOT)

CONSENSUS REPORTS: Zinc and its compounds are on the Community Right-To-Know List. Reported in EPA TSCA Inventory. EPA Genetic Toxicology Program.

DOT Classification: Flammable Solid; Label: Dangerous When Wet, non-pyrophoric; Flammable Solid; Label: Spontaneously Combustible, pyrophoric.

SAFETY PROFILE: Human systemic effects by ingestion: cough, dyspnea, and sweating. A human skin irritant. Pure zinc powder, dust, fume is relatively nontoxic to humans by inhalation. The difficulty arises from oxidation of zinc fumes immediately prior to inhalation or presence of impurities such as Cd, Sb, As, Pb. Inhalation may cause sweet taste, throat dryness, cough, weakness, generalized aches, chills, fever, nausea, vomiting.
    Flammable in the form of dust when exposed to heat or flame. May ignite spontaneously in air when dry. Explosive in the form of dust when reacted with acids. Incompatible with NH_4NO_3, BaO_2, $Ba(NO_3)_2$, Cd, CS_2, chlorates, Cl_2, ClF_3, CrO_3, (ethyl acetoacetate + tribromoneopentyl alcohol), F_2, hydrazine mononitrate, hydroxylamine, $Pb(N_3)_2$, (Mg + $Ba(NO_3)_2$ + BaO_2), $MnCl_2$, HNO_3, performic acid, $KClO_3$, KNO_3, K_2O_2, Se, $NaClO_3$, Na_2O_2, S, Te, H_2O, $(NH_4)_2S$, As_2O_3, CS_2, $CaCl_2$, NaOH, chlorinated rubber, catalytic metals, halocarbons, o-nitroanisole, nitrobenzene, non-metals, oxidants, paint primer base, pentacarbonyliron, transition metal halides, seleninyl bromide. To fight fire, use special mixtures of dry chemical. When heated to decomposition it emits toxic fumes of ZnO.

ZBS000   CAS: 557-34-6   **HR: 3**
ZINC ACETATE

DOT: NA 9153
mf: $C_4H_6O_4 \cdot Zn$ mw: 183.47

PROP: Crystals; astringent taste. D: 1.735, mp: 237°. Very sol in water; somewhat sol in alc.

SYNS: ACETIC ACID, ZINC SALT * DICARBOMETHOXYZINC * ZINC DIACETATE

CONSENSUS REPORTS: Zinc and its compounds are on the Community Right-To-Know List. Reported in EPA TSCA Inventory.

DOT Classification: ORM-E; Label: None.

SAFETY PROFILE: Poison by intraperitoneal and intravenous routes. Moderately toxic by ingestion. Experimental reproductive effects. Incompatible with zinc salts, alkalies and their carbonates, oxalates, phosphates, sulfides. When heated to decomposition it emits toxic fumes of ZnO.

ZDA000   CAS: 63885-01-8   **HR: 2**
ZINC AMMONIUM NITRITE

DOT: UN 1512

PROP: Solid.

CONSENSUS REPORTS: Zinc and its compounds are on the Community Right-To-Know List.

DOT Classification: Oxidizer; Label: Oxidizer.

SAFETY PROFILE: Flammable by spontaneous chemical reaction. A powerful oxidizing agent. When heated to decomposition it emits toxic fumes of NO_x, NH_3, and ZnO.

ZDJ000   CAS: 1303-39-5   **HR: 3**
ZINC ARSENATE

DOT: UN 1712
mf: $As_4O_{15} \cdot 5Zn$ mw: 866.53

PROP: White, odorless powder.

SYNS: ARSENIC ACID, ZINC SALT * ZINC ARSENATE, BASIC * ZINC ARSENATE, solid (DOT)

CONSENSUS REPORTS: Arsenic and its compounds, as well as zinc and its compounds, are on the Community Right-To-Know List.

OSHA PEL: TWA 0.01 mg(As)/m^3; Cancer Hazard
NIOSH REL: CL 0.002 mg(As)/m^3/15M
DOT Classification: Poison B; Label: Poison.

SAFETY PROFILE: Confirmed human carcinogen. A poison. When heated to decomposition it emits toxic fumes of As and ZnO.

ZDS000   CAS: 10326-24-6   **HR: 3**
ZINC-m-ARSENITE

DOT: UN 1712
mf: AsHO$_2$ • 1/2Zn   mw: 140.61

PROP: A white powder.

SYNS: ARSENIOUS ACID, ZINC SALT * ZINC AR-SENITE, solid (DOT) * ZINC METAARSENITE * ZINC METHARSENITE * ZMA

CONSENSUS REPORTS: Arsenic and its compounds, as well as zinc and its compounds, are on the Community Right-To-Know List.

OSHA PEL: TWA 0.01 mg(As)/m^3: Cancer Hazard
ACGIH TLV: TWA 0.2 mg(As)/m^3
NIOSH REL: (Inorganic Arsenic) CL 0.002 mg(As)/m^3/15M
DOT Classification: Poison B; Label: Poison.

SAFETY PROFILE: Confirmed human carcinogen. A poison. When heated to decomposition it emits toxic fumes of As and ZnO.

ZES000   CAS: 10361-95-2   **HR: 3**
ZINC CHLORATE

DOT: UN 1513
mf: Cl$_2$O$_6$ • Zn   mw: 232.27

PROP: Colorless, very deliquescent crystals.

CONSENSUS REPORTS: Zinc and its compounds are on the Community Right-To-Know List. Reported in EPA TSCA Inventory.

DOT Classification: Oxidizer; Label: Oxidizer.

SAFETY PROFILE: A powerful oxidizer. Probably a skin, eye, and mucous membrane irritant. The tetrahydrated salt explodes at 60°C. Explosive reaction with copper(II) sulfide. Can react violently with Al, Sb$_2$S$_3$, As, C, charcoal, Cu, MnO$_2$, metal sulfides, dibasic organic acids, organic matter, P, S, H$_2$SO$_4$. Incandescent reac-

tion with antimony(III) sulfide, arsenic (III) sulfide, tin(II) sulfide, tin(IV) sulfide. When heated to decomposition it emits toxic fumes of Cl$^-$ and ZnO.

ZFA000   CAS: 7646-85-7   **HR: 3**
ZINC CHLORIDE

DOT: UN 1840/UN 2331
mf: Cl$_2$Zn   mw: 136.27

PROP: Odorless, cubic, white, deliquescent crystals. Mp: 290°, bp: 732°, d: 2.91 @ 25°, vap press: 1 mm @ 428°.

SYNS: BUTTER of ZINC * CHLORURE de ZINC (FRENCH) * TINNING GLUX (DOT) * ZINC CHLO-RIDE, anhydrous (DOT) * ZINC CHLORIDE, solid (DOT) * ZINC CHLORIDE, solution (DOT) * ZINC (CHLO-RURE de) (FRENCH) * ZINC DICHLORIDE * ZINC MURIATE, solution (DOT) * ZINCO (CLORURO di) (ITALIAN) * ZINKCHLORID (GERMAN) * ZINK-CHLORIDE (DUTCH)

CONSENSUS REPORTS: Zinc and its compounds are on the Community Right-To-Know List. Reported in EPA TSCA Inventory. EPA Genetic Toxicology Program.

OSHA PEL: Fume: (Transitional: TWA 1 mg/m^3) TWA 1 mg/m^3; STEL 2 mg/m^3
ACGIH TLV: TWA 1 mg/m^3; STEL 2 mg/m^3 (fume)
DOT Classification: Corrosive Material; Label: Corrosive and Corrosive, solution; ORM-E; Label: None, solid.

SAFETY PROFILE: Poison by ingestion, intravenous, subcutaneous, and intraperitoneal routes. Human systemic effects by inhalation: pulmonary changes. Experimental teratogenic and reproductive effects. Questionable carcinogen with experimental tumorigenic data. Human mutation data reported. A corrosive irritant to skin, eyes, and mucous membranes. Exposure to ZnCl$_2$ fumes or dusts can cause dermatitis, boils, conjunctivitis, gastrointestinal tract upsets. The fumes are highly toxic. Incompatible with potassium. Mixtures of the powdered chloride and powdered zinc are flammable. When heated to decomposition it emits toxic fumes of Cl$^-$ and ZnO.

ZFJ100   CAS: 13530-65-9   **HR: 3**
ZINC CHROMATE
mf: CrH$_2$O$_4$ • Zn   mw: 183.39

SYNS: BASIC ZINC CHROMATE * BUTTERCUP YEL-LOW * CHROMIC ACID, ZINC SALT * CHROMIUM

ZINC OXIDE * C.I. 77955 * C.I. PIGMENT YELLOW 36 * CITRON YELLOW * C.P. ZINC YELLOW X-883 * PRIMROSE YELLOW * PURE ZINC CHROME * ZINC CHROMATE(VI) HYDROXIDE * ZINC CHROME YELLOW * ZINC CHROMIUM OXIDE * ZINC HYDROXYCHROMATE * ZINC TETRAOX-YCHROMATE 76A * ZINC YELLOW

CONSENSUS REPORTS: IARC Cancer Review: GROUP 1 IMEMDT 7,165,87; Human Sufficient Evidence IMEMDT 23,205,80; Animal Sufficient Evidence IMEMDT 23,205,80. NTP Fourth Annual Report On Carcinogens, 1984. EPA Genetic Toxicology Program. Reported in EPA TSCA Inventory. Zinc and chromium and their compounds, are on the Community Right-To-Know List.

OSHA PEL: (Transitional: 1 mg(CrO_3)/10m^3) CL 0.1 mg(CrO_3)/m^3
ACGIH TLV: TWA 0.01 mg(Cr)/M^3; Confirmed Human Carcinogen
DFG TRK: 0.1 mg/m^3; Human Carcinogen.
NIOSH REL: (Chromium (VI)) TWA 0.001 mg(Cr(VI))/m^3

SAFETY PROFILE: Confirmed human carcinogen producing lung tumors. A poison via intravenous route. Human mutation data reported.

***

ZFJ120　　　CAS: 37300-23-5　　**HR: 3**
ZINC CHROMATE with ZINC HYDROXIDE and CHROMIUM OXIDE (9:1)
mf: $CrO_4 \cdot Zn \cdot H_4O_2Zn \cdot CrO_3$　　mw: 183.39

SYN: ZINC YELLOW

CONSENSUS REPORTS: Reported in EPA TSCA Inventory.

OSHA PEL: (Transitional: 1 mg(CrO_3)/10m^3) CL 0.1 mg(CrO_3)/m^3
ACGIH TLV: TWA 0.01 mg(Cr)/M^3; Confirmed Human Carcinogen
DFG TRK: 0.1 mg/m^3; Human Carcinogen.
NIOSH REL: (Chromium (VI)) TWA 0.001 mg(Cr(VI))/m^3

SAFETY PROFILE: Confirmed human carcinogen producing lung tumors. Mutation data reported.

***

ZFJ150　　　　　　　　　　　**HR: 3**
ZINC CHROMATE, POTASSIUM DICHROMATE and ZINC HYDROXIDE (3:1:1)
mf:　　$CrK_2O_4 \cdot 3CrO_4Zn \cdot H_2O_2Zn$　　mw: 837.70

SYN: POTASSIUM DICHROMATE, ZINC CHROMATE and ZINC HYDROXIDE (1:3:1)

OSHA PEL: (Transitional: 1 mg(CrO_3)/10m^3) CL 0.1 mg(CrO_3)/m^3
ACGIH TLV: TWA 0.01 mg(Cr)/M^3; Confirmed Human Carcinogen
DFG TRK: 0.1 mg/m^3; Human Carcinogen.
NIOSH REL: (Chromium (VI)) TWA 0.001 mg(Cr(VI))/m^3

SAFETY PROFILE: Confirmed human carcinogen with experimental carcinogenic data.

***

ZFS000　　　　　　　　　　　**HR: 3**
ZINC COMPOUNDS
CONSENSUS REPORTS: Zinc and its compounds are on the Community Right-To-Know List.

SAFETY PROFILE: Variable toxicity, but generally of low toxicity. However, zinc salts, such as chromates and arsenates, are experimental carcinogens. Zinc is not inherently a toxic element. However, when heated, it evolves a fume of zinc oxide which, when inhaled fresh, can cause a disease known as "brass founders" "ague," or "brass chills," sweet taste, throat dryness, cough, weakness, generalized aching, fever, nausea, and vomiting. It is possible for people to become immune to it, but this immunity can be broken by cessation of exposure of only a few days. Zinc oxide dust which is not freshly formed is virtually innocuous. There is no cumulative effect from the inhalation of zinc fumes. Exposure to zinc chloride fumes can cause damage to the mucous membranes of the nasopharynx and respiratory tract, and give rise to a pale gray cyanosis; fatalities have resulted. Soluble salts of zinc have a harsh metallic taste; small doses can cause nausea and vomiting, while larger doses cause violent vomiting and purging. Some cases of intoxication have been reported due to drinking liquids stored in galvanized containers and in dialysis patients using a dialyzate prepared with water that had been stored in a galvanized tank. In general, the continued administration of zinc salts in small doses has no effect in humans except those of disordered digestion and constipation. Workers in zinc refining have been reported to suffer from a variety of non-specific intestinal, respiratory and nervous symptoms. Ulceration of the nasal septum and eczematous dermatosis are also reported. It has been stated that zinc oxide or zinc stearate dust can block the ducts of the sebaceous glands and give rise to a papular, pustular eczema in workers engaged in packing these compounds into barrels. Sensitivity to zinc oxide in workers is extremely rare. Zinc chloride and zinc sulfate, because of caustic action, can cause ulceration of the fingers, hands and forearms of those who use them as a flux in soldering or other industrial use. This condition has even been observed in men who handle railway ties which have been impregnated with this material. Common air contaminants. When heated to decomposition it emits toxic fumes of ZnO.

ZGA000   CAS: 557-21-1   HR: 3
ZINC CYANIDE

DOT: UN 1713
mf: C_2N_2Zn mw: 117.41

PROP: Rhombic, colorless crystals. Mp: decomp @ 800°. Insol in water; sol in solns of alkali cyanides; decomp by dil mineral acid.

SYNS: CYANURE de ZINC (FRENCH) * RCRA WASTE NUMBER P121 * ZINC DICYANIDE

CONSENSUS REPORTS: Zinc and its compounds, as well as cyanide and its compounds, are on the Community Right-To-Know List. Reported in EPA TSCA Inventory.

DOT Classification: Poison B; Label: Poison.

SAFETY PROFILE: Poison by intraperitoneal route. Can react violently with Mg. When heated to decomposition it emits toxic fumes of CN^-, ZnO, and NO_x. Used in electroplating operations.

ZHS000   CAS: 7783-49-5   HR: 3
ZINC FLUORIDE

DOT: NA 9158
mf: F_2Zn mw: 103.37

PROP: Tetragonal needles or white crystalline mass. D: 5.00 @ 25°, mp: 872°, bp: 1500°, vap press: 1 mm @ 970°. Sltly sol in aq HF; sol in HCl, HNO_3, and NH_4OH.

SYN: ZINC FLUORURE (FRENCH)

CONSENSUS REPORTS: Zinc and its compounds are on the Community Right-To-Know List. Reported in EPA TSCA Inventory.

OSHA PEL: TWA 2.5 mg(F)/m^3
ACGIH TLV: TWA 2.5 mg(F)/m^3

NIOSH REL: TWA 2.5 mg(F)/m^3
DOT Classification: ORM-E; Label: None.

SAFETY PROFILE: Poison by subcutaneous route. Can react violently with potassium. A fluorination agent. When heated to decomposition it emits toxic fumes of F^- and ZnO.

ZIA000   CAS: 16871-71-9   HR: 3
ZINC FLUOSILICATE

DOT: UN 2855
mf: $F_6Si \cdot Zn$ mw: 207.46

SYNS: FLUOSILICATE de ZINC * ZINC HEXAFLUOROSILICATE

CONSENSUS REPORTS: Zinc and its compounds are on the Community Right-To-Know List. Reported in EPA TSCA Inventory.

NIOSH REL: TWA 2.5 mg(F)/m^3
DOT Classification: Poison B; Label: St. Andrews Cross

SAFETY PROFILE: Poison by ingestion and subcutaneous routes. When heated to decomposition it emits toxic fumes of F^- and ZnO.

ZJA000   CAS: 22323-45-1   HR: 2
ZINC MERCURY CHROMATE COMPLEX
mf: $7ZnO \cdot 2HgO \cdot 2CrO_3 \cdot 7H_2O$ mw: 1328.91

SYNS: CHROMIC ACID, MERCURY ZINC COMPLEX * EXPERIMENTAL FUNGICIDE 224 (UNION CARBIDE) * MERCURY ZINC CHROMATE COMPLEX

CONSENSUS REPORTS: Zinc, mercury, chromium, and their compounds are on the Community Right-To-Know List.

OSHA PEL: (Transitional: 1 mg/10m^3) CL 0.1 mg(CrO_3)/m^3
ACGIH TLV: TWA 0.05 mg(Cr)/m^3, Confirmed Human Carcinogen.
DFG MAK: Animal Carcinogen, Suspected Human Carcinogen.
NIOSH REL: TWA 0.025 mg(Cr(VI))/m^3; CL 0.05/15M; 0.05 mg(Hg)/m^3

SAFETY PROFILE: Confirmed carcinogen. Moderately toxic by ingestion. When heated to decomposition it emits very toxic fumes of Hg and ZnO.

ZJJ000   CAS: 7779-88-6   HR: 3
ZINC NITRATE

DOT: UN 1514
mf: $N_2O_6 \cdot Zn$ mw: 189.39

PROP: A: needles; B: tetragonal, colorless crystals; A: trihydrate; B: hexahydrate; d: (B) 2.065 @ 14°; mp: (A) 42.5°; mp: (B): 36.4°; bp: (B): loses $6H_2O$ @ 105-131°. Very sol in alc; sol in water.

SYNS: NITRATE de ZINC (FRENCH) * NITRIC ACID, ZINC SALT

CONSENSUS REPORTS: Zinc and its compounds are on the Community Right-To-Know List. Reported in EPA TSCA Inventory.

DOT Classification: Oxidizer; Label: Oxidizer.

SAFETY PROFILE: A powerful oxidizer. Can react violently with C, Cu, metal sulfides, organic matter, P, S. When heated to decomposition it emits toxic fumes of NO_x and ZnO.

ZKA000   CAS: 1314-13-2   **HR: 3**
ZINC OXIDE
mf: OZn   mw: 81.37

PROP: Odorless, white or yellowish powder. Mp: >1800°, d: 5.47. Insol in water, alc; sol in dil acetic or mineral acids, ammonia.

SYNS: AKRO-ZINC BAR 85 * AKRO-ZINC BAR 90 * AMALOX * AZO-33 * AZO-55 * AZO-66 * AZO-77 * AZODOX-55 * AZODOX-55TT * CALAMINE (spray) * CHINESE WHITE * C.I. 77947 * C.I. PIGMENT WHITE 4 * CYNKU TLENEK (POLISH) * EMANAY ZINC OXIDE * EMAR * FELLING ZINC OXIDE * FLOWERS of ZINC * GREEN SEAL-8 * HUBBUCK'S WHITE * KADOX-25 * K-ZINC * OZIDE * OZLO * PASCO * PERMANENT WHITE * PHILOSOPHER'S WOOL * PROTOX TYPE 166 * PROTOX TYPE 167 * PROTOX TYPE 168 * PROTOX TYPE 169 * PROTOX TYPE 267 * PROTOX TYPE 268 * RED-SEAL-9 * SNOW WHITE * WHITE SEAL-7 * ZINCITE * ZINCOID * ZINC OXIDE FUME (MAK) * ZINC WHITE

CONSENSUS REPORTS: Zinc and its compounds are on the Community Right-To-Know List. Reported in EPA TSCA Inventory.

OSHA PEL: Fume: (Transitional: TWA 5 mg/m^3) TWA 5 mg/m^3; STEL 10 mg/m^3; Dust: (Transitional: Total Dust: 15 mg/m^3; Respirable Fraction: 5 mg/m^3;) TWA Total Dust: 10 mg/m^3; Respirable Fraction: 5 mg/m^3
ACGIH TLV: Fume: TWA 5 mg/m^3; STEL 10 mg/m^3; Dust: 10 mg/m^3 of total dust (when toxic impurities are not present, e.g., quartz < 1%).

DFG MAK: 5 mg/m^3
NIOSH REL: TWA (Zinc Oxide) 5 mg/m^3; CL 15 mg/m^3/15M

SAFETY PROFILE: Poison by intraperitoneal route. An experimental teratogen. Human systemic effects by inhalation of freshly formed fumes: metal fume fever with chills, fever, tightness of chest, cough, dyspnea, and other pulmonary changes. Mutation data reported. A skin and eye irritant. Has exploded when mixed with chlorinated rubber. Violent reaction with Mg, linseed oil. When heated to decomposition it emits toxic fumes of ZnO.

ZLA000   CAS: 23414-72-4   **HR: 1**
ZINC PERMANGANATE
DOT: UN 1515
mf: $Mn_2O_8 \cdot Zn$ mw: 303.25

PROP: Violet-brown or black, hygroscopic crystals.

CONSENSUS REPORTS: Zinc, manganese, and their compounds are on the Community Right-To-Know List.

DOT Classification: Oxidizer; Label: Oxidizer.

SAFETY PROFILE: Probably a skin, eye, and mucous membrane irritant. Flammable by chemical reaction with reducing agents. A powerful oxidizing agent. When heated to decomposition it emits toxic fumes of ZnO. Used as an antiseptic.

ZLJ000   CAS: 1314-22-3   **HR: 3**
ZINC PEROXIDE
DOT: UN 1516
mf: O_2Zn mw: 97.37

PROP: Odorless, yellow-white powder. D: 1.571 (theoretical). Decomp >150°. Sol in dil acids.

SYN: ZINC SUPEROXIDE

CONSENSUS REPORTS: Zinc and its compounds are on the Community Right-To-Know List. Reported in EPA TSCA Inventory.

DOT Classification: Oxidizer; Label: Oxidizer.

SAFETY PROFILE: Systemic toxicity is similar to zinc oxide. Flammable when exposed to heat or by chemical reaction with reducing materials. Finely divided powder is slightly soluble in water, decomposes rapidly at 150°. A powerful oxidizer and dangerous when mixed with highly combustible materials. A very dangerous explosion hazard when exposed to heat. Explodes at 212°. Can react violently with Al, Zn. Very dangerous, will react with water or steam to produce heat. Vigorous reaction with reducing materials. When heated to decomposition it emits toxic fumes of ZnO.

ZLS000 CAS: 1314-84-7 **HR: 3**
ZINC PHOSPHIDE

DOT: UN 1714
mf: P_2Zn_3 mw: 258.05

PROP: Cubic, dark gray crystals or powder. Mp: 420°, bp: 1100°, d: 4.55 @ 13°. Insol in water, alc; sol in benzene, carbon disulfide.

SYNS: BLUE-OX * KILRAT * MOUS-CON * PHOSPHURE de ZINC (FRENCH) * PHOSVIN * RCRA WASTE NUMBER P122 * RUMETAN * ZINCO(FOSFURO di) (ITALIAN) * ZINC(PHOSPHURE de) (FRENCH) * ZINC-TOX * ZINKFOSFIDE (DUTCH) * ZINKPHOSPHID (GERMAN) * ZP

CONSENSUS REPORTS: Zinc and its compounds are on the Community Right-To-Know List. Reported in EPA TSCA Inventory.

DOT Classification: Flammable Solid; Label: Flammable Solid and Dangerous When Wet; Poison B; Label: Poison.

SAFETY PROFILE: Human poison by ingestion causing nausea, vomiting, death. Flammable when exposed to heat or flame. This material is stable while kept dry. In moist air, it decomposes slowly. Reacts violently with acids or acid fumes to emit the highly toxic and flammable phosphine. Violent reaction with concentrated sulfuric acid, nitric acid, and oxidizing materials. Incompatible with HCl, H_2SO_4. When heated to decomposition it emits toxic fumes of PO_x and ZnO. Used as an acute rodenticide.

ZMS000 CAS: 557-05-1 **HR: 3**
ZINC STEARATE
mf: $Zn(C_{18}H_{35}O_2)_2$ mw: 632.30

PROP: White powder. Mp: 130°, flash p: 530°F (OC), autoign temp: 790°F. Insol in water, alc, ether; sol in benzene. Decomp by dil acids.

SYNS: DIBASIC ZINC STEARATE * OCTADECANOIC ACID, ZINC SALT * STEARIC ACID, ZINC SALT * ZINC DISTERATE * ZINC OCTADECANOATE

CONSENSUS REPORTS: Zinc and its compounds are on the Community Right-To-Know List. Reported in EPA TSCA Inventory.

OSHA PEL: (Transitional: TWA Total Dust: 15 mg/m³; Respirable Fraction: 5 mg/m³) TWA Total Dust: 10 mg/m³; Respirable Fraction: 5 mg/m³
ACGIH TLV: TWA 10 mg/m³ of total dust when toxic impurities are not present, e.g., quartz < 1%

SAFETY PROFILE: Poison by intratracheal route. Inhalation of zinc stearate has been reported as causing pulmonary fibrosis. A nuisance dust. Combustible when exposed to heat or flame. To fight fire, use water, foam, CO_2, dry chemical. When heated to decomposition it emits toxic fumes of ZnO.

ZNA000 CAS: 7733-02-0 **HR: 3**
ZINC SULFATE

DOT: NA 9161
mf: $O_4S \cdot Zn$ mw: 161.43

PROP: Rhombic, colorless crystals or crystalline powder. Mp: decomp @ 740°, d: 3.74 @ 15°. Sol in water; almost insol in alc.

SYNS: BONAZEN * BUFOPTO ZINC SULFATE * OP-THAL-ZIN * SULFATE de ZINC (FRENCH) * SULFURIC ACID, ZINC SALT (1:1) * VERAZINC * WHITE COPPERAS * WHITE VITRIOL * ZINC SULPHATE * ZINC VITRIOL * ZINKOSITE

CONSENSUS REPORTS: Zinc and its compounds are on the Community Right-To-Know List. Reported in EPA TSCA Inventory. EPA Genetic Toxicology Program.

DOT Classification: ORM-E; Label: None.

SAFETY PROFILE: Poison by intraperitoneal, subcutaneous, and intravenous routes. Moderately toxic by ingestion. Human systemic effects by ingestion: increased pulse rate without blood pressure decrease, blood pressure decrease, acute pulmonary edema, normocytic anemia, hypermotility, diarrhea, and other gastrointestinal changes. Experimental teratogenic and reproductive effects. Questionable carcinogen with experimental tumorigenic data. Human mutation data reported. An eye irritant. When heated to decomposition it emits toxic fumes of SO_x and ZnO.

ZNJ000   CAS: 7446-20-0   **HR: 3**
ZINC SULFATE HEPTAHYDRATE
(1:1:7)
mf: $O_4SZn \cdot 7H_2O$ mw: 287.57

PROP: Colorless crystals or crystalline powder; odorless. D: 1.97; mp: 100°. Decomp >500°. Insol in alc; glycerin.

SYNS: SULFURIC ACID, ZINC SALT (1:1), HEPTAHY-DRATE * WHITE VITRIOL * ZINC SULFATE * ZINC SULFATE (1:1) HEPTAHYDRATE * ZINC VI-TRIOL

CONSENSUS REPORTS: Zinc and its compounds are on the Community Right-To-Know List.

SAFETY PROFILE: Human poison by an unspecified route. Poison experimentally by subcutaneous, intravenous, and intraperitoneal routes. Moderately toxic by ingestion. When heated to decomposition it emits toxic fumes of SO_x and ZnO.

ZOA000   CAS: 7440-67-7   **HR: 3**
ZIRCONIUM

DOT: UN 1308/UN 1358/UN 2008/UN 2009/ UN 2858
af: Zr   aw: 91.224

PROP: A grayish-white, lustrous, metallic element; very sltly radioactive. Mp: 1852°, bp: 3577°, d: 6.506 @ 20°.

SYNS: ZIRCAT * ZIRCONIUM METAL * ZIRCO-NIUM METAL, DRY (DOT) * ZIRCONIUM SHAVINGS * ZIRCONIUM SHEETS (DOT) * ZIRCONIUM TURN-INGS

CONSENSUS REPORTS: Reported in EPA TSCA Inventory.

OSHA PEL: (Transitional: TWA 5 mg(Zr)/m^3) TWA 5 mg(Zr)/m^3; STEL 10 mg(Zr)/m^3
ACGIH TLV: TWA 5 mg(Zr)/m^3; STEL 10 mg(Zr)/m^3
DFG MAK: 5 mg(Zr)/m^3
DOT Classification: Flammable Solid; Label: Flammable Solid; Flammable Solid; Label: Spontaneously Combustible; Flammable Liquid; Label: Flammable Liquid.

SAFETY PROFILE: A very dangerous fire hazard in the form of dust when exposed to heat or flame or by chemical reaction with oxidizers. May ignite spontaneously. A dangerous explosion hazard in the form of dust by chemical reaction with air, alkali hydroxides, alkali metal chromates, dichromates, molybdates, sulfates, tungstates, borax, CCl_4, CuO, Pb, PbO, P, $KClO_3$, KNO_3, nitrylfluoride. Explosive range: 0.16 g/L in air. To fight fire, use special mixtures, dry chemical, salt, or dry sand.

ZPA000   CAS: 10026-11-6   **HR: 3**
ZIRCONIUM CHLORIDE

DOT: UN 2503
mf: Cl_4Zr mw: 233.02

PROP: White, lustrous crystals. Mp: sublimes @ 300°, bp: 331°, d: 2.80, vap press: 1 mm @ 190°.

SYNS: ZIRCONIUM(IV) CHLORIDE (1:4) * ZIRCO-NIUM TETRACHLORIDE (DOT) * ZIRCONIUM TETRA-CHLORIDE, solid (DOT)

CONSENSUS REPORTS: Reported in EPA TSCA Inventory.

OSHA PEL: (Transitional: TWA 5 mg(Zr)/m^3) TWA 5 mg(Zr)/m^3; STEL 10 mg(Zr)/m^3
ACGIH TLV: TWA 5 mg(Zr)/m^3; STEL 10 mg(Zr)/m^3
DFG MAK: 5 mg(Zr)/m^3
DOT Classification: Corrosive Material; Label: Corrosive.

SAFETY PROFILE: Moderately toxic by ingestion. A corrosive irritant to skin, eyes, and mucous membranes. Ignites spontaneously in air. When heated to decomposition it emits toxic fumes of Cl^-.

ZQA000   **HR: 2**
ZIRCONIUM COMPOUNDS

OSHA PEL: (Transitional: TWA 5 mg(Zr)/m^3) TWA 5 mg(Zr)/m^3; STEL 10 mg(Zr)/m^3
ACGIH TLV: TWA 5 mg(Zr)/m^3; STEL 10 mg(Zr)/m^3
DFG MAK: 5 mg(Zr)/m^3

SAFETY PROFILE: Zirconium is not an important industrial poison, however, poisoning may occur due to excessive exposure to zirconium salts. Deaths in rabbits have been caused by intravenous injection of 150 mg/kg of body weight. Inhalation of $ZrCl_4$ (6 mg Zr/m^3) for 60 days produces slight decreases in hemoglobin and red blood cell count in dogs and increases mortality in rats and guinea pigs. Most zirconium compounds in common use are insoluble and considered inert. Pulmonary granuloma in zirconium workers has been reported and sodium zirconium lactate has been held responsible for skin granulomas. Avoid inhalation of Zr-containing aerosols, which can cause lung granulomas. Zirconium-containing drugs or cosmetic products are being controlled by the FDA.

ZQS000   CAS: 7783-64-4   **HR: 3**
ZIRCONIUM FLUORIDE
mf: F_4Zr mw: 167.22

PROP: Refractive crystals. Water-sol. D: 4.6 @ 16°, sublimes @ 600°. Very sol in HF.

SYN: ZIRCONIUM TETRAFLUORIDE

CONSENSUS REPORTS: Reported in EPA TSCA Inventory.

OSHA PEL: (Transitional: TWA 5 mg(Zr)/m^3) TWA 5 mg(Zr)/m^3; STEL 10 mg(Zr)/m^3
ACGIH TLV: TWA 5 mg(Zr)/m^3; STEL 10 mg(Zr)/m^3
DFG MAK: 5 mg(Zr)/m^3

SAFETY PROFILE: Poison by intravenous route. When heated to decomposition it emits toxic fumes of F$^-$.

ZRA000   CAS: 7704-99-6   **HR: 3**
ZIRCONIUM HYDRIDE
DOT: UN 1437
mf: H_2Zr mw: 93.24

PROP: Metallic dark gray to black powder. D: 5.6, autoign temp: 270° (in air).

CONSENSUS REPORTS: Reported in EPA TSCA Inventory.

OSHA PEL: (Transitional: TWA 5 mg(Zr)/m^3) TWA 5 mg(Zr)/m^3; STEL 10 mg(Zr)/m^3
ACGIH TLV: TWA 5 mg(Zr)/m^3; STEL 10 mg(Zr)/m^3
DFG MAK: 5 mg(Zr)/m^3
DOT Classification: Flammable Solid; Label: Flammable Solid and Dangerous When Wet; Flammable Solid; Label: Flammable Solid.

SAFETY PROFILE: A powerful reducing agent. Flammable when dry or wet. Very dangerous to handle; can explode. Incandesces when heated in air.

ZSA000   CAS: 13746-89-9   **HR: 2**
ZIRCONIUM NITRATE
DOT: UN 2728
mf: $N_4O_{12} \cdot Zr$ mw: 339.26

PROP: White crystals.

SYN: DUSICNAN ZIRKONICITY (CZECH)

CONSENSUS REPORTS: Reported in EPA TSCA Inventory.

OSHA PEL: (Transitional: TWA 5 mg(Zr)/m^3) TWA 5 mg(Zr)/m^3; STEL 10 mg(Zr)/m^3
ACGIH TLV: TWA 5 mg(Zr)/m^3; STEL 10 mg(Zr)/m^3
DFG MAK: 5 mg(Zr)/m^3
DOT Classification: Oxidizer; Label: Oxidizer.

SAFETY PROFILE: Moderately toxic by inhalation and ingestion. A powerful oxidizer. When heated to decomposition it emits toxic fumes of NO$_x$.

ZSJ000   CAS: 7699-43-6   **HR: 3**
ZIRCONIUM OXYCHLORIDE
mf: Cl_2OZr mw: 178.12

PROP: Crystals. D: 1.91. Very sol in water, alc.

SYNS: BASIC ZIRCONIUM CHLORIDE * CHLORO-ZIRCONYL * DICHLOROOXOZIRCONIUM * NCI-C60811 * ZIRCONYL CHLORIDE

CONSENSUS REPORTS: Reported in EPA TSCA Inventory.

OSHA PEL: (Transitional: TWA 5 mg(Zr)/m^3) TWA 5 mg(Zr)/m^3; STEL 10 mg(Zr)/m^3
ACGIH TLV: TWA 5 mg(Zr)/m^3; STEL 10 mg(Zr)/m^3
DFG MAK: 5 mg(Zr)/m^3

SAFETY PROFILE: Poison by intraperitoneal route. Moderately toxic by ingestion and subcutaneous routes. Questionable carcinogen with experimental neoplastigenic data. When heated to decomposition it emits toxic fumes of Cl$^-$. Used as an antiperspirant.

ZTJ000   CAS: 14644-61-2   **HR: 3**
ZIRCONIUM(IV) SULFATE (1:2)
DOT: NA 9163
mf: $O_8S_2 \cdot Zr$ mw: 283.34

PROP: Tetrahydrate, crystalline solid.

SYNS: DISULFATOZIRCONIC ACID * SULFURIC ACID, ZIRCONIUM(4+) SALT (2:1) * ZIRCONYL SULFATE

CONSENSUS REPORTS: Reported in EPA TSCA Inventory.

OSHA PEL: (Transitional: TWA 5 mg(Zr)/m^3) TWA 5 mg(Zr)/m^3; STEL 10 mg(Zr)/m^3
ACGIH TLV: TWA 5 mg(Zr)/m^3; STEL 10 mg(Zr)/m^3

DFG MAK: 5 mg(Zr)/m^3
DOT Classification: ORM-B; Label: None

SAFETY PROFILE: Poison by intraperitoneal route. Moderately toxic by ingestion and subcutaneous routes. When heated to decomposition it emits toxic fumes of SO$_x$.

ZUS000   CAS: 22144-77-0   **HR: 3**
ZYGOSPORIN A
mf: C$_{30}$H$_{37}$NO$_6$ mw: 507.68

SYNS: 3-BENZYL-3,3-α,4,5,6,6-α,9,10,12,15-DE-CAHYDRO-6,12,15-TRIHYDROXY-4,10,12-TRIMETHYL-5-METHYLENE-1H-CYCLOUNDEC(d)ISOINDOLE-1,11(2H)-DIONE, 15-ACETATE * CYTOCHALA-SIN D

SAFETY PROFILE: Poison by ingestion, subcutaneous, and intraperitoneal routes. Experimental teratogenic and reproductive effects. Human mutation data reported. When heated to decomposition it emits toxic fumes of NO$_x$.

II. Synonym Cross-Index

1080 see SHG500
A 21 see DMV600
A-36 see DAE600
A 71 see CHG000
A 100 (pharmaceutical) see IGS000
A-139 see BDC750
A-20D see GLU000
A 361 see ARQ725
A 363 see DOR400
A 3-80 see SMQ500
A 468 see DLS800
A-502 see SNJ000
688A see DDG800
A 688 see PDT250
A 884 see DBA800
A-980 see CEW500
A 1141 see GGS000
A-1981 see DTO200
A 2079 see BJP000
A 2297 see TLP750
A-2371 see MQW750
A 3322 see TEO250
A-4760 see EDM000
A-4828 see TNT500
A 4942 see IMH000
A 10846 see DUD800
A 11032 see UVA000
A 3823A see MRE225
A-91033 see DQA400
A 1 (SORBENT) see AHE250
9AA see AHS500
AA-9 see DXW200
AAB see PEI000
AACAPTAN see CBG000
AACIFEMINE see EDU500
AAF see FDR000
2-AAF see FDR000
AAFERTIS see FAS000
AALINDAN see BBQ500
AAMANGAN see MAS500
AAPROTECT see BJK500
AAT see AIC250, PAK000
o-AAT see AIC250
AATACK see TFS350
AATP see PAK000
AATREX see ARQ725
AATREX 4L see ARQ725
AATREX 80W see ARQ725
AATREX NINE-O see ARQ725
AAVOLEX see BJK500
AAZIRA see BJK500
2-AB see BPY000
AB-42 see COH250
ABAR see LEN000
ABASIN see ACE000
ABATE see TAL250
ABATHION see TAL250
ABAVIT see PFP500
ABBOCILLIN see BDY669
ABBOTT-28440 see DEW400
ABBOTT-30360 see FIW000
ABBOTT 40566 see TCM250
ABCID see SNN300

ABENSANIL see HIM000
ABESON NAM see DXW200
ABIES ALBA OIL see AAC250
ABIETIC ACID, METHYL ESTER see MFT500
ABIOL see HJL500
"A" BLASTING POWDER see ERF500
ABMINTHIC see DJT800
ABRACOL S.L.G see OAV000
ABRAMYCIN see TBX000
ABRAREX see AHE250
ABRICYCLINE see TBX000
ABRIN see AAD000
ABRINS see AAD000
ABROVAL see BNP750
ABSENTOL see TLP750
ABSETIL see TLP750
ABSIN see ACE000
ABSINTHIUM see ARL250
ABSOLUTE ETHANOL see EFU000
ABSTENSIL see DXH250
ABSTINYL see DXH250
AC 3422 see EEH600, PAK000
AC 5223 see DXX400
AC 5230 see ADA725
AC-12682 see DSP400
AC 18133 see EPC500
AC 18682 see IOT000
AC-18,737 see EAS000
AC-43064 see DXN600
AC 47031 see PGW750
AC 47470 see DHH400
AC 52160 see TAL250
ACACIA see AQQ500
ACACIA DEALBATA GUM see AQQ500
ACACIA GUM see AQQ500
ACACIA MOLLISSIMA TANNIN see MQV250
ACACIA SENEGAL see AQQ500
ACACIA SYRUP see AQQ500
ACADYL see BJZ000
ACAMOL see HIM000
ACAR see DER000
ACARABEN 4E see DER000
ACARACIDE see SOP500
ACARICYDOL E 20 see CJT750
ACARIN see BIO750
ACARITHION see TNP250
ACARON see CJJ250
ACAVYL see BJZ000
ACC 3422 see PAK000
ACCELERATE see DXD000
ACCELERATOR L see BJK500
ACCELERATOR THIURAM see TFS350
ACCELERINE see DSY600
ACCEL R see BKU500
ACCENT see MRL500
ACCO FAST RED KB BASE see CLK225
ACCOTHION see DSQ000
AC-DI-SOL NF see SFO500
ACECARBROMAL see ACE000
ACECLIDINE see AAE250
ACECOLINE see ABO000
ACEDE CRESYLIQUE (FRENCH) see CNW500
ACEDOXIN see DKL800

ACEDRON see BBK500
ACENTERINE see ADA725
ACEOTHION see DSQ000
ACEPHAT (GERMAN) see DOP600
ACEPHATE see DOP600
ACEPHATE-MET see DTQ400
ACEPROMAZINA see ABH500
ACEPROMAZINE see ABH500
ACEPROMAZINE MALEATE see AAF750
ACEPROMIZINA see ABH500
ACESAL see ADA725
ACETAAL (DUTCH) see AAG000
ACETACID RED B see HJF500
ACETAGESIC see HIM000
ACETAL see AAG000, ADA725
ACETALDEHYD (GERMAN) see AAG250
ACETALDEHYDE see AAG250
ACETALDEHYDE, AMINE SALT see AAG500
ACETALDEHYDE AMMONIA see AAG500
ACETALDEHYDE DIMETHYL ACETAL see DOO600
ACETALDEHYDE-N-FORMYL-N-METHYLHYDRAZONE see
  AAH000
ACETALDEHYDE-N-METHYL-N-FORMYLHYDRAZONE see
  AAH000
ACETALDEHYDE OXIME see AAH250
ACETALDEHYDE, TETRAMER see TDW500
ACETALDEHYDE, TRIMER see PAI250
ACETAL DIETHYLIQUE (FRENCH) see AAG000
ACETALDOL see AAH750
ACETALDOXIME see AAH250
ACETALE (ITALIAN) see AAG000
ACETALGIN see HIM000
ACETAMIDE see AAI000
5-ACETAMIDE-1,3,4-THIADIAZOLE-2-SULFONAMIDE see
  AAI250
ACETAMIDOBENZENE see AAQ500
4-ACETAMIDOBIPHENYL see PDY500
1-ACETAMIDO-4-ETHOXYBENZENE see ABG750
2-ACETAMIDOFLUORENE see FDR000
1-(N-ACETAMIDOFLUOROMETHYL)-NAPHTHALENE see
  MME809
3-ACETAMIDO-4-HYDROXYBENZENEARSONIC ACID see
  ABX500
3-ACETAMIDO-4-HYDROXY-PHENYLARSONIC ACID see
  ABX500
l-α-ACETAMIDO-β-MERCAPTOPROPIONIC ACID see
  ACH000
2-ACETAMIDO-4-(5-NITRO-2-FURYL)THIAZOLE see
  AAL750
ACETAMIDO-5-NITROTHIAZOLE see ABY900
p-ACETAMIDOPHENACYL CHLORIDE see CEC000
4-ACETAMIDOPHENOL see HIM000
p-ACETAMIDOPHENOL see HIM000
2-ACETAMIDO-5-SULFONAMIDO-1,3,4-THIADIAZOLE see
  AAI250
ACETAMIDOTHIADIAZOLESULFONAMIDE see AAI250
p-ACETAMIDOTOLUENE see ABJ250
ACETAMINE DIAZO BLACK RD see DCJ200
ACETAMINE YELLOW CG see AAQ250
2-ACETAMINOFLUORENE see FDR000
2-ACETAMINO-4-(5-NITRO-2-FURYL)THIAZOLE see
  AAL750
4-ACETAMINO-2-NITROPHENETOLE see NEL000
ACETAMINOPHEN see HIM000
p-ACETAMINOPHENOL see HIM000
ACETAMOX see AAI250
ACETANIL see AAQ500
ACETANILIDE see AAQ500
ACETANISOLE (FCC) see MDW750
ACETARSOL see ABX500
ACETARSONE see ABX500
ACETATE d'AMYLE (FRENCH) see AOD725
ACETATE de BUTYLE (FRENCH) see BPU750
ACETATE de BUTYLE SECONDAIRE (FRENCH) see
  BPV000

ACETATE C-8 see OEG000
ACETATE de CELLOSOLVE (FRENCH) see EES400
ACETATE de CUIVRE (FRENCH) see CNI250
ACETATE de l'ETHER MONOETHYLIQUE DE L'ETH-
  YLENE-GLYCOL (FRENCH) see EES400
ACETATE d'ETHYLGLYCOL (FRENCH) see EES400
ACETATE FAST ORANGE R see AKP750
ACETATE d'ISOBUTYLE (FRENCH) see IIJ000
ACETATE d'ISOPROPYLE (FRENCH) see INE100
ACETATE de L'ETHER MONOMETHYLIQUE de L'ETH-
  YLENE-GLYCOL (FRENCH) see EJJ500
ACETATE of LIME see CAL750
ACETATE de METHYLE (FRENCH) see MFW100
ACETATE de METHYLE GLYCOL (FRENCH) see EJJ500
ACETATE P.A. see AGQ750
ACETATE PHENYLMERCURIQUE (FRENCH) see ABU500
ACETATE de PLOMB (FRENCH) see LCG000
ACETATE de PROPYLE NORMAL (FRENCH) see PNC250
ACETATE de TRIPHENYL-ETAIN (FRENCH) see ABX250
ACETATO(2-AMINO-5-NITROPHENYL)MERCURY see
  ABQ250
(ACETATO)(p-AMINOPHENYL)MERCURY see ABQ000
(ACETATO) di CELLOSOLVE (ITALIAN) see EES400
(ACETATO)(DIETHOXYPHOSPHINYL)MERCURY see
  AAS250
ACETATO DI METIL CELLOSOLVE (ITALIAN) see EJJ500
ACETATO(2-METHOXYETHYL)MERCURY see MEO750
(ACETATO)PHENYLMERCURY see ABU500
ACETATO di STAGNO TRIFENILE (ITALIAN) see ABX250
(ACETATO)(2,3,5,6-TETRAMETHYLPHENYL)MERCURY
  see AAS500
(ACETATO)(TRIMETAARSENITO)DICOPPER see COF500
ACETATOTRIPHENYLSTANNANE see ABX250
ACETAZINE see ABH500
ACETAZOLAMID see AAI250
ACETAZOLAMIDE see AAI250
ACETAZOLAMIDE SODIUM see AAS750
ACETAZOLAMIDE SODIUM SALT see AAS750
ACETAZOLEAMIDE see AAI250
ACETCARBROMAL see ACE000
ACETDIMETHYLAMIDE see DOO800
ACETEIN see ACH000
ACETENE see EIO000
ACETETHYLANILIDE see EFQ500
ACETEUGENOL see EQS000
ACETHYDRAZIDE see ACM750
ACETHYLPROMAZIN see ABH500
ACETIC ACID see AAT250
ACETIC ACID (aqueous solution) (DOT) see AAT250
ACETIC ACID (N-ACETYL-N-(2-FLUORENYL)AMINO) ES-
  TER see ABL000
ACETIC ACID ALLYL ESTER see AFU750
ACETIC ACID AMIDE see AAI000
ACETIC ACID, AMMONIUM SALT see ANA000
ACETIC ACID, AMYL ESTER see AOD725, AOD750
ACETIC ACID, ANHYDRIDE see AAX500
ACETIC ACID, ANHYDRIDE with NITRIC ACID (1:1) see
  ACS750
ACETIC ACID ANILIDE see AAQ500
ACETIC ACID BENZYL ESTER see BDX000
ACETIC ACID-1,3-BUTADIENYL ESTER see ABM250
ACETIC ACID-2-BUTOXY ESTER see BPV000
ACETIC ACID n-BUTYL ESTER see BPU750
ACETIC ACID-tert-BUTYL ESTER see BPV100
ACETIC ACID, CADMIUM SALT see CAD250
ACETIC ACID CHLORIDE see ACF750
ACETIC ACID, CINNAMYL ESTER see CMQ730
ACETIC ACID, CITRONELLYL ESTER see AAU000
ACETIC ACID, COBALT(2+) SALT, TETRAHYDRATE see
  CNA500
ACETIC ACID, CUPRIC SALT see CNI250
ACETIC ACID DIMETHYLAMIDE see DOO800
ACETIC ACID-1,3-DIMETHYLBUTYL ESTER see HFJ000
ACETIC ACID-2,6-DIMETHYL-m-DIOXAN-4-YL ESTER see
  ABC250

ACETIC ACID-1,1-DIMETHYLETHYL ESTER see BPV100
ACETIC ACID-3,7-DIMETHYL-6-OCTEN-1-YL ESTER see AAU000
ACETIC ACID-(2,4-DINITRO-6-sec-BUTYLPHENYL) ESTER see ACE500
ACETIC ACID-(4,6-DINITRO-2-sec-BUTYLPHENYL) ESTER see ACE500
ACETIC ACID-4,6-DINITRO-o-CRESYL ESTER see AAU250
ACETIC ACID ETHENYL ESTER see VLU250
ACETIC ACID ETHENYL ESTER HOMOPOLYMER see AAX250
ACETIC ACID ETHENYL ESTER POLYMER with CHLOR-ETHENE (9CI) see AAX175
ACETIC ACID-2-ETHOXYETHYL ESTER see EES400
ACETIC ACID GERANIOL ESTER see DTD800
ACETIC ACID, GLACIAL (DOT) see AAT250
ACETIC ACID HEXYL ESTER see HFI500
ACETIC ACID, ISOBUTYL ESTER see IIJ000
ACETIC ACID, ISOPENTYL ESTER see IHO850
ACETIC ACID ISOPROPYL ESTER see INE100
ACETIC ACID LEAD (2+) SALT see LCG000
ACETIC ACID, LEAD(+2) SALT TRIHYDRATE see LCJ000
ACETIC ACID LINALOOL ESTER see LFY100
ACETIC ACID, MAGNESIUM SALT see MAD000
ACETIC ACID, MERCURY(2+) SALT see MCS750
ACETIC ACID-3-METHOXYBUTYL ESTER see MHV750
ACETIC ACID METHYL ESTER see MFW100
ACETIC ACID-1-METHYLETHYL ESTER (9CI) see INE100
ACETIC ACID-2-METHYL-6-METHYLENE-7-OCTEN-2-YL ESTER see AAW500
ACETIC ACID METHYLNITROSAMINOMETHYL ESTER see AAW000
ACETIC ACID-4-METHYLPHENYL ESTER see MNR250
ACETIC ACID-2-METHYLPROPYL ESTER see IIJ000
ACETIC ACID-1-METHYLPROPYL ESTER (9CI) see BPV000
ACETIC ACID MYRCENYL ESTER see AAW500
ACETIC ACID, NICKEL(2+) SALT see NCX000
ACETIC ACID, OCTYL ESTER see OEG000
ACETIC ACID-2-PHENYLETHYL ESTER see PFB250
ACETIC ACID, PHENYLMERCURY DERIV. see ABU500
ACETIC ACID PHENYLMETHYL ESTER see BDX000
ACETIC ACID-2-PROPENYL ESTER see AFU750
ACETIC ACID, n-PROPYL ESTER see PNC250
ACETIC ACID-1-(PROPYLNITROSAMINO)PROPYL ESTER see ABT750
ACETIC ACID, SAMARIUM SALT see SAR000
ACETIC ACID, SODIUM SALT see SEG500
ACETIC ACID VINYL ESTER see VLU250
ACETIC ACID, VINYL ESTER, POLYMER with CHLORO-ETHYLENE see AAX175
ACETIC ACID VINYL ESTER POLYMERS see AAX250
ACETIC ACID, ZINC SALT see ZBS000
ACETIC ALDEHYDE see AAG250
ACETIC ANHYDRIDE see AAX500
ACETIC CHLORIDE see ACF750
ACETIC ETHER see EFR000
ACETIC OXIDE see AAX500
ACETICYL see ADA725
ACETIDIN see EFR000
ACETILSALICILICO see ADA725
ACETILUM ACIDULATUM see ADA725
ACETIMIDIC ACID see AAI000
ACETISAL see ADA725
ACETISOEUGENOL see AAX750
ACETKARBROMAL see ACE000
ACETOACETAMIDOBENZENE see AAY000
ACETOACETANILIDE see AAY000
o-ACETOACETANISIDE see ABA500
ACETOACET-o-ANISIDIN (CZECH) see ABA500
ACETOACETIC ACID ANILIDE see AAY000
ACETOACETIC ACID-o-ANISIDIDE see ABA500
ACETOACETIC ACID BUTYL ESTER see BPV250
ACETOACETIC ACID, ETHYL ESTER see EFS000

ACETOACETIC ANILIDE see AAY000
ACETOACETIC ESTER see EFS000
ACETOACETIC METHYL ESTER see MFX250
ACETOACETONE see ABX750
ACETOACET-o-TOLUIDIDE see ABA000
2-ACETOACETYLAMINOANISOLE see ABA500
((ACETOACETYL)AMINO)BENZENE see AAY000
2-ACETOACETYLAMINOTOLUENE see ABA000
ACETOACETYLANILINE see AAY000
ACETOACETYL-o-ANISIDE see ABA500
ACETOACETYL-o-ANISIDINE see ABA500
ACETOACETYL-o-ANISINE see ABA500
ACETOACETYL-2-METHYLANILIDE see ABA000
ACETOAMINOFLUORENE see FDR000
ACETOANILIDE see AAQ500
ACETOARSENITE de CUIVRE (FRENCH) see COF500
ACETO AZIB see ASL750
ACETO-CAUSTIN see TII250
ACETO DIPP see NBL000
ACETO HMT see HEI500
ACETOHYDRAZIDE see ACM750
ACETOIN see ABB500
ACETOL see ADA725
ACETOL (1) see ABC000
ACETOMETHOXAN see ABC250
ACETOMETHOXANE see ABC250
ACETOMORFINE see HBT500
ACETOMORPHINE see HBT500
β-ACETONAPHTHALENE see ABC500
ACETONAPHTHONE see ABC500
2-ACETONAPHTHONE see ABC500
β-ACETONAPHTHONE see ABC500
2'-ACETONAPHTHONE see ABC500
ACETONCIANHIDRINEI (ROUMANIAN) see MLC750
ACETONCIANIDRINA (ITALIAN) see MLC750
ACETONCYAANHYDRINE (DUTCH) see MLC750
ACETONCYANHYDRIN (GERMAN) see MLC750
ACETON (GERMAN, DUTCH, POLISH) see ABC750
ACETONE see ABC750
ACETONE ANIL see TLP500
ACETONE CHLOROFORM see ABD000
ACETONECYANHYDRINE (FRENCH) see MLC750
ACETONE CYANOHYDRIN (DOT) see MLC750
ACETONE OIL see ABD750
ACETONE PEROXIDE see ABE000
ACETONE SEMICARBAZONE see ABE250
ACETONIC ACID see LAG000
ACETONITRIL (GERMAN, DUTCH) see ABE500
ACETONITRILE see ABE500
ACETONKYANHYDRIN (CZECH) see MLC750
ACETONYL see ADA725
3-(α-ACETONYLBENZYL)-4-HYDROXYCOUMARIN see WAT200
ACETONYL BROMIDE see BNZ000
ACETONYL CHLORIDE see CDN200
3-(α-ACETONYLFURFURYL)-4-HYDROXYCOUMARIN see ABF500
ACETO PBN see PFT500
ACETOPHEN see ADA725
ACETO-p-PHENALIDE see ABG750
ACETOPHENAZINE see ABG000
ACETOPHENAZINE MALEATE see ABG000
p-ACETOPHENETIDE see ABG750
p-ACETOPHENETIDIDE see ABG750
ACETO-p-PHENETIDIDE see ABG750
ACETOPHENETIDIN see ABG750
ACETOPHENETIDINE see ABG750
ACETO-4-PHENETIDINE see ABG750
ACETOPHENETIN see ABG750
ACETOPHENONE see ABH000
ACETOPHOS see DIW600
ACETOPROMAZINE see ABH500
ACETOQUAT CPC see CCX000
ACETOQUAT CTAB see HCQ500

ACETOQUINONE LIGHT ORANGE JL see AKP750
ACETOSAL see ADA725
ACETOSALIC ACID see ADA725
ACETOSALIN see ADA725
ACETO SDD 40 see SGM500
ACETO TETD see TFS350
ACETOTHIOAMIDE see TFA000
ACETO TMTM see BJL600
4-ACETOTOLUIDE see ABJ250
p-ACETOTOLUIDE see ABJ250
p-ACETOTOLUIDIDE see ABJ250
ACETOXON see DIW600
N-ACETOXY-2-ACETAMIDOFLUORENE see ABL000
N-ACETOXY-2-ACETYLAMINOFLUORENE see ABL000
N-ACETOXY-N-ACETYL-2-AMINOFLUORENE see ABL000
2-ACETOXYACRYLONITRILE see ABL500
α-ACETOXYACRYLONITRILE see ABL500
2-ACETOXYBENZOIC ACID see ADA725
o-ACETOXYBENZOIC ACID see ADA725
1-ACETOXY-1,3-BUTADIENE see ABM250
17-ACETOXY-6-CHLORO-6-DEHYDROPROGESTERONE see
  CBF250
17-α-ACETOXY-6-CHLORO-6-DEHYDROPROGESTERONE
  see CBF250
17-α-ACETOXY-6-CHLORO-6,7-DEHYDROPROGESTERONE
  see CBF250
2-ACETOXY-2′-CHLORO-N-METHYL-DIETHYLAMINE see
  MFW750
17-α-ACETOXY-6-CHLORO-4,6-PREGNADIENE-3,20-DIONE
  see CBF250
17-α-ACETOXY-6-CHLOROPREGNA-4,6-DIENE-3,20-DIONE
  see CBF250
ACETOXYCYCLOHEXIMIDE see ABN000
17-α-ACETOXY-6-DEHYDRO-6-METHYLPROGESTERONE
  see VTF000
2-ACETOXY-3-DIETHYLCARBAMYL-9,10-DIMETHOXY-
  1,2,3,4,6,7-HEXAHYDRO-11B-BENZO(a)QUINOLIZINE
  see BCL250
ACETOXYDIETHYLPHENYLSTANNANE see DJV800
6-ACETOXY-2,4-DIMETHYL-m-DIOXANE see ABC250
α-ACETOXY DIMETHYLNITROSAMINE see AAW000
3-(2-(5-ACETOXY-3,5-DIMETHYL-2-OXOCYCLOHEXYL)-2-
  HYDROXYETHYL)GLUTARIMIDE see ABN000
ACETOXYETHANE see EFR000
N-ACETOXYETHYL-N-CHLOROETHYLMETHYLAMINE see
  MFW750
1-ACETOXYETHYLENE see VLU250
2-ACETOXYETHYLTRIMETHYLAMMONIUM CHLORIDE
  see ABO000
N-ACETOXY-2-FLUORENYLACETAMIDE see ABL000
2-ACETOXYISOSUCCINODINITRILE see MFX000
ACETOXYL see BDS000
p-(ACETOXYMERCURI)ANILINE see ABQ000
(ACETOXYMERCURI)BENZENE see ABU500
2-(ACETOXYMERCURI)-4-NITROANILINE see ABQ250
1-ACETOXY-2-METHOXY-4-ALLYLBENZENE see EQS000
4-ACETOXY-3-METHOXY-1-PROPENYLBENZENE see
  AAX750
ACETOXYMETHYLBUTYLNITROSAMINE see BRX500
N-(ACETOXY)METHYL-N,N-BUTYLNITROSAMINE see
  BRX500
ACETOXYMETHYLETHYLNITROSAMINE see ENR500
N-(ACETOXY)METHYL-N-ETHYLNITROSAMINE see
  ENR500
ACETOXYMETHYL-METHYL-NITROSAMIN (GERMAN) see
  AAW000
ACETOXYMETHYL METHYLNITROSAMINE see AAW000
N-α-ACETOXYMETHYL-N-METHYLNITROSAMINE see
  AAW000
N-ACETOXYMETHYL-N-NITROSOETHYLAMINE see
  ENR500
N-(1-ACETOXYMETHYL)-N-NITROSOETHYL AMINE see
  ENR500
17-ACETOXY-6-METHYLPREGNA-4,6-DIENE-3,20-DIONE
  see VTF000

17-α-ACETOXY-6-METHYLPREGNA-4,6-DIENE-3,20-DIONE
  see VTF000
17-α-ACETOXY-6-METHYL-4,6-PREGNADIENE-3,20-DIONE
  see VTF000
17-α-ACETOXY-6-α-METHYLPREGN-4-ENE-3,20-DIONE
  see MCA000
17-ACETOXY-6-α-METHYLPROGESTERONE see MCA000
ACETOXYMETHYLPROPYLNITROSAMINE see PNR250
N-(ACETOXY)METHYL-N-n-PROPYLNITROSAMINE see
  PNR250
p-ACETOXYNITROBENZENE see ABS750
1-ACETOXY-N-NITROSODIMETHYLAMINE see AAW000
1-ACETOXY-N-NITROSODIPROPYLAMINE see ABT750
17-ACETOXY-19-NOR-17-α-PREGN-4-EN-20-YN-3-ONE see
  ABU000
17-β-ACETOXY-19-NOR-17-α-PREGN-4-EN-20-YN-3-ONE see
  ABU000
2-ACETOXYPENTANE see AOD735
ACETOXYPHENYLMERCURY see ABU500
3-ACETOXYPHENYLTRIMETHYLAMMONIUM IODIDE see
  ABV250
1-ACETOXYPROPANE see PNC250
2-ACETOXYPROPANE see INE100
3-ACETOXYPROPENE see AFU750
N-(α-ACETOXY)PROPYL-N-N-PROPYLNITROSAMINE see
  ABT750
3-ACETOXYQUINUCLIDINE GLAUCOSTAT see AAE250
4-ACETOXYTOLUENE see MNR250
p-ACETOXYTOLUENE see MNR250
α-ACETOXYTOLUENE see BDX000
ACETOXYTRIETHYLSTANNANE see ABW750
ACETOXYTRIETHYLTIN see ABW750
ACETOXYTRIHEXYLSTANNANE see ABX000
ACETOXYTRIHEXYLTIN see ABX000
ACETOXYTRIISOPROPYLSTANNANE see TKT750
ACETOXYTRIMETHYLSTANNANE see TMI000
ACETOXY-TRIPHENYL-STANNAN (GERMAN) see
  ABX250
ACETOXYTRIPHENYLSTANNANE see ABX250
ACETOXY-TRIPHENYL-STANNANE see ABX250
ACETOXYTRIPHENYLTIN see ABX250
ACETOZALAMIDE see AAI250
ACETO ZDBD see BIX000
ACETO ZDED see BJK500
ACETO ZDMD see BJK500
ACET-p-PHENALIDE see ABG750
ACETPHENARSINE see ABX500
ACETPHENETIDIN see ABG750
ACET-p-PHENETIDIN see ABG750
p-ACETPHENETIDIN see ABG750
ACET-THEOCIN see TEP000
ACETYLACETANILIDE see AAY000
α-ACETYLACETANILIDE see AAY000
ACETYL ACETONE see ABX750
N-(ACETYLACETYL)ANILINE see AAY000
ACETYL ADALIN see ACE000
ACETYLADRIAMYCIN see DAC000
ACETYLAMINOBENZENE see AAQ500
4-ACETYLAMINOBIPHENYL see PDY500
N-((ACETYLAMINO)CARBONYL)-2-BROMO-2-ETHYLBU-
  TANAMIDE see ACE000
S-(2-(ACETYLAMINO)ETHYL)-O,O-DIMETHYL PHOSPHO-
  RODITHIOATE see DOP200
2-ACETYLAMINO-FLUOREN (GERMAN) see FDR000
4-ACETYLAMINOFLUOREN (GERMAN) see ABY000
4-ACETYLAMINOFLUORENE see ABY000
N-ACETYL-2-AMINOFLUORENE see FDR000
2-ACETYLAMINOFLUORENE (OSHA) see FDR000
3-ACETYLAMINO-4-HYDROXYPHENYLARSONIC ACID see
  ABX500
(3-(ACETYLAMINO)-4-HYDROXYPHENYL)ARSONINE (9CI)
  see ABX500
2-ACETYLAMINO-4-(5-NITRO-2- FURYL)THIAZOLE see
  AAL750

2-ACETYLAMINO-5-NITROTHIAZOLE see ABY900
p-(ACETYLAMINO)PHENACYL CHLORIDE see CEC000
p-ACETYLAMINOPHENOL see HIM000
N-ACETYL-p-AMINOPHENOL see HIM000
p-ACETYLAMINOPHENYL DERIVATIVE of NITROGEN
     MUSTARD see BHO500
2-ACETYLAMINO-1,3,4-THIADIAZOLE-5-SULFONAMIDE
     see AAI250
4-(ACETYLAMINO)TOLUENE see ABJ250
ACETYL ANHYDRIDE see AAX500
ACETYLANILINE see AAQ500
3-ACETYLANILINE see AHR500
4-ACETYLANILINE see AHR750
m-ACETYLANILINE see AHR500
N-ACETYLANILINE see AAQ500
4-ACETYLANISOLE see MDW750
p-ACETYLANISOLE see MDW750
1-ACETYLAZIRIDINE see ACB250
ACETYLBENZENE see ABH000
ACETYL BENZOYL PEROXIDE (solid) see ACC250
ACETYL BENZOYLPEROXIDE (solution) see ACC500
N-ACETYL-4-BIPHENYLHYDROXYLAMINE see ACD000
N-(N-ACETYL-3-(p-(BIS(2-CHLOROETHYL)AMINO)PHE-
     NYL)ALANYL-3-PHENYLALANINE ETHYL ESTER see
     ACD250
ACETYL BROMIDE see ACD750
ACETYLBROMODIETHYLACETYLCARBAMIDE see
     ACE000
N-ACETYL-N-BROMODIETHYLACETYLCARBAMIDE see
     ACE000
N-ACETYL-N-BROMODIETHYLACETYLUREA see ACE000
N-ACETYL-N'-α-BROMO-α-ETHYLBUTYRYLCARBAMIDE
     see ACE000
1-ACETYL-3-(2-BROMO-2-ETHYLBUTYRYL)UREA see
     ACE000
1-ACETYL-3-(α-BROMO-α-ETHYLBUTYRYL)UREA see
     ACE000
o-ACETYL-2-sec-BUTYL-4,6-DINITROPHENOL see ACE500
3'-o-ACETYLCALOTROPIN see ACF000
ACETYLCARBROMAL see ACE000
ACETYL CHLORIDE see ACF750
ACETYLCHOLINE CHLORIDE see ABO000
ACETYLCHOLINE HYDROCHLORIDE see ABO000
ACETYLCHOLINIUM CHLORIDE see ABO000
ACETYLCYSTEINE see ACH000
N-ACETYLCYSTEINE see ACH000
N-ACETYL-l-CYSTEINE see ACH000
N-ACETYL-N-CYSTEINE see ACH000
N-ACETYL-l-CYSTEINE (9CI) see ACH000
1-ACETYL-9,10-DIDEHYDRO-N,N-DIETHYL-6-METHYLER-
     GOLINE-8-β-CARBOXAMIDE BITARTRATE see ACP500
1-ACETYL-9,10-DIDEHYDRO-N-ETHYL-6-METHYLERGO-
     LINE-8-β-CARBOXAMIDE see ACP750
ACETYLDIGITOXIN-α see ACH500
α-ACETYLDIGITOXIN see ACH500
ACETYLDIGOXIN-α see ACI000
α-ACETYLDIGOXIN see ACI000
ACETYLDIGOXIN-β see ACI250
β-ACETYLDIGOXIN see ACI250
ε-ACETYLDIGOXIN (GERMAN) see DKN600
3-ACETYL-10-(3-DIMETHYLAMINOPROPYL)PHENO-
     THIAZINE see ABH500
2-ACETYL-10-(3-(DIMETHYLAMINO)PROPYL)PHENO-
     THIAZINE, MALEATE see AAF750
ACETYLEN see ACI750
ACETYLENE see ACI750
ACETYLENE, dissolved (DOT) see ACI750
ACETYLENE BLACK see CBT750
ACETYLENE CHLORIDE see ACJ000
ACETYLENEDICARBOXAMIDE see ACJ250
ACETYLENEDICARBOXYLIC ACID DIAMIDE see ACJ250
ACETYLENEDICARBOXYLIC ACID MONOPOTASSIUM
     SALT see ACJ500

ACETYLENE DICHLORIDE see DFI100
trans-ACETYLENE DICHLORIDE see ACK000
ACETYLENE TETRABROMIDE see ACK250
ACETYLENE TETRACHLORIDE see TBQ100
ACETYLENE TRICHLORIDE see TIO750
ACETYL ETHER see AAX500
ACETYL ETHYLENE see BOY500
ACETYLETHYLENEIMINE see ACB250
ACETYL ETHYL TETRAMETHYL TETRALIN see ACL750
ACETYLETHYL TETRAMETHYLTETRALIN see ACL750
ACETYLEUGENOL see EQS000
N-ACETYL-N-9H-FLUOREN-2-YL-ACETAMIDE see
     DBF200
ACETYL FLUORIDE see ACM000
ACETYL HYDRAZIDE see ACM750
N-ACETYLHYDRAZINE see ACM750
ACETYL HYDROPEROXIDE see PCL500
N-ACETYL-4-HYDROXY-m-ARSANILIC ACID see ABX500
N-ACETYL-4-HYDROXY-m-ARSANILIC ACID, CALCIUM
     SALT see CAL500
2-ACETYL-10-(3-(4-(β-HYDROXYETHYL)PIPERAZINYL)
     PROPYL)PHENOTHIAZINE see ABG000
2-ACETYL-5-HYDROXY-3-OXO-4-HEXENOIC ACID Δ-LAC-
     TONE see MFW750
ACETYLIDES see ACO000
ACETYLIN see ADA725
ACETYL IODIDE see ACO500
ACETYLISOEUGENOL see AAX750
ACETYLKIDAMYCIN see ACP000
1-ACETYLLYSERGIC ACID DIETHYLAMIDE BITARTRATE
     see ACP500
1-ACETYLLYSERGIC ACID ETHYLAMIDE see ACP750
d-1-ACETYL LYSERGIC ACID MONOETHYLAMIDE see
     ACP750
ACETYL MERCAPTAN see TFA500
N-ACETYL-3-MERCAPTOALANINE see ACH000
4-(N-ACETYL-N-METHYL)AMINO-4'-(N',N'-DIMETHYLAM-
     INO)AZOBENZENE see DPQ200
N'-ACETYL-N'-METHYL-4'-AMINO-N,N-DIMETHYL-4-AMI-
     NOAZOBENZENE see DPQ200
ACETYL METHYL BROMIDE see BNZ000
ACETYL METHYL CARBINOL see ABB500
N-ACETYL-N-(2-METHYL-4-((2-METHYLPHENYL)AZO)
     PHENYL)ACETAMIDE see ACR300
ACETYL-METHYL-NITROSO-HARNSTOFF (GERMAN) see
     ACR400
ACETYLMETHYLNITROSOUREA see ACR400
N'-ACETYL-METHYLNITROSOUREA see ACR400
3-ACETYL-6-METHYL-2,4-PYRANDIONE see MFW500
3-ACETYL-6-METHYLPYRANDIONE-2,4 see MFW500
3-ACETYL-6-METHYL-2H-PYRAN-2,4(3H)-DIONE see
     MFW500
2-ACETYLNAPHTHALENE see ABC500
β-ACETYLNAPHTHALENE see ABC500
ACETYL NITRATE see ACS750
ACETYL OXIDE see AAX500
2-(ACETYLOXY)BENZOIC ACID see ADA725
2-(ACETYLOXY)BENZOIC ACID, mixed with 3,7-DIHYDRO-
     1,3,7-TRIMETHYL-1H-PURINE-2,6-DIONE and N-(4-
     ETHOXYPHENYL)ACETAMIDE see ARP250
17-(ACETYLOXY)-6-CHLOROPREGNA-4,6-DIENE-3,20-
     DIONE see CBF250
ACETYLOXYCYCLOHEXIMIDE see ABN000
4-ACETYLOXY-12,13-EPOXY-3,7,15-TRIHYDROXY-(3-α,4-
     β,7-β)-TRICHOTHEC-9-EN-8-ONE see FQR000
17-(ACETYLOXY)-6-METHYL-16-METHYLENEPREGNA-4,6-
     DIENE-3,20-DIONE (9CI) see MCB380
(6-α)-17-(ACETYLOXY)-6-METHYLPREG-4-ENE-3,20-DIONE
     see MCA000
(17-α)-17-(ACETYLOXY)-19-NORPREGN-4-EN-20-YN-3-ONE
     see ABU000
17-ACETYLOXY(17-α)-19-NORPREGN-4-ESTREN-17-β-OL-
     ACETATE-3-ONE see ABU000

2-(ACETYLOXY)-N,N,N-TRIMETHYLETHANAMINIUM
CHLORIDE see ABO000
(ACETYLOXY)TRIPHENYL-STANNANE (9CI) see ABX250
ACETYL PEROXIDE see ACV500
ACETYLPHENETIDIN see ABG750
N-ACETYL-p-PHENETIDINE see ABG750
ACETYLPHOSPHORAMIDOTHIOIC ACID-O,S-DIMETHYL
ESTER see DOP600
ACETYLPROMAZINE see ABH500
ACETYLPROMAZINE MALEATE (1:1) see AAF750
ACETYLSAL see ADA725
ACETYLSALICYLIC ACID see ADA725
ACETYLSALICYLSAURE (GERMAN) see ADA725
6-ACETYL-1,1,4,4-TETRAMETHYL-7-ETHYL-1,2,3,4,-
TETRALIN see ACL750
7-ACETYL-1,1,4,4-TETRAMETHYL-1,2,3,4-TETRAHYDRO-
NAPHTHALENE see ACL750
1-ACETYL-2-THIOHYDANTOIN see ADC750
p-ACETYLTOLUENE see MFW250
N-ACETYL-p-TOLUIDIDE see ABJ250
ACETYL-p-TOLUIDINE see ABJ250
ACETYL TRIETHYL CITRATE see ADD750
ACH CHLORIDE see ABO000
ACHIOTE see APE100
ACHROCIDIN see ABG750
ACHROMYCIN see TBX000, TBX250
ACHROMYCIN HYDROCHLORIDE see TBX250
ACID see DJO000
ACIDAL FAST ORANGE see HGC000
ACIDAL LIGHT GREEN SF see FAF000
ACID AMIDE see NCR000
ACID AMMONIUM CARBONATE see ANB250
ACID BLUE 9 see FMU059
ACID BLUE W see FAE100
ACID BRILLIANT GREEN SF see FAF000
ACID BRILLIANT PINK B see FAG070
ACID BRILLIANT RUBINE 2G see HJF500
ACID BUTYL PHOSPHATE see ADF250
ACID CARBOYS, EMPTY see ADF500
ACID CHROME BLUE BA see HJF500
ACID COPPER ARSENITE see CNN500
ACIDE ACETIQUE (FRENCH) see AAT250
ACIDE ACETYLSALICYLIQUE (FRENCH) see ADA725
ACIDE ANISIQUE (FRENCH) see MPI000
ACIDE ARSENIEUX (FRENCH) see ARI750
ACIDE ARSENIQUE LIQUIDE (FRENCH) see ARB250
ACIDE BENZOIQUE (FRENCH) see BCL750
ACIDE BROMACETIQUE (FRENCH) see BMR750
ACIDE BROMHYDRIQUE (FRENCH) see HHJ000
ACIDE CACODYLIQUE (FRENCH) see HKC000
ACIDE CARBOLIQUE (FRENCH) see PDN750
ACIDE CHLORACETIQUE (FRENCH) see CEA000
ACIDE CHLORHYDRIQUE (FRENCH) see HHL000
ACIDE 2-(4-CHLORO-2-METHYL-PHENOXY)PROPIONIQUE
(FRENCH) see CIR500
ACIDE CHROMIQUE (FRENCH) see CMH250
ACIDE CYANHYDRIQUE (FRENCH) see HHS000
ACIDE-2,4-DICHLORO PHENOXYACETIQUE (FRENCH) see
DAA800
ACIDE-2-(2,4-DICHLORO-PHENOXY) PROPIONIQUE
(FRENCH) see DGB000
ACIDE DIMETHYLARSINIQUE (FRENCH) see HKC000
ACIDE ETHYLENEDIAMINETETRACETIQUE (FRENCH) see
EIX000
ACIDE 1-ETIL-7-METIL-1,8-NAFTIRIDIN-4-ONE-3-CARBOS-
SILICO (ITALIAN) see EID000
ACIDE FLUORHYDRIQUE (FRENCH) see HHU500
ACIDE FLUOROSILICIQUE (FRENCH) see SCO500
ACIDE FLUOSILICIQUE (FRENCH) see SCO500
ACIDE FORMIQUE (FRENCH) see FNA000
ACIDE METHYL-o-BENZOIQUE (FRENCH) see MPI000
ACIDE MONOCHLORACETIQUE (FRENCH) see CEA000
ACIDE-MONOFLUORACETIQUE (FRENCH) see FIC000
ACIDE NALIDIXICO (ITALIAN) see EID000
ACIDE NALIDIXIQUE (FRENCH) see EID000

ACIDE NICOTINIQUE (FRENCH) see NCQ900
ACIDE NITRIQUE (FRENCH) see NED500
l'ACIDE OLEIQUE (FRENCH) see OHU000
ACIDE OXALIQUE (FRENCH) see OLA000
ACIDE PÉRACETIQUE (FRENCH) see PCL500
ACIDE PHOSPHORIQUE (FRENCH) see PHB250
ACIDE PHTALIQUE (FRENCH) see PHW250
ACIDE PICRIQUE (FRENCH) see PID000
ACIDE PROPIONIQUE (FRENCH) see PMU750
ACIDE SULFHYDRIQUE (FRENCH) see HIC500
ACIDE SULFURIQUE (FRENCH) see SOI500
d'ACIDE TANNIQUE (FRENCH) see TAD750
ACIDE THIOGLYCOLIQUE (FRENCH) see TFJ100
ACIDE TRICHLORACETIQUE (FRENCH) see TII250
ACIDE 2,4,5-TRICHLORO PHENOXYACETIQUE (FRENCH)
see TAA100
ACIDE 2-(2,4,5-TRICHLORO-PHENOXY) PROPIONIQUE
(FRENCH) see TIX500
ACID FAST ORANGE EGG see HGC000
ACID FAST RED FB see HJF500
ACID GREEN A see FAF000
ACID LEAD ARSENATE see LCK000
ACID LEAD ORTHOARSENATE see LCK000
ACID LEATHER BLUE IC see FAE100
ACID LEATHER BROWN 2G see XMA000
ACID LEATHER ORANGE PGW see HGC000
ACID LEATHER YELLOW T see FAG140
ACID LIGHT ORANGE G see HGC000
ACIDO ACETICO (ITALIAN) see AAT250
ACIDO o-ACETIL-BENZOICO (ITALIAN) see ADA725
ACIDO ACETILSALICILICO (ITALIAN) see ADA725
ACIDO BROMIDRICO (ITALIAN) see HHJ000
ACIDO CIANIDRICO (ITALIAN) see HHS000
ACIDO CLORIDRICO (ITALIAN) see HHL000
ACIDO 2-(4-CLORO-2-METIL-FENOSSI)-PROPIONICO
(ITALIAN) see CIR500
ACIDO (2,4-DICLORO-FENOSSI)-ACETICO (ITALIAN) see
DAA800
ACIDO-2-(2,4-DICLORO-FENOSSI)-PROPIONICO (ITALIAN)
see DGB000
ACIDO (3,6-DICLORO-2-METOSSI)-BENZOICO (ITALIAN)
see MEL500
ACIDO-5-FENIL-5-ETILBARBITURICO (ITALIAN) see
EOK000
ACIDO FLUORIDRICO (ITALIAN) see HHU500
ACIDO FLUOSILICICO (ITALIAN) see SCO500
ACIDO FORMICO (ITALIAN) see FNA000
ACIDO FOSFORICO (ITALIAN) see PHB250
ACIDOMONOCLOROACETICO (ITALIAN) see CEA000
ACIDO MONOFLUOROACETIO (ITALIAN) see FIC000
ACIDO NITRICO (ITALIAN) see NED500
ACIDO OSSALICO (ITALIAN) see OLA000
ACIDO PICRICO (ITALIAN) see PID000
ACID ORANGE 10 see HGC000
ACID ORANGE 24 see XMA000
ACIDO SALICILICO (ITALIAN) see SAI000
ACIDO SOLFORICO (ITALIAN) see SOI500
ACIDO TRICLOROACETICO (ITALIAN) see TII250
ACIDO (2,4,5-TRICLORO-FENOSSI)-ACETICO (ITALIAN)
see TAA100
ACIDO 2-(2,4,5-TRICLORO-FENOSSI)-PROPIONICO (ITAL-
IAN) see TIX500
ACID POTASSIUM SULFATE see PKX750
ACID QUININE HYDROCHLORIDE see QIJ000
ACID RUBINE see HJF500
ACID SKY BLUE A see FAE000
ACID, SLUDGE (DOT) see SEA000
ACID-TREATED HEAVY NAPHTHENIC DISTILLATE see
MQV760
ACID-TREATED HEAVY PARAFFINIC DISTILLATE see
MQV765
ACID-TREATED LIGHT NAPHTHENIC DISTILLATE see
MQV770
ACID-TREATED LIGHT PARAFFINIC DISTILLATE see
MQV775

ACID-TREATED RESIDUAL OIL see MQV872
ACIDUM ACETYLSALICYLICUM see ADA725
ACIDUM NICOTINICUM see NCQ900
ACID VIOLET see FAG120
ACID YELLOW TRA see FAG150
ACIFLOCTIN see AEN250
ACIGENA see HCL000
ACILAN GREEN SFG see FAF000
ACILAN ORANGE GX see HGC000
ACILAN TURQUOISE BLUE AE see FMU059
ACILAN YELLOW GG see FAG140
ACILETTEN see CMS750
ACILLIN see AIV500
ACIMETION see MDT740
ACIMETTEN see ADA725
ACINETTEN see AEN250
ACINITRAZOLE see ABY900
ACINTENE A see PIH250
ACINTENE DP see MCC250
ACINTENE DP DIPENTENE see MCC250
ACINTENE O see PMQ750
ACISAL see ADA725
ACKET see SAH000
ACL 60 see SGG500
ACL 70 see DGN200
ACL 85 see TIQ750
ACNEGEL see BDS000
ACNESTROL see DKA600
ACOCANTHERIN see OKS000
ACOINE see PDN500
ACON see VSK600
ACONITUM CARMICHAELI see ADI250
ACQUINITE see ADR000, CKN500
ACRALDEHYDE see ADR000
ACRAMINE RED see DBN000
9-ACRIDINAMINE see AHS500
ACRIDINE see ADJ500
2,6-ACRIDINEDIAMINE see DBN000
3,6-ACRIDINEDIAMINE see DBN600
3,6-ACRIDINEDIAMINE, MONOHYDROCHLORIDE (9CI)
   see PMH250
3,6-ACRIDINEDIAMINE SULFATE (2:1) see DBN400
3,6-ACRIDINEDIAMINE SULPHATE see DBN400
ACRIDINE ORANGE see BJF000
ACRIDINE ORANGE FREE BASE see BJF000
ACRIDINE YELLOW BASE see DBT200
ACRIFLAVIN see DBX400
ACRIFLAVINE mixture with PROFLAVINE see DBX400
ACRIFLAVINIUM CHLORIDE see DBX400
ACRIFLAVINIUM CHLORIDUM see DBX400
ACRIFLAVON see DBX400
ACRILAFIL see ADY500
ACROLEIC ACID see ADS750
ACROLEIN see ADR000
ACROLEINA (ITALIAN) see ADR000
ACROLEIN CYANOHYDRIN see HJQ000
ACROLEIN DIMER see ADR500
ACROLEINE (DUTCH, FRENCH) see ADR000
ACROMONA see MMN250
ACRONIZE see CMA750
ACRYLALDEHYD (GERMAN) see ADR000
ACRYLALDEHYDE see ADR000
ACRYLAMIDE see ADS250
ACRYLATE d'ETHYLE (FRENCH) see EFT000
ACRYLATE de METHYLE (FRENCH) see MGA500
ACRYLIC ACID see ADS750
ACRYLIC ACID, inhibited (DOT) see ADS750
ACRYLIC ACID (ACGIH,DOT,OSHA) see ADS750
ACRYLIC ACID BUTYL ESTER see BPW100
ACRYLIC ACID n-BUTYL ESTER (MAK) see BPW100
ACRYLIC ACID, ETHYLENE ESTER see EIP000
ACRYLIC ACID, ETHYLENE GLYCOL DIESTER see
   EIP000
ACRYLIC ACID ETHYL ESTER see EFT000

ACRYLIC ACID, GLACIAL see ADS750
ACRYLIC ACID-2-HYDROXYPROPYL ESTER see HNT600
ACRYLIC ACID ISOBUTYL ESTER see IIK000
ACRYLIC ACID-2-METHOXYETHYL ESTER see MIF750
ACRYLIC ACID METHYL ESTER (MAK) see MGA500
ACRYLIC ALDEHYDE see ADR000
ACRYLIC AMIDE see ADS250
ACRYLITE see PKB500
ACRYLNITRIL (GERMAN, DUTCH) see ADX500
ACRYLONITRILE see ADX500
ACRYLONITRILE MONOMER see ADX500
ACRYLONITRILE POLYMER with STYRENE see ADY500
ACRYLONITRILE-STYRENE COPOLYMER see ADY500
ACRYLONITRILE-STYRENE POLYMER see ADY500
ACRYLONITRILE-STYRENE RESIN see ADY500
ACRYLSAEUREAETHYLESTER (GERMAN) see EFT000
ACRYLSAEUREMETHYLESTER (GERMAN) see MGA500
ACRYPET see PKB500
ACS see ADY500
ACTEDRON see BBK000
ACTELIC see DIN800
ACTELLIC see DIN800
ACTELLIFOG see DIN800
ACTICEL see SCH000
ACTI-CHLORE see CDP000
ACTINE see EHP000
ACTINIC RADIATION see AEA000
ACTINOLITE ASBESTOS see ARM260
ACTINOMYCIN 1048A see AEC000
ACTINOMYCIN 2104L see AEB750
ACTINOMYCIN L see AEB750
ACTINOMYCIN S see AEC000
ACTIOQUINONE LIGHT YELLOW see AAQ250
ACTIVATED ALUMINUM OXIDE see AHE250
ACTIVATED CARBON see CDI000
ACTIVE ACETYL ACETATE see EFS000
ACTIVE DICUMYL PEROXIDE see DGR600
ACTOR Q see DVR200
ACTOZINE see BCA000, PGA750
ACTRIL see HKB500
ACTYBARYTE see BAP000
ACTYLOL see LAJ000
ACUPAN see NBS500
ACYLANID see ACH500
ACYLPYRIN see ADA725
ACYTOL see LAJ000
AD1M see AGX000
ADALIN see BNK000
ADAMSITE see PDB000
ADANON see MDO750
ADANON HYDROCHLORIDE see MDP000
ADC AURAMINE O see IBA000
ADC BRILLIANT GREEN CRYSTALS see BAY750
ADC RHODAMINE B see FAG070
ADDISOMNOL see BNK000
ADEPSINE OIL see MQV750
ADERMINE see PPK250
ADERMINE HYDROCHLORIDE see PPK500
ADHERE see MIQ075
ADILACTETTEN see AEN250
ADIPAMIDE see AEN000
ADIPAN see BBK000
ADIPEX see MDQ500
ADIPHENIN see DHX800
ADIPHENINE see DHX800
ADIPIC ACID see AEN250
ADIPIC ACID DIAMIDE see AEN000
ADIPIC ACID DINITRILE see AER250
ADIPIC ACID NITRILE see AER250
ADIPIC ACID, POLYMER with 1,4-BUTANEDIOL and
   METHYLENEDI-p-PHENYLENE ISOCYANATE see
   PKM250
ADIPIC ACID, POLYMER with 1,4-BUTANEDIOL, METHY-
   LENEDI-p-PHENYLENE ISOCYANATE and 2,2'-(p-PHE-
   NYLENEDIOXY)DIETHANOL see PKM500

ADIPIC ACID, POLYMER with ETHYLENE GLYCOL and
　METHYLENEDI-p-PHENYLENE ISOCYANATE see
　PKL750
ADIPIC DIAMIDE see AEN000
ADIPIC KETONE see CPW500
ADIPINIC ACID see AEN250
ADIPODINITRILE see AER250
ADIPONITRILE see AER250
ADITYL see ACE000
ADJUDETS see BBK500
ADM see AES750
ADMER PB 02 see PMP500
ADMUL see OAV000
ADNEPHRINE see VGP000
2-ADO see DDB600
ADOBACILLIN see AIV500
ADOL see HCP000, OAX000
ADOL 68 see OAX000
ADONAL see EOK000
ADORM see TDA500
ADPHEN see DKE800
ADRENAL see VGP000
1-ADRENALIN see VGP000
d-ADRENALINE see AES250
l-(+)-ADRENALINE see AES250
ADRENALIN-MEDIHALER see VGP000
ADRENAMINE see VGP000
ADRENAN see VGP000
ADRENAPAX see VGP000
ADRENASOL see VGP000
ADRENATRATE see VGP000
ADRENODIS see VGP000
ADRENOHORMA see VGP000
ADRENOR see NNO500
ADRENUTOL see VGP000
ADRIAMICINA see MDO250
ADRIAMYCIN see AES750
ADRIAMYCIN-HCl see AES750
ADRIAMYCIN SEMIQUINONE see AES750
ADRIBLASTINA see AES750
ADRINE see VGP000
ADRIXINE see BBK500
ADROIDIN see PAN100
ADRONAL see CPB750
ADROYD see PAN100
ADRUCIL see FMM000
ADVASTAB 401 see BFW750
ADVASTAB 17 MO see BKK750
ADVAWAX 140 see OAV000
AENH (GERMAN) see ENV000
AEORLIN see BQF500
AEPHENAL see EOK000
AERO see MCB000
AERO-CYANAMID see CAQ250
AERO CYANAMID GRANULAR see CAQ250
AERO CYANAMID SPECIAL GRADE see CAQ250
AERO CYANATE see PLC250
AERO liquid HCN see HHS000
AEROL 1 (pesticide) see TIQ250
AEROSEB-DEX see SOW000
AEROSIL see SCH000
AEROSOL GPG see DJL000
AEROSOL of THERMOVACUUM CADMIUM see CAK000
AEROTEX GLYOXAL 40 see GIK000
AEROTHENE MM see MJP450
AEROTHENE TT see MIH275
AESCIN SODIUM SALT see EDM000
AESCULETIN DIMETHYL ETHER see DRS800
AESCUSAN SODIUM SALT see EDM000
AETHALDIAMIN (GERMAN) see EEA500
AETHANETHIOL (GERMAN) see EMB100
AETHANOL (GERMAN) see EFU000
AETHANOLAMIN (GERMAN) see EEC600
AETHER see EJU000

AETHIONIN see EEI000
AETHON see ENY500
AETHOPROPROPAZIN see DIR000
AETHOSUXIMIDE (GERMAN) see ENG500
2-AETHOXY-AETHYLACETAT (GERMAN) see EES400
p-AETHOXYPHYLHARNSTOFF (GERMAN) see EFE000
AETHYLACETAT (GERMAN) see EFR000
AETHYL ACETOXYMETHYLNITROSAMIN (GERMAN) see
　ENR500
AETHYLACRYLAT (GERMAN) see EFT000
AETHYL-AETHANOL-NITROSOAMIN (GERMAN) see
　ELG500
AETHYLALKOHOL (GERMAN) see EFU000
AETHYLAMINE (GERMAN) see EFU400
2-AETHYLAMINO-4-CHLOR-6-ISOPROPYLAMINO-1,3,5-
　TRIAZIN (GERMAN) see ARQ725
2-AETHYLAMINO-4-ISOPROPYLAMINO-6-CHLOR-1,3,5-
　TRIAZIN (GERMAN) see ARQ725
AETHYLANILIN (GERMAN) see EGK000
AETHYLBENZOL (GERMAN) see EGP500
AETHYLBUTYLKETON (GERMAN) see EHA600
AETHYL-N-BUTYL-NITROSOAMIN (GERMAN) see
　EHC000
AETHYLCARBAMAT (GERMAN) see UVA000
AETHYLCHLORID (GERMAN) see EHH000
AETHYL-CHLORVYNOL see CHG000
O-AETHYL-S-(2-DIMETHYLAMINOAETHYL)-METHYL-
　PHOSPHONOTHIOATE (GERMAN) see EIF500
O-AETHYL-S,S-DIPHENYL-DITHIOPHOSPHAT (GERMAN)
　see EIM000
S-AETHYL-N,N-DIPROPYLTHIOLCARBAMAT (GERMAN)
　see EIN500
AETHYLEN-BIS-THIURAMMONOSULFID (GERMAN) see
　ISK000
AETHYLENBROMID (GERMAN) see EIY500
AETHYLENCHLORID (GERMAN) see EIY600
AETHYLENECHLORHYDRIN (GERMAN) see EIU800
AETHYLENEDIAMIN (GERMAN) see EEA500
AETHYLENGLYKOLAETHERACETAT (GERMAN) see
　EES400
AETHYLENGLYKOLMETHYLAETHERACETAT (GERMAN)
　see EJJ500
AETHYLENGLYKOL-MONOMETHYLAETHER (GERMAN)
　see EJH500
AETHYLENIMIN (GERMAN) see EJM900
AETHYLENOXID (GERMAN) see EJN500
AETHYLENSULFID (GERMAN) see EJP500
AETHYLFORMIAT (GERMAN) see EKL000
AETHYLHARNSTOFF und NATRIUMNITRIT (GERMAN) see
　EQE000
AETHYLHARNSTOFF und NITRIT (GERMAN) see EQE000
AETHYLIDENCHLORID (GERMAN) see DFF809
AETHYLIS see EHH000
AETHYLIS CHLORIDUM see EHH000
2-(O-AETHYL-N-ISOPROPYLAMINDOTHIOPHOS-
　PHORYLOXY)-BENZOSAEURE-ISOPROPYLESTER
　(GERMAN) see IMF300
AETHYL-ISOPROPYL-NITROSOAMIN (GERMAN) see
　ELX500
AETHYLMERCAPTAN (GERMAN) see EMB100
AETHYLMETHYLKETON (GERMAN) see MKA400
O-AETHYL-O-(3-METHYL-4-METHYLTHIOPHENYL)-ISO-
　PROPYLAMIDO-PHOSPHORSAEURE ESTER (GERMAN)
　see FAK000
3-β-AETHYL-1-METHYL-4-PHENYL-4-α-PIPERIDYLPRO-
　PIONAT HYDROCHLORID (GERMAN) see NOD500
3-β-AETHYL-1-METHYL-4-PHENYL-4-α-PROPIONYLOXYPI-
　PERIDIN HYDROCHLORID (GERMAN) see NOD500
N-AETHYL-N'-NITRO-N-NITROSOGUANIDIN (GERMAN)
　see ENU000
O-AETHYL-O-n(4-NITROPHENYL)-PHENYL-MONOTHIO-
　PHOSPHONAT (GERMAN) see EBD700
AETHYLNITROSO-HARNSTOFF (GERMAN) see ENV000
AETHYLNITROSOURETHAN (GERMAN) see NKE500

5-AETHYL-5-PENTYL-(2')-BARBITURSAEURE (GERMAN) see PBS250
O-AETHYL-S-PHENYL-AETHYL-DITHIOPHOSPHONAT (GERMAN) see FMU045
5-AETHYL-5-PHENYL-HEXAHYDROPYRIMIDIN-4,6-DION (GERMAN) see DBB200
AETHYL-4-PICOLYLNITROSAMIN (GERMAN) see NLH000
N-AETHYLPIPERIDIN (GERMAN) see EOS500
AETHYLRHODANID (GERMAN) see EPP000
AETHYLSENFOEL (GERMAN) see ISK000
S-2-AETHYLSULFINYL-1-METHYL AETHYL-O,O DI-METHYL-MONOTHIOPHOSPHAT see DSK600
O-AETHYL-O-(2,4,5-TRICHLORPHENYL)-AETHYLTHIONO-PHOSPHONAT (GERMAN) see EPY000
AETHYLURETHAN (GERMAN) see UVA000
AETHYL-VINYL-NITROSOAMIN (GERMAN) see NKF000
AETINA see EPQ000
AETIVA see EPQ000
AETM (GERMAN) see ISK000
AETT see ACL750
AF-2 (preservative) see FQN000
AF 101 see DXQ500
AF 864 see BBW500
AFASTOGEN BLUE 5040 see DNE400
AFATIN see BBK500
AFAXIN see VSK600
AFBI see AEU250
A.F. BLUE No. 1 see FMU059
A.F. BLUE No. 2 see FAE100
AFCOLAC B 101 see SMR000
AFCOLENE see SMQ500
AFESIN see CKD500
A.F. GREEN No. 2 see FAF000
AFICIDE see BBQ500
A-FIL CREAM see TGG760
AFI-TIAZIN see PDP250
AFKO-HIST see WAK000
AFKO-SAL see SAH000
AFL 1081 see FFF000
AFL 1082 see FFH000
AFLATOXICOL see AEW500
AFLATOXIN see AET750
AFLATOXIN B see AEU250
AFLATOXIN B1 see AEU250
AFLATOXIN B2 see AEU750
AFLATOXIN G1 see AEV000
AFLATOXIN G2 see AEV500
AFLATOXIN G1 mixed with AFLATOXIN B1 see AEV250
AFLATOXIN M1 see AEW000
AFLATOXIN Ro see AEW500
AFLIX see DRR200
AFLON see TAI250
AFNOR see CJJ000
A.F.ORANGE No. 2 see TGW000
AFOS see DJI000
A.F. RED No. 1 see FAG018
A.F. RED No. 5 see XRA000
AFRICAN COFFEE TREE see CCP000
A.F. VIOLET No 1 see FAG120
A.F YELLOW No. 2 see FAG130
A.F. YELLOW No. 3 see FAG135
AGALITE see TAB750
AGALLOL see MEP250
AGALLOLAT see MEP250
AGAR see AEX250
AGAR-AGAR see AEX250
AGAR AGAR FLAKE see AEX250
AGAR-AGAR GUM see AEX250
AGATE see SCI500, SCJ500
AGC see GFA000
AGE see AGH150
AGEDAL see DPH600
AGEFLEX BGE see BRK750
AGEFLEX CGE see GGS000

AGEFLEX FM-1 see DPG600
AGEFLEX FM-4 see BQD250
AGENAP see NAR000
AGENT 504 see DAI600
AGENT AT 717 see PKQ250
AGENT BLUE see HKC000
AGERITE see BLE500
AGERITEDPPD see BLE500
AGERITE POWDER see PFT500
AGERITE WHITE see NBL000
AGGLUTININ see AAD000
AGIDOL see BFW750
AGILENE see DRK600, PJS750
AGIOLAN see VSK600
AGI TALC, BC 1615 see TAB750
AGLICID see BSQ000
AGOFOLLIN see EDR000
AGOSTILBEN see DKA600
AGOVIRIN see TBG000
AGRAZINE see PDP250
AGREFLAN see DUV600
AGRIA 1050 see DSQ000
AGRIBON see SNN300
AGRICIDE MAGGOT KILLER (F) see CDV100
AGRICULTURAL LIMESTONE see CAO000
AGRIDIP see CNU750
AGRIFLAN 24 see DUV600
AGRIMYCIN 17 see SLW500
AGRION see SGH500
AGRISIL see EPY000
AGRISOL G-20 see BBQ500
AGRITAN see DAD200
AGRITOX see CIR250, EPY000
AGRIYA 1050 see DSQ000
A-GRO see MNH000
AGROCERES see HAR000
AGROCIDE see BBQ500
AGROFOROTOX see TIQ250
AGROMICINA see TBX000
AGRONEXIT see BBQ500
AGROSAN see ABU500
AGROSOL see MLF250
AGROSOL S see CBG000
AGROTECT see DAA800
AGROTHION see DSQ000
AGROXONE see CIR250
AGROX 2-WAY and 3-WAY see CBG000
AGRYPNAL see EOK000
AGSTONE see CAO000
AGUATHOL see DXD000
AH see DBM800
AH-42 see TEO250
AH 289 see CDR000
AH 3365 see BQF500
AHCO DIRECT BLACK GX see AQP000
AH-289 HYDROCHLORIDE see CDR250
AHR-1680 see DMB000
A 66 HYDROCHLORIDE see MNV750
A-HYDROCORT see HHR000
AHYPNON see MKA250
AI 318284 see MRQ750
AI 3-22542 see DKC800
AI3-29158 see AHJ750
AI3-35966 see ISZ000
AIBN see ASL750
AIMAX see MLJ500
AIMSAN see DRR400
AIP see AHE750
AIR, compressed see AFG250
AIRBRON see ACH000
AIRDALE BLUE IN see FAE100
AIREDALE BLACK ED see AQP000
AIREDALE BLUE 2BD see CMO000
AIREDALE CARMOISINE see HJF500

AIREDALE YELLOW T see FAG140
AIR-FLO GREEN see CNN500
AISELAZINE see HGP500
AISEMIDE see CHJ750
AITC see AGJ250
AIZEN AURAMINE see IBA000
AIZEN BRILLIANT BLUE FCF see FMU059
AIZEN CRYSTAL VIOLET EXTRA PURE see AOR500
AIZEN DIAMOND GREEN GH see BAY750
AIZEN DIRECT BLUE 2BH see CMO000
AIZEN DIRECT DEEP BLACK GH see AQP000
AIZEN EOSINE GH see BNH500
AIZEN ERYTHROSINE see FAG040
AIZEN FOOD BLUE No. 2 see FAE000
AIZEN FOOD GREEN No. 3 see FAG000
AIZEN FOOD ORANGE No. 2 see TGW000
AIZEN FOOD RED No. 5 see XRA000
AIZEN FOOD VIOLET No 1 see FAG120
AIZEN FOOD YELLOW No. 5 see FAG150
AIZEN METHYLENE BLUE BH see BJI250
AIZEN PRIMULA BROWN BRLH see CMO750
AIZEN RHODAMINE BH see FAG070
AIZEN TARTRAZINE see FAG140
AIZEN URANINE see FEW000
AJAN see NBS500
AJINOMOTO see MRL500
AKAR see DER000
AKARITHION see TNP250
AKARITOX see CKM000
AKIRIKU RHODAMINE B see FAG070
AKLOMIX-3 see HMY000
AKOIN HYDROCHLORID (GERMAN) see PDN500
AKOTIN see NCQ900
AKROLEIN (CZECH) see ADR000
AKROLEINA (POLISH) see ADR000
AKRO-ZINC BAR 85 see ZKA000
AKRO-ZINC BAR 90 see ZKA000
AKRYLAMID (CZECH) see ADS250
AKRYLONITRYL (POLISH) see ADX500
AKTAMIN see NNO500
AKTIKON see ARQ725
AKTIKON PK see ARQ725
AKTINIT A see ARQ725
AKTINIT PK see ARQ725
AKTINIT S see BJP000
AKTIVEX see CCX000
AKTIVIN see CDP000
AKTON see DIX600
AKULON see PJY500
AK-33X see MAV750
AKZO CHEMIE MANEB see MAS500
AL-1021 see FGV000
ALABASTER see CAX750
ALACHLOR (USDA) see CFX000
ALACINE see PFC750
ALAMINE 6 see HCO500
ALANE see AHB500
ALANEX see CFX000
ALANINE MUSTARD see BHN500
ALANINE NITROGEN MUSTARD see BHV250
ALANINE NITROGEN MUSTARD see PED750
ALAR see DQD400
ALAR-85 see DQD400
ALAUN (GERMAN) see AGX000
ALAZIN see PFC750
ALAZINE see PFC750
ALBAGEL PREMIUM USP 4444 see BAV750
ALBAMYCIN see NOB000
ALBAMYCIN SODIUM see NOB000
ALBEMAP see BBK500
ALBENDAZOLE (USDA) see VAD000
ALBEXAN see SNM500
ALBIGEN A see PKQ250
ALBIOTIC see LGD000

ALBOLINE see MQV750
ALBON see SNN300
ALBONE see HIB000
ALBOSAL see SNM500
ALBSAPOGENIN see GMG000
ALBUMIN see AFI850
ALBUMIN MACRO AGGREGATES see AFI850
ALBUTEROL see BQF500
ALCALASE see BAC000
ALCANFOR see CBB250
ALCHLOQUIN see CHR500
ALCIDE see CDW450
ALCOA F 1 see AHE250
ALCOA SODIUM FLUORIDE see SHF500
ALCOBAM NM see SGM500
ALCOBAM ZM see BJK500
ALCOHOL see EFU000
ALCOHOL, anhydrous see EFU000
ALCOHOL, dehydrated see EFU000
ALCOHOL C-8 see OEI000
ALCOHOL C-9 see NNB500
ALCOHOL C-10 see DAI600
ALCOHOL C-11 see UNA000
ALCOHOL C-12 see DXV600
ALCOHOL C-16 see HCP000
ALCOHOL, DENATURED see AFJ000
ALCOHOLS, N.O.S. see AFJ250
ALCOOL ALLILCO (ITALIAN) see AFV500
ALCOOL ALLYLIQUE (FRENCH) see AFV500
ALCOOL AMILICO (ITALIAN) see IHP000
ALCOOL AMYLIQUE (FRENCH) see AOE000
ALCOOL BUTYLIQUE (FRENCH) see BPW500
ALCOOL BUTYLIQUE SECONDAIRE (FRENCH) see
   BPW750
ALCOOL BUTYLIQUE TERTIAIRE (FRENCH) see BPX000
ALCOOL ETHYLIQUE (FRENCH) see EFU000
ALCOOL ETILICO (ITALIAN) see EFU000
l'ALCOOL n-HEPTYLIQUE PRIMAIRE (FRENCH) see
   HBL500
ALCOOL ISOAMYLIQUE (FRENCH) see IHP000
ALCOOL ISOBUTYLIQUE (FRENCH) see IIL000
ALCOOL ISOPROPILICO (ITALIAN) see INJ000
ALCOOL ISOPROPYLIQUE (FRENCH) see INJ000
ALCOOL METHYL AMYLIQUE (FRENCH) see MKW600
ALCOOL METHYLIQUE (FRENCH) see MGB150
ALCOOL METILICO (ITALIAN) see MGB150
ALCOOL PROPILICO (ITALIAN) see PND000
ALCOOL PROPYLIQUE (FRENCH) see PND000
ALCOPAN-250 see PAG200
ALCOPHOBIN see DXH250
ALCOPOL O see DJL000
ALDACOL Q see PKQ250
ALDECARB see CBM500
ALDEHYDE-14 see UJJ000
ALDEHYDE ACETIQUE (FRENCH) see AAG250
ALDEHYDE ACRYLIQUE (FRENCH) see ADR000
ALDEHYDE AMMONIA see AAG500
ALDEHYDE B see COU500
ALDEHYDE BUTYRIQUE (FRENCH) see BSU250
ALDEHYDE C-6 see HEM000
ALDEHYDE C-8 see OCO000
ALDEHYDE C-9 see NMW500
ALDEHYDE C10 see DAG000
ALDEHYDE C-18 see CNF250
ALDEHYDECOLLIDINE see EOS000
ALDEHYDE C-14 PURE see UJA800
ALDEHYDE CROTONIQUE (FRENCH) see COB260
ALDEHYDE C-11, UNDECYLENIC see ULJ000
ALDEHYDE-2-ETHYLBUTYRIQUE (FRENCH) see DHI000
ALDEHYDE FORMIQUE (FRENCH) see FMV000
ALDEHYDE PROPIONIQUE (FRENCH) see PMT750
ALDEHYDES see AFJ800
ALDEHYDINE see EOS000
ALDEIDE ACETICA (ITALIAN) see AAG250

ALDEIDE ACRILICA (ITALIAN) see ADR000
ALDEIDE BUTIRRICA (ITALIAN) see BSU250
ALDEIDE FORMICA (ITALIAN) see FMV000
ALDERLIN HYDROCHLORIDE see INT000
ALDICARB (USDA) see CBM500
ALDICARBE (FRENCH) see CBM500
ALDIFEN see DUZ000
ALDINAMID see POL500
ALDO-28 see OAV000
ALDO-72 see OAV000
ALDO HMS see OAV000
ALDOL see AAH750
ALDOMET see DNA800
ALDOMETIL see DNA800
ALDOMIN see DNA800
ALDO MS see OAV000
ALDO MSA see OAV000
ALDO MSLG see OAV000
ALDOMYCIN see NGE500
ALDOXIME see AAH250
ALDREX see AFK250
ALDREX 30 see AFK250
ALDRICH see DOJ200
ALDRIN see AFK250
ALDRIN, cast solid (DOT) see AFK250
ALDRINE (FRENCH) see AFK250
ALDRITE see AFK250
ALDROSOL see AFK250
ALENTIN see BSM000
ALEPSIN see DNU000
ALERYL see BBV500
ALEUDRIN see DMV600
ALEVIATIN see DKQ000
ALFACALCIDOL see HJV000
ALFALFA MEAL see AFK750
ALFAMAT see GFA000
ALFANAFTILAMINA (ITALIAN) see NBE000
ALFA-NAFTYLOAMINA (POLISH) see NBE000
ALFA-TOX see DCM750
ALFENOL 3 see PKF500
ALFENOL 9 see PKF500
ALFICETYN see CDP250
ALFIMID see DYC800
ALFOL 8 see OEI000
ALFOL 12 see DXV600
ALFUCIN see NGE500
ALGAFAN see PNA500
ALGAMON see SAH000
ALGAROBA see LIA000
ALGIAMIDA see SAH000
ALGIL see DAM700
ALGIMYCIN see ABU500
ALGIN see SEH000
ALGINATE KMF see SEH000
ALGINIC ACID see AFL000
ALGIN (POLYSACCHARIDE) see SEH000
ALGIPON L-1168 see SEH000
ALGISTAT see DFT000
ALGLOFLON see TAI250
ALGO-DEX see BBK500
ALGOFLON SV see TAI250
ALGOFRENE TYPE 1 see TIP500
ALGOFRENE TYPE 2 see DFA600
ALGOFRENE TYPE 5 see DFL000
ALGOFRENE TYPE 6 see CFX500
ALGOFRENE TYPE 67 see ELN500
ALGOSEDIV see TEH500
ALGOTROPYL see HIM000
ALGRAIN see EFU000
ALGYLEN see TIO750
ALICYANATE see PLC250
ALIDOCHLOR see CFK000
ALINDOR see BRF500
ALIPHATIC and AROMATIC EPOXIDES see AFM250

ALIPHATIC CHLORINATED HYDROCARBONS see
    CDV250
ALIQUAT 336 see MQH000
ALIZARINE CYANINE GREEN BASE see BLK000
ALKABUTAZONA see BRF500
ALKALIES see AFM500
ALKALOID H 3, from COLCHICUM ANTUMNALE see
    MIW500
ALKALOIDS see AFM750
ALKALOID SALTS see AFM750
ALKALOIDS, VERATRUM see VIZ000
ALKAMID see PJY500
ALKANES see AFN250
ALKAPOL PEG-200 see PJT000
ALKAPOL PPG-1200 see PKI500
ALKARSODYL see HKC500
ALKATHENE see PJS750
ALKATHENE RXDG33 see TAI250
ALK-AUBS see DXH250
ALKAVERVIR see VIZ000
ALK-ENZYME see BAC000
ALKERAN see PED750
ALKERAN (RUSSIAN) see BHV000
ALKIRON see MPW500
ALKOHOL (GERMAN) see EFU000
ALKOHOLU ETYLOWEGO (POLISH) see EFU000
ALKOVERT see PIB250
ALKYL(C$_8$H$_{17}$ to C$_{18}$H$_{37}$) DIMETHYL-3,4-DICHLOROBEN-
    ZYL AMMONIUM CHLORIDE see AFP750
ALKYL(C9-15)TOLYL METHYLTRIMETHYL AMMONIUM
    CHLORIDE see DYA600
ALKYL(C$_8$C$_{18}$)DIMETHYL-3,4-DICHLOROBENZYLAMMO-
    NIUM CHLORIDE see AFP750
ALLBRI NATURAL COPPER see CNI000
(+)-ALLELRETHONYL (+)-cis,trans-CHRYSANTHEMATE
    see AFR250
ALLEOSIDE A DIHYDRATE see HAO000
ALLERCLOR see TAI500
ALLERGAN 211 see DAS000
ALLERGAN B see BBV500
ALLERGEN see LJR000
ALLERGEVAL see BBV500
ALLERGICAL see BBV500
ALLERGIN see BBV500, TAI500
ALLERGINA see BBV500
ALLERGISAN see TAI500
ALLERGIVAL see BBV500
ALLERON see PAK000
ALLETHRIN see AFR250
d-ALLETHRIN see AFR250
(+)-cis-ALLETHRIN see AFR500
ALLETHRIN I see AFR250
ALLETHRIN RACEMIC MIXTURE see AFS000
ALLIDOCHLOR see CFK000
ALLILE (CLORURO DI) (ITALIAN) see AGB250
ALLIL-GLICIDIL-ETERE (ITALIAN) see AGH150
1-ALLILOSSI-2,3 EPOSSIPROPANO (ITALIAN) see AGH150
ALLILOWY ALKOHOL (POLISH) see AFV500
ALLOCAINE see AIL750, AIT250
ALLODENE see BBK000
ALLOMALEIC ACID see FOU000
ALLO-OCIMENOL see LFX000
ALLOXAN MONOHYDRATE see MDL500
ALLSPICE see PIG740
ALLTEX see CDV100
ALLTOX see CDV100
ALLUMINIO(CLORURO DI) (ITALIAN) see AGY750
ALLUMINIO DIISOBUTIL-MONOCLORURO (ITALIAN) see
    CGB500
ALLURA RED AC see FAG100
ALLUVAL see BNP750
ALLYL ACETATE see AFU750
ALLYL AL see AFV500

ALLYL ALCOHOL see AFV500
ALLYL ALDEHYDE see ADR000
ALLYLALKOHOL (GERMAN) see AFV500
ALLYLAMINE see AFW000
p-ALLYLANISOLE see AFW750
5-ALLYL-1,3-BENZODIOXOLE see SAD000
ALLYL BROMIDE see AFY000
ALLYL CAPROATE see AGA500
ALLYL CAPRYLATE see AGM500
ALLYLCATECHOL METHYLENE ETHER see SAD000
ALLYLCHLORID (GERMAN) see AGB250
ALLYL CHLORIDE see AGB250
ALLYL CHLOROCARBONATE see AGB500
ALLYL CHLOROFORMATE see AGB500
ALLYL CHLOROFORMATE (DOT) see AGB500
ALLYL CINERIN see AFR250
ALLYL CINNAMATE see AGC000
ALLYL CYCLOHEXANEPROPIONATE see AGC500
3-ALLYLCYCLOHEXYL PROPIONATE see AGC500
1-ALLYL-3,4-DIMETHOXYBENZENE see AGE250
4-ALLYL-1,2-DIMETHOXYBENZENE see AGE250
ALLYLDIOXYBENZENE METHYLENE ETHER see SAD000
ALLYLE (CHLORURE D') (FRENCH) see AGB250
ALLYL ENANTHATE see AGH250
ALLYL-2,3-EPOXYPROPYL ETHER see AGH150
ALLYLETHER see DBK000
ALLYL FLUORIDE see AGG500
ALLYL FORMATE see AGH000
ALLYLGLYCIDAETHER (GERMAN) see AGH150
ALLYL GLYCIDYL ETHER see AGH150
4-ALLYLGUAIACOL see EQR500
ALLYL HEPTANOATE see AGH250
ALLYL HEPTOATE see AGH250
ALLYL HEPTYLATE see AGH250
ALLYL HEXAHYDROPHENYLPROPIONATE see AGC500
ALLYL HEXANOATE (FCC) see AGA500
ALLYL HOMOLOG of CINERIN I see AFR250
ALLYL HYDROPEROXIDE see AGH750
4-ALLYL-1-HYDROXY-2-METHOXYBENZENE see EQR500
d,l-2-ALLYL-4-HYDROXY-3-METHYL-2-CYCLOPENTEN-1-
  ONE-d,l-CHRYSANTHEMUMMONOCARBOXYLATE see
  AFR250
ALLYLIC ALCOHOL see AFV500
ALLYL IODIDE see AGI250
ALLYL-α-IONONE see AGI500
ALLYL ISORHODANIDE see AGJ250
ALLYL ISOSULFOCYANATE see AGJ250
ALLYL ISOTHIOCYANATE see AGJ250
ALLYL ISOTHIOCYANATE, stabilized (DOT) see AGJ250
ALLYL ISOVALERATE see ISV000
ALLYL ISOVALERIANATE see ISV000
3-ALLYL-4-KETO-2-METHYLCYCLOPENTENYL CHRYSAN-
  THEMUMMONOCARBOXYLATE see AFR250
ALLYL MERCAPTAN see AGJ500
4-ALLYL-1-METHOXYBENZENE see AFW750
4-ALLYL-2-METHOXYPHENOL see EQR500
4-ALLYL-2-METHOXYPHENOL ACETATE see EQS000
5-ALLYL-5-(1-METHYLBUTYL)BARBITURIC ACID see
  SBM500
5-ALLYL-5-(1-METHYLBUTYL)MALONYLUREA see
  SBM500
5-ALLYL-5-(1-METHYLBUTYL)-2-THIOBARBITURATE SO-
  DIUM see SOX500
5-ALLYL-5-(1-METHYLBUTYL)-2-THIO-BARBITURIC ACID
  SODIUM SALT see SOX500
ALLYL 3-METHYLBUTYRATE see ISV000
1-ALLYL-3,4-METHYLENEDIOXYBENZENE see SAD000
4-ALLYL-1,2-METHYLENEDIOXYBENZENE see SAD000
5-ALLYL-1-METHYL-5-(1-METHYL-2-PENTYNYL)BARBI-
  TURIC ACID SODIUM SALT see MDU500
3-ALLYL-2-METHYL-4-OXO-2-CYCLOPENTEN-1-YL CHRY-
  SANTHEMATE see AFR250
dl-3-ALLYL-2-METHYL-4-OXOCYCLOPENT-2-ENYL-dl-cis
  trans CHRYSANTHEMATE see AFR250

ALLYL MUSTARD OIL see AGJ250
ALLYL OCTANOATE see AGM500
(±)-1-(β-(ALLYLOXY)-2,4-DICHLOROPHENETHYL)IMID-
  AZOLE see FPB875
1-ALLYLOXY-2,3-EPOXY-PROPAAN (DUTCH) see
  AGH150
1-ALLYLOXY-2,3-EPOXYPROPAN (GERMAN) see AGH150
1-(ALLYLOXY)-2,3-EPOXYPROPANE see AGH150
ALLYL PHENOXYACETATE see AGQ750
ALLYL PHENYLACETATE see PMS500
ALLYL-3-PHENYLACRYLATE see AGC000
ALLYL PROPYL DISULFIDE see AGR500
m-ALLYLPYROCATECHIN METHYLENE ETHER see
  SAD000
4-ALLYLPYROCATECHOL FORMALDEHYDE ACETAL see
  SAD000
ALLYLPYROCATECHOL METHYLENE ETHER see
  SAD000
ALLYLRETHRONYL dl-cis-trans-CHRYSANTHEMATE see
  AFR250
ALLYLSENFOEL (GERMAN) see AGJ250
ALLYL SEVENOLUM see AGJ250
ALLYL THIOCARBONIMIDE see AGJ250
ALLYL TRICHLORIDE see TJB600
ALLYL TRICHLOROSILANE see AGU250
4-ALLYLVERATROLE see AGE250
ALMEDERM see HCL000
ALMEFROL see VSZ450
ALMITE see AHE250
ALMOCARPINE see PIF000
ALMOND ARTIFICIAL ESSENTIAL OIL see BAY500
ALMOND OIL BITTER, FFPA (FCC) see BLV500
ALOCHLOR see CFX000
ALODAN (GEROT) see DAM700
ALON see AHE250
ALPEN see AIV500
ALPEROX C see LBR000
ALPHALIN see VSK600
ALPHA MEDOPA see DNA800
ALPHANAPHTHYL THIOUREA see AQN635
ALPHANAPHTYL THIOUREE (FRENCH) see AQN635
ALPHASOL OT see DJL000
ALPHASTEROL see VSK600
ALPHAZURINE see FMU059
AL-PHOS see AHE750
ALPINE TALC USP, BC 127 see TAB750
ALPINE TALC USP, BC 141 see TAB750
ALPINE TALC USP, BC 662 see TAB750
ALPINYL see HIM000
ALQOVERIN see BRF500
ALRATO see AQN635
ALTADIOL see EDS100
ALTAX see BDE750
ALTHOSE HYDROCHLORIDE see MDP000
ALTOSID see KAJ000
ALTOSID IGR see KAJ000
ALTOSID SR 10 see KAJ000
ALTOX see AFK250
ALTRAD see EDO000
ALTRETAMINE see HEJ500
ALUDRINE see DMV600
ALUM see AHG750
ALUMINA see AHE250
α-ALUMINA (OSHA) see AHE250
β-ALUMINA see AHE250, AHG000
γ-ALUMINA see AHE250
β''-ALUMINA see AHG000
ALUMINA FIBRE see AGX000
ALUMINUM see AGX000
ALUMINUM ALLOY, Al,Be see BFP250
ALUMINUM AMMONIUM SULFATE see AGX250
ALUMINUM BERYLLIUM ALLOY see BFP250
ALUMINUM BOROHYDRIDE see AGX500
ALUMINUM BROMIDE see AGX750

ALUMINUM BROMIDE, anhydrous see AGX750
ALUMINUM BROMIDE, solution (DOT) see AGX750
ALUMINUMCHLORID (GERMAN) see AGY750
ALUMINUM CHLORIDE see AGY750
ALUMINUM CHLORIDE (1:3) see AGY750
ALUMINUM CHLORIDE, anhydrous (DOT) see AGY750
ALUMINUM CHLORIDE, solution (DOT) see AGY750
ALUMINUM DEHYDRATED see AGX000
ALUMINUM DEXTRAN see AHA250
ALUMINUM ETHYLATE see AHA750
ALUMINUM FLAKE see AGX000
ALUMINUM FOSFIDE (DUTCH) see AHE750
ALUMINUM HYDRIDE see AHB500
ALUMINUM IODIDE see AHC500
ALUMINUM LITHIUM HYDRIDE see LHS000
ALUMINUM MAGNESIUM PHOSPHIDE see AHD250
ALUMINUM, METALLIC, POWDER (DOT) see AGX000
ALUMINUM METHYL see AHD500
ALUMINUM MONOPHOSPHIDE see AHE750
ALUMINUM MONOSTEARATE see AHA250
ALUMINUM NITRATE (DOT) see AHD750
ALUMINUM(III) NITRATE (1:3) see AHD750
ALUMINUM OXIDE see AHE250, EAL100
α-ALUMINUM OXIDE see AHE250
β-ALUMINUM OXIDE see AHE250
γ-ALUMINUM OXIDE see AHE250
ALUMINUM OXIDE (2:3) see AHE250
ALUMINUM PHOSPHIDE see AHE750
ALUMINUM PICRATE see AHF000
ALUMINUM POWDER see AGX000
ALUMINUM POWDER, UNCOATED, NON-PYROPHORIC
 (DOT) see AGX000
ALUMINUM SESQUIOXIDE see AHE250
ALUMINUM SODIUM HYDRIDE see SEM500
ALUMINUM SODIUM OXIDE see AHG000
ALUMINUM SODIUM SULFATE see AHG500
ALUMINUM SULFATE (2:3) see AHG750
ALUMINUM TETRAHYDROBORATE see AGX500,
 AHG875
ALUMINUM THALLIUM SULFATE see AHH000
ALUMINUM TRIBROMIDE see AGX750
ALUMINUM TRICHLORIDE see AGY750
ALUMINUM TRIHYDRIDE see AHB500
α-ALUMINUM TRIHYDRIDE see AHB500
ALUMINUM TRINITRATE see AHD750
ALUMINUM TRIPROPYL see AHH750
ALUMINUM TRISULFATE see AHG750
ALUMITE see AHE250
ALUNDUM see AHE250
ALUNEX see TAI500
ALURAL see BNP750
ALUTOR M 70 see PKB500
ALUZINE see CHJ750
ALVEDON see HIM000
ALVINOL see CHG000
ALVIT see DHB400
ALYSINE see SJO000
ALZODEF see CAQ250
AMABEVAN see CBJ000
AMACEL DEVELOPED NAVY SD see DCJ200
AMACEL YELLOW G see AAQ250
AMACID BLUE FG CONC see FMU059
AMACID BRILLIANT BLUE see FAE100
AMACID CHROME BLUE R see HJF500
AMACID GREEN G see FAF000
AMADIL see HIM000
AMALOX see ZKA000
AMANIL BLACK GL see AQP000
AMANIL BLUE 2BX see CMO000
AMANIL SKY BLUE see CMO250
AMANIL SUPRA BROWN LBL see CMO750
AMAPLAST GREEN OZ see BLK000
AMARSAN see ABX500
AMARTHOL FAST RED TR BASE see CLK220, CLK235

AMARTHOL FAST RED TR SALT see CLK235
AMATIN see HCC500
AMATOL see AHI750
AMAX see BDG000
AMAZE see IMF300
AMBENYL see BAU750
AMBER see AHJ000
AMBER ACID see SMY000
AMBERGRIS TINCTURE see AHJ000
AMBESIDE see SNM500
AMBILHAR see NML000
AMBLOSIN see AIV500
AMBOCHLORIN see CDO500
AMBOCLORIN see CDO500
AMBOFEN see CDP250
AMBRA see AHJ000
AMBRACYN see TBX250
AMBRAMICINA see TBX000
AMBRAMYCIN see TBX000
AMBUSH see AHJ750, CBM500
AMCAP see AOD125
AMCHEM 68-250 see CDS125
AMCHEM GRASS KILLER see TII250
AMCHEM R 14 see PKL750
AMCHEM 2,4,5-TP see TIX500
AMCIDE see ANU650
AMCILL see AIV500, AOD125
AMCO see PMP500
AMD see DNA800
AMDEX see BBK500
AMDON GRAZON see PIB900
AMDRAM see DBA800
AMEBAN see CBJ000
AMEBARSONE see CBJ000
AMEBICIDE see EAN000
AMEBIL see CHR500
AMEDRINE see DBA800
AMEISENATOD see BBQ500
AMEISENMITTEL MERCK see BBQ500
AMEISENSAEURE (GERMAN) see FNA000
AMEPROMAT see MQU750
AMERCIAN CYANAMID 18133 see EPC500
AMERCIDE see CBG000
AMERFIL see PMP500
AMERICAINE see EFX000
AMERICAN CYANAMID 4,049 see MAK700
AMERICAN CYANAMID 5223 see DXX400
AMERICAN CYANAMID 12,008 see DJN600
AMERICAN CYANAMID 12,503 see POP000
AMERICAN CYANAMID 12880 see DSP400
AMERICAN CYANAMID 18682 see IOT000
AMERICAN CYANAMID 18706 see DNX600
AMERICAN CYANAMID-38023 see FAB600
AMERICAN CYANAMID-43073 see MJG500
AMERICAN CYANAMID 47031 see PGW750
AMERICAN CYANAMID AC 43,064 see DXN600
AMERICAN CYANAMID AC 52,160 see TAL250
AMERICAN CYANAMID CL-38,023 see FAB800
AMERICAN CYANAMID CL-47,300 see DSQ000
AMERICAN CYANAMID CL-47470 see DHH400
AMERICAN HELLEBORE see VIZ000
AMERICAN PENICILLIN see BFD250
AMERICAN PENNYROYAL OIL see PAR500
AMERICAN VERATRUM see VIZ000
AMERICIUM see AHK000
AMERICIUM TRICHLORIDE see AHK250
AMERIZOL see TOA000
AMEROL see AMY050
AMETHOCAINE see BQA010
AMETHOPTERIN see MDV500
AMETHYST see SCI500, SCJ500
AMETOX see SKI500
AMETYCIN see AHK500
d-AMFETASUL see BBK500

AMFIPEN see AIV500
AMIANTHUS see ARM250
AMIBIARSON see CBJ000
AMICIDE see ANU650
AMIDAZIN see EPQ000
AMIDAZOPHEN see DOT000
AMIDE PP see NCR000
AMIDES see AHL750
AMIDINE BLUE 4B see CMO250
4-AMIDINO-1-(NITROSAMINOAMIDINO)-1-TETRAZENE see
  TEF500
AMID KYSELINY OCTOVE see AAI000
AMID KYSELINY STAVELOVE (CZECH) see OLO000
o-AMIDOAZOTOLUOL (GERMAN) see AIC250
o-AMIDOBENZOIC ACID see API500
AMIDOCYANOGEN see COH500
AMIDOFEBRIN see DOT000
AMIDOFOS see COD850
AMIDON see ILD000
AMIDONE see MDO750
AMIDONE HYDROCHLORIDE see MDP000
AMIDOPHEN see DOT000
AMIDOPHENAZONE see DOT000
AMIDOPHOS see COD850
AMIDOPYRAZOLINE see DOT000
AMIDOPYRIN see DOT000
AMIDOSAL see SAH000
AMIDOSULFONIC ACID see SNK500
AMIDOSULFURIC ACID see SNK500
AMIDOUREA HYDROCHLORIDE see SBW500
AMIDOX see DAA800
AMIDRINE see ILM000
AMIDRYL see BBV500
AMID-SAL see SAH000
AMIFUR see NGE500
AMILAN see NOH000
AMILAN CM 1001 see PJY500
AMINACRINE see AHS500
AMINARSON see CBJ000
AMINARSONE see CBJ000
AMINES see AHP750
AMINE 2,4,5-T FOR RICE see TAA100
AMINIC ACID see FNA000
AMINICOTIN see NCR000
AMINITROZOLE see ABY900
AMINOACETIC ACID see GHA000
2-AMINOACETOPHENONE see AHR250
m-AMINOACETOPHENONE see AHR500
p-AMINO ACETOPHENONE see AHR750
β-AMINOACETOPHENONE see AHR500
3'-AMINOACETOPHENONE see AHR500
4'-AMINOACETOPHENONE see AHR750
φ-AMINOACETOPHENONE see AHR250
m-AMINOACETYLBENZENE see AHR500
p-AMINOACETYLBENZENE see AHR750
5-AMINOACRIDINE see AHS500
9-AMINOACRIDINE see AHS500
2-AMINOAETHANOL (GERMAN) see EEC600
6-AMINO-4-((3-AMINO-4-(((4-((1-METHYLPYRIDINIUM-4-
  YL)AMINO)PHENYL)AMINO)CARBONYL)PHENYL)
  AMINO)-1-METHYLQUINOLINIUM),DIIODIDE see
  AHT850
2-AMINOANILINE see PEY250
3-AMINOANILINE see PEY000
4-AMINOANILINE see PEY500
m-AMINOANILINE see PEY000
p-AMINOANILINE see PEY500
3-AMINOANILINE DIHYDROCHLORIDE see PEY750
m-AMINOANILINE DIHYDROCHLORIDE see PEY750
2-AMINOANISOLE see AOV900
4-AMINOANISOLE see AOW000
o-AMINOANISOLE see AOV900
p-AMINOANISOLE see AOW000
2-AMINOANTHRACENE see APG000
β-AMINOANTHRACENE see APG000

1-AMINO-9,10-ANTHRACENEDIONE see AIA750
2-AMINO-9,10-ANTHRACENEDIONE see AIB000
1-AMINOANTHRACHINON (CZECH) see AIA750
1-AMINOANTHRAQUINONE see AIA750
2-AMINOANTHRAQUINONE see AIB000
α-AMINOANTHRAQUINONE see AIA750
β-AMINOANTHRAQUINONE see AIB000
1-AMINO-9,10-ANTHRAQUINONE see AIA750
2-AMINO-9,10-ANTRAQUINONE see AIB000
AMINOARSON see CBJ000
AMINOAZOBENZENE see PEI000
4-AMINOAZOBENZENE see PEI000
p-AMINOAZOBENZENE see PEI000
4-AMINO-1,1'-AZOBENZENE see PEI000
4-AMINOAZOBENZOL see PEI000
p-AMINOAZOBENZOL see PEI000
2-AMINO-5-AZOTOLUENE see AIC250
4'-AMINO-2,3'-AZOTOLUENE see AIC250
4'-AMINO-2:3'-AZOTOLUENE see AIC250
AMINOAZOTOLUENE (indicator) see AIC250
o-AMINOAZOTOLUENE (MAK) see AIC250
o-AMINOAZOTOLUENO (SPANISH) see AIC250
o-AMINOAZOTOLUOL see AIC250
AMINOBENZ see AKF000
m-AMINOBENZAL FLUORIDE see AID500
AMINOBENZENE see AOQ000
4-AMINOBENZENEARSONIC ACID see ARA250
p-AMINOBENZENEARSONIC ACID see ARA250
(S)-α-AMINOBENZENEPROPANOIC ACID see PEC750
p-AMINOBENZENESULFAMIDE see SNM500
4-AMINOBENZENESULFONAMIDE see SNM500
p-AMINOBENZENESULFONAMIDE see SNM500
2-(p-AMINOBENZENESULFONAMIDO)-4,6-DIMETHYLPY-
  RIMIDINE see SNJ000
2-p-AMINOBENZENESULFONAMIDOQUINOXALINE see
  QTS000
2-(p-AMINOBENZENESULFONAMIDO)THIAZOLE see
  TEX250
N-(4-AMINOBENZENESULFONYL)-N'-BUTYLUREA see
  BSM000
2-p-AMINOBENZENESULPHONAMIDOQUINOXALINE see
  QTS000
2-(p-AMINOBENZENESULPHONAMIDO)THIAZOLE see
  TEX250
2-AMINOBENZENETHIOL see AIF500
2-AMINOBENZIMIDAZOLE see AIG000
2-AMINOBENZOIC ACID see API500
4-AMINOBENZOIC ACID see AIH600
o-AMINOBENZOIC ACID see API500
p-AMINOBENZOIC ACID see AIH600
γ-AMINOBENZOIC ACID see AIH600
p-AMINOBENZOIC ACID-2-N-AMYLAMINOETHYL ESTER
  see PBV750
p-AMINOBENZOIC ACID-3-(DIBUTYLAMINO)PROPYL ES-
  TER, HYDROCHLORIDE see AIT000
p-AMINOBENZOIC ACID-3-(β-DIETHYLAMINO)ETHOXY)
  PROPYL ESTER see AIL500
4-AMINOBENZOIC ACID DIETHYLAMINOETHYL ESTER
  see AIL750
p-AMINOBENZOIC ACID-2-DIETHYLAMINOETHYL ESTER
  see AIL750
4-AMINOBENZOIC ACID 2-(DIETHYLAMINO)ETHYL ES-
  TER, HYDROCHLORIDE see AIT250
p-AMINOBENZOIC ACID-2-DIETHYLAMINOETHYL ESTER,
  HYDROCHLORIDE see AIT250
p-AMINOBENZOIC ACID 3-(DIMETHYLAMINO)-1,2-DI-
  METHYLPROPYL ESTER, HYDROCHLORIDE see
  AIT750
4-AMINOBENZOIC ACID ETHYL ESTER see EFX000
o-AMINOBENZOIC ACID, ETHYL ESTER see EGM000
p-AMINOBENZOIC ACID ETHYL ESTER see EFX000
2-AMINOBENZOIC ACID METHYL ESTER see APJ250
o-AMINOBENZOIC ACID METHYL ESTER see APJ250
2-AMINOBENZOIC ACID-3-PHENYL-2-PROPENYL ESTER
  see API750

6-(4'-AMINOBENZOL-SULFONAMIDO)-2,4-DIMETHYLPY-
RIMIDIN (GERMAN) see SNJ000
(p-AMINOBENZOLSULFONYL)-2-AMINO-4,6-DIMETHYL-
PYRIMIDIN (GERMAN) see SNJ000
p-AMINOBENZOPHENONE see AIR250
3-AMINOBENZOTRIFLUORIDE see AID500
m-AMINOBENZOTRIFLUORIDE see AID500
2-AMINOBENZOXAZOLE see AIS600
3-(p-AMINOBENZOXY)-1-DI-n-BUTYLAMINOPROPANE see
BOO750
3-(p-AMINOBENZOXY)-1-DI-n-BUTYLAMINOPROPANE
SULFATE see BOP000
p-AMINOBENZOYLDIBUTYLAMINOPROPANOL see
BOO750
AMINOBENZOYLDIBUTYLAMINOPROPANOL HYDRO-
CHLORIDE see AIT000
p-AMINOBENZOYLDIBUTYLAMINOPROPANOL SULFATE
see BOP000
p-AMINOBENZOYLDIETHYLAMINOETHANOL see AIL750
p-AMINOBENZOYLDIETHYLAMINOETHANOL HYDRO-
CHLORIDE see AIT250
o-AMINOBENZOYL DI(ISOPROPYLAMINO)ETHANOL HY-
DROCHLORIDE see IJZ000
p-AMINOBENZOYLDIMETHYLAMINO-1,2-DIMETHYLPRO-
PANOL HYDROCHLORIDE see AIT750
4-(4-AMINOBENZYL)ANILINE see MJQ000
AMINOBENZYLPENICILLIN see AIV500
d-(−)-α-AMINOBENZYLPENICILLIN see AIV500
AMINOBENZYLPENICILLIN TRIHYDRATE see AOD125
α-AMINOBENZYLPENICILLIN TRIHYDRATE see AOD125
4-AMINOBIPHENYL see AJS100
p-AMINOBIPHENYL see AJS100
5-AMINO-1-BIS(DIMETHYLAMIDE)PHOSPHORYL-3-PHE-
NYL-1,2,4-TRIAZOLE see AIX000
5-AMINO-1-BIS(DIMETHYLAMIDO)PHOSPHORYL-3-PHE-
NYL-1,2,4-TRIAZOLE see AIX000
5-AMINO-1-(BIS(DIMETHYLAMINO)PHOSPHINYL)-3-PHE-
NYL-1,2,4-TRIAZOLE see AIX000
AMINOBIS(PROPYLAMINE) see AIX250
1-AMINO-BUTAAN (DUTCH) see BPX750
1-AMINOBUTAN (GERMAN) see BPX750
1-AMINOBUTANE see BPX750
2-AMINOBUTANE see BPY000
2-AMINOBUTAN-1-OL see AJA250
2-AMINO-1-BUTANOL see AJA250
2-AMINO-n-BUTYL ALCOHOL see AJA250
4-AMINO-N-((BUTYLAMINO)CARBONYL)BENZENE-
SULFONAMIDE see BSM000
3-(2-AMINOBUTYL)INDOLE ACETATE see AJB250
4-AMINO-6-tert-BUTYL-3-METHYLTHIO-as-TRIAZIN-5-ONE
see MQR275
4-AMINO-6-tert-BUTYL-3-(METHYLTHIO)-1,2,4-TRIAZIN-5-
ONE see MQR275
γ-AMINOBUTYRIC ACID CETYL ESTER see AJC500
AMINOCAINE see AIT250
AMINOCAPROIC LACTAM see CBF700
AMINOCARB see DOR400
AMINOCARBE (FRENCH) see DOR400
AMINO-α-CARBOLINE see AJD750
2-AMINO-α-CARBOLINE see AJD750
2,2'-((4-((AMINOCARBONYL)AMINO)PHENYL)
ARSINIDENE)BIS(THIO)BISACETIC ACID see CBI250
2,2'-(((4-((AMINOCARBONYL)AMINO)PHENYL)
ARSINIDENEBIS(THIO))BIS) BENZOIC ACID see
TFD750
(4-((AMINOCARBONYL)AMINO)PHENYL)ARSONIC ACID
see CBJ000
N-(AMINOCARBONYL)-2-BROMO-2-ETHYLBUTANAMIDE
see BNK000
N-(AMINOCARBONYL)-2-BROMO-3-METHYLBUTANAM-
IDE see BNP750
(Z)-N-(AMINOCARBONYL)-2-ETHYL-2-BUTENAMIDE see
EHP000
N-(AMINOCARBONYL)HYDROXYLAMINE see HOO500

2-(((AMINOCARBONYL)OXY)METHYL)-2-METHYLPENTYL
ESTER BUTYL CARBAMIC ACID see MOV500
1-AMINO-2-CARBOXYBENZENE see API500
1-AMINO-4-CARBOXYBENZENE see AIH600
3-AMINO-N-(α-CARBOXYPHENETHYL)SUCCINAMIC ACID
N-METHYL ESTER, stereoisomer see ARN825
m-AMINOCHLOROBENZENE see CEH675
1-AMINO-2-CHLOROBENZENE see CEH670
1-AMINO-3-CHLOROBENZENE see CEH675
1-AMINO-4-CHLOROBENZENE see CEH680
5-AMINO-4-CHLORO-2,3-DIHYDRO-3-OXO-2-PHENYLPYRI-
DAZINE see PEE750
1-AMINO-3-CHLORO-6-METHYLBENZENE see CLK225
2-AMINO-4-CHLOROPHENOL (DOT) see CEH250
5-AMINO-4-CHLORO-2-PHENYL-3(2H)-PYRIDAZINONE see
PEE750
2-AMINO-4-CHLOROTOLUENE see CLK225
2-AMINO-5-CHLOROTOLUENE see CLK220
2-AMINO-5-CHLOROTOLUENE HYDROCHLORIDE see
CLK235
3-AMINO-p-CRESOL METHYL ESTER see MGO750
m-AMINO-p-CRESOL, METHYL ESTER see MGO750
AMINOCYCLOHEXANE see CPF500
1-AMINODECANE see DAG600
4-AMINO-4-DEOXY-N[10]-METHYLPTEROYLGLUTAMATE
see MDV500
4-AMINO-4-DEOXY-N[10]-METHYLPTEROYLGLUTAMIC
ACID see MDV500
4-AMINO-4-DEOXYPTEROYLGLUTAMATE see AMG750
2-AMINO-9-(2-DEOXY-β-d-RIBOFURANOSYL)-9H-PURINE-
6-THIOL HYDRATE see TFJ250
3-AMINODIBENZOFURAN see DDB600
2-AMINO-4-DICHLOROARSINOPHENOL HYDROCHLORIDE
see DFX400
1-AMINO-3-(DIETHYLAMINO)PROPANE see DIY800
4-AMINODIFENIL (SPANISH) see AJS100
6-AMINO-1,2-DIHYDRO-1-HYDROXY-2-IMINO-4-PIPERIDI-
NOPYRIMIDINE see DCB000
(s)-7-AMINO-6,7-DIHYDRO-10-HYDROXY-1,2,3-TRIME-
THOXYBENZO(a)HEPTALEN-9(5H)-ONE see TLN750
l-2-AMINO-1-(3,4-DIHYDROXYPHENYL)ETHANOL see
NNO500
2-AMINO-3-(3,4-DIHYDROXYPHENYL)PROPANOIC ACID
see DNA200
2-AMINO-1-(2,5-DIMETHOXYPHENYL)-1-PROPANOL HY-
DROCHLORIDE see MDW000
4-AMINO-N-(2,6-DIMETHOXY-4-PYRIMIDINYL)BENZENE-
SULFONAMIDE see SNN300
1-AMINO-3-DIMETHYLAMINOPROPANE see AJP750
p-AMINODIMETHYLANILINE see DTL800
4-AMINO-2',3-DIMETHYLAZOBENZENE see AIC250
4'-AMINO-2,3'-DIMETHYLAZOBENZENE see AIC250
AMINODIMETHYLBENZENE see XMA000
1-AMINO-2,4-DIMETHYLBENZENE see XMS000
1-AMINO-2,5-DIMETHYLBENZENE see XNA000
3-AMINO-1,4-DIMETHYLBENZENE see XNA000
4-AMINO-1,3-DIMETHYLBENZENE see XMS000
1-AMINO-2,4-DIMETHYLBENZENE HYDROCHLORIDE see
XOJ000
1-AMINO-2,5-DIMETHYLBENZENE HYDROCHLORIDE see
XOS000
3-AMINO-1,4-DIMETHYLBENZENE HYDROCHLORIDE see
XOS000
4-AMINO-1,3-DIMETHYLBENZENE HYDROCHLORIDE see
XOJ000
5-AMINO-1,4-DIMETHYLBENZENE HYDROCHLORIDE see
XOS000
3-AMINO-1,4-DIMETHYL-γ-CARBOLINE see TNX275
4-AMINO-6-(1,1-DIMETHYLETHYL)-3-(METHYLTHIO)-
1,2,4-TRIAZIN-5(4H)-ONE see MQR275
2-AMINO-3,8-DIMETHYLIMIDAZO(4,5-f)QUINOXALINE see
AJQ675
2-AMINO-3,8-DIMETHYL-3H-IMIDAZO(4,5-f)QUINOXALINE
see AJQ675

3-AMINO-1,4-DIMETHYL-5H-PYRIDO(4,3-b)INDOLE see
  TNX275
3-AMINO-1,4-DIMETHYL-5H-PYRIDO(4,3-b)INDOLE ACE-
  TATE see AJR500
4-AMINODIPHENYL see AJS100
p-AMINODIPHENYL see AJS100
2-AMINODIPHENYLENE OXIDE see DDB600
p-AMINODIPHENYLIMIDE see PEI000
2-AMINODIPYRIDO(1,2-a:3′,2′-d)-IMIDAZOLE see
  DWW700
2-AMINOETANOLO (ITALIAN) see EEC600
AMINOETHANE see EFU400
1-AMINOETHANE see EFU400
2-AMINOETHANETHIOL see AJT250
1-AMINOETHANOL see AAG500
2-AMINOETHANOL (MAK) see EEC600
2-AMINOETHOXYETHANOL see AJU250
2-(2-AMINOETHOXY)ETHANOL see AJU250
α-AMINOETHYL ALCOHOL see AAG500
β-AMINOETHYL ALCOHOL see EEC600
o-AMINOETHYLBENZENE see EGK500
β-AMINOETHYLBENZENE see PDE250
1-AMINO-4-ETHYLBENZENE see EGL000
α-(1-AMINOETHYL)-BENZYL ALCOHOL see NNM000
3-(2-AMINOETHYL)-1-BENZYL-5-METHOXY-2-METHYLIN-
  DOLE HYDROCHLORIDE see BEM750
3-AMINO-9-ETHYLCARBAZOLE see AJV000
3-AMINO-N-ETHYLCARBAZOLE see AJV000
3-AMINO-9-ETHYLCARBAZOLEHYDROCHLORIDE see
  AJV250
α-(1-AMINOETHYL)-2,5-DIMETHOXYBENZYL ALCOHOL
  HYDROCHLORIDE see MDW000
AMINOETHYLENE see EJM900
AMINOETHYLETHANDIAMINE see DJG600
N-(2-AMINOETHYL)ETHYLENEDIAMINE see DJG600
β-AMINOETHYLGLYOXALINE see HGD000
1-AMINO-2-ETHYLHEXAN (CZECH) see EKS500
1-α-(1-AMINOETHYL)-m-HYDROXYBENZYL ALCOHOL see
  HNB875
β-AMINOETHYLIMIDAZOLE see HGD000
4-(2-AMINOETHYL)IMIDAZOLE see HGD000
2-AMINOETHYL MERCAPTAN see AJT250
3-(2-AMINOETHYL)-6-METHOXYINDOLE HYDROCHLO-
  RIDE see MFS500
AMINOETHYLPIPERAZINE see AKB000
N-AMINOETHYLPIPERAZINE see AKB000
1-(2-AMINOETHYL)PIPERAZINE see AKB000
N-(2-AMINOETHYL)PIPERAZINE see AKB000
N-(β-AMINOETHYL)PIPERAZINE see AKB000
4-(2-AMINOETHYL)PYROCATECHOL see DYC400
2-AMINO-4-(ETHYLTHIO)BUTYRIC ACID see EEI000
dl-2-AMINO-4-(ETHYLTHIO)BUTYRIC ACID see EEI000
AMINOFENAZONE (ITALIAN) see DOT000
5-AMINO-3-FENIL-1-BIS(-DIMETILAMINO)-FOSFORIL-1,2,4-
  TRIAZOLO (ITALIAN) see AIX000
m-AMINOFENOL (CZECH) see ALT500
5-AMINO-3-FENYL-1-BIS(DIMETHYL-AMINO)-FOSFORYL-
  1,2,4-TRIAZOOL (DUTCH) see AIX000
AMINOFLUOREN (GERMAN) see FDI000
2-AMINOFLUORENE see FDI000
AMINOFORM see HEI500
AMINOFORMAMIDINE see GKW000
2-AMINOGLUTARAMIC ACID see GFO050
l-2-AMINOGLUTARAMIDIC ACID see GFO050
α-AMINOGLUTARIC ACID see GFO000
l-2-AMINOGLUTARIC ACID see GFO000
AMINOHEXAHYDROBENZENE see CPF500
1-AMINOHEXANE see HFK000
6-AMINOHEXANOIC ACID CYCLIC LACTAM see CBF700
6-AMINOHEXANOIC ACID HOMOPOLYMER see PJY500
α-AMINOHYDROCINNAMIC ACID see PEC000
1-AMINO-4-HYDROXYANTHRAQUINONE see AKE250
2-AMINO-1-HYDROXYBENZENE see ALT000
3-AMINO-1-HYDROXYBENZENE see ALT500

3-AMINO-4-HYDROXYBENZOIC ACID METHYL ESTER see
  AKF000
α-AMINO-p-HYDROXYBENZYLPENICILLIN TRIHYDRATE
  see AOA100
l-2-AMINO-3-HYDROXYBUTYRIC ACID see TFU750
(R)-4-(2-AMINO-1-HYDROXYETHYL)-1,2-BENZENEDIOL
  see NNO500
((5-(3-AMINO-4-HYDROXYPHENYL)ARSENO)-2-HYDROXY-
  ANILINO)METHANOL SULFOXYLATE SODIUM see
  NCJ500
(3-AMINO-4-HYDROXYPHENYL)ARSONOUS DICHLORIDE
  MONOHYDROCHLORIDE see DFX400
3-AMINO-4-HYDROXYPHENYL DICHLORARSINE HYDRO-
  CHLORIDE see DFX400
(3-AMINO-4-HYDROXYPHENYL)DICHLOROARSINE HY-
  DROCHLORIDE see DFX400
l-N-(p-(((-2-AMINO-4-HYDROXY-6-PTERIDINYL)
  METHYL)AMINO)BENZOYL)GLUTAMIC ACID see
  FMT000
(2S-(2-α,5-α,6-β(S*)))-6-((AMINO(4-HYEROXYPHENYL)ACE-
  TYL)AMINO)-3,3-DIMETHYL-7-OXO-4-THIA-1-
  AZABICYCLO(3.2.0)HEPTANE-2-CARBOXYLIC ACID
  TRIHYDRATE see AOA100
l-α-AMINO-4(OR 5)-IMIDAZOLEPROPIONIC ACID see
  HGE700
α-AMINOIMIDAZOLE-4-PROPIONIC ACID, COBALT(2+)
  SALT see BJY000
α-AMINO-INDOLE-3-PROPRIONIC ACID see TNX000
l-α-AMINO-3-INDOLEPROPRIONIC ACID see TNX000
α′-AMINO-3-INDOLEPROPRIONIC ACID see TNX000
2-AMINO-3-INDOL-3-YL-PROPRIONIC ACID see TNX000
2-AMINOISOBUTANE see BPY250
α-AMINOISOCAPROIC ACID see LES000
2-AMINOISOPROPYLBENZENE see INX000
o-AMINOISOPROPYLBENZENE see INX000
l-(+)-α-AMINOISOVALERIC ACID see VBP000
AMINOMERCURIC CHLORIDE see MCW500
AMINOMESITYLENE see TLG500
2-AMINOMESITYLENE see TLG500
AMINOMESITYLENE HYDROCHLORIDE see TLH000
2-AMINOMESITYLENE HYDROCHLORIDE see TLH000
AMINOMETHANAMIDINE see GKW000
AMINOMETHANE see MGC250
1-AMINO-2-METHOXYBENZENE see AOV900
1-AMINO-4-METHOXYBENZENE see AOW000
1-AMINO-2-METHOXY-5-METHYLBENZENE see MGO750
7-AMINO-9-α-METHOXYMITOSANE see AHK500
3-AMINO-4-METHOXYNITROBENZENE see NEQ500
2-AMINO-1-METHOXY-4-NITROBENZENE see NEQ500
3-AMINO-4-METHOXYTOLUENE see MGO750
4-AMINO-2-METHYLANILINE see TGM000
2-AMINO-4-METHYLANISOLE see MGO750
1-AMINO-2-METHYL-9,10-ANTHRACENEDIONE see
  AKP750
1-AMINO-2-METHYLANTHRAQUINONE see AKP750
1-AMINO-2-METHYLBENZENE see TGQ750
2-AMINO-1-METHYLBENZENE see TGQ750
3-AMINO-1-METHYLBENZENE see TGQ500
4-AMINO-1-METHYLBENZENE see TGR000
1-AMINO-2-METHYLBENZENE HYDROCHLORIDE see
  TGS500
2-AMINO-1-METHYLBENZENE HYDROCHLORIDE see
  TGS500
β-(AMINOMETHYL)-BENZENEPROPANOIC ACID HYDRO-
  CHLORIDE see GAD000
2-AMINO-3-METHYL-α-CARBOLINE see ALD750
3-AMINO-1-METHYL-γ-CARBOLINE see ALD500
l-α-(AMINOMETHYL)-3,4-DIHYDROXYBENZYL ALCOHOL
  see NNO500
2-AMINO-6-METHYLDIPYRIDO(1,2-a:3′,2′-d)IMIDAZOLE
  see AKS250
4-AMINO-10-METHYLFOLIC ACID see MDV500
2-AMINO-6-METHYLHEPTANE see ILM000
6-AMINO-2-METHYLHEPTANE see ILM000

AMINOPTERIDINE see AMG750
AMINOPTERIN see AMG750
4-AMINOPTEROYLGLUTAMIC ACID see AMG750
2-AMINOPYRIDINE see AMI000
AMINO-2-PYRIDINE see AMI000
3-AMINOPYRIDINE see AMI250
AMINO-3-PYRIDINE see AMI250
o-AMINOPYRIDINE see AMI000
α-AMINOPYRIDINE see AMI000
m-AMINOPYRIDINE (DOT) see AMI250
2-AMINO-9H-PYRIDO(2,3-B)INDOLE see AJD750
AMINOPYRINE see DOT000
AMINOPYRINE SODIUM SULFONATE see AMK500
AMINOREXFUMARATE see ALX250
p-AMINOSALICYLSAURES SALZ (GERMAN) see TMJ750
4-AMINOSEMICARBAZIDE see CBS500
AMINOSULFONIC ACID see SNK500
5-(AMINOSULFONYL)-4-CHLORO-2-((2-FURNAYL-
   METHYL)AMINO)BENZOIC ACID see CHJ750
N-(5-(AMINOSULFONYL)-1,3,4-THIADIAZOL-2-YL)ACE-
   TAMIDE see AAI250
5-AMINO-2,2,4,4-TETRAKIS(TRIFLUOROMETHYL)
   IMIDAZOLIDINE see AMQ500
4-AMINO-2,2,5,5-TETRAKIS(TRIFLUOROMETHYL)-3-IMI-
   DAZOLINE see AMQ500
4-AMINO-N-2-THIAZOLYLBENZENESULFONAMIDE see
   TEX250
2-AMINOTHIOPHENOL see AIF500
o-AMINOTHIOPHENOL see AIF500
N-AMINOTHIOUREA see TFQ000
3-AMINOTOLUEN (CZECH) see TGQ500
4-AMINOTOLUEN (CZECH) see TGR000
2-AMINOTOLUENE see TGQ750
3-AMINOTOLUENE see TGQ500
4-AMINOTOLUENE see TGR000
m-AMINOTOLUENE see TGQ500
o-AMINOTOLUENE see TGQ750
p-AMINOTOLUENE see TGR000
2-AMINOTOLUENE HYDROCHLORIDE see TGS500
4-AMINOTOLUENE HYDROCHLORIDE see TGS750
o-AMINOTOLUENE HYDROCHLORIDE see TGS500
3-AMINO-p-TOLUIDINE see TGL750
5-AMINO-o-TOLUIDINE see TGL750
AMINOTRIACETIC ACID see AMT500
AMINOTRIAZOLE see AMY050
2-AMINOTRIAZOLE see AMY050
3-AMINOTRIAZOLE see AMY050
3-AMINO-s-TRIAZOLE see AMY050
2-AMINO-1,3,4-TRIAZOLE see AMY050
3-AMINO-1,2,4-TRIAZOLE see AMY050
3-AMINO-1H-1,2,4-TRIAZOLE see AMY050
AMINOTRIAZOLE (plant regulator) see AMY050
AMINO TRIAZOLE WEEDKILLER 90 see AMY050
AMINOTRIAZOL-SPRITZPULVER see AMY050
4-AMINO-3,5,6-TRICHLOROPICOLINIC ACID see PIB900
4-AMINO-3,5,6-TRICHLORO-2-PICOLINIC ACID see PIB900
4-AMINO-3,5,6-TRICHLORPICOLINSAEURE (GERMAN) see
   PIB900
1-AMINO-2,4,6-TRIMETHYLBENZEN (CZECH) see TLG500
1-AMINO-2,4,5-TRIMETHYLBENZENE see TLG250
2-AMINO-1,3,5-TRIMETHYLBENZENE see TLG500
1-AMINO-2,4,5-TRIMETHYLBENZENE HYDROCHLORIDE
   see TLG750
2-AMINO-1,3-5-TRIMETHYLBENZENE HYDROCHLORIDE
   see TLH000
AMINOUNDECANOIC ACID see AMW000
11-AMINOUNDECANOIC ACID see AMW000
11-AMINOUNDECYLIC ACID see AMW000
AMINOURACIL MUSTARD see BIA250
AMINOUREA see HGU000
AMINOUREA HYDROCHLORIDE see SBW500
2-AMINO-1,4-XYLENE see XNA000
4-AMINO-1,3-XYLENE see XMS000

2-AMINO-1,4-XYLENE HYDROCHLORIDE see XOS000
4-AMINO-1,3-XYLENE HYDROCHLORIDE see XOJ000
AMINOZIDE see DQD400
1,2,-AMINOZOPHENYLENE see BDH250
AMIOYL see BCA000
AMIP 15m see TAI250
AMIPAN T see DUO400
AMIPENIX S see AIV500
AMIPHOS see DOP200
AMIRAL see CJO250
AMISYL see BCA000
AMITAKON see BCA000
AMITAL see AMX750
AMITID see EAI000
AMITON see DJA400
AMITOL see AMY050
AMITRENE see BBK250, BBK500
AMITRIL see AMY050, EAI000
AMITRIL T.L. see AMY050
AMITRIPTILINE see EAH500
AMITRIPTYLIN (GERMAN) see EAH500
AMITRIPTYLINE see EAH500
AMITRIPTYLINE CHLORIDE see EAI000
AMITROL see AMY050
AMITROL 90 see AMY050
AMITROLE see AMY050
AMITROL-T see AMY050
AMITRYPTYLINE HYDROCHLORIDE see EAI000
AMIXICOTYN see NCR000
AMIZIL HYDROCHLORIDE see BCA000
AMIZOL see AMY050
AMMAT see ANU650
AMMATE see ANU650
AMMN see AAW000
AMMOFORM see HEI500
AMMOIDIN see XDJ000
AMMONIA see AMY500
AMMONIA, solution (DOT) see ANK250
AMMONIAC (FRENCH) see AMY500
AMMONIACA (ITALIAN) see AMY500
AMMONIA GAS see AMY500
AMMONIAK (GERMAN) see AMY500
AMMONIATED MERCURY see MCW500
AMMONIO (DICROMATO DI) (ITALIAN) see ANB500
AMMONIOFORMALDEHYDE see HEI500
AMMONIUM ACETATE see ANA000
AMMONIUM ACID ARSENATE see DCG800
AMMONIUM AMIDOSULFONATE see ANU650
AMMONIUM AMIDOSULPHATE see ANU650
AMMONIUM AMINOFORMATE see AND750
AMMONIUM ARSENATE, solid (DOT) see DCG800
AMMONIUM AZIDE see ANA750
AMMONIUM BICARBONATE (1:1) see ANB250
AMMONIUMBICHROMAAT (DUTCH) see ANB500
AMMONIUM BICHROMATE see ANB500
AMMONIUM BIFLUORIDE see ANJ000
AMMONIUM BISULFIDE see ANJ750
AMMONIUM BOROFLUORIDE see ANH000
AMMONIUM BROMATE see ANC000
AMMONIUM BROMIDE see ANC250
AMMONIUM CADMIUM CHLORIDE see AND250
AMMONIUM CALCIUM ARSENATE see AND500
AMMONIUMCARBONAT (GERMAN) see ANE000
AMMONIUM CARBONATE see ANB250, AND750,
   ANE000
AMMONIUM CHLORATE see ANE250
AMMONIUMCHLORID (GERMAN) see ANE500
AMMONIUM CHLORIDE see ANE500
AMMONIUM CHLOROPALLADATE(IV) see ANF000
AMMONIUM CHLOROPLATINATE see ANF250
AMMONIUM CHROMATE see ANF500
AMMONIUM CITRATE see ANF800
AMMONIUM CITRATE, DIBASIC (DOT) see ANF800
AMMONIUM CYANIDE see ANG000

AMMONIUMDICHROMAAT (DUTCH) see ANB500
AMMONIUMDICHROMAT (GERMAN) see ANB500
AMMONIUM DICHROMATE see ANB500
AMMONIUM DICHROMATE(VI) see ANB500
AMMONIUM DIFLUORIDE mixed with HYDROCHLORIC
  ACID see ANG250
AMMONIUM FLUOBORATE see ANH000
AMMONIUM FLUORIDE see ANH250
AMMONIUM FLUOROBORATE see ANH000
AMMONIUM FLUORURE (FRENCH) see ANH250
AMMONIUM FLUOSILICATE see COE000
AMMONIUMGLUTAMINAT (GERMAN) see MRF000
AMMONIUM HEXACHLOROPALLADATE see ANF000
AMMONIUM HEXACHLOROPLATINATE(IV) see ANF250
AMMONIUM HEXAFLUOROSILICATE see COE000
AMMONIUM HEXAFLUOROTITANATE see ANI250
AMMONIUM HEXAFLUOROVANADATE see ANI500
AMMONIUM HYDROGEN CARBONATE see ANB250
AMMONIUM HYDROGEN FLUORIDE see ANJ000
AMMONIUM HYDROGEN FLUORIDE, solid see ANJ000
AMMONIUM HYDROGEN FLUORIDE (solution) see ANJ250
AMMONIUM HYDROGEN FLUORIDE, solution (DOT) see
  ANJ250
AMMONIUM HYDROGEN SULFIDE see ANJ750
AMMONIUM HYDROSULFIDE see ANJ750
AMMONIUM HYDROSULFIDE (solution) see ANK000
AMMONIUM HYDROSULFIDE, solution (DOT) see ANJ750,
  ANK000
AMMONIUM HYDROXIDE see ANK250
AMMONIUM IODATE see ANK750
AMMONIUM IODIDE see ANL000
AMMONIUM MAGNESIUM ARSENATE see ANL750
AMMONIUM MAGNESIUM CHROMATE see ANM000
AMMONIUM MERCAPTAN see ANJ750
AMMONIUM MERCAPTOACETATE see ANM500
AMMONIUM METAVANADATE (DOT) see ANY250
AMMONIUM MOLYBDATE see ANM750
AMMONIUM MURIATE see ANE500
AMMONIUM NITRATE see ANN000
AMMONIUM NITRATE (DOT) see ANN000
AMMONIUM(I) NITRATE(1:1) see ANN000
AMMONIUM-N-NITROSOPHENYLHYDROXYLAMINE see
  ANO500
AMMONIUM ORTHOPHOSPHITE see ANS250
AMMONIUM OXALATE see ANO750
AMMONIUM PARAMOLYBDATE see ANM750
AMMONIUM PENTADECAFLUOROOCTANATE see
  ANP625
AMMONIUM PERCHLORATE see ANP250
AMMONIUM PERCHLORATE (DOT) see PCD500
AMMONIUM PERCHLORYL AMIDE: see ANP500
AMMONIUM PERFLUOROCAPRILATE see ANP625
AMMONIUM PERFLUOROCAPRYLATE see ANP625
AMMONIUM PERFLUOROOCTANOATE see ANP625
AMMONIUM-m-PERIODATE see ANP750
AMMONIUM PERMANGANATE see ANQ000, PCJ750
AMMONIUM PEROXO BORATE see ANQ250
AMMONIUM PEROXY CHROMATE see ANQ750
AMMONIUM PEROXYDISULFATE see ANR000
AMMONIUM PERSULFATE see ANR000
AMMONIUM PERSULFATE (DOT) see ANR000
AMMONIUM PHOSPHATE see ANR500
AMMONIUM PHOSPHATE DIBASIC see ANR500
AMMONIUM PHOSPHATE, MONOBASIC see ANR750
AMMONIUM PHOSPHIDE see ANS000
AMMONIUM PHOSPHITE see ANS250
AMMONIUM PICRATE see ANS500
AMMONIUM PICRATE, wet see ANS750
AMMONIUM PICRATE, wet with 10% or more water (DOT)
  see ANS750
AMMONIUM PICRATE, wet with 10% or more water, over 16
  oz in one outside packaging (DOT) see ANS750
AMMONIUM PICRONITRATE see ANS500
AMMONIUM PLATINIC CHLORIDE see ANF250

AMMONIUM RHODANATE see ANW750
AMMONIUM RHODANIDE see ANW750
AMMONIUMSALZ der AMIDOSULFONSAURE (GERMAN)
  see ANU650
AMMONIUM SILICOFLUORIDE (DOT) see COE000
AMMONIUM SULFAMATE see ANU650
AMMONIUM SULFATE (2:1) see ANU750
AMMONIUM SULFHYDRATE see ANJ750
AMMONIUM SULFOCYANATE see ANW750
AMMONIUM SULFOCYANIDE see ANW750
AMMONIUM SULPHAMATE see ANU650
AMMONIUM SULPHATE see ANU750
AMMONIUM-d-TARTRATE see DCH000
AMMONIUM TARTRATE (DOT) see DCH000
AMMONIUM TETRAFLUOROBORATE see ANH000
AMMONIUM TETRAFLUOROBORATE(1-) see ANH000
AMMONIUM THIOCYANATE see ANW750
AMMONIUM THIOGLYCOLATE see ANM500
AMMONIUM THIOGLYCOLLATE see ANM500
AMMONIUM TRICHLOROACETATE see ANX750
AMMONIUM VANADATE see ANY250
AMMONYX 4 see DTC600
AMMONYX CA SPECIAL see DTC600
AMMONYX CPC see CCX000
AMN see AOL000
AMNESTROGEN see ECU750
AMNICOTIN see NCR000
AMNOSED see TDA500
AMNUCOL see SEH000
AMOBARBITAL see AMX750
AMOCO 1010 see PMP500
AMOEBAL see ABX500
AMOENOL see CHR500
AMOGLANDIN see DMU800
AMONIAK (POLISH) see AMY500
A1-MORIN see MRN500
AMORPHOUS CROCIDOLITE ASBESTOS see ARM275
AMORPHOUS FUSED SILICA see SCK600
AMORPHOUS SILICA DUST see SCH000
AMOSENE see MQU750
AMOSITE ASBESTOS see ARM262
AMOSITE (OBS.) see ARM250
AMOSYT see DYE600
AMOX see NGS500
AMOXICILLIN TRIHYDDRATE see AOA100
AMOXONE see DAA800
AMPAZINE see DQA600
AMPERIL see AIV500, AOD125
AMPHAETEX see BBK500
AMPHEDRINE see BBK500
AMPHEDROXY see DBA800
AMPHEDROXYN see DBA800
AMPHENICOL see CDP250
AMPHEREX see BBK500
dl-AMPHETAMINE see BBK000
AMPHETAMINE HYDROCHLORIDE see AOA750
(−)-AMPHETAMINE SULFATE see BBK750
(+)-AMPHETAMINE SULFATE see BBK500
d-AMPHETAMINE SULFATE see BBK500
l-AMPHETAMINE SULFATE see BBK750
AMPHETASUL see BBK500
AMPHEX see BBK500
AMPHIBOLE see ARM250
AMPHICOL see CDP250
AMPHOIDS S see BBK250
AMPHOMORONAL see AOC500
AMPHORDS S see BBK250
AMPHOTERICIN B see AOC500
AMPHOTERICINE B see AOC500
AMPI-BOL see AIV500
AMPICHEL see AOD125
d-AMPICILLIN see AIV500
d-(−)-AMPICILLIN see AIV500
AMPICILLIN A see AIV500

AMPICILLIN ACID see AIV500
AMPICILLIN ANHYDRATE see AIV500
AMPICILLIN TRIHYDRATE see AOD125
AMPICILLIN (USDA) see AIV500
AMPICIN see AIV500
AMPIKEL see AIV500, AOD125
AMPIMED see AIV500
AMPINOVA see AOD125
AMPIPENIN see AIV500
AMPLIN see AOD125
AMPLISOM see AIV500
AMPLITAL see AIV500
AMPROLENE see EJN500
AMPTREREX see BBK500
AMPY-PENYL see AIV500
AMS see ANU650
AMSCO TETRAMER see PMP750
AMSECLOR see CDP250
AMSUSTAIN see BBK500
AMTHIO see ANW750
AMUDANE see GKE000
AMUNO see IDA000
AMYGDALIC ACID see MAP000
AMYGDALINIC ACID see MAP000
AMYGDALONITRILE see MAP250
n-AMYL ACETATE see AOD725
AMYL ACETATE (DOT) see AOD725
sec-AMYL ACETATE see AOD735
AMYL ACETATE (mixed isomers) see AOD750
AMYL ACETIC ESTER see AOD725
AMYL ACETIC ETHER see AOD725
AMYL ACID PHOSPHATE (DOT) see PBW750
AMYL ALCOHOL see AOE000
N-AMYL ALCOHOL see AOE000
sec-AMYL ALCOHOL (DOT) see PBM750
tert-AMYL ALCOHOL (DOT) see PBV000
AMYL ALCOHOL, NORMAL see AOE000
AMYL ALDEHYDE see VAG000
N-AMYLALKOHOL (CZECH) see AOE000
AMYLAMINE (mixed isomers) (DOT) see PBV500
2-N-AMYLAMINOETHYL-p-AMINOBENZOATE see
   PBV750
AMYLAZETAT (GERMAN) see AOD725
AMYLBARBITONE see AMX750
AMYL BENZOATE see IHP100
4-AMYL-N-BENZOHYDRYLPYRIDINIUM BROMIDE see
   AOF250
d-AMYL BROMIDE see AOF750
AMYL BUTYRATE see IHP400
γ-N-AMYLBUTYROLACTONE see CNF250
AMYLCAINE see PBV750
AMYLCARBINOL see HFJ500
n-AMYL CHLORIDE see PBW500
AMYL CHLORIDE (DOT) see PBW500
α-AMYL CINNAMALDEHYDE see AOG500
AMYL CINNAMATE see AOG600
α-AMYL CINNAMIC ALDEHYDE see AOG500
AMYLEINE see AOM000
α,η-AMYLENE see AOI750
AMYLENE HYDRATE see PBV000
AMYLENES, MIXED see AOJ000
AMYL ETHER see PBX000
N-AMYL ETHER see PBX000
AMYLETHYLCARBINOL see OCY100
AMYL ETHYL KETONE see EGI750, EGI755
n-AMYL FORMATE see AOJ500
AMYL FORMATE (DOT) see AOJ500
AMYL HEXANOATE see IHU100
AMYL HYDRIDE (DOT) see PBK250
AMYL HYDROSULFIDE see PBM000
AMYL LACTATE see AOK250
AMYL LAURATE see AOK500
n-AMYL MERCAPTAN see PBM000
AMYL MERCAPTAN (DOT) see PBM000

AMYL METHYL ALCOHOL see AOK750
AMYL METHYL CARBINOL see HBE500
AMYL-METHYL-CETONE (FRENCH) see MGN500
n-AMYL METHYL KETONE see MGN500
AMYL METHYL KETONE (DOT) see MGN500
n-AMYL-N-METHYLNITROSAMINE see AOL000
AMYL NITRATE see AOL250
n-AMYL NITRITE see AOL500
AMYL NITRITE (DOT) see AOL500
n-AMYLNITROSOUREA see PBX500
1-AMYL-1-NITROSOUREA see PBX500
AMYLOBARBITAL see AMX750
AMYLOBARBITONE see AMX750
AMYLOCAINE see AOM000
AMYLOFENE see EOK000
AMYLOWY ALKOHOL (POLISH) see IHP000
4-n-AMYLPHENOL see AOM250
2-sec-AMYLPHENOL see AOM500
α-AMYL-β-PHENYLACROLEIN see AOG500
AMYL PROPIONATE see AON350
AMYLSINE see PBV750
AMYL SULFHYDRATE see PBM000
AMYL THIOALCOHOL see PBM000
n-AMYL THIOCYANATE see AON500
AMYL TRICHLOROSILANE see PBY750
AMYL-Δ-VALEROLACTONE see DAF200
AMYLVINYLCARBINOL see ODW000
AMYL ZIMATE see BJK500
AMYTAL see AMX750
AMYTAL SODIUM see AON750
AN see AAQ500
AN 1041 see LJR000
ANAC 110 see CNI000
ANACARDONE see DJS200
ANACETIN see CDP250
ANACOBIN see VSZ000
ANACORDONE see DJS200
ANADOLOR see AIT250
ANADOMIS GREEN see CMJ900
ANADREX see DBB000
ANADROL see PAN100
ANADROYD see PAN100
ANAESTHETIC ETHER see EJU000
ANAFEBRINA see DOT000
ANAFLON see HIM000
ANAGIARDIL see MMN250
ANALGIZER see DFA400
O-ANALOG of DIMETHOATE see DNX800
ANAMENTH see TIO750
ANAMID see SAH000
ANANASE see BMO000
ANANSIOL see MNM500
ANAPAC see ABG750
ANAPOLON see PAN100
ANASTERON see PAN100
ANASTERONAL see PAN100
ANASTERONE see PAN100
ANASTRESS see MQU750
ANATENSIN see GGS000
ANATHYLMON see MQU750
ANATOLA see VSK600
ANATRAN see ABH500
ANAUTINE see DYE600
ANAYODIN see SHW000
ANCHRED STANDARD see IHD000
ANCILLIN see AOD125
ANCOLAN see HGC500
ANCOR EN 80/150 see IGK800
ANCORTONE see PLZ000
ANCYLOL see DNG000
ANDAKSIN see MQU750
ANDAXIN see MQU750
ANDERE see BQL000
ANDRAMINE see DYE600

ANDRAZIDE see ILD000
ANDREZ see SMR000
ANDROGEN see TBG000
ANDROLIN see TBF500
ANDROMETH see MPN500
ANDRONAQ see TBF500
ANDROSAN see MPN500, TBG000
ANDROSAN (tablets) see MPN500
ANDROSTEN see MPN500
4-ANDROSTENE-17-α-METHYL-17-β-OL-3-ONE see
   MPN500
Δ⁴-ANDROSTENE-17-β-PROPIONATE-3-ONE see TBG000
ANDROST-4-EN-17β-OL-3-ONE see TBF500
Δ⁴-ANDROSTEN-17(β)-OL-3-ONE see TBF500
ANDROTESTON see TBG000
ANDROTEST P see TBG000
ANDRUSOL see TBF500
ANDRUSOL-P see TBG000
ANECOTAN see DFA400
ANECTINE see CMG250, HLC500
ANECTINE CHLORIDE see HLC500
ANELIX see HIM000
ANELMID see DJT800
ANERGAN see ABH500
ANERTAN see MPN500, TBG000
ANERTAN (tablets) see MPN500
ANERVAL see BRF500
ANESTACON see DHK400
ANESTACON HYDROCHLORIDE see DHK600
ANESTHENYL see MGA850
ANESTHESIA ETHER see EJU000
ANESTHESIN see EFX000
ANESTHESOL see AIT250
ANESTHETIC COMPOUND No. 347 see EAT900
ANESTHETIC ETHER see EJU000
ANESTHONE see EFX000
ANESTIL see AIT250
ANETAIN see BQA010
ANETHOLE (FCC) see PMQ750
ANEURAL see MQU750
ANEUXRAL see MQU750
ANEXOL see CDP000
ANFRAM 3PB see TNC500
ANG 66 see AOP250
ANGEL TULIP see SLV500
ANGICAP see PBC250
ANGIFLAN see DBX400
ANGININE see NGY000
ANGITET see PBC250
ANGLISLITE see LDY000
ANGUIDIN see AOP250
ANGUIDINE see AOP250
ANGUIFUGAN see DJT800
ANHIBA see HIM000
ANHISTABS see WAK000
ANHISTOL see WAK000
ANHYDRIDE ACETIQUE (FRENCH) see AAX500
ANHYDRIDE ARSENIEUX (FRENCH) see ARI750
ANHYDRIDE ARSENIQUE (FRENCH) see ARH500
ANHYDRIDE CARBONIQUE (FRENCH) see CBU250
ANHYDRIDE CARBONIQUE et OXYDE d'ETHYLENE
   MELANGES (FRENCH) see EJO000
ANHYDRIDE CHROMIQUE (FRENCH) see CMK000
ANHYDRIDE PHTALIQUE (FRENCH) see PHW750
ANHYDRIDES see AOP500
ANHYDRIDE VANADIQUE (FRENCH) see VDU000
ANHYDRID KYSELINY TETRAHYDROFTALOVE (CZECH)
   see TDB000
3,6-ANHYDRO-d-GALACTAN see CCL250
ANHYDROGITALIN see GEU000
ANHYDRO-d-GLUCITOL MONOOCTADECANOATE see
   SKV150
ANHYDROGLUCOCHLORAL see GFA000
ANHYDROL see EFU000

ANHYDRONE see PCE000
ANHYDROSORBITOL STEARATE see SKV150
ANHYDRO-o-SULFAMINE BENZOIC ACID see BCE500
ANHYDROTRIMELLIC ACID see TKV000
ANHYDROUS AMMONIA see AMY500
ANHYDROUS CHLOROBUTANOL see ABD000
ANHYDROUS IRON OXIDE see IHD000
ANHYDROUS OXIDE of IRON see IHD000
ANICON KOMBI see CIR250
ANICON M see CIR250
ANIDRIDE ACETICA (ITALIAN) see AAX500
ANIDRIDE CROMICA (ITALIAN) see CMK000
ANIDRIDE CROMIQUE (FRENCH) see CMJ900
ANIDRIDE FTALICA (ITALIAN) see PHW750
ANILIN (CZECH) see AOQ000
ANILINA (ITALIAN, POLISH) see AOQ000
ANILINE see AOQ000
p-ANILINEARSONIC ACID see ARA250
ANILINE CARMINE POWDER see FAE100
ANILINE CHLORIDE see BBL000
ANILINE DYES see AOQ500
ANILINE GREEN see BAY750
ANILINE HYDROCHLORIDE (DOT) see BBL000
ANILINE OIL see AOQ000
ANILINE OIL DRUMS, EMPTY see AOR000
''ANILINE SALT'' see BBL000
p-ANILINESULFONAMIDE see SNM500
ANILINE-p-SULFONIC AMIDE see SNM500
ANILINE VIOLET see AOR500
ANILINE YELLOW see PEI000
ANILINIUM CHLORIDE see BBL000
ANILINOBENZENE see DVX800
2-ANILINOBENZOIC ACID see PEG500
o-ANILINOBENZOIC ACID see PEG500
ANILINOETHANE see EGK000
2-ANILINOETHANOL see AOR750
ANILINOMETHANE see MGN750
ANILINONAPHTHALENE see PFT500
2-ANILINONAPHTHALENE see PFT500
ANILITE see AOT250
ANIMAL CONIINE see PBK500
ANIMAL OIL see BMA750
p-ANISALDEHYDE see AOT500
ANISE ALCOHOL see MED500
ANISE CAMPHOR see PMQ750
ANISEED OIL see AOU250
ANISENE see CLO750
ANISE OIL see AOU250
m-ANISIC ACID see AOU500
o-ANISIC ACID see MPI000
p-ANISIC ACID, ETHYL ESTER see AOV000
ANISIC ACID HYDRAZIDE see AOV500
p-ANISIC ACID, HYDRAZIDE see AOV500
ANISIC ALCOHOL see MED500
ANISIC ALDEHYDE see AOT500
ANISIC HYDRAZIDE see AOV500
2-ANISIDINE see AOV900
4-ANISIDINE see AOW000
o-ANISIDINE see AOV900
p-ANISIDINE see AOW000
o-ANISIDINE HYDROCHLORIDE see AOX250
p-ANISIDINE HYDROCHLORIDE see AOX500
o-ANISIDINE NITRATE see NEQ500
ANIS OEL (GERMAN) see AOU250
p-ANISOL ALCOHOL see MED500
ANISOLE see AOX750
ANISOPIROL see HAH000
ANISOPYRADAMINE see DBM800
ANISOYL CHLORIDE see AOY250
ANISOYLHYDRAZINE see AOV500
p-ANISOYLHYDRAZINE see AOV500
ANISYL ACETATE see AOY400
2-(p-ANISYL)ACETIC ACID see MFE250
ANISYLACETONE see MFF580

ANISYL ALCOHOL (FCC) see MED500
o-ANISYLAMINE see AOV900
p-ANISYLAMINE see AOW000
p-ANISYL CHLORIDE see AOY250
ANISYL FORMATE see MFE250
ANKILOSTIN see PCF275
ANN (GERMAN) see AAW000
ANNALINE see CAX750
ANNATTO EXTRACT see APE100
(6)ANNULENE see BBL250
ANODYNON see EHH000
ANOFEX see DAD200
ANOL see CPB750
ANOREXIDE see BBK000
ANOVLAR 21 see EEH520
ANOZOL see DJX000
ANPROLENE see EJN500
ANPROLINE see EJN500
ANPUZONE see BRF500
ANSAR see HKC000
ANSAR 160 see HKC500
ANSAR 170 see MRL750
ANSAR 184 see DXE600
ANSAR DSMA LIQUID see DXE600
ANSEPRON see PGA750
ANSIACAL see MDQ250
ANSIATAN see MQU750
ANSIBASE RED KB see CLK225
ANSIL see MQU750
ANSIOWAS see MQU750
ANTABUS see DXH250
ANTABUSE see DXH250
ANTADIX see DXH250
ANTADOL see BRF500
ANTAENYL see DXH250
ANTAETHAN see DXH250
ANTAETHYL see DXH250
ANTAETIL see DXH250
ANTAGONATE see TAI500
ANTAGOTHYROID see TFR250
ANTAGOTHYROIL see TFR250
ANTAK see DAI600
ANTALCOL see DXH250
ANTALERGAN see WAK000
ANTALVIC see PNA500
ANTAMINE see WAK000
ANTAN see NAH500
ANTAROX A-200 see PKF500
ANTASTEN see PDC000
ANTAZOLINE see PDC000
ANTENE see BJK500
ANTETAN see DXH250
ANTETHYL see DXH250
ANTETIL see DXH250
ANTEYL see DXH250
ANTHANTHREN (GERMAN) see APE750
ANTHANTHRENE see APE750
ANTHIO see DRR200
ANTHIOLIMINE see LGU000
ANTHIOMALINE see LGU000
ANTHION see DWQ000
ANTHISAN see WAK000
ANTHISAN MALEATE see DBM800
ANTHIUM DIOXCIDE see CDW450
ANTHON see TIQ250
ANTHOPHYLITE see ARM264
ANTHRACEN (GERMAN) see APG500
2-ANTHRACENAMINE see APG000
ANTHRACENE see APG500
9,10-ANTHRACENEDIONE see APK250
1,8,9-ANTHRACENETRIOL see APH250
ANTHRACIN see APG500
ANTHRACITE PARTICLES see CMY635
2-ANTHRACYLAMINE see APG000

ANTHRADIONE see APK250
ANTHRALIN see APH250
2-ANTHRAMINE see APG000
ANTHRANILIC ACID see API500
ANTHRANILIC ACID, CINNAMYL ESTER see API750
ANTHRANILIC ACID, METHYL ESTER see APJ250
ANTHRANILIC ACID, PHENETHYL ESTER see APJ500
ANTHRANTHRENE see APE750
ANTHRAPOLE AZ see BQK250
ANTHRAQUINONE see APK250
9,10-ANTHRAQUINONE see APK250
α-ANTHRAQUINONYLAMINE see AIA750
β-ANTHRAQUINONYLAMINE see AIB000
1,8,9-ANTHRATRIOL see APH250
2-ANTHRYLAMINE see APG000
ANTIAETHAN see DXH250
ANTIBASON see MPW500
ANTIBIOTIC No. 899 see VRF000
ANTIBIOTIC A-5283 see PIF750
ANTIBIOTIC FN 1636 see PEC750
ANTIBIOTIC LA 7017 see MQW750
ANTIBIOTIC X 537 see LBF500
ANTIBULIT see SHF500
ANTICARIE see HCC500
ANTICHLOR see SKI500
ANTIDEPRIN see DLH600
ANTIDEPRIN HYDROCHLORIDE see DLH630
ANTIDUROL see DAM700
ANTIEGENE MB see BCC500
ANTIETANOL see DXH250
ANTI-ETHYL see DXH250
ANTIETIL see DXH250
ANTIFEBRIN see AAQ500
ANTIFOLAN see MDV500
ANTIFORMIN see SHU500
ANTI-GERM 77 see BEN000
ANTIGESTIL see DKA600
ANTIHELMYCIN see AQB000
ANTIHIST see DBM800
ANTIHISTAL see PDC000
ANTI-INFECTIVE VITAMIN see VSK600
ANTIKNOCK-33 see MAV750
ANTIKOL see DXH250
ANTILEPSIN see DNU000
ANTIMICINA see ILD000
ANTIMIGRANT C 45 see SEH000
ANTIMILACE see TDW500
ANTIMIT see BIE500
ANTIMOINE FLUORURE (FRENCH) see AQE000
ANTIMOINE (TRICHLORURE D') see AQC500
ANTIMOL see SFB000
ANTIMONIC CHLORIDE see AQD000
ANTIMONIO (PENTACLORURO DI) (ITALIAN) see AQD000
ANTIMONIO (TRICLORURO DI) see AQC500
ANTIMONIOUS OXIDE see AQF000
ANTIMONOUS CHLORIDE see AQC500
ANTIMONOUS CHLORIDE (DOT) see AQC500
ANTIMONOUS FLUORIDE see AQE000
ANTIMONOUS SULFIDE see AQL500
ANTIMONPENTACHLORID (GERMAN) see AQD000
ANTIMONTRICHLORID see AQC500
ANTIMONWASSERSTOFFES (GERMAN) see SLQ000
ANTIMONY see AQB750
ANTIMONY BLACK see AQB750
ANTIMONY BUTTER see AQC500
ANTIMONY CHLORIDE see AQC500
ANTIMONY(V) CHLORIDE see AQD000
ANTIMONY CHLORIDE (DOT) see AQC500
ANTIMONY(III) CHLORIDE see AQC500
ANTIMONY COMPOUNDS see AQD500
ANTIMONY EMETINE IODIDE see EAM000
ANTIMONY FLUORIDE see AQF250
ANTIMONY(V) FLUORIDE see AQF250

ANTIMONY(III) FLUORIDE (1:3) see AQE000
ANTIMONY GLANCE see AQL500
ANTIMONY HYDRIDE see SLQ000
ANTIMONY LACTATE see AQE250
ANTIMONY LACTATE, solid (DOT) see AQE250
ANTIMONYL POTASSIUM TARTRATE see AQG250
ANTIMONY NITRIDE see AQE750
ANTIMONY ORANGE see AQL500
ANTIMONY OXIDE see AQF000
ANTIMONY PENTACHLORIDE see AQD000
ANTIMONY PENTACHLORIDE (DOT) see AQD000
ANTIMONY(V) PENTAFLUORIDE see AQF250
ANTIMONY PERCHLORIDE see AQD000
ANTIMONY PEROXIDE see AQF000
ANTIMONY POTASSIUM TARTRATE see AQG250
ANTIMONY REGULUS see AQB750
ANTIMONY SESQUIOXIDE see AQF000
ANTIMONY SODIUM OXIDE-l-(+)-TARTRATE see AQI750
ANTIMONY SODIUM TARTRATE see AQI750
ANTIMONY SULFIDE see AQL500
ANTIMONY TELLURIDE see AQL750
ANTIMONY TRICHLORIDE see AQC500
ANTIMONY TRICHLORIDE, liquid (DOT) see AQC500
ANTIMONY TRICHLORIDE, solid (DOT) see AQC500
ANTIMONY TRICHLORIDE, solution (DOT) see AQC500
ANTIMONY TRIETHYL see AQK500
ANTIMONY TRIFLUORIDE see AQE000
ANTIMONY TRIHYDRIDE see SLQ000
ANTIMONY TRIIODIDE see AQK750
ANTIMONY TRIMETHYL see AQL000
ANTIMONY TRIOXIDE (MAK) see AQF000
ANTIMONY TRISULFIDE see AQL500
ANTIMONY TRITELLURIDE see AQL750
ANTIMONY WHITE see AQF000
ANTIMOONPENTACHLORIDE (DUTCH) see AQD000
ANTIMOONTRICHLRIDE see AQC500
ANTIMUCIN WDR see ABU500
ANTINONIN see DUS700
ANTIO see DRR200
ANTIOXIDANT 29 see BFW750
ANTIOXIDANT 116 see PFT500
ANTIOXIDANT DBPC see BFW750
ANTIOXIDANT MB (CZECH) see BCC500
ANTIOXIDANT PBN see PFT500
ANTI-PELLAGRA VITAMIN see NCQ900
ANTIPHEN see MJM500
ANTI-PICA see FDA880
ANTIPRESSINE DIHYDROCHLORIDE see DQB800
ANTIPYONIN see SFF000
ANTIPYRINE see AQN000
(ANTIPYRINYLMETHYLAMINO)METHANESULFONIC
    ACID SODIUM SALT see AMK500
ANTIREN see PIJ000
ANTI-RUST see SIQ500
ANTISACER see DKQ000, DNU000
ANTISAL 1a see TGK750
ANTISEPTOL see BEN000
ANTISOL 1 see PCF275
ANTISTERILITY VITAMIN see VSZ450
ANTISTINE see PDC000
ANTISTOMINUM see BBV500
ANTISTREPT see SNM500
ANTITROMBOSIN see BJZ000
ANTITUBERKULOSUM see ILD000
ANTIVERM see PDP250
ANTIVITIUM see DXH250
ANTIXEROPHTHALMIC VITAMIN see VSK600
ANTOMIN see BBV500
ANTORA see PBC250
ANTOXYLIC ACID see ARA250
ANTRANCINE 12 see BQI000
ANTRENIL see ORQ000
ANTRENYL see ORQ000
ANTRENYL BROMIDE see ORQ000

ANTU see AQN635
ANTURAT see AQN635
ANTYMON (POLISH) see AQB750
ANTYMONOWODOR (POLISH) see SLQ000
ANU see PBX500
ANURAL see MQU750
ANUSPIRAMIN see BRF500
ANXIETIL see MQU750
ANXINE see GGS000
AO 29 see BFW750
AO 4K see BFW750
AOM see ASP250
AOMB see BCC500
A OO see AGX000
AORAL see VSK600
AP see PEK250
A1-0109 P see AHE250
APACHLOR see CDS750
APADODINE see DXX400
APADON see HIM000
APADRIN see MRH209
APAMIDE see HIM000
APAMINE see DBA800
APAP see HIM000
APARKAN see BBV000
APARSIN see BBQ500
APASCIL see MQU750
APAVAP see DGP900
APAVINPHOS see MQR750
APC see ABG750
APC (pharmaceutical) see ARP250
APCO 2330 see PEY000
APELAGRIN see NCQ900
APESAN see IPU000
APETAIN see BBK500
APEX 462-5 see TNC500
APEXOL see VSK600
APFO see ANP625
APGA see AMG750
APHAMITE see PAK000
APHENYLBARBIT see EOK000
APHENYLETTEN see EOK000
APHOSAL see GFA000
APHOXIDE see TND250
APHRODINE see YBJ000
APHROSOL see YBJ000
APHTIRIA see BBQ500
APIGENIN see CDH250
APIGENINE see CDH250
APIGENOL see CDH250
APLIDAL see BBQ500
APL-LUSTER see TEX000
APO see TND250
APOCID ORANGE 2G see HGC000
APOCODEINE see AQO750
A 15 (polymer) see AAX175
APOMINE BLACK GX see AQP000
A-POXIDE see MDQ250
APPA see PHX250
APPRESINUM see HGP500
APPRESSIN see PHW000
APRELAZINE see HGP500
APRESAZIDE see HGP500
APRESINE see HGP500
APRESOLIN see HGP500
APRESOLIN see PHW000
APRESOLINE-ESIDRIX see HGP500
APRESOLINE HYDROCHLORIDE see HGP500
APREZOLIN see HGP500, PHW000
APROBARBITAL SODIUM see BOQ750
APROBARBITONE SODIUM see BOQ750
APROBIT see DQA400
APROCARB see PMY300
APTAL see CFE250

APV see CBR000
APYONINE AURAMINE BASE see IBB000
AQUA AMMONIA see ANK250
AQUACAT see CNA250
AQUA CERA see HKJ000
AQUACHLORAL see CDO000
AQUACIDE see DWX800
AQUACRINE see EDV000
AQUAFIL see SCH000
AQUA FORTIS see NED500
AQUAKAY see MMD500
AQUA-KLEEN see DAA800
AQUALINE see ADR000
AQUAMOLLIN see EIV000
AQUAMYCETIN see CDP250
AQUAMYCIN see ACJ250
AQUAPLAST see SFO500
AQUA REGIA see HHM000
AQUAREX METHYL see SIB600
AQUARILLS see CFY000
AQUARIUS see CFY000
AQUASOL see VSP000
AQUASYNTH see VSK600
AQUATAG see BDE250
AQUATHOL see EAR000
AQUATIN see CLU000
AQUA-VEX see TIX500
AQUAVIRON see TBG000
AQUAZINE see BJP000
AQUEOUS AMMONIA see ANK250
AQUINONE see MMD500
AR-32 see DQY909
AR 12008 see DIO200
ARABIC GUM see AQQ500
ARAB RAT DETH see WAT200
ARACHIC ACID see EAF000
ARACHIDIC ACID see EAF000
ARACHIS OIL see PAO000
ARACID see CJR500
ARACIDE see SOP500
ARAGONITE see CAO000
ARALDITE ACCELERATOR 062 see DQP800
ARALDITE ERE 1359 see REF000
ARALDITE HARDENER HY 951 see TJR000
ARALDITE HY 951 see TJR000
ARALO see PAK000
ARAMINE see HNB875
ARAMITE see SOP500
ARAMITEARARAMITE-15W see SOP500
ARANCIO CROMO (ITALIAN) see LCS000
ARASAN see TFS350
ARATAN see MEP250
ARATHANE see AQT500
ARATRON see SOP500
ARBITEX see BBQ500
ARBOGAL see DSQ000
ARBORICID see BSQ750
ARBOROL see DUS700
ARBOTECT see TEX000
ARBUZ see PAG500
ARCADINE see BCA000
ARCHIDYN see RKP000
ARCOBAN see MQU750
ARCOSOLV see DWT200
ARCOTRATE see PBC250
ARCTON see CBY750
ARCTON 3 see CLR250
ARCTON 4 see CFX500
ARCTON 6 see DFA600
ARCTON 7 see DFL000
ARCTON 9 see TIP500
ARCTON 33 see FOO509
ARCTON 63 see FOO000
ARCTON 114 see FOO509

ARCTON O see CBY250
ARCTUVIN see HIH000
ARDALL see SJO000
ARDEX see BBK500
ARDUAN see PII250
ARECA/CATECHU see BFW000
ARECA CATECHU Linn., fruit extract see BFW000
ARECA CATECHU Linn., nut extract see BFW000
ARECAIDINE METHYL ESTER see AQT750
ARECOLINE see AQT750
ARECOLINE BASE see AQT750
AREDION see CKM000
AREGINAL see EKL000
ARESIN see CKD500
ARETIT see ACE500, BRE500
AREZIN see CKD500
AREZINE see CKD500
ARFICIN see RKP000
ARGAMINE see AQW000
ARGENT FLUORURE (FRENCH) see SDQ500
ARGENTIC FLUORIDE see SDQ500
ARGENTUM see SDI500
ARGEZIN see ARQ725
ARGININE HYDROCHLORIDE see AQW000
l-ARGININE HYDROCHLORIDE see AQW000
ARGININE MONOHYDROCHLORIDE see AQW000
l-ARGININE MONOHYDROCHLORIDE see AQW000
ARGIVENE see AQW000
ARGON see AQW250
ARILATE see BAV575
ARIOTOX see TDW500
ARISAN see CKF750
ARISTOLOCHIC ACID see AQY250
ARISTOLOCHINE see AQY250
ARIZOLE see PMQ750
ARKLONE P see FOO000
ARKOPAL N-090 see PKF000
ARKOTINE see DAD200
ARKOZAL see BSQ000
ARLACEL 60 see SKV150
ARLACEL 161 see OAV000
ARLACEL 169 see OAV000
ARLIDIN HYDROCHLORIDE see DNU200
ARLOSOL GREEN B see BLK000
ARMACIDE see ILD000
ARMAZAL see COH250
ARMCO IRON see IGK800
ARMEEN 16D see HCO500
ARMENIAN BOLE see IHD000
ARMODOUR see PKQ059
ARMOFOS see SKN000
ARMOSTAT 801 see OAV000
ARMOTAN MS see SKV150
ARMOTAN PMO-20 see PKL100
ARMYL see MRV250
ARNICA see AQY500
ARNOSULFAN see SNN300
AROALL see SJO000
AROCHLOR 1221 see PJM000
AROCHLOR 1242 see PJM500
AROCHLOR 1254 see PJN000
AROCHLOR 1260 see PJN250
AROCLOR see PJL750
AROCLOR 1232 see PJM250
AROCLOR 1242 see PJM500
AROCLOR 1248 see PJM750
AROCLOR 1254 see PJN000
AROCLOR 1260 see PJN250
AROCLOR 1262 see PJN500
AROCLOR 1268 see PJN750
AROCLOR 2565 see PJO000
AROCLOR 4465 see PJO250
AROMATIC AMINES see AQY750
AROMATIC CASTOR OIL see CCP250

AROMATIC SOLVENT see NAI500
AROMATIC SPIRITS of AMMONIA see AQZ000
ARON COMPOUND HW see PKQ059
AROSOL see PER000
AROVIT see VSP000
ARQUAD DM14B-90 see TCA500
ARQUAD DM18B-90 see DTC600
ARRESIN see CKD500
ARRHENAL see DXE600
ARSAMBIDE see CBJ000
ARSAN see HKC000
ARSANILIC ACID see ARA250
4-ARSANILIC ACID see ARA250
p-ARSANILIC ACID see ARA250
ARSANILIC ACID, MONOSODIUM SALT see ARA500
ARSANILIC ACID SODIUM SALT see ARA500
ARSECLOR see DFX400
ARSECODILE see HKC500
ARSEN (GERMAN, POLISH) see ARA750
ARSENATE see ARB250
ARSENATE of IRON, FERRIC see IGN000
ARSENATE of IRON, FERROUS see IGM000
ARSENATE of LEAD see LCK000
ARSENENOUS ACID, POTASSIUM SALT see PKV500
ARSENENOUS ACID, SODIUM SALT (9CI) see SEY500
ARSENIATE de CALCIUM (FRENCH) see ARB750
ARSENIATE de MAGNESIUM (FRENCH) see ARD000
ARSENIATE de PLOMB (FRENCH) see ARC750
ARSENIC see ARA750
ARSENIC-75 see ARA750
ARSENIC ACID see ARH500
m-ARSENIC ACID see ARB000
o-ARSENIC ACID see ARB250
ARSENIC ACID, liquid (DOT) see ARB250
ARSENIC ACID, solid (DOT) see ARB250, ARC500
ARSENIC ACID ANHYDRIDE see ARH500
ARSENIC ACID, CALCIUM SALT (2:3) see ARB750
ARSENIC ACID, DISODIUM SALT see ARC000
ARSENIC ACID, DISODIUM SALT, HEPTAHYDRATE see
   ARC250
o-ARSENIC ACID, HEMIHYDRATE see ARC500
ARSENIC ACID, LEAD SALT see ARC750
ARSENIC ACID, MAGNESIUM SALT see ARD000
ARSENIC ACID, MONOPOTASSIUM SALT see ARD250
ARSENIC ACID, MONOSODIUM SALT see ARD500,
   ARD600
ARSENIC ACID, SODIUM SALT see ARD750
ARSENIC ACID, SODIUM SALT (9CI) see ARD500
ARSENIC(V) ACID, TRISODIUM SALT, HEPTAHYDRATE
   (1:3:7) see ARE000
ARSENIC ACID, ZINC SALT see ZDJ000
ARSENICAL solution see FOM050
ARSENICAL DIP see ARE250
ARSENICAL DIP, LIQUID (DOT) see ARE250
ARSENICAL DUST see ARE500
ARSENICAL FLUE DUST see ARE500
ARSENICAL FLUE DUST (DOT) see ARE750
ARSENICALS see ARA750, ARF750
ARSENIC ANHYDRIDE see ARH500
ARSENIC BISULFIDE see ARF000
ARSENIC BLACK see ARA750
ARSENIC BLANC (FRENCH) see ARI750
ARSENIC(III) BROMIDE see ARF250
ARSENIC BUTTER see ARF500
ARSENIC CHLORIDE see ARF500
ARSENIC(III) CHLORIDE see ARF500
ARSENIC COMPOUNDS see ABX750
ARSENIC DICHLOROETHANE see DFH200
ARSENIC DIETHYL see ARG000
ARSENIC DIMETHYL see ARG250
ARSENIC FLUORIDE see ARI250
ARSENIC HEMISELENIDE see ARG500
ARSENIC HYDRIDE see ARK250
ARSENIC IODIDE see ARG750

ARSENIC OXIDE see ARH500, ARI750
ARSENIC(V) OXIDE see ARH500
ARSENIC(III) OXIDE see ARI750
ARSENIC PENTASULFIDE see ARH250
ARSENIC PENTOXIDE see ARH500
ARSENIC PHOSPHIDE see ARH750
ARSENIC SESQUIOXIDE see ARI750
ARSENIC SESQUISULFIDE see ARI000
ARSENIC SULFIDE see ARI000
ARSENIC SULFIDE YELLOW see ARI000
ARSENIC SULPHIDE see ARI000
ARSENIC TRIBROMIDE see ARF250
ARSENIC TRIFLUORIDE see ARI250
ARSENIC TRIHYDRIDE see ARK250
ARSENIC TRIIODIDE see ARG750
ARSENIC TRIIODIDE mixed with MERCURIC IODIDE see
   ARI500
ARSENIC TRIOXIDE see ARI750
ARSENIC TRIOXIDE mixed with SELENIUM DIOXIDE (1:1)
   see ARJ000
ARSENIC TRISULFIDE see ARI000
ARSENIC YELLOW see ARI000
ARSENIDES see ARJ250
ARSENIGEN SAURE (GERMAN) see ARI750
ARSENIOUS ACID, CALCIUM SALT see CAM500
ARSENIOUS ACID DISODIUM SALT see DXB200
ARSENIOUS ACID (MAK) see ARI750
ARSENIOUS ACID, SODIUM SALT see ARJ500, SEY500
ARSENIOUS ACID, SODIUM SALT POLYMERS see
   ARJ500
ARSENIOUS ACID, STRONTIUM SALT see SME500
ARSENIOUS ACID, ZINC SALT see ZDS000
ARSENIOUS CHLORIDE see ARF500
ARSENIOUS and MERCURIC IODIDE, solution (DOT) see
   ARI500
ARSENIOUS OXIDE see ARI750
ARSENIOUS SULPHIDE see ARI000
ARSENIOUS TRIOXIDE see ARI750
ARSENITE de POTASSIUM (FRENCH) see PKV500
ARSENITE de SODIUM (FRENCH) see SEY500
ARSENIURETTED HYDROGEN see ARK250
ARSENOMARCASITE see ARJ750
ARSENOPYRITE see ARJ750
ARSENOUS ACID see ARI750
ARSENOUS ACID ANHYDRIDE see ARI750
ARSENOUS ANHYDRIDE see ARI750
ARSENOUS BROMIDE see ARF250
ARSENOUS CHLORIDE see ARF500
ARSENOUS FLUORIDE see ARI250
ARSENOUS HYDRIDE see ARK250
ARSENOUS IODIDE see ARG750
ARSENOUS OXIDE see ARI750
ARSENOUS OXIDE ANHYDRIDE see ARI750
ARSENOUS SULFIDE see ARI000
ARSENOUS TRIBROMIDE see ARF250
ARSENOUS TRICHLORIDE (9CI) see ARF500
ARSENOUS TRIIODIDE (9CI) see ARG750
ARSENOWODOR (POLISH) see ARK250
ARSENPHENOLAMINE HYDROCHLORIDE see SAP500
ARSENWASSERSTOFF (GERMAN) see ARK250
ARSEVAN see NCJ500
ARSINE see ARK250
ARSINE BORON TRIBROMIDE see ARK500
ARSINETTE see LCK000
ARSINYL see DXE600
ARSONATE liquid see MRL750
ARSONIC ACID see ABX500
ARSONIC ACID, COPPER(2+) SALT (1:1) (9CI) see
   CNN500
ARSONIC ACID, POTASSIUM SALT see PKV500
ARSONIC ACID, SODIUM SALT (9CI) see ARJ500
4-ARSONOPHENYLGLYCINAMIDE see CBJ750
p-ARSONOPHENYLUREA see CBJ000
ARSPHEN see ABX500

ARSPHENAMINE see SAP500
ARSPHENAMINE METHYLENESULFOXYLIC ACID SO-
    DIUM SALT see NCJ500
ARSYCODILE see HKC500
ARSYNAL see DXE600
ARTANE see BBV000
ARTANE HYDROCHLORIDE see BBV000
ARTANE TRIHEXYPHENIDYL see BBV000
ARTEMISIA OIL see ARL250
ARTEMISIA OIL (WORMWOOD) see ARL250
ARTERENOL see NNO500
l-ARTERENOL see NNO500
ARTEROCOLINE see ABO000
ARTERODY see BBJ750
d-ARTHIN see VSZ100
ARTHODIBROM see NAG400
ARTHO LM see MLH000
ARTIC see MIF765
ARTIFICIAL ALMOND OIL see BAY500
ARTIFICIAL ANT OIL see FPQ875
ARTIFICIAL BARITE see BAP000
ARTIFICIAL CINNAMON OIL see CCO750
ARTIFICIAL GUM see DBD800
ARTIFICIAL HEAVY SPAR see BAP000
ARTIFICIAL MUSTARD OIL see AGJ250
ARTIFICIAL SWEETENING SUBSTANZ GENDORF 450 see
    SJN700
ARTISIL ORANGE 3RP see AKP750
ARTIZIN see BRF500
ARTOLON see MQU750
ARTOMYCIN see TBX250
ARTOSIN see BSQ000
ARTOZIN see BSQ000
ARTRACIN see IDA000
ARTRINOVO see IDA000
ARTRIVIA see IDA000
ARTRIZONE see BRF500
ARTROFLOG see HNI500
ARTROPAN see BRF500
ARUMEL see FMM000
ARUSAL see IPU000
ARVYNOL see CHG000
ARWOOD COPPER see CNI000
AS-17665 see NDY500
ASA see ADA725
A.S.A. see ADA725
ASABAINE see XCJ000
ASA COMPOUND see ABG750
A.S.A. EMPIRIN see ADA725
ASAGRAEA OFFICINALIS see VHZ000
ASAGRAN see ADA725
ASAHISOL 1527 see AAX250
ASALIN see ARM000
ASAMEDOL see DHS200
ASAMID see ENG500
ASATARD see ADA725
ASAZOL see MRL750
ASB 516 see AAX250
ASBEST (GERMAN) see ARM250
ASBESTINE see TAB750
ASBESTOS see ARM250, ARM260, ARM262, ARM264,
    ARM268, ARM275, ARM280
ASBESTOS, ACTINOLITE see ARM260
ASBESTOS, AMOSITE see ARM262
ASBESTOS, ANTHOPHYLITE see ARM264
ASBESTOS, ANTHOPHYLLITE see ARM266
ASBESTOS, CHRYSOTILE see ARM268
ASBESTOS, CROCIDOLITE see ARM275
ASBESTOS FIBER see ARM250
ASBESTOS, TREMOLITE see ARM280
ASBESTOS, WHITE (DOT) see ARM268
ASCABIN see BCM000
ASCABIOL see BCM000
ASCARIDOLE see ARM500

ASCARISIN see ARM500
ASCARYL see HFV500
ASCEPTICHROME see MCV000
ASCLEPIN see ACF000
ASCOPHEN see ARP250
ASCORBIC ACID see ARN000
l-ASCORBIC ACID see ARN000
l(+)-ASCORBIC ACID see ARN000
ASCORBIC ACID SODIUM SALT see ARN125
l-ASCORBIC ACID SODIUM SALT see ARN125
ASCORBICIN see ARN125
ASCORBIN see ARN125
ASCORBUTINA see ARN000
ASCURON see BJI000
ASECRYL see GIC000
ASEPTICHROME see MCV000
ASEPTOFORM see HJL500
ASEPTOFORM E see HJL000
ASEPTOFORM P see HNU500
ASEX see SFS000
ASHLENE see NOH000
ASIDON 3 see TEH500
ASIPRENOL see DMV600
ASL-603 see BMV750
AS 61CL see ADY500
ASMADION see TEH500
ASMALAR see DMV600
ASMATANE MIST see VGP000
ASMAVAL see TEH500
ASM MB see BCC500
ASOZIN see MGQ750
ASPALON see ADA725
ASPARAGINASE see ARN800
l-ASPARAGINASE see ARN800
l-ASPARAGINASE X see ARN800
l-ASPARAGINASI (ITALIAN) see ARN800
l-ASPARAGINE AMIDOHYDROLASE see ARN800
ASPARTAME see ARN825
ASPARTYLPHENYLALANINE METHYL ESTER see
    ARN825
N-l-α-ASPARTYL-l-PHENYLALANINE l-METHYL ESTER
    (9CI) see ARN825
ASPERGUM see ADA725
ASPHALT see ARO500
ASPHALT (CUT BACK) see ARO750
ASPHALT, PETROLEUM see PCR500
ASPHALTUM see ARO500
ASPIRDROPS see ADA725
ASPIRIN see ADA725
ASPIRINE see ADA725
ASPIRIN, PHENACETIN and CAFFEINE see ARP250
ASPON-CHLORDANE see CDR750
ASPOR see EIR000
ASPORUM see EIR000
ASPRO see ADA725
ASPRON see HBT500
ASSIFLAVINE see DBX400
ASSIPRENOL see DMV600
ASSUGRIN see SGC000
ASSUGRIN FEINUSS see SGC000
ASSUGRIN VOLLSUSS see SGC000
ASTA see EAS500
ASTA B518 see EAS500
ASTA Z 4828 see TNT500
ASTA Z 4942 see IMH000
ASTERIC see ADA725
ASTHENTHILO see DKL800
ASTHMA METER MIST see VGP000
ASTMAHALIN see VGP000
A-STOFF see CDN200
ASTOMIN see MLP250
ASTRA CHRYSOIDINE R see PEK000
ASTRA DIAMOND GREEN GX see BAY750
ASTRAFER see IGU000

ASTRALON see PKQ059
ASTRANTIAGENIN D see GMG000
ASTROBAIN see OKS000
ASTROBOT see DGP900
ASTROCAR see DJS200
ASTYN see EHP000
ASUGRYN see SGC000
ASUNTHOL see CNU750
ASYMMETRICAL TRIMETHYL BENZENE see TLL750
ASYMMETRIN see FNO000
AT see AMY050
AT 7 see HCL000
o-AT see AIC250
AT-290 see PED750
AT 327 see BLV000
AT 717 see PKQ250
ATA see AMY050
ATABRINE see ARQ250
ATACTIC POLYPROPYLENE see PMP500
ATACTIC POLYSTYRENE see SMQ500
ATACTIC POLY(VINYL CHLORIDE) see PKQ059
ATALCO C see HCP000
ATALCO S see OAX000
ATARA see CJR909
ATARAX see CJR909
ATARAXOID see CJR909
ATARAZOID see CJR909
ATAZINA see CJR909
ATAZINAX see ARQ725
ATCOTIBINE see ILD000
ATCP see PIB900
ATEM see IGG000
ATENSIN see GGS000
ATERAX see CJR909
ATGARD see DGP900
ATHAPROPAZINE see DIR000
ATHOPROPAZIN see DIR000
ATHROMBINE-K see WAT200
ATHYLEN (GERMAN) see EIO000
ATHYLENGLYKOL (GERMAN) see EJC500
ATHYLENGLYKOL-MONOATHYLATHER (GERMAN) see
   EES350
ATHYL-GUSATHION see EKN000
ATIRAN see MEP250
ATLACIDE see SFS000
ATLANTIC BLACK BD see AQP000
ATLANTIC BLUE 2B see CMO000
ATLANTIC RESIN FAST BROWN BRL see CMO750
ATLAS "A" see SEY500
ATLAS G 2146 see HKJ000
ATLAS WHITE TITANIUM DIOXIDE see TGG760
AT LIQUID see AMY050
ATLOX 1087 see PKL100
ATMOS 150 see OAV000
ATMUL 67 see OAV000
ATMUL 84 see OAV000
ATMUL 124 see OAV000
ATOMIT see CAO000
ATOSIL see DQA400
ATOXICOCAINE see AIT250
ATOXYL see ARA500
ATOXYLIC ACID see ARA250
AT-17 PHOSPHATE see MLP250
ATRANEX see ARQ725
ATRASINE see ARQ725
ATRATOL see SFS000
ATRATOL A see ARQ725
ATRAVET see AAF750, ABH500
ATRAXINE see MQU750
ATRAZIN see ARQ725
ATRAZINE see ARQ725
ATRED see ARQ725
ATREX see ARQ725
ATRIVYL see MMN250

ATROCHIN see SBG000
ATROPIN (GERMAN) see ARR000
ATROPINE see ARR000
(−)-ATROPINE see HOU000
ATROPINE SULFATE (1 : 1) see ARR250
ATROPINE SULFATE (2 : 1) see ARR500
ATROPIN SIRAN (CZECH) see ARR500
ATROPINSULFAT (GERMAN) see ARR500
ATROQUIN see SBG000
ATROVENT see IGG000
ATSETOZIN see ABH500
ATTAC 6 see CDV100
ATTAC 6-3 see CDV100
ATUL ACID CRYSTAL ORANGE G see HGC000
ATUL CRYSTAL RED F see HJF500
ATUL DIRECT BLACK E see AQP000
ATUL DIRECT BLUE 2B see CMO000
ATUL FAST YELLOW R see DOT300
ATUL INDIGO CARMINE see FAE100
ATUL OIL ORANGE T see TGW000
ATUL ORANGE R see PEJ500
ATUL TARTRAZINE see FAG140
ATX II see ARS000
ATYSMAL see ENG500
AUBYGEL GS see CCL250
AUBYGUM DM see CCL250
AULES see TFS350
AULIGEN see BJU000
AURAMINE (MAK) see IBA000, IBB000
AURAMINE BASE see IBB000
AURAMINE HYDROCHLORIDE see IBA000
AURAMINE O (BIOLOGICAL STAIN) see IBA000
AURAMINE YELLOW see IBA000
AURANILE see DKQ000, DNU000
AURANTICA see MRN500
AUREMETINE see ARS750
AUREOCINA see CMA750
AUREOLIC ACID see MQW750
AUREOMYCIN see CMA750
AUREOMYCIN A-377 see CMA750
AUREOMYKOIN see CMA750
AUREOTAN see ART250
AURICIDINE see GJG000
AURLELIC ACID see MQW750
AUROCIDIN see GJG000
AUROLIN see GJG000
AUROMYOSE see ART250
AUROPAN see NBU000
AUROPEX see GJG000
AUROPIN see GJG000
AURORA YELLOW see CAJ750
AUROSAN see GJG000
AUROTAN see ART250
1-AUROTHIO-d-GLUCOPYRANOSE see ART250
AUROTHIOGLUCOSE see ART250
AUROTHION see GJG000
AURUMINE see ART250
AUSTIOX see TGG760
AUSTRACIL see CDP250
AUSTRACOL see CDP250
AUSTRALIAN GUM see AQQ500
AUSTRALOL see IQZ000
AUSTRAPEN see AIV500
AUSTRIAN CINNABAR see LCS000
AUSTROMINAL see EOK000
AUSTROVIT PP see NCR000
AUTAN see DKC800
AUTHRON see ART250
AuTM see GJC000
AUTOMIN see BBV500
AUXINUTRIL see PBB750
AVADEX see DBI200
AVAGAL see XCJ000
AVERSAN see DXH250

AVERTIN see ARW250, THV000
AVERZAN see DXH250
AVESYL see GGS000
AVIBEST C see ARM268
AVIBON see VSK600
AVICOL see PAX000
AVIOMARIN see DYE600
AVIROL 118 CONC see SIB600
AVISUN see PMP500
AVITA see VSK600
AVITOL see VSK600
AVLON see DBX400
AVLOSULPHONE see SOA500
AVLOTANE see HCI000
AVOLIN see DTR200
AVOMINE see DQA400
AVON GREEN A-4379 see BAY750
AVOXYL see GGS000
AWPA #1 see CMY825
AXIOM see DIX600
AXM see ABN000
AY-5406 see BCA000
AY-6108 see AIV500
AY 9944 see BHN000
AY-57,062 see ABH500
AY-61122 see MLJ500
AY 64043 see ICB000
AYAA see AAX250
AYAF see AAX250
AYERMATE see MQU750
AYFIVIN see BAC250
9-AZAANTHRACENE see ADJ500
10-AZAANTHRACENE see ADJ500
5-AZA-10-ARSENAANTHRACENE CHLORIDE see PDB000
12-AZABENZ(a)ANTHRACENE see BAW750
AZABENZENE see POP250
AZACYCLOHEPTANE see HDG000
1-AZACYCLOHEPTANE see HDG000
2-AZACYCLOHEPTANONE see CBF700
AZACYCLOHEXANE see PIL500
1-AZA-2,4-CYCLOPENTADIENE see PPS250
AZACYCLOPENTANE see PPS500
AZACYCLOPROPANE see EJM900
7-AZADIBENZ(a,h)ANTHRACENE see DCS400
7-AZADIBENZ(a,j)ANTHRACENE see DCS600
14-AZADIBENZ(a,j)ANTHRACENE see DCS800
7-AZA-7H-DIBENZO(c,g)FLUORENE see DCY000
9-AZAFLUORENE see CBN000
1-AZAINDENE see ICM000
3-AZAINDOLE see BCB750
AZALONE see PDC000
1-AZANAPHTHALENE see QMJ000
2-AZANAPHTHALENE see IRX000
AZANIL RED SALT TRD see CLK235
AZANIN see ASB250
3-AZAPENTANE-1,5-DIAMINE see DJG600
AZAPERONE (USDA) see FLU000
AZAPLANT see AMY050
AZASERIN see ASA500
AZASERINE see ASA500
l-AZASERINE see ASA500
AZATADINE DIMALEATE see DLV800
AZATADINE MELEATE see DLV800
AZATHIOPRINE see ASB250
8-AZATHIOXANTHINE see MLY000
AZATIOPRIN see ASB250
AZBLLEN ASBESTOS see ARM266
AZBOLEN ASBESTOS see ARM264
AZDEL see PMP500
AZDID see BRF500
AZEPERONE see FLU000
AZEPROMAZINE see ABH500
AZETYLAMINOFLUOREN (GERMAN) see FDR000
AZIDE see SFA000

AZIDES see ASC750
AZIDINE BLUE 3B see CMO250
AZIJNZUUR (DUTCH) see AAT250
AZIJNZUURANHYDRIDE (DUTCH) see AAX500
AZIMETHYLENE see DCP800
AZIMIDOBENZENE see BDH250
AZIMINOBENZENE see BDH250
AZINDOLE see BCB750
AZINE see POP250
AZINE DEEP BLACK EW see AQP000
AZINFOS-ETHYL (DUTCH) see EKN000
AZINFOS-METHYL (DUTCH) see ASH500
AZINOS see EKN000
AZINPHOS-AETHYL (GERMAN) see EKN000
AZINPHOS ETHYL see EKN000
AZINPHOS-ETILE (ITALIAN) see EKN000
AZINPHOS METHYL see ASH500
AZINPHOS METHYL, liquid (DOT) see ASH500
AZINPHOS-METILE (ITALIAN) see ASH500
AZIRANE see EJM900
AZIRIDIN (GERMAN) see EJM900
AZIRIDINE see EJM900
AZIRIDINE CARBOXYLIC ACID ETHYL ESTER see
    ASH750
1-AZIRIDINE ETHANOL see ASI000
2-(1-AZIRIDINYL)ETHANOL see ASI000
1-(1-AZIRIDINYL)-N-(p-METHOXYPHENYL)FORMAMIDE
    see MFF250
1-AZIRIDINYL PHOSPHINE OXIDE (TRIS) (DOT) see
    TND250
AZIRIDYL BENZOQUINONE see BDC750
AZIUM see SFA000, SOW000
AZO-33 see ZKA000
AZO-55 see ZKA000
AZO-66 see ZKA000
AZO-77 see ZKA000
AZOAETHAN (GERMAN) see ASN250
AZOAMINE SCARLET see NEQ500
AZOBASE MNA see NEN500
AZOBENZEEN (DUTCH) see ASL250
AZOBENZENE see ASL250
AZOBENZENE OXIDE see ASO750
AZOBENZIDE see ASL250
AZOBENZOL see ASL250
AZOBISBENZENE see ASL250
1,1'-AZOBISCARBAMIDE see ASM270
AZOBISCARBONAMIDE see ASM270
AZOBISCARBOXAMIDE see ASM270
1,1'-AZOBIS(FORMAMIDE) see ASM270
AZOBISISOBUTYLONITRILE see ASL750
α,α'-AZOBISISOBUTYLONITRILE see ASL750
AZOBISISOBUTYRONITRILE see ASL750
2,2'-AZOBIS(ISOBUTYRONITRILE) see ASL750
2,2'-AZOBIS(2-METHYLPROPIONITRILE) see ASL750
AZOBUTYL see BRF500
AZOCARD BLACK EW see AQP000
AZOCARD BLUE 2B see CMO000
AZODIBENZENE see ASL250
AZODIBENZENEAZOFUME see ASL250
AZODICARBAMIDE see ASM270
AZODICARBOAMIDE see ASM270
AZODICARBONAMIDE see ASM300, ASM270
AZODICARBOXAMIDE see ASM270
AZODICARBOXYLIC ACID DIAMIDE see ASM270
AZODIISOBUTYRONITRILE see ASL750
2,2'-AZODIISOBUTYRONITRILE see ASL750
AZODIISOBUTYRONITRILE (DOT) see ASL750
α,α'-AZODIISOBUTYRONITRILE see ASL750
AZODINE see PDC250
AZODIUM see PDC250
AZODOX-55 see ZKA000
AZODOX-55TT see ZKA000
AZODRIN see MRH209
"AZODRIN" see ASN000

AZODYNE see PDC250
AZOENE FAST BLUE BASE see DCJ200
AZOENE FAST ORANGE GR SALT see NEO000
AZOENE FAST RED KB BASE see CLK225
AZOENE FAST RED TR BASE see CLK220
AZOENE FAST RED TR SALT see CLK235
AZO ETHANE see ASN250
AZOFENE see BDJ250
AZOFIX BLUE B SALT see DCJ200
AZOFIX RED GG SALT see NEO500
AZOFIX SCARLET G SALT see NMP500
AZOFOS see MNH000
AZO-GANTANOL see SNK000
AZO GANTRISIN see PDC250
AZO GASTANOL see PDC250
AZOGEN DEVELOPER A see NAX000
AZOGEN DEVELOPER H see TGL750
AZOGENE ECARLATE R see NEQ500
AZOGENE FAST RED TR see CLK220, CLK235
AZOGENE FAST SCARLET G see NMP500
AZOGNE FAST BLUE B see DCJ200
AZOIC DIAZO COMPONENT 32 see CLK225
AZOIC DIAZO COMPONENT 37 see NEO500
AZOIC DIAZO COMPONENT 11 BASE see CLK220,
   CLK235
AZOIC DIAZO COMPONENT 13 BASE see NEQ500
AZOIC RED 36 see MGO750
AZOIMIDE see HHG500
AZOLAN see AMY050
AZOLE see AMY050, PPS250
AZOLID see BRF500
AZOLMETAZIN see SNJ000
AZO-MANDELAMINE see PDC250
AZOMINE see PDC250
AZOMINE BLACK EWO see AQP000
AZOMINE BLUE 2B see CMO000
AZOPHENYLENE see PDB500
AZOPHOS see MNH000
AZORUBIN see HJF500
AZOSEPTALE see TEX250
AZOSSIBENZENE (ITALIAN) see ASO750
AZO-STANDARD see PDC250
AZO-STAT see PDC250
AZOTE (FRENCH) see NGR500
AZOTHIOPRINE see ASB250
AZOTIC ACID see NED500
AZOTO (ITALIAN) see NGR500
AZOTOWY KWAS (POLISH) see NED500
AZOTOX see DAD200
AZOTOYPERITE see BIE500
AZOTREX see PDC250
AZOTURE de SODIUM (FRENCH) see SFA000
AZOXYAETHAN (GERMAN) see ASP000
AZOXYBENZEEN (DUTCH) see ASO750
AZOXYBENZENE see ASO750
AZOXYBENZIDE see ASO750
AZOXYBENZOL (GERMAN) see ASO750
AZOXYDIBENZENE see ASO750
AZOXYETHANE see ASP000
AZOXYMETHANE see ASP250
AZS see ASA500
AZTEC BPO see BDS000

B10 see SFO500
B 10 (polysaccharide) see SFO500
B-12 see VSZ000
B32 see HCL000
B-45 see AOF250
B-71 see PKC000
B/77 see DNX600
B 404 see PAK000
B 518 see EAS500
B 995 see DQD400
B-1,776 see BSH250

B-1843 see BLG500
B-9002 see HMV000
B 77488 see BAT750
BA see BBC250
BA 2726 see DBF800
Ba 2797 see CAY500
BA 5968 see HGP500
BA5968 see PHW000
BA 30,803 see BCH750
BA 32644 see NML000
BA 51-090462 see DCY400
BABROCID see NGE500
B(c)AC see BAW750
BACARATE see DKE800
BA 32644 CIBA see NML000
BACIGUENT see BAC250
BACI-JEL see BAC250
BACILIQUIN see BAC250
BACILLIN see ILD000
BACILLOL see CNW500
BACILLOMYCIN (8CI, 9CI) see BAB750
BACILLOMYCIN R see BAB750
BACILLOPEPTIDASE A see BAC000
BACILLOPEPTIDASE B see BAC000
BACILLUS SUBTILIS BPN see BAB750
BACILLUS SUBTILIS CARLSBERG see BAC000
BACITEK OINTMENT see BAC250
BACITRACIN see BAC250
BACTERAMID see SNM500
BACTERIAL VITAMIN H1 see AIH600
BACTOL see CHR500
BACTOLATEX see SMQ500
BACTOPEN see SLJ000
BACTRIM see SNK000
BACTROL see BGJ750
BACTROVET see SNN300
BA-1,2-DIHYDRODIOL see BBD250
BA-3,4-DIHYDRODIOL see BBD500
BA-5,6-DIHYDRODIOL see BBE250
BA-8,9-DIHYDRODIOL see BBE750
BA-10,11-DIHYDRODIOL see BBF000
BA-5,6-trans-DIHYDRODIOL see BBE250
BA-3,4-DIOL-1,2-EPOXIDE-1 see DLE000
BA-3,4-DIOL-1,2-EPOXIDE-2 see DLE000
BAGAODRYL see BBV500
BAGASSE DUST see BAD250
BAKELITE AYAA see AAX250
BAKELITE DYNH see PJS750
BAKELITE LP 70 see AAX175
BAKELITE LP 90 see AAX250
BAKELITE RMD 4511 see ADY500
BAKELITE SMD 3500 see SMQ500
BAKELITE VLFV see AAX175
BAKELITE VMCC see AAX175
BAKELITE VYNS see AAX175
BAKER'S ANTIFOL see DKC800
BAKER'S P AND S LIQUID and OINTMENT see PDN750
BAKONTAL see BAP000
BAKTOL see CFE250
BAKTOLAN see CFE250
BAKUCHIOL see BAD625
BAL see BAD750
BALFON 7000 see TAI250
BALMADREN see VGP000
BALSAM CAPTIVI see CNH792
BALSAM of PERU see BAE750
BALSAM PERU OIL (FCC) see BAE750
BALSAMS, COPAIBA see CNH792
BAMBERMYCIN see MRA250
BAMD 400 see MQU750
BAMIFYLLINE HYDROCHLORIDE see BEO750, THL750
BAMIPHYLLINE HYDROCHLORIDE see BEO750, THL750
BAMN see BRX500
BANANA OIL see IHO850

BANANOTE see MDW750
BANEX see MEL500
BANGTON see CBG000
BANISTERINE see HAI500
BANLEN see MEL500
BANMINTH see TCW750
BANOCIDE see DIW200
BANTHIN see XCJ000
BANTHINE see XCJ000
BANTHINE BROMIDE see XCJ000
BANTHIONINE see MDT740
BANTROL see HKB500
BANVEL see MEL500
BANVEL HERBICIDE see MEL500
BAP see NJM500
BAPN see AMB500
BARACOUMIN see BJZ000
BARAMINE see BBV500
BARBAMATE see CEW500
BARBAN see CEW500
BARBANE see CEW500
BARBAPIL see EOK000
BARBASCO see RNZ000
BARBELLON see EOK000
BARBENYL see EOK000
BARBIDORM see ERD500
BARBILEHAE (BARBILETTAE) see EOK000
BARBINAL see EOK000
BARBIPHENYL see EOK000
BARBITA see EOK000
BARBITAL see BAG500
BARBITAL Na see BAG250
BARBITAL SODIUM see BAG250, BAG500
BARBITAL SOLUBLE see BAG250
BARBITONE see BAG500
BARBITONE SODIUM see BAG250
BARBITURATES see BAG500
BARBIVIS see EOK000
BARBONAL see EOK000
BARBOPHEN see EOK000
BARBOSEC see SBM500
BARDIOL see EDO000
BARDORM see EOK000
BARIDIUM see PDC250
BARIDOL see BAP000
BARIO (PEROSSIDO di) (ITALIAN) see BAO250
BARITE see BAP000
BARITOP see BAP000
BARITRATE see PBC250
BARIUM see BAH250
BARIUM AZIDE see BAI000
BARIUM AZIDE, dry or containing less than 50% water (DOT)
    see BAI000
BARIUM BENZOATE see BAI500
BARIUM BICHROMATE see BAL500
BARIUM BINOXIDE see BAO250
BARIUM BROMATE see BAI750
BARIUM CARBIDE see BAJ000
BARIUM CHLORATE see BAJ500
BARIUM CHROMATE (1:1) see BAK250
BARIUM CHROMATE(VI) see BAK250
BARIUM CHROMATE OXIDE see BAK250
BARIUM COMPOUNDS (soluble) see BAK500
BARIUM CYANIDE see BAK750
BARIUM CYANIDE, solid (DOT) see BAK750
BARIUM CYANOPLATINITE see BAL000
BARIUM DICHROMATE see BAL500
BARIUM DICYANIDE see BAK750
BARIUM DINITRATE see BAN250
BARIUM DIOXIDE see BAO250
BARIUM HYDRIDE see BAM250
BARIUM HYPOPHOSPHITE see BAM750
BARIUM IODATE see BAN000
BARIUM MONOXIDE see BAO000

BARIUM NITRATE (DOT) see BAN250
BARIUM(II) NITRATE (1:2) see BAN250
BARIUM NITRIDE see BAN500
BARIUM OXIDE see BAO000
BARIUM PERCHLORATE (DOT) see PCD750
BARIUM PERMANGANATE see PCK000
BARIUMPEROXID (GERMAN) see BAO250
BARIUM PEROXIDE see BAO250
BARIUMPEROXYDE (DUTCH) see BAO250
BARIUM PROTOXIDE see BAO000
BARIUM SULFATE see BAP000
BARIUM SULFIDE see BAP250
BARIUM SUPEROXIDE see BAO250
BAROS CAMPHOR see BMD000
BAROSPERSE see BAP000
BAROTRAST see BAP000
BARPENTAL see NBU000
BARQUAT SB-25 see DTC600
BARQUINOL see CHR500
BAR-TIME see BBK250
BARTOL see EOK000
BARYTA see BAO000
BARYTA WHITE see BAP000
BARYTA YELLOW see BAK250
BARYTES see BAP000
BAS see BEM750
BASAGRAN see MJY500
BASAMID see DSB200
BASAMID G see DSB200
BASAMID-GRANULAR see DSB200
BASAMID P see DSB200
BASAMID-PUDER see DSB200
BASANITE see BRE500
BASCOREZ see AAX250
BASE 661 see SMR000
BASECIL see MPW500
BASE OIL see PCR250
BASERGIN see LJL000
BASETHYRIN see MPW500
BAS 352 F see RMA000
BASFAPON see DGI400
BASFAPON B see DGI400, DGI600
BASFAPON/BASFAPON N see DGI400
BASF III see SMQ500
BASF-MANEB SPRITZPULVER see MAS500
BASF URSOL 3GA see ALT000
BASF URSOL D see PEY500
BASF URSOL EG see ALT500
BASF URSOL ERN see NAW500
BASF URSOL SLA see DBO400
BAS 351-H see MJY500
BASIC BLUE 9 see BJI250
BASIC BRIGHT GREEN see BAY750
BASIC CHROMIC SULFATE see NBW000
BASIC CHROMIC SULPHATE see NBW000
BASIC CHROMIUM SULFATE see NBW000
BASIC CHROMIUM SULPHATE see NBW000
BASIC LEAD ACETATE see LCH000
BASIC LEAD CHROMATE see LCS000
BASIC MERCURIC SULFATE see MDG000
BASIC NICKEL CARBONATE see NCY500
BASIC ORANGE 3RN see BJF000
BASIC PARAFUCHSINE see RMK020
BASIC VIOLET 10 see FAG070
BASIC ZINC CHROMATE see ZFJ100
BASIC ZIRCONIUM CHLORIDE see ZSJ000
BASIL OIL see BAR250
BASIL OIL, EUROPEAN TYPE (FCC) see BAR250
BASINEX see DGI400
BASLE GREEN see COF500
BASOFORTINA see PAM000
BASOLAN see MCO500
BASSA see MOV000
BASUDIN see DCM750

BASUDIN 10 G see DCM750
BATASAN see ABX250
BATAZINA see BJP000
BATRILEX see PAX000
BAUXITE RESIDUE see IHD000
BAX see BAU750
BAX 2793Z see BEO750, THL750
BAXACOR see DHS200
BAY 1470 see DMW000
BAY 1521 see DPH600
BAY 2353 see DFV400
BAY 4934 see MGQ750
BAY 5621 see BAT750
BAY 5821 see DJY200
BAY 9010 see PMY300
BAY 9015 see DFD000
BAY 9026 see DST000
BAY 9027 see ASH500
BAY 10756 see DAO600
BAY 11405 see MNH000
BAY 15203 see DAO800, MIW100
BAY 15922 see TIQ250
BAY 16225 see EKN000
BAY 18436 see DAP400
BAY 19149 see DGP900
BAY 21097 see DAP000
BAY 23129 see PHI500
BAY 23323 see OQS000
BAY 23655 see DSK600
BAY 25141 see FAQ800
BAY 29493 see FAQ999
BAY 30130 see DGI000
BAY 33051 see DRR400
BAY 37342 see DST200
BAY 41831 see DSQ000
BAY 42696 see DQE800
BAY 44646 see DOR400
BAY 45432 see DNX800
BAY 48130 see DLS800
BAY 61597 see MQR275
BAY 62863 see DLS800
BAY 68138 see FAK000
BAY 70143 see CBS275
BAY 71628 see DTQ400
BAY 77049 see DJY200
BAY 77488 see BAT750
BAY-92114 see IMF300
BAY 105807 see MIA250
BAYCARD see MOV000
BAYCID see FAQ999
BAYCOVIN see DIZ100
BAY DIC 1468 see MQR275
BAY E-601 see MNH000
BAY E-605 see PAK000
BAYER 73 see DFV400, DFV600
BAYER 1440 L see DNA800
BAYER 2353 see DFV400
BAYER 3231 see TND000
BAYER 5072 see DOU600
BAYER 5080 see DOR400
BAYER 5081 see EPY000
BAYER 5312 see EPQ000
BAYER 5360 see MMN250
BAYER 8169 see DAO600
BAYER 9007 see FAQ999
BAYER 9013 see DST200
BAYER 9015 see DFD000
BAYER 15080 see BDD000
BAYER 15922 see TIQ250
BAYER 16259 see EKN000
BAYER 17147 see ASH500
BAYER 18510 see DRR400
BAYER 19639 see DXH325
BAYER 20315 see DAP600

BAYER 23655 see DSK600
BAYER 25648 see DFV600
BAYER 25820 see HMV000
BAYER 37289 see EPY000
BAYER 37342 see DST200
BAYER 37344 see DST000
BAYER 39007 see PMY300
BAYER 41831 see DSQ000
BAYER 44646 see DOR400
BAYER 45,432 see DNX800
BAYER 6159H see MQR275
BAYER 62863 see DLS800
BAYER 6443H see MQR275
BAYER 71628 see DTQ400
BAYER 78418 see EIM000
BAYER 94337 see MQR275
BAYER 21/116 see MIW100
BAYER 21/199 see CNU750
BAYER 25/154 see DAP400
BAYER A 139 see BDC750
BAYER-E 393 see SOD100
BAYER E-605 see PAK000
BAYERITIAN see TGG760
BAYER L 13/59 see TIQ250
BAYER R39 SOLUBLE see BDC750
BAYER S767 see FAQ800
BAYER S 4400 see EPY000
BAYER S 5660 see DSQ000
BAYERTITAN see TGG760
BAY 6681 F see CJO250
BAYGON see PMY300
BAY LEAF OIL see BAT500, LBK000
BAYLETON see CJO250
BAYLUSCID see DFV400, DFV600
BAYLUSCIDE see DFV600
BAY-MEB-6447 see CJO250
BAYMIX 50 see CNU750
BAY-NTN-9306 see SOU625
BAY OIL see BAT500
BAYOL F see MQV750
BAYPRESOL see DNA800
BAYRE 77488 see BAT750
BAYRITES see BAP000
BAYRUSIL see DJY200
BAY-SRA-12869 see IMF300
BAYTAN see MEP250
BAYTEX see FAQ999
BAYTHION see BAT750
BAYTITAN see TGG760
BAY VA 1470 see DMW000
BAZUDEN see DCM750
BB-8 see OAH000
BBC see BAV575, BMW250
BBC 12 see DDL800
BBCE see BIQ500
BBH see BBQ500
"B" BLASTING POWDER see ERF500
BBN see BMW250, HJQ350
BBNOH see HJQ350
BBP see BEC500
BCF-BUSHKILLER see TAA100
BCME see BIK000
BCNU see BIF750
BCPN see BQQ250
BCS COPPER FUNGICIDE see CNP250
BCTB see DIG400
BDCM see BND500
BDH 312 see GGS000
BDH 1298 see VTF000
BDH 29-801 see TAI250
BDMA see DQP800
BD(a,h)P see DCY200
BDU see BNC750
5-BDU see BNC750

BE see BNI500
BEAMETTE see PMP500
BEAN SEED PROTECTANT see CBG000
BECAPTAN see AJT250
BECAPTAN DISULFURE (FRENCH) see MCN500
BECILAN see PPK500
BECOREL see PAN100
BEESIX see PPK250
BEESWAX see BAU000
BEESWAX, WHITE see BAU000
BEESWAX, YELLOW see BAU000
BEET-KLEEN see CKC000
BEET SUGAR see SNH000
BEFLAVINE see RIK000
BEHP see DVL700
BEK see BJU000
BELAMINE BLACK GX see AQP000
BELAMINE BLUE 2B see CMO000
BELDAVRIN see HOT500
BELFENE see LJR000
BELLADONNA see BAU500
BELLASTHMAN see DMV600
BELL MINE see CAT250
BELL MINE PULVERIZED LIMESTONE see CAO000
BELMARK see FAR100
BELT see CDR750
BELUSTINE see CGV250
BEMEGRIDE see MKA250
BENA see BAU750, BBV500
BENACHLOR see BBV500
BENACTIZINE HYDROCHLORIDE see BCA000
BENACTYZIN (CZECH) see BCA000
BENACTYZINE CHLORIDE see BCA000
BENACTYZINE HYDROCHLORIDE see BCA000
BENADON see BBV500, PPK500
BENADRIN see BBV500
BENADRYL see BAU750, BBV500
BENADRYL HYDROCHLORIDE see BAU750
BEN-A-HIST see PDC000
BENAKTIN see BCA000
BENALGIN see BBW500
BEN-ALLERGIN see BBV500
BENANSERIN HYDROCHLORIDE see BEM750
BENAPON see BBV500
BENASPIR see ADA725
BENAZIDE see IKC000
BENCARBATE see DQM600
BENCHINOX see BDD000
BENCIDAL BLACK E see AQP000
BENCIDAL BLUE 2B see CMO000
BENCIDAL BLUE 3B see CMO250
BENDEX see BLU000
BENDIOCARB see DQM600, MHZ000
BENDIOXIDE see MJY500
BENDOPA see DNA200
BENDYLATE see BAU750
BENESAL see SAH000
BENFOS see DGP900
BENGAL GELATIN see AEX250
BENGAL ISINGLASS see AEX250
BENGUINOX see BDD000
BEN-HEX see BBQ500
BENICOT see NCR000
BENKFURAN see NGE000
BENLATE 50 see BAV575
BENOCTEN see BAU750
BENODAINE HYDROCHLORIDE see BCI500
BENODIN see BBV500
BENODINE see BBV500
BENOMYL see BAV575
BENOMYL 50W see BAV575
BENOVOCYLIN see EDP000
BENOXYL see BDS000
BENOZIL see DAB800

BENQUINOX see BDD000
BENSULFOID see SOD500
BENSYLYTE see PDT250
BENSYLYT NEN see DDG800
BENT see MDQ250
BENTAZON see MJY500
BENTONITE see BAV750
BENTONITE 2073 see BAV750
BENTONITE MAGMA see BAV750
BENTOX 10 see BBQ500
BENTROL see HKB500
BEN-U-RON see HIM000
BENVIL see MOV500
(5R,6R)-BENXYLPENICILLIN see BDY669
BENYLAN see BBV500
BENYLATE see BCM000
BENZAC see BDS000
1,2-BENZACENAPHTHENE see FDF000
BENZ(e)ACEPHENANTHRYLENE see BAW250
3,4-BENZ(e)ACEPHENANTHRYLENE see BAW250
BENZACIN see BAW500
BENZACINE see BAW500
BENZACINE HYDROCHLORIDE see BAW500
BENZACIN HYDROCHLORIDE see BAW500
BENZACONINE see PIC250
BENZ(c)ACRIDINE see BAW750
3,4-BENZACRIDINE see BAW750
7,8-BENZACRIDINE (FRENCH) see BAW750
BENZAHEX see BBQ750
BENZAKNEW see BDS000
BENZAL ALCOHOL see BDX500
BENZAL-(BENZYL-CYANID) (GERMAN) see DVX600
BENZAL CHLORIDE see BAY300
BENZALDEHYDE see BAY500
BENZALDEHYDE CYANOHYDRIN see MAP250
BENZALDEHYDE GLYCERYL ACETAL (FCC) see BBA000
BENZALDEHYDE GREEN see BAY750
BENZALDEHYDKYANHYDRIN (CZECH) see MAP250
BENZAL GLYCERYL ACETAL see BBA000
BENZALIN see DLY000
1-BENZAMIDO-1-PHENYL-3-PIPERIDINOPROPANE HY-
    DROCHLORIDE see DKK800
N-(3-BENZAMIDO-3-PHENYL)PROPYL PIPERIDINE HY-
    DROCHLORIDE see DKK800
BENZAMIL BLACK E see AQP000
BENZAMIL SUPRA BROWN BRLL see CMO750
BENZAMINE BLUE see CMO250
BENZANIL BLUE 2B see CMO000
BENZANTHRACENE see BBC250
BENZ(a)ANTHRACENE see BBC250
1,2-BENZANTHRACENE see BBC250
1,2-BENZ(a)ANTHRACENE see BBC250
1,2:5,6-BENZANTHRACENE see DCT400
BENZ(a)ANTHRACENE-1,2-DIHYDRODIOL see BBD250
BENZ(a)ANTHRACENE-3,4-DIHYDRODIOL see BBD500
BENZ(a)ANTHRACENE-5,6-DIHYDRODIOL see BBE250
BENZ(a)ANTHRACENE-10,11-DIHYDRODIOL see BBF000
BENZ(a)ANTHRACENE-5,6-trans-DIHYDRODIOL see
    BBE250
trans-BENZ(a)ANTHRACENE-8,9-DIHYDRODIOL see
    BBE750
BENZ(a)ANTHRACENE 3,4-DIOL-1,2-EPOXIDE-2 see
    DLE000
BENZ(a)ANTHRACENE-7-METHANOL see BBH250
1,2-BENZANTHRAZEN (GERMAN) see BBC250
BENZANTHRENE see BBC250
1,2-BENZANTHRENE see BBC250
BENZANTINE see BBV500
BENZATROPINE METHANESULFONATE see TNU000
BENZAZIDE, BENZOIC ACID AZIDE see BDL750
1-BENZAZINE see QMJ000
2-BENZAZINE see IRX000
1-BENZAZOLE see ICM000
BENZAZOLINE HYDROCHLORIDE see BBJ750

BENZ-o-CHLOR see DER000
BENZCURINE IODIDE see PDD300
15,16-BENZDEHYDROCHOLANTHRENE see DCR400
BENZEDRINE see BBK000
(±)-BENZEDRINE see BBK000
dl-BENZEDRINE see BBK000
BENZEDRINE SULFATE see BBK250
d-BENZEDRINE SULFATE see BBK500
l-BENZEDRINE SULFATE see BBK750
BENZEEN (DUTCH) see BBL250
BENZEHIST see BAU750
BENZEN (POLISH) see BBL250
BENZENACETIC ACID see PDY850
BENZENAMINE see AOQ000
BENZENAMINE, 4-FLUORO-(9CI) see FFY000
BENZENAMINE HYDROCHLORIDE see BBL000
BENZENAMINIUM, 3-(((DIMETHYLAMINO)CAR-
    BONYL)OXY)-N,N,N-TRIMETHYL-, BROMIDE
    (9CI) see DQY800
BENZENE see BBL250
BENZENEACETALDEHYDE see BBL500
BENZENEACETIC ACID see PDY850
BENZENEACETIC ACID, ETHYL ESTER (9CI) see EOH000
BENZENEACETIC ACID, METHYL ESTER see MHA500
BENZENEACETIC ACID, α-PHENYL-, 2-(DIETHYL-
    AMINO)ETHYL ESTER, (9CI) see DHX800
BENZENEACETIC ACID, 2-PHENYLETHYL ESTER see
    PDI000
BENZENEACETIC ACID, 2-PROPENYL ESTER see PMS500
BENZENEACETONITRILE see PEA750
BENZENE AZIMIDE see BDH250
4-BENZENEAZOANILINE see PEI000
BENZENEAZOBENZENE see ASL250
BENZENEAZOBENZENEAZO-β-NAPHTHOL see OHA000
BENZENEAZODIMETHYLANILINE see DOT300
BENZENEAZO-β-NAPHTHOL see PEJ500
BENZENE-1-AZO-2-NAPHTHOL see PEJ500
1-BENZENEAZO-2-NAPHTHYLAMINE see FAG130
1-BENZENE-AZO-β-NAPHTHYLAMINE see FAG130
p-BENZENEAZOPHENOL see HJF000
BENZENECARBALDEHYDE see BAY500
BENZENECARBINOL see BDX500
BENZENECARBONAL see BAY500
BENZENECARBONYL CHLORIDE see BDM500
BENZENECARBOPEROXOIC ACID (9CI) see PCM000
BENZENECARBOXYLIC ACID see BCL750
BENZENE CHLORIDE see CEJ125
m-BENZENEDIAMINE see PEY000
o-BENZENEDIAMINE see PEY250
p-BENZENEDIAMINE see PEY500
1,2-BENZENEDIAMINE see PEY250
1,3-BENZENEDIAMINE see PEY000
1,4-BENZENEDIAMINE see PEY500
m-BENZENEDIAMINE DIHYDROCHLORIDE see PEY750
1,3-BENZENEDIAMINE HYDROCHLORIDE see PEY750
BENZENE, 1,4-DIAZIDO- see DCL125
BENZENEDIAZONIUM, 4-(HYDROXYMETHYL)-,
    TETRAFLUOROBORATE(1-) see HLX925
1,3-BENZENEDICARBONITRILE see PHX550
o-BENZENEDICARBOXYLIC ACID see PHW250
BENZENE-1,2-DICARBOXYLIC ACID see PHW250
1,2-BENZENEDICARBOXYLIC ACID see PHW250
1,2-BENZENEDICARBOXYLIC ACID ANHYDRIDE see
    PHW750
1,2-BENZENEDICARBOXYLIC ACID BI(2-METHOXY-
    ETHYL)ESTER (9CI) see DOF400
1,2-BENZENEDICARBOXYLIC ACID, BUTYL PHENYL-
    METHYL ESTER see BEC500
o-BENZENEDICARBOXYLIC ACID, DIBUTYL ESTER see
    DEH200
BENZENE-o-DICARBOXYLIC ACID DI-n-BUTYL ESTER see
    DEH200
1,2-BENZENEDICARBOXYLIC ACID, DIETHYL ESTER see
    DJX000

1,2-BENZENEDICARBOXYLIC ACID, DIISOOCTYL ESTER
    see ILR100
1,2-BENZENEDICARBOXYLIC ACID DIMETHYL ESTER see
    DTR200
1,4-BENZENE DICARBOXYLIC ACID DIMETHYL ESTER
    (9CI) see DUE000
o-BENZENEDICARBOXYLIC ACID DIOCTYL ESTER see
    DVL600
1,2-BENZENEDICARBOXYLIC ACID DIOCYTL ESTER see
    DVL600
BENZENE-1,3-DIISOCYANATE see BBP000
BENZENE-, 1,3-DIISOCYANATOMETHYL- see TGM740
p-BENZENEDINITRILE see BBP250
m-BENZENEDIOL see REA000
o-BENZENEDIOL see CCP850
p-BENZENEDIOL see HIH000
1,2-BENZENEDIOL see CCP850
1,3-BENZENEDIOL see REA000
1,4-BENZENEDIOL see HIH000
BENZENEFORMIC ACID see BCL750
BENZENE HEXACHLORIDE see BBP750
α-BENZENEHEXACHLORIDE see BBQ000
β-BENZENEHEXACHLORIDE see BBR000
γ-BENZENE HEXACHLORIDE see BBQ500
trans-α-BENZENEHEXACHLORIDE see BBR000
Δ-BENZENEHEXACHLORIDE see BFW500
BENZENE HEXACHLORIDE-α-isomer see BBQ000
BENZENE HEXACHLORIDE-γ isomer see BBQ500
BENZENEHEXACHLORIDE (mixed isomers) see BBQ750
BENZENE ISOPROPYL see COE750
BENZENE-1-ISOTHIOCYANATE see ISQ000
BENZENEMETHANOIC ACID see BCL750
BENZENEMETHANOL see BDX500
BENZENEMETHANOL, α-(1-AMINOETHYL)-3-HYDROXY-,
    (R-(R*,S*))-(9CI) see HNB875
BENZENEMETHANOL, α-METHYL- see PDE000
BENZENE, METHYL- see TGK750
BENZENENITRILE see BCQ250
BENZENE PHOSPHORUS DICHLORIDE (DOT) see DGE400
BENZENE PHOSPHORUS THIODICHLORIDE see PFW500
BENZENEPROPANAL see HHP000
BENZENEPROPANOL see HHP050
3-BENZENEPROPANOL see HHP050
BENZENEPROPANOL CARBAMATE see PGA750
BENZENESULFONAMIDE, N-CHLORO-4-METHYL-, SO-
    DIUM SALT (9CI) see CDP000
BENZENESULFONATE de 4-CHLOROPHENYLE (FRENCH)
    see CJR500
BENZENE SULFONCHLORIDE see BBS750
BENZENESULFONIC ACID see BBS250
BENZENESULFONIC (ACID) CHLORIDE see BBS750
BENZENESULFONIC ACID, 4-CHLOROPHENYL ESTER see
    CJR500
BENZENESULFONYL CHLORIDE see BBS750
BENZENE SULPHONYL CHLORIDE (DOT) see BBS750
BENZENESULPHONYL FLUORIDE see BBT250
BENZENETETRAHYDRIDE see CPC579
BENZENETHIOL (DOT) see PFL850
1,2,4-BENZENETRICARBOXYLIC ACID ANHYDRIDE see
    TKV000
1,2,4-BENZENETRICARBOXYLIC ACID, CYCLIC 1,2-AN-
    HYDRIDE see TKV000
1,2,4-BENZENETRICARBOXYLIC ANHYDRIDE see
    TKV000
BENZENE-s-TRIOL see PGR000
1,2,3-BENZENETRIOL see PPQ500
1,2,4-BENZENETRIOL see BBU250
BENZENE-1,3,5-TRIOL see PGR000
1,3,5-BENZENETRIOL see PGR000
BENZENOL see PDN750
BENZENOSULFOCHLOREK (POLISH) see BBS750
BENZENOSULPHOCHLORIDE see BBS750
BENZENYL CHLORIDE see BFL250
BENZENYL FLUORIDE see BDH500
BENZENYL TRICHLORIDE see BFL250

BENZETAMOPHYLLINE HYDROCHLORIDE see BEO750, THL750
BENZETHONIUM CHLORIDE see BEN000
BENZETONIUM CHLORIDE see BEN000
BENZEX see BBQ750
2,3-BENZFLUORANTHENE see BAW250
3,4-BENZFLUORANTHENE see BAW250
10,11-BENZFLUORANTHENE see BCJ500
BENZ(j)FLUOROCANTHRENE see BCJ500
BENZHEXOL CHLORIDE see BBV000
BENZHEXOL HYDROCHLORIDE see BBV000
BENZHORMOVARINE see EDP000
BENZHYDRAMINE see BBV500
BENZHYDRAMINE HYDROCHLORIDE see BAU750
BENZHYDRAMINUM see BBV500
BENZHYDRAZIDE see BBV250
BENZHYDRIL see BBV500
BENZHYDRYL see BBV500
o-BENZHYDRYLDIMETHYLAMINOETHANOL see BBV500
o-BENZHYDRYLDIMETHYLAMINOETHANOL-8-CHLORO-THEOPHYLLINATE see DYE600
2-(BENZHYDRYLOXY)-N,N-DIMETHYLETHYLAMINE see BBV500
2-(BENZHYDRYLOXY)-N,N-DIMETHYLETHYLAMINE with 8-CHLOROTHEOPHYLLINE see DYE600
2-(BENZHYDRYLOXY)-N,N-DIMETHYLETHYLAMINE-HYDROCHLORIDE see BAU750
4-(BENZHYDRYLOXY)-1-METHYLPIPERIDINE see LJR000
BENZIDAMINE HYDROCHLORIDE see BBW500
BENZIDIN (CZECH) see BBX000
BENZIDINA (ITALIAN) see BBX000
BENZIDINE see BBX000
3,3'-BENZIDINEDICARBOXYLIC ACID see BFX250
BENZIDINE HYDROCHLORIDE see BBX750
BENZIDINE SULFATE see BBY000
BENZIDINE SULPHATE and HYDRAZINE-BENZENE see BBY300
BENZILAN see DER000
BENZILATE DU DIETHYLAMINO-ETHANOL CHLORHY-DRATE (FRENCH) see BCA000
BENZILE (CLORURO di) (ITALIAN) see BEE375
BENZILIC ACID-β-DIETHYLAMINOETHYL ESTER HYDRO-CHLORIDE see BCA000
BENZILIC ACID ESTER with 1-ETHYL-3-HYDROXY-1-METHYLPIPERIDINIUM BROMIDE see PJA000
BENZILIC ACID-1-METHYL-3-PIPERIDYL ESTER see MON250
BENZIMIDAZOLE see BCB750
o-BENZIMIDAZOLE see BCB750
1H-BENZIMIDAZOLE (9CI) see BCB750
BENZIMIDAZOLE METHYLENE MUSTARD see BCC250
BENZIMIDAZOLE MUSTARD see BCC250
2-BENZIMIDAZOLETHIOL see BCC500
4-(2-BENZIMIDAZOLYL)THIAZOLE see TEX000
BENZIMINAZOLE see BCB750
BENZIN see NAI500
BENZIN (OBS.) see BBL250
BENZINDAMINE HYDROCHLORIDE see BBW500
BENZINE see PCT250
1-BENZINE see QMJ000
BENZINE (OBS.) see BBL250
BENZINOFORM see CBY000
BENZINOL see TIO750
3-BENZISOTHIAZOLINONE-1,1-DIOXIDE see BCE500
1,2-BENZISOTHIAZOL-3(2H)-ONE-1,1-DIOXIDE see BCE500
1,2-BENZISOTHIAZOL-3(2H)-ONE-1,1-DIOXIDE, CALCIUM SALT see CAM750
BENZISOTRIAZOLE see BDH250
3,4-BENZOACRIDINE see BAW750
BENZOANTHRACENE see BBC250
BENZO(a)ANTHRACENE see BBC250
1,2-BENZOANTHRACENE see BBC250

BENZOATE see BCL750
BENZOATE d'OESTRADIOL (FRENCH) see EDP000
BENZOATE d'OESTRONE (FRENCH) see EDV500
BENZOATE of SODA see SFB000
BENZOATE SODIUM see SFB000
1-BENZOAZO-2-NAPHTHOL see PEJ500
BENZO BLUE see CMO250
BENZO BLUE GS see CMO000
BENZOCAINE see EFX000
BENZOCHINAMIDE see BCL250
BENZO-CHINON (GERMAN) see QQS200
BENZO(b)CHRYSENE see BCG250
BENZO(d,e,f)CHRYSENE see BCS750
2,3-BENZOCHRYSENE see BCG500
BENZOCTAMINE HYDROCHLORIDE see BCH750
BENZO DEEP BLACK E see AQP000
BENZODIAPIN see MDQ250
1,3-BENZODIAZOLE see BCB750
1,2-BENZODIHYDROPYRONE (FCC) see HHR500
BENZODIOXANE HYDROCHLORIDE see BCI500
1-(1,4-BENZODIOXAN-2-YLMETHYL)PIPERIDINEHY-DROCHLORIDE see BCI500
3,4-BENZODIOXOLE-5-CARBOXALDEHYDE see PIW250
1,3-BENZODIOXOLE-5-(2-PROPEN-1-OL) see BCJ000
BENZOEPIN see EAQ750
BENZOESAEURE (GERMAN) see BCL750
BENZOESAEURE (NA-SALZ) (GERMAN) see SFB000
BENZOESTROFOL see EDP000
BENZO(1)FLUORANTHENE see BCJ500
BENZO(b)FLUORANTHENE see BAW250
BENZO(e)FLUORANTHENE see BAW250
BENZO(j)FLUORANTHENE see BCJ500
BENZO(k)FLUORANTHENE see BCJ750
2,3-BENZOFLUORANTHENE see BAW250
3,4-BENZOFLUORANTHENE see BAW250
7,8-BENZOFLUORANTHENE see BCJ500
8,9-BENZOFLUORANTHENE see BCJ750
11,12-BENZOFLUORANTHENE see BCJ750
11,12-BENZO(k)FLUORANTHENE see BCJ750
2,3-BENZOFLUORANTHRENE see BAW250
BENZO(jk)FLUORENE see FDF000
BENZOFOLINE see EDP000
BENZOFORM BLACK BCN-CF see AQP000
BENZOFUR D see PEY500
BENZOFUR GG see ALT000
BENZOFUR MT see TGL750
BENZOFUROLINE see BEP500
BENZOGUANAMINE see BCL250
BENZO-GYNOESTRYL see EDP000
BENZOHYDRAZIDE see BBV250
BENZOHYDRAZINE see BBV250
BENZOHYDROQUINONE see HIH000
2-(BENZOHYDRYLOXY)-N,N-DIMETHYLETHYLAMINE see BBV500
BENZOIC ACID see BCL750
BENZOIC ACID (DOT) see BCL750
BENZOIC ACID, BENZYL ESTER see BCM000
BENZOIC ACID-n-BUTYL ESTER see BQK250
BENZOIC ACID, CHLORIDE see BDM500
BENZOIC ACID ESTRADIOL see EDP000
BENZOIC ACID(4-(HYDROXYIMINO)-2,5-CYCLOHEXA-DIEN-1-YLIDENE) HYDRAZIDE see BDD000
BENZOIC ACID, 1-(3-METHYL)BUTYL ESTER see IHP100
BENZOIC ACID NITRILE see BCQ250
BENZOIC ACID, PEROXIDE see BDS000
BENZOIC ACID, PHENYLMETHYL ESTER see BCM000
BENZOIC ACID, SODIUM SALT see SFB000
BENZOIC ALDEHYDE see BAY500
BENZOIC ETHER see EGR000
BENZOIC HYDRAZIDE see BBV250
o-BENZOIC SULPHIMIDE see BCE500
BENZOIC TRICHLORIDE see BFL250
BENZOIMIDAZOLE see BCB750
BENZOIN see BCP250

BENZOL (DOT) see BBL250
BENZOLE see BBL250
BENZOLENE see BBL250
BENZOLIN see NOF500
BENZOLINE see PCT250
BENZOLIN HYDROCHLORIDE see NOF500
BENZOLO (ITALIAN) see BBL250
BENZONE see BRF500
BENZONITRILE see BCQ250
BENZONITRILE (DOT) see BCQ250
BENZOPENICILLIN see BDY669
BENZO(rst)PENTAPHENE see BCQ500
BENZOPEROXIDE see BDS000
BENZO(a)PHENANTHRENE see BBC250, CML810
BENZO(b)PHENANTHRENE see BBC250
BENZO(c)PHENANTHRENE see BCR750
1,2-BENZOPHENANTHRENE see CML810
2,3-BENZOPHENANTHRENE see BBC250
3,4-BENZOPHENANTHRENE see BCR750
BENZO(def)PHENANTHRENE see PON250
BENZOPHENONE see BCS250
BENZOPHOSPHATE see BDJ250
3,4-BENZOPIRENE (ITALIAN) see BCS750
2H-1-BENZOPYRAN-2-ONE see CNV000
BENZO(a)PYRENE see BCS750
BENZO(e)PYRENE see BCT000
1,2-BENZOPYRENE see BCT000
3,4-BENZOPYRENE see BCS750
4,5-BENZOPYRENE see BCT000
6,7-BENZOPYRENE see BCS750
(E)-BENZO(a)PYRENE-4,5-DIHYDRODIOL see DLC600
BENZO(a)PYRENE-7,8-DIHYDRODIOL-9,10-EPOXIDE (anti)
    see BCU000
anti-BENZO(a)PYRENE-7,8-DIHYDRODIOL-9,10-OXIDE see
    BCU000
BENZO(a)PYRENE-6-METHANOL see BCV250
BENZO(a)PYREN-2-OL see BCX000
BENZO(a)PYREN-3-OL see BCX250
BENZO(a)PYREN-7-OL see BCY000
BENZO(b)PYRIDINE see QMJ000
BENZO(c)PYRIDINE see IRX000
1,2-BENZOPYRONE see CNV000
BENZOPYRROLE see ICM000
2,3-BENZOPYRROLE see ICM000
BENZOQUINAMIDE see BCL250
1,4-BENZOQUINE see QQS200
BENZOQUINOL see HIH000
BENZO(b)QUINOLINE see ADJ500
2,3-BENZOQUINOLINE see ADJ500
o-BENZOQUINONE see BDC250
p-BENZOQUINONE see QQS200
1,2-BENZOQUINONE see BDC250
1,4-BENZOQUINONE see QQS200
BENZOQUINONE (DOT) see BDC250, QQS200
BENZOQUINONE AZIRIDINE see BDC750
1,4-BENZOQUINONE-N'-BENZOYLHYDRAZONE OXIME
    see BDD000
1,4-BENZOQUINONE DIOXINE see DVR200
p-BENZOQUINONE OXIME BENZOYLHYDRAZONE see
    BDD000
o-BENZOSULFIMIDE see BCE500
BENZOSULFONAZOLE see BDE500
BENZOSULPHIMIDE see BCE500
BENZO-2-SULPHIMIDE see BCE500
3,4-BENZOTETRACENE see BCG500
BENZO(c)TETRAPHENE see BCG500
3,4-BENZOTETRAPHENE see BCG500
BENZOTHIAZIDE see BDE250
BENZOTHIAZOLE see BDE500
BENZOTHIAZOLE DISULFIDE see BDE750
2-BENZOTHIAZOLETHIOL see BDF000
2-BENZOTHIAZOLETHIOL, ZINC SALT (2:1) see BHA750
BENZOTHIAZOLYL DISULFIDE see BDE750
2-BENZOTHIAZOLYL DISULFIDE see BDE750

2-BENZOTHIAZOLYL-N-MORPHOLINOSULFIDE see
    BDG000
2-BENZOTHIAZOLYLSULFENYL MORPHOLINE see
    BDG000
4-(2-BENZOTHIAZOLYLTHIO)MORPHOLINE see BDG000
BENZOTRIAZINEDITHIOPHOSPHORIC ACID DIMETHOXY
    ESTER see ASH500
BENZOTRIAZINE derivative of an ETHYL DITHIOPHOS-
    PHATE see EKN000
BENZOTRIAZINE derivative of a METHYL DITHIOPHOS-
    PHATE see ASH500
1H-BENZOTRIAZOLE see BDH250
1,2,3-BENZOTRIAZOLE see BDH250
BENZOTRICHLORIDE (DOT, MAK) see BFL250
BENZOTRIFLUORIDE see BDH500
BENZO(b)TRIPHENYLENE see BDH750
BENZOTROPINE MESYLATE see TNU000
BENZOTROPINE METHANESULFONATE see TNU000
BENZOXALE see TMP750
S-((3-BENZOXAZOLINYL-6-CHLORO-2-OXO)METHYL) O,O-
    DIETHYLPHOSPHORODITHIOATE see BDJ250
3-BENZOXY-1-(2-METHYLPIPERIDINO)PROPANE see
    PIV750
3-BENZOXY-1-(2-METHYLPIPERIDINO)PROPANE HYDRO-
    CHLORIDE see IJZ000
dl-3-BENZOXY-1-(2-METHYLPIPERIDINO)PROPANE HY-
    DROCHLORIDE see IJZ000
BENZOYL see BDS000
BENZOYLACONINE see PIC250
BENZOYL ALCOHOL see BDX500
BENZOYL AZIDE see BDL750
BENZOYLBENZENE see BCS250
N-2 (5-BENZOYL-BENZIMIDAZOLE) CARBAMATE de
    METHYLE (FRENCH) see MHL000
5-BENZOYL-2-BENZIMIDAZOLECARBAMIC ACID
    METHYL ESTER see MHL000
N-(BENZOYL-5-BENZIMIDAZOLYL)-2, CARBAMATE de
    METHYLE (FRENCH) see MHL000
BENZOYL CHLORIDE see BDM500
BENZOYL CHLORIDE (DOT) see BDM500
BENZOYL CYANIDE-o-(DIETHOXYPHOSPHINO-
    THIOYL)OXIME see BAT750
BENZOYL HYDRAZIDE see BBV250
BENZOYLHYDROGEN PEROXIDE see PCM000
BENZOYL HYDROPEROXIDE see PCM000
BENZOYL METHIDE see ABH000
BENZOYL-γ-(2-METHYLPIPERIDINE)PROPANOL HYDRO-
    CHLORIDE see IJZ000
BENZOYL-γ-(2-METHYLPIPERIDINO)PROPANOL see
    PIV750
N-BENZOYLOXY-ACETYLAMINOFLUORENE see FDZ000
3-(BENZOYLOXY)ESTRA-1,3,5(10)-TRIEN-17-ONE see
    EDV500
3-(BENZOYLOXY)-8-METHYL-8-AZABICYCLO(3.2.1)
    OCTANE-2-CARBOXYLIC ACID PROPYL ESTER, HY-
    DROCHLORIDE (1R-(2-ENDO,3-EXO)) see NCJ000
BENZOYLOXYTRIBUTYLSTANNANE see BDR750
BENZOYLPEROXID (GERMAN) see BDS000
BENZOYL PEROXIDE see BDS000
BENZOYL PEROXIDE, WET see BDS250
BENZOYLPEROXYDE (DUTCH) see BDS000
BENZOYLPHENYLCARBINOL see BCP250
o-BENZOYL SULFIMIDE see BCE500
o-BENZOYL SULPHIMIDE see BCE500
BENZOYL SUPEROXIDE see BDS000
BENZ(a)PHENANTHRENE see CML810
1,2-BENZPHENANTHRENE see CML810
2,3-BENZPHENANTHRENE see BBC250
3,4-BENZPHENANTHRENE see BCR750
BENZPHOS see BDJ250
3,4-BENZPYREN (GERMAN) see BCS750
BENZ(a)PYRENE see BCS750
1,2-BENZPYRENE see BCT000
3,4-BENZ(a)PYRENE see BCS750

BENZQUINAMIDE see BCL250
BENZQUINAMIDU (POLISH) see BCL250
BENZTROPINE MESYLATE see TNU000
BENZTROPINE METHANESULFONATE see TNU000
BENZYDAMINE HYDROCHLORIDE see BBW500
BENZYDYNA (POLISH) see BBX000
BENZYFUROLINE see BEP500
BENZYHYDRYLCYANIDE see DVX200
BENZYLACETALDEHYDE see HHP000
BENZYL ACETATE see BDX000
BENZYL ALCOHOL see BDX500
BENZYL ALCOHOL BENZOIC ESTER see BCM000
BENZYL ALCOHOL CINNAMIC ESTER see BEG750
BENZYL ALCOHOL FORMATE see BEP250
BENZYL-6-AMINOPENICILLINIC ACID see BDY669
2-(N-BENZYLANILINOMETHYL)-2-IMIDAZOLINE see
    PDC000
BENZYL ANTISEROTONIN see BEM750
BENZYLBARBITAL see BEA500
BENZYL BENZENECARBOXYLATE see BCM000
BENZYL BENZOATE (FCC) see BCM000
BENZYLBIS(β-CHLOROETHYL)AMINE see BIA750
BENZYL BROMIDE see BEC000
BENZYL n-BUTANOATE see BED000
BENZYL BUTYL PHTHALATE see BEC500
BENZYL n-BUTYRATE see BED000
BENZYL CARBINOL see PDD750
BENZYLCARBINOL ISOBUTYRATE see PDF750
BENZYLCARBINYL ACETATE see PFB250
BENZYLCARBINYL ANTHRANILATE see APJ500
BENZYLCARBINYL ISOBUTYRATE see PDF750
BENZYLCARBINYL-α-TOLUATE see PDI000
BENZYLCARBONYL CHLORIDE see BEF500
BENZYLCHLORID (GERMAN) see BEE375
BENZYL CHLORIDE see BEE375
BENZYL CHLOROCARBONATE (DOT) see BEF500
2-(N-BENZYL-2-CHLOROETHYLAMINO)-1-PHENOXYPRO-
    PANE see PDT250
2-(N-BENZYL-2-CHLOROETHYLAMINO)-1-PHENOXYPRO-
    PANE HYDROCHLORIDE see DDG800
BENZYL(2-CHLOROETHYL)-(1-METHYL-2-PHENOXY-
    ETHYL)AMINE see PDT250
BENZYL(2-CHLOROETHYL)(1-METHYL-2-PHENOXY-
    ETHYL)AMINE HYDROCHLORIDE see DDG800
BENZYL CHLOROFORMATE see BEF500
BENZYL CHLOROFORMATE (DOT) see BEF500
BENZYL CINNAMATE see BEG750
BENZYL CYANIDE see PEA750
3-BENZYL-3,3-α,4,5,6,6-α,9,10,12,15-DECAHYDRO-6,12,15-
    TRIHYDROXY-4,10,12-TRIMETHYL-5-METHYLENE-1H-
    CYCLOUNDEC(d)ISOINDOLE-1,11(2H)-DIONE, 15-ACE-
    TATE see ZUS000
BENZYL DICHLORIDE see BAY300
BENZYLDIMETHYLAMINE see DQP800
N-BENZYLDIMETHYLAMINE see DQP800
BENZYL-N,N-DIMETHYLAMINE see DQP800
BENZYLDIMETHYLAMINE METHIODIDE see BFM750
2-(BENZYL(2-DIMETHYL AMINOETHYL)AMINO)PYRIDINE
    see TMP750
1-BENZYL-3-(3-(DIMETHYLAMINO)PROPOXY)-1H-INDA-
    ZOLE HYDROCHLORIDE see BBW500
1-BENZYL-3-γ-DIMETHYLAMINOPROPOXY-1H-INDAZOLE
    HYDROCHLORIDE see BBW500
BENZYL DIMETHYL CARBINOL see DQQ200
BENZYLDIMETHYLDODECYLAMMONIUM CHLORIDE see
    BEM000
N-BENZYL-N′,N′-DIMETHYL-N-2-PYRIDYLETHYLENE DI-
    AMINE see TMP750
1-BENZYL-2,5-DIMETHYL SEROTONIN HYDROCHLORIDE
    see BEM750
BENZYLDIMETHYLSTEARYLAMMONIUM CHLORIDE see
    DTC600
BENZYLDIMETHYL-p-(1,1,3,3-TETRAMETHYLBUTYL)-
    PHENOXYETHOXY-ETHYLAMMONIUM CHLORIDE
    see BEN000

BENZYLDIMETHYL(2-(2-(p-(1,1,3,3-TETRAMETHYL-
    BUTYL)PHENOXY)ETHOXY)ETHYL) AMMONIUM
    CHLORIDE see BEN000
BENZYLE (CHLORURE de) (FRENCH) see BEE375
BENZYLENE CHLORIDE see BAY300
BENZYL ETHANOATE see BDX000
BENZYL ETHER see BEO250
5-BENZYL-5-ETHYLBARBITURIC ACID see BEA500
8-BENZYL-7-(2-(ETHYL(2-HYDROXYETHYL)AMINO)
    ETHYL)THEOPHYLLINE, HYDROCHLORIDE see
    THL750
8′-BENZYL-7(2-(ETHYL(2-HYDROXYETHYL)AMINO)
    ETHYL) THEOPHYLLINE HYDROCHLORIDE see
    BEO750
8-BENZYL-7-(N-ETHYL-N-(β-HYDROXYETHYL)AMINO-
    ETHYL)THEOPHYLLINE HYDROCHLORIDE see
    BEO750
BENZYLETS see BCM000
BENZYL FORMATE see BEP250
5-BENZYL-3-FURYL METHYL(±)-cis,trans-CHRYSANTHE-
    MATE see BEP500
(5-BENZYL-3-FURYL) METHYL-2,2-DIMETHYL-3-(2-
    METHYLPROPENYL)-CYCLOPROPANECARBOXYLATE
    see BEP500
BENZYL-o-HYDROXYBENZOATE see BFJ750
BENZYLIDENEACETALDEHYDE see CMP969
BENZYLIDENE CHLORIDE (DOT) see BAY300
BENZYLIDENE GLYCEROL see BBA000
BENZYLIDENEPHENYLACETONITRILE see DVX600
BENZYLIDYNE CHLORIDE see BFL250
BENZYLIDYNE FLUORIDE see BDH500
BENZYLIMIDAZOLINE HYDROCHLORIDE see BBJ750
2-BENZYL-2-IMIDAZOLINE MONOHYDROCHLORIDE see
    BBJ750
BENZYL ISOBUTYRATE (FCC) see IJV000
BENZYL-ISOTHIOCYANATE see BEU250
BENZYLISOTHIOUREA HYDROCHLORIDE see BEU500
BENZYLISOTHIOURONIUM CHLORIDE see BEU500
2-BENZYLISOTHIOURONIUM CHLORIDE see BEU500
BENZYL ISOVALERATE (FCC) see ISW000
BENZYL MERCAPTAN see TGO750
BENZYL METHANOATE see BEP250
1-BENZYL-2-METHYL-3-(2-AMINOETHYL)-5-METHOXYIN-
    DOLE HYDROCHLORIDE see BEM750
BENZYL-3-METHYLBUTANOATE see ISW000
BENZYL-3-METHYL BUTYRATE see ISW000
1-BENZYL-2-METHYLHYDRAZINE see MHN750
1-BENZYL-1-(5-METHYL-3-ISOXAZOIYLCARBONYL)HY-
    DRAZINE see IKC000
1-BENZYL-2-(5-METHYL-3-ISOXAZOIYL-CARBONYL)HY-
    DRAZINE see IKC000
N′-BENZYL N-METHYL-5-ISOXAZOLECARBOXYLHYDRA-
    ZIDE-3 see IKC000
1-BENZYL-2-METHYL-5-METHOXYTRYPTAMINE HYDRO-
    CHLORIDE see BEM750
BENZYL-2-METHYL PROPIONATE see IJV000
BENZYL MUSTARD OIL see BEU250, BFL000
BENZYL NITRILE see PEA750
BENZYL NORMECHLORETHAMINE see BIA750
BENZYL OXIDE (CZECH) see BEO250
BENZYLOXYCARBONYL CHLORIDE see BEF500
BENZYLPENICILLIN see BDY669
BENZYLPENICILLIN G see BDY669
BENZYLPENICILLINIC ACID see BDY669
BENZYL PENICILLINIC ACID SODIUM SALT see BFD250
BENZYLPENICILLIN SODIUM see BFD250
N-BENZYL-N-PHENOXYISOPROPYL-β-CHLORETHYLA-
    MINE HYDROCHLORIDE see DDG800
BENZYL PHENYLACETATE see BFD400
BENZYL γ-PHENYLACRYLATE see BEG750
BENZYL PHENYLFORMATE see BCM000
BENZYL-(α-PYRIDYL)-DIMETHYLAETHYLENDIAMIN
    (GERMAN) see TMP750

BENZYL SALICYLATE see BFJ750
BENZYLSENFOEL (GERMAN) see BEU250
BENZYLSTEARYLDIMETHYLAMMONIUM CHLORIDE see DTC600
BENZYLT see PDT250
BENZYL THIOCYANATE see BFL000
BENZYLTHIOL see TGO750
3-((BENZYLTHIO)METHYL)-6-CHLORO-1,2,4-BENZOTHIA-DIAZINE-7-SULFONAMIDE-1,1-DIOXIDE see BDE250
3-BENZYLTHIOMETHYL-6-CHLORO-2H-1,2,4-BENZOTHIA-DIAZINE-7-SULFONAMIDE-1,1-DIOXIDE see BDE250
3-BENZYLTHIOMETHYL-6-CHLORO-7-SULFAMOYL-1,2,4-BENZOTHIADIAZINE-1,1-DIOXIDE see BDE250
3-BENZYLTHIOMETHYL-6-CHLORO-7-SULFAMYL-1,2,4-BENZOTHIADIAZINE-1,1-DIOXIDE see BDE250
3-BENZYLTHIOMETHYL-6-CHLORO-7-SULFAMYL-2H-1,2,4-BENZOTHIADIAZINE-1,1-DIOXIDE see BDE250
BENZYL THIOPSEUDOUREA HYDROCHLORIDE see BEU500
2-BENZYL-2-THIO-PSEUDOUREA HYDROCHLORIDE see BEU500
BENZYLTHIURONIUM CHLORIDE see BEU500
S-BENZYLTHIURONIUM CHLORIDE see BEU500
BENZYL TRICHLORIDE see BFL250
BENZYL TRIMETHYL AMMONIUM IODIDE see BFM750
BENZYL VIOLET see FAG120
BENZYL VIOLET 3B see FAG120
BENZYLYT see DDG800
3,4-BENZYPYRENE see BCS750
BENZYRIN see BBW500
BENZYTOL see CLW000
BEOSIT see EAQ750
BEPANTHEN see PAG200
BEPANTHENE see PAG200
BEPANTOL see PAG200
BERBERIN see BFN500
BERBERINE see BFN500
BERBERINE SULFATE TRIHYDRATE see BFN750
BERCEMA see EIR000
BERCEMA FERTAM 50 see FAS000
BERELEX see GEM000
BERGAMIOL see LFY100
BERGAMOT OIL RECTIFIED see BFO000
BERGAMOTTE OEL (GERMAN) see BFO000
BERGAPTEN see MFN275
BERKENDYL see CDP000
BERKFURIN see NGE000
BERKOMINE see DLH600, DLH630
BERMAT see CJJ250
BERNARENIN see VGP000
BERNOCAINE see AIT250
BERNSTEINSAEURE-2,2-DIMETHYLHYDRAZID (GERMAN) see DQD400
BERNSTEINSAURE (GERMAN) see SMY000
BERNSTEINSAURE-ANHYDRID (GERMAN) see SNC000
BEROL 478 see DJL000
BERONALD see CHJ750
BERTHOLITE see CDV750
BERTHOLLET SALT see PLA250
BERTRANDITE see BFO250
BERUBIGEN see VSZ000
BERYL see BFO500
BERYLLIA see BFT250
BERYLLIUM see BFO750
BERYLLIUM-9 see BFO750
BERYLLIUM, metal powder (DOT) see BFO750
BERYLLIUM ACETATE see BFP000
BERYLLIUM ACETATE, BASIC see BFT500
BERYLLIUM ACETATE, NORMAL see BFP000
BERYLLIUM ALUMINOSILICATE see BFO500
BERYLLIUM ALUMINUM ALLOY see BFP250
BERYLLIUM ALUMINUM SILICATE see BFO500
BERYLLIUM CARBONATE see BFP500
BERYLLIUM CARBONATE (1:1) see BFP750

BERYLLIUM CARBONATE, BASIC see BFP500
BERYLLIUM CHLORIDE see BFQ000
BERYLLIUM CHLORIDE TETRAHYDRATE see BFQ250
BERYLLIUM COMPOUND with NIOBIUM (12:1) see BFQ750
BERYLLIUM COMPOUNDS see BFQ500
BERYLLIUM COMPOUND with TITANIUM (12:1) see BFR000
BERYLLIUM COMPOUND with VANADIUM (12:1) see BFR250
BERYLLIUM-COPPER-COBALT ALLOY see CNK700
BERYLLIUM DICHLORIDE see BFQ000
BERYLLIUM DIFLUORIDE see BFR500
BERYLLIUM DIHYDROXIDE see BFS250
BERYLLIUM DINITRATE see BFT000
BERYLLIUM FLUORIDE see BFR500
BERYLLIUM HYDRATE see BFS250
BERYLLIUM HYDRIDE see BFR750
BERYLLIUM HYDROGEN PHOSPHATE (1:1) see BFS000
BERYLLIUM HYDROXIDE see BFS250
BERYLLIUM LACTATE see LAH000
BERYLLIUM MANGANESE ZINC SILICATE see BFS750
BERYLLIUM MONOXIDE see BFT250
BERYLLIUM NITRATE see BFT000
BERYLLIUM ORTHOSILICATE see SCN500
BERYLLIUM OXIDE see BFT250
BERYLLIUM OXIDE ACETATE see BFT500
BERYLLIUMOXIDE CARBONATE see BFP500
BERYLLIUM OXYACETATE see BFT500
BERYLLIUM OXYFLUORIDE see BFT750
BERYLLIUM PERCHLORATE see BFU000
BERYLLIUM PHOSPHATE see BFS000
BERYLLIUM SILICATE see SCN500
BERYLLIUM SILICATE HYDRATE see BFO250
BERYLLIUM SILICIC ACID see SCN500
BERYLLIUM SULFATE (1:1) see BFU250
BERYLLIUM SULFATE TETRAHYDRATE (1:1:4) see BFU500
BERYLLIUM SULPHATE TETRAHYDRATE see BFU500
BERYLLIUM TETRAHYDROBORATE see BFU750
BERYLLIUM TETRAHYDROBORATETRIMETHYLAMINE see BFV000
BERYLLIUM ZINC SILICATE see BFV250
BERYL ORE see BFO500
BETACIDE P see HNU500
BETAFEDRINA see BBK500
BETAFEDRINE see BBK500
BETALGIL see DTL200
BETALIN 12 CRYSTALLINE see VSZ000
BETAMETHASONE see BFV750
d-BETAPHEDRINE see BBK500
BETAPRONE see PMT100
BETAPYRIMIDUM see DJS200
BETAXINA see EID000
BETAZED see BRF500
BETEL NUT see BFW000
BETEL QUID EXTRACT see BFW125
BETEL TOBACCO EXTRACT see BFW135
BETNELAN see BFV750
BETRAMIN see BBV500
BETSOLAN see BFV750
BETULA OIL see MPI000
BEVATINE-12 see VSZ000
BEVIDOX see VSZ000
BEXIDE see BJU000
BEXOL see BBQ500
BEXON see MMN250
BEXT see BJU000, DKE400
BEXTON see CHS500
BEXTRENE XL 750 see SMQ500
B(b)F see BAW250
B(j)F see BCJ500
BF 5930 see OJW000

BFP see BJE750
BFPO see BJE750
BFV see FMV000
BG 5930 see OJW000
BGE see BRK750
BHA (FCC) see BQI000
BHBN see HJQ350
BHC see BBQ500, BJZ000
α-BHC see BBQ000
β-BHC see BBR000
γ-BHC see BBQ500
Δ-BHC see BFW500
BHC (USDA) see BBP750
BH 2,4-D see DAA800
BH DALAPON see DGI400
B-HERBATOX see SFS000
BHIMSAIM CAMPHOR see BMD000
BH MCPA see CIR250
BH MECOPROP see CIR500
BHP see DNB200
BHT (food grade) see BFW750
BI-58 see DSP400
Bi 3411 see CDO000
4',4'''-BIACETANILIDE see BFX000
BIACETYL see BOT500
BIALFLAVINA see DBX400
BIALLYL see HCR500
BIALMINAL see EOK000
BIALPIRINIA see ADA725
p,p-BIANILINE see BBX000
4,4'-BIANILINE see BBX000
N,N'-BIANILINE see HHG000
BIANISIDINE see TGJ750
5,5'-BIANTHRANILIC ACID see BFX250
BIBENZENE see BGE000
BIBENZYL see BFX500
BIBESOL see DGP900
BIC see BRQ500, IAN000
BICAM ULV see DQM600
BICARBURET of HYDROGEN see BBL250
BICARBURRETTED HYDROGEN see EIO000
BICHLORACETIC ACID see DEL000
BICHLORENDO see MQW500
BICHLORIDE of MERCURY see MCY475
BICHLORURE d'ETHYLENE (FRENCH) see EIY600
BICHLORURE de MERCURE (FRENCH) see MCY475
BICHLORURE de PROPYLENE (FRENCH) see PNJ400
BICHROMATE d'AMMONIUM (FRENCH) see ANB500
BICHROMATE OF POTASH see PKX250
BICHROMATE of SODA see SGI000
BICHROMATE de SODIUM (FRENCH) see SGI000
BICKIE-MOL see HIM000
BiCNU see BIF750
BICOLASTIC A 75 see SMQ500
BICOLENE P see PMP500
BICORTONE see PLZ000
BICYCLO(4.4.0)DECANE see DAE800
BICYCLO(2.2.1)HEPTENE-2-DICARBOXYLIC ACID, 2-
  ETHYLHEXYLIMIDE see OES000
BICYCLOPENTADIENE see DGW000
BIDIRL see DGQ875
BIDRIN see DGQ875
BIETHYLENE see BOP500
1,1'-BI(ETHYLENE OXIDE) see BGA750
BIETHYLXANTHOGENTRISULFIDE see BJU000
BIFEX see PMY300
BIFLORINE see FOW000
BIFLUORIDEN (DUTCH) see FEZ000
BIFLUORURE de POTASSIUM (FRENCH) see PKU250
BIFORMAL see GIK000
BIFORMYL see GIK000
BIFORON see BQL000
BIFURON see NGG500
BIG DIPPER see DVX800

BIGITALIN see GEU000
BIGUMAL see CKB250
BIGUNAL see BQL000
BILARCIL see TIQ250
BILEVON see HCL000
BILEVON M see DFD000
BILOBRAN see MRH209
BIMETHYL see EDZ000
2,3,1',8'-BINAPHTHYLENE see BCJ750
BINDON see ONY000
BINDON ATHYLATHER see BGC250
BINDON ETHYL ETHER see BGC250
BINITROBENZENE see DUQ200
BINOTAL see AIV500
BIO 5,462 see EAQ750
BIOACRIDIN see DBX400
BIOALLETHRIN see AFR250
BIOBAMAT see MQU750
BIOCETIN see CDP250
BIOCIDE see ADR000
BIOCOLINA see CMF750
BIO-DES see DKA600
BIODOPA see DNA200
BIOFANAL see NOH500
BIOFUREA see NGE500
BIOGRISIN-FP see GKE000
BIOMET TBTO see BLL750
BIOMITSIN see CMA750
BIOMYCIN see CMA750
BIONIC see NCQ900
BIOPHEDRIN see EAW000
BIOPHENICOL see CDP250
BIOPHYLL see CKN000
BIOPRASE see BAC000
BIOQUIN see BLC250, QPA000
BIOQUIN 1 see BLC250
BIORENINE see VGP000
BIOSEDAN see NBU000
BIOSEPT see CCX000
BIO-SOFT D-40 see DXW200
BIOSOL VETERINARY see NCG000
BIOSTAT see HOH500
BIOSTEROL see VSK600
BIOSUPRESSIN see HOO500
BIO-TESTICULINA see TBG000
BIO-TETRA see TBX000
BIOTHION see TAL250
BIOXIRANE see BGA750
2,2'-BIOXIRANE see BGA750
(R*,S*)-2,2'-BIOXIRANE see DHB800
(S-(R*,R*))-2,2'-BIOXIRANE see BOP750
BIOXYDE d'AZOTE (FRENCH) see NEG100
BIOXYDE de PLOMB (FRENCH) see LCX000
BIPHENYL see BGE000
1,1'-BIPHENYL see BGE000
4-BIPHENYLACETAMIDE see PDY500
N-4-BIPHENYLACETAMIDE see PDY500
4-BIPHENYLACETHYDROXAMIC ACID see ACD000
BIPHENYLAMINE see AJS100
4-BIPHENYLAMINE see AJS100
p-BIPHENYLAMINE see AJS100
(1,1'-BIPHENYL)-4-AMINE see AJS100
BIPHENYL, mixed with BIPHENYL OXIDE (3:7) see PFA860
4,4'-BIPHENYLDIAMINE see BBX000
(1,1'-BIPHENYL)-4,4'-DIAMINE (9CI) see BBX000
(1,1'-BIPHENYL)-4,4'-DIAMINE, DIHYDROCHLORIDE see
  BBX750
(1,1'-BIPHENYL)-4,4'-DIAMINE SULFATE (1:1) see
  BBY000
BIPHENYL-DIPHENYL ETHER mixture see PFA860
N,N'-(1,1'-BIPHENYL)-4,4'-DIYLBIS-ACETAMIDE 4',4'''-
  BIACETANILIDE see BFX000
2,2'-(1,1'-BIPHENYL-4,4'-DIYLBIS(2-HYDROXY-4,4-DI-
  METHYL-MORPHOLINIUM DIBROMIDE see HAQ000

4,4'-BIPHENYLENEDIAMINE see BBX000
N-(p-BIPHENYLMETHYL)-ATROPINIUM BROMIDE see PEM750
N,4-BIPHENYL-METHYL-dl-TROPEYL-α-TROPINIUMBRO-MIDS (GERMAN) see PEM750
p-BIPHENYLMETHYL-(dl-TROPYL-α-TROPINIUM)BROMIDE see PEM750
4-BIPHENYLOL see BGJ500
2-BIPHENYLOL, SODIUM SALT see BGJ750
BIPHENYL OXIDE see PFA850
1,1'-BIPHENYL, mixed with 1,1'-OXYBIS(BENZENE) see PFA860
N-(4-BIPHENYLYL)ACETAMIDE see PDY500
N,N'-4,4'-BIPHENYLYLENEBISACETAMIDE see BFX000
BIPOTASSIUM CHROMATE see PLB250
BIRCH TAR OIL see BGO750
BIRCH TAR OIL, RECTIFIED (FCC) see BGO750
BIRLANE see CDS750
BIRNENOEL see AOD725
BIRTHWORT see AQY250
2,7-BIS(ACETAMIDO)FLUORENE see BGP250
BIS(ACETATO)TETRAHYDROXYTRILEAD see LCH000
BIS(ACETATO)TRIHYDROXYTRILEAD see LCJ000
BIS(ACETO)DIHYDROXYTRILEAD see LCH000
BIS(ACETOXY)CADMIUM see CAD250
BIS(ACETYLACETONATO) TITANIUM OXIDE see BGQ750
BIS(ACETYLOXY)DIBUTYLSTANNANE see DBF800
BIS(ACETYLOXY)MERCURY see MCS750
S-(1,2-BIS(AETHOXY-CARBONYL)-AETHYL)-O,O-DI-METHYL-DITHIOPHASPHAT (GERMAN) see MAK700
2,4-BIS(AETHYLAMINO)-6-CHLOR-1,3,5-TRIAZIN (GER-MAN) see BJP000
BIS AMINE see MJM200
BIS(4-AMINO-3-CHLOROPHENYL) ETHER see BGT000
BIS(2-AMINOETHYL)AMINE see DJG600
BIS(β-AMINOETHYL)AMINE see DJG600
N,N'-BIS(2-AMINOETHYL)-1,2-DIAMINOETHANE see TJR000
BIS(β-AMINOETHYL)DISULFIDE see MCN500
N,N'-BIS(2-AMINOETHYL)ETHYLENEDIAMINE see TJR000
N,N'-BIS(2-AMINOETHYL)-1,2-ETHYLENEDIAMINE see TJR000
BIS-p-AMINOFENYLMETHAN (CZECH) see MJQ000
1,3-BIS-AMINOMETHYLBENZEN (CZECH) see XHS800
BIS-4-AMINO-3-METHYLFENYLMETHAN (CZECH) see MJO250
BIS(2-AMINOPHENYL)DISULFIDE see DXJ800
BIS(o-AMINOPHENYL)DISULFIDE see DXJ800
1,1'-BIS(2-AMINOPHENYL)DISULFIDE see DXJ800
BIS(4-AMINOPHENYL)ETHER see OPM000
BIS(p-AMINOPHENYL)ETHER see OPM000
BIS(4-AMINOPHENYL)METHANE see MJQ000
BIS(p-AMINOPHENYL)METHANE see MJQ000
2',4-BIS(AMINOPHENYL)METHANE see MJP750
BIS(4-AMINOPHENYL) SULFIDE see TFI000
BIS(p-AMINOPHENYL)SULFIDE see TFI000
BIS(4-AMINOPHENYL) SULFONE see SOA500
BIS(p-AMINOPHENYL) SULFONE see SOA500
BIS(4-AMINOPHENYL) SULPHIDE see TFI000
BIS(p-AMINOPHENYL)SULPHIDE see TFI000
BIS(4-AMINOPHENYL)SULPHONE see SOA500
BIS(p-AMINOPHENYL)SULPHONE see SOA500
BIS-(3-AMINOPROPYL)AMINE see AIX250
BIS(3-AMINOPROPYL)METHYLAMINE see BGU750
BIS(γ-AMINOPROPYL)METHYLAMINE see BGU750
N,N-BIS(3-AMINOPROPYL)METHYLAMINE see BGU750
N,N-BIS-(γ-AMINOPROPYL)METHYLAMINE see BGU750
BIS(φ-AMINOPROPYL)METHYLAMINE see BGU750
1,4-BIS(AMINOPROPYL)PIPERAZINE see BGV000
BIS(AMINOPROPYL)PIPERAZINE (DOT) see BGV000
2,2-BIS(p-ANISYL)-1,1,1-TRICHLOROETHANE see MEI450
2,5-BIS(1-AZIRIDINYL)-3,6-BIS(2-METHOXYETHOXY)-p-BENZOQUINONE see BDC750

2,5-BIS(1-AZIRIDINYL)-3,6-BIS(2-METHOXYETHOXY)-2,5-CYCLOHEXADIENE-1,4-DIONE see BDC750
BIS(1-AZIRIDINYL)(2-METHYL-3-THIAZOLIDINYL)PHOS-PHINE OXIDE see BGY000
(BIS(1-AZIRIDINYL)PHOSPHINYL)CARBAMIC ACID, ETHYL ESTER see EHV500
BIS(BENZOTHIAZOLYL)DISULFIDE see BDE750
BIS(2-BENZOTHIAZOLYLTHIO)ZINC see BHA750
BIS(2-BENZOTHIAZYL) DISULFIDE see BDE750
BIS(2-BENZOYLBENZOATO)BIS(3-(1-METHYL-2-PYRROLI-DINYL)PYRIDINE) NICKEL TRIHYDRATE see BHB000
1,4-BIS(BIS(1-AZIRIDINYL)PHOSPHINYL)PIPERAZINE see BJC250
BIS((4-(BIS(2-CHLOROETHYL)AMINO)BENZENE)
ACETATE)ESTRA-1,3,5(10)-TRIENE-3,17-DIOL(17-β) see EDR500
BIS((4-(BIS(2-CHLOROETHYL)AMINO)BENZENE)
ACETATE)OESTRA-1,3,5(10)-TRIENE-3,17-DIOL(17-β) see EDR500
2,5-BIS(BIS(2-CHLOROETHYL)AMINOMETHYL)
HYDROQUINONE see BHB750
BIS((p-(BIS(2-CHLOROETHYL)AMINO)PHENYL)ACE-TATE)ESTRADIOL see EDR500
BIS((p-(BIS(2-CHLOROETHYL)AMINO)PHENYL)
ACETATE)ESTRA-1,3,5(10)-TRIENE-3,17-β-DIOL see EDR500
BIS((p-(BIS(2-CHLOROETHYL)AMINO)PHENYL)ACE-TATE)OESTRADIOL see EDR500
BIS((p-BIS(2-CHLOROETHYL)AMINOPHENYL)ACE-TATE)OESTRA-1,3,5(10)-TRIENE-3,17-β-DIOL see EDR500
BIS(BISDIMETHYLAMINOPHOSPHONOUS)ANHYDRIDE see OCM000
1,2-BIS(BROMOACETOXY)ETHANE see BHD250
BIS(BUTOXYMALEOYLOXY)DIBUTYLSTANNANE see BHK250
BIS(BUTOXYMALEOYLOXY)DIOCTYLSTANNANE see BHK500
BIS(3-tert-BUTYL-4-HYDROXY-6-METHYLPHENYL) SUL-FIDE see TFC600
BIS(n-BUTYL)SEBACATE see DEH600
S-(1,2-BIS(CARBETHOXY)ETHYL)-O,O-DIMETHYL DITHIO-PHOSPHATE see MAK700
BIS(CARBONATO(2-))DIHYDROXYTRIBERYLLIUM see BFP500
BIS(2-CARBOXYETHYL) SULFIDE see BHM000
3,6-BIS(CARBOXYMETHYL)-3,5-DIAZOOCTANEDIOIC ACID see EIX000
N,N-BIS(CARBOXYMETHYL)GLYCINE TRISODIUM SALT MONOHYDRATE see NEI000
N,N-BIS(CARBOXYMETHYL)GLYSINE see AMT500
BIS(CARBOXYMETHYLMERCAPTO)(p-UREIDOPHENYL) ARSINE see CBI250
BIS(CARBOXYMETHYLTHIO)(p-UREIDOPHENYL)ARSINE see CBI250
p-(BIS(o-CARBOXYPHENYLMERCAPTO)-ARSINO)-PHEN-YLUREA see TFD750
BIS(3-CARBOXYPROPIONYL) PEROXIDE see SNC500
N,N-BIS-(β-CHLORAETHYL)-AMIN (GERMAN) see BHN750
N,N-BIS-(β-CHLORAETHYL)-N',O-PROPYLEN-PHOSPHOR-SAEURE-ESTER-DIAMID (GERMAN) see EAS500
BIS(5-CHLOR-2-HYDROXYPHENYL)-METHAN (GERMAN) see MJM500
BIS(p-CHLOROBENZOYL) PEROXIDE see BHM750
trans-N,N'-BIS(2-CHLOROBENZYL)-1,4-CYCLOHEXANEBIS-(METHYLAMINE) DIHYDROCHLORIDE see BHN000
1,3-BIS((p-CHLOROBENZYLIDENE)AMINO)GUANIDINE see RLK890
BIS(2-CHLOROETHOXY)METHANE see BID750
N,N-BIS(β-CHLOROETHYL)-dl-ALANINE HYDROCHLO-RIDE see BHN500
BIS-β-CHLOROETHYLAMINE see BHN750

BIS(p-DIMETHYLAMINOPHENYL)METHYLENEIMINE see IBB000

1,1-BIS(p-DIMETHYLAMINOPHENYL)METHYLENIMINE-HYDROCHLORIDE see IBA000

BIS(DIMETHYLAMINO)PHOSPHONOUS ANYHYDRIDE see OCM000

BIS(DIMETHYLAMINO)PHOSPHORIC ANHYDRIDE see OCM000

BIS(α,α-DIMETHYLBENZYL)PEROXIDE see DGR600

BIS(DIMETHYLCARBAMODITHIOATO-S,S')LEAD see LCW000

BIS(DIMETHYLCARBAMODITHIOATO-S,S')ZINC see BJK500

BIS(DIMETHYLDITHIOCARBAMATE de ZINC) (FRENCH) see BJK500

BIS(DIMETHYLDITHIOCARBAMATO)ZINC see BJK500

BIS(DIMETHYLDITHIOCARBAMIATO)LEAD see LCW000

2,6-BIS(1,1-DIMETHYLETHYL)-4-METHYLPHENOL see BFW750

1,1'-BIS(3,5-DIMETHYLMORPHOLINOCARBONYL-METHYL)-4,4'-BIPYRIDINIUM-DICHLORID (GERMAN) see BJK750

1,1'-BIS(3,5-DIMETHYLMORPHOLINOCARBONYL-METHYL)-4,4'-BIPYRIDYNIUM DICHLORIDE see BJK750

1,1'-BIS(2-(3,5-DIMETHYL-4-MORPHOLINYL)-2-OXO-ETHYL)-4,4'-BIPYRIDINIUM DICHLORIDE see BJK750

BIS(DIMETHYL-THIOCARBAMOYL)-DISULFID (GERMAN) see TFS350

BIS(DIMETHYLTHIOCARBAMOYL) DISULFIDE see TFS350

BIS(DIMETHYLTHIOCARBAMOYL)SULFIDE see BJL600

BIS(DIMETHYLTHIOCARBAMYL) MONOSULFIDE see BJL600

BIS(N,N-DIMETIL-DITIOCARBAMMATO) DI ZINCO (ITALIAN) see BJK500

(±)-1,2-BIS(3,5-DIOXOPIPERAZINE-1-YL)PROPANE see PIK250

(±)-1,2-BIS(3,5-DIOXOPIPERAZINYL)PROPANE see PIK250

BIS(DITHIOPHOSPHATE de O,O-DIETHYLE) de S,S'-(1,4-DI-OXANNE-2,3-DIYLE) (FRENCH) see DVQ709

BIS(DODECANOYLOXY)DI-n-BUTYLSTANNANE see DDV600

BIS(DODECYLOXYCARBONYLETHYL) SULFIDE see TFD500

BIS(2,3-EPOXYCYCLOPENTYL) ETHER see BJN250

m-BIS(2,3-EPOXYPROPOXY)BENZENE see REF000

1,3-BIS(2,3-EPOXYPROPOXY)BENZENE see REF000

BIS(2,3-EPOXYPROPYL)ETHER see DKM200

2,2-BIS(4-(2,3-EPOXYPROPYLOXY)PHENYL)PROPANE see BLD750

S-(1,2-BIS(ETHOXY-CARBONYL)-ETHYL)-O,O-DIMETHYL-DITHIOFOSFAAT (DUTCH) see MAK700

S-(1,2-BIS(ETHOXYCARBONYL)ETHYL)-O,O-DIMETHYL PHOSPHORODITHIOATE see MAK700

S-1,2-BIS(ETHOXYCARBONYL)ETHYL-O,O-DIMETHYL THIOPHOSPHATE see MAK700

BIS(ETHOXYTHIOCARBONYL)TRISULFIDE see DKE400

2,4-BIS(ETHYLAMINO)-6-CHLORO-s-TRIAZINE see BJP000

BIS(ETHYLENIMIDO)PHOSPHORYLURETHAN see EHV500

2,6-BIS(ETHYLEN-IMINO)-4-AMINO-s-TRIAZINE see BJP500

BIS(2-ETHYLHEXANOYLOXY)DIBUTYL STANNANE see BJQ250

BIS(2-ETHYLHEXYL)-1,2-BENZENEDICARBOXYLATE see DVL700

BIS(2-ETHYLHEXYL) ESTER, PEROXYDICARBONIC ACID see DJK800

BIS(2-ETHYLHEXYL) ESTER PHOSPHORUS ACID CAD-MIUM SALT see CAD500

BIS(ETHYLHEXYL) ESTER of SODIUM SULFOSUCCINIC ACID see DJL000

BIS(2-ETHYLHEXYL) FUMARATE see DVK600

BIS(2-ETHYLHEXYL)HYDROGEN PHOSPHATE see BJR750

BIS(2-ETHYLHEXYL)ORTHOPHOSPHORIC ACID see BJR750

BIS(2-ETHYLHEXYL)PHOSPHATE see BJR750

BIS(2-ETHYLHEXYL)PHOSPHORIC ACID see BJR750

BIS(2-ETHYLHEXYL)PHTHALATE see DVL700

BIS(2-ETHYLHEXYL)SEBACATE see BJS250

BIS(2-ETHYLHEXYL)SODIUM SULFOSUCCINATE see DJL000

BIS(2-ETHYLHEXYL)-S-SODIUM SULFOSUCCINATE see DJL000

1,4-BIS(2-ETHYLHEXYL) SODIUM SULFOSUCCINATE see DJL000

1,4-BIS(2-ETHYLHEXYL)SULFOBUTANEDIOIC ACID ES-TER, SODIUM SALT see DJL000

BIS(ETHYLMERCURI)PHOSPHATE see BJT250

1,1-BIS(p-ETHYLPHENYL)-2,2-DICHLOROETHANE see DJC000

2,2-BIS(p-ETHYLPHENYL)-1,1-DICHLOROETHANE see DJC000

BIS(N-ETHYL-N-PHENYL)UREA see DJC400

BIS(ETHYLXANTHIC)DISULFIDE see BJU000

BISETHYL XANTHOGEN DISULFIDE see BJU000

BIS(ETHYLXANTHOGEN) TETRASULFIDE see BJU250

BIS(ETHYLXANTHOGEN) TRISULFIDE see DKE400

S-(1,2-BIS(ETOSSI-CARBONIL)-ETIL)-O,O-DIMETIL-DITIO-FOSFATO (ITALIAN) see MAK700

BISFEROL A (GERMAN) see BLD500

1,6-BIS(9 FLUORENYLDIMETHYL-AMMONIUM)HEXANE BROMIDE see HEG000

BIS(3-FLUOROSALICYLALDEHYDE)-ETHYLENEDIIMINE-COBALT see EIS000

2-(BIS(FURFURYLIDENAMINO))METHYLFURAN see FPS000

m-BIS(GLYCIDYLOXY)BENZENE see REF000

BIS(4-GLYCIDYLOXYPHENYL)DIMETHYAMETHANE see BLD750

2,2-BIS(p-GLYCIDYLOXYPHENYL)PROPANE see BLD750

BIS(l-HISTIDINATO)COBALT see BJY000

BIS(l-HISTIDINE)COBALT see BJY000

2,2-BIS(HYDROPEROXY)PROPANE see BJY825

BIS(HYDROXYAETHYL)-AETHER-DINITRAT (GERMAN) see DJE400

BIS(β-HYDROXYAETHYL)NITROSAMIN (GERMAN) see NKM000

BIS(4-HYDROXY-5-tert-BUTYL-2-METHYLPHENYL) SUL-FIDE see TFC600

BIS-2-HYDROXY-5-CHLORFENYLMETHAN (CZECH) see MJM500

BIS(2-HYDROXY-5-CHLOROPHENYL)METHANE see MJM500

BISHYDROXYCOUMARIN see BJZ000

BIS(4-HYDROXY-3-COUMARIN) ACETIC ACID ETHYL ES-TER see BKA000

BIS-3,3'-(4-HYDROXYCOUMARINYL)ACETIC ACID ETHYL ESTER see BKA000

BIS-(4-HYDROXY-3-COUMARINYL)ETHYL ACETATE see BKA000

BIS(4-HYDROXYCOUMARIN-3-YL)METHANE see BJZ000

BIS(2-HYDROXY ETHYL)AMINE see DHF000

BIS(2-HYDROXYETHYL)CARBAMODITHIOIC ACID, MONOPOTASSIUM SALT see PKX500

BIS(2-HYDROXYETHYL)-2-(2-CHLORO ETHYL THIO) ETHYL SULFONIUM) CHLORIDE see BKD750

BIS(2-HYDROXYETHYL)DITHIOCARBAMIC ACID, MONO-POTASSIUM SALT see PKX500

BIS(2-HYDROXYETHYL)DITHIOCARBAMIC ACID, POTAS-SIUM SALT see PKX500

N,N-BIS(2-HYDROXYETHYL)DODECAN AMIDE see BKE500

BIS(2-HYDROXYETHYL) ETHER see DJD600

1,1-BIS(2-HYDROXYETHYL)HYDRAZINE see HHH000

BIS(2-HYDROXYETHYL)LAURAMIDE see BKE500

N,N-BIS(HYDROXYETHYL)LAURAMIDE see BKE500
N,N-BIS(2-HYDROXYETHYL)LAURAMIDE see BKE500
N,N-BIS(β-HYDROXYETHYL)LAURAMIDE see BKE500
N′,N′-BIS(2-HYDROXYETHYL)-N-METHYL-2-NITRO-p-PHE-NYLENEDIAMINE see BKF250
BIS(β-HYDROXYETHYL)NITROSAMINE see NKM000
2,2-BIS-4′-HYDROXYFENYLPROPAN (CZECH) see BLD500
BIS(HYDROXYLAMINE) SULFATE see OLS000
3-BIS(HYDROXYMETHYL)AMINO-6-(5-NITRO-2-FURYL-ETHENYL)-1,2,4-TRIAZINE see BKH500
BIS(HYDROXYMETHYL)FURATRIZINE see BKH500
2,2-BIS(HYDROXYMETHYL)-1,3-PROPANEDIOL see PBB750
2,2-BIS(HYDROXYMETHYL)-1,3-PROPANEDIOL TETRANI-TRATE see PBC250
BIS(4-HYDROXY-2-OXO-2H-1-BENZOPYRAN-3-YL)ACETIC ACID ETHYL ESTER see BKA000
BIS(4-HYDROXYPHENYL) DIMETHYLMETHANE see BLD500
BIS(4-HYDROXYPHENYL)DIMETHYLMETHANE DIGLYCI-DYL ETHER see BLD750
3,4-BIS(4-HYDROXYPHENYL)-2,4-HEXADIENE see DAL600
3,4-BIS(p-HYDROXYPHENYL)-2,4-HEXADIENE see DAL600
3,4-BIS(p-HYDROXYPHENYL)-3-HEXENE see DKA600
BIS(4-HYDROXYPHENYL)PROPANE see BLD500
2,2-BIS(4-HYDROXYPHENYL)PROPANE see BLD500
2,2-BIS(p-HYDROXYPHENYL)PROPANE see BLD500
2,2-BIS(4-HYDROXYPHENYL)PROPANE, DIGLYCIDYL ETHER see BLD750
2,2-BIS(p-HYDROXYPHENYL)PROPANE, DIGLYCIDYL ETHER see BLD750
N-BIS(2-HYDROXYPROPYL)NITROSAMINE see DNB200
2,2′-BISHYDROXYPROPYLNITROSAMINE see DNB200
BIS(2-HYDROXY-3,5,6-TRICHLOROPHENYL)METHANE see HCL000
BIS(ISOBUTYL)ALUMINUM CHLORIDE see CGB500
BIS(4-ISOCYANATOCYCLOHEXYL)METHANE see MJM600
BIS(4-ISOCYANATOPHENYL)METHANE see MJP400
BIS(p-ISOCYANATOPHENYL)METHANE see MJP400
BIS(1,4-ISOCYANATOPHENYL)METHANE see MJP400
BIS(ISOOCTYLOXYCARBONYLMETHYLTHIO)DIOCTYL STANNANE see BKK750
BIS(ISOPROPYLAMIDO) FLUOROPHOSPHATE see PHF750
2,4-BIS(ISOPROPYLAMINO)-6-CHLORO-s-TRIAZINE see PMN850
2,4-BIS(ISOPROPYLAMINO)-6-METHYLMERCAPTO-s-TRIAZINE see BKL250
4,6-BIS(ISOPROPYLAMINO)-2-METHYLMERCAPTO-s-TRIAZINE see BKL250
2,4-BIS(ISOPROPYLAMINO)-6-METHYLTHIO-s-TRIAZINE see BKL250
2,4-BIS(ISOPROPYLAMINO)-6-METHYLTHIO-1,3,5-TRI-AZINE see BKL250
BIS(LAUROYLOXY)DIBUTYLSTANNANE see DDV600
BIS(LAUROYLOXY)DI(n-BUTYL)STANNANE see DDV600
1,3-BISMALEIMIDO BENZENE see BKL750
BISMARSEN see BKV250
BISMATE see BKW000
BIS(MERCAPTOACETATE)DIOCTYL-TIN BIS(ISOOCTYL) ESTER see BKK750
BIS(MERCAPTOBENZOTHIAZOLATO)ZINC see BHA750
1,4-BIS(METHANESULFONOXY)BUTANE see BOT250
BIS(METHANE SULFONYL)-d-MANNITOL see BKM500
(1,4-BIS(METHANESULFONYLOXY)BUTANE) see BOT250
2,5-BISMETHOXYETHOXY-3,6-BISETHYLENEIMINO-1,4-BENZOQUINONE see BDC750
3,6-BIS (β-METHOXYETHOXY)-2,5-BIS(ETHYLENEIMINO)-p-BENZOQUINONE see BDC750
3,6-BIS(β-METHOXYETHOXY)-2,5-BIS(ETHYLENIMINO)-p-BENZOQUINONE see BDC750
BIS(METHOXYETHYL) PHTHALATE see DOF400

BIS(2-METHOXYETHYL) PHTHALATE see DOF400
N,N′-BIS(4-METHOXYPHENYL)-N″-(4-ETHOXYPHENYL) GUANIDINE HYDROCHLORIDE see PDN500
3,4-BIS(p-METHOXYPHENYL)-3-HEXENE see DJB200
1,1-BIS(p-METHOXYPHENYL)-2,2,2-TRICHLOROETHANE see MEI450
2,2-BIS(p-METHOXYPHENYL)-1,1,1-TRICHLOROETHANE see MEI450
N,N′-BIS(1-METHYLETHYL)-6-METHYL-THIO-1,3,5-TRI-AZINE-2,4-DIAMINE see BKL250
BIS(6-METHYLHEPTYL)ESTER of PHTHALIC ACID see ILR100
N,N-BIS(N-METHYL-N-PHENYL-tert-BUTYLACETAMIDO)-β-HYDROXYETHYLAMINE see DTL200
N,N′-BIS(2-METHYLPHENYLTHIOUREA see DXP600
N-BISMETHYLPTEROYLGLUTAMIC ACID see MDV500
N,N-BIS(METHYLSULFONEPROPOXY)AMINE HYDRO-CHLORIDE see YCJ000
1,6-BIS-o-METHYLSULFONYL-d-MANNITOL see BKM500
BIS(MONOISOPROPYLAMINO)FLUOROPHOSPHATE see PHF750
BIS(MONOISOPROPYLAMINO)FLUOROPHOSPHINE OXIDE see PHF750
N,N′-BISMORPHOLINE DISULFIDE see BKU500
BISMORPHOLINO DISULFIDE see BKU500
BIS(MORPHOLINO-)METHAN (GERMAN) see MJQ750
BISMORPHOLINO METHANE see MJQ750
BISMUTH see BKU750
BISMUTH-209 see BKU750
BISMUTH ARSPHENAMINE SULFONATE see BKV250
BISMUTH COMPOUNDS see BKV750
BISMUTH DIMETHYL DITHIOCARBAMATE see BKW000
BISMUTH EMETINE IODIDE see EAM500
BISMUTH NITRATE see BKW250
BISMUTH PENTAFLUORIDE see BKW750
BISMUTH SODIUM THIOGLYCOLLATE see BKX750
BIS(NITRATO-O,O′)DIOXO URANIUM (solid) see USA000
BIS(NITRATO)DIOXOURANIUM HEXAHYDRATE see URS000
BIS(OCTANOYLOXY)DI-n-BUTYL STANNANE see BLB250
BIS(OCTANOYLOXY)DI-n-BUTYLTIN see BLB250
BISODIUM TARTRATE see BLC000
BISOFLEX 81 see DVL700
BISOFLEX DOP see DVL700
BISOFLEX DOS see BJS250
BISOLVOMYCIN see HOI000
BIS(1-OXODODECYL)PEROXIDE see LBR000
BIS-(2-OXOPROPYL)-N-NITROSAMINE see NJN000
BIS(1-OXOPROPYL)PEROXIDE see DWQ800
BIS(8-OXYQUINOLINE)COPPER see BLC250
BISPENTAFLUOROSULFUR OXIDE see BLD000
BIS(2,4-PENTANEDIONATO)TITANIUM OXIDE see BGQ750
BISPHENOL A see BLD500
BISPHENOL A DIGLYCIDYL ETHER see BLD750
1,4-BIS(PHENYL AMINO)BENZENE see BLE500
BIS(PHENYLMERCURI)METHYLENEDINAPHTHALENE-SULFONATE see PFN000
2,4-BIS(PROPYLAMINO)-6-CHLOR-1,3,5-TRIAZIN (GER-MAN) see PMN850
trans-1,2-BIS(n-PROPYLSULFONYL)ETHYLENE see BLG500
BIS(8-QUINOLINATO)COPPER see BLC250
BIS(8-QUINOLINOLATO)COPPER see BLC250
BIS(8-QUINOLINOLATO-N(^{1}),O(^{8}))-COPPER see BLC250
BIS(SUCCINYLDICHLOROCHOLINE) see HLC500
BISTERIL see PDC250
BIS-N,N,N′,N′-TETRAMETHYLPHOSPHORODIAMIDIC AN-HYDRIDE see OCM000
BIS(THIOCYANATO)-MERCURY see MCU250
1,4-BIS(p-TOLYLAMINO)ANTHRAQUINONE see BLK000
BIS-1,4-p-TOLYLAMINOANTHRCHINON (CZECH) see BLK000
1,3-BIS(o-TOLYL)-2-THIOUREA see DXP600

BISTON see DCV200
BIS-(TRI-N-BUTYLCIN)OXID (CZECH) see BLL750
BIS(TRIBUTYLOXIDE) of TIN see BLL750
BIS(TRI-N-BUTYLPHOSPHINE)DICHLORONICKEL see BLS250
BIS(TRIBUTYLSTANNYL)OXIDE see BLL750
BIS(TRIBUTYL TIN)OXIDE see BLL750
BIS(TRI-N-BUTYLZINN)-OXYD (GERMAN) see BLL750
BIS-2,3,5-TRICHLOR-6-HYDROXYFENYLMETHAN (CZECH) see HCL000
BIS(3,5,6-TRICHLORO-2-HYDROXYPHENYL)METHANE see HCL000
BISTRICHLOROMETHYLTRISULFID (CZECH) see BLM750
BIS(TRICHLORO METHYL)TRISULFIDE see BLM750
BIS(TRIETHYLTIN) SULFATE see BLN250
BIS(TRIFLUOROETHYL)ETHER see HDC000
BIS(2,2,2-TRIFLUOROETHYL)ETHER see HDC000
2,2'-BIS(1,6,7-TRIHYDROXY-3-METHYL-5-ISOPROPYL-8-ALDEHYDONAPHTHALENE see GJM000
BISTRIMATE see BKX750
α,omega-BIS(TRIMETHYL AMMONIUM)HEXANE DIBROMIDE see HEA000
BIS(TRINITROPHENYL)SULFIDE see BLR750
BIS(TRIPHENYLPHOSPHINE)DICHLORONICKEL see BLS250
BIS(TRIPHENYL PHOSPHINE)NICKEL DITHIOCYANATE see BLS500
BIS(TRIPHENYL SILYL)CHROMATE see BLS750
BIS(TRIPHENYL TIN)SULFIDE see BLT250
BIS(TRIS(β,β-DIMETHYLPHENETHYL)TIN)OXIDE see BLU000
BIS(TRIS(2-METHYL-2-PHENYLPROPYL)TIN)OXIDE see BLU000
BISULFAN see BOT250
BISULFITE see SOH500
BISULFITE de SODIUM (FRENCH) see SFE000
BISULPHANE see BOT250
BITEMOL see BJP000
BITEMOL S 50 see BJP000
BITHION see TAL250
BITIODIN see BLV000
4,4'-BI-o-TOLUIDINE see TGJ750
(m,o'-BITOLYL)-4-AMINE see BLV250
BITOSCANATE see PFA500
BITTER ALMOND OIL see BLV500
BITTER ALMOND OIL CAMPHOR see BCP250
BITTER FENNEL OIL see FAP000
BITTER ORANGE OIL see BLV750
BITTER SALTS see MAJ500
BITUMEN (MAK) see ARO500
Δ(1,1')-BIUREA see ASM270
BIVERM see PDP250
BIVINYL see BOP500
BIXA ORELLANA see APE100
BIZOLIN 200 see BRF500
BK see BML500
B-K LIQUID see SHU500
B-K POWDER see HOV500
γ-BL see BOV000
BL 139 see DOY400
BLA see LCH000
BLACAR 1716 see PKQ059
BLACK AND WHITE BLEACHING CREAM see HIH000
BLACK BLASTING POWDER see ERF500
BLACK 2EMBL see AQP000
BLACK LEAF see NDN000
BLACK MANGANESE OXIDE see MAS000
BLACK OXIDE of IRON see IHD000
BLACK PEARLS see CBT500
BLACK PEPPER OIL see BLW250
BLACOSOLV see TIO750
BLADAFUME see SOD100
BLADAFUN see SOD100
BLADAN see EEH600, HCY000, PAK000, TCF250

BLADAN BASE see HCY000
BLADAN F see PAK000
BLADAN-M see MNH000
BLADEX see BLW750
BLADEX 80WP see BLW750
BLANC FIXE see BAP000
BLANDLUBE see MQV750
BLANOSE BWM see SFO500
BLASTING GELATIN (DOT) see NGY000
BLASTING OIL see NGY000
BLATTANEX see PMY300
BLATTERALKOHOL see HFE000
L-BLAU 2 (GERMAN) see FAE100
BLAUSAEURE (GERMAN) see HHS000
BLAUWZUUR (DUTCH) see HHS000
BLEACHING POWDER see HOV500
BLEACHING POWDER, CONTAINING 39% OR LESS CHLORINE (DOT) see HOV500
BLEIACETAT (GERMAN) see LCG000
BLEIAZETAT (GERMAN) see LCJ000
BLEIPHOSPHAT (GERMAN) see LDU000
BLEISULFAT (GERMAN) see LDY000
BLENDED RED OXIDES of IRON see IHD000
BLENOXANE see BLY000
BLEO see BLY000
BLEOCIN see BLY000
BLEOMYCIN see BLY000
BLEOMYCIN A2 see BLY250
BLEU BRILLIANT FCF see FMU059
BLEU DIAMINE see CMO250
BLEX see DIN800
BLIGHTOX see EIR000
BLISTERING BEETLES see CBE250
BLISTERING FLIES see CBE250
BLITEX see EIR000
BLIZENE see EIR000
BLM see BLY000
BLO see BOV000
BLOC see FAK100
BLOCADREN see DDG800
BLON see BOV000
BLOOD STONE see HAO875
BLOTIC see MKA000
BLUE 2B see CMO000
1206 BLUE see FAE000
1311 BLUE see FAE100
11388 BLUE see FMU059
12070 BLUE see FAE100
BLUE ASBESTOS (DOT) see ARM275
BLUE BN BALSE see DCJ200
BLUE COPPER see CNP250
BLUE CROSS see CGN000
BLUE EMB see CMO250
BLUE OIL see AOQ000, COD750
BLUE-OX see ZLS000
BLUE POWDER see ZBJ000
BLUE STONE see CNP250
BLUE VITRIOL see CNP250
BLU-PHEN see EOK000
BM 1 see HNI500
7-BMBA see BNO750
BMC see BRS750
BMIH see IKC000
BMOO see BRT000
BN see BFW000
B-NINE see DQD400
BNM see BAV575
BNP 30 see BRE500
BNU see BSA250
BO-ANA see FAB600
BOEA see BKA000
B.O.E.A. see BKA000
BOG MANGANESE see MAS000
BOH see HHC000

BOIS D'ARC (FRENCH) see MRN500
BOIS d'INDE see BAT500
BOL see BNM250
BOL-148 see BNM250
BOLATRON see PKQ059
BOLETIC ACID see FOU000
BOLINAN see PKQ250
BOLLS-EYE see HKC000, HKC500
BOLSTAR see SOU625
BOMBITA see DBA800
BOMYL see SOY000
BONADETTES see HGC500
BONADOXIN see HGC500
BONAMID see PJY500
BONAMINE see HGC500
BONAPICILLIN see AIV500
BONAZEN see ZNA000
BONBRAIN see TEH500
BOND CH 18 see AAX250
BONE OIL see BMA750
BONIBAL see DXH250
BONIDE BLUE DEATH RAT KILLER see PHO750
BONIDE KRAB CRABGRASS KILLER see PLC250
BONIDE RYATOX see RSZ000
BONIDE TOPZOL RAT BAITS and KILLING SYRUP see RCF000
BONINE see MBX500
BONLOID see PKQ059
BONOFORM see TBQ100
BONOMOLD OE see HJL000
BONOMOLD OP see HNU500
BOOKSAVER see AAX250
BOP see NJN000
BORACIC ACID see BMC000
BORACSU see SFF000
BORANE with DIMETHYLAMINE (1:1) see DOR200
BORATES, TETRA, SODIUM SALT, anhydrous (OSHA, ACGIH) see SFE500, SFF000
BORAX (8CI) see SFF000
BORAX DECAHYDRATE see SFF000
BORAZINE see BMB500
BORAZOLE see BMB500
BORDEAU ARSENITE, liquid or solid (DOT) see BMB750
BORDEAUX ARSENITE see BMB750
BORDEN 2123 see AAX250
BORDERMASTER see CIR250
BOREA see BMM650
BORER SOL see EIY600
BORESTER 2 see THX750
BORESTER O see TLN000
BORIC ACID see BMC000
BORIC ACID, ETHYL ESTER see BMC250
BORIC ACID, TRIBUTYL ESTER see THX500
BORIC ACID, TRI-sec-BUTYL ESTER see THX750
BORIC ACID, TRI-n-OCTYL ESTER see TMO250
BORIC ANHYDRIDE see BMG000
BORICIN see SFF000
1-2-BORNANOL see NCQ820
2-BORNANONE see CBA750
(+)-2-BORNANONE see CBB250
d-2-BORNANONE see CBB250
BORNATE see IHZ000
BORNEO CAMPHOR see BMD000
BORNEOL see BMD000
(−)-BORNEOL see NCQ820
BORNEOL (DOT) see BMD000
trans-BORNEOL see BMD000
(1S,2R,4S)-(−)-1-BORNEOL see NCQ820
BORNYL ACETATE see BMD100
l-BORNYL ACETATE see BMD100
BORNYL ALCOHOL see BMD000
1-BORNYL ALCOHOL see NCQ820
BOROETHANE see DDI450
BOROFAX see BMC000

BOROHYDRURE de POTASSIUM (FRENCH) see PKY250
BOROHYDRURE de SODIUM (FRENCH) see SFF500
BOROLIN see PIB900
BORON see BMD500
BORON BROMIDE see BMG400
BORON CHLORIDE see BMG500
BORON COMPOUNDS see BME500
BORON FLUORIDE see BMG700
BORON HYDRIDE see DDI450
BORON OXIDE see BMG000
BORON PHOSPHIDE see BMG250
BORON SESQUIOXIDE see BMG000
BORON TRIBROMIDE see BMG400
BORON TRICHLORIDE see BMG500
BORON TRIFLUORIDE see BMG700
BORON TRIFLUORIDE-ACETIC ACID COMPLEX see BMG750
BORON TRIFLUORIDE-ACETIC ACID COMPLEX (DOT) see BMG750
BORON TRIFLUORIDE-DIMETHYL ETHER see BMH000
BORON TRIFLUORIDE DIMETHYL ETHERATE (DOT) see BMH000
BORON TRIIODIDE see BMH500
BORON TRIOXIDE see BMG000
BORSAURE (GERMAN) see BMC000
BOSAN SUPRA see DAD200
BOSMIN see VGP000
BOURBONAL see EQF000
BOV see SOI500
BOVIDERMOL see DAD200
BOVINOX see TIQ250
BOVIZOLE see TEX000
BOVOFLAVIN see DBX400
BOY see HBT500
BOYGON see PMY300
B(e)P see BCT000
BP 400 see MOO750
BP-7,8-DIHYDRODIOL see DML200
BP-7,8-DIHYDRODIOL-9,10-EPOXIDE (anti) see BCU000
anti-BP-7,8-DIHYDRODIOL-9,10-OXIDE see BCU000
BP-3-HYDROXY see BCX250
BPL see PMT100
BPMC see MOV000
BPPS see SOP000
BR-931 see CLW500
BRACKEN FERN, DRIED see BML000
BRACKEN FERN TOXIC COMPONENT see SCE000
BRADYKININ see BML500
BRADYKININ (SYNTHETIC) see BML500
BRASILAMINA BLACK GN see AQP000
BRASILAMINA BLUE 2B see CMO000
BRASILAMINA BLUE 3B see CMO250
BRASILAN AZO RUBINE 2NS see HJF500
BRASILAN ORANGE 2G see HGC000
BRASILAZINA OIL RED B see SBC500
BRASILAZINA OIL SCARLET 6G see XRA000
BRASILAZINA OIL YELLOW G see PEI000
BRASILAZINA OIL YELLOW R see AIC250
BRASILAZINA ORANGE Y see PEK000
BRASSICOL see PAX000
BRAUNSTEIN (GERMAN) see MAS000
BRAVO see TBQ750
BRAVO 6F see TBQ750
BRAVO-W-75 see TBQ750
BRECOLANE NDG see DJD600
BRELLIN see GEM000
BREMIL see CFY000
BRENOL see ART250
BRENTAMINE FAST BLUE B BASE see DCJ200
BRENTAMINE FAST ORANGE GR BASE see NEO000
BRENTAMINE FAST RED TR BASE see CLK220
BRENTAMINE FAST RED TR SALT see CLK235
BREON see PKQ059
BREON 351 see AAX175

BRESTAN see ABX250
BRESTANOL see CLU000
BRETYLAN see BMV750
BRETYLATE see BMV750
BRETYLIUM-p-TOLUENESULFONATE see BMV750
BRETYLIUM TOSYLATE see BMV750
BRETYLOL see BMV750
BREVIMYTAL see MDU500
BREVINYL see DGP900
BREVIRENIN see VGP000
BREVITAL SODIUM see MDU500
BRIANIL see IPU000
BRICK OIL see CMY825
BRIETAL SODIUM see MDU500
BRIGHT RED see CHP500
BRILLIANT ACRIDINE ORANGE E see BJF000
BRILLIANT BLUE see FMU059
BRILLIANT BLUE FCD No. 1 see FAE000
BRILLIANT BLUE FCF see FAE000
BRILLIANT BLUE R see BMM500
BRILLIANT CRIMSON RED see HJF500
BRILLIANT FAST YELLOW see DOT300
BRILLIANT GREEN SULFATE see BAY750
BRILLIANT OIL ORANGE R see PEJ500
BRILLIANT OIL ORANGE Y BASE see PEK000
BRILLIANT OIL SCARLET B see XRA000
BRILLIANT OIL YELLOW see IBB000
BRILLIANT PINK B see FAG070
BRILLIANT RED see CHP500
BRILLIANT SCARLET see CHP500
BRILLIANT TONER Z see CHP500
BRIMSTONE see SOD500
BRISTACICLIN α see TBX000
BRISTACYCLINE see TBX000, TBX250
BRISTAMIN HYDROCHLORIDE see DTO800
BRITACIL see AIV500
BRITISH ANTILEWISITE see BAD750
BRITISH EAST INDIAN LEMONGRASS OIL see LEG000
BRITON see TIQ250
BRITTEN see TIQ250
BRITTOX see DDP000
BRL see AIV500
BRL 1341 see AIV500
BRL-1621 see SLJ000
BRL 2333 TRIHYDRATE see AOA100
BROBAMATE see MQU750
BROCADISIPAL see OJW000
BROCADOPA see DNA200
BROCASIPAL see OJW000
BROCIDE see EIY600
BROCKMANN, ALUMINUM OXIDE see AHE250
BRODAN see CMA100
BROGDEX 555 see SGM500
BROM (GERMAN) see BMP000
BROMACETOCARBAMIDE see BNK000
BROMACETYLENE see BMS500
BROMACIL see BMM650
BROMADAL see BNK000
BROMADEL see BNK000
BROMADRYL see BMN250
BROMALLYLENE see AFY000
BROMARAL see BNP750
BROMAT see HCQ500
BROMATES see BMN500
BROMATE de SODIUM (FRENCH) see SFG000
BROMAZIL see BMM650
BROMBENZYL CYANIDE see BMW250
BROMCARBAMIDE see BNP750
BROMCHLOPHOS see NAG400
d-2-BROM-DIETHYLAMIDE of LYSERGIC ACID see
  BNM250
BROME (FRENCH) see BMP000
BROMELAIN see BMO000
BROMELAINS see BMO000

BROMELIN see BMO000
BROMETHOL see THV000
BROMEX see DFK600, NAG400
BROMIC ACID, POTASSIUM SALT see PKY300
BROMIC ACID, SODIUM SALT see SFG000
BROMIDES see BMO750
BROMIDE SALT OF POTASSIUM see PKY500
BROMINAL see DDP000
BROMINAL M & PLUS see CIR250
BROMINATED VEGETABLE (SOYBEAN) OIL see BMO825
BROMINE see BMP000
BROMINE, solution (DOT) see BMP000
BROMINE AZIDE see BMP250
BROMINE CYANIDE see COO500
BROMINE DIOXIDE see BMP500
BROMINE FLUORIDE see BMP750
BROMINE PENTAFLUORIDE see BMQ000
BROMINE TRIFLUORIDE see BMQ325
BROMINEX see DDP000
BROMINIL see DDP000
BROMISOVAL see BNP750
BROMISOVALERYLUREA see BNP750
α-BROMISOVALERYLUREA see BNP750
BROMISOVALUM see BNP750
BROMIZOVAL see BNP750
BROMKAL 80 see OAH000
BROMKAL P 67-6HP see TNC500
BROM LSD see BNM250
BROMLYSERGAMIDE see BNM250
2-BROM-d-LYSERGIC ACID DIETHYLAMINE see BNM250
BROM-METHAN (GERMAN) see MHR200
BROMO (ITALIAN) see BMP000
α-BROMOACETIC ACID see BMR750
BROMOACETIC ACID, solid (DOT) see BMR750
BROMOACETIC ACID, solution (DOT) see BMR750
BROMOACETIC ACID ETHYLENE ESTER see BHD250
BROMOACETIC ACID, ETHYL ESTER see EGV000
BROMOACETIC ACID METHYL ESTER see MHR250
BROMOACETONE see BNZ000
BROMOACETONE (DOT) see BNZ000
BROMOACETONE, liquid (DOT) see BNZ000
BROMOACETYLENE see BMS500
BROMO ACID see BNH500
5-(2-BROMOALLYL)-5-sec-BUTYLBARBITURIC ACID see
  BOR000
γ-BROMOALLYLENE see PMN500
5-(2'-BROMOALLYL)-5-(1'-METHYL-N-PROPYL)BARBI-
  TURIC ACID see BOR000
BROMOAZIDE see BMP250
BROMOBENZENE (DOT) see PEO500
4-BROMOBENZENEACETONITRILE see BNV750
β-(p-BROMOBENZHYDRYLOXY)ETHYLDIMETHYLAMINE
  HYDROCHLORIDE see BNW500
2-(4-BROMOBENZOHYDRYLOXY)ETHYLDIMETHYL-
  AMINE HYDROCHLORIDE see BNW500
4-BROMOBENZYLCYANIDE see BNV750
p-BROMOBENZYL CYANIDE see BNV750
α-BROMOBENZYL CYANIDE see BMW250
(o-BROMOBENZYL)ETHYLDIMETHYLAMMONIUM-p-TO-
  LUENESULFONATE see BMV750
BROMOBENZYLNITRILE see BMW250
α-BROMOBENZYLNITRILE see BMW250
1-BROMOBUTANE see BMX500
2-BROMOBUTANE see BMX750
5-BROMO-3-sec-BUTYL-6-METHYLURACIL see
  BMM650
BROMOCARBAMIDE see BNP750
BROMOCHLORODIFLUOROMETHANE see BNA250
3-BROMO-N-(2-CHLOROMERCURICYCLOHEXYL)PRO-
  PIONAMIDE see CET000
BROMOCHLOROMETHANE see CES650
O-(4-BROMO-2-CHLOROPHENYL)-O-ETHYL-S-PROPYL
  PHOSPHOROTHIOATE see BNA750

1-BROMO-3-CHLOROPROPANE see BNA825
BROMOCHLOROTRIFLUOROETHANE see HAG500
2-BROMO-2-CHLORO-1,1,1-TRIFLUOROETHANE see HAG500
BROMOCYAN see COO500
BROMOCYANOGEN see COO500
BROMODEOXYURIDINE see BNC750
5-BROMODEOXYURIDINE see BNC750
5-BROMO-2-DEOXYURIDINE see BNC750
5-BROMO-2'-DEOXY URIDINE see BNC750
5-BROMODESOXYURIDINE see BNC750
BROMODICHLOROMETHANE see BND500
4-BROMO-2,5-DICHLOROPHENOL-o-ESTER with O,O-DI-ETHYL PHOSPHOROTHIOATE see EGV500
O-(4-BROMO-2,5-DICHLOROPHENYL)-O,O-DIETHYL PHOS-PHOROTHIOATE see EGV500
O-(4-BROMO-2,5 DICHLOROPHENYL)-O,O-DIETHYLPHOS-PHOROTHIONATE see EGV500
O-(4-BROMO-2,5-DICHLOROPHENYL)-O-METHYL PHE-NYLPHOSPHONOTHIOATE see LEN000
2-BROMO-9,10-DIDEHYDRO-N,N-DIETHYL-6-METHYL-ERGOLINE-8-β-CARBOXAMIDE see BNM250
BROMODIETHYLACETYLCARBAMIDE see BNK000
BROMODIETHYLACETYLUREA see BNK000
α-BROMO-β-DIMETHYLPROPANOYLUREA see BNP750
2-(1-(4-BROMODIPHENYL)ETHOXY)-N,N-DIMETHYL-ETHYLAMINE HYDROCHLORIDE see BMN250
BROMODIPHENYLMETHANE see BNG750
BROMOEOSINE see BNH500
3-BROMO-1,2-EPOXYPROPANE see BNI000
BROMOETHANE see EGV400
BROMOETHANIOC ACID see BMR750
α-BROMOETHANIOC ACID see BMR750
BROMOETHANOL see BNI500
2-BROMO ETHANOL see BNI500
BROMOETHENE see VMP000
2-BROMO-2-ETHYLBUTYRLUREA see BNK000
(α-BROMO-α-ETHYLBUTYRYL)CARBAMIDE see BNK000
1-BROMO-ETHYL-BUTYRYL-UREA see BNK000
2-BROMO-2-ETHYLBUTYRYLUREA see BNK000
(α-BROMO-α-ETHYLBUTYRYL)UREA see BNK000
2-BROMO-N-ETHYL-N,N-DIMETHYLBENZENEMETHAN-AMINIUM 4-METHYLBENZENESULFONATE see BMV750
BROMOETHYLENE see VMP000
BROMOETHYLENE POLYMER see PKQ000
2-BROMO ETHYL ETHYL ETHER see BNK250
BROMOETHYNE see BMS500
BROMOFLOR see CDS125
BROMOFLUORESCEIC ACID see BNH500
BROMO FLUORESCEIN see BNH500
BROMOFLUOROFORM see TJY100
BROMOFORM see BNL000
BROMOFORME (FRENCH) see BNL000
BROMOFORMIO (ITALIAN) see BNL000
BROMOFOS-ETHYL see EGV500
BROMOFUME see EIY500
BROMO-O-GAS see MHR200
2-BROMOISOBUTANE see BQM250
α-BROMOISOVALERIC ACID UREIDE see BNP750
α-BROMOISOVALEROYLUREA see BNP750
(α-BROMOISOVALERYL)UREA see BNP750
2-BROMO-d-LYSERGIC ACID DIETHYLAMIDE see BNM250
BROMOLYSERGIDE see BNM250
BROMOMETANO (ITALIAN) see MHR200
BROMO METHANE see MHR200
7-BROMO METHYL BENZ(a)ANTHRACENE see BNO750
(BROMOMETHYL)BENZENE see BEC000
p-BROMO-α-METHYLBENZHYDRYL-2-DIMETHYLAMINO-ETHYL ETHER HYDROCHLORIDE see BMN250
1-BROMO-3-METHYL BUTANE see BNP250
2-BROMO-3-METHYLBUTYRYLUREA see BNP750
BROMOMETHYL METHYL KETONE see BNZ000

5-BROMO-6-METHYL-3-(1-METHYLPROPYL)-2,4(1H,3H)-PYRIMIDINEDIONE see BMM650
5-BROMO-6-METHYL-3-(1-METHYLPROPYL)URACIL see BMM650
p-(BROMOMETHYL)NITROBENZENE see BEC000
2-((p-BROMO-α-METHYL-α-PHENYLBENZYL)OXY)-N,N-DI-METHYLETHYLAMINE HYDROCHLORIDE see BMN250
1-BROMO-2-METHYL PROPANE see BNR750
2-BROMO-2-METHYLPROPANE (DOT) see BQM250
8-β-((5-BROMONICOTINOYLOXY)METHYL)-1,6-DI-METHYL-10-α-METHOXYERGOLINE see NDM000
2-BROMO-2-NITROPANE-1,3-DIOL see BNT250
2-BROMO-2-NITROPROPAN-1,3-DIOL see BNT250
2-BROMO-2-NITRO-1,3-PROPANEDIOL see BNT250
β-BROMO-β-NITROTRIMETHYLENEGLYCOL see BNT250
2-BROMOPENTANE see BNU500
BROMO PHENOLS see BNV250
4-BROMOPHENYLACETONITRILE see BNV750
p-BROMOPHENYLACETONITRILE see BNV750
α-BROMOPHENYLACETONITRILE see BMW250
2-(4-BROMOPHENYL)ACETONITRILE see BNV750
BROMO PHENYL HYDRAMINE HYDROCHLORIDE see BNW500
BROMOPHENYLMETHANE see BEC000
1-(p-BROMOPHENYL)-1-PHENYL-1-(2-DIMETHYLAMINO-ETHOXY)ETHANE HYDROCHLORIDE see BMN250
2-(1-(4-BROMOPHENYL)-1-PHENYLETHOXY)-N,N-DI-METHYLETHANAMINE HYDROCHLORIDE see BMN250
(2-(1-p-BROMOPHENYL-1-PHENYLETHOXY)ETHYL)DIMETHYLETHYLAMINE HYDROCHLORIDE see BMN250
BROMOPHOSETHYL see EGV500
BROMOPICRIN see NMQ000
1-BROMOPROPANE see BNX750
1-BROMOPROPANE (DOT) see BNX750
BROMO-2-PROPANONE see BNZ000
1-BROMO-2-PROPANONE see BNZ000
3-BROMOPROPENE see AFY000
3-BROMOPROPIONIC ACID see BOB250
β-BROMOPROPIONIC ACID see BOB250
3-BROMOPROPYL CHLORIDE see BNA825
3-BROMOPROPYLENE see AFY000
3-BROMO-1-PROPYNE see PMN500
3-BROMOPROPYNE (DOT) see PMN500
2-(6-(5-BROMO-2-PYRIDYL OXY)HEXYL)AMINOETHANE THIOL HYDROCHLORIDE see BOD500
1-BROMO-2,5-PYRROLIDINEDIONE see BOF500
BROMO SELTZER see ABG750
BROMO SILANE see BOE750
N-BROMOSUCCIMIDE see BOF500
N-BROMO SUCCINIMIDE see BOF500
α-BROMOTOLUENE (DOT) see BEC000
φ-BROMOTOLUENE see BEC000
α-BROMO-α-TOLUNITRILE see BMW250
BROMOTRICHLOROMETHANE see BOH750
BROMOTRIFLUOROETHENE see BOJ000
BROMO TRIFLUOROETHYLENE see BOJ000
BROMOTRIFLUOROMETHANE see TJY100
BROMOURACIL DEOXYRIBOSIDE see BNC750
5-BROMOURACIL DEOXYRIBOSIDE see BNC750
5-BROMOURACIL-2-DEOXYRIBOSIDE see BNC750
BROMOVAL see BNP750
BROMOVALEROCARBAMIDE see BNP750
BROMOVALERYLUREA see BNP750
BROMOWODOR (POLISH) see HHJ000
BROMOXIL see BNP750
BROMOXYNIL see DDP000
BROMURAL see BNP750
BROMURE de CYANOGEN (FRENCH) see COO500
BROMURE d'ETHYLE see EGV400
BROMURE de METHYLE (FRENCH) see MHR200
BROMURE de VINYLE (FRENCH) see VMP000
BROMURE de XYLYLE (FRENCH) see XRS000
BROMURO di ETILE (ITALIAN) see EIY500

BROMURO di METILE (ITALIAN) see MHR200
BROMUVAN see BNP750
BROMVALERYLUREA see BNP750
BROMVALETONE see BNP750
BROMVALETONUM see BNP750
BROMVALUREA see BNP750
BROMWASSERSTOFF (GERMAN) see HHJ000
BROMYL see BNP750
BRONCHIOCAIN see BQH250
BRONCHOCAIN see BQH250
BRONCHOCAINE see BQH250
BRONCHODIL see DNA600
BRONCHOLYSIN see ACH000
BRONCHOSPASMIN see DNA600
BRONCOVALEAS see BQF500
BRONKAID MIST see VGP000
BRONKEPHRINE see DMV600
BRONKEPHRINE HYDROCHLORIDE see ENX500
BRONOCOT see BNT250
BRONOPOL see BNT250
BRONOSOL see BNT250
BRONZE BROMO see BNH500
BRONZE POWDER see CNI000
BRONZE RED RO see CHP500
BRONZE SCARLET see CHP500
BROOM (DUTCH) see BMP000
BROOMMETHAAN (DUTCH) see MHR200
BROOMWATERSTOF (DUTCH) see HHJ000
BROVALIN see BNP750
BROVALUREA see BNP750
BROVARIN see BNP750
11460 BROWN see XMA000
BROWN ACETATE see CAL750
BROWN HEMATITE see LFW000
BROWN IRON ORE see LFW000
BROWN IRONSTONE CLAY see LFW000
BROXURIDINE see BNC750
BROXYNIL see DDP000
BRS 640 see BML500
BRUCIN (GERMAN) see BOL750
BRUCINA (ITALIAN) see BOL750
BRUCINE see BOL750, BOL750
BRUCINE, solid (DOT) see BOL750
BRUCINE ALKALOID see BOL750
BRUCINE IODOMETHYLATE see BOM000
BRUCINE IODOMETHYLE (FRENCH) see BOM000
BRUCINE METHIODIDE see BOM000
BRUDR see BNC750
BRUFANEUXOL see DOT000
BRUINSTEEN (DUTCH) see MAS000
BRUMIN see WAT200
BRUSH BUSTER see MEL500
BRUSH-OFF 445 MLD VOLATILE BRUSH KILLER see
    TAA100
BRUSH RHAP see TAA100
BRUSHTOX see TAA100
BS 5930 see OJW000
BSC-REFINE D see BBS750
B-SELEKTONON M see CIR250
BT see BPU000
BTKH see BEU500
BTO see BLL750
B.T.Z. see BRF500
BU2AE see DDU600
BUCACID AZURE BLUE see FMU059
BUCACID FAST ORANGE G see HGC000
BUCACID INDIGOTINE B see FAE100
BUCACID TARTRAZINE see FAG140
BUCARBAN see BSM000
BUCCALSONE see HHR000
BUCROL see BSM000
BUCS see BPJ850
BUCTRIL see DDP000
BUCTRIL INDUSTRIAL see DDP000

BUD-NIP see CKC000
BUDOFORM see CHR500
BUDORM see BPF500
BUDR see BNC750
5-BUDR see BNC750
BUENO see MRL750
BUFEN see ABU500
BUFENCARB see BTA250
BUFEXAMIC ACID see BPP750
BUFF-A-COMP see ABG750
BUFON see DKA600
BUFONAMIN see BOM750, BQL000
BUFOPTO ZINC SULFATE see ZNA000
BUFORMIN HYDROCHLORIDE see BOM750, BQL000
BUFOTALIN see BON000
BUFOTALINE see BON000
BUFOTENIN see DPG109
BUHACH see POO250
BUKARBAN see BSM000
BUKS see BFW750
BULBONIN see BQL000
BULPUR see PLC250
bu-MDI see BRC500
BUNSENITE see NDF500
BUNT-CURE see HCC500
BUNT-NO-MORE see HCC500
BUPHENINE HYDROCHLORIDE see DNU200
BUPICAINE HYDROCHLORIDE (+) see BON750
BUPIVACAINE see BSI250
dl-BUPIVACAINE see BSI250
BURCOL see BSM000
BUREX (CZECH) see PEE750
BURITAL SODIUM see SOX500
BURNISH GOLD see GIS000
BURNOL see DBX400
BURNTISLAND RED see IHD000
BURNT LIME see CAU500
BURNT SIENNA see IHD000
BURNT UMBER see IHD000
BUROFLAVIN see DBX400
BURONIL see FKI000
BURTONITE 44 see FPQ000
BURTONITE V-7-E see GLU000
BURTONITE-V-40-E see CCL250
BUSONE see BRF500
BUSTREN see SMQ500
BUTABARB see BPF000
BUTABARBITAL see BPF000
BUTABARBITONE see BPF000
BUTACAINE see BOO750
BUTACAINE SULFATE see BOP000
BUTACARB see DEG400
BUTACARBE (FRENCH) see DEG400
BUTACIDE see PIX250
BUTACOMPREN see BRF500
BUTACOTE see BRF500
BUTADIEEN (DUTCH) see BOP500
BUTA-1,3-DIEEN (DUTCH) see BOP500
BUTADIEN (POLISH) see BOP500
BUTA-1,3-DIEN (GERMAN) see BOP500
BUTADIENDIOXYD (GERMAN) see BGA750
1,3-BUTADIENE see BOP500
BUTA-1,3-DIENE see BOP500
α-γ-BUTADIENE see BOP500
BUTADIENE DIEPOXIDE see BGA750
l-BUTADIENE DIEPOXIDE see BOP750
1,3-BUTADIENE DIEPOXIDE see BGA750
BUTADIENE DIMER see CPD750
BUTADIENE DIOXIDE see BGA750
dl-BUTADIENE DIOXIDE see DHB600
BUTADIENE MONOXIDE see EBJ500
1,3-BUTADIENE-STYRENE COPOLYMER see SMR000
BUTADIENE-STYRENE POLYMER see SMR000
1,3-BUTADIENE-STYRENE POLYMER see SMR000

BUTADIENE-STYRENE RESIN see SMR000
BUTADIENE-STYRENE RUBBER (FCC) see SMR000
BUTADION see BRF500
BUTADIONA see BRF500
BUTAFLOGIN see HNI500
BUTAFUME see BPY000
BUTAGESIC see BRF500
BUTAKON 85-71 see SMR000
BUTAL see BSU250
BUTALAN see BRF500
BUTALBITAL SODIUM see BOQ750
BUTALDEHYDE see BSU250
BUTALGINA see BRF500
BUTALIDON see BRF500
BUTALLYLONAL see BOR000
BUTALLYLONAL SODIUM see BOR250
BUTALUY see BRF500
BUTALYDE see BSU250
BUTAMID see BSQ000
BUTAMIN see AIT750
BUTANAL see BSU250
n-BUTANAL (CZECH) see BSU250
BUTANAL OXIME see BSU500
1-BUTANAMINE see BPX750
2-BUTANAMINE see BPY000
1,3-BUTANDIOL (GERMAN) see BOS500
BUTANE see BOR500
n-BUTANE (DOT) see BOR500
BUTANECARBOXYLIC ACID see VAQ000
1-BUTANECARBOXYLIC ACID see VAQ000
1,3-BUTANEDIAMINE see BOR750
1,4-BUTANEDIAMINE, 2-METHYL-, POLYMER with α-HY-
   DRO-omega-HYDROXYPOLY(OXY-1,4-BUTANEDIYL) and
   1,1'-METHYLENEBIS(4-ISOCYANATOCYCLOHEXANE)
   see PKN500
1,4-BUTANEDICARBOXAMIDE see AEN000
1,4-BUTANEDICARBOXYLIC ACID see AEN250
BUTANE DIEPOXIDE see BGA750
1,4-BUTANEDINITRILE see SNE000
BUTANEDIOIC ACID see SMY000
BUTANEDIOIC ACID, DIETHYL ESTER see SNB000
BUTANEDIOIC ACID MONO(2,2-DIMETHYLHYDRAZIDE)
   see DQD400
BUTANEDIOIC ANHYDRIDE see SNC000
1,2-BUTANEDIOL see BOS250
1,3-BUTANEDIOL see BOS500
BUTANE-1,3-DIOL see BOS500
1,4-BUTANEDIOL see BOS750
BUTANE-1,4-DIOL see BOS750
2,3-BUTANEDIOL see BOT000
1,3-BUTANEDIOL, CYCLIC SULFITE see MQG500
1,4-BUTANEDIOL DIMETHANESULPHONATE see BOT250
1,4-BUTANEDIOL DIMETHYL SULFONATE see BOT250
1,4-BUTANEDIOL POLYMER with 1,1'-METHYLENEBIS(4-
   ISOCYANATOBENZENE) see PKP000
2,3-BUTANEDIONE see BOT500
BUTANEFRINE HYDROCHLORIDE see ENX500
BUTANEN (DUTCH) see BOR500
BUTANENITRILE see BSX250
n-BUTANENITRILE see BSX250
BUTANESULFONE see BOU250
BUTANE SULTONE see BOU250
Δ-BUTANE SULTONE see BOU250
1,4-BUTANESULTONE (MAK) see BOU250
5H,6H-6,5A,13A,14-(1,2,3,4)BUTANETETRAYCYCLO-
   OCTA(1,2-B:5,6-B')DINAPHTHALENE see LIV000
BUTANETHIOL see BRR900
n-BUTANETHIOL see BRR900
tert-BUTANETHIOL see MOS000
BUTANI (ITALIAN) see BOR500
BUTANOIC ACID see BSW000
BUTANOIC ACID-2-BUTOXY-1-METHYL-2-OXOETHYL ES-
   TER (9CI) see BQP000

BUTANOIC ACID ETHYL ESTER see EHE000
BUTANOIC ACID, 1,2,3-PROPANETRIYL ESTER see
   TIG750
BUTANOIC ACID 2,2,2-TRICHLORO-1-(DIMETHOXYPHOS-
   PHINYL)ETHYL ESTER see BPG000
1-BUTANOL see BPW500
BUTAN-1-OL see BPW500
BUTAN-2-OL see BPW750
2-BUTANOL see BPW750
n-BUTANOL see BPW500
BUTANOL (DOT) see BPW500
tert-BUTANOL see BPX000
BUTANOL (FRENCH) see BPW500
sec-BUTANOL (DOT) see BPW750
2-BUTANOL ACETATE see BPV000
3-BUTANOLAL see AAH750
BUTANOL-2-AMINE see AJA250
BUTANOL (4)-BUTYL-NITROSAMINE see HJQ350
BUTANOLEN (DUTCH) see BPW500
4-BUTANOLIDE see BOV000
BUTANOLO (ITALIAN) see BPW500
2-BUTANOL-3-ONE see ABB500
BUTANOL SECONDAIRE (FRENCH) see BPW750
BUTANOL TERTIAIRE (FRENCH) see BPX000
BUTANONE 2 (FRENCH) see MKA400
2-BUTANONE (OSHA) see MKA400
2-BUTANONE, SEMICARBAZONE see MKA750
BUTANOVA see HNI500
BUTAPHEN see BRF500
BUTAPHENE see BRE500
BUTAPIRAZOL see BRF500
BUTAPIRONE see HNI500
BUTAPYRAZOLE see BRF500
BUTARECBON see BRF500
BUTARTRIL see BRF500
BUTARTRINA see BRF500
BUTATAB see BPF000
BUTATAL see BPF000
BUTAZATE see BIX000
BUTAZATE 50-D see BIX000
BUTAZINA see BRF500
BUTAZOLIDIN see BRF500
BUTAZOLIDINE SODIUM see BOV750
BUTAZONA see BRF500
BUTAZONE see BRF500
BUTE see BRF500
BUTELLINE see BOP000
2-BUTENAL see COB250
(E)-2-BUTENAL see COB260
trans-2-BUTENAL see COB260
1-BUTENE see BOW250
cis-2-BUTENE see BOW500
trans-2-BUTENE see BOW750
(E)-BUTENEDIOIC ACID see FOU000
(Z)-BUTENEDIOIC ACID see MAK900
cis-BUTENEDIOIC ACID see MAK900
trans-BUTENEDIOIC ACID see FOU000
2-BUTENEDIOIC ACID BIS(2-ETHYLHEXYL) ESTER see
   DVK600
2-BUTENEDIOIC ACID, DIBUTYL ESTER see DED600
cis-BUTENEDIOIC ANHYDRIDE see MAM000
3-BUTENE-2-ONE see BOY500
1-BUTENE, 2,3,4-TRICHLORO- see TIL360
2-BUTENOIC ACID see COB500
α-BUTENOIC ACID see COB500
2-BUTENOL see BOY000
2-BUTEN-1-OL see BOY000
3-BUTENO-β-LACTONE see KFA000
3-BUTEN-2-ONE see BOY500
2-BUTENYL ALCOHOL see BOY000
(2-BUTENYLIDENE)ACETIC ACID see SKU000
BUTEN-3-YNE see BPE109
BUTETHAL see BPF500
BUTETHAMINE HYDROCHLORIDE see IAC000

BUTICAPS see BPF000
BUTIDIONA see BRF500
BUTIFOS see BSH250
n-BUTILAMINA (ITALIAN) see BPX750
BUTILCHLOROFOS see BPG000, DDP000
BUTILE (ACETATI di) (ITALIAN) see BPU750
BUTILENE see HNI500
BUTIL METACRILATO (ITALIAN) see MHU750
BUTINOX see BLL750
BUTIPHOS see BSH250
BUTISOL see BPF000
BUTISULFINA see BSM000
BUTIWAS-SIMPLE see BRF500
BUTOBARBITAL see BPF500
BUTOBARBITONE see BPF500
BUTOBARBITURAL see BPF500
BUTOBEN see DTC800
BUTOCIDE see PIX250
BUTOKSYETYLOWY ALKOHOL (POLISH) see BPJ850
BUTONATE see BPG000
BUTONE see BRF500
BUTOPHEN see BPG250
BUTOPYRONOXYL see BRT000
2-BUTOSSI-ETANOLO (ITALIAN) see BPJ850
BUTOXIDE see PIX250
2-BUTOXY-AETHANOL (GERMAN) see BPJ850
1-(2-(4-BUTOXYBENZOYL)ETHYL)PIPERIDINE HYDRO-
   CHLORIDE see BPR500
1-BUTOXYBUTANE see BRH750
tert-BUTOXY CARBONYL AZIDE see BQI250
BUTOXYCINCHONINIC ACID DIETHYLETHYLENEDIAM-
   IDE HYDROCHLORIDE see NOF500
2-BUTOXY-N-(2-(DIETHYLAMINO)ETHYL)CINCHONIN-
   AMIDE see DDT200
2-BUTOXY-N-(β-DIETHYLAMINOETHYL)CINCHONIN-
   AMIDE see DDT200
2-BUTOXY-N-(2-DIETHYLAMINOETHYL)CINCHONIN-
   AMIDE HYDROCHLORIDE see NOF500
2-N-BUTOXY-N-(2-DIETHYLAMINOETHYL)CINCHONIN-
   AMIDE HYDROCHLORIDE see NOF500
2-BUTOXY-N-(2-DIETHYLAMINOETHYL)CINCHONINIC
   ACID AMIDE HYDROCHLORIDE see NOF500
BUTOXYETHANOL see BPJ850
2-BUTOXYETHANOL see BPJ850
n-BUTOXYETHANOL see BPJ850
2-BUTOXY-1-ETHANOL see BPJ850
2-BUTOXYETHANOL ACETATE see BPM000
2-BUTOXYETHANOL PHOSPHATE see BPK250
BUTOXYETHENE see VMZ000
2-(2-BUTOXYETHOXY)ETHANOL ACETATE see BQP500
α-(2-(2-BUTOXYETHOXY)ETHOXY)-4,5-METHYLENEDI-
   OXY-2-PROPYLTOLUENE see PIX250
α-(2-(2-n-BUTOXYETHOXY)-ETHOXY)-4,5-METHYLENEDI-
   OXY-2-PROPYLTOLUENE see PIX250
5-((2-(2-BUTOXYETHOXY)ETHOXY)METHYL)-6-PROPYL-
   1,3-BENZODIOXOLE see PIX250
2-(2-BUTOXYETHOXY)ETHYL ACETATE see BQP500
2-(2-BUTOXY ETHOXY)ETHYL THIOCYANATE see
   BPL250
2-(2-(BUTOXY)ETHOXY)ETHYL THIOCYANIC ACID ESTER
   see BPL250
1-BUTOXY ETHOXY-2-PROPANOL see BPL500
1-(2-BUTOXYETHOXY)-2-PROPANOL see BPL500
2-BUTOXYETHYL ACETATE see BPM000
2-BUTOXYETHYL ESTER ACETIC ACID see BPM000
BUTOXYL see MHV750
BUTOXYPHENYL see BSF750
4-BUTOXYPHENYLACETOHYDROXAMIC ACID see
   BPP750
p-BUTOXY PHENYL ACETOHYDROXAMIC ACID see
   BPP750
4'-BUTOXY-3-PIPERIDINO PROPIOPHENONE HYDRO-
   CHLORIDE see BPR500

4-n-BUTOXY-β-(1-PIPERIDYL)PROPIOPHENONE HYDRO-
   CHLORIDE see BPR500
2-BUTOXYQUINOLINE-4-CARBOXYLIC ACID DIETHYL-
   AMINOETHYLAMIDE see DDT200
BUTOXYRHODANODIETHYL ETHER see BPL250
1-BUTOXY-2-(2-THIOCYANATOETHYXY)ENTHANE see
   BPL250
2-BUTOXY-2'-THIOCYANODIETHYL ETHER see BPL250
β-BUTOXY-β'-THIOCYANODIETHYL ETHER see BPL250
1-BUTOXY-2-(2-THIOCYANOETHOXY)ETHANE see
   BPL250
BUTOZ see BRF500
BUTRATE see BPF000
BUTRIZOL see BPU000
BUTTER of ANTIMONY see AQC500, AQD000
BUTTERCUP YELLOW see CMK500, ZFJ100, PLW500
BUTTERSAEURE (GERMAN) see BSW000
BUTTER YELLOW see AIC250, DOT300
BUTTER of ZINC see ZFA000
BUTURON see CKF750
BUTYLACETAT (GERMAN) see BPU750
BUTYL ACETATE see BPU750
1-BUTYL ACETATE see BPU750
2-BUTYL ACETATE see BPV000
n-BUTYL ACETATE see BPU750
sec-BUTYL ACETATE see BPV000
tert-BUTYL ACETATE see BPV100
BUTYLACETATEN (DUTCH) see BPU750
BUTYLACETIC ACID see HEU000
BUTYL ACETOACETATE see BPV250
BUTYL ACETOXYMETHYLNITROSAMINE see BRX500
N-BUTYL-N-(ACETOXYMETHYL)NITROSAMINE see
   BRX500
n-BUTYL ACID PHOSPHATE see ADF250
BUTYL ACRYLATE see BPW100
n-BUTYL ACRYLATE see BPW100
BUTYLACRYLATE, INHIBITED (DOT) see BPW100
2-BUTYL ALCOHOL see BPW750
n-BUTYL ALCOHOL see BPW500
BUTYL ALCOHOL (DOT) see BPW500
sec-BUTYL ALCOHOL see BPW750
tert-BUTYL ALCOHOL see BPX000
sec-BUTYL ALCOHOL ACETATE see BPV000
n-BUTYL ALDEHYDE see BSU250
BUTYLALYLONAL see BOR000
n-BUTYLAMIN (GERMAN) see BPX750
n-BUTYLAMINE see BPX750
sec-BUTYLAMINE see BPY000
tert-BUTYLAMINE see BPY250
BUTYLAMINE, tertiary see BPY250
p-(BUTYLAMINO)BENZOIC ACID-2-(DIMETHYLAMINO)
   ETHYL ESTER see BQA010
p-BUTYLAMINOBENZOYL-2-DIMETHYLAMINOETHANOL
   see BQA010
N-((BUTYLAMINO)CARBONYL)-4-METHYLBENZENESUL-
   FONAMIDE see BSQ000
1-(BUTYLAMINO)CYCLOHEXYLPHOSPHONIC ACID DI-
   BUTYL ESTER see ALZ000
2-BUTYLAMINOETHANOL see BQC000
tert-BUTYL AMINO ETHYL METHACRYLATE see
   BQD250
2-(tert-BUTYLAMINO)ETHYL METHACRYLATE see
   BQD250
2-(tert-BUTYLAMINO)-1-(4-HYDROXY-3-HYDROXY-
   METHYLPHENYL)ETHANOL see BQF500
α-1-((tert-BUTYLAMINO)METHYL)-4-HYDROXY-m-
   XYLENE-α,α-DIOL see BQF500
α'-((tert-BUTYL AMINO)METHYL)-4-HYDROXY-m-
   XYLENE-α,α'-DIOL see BQF500
4-(BUTYLAMINO)SALICYLIC ACID 2-(DIETHYLAMINO)
   ETHYL ESTER HYDROCHLORIDE see BQH250
p-BUTYLAMINO SALICYLIC ACID-2-(DIETHYLAMINO)
   ETHYL ESTER HYDROCHLORIDE see BQH250

4-(BUTYLAMINO)-SALICYLIC ACID 2-(DIETHYLAMINO) ETHYL ESTER MONOHYDROCHLORIDE see BQH250
BUTYLAMYLNITROSAMIN (GERMAN) see BRY250
N-BUTYLANILINE see BQH850
N-(n-BUTYL)ANILINE see BQH850
N-n-BUTYLANILINE (DOT) see BQH850
BUTYLATED HYDROXYANISOLE see BQI000
BUTYLATED HYDROXYTOLUENE see BFW750
BUTYLATE-2,4,5-T see BSQ750
tert-BUTYL AZIDO FORMATE see BQI250
N-BUTYLBENZENAMINE (9CI) see BQH850
n-BUTYLBENZENE see BQI750
sec-BUTYLBENZENE see BQJ000
tert-BUTYLBENZENE see BQJ250
5-BUTYL-2-BENZIMIDAZOLECARBAMIC ACID METHYL ESTER see BQK000
N-(BUTYL-5-BENZIMIDAZOLYL)-2-CARBAMATE de METHYLE (FRENCH) see BQK000
(4-BUTYL-1H-BENZIMIDAZOL-2-YL)-CARBAMIC ACID METHYL ESTER see BQK000
BUTYL BENZOATE see BQK250
n-BUTYL BENZOATE see BQK250
p-tert-BUTYL BENZOIC ACID see BQK500
BUTYL BENZYL PHTHALATE see BEC500
n-BUTYL BENZYL PHTHALATE see BEC500
1-BUTYLBIGUANIDE HYDROCHLORIDE see BQL000
N-BUTYLBIGUANIDE HYDROCHLORIDE see BQL000
N-BUTYL-N,N-BIS(HYDROXY ETHYL)AMINE see BQM000
BUTYL BORATE see THX750
n-BUTYL BORATE see THX750
sec-BUTYL-BROM-ALLYL BARBITURIC ACID SODIUM SALT see BOR250
1-BUTYL BROMIDE see BNR750
N-BUTYL BROMIDE see BMX500
i-BUTYL BROMIDE see BNR750
BUTYL BROMIDE (DOT) see BMX500
sec-BUTYL BROMIDE see BMX750
tert-BUTYL BROMIDE see BQM250
BUTYL BROMIDE, NORMAL (DOT) see BMX500
3-sek.BUTYL-5-BROM-6-METHYLURACIL (GERMAN) see BMM650
5-sec-BUTYL-5-(β-BROMOALLYL)BARBITURIC ACID see BOR000
N-BUTYL-1-BUTANAMINE see DDT800
n-BUTYL n-BUTANOATE see BQM500
BUTYL-BUTANOL(4)-NITROSAMIN see HJQ350
BUTYL-BUTANOL-NITROSAMINE see HJQ350
n-BUTYL BUTYRATE see BQM500
n-BUTYL n-BUTYRATE see BQM500
BUTYL BUTYRATE (FCC) see BQM500
γ-n-BUTYL-γ-BUTYROLACTONE see OCE000
BUTYL BUTYROLLACTATE see BQP000
BUTYL BUTYRYL LACTATE see BQP000
BUTYL CARBAMATE see BQP250
1-(BUTYLCARBAMOYL)-2-BENZIMIDAZOLECARBAMIC ACID, METHYL ESTER see BAV575
1-(BUTYLCARBAMOYL)-2-BENZIMIDAZOL-METHYLCAR-BAMAT (GERMAN) see BAV575
1-(N-BUTYLCARBAMOYL)-2-(METHOXY-CARBOXAMIDO)-BENZIMIDAZOL (GERMAN) see BAV575
N'-(BUTYLCARBAMOYL)SULFANILAMIDE see BSM000
N¹-(BUTYLCARBAMOYL)SULFANILAMIDE see BSM000
N-BUTYLCARBINOL see AOE000
dl-sec-BUTYLCARBINOL see MHS750
BUTYL CARBITOL ACETATE see BQP500
BUTYL CARBITOL 6-PROPYLPIPERONYL ETHER see PIX250
BUTYL CARBITOL RHODANATE see BPL250
BUTYL CARBITOL THIOCYANATE see BPL250
BUTYL-CARBITYL (6-PROPYLPIPERONYL) ETHER see PIX250
BUTYL CARBOBUTOXYMETHYL PHTHALATE see BQP750

5-BUTYL-2-(CARBOMETHOXYAMINO)BENZIMIDAZOLE see BQK000
1-((tert-BUTYLCARBONYL-4-CHLOROPHENOXY)METHYL)-1H-1,2,4-TRIAZOLE see CJO250
N-BUTYL-(3-CARBOXY PROPYL)NITROSAMINE see BQQ250
4-tert-BUTYLCATECHOL see BSK000
BUTYL CELLOSOLVE see BPJ850
BUTYL CELLOSOLVE ACETATE see BPM000
o-(4-tert BUTYL-2-CHLOOR-FENYL)-o-METHYL-FOS-FORZUUR-N-METHYL-AMIDE (DUTCH) see COD850
BUTYL CHLORIDE (DOT) see BQQ750
n-BUTYL CHLORIDE see BQQ750
sec-BUTYL CHLORIDE see CEU250
tert-BUTYL CHLORIDE see BQR000
4-tert-BUTYL-2-CHLORO PHENYL METHYL METHYL PHOSPHORAMIDATE see COD850
4-tert. BUTYL 2-CHLOROPHENYL METHYLPHOSPHORAMI-DATE de METHYLE (FRENCH) see COD850
o-(4-tert-BUTYL-2-CHLOR-PHENYL)-o-METHYL-PHOSPHOR-SAEURE-N-METHYL AMID (GERMAN) see COD850
tert-BUTYL CHROMATE see BQV000
2-tert-BUTYL-p-CRESOL see BQV750
N-BUTYL CYCLOHEXYL AMINE see BQW750
N-(4-tert-BUTYL CYCLOHEXYL)-3,3-DIPHENYL PROPYL-AMINE HYDROCHLORIDE see BQX000
BUTYL 2,4-D see BQZ000
BUTYL DICHLOROPHENOXYACETATE see BQZ000
BUTYL (2,4-DICHLOROPHENOXY)ACETATE see BQZ000
N-BUTYLDIETHANOLAMINE see BQM000
1-BUTYLDIGUANIDE HYDROCHLORIDE see BQL000
BUTYL-3,4-DIHYDRO-2,2-DIMETHYL-4-OXO-2H-PYRAN-6-CARBOXYLATE see BRT000
3-BUTYL-1-(2-(DIMETHYLAMINO)ETHOXY)ISOQUINOLINE HYDROCHLORIDE see DNX400
6-BUTYL-5-DIMETHYLAMINO-5H-INDENO(5,6-d)-1,3-DIOX-OLE HYDROCHLORIDE see BRC500
2-n-BUTYL-3-DIMETHYLAMINO-5,6-METHYLENEDIOXY-INDENE HYDROCHLORIDE see BRC500
2-sec-BUTYL-4,6-DINITROPHENOL see BRE500
2-sec-BUTYL-4,6-DINITROPHENOL AMMONIUM SALT see BPG250
2-sec-BUTYL-4,6-DINITROPHENYLACETATE see ACE500
6-sec-BUTYL-2,4-DINITROPHENYLACETATE see ACE500
4-BUTYL-1,2-DIPHENYL-3,5-DIOXO PYRAZOLIDINE see BRF500
4-BUTYL-1,2-DIPHENYLPYRAZOLIDINE-3,5-DIONE see BRF500
4-BUTYL-1,2-DIPHENYL-3,5-PYRAZOLIDINEDIONE see BRF500
4-BUTYL-1,2-DIPHENYL-3,5-PYRAZOLIDINEDIONE SO-DIUM SALT see BOV750
BUTYLE (ACETATE de) (FRENCH) see BPU750
BUTYLENE see BOW250
α-BUTYLENE see BOW250
γ-BUTYLENE see IIC000
β-BUTYLENE GLYCOL see BOS500
1,2-BUTYLENE GLYCOL see BOS250
1,3-BUTYLENE GLYCOL (FCC) see BOS500
1,4-BUTYLENE GLYCOL see BOS750
2,3-BUTYLENE GLYCOL see BOT000
BUTYLENE HYDRATE see BPW750
BUTYLENE OXIDE see TCR750
1,4-BUTYLENE SULFONE see BOU250
2,4,5-T-N-BUTYL ESTER see BSQ750
BUTYL ESTER 2,4-D see BQZ000
n-BUTYL ESTER of 3,4-DIHYDRO-2,2-DIMETHYL-4-OXO-2H-PYRAN-6-CARBOXYLIC ACID see BRT000
N-BUTYLESTER KYSELINI-2,4,5-TRICHLORFENOXYOC-TOVE (CZECH) see BSQ750
BUTYLESTER KYSELINY MRAVENCI see BRK000
BUTYL ETHANOATE see BPU750
n-BUTYL ETHER see BRH750
BUTYL ETHER (DOT) see BRH750

BUTYL ETHYL ACETALDEHYDE see BRI000
BUTYL ETHYL ACETIC ACID see BRI250
5-BUTYL-5-ETHYLBARBITURIC ACID see BPF500
5-sec-BUTYL-5-ETHYLBARBITURIC ACID see BPF000
BUTYL ETHYLENE see HFB000
o-BUTYL ETHYLENE GLYCOL see BPJ850
n-BUTYL ETHYL KETONE see EHA600
5-sec-BUTYL-5-ETHYLMALONYL UREA see BPF000
5-BUTYL-5-ETHYL-2,4,6(1H,3H,5H)-PYRIMIDINETRIONE
    (9CI) see BPF500
BUTYLETHYLTHIOCARBAMIC ACID S-PROPYL ESTER see
    PNF500
2-sec.-BUTYLFENOL (CZECH) see BSE000
p-tert-BUTYLFENOL (CZECH) see BSE500
BUTYL FORMAL see VAG000
tert-BUTYL FORMAMIDE see BRJ750
n-BUTYL FORMATE see BRK000
BUTYL FORMATE (DOT) see BRK000
n-BUTYL GLYCIDYL ETHER see BRK750
BUTYL GLYCOL see BPJ850
BUTYLGLYCOL (FRENCH, GERMAN) see BPJ850
terc. BUTYLHYDROPEROXID (CZECH) see BRM250
tert-BUTYLHYDROPEROXIDE see BRM250
tert-BUTYLHYDROQUINONE see BRM500
BUTYL HYDROXIDE see BPW500
tert-BUTYL HYDROXIDE see BPX000
BUTYLHYDROXYANISOLE see BQI000
tert-BUTYLHYDROXYANISOLE see BQI000
tert-BUTYL-4-HYDROXYANISOLE see BQI000
2(3)-tert-BUTYL-4-HYDROXYANISOLE see BQI000
BUTYL p-HYDROXYBENZOATE see DTC800
n-BUTYL-(4-HYDROXYBUTYL)NITROSAMINE see HJQ350
N-BUTYL-N-(4-HYDROXYBUTYL)NITROSAMINE see
    HJQ350
BUTYLHYDROXYOXOSTANNANE see BSL500
4-BUTYL-2-(4-HYDROXYPHENYL)-1-PHENYL-3,5-DIOXO-
    PYRAZOLIDINE see HNI500
4-BUTYL-1-(4-HYDROXYPHENYL)-2-PHENYL-3,5-PYRAZO-
    LIDINEDIONE see HNI500
4-BUTYL-1-(p-HYDROXYPHENYL)-2-PHENYL-3,5-PYRAZO-
    LIDINEDIONE see HNI500
4-BUTYL-2-(p-HYDROXYPHENYL)-1-PHENYL-3,5-PYRAZO-
    LIDINEDIONE see HNI500
BUTYL α-HYDROXYPROPIONATE see BRR600
BUTYLHYDROXYTOLUENE see BFW750
N-BUTYLIMIDODICARBONIMIDIC DIAMIDE MONOHY-
    DROCHLORIDE (9CI) see BQL000
N-BUTYL-2,2'-IMINODIETHANOL see BQM000
sec-BUTYL IODIDE see IEH000
BUTYL ISOBUTYRATE see BRQ350
n-BUTYL ISOCYANATE see BRQ500
n-BUTYL ISOPENTANOATE see ISX000
tert-BUTYL ISOPROPYL BENZENE HYDROPEROXIDE see
    BRR250
tert-BUTYL ISOPROPYL BENZENE HYDROPEROXIDE
    (DOT) see BRR250
2-sec-BUTYL-6-ISOPROPYLPHENOL see BRR500
2-((3-BUTYL-1-ISOQUINOLINYL)OXY)-N,N-DIMETHYL-
    ETHANAMINE MONOHYDROCHLORIDE see DNX400
1-BUTYL ISOVALERATE see ISX000
n-BUTYL ISOVALERATE see ISX000
BUTYL ISOVALERIANATE see ISX000
2-tert-BUTYL-p-KRESOL (CZECH) see BQV750
BUTYL LACTATE see BRR600
n-BUTYL LACTATE see BRR600
BUTYL MERCAPTAN see BRR900
n-BUTYL MERCAPTAN see BRR900
BUTYL MERCAPTAN (DOT) see BRR900
tert-BUTYL MERCAPTAN see MOS000
n-BUTYLMERCURIC CHLORIDE see BRS750
n-BUTYL MESITYL OXIDE OXALATE see BRT000
n-BUTYLMESITYLOXID OXALATE see BRT000
BUTYLMETHACRYLAAT (DUTCH) see MHU750
BUTYL-2-METHACRYLATE see MHU750

N-BUTYL METHACRYLATE see MHU750
BUTYL 3-METHYLBUTYRATE see ISX000
6-tert-BUTYL-3-METHYL-2,4-DINITRO ANISOLE see
    BRU500
BUTYL METHYL KETONE see HEV000
n-BUTYL METHYL KETONE see HEV000
2-tert-BUTYL-4-METHYLPHENOL see BQV750
1-BUTYL-3-(p-METHYLPHENYLSULFONYL)UREA see
    BSQ000
BUTYL-2-METHYL-2-PROPENOATE see MHU750
N-BUTYL-2-METHYL-2-PROPYL-1,3-PROPANEDIOL DI-
    CARBAMATE see MOV500
N-N-BUTYL-2-METHYL-2-PROPYL-1,3-PROPANEDIOL DI-
    CARBAMATE see MOV500
BUTYL NAMATE see SGF500
n-BUTYL NITRITE see BRV500
BUTYL NITRITE (DOT) see BRV500
sec-BUTYL NITRITE see BRV750
4-(BUTYLNITROSAMINO)-1-BUTANOL see HJQ350
4-(n-BUTYLNITROSAMINO)-1-BUTANOL see HJQ350
4-(BUTYLNITROSOAMINO)BUTANOIC ACID see BQQ250
BUTYLNITROSOAMINOMETHYL ACETATE see BRX500
N-BUTYL-N-NITROSO AMYL AMINE see BRY250
n-BUTYL-N-NITROSO-1-BUTAMINE see BRY500
N-BUTYL-N-NITROSO ETHYL CARBAMATE see BRZ000
BUTYLNITROSOHARNSTOFF (GERMAN) see BSA250
N-BUTYL-N-NITROSOPENTYLAMINE see BRY250
n-BUTYLNITROSOUREA see BSA250
1-BUTYL-1-NITROSOUREA see BSA250
N-n-BUTYL-N-NITROSOUREA see BSA250
1-BUTYL-1-NITROSOURETHAN see BRZ000
N-BUTYL-N-NITROSOURETHAN see BRZ000
2-BUTYL-1-OCTANOL see BSA500
2-BUTYLOCTYL ALCOHOL see BSA500
BUTYLOHYDROKSYANIZOL (POLISH) see BQI000
BUTYL OLEATE see BSB000
BUTYLONE see NBU000
BUTYLOWY ALKOHOL (POLISH) see BPW500
BUTYL OXITOL see BPJ850
N-BUTYL-N-(2-OXOBUTYL)NITROSAMINE see BSB500
tert-BUTYLOXYCARBONYL AZIDE see BQI250
α-BUTYLOXYCINCHONINIC ACID DIETHYLETHYLENE-
    DIAMIDE see DDT200
N-BUTYL-N-PENTYLINITROSAMINE see BRY250
tert-BUTYL PERACETATE see BSC250
tert-BUTYLPERBENZOAN (CZECH) see BSC500
tert-BUTYL PERBENZOATE see BSC500
tert-BUTYL PEROXIDE see BSC750
tert-BUTYL PEROXYACETATE see BSC250
tert-BUTYL PEROXYACETATE, more than 76% in solution
    (DOT) see BSC250
tert-BUTYL PEROXY BENZOATE see BSC500
tert-BUTYL PEROXYBENZOATE, technical pure or in concen-
    tration of more than 75% (DOT) see BSC500
sec-BUTYL PEROXYDICARBONATE see BSD000
tert-BUTYL PEROXYPIVALATE see BSD250
BUTYLPHEN see BSE500
2-n-BUTYLPHENOL see BSD500
4-n-BUTYLPHENOL see BSD750
4-sec BUTYL PHENOL see BSE250
o-sec-BUTYLPHENOL see BSE000
2-(p-tert-BUTYLPHENOXY)CYCLOHEXYL PROPARGYL
    SULFITE see SOP000
2-(p-tert-BUTYLPHENOXY)CYCLOHEXYL 2-PROPYNYL
    SULFITE see SOP000
BUTYLPHENOXYISOPROPYL CHLOROETHYL SULFITE
    see SOP500
2-(p-BUTYLPHENOXY)ISOPROPYL 2-CHLOROETHYL SUL-
    FITE see SOP500
2-(4-tert-BUTYLPHENOXY)ISOPROPYL-2-CHLOROETHYL
    SULFITE see SOP500
2-(p-tert-BUTYLPHENOXY)ISOPROPYL 2'-CHLOROETHYL
    SULPHITE see SOP500
2-(p-tert-BUTYLPHENOXY)-1-METHYLETHYL 2-CHLORO-
    ETHYL ESTER of SULPHUROUS ACID see SOP500

2-(p-BUTYLPHENOXY)-1-METHYLETHYL 2-CHLORO-
ETHYL SULFITE see SOP500
2-(p-tert-BUTYLPHENOXY)-1-METHYLETHYL-2-CHLORO-
ETHYL SULFITE ESTER see SOP500
2-(p-tert-BUTYLPHENOXY)-1-METHYLETHYL 2'-CHLORO-
ETHYL SULPHITE see SOP500
2-(p-tert-BUTYLPHENOXY)-1-METHYLETHYL SULPHITE of
2-CHLOROETHANOL see SOP500
1-(p-tert-BUTYLPHENOXY)-2-PROPANOL-2-CHLOROETHYL
SULFITE see SOP500
BUTYL PHENYL ACETATE see BBA000
o-sec-BUTYLPHENYL CARBAMATE see BSF250
BUTYL PHENYL ETHER see BSF750
o-sec-BUTYLPHENYL METHYLCARBAMATE see MOV000
2-sec-BUTYLPHENYL-N-METHYLCARBAMATE see
MOV000
BUTYL PHOSPHORIC ACID see ADF250
BUTYL PHOSPHOROTRITHIOATE see BSH250
n-BUTYL PHTHALATE (DOT) see DEH200
BUTYL PHTHALATE BUTYL GLYCOLATE see BQP750
BUTYL PHTHALYL BUTYL GLYCOLATE see BQP750
5-BUTYL PICOLINIC ACID see BSI000
1-BUTYL-2',6'-PIPECOLOXYLIDIDE see BSI250
1-BUTYL-2',6'-PIPECOLOXYLIDIDE HYDROCHLORIDE (+)
see BON750
BUTYL PROPANOATE see BSJ500
BUTYL-2-PROPENOATE see BPW100
BUTYL PROPIONATE see BSJ500
n-BUTYL PROPIONATE see BSJ500
5-BUTYL-2-PYRIDINECARBOXYLIC ACID see BSI000
BUTYLPYRIN see BRF500
4-tert-BUTYLPYROCATECHOL see BSK000
p-tert-BUTYLPYROCATECHOL see BSK000
4-tert-BUTYLPYROKATECHIN (CZECH) see BSK000
n-BUTYL RHODANATE see BSN500
BUTYL STANNOIC ACID see BSL500
N-BUTYLSULFANILYLUREA see BSM000
1-BUTYL-3-SULFANILYL UREA see BSM000
BUTYL-2,4,5-T see BSQ750
n-BUTYL THIOCYANATE see BSN500
n-BUTYL THIOUREA see BSO500
BUTYL TITANATE see BSP250
p-tert-BUTYLTOLUENE see BSP500
n-BUTYL-N'-p-TOLUENESULFONYLUREA see BSQ000
1-BUTYL-3-(p-TOLYL SULFONYL)UREA see BSQ000
1-BUTYL-3-TOSYLUREA see BSQ000
N-n-BUTYL-N'-TOSYLUREA see BSQ000
4-BUTYL-s-TRIAZOLE see BPU000
4-N-BUTYL-4H-1,2,4-TRIAZOLE see BPU000
BUTYL-2,4,5-TRICHLOROPHENOXYACETATE see BSQ750
N-BUTYL (2,4,5-TRICHLOROPHENOXY)ACETATE see
BSQ750
BUTYL TRICHLORO SILANE see BSR000
N-BUTYLUREA see BSS250
1-BUTYLUREA and SODIUM NITRITE (2:1) see BSS500
BUTYL VINYL ETHER see VMZ000
BUTYL VINYL ETHER (inhibited) see VMZ000
BUTYL ZIMATE see BIX000
BUTYL ZIRAM see BIX000
BUTYN see BOO750
1-BUTYNE see EFS500
2-BUTYNE see COC500
2-BUTYNEDIAMIDE see ACJ250
2-BUTYNE-1,4-DIOL see BST500
1,4-BUTYNEDIOL (DOT) see BST500
2-BUTYNE-1-THIOL see BST750
BUTYNORATE see DDV600
BUTYN SULFATE see BOP000
2-BUTYNYL-4-CHLORO-m-CHLOROCARBANILATE see
CEW500
2-BUTYOXY-N-(2-(DIETHYLAMINO)ETHYL)-4-QUINOLINE-
CARBOXAMIDE see DDT200
BUTYRAL see BSU250
BUTYRALDEHYD (GERMAN) see BSU250

BUTYRALDEHYDE (CZECH) see BSU250
n-BUTYRALDEHYDE see BSU250
m-BUTYRALDEHYDE OXIME see BSU500
N-BUTYRALDOXIME see BSU500
BUTYRALDOXIME (DOT) see BSU500
BUTYRHODANID (GERMAN) see BSN500
n-BUTYRIC ACID see BSW000
BUTYRIC ACID ESTER with BUTYL LACTATE see BQP000
BUTYRIC ACID ISOBUTYL ESTER see BSW500
BUTYRIC ACID LACTONE see BOV000
BUTYRIC ACID NITRILE see BSX250
BUTYRIC ACID TRIESTER with GLYCERIN see TIG750
BUTYRIC ACID, VINYL ESTER see VNF000
BUTYRIC ALDEHYDE see BSU250
BUTYRIC ETHER see EHE000
BUTYRIC or NORMAL PRIMARY BUTYL ALCOHOL see
BPW500
α-BUTYROLACTONE see BOV000
β-BUTYROLACTONE see BSX000
γ-BUTYROLACTONE (FCC) see BOV000
BUTYRON see CKF750
BUTYRONE (DOT) see DWT600
BUTYRONITRILE see BSX250
BUTYRONITRILE (DOT) see BSX250
1-BUTYRYLAZIRIDINE see BSY000
1-n-BUTYRYLAZIRIDINE see BSY000
BUTYRYL CHLORIDE see BSY250
BUTYRYLETHYLENEIMINE see BSY000
BUTYRYLETHYLENIMINE see BSY000
BUTYRYL LACTONE see BOV000
BUTYRYL TRIGLYCERIDE see TIG750
BUVETZONE see BRF500
BUX see BTA250
BUX-TEN see BTA250
2-n-BUYTLAMINOETHANOL see BQC000
BUZON see BRF500
BUZULFAN see BOT250
BVU see BNP750
B-W see SJU000
BW 57-322 see ASB250
BW-21-Z see AHJ750
BYLADOCE see VSZ000
BZ 55 see BSQ000
BZCF see BEF500
BZF-60 see BDS000
BZI see BCB750
BZL see BTA500
BZQ see BCL250

C 6 see HEA000
C-56 see HCE500
C-272 see BLG500
C 709 see DGQ875
C-847 see CEW500
C-854 see CJT750
C 1,006 see CJT750
C 1120 see DLS800
C 2059 see DUK800
C 3067 see CIK500
C 3172 see TNK250
C 3235 see DII400
C 4208 see BQH250
C-5068 see PHW250
C 5420 see CIW250
C 5968 see PHW000
C 6379 see DBA800
C 6866 see BIE500
C 8514 see CJJ250
C-10015 see CDS750
CA see BMW250
CABADON M see VSZ000
CAB-O-GRIP see AHE250
CAB-O-GRIP II see SCH000
CAB-O-SIL see SCH000

CAB-O-SPERSE see SCH000
CABRONAL see EOK000
CACHALOT C-50 see HCP000
CACHALOT L-50 see DXV600
C-8 ACID see OCY000
CACODYLATE de SODIUM (FRENCH) see HKC500
CACODYL HYDRIDE see DQG600
CACODYLIC ACID (DOT) see HKC000
CACODYLIC ACID SODIUM SALT see HKC500
CACODYL NEW see DXE600
CACODYL SULFIDE see CAC250
CACP see PJD000
CADAVERIN see PBK500
CADAVERINE see PBK500
CADCO 0115 see SMQ500
CADDY see CAE250
CADET see BDS000
CADMINATE see CAI750
CADMIUM see CAD000
CADMIUM(II) ACETATE see CAD250
CADMIUM ACETATE (DOT) see CAD250
CADMIUM AMIDE see CAD325
CADMIUM AZIDE see CAD350
CADMIUM BIS(2-ETHYLHEXYL) PHOSPHITE see CAD500
CADMIUM CAPRYLATE see CAD750
CADMIUM CHLORATE see CAE000
CADMIUM CHLORIDE see CAE250
CADMIUM CHLORIDE, DIHYDRATE see CAE375
CADMIUM CHLORIDE, HYDRATE (2:5) see CAE425
CADMIUM CHLORIDE, MONOHYDRATE see CAE500
CADMIUM COMPOUNDS see CAE750
CADMIUM DIACETATE see CAD250
CADMIUM DIAMIDE see CAD325
CADMIUM DIAZIDE see CAD350
CADMIUM DICHLORIDE see CAE250
CADMIUM DICYANIDE see CAF500
CADMIUM DIETHYL DITHIOCARBAMATE see BJB500
CADMIUM DINITRATE see CAH000
CADMIUM(II) EDTA COMPLEX see CAF750
CADMIUM FLUOBORATE see CAG000
CADMIUM FLUORIDE see CAG250
CADMIUM FLUOROBORATE see CAG000
CADMIUM FLUORURE (FRENCH) see CAG250
CADMIUM FLUOSILICATE see CAG500
CADMIUM FUME see CAH750
CADMIUM GOLDEN 366 see CAJ750
CADMIUM LACTATE see CAG750
CADMIUM LEMON YELLOW 527 see CAJ750
CADMIUM NITRATE see CAH000
CADMIUM(II) NITRATE TETRAHYDRATE (1:2:4) see CAH250
CADMIUM NITRIDE see TIH000
CADMIUM ORANGE see CAJ750
CADMIUM OXIDE see CAH500
CADMIUM OXIDE FUME see CAH750
CADMIUM PHOSPHATE see CAI000
CADMIUM PHOSPHIDE see CAI125
CADMIUM PRIMROSE 819 see CAJ750
CADMIUM PROPIONATE see CAI250
CADMIUM SELENIDE see CAI500
CADMIUM STEARATE see OAT000
CADMIUM SUCCINATE see CAI750
CADMIUM SULFATE see CAJ000
CADMIUM SULFATE (1:1) see CAJ000
CADMIUM SULFATE (1:1) HYDRATE (3:8) see CAJ250
CADMIUM SULFATE OCTAHYDRATE see CAJ250
CADMIUM SULFATE TETRAHYDRATE see CAJ500
CADMIUM SULFIDE see CAJ750
CADMIUM SULPHATE see CAJ000
CADMIUM SULPHIDE see CAJ750
CADMIUM THERMOVACUUM AEROSOL see CAK000
CADMIUM-THIONEIN see CAK250
CADMIUM salt of 2,4-5-TRIBROMOIMIDAZOLE see THV500

CADMIUM YELLOW see CAJ750
CADMOPUR YELLOW see CAJ750
CADOX see BDS000, BSC750
CADOX TBH see BRM250
CADPX PS see BHM750
CAF see CDP250, CEA750
CAFFEIN see CAK500
CAFFEINE see CAK500
CAFFEINE BROMIDE see CAK750
CAFFEINE HYDROBROMIDE see CAK750
CAFRON see BCA000
CAID see CJJ000
CAIROX see PLP000
CAJEPUTENE see MCC250
CAJEPUTOL see CAL000
CAKE ALUM see AHG750
CALAMINE (spray) see ZKA000
CALAMUS OIL see OGK000
CAL CHEM 5655 see MOU750
CALCIA see CAU500
CALCICAT see CAL250
CALCIC LIVER of SULFUR see CAY000
CALCIFEROL see VSZ100
CALCIFERON 2 see VSZ100
CALCINED BARYTA see BAO000
CALCINED BRUCITE see MAH500
CALCINED DIATOMITE see SCJ000
CALCINED MAGNESIA see MAH500
CALCINED MAGNESITE see MAH500
CALCINED SODA see SIN500
CALCITE see CAO000
CALCITRIOL see DMJ400
CALCIUM see CAL250
CALCIUM, non-pyrophoric (DOT) see CAL250
CALCIUM, pyrophoric (DOT) see CAL250
CALCIUM ACETARSONE see CAL500
CALCIUM ACETATE see CAL750
CALCIUM ACETYLIDE see CAN750
CALCIUMARSENAT see ARB750
CALCIUM ARSENATE (MAK) see ARB750
CALCIUM ARSENITE see CAM500
CALCIUM ARSENITE, solid (DOT) see CAM500
CALCIUM BENZOATE see CAM680
CALCIUM-o-BENZOSULFIMIDE see CAM750
CALCIUM-2-BENZOSULPHIMIDE see CAM750
CALCIUM-o-BENZOSULPHIMIDE see CAM750
CALCIUM BISULFITE (solution) see CAN000
CALCIUM BISULFITE, solution (DOT) see CAN000
CALCIUM CARBIDE see CAN750
CALCIUM CARBIMIDE see CAQ250
CALCIUM CARBONATE see CAO000
CALCIUM CHEL-330 see CAY500
CALCIUM CHLORATE see CAO500
CALCIUM CHLORATE, aqueous solution (DOT) see CAO500
CALCIUM CHLORIDE see CAO750
CALCIUM CHLORIDE, anhydrous see CAO750
CALCIUM CHLORITE see CAP000
CALCIUM CHLOROHYDROCHLORITE see HOV500
CALCIUM CHROMATE see CAP500
CALCIUM CHROMATE (VI) see CAP500
CALCIUM CHROMATE(VI) DIHYDRATE see CAP750
CALCIUM CHROME YELLOW see CAP500, CAP750
CALCIUM CHROMIUM OXIDE (CaCrO₄) see CAP500
CALCIUM COMPOUNDS see CAQ000
CALCIUM CYANAMID see CAQ250
CALCIUM CYANAMIDE see CAQ250
CALCIUM CYANIDE see CAQ500
CALCIUM CYANIDE (mixture) see CAQ750
CALCIUM CYANIDE, solid (DOT) see CAQ500
CALCIUM CYANIDE MIXTURE, solid (DOT) see CAQ750
CALCIUM CYCLAMATE see CAR000
CALCIUM CYCLOHEXANESULFAMATE see CAR000
CALCIUM CYCLOHEXANE SULPHAMATE see CAR000
CALCIUM CYCLOHEXYLSULFAMATE see CAR000

CALCIUM CYCLOHEXYLSULPHAMATE see CAR000
CALCIUM DIACETATE see CAL750
CALCIUM d(+)-N-(α,γ-DIHYDROXY-β,β-DIMETHYLBUTY-
 RYL)-β-ALANINATE see CAU750
CALCIUM DIOXIDE see CAV500
CALCIUM-DTPA see CAY500
CALCIUM FLUOSILICATE see CAX250
CALCIUM FORMATE see CAS250
CALCIUM GLUCONATE see CAS750
CALCIUM HEXAFLUOROSILICATE see CAX250
CALCIUM HYDRATE see CAT250
CALCIUM HYDROGEN SULFITE, solution (DOT) see
 CAN000
CALCIUM HYDROXIDE see CAT250
CALCIUM HYPOCHLORIDE see HOV500
CALCIUM HYPOCHLORITE see HOV500
CALCIUM HYPOCHLORITE MIXTURE, DRY (DOT) see
 HOW000
CALCIUM, METAL (DOT) see CAL250
CALCIUM, METAL, CRYSTALLINE (DOT) see CAL250
CALCIUM MONOCHROMATE see CAP500
CALCIUM NEMBUTAL see CAV000
CALCIUM NITRATE (DOT) see CAU000
CALCIUM(II) NITRATE (1:2) see CAU000
CALCIUM ORTHOARSENATE see ARB750
CALCIUM OXIDE see CAU500
CALCIUM OXYCHLORIDE see HOV500
CALCIUM PANTHOTHENATE (FCC) see CAU750
CALCIUM PANTOTHENATE see CAU750
CALCIUM-d-PANTOTHENATE see CAU750
d-CALCIUM PANTOTHENATE see CAU750
CALCIUM PANTOTHENATE, CALCIUM CHLORIDE DOU-
 BLE SALT see CAU780
CALCIUM PENTOBARBITAL see CAV000
CALCIUM PERMANGANATE see CAV250
CALCIUM PEROXIDE see CAV500
CALCIUM PHOSPHATE, DIBASIC see CAW100
CALCIUM PHOSPHATE, TRIBASIC see CAW120
CALCIUM PHOSPHIDE see CAW250
CALCIUM RESINATE see CAW500
CALCIUM RESINATE, fused (DOT) see CAW500
CALCIUM RESINATE, technically pure (DOT) see CAW500
CALCIUM RHODANID (GERMAN) see CAY250
CALCIUM SACCHARIN see CAM750
CALCIUM SACCHARINA see CAM750
CALCIUM SACCHARINATE see CAM750
CALCIUM SILICATE see CAW850
CALCIUM SILICOFLUORIDE see CAX250
CALCIUM SULFATE see CAX500
CALCIUM(II) SULFATE DIHYDRATE (1:1:2) see CAX750
CALCIUM SULFIDE see CAY000
CALCIUM SUPEROXIDE see CAV500
CALCIUM THIOCYANATE see CAY250
CALCIUM TRISODIUM CHEL 330 see CAY500
CALCIUM TRISODIUM DIETHYLENE TRIAMINE PENTA-
 ACETATE see CAY500
CALCIUM TRISODIUM DTPA see CAY500
CALCIUM TRISODIUM PENTETATE see CAY500
CALCIUM TRISODIUM SALT of DIETHYLENETRIAMINE-
 PENTAACETIC ACID see CAY500
CALCOCID BLUE EG see FMU059
CALCOCID ERYTHROSINE N see FAG040
CALCOCID FAST LIGHT ORANGE 2G see HGC000
CALCOCID URANINE B4315 see FEW000
CALCOCID VIOLET 4BNS see FAG120
CALCOCID YELLOW XX see FAG140
CALCODUR BROWN BRL see CMO750
CALCOGAS ORANGE NC see PEJ500
C 10 ALCOHOL see DAI600
CALCOMINE BLACK see AQP000
CALCOMINE BLUE 2B see CMO000
CALCO OIL ORANGE 7078 see PEJ500
CALCO OIL RED D see SBC500

CALCO OIL SCARLET BL see XRA000
CALCOSYN YELLOW GC see AAQ250
CALCOTONE RED see IHD000
CALCOTONE WHITE T see TGG760
CALCOZINE BLUE ZF see BJI250
CALCOZINE BRILLIANT GREEN G see BAY750
CALCOZINE CHRYSOIDINE Y see PEK000
CALCOZINE MAGENTA N see RMK020
CALCOZINE ORANGE YS see PEK000
CALCOZINE RED BX see FAG070
CALCOZINE RHODAMINE BX see FAG070
CALCOZINE YELLOW OX see IBA000
CALCYANIDE see CAQ500
C-8 ALDEHYDE see OCO000
C-9 ALDEHYDE see NMW500
C-10 ALDEHYDE see DAG000
C-16 ALDEHYDE see ENC000
C-12 ALDEHYDE, LAURIC see DXT000
CALDON see BRE500
CALGON see SHM500
CALICO YELLOW see MRN500
CALIDRIA RG 100 see ARM268
CALIDRIA RG 144 see ARM268
CALIDRIA RG 600 see ARM268
CALMADIN see MQU750
CALMATHION see MAK700
CALMAX see MQU750
CALMETTEN see EOK000
CALMINAL see EOK000
CALMINOL see MNM500
CALMIREN see MQU750
CALMIXENE see MOO750
CALMODEN see MDQ250
CALMONAL see HGC500
CALMORE see TEH500
CALMOREX see TEH500
CALMOTIN see BNP750
CALOCAIN see MNQ000
CALOCHLOR see MCY475
CALOGREEN see MCW000
CALOMEL see MCW000
CALOMELANO (ITALIAN) see MCW000
CALOSAN see MCW000
CALPANATE see CAU750
CALPLUS see CAO750
CALPOL see HIM000
CALSMIN see DLY000
CALSOFT F-90 see DXW200
CALSOL see EIV000
CALTAC see CAO750
CALX see CAU500
CAM see CDP250
CAMCOLIT see LGZ000
CAMITE see BMW250
CAMOMILE OIL, ENGLISH TYPE (FCC) see CDH750
CAMOMILE OIL GERMAN see CDH500
CAMPAPRIM A 1544 see AMY050
CAMPBELLINE OIL ORANGE see PEJ500
2-CAMPHANOL see BMD000
1-2-CAMPHANOL see NCQ820
2-CAMPHANONE see CBA750
d-2-CAMPHANONE see CBB250
CAMPHECHLOR see CDV100
CAMPHENE see CBA500
CAMPHOCHLOR see CDV100
CAMPHOCLOR see CDV100
CAMPHOFENE HUILEUX see CDV100
CAMPHOGEN see CQI000
CAMPHOR see CBA750
(+)-CAMPHOR see CBB250
d-CAMPHOR see CBB250
d-(+)-CAMPHOR see CBB250
(1R,4R)-(+)-CAMPHOR see CBB250
CAMPHOR, synthetic (ACGIH, DOT) see CBA750

CAMPHOR-NATURAL see CBA750
CAMPHOR OIL see CBB500
CAMPHOR OIL, RECTIFIED see CBB500
CAMPHOR OIL WHITE see CBB500
CAMPHOR OIL YELLOW see CBB500
CAMPHOR TAR see NAJ500
CAMPHOR USP see CBB250
CAMPHOZONE see DJS200
CAMPILIT see COO500
CAMPOSAN see CDS125
CAMPOVITON 6 see PPK500
CANACERT BRILLIANT BLUE FCF see FAE000
CANACERT ERYTHROSINE BS see FAG040
CANACERT INDIGO CARMINE see FAE100
CANACERT SUNSET YELLOW FCF see FAG150
CANACERT TARTRAZINE see FAG140
CANADOL see PCT250
CANARY CHROME YELLOW 40-2250 see LCR000
CANDAMIDE see LGZ000
CANDASETPIC see CFE250
CANDEPTIN see CBC250, LFF000
CANDEREL see ARN825
CANDEX see ARQ725, NOH500
CANDICIDIN see CBC250
CANDIMON see CBC250, LFF000
CANDIO-HERMAL see NOH500
CANE SUGAR see SNH000
CANOGARD see DGP900
CANQUIL-400 see MQU750
CANTABILINE SODIUM see HMB000
CANTHARIDES see CBE250
CANTREX see KAL000
CAO 1 see BFW750
CAO 3 see BFW750
CAP see CBF250, CDP250, CEA750
CAPAROL see BKL250
CAPITUS see ENG500
CAPMUL see PKL030
CAPMUL POE-O see PKL100
CAPORIT see HOV500
CAP-O-TRAN see MQU750
CAP-P see CDP700
CAP-PALMITATE see CDP700
CAPRALDEHYDE see DAG000
CAPRAN 80 see PJY500
CAPRIC ACID see DAH400
n-CAPRIC ACID see DAH400
CAPRIC ACID ETHYL ESTER see EHE500
CAPRIC ALCOHOL see DAI600
CAPRIN see ADA725
CAPRINIC ACID see DAH400
CAPRINIC ALCOHOL see DAI600
CAPROALDEHYDE see HEM000
CAPROAMIDE see HEM500
CAPROAMIDE POLYMER see PJY500
CAPRODAT see IPU000
CAPROIC ACID see HEU000
n-CAPROIC ACID see HEU000
CAPROIC ALDEHYDE see HEM000
CAPROKOL see HFV500
CAPROLACTAM see CBF700
6-CAPROLACTAM see CBF700
omega-CAPROLACTAM (MAK) see CBF700
CAPROLACTAM OLIGOMER see PJY500
epsilon-CAPROLACTAM POLYMERE (GERMAN) see
    PJY500
CAPROLATTAME (FRENCH) see CBF700
CAPROLON see NOH000
CAPRON see HNT500, PJY500
CAPRONALDEHYDE see HEM000
CAPRONAMIDE see HEM500
CAPRONIC ACID see HEU000
CAPRONITRILE see HER500
CAPROYL ALCOHOL see HFJ500

n-CAPROYLALDEHYDE see HEM000
1-CAPROYLAZIRIDINE see HEW000
CAPROYLETHYLENEIMINE see HEW000
CAPRYL ALCOHOL see OEI000
CAPRYLDINITROPHENYL CROTONATE see AQT500
2-CAPRYL-4,6-DINITROPHENYL CROTONATE see
    AQT500
CAPRYLIC ACID see OCY000
n-CAPRYLIC ACID see OCY000
CAPRYLIC ALCOHOL see OEI000
CAPRYLYL ACETATE see OEG000
CAPRYLYL PEROXIDE, solution (DOT) see OFI000
CAPRYNIC ACID see DAH400
CAPSEBON see CAJ750
CAPSINE see DUS700
CAPTAF see CBG000
CAPTAFOL see CBF800
CAPTAN see CBG000
CAPTANCAPTENEET 26,538 see CBG000
CAPTANE see CBG000
CAPTAN-STREPTOMYCIN 7.5-0.1 POTATO SEED PIECE
    PROTECTANT see CBG000
CAPTAX see BDF000
CAPTEX see CBG000
CAPTOFOL see CBF800
CAPUT MORTUUM see IHD000
CARADATE 30 see MJP400
CARAMEL see CBG125
CARAMEL COLOR see CBG125
CARASTAY see CCL250
CARASTAY G see CCL250
CARAWAY OIL see CBG500
CARBACRYL see ADX500
CARBADINE see EIR000
CARBADOX (USDA) see FOI000
CARBAETHOXYDIGOXIN (GERMAN) see EEP000
CARBAMALDEHYDE see FMY000
CARBAMAMIDINE see GKW000
CARBAMATE see FAS000
CARBAMATE de METHYLPENTINOL (FRENCH) see
    MNM500
CARBAMATES see CBH750
CARBAMAZEPEN see DCV200
CARBAMAZEPINE see DCV200
CARBAMEZEPINE see DCV200
CARBAMIC ACID, BUTYL ESTER see BQP250
CARBAMIC ACID, DIMETHYLDITHIO-, ANHYDROSUL-
    FIDE see BJL600
CARBAMIC ACID, DIMETHYLDITHIO-, ZINC SALT (2:)
    see BJK500
CARBAMIC ACID, DIMETHYL-, ester with (m-HYDROXY-
    PHENYL)TRIMETHYLAMMONIUM BROMIDE see
    DQY800
CARBAMIC ACID, ESTER with 2-(HDYROXYMETHYL)-1-
    METHYLPENTYLISOPROPYLCARBAMATE see IPU000
CARBAMIC ACID, ESTER with 2-(HYDROXYMETHYL)-2-
    METHYLPENTYL BUTYLCARBAMATE see MOV500
CARBAMIC ACID, ESTER with 2-METHYL-2-PROPYL-1,3-
    PROPANEDIOL BUTYLCARBAMATE see MOV500
CARBAMIC ACID, ESTER with 2-METHYL-2-PROPYL-1,3-
    PROPANEDIOL ISOPROPYLCARBAMATE see IPU000
CARBAMIC ACID, ETHYL ESTER see UVA000
CARBAMIC ACID-1-ETHYL-1-METHYL-2-PROPYNYL ES-
    TER see MNM500
CARBAMIC ACID-2-ETHYNYL-2-BUTYL ESTER see
    MNM500
CARBAMIC ACID HYDRAZIDE see HGU000
CARBAMIC ACID, ISOPROPYL ESTER see IOJ000
CARBAMIC ACID-N-METHYL-3-DIMETHYLAMINOPHENYL
    ESTER METHIODIDE see HNO500
CARBAMIC ACID-1-METHYLETHYL ESTER see IOJ000
CARBAMIC ACID-3-PHENYLPROPYL ESTER see PGA750
CARBAMIC ACID, PROPYL ESTER see PNG250
CARBAMIC ACID, VINYL ESTER see VNK000

CARBAMIDAL see DJS200
CARBAMIDE see USS000
CARBAMIDE PEROXIDE see HIB500
CARBAMIDE RESIN see USS000
CARBAMIDINE see GKW000
p-CARBAMIDOBENZENEARSONIC ACID see CBJ000
4-CARBAMIDOPHENYL BIS(CARBOXYMETHYLTHIO)AR-
    SENITE see CBI250
p-CARBAMIDOPHENYL-BIS(2-CARBOXYPHENYLMER-
    CAPTO)ARSINE see TFD750
4-CARBAMIDOPHENYL BIS(o-CARBOXYPHENYLTHIO)AR-
    SENITE see TFD750
p-CARBAMIDOPHENYL-DI(1'-CARBOXYPHENYL-2')
    THIOARSENITE see TFD750
CARBAMIDSAEURE-AETHYLESTER (GERMAN) see
    UVA000
CARBAMIMIDIC ACID see USS000
p-CARBAMINO PHENYL ARSONIC ACID see CBJ000
CARBAMINOPHENYL-p-ARSONIC ACID see CBJ000
CARBAMOHYDROXAMIC ACID see HOO500
CARBAMOHYDROXIMIC ACID see HOO500
CARBAMOHYDROXYAMIC ACID see HOO500
CARBAMONITRILE see COH500
(p-CARBAMOYLAMINO)PHENYLARSINOBIS(2-THIO-ACE-
    TIC ACID) see CBI250
N-CARBAMOYLARSANILIC ACID see CBJ000
5-CARBAMOYL-5H-DIBENZ(b,f)AZEPINE see DCV200
5-CARBAMOYLDIBENZO(b,f)AZEPINE see DCV200
5-CARBAMOYL-5H-DIBENZO(b,f)AZEPINE see DCV200
1-CARBAMOYLFORMIMIDIC ACID see OLO000
CARBAMOYLHYDRAZINE see HGU000
N-CARBAMOYLHYDROXYLAMINE see HOO500
p-((CARBAMOYLMETHYL)AMINO)-BENXENEARSONIC
    ACID see CBJ750
N-(CARBAMOYLMETHYL)ARSANILIC ACID see CBJ750
CARBAMOYL OXIME see HOO500
3-CARBAMOYLOXY-3-METHYL-4-PENTYNE see MNM500
1-CARBAMOYLOXY-3-PHENYLPROPANE see PGA750
10-(3-(4-CARBAMOYLPIPERIDINE)PROPYL)-2-(METHANE-
    SULFONYL)PHENOTHIAZINE see MQR000
4-CARBAMYLAMINOPHENYLARSONIC ACID see CBJ000
N-CARBAMYL ARSANILIC ACID see CBJ000
5-CARBAMYLDIBENZO(b,f)AZEPINE see DCV200
5-CARBAMYL-5H-DIBENZO(b,f)AZEPINE see DCV200
CARBAMYLHYDRAZINE see HGU000
CARBAMYLHYDRAZINE HYDROCHLORIDE see SBW500
CARBAMYL HYDROXAMATE see HOO500
2-CARBAMYL PYRAZINE see POL500
CARBANIL see PFK250
CARBANOLATE see CBM500
CARBARSONE (USDA) see CBJ000
CARBARYL see CBM750
CARBASED see ACE000
CARBASONE see CBJ000
CARBATOX-60 see CBM750
CARBAX see BIO750
CARBAZAMIDE see HGU000
CARBAZEPINE see DCV200
CARBAZIC ACID HYDRAZIDE see CBS500
CARBAZIDE see CBS500
CARBAZIDE (DOT) see CBS500
CARBAZINC see BJK500
CARBAZOLE see CBN000
9H-CARBAZOLE see CBN000
CARBAZOTIC ACID see PID000
CARBENDAZIM and SODIUM NITRITE (5:1) see CBN375
CARBETHOXYACETIC ESTER see EMA500
N-CARBETHOXYETHYLENIMINE see ASH750
CARBETHOXY MALATHION see MAK700
p-CARBETHOXYPHENOL see HJL000
CARBETOVUR see MAK700
CARBETOX see MAK700
CARBICRON see DGQ875

CARBIDE 6-12 see EKV000
CARBIDE BLACK E see AQP000
CARBIMIDE see COH500
CARBIN see CEW500
CARBINAMINE see MGC250
CARBINOL see MGB150
CARBINOXAMINE MALEATE see TAI500
CARBITOL see CBR000, DJD600
CARBITOL ACETATE see CBQ750
CARBITOL CELLOSOLVE see CBR000
CARBITOL SOLVENT see CBR000
CARBOBENZOXY CHLORIDE see BEF500
CARBOBENZYLOXY CHLORIDE see BEF500
2-CARBO-n-BUTOXY-6,6-DIMETHYL-5,6-DIHYDRO-1,4-PY-
    RONE see BRT000
CARBO-CORT see CMY800
CARBODIHYDRAZIDE see CBS500
CARBOFENOTHION see TNP250
CARBOFENOTHION (DUTCH) see TNP250
CARBOFOS see MAK700
CARBOFURAN see CBS275
CARBOHYDRAZIDE see CBS500
CARBOLIC ACID see PDN750
CARBOLITH see LGZ000
CARBOLON see SCQ000
CARBOLSAURE (GERMAN) see PDN750
CARBOMAL see BNK000
CARBOMETHENE see KEU000
2-CARBOMETHOXYANILINE see APJ250
o-CARBOMETHOXYANILINE see APJ250
4'-CARBOMETHOXY-2,3'-DIMETHYLAZOBENZENE see
    CBS750
4'-CARBOMETHOXY-2,3'-DIMETHYLAZOBENZOL see
    CBS750
α-2-CARBOMETHOXY-1-METHYLVINYL DIMETHYL
    PHOSPHATE see MQR750
2-CARBOMETHOXY-1-PROPEN-2-YL DIMETHYL PHOS-
    PHATE see MQR750
CARBOMYCIN see CBT250
CARBOMYCIN A see CBT250
CARBON see CBT500
CARBONA see CBY000
CARBON, ACTIVATED see CDI000
CARBONATE MAGNESIUM see MAC650
CARBONAZIDIC ACID, 1,1-DIMETHYLETHYL ESTER see
    BQI250
CARBON BICHLORIDE see PCF275
CARBON BISULFIDE (DOT) see CBV500
CARBON BISULPHIDE see CBV500
CARBON BLACK see CBT750
CARBON BROMIDE see CBX750
CARBON CHLORIDE see CBY000
CARBON CHLOROSULFIDE see TFN500
CARBON D see DXD200
CARBON DICHLORIDE see PCF275
CARBON DIFLUORIDE OXIDE see CCA500
CARBON DIOXIDE see CBU250
CARBON DIOXIDE (liquefied) see CBU500
CARBON DIOXIDE, liquefied (DOT) see CBU500
CARBON DIOXIDE, refrigerated liquid (DOT) see CBU500
CARBON DIOXIDE and ETHYLENE OXIDE MIXTURES, with
    more than 6% ETHYLENE OXIDE (DOT) see EJO000
CARBON DIOXIDE, mixture with NITROGEN OXIDE (N_2O)
    see CBV000
CARBON DIOXIDE mixed with NITROUS OXIDE see
    CBV000
CARBON DIOXIDE-NITROUS OXIDE mixture (DOT) see
    CBV000
CARBON DISULFIDE see CBV500
CARBON DISULPHIDE see CBV500
CARBONE (ITALIAN) see CBT500
CARBONE (OXYCHLORURE de) (FRENCH) see PGX000
CARBONE (OXYDE de) (FRENCH) see CBW750
CARBONE (SUFURE de) (FRENCH) see CBV500

CARBON FERROCHROMIUM see FBD000
CARBON FLUORIDE see CBY250
CARBON FLUORIDE OXIDE see CCA500
CARBON HEXACHLORIDE see HCI000
CARBONIC ACID, AMMONIUM SALT see ANE000
CARBONIC ACID BERYLLIUM SALT (1:1) see BFP750
CARBONIC ACID, CALCIUM SALT (1:1) see CAO000
CARBONIC ACID, DIAMMONIUM SALT see ANE000
CARBONIC ACID, DICESIUM SALT see CDC750
CARBONIC ACID DIHYDRAZIDE see CBS500
CARBONIC ACID, DILITHIUM SALT see LGZ000
CARBONIC ACID, DIPHENYL ESTER see DVZ000
CARBONIC ACID, DIPOTASSIUM SALT see PLA000
CARBONIC ACID, DISODIUM SALT see SFO000
CARBONIC ACID GAS see CBU250
CARBONIC ACID, LEAD(2+) SALT (1:1) see LCP000
CARBONIC ACID LITHIUM SALT see LGZ000
CARBONIC ACID, MAGNESIUM SALT see MAC650
CARBONIC ACID METHYL-4-(o-TOLYLAZO)-o-TOLYL ES-
   TER see CBS750
CARBONIC ACID, MONOAMMONIUM SALT see ANB250
CARBONIC ACID, NICKEL SALT (1:1) see NCY500
CARBONIC ANHYDRASE INHIBITOR No. 6063 see AAI250
CARBONIC ANHYDRIDE see CBU250
CARBONIC DIFLUORIDE see CCA500
CARBONIC DIHYDRAZIDE see CBS500
CARBONIC OXIDE see CBW750
CARBONIMIDIC DIHYDRAZIDE, BIS((4-CHLOROPHENYL)
   METHYLENE)- see RLK890
4,4'-CARBONIMIDOYLBIS(N,N-DIMETHYLBENZENAMINE)
   see IBB000
4,4'-CARBONIMIDOYLBIS(N,N-DIMETHYLBENZEN-
   AMINE)MONOHYDROCHLORIDE see IBA000
CARBONIO (OSSICLORURO di) (ITALIAN) see PGX000
CARBONIO (OSSIDO di) (ITALIAN) see CBW750
CARBONIO (SOLFURO di) (ITALIAN) see CBV500
CARBON MONOXIDE see CBW750
CARBON MONOXIDE, CRYOGENIC liquid (DOT) see
   CBW750
CARBON NITRIDE see COO000
CARBON NITRIDE ION (CN1-) see COI500
CARBONOCHLORIDE ACID-1-METHYL ESTER see IOL000
CARBONOCHLORIDIC ACID PHENYL ESTER see CBX109
CARBONOCHLORIDIC ACID, PROPYL ESTER see PNH000
CARBONOHYDRAZIDE see CBS500
CARBON OIL see BBL250
CARBONOTHIOIC DICHLORIDE see TFN500
CARBON OXIDE (CO) see CBW750
CARBON OXIDE SULFIDE see CCC000
CARBON OXYCHLORIDE see PGX000
CARBON OXYFLUORIDE see CCA500
CARBON OXYSULFIDE see CCC000
CARBON REMOVER (liquid) see CBX250
CARBON S see SGM500
CARBON SILICIDE see SCQ000
CARBON SULFIDE see CBV500
CARBON SULPHIDE (DOT) see CBV500
CARBON TET see CBY000
CARBON TETRABROMIDE see CBX750
CARBON TETRACHLORIDE see CBY000
CARBON TETRAFLUORIDE see CBY250
CARBON TRIFLUORIDE see CBY750
CARBONYLCHLORID (GERMAN) see PGX000
CARBONYL CHLORIDE see PGX000
CARBONYL DIAMIDE see USS000
CARBONYLDIAMINE see USS000
CARBONYL DIFLUORIDE see CCA500
CARBONYLDIHYDRAZINE see CBS500
CARBONYL FLUORIDE see CCA500
CARBONYL IRON see IGK800
CARBONYLS see CCB609
CARBONYL SULFIDE see CCC000
CARBONYL SULFIDE-^{32}s see CCC000

CARBOPHOS see MAK700
CARBORAFFIN see CDI000
CARBORAFINE see CDI000
CARBORUNDEUM see SCQ000
CARBORUNDUM see SCQ000
CARBOSPOL see AGJ250
CARBOTHIALDIN see DSB200
CARBOTHIALDINE see DSB200
CARBOWAX see PJT000, PJT200
CARBOWAX 1000 see PJT250
CARBOWAX 1500 see PJT500
CARBOWAX 4000 see PJT750
CARBOWAX 6000 see PJU000
5-CARBOXANILIDO-2,3-DIHYDRO-6-METHYL-1,4-OXA-
   THIIN see CCC500
CARBOXIN (USDA) see CCC500
CARBOXINE see CCC500
CARBOXYANILINE see API500
2-CARBOXYANILINE see API500
4-CARBOXYANILINE see AIH600
o-CARBOXYANILINE see API500
3-(α-CARBOXY-o-ANISAMIDO)-2-METHOXYPROPYL HY-
   DROXYMERCURY, MONOSODIUM SALT see SIH500
CARBOXYBENZENE see BCL750
2-CARBOXY-4'-(DIMETHYLAMINO)AZOBENZENE see
   CCE500
2-CARBOXYDIPHENYLAMINE see PEG500
CARBOXYETHANE see PMU750
3-CARBOXY-1-ETHYL-7-METHYL-1,8-NAPHTHIDIN-4-ONE
   see EID000
2'-CARBOXY-2-HYDROXY-4-METHOXYBENZOPHENONE
   (o-(2-HYDROXY-p-ANISOYL)BENZOIC ACID) see
   HLS500
3-CARBOXY-5-HYDROXY-1-p-SULFOPHENYL-4-o-SULFO-
   PHENYLAZOPYRAZOLE TRISODIUM SALT see FAG140
(4-(CARBOXY METHOXY)-3-CHLOROPHENYL)(5,5-DI-
   ETHYL-2,4,6(1H,3H,5H)-PYRIMIDINETRIONATO-O^2-
   MERCURY, MONOSODIUM SALT see CCG500
CARBOXYMETHYL CELLULOSE see SFO500
CARBOXYMETHYL CELLULOSE, SODIUM see SFO500
CARBOXYMETHYL CELLULOSE, SODIUM SALT see
   SFO500
3,3'-(CARBOXYMETHYLENE)BIS(4-HYDROXYCOUMARIN)
   ETHYL ESTER see BKA000
((CARBOXYMETHYLIMINO)BIS(ETHYLENENITRILO))
   TETRAACETIC ACID see DJG800
N-(γ-CARBOXYMETHYLMERCAPTOMERCURI-β-METHOX-
   Y)PROPYLCAMPHORAMIC ACID DISODIUM SALT see
   DXC000
(CARBOXYMETHYLTHIO)ACETIC ACID see MCM750
o-CARBOXYPHENYL ACETATE see ADA725
p-CARBOXYPHENYLAMINE see AIH600
(p-CARBOXYPHENYL)CHLOROMERCURY see CHU500
9-o-CARBOXYPHENYL-6-DIETHYLAMINO-3-ETHYLIMINO-
   3-ISOXANTHENE, 3-ETHOCHLORIDE see FAG070
(9-(o-CARBOXYPHENYL)-6-(DIETHYLAMINO)-3H-XAN-
   THEN-3-YLIDENE) DIETHYLAMMONIUM CHLORIDE
   see FAG070
9-(o-CARBOXYPHENYL)-6-HYDROXY-3-ISOXANTHENONE
   see FEV000
9-o-CARBOXYPHENYL-6-HYDROXY-3-ISOXANTHONE, DI-
   SODIUM SALT see FEW000
9-(o-CARBOXYPHENYL)-6-HYDROXY-2,4,5,7-TETRAIODO-
   3-ISOXANTHONE see FAG040
9-(o-CARBOXYPHENYL)-6-HYDROXY-3H-XANTHEN-3-ONE
   see FEV000
((o-CARBOXYPHENYL)THIO)ETHYLMERCURY SODIUM
   SALT see MDI000
4-CARBOXYPHTHALIC ANHYDRIDE see TKV000
3-CARBOXYPYRIDINE see NCQ900
CARBRITAL see NBU000
CARBUTAMID see BSM000
CARBUTAMIDE see BSM000
CARBYNE see CEW500

CB 2511 see BKM500
CB 2562 see TFU500
CB 3008 see BHV000
CB 3025 see BHV250, PED750
CB-3307 see BHT750
CB 4564 see EAS500, CQC675
CB-4835 see BIA250
CB 8019 see IPU000
CBC 806495 see TFQ750
CBC 906288 see TND250
CBD 90 see TIQ750
8102 CB HYDROCHLORIDE see BEO750
CBN see CEW500
CC 914 see CBI250
CC 11511 see DYC800
CCC see CAQ250
CCH see HOV500
CCHO see CPD000
C.C. No. 914 see CBI250
CCNU see CGV250
CCS see CJT750
CCS 203 see BPW500
CCS 301 see BPW750
CCUCOL see ASB250
CD see TGD000
CD 2 see LFK000
CD 68 see CDR750
CDA 101 see CNI000
CDA 102 see CNI000
CDA 110 see CNI000
CDA 122 see CNI000
CDAA see CFK000
CDAAT see CFK000
CDB 63 see SGG500
CDBM see CFK500
CDDP see PJD000
CDEC see CDO250
CDHA see DGT600
CDM see CJJ250
CDP see LFK000
CDT see BJP000
CEBETOX see MIW250
CEBITATE see ARN125
CEBROGEN see GFO050
CEBRUM see MDQ250
CECALGINE TBV see SEH000
CECENU see CGV250
CECOLENE see TIO750
CEDAD see BCA000
CEDAR LEAF OIL see CCQ500
CEDILANID see LAU000
CEDIN see ILD000
CEDRO OIL see LEI000
CEE see PMB000
CEE DEE see HCQ500
CEENU see CGV250
CEEPRYN see CCX000, CDF750
CEEPRYN CHLORIDE see CCX000
CEFAPIRIN (GERMAN) see CCX500
CEFATIN see OAV000
CEFOXITIN see CCS500
CEFRACYCLINE SUSPENSION see TBX000
CEFRACYCLINE TABLETS see TBX250
CEGLUTION see LGZ000
CEKIURON see DXQ500
CEKUDIFOL see BIO750
CEKUFON see TIQ250
CEKUGIB see GEM000
CEKUMETA see TDW500
CEKUMETHION see MNH000
CEKUQUAT see PAJ000
CEKUSAN see BJP000, DGP900
CEKUSIL see ABU500
CEKUSIL UNIVERSAL A see MEO750

CEKUSIL UNIVERSAL C see MEP250
CEKUTHOATE see DSP400
CEKUTROTHION see DSQ000
CEKUZINA-S see BJP000
CEKUZINA-T see ARQ725
CELA A-36 see DAE600
CELANEX see BBQ500
CELANOL DOS 75 see DJL000
CELA S-2225 see EGV500
CELESTONE see BFV750
CELGARD 2500 see PMP500
CELINHOL -A see OAV000
CELLITAZOL B see DCJ200
CELLITON FAST YELLOW G see AAQ250
CELLITON ORANGE R see AKP750
CELLOCIDIN see ACJ250
CELLOFAS see SFO500
CELLOFOR (CZECH) see DWO800
CELLOGEL C see SFO500
CELLOIDIN see CCU250
CELLON see TBQ100
CELLOPHANE see CCT250
CELLOSOLVE (DOT) see EES350
CELLOSOLVE ACETATE (DOT) see EES400
CELLOSOLVE SOLVENT see EES350
CELLPRO see SFO500
CELLUFIX FF 100 see SFO500
CELLUFLEX see CGO500
CELLUFLEX 179C see TNP500
CELLUFLEX DOP see DVL600
CELLUFLEX DPB see DEH200
CELLUFLEX TPP see TMT750
CELLUGEL see SFO500
"CELLULOID" see CCU000
CELLULOID, IN BLOCKS, RODS, ROLLS, SHEETS, TUBES
   (DOT) see CCU000
CELLULOID SCRAP (DOT) see CCU000
CELLULOSE GLYCOLIC ACID, SODIUM SALT see SFO500
CELLULOSE GUM see SFO500
CELLULOSE NITRATE see CCU250
CELLULOSE, POWDERED see CCU150
CELLULOSE SODIUM GLYCOLATE see SFO500
CELLULOSE TETRANITRATE see CCU250
CELLUPHOS 4 see TIA250
CELLU-QUIN see BLC250
CELMER see ABU500, MEP250
CELMIDE see EIY500
CELOCURINE see BJI000
CELON A see EIX000
CELON ATH see EIX000
CELON E see EIV000
CELON H see EIV000
CELON IS see EIV000
CELOSEN AZ see ASM270
CELPHIDE see AHE750
CELPHOS see AHE750, PGY000
CELTHIGN see MAK700
CEMENT, adhesive (DOT) see CCV250
CEMENT, leather see CCV000
CEMENT (liquid) see CCV250
CEMENT (pyroxylin) see CCV750
CEMENT (roofing liquid) see CCW000
CEMENT (rubber) see CCW250
CEMENT BLACK see MAS000
CEMENT, PORTLAND see PKS750
CEMENT, PYROXYLIN (DOT) see CCV750
CEMENT, ROOFING, liquid (DOT) see CCW000
CEMENT, RUBBER (DOT) see CCW250
CEMIDON see ILD000
CENOLATE see ARN125
CENOL GARDEN DUST see RNZ000
CENSTIM see DLH630
CENSTIN see DLH600, DLH630
CENTEDEIN see MNQ000

CENTIMIDE see HCQ500
CENTRALGIN see DAM700
CENTRALINE BLUE 3B see CMO250
CENTREDIN see MNQ000
CENTRINE see DOY400
CENTURY 1240 see SLK000
CENTURY CD FATTY ACID see OHU000
CEP see CDS125
2-CEPA see CDS125
CEPACOL see CDF750
CEPACOL CHLORIDE see CCX000
CEPHA see CDS125
CEPHAELINE METHYL ETHER see EAL500
CEPHA 10LS see CDS125
CEPHAPIRIN see CCX500
CEPHOXITIN see CCS500
CEPHROL see CMT250
CEPRIM see CCX000
CERASINE YELLOW GG see DOT300
CERASINROT see OHA000
CERASYNT see HKJ000
CERASYNT 1000-D see OAV000
CERASYNT S see OAV000
CERASYNT SD see OAV000
CERASYNT SE see OAV000
CERASYNT WM see OAV000
CERAZOL (suspension) see TEX250
CEREDON see BDD000
CERELINE see BDD000
CERELOSE see GFG000
CERENOX see BDD000
CEREPAP see DNA200
CERESAN see ABU500, CHC500
CERESAN M see EME500
CERESAN UNIVERSAL NAZBEIZE see MEP250
CERES ORANGE R see PEJ500
CERES ORANGES RR see XRA000
CERES RED 7B see EOJ500
CERES YELLOW R see PEI000
CERIC OXIDE see CCY000
CERISE TONER X1127 see FAG070
CERISOL SCARLET G see XRA000
CERISOL YELLOW AB see FAG130
CERISOL YELLOW TB see FAG135
CERIUM see CCY250
CERIUM ACETATE see CCY500
CERIUM AZIDE see CCY699
CERIUM CHLORIDE see CCY750
CERIUM(III) CHLORIDE see CCY750
CERIUM CITRATE see CCZ000
CERIUM(III) CITRATE see CCZ000
CERIUM 2COMPOUNDS see CDA250
CERIUM DIOXIDE see CCY000
CERIUM EDETATE see CDA500
CERIUM FLUORIDE see CDA750
CERIUM FLUORURE (FRENCH) see CDA750
CERIUM NITRATE, HEXAHYDRATE see CDB250
CERIUM(III) NITRATE, HEXAHYDRATE (1:3:6) see
　　CDB250
CERIUM(III) TETRAHYDROALUMINATE see CDB500
CERIUM TRIACETATE see CCY500
CERIUM TRICHLORIDE see CCY750
CERIUM TRIFLUORIDE see CDA750
CERIUM TRINITRATE HEXAHYDRATE see CDB250
CERN PRIMA 38 see AQP000
CEROTINORANGE G see PEJ500
CEROTINSCHARLACH G see XRA000
CEROUS ACETATE see CCY500
CEROUS CHLORIDE see CCY750
CEROUS CITRATE see CCZ000
CEROUS FLUORIDE see CDA750
CEROUS NITRATE HEXAHYDRATE see CDB250
CEROXONE see BJK750
CERTICOL CARMOISINE S see HJF500

CERTICOL ORANGE GS see HGC000
CERTIQUAL EOSINE see BNH500
CERTIQUAL FLUORESCEINE see FEW000
CERTIQUAL RHODAMIEN see FAG070
CERTOX see SMN500
CERTROL see HKB500
CERUBIDIN see DAC000
CERUSSETE see LCP000
CERVEN KUMIDINOVA see FAG018
CERVICUNDIN see EQJ500
CES see ECU750, SOP500
CESIUM see CDC000
CESIUM-133 see CDC000
CESIUM BROMIDE see CDC500
CESIUM BROMOXENATE see CDC699
CESIUM CARBONATE see CDC750
CESIUM CHLORIDE see CDD000
CESIUM FLUORIDE see CDD500
CESIUM HYDRATE see CDD750
CESIUM HYDROXIDE see CDD750
CESIUM HYDROXIDE, solid (DOT) see CDD750
CESIUM HYDROXIDE, solution (DOT) see CDD750
CESIUM HYDROXIDE DIMER see CDD750
CESIUM IODIDE see CDE000
CESIUM METAL (DOT) see CDC000
CESIUM MONOCHLORIDE see CDD000
CESIUM MONOFLUORIDE see CDD500
CESIUM NITRATE (DOT) see CDE250
CESIUM(I) NITRATE (1:1) see CDE250
CESIUM, POWDERED (DOT) see CDC000
CET see BJP000
CETAB see HCQ500
CETADOL see HIM000
CETAFFINE see HCP000
CETAIN see AIT250
CETAL see HCP000
CETALOL CA see HCP000
CETAMIUM see CCX000
CETARIN see MKR250
CETAROL see HCQ500
CETAVLON see HCQ500
CETIL LIGHT ORANGE GG see HGC000
CETONE V see AGI500
CETRIMIDE see HCQ500
CETRIMONIUM BROMIDE see HCQ500
CETYL ALCOHOL see HCP000
CETYLAMIN (GERMAN) see HCO500
CETYLAMINE see HCO500, HCQ500
CETYL-γ-AMINOBUTYRATE see AJC500
CETYL GABA see AJC500
CETYLIC ACID see PAE250
CETYLIC ALCOHOL see HCP000
CETYLOL see HCP000
CETYLPYRIDINIUM CHLORIDE see CCX000
1-CETYLPYRIDINIUM CHLORIDE see CCX000
N-CETYLPYRIDINIUM CHLORIDE see CCX000
CETYLPYRIDINIUM CHLORIDE MONOHYDRATE see
　　CDF750
CETYLTRIMETHYLAMMONIUM BROMIDE see HCQ500
N-CETYLTRIMETHYLAMMONIUM BROMIDE see HCQ500
CEVADILLA see VHZ000
CEVADINE see VHZ000
CEVANOL see BCA000
CEVIAN A 678 see AAX250
CEVIAN HL see ADY500
CEVITAMIC ACID see ARN000
CEVITAMIN see ARN000
CEYLON ISINGLASS see AEX250
CFC see PGE000
CFC 31 see CHI900
C.F.S. see TEM000
CFV see CDS750
CFX see CCS500
CG 315 see THJ500

CG-1283 see MQW500
CGA 15324 see BNA750
CGA 26351 see CDS750
C-GREEN 10 see BLK000
CH see SBW500
CHA see CPF500
CHALCEDONY see SCI500, SCJ500
CHALK see CAO000
CHALOTHANE see HAG500
CHAMBER CRYSTALS see NMJ000
CHAMELEON MINERAL see PLP000
CHAMOMILE see CDH250
CHAMOMILE-GERMAN OIL see CDH500
CHAMOMILE OIL see CDH500
CHAMOMILE OIL (ROMAN) see CDH750
CHANNEL BLACK see CBT750
CHANNING'S SOLUTION see NCP500
CHARCOAL see CDI250
CHARCOAL, ACTIVATED (DOT) see CDI000
CHARCOAL BLACK see CBT500
CHARCOAL (BRIQUETTES) see CDI250
CHARCOAL SCREENINGS, MADE from ''PINON'' WOOD
   (DOT) see CDI500
CHARCOAL (SHELL) see CDJ000
CHARCOAL, SHELL (DOT) see CDJ000
CHARCOAL (wood, ground, crushed, granulated or pulverized)
   see CDJ500
CHARCOAL WOOD SCREENINGS, OTHER THAN ''PINON''
   WOOD SCREENINGS (DOT) see CDK000
CHAVICOL METHYL ETHER see AFW750
1,3-CHBP see BNA825
CHEELOX BF see EIV000
CHEELOX BF ACID see EIX000
CHEELOX BR-33 see EIV000
CHEL 330 see DJG800
CHEL 330 ACID see DJG800
CHELADRATE see EIX500
CHELAFER see FBC100
CHELAFRIN see VGP000
CHELAPLEX III see EIX500
CHELATON III see EIX500
CHEL DTPA see DJG800
CHELEN see EHH000
CHELIDONINE see CDL000
CHEL-IRON see FBC100
CHELON 100 see EIV000
CHEMAGRO 1,776 see BSH250
CHEMAGRO 2353 see DFV400
CHEMAGRO 5461 see BJD000
CHEMAGRO 25141 see FAQ800
CHEMAGRO 37289 see EPY000
CHEMAGRO B-1776 see BSH250, TIG250
CHEMAGRO B-1843 see BLG500
CHEMAGRO B-9002 see HMV000
CHEMAGRO R-5461 see BJD000
CHEMAID see HKC500
CHEMANOX 11 see BFW750
CHEMATHION see MAK700
CHEM BAM see DXD200
CHEMBUTAZONE see BRF500
CHEMCOCCIDE see RLK890
CHEMCOLOX 200 see EIV000
CHEMCOLOX 340 see EIX000
CHEM FISH see RNZ000
CHEMFORM see DKC800, MEI450, SLW500
CHEMIAZID see ILD000
CHEMICAL 109 see AQN635
CHEMICAL MACE see CEA750
CHEMICETIN see CDP250
CHEMICETINA see CDP250
CHEMI-CHARL see SHM500
CHEMIDON see ILD000
CHEMIFLUOR see SHF500
CHEMIOFURAN see NGE000

CHEMLON see PJY500
CHEM-MITE see RNZ000
CHEM NEB see MAS500
CHEMOCCIDE see RLK890
CHEMOFURAN see NGE500
CHEMOSEPT see TEX250
CHEMOX GENERAL see BRE500
CHEMOX P.E. see BRE500
CHEMOX PE see DUZ000
CHEMOX SELECTIVE see BPG250
CHEM PELS C see SEY500
CHEM-PHENE see CDV100
CHEMRAT see PIH175
CHEM RICE see DGI000
CHEMSECT DNOC see DUS700
CHEM-SEN 56 see SEY500
CHEM-TOL see PAX250
CHEM ZINEB see EIR000
CHENOPODIUM OIL see CDL500
CHEQUE see MQS225
CHERRY LAUREL OIL see CDM000
CHERTS see SCI500, SCJ500
CHESTNUT TANNIN see CDM250
CHEVRON 9006 see DTQ400
CHEVRON ORTHO 9006 see DTQ400
CHEVRON RE 5655 see MOU750
CHEVRON RE 12,420 see DOP600
CHEWING TOBACCO see SED400
CHEXMATE see HKC000
CHICLIDA see HGC500
CHILE SALTPETER see SIO900
CHIMCOCCIDE see RLK890
CHIMOREPTIN see DLH630
CHINA CLAY see KBB600
CHINALPHOS see DJY200
CHINAWOOD OIL see TOA510
CHINESE ISINGLASS see AEX250
CHINESE RED see LCS000
CHINESE SEASONING see MRL500
CHINESE WHITE see ZKA000
CHINIDIN (GERMAN) see QFS000
CHININ (GERMAN) see QHJ000
CHININDIHYDROCHLORID (GERMAN) see QIJ000
CHINOFER see IGS000
CHINOFORM see CHR500
CHINOIN see EID000
CHINOLEINE see QMJ000
CHINOLIN (CZECH) see QMJ000
CHINOLINE see QMJ000
p-CHINON (GERMAN) see QQS200
CHINON (DUTCH, GERMAN) see QQS200
CHINONE see QQS200
CHINONOXIM-BENZOYLHYDRAZON (GERMAN) see
   BDD000
CHINONOXIME-BENZOYLHYDRAZONE see BDD000
CHINORTA see NIM500
CHIPCO 26019 see GIA000
CHIPCO BUCTRIL see DDP000
CHIPCO CRAB-KLEEN see DDP000, DXE600
CHIPCO THIRAM 75 see TFS350
CHIPCO TURF HERBICIDE ''D'' see DAA800
CHIPCO TURF HERBICIDE MCPP see CIR500
CHIPMAN 6200 see DJA400
CHIPMAN 11974 see BDJ250
CHIPTOX see CIR250
CHISSO 507B see PMP500
CHISSONOX 206 see VOA000
ChKhZ 21 see ASM270
ChKhZ 21R see ASM270
CHLODITAN see CDN000
CHLODITHANE see CDN000
CHLOFENVINPHOS see CDS750
CHLOMIN see CDP250
CHLOMYCOL see CDP250

CHLOOR (DUTCH) see CDV750
3-CHLOORANILINEN (DUTCH) see CEH675
2-CHLOORBENZALDEHYDE (DUTCH) see CEI500
o-CHLOORBENZALDEHYDE (DUTCH) see CEI500
CHLOORBENZEEN (DUTCH) see CEJ125
CHLOORBENZIDE (DUTCH) see CEP000
(4-CHLOOR-BENZYL)-(4-CHLOOR-FENYL)-SULFIDE
    (DUTCH) see CEP000
2-CHLOOR-1,3-BUTADIEEN (DUTCH) see NCI500
(4-CHLOOR-BUT-2-YN-YL)-N-(3-CHLOOR-FENYL)-CARBA-
    MAAT (DUTCH) see CEW500
CHLOORDAAN (DUTCH) see CDR750
O-2-CHLOOR-1-(2,4-DICHLOOR-FENYL)-VINYL-O,O-DI-
    ETHYLFOSFAAT (DUTCH) see CDS750
2-CHLOOR-4-DIMETHYLAMINO-6-METHYL-PYRIMIDINE
    (DUTCH) see CCP500
1-CHLOOR-2,4-DINITROBENZEEN (DUTCH) see CGM000
1-CHLOOR-2,3-EPOXY-PROPAAN (DUTCH) see EAZ500
CHLOORETHAAN (DUTCH) see EHH000
2-CHLOORETHANOL (DUTCH) see EIU800
CHLOORFACINON (DUTCH) see CJJ000
CHLOORFENSON (DUTCH) see CJT750
(4-CHLOOR-FENYL)-BENZEEN-SULFONAAT (DUTCH) see
    CJR500
(4-CHLOOR-FENYL)-4-CHLOOR-BENZEEN-SULFONAAT
    (DUTCH) see CJT750
3-(4-CHLOOR-FENYL)-1,1-DIMETHYLUREUM (DUTCH) see
    CJX750
2(2-(4-CHLOOR-FENYL-2-FENYL)-ACETYL)-INDAAN-1,3-
    DION (DUTCH) see CJJ000
N-(3-CHLOOR-FENYL)-ISOPROPYL CARBAMAAT (DUTCH)
    see CKC000
CHLOOR-METHAAN (DUTCH) see MIF765
2-(4-CHLOOR-2-METHYL-FENOXY)-PROPIONZUUR
    (DUTCH) see CIR500
1-CHLOOR-4-NITROBENZEEN (DUTCH) see NFS525
O-(3-CHLOOR-4-NITRO-FENYL)-O,O-DIMETHYL-MONO-
    THIOFOSFAAT (DUTCH) see MIJ250
O-(4-CHLOOR-3-NITRO-FENYL)-O,O-DIMETHYLMONO-
    THIOFOSFAAT (DUTCH) see NFT000
CHLOORPIKRINE (DUTCH) see CKN500
CHLOORTHION (DUTCH) see MIJ250
CHLOORWATERSTOF (DUTCH) see HHL000
CHLOPHEN see PJL750
CHLOR (GERMAN) see CDV750
CHLORACETAMID (GERMAN) see CDY850
CHLORACETIC ACID see CEA000
CHLORACETONE see CDN200, CDN200
CHLORACETONITRILE see CDN500
CHLORACETYL CHLORIDE see CEC250
2-CHLORAETHANOL (GERMAN) see EIU800
N-(2-CHLORAETHYL)-N′-(2 CHLOROETHYL)-N′-o-PRO-
    PYLEN-PHOSPHORSAUREESTER-DIAMID (GERMAN)
    see IMH000
2-CHLORAETHYL-PHOSPHONSAEURE (GERMAN) see
    CDS125
CHLORAK see TIQ250
CHLORALDURAT see CDO000
CHLORAL HYDRATE see CDO000
CHLORALLYL DIETHYLDITHIOCARBAMATE see CDO250
2-CHLORALLYL DIETHYLDITHIOCARBAMATE see
    CDO250
CHLORALLYLENE see AGB250
CHLORALONE see CDP000
CHLORALOSANE see GFA000
α-CHLORALOSE see GFA000
CHLORAMBUCIL see CDO500
CHLORAMEISENSAEURE METHYLESTER (GERMAN) see
    MIG000
CHLORAMEX see CDP250
CHLORAMFICIN see CDP250
CHLORAMFILIN see CDP250
CHLORAMIN see BIE500
CHLORAMINE see BIE500

CHLORAMINE BLACK C see AQP000
CHLORAMINE BLUE see CMO250
CHLORAMINE BLUE 2B see CMO000
CHLORAMINE FAST BROWN BRL see CMO750
CHLORAMINE T see CDP000
CHLORAMIN HYDROCHLORIDE see BIE500
CHLORAMINOPHEN see CDO500
CHLORAMINOPHENE see CDO500
CHLORAMIPHENE see CMX700
CHLORAMIPHENE CITRATE see CMX700
CHLORAMP (RUSSIAN) see PIB900
CHLORAMPHENICOL see CDP250
d-CHLORAMPHENICOL see CDP250
d-threo-CHLORAMPHENICOL see CDP250
CHLORAMPHENICOL MONOPALMITATE see CDP700
CHLORAMPHENICOL PALMITATE see CDP700
CHLORAMSAAR see CDP250
CHLORANAUTINE see DYE600
4-CHLORANILIN (CZECH) see CEH680
m-CHLORANILINE see CEH675
o-CHLORANILINE see CEH670
p-CHLORANILINE see CEH680
1-CHLORANTHRACHINON (CZECH) see CEI000
CHLORARSOL see DFX400
CHLORASAN see CDP000
CHLORASEN see DFX400
CHLORASEPTINE see CDP000
CHLORASOL see CDP250
CHLORA-TABS see CDP250
CHLORATE de CALCIUM (FRENCH) see CAO500
CHLORATE OF POTASH (DOT) see PLA250
CHLORATE de POTASSIUM (FRENCH) see PLA250
CHLORATES see CDQ000
CHLORATE SALT of MAGNESIUM see MAE000
CHLORATE SALT of SODIUM see SFS000
CHLORATE of SODA (DOT) see SFS000
CHLORAX see SFS000
CHLORAZAN see CDP000
CHLORAZENE see CDP000
CHLORAZOL BLACK E (biological stain) see AQP000
CHLORAZOL BLACK EA see AQP000
CHLORAZOL BLACK EN see AQP000
CHLORAZOL BLUE 3B see CMO250
CHLORAZOL BLUE B see CMO000
CHLORAZONE see CDP000
CHLORBENSID (GERMAN) see CEP000
CHLORBENSIDE see CEP000
CHLORBENXIDE see CEP000
2-CHLORBENZALDEHYD (GERMAN) see CEI500
CHLORBENZENE see CEJ125
CHLORBENZIDE see CEP000
CHLORBENZILATE see DER000
CHLORBENZOL see CEJ125
o-CHLORBENZONITRIL (CZECH) see CEM000
N-p-CHLORBENZOYL-5-METHOXY-2-METHYLINDOLE-3-
    ACETIC ACID see IDA000
(4-CHLOR-BENZYL)-(4-CHLOR-PHENYL)-SULFID (GER-
    MAN) see CEP000
CHLORBICYCLENE (FRENCH) see DAM700
2-CHLOR-1,3-BUTADIEN (GERMAN) see NCI500
CHLORBUTANOL see ABD000
4-CHLORBUTAN-1-OL (GERMAN) see CEU500
(4-CHLOR-BUT-2-IN-YL)-N-(3-CHLOR-PHENYL)-CARBA-
    MAT (GERMAN) see CEW500
CHLORBUTOL see ABD000
p-CHLOR-m-CRESOL see CFE250
CHLORCYAN see COO750
CHLORCYCLINE see CFF500
CHLORCYCLIZINE see CFF500
CHLORCYCLIZINE DIHYDROCHLORIDE see CDR000
CHLORCYCLIZINE HYDROCHLORIDE see CDR250
CHLORCYCLIZINIUM CHLORIDE see CDR250
CHLORDAN see CDR750
γ-CHLORDAN see CDR750

CHLORDANE see CDR750
CHLORDANE, liquid (DOT) see CDR750
CHLORDECONE see KEA000
CHLORDIAZACHEL see MDQ250
CHLORDIAZEPOXIDE see LFK000
CHLORDIAZEPOXIDE HYDROCHLORIDE see MDQ250
CHLORDIAZEPOXIDE MONOHYDROCHLORIDE see MDQ250
O-2-CHLOR-1-(2,4-DICHLOR-PHENYL)-VINYL-O,O-DIA-ETHYLPHOSPHAT (GERMAN) see CDS750
CHLORDIMEFORM see CJJ250
2-CHLOR-4-DIMETHYLAMINO-6-METHYLPYRIMIDIN (GERMAN) see CCP500
CHLORDIMETHYLETHER (CZECH) see CIO250
1-CHLOR-2,4-DINITROBENZENE see CGM000
CHLORE (FRENCH) see CDV750
CHLOREFENIZON (FRENCH) see CJT750
CHLORENDIC ACID see CDS000
CHLOREPIN see CIR750
1-CHLOR-2,3-EPOXY-PROPAN (GERMAN) see EAZ500
CHLORESENE see BBQ500
CHLORESSIGSAEURE-N-ISOPROPYLANILID (GERMAN) see CHS500
CHLORESSIGSAEURE-N-(METHOXYMETHYL)-2,6-DIA-ETHYLANILID (GERMAN) see CFX000
CHLORESTROLO see CLO750
CHLORETHAMINACIL see BIA250
CHLORETHAMINE see BIE500
2-CHLORETHANOL (GERMAN) see EIU800
CHLORETHAZINE see BIE500
CHLORETHENE see VNP000
CHLORETHEPHON see CDS125
CHLORETHYL see EHH000
CHLORETHYLENE see VNP000
2-CHLORETHYLPHOSPHONIC ACID see CDS125
2-CHLORETHYL VINYL ETHER see CHI250
CHLORETONE see ABD000
CHLOREX see DFJ050
CHLOREXTOL see PJL750
CHLORFACINON (GERMAN) see CJJ000
CHLORFENAMIDINE see CJJ250
CHLORFENIDIM see CJX750
p-CHLORFENOL (CZECH) see CJK750
CHLORFENSON see CJT750
CHLORFENSONE see CJT750
CHLORFENVINFOS see CDS750, CDS750
CHLORFENVINPHOS see CDS750
p-CHLORFENYLISOKYANAT (CZECH) see CKB000
CHLORFOS see TIQ250
CHLOR-N-(2-FURYLMETHYL)-5-SULFAMYLANTHRANIL-SAEURE (GERMAN) see CHJ750
CHLORGUANIDE see CKB250
CHLORHEXIDIN (CZECH) see BIM250
CHLORHEXIDINE see BIM250
CHLORHYDRATE d'ANILINE (FRENCH) see BBL000
CHLORHYDRATE de 4-CHLOROORTHOTOLUIDINE (FRENCH) see CLK235
CHLORHYDRATE de NICOTINE (FRENCH) see NDP400
CHLORHYDRATE de TETRACYCLINE (FRENCH) see TBX250
CHLORHYDRIN see CDT750
α-CHLORHYDRIN see CDT750
CHLORIC ACID see CDU000
CHLORIC ACID, solution, containing not more than 10% acid (DOT) see CDU000
CHLORIC ACID, BARIUM SALT see BAJ500
CHLORICOL see CDP250
CHLORID AMONNY (CZECH) see ANE500
CHLORID ANILINU (CZECH) see BBL000
CHLORID ANTIMONITY see AQC500
CHLORIDAZON see PEE750
CHLORID DI-n-BUTYLCINICITY (CZECH) see DDY200
CHLORID DRASELNY (CZECH) see PLA500
CHLORIDEAZEPOXIDE HYDROCHLORIDE see MDQ250

CHLORIDE de CHOLINE (FRENCH) see CMF750
CHLORIDE of LIME (DOT) see HOV500
CHLORIDE of PHOSPHORUS see PHT275
CHLORIDES see CDU250
CHLORIDE of SULFUR (DOT) see SOG500, SON510
CHLORID FENYLRTUTNATY (CZECH) see PFM500
CHLORIDIAZEPIDE see LFK000
CHLORIDIAZEPOXIDE see LFK000
CHLORIDIN see TGD000
CHLORIDINE see TGD000
CHLORID KREMICITY (CZECH) see SCQ500
CHLORID KYSELINY CHLORMETHANSULFONOVE (CZECH) see CHY000
CHLORID RTUTNATY (CZECH) see MCY475
CHLORID TRI-n-BUTYLCINICITY (CZECH) see CLP500
CHLORIDUM see EHH000
CHLORIERTE BIPHENYLE, CHLORGEHALT 42% (GER-MAN) see PJM500
CHLORIERTE BIPHENYLE, CHLORGEHALT 54% (GER-MAN) see PJN000
CHLOR-IFC see CKC000
CHLORINAT see CEW500
CHLORINATED BIPHENYL see PJL750
CHLORINATED CAMPHENE see CDV100
CHLORINATED DIPHENYL see PJL750
CHLORINATED DIPHENYLENE see PJL750
CHLORINATED DIPHENYL OXIDE see CDV175
CHLORINATED HC, ALIPHATIC see CDV250
CHLORINATED HC AROMATIC see CDV500
CHLORINATED HYDROCARBONS, ALIPHATIC see CDV250
CHLORINATED HYDROCARBONS, AROMATIC see CDV500
CHLORINATED HYDROCHLORIC ETHER see DFF809
CHLORINATED LIME (DOT) see HOV500
CHLORINATED PARAFFINS (C12, 60% CHLORINE) see PAH800
CHLORINATED POLYETHER POLYURETHAN see CDV625
CHLORINDAN see CDR750
CHLORINE see CDV750
CHLORINE AZIDE see CDW000
CHLORINE CYANIDE see COO750
CHLORINE DIOXIDE see CDW450
CHLORINE DIOXIDE, not hydrated (DOT) see CDW450
CHLORINE FLUORIDE see CDX750
CHLORINE FLUORIDE (ClF$_5$) see CDX250
CHLORINE FLUORIDE OXIDE see PCF750
CHLORINE MOL. see CDV750
CHLORINE NITRIDE (NITROGEN) TRICHLORIDE see NGQ500
CHLORINE OXIDE see CDW450
CHLORINE(IV) OXIDE see CDW450
CHLORINE OXYFLUORIDE see PCF750
CHLORINE PENTAFLUORIDE see CDX250
CHLORINE PENTAFLUORIDE (DOT) see CDX250
CHLORINE PEROXIDE see CDW450
CHLORINE SULFIDE see SOG500
CHLORINE TETROXYFLUORIDE see FFD000
CHLORINE TRIFLUORIDE see CDX750
CHLOR-IPC see CKC000
CHLORITES see CDY250
5-CHLOR-7-JOD-8-8HYDROXY-CHINOLIN (GERMAN) see CHR500
CHLOR KIL see CDR750
CHLORKU LITU (POLISH) see LHB000
CHLORMADINON ACETATE see CBF250
CHLORMADINONE ACETATE see CBF250
CHLORMADINONU (POLISH) see CBF250
CHLORMENE see TAI500
CHLOR-METHAN (GERMAN) see MIF765
CHLORMETHANSULFOCHLORID (CZECH) see CHY000
CHLORMETHINE see BIE250
CHLORMETHINE HYDROCHLORIDE see BIE500
CHLORMETHINE-N-OXIDE HYDROCHLORIDE see CFA750

CHLORMETHINUM see BIE500
2-(4-CHLOR-2-METHYL-PHENOXY)-PROPIONSAEURE
(GERMAN) see CIR500
3-CHLOR-2-METHYL-PROP-1-EN (GERMAN) see CIU750
CHLORNAFTINA see BIF250
CHLORNAPHAZIN see BIF250
CHLORNAPHTHIN see BIF250
1-CHLOR-4-NITROBENZOL (GERMAN) see NFS525
CHLORNITROMYCIN see CDP250
O-(3-CHLOR-4-NITRO-PHENYL)-O,O-DIMETHYL-MONO-
THIOPHOSPHAT (GERMAN) see MIJ250
O-(4-CHLOR-3-NITRO-PHENYL)-O,O-DIMETHYL-MONO-
THIOPHOSPHAT (GERMAN) see NFT000
CHLOROACETALDEHYDE see CDY500
2-CHLOROACETALDEHYDE see CDY500
CHLOROACETALDEHYDE MONOMER see CDY500
CHLOROACETAMIDE see CDY850
2-CHLORO ACETAMIDE see CDY850
α-CHLOROACETAMIDE see CDY850
CHLOROACETIC ACID see CEA000
α-CHLOROACETIC ACID see CEA000
CHLOROACETIC ACID, liquid (DOT) see CEA000
CHLOROACETIC ACID, solid (DOT) see CEA000
CHLOROACETIC ACID CHLORIDE see CEC250
CHLOROACETIC ACID, ETHYL ESTER see EHG500
CHLOROACETIC ACID SODIUM SALT see SFU500
CHLOROACETIC CHLORIDE see CEC250
CHLOROACETONE see CDN200
CHLOROACETONE, stabilized (DOT) see CDN200
2-CHLOROACETONITRILE see CDN500
α-CHLOROACETONITRILE see CDN500
CHLOROACETONITRILE (DOT) see CDN500
1-CHLOROACETOPHENONE see CEA750
4-CHLOROACETOPHENONE see CEB250
p-CHLOROACETOPHENONE see CEB250
α-CHLOROACETOPHENONE see CEA750
4′-CHLOROACETOPHENONE see CEB250
omega-CHLOROACETOPHENONE see CEA750
CHLOROACETOPHENONE, gas, liquid or solid (DOT) see
CEA750
6-CHLORO-17-α-ACETOXY-4,6-PREGNADIENE-3,20-DIONE
see CBF250
6-CHLORO-Δ⁶-17-ACETOXYPROGESTERONE see CBF250
Δ⁶-6-CHLORO-17-α-ACETOXYPROGESTERONE see
CBF250
6-CHLORO-Δ⁶-(17-α)ACETOXYPROGESTERONE see
CBF250
4′-CHLOROACETYL ACETANILIDE see CEC000
4′-(CHLOROACETYL)ACETANILIDE see CEC000
CHLOROACETYL CHLORIDE see CEC250
CHLOROAETHAN (GERMAN) see EHH000
α-CHLOROALLYL CHLORIDE see DGG950
γ-CHLOROALLYL CHLORIDE see DGG950
2-CHLOROALLYL DIETHYLDITHIOCARBAMATE see
CDO250
2-CHLOROALLYL-N,N-DIETHYLDITHIOCARBAMATE see
CDO250
CHLOROALLYLENE see AGB250
CHLOROALONIL see TBQ750
CHLOROALOSANE see GFA000
CHLOROAMBUCIL see CDO500
p-CHLORO-o-AMINOPHENOL see CEH250
4-CHLORO-2-AMINOTOLUENE see CLK225
5-CHLORO-2-AMINOTOLUENE see CLK220
5-CHLORO-2-AMINOTOLUENE HYDROCHLORIDE see
CLK235
2-CHLOROANILINE see CEH670
3-CHLOROANILINE see CEH675
4-CHLOROANILINE see CEH680
m-CHLOROANILINE see CEH675
o-CHLOROANILINE see CEH670
p-CHLOROANILINE see CEH680
3-CHLOROANILINE (ITALIAN) see CEH675
m-CHLOROANILINE, liquid (DOT) see CEH675

m-CHLOROANILINE, solid (DOT) see CEH675
o-CHLOROANILINE, liquid (DOT) see CEH670
o-CHLOROANILINE, solid (DOT) see CEH670
p-CHLOROANILINE, liquid (DOT) see CEH680
p-CHLOROANILINE, solid (DOT) see CEH680
1-CHLORO-9,10-ANTHRACENEDIONE see CEI000
1-CHLOROANTHRAQUINONE see CEI000
α-CHLOROANTHRAQUINONE see CEI000
1-CHLORO-9,10-ANTHRAQUINONE see CEI000
CHLOR(O)AZIDE see CDW000
CHLOROBEN see DEP600
2-CHLOROBENZALDEHYDE see CEI500
o-CHLOROBENZALDEHYDE see CEI500
α-CHLOROBENZALDEHYDE see BDM500
2-CHLOROBENZAL MALONONITRILE see CEQ600
o-CHLOROBENZAL MALONONITRILE see CEQ600
CHLOROBENZEN (POLISH) see CEJ125
3-CHLOROBENZENAMINE see CEH675
4-CHLOROBENZENAMINE see CEH680
2-CHLORO-BENZENAMINE (9CI) see CEH670
CHLOROBENZENE see CEJ125
4-CHLORO BENZENEAMINE see CEH680
o-CHLOROBENZENECARBOXALDEHYDE see CEI500
4-CHLORO-1,3-BENZENEDIAMINE see CJY120
4-CHLOROBENZENESULFONATE de 4-CHLOROPHENYLE
(FRENCH) see CJT750
p-CHLOROBENZENESULFONIC ACID-p-CHLOROPHENYL
ESTER see CJT750
1-(p-CHLOROBENZHYDRYL)-4-(2-(2-HYDROXYETHOXY)
ETHYL)DIETHYLENEDIAMINE see CJR909
N-(4-CHLOROBENZHYDRYL)-N′-(HYDROXYETHOXY-
ETHYL)PIPERAZINE see CJR909
1-(p-CHLOROBENZHYDRYL)-4-(2-(2-HYDROXYETHOXY)
ETHYL)PIPERAZINE see CJR909
1-(p-CHLOROBENZHYDRYL)-4-(m-METHYLBENZYL)DI-
ETHYLENEDIAMINE see HGC500
1-p-CHLOROBENZHYDRYL-4-m-METHYLBENZYLPIPER-
AZINE see HGC500
1-(4-CHLOROBENZHYDRYL)-4-METHYLPIPERAZINE see
CFF500
1-(4-CHLOROBENZHYDRYL)-4-METHYLPIPERAZINE DI-
HYDROCHLORIDE see CDR000
1-(p-CHLOROBENZHYDRYL)-4-METHYLPIPERAZINE HY-
DROCHLORIDE see CDR250
CHLOROBENZOL (DOT) see CEJ125
o-CHLOROBENZONITRILE see CEM000
p-CHLOROBENZOTRIFLUORIDE see CEM825
1-(p-CHLOROBENZOYL)-5-METHOXY-2-METHYLINDOLE-
3-ACETIC ACID see IDA000
1-(p-CHLOROBENZOYL)-2-METHYL-5-METHOXYINDOLE-
3-ACETIC ACID see IDA000
1-(p-CHLOROBENZOYL)-2-METHYL-5-METHOXY-3-IN-
DOLE-ACETIC ACID see IDA000
α-(1-(p-CHLOROBENZOYL)-2-METHYL-5-METHOXY-3-IN-
DOLYL)ACETIC ACID see IDA000
p-CHLOROBENZOYL PEROXIDE see BHM750
p-CHLOROBENZOYL PEROXIDE (DOT) see BHM750
CHLOROBENZYLATE see DER000
p-CHLOROBENZYL-p-CHLOROPHENYL SULFIDE see
CEP000
4-CHLOROBENZYL-4-CHLOROPHENYL SULPHIDE see
CEP000
p-CHLOROBENZYL-p-CHLOROPHENYL SULPHIDE see
CEP000
o-CHLOROBENZYLIDENE MALONITRILE see CEQ600
2-CHLOROBENZYLIDENE MALONONITRILE see CEQ600
o-CHLOROBENZYLIDENE MALONONITRILE see CEQ600
CHLORO BIPHENYL see PJL750
CHLORO-1,1-BIPHENYL see PJL750
2-CHLORO-4,6-BIS(ETHYLAMINO)-s-TRIAZINE see BJP000
1-CHLORO-3,5-BISETHYLAMINO-2,4,6-TRIAZINE see
BJP000
2-CHLORO-4,6-BIS(ETHYLAMINO)-1,3,5-TRIAZINE see
BJP000

CHLOROBIS(2-METHYLPROPYL)ALUMINUM see CGB500
CHLOROBLE M see MAS500
2-CHLOROBMN see CEQ600
CHLOROBROMOMETHANE see CES650
1-CHLORO-3-BROMOPROPANE (DOT) see BNA825
omega-CHLOROBROMOPROPANE see BNA825
trans-CHLORO(2-(3-BROMOPROPIONAMIDO)CYCLO-
   HEXYL)MERCURY see CET000
CHLOROBUTADIENE see NCI500
1-CHLOROBUTADIENE see CET250
1-CHLORO-1,3-BUTADIENE see CET250
2-CHLOROBUTA-1,3-DIENE see NCI500
2-CHLORO-1,3-BUTADIENE see NCI500
2-CHLOROBUTANE see CEU250
1-CHLOROBUTANE (DOT) see BQQ750
4-CHLORO-1-BUTANE-OL see CEU500
CHLOROBUTANOL see ABD000
4-CHLOROBUTANOL see CEU500
4-CHLORO-1-BUTANOL see CEU500
CHLOROBUTIN see CDO500
CHLOROBUTINE see CDO500
CHLORO-2-BUTYNYL-m-CHLOROCARBAMATE see
   CEW500
4-CHLOROBUT-2-YNYL-m-CHLOROCARBANILATE see
   CEW500
4-CHLORO-2-BUTYNYL-m-CHLOROCARBANILATE see
   CEW500
4-CHLOROBUT-2-YNYL-3-CHLOROPHENYLCARBAMATE
   see CEW500
4-CHLORO-2-BUTYNYL-N-(3-CHLOROPHENYL)CARBA-
   MATE see CEW500
CHLOROCAINE see AIT250
CHLOROCAMPHENE see CDV100
CHLOROCAPS see CDP250
m-CHLORO CARBANILIC ACID-4-CHLORO-2-BUTYNYL
   ESTER see CEW500
3-CHLOROCARBANILIC ACID, ISOPROPYL ESTER see
   CKC000
m-CHLOROCARBANILIC ACID, ISOPROPYL ESTER see
   CKC000
CHLOROCARBONATE D'ETHYLE (FRENCH) see EHK500
CHLOROCARBONATE de METHYLE (FRENCH) see
   MIG000
CHLOROCARBONIC ACID METHYL ESTER see MIG000
3-CHLOROCHLORDENE see HAR000
1-CHLORO-2-(β-CHLOROETHOXY)ETHANE see DFJ050
2-CHLORO-N-(2-CHLOROETHYL)ETHANAMINE HYDRO-
   CHLORIDE see BHO250
2-CHLORO-N-(2-CHLOROETHYL)-N-METHYLETHAN-
   AMINE HYDROCHLORIDE see BIE500
2-CHLORO-N-(2-CHLOROETHYL)-N-METHYL ETHAN-
   AMINE-N-OXIDE see CFA500
2-CHLORO-N-(2-CHLOROETHYL)-N-METHYLETHAN-
   AMINE-N-OXIDE HYDROCHLORIDE see CFA750
1-CHLORO-2-(β-CHLOROETHYLTHIO)ETHANE see
   BIH250
CHLORO(CHLOROMETHOXY)METHANE see BIK000
5-CHLORO-N-(2-CHLORO-4-NITROPHENYL)-2-HYDROXY-
   BENZAMIDE see DFV400
5-CHLORO-N-(2-CHLORO-4-NITROPHENYL)-2-HYDROXY-
   BENZAMIDE with 2-AMINOETHANOL (1:1) see DFV600
5-CHLORO-2'-CHLORO-4'-NITROSALICYLANILIDE see
   DFV400
1-CHLORO-4-(((4-CHLOROPHENYL)METHYL)THIO)BEN-
   ZENE see CEP000
4-CHLORO-α-(4-CHLOROPHENYL)-α-(TRICHLORO-
   METHYL)BENZENEMETHANOL see BIO750
CHLOROCID see CDP250
CHLOROCIDE see CEP000
CHLOROCIDIN C TETRAN see CDP250
CHLOROCOL see CDP250
CHLOROCRESOL see CFE250
p-CHLOROCRESOL see CFE250

4-CHLORO-m-CRESOL see CFE250
6-CHLORO-m-CRESOL see CFE250
p-CHLORO-m-CRESOL see CFE250
4-CHLORO-o-CRESOXYACETIC ACID see CIR250
CHLOROCTAN SODNY (CZECH) see SFU500
CHLOROCYAN see COO750
CHLOROCYANIDE see COO750
CHLOROCYANOGEN see COO750
2-CHLORO-4-(1-CYANO-1-METHYLETHYLAMINO)-6-
   ETHYLAMINO-1,3,5-TRIAZINE see BLW750
CHLOROCYCLINE see CFF500
CHLOROCYCLIZINE see CFF500
CHLORODANE see CDR750
6-CHLORO-6-DEHYDRO-17-α-ACETOXYPROGESTERONE
   see CBF250
6-CHLORO-Δ⁶-DEHYDRO-17-ACETOXYPROGESTERONE
   see CBF250
6-CHLORO-6-DEHYDRO-17-α-HYDROXYPROGESTERONE
   ACETATE see CBF250
7-CHLORO-6-DEMETHYLTETRACYCLINE DEMETHYL-
   CHLOROTETRACYCLINE see MIJ500
CHLORODEN see DEP600
CHLORODEOXYGLYCEROL see CDT750
2-CHLORO-N,N-DIALLYLACETAMIDE see CFK000
α-CHLORO-N,N-DIALLYLACETAMIDE see CFK000
1-CHLORO-2,4-DIAMINOBENZENE see CJY120
4-CHLORO-1,2-DIAMINOBENZENE see CFK125
CHLORODIAZEPOXIDE see LFK000
CHLORODIBROMOMETHANE see CFK500
1-CHLORO-2,3-DIBROMOPROPANE see DDL800
3-CHLORO-1,2-DIBROMOPROPANE see DDL800
1-CHLORO-2-(2,2-DICHLORO-1-(4-CHLOROPHENYL)
   ETHYL)BENZENE see CDN000
1-CHLORO-2,2-DICHLOROETHYLENE see TIO750
O-(2-CHLORO-1-(2,5-DICHLOROPHENYL)-O,O-DIETHYL
   ESTER PHOSPHOROTHIOIC ACID see DIX600
2-CHLORO-1-(2,4-DICHLOROPHENYL)VINYL DIETHYL
   PHOSPHATE see CDS750
β-2-CHLORO-1-(2',4'-DICHLOROPHENYL) VINYL DI-
   ETHYLPHOSPHATE see CDS750
O-(2-CHLORO-1-(2,5-DICHLOROPHENYL)VINYL)-O,O-DI-
   ETHYL PHOSPHOROTHIOATE see DIX600
2-CHLORO-1-(p-(β-DIETHYLAMINOETHOXY)PHENYL)-1,2-
   DIPHENYLETHYLENE see CMX700
7-CHLORO-1-(2-(DIETHYLAMINO)ETHYL)-5-(2-FLUORO-
   PHENYL)-1H-1,4-BENZODIAZEPIN-2(3H)-ONE see
   FMQ000
6-CHLORO-9-((4-(DIETHYL AMINO)-1-METHYL BUTYL)
   AMINO)-2-METHOXYACRIDINE see ARQ250
2-CHLORO-10-(3'-DIETHYLAMINOPROPYL)PHENO-
   THIAZINE HYDROCHLORIDE see CLZ000
2-CHLORO-2',6'-DIETHYL-N-(METHOXYMETHYL)AC-
   ETANILIDE see CFX000
2-CHLORO-N-(2,6-DIETHYLPHENYL)-N-(METHOXY-
   METHYL)ACETAMIDE see CFX000
CHLORODIFLUOROBROMOMETHANE (DOT) see BNA250
1-CHLORO-1,1-DIFLUOROETHANE see CFX250
CHLORODIFLUOROETHANE (DOT) see CFX250
CHLORODIFLUOROMETHANE see CFX500
2-CHLORO-1-(DIFLUOROMETHOXY)-1,1,2-TRIFLUORO-
   ETHANE see EAT900
CHLORODIFLUOROMONOBROMOMETHANE see BNA250
10-CHLORO-5,10-DIHYDROARSACRIDINE see PDB000
6-CHLORO-3,4-DIHYDRO-2H-1,2,4-BENZOTHIADIAZINE-7-
   SULFONAMIDE- 1,1-DIOXIDE see CFY000
S-(2-CHLORO-1-(1,3-DIHYDRO-1,3-DIOXO-2H-ISOINDOL-2-
   YL)ETHYL)-O,O-DIETHYL PHOSPHORODITHIOATE see
   DBI099
(R)N-((5-CHLORO-3,4-DIHYDRO-8-HYDROXY-3-METHYL-1-
   OXO-1H-2-BENZOPYRAN-7-YL)PHENYLALANINE see
   CHP250
10-CHLORO-5,10-DIHYDROPHENARSAZINE see PDB000
7-CHLORO-1,3-DIHYDRO-5-PHENYL-1-TRIMETHYLSILYL-
   2H-1,4-BENZODIAZEPIN-2-ONE see CGB000

6-CHLORO-3,4-DIHYDRO-7-SULFAMOYL-2H-1,2,4-BENZO-THIADIAZINE-1,1-DIOXIDE see CFY000
1-CHLORO-2,3-DIHYDROXYPROPANE see CDT750
3-CHLORO-1,2-DIHYDROXYPROPANE see CDT750
CHLORO DIISOBUTYL ALUMINUM see CGB500
p-CHLORO DIMETHYLAMINOAZOBENZENE see CGD250
4'-CHLORO-4-DIMETHYLAMINOAZOBENZENE see CGD250
CHLORO(DIMETHYLAMINO)ETHANE see CGW000
dl-2(-p-CHLORO-α-2-(DIMETHYLAMINO)ETHYLBENZYL) PYRIDINE BIMALEATE see TAI500
(+)-2-(p-CHLORO-α-(2-(DIMETHYLAMINO)ETHYL) BENZYL)PYRIDINE MALEATE see PJJ325
2-CHLORO-4-DIMETHYLAMINO-6-METHYL-PYRIMIDINE see CCP500
7-CHLORO-4-(DIMETHYLAMINO)-1,4,4a,5,5a,6,11,12a-OC-TAHYDRO-2-NAPHTHACENECARBOXAMIDE see CMA750
2-CHLORO-N,N-DIMETHYLETHYLAMINE HYDROCHLO-RIDE see DRC000
p-CHLORO-N-α-DIMETHYLPHENETHYLAMINE see CIF250
4-CHLORO-3,5-DIMETHYLPHENOL see CLW000
((4-CHLORO-6-((2,3-DIMETHYLPHENYL)AMINO)-2-PYRIMI-DINYL)THIO)ACETIC ACID see CLW250
CHLORODINITROBENZENE see CGL750
1-CHLORO-2,4-DINITROBENZENE see CGM000
4-CHLORO-1,3-DINITROBENZENE see CGM000
6-CHLORO-1,3-DINITROBENZENE see CGM000
CHLORODINITROBENZENE (DOT) see CGL750
CHLORODINITRO BENZENE (mixed isomers) see CGL750
1-CHLORO-2,4-DINITROBENZOL (GERMAN) see CGM000
1-CHLORO-2,4-DINITRONAPHTHALENE see DUS600
2-CHLORO-2,2-DIPHENYLACETIC ACID-2-(DIETHYLAMI-NO)ETHYL ESTER HYDROCHLORIDE see DHW200
CHLORODIPHENYLARSINE see CGN000
CHLORODIPHENYL (21% Cl) see PJM000
CHLORODIPHENYL (32% Cl) see PJM250
CHLORODIPHENYL (48% Cl) see PJM750
CHLORODIPHENYL (60% Cl) see PJN250
CHLORODIPHENYL (62% Cl) see PJN500
CHLORODIPHENYL (68% Cl) see PJN750
CHLORODIPHENYL (42% Cl) (OSHA) see PJM500
CHLORODIPHENYL (54% Cl) (OSHA) see PJN000
1-(p-CHLORODIPHENYLMETHYL)-4-(2-(2-HYDROXY-ETHOXY)ETHYL)PIPERAZINE see CJR909
2-(p-(2-CHLORO-1,2-DIPHENYL VINYL)PHENOXY)TRI-ETHYLAMINE CITRATE (1:1) see CMX700
2-CHLORO-N,N-DI-2-PROPENYLACETAMIDE see CFK000
1-CHLORO-2,3-EPOXYPROPANE see EAZ500
3-CHLORO-1,2-EPOXYPROPANE see EAZ500
CHLOROETENE see MIH275
2-CHLORO-1-ETHANAL see CDY500
2-CHLOROETHANAMIDE see CDY850
CHLOROETHANE see EHH000
2-CHLOROETHANEPHOSPHONIC ACID see CDS125
CHLOROETHANOIC ACID see CEA000
2-CHLOROETHANOL (MAK) see EIU800
Δ-CHLOROETHANOL see EIU800
2-CHLOROETHANOL-2-(p-tert-BUTYLPHENOXY)-1-METH-YLETHYL SULFITE see SOP500
2-CHLOROETHANOL ESTER with 2-(p-tert-BUTYLPHEN-OXY)-1-METHYLETHYL SULFITE see SOP500
2-CHLOROETHANOL HYDROGEN PHOSPHATE ESTER with 3-CHLORO-7-HYDROXY-4-METHYLCOUMARIN see DFH600
2-CHLOROETHANOL PHOSPHATE see CGO500
2-CHLOROETHANOL PHOSPHATE DIESTER ESTER with 3-CHLORO-7-HYDROXY-4-METHYLCOUMARIN see DFH600
CHLOROETHENE see MIH275, VNP000
CHLOROETHENE HOMOPOLYMER see PKQ059
(2-CHLOROETHENYL) ARSONOUS DICHLORIDE see CLV000
1,1',1''-(1-CHLORO-1-ETHENYL-2-YLIDENE)-TRIS(4-ME-THOXYBENZENE) see CLO750

(2-CHLOROETHOXY)ETHENE see CHI250
2-CHLOROETHYL ALCOHOL see EIU800
β-CHLOROETHYL ALCOHOL see EIU800
2-CHLORO-4-ETHYLAMINEISOPROPYLAMINE-s-TRIAZINE see ARQ725
2-CHLORO-4-ETHYLAMINO-6-(1-CYANO-1-METHYL) ETHYLAMINO-s-TRIAZINE see BLW750
1-CHLORO-3-ETHYLAMINO-5-ISOPROPYLAMINO-s-TRI-AZINE see ARQ725
2-CHLORO-4-ETHYLAMINO-6-ISOPROPYLAMINO-s-TRI-AZINE see ARQ725
1-CHLORO-3-ETHYLAMINO-5-ISOPROPYLAMINO-2,4,6-TRIAZINE see ARQ725
2-CHLORO-4-ETHYLAMINO-6-ISOPROPYLAMINO-1,3,5-TRIAZINE see ARQ725
2-(4-CHLORO-6-ETHYLAMINO-s-TRIAZINE-2-YLAMINO)-2-METHYL-PROPIONITRILE see BLW750
2-(4-CHLORO-6-ETHYLAMINO-1,3,5-TRIAZINE-2-YL-AMINO)-2-METHYLPROPIONITRILE see BLW750
2-((4-CHLORO-6-(ETHYLAMINO)-1,3,5-TRIAZIN-2-YL) AMINO)-2-METHYL-PROPANENITRILE see BLW750
2-((4-CHLORO-6-(ETHYLAMINO)-s-TRIAZIN-2-YL)AMINO)-2-METHYLPROPIONITRILE see BLW750
CHLOROETHYLBENZENE see EHH500
3-(2-CHLOROETHYL)-2-(BIS(2-CHLOROETHYL)AMINO) PERHYDRO-2H-1,3,2-OXAZAPHOSPHORINE-2-OXIDE see TNT500
β-CHLOROETHYL-β-(BIS(β-HYDROXYETHYL)SUL-FONIUM)ETHYL SULFIDE CHLORIDE see BKD750
β-CHLOROETHYL-β'-(p-tert-BUTYLPHENOXY)-α'- METH-YLETHYL SULFITE see SOP500
β-CHLOROETHYL-β-(p-tert-BUTYLPHENOXY)-α-METHYL-ETHYL SULPHITE see SOP500
3-(2-CHLOROETHYL)-2-((2-CHLOROETHYL)AMINO) PERHYDRO-2H-1,3,2-OXAZAPHOSPHORINE OXIDE see IMH000
3-(2-CHLOROETHYL)-2-((2-CHLOROETHYL)AMINO)TET-RAHYDRO-2H-1,3,2-OXAZAPHOSPHORINE-2-OXIDE see IMH000
N-(2-CHLOROETHYL)-N'-(2-CHLOROETHYL)-N',O-PRO-PYLENEPHOSPHORIC ACID DIAMIDE see IMH000
N-(2-CHLOROETHYL)-N'-(2-CHLOROETHYL)-N',O-PRO-PYLENEPHOSPHORIC ACID ESTER DIAMIDE see IMH000
CHLOROETHYLCYCLOHEXYLNITROSOUREA see CGV250
((CHLORO-2-ETHYL)-1-CYCLOHEXYL-3-NITROSOUREA see CGV250
1-(2-CHLOROETHYL)-3-CYCLOHEXYL-1-NITROSOUREA see CGV250
N-(2-CHLOROETHYL)-N'-CYCLOHEXYL-N-NITROSOUREA see CGV250
N-(2-CHLOROETHYL)DIBENZYLAMINE HYDROCHLORIDE see DCR200
(2-CHLOROETHYL)DIETHYLAMINE see CGV500
N-(2-CHLORO ETHYL)DIETHYLAMINE see CGV500
(2-CHLOROETHYL)DIMETHYLAMINE see CGW000
β-CHLOROETHYLDIMETHYLAMINE see CGW000
N-(2-CHLOROETHYL)DIMETHYLAMINE see CGW000
CHLOROETHYLENE see VNP000
CHLOROETHYLENE POLYMER see PKQ059
CHLOROETHYLENEVINYL ACETATE POLYMER see AAX175
CHLOROETHYL ETHER see DFJ050
CHLOROETHYL ETHYL SULFIDE see CGY750
2-CHLOROETHYL ETHYL SULFIDE see CGY750
2-CHLOROETHYL ETHYL THIOETHER see CGY750
α-CHLOROETHYLIDENE FLUORIDE see CFX250
CHLOROETHYL MERCURY see CHC500
2-CHLOROETHYL METHANESULFONATE see CHC750
β-CHLOROETHYLMETHANESULFONATE see CHC750
CHLOROETHYL METHANESULPHONATE see CHC750
2-((2-CHLOROETHYL)METHYLAMINO)ETHANOL ACE-TATE see MFW750

2-CHLOROETHYL 1-METHYL-2-(p-tert-BUTYLPHENOXY)
ETHYL SULPHATE see SOP500

1-(2-CHLOROETHYL)-3-(4-METHYL-CYCLOHEXYL)-1-NI-
TROSOUREA see CHD250

1-(2-CHLOROETHYL)-3-(trans-4-METHYL-CYCLOHEXYL)-1-
NITROSOUREA see CHD250

N-(2-CHLOROETHYL)-N′-(trans-4-METHYLCYCLOHEXYL)-
N-NITROSOUREA see CHD250

6-CHLORO-N-ETHYL-N′-(1-METHYLETHYL)-1,3,5-TRI-
AZINE-2,4-DIAMINE (9CI) see ARQ725

N-(2-CHLOROETHYL)-N-(1-METHYL-2-PHENOXYETHYL)
BENZENEMETHANAMINE see PDT250

N-(2-CHLOROETHYL)-N-(1-METHYL-2-PHENOXYETHYL)
BENZENEMETHANAMINE HYDROCHLORIDE see
DDG800

N-(2-CHLOROETHYL)-N-(1-METHYL-2-PHENOXYETHYL)
BENZYLAMINE see PDT250

N-(2-CHLOROETHYL)-N-(1-METHYL-2-PHENOXYETHYL)
BENZYLAMINE HYDROCHLORIDE see DDG800

N-(2-CHLOROETHYL)-N-NITROSOETHYLCARBAMATE see
CHF500

N-(β-CHLOROETHYL)-N-NITROSOURETHAN see CHF500

2-CHLOROETHYL-N-NITROSOURETHANE see CHF500

CHLOROETHYLOWY ALKOHOL (POLISH) see EIU800

1-CHLORO-3-ETHYL-1-PENTEN-4-YN-3-OL see CHG000

2-CHLOROETHYL SULFUROUS ACID-2-(4-(1,1-DIMETH-
YLETHYL)PHENOXY)-1-METHYLETHYL ESTER see
SOP500

2-CHLOROETHYL SULPHITE of 1-(p-tert-BUTYLPHENOXY)-
2-PROPANOL see SOP500

1-CHLORO-2-(ETHYLTHIO)ETHANE see CGY750

2-(2-CHLOROETHYL)THIOETHYLBIS(2-HYDROXYETHYL)-
CHLORIDE see BKD750

2-CHLOROETHYL VINYL ETHER see CHI250

CHLOROETHYNE see ACJ000

CHLOROFENIZON see CJT750

CHLOROFENVINPHOS see CDS750

p-CHLOROFENYLESTER KYSELINY BENZENSULFONOVE
(CZECH) see CJR500

CHLOROFLUOROMETHANE see CHI900

CHLOROFORM see CHJ500

CHLOROFORME (FRENCH) see CHJ500

CHLOROFORMIC ACID BENZYL ESTER see BEF500

CHLOROFORMIC ACID DIMETHYLAMIDE see DQY950

CHLOROFORMIC ACID ETHYL ESTER see EHK500

CHLOROFORMIC ACID ISOPROPYL ESTER see IOL000

CHLOROFORMIC ACID METHYL ESTER see MIG000

CHLOROFORMIC ACID PHENYL ESTER see CBX109

CHLOROFORMIC ACID PROPYL ESTER see PNH000

CHLOROFORMIC DIGITALIN see DKN400

CHLOROFORMYL CHLORIDE see PGX000

CHLOROFOS see TIQ250

CHLOROFTALM see TIQ250

4-CHLORO-N-FURFURYL-5-SULFAMOYLANTHRANILIC
ACID see CHJ750

4-CHLORO-N-(2-FURYLMETHYL)-5-SULFAMOYLAN-
THRANILIC ACID see CHJ750

CHLOROGUANIDE see CKB250

CHLOROHYDRIC ACID see HHL000

α-CHLOROHYDRIN see CDT750

epi-CHLOROHYDRIN see EAZ500

CHLOROHYDROQUINONE see CHM000

2-CHLORO-10-3-(1-(2-HYDROXYETHYL)-4-PIPERAZINYL)
PROPYL PHENOTHIAZINE see CJM250

5-CHLORO-8-HYDROXY-7-IODOQUINOLINE see CHR500

3-CHLORO-7-HYDROXY-4-METHYLCOUMARIN BIS(2-
CHLOROETHYL)PHOSPHATE see DFH600

3-CHLORO-7-HYDROXY-4-METHYL-COUMARIN-O,O-DI-
ETHYL PHOSPHOROTHIOATE see CNU750

3-CHLORO-7-HYDROXY-4-METHYL-COUMARIN-O-ESTER
with O,O-DIETHYL PHOSPHOROTHIOATE see CNU750

(−)-N-((5-CHLORO-8-HYDROXY-3-METHYL-1-OXO-7-ISO-
CHROMANYL)CARBONYL)-3-PHENYLALANINE see
CHP250

5-CHLORO-2-((2-HYDROXY-1-NAPHTHALENYL)AZO)-4-
METHYLBENZENE SULFONIC ACID, BARIUM SALT
(2:1) see CHP500

5-CHLORO-2-((2-HYDROXY-1-NAPHTHALENYL)AZO)-4-
METHYLBENZENE SULPHONIC ACID, BARIUM SALT
see CHP500

5-CHLORO-2-((2-HYDROXY-1-NAPHTHYL)AZO)-p-TOL-
UENE SULFONIC ACID, BARIUM SALT see CHP500

6-CHLORO-17-α-HYDROXYPREGNA-4,6-DIENE-3,20-DIONE
ACETATE see CBF250

6-CHLORO-17-α-HYDROXY-Δ⁶-PROGESTERONE ACETATE
see CBF250

2-CHLORO-HYDROXYTOLUENE see CFE250

6-CHLORO-3-HYDROXYTOLUENE see CFE250

5-CHLORO-7-IODO-8-HYDROXYQUINOLINE see CHR500

CHLOROIODOQUINE see CHR500

5-CHLORO-7-IODO-8-QUINOLINOL see CHR500

2-CHLOROISOBUTANE see BQR000

γ-CHLOROISOBUTYLENE see CIU750

2-CHLORO-N-ISOPROPYLACETANILIDE see CHS500

α-CHLORO-N-ISOPROPYLACETANILIDE see CHS500

2-CHLORO-N-ISOPROPYL-N-PHENYLACETAMIDE see
CHS500

CHLOROJECT L see CDP250

CHLOROMADINONE ACETATE see CBF250

CHLOROMAX see CDP250

S-(6-CHLORO-3-(MERCAPTOMETHYL)-2-BENZOXAZOLI-
NONE)-O,O-DIETHYL PHOSPHORODITHIOATE see
BDJ250

(CHLOROMERCURI)BENZENE see PFM500

p-(CHLOROMERCURI)BENZOIC ACID see CHU500

p-CHLOROMERCURIC BENZOIC ACID see CHU500

CHLOROMETHANE see MIF765

CHLOROMETHANE mixed with DICHLOROMETHANE see
CHX750

CHLOROMETHANE SULFONATE d′ETHYLE (FRENCH) see
CHC750

CHLOROMETHANE SULFONYL CHLORIDE see CHY000

CHLOROMETHAPYRILENE see CHY250

CHLOROMETHOXY ETHANE see CIM000

CHLORO(2-METHOXYETHYL)MERCURY see MEP250

3-CHLORO-7-METHOXY-9-(1-METHYL-4-DIETHYL-
AMINOBUTYLAMINO)ACRIDINE see ARQ250

7-CHLORO-2-METHYLAMINO-5-PHENYL-3H-1,4-BENZO-
DIAZEPINE 4-OXIDE see LFK000

7-CHLORO-2-METHYLAMINO-5-PHENYL-3H-1,4-BENZO-
DIAZEPIN 4-OXIDE see LFK000

7-CHLORO-2-METHYLAMINO-5-PHENYL-3H-1,4-BENZO-
DIAZEPIN, 4-OXIDE, HYDROCHLORIDE see MDQ250

p-CHLORO-N-METHYLAMPHETAMINE see CIF250

d-1-p-CHLORO-METHYLAMPHETAMINE (FRENCH) see
CIF250

3-CHLORO-6-METHYLANILINE see CLK225

4-CHLORO-2-METHYLANILINE see CLK220

4-CHLORO-6-METHYLANILINE see CLK220

5-CHLORO-2-METHYLANILINE see CLK225

4-CHLORO-2-METHYLANILINE HYDROCHLORIDE see
CLK235

4-CHLORO-6-METHYLANILINE HYDROCHLORIDE see
CLK235

7-CHLOROMETHYL BENZ(a)ANTHRACENE see CIG250

CHLOROMETHYLBENZENE see BEE375

4-CHLORO-1-METHYLBENZENE see TGY075

2-CHLORO-1-METHYLBENZENE (9CI) see CLK100

4-CHLORO-2-METHYLBENZENEAMINE see CLK220

4-CHLORO-2-METHYLBENZENEAMINE HYDROCHLORIDE
see CLK235

6-CHLOROMETHYL BENZO(a)PYRENE see CIH000

2-CHLORO-6-METHYLCARBANILIC ACID-2-(PYRROLIDI-
NYL)ETHYL ESTER HYDROCHLORIDE see CIK500

3-CHLORO-4-METHYL-7-COUMARINYL DIETHYLPHOS-
PHATE see CIK750

3-CHLORO-4-METHYL-7-COUMARINYL DIETHYL PHOS-
PHOROTHIOATE see CNU750

O-3-CHLORO-4-METHYL-7-COUMARINYL-O,O-DIETHYL
  PHOSPHOROTHIOATE see CNU750
CHLOROMETHYL CYANIDE see CDN500
2-CHLORO-4-METHYL-6-DIMETHYLAMINOPYRIMIDINE
  see CCP500
d-2-CHLORO-6-METHYLERGOLINE-8-β-ACETONITRILE
  METHANESULFONIC ACID SALT see LEP000
(CHLOROMETHYL)ETHYLENE OXIDE see EAZ500
CHLOROMETHYL ETHYL ETHER see CIM000
(2-CHLORO-1-METHYLETHYL) ETHER see BII250
2-CHLORO-N-(1-METHYLETHYL)-N-PHENYLACETAMIDE
  see CHS500
3-CHLORO-4-METHYL-7-HYDROXYCOUMARIN DIETHYL
  THIOPHOSPHORIC ACID ESTER see CNU750
CHLOROMETHYLMERCURY see MDD750
4-CHLORO-N-METHYL-3-((METHYLAMINO)SULFONYL)
  BENZAMIDE see CIP500
7-CHLOROMETHYL-12-METHYL BENZ(a)ANTHRACENE
  see CIN750
CHLOROMETHYL METHYL ETHER see CIO250
4-CHLORO-N-METHYL-3-(METHYLSULFAMOYL)BENZAM-
  IDE see CIP500
2-CHLORO-2-METHYL-N-NITROSOETHANAMINE see
  CIQ500
2-CHLORO-N-METHYL-N-NITROSOETHYLAMINE see
  CIQ500
CHLOROMETHYLOXIRANE see EAZ500
2-(CHLOROMETHYL)OXIRANE see EAZ500
4-CHLORO-3-METHYLPHENOL see CFE250
(4-CHLORO-2-METHYLPHENOXY)ACETIC ACID see
  CIR250
2-(4-CHLORO-2-METHYLPHENOXY)PROPIONIC ACID see
  CIR500
4-CHLORO-2-METHYLPHENOXY-α-PROPIONIC ACID see
  CIR500
(+)-α-(4-CHLORO-2-METHYLPHENOXY) PROPIONIC ACID
  see CIR500
7-CHLORO-N-METHYL-5-PHENYL-3H-1,4-BENZODIAZE-
  PIN-2-AMINE-4-OXIDE see LFK000
7-CHLORO-1-METHYL-5-PHENYL-1H-1,5-BENZODIAZE-
  PINE-2,4(3H,5H)-DIONE see CIR750
N′-(4-CHLORO-2-METHYLPHENYL)-N,N-DIMETHYLMETH-
  ANIMIDAMIDE see CJJ250
7-CHLORO-N-METHYL-5-PHENYL-EH-1,4-BENZODIAZE-
  PIN-2-AMINE-4-OXIDE, MONOHYDROCHLORIDE see
  MDQ250
CHLOROMETHYL PHENYL KETONE see CEA750
2-CHLORO-11-(4-METHYLPIPERAZINO)DIBENZO(b,f)(1,4)
  THIAZEPINE see CIS750
2-CHLORO-11-(4-METHYL-1-PIPERAZINYL)-DIBENZO
  (b,f)(1,4)OXAZEPINE see DCS200
2-CHLORO-11-(4-METHYL-1-PIPERAZINYL)-DIBENZO
  (b,f)(1,4)OXOAZEPINE see DCS200
2-CHLORO-11-(4-METHYL-1-PIPERAZINYL)DIBENZO
  (b,f)(1,4)THIAZEPINE see CIS750
2-CHLORO-10-(3-(1-METHYL-4-PIPERAZINYL)-PROPYL)-
  PHENOTHIAZINE see PMF500
2-CHLORO-10-(3-(4-METHYL-1-PIPERAZINYL)PROPYL)
  PHENOTHIAZINE see PMF500
3-CHLORO-10-(3-(1-METHYL-4-PIPERAZINYL)PROPYL)
  PHENOTHIAZINE see PMF500
CHLORO-3 (N-METHYLPIPERAZINYL-3 PROPYL)-10 PHE-
  NOTHIAZINE (FRENCH) see PMF500
2-CHLORO-10-(3-(4-METHYL-1-PIPERAZINYL)-PROPYL-
  10H-PHENOTHIAZINE-(Z)-2-BUTENEDIOATE (1:2) see
  PMF250
2-CHLORO-10-(3-(1-METHYL-4-PIPERAZINYL)PROPYL)
  PHENOTHIAZINE, DIMALEATE see PMF250
2-CHLORO-10-(3-(4-METHYL-1-PIPERAZINYL)PROPYL)
  PHENOTHIAZINE DIMALEATE see PMF250
2-CHLORO-10-(3-(4-METHYL-1-PIPERAZINYL)PROPYL)
  PHENOTHIAZINE MALEATE see PMF250
2-CHLORO-2-METHYLPROPANE see BQR000

3-CHLORO-2-METHYLPROPENE see CIU750
3-CHLORO-2-METHYL-1-PROPENE see CIU750
2-CHLORO-5-(1-METHYLPROPYL)PHENYL METHYLCAR-
  BAMATE see MOU750
3-(CHLOROMETHYL) PYRIDINE HYDROCHLORIDE see
  CIV000
5′-CHLORO-2-(METHYL(2-(PYRROLIDINYL)ETHYL)
  AMINO)-o-ACETOTOLUIDIDE DIHYDROCHLORIDE
  see CIW250
2-CHLOROMETHYLTHIOPHENE see CIY250
3-CHLORO-4-METHYL-UMBELLIFERONE BIS(2-CHLORO-
  ETHYL)PHOSPHATE see DFH600
3-CHLORO-4-METHYLUMBELLIFERONE-O-ESTER with
  O,O-DIETHYL PHOSPHOROTHIOATE see CNU750
CHLOROMYCETIN see CDP250
CHLORONAFTINA see BIF250
CHLORONAPHTHINE see BIF250
CHLORONITRIN see CDP250
2-CHLORONITROBENZENE see CJB750
4-CHLORONITROBENZENE see NFS525
CHLORO-m-NITROBENZENE see CJB250
m-CHLORONITROBENZENE see CJB250
CHLORO-o-NITROBENZENE see CJB750
o-CHLORONITROBENZENE see CJB750
p-CHLORONITROBENZENE see NFS525
1-CHLORO-2-NITROBENZENE see CJB750
1-CHLORO-3-NITROBENZENE see CJB250
1-CHLORO-4-NITROBENZENE see NFS525
2-CHLORO-1-NITROBENZENE see CJB750
4-CHLORO-1-NITROBENZENE see NFS525
m-CHLORONITROBENZENE (DOT) see CJB250
o-CHLORONITROBENZENE (DOT) see CJB750
2-((o-CHLORO-α-(NITROMETHYL)BENZYL)THIO)
  ETHYLAMINE HYDROCHLORIDE see NEC000
2-CHLORO-4-NITROPHENYLAMIDE-6-CHLOROSALICYLIC
  ACID see DFV400
N-(2-CHLORO-4-NITROPHENYL)-5-CHLOROSALICYLAM-
  IDE see DFV400
O-(2-CHLORO-4-NITROPHENYL) O,O-DIMETHYL PHOS-
  PHOROTHIOATE see NFT000
O-(3-CHLORO-4-NITROPHENYL) O,O-DIMETHYL PHOS-
  PHOROTHIOATE see MIJ250
CHLORONITROPROPAN (POLISH) see CJD750
CHLORONITROPROPANE see CJD750, CJE000
1-CHLORO-1-NITROPROPANE see CJE000
1-CHLORO-2-NITROPROPANE see CJD750
2-CHLORO-2-NITROPROPANE see CJE250
4-CHLORO-3-NITRO-α,α,α-TRIFLUOROTOLUENE see
  NFS525
3-(6-CHLORO-2-OXOBENZOXAZOLIN-3-YL)METHYL-
  O,O-DIETHYL PHOSPHOROTHIOLOTHIONATE see
  BDJ250
CHLOROPARACIDE see CEP000
CHLOROPENTAFLUOROETHANE see CJI500
1-CHLOROPENTANE see PBW500
CHLOROPEROXYL see CDW450
CHLOROPHACINONE see CJJ000
CHLOROPHEN see PAX250
CHLOROPHENAMADIN see CJJ250
CHLOROPHENAMIDINE see CJJ250
4-CHLOROPHENE-1,3-DIAMINE see CJY120
2-CHLOROPHENOL see CJK250
3-CHLOROPHENOL see CJK500
4-CHLOROPHENOL see CJK750
m-CHLOROPHENOL see CJK500
o-CHLOROPHENOL see CJK250
p-CHLOROPHENOL see CJK750
m-CHLOROPHENOL, liquid (DOT) see CJK500
m-CHLOROPHENOL, solid (DOT) see CJK500
o-CHLOROPHENOL, liquid (DOT) see CJK250
o-CHLOROPHENOL, solid (DOT) see CJK250
p-CHLOROPHENOL, liquid (DOT) see CJK750
p-CHLOROPHENOL, solid (DOT) see CJK750

S-(((p-CHLOROPHENYL)THIO)METHYL) O,O-DIMETHYL
  PHOSPHORODITHIOATE see MQH750
4-CHLOROPHENYL-2,4,5-TRICHLOROPHENYL SULFONE
  see CKM000
p-CHLOROPHENYL-2,4,5-TRICHLOROPHENYL SULFONE
  see CKM000
p-CHLOROPHENYL-2,4,5-TRICHLOROPHENYL SULPHONE
  see CKM000
CHLOROPHENYLTRICHLOROSILANE see CKM250
(p-CHLOROPHENYL)TRIFLUOROMETHANE see CEM825
CHLOROPHOS see TIQ250
CHLORO-PHOSPHONOTHIOIC ACID-O,O-DIETHYL ESTER
  see DJW600
CHLOROPHOSPHONOTHIOIC ACID-O,O-DIMETHYL ESTER
  see DTQ600
CHLOROPHOSPHORIC ACID DIETHYL ESTER see DIY000
S-(2-CHLORO-1-PHTHALIMIDOETHYL)-O,O-DIETHYL
  PHOSPHORODITHIOATE see DBI099
CHLOROPHTHALM see TIQ250
CHLOROPHYL, GREEN see CKN000
CHLOROPHYLL see CKN000
CHLOROPICRIN see CKN500
CHLOROPICRIN, liquid (DOT) see CKN500
CHLOROPICRIN, ABSORBED (DOT) see CKN500
CHLOROPICRINE see CKN510
CHLOROPICRINE (FRENCH) see CKN500
CHLOROPICRIN MIXTURE (flammable) see CKN510
CHLOROPLATINIC ACID see CKO750
CHLOROPLATINIC(IV) ACID see CKO750
CHLOROPOTASSURIL see PLA500
CHLOROPREEN (DUTCH) see NCI500
6-CHLORO-Δ⁴,⁶-PREGNADIENE-17-α-OL-3,20-DIONE-17-
  ACETATE see CBF250
6-CHLORO-PREGNA-4,6-DIEN-17-α-OL-3,20-DIONE ACE-
  TATE see CBF250
CHLOROPREN (GERMAN, POLISH) see NCI500
CHLOROPRENE see NCI500
3-CHLOROPRENE see AGB250
β-CHLOROPRENE (OSHA, MAK) see NCI500
CHLOROPRENE, inhibited (DOT) see NCI500
CHLOROPRENE, uninhibited (DOT) see NCI500
CHLOROPROMURITE see DEQ000
1-CHLOROPROPANE see CKP750
1-CHLOROPROPANE-2,3-DIOL see CDT750
1-CHLORO-2,3-PROPANEDIOL see CDT750
3-CHLOROPROPANE-1,2-DIOL see CDT750
3-CHLORO-1,2-PROPANEDIOL see CDT750
2-CHLORO-1-PROPANOL see CKR500
CHLOROPROPANONE see CDN200
1-CHLORO-2-PROPANONE see CDN200
1-CHLOROPROPENE see PMR750
3-CHLOROPROPENE see AGB250
1-CHLORO-1-PROPENE see PMR750
1-CHLORO PROPENE-2 see AGB250
1-CHLORO-2-PROPENE see AGB250
2-CHLORO-1-PROPENE see CKS000
3-CHLORO-1-PROPENE see AGB250
2-CHLOROPROPENE (DOT) see CKS000
2-CHLORO-2-PROPENE-1-THIOL DIETHYLDITHIOCARBA-
  MATE see CDO250
2-CHLORO-2-PROPENYL DIETHYLCARBAMODITHIOATE
  see CDO250
CHLOROPROPHAM see CKC000
CHLOROPROPHENYPYRIDAMINE MALEATE see TAI500
2-CHLOROPROPIONIC ACID METHYL ESTER see CKT000
p-CHLOROPROPIOPHENONE see CKT500
2-CHLOROPROPYL ALCOHOL see CKR500
2-CHLORO-4-(2-PROPYLAMINO)-6-ETHYLAMINO-s-TRI-
  AZINE see ARQ725
3-CHLOROPROPYL BROMIDE see BNA825
3-CHLOROPROPYLENE see AGB250
α-CHLOROPROPYLENE see AGB250
3-CHLORO-1-PROPYLENE see AGB250
3-CHLOROPROPYLENE GYLCOL see CDT750

CHLOROPROPYLENE OXIDE see EAZ500
γ-CHLOROPROPYLENE OXIDE see EAZ500
3-CHLORO-1,2-PROPYLENE OXIDE see EAZ500
CHLOROPTIC see CDP250
2-CHLOROPYRIDINE see CKW000
o-CHLOROPYRIDINE see CKW000
α-CHLOROPYRIDINE see CKW000
CHLOROPYRILENE see CHY250
CHLOROS see SHU500
CHLOROSILANES see CLE250
CHLOROSTOP see PKQ059
o-CHLOROSTYRENE see CLE750
CHLOROSULFACIDE see CEP000
6-CHLORO-7-SULFAMOYL-3,4-DIHYDRO-2H-1,2,4-BENZO-
  THIADIAZINE-1,1-DIOXIDE see CFY000
CHLOROSULFONIC ACID (DOT) see CLG500
CHLOROSULFONIC ANHYDRIDE see PPR500
1-CHLOROSULFONYL-5-DIMETHYLAMINONAPHTHALENE
  see DPN200
1-(4-CHLORO-o-SULFO-5-TOLYLAZO)-2-NAPHTHOL,
  BARIUM SALT see CHP500
CHLOROSULFURIC ACID see CLG500
CHLOROSULTHIADIL see CFY000
7-CHLOROTETRACYCLINE see CMA750
CHLOROTETRAFLUOROETHANE see CLH000
CHLOROTHALONIL see TBQ750
CHLOROTHANE NU see MIH275
CHLOROTHEN see CHY250
CHLOROTHENE see MIH275
CHLOROTHENE (INHIBITED) see MIH275
CHLOROTHENE NU see MIH275
CHLOROTHENE VG see MIH275
2-((5-CHLORO-2-THENYL)(2-DIMETHYLAMINOETHYL)
  AMINO)PYRIDINE see CHY250
CHLOROTHENYLPYRAMINE see CHY250
CHLOROTHIAMIDE see DGM600
CHLOROTHIOFORMIC ACID ETHYL ESTER see CLJ750
2,3,4,5-CHLOROTHIOPHENE see TBV750
4-CHLORO-o-TOLOXYACETIC ACID see CIR250
2-CHLOROTOLUENE see CLK100
4-CHLOROTOLUENE see TGY075
o-CHLOROTOLUENE see CLK100
α-CHLOROTOLUENE see BEE375
p-CHLOROTOLUENE (DOT) see TGY075
φ-CHLOROTOLUENE see BEE375
4-CHLORO-2-TOLUIDINE see CLK220
4-CHLORO-o-TOLUIDINE see CLK220
5-CHLORO-o-TOLUIDINE see CLK225
4-CHLORO-2-TOLUIDINE HYDROCHLORIDE see CLK235
4-CHLORO-o-TOLUIDINE HYDROCHLORIDE see CLK235
4-CHLORO-o-TOLUIDINE HYDROCHLORIDE (DOT) see
  CLK235
N'-(4-CHLORO-o-TOLYL)-N,N-DIMETHYLFORMAMIDINE
  see CJJ250
((4-CHLORO-o-TOLYL)OXY)ACETIC ACID see CIR250
2-(p-CHLORO-o-TOLYLOXY)PROPIONIC ACID see CIR500
CHLOROTRIANISENE see CLO750
CHLOROTRIANIZEN see CLO750
CHLOROTRIBUTYLSTANNANE see CLP500
2-CHLORO-6-(TRICHLOROMETHYL)PYRIDINE see CLP750
2-CHLORO-1-(2,4,5-TRICHLOROPHENYL)VINYL DI-
  METHYL PHOSPHATE see TBW100
2-CHLORO-1-(2,4,5-TRICHLOROPHENYL(VINYL PHOS-
  PHORIC ACID DIMETHYL ESTER see TBW100
2-CHLOROTRIETHYLAMINE see CGV500
β-CHLOROTRIETHYLAMINE see CGV500
CHLORO(TRIETHYLPHOSPHINE)GOLD see CLQ500
CHLOROTRIFLUORIDE see CDX750
2-CHLORO-1,1,2-TRIFLUOROETHYL DIFLUOROMETHYL
  ETHER see EAT900
CHLOROTRIFLUOROETHYLENE see CLQ750
1-CHLORO-1,2,2-TRIFLUOROETHYLENE see CLQ750
2-CHLORO-1,1,2-TRIFLUOROETHYLENE see CLQ750

CHLOROTRIFLUOROMETHANE see CLR250
4-CHLOROTRIFLUOROMETHYLBENZENE see CEM825
p-CHLOROTRIFLUOROMETHYLBENZENE see CEM825
4-(4-CHLORO-α,α,α-TRIFLUORO-m-TOLYL)-1-(4,4-BIS(p-FLUOROPHENYL)BUTYL)-4-PIPERIDINOL see PAP250
4-(4-(4-CHLORO-α,α,α-TRIFLUORO-m-TOLYL)-4-HY-DROXYPIPERIDINO)BUTYROPHENONE-4'-FLUOROHY-DROCHLORIDE see CLS250
7-CHLORO-4,6,2'-TRIMETHOXY-6'-METHYLGRIS-2'-EN-3,4'-DIONE see GKE000
1-CHLORO-4-(TRIMETHYL)-BENZENE (9CI) see CEM825
CHLOROTRIMETHYLSILICANE see TLN250
CHLOROTRIMETHYLSTANNANE see CLT000
CHLOROTRIMETHYLTIN see CLT000
CHLOROTRINITROMETHANE see CLT250
CHLOROTRIPHENYLSTANNANE see CLU000
CHLOROTRIPHENYLTIN see CLU000
CHLOROTRIPROPYLSTANNANE see CLU250
CHLOROTRISIN see CLO750
CHLOROTRIS(p-METHOXYPHENYL)ETHYLENE see CLO750
CHLORO(TRIVINYL)STANNANE see CLU500
CHLOROVINYLARSINE DICHLORIDE see CLV000
β-CHLOROVINYLBICHLOROARSINE see CLV000
2-CHLOROVINYLDICHLOROARSINE see CLV000
(2-CHLOROVINYL)DICHLOROARSINE see CLV000
β-CHLOROVINYL ETHYLETHYNYL CARBINOL see CHG000
3-(β-CHLOROVINYL)-1-PENTYN-3-OL see CHG000
CHLOROVULES see CDP250
CHLOROWODOR (POLISH) see HHL000
CHLOROX see SHU500
CHLOROXONE see DAA800
CHLORO-XYLENOL see CLW000
p-CHLORO-m-XYLENOL see CLW000
4-CHLORO-3,5-XYLENOL see CLW000
(4-CHLORO-6-(2,3-XYLIDINO)-2-PYRIMIDINYLTHIO)-ACETIC ACID see CLW250
2-((4-CHLORO-6-(2,3-XYLIDINO)-2-PYRIMIDINYL)THIO)-N-(2-HYDROXYETHYL)ACETAMIDE see CLW500
CHLOROXYPHOS see TIQ750
CHLOROZIRCONYL see ZSJ000
CHLOROZONE see CDP000
CHLORPARACIDE see CEP000
CHLORPERAZINE see PMF500
CHLORPHACINON (ITALIAN) see CJJ000
CHLORPHENAMIDINE see CJJ250
CHLORPHENIRAMINE MALEATE see TAI500
(+)-CHLORPHENIRAMINE MALEATE see PJJ325
d-CHLORPHENIRAMINE MALEATE see PJJ325
S-(+)-CHLORPHENIRAMINE MALEATE see PJJ325
o-CHLORPHENOL (GERMAN) see CJK250
CHLORPHENVINFOS see CDS750
CHLORPHENVINPHOS see CDS750
(4-CHLOR-PHENYL)-BENZOLSULFONAT (GERMAN) see CJR500
4-CHLORPHENYL-4'-CHLORBENZOLSULFONAT (GER-MAN) see CJT750
(4-CHLOR-PHENYL)-4-CHLOR-BENZOL-SULFONATE (GER-MAN) see CJT750
3-(4-CHLOR-PHENYL)-1,1-DIMETHYL-HARNSTOFF (GER-MAN) see CJX750
N-(3-CHLOR-PHENYL)-ISOPROPYL-CARBAMAT (GERMAN) see CKC000
3-(4-CHLORPHENYL)-1-METHOXY-1-METHYLHARNSTOFF (GERMAN) see CKD500
3-(4-CHLORPHENYL)-1-METHYL-1-ISOBUTINYLHARNS-TOFF (GERMAN) see CKF750
N-(4-CHLORPHENYL)-N'-METHYL-N'-ISOBUTINYLHARNS-TOFF (GERMAN) see CKF750
((4-CHLORPHENYL)-1-PHENYL)-ACETYL-1,3-INDANDION (GERMAN) see CJJ000
1-(4-CHLORPHENYL)-1-PHENYL-ACETYL-INDAN-1,3-DION (GERMAN) see CJJ000

2(2-(4-CHLOR-PHENYL-2-PHENYL)ACETYL)INDAN-1,3-DION (GERMAN) see CJJ000
CHLOR-O-PIC see CKN500
CHLORPIKRIN (GERMAN) see CKN500
CHLORPROETHAZINE HYDROCHLORIDE see CLZ000
3-CHLORPROPEN (GERMAN) see AGB250
CHLORPROPHAM see CKC000
CHLORPROPHAME (FRENCH) see CKC000
CHLORPYRIFOS see CMA100
CHLORPYRIFOS-METHYL see CMA250
CHLORSAURE (GERMAN) see SFS000
CHLORSEPTOL see CDP000
CHLORSUCCINYLCHOLIN (GERMAN) see HLC500
CHLORSULFONAMIDO DIHYDROBENZOTHIADIAZINE DI-OXIDE see CFY000
CHLORSULPHACIDE see CEP000
CHLORTEN see MIH275
CHLORTETRACYCLINE see CMA750
CHLORTHALONIL (GERMAN) see TBQ750
CHLORTHIEPIN see EAQ750
CHLORTHION METHYL see MIJ250
CHLORTION (CZECH) see MIJ250
α-CHLORTOLUOL (GERMAN) see BEE375
N'-(4-CHLOR-o-TOLYL)-N,N-DIMETHYLFORMAMIDIN (GERMAN) see CJJ250
CHLORTOX see CDR750
CHLORTRIANISEN see CLO750
CHLORTRIFLUORAETHYLEN (GERMAN) see CLQ750
CHLOR-TRIMETON see TAI500
CHLOR-TRIMETON MALEATE see TAI500
CHLOR-TRIPOLON see TAI500
CHLORURE d'ALUMINUM (FRENCH) see AGY750
CHLORURE ANTIMONIEUX (FRENCH) see AQC500
CHLORURE d'ARSENIC (FRENCH) see ARF500
CHLORURE ARSENIEUX (FRENCH) see ARF500
CHLORURE de BENZENYLE (FRENCH) see BFL250
CHLORURE de BENZYLE (FRENCH) see BEE375
CHLORURE de BENZYLIDENE (FRENCH) see BAY300
CHLORURE de BORE (FRENCH) see BMG500
CHLORURE de BUTYLE (FRENCH) see BQQ750
CHLORURE de CHLORACETYLE (FRENCH) see CEC250
CHLORURE de CHROMYLE (FRENCH) see CML125
CHLORURE de CYANOGENE (FRENCH) see COO750
CHLORURE de DICHLORACETYLE (FRENCH) see DEN400
CHLORURE d'ETHYLE (FRENCH) see EHH000
CHLORURE d'ETHYLENE (FRENCH) see EIY600
CHLORURE d'ETHYLIDENE (FRENCH) see DFF809
CHLORURE de FUMARYLE (FRENCH) see FOY000
CHLORURE de LITHIUM (FRENCH) see LHB000
CHLORURE MERCUREUX (FRENCH) see MCW000
CHLORURE MERCURIQUE (FRENCH) see MCY475
CHLORURE de METHALLYLE (FRENCH) see CIU750
CHLORURE de METHYLE (FRENCH) see MIF765
CHLORURE de METHYLENE (FRENCH) see MJP450
CHLORURE PERRIQUE see FAU000
CHLORURE de SUCCINILCOLINE (FRENCH) see HLC500
CHLORURE de VINYLE (FRENCH) see VNP000
CHLORURE de VINYLIDENE (FRENCH) see VPK000
CHLORURE de ZINC (FRENCH) see ZFA000
CHLORVINPHOS see DGP900
CHLORWASSERSTOFF (GERMAN) see HHL000
CHLORYL see EHH000
CHLORYL ANESTHETIC see EHH000
CHLORYLEA see TIO750
CHLORYL RADICAL see CDW450
CHLORZIDE see CFY000
CHOCOLA A see VSK600
CHOLAXINE see SKV200
CHOLECALCIFEROL see CMC750
CHOLEIC ACID see DAQ400
CHOLEREBIC see DAQ400
3-β-CHOLESTANOL see DKW000
(3-β,5-β)-CHOLESTAN-3-OL see DKW000
CHOLEST-5-EN-3-β-OL see CMD750

5-CHOLESTEN-3-β-OL see CMD750
5:6-CHOLESTEN-3-β-OL see CMD750
Δ⁵-CHOLESTEN-3-β-OL see CMD750
5-CHOLESTEN-3-β-OL 3-(p-(BIS(2-CHLOROETHYL)AMINO)
PHENYL)ACETATE see CME250
CHOLESTERIN see CMD750
CHOLESTEROL see CMD750
CHOLESTEROL BASE H see CMD750
CHOLESTEROL-α-EPOXIDE see EBM000
CHOLESTEROL-5-α,6-α-EPOXIDE see EBM000
CHOLESTEROL OXIDE see EBM000
CHOLESTEROL-α-OXIDE see EBM000
CHOLESTERYL ALCOHOL see CMD750
CHOLESTERYL-p-BIS(2-CHLOROETHYL)AMINO PHE-
NYLACETATE see CME250
CHOLESTRIN see CMD750
CHOLESTROL see CMD750
CHOLIC ACID, MONOSODIUM SALT see SFW000
CHOLIC ACID, SODIUM SALT see SFW000
CHOLIFLAVIN see DBX400
CHOLINE CHLORHYDRATE see CMF750
CHOLINE CHLORIDE ACETATE see ABO000
CHOLINE CHLORIDE (FCC) see CMF750
CHOLINE HYDROCHLORIDE see CMF750
CHOLINE IODIDE SUCCINATE (2:1) see BJI000
CHOLINE SUCCINATE (ester) see CMG250
CHOLINE SUCCINATE DICHLORIDE see HLC500
CHOLINE SUCCINATE (2:1) (ESTER) see CMG250
CHOLINIUM CHLORIDE see CMF750
CHOLOREBIC see DAQ400
CHONDRUS see CCL250
CHONDRUS EXTRACT see CCL250
CHORYLEN see TIO750
CHOT see PBC250
2-CHROMANONE see HHR500
CHROMAR see XHS000
CHROMARGYRE see MCV000
CHROMATE OF POTASSIUM see PLB250
CHROMATE de PLOMB (FRENCH) see LCR000
CHROMATE of SODA see DXC200
CHROME see CMI750
CHROME ALUM see PLB500
CHROME FAST BLUE 2R see HJF500
CHROME FERROALLOY see FBD000
CHROME GREEN see CMJ900, LCR000
CHROME LEATHER BLACK EM see AQP000
CHROME LEATHER BLUE 2B see CMO000
CHROME LEATHER BLUE 3B see CMO250
CHROME LEATHER BROWN BRLL see CMO750
CHROME LEMON see LCR000
CHROME OCHER see CMJ900
CHROME ORANGE see LCS000
CHROME ORE see CMI500
CHROME OXIDE see CMJ900
CHROME OXIDE GREEN see CMJ900
CHROME POTASH ALUM see PLB500
CHROME (TRIOXYDE de) (FRENCH) see CMK000
CHROME YELLOW see LCR000
CHROMIA see CMJ900
CHROMIC ACETATE see CMH000
CHROMIC ACETATE(III) see CMH000
CHROMIC ACID see CMH250, CMJ900, CMK000
CHROMIC(VI) ACID see CMH250, CMK000
CHROMIC ACID (mixture) see CMH500
CHROMIC ACID, solid (DOT) see CMK000
CHROMIC ACID (solution) see CMH750
CHROMIC ACID, solution (DOT) see CMK000
CHROMIC ACID, BARIUM SALT (1:1) see BAK250
CHROMIC ACID, BIS(TRIPHENYLSILYL) ESTER see
BLS750
CHROMIC ACID, CALCIUM SALT (1:1) see CAP500
CHROMIC ACID, CALCIUM SALT (1:1), DIHYDRATE see
CAP750
CHROMIC ACID, CHROMIUM(3+) SALT (3:2) see CMI250

CHROMIC ACID, DI-tert-BUTYL ESTER see BQV000
CHROMIC ACID, DILITHIUM SALT see LHD000
CHROMIC ACID, DIPOTASSIUM SALT see PKX250
CHROMIC ACID, DISODIUM SALT see SGI000
CHROMIC ACID, DISODIUM SALT, DECAHYDRATE see
SFW500
CHROMIC ACID GREEN see CMJ900
CHROMIC ACID, LEAD and MOLYBDENUM SALT see
LDM000
CHROMIC ACID, LEAD(2+) SALT (1:1) see LCR000
CHROMIC ACID LEAD SALT with LEAD MOLYBDATE see
LDM000
CHROMIC ACID, MERCURY ZINC COMPLEX see ZJA000
CHROMIC ACID MIXTURE, DRY (DOT) see CMH500
CHROMIC ACID, POTASSIUM ZINC SALT (2:2:1) see
PLW500
CHROMIC ACID, STRONTIUM SALT (1:1) see SMH000
CHROMIC ACID, ZINC SALT see ZFJ100
CHROMIC ACID, ZINC SALT (1:2) see CMK500
CHROMIC ANHYDRIDE (DOT) see CMK000
CHROMIC CHLORIDE see CMJ250
CHROMIC CHROMATE see CMI250
CHROMIC(III) HYDROXIDE see CMH750
CHROMIC OXIDE see CMJ900
CHROMIC OXYCHLORIDE see CML125
CHROMIC POTASSIUM SULFATE see PLB500
CHROMIC POTASSIUM SULPHATE see PLB500
CHROMIC TRIOXIDE (DOT) see CMK000
CHROMITE see CMI500
CHROMITE (MINERAL) see CMI500
CHROMITE ORE see CMI500
CHROMIUM see CMI750
CHROMIUM ACETATE see CMH000
CHROMIUM(III) ACETATE see CMH000
CHROMIUM ALLOY, BASE, Cr,C,Fe,N,Si (FERROCHRO-
MIUM) see FBD000
CHROMIUM ALLOY, Cr,C,Fe,N,Si see FBD000
CHROMIUM CARBONYL (MAK) see HCB000
CHROMIUM CARBONYL (OC-6-11) (9CI) see HCB000
CHROMIUM CHLORIDE see CMJ250
CHROMIUM(III) CHLORIDE (1:3) see CMJ250
CHROMIUM CHLORIDE, anhydrous see CMJ250
CHROMIUM CHLORIDE OXIDE see CML125
CHROMIUM CHROMATE (MAK) see CMI250
CHROMIUM-COBALT ALLOY see CNA750
CHROMIUM COMPOUNDS see CMJ500
CHROMIUM DICHLORIDE DIOXIDE see CML125
CHROMIUM DIOXIDE DICHLORIDE see CML125
CHROMIUM(VI) DIOXYCHLORIDE see CML125
CHROMIUM DISODIUM OXIDE see DXC200
CHROMIUM HEXACARBONYL see HCB000
CHROMIUM HYDROXIDE SULFATE see NBW000
CHROMIUM LEAD OXIDE see LCS000
CHROMIUM LITHIUM OXIDE see LHD000
CHROMIUM OXIDE see CMJ900, CMK000
CHROMIUM(3+) OXIDE see CMJ900
CHROMIUM(VI) OXIDE see CMK000
CHROMIUM(III) OXIDE see CMJ900
CHROMIUM(VI) OXIDE (1:3) see CMK000
CHROMIUM(III) OXIDE (2:3) see CMJ900
CHROMIUM OXIDE, NICKEL OXIDE, and IRON OXIDE
FUME see IHE000
CHROMIUM OXYCHLORIDE see CML125
CHROMIUM POTASSIUM SULFATE (1:1:2) see PLB500
CHROMIUM POTASSIUM SULPHATE see PLB500
CHROMIUM POTASSIUM ZINC OXIDE see CMK400
CHROMIUM SESQUIOXIDE see CMJ900
CHROMIUM SODIUM OXIDE see DXC200, SGI000
CHROMIUM SULFATE see NBW000
CHROMIUM SULFATE, BASIC see NBW000
CHROMIUM SULPHATE see NBW000
CHROMIUM TRIACETATE see CMH000
CHROMIUM TRICHLORIDE see CMJ250
CHROMIUM TRIOXIDE see CMK000

CHROMIUM(3+) TRIOXIDE see CMJ900
CHROMIUM(6+) TRIOXIDE see CMK000
CHROMIUM TRIOXIDE, anhydrous (DOT) see CMK000
CHROMIUM YELLOW see LCR000
CHROMIUM ZINC OXIDE see ZFJ100
CHROMIUM(6+)ZINC OXIDE HYDRATE (1:2:6:1) see CMK500
CHROMOFLAVINE see DBX400
CHROMOSMON see BJI250
CHROMOSORB T see TAI250
CHROMOTRICHIA FACTOR see AIH600
ANTI-CHROMOTRICHIA FACTOR see AIH600
CHROMO (TRIOSSIDO di) (ITALIAN) see CMK000
CHROMOXYCHLORID (GERMAN) see CML125
CHROMSAEUREANHYDRID (GERMAN) see CMK000
CHROMTRIOXID (GERMAN) see CMK000
CHROMYLCHLORID (GERMAN) see CML125
CHROMYL CHLORIDE see CML125
CHROOMOXYLCHLORIDE (DUTCH) see CML125
CHROOMTRIOXYDE (DUTCH) see CMK000
CHROOMZUURANHYDRIDE (DUTCH) see CMK000
CHRYSANTHEMUM CINERAREAEFOLIUM see POO250
CHRYSANTHEMUMDICARBOXYLIC ACID MONOMETHYL ESTER PYRETHROLONE ESTER see POO100
CHRYSAROBIN see CML750
CHRYSENE see CML810
α-CHRYSIDINE see BAW750
CHRYSOIDIN see PEK000
CHRYSOIDINE see PEK000
CHRYSOIDINE(II) see PEK000
CHRYSOIDINE A see PEK000
CHRYSOIDINE B see PEK000
CHRYSOIDINE C CRYSTALS see PEK000
CHRYSOIDINE G see PEK000
CHRYSOIDINE GN see PEK000
CHRYSOIDINE HR see PEK000
CHRYSOIDINE J see PEK000
CHRYSOIDINE M see PEK000
CHRYSOIDINE ORANGE see PEK000
CHRYSOIDINE PRL see PEK000
CHRYSOIDINE PRR see PEK000
CHRYSOIDINE SL see PEK000
CHRYSOIDINE SPECIAL (biological stain and indicator) see PEK000
CHRYSOIDINE SS see PEK000
CHRYSOIDINE Y see PEK000
CHRYSOIDINE Y BASE NEW see PEK000
CHRYSOIDINE Y CRYSTALS see PEK000
CHRYSOIDINE Y EX see PEK000
CHRYSOIDINE YGH see PEK000
CHRYSOIDINE YL see PEK000
CHRYSOIDINE YN see PEK000
CHRYSOIDINE Y SPECIAL see PEK000
CHRYSOIDIN FB see PEK000
CHRYSOIDIN Y see PEK000
CHRYSOIDIN YN see PEK000
CHRYSOMYKINE see CMA750
CHRYSON see BEP500
CHRYSOPHANIC ACID ANTHRANOL see CML750
CHRYSOTILE (DOT) see ARM268
CHRYSOTILE ASBESTOS see ARM268
CHRYSRON see BEP500
CHRYTEMIN see DLH630
CHRYZOIDYNA F.B. (POLISH) see PEK000
CHWASTOX see CIR250
CI-2 see MAV750
C.I. 27 see HGC000
C.I. 258 see SBC500
CI-337 see ASA500
CI 366 see ENG500
CI-406 see PAN100
C.I. 456 see CIP500
CI 581 see CKD750
CI-628 see NHP500

C.I. 671 see FMU059
CI-705 see QAK000
C.I. 749 see FAG070
C.I. 766 see FEW000
C.I. 1956 see CKN000
C.I. 7581 see FAE100
C.I. 10305 see PID000
C.I. 10355 see DVX800
C.I. 11000 see PEI000
C.I. 11020 see DOT300
C.I. 11160 see AIC250
C.I. 11270 see PEK000
C.I. 11380 see FAG130
C.I. 11390 see FAG135
C.I. 11855 see AAQ250
C.I. 12055 see PEJ500
C.I. 12100 see TGW000
C.I. 12140 see XRA000
C.I. 12156 see DOK200
C.I. 13020 see CCE500
C.I. 14720 see HJF500
C.I. 15985 see FAG150
C.I. 16035 see FAG100
C.I. 16155 see FAG018
C.I. 19140 see FAG140
C.I. 20285 see CMP500
C.I. 22610 see CMO000
C.I. 23060 see DEQ600
C.I. 23850 see CMO250
C.I. 24110 see DCJ200
C.I. 26050 see EOJ500
C.I. 30145 see CMO750
C.I. 30235 see AQP000
C.I. 37025 see NEO000
C.I. 37030 see NEN500
C.I. 37035 see NEO500
C.I. 37077 see TGQ750
C.I. 37085 see CLK235
C.I. 37105 see NMP500
C.I. 37107 see TGR000
C.I. 37115 see AOX250
C.I. 37130 see NEQ500
C.I. 37225 see BBX000
C.I. 37230 see TGJ750
C.I. 37270 see NBE500
C.I. 37275 see AIA750
C.I. 37500 see NAX000
C.I. 41000 see IBA000
C.I. 42040 see BAY750
C.I. 42053 see FAG000
C.I. 42090 see FAE000, FMU059
C.I. 42095 see FAF000
C.I. 42500 see RMK020
C.I. 42640 see FAG120
C.I. 45160 see RGW000
C.I. 45330 see FEV000
C.I. 45380 see BNH500
C.I. 45430 see FAG040
C.I. 46005 see BJF000
C.I. 47031 see PGW750
C.I. 59825 see JAT000
C.I. 60700 see AKP750
C.I. 60710 see AKE250
C.I. 61200 see BMM500
C.I. 61565 see BLK000
C.I. 63340 see TGS000
C.I. 73015 see FAE100
C.I. 75300 see COG000
C.I. 75500 see WAT000
C.I. 75670 see QCA000
C.I. 76000 see AOQ000
C.I. 76010 see PEY250
C.I. 76025 see PEY000
C.I. 76027 see CJY120

C.I. 76035 see TGL750
C.I. 76042 see TGM000
C.I. 76043 see DCE600
C.I. 76050 see DBO000
C.I. 76051 see DBO400
C.I. 76060 see PEY500
C.I. 76070 see ALL750
C.I. 76075 see DTL800
C.I. 76500 see CCP850
C.I. 76505 see REA000
C.I. 76515 see PPQ500
C.I. 76520 see ALT000
C.I. 76545 see ALT500
C.I. 76555 see NEM480
C.I. 76605 see NAW500
C.I. 77000 see AGX000
C.I. 77050 see AQB750
C.I. 77056 see AQC500
C.I. 77060 see AQL500
C.I. 77086 see ARI000
C.I. 77103 see BAK250
C.I. 77120 see BAP000
C.I. 77180 see CAD000
C.I. 77185 see CAD250
C.I. 77199 see CAJ750
C.I. 77223 see CAP500, CAP750
C.I. 77231 see CAX750
C.I. 77266 see CBT500
C.I. 77288 see CMJ900
C.I. 77295 see CMJ250
C.I. 77320 see CNA250
C.I. 77400 see CNI000
C.I. 77410 see COF500
C.I. 77491 see IHD000
C.I. 77575 see LCF000
C.I. 77577 see LDN000
C.I. 77578 see LDS000
C.I. 77580 see LCX000
C.I. 77600 see LCR000
C.I. 77601 see LCS000
C.I. 77610 see LCU000
C.I. 77622 see LDU000
C.I. 77630 see LDY000
C.I. 77640 see LDZ000
C.I. 77713 see MAC650
C.I. 77718 see TAB750
C.I. 77726 see MAT250
C.I. 77727 see MAT500
C.I. 77728 see MAS000
C.I. 77755 see PLP000
C.I. 77760 see MCT500
C.I. 77764 see MCW000
C.I. 77775 see NCW500
C.I. 77777 see NDF500
C.I. 77779 see NCY500
C.I. 77795 see PJD500
C.I. 77805 see SBO500
C.I. 77820 see SDI500
C.I. 77847 see SMM000
C.I. 77864 see TGC000
C.I. 77891 see TGG760
C.I. 77938 see VDU000
C.I. 77940 see VEZ000
C.I. 77945 see ZBJ000
C.I. 77947 see ZKA000
C.I. 77955 see ZFJ100
C.I. 11160B see AIC250
C.I. 52 015 (CZECH) see BJI250
C.I. ACID BLUE 74 see FAE100
C.I. ACID BLUE 9, DIAMMONIUM SALT see FMU059
C.I. ACID BLUE 9, DISODIUM SALT see FAE000
C.I. ACID GREEN 5 see FAF000
C.I. ACID GREEN 5, DISODIUM SALT see FAF000
C.I. ACID ORANGE 10 see HGC000

C.I. ACID RED 2 see CCE500
C.I. ACID RED 51 see FAG040
C.I. ACID RED 14, DISODIUM SALT see HJF500
C.I. ACID YELLOW 73 see FEW000
CIANAZIL see COH250
CIANURO di SODIO (ITALIAN) see SGA500
CIANURO di VINILE (ITALIAN) see ADX500
C.I. AZOIC COUPLING COMPONENT 1 see NAX000
C.I. AZOIC COUPLING COMPONENT 107 see TGR000
C.I. AZOIC DIAZO COMPONENT 6 see NEO000
C.I. AZOIC DIAZO COMPONENT 7 see NEN500
C.I. AZOIC DIAZO COMPONENT 11 see CLK235
C.I. AZOIC DIAZO COMPONENT 12 see NMP500
C.I. AZOIC DIAZO COMPONENT 13 see NEQ500
C.I. AZOIC DIAZO COMPONENT 37 see NEO500
C.I. AZOIC DIAZO COMPONENT 48 see DCJ200
C.I. AZOIC DIAZO COMPONENT 112 see BBX000
C.I. AZOIC DIAZO COMPONENT 113 see TGJ750
C.I. AZOIC DIAZO COMPONENT 114 see NBE000
C.I. AZOIC RED 83 see MGO750
C.I. 41000B see IBB000
CIBA 709 see DGQ875
CIBA 2059 see DUK800
CIBA 5968 see HGP500, PHW000
CIBA 8353 see DVS000
CIBA 8514 see CJJ250, KEA000
CIBA 32644 see NML000
CIBA 12669A see MIW500
CIBA 32644-BA see NML000
CIBA C-768 see DTP800
CIBA C-2307 see DOL800
CIBA C-7824 see MPG250
CIBACETE DIAZO NAVY BLUE 2B see DCJ200
CIBACET YELLOW GBA see AAQ250
CIBA CO. 2825 see PIT250
CIBA-GEIGY GS 13005 see DSO000
C.I. BASIC BLUE 9 see BJI250
C.I. BASIC GREEN 1, SULFATE (1:1) see BAY750
C.I. BASIC ORANGE 2 see PEK000
C.I. BASIC ORANGE 3 see PEK000
C.I. BASIC ORANGE 14 see BJF000
C.I. BASIC ORANGE 2, MONOHYDROCHLORIDE see
    PEK000
C.I. BASIC RED 1, MONOHYDROCHLORIDE see RGW000
C.I. BASIC RED 9, MONOHYDROCHLORIDE see RMK020
C.I. BASIC VIOLET 10 see FAG070
C.I. BASIC YELLOW 2 see IBA000
C.I. BASIC YELLOW 2, FREE BASE see IBB000
C.I. BASIC YELLOW 2, MONOHYDROCHLORIDE see
    IBA000
CI-628 CITRATE see NHP500
CICLOBIOTIC see MDO250
CICLOESANO (ITALIAN) see CPB000
CICLOESANOLO (ITALIAN) see CPB750
CICLOESANONE (ITALIAN) see CPC000
6-CICLOESIL-2,4-DINITR-FENOLO (ITALIAN) see CPK500
CICLORAL see BSM000
CICLOSOM see TIQ250
CICP see CKC000
CICUTOXIN see CMN000
CIDAL see SAH000
CIDALON see IHZ000
CIDAMEX see AAI250
CIDANDOPA see DNA200
CIDEMUL see DRR400
C.I. DEVELOPER 4 see REA000
C.I. DEVELOPER 5 see NAX000
C.I. DEVELOPER 13 see PEY500
C.I. DEVELOPER 17 see NEO500
CIDEX see GFQ000
CIDIAL see DRR400
C.I. DIRECT BLACK 38 see AQP000
C.I. DIRECT BLUE 14 see CMO250
C.I. DIRECT BLUE 6, TETRASODIUM SALT see CMO000

C.I. DIRECT BLUE 14, TETRASODIUM SALT see CMO250
C.I. DIRECT BROWN see CMO750
C.I. DIRECT BROWN 78, DIAMMONIUM SALT see FMU059
C.I. 45350 DISODIUM SALT see FEW000
C.I. DISPERSE BLACK 6 see DCJ200
C.I. DISPERSE BLACK-6-DIHYDROCHLORIDE see DOA800
C.I. DISPERSE ORANGE 11 see AKP750
C.I. DISPERSE YELLOW 3 see AAQ250
CIDOCETINE see CDP250
CIDREX see CFY000
C.I. FOOD BLUE 1 see FAE100
C.I. FOOD BLUE 2 see FAE000, FMU059
C.I. FOOD BROWN 3, DISODIUM SALT see CMP500
C.I. FOOD GREEN 2 see FAF000
C.I. FOOD GREEN 3 see FAG000
C.I. FOOD ORANGE 4 see HGC000
C.I. FOOD RED 3 see HJF500
C.I. FOOD RED 6 see FAG018
C.I. FOOD RED 15 see FAG070
C.I. FOOD RED 6, DISODIUM SALT see FAG018
C.I. FOOD VIOLET 2 see FAG120
C.I. FOOD YELLOW 4 see FAG140
C.I. FOOD YELLOW 10 see FAG130
C.I. FOOD YELLOW 11 see FAG135
C.I. 45350 (FREE ACID) see FEV000
CIGARETTE REFINED TAR see CMP800
CIGARETTE SMOKE CONDENSATE see SEC000
CIGARETTE TAR see CMP800
CI-IPC see CKC000
CILAG 61 see HDY000
CILEFA PINK B see FAG040
CILLA ORANGE R see AKP750
CILLORAL see BDY669
CILOPEN see BDY669
CIMEXAN see MAK700
C.I. NATURAL BROWN 7 see WAT000
C.I. NATURAL BROWN 8 see MAT500
C.I. NATURAL RED 1 see QCA000
C.I. NATURAL YELLOW 1 see CDH250
C.I. NATURAL YELLOW 8 see MRN500
C.I. NATURAL YELLOW 10 see QCA000
CINCAINE HYDROCHLORIDE see NOF500
CINCHOCAINE see DDT200
CINCHOCAINE HYDROCHLORIDE see NOF500
CINCHOCAINIUM CHLORIDE see NOF500
CINEB see EIR000
CINENE see MCC250
1,8-CINEOL see CAL000
CINEOLE see CAL000
1,8-CINEOLE see CAL000
CINERIN I ALLYL HOMOLOG see AFR250
CINERIN I or II see POO250
CINNAMAL see CMP969
CINNAMALDEHYDE see CMP969
CINNAMEIN see BEG750
CINNAMENE see SMQ000
CINNAMENOL see SMQ000
CINNAMIC ACID see CMP975
trans-CINNAMIC ACID BENZYL ESTER see BEG750
CINNAMIC ACID, ISOBUTYL ESTER see IIQ000
CINNAMIC ALCOHOL see CMQ740
CINNAMON BARK OIL see CCO750
CINNAMON BARK OIL, CEYLON TYPE (FCC) see CCO750
CINNAMON OIL see CCO750
CINNAMYL ACETATE see CMQ730
CINNAMYL ALCOHOL see CMQ740
CINNAMYL ALCOHOL ANTHRANILATE see API750
CINNAMYL ALCOHOL, FORMATE see CMR500
CINNAMYL ALCOHOL, SYNTHETIC see CMQ740
CINNAMYL ALDEHYDE see CMP969
CINNAMYL-2-AMINOBENZOATE see API750
CINNAMYL-o-AMINOBENZOATE see API750

CINNAMYL ANTHRANILATE (FCC) see API750
d-CINNAMYLEPHEDRINE HYDROCHLORIDE see CMR250
CINNAMYLEPHEDRINE HYDROCHLORIDE, DEXTRO see CMR250
CINNAMYL FORMATE see CMR500
CINNAMYL ISOVALERATE see CMR800
CINNAMYL METHANOATE see CMR500
CINNAMYL PROPIONATE see CMR850
CINNIMIC ALDEHYDE see CMP969
C.I. No. 77278 see CMJ900
C.I. No. 46005:1 see BJF000
CIN-QUIN see QFS000
CINU see CGV250
C.I. OXIDATION BASE see TGL750
C.I. OXIDATION BASE 7 see ALT500
C.I. OXIDATION BASE 10 see PEY500
C.I. OXIDATION BASE 12 see DBO000
C.I. OXIDATION BASE 12A see DBO400
C.I. OXIDATION BASE 16 see PEY250
C.I. OXIDATION BASE 17 see ALT000
C.I. OXIDATION BASE 22 see ALL750
C.I. OXIDATION BASE 26 see CCP850
C.I. OXIDATION BASE 31 see REA000
C.I. OXIDATION BASE 32 see PPQ500
C.I. OXIDATION BASE 33 see NAW500
CIPC see CKC000
C.I. PIGMENT BLACK 14 see MAS000
C.I. PIGMENT BLACK 16 see ZBJ000
C.I. PIGMENT BROWN 8 see MAS000
C.I. PIGMENT GREEN 17 see CMJ900
C.I. PIGMENT GREEN 21 (9CI) see COF500
C.I. PIGMENT METAL 2 see CNI000
C.I. PIGMENT METAL 4 see LCF000
C.I. PIGMENT METAL 6 see ZBJ000
C.I. PIGMENT ORANGE 20 see CAJ750
C.I. PIGMENT ORANGE 21 see LCS000
C.I. PIGMENT RED see CHP500, LCS000
C.I. PIGMENT RED 101 see IHD000
C.I. PIGMENT RED 104 see LDM000
C.I. PIGMENT RED 105 see LDS000
C.I. PIGMENT WHITE 3 see LDY000
C.I. PIGMENT WHITE 4 see ZKA000
C.I. PIGMENT WHITE 6 see TGG760
C.I. PIGMENT WHITE 11 see AQF000
C.I. PIGMENT WHITE 21 see BAP000
C.I. PIGMENT WHITE 25 see CAX750
C.I. PIGMENT YELLOW 31 see BAK250
C.I. PIGMENT YELLOW 32 see SMH000
C.I. PIGMENT YELLOW 33 see CAP500, CAP750
C.I. PIGMENT YELLOW 34 see LCR000
C.I. PIGMENT YELLOW 36 see ZFJ100
C.I. PIGMENT YELLOW 37 see CAJ750
C.I. PIGMENT YELLOW 46 see LDN000
C.I. PIGMENT YELLOW 48 see LCU000
CIPLAMYCETIN see CDP250
CIPROMID see CQJ250
CIRAM see BJK500
CIRCOSOLV see TIO750
C.I. REACTIVE BLUE 19 see BMM500
C.I. REACTIVE BLUE 19, DISODIUM SALT see BMM500
CIRPONYL see MQU750
CIRRASOL 185A see NMY000
CIRRASOL-OD see HCQ500
C.I. SOLVENT BLUE 7 see PEI000
C.I. SOLVENT GREEN 3 see BLK000
C.I. SOLVENT ORANGE 2 see TGW000
C.I. S0
CO CAP IMIPRAMINE 25 see DLH630
COCCIDINE A see DUP300
COCCIDIOSTAT C see CMX850
COCCIDOT see DUP300
COCCULIN see PIE500
COCCULUS see PIE500
COCCULUS solid (DOT) see PIE500

COCHIN see LEG000
COCO DIETHANOLAMIDE see BKE500
COCONUT ALDEHYDE see CNF250
COCONUT BUTTER see CNR000
COCONUT MEAL PELLETS, containing 6-13% moisture and no
  more than 10% residual fat (DOT) see CNR000
COCONUT OIL AMIDE of DIETHANOLAMINE see BKE500
COCONUT OIL (FCC) see CNR000
COCONUT PALM OIL see CNR000
CODECHINE see BBQ500
CODELCORTONE see PMA000
CODEMPIRAL e ENK000
CODHYDRINE see DKW800
CODIBARBITA see EOK000
CO-ESTRO see ECU750
COFFEIN (GERMAN) see CAK500
COFFEINE see CAK500
COGESIC see DTO200
COGILOR BLUE 512.12 see FAE000
COGILOR RED 321.10 see FAG070
COHASAL-1H see SEH000
CO-HYDELTRA see PMA000
COHYDRIN see DKW800
COIR DEEP BLACK C see AQP000
COLACE see DJL000
COLAMINE see EEC600
COLCEMIDE see MIW500
COLCHAMINE see MIW500
COLCHINE, N-DEACETYL-N-METHYL see MIW500
COLCOTHAR see IHD000
COLEBENZ see BCM000
COLECALCIFEROL see CMC750
COLEMID see MIW500
COLFARIT see ADA725
COLISONE see PLZ000
COLISTINASE see BAC000
COLLIDINE, ALDEHYDECOLLIDINE see EOS000
COLLIRON I.V. see IHG000
COLLODION see CNH000
COLLODION COTTON see CCU250
COLLOID 775 see CCL250
COLLOIDAL ARSENIC see ARA750
COLLOIDAL CADMIUM see CAD000
COLLOIDAL FERRIC OXIDE LYL ACETATE see RHA000
CITRONELLYL ACETATE (FCC) see AAU000
CITRONELLYL BUTYRATE see CMT600
CITRONELLYL FORMATE see CMT750
CITRONELLYL ISOBUTYRATE see CMT900
CITRONELLYL PROPIONATE see CMU100
CITRON YELLOW see ZFJ100, PLW500
CITRULLAMON see DKQ000, DNU000
CITRUS RED No. 2 see DOK200
CIZARON see TMP750
CL 337 see ASA500
CL 369 see CKD750
CL 12503 see POP000
CL 12,625 see PIF750
CL 12880 see DSP400
CL-14377 see MDV500
CL 18133 see EPC500
CL-38023 see FAB600
CL-43,064 see DXN600
CL 47300 see DSQ000
CL-47,470 see DHH400
CL 52160 see TAL250
CL-62362 see DCS200
CL-71563 see DCS200
CL 19217 4090L 7-5525 see EGI000
CLAFEN see EAS500, CQC675
CLAIRSIT see PCF300
CLAPHENE see EAS500
CLARK I see CGN000
CLEARASIL BENZOYL PEROXIDE LOTION see BDS000
CLEARASIL BP ACNE TREATMENT see BDS000

CLESTOL see DJL000
CLHORAMEISENSAEUREAETHYLESTER (GERMAN) see
  EHK500
CLIFT see MCI750
CLIMATERINE see DKA600
CLIMESTRONE see ECU750
CLIN see SJO000
CLINDROL 101CG see BKE500
CLINDROL SDG see HKJ000
CLINDROL SEG see EJM500
CLINDROL SUPERAMIDE 100L see BKE500
CLINESTROL see DKB000
CLIOQUINOL see CHR500
CLIQUINOL see CHR500
CLIRADON HYDROCHLORIDE see KFK000
CLIXODYNE see HIM000
4-Cl-M-PD see CJY120
CLOAZEPAM see CMW000, CMW000
CLOBAZAM see CIR750
CLOBBER see CQJ250
CLOFENOTANE see DAD200
CLOFLUPEROL HYDROCHLORIDE see CLS250
CLOMID see CMX700
CLOMIFEN CITRATE see CMX700
CLOMIFENO see CMX700
CLOMIPHENE CITRATE see CMX700
racemic-CLOMIPHENE CITRATE see CMX700
CLOMIPHENE DIHYDROGEN CITRATE see CMX700
CLOMIPHENE-R see CMX700
CLOMIPHINE see CMX700
CLOMIVID see CMX700
CLOMPHID see CMX700
CLONAZEPAM see CMW000
CLONITARLID see DFV600
CLONITRALID see DFV400
CLONT see MMN250
CLOPHEN see PJL750
CLOPHEN A60 see PJN250
CLOPIDOL see CMX850
CLOPOXIDE see LFK000
CLOPROSTENOL see CMX880
CLORAMIDINA see CDP250
CLORAMIN see BIE250
CLORARSEN see DFX400
CLOR CHEM T-590 see CDV100
CLORDAN (ITALIAN) see CDR750
CLORDIAZEPOSSIDO (ITALIAN) see LFK000
CLORDION see CBF250
CLOREPIN see CIR750
CLORESTROLO see CLO750
CLOREX see DFJ050
CLORGYLINE HYDROCHLORIDE see CMY000
CLORINA see CDP000
CLORNAPHAZINE see BIF250
CLORO (ITALIAN) see CDV750
CLOROAMFENICOLO (ITALIAN) see CDP250
CLOROBEN see DEP600
2-CLOROBENZALDEIDE (ITALIAN) see CEI500
CLOROBENZENE (ITALIAN) see CEJ125
(4-CLORO-BENZIL)-(4-CLORO-FENIL)-SOLFURO (ITALIAN)
  see CEP000
1-p-CLORO-BENZOIL-5-METOXI-2-METILINDOL-3-ACIDO
  ACETICO (SPANISH) see IDA000
2-CLORO-1,3-BUTADIENE (ITALIAN) see NCI500
(4-CLORO-BUT-2-IN-IL)-N-(3-CLORO-FENIL)-CARBAM-
  MATO (ITALIAN) see CEW500
O-2-CLORO-1-(2,4-DICLORO-FENIL)-VINYL-O,O-DIETIL-
  FOSFATO (ITALIAN) see CDS750
CLORODIFENILI, CLORO 42% (ITALIAN) see PJM500
CLORODIFENILI, CLORO 54% (ITALIAN) see PJN000
2-CLORO-4-DIMETILAMINO-6-METIL-PIRIMIDINA (ITAL-
  IAN) see CCP500
1-CLORO-2,4-DINITROBENZENE (ITALIAN) see CGM000
1-CLORO-2,3-EPOSSIPROPANO (ITALIAN) see EAZ500

CLOROETANO (ITALIAN) see EHH000
2-CLOROETANOLO (ITALIAN) see EIU800
(CLORO-2-ETIL)-1-CICLOESIL-3-NITROSOUREA (ITALIAN) see CGV250
(4-CLORO-FENIL)-BENZOL-SOLFONATO (ITALIAN) see CJR500
(4-CLORO-FENIL)-4-CLORO-VENZOL-SOLFONATO (ITALIAN) see CJT750
3-(4-CLORO-FENIL)-1,1-DIMETIL-UREA (ITALIAN) see CJX750
2(2-(4-CLORO-FENIL-2-FENIL)-ACETIL)INDAN-1,3-DIONE (ITALIAN) see CJJ000
N-(3-CLORO-FENIL)-ISOPROPIL-CARBAMMATO (ITALIAN) see CKC000
CLOROFORMIO (ITALIAN) see CHJ500
CLOROFOS (RUSSIAN) see TIQ250
CLOROMETANO (ITALIAN) see MIF765
7-CLORO-2-METILAMINO-5-FENIL-3H-1,4-BENZOIDI-AZEPINA 4-OSSIDO (ITALIAN) see LFK000
3-CLORO-2-METIL-PROP-1-ENE (ITALIAN) see CIU750
CLOROMISAN see CDP250
1-CLORO-4-NITROBENZENE (ITALIAN) see NFS525
O-(4-CLORO-3-NITRO-FENIL)-O,O-DIMETIL-MONOIIOFOS-FATO (ITALIAN) see NFT000
O-(3-CLORO-4-NITRO-FENIL)-O,O-DIMETIL-MONOTIOFOS-FATO (ITALIAN) see MIJ250
CLOROPICRINA (ITALIAN) see CKN500
CLOROPIRIL see TAI500
CLOROPRENE (ITALIAN) see NCI500
CLOROSAN see CDP000
CLOROSINTEX see CDP250
CLOROTRISIN see CLO750
CLOROX see SHU500
CLORTRAN see ABD000
CLORURO DI ETILE (ITALIAN) see EHH000
CLORURO di ETHENE (ITALIAN) see EIY600
CLORURO di ETILIDENE (ITALIAN) see DFF809
CLORURO di MERCURIO (ITALIAN) see MCY475
CLORURO MERCUROSO (ITALIAN) see MCW000
CLORURO di METALLILE (ITALIAN) see CIU750
CLORURO di METILE (ITALIAN) see MIF765
CLORURO di SUCCINILCOLINA (ITALIAN) see HLC500
CLORURO di VINILE (ITALIAN) see VNP000
SYMCLOSEN see TIQ750
SYMCLOSENE see TIQ750
CLOUT see DXE600
CLOVE LEAF OIL see CMY100
CLOVE LEAF OIL MADAGASCAR see CMY100
CLOXACILLIN SODIUM MONOHYDRATE see SLJ000
CLOXAPEN see SLJ000
CLOXAZEPINE see DCS200
CLOXYPEN see SLJ000
4-Cl-o-PD see CFK125
CLYSAR see PMP500
CMA see CBF250, CIF250
CMC see SFO500
CMC 7H see SFO500
CM-CELLULOSE Na SALT see SFO500
CMC SODIUM SALT see SFO500
CMDP see MQR750
C-METON see TAI500
CMME see CIO250
CMPP see CIR500
CMU see CJX750
CMW BONE CEMENT see PKB500
CN see CEA750
CN 8676 see EEI000
CN-15,757 see ASA500
CN-36337 see CIP500
CN 38703 see QAK000
CN-52,372-2 see CKD750
CN-55945-27 see NHP500
CO 12 see DXV600
CO-1214 see DXV600

CO-1670 see HCP000
CO-1895 see OAX000
CO-1897 see OAX000
COAL CONVERSION MATERIALS, SRC-II HEAVY DISTIL-LATE see CMY625
COAL DUST see CMY635
COAL FACINGS see CMY635
COAL GAS see HHJ500
COAL, GROUND BITUMINOUS (DOT) see CMY635
COAL LIQUID see PCR250
COAL-MILLED see CMY635
COAL NAPHTHA see BBL250
COAL OIL see KEK000, PCR250
COAL OIL (EXPORT SHIPMENT ONLY) (DOT) see KEK000
COAL SLAG-MILLED see CMY635
COAL TAR see CMY800
COAL TAR CREOSOTE see CMY825
COAL TAR DYE, liquid (DOT) see CMY840
COAL TAR NAPHTHA see NAI500
COAL TAR OIL see CMY825
COAL TAR OIL (DOT) see CMY825
COAL TAR PITCH VOLATILES see CMZ100
COAPT see MIQ075
COATHYLENE PF 0548 see PMP500
COBADOCE FORTE see VSZ000
COBALIN see VSZ000
COBALT see CNA250
COBALT-59 see CNA250
COBALT ACETATE TETRAHYDRATE see CNA500
COBALT ALLOY, Co, Cr see CNA750
COBALT CARBONYL see CNB500
COBALT(II) CHLORIDE see CNB599
COBALT-CHROMIUM ALLOY see CNA750
COBALT COMPOUNDS see CNB850
COBALT DIACETATE TETRAHYDRATE see CNA500
COBALT DICHLORIDE see CNB599
COBALT-HISTIDINE see BJY000
COBALT HYDROCARBONYL see CNC230
COBALT MURIATE see CNB599
COBALT NAPHTHENATE, POWDER (DOT) see NAR500
COBALT OCTACARBONYL see CNB500
COBALTOUS ACETATE TETRAHYDRATE see CNA500
COBALTOUS CHLORIDE see CNB599
COBALTOUS DICHLORIDE see CNB599
COBALTOUS SULFATE see CNE125
COBALT SULFATE see CNE125
COBALT SULFATE (1:1) see CNE125
COBALT (2+) SULFATE see CNE125
COBALT(II) SULFATE (1:1) see CNE125
COBALT(II) SULPHATE see CNE125
COBALT TETRACARBONYL see CNB500
COBALT TETRACARBONYL DIMER see CNB500
COBAMIN see VSZ000
COBEX (polymer) see PKQ059
COBH see BDD000
COBIONE see VSZ000
COBRATEC #99 see BDH250
COCAFURIN see NGE500
CO CAP IMIPRAMINE 25 see DLH630
COCCIDINE A see DUP300
COCCIDIOSTAT C see CMX850
COCCIDOT see DUP300
COCCULIN see PIE500
COCCULUS see PIE500
COCCULUS solid (DOT) see PIE500
COCHIN see LEG000
COCO DIETHANOLAMIDE see BKE500
COCONUT ALDEHYDE see CNF250
COCONUT BUTTER see CNR000
COCONUT MEAL PELLETS, containing 6-13% moisture and no more than 10% residual fat (DOT) see CNR000
COCONUT OIL AMIDE of DIETHANOLAMINE see BKE500
COCONUT OIL (FCC) see CNR000
COCONUT PALM OIL see CNR000

CODECHINE see BBQ500
CODELCORTONE see PMA000
CODEMPIRAL see ABG750
CODETHYLINE see ENK000
CODHYDRINE see DKW800
CODIBARBITA see EOK000
CO-ESTRO see ECU750
COFFEIN (GERMAN) see CAK500
COFFEINE see CAK500
COGESIC see DTO200
COGILOR BLUE 512.12 see FAE000
COGILOR RED 321.10 see FAG070
COHASAL-1H see SEH000
CO-HYDELTRA see PMA000
COHYDRIN see DKW800
COIR DEEP BLACK C see AQP000
COLACE see DJL000
COLAMINE see EEC600
COLCEMIDE see MIW500
COLCHAMINE see MIW500
COLCHINE, N-DEACETYL-N-METHYL see MIW500
COLCOTHAR see IHD000
COLEBENZ see BCM000
COLECALCIFEROL see CMC750
COLEMID see MIW500
COLFARIT see ADA725
COLISONE see PLZ000
COLISTINASE see BAC000
COLLIDINE, ALDEHYDECOLLIDINE see EOS000
COLLIRON I.V. see IHG000
COLLODION see CNH000
COLLODION COTTON see CCU250
COLLOID 775 see CCL250
COLLOIDAL ARSENIC see ARA750
COLLOIDAL CADMIUM see CAD000
COLLOIDAL FERRIC OXIDE see IHD000
COLLOIDAL GOLD see GIS000
COLLOIDAL MANGANESE see MAP750
COLLOIDAL MERCURY see MCW250
COLLOIDAL SELENIUM see SBO500
COLLOIDAL SILICA see SCH000
COLLOIDAL SILICON DIOXIDE see SCH000
COLLOIDAL SULFUR see SOD500
COLLOKIT see SOD500
COLLOMIDE see SNM500
COLLOWELL see SFO500
COLLOXYLIN see CCU250
COLLUNOSOL see TIV750
COLLUNOVAR see NCJ500
COLLUNOVER see NCJ500
COLOGNE EARTH see MAT500
COLOGNE SPIRIT see EFU000
COLOGNE SPIRITS (ALCOHOL) (DOT) see EFU000
COLOGNE UMBER see MAT500
COLOGNE YELLOW see LCR000
COLOMBIAN BLACK TOBACCO CIGARETTE REFINED
    TAR see CMP800
COLONATRAST see BAP000
COLONIAL SPIRIT see MGB150
COLOR-SET see TIX500
COLPOVISTER see EDU500
COLSUL see SOD500
COLSULANYDE see SNM500
COLTS FOOT see PCR000
COLUMBIA BLACK EP see AQP000
COLUMBIAN CARBON see CBT500
COLUMBIAN SPIRITS (DOT) see MGB150
COLUMBIUM PENTACHLORIDE see NEA000
COMBOT EQUINE see TIQ250
COMBUSTION IMPROVER -2 see MAV750
COMESA see MNM500
COMESTROL see DKA600
COMESTROL ESTROBENE see DKA600
COMFREY, RUSSIAN see RRP000

COMITAL see DKQ000
COMITE see SOP000
COMMON SALT see SFT000
COMMOTIONAL see ABG750
COMPALOX see AHE250
COMPAZINE see PMF250, PMF500
COMPERLAN LD see BKE500
COMPITOX see CIR500
COMPLEMIX see DJL000
COMPLEXON II see EIX000
COMPLEXON III see EIX500
COMPLEXONE see EIV000
COMPOCILLIN G see BDY669
COMPOUND 42 see WAT200
COMPOUND 118 see AFK250
COMPOUND 269 see EAT500
COMPOUND 338 see DER000
COMPOUND 347 see EAT900
COMPOUND 497 see DHB400
COMPOUND 604 see DFT000
COMPOUND-666 see BBP750
COMPOUND 711 see IKO000
COMPOUND 864 see YCJ000
COMPOUND 889 see DVL700
COMPOUND 88R see SOP500
COMPOUND 923 see DFY400
COMPOUND 1081 see FFF000
COMPOUND 1189 see KEA000
COMPOUND 2046 see MQR750
COMPOUND 3422 see PAK000
COMPOUND 3956 see CDV100
COMPOUND 4049 see MAK700
COMPOUND 4072 see CDS750
COMPOUND 01748 see DJT800
COMPOUND 33355 see MKB750
COMPOUND 33,828 see MLJ500
COMPOUND B DICAMBA see MEL500
COMPOUND-1452-F see EME500
COMPOUND G-11 see HCL000
COMPOUND 6-12 INSECT REPELLENT see EKV000
COMPOUND M-81 see PHI500
COMPOUND No. 1080 see SHG500
COMYCETIN see CDP250
d-CON see WAT200
CONCHININ see QFS000
CONCO AAS-35 see DXW200
CONCO NI-90 see PKF000
CONCO NIX-100 see PKF500
CONCO SULFATE WA see SIB600
CONDACAPS see VSZ100
CONDENSATE PL see BKE500
CONDENSATES (PETROLEUM), VACUUM TOWER (9CI)
    see MQV755
CONDOCAPS see VSZ100
CONDOL see VSZ100
CONDY'S CRYSTALS see PLP000
CONESSINE DIHYDROBROMIDE see DOX000
CONEST see ECU750
CONESTRON see ECU750
CONFECTIONER'S SUGAR see SNH000
CONFORTID see IDA000
CONGOBLAU 3B see CMO250
CONGO BLUE see CMO250
CONIGON BC see EIV000
CONJES see ECU750
CONJUGATED EQUINE ESTROGEN see PMB000
CONJUGATED ESTROGENS see ECU750
CONJUTABS see ECU750
CONOCO C-50 see DXW200
CONOTRANE see PFN000
CONOVID see EAP000
CONOVID E see EAP000
CONQUININE see QFS000
CONSTONATE see DJL000

CONT see MMN250
CONTAVERM see PDP250
CONTERGAN see TEH500
CONTIMET 30 see TGF250
CONTINAL see NBU000
CONTIZELL see PKQ059
CONTRA CREME see ABU500
CONTRADOL see ABG750
CONTRALGIN see BQA010
CONTRALIN see DXH250
CONTRAPOT see DXH250
CONTRHEUMA RETARD see ADA725
CONTROVLAR see EEH520
CONVENIXA see TLP750
CONVUL see DKQ000
COOMASSIE VIOLET see FAG120
CO-OP HEXA see HCC500
COPAGEL PB 25 see SFO500
COPAIBA BALSAM see CNH792
COPAIBA OIL see CNH792
COPAIBA OLEORESIN see CNH792
COPAL Z see SMQ500
COPHARCILIN see AIV500
COPPER see CNI000
COPPER-8 see BLC250
COPPER ACETATE see CNI250
COPPER(2+) ACETATE see CNI250
COPPER(II) ACETATE see CNI250
COPPER ACETOARSENITE (DOT) see COF500
COPPER ACETOARSENITE, solid (DOT) see COF500
COPPER(II) ACETYLIDE see CNI500
COPPER-AIRBORNE see CNI000
COPPER ALLOY, Cu, Be see CNI750
COPPER ALLOY, Cu, Be, Co see CNK700
COPPER ARSENATE (BASIC) see CNI900
COPPER ARSENATE HYDROXIDE see CNI900
COPPER ARSENITE, solid (DOT) see CNN500
COPPERAS see FBN100, FBO000
COPPER-BERYLLIUM ALLOY see CNI750
COPPER BRONZE see CNI000
COPPER CHLORIDE (DOT) see CNK500
COPPER(II) CHLORIDE (1:2) see CNK500
COPPER-COBALT-BERYLLIUM see CNK700
COPPER COMPOUNDS see CNK750
COPPER CYANIDE see CNL000
COPPER(II) CYANIDE see CNL250
COPPER CYANIDE (DOT) see CNL250
COPPER CYNANAMIDE see CNL250
COPPER DIACETATE see CNI250
COPPER(2+) DIACETATE see CNI250
COPPER DINITRATE see CNM750
COPPER HYDROXYQUINOLATE see BLC250
COPPER-8-HYDROXYQUINOLATE see BLC250
COPPER-8-HYDROXYQUINOLINATE see BLC250
COPPER-8-HYDROXYQUINOLINE see BLC250
COPPER-MILLED see CNI000
COPPER MONOSULFATE see CNP250
COPPER NAPHTHENATE see NAS000
COPPER(2+) NITRATE see CNM750
COPPER(II) NITRATE see CNM750
COPPER ORTHOARSENITE see CNN500
COPPER OXINATE see BLC250
COPPER (2+) OXINATE see BLC250
COPPER OXINE see BLC250
COPPER OXYQUINOLATE see BLC250
COPPER OXYQUINOLINE see BLC250
COPPER QUINOLATE see BLC250
COPPER-8-QUINOLATE see BLC250
COPPER-8-QUINOLINOL see BLC250
COPPER QUINOLINOLATE see BLC250
COPPER-8-QUINOLINOLATE see BLC250
COPPER SLAG-AIRBORNE see CNI000
COPPER SLAG-MILLED see CNI000
COPPER SULFATE see CNP250

COPPER(II) SULFATE (1:1) see CNP250
COPPER UVERSOL see NAS000
COPRA (DOT) see CNR000
COPRA (OIL) see CNR000
COPRA PELLETS (DOT) see CNR000
COPROL see DJL000
COPROSTANOL see DKW000
COPROSTAN-3-β-OL see DKW000
COPROSTEROL see DKW000
COPSAMINE see WAK000
COPTICIDE see SNM500
COQUES DU LEVANT (FRENCH) see PIE500
CORACON see DJS200
CORADON see WAK000
CORAETHAMIDE see DJS200
CORAETHAMIDUM see DJS200
CORALEPT see DJS200
CORAMINE see DJS200
CORAVITA see DJS200
CORAX see MDQ250, WAT200
CORAZONE see DJS200
CORDIAMID see DJS200
CORDIAMIN see DJS200
CORDIAMINE see DJS200
CORDITON see DJS200
CORDOVAL see GEW000
CORDULAN see CMD750
CORDYNIL see DJS200
COREDIOL see DJS200
COREINE see CCL250
CORESPIN see DJS200
CORETHAMIDE see DJS200
CORETONE see DJS200
CORFLEX 880 see ILR100
CORIANDER OIL see CNR735
CORICIDIN see ABG750
CORIFORTE see ABG750
CORISOL see VGP000
CORIZIUM see NGG500
CORLAN see HHR000
CORLUTIN see PMH500
CORLUTIN L.A. see HNT500
CORLUVITE see PMH500
CORMED see DJS200
CORMID see DJS200
CORMOTYL see DJS200
CORNMINT OIL, PARTIALLY DEMENTHOLIZED see
   MCB625
CORNOCENTIN see EDB500, LJL000
CORN OIL see CNS000
CORNOTONE see DJS200
CORNOX-M see CIR250
CORNOX RD see DGB000
CORNOX RK see DGB000
CORN SUGAR see GFG000
CORODANE see CDR750
CORODILAN see DHS200
CORODINOC see DUU600
CORONA COROZATE see BJK500
CORONALETTA see EOK000
COROSUL D AND S see SOD500
COROTHION see PAK000
COROTONIN see DJS200
COROTRAN see CJT750
COROVIT see DJS200
COROXON see CIK750
COROZATE see BJK500
CORPORIN see PMH500
CORPS PRALINE see MAO350
CORPUS LUTEUM HORMONE see PMH500
CORRIGEN see OHQ000
CORRONAROBETIN see TEH500
CORROSIVE MERCURY CHLORIDE see MCY475
CORROSIVE SUBLIMATE see MCY475

CORRY'S SLUG DEATH see TDW500
CORSONE see SOW000
CORTAN see PLZ000
CORTANCYL see PLZ000
Δ-CORTELAN see PLZ000
CORTHION see PAK000
CORTHIONE see PAK000
CORTIDELT see PLZ000
CORTILAN-NEU see CDR750
CORTINAZINE see ILD000
Δ¹-CORTISOL see PMA000
CORTISOL HEMISUCCINATE SODIUM SALT see HHR000
CORTISOL SODIUM HEMISUCCINATE see HHR000
CORTISOL SODIUM SUCCINATE see HHR000
CORTISOL-21-SODIUM SUCCINATE see HHR000
CORTISOL SUCCINATE, SODIUM SALT see HHR000
Δ-CORTISONE see PLZ000
Δ¹-CORTISONE see PLZ000
Δ-CORTONE see PLZ000
CORUNDUM see EAL100
CORUNDUM FUME see CNT250
CORVIC 55/9 see PKQ059
CORVIC 236581 see AAX175
CORVITAN see DJS200
CORVITIN see DBA800
CORVITOL see DJS200
CORVITONE see DJS200
CORYBAN-D see ABG750
CORYDININE see FOW000
CORYLON see HMB500
CORYLONE see HMB500
CORYNINE see YBJ000
CORYWAS see DJS200
COSAN see SOD500
COSDEN 550 see SMQ500
COSMETIC BLUE LAKE see FAE000
COSMETIC BRILLIANT PINK BLUISH D CONC see FAG070
COSMETIC CORAL RED KO BLUISH see CHP500
COSMETIC WHITE C47-5175 see TGG760
COSMETOL see CCP250
COSMOPEN see BDY669
COTEL see VSZ000
COTINAZIN see ILD000
COTINIZIN see ILD000
COTNION-ETHYL see EKN000
COTNION METHYL see ASH500
COTOFILM see HCL000
COTONE see PLZ000
COTORAN see DUK800
COTORAN MULTI 50WP see DUK800
CO-TRIMOXAZOLE see SNK000
COTTON DUST see CNT750
COTTONEX see DUK800
COTTONSEED OIL (unhydrogenated) see CNU000
COUMADIN see WAT200
COUMAFENE see WAT200
COUMAFURYL see ABF500
COUMAPHOS see CNU750
COUMAPHOS-O-ANALOG see CIK750
COUMAPHOS OXYGEN ANALOG (USDA) see CIK750
COUMARIN see CNV000
cis-o-COUMARINIC ACID LACTONE see CNV000
COUMARINIC ANHYDRIDE see CNV000
COURLOSE A 590 see SFO500
COVI-OX see VSZ450
COVIT see VSZ000
COXISTAT see NGE500
COYDEN see CMX850
COZYME see PAG200
CP see EAS500
4-CP see CJN000
CP 34 see TFM250
CP 4572 see CDO250

CP 6,343 see CFK000
CP 14,957 see OAN000
CP 15,336 see DBI200
CP 31393 see CHS500
CP 47114 see DSQ000
CP 49674 see DOP200
CP 50144 see CFX000
CP 53926 see DRR200
CP-16533-1 see IRV000
CP 10423-18 see TCW750
CP-15467-61 see LGZ000
CPA see EAS500, CJN000
CPB see CJR500
CP BASIC SULFATE see CNP250
CPBS see CJR500
CPCA see BIO750
CPCBS see CJT750
C.P. CHROME LIGHT 2010 see LCS000
C.P. CHROME ORANGE DARK 2030 see LCS000
C.P. CHROME ORANGE MEDIUM 2020 see LCS000
C.P. CHROME YELLOW LIGHT see LCR000
CPDC see PJD000
CPDD see PJD000
CPH see CDP250
CPIRON see FBJ100
CP 1044 J3 see BPP750
CPMC see CKF000
C.P. TITANIUM see TGF250
C.P. ZINC YELLOW X-883 see ZFJ100
C-QUENS see CBF250
CR see DDE200
CR 409 see BJE750
CR/662 see BLV000
CR 3029 see MAS500
CRAB-E-RAD see DXE600
CRAB'S EYES see AAD000
CRADEX see BBK500
CRAG 85W see DSB200
CRAG 974 see DSB200
CRAG FUNGICIDE 974 see DSB200
CRAG HERBICIDE see CNW000
CRAG HERBICIDE 1 see CNW000
CRAG NEMACIDE see DSB200
CRAG SESONE see CNW000
CRAG SEVIN see CBM750
CRALO-E-RAD see DXE600
CRATECIL see EOK000
CRAWHASPOL see TIO750
CREDO see SHF500
CREMOMETHAZINE see SNJ000
CREOSOTE see CMY825
CREOSOTE, from COAL TAR see CMY825
CREOSOTE OIL see CMY825
CREOSOTE P1 see CMY825
CREOSOTUM see CMY825
CRESIDINE see MGO750
m-CRESIDINE see MGO500
p-CRESIDINE see MGO750
CRESODIOL see GGS000
CRESOL see CNW500
2-CRESOL see CNX000
3-CRESOL see CNW750
4-CRESOL see CNX250
m-CRESOL see CNW750
o-CRESOL see CNX000
p-CRESOL see CNX250
p-CRESOL ACETATE see MNR250
o-CRESOL GLYCERYL ETHER see GGS000
CRESOLI (ITALIAN) see CNW500
p-CRESOL METHYL ETHER see MGP000
CRESORCINOL DIISOCYANATE see TGM750
CRESOSSIDIOLO see GGS000
CRESOSSIPROPANDIOLO see GGS000
CRESOTINE BLUE 2B see CMO000

CRESOTINE BLUE 3B see CMO250
CRESOTOL see DUU600
CRESOXYDIOL see GGS000
CRESOXYPROPANEDIOL see GGS000
CRESTANIL see MQU750
CRESTOXO see CDV100
p-CRESYL ACETATE (FCC) see MNR250
m-CRESYL ESTER of N-METHYLCARBAMIC ACID see MIB750
o-CRESYL-α-GLYCERYL ETHER see GGS000
CRESYLIC ACID see CNW500
m-CRESYLIC ACID see CNW750
o-CRESYLIC ACID see CNX000
p-CRESYLIC ACID see CNX250
CRESYLIC CREOSOTE see CMY825
p-CRESYL ISOBUTYRATE see THA250
CRESYLITE see TML500
m-CRESYL METHYLCARBAMATE see MIB750
p-CRESYL METHYL ETHER see MGP000
CRESYL PHOSPHATE see TNP500
o-CRESYL PHOSPHATE see TMO600
CRILL 3 see SKV150
CRILL 10 see PKL100
CRILL K 3 see SKV150
CRILLON L.D.E. see BKE500
CRIMIDIN (GERMAN) see CCP500
CRIMIDINA (ITALIAN) see CCP500
CRIMIDINE see CCP500
CRIMSON ANTIMONY see AQL500
CRIMSON EMBL see HJF500
CRINOTHENE see PKB500
CRINOVARYL see EDV000
CRINURYL see DFP600
CRISALBINE see GJG000
CRISALIN see DUV600
CRISAPON see DGI400
CRISATRINA see ARQ725
CRISAZINE see ARQ725
CRISEOCICLINE see TBX000
CRISODIN see MRH209
CRISODRIN see MRH209
CRISPATINE see FOT000
CRISQUAT see PAJ000
CRISTALLOSE see SJN700
CRISTALLOVAR see EDV000
CRISTAPURAT see DKL800
CRISTERONE T see TBF500
CRISTOBALITE see SCI500, SCJ000
CRISTOXO 90 see CDV100
CRISULFAN see EAQ750
CRISURON see DXQ500
CRITTOX see EIR000
CROCIDOLITE (DOT) see ARM275
CROCIDOLITE ASBESTOS see ARM275
CROCOITE see LCR000
CROCUS MARTIS ADSTRINGENS see IHD000
CRODACID see MSA250
CRODACOL-CAS see HCP000
CRODACOL-S see OAX000
CROLEAN see ADR000
CROMILE, CLORURO di (ITALIAN) see CML125
CROMO, OSSICLORURO di (ITALIAN) see CML125
CRONETAL see DXH250
CRONIL see EHP000
CROP RIDER see DAA800
CROTALINE see MRH000
CROTILIN see DAA800
CROTONAL see COB260
CROTONALDEHYDE see COB250, COB260
(E)-CROTONALDEHYDE see COB260
CROTONATE de 2,4-DINITRO 6-(1-METHYL-HEPTYL)-PHENYLE (FRENCH) see AQT500
CROTONATE d'ETHYLE (FRENCH) see COB750
CROTONIC ACID see COB500

α-CROTONIC ACID see COB500
CROTONIC ACID, solid see COB500
(E)-CROTONIC ACID, ETHYL ESTER see COB750
α-CROTONIC ACID ETHYL ESTER see COB750
CROTONIC ALDEHYDE see COB250, COB260
CROTONYL ALCOHOL see BOY000
CROTONYLENE see COC500
CROTURAL see EHP000
CROTYL ALCOHOL see BOY000
CROTYLIDENE ACETIC ACID see SKU000
CROVARIL see HNI500
CROYSULFONE see SOA500
CRTRON see VSZ100
CRUDE ARSENIC see ARI750
CRUDE COAL TAR see CMY800
CRUDE ERGOT see EDB500
CRUDE OIL see PCR250
CRUDE SHALE OILS see COD750
CRUFOMATE see COD850
CRUFOMATE A see COD850
CRUFORMATE see COD850
CRYOFLUORAN see FOO509
CRYOFLUORANE see FOO509
CRYPTOGIL OL see PAX250
CRYPTOHALITE see COE000
CRYSTAL CHROME ALUM see PLB500
CRYSTALETS see VSK900
CRYSTALLINA see VSZ100
CRYSTALLINE DIGITALIN see DKL800
CRYSTALLIZED VERDIGRIS see CNI250
CRYSTALLOSE see SJN700
CRYSTAL O see CCP250
CRYSTAL ORANGE 2G see HGC000
CRYSTAL PROPANIL-4 see DGI000
CRYSTALS of VENUS see CNI250
CRYSTAMET see SJU000
CRYSTAMIN see VSZ000
CRYSTAPEN see BFD250
CRYSTAR see ADA725
CRYSTEX see SOD500
CRYSTHION 2L see ASH500
CRYSTHYON see ASH500
CRYSTODIGIN see DKL800
CRYSTOGEN see EDV000
CRYSTOIDS see HFV500
CRYSTOL CARBONATE see SFO000
CRYSTOSOL see MQV750
CRYSTWEL see VSZ000
CS see CEQ600
CS-847 see CEW500
50-CS-46 see EME050
CSAC see EES400
CSC see SEC000
CSF-GIFTWEIZEN see TEM000
C-Sn-9 see BLL750
CT 4436 see MFO250
CTA see CLO750
CTAB see HCQ500
CTC see CMA750
CTFE see CLQ750
CTX see EAS500
CUBE see RNZ000
CUBE EXTRACT see RNZ000
CUBE-PULVER see RNZ000
CUBE ROOT see RNZ000
CUBES see DJO000
CUBIC NITER see SIO900
CUBOR see RNZ000
CUCUMBER ALDEHYDE see NMV760
CULLEN EARTH see MAT500
CUM see COE750
CUMA see BJZ000
CUMAFOS (DUTCH) see CNU750
CUMAFURYL (GERMAN) see ABF500

CUMALDEHYDE see COE500
CUMAN see BJK500
CUMAN L see BJK500
CUMEEN (DUTCH) see COE750
CUMEENHYDROPEROXYDE (DUTCH) see IOB000
CUMENE see COE750
psi-CUMENE see TLL750
CUMENE ALDEHYDE see COF000
CUMENE HYDROPEROXIDE (DOT) see IOB000
CUMENE HYDROPEROXIDE, TECHNICALLY PURE (DOT)
  see IOB000
CUMENE PEROXIDE see DGR600
p-CUMENOL see IQZ000
CUMENT HYDROPEROXIDE see IOB000
CUMENYL HYDROPEROXIDE see IOB000
p-CUMIC ALDEHYDE see COE500
CUMID see BJZ000
o-CUMIDINE see INX000
psi-CUMIDINE see TLG250
psi-CUMIDINE HYDROCHLORIDE see TLG750
CUMINALDEHYDE see COE500
CUMINIC ALDEHYDE (FCC) see COE500
CUMIN OIL see COF325
CUMINYL ALDEHYDE see COE500
CUMMIN see COF325
CUMOLHYDROPEROXID (GERMAN) see IOB000
CUMYL HYDROPEROXIDE see IOB000
α-CUMYL HYDROPEROXIDE see IOB000
CUMYL HYDROPEROXIDE, TECHNICAL PURE (DOT) see
  IOB000
CUMYL PEROXIDE see DGR600
CUNILATE see BLC250
CUNILATE 2472 see BLC250
CUPFERRON see ANO500
CUPRAL see SGJ000
CUPRENIL see MCR750
CUPRIC ACETATE see CNI250
CUPRIC ACETOARSENITE see COF500
CUPRIC ARSENITE see CNN500
CUPRIC CHLORIDE see CNK500
CUPRIC CYANIDE (DOT) see CNL250
CUPRIC DIACETATE see CNI250
CUPRIC DINITRATE see CNM750
CUPRIC GREEN see CNN500
CUPRIC-8-HYDROXYQUINOLATE see BLC250
CUPRICIN see CNL000
CUPRIC NITRATE (DOT) see CNM750
CUPRIC-8-QUINOLINOLATE see BLC250
CUPRIC SULFATE see CNP250
CUPRIETHYLENE DIAMINE see DBU800
CUPRIETHYLENEDIAMINE, solution (DOT) see DBU800
CUPRIMINE see MCR750
CUPRINOL see NAS000
CUPROUS ARSENATE, BASIC see CNI900
CUPROUS CYANIDE see CNL000
CURACIT see BJI000
CURACRON see BNA750
CURALIN M see MJM200
CURARIL see GGS000
CURARIN-HAF see TOA000
CURARYTHAN see GGS000
CURATERR see CBS275
CURCUMA OIL see COG000
CURCUMIN see COG000
CURCUMINE see COG000
CURENE 442 see MJM200
CURETARD A see DWI000
CUREX FLEA DUSTER see RNZ000
CURITAN see DXX400
CURITHANE see MJQ000
CURITHANE 103 see MGA500
CURITHANE C126 see DEQ600
CURON FAST YELLOW 5G see FAG140
CUTICURA ACNE CREAM see BDS000

CUTTING OILS see COH000
CVP see CDS750
C-WEISS 7 (GERMAN) see TGG760
CX-59 see HKE000
CY see EAS500
CY-39 see PHU500
CYAANWATERSTOF (DUTCH) see HHS000
CYACETACID see COH250
CYACETACIDE see COH250
CYACETAZID see COH250
CYACETAZIDE see COH250
CYALANE see DXN600
CYAMOPSIS GUM see GLU000
CYANACETATE ETHYLE (GERMAN) see EHP500
CYANACETHYDRAZIDE see COH250
CYANACETIC ACID HYDRAZIDE see COH250, COH250
CYANACETOHYDRAZIDE see COH250
CYANACETYLHYDRAZIDE see COH250
CYANAMIDE see CAQ250, COH500
CYANAMIDE CALCIQUE (FRENCH) see CAQ250
CYANAMIDE, CALCIUM SALT (1:1) see CAQ250
CYANAMID GRANULAR see CAQ250
CYANAMID SPECIAL GRADE see CAQ250
CYANASET see MJM200
CYANAZIDE see COH250
CYANAZINE see BLW750
CYANHYDRINE d'ACETONE (FRENCH) see MLC750
CYANIC ACID, POTASSIUM SALT see PLC250
CYANIDE see COI500
CYANIDE ANION see COI500
CYANIDELONON 1522 see QCA000
CYANIDE of POTASSIUM see PLC500
CYANIDE of SODIUM see SGA500
CYANINE GREEN G BASE see BLK000
CYANIZIDE see COH250
CYANOACETHYDRAZIDE see COH250
CYANOACETIC ACID ETHYL ESTER see EHP500
CYANOACETIC ACID HYDRAZIDE see COH250
CYANOACETIC ESTER see EHP500
CYANOACETOHYDRAZIDE see COH250
α-CYANOACETOHYDRAZIDE see COH250
CYANOACETONITRILE see MAO250
CYANOACETYLHYDRAZIDE see COH250
2-CYANOACRYLATE ACID METHYL ESTER see MIQ075
α-CYANOACRYLATE ACID METHYL ESTER see MIQ075
CYANOAMINE see COH500
CYANO-B12 see VSZ000
CYANOBENZENE see BCQ250
4-CYANOBENZONITRILE see BBP250
CYANOBRIK see SGA500
CYANOBROMIDE see COO500
CYANOCOBALAMIN see VSZ000
2-CYANO-N-(2-CYANOETHYL)ETHANAMINE see BIQ500
4-CYANO-2,6-DIIODOPHENOL see HKB500
4-CYANO-2,6-DIJODPHENOL (GERMAN) see HKB500
4-CYANO-2,6-DIJODPHENOL CAPRYSAEUREESTER (GER-
  MAN) see DNG200
α-CYANODIPHENYLMETHANE see DVX200
CYANOETHANE see PMV750
2-CYANOETHANOL see HGP000
CYANOETHYDRAZIDE see COH250
2-CYANOETHYL ALCOHOL see HGP000
β-CYANOETHYLAMINE see AMB500
CYANOETHYLENE see ADX500
2-CYANOETHYL-2'-FLUOROETHYLETHER see CON500
2-CYANO-2'-FLUORODIETHYL ETHER see CON500
CYANOGAS see CAQ500
CYANOGEN see COO000
CYANOGENAMIDE see COH500
CYANOGEN BROMIDE see COO500
CYANOGEN CHLORIDE see COO750
CYANOGEN CHLORIDE, inhibited (DOT) see COO750
CYANOGEN CHLORIDE, containing less than 0.9% water
  (DOT) see COO750

CYANOGENE (FRENCH) see COO000
CYANOGEN GAS (DOT) see COO000
CYANOGEN MONOBROMIDE see COO500
CYANOGEN NITRIDE see COH500
CYANOGRAN see SGA500
CYANOGUANIDINE METHYLMERCURY DERIV. see
MLF250
2-CYANO-10-(3-(4-HYDROXYPIPERIDINO)PROPYL)
PHENOTHIAZINE see PIW000
2-CYANO-10-(3-(4-HYDROXY-1-PIPERIDYL)PROPYL)PHE-
NOTHIAZINE see PIW000
CYANO-3 ((HYDROXY-4 PIPERIDYL-1)-3 PROPYL)-10-PHE-
NOTHIAZINE (FRENCH) see PIW000
α-CYANOISOPROPYLAMIDE OF THE O,O-DIETHYLTHIO-
PHOSPHORYL ACETIC ACID see PHK250
CYANOLYT see MIQ075
CYANOMETHANE see ABE500
CYANOMETHANOL see HIM500
(CYANOMETHYL)BENZENE see PEA750
N-(CYANOMETHYL)DIMETHYLAMINE see DOS200
S-N-(1-CYANO-1-METHYLETHYL)CARBAMOYLMETHYL
DIETHYL PHOSPHOROTHIOLATE see PHK250
S-(((1-CYANO-1-METHYL-ETHYL)CARBAMOYL)METHYL)
O,O-DIETHYL PHOSPHOROTHIOATE see PHK250
3-(CYANOMETHYL)INDOLE see ICW000
CYANO(METHYLMERCURI)GUANIDINE see MLF250
α-CYANO-3-PHENOXYBENZYL-2-(4-CHLOROPHENYL)ISO-
VALERATE PYDRIN see FAR100
α-CYANO-3-PHENOXYBENZYL-2-(4-CHLOROPHENYL)-3-
METHYLBUTYRATE see FAR100
CYANO(3-PHENOXYPHENYL)METHYL 4-CHLORO-α-(1-
METHYLETHYL)BENZENEACETATE see FAR100
CYANOPHENYLMETHYL-β-d-GLUCOPYRANOSIDURONIC
ACID see LAS000
CYANOPHOS see DGP900
1-CYANOPROPANE see BSX250
2-CYANOPROPENE-1 see MGA750
1-CYANO-2-PROPEN-1-OL see HJQ000
α-CYANOSTILBENE see DVX600
α-CYANOTOLUENE see PEA750
φ-CYANOTOLUENE see PEA750
α-CYANOVINYL ACETATE see ABL500
CYANTIN see NGE000
CYANURAMIDE see MCB000
CYANURE (FRENCH) see COI500
CYANURE d'ARGENT (FRENCH) see SDP000
CYANURE de CALCIUM (FRENCH) see CAQ500
CYANURE de CUIVRE (FRENCH) see CNL250
CYANURE de MERCURE (FRENCH) see MDA250
CYANURE de METHYL (FRENCH) see ABE500
CYANURE de PLOMB (FRENCH) see LCU000
CYANURE de POTASSIUM (FRENCH) see PLC500
CYANURE de SODIUM (FRENCH) see SGA500
CYANURE de VINYLE (FRENCH) see ADX500
CYANURE de ZINC (FRENCH) see ZGA000
CYANUROTRIAMIDE see MCB000
CYANUROTRIAMINE see MCB000
CYANWASSERSTOFF (GERMAN) see HHS000
CYAZID see COH250
CYAZIDE see COH250
CYAZIN see ARQ725
CYBIS see EID000
CYCASIN see COU000
CYCAS REVOLUTA GLUCOSIDE see COU000
CYCLADIENE see DAL600
CYCLAL CETYL ALCOHOL see HCP000
CYCLALIA see CMS850
CYCLAMAL see COU500
CYCLAMATE see CPQ625, SGC000
CYCLAMATE CALCIUM see CAR000
CYCLAMATE, CALCIUM SALT see CAR000

CYCLAMATE SODIUM see SGC000
CYCLAMEN ALDEHYDE see COU500
CYCLAMIC ACID see CPQ625
CYCLAMIC ACID SODIUM SALT see SGC000
CYCLAN see CAR000
CYCLIC ETHYLENE (DIETHOXYPHOSPHINOTHIOYL)
DITHIOIMIDOCARBONATE see DXN600
CYCLIC ETHYLENE(DIETHOXYPHOSPHINOTHIOYL)
DITHIOIMIDOCARBONATE see PGW750
CYCLIC ETHYLENE P,P-DIETHYL PHOSPHONODITHIOIMI-
DOCARBONATE see PGW750
CYCLIC ETHYLENE ESTER of (DIETHOXYPHOSPHINO-
THIOYL)DITHIOIMIDOCARBONIC ACID see DXN600
CYCLIC (HYDROXYMETHYL)ETHYLENE ACETAL ACE-
TONE see DVR600
CYCLIC PROPYLENE (DIETHOXYPHOSPHINYL)
DITHIOIMIDOCARBONATE see DHH400
CYCLIC N',O-PROPYLENE ESTER of N,N-BIS(2-CHLORO-
ETHYL)PHOSPHORODIAMIDIC ACID MONOHYDRATE
see CQC675
CYCLOBARBITAL see TDA500
CYCLOBARBITOL see TDA500
CYCLOBARBITONE see TDA500
CYCLOBUTANE see COW000
CYCLOBUTENE see COW250
CYCLOBUTYLENE see COW250
CYCLOCHEM GMS see OAV000
α-CYCLOCITRYLIDENEACETONE see IFW000
β-CYCLOCITRYLIDENEACETONE see IFX000
CYCLODAN see EAQ750
CYCLODOL see BBV000
CYCLODORM see TDA500
CYCLOGEST see PMH500
CYCLOHEPTANE see COX500
1,3,5-CYCLOHEPTATRIENE see COY000
CYCLOHEPTATRIENE (DOT) see COY000
CYCLOHEXAAN (DUTCH) see CPB000
CYCLOHEXADEINEDIONE see QQS200
1,4-CYCLOHEXADIENEDIONE see QQS200
2,5-CYCLOHEXADIENE-1,4-DIONE see QQS200
3,5-CYCLOHEXADIENE-1,2-DIONE see BDC250
2,5-CYCLOHEXADIENE-1,4-DIONE DIOXIME see
DVR200
1,4-CYCLOHEXADIENE DIOXIDE see QQS200
CYCLOHEXAMETHYLENIMINE see HDG000
CYCLOHEXAN (GERMAN) see CPB000
CYCLOHEXANAMINE see CPF500
CYCLOHEXANE see CPB000
CYCLOHEXANEACETIC ACID, 1-(HYDROXYMETHYL)-,
MONOSODIUM SALT (9CI) see SHL250
CYCLOHEXANECARBOXYLIC ACID, LEAD SALT see
NAS500
CYCLOHEXANE OXIDE see CPD000
CYCLOHEXANESULFAMIC ACID, CALCIUM SALT see
CAR000
CYCLOHEXANESULFAMIC ACID, MONOSODIUM SALT
see SGC000
CYCLOHEXANESULPHAMIC ACID see CPQ625
CYCLOHEXANESULPHAMIC ACID, MONOSODIUM SALT
see SGC000
CYCLOHEXANETHIOL see CPB625
CYCLOHEXANOL see CPB750
CYCLOHEXANOL ACETATE see CPF000
CYCLOHEXANOLAZETAT (GERMAN) see CPF000
CYCLOHEXANON (DUTCH) see CPC000
CYCLOHEXANONE see CPC000
CYCLOHEXANONE-Δ see CPC250
CYCLOHEXANONE ISO-OXIME see CBF700
CYCLOHEXANONE PEROXIDE and BIS(1-HYDROXYCY-
CLOHEXYL)PEROXIDE MIXTURE see CPC500
CYCLOHEXANYL ACETATE see CPF000
CYCLOHEXATRIENE see BBL250
CYCLOHEXENE see CPC579

CYCLOHEXENECARBOXALDEHYDE see TCJ100
3-CYCLOHEXENE-1-CARBOXALDEHYDE see FNK025
1-CYCLOHEXENE-1-CARBOXYLIC ACID, 3,4,5 see
BML000
CYCLOHEXENE EPOXIDE see CPD000
CYCLOHEXENE OXIDE see CPD000
CYCLOHEXENE-1-OXIDE see CPD000
1,2-CYCLOHEXENE OXIDE see CPD000
CYCLOHEXENONE see CPD250
2-CYCLOHEXEN-1-ONE see CPD250
5-(1-CYCLOHEXEN-1-YL)-1,5-DIMETHYLBARBITURIC
ACID see ERD500
5-(1-CYCLOHEXEN-1-YL)-1,5-DIMETHYL-2,4,6(1H,3H,5H)-
PYRIMIDINETRIONE see ERD500
CYCLOHEXENYL-ETHYL BARBITURIC ACID see TDA500
5-(1-CYCLOHEXENYL)-5-ETHYLBARBITURIC ACID see
TDA500
5-(1-CYCLOHEXEN-1-YL)-5-ETHYLBARBITURIC ACID see
TDA500
CYCLOHEXENYLETHYLENE see CPD750
5-(1-CYCLOHEXEN-1-YL)-5-ETHYL-2,4,6(1H,3H,5H)-PYRI-
MIDINETRIONE see TDA500
5-(1-CYCLOHEXENYL-1)-1-METHYL-5-METHYLBARBI-
TURIC ACID see ERD500
5-(Δ-1,2-CYCLOHEXENYL)-5-METHYL-N-METHYL-BARBI-
TURSAEURE (GERMAN) see ERD500
CYCLOHEXENYL TRICHLOROSILANE see CPE500
CYCLOHEXYL ACETATE see CPF000
CYCLOHEXYL ALCOHOL see CPB750
CYCLOHEXYLAMIDOSULPHURIC ACID see CPQ625
CYCLOHEXYLAMINE see CPF500
CYCLOHEXYLAMINESULPHONIC ACID see CPQ625
N-CYCLOHEXYLCYCLOHEXANAMINE see DGT600
CYCLOHEXYLDIMETHYLAMINE see DRF709
N-CYCLOHEXYLDIMETHYLAMINE see DRF709
3-CYCLOHEXYL-6-(DIMETHYLAMINO)-1-METHYL-s-
TRIAZINE-2,4(1H,3H)-DIONE see HFA300
3-CYCLOHEXYL-6-(DIMETHYLAMINO)-1-METHYL-1,3,5-
TRIAZINE-2,4(1H,3H)-DIONE see HFA300
2-CYCLOHEXYL-4,6-DINITROFENOL (DUTCH) see
CPK500
2-CYCLOHEXYL-4,6-DINITROPHENOL see CPK500
6-CYCLOHEXYL-2,4-DINITROPHENOL see CPK500
trans-N,N'-(1,4-CYCLOHEXYLENEDIMETHYLENE)BIS(2-
CHLOROBENZYLAMINE) DIHYDROCHLORIDE see
BHN000
CYCLOHEXYLENE OXIDE see CPD000
CYCLOHEXYL ISOCYANATE see CPN500
CYCLOHEXYLMETHANE see MIQ740
α-CYCLOHEXYL-α-PHENYL-1-PIPERIDINEPROPANOL HY-
DROCHLORIDE see BBV000
CYCLOHEXYLSULFAMIC ACID (9CI) see CPQ625
CYCLOHEXYL SULPHAMATE SODIUM see SGC000
CYCLOHEXYLSULPHAMIC ACID see CPQ625
N-CYCLOHEXYLSULPHAMIC ACID see CPQ625
CYCLOHEXYLSULPHAMIC ACID, CALCIUM SALT see
CAR000
CYCLOHEXYLTRICHLOROSILANE see CPR250
CYCLOMYCIN see TBX000
CYCLON see HHS000
CYCLONAL see ERD500
CYCLONE B see HHS000
CYCLONITE see CPR800
CYCLOOCTAFLUOROBUTANE see CPS000
1,3,5,7-CYCLOOCTATETRAENE see CPS500
CYCLOPAN see ERD500
CYCLOPAR see TBX250
CYCLOPENTADIENE see CPU500
1,3-CYCLOPENTADIENE see CPU500
1,3-CYCLOPENTADIENE, DIMER see DGW000
pi-CYCLOPENTADIENYL COMPOUND with NICKEL see
NDA500
CYCLOPENTADIENYLMANGANESE TRICARBONYL see
CPV000

α,β-CYCLOPENTAMETHYLENETETRAZOLE see PBI500
CYCLOPENTANE see CPV750
CYCLOPENTANONE see CPW500
CYCLOPENTENE see CPX750
2-CYCLOPENTENYL-4-HYDROXY-3-METHYL-2-CYCLO-
PENTEN-1-ONE CHRYSANTHEMATE see POO000
3-(2-CYCLOPENTEN-1-YL)-2-METHYL-4-OXO-2-CYCLO-
PENTEN-1-YL CHRYSANTHEMUMATE see POO000
3-(2-CYCLOPENTENYL)-2-METHYL-4-OXO-2-CYCLOPEN-
TENYL CHRYSANTHEMUMMONOCARBOXYLATE see
POO000
CYCLOPENTENYLRETHONYL CHRYSANTHEMATE see
POO000
CYCLOPENTIMINE see PIL500
α-CYCLOPENTYL-2-THIOPHENEGLYCOLATE DIETHYL(2-
HYDROXYETHYL)METHYLAMMONIUM BROMIDE see
PBS000
CYCLOPHOSPHAMIDE see EAS500
CYCLOPHOSPHAMIDE HYDRATE see CQC675
CYCLOPHOSPHAMIDE and MNU (1:2) see CQC750
CYCLOPHOSPHAMIDE-N-MONOCHLOROETHYL derivative
see TNT500
CYCLOPHOSPHAMIDE MONOHYDRATE see CQC675
CYCLOPHOSPHAMIDUM see EAS500, CQC675
CYCLOPHOSPHAN see EAS500, CQC675
CYCLOPHOSPHANE see CQC675
CYCLOPHOSPHANUM see CQC675
CYCLOPHOSPHORAMIDE see EAS500
CYCLOPROPANAMINE, 2-PHENYL-, trans-(+-)-, SULFATE
(2:1) see PET500
CYCLOPROPANE see CQD750
CYCLOPROPANE, liquefied (DOT) see CQD750
CYCLORYL 21 see SIB600
CYCLOSAN see MCW000
CYCLOSIA see CMS850
CYCLOSTIN see EAS500
CYCLOTEN see HMB500
CYCLOTETRAMETHYLENE OXIDE see TCR750
CYCLOTETRAMETHYLENE TETRANITRAMINE see
CQH250
CYCLOTETRAMETHYLENE TETRANITRAMINE, dry (DOT)
see CQH250
CYCLOTON V see HCQ500
CYCLOTRIMETHYLENENITRAMINE see CPR800
CYCLOTRIMETHYLENETRINITRAMINE see CPR800
CYCLOTRIMETHYLENETRINITRAMINE, containing at least
10%-25% water (DOT) see CPR800
CYCLOTRIMETHYLENETRINITRAMINE, desensitized (DOT)
see CPR800
CYCOLAMIN see VSZ000
CYCTEINAMINE see AJT250
CYFEN see DSQ000
CYFLEE see FAB600
CYFOS see IMH000
CYGON see DSP400
CYGON INSECTICIDE see DSP400
CYHEXATIN see CQH650
CYJANOWODOR (POLISH) see HHS000
CYKAZINE see COU000
CYKLOHEKSAN (POLISH) see CPB000
CYKLOHEKSANOL (POLISH) see CPB750
CYKLOHEKSANON (POLISH) see CPC000
CYKLOHEKSEN (POLISH) see CPC579
CYKLOHEXANTHIOL see CPB625
CYKLOHEXYLMERKATPAN (CZECH) see CPB625
CYKOBEMINET see VSZ000
CY-L 500 see CAQ250
CYLAN see CAR000, DXN600, PGW75
CYLPHENICOL see CDP250
CYMAG see SGA500
CYMATE see BJK500
CYMBI see AIV500, AOD125
CYMEL see MCB000
CYMENE see CQI000

p-CYMENE see CQI000
2-p-CYMENOL see CCM000
3-p-CYMENOL see TFX810
p-CYMEN-3-OL see TFX810
CYMETHION see MDT750
CYMETOX see MIW250
CYMOL see CQI000
CYMONIC ACID see FIC000
CYNARON see MDT740
CYNEM see EPC500
CYNKOTOX see EIR000
CYNKU TLENEK (POLISH) see ZKA000
CYNOGAN see BMM650
CYOLAN see DXN600
CYOLANE see PGW750
CYOLANE INSECTICIDE see DXN600, PGW750
CYPENTIL see PIL500
CYPONA see DGP900
CYPREX see DXX400
CYPREX 65W see DXX400
CYPROMID see CQJ250
CYPRON see MQU750
CYRAL see DBB200
CYREDIN see VSZ000
CYREN see DKA600
CYREN B see DKB000
CYRSTHION see EKN000
CYSTAMIN see HEI500
CYSTAMINE see MCN500
CYSTAMINE "MCCLUNG" see PDC250
CYSTEAMIDE see AJT250
CYSTEAMINE see AJT250
CYSTEIN see CQK000
CYSTEINAMINE DISULFIDE see MCN500
CYSTEINE see CQK000
l-CYSTEINE see CQK000
l-(+)-CYSTEINE see CQK000
CYSTEINE CHLORHYDRATE see CQK250
CYSTEINE DISULFIDE see CQK325
CYSTEINE HYDROCHLORIDE see CQK250
l-CYSTEINE HYDROCHLORIDE see CQK250
l-CYSTEINE MONOHYDROCHLORIDE (FCC) see CQK250
l-CYSTEIN HYDROCHLORIDE see CQK250
CYSTIN see CQK325
CYSTINAMIN (GERMAN) see MCN500
(−)-CYSTINE see CQK325
l-CYSTINE see CQK325
CYSTINE ACID see CQK325
CYSTINEAMINE see MCN500
CYSTOGEN see HEI500
CYSTOIDS ANTHELMINTIC see HFV500
CYSTOPYRIN see PDC250
CYSTURAL see PDC250
CYTACON see VSZ000
CYTAMEN see VSZ000
CYTEL see DSQ000
CYTEN see DSQ000
CYTHIOATE see CQL250
CYTHION see MAK700
CYTOBION see VSZ000
CYTOCHALASIN D see ZUS000
CYTOCHALASIN E see CQM250
CYTOPHOSPHAN see EAS500, CQC675
CYTOXAL ALCOHOL see CQN000
CYTOXAN see EAS500, CQC675
CYTOXYL ALCOHOL CYCLOHEXYLAMMONIUM SALT
    see CQN000
CYTROL see AMY050
CYTROLANE see DHH400
CYURAM DS see TFS350
CZTEROCHLOREK WEGLA (POLISH) see CBY000
2,3,7,8-CZTEROCHLORODWUBENZO-p-DWUOKSYNY
    (POLISH) see TAI000

1,1,2,2-CZTEROCHLOROETAN (POLISH) see TBQ100
CZTEROCHLOROETYLEN (POLISH) see PCF275

D₂ see DBB800
2,4-D see DAA800
D 50 see AAX250, DAA800
D-50 see POL500
D-365 see IRV000
D 735 see CCC500
D 854 see CJT750
D 860 see BSQ000
D 1221 see CBS275
D-1410 see DSP600
D 1593 see CIP500
DA see CGN000, DNA200
2,4-DAA see DBO000
DAAB see DWO800
DAAE see DCN800
2,4-DAA SULFATE see DBO400
DAB see DOT300
DABI see DOT600
DAB-N-OXIDE see DTK600
DABYLEN see BAU750, BBV500
DAC 2797 see TBQ750
DACAMINE see DAA800, TAA100
DACARBAZINE see DAB600
2,4-D ACID see DAA800
DACONATE 6 see MRL750
DACONIL see TBQ750
DACONIL 2787 FLOWABLE FUNGICIDE see TBQ750
DACORTIN see PLZ000
DACOSOIL see TBQ750
DACOVIN see PKQ059
DACTIN see DFE200
DACTINOL see RNZ000
DAD see DCI600
DADEX see BBK500
DADOX d-CITRAMINE see BBK500
DADPE see OPM000
DADPS see SOA500
DAEP see DOP200
DAF 68 see DVL700
DAFEN see LJR000
DAG see DCI600
DAGADIP see TNP250
DAGUTAN see SJN700
DAI CARI XBN see BQK250
DAICEL 1150 see SFO500
DAIFLON see CLQ750
DAIFLON S 3 see FOO000
DAILON see DXQ500
DAINICHI CHROME ORANGE R see LCS000
DAINICHI CHROME YELLOW G see LCR000
DAINICHI FAST SCARLET G BASE see NMP500
DAINICHI LAKE RED C see CHP500
DAISEN see EIR000
DAITO ORANGE BASE R see NEN500
DAITO RED BASE TR see CLK220
DAITO RED SALT TR see CLK235
DAITO SCARLET BASE G see NMP500
DAKINS SOLUTION see SHU500
DAKTIN see DFE200
DALAPON see DGI600
DALAPON (USDA) see DGI400
DALAPON 85 see DGI400
DALAPON SODIUM see DGI600
DALAPON SODIUM SALT see DGI600
DAL-E-RAD see MRL750
DAL-E-RAD 100 see DXE600
DALF see MNH000
DALMADORM see DAB800
DALMADORM HYDROCHLORIDE see DAB800
DALMANE see DAB800
DALMATE see DAB800

DALMATIAN SAGE OIL see SAE500
DALMATION INSECT FLOWERS see POO250
DALTOGEN see TKP500
DAM-57 see LJH000
DAMILAN see EAH500
DAMILEN HYDROCHLORIDE see EAI000
DAMINOZIDE (USDA) see DQD400
2,4-D AMMONIUM SALT see DAB020
DAMORAL see EOK000
DAMS see BBK500
DAN see DSU600
DANA see NJW500
DANAMID see PJY500
DANAMINE see DJS200
DANANTIZOL see MCO500
DANEX see TIQ250
DANFIRM see AAX250
DANIFOS see DIX800
DANIZOL see MMN250
DANSYL see DPN200
DANSYL CHLORIDE see DPN200
DANTAFUR see NGE000
DANTEN see DKQ000, DNU000
DANTHION see PAK000
DANTINAL see DKQ000
DANTOIN see DFE200, DNU000
DANTOINAL KLINOS see DKQ000
DANTOINE see DKQ000
DANUVIL 70 see PKQ059
DAP see DOT000
DAPA see DOU600
DAPAZ see MQU750
DAPHENE see DSP400
DAPLEN AD see PMP500
DAPON 35 see DBL200
DAPON R see DBL200
DAPRISAL see ABG750
DAPSONE see SOA500
DARACLOR see TGD000
DARAL see VSZ100
DARAMIN see CAM750
DARAPRAM see TGD000
DARAPRIM see TGD000
DARAPRIME see TGD000
DARATAK see AAX250
DAR-CHEM 14 see SLK000
DARENTHIN see BMV750
DARID QH see SEH000
DARILOID QH see SEH000
DAROLON see ACE000
DAROPERVAMIN see DBA800
DAROTOL see CKN000
DARVIC 110 see PKQ059
DARVIS CLEAR 025 see PKQ059
DARVON see DAB879
DARVON COMPOUND see ABG750
DARVON HYDROCHLORIDE see PNA500
DAS see AOP250, BBK500, DOU600
DASANIT see FAQ800
DASERD see GGS000
DASEROL see GGS000
DASIKON see ABG750
DASKIL see NCQ900
DATC see DBI200
DATRIL see HIM000
DATURA STRAMONIUM see SLV500
DATURINE see HOU000
DAUNAMYCIN see DAC000
DAUNOMYCIN see DAC000
DAUNORUBICIN see DAC000
DAUNORUBICINE see DAC000
DAVISON SG-67 see SCH000
DAVITAMON D see VSZ100
DAVITAMON PP see NCQ900

DAVITIN see VSZ100
DAWE'S DESTROL see DKA600
DAWSON 100 see MHR200
DAYFEN see LJR000
DAZOMET see DSB200
DAZZEL see DCM750
DBA see DCT400, DQJ200
DB(a,c)A see BDH750
DB(a,h)A see DCT400
1,2,5,6-DBA see DCT400
DB(a,h)AC see DCS400
DB(a,j)AC see DCS600
7H-DB(c,g)C see DCY000
DBCP see DDL800
DBD see ASH500, DDJ000
DBE see EIY500
DBH see BBP750, BBQ500
DBM see DDP600, DED600
DBMP see BFW750
DBN see BRY500
DBNA see BRY500
DBNPA see DDM000
DBOT see DEF400
DBP see DEH200
DB(a,e)P see NAT500
DB(a,i)P see BCQ500
DB(a,l)P see DCY400
DBPC (technical grade) see BFW750
D.B.T.C. see DDY200
DBTL see DDV600
2,4-D BUTYL ESTER see BQZ000
DCA see DEL000, DFE200
DCA 70 see AAX250
DCB see DEP600, DEQ600, DEV000
1,4-DCB see DEV000
DCBA see BIA750
D&C BLUE No. 4 see FAE000, FMU059
D&C BLUE NUMBER 1 see BJI250
DCBN see DGM600
DCDD see DAC800
2,3-DCDT see DBI200
1-1-DCE see VPK000
1,2-DCE see EIY600
DCEE see DFJ050
D&C GREEN No. 4 see FAF000
D&C GREEN No. 6 see BLK000
DCHFB see DFM000
DCI LIGHT MAGNESIUM CARBONATE see MAC650
DCM see DFO000, MJP450
DCMO see CCC500
DCMU see DXQ500
D&C ORANGE No. 2 see TGW000
D&C ORANGE No. 3 see HGC000
DCP see DFX800
2,4-DCP see DFX800
DCPA see DGI000
D&C RED No. 3 see FAG040
D&C RED No. 9 see CHP500
D&C RED No. 17 see OHA000
D&C RED No. 19 see FAG070
D&C RED No. 22 see BNH500
D.C.S. see BGJ750
D&C YELLOW No. 5 see FAG140
D&C YELLOW No. 7 see FEV000
D&C YELLOW No. 8 see FEW000
D-D see DGG000
DDC see DQY950, SGJ000
DDD see BIM500
2,4'-DDD see CDN000
o,p'-DDD see CDN000
p,p'-DDD see BIM500
DDE see BIM750
p,p'-DDE see BIM750
DDETA see HMQ500

DDM see DSU000, MJQ000
DD MIXTURE see DGG000
DDNP see DUR800
DDOA see ABC250
DDP see PJD000
cis-DDP see PJD000
DDS see SOA500
DD SOIL FUMIGANT see DGG000
DDT see DAD200
p,p'-DDT see DAD200
DDT DEHYDROCHLORIDE see BIM750
DDVF see DGP900
DDVP see DGP900
D.E. see DCJ800
DEA see DHF000
DEACETYLCHOLCHICEINE see TLN750
N-DEACETYLCHOLCHICEINE see TLN750
DEACETYLMETHYLCOLCHICINE see MIW500
DEACETYL-N-METHYLCOLCHICINE see MIW500
N-DEACETYL-N-METHYLCOLCHICINE see MIW500
DEACTIVATOR E see DJD600
DEACTIVATOR H see DJD600
DEADLY NIGHTSHADE see BAU500
DEADOPA see DNA200
DEAE see DHO500
DEALCA TP1 see GLU000
DEANOL see DOY800
DEANOX see IHD000
DEA OXO-5 see DBA800
DEB see BGA750, DKA600
DEBECACIN see DCQ800
DEBENDRIN see BBV500
DEBRICIN see FBS000
DEBROUSSAILLANT 600 see DAA800
DEBROUSSAILLANT CONCENTRE see TAA100
DEBROXIDE see BDS000
DEC see DAE800, DIX200
DECABORANE see DAE400
DECABORANE(14) see DAE400
1,2,3,5,6,7,8,9,10,10-DECACHLORO(5.2.1.0$(^{2,6})$.0$(^{3,9})$
.0$(^{5,8})$)DECANO-4-ONE see KEA000
DECACHLOROKETONE see KEA000
DECACHLORO-1,3,4-METHENO-2H-CYCLOBUTA(cd)PEN-
TALEN-2-ONE see KEA000
DECACHLOROOCTAHYDROKEPONE-2-ONE see KEA000
DECACHLOROOCTAHYDRO-1,3,4-METHENO-2H-CYCLO-
BUTA(cd)PENTALEN-2-ONE see KEA000
1,1a,3,3a,4,5,5,5a,5b,6-DECACHLOROOCTAHYDRO-1,3,4-
METHENO-2H-CYCLOBUTA(cd)PENTALEN-2-ONE see
KEA000
DECACHLOROPENTACYCLO(5.2.1.0$(^{2,6})$.0$(^{3,9})$.0$(^{5,8})$)
DECAN-4-ONE see KEA000
DECACHLOROPENTACYCLO(5.3.0.0$(^{2,6})$.0$(^{4,10})$.0$(^{5,9})$)
DECAN-3-ONE see KEA000
DECACHLOROTETRACYCLODECANONE see KEA000
DECACHLOROTETRAHYDRO-4,7-METHANOINDENEONE
see KEA000
DECACIL see LFK000
DECADERM see SOW000
trans,trans-2,4-DECADIENAL see DAE450
DECADRON see SOW000
DECAFENTIN see DAE600
DECAHYDRONAPHTHALENE see DAE800
DECAHYDRO-4a,7,9-TRIHYDROXY-2-METHYL-6,8-BIS
(METHYLAMINO)-4H-PYRANO(2,3-b)(1,4)BENZODIOXIN-
4-ONE DIHYDROCHLORIDE, (2R-(2-α,4a-β,5a-β,6-β,7-
β,8-β,9-α,9a-α,10a-β))- see SLI325
Δ-DECALACTONE see DAF200
DECALIN see DAE800
DECALIN (DOT) see DAE800
DECALIN SOLVENT see DAE800
DECAMINE see DAA800
DECAMINE 4T see TAA100
1-DECANAL see DAG000

1-DECANAL (mixed isomers) see DAG200
DECANAL DIMETHYL ACETAL see DAI600
DECANE see DAG400
n-DECANE (DOT) see DAG400
1-DECANEAMINE see DAG600
DECANEDIOIC ACID, BIS(2-ETHYLHEXYL) ESTER see
BJS250
DECANEDIOIC ACID, DIBUTYL ESTER see DEH600
DECANOIC ACID see DAH400
n-DECANOIC ACID see DAH400
DECANOIC ACID, ETHYL ESTER see EHE500
DECANOL see DAI600
n-DECANOL see DAI600
1-DECANOL (FCC) see DAI600
DECANOLIDE-1,5 see DAF200
DECAPS see VSZ100
DECARBOFURAN see DLS800
DECARBOXYCYSTEINE see AJT250
DECARBOXYCYSTINE see MCN500
DECARIS see LFA020
DECASONE see SOW000
DECASPRAY see SOW000
n-DECATYL ALCOHOL see DAI600
DECCOTANE see BPY000
DECELITH H see PKQ059
DECEMTHION P-6 see PHX250
2-DECENAL see DAI350
trans-2-DECEN-1-AL see DAI350
DECENALDEHYDE see DAI350
DECENTAN see CJM250
DECHLORANE 4070 see MQW500
DECHLORANE-A-O see AQF000
2,4-DECHLOROPHENYL-p-NITROPHENYL ETHER see
DFT800
DECLOMYCIN see MIJ500
DECOFOL see BIO750
n-DECOIC ACID see DAH400
DECONTRACTIL see GGS000
DECORPA see GLU000
DECORTANCYL see PLZ000
DECORTIN see PLZ000
DECORTIN H see PMA000
DECORTISYL see PLZ000
DECTANCYL see SOW000
DECYL ALCOHOL see DAI600
n-DECYL ALCOHOL see DAI600
1-DECYL ALDEHYDE see DAG000, UJJ000
DECYLAMINE see DAG600
DECYL BENZENE SODIUM SULFONATE see DAJ000
DECYLIC ACID see DAH400
n-DECYLIC ACID see DAH400
DECYLIC ALCOHOL see DAI600
DECYL OCTYL ALCOHOL see OAX000
DECYLTRIPHENYLPHOSPHONIUM BROMOCHLOROTRI-
PHENYLSTANNATE see DAE600
(DECYL-TRIPHENYL-PHOSPHONIUM)-TRIPHENYL-BROM-
CHLOR-STANNAT (GERMAN) see DAE600
DEDC see SGJ000
DEDELO see DAD200
DEDEVAP see DGP900
DEDK see SGJ000
DED-WEED see CIR250, DAA800, DGI400, TIX500
DED-WEED BRUSH KILLER see TAA100
DED-WEED CRABGRASS KILLER see PLC250
DED-WEED LV-69 see DAA800
DED-WEED LV-6 BRUSH KIL and T-5 BRUSH KIL see
TAA100
DEE-OSTEROL see VSZ100
DEEP LEMON YELLOW see SMH000
DEE-RON see VSZ100
DEE-RONAL see VSZ100
DEE-ROUAL see VSZ100
DEER'S TONGUE see DAJ800
DEERTONGUE INCOLORE see DAJ800

DEET see DKC800
DEF see BSH250
DEF DEFOLIANT see BSH250
DE-FEND see DSP400
DEFILIN see DJL000
DEFILTRAN see AAI250
DEFLAMON-WIRKSTOFF see MMN250
DEFLOGIN see HNI500
DE-FOL-ATE see MAE000, SFS000
DEFOLIANT 713 see DKE400
DEFOLIT see TEX600
DEFONIN see ILD000
DEG see DJD600
DEGALAN S 85 see PKB500
DEGALOL see DAQ400
DEGRANOL see MAW500
DEGRASSAN see DUS700
DE-GREEN see BSH250
DEGUELIA ROOT see DBA000
DEH see HHH000
D.E.H. 20 see DJG600
DEH 24 see TJR000
D.E.H. 26 see TCE500
DEHACODIN see DKW800
DEHIDROBENZPERIDOL see DYF200
DEHISTIN see TMP750
DEHP see DVL700
DEHPA EXTRACTANT see BJR750
DEHYDRACETIC ACID see MFW500
DEHYDRATIN see AAI250
DEHYDRITE see PCE000
DEHYDROACETIC ACID (FCC) see MFW500
DEHYDROACETIC ACID, SODIUM SALT see SGD000
DEHYDROBENZPERIDOL see DYF200
6-DEHYDRO-6-CHLORO-17-α-ACETOXYPROGESTERONE
   see CBF250
7-DEHYDROCHOLESTROL, ACTIVATED see CMC750
Δ¹-DEHYDROCORTISOL see PMA000
1-DEHYDROCORTISONE see PLZ000
Δ-1-DEHYDROCORTISONE see PLZ000
DEHYDROERGOTAMINE see DLK800
1-DEHYDROHYDROCORTISONE see PMA000
Δ¹-DEHYDROHYDROCORTISONE see PMA000
6-DEHYDRO-6-METHYL-17-α-ACETOXYPROGESTERONE
   see VTF000
6-DEHYDRO-16-METHYLENE-6-METHYL-17-ACETOXY-
   PROGESTERONE see MCB380
1-DEHYDRO-16-α-METHYL-9-α-FLUOROHYDROCORTI-
   SONE see SOW000
DEHYDRORETRONECINE see DAL400
DEHYDROSTILBESTROL see DAL600
DEHYDROSTILBOESTROL see DAL600
DEHYQUART STC-25 see DTC600
DEIDROBENZPERIDOLO see DYF200
DEINAIT see CJR909
DEIQUAT see DWX800
DEJO see DJT800
DEK see DJN750
DE-KALIN see DAE800
DEKALINA (POLISH) see DAE800
DEKORTIN see PLZ000
DEKRYSIL see DUS700
DEKSONAL see DOU600
DELAC J see DWI000
DELACURARINE see TOA000
DELADIOL see EDS100
DELAHORMONE UNIMATIC see EDS100
DELALUTIN see HNT500
DELAN see DLK200
DELAN-COL see DLK200
DELATESTRYL see MPN500
DELCORTOL see PMA000
DELEAF DEFOLIANT see TIG250
DELESTROGEN see EDS100

DELESTROGEN 4X see EDS100
DELGESIC see ADA725
DELICIA see AHE750, PGY000
DELLIPSOIDS see BBK500
DELMOFULVINA see GKE000
DELNAV see DVQ709
DELONIN AMIDE see NCR000
DELOWAS S see TBD000
DELOWAX OM see TBD000
DELPET 50M see PKB500
DELPHENE see DKC800
m-DELPHENE see DKC800
DELPHINIC ACID see ISU000
DELSTEROL see CMC750
DELTA see CJJ000
DELTA-CORTEF see PMA000
DELTACORTELAN see PLZ000
DELTACORTENOL see PMA000
DELTACORTISONE see PLZ000
DELTACORTONE see PLZ000
DELTACORTRIL see PMA000
DELTA-DOME see PLZ000
DELTA F see PMA000
DELTAFLUORENE see SOW000
DELTALIN see VSZ100
DELTA-MVE see MKB750
DELTAMYCIN A see CBT250
DELTAN see DUD800
DELTA-STAB see PMA000
DELTISONE see PLZ000
DELTRATE-20 see PBC250
DELVEX see DJT800
DELVINAL SODIUM see VKP000
DELYSID see DJO000
DEMA see BIE500
DEMAROL see DAM600
DEMASORB see DUD800
DEMAVET see DUD800
DEMECLOCYCLINE see MIJ500
DEMECOLCINE see MIW500
DEMEPHION see MIW250
DEMEROL see DAM600, DAM700
DEMEROL HYDROCHLORIDE see DAM700
DEMESO see DUD800
DEMETHON-METHYL (MAK) see MIW100
4-DEMETHOXYDAUNOMYCIN see DAN000
4-DEMETHOXYDAUNORUBICIN see DAN000
DEMETHYLAMITRIPTYLENE see NNY000
DEMETHYLCHLOROTETRACYCLIN see MIJ500
DEMETHYLCHLOROTETRACYCLINE see MIJ500
6-DEMETHYLCHLOROTETRACYCLINE see MIJ500
6-DEMETHYL-7-CHLOROTETRACYCLINE DEMETHYL-
   CHLORTETRACYCLINE see MIJ500
DEMETHYLCHLORTETRACYCLINE see MIJ500
6-DEMETHYLCHLORTETRACYCLINE see MIJ500
6-DEMETHYL-7-CHLORTETRACYCLINE see MIJ500,
   MIJ500
DEMETHYLCHLORTETRACYCLINE, BASE see MIJ500
o-DEMETHYLDAUNOMYCIN see KBU000
DEMETHYLDOPAN see BIA250
DEMETHYL-EPIODOPHYLLOTOXIN ETHYLIDENE GLUCO-
   SIDE see EAV500
4-DEMETHYLEPIODOPHYLLOTOXIN-β,d-ETHYLIDENE-
   GLUCOSIDE see EAV500
4'-DEMETHYLEPIPODOPHYLLOTOXIN-9-(4,6-O-ETHYLI-
   DENE-β-d-GLUCOPYRANOSIDE see EAV500
4'-DEMETHYLEPIPODOPHYLLOTOXIN ETHYLIDENE-β,d-
   GLUCOSIDE see EAV500
4-DEMETHYL-EPIPODOPHYLLOTOXIN-β,d-ETHYLIDEN-
   GLUCOSIDE see EAV500
4'-DEMETHYLEPIPODOPHYLLOTOXIN-9-(4,6-O-2-THENYL-
   IDENE-β-d-GLUCOPYRANOSIDE see EQP000
4'-DEMETHYL-EPIPODOPHYLLOTOXIN-β-d-THENYLI-
   DENE-GLUCOSIDE see EQP000

4'-O-DEMETHYL-1-O-(4,6-O-ETHYLIDENE-β,d-
GLUCOPYRANOSYL)EPIPODOPHYLLOTOXIN see
EAV500
DEMETHYLIMIPRAMINE see DSI709
4'-DEMETHYL 1-O-(4,6-O,O-(2-THENYLIDENE)-β-d-
GLUCOPYRANOSYL)EPIPODOPHYLLOTOXIN see
EQP000
DEMETON see DAO600
DEMETON METHYL see MIW100
DEMETON-O-METHYL see DAO800
DEMETON-S-METHYL see DAP400
DEMETON-S-METHYLSULFON (GERMAN) see DAP600
DEMETON-S-METHYLSULFONE see DAP600
DEMETON-S-METHYL-SULFOXID (GERMAN) see DAP000
DEMETON-O-METHYL SULFOXIDE see DAP000
DEMETON-S-METHYL SULFOXIDE see DAP000
DEMETON-S-METHYL-SULPHONE see DAP600
DEMETON-METHYL SULPHOXIDE see DAP000
DEMETON-O-METILE (ITALIAN) see DAO800
DEMETON-S-METILE (ITALIAN) see DAP400
DEMETON-O + DEMETON-S see DAO600
DEMETON-S see DAP200
DEMOCRACIN see TBX000
DEMOS-L40 see DSP400
DEMOX see DAO600
DEMSODROX see DUD800
DEN see NJW500
DENA see NJW500
DENAMONE see VSZ450
DENAPON, NITROSATED (JAPANESE) see NBJ500
DENATURED SPIRITS see AFJ000
DENDREPAR see PEM750
DENDRID see DAS000
DENDRITIS see SFT000
DENKALAC 61 see AAX175
DENKA QP3 see SMQ500
DENKA VINYL SS 80 see PKQ059
DENSINFLUAT see TIO750
DENYL see DKQ000, DNU000
DENYLSODIUM see DNU000
DEODOPHYLL see CKN000
DEODORIZED WINTERIZED COTTONSEED OIL see
CNU000
DEOFED see DBA800
DEORLENE GREEN JJO see BAY750
DEOVAL see DAD200
2-DEOXY-3-ARABINO-HEXOSE see DAR600
2-DEOXY-d-ARABINO-HEXOSE see DAR600
DEOXYCHOLATIC ACID see DAQ400
7-α-DEOXYCHOLIC ACID see DAQ400
DEOXYCHOLIC ACID (FCC) see DAQ400
14-DEOXY-14-((2-DIETHYLAMINOETHYL)MERCAPTO-
ACETOXY)-MUTILIN HYDROGEN FUMARATE see
DAR000
14-DEOXY-14-((2-DIETHYLAMINOETHYL-THIO)-ACE-
TOXY)MUTILINE see TET800
DEOXYEPHEDRINE see DBA800, DBB000
9-DEOXY-12,13-EPOXY-9-OXOLEUCOMYCIN V 3-ACE-
TATE 4ᴮ-(3-METHYLBUTANOATE) see CBT250
DEOXYFLUOROURIDINE see DAR400
2'-DEOXY-5-FLUOROURIDINE see DAR400
2-DEOXYGLUCOSE see DAR600
2-DEOXY-d-GLUCOSE see DAR600
d-2-DEOXYGLUCOSE see DAR600
2'-DEOXY-5-IODOURIDINE see DAS000
2-DEOXY-2-(((METHYLNITROSOAMINO)CARBONYL)
AMINO)-d-GLUCOPYRANOSE see SMD000
2-DEOXY-2-(3-METHYL-3-NITROSOUREIDO)-d-GLUCOPY-
RANOSE see SMD000
2-DEOXY-2-(3-METHYL-3-NITROSOUREIDO)-α(and β)-d-
GLUCOPYRANOSE see SMD000
DEOXYNOREPHEDRINE see BBK000
2-DEOXYPHENOBARBITAL see DBB200

1-β-d-2'-DEOXYRIBOFURANOSYL-5-FLUROURACIL see
DAR400
1-(2-DEOXY-β-d-RIBOFURANOSYL)-5-IODOURACIL see
DAS000
1-β-d-2'-DEOXYRIBOFURANOSYL-5-IODOURACIL see
DAS000
4-DEOXYTETRONIC ACID see BOV000
β-DEOXYTHIOGUANOSINE see TFJ250
β-2'-DEOXY-6-THIOGUANOSINE MONOHYDRATE see
TFJ250
DEP (pesticide) see TIQ250
DEPARAL see CMC750
DEPC see DIZ100
DEPEN see MCR750
DEPHADREN see BBK500
DEPINAR see VSZ000
DEPO-PROLUTON see HNT500
DEPO-PROVERA see MCA000
DEPOSUL see SNN300
DEPOT-OESTROMENINE see DJB200
DEPOT-OESTROMON see DJB200
DEPOXIN see DBA800
DEPRANCOL see PNA500
DEPRELIN see SNE000
DEPREX see EAI000
DEPRINOL see DLH630
DEPROMIC see PNA500
DEPTHON see TIQ250
D.E.R. 332 see BLD750
DERACIL see TFR250
DERATOL see VSZ100
DEREUMA see DOT000
DERGRAMIN see SOW000
DERIBAN see DGP900
DERIL see RNZ000
DERIZENE see DNU000
DERMACAINE see DDT200
DERMADEX see HCL000
DERMA FAST BROWN W-GL see CMO750
DERMAFOSU (POLISH) see RMA500
DERMAGAN see ACR300
DERMAGEN see ACR300
DERMAGINE see OAV000
DERMAPHOS see RMA500
DERMASORB see DUD800
DERMATON see CDS750
DERMISTINE see BBV500
DERMODRIN see BBV500
DERMOFURAL see NGE500
DERONIL see SOW000
DERRIBANTE see DGP900
DERRIN see RNZ000
DERRIS see RNZ000
DERRIS ELLIPTICA, root see DBA000
DERRIS RESINS see DBA000
DERRIS ROOT see DBA000
DES (synthetic estrogen) see DKA600
DESACETYLCHOLCHICEINE see TLN750
N-DESACETYL-N-METHYLCOLCHICINE see MIW500
DESADRENE see SOW000
DESAMETASONE see SOW000
DESAMINE see DBA800
DESCHLOROBIOMYCIN see TBX000
DESD see DKB000
DESDEMIN see CHJ750
DES DISODIUM SALT see DKA400
DESENTOL see BBV500
DESERTALC 57 see TAB750
DESERT RED see CHP500
DESFEDRIN see DBA800
DES-I-CATE see DXD000
DESICCANT L-10 see ARB250
DESIMIPRAMINE see DSI709
DESINFECT see CDP000

DESIPRAMIN see DSI709
DESIPRAMINE (D4) see DSI709
DESIPRAMINE HYDROCHLORIDE see DLS600
DESMA see DKA600
DESMECOLCINE see MIW500
DESMETHYLAMITRIPTYLINE see NNY000
DESMETHYLDOPAN see BIA250
DESMETHYLDOXEPIN see DBA600
DESMETHYLIMIPRAMINE see DSI709
DESMETHYLIMIPRAMINE HYDROCHLORIDE see DLS600
DESMODUR 44 see MJP400
DESMODUR T80 see TGM750
DESMODUR T100 see TGM740
2,4-DES-Na see CNW000
2,4-DES-NATRIUM (GERMAN) see CNW000
DESOLET see SFS000
DESORMONE see DAA800, DGB000
14-DESOSSI-14-((2-DIETILAMINOETIL)MERCAPTO-ACE-
    TOSSI)MUTILIN IDROGENO FUMARATO (ITALIAN) see
    TET800
DESOSSIEFEDRINA see DBA800
DES-OXA-D see DBA800
DESOXEDRINE see DBA800
DESOXIN see DBA800
DESOXO-5 see DBA800
DESOXYCHOLIC ACID see DAQ400
DESOXYCHOLSAEURE (GERMAN) see DAQ400
14-DESOXY-14-((DIETHYLAMINOETHYL)-MERCAPTO
    ACETOXYL)-MUTILIN HYDROGEN FUMARATE see
    TET800
DESOXYEPHEDRINE see DBB000
DESOXYEPHEDRINE HYDROCHLORIDE see DBA800
l-DESOXYEPHEDRINE HYDROCHLORIDE see MDQ500
DESOXYFED see DBA800
2-DESOXY-d-GLUCOSE (FRENCH) see DAR600
DESOXYN see BBK500, DBA800, DBB000
(±)-DESOXYNOREPHEDRINE see BBK000
racemic-DESOXYNOR-EPHEDRINE see BBK000
DESOXYPHED see DBA800
2-DESOXYPHENOBARBITAL see DBB200
DESOXYPHENOBARBITONE see DBB200
DESPHEN see CDP250
DESSON see CLW000
DESTENDO see BCA000
DESTIM see DBA800
DESTRIOL see EDU500
DESTROL see DKA600
DESTRONE see EDV000
DESTRUXOL APPLEX see DXE000
DESTRUXOL BORER-SOL see EIY600
DESTRUXOL ORCHID SPRAY see NDN000
DET see DKC800
m-DET see DKC800
DETA see DJG600
m-DETA see DKC800
DETAL see DUS700
DETALUP see VSZ100
DETAMIDE see DKC800
DETARIL see DWK400
DETERGENT 66 see SIB600
DETERGENT HD-90 see DXW200
DETF see TIQ250
DETHMORE see WAT200
DETIA GAS EX-B see AHE750, PGY000
DETICENE see DAB600
DETMOL-EXTRAKT see BBQ500
DETMOL MA see MAK700
DETMOL MA 96% see MAK700
DETOX see DAD200
DETOX 25 see BBQ500
DETOXAN see DAD200
DETOXARGIN see AQW000
DETREOMYCINE see CDP250
DETREOPAL see CDP700

DETREX see DBA800
DETTOL see CLW000
DEUSLON-A see EDU500
DEUTERIOMORPHINE see DBB600
DEUTERIUM see DBB800
DEUTERIUM FLUORIDE see DBC000
DEUTERIUM OXIDE see HAK000
DEVAL RED K see CLK220
DEVAL RED TR see CLK220
DEVEGAN see ABX500
DEVELIN see PNA500
DEVELOPER 11 see PEY000
DEVELOPER 13 see PEY500
DEVELOPER A see NAX000
DEVELOPER AMS see NAX000
DEVELOPER BN see NAX000
DEVELOPER H see TGL750
DEVELOPER P see NEO500
DEVELOPER PF see PEY500
DEVELOPER R see REA000
DEVELOPER SODIUM see NAX000
DEVIGON see DSP400
DEVIKOL see DGP900
DEVIL'S APPLE see SLV500
DEVIPON see DGI400
DEVISULPHAN see EAQ750
DEVITHION see MNH000
DEVOL ORANGE B see NEO000
DEVOL ORANGE R see NEN500
DEVOL RED K see CLK235
DEVOL RED TA SALT see CLK235
DEVOL RED TR see CLK235
DEVOL SCARLET B see NMP500
DEVORAN see BBQ500
DEVOTON see MFW100
DEX see BJU000
DEXA see SOW000
DEXACORT see SOW000
DEXA-CORTIDELT see SOW000
DEXA-CORTIDELT HOSTACORTIN H see PMA000
DEXADELTONE see SOW000
DEXAIME see BBK500
DEXALINE see BBK500
DEXALME see BBK500
DEXALONE see BBK500
DEXAMED see BBK500
DEXAMETH see SOW000
DEXAMETHASONE ALCOHOL see SOW000
DEXAMINE see BBK500
DEXAMPHAMINE see BBK500
DEXAMPHETAMINE see BBK500
DEXAMPHETAMINE SULFATE see BBK500
DEXAMYL see BBK500
DEXCHLOROPHENIRAMINE MALEATE see PJJ325
DEXCHLORPHENIRAMINE MALEATE see PJJ325
DEXEDRINA see BBK500
DEXEDRINE SULFATE see BBK500
DEXIES see BBK500
DEXON see DOU600
DEXONE see SOW000
DEXON E 117 see PMP500
DEXOPHRINE see DBA800
DEXOVAL see DBA800
DEXPANTHENOL (FCC) see PAG200
DEXTELAN see SOW000
DEXTRAN 1 see DBC800
DEXTRAN 2 see DBD000
DEXTRAN 5 see DBD200
DEXTRAN 10 see DBD400
DEXTRAN 11 see DBD600
DEXTRAN ION COMPLEX see IGS000
DEXTRANS see DBD800
DEXTRIFERRON see IGU000
DEXTRIFERRON INJECTION see IGU000

DEXTRINS see DBD800
DEXTROAMPHETAMINE SULFATE see BBK500
DEXTRO CALCIUM PANTOTHENATE see CAU750
DEXTROCHLORPHENIRAMINE MALEATE see PJJ325
DEXTROFER 75 see IGS000
DEXTRO-α-METHYLPHENETHYLAMINE SULFATE see
   BBK500
DEXTROMYCETIN see CDP250
DEXTRONE see DWX800, PAJ000
DEXTRO-1-PHENYL-2-AMINOPROPANE SULFATE see
   BBK500
DEXTRO-β-PHENYLISOPROPYLAMINE SULFATE see
   BBK500
DEXTROPROPOXYPHENE see DAB879
DEXTROPROPOXYPHENE HYDROCHLORIDE see PNA500
DEXTROPROXYPHEN HYDROCHLORIDE see PNA500
DEXTROPUR see GFG000
DEXTROSE, anhydrous see GFG000
DEXTROSE (FCC) see GFG000
DEXTROSOL see GFG000
DEXTROTUBOCURARINE CHLORIDE see TOA000
DEZIBARBITUR see EOK000
DEZONE see SOW000
DF 118 see DKW800, DKX000
DFA see DVX800
DFP see IRF000
DFT see DWN800
2-DG see DAR600
DGE see DKM200
DHA see MFW500
DHA-SODIUM see SGD000
DHBP see DYF200
DHMS see DME000
DHNT see BKH500
DHPN see DNB200
DHS see MFW500
DHUTRA see SLV500
DIABASE SCARLET G see NMP500
DIABASIC RHODAMINE B see FAG070
DIABEN see BSQ000
DIABENYL see BBV500
DIABETAMID see BSQ000
DIABETOL see BSQ000
DIABORAL see BSM000
DIABRIN see BOM750, BQL000
DIABUTAL see NBU000
DIABUTON see BSQ000
DIABYLEN see BBV500
DIACARB see AAI250
DIACELLITON FAST GREY G see DCJ200
DIACEL NAVY DC see DCJ200
DIACEPHIN see HBT500
2-DIACETAMIDOFLUORENE see DBF200
2,7-DIACETAMIDOFLUORENE see BGP250
1,3-DIACETATE GLYCEROL see DBF600
1,2-DIACETATE 1,2,3-PROPANETRIOL see DBF600
DIACETAZOTOL see ACR300
DIACETIC ETHER see EFS000
DIACETIN see DBF600
1,2-DI-ACETIN see DBF600
1,3-DIACETIN see DBF600
2,3-DIACETIN see DBF600
DIACETONALCOHOL (DUTCH) see DBF750
DIACETONALCOOL (ITALIAN) see DBF750
DIACETONALKOHOL (GERMAN) see DBF750
DIACETONE see DBF750
DIACETONE ALCOHOL see DBF750
DIACETONE-ALCOOL (FRENCH) see DBF750
DIACETOTOLUIDE see ACR300
o-DIACETOTOLUIDIDE, 4''-(o-TOLYLAZO)-(8CI) see
   ACR300
DIACETOXYBUTYLTIN see DBF800
DIACETOXYDIBUTYLPLUMBANE see DED400
DIACETOXYDIBUTYL STANNANE see DBF800

DIACETOXYDIBUTYLTIN see DBF800
3-β,17-β-DIACETOXY-17-α-ETHYNYL-4-OESTRENE see
   EQJ500
4-β,15-DIACETOXY-3-α-HYDROXY-12,13-EPOXYTRICHO-
   THEC-9-ENE see AOP250
DIACETOXYMERCURY see MCS750
4,15-DIACETOXY-8-(3-METHYLBUTYRYLOXY)-12,13-
   EPOXY-Δ-9-TRICHOTHECEN-3-OL see FQS000
4-β,15-DIACETOXY-8-α-(3-METHYLBUTYRYLOXY)-3-α-
   HYDROXY-12,13-EPOXYTRICHOTHEC-9-ENE see
   FQS000
3-β,17-β-DIACETOXY-19-NOR-17-α-PREGN-4-EN-20-YNE
   see EQJ500
DIACETOXYSCIRPENOL see AOP250
4,15-DIACETOXYSCIRPEN-3-OL see AOP250
DIACETYL (FCC) see BOT500
DIACETYLAMINOAZOTOLUENE see ACR300
4,4'-DIACETYLAMINOBIPHENYL see BFX000
2-DIACETYLAMINOFLUORENE see DBF200
2,7-DIACETYLAMINOFLUORENE see BGP250
N-DIACETYL-2-AMINOFLUORENE see DBF200
N,N-DIACETYL-2-AMINOFLUORENE see DBF200
4,4'-DIACETYLBENZIDINE see BFX000
N,N'-DIACETYL BENZIDINE see BFX000
DIACETYLCHOLINE see CMG250
DIACETYLCHOLINE CHLORIDE see HLC500
DIACETYLCHOLINE DICHLORIDE see HLC500
DIACETYLCHOLINE DIIODIDE see BJI000
DIACETYL DIOXIME see DBH000
N,N-DIACETYL-2-FLUORENAMINE see DBF200
DIACETYL GLYCERINE see DBF600
DIACETYLMETHANE see ABX750
DIACETYLMORFIN see HBT500
DIACETYLMORPHINE see HBT500
DIACETYL PEROXIDE (MAK) see ACV500
3,4-DI(ACETYLTHIOMETHYL)-5-HYDROXY-6-METHYLPY-
   RIDINE HYDROBROMIDE see DBH800
N,N-DIACETYL-o-TOLYLAZO-o-TOLUIDINE see ACR300
DIACID see BNK000
DIACOTTON BLUE BB see CMO000
DIACOTTON DEEP BLACK see AQP000
DIACRID see DBX400
DIACTOL see VSZ100
DIACYCINE see TBX250
DIADEM CHROME BLUE R see HJF500
DI-ADRESON F see PMA000
DIAETHANOLAMIN (GERMAN) see DHF000
DIAETHANOLNITROSAMIN (GERMAN) see NKM000
1,1-DIAETHOXY-AETHAN (GERMAN) see AAG000
DIAETHYLACETAL (GERMAN) see AAG000
DIAETHYLAETHER (GERMAN) see EJU000
O,O-DIAETHYL-S-(2-AETHYLTHIO-AETHYL)-DITHIO-
   PHOSPHAT (GERMAN) see DXH325
O,O-DIAETHYL-S-(2-AETHYLTHIO-AETHYL)-MONOTHIO-
   PHOSPHAT (GERMAN) see DAP200
O,O-DIAETHYL-S-(AETHYLTHIO-METHYL)-DITHIOPHOS-
   PHAT (GERMAN) see PGS000
DIAETHYLALLYLACETAMIDE (GERMAN) see DJU200
DIAETHYLAMIN (GERMAN) see DHJ200
DIAETHYLAMINOAETHANOL (GERMAN) see DHO500
o-DIAETHYLAMINOAETHOXY-BENZANILID (GERMAN)
   see DHP200
DIAETHYLANILIN (GERMAN) see DIS700
O,O-DIAETHYL-O-(4-BROM-2,5-DICHLOR)-PHENYL-MO-
   NOTHIOPHOSPHAT (GERMAN) see EGV500
DIAETHYLCARBONAT (GERMAN) see DIX200
O,O-DIAETHYL-O-(CHINOXALYL-(2))-MONOTHIOPHOS-
   PHAT (GERMAN) see DJY200
O,O-DIAETHYL-O-(3-CHLOR-4-METHYL-CUMARIN-7-YL)-
   MONOTHIOPHOSPHAT (GERMAN) see CNU750
O,O-DIAETHYL-S-(6-CHLOR-2-OXO-BEN(b)-1,3-OXALIN-3-
   YL)-METHYL-DIT HIOPHOSPHAT (GERMAN) see
   BDJ250
O,O-DIAETHYL-S-((4-CHLOR-PHENYL-THIO)-METHYL)DI-
   THIOPHOSPHAT (GERMAN) see TNP250

O,O-DIAETHYL-o-(α-CYANBENZYLIDEN-AMINO)-THION-PHOSPHAT (GERMAN) see BAT750
O,O-DIAETHYL-o-(α-CYANO-BENZYLIDENAMINO)-MONO-THIOPHOSPHAT (GERMAN) see BAT750
N-DIAETHYL CYSTEAMIN (GERMAN) see DIY600
O,O-DIAETHYL-O-(2,5-DICHLOR-4-BROMPHENYL)-THION-OPHOSPHAT (GERMAN) see EGV500
O,O-DIAETHYL-O-1-(4,5-DICHLORPHENYL)-2-CHLOR-VI-NYL-PHOSPHAT (GERMAN) see CDS750
O,O-DIAETHYL-O-2,4-DICHLOR-PHENYL-MONOTHIO-PHOSPHAT (GERMAN) see DFK600
O,O-DIAETHYL-S((2,5-DICHLOR-PHENYL-THIO)-METHYL)-DITHIOPHOSPHAT (GERMAN) see PDC750
O,O-DIAETHYL-O-2,4-DICHLORPHENYL-THIONOPHOS-PHAT (GERMAN) see DFK600
1,2-DIAETHYLHYDRAZINE (GERMAN) see DJL400
O,O-DIAETHYL-O-(2-ISOPROPYL-4-METHYL-PYRIMIDIN-6-YL)-MONOTHIOPHOSPHAT (GERMAN) see DCM750
O,O-DIAETHYL-O-(2-ISOPROPYL-4-METHYL)-6-PYRIMI-DYL-THIONOPHOSPHAT (GERMAN) see DCM750
O,O-DIAETHYL-S-1-METHYL)AETHYL)-CARBAMOYL-METHYL-MONOTHIOPHOSPHAT (GERMAN) see PHK250
O,O-DIAETHYL-S-(3-METHYL-2,4-DIOXO-5-OXA-3-AZA-HEPTYL)-DITHIOPHOSPHAT (GERMAN) see DJI000
O,O-DIAETHYL-O-(3-METHYL-1H-PYRAZOL-5-YL)-PHOS-PHAT (GERMAN) see MOX250
O,O-DIAETHYL-O-4-METHYLSULFINYL-PHENYL-MONO-THIOPHOSPHAT (GERMAN) see FAQ800
DIAETHYL-NICOTINAMID (GERMAN) see DJS200
O,O-DIAETHYL-O-(4-NITROPHENYL)-MONOTHIOPHOS-PHAT (GERMAN) see PAK000
O,O'-DIAETHYL-p-NITROPHENYLPHOSPHAT (GERMAN) see NIM500
DIAETHYL-p-NITROPHENYLPHOSPHORSAEUREESTER (GERMAN) see NIM500
DIAETHYLNITROSAMIN (GERMAN) see NJW500
O,O-DIAETHYL-S-(4-OXOBENZOTRIAZIN-3-METHYL)-DI-THIOPHOSPHAT (GERMAN) see EKN000
O,O-DIAETHYL-S-((4-OXO-3H-1,2,3-BENZOTRIAZIN-3-YL)-METHYL)-DITHIOPHOSPHAT (GERMAN) see EKN000
O,O-DIAETHYL-O-(PYRAZIN-2YL)-MONOTHIOPHOSPHAT (GERMAN) see EPC500
O,O-DIAETHYL-O-(2-PYRAZINYL)-THIONOPHOSPHAT (GERMAN) see EPC500
DIAETHYLSULFAT (GERMAN) see DKB110
O,O-DIAETHYL-S-(3-THIA-PENTYL)-DITHIOPHOSPHAT (GERMAN) see DXH325
DIAETHYLTHIOPHOSPHORSAEUREESTER des AETHYL-THIOGLYKOL (GERMAN) see DAP200
O,O-DIAETHYL-O-3,5,6-TRICHLOR-2-PYRIDYLMONOTHIO-PHOSPHAT (GERMAN) see CMA100
DIAETHYLZINNDICHLORID (GERMAN) see DEZ000
DIAFEN see LJR000
DIAFURON see NGG500
DIAGRABROMYL see BNP750
DIAKARB see AAI250
DIAKARMON see SKV200
DIAKON see MLH750, PKB500
DIAL-A-GESIC see HIM000
DIALICOR see DHS200
DIALIFOR see DBI099
DIALLAAT (DUTCH) see DBI200
DIALLAT (GERMAN) see DBI200
DIALLATE see DBI200
DIALLYL see HCR500
DIALLYLAMINE see DBI600
DIALLYLCHLOROACETAMIDE see CFK000
N,N-DIALLYLCHLOROACETAMIDE see CFK000
N,N-DIALLYL-2-CHLOROACETAMIDE see CFK000
N,N-DIALLYL-α-CHLOROACETAMIDE see CFK000
DIALLYLDIBROMO STANNANE see DBJ400
DIALLYL ETHER see DBK000
DIALLYL MALEATE see DBK200
DIALLYL PHTHALATE see DBL200

DIALLYLTIN DIBROMIDE see DBJ400
DIALUMINUM SULPHATE see AHG750
DIALUMINUM TRIOXIDE see AHE250
DIALUMINUM TRISULFATE see AHG750
DIALUX see ADY500
DIAMARIN see DYE600
DIAMAZO see ACR300
DIAMIDAFOS see PEV500
DIAMIDE see DRP800, HGS000
DIAMIDFOS see PEV500
DIAMIDINE see DBL800
4,4'-DIAMIDINODIPHENOXYPENTANE DI(β-HYDROXY-ETHANESULFONATE see DBL800
4,4'-DIAMIDINO-α,omega-DIPHENOXYPENTANE ISETH-IONATE see DBL800
DIAMINE see HGS000
2,4-DIAMINEANISOLE see DBO000
DIAMINE BLUE 2B see CMO000
DIAMINE BLUE 3B see CMO250
DIAMINE DEEP BLACK EC see AQP000
cis-DIAMINEDICHLOROPLATINUM see PJD000
(4,4'-DIAMINE-3,3'-DIMETHYL(1,1'-BIPHENYL) see TGJ750
DIAMINIDE MALEATE see DBM800
2,6-DIAMINOACRIDINE see DBN000
2,8-DIAMINOACRIDINE see DBN600
3,6-DIAMINOACRIDINE see DBN600
3,7-DIAMINOACRIDINE see DBN000
3,6-DIAMINOACRIDINE BISULPHATE see DBN400
3,6-DIAMINOACRIDINE mixture with 3,6-DIAMINO-10-METHYLACRIDINIUM CHLORIDE see DBX400
3,6-DIAMINOACRIDINE MONOHYDROCHLORIDE see PMH250
3,6-DIAMINOACRIDINE SULFATE (1:1) see DBN400
3,6-DIAMINOACRIDINE SULPHATE (1:1) see DBN400
2,8-DIAMINOACRIDINIUM see DBN600
3,6-DIAMINOACRIDINIUM see DBN600
3,6-DIAMINOACRIDINIUM CHLORIDE see PMH250
3,6-DIAMINOACRIDINIUM CHLORIDE HYDROCHLORIDE see PMH250
2,8-DIAMINOACRIDINIUM CHLORIDE MONOHYDRO-CHLORIDE see PMH250
3,6-DIAMINOACRIDINIUM MONOHYDROGEN SULPHATE see DBN400
2,8-DIAMINOACRIDINIUM SULPHATE see DBN400
1,2-DIAMINOAETHAN (GERMAN) see EEA500
2,4-DIAMINOANISOL see DBO000
2,4-DIAMINOANISOLE see DBO000
2,4-DIAMINOANISOLE BASE see DBO000
m-DIAMINOANISOLE 1,3-DIAMINO-4-METHOXYBENZENE see DBO000
2,4-DIAMINOANISOLE SULFATE see DBO400
2,4-DIAMINOANISOLE SULPHATE see DBO400
2,4-DIAMINO-ANISOL SULPHATE see DBO400
1,5-DIAMINOANTHRARUFIN see DBP909
4,8-DIAMINOANTHRARUFIN see DBP909
3,7-DIAMINO-5-AZAANTHRACENE see DBN600
2,4-DIAMINOAZOBENZENE HYDROCHLORIDE see PEK000
m-DIAMINOBENZENE see PEY000
o-DIAMINOBENZENE see PEY250
p-DIAMINOBENZENE see PEY500
1,2-DIAMINOBENZENE see PEY250
1,3-DIAMINOBENZENE see PEY000
1,4-DIAMINOBENZENE see PEY500
m-DIAMINOBENZENE DIHYDROCHLORIDE see PEY750
1,3-DIAMINOBENZENE DIHYDROCHLORIDE see PEY750
6,6'-DIAMINO-m,m'-BIPHENOL see DMI400
4,4'-DIAMINOBIPHENYL see BBX000
p,p'-DIAMINOBIPHENYL see BBX000
4,4'-DIAMINO-1,1'-BIPHENYL see BBX000
4,4'-DIAMINO-3,3'-BIPHENYLDICARBOXYLIC ACID see BFX250
4,4'-DIAMINOBIPHENYL-3,3'-DICARBOXYLIC ACID see BFX250

4,4′-DIAMINO-3,3′-BIPHENYLDIOL see DMI400
4,4′-DIAMINOBIPHENYLOXIDE see OPM000
1,3-DIAMINOBUTANE see BOR750
3,4-DIAMINOCHLOROBENZENE see CFK125
1,2-DIAMINO -4-CHLOROBENZENE see CFK125
3,4-DIAMINO-1-CHLOROBENZENE see CFK125
2,4-DIAMINO-5-(4-CHLOROPHENYL)-6-ETHYLPYRIMIDINE
    see TGD000
2,4-DIAMINO-5-p-CHLOROPHENYL-6-ETHYLPYRIMIDINE
    see TGD000
DI(-4-AMINO-3-CHLOROPHENYL)METHANE see MJM200
DI-(4-AMINO-3-CLOROFENIL)METANO (ITALIAN) see
    MJM200
4,4′-DIAMINO-3,3′-DICHLOROBIPHENYL see DEQ600
4,4′-DIAMINO-3,3′-DICHLORODIPHENYL see DEQ600
4,4′-DIAMINO-3,3′-DICHLORODIPHENYLMETHANE see
    MJM200
2,2′-DIAMINODIETHYLAMINE see DJG600
β,β′-DIAMINODIETHYL DISULFIDE see MCN500
DIAMINODIFENILSULFONA (SPANISH) see SOA500
p,p′-DIAMINODIFENYLMETHAN (CZECH) see MJQ000
1,5-DIAMINO-4,8-DIHYDROXY-9,10-ANTHRACENEDIONE
    see DBP909
1,5-DIAMINO-4,8-DIHYDROXYANTHRAQUINONE see
    DBP909
4,8-DIAMINO-1,5-DIHYDROXYANTHRAQUINONE see
    DBP909
leuco-1,5-DIAMINO-4,8-DIHYDROXYANTHRAQUINONE see
    DBP909
3,3′-DIAMINO-4,4′-DIHYDROXYARSENOBENZENE DIHY-
    DROCHLORIDE see SAP500
3,3′-DIAMINO-4,4′-DIHYDROXY ARSENOBENZENE
    METHYLENESULFOXYLATE SODIUM see NCJ500
2,8-DIAMINO-3,7-DIMETHYLACRIDINE see DBT200
3,6-DIAMINO-2,7-DIMETHYLACRIDINE see DBT200
4,4′-DIAMINO-3,3′-DIMETHYLBIPHENYL see TGJ750
4,4′-DIAMINO-3,3′-DIMETHYLDIPHENYL see TGJ750
p-DIAMINODIPHENYL see BBX000
4,4′-DIAMINODIPHENYL see BBX000
O,O′-DIAMINO DIPHENYL DISULFIDE see DXJ800
DIAMINODIPHENYL ETHER see OPM000
4,4-DIAMINODIPHENYL ETHER see OPM000
p,p′-DIAMINODIPHENYL ETHER see OPM000
2,4′-DIAMINODIPHENYLMETHAN (GERMAN) see MJP750
4,4′-DIAMINODIPHENYLMETHAN (GERMAN) see MJQ000
DIAMINODIPHENYLMETHANE see MJQ000
2,4′-DIAMINODIPHENYLMETHANE see MJP750
4,4′-DIAMINODIPHENYLMETHANE see MJQ000
o,p′-DIAMINODIPHENYLMETHANE see MJP750
p,p′-DIAMINODIPHENYLMETHANE see MJQ000
4,4′-DIAMINODIPHENYL OXIDE see OPM000
4,4′-DIAMINODIPHENYL SULFIDE see TFI000
p,p′-DIAMINODIPHENYL SULFIDE see TFI000
DIAMINO-4,4′-DIPHENYL SULFONE see SOA500
4,4′-DIAMINODIPHENYL SULFONE see SOA500
p,p′-DIAMINODIPHENYL SULFONE see SOA500
p,p′-DIAMINODIPHENYL SULPHIDE see TFI000
p,p-DIAMINODIPHENYL SULPHONE see SOA500
DIAMINO-4,4′-DIPHENYL SULPHONE see SOA500
3,3-DIAMINODIPROPYLAMINE see AIX250
3,3′-DIAMINODIPROPYLAMINE see AIX250
DIAMINODITOLYL see TGJ750
1,2-DIAMINO-ETHAAN (DUTCH) see EEA500
1,2-DIAMINOETHANE see EEA500
1,2-DIAMINOETHANE COPPER COMPLEX see DBU800
1,2-DIAMINO-ETHANO (ITALIAN) see EEA500
3,8-DIAMINO-5-ETHYL-6-PHENYLPHENANTHRIDINIUM
    BROMIDE see DBV400
2,7-DIAMINO-10-ETHYL-9-PHENYLPHENANTHRIDINIUM
    BROMIDE see DBV400
1,6-DIAMINOHEXANE see HEO000
2,6-DIAMINOHEXANOIC ACID HYDROCHLORIDE see
    LJO000
4,6-DIAMINO-2-HYDROXY-1,3-CYCLOHEXANE-3,6′-DI-
    AMINO-3,6′-DIDEOXYDI-α-d-GLUCOSIDE see KAL000

4,6-DIAMINO-2-HYDROXY-1,3-CYCLOHEXYLENE 3,6′-
    DIAMINO-3,6′-DIDEOXYDI-d-GLUCOPYRANOSIDE see
    KAL000
2,4-DIAMINO-1-METHOXYBENZENE see DBO000,
    DBO400
1,3-DIAMINO-4-METHOXYBENZENE SULPHATE see
    DBO400
2,4-DIAMINO-1-METHOXYBENZENE SULPHATE see
    DBO400
3,6-DIAMINO-10-METHYLACRIDINIUM CHLORIDE with
    3,6-ACRIDINEDIAMINE see DBX400
2,8-DIAMINO-10-METHYLACRIDINIUM CHLORIDE mixture
    with 2,8-DIAMINOACRIDINE see DBX400
1,3-DIAMINO-4-METHYLBENZENE see TGL750
2,4-DIAMINO-1-METHYLBENZENE see TGL750
3,7′-DIAMINO-N-METHYLDIPROPYLAMINE see BGU750
DIAMINON see MDO750
1,5-DIAMINONAPHTHALENE see NAM000
DIAMINON HYDROCHLORIDE see MDP000
1,4-DIAMINO-2-NITROBENZENE see ALL750
4,6-DIAMINO-2-(5-NITRO-2-FURYL)-S-TRIAZINE see
    DBY800
1,5-DIAMINOPENTANE see PBK500
DIAMINOPHEN see DHW200
2,6-DIAMINO-3-PHENYLAZOPYRIDINE see PEK250
2,6-DIAMINO-3-PHENYLAZOPYRIDINE HYDROCHLORIDE
    see PDC250
2,6-DIAMINO-3-(PHENYLAZO)PYRIDINE MONOHYDRO-
    CHLORIDE see PDC250
4,4′-DIAMINOPHENYL ETHER see OPM000
2,7-DIAMINO-9-PHENYL-10-ETHYLPHENANTHRIDINIUM
    BROMIDE see DBV400
DI-(4-AMINOPHENYL)METHANE see MJQ000
2,7-DIAMINO-9-PHENYLPHENANTHRIDINE ETHOBRO-
    MIDE see DBV400
DI(p-AMINOPHENYL) SULFIDE see TFI000
DI(4-AMINOPHENYL)SULFONE see SOA500
DI(p-AMINOPHENYL) SULFONE see SOA500
DI(p-AMINOPHENYL)SULPHIDE see TFI000
DI(4-AMINOPHENYL)SULPHONE see SOA500
DI(p-AMINOPHENYL)SULPHONE see SOA500
2,4-DIAMINO-6-PIPERIDINILPIRIMIDINA-3-OSSIDO (ITAL-
    IAN) see DCB000
2,4-DIAMINO-6-PIPERIDINOPYRIMIDINE-3-OXIDE see
    DCB000
1,2-DIAMINOPROPANE see PMK250
l-(+)-N-(p-(((2,4-DIAMINO-6-PTERIDINYL)METHYL)
    METHYLAMINO)BENZOYL)GLUTAMIC ACID see
    MDV500
2,6-DIAMINOPYRIDINE see DCC800
DIAMINOPYRITAMIN see TGD000
2,4-DIAMINOSOLE SULPHATE see DBO400
2,4-DIAMINOTOLUEN (CZECH) see TGL750
DIAMINOTOLUENE see TGL500, TGL750
2,4-DIAMINOTOLUENE see TGL750
2,5-DIAMINOTOLUENE see TGM000
2,4-DIAMINO-1-TOLUENE see TGL750
2,4-DIAMINOTOLUENE DIHYDROCHLORIDE see DCE000
2,5-DIAMINOTOLUENE DIHYDROCHLORIDE see DCE200
2,6-DIAMINOTOLUENE DIHYDROCHLORIDE see DCE400
p-DIAMINOTOLUENE SULFATE see DCE600
2,5-DIAMINOTOLUENE SULFATE see DCE600
2,5-DIAMINOTOLUENE SULPHATE see DCE600
2,4-DIAMINOTOLUOL see TGL750
3,5-DIAMINO-s-TRIAZOLE see DCF200
1,3-DIAMINOUREA see CBS500
trans-DIAMMINEDICHLOROPLATINUM(II) see DEX000
DIAMMINEMALONATO PLATINUM (II) see DCG000
DIAMMONIUM ARSENATE see DCG800
DIAMMONIUM CARBONATE see ANE000
DIAMMONIUM CITRATE see ANF800
DIAMMONIUM HEXACHLOROPALLADATE see ANF000
DIAMMONIUM HEXACHLOROPLATINATE (2-) see
    ANF250

DIAMMONIUM HEXAFLUOROSILICATE see COE000
DIAMMONIUM HYDROGEN ARSENATE see DCG800
DIAMMONIUM HYDROGEN PHOSPHATE see ANR500
DIAMMONIUM MOLYBDATE see ANM750
DIAMMONIUM MONOHYDROGEN ARSENATE see
   DCG800
DIAMMONIUM SULFATE see ANU750
DIAMMONIUM TARTRATE see DCH000
DIAMOND GREEN G see BAY750
DIAMOND SHAMROCK 40 see PKQ059
DIAMOND SHAMROCK 744 see AAX175
DIAMORFINA see HBT500
DIAMORPHINE see HBT500
DIAMOX see AAI250
DIAMPHETAMINE SULFATE see BBK250
DIAMYL AMINE see DCH200
DI-n-AMYLAMINE (DOT) see DCH200
DIAMYLNITROSAMIN (GERMAN) see DCH600
DI-n-AMYLNITROSAMINE see DCH600
DIAN see BLD500
DIANABOL see MPN500
DIANALINEMETHANE see MJQ000
DIANAT (RUSSIAN) see MEL500
DIANATE see MEL500
DIANEMYCIN see DCI400
DIANHYDROCULCITOL see DCI600
1,2:5,6-DIANHYDRODULCITOL see DCI600
1,2:3,4-DIANHYDROERYTHRITOL see DHB800
DIANHYDROGALACTITOL see DCI600
1,2:5,6-DIANHYDROGALACTITOL see DCI600
1,2:3,4-DIANHYDRO-dl-THREITOL see DHB600
DIANILBLAU see CMO250
DIANIL BLUE see CMO250
p,p'-DIANILINE see BBX000
o-DIANISIDIN (CZECH, GERMAN) see DCJ200
o-DIANISIDINA (ITALIAN) see DCJ200
o-DIANISIDINE see DCJ200
O,O'DIANISIDINE see DCJ200
3,3'-DIANISIDINE see DCJ200
o-DIANISIDINE DIHYDROCHLORIDE see DOA800
DIANISIDINE DIISOCYANATE see DCJ400
α,γ-DI-p-ANISYL-β-(ETHOXYPHENYL)GUANIDINE HY-
   DROCHLORIDE see PDN500
3,4-DIANISYL-3-HEXENE see DJB200
DIANISYL-MONOPHENETHYLGUANIDINE HYDROCHLO-
   RIDE see PDN500
DIANISYLTRICHLORETHANE see MEI450
2,2-DI-p-ANISYL-1,1,1-TRICHLOROETHANE see MEI450
DIANON see DCM750
DIANTIMONY TRIOXIDE see AQF000
DIAPADRIN see DGQ875
DIAPAMIDE see CIP500
DIAPHEN (NEUROPLEGIC) see DHW200
DIAPHENYLSULFONE see SOA500
DIAPHENYLSULPHON see SOA500
DIAPHENYLSULPHONE see SOA500
DIAPHORM see HBT500
DIAPHTAMINE BLACK V see AQP000
DIAPHTAMINE BLUE BB see CMO000
DIAPP see BEN000
DIAQUODIAMMINEPLATINUM DINITRATE see DCJ600
cis-DIAQUODIAMMINEPLATINUM(II) DINITRATE see
   DCJ600
DIAREX 43G see SMQ500
DIAREX 600 see SMR000
DIAREX HF 77 see SMQ000
DIARSEN see DXE600
DIARSENIC PENTOXIDE see ARH500
DIARSENIC TRIOXIDE see ARI750
DIARSENIC TRISULFIDE see ARI000
DIASETIELMORFIEN see HBT500
DIASETILMORFIN see HBT500
DIASETYLMORFIIMI see HBT500
DIASONE HYDROCHLORIDE see MDP000

DIASTATIN see NOH500
DIASTYL see DKA600
DIASULFA see SNN300
DIASULFYL see SNN300
DIATER see DXQ500
DIATERR-FOS see DCM750
DIATO BLUE BASE B see DCJ200
DIATOMACEOUS EARTH see DCJ800
DIATOMACEOUS SILICA see DCJ800
DIATOMITE see DCJ800
DIAZACHEL (OBS.) see MDQ250
1,3-DIAZA-2,4-CYCLOPENTADIENE see IAL000
1,6-DIAZA-3,4,8,9,12,13-HEXAOXABICYCLO(4.4.4)
   TETRADECANE see DCK700
1,3-DIAZAINDENE see BCB750
2,3-DIAZAINDOLE see BDH250
DIAZAJET see DCM750
3,6-DIAZAOCTANE-1,8-DIAMINE see TJR000
DIAZATOL see DCM750
DIAZENEDICARBOXAMIDE see ASM270
DIAZETOXYSKIRPENOL (GERMAN) see AOP250
DIAZETYLMORPHINE see HBT500
DIAZIDE see DCM750
1,4-DIAZIDOBENZENE see DCL125, DCL125
p-DIAZIDOBENZENE (DOT) see DCL125
1,2-DIAZIDOETHANE see DCL600
DIAZINE BLACK E see AQP000
DIAZINE BLUE 2B see CMO000
DIAZINE BLUE 3B see CMO250
DIAZINON see DCM750
DIAZINONE see DCM750
DIAZIRINE see DCP800
DIAZITOL see DCM750
DIAZO see DUR800
DI-AZO see PDC250
DIAZOACETATE (ESTER)-l-SERINE see ASA500
l-DIAZOACETATE (ESTER) SERINE see ASA500
DIAZO-ACETIC ACID ESTER with SERINE see ASA500
DIAZOACETIC ACID, ETHYL ESTER see DCN800
DIAZOACETIC ESTER see DCN800
N-DIAZOACETILGLICINA-IDRAZIDE (ITALIAN) see
   DCO800
DIAZOACETYLGLYCINE HYDRAZIDE see DCO800
N-(DIAZOACETYL)GLYCINE HYDRAZINE see DCO800
N-DIAZOACETYL GLYCYLHYDRAZIDE see DCO800
o-DIAZOACETYL-l-SERINE see ASA500
DIAZOAMINOBENZEN (CZECH) see DWO800
DIAZOAMINOBENZENE see DWO800
p-DIAZOAMINOBENZENE see DWO800
DIAZOAMINOBENZOL (GERMAN) see DWO800
DIAZOBENZENE see ASL250
DIAZOCARD CHRYSOIDINE G see PEK000
2-DIAZO-4,6-DINITROBENZENE-1-OXIDE see DUR800
DIAZODINITROPHENOL (DOT) see DUR800
DIAZODINITROPHENOL, containing, by weight, at least 40%
   water (DOT) see DUR800
DIAZODINITROPHENOL, DRY (DOT) see DUR800
DIAZOESSIGSAEURE-AETHYLESTER (GERMAN) see
   DCN800
DIAZO FAST ORANGE R see NEN500
DIAZO FAST RED AL see AIA750
DIAZO FAST RED GG see NEO500
DIAZO FAST RED TR see CLK235
DIAZO FAST RED TRA see CLK220, CLK235
DIAZO FAST SCARLET G see NMP500
DIAZOIMIDE see HHG500
DIAZOL see DCM750
DIAZOL BLACK 2V see AQP000
DIAZOL BLUE 2B see CMO000
1,3-DIAZOLE see IAL000
DIAZOMETHANE see DCP800
5-DIAZOPYRIMIDINE-2,4(3H)-DIONE see DCQ600
5-DIAZO-2,4(1H,3H)-PYRIMIDINEDIONE see DCQ600
DIAZOTIZING SALTS see SIQ500

DIAZOURACIL see DCQ600
5-DIAZOURACIL see DCQ600
DIAZYL see SNJ000
DIBA see DNH125
DIBAM see SGM500
DIBASIC AMMONIUM ARSENATE see DCG800
DIBASIC AMMONIUM PHOSPHATE see ANR500
DIBASIC LEAD ACETATE see LCG000
DIBASIC LEAD ARSENATE see LCK000
DIBASIC LEAD CARBONATE see LCP000
DIBASIC SODIUM PHOSPHATE see SJH090
DIBASIC ZINC STEARATE see ZMS000
DIBEKACIN see DCQ800
DIBENAMINE see DCR200
DIBENAMINE HYDROCHLORIDE see DCR200
DIBENYLIN see PDT250
DIBENYLINE see PDT250
DIBENZ(a,j)ACEANTHRYLENE see DCR400
DIBENZACEPIN see DCS200
DIBENZ(a,d)ACRIDINE see DCS400
DIBENZ(a,f)ACRIDINE see DCS600
DIBENZ(a,h)ACRIDINE see DCS400
DIBENZ(a,j)ACRIDINE see DCS600
DIBENZ(c,h)ACRIDINE see DCS800
1,2,5,6-DIBENZACRIDINE see DCS400
1,2,7,8-DIBENZACRIDINE see DCS600
3,4,5,6-DIBENZACRIDINE see DCS600
3,4:5,6-DIBENZACRIDINE see DCS800
1,2,7,8-DIBENZACRIDINE (FRENCH) see DCS800
1,2,5,6-DIBENZANTHRACEEN (DUTCH) see DCT400
DIBENZ(a,c)ANTHRACENE see BDH750
DIBENZ(a,h)ANTHRACENE see DCT400
DIBENZ(a,j)ANTHRACENE see DCT600
1,2:3,4-DIBENZANTHRACENE see BDH750
1,2:5,6-DIBENZANTHRACENE see DCT400
1,2:7,8-DIBENZANTHRACENE see DCT600
1,2:5,6-DIBENZ(a)ANTHRACENE see DCT400
5H-DIBENZ(b,f)AZEPINE-5-CARBOXAMIDE see DCV200
4-(3-(5H-DIBENZ(b,f)AZEPIN-5-YL)PROPYL)-1-PIPER-
    AZINEETHANOL DIHYDROCHLORIDE see IDF000
3,4,5,6-DIBENZCARBAZOL see DCY000
3,4,5,6-DIBENZCARBAZOLE see DCY000
DIBENZEPIN see DCW600
DIBENZEPINE see DCW600
DIBENZEPINE HYDROCHLORIDE see DCW800
DIBENZEPIN HYDROCHLORIDE see DCW800
DIBENZO(a,j)ACRIDINE see DCS600
1,2,5,6-DIBENZOACRIDINE see DCS400
DIBENZO(a,c)ANTHRACENE see BDH750
DIBENZO(a,h)ANTHRACENE see DCT400
1,2:3,4-DIBENZOANTHRACENE see BDH750
1,2:5,6-DIBENZOANTHRACENE see DCT400
DIBENZOAZEPINE see DCS200
3,4,5,6-DIBENZOCARBAZOLE see DCY000
7H-DIBENZO(c,g)CARBAZOLE see DCY000
DIBENZO(b,def)CHRYSENE see DCY200
DIBENZO(def,p)CHRYSENE see DCY400
DIBENZO-(drf,mno)CHRYSENE see APE750
DIBENZODIOXIN see DDA800
DIBENZO-p-DIOXIN see DDA800
DIBENZO(1,4)DIOXIN see DDA800
DIBENZO(e)(1,4)DIOXIN see DDA800
DIBENZO(a,jk)FLUORENE see BCJ500
DIBENZO(b,jk)FLUORENE see BCJ750
2-DIBENZOFURANAMINE see DDB600
1,2,5,6-DIBENZONAPHTHALENE see CML810
DIBENZOPARADIAZINE see PDB500
DIBENZOPARATHIAZINE see PDP250
1,2:6,7-DIBENZOPHENANTHRENE see BCG500
2,3:7,8-DIBENZOPHENANTHRENE see BCG500
DIBENZO-2,3,7,8-PHENANTHRENE see BCG500
DIBENZO PQD see DVR200
DIBENZOPYRAZINE see PDB500
DIBENZO(a,d)PYRENE see DCY400

DIBENZO(a,e)PYRENE see NAT500
DIBENZO(a,h)PYRENE see DCY200
DIBENZO(a,i)PYRENE see BCQ500
DIBENZO(a,l)PYRENE see DCY400
DIBENZO(b,h)PYRENE see BCQ500
1,2:3,4-DIBENZOPYRENE see DCY400
1,2,4,5-DIBENZOPYRENE see NAT500
1,2,6,7-DIBENZOPYRENE see DCY200
1,2,7,8-DIBENZOPYRENE see BCQ500
2,3:4,5-DIBENZOPYRENE see DCY400
3,4,8,9-DIBENZOPYRENE see DCY200
DIBENZO(cd,mk)PYRENE see APE750
1,2,9,10-DIBENZOPYRENE see DCY400
3,4:9,10-DIBENZOPYRENE see BCQ500
DIBENZO(b,e)PYRIDINE see ADJ500
DIBENZOPYRROLE see CBN000
DIBENZO(b,d)PYRROLE see CBN000
DIBENZOSUBERONE OXIME see DDD000
DIBENZOTHIAZEPINE see CIS750
DIBENZOTHIAZINE see PDP250
DIBENZO-1,4-THIAZINE see PDP250
DI-2-BENZOTHIAZOLYLDISULFIDE see BDE750
DIBENZOTHIAZYL DISULFIDE see BDE750
2,2'-DIBENZOTHIAZYLDISULFIDE see BDE750
DIBENZ(b,f)(1,4)OXAZEPINE see DDE200
DIBENZ(b,e)OXEPIN-Δ$^{11(6H)}$,γ-PROPYLAMINE see DBA600
DIBENZOYLPEROXID (GERMAN) see BDS000
DIBENZOYL PEROXIDE (MAK) see BDS000
DIBENZOYLPEROXYDE (DUTCH) see BDS000
DIBENZOYLTHIAZYL DISULFIDE see BDE750
DIBENZ(a,i)PYRENE see BCQ500
1,2,3,4-DIBENZPYRENE see DCY400
1,2:7,8-DIBENZPYRENE see BCQ500
3,4,8,9-DIBENZPYRENE see DCY200
4,5,6,7-DIBENZPYRENE see DCY400
3,4:9,10-DIBENZPYRENE see BCQ500
DIBENZTHIAZYL DISULFIDE see BDE750
DIBENZYL see BFX500
N,N-DIBENZYLAMINOETHYL CHLORIDE HYDROCHLO-
    RIDE see DCR200
DIBENZYLCHLORETHAMINE HYDROCHLORIDE see
    DCR200
DIBENZYLCHLORETHYLAMINE HYDROCHLORIDE see
    DCR200
N,N-DIBENZYL-2-CHLOROETHYLAMINE HYDROCHLO-
    RIDE see DCR200
N,N-DIBENZYL-β-CHLOROETHYLAMINE HYDROCHLO-
    RIDE see DCR200
DIBENZYLENE see DDG800
DIBENZYLETHER (CZECH) see BEO250
DIBENZYLIN see DDG800
DIBENZYLINE see PDT250
DIBENZYLINE HYDROCHLORIDE see DDG800
DIBENZYLMERCURY see DDH000
DIBENZYRAN see DDG800
DIBESTIL see DKB000
DIBESTROL see DKA600
DIBETOS see BQL000
DIBONDRIN see BBV500
DIBORANE see DDI450
DIBORANE(6) see DDI450
DIBORON HEXAHYDRIDE see DDI450
DIBOVAN see DAD200
DIBP see DNJ400
DIBROLUUR see BNP750
DIBROM see NAG400
1,2-DIBROMAETHAN (GERMAN) see EIY500
DIBROMANNIT see DDP600
DIBROMANNITOL see DDP600
d-DIBROMANNITOL see DDP600
DIBROMCHLORPROPAN (GERMAN) see DDL800
1,2-DIBROM-3-CHLOR-PROPAN (GERMAN) see DDL800
O-(1,2-DIBROM-2,2-DICHLORAETHYL)-O,O-DIMETHYL-
    PHOSPHAT (GERMAN) see NAG400

DIBROMDULCITOL see DDJ000
DIBROMOACETYLENE see DDJ800
2,2'-DIBROMOBIACETYL see DDK600
α,α'-DIBROMOBIACETYL see DDK600
1,4-DIBROMO-2-BUTENE see DDL400
DIBROMOCHLOROMETHANE see CFK500
DIBROMOCHLOROPROPANE see DDL800
1,2-DIBROMO-3-CHLOROPROPANE see DDL800
1,2-DIBROMO-3-CLORO-PROPANO (ITALIAN) see DDL800
DIBROMOCYANOACETAMIDE see DDM000
α,α-DIBROMO-α-CYANOACETAMIDE see DDM000
2,6-DIBROMO-4-CYANOPHENOL see DDP000
DIBROMODIBUTYLSTANNANE see DDM400
DIBROMODIBUTYLTIN see DDM400
1,2-DIBROMO-2,2-DICHLOROETHYL DIMETHYL PHOS-
  PHATE see NAG400
O-(1,2-DIBROMO-2,2-DICLORO-ETIL)-O,O-DIMETIL-FOS-
  TATO (ITALIAN) see NAG400
1,6-DIBROMODIDEOXYDULCITOL see DDJ000
1,6-DIBROMO-1,6-DIDEOXYDULCITOL see DDJ000
1,6-DIBROMO-1,6-DIDEOXYGALACTITOL see DDJ000
1,6-DIBROMO-1,6-DIDEOXY-d-GALACTITOL see DDJ000
1,6-DIBROMO-1,6-DIDEOXY-d-MANNITOL see DDP600
1,6-DIBROMO-1,6-d-DIDESOXYMANNITOL see DDP600
DIBROMODIFLUOROMETHANE see DKG850
DIBROMODIMETHYL STANNANE see DUG800
DIBROMODULCITOL see DDJ000
1,6-DIBROMODULCITOL see DDJ000
1,2-DIBROMOETANO (ITALIAN) see EIY500
α,β-DIBROMOETHANE see EIY500
sym-DIBROMOETHANE see EIY500
1,2-DIBROMOETHANE (MAK) see EIY500
3,5-DIBROMO-4-HYDROXYBENZONITRILE see DDP000
2,7-DIBROMO-4-HYDROXYMERCURIFLUORESCEINE DI-
  SODIUM SALT see MCV000
3,5-DIBROMO-4-HYDROXYPHENYLCYANIDE see DDP000
1,6-DIBROMOMANNITOL see DDP600
DIBROMOMETHANE see DDP800
2,2-DIBROMO-3-NITRILOPROPIONAMIDE see DDM000
3,4-DIBROMONITROSOPIPERIDINE see DDQ800
DIBROMOPHENYLARSINE see DDR200
DIBROMOPROPANAL see DDS400
1,3-DIBROMOPROPANE polymer with N,N,N',N'-TETRA-
  METHYL-1,6-HEXANEDIAMINE see HCV500
2,3-DIBROMO-1-PROPANOL PHOSPHATE see TNC500
2,3-DIBROMO-1-PROPANOL, PHOSPHATE (3:1) see
  TNC500
2,3-DIBROMOPROPIONALDEHYDE see DDS400
(2,3-DIBROMOPROPYL) PHOSPHATE see TNC500
DIBROMURE d'ETHYLENE (FRENCH) see EIY500
1,2-DIBROOM-3-CHLOORPROPAAN (DUTCH) see DDL800
O-(1,2-DIBROOM-2,2-DICHLOOR-ETHYL)-O,O-DIMETHYL-
  FOSFAAT (DUTCH) see NAG400
1,2-DIBROOMETHAAN (DUTCH) see EIY500
DIBUCAIN see NOF500
DIBUCAINE see DDT200
DIBUCAINE HYDROCHLORIDE see NOF500
DIBULINESULFAT see DDW000
DIBULINE SULFATE see DDW000
DIBUTIL see DIR000
DIBUTIN see ILD000
DIBUTOLINE see DDW000
DIBUTOLINE SULFATE see DDW000
n-DIBUTYLAMINE see DDT800
DI-n-BUTYLAMINE see DDT800
DI(n-BUTYL)AMINE (DOT) see DDT800
DIBUTYLAMINE, 4-HYDROXY-N-NITROSO- see HJQ350
1-(((DIBUTYLAMINO)CARBONYL)OXY)-N-ETHYL-N,N-DI-
  METHYLETHANAMINIUM SULFATE (2:1) see DDW000
DIBUTYLAMINOETHANOL see DDU600
2-DIBUTYLAMINOETHANOL see DDU600
2-N-DIBUTYLAMINOETHANOL see DDU600
2-DI-n-BUTYLAMINOETHANOL see DDU600
N,N-DI-n-BUTYLAMINOETHANOL (DOT) see DDU600

β-N-DIBUTYLAMINOETHYL ALCOHOL see DDU600
3-(DIBUTYLAMINO)-1-PROPANOL-p-AMINOBENZOATE see
  BOO750
3-(DIBUTYLAMINO)-1-PROPANOL-p-AMINOBENZOATE
  (ESTER) SULFATE (2:1) see BOP000
3-DIBUTYLAMINO-1-PROPANOL-4-AMINOBENZOATE (ES-
  TER) SULFATE (SALT) (2:1) see BOP000
3-DIBUTYLAMINOPROPYL-p-AMINOBENZOATE see
  BOO750
DIBUTYLAMINOPROPYL-p-AMINOBENZOATE SULFATE
  see BOP000
3'-DIBUTYLAMINOPROPYL-4-AMINOBENZOATE SULFATE
  see BOP000
DIBUTYLATED HYDROXYTOLUENE see BFW750
DIBUTYL-1,2-BENZENEDICARBOXYLATE see DEH200
DIBUTYLBIS((2-ETHYLHEXANOYL)OXY)-STANNANE see
  BJQ250
DIBUTYLBIS((2-ETHYL-1-OXOHEXYL)OXY)-STANNANE
  (9CI) see BJQ250
DIBUTYLBIS(LAUROYLOXY)STANNANE see DDV600
DIBUTYLBIS(LAUROYLOXY)TIN see DDV600
DIBUTYLBIS(OCTANOYLOXY)STANNANE see BLB250
DIBUTYLBIS((1-OXOOCTYL)OXY)STANNANE see BLB250
DIBUTYL BUTANEPHOSPHONATE see DDV800
O,O-DIBUTYL-1-BUTYLAMINO-CYCLOHEXYLPHOSPHO-
  NATE see ALZ000
DIBUTYL BUTYLPHOSPHONATE see DDV800
DI-n-BUTYL-CARBAMYLCHOLINE SULPHATE see
  DDW000
(2-DIBUTYLCARBAMYLOXYETHYL)-DIMETHYLETHYL-
  AMMONIUM SULFATE see DDW000
DIBUTYL-o-(o-CARBOXYBENZOYL) GLYCOLATE see
  BQP750
DIBUTYL-o-CARBOXYBENZOYLOXYACETATE see
  BQP750
2,6-DI-tert-BUTYL-p-CRESOL (OSHA, ACGIH) see BFW750
N,N'-DIBUTYL-N,N'-DICARBOXYETHYLENE DIAMINE-
  MORPHOLIDE see DUO400
N,N'-DIBUTYL-N,N'-DICARBOXYMORPHOLIDE-ETHYL-
  ENEDIAMINE see DUO400
DIBUTYLDICHLOROGERMANE see DDY000
DIBUTYLDICHLOROSTANNANE see DDY200
DIBUTYLDICHLOROTIN see DDY200
DIBUTYL(DIFORMYLOXY)STANNANE see DDZ000
DIBUTYLDIPENTANOYLOXYSTANNANE see DEA600
DIBUTYLDITHIOCARBAMIC ACID, NICKEL SALT see
  BIW750
DIBUTYLDITHIOCARBAMIC ACID SODIUM SALT see
  SGF500
DIBUTYLDITHIO-CARBAMIC ACID ZINC COMPLEX see
  BIX000
DIBUTYLDITHIOCARBAMIC ACID ZINC SALT see BIX000
N,N-DI-sec-BUTYL DITHIOOXAMIDE see DEB800
DI-sec-BUTYL ESTER PHOSPHOROFLUORIDIC ACID see
  DEC200
DIBUTYL ESTER SULFURIC ACID see DEC000
N,N-DIBUTYLETHANOLAMINE see DDU600
DI-n-BUTYL ETHER (DOT) see BRH750
N,N'-DI-n-BUTYLETHYLENEDIAMINE-N,N'-DICARBOXY-
  BISMORPHOLIDE see DUO400
2,6-DI-sec-BUTYLFENOL (CZECH) see DEF800
DI-sec-BUTYLFLUOROPHOSPHATE see DEC200
DI-sec-BUTYL FLUOROPHOSPHONATE see DEC200
DIBUTYL FUMARATE see DEC600
DI-n-BUTYLGERMANEDICHLORIDE see DDY000
N,N-DIBUTYL-N-(2-HYDROXYETHYL)AMINE see DDU600
2,6-DI-tert-BUTYL-1-HYDROXY-4-METHYLBENZENE see
  BFW750
3,5-DI-tert-BUTYL-4-HYDROXYTOLUENE see BFW750
2,6-DI-terc. BUTYL-p-KRESOL (CZECH) see BFW750
DIBUTYL LEAD DIACETATE see DED400
DIBUTYL MALEATE see DED600
DIBUTYLMERCURY see DEE000
DI-sec-BUTYLMERCURY see DEE200

2,6-DI-tert-BUTYL-4-METHYLPHENOL see BFW750
2,6-DI-tert-BUTYL-p-METHYLPHENOL see BFW750
DI-n-BUTYLNITROSAMIN (GERMAN) see BRY500
DI-n-BUTYLNITROSAMINE see BRY500
N,N-DI-n-BUTYLNITROSAMINE see BRY500
DIBUTYLNITROSOAMINE see BRY500
N,N-DIBUTYLNITROSOAMINE see BRY500
DIBUTYL OXIDE see BRH750
DIBUTYLOXIDE of TIN see DEF400
DIBUTYLOXOSTANNANE see DEF400
DIBUTYLOXOTIN see DEF400
DI-tert-BUTYLPEROXID (GERMAN) see BSC750
DI-tert-BUTYL PEROXIDE (MAK) see BSC750
DI-tert-BUTYL PEROXYDE (DUTCH) see BSC750
DI-sec-BUTYL PEROXYDICARBONATE see BSD000
DI-sec-BUTYL PEROXYDICARBONATE, not more than 52%
   in solution (DOT) see BSD000
DI-sec-BUTYL PEROXYDICARBONATE, technically pure
   (DOT) see BSD000
2,6-DI-sec-BUTYLPHENOL see DEF800
N,N'-DI-sec-BUTYL-p-PHENYLENEDIAMINE see DEG200
3,5-DI-tert-BUTYLPHENYLMETHYLCARBAMATE see
   DEG400
DIBUTYL PHENYL PHOSPHATE see DEG600
DIBUTYL PHOSPHATE see DEG700
DIBUTYL-PHOSPHINIC ACID, 4-NITROPHENYL ESTER see
   NIM000
DIBUTYL PHTHALATE see DEH200
DI-n-BUTYL PHTHALATE see DEH200
DIBUTYL SEBACATE see DEH600
DI-n-BUTYL SEBACATE see DEH600
DIBUTYLSTANNANE OXIDE see DEF400
DI-n-BUTYLSULFAT (GERMAN) see DEC000
DIBUTYL SULFATE see DEC000
DIBUTYLTIN BIS(2-ETHYLHEXANOATE) see BJQ250
DIBUTYLTIN BIS(α-ETHYLHEXANOATE) see BJQ250
DIBUTYLTIN CHLORIDE see DDY200
DIBUTYL TIN DIACETATE see DBF800
DIBUTYL TIN DIBROMIDE see DDM400
DIBUTYLTIN DICAPRYLATE see BLB250
DIBUTYLTIN DICHLORIDE see DDY200
DI-n-BUTYLTIN DICHLORIDE see DDY200
DI-n-BUTYLTIN DI(DODECANOATE) see DDV600
DIBUTYLTIN DI(2-ETHYLHEXANOATE) see BJQ250
DI-n-BUTYLTIN DI-2-ETHYLHEXANOATE see BJQ250
DIBUTYLTIN DI(2-ETHYLHEXOATE) see BJQ250
DI-n-BUTYLTIN DIFORMATE see DDZ000
DIBUTYLTIN DILAURATE (USDA) see DDV600
DI-N-BUTYLTIN DI(MONOBUTYL)MALEATE see BHK250
DIBUTYLTIN DIOCTANOATE see BLB250
DIBUTYLTIN DIOCTATE see BLB250
DI-n-BUTYLTIN DIPENTANOATE see DEA600
DIBUTYLTIN LAURATE see DDV600
DIBUTYLTIN OCTANOATE see BLB250
DIBUTYLTIN OXIDE see DEF400
DI-n-BUTYLTIN OXIDE see DEF400
DI-n-BUTYL-ZINN-DICHLORID (GERMAN) see DDY200
DIBUTYL-ZINN-DILAURAT (GERMAN) see DDV600
DI-N-BUTYL-ZINN-DI(MONOBUTYL)MALEINAT (GER-
   MAN) see BHK250
DI-n-BUTYL-ZINN-OXYD (GERMAN) see DEF400
DIC see DAB600
DIC 1468 see MQR275
DICACODYL SULFIDE see CAC250
DICAIN see BQA010
DICAINE see BQA010
DICALCIUM PHOSPHATE see CAW100
DICALITE see SCH000
DICAMBA (DOT) see MEL500
DICANDIOL see MQU750
DICAPTOL see BAD750
2,2-DI(CARBAMOYLOXYMETHYL)PENTANE see MQU750
S-(1,2-DICARBETHOXYETHYL)-O,O-DIMETHYLDITHIO-
   PHOSPHATE see MAK700

DICARBETHOXYMETHANE see EMA500
DICARBOETHOXYETHYL-O,O-DIMETHYL PHOSPHORO-
   DITHIOATE see MAK700
DICARBOMETHOXYZINC see ZBS000
DICARBONIC ACID DIETHYL ESTER see DIZ100
2,3-DICARBONITRILO-1,4-DIATHIAANTHRACHINON
   (GERMAN) see DLK200
DI-mu-CARBONYLHEXACARBONYLDICOBALT see
   CNB500
o-DICARBOXYBENZENE see PHW250
3,3'-DICARBOXYBENZIDINE see BFX250
((1,2-DICARBOXYETHYL)THIO)GOLD DISODIUM SALT see
   GJC000
DICAROCIDE see DIW200
DICARZOL see DSO200
DICATRON see PDN000
DICESIUM CARBONATE see CDC750
DICESIUM DICHLORIDE see CDD000
DICESIUM DIFLUORIDE see CDD500
DICESTAL see MJM500
DICHAPETULUM CYMOSUM (HOOK) ENGL see PLG000
DICHLOFENTHION see DFK600
DICHLOFENTION see DFK600
DICHLONE (DOT) see DFT000
p-DICHLOORBENZEEN (DUTCH) see DEP800
1,4-DICHLOORBENZEEN (DUTCH) see DEP800
1,1-DICHLOOR-2,2-BIS(4-CHLOOR FENYL)-ETHAAN
   (DUTCH) see BIM500
1,1-DICHLOORETHAAN (DUTCH) see DFF809
1,2-DICHLOORETHAAN (DUTCH) see EIY600
2,2'-DICHLOORETHYLETHER (DUTCH) see DFJ050
DICHLOORFEEN (DUTCH) see MJM500
(2,4-DICHLOOR-FENOXY)-AZIJNZUUR (DUTCH) see
   DAA800
2-(2,4-DICHLOOR-FENOXY)-PROPIONZUUR (DUTCH) see
   DGB000
(3,4-DICHLOOR-FENYL-AZO)-THIOUREUM (DUTCH) see
   DEQ000
3-(3,4-DICHLOOR-FENYL)-1,1-DIMETHYLUREUM (DUTCH)
   see DXQ500
3-(3,4-DICHLOOR-FENYL)-1-METHOXY-1-METHYLUREUM
   (DUTCH) see DGD600
3,6-DICHLOOR-2-METHOXY-BENZOEIZUUR (DUTCH) see
   MEL500
1,1-DICHLOOR-1-NITROETHAAN (DUTCH) see DFU000
(2,2-DICHLOOR-VINYL)-DIMETHYL-FOSFAAT (DUTCH)
   see DGP900
DICHLOORVO (DUTCH) see DGP900
DICHLORACETIC ACID see DEL000
DICHLORACETYL CHLORIDE see DEN400
1,1-DICHLORAETHAN (GERMAN) see DFF809
1,2-DICHLOR-AETHAN (GERMAN) see EIY600
1,2-DICHLOR-AETHEN (GERMAN) see DFI100
p-DI-(2-CHLORAETHYL)-AMINO-dl-PHENYL-ALANIN
   (GERMAN) see BHT750
S-(2,3-DICHLOR-ALLYL)-N,N-DIISOPROPYL-MONOTHIO-
   CARBAMAAT (DUTCH) see DBI200
2,3-DICHLORALLYL-N,N-(DIISOPROPYL)-THIOCARBAMAT
   (GERMAN) see DBI200
DICHLOR AMINE see BIE250
DICHLORANTIN see DFE200
o-DICHLORBENZENE see DEP600
3,3'-DICHLORBENZIDIN (CZECH) see DEQ600
4,4'-DICHLORBENZILSAEUREAETHYLESTER (GERMAN)
   see DER000
o-DICHLOR BENZOL see DEP600
p-DICHLORBENZOL (GERMAN) see DEP800
1,4-DICHLOR-BENZOL (GERMAN) see DEP800
1,1-DICHLOR-2,2-BIS(4-CHLOR-PHENYL)-AETHAN (GER-
   MAN) see BIM500
2,2'-DICHLOR-DIAETHYLAETHER (GERMAN) see DFJ050
3,3'-DICHLOR-4,4'-DIAMINO-DIPHENYLAETHER (GER-
   MAN) see BGT000
3,3'-DICHLOR-4,4'-DIAMINODIPHENYLMETHAN (GER-
   MAN) see MJM200

DICHLORDIMETHYLAETHER (GERMAN) see BIK000
DICHLOREMULSION see EIY600
DICHLOREN see BIE500
DICHLOREN (GERMAN) see BIE250
DICHLOREN HYDROCHLORIDE see BIE500
DICHLORETHANOIC ACID see DEL000
2,2'-DICHLORETHYL ETHER see DFJ050
β,β-DICHLOR-ETHYL-SULPHIDE see BIH250
DICHLORFENIDIM see DXQ500
2,6-DICHLORFENOL (CZECH) see DFY000
DICHLORFOS (POLISH) see DGP900
DI-CHLORICIDE see DEP800
DICHLORICIDE MOTHPROOFER see TBC500
DICHLORID KYSELINY FUMAROVE (CZECH) see FOY000
DICHLORMETHAZANONE see DEM000
3,6-DICHLOR-3-METHOXY-BENZOESAEURE (GERMAN)
    see MEL500
DICHLORMEZANONE see DEM000
DI-CHLOR-MULSION see EIY600
2,3-DICHLOR-1,4-NAPHTHOCHINON (GERMAN) see
    DFT000
1,1-DICHLOR-1-NITROAETHAN (GERMAN) see DFU000
2,4-DICHLOR-6-NITROFENOL (CZECH) see DFU600
2',5-DICHLOR-4'-NITRO-SALIZYLSAEUREANILID (GER-
    MAN) see DFV400
d-(−)-threo-2-DICHLOROACETAMIDO-1-p-NITROPHENYL-
    1,3-PROPANEDIOL see CDP250
DICHLOROACETATE SODIUM SALT see SGG000
2,2-DICHLOROACETIC ACID see DEL000
DICHLOROACETIC ACID METHYL ESTER see DEM800
DICHLOROACETIC ACID SODIUM SALT see SGG000
1,3-DICHLOROACETONE see BIK250
α,γ-DICHLOROACETONE see BIK250
sym-DICHLOROACETONE see BIK250
α,α'-DICHLOROACETONE see BIK250
1,3-DICHLOROACETONE (DOT) see BIK250
DICHLOROACETYL CHLORIDE see DEN400
2,2-DICHLOROACETYL CHLORIDE see DEN400
α,α-DICHLOROACETYL CHLORIDE see DEN400
DICHLOROACETYL CHLORIDE (DOT) see DEN400
DICHLOROACETYLENE see DEN600
d-threo-N-DICHLOROACETYL-1-p-NITROPHENYL-2-AMINO-
    1,3-PROPANEDIOL see CDP250
S-(2,3-DICHLORO-ALLIL)-N,N-DIISOPROPIL-MONOTIO-
    CARBAMMATO (ITALIAN) see DBI200
DICHLOROALLYL DIISOPROPYLTHIOCARBAMATE see
    DBI200
S-2,3-DICHLOROALLYL DIISOPROPYLTHIOCARBAMATE
    see DBI200
2,3-DICHLOROALLYL-N,N-DIISOPROPYLTHIOLCARBA-
    MATE see DBI200
DICHLOROAMETHOPTERIN see DFO000
3',5'-DICHLOROAMETHOPTERIN see DFO000
3',5'-DICHLORO-4-AMINO-4-DEOXY-N₁₀-METHYLPTERO-
    GLUTAMIC ACID see DFO000
3,6-DICHLORO-o-ANISIC ACID see MEL500
DICHLOROBENZALKONIUM CHLORIDE see AFP750
m-DICHLOROBENZENE see DEP699
o-DICHLOROBENZENE see DEP600
p-DICHLOROBENZENE see DEP800
1,3-DICHLOROBENZENE see DEP699
2,6-DICHLOROBENZENECARBOTHIOAMIDE see DGM600
1-(3',4'-DICHLOROBENZENEDIAZOL)-2-THIOUREA see
    DEQ000
3,4-DICHLOROBENZENE DIAZOTHIOCARBAMID see
    DEQ000
3,4-DICHLOROBENZENE DIAZOTHIOUREA see DEQ000
1,2-DICHLOROBENZENE (MAK) see DEP600
1,4-DICHLOROBENZENE (MAK) see DEP800
DICHLOROBENZENE, ORTHO, liquid (DOT) see DEP600
DICHLOROBENZENE, PARA, solid (DOT) see DEP800
3,3'-DICHLOROBENZIDENE see DEQ600
3,3'-DICHLOROBENZIDINA (SPANISH) see DEQ600
DICHLOROBENZIDINE see DEQ600

o,o'-DICHLOROBENZIDINE see DEQ600
3',3'-DICHLOROBENZIDINE see DEQ600
DICHLOROBENZIDINE BASE see DEQ600
3,3'-DICHLOROBENZIDINE DIHYDROCHLORIDE see
    DEQ800
4,4'-DICHLOROBENZILATE see DER000
4,4'-DICHLOROBENZILIC ACID ETHYL ESTER see
    DER000
3,4-DICHLOROBENZOIC ACID see DER600
p-DICHLOROBENZOL see DEP800
DI-(4-CHLOROBENZOYL) PEROXIDE see BHM750
p,p'-DICHLOROBENZOYL PEROXIDE see BHM750
3,3'-DICHLORO-4,4'-BIPHENYLDIAMINE see DEQ600
3,3'-DICHLOROBIPHENYL-4,4'-DIAMINE see DEQ600
3,3'-DICHLORO-(1,1'-BIPHENYL)-4,4'-DIAMINE DIHYDRO-
    CHLORIDE see DEQ800
1,1-DICHLORO-2,2-BIS(p-CHLOROPHENYL)ETHANE see
    BIM500
1,1-DICHLORO-2,2-BIS(p-CHLOROPHENYL)ETHANE (DOT)
    see BIM500
1,1-DICHLORO-2,2-BIS(4-CHLOROPHENYL)-ETHANE
    (FRENCH) see BIM500
1,1-DICHLORO-2,2-BIS(p-CHLOROPHENYL)ETHYLENE see
    BIM750
cis-DICHLOROBIS(CYCLOBUTYLAMMINE)PLATINUM(II)
    see DGT200
DICHLOROBIS(eta-CYCLOPENTADIENYL)HAFNIUM see
    HAE500
cis-DICHLOROBIS(CYCLOPENTYLAMMINE)PLATINUM(II)
    see BIS250
1,1-DICHLORO-2,2-BIS(2,4'-DICHLOROPHENYL)ETHANE
    see CDN000
1,1-DICHLORO-2,2-BIS(4-ETHYLPHENYL)ETHANE see
    DJC000
1,1-DICHLORO-2,2-BIS(p-ETHYLPHENYL)ETHANE see
    DJC000
2,2-DICHLORO-1,1-BIS(p-ETHYLPHENYL)ETHANE see
    DJC000
α,α-DICHLORO-2,2-BIS(p-ETHYLPHENYL)ETHANE see
    DJC000
1,1-DICHLORO-2,2-BIS(PARACHLOROPHENYL)ETHANE
    (DOT) see BIM500
cis-DICHLOROBIS(PYRROLIDINE)PLATINUM(II) see
    DEU200
DICHLOROBROMOMETHANE see BND500
O-(2,5-DICHLORO-4-BROMOPHENYL)-O-METHYL PHE-
    NYLTHIOPHOSPHONATE see LEN000
1,4-DICHLORO-2-BUTENE see DEV000
1,4-DICHLOROBUTENE-2 (MAK) see DEV000
1,4-DICHLOROBUTYNE see DEV400
1,4-DICHLORO-2-BUTYNE see DEV400
DICHLOROCHLORDENE see CDR750
1,1-DICHLORO-2-CHLOROETHYLENE see TIO750
2,4-DICHLORO-α-(CHLOROMETHYLENE)BENZYL ALCO-
    HOL DIETHYL PHOSPHATE see CDS750
1,1-DICHLORO-2-(o-CHLOROPHENYL)-2-(p-CHLOROPHE-
    NYL)ETHANE see CDN000
DICHLORO(2-CHLOROVINYL)ARSINE see CLV000
DICHLORO(2-CHLOROVINYL)ARSINE OXIDE see
    DEW000
DICHLOROCTAN SODNY (CZECH) see SGG000
3,4'-DICHLOROCYCLOPROPANECARBOXANILIDE see
    CQJ250
2,6-DICHLORO-N-CYCLOPROPYL-N-ETHYL ISONICOTI-
    NAMIDE see DEW400
3,3'-DICHLORO-4,4'-DIAMINOBIPHENYL see DEQ600
3,3'-DICHLORO-4,4'-DIAMINO(1,1-BIPHENYL) see
    DEQ600
3,3'-DICHLORO-4,4'-DIAMINODIPHENYL ETHER see
    BGT000
3,3'-DICHLORO-4,4'-DIAMINODIPHENYLMETHANE see
    MJM200
N-(3,5-DICHLORO-4-((2,4-DIAMINO-6-PTERIDINYL
    METHYL)METHYLAMINO)BENZOYL)GLUTAMIC ACID
    see DFO000

DI-(2-CHLOROETHYL)-3-CHLORO-4-METHYLCOUMARIN-7-YL PHOSPHATE see DFH600

O,O-DI(2-CHLOROETHYL)-7-(3-CHLORO-4-METHYLCOUMARINYL)PHOSPHATE see CIK750

O,O-DI(2-CHLOROETHYL)-O-(3-CHLORO-4-METHYLCOUMARIN-7-YL) PHOSPHATE see DFH600

DICHLOROETHYLENE see DFH800, EIY600

1,1-DICHLOROETHYLENE see VPK000

1,2-DICHLOROETHYLENE see DFI100, DFI200

cis-DICHLOROETHYLENE see DFI200

sym-DICHLOROETHYLENE see DFI100

trans-DICHLOROETHYLENE see ACK000

DICHLORO-1,2-ETHYLENE (FRENCH) see DFI100

DICHLORO(ETHYLENEDIAMMINE)PLATINUM(II) see DFJ000

trans-1,2-DICHLOROETHYLENE (MAK) see ACK000

DI(2-CHLOROETHYL) ESTER, MALEIC ACID see DFJ200

DICHLOROETHYL ETHER see DFJ050

DI(β-CHLOROETHYL)ETHER see DFJ050

sym-DICHLOROETHYL ETHER see DFJ050

β,β'-DICHLOROETHYL ETHER see DFJ050

2,2'-DICHLOROETHYL ETHER (MAK) see DFJ050

DICHLOROETHYL FORMAL see BID750

DI-2-CHLOROETHYL FORMAL see BID750

DI-2-CHLOROETHYL MALEATE see DFJ200

2,3-DICHLORO-N-ETHYLMALEINIMIDE see DFJ400

DI(2-CHLOROETHYL)METHYLAMINE see BIE250

DI(2-CHLOROETHYL)METHYLAMINE HYDROCHLORIDE see BIE500

DICHLOROETHYL-β-NAPHTHYLAMINE see BIF250

DI(2-CHLOROETHYL)-β-NAPHTHYLAMINE see BIF250

2-N,N-DI(2-CHLOROETHYL)NAPHTHYLAMINE see BIF250

N,N-DI(2-CHLOROETHYL)-β-NAPHTHYLAMINE see BIF250

DICHLOROETHYL OXIDE see DFJ050

DICHLOROETHYLPHENYLSILANE see DFJ800

DICHLOROETHYLPHOSPHINE see EOQ000

N,N-DI(2-CHLOROETHYL)-N,o-PROPYLENE-PHOSPHORIC ACID ESTER DIAMIDE see EAS500

DICHLOROETHYLSILANE see DFK000

DI-2-CHLOROETHYL SULFIDE see BIH250

β,β'-DICHLOROETHYL SULFIDE see BIH250

2,2'-DICHLOROETHYL SULPHIDE (MAK) see BIH250

DICHLOROETHYNE see DEN600

3-(3,4-DICHLORO-FENIL)-1-METOSSI-1-METIL-UREA (ITALIAN) see DGD600

DICHLOROFENTHION see DFK600

DICHLOROFLUOROMETHANE see DFL000

2,3-DICHLOROHEXAFLUOROBUTENE-2 see DFM000

2,3-DICHLOROHEXAFLUORO-2-BUTENE see DFM000

2,3-DICHLORO-1,1,1,4,4,4-HEXAFLUOROBUTENE-2 see DFM000

4,5-DICHLORO-3,3,4,5,6,6-HEXAFLUORO-1,2-DIOXANE see DFM099

DICHLOROHYDRIN see DGG400

α-DICHLOROHYDRIN see DGG400

6,7-DICHLORO-10-(3-(N-(2-HYDROXYETHYL)METHYLAMINO)PROPYL) ISOALLOXAZINE SULFATE see DFM800

d-(−)-threo-2,2-DICHLORO-N-(β-HYDROXY-α-(HYDROXYMETHYL))-p-NITROPHENETHYLACETAMIDE see CDP250

d-(−)-2,2-DICHLORO-N-(β-HYDROXY-α-(HYDROXYMETHYL)-p-NITROPHENYLETHYL)ACETAMIDE see CDP250

DI-(5-CHLORO-2-HYDROXYPHENYL)METHANE see MJM500

DICHLOROISOCYANURIC ACID see DGN200

DICHLOROISOCYANURIC ACID, dry (DOT) see DGN200

DICHLOROISOCYANURIC ACID POTASSIUM SALT see PLD000

DICHLOROISOCYANURIC ACID SODIUM SALT (DOT) see SGG500

sym-DICHLOROISOPROPYL ALCOHOL see DGG400

2,2'-DICHLORO-N-ISOPROPYLDIETHYLAMINE HYDROCHLORIDE see IPG000

2,2'-DICHLOROISOPROPYL ETHER see BII250

DICHLOROISOPROPYL ETHER (DOT) see BII250

DICHLOROKELTHANE see BIO750

DICHLOROMALEIMIDE see DFN800

DICHLOROMALEINIMIDE see DFN800

DICHLOROMAPHARSEN see DFX400

DICHLOROMETHANE (MAK, DOT) see MJP450

DICHLOROMETHOTREXATE see DFO000

3'5'-DICHLOROMETHOTREXATE see DFO000

2,5-DICHLORO-6-METHOXYBENZOIC ACID see MEL500

3,6-DICHLORO-2-METHOXYBENZOIC ACID see MEL500

DICHLOROMETHYLARSINE see DFP200

1,5-DICHLORO-3-METHYL-3-AZAPENTANE HYDROCHLORIDE see BIE500

2,2'-DICHLORO-N-METHYLDIETHYLAMINE see BIE250

2,2'-DICHLORO-N-METHYLDIETHYLAMINE HYDROCHLORIDE see BIE500

2,2'-DICHLORO-N-METHYLDIETHYLAMINE-N-OXIDE see CFA500

2,2'-DICHLORO-N-METHYLDIETHYLAMINE N-OXIDE HYDROCHLORIDE see CFA750

2,3-DICHLORO-4-(2-METHYLENEBUTYRL)PHENOXY ACETIC ACID see DFP600

(2,3-DICHLORO-4-(2-METHYLENEBUTYRYL)PHENOXY) ACETIC ACID see DFP600

(2,3-DICHLORO-4-(2-METHYLENE-1-OXOBUTYL)PHENOXY)ACETIC ACID see DFP600

sym-DICHLOROMETHYL ETHER see BIK000

1,3-DICHLORO-5,5'-METHYLHYDANTOIN see DFE200

DICHLORO-N-METHYLMALEIMIDE see DFP800

2,3-DICHLORO-N-METHYLMALEIMIDE see DFP800

DICHLOROMETHYLPHENYLSILANE see DFQ800

DICHLOROMETHYL PHOSPHINE see EOQ000

DICHLOROMETHYLSILANE see DFS000

2,2'-DICHLORO-1''-METHYLTRIETHYLAMINE HYDROCHLORIDE see IPG000

DICHLOROMONOFLUOROMETHANE (OSHA, DOT) see DFL000

2,3-DICHLORO-1,4-NAPHTHALENEDIONE see DFT000

2,3-DICHLORO-1,4-NAPHTHAQUINONE see DFT000

DICHLORONAPHTHOQUINONE see DFT000

2,3-DICHLORONAPHTHOQUINONE see DFT000

2,3-DICHLORO-α-NAPHTHOQUINONE see DFT000

2,3-DICHLORO-1,4-NAPHTHOQUINONE see DFT000

2,3-DICHLORONAPHTHOQUINONE-1,4 see DFT000

2',4'-DICHLORO-4-NITROBIPHENYL ETHER see DFT800

2,4-DICHLORO-4'-NITRODIPHENYL ETHER see DFT800

DICHLORONITROETHANE see DFU000

1,1-DICHLORO-1-NITROETHANE see DFU000

2,4-DICHLORO-6-NITROPHENOL see DFU600

2,4-DICHLORO-1-(4-NITROPHENOXY)BENZENE see DFT800

2',5-DICHLORO-4'-NITROSALICYLANILIDE see DFV400

2',5-DICHLORO-4'-NITROSALICYLANILIDE-2-AMINOETHANOL SALT see DFV600

5,2'-DICHLORO-4'-NITROSALICYLANILIDE ETHANOLAMINE SALT see DFV600

5,2-DICHLORO-4'-NITROSALICYLIC ANILIDE-2-AMINOETHANOL SALT see DFV600

2',5-DICHLORO-4'-NITROSALICYLOYLANILIDE ETHANOLAMINE SALT see DFV600

2,2'-DICHLORO-N-NITROSODIPROPYLAMINE see DFW000

3,4-DICHLORONITROSOPIPERIDINE see DFW200

3,4-DICHLORO-N-NITROSOPYRROLIDINE see DFW600

DICHLOROOXOZIRCONIUM see ZSJ000

DICHLOROPENTANE see DFX000

DICHLOROPENTANES (DOT) see DFX000

DICHLOROPHENARSINE HYDROCHLORIDE see DFX400

2,4-DICHLOROPHENOL see DFX800

2,6-DICHLOROPHENOL see DFY000

2,4-DICHLOROPHENOL BENZENESULFONATE see DFY400

3-(3,4-DICHLOROPHENOL)-1,1-DIMETHYLUREA see DXQ500

2,4-DICHLORO-PHENOL-O-ESTER with O,O-DIETHYL PHOSPHOROTHIOATE see DFK600

DICHLOROPHENOXYACETIC ACID see DAA800

2,4-DICHLOROPHENOXYACETIC ACID (DOT) see DAA800

(2,4-DICHLOROPHENOXY)ACETIC ACID, BUTYL ESTER see BQZ000

2,4-DICHLOROPHENOXYACETIC ACID ISOOCTYL ESTER see ILO000

(2,4-DICHLOROPHENOXY)ACETIC ACID, ISOPROPYL ESTER see IOY000

(2-4-DICHLOROPHENOXY)ACETIC ACID-1-METHYLETHYL ESTER (9CI) see IOY000

2,4-DICHLOROPHENOXYACETIC ACID, SODIUM SALT see SGH500

5,6-DICHLORO-1-PHENOXYCARBONYL-2-TRIFLUORO-METHYLBENZIMIDAZOLE see DGA200

2-(2,4-DICHLOROPHENOXY)ETHANOL HYDROGEN SULFATE SODIUM SALT see CNW000

2,4-DICHLOROPHENOXYETHYL SULFATE, SODIUM SALT see CNW000

4-(2,4-DICHLOROPHENOXY)NITROBENZENE see DFT800

2-(2,4-DICHLOROPHENOXY) PROPIONIC ACID see DGB000

α-(2,4-DICHLOROPHENOXY) PROPIONIC ACID see DGB000

DICHLOROPHENYLARSINE see DGB600

3,4-DICHLOROPHENYLAZOTHIOUREA see DEQ000

3,4-DICHLOROPHENYL-AZOTHIOUREE (FRENCH) see DEQ000

2,4-DICHLOROPHENYL BENZENESULFONATE see DFY400

2,4-DICHLOROPHENYL BENZENESULPHONATE see DFY400

N-(3,4-DICHLOROPHENYL)CYCLOPROPANECARBOX-AMIDE see CQJ250

2,4′-DICHLOROPHENYLDICHLOROETHANE see CDN000

O-2,4-DICHLOROPHENYL-O,O-DIETHYL PHOSPHORO-THIOATE see DFK600

2,4-DICHLORO-PHENYL DIETHYL PHOSPHOROTHIONATE see DFK600

1,6-DI(4′-CHLOROPHENYLDIGUANIDO)HEXANE see BIM250

N′-(3,4-DICHLOROPHENYL)-N,N-DIMETHYLUREA see DXQ500

1-(3,4-DICHLOROPHENYL)-3,3-DIMETHYLUREE (FRENCH) see DXQ500

cis-DICHLORO(o-PHENYLENEDIAMINE)PLATINUM(II) see DGC600

DICHLORO(1,2-PHENYLENEDIAMMINE)PLATINUM(II) see DGC600

2,4-DICHLOROPHENYL ESTER of BENZENESULFONIC ACID see DFY400

2,4-DICHLOROPHENYL ESTER BENZENESULPHONIC ACID see DFY400

3-(3,5-DICHLOROPHENYL)-5-ETHENYL-5-METHYL-2,4-OXAZOLIDINEDIONE see RMA000

DICHLOROPHENYL ETHER see DFE800

3-(3,4-DICHLOROPHENYL)-1-METHOXYMETHYLUREA see DGD600

3-(3,4-DICHLOROPHENYL)-1-METHOXY-1-METHYLUREA see DGD600

N′-(3,4-DICHLOROPHENYL)-N-METHOXY-N-METHYL-UREA see DGD600

1-(3,4-DICHLOROPHENYL)3-METHOXY-3-METHYLUREE (FRENCH) see DGD600

3-(3,5-DICHLOROPHENYL)-N-(1-METHYLETHYL)-2,4-DI-OXO-1-IMIDAZOLIDINECARBOXAMIDE see GIA000

O-(2,4-DICHLOROPHENYL)-O-METHYLISOPROPYL-PHOSPHORAMIDOTHIOATE see DGD800

O-(2,4-DICHLOROPHENYL)-O-METHYL-N-ISOPROPYL-PHOSPHORAMIDOTHIOATE see DGD800

2-(3,4-DICHLOROPHENYL)-3-METHYL-4-METATHIAZA-NONE-1,1-DIOXIDE see DEM000

N-(3,4-DICHLOROPHENYL)-N′-METHYL-N′-METHOXY-UREA see DGD600

3-(3,5-DICHLOROPHENYL)-5-METHYL-5-VINYL-2,4-OXA-ZOLIDINEDIONE see RMA000

2,4-DICHLOROPHENYL-4-NITROPHENYL ETHER see DFT800

2,4-DICHLOROPHENYL-p-NITROPHENYL ETHER see DFT800

DICHLOROPHENYLPHOSPHINE see DGE400

N-(3,4-DICHLOROPHENYL)PROPANAMIDE see DGI000

1-(2-(2,4-DICHLOROPHENYL)-2-(2-PROPENYLOXY)ETHYL)-1H-IMIDAZOLE see FPB875

N-(3,4-DICHLOROPHENYL)PROPIONAMIDE see DGI000

1-(3,5-DICHLOROPHENYL)-2,5-PYRROLIDINEDIONE see DGF000

N-(3,5-DICHLOROPHENYL)SUCCINIMIDE see DGF000

2-(3,4-DICHLOROPHENYL)TETRAHYDRO-3-METHYL-4H-1,3-THIAZIN-4-ONE-1,1-DIOXIDE see DEM000

2,5-DICHLOROPHENYLTHIOMETHYL O,O-DIETHYL PHOS-PHORODITHIOATE see PDC750

S-(2,5-DICHLOROPHENYLTHIOMETHYL) O,O-DIETHYL PHOSPHORODITHIOATE see PDC750

S-((3,4-DICHLOROPHENYLTHIO)METHYL)-O,O-DIETHYL PHOSPHORODITHIOATE see DJA200

S-(2,5-DICHLOROPHENYLTHIOMETHYL) DIETHYL PHOS-PHOROTHIOLOTHIONATE see PDC750

DI-(p-CHLOROPHENYL)TRICHLOROMETHYLCARBINOL see BIO750

(DICHLOROPHENYL)TRICHLOROSILANE see DGF200

DICHLOROPHENYLTRICHLOROSILANE (DOT) see DGF200

DICHLOROPHOS see DGP900

DICHLOROPHOSPHORIC ACID, ETHYL ESTER see EOR000

DICHLOROPROP see DGB000

1,2-DICHLOROPROPANE see PNJ400

1,3-DICHLOROPROPANE see DGF800

α,β-DICHLOROPROPANE see PNJ400

1,2-DICHLOROPROPANE mixed with DICHLOROPROPENE see DGG200

DICHLOROPROPANE-DICHLOROPROPENE MIXTURE see DGG000

2,3-DICHLOROPROPANOL see DGG600

1,2-DICHLORO-3-PROPANOL see DGG600

1,2-DICHLOROPROPANOL-3 see DGG600

1,3-DICHLORO-2-PROPANOL see DGG400

2,3-DICHLORO-1-PROPANOL see DGG600

1,3-DICHLOROPROPANOL-2 (DOT) see DGG400

1,3-DICHLORO-2-PROPANONE see BIK250

1,3-DICHLOROPROPENE see DGG950

2,3-DICHLOROPROPENE see DGH400

1,3-DICHLOROPROPENE-1 see DGG950

2,3-DICHLORO-1-PROPENE see DGH400

(E)-1,3-DICHLOROPROPENE see DGH000

(Z)-1,3-DICHLOROPROPENE see DGH200

cis-1,3-DICHLOROPROPENE see DGH200

trans-1,3-DICHLOROPROPENE see DGH000

1,3-DICHLOROPROPENE and 1,2-DICHLOROPROPANE MIXTURE see DGG000

DICHLOROPROPENE and PROPYLENE DICHLORIDE MIX-TURE (DOT) see DGG200

2,3-DICHLORO-2-PROPENE-1-THIOL DIISOPROPYLCARBA-MATE see DBI200

S-(2,3-DICHLORO-2-PROPENYL)ESTER, BIS(1-METHYL-ETHYL) CARBAMOTHIOIC ACID see DBI200

1,2-DICHLORO-3-PROPIONAL see DGH800

2,3-DICHLORO PROPIONALDEHYDE see DGH800

α,β-DICHLOROPROPIONALDEHYDE see DGH800

DICHLOROPROPIONANILIDE see DGI000

3,4-DICHLOROPROPIONANILIDE see DGI000

3′,4′-DICHLOROPROPIONANILIDE see DGI000

α-DICHLOROPROPIONIC ACID see DGI400

2,2-DICHLOROPROPIONIC ACID see DGI400
α,α-DICHLOROPROPIONIC ACID see DGI400
2,2-DICHLOROPROPIONIC ACID, SODIUM SALT see DGI600
α,α-DICHLOROPROPIONIC ACID SODIUM SALT see DGI600
1,3-DICHLOROPROPYLENE see DGG950
2,3-DICHLOROPROPYLENE see DGH400
α,γ-DICHLOROPROPYLENE see DGG950
cis-1,3-DICHLOROPROPYLENE see DGH200
trans-1,3-DICHLOROPROPYLENE see DGH000
3,4-DICHLORO-2,5-PYRROLIDINEDIONE see DFN800
DICHLOROSAL see CFY000
3,5-DICHLOROSALICYLIC ACID see DGK200
DICHLOROSILANE see DGK300
DICHLOROSULFANE see SOG500
DICHLOROTETRAFLUOROETHANE see DGL600
DICHLOROTETRAFLUOROETHANE (OSHA, ACGIH) see FOO509
sym-DICHLOROTETRAFLUOROETHANE see FOO509
1,2-DICHLORO-1,1,2,2-TETRAFLUOROETHANE (MAK) see FOO509
2,6-DICHLOROTHIOBENZAMIDE see DGM600
DICHLOROTHIOCARBONYL see TFN500
α,α-DICHLOROTOLUENE see BAY300
1,3-DICHLORO-s-TRIAZINE-2,4,6(1H,3H,5H)-TRIONE see DGN200
DICHLORO-s-TRIAZINE-2,4,6(1H,3H,5H)-TRIONE POTASSIUM DERIV see PLD000
4,4′-DICHLORO-α-(TRICHLOROMETHYL)BENZHYDROL see BIO750
2,2′-DICHLOROTRIETHYLAMINE see BID250
5,6-DICHLORO-2-TRIFLUOROMETHYLBENZIMIDAZOLE-1-CARBOXYLATE see DGA200
5,6-DICHLORO-2-(TRIFLUOROMETHYL)-1H-BENZIMIDAZOLE-1-CARBOXYLIC ACID PHENYL ESTER see DGA200
DICHLOROVAS see DGP900
(2,2-DICHLORO-VINIL)DIMETILFOSFATO (ITALIAN) see DGP900
2,2-DICHLOROVINYL ALCOHOL, DIMETHYL PHOSPHATE see DGP900
2,2-DICHLOROVINYL DIMETHYL PHOSPHATE see DGP900
2,2-DICHLOROVINYL DIMETHYL PHOSPHORIC ACID ESTER see DGP900
DICHLOROVOS see DGP900
α,α′-DICHLORO-o-XYLENE see DGP400
2,4-DICHLORPHENOXYACETIC ACID see DAA800
(2,4-DICHLOR-PHENOXY)-ESSIGSAEURE (GERMAN) see DAA800
2-(2,4-DICHLOR-PHENOXY)-PROPIONSAEURE (GERMAN) see DGB000
(3,4-DICHLOR-PHENYL-AZO)-THIOHARNSTOFF (GERMAN) see DEQ000
3-(3,4-DICHLOR-PHENYL)-1,1-DIMETHYL-HARNSTOFF (GERMAN) see DXQ500
3-(3,4-DICHLOR-PHENYL)-1-METHOXY-1-METHYL-HARNSTOFF (GERMAN) see DGD600
3-(4,5-DICHLORPHENYL)-1-METHOXY-1-METHYLHARNSTOFF (GERMAN) see DGD600
2,4,-DICHLORPHENYL-4-NITROPHENYLAETHER (GERMAN) see DFT800
1-(2-(2,4-DICHLORPHENYL)-2-PROPENYLOXY)AETHYL)-1H-IMIDAZOLE see FPB875
DICHLORPHOS see DGP900
DICHLORPROP see DGB000
DICHLORPROPAN-DICHLORPROPENGEMISCH (GERMAN) see DGG000
DICHLOR-s-TRIAZIN-2,4,6(1H,3H,5H)TRIONE POTASSIUM see PLD000
(2,2-DICHLOR-VINYL)-DIMETHYL-PHOSPHAT (GERMAN) see DGP900
O-(2,2-DICHLORVINYL)-O,O-DIMETHYLPHOSPHAT (GERMAN) see DGP900

DICHLORVOS see DGP900
DICHLOSALE see DFV400
DICHLOTIAZID see CFY000
DICHLOTRIDE see CFY000
DICHOLINE SUCCINATE see CMG250
DICHROMIUM TRIOXIDE see CMJ900
DI(2-CIANOETIL)AMMINA (ITALIAN) see BIQ500
DICK (GERMAN) see DFH200
DICLONIA see BPR500
p-DICLOROBENZENE (ITALIAN) see DEP800
1,4-DICLOROBENZENE (ITALIAN) see DEP800
1,1-DICLORO-2,2-BIS(4-CLORO-FENIL)-ETANO (ITALIAN) see BIM500
3,3′-DICLORO-4,4′-DIAMINODIFENILMETANO (ITALIAN) see MJM200
1,1-DICLOROETANO (ITALIAN) see DFF809
1,2-DICLOROETANO (ITALIAN) see EIY600
2,2′-DICLOROETILETERE (ITALIAN) see DFJ050
(3,4-DICLORO-FENIL-AZO)-TIOUREA (ITALIAN) see DEQ000
3-(3,4-DICLORO-FENYL)-1,1-DIMETIL-UREA (ITALIAN) see DXQ500
1,1-DICLORO-1-NITROETANO (ITALIAN) see DFU000
DICLOTRIDE see CFY000
DICOBALT CARBONYL see CNB500
DICOBALT OCTACARBONYL see CNB500
DICOFOL see BIO750
DICOL see DJD600
DICONIRT D see SGH500
DICOPHANE see DAD200
DICOPUR see DAA800
DICOPUR-M see CIR250
DICORTOL see PMA000
DICORVIN see DKA600
DICOTEX see CIR250
DICOTOX see DAA800
DICOUMARIN see BJZ000
DICOUMAROL see BJZ000
DICPUR see DAA800
DICRESYL see MIB750
DICROTOFOS (DUTCH) see DGQ875
DICROTOPHOS see DGQ875
DICTYCIDE see COH250
DICTYZIDE see COH250
DICUMACYL see BKA000
DICUMAN see BJZ000
DICUMARINE see BJZ000
DI-α-CUMYL PEROXIDE see DGR600
DICUMYL PEROXIDE (DOT) see DGR600
DI-CUP see DGR600
DI-CUP 40 KF see DGR600
DI-CUPR see DGR600
DICUPRAL see DXH250
2,2′-DICYANO-2,2′-AZOPROPANE see ASL750
m-DICYANOBENZENE see PHX550
o-DICYANOBENZENE see PHY000
p-DICYANOBENZENE see BBP250
1,2-DICYANOBENZENE see PHY000
1,3-DICYANOBENZENE see PHX550
1,4-DICYANOBENZENE see BBP250
1,4-DICYANOBUTANE see AER250
β,β-DICYANO-o-CHLOROSTYRENE see CEQ600
2,2′-DICYANODIETHYLAMINE see BIQ500
2,3-DICYANO-1,4-DITHIA-ANTHRAQUINONE see DLK200
s-DICYANOETHANE see SNE000
1,2-DICYANOETHANE see SNE000
DI-(2-CYANOETHYL)AMINE see BIQ500
DICYANOGEN see COO000
DICYANOMETHANE see MAO250
1,3-DICYANOPROPANE see TLR500
1,3-DICYANOTETRACHLOROBENZENE see TBQ750
cis-DICYCLOBUTYLAMMINEDICHLOROPLATINUM(II) see DGT200
N,N-DICYCLOHEXYLAMINE see DGT600
DICYCLOHEXYLAMINE (DOT) see DGT600

DICYCLOPENTADIENE see DGW000
DICYCLOPENTADIENYLHAFNIUM DICHLORIDE see
    HAE500
DI-2,4-CYCLOPENTADIEN-1-YL IRON see FBC000
DICYCLOPENTADIENYL IRON (OSHA, ACGIH) see
    FBC000
DI-pi-CYCLOPENTADIENYLNICKEL see NDA500
cis-DICYCLOPENTYLAMMINEDICHLOROPLATINUM(II) see
    BIS250
DICYKLOHEXYLAMIN (CZECH) see DGT600
DICYKLOPENTADIEN (CZECH) see DGW000
DICYSTEINE see CQK325
DIDAKENE see PCF275
DIDANDIN see DVV600
DIDAN-TDC-250 see DKQ000
9,10-DIDEHYDRO-N,N-DIETHYL-2-BROMO-6-METHYL-
    ERGOLINE-8-β-CARBOXAMIDE see BNM250
9,10-DIDEHYDRO-N,N-DIETHYL-6-METHYL-ERGOLINE-8-
    β-CARBOXAMIDE see DJO000
9,10-DIDEHYDRO-N,N-DIETHYL-6-METHYL-ERGOLINE-8-
    β-CARBOXAMIDE-d- TARTRATE with METHANOL (1:2)
    see LJG000
(5-α,6-α)-7,8-DIDEHYDRO-4,5-EPOXY-3-ETHOXY-17-
    METHYLMORPHINAN-6-OL see ENK000
7,8-DIDEHYDRO-4,5-α-EPOXY-3-ETHOXY-17-METHYL-
    MORPHINAN-6-α-OL HYDROCHLORIDE DIHYDRATE
    see ENK500
7,8-DIDEHYDRO-4,5-α-EPOXY-17-METHYLMORPHINAN-
    3,6-α-DIOL HYDROCHLORIDE see MRO750
7,8-DIDEHYDRO-4,5-α-EPOXY-17-METHYLMORPHINE HY-
    DROCHLORIDE see MRO750
9,10-DIDEHYDRO-N-ETHYL-1,6-DIMETHYLERGOLINE-8-β-
    CARBOXAMIDE see MLD500
9,10-DIDEHYDRO-N-ETHYL-6-METHYLERGOLINE-8-β-
    CARBOXAMIDE, N-ETHYLLYSERGAMIDE see LJI000
9,10-DIDEHYDRO-N-(α-(HYDROXYMETHYL)ETHYL)-6-
    METHYLERGOLINE-8-β-CARBOXAMIDE see LJL000
9,10-DIDEHYDRO-N-(α-(HYDROXYMETHYL)PROPYL)-6-
    METHYL-ERGOLINE-8-β-CARBOXAMIDE see PAM000
13,19-DIDEHYDRO-12-HYDROXY-SENECIONAN-11,16-
    DIONE see SBX500
3,8-DIDEHYDRORETRONECINE see DAL400
9,10-DIDEHYDRO-N,N,6-TRIMETHYLERGOLINE-8-β-CAR-
    BOXAMIDE see LJH000
1,6-DIDEOXY-1,6-DI(2-CHLOROETHYLAMINO)-d-MANNI-
    TOLDIHYDROCHLORIDE see MAW750
DIDEOXYKANAMYCIN B see DCQ800
3',4'-DIDEOXYKANAMYCIN B see DCQ800
3',4'-DIDEOXYKANAMYCIN B SULFATE see DHA400
DIDEUTERIUM OXIDE see HAK000
DIDIGAM see DAD200
DIDIMAC see DAD200
3,6-DI(DIMETHYLAMINO)ACRIDINE see BJF000
1,2-DI-(DIMETHYLAMINO)ETHANE (DOT) see TDQ750
DIDOC see AAI250
DIDODECYL-3,3'-THIODIPROPIONATE see TFD500
DIDRATE see DKW800
DIELDREX see DHB400
DIELDRIN see DHB400
DIELDRINE (FRENCH) see DHB400
DIELDRITE see DHB400
DIELTAMID see DKC800
DIENESTROL see DAL600
DIENOESTROL see DAL600
β-DIENOESTROL see DAL600
DIENOL see DAL600
DIENOL S see SMR000
DIEPOXYBUTANE see BGA750
l-DIEPOXYBUTANE see BOP750
2,4-DIEPOXYBUTANE see BGA750
dl-DIEPOXYBUTANE see DHB600
1,2:3,4-DIEPOXYBUTANE see BGA750
(2S,3S)-DIEPOXYBUTANE see BOP750
meso-DIEPOXYBUTANE see DHB800

(R*,S*)-DIEPOXYBUTANE see DHB800
l-1,2:3,4-DIEPOXYBUTANE see BOP750
(±)-1,2:3,4-DIEPOXYBUTANE see DHB600
dl-1,2:3,4-DIEPOXYBUTANE see DHB600
(2S,3S)-1,2:3,4-DIEPOXYBUTANE see BOP750
meso-1,2,3,4-DIEPOXYBUTANE see DHB800
1,2:5,6-DIEPOXYDULCITOL see DCI600
1,2,8,9-DIEPOXYLIMONENE see LFV000
1,2:8,9-DIEPOXYMENTHANE see LFV000
1,2:8,9-DIEPOXY-p-MENTHANE see LFV000
DI(2,3-EPOXYPROPYL) ETHER see DKM200
DIESEL FUEL (DOT) see FOP000
S,S-DIESTER with DITHIO-p-UREIDOBENZENEARSONOUS
    ACID o-MERCAPTOBENZOIC ACID see TFD750
DIESTER with o-MERCAPTOBENZOIC ACID DITHIO-p-
    UREIDOBENZENEARSONOUS ACID see TFD750
DI-ESTRYL see DKA600
DIETADIONE (ITALIAN) see DJT400
DIETHADION see DJT400
DIETHADIONE see DJT400
DIETHANOLAMIN (CZECH) see DHF000
DIETHANOLAMINE see DHF000
DIETHANOLAMMONIUM MALEIC HYDRAZIDE see
    DHF200
1,1-DIETHANOLHYDRAZINE see HHH000
DIETHANOLLAURAMIDE see BKE500
N,N-DIETHANOLLAURAMIDE see BKE500
N,N-DIETHANOLLAURIC ACID AMIDE see BKE500
DIETHANOLNITROSOAMINE see NKM000
DIETHION see EEH600
1,2-DI(ETHOXYCARBONYL)ETHYL-O,O-DIMETHYL PHOS-
    PHORODITHIOATE see MAK700
S-(1,2-DI(ETHOXYCARBONYL)ETHYL DIMETHYL PHOS-
    PHOROTHIOLOTHIONATE see MAK700
DIETHOXYDIMETHYLSILANE see DHG000
1,1-DIETHOXY-ETHAAN (DUTCH) see AAG000
1,1-DIETHOXYETHANE see AAG000
1,2-DIETHOXYETHANE see EJE500
DIETHOXYMETHANE (DOT) see EFT500
α-(((DIETHOXYPHOSPHINOTHIOYL)OXY)IMINO)BEN-
    ZENEACETONITRILE see BAT750
(DIETHOXYPHOSPHINYL)DITHIOIMIDOCARBONIC ACID
    CYCLIC ETHYLENE ESTER see PGW750
2-(DIETHOXYPHOSPHINYLIMINO)-1,3-DITHIOLANE see
    DXN600
2-(DIETHOXYPHOSPHINYLIMINO)-1,3-DITHIOLANE see
    PGW750
2-(DIETHOXYPHOSPHINYLIMINO)-4-METHYL-1,3-DI-
    THIOLANE see DHH400
(DIETHOXY-PHOSPHINYL)MERCURY ACETATE see
    AAS250
2-((DIETHOXYPHOSPHINYL)OXY)-1H-BENZ(de)ISOQUINO-
    LINE-1,3(2H)-DIONE see HMV000
DIETHOXYPHOSPHORUS OXYCHLORIDE see DIY000
DIETHOXYPHOSPHORYL CYANIDE see DJW800
DI-ETHOXYTHIOKARBONYL-TRISULFID see DKE400
DIETHOXY THIOPHOSPHORIC ACID ESTER of 2-ETHYL-
    MERCAPTOETHANOL see DAO600
(DIETHOXY-THIOPHOSPHORYLOXYIMINO)-PHENYL ACE-
    TONITRILE see BAT750
DIETHQUINALPHION see DJY200
DIETHQUINALPHIONE see DJY200
DIETHYL see BOR500
DIETHYL ACETAL see AAG000
DIETHYL ACETALDEHYDE see DHI000
N,N-DIETHYLACETAMIDE see DHI200
DIETHYLACETIC ACID see DHI400
1-DIETHYLACETYLAZIRIDINE see DHI800
DIETHYLACETYLETHYLENEIMINE see DHI800
DIETHYLAMIDE de VANILLIQUE see DKE200
DIETHYLAMINE see DHJ200
N,N-DIETHYLAMINE see DHJ200
DIETHYLAMINE, 2,2'-DICHLORO-N-METHYL-, OXIDE see
    CFA500

2-DIETHYLAMINO-6-METHYLPYRIMIDIN-4-YL DIMETHYL PHOSPHOROTHIONATE see DIN800

O-(2-(DIETHYLAMINO)-6-METHYL-4-PYRIMIDINYL)-O,O-DIMETHYL PHOSPHOROTHIOATE see DIN800

O-(2-DIETHYLAMINO-6-METHYLPYRIMIDIN-4-YL)-O,O-DI-METHYL PHOSPHOROTHIOATE see DIN800

7-DIETHYLAMINO-5-METHYL-s-TRIAZOLO(1,5-a)PYRIMI-DINE see DIO200

DIETHYL-m-AMINO-PHENOLPHTHALEIN HYDROCHLO-RIDE see FAG070

10-DIETHYLAMINOPROPIONYL-3-TRIFLUOROMETHYL PHENOTHIAZINE HYDROCHLORIDE see FDE000

N-(3-DIETHYLAMINOPROPYL)AMINE see DIY800

N,N-DIETHYLAMINOPROPYLAMINE see DIY800

3-(DIETHYLAMINO)PROPYLAMINE (DOT) see DIY800

2-DIETHYLAMINO-1-PROPYL-N-DIBENZOPARATHIAZINE see DIR000

10-(2-DIETHYLAMINOPROPYL)PHENOTHIAZINE see DIR000

2-DIETHYLAMINO-2',4',6'-TRIMETHYLACETANILIDE HY-DROCHLORIDE see DHL800

2-(DIETHYLAMINO)-2',4',6'-TRIMETHYLACETANILIDE MONOHYDROCHLORIDE see DHL800

DIETHYLAMINOTRIMETHYLENAMINE see DIY800

2-(DIETHYLAMINO)-N-(2,4,6-TRIMETHYLPHENYL) ACETAMIDE MONOHYDROCHLORIDE see DHL800

N,N-DIETHYLANILIN (CZECH) see DIS700

DIETHYLANILINE see DIS700

N,N-DIETHYLANILINE see DIS700

DIETHYLBARBITURATE MONOSODIUM see BAG250

5,5-DIETHYLBARBITURIC ACID SODIUM deriv. see BAG250

DIETHYL BENZENE see DIU000

m-DIETHYLBENZENE see DIU200

N,N-DIETHYLBENZENESULFONAMIDE see DIU400

DIETHYLBERYLLIUM see DIV000

DIETHYLBIS(OCTANOYLOXY)STANNANE see DIV600

DIETHYLBIS(1-OXOOCTYL)OXY)STANNANE see DIV600

O,O-DIETHYL-O-(4-BROOM-2,5-DICHLOOR-FENYL)-MONOTHIOFOSFAAT (DUTCH) see EGV500

DIETHYLCADMIUM see DIV800

DIETHYLCARBAMAZANE CITRATE see DIW200

DIETHYLCARBAMAZINE ACID CITRATE see DIW200

DIETHYLCARBAMAZINE CITRATE see DIW200

DIETHYLCARBAMAZINE HYDROGEN CITRATE see DIW200

DIETHYLCARBAMIC CHLORIDE see DIW400

DIETHYLCARBAMIDOYL CHLORIDE see DIW400

DIETHYLCARBAMODITHIOIC ACID 2-CHLORO-2-PROPE-NYL ESTER see CDO250

DIETHYLCARBAMODITHIOIC ACID, SODIUM SALT see SGJ000

DIETHYLCARBAMOYL CHLORIDE see DIW400

N,N-DIETHYLCARBAMOYL CHLORIDE see DIW400

1-DIETHYLCARBAMOYL-4-METHYLPIPERAZINE DIHY-DROGEN CITRATE see DIW200

DIETHYLCARBAMYL CHLORIDE see DIW400

N,N-DIETHYLCARBANILIDE see DJC400

O,O-DIETHYL-S-(CARBETHOXY)METHYL PHOSPHORO-THIOLATE see DIW600

DIETHYL CARBINOL see IHP010

DIETHYLCARBINOL (DOT) see IHP010

O,O-DIETHYL-S-CARBOETHOXYMETHYL PHOSPHORO-THIOATE see DIW600

O,O-DIETHYL-S-CARBOETHOXYMETHYL THIOPHOS-PHATE see DIW600

DIETHYL CARBONATE see DIX200

DIETHYL CARBONATE (DOT) see DIX200

DIETHYL CELLOSOLVE (DOT) see EJE500

DIETHYLCETONE (FRENCH) see DJN750

O,O-DIETHYL-O-(2-CHINOXALYL)PHOSPHOROTHIOATE see DJY200

O,O-DIETHYL-S-(4-CHLOOR-FENYL-THIO)-METHYL)-DI-THIOFOSFAAT (DUTCH) see TNP250

O,O-DIETHYL-O-(3-CHLOOR-4-METHYL-CUMARIN-7-YL) MONOTHIOFOSFAAT (DUTCH) see CNU750

O,O-DIETHYL-S-((6-CHLOOR-2-OXO-BENZOXAZOLIN-3-YL)-METHYL)-DITHIO FOSFAAT (DUTCH) see BDJ250

O,O-DIETHYL-S-p-CHLORFENYLTHIOMETHYLESTER KY-SELINY DITHIOFOSFORECNE (CZECH) see TNP250

O,O-DIETHYL-S-p-CHLORLPHENYLTHIOMETHYL DITHIO-PHOSPHATE see TNP250

O,O-DIETHYL-S-(6-CHLOROBENZOXAZOLINYL-3-METHYL)DITHIOPHOSPHATE see BDJ250

O,O-DIETHYL-O-(2-CHLORO-1-(2',4'-DICHLOROPHENYL) VINYL) PHOSPHATE see CDS750

O,O-DIETHYL-O-(2-CHLORO-1,2,5-DICHLOROPHENYLVI-NYL) PHOSPHOROTHIOATE see CDS750

DIETHYL(2-CHLOROETHYL)AMINE see CGV500

DIETHYL-3-CHLORO-4-METHYL-7-COUMARINYL PHOS-PHATE see CIK750

O,O-DIETHYL-O-(3-CHLORO-4-METHYLCOUMARIN-7-YL) PHOSPHATE see CIK750

O,O-DIETHYL-O-(3-CHLORO-4-METHYL-7-COUMARINYL) PHOSPHOROTHIOATE see CNU750

O,O-DIETHYL-O-(3-CHLORO-4-METHYLCOUMARINYL-7) THIOPHOSPHATE see CNU750

O,O-DIETHYL-O-(3-CHLORO-4-METHYL-2-OXO-2H-BENZO-PYRAN-7-YL)PHOSPHOROTHIOATE see CNU750

S,S-DIETHYL(CHLOROMETHYL)PHOSPHONODITHIOATE see BJD000

O,O-DIETHYL-3-CHLORO-4-METHYL-7-UMBELLIFERONE THIOPHOSPHATE see CNU750

O,O-DIETHYL-O-(3-CHLORO-4-METHYLUMBEL-LIFERYL)PHOSPHOROTHIOATE see CNU750

DIETHYL-3-CHLORO-4-METHYLUMBELLIFERYL THIONO-PHOSPHATE see CNU750

O,O-DIETHYL-S-((6-CHLORO-2-OXOBENZOXAZOLIN-3-YL) METHYL) PHOSPHORODITHIOATE see BDJ250

O,O-DIETHYL-S-(6-CHLORO-2-OXO-BENZOXAZOLIN-3-YL) METHYL-PHOSPHORO THIOLOTHIONATE see BDJ250

O,O-DIETHYL-P-CHLOROPHENYLMERCAPTOMETHYL DI-THIOPHOSPHATE see TNP250

O,O-DIETHYL-S-(4-CHLOROPHENYLTHIOMETHYL) DI-THIOPHOSPHATE see TNP250

O,O-DIETHYL-S-(p-CHLOROPHENYLTHIOMETHYL) PHOS-PHORODITHIOATE see TNP250

O,O-DIETHYL-S-p-CHLOROPHENYL THIOMETHYLPHOS-PHOROTHIOATE see DIX800

DIETHYL CHLOROPHOSPHATE see DIY000

O,O-DIETHYL-S-(2-CHLORO-1-PHTHALIMIDOETHYL)-PHOSPHORODITHIOATE see DBI099

DIETHYLCHLOROTHIOPHOSPHATE see DJW600

DIETHYLCHLORTHIOFOSFAT (CZECH) see DJW600

O,O-DIETHYL-S-((2-CYAAN-2-METHYL-ETHYL)-CARBA-MOYL)-METHYL-MONOTHIOFOSFAAT (DUTCH) see PHK250

O,O-DIETHYL-S-N-(A-CYANOISOPROPYL)CARBOMOYL-METHYL PHOSPHOROTHIOATE see PHK250

DIETHYLCYANOPHOSPHATE see DJW800

DIETHYL CYANOPHOSPHONATE see DJW800

P,P-DIETHYL CYCLIC ETHYLENE ESTER OF PHOSPHONO-DITHIOIMIDOCARBONIC ACID see PGW750

p,p-DIETHYL CYCLIC PROPYLENE ESTER of PHOSPHONO-DITHIOIMIDOCARBONIC ACID see DHH400

DIETHYLCYSTEAMIN see DIY600

DIETHYLCYSTEAMINE see DIY600

N-DIETHYL CYSTEAMINE see DIY600

N,N-DIETHYL CYSTEAMINE see DIY600

DIETHYL DECANEDIOATE see DJY600

DIETHYL-1,10-DECANEDIOATE see DJY600

N,N-DIETHYL-1,3-DIAMINOPROPANE see DIY800

DIETHYLDIAZENE-1-OXIDE see ASP000

DIETHYL DICARBONATE see DIZ100

O,O-DIETHYL-O-(2,4-DICHLOOR-FENYL)-MONOTHIOFOS-FAAT (DUTCH) see DFK600

DIETHYL-S-(2-ETHTHIONYLETHYL) THIOPHOSPHATE see ISD000

O,O-DIETHYL-S-ETHYL-2-ETHYLMERCAPTOPHOS-PHOROTHIOLATE see DAP200

O,O-DIETHYL-S-ETHYL-2-ETHYLMERCAPTO PHOS-PHOROTHIOLATE SULFOXIDE see ISD000

O,O-DIETHYL-S-(2-ETHYLMERCAPTOETHYL) DITHIO-PHOSPHATE see DXH325

O,O-DIETHYL 2-ETHYLMERCAPTOETHYL THIOPHOS-PHATE see DAO600

O,O-DIETHYL-2-ETHYLMERCAPTOETHYL THIOPHOS-PHATE, THIONO ISOMER see SPF000

O,O-DIETHYL-S-ETHYLMERCAPTOMETHYL DITHIOPHOS-PHONATE see PGS000

O,O-DIETHYL-S-((ETHYLSULFINYL)ETHYL)PHOS-PHORODITHIOATE see OQS000

O,O-DIETHYL S-(2-(ETHYLSULFINYL)ETHYL) PHOSPHO-RODITHIOATE see OQS000

O,O-DIETHYL-O-(2-ETHYLSULFONYLETHYL)PHOS-PHOROTHIOATE see SPF000

DIETHYL-2-ETHYLSULFONYLETHYL THIONOPHOSPHATE see SPF000

O,O-DIETHYL-S-(2-ETHYLTHIO-ETHYL)-DITHIOFOSFAAT (DUTCH) see DXH325

O,O-DIETHYL-S-(2-ETHYLTHIO-ETHYL)-MONOTHIOFOS-FAAT (DUTCH) see DAP200

O,O-DIETHYL-2-ETHYLTHIOETHYL PHOSPHORODI-THIOATE see DXH325

O,O-DIETHYL-S-2-(ETHYLTHIO)ETHYL PHOSPHORODI-THIOATE see DXH325

O,O-DIETHYL-S-2-(ETHYLTHIO)ETHYL PHOSPHORO-THIOATE see DAP200

O,O-DIETHYL O(and S)-2-(ETHYLTHIO)ETHYL PHOSPHO-ROTHIOATE MIXTURE see DAO600

O,O-DIETHYL-S-(2-(ETHYLTHIO)ETHYL) PHOSPHORO-THIOLATE (USDA) see DAP200

O,O-DIETHYL-S-ETHYLTHIOMETHYL DITHIOPHOSPHO-NATE see PGS000

O,O-DIETHYL-ETHYLTHIOMETHYL PHOSPHORODI-THIOATE see PGS000

O,O-DIETHYL-S-ETHYLTHIOMETHYL THIOTHIONO-PHOSPHATE see PGS000

DIETHYL FLUOROPHOSPHATE see DJJ400

N,N-DIETHYLGLYCINONITRILE see DHJ600

DIETHYL GOLD BROMIDE see DJJ850

DI-(2-ETHYLHEXYL) ESTER, PEROXYDICARBONIC ACID see DJK800

DI(2-ETHYLHEXYL) FUMARATE see DVK600

DI(2-ETHYLHEXYL)ORTHOPHTHALATE see DVL700

DI(2-ETHYLHEXYL) PEROXYDICARBONATE see DJK800

DI-(2-ETHYLHEXYL)PEROXYDICARBONATE, technical pure (DOT) see DJK800

DI(2-ETHYLHEXYL)PHOSPHATE see BJR750

DI-2(ETHYLHEXYL)PHOSPHORIC ACID see BJR750

DI-(2-ETHYLHEXYL)PHOSPHORIC ACID (DOT) see BJR750

DI(2-ETHYLHEXYL)PHTHALATE see DVL700

DI(2-ETHYLHEXYL)SEBACATE see BJS250

DI-(2-ETHYLHEXYL) SODIUM SULFOSUCCINATE see DJL000

1,2-DIETHYLHYDRAZINE see DJL400

N-N'-DIETHYLHYDRAZINE see DJL400

sym-DIETHYLHYDRAZINE see DJL400

1,2-DIETHYLHYDRAZINE DIHYDROCHLORIDE see DJL600

N,N-DIETHYL-N-(β-HYDROXYETHYL)AMINE see DHO500

DIETHYL(2-HYDROXYETHYL)METHYLAMMONIUM BROMIDE, α-CYCLOPENTYL-2-THIOPHENEGLYCOLATE see PBS000

DIETHYL(2-HYDROXYETHYL) METHYLAMMONIUM-BROMIDE ESTER see VBK000

DIETHYL(2-HYDROXYETHYL)METHYL-AMMONIUM

BROMIDE 3-METHYL-2-PHENYLVALERATE see VBK000

DIETHYL(2-HYDROXYETHYL)METHYLAMMONIUM BROMIDE α-PHENYLCYCLOHEXANEGLYCOLATE see ORQ000

DIETHYL(2-HYDROXYETHYL)METHYLAMMONIUM-BROMIDE XANTHENE-9-CARBOXYLATE see XCJ000

DIETHYL(2-HYDROXYETHYL)METHYLAMMONIUM 3-METHYL-2-PHENYLVALERATE BROMIDE see VBK000

O,O-DIETHYL N-HYDROXYNAPHTHALIMIDE PHOSPHATE see HMV000

O,O-DIETHYL-7-HYDROXY-3,4-TETRAMETHYLENE COUMARINYL PHOSPHOROTHIOATE see DXO000

DIETHYL HYDROXYTIN HYDROPEROXIDE see DJN489

4,4'-(1,2-DIETHYLIDENE-1,2-ETHANEDIYL)BISPHENOL see DAL600

4,4'-(DIETHYLIDENEETHYLENE)DIPHENOL see DAL600

p,p'-(DIETHYLIDENEETHYLENE)DIPHENOL see DAL600

O,O-DIETHYL-S-(N-ISOPROPYLCARBAMOYLMETHYL) DI-THIOPHOSPHATE see IOT000

O,O-DIETHYL-S-ISOPROPYLCARBAMOYLMETHYL PHOS-PHORODITHIOATE see IOT000

O,O-DIETHYL-S-(N-ISOPROPYLCARBAMOYLMETHYL) PHOSPHORODITHIOATE see IOT000

O,O-DIETHYL-S-2-ISOPROPYLMERCAPTOMETHYL-DITHIOPHOSPHATE see DJN600

O,O-DIETHYL-S-(ISOPROPYLMERCAPTOMETHYL) PHOS-PHORODITHIOATE see DJN600

O,O-DIETHYL-O-(2-ISOPROPYL-4-METHYL-PYRIMIDIN-6-YL)MONOTHIOFOSFAAT (DUTCH) see DCM750

O,O-DIETHYL-O-2-ISOPROPYL-4-METHYL-6-PYRIMIDI-NYL)PHOSPHOROTHIOATE see DCM750

O,O-DIETHYL-O-(2-ISOPROPYL-6-METHYL-4-PYRIMIDI-NYL) PHOSPHOROTHIOATE see DCM750

DIETHYL 4-(2-ISOPROPYL-6-METHYLPYRIMIDINYL) PHOSPHOROTHIONATE see DCM750

O,O-DIETHYL-O-(2-ISOPROPYL-4-METHYL-6-PYRIMIDYL) PHOSPHOROTHIOATE see DCM750

O,O-DIETHYL-O-(2-ISOPROPYL-4-METHYL-6-PYRIMIDYL) THIONOPHOSPHATE see DCM750

O,O-DIETHYL-2-ISOPROPYL-4-METHYLPYRIMIDYL-6-THIOPHOSPHATE see DCM750

O,O-DIETHYL-S-(ISOPROPYLTHIOMETHYL) PHOSPHORO-DITHIOATE see DJN600

DIETHYL KETONE see DJN750

DIETHYL LEAD DIACETATE see DJN800

N,N-DIETHYLLYSERGAMIDE see DJO000

DIETHYL MAGNESIUM see DJO100

DIETHYL MALONATE (FCC) see EMA500

DIETHYLMALONYLUREA SODIUM see BAG250

DIETHYL(2-MERCAPTOETHYL)AMINE see DIY600

DIETHYL MERCAPTOSUCCINATE-O,O-DIMETHYL DI-THIOPHOSPHATE, S-ESTER see MAK700

DIETHYL MERCAPTOSUCCINATE-O,O-DIMETHYL PHOS-PHORODITHIOATE see MAK700

DIETHYL MERCAPTOSUCCINATE-O,O-DIMETHYL THIO-PHOSPHATE see MAK700

DIETHYL MERCAPTOSUCCINATE-S-ESTER with O,O-DI-METHYLPHOSPHORODITHIOATE see MAK700

DIETHYL MERCAPTOSUCCINIC ACID O,O-DIMETHYL PHOSPHORODITHIOATE see MAK700

DIETHYL MERCURY see DJO400

N,N-DIETHYL-3-METHYLBENZAMIDE see DKC800

O,O-DIETHYL S-(N-METHYL-N-CARBOETHOXYCARBA-MOYLMETHYL) DITHIOPHOSPHATE see DJI000

O,O-DIETHYL-S-(3-METHYL-2,4-DIOXO-5-OXA-3-AZA-HEP-TYL)-DITHIOFOSFAAT (DUTCH) see DJI000

DIETHYL (4-METHYL-1,3-DITHIOLAN-2-YLIDENE)PHOS-PHOROAMIDATE see DHH400

O,O-DIETHYL-O-6-METHYL-2-ISOPROPYL-4-PYRIMIDINYL PHOSPHOROTHIOATE see DCM750

DIETHYLMETHYL METHANE see MNI500

1,1-DIETHYL-3-METHYL-3-NITROSOUREA see DJP600

N,N-DIETHYL-α-METHYL-10H-PHENOTHIAZINE-10-ETH-ANAMINE see DIR000

N,N-DIETHYL-4-METHYL-1-PIPERAZINE CARBOXAMIDE CITRATE see DIW200

N,N-DIETHYL-4-METHYL-1-PIPERAZINECARBOXAMIDE DIHYDROGEN CITRATE see DIW200

N,N-DIETHYL-4-METHYL-1-PIPERAZINECARBOXAMIDE-2-HYDROXY-1,2,3-PROPANETIRCARBOXYLATE see DIW200

O,O-DIETHYL-O-(3-METHYL-1H-PYRAZOL-5-YL)-FOSFAAT (DUTCH) see MOX250

DIETHYL-3-METHYL-5-PYRAZOLYL PHOSPHATE see MOX250

O,O-DIETHYL-O-(3-METHYL-5-PYRAZOLYL) PHOSPHATE see MOX250

O,O-DIETHYL-O-(p-(METHYLSULFINYL)PHENYL) PHOS-PHOROTHIOATE see FAQ800

O,O-DIETHYL-O-p-(METHYLSULFINYL)PHENYL THIO-PHOSPHATE see FAQ800

N,N-DIETHYL-5-METHYL-(1,2,4)TRIAZOLO(1,5-a)PYRIMI-DINE-7-AMINE see DIO200

O,O-DIETHYL-o-NAPHTHALIMIDE PHOSPHOROTHIOATE see NAQ500

O,O-DIETHYL-o-NAPHTHALOXIMIDO PHOSPHORO-THIOATE see NAQ500

O,O-DIETHYL-o-NAPHTHALOXIMIDOPHOSPHORO-THIONATE see NAQ500

O,O-DIETHYL-o-NAPHTHYLAMIDOPHOSPHOROTHIOATE see NAQ500

DIETHYL-NICOTAMIDE see DJS200

N,N-DIETHYLNICOTINAMIDE see DJS200

O,O-DIETHYL-O-(4-NITRO-FENIL)-MONOTHIOFOSFAAT (DUTCH) see PAK000

DIETHYL-p-NITROFENYL ESTER KYSELINY FOSFORECNE (CZECH) see NIM500

O,O-DIETHYL-O-p-NITROFENYLESTER KYSELINYTHIO-FOSFORECNE (CZECH) see PAK000

O,O-DIETHYL-O-p-NITROFENYLTIOFOSFAT (CZECH) see PAK000

O,O-DIETHYL-O-(p-NITROPHENYL) ESTER (DRY MIX-TURE) PHOSPHOROTHIOIC ACID see PAK250

DIETHYL p-NITROPHENYL PHOSPHATE see NIM500

O,O-DIETHYL-O-p-NITROPHENYL PHOSPHATE see NIM500

O,O-DIETHYL-O-4-NITROPHENYLPHOSPHOROTHIOATE see PAK000

O,O-DIETHYL-O-(4-NITROPHENYL) PHOSPHOROTHIOATE see PAK000

O,O-DIETHYL-O-(p-NITROPHENYL) PHOSPHOROTHIOATE see PAK000

O,S-DIETHYL-O-(4-NITROPHENYL)PHOSPHOROTHIOATE see DJT000

O,S-DIETHYL-O-(p-NITROPHENYL) PHOSPHOROTHIOATE see DJT000

O,S-DIETHYL-O-(4-NITROPHENYL)PHOSPHOROTHIOIC ACID ESTER see DJT000

O,S-DIETHYL-O-(p-NITROPHENYL)PHOSPHOROTHIOIC ACID ESTER see DJT000

DIETHYL-4-NITROPHENYL PHOSPHOROTHIONATE see PAK000

DIETHYL-p-NITROPHENYLTHIONOPHOSPHATE see PAK000

O,O-DIETHYL-O-(p-NITROPHENYL)THIONOPHOSPHATE see PAK000

DIETHYL-p-NITROPHENYLTHIOPHOSPHATE see PAK000

O,O-DIETHYL-O-4-NITROPHENYL THIOPHOSPHATE see PAK000

O,O-DIETHYL-O-p-NITROPHENYL THIOPHOSPHATE see PAK000

O,S-DIETHYL-O-(4-NITROPHENYL)THIOPHOSPHATE see DJT000

DIETHYLNITROSAMINE see NJW500

N,N-DIETHYLNITROSAMINE see NJW500

DIETHYLNITROSOAMINE see NJW500

DIETHYLOLAMINE see DHF000

O,O-DIETHYL-O,2-PYRAZINYL PHOSPHOROTHIOATE see EPC500

DIETHYL OXALATE see DJT200

5,5-DIETHYL-1,3-OXAZIN-2,4-DIONE see DJT400

5,5-DIETHYL-1,3-OXAZINE-2,4-DIONE see DJT400

DIETHYL OXIDE see EJU000

O,O-DIETHYL-S-(4-OXO-3H-1,2,3-BENZOTRIAZINE-3-YL)-METHYL-DITHIOPHOSPHATE see EKN000

O,O-DIETHYL-S-(4-OXOBENZOTRIAZINO-3-METHYL) PHOSPHORODITHIOATE see EKN000

O,O-DIETHYL-S-((4-OXO-3H-1,2,3-BENZOTRIAZIN-3-YL)-METHYL)-DITHIO FOSFAAT (DUTCH) see EKN000

DIETHYL OXYDIFORMATE see DIZ100

DIETHYL PARAOXON see NIM500

DIETHYLPARATHION see PAK000

3,3′-DIETHYLPENTAMETHINETHIACYANINE IODIDE see DJT800

2,2-DIETHYL-4-PENTENAMIDE see DJU200

DIETHYL PEROXYDICARBONATE see DJU600

DIETHYLPHENYLAMINE see DIS700

DI(p-ETHYLPHENYL)DICHLOROETHANE see DJC000

DIETHYL-p-PHENYLENEDIAMINE see DJV200

N,N-DIETHYL-p-PHENYLENEDIAMINE see DJV200

DIETHYL PHENYLTIN ACETATE see DJV800

3,3-DIETHYL-1-PHENYLTRIAZENE see PEU500

O,O-DIETHYL PHOSPHORIC ACID O-p-NITROPHENYL ES-TER see NIM500

O,O-DIETHYLPHOSPHOROCHLORIDOTHIOATE see DJW600

DIETHYL PHOSPHOROCYANIDATE see DJW800

O,O-DIETHYL PHOSPHORODITHIOATE S-ester with 3-(MER-CAPTOMETHYL)-1,2,3-BENZOTRIAZIN-4(3H)-ONE see EKN000

O,O-DIETHYLPHOSPHOROTHIOATE, O-ESTER with 6-HY-DROXY-2-PHENYL-3(2H)-PYRIDAZINONE see POP000

O,O-DIETHYL PHOSPHOROTHIOATE, o-ESTER with PHE-NYLGLYOXYLONITRILE OXIME see BAT750

DIETHYL PHTHALATE see DJX000

DIETHYL-o-PHTHALATE see DJX000

DIETHYL PROPANEDIOATE see EMA500

DIETHYL-O-2-PYRAZINYL PHOSPHOROTHIONATE see EPC500

O,O-DIETHYL-O-2-PYRAZINYL PHOSPHOTHIONATE see EPC500

O,O-DIETHYL-O-PYRAZINYL THIOPHOSPHATE see EPC500

N,N-DIETHYL-3-PYRIDINECARBOXAMIDE see DJS200

3,3-DIETHYL-1-(m-PYRIDYL)TRIAZENE see DJY000

DIETHYL PYROCARBONATE see DIZ100

DIETHYL PYROCARBONIC ACID see DIZ100

O,O-DIETHYL-O-QUINOXALIN-2-YL PHOSPHOROTHIOATE see DJY200

O,O-DIETHYL-O-(2-QUINOXALINYL) PHOSPHORO-THIOATE see DJY200

O,O-DIETHYL-O-(2-QUINOXALYL) PHOSPHOROTHIOATE see DJY200

O,O-DIETHYL-O-2-QUINOXALYLTHIOPHOSPHATE see DJY200

DIETHYL SEBACATE see DJY600

N,N-DIETHYLSELENOUREA see DJY800

1,1-DIETHYL-2-SELENOUREA see DJY800

DIETHYL SODIUM DITHIOCARBAMATE see SGJ000

DIETHYLSTANNYL DICHLORIDE see DEZ000

α,α′-DIETHYLSTILBENEDIOL see DKA600

2,2′-DIETHYL-4,4′-STILBENEDIOL see DKA600

α,α′-DIETHYL-4,4′-STILBENEDIOL see DKA600

α,α′-DIETHYL-(E)-4,4′-STILBENEDIOL see DKA600

trans-α,α′-DIETHYL-4,4′-STILBENEDIOL see DKA600

α,α′-DIETHYL-4,4′-STILBENEDIOL DIPALMITATE see DKA800

α,α′-DIETHYL-4,4′-STILBENEDIOL, DIPROPIONATE see DKB000

trans-α,α′-DIETHYL-4,4′-STILBENEDIOL DIPROPIONATE see DKB000

α,α′-DIETHYL-4,4′-STILBENEDIOL trans-DIPROPIONATE see DKB000

α,α′-DIETHYL-4,4′-STILBENEDIOL DIPROPIONYL ESTER see DKB000

α,α′-DIETHYL-4,4′-STILBENEDIOL DISODIUM SALT see DKA400

DIETHYLSTILBENE DIPROPIONATE see DKB000

DIETHYLSTILBESTEROL see DKA600

trans-DIETHYLSTILBESTEROL see DKA600

DIETHYLSTILBESTEROL DIPROPIONATE see DKB000

DIETHYLSTILBESTROL see DKA600

trans-DIETHYLSTILBESTROL see DKA600

DIETHYLSTILBESTROL DIMETHYL ETHER see DJB200

DIETHYLSTILBESTROL DIPALMITATE see DKA800

DIETHYLSTILBESTROL DIPROPIONATE see DKB000

DIETHYLSTILBESTROL DISODIUM SALT see DKA400

DIETHYLSTILBESTROL PROPIONATE see DKB000

DIETHYLSTILBOESTEROL see DKA600

trans-DIETHYLSTILBOESTEROL see DKA600

DIETHYL SUCCINATE (FCC) see SNB000

DIETHYL SULFATE see DKB110

DIETHYLSULFID (CZECH) see EPH000

DIETHYL SULFIDE (DOT) see EPH000

DIETHYL SULFIDE-2,2′-DICARBOXYLIC ACID see BHM000

5,5-DIETHYLTETRAHYDRO-2H-1,3-OXAZINE-2,4(3H)-DIONE see DJT400

O,O-DIETHYL-O-(7,8,9,10-TETRAHYDRO-6-OXOBENZO(C)CHROMAN-3-YL)PHOSPHOROTHIOATE see DXO000

O,O-DIETHYL-O-(7,8,9,10-TETRAHYDRO-6-OXO-6H-DIBENZO(b,d)PYRAN-3-YL)PHOSPHOROTHIOATE see DXO000

O,O-DIETHYL-O-(3,4-TETRAMETHYLENECOUMARINYL-7)THIOPHOSPHATE see DXO000

DIETHYLTHIADICARBOCYANINE IODIDE see DJT800

3,3′-DIETHYLTHIADICARBOCYANINE IODIDE see DJT800

N,N′-DIETHYLTHIOCARBAMIDE see DKC400

N,N-DIETHYLTHIOCARBAMYL-O,O-DIISOPROPYLDITHIOPHOSPHATE see DKB600

DIETHYLTHIOETHER see EPH000

2,2-DIETHYL-3-THIOMORPHOLINONE see DKC200

DIETHYL THIOPHOSPHORIC ACIDESTER of 3-CHLORO-4-METHYL-7-HYDROXYCOUMARIN see CNU750

DIETHYLTHIOPHOSPHORYL CHLORIDE (DOT) see DJW600

1,3-DIETHYLTHIOUREA see DKC400

1,3-DIETHYL-2-THIOUREA see DKC400

N,N′-DIETHYLTHIOUREA see DKC400

DIETHYLTIN CHLORIDE see DEZ000

DIETHYLTIN DICAPRYLATE see DIV600

DIETHYLTIN DICHLORIDE see DEZ000

DIETHYLTIN DIIODIDE see DJB000

DIETHYLTIN DIOCTANOATE see DIV600

DIETHYLTOLUAMIDE see DKC800

DIETHYL-m-TOLUAMIDE see DKC800

N,N-DIETHYL-m-TOLUAMIDE see DKC800

DIETHYL TRIAZENE see DKD200

O,O-DIETHYL-O-3,5,6-TRICHLORO-2-PYRIDYL PHOSPHOROTHIOATE see CMA100

N,N-DIETHYLVANILLAMIDE see DKE200

DIETHYL XANTHOGENATE see BJU000

DIETHYLXANTHOGEN DISULFIDE see BJU000

DI(ETHYLXANTHOGEN)TRISULFIDE see DKE400

DIETHYLZINC see DKE600

DIETIL see CAR000

DIETILAMIDE-CARBOPIRIDINA see DJS200

DIETILAMINA (ITALIAN) see DHJ200

α-DIETILAMINO-2,6-DIMETILACETANILIDE (ITALIAN) see DHK400

O,O-DIETIL-O-(4-BROMO-2,5 DICLORO-FENIL)-MONOTIOFOSFATO (ITALIAN) see EGV500

O,O-DIETIL-S-((2-CIAN-2-METIL-ETIL)-CARBAMOIL)-METIL-MONOTIOFOSFATO (ITALIAN) see PHK250

O,O-DIETIL-S-((4-CLORO-FENIL-TIO)-METILE)-DITIOFOSFATO (ITALIAN) see TNP250

O,O-DIETIL-S-(3-CLORO-4-METIL-CUMARIN-7-IL-MONO-TIOFOSFATO) (ITALIAN) see CNU750

O,O-DIETIL-S-((6-CLORO-2-OXO-BENZOSSAZOLIN-3-IL)-METIL)-DITIOFOSFATO (ITALIAN) see BDJ250

O,O-DIETIL-O-(2,4-DICLORO-FENIL)-MONOTIOFOSFATO (ITALIAN) see DFK600

5,5-DIETILDIIDRO-1,3-OSSAZIN-2,4-DIONE (ITALIAN) see DJT400

DIETILESTILBESTROL (SPANISH) see DKA600

O,O-DIETIL-S-(2-ETILTIO-ETIL)-DITIOFOSFATO (ITALIAN) see DXH325

O,O-DIETIL-S-(2-ETILTIO-ETIL)-MONOTIOFOSFATO (ITALIAN) see DAP200

O,O-DIETIL-S-(ETILTIO-METIL)-DITIOFOSFATO (ITALIAN) see PGS000

O,O-DIETIL-S-(N-ETOSSI-CARBONIL-N-METIL-CARBAMOIL-METIL)-DITIOFOSFATO (ITALIAN) see DJI000

O,O-DIETIL-O-(2-ISOPROPIL-4-METIL-PIRIMIDIN-6-IL)-MONOTIOFOSFATO (ITALIAN) see DCM750

O,O-DIETIL-O-(3-METIL-1H-PIRAZOL-5-IL)-FOSFATO (ITALIAN) see MOX250

O,O-DIETIL-O-(4-NITRO-FENIL)-MONOTIOFOSFATO (ITALIAN) see PAK000

O,O-DIETIL-S-((4-OXO-3H-1,2,3-BENZOTRIAZIN-3-IL)-METIL)-DITIOFOSFATO (ITALIAN) see EKN000

1,1-DIETOSSIETANO (ITALIAN) see AAG000

DIETROL see DKE800

DIETROXINE see DJT400

O,O-DIETYL-S-2-ETYLMERKAPTOETYLTIOFOSFAT (CZECH) see DAP200

O,O-DIETYL-o-p-NITROFENYLFOSFAT (CZECH) see NIM500

DIF 4 see DRP800

DIFACIL see DHX800

DIFEDRYL see BBV500

DIFENHYDRAMIN see BBV500

DIFENHYDRAMINE HYDROCHLORIDE see BAU750

DIFENIDRAMINA (ITALIAN) see BBV500

DIFENILHIDANTOINA (SPANISH) see DKQ000

DIFENIL-METAN-DIISOCIANATO (ITALIAN) see MJP400

DIFENIN see DKQ000, DNU000

DIFENSON see CJT750

DIFENTHOS see TAL250

N,N′-DIFENYL-p-FENYLENDIAMIN (CZECH) see BLE500

2-(DIFENYL-HYDROXYACETOXY)ETHYL-DIETHYLAMMONIUMCHLORID (CZECH) see BCA000

DIFENYLMETHAAN-DISSOCYANAAT (DUTCH) see MJP400

DIFETOIN see DNU000

DIFFLAM see BBW500

DIFFOLLISTEROL see EDP000

DIFHYDAN see DKQ000, DNU000

DIFLAVINE (ACRIDINE) see DBN000

DIFLUBENZURON see CJV250

DIFLUNISAL see DKI600

2,10-DIFLUOROBENZO(rst)PENTAPHENE see DKG400

1,1-DIFLUORO-1-CHLOROETHANE see CFX250

DIFLUOROCHLOROMETHANE see CFX500

2,10-DIFLUORODIBENZO(a,i)PYRENE see DKG400

DIFLUORODIBROMOMETHANE see DKG850

DIFLUORODICHLOROMETHANE see DFA600

DIFLUORODIMETHYLSTANNANE see DKH200

DIFLUOROETHANE see ELN500

1,1-DIFLUOROETHANE (DOT) see ELN500

1,1-DIFLUOROETHYLENE (DOT, MAK) see VPP000

DIFLUOROFORMALDEHYDE see CCA500

2′,4′-DIFLUORO-4-HYDROXY-3-BIPHENYLCARBOXYLIC ACID see DKI600

2′,4′-DIFLUORO-4-HYDROXY-(1,1′-BIPHENYL)-3-CARBOXYLIC ACID see DKI600

2′,4′-DIFLUORO-4-HYDROXY-(1′,1-DIPHENYL)-3-CARBOX-YLIC ACID see DKI600
DIFLUOROMONOCHLOROETHANE (DOT) see CFX250
DIFLUOROMONOCHLOROMETHANE see CFX500
DIFLUOROPHENYLARSINE see DKI400
5-(2,4-DIFLUOROPHENYL)SALICYLIC ACID see DKI600
DIFLUOROPHOSPHORIC ACID, anhydrous (DOT) see PHF500
1,2-DIFLUORO-1,1,2,2-TETRACHLOROETHANE see TBP050
3,3-DIFLUORO-2-(TRIFLUOROMETHYL)ACRYLIC ACID, METHYL ESTER see MNN000
3,3-DIFLUORO-2-(TRIFLUOROMETHYL)-2-PROPENOIC ACID, METHYL ESTER see MNN000
DIFLUPYL see IRF000
DIFLURON see CJV250
DIFLUROPHATE see IRF000
DIFO see BJE750
DIFOLATAN see CBF800
DIFOLLICULINE see EDP000
DIFONATE see FMU045
DIFORIN see ILD000
DIFORMAL see GIK000
DIFOSAN see CBF800
DIFUMARATE see DKK000
DIGACIN see DKN400
DIGAMMACAINE see DKK800
DIGENEA SIMPLEX MUCILAGE see AEX250
DIGERMIN see DUV600
DIGIBUTINA see BRF500
DIGILANID C see LAU000
DIGILONG see DKL800
DIGIMED see DKL800
DIGIMERCK see DKL800
DIGISIDIN see DKL800
DIGITALIN see DKL800
DIGITALINE (FRENCH) see DKL800
DIGITALINE CRISTALLISEE see DKL800
DIGITALINE NATIVELLE see DKL800
DIGITALINUM VERUM see DKL800
DIGITALIS see DKL200
DIGITALIS GLYCOSIDE see DKN400
DIGITALIS PURPUREA, LEAF see DKL200
DIGITANNOID see DKL200
DIGITIN see DKL400
DIGITONIN see DKL400
DIGITOPHYLLIN see DKL800
DIGITOXIGENIN-TRIDIGITOXOSID (GERMAN) see DKL800
DIGITOXIGENIN TRIDIGITOXOSIDE see DKL800
DIGITOXIN see DKL800
DIGLYCERIDE ACETIC ACID see DBF600
DIGLYCIDYL BISPHENOL A ETHER see BLD750
DIGLYCIDYL ETHER see DKM200
DIGLYCIDYL ETHER of 2,2-BIS(4-HYDROXYPHENYL)PRO-PANE see BLD750
DIGLYCIDYL ETHER of 2,2-BIS(p-HYDROXYPHENYL)PRO-PANE see BLD750
DIGLYCIDYL ETHER of BISPHENOL A see BLD750
DIGLYCIDYL ETHER of 4,4′-ISOPROPYLIDENEDIPHENOL see BLD750
1,3-DIGLYCIDYLOXYBENZENE see REF000
DIGLYCIDYL RESORCINOL ETHER see REF000
DIGLYCOL see DJD600
DIGLYCOLAMINE see AJU250
DIGLYCOLDINITRAAT (DUTCH) see DJE400
DIGLYCOL (DINITRATE de) (FRENCH) see DJE400
DIGLYCOL MONOBUTYL ETHER ACETATE see BQP500
DIGLYCOL MONOETHYL ETHER see CBR000
DIGLYCOL MONOETHYL ETHER ACETATE see CBQ750
DIGLYCOL MONOSTEARATE see HKJ000
DIGLYCOL STEARATE see HKJ000
DIGLYKOLDINITRAT (GERMAN) see DJE400
DIGORID A see ACI000

DIGORID B see ACI250
DIGOXIGENIN-TRIDIGITOXOSID (GERMAN) see DKN400
DIGOXIGENIN + ZUCKERKETTE WIE BEI ACETYL-DIGI-TOXIN-α (GERMAN) see ACI250
DIGOXIGENIN + ZUCKERKETTE WIE BIE ACETYL-DIGI-TOXIN A (GERMAN) see ACI000
DIGOXIN see DKN400
ε-DIGOXIN ACETATE see DKN600
DIGOXINE see DKN400
DIHDYROPYRONE see BRT000
DIHEPTYLMERCURY see DKO000
DIHEXYLAMINE see DKO600
DI-N-HEXYLAMINE see DKO600
DIHEXYLTIN DICHLORIDE see DFC200
DIHIDRAL see BBV500
DIHIDROBENZPERIDOL see DYF200
DIHIDROCLORURO de BENZIDINA (SPANISH) see BBX750
DIHYCON see DKQ000
DI-HYDAN see DKQ000, DNU000
DIHYDANTOIN see DKQ000, DNU000
DIHYDRIN see DKW800
DIHYDROAFLATOXIN B1 see AEU750
DIHYDROANETHOLE see PNE250
22,23-DIHYDROAVERMECTIN B1 see ITD875
DIHYDROAZIRENE see EJM900
DIHYDRO-1H-AZIRINE see EJM900
6,13-DIHYDROBENZO(e)(1)BENZOTHIOPYRANO(4,3-b)IN-DOLE see DKS800
7,8-DIHYDROBENZO(a)PYRENE see DKU000
9,10-DIHYDROBENZO(e)PYRENE see DKU400
9,10-DIHYDROBENZO(a)PYRENE-9,10-DIOL see DLC000
trans-4,5-DIHYDROBENZO(a)PYRENE-4,5-DIOL see DLC600
2,3-DIHYDRO-5-CARBOXANILIDO-6-METHYL-1,4-OXA-THIIN see CCC500
DIHYDROCARVEOL see DKV150
1,6-DIHYDROCARVEOL see DKV150
d-DIHYDROCARVONE see DKV175
DIHYDROCHLORIDE SALT OF DIETHYLENEDIAMINE see PIK000
3,4-DIHYDRO-6-CHLORO-7-SULFAMYL-1,2,4-BENZOTHIA-DIAZINE-1,1-DIOXIDE see CFY000
DIHYDROCHLOROTHIAZID see CFY000
DIHYDROCHLOROTHIAZIDE see CFY000
3,4-DIHYDROCHLOROTHIAZIDE see CFY000
DIHYDROCHOLESTEROL see DKW000
DIHYDROCINNAMALDEHYDE see HHP000
DIHYDROCITRONELLOL see DTE600
DIHYDROCODEINE see DKW800
7,8-DIHYDROCODEINE see DKW800
DIHYDROCODEINE ACID TARTRATE see DKX000
DIHYDROCODEINE BITARTRATE see DKX000
DIHYDROCODEINE TARTRATE see DKX000
DIHYDROCODEINE TARTRATE (1:1) see DKX000
DIHYDROCOUMARIN see HHR500
3,4-DIHYDROCOUMARIN see HHR500
9,10-DIHYDRO-8a,10,-DIAZONIAPHENANTHRENE DI-BROMIDE see DWX800
9,10-DIHYDRO-8a,10a-DIAZONIAPHENANTHRENE(1,1′-ETHYLENE-2,2′-BIPYRIDYLIUM)DIBROMIDE see DWX800
10,11-DIHYDRO-5H-DIBENZO(a,d)CYCLOHEPTEN-5-ONE OXIME see DDD000
3,10-DIHYDRO-5H-DIBENZO(a,d)CYCLOHEPTEN-5-YLI-DENE-N,N-DIMETHYL-1-PROPANAMINE see EAH500
DIHYDRO-5,5-DIETHYL-2H-1,3-OXAZINE-2,4(3H)-DIONE see DJT400
trans-3,4-DIHYDRO-3,4-DIHYDROXYBENZO(a)ANTHRA-CENE see BBD500
9,10-DIHYDRO-9,10-DIHYDROXYBENZO(a)PYRENE see DLC000
trans-4,5-DIHYDRO-4,5-DIHYDROXYBENZO(a)PYRENE see DLC600

(E)-8,9-DIHYDRO-8,9-DIHYDROXY-7,12-DIMETHYL-BENZ(a)ANTHRACENE see DLD800

trans-3,4-DIHYDRO-3,4-DIHYDROXY-7,12-DIMETHYL-BENZ(a)ANTHRACENE see DLD600

trans-8,9-DIHYDRO-8,9-DIHYDROXY-7,12-DIMETHYL-BENZ(a)ANTHRACENE see DLD800

trans-3,4-DIHYDRO-3,4-DIHYDROXY DMBA see DLD600

trans-8,9-DIHYDRO-8,9-DIHYDROXY DMBA see DLD800

(±)-(1R,2S,3R,4R)-3,4-DIHYDRO-3,4-DIHYDROXY-1,2-EPOXYBENZ(a)ANTHRACENE see DLE000

trans-8,9-DIHYDRO-8,9-DIHYDROXY-7-METHYLBENZ(a)ANTHRACENE see DLF200

7,8-DIHYDRO-7,8-DIHYDROXY-5-METHYLCHRYSENE see DLF600

14,19-DIHYDRO-12,13-DIHYDROXY(13-α,14-α)-20-NORCROTALANAN-11,15-DIONE see MRH000

1,4-DIHYDRO-1,4-DIKETONAPHTHALENE see NBA500

5,6-DIHYDRO-9,10-DIMETHOXYBENZO(g)-1,3-BENZODIOXOLO(5,6-a)QUINOLIZINIUM SULFATE TRIHYDRATE see BFN750

5,10-DIHYDRO-10-(2-(DIMETHYLAMINO)ETHYL)-5-METHYL-11H-DIBENZO(b,e)(1,4)DIAZEPIN-11-ONE see DCW600

5,6-DIHYDRO-N-(3-(DIMETHYLAMINO)PROPYL)-11H-DIBENZ(b,e)AZEPINE see DLH600

10,11-DIHYDRO-5-(3-(DIMETHYLAMINO)PROPYL)-5H-DIBENZ(b,f)AZEPINE see DLH600

10,11-DIHYDRO-5-(3-(DIMETHYLAMINO)PROPYL)-5H-DIBENZ(b,f)AZEPINE HYDROCHLORIDE see DLH630

10,11-DIHYDRO-5-(γ-DIMETHYLAMINOPROPYLIDENE)-5H-DIBENZO(a,d)CYCLOHEPTENE see EAH500

2,3-DIHYDRO-2,2-DIMETHYL-7-BENZOFURANYL METHYLCARBAMATE see CBS275

2,3-DIHYDRO-2,2-DIMETHYLBENZOFURANYL-7-N-METHYLCARBAMATE see CBS275

3,4-DIHYDRO-1,11-DIMETHYLCHRYSENE see DLI000

16,17-DIHYDRO-11,17-DIMETHYLCYCLOPENTA(a)PHENANTHRENE see DLI200

10,11-DIHYDRO-N,N-DIMETHYL-5H-DIBENZ(b,f)AZEPINE-5-PROPANAMINE MONOHYDROCHLORIDE see DLH630

10,11-DIHYDRO-N,N-DIMETHYL-5H-DIBENZO(a,d)-CYCLOHEPTENE-Δ⁵-γ-PROPYLAMINE HCL see EAI000

10,11-DIHYDRO-N,N-DIMETHYL-5H-DIBENZO(a,d)HEPTALENE-Δ⁵-γ-PROPYLAMINE see EAH500

10,11-DIHYDRO-N,N-DIMETHYL-5,10-METHANO-5H-DIBENZO(a,d)CYCLOHEPTENE-12-METHANAMINE HCl see DPK000

1,2-DIHYDRO-1,5-DIMETHYL-4-((1-METHYLETHYL)AMINO)-2-PHENYL-3H-PYRAZOL-3-ONE see INY000

3,4-DIHYDRO-2,2-DIMETHYL-4-OXO-2H-PYRAN-6-CARBOXYLIC ACID-n-BUTYL ESTER see BRT000

3,7-DIHYDRO-1,3-DIMETHYL-1H-PURINE-2,6-DIONE see TEP000

3,7-DIHYDRO-3,7-DIMETHYL-1H-PURINE-2,6-DIONE see TEO500

1,3-DIHYDRO-1,3-DIOXO-5-ISOBENZOFURANCARBOXYLIC ACID see TKV000

5,10-DIHYDRO-5,10-DIOXONAPHTHO(2,3-b)-p-DITHIIN-2,3-DICARBONITRILE see DLK200

5,6-DIHYDRO-DIPYRIDO(1,2a;2,1c)PYRAZINIUM DIBROMIDE see DWX800

7,8-DIHYDRO-4,5-α-EPOXY-17-METHYLMORPHINAN-3,6-α-DIOL DIACETATE see HBT500

DIHYDROERGOTAMINE see DLK800

DIHYDROERGOTAMINE TARTRATE (2:1) see DLL000

DIHYDRO-β-ERYTHROIDINE HYDROBROMIDE see DLL600

DIHYDROESTRIN BENZOATE see EDP000

1,2-DIHYDRO-6-ETHOXY-2,2,4-TRIMETHYLQUINOLINE see SAV000

1,4-DIHYDRO-1-ETHYL-7-METHYL-4-OXO-1,8-NAPHTHYRIDINE-3-CARBOXYLIC ACID see EID000

6,11-DIHYDRO-4-FLUORO(1)BENZOTHIOPYRANO(4,3-b)INDOLE see DLO200

DIHYDROFOLLICULAR HORMONE see EDO000

DIHYDROFOLLICULIN see EDO000

DIHYDROFOLLICULIN BENZOATE see EDP000

DIHYDRO-2,5-FURANDIONE see SNC000

2,5-DIHYDROFURAN-2,5-DIONE see MAM000

DIHYDRO-2(3H)-FURANONE see BOV000

DIHYDROGEN DIOXIDE see HIB000

DIHYDROGEN HEXACHLOROPLATINATE see CKO750

DIHYDROGEN HEXACHLOROPLATINATE(2-) see CKO750

(DIHYDROGEN MERCAPTOSUCCINATO)GOLD DISODIUM SALT see GJC000

DIHYDROHYDROXYCODEINONE see PCG500

DIHYDRO-14-HYDROXYCODEINONE see PCG500

6,7-DIHYDRO-6-(2-HYDROXYETHYL)-5H-DIBENZ(c,e)AZEPINE see DLP000

2,3-DIHYDRO-3-HYDROXY-2-IMINO-6-(1-PIPERIDINYL)-4-PYRIMIDINAMINE see DCB000

3a,12c-DIHYDRO-8-HYDROXY-6-METHOXY-7H-FURO(3′,2′:4,5)FURO(2,3-C)XANTHEN-7-ONE see SLP000

4,5-DIHYDRO-2-HYDROXYMETHYLENE-17-α-METHYL-TESTOSTERONE see PAN100

3,7-DIHYDRO-1-(3-HYDROXYPROPYL)-3,7-DIMETHYL-1H-PURINE-2,6-DIONE see HNZ000

3,7-DIHYDRO-3-HYDROXY-1H-PURINE-2,6-DIONE see HOP000

(R)-2,3-DIHYDRO-1-HYDROXY-1H-PYRROLIZINE-7-METHANOL see DAL400

4,5-DIHYDROIMIDAZOLE-2(3H)-THIONE see IAQ000

DIHYDROISOCODEINE ACID TARTRATE see DLQ000

DIHYDROISOCODEINE TARTRATE see DLQ000

DIHYDRO ISOCODEINE TARTRATE (1:1) see DLQ000

2,3-DIHYDRO-9H-ISOXAZOLO(3,2-b)QUINAZOLIN-9-ONE see DLQ400

12,β,13,α-DIHYDROJERVINE see DLQ800

1,2-DIHYDRO-2-KETOBENZISOSULFONAZOLE see BCE500

1,2-DIHYDRO-2-KETOBENZISOSULPHONAZOLE see BCE500

DIHYDROMENFORMON see EDO000

10,11-DIHYDRO-6′-METHOXYCINCHONAN-9-OL see HIG500

S-(2,3-DIHYDRO-5-METHOXY-2-OXO-1,3,4-THIADIAZOL-3-METHYL) see DSO000

10,11-DIHYDRO-5-(3-(METHYLAMINO)PROPYL)-5H-DIBENZ(b,f)AZEPINE HYDROCHLORIDE see DLS600

1,2-DIHYDRO-3-METHYL-BENZ(j)ACEANTHRYLENE see MIJ750

2,3-DIHYDRO-2-METHYLBENZOPYRANYL-7,N-METHYL-CARBAMATE see DLS800

4,5-DIHYDRO-2-((2-METHYLBENZO(b)THIEN-3-YL)METHYL)-1H-IMIDAZOLE HYDROCHLORIDE see MHJ500

5,6-DIHYDRO-2-METHYL-3-CARBOXANILIDO-1,4-OXATHIIN (GERMAN) see CCC500

7,8-DIHYDRO-5-METHYL-7,8-CHRYSENEDIOL see DLF600

15,16-DIHYDRO-11-METHYLCYCLOPENTA(a)PHENANTHREN-17-ONE see MJE500

15,16-DIHYDRO-11-METHYL-17H-CYCLOPENTA(a)PHENANTHREN-17-ONE see MJE500

10,11-DIHYDRO-N-METHYL-5H-DIBENZO(a,d)CYCLOHEPTANE-Δ,γ-PROPYLAMINE see NNY000

2,3-DIHYDRO-6-METHYL-1,4-OXATHIIN-5-CARBOXANILIDE see CCC500

5,6-DIHYDRO-2-METHYL-1,4-OXATHIIN-3-CARBOXANILIDE see CCC500

3,4-DIHYDRO-2-METHYL-4-OXO-3-o-TOLYLQUINAZOLINE see QAK000

5,6-DIHYDRO-2-METHYL-N-PHENYL-1,4-OXATHIIN-3-CARBOXAMIDE see CCC500

6,11-DIHYDRO-11-(1-METHYL-4-PIPERIDYLIDENE)-5H-BENZO(5,6)CYCLOHEPTA (1,2-b) PYRIDINE DIMALEATE see DLV800

3,4-DIHYDROMETHYL-2H-PYRAN see MJE750
(S)-(+)-5,6-DIHYDRO-6-METHYL-2H-PYRAN-2-ONE see
    PAJ500
2,3-DIHYDRO-6-METHYL-2-THIOXO-4(1H)-PYRIMIDINONE
    see MPW500
DIHYDROMORPHINE HYDROCHLORIDE see DNU310
DIHYDROMORPHINONE see DLW600
DIHYDROMORPHINONE HYDROCHLORIDE see DNU300
DIHYDRONE HYDROCHLORIDE see DLX400
DIHYDRONEOPINE see DKW800
1,2-DIHYDRO-5-NITRO-ACENAPHTHYLENE see NEJ500
1,3-DIHYDRO-7-NITRO-5-PHENYL-2H-1,4-BENZODIAZE-
    PIN-2-ONE see DLY000
DIHYDRO-1-NITROSO-2,4(1H,3H)-PYRIMIDINEDIONE see
    NJY000
5,6-DIHYDRO-1-NITROSOURACIL see NJY000
DIHYDRONORGUAIARETIC ACID see NBR000
DIHYDROOXIRENE see EJN500
2,3-DIHYDRO-3-OXOBENZISOSULFONAZOLE see BCE500
2,3-DIHYDRO-3-OXOBENZISOSULPHONAZOLE see
    BCE500
3,4-DIHYDRO-4-OXO-3-BENZOTRIAZINYLMETHYL O,O-
    DIETHYL PHOSPHORODITHIOATE see EKN000
S-(3,4-DIHYDRO-4-OXO-1,2,3-BENZOTRIAZIN-3-YL-
    METHYL) O,O-DIETHYL PHOSPHORODITHIOATE see
    EKN000
S-(3,4-DIHYDRO-4-OXO-1,2,3-BENZOTRIAZIN-3-YL-
    METHYL)- O,O-DIMETHYL PHOSPHORODITHIOATE see
    ASH500
S-(3,4-DIHYDRO-4-OXO-BENZO(α)(1,2,3)TRIAZIN-3-YL-
    METHYL)-O,O-DIMETHYL PHOSPHORODITHIOATE see
    ASH500
O-(1,6)-DIHYDRO-6-OXO-1-PHENYLPYRIDAZIN-3-LY),
    O,O-DIETHYL PHOSPHOROTHIOATE see POP000
DIHYDROOXYCODEINONE HYDROCHLORIDE see
    DLX400
4,5-DIHYDRO-N-PHENYL-N-PHENYLMETHYL-1H-IMID-
    AZOLE-2-METHANAMINE see PDC000
5-(2-(3,6-DIHYDRO-4-PHENYL-1(2H)-PYRIDYL)ETHYL)-3-
    METHYL-2-OXAZOL IDINONE see DMB000
2,3-DIHYDRO-6-PROPYL-2-THIOXO-4(1H)-PYRIMIDINONE
    see PNX000
3,7-DIHYDRO-1H-PURINE-2,6-DIONE see XCA000
1,7-DIHYDRO-6H-PURINE-6-THIONE see POK000
DIHYDROPYRAN see DMC200
3,4-DIHYDROPYRAN see DMC200
2H-3,4-DIHYDROPYRAN see DMC200
Δ²-DIHYDROPYRAN see DMC200
3,4-DIHYDRO-2H-PYRAN-2-CARBOXALDEHYDE see
    ADR500
1,2-DIHYDROPYRIDAZINE-3,6-DIONE see DMC600
1,2-DIHYDRO-3,6-PYRIDAZINEDIONE see DMC600
6,7-DIHYDROPYRIDO(1,2a;2′,1′-C)PYRAZINEDIUM DI-
    BROMIDE see DWX800
DIHYDROQUINIDINE see HIG500
10,11-DIHYDROQUINIDINE see HIG500
DIHYDROSAFROLE see DMD600
DIHYDROSTREPTOMYCIN see DME000
2,3-DIHYDROSUCCINIC ACID see TAF750
DIHYDROTACHY STEROL see IGF000
6,7-DIHYDRO-1,2,3,10-TETRAMETHOXY-7-(METHYL-
    AMINO)-BENZO(α)HEPTALEN-9(5H)-ONE see MIW500
(S)-6,7-DIHYDRO-1,2,3,10-TETRAMETHOXY-7-(METHYL-
    AMINO)-BENZO(a)HEPTALEN-9(5H)-ONE see MIW500
DIHYDROTHEELIN see EDO000
N-(5,6-DIHYDRO-4H-1,3-THIAZINYL)-2,6-XYLIDINE see
    DMW000
2,3-DIHYDROTHIIRENE see EJP500
2,3-DIHYDRO-2-THIOXO-4(1H)-PYRIMIDINONE see
    TFR250
1,2-DIHYDRO-2,2,4-TRIMETHYL-6-ETHOXYQUINOLINE
    see SAV000
3,7-DIHYDRO-1,3,7-TRIMETHYL-1H-PURINE-2,6-DIONE see
    CAK500

3,7-DIHYDRO-1,3,7-TRIMETHYL-1H-PURINE-2,6-DIONE
    MONOHYDROBROMIDE see CAK750
1,2-DIHYDRO-2,2,4-TRIMETHYLQUINOLINE see TLP500
1,4-DIHYDROXYANTHRACHINON (CZECH) see DMH000
DIHYDROXYANTHRANOL see APH250
1,8-DIHYDROXYANTHRANOL see APH250
1,8-DIHYDROXY-9-ANTHRANOL see APH250
1,4-DIHYDROXYANTHRAQUINONE see DMH000
1,4-DIHYDROXY-9,10-ANTHRAQUINONE see DMH000
6,6′-((4,8-DIHYDROXY-1,5-ANTHRAQUINONYLENE)DIIM-
    INO) DI-m-TOLUENE SULFONIC ACID DISODIUM SALT
    see TGS000
1,8-DIHYDROXY-9-ANTHRONE see APH250
3,4-DIHYDROXYBENZALDEHYDE METHYLENE KETAL
    see PIW250
1,4-DIHYDROXY-BENZEEN (DUTCH) see HIH000
1,4-DIHYDROXYBENZEN (CZECH) see HIH000
DIHYDROXYBENZENE see HIH000
m-DIHYDROXYBENZENE see REA000
o-DIHYDROXYBENZENE see CCP850
p-DIHYDROXYBENZENE see HIH000
1,2-DIHYDROXYBENZENE see CCP850
1,3-DIHYDROXYBENZENE see REA000
1,4-DIHYDROXYBENZENE see HIH000
3,3′-DIHYDROXYBENZIDINE see DMI400
2,4-DIHYDROXYBENZOFENON (CZECH) see DMI600
1,4-DIHYDROXY-BENZOL (GERMAN) see HIH000
2,4-DIHYDROXYBENZOPHENONE see DMI600
2,6-DIHYDROXY-5-BIS(2-CHLOROETHYL)AMINOPYRAMI-
    DINE see BIA250
1,3-DIHYDROXYBUTANE see BOS500
1,4-DIHYDROXYBUTANE see BOS750
2,3-DIHYDROXYBUTANE see BOT000
2,3-DIHYDROXYBUTANEDIOC ACID see TAF750
2,3-DIHYDROXYBUTANEDIOIC ACID, DIAMMONIUM
    SALT see DCH000
2,3-DIHYDROXY-(R-(R*,R*))-BUTANEDIOIC ACID DISO-
    DIUM SALT (9CI) see BLC000
DIHYDROXYCHLOROTHIAZIDUM see CFY000
3,12-DIHYDROXYCHOLANIC ACID see DAQ400
3-α,12-α-DIHYDROXYCHOLANIC ACID see DAQ400
3-α,12-α-DIHYDROXY-5-β-CHOLANOIC ACID see DAQ400
3-α,12-α-DIHYDROXY-5-β-CHOLAN-24-OIC ACID see
    DAQ400
3-α,12-α-DIHYDROXYCHOLANSAEURE (GERMAN) see
    DAQ400
1,25-DIHYDROXYCHOLECALCIFEROL see DMJ400
1a,25-DIHYDROXYCHOLECALCIFEROL see DMJ400
1-α,25-DIHYDROXYCHOLECALCIFEROL see DMJ400
DIHYDROXYCODEINONE HYDROCHLORIDE see DLX400
DI-(4-HYDROXY-3-COUMARINYL)METHANE see BJZ000
1,5-DIHYDROXY-4,8-DIAMINOANTHRACHINON (CZECH)
    see DBP909
1,5-DIHYDROXY-4,8-DIAMINOANTHRAQUINONE see
    DBP909
2,2′-DIHYDROXY-5,5′-DICHLORODIPHENYLMETHANE see
    MJM500
2,2′-DIHYDROXYDIETHYLAMINE see DHF000
DIHYDROXYDIETHYL ETHER see DJD600
β,β′-DIHYDROXYDIETHYL ETHER see DJD600
4,4′-DIHYDROXYDIETHYLSTILBENE see DKA600
4,4′-DIHYDROXY-α,β-DIETHYLSTILBENE see DKA600
DIHYDROXYDIETHYLSTILBENE DIPROPIONATE see
    DKB000
4,4′-DIHYDROXY-α,β-DIETHYLSTILBENE DIPROPIONATE
    see DKB000
4,4′-DIHYDROXY-α,β-DIETHYLSTILBENE PALMITATE
    see DKA800
trans-1,2-DIHYDROXY-1,2-DIHYDROBENZ(a)ANTHRACENE
    see BBD250
trans-3,4-DIHYDROXY-3,4-DIHYDROBENZ(a)ANTHRACENE
    see BBD500
trans-5,6-DIHYDROXY-5,6-DIHYDROBENZ(a)ANTHRACENE
    see BBE250

trans-8,9-DIHYDROXY-8,9-DIHYDROBENZ(a)ANTHRACENE
see BBE750

trans-10,11-DIHYDROXY-10,11-DIHYDROBENZ(a)ANTHRA-
CENE see BBF000

trans-4,5-DIHYDROXY-4,5-DIHYDROBENZO(a)PYRENE see
DLC600

(−)-trans-7,8-DIHYDROXY-7,8-DIHYDROBENZO(a)PYRENE
see DML200

d-7′,12′-DIHYDROXY-6,6′-DIMETHOXY-2,2′,2′-TRI-
METHYLTUBOCURARANIUM CHLORIDE see TOA000

3,4-DIHYDROXY-α-(DIMETHYLAMINOMETHYL) BENZYL
ALCOHOL see MJV000

2,5-DIHYDROXY-3-DIMETHYLAMINO-5-METHYL-2-CY-
CLOPENTEN-1-ONE see DXS200

N-(2,4-DIHYDROXY-3,3-DIMETHYLBUTYRYL)-β-ALANINE
CALCIUM see CAU750

4,4′-DIHYDROXYDIPHENYLDIMETHYLMETHANE see
BLD500

p,p′-DIHYDROXYDIPHENYLDIMETHYLMETHANE see
BLD500

4,4′-DIHYDROXYDIPHENYLDIMETHYLMETHANE DIGLY-
CIDYL ETHER see BLD750

p,p′-DIHYDROXYDIPHENYLDIMETHYLMETHANE DIGLY-
CIDYL ETHER see BLD750

4,4′-DIHYDROXYDIPHENYLPROPANE see BLD500

p,p′-DIHYDROXYDIPHENYLPROPANE see BLD500

2,2-(4,4′-DIHYDROXYDIPHENYL)PROPANE see BLD500

4,4′-DIHYDROXYDIPHENYL-2,2-PROPANE see BLD500

4,4′-DIHYDROXY-2,2-DIPHENYLPROPANE see BLD500

2,2′-DIHYDROXY-DI-n-PROPYLNITROSOAMINE see
DNB200

2,4-DIHYDROXY-3,5-DI(4-SULPHO-1-NAPHTHYLAZO)BEN-
ZYL ALCOHOL, DISODIUM SALT see CMP500

(±)-3-α,4-β-DIHYDROXY-1-α,2-α-EPOXY-1,2,3,4-TETRAHY-
DROBENZ(a)ANTHRACENE see DLE000

3,17-β-DIHYDROXYESTRA-1,3,5(10)-TRIENE see EDO000

3,17-β-DIHYDROXY-1,3,5(10)-ESTRATRIENE see EDO000

DIHYDROXYESTRIN see EDO000

1,2-DIHYDROXYETHANE see EJC500

DI-β-HYDROXYETHOXYETHANE see TJQ000

DI(2-HYDROXYETHYL)AMINE see DHF000

2,2′-DIHYDROXYETHYL ETHER see DJD600

DI(HYDROXYETHYL) ETHER DINITRATE see DJE400

3,17-β-DIHYDROXY-17-α-ETHYNYL-1,3,5(10)-ESTRA-
TRIENE see EEH500

3,17-β-DIHYDROXY-17-α-ETHYNYL-1,3,5(10)-OESTRA-
TRIENE see EEH500

3′,6′-DIHYDROXYFLUORAN see FEV000

DIHYDROXYFLUORANE see FEV000

2,2′-DIHYDROXY-3,3′,5,5′,6,6′-HEXACHLORODIPHENYL-
METHANE see HCL000

2,2′-DIHYDROXY-3,5,6,3′,5′,6′-HEXACHLORODIPHENYL-
METHANE see HCL000

7-(3,5-DIHYDROXY-2-(3-HYDROXY-1-OCTENYL)CYCLO-
PENTYL)-5-HEPTENOIC ACID see DMU800

7-(3,5-DIHYDROXY-2-(3-HYDROXY-1-OCTENYL)CYCLO-
PENTYL)-5-HEPTENOIC ACID, THAM see POC750

7-(3,5-DIHYDROXY-2-(3-HYDROXY-1-OCTENYL)CYCLO-
PENTYL)-5-HEPTENOIC ACID, TRIMETHAMINE SALT
see POC750

5,7-DIHYDROXY-2-(4-HYDROXYPHENYL)-4H-1-BENZOPY-
RAN-4-ONE see CDH250

2,3-DIHYDROXY-2-((4-HYDROXYPHENYL)METHYL)
BUTANEDIOIC ACID see HJO500

d-(+)-2,4-DIHYDROXY-N-(3-HYDROXYPROPYL)-3,3-DI-
METHYLBUTYRAMIDE see PAG200

3,4-DIHYDROXY-α-((ISOPROPYLAMINO)METHYL)BENZYL
ALCOHOL see DMV600

β,β′-DIHYDROXYISOPROPYL CHLORIDE see CDT750

5,6-DIHYDRO-2-(2,6-XYLIDINO)-4H-1,3-THIAZINE see
DMW000

3,4-DIHYDROXY-α-((METHYLAMINO)METHYL)BENZYL
ALCOHOL see VGP000

3-DI(HYDROXYMETHYL)AMINO-6-(5-NITRO-2-FURYLETH-
NEYL)-1,2,4-TRIAZINE see BKH500

3-DI(HYDROXYMETHYL)AMINO-6-(2-(5-NITRO-2-FURYL)
VINYL)-1,2,4-TRIAZINE see BKH500

1,3-DIHYDROXY-5-METHYLBENZENE see MPH500

2,12-DIHYDROXY-4-METHYL-11,16-DIOXOSENECIONA-
NIUM see DMX200

DI-4-HYDROXY-3,3′-METHYLENEDICOUMARIN see
BJZ000

DIHYDROXYMETHYL FURATRIZINE see BKH500

2,4-DIHYDROXY-2-METHYLPENTANE see HFP875

1,2-DIHYDROXY-3-(2-METHYLPHENOXY)PROPANE see
GGS000

α,β-DIHYDROXY-γ-(2-METHYLPHENOXY)PROPANE see
GGS000

d-threo-N-(1,1′-DIHYDROXY-1-p-NITROPHENYLISOPRO-
PYL)DICHLOROACETAMIDE see CDP250

2,2′-DIHYDROXY-N-NITROSODIETHYLAMINE see
NKM000

3,4-DIHYDROXYNOREPHEDRINE HYDROCHLORIDE see
AMB000

3,17-β-DIHYDROXYOESTRA-1,3,5-TRIENE see EDO000

3,17-β-DIHYDROXY-1,3,5(10)-OESTRATRIENE see EDO000

DIHYDROXYOESTRIN see EDO000

(5Z,11-α,13E,15S)-11,15-DIHYDROXY-9-OXOPROSTA-5,13-
DIEN-1-OIC ACID see DVJ200

3,5-DIHYDROXYPHENOL see PGR000

DIHYDROXY-l-PHENYLALANINE see DNA200

3,4-DIHYDROXYPHENYLALANINE see DNA200

(−)-3,4-DIHYDROXYPHENYLALANINE see DNA200

l-DIHYDROXYPHENYL-l-ALANINE see DNA200

l-α-DIHYDROXYPHENYLALANINE see DNA200

3,4-DIHYDROXYPHENYL-l-ALANINE see DNA200

3,4-DIHYDROXY-l-PHENYLALANINE see DNA200

l-3,4-DIHYDROXYPHENYLALANINE see DNA200

(−)-3-(3,4-DIHYDROXYPHENYL)-l-ALANINE see DNA200

3-(3,4-DIHYDROXYPHENYL)-l-ALANINE see DNA200

l-3-(3,4-DIHYDROXYPHENYL)ALANINE see DYC200

l-3-(3,4-DIHYDROXYPHENYL-α-ALANINE see DNA200

l-β-(3,4-DIHYDROXYPHENYL)ALANINE see DNA200

β-(3,4-DIHYDROXYPHENYL)-l-ALANINE see DNA200

β-(3,4-DIHYDROXYPHENYL)-α-ALANINE see DNA200

1-(3,4-DIHYDROXYPHENYL)-2-AMINO-1-BUTANOL HY-
DROCHLORIDE see ENX500

l-1-(3,4-DIHYDROXYPHENYL)-2-AMINOETHANOL see
NNO500

3,4-DIHYDROXYPHENYLAMINOPROPANOL HYDROCHLO-
RIDE see AMB000

α-(3,4-DIHYDROXYPHENYL)-β-DIMETHYLAMINOETHA-
NOL see MJV000

l-3,4-DIHYDROXYPHENYLETHANOLAMINE see NNO500

DIHYDROXYPHENYLETHANOLISOPROPYLAMINE see
DMV600

3,4-DIHYDROXYPHENYLETHYLMETHYLAMINE HYDRO-
CHLORIDE see EAZ000

3,4′(4,4′-DIHYDROXYPHENYL)HEX-3-ENE see DKA600

1-(3,4-DIHYDROXYPHENYL)-1-HYDROXY-2-AMINOBU-
TANE HYDROCHLORIDE see ENX500

α-(3,4-DIHYDROXYPHENYL)-α-HYDROXY-β-DIMETHYL-
AMINOETHANE see MJV000

7-(3-(2-(3,5-DIHYDROXYPHENYL-2-HYDROXY-ETHYLAMI-
NO)PROPYL)THEOPHYLLINE HYDROCHLORIDE see
DNA600

1-(3,4-DIHYDROXYPHENYL)-2-ISOPROPYLAMINOETHA-
NOL see DMV600

dl-α-3,4-DIHYDROXYPHENYL-β-ISOPROPYLAMINOETHA-
NOL SULFATE see IRU000

l-(−)-3-(3,4-DIHYDROXYPHENYL)-2-METHYLALANINE see
DNA800

l(−)-β-(3,4-DIHYDROXYPHENYL)-α-METHYLALANINE
see DNA800

3,4-DIHYDROXYPHENYL-1-METHYLAMINO-2-ETHANE
HYDROCHLORIDE see EAZ000

1-1-(3,4-DIHYDROXYPHENYL)-2-METHYLAMINOETHA-
NOL see VGP000

β-DI-p-HYDROXYPHENYLPROPANE see BLD500

2,2-DI(4-HYDROXYPHENYL)PROPANE see BLD500

3,4-DIHYDROXYPHENYLPROPANOLAMINE HYDROCHLO-
RIDE see AMB000

2-(3,4-DIHYDROXYPHENYL)-3,5,7-TRIHYDROXY-4H-1-
BENZOPYRAN-4-ONE see QCA000

2-2-(2,4-DIHYDROXYPHENYL)-3,5,7-TRIHYDROXY-4H-1-
BENZOPYRAN-4-ONE see MRN500

17,21-DIHYDROXYPREGNA-1,4-DIENE-3,11,20-TRIONE see
PLZ000

1,2-DIHYDROXYPROPANE see PML000

17R,21-α-DIHYDROXY-4-PROPYLAJMALANIUM HYDRO-
GEN TARTRATE see DNB000

DI(2-HYDROXY-n-PROPYL)AMINE see DNB200

2,3-DIHYDROXYPROPYL CHLORIDE see CDT750

N,N-DI-(2-HYDROXYPROPYL)NITROSAMINE see DNB200

4,8-DIHYDROXYQUINALDIC ACID see DNC200

4,8-DIHYDROXYQUINALDINIC ACID see DNC200

4,8-DIHYDROXYQUINOLINE-2-CARBOXYLIC ACID see
DNC200

8,8'-DIHYDROXY-RUGULOSIN see LIV000

12,18-DIHYDROXY-SENECIONAN-11,16-DIONE see
RFP000

3',6'-DIHYDROXYSPIRO(ISOBENZOFURAN-1(3H),9'(9H)-
XANTHEN)-3-ONE see FEV000

2,3-DIHYDROXYTOLUENE see DNE000

3,5-DIHYDROXYTOLUENE see MPH500

1,3-DIHYDROXY-2,4,6-TRINITROBENZENE see SMP500

2,4-DIHYDROXY-1,3,5-TRINITROBENZENE see SMP500

6-(6,10-DIHYDROXYUNDECYL)-β-RESORCYLIC ACID-mu-
LACTONE see RBF100

DIHYDROXYVITAMIN D3 see DMJ400

1-α-DIHYDROXYVITAMIN D3 see HJV000

1-α,25-DIHYDROXYVITAMIN D3 see DMJ400

1,4-DIIDROBENZENE (ITALIAN) see HIH000

DIIDRO-5,5-DIETIL-2H-1,3-OSSAZIN-2,4(3H)-DIONE (ITAL-
IAN) see DJT400

1,3-DIIMINOISOINDOLIN (CZECH) see DNE400

1,3-DIIMINOISOINDOLINE see DNE400

DIIODOACETYLENE see DNE500

DIIODOETHYNE see DNE500

3,5-DIIODO-4-HYDROXYBENZONITRILE see HKB500

3,5-DIIODO-4-HYDROXYBENZONITRILE OCTANOATE see
DNG100

DIIODOHYDROXYQUIN see DNF600

DIIODOHYDROXYQUINOLINE see DNF600

5,7-DIIODO-8-HYDROXYQUINOLINE see DNF600

2,6-DIIODO-4-NITROPHENOL see DNG000

3,5-DIIODO-4-OCTANOYLOXYBENZONITRILE see
DNG200

5,7-DIIODO-OXINE see DNF600

5,7-DIIODO-8-QUINOLINOL see DNF600

O,O-DIIOSPROPYL DITHIOPHOSPHORIC ACID ESTER OF-
N,N-S-DIETHYLTHIOCARBAMOYL-O,O-DIISOPROPYL
PHOSPHOROTHIOATE see DKB600

DIIRON TRISULFATE see FBA000

DIISOBUTILCHETONE (ITALIAN) see DNI800

DIISOBUTYL ADIPATE see DNH125

DIISOBUTYLALUMINUM CHLORIDE see CGB500

DIISOBUTYLALUMINUM MONOCHLORIDE see CGB500

DIISOBUTYLAMINE see DNH400

DIISOBUTYL CARBINOL see DNH800

DI-ISOBUTYLCETONE (FRENCH) see DNI800

DIISOBUTYLCHLOROALUMINUM see CGB500

DIISOBUTYLKETON (DUTCH, GERMAN) see DNI800

DIISOBUTYL KETONE see DNI800

DI-ISO-BUTYLNITROSAMINE see DRQ200

DIISOBUTYLOXOSTANNANE see DNJ000

DIISOBUTYLPHENOXYETHOXYETHYLDIMETHYL BEN-
ZYL AMMONIUM CHLORIDE see BEN000

DIISOBUTYL PHTHALATE see DNJ400

DIISOBUTYLTIN OXIDE see DNJ000

4-4'-DIISOCYANATE de DIPHENYLMETHANE (FRENCH)
see MJP400

DI-ISOCYANATE de TOLUYLENE see TGM750

1,3-DIISOCYANATOBENZENE see BBP000

4,4'-DIISOCYANATO-3,3'-DIMETHOXY-1,1'-BIPHENYL
see DCJ400

4,4'-DIISOCYANATODIPHENYLMETHANE see MJP400

1,6-DIISOCYANATOHEXANE see DNJ800

DI-ISO-CYANATOLUENE see TGM750

DIISOCYANATOMETHYLBENZENE see TGM740

2,6-DIISOCYANATO-1-METHYLBENZENE see TGM800

2,4-DIISOCYANATO-1-METHYLBENZENE (9CI) see
TGM750

1,5-DIISOCYANATONAPHTHALENE see NAM500

DIISOCYANATOTOLUENE see TGM740

2,4-DIISOCYANATOTOLUENE see TGM750

2,6-DIISOCYANATOTOLUENE see TGM800

DIISOCYANAT-TOLUOL see TGM750

2,3-DIISONITROSOBUTANE see DBH000

DIISOOCTYL ACID PHOSPHATE see DNK800

DIISOOCTYL ((DIOCTYLSTANNYLENE)DITHIO)DIACE-
TATE see BKK750

DIISOOCTYL PHOSPHATE (DOT) see DNK800

DIISOOCTYL PHTHALATE see ILR100

DIISOPHENOL see DNG000

DIISOPROPANOLNITROSAMINE see DNB200

N,N'-DIISOPROPIL-FOSFORODIAMMIDO-FLUORURO
(ITALIAN) see PHF750

DIISOPROPOXYPHOSPHORYL FLUORIDE see IRF000

s-DIISOPROPYLACETONE see DNI800

DI(ISOPROPYLAMIDO)PHOSPHORYLFLUORIDE see
PHF750

DIISOPROPYLAMINE see DNM200

DIISOPROPYLBENZENE PEROXIDE see DGR600

DIISOPROPYLBERYLLIUM see DNO200

N,N'-DIISOPROPYL-DIAMIDO-FOSFORZUUR-FLUORIDE
(DUTCH) see PHF750

N,N'-DIISOPROPYL-DIAMIDO-PHOSPHORSAEURE-
FLUORID (GERMAN) see PHF750

N,N'-DIISOPROPYLDIAMIDOPHOSPHORYL FLUORIDE see
PHF750

O,O-DIISOPROPYL-S-DIETHYLDITHIOCARBAMOYL-
PHOSPHORODITHIOATE see DKB600

DIISOPROPYL ESTER OF DITHIOCARBAMYL PHOSPHO-
ROTHIOIC ACID see DKB600

DIISOPROPYL ESTER SULFURIC ACID see DNO900

DIISOPROPYL ETHER see IOZ750

DIISOPROPYL FLUOROPHOSPHATE see IRF000

O,O-DIISOPROPYL FLUOROPHOSPHATE see IRF000

DIISOPROPYL FLUOROPHOSPHONATE see IRF000

DIISOPROPYLFLUOROPHOSPHORIC ACID ESTER see
IRF000

DIISOPROPYLFLUORPHOSPHORSAEUREESTER (GERMAN)
see IRF000

DIISOPROPYLIDENE ACETONE see PGW250

sym-DIISOPROPYLIDENE ACETONE see PGW250

DIISOPROPYLMERCURY see DNQ800

DIISOPROPYL-p-NITROPHENYL PHOSPHATE see DNR309

O,O-DIISOPROPYL-o,p-NITROPHENYL PHOSPHATE see
DNR309

DIISOPROPYLNITROSAMIN (GERMAN) see NKA000

DIISOPROPYL OXIDE see IOZ750

DIISOPROPYLOXOSTANNANE see DNR200

DIISOPROPYL PARAOXON see DNR309

DIISOPROPYL PERDICARBONATE see DNR400

DIISOPROPYL PEROXYDICARBONATE see DNR400

2,6-DIISOPROPYLPHENOL see DNR800

DIISOPROPYL PHOSPHOFLUORIDATE see IRF000

N,N'-DIISOPROPYLPHOSPHORODIAMIDIC FLUORIDE see
PHF750

DIISOPROPYL PHOSPHOROFLUORIDATE see IRF000

O,O'-DIISOPROPYL PHOSPHORYL FLUORIDE see
IRF000

DI-ISOPROPYLSULFAT (GERMAN) see DNO900
DI-ISOPROPYLSULFATE see DNO900
DI-ISOPROPYLTHIOLOCARBAMATE de S-(2,3-DICHLORO-
   ALLYLE) (FRENCH) see DBI200
DIISOPROPYLTIN DICHLORIDE see DNT000
DIISOPROPYLTIN OXIDE see DNR200
1,4-DIISOTHIOCYANATOBENZENE see PFA500
1,2-DIISOTHIOCYANATOETHANE see ISK000
3,5-DIJOD-4-HYDROXY-BENZONITRIL (GERMAN) see
   HKB500
3,5-DIJOD-4-HYDROXY-BENZONITRIL CAPRYSAEUREES-
   TER (GERMAN) see DNG200
DIKAIN see BQA010
DIKETENE see KFA000
DIKETENE, INHIBITED (DOT) see KFA000
2,3-DIKETOBUTANE see BOT500
1,3-DIKETOHYDRINDENE see IBS000
DIKETONE ALCOHOL see DBF750
2,5-DIKETOTETRAHYDROFURAN see SNC000
DIKONIT see SGG500
DILACORAN see IRV000
DILANGIL see MAW250
DILANTIN see DKQ000, DNU000
DILANTIN DB see DEP600
DILANTINE see DKQ000
DILANTIN SODIUM see DNU000
DILATIN DB see DEP600
DILATOL HYDROCHLORIDE see DNU200
DILATYL see DNU200
DILAUDID see DNU300
DILAUDID HYDROCHLORIDE see DNU300, DNU310
DILAUROYL PEROXIDE see LBR000
DILAUROYL PEROXIDE, TECHNICAL PURE (DOT) see
   LBR000
DILAURYLESTER KYSELINY β′,β′-THIODIPROPIONOVE
   (CZECH) see TFD500
DILAURYL THIODIPROPIONATE see TFD500
DILAURYL-β-THIODIPROPIONATE see TFD500
DILAURYL-3,3′-THIODIPROPIONATE see TFD500
DILAURYL-β′,β′-THIODIPROPIONATE see TFD500
DILEAD(II) LEAD(IV) OXIDE see LDS000
DI-LEN see DNU000
DILENE see BIM500
DILIC see HKC000
DILITHIUM CARBONATE see LGZ000
DILITHIUM CHROMATE see LHD000
DILL FRUIT OIL see DNU400
DILL HERB OIL see DNU400
DILL OIL see DNU400
DILL SEED OIL see DNU400
DILL SEED OIL, EUROPEAN TYPE see DNU400
DILL WEED OIL see DNU400
DILOMBRIN see DJT800
DILOSPAN S see PGR000
DILOSYN see MPE250
DILOXOL see GGS000
DILURAN see AAI250
DILVASENE see FMX000
1,3-DIMALEIMIDOBENZENE see BKL750
DIMANGANESE TRIOXIDE see MAT500
DIMANIN C see SGG500
DIMAPP see DQA400
DIMAPYRIN see DOT000
DIMAS see DQD400
DIMATE 267 see DSP400
DIMAZ see DXH325
DIMAZINE see DSF400
DIMAZON see ACR300
DIMEDROL see BBV500
DIMEDRYL see BBV500
DIMEFLINE see DNV000
DIMEFOX see BJE750
DIMEMORFAN PHOSPHATE see MLP250
DIMENFORMON see EDO000

DIMENFORMON BENZOATE see EDP000
DIMENFORMON DIPROPIONATE see EDR000
DIMENFORMONE see EDP000
DIMENFORMON PROLONGATUM see EDO000
DIMENHYDRINATE see DYE600
DIMEPHENTHIOATE see DRR400
DIMEPHENTHOATE see DRR400
DIMERCAPROL PROPANOL see BAD750
1,2-DIMERCAPTOETHANE see EEB000
DIMERCAPTOL see BAD750
2,3-DIMERCAPTOL-1-PROPANOL see BAD750
4,5-DI(MERCAPTOMETHYL)-2-METHYL-3-PYRIDINOL DI-
   THIOACETATE HYDROBROMIDE see DBH800
DIMERCAPTOPROPANOL see BAD750
2,3-DIMERCAPTOPROPANOL see BAD750
2,3-DIMERCAPTOPROPAN-1-OL see BAD750
4,5-DIMERCAPTOPYRIDOXINDI-THIOACETAT HYDRO-
   BROMID (GERMAN) see DBH800
DIMERCUROUS METHANE ARSONATE see DNW000
DIMER CYKLOPENTADIENU (CZECH) see DGW000
DIMESTROL see DJB200
1,6-DIMESYL-d-MANNITOL see BKM500
1,4-DIMESYLOXYBUTANE see BOT250
DIMET see DXE600
DIMETACRINE BITARTRATE see DRM000
DIMETACRIN HYDROGENTARTRATE see DRM000
DIMETAN see DRL200
DIMETATE see DSP400
DIMETAZINA see SNN300
DIMETHACHLON see DGF000
DIMETHACRINE TARTRATE see DRM000
2,5-DIMETHANESULFOMYLOXYHEXANE see DSU000
1,6-DIMETHANESULFONATE-d-MANNITOL see BKM500
1,4-DIMETHANESULFONATE THREITOL see TFU500
(2s,3s)-1,4-DIMETHANESULFONATE TREITOL see TFU500
1,4-DIMETHANESULFONOXYBUTANE see BOT250
1,4-DIMETHANESULFONOXY-1,4-DIMETHYLBUTANE see
   DSU000
1,6-DIMETHANE-SULFONOXY-d-MANNITOL see BKM500
1,4-DI(METHANESULFONYLOXY)BUTANE see BOT250
1,6-DIMETHANESULPHONOXY-1,6-DIDEOXY-d-MANNI-
   TOL see BKM500
1,4-DIMETHANESULPHONYLOXYBUTANE see BOT250
DIMETHICONE 350 see PJR000
DIMETHISOQUIN HYDROCHLORIDE see DNX400
DIMETHISTERONE and ETHINYL ESTRADIOL see
   DNX500
DIMETHOAAT (DUTCH) see DSP400
DIMETHOAT (GERMAN) see DSP400
DIMETHOATE O-ANALOG see DNX800
DIMETHOATE-ETHYL see DNX600
DIMETHOATE OXYGEN ANALOG see DNX800
DIMETHOATE PO ISOLOGUE see DNX800
DIMETHOATE (USDA) see DSP400
DIMETHOAT TECHNISCH 95% see DSP400
DIMETHOGEN see DSP400
DIMETHOXANE see ABC250
DIMETHOXON see DNX800
1,2-DIMETHOXY-4-ALLYLBENZENE see AGE250
2,6-DIMETHOXY-4-(p-AMINOBENZENESULFONAMIDO)-
   PYRIMIDINE see SNN300
(trans)-2,5-DIMETHOXY-4′-AMINOSTILBENE see DON400
2,5-DIMETHOXYAMPHETAMINE HYDROCHLORIDE see
   DOJ800
3,4-DIMETHOXYAMPHETAMINE HYDROCHLORIDE see
   DOK000
1-5,6-DIMETHOXYAPORPHINE see DNZ000, NOE500
(R)-1,2-DIMETHOXYAPORPHINE see DNZ000, NOE500
1,2-DIMETHOXY-6a-β-APORPHINE see DNZ000, NOE500
2,5-DIMETHOXYBENZENEAZO-β-NAPHTHOL see DOK200
3,3′-DIMETHOXYBENZIDIN (CZECH) see DCJ200
3,3′-DIMETHOXYBENZIDINE see DCJ200
3,3′-DIMETHOXYBENZIDINE DIHYDROCHLORIDE see
   DOA800

3,3'-DIMETHOXYBENZIDINE-4,4'-DIISOCYANATE see
DCJ400
6,7-DIMETHOXYBENZOPYRAN-2-ONE see DRS800
1-(3,4)-DIMETHOXYBENZYL-6,7-DIMETHOXYISOQUINO-
LINE-3-CARBOXYLIC ACID, SODIUM SALT see
PAG750
3,3-DIMETHOXY-(1,1'-BIPHENYL)-4,4'-DIAMINE DIHY-
DROCHLORIDE see DOA800
3,3'-DIMETHOXY-4,4'-BIPHENYLENE DIISOCYANATE see
DCJ400
6,7-DIMETHOXYCOUMARIN see DRS800
DIMETHOXY-DDT see MEI450
4,4'-DIMETHOXY-α,β-DIETHYLSTILBENE see DJB200
2,5-DIMETHOXY-α,4-DIMETHYLPHENETHYLAMINE HY-
DROCHLORIDE see DOG600
p,p'-DIMETHOXYDIPHENYLTRICHLOROETHANE see
MEI450
3,4-DIMETHOXYDOPAMINE see DOE200
DIMETHOXY-DT see MEI450
DIMETHOXYETHANE see DOE600
1,2-DIMETHOXYETHANE see DOE600
α,β-DIMETHOXYETHANE see DOE600
1,1-DIMETHOXYETHANE (DOT) see DOO600
1,2-DIMETHOXYETHANE (DOT) see DOE600
(2,2-DIMETHOXYETHYL)-BENZENE (9CI) see PDX000
DIMETHOXY ETHYL PHTHALATE see DOF400
DI(2-METHOXYETHYL)PHTHALATE see DOF400
DIMETHOXYMETHANE see MGA850
2,5-DIMETHOXY-4-METHYLAMPHETAMINE HYDRO-
CHLORIDE see DOG600
2,5-DIMETHOXY-α-METHYLBENZENEETHANAMINE HY-
DROCHLORIDE see DOJ800
9,10-DIMETHOXY-2,3-(METHYLENEDIOXY)-7,8,13,13A-
TETRAHYDROBERBINIUM see BFN500
2,5-DIMETHOXY-α-METHYLPHENETHYLAMINE HYDRO-
CHLORIDE see DOJ800
1-(2,5-DIMETHOXY-4-METHYLPHENYL)-2-AMINOPRO-
PANE see DOG600
2,5-DIMETHOXY-α-METHYL-β-PHENYLETHYLAMINE HY-
DROCHLORIDE see DOJ800
3,4-DIMETHOXY-α-METHYL-β-PHENYLETHYLAMINEHY-
DROCHLORIDE see DOK000
3,4-DIMETHOXYPHENETHYLAMINE see DOE200
3,4-DIMETHOXY-β-PHENETHYLAMINE see DOE200
3,4-DIMETHOXYPHENETHYLAMINE HYDROCHLORIDE
see DOI400
4-(2,5-DIMETHOXYPHENETHYL)ANILINE see DON400
5-((3,4-DIMETHOXYPHENETHYL)METHYLAMINO)-2-(3,4-
DIMETHOXYPHENYL)-2-ISOPROPYLVALERONITRILE
see IRV000
2,6-DIMETHOXYPHENOL see DOJ200
1-(2,5-DIMETHOXYPHENYL)-2-AMINOPROPANE see
DOJ800
1-(3,4-DIMETHOXYPHENYL)-2-AMINOPROPANE see
DOK000
1-((2,5-DIMETHOXYPHENYL)AZO)-2-NAPHTHALENOL see
DOK200
1-((2,5-DIMETHOXYPHENYL)AZO)-2-NAPHTHOL see
DOK200
2,5-DIMETHOXY-1-(PHENYLAZO)-2-NAPHTHOL see
DOK200
1-(1-(2,5-DIMETHOXYPHENYL)AZO)-2-NAPHTHOL see
DOK200
1,1-DIMETHOXY-2-PHENYLETHANE see PDX000
DIMETHOXYPHENYLETHYLAMINE see DOE200
3,4-DIMETHOXYPHENYLETHYLAMINE see DOE200
3,4-DIMETHOXYPHENYLETHYLAMINE (base) see DOE200
2-(3,4-DIMETHOXYPHENYL)ETHYLAMINE see DOE200
3,4-DIMETHOXY-β-PHENYLETHYLAMINE see DOE200
β-(3,4-DIMETHOXYPHENYL)ETHYLAMINE see DOE200
3,4-DIMETHOXY-β-PHENYLETHYLAMINE HYDROCHLO-
RIDE see DOI400
4-(2-(2,5-DIMETHOXYPHENYL)ETHYL)BENZENAMINE see
DON400

β-(2,5-DIMETHOXYPHENYL)-β-HYDROXYISOPROPYL-
AMINE HYDROCHLORIDE see MDW000
β-(2,5-DIMETHOXYPHENYL)ISOPROPYLAMINE HYDRO-
CHLORIDE see DOJ800
1-((3,4-DIMETHOXYPHENYL)METHYL)-6,7-DIMETHOXY-
ISOQUINOLINE see PAH000
1-(3,4-DIMETHOXYPHENYL)-2-PROPENE see AGE250
2,2-DI-(p-METHOXYPHENYL)-1,1,1-TRICHLOROETHANE
see MEI450
DI(p-METHOXYPHENYL)-TRICHLOROMETHYL METHANE
see MEI450
(DIMETHOXYPHOSPHINOTHIOYL)THIO)BUTANEDIOIC
ACID DIETHYL ESTER see MAK700
3-((DIMETHOXYPHOSPHINYL)OXY)-2-BUTENOIC ACID
METHYL ESTER see MQR750
3-(DIMETHOXYPHOSPHINYLOXY)-N,N-DIMETHYL-cis-
CROTONAMIDE see DGQ875
3-(DIMETHOXYPHOSPHINYLOXY)-N,N-DIMETHYLISO-
CROTONAMIDE see DGQ875
3-(DIMETHOXYPHOSPHINYLOXY)-N-METHYL-N-ME-
THOXY-cis-CROTONAMIDE see DOL800
3-(DIMETHOXYPHOSPHINYLOXY)N-METHYL-cis-CROTO-
NAMIDE see MRH209
1,2-DIMETHOXY-4-PROPENYLBENZENE see IKR000
N^1-(2,6-DIMETHOXY-4-PYRIMIDINYL)SULFANILAMIDE
see SNN300
4-(2,5-DIMETHOXY)STILBENAMINE see DON400
2',5'-DIMETHOXYSTILBENAMINE see DON400
2,5-DIMETHOXY-4'-STILBENAMINE see DON400
DIMETHOXY STRYCHNINE (DOT) see BOL750
2,3-DIMETHOXYSTRYCHNINE see BOL750
DIMETHOXYSULFADIAZINE see SNN300
2,4-DIMETHOXY-6-SULFANILAMIDO-1,3-DIAZINE see
SNN300
2,6-DIMETHOXY-4-SULFANILAMIDOPYRIMIDINE see
SNN300
DIMETHOXY-2,2,2-TRICHLORO-1-N-BUTYRYLOXY-
ETHYLPHOSPHINE OXIDE see BPG000
DIMETHOXY-2,2,2-TRICHLORO-1-HYDROXY-ETHYL-
PHOSPHINE OXIDE see TIQ250
6,7-DIMETHOXY-1-VERATRYLISOQUINOLINE see
PAH000
6,7-DIMETHOXY-1-VERATRYLISOQUINOLINE-3-CARBOX-
YLIC ACID SODIUM SALT see PAG750
DIMETHOXYVIOLANTHRONE see JAT000
16,17-DIMETHOXYVIOLANTHRONE see JAT000
DIMETHYL see EDZ000
DIMETHYLACETAL see DOO600
DIMETHYLACETAMIDE see DOO800
N,N-DIMETHYLACETAMIDE see DOO800
O,O-DIMETHYL-S-(2-ACETAMIDOETHYL) ESTER PHOS-
PHORODITHIOIC ACID see DOP200
DIMETHYLACETIC ACID see IJU000
DIMETHYLACETONE see DJN750
DIMETHYLACETONE AMIDE see DOO800
O,O-DIMETHYL-S-(2-(ACETYLAMINO)ETHYL) DITHIO-
PHOSPHATE see DOP200
O,O-DIMETHYL-S-(2-ACETYLAMINOETHYL) PHOSPHORO-
DITHIOATE see DOP200
DIMETHYLACETYLENE see COC500
O,S-DIMETHYLACETYLPHOSPHOROAMIDOTHIOATE see
DOP600
DIMETHYLAETHANOLAMIN (GERMAN) see DOY800
O,O-DIMETHYL-S-(2-AETHYLSULFINYL-AETHYL)-THIOL-
PHOSPHAT (GERMAN) see DAP000
O,O-DIMETHYL-S-(2-AETHYLSULFONYL-AETHYL)-THIOL-
PHOSPHAT (GERMAN) see DAP600
O,O-DIMETHYL-S-(2-AETHYLTHIO-AETHYL)-DITHIO
PHOSPHAT (GERMAN) see PHI500
O,O-DIMETHYL-O-(2-AETHYLTHIO-AETHYL MONOTHIO-
PHOSPHAT (GERMAN) see DAO800
O,O-DIMETHYL-S-(2-AETHYLTHIO-AETHYL)-MONOTHIO-
PHOSPHAT (GERMAN) see DAP400
DIMETHYL ALDEHYDE see DOO600

2-(3,3-DIMETHYLALLYL)CYCLAZOCINE see DOQ400
2-DIMETHYLALLYL-5,9-DIMETHYL-2′-HYDORXYBENZO-
MORPHAN see DOQ400
2-(3,3-DIMETHYLALLYL)-5-ETHYL-2′-HYDROXY-9-
METHYL-6,7-BENZOMORPHAN see DOQ600
2-(3,3-DIMETHYLALLYL)-2′,2′-HYDROXY-5,9-DIMETHYL-
6,7-BENZOMORPHAN see DOQ400
DIMETHYLAMIDE ACETATE see DOO800
DIMETHYLAMIDOETHOXYPHOSPHORYL CYANIDE see
EIF000
DIMETHYLAMINE see DOQ800
DIMETHYLAMINE (anhydrous) see DOR000
DIMETHYLAMINE, anhydrous (DOT) see DOQ800
DIMETHYLAMINE, aqueous solution (DOT) see DOQ800
DIMETHYLAMINE, solution (DOT) see DOQ800
DIMETHYLAMINE BENZHYDRYL ESTER HYDROCHLO-
RIDE see BAU750
DIMETHYLAMINE BORANE see DOR200
4-DIMETHYLAMINE m-CRESYL METHYLCARBAMATE see
DOR400
4-(DIMETHYLAMINE)-3,5-XYLYL-N-METHYLCARBAMATE
see DOS000
DIMETHYLAMINOACETONITRILE see DOS200
N′,N′-DIMETHYL-4′-AMINO-N-ACETYL-N-MONOMETHYL-
4-AMINOAZOBENZENE see DPQ200
DIMETHYLAMINOAETHANOL (GERMAN) see DOY800
β-DIMETHYLAMINO-AETHYL-BENZHYDRYL-AETHER
(GERMAN) see BBV500
N-DIMETHYLAMINO-AETHYL-N-p-METHOXY-BENZYL-α-
AMINO-PYRIDIN-MALEAT (GERMAN) see WAK000
5-(DIMETHYLAMINOAETHYL-OXYIMINO)-5H-
DIBENZO(a,d)CYCLOHEPTA-1,4-DIENHYDROCHLORID
(GERMAN) see DPH600
DIMETHYLAMINO-ANALGESINE see DOT000
DIMETHYLAMINOANTIPYRINE see DOT000
4-(DIMETHYLAMINO)ANTIPYRINE see DOT000
p-DIMETHYLAMINOAZOBENZEN (CZECH) see DOT300
DIMETHYLAMINOAZOBENZENE see DOT300
4-DIMETHYLAMINOAZOBENZENE see DOT300
p-DIMETHYLAMINOAZOBENZENE see DOT300
4-(N,N-DIMETHYLAMINO)AZOBENZENE see DOT300
N,N-DIMETHYL-4-AMINOAZOBENZENE see DOT300
N,N-DIMETHYL-p-AMINOAZOBENZENE see DOT300
2′,3-DIMETHYL-4-AMINOAZOBENZENE see AIC250
4-DIMETHYLAMINOAZOBENZENE AMINE-N-OXIDE see
DTK600
p-(DIMETHYLAMINO)AZOBENZENE-o-CARBOXYLIC ACID
see CCE500
4′-DIMETHYLAMINOAZOBENZENE-2-CARBOXYLIC ACID
see CCE500
N,N-DIMETHYLAMINOAZOBENZENE-N-OXIDE see
DTK600
DIMETHYLAMINOAZOBENZOL see DOT300
4-DIMETHYLAMINOAZOBENZOL see DOT300
p-DIMETHYLAMINO-AZOBENZOL (GERMAN) see DOT300
DIMETHYLAMINOAZOPHENE see DOT000
1-(4-DIMETHYLAMINOBENZAL)INDENE see DOT600
p-DIMETHYLAMINOBENZALRHODANINE see DOT800
5-(p-DIMETHYLAMINOBENZAL)RHODANINE see DOT800
p-(DIMETHYLAMINO)BENZAL-5-RHODANINE see
DOT800
(DIMETHYLAMINO)BENZENE see DQF800
p-DIMETHYLAMINOBENZENEAZO-1-NAPHTHALENE see
DSU600
p-DIMETHYLAMINOBENZENE-1-AZO-1-NAPHTHALENE
see DSU600
p-DIMETHYLAMINOBENZENE DIAZO SODIUM SULFO-
NATE see DOU600
p-DIMETHYLAMINOBENZENEDIAZOSODIUM SULPHO-
NATE see DOU600
p-(DIMETHYLAMINO)BENZENEDIAZOSULFONATE see
DOU600
4-DIMETHYLAMINOBENZENEDIAZOSULFONIC ACID, SO-
DIUM SALT see DOU600

p-DIMETHYLAMINOBENZENEDIAZOSULFONIC ACID, SO-
DIUM SALT see DOU600
p-(DIMETHYLAMINO)BENZENEDIAZOSULPHONATE see
DOU600
4-DIMETHYLAMINOBENZENEDIAZOSULPHONIC ACID,
SODIUM SALT see DOU600
p-(DIMETHYLAMINO)BENZENEDIAZOSULPHONIC ACID,
SODIUM SALT see DOU600
p-DIMETHYLAMINOBENZOLDIAZOSULFONAT (NATRIUM-
SALZ) (GERMAN) see DOU600
4,4′-DIMETHYLAMINOBENZOPHENONIMIDE see IBB000
5-(p-DIMETHYLAMINOBENZOYLIDENE)RHODANINE see
DOT800
(4-DIMETHYLAMINOBENZYLIDENE)INDENE see DOT600
p-DIMETHYLAMINOBENZYLIDENE RHODAMINE see
DOT800
2′,3-DIMETHYL-4-AMINOBIPHENYL see BLV250
3,2′-DIMETHYL-4-AMINOBIPHENYL see BLV250
N-DIMETHYL AMINO-β-CARBAMYL PROPIONIC ACID see
DQD400
(DIMETHYLAMINO)CARBONYL CHLORIDE see DQY950
1-(N,N-DIMETHYLAMINO)-3-(p-CHLOROPHENYL-3-α-PYRI-
DYL)PROPANE MALEATE see TAI500
3-β-(DIMETHYLAMINO)CON-5-ENINE-DIHYDROBROMIDE
see DOX000
4-DIMETHYLAMINO-3-CRESYL METHYLCARBAMATE see
DOR400
DIMETHYLAMINOCYANPHOSPHORSAEUREAETHYL-
ESTER (GERMAN) see EIF000
(DIMETHYLAMINO)CYCLOHEXANE see DRF709
N,N-DIMETHYLAMINOCYCLOHEXANE see DRF709
4-(DIMETHYLAMINO)-1,2-DIHYDRO-1,5-DIMETHYL-2-PHE-
NYL-3H-PYRAZOL-3-ONE see DOT000
4-(DIMETHYLAMINO)-3,5-DIMETHYLPHENOL METHYL-
CARBAMATE (ESTER) see DOS000
4-(DIMETHYLAMINO)-3,5-DIMETHYLPHENYL ESTER,
METHYLCARBAMIC ACID see DOS000
4-(DIMETHYLAMINO)-3,5-DIMETHYLPHENYL-N-METHYL-
CARBAMATE see DOS000
4-DIMETHYLAMINO-2,3-DIMETHYL-1-PHENYL-3-PYRAZO-
LIN-5-ONE see DOT000
4-DIMETHYLAMINO-2,3-DIMETHYL-1-PHENYL-5-PYRAZO-
LONE see DOT000
3-DIMETHYLAMINO-1,2-DIMETHYLPROPYL p-AMINOBEN-
ZOATE HYDROCHLORIDE see AIT750
3,2′-DIMETHYL-4-AMINODIPHENYL see BLV250
6-DIMETHYLAMINO-4,4-DIPHENYL-3-HEPTANONE HY-
DROCHLORIDE see MDP000
1-6-DIMETHYLAMINO-4,4-DIPHENYL-3-HEPTANONE HY-
DROCHLORIDE see MDP250
p,p-DIMETHYLAMINODIPHENYLMETHANE see MJN000
α-(+)-4-DIMETHYLAMINO-1,2-DIPHENYL-3-METHYL-2-
BUTANOL PROPIONATE ESTER see DAB879
4-(DIMETHYLAMINO)-2,2-DIPHENYLVALERAMIDE see
DOY400
DIMETHYLAMINOETHANOL see DOY800
2-(DIMETHYLAMINO)ETHANOL see DOY800
N-DIMETHYLAMINOETHANOL see DOY800
β-DIMETHYLAMINOETHANOL see DOY800
N,N-DIMETHYLAMINOETHANOL see DOY800
β-DIMETHYLAMINOETHANOL DIPHENYLMETHYL ETHER
see BBV500
2-(DIMETHYLAMINO)ETHANOL METHACRYLATE see
DPG600
1-(β-DIMETHYLAMINOETHOXY)-3-N-BUTYLISOQUINO-
LINE HYDROCHLORIDE see DNX400
1-(β-DIMETHYLAMINOETHOXY)-3-N-BUTYLISOQUINO-
LINE MONOHYDROCHLORIDE see DNX400
α-(2-DIMETHYLAMINOETHOXY)DIPHENYLMETHANE see
BBV500
β-DIMETHYLAMINOETHYL ALCOHOL see DOY800
β-DIMETHYLAMINOETHYLBENZHYDRYLETHER see
BBV500
β-DIMETHYLAMINOETHYL BENZHYDRYL ETHER HY-
DROCHLORIDE see BAU750

DIMETHYLAMINOETHYL BENZILATE, HYDROCHLORIDE
see BAW500
2-(DIMETHYLAMINO)ETHYL BENZILATE HYDROCHLO-
RIDE see BAW500
β-DIMETHYLAMINOETHYL BENZILATE HYDROCHLO-
RIDE see BAW500
DIMETHYLAMINOETHYL BENZYLATE HYDROCHLORIDE
see BAW500
2-(α-(2-DIMETHYLAMINOETHYL)BENZYL)PYRIDINE see
TMJ750
β-DIMETHYLAMINOETHYL-p-BROMO-α-METHYLBENZ-
HYDRYL ETHER HYDROCHLORIDE see BMN250
DIMETHYLAMINOETHYL-p-BUTYL-AMINOBENZOATE see
BQA010
2-DIMETHYLAMINOETHYL-p-BUTYLAMINOBENZOATE
see BQA010
DIMETHYLAMINOETHYL CHLORIDE see CGW000
2-DIMETHYLAMINOETHYLCHLORIDE see CGW000
β-(DIMETHYLAMINO)ETHYL CHLORIDE see CGW000
10-(2-(DIMETHYLAMINO)ETHYL)-5,10-DIHYDRO-5-
METHYL-11H-DIBENZO(B,E)(1,4)DIAZEPIN-11-ONE see
DCW600
DIMETHYLAMINOETHYL DIPHENYLHYDROXYACETATE
HYDROCHLORIDE see BAW500
2-(DIMETHYLAMINO)ETHYL ESTER METHACRYLIC ACID
see DPG600
3-(β-DIMETHYLAMINOETHYL)-5-HYDROXYINDOLE see
DPG109
3-(2-(DIMETHYLAMINO)ETHYL)INDOLE see DPF600
3-(2-DIMETHYLAMINOETHYL)INDOL-4-OL see HKE000
3-(2-DIMETHYLAMINOETHYL)-5-INDOLOL see DPG109
3-(2-(DIMETHYLAMINO)ETHYL)-1H-INDOL-4-OL DIHY-
DROGEN PHOSPHATE ESTER see PHU500
3-2′-DIMETHYLAMINOETHYLINDOL-4-PHOSPHATE see
PHU500
3-(2-DIMETHYLAMINOETHYL)INDOL-4-YL DIHYDROGEN
PHOSPHATE see PHU500
DIMETHYLAMINOETHYL METHACRYLATE see DPG600
2-(DIMETHYLAMINO)ETHYL METHACRYLATE see
DPG600
β-DIMETHYLAMINOETHYL METHACRYLATE see
DPG600
N,N-DIMETHYLAMINOETHYL METHACRYLATE see
DPG600
N-DIMETHYLAMINOETHYL-N-p-METHOXY-α-AMINOPY-
RIDINE MALEATE see DBM800
2-((2-(DIMETHYLAMINO)ETHYL)-(p-METHOXYBENZYL)
AMINO)PYRIDINE see WAK000
2-((2-(DIMETHYLAMINO)ETHYL)(p-METHOXYBENZYL)
AMINO)PYRIDINE BIMALEATE see DBM800
2-((2-(DIMETHYLAMINO)ETHYL)(p-METHOXYBENZYL)
AMINO)PYRIDINE MALEATE see DBM800
2-DIMETHYLAMINOETHYL-2-METHYL-BENZHYDRYL
ETHER CITRATE see DPH000
2-DIMETHYLAMINO-2-METHYLBENZHYDRYL
ETHERHYDROCHLORIDE see OJW000
10-(2-(DIMETHYLAMINO)ETHYL)-5-METHYL-5H-DI-
BENZO(b,e)(1,4)DIAZEPIN-11(10H)-ONE see DCW600
5-DIMETHYLAMINOETHYLOXYIMINO-5H-DIBENZO
(a,d)CYCLOHEPTA-1,4-DIENE HYDROCHLORIDE see
DPH600
β-DIMETHYLAMINO ETHYL-2-PYRIDYLAMINOTOLUENE
see TMP750
β-DIMETHYLAMINOETHYL-2-PYRIDYLBENZYLAMINE see
TMP750
2-DIMETHYLAMINOETHYL SUCCINATE DIMETHOCHLO-
RIDE see HLC500
2-((2-(DIMETHYLAMINO)ETHYL)-2-THENYLAMINO)PYRI-
DINE see TEO250
4-(DIMETHYLAMINO)-4′-FLUOROAZOBENZENE see
DSA000
DIMETHYLAMINO HEXOSE REDUCTIONE see DXS200
DIMETHYLAMINO-ISOPROPYL-PHENTHIAZIN (GERMAN)
see DQA400

2-(DIMETHYLAMINO)-N-(((METHYLAMINO)CARBONYL)
OXY)-2-OXOETHANIMIDOTHIOIC ACID METHYL ESTER
see DSP600
4-(N,N-DIMETHYLAMINO)-3′-METHYLAZOBENZENE see
DUH600
4-(DIMETHYLAMINO)-3-METHYL-2-BUTANOL 4-AMINO-
BENZOATE (ester) HYDROCHLORIDE see AIT750
1-(DIMETHYLAMINO)-2-METHYL-2-BUTANOL BENZOATE
(ESTER) see AOM000
anti-8-(N,N-DIMETHYLAMINOMETHYL)DIBENZOBICYCLO
(3.2.1)OCTADIENE HYDROCHLORIDE see DPK000
d-4-DIMETHYLAMINO-3-METHYL-1,2-DIPHENYL-2-BUTA-
NOL PROPIONATE HYDROCHLORIDE see PNA500
6-DIMETHYLAMINO-5-METHYL-4,4-DIPHENYL-3-HEXA-
NONE see IKZ000
m-(((DIMETHYLAMINO)METHYLENE)AMINO)
PHENYLMETHYL CARBAMATE,HYDROCHLORIDE see
DSO200
3-DIMETHYLAMINOMETHYLENEIMINOPHENYL-N-
METHYLCARBAMATE, HYDROCHLORIDE see
DSO200
(2-(DIMETHYLAMINO-2-METHYL)ETHYL-N-DIBENZO-
PARATHIAZINE see DQA400
10-(2-(DIMETHYLAMINO)-2-METHYLETHYL)PHENO-
THIAZINE see DQA400
N-(2′-DIMETHYLAMINO-2′-METHYL)ETHYLPHENOTHI-
AZINE see DQA400
s-α-(2-(DIMETHYLAMINO)-1-METHYLETHYL)-α-PHENYL-
BENZENEETHANOL PROPIOATE HYDROCHLORIDE see
PNA500
N-DIMETHYLAMINO-2-METHYLETHYL THIODIPHENYL-
AMINE see DQA400
trans-2-((DIMETHYLAMINO)METHYLIMINO)-5-(2-(5-NITRO-
2-FURYL)VINYL)-1,3,4-OXADIAZOLE see DPL000
8-(DIMETHYLAMINOMETHYL)-7-METHOXY-3-METHYL-
FLAVONE see DNV000
8-((DIMETHYLAMINO)METHYL)-7-METHOXY-3-METHYL-
2-PHENYLFLAVONE see DNV000
2-DIMETHYLAMINOMETHYL-1-(m-METHOXYPHENYL)CY-
CLOHEXANOL see DPL200
(±)-trans-2-((DIMETHYLAMINO)METHYL-1-(m-METHOXY-
PHENYL)CYCLOHEXANOL see THJ500
3-(DIMETHYLAMINO)-1-METHYL-3-OXO-1-PROPENYL DI-
METHYL PHOSPHATE see DGQ875
4-(DIMETHYLAMINO)-3-METHYLPHENOL METHYL CAR-
BAMATE (ester) see DOR400
(4-DIMETHYLAMINO-3-METHYL-PHENYL)N-METHYL-
CARBAMAAT (DUTCH) see DOR400
(4-DIMETHYLAMINO-3-METHYL-PHENYL)N-METHYL-
CARBAMAT (GERMAN) see DOR400
(4-DIMETHYLAMINO-3-METHYL-PHENYL)N-METHYL-
CARBAMATE see DOR400
α-(DIMETHYLAMINOMETHYL)PROTOCATECHUYL ALCO-
HOL see MJV000
2-DIMETHYLAMINO-1-(METHYLTHIO)GLYOXAL-o-
METHYLCARBAMOYLMONOXIME see DSP600
1-DIMETHYLAMINONAPHTHALENE see DSU400
DIMETHYLAMINONAPHTHALENESULFONYL CHLORIDE
see DPN200
1-DIMETHYLAMINONAPHTHALENE-5-SULFONYL CHLO-
RIDE see DPN200
1-(DIMETHYLAMINO)-5-NAPHTHALENESULFONYLCHLO-
RIDE see DPN200
5-(DIMETHYLAMINO)-1-NAPHTHALENESULFONYL CHLO-
RIDE see DPN200
5-DIMETHYLAMINONAPHTHYL-5-SULFONYL CHLORIDE
see DPN200
4-(DIMETHYLAMINO)NITROSOBENZENE see DSY600
p-(DIMETHYLAMINO)NITROSOBENZENE see DSY600
5-(DIMETHYLAMINOOXYIMINO)-5H-DIBENZO(a,b)CY-
CLOHEPTA-1,4-DIENE HYDROCHLORIDE see DPH600
DIMETHYLAMINOPHENAZON (GERMAN) see DOT000
DIMETHYLAMINOPHENAZONE see DOT000
4-DIMETHYLAMINOPHENAZONE see DOT000

p-DIMETHYLAMINOPHENYLAMINE see DTL800
4-DIMETHYLAMINOPHENYLAZOBENZENE see DOT300
2-((4-DIMETHYLAMINO)PHENYLAZO)BENZOIC ACID see
CCE500
o-((p-(DIMETHYLAMINO)PHENYL)AZO)BENZOIC ACID see
CCE500
4-(p-(DIMETHYLAMINO)PHENYL)AZO)-N-METHYLACE-
TANILIDE see DPQ200
N-(4-((4-(DIMETHYLAMINO)PHENYL)AZO)PHENYL)-N-
METHYLACETAMIDE see DPQ200
6-((p-(DIMETHYLAMINO)PHENYL)AZO)QUINOLINE see
DPR000
(4-(DIMETHYLAMINO)PHENYL)DIAZENESULFONIC ACID,
SODIUM SALT see DOU600
4-((DIMETHYLAMINO)PHENYL)DIAZENESULFONIC ACID,
SODIUM SALT see DOU600
p-(DIMETHYLAMINO)-PHENYLDIAZO-NATRIUMSULFO-
NAT (GERMAN) see DOU600
DIMETHYLAMINOPHENYLDIMETHYLPYRAZOLIN see
DOT000
4-DIMETHYLAMINO-1-PHENYL-2,3-DIMETHYLPYRAZO-
LONE see DOT000
2-(p-DIMETHYLAMINOPHENYL)-1,6-DIMETHYLQUINOLI-
NIUM CHLORIDE see DPS200
2-(3-DIMETHYLAMINO-1-PHENYLPROPYL)PYRIDINE see
TMJ750
3-(DIMETHYLAMINO)PROPIONITRILE see DPU000
β-DIMETHYLAMINOPROPIONITRILE see DPU000
3-(DIMETHYLAMINO)PROPYLAMINE see AJP750
N,N-DIMETHYL-N-(3-AMINOPROPYL)AMINE see AJP750
2,2′-(3-DIMETHYLAMINOPROPYLAMINO)BIBENZYL see
DLH600
5-(3-(DIMETHYLAMINO)PROPYL)-5H-DIBENZ(b,f)AZEPINE
see DPW600
1-(3-DIMETHYLAMINOPROPYL)-4,5-DIHYDRO-2,3,6,7-DI-
BENZAZEPINE see DLH600
5-(3-(DIMETHYLAMINO)PROPYL)-10,11-DIHYDRO-5H-DI-
BENZ(b,f)AZEPINE HYDROCHLORIDE see DLH630
5-(3-(DIMETHYLAMINO)PROPYL)-10,11-DIHYDRO-5H-DI-
BENZ(b,f)AZEPINE-5-OXIDE see IBP309
5-(3-(DIMETHYLAMINO)PROPYL)-10,11-DIHYDRO-5H-DI-
BENZO(b,f)AZEPINE see DLH600
10-(3-(DIMETHYLAMINO)PROPYL)-9,9-DIMETHYLACRI-
DAN TARTRATE (1:1) see DRM000
5-(3′-DIMETHYLAMINOPROPYLIDENE)-DIBENZO-
(a,d)(1,4)-CYCLOHEPTADIENE see EAH500
3-(3-DIMETHYLAMINOPROPYLIDENE)-1:2-4:5-DIBENZO-
CYCLOHEPTA-1:4-DIENE see EAI000
5-(3-DIMETHYLAMINOPROPYLIDENE)DIBENZO(a,d)
(1,4)CYCLOHEPTADIENE HYDROCHLORIDE see
EAI000
5-(γ-DIMETHYLAMINOPROPYLIDENE)-5H-DIBENZO(a,d)-
10,11-DIHYDROCYCLOHEPTENE see EAH500
5-(3-DIMETHYLAMINOPROPYLIDENE)-10,11-DIHYDRO-5H-
DIBENZO(a,d)CYCLOHEPTENE see EAH500
5-(γ-DIMETHYLAMINOPROPYLIDENE)-10,11-DIHYDRO-5H-
DIBENZO(A,D)CYCLOHEPTENE see EAH500
11-(3-DIMETHYLAMINOPROPYLIDENE)-6,11-DIHYDRODI-
BENZ(b,e)OXIPIN see DYE409
N-(γ-DIMETHYLAMINOPROPYL)IMINODIBENZYL see
DLH600
2,2′-(3-DIMETHYLAMINOPROPYLIMINO)DIBENZYL see
DLH600
N-(3-DIMETHYLAMINOPROPYL)IMINODIBENZYL HYDRO-
CHLORIDE see DLH630
5-DIMETHYLAMINO-6-PROPYL-5H-INDENO(5,6-d)-1,3-DI-
OXOLE HYDROCHLORIDE see DPY600
10-(3-DIMETHYLAMINOPROPYL)-2-METHOXYPHENO-
THIAZINE see MFK500
10-(3-(DIMETHYLAMINO)PROPYL)-2-METHOXY)PHENO-
THIAZINE, MALEATE see MFK750
10-(2-(DIMETHYLAMINO)PROPYL)PHENOTHIAZINE see
DQA400
10-(3-(DIMETHYLAMINO)PROPYL)PHENOTHIAZINE see
DQA600

10-(3-DIMETHYLAMINOPROPYL)PHENOTHIAZINE-3-
ETHYLONE see ABH500
10-(3-(DIMETHYLAMINO)PROPYL)PHENOTHIAZINE HY-
DROCHLORIDE see PMI500
10-(γ-DIMETHYLAMINO-N-PROPYL)PHENOTHIAZINE HY-
DROCHLORIDE see PMI500
(DIMETHYLAMINO-2-PROPYL-10-PHENOTHIAZINE HY-
DROCHLORIDE (FRENCH) see DQA400
1-(10-(3-(DIMETHYLAMINO)PROPYL)-10H-PHENOTHIAZIN-
2-YL)ETHANONE see ABH500
10-(3-DIMETHYLAMINOPROPYL)PHENOTHIAZIN-3-YL-
METHYL KETONE see ABH500
10-(3-(DIMETHYLAMINO)PROPYL)PHENOTHIAZIN-2-YL
METHYL KETONE MALEATE (1:1) see AAF750
α-(2-(DIMETHYLAMINO)PROPYL)-α-PHENYLBENZENE-
ACETAMIDE see DOY400
10-(3-(DIMETHYLAMINO)PROPYL-2-(TRIFLUOROMETHYL)
PHENOTHIAZINE see TKL000
2-(DIMETHYLAMINO) RESERPILINATE see DQB800
2-(DIMETHYLAMINO) RESERPILIN-24-OIC ACID ETHYL
ESTER see DQB800
4-DIMETHYLAMINOSTILBEN (GERMAN) see DUB800
N,N-DIMETHYL-4-AMINOSTILBENE see DUB800
4-DIMETHYLAMINO-trans-STILBENE see DUC000
trans-4-DIMETHYLAMINOSTILBENE see DUC000
trans-p-(DIMETHYLAMINO)STILBENE see DUC000
DIMETHYLAMINOSUCCINAMIC ACID see DQD400
N-(DIMETHYLAMINO)SUCCINAMIC ACID see DQD400
N-DIMETHYLAMINO-SUCCINAMIDSAEURE (GERMAN) see
DQD400
O-(4-((DIMETHYLAMINO)SULFONYL)PHENYL) O,O-DI-
METHYL PHOSPHOROTHIOATE see FAB600
4-(DIMETHYLAMINO)-m-TOLYL METHYLCARBAMATE
see DOR400
5-DIMETHYLAMINO-4-TOLYL METHYLCARBAMATE see
DQE800
4-(DIMETHYLAMINO)-3,5-XYLENOL METHYLCARBA-
MATE (ESTER) see DOS000
4-(DIMETHYLAMINO)-3,5-XYLYL ESTER METHYLCAR-
BAMIC ACID see DOS000
4-DIMETHYLAMINO-3,5-XYLYL METHYLCARBAMATE
see DOS000
4-DIMETHYLAMINO-3,5-XYLYL-N-METHYLCARBAMATE
see DOS000
4-(N,N-DIMETHYLAMINO)-3,5-XYLYL N-METHYLCARBA-
MATE see DOS000
DIMETHYLANILINE see XMA000
2,3-DIMETHYLANILINE see XMJ000
2,4-DIMETHYLANILINE see XMS000
2,5-DIMETHYLANILINE see XNA000
2,6-DIMETHYLANILINE see XNJ000
3,5-DIMETHYLANILINE see XOA000
N,N-DIMETHYLANILINE see DQF800
N,N-DIMETHYL-p-ANILINEDIAZOSULFONIC ACID SO-
DIUM SALT see DOU600
2,4-DIMETHYLANILINE HYDROCHLORIDE see XOJ000
2,5-DIMETHYLANILINE HYDROCHLORIDE see XOS000
2-(2,6-DIMETHYLANILINO)-5,6-DIHYDRO-4H-1,3-THIAZINE
see DMW000
9,10-DIMETHYLANTHRACENE see DQG200
DIMETHYL ANTHRANILATE (FCC) see MGQ250
DIMETHYLARSENIC ACID see HKC000
DIMETHYLARSINE see DQG600
DIMETHYLARSINIC ACID see HKC000
((DIMETHYLARSINO)OXY)SODIUM-As-OXIDE see
HKC500
N,N-DIMETHYL-p-AZOANILINE see DOT300
2,3′-DIMETHYLAZOBENZENE-4′-METHYLCARBONATE
see CBS750
7,9-DIMETHYLBENZ(c)ACRIDINE see DQI200
5,7-DIMETHYL-1,2-BENZACRIDINE see DQI600
6,9-DIMETHYL-1,2-BENZACRIDINE see DQI800
7,10-DIMETHYLBENZ(c)ACRIDINE see DQI800
8,10-DIMETHYL-BENZ(a)ACRIDINE see DQI600

2,10-DIMETHYL-7,8-BENZACRIDINE (FRENCH) see DQI800

3,10-DIMETHYL-7,8-BENZACRIDINE (FRENCH) see DQI200

DIMETHYLBENZANTHRACENE see DQJ200

DIMETHYLBENZ(a)ANTHRACENE see DQJ200

7,12-DIMETHYLBENZANTHRACENE see DQJ200

9,10-DIMETHYL-BENZANTHRACENE see DQJ200

6,7-DIMETHYL-1,2-BENZANTHRACENE see DQL200

7,12-DIMETHYLBENZ(a)ANTHRACENE see DQJ200

9,10-DIMETHYLBENZ(a)ANTHRACENE see DQJ200, DQL200

9,10-DIMETHYL-1,2-BENZANTHRACENE see DQJ200

9,10-DIMETHYL-1,2-BENZANTHRAZEN (GERMAN) see DQJ200

DIMETHYLBENZANTHRENE see DQJ200

O,O-DIMETHYL-S-(BENZAZIMINOMETHYL) DITHIO-PHOSPHATE see ASH500

2,2-DIMETHYL-1,3-BENZDIOXOL-4-YL-N-METHYLCARBA-MATE see DQM600

α,α-DIMETHYLBENZEETHANAMINE see DTJ400

2,3-DIMETHYLBENZENAMINE see XMJ000

2,4-DIMETHYLBENZENAMINE see XMS000

2,5-DIMETHYLBENZENAMINE see XNA000

2,6-DIMETHYLBENZENAMINE see XNJ000

3,5-DIMETHYLBENZENAMINE see XOA000

2,4-DIMETHYLBENZENAMINE HYDROCHLORIDE see XOJ000

2,5-DIMETHYLBENZENAMINE HYDROCHLORIDE see XOS000

DIMETHYLBENZENE see XGS000

m-DIMETHYLBENZENE see XHA000

o-DIMETHYLBENZENE see XHJ000

p-DIMETHYLBENZENE see XHS000

1,2-DIMETHYLBENZENE see XHJ000

1,3-DIMETHYLBENZENE see XHA000

1,4-DIMETHYLBENZENE see XHS000

N,N-DIMETHYLBENZENEAMINE see DQF800

N,N-DIMETHYL-1,4-BENZENEDIAMINE see DTL800

DIMETHYL-1,2-BENZENEDICARBOXYLATE see DTR200

DIMETHYL-1,4-BENZENE DICARBOXYLATE see DUE000

N,N-DIMETHYLBENZENEMETHANAMINE see DQP800

DIMETHYL BENZENEORTHODICARBOXYLATE see DTR200

3,3′-DIMETHYLBENZIDIN see TGJ750

3,3′-DIMETHYLBENZIDINE see TGJ750

DIMETHYLBENZIMIDAZOLYCOBAMIDE see VSZ000

5,6-DIMETHYLBENZIMIDAZOLYCOBAMIDE CYANIDE see VSZ000

7,12-DIMETHYLBENZO(a)ANTHRACENE see DQJ200

2,2-DIMETHYL-1,3-BENZODIOXOL-4-OL METHYLCARBA-MATE see DQM600

2,2-DIMETHYLBENZO-1,3-DIOXOL-4-YL METHYLCARBA-MATE see DQM600

AR,AR-DIMETHYLBENZOPHENONE see PGP750

O,O-DIMETHYL-S-(1,2,3-BENZOTRIAZINYL-4-KETO) METHYL PHOSPHORODITHIOATE see ASH500

1,4-DIMETHYL-2,3-BENZPHENANTHRENE see DQJ200

N,N-DIMETHYLBENZYLAMINE see DQP800

DIMETHYLBENZYLAMINE HYDROCHLORIDE see DQQ000

DIMETHYLBENZYLAMMONIUM CHLORIDE see DQQ000

DIMETHYL BENZYL CARBINOL see DQQ200

DIMETHYL BENZYL CARBINYL ACETATE see DQQ375

DIMETHYL BENZYL CARBINYL BUTYRATE see DQQ380

α,α-DIMETHYLBENZYL HYDROPEROXIDE (MAK) see IOB000

DIMETHYLBENZYLOCTADECYLAMMONIUM CHLORIDE see DTC600

N,N-DIMETHYL-N′-BENZYL-N′-(α-PYRIDYL)ETHYLENEDI-AMINE see TMP750

DIMETHYL BERYLLIUM see DQR200

DIMETHYLBERYLLIUM-1,2-DIMETHOXYETHANE see DQR289

1,1-DIMETHYLBIGUANIDE see DQR600

N,N-DIMETHYLBIGUANIDE see DQR600

3,2′-DIMETHYL-4-BIPHENYLAMINE see BLV250

3,3′-DIMETHYL-4,4′-BIPHENYLDIAMINE see TGJ750

3,3′-DIMETHYLBIPHENYL-4,4′-DIAMINE see TGJ750

3,3′-DIMETHYL-(1,1′-BIPHENYL)-4,4′-DIAMINE see TGJ750

N,N′-DIMETHYL-4,4′-BIPYRIDINIUM DICHLORIDE see PAJ000

DIMETHYL 1,3-BIS(CARBOMETHOXY)-1-PROPEN-2-YL PHOSPHATE see SOY000

β,γ-DIMETHYL-α,Δ-BIS(3,4-DIHYDROXYPHENYL)BUTANE see NBR000

O,O-DIMETHYL-S-(1,2-BIS(ETHOXYCARBONYL)ETHYL) DITHIOPHOSPHATE see MAK700

DIMETHYL BIS(p-HYDROXYPHENYL)METHANE see BLD500

2,2-DIMETHYLBUTANE see DQT200

2,3-DIMETHYLBUTANE see DQT400

1,3-DIMETHYL BUTANOL see AOK750

3,3-DIMETHYL-2-BUTANOL METHYLPHOSPHONOFLUORI-DATE see SKS500

5-(1,3-DIMETHYL-2-BUTENYL)-5-ETHYL BARBITURIC ACID see DQU200

5-(1,3-DIMETHYL-2-BUTENYL)-5-ETHYL-2,4,6(1H,3H,5H) PYRIMIDINETRIONE see DQU200

1,3-DIMETHYLBUTYL ACETATE see HFJ000

1,3-DIMETHYL BUTYLAMINE see DQU600

3,3-DIMETHYL-2-BUTYL METHYLPHOSPHONOFLUORI-DATE see SKS500

3,3-DIMETHYL-n-BUT-2-YL METHYLPHOSPHONOFLUORI-DATE see SKS500

O,O-DIMETHYL-(1-BUTYRYLOXY-2,2,2-TRICHLORO-ETHYL) PHOSPHONATE see BPG000

DIMETHYLCADMIUM see DQW800

DIMETHYLCARBAMATE de 5,5-DIMETHYL DIHYDRO-RESORCINOL (FRENCH) see DRL200

DIMETHYLCARBAMATE-d′l-ISOPROPYL-3-METHYL-5-PY-RAZOLYLE (FRENCH) see DSK200

DIMETHYLCARBAMIC ACID CHLORIDE see DQY950

N,N-DIMETHYLCARBAMIC ACID-3-DIMETHYLAMINO-PHENYL ESTER METHOSULFATE see DQY909

DIMETHYLCARBAMIC ACID ESTER with (m-HYDROXYPHENYL)TRIMETHYLAMMONIUM METHYL SULFATE see DQY909

DIMETHYLCARBAMIC ACID ester with 3-HYDROXY-5,5-DI-METHYL-2-CYCLOHEXEN-1-ONE see DRL200

DIMETHYLCARBAMIC ACID 3-METHYL-1-(1-METHYL-ETHYL)-1H-PYRAZOL-5-YL ESTER see DSK200

N,N-DIMETHYLCARBAMIC ACID-8-QUINOLINYL ESTER METHOSULFATE see DQY400

N,N-DIMETHYLCARBAMIC ACID-3-(TRIMETHYLAMMO-NIO)PHENYL ESTER METHYLSULFATE see DQY909

DIMETHYLCARBAMIC CHLORIDE see DQY950

DIMETHYLCARBAMIC ESTER of 8-OXYMETHYLQUINOLI-NIUM METHYLSULFATE see DQY400

DIMETHYLCARBAMIC ESTER of 3-OXYPHENYLTRI-METHYLAMMONIUM METHYLSULFATE see DQY909

DIMETHYLCARBAMIDOYL CHLORIDE see DQY950

DIMETHYLCARBAMODITHIOIC ACID, IRON COMPLEX see FAS000

DIMETHYLCARBAMODITHIOIC ACID, IRON(3+) SALT see FAS000

DIMETHYLCARBAMODITHIOIC ACID, ZINC COMPLEX see BJK500

DIMETHYLCARBAMODITHIOIC ACID, ZINC SALT see BJK500

3-DIMETHYLCARBAMOXYPHENYL TRIMETHYL AMMO-NIUM BROMIDE see DQY800

3-DIMETHYLCARBAMOXYPHENYLTRIMETHYLAM-MONIUM BROMIDE see POD000

3-(DIMETHYLCARBAMOXY)PHENYL TRIMETHYLAMMO-NIUM METHYL SULFATE see DQY909

DIMETHYL CARBAMOYL CHLORIDE see DQY950
N,N-DIMETHYLCARBAMOYL CHLORIDE (DOT) see DQY950
cis-2-DIMETHYLCARBAMOYL-1-METHYLVINYL DI-METHYLPHOSPHATE see DGQ875
(3-(DIMETHYLCARBAMOYLOXY)PHENYL)TRIMETHYL-AMMONIUM METHYLSULFATE see DQY909
DIMETHYLCARBAMYL CHLORIDE see DQY950
N,N-DIMETHYLCARBAMYL CHLORIDE see DQY950
DIMETHYLCARBINOL see INJ000
α,α-DIMETHYL-α'-CARBOBUTOXY-DIHYDRO-γ-PYRONE see BRT000
2,2-DIMETHYL-6-CARBOBUTOXY-2,3-DIHYDRO-4-PY-RONE see BRT000
O,O-DIMETHYL-S-(1-CARBOETHOXYBENZYL) DITHIO-PHOSPHATE see DRR400
O,O-DIMETHYL-O-(2-CARBOMETHOXY-1-METHYLVINYL) PHOSPHATE see MQR750
DIMETHYL-1-CARBOMETHOXY-1-PROPEN-2-YL PHOS-PHATE see MQR750
DIMETHYL CARBONATE see MIF000
O,O-DIMETHYL-S-(CARBONYLMETHYLMORPHOLINO) PHOSPHORODITHIOATE see PHI500
DIMETHYLCELLOSOLVE see DOE600
O,O-DIMETHYL-O-3-CHLOR-4-NITROFENYLTIOFOSFAT (CZECH) see MIJ250
O,O-DIMETHYL-O-(3-CHLOR-4-NITROPHENYL)-MONO-THIOPHOSPHAT (GERMAN) see MIJ250
DIMETHYLCHLOROETHER see CIO250
DIMETHYL(2-CHLOROETHYL)AMINE see CGW000
DIMETHYL(2-CHLOROETHYL)AMINE HYDROCHLORIDE see DRC000
DIMETHYL-β-CHLOROETHYLAMINE HYDROCHLORIDE see DRC000
O,O-DIMETHYL O-2-CHLORO-4-NITROPHENYL PHOSPHO-ROTHIOATE see NFT000
O,O-DIMETHYL-O-(3-CHLORO-4-NITROPHENYL) PHOS-PHOROTHIOATE see MIJ250
DIMETHYL-3-CHLORO-4-NITROPHENYL THIONOPHOS-PHATE see MIJ250
DIMETHYL-2-CHLORONITROPHENYL THIOPHOSPHATE see NFT000
O,O-DIMETHYL-O-(3-CHLORO-4-NITROPHENYL) THIO-PHOSPHATE see MIJ250
N,N-DIMETHYL-p-((p-CHLOROPHENYL)AZO)ANILINE see CGD250
O,O-DIMETHYL-S-p-CHLOROPHENYL PHOSPHORO-THIOATE see FOR000
DIMETHYL-p-CHLOROPHENYLTHIOMETHYL DITHIO-PHOSPHATE see MQH750
O,O-DIMETHYL-S-(p-CHLOROPHENYLTHIOMETHYL) PHOSPHORODITHIOATE see MQH750
1,1-DIMETHYL-3-(p-CHLOROPHENYL)UREA see CJX750
N,N-DIMETHYL-N'-(4-CHLOROPHENYL)UREA see CJX750
DIMETHYL CHLOROTHIOPHOSPHATE (DOT) see DTQ600
DIMETHYLCHLORTHIOFOSAT (CZECH) see DTQ600
O,O-DIMETHYL-O-2-CHLOR-1-(2,4,5-TRICHLORPHENYL)-VINYL-PHOSPHAT (GERMAN) see TBW100
5,7-DIMETHYLCHRYSENE see DRE400
1,11-DIMETHYLCHRYSENE see DRE400
2,2-DIMETHYL-7-COUMARANYL N-METHYLCARBAMATE see CBS275
DIMETHYLCYANAMIDE see DRF600
N,N-DIMETHYLCYCLOHEXANAMINE see DRF709
m-DIMETHYLCYCLOHEXANE see DRG000
o-DIMETHYLCYCLOHEXANE see DRF800
1,3-DIMETHYLCYCLOHEXANE see DRG000
1,4-DIMETHYLCYCLOHEXANE see DRG200
1,2-DIMETHYLCYCLOHEXANE (DOT) see DRF800
cis-1,2-DIMETHYLCYCLOHEXANE see DRF800
1,5-DIMETHYL-5-(1-CYCLOHEXENYL)BARBITURIC ACID see ERD500
DIMETHYLCYCLOHEXYLAMINE see DRF709
N,N-DIMETHYLCYCLOHEXYLAMINE (DOT) see DRF709

DIMETHYLCYSTEINE see MCR750
β,β-DIMETHYLCYSTEINE see MCR750
3,3'-DIMETHYL-4,4'-DIAMINODIPHENYLMETHANE see MJO250
N,N-DIMETHYL-1,3-DIAMINOPROPANE see AJP750
1,5-DIMETHYL-1,5-d-DIAZAUNDECAMETHYLENE POLY-METHOBROMIDE see HCV500
2,3-DIMETHYL-1,4-DIAZINE see DTU400
2,5-DIMETHYL-1,4-DIAZINE see DTU600
N,N-DIMETHYLDIBENZ(b,e)OXEPIN-Δ$^{11(6H)}$-γ-PROPYLA-MINE see DYE409
O,O-DIMETHYL-O-(1,2-DIBROMO-2,2-DICHLOROETHYL) PHOSPHATE see NAG400
DIMETHYL-1,2-DIBROMO-2,2-DICHLOROETHYL PHOS-PHATE (OSHA) see NAG400
2,5-DIMETHYL-2,5-DI-(tert-BUTYLPEROXY)HEXANE see DRJ800
2,5-DIMETHYL-2,5-DI-(tert-BUTYLPEROXY)HEXANE, tech-nically pure (DOT) see DRJ800
O,O-DIMETHYL-S-1,2-(DICARBAETHOXYAETHYL)-DI-THIOPHOSPHAT (GERMAN) see MAK700
O,O-DIMETHYL-S-(1,2-DICARBETHOXYETHYL) DITHIO-PHOSPHATE see MAK700
O,O-DIMETHYL-S-(1,2-DICARBETHOXYETHYL)PHOS-PHORODITHIOATE see MAK700
O,O-DIMETHYL-S-(1,2-DICARBETHOXYETHYL) THIOTH-IONOPHOSPHATE see MAK700
DIMETHYL-1,3-DI(CARBOMETHOXY)-1-PROPEN-2-YL PHOSPHATE see SOY000
O,O-DIMETHYL-O-2,2-DICHLORO-1,2-DIBROMOETHYL PHOSPHATE see NAG400
DIMETHYL-2,2-DICHLOROETHENYL PHOSPHATE see DGP900
DIMETHYL-1,1'-DICHLOROETHER see BIK000
1,1-DIMETHYL-3-(3,4-DICHLOROPHENYL)UREA see DXQ500
DIMETHYL DICHLOROVINYL PHOSPHATE see DGP900
DIMETHYL-2,2-DICHLOROVINYL PHOSPHATE see DGP900
O,O-DIMETHYL DICHLOROVINYL PHOSPHATE see DGP900
O,O-DIMETHYL-O-2,2-DICHLOROVINYL PHOSPHATE see DGP900
O,O-DIMETHYL-O-(2,2-DICHLOR-VINYL)-PHOSPHAT (GERMAN) see DGP900
O,O-DIMETHYL-S-1,2-DI(ETHOXYCARBAMYL)ETHYL PHOSPHORODITHIOATE see MAK700
DIMETHYL-DIETHOXYSILAN (CZECH) see DHG000
DIMETHYLDIETHOXYSILANE (DOT) see DHG000
2,6-DIMETHYL-1,1-DIETHYLPIPERIDINIUM BROMIDE see DRK600
N,N-DIMETHYLDIGUANIDE see DQR600
2,2-DIMETHYL-2,3-DIHYDROBENZOFURAN-7-YL ESTER, METHYLCARBAMIC ACID see CBS275
2,2-DIMETHYL-2,3-DIHYDRO-7-BENZOFURANYL-N-METHYLCARBAMATE see CBS275
11,17-DIMETHYL-16,17-DIHYDRO-15H-CYCLOPENTA(a) PHENANTHRENE see DLI200
O,O-DIMETHYL-S-(3,4-DIHYDRO-4-KETO-1,2,3-BENZO-TRIAZINYL-3-METHYL) DITHIOPHOSPHATE see ASH500
5,5-DIMETHYL-DIHYDRORESORCINOL-N,N-DIMETHYL-CARBAMAT (GERMAN) see DRL200
5,5-DIMETHYLDIHYDRORESORCINOL DIMETHYLCARBA-MATE see DRL200
5,5-DIMETHYL-4,5-DIHYDRO-3-RESORCYL-DIMETHYL-CARBAMAT (GERMAN) see DRL200
O,O-DIMETHYL-S-1,2-DIKARBETOXYLETHYLDITIOFOS-FAT (CZECH) see MAK700
DIMETHYL DIKETONE see BOT500
DIMETHYL 3-(DIMETHOXYPHOSPHINYLOXY)GLUTA-CONATE see SOY000

O,O-DIMETHYL-S-(2-ETHYLSULFINYL)ETHYL THIO-
PHOSPHATE see DAP000
O,O-DIMETHYL-S-ETHYL-2-SULFONYLETHYL PHOSPHO-
ROTHIOLATE see DAP600
O,O-DIMETHYL-S-ETHYLSULPHINYLETHYL PHOSPHORO-
THIOLATE see DAP000
O,O-DIMETHYL-S-ETHYLSULPHONYLETHYL PHOSPHO-
ROTHIOLATE see DAP600
O,O-DIMETHYL-S-(2-ETHYLTHIO-ETHYL)-DITHIOFOS-
FAAT (DUTCH) see PHI500
O,O-DIMETHYL-O-(2-ETHYL-THIO-ETHYL)-MONOTHIO-
FOSFAAT (DUTCH) see DAO800
O,O-DIMETHYL-S-(2-ETHYLTHIO-ETHYL)-MONOTHIOFOS-
FAAT (DUTCH) see DAP400
O,O-DIMETHYL S-(2-(ETHYLTHIO)ETHYL) PHOSPHORO-
DITHIOATE see PHI500
O,O-DIMETHYL-O-2-(ETHYLTHIO)ETHYL PHOSPHORO-
THIOATE see DAO800
O,O-DIMETHYL-S-2-(ETHYLTHIO)ETHYL)PHOSPHORO-
THIOATE see DAP400
7,12-DIMETHYL-5-FLUOROBENZ(a)ANTHRACENE see
DRY600
7,12-DIMETHYL-11-FLUOROBENZ(a)ANTHRACENE see
DRZ000
N,N-DIMETHYL-p-((p-FLUOROPHENYL)AZO)ANILINE see
DSA000
DIMETHYL FLUOROPHOSPHATE see DSA800
DIMETHYL FORMAL see MGA850
DIMETHYLFORMALDEHYDE see ABC750
DIMETHYLFORMAMID (GERMAN) see DSB000
DIMETHYLFORMAMIDE see DSB000
N,N-DIMETHYL FORMAMIDE see DSB000
N,N-DIMETHYLFORMAMIDE (DOT) see DSB000
DIMETHYLFORMOCARBOTHIALDINE see DSB200
O,O-DIMETHYL-S-(N-FORMYL-N-METHYLCARBAMOYL-
METHYL) PHOSPHORODITHIOATE see DRR200
N,N-DIMETHYLGLYCINONITRILE see DOS200
DIMETHYLGLYOXAL see BOT500
DIMETHYLGLYOXIME see DBH000
N,N'-DIMETHYLHARNSTOFF (GERMAN) see DUM200
2,6-DIMETHYL-2,5-HEPTADIEN-4-ONE see PGW250
2,6-DIMETHYL-4-HEPTANOL see DNH800
2,6-DIMETHYL HEPTANOL-4 see DNH800
2,6-DIMETHYL-HEPTAN-4-ON (DUTCH, GERMAN) see
DNI800
2,6-DIMETHYLHEPTAN-4-ONE see DNI800
2,6-DIMETHYL-4-HEPTANONE see DNI800
2,6-DIMETHYL-5-HEPTENAL see DSD775
1,5-DIMETHYLHEXYLAMINE see ILM000
α,ε-DIMETHYLHEXYLAMINE see ILM000
1,2-DIMETHYLHYDRAZIN (GERMAN) see DSF600
DIMETHYLHYDRAZINE see DSF400
1,1-DIMETHYLHYDRAZINE see DSF400
1,2-DIMETHYLHYDRAZINE see DSF600
N,N-DIMETHYLHYDRAZINE see DSF400
N,N'-DIMETHYLHYDRAZINE see DSF600
sym-DIMETHYLHYDRAZINE see DSF600
asym-DIMETHYLHYDRAZINE see DSF400
unsym-DIMETHYLHYDRAZINE see DSF400
1,1-DIMETHYLHYDRAZINE (GERMAN) see DSF400
DIMETHYLHYDRAZINE, symmetrical (DOT) see DSF600
DIMETHYLHYDRAZINE, unsymmetrical (DOT) see DSF400
1,2-DIMETHYLHYDRAZINE DIHYDROCHLORIDE see
DSF800
N,N'-DIMETHYLHYDRAZINE DIHYDROCHLORIDE see
DSF800
sym-DIMETHYLHYDRAZINE DIHYDROCHLORIDE see
DSF800
1,1-DIMETHYLHYDRAZINE HYDROCHLORIDE see
DSG000
1,2-DIMETHYLHYDRAZINE HYDROCHLORIDE see
DSG200
sym-DIMETHYLHYDRAZINE HYDROCHLORIDE see
DSG200

2-(2,2-DIMETHYLHYDRAZINO)-4-(5-NITRO-2-FURYL)THI-
AZOLE see DSG400
N,N-DIMETHYL-2-HYDROXYETHYLAMINE see DOY800
N,N-DIMETHYL-N-(2-HYDROXYETHYL)AMINE see
DOY800
DIMETHYL 3-HYDROXYGLUTACONATE DIMETHYL
PHOSPHATE see SOY000
2,2-DIMETHYL-5-HYDROXYMETHYL-1,3-DIOXOLANE see
DVR600
2',3-DIMETHYL-4-(2-HYDROXYNAPHTHYLAZO)AZOBEN-
ZENE see SBC500
3,7-DIMETHYL-7-HYDROXYOCTANAL see CMS850
1,1-DIMETHYL-3-HYDROXYPYRROLIDINIUM BROMIDE-α-
CYCLOPENTYLMANDELATE see GIC000
O,O-DIMETHYL-(1-HYDROXY-2,2,2-TRICHLORAETHYL)
PHOSPHONSAEURE ESTER (GERMAN) see TIQ250
O,O-DIMETHYL-(1-HYDROXY-2,2,2-TRICHLORATHYL)-
PHOSPHAT (GERMAN) see TIQ250
O,O-DIMETHYL-(1-HYDROXY-2,2,2-TRICHLORO)ETHYL
PHOSPHATE see TIQ250
DIMETHYL-1-HYDROXY-2,2,2-TRICHLOROETHYL PHOS-
PHONATE see TIQ250
O,O-DIMETHYL-(1-HYDROXY-2,2,2-TRICHLOROETHYL)
PHOSPHONATE see TIQ250
N,N-DIMETHYL-5-HYDROXYTRYPTAMINE see DPG109
3,8-DIMETHYL-3H-IMIDAZO(4,5-f)QUINOXALIN-2-AMINE
see AJQ675
DIMETHYLIMIPRAMINE see DSI709
DIMETHYLIMIPRAMINE HYDROCHLORIDE see DLS600
N,N-DIMETHYL-α-INDOLYLIDENE-p-TOLUIDINE see
DOT600
DIMETHYL-5-(1-ISOPROPYL-3-METHYLPYRAZOLYL)CAR-
BAMATE see DSK200
O,O-DIMETHYL-S-ISOPROPYL-2-SULFINYLETHYLPHOS-
PHOROTHIOATE see DSK600
O,O-DIMETHYL-S-2-(ISOPROPYLTHIO)ETHYLPHOS-
PHORODITHIOATE see DSK800
DIMETHYLKETAL see ABC750
DIMETHYLKETOL see ABB500
DIMETHYL KETONE see ABC750
DIMETHYLMAGNESIUM see DSL600
DIMETHYLMALEIC ANHYDRIDE see DSM000
α,β-DIMETHYLMALEIC ANHYDRIDE see DSM000
DIMETHYLMESCALINE see DOE200
N,N-DIMETHYLMETHANAMINE see TLD500
DIMETHYLMETHANE see PMJ750
N,N-DIMETHYL-N'-(4-METHOXYBENZYL)-N'-(2-PYRI-
DYL)ETHYLENEDIAMINE MALEATE see DBM800
O,O-DIMETHYL-O-2-METHOXYCARBONYL-1-METHYL-VI-
NYL-PHOSPHAT (GERMAN) see MQR750
DIMETHYL 2-METHOXYCARBONYL-1-METHYLVINYL
PHOSPHATE see MQR750
DIMETHYL METHOXYCARBONYLPROPENYL PHOSPHATE
see MQR750
DIMETHYL (1-METHOXYCARBOXYPROPEN-2-YL)PHOS-
PHATE see MQR750
O,O-DIMETHYL-S-(5-METHOXY-4-OXO-4H-PYRAN-2-YL)
PHOSPHOROTHIOATE see EAS000
3,3-DIMETHYL-1-p-METHOXYPHENYLTRIAZENE see
DSN600
O,O-DIMETHYL-S-(5-METHOXY-PYRON-2-YL)-METHYL)-
THIOLPHOSPHAT (GERMAN) see EAS000
O,O-DIMETHYL-S-(5-METHOXYPYRONYL-2-METHYL)
THIOPHOSPHATE see EAS000
(O,O-DIMETHYL)-S-(-2-METHOXY-Δ²-1,3,4-THIADIAZOLIN-
5-ON-4-YLMETHYL)DITHIOPHOSPHATE DIMETHYL
PHOSPHOROTHIOLOTHIONATE see DSO000
O,O-DIMETHYL-S-(5-METHOXY-1,3,4-THIADIAZOLINYL-3-
METHYL) DITHIOPHOSPHATE see DSO000
O,O-DIMETHYL-S-(2-METHOXY-1,3,4-THIADIAZOL-5-(4H)-
ONYL-(4)-METHYL)-DITHIOPHOSPHAT (GERMAN) see
DSO000
O,O-DIMETHYL-S-(2-METHOXY-1,3,4-THIADIAZOL-5(4H)-

ONYL-(4)-METHYL) PHOSPHORODITHIOATE see
DSO000

O,O-DIMETHYL-S-((2-METHOXY-1,3,4 (4H)-THIODIAZOL-5-
ON-4-YL)-METHYL)DITHIOFOSFAAT (DUTCH) see
DSO000

N,N-DIMETHYL-N'-(((METHYLAMINO)CARBONYL)
OXY)PHENYLMETHANIMIDAMIDE MONOHYDRO-
CHLORIDE see DSO200

2,2-DIMETHYL-4-(N-METHYLAMINOCARBOXYLATO)-1,3-
BENZODIOXOLE see DQM600

O,O-DIMETHYL-S-(2-(METHYLAMINO)-2-OXOETHYL)
PHOSPHORODITHIOATE see DSP400

2,2-DIMETHYL-4-(N-METHYLCARBAMATO)-1,3-BENZODI-
OXOLE see DQM600

O,O-DIMETHYL-S-2-(1-N-METHYLCARBAMOYLETHYL-
MERCAPTO)ETHYL THIOPHOSPHATE see MJG500

DIMETHYL-S-(2-(1-METHYLCARBAMOYLETHYLTHIO
ETHYL) PHOSPHOROTHIOLATE see MJG500

O,O-DIMETHYL-S-(2-(1-METHYLCARBAMOYLETHYL-
THIO)ETHYL) PHOSPHOROTHIOATE see MJG500

O,O-DIMETHYL-S-(N-METHYL-CARBAMOYL)-METHYL-DI-
THIOFOSFAAT (DUTCH) see DSP400

(O,O-DIMETHYL-S-(N-METHYL-CARBAMOYL-METHYL)-
DITHIOPHOSPHAT) (GERMAN) see DSP400

O,O-DIMETHYL-S-(N-METHYLCARBAMOYLMETHYL) DI-
THIOPHOSPHATE see DSP400

O,O-DIMETHYL-S-((N-METHYL-CARBAMOYL)-METHYL)
MONOTHIOFOSFAAT (DUTCH) see DNX800

O,O-DIMETHYL-S-(N-METHYL-CARBAMOYL)-METHYL-
MONOTHIOPHOSPHAT (GERMAN) see DNX800

O,O-DIMETHYL METHYLCARBAMOYLMETHYL PHOS-
PHORODITHIOATE see DSP400

O,O-DIMETHYL-S-(N-METHYLCARBAMOYLMETHYL)
PHOSPHORODITHIOATE see DSP400

O,O-DIMETHYL-S-((METHYLCARBAMOYL)METHYL)
PHOSPHOROTHIOATE see DNX800

O,O-DIMETHYL-S-(N-METHYLCARBAMOYLMETHYL)
PHOSPHOROTHIOATE see DNX800

DIMETHYL-S-(N-METHYL-CARBAMOYL-METHYL)PHOS-
PHOROTHIOLATE see DNX800

O,O-DIMETHYL-S-(N-METHYLCARBAMOYLMETHYL)
PHOSPHOROTHIOLATE see DNX800

O,O-DIMETHYL-S-(N-METHYLCARBAMOYLMETHYL)
THIOPHOSPHATE see DNX800

O,O-DIMETHYL-O-(2-N-METHYLCARBAMOYL-1-METHYL-
VINYL)-FOSFAAT (DUTCH) see MRH209

O,O-DIMETHYL-O-(2-N-METHYLCARBAMOYL-1-
METHYL)-VINYL-PHOSPHAT (GERMAN) see
MRH209

O,O-DIMETHYL-O-(2-N-METHYLCARBAMOYL-1-METHYL-
VINYL) PHOSPHATE see MRH209

N,N-DIMETHYL-α-METHYLCARBAMOYLOXYIMINO-α-
(METHYLTHIO)ACETAMIDE see DSP600

N',N'-DIMETHYL-N-((METHYLCARBAMOYL)OXY)-1-
METHYLTHIOOXAMIMIDIC ACID see DSP600

N',N'-DIMETHYL-N-((METHYLCARBAMOYL)OXY)-1-
THIOOXAMIMIDIC ACID METHYL ESTER see
DSP600

O,O-DIMETHYL-S-(N-METHYLCARBAMYLMETHYL)
THIOTHIONOPHOSPHATE see DSP400

O,O-DIMETHYL O-(1-METHYL-2-CARBOXYVINYL) PHOS-
PHATE see MQR750

N,N-DIMETHYL-N'-(2-METHYL-4-CHLOROPHENYL)-FOR-
MAMIDINE see CJJ250

N,N-DIMETHYL-N'-(2-METHYL-4-CHLORPHENYL)-FOR-
MADIN (GERMAN) see CJJ250

O,O-DIMETHYL-S-(3-METHYL-2,4-DIOXO-3-AZA-BUTYL)-
DITHIOFOSFAAT (DUTCH) see DRR200

O,O-DIMETHYL-S-(3-METHYL-2,4-DIOXO-3-AZA-BUTYL)-
DITHIOPHOSPHAT (GERMAN) see DRR200

6,6-DIMETHYL-2-METHYLENEBICYCLO(3.1.1)HEPTANE
see POH750

DIMETHYLMETHYLENE-p,p'-DIPHENOL see BLD500

O,O-DIMETHYL-S-(N-METHYL-N-FORMYL-CARBAMOYL-
METHYL)-DITHIOPHOSPHAT see DRR200

O,O-DIMETHYL-S-(N-METHYL-N-FORMYLCARBAMOYL-
METHYL)PHOSPHORODITHIOATE see DRR200

O,O-DIMETHYL-O-4-(METHYLMERCAPTO)-3-METHYLPHE-
NYL PHOSPHOROTHIOATE see FAQ999

O,O-DIMETHYL-p-4-(METHYLMERCAPTO)-3-METHYLPHE-
NYL THIOPHOSPHATE see FAQ999

O,O-DIMETHYL O-(4-METHYLMERCAPTOPHENYL)PHOS-
PHATE see PHD250

(E)-DIMETHYL 1-METHYL-3-(METHYLAMINO)-3-OXO-1-
PROPENYL PHOSPHATE see MRH209

DIMETHYL-1-METHYL-2-(METHYLCARBAMOYL)VINYL-
PHOSPHATE, cis see MRH209

O,O-DIMETHYL-O-(3-METHYL-4-METHYLMERCAPTO-
PHENYL)PHOSPHOROTHIOATE see FAQ999

O,O-DIMETHYL-O-(3-METHYL-4-METHYLTHIO-FENYL)-
MONOTHIOFOSFAAT (DUTCH) see FAQ999

O,O-DIMETHYL-O-(3-METHYL-4-METHYLTHIOPHENYL)-
MONOTHIOPHOSPHAT (GERMAN) see FAQ999

O,O-DIMETHYL-O-3-METHYL-4-METHYLTHIOPHENYL
PHOSPHOROTHIOATE see FAQ999

O,O-DIMETHYL-O-(3-METHYL-4-METHYLTHIO-PHENYL)-
THIONOPHOSPHAT (GERMAN) see FAQ999

O,O-DIMETHYL-O-(3-METHYL-4-NITROFENYL)-MONO-
THIOFOSFAAT (DUTCH) see DSQ000

O,O-DIMETHYL-O-(3-METHYL-4-NITRO-PHENYL)-MONO-
THIOPHOSPHAT (GERMAN) see DSQ000

O,O-DIMETHYL-O-(3-METHYL-4-NITROPHENYL)PHOS-
PHORATE see PHD750

O,O-DIMETHYL-O-(3-METHYL-4-NITROPHENYL) PHOS-
PHOROTHIOATE see DSQ000

DIMETHYL-3-METHYL-4-NITROPHENYLPHOSPHORO-
THIONATE see DSQ000

O,O-DIMETHYL-O-(3-METHYL-4-NITROPHENYL) THIO-
PHOSPHATE see DSQ000

N,N-DIMETHYL-p-(3'-METHYLPHENYLAZO)ANILINE see
DUH600

N,N-DIMETHYL-4-((3-METHYLPHENYL)AZO)BENZEN-
AMINE see DUH600

N,N-DIMETHYL-2-((o-METHYL-α-PHENYL-BENZYL)OXY)-
ETHYLAMINE CITRATE see DPH000

N,N-DIMETHYL-2-(o-METHYL-α-PHENYLBENZYLOXY)
ETHYLAMINE HYDROCHLORIDE see OJW000

3,3-DIMETHYL-1-(m-METHYLPHENYL)TRIAZENE see
DSR200

3,3-DIMETHYL-1-(o-METHYLPHENYL)TRIAZENE see
MNT500

O,O-DIMETHYL-O-(3-METHYL) PHOSPHOROTHIOATE see
DSQ000

DIMETHYL-3-(2-METHYL-1-PROPENYL)CYCLOPRO-
PANECARBOXYLATE see BEP500

(+)-(Z)-2,2-DIMETHYL-3-(2-METHYLPROPENYL)-CYCLO-
PROPANECARBOXYLIC ACID ESTER with 2-ALLYL-4-
HYDROXY-3-METHYL-2-CYCLOPENTEN-ONE see
AFR500

O,O-DIMETHYL-o-(4-(METHYLSULFONYL)-m-TOLYL)
PHOSPHOROTHIOATE see DSS800

O,O-DIMETHYL-O-(4-METHYLTHIO-3-METHYLPHENYL)
PHOSPHOROTHIOATE see FAQ999

3,5-DIMETHYL-4-(METHYLTHIO)PHENOL METHYLCAR-
BAMATE see DST000

3,5-DIMETHYL-4-METHYL-THIOPHENYL-N-CARBAMAT
(GERMAN) see DST000

O-(3,5-DIMETHYL-4-(METHYLTHIO)PHENYL)-O,O-DI-
METHYL PHOSPHOROTHIOATE see DST200

3,5-DIMETHYL-4-METHYLTHIOPHENYL-N-METHYLCAR-
BAMATE see DST000

DIMETHYL-p-(METHYLTHIO)PHENYL PHOSPHATE see
PHD250

O,O-DIMETHYL-O-(4-(METHYLTHIO)-m-TOLYL) PHOS-
PHOROTHIOATE see FAQ999

O,O-DIMETHYL-o-((4-METHYLTHIO)-m-TOLYL) PHOSPHO-
ROTHIOATE SULFONE see DSS800

O,O-DIMETHYL-O-4-(METHYLTHIO)-3,5-XYLYL PHOS-
PHOROTHIOATE see DST200
O,O-DIMETHYL-S-(N-MONOMETHYL)-CARBAMYL
METHYLDITHIOPHOSPHATE see DSP400
DIMETHYL MONOSULFATE see DUD100
O,O-DIMETHYL-S-((MORFOLINO-CARBONYL)-METHYL)-
DITHIOFOSFAAT (DUTCH) see MRU250
3,17-DIMETHYL-9-α,13-α,14-α-MORPHINAN PHOSPHATE
see MLP250
(9-α,13-α,14-α)-3,17-DIMETHYLMORPHINAN PHOSPHATE
see MLP250
O,O-DIMETHYL-S-(MORPHOLINOCARBAMOYLMETHYL)
DITHIOPHOSPHATE see MRU250
O,O-DIMETHYL-S-((MORPHOLINO-CARBONYL)-METHYL)-
DITHIOPHOSPHAT (GERMAN) see MRU250
O,O-DIMETHYL MORPHOLINOCARBONYLMETHYL PHOS-
PHORODITHIOATE see MRU250
O,O-DIMETHYL-S-(MORPHOLINOCARBONYLMETHYL)
PHOSPHORODITHIOATE see MRU250
DIMETHYL S-(MORPHOLINOCARBONYLMETHYL) PHOS-
PHOROTHIOLOTHIONATE see MRU250
DIMETHYLMYLERAN see DSU000
DIMETHYL-α-NAPHTHYLAMINE see DSU400
α-DIMETHYLNAPHTHYLAMINE see DSU400
N,N-DIMETHYL-1-NAPHTHYLAMINE see DSU400
N,N-DIMETHYL-α-NAPHTHYLAMINE see DSU400
N,N-DIMETHYL-p-(1-NAPHTHYLAZO)ANILINE see
DSU600
DIMETHYLNITRAMIN (GERMAN) see DSV200
DIMETHYLNITRAMINE see DSV200
DIMETHYLNITROAMINE see DSV200
O,O-DIMETHYL-p-NITRO-m-CHLOROPHENYL THIOPHOS-
PHATE see MIJ250
O,O-DIMETHYL-O-4-NITRO-3-CHLOROPHENYL THIO-
PHOSPHATE see MIJ250
O,O-DIMETHYL-S-p-NITROFENYL ESTER KYSELINY
THIOFOSFORECEN (CZECH) see DTH800
O,O-DIMETHYL-O-p-NITROFENYLESTER KYSELINY THIO-
FOSFORECNE (CZECH) see MNH000
O,O-DIMETHYL-O-(4-NITROFENYL)-MONOTHIOFOSFAAT
(DUTCH) see MNH000
1,2-DIMETHYL-5-NITROIMIDAZOLE see DSV800
1,2-DIMETHYL-5-NITRO-1H-IMIDAZOLE see DSV800
DIMETHYLNITROMETHANE see NIY000
O,O-DIMETHYL-O-(4-NITRO-3-METHYLPHENYL)THIO-
PHOSPHATE see DSQ000
O,O-DIMETHYL-O-(4-NITRO-PHENYL)-MONOTHIOPHOS-
PHAT (GERMAN) see MNH000
DIMETHYL p-NITROPHENYL MONOTHIOPHOSPHATE see
MNH000
O,O-DIMETHYL-O-(4-NITROPHENYL) PHOSPHORO-
THIOATE see MNH000
O,O-DIMETHYL-O-(p-NITROPHENYL) PHOSPHORO-
THIOATE see MNH000
O,O-DIMETHYL-S-(p-NITROPHENYL) PHOSPHORO-
THIOATE see DTH800
DIMETHYL 4-NITROPHENYL PHOSPHOROTHIONATE see
MNH000
O,O-DIMETHYL-O-(4-NITROPHENYL)-THIONOPHOS see
MNH000
DIMETHYLPHOSPHORAMIDOCYANIDIC ACID, ETHYL ES-
TER see EIF000
O,S-DIMETHYL PHOSPHORAMIDOTHIOATE see DTQ400
O,O-DIMETHYLPHOSPHOROCHLORIDOTHIOATE see
DTQ600
DIMETHYL PHOSPHOROCHLORIDOTHIOATE (DOT) see
DTQ600
O,O-DIMETHYL PHOSPHORODITHIOATE N-FORMYL-2-
MERCAPTO-N-METHYLACETAMIDE-S-ESTER see
DRR200
N-((O,O-DIMETHYLPHOSPHORODITHIOYL)ETHYL)
ACETAMIDE see DOP200
O,O-DIMETHYL PHOSPHOROTHIOATE-O,O-DIESTER with
4,4'-THIODIPHENOL see TAL250

DIMETHYLPHTHALATE see DTR200
(O,O-DIMETHYL-PHTHALIMIDIOROSOANILINE see
DSY600
DIMETHYL-p-NITROSOANILINE (DOT) see DSY600
N,N-DIMETHYL-4-NITROSOBENZENAMINE see DSY600
DIMETHYLNITROSOHARNSTOFF (GERMAN) see DTB200
DIMETHYLNITROSOMORPHOLINE see DTA000
2,6-DIMETHYLNITROSOMORPHOLINE see DTA000
2,6-DIMETHYL-N-NITROSOMORPHOLINE see DTA000
DIMETHYL(p-NITROSOPHENYL)AMINE see DSY600
1,3-DIMETHYLNITROSOUREA see DTB200
1,3-DIMETHYL-N-NITROSOUREA see DTB200
N,N'-DIMETHYLNITROSOUREA see DTB200
O,O-DIMETHYL-O-4-NITRO-m-TOLYL PHOSPHORO-
THIOATE see DSQ000
3,7-DIMETHYL-2,6-OCTADIEN-1-YL PROPIONATE see
GDM450
DIMETHYLOCTADECYLBENZYLAMMONIUM CHLORIDE
see DTC600
3,7-DIMETHYL-2,6-OCTADIENAL see DTC800
2,6-DIMETHYL-2,7-OCTADIENE-6-OL see LFX000
3,7-DIMETHYL-2,6-OCTADIENE-1-YL BUTYRATE see
GDE825
2,6-DIMETHYLOCTA-2,7-DIEN-6-OL see LFX000
3,7-DIMETHYLOCTA-1,6-DIEN-3-OL see LFX000
3,7-DIMETHYL-1,6-OCTADIEN-3-OL see LFX000
3,7-DIMETHYL-(E)-2,6-OCTADIEN-1-OL see DTD000
3,7-DIMETHYL-(Z)-2,6-OCTADIEN-1-OL see DTD200
2-cis-3,7-DIMETHYL-2,6-OCTADIEN-1-OL see DTD200
2,6-DIMETHYL-trans-2,6-OCTADIEN-8-OL see DTD000
3,7-DIMETHYL-trans-2,6-OCTADIEN-1-OL see DTD000
3,7-DIMETHYL-1,6-OCTADIEN-3-OL ACETATE see
LFY100
trans-3,7-DIMETHYL-2,6-OCTADIEN-1-OL ACETATE see
DTD800
3,7-DIMETHYL-1,6-OCTADIEN-3-OL BENZOATE see
LFZ000
trans-3,7-DIMETHYL-2,6-OCTADIEN-1-OL FORMATE see
GCY000
3,7-DIMETHYL-1,6-OCTADIEN-3-OL ISOBUTYRATE see
LGB000
3,7-DIMETHYL-1,6-OCTADIEN-3-YL ACETATE see
LFY100
3,7-DIMETHYL-2-trans-6-OCTADIENYL ACETATE see
DTD800
trans-3,7-DIMETHYL-2,6-OCTADIEN-1-YL ACETATE see
DTD800
3,7-DIMETHYL-1,6-OCTADIEN-3-YL BENZOATE see
LFZ000
3,7-DIMETHYL-2,6-OCTADIEN-1-YL BENZOATE see
GDE800
3,7-DIMETHYL-2,6-OCTADIENYL ESTER FORMIC ACID (E)
see GCY000
trans-2,6-DIMETHYL-2,6-OCTADIEN-8-YL ETHANOATE see
DTD800
3,7-DIMETHYL-1,6-OCTADIEN-3-YL FORMATE see
LGA050
trans-3,7-DIMETHYL-2,6-OCTADIEN-1-YL FORMATE see
GCY000
3,7-DIMETHYL-1,6-OCTADIEN-3-YL ISOBUTYRATE see
LGB000
trans-3,7-DIMETHYL-2,6-OCTADIENYL ISOPENTANOATE
see GDK000
3,7-DIMETHYL-2,6-OCTADIEN-1-YL PHENYLACETATE see
GDM400
4,4-DIMETHYLOCTANOIC ACID, TRIBUTYLSTANNYL ES-
TER see TIF250
DIMETHYLOCTANOL see DTE600
2,6-DIMETHYL-8-OCTANOL see DTE600
3,7-DIMETHYL-3-OCTANOL see TCU600
3,7-DIMETHYL-1-OCTANOL (FCC) see DTE600
(4,4-DIMETHYLOCTANOYLOXY)TRIBUTYLSTANNANE see
TIF250

DIMETHYL PHOSPHATE ESTER OF 3-HYDROXY-N-METHYL-cis-CROTONAMIDE see MRH209

DIMETHYL PHOSPHATE of 3-HYDROXY-N,N-DIMETHYL-cis-CROTONAMIDE see DGQ875

DIMETHYL PHOSPHATE OF 3-HYDROXY-N-METHYL-cis-CROTONAMINE see MRH209

DIMETHYLPHOSPHORAMIDOCYANIDIC ACID, ETHYL ESTER see EIF000

O,S-DIMETHYL PHOSPHORAMIDOTHIOATE see DTQ400

O,O-DIMETHYLPHOSPHOROCHLORIDOTHIOATE see DTQ600

DIMETHYL PHOSPHOROCHLORIDOTHIOATE (DOT) see DTQ600

O,O-DIMETHYL PHOSPHORODITHIOATE N-FORMYL-2-MERCAPTO-N-METHYLACETAMIDE-S-ESTER see DRR200

N-((O,O-DIMETHYLPHOSPHORODITHIOYL)ETHYL) ACETAMIDE see DOP200

O,O-DIMETHYL PHOSPHOROTHIOATE-O,O-DIESTER with 4,4'-THIODIPHENOL see TAL250

DIMETHYLPHTHALATE see DTR200

(O,O-DIMETHYL-PHTHALIMIDIOMETHYL-DITHIOPHOSPHATE) see PHX250

O,O-DIMETHYL S-(N-PHTHALIMIDOMETHYL) DITHIOPHOSPHATE see PHX250

O,O-DIMETHYL S-PHTHALIMIDOMETHYL PHOSPHORODITHIOATE see PHX250

2-β,16-β-(4'-DIMETHYL-1'-PIPERAZINO)-3-α,17-β-DIACETOXY-5-α-ANDROSTANE 2BR see PII250

DIMETHYLPOLYSILOXANE see DTR850

2,2-DIMETHYLPROPANE (DOT) see NCH000

N,N-DIMETHYL-1,3-PROPANEDIAMINE see AJP750

2,2-DIMETHYLPROPANOIC ACID see PJA500

2,2-DIMETHYLPROPANOYL CHLORIDE see DTS400

DIMETHYL PROPIOLACTONE see DTH000

3,3-DIMETHYL-β-PROPIOLACTONE see DTH000

2,2-DIMETHYLPROPIONIC ACID see PJA500

α,α-DIMETHYLPROPIONIC ACID see PJA500

2,2-DIMETHYLPROPIONYL CHLORIDE see DTS400

N,N-DIMETHYL-1,3-PROPYLENEDIAMINE see AJP750

5-(3-DIMETHYLPROPYLIDENE)DIBENZO(a,d)(1,4) CYCLOHEPTADIENE see EAH500

N-(1,1-DIMETHYLPROPYNYL)-3,5-DICHLOROBENZAMIDE see DTT600

2,3-DIMETHYLPYRAZINE see DTU400

2,5-DIMETHYLPYRAZINE see DTU600

2,6-DIMETHYLPYRAZINE see DTU800

3,4-DIMETHYLPYRIDINE see LJB000

N,N-DIMETHYL-N'-2-PYRIDINYL-N'-(2-THIENYLMETHYL)-1,2-ETHANEDIAMIDE see TEO250

1,4-DIMETHYL-5H-PYRIDO(4,3-b)INDOL-3-AMINE see TNX275

1,4-DIMETHYL-5H-PYRIDO(4,3-b)INDOL-3-AMINE ACETATE see AJR500

1,4-DIMETHYL-5H-PYRIDO(4,3-b)INDOL-3-AMINE MONOACETATE see AJR500

N,N-DIMETHYL-N'-(2-PYRIDYL)-N'-(5-CHLORO-2-THENYL)ETHYLENEDIAMINE see CHY250

(3,3-DIMETHYL-1-(m-PYRIDYL-N-OXIDE))TRIAZENE see DTV200

N,N-DIMETHYL-N'-PYRID-2-YL-N'-2-THENYLETHYLENEDIAMINE see TEO250

S-(4,6-DIMETHYL-2-PYRIMIDINYL)-O,O-DIETHYL PHOSPHORODITHIOATE see DTV400

N¹-(4,6-DIMETHYL-2-PYRIMIDINYL)SULFANILAMIDE see SNJ000

N-(4,6-DIMETHYL-2-PYRIMIDYL)SULFANILAMIDE see SNJ000

1,3-DIMETHYL PYROGALLATE see DOJ200

N,N-DIMETHYL-4-(6'-QUINOLYLAZO)ANILINE see DPR000

6,7-DIMETHYL-9-d-RIBITYLISOALLOXAZINE see RIK000

7,8-DIMETHYL-10-d-RIBITYLISOALLOXAZINE see RIK000

7,8-DIMETHYL-10-(d-RIBO-2,3,4,5-TETRAHYDROXY-PENTYL)ISOALLOXAZINE see RIK000

N,N-DIMETHYLSEROTONIN see DPG109

DIMETHYL SILICONE see DTR850

N,N-DIMETHYL-4-STILBENAMINE see DUB800

(E)-N,N-DIMETHYL-4-STILBENAMINE see DUC000

trans-N,N-DIMETHYL-4-STILBENAMINE see DUC000

N,N-DIMETHYL-p-STYRYLANILINE see DUB800

DIMETHYLSULFAAT (DUTCH) see DUD100

O,O-DIMETHYL-O,p-SULFAMOYLPHENYL PHOSPHOROTHIOATE see CQL250

4,6-DIMETHYL-2-SULFANILAMIDOPYRIMIDINE see SNJ000

DIMETHYLSULFAT (CZECH) see DUD100

DIMETHYL SULFATE see DUD100

DIMETHYLSULFID (CZECH) see TFP000

DIMETHYL SULFIDE (DOT) see TFP000

DIMETHYLSULFIDE-α,α'-DICARBOXYLIC ACID see MCM750

2,4-DIMETHYL SULFOLANE see DUD400

1,4-DIMETHYLSULFONOXYBUTANE see BOT250

DIMETHYL SULFOXIDE see DUD800

as-DIMETHYL SULPHATE see MLH500

DIMETHYL SULPHIDE see TFP000

DIMETHYL SULPHOXIDE see DUD800

DIMETHYL TEREPHTHALATE see DUE000

N,N-DIMETHYL-N-TETRADECYLBENZENE-METHANAMINIUM, CHLORIDE (9CI) see TCA500

3,5-DIMETHYLTETRAHYDRO-1,3,5-THIADIAZINE-2-THIONE see DSB200

3,5-DIMETHYLTETRAHYDRO-1,3,5-2H-THIADIAZINE-2-THIONE see DSB200

3,5-DIMETHYL-1,3,5-2H-TETRAHYDROTHIADIAZINE-2-THIONE see DSB200

3,5-DIMETHYLTETRAHYDRO-2H-1,3,5-THIADIAZINE-2-THIONE see DSB200

3,5-DIMETHYL-1,2,3,5-TETRAHYDRO-1,3,5-THIADIAZINETHIONE-2 see DSB200

4,4'-(2,3-DIMETHYLTETRAMETHYLENE)DIPYROCATECHOL see NBR000

O,O-DIMETHYL-S-(3-THIA-PENTYL)-MONOTHIOPHOSPHAT (GERMAN) see DAP400

2,2'-DIMETHYLTHIOCARBANILIDE see DXP600

2,2-DIMETHYL-3-THIOMORPHOLINONE see DUG600

2,2-DIMETHYL-3-THIOMORPHOLONE see DUG600

3,5-DIMETHYL-2-THIONOTETRAHYDRO-1,3,5-THIADIAZINE see DSB200

O,O-DIMETHYLTHIOPHOSPHORIC ACID, p-CHLOROPHENYL ESTER see MQH750

DIMETHYLTIN DIBROMIDE see DUG800

DIMETHYLTIN DIFLUORIDE see DKH200

DIMETHYLTIN FLUORIDE see DKH200

DIMETHYLTIN OXIDE see DTH400

N,N-DIMETHYL-p-(m-TOLYLAZO)ANILINE see DUH600

3,3-DIMETHYL-1-(m-TOLYL)TRIAZENE see DSR200

3,3-DIMETHYL-1-(o-TOLYL)TRIAZENE see MNT500

(DIMETHYLTRIAZENO)IMIDAZOLECARBOXAMIDE see DAB600

4-(DIMETHYLTRIAZENO)IMIDAZOLE-5-CARBOXAMIDE see DAB600

5-(DIMETHYLTRIAZENO)IMIDAZOLE-4-CARBOXAMIDE see DAB600

5-(3,3-DIMETHYLTRIAZENO)IMIDAZOLE-4-CARBOXAMIDE see DAB600

4-(3,3-DIMETHYL-1-TRIAZENO)IMIDAZOLE-5-CARBOXAMIDE see DAB600

5-(3,3-DIMETHYL-1-TRIAZENO)IMIDAZOLE-4-CARBOXAMIDE see DAB600

4-(5)-(3,3-DIMETHYL-1-TRIAZENO)IMIDAZOLE-5(4)-CARBOXAMIDE see DAB600

5-(3,3-DIMETHYL-1-TRIAZENO)IMIDAZOLE-4-CARBOXAMIDE CITRATE see DUI400

3-(3′,3′-DIMETHYLTRIAZENO)PYRIDINE-N-OXIDE see DTV200

3-(3′,3′-DIMETHYLTRIAZENO)-PYRIDIN-N-OXID (GERMAN) see DTV200

5-(3,3-DIMETHYL-1-TRIAZENYL)-1H-IMIDAZOLE-4-CARBOXAMIDE see DAB600

O,O-DIMETHYL-(2,2,2-TRICHLOOR-1-HYDROXY-ETHYL)-FOSFONAAT (DUTCH) see TIQ250

O,O-DIMETHYL-(2,2,2-TRICHLOR-1-HYDROXY-AETHYL) PHOSPHONAT (GERMAN) see TIQ250

O,O-DIMETHYL 2,2,2-TRICHLORO-1-(N-BUTYRYLOXY) ETHYLPHOSPHONATE see BPG000

DIMETHYLTRICHLOROHYDROXYETHYL PHOSPHONATE see TIQ250

DIMETHYL-2,2,2-TRICHLORO-1-HYDROXYETHYLPHOSPHONATE see TIQ250

O,O-DIMETHYL-2,2,2-TRICHLORO-1-HYDROXYETHYL PHOSPHONATE see TIQ250

O,O-DIMETHYL-O-2,4,5-TRICHLOROPHENYL PHOSPHOROTHIOATE see RMA500

DIMETHYL TRICHLOROPHENYL THIOPHOSPHATE see RMA500

O,O-DIMETHYL-O-(2,4,5-TRICHLOROPHENYL)THIOPHOSPHATE see RMA500

O,O-DIMETHYL-O-(3,5,6-TRICHLORO-2-PYRIDYL)PHOSPHOROTHIOATE see CMA250

O,O-DIMETHYL-O-(2,4,5-TRICHLORPHENYL)-THIONOPHOSPHAT(GERMAN) see RMA500

5-(2,3-DIMETHYLTRICYCLO(2.2.1.0^{2,6})HEPT-3-YL)-2-METHYL-2-PENTEN-1-OL see OHG000

N,N-DIMETHYL-2-(TRIFLUOROMETHYL)-10H-PHENOTHIAZINE-10-PROPANAMINE see TKL000

1,1-DIMETHYL-3-(3-TRIFLUOROMETHYLPHENYL)UREA see DUK800

N,N-DIMETHYL-N′-(3-TRIFLUOROMETHYLPHENYL)UREA see DUK800

1,1-DIMETHYL-3-(α,α,α-TRIFLUORO-m-TOLYL) UREA see DUK800

3,7-DIMETHYL-9-(2,6,6-TRIMETHYL-1-CYCLOHEXEN-1-YL)-2,4,6,8-NONATETRAEN-1-OL see VSK600

N,N-DIMETHYLTRYPTAMINE see DPF600

DIMETHYL TUBOCURARINE see DUL800

o,o-DIMETHYLTUBOCURARINE see DUL800

o,o′-DIMETHYLTUBOCURARINE see DUL800

1,3-DIMETHYLUREA see DUM200

N,N′-DIMETHYLUREA see DUM200

sym-DIMETHYLUREA see DUM200

DIMETHYL VICLOGEN CHLORIDE see PAJ000

1,5-DIMETHYL-1-VINYL-4-HEXEN-1-OL BENZOATE see LFZ000

1,5-DIMETHYL-1-VINYL-4-HEXEN-1-YL BENZOATE see LFZ000

1,5-DIMETHYL-1-VINYL-4-HEXENYL ESTER, ISOBUTYRIC ACID see LGB000

DIMETHYL VIOLOGEN see PAI990

1,3-DIMETHYLXANTHINE see TEP000

3,7-DIMETHYLXANTHINE see TEO500

DIMETHYL YELLOW see DOT300

DIMETHYL YELLOW-N,N-DIMETHYLANILINE see DOT300

DIMETHYOXYDOPAMINE see DOE200

10,11-DIMETHYSTRYCHNINE see BOL750

5-(DIMETILAMINOETILOXIMINO-5H-DIBENZO(a,d) CICLOEPTA-1,4-DIENE) CLORIDRATO (ITALIAN) see DPH600

(4-DIMETILAMINO-3-METIL-FENIL)-N-METIL-CARBAMMATO (ITALIAN) see DOR400

N-(γ-DIMETILAMINOPROPIL)-IMINODIBENZILE CLORIDRATO (ITALIAN) see DLH630

O,O-DIMETIL-O-(1,4-DIMETIL-3-OXO-4-AZA-PENT-1-ENIL)-FOSFATO (ITALIAN) see DGQ875

2,6-DIMETIL-EPTAN-4-ONE (ITALIAN) see DNI800

O,O-DIMETIL-S-(2-ETILTIO-ETIL)-MONOTIOFOSFATO (ITALIAN) see DAP400

O,O-DIMETIL-S-(2-ETIL-SOLFINIL-ETIL)-MONOTIOFOSFATO (ITALIAN) see DAP000

O,O-DIMETIL-S-(ETILTIO-ETIL)-DITIOFOSFATO (ITALIAN) see PHI500

O,O-DIMETIL-O-(2-ETILTIO-ETIL)-MONOTIOFOSFATO (ITALIAN) see DAO800

DIMETILFORMAMIDE (ITALIAN) see DSB000

O,O-DIMETIL-S-(N-FORMIL-N-METIL-CARBAMOIL-METIL)-DITIOFOSFATO (ITALIAN) see DRR200

O,O-DIMETIL-S-(N-METIL-CARBAMOIL-METIL)-DITIOFOSFATO (ITALIAN) see DSP400

O,O-DIMETIL-S-(N-METIL-CARBAMOIL)-METIL-MONOTIOFOSFATO (ITALIAN) see DNX800

O,O-DIMETIL-O-(2-N-METILCARBAMOIL-1-METIL-VINIL)-FOSFATO (ITALIAN) see MRH209

O,O-DIMETIL-O-(3-METIL-4-METILTIO-FENIL)-MONOTIOFOSFATO (ITALIAN) see FAQ999

O,O-DIMETIL-O-(3-METIL-4-NITRO-FENIL)-MONOTIOFOSFATO (ITALIAN) see DSQ000

O,O-DIMETIL-S-((2-METOSSI-1,3,4-(4H)-TIADIZAOL-5-ON-4-IL)-METIL)-DITIFOSFATO (ITALIAN) see DSO000

O,O-DIMETIL-S-((MORFOLINO-CARBONIL)-METIL)-DITIOFOSFATO (ITALIAN) see MRU250

O,O-DIMETIL-O-(4-NITRO-FENIL)-MONOTIOFOSFATO (ITALIAN) see MNH000

O,O-DIMETIL-S-((4-OXO-3H-1,2,3-BENZOTRIAZIN-3-IL)-METIL)-DITIOFOSFATO (ITALIAN) see ASH500

(5,5-DIMETIL-3-OXO-CICLOES-1-EN-IL)-N,N-DIMETIL-CARBAMMATO (ITALIAN) see DRL200

3,5-DIMETIL-PERIDRO-1,3,5-TIHADIAZIN-2-TIONE (ITALIAN) see DSB200

DIMETILSOLFATO (ITALIAN) see DUD100

O,O-DIMETIL-(2,2,2-TRICLORO-1-IDROSSI-ETIL)-FOSFONATO (ITALIAN) see TIQ250

DIMETON see DSP400

3,3′-DIMETOSSIBENZODINA (ITALIAN) see DCJ200

DIMETOX see TIQ250

DIMETRIDAZOLE see DSV800

DIMETYLFORMAMIDU (CZECH) see DSB000

DIMEVAMIDE see DOY400

DIMEVUR see DSP400

DIMEXIDE see DUD800

DIMEZATHINE see SNJ000

DIMID see DRP800

DIMILIN see CJV250

DIMIPRESSIN see DLH600, DLH630

DIMITAN see BIE500

DIMO see DLW600

DIMORPHOLAMINE see DUO400

DIMORPHOLINE DISULFIDE see BKU500

DIMORPHOLINO DISULFIDE see BKU500

DIMPEA see DOE200

DIMPYLATE see DCM750

DIN 2.4602 see CNA750

DIN 2.4964 see CNA750

DINACORYL see DJS200

DINACRIN see ILD000

3,4,5,6-DINAPHTHACARBAZOLE see DCY000

1,2,5,6-DINAPHTHACRIDINE see DCS400

3,4,6,7-DINAPHTHACRIDINE see DCS600

DI-β-NAPHTHYL-p-PHENYLDIAMINE see NBL000

DI-β-NAPHTHYL-p-PHENYLENEDIAMINE see NBL000

N,N′-DI-β-NAPHTHYL-p-PHENYLENEDIAMINE see NBL000

sym-DI-β-NAPHTHYL-p-PHENYLENEDIAMINE see NBL000

DINARKON see DLX400

DINATE see DXE600

DINATRIUM-AETHYLENBISDITHIOCARBAMAT (GERMAN) see DXD200

DINATRIUM-(N,N′-AETHYLEN-BIS(DITHIOCARBAMAT)) (GERMAN) see DXD200

DINATRIUM-(3,6-EPOXY-CYCLOHEXAAN-1,2-DICARBOXYLAAT) (DUTCH) see DXD000

DINATRIUM-(3,6-EPOXY-CYCLOHEXAN-1,2-DICARBOXYLAT) (GERMAN) see DXD000

DINATRIUM-(N,N'-ETHYLEEN-BIS(DITHIOCARBAMAAT)) (DUTCH) see DXD200
DINATRIUMPYROPHOSPHAT (GERMAN) see DXF800
DINEX see CPK500
DINICKEL TRIOXIDE see NDH500
DINIL see PFA860
DINILE see SNE000
DINITOLMID see DUP300
DINITOLMIDE see DUP300
2,4-DINITRANILINE see DUP600
DINITRATE de DIETHYLENE-GLYCOL (FRENCH) see DJE400
1,3-DINITRATO-2,2-BIS(NITRATOMETHYL)PROPANE see PBC250
DINITRATODIOXOURANIUM, HEXAHYDRATE see URS000
DINITRILE of ISOPHTHALIC ACID see PHX550
2,3-DINITRILO-1,4-DITHIA-ANTHRAQUINONE see DLK200
2,3-DINITRILO-1,4-DITHIOANTHRACHINON (GERMAN) see DLK200
DINITRO see BRE500
DINITRO-3 see BRE500
2,4-DINITROANILIN (GERMAN) see DUP600
2,4-DINITROANILINA (ITALIAN) see DUP600
2,4-DINITROANILINE see DUP600
2,4-DINITROANISOL see DUP800
α-DINITROANISOLE see DUP800
2,4-DINITROANISOLE see DUP800
2,4-DINITROBENZENAMIME see DUP600
DINITROBENZENE see DUQ180
m-DINITROBENZENE see DUQ200
o-DINITROBENZENE see DUQ400
p-DINITROBENZENE see DUQ600
1,2-DINITROBENZENE see DUQ400
1,3-DINITROBENZENE see DUQ200
2,4-DINITROBENZENE see DUQ200
DINITROBENZENE, solution (DOT) see DUQ180
1,3-DINITROBENZOL see DUQ200
DINITROBENZOL, solid (DOT) see DUQ180
5,7-DINITRO-1,2,3-BENZOXADIAZOLE see DUR800
4,4'-DINITROBIFENYL (CZECH) see DUS000
4,4'-DINITROBIPHENYL see DUS000
4,6-DINITRO-2-sec.BUTYLFENOL (CZECH) see BRE500
4,6-DINITRO-2-sec.BUTYLFENOLATE AMMONY (CZECH) see BPG250
2,4-DINITRO-6-sec-BUTYLFENYLESTER KYSELINY OC-TOVE (CZECH) see ACE500
DINITROBUTYLPHENOL see BRE500
2,4-DINITRO-6-sec-BUTYLPHENOL see BRE500
4,6-DINITRO-2-sec-BUTYLPHENOL see BRE500
4,6-DINITRO-o-sec-BUTYLPHENOL see BRE500
4,6-DINITRO-2-sec-BUTYLPHENOL AMMONIUM SALT see BPG250
4,6-DINITRO-o-sec-BUTYLPHENOL AMMONIUM SALT see BPG250
2,4-DINITRO-6-sek.BUTYL-PHENYLACETAT (GERMAN) see ACE500
4,6-DINITRO-2-sec-BUTYLPHENYL ACETATE see ACE500
4,6-DINITRO-2-CAPRYLPHENYL CROTONATE see AQT500
4,6-DINITRO-2-(2-CAPRYL)PHENYL CROTONATE see AQT500
DINITROCHLOROBENZENE see CGL750
DINITROCHLOROBENZENE (DOT) see CGL750
2,4-DINITROCHLOROBENZENE see CGM000
1,3-DINITRO-4-CHLOROBENZENE see CGM000
2,4-DINITRO-1-CHLOROBENZENE see CGM000
DINITROCHLOROBENZOL see CGM000
DINITROCHLOROBENZOL (DOT) see CGM000
2,4-DINITRO-1-CHLORO-NAPHTHALENE see DUS600
DINITROCRESOL see DUS700
DINITRO-o-CRESOL see DUS800, DUS00
DINITRO-p-CRESOL see DUT600

2,4-DINITRO-o-CRESOL see DUS700
2,6-DINITRO-p-CRESOL see DUT600
3,5-DINITRO-o-CRESOL see DUT000
3,5-DINITRO-p-CRESOL see DUT200
4,6-DINITRO-o-CRESOL see DUS700
DINITRO-o-CRESOL, liquid (DOT) see DUS800
DINITRO-o-CRESOL, solid (DOT) see DUS800
4,6-DINITRO-o-CRESOLO (ITALIAN) see DUS700
DINITRO-o-CRESOL SODIUM SALT see DUU600
3,5-DINITRO-o-CRESOL SODIUM SALT see DUU600
4,6-DINITRO-o-CRESOL SODIUM SALT see DUU600
DINITROCYCLOHEXYLPHENOL see CPK500
DINITROCYCLOHEXYLPHENOL (DOT) see CPK500
DINITRO-o-CYCLOHEXYLPHENOL see CPK500
2,4-DINITRO-6-CYCLOHEXYLPHENOL see CPK500
4,6-DINITRO-o-CYCLOHEXYLPHENOL see CPK500
DINITRODENDTROXAL see DUS700
DINITRODIGLICOL (ITALIAN) see DJE400
DINITRODIGLYKOL (CZECH) see DJE400
2,6-DINITRO-N,N-DIPROPYL-4-(TRIFLUOROMETHYL) BENZENAMINE see DUV600
2,6-DINITRO-N,N-DI-N-PROPYL-α,α,α-TRIFLURO-p-TOLU-IDINE see DUV600
2,4-DINITROFENOL (DUTCH) see DUZ000
DINITROFENOLO (ITALIAN) see DUZ000
2,4-DINITROFLUOROBENZENE see DUW400
2,4-DINITRO-1-FLUOROBENZENE see DUW400
DINITROGEN DIOXIDE see NGU500
DINITROGEN MONOXIDE see NGU000
DINITROGEN TETRAFLUORIDE see TCI000
DINITROGEN TETROXIDE (DOT) see NGU500
DINITROGLICOL (ITALIAN) see EJG000
DINITROGLYCOL see EJG000
3,5-DINITRO-2-HYDROXYTOLUENE see DUS700
4,6-DINITROKRESOL (DUTCH) see DUS700
4,6-DINITRO-o-KRESOL (CZECH) see DUS700
4,6-DINITRO-o-KRESYLESTER KYSELINY OCTOVE (CZECH) see AAU250
DINITROL see DUS700
2,6-DINITRO-3-METHOXY-4-tert-BUTYLTOLUENE see BRU500
DINITROMETHYL CYCLOHEXYLTRIENOL see DUS700
DINITRO(1-METHYLHEPTYL)PHENYL CROTONATE see AQT500
2,4-DINITRO-6-(1-METHYLHEPTYL)PHENYL CROTONATE see AQT500
2,4-DINITRO-6-METHYLPHENOL see DUS700
2,4-DINITRO-6-METHYLPHENOL SODIUM SALT see DUU600
4,6-DINITRO-2-(1-METHYL-N-PROPYL)PHENOL see BRE500
2,4-DINITRO-6-(1-METHYL-PROPYL)PHENOL (FRENCH) see BRE500
1,5-DINITRONAPHTHALENE see DUX700
2,4-DINITRO-1-NAPHTHOL see DUX800
2-4 DINITRO-α-NAPHTOL (FRENCH) see DUX800
2,4-DINITRO-6-(2-OCTYL)PHENYL CROTONATE see AQT500
DINITROPHENOL see DUY600
α-DINITROPHENOL see DUZ000
β-DINITROPHENOL see DVA200
2,3-DINITROPHENOL see DUY800
2,4-DINITROPHENOL see DUZ000
2,6-DINITROPHENOL see DVA200
3,4-DINITROPHENOL see DVA400
3,5-DINITROPHENOL see DVA600
2,4-DINITROPHENYL ETHER of MORPHINE see DVC800
DINITROPHENYLMETHANE see DVG600
2,4-DINITROPHENYLMETHYL ETHER see DUP800
2,4-DINITROPHENYLMORPHINE HYDROCHLORIDE see DVC800
DINITROPYRENE see DVD400, DVD600, DVD800
1,3-DINITROPYRENE see DVD400
1,6-DINITROPYRENE see DVD600

1,8-DINITROPYRENE see DVD800
4,6-DINITROQUINOLINE-1-OXIDE see DVE000
DINITROSO-2,5-DIMETHYLPIPERAZINE see DRN800
DINITROSO-2,6-DIMETHYLPIPERAZINE see DRO000
1,4-DINITROSO-2,6-DIMETHYLPIPERAZINE see DRO000
N,N'-DINITROSO-2,6-DIMETHYLPIPERAZINE see DRO000
DINITROSODIMETHYLPROPANEDIAMINE see DRO200
N,N'-DINITROSO-N,N'-DIMETHYL-1,3-PROPANEDIAMINE
   see DRO200
DINITROSOPIPERAZIN (GERMAN) see DVF200
DINITROSOPIPERAZINE see DVF200
1,4-DINITROSOPIPERAZINE see DVF200
N,N'-DINITROSOPIPERAZINE see DVF200
3,5-DINITRO-o-TOLUAMIDE see DUP300
DINITROTOLUENE see DVG600
2,4-DINITROTOLUENE see DVH000
ar,ar-DINITROTOLUENE see DVG600
DINITROTOLUENE, liquid (DOT) see DVG600
DINITROTOLUENE, molten (DOT) see DVG600
DINITROTOLUENE, solid (DOT) see DVG600
2,4-DINITROTOLUOL see DVH000
4,6-DINITRO-1,2,3-TRICHLOROBENZENE see DVI600
2,6-DINITRO-4-TRIFLUORMETHYL-N,N-DIPROPYLANILIN
   (GERMAN) see DUV600
DINKUM OIL see EQQ000
DINOC see DUS700, DUU600
DINOPOL NOP see DVL600
DINOPROST see DMU800
DINOPROSTONE see DVJ200
DINOPROST TROMETHAMINE (USDA) see POC750
DINOSEB see BRE500
DINOSEB-ACETATE see ACE500
DINOSEB (AMINE) see BPG250
DINOSEBE (FRENCH) see BRE500
DINOSOL see SNN300
DINOVEX see DAL600
DINOXOL see DAA800, TAA100
DINTOIN see DKQ000
DINTOINA see DNU000
DINURANIA see DUS700
DINYL see PFA860
DIOCTLYN see DJL000
DIOCTYLAL see DJL000
DIOCTYL-o-BENZENEDICARBOXYLATE see DVL600
2,2-DIOCTYL-1,3-DIOXA-2-STANNA-7-THIADECAN-4,10-
   DIONE see DVN909
DIOCTYL ESTER of SODIUM SULFOSUCCINATE see
   DJL000
DIOCTYL ESTER of SODIUM SULFOSUCCINIC ACID see
   DJL000
DIOCTYL FUMARATE see DVK600
DIOCTYL-MEDO FORTE see DJL000
DIOCTYL PHTHALATE see DVL600, DVL700
n-DIOCTYL PHTHALATE see DVL600
DI-sec-OCTYL PHTHALATE see DVL700
DIOCTYL(1,2-PROPYLENEDIOXYBIS(MALEOYLDIOXY))
   STANNANE see DVL800
DIOCTYL SEBACATE see BJS250
DIOCTYL SODIUM SULFOSUCCINATE (FCC) see DJL000
DIOCTYL SULFOSUCCINATE SODIUM SALT see DJL000
DI-N-OCTYLTIN BIS(BUTYL MALEATE) see BHK500
DIOCTYLTIN BIS(ISOOCTYL MERCAPTOACETATE) see
   BKK750
DIOCTYLTIN-S,S'-BIS(ISOOCTYL MERCAPTOACETATE)
   see BKK750
DIOCTYLTIN BIS(ISOOCTYL THIOGLYCOLATE) see
   BKK750
DIOCTYL-TIN BIS(ISOOCTYLTHIOGLYCOLLATE) see
   BKK750
DI-n-OCTYLTIN DIISOOCTYL THIOGLYCOLATE see
   BKK750
DI-N-OCTYLTIN DIMONOBUTYLMALEATE see BHK500
DI-n-OCTYLTIN DI(1,2-PROPYLENEGLYCOLMALEATE) see
   DVL800

DIOCTYLTIN-3,3'-THIODIPROPIONATE see DVN909
DI-n-OCTYL-ZINN-DI-ISOOCTYLTHIOGLYKOLAT (GER-
   MAN) see BKK750
DI-N-OCTYLZINN-DIMONOBUTYLMALEINAT (GERMAN)
   see BHK500
DI-n-OCTYL-ZINN-DI-(1,2-PROPYLENGLYKOLMALEINAT)
   (GERMAN) see DVL800
DIODOHYDROXYQUIN see DNF600
DIOFORM see DFI100
DIOGYN see EDO000
DIOGYN B see EDP000
DIOGYNETS see EDO000
DIOKAN see DVQ000
DIOKSAN (POLISH) see DVQ000
DIOKSYNY (POLISH) see TAI000
DIOLAMINE see DHF000
DIOLANE see HFP875
DIOLENE see IPU000
DIOLICE see CNU750
DIOMEDICONE see DJL000
DI-ON see DXQ500
DIONE 21-ACETATE see RKP000
DIONIN see ENK000, ENK500
DIONINE see ENK000
DIORTHOTOLYLGUANIDINE see DXP200
1,4-DIOSSAN-2,3-DIYL-BIS(O,O-DIETIL-DITIOFOSFATO)
   (ITALIAN) see DVQ709
DIOSSANO-1,4 (ITALIAN) see DVQ000
1,4-DIOSSIBENZENE (ITALIAN) see QQS200
2,4-DIOSSI-5-DIAZOPIRIMIDINA (ITALIAN) see DCQ600
DIOSSIDONE see BRF500
DIOSUCCIN see DJL000
DIOTHENE see PJS750
DIOTILAN see DJL000
DIOVAC see DJL000
DIOVOCYCLIN see EDR000
DIOVOCYLIN see EDR000
DIOXAAN-1,4 (DUTCH) see DVQ000
1,4-DIOXAAN-2,3-DIYL-BIS(O,O-DIETHYL-DITHIOFOS-
   FAAT) (DUTCH) see DVQ709
DIOXACARB see DVS000
1,4-DIOXACYCLOHEXANE see DVQ000
2,5-DIOXAHEXANE see DOE600
p-DIOXAN (CZECH) see DVQ000
DIOXAN-1,4 (GERMAN) see DVQ000
2,3-p-DIOXANDITHIOL S,S-BIS(O,O-DIETHYL PHOSPHO-
   RODITHIOATE) see DVQ709
1,4-DIOXAN-2,3-DIYL-BIS(O,O-DIAETHYL-DITHIOPHOS-
   PHAT) (GERMAN) see DVQ709
1,4-DIOXAN-2,3-DIYL-BIS(O,O-DIETHYLPHOSPHORO-
   THIOLOTHIONATE) see DVQ709
1,4-DIOXAN-2,3-DIYL-O,O,O',O'-TETRAETHYL DI(PHOS-
   PHOROMITHIOATE) see DVQ709
DIOXANE see DVQ000
p-DIOXANE see DVQ000
2,3-p-DIOXANE-S,S-BIS(O,O-DIETHYLPHOSPHOROI-
   THIOATE) see DVQ709
p-DIOXANE-2,3-DITHIOL-S,S-DIESTER with O,O-DIETHYL
   PHOSPHORODITHIOATE see DVQ709
p-DIOXANE-2,3-DIYL ETHYL PHOSPHORODITHIOATE see
   DVQ709
1,4-DIOXANE (MAK) see DVQ000
DIOXANNE (FRENCH) see DVQ000
3,6-DIOXAOCTANE-1,8-DIOL see TJQ000
DIOXATHION see DVQ709
2,2-DIOXIDE-1,3,2-DIOXATHIOLANE see EJP000
DIOXIME-p-BENZOQUINONE see DVR200
DIOXIME-1,4-CYCLOHEXADIENEDIONE see DVR200
DIOXIME-2,5-CYCLOHEXADIENE-1,4-DIONE see DVR200
DIOXIN (bactericide) (OBS.) see ABC250
DIOXIN (herbicide contaminant) see TAI000
DIOXINE see TAI000
DIOXITOL see CBR000
9,10-DIOXOANTHRACENE see APK250

p-DIOXOBENZENE see HIH000
2,2'-((1,4-DIOXO-1,4-BUTANEDIYL)BIS(OXY))BIS(N,N,N-TRIMETHYLETHANAMINIUM see CMG250
2,2'-((1,4-DIOXO-1,4-BUTANEDIYL)BIS(OXY))BIS(N,N,N-TRIMETHYLETHANAMINIUM DICHLORIDE see HLC500
1,1',1''-(3,6-DIOXO-1,4-CYCLOHEXADIENE-1,2,4-TRIYL) TRISAZIRIDINE see TND000
2,6-DIOXO-5-DIAZOPYRIMIDINE see DCQ600
DIOXODICHLOROCHROMIUM see CML125
5,5-DIOXO-10-(2-(DIMETHYLAMINO)PROPYL)PHENO-THIAZINE HYDROCHLORIDE see DVT400
3,5-DIOXO-1,2-DIPHENYL-4-N-BUTYLPYRAZOLIDENE see BRF500
3,5-DIOXO-1,2-DIPHENYL-4-N-BUTYL-PYRAZOLIDIN see BRF500
3,5-DIOXO-1,2-DIPHENYL-4-N-BUTYL-PYRAZOLIDINE see BRF500
3,5-DIOXO-1,2-DIPHENYL-4-N-BUTYLPYRAZOLIDIN SO-DIUM see BOV750
DI-OXO-DI-N-PROPYLNITROSAMINE see NJN000
2,2'-DIOXO-DI-N-PROPYLNITROSAMINE see NJN000
DIOXOLAN see DVR600
DIOXOLANE (DOT) see DVR600
2-(1,3-DIOXOLANE-2-YL)PHENYL N-METHYLCARBAMATE see DVS000
((1,3-DIOXOLAN-4-YL)METHYL)TRIMETHYLAMMONIUM IODIDE see FMX000
2-(1,3-DIOXOLAN-2-YL)PHENYL-N-METHYLCARBAMAT see DVS000
o-(1,3-DIOXOLAN-2-YL)PHENYL METHYLCARBAMATE see DVS000
2,6-DIOXO-4-METHYL-4-ETHYLPIPERIDINE see MKA250
DIOXONE see DJT400
2,2'-DIOXO-N-NITROSODIPROPYLAMINE see NJN000
3,5-DIOXO-1-PHENYL-2-(p-HYDROXYPHENYL)-4-N-BU-TYLPYRAZOLIDENE see HNI500
1,3-DIOXOPHTHALAN see PHW750
1,3-DIOXO-5-PHTHALANCARBOXYLIC ACID see TKV000
2,6-DIOXO-3-PHTHALIMIDOPIPERIDINE see TEH500
2-(2-6-DIOXO-3-PIPERIDINYL)1H-ISOINDOLE-1,3(2H)-DIONE see TEH500
N-(2,6-DIOXO-3-PIPERIDYL)PHTHALIMIDE see TEH500
DIOXOPROMETHAZINE HYDROCHLORIDE see DVT400
N,N-DI(2-OXOPROPYL)NITROSAMINE see NJN000
2,2'-DIOXOPROPYL-N-PROPYLNITROSAMINE see NJN000
2,6-DIOXOPURINE see XCA000
DIOXYANTHRANOL see APH250
1,4-DIOXYANTHRAQUINONE (RUSSIAN) see DMH000
m-DIOXYBENZENE see REA000
o-DIOXYBENZENE see CCP850
1,4-DIOXYBENZENE see QQS200
3,3'-DIOXYBENZIDINE see DMI400
1,4-DIOXY-BENZOL (GERMAN) see QQS200
DIOXYBUTADIENE see BGA750
DIOXYDE de BARYUM (FRENCH) see BAO250
DIOXYDEMETON-S-METHYL see DAP600
3,3'-(DIOXYDICARBONYL)DIPROPIONIC ACID see SNC500
DIOXYETHYLENE ETHER see DVQ000
DIOXYMETHYLENE-PROTOCATECHUIC ALDEHYDE see PIW250
DI(p-OXYPHENYL)-2,4-HEXADIENE see DAL600
DIOZOL see BRF500
DIPA see DNM200
DIPAN see DVX200
DIPANE see WAK000
DIPANOL see MCC250
DIPARAANISYL-MONOPHENETHYL-GUANIDIN-HYDRO-CHLORID (GERMAN) see PDN500
DI-PARALEN see CFF500
DIPARALENE see CFF500
DIPARALENE HYDROCHLORIDE see CDR250

DI-PARALENE-2-HYDROCHLORIDE see CDR000
DIPAXIN see DVV600
DIPEGYL see NCR000
DI(PENTANOYLOXY)DIBUTYLSTANNANE see DEA600
DIPENTENE see MCC250
DIPENTENE DIOXIDE see LFV000
DIPENTYLAMINE see DCH200
DIPENTYL ETHER see PBX000
DIPENTYLNITROSAMINE see DCH600
DI-n-PENTYLNITROSAMINE see DCH600
DIPENTYLTIN DICHLORIDE see DVV200
DIPEPTIDE SWEETENER see ARN825
DIPHACIL see DHX800
DIPHACIN see DVV600
DIPHACINONE see DVV600
DIPHACYL see DHX800
DIPHANTINE see BBV500
DIPHANTOIN see DKQ000
DIPHANTOINE SODIUM see DNU000
DIPHEBUZOL see BRF500
DIPHEDAL see DKQ000
DIPHEDAN see DNU000
DIPHENACIN see DVV600
DIPHENADIONE see DVV600
DIPHENAMID see DRP800
DIPHENAMIDE see DRP800
DIPHENATE see DNU000
DIPHENATRILE see DVX200
DIPHENHYDRINATE see DYE600
DIPHENIN see DNU000
DIPHENINE see DKQ000
DIPHENINE SODIUM see DNU000
o-DIPHENOL see CCP850
DIPHENTOIN see DKQ000, DNU000
DIPHENYL (OSHA) see BGE000
DIPHENYLACETIC ACID DIETHYLAMINOETHYL ESTER see DHX800
DIPHENYLACETIC ACID, 2-(DIETHYLAMINO)ETHYL ES-TER see DHX800
DIPHENYLACETONITRILE see DVX200
DIPHENYLACETYLDIETHYLAMINOETHANOL see DHX800
2-DIPHENYLACETYL-1,3-DIKETOHYDRINDENE see DVV600
2-DIPHENYLACETYL-1,3-INDANDIONE see DVV600
2-(DIPHENYLACETYL)INDAN-1,3-DIONE see DVV600
2-(DIPHENYLACETYL)-1H-INDENE-1,3(2H)-DIONE see DVV600
2,3-DIPHENYLACRYLONITRILE see DVX600
α,β-DIPHENYLACRYLONITRILE see DVX600
DIPHENYLAMIDE see DRP800
DIPHENYLAMINE see DVX800
N,N-DIPHENYLAMINE see DVX800
DIPHENYLAMINE-2-CARBOXYLIC ACID see PEG500
DIPHENYLAMINECHLORARSINE see PDB000
DIPHENYLAMINECHLOROARSINE (DOT) see PDB000
DIPHENYLAN see DKQ000
N,N-DIPHENYLANILINE see TMQ500
DIPHENYLAN SODIUM see DNU000
DIPHENYLARSINOUS CHLORIDE see CGN000
DIPHENYLBENZENE see TBD000
m-DIPHENYLBENZENE see TBC620
p-DIPHENYLBENZENE see TBC750
1,2-DIPHENYLBENZENE see TBC640
1,4-DIPHENYLBENZENE see TBC750
DIPHENYL BLUE 2B see CMO000
DIPHENYL BLUE 3B see CMO250
DIPHENYLBUTAZONE see BRF500
1,2-DIPHENYL-4-BUTYL-3,5-DIOXOPYRAZOLIDINE see BRF500
1,2-DIPHENYL-4-BUTYL-3,5-PYRAZOLIDINEDIONE see BRF500
DIPHENYL CARBONATE see DVZ000
DIPHENYLCHLOORARSINE (DUTCH) see CGN000

DIPHENYLCHLOROARSINE (DOT) see CGN000
α,β-DIPHENYLCINNAMONITRILE see TMQ250
DIPHENYL-α-CYANOMETHANE see DVX200
DIPHENYL DEEP BLACK G see AQP000
DIPHENYLDIAZENE see ASL250
1,2-DIPHENYLDIAZENE see ASL250
DIPHENYL DICHLOROSILANE (DOT) see DFF000
DIPHENYLDIIMIDE see ASL250
2,2-DIPHENYL-N,N-DIMETHYLACETAMIDE see DRP800
1,2-DIPHENYL-1-(DIMETHYLAMINO)ETHANE see DWA600
1,1-DIPHENYL-1-(DIMETHYLAMINOISOPROPYL) BUTANONE-2 see IKZ000
1,1-DIPHENYL-1-(β-DIMETHYLAMINOPROPYL)BUTANONE-2 HYDROCHLORIDE see MDP000
α,α-DIPHENYL-γ-DIMETHYLAMINOVALERAMIDE see DOY400
1,2-DIPHENYL-3,5-DIOXO-4-BUTYLPYRAZOLIDINE see BRF500
DIPHENYLDIOXOBUTYLPYRAZOLIDINE-BUTAZOLIDINE-SODIUM see BOV750
1,2-DIPHENYL-2,3-DIOXO-4-N-BUTYLPYRAZOLINE see BRF500
DIPHENYL mixed with DIPHENYL OXIDE see PFA860
DIPHENYLE CHLORE, 42% de CHLORE (FRENCH) see PJM500
DIPHENYLE CHLORE, 54% de CHLORE (FRENCH) see PJN000
4,4′-DIPHENYLENEDIAMINE see BBX000
DIPHENYLENE DIOXIDE see DDA800
DIPHENYLENEIMINE see CBN000
DIPHENYLENIMIDE see CBN000
DIPHENYLENIMINE see CBN000
1,2-DIPHENYLETHANE see BFX500
DIPHENYL ETHER see PFA850
DIPHENYLETHYLENE see SLR000
DIPHENYL-2-ETHYLHEXYL PHOSPHATE see DWB800
DIPHENYL FAST BROWN BRL see CMO750
DIPHENYLGLYCOLLIC ACID-2-(DIETHYLAMINO)ETHYL ESTER HYDROCHLORIDE see BCA000
DIPHENYLGLYOXAL PEROXIDE see BDS000
DIPHENYLGUANIDINE see DWC600
1,3-DIPHENYLGUANIDINE see DWC600
N,N′-DIPHENYLGUANIDINE see DWC600
DIPHENYLHYDANTOIN see DKQ000
5,5-DIPHENYLHYDANTOIN see DKQ000
DIPHENYLHYDANTOINE (FRENCH) see DKQ000
DIPHENYLHYDANTOIN SODIUM see DNU000
5,5-DIPHENYLHYDANTOIN SODIUM see DNU000
DIPHENYLHYDRAMINE see BBV500
DIPHENYLHYDRAMINE HYDROCHLORIDE see BAU750
1,2-DIPHENYLHYDRAZINE see HHG000
sym-DIPHENYLHYDRAZINE see HHG000
2,2-DIPHENYL-3-HYDROXYPROPIONIC ACID LACTONE see DWI400
5,5-DIPHENYLIMIDAZOLIDIN-2,4-DIONE see DKQ000
5,5-DIPHENYL-2,4-IMIDAZOLIDINEDIONE see DKQ000
5,5-DIPHENYL-2,4-IMIDAZOLIDINE-DIONE, MONOSODIUM SALT see DNU000
DIPHENYL KETONE see BCS250
DIPHENYLMERCURIDINAPHTHYLMETHANEDISULFONATE see PFM750
DIPHENYLMERCURY see DWD800
DIPHENYLMETHAN-4,4′-DIISOCYANAT (GERMAN) see MJP400
2,4′-DIPHENYLMETHANEDIAMINE see MJP750
4,4′-DIPHENYLMETHANEDIAMINE see MJQ000
DIPHENYL METHANE DIISOCYANATE see MJP400
4,4′-DIPHENYLMETHANE DIISOCYANATE see MJP400
p.p′-DIPHENYLMETHANE DIISOCYANATE see MJP400
DIPHENYLMETHANE 4,4′-DIISOCYANATE (DOT) see MJP400
DIPHENYLMETHANE-4,4′-DIISOCYANATE-TRIMELLIC ANHYDRIDE-ETHOMID HT POLYMER see TKV000

DIPHENYLMETHANONE see BCS250
2-(DIPHENYLMETHOXY)-N,N-DIMETHYLETHYLAMINE see BBV500
2-DIPHENYLMETHOXY-N,N-DIMETHYLETHYLAMINE HYDROCHLORIDE see BAU750
4-(DIPHENYLMETHOXY)-1-METHYLPIPERIDINE see LJR000
4-(DIPHENYLMETHOXY)-1-METHYLPIPERIDINE CHLOROTHEOPHYLLINE see DWF000
3-DIPHENYLMETHOXYTROPANE MESYLATE see TNU000
3-DIPHENYLMETHOXYTROPANE METHANESULFONATE see TNU000
DIPHENYLMETHYL BROMIDE (DOT) see BNG750
DIPHENYL METHYL BROMIDE, solid (DOT) see BNG750
DIPHENYL METHYL BROMIDE, solution (DOT) see BNG750
DIPHENYLMETHYLCYANIDE see DVX200
α-(DIPHENYLMETHYLENE)BENZENEACETIC ACID see TMQ250
1-(3-(4-(DIPHENYLMETHYL)-1-PIPERAZINYL)PROPYL)-2-BENZIMIDAZOLINONE see OMG000
1-(3-(4-(DIPHENYLMETHYL)-1-PIPERAZINYL)PROPYL)-1,3-DIHYDRO-2H-BENZ IMIDAZOL-2-ONE see DWF790
1-(3-(4-(DIPHENYLMETHYL)-1-PIPERAZINYL)PROPYL)-1,3-DIHYDRO-2H-BENZ IMIDAZOL-2-ONE see OMG000
4-DIPHENYLMETHYLTROPYLTROPINIUM BROMIDE see PEM750
4-DIPHENYLMETHYL-dl-TROPYLTROPINIUM BROMIDE see PEM750
DIPHENYLNITROSAMIN (GERMAN) see DWI000
DIPHENYLNITROSAMINE see DWI000
N,N-DIPHENYLNITROSAMINE see DWI000
DIPHENYL N-NITROSOAMINE see DWI000
2,2-DI(4-PHENYLOL)PROPANE see BLD500
3,3-DIPHENYL-2-OXETANONE see DWI400
DIPHENYL OXIDE see PFA850
DIPHENYL-p-PHENYLENEDIAMINE see BLE500
N,N′-DIPHENYL-p-PHENYLENEDIAMINE see BLE500
1,2-DIPHENYL-4-(2′-PHENYLSULFINETHYL)-3,5-PYRAZO-LIDINEDIONE see DWM000
α,α-DIPHENYL-2-PIPERIDINEMETHANOL see DWK400
α,α-DIPHENYL-2-PIPERIDINEMETHANOL HYDROCHLORIDE see PII750
α,α-DIPHENYL-1-PIPERIDINEPROPANOL HYDROCHLORIDE see PMC250
1,1-DIPHENYL-3-PIPERIDINO-1-PROPANOL HYDROCHLORIDE see PMC250
1,1-DIPHENYL-3-(1-PIPERIDYL)-1-PROPANOL HYDROCHLORIDE see PMC250
α,α-DIPHENYL-β-PROPIOLACTONE see DWI400
(+)-1,2-DIPHENYL-2-PROPIONOXY-3-METHYL-4-DIMETHYLAMINOBUTANE HYDROCHLORIDE see PNA500
1,1-DIPHENYL-2-PROPYN-1-OL CYCLOHEXANECARBAMATE see DWL400
1,1-DIPHENYL-2-PROPYNYL-N-CYCLOHEXYLCARBAMATE see DWL400
1,1-DIPHENYL-2-PROPYNYL ESTER CYCLOHEXANECARBAMIC ACID see DWL400
DIPHENYLPYRALINE see LJR000
DIPHENYLPYRAZONE see DWM000
DIPHENYLPYRILENE see LJR000
N,N′-DIPHENYLTHIOCARBAMIDE see DWN800
sym-DIPHENYLTHIOCARBAMIDE see DWN800
DIPHENYLTHIOCARBAZONE see DWN200
DIPHENYLTHIOUREA see DWN800
1,3-DIPHENYLTHIOUREA see DWN800
1,3-DIPHENYL-2-THIOUREA see DWN800
N,N′-DIPHENYLTHIOUREA see DWN800
sym-DIPHENYLTHIOUREA see DWN800
1,3-DIPHENYLTRIAZENE see DWO800
DIPHENYLTRICHLOROETHANE see DAD200
2,2-DIPHENYL-VALERIC ACID-2-(DIETHYLAMINO) ETHYL)ESTER see DIG400
DIPHER see EIR000

DI-PHETINE see DKQ000, DNU000
DIPHONE see SOA500
DIPHOSGENE see PGX000
DIPHOSPHORIC ACID, DISODIUM SALT see DXF800
DIPHOSPHORIC ACID THETRAETHYL ESTER see TCF250
DIPHOSPHORUS PENTOXIDE see PHS250
DIPHOSPHORUS TRIOXIDE see PHT500
DIPHYL see PFA860
2-(DIPHYLMETHOXY)-N,N-DIMETHYL-ETHANAMINE HYDROCHLORIDE see BAU750
DIPIGYL see NCR000
DIPIN see BJC250
DIPINE see BJC250
DIPIPERAL see FHG000
DIPIPERON see FHG000
DIPIPERONE see FHG000
DIPIRARTRIL-TROPICO see DUD800
DIPIRIN see DOT000
DIPN see DNB200
DIPOFENE see DCM750
DIPOTASSIUM CHROMATE see PLB250
DIPOTASSIUM DICHLORIDE see PLA500
DIPOTASSIUM DICHROMATE see PKX250
DIPOTASSIUM MONOCHROMATE see PLB250
DIPOTASSIUM NICKEL TETRACYANIDE see NDI000
DIPOTASSIUM PERSULFATE see DWQ000
DIPOTASSIUM TETRACYANONICKELATE see NDI000
DIPPEL'S OIL see BMA750
DIPPING ACID see SOI500
DIPRAM see DGI000
DIPRAZINE see DQA400
DIPRON see SNM500
DI-2-PROPENYLAMINE see DBI600
DI-2-PROPENYL ESTER, 1,2-BENZENEDICARBOXYLIC ACID see DBL200
N-N-DI-2-PROPENYL-2-PROPEN-1-AMINE see THN000
DIPROPIONATE d'OESTRADIOL (FRENCH) see EDR000
DIPROPIONATO de ESTILBENE (SPANISH) see DKB000
p,p'-DIPROPIONOXY-trans-α,β-DIETHYLSTILBENE see DKB000
DIPROPIONYL PEROXIDE see DWQ800
DIPROPYLAMINE see DWR000
DI-n-PROPYLAMINE see DWR000
n-DIPROPYLAMINE see DWR000
4-(DI-N-PROPYLAMINO)-3,5-DINITRO-1-TRIFLUOROMETHYLBENZENE see DUV600
DIPROPYLCARBAMOTHIOIC ACID-S-ETHYL ESTER see EIN500
N,N-DI-N-PROPYL-2,6-DINITRO-4-TRIFLUOROMETHYL-ANILINE see DUV600
DIPROPYLENE GLYCOL METHYL ETHER see DWT200
DIPROPYLENE GLYCOL MONOMETHYL ETHER see DWT200
DIPROPYLENETRIAMINE see AIX250
DIPROPYL ETHER see PNM000
DI-n-PROPYLETHER see PNM000
DIPROPYL KETONE see DWT600
DIPROPYL MERCURY see DWU000
DIPROPYL METHANE see HBC500
DI-n-PROPYLNITROSAMINE see NKB700
DIPROPYLNITROSOAMINE see NKB700
DIPROPYL OXIDE see PNM000
DIPROPYLOXOSTANNANE see DWV000
DI-n-PROPYL PEROXYDICARBONATE see DWV400
N,N-DIPROPYL-1-PROPANAMINE see TMY250
p-(DIPROPYLSULFAMOYL)BENZOIC ACID SODIUM SALT see DWW200
p-(DI-N-PROPYLSULFAMYL)BENZOIC ACID SODIUM SALT see DWW200
N,N-DIPROPYLTHIOCARBAMIC ACID-S-ETHYL ESTER see EIN500
DIPROPYLTIN CHLORIDE see DFF400
DIPROPYLTIN DICHLORIDE see DFF400
DI-n-PROPYLTIN DICHLORIDE see DFF400

DIPROPYLTIN OXIDE see DWV000
N,N-DIPROPYL-4-TRIFLUOROMETHYL-2,6-DINITROANILINE see DUV600
DIPROSTRON see EDR000
DIPROZIN see DQA400
DIPTERAX see TIQ250
DIPTEREX see TIQ250
DIPTEREX 50 see TIQ250
DIPTEVUR see TIQ250
DIPYRIDO(1,2-a:3',2'-d)IMIDAZOL-2-AMINE see DWW700
DIPYRIDYL HYDROGEN PHOSPHATE see DWX000
DI-3-PYRIDYLMERCURY see DWX200
DIPYRIDYL PHOSPHATE see DWX000
DIPYRIN see DOT000
cis-DIPYRROLIDINEDICHLOROPLATINUM(II) see DEU200
DIQUAT see DWX800
DIQUAT DIBROMIDE see DWX800
DIQUAT DICHLORIDE see DWY000
DIRAX see AQN635
DIRECT BLACK A see AQP000
DIRECT BLACK META see AQP000
DIRECT BLUE 6 see CMO000
DIRECT BLUE 14 see CMO250
DIRECT BROWN 95 see CMO750
DIRECT BROWN BR see PEY000
DIREKTAN see NCQ900
DIREMA see CFY000
DIREX 4L see DXQ500
DIRIDONE see PDC250, PEK250
DIROX see HIM000
DISALUNIL see CFY000
DISATABS TABS see VSK600
3,3'-DISELENODIALANINE see DWY800
DISETIL see DXH250
DISFLAMOLL TKP see TNP500
DISILANE see DXA000
DISILYN see BEN000
DISIPAL HYDROCHLORIDE see OJW000
DISODIUM ARSENATE see ARC000
DISODIUM ARSENATE, HEPTAHYDRATE see ARC250
DISODIUM ARSENIC ACID see ARC000
DISODIUM ARSENITE see DXB200
DISODIUM AUROTHIOMALATE see GJC000
DISODIUM CARBONATE see SFO000
DISODIUM-N-(3-(CARBOXYMETHYLTHIOMERCURI)-2-METHOXYPROPYL)-α-CAMPHORAMATE see DXC000
DISODIUM CHROMATE see DXC200
DISODIUM CINNAMYLIDENE BISULFITE derivative of SULFAPYRIDINE see DXF400
DISODIUM CITRATE see DXC400
DISODIUM DIACID ETHYLENEDIAMINETETRAACETATE see EIX500
DISODIUM 2,7-DIBROM-4-HYDROXY-MERCURI-FLUORESCEIN see MCV000
DISODIUM 2',7'-DIBROMO-4'-(HYDROXYMERCURY) FLUORESCEIN see MCV000
DISODIUM DICHROMATE see SGI000
DISODIUM DIFLUORIDE see SHF500
DISODIUM DIHYDROGEN ETHYLENEDIAMINETETRA-ACETATE see EIX500
DISODIUM DIHYDROGEN(ETHYLENEDINITRILO)TETRA-ACETATE see EIX500
DISODIUM DIHYDROGEN PYROPHOSPHATE see DXF800
DISODIUM DIOXIDE see SJC500
DISODIUM DIPHOSPHATE see DXF800
DISODIUM EDATHAMIL see EIX500
DISODIUM EDETATE see EIX500
DISODIUM EDTA (FCC) see EIX500
DISODIUM-3,6-ENDOXOHEXAHYDROPHTHALATE see DXD000
DISODIUM EOSIN see BNH500
DISODIUM 3,6-EPOXYCYCLOHEXANE-1,2-DICARBOXYLATE see DXD000

DISODIUM ETHYLENEBIS(DITHIOCARBAMATE) see DXD200

DISODIUM ETHYLENE-1,2-BISDITHIOCARBAMATE see DXD200

DISODIUM ETHYLENEDIAMINETETRAACETATE see EIX500

DISODIUM ETHYLENEDIAMINETETRAACETIC ACID see EIX500

DISODIUM (ETHYLENEDINITRILO)TETRAACETATE see EIX500

DISODIUM (ETHYLENEDINITRILO)TETRAACETIC ACID see EIX500

DISODIUM FUMARATE see DXD800

DISODIUM GMP see GLS800

DISODIUM-5'-GMP see GLS800

DISODIUM-5'-GUANYLATE see GLS800

DISODIUM GUANYLATE (FCC) see GLS800

DISODIUM HEXAFLUOROSILICATE see DXE000

(2-)-DISODIUM HEXAFLUOROSILICATE see DXE000

DISODIUM HYDROGEN ARSENATE see ARC000

DISODIUM HYDROGEN CITRATE see DXC400

DISODIUM HYDROGEN ORTHOARSENATE see ARC000

DISODIUM HYDROGEN PHOSPHATE see SJH090

DISODIUM-6-HYDROXY-3-OXO-9-XANTHENE-o-BEN-ZOATE see FEW000

DISODIUM 3-HYDROXY-4-((2,4,5-TRIMETHYLPHE-NYL)AZO)-2,7-NAPHTHALENEDISULFONATE see FAG018

DISODIUM 3-HYDROXY-4-((2,4,5-TRIMETHYLPHE-NYL)AZO)-2,7-NAPHTHALENEDISULFONIC ACID see FAG018

DISODIUM 3-HYDROXY-4-((2,4,5-TRIMETHYLPHE-NYL)AZO)-2,7-NAPHTHALENEDISULPHONATE see FAG018

DISODIUM 3-HYDROXY-4-((2,4,5-TRIMETHYLPHE-NYL)AZO)-2,7-NAPHTHALENEDISULPHONIC ACID see FAG018

DISODIUM IMP see DXE500

DISODIUM INDIGO-5,5-DISULFONATE see FAE100

DISODIUM INOSINATE see DXE500

DISODIUM-5'-INOSINATE see DXE500

DISODIUM INOSINE-5'-MONOPHOSPHATE see DXE500

DISODIUM INOSINE-5'-PHOSPHATE see DXE500

DISODIUM METASILICATE see SJU000

DISODIUM METHANEARSENATE see DXE600

DISODIUM METHANEARSONATE see DXE600

DISODIUM METHYLARSENATE see DXE600

DISODIUM METHYLARSONATE see DXE600

DISODIUM MOLYBDATE see DXE800

DISODIUM MONOHYDROGEN ARSENATE see ARC000

DISODIUM MONOHYDROGEN PHOSPHATE see SJH090

DISODIUM MONOMETHYLARSONATE see DXE600

DISODIUM MONOSILICATE see SJU000

DISODIUM MONOXIDE see SIN500

DISODIUM NITROSYLPENTACYANOFERRATE see SIU500

DISODIUM ORTHOPHOSPHATE see SJH090

DISODIUM 7-OXABICYCLO(2.2.1)HEPTANE-2,3-DICAR-BOXYLATE see DXD000

DISODIUM OXIDE see SIN500

DISODIUM PEROXIDE see SJC500

DISODIUM-2-(p-(γ-PHENYLPROPYLAMINO)BENZENE-SULFONAMIDO) PYRIDINE see DXF400

DISODIUM PHOSPHATE see SJH090

DISODIUM PHOSPHORIC ACID see SJH090

DISODIUM PYROPHOSPHATE see DXF800

DISODIUM PYROSULFITE see SII000

DISODIUM SALT of EDTA see EIX500

DISODIUM SALT of ENDOTHALL see DXD000

DISODIUM SALT of 1-INDIGOTIN-S,S'-DISULPHONIC ACID see FAE100

DISODIUM SALT of 7-OXABICYCLO(2.2.1)HEPTANE-2,3-DICARBOXYLIC ACID see DXD000

DISODIUM SALT of 2-(4-SULPHO-1-NAPHTHYLAZO)-1-NAPHTHOL-4-SULPHONIC ACID see HJF500

DISODIUM SELENATE see DXG000

DISODIUM SELENITE see SJT500

DISODIUM SEQUESTRENE see EIX500

DISODIUM SILICOFLUORIDE see DXE000

DISODIUM SULFATE see SJY000

DISODIUM SULFITE see SJZ000

DISODIUM 2-(4-SULFO-1-NAPHTHYLAZO)-1-NAPHTHOL-4-SULFONATE see HJF500

DISODIUM 2-(4-SULPHO-1-NAPHTHYLAZO)-1-NAPHTHOL-4-SULPHONATE see HJF500

DISODIUM TARTRATE see BLC000

DISODIUM 1-(+)-TARTRATE see BLC000

DISODIUM TETRACEMATE see EIX500

DISODIUM VERSENATE see EIX500

DISODIUM VERSENE see EIX500

DISOFEN see DNG000

DISOLFURO DI TETRAMETILTIOURAME (ITALIAN) see TFS350

DISOMAR see DXE600

2,4-D ISOOCTYL ESTER see ILO000

DISOPHENOL see DNG000

2,4-D ISOPROPYL ESTER see IOY000

DISPADOL see DAM700

DISPAL see AHE250

DISPARICIDA see ABX500

DISPASOL M see PKB500

DISPERMINE see PIJ000

DISPERSED BLUE 12195 see FAE000

DISPERSED VIOLET 12197 see FAG120

DISPERSE MB-61 see TFC600

DISPERSE ORANGE see AKP750

DISPERSOL YELLOW PP see PEJ500

DISSOLVANT APV see DJD600

DISTAMINE see PAP550

DISTAVAL see TEH500

DISTAXAL see TEH500

DISTEARIN see OAV000

DISTESOL see EHP000

DISTESSOL see EHP000

DISTIGMINE BROMIDE see DXG800

DISTILBENE see DKA600, DKB000

DISTILLATES (PETROLEUM), ACID-TREATED HEAVY NAPHTHENIC (9CI) see MQV760

DISTILLATES (PETROLEUM), ACID-TREATED HEAVY PARAFFINIC (9CI) see MQV765

DISTILLATES (PETROLEUM), ACID-TREATED LIGHT NAPHTHENIC (9CI) see MQV770

DISTILLATES (PETROLEUM), ACID-TREATED LIGHT PARAFFINIC (9CI) see MQV775

DISTILLATES (PETROLEUM), HEAVY NAPHTHENIC (9CI) see MQV780

DISTILLATES (PETROLEUM), HEAVY PARAFFINIC (9CI) see MQV785

DISTILLATES (PETROLEUM), HYDROTREATED HEAVY NAPHTHENIC (9CI) see MQV790

DISTILLATES (PETROLEUM), HYDROTREATED HEAVY PARAFFINIC (9CI) see MQV795

DISTILLATES (PETROLEUM), HYDROTREATED LIGHT NAPHTHENIC (9CI) see MQV800

DISTILLATES (PETROLEUM), HYDROTREATED LIGHT PARAFFINIC (9CI) see MQV805

DISTILLATES (PETROLEUM), LIGHT NAPHTHENIC (9CI) see MQV810

DISTILLATES (PETROLEUM), LIGHT PARAFFINIC (9CI) see MQV815

DISTILLATES (PETROLEUM), SOLVENT-DEWAXED LIGHT NAPHTHENIC (9CI) see MQV835

DISTILLATES (PETROLEUM), SOLVENT-DEWAXED LIGHT PARAFFINIC (9CI) see MQV840

DISTILLATES (PETROLEUM), SOLVENT-REFINED LIGHT NAPHTHENIC (9CI) see MQV852

DISTILLATES (PETROLEUM), SOLVENT-REFINED LIGHT PARAFFINIC (9CI) see MQV855

DISTILLED LIME OIL see OGO000

DISTILLED MUSTARD see BIH250
DISTIVIT (B12 PEPTIDE) see VSZ000
DISTOKAL see HCI000
DISTOL 8 see EIV000
DISTOPAN see HCI000
DISTOPIN see HCI000
DISTOVAL see TEH500
DISUL see CNW000
DISULFAN see DXH250
DISULFATON see DXH325
DISULFATOZIRCONIC ACID see ZTJ000
DISULFIRAM see DXH250
DISULFOTON see DXH325
DISULFOTON DISULIDE see OQS000
DISULFOTON SULFOXIDE see OQS000
DISULFURAM see DXH250
DISULFUR DICHLORIDE see SON510
DISULFURE de TETRAMETHYLTHIOURAME (FRENCH) see
    TFS350
DISULFUR PENTOXYDICHLORIDE see PPR500
DISULFURYL CHLORIDE see PPR500
DISULFURYL DICHLORIDE see PPR500
DISUL-Na see CNW000
DISULONE see SOA500
DISULPHINE LAKE BLUE EG see FMU059
DISULPHURAM see DXH250
DISULPHURIC ACID see SOI520
DISUL-SODIUM see CNW000
DISYNCRAM see MPE250
DISYNCRAN see MPE250
DISYNFORMON see EDV000
DI-SYSTON see DXH325
DISYSTON SULFOXIDE see OQS000
DISYSTOX see DXH325
DI-TAC see DXE600
DITAVEN see DKL800
DITHALLIUM SULFATE see TEM000
DITHALLIUM(1+) SULFATE see TEM000
DITHALLIUM TRIOXIDE see TEL050
DITHANE D-14 see DXD200
DITHANE A-4 see DUQ600
DITHANE A-40 see DXD200
DITHANE M 22 SPECIAL see MAS500
DITHANE R-24 see BPU000
DITHANE Z see EIR000
1,4-DITHIAANTHRAQUINONE-2,3-DICARBONITRILE see
    DLK200
1,4-DITHIAANTHRAQUINONE-2,3-DINITRILE see DLK200
DITHIANON see DLK200
DITHIANONE see DLK200
DITHIAZANINE see DXI600
DITHIAZANINE IODIDE see DJT800
DITHIAZANIN IODIDE see DJT800
DITHIAZININE see DJT800
3-(DI-2-THIENYLMETHYLENE)-1-METHYLPIPERIDINE see
    BLV000
DITHIO see SOD100
O,O-DITHIO-BIS-ANILINE see DXJ800
2,2'-DITHIOBISANILINE see DXJ800
2,2'-DITHIOBIS(BENZOTHIAZOLE) see BDE750
1,1'-DITHIOBIS(N,N-DIETHYLTHIOFORMAMIDE) see
    DXH250
α,α'-DITHIOBIS(DIMETHYLTHIO)FORMAMIDE see
    TFS350
1,1'-DITHIOBIS(N,N-DIMETHYLTHIO)FORMAMIDE see
    TFS350
2,2'-DITHIOBIS(ETHYLAMINE) see MCN500
DITHIOBISMORPHOLINE see BKU500
4,4'-DITHIOBIS(MORPHOLINE) see BKU500
2,2'-DITHIOBIS(PYRIDINE-1-OXIDE)MAGNESIUM SUL-
    FATE TRIHYDRATE see DXL400
DITHIOBIS(THIOFORMIC ACID)-o,o-DIETHYL ESTER see
    BJU000
DITHIOBIURET see DXL800

DITHIOCARB see SGJ000
DITHIOCARBAMATE see SGJ000
DITHIOCARBAMOYLHYDRAZINE see MLJ500
DITHIOCARBONIC ANHYDRIDE see CBV500
DITHIODEMETON see DXH325
β,β'-DITHIODIALANINE see CQK325
2,2'-DITHIODIANILINE see DXJ800
N,N'-(DITHIODICARBONOTHIOYL)BIS(N-METHYLMETHA-
    NAMINE) see TFS350
2,2-DITHIODIETHANOL see DXM600
DITHIODIGLYCOL see DXM600
N,N-DITHIODIMORPHOLINE see BKU500
4,4'-DITHIODIMORPHOLINE see BKU500
DITHIODIPHOSPHORIC ACID, TETRAETHYL ESTER see
    SOD100
DITHIOETHYLENEGLYCOL see EEB000
DITHIOFOS see SOD100
DITHIOGLYCEROL see BAD750
1,2-DITHIOGLYCEROL see BAD750
DITHIOGLYCOL see EEB000
DITHIOLANE see DXN600
DITHIOLANE IMINOPHOSPHATE see DXN600
1,3-DITHIOLAN-2-YLIDENE-PHOSPHORAMIDOTHIOIC
    ACID DIETHYL ESTER see DXN600
1,3-DITHIOLAN-2-YLIDENE-PHOSPHORAMIDOTHIOIC
    ACID-O,O-DIETHYL ESTER see DXN600
DITHIOMETON (FRENCH) see PHI500
4,4'-DITHIOMORPHOLINE see BKU500
DITHION see DXO000
DITHIONE see DXO000, SOD100
DITHIONIC ACID see SOI520
DITHIOOXALDIIMIDIC ACID see DXO200
DITHIOOXAMIDE see DXO200
DITHIOPHOSPHATE de O,O-DIETHYLE et de (4-CHLORO-
    PHENYL) THIOMETHYLE (FRENCH) see TNP250
DITHIOPHOSPHATE de-O,O-DIETHYLE et de S(2,5-DICHLO-
    ROPHENYL) THIOMETHYLE (FRENCH) see PDC750
DITHIOPHOSPHATE de O,O-DIETHYLE et de S-(2-ETHYL-
    THIO-ETHYLE) (FRENCH) see DXH325
DITHIOPHOSPHATE de O,O-DIETHYLE et d'ETHYLTHIO-
    METHYLE (FRENCH) see PGS000
DITHIOPHOSPHATE de O,O-DIETHYLE et de S-N-METHYL-
    N-CARBOETHOXY CARBAMOYLMETHYLE (FRENCH)
    see DJI000
DITHIOPHOSPHATE de O,O-DIMETHYLE et de S-(1,2-DI-
    CARBOETHOXYETHYLE) (FRENCH) see MAK700
DITHIOPHOSPHATE de O,O-DIMETHYLE et de S-(2-ETHYL-
    THIO-ETHYLE) (FRENCH) see PHI500
DITHIOPHOSPHATE de O,O-DIMETHYLE et de S(-N-
    METHYLCARBAMOYL-METHYLE) (FRENCH) see
    DSP400
DITHIOPHOSPHATE de O,O-DIMETHYLE et de S-((MOR-
    PHOLINOCARBONYL)-METHYLE) (FRENCH) see
    MRU250
DI(THIOPHOSPHORIC) ACID, TETRAETHYL ESTER see
    SOD100
DITHIOPHOSPHORSAEURE-O-AETHYL-S,S-DIPHENYLES-
    TER (GERMAN) see EIM000
2,3-DITHIOPROPANOL see BAD750
DITHIOPYROPHOSPHATE de TETRAETHYLE (FRENCH) see
    SOD100
DITHIOSYSTOX see DXH325
DITHIOTEP see SOD100
DITHIOXAMIDE see DXO200
DITHIZON see DWN200
DITHIZONE see DWN200
DITIAMINA see EIR000
DITILIN see CMG250, HLC500
DITILINE see CMG250, HLC500
DITILIN IODIDE see BJI000
DITOIN see DNU000
DITOINATE see DKQ000
4,4'-DI-o-TOLUIDINE see TGJ750
1,4-DI-p-TOLUIDINOANTHRAQUINONE see BLK000
DI-o-TOLUYLTHIOUREA see DXP600

DITOLYLETHANE see DXP000
DI-o-TOLYLGUANIDINE see DXP200
1,3-DI-o-TOLYLGUANIDINE see DXP200
DI-o-TOLYLTHIOUREA see DXP600
DITRAN see DXP800
DITRAZIN see DIW200
DITRAZIN CITRATE see DIW200
DITRAZINE see DIW200
DITRAZINE CITRATE see DIW200
DI(TRI-(2,2-DIMETHYL-2-PHENYLETHYL)TIN)OXIDE see
   BLU000
DITRIFON see TIQ250
DITRIPENTAT see CAY500
DITROSOL see DUS700
DITUBIN see ILD000
DIUCARDYN SODIUM see DXC000
DIURAL see CHJ750
DIURAMID see AAI250
1,1-DIUREIDISOBUTANE see IIV000
DIUREIDOISOBUTANE see IIV000
DIURETICUM-HOLZINGER see AAI250
DIURETIN see SJO000
DIUREX see DXQ500
DIUROBROMINE see TEO500
DIUROL see AMY050, DXQ500
DIURON see DXQ500
DIURON 4L see DXQ500
DIUTAZOL see AAI250
DIVERCILLIN see AIV500, AOD125
DIVERON see MQU750
DIVINYL see BOP500
DIVINYLBENZENE see DXQ745
DIVINYLENE OXIDE see FPK000
DIVINYLENE SULFIDE see TFM250
DIVINYLENIMINE see PPS250
DIVINYL ETHER (DOT) see VOP000
DIVINYL ETHER, inhibited (DOT) see VOP000
DIVINYL SULFONE see DXR200
DIVIPAN see DGP900
DIVIT URTO see VSZ100
DIVULSAN see DNU000
DIVYNYL OXIDE see VOP000
DIXANTHOGEN see BJU000
DIXIBEN see EID000
DIXON 164 see TAI250
DIXYRAZINE DIHYDROCHLORIDE see DXR800
DIZENE see DEP600
DIZINON see DCM750
DKB see DCQ800
DKB SULFATE see DHA400
DKD see DIZ100
DLX-6000 see TAI250
DM see DAC000
DM see PDB000
DMA see DOO800, DOQ800, DXE600, DXS200
DMA-4 see DAA800
DMAA see HKC000
DMAB see DOR200, DOT300
3,2′-DMAB see BLV250
DMAC see DOO800
DMAE see DOY800
DMAP see VTF000
DMASA see DQD400
DMBA see DQJ200
7,12-DMBA see DQJ200
DMBC see DQQ200
DMCC see DQY950
DMCT see MIJ500
DMDK see SGM500
DMDPN see DRQ200
DMDT see MEI450
p,p′-DMDT see MEI450
DMEP see DOF400
DMF see BJE750, DSB000

DMFA see DSB000
DMH see DSF400, DSF600, DSF800, DSG200
DMI see DSI709
DMI 50475 see DSI709
DMI HYDROCHLORIDE see DLS600
DMM see BKM500
DMN see NKA600
DMNA see NKA600
DMNM see DSV200, DTA000
DMNO see DSV200
DMN-OAC see AAW000
DMNT see DSG400
DMP see DTR200
2,5-DMP see XKS000
3,5-DMP see XLS000
DMPA see DGD800
DMPD see DTL800
DMPE see DOE200
DMPEA see DOE200
DMPP see DTO000
DMPP IODIDE see DTO000
DMPT see DTP000
DMS see DUD100, DUD400, TFP000
DMS-70 see DUD800
DMS-90 see DUD800
DMSA see DQD400
DMS(METHYL SULFATE) see DUD100
DMSO see DUD800
DMSP see FAQ800
DMT see DPF600
DMTP see FAQ999
DMTP (JAPAN) see DSO000
DMTT see DSB200
DMU see DXQ500
DN 289 see BRE500
DNA see DUP600
DNBP see BRE500
DNBP AMMONIUM SALT see BPG250
DNCB see CGM000
DNDMP see DRO000
DN DRY MIX No. 1 see CPK500
DN-DRY MIX No.2 see DUS700
DN DUST No. 12 see CPK500
2,4-DNFB see DUW400
DNOCHP see CPK500
DNOC SOLDIUM SALT see DUU600
DNOK (CZECH) see DUS700
DNOK-ACETAT (CZECH) see AAU250
DNOP see DVL600
DNOSBP see BRE500
DNP see DNG000
2,4-DNP see DUZ000
DNPC see DUT600
DNPD see NBL000
DNPZ see DVF200
DNSBP see BRE500
DNT see DVH000
2,4-DNT see DVH000
DNTP see PAK000
DO 14 see SOP000
DOBENDAN see CCX000
DOBETIN see VSZ000
DOCEMINE see VSZ000
DOCIBIN see VSZ000
DOCIGRAM see VSZ000
DOCITON see ICB000
DOCTAMICINA see CDP250
DOCUSATE SODIUM see DJL000
DODAT see DAD200
DODECABEE see VSZ000
DODECACHLOROOCTAHYDRO-1,3,4-METHENO-2H-CY-
   CLOBUTA(c,d)PENTALENE see MQW500
1,1a,2,2,3,3a,4,5,5,5a,5b,6-DODECACHLOROOCTAHYDRO-
   1,3,4-METHENO-1H-CYCLOBUTA(c,d)PENTALENE see
   MQW500

DODECACHLOROPENTACYCLODECANE see MQW500
DODECACHLOROPENTACYCLO(3,2,2,0^{2,6},0^{3,9},0^{5,10})
  DECANE see MQW500
DODECAHYDRODIPHENYLAMINE see DGT600
Δ-DODECALACTONE see DXS700
1-DODECANAL see DXT000
1-DODECANETHIOL see LBX000
tert-DODECANETHIOL see DXT800
DODECANOIC ACID see LBL000
1-DODECANOL see DXV600
n-DODECANOL see DXV600
DODECANOYL PEROXIDE see LBR000
DODECATRIETHYLAMMONIUM BROMIDE see DXU200
DODECAVITE see VSZ000
DODECENE see PMP750
DODECENE EPOXIDE see DXU400
DODECENYLSUCCINIC ANHYDRIDE see DXV000
DODECOIC ACID see LBL000
DODECYL ALCOHOL see DXV600
n-DODECYL ALCOHOL see DXV600
DODECYL ALCOHOL, HYDROGEN SULFATE, SODIUM
  SALT see SIB600
1-DODECYL ALDEHYDE see DXT000
DODECYLAMINE see DXW000
DODECYL BENZENE SODIUM SULFONATE see DXW200
DODECYLBENZENESULFONIC ACID SODIUM SALT see
  DXW200
DODECYLBENZENESULPHONATE, SODIUM SALT see
  DXW200
DODECYLBENZENSULFONAN SODNY (CZECH) see
  DXW200
DODECYL DIMETHYL BENZYLAMMONIUM CHLORIDE
  see BEM000
DODECYLDIMETHYL(2-PHENOXYETHYL)AMMONIUM
  BROMIDE see DXX000
DODECYLENE see PMP750
DODECYL GALLATE see DXX200
N-DODECYLGUANIDINACETAT (GERMAN) see DXX400
DODECYLGUANIDINE ACETATE see DXX400
N-DODECYLGUANIDINE ACETATE see DXX400
2-DODECYLISOQUINOLINIUM BROMIDE see LBW000
DODECYL MERCAPTAN see LBX000
1-DODECYL MERCAPTAN see LBX000
m-DODECYL MERCAPTAN see LBX000
tert-DODECYLMERCAPTAN see DXT800
terc.DODECYLMERKAPTAN (CZECH) see DXT800
DODECYLPHENOL see DXY600
N-DODECYLSARCOSINE SODIUM SALT see DXZ000
DODECYL SODIUM SULFATE see SIB600
DODECYL SULFATE, SODIUM SALT see SIB600
tert-DODECYLTHIOL see DXT800
DODECYL-p-TOLYL TRIMETHYL AMMONIUM CHLORIDE
  see DYA600
DODECYLTRICHLOROSILANE see DYA800
DODEX see VSZ000
DODGUADINE see DXX400
DODINE see DXX400
DODINE ACETATE see DXX400
DODINE, mixture with GLYODIN see DXX400
DOF see DVK600
DOFSOL see VSK600
DOGQUADINE see DXX400
DOJYOPICRIN see CKN500
DOKIRIN see BLC250
DOKTACILLIN see AIV500
DOL see BBQ750
DOLADENE see XCJ000
DOLANTAL see DAM700
DOLANTIN see DAM700
DOLANTIN HYDROCHLORIDE see DAM700
DOLANTOL see DAM700
DOLAREN see DAM700
DOLARGAN see DAM700
DOLCO MOUSE CEREAL see SMN500

DOLCONTRAL see DAM600, DAM700
DOLCYMENE see CQI000
DOLEAN pH 8 see ADA725
DOLENAL see DAM700
DOLENE see DAB879
DOLENE see PNA500
DOLENOL see DAM700
DOLEN-PUR see HCD250
DOLESTAN see BAU750
DOLESTINE see DAM700
DOL GRANULE see BBQ500
DOLICUR see DUD800
DOLIGUR see DUD800
DOLIN see DAM700
DOLIPOL see BSQ000
DOLIPRANE see HIM000
DOLKWAL BRILLIANT BLUE see FAE000
DOLKWAL ERYTHROSINE see FAG040
DOLKWAL INDIGO CARMINE see FAE100
DOLKWAL ORANGE SS see TGW000
DOLKWAL PONCEAU 3R see FAG018
DOLKWAL TARTRAZINE see FAG140
DOLKWAL YELLOW AB see FAG130
DOLKWAL YELLOW OB see FAG135
DOLMIX see BBQ750
DOLOBID see DKI600
DOLOBIL see DKI600
DOLOBIS see DKI600
DOLOCAP see PNA500
DOLOCHLOR see CKN500
DOLOGAL see DAM700
DOLOMIDE see SAH000
DOLOMITE see CAO000
DOLONEURINE see DAM700
DOLONIL see PDC250
DOLOPETHIN see DAM700
DOLOPHINE see MDO750, MDP000
DOLOPHINE HYDROCHLORIDE see MDP000
d-DOLOPHINE HYDROCHLORIDE see MDP500
DOLOSAL see DAM600, DAM700
DOLOVIN see IDA000
DOLOXENE see DAB879
DOLOXENE see PNA500
DOLPHINE see MDP000
DOLSIN see DAM600
DOLVANOL see DAM700
DOMARAX see IPU000
DOMATOL see AMY050
DOMESTROL see DKA600
DOMF see MCV000
DOMICAL see EAI000
DOMOSO see DUD800
DONMOX see AAI250
DONOVAN'S SOLUTION see ARI500
DOOJE see HBT500
DOP see DVL700
(−)-DOPA see DNA200
l-DOPA see DNA200
DOPAFLEX see DNA200
l-DOPA HYDROCHLORIDE see DYC200
DOPAL see DNA200
DOPAMET see DNA800
DOPAMINE see DYC400
DOPARKINE see DNA200
DOPASOL see DNA200
DOPEGYT see DNA800
DOPIDRIN see DBA800
DOPN see NJN000
DOPRIN see DNA200
DOPTAEC see DNA800
DORAL see VSZ100
DORANTAMIN see WAK000
DORAPHEN see PNA500
DORICO see ERD500

DORIDEN see DYC800
DORIDEN-SED see DYC800
DORINAMIN see BBW500
DORISUL see SNN300
DORLYL see PKQ059
DORMABROL see MQU750
DORMAL see CDO000
DORMIGENE see BNP750
DORMIGOA see QAK000
DORMINA see EOK000
DORMIRAL see EOK000
DORMITURIN see BNK000
DORMODOR see DAB800
DORMOGEN see QAK000
DORMONE see DAA800
DORMUTIL see QAK000
DORSEDIN see QAK000
DORSITAL see NBT500
DORVICIDE A see BGJ750
DORVON see SMQ500
DOS see BJS250
DOSCALUN see EOK000
D.O.T. see DUP300
DOTG see BKK750
DOTG ACCELERATOR see DXP200
DOTMENT 324 see AHE250
DOTYCIN see EDH500
DOUBLE STRENGTH see TIX500
DOVENIX see HLJ500
DOVIP see FAB600
DOW 209 see SMR000
DOW 860 see SMQ500
DOW 1329 see DGD800
DOWANOL see CBR000
DOWANOL 33B see PNL250
DOWANOL-50B see DWT200
DOWANOL DE see CBR000
DOWANOL DPM see DWT200
DOWANOL EB see BPJ850
DOWANOL EE see EES350
DOWANOL EIPAT see INA500
DOWANOL EM see EJH500
DOWANOL EP see PER000
DOWANOL EPH see PER000
DOWANOL TE see EFL000
DOWCC 132 see COD850
DOWCHLOR see CDR750
DOWCIDE 7 see PAX250
DOWCO 118 see DGD800
DOWCO 139 see DOS000
DOWCO-163 see CLP750
DOWCO 169 see PEV500
DOWCO 179 see CMA100
DOWCO 186 see HON000
DOWCO-213 see CQH650
DOWCO 217 see CMA250
DOW CORNING 346 see PJR000
DOW DEFOLIANT see SFU500
DOW DORMANT FUNGICIDE see SJA000
DOW ET 14 see RMA500
DOW ET 57 see RMA500
DOWFLAKE see CAO750
DOWFROST see PML000
DOWFUME see MHR200
DOWFUME 40 see EIY500
DOWFUME EB-5 see DYE400
DOWFUME EDB see EIY500
DOWFUME MC-2 SOIL FUMIGANT see MHR200
DOWFUME N see DGG000
DOWFUME W-8 see EIY500
DOW GENERAL see BRE500
DOW GENERAL WEED KILLER see BRE500
DOWICIDE see BGJ750
DOWICIDE 2 see TIV750

DOWICIDE 6 see TBT000
DOWICIDE 7 see PAX250
DOWICIDE 2S see TIW000
DOWICIDE B see TIV750
DOWICIDE EC-7 see PAX250
DOWICIDE G see PAX250
DOWICIDE G-ST see SJA000
DOW LATEX 612 see SMR000
DOW MCP AMINE WEED KILLER see CIR250
DOW PENTACHLOROPHENOL DP-2 ANTIMICROBIAL see
    PAX250
DOW-PER see PCF275
DOWPON see DGI400, DGI600
DOWPON M see DGI400
DOW SELECTIVE see BPG250
DOW SELECTIVE WEED KILLER see BRE500
DOW SODIUM TCA INHIBITED see TII250
DOWSPRAY 17 see CPK500
DOWTHERM see PFA860
DOWTHERM 209 see PNL250
DOWTHERM A see PFA860
DOWTHERM E see DEP600
DOWTHERM SR 1 see EJC500
DOW-TRI see TIO750
DOWZENE DHC see PIK000
D-OX see SHR500
DOXCIDE 50 see CDW450
DOXEPHRIN see DBA800
DOXEPIN see DYE409
DOXINATE see DJL000
DOXOL see DJL000
DOXORUBICIN see AES750
DOXYFED see DBA800
2,4-DP see DGB000
2-(2,4-DP) see DGB000
DPA see DGI000, DVX800
2,2-DPA see DGI600
D-P-A INJECTION see PAG200
DPBS see DFY400
D & P DOUBLE O CRABGRASS KILLER see PLC250
DPG see DWC600
DPG ACCELERATOR see DWC600
DPH see DKQ000, DNU000
DPID see DLH600
DPN see NKB700
DPNA see NKB700
DPP see PAK000, PEK250
DPPD see BLE500
DPX 1410 see DSP600
DP X 1410 see MME809
DPX 3674 see HFA300
DQUIGARD see DGP900
DRABET see BSQ000
DRACYLIC ACID see BCL750
DRAKEOL see MQV750
DRALZINE see HGP500
DRAMAMIN see DYE600
DRAMAMINE see DYE600
DRAMARIN see DYE600
DRAMYL see DYE600
DRAT see CJJ000
DRAZA see DST000
DRAZOXOLON see MLC250
DRAZOXOLONE see MLC250
DRC 3340 see DTN200
DRC 3341 see MIB750
DREFT see SIB600
DRENAMIST see VGP000
DRENOL see CFY000
DREWMULSE POE-SMO see PKL100
DREWMULSE TP see OAV000
DREWMULSE V see OAV000
DREWSORB 60 see SKV150
DREXEL see DXQ500

DREXEL DEFOL see SFS000
DREXEL DIURON 4L see DXQ500
DREXEL DSMA LIQUID see DXE600
DREXEL METHYL PARATHION 4E see MNH000
DREXEL PARATHION 8E see PAK000
DRI-DIE INSECTICIDE 67 see SCH000
DRIDOL see DYF200
DRI-KIL see RNZ000
DRILL TOX-SPEZIAL AGLUKON see BBQ500
DRINALFA see DBA800
DRINOX see AFK250, HAR000
DRISDOL see VSZ100
DRI-TRI see SJH200
DROCODE see DKW800
DROLEPTAN see DYF200
DROMILAC see DBV400
DROMISOL see DUD800
DROMORAN see MKR250
levo-DROMORAN see LFG000
racemic DROMORAN see MKR250
DROMORAN HYDROBROMIDE see MDV250
DROMYL see DYE600
DROPCILLIN see BDY669
DROPERIDOL see DYF200
DROP LEAF see SFS000
DROPP see TEX600
DROPSPRIN see SAH000
DROXAROL see BPP750
DROXARYL see BPP750
DROXOL see MDQ250
DROXOLAN see DAQ400
DRP 859025 see TND500
DRUMULSE AA see OAV000
DRUPINA 90 see BJK500
DRY AND CLEAR see BDS000
DRYISTAN see BBV500
DRYLISTAN see BBV500
DRY MIX No. 1 see CPK500
DRYOBALANOPS CAMPHOR see BMD000
DRYPTAL see CHJ750
DSDP see DJA400
DSE see DXD200
DSMA LIQUID see DXE600
DS-Na see SFW000
2,4-D SODIUM SALT see SGH500
DSP see SJH090
DSS see DJL000, SOA500
DST see DME000
DST 50 see SMR000
DTA see DLK200
DTB see DXL800
DTBP see BSC750
DTIC see DAB600
DTIC CITRATE see DUI400
DTIC-DOME see DAB600
DTMC see BIO750
DTPA see DJG800
DTPA CALCIUM TRISODIUM SALT see CAY500
DU see DCQ600
DU 112307 see CJV250
DUATOK see TEX250
DUBRONAX see SOA500
DUCKALGIN see SEH000
DUCOBEE see VSZ000
DUFALONE see BJZ000
DUGERASE see AIT250
DUKERON see TIO750
DULCIDOR see GFA000
DULCINE see EFE000
DULCITOLDIEPOXIDE see DCI600
DULL 704 see PJY500
DULSIVAC see DJL000
DULZOR-ETAS see SGC000
DUMASIN see CPW500

DUMITONE see SOA500
DUMOCYCIN see TBX250
DUMOGRAN see MPN500
DUNCAINE see DHK400
DUNCAINE HYDROCHLORIDE see DHK600
DUNERYL see EOK000
DUNKELGELB see PEJ500
DUODECIBIN see VSZ000
DUODECYL ALCOHOL see DXV600
DUODECYLIC ACID see LBL000
DUODECYLIC ALDEHYDE see DXT000
DUO-KILL see DGP900
DUOMYCIN see CMA750
DUOSOL see DJL000
DUOTRATE see PBC250
DUPHAR see CKM000
DUPONOL see SIB600
DU PONT 326 see DGD600
DU PONT 1991 see BAV575
DUPONT HERBICIDE 326 see DGD600
DU PONT HERBICIDE 976 see BMM650
DU PONT INSECTICIDE 1179 see MDU600
DU PONT INSECTICIDE 1519 see DVS000
DUPONT PC CRABGRASS KILLER see PLC250
DURAD see TNP500
DURA-ESTRADIOL see EDS100
DURAFUR BLACK R see PEY500
DURAFUR BROWN see ALL750
DURAFUR BROWN 2R see ALL750
DURAFUR BROWN MN see DBO400
DURAFUR DEVELOPER C see CCP850
DURAFUR DEVELOPER D see NAW500
DURAFUR DEVELOPER G see REA000
DURALUTON see HNT500
DURAMAX see ADA725
DURAN see DXQ500
DURANIT see SMR000
DURANOL ORANGE G see AKP750
DURAPHOS see MQR750
DURASORB see DUD800
DURATOX see DAP400, MIW100
DURAVOS see DGP900
DURETHAN BK see PJY500
DURETTER see FBN100
DURFAX 80 see PKL100
DUROFERON see FBN100
DUROFOL P see PKQ059
DUROID 5870 see TAI250
DUROMINE see DTJ400
DUROTOX see PAX250
DURSBAN see CMA100
DURSBAN F see CMA100
DURSBAN METHYL see CMA250
DURTAN 60 see SKV150
DUSICNAN BARNATY (CZECH) see BAN250
DUSICNAN KADEMNATY (CZECH) see CAH250
DUSICNAN ZIRKONICITY (CZECH) see ZSA000
DUSITAN SODNY (CZECH) see SIQ500
DUSOLINE see CMD750
DUSORAN see CMD750
DUS-TOP see MAE250
DUTCH LIQUID see EIY600
DUTCH OIL see EIY600
DUTCH-TREAT see HKC500
DU-TER see HON000
DUVILAX BD 20 see AAX250
DV see DAL600
DV 400 see PKB500
D3-VIGANTOL see CMC750
DW 62 see DNV000
DW3418 see BLW750
DWARF PINE NEEDLE OIL see PIH400
DWUBROMOETAN (POLISH) see EIY500
DWUCHLOROCZTEROFLUOROETAN (POLISH) see DGL600

DWUCHLORODWUETYLOWY ETER (POLISH) see DFJ050
DWUCHLORODWUFLUOROMETAN (POLISH) see DFA600
2,4-DWUCHLOROFENOKSYOCTOSY KWAS (POLISH) see DAA800
DWUCHLOROFLUOROMETAN (POLISH) see DFL000
DWUCHLOROPROPAN (POLISH) see PNJ400
DWUETYLOAMINA (POLISH) see DHJ200
DWUETYLOWY ETER (POLISH) see EJU000
DWUFENYLOGUANIDYNA (POLISH) see DWC600
DWUMETHYLOFORMAMID (POLISH) see DSB000
DWUMETYLOANILINA (POLISH) see DQF800
symetryczna DWUMETYLOHYDRAZYNA (POLISH) see DSF600
DWUMETYLOSULFOTLENKU (POLISH) see MAO250
DWUMETYLOWY SIARCZAN (POLISH) see DUD100
DWU-β-NAFTYLO-p-FENYLODWUAMINA (POLISH) see NBL000
DWUNITROBENZEN (POLISH) see DUQ200
DWUNITRO-o-KREZOL (POLISH) see DUS700
3,3'-DWUOKSYBENZYDYNA (POLISH) see DMI400
DWUSIARCZEK DWUBENZOTIAZYLU (POLISH) see BDE750
DX see AES750
DXMS see SOW000
DYANACIDE see ABU500
DYAZIDE see CFY000
DYCARB see DQM600
DYCLOCAINUM see BPR500
DYCLONE HYDROCHLORIDE see BPR500
DYCLONINE HYDROCHLORIDE see BPR500
DYCLOTHANE see BPR500
DYDELTRONE see PMA000
DYE FD&C RED No. 3 see FAG040
DYE GS see ALL750
DYESTROL see DKA600
DYETONE see SFG000
DYFLOS see IRF000
DYFONATE see FMU045
DYKANOL see PJL750
DYKOL see DAD200
DYLAMON see BBV500
DYLENE see SMQ500
DYLEPHRIN see VGP000
DYLITE F 40 see SMQ500
DYLOX see TIQ250
DYLOX-METASYSTOX-R see TIQ250
DYMADON see HIM000
DYMEX see ABH000
DYMID see DRP800
DYNACORYL see DJS200
DYNADUR see PKQ059
DYNALIN INJECTABLE see TET800
DYNAMICARDE see DJS200
DYNAMITE see DYG000
DYNAMUTILIN see TET800
DYNAPRIN see DLH600
DYNARSAN see ABX500
DYNASTEN see PAN100
DYNA-ZINA see DLH600, DLH630
DYNAZONE see NGE500
DYNERIC see CMX700
DYNEX see DXQ500
DYNOSOL see DUU600
DYP-97 F see LBR000
DYPHONATE see FMU045
DYPRIN see MDT740
DYREX see TIQ250
DYSPNE-INHAL see VGP000
DYSPROSIUM see DYG400
DYSPROSIUM CHLORIDE see DYG600
DYSPROSIUM CITRATE see DYG800
DYSPROSIUM(III) NITRATE HEXAHYDRATE (1:3:6) see DYH000
DYTHOL see CMD750

DYTOL S-91 see DAI600
DYTOL E-46 see OAX000
DYTOL F-11 see HCP000
DYTOL J-68 see DXV600
DYTOL M-83 see ŒEI000

E^1 see EDV000
E^2 see EDO000
E6 see PKO500
E 62 see PKQ059
E 127 see FAG040
E 132 see FAE100
E 140 see CKN000
E 158 see DAP600
E393 see SOD100
583E see POC750
E 600 see NIM500
E 605 see PAK000
E 66P see PKQ059
E 1059 see DAO600
E 3314 see HAR000
EA 3547 see DDE200
EAA see EFS000
E-73 ACETATE see ABN000
EAK see EGI755
EAMN see ENR500
EARTHCIDE see PAX000
EARTHNUT OIL see PAO000
EASEPTOL see HJL000
EASTBOND M 5 see PMP500
EASTERN STATES DUOCIDE see WAT200
EAST INDIAN LEMONGRASS OIL see LEG000
EASTMAN 910 see MIQ075
EASTMAN 7663 see DJT800
EASTMAN INHIBITOR DHPB see DMI600
EASY OFF-D see TIG250
EATAN see DLY000
EB see EGP500
EBI see ISK000
EBIDENE see ILD000
EBIS see ISK000
EBS see FAG040
EBZ see EDP000
E.C. 3.4.4.16 see BAC000
E.C. 3.4.4.24 see BMO000
E.C. 3.4.21.14 see BAC000
ECATOX see PAK000
ECCOTHAL see TEM000
ECF see EHK500
ECH see EAZ500
ECIPHIN see EAW000
ECLORIL see CDO500
ECLORION see HBT500
ECM see ADA725
ECOBUTAZONE see BRF500
ECONOCHLOR see CDP250
ECOPRO see TAL250
ECOTRIN see ADA725
ECP see DFK600
ECTIBAN see AHJ750
ECTIDA see EHP000
ECTILURAN see TEH500
ECTILUREA see EHP000
ECTON see EHP000
ECTORAL see RMA500
ECTRIN see FAR100
ECTYDA see EHP000
ECTYLCARBAMIDE see EHP000
ECTYLUREA see EHP000
ECTYN see EHP000
ECUANIL see MQU750
ECZECIDIN see CHR500
ED see DFH200
EDA see DCN800

EDATHAMIL see EIX000
EDATHAMIL DISODIUM see EIX500
EDATHANIL TETRASODIUM see EIV000
EDB see EIY500
EDB-85 see EIY500
E-D-BEE see EIY500
EDC see EIY600
EDCO see MHR200
EDDP see EIM000
EDECRIL see DFP600
EDECRIN see DFP600
EDECRINA see DFP600
EDEMEX see BDE250
EDEMOX see AAI250
EDEN see LFK000
EDENAL see MQU750
EDETATE DISODIUM see EIX500
EDETATE SODIUM see EIV000
EDETATE TRISODIUM see TNL250
EDETIC ACID see EIX000
EDETIC ACID TETRASODIUM SALT see EIV000
EDICOL BLUE CL 2 see FAE000
EDICOL SUPRA BLUE E6 see FMU059
EDICOL SUPRA CARMOISINE WS see HJF500
EDICOL SUPRA ERYTHROSINE A see FAG040
EDICOL SUPRA ROSE B see FAG070
EDICOL SUPRA TARTRAZINE N see FAG140
EDIFENPHOS see EIM000
EDION see TLP750
EDIPHENPHOS see EIM000
EDISTIR RB 268 see SMR000
EDTA ACID see EIX000
EDTA (CHELATING AGENT) see EIX000
d'E.D.T.A. DISODIQUE (FRENCH) see EIX500
EDTA, DISODIUM SALT see EIX500
EDTA, SODIUM SALT see EIV000
EDTA TETRASODIUM SALT see EIV000
EDTA TRISODIUM SALT see TNL250
EENA see ELG500
EENKAPTON (DUTCH) see PDC750
EEREX GRANULAR WEED KILLER see BMM650
EEREX WATER SOLUBLE CONCENTRATE WEED KILLER
    see BMM650
EFACIN see NCQ900
EFED see TMU250
EFEDRIN see EAW000
EFFISAX see MOV500
EFFLUDERM (FREE BASE) see FMM000
EFFROXINE see DBA800
EFFUSAN see DUS700
EFLORAN see MMN250
EFUDEX see FMM000
EFUDIX see FMM000
EFURANOL see DLH630
EGDME see DOE600
EGDN see EJG000
EGG YELLOW A see FAG140
EGITOL see HCI000
EGM see EJH500
EGME see EJH500
EH 121 see TNX000
EHEN see ELG500
EHRLICH 594 see ABX500
EHRLICH 606 see SAP500
EI see EJM900
EI-12880 see DSP400
EI-18706 see DNX600
EI 47031 see PGW750
EI 47300 see DSQ000
EI-47470 see DHH400
EI 52160 see TAL250
EICOSANOIC ACID see EAF000
EIRENAL see PGA750
EISENDEXTRAN (GERMAN) see IGS000

EISENDIMETHYLDITHIOCARBAMAT (GERMAN) see
    FAS000
EISENOXYD see IHD000
EISEN(III)-TRIS(N,N-DIMETHYLDITHIOCARBAMAT) (GER-
    MAN) see FAS000
EK 54 see DUU600
EK 1700 see PEY250
EKAGOM TB see TFS350
EKAGOM TEDS see DXH250
EKALUX see DJY200
EKATIN see PHI500
EKATIN AEROSOL see PHI500
EKATINE-25 see PHI500
EKATIN ULV see PHI500
EKATIN WF & WF ULV see PAK000
EKATOX see PAK000
EKAVYL SD 2 see PKQ059
EKKO CAPSULES see DKQ000
EKTAFOS see DGQ875
EKTASOLVE de ACETATE see CBQ750
EKTASOLVE DB ACETATE see BQP500
EKTASOLVE EB see BPJ850
EKTASOLVE EB ACETATE see BPM000
EKTASOLVE EE see EES350
EKTASOLVE EE ACETATE SOLVENT see EES400
EKTASOLVE EIB see IIP000
EKTYLCARBAMID see EHP000
EKVACILLIN see SLJ000
EK 1108GY-A see TAI250
EL 222 see FAK100
EL 4049 see MAK700
ELAIOMYCIN see EAG000
ELALDEHYDE see PAI250
ELANCOBAN see MRE225
ELANCOLAN see DUV600
ELANIL see EAH500
ELAOL see DEH200
ELASTONON see BBK000
ELAVIL see EAH500, EAI000
ELAVIL HYDROCHLORIDE see EAI000
ELAYL see EIO000
ELBANIL see CKC000
ELCIDE 75 see MDI000
ELCORIL see CDO500
EL-CORTELAN SOLUBLE see HHR000
ELCOZINE CHRYSOIDINE Y see PEK000
ELCOZINE RHODAMINE B see FAG070
ELDADRYL see BAU750
ELDEZOL see NGE500
ELDIATRIC C see CAK500
ELDODRAM see DYE600
ELDOPAL see DNA200
ELDOPAQUE see HIH000
ELDOQUIN see HIH000
ELECTRO-CF 11 see TIP500
ELECTRO-CF 12 see DFA600
ELECTRO-CF 22 see CFX500
ELECTROCORUNDUM see EAL100
ELECTRONIC E-2 see HIC000
ELEMENTAL SELENIUM see SBO500
ELENIUM see LFK000, MDQ250
ELEPSINDON see DKQ000
ELEUDRON see TEX250
ELGETOL see BRE500, DUS700, DUU600
ELGETOL 318 see BRE500
ELICIDE see MDI000
ELIMOCLAVIN see EAJ000
ELIPOL see DUS700
ELITONE see DJS200
ELIXICON see TEP000
ELIXOPHYLLIN see TEP000
ELIXOPHYLLINE see TEP000
ELJON LAKE RED C see CHP500
ELLSYL see MHJ500

ELMASIL see AMY050
ELMEDAL see BRF500
ELMER'S GLUE ALL see AAX250
ELOBROMOL see DDJ000
ELOCRON see DVS000
EL PETN see PBC250
ELPON see PMP500
ELRODORM see DYC800
ELSAN see DRR400
ELSYL see MHJ500
ELVACITE see PKB500
ELVANOL see PKP750
ELYMOCLAVIN see EAJ000
ELYMOCLAVINE see EAJ000
ELYZOL see MMN250
EM see EDH500
EM 923 see DFY400
EMAFORM see CHR500
EMANAY ATOMIZED ALUMINUM POWDER see AGX000
EMANAY ZINC DUST see ZBJ000
EMANAY ZINC OXIDE see ZKA000
EMANIL see DAS000
EMAR see ZKA000
EMAZOL RED B see EAJ500
EMBACETIN see CDP250
EMBAFUME see MHR200
EMBANOX see BQI000
EMBATHION see EEH600
EMBECHINE see BIE500
EMBICHIN see BIE250, BIE500
EMBICHIN HYDROCHLORIDE see BIE500
EMBIKHINE see BIE500
EMBINAL see BAG250
EMBIOL see VSZ000
EMBRAMINE HYDROCHLORIDE see BMN250
EMBUTAL see NBU000
EMC see CHC500
EMCEPAN see CIR250
EMCOL CA see OAV000
EMCOL DS-50 CAD see HKJ000
EMCOL MSK see OAV000
EMEDAN see BSM000
EMERALD GREEN see BAY750, COF500
EMERESSENCE 1160 see PER000
EMEREST 2301 see OHW000
EMEREST 2350 see EJM500
EMEREST 2400 see OAV000
EMEREST 2401 see OAV000
EMEREST 2801 see OHW000
EMERGIL see FMO129
EMERSAL 6400 see SIB600
EMERSAL 6465 see TAV750
EMERSOL 120 see SLK000
EMERSOL 140 see PAE250
EMERSOL 143 see PAE250
EMERSOL 210 see OHU000
EMERSOL 213 see OHU000
EMERSOL 6321 see OHU000
EMERSOL 233LL see OHU000
EMERSOL 221 LOW TITER WHITE OLEIC ACID see
    OHU000
EMERSOL 220 WHITE OLEIC ACID see OHU000
EMERY see EAL100
EMERY 655 see MSA250
EMERY 2218 see MJW000
EMERY 2219 see OHW000
EMERY 2310 see OHW000
EMERY 5791 see MCN250
EMERY 6705 see PER000
EMERY OLEIC ACID ESTER 2301 see OHW000
EMESIDE see ENG500
EMETHIBUTIN HYDROCHLORIDE see EIJ000
EMETINE see EAL500, EAL500
EMETINE ANTIMONY IODIDE see EAM000

EMETINE BISMUTH IODIDE see EAM500
EMETINE with BISMUTH(III) TRIIODIDE see EAM500
EMETINE, DIHYDROCHLORIDE see EAN000
(−)-EMETINE DIHYDROCHLORIDE see EAN000
1-EMETINE DIHYDROCHLORIDE see EAN000
EMETINE HYDROCHLORIDE see EAN000
EMETINE TRIIODOBISMUTH(III) see EAM500
EMETIQUE (FRENCH) see AQG250
EMETIRAL see PMF250
EMETREN see CDP250
EMFAC 1202 see NMY000
EMI-CORLIN see HHR000
EMID 6511 see BKE500
EMID 6541 see BKE500
EMIPHEROL see VSZ450
EMISAN 6 see MEP250
EMISOL see AMY050
EMMATOS see MAK700
EMMATOS EXTRA see MAK700
EMMI see EME050
EMOCICLINA see VSZ000
EMO-NIK see NDN000
EMOREN see DTL200
EMPAL see CIR250
EMPG see ENC000
EMPILAN 2848 see EJM500
EMPIRIN see ADA725
EMPIRIN COMPOUND see ABG750, ARP250
EMPLETS POTASSIUM CHLORIDE see PLA500
EMQ see SAV000
EMS see EMF500
EMSORB 2505 see SKV150
EMSORB 6900 see PKL100
EMT 25,299 see MDV500
EMTAL 596 see TAB750
EMTEXATE see MDV500
EMTRYL see DSV800
EMTRYLVET see DSV800
EMTRYMIX see DSV800
EMTS see EME500
EMUL P.7 see OAV000
EMULSAMINE BK see DAA800
EMULSAMINE E-3 see DAA800
EMULSIFIER No. 104 see SIB600
EMULSIPHOS 440/660 see SJH200
E-MYCIN see EDH500
EN 237 see EAL100
EN 18133 see EPC500
ENALLYNYMAL SODIUM see MDU500
ENAMEL WHITE see BAP000
ENANTHAL see HBB500
ENANTHALDEHYDE see HBB500
ENANTHIC ALCOHOL see HBL500
ENANTHOLE see HBB500
ENARMON see TBG000
ENAVID see EAP000
ENBU see ENT000
ENCORTON see PLZ000
ENDECRIL see DFP600
ENDEP see EAI000
ENDOBION see NCR000
ENDOCEL see EAQ750
ENDOCID see EAS000
ENDOCIDE see EAS000
ENDO E see VSZ450
ENDOFOLLICOLINA D.P. see EDR000
ENDOFOLLICULINA see EDV000
ENDOLAT see DAM700
ENDOMETHYLENETETRAHYDROPHTHALIC ACID, N-2-
    ETHYLHEXYL IMIDE see OES000
3,6-ENDOOXOHEXAHYDROPHTHALIC ACID see EAR000
ENDOSOL see EAQ750
ENDOSULFAN see EAQ750
ENDOSULPHAN see EAQ750

ENDOTAL see DXD000
ENDOTHAL see DXD000, EAR000
ENDOTHALL see EAR000
ENDOTHAL-NATRIUM (DUTCH) see DXD000
ENDOTHAL-SODIUM see DXD000
ENDOTHAL TECHNICAL see EAR000
ENDOTHAL WEED KILLER see DXD000
ENDOTHION see EAS000
ENDOXAN see EAS500
ENDOXANA see CQC675
ENDOXANAL see EAS500
ENDOXAN-ASTA see CQC675, EAS500
ENDOXAN MONOHYDRATE see CQC675
ENDOXAN R see CQC675, EAS500
ENDUXAN see CQC675
3,6-ENDOXOHEXAHYDROPHTHALIC ACID see EAR000
3,6-ENDOXOHEXAHYDROPHTHALIC ACID DISODIUM
   SALT see DXD000
ENDRATE see EIX000
ENDRATE DISODIUM see EIX500
ENDRATE TETRASODIUM see EIV000
ENDREX see EAT500
ENDRIN see EAT500
ENDRINE (FRENCH) see EAT500
ENDUXAN see CQC675
ENDYDOL see ADA725
ENDYL see TNP250
E.N.E. see ENX500
ENELFA see HIM000
ENERIL see HIM000
ENERZER see IKC000
ENFENEMAL see ENB500
ENFLURANE see EAT900
ENGLISH RED see IHD000
ENHEPTIN-A see ABY900
ENHEXYMAL see ERD500
ENIACID BRILLIANT RUBINE 3B see HJF500
ENIACID LIGHT ORANGE G see HGC000
ENIAL ORANGE I see PEJ500
ENIAL YELLOW 2G see DOT300
ENIANIL BLACK CN see AQP000
ENIANIL BLUE 2BN see CMO000
ENICOL see CDP250
ENIDE see DRP800
ENIDREL see EAP000
ENILOCONAZOL (SP) see FPB875
ENJAY CD 460 see PMP500
ENKALON see NOH000
ENKEFAL see DNU000
ENKELFEL see DKQ000
ENNG see ENU000
ENORDEN see MQU750
ENOVID see EAP000
ENOVID-E see EAP000
ENPHENEMAL see ENB500
ENPROMATE see DWL400
ENRUMAY see WAK000
E.N.S. see ENX500
ENS see EPI300
ENSEAL see PLA500
ENSOBARB see EOK000
ENSODORM see EOK000
ENSURE see EAQ750
ENS-ZEM WEEVIL BAIT see DXE000
ENT 6 see BPL250
ENT 9 see BRT000
17-ENT see ABU000
ENT 38 see PDP250
ENT 54 see ADX500
ENT 92 see IHZ000
ENT 123 see VHZ000
ENT 133 see RNZ000
ENT 154 see DUS700
ENT 157 see CPK500
ENT 262 see DTR200

ENT 375 see EKV000
ENT 884 see COF500
ENT 988 see BJK500
ENT 1,122 see BRE500
ENT 1,501 see DXE000
ENT 1,506 see DAD200
ENT 1,656 see EIY600
ENT 1,716 see MEI450
ENT 1,860 see PCF275
ENT 3,424 see NDN000
ENT 3,776 see DFT000
ENT 4,225 see BIM500
ENT 4,504 see DFJ050
ENT 4,585 see CJR500
ENT 4,705 see CBY000
ENT 7,543 see POO100
ENT 7,796 see BBQ500
ENT 8,184 see OES000
ENT 8,420 see DGG000
ENT 8,538 see DAA800
ENT 8,601 see BBP750
ENT 9,232 see BBQ000
ENT 9,233 see BBR000
ENT 9,234 see BFW500
ENT 9,735 see CDV100
ENT 9,932 see CDR750
ENT 14,250 see PIX250
ENT 14,611 see ASL250
ENT 14,689 see FAS000
ENT 14,874 see EIR000
ENT 14,875 see MAS500
ENT 15,108 see PAK000
ENT 15,152 see HAR000
ENT 15,349 see EIY500
ENT 15,406 see PNJ400
ENT 15,949 see AFK250
ENT 16,087 see NIM500
ENT 16,225 see DHB400
ENT 16,273 see SOD100
ENT 16,358 see CJT750
ENT 16,391 see KEA000
ENT 16,436 see DXX400
ENT 16,519 see SOP500
ENT 16,634 see ISA000
ENT 17,034 see MAK700
ENT 17,035 see NFT000
ENT 17,251 see EAT500
ENT 17,291 see OCM000
ENT 17,292 see MNH000
ENT 17,295 see DAO600
ENT 17,470 see DFK600
ENT 17,510 see AFR250
ENT 17,798 see EBD700
ENT 17,956 see CNU750
ENT 18,060 see CKC000
ENT 18,596 see DER000
ENT 18,771 see TCF250
ENT 18,861 see MIJ250
ENT 18,862 see DAO800, MIW100
ENT 18,870 see DMC600
ENT 19,060 see DSK200
ENT 19,109 see BJE750
ENT 19,244 see IKO000
ENT 19,442 see TBC500
ENT 19,507 see DCM750
ENT 19,763 see TIQ250
ENT 20,218 see DKC800
ENT 20,696 see CEP000
ENT 20,738 see DGP900
ENT 20,852 see BPG000, DDP000
ENT 21,040 see AGE250
ENT 22,014 see EKN000
ENT 22,335 see TBR250
ENT 22,374 see MQR750

ENT 22,542 see DKC800
ENT 22,865 see DJN600
ENT 22,897 see DVQ709
ENT 22,952 see POO000
ENT 23,233 see ASH500
ENT 23,284 see RMA500
ENT 23,437 see DXH325
ENT 23,438 see DRR400
ENT 23,648 see BIO750
ENT 23,708 see TNP250
ENT 23,737 see CKM000
ENT 23,968 see POP000
ENT 23,969 see CBM750
ENT 23,979 see EAQ750
ENT 24,042 see PGS000
ENT 24,105 see EEH600
ENT 24,482 see DGQ875
ENT 24,650 see DSP400
ENT 24,652 see IOT000
ENT 24,653 see EAS000
ENT 24,723 see MOX250
ENT 24,725 see DKB600
ENT 24,727 see AQT500
ENT 24,738 see DRL200
ENT 24,833 see SOY000
ENT 24,915 see TND250
ENT 24,945 see FAQ800
ENT 24,964 see DAP000
ENT 24,969 see CDS750
ENT 24,970 see NAQ500
ENT 24,979 see BLL750
ENT 24,986 see DXO000
ENT 24,988 see NAG400
ENT 25,208 see ABX250
ENT 25,294 see BIE250
ENT 25,296 see TND500
ENT 25,445 see AMY050
ENT 25,506 see DNX600
ENT 25,540 see FAQ999
ENT 25,545 see OAN000
ENT 25,550 see SCH000
ENT 25,567 see HMV000
ENT 25,580 see EPC500
ENT 25,584 see EBW500
ENT 25,585 see PDC750
ENT 25,599 see MQH750
ENT 25,640 see CQL250
ENT 25,644 see FAB600
ENT 25,647 see DGD800
ENT 25,670 see MIA250
ENT 25,671 see PMY300
ENT 25,674 see DSK600
ENT 25,684 see DST200
ENT 25,705 see PHX250
ENT 25,712 see EPY000
ENT 25,715 see DSQ000
ENT 25,719 see MQW500
ENT 25,726 see DST000
ENT 25,734 see PHD250
ENT 25,737 see DTV400
ENT 25,764 see TBV750
ENT 25,766 see DOS000
ENT 25,776 see DNX800
ENT 25,784 see DOR400
ENT 25,796 see FMU045
ENT 25,809 see DXN600
ENT 25,823 see DFV400
ENT 25,830 see PGW750
ENT 25,843 see TMD000
ENT 25,991 see DHH400
ENT 26,079 see AMG750
ENT 26,263 see EJN500
ENT 26,396 see EMF500
ENT 26,538 see CBG000

ENT 26,592 see BGA750
ENT 26,613 see MJG500
ENT 27,093 see CBM500
ENT 27,102 see DIX600
ENT 27,128 see MOU750
ENT 27,129 see MRH209
ENT 27,163 see BDJ250
ENT 27,164 see CBS275
ENT 27,165 see TAL250
ENT 27,180 see MOB250
ENT 27,193 see DSO000
ENT 27,223 see AIX000
ENT 27,226 see SOP000
ENT 27,257 see DRR200
ENT 27,258 see EGV500
ENT 27,267 see BJD000
ENT 27,311 see CMA100
ENT 27,318 see EIN000
ENT 27,320 see DBI099
ENT 27,324 see DLS800
ENT 27,335 see CJJ250
ENT 27,341 see MDU600
ENT 27,346 see DOP200
ENT 27,357 see DTP800
ENT 27,389 see DVS000
ENT 27,394 see DJY200
ENT 27,396 see DTQ400
ENT 27,407 see MPG250
ENT 27,438 see DGA200
ENT 27,474 see BEP500
ENT 27,488 see BAT750
ENT 27,520 see CMA250
ENT 27,566 see DSO200
ENT 27,567 see CJJ250
ENT 27,572 see FAK000
ENT 27,625 see DOL800
ENT 27,738 see BLU000
ENT 27,822 see DOP600
ENT 27,989 see MKA000
ENT 28,009 see HON000
ENT 29,054 see CJV250
ENT 50,003 see TNK250
ENT 50,107 see BJC250
ENT 50,146 see RDK000
ENT 50,324 see EJM900
ENT 50,434 see AQG250
ENT 50,439 see BIA250
ENT 50,825 see MLY000
ENT 50,852 see HEJ500
ENT 50,882 see HEK000
ENT 51,799 see MJQ500
ENT 61,241 see ACM750
ENT 70,460 see KAJ000
ENT 27,386GC see DRR400
ENT 27,699GC see DIN800
ENTERICIN see ADA725
ENTERO-BIO FORM see CHR500
ENTEROMYCETIN see CDP250
ENTEROPHEN see ADA725
ENTEROQUINOL see CHR500
ENTEROSALICYL see SJO000
ENTEROSALIL see SJO000
ENTEROSARINE see ADA725
ENTEROSEDIV see TEH500
ENTEROSEPTOL see CHR500
ENTEROTOXON see NGG500
ENTERO-VIOFORM see CHR500
ENTEROZOL see CHR500
ENTERUM LOCORTEN see CHR500
ENTEX see FAQ999
ENTIZOL see MMN250
ENTOMOXAN see BBQ500
ENTROKIN see CHR500
ENTROPHEN see ADA725

ENTSUFON see TMN490
ENT 24,980-X see DJA400
ENT 25,545-X see OAN000
ENT 25,552-X see CDR750
ENT 25,555-X see DJA200
ENT 25,602-X see COD850
ENT 27,395-X see CQH650
ENU see ENV000, NKE500
ENVERT 171 see DAA800
ENVERT DT see DAA800
ENVERT-T see TAA100
ENZACTIN see THM500
ENZAMIN see BBW500
ENZAPROST see DMU800
ENZAPROST F see DMU800
E.O. see EJN500
EOSINE see BNH500
EOSINE SODIUM SALT see BNH500
EOSINE YELLOWISH see BNH500
EOSIN GELBLICH (GERMAN) see BNH500
EP 30 see PAX250
EP-205 see BJN250
EP-206 see VOA000
EP-332 see DSO200
EP-333 see CJJ250
EP-411 see PFC750
EP 1463 see AAX250
EP-161E see ISE000
EPAL 6 see HFJ500
EPAL 8 see OEI000
EPAL 10 see DAI600
EPAL 12 see DXV600
EPAL 16NF see HCP000
EPAMIN see DKQ000, DNU000
EPANAL see EOK000
EPANUTIN see DKQ000, DNU000
EPASMIR '5' see DKQ000
EPDANTOINE SIMPLE see DKQ000
EPE see EAV500
EPELIN see DKQ000, DNU000
EPHEDRAL see EAW000
EPHEDRATE see EAW000
EPHEDREMAL see EAW000
EPHEDRIN see EAW000
EPHEDRINE see EAW000
l-EPHEDRINE see EAW000
l(−)-EPHEDRINE see EAW000
EPHEDRINE HYDROCHLORIDE see EAW500, EAY000
(−)-EPHEDRINE HYDROCHLORIDE see EAY000
d-EPHEDRINE HYDROCHLORIDE see EAX000
l-EPHEDRINE HYDROCHLORIDE see EAY000
dl-EPHEDRINE HYDROCHLORIDE see EAX500
1-EPHEDRINE SULFATE see EAY500
EPHEDRITAL see EAW000
EPHEDROL see EAW000
EPHEDROSAN see EAW000
EPHEDROTAL see EAW000
EPHEDSOL see EAW000
EPHENDRONAL see EAW000
EPHETONIN see EAX500
EPHETONINE see EAX500
EPHININE HYDROCHLORIDE see EAZ000
EPHIRSULPHONATE see CJT750
EPHORRAN see DXH250
EPHOXAMIN see EAW000
EPHYNAL see VSZ450
EPIBENZALIN see DLY000
EPIBLOC see CDT750
EPIBROMHYDRIN see BNI000
EPIBROMOHYDRIN (DOT) see BNI000
EPIBROMOHYDRINE see BNI000
EPICHLOORHYDRIN (DUTCH) see EAZ500
EPICHLORHYDRIN (GERMAN) see EAZ500
EPICHLORHYDRINE (FRENCH) see EAZ500

EPICHLOROHYDRIN see EAZ500
α-EPICHLOROHYDRIN see EAZ500
(dl)-α-EPICHLOROHYDRIN see EAZ500
EPICHLOROHYDRYNA (POLISH) see EAZ500
EPICHLOROPHYDRIN see EAZ500
EPI-CLEAR see BDS000
EPICLORIDRINA (ITALIAN) see EAZ500
EPICUR see MQU750
EPICURE DDM see MJQ000
EPIDERMOL see ACR300
3,17-EPIDIHYDROXYESTRATRIENE see EDO000
3,17-EPIDIHYDROXYOESTRATRIENE see EDO000
EPIDIONE see TLP750
EPIDONE see TLP750
EPIDORM see EOK000
EPIFENYL see DKQ000, DNU000
EPIFRIN see VGP000
EPIHYDAN see DKQ000, DNU000
EPIHYDRIN ALCOHOL see GGW500
EPIHYDRINALDEHYDE see GGW000
EPIHYDRINE ALDEHYDE see GGW000
EPILAN see DKQ000, MKB250
EPILAN-D see DNU000
EPILANTIN see DKQ000, DNU000
EPILEO PETIT MAL see ENG500
EPILOL see EOK000
EPINAT see DKQ000, DNU000
EPINELBON see DLY000
EPINEPHRAN see VGP000
EPINEPHRINE see VGP000, VGP000
1-EPINEPHRINE see VGP000
d-EPINEPHRINE see AES250
(R)-EPINEPHRINE see VGP000
dl-EPINEPHRINE see EBB500
EPINEPHRINE racemic see EBB500
EPINEPHRINE ISOPROPYL HOMOLOG see DMV600
1-EPINEPHRINE (SYNTHETIC) see VGP000
EPINOVAL see DJU200
EPIRENAMINE see VGP000
EPIRENAN see VGP000
EPI-REZ 508 see BLD750
EPI-REZ 510 see BLD750
EPIROTIN see BBW500
EPISED see DKQ000
EPISEDAL see EOK000
EPITELIOL see VSK600
EPITHELONE see ACR300
EPITRATE see VGP000
EPN see EBD700
EPOLENE M 5K see PMP500
EPON 828 see BLD750
EPORAL see SOA500
(3,6-EPOSSI-CICLOESAN-1,2-DICARBOSSILATO) DISODICO (ITALIAN) see DXD000
EPOXIDE 269 see LFV000
EPOXIDE A see BLD750
1,2-EPOXYAETHAN (GERMAN) see EJN500
1,4-EPOXYBUTANE see TCR750
1,2-EPOXYBUTENE-3 see EBJ500
3,4-EPOXY-1-BUTENE see EBJ500
4,9-EPOXYCEVANE-3,4,12,14,16,17,20-HEPTOL 3-(3,4-DI-METHOXYBENZOATE) see VHU000
1,2-EPOXY-3-CHLOROPROPANE see EAZ500
5-α,6-α-EPOXYCHOLESTANOL see EBM000
5,6-α-EPOXY-5-α-CHOLESTAN-3-β-OL see EBM000
EPOXYCHOLESTEROL see EBM000
1,2-EPOXYCYCLOHEXANE see CPD000
3,6-EPOXY-CYCLOHEXANE 1,2-CARBOXYLATE DISO-DIQUE (FRENCH) see DXD000
3,6-endo-EPOXY-1,2-CYCLOHEXANEDICARBOXYLIC ACID see EAR000
(3-α,4-β)-12,13-EPOXY-4,15-DIACETATE-TRICHOTHEC-9-ENE-3,4,15-TRIOL see AOP250
12,13-EPOXY-4-β,15-DIAZETOXY-3-α-HYDROXY-TRICHO-THEC-9-ENE see AOP250

15,20-EPOXY-15,30-DIHYDRO-12-HYDROXYSENECIONAN-11,16-DIONE see JAK000
5,6-EPOXY-5,6-DIHYDRO-7-METHYLBENZ(A) ANTHRA-CENE see MGZ000
1,2-EPOXYDODECANE see DXU400
1,2-EPOXY-4-(EPOXYETHYL)CYCLOHEXANE see VOA000
EPOXYETHANE see EJN500
1,2-EPOXYETHANE see EJN500
1,2-EPOXYETHYLBENZENE see EBR000
EPOXYETHYLBENZENE (8CI) see EBR000
1-EPOXYETHYL-3,4-EPOXYCYCLOHEXANE see VOA000
3-(EPOXYETHYL)-7-OXABICYCLO(4.1.0)HEPTANE see VOA000
4-(EPOXYETHYL)-7-OXABICYCLO(4.1.0)HEPTANE see VOA000
3-(1,2-EPOXYETHYL)-7-OXABICYCLO(4.1.0)HEPTANE see VOA000
4-(1,2-EPOXYETHYL)-7-OXABICYCLO(4.1.0)HEPTANE see VOA000
EPOXYHEPTACHLOR see EBW500
1,2-EPOXYHEXADECANE see EBX500
4,5-α-EPOXY-3-HYDROXY-17-METHYLMORPHINAN-6-ONE HYDROCHLORIDE see DNU300
12,13-EPOXY-4-HYDROXYTRICHOTHEC-9-EN-8-ONE CRO-TONATE see TJE750
1,8-EPOXY-p-MENTHANE see CAL000
4-(1,2-EPOXY-1-METHYLETHYL)-1-METHYL-7-OXABI-CYCLO(4.1.0)HEPTANE see LFV000
α-β-EPOXY-β-METHYLHYDROCINNAMIC ACID, ETHYL ESTER see ENC000
cis-9,10-EPOXYOCTADECANOATE see ECD500
cis-9,10-EPOXYOCTADECANOIC ACID see ECD500
EPOXYOLEIC ACID see ECD500
12,13-EPOXY-4-((1-OXO-2-BUTENYL)OXY)TRICHOTHEC-9-EN-8-ONE see TJE750
1,2-EPOXY-3-PHENOXYPROPANE see PFH000
2,3-EPOXYPROPANAL see GGW000
2,3-EPOXY-1-PROPANAL see GGW000
EPOXYPROPANE see PNL600
1,2-EPOXYPROPANE see PNL600
2,3-EPOXYPROPANE see PNL600
2,3-EPOXYPROPANOL see GGW500
2,3-EPOXY-1-PROPANOL see GGW500
2,3-EPOXY-1-PROPANOL ACRYLATE see ECH500
2,3-EPOXY-1-PROPANOL OLEATE see ECJ000
2,3-EPOXY-1-PROPANOL STEARATE see SLK500
2,3-EPOXYPROPIONALDEHYDE see GGW000
2,3-EPOXYPROPYL ACRYLATE see ECH500
2,3-EPOXYPROPYL BUTYL ETHER see BRK750
2,3-EPOXYPROPYL CHLORIDE see EAZ500
2,3-EPOXYPROPYL ESTER ACRYLIC ACID see ECH500
2,3-EPOXYPROPYL ESTER of OLEIC ACID see ECJ000
2,3-EPOXYPROPYL ESTER of STEARIC ACID see SLK500
2,3-EPOXYPROPYL OLEATE see ECJ000
2,3-EPOXYPROPYLPHENYL ETHER see PFH000
2,3-EPOXYPROPYL STEARATE see SLK500
EPOXY RESINS, CURED see ECK500
EPOXY RESINS, UNCURED see ECL000
9,10-EPOXYSTEARIC ACID see ECD500
cis-9,10-EPOXYSTEARIC ACID see ECD500
EPOXYSTYRENE see EBR000
α,β-EPOXYSTYRENE see EBR000
6-β,7-β-EPOXY-3-α-TROPANYL S-(−)-TROPATE see SBG000
EPOXYTROPINE TROPATE see SBG000
EPRAZIN see POL500
EPROFIL see TEX000
EPROLIN see VSZ450
EPSILAN see VSZ450
EPSOM SALTS see MAJ250, MAJ500
EPSYLONE see EOK000
EPSYLON KAPROLAKTAM (POLISH) see CBF700
EPT see EQP000
EPTAC 1 see BJK500

EPTACLORO (ITALIAN) see HAR000
1,4,5,6,7,8,8-EPTACLORO-3a,4,7,7a-TETRAIDRO-4,7-endo-METANO-INDENE (ITALIAN) see HAR000
EPTAL see DKQ000
EPTAM see EIN500
EPTANI (ITALIAN) see HBC500
EPTAN-3-ONE (ITALIAN) see EHA600
EPTAPUR see CKF750
EPTC see EIN500
EPTOIN see DKQ000, DNU000
E-PVC see PKQ059
EQ see SAV000
EQUAL see ARN825
EQUANIL SUSPENSION see MQU750
EQUIBRAL see MDQ250
EQUI BUTE see BRF500
EQUIGEL see DGP900
EQUIGYNE see ECU750, ECU750
EQUILENIN see ECV000
EQUILENINA (SPANISH) see ECV000
EQUILENINE see ECV000
EQUILIN see ECW000
EQUILIUM see MQU750, PGE000
EQUINIL see MQU750
EQUINO-ACID see TIQ250
EQUINO-AID see TIQ250
EQUIPOISE see CJR909
EQUIZOLE see TEX000
ERADEX see CMA100
ERALON see ILD000
ERAMIDE see CDR250
ERAMIN see HGD000
ERANTIN see PNA500
ERASE see HKC000
ERASOL see BIE500
ERASOL HYDROCHLORIDE see BIE500
ERASOL-IDO see BIE500
ERBAPLAST see CDP250
ERBAPRELINA see TGD000
ERBIUM CHLORIDE see ECX500
ERBIUM(III) NITRATE (1:3) see ECY500
ERBIUM(III) NITRATE, HEXAHYDRATE (1:3:6) see ECZ000
ERBIUM TRICHLORIDE see ECX500
ERCO-FER see FBJ100
ERCOFERRO see FBJ100
ERE 1359 see REF000
ERGAM see EDC500
ERGAMINE see HGD000
ERGATE see EDC500
ERGOATETRINE see LJL000
ERGOBASINE see LJL000
ERGOCALCIFEROL see VSZ100
ERGOKLININE see LJL000
ERGOMAR see EDC500
ERGOMETRINE see LJL000
ERGOMETRINE ACID MALEATE see EDB500
ERGOMETRINE MALEATE see EDB500
ERGONOVINE see LJL000
ERGONOVINE, MALEATE (1:1) (SALT) see EDB500
ERGOPLAST FDO see DVL700
ERGORONE see VSZ100
ERGOSTAT see EDC500
ERGOSTEROL, ACTIVATED see VSZ100
ERGOSTEROL, IRRADIATED see VSZ100
ERGOT see EDB500
ERGOTAMINE see EDC000
ERGOTAMINE BITARTRATE see EDC500
ERGOTAMINE TARTRATE see EDC500
ERGOTARTRATE see EDC500
ERGOTIDINE see HGD000
ERGOTOCINE see LJL000
ERGOTRATE see EDB500, LJL000
ERGOTRATE MALEATE see EDB500

ERIBUTAZONE see BRF500
ERIE BLACK B see AQP000
ERINA see MQU750
ERINIT see PBC250
ERINITRIT see SIQ500
ERIO FAST ORANGE AS see HGC000
ERIOGLAUCINE see FMU059
ERIOGLAUCINE G see FAE000
ERIONITE see EDC650
ERIOSIN RHODAMINE B see FAG070
ERIOSKY BLUE see FMU059
ERISIMIN DIHYDRATE see HAO000
ERITRONE see VSZ000
ERL-2774 see BLD750
ERLA-2270 see VOA000
ERLA-2271 see VOA000
ERMETRINE see LJL000
EROINA see HBT500
ERR 4205 see BJN250
ERROLON see CHJ750
ERSERINE see PIA500
ERTALON 6SA see PJY500
ERTILEN see CDP250
ERTRON see VSZ100
ERTUBAN see ILD000
ERYCIN see EDH500
ERYCYTOL see VSZ000
ERYSAN see BIF250
ERYSIMIN DIHYDRATE see HAO000
ERYTHRENE see BOP500
ERYTHRITOL ANHYDRIDE see BGA750, DHB800
ERYTHROCIN see EDH500
ERYTHROGRAN see EDH500
ERYTHROGUENT see EDH500
ERYTHROMYCIN see EDH500
ERYTHROMYCIN A see EDH500
ERYTHROSIN see FAG040
ERYTHROSINE B-FO (BIOLOGICAL STAIN) see FAG040
ERYTHROTIN see VSZ000
ESACHLOROBENZENE (ITALIAN) see HCC500
ESACICLONATO see SHL500
ESAIDRO-1,3,5-TRINITRO-1,3,5-TRIAZINA (ITALIAN) see CPR800
ESAMETILENTETRAMINA (ITALIAN) see HEI500
ESAMETINA see HEA000
ESAMETONIO IODURO (ITALIAN) see HEB000
ESANI (ITALIAN) see HEN000
ESBRITE see SMQ500
ESCAMBIA 2160 see PKQ059
ESCIN, SODIUM SALT see EDM000
ESCOPARONE see DRS800
ESCOREZ 7404 see SMQ500
ESCULETIN DIMETHYL ETHER see DRS800
ESDRAGOL see AFW750
ESEN see PHW750
ESERINE see PIA500
ESERINE SALICYLATE see PIA750
ESEROLEIN, METHYLCARBAMATE (ESTER) see PIA500
ESGRAM see PAJ000
ESIDREX see CFY000
ESIDRIX see CFY000
ESKABARB see EOK000
ESKALIN V see VRF000
ESKALITH see LGZ000
ESKIMON 11 see TIP500
ESKIMON 12 see DFA600
ESKIMON 22 see CFX500
ESOBARBITALE (ITALIAN) see ERD500
E 39 SOLUBLE see BDC750
ESOPHOTRAST see BAP000
ESORB see VSZ450
ESP see DSK600
ESPADOL see CLW000
ESPARIN see DQA600

ESPENAL see DXH250
ESPERAL see DXH250
ESPEROX 10 see BSC500
ESPEROX 31M see BSD250
ESPHYGMOGENINA see VGP000
ESSENCE of MIRBANE see NEX000
ESSENCE of MYRBANE see NEX000
ESSENCE of NIOBE see EGR000
ESSIGESTER (GERMAN) see EFR000
ESSIGSAEURE (GERMAN) see AAT250
ESSIGSAEUREANHYDRID (GERMAN) see AAX500
ESSO FUNGICIDE 406 see CBG000
ESSO HERBICIDE 10 see BQZ000
ESTANE 5703 see UVA000
ESTAR see CMY800
ESTASIL see MQU750
ESTER 25 see NIM500
ESTERCIDE T-2 and T-245 see TAA100
S-ESTER with O,O-DIMETHYL PHOSPHOROTHIOATE see MAK700
ESTER DWUETYLOAMINOETYLOWY KWASU DWU-FENYLOOCTOWEGO see DHX800
O-ESTER-p-NITROPHENOL with O-ETHYL PHENYL PHOS-PHONOTHIOATE see EBD700
ESTERON see DAA800
ESTERON 44 see IOY000
ESTERON 99 see DAA800
ESTERON 76 BE see DAA800
ESTERON 245 BE see TAA100
ESTERON BRUSH KILLER see DAA800, TAA100
ESTERON 99 CONCENTRATE see DAA800
ESTERONE see EDV000
ESTERONE FOUR see DAA800
ESTERON 44 WEED KILLER see DAA800
α-ESTER PALMITIC ACID with D-threo-(−)-2,2-DICHLORO-N-(β-HYDROXY-α-(HYDROXYMETHYL)-p-NITROPHEN-ETHYL)ACETAMIDE see CDP700
ESTERS see EDN500
ESTER SULFONATE see CJT750
ESTEVE see BRF500
ESTILBEN see DKA600, DKB000
ESTILBIN see DKB000
ESTIMULEX see DBA800
ESTINERVAL see PFC750
ESTOL 603 see OAV000
ESTOL 1550 see DJX000
ESTON see DSK600
ESTONATE see DAD200
ESTON-B see EDP000
ESTONMITE see CJT750
ESTONOX see CDV100
ESTOX see DSK600
ESTRADIOL see EDO000
d-ESTRADIOL see EDO000
α-ESTRADIOL see EDO000
β-ESTRADIOL see EDO000
ESTRADIOL-17-β see EDO000
17-β-ESTRADIOL see EDO000
cis-ESTRADIOL see EDO000
3,17-β-ESTRADIOL see EDO000
d-3,17-β-ESTRADIOL see EDO000
ESTRADIOL BENZOATE see EDP000
ESTRADIOL-3-BENZOATE see EDP000
β-ESTRADIOL BENZOATE see EDP000
β-ESTRADIOL-3-BENZOATE see EDP000
ESTRADIOL-17-β-BENZOATE see EDP000
17-β-ESTRADIOL BENZOATE see EDP000
ESTRADIOL-17-β-3-BENZOATE see EDP000
17-β-ESTRADIOL-3-BENZOATE see EDP000
ESTRADIOL DIPROPIONATE see EDR000
β-ESTRADIOL DIPROPIONATE see EDR000
17-β-ESTRADIOL DIPROPIONATE see EDR000
ESTRADIOL-3,17-DIPROPIONATE see EDR000
3,17-β-ESTRADIOL DIPROPIONATE see EDR000

β-ESTRADIOL-3,17-DIPROPIONATE see EDR000
ESTRADIOL MONOBENZOATE see EDP000
17-β-ESTRADIOL MONOBENZOATE see EDP000
ESTRADIOL MUSTARD see EDR500
ESTRADIOL PHOSPHATE POLYMER see EDS000
ESTRADIOL POLYESTER with PHOSPHORIC ACID see EDS000
ESTRADIOL VALERATE see EDS100
ESTRADIOL-17-VALERATE see EDS100
ESTRADIOL 17-β-VALERATE see EDS100
ESTRADIOL VALERIANATE see EDS100
ESTRADURIN see EDS000
ESTRAGARD see DAL600
ESTRAGON OIL see TAF700
ESTRALDINE see EDO000
ESTRALUTIN see HNT500
ESTRATAB see ECU750
1,3,5,7-ESTRATETRAEN-3-OL-17-ONE see ECW000
1,3,5-ESTRATRIENE-3,17-β-DIOL see EDO000
ESTRA-1,3,5(10)-TRIENE-3,17-β-DIOL see EDO000
17-β-ESTRA-1,3,5(10)-TRIENE-3,17-DIOL see EDO000
ESTRA-1,3,5(10)-TRIENE-3,17-β-DIOL, 3-BENZOATE see EDP000
1,3,5(10)-ESTRATRIENE-3,17-β-DIOL 3-BENZOATE see EDP000
ESTRA-1,3,5(10)-TRIENE-3,17-DIOL (17-β)-3-BENZOATE see EDP000
1,3,5(10)-ESTRATRIENE-3,17-β-DIOL DIPROPIONATE see EDR000
ESTRA-1,3,5(10)-TRIENE-3,17-DIOL (17-β)-DIPROPIONATE see EDR000
(17-β)-ESTRA-1,3,5(10)-TRIENE-3,17-DIOL-17-PENTANOATE (9CI) see EDS100
(17-β)-ESTRA-1,3,5(10)-TRIENE-3,17-DIOL POLYMER with PHOSPHORIC ACID see EDS000
1,3,5-ESTRATRIENE-3-β,16-α,17-β-TRIOL see EDU500
ESTRA-1,3,5(10)-TRIENE-3,16-α,17-β-TRIOL see EDU500
(16-α,17-β)-ESTRA-1,3,5(10)-TRIENE-3,16,17-TRIOL see EDU500
1,3,5-ESTRATRIEN-3-OL-17-ONE see EDV000
1,3,5(10)-ESTRATRIEN-3-OL-17-ONE see EDV000
Δ-1,3,5-ESTRATRIEN-3-β-OL-17-ONE see EDV000
ESTRATRIOL see EDU500
ESTRAVEL see EDS100
ESTREPTOCIDA see SNM500
ESTRIFOL see ECU750
ESTRIL see DKA600
ESTRIN see EDV000
ESTRIOL see EDU500
16-α,17-β-ESTRIOL see EDU500
3,16-α,17-β-ESTRIOL see EDU500
ESTRIOLO (ITALIAN) see EDU500
ESTROATE see ECU750
ESTROBEN see DKB000
ESTROBENE see DKA600, DKB000
ESTROCON see ECU750
ESTRODIENOL see DAL600
ESTROGEN see DKA600, EEH500
ESTROGENIN see DKB000
ESTROGENS, CONJUGATES see PMB000
ESTROICI see EDR000
ESTROL see EDV000
ESTROMED see ECU750
ESTROMENIN see DKA600
ESTRON see EDV000
ESTRONA (SPANISH) see EDV000
ESTRONE see EDV000
ESTRONE-A see EDV000
ESTRONE BENZOATE see EDV500
ESTRONEX see EDR000
ESTROPAN see ECU750
ESTRORAL see DAL600
ESTROSEL see DGP900
ESTROSOL see DGP900

ESTROSTILBEN see DKB000
ESTROSYN see DKA600
ESTROVITE see EDO000
ESTRUGENONE see EDV000
ESTRUMATE see CMX880
ESTRUSOL see EDV000
ESTYRENE AS see ADY500
ESTYRENE G 20 see SMQ500
ESZ see IHL000
ET 14 see RMA500
ET 57 see RMA500
ETABUS see DXH250
ETACRINIC ACID see DFP600
ETAFENONE HYDROCHLORIDE see DHS200
ETAIN (TETRACHLORURE d') (FRENCH) see TGC250
ETAKRINIC ACID see DFP600
ETAMICAN see VSZ450
ETAMINAL SODIUM see NBU000
ETANAUTINE see BBV500
ETANOLAMINA (ITALIAN) see EEC600
ETANOLO (ITALIAN) see EFU000
ETANTIOLO (ITALIAN) see EMB100
ETAPERAZIN see CJM250
ETAPERAZINE see CJM250
ETAVIT see VSZ450
ETCHLORVINOLO see CHG000
ETERE ETILICO (ITALIAN) see EJU000
ETH see EEI000, EPQ000
ETHAANTHIOL (DUTCH) see EMB100
ETHACRYNIC ACID see DFP600
ETHAL see HCP000
ETHAMINAL see NBT500
ETHAMINAL SODIUM see NBU000
ETHAMOXYTRIPHETOL see DHS000
ETHANAL see AAG250
ETHANAL OXIME see AAH250
ETHANAMIDE see AAI000
ETHANAMINE see EFU400
ETHANAMINE (anhydrous) see EDY000
ETHANAMINE, aqueous solution see EDY500
ETHANAMINIUM, 2-((CYCLOHEXYLHYDROXYPHENYL-ACETYL)OXY)-N,N-DIETHYL-N-METHYL-, BROMIDE (9CI) see ORQ000
ETHANAMINIUM, N,N-DIETHYL-N-METHYL-2-((9H-XAN-THEN-9-YLCARBONYL)OXY)-, BROMIDE (9CI) see XCJ000
ETHANDIAL see GIK000
ETHANE see EDZ000
ETHANE, compressed (DOT) see EDZ000
ETHANE, refrigerated liquid (DOT) see EDZ000
ETHANECARBOXYLIC ACID see PMU750
ETHANEDIAL see GIK000
ETHANEDIAMIDE see OLO000
1,2-ETHANEDIAMINE see EEA500
1,2-ETHANEDICARBOXYLIC ACID see SMY000
ETHANE DICHLORIDE see EIY600
ETHANEDINITRILE see COO000
ETHANEDIOIC ACID see OLA000
1,2-ETHANEDIOL see EJC500
1,2-ETHANEDIOL DIACETATE see EJD759
ETHANEDIOL DINITRATE see EJG000
1,2-ETHANEDIOL DIPROPANOATE (9CI) see COB260
1,2-ETHANEDIONE see GIK000
ETHANEDIONIC ACID see OLA000
ETHANEDITHIOAMIDE see DXO200
1,2-ETHANEDITHIOL see EEB000
1,2-ETHANEDITHIOL, CYCLIC ESTER with P,P-DIETHYL PHOSPHONODITHIOIMIDOCARBONATE see PGW750
1,2-ETHANEDITHIOL, CYCLIC S,S-ESTER with PHOSPHO-NODITHIOIMIDOCARBONIC ACID P,P-DIETHYL ESTER see PGW750
N,N'-1,2-ETHANEDIYLBIS(N-BUTYL-4-MORPHOLINECAR-BOXAMIDE) see DUO400

1,2-ETHANEDIYLBIS(CARBAMODITHIOATO)(2−)-MANGA-
NESE see MAS500
((1,2-ETHANEDIYLBIS(CARBAMODITHIOATO))(2-)ZINC
see EIR000
1,2-ETHANEDIYLBIS(CARBAMODITHIOATO) (2-)-S,S′-
ZINC see EIR000
1,2-ETHANEDIYLBISCARBAMODITHIOIC ACID DISODIUM
SALT see DXD200
1,2-ETHANEDIYLBISCARBAMODITHIOIC ACID MANGA-
NESE COMPLEX see MAS500
1,2-ETHANEDIYLBISCARBAMODITHIOIC ACID,
MANGANESE(2+) SALT (1:1) see MAS500
1,2-ETHANEDIYLBISCARBAMODITHIOIC ACID, ZINC
COMPLEX see EIR000
1,2-ETHANEDIYLBISCARBAMOTHIOIC ACID, ZINC SALT
see EIR000
N,N′-1,2-ETHANEDIYLBIS(N-(CARBOXYMETHYL)
GLYCINE see EIX000
N,N′-1,2-ETHANEDIYLBIS(N-(CARBOXYMETHYL)
GLYCINE) DISODIUM SALT see EIX500
N,N′-1,2-ETHANEDIYLBIS(N-(CARBOXYMETHYL)
GLYCINE TETRASODIUM SALT see EIV000
N,N′-1,2-ETHANEDIYLBIS(N-CARBOXYMETHYL)GLY-
CINE, TRISODIUM SALT see TNL250
1,2-ETHANEDIYLBISMANEB, MANGANESE (2+) SALT
(1:1) see MAS500
2,2′-(1,2-ETHANEDIYLBIS(OXY))BISETHANOL see TJQ000
1,2-ETHANEDIYL ESTER CARBAMIMIDOTHIOIC ACID
DIHYDROBROMIDE see EJA000
ETHANE HEXACHLORIDE see HCI000
ETHANEHYDRAZONIC ACID see ACM750
ETHANENITRILE see ABE500
ETHANE PENTACHLORIDE see PAW500
ETHANEPEROXOIC ACID see PCL500
ETHANEPEROXOIC ACID-1,1-DIMETHYLETHYL ESTER
see BSC250
ETHANETHIOAMIDE see TFA000
ETHANETHIOIC ACID see TFA500
ETHANETHIOL see EMB100
ETHANETHIOLIC ACID see TFA500
ETHANE TRICHLORIDE see TIN000
ETHANOIC ACID see AAT250
ETHANOIC ANHYDRATE see AAX500
ETHANOL, solution (DOT) see EFU000
ETHANOLAMINE see EEC600
β-ETHANOLAMINE see EEC600
ETHANOLAMINE, solution (DOT) see EEC600
ETHANOLAMINE SALT of 5,2′-DICHLORO-4′-NITROSALI-
CYCLICANILIDE see DFV600
ETHANOL-2-(2-BUTOXYETHOXY) THIOCYANATE see
BPL250
ETHANOL (MAK) see EFU000
ETHANOL, 1-PHENYL- see PDE000
ETHANOL 200 PROOF see EFU000
4-ETHANOLPYRIDINE see POR500
1-ETHANOL-2-THIOL see MCN250
ETHANONE-1-(3-ETHYL-5,6,7,8-TETRAHYDRO-5,5,8,8-TET-
RAMETHYL-2-NAPHTHALENYL)(9CI) see ACL750
ETHANOX see EEH600
ETHANOXYTRIPHETOL see DHS000
ETHANOYL CHLORIDE see ACF750
ETHAPERAZINE see CJM250
ETHAVAN see EQF000
ETHAZATE see BJC000
ETHCHLOROVYNOL see CHG000
ETHCHLORVINYL see CHG000
ETHCLORVYNOL see CHG000
ETHEFON see CDS125
ETHEL see CDS125
ETHENE see EIO000
ETHENE OXIDE see EJN500
ETHENE POLYMER see PJS750
ETHENOL HOMOPOLYMER (9CI) see PKP750

ETHENONE see KEU000
ETHENYL ACETATE see VLU250
ETHENYLBENZENE see SMQ000
ETHENYLBENZENE HOMOPOLYMER see SMQ500
ETHENYLBENZENE POLYMER with 1,3-BUTADIENE see
SMR000
4-ETHENYL-1-CYCLOHEXENE see CPD750
S-ETHENYL-dl-HOMOCYSTEINE see VLU200
1-(ETHENYLOXY) BUTANE see VMZ000
ETHENYLOXYETHENE see VOP000
1-ETHENYL PYRENE see EEF000
1-ETHENYL-2-PYRROLIDINONE see EEG000
1-ETHENYL-2-PYRROLIDINONE HOMOPOLYMER see
PKQ250
1-ETHENYL-2-PYRROLIDINONE POLYMERS see PKQ250
ETHEPHON see CDS125
ETHER see EJU000
ETHER BUTYLIQUE (FRENCH) see BRH750
ETHER CHLORATUS see EHH000
ETHER CYANATUS see PMV750
ETHER DICHLORE (FRENCH) see DFJ050
ETHER ETHYLBUTYLIQUE (FRENCH) see EHA500
ETHER ETHYLIQUE (FRENCH) see EJU000
ETHER HYDROCHLORIC see EHH000
ETHER ISOPROPYLIQUE (FRENCH) see IOZ750
ETHER METHYLIQUE MONOCHLORE (FRENCH) see
CIO250
ETHER MONOETHYLIQUE de l'ETHYLENE-GLYCOL
(FRENCH) see EES350
ETHER MONOMETHYLIQUE de l'ETHYLENE-GLYCOL
(FRENCH) see EJH500
ETHER MURIATIC see EHH000
ETHERON SPONGE see PKL500
ETHER, PENTACHLOROPHENYL see PAW250
ETHERS see EEG500
ETHERSULFONATE see CJT750
ETHEVERSE see CDS125
ETHICON PTFE see TAI250
ETHIDE see DFU000
ETHIDIUM BROMIDE see DBV400
ETHIMIDE see EPQ000
ETHINA see EPQ000
ETHINAMIDE see EPQ000
ETHINE see ACI750
ETHINODIOL DIACETATE see EQJ500
17-α-ETHINYL-3,17-DIHYDROXY-Δ^{1,3,5}-ESTRATRIENE see
EEH500
17-α-ETHINYL-3,17-DIHYDROXY-Δ^{1,3,5}-OESTRATRIENE see
EEH500
ETHINYL ESTRADIOL see EEH500
17-ETHINYLESTRADIOL see EEH500
17-α-ETHINYLESTRADIOL see EEH500
17-ETHINYL-3,17-ESTRADIOL see EEH500
17-α-ETHINYL-17-β-ESTRADIOL see EEH500
ETHINYL ESTRADIOL and DIMETHISTERONE see
DNX500
ETHINYLESTRADIOL-3-METHYL ETHER see MKB750
17-α-ETHINYL ESTRADIOL 3-METHYL ETHER see
MKB750
ETHINYLESTRADIOL-3-METHYL ETHER and NORETHY-
NODRED (1:50) see EAP000
ETHINYLESTRADIOL and NORETHINDRONE ACETATE see
EEH520
17-ETHINYL-5(10)-ESTRAENEOLONE see EEH550
17-α-ETHINYL-ESTRA(5,10)ENEOLONE see EEH550
17-α-ETHINYLESTRA-4-EN-17-β-OL-3-ONE see NNP500
17-α-ETHINYLESTRA-1,3,5(10)-TRIENE-3,17-β-DIOL see
EEH500
17-α-ETHINYL-5,10-ESTRENOLONE see EEH550
ETHINYLESTRIOL see EEH500
17-α-ETHINYL-17-β-HYDROXY-Δ:4-ESTREN-3-ONE see
NNP500
17-α-ETHINYL-17-β-HYDROXY-Δ$^{5(10)}$-ESTREN-3-ONE see
EEH550

17-α-ETHINYL-19-NORTESTOSTERONE see NNP500
17-α-ETHINYL-Δ^{5,10-19}-NORTESTOSTERONE see EEH550
17-α-ETHINYL-19-NORTESTOSTERONE ACETATE see
  ABU000
17-α-ETHINYL-19-NORTESTOSTERONE-17-β-ACETATE see
  ABU000
ETHINYLOESTRADIOL see EEH500
17-ETHINYL-3,17-OESTRADIOL see EEH500
ETHINYLOESTRADIOL-3-METHYL ETHER see MKB750
17-α-ETHINYL OESTRADIOL-3-METHYL ETHER see
  MKB750
ETHINYL OESTRADIOL mixed with NORETHISTERONE
  ACETATE see EEH520
ETHINYL-OESTRANOL see EEH500
17-α-ETHINYLOESTRA-1,3,5(10)-TRIENE-3,17-β-DIOL see
  EEH500
17-α-ETHINYL-Δ(SUP 1,3,5(10))OESTRATRIENE-3,17-β-
  DIOL see EEH500
ETHINYLOESTRIOL see EEH500
ETHINYL TRICHLORIDE see TIO750
ETHIOL see EEH600
ETHIOLACAR see MAK700
ETHION see EEH600
ETHIONIAMIDE see EPQ000
ETHIONIN see EEI000
ETHIONINE see EEI000
(±)-ETHIONINE see EEI000
dl-ETHIONINE see EEI000
ETHLON see PAK000
ETHOATE METHYL see DNX600
ETHOBROM see THV000
ETHOCAINE see AIT250
ETHOCHLORVYNOL see CHG000
ETHODAN see EEH600
ETHODRYL CITRATE see DIW200
ETHOHEXADIOL see EKV000
ETHOL see HCP000
ETHOMEEN C/15 see EEJ000
ETHONE see ENY500
ETHOPROMAZINE see DIR000
ETHOPROP see EIN000
ETHOPROPHOS see EIN000
ETHOSUCCIMIDE see ENG500
ETHOSUCCINIMIDE see ENG500
ETHOSUXIDE see ENG500
ETHOSUXIMIDE see ENG500
ETHOVAN see EQF000
4-ETHOXYACETANILIDE see ABG750
p-ETHOXYACETANILIDE see ABG750
ETHOXY ACETATE see EES400
ETHOXY ACETYLENE see EEL000
4-ETHOXYANILINE see PDD500
p-ETHOXYANILINE see PDD500
6-ETHOXY-m-ANOL see IRY000
N-(ETHOXYCARBONYL)AZIRIDINE see ASH750
S-α-ETHOXYCARBONYLBENZYL-O,O-DIMETHYL PHOS-
  PHORODITHIOATE see DRR400
S-α-ETHOXYCARBONYLBENZYL DIMETHYL PHOSPHO-
  ROTHIOLOTHIONATE see DRR400
ETHOXYCARBONYLDIAZOMETHANE see DCN800
ETHOXY CARBONYL DIGOXIN see EEP000
ETHOXYCARBONYLETHYLENE see EFT000
N-ETHOXYCARBONYLETHYLENEIMINE see ASH750
ETHOXYCARBONYL-1-ETHYLENIMINE see ASH750
ETHOXYCARBONYLMETHYL BROMIDE see EGV000
S-(N-ETHOXYCARBONYL-N-METHYLCARBAMOYL-
  METHYL)-DIETHYL PHOSPHORODITHIOATE see
  DJI000
S-((ETHOXYCARBONYL)METHYLCARBAMOYL)METHYL-
  O,O-DIETHYL PHOSPHORODITHIOATE see DJI000
N-ETHOXYCARBONYL-N-METHYLCARBAMOYLMETHYL-
  O,O-DIETHYL PHOSPHORODITHIOATE see DJI000
ETHOXY CHLOROMETHANE see CIM000

ETHOXY DIGLYCOL see CBR000
2-ETHOXY DIHYDROPYRAN see EER500
2-ETHOXY-2,3-DIHYDRO-γ-PYRAN see EER500
2-ETHOXY-3,4-DIHYDRO-1,2-PYRAN see EER500
2-ETHOXY-3,4-DIHYDRO-2H-PYRAN see EER500
6-ETHOXY-1,2-DIHYDRO-2,2,4-TRIMETHYLQUINOLINE
  see SAV000
ETHOXYETHANE see EJU000
2-ETHOXYETHANOL see EES350
2-ETHOXYETHANOL ACETATE see EES400
2-ETHOXYETHANOL, ESTER with ACETIC ACID see
  EES400
ETHOXY ETHENE see EQF500
2-(2-ETHOXYETHOXY)ETHANOL see CBR000
2-(2-ETHOXYETHOXY)ETHANOL ACETATE see CBQ750
2-(2-(2-ETHOXYETHOXY)ETHOXY)ETHANOL see EFL000
2-ETHOXY-ETHYLACETAAT (DUTCH) see EES400
ETHOXYETHYL ACETATE see EES400
2-ETHOXYETHYL ACETATE see EES400
2-ETHOXYETHYLACETATE see EES400
β-ETHOXYETHYL ACETATE see EES400
2-ETHOXYETHYLE, ACETATE de (FRENCH) see EES400
ETHOXYETHYNE see EEL000
ETHOXYFORMIC ANHYDRIDE see DIX200
3-ETHOXY-4-HYDROXYBENZALDEHYDE see EQF000
1-ETHOXY-2-HYDROXY-4-PROPENYLBENZENE see
  IRY000
2-(3-ETHOXY-1-INDANYLIDENE)-1,3-DINDANDIONE see
  BGC250
ETHOXYLATED SORBITAN MONOOLEATE see PKL100
ETHOXYMETHANE see EMT000
ETHOXY METHYL CHLORIDE see CIM000
2-((ETHOXY((1-METHYLETHYL)AMINO)PHOSPHINO-
  THIOYL)OXY)BENZOIC ACID 1-METHYLETHYL ESTER
  see IMF300
N-ETHOXYMORPHOLINO DIAZENIUM FLUOROBORATE
  see EEX500
ETHOXY-4-NITROPHENOXYPHENYLPHOSPHINE SULFIDE
  see EBD700
N-(4-ETHOXY-3-NITRO)PHENYLACETAMIDE see NEL000
N-(4-ETHOXYPHENYL)ACETAMIDE see ABG750
N-p-ETHOXYPHENYLACETAMIDE see ABG750
2-(4-ETHOXYPHENYL)-1,3-BIS(4-METHOXYPHENYL)
  GUANIDINE HYDROCHLORIDE see PDN500
4-ETHOXY-7-PHENYL-3,5-DIOXA-6-AZA-4-PHOSPHAOCT-6-
  ENE-8-NITRILE 4 SULFIDE see BAT750
N-(4-ETHOXYPHENYL)-3'-NITROACETAMIDE see NEL000
4-ETHOXYPHENYLUREA see EFE000
p-ETHOXYPHENYLUREA see EFE000
N-(4-ETHOXYPHENYL)UREA see EFE000
4-ETHOXY-β-(1-PIPERIDYL)PROPIOPHENONE HYDRO-
  CHLORIDE see EFE500
1-ETHOXYPROPANE see EPC125
ETHOXY PROPIONALDEHYDE see EFG500
ETHOXYPROPIONIC ACID see EFH000
ETHOXYQUINE see SAV000
ETHOXYQUIN (FCC) see SAV000
ETHOXYTRIETHYLENE GLYCOL see EFL000
ETHOXYTRIGLYCOL see EFL000
6-ETHOXY-2,2,4-TRIMETHYL-1,2-DIHYDROQUINOLINE
  see SAV000
ETHRANE see EAT900
ETHREL see CDS125
ETHYLACETAAT (DUTCH) see EFR000
ETHYLACETANILIDE see EFQ500
N-ETHYLACETANILIDE see EFQ500
ETHYL ACETATE see EFR000
ETHYLACETIC ACID see BSW000
ETHYL ACETIC ESTER see EFR000
ETHYL ACETOACETATE (FCC) see EFS000
ETHYL ACETONE see PBN250
ETHYL ACETOXYMETHYLNITROSAMINE see ENR500
N-ETHYL-N-(ACETOXYMETHYL)NITROSAMINE see
  ENR500

ETHYL ACETYL ACETATE see EFS000
ETHYL ACETYLACETONATE see EFS000
ETHYL ACETYLENE see EFS500
ETHYL ACETYLENE, INHIBITED (DOT) see EFS500
ETHYLACRYLAAT (DUTCH) see EFT000
ETHYL ACRYLATE see EFT000
ETHYLAKRYLAT (CZECH) see EFT000
ETHYLAL see EFT500
ETHYL ALCOHOL see EFU000
ETHYLALCOHOL (DUTCH) see EFU000
ETHYL ALCOHOL, anhydrous see EFU000
ETHYL ALDEHYDE see AAG250
ETHYLAMINE see EFU400
ETHYLAMINE, solution, in water, concentrations up to 70%
    (DOT) see EDY500
ETHYLAMINE-2-(DIPHENYLMETHOXY)-N,N-DIMETHYL,
    compound with 8-CHLOROTHEOPHYLLINE (1:1) see
    DYE600
N-ETHYLAMINOBENZENE see EGK000
ETHYL AMINOBENZOATE see EFX000
ETHYL-4-AMINOBENZOATE see EFX000
ETHYL-o-AMINOBENZOATE see EGM000
ETHYL-p-AMINOBENZOATE see EFX000
2-ETHYLAMINOETHANOL see EGA500
2-(ETHYLAMINO)ETHANOL see EGA500
S-(2-(ETHYLAMINO-2-OXOETHYL)-O,O-DIMETHYL PHOS-
    PHORODITHIOATE see DNX600
ETHYL-1-(p-AMINOPHENETHYL)-4-PHENYLISONIPECO-
    TATE see ALW750
2-ETHYLAMINOTHIADIAZOLE see EGI000
2-ETHYLAMINO-1,3,4-THIADIAZOLE see EGI000
ETHYLAMYLCARBINOL see OCY100
ETHYL-n-AMYLCARBINOL see OCY100
ETHYL AMYL KETONE see EGI750, EGI755
ETHYLAN see DJC000
ETHYLANILINE see EGK000
2-ETHYLANILINE see EGK500
4-ETHYLANILINE see EGL000
N-ETHYLANILINE see EGK000
o-ETHYLANILINE see EGK500
p-ETHYLANILINE see EGL000
ETHYL ANISATE see AOV000
ETHYL-p-ANISATE (FCC) see AOV000
ETHYLAN MLD see BKE500
ETHYL ANTHRANILATE see EGM000
ETHYLARSONOUS DICHLORIDE see DFH200
ETHYL AZIRIDINECARBOXYLATE see ASH750
ETHYL-1-AZIRIDINECARBOXYLATE see ASH750
ETHYL AZIRIDINOCARBOXYLATE see ASH750
ETHYL-1-AZIRIDINYLCARBOXYLATE see ASH750
ETHYL AZIRIDINYLFORMATE see ASH750
ETHYLBENZEEN (DUTCH) see EGP500
2-ETHYLBENZENAMINE see EGK500
N-ETHYLBENZENAMINE see EGK000
N-ETHYLBENZENAMINO see EGK000
ETHYL BENZENE see EGP500
ETHYL BENZENEACETATE see EOH000
ETHYL BENZOATE see EGR000
ETHYLBENZOL see EGP500
ETHYL BENZYL ACETOACETATE see EFS000
ETHYLBENZYLBARBITURIC ACID see BEA500
ETHYL (BIS(1-AZIRIDINYL)PHOSPHINYL)CARBAMATE
    see EHV500
ETHYLBIS(2-CHLOROETHYL)AMINE see BID250
ETHYLBIS(β-CHLOROETHYL)AMINE see BID250
ETHYL BISCOUMACETATE see BKA000
ETHYL BIS(4-HYDROXYCOUMARINYL)ACETATE see
    BKA000
ETHYL BIS(4-HYDROXY-3-COUMARINYL)ACETATE see
    BKA000
ETHYL BORATE (DOT) see BMC250
ETHYL BROMACETATE see EGV000
ETHYL BROMIDE see EGV400
ETHYL BROMOACETATE see EGV000

ETHYL-α-BROMOACETATE see EGV000
N-ETHYL-N-o-BROMOBENZYL-N,N-DIMETHYLAMMO-
    NIUM TOSYLATE see BMV750
ETHYL BROMOPHOS see EGV500
2-ETHYLBUTANAL see DHI000
2-ETHYL-1-BUTANAMINE see EHA000
ETHYL BUTANOATE see EHE000
2-ETHYL BUTANOIC ACID see DHI400
2-ETHYLBUTANOL see EGW000
2-ETHYLBUTANOL-1 see EGW000
2-ETHYL-1-BUTANOL see EGW000
2-ETHYL-1-BUTENE see EGW500
2-ETHYLBUTRIC ALDEHYDE see DHI000
ETHYLBUTYLACETALDEHYDE see BRI000
ETHYL BUTYLACETATE (DOT) see EHF000
2-ETHYLBUTYLACRYLATE see EGZ000
2-ETHYLBUTYL ALCOHOL see EGW000
2-ETHYLBUTYLAMINE see EHA000
5-ETHYL-5-N-BUTYLBARBITURIC ACID see BPF500
ETHYLBUTYLCETONE (FRENCH) see EHA600
2-ETHYLBUTYL ESTER, ACRYLIC ACID see EGZ000
ETHYL BUTYL ETHER see EHA500
ETHYLBUTYLKETON (DUTCH) see EHA600
ETHYL BUTYL KETONE see EHA600
ETHYL-N-BUTYLNITROSAMINE see EHC000
ETHYL BUTYRALDEHYDE see DHI000
α-ETHYLBUTYRALDEHYDE see DHI000
ETHYL BUTYRALDEHYDE (DOT) see DHI000
2-ETHYLBUTYRALDEHYDE (DOT,FCC) see DHI000
ETHYL n-BUTYRATE see EHE000
ETHYL BUTYRATE (DOT,FCC) see EHE000
α-ETHYLBUTYRIC ACID see DHI400
2-ETHYLBUTYRIC ACID (FCC) see DHI400
ETHYL CADMATE see BJB500
ETHYL CAPRATE see EHE500
ETHYL CAPRINATE see EHE500
α-ETHYLCAPROALDEHYDE see BRI000
ETHYL CAPROATE see EHF000
α-ETHYLCAPROIC ACID see BRI250
ETHYL CAPRYLATE see ENY000
ETHYL CARBAMATE see UVA000
2-(N-ETHYL CARBAMOYL HYDROXYMETHYL)FURAN see
    EHF500
S-(N-ETHYLCARBAMOYLMETHYL) DIMETHYL PHOSPHO-
    RODITHIOATE see DNX600
ETHYL CARBINOL see PND000
ETHYL CARBITOL see CBR000
ETHYL CARBONATE see DIX200
ETHYL CELLOSOLVE see EES350
ETHYL CELLOSOLVE ACETAAT (DUTCH) see EES400
ETHYLCHLOORFORMIAAT (DUTCH) see EHK500
ETHYL CHLORACETATE see EHG500
ETHYL CHLORIDE see EHH000
ETHYL CHLOROACETATE see EHG500
ETHYL-α-CHLOROACETATE see EHG500
ETHYL CHLORO BENZENE see EHH500
ETHYL CHLOROCARBONATE (DOT) see EHK500
ETHYL 4-CHLORO-α-(4-CHLOROPHENYL)-α-HYDROXY-
    BENZENEACETATE see DER000
ETHYL CHLOROETHANOATE see EHG500
9-(3-ETHYL-2-CHLOROETHYL)AMINOPROPYLAMINO)-4-
    METHOXYACRIDINE DIHYDROCHLORIDE see EHJ500
ETHYL-N-(β-CHLOROETHYL)-N-NITROSOCARBAMATE
    see CHF500
ETHYL-2-CHLOROETHYL SULFIDE see CGY750
ETHYL-β-CHLOROETHYL SULFIDE see CGY750
ETHYL CHLOROFORMATE see EHK500
ETHYL CHLOROTHIOFORMATE (DOT) see CLJ750
ETHYL-β-CHLOROVINYLETHYNYL CARBINOL see
    CHG000
ETHYLCHLORVYNOL see CHG000
ETHYL-trans-CINNAMATE see EHN000
ETHYL CINNAMATE (FCC) see EHN000
ETHYL CITRATE see TJP750

ETHYLCROTONATE see COB750
(α-ETHYL-cis-CROTONYL)CARBAMIDE see EHP000
2-ETHYLCROTONYLUREA see EHP000
2-ETHYL-cis-CROTONYLUREA see EHP000
cis-(2-ETHYLCROTONYL) UREA see EHP000
ETHYL CYANIDE see PMV750
ETHYL CYANOACETATE see EHP500
ETHYL CYANOETHANOATE see EHP500
5-ETHYL-5-CYCLOHEXENYLBARBITURIC ACID see
    TDA500
N-ETHYL(CYCLOHEXYL)AMINE see EHT000
N-ETHYL-CYCLOHEXYLAMINE see EHT000
ETHYL CYMATE see BJC000
p,p-ETHYL DDD see DJC000
p,p'-ETHYL-DDD see DJC000
ETHYL DECABORANE see EHV000
ETHYL DECANOATE (FCC) see EHE500
ETHYL DECYLATE see EHE500
ETHYL(DI-(1-AZIRIDINYL)PHOSPHINYL)CARBAMATE see
    EHV500
ETHYL DIAZOACETATE see DCN800
ETHYL-4,4'-DICHLOROBENZILATE see DER000
ETHYL-ρ,ρ'-DICHLOROBENZILATE see DER000
ETHYL-4,4'-DICHLORODIPHENYL GLYCOLLATE see
    DER000
N-ETHYL-DICHLOROMALEINIMIDE see DFJ400
ETHYL-4,4'-DICHLOROPHENYL GLYCOLLATE see
    DER000
O-ETHYL-O-2,4-DICHLOROPHENYL THIONOBENZENE-
    PHOSPHONATE see SCC000
ETHYL DICHLOROSILANE (DOT) see DFK000
ETHYLDICOUMAROL see BKA000
ETHYLDICOUMAROL ACETATE see BKA000
ETHYL DIETHYLENE GLYCOL see CBR000
5-ETHYLDIHYDRO-5-(1-METHYLBUTYL)-2-THIOXO-4,-
    6(1H,5H)-PYRIMIDINEDIONE, MONOSODIUM SALT see
    PBT500
(3S-cis)-3-ETHYLDIHYDRO-4-((1-METHYL-1H-IMIDAZOL-5-
    YL)METHYL)-2(3H)-FURANONE see PIF000
1-ETHYL-1,4-DIHYDRO-7-METHYL-4-OXO-1,8-NAPHTHY-
    RIDINE-3-CARBOXYLIC ACID see EID000
5-ETHYLDIHYDRO-5-PHENYL-4,6(1H,5H)-PYRIMIDINE-
    DIONE see DBB200
ETHYL-4,4'-DIHYDROXYDICOUMARINYL-3,3'-ACETATE
    see BKA000
ETHYL-α-((DIMETHOXYPHOSPHENOTHIOYL)THIO)
    BENZENEACETATE see DRR400
ETHYL DIMETHYLAMIDOCYANOPHOSPHATE see EIF000
ETHYL N,N-DIMETHYLAMINO CYANOPHOSPHATE see
    EIF000
O-ETHYL-S-(2-DIMETHYL AMINO ETHYL)-METHYLPHOS-
    PHONOTHIOATE see EIF500
N-ETHYL-N-1-DIMETHYL-3,3-DI-2-THIENYLALLYLAMINE
    HYDROCHLORIDE see EIJ000
N-ETHYL-N-1-DIMETHYL-3,3-DI-2-THIENYL-2-PROPEN-
    AMINE HYDROCHLORIDE see EIJ000
ETHYLDIMETHYLMETHANE see EIK000
ETHYL DIMETHYLPHOSPHORAMIDOCYANIDATE see
    EIF000
ETHYL-N,N-DIMETHYLPHOSPHORAMIDOCYANIDATE see
    EIF000
ETHYL-O,O-DIMETHYL PHOSPHORODITHIOYLPHENYL
    ACETATE see DRR400
2-ETHYL-3,5(6)-DIMETHYLPYRAZINE see EIL100
ETHYLDIOL ACRILATE (RUSSIAN) see EIP000
O-ETHYL-S,S-DIPHENYL DITHIOPHOSPHATE see EIM000
O-ETHYL-S,S-DIPHENYL PHOSPHORODITHIOATE see
    EIM000
O-ETHYL-S,S-DIPROPYL ESTER, PHOSPHORODITHIOIC
    ACID see EIN000
O-ETHYL-S,S-DIPROPYLPHOSPHORODITHIOATE see
    EIN000
S-ETHYL-N,N-DIPROPYLTHIOCARBAMATE see EIN500

S-ETHYL-N,N-DI-N-PROPYLTHIOCARBAMATE see EIN500
ETHYL DI-N-PROPYLTHIOLCARBAMATE see EIN500
ETHYL-N,N-DIPROPYLTHIOLCARBAMATE see EIN500
ETHYL-N,N-DI-N-PROPYLTHIOLCARBAMATE see EIN500
ETHYLDITHIOURAME see DXH250
ETHYLDITHIURAME see DXH250
ETHYL DODECANOATE see ELY700
ETHYLE (ACETATE d') (FRENCH) see EFR000
ETHYLE, CHLOROFORMIAT D' (FRENCH) see EHK500
ETHYLEEN-CHLOORHYDRINE (DUTCH) see EIU800
ETHYLEENDIAMINE (DUTCH) see EEA500
ETHYLEENDICHLORIDE (DUTCH) see EIY600
ETHYLEENIMINE (DUTCH) see EJM900
ETHYLEENOXIDE (DUTCH) see EJN500
ETHYLE (FORMIATE d') (FRENCH) see EKL000
ETHYLENE see EIO000
ETHYLENE, compressed (DOT) see EIO000
ETHYLENE, refrigerated liquid (DOT) see EIO000
ETHYLENE ACETATE see EJD759
ETHYLENE ACRYLATE see EIP000
ETHYLENE ALCOHOL see EJC500
ETHYLENE ALDEHYDE see ADR000
1,1'-ETHYLENE-2,2'-BIPYRIDYLIUM DIBROMIDE see
    DWX800
ETHYLENE BIS(BROMOACETATE) see BHD250
N,N'-ETHYLENEBIS(N-BUTYL-4-MORPHOLINECARBOX-
    AMIDE) see DUO400
ETHYLENEBIS(DITHIOCARBAMATE) DISODIUM SALT see
    DXD200
ETHYLENEBISDITHIOCARBAMATE MANGANESE see
    MAS500
N,N'-ETHYLENE BIS(DITHIOCARBAMATE MANGANEUX)
    (FRENCH) see MAS500
N,N'-ETHYLENE BIS(DITHIOCARBAMATE de SODIUM)
    (FRENCH) see DXD200
ETHYLENEBIS(DITHIOCARBAMATO) MANGANESE see
    MAS500
ETHYLENEBIS(DITHIOCARBAMATO)MANGANESE and
    ZINC ACETATE (50:1) see EIQ500
ETHYLENE BIS(DITHIOCARBAMATO)ZINC see EIR000
ETHYLENEBIS(DITHIOCARBAMIC ACID) DISODIUM SALT
    see DXD200
ETHYLENEBIS(DITHIOCARBAMIC ACID) MANGANESE
    SALT see MAS500
ETHYLENEBIS(DITHIOCARBAMIC ACID) MANGANOUS
    SALT see MAS500
ETHYLENEBIS(DITHIOCARBAMIC ACID), ZINC SALT see
    EIR000
N,N'-ETHYLENE BIS(3-FLUOROSALICYLIDENEI-
    MINATO)COBALT(II) see EIS000
ETHYLENEBIS(IMINODIACETIC ACID) DISODIUM SALT
    see EIX500
ETHYLENEBIS(IMINODIACETIC ACID) TETRASODIUM
    SALT see EIV000
ETHYLENEBISISOTHIOCYANATE see ISK000
2,2'-(ETHYLENEBIS(NITROSOIMINO))BISBUTANOL see
    HMQ500
(ETHYLENEBIS(OXYETHYLENENITRILO))TETRAACETIC
    ACID see EIT000
2,2'-ETHYLENE-BIS-(2-THIOPSEUDOUREA), DIHYDRO-
    BROMIDE see EJA000
ETHYLENE-BIS-THIURAMMONO-SULFIDE see ISK000
ETHYLENE BROMIDE see EIY500
ETHYLENE BROMOACETATE see BHD250
ETHYLENEBROMOHYDRIN see BNI500
ETHYLENECARBOXAMIDE see ADS250
ETHYLENECARBOXYLIC ACID see ADS750
ETHYLENE CHLORIDE see EIY600
ETHYLENE CHLOROHYDRIN see EIU800
ETHYLENE CYANIDE see SNE000
ETHYLENE CYANOHYDRIN see HGP000
ETHYLENE DIACRYLATE see EIP000
ETHYLENEDIAMINE (OSHA) see EEA500

ETHYLENE THIOUREA see IAQ000
1,3-ETHYLENE-2-THIOUREA see IAQ000
N,N'-ETHYLENETHIOUREA see IAQ000
ETHYLENETHIOUREA mixed with SODIUM NITRITE see
IAR000
l'ETHYLENE THIOUREE (FRENCH) see IAQ000
ETHYLENE TRICHLORIDE see TIO750
ETHYLENGLYKOLDINITRAT (CZECH) see EJG000
ETHYLENIMINE see EJM900
ETHYL-α,β-EPOXYHYDROCINNAMATE see EOK600
ETHYL α,β-EPOXY-β-METHYLHYDROCINNAMATE see
ENC000
ETHYL 2,3-EPOXY-3-METHYL-3-PHENYLPROPIONATE see
ENC000
ETHYL-α,β-EPOXY-α-PHENYLPROPIONATE see EOK600
ETHYL ESTER of N-ACETYL-dl-SARCOLYSYL-l-PHENYL-
ALANINE see ACD250
ETHYL ESTER of N-ACETYL-dl-SARCOSYLYL-dl-VALINE
see ARM000
ETHYL ESTER of 4,4'-DICHLOROBENZILIC ACID see
DER000
ETHYL ESTER of O,O-DIMETHYLDITHIOPHOSPHORYL α-
PHENYL ACETATE ACID see DRR400
ETHYL ESTER of 2,3-EPOXY-3-PHENYLBUTANOIC ACID
see ENC000
ETHYLESTER KYSELINY ORTHOMRAVENCI (CZECH) see
ENY500
ETHYL ESTER of METHANESULFONIC ACID see EMF500
ETHYL ESTER of METHYLNITROSO-CARBAMIC ACID see
MMX250
ETHYL ESTER of METHYLSULFONIC ACID see EMF500
ETHYL ESTER of METHYLSULPHONIC ACID see EMF500
N-ETHYL-ETHANAMINE see DHJ200
ETHYL ETHANOATE see EFR000
ETHYL ETHER see EJU000
3-ETHYL-2-(5-(3-ETHYL-2-BENZOTHIAZOLINYLIDENE)-
1,3-PENTADIENYL)BENZOTHIAZOLIUM IODIDE see
DJT800
5-ETHYL-5-(1-ETHYLPROPYL)BARBITURIC ACID see
EJY000
5-ETHYL-5-(1-ETHYLPROPYL)2,4,6(1H,3H,5H)-PYRIMIDI-
NETRIONE see EJY000
O,O-ETHYL-S-2(ETHYLTHIO)ETHYL PHOSPHORODI-
THIOATE see DXH325
ETHYLETHYNE see EFS500
13-ETHYL-17-α-ETHYNYLGON-4-EN-17-β-OL-3-ONE see
NNQ500
13-ETHYL-17-α-ETHYNYL-17-β-HYDROXY-4-GONEN-3-
ONE see NNQ500
(±)-13-ETHYL-17-α-ETHYNYL-17-HYDROXYGON-4-EN-3-
ONE see NNQ500
dl-13-β-ETHYL-17-α-ETHYNYL-17-β-HYDROXYGON-4-EN-
3-ONE see NNQ500
dl-13-β-ETHYL-17-α-ETHYNYL-19-NORTESTOSTERONE
see NNQ500
ETHYL FLUORIDE (DOT) see FIB000
ETHYL-10-FLUORODECANOATE see EKI000
ETHYL-φ-FLUORODECANOATE see EKI000
ETHYL-9-FLUORONONANECARBOXYLATE see EKI000
ETHYL-8-FLUORO OCTANOATE see EKK500
ETHYL-φ-FLUOROOCTANOATE see EKK500
ETHYL FORMATE see EKL000
ETHYLFORMIAAT (DUTCH) see EKL000
ETHYLFORMIC ACID see PMU750
ETHYL FORMIC ESTER see EKL000
ETHYLGLYKOLACETAT (GERMAN) see EES400
ETHYL GLYME see EJE500
ETHYL GREEN see BAY750
ETHYL GUSATHION see EKN000
ETHYL GUTHION see EKN000
ETHYL HEPTANOATE see EKN050
ETHYL HEPTOATE see EKN050
ETHYLHEXABITAL see TDA500
5-ETHYLHEXAHYDRO-4,6-DIOXO-5-PHENYLPHRIMIDINE
see DBB200

5-ETHYLHEXAHYDRO-5-PHENYLPYRIMIDINE-4,6-DIONE
see DBB200
2-ETHYLHEXALDEHYDE see BRI000
ETHYLHEXALDEHYDE (DOT) see BRI000
2-ETHYLHEXANAL see BRI000
ETHYL HEXANEDIOL see EKV000
2-ETHYL-1,3-HEXANEDIOL see EKV000
2-ETHYLHEXANE-1,3-DIOL see EKV000
2-ETHYLHEXANEDIOL-1,3 see EKV000
ETHYL HEXANOATE (FCC) see EHF000
2-ETHYLHEXANOIC ACID see BRI250
2-ETHYL-1-HEXANOL ESTER with DIPHENYL PHOSPHATE
see DWB800
2-ETHYL-1-HEXANOL HYDROGEN PHOSPHATE see
BJR750
2-ETHYL-1-HEXANOL HYDROGEN SULFATE, SODIUM
SALT see TAV750
2-ETHYL-1-HEXANOL SULFATE SODIUM SALT see
TAV750
2-ETHYL-1-HEXENE see EKR500
2-ETHYL HEXENE-1 see EKR500
2-ETHYLHEXOIC ACID see BRI250
2-ETHYL HEXYLAMINE see EKS500
N-(2-ETHYLHEXYL)BICYCLO-(2,2,1)-HEPT-5-ENE-2,3-DI-
CARBOXIMIDE see OES000
2-ETHYLHEXYL DIPHENYL ESTER PHOSPHORIC ACID see
DWB800
2-ETHYLHEXYL DIPHENYLPHOSPHATE see DWB800
ETHYL HEXYLENE GLYCOL see EKV000
2-ETHYLHEXYL FUMARATE see DVK600
N-2-ETHYLHEXYLIMIDEENDOMETHYLENETETRAHY-
DROPHTHALIC ACID see OES000
N-(2-ETHYLHEXYL)-5-NORBORNENE-2,3-DICARBOXIMIDE
see OES000
ETHYLHEXYL PHTHALATE see DVL700
2-ETHYLHEXYL PHTHALATE see DVL700
2-ETHYLHEXYL SEBACATE see BJS250
2-ETHYLHEXYL SODIUM SULFATE see TAV750
2-ETHYLHEXYL SULFOSUCCINATE SODIUM see DJL000
2-(2-ETHYLHEXYL)-3a,4,7,7a-TETRAHYDRO-4,7-METH-
ANO-1H-ISOINDOLE-1,3(2H)-DIONE see OES000
S-ETHYL-HOMOCYSTEINE see EEI000
S-ETHYL-dl-HOMOCYSTEINE see EEI000
ETHYL HYDRATE see EFU000
ETHYL HYDRIDE see EDZ000
ETHYLHYDROCUPREINE HYDROCHLORIDE see ELC500
ETHYL HYDROGEN PEROXIDE see ELD000
ETHYL HYDROPEROXIDE see ELD000
ETHYL HYDROPERSULFIDE see EEB000
ETHYL HYDROSULFIDE see EMB100
ETHYL HYDROXIDE see EFU000
ETHYL-o-HYDROXYBENZOATE see SAL000
ETHYL-p-HYDROXYBENZOATE see HJL000
ETHYL-2-HYDROXY-2,2-BIS(4-CHLOROPHENYL)ACETATE
see DER000
(±)-13-ETHYL-17-HYDROXY-18,19-DINOR-17-α-PREGN-4-
EN-20-YN-3-ONE see NNQ500
ETHYL(2-HYDROXYETHYL)DIMETHYL-AMMONIUM SUL-
FATE (SALT), BIS(DIBUTYLCARBAMATE) see DDW000
ETHYL-2-HYDROXYETHYLNITROSAMINE see ELG500
N-ETHYL-N-HYDROXYETHYLNITROSAMINE see ELG500
(8-α,9R)-1-ETHYL-9-HYDROXY-6'-METHOXYCINCHONAN-
1-IUM IODIDE see QIS000
5-ETHYL-2'-HYDROXY-2(N)-(3-METHYL-2-BUTENYL)-9-
METHYL-6,7-BENZOMORPHAN see DOQ600
α-ETHYL-β-(HYDROXYMETHYL)-1-METHYL-IMIDAZOLE-
5-BUTYRIC ACID, γ-LACTONE see PIF000
1-ETHYL-3-HYDROXY-1-METHYL-PIPERIDINIUM BRO-
MIDE BENZILATE see PJA000
ETHYL-p-HYDROXYPHENYL KETONE see ELL500
ETHYL 2-HYDROXYPROPIONATE see LAJ000
ETHYL α-HYDROXYPROPIONATE see LAJ000
2-ETHYL-3-HYDROXY-4H-PYRAN-4-ONE see EMA600

ETHYLIC ACID see AAT250
5-ETHYLIDENEBICYCLO(2.2.1)HEPT-2-ENE see ELO500
ETHYLIDENE CHLORIDE see DFF809
ETHYLIDENE DICHLORIDE see DFF809
ETHYLIDENE DIETHYL ETHER see AAG000
ETHYLIDENE DIFLUORIDE see ELN500
ETHYLIDENE DIMETHYL ETHER see DOO600
ETHYLIDENE FLUORIDE see ELN500
ETHYLIDENE GYROMITRIN see AAH000
ETHYLIDENEHYDROXYLAMINE see AAH250
trans-15-ETHYLIDENE-12-β-HYDROXY-4,12-α,13-β-TRI-
    METHYL 8-OXO-4,8 SECOSENEC-1-ENINE see
    DMX200
ETHYLIDENELACTIC ACID see LAG000
ETHYLIDENE NORBORNENE see ELO500
5-ETHYLIDENE-2-NORBORNENE see ELO500
ETHYLIDICHLORARSINE see DFH200
ETHYLIDICHLOROARSINE (DOT) see DFH200
ETHYLIMINE see EJM900
5-ETHYL-5-ISOAMYLBARBITURIC ACID see AMX750
5-ETHYL-5-ISOAMYLMALONYL UREA see AMX750
ETHYL ISOBUTANOATE see ELS000
ETHYL ISOBUTYRATE see ELS000
ETHYLISOBUTYRATE (DOT) see ELS000
ETHYL ISOCYANATE see ELS500
ETHYL ISOCYANATE (DOT) see ELS500
ETHYL ISONICOTINATE see ELU000
2-ETHYLISONICOTINIC ACID THIOAMIDE see EPQ000
α-ETHYLISONICOTINIC ACID THIOAMIDE see EPQ000
2-ETHYLISONICOTINIC THIOAMIDE see EPQ000
α-ETHYLISONICOTINOYLTHIOAMIDE see EPQ000
ETHYLISOPENTYLBARBITURIC ACID see AMX750
5-ETHYL-5-ISOPENTYLBARBITURIC ACID see AMX750
5-ETHYL-5-ISOPENTYLBARBITURIC ACID SODIUM SALT
    see AON750
O-ETHYL-O-(2-ISOPROPOXY-CARBONYL)-PHENYL ISO-
    PROPYLPHOSPHORAMIDOTHIOATE see IMF300
ETHYLISOPROPYLBARBITURIC ACID see ELX000
5-ETHYL-5-ISOPROPYLBARBITURIC ACID see ELX000
ETHYLISOPROPYLNITROSOAMINE see ELX500
ETHYLISOTHIAMIDE see EPQ000
2-ETHYLISOTHIONICOTINAMIDE see EPQ000
α-ETHYLISOTHIONICOTINAMIDE see EPQ000
ETHYL ISOVALERATE (FCC) see ISY000
ETHYL LACTATE (DOT,FCC) see LAJ000
ETHYL LAURATE see ELY700
N-ETHYLMALEIMIDE see MAL250
ETHYL MALONATE see EMA500
ETHYL MALTOL see EMA600
ETHYLMERCAPTAAN (DUTCH) see EMB100
ETHYL MERCAPTAN see EMB100
β-ETHYLMERCAPTOETHYL DIMETHYL THIONOPHOS-
    PHATE see DAO800
ETHYL MERCAPTOPHENYLACETATE-O,O-DIMETHYL
    PHOSPHOROCITHIOATE see DRR400
ETHYLMERCURIC CHLORIDE see CHC500
ETHYLMERCURICHLORENDIMIDE see EME050
ETHYLMERCURIC PHOSPHATE see BJT250
N-(ETHYLMERCURI)-1,4,5,6,7,7-HEXACHLOROBICYCLO
    (2.2.1)HEPT-5-ENE- 2,3-DICARBOXIMIDE see EME050
N-ETHYLMERCURI-3,4,5,6,7,7-HEXACHLORO-3,6-ENDO-
    METHYLENE-1,2,3,6- TETRAHYDROPHTHALIMIDE see
    EME050
N-ETHYLMERCURI-N-PHENYL-p-TOLUENESULFONAMIDE
    see EME500
N-ETHYLMERCURI-1,2,3,6-TETRAHYDRO-3,6-ENDOMETH-
    ANO- 3,4,5,6,7,7-HEXACHLOROPHTHALIMIDE see
    EME050
o-(ETHYLMERCURITHIO)BENZOIC ACID SODIUM SALT
    see MDI000
ETHYLMERCURITHIOSALICYLIC ACID SODIUM SALT see
    MDI000
N-(ETHYLMERCURI)-p-TOLUENESULFONANILIDE see
    EME500

N-(ETHYLMERCURI)-p-TOLUENESULPHONANILIDE see
    EME500
ETHYLMERCURY CHLORIDE see CHC500
ETHYLMERCURY PHOSPHATE see BJT250
ETHYLMERCURY p-TOLUENESULFANILIDE see EME500
ETHYLMERCURY-p-TOLUENE SULFONAMIDE see
    EME500
ETHYLMERCURY-p-TOLUENESULFONANILIDE see
    EME500
ETHYLMERKAPTAN (CZECH) see EMB100
β-ETHYLMERKAPTOETHYLCHLORID (CZECH) see
    CGY750
ETHYL METHACRYLATE see EMF000
ETHYL METHACRYLATE, INHIBITED (DOT) see EMF000
ETHYL METHANESULFONATE see EMF500
ETHYL METHANESULPHONATE see EMF500
ETHYL METHANOATE see EKL000
ETHYL METHANSULFONATE see EMF500
ETHYL METHANSULPHONATE see EMF500
ETHYLMETHIAMBUTENE HYDROCHLORIDE see EIJ000
ETHYL-4-METHOXYBENZOATE see AOV000
ETHYL-p-METHOXYBENZOATE see AOV000
ETHYL-2-METHYLACRYLATE see EMF000
ETHYL-α-METHYL ACRYLATE see EMF000
4-ETHYLMETHYLAMINOAZOBENZENE see ENB000
p-ETHYLMETHYLAMINOAZOBENZENE see ENB000
N-ETHYL-N-METHYL-p-AMINOAZOBENZENE see ENB000
3-ETHYLMETHYLAMINO-1,1-DI(2′-THIENYL)BUT-1-ENE
    HYDROCHLORIDE see EIJ000
ETHYL-p-METHYL BENZENESULFONATE see EPW500
5-ETHYL-5-(1-METHYL-1-BUTENYL)BARBITURATE see
    EMO500
5-ETHYL-5-(1-METHYL-1-BUTENYL)BARBITURIC ACID
    see EMO500
5-ETHYL-5-(1-METHYL-1-BUTENYL)BARBITURIC ACID
    SODIUM SALT see VKP000
5-ETHYL-5-(1-METHYL-1-BUTENYL)-2,4,6(1H,3H,5H)-PY-
    RIMIDINETRIONE see EMO500
5-ETHYL-5-(1-METHYL-1-BUTENYL)-2,4,6(1H,3H,5H)-PY-
    RIMIDINETRIONE SODIUM SALT see VKP000
5-ETHYL-5-(1-METHYLBUTYL)BARBITURIC ACID see
    NBT500
5-ETHYL-5-(3-METHYLBUTYL)BARBITURIC ACID see
    AMX750
5-ETHYL-5-(3-METHYLBUTYL)BARBITURIC ACID SODIUM
    DERIVATIVE see AON750
5-ETHYL-5-(1-METHYLBUTYL)BARBITURIC ACID SODIUM
    SALT see NBU000
5-ETHYL-5-(1-METHYLBUTYL)MALONYLUREA see
    NBT500
5-ETHYL-5-(1-METHYLBUTYL)-2,4,6(1H,3H,5H)-PYRIMI-
    DINETRIONE (9CI) see NBT500
5-ETHYL-5-(1-METHYLBUTYL)-2,4,6(1H,3H,5H)-PYRIMI-
    DINETRIONE MONOSODIUM SALT (9CI) see NBU000
5-ETHYL-5-(1-METHYLBUTYL)-2-THIOBARBITURIC ACID
    MONOSODIUM see PBT500
ETHYL 2-METHYLBUTYRATE see EMP600
ETHYL-N-METHYL CARBAMATE see EMQ500
ETHYLMETHYL CARBINOL see BPW750
ETHYL METHYL CETONE (FRENCH) see MKA400
1-ETHYL-7-METHYL-1,4-DIHYDRO-1,8-NAPHTHYRIDINE-4-
    ONE-3-CARBOXYLIC ACID see EID000
1-ETHYL-7-METHYL-1,4-DIHYDRO-1,8-NAPHTHYRIDIN-4-
    ONE-3-CARBOXYLIC ACID see EID000
4-ETHYL-4-METHYL-2,6-DIOXOPIPERIDINE see MKA250
ETHYL METHYLENE PHOSPHORODITHIOATE see
    EEH600
ETHYL METHYL ETHER see EMT000
ETHYL METHYL ETHER (DOT) see EMT000
5-ETHYL-5-(1-METHYLETHYL)-2,4,6(1H,3H,5H)-PYRIMI-
    DINETRIONE see ELX000
3-ETHYL-3-METHYLGLUTARIMIDE see MKA250
β-ETHYL-β-METHYLGLUTARIMIDE see MKA250
ETHYLMETHYLKETON (DUTCH) see MKA400

ETHYL METHYL KETONE (DOT) see MKA400

O-ETHYL-O-(4-METHYLMERCAPTO)PHENYL)-S-N-PRO-PYLPHOSPHOROTHIONOTHIOLATE see SOU625

ETHYL-3-METHYL-4-(METHYLTHIO)PHENYL(1-METHYL-ETHYL)PHOSPHORAMIDATE see FAK000

ETHYLMETHYLNITROSAMINE see MKB000

1-ETHYL-7-METHYL-4-OXO-1,4-DIHYDRO-1,8-NAPH-THYRIDINE-3-CARBOXYLIC ACID see EID000

N-ETHYL-N-METHYL-p-(PHENYLAZO)ANILINE see ENB000

N-ETHYLMETHYLPHENYLBARBITURIC ACID see ENB500

5-ETHYL-1-METHYL-5-PHENYLBARBITURIC ACID see ENB500

5-ETHYL-N-METHYL-5-PHENYLBARBITURIC ACID see ENB500

ETHYL METHYLPHENYLGLYCIDATE see ENC000

5-ETHYL-3-METHYL-5-PHENYLHYDANTOIN see MKB250

5-ETHYL-3-METHYL-5-PHENYL-2,4(3H,5H)-IMIDAZOLE-DIONE see MKB250

5-ETHYL-3-METHYL-5-PHENYLIMIDAZOLIDIN-2,4-DIONE see MKB250

ETHYL-1-METHYL-4-PHENYLISONIPECOTATE see DAM600

ETHYL-1-METHYL-4-PHENYLISONIPECOTATE HYDRO-CHLORIDE see DAM700

ETHYL-1-METHYL-4-PHENYLPIPERIDINE-4-CARBOXYL-ATE see DAM600

ETHYL-1-METHYL-4-PHENYLPIPERIDINE-4-CARBOXYL-ATE HYDROCHLORIDE see DAM700

ETHYL-1-METHYL-4-PHENYLPIPERIDYL-4-CARBOXYL-ATE HYDROCHLORIDE see DAM700

5-ETHYL-1-METHYL-5-PHENYL-2,4,6(1H,3H,5H)-PYRIMI-DINETRIONE see ENB500

4-ETHYL-4-METHYL-2,6-PIPERIDINEDIONE see MKA250

ETHYL-2-METHYLPROPANOATE see ELS000

ETHYL-2-METHYL-2-PROPENOATE see EMF000

ETHYL-2-METHYLPROPIONATE see ELS000

5-ETHYL-5-(1-METHYLPROPYL)BARBITURATE see BPF000

5-ETHYL-5-(1-METHYLPROPYL)BARBITURIC ACID see BPF000

5-ETHYL-5-(1-METHYLPROPYL)-2,4,6(1H,3H,5H)-PYRIMI-DINETRIONE (9CI) see BPF000

1-ETHYL-1-METHYL-2-PROPYNYL CARBAMATE see MNM500

3-ETHYL-6-METHYLPYRIDINE see EOS000

5-ETHYL-2-METHYLPYRIDINE see EOS000

3-ETHYL-3-METHYLPYRROLIDINE-2,5-DIONE see ENG500

3-ETHYL-3-METHYL-2,5-PYRROLIDINE-DIONE see ENG500

2-ETHYL-2-METHYLSUCCINIMIDE see ENG500

α-ETHYL-α-METHYLSUCCINIMIDE see ENG500

ETHYLMETHYLTHIAMBUTENE HYDROCHLORIDE see EIJ000

O-ETHYL-O-(4-(METHYLTHIO)PHENYL)PHOSPHORO-DITHIOIC ACID-S-PROPYL ESTER see SOU625

O-ETHYL-O-(4-(METHYLTHIO)PHENYL) S-PROPYL PHOS-PHORODITHIOATE see SOU625

ETHYL-4-(METHYLTHIO)-m-TOLYL ISOPROPYL PHOS-PHOR AMIDATE see FAK000

N-ETHYL-α-METHYL-m-(TRIFLUOROMETHYL)PHEN-ETHYLAMINE see ENJ000

N-ETHYL-α-METHYL-m-TRIFLUOROMETHYLPHENETHYL-AMINE see PDM250

N-ETHYL-α-METHYL-m-(TRIFLUOROMETHYL)PHEN-ETHYLAMINE HYDROCHLORIDE see PDM250

ETHYL MONOBROMOACETATE see EGV000

ETHYL MONOCHLORACETATE see EHG500

ETHYL MONOCHLOROACETATE see EHG500

ETHYL MONOSULFIDE see EPH000

ETHYLMORPHINE see ENK000

3-o-ETHYLMORPHINE see ENK000

ETHYLMORPHINE HYDROCHLORIDE see ENK500

ETHYL MORPHINE HYDROCHLORIDE DIHYDRATE see ENK500

4-ETHYLMORPHOLINE see ENL000

N-ETHYLMORPHOLINE see ENL000

ETHYL MYRISTATE see ENL850

3-ETHYLNIRVANOL see MKB250

ETHYL NITRATE see ENM500

ETHYL NITRILE see ABE500

ETHYL NITRITE see ENN000

ETHYL NITRITE (DOT) see ENN000

ETHYL NITRITE, solution (DOT) see ENN000

O-ETHYL-O-((4-NITROFENYL)-FENYL)-MONOTHIOFOS-FONAAT (DUTCH) see EBD700

N-ETHYL-N′-NITRO-N-NITROSOGUANIDINE see ENU000

ETHYL-p-NITROPHENYL BENZENETHIONOPHOSPHONATE see EBD700

O-ETHYL O-(4-NITROPHENYL)BENZENETHIONOPHOS-PHONATE see EBD700

ETHYL-p-NITROPHENYL BENZENETHIOPHOSPHATE see EBD700

ETHYL-p-NITROPHENYL BENZENETHIOPHOSPHONATE see EBD700

ETHYL p-NITROPHENYL ETHYLPHOSPHATE see NIM500

ETHYL-p-NITROPHENYL PHENYLPHOSPHONOTHIOATE see EBD700

O-ETHYL-O-(4-NITROPHENYL) PHENYLPHOSPHONO-THIOATE see EBD700

O-ETHYL-O-p-NITROPHENYL PHENYLPHOSPHONO-THIOLATE see EBD700

O-ETHYL-O-p-NITROPHENYL PHENYLPHOSPHORO-THIOATE see EBD700

ETHYL-p-NITROPHENYL THIONOBENZENEPHOSPHATE see EBD700

ETHYL-p-NITROPHENYL THIONOBENZENEPHOSPHONATE see EBD700

2-(ETHYLNITROSAMINO)ETHANOL see ELG500

(ETHYLNITROSAMINO)METHYL ACETATE see ENR500

4-((ETHYLNITROSAMINO)METHYL)PYRIDINE see NLH000

ETHYLNITROSOANILINE see NKD000

N-ETHYL-N-NITROSOBENZENAMINE see NKD000

ETHYLNITROSOBIURET see ENT000

N-ETHYL-N-NITROSOBIURET see ENT000

N-ETHYL-N-NITROSOBUTYLAMINE see EHC000

ETHYLNITROSOCARBAMIC ACID, ETHYL ESTER see NKE500

N-ETHYL-N-NITROSOCARBAMIC ACID ETHYL ESTER see NKE500

N-ETHYL-N-NITROSOCARBAMIDE see ENV000

N-ETHYL-N-NITROSO-ETHANAMINE see NJW500

N-ETHYL-N-NITROSOETHENAMINE see NKF000

N-ETHYL-N-NITROSOETHENYLAMINE see NKF000

N-ETHYL-N-NITROSO-N′-NITROGUANIDINE see ENU000

ETHYLNITROSOUREA see ENV000

1-ETHYL-1-NITROSOUREA see ENV000

N-ETHYL-N-NITROSO-UREA see ENV000

N-ETHYL-N-NITROSOURETHANE see NKE500

N-ETHYL-N-NITROSOVINYLAMINE see NKF000

1-ETHYL-3-(5-NITRO-2-THIAZOLYL) UREA see ENV500

N-ETHYL-N′-(5-NITRO-2-THIAZOLYL)UREA see ENV500

ETHYL NONANOATE see ENW000

ETHYL NONYLATE see ENW000

ETHYLNORADRENALINE HYDROCHLORIDE see ENX500

ETHYL NOREPINEPHRINE HYDROCHLORIDE see ENX500

α-ETHYLNOREPINEPHRINE HYDROCHLORIDE see ENX500

ETHYLNORSUPRARENIN HYDROCHLORIDE see ENX500

ETHYL OCTANOATE see ENY000

ETHYL OCTYLATE see ENY000

ETHYLOLAMINE see EEC600
1-(β-ETHYLOL)-2-METHYL-5-NITRO-3-AZAPYRROLE see MMN250
ETHYL ORTHOFORMATE see ENY500
ETHYL ORTHOSILICATE see EPF550
ETHYL OXALATE see DJT200
ETHYL OXALATE (DOT) see DJT200
ETHYL-3-OXOBUTANOATE see EFS000
ETHYL-3-OXOBUTYRATE see EFS000
ETHYL PARABEN see HJL000
ETHYL PARAOXON see NIM500
ETHYL PARASEPT see HJL000
ETHYL PARATHION see PAK000
S-ETHYL PARATHION see DJT000
ETHYL PELARGONATE see ENW000
5-ETHYL-5-PENTYLBARBITURIC ACID see PBS250
ETHYL PERCHLORATE see EOD000
ETHYL PHENACETATE see EOH000
ETHYL PHENYLACETATE see EOH000
ETHYL-β-PHENYLACRYLATE see EHN000
ETHYLPHENYLAMINE see EGK000
N-ETHYL-1-((4-(PHENYLAZO)PHENYL)AZO)-2-NAPH-THALENAMINE see EOJ500
N-ETHYL-1-((p-(PHENYLAZO)PHENYL)AZO)-2-NAPH-THALENAMINE see EOJ500
N-ETHYL-1-((4-(PHENYLAZO)PHENYL)AZO)-2-NAPH-THYLAMINE see EOJ500
N-ETHYL-1-((p-(PHENYLAZO)PHENYL)AZO)-2-NAPH-THYLAMINE see EOJ500
5-ETHYL-5-PHENYLBARBITURIC ACID see EOK000
5-ETHYL-5-PHENYLBARBITURIC ACID SODIUM see SID000
5-ETHYL-5-PHENYLBARBITURIC ACID SODIUM SALT see SID000
ETHYL PHENYL DICHLOROSILANE (DOT) see DFJ800
3-ETHYL-3-PHENYL-2,6-DIKETOPIPERIDINE see DYC800
3-ETHYL-3-PHENYL-2,6-DIOXOPIPERIDINE see DYC800
ETHYL-2-PHENYLETHANOATE see EOH000
O-ETHYL-S-PHENYL ETHYLDITHIOPHOSPHONATE see FMU045
O-ETHYL-S-PHENYL ETHYLPHOSPHONODITHIOATE see FMU045
2-ETHYL-2-PHENYLGLUTARIMIDE see DYC800
α-ETHYL-α-PHENYLGLUTARIMIDE see DYC800
ETHYL PHENYLGLYCIDATE see EOK600
ETHYL-3-PHENYLGLYCIDATE see EOK600
5-ETHYL-5-PHENYLHEXAHYDROPYRIMIDINE-4,6-DIONE see DBB200
ETHYLPHENYLHYDANTOIN see EOL000
5-ETHYL-5-PHENYLHYDANTOIN see EOL000
5-ETHYL-5-PHENYL-2,4-IMIDAZOLIDINEDIONE see EOL000
5-ETHYL-5-PHENYL-N-METHYLBARBITURIC ACID see ENB500
5-ETHYL-5-(PHENYLMETHYL)-2,4,6(1H,3H,5H)-PYRIMI-DINETRIONE (9CI) see BEA500
O-ETHYL PHENYL-p-NITROPHENYL THIOPHOSPHONATE see EBD700
3-ETHYL-3-PHENYL-2,6-PIPERIDINEDIONE see DYC800
ETHYL-3-PHENYLPROPENOATE see EHN000
5-ETHYL-5-PHENYL-2,4,6-(1H,3H,5H)PYRIMIDINETRIONE see EOK000
5-ETHYL-5-PHENYL-2,4,6-(1H,3H,5H)PYRIMIDINETRIONE MONOSODIUM SALT see SID000
ETHYL(N-PHENYL-p-TOLUENESULFONAMIDATO)MER-CURY see EME500
ETHYL(N-PHENYL-p-TOLUENESULFONAMIDO)MERCURY see EME500
ETHYLPHOSPHONOUS DICHLORIDE see EOQ000
ETHYL PHOSPHONOUS DICHLORIDE, anhydrous (DOT) see EOQ000
ETHYL PHOSPHORODICHLORIDATE see EOR000
ETHYL PHTHALATE see DJX000
5-ETHYL-2-PICOLINE see EOS000

5-ETHYL-α-PICOLINE see EOS000
1-ETHYLPIPERIDINE see EOS500
N-ETHYL-3-PIPERIDYLBENZILATE METHOBROMIDE see PJA000
1-ETHYL-3-PIPERIDYL BENZILATE METHYLBROMIDE see PJA000
ETHYL PROPENOATE see EFT000
ETHYL-2-PROPENOATE see EFT000
ETHYL PROPIONATE see EPB500
5-ETHYL-5-(1-PROPYL-1-BUTENYL)BARBITURIC ACID see EPC000
ETHYL PROPYL ETHER see EPC125
2-ETHYL-3-PROPYL-1,3-PROPANEDIOL see EKV000
ETHYLPROTAL see EQF000
ETHYL PTS see EPW500
ETHYL PYRAZINYL PHOSPHOROTHIOATE see EPC500
2-ETHYL-4-PYRIDINECARBOTHIOAMIDE see EPQ000
ETHYL PYROCARBONATE see DIZ100
2-ETHYL PYROMECONIC ACID see EMA600
ETHYL RHODANATE see EPP000
ETHYL-S see BID250
ETHYL SALICYLATE (FCC) see SAL000
ETHYL SEBACATE see DJY600
ETHYL SELENAC see DJD400
ETHYL SILICATE see EPF550
ETHYL SILICON TRICHLORIDE see EPY500
ETHYL SUCCINATE see SNB000
ETHYL SULFATE see DKB110
ETHYL SULFHYDRATE see EMB100
ETHYL SULFIDE see EPH000
S-(2-(ETHYLSULFINYL)ETHYL)-O,O-DIMETHYL PHOS-PHOROTHIOATE see DAP000
S-2-ETHYL-SULFINYL-1-METHYL-ETHYL-O,O-DIMETHYL-MONOTHIOFOSFAAT see DSK600
ETHYL SULFOCYANATE see EPP000
4-(ETHYLSULFONYL)-1-NAPHTHALENE SULFONAMIDE see EPI300
S-2-ETHYL-SULPHINYL-1-METHYL-ETHYL-O,O-DI-METHYL PHOSPHOROTHIOLATE see DSK600
4-ETHYLSULPHONYLNAPHTHALENE-1-SULFONAMIDE see EPI300
4-ETHYLSULPHONYLNAPHTHALENE-1-SULPHONAMIDE see EPI300
ETHYL TELLURAC see EPJ000
3'-ETHYL-5',6',7',8'-TETRAHYDRO-5',5',8'-TETRA-METHYL-2'-ACETONAPHTHONE see ACL750
1-(3-ETHYL-5,6,7,8-TETRAHYDRO-5,5,8,8-TETRAMETHYL-2-NAPHTHALENYL)-ETHANONE see ACL750
ETHYL TETRAPHOSPHATE see HCY000
ETHYL THIOALCOHOL see EMB100
2-ETHYL-4-THIOAMIDYLPYRIDINE see EPQ000
2-ETHYL-4-THIOCARBAMOYLPYRIDINE see EPQ000
2-(ETHYLTHIO)CHLOROETHANE see CGY750
ETHYL THIOCYANATE see EPP000
ETHYLTHIOETHANE see EPH000
2-(ETHYLTHIO)-ETHANETHIOL S-ESTER with O,O-DI-ETHYL PHOSPHOROTHIOATE see DAP200
ETHYL THIOETHER see EPH000
2-ETHYLTHIOETHYL CHLORIDE see CGY750
S-2-(ETHYLTHIO)ETHYL O,O-DIETHYL ESTER of PHOS-PHORODITHIOIC ACID see DXH325
2-ETHYLTHIOETHYL O,O-DIMETHYL PHOSPHORODI-THIOATE see PHI500
S-(2-(ETHYLTHIO)ETHYL) O,O-DIMETHYLPHOSPHORO-DITHIONATE see PHI500
O-(2-(ETHYLTHIO)ETHYL)-O,O-DIMETHYL PHOSPHORO-THIOATE see DAO800
S-(2-(ETHYLTHIO)ETHYL)-O,O-DIMETHYL PHOSPHORO-THIOATE see DAP400
S(and O)-2-(ETHYLTHIO)ETHYL-O,O-DIMETHYL PHOS-PHOROTHIOATE see MIW100
S-(2-(ETHYLTHIO)ETHYL)DIMETHYL PHOSPHOROTHIO-LATE see DAP400
S-(2-(ETHYLTHIO)ETHYL)DIMETHYL PHOSPHOROTHIO-LOTHIONATE see PHI500

2-(ETHYLTHIO)ETHYL DIMETHYL PHOSPHOROTHIONATE see DAO800

S-(2-(ETHYLTHIO)ETHYL)-O,O-DIMETHYL THIOPHOS-PHATE see DAP400

2-ETHYLTHIOISONICOTINAMIDE see EPQ000

α-ETHYLTHIOISONICOTINAMIDE see EPQ000

ETHYLTHIOMETON SULFOXIDE see OQS000

ETHYL THIOPYROPHOSPHATE see SOD100

ETHYL THIRAM see DXH250

ETHYL THIUDAD see DXH250

ETHYL THIURAD see DXH250

ETHYL-α-TOLUATE see EOH000

ETHYL(p-TOLUENESULFONANILIDATO)MERCURY see EME500

ETHYL-p-TOLUENESULFONATE see EPW500

ETHYL TOSYLATE see EPW500

ETHYL-p-TOSYLATE see EPW500

ETHYL TRICHLOROPHENYLETHYLPHOSPHONOTHIOATE see EPY000

O-ETHYL-O-2,4,5-TRICHLOROPHENYL ETHYLPHOS-PHONOTHIOATE see EPY000

ETHYL TRICHLOROSILANE see EPY500

ETHYL-γ-TRIMETHYLAMMONIUM PROPANEDIOL IODIDE see MJH250

α-ETHYLTRYPTAMINE ACETATE see AJB250

dl-α-ETHYLTRYPTAMINE ACETATE see AJB250

ETHYL TUADS see BJB500, DXH250

ETHYL TUEX see DXH250

ETHYLUREA and SODIUM NITRITE (2:1) see EQE000

ETHYLURETHAN see UVA000

ETHYL URETHANE see UVA000

o-ETHYLURETHANE see UVA000

ETHYL VANILLIN see EQF000

ETHYL VINYL ETHER see EQF500

ETHYL VINYL ETHER (inhibited) see EQG000

ETHYLVINYLNITROSAMINE see NKF000

ETHYL XANTHOGEN DISULFIDE see BJU000

ETHYL ZIMATE see BJC000, EIR000

ETHYL ZIRUM see BJC000

ETHYMAL see ENG500

ETHYNE see ACI750

ETHYNODIOL ACETATE see EQJ500

ETHYNODIOL DIACETATE see EQJ500

β-ETHYNODIOL DIACETATE see EQJ500

17-α-ETHYNYL-17-β-ACETOXY-19-NORANDROST-4-EN-3-ONE see ABU000

α-ETHYNYLBENZYL CARBAMATE see PGE000

2-ETHYNYL-2-BUTYL CARBAMATE see MNM500

ETHYNYLCARBINOL see PMN450

17-α-ETHYNYL-3,17-DIHYDROXY-4-ESTRENE DIACETATE see EQJ500

17-ETHYNYL-3,17-DIHYDROXY-1,3,5-OESTRATRIENE see EEH500

ETHYNYLESTRADIOL see EEH500

17-α-ETHYNYLESTRADIOL see EEH500

17-α-ETHYNYLESTRADIOL-17-β see EEH500

ETHYNYLESTRADIOL-3-METHYL ETHER see MKB750

17-ETHYNYLESTRADIOL-3-METHYL ETHER see MKB750

17-α-ETHYNYLESTRADIOL-3-METHYL ETHER see MKB750

17-α-ETHYNYL-1,3,5(10)-ESTRATRIENE-3,17-β-DIOL see EEH500

17-α-ETHYNYLESTRA-1,3,5(10)-TRIENE-3,17-β-DIOL see EEH500

17-α-ETHYNYLESTR-4-ENE-3-β,17-β-DIOL ACETATE see EQJ500

17-α-ETHYNYL-4-ESTRENE-3-β,17-β-DIOL DIACETATE see EQJ500

17-α-ETHYNYL-4-ESTRENE-3-β,17-β-DIOL DIACETATE see EQJ500

17-α-ETHYNYL-4-ESTREN-17-OL-3-ONE see NNP500

17-α-ETHYNYL-5(10)-ESTREN-17-OL-3-ONE see EEH550

17-α-ETHYNYLESTR-5(10)-EN-17-β-OL-3-ONE see EEH550

17-α-ETHYNYL-ESTR-5(10)-EN-3-ON-17-β-OL see EEH550

17-α-ETHYNYL-17-HYDROXY-4-ESTREN-3-ONE see NNP500

17-α-ETHYNYL-17-HYDROXYESTR-5(10)-EN-3-ONE see EEH550

17-α-ETHYNYL-17-HYDROXY-5(10)-ESTREN-3-ONE see EEH550

17-α-ETHYNYL-17-β-HYDROXY-5(10)-ESTREN-3-ONE see EEH550

17-α-ETHYNYL-17-β-HYDROXYESTR-5(10)-EN-3-ONE see EEH550

17-α-ETHYNYL-17-β-HYDROXY-Δ$^{-5(10)}$-ESTREN-3-ONE see EEH550

17-α-ETHYNYL-17-HYDROXYESTR-4-EN-3-ONE ACETATE see ABU000

(+)-17-α-ETHYNYL-17-β-HYDROXY-3-METHOXY-1,3,5(10)-ESTRATRIENE see MKB750

(+)-17-α-ETHYNYL-17-β-HYDROXY-3-METHOXY-1,3,5(10)-OESTRATRIENE see MKB750

17-α-ETHYNYL-17-β-HYDROXY-19-NORANDROST-4-EN-3-ONE see NNP500

17-α-ETHYNYL-17-β-HYDROXY-3-OXO-Δ$^{5(10)}$-ESTRENE see EEH550

ETHYNYLMETHANOL see PMN450

17-ETHYNYL-3-METHOXY-1,3,5(10)-ESTRATRIEN-17-β-OL see MKB750

17-α-ETHYNYL-3-METHOXY-1,3,5(10)-ESTRATRIEN-17-β-OL see MKB750

17-α-ETHYNYL-3-METHOXY-17-β-HYDROXY-Δ-1,3,5(10)-ESTRATRIENE see MKB750

17-α-ETHYNYL-3-METHOXY-17-β-HYDROXY-Δ-1,3,5(10)-OESTRATRIENE see MKB750

17-ETHYNYL-3-METHOXY-1,3,5(10)-OESTRATIEN-17-β-OL see MKB750

17-ETHYNYL-18-METHYL-19-NORTESTOSTERONE see NNQ500

17-α-ETHYNYL-19-NORANDROST-4-ENE-3-β,17-β-DIOL DI-ACETATE see EQJ500

17-α-ETHYNYL-19-NORANDROST-4-EN-17-β-OL-3-ONE see NNP500

17-α-ETHYNYL-19-NOR-4-ANDROSTEN-17-β-OL-3-ONE see NNP500

17-α-ETHYNYL-19-NOR-5(10)-ANDROSTEN-17-β-OL-3-ONE see EEH550

17-α-ETHYNYL-19-NORTESTOSTERONE see NNP500

17-α-ETHYNYL-19-NORTESTOSTERONE ACETATE see ABU000

ETHYNYLOESTRADIOL see EEH500

17-ETHYNYLOESTRADIOL see EEH500

17-α-ETHYNYLOESTRADIOL see EEH500

17-α-ETHYNYL-17-β-OESTRADIOL see EEH500

17-α-ETHYNYLOESTRADIOL-17-β see EEH500

ETHYNYLOESTRADIOL METHYL ETHER see MKB750

17-ETHYNYLOESTRADIOL-3-METHYL ETHER see MKB750

17-α-ETHYNYLOESTRADIOL METHYL ETHER see MKB750

17-α-ETHYNYLOESTRADIOL-3-METHYL ETHER see MKB750

17-α-ETHYNYL-1,3,5-OESTRATRIENE-3,17-β-DIOL see EEH500

17-ETHYNYLOESTRA-1,3,5(10)-TRIENE-3,17-β-DIOL see EEH500

17-α-ETHYNYL-1,3,5(10)-OESTRATRIENE-3,17-β-DIOL see EEH500

17-α-ETHYNYLOESTRA-1,3,5(10)-TRIENE-3,17-β-DIOL see EEH500

ETHYONOMIDE see EPQ000

ETICOL see NIM500

ETIL ACRILATO (ITALIAN) see EFT000

ETILACRILATULUI (ROMANIAN) see EFT000

ETILAMINA (ITALIAN) see EFU400

ETILBENZENE (ITALIAN) see EGP500

ETILBUTILCHETONE (ITALIAN) see EHA600

ETIL CLOROCARBONATO (ITALIAN) see EHK500

ETIL CLOROFORMIATO (ITALIAN) see EHK500
ETILE (ACETATO di) (ITALIAN) see EFR000
ETILE (FORMIATO di) (ITALIAN) see EKL000
N,N'-ETILEN-BIS(DITIOCARBAMMATO) di MANGANESE (ITALIAN) see MAS500
N,N'-ETILEN-BIS(DITIOCARBAMMATO) di SODIO (ITALIAN) see DXD200
ETILENE (OSSIDO di) (ITALIAN) see EJN500
ETILENIMINA (ITALIAN) see EJM900
ETILFEN see EOK000
ETILMERCAPTANO (ITALIAN) see EMB100
O-ETIL-O-((4-NTIRO-FENIL)-FENIL)-MONOTIOFOSFONATO (ITALIAN) see EBD700
S-2-ETIL-SULFINIL-1-METIL-ETIL-O,O-DIMETIL-MONOTIO-FOSFATO see DSK600
ETIMID see EPQ000
ETIN see EDC500
ETIOCIDAN see EPQ000
ETIOL see MAK700
ETIONAMID see EPQ000
ETIONIZINA see EPQ000
ETO see EJN500
ETOKSYETYLOWY ALKOHOL (POLISH) see EES350
ETOMAL see ENG500
ETOPOSIDE see EAV500
ETOPROPEZINA see DIR000
2-ETOSSIETIL-ACETATO (ITALIAN) see EES400
ETOSUXIMIDA see ENG500
ETOVAL see BPF500
ETP see EPQ000, EQP000
ETRENOL see HGO500
ETROFOL see CKF000
ETROFOLAN see MIA250
ETROLENE see RMA500
ETROZOLIDINA see HNI500
ETRYPTAMINE ACETATE see AJB250
ETU see IAQ000
ETYDION see TLP750
ETYLENU TLENEK (POLISH) see EJN500
ETYLOAMINA (POLISH) see EFU400
ETYLOBENZEN (POLISH) see EGP500
ETYLOWY ALKOHOL (POLISH) see EFU000
ETYLU BROMEK (POLISH) see EGV400
ETYLU CHLOREK (POLISH) see EHH000
ETYLU KRZEMIAN (POLISH) see EPF550
EUBINE see DLX400
EUCALMYL see FLU000
EUCALYPTOLE see CAL000
EUCALYPTOL (FCC) see CAL000
EUCALYPTUS OIL see EQQ000
EUCANINE GB see TGL750
EUCHEUMA SPINOSUM GUM see CCL250
EUCHRYSINE see BJF000
EUCISTEN see EID000
EUCISTIN see PDC250
EUCLORINA see CDP000
EUCODAL see DLX400
EUCORAN see DJS200
EUFIN see DIX200
EUFLAVINE see DBX400
EUGENIC ACID see EQR500
EUGENOL see EQR500
EUGENOL ACETATE see EQS000
1,3,4-EUGENOL ACETATE see EQS000
1,3,4-EUGENOL METHYL ETHER see AGE250
EUGENYL ACETATE see EQS000
EUGENYL METHYL ETHER see AGE250
EUHAEMON see VSZ000
EUKALYPTUS OEL (GERMAN) see EQQ000
EUKODAL see DLX400
EUKRATON see MKA250
EUMICTON see AAI250
EUMIN see MMN250
EUNATROL see OIA000

EUNERPAN see FKI000
EUNERYL see EOK000
EUNOCTIN see DLY000
EUPHODRIN see DBA800
EUPHOZID see ILE000
EUPLACID see EHP000
EUPRAMIN see DLH600, DLH630
EUROCERT AZORUBINE see HJF500
EUROCERT TARTRAZINE see FAG140
EURODOPA see DNA200
EUROPHAN see PKQ059
EUROPIC CHLORIDE see ERA500
EUROPIUM CHLORIDE see ERA500
EUROPIUM(III) NITRATE, HEXAHYDRATE (1:3:6) see ERC000
EUSAPRIM see SNK000
EUSCOPOL see HOT500
EUSTIDIL see DFH600
EUSTIGMIN BROMIDE see DQY800
EUSTIGMIN METHYLSULFATE see DQY909
EUTAGEN see DLX400
EUTENSIN see CHJ750
EUTHATAL see NBU000
EUTIZON see ILD000
EUVESTIN see DKB000
EVALON see ILD000
EVAU-SUPER see SFS000
EVE see EQF500
EVEX see ECU750
EVION see VSZ450
EVIPAL see ERD500
EVIPAN see ERD500
EVIPLAST 80 see DVL700
EVIPLAST 81 see DVL700
EVITAMINUM see VSZ450
EVOLA see DEP800
EVRONAL see SBM500
EWEISS see BAP000
EX 4355 see DLS600
EX 10-781 see MHJ500
EXACT-S see TFP000
EXADRIN see VGP000
EXAGAMA see BBQ500
EXAL see VLA000
EXD see BJU000
EXDOL see HIM000
EXHAUST GAS see CBW750
EXHORAN see DXH250
EXHORRAN see DXH250
EXMIGRA see EDC500
EXMIN see AHJ750
EXNA see BDE250
EXOFENE see HCL000
EXON 450 see AAX175
EXON 454 see AAX175
EXON 605 see PKQ059
EXOSALT see BDE250
EXOTHERM see TBQ750
exothermic FERROCHROME (DOT) see FBD000
exothermic FERROMANGANESE (DOT) see FBE000
EXOTHERM TERMIL see TBQ750
EXOTHION see EAS000
EXP 338 see AMQ500
EXP 999 see MQR000
EXPERIMENTAL FUNGICIDE 5223 see DXX400
EXPERIMENTAL FUNGICIDE 224 (UNION CARBIDE) see ZJA000
EXPERIMENTAL INSECTICIDE 711 see IKO000
EXPERIMENTAL INSECTICIDE 4049 see MAK700
EXPERIMENTAL INSECTICIDE 4124 see NFT000
EXPERIMENTAL INSECTICIDE 7744 see CBM750
EXPERIMENTAL INSECTICIDE 12008 see DJN600
EXPERIMENTAL INSECTICIDE 12,880 see DSP400
EXPERIMENTAL INSECTICIDE 52160 see TAL250

EXPERIMENTAL NEMATOCIDE 18,133 see EPC500
EXPLOSIVE D see ANS750
EXPLOSIVES, HIGH see ERF000
EXPLOSIVES, LOW see ERF500
EXPLOSIVES, PERMITTED see ERG000
EXSEL see SBR000
EXSICATED SODIUM SULFITE see SJZ000
EXSICCATED FERROUS SULFATE see FBN100
EXSICCATED FERROUS SULPHATE see FBN100
EXSICCATED SODIUM PHOSPHATE see SJH090
EXTACOL see PGA750
EXT. D&C RED No. 15 see FAG018
EXT. D&C YELLOW No. 9 see FAG130
EXT. D&C YELLOW No. 10 see FAG135
EXTERMATHION see MAK700
EXTERNAL BLUE 1 see BJI250
EXTHRIN see AFR250
EXTRACT D&C ORANGE No. 4 see TGW000
EXTRACT D&C RED No. 10 see HJF500
EXTRACT D&C RED No. 14 see XRA000
EXTRACTS (PETROLEUM), HEAVY NAPHTHENIC DISTIL-
    LATE SOLVENT (9CI) see MQV857
EXTRACTS (PETROLEUM), HEAVY PARAFFINIC DISTIL-
    LATE SOLVENT (9CI) see MQV859
EXTRACTS (PETROLEUM), LIGHT NAPHTHENIC DISTIL-
    LATE SOLVENT (9CI) see MQV860
EXTRACTS (PETROLEUM), LIGHT PARAFFINIC DISTIL-
    LATE SOLVENT (9CI) see MQV862
EXTRACTS (PETROLEUM), RESIDUAL OIL SOLVENT (9CI)
    see MQV863
EXTRA FINE 200 SALT see SFT000
EXTRA FINE 325 SALT see SFT000
EXTRANASE see BMO000
EXTRAR see DUS700
EXTRAX see RNZ000
EXTREMA see EPF550, GDY000, SCQ500
EXTREN see ADA725
EXTRINSIC FACTOR see VSZ000
E-Z-OFF see MAE000
E-Z-OFF D see BSH250

F 1 (complexon) see EIX500
F 10 (pesticide) see MAS500
F 12 see DFA600
F 13 see CLR250
16 F see ARM266
F 22 see CFX500
F-33 see FLJ000
F-112 see TBP050
F 114 see FOO509
F-115 see CJI500
F-139 see BPG000
F 156 see HEL500
190 F see ABX500
F 190 see ABX500
F 735 see CCC500
F 933 see BCI500
F III (sugar fraction) see FAB000
F 1162 see SNM500
1358F see SOA500
F 1358 see SOA500
F-13B1 see TJY100
2249F see FMX000
F 2387 see DVX600
F 2559 see PDD300
FA see FMV000
Fa 100 see EQR500
FA-192 see BRF500
FAA see FDR000, FFF000, FIC000
2-FAA see FDR000
2,7-FAA see BGP250
F-diAA see DBF200
FAC see IOT000
FAC 20 see IOT000

FAC 5273 see PIX250
FACTITIOUS AIR see NGU000
FACTOR II (VITAMIN) see VSZ000
FACTOR PP see NCR000
FADORMIR see QAK000
FALISAN see MEP250
FALITHION see DSQ000
FALITIRAM see TFS350
FALKITOL see HCI000
FALL see SFS000
FAM see MME809
FAMFOS see FAB600
FAMID see DVS000
FAMOPHOS see FAB600
FAMOPHOS WARBEX see FAB600
FAMPHOS see FAB600
FAMPHUR see FAB600
FANFOS see FAB600
FANFT see NGM500
FANNOFORM see FMV000
FANODORMO see TDA500
FARGAN see DQA400
FARLUTIN see MCA000
FARMCO see DAA800
FARMCO ATRAZINE see ARQ725
FARMCO DIURON see DXQ500
FARMCO FENCE RIDER see TAA100
FARMCO PROPANIL see DGI000
FARMICETINA see CDP250
FARMOTAL see PBT500
FARNESOL see FAB800
FARNESYL ALCOHOL see FAB800
FARTOX see PAX000
FAS-CILE see MQU750
FASCIOLIN see CBY000, HCI000
FASCO-TERPENE see CDV100
FASCO WY-HOE see CKC000
FASERTON see AHE250
FAST ACID GREEN N see FAF000
FASTBALLS see BBK500
FAST BLUE B BASE see DCJ200
FAST CORINTH BASE B see BBX000
FAST DARK BLUE BASE R see TGJ750
FAST GARNET GBC BASE see AIC250
FAST GREEN FCF see FAG000
FAST GREEN JJO see BAY750
FAST LIGHT ORANGE GA see HGC000
FASTOGEN BLUE FP-3100 see DNE400
FASTOGEN BLUE SH-100 see DNE400
FAST OIL ORANGE see PEJ500
FAST OIL ORANGE II see XRA000
FAST OIL RED B see SBC500
FAST OIL YELLOW see AIC250
FAST OIL YELLOW B see DOT300
FAST ORANGE see PEJ500
FAST ORANGE BASE GR see NEO000
FAST ORANGE GC BASE see CEH675
FAST ORANGE R SALT see NEN500
FAST RED BASE GG see NEO500
FAST RED BASE TR see CLK220
FAST RED 5CT BASE see CLK220
FAST RED 5CT SALT see CLK235
FAST RED KB AMINE see CLK225
FAST RED KB BASE see CLK225
FAST RED KB SALT see CLK225
FAST RED KB SALT SUPRA see CLK225
FAST RED KBS SALT see CLK225
FAST RED 2G SALT see NEO500
FAST RED SALT TR see CLK235
FAST RED SALT TRA see CLK235
FAST RED SALT TRN see CLK235
FAST RED SG BASE see NMP500
FAST RED TR see CLK220
FAST RED TR11 see CLK220

FAST RED TR BASE see CLK220
FAST RED TRO BASE see CLK220
FAST RED TR SALT see CLK235
FAST SCARLET BASE B see NBE500
FAST SCARLET G see NMP500
FAST SCARLET R see NEQ500
FAST SPIRIT YELLOW AAB see PEI000
FAST WHITE see LDY000
FAST YELLOW AT see AIC250
FAST YELLOW B see AIC250
FAST YELLOW GC BASE see CEH670
FATOLIAMID see EPQ000
FAT ORANGE II see TGW000
FAT RED 7B see EOJ500
FAT RED B see SBC500
FAT RED (YELLOWISH) see XRA000
FAT SCARLET 2G see XRA000
FAT SOLUBLE GREEN ANTHRAQUINONE see BLK000
FAT YELLOW see DOT300
FAVISTAN see MCO500
FB/2 see DWX800
FBC CMPP see CIR500
FBHC see BBQ750
FBZ see BRF500
FC 12 see DFA600
FC 14 see CBY250
FC 31 see CHI900
FC 114 see FOO509
FC-143 see ANP625
FC-1318 see OBO000
FC142b see CFX250
FC 152a see ELN500
FC 4648 see PKQ059
FC-C 318 see CPS000
FCDR see FHO000
FCdR see FHO000
FDA see DGB600
FDA 0101 see SHF500
FDA 0109 see BHT250
FDA 0345 see BIF750
FDA 1446 see AFR250
FDA 1541 see EIN500
FD&C BLUE No. 1 see FAE000
FD&C BLUE No. 2 see FAE100
FD&C GREEN No. 2 see FAF000
FD&C GREEN No. 3 see FAG000
FD&C GREEN No. 2-ALUMINUM LAKE see FAF000
FD&C RED No. 1 see FAG018
FD&C RED No. 3 see FAG040
FD&C RED No. 19 see FAG070
FD&C RED No. 40 see FAG100
FD&C VIOLET No. 1 see FAG120
FD&C YELLOW No. 3 see FAG130
FD&C YELLOW No. 4 see FAG135
FD&C YELLOW No. 5 see FAG140
FD&C YELLOW No. 6 see FAG150
FDN see DRP800
FDUR see DAR400
FEBRILIX see HIM000
FEBRININA see DOT000
FEBRO-GESIC see HIM000
FEBROLIN see HIM000
FEBRON see DOT000
FEBUZINA see BRF500
FECAMA see DGP900
FECTRIM see SNK000
FEDACIN see NGE500
FEDAL-UN see PCF275
Fe-DEXTRAN see IGS000
FEDRIN see EAW000
FEENO see PDP250
FEGLOX see DWX800
FEINALMIN see DLH630
FEKABIT see PLA250

FELAZINE see PFC750
FELBEN see BAU750
FELISON see DAB800
α-FELLANDRENE see MCC000
FELLING ZINC OXIDE see ZKA000
FELMANE see FMQ000
FELSULES see CDO000
FEMACOID see ECU750
FEMADOL see PNA500
FEMA No. 2003 see AAG250
FEMA No. 2005 see MDW750
FEMA No. 2006 see AAT250
FEMA No. 2007 see THM500
FEMA No. 2008 see ABB500
FEMA No. 2009 see ABH000
FEMA No. 2011 see AEN250
FEMA No. 2026 see AGC500
FEMA No. 2031 see AGH250
FEMA No. 2032 see AGA500
FEMA No. 2033 see AGI500
FEMA No. 2034 see AGJ250
FEMA No. 2037 see AGM500
FEMA No. 2045 see ISV000
FEMA No. 2055 see IHO850
FEMA No. 2058 see IHP100
FEMA No. 2060 see IHP400
FEMA No. 2061 see AOG500
FEMA No. 2063 see AOG600
FEMA No. 2069 see IHS000
FEMA No. 2075 see IHU100
FEMA No. 2082 see AON350
FEMA No. 2084 see IME000
FEMA No. 2085 see ITB000
FEMA No. 2086 see PMQ750
FEMA No. 2097 see AOX750
FEMA No. 2098 see AOY400
FEMA No. 2099 see MED500
FEMA No. 2109 see ARN000
FEMA No. 2127 see BAY500
FEMA No. 2134 see BCS250
FEMA No. 2135 see BDX000
FEMA No. 2137 see BDX500
FEMA No. 2138 see BCM000
FEMA No. 2140 see BED000
FEMA No. 2141 see IJV000
FEMA No. 2142 see BEG750
FEMA No. 2149 see BFD400
FEMA No. 2151 see BFJ750
FEMA No. 2152 see ISW000
FEMA No. 2159 see BMD100
FEMA No. 2160 see IHX600
FEMA No. 2170 see MKA400
FEMA No. 2174 see BPU750
FEMA No. 2175 see IIJ000
FEMA No. 2178 see BPW500
FEMA No. 2179 see IIL000
FEMA No. 2183 see BQI000
FEMA No. 2184 see BFW750
FEMA No. 2186 see BQM500
FEMA No. 2187 see BSW500
FEMA No. 2188 see BRQ350
FEMA No. 2190 see BQP000
FEMA No. 2193 see IIQ000
FEMA No. 2203 see DTC800
FEMA No. 2209 see BBA000
FEMA No. 2210 see IJF400
FEMA No. 2213 see IJN000
FEMA No. 2218 see ISX000
FEMA No. 2219 see BSU250
FEMA No. 2220 see IJS000
FEMA No. 2221 see BSW000
FEMA No. 2222 see IJU000
FEMA No. 2223 see TIG750
FEMA No. 2224 see CAK500

FEMA No. 2856 see MCC000
FEMA No. 2857 see PFB250
FEMA No. 2858 see PDD750
FEMA No. 2862 see PDF750
FEMA No. 2866 see PDI000
FEMA No. 2868 see PDK200
FEMA No. 2871 see PDF775
FEMA No. 2873 see PDS900
FEMA No. 2874 see BBL500
FEMA No. 2876 see PDX000
FEMA No. 2878 see PDY850
FEMA No. 2885 see HHP050
FEMA No. 2886 see COF000
FEMA No. 2887 see HHP000
FEMA No. 2890 see HHP500
FEMA No. 2902 see PIH250
FEMA No. 2903 see POH750
FEMA No. 2911 see PIW250
FEMA No. 2922 see IRY000
FEMA No. 2923 see PMT750
FEMA No. 2926 see INE100
FEMA No. 2930 see PNE250
FEMA No. 2962 see MCE750
FEMA No. 2980 see CMT250
FEMA No. 2981 see DTF400, RHA000
FEMA No. 3006 see OHG000
FEMA No. 3007 see SAU400
FEMA No. 3045 see TBD500
FEMA No. 3047 see TBE250
FEMA No. 3053 see TBE600
FEMA No. 3060 see TCU600
FEMA No. 3073 see MNR250
FEMA No. 3075 see THA250
FEMA No. 3091 see UJA800
FEMA No. 3092 see UJJ000
FEMA No. 3095 see ULJ000
FEMA No. 3097 see UNA000
FEMA No. 3101 see VAQ000
FEMA No. 3102 see ISU000
FEMA No. 3103 see VAV000
FEMA No. 3107 see VFK000
FEMA No. 3135 see DAE450
FEMA No. 3149 see EIL100
FEMA No. 3164 see HAV450
FEMA No. 3183 see MEX350
FEMA No. 3213 see NNA300
FEMA No. 3237 see TDV725
FEMA No. 3244 see TME270
FEMA No. 3271 see DTU400
FEMA No. 3272 see DTU600
FEMA No. 3273 see DTU800
FEMA No. 3289 see HBI800
FEMA No. 3291 see BOV000
FEMA No. 3302 see MFN285
FEMA No. 3309 see MOW750
FEMA No. 3317 see NMV760
FEMA No. 3326 see ABC750
FEMA No. 3386 see PPS250
FEMA No. 3497 see HFE550
FEMA No. 3498 see ISZ000
FEMA No. 3499 see HFR200
FEMA No. 3558 see MLA250
FEMA No. 3559 see MCB750
FEMA No. 3565 see DKV175
FEMA No. 3581 see OCY100
FEMA No. 3583 see OEG100
FEMA No. 3632 see PDF790
FEMERGIN see EDC500
FEMEST see ECU750
FEMESTRAL see EDO000
FEMESTRONE see EDP000
FEMESTRONE INJECTION see EDV000
FEM H see ECU750
FEMIDYN see EDV000

FEMMA see ABU500
FEMOGEN see ECU750, EDO000
FEMOGEX see EDS100
FEMPROPAZINE see DIR000
FEMULEN see EQJ500
FENACETINA see ABG750
FENAM see DRP800
FENAMIC ACID see PEG500
FENAMIN see ARQ725
FENAMIN BLACK E see AQP000
FENAMIN BLUE 2B see CMO000
FENAMINE see AMY050, ARQ725
FENAMINOSULF see DOU600
FENAMIPHOS see FAK000
FENANTOIN see DKQ000, DNU000
FENARIMOL see FAK100
FENAROL see CJR909
FENARSONE see CBJ000
FENARTIL see BRF500
FENASAL see DFV400
FENATE see IGS000
FENATROL see ARQ725
FENAVAR see AMY050
FENAZAFLOR see DGA200
FENAZIL see DQA400
FENAZO BLUE XI see FAE000
FENAZO BLUE XR see FMU059
FENAZO EOSINE XG see BNH500
FENAZO GREEN 7G see FAF000
FENAZO RED C see HJF500
FENAZOXINE see FAL000
FENAZOXINE HYDROCHLORIDE see NBS500
FENAZO YELLOW T see FAG140
FENBENDAZOL see FAL100
FENBENDAZOLE see FAL100
FENBITAL see EOK000
FENBUTATIN OXIDE see BLU000
FENCHEL OEL (GERMAN) see FAP000
FENCHLOORFOS (DUTCH) see RMA500
FENCHLORFOS see RMA500
FENCHLORFOSU (POLISH) see RMA500
FENCHLOROPHOS see RMA500
FENCHLORPHOS see RMA500
FENCLOR see PJL750
FENDON see HIM000
FENELZIN see PFC750
FENEMAL see EOK000
FENERGAN see DQA400
FENESTERIN see CME250
FENESTRIN see CME250
FENETAZINA see DQA400
FENFLURAMINE see ENJ000
FENFLURAMINE HYDROCHLORIDE see PDM250
FENIBUTASAN see BRF500
FENIBUTAZONA see BRF500
FENIBUT HYDROCHLORIDE see GAD000
FENIBUTOL see BRF500
FENICOL see CDP250
FENIDANTOIN 'S' see DKQ000
FENIGAM HYDROCHLORIDE see GAD000
FENILBUTAZONA see BRF500
FENILBUTINE see BRF500
FENILDICLOROARSINA (ITALIAN) see DGB600
FENILIDINA see BRF500
FENILIDRAZINA (ITALIAN) see PFI000
FENILISOPROPILIDRAZINA see PDN000
2-FENILPROPANO (ITALIAN) see COE750
FENILPROPANOLAMINA (ITALIAN) see NNM000
FENITOIN see DNU000
FENITOX see DSQ000
FENITROTHION see DSQ000
FENITROTION (HUNGARIAN) see DSQ000
FENITROXON see PHD750
FENIZON (FRENCH) see CJR500

FENNEL OIL see FAP000
FENNOSAN see QPA000
FENNOSAN B 100 see DSB200
FENOBARBITAL see EOK000
FENOFLURAZOLE see DGA200
FENOL (DUTCH, POLISH) see PDN750
FENOLO (ITALIAN) see PDN750
FENOLOVO see HON000
FENOLOVO ACETATE see ABX250
FENOPHOSPHON see EPY000
FENOPROP see TIX500
FENORMONE see TIX500
FENOSED see EOK000
FENOTHIAZINE (DUTCH) see PDP250
FENOTIAZINA (ITALIAN) see PDP250
FENOTONE see BRF500
FENOVERM see PDP250
FENOXYBENZAMIN see DDG800
2-FENOXYETHANOL (CZECH) see PER000
FENOXYL CARBON N see DUZ000
FENOZAFLOR see DGA200
FENPROBAMATO see PGA750
FENPROPAZINA see DIR000
FENSON see CJR500
FENSULFOTHION see FAQ800
FENTANEST see PDW500, PDW750
FENTANIL see PDW500
FENTANYL see PDW500
FENTANYL CITRATE see PDW750
FENTAZIN see CJM250
FENTHION see FAQ999
FENTHIONE SULFONE see DSS800
FENTHOATE see DRR400
FENTIAZIN see PDP250
FENTIN ACETAAT (DUTCH) see ABX250
FENTIN ACETAT (GERMAN) see ABX250
FENTIN ACETATE see ABX250
FENTIN CHLORIDE see CLU000
FENTINE ACETATE (FRENCH) see ABX250
FENTIN HYDROXIDE see HON000
FENVALERATE see FAR100
1-FENYL-4-AMINO-5-CHLOR-6-PYRIDAZINON (CZECH) see PEE750
FENYLBUTAZON see BRF500
FENYL-CELLOSOLVE (CZECH) see PER000
1-FENYL-3,3-DIETHYLTRIAZEN (CZECH) see PEU500
1-FENYL-3,3-DIMETHYLTRIAZIN see DTP000
m-FENYLENDIAMIN (CZECH) see PEY000
FENYLENODWUAMINA (POLISH) see PEY500
FENYLEPSIN see DKQ000
FENYLESTER KYSELINY CHLORMRAVENCI (CZECH) see CBX109
1-FENYLETHANOL see PDE000
FENYLETTAE see EOK000
FENYL-GLYCIDYLETHER (CZECH) see PFH000
FENYLHIST see BAU750
FENYLHYDRAZINE (DUTCH) see PFI000
FENYLMERCURIACETAT (CZECH) see ABU500
FENYLMERCURICHLORID (CZECH) see PFM500
FENYL-METHYLKARBINOL see PDE000
2-FENYLOTIOMOCZNIK (POLISH) see DWN800
2-FENYL-PROPAAN (DUTCH) see COE750
FENYPRIN see DBA800
FENYTOINE see DKQ000, DNU000
FENZAFLOR see DGA200
FENZEN (CZECH) see BBL250
FEOJECTIN see IHG000
FEOSOL see FBN100, FBO000
FEOSPAN see FBN100
FEOSTAT see FBJ100
FERBAM see FAS000
FERBAM 50 see FAS000
FERBAM, IRON SALT see FAS000
FERBECK see FAS000

FERDEX 100 see IGS000
FERGON see FBK000
FERGON PREPARATIONS see FBK000
FER-IN-SOL see FBN100, FBO000
FERKETHION see DSP400
FERLUCON see FBK000
FERMATE FERBAM FUNGICIDE see FAS000
FERMENICIDE LIQUID see SOH500
FERMENICIDE POWDER see SOH500
FERMENTATION ALCOHOL see EFU000
FERMENTATION AMYL ALCOHOL see IHP000
FERMENTATION BUTYL ALCOHOL see IIL000
FERMIDE see TFS350
FERMINE see DTR200
FERMOCIDE see FAS000
FERNACOL see TFS350
FERNASAN see TFS350
FERNESTA see BQZ000, DAA800
FERMINE see DAA800
FERNIDE see TFS350
FERNIMINE see DAA800
FERNISOLONE see PMA000
FERNOXENE see SGH500
FERNOXONE see DAA800
FERO-GRADUMET see FBN100, FBO000
FEROTON see FBJ100
FER PENTACARBONYLE (FRENCH) see IHG500
FERRADOW see FAS000
FERRALYN see FBN100
FERRIAMICIDE see MQW500
FERRIC ARSENATE, solid (DOT) see IGN000
FERRIC ARSENITE, solid (DOT) see IGO000
FERRIC ARSENITE, BASIC see IGO000
FERRIC CHLORIDE see FAU000
FERRIC CHLORIDE, anhydrous (DOT) see FAU000
FERRIC CHLORIDE, solid (DOT) see FAU000
FERRIC CHLORIDE, solid, anhydrous (DOT) see FAU000
FERRIC CHLORIDE (solution) see FAW000
FERRIC CHLORIDE, solution (DOT) see FAU000
FERRIC CHOLINE CITRATE see FBC100
FERRIC DEXTRAN see IGS000
FERRIC DIMETHYLDITHIOCARBAMATE see FAS000
FERRIC FLUORIDE see FAX000
FERRIC OXIDE see IHD000
FERRIC OXIDE, SACCHARATED see IHG000
FERRIC SACCHARATE IRON OXIDE (MIX.) see IHG000
FERRIC SULFATE see FBA000
FERRIDEXTRAN see IGS000
FERRIGEN see IGU000
FERRIVENIN see IHG000
FERROANTHOPHYLLITE see ARM264
FERROCENE see FBC000
FERROCHOLINATE see FBC100
FERROCHROME see FBD000
FERROCHROME (exothermic) see FBD000
FERROCHROME, exothermic (DOT) see FBD000
FERROCHROMIUM see FBD000
FERRODEXTRAN see IGS000
FERROFLUKIN 75 see IGS000
FERROFUME see FBJ100
FERROGLUCIN see IGS000
FERROGLUKIN 75 see IGS000
FERRO-GRADUMET see FBN100
FERROLIP see FBC100
FERROMANGANESE (EXOTHERMIC) see FBE000
FERRONAT see FBJ100
FERRONE see FBJ100
FERRONICUM see FBK000
FERROPHOSPHORUS see FBF000
FERROSILICON see FBG000
FERROSILICON, containing more than 30% but less than 90% SILICON (DOT) see FBG000
FERROSULFAT (GERMAN) see FBN100
FERROSULFATE see FBN100
FERROTEMP see FBJ100

FERRO-THERON see FBN100
FERROUS ARSENATE (DOT) see IGM000
FERROUS ARSENATE, solid (DOT) see IGM000
FERROUS CHLORIDE see FBI000
FERROUS FUMARATE see FBJ100
FERROUS GLUCONATE see FBK000
FERROUS LACTATE see LAL000
FERROUS SULFATE see FBN100
FERROUS SULFATE (FCC) see FBO000
FERROUS SULFATE HEPTAHYDRATE see FBO000
FERROVANADIUM DUST see FBP000
FERRO YELLOW see CAJ750
FERRUGO see IHD000
FERRUM see FBJ100
FERSAMAL see FBJ100
FERSOLATE see FBN100
FESOFOR see FBO000
FESOTYME see FBO000
FETTORANGE B see XRA000
FETTORANGE R see PEJ500
FETTSCHARLACH see OHA000
FF see FQN000
FG 5111 see FKI000
FH 099 see DTL200
FH 122-A see NNQ500
FHCH see BBQ750
F.I 106 see AES750
F.I. 58-30 see EPQ000
F.I. 6145 see PIW000
FI6339 see DAC000
FI 6714 see NDM000
FIBERGLASS see FBQ000
FIBRE BLACK VF see AQP000
FIBRENE C 400 see TAB750
FIBROTAN see PFN000
FIBROUS CROCIDOLITE ASBESTOS see ARM275
FIBROUS GLASS see FBQ000
FIBROUS GLASS DUST (ACGIH) see FBQ000
FIBROUS GRUNERITE see ARM250
FIBROUS TREMOLITE see ARM280
FICAM see DQM600
FICHLOR 91 see TIQ750
FICIN see FBS000
FI CLOR 91 see TIQ750
FI CLOR 60S see SGG500
FICUS PROTEASE see FBS000
FICUS PROTEINASE see FBS000
FILARIOL see EGV500
FILARSEN see DFX400
FILMERINE see SIQ500
FIMALENE see ILD000
FINDOLAR see GGS000
FINE GUM HES see SFO500
FINEMEAL see BAP000
FINIMAL see HIM000
FINLEPSIN see DCV200
FINTIN ACETATO (ITALIAN) see ABX250
FINTINE HYDROXYDE (FRENCH) see HON000
FINTIN HYDROXID (GERMAN) see HON000
FINTIN HYDROXYDE (DUTCH) see HON000
FINTIN IDROSSIDO (ITALIAN) see HON000
FIORINAL see ABG750
FIRE DAMP see MDQ750
FIREMASTER BP-6 see FBU000
FIREMASTER FF-1 see FBU509
FIREMASTER T23P-LV see TNC500
FIRMOTOX see POO250
FIR NEEDLE OIL, SIBERIAN see FBV000
FIRON see FBJ100
FISH BERRY see PIE500
FISH-TOX see RNZ000
FISONS B25 see CEW500
FISONS NC 2964 see DSO000
FISONS NC 5016 see DGA200

FISSUCAIN see BQA010
FITIOS see DNX600
FITIOS B/77 see DNX600
FIXANOL BLACK E see AQP000
FIXANOL BLUE 2B see CMO000
FIXOL see CMS850
FLACAVON R see TNC500
FLAGEMONA see MMN250
FLAGESOL see MMN250
FLAGIL see MMN250
FLAGYL see MMN250
FLAMARIL see HNI500
FLAMENCO see TGG760
FLAMMEX AP see TNC500
FLAMYCIN see CMA750
FLANARIL see HNI500
FLANOGEN ELA see CCL250
FLAVACRIDINUM HYDROCHLORICUM see DBX400
FLAVAXIN see RIK000
FLAVAZONE see NGE500
FLAVINE see DBN400, DBX400
FLAVIN SULPHATE see DBN400
FLAVIOFORM see DBX400
FLAVIPIN see DBX400
FLAVISEPT see DBX400
FLAVOMYCELIN see LIV000
FLAVOMYCIN see MRA250
FLAVOPHOSPHOLIPOL see MRA250
FLAVUROL see MCV000
FLAXEDIL see PDD300
FLEBOCORTID see HHR000
FLECK-FLIP see TIO750
FLECTOL H see TLP500
FLEET-X see TLM050
FLEXAL see IPU000
FLEXAMINE G see BLE500
FLEXARTAL see IPU000
FLEXARTEL see IPU000
FLEXAZONE see BRF500
FLEXIMEL see DVL700
FLEXOL DOP see DVL700
FLEXOL PLASTICIZER DIP see ILR100
FLEXOL PLASTICIZER DOP see DVL700
FLEXOL PLASTICIZER TCP see TNP500
FLIBOL E see TIQ250
FLIEGENTELLER see TIQ250
FLINT see SCI500, SCJ500
FLIT 406 see CBG000
FLOCOOL 180 see SJC500
FLOCOR see PKQ059
FLO-GARD see SCH000
FLOGHENE see HNI500
FLOGICID see BPP750
FLOGISTIN see HNI500
FLOGITOLO see HNI500
FLOGOCID N PLASTIGEL see BPP750
FLOGODIN see HNI500
FLOGORIL see HNI500
FLOGOSTOP see HNI500
FLO-MOR see PAI000
FLOMORE see BSQ750
FLOPIRINA see HNI500
FLO PRO T SEED PROTECTANT see TFS350
FLO PRO V SEED PROTECTANT see CCC500
FLORALTONE see GEM000
FLORDIMEX see CDS125
FLOREL see CDS125
FLORES MARTIS see FAU000
FLORIDINE see SHF500
FLOROCID see SHF500
FLOROPIPAMIDE see FHG000
FLOROPRYL see IRF000
FLOROXENE see TKB250
FLOVACIL see DKI600

FLOWERS of ANTIMONY see AQF000
FLOWERS of SULPHUR (DOT) see SOD500
FLOWERS of ZINC see ZKA000
FLOXURIDIN see DAR400
FLOXURIDINE see DAR400
FLOZENGES see SHF500
FLUANISONE HYDROCHLORIDE see FDA880
FLUANXOL see FMO129
FLUATE see TIO750
FLUE DUST, ARSENIC containing see ARE500
FLUE GAS see CBW750
FLUIMUCETIN see ACH000
FLUIMUCIL see ACH000
FLUKOIDS see CBY000
FLUMAMINE see DQR600
FLUMICIL see ACH000
FLUNIGET see DKI600
FLUO-KEM see TAI250
FLUOMETURON see DUK800
FLUOMINE see EIS000
FLUOMINE DUST see EIS000
FLUON see TAI250
FLUOOXENE see TKB250
FLUOPERIDOL see FLU000
FLUOPHOSGENE see CCA500
FLUOPHOSPHORIC ACID DI(DIMETHYLAMIDE) see
  BJE750
FLUOPHOSPHORIC ACID, DIETHYL ESTER see DJJ400
FLUOPHOSPHORIC ACID, DIISOPROPYL ESTER see
  IRF000
FLUOPHOSPHORIC ACID, DIMETHYL ESTER see DSA800
FLUOR (DUTCH, FRENCH, GERMAN, POLISH) see
  FEZ000
FLUORACETATO di (ITALIAN) see SHG500
5-FLUORACIL (GERMAN) see FMM000
FLUORACIZINE see FDE000
FLUORAKIL 100 see FFF000
FLUORAL see SHF500
3,6-FLUORANDIOL see FEV000
3',6'-FLUORANDIOL see FEV000
FLUORANE 114 see FOO509
4-FLUORANILIN see FFY000
FLUORANTHENE see FDF000
5-FLUOR-DESOXYCYTIDIN (GERMAN) see FHO000
FLUOREN-2-AMINE see FDI000
2-FLUORENAMINE see FDI000
2-FLUORENEAMINE see FDI000
2-FLUORENYLACETAMIDE see FDR000
N-FLUOREN-2-YL ACETAMIDE see FDR000
N-2-FLUORENYLACETAMIDE see FDR000
N-FLUOREN-4-YLACETAMIDE see ABY000
N-4-FLUORENYLACETAMIDE see ABY000
FLUORENYL-2-ACETHYDROXAMIC ACID see HIP000
N-(FLUOREN-2-YL)ACETOHYDROXAMIC ACETAMIDE see
  ABL000
N-FLUOREN-2-YL ACETOHYDROXAMIC ACID see HIP000
N-2-FLUORENYL ACETOHYDROXAMIC ACID see HIP000
N-FLUOREN-2-YL BENZOHYDROXAMIC ACID see
  FDZ000
N-(2-FLUORENYL)BENZOHYDROXAMIC ACID see
  FDZ000
2,7-FLUORENYLBISACETAMIDE see BGP250
N,N'-FLUOREN-2,7-YLBISACETAMIDE see BGP250
2-FLUORENYLDIACETAMIDE see DBF200
N-FLUOREN-2-YLDIACETAMIDE see DBF200
N-2-FLUORENYLDIACETAMIDE see DBF200
N,N'-FLUOREN-2,7-YLENEBISACETAMIDE see BGP250
N,N'-2,7-FLUORENYLENEBISACETAMIDE see BGP250
N,N'-(FLUOREN-2,7-YLENE)BIS(ACETYLAMINE) see
  BGP250
N,N'-2,7-FLUORENYLENEDIACETAMIDE see BGP250
N-9H-FLUOREN-2-YL-N-HYDROXYBENZAMIDE see
  FDZ000
FLUORESCEIN see FEV000

FLUORESCEINE see FEV000
FLUORESCEIN SODIUM see FEW000
FLUORESCEIN SODIUM B.P see FEW000
FLUORESCEIN, SOLUBLE see FEW000
FLUORESSIGAEURE (GERMAN) see SHG500
FLUORIC ACID see HHV500
FLUORIDENT see SHF500
FLUORIDES see FEY000
FLUORID SODNY (CZECH) see SHF500
FLUORIGARD see SHF500
FLUORINE see FEZ000
FLUORINE, compressed (DOT) see FEZ000
FLUORINEED see SHF500
FLUORINE MONOXIDE see ORA000
FLUORINE OXIDE see ORA000
FLUORINE PERCHLORATE see FFD000
FLUORINSE see SHF500
FLUOR-I-STRIP A.T. see FEW000
FLUORITAB see SHF500
FLUOR-O-KOTE see SHF500
FLUORO (ITALIAN) see FEZ000
FLUOROACETALDEHYDE see FFE000
FLUOROACETAMIDE see FFF000
2-FLUOROACETAMIDE see FFF000
7-FLUORO-2-ACETAMIDO-FLUORENE see FFG000
FLUOROACETANILIDE see FFH000
2-FLUOROACETANILIDE see FFH000
FLUOROACETATE see FIC000
FLUOROACETIC ACID see FIC000
2-FLUOROACETIC ACID see FIC000
FLUOROACETIC ACID (DOT) see FIC000
FLUOROACETIC ACID AMIDE see FFF000
FLUOROACETIC ACID METHYL ESTER see MKD000
FLUOROACETIC ACID, SODIUM SALT see SHG500
FLUOROACETIC ACID, TRIETHYLLEAD SALT see TJS500
FLUOROACETONITRILE see FFJ000
7-FLUORO-2-ACETYLAMINOFLUORENE see FFG000
p-FLUOROACETYLAMINOPHENYL DERIVATIVE of NITRO-
  GEN MUSTARD see BHP750
FLUOROACETYL CHLORIDE see FFR000
o-(FLUOROACETYL)SALICYLIC ACID see FFT000
5-FLUORO AMYLAMINE see FFV000
5-FLUOROAMYL THIOCYANATE see FFX000
4-FLUOROANILINE see FFY000
p-FLUOROANILINE see FFY000
4-FLUOROBENZENAMINE see FFY000
FLUOROBENZENE see FGA000
1-(4'-FLUOROBENZOIL)-3-PIRROLIDINOPROPANO
  MALEATO (ITALIAN) see FGW000
6-FLUOROBENZO(a)PYRENE see FGI100
1'-(3-(p-FLUOROBENZOYL)PROPYL)(1,4'-BIPIPERIDINE
  1-4'-CARBOXAMIDE see FHG000
1-(3-(p-FLUOROBENZOYL)PROPYL)-4-PIPERIDINOISONIPA-
  COTAMIDE see FHG000
1-(3-(4-FLUOROBENZOYL)PROPYL)-4-PIPERIDYL-N-ISO-
  PROPYL CARBAMATE see FGV000
1-(3-(4-FLUOROBENZOYL)PROPYL)-4-(2-PYRIDYL)PIPER-
  AZINE see FLU000
1-(1-(3-(p-FLUOROBENZOYL)PROPYL)-1,2,3,6-TETRAHY-
  DRO-4-PYRIDYL)-2 -BENZIMIDAZOLINONE see
  DYF200
1-(4'-FLUOROBENZOYL)-3-PYRROLIDINYLPROPANE MA-
  LEATE see FGW000
FLUOROBISISOPROPYLAMINO- PHOSPHINE OXIDE see
  PHF750
FLUOROBLASTIN see FMM000
4-FLUOROBUTYL THIOCYANATE see FHD000
4-FLUOROBUTYRIC ACID METHYL ESTER see MKE000
FLUOROBUTYROPHENONE see FHG000
FLUOROCARBON-12 see DFA600
FLUOROCARBON-22 see CFX500
FLUOROCARBON 113 see FOO000
FLUOROCARBON 114 see FOO509
FLUOROCARBON-115 see CJI500

FLUOROCARBON FC142b see CFX250
FLUOROCARBON No. 11 see TIP500
FLUOROCHROME see MCV000
4-FLUORO-CROTONIC ACID METHYL ESTER see MKE250
2-FLUORO-2'-CYANODIETHYL ETHER see CON500
5-FLUORODEOXYCYTIDINE see FHO000
5-FLUORO-2'-DEOXYCYTIDINE see FHO000
FLUORODEOXYURIDINE see DAR400
5-FLUORODEOXYURIDINE see DAR400
5-FLUORO-2-DEOXYURIDINE see DAR400
5-FLUORO-2'-DEOXYURIDINE see DAR400
β-5-FLUORO-2'-DEOXYURIDINE see DAR400
FLUORODICHLOROMETHANE see DFL000
FLUORODIISOPROPYL PHOSPHATE see IRF000
4'-FLUORO-4-DIMETHYLAMINOAZOBENZENE see DSA000
4'-FLUORO-p-DIMETHYLAMINOAZOBENZENE see DSA000
4'-FLUORO-N,N-DIMETHYL-4-AMINOAZOBENZENE see DSA000
5-FLUORO-7,12-DIMETHYLBENZ(a)ANTHRACENE see DRY600
11-FLUORO-7,12-DIMETHYLBENZ(a)ANTHRACENE see DRZ000
4'-FLUORO-N,N-DIMETHYL-p-PHENYLAZOANILINE see DSA000
1,2,4-FLUORODINITROBENZENE see DUW400
1-FLUORO-2,4-DINITROBENZENE see DUW400
FLUOROETHANE see FIB000
FLUOROETHANOIC ACID see FIC000
FLUOROETHANOL see FID000
2-FLUOROETHANOL see FIE000
β-FLUOROETHANOL see FIE000
2-FLUOROETHANOL, PHOSPHITE (3:1) see PHO250
FLUOROETHENE see VPA000
FLUOROETHYL see HDC000
FLUOROETHYLENE see VPA000
2-FLUOROETHYL ESTER DIPHENYLACETIC ACID see FIP999
2-FLUOROETHYL FLUOROACETATE see FIM000
β-FLUOROETHYL FLUOROACETATE see FIM000
2-FLUORO ETHYL-γ-FLUORO BUTYRATE see FIN000
β-FLUOROETHYL-γ-FLUOROBUTYRATE see FIN000
2-FLUOROETHYL-5-FLUOROHEXOATE see FIO000
β-FLUOROETHYLIC ESTER of XENYLACETIC ACID see FIP999
FLUOROFLEX see TAI250
N-(7-FLUOROFLUORENE-2-YL)ACETAMIDE see FFG000
4'-FLUORO-4-(8-FLUORO-2,3,4,5-TETRAHYDRO-1H-PYRIDO(4,3-b)INDOL-2-YL)BUTYROPHENONE HYDROCHLORIDE see FIW000
FLUOROFORM see CBY750
FLUOROFORMYL FLUORIDE see CCA500
4-FLUORO-3-HYDROXY-BUTANETHIOIC ACID METHYL ESTER see MKF000
γ-FLUORO-β-HYDROXY-BUTYRIC ACID METHYL ESTER see MKE750
4'-FLUORO-4-(4-HYDROXY-4-p-TOLYLPIPERIDINO)BUTYROPHENONE, HYDROCHLORIDE see MNN250
4'-FLUORO-4-(4-HYDROXY-4-(α,α,α-TRIFLUORO-m-TOLYL)PIPERIDINO)BUTYROPHENONE see TKK500
4-FLUORO-4,4-IDROSSI-4-(m-TRIFLUOROMETIL-FENIL)-PIPERIDINO-BUTIRROFENONE (ITALIAN) see TKK500
FLUOROISOPROPOXYMETHYLPHOSPINE OXIDE see IPX000
FLUOROLON 4 see TAI250
FLUOROMAR see TKB250
FLUOROMETHANE see FJK000
4'-FLUORO-4-(4-(o-METHOXYPHENYL)-1-PIPERAZINYL)BUTYROPHENONE HYDROCHLORIDE see FDA880
Δ¹-9-α-FLUORO-16-α-METHYLCORTISOL see SOW000
FLUOROMETHYL CYANIDE see FFJ000
2-FLUORO-N-METHYL-N-1-NAPHTHALENYLACETAMIDE see MME809

2-FLUORO-N-METHYL-N-1-NAPHTHYLACETAMIDE see MME809
4'-FLUORO-4-(4-METHYLPIPERIDINO)BUTYROPHENONE HYDROCHLORIDE see FKI000
9-α-FLUORO-16-α-METHYLPREDNISOLONE see SOW000
9-α-FLUORO-16-β-METHYLPREDNISOLONE see BFV750
9-α-FLUORO-16-α-METHYL-1,4-PREGNADIENE-11-β,17-α,21-TRIOL-3,20-DIONE see SOW000
9-α-FLUORO-16-β-METHYL- 1,4-PREGNADIENE-11-β,17-α,21-TRIOL-3,20-DIONE see BFV750
4-α-FLUORO-16-α-METHYL-11-β,17,21-TRIHYDROXY-PREGNA-1,4-DIENE-3,20-DIONE see SOW000
FLUOROMETHYL(1,2,2-TRIMETHYLPROPOXY)PHOSPHINE OXIDE see SKS500
8-FLUOROOCTANOIC ACID, ETHYL ESTER see EKK500
FLUOROPAK 80 see TAI250
5-FLUOROPENTYLAMINE see FFV000
5-FLUOROPENTYL THIOCYANATE see FFX000
2-FLUORO-N-PHENYLACETAMIDE see FFH000
p-FLUOROPHENYLAMINE see FFY000
p-((p-FLUOROPHENYL)AZO)-N,N-DIMETHYLANILINE see DSA000
4-((4-FLUOROPHENYL)AZO)-N,N-DIMETHYLBENZEN-AMINE see DSA000
1-(4-FLUOROPHENYL)-4-(4-HYDROXY-4-(4-METHYLPHE-NYL)-1-PIPERIDINYL)-1-BUTANONE HYDROCHLORIDE see MNN250
1-(4-FLUOROPHENYL)-4-(4-HYDROXY-4-(3-(TRIFLUORO-METHYL)PHENYL)-1-PIPERIDINYL)-1-BUTANONE see TKK500
(+-)-α-(p-FLUOROPHENYL)-4-(o-METHOXYPHENYL)-1-PI-PERAZINEBUTANOL see HAH000
dl-1-(4-FLUOROPHENYL)-4-(1-(4-(2-METHOXY-PHENYL))-PIPERAZINYL)BUTANOL see HAH000
8-(4-p-FLUORO PHENYL-4-OXOBUTYL)-2-METHYL-2,8-DIAZASPIRO(4.5)DECANE-1,3-DIONE see FLJ000
1-(1-(4-(p-FLUOROPHENYL-4-OXOBUTYL)-1,2,3,6-TETRA-HYDRO-4-PYRIDYL)-2-BENZIMIDAZOLINONE see DYF200
1-(4-FLUOROPHENYL)-4-(4-(2-PYRIDINYL)-1-PIPERAZI-NYL)-1-BUTANONE see FLU000
FLUOROPHOSGENE see CCA500
FLUOROPHOSPHORIC ACID, anhydrous see PHJ250
4'-FLUORO-4-N-PIPERIDINO-4-CARBAMIDO-PIPERIDINO)BUTYROPHENONE see FHG000
p-FLUORO-γ-(4-PIPERIDINO-4-CARBAMOYLPIPERI-DINO)BUTYROPHENONE see FHG000
FLUOROPLAST 3 see CLQ750
FLUOROPLAST 4 see TCH500
FLUOROPLEX see FMM000
3-FLUOROPROPENE see AGG500
2-FLUORO-2-PROPEN-1-OL see FLQ000
FLUOROPRYL see IRF000
4'-FLUORO-4-(4-(2-PYRIDYL)-1-PIPERAZINYL)BUTYRO-PHENONE see FLU000
5-FLUORO-2,4-PYRIMIDINEDIONE see FMM000
5-FLUORO-2,4(1H,3H)-PYRIMIDINEDIONE see FMM000
4'-FLUORO-4-(n-(4-PYRROLIDINAMIDO-4-m-TOLYPIPERI-DINO)BUTYROPHENONE see FLV000
4'-FLUORO-4-(1-PYRROLIDINYL)BUTYROPHENONE MA-LEATE see FGW000
FLUOROSILICIC ACID see SCO500
FLUOROSUFONIC ACID (DOT) see FLZ000
FLUOROSULFURIC ACID see FLZ000
FLUOROTANE see HAG500
FLUOROTHYL see HDC000
p-FLUOROTOLUENE see FMC000
FLUOROTRICHLOROMETHANE (OSHA) see TIP500
4-FLUORO-4'-TRIFLUOROMETHYLBENZOPHENONE GUANYLHYDRAZONE HYDROCHLORIDE see FMH000
9-FLUORO-11-β,17,21-TRIHYDROXY-16-α-METHYL-PREGNA-1,4-DIENE-3,20-DIONE see SOW000

9-FLUORO-11-β,17,21-TRIHYDROXY-16-β-METHYL-
PREGNA-1,4-DIENE-3,20-DIONE see BFV750
9-α-FLUORO-11-β,17,21-TRIHYDROXY-16-β-METHYL-
PREGNA-1,4-DIENE- 3,20-DIONE see BFV750
9-α-FLUORO-11-β,17-α,21-TRIHYDROXY-16-α-METHYL-
PREGNA-1,4-DIENE-3,20-DIONE see SOW000
FLUOROTROJCHLOROMETAN (POLISH) see TIP500
3-FLUOROTYROSIN see FMJ000
3-FLUOROTYROSINE see FMJ000
m-FLUOROTYROSINE see FMJ000
FLUOROURACIL see FMM000
5-FLUOROURACIL see FMM000
5-FLUOROURACIL DEOXYRIBOSIDE see DAR400
5-FLUOROURACIL-2'-DEOXYRIBOSIDE see DAR400
5-FLUOROURIDINE see FMN000
FLUOROWODOR (POLISH) see HHU500
FLUOROXENE see TKB250
FLUORPLAST 4 see TAI250
5-FLUORPROPYRIMIDINE-2,4-DIONE see FMM000
FLUORTHYRIN see FMJ000
3-FLUORTYROSIN (GERMAN) see FMJ000
5-FLUORURACIL (GERMAN) see FMM000
FLUORURE de BORE (FRENCH) see BMG700
FLUORURE de N,N'-DIISOPROPYLE PHOSPHORODIAMIDE
(FRENCH) see PHF750
FLUORURE de POTASSIUM (FRENCH) see PLF500
FLUORURES ACIDE (FRENCH) see FEZ000
FLUORURE de SODIUM (FRENCH) see SHF500
FLUORURE de SULFURYLE (FRENCH) see SOU500
FLUORURE de N,N,N',N'-TETRAMETHYLE PHOSPHORO-
DIAMIDE (FRENCH) see BJE750
FLUORURE de THIONYLE (FRENCH) see TFL250
FLUORURI ACIDI (ITALIAN) see FEZ000
FLUORURIDINE DEOXYRIBOSE see DAR400
FLUORWASSERSTOFF (GERMAN) see HHU500
FLUORWATERSTOF (DUTCH) see HHU500
FLUORXENE see TKB250
FLUORYL see CBY750
FLUOSILICATE de AMMONIUM (FRENCH) see COE000
FLUOSILICATE de MAGNESIUM (FRENCH) see MAG250
FLUOSILICATE de SODIUM see DXE000
FLUOSILICATE de ZINC see ZIA000
FLUOSILICIC ACID see SCO500
FLUOSTIGMINE see IRF000
FLUOSULFONIC ACID (DOT) see FLZ000
FLUOTHANE see HAG500
FLUOTITANATE de POTASSIUM (FRENCH) see PLI000
FLUPENTHIXOL see FMO129
(α,β)-FLUPENTHIXOL see FMO129
cis-(Z)-FLUPENTHIXOL see FMO129
FLUPENTHIXOLE see FMO129
FLUPENTIXOL see FMO129
FLURACIL see FMM000
FLURA-GEL see SHF500
FLURAZEPAM see FMQ000
FLURAZEPAM HYDROCHLORIDE see DAB800
FLURAZEPAN DIHYDROCHLORIDE see DAB800
FLURCARE see SHF500
FLURENTIXOL see FMO129
FLURI see FMM000
FLURIL see FMM000
FLUROTHYL see HDC000
FLUROXENE see TKB250
FLUVIN see CFY000
FLUXANXOL see FMO129
FLUX MAAG see NDN000
FLY BAIT GRITS see SOY000
FLY-DIE see DGP900
FLY FIGHTER see DGP900
FLYPEL see DKC800
FMA see ABU500
FMC 249 see AFR250
FMC-1240 see EEH600
FMC 5273 see PIX250

FMC 5462 see EAQ750
FMC 5488 see CKM000
FMC 10242 see CBS275
FMC 17370 see BEP500
FMC 33297 see AHJ750
FMC 41655 see AHJ750
FNT see NDY500
FOBEX see BCA000
FOLACIN see FMT000
FOLATE see FMT000
FOLBEX see DER000
FOLBEX SMOKE-STRIPS see DER000
FOLCID see CBF800
FOLCYSTEINE see FMT000
FOLETHION see DSQ000
FOLEX see TIG250
FOLIANDRIN see OHQ000
FOLIC ACID see FMT000
FOLIC ACID, 4-AMINO- see AMG750
FOLIDOL see PAK000
FOLIDOL E605 see PAK000
FOLIDOL E & E 605 see PAK000
FOLIDOL M see MNH000
FOLIKRIN see EDV000
FOLIMAT see DNX800
FOLINERIN see OHQ000
FOLINEVIN see OHQ000
FOLIONE see MND275
FOLIPEX see EDV000
FOLISAN see EDV000
FOLLESTRINE see EDV000
FOLLICORMON see EDP000
FOLLICULAR HORMONE see EDV000
FOLLICULAR HORMONE HYDRATE see EDU500
FOLLICULIN see EDV000
FOLLICULINE BENZOATE see EDV000
FOLLICUNODIS see EDV000
FOLLICYCLIN P see EDR000
FOLLIDIENE see DAL600, DKA600
FOLLIDRIN see EDP000, EDV000
FOLLORMON see DAL600
FOLOSAN see PAX000
FOMAC see HCL000
FOMAC 2 see PAX000
FOMREZ SUL-3 see DBF800
FOMREZ SUL-4 see DDV600
FONATOL see DKA600
FONOFOS see FMU045
FONOLINE see MQV750
FONTARSOL see DFX400
FONURIT see AAI250
FOOD BLUE 1 see FMU059
FOOD BLUE 2 see FAE000
FOOD BLUE DYE No. 1 see FAE000
FOOD GREEN 2 see FAF000
FOOD RED 5 see HJF500
FOOD RED 14 see FAG040
FOOD RED 15 see FAG070
FOOD YELLOW No. 4 see FAG140
FORAAT (DUTCH) see PGS000
FOREDEX 75 see DAA800
FORLIN see BBQ500
FORMAGENE see PAI000
FORMAL see MAK700, MGA850
FORMALDEHYD (CZECH, POLISH) see FMV000
FORMALDEHYDE see FMV000
FORMALDEHYDE, solution (DOT) see FMV000
FORMALDEHYDE BIS(β-CHLOROETHYL) ACETAL see
BID750
FORMALDEHYDE CYANOHYDRIN see HIM500
FORMALDEHYDE DIMETHYLACETAL see MGA850
FORMALIN see FMV000
FORMALIN 40 see FMV000
FORMALIN (DOT) see FMV000

FORMALINA (ITALIAN) see FMV000
FORMALINE (GERMAN) see FMV000
FORMALINE BLACK C see AQP000
FORMALIN-LOESUNGEN (GERMAN) see FMV000
FORMALITH see FMV000
FORMAL-γ-TRIMETHYLAMMONIUM PROPANEDIOL see FMX000
FORMAMIDE see FMY000
FORMAMINE see HEI500
FORMARIN see MNM500
FORMATRIX see ECU750
FORMETANATE HYDROCHLORIDE see DSO200
FORMIATE de METHYLE (FRENCH) see MKG750
FORMIATE de PROPYLE (FRENCH) see PNM500
FORMIC ACID see FNA000
FORMIC ACID, ALLYL ESTER see AGH000
FORMIC ACID, CALCIUM SALT see CAS250
FORMIC ACID, CINNAMYL ESTER see CMR500
FORMIC ACID, CITRONELLYL ESTER see CMT750
FORMIC ACID-3,7-DIMETHYL-6-OCTEN-1-YL ESTER see CMT750
FORMIC ACID, ETHYL ESTER see EKL000
FORMIC ACID, GERANIOL ESTER see GCY000
FORMIC ACID, HEPTYL ESTER see HBO500
FORMIC ACID, ISOBUTYL ESTER see IIR000
FORMIC ACID, ISOPENTYL ESTER see IHS000
FORMIC ACID, ISOPROPYL ESTER see IPC000
FORMIC ACID, METHYLHYDRAZIDE see FNW000
FORMIC ALDEHYDE see FMV000
FORMIC BLACK C see AQP000
FORMIC ETHER see EKL000
FORMIC 2-(4-(5-NITROFURYL)-2-THIAZOLYL)HYDRAZIDE see NDY500
FORMIN see HEI500
FORMOL see FMV000
FORMOLA 40 see DAA800
FORMOSA CAMPHOR see CBA750
FORMOSA CAMPHOR OIL see CBB500
FORMOSE OIL of CAMPHOR see CBB500
FORMOSULFATHIAZOLE see TEX250
FORMOTHION see DRR200
FORMVAR 1285 see AAX250
2-FORMYLAMINO-4-(5-NITRO-2-FURYL)THIAZOLE see NGM500
4-FORMYLCYCLOHEXENE see FNK025
2-FORMYL-3,4-DIHYDRO-2H-PYRAN see ADR500
N-FORMYLDIMETHYLAMINE see DSB000
α-FORMYLETHYLBENZENE see COF000
2-(2-FORMYLHYDRAZINO)-4-(5-NITRO-2-FURYL)THI-AZOLE see NDY500
N-FORMYL HYDROXYAMINOACETIC ACID see FNO000
N-FORMYL-N-HYDROXYGLYCINE see FNO000
FORMYLIC ACID see FNA000
S-(2-(FORMYLMETHYLAMINO)-2-OXOETHYL)-O,O-DI-METHYLPHOSPHORODITHIOATE see DRR200
N-FORMYL-N-METHYLCARBAMOYLMETHYL-O,O-DI-METHYL PHOSPHORODITHIOATE see DRR200
S-(N-FORMYL-N-METHYLCARBAMOYLMETHYL)-O,O-DI-METHYL PHOSPHORODITHIOATE see DRR200
S-(N-FORMYL-N-METHYLCARBAMOYLMETHYL) DI-METHYL PHOSPHOROTHIOLOTHIONATE see DRR200
1-FORMYL-1-METHYLHYDRAZINE see FNW000
N-FORMYL-N-METHYLHYDRAZINE see FNW000
2-FORMYL-1-METHYLPYRIDINIUM CHLORIDE OXIME see FNZ000
2-FORMYL-1-METHYLPYRIDINIUM IODIDE OXIME see POS750
2-FORMYL-N-METHYLPYRIDINIUM OXIME CHLORIDE see FNZ000
2-FORMYL-N-METHYLPYRIDINIUM OXIME IODIDE see POS750
2-FORMYLQUINOXALINE-1,4-DIOXIDE CARBOMETHOX-YHYDRAZONE see FOI000
N-FORMYL-l-p-SARCOLYSIN see BHX250

FORMYL TRICHLORIDE see CHJ500
8-FORMYL-1,6,7-TRIHYDROXY-5-ISOPROPYL-3-METHYL-2,2'-BISNAPHTHALENE see GJM000
FORMYL VIOLET S4BN see FAG120
FOROTOX see TIQ250
FORRON see TAA100
FORST U 46 see TAA100
FOR-SYN see BEP500
FORTALGESIC see DOQ400
FORTALIN see DOQ400
FORTECORTIN see SOW000
FORTEX see TAA100
FORTHION see MAK700
FORTIGRO see FOI000
FORTION NM see DSP400
FORTODYL see VSZ100
FORTRACIN see BAC250
FORTRAL see DOQ400
FORTROL see BLW750
FORTURF see TBQ750
FOSCHLOR see TIQ250
FOSCHLOREM (POLISH) see TIQ250
FOSCHLOR R-50 see TIQ250
FOSDRIN see MQR750
FOSFAKOL see NIM500
FOS-FALL ''A'' see BSH250
FOSFAMID see DSP400
FOSFAZIDE see ILE000
FOSFERMO see PAK000
FOSFERNO see PAK000
FOSFEX see PAK000
FOSFIVE see PAK000
FOSFONO 50 see EEH600
FOSFORO BIANCO (ITALIAN) see PHO750
FOSFORO(PENTACHLORURO di) (ITALIAN) see PHR5(
FOSFORO(TRICLORURO di) (ITALIAN) see PHT275
FOSFOROWODOR (POLISH) see PGY000
FOSFORPENTACHLORIDE (DUTCH) see PHR500
FOSFORTRICHLORIDE (DUTCH) see PHT275
FOSFORYN TROJETYLOWY (CZECH) see TJT800
FOSFORYN TROJMETYLOWY (CZECH) see TMD500
FOSFORZUUROPLOSSINGEN (DUTCH) see PHB250
FOSFOTHION see MAK700
FOSFOTION see MAK700
FOSFOTOX see DSP400
FOSFURI di ALLUMINIO (ITALIAN) see AHE750
FOSFURI di MAGNESIO (ITALIAN) see MAI000
FOSGEEN (DUTCH) see PGX000
FOSGEN (POLISH) see PGX000
FOSGENE (ITALIAN) see PGX000
FOSOVA see PAK000
FOSSIL FLOUR see SCH000
FOSTER GRANT 834 see SMQ500
FOSTERN see PAK000
FOSTEX see BDS000
FOSTION see IOT000
FOSTION MM see DSP400
FOSTOX see PAK000
FOSTRIL see HCL000
FOSVEL see LEN000
FOSVEX see TCF250, TCF280
FOUMARIN see ABF500
FOURAMIEN 2R see ALL750
FOURAMINE see TGL750
FOURAMINE BA see DBO400
FOURAMINE BROWN AP see PPQ500
FOURAMINE D see PEY500
FOURAMINE EG see ALT500
FOURAMINE ERN see NAW500
FOURAMINE OP see ALT000
FOURAMINE PCH see CCP850
FOURAMINE RS see REA000
FOURNEAU 190 see ABX500
FOURNEAU 933 see BCI500

FOURNEAU 1162 see SNM500
FOURNEAU 2268 see MJH250
FOURRINE 1 see PEY500
FOURRINE 36 see ALL750
FOURRINE 57 see NEM480
FOURRINE 65 see ALT500
FOURRINE 68 see CCP850
FOURRINE 79 see REA000
FOURRINE 99 see NAW500
FOURRINE BROWN 2R see ALL750
FOURRINE BROWN PR see NEM480
FOURRINE BROWN PROPYL see NEM480
FOURRINE D see PEY500
FOURRINE EG see ALT500
FOURRINE ERN see NAW500
FOURRINE M see TGL750
FOURRINE PG see PPQ500
FOURRINE SLA see DBO400
FOUR THOUSAND FORTY-NINE see MAK700
FOVANE see BDE250
FOWLER'S SOLUTION see FOM050
FOXGLOVE see DKL200
FOZALON see BDJ250
FPA see FHG000
FR-33 see FLJ000
FRABEL see HNI500
FRACINE see NGE500
FRADIOMYCIN SULFATE see NCG000
FRAESEOL see ENC000
FRAMBINONE see RBU000
FRAMED see BJP000
FRANKINCENSE GUM see OIM000
FRANKLIN see CAO000
FRANOCIDE see DIW200
FRANOZAN see DIW200
FRATOL see SHG500
FREE BENZYLPENICILLIN see BDY669
FREE COCONUT OIL see CNR000
FREE HISTAMINE see HGD000
FREEMANS WHITE LEAD see LDY000
FREEURIL see BDE250
FRENANTOL see ELL500
FRENCH GREEN see COF500
FRENOGASTRICO see XCJ000
FRENOHYPON see ELL500
FRENOLON DIFUMARATE see MDU750
FRENTIROX see MCO500
FREON see CFX500
FREON 11 see TIP500
FREON 13 see CLR250
FREON 14 see CBY250
FREON 21 see DFL000
FREON 22 see CFX500
FREON 23 see CBY750
FREON 30 see MJP450
FREON 31 see CHI900
FREON 41 see FJK000
FREON 112 see TBP050
FREON 113 see FOO000
FREON 114 see FOO509
FREON 115 see CJI500
FREON 142 see CFX250
FREON 152 see ELN500
FREON 500 see DFB400
FREON 12-B2 see DKG850
FREON 13B1 see TJY100
FREON 142b see CFX250
FREON C-318 see CPS000
FREON F-12 see DFA600
FREON F-23 see CBY750
FREON MF see TIP500
FREON 113TR-T see FOO000
FRESMIN see VSZ000
FRIDERON see RBF100

FRIGEN see CFX500
FRIGEN 11 see TIP500
FRIGEN 12 see DFA600
FRIGEN 114 see FOO509
FRIGEN 113a see FOO000
FRIGIDERM see FOO509
FRISIUM see CIR750
FRUCOTE see BPY000
FRUCTOSE (FCC) see LFI000
FRUITDO see BLC250
FRUITONE A see TAA100
FRUITONE T see TIX500
FRUIT RED A EXTRA YELLOWISH GEIGY see HJF500
FRUIT SUGAR see LFI000
FRUMIN AL see DXH325
FRUSEMIDE see CHJ750
FRUSEMIN see CHJ750
FRUSID see CHJ750
FRUTABS see LFI000
FTAALZUURANHYDRIDE (DUTCH) see PHW750
F1-TABS see SHF500
FTAFLEX DIBA see DNH125
FTALOPHOS see PHX250
FTALOWY BEZWODNIK (POLISH) see PHW750
FTBG see FMH000
FTORLON 4 see TAI250
FTOROPLAST 4 see TAI250
FTOROTAN (RUSSIAN) see HAG500
5-FU see FMM000
p-FUCHSIN see RMK020
FUCHSINE BASE see MAC500
FUCHSINE DR-001 see RMK020
FUCHSINE SPC see RMK020
FUCLASIN see BJK500
FUCLASIN ULTRA see BJK500
FUDR see DAR400
5-FUDR see DAR400
FUEL OIL see FOP000
FUGU POISON see FOQ000
FUJITHION see FOR000
FUKI-NO-TOH (JAPANESE) see PCR000
FUKLASIN see BJK500
FUKLASIN ULTRA see FAS000
FULCIN see GKE000
FULCINE see GKE000
FUL-GLO see FEW000
FULMINATE of MERCURY, DRY (DOT) see MDC000
FULMINATE of MERCURY, WET (DOT) see MDC250
FULMINATES see FOS000
FULMINIC ACID see FOS050
FULSIX see CHJ750
FULUVAMIDE see CHJ750
FULVICAN GRISACTIN see GKE000
FULVICIN see GKE000
FULVINA see GKE000
FULVINE see FOT000
FULVISTATIN see GKE000
FUMAFER see FBJ100
FUMAGON see DDL800
FUMAR-F see FBJ100
FUMARIC ACID see FOU000
FUMARIC ACID, DIBUTYL ESTER see DEC600
FUMARINE see FOW000
FUMAROYL CHLORIDE see FOY000
FUMARYLCHLORID (CZECH) see FOY000
FUMARYL CHLORIDE see FOY000
FUMAZONE see DDL800
FUMED SILICA see SCH000
FUMED SILICON DIOXIDE see SCH000
FUMETOBAC see NDN000
FUMIGANT-1 (OBS.) see MHR200
FUMIGRAIN see ADX500
FUMING LIQUID ARSENIC see ARF500
FUMING SULFURIC ACID (DOT) see SOI520

FUMIRON see FBJ100
FUMITOXIN see AHE750
FUMO-GAS see EIY500
FUNDAL see CJJ250
FUNDAL 500 see CJJ250
FUNDASOL see BAV575
FUNDEX see CJJ250
FUNDUSCEIN see FEW000
FUNGACETIN see THM500
FUNGAFLOR see FPB875
FUNGICIDE 1991 see BAV575
FUNGICLOR see PAX000
FUNGIFEN see PAX250
FUNGILIN see AOC500
FUNGISONE see AOC500
FUNGITOX OR see ABU500
FUNGIVIN see GKE000
FUNGOCIN see BAB750
FUNGOL B see SHF500
FUNGOSTOP see BJK500
FUNGUS BAN TYPE II see CBG000
FUNICOLOSIN see FPD000
FUR see FMN000
5-FUR see FMN000
FURACILLIN see NGE500
FURACINETTEN see NGE500
FURACOCCID see NGE500
FURACORT see NGE500
FURACYCLINE see NGE500
FURADAN see CBS275
FURADANTIN see NGE000
FURADONIN see NGE000
FURAL see FPQ875
2-FURALDEHYDE see FPQ875
FURALDON see NGE500
FURALE see FPQ875
l-FURALTADONE HYDROCHLORIDE see FPI150
FURAN see FPK000
2-FURANALDEHYDE see FPQ875
2-FURANCARBINOL see FPU000
2-FURANCARBONAL see FPQ875
2-FURANCARBOXALDEHYDE see FPQ875
2,5-FURANDIONE see MAM000
FURANIDINE see TCR750
FURANIUM see FEW000
2-FURANMETHANOL see FPU000
2-FURANMETHYLAMINE see FPW000
FURAN-OFTENO see NGE500
FURANTHRIL see CHJ750
FURANTHRYL see CHJ750
FURANTOIN see NGE000
FURANTRIL see CHJ750
1-(3-FURANYL)-4-METHYL-1-PENTANONE see PCI750
FURAPLAST see NGE500
FURASEPTYL see NGE500
FURATOL see SHG500
FURATONE see BKH500
FURATONE-S see BKH500
FURAXONE see NGG500
FURAZOL see NGG500
FURAZOLIDON see NGG500
FURAZOLIDONE (USDA) see NGG500
FURAZON see NGG500
FURAZONE see NGE500
FUR BLACK 41867 see PEY500
FUR BROWN 41866 see PEY500
FURCELLERAN GUM see FPQ000
FURESIS see CHJ750
FURESOL see NGE500
FURFURAL see FPQ875
2-FURFURAL see FPQ875
FURFURAL ALCOHOL see FPU000
FURFURALDEHYDE see FPQ875
FURFURALE (ITALIAN) see FPQ875

FURFURAMIDE see FPS000
FURFURAN see FPK000
FURFURIN see NGE500
FURFUROL see FPQ875
FURFUROLE see FPQ875
FURFURYL ALCOHOL see FPU000
2-FURFURYLALKOHOL (CZECH) see FPU000
FURFURYLAMINE see FPW000
FURIDIAZINE see NGI500
FURIDON see NGG500
2-FURIL-METANALE (ITALIAN) see FPQ875
FURLOE see CKC000
FURLOE 4EC see CKC000
FURMETHONOL see FPI150
FURNACE BLACK see CBT750
FUROBACTINA see NGE000
FURODAN see CBS275
FUROLE see FPQ875
α-FUROLE see FPQ875
FUROSEDON see CHJ750
FUROSEMID see CHJ750
FUROSEMIDE see CHJ750
FUROSEMIDE "MITA" see CHJ750
FUROTHIAZOLE see FQJ000
FUROVAG see NGG500
FUROX see NGG500
FUROXAL see NGG500
FUROXANE see NGG500
FUROXONE SWINE MIX see NGG500
FUROZOLIDINE see NGG500
FURRO D see PEY500
FURRO EG see ALT500
FURRO ER see NAW500
FURRO L see DBO000
FURRO SLA see DBO400
FURSEMID see CHJ750
FURSEMIDE see CHJ750
FUR YELLOW see PEY500
3-(α-FURYL-β-ACETYLAETHYL)-4-HYDROXYCUMARIN
    (GERMAN) see ABF500
3-(1-FURYL-3-ACETYLETHYL)-4-HYDROXYCOUMARIN
    see ABF500
FURYL ALCOHOL see FPU000
FURYLAMIDE see FQN000
2-FURYLCARBINOL see FPU000
α-FURYLCARBINOL see FPU000
FURYLFURAMIDE see FQN000
β-FURYL ISOAMYL KETONE see PCI750
2-FURYL-METHANAL see FPQ875
(2-FURYL)METHANOL see FPU000
1-(2-FURYL)METHYLAMINE see FPW000
1-(3-FURYL)-4-METHYL-1-PENTANONE see PCI750
α-2-FURYL-5-NITRO-2-FURANACYRLAMIDE see FQN000
2-(2-FURYL)-3-(5-NITRO-2-FURYL)ACRYLAMIDE see
    FQN000
2-(2-FURYL)-3-(5-NITRO-2-FURYL)ACRYLIC ACID AMIDE
    see FQN000
α-(FURYL)-β-(5-NITRO-2-FURYL)ACRYLIC AMIDE see
    FQN000
FUSARENONE X see FQR000
FUSARIC ACID see BSI000
FUSARINIC ACID see BSI000
FUSARIOTOXIN T 2 see FQS000
FUSED BORIC ACID see BMG000
FUSED QUARTZ see SCK600
FUSED SILICA (ACGIH) see SCK600
FUSELOEL (GERMAN) see FQT000
FUSEL OIL see FQT000
FUSEL OIL, REFINED (FCC) see FQT000
FUSID see CHJ750
FUSSOL see FFF000
FUTRAMINE D see PEY500
FUTRAMINE EG see ALT500
FUVACILLIN see NGE500

FUXAL see SNN300
FW 293 see BIO750
FW 734 see DGI000
FW 925 see DFT800
FYDALIN see BNK000
FYDE see FMV000
FYFANON see MAK700
FYROL CEF see CGO500
FYROL HB32 see TNC500
FYRQUEL 150 see TNP500

G 0 see HDG000
G 1 see HIM000
G-11 see HCL000
G 301 see DCM750
G 338 see DER000
G 347 see DST200
G 996 see CDS125
G 22150 see DLH630
G 22355 see DLH600, DLH630
G 23992 see DER000
G-24480 see DCM750
G 27202 see HNI500
G 27365 see DJA200
G 27692 see BJP000
G-29288 see MQH750
G 30027 see ARQ725
G 32883 see DCV200
G 34161 see BKL250
G 35020 see DLS600
GA see EIF000, GEM000
p-GABA HYDROCHLORIDE see GAD000
GADEXYL see MQU750
GADOLINIUM see GAF000
GADOLINIUM CHLORIDE see GAH000
GADOLINIUM CITRATE see GAJ000
GADOLINIUM(III) NITRATE (1:3) see GAL000
GADOLINIUM TRICHLORIDE see GAH000
GAFCOL EB see BPJ850
GAFCOTE see PKQ059
GALACTASOL see GLU000
GALACTICOL see DDJ000
GALACTOSE see GAV000
d-GALACTOSE see GAV000
4-(β-d-GALACTOSIDO)-d-GLUCOSE see LAR000
GALECRON see CJJ250
GALENA see LDZ000
GALFER see FBJ100
GALLAMINE see PDD300
GALLIC ACID see GBE000
GALLIC ACID, PROPYL ESTER see PNM750
GALLIUM see GBG000
GALLIUM CHLORIDE see GBM000
GALLIUM (3+) CHLORIDE see GBM000
GALLIUM CITRATE see GBO000
GALLIUM METAL, liquid (DOT) see GBG000
GALLIUM METAL, solid (DOT) see GBG000
GALLIUM-NICKEL ALLOY see NDD500
GALLIUM NITRATE see GBS000
GALLIUM(III) NITRATE (1:3) see GBS000
GALLOCHROME see MCV000
GALLOGAMA see BBQ500
GALLOTANNIC ACID see TAD750
GALLOTANNIN see TAD750
GALLOTOX see ABU500
GALLOXON see DFH600
GALOFAK see BDY669
GALOXANE see DFH600
GALOZONE see CCL250
GAMACID see BBQ500
GAMAPHEX see BBQ500
GAMAQUIL see PGA750
GAMASOL 90 see DUD800
GAMENE see BBQ500

GAMISO see BBQ500
GAMMA-COL see BBQ500
GAMMACORTEN see SOW000
GAMMAHEXA see BBQ500
GAMMAHEXANE see BBQ500
GAMMALIN see BBQ500
GAMMEXANE see BBP750
GAMMOPAZ see BBQ500
GAMOPHENE see HCL000
GANEAKE see ECU750
GANEX P 804 see PKQ250
GANGLIOSTAT see HEA000
GANOCIDE see MLC250
GANOZAN see CHC500
GANSIL see CDP000
GANTANOL see SNK000
GARAMYCIN see GCO000, GCS000
GARANTOSE see BCE500
GARDENAL see EOK000
GARDENAL SODIUM see SID000
GARDENTOX see DCM750
GARDEPANYL see EOK000
GARLIC OIL see GBU800
GARNITAN see DGD600
GAROX see BDS000
GARRATHION see TNP250
GARVOX see DQM600
GASOLINE see GBY000
GASOLINE, UNLEADED see GCE100
GASTRACID see PEK250
GASTRINIDE see TEH500
GASTRIPON see PEM750
GASTRODYN see GIC000
GASTRON see XCJ000
GASTROSEDAN see XCJ000
GASTROTEST see PEK250
GAULTHERIA OIL, ARTIFICIAL see MPI000
GB see IPX000
GBH see BDD000
GBL see DWT600
GBS see SEG800
GC 928 see CJR500
GC-1106 see HCL500
GC 3707 see SOY000
GC 4072 see CDS750
GC 6936 see ABX250
GC 7787 see HDA500
GC 8993 see CLU000
GC 3944-3-4 see PAX000
GD see SKS500
GEARPHOS see MNH000, PAK000
GEBUTOX see BRE500
GECHLOREERDEDIFENYL (DUTCH) see PJM500
GEDEX see SMQ500
GEIGY 338 see DER000
GEIGY 13005 see DSO000
GEIGY 19258 see DRL200
GEIGY 24480 see DCM750
GEIGY 27,692 see BJP000
GEIGY 30,027 see ARQ725
GEIGY 32883 see DCV200
GEIGY G-23611 see DSK200
GEIGY G-27365 see DJA200
GEIGY G-28029 see PDC750
GEIGY G-29288 see MQH750
GEL II see SHF500
GELACILLIN see BDY669
GELATINE DYNAMITE see DYG000
GELBER PHOSPHOR (GERMAN) see PHO750
GELBIN see CAP500
GELBIN YELLOW ULTRAMARINE see CAP750
GELBORANGE-S (GERMAN) see FAG150
GELCARIN see CCL250
GELCARIN HMR see CCL250

G-ELEVEN see HCL000
GELOCATIL see HIM000
GELOSE see AEX250
GELOZONE see CCL250
GELSEMIN see GCK000
GELSEMINE see GCK000
GELSTAPH see SLJ000
GELTABS see VSZ100
GELUCYSTINE see CQK325
GELUTION see SHF500
GELVA CSV 16 see AAX250
GELVATOLS see PKP750
GENAZO RED KB SOLN see CLK225
GENDRIV 162 see GLU000
GENEP EPTC see EIN500
GENERAL CHEMICALS 1189 see KEA000
GENERAL CHEMICALS 3707 see SOY000
GENERAL CHEMICALS 8993 see CLU000
GENETRON 11 see TIP500
GENETRON 12 see DFA600
GENETRON 13 see CLR250
GENETRON 21 see DFL000
GENETRON 22 see CFX500
GENETRON-23 see CBY750
GENETRON 100 see ELN500
GENETRON 101 see CFX250
GENETRON 112 see TBP050
GENETRON 113 see FOO000
GENETRON 114 see FOO509
GENETRON 115 see CJI500
GENETRON 316 see FOO509
GENETRON 1113 see CLQ750
GENETRON 142b see CFX250
GENIPHENE see CDV100
GENISIS see ECU750
GENITE see DFY400
GENITE 883 see CJT750
GENITHION see PAK000
GENITOL see DFY400
GENITOX see DAD200
GENITRON AC see ASM270
GENITRON AC 2 see ASM270
GENITRON AC 4 see ASM270
GENO-CRISTAUZ GREMY see TBF500
GENOPTIC see GCS000
GENOPTIC S.O.P. see GCS000
GENOTHERM see PKQ059
GENOXAL see CQC675, EAS500
GENOZYM see CMX700
GENTAMICIN see GCO000
GENTAMYCIN see GCO000
GENTAMYCIN-CREME (GERMAN) see GCO000
GENTAMYCIN SULFATE see GCS000
GENTIAN VIOLET see AOR500
GENU see CCL250
GENUGEL see CCL250
GENUGEL CJ see CCL250
GENUGOL RLV see CCL250
GENUINE ACETATE CHROME ORANGE see LCS000
GENUINE ORANGE CHROME see LCS000
GENUINE PARIS GREEN see COF500
GENUVISCO J see CCL250
GEON see PJR000, PKQ059
GEON 135 see AAX175
GEON LATEX 151 see PKQ059
GERANIOL ACETATE see DTD800
GERANIOL ALCOHOL see DTD000
GERANIOL EXTRA see DTD000
GERANIOL (FCC) see DTD000
GERANIOL FORMATE see GCY000
GERANIOL TETRAHYDRIDE see DTE600
GERANIUM CRYSTALS see PFA850
GERANIUM LAKE N see FAG070
GERANIUM OIL see GDA000

GERANIUM OIL ALGERIAN TYPE see GDA000
GERANIUM OIL, EAST INDIAN TYPE see PAE000
GERANIUM OIL, TURKISH TYPE see PAE000
GERANYL ACETATE (FCC) see DTD800
GERANYL ALCOHOL see DTD000
GERANYL BENZOATE see GDE800
GERANYL BUTYRATE see GDE825
GERANYL FORMATE (FCC) see GCY000
GERANYL ISOVALERATE see GDK000
GERANYL PHENYLACETATE see GDM400
GERANYL PROPIONATE see GDM450
GERFIL see PMP500
GERISON see SNM500
GERMALGENE see TIO750
GERMA-MEDICA see HCL000
GERMAN CHAMOMILE OIL see CDH500
GERMANE (DOT) see GEI100
GERMANIC OXIDE (crystalline) see GDS000
GERMANIUM BROMIDE see GDW000
GERMANIUM CHLORIDE see GDY000
GERMANIUM COMPOUNDS see GEA000
GERMANIUM HYDRIDE see GEI100
GERMANIUM TETRABROMIDE see GDW000
GERMANIUM TETRACHLORIDE see GDY000
GERMANIUM TETRAHYDRIDE see GEI100
GERMICICLIN see MDO250
GEROBIT see DBA800
GERODYL see DWK400
GEROT-EPILAN see MKB250
GEROT-EPILAN-D see DKQ000
GEROVIT see DBA800
GEROVITAL see AIL750
GEROX see SLW500
GERTLEY BORATE see SFF000
GESAFID see DAD200
GESAGARD see BKL250
GESAMIL see PMN850
GESAPON see DAD200
GESAPRIM see ARQ725
GESARAN see BJP000
GESAREX see DAD200
GESAROL see DAD200
GESATOP see BJP000
GESATOP 50 see BJP000
GESFID see MQR750
GESOPRIM see ARQ725
GESTEROL L.A. see HNT500
GESTID see MQR750
GETTYSOLVE-B see HEN000
GETTYSOLVE-C see HBC500
GEVILON see SHL500
GIALLO CROMO (ITALIAN) see LCR000
GIARDIL see NGG500
GIARLAM see NGG500
GIATRICOL see MMN250
GIBBERELLIC ACID see GEM000
GIBBERELLIN see GEM000
GIBBREL see GEM000
GIB-SOL see GEM000
GIB-TABS see GEM000
GIE see DVR600
GIEGY GS-13798 see MPG250
GIFBLAAR see PLG000
GIFBLAAR POISON see FIC000
GILOTHERM OM 2 see TBD000
GILSONITE see GEO000
GIMID see DYC800
GINARSOL see ABX500
GINEFLAVIR see MMN250
GINGER OIL see GEQ000
GIRACID see PDC250
GIROSTAN see TFQ750
GITHAGENIN see GMG000
GITOXIGENIN-TRIDIGITOXOSID (GERMAN) see GEU000

GITOXIN see GEU000
GITOXIN PENTAACETATE see GEW000
GLACIAL ACETIC ACID see AAT250
GLACIAL ACRYLIC ACID see ADS750
GLANDUBOLIN see EDV000
GLANDUCORPIN see PMH500
GLASS see FBQ000
GLASS FIBERS see FBQ000
GLAUPAX see AAI250
GLAURAMINE see IBB000
GLAZD PENTA see PAX250
GLEEM see SHF500
GLIBUTIDE see BQL000
GLICOL MONOCLORIDRINA (ITALIAN) see EIU800
GLIKOCEL TA see SFO500
GLIMID see DYC800
GLIPORAL see BQL000
GLOBENICOL see CDP250
GLOBOCICLINA see MDO250
GLOBOID see ADA725
GLOGAL see HNI500
GLONOIN see NGY000
GLONSEN see SHK800
GLOROUS see CDP250
GLOSSO STERANDRYL see MPN500
GLOVER see LCF000
GLUCID see BCE500
GLUCIDORAL see BSM000
GLUCINUM see BFO750
GLUCITOL see SKV200
d-GLUCITOL see SKV200
GLUCOCHLORAL see GFA000
GLUCOCHLORALOSE see GFA000
α-d-GLUCOCHLORALOSE see GFA000
GLUCODIGIN see DKL800
GLUCO-FERRUM see FBK000
GLUCOFREN see BSM000
GLUCOLIN see GFG000
GLUCONATE de CALCIUM (FRENCH) see CAS750
GLUCONATO di SODIO (ITALIAN) see SHK800
d-GLUCONIC ACID, MONOPOTASSIUM SALT (9CI) see
    PLG800
GLUCONIC ACID POTASSIUM SALT see PLG800
GLUCONIC ACID SODIUM SALT see SHK800
GLUCONSAN K see PLG800
GLUCOPHAGE see DQR600
GLUCOPHAGE LA 6023 see DQR600
GLUCOPROSCILLARIDIN A see GFC000
4-(α-d-GLUCOPYRANOSIDO)-α-GLUCOPYRANOSE see
    MAO500
α-d-GLUCOPYRANOSYL β-d-FRUCTOFURANOSIDE see
    SNH000
(d-GLUCOPYRANOSYLTHIO)GOLD see ART250
GLUCOSE see GFG000
d-GLUCOSE see GFG000
d-GLUCOSE, anhydrous see GFG000
GLUCOSE LIQUID see GFG000
(α-d-GLUCOSIDO)-β-d-FRUCTOFURANOSIDE see SNH000
4-(α-d-GLUCOSIDO)-d-GLUCOSE see MAO500
N-d-GLUCOSYL-(2)-N'-NITROSOMETHYLHARNSTOFF
    (GERMAN) see SMD000
N-d-GLUCOSYL-(2)-N'-NITROSOMETHYLUREA see
    SMD000
β-d-GLUCOSYLOXYAZOXYMETHANE see COU000
(1-d-GLUCOSYLTHIO)GOLD see ART250
GLUEOPHOGE see DQR600
GLUKRESIN see GGS000
GLUMIN see GFO050
GLU-P-2 see DWW700
GLUPAN see TEH500
GLUPAX see AAI250
GLU-P-I see AKS250
GLUSATE see GFO000
GLUSIDE see BCE500

GLUTACID see GFO000
GLUTACYL see MRL500
GLUTAMIC ACID see GFO000
l-GLUTAMIC ACID see GFO000
α-GLUTAMIC ACID see GFO000
GLUTAMIC ACID AMIDE see GFO050
GLUTAMIC ACID-5-AMIDE see GFO050
l-GLUTAMIC ACID, MONOPOTASSIUM SALT see
    MRK500
GLUTAMIC ACID, SODIUM SALT see MRL500
d-GLUTAMIENSUUR see GFO000
GLUTAMINE see GFO050
γ-GLUTAMINE see GFO050
l-GLUTAMINE (9CI, FCC) see GFO050
GLUTAMINIC ACID see GFO000
l-GLUTAMINIC ACID see GFO000
GLUTAMINOL see GFO000
GLUTAMMATO MONOSODICO (ITALIAN) see MRL500
GLUTANON see TEH500
GLUTARAL see GFQ000
GLUTARALDEHYD (CZECH) see GFQ000
GLUTARALDEHYDE see GFQ000
GLUTARDIALDEHYDE see GFQ000
GLUTARIC ACID DINITRILE see TLR500
GLUTARIC DIALDEHYDE see GFQ000
GLUTARODINITRILE see TLR500
GLUTARONITRILE see TLR500
GLUTATHIMID see DYC800
GLUTATON see GFO000
GLUTAVENE see MRL500
GLUTETHIMID see DYC800
GLUTETHIMIDE see DYC800
GLUTETIMIDE see DYC800
GLYBUTAMIDE see BSM000
GLYCERIN see GGA000
GLYCERIN, anhydrous see GGA000
GLYCERINE see GGA000
GLYCERINE TRIACETATE see THM500
GLYCERIN-α-MONOCHLORHYDRIN see CDT750
GLYCERIN MONOSTEARATE see OAV000
GLYCERIN, SYNTHETIC see GGA000
GLYCERINTRINITRATE (CZECH) see NGY000
GLYCERITE see TAD750
GLYCERITOL see GGA000
GLYCEROL see GGA000
GLYCEROLACETONE see DVR600
GLYCEROL CHLOROHYDRIN see CDT750
GLYCEROL-α-CHLOROHYDRIN see CDT750
GLYCEROL DIACETATE see DBF600
GLYCEROL-α,β-DICHLOROHYDRIN see DGG600
GLYCEROL α,γ-DICHLOROHYDRIN see DGG400
sym-GLYCEROL DICHLOROHYDRIN see DGG400
GLYCEROL DIMETHYLKETAL see DVR600
GLYCEROL EPICHLORHYDRIN see EAZ500
GLYCEROL-α-MONOCHLOROHYDRIN (DOT) see CDT750
GLYCEROL MONOSTEARATE see OAV000
GLYCEROL, NITRIC ACID TRIESTER see NGY000
GLYCEROL TRIACETATE see THM500
GLYCEROL TRIBUTYRATE see TIG750
GLYCEROL TRICHLOROHYDRIN see TJB600
GLYCEROL (TRI(CHLOROMETHYL))ETHER see GGI000
GLYCEROL TRINITRATE see NGY000
GLYCEROL(TRINITRATE de) (FRENCH) see NGY000
GLYCEROLTRINTRAAT (DUTCH) see NGY000
GLYCERYL-α-CHLOROHYDRIN see CDT750
GLYCERYL-1,3-DIACETATE see DBF600
GLYCERYL MONOSTEARATE see OAV000
GLYCERYL NITRATE see NGY000
GLYCERYL-o-TOLYL ETHER see GGS000
GLYCERYL TRIACETATE see THM500
GLYCERYL TRICHLOROHYDRIN see TJB600
GLYCERYL TRINITRATE see NGY000
GLYCERYL TRINITRATE, solution see SLD000
GLYCERYL TRINITRATE, solution up to 1% in alcohol (DOT)
    see NGY000

GLYCIDAL see GGW000
GLYCIDALDEHYDE see GGW000
GLYCIDE see GGW500
GLYCIDOL see GGW500
GLYCIDOL OLEATE see ECJ000
GLYCIDOL STEARATE see SLK500
GLYCIDYL ACRYLATE see ECH500
GLYCIDYL ALCOHOL see GGW500
GLYCIDYLALDEHDYE see GGW000
GLYCIDYL BUTYL ETHER see BRK750
GLYCIDYL OCTADECANOATE see SLK500
GLYCIDYL OCTADECENOATE see ECJ000
GLYCIDYL OLEATE see ECJ000
GLYCIDYL PHENYL ETHER see PFH000
GLYCIDYL PROPENATE see ECH500
GLYCIDYL STEARATE see SLK500
GLYCINE see GHA000
GLYCINE MUSTARD see GHE000
GLYCINE NITROGEN MUSTARD see GHE000
GLYCINOL see EEC600
GLYCIRENAN see VGP000
GLYCO-FLAVINE see DBX400
GLYCOL see EJC500
GLYCOL ALCOHOL see EJC500
GLYCOL BIS(HYDROXYETHYL) ETHER see TJQ000
GLYCOL BROMIDE see EIY500
GLYCOL BROMOHYDRIN see BNI500
GLYCOL BUTYL ETHER see BPJ850
GLYCOL CHLOROHYDRIN see EIU800
GLYCOL CYANOHYDRIN see HGP000
GLYCOL DIACETATE see EJD759
GLYCOL DIBROMIDE see EIY500
GLYCOL DICHLORIDE see EIY600
GLYCOL DIFORMATE see EJF000
GLYCOL DIMETHYL ETHER see DOE600
GLYCOLDINITRAAT (DUTCH) see EJG000
GLYCOL DINITRATE see EJG000
GLYCOL (DINITRATE DE) (FRENCH) see EJG000
GLYCOL ETHER see DJD600
GLYCOL ETHER de ACETATE see CBQ750
GLYCOL ETHER DB ACEATATE see BQP500
GLYCOL-ETHERDIAMINETETRAACETIC ACID see EIT000
GLYCOL ETHER EB see BPJ850
GLYCOL ETHER EB ACETATE see BPJ850
GLYCOL ETHER EE see EES350
GLYCOL ETHER EE ACETATE see EES400
GLYCOL ETHER EM see EJH500
GLYCOL ETHER EM ACETATE see EJJ500
GLYCOL ETHYLENE ETHER see DVQ000
GLYCOL ETHYL ETHER see DJD600, EES350
GLYCOLIC NITRILE see HIM500
GLYCOLIXIR see GHA000
GLYCOLMETHYL ETHER see EJH500
GLYCOL MONOBUTYL ETHER see BPJ850
GLYCOL MONOBUTYL ETHERACETATE see BPM000
GLYCOLMONOCHLOORHYDRINE (DUTCH) see EIU800
GLYCOL MONOCHLOROHYDRIN see EIU800
GLYCOL MONOETHYL ETHER see EES350
GLYCOL MONOETHYL ETHER ACETATE see EES400
GLYCOL MONOMETHYL ETHER see EJH500
GLYCOL MONOMETHYL ETHER ACETATE see EJJ500
GLYCOL MONOMETHYL ETHER ACRYLATE see MIF750
GLYCOL MONOPHENYL ETHER see PER000
GLYCOL MONOSTEARATE see EJM500
GLYCOLONITRILE see HIM500
GLYCOL STEARATE see EJM500
GLYCOL SULFATE see EJP000
GLYCOMONOCHLORHYDRIN see EIU800
GLYCOMUL S see SKV150
GLYCON S-70 see SLK000
GLYCON DP see SLK000
GLYCONITRILE see HIM500
GLYCON RO see OHU000
GLYCON TP see SLK000

GLYCON WO see OHU000
GLYCOPHEN see GIA000
GLYCOPHENE see GIA000
GLYCOPYRROLATE see GIC000
GLYCOPYRROLATE BROMIDE see GIC000
GLYCOPYRRONIUM BROMIDE see GIC000
GLYCOSPERSE O-20 see PKL100
GLYCOSPERSE L-20X see PKL000
GLYCO STEARIN see HKJ000
GLYCYL ALCOHOL see GGA000
GLYCYRRHIZA see LFN300
GLYCYRRHIZAE (LATIN) see LFN300
GLYCYRRHIZA EXTRACT see LFN300
GLYCYRRHIZINA see LFN300
GLYESTRIN see ECU750
GLYKOLDINITRAT (GERMAN) see EJG000
GLYME see DOE600
GLYMOL see MQV750
GLYODEX 3722 see CBG000
GLYOTOL see GGS000
GLYOXAL see GIK000
GLYOXALIN see IAL000
GLYOXALINE see IAL000
GLYOXALINE-5-ALANINE see HGE700
GLY-OXIDE see HIB500
GLYOXYLALDEHYDE see GIK000
GLYPED see THM500
GLYPHOSATE see PHA500
GLYSANOL B see ART250
GLYSOLETTEN see EOK000
G-M-F see DVR200
GMI see DLS600
GMP DISODIUM SALT see GLS800
5'-GMP DISODIUM SALT see GLS800
GMP SODIUM SALT see GLS800
GM SULFATE see GCS000
GO 186 see SHL500
GOHSENOLS see PKP750
GOHSENYL E 50 Y see AAX250
GOLD see GIS000
1721 GOLD see CNI000
GOLD BOND see CCP250
GOLD BRONZE see CNI000
GOLDEN YELLOW see DUX800
GOLD FLAKE see GIS000
GOLD LEAF see GIS000
GOLD ORANGE MP see DOU600
GOLD POWDER see GIS000
GOLD SATINOBRE see LDS000
GOLD SODIUM THIOMALATE see GJC000
GOLD SODIUM THIOSULFATE DIHYDRATE see GJG000
GOLD THIOGLUCOSE see ART250
GOMBARDOL see SNM500
GONACRINE see DBX400
GOOD-RITE see PJR000
GOODRITE 1800X73 see SMR000
GOOD-RITE GP 264 see DVL700
GORE-TEX see TAI250
GOSSYPOL see GJM000
GOTAMINE TARTRATE see EDC500
GOTHNION see ASH500
G 2 (OXIDE) see AHE250
GOYL see ABX500
GP 38383 see IBP309
GP-40-66:120 see HCD250
GPKh see HAR000
GRAAFINA see EDP000, EDP000
GRAFESTROL see DKA600
GRAHAM'S SALT see SII500
GRAIN ALCOHOL see EFU000
GRAIN SORGHUM HARVEST-AID see SFS000
GRAMEVIN see DGI400, DGI600
GRAMICIDIN see GJO000
GRAMISAN see MEP250

GRAMOXONE S see PAI990
GRAMOZONE see PAJ000
GRAMPENIL see AIV500
GRANEX O see SFS000
GRANOSAN see CHC500
GRANOSAN M see EME500
GRANOX NM see HCC500
GRANOX PPM see CBG000
GRANULAR ZINC see ZBJ000
GRANULATED SUGAR see SNH000
GRANULIN see ACR300
GRANUTOX see PGS000
GRAPE BLUE A GEIGY see FAE100
GRAPEFRUIT OIL see GJU000
GRAPEFRUIT OIL, COLDPRESSED see GJU000
GRAPEFRUIT OIL, EXPRESSED see GJU000
GRAPE SUGAR see GFG000
GRAPHITE (MAK) see CBT500
GRAPHITE, NATURAL (ACGIH) see CBT500
GRAPHITE, SYNTHETIC see CBT500
GRASAL BRILLIANT YELLOW see DOT300
GRASAL YELLOW see FAG130
GRASAN ORANGE 3R see XRA000
GRASAN ORANGE R see PEJ500
GRASCIDE see DGI000
GRATIBAIN see OKS000
GRATUS STROPHANTHIN see OKS000
GRAVIDOX see PPK250
GRAVINOL see DYE600
GRAVOCAIN see DHK400
GRAVOCAIN HYDROCHLORIDE see DHK600
GRAVOL see DYE600
GRAY ACETATE see CAL750
GRAY AMBER see AHJ000
1724 GREEN see FAG000
11091 GREEN see BLK000
11661 GREEN see CMJ900
GREEN CHLOROPHYL see CKN000
GREEN CHROME OXIDE see CMJ900
GREEN CHROMIC OXIDE see CMJ900
GREEN CINNABAR see CMJ900
GREEN CROSS CRABGRASS KILLER see PLC250
GREEN CROSS WARBLE POWDER see RNZ000
GREEN HELLEBORE see VIZ000
GREEN NICKEL OXIDE see NDF500
GREEN No. 2 see BLK000
GREEN No. 203 see FAF000
GREENOCKITE see CAJ750
GREEN OIL see APG500, COD750
GREEN ROUGE see CMJ900
GREEN SEAL-8 see ZKA000
GREEN VITRIOL see FBN100
GREEN VITROL see FBO000
GREOSIN see GKE000
GRESFEED see GKE000
GREY ARSENIC see ARA750
GRICIN see GKE000
GRIFFEX see ARQ725
GRIFFIN MANEX see MAS500
GRIFULVIN see GKE000
GRILON see NOH000, PJY500
GRIPPEX see TEH500
GRISACTIN see GKE000
GRISCOFULVIN see GKE000
GRISEFULINE see GKE000
GRISEO see GKE000
(+)-GRISEOFULVIN see GKE000
GRISEOFULVIN-FORTE see GKE000
GRISEOFULVINUM see GKE000
GRISEOVIRIDIN see GKC000
GRISETIN see GKE000
GRISOFULVIN see GKE000
GRISOL see TCF250
GRISOVIN see GKE000

GRIS-PEG see GKE000
GROCEL see GEM000
GROCO see LGK000
GROCO 2 see OHU000
GROCO 4 see OHU000
GROCO 54 see SLK000
GROCO 5L see OHU000
GROCOLENE see GGA000
GROCOR 5500 see OAV000
GROCOR 6000 see OAV000
GROUNDNUT OIL see PAO000
GROUND RYANIA SPECISA(VAHL) STEMWOOD (ALKO-
   LOID RYANODINE) see RSZ000
GROUND VOCLE SULPHUR see SOD500
GRUNDIER ARBEZOL see PAX250
GRYSIO see GKE000
GS 6244 see FOI000
GS-13,798 see MPG250
G-STROPHANTHIN see OKS000
GT41 see BOT250
GT-1012 see DNB000
GT 2041 see BOT250
GTG see ART250
GTN see NGY000
GUAIAC GUM see GLY100
GUAIACOL see GKI000
GUAICOL see GKI000
GUANAZOLE see DCF200
GUANETHIDINE MONOSULFATE see GKU000
GUANICAINE see PDN500
GUANIDINE see GKW000
GUANIDINE MONONITRATE see GLA000
GUANIDINE NITRATE (DOT) see GLA000
N-(2-GUANIDINO ETHYL)HEPTAMETHYLENIMINE SUL-
   FATE see GKU000
GUANIOL see DTD000
GUANYLIC ACID SODIUM SALT see GLS800
1-GUANYL-4-NITROSAMINOGUANYLTETRAZENE see
   TEF500
GUANYL NITROSAMINO GUANYL TETRAZENE (DOT) see
   TEF500
GUANYL NITROSAMINO GUANYL TETRAZENE, contain-
   ing, by weight, at least 30% water (DOT) see TEF500
GUAR see GLU000
GUARANINE see CAK500
GUAR FLOUR see GLU000
GUAR GUM see GLU000
GUATEMALA LEMONGRASS OIL see LEH000
GUDAKHU (INDIA) see SED400
GUESAPON see DAD200
GUESAROL see DAD200
GUICITRINA see AIV500
GUICITRINE see AIV500
GUIGNER'S GREEN see CMJ900
GULITOL see SKV200
l-GULITOL see SKV200
GUM see PJR000
GUM ARABIC see AQQ500
GUM CAMPHOR see CBA750
GUM CARRAGEENAN see CCL250
GUM CHON 2 see CCL250
GUM CHROND see CCL250
GUM CYAMOPSIS see GLU000
GUM GHATTI see GLY000
GUM GUAIAC see GLY100
GUM GUAR see GLU000
GUM OPIUM see OJG000
GUM OVALINE see AQQ500
GUM SENEGAL see AQQ500
GUM TRAGACANTH see THJ250
GUNCOTTON see CCU250
GUNPOWDER see ERF500
GUSATHION see ASH500
GUSATHION A see EKN000

GUSERVIN see GKE000
GUSTAFSON CAPTAN 30-DD see CBG000
GUTHION (DOT) see ASH500
GUTHION, liquid (DOT) see ASH500
GUTHION (ETHYL) see EKN000
GUTTAGENA see PKQ059
G1V GARD DXN see ABC250
GYNAESAN see EDU500
GYN-ANOVLAR see EEH520
GYNECLORINA see CDP000
GYNECORMONE see EDP000
GYNEFOLLIN see DAL600
GYNERGEN see EDC500
GYNERGON see EDO000
GYNESTREL see EDO000
GYNFORMONE see EDP000
GYNOCHROME see MCV000
GYNOESTRYL see EDO000
GYNOFON see ABY900
GYNOLETT see DKB000
GYNONLAR 21 see EEH520
GYNOPHARM see DKA600
GYNOPLIX see ABX500
GY-PHENE see CDV100
GYPSINE see LCK000
GYPSOGENIN see GMG000
GYPSOPHILASAPOGENIN see GMG000
GYPSOPHILASAPONIN see GMG000
GYPSUM see CAX500, CAX750
GYPSUM STONE see CAX750
"H" see HBT500
H-34 see HAR000
H-69 see DTN200
H 95 see CKF750
H 321 see DST000
H-365 see ELL500
H-490 see ENG500
H 940 see ENG500
H 1032 see MHJ500
H 1803 see BJP000
H4 099 see DTL200
H-4723 see CIR750
96H60 see DFH600
HACHI-SUGAR see SGC000
HADACIDIN see FNO000
HADACIDINE see FNO000
HADACIN see FNO000
HAEMATITE see HAO875
HAEMOFORT see FBO000
HAEMOSTASIN see VGP000
HAFNIUM see HAC000
HAFNIUM, wet with not less than 25% water (DOT) see
 HAC000
HAFNIUM CHLORIDE OXIDE see HAD500
HAFNIUM DICYCLOPENTADIENE DICHLORIDE see
 HAE500
HAFNIUM METAL, dry (DOT) see HAC000
HAFNIUM METAL, wet (DOT) see HAC000
HAFNIUM OXYCHLORIDE see HAD500
HAFNOCENE DICHLORIDE see HAE500
HAIARI see RNZ000
HAIMASED see SIA500
HAIRY see HBT500
HAITIN see HON000
HALAMID see CDP000
HALANE see DFE200
HALARSOL see DFX400
HALBMOND see BAU750
HALF-CYSTEINE see CQK000
HALF-CYSTINE see CQK000
HALF-MUSTARD GAS see CGY750
HALF-MYLERAN see EMF500
HALITE see SFT000
HALIZAN see TDW500

HALKAN see DYF200
HALLTEX see SEH000
HALOANISONE COMPOSITUM see FDA880
HALOCARBON 11 see TIP500
HALOCARBON 14 see CBY250
HALOCARBON 23 see CBY750
HALOCARBON 112 see TBP050
HALOCARBON 113 see FOO000
HALOCARBON 114 see FOO509
HALOCARBON 115 see CJI500
HALOCARBON 112a see TBP000
HALOCARBON 152A see ELN500
HALOCARBON 1132A see VPP000
HALOCARBON C-138 see CPS000
HALOCARBON 13/UCON 13 see CLR250
HALOMYCETIN see CDP250
HALON see DFA600
HALON 14 see CBY250
HALON 1001 see MHR200
HALON 1011 see CES650
HALON 1202 see DKG850
HALON 1211 see BNA250
HALON 1301 see TJY100
HALON 2001 see EGV400
HALON TFEG 180 see TAI250
HALOTAN see HAG500
HALOTHANE see HAG500
HALOWAX see TBR000, TIT500
HALOXON see DFH600
HALSAN see HAG500
HALSO 99 see CLK100
HALVIC 223 see PKQ059
HALVISOL see HAH000
HAMIDOP see DTQ400
HAMILTON RED see CHP500
HAMP-ENE 100 see EIV000
HAMP-ENE 215 see EIV000
HAMP-ENE 220 see EIV000
HAMP-ENE ACID see EIX000
HAMP-ENE Na4 see EIV000
HAMP-EX ACID see DJG800
HAMPSHIRE GLYCINE see GHA000
HAMPSHIRE NTA see SIP500
HANANE see BJE750
HANSACOR see DJS200
HANSAMID see NCR000
HAPLOPAN see EOK000
HAPLOS see EOK000
4HAQO see HIY500
HARMAR see PNA500
HARMINE see HAI500
HARMONIN see MQU750
HARRY see HBT500
HARTOL see MQU750
HARVAMINE see WAK000
HARVEN see SGD000
HARVEST-AID see SFS000
HASETHROL see PBC250
HASTELLOY C see CNA750
HATCOL DOP see DVL700
HAVERO-EXTRA see DAD200
HAVIDOTE see EIX000
HAYNES STELLITE 21 see CNA750
HAZODRIN see MRH209
HB see HEA000
HBBN see HJQ350
9,10-H2 B(e)P see DKU400
HC-3 see HAQ000
HC 2072 see NIM500
8057HC see DSQ000
HCA see HCL500
HCB see HCC500
HCBD see HCD250
HC BLUE 1 see BKF250

α-HCC see HJV000
HCCH see BBP750, BBQ500
HCCPD see HCE500
HCDD see HAJ500
HCE see EBW500
HCH see BBQ500
α-HCH see BBQ000
β-HCH see BBR000
γ-HCH see BBQ500
HCl SALZ DES p,N,N-BUTYLAMINOSALICYLSAEURE-
    DIAETHYLAMINOAETHYLESTER (GERMAN) see
    BQH250
HCN see HHS000
HCP see ABD000, HCL000
HC RED No. 3 see ALO750
HCS 3260 see CDR750
HCT see HGO500
HCTZ see CFY000
HCZ see CFY000
HDEHP see BJR750
HDMTX see MDV500
HEARTS see BBK500
HEAVENLY BLUE see DJO000
HEAVY NAPHTHENIC DISTILLATE see MQV780
HEAVY NAPHTHENIC DISTILLATE SOLVENT EXTRACT
    see MQV857
HEAVY NAPHTHENIC DISTILLATES (PETROLEUM) see
    MQV780
HEAVY OIL see CMY825
HEAVY PARAFFINIC DISTILLATE see MQV785
HEAVY PARAFFINIC DISTILLATE, SOLVENT EXTRACT
    see MQV859
HEAVY WATER see HAK000
HEAVY WATER-d2 see HAK000
HEAZLEWOODITE see NDJ500
HEBABIONE HYDROCHLORIDE see PPK500
HECLOTOX see BBQ500
HEDAPUR M 52 see CIR250
HEDEX see HIM000
HEDOLIT see DUS700
HEDONAL see DGB000
HEDONAL (The herbicide) see DAA800
HEDONAL DP see DGB000
HEDONAL MCPP see CIR500
HEF-2 see JDA100
HEKSAN (POLISH) see HEN000
HEKSOGEN (POLISH) see CPR800
HEKTALIN see VGP000
HELFO DOPA see DNA200
HELICON see ADA725
HELIOGEN see CDP000
HELIONAL see EOK000
HELIO RED TONER LCLL see CHP500
HELIOTRIDINE ESTER with LASIOCARPUM and ANGELIC
    ACID see LBG000
HELIOTRINE see HAL500
HELIOTRON see HAL500
HELIOTROPIN see PIW250
HELIOTROPYL ACETATE see PIX000
HELIUM see HAM500
HELIUM, compressed (DOT) see HAM500
HELIUM, refrigerated liquid (DOT) see HAM500
HELIUM-OXYGEN (MIXTURE) see HAN000
HELIUM-OXYGEN MIXTURE (DOT) see HAN000
HELMATAC see BQK000
HELMETINA see PDP250
HELMIRANE see DFH600
HELMIRON see DFH600
HELMIRONE see DFH600
HELMOX see COH250
HELOTHION see SOU625
HELVETICOSIDE DIHYDRATE see HAO000
HEMATITE see HAO875

HEMEL see HEJ500
HEMICHOLINE see HAQ000
HEMICHOLINIUM-3 see HAQ000
HEMICHOLINIUM BROMIDE see HAQ000
HEMICHOLINIUM-3-BROMIDE see HAQ000
HEMICHOLINIUM DIBROMIDE see HAQ000
HEMICHOLINIUM-3-DIBROMIDE see HAQ000
HEMISINE see VGP000
HEMO-B-DOZE see VSZ000
HEMODAL see MMD500
HEMODESIS see PKQ250
HEMODEZ see PKQ250
HEMOFURAN see NGE500
HEMOMIN see VSZ000
HEMOSTASIN see VGP000
HEMOSTYPTANON see EDU500
HEMOTON see FBJ100
HENDECANAL see UJJ000
HENDECANALDEHYDE see UJJ000
HENDECANE see UJS000
HENDECANOIC ALCOHOL see UNA000
1-HENDECANOL see UNA000
2-HENDECANONE see UKS000
HENDECENAL see ULJ000
HENDECYL ALCOHOL see UNA000
n-HENDECYLENIC ALCOHOL see UNA000
HENNOLETTEN see EOK000
HENU see HKW500
HEOD see DHB400
HEPACHOLINE see CMF750
HEPAGON see VSZ000
HEPAR CALCIS see CAY000
HEPARIN see HAQ500
α-HEPARIN see HAQ500
HEPARINATE see HAQ500
HEPARINIC ACID see HAQ500
HEPARIN SULFATE see HAQ500
HEPAR SULFUROUS see PLT250
HEPAVIS see VSZ000
HEPCOVITE see VSZ000
HEPT see TCF250
HEPTACAINE see PIO750
HEPTACHLOOR (DUTCH) see HAR000
1,4,5,6,7,8,8-HEPTACHLOOR-3a,4,7,7a-TETRAHYDRO-4,7-
    endo-METHANO-INDEEN (DUTCH) see HAR000
HEPTACHLOR see HAR000
HEPTACHLORE (FRENCH) see HAR000
HEPTACHLOR EPOXIDE (USDA) see EBW500
3,4,5,6,7,8,8-HEPTACHLORODICYCLOPENTADIENE see
    HAR000
3,4,5,6,7,8,8a-HEPTACHLORODICYCLOPENTADIENE see
    HAR000
1,4,5,6,7,8,8-HEPTACHLORO-2,3-EPOXY-2,3,3a,4,7,7a-
    HEXAHYDRO-4,7-METHANOINDAN see EBW500
1,4,5,6,7,8,8-HEPTACHLORO-2,3-EPOXY-3a,4,7,7a-TETRA-
    HYDRO-4,7-METHANOINDAN see EBW500
2,3,4,5,6,7,7-HEPTACHLORO-1a,1b,5,5a,6,6a-HEXAHYDRO-
    2,5-METHANO-2H-INDENO(1,2-b)OXIRENE see EBW500
1,4,5,6,7,8,8-HEPTACHLORO-3a,4,7,7a-TETRAHYDRO-4,7-
    ENDOMETHANOINDENE see HAR000
1,4,5,6,7,10,10-HEPTACHLORO-4,7,8,9,-TETRAHYDRO-4,7-
    ENDOMETHYLENEINDENE see HAR000
1,4,5,6,7,8,8a-HEPTACHLORO-3a,4,7,7a-TETRAHYDRO-4,7-
    METHANOINDANE see HAR000
1(3a),4,5,6,7,8,8-HEPTACHLORO-3a(1),4,7,7a-TETRAHY-
    DRO-4,7-METHANOINDENE see HAR000
1,4,5,6,7,8,8-HEPTACHLORO-3a,4,7,7a-TETRAHYDRO-4,7-
    METHANOINDENE see HAR000
1,4,5,6,7,8,8-HEPTACHLORO-3a,4,7,7a-TETRAHYDRO-4,7-
    METHANOL-1H-INDENE see HAR000
1,4,5,6,7,8,8-HEPTACHLORO-3a,4,7,7,7a-TETRAHYDRO-4,7-
    METHYLENE INDENE see HAR000
1,4,5,6,7,8,8-HEPTACHLOR-3a,4,7,7,7a-TETRAHYDRO-4,7-
    endo-METHANO-INDEN (GERMAN) see HAR000

1-HEPTADECANECARBOXYLIC ACID see SLK000
8,10,12-HEPTADECATRIENE-4,6-DIYNE-1,14-DIOL,
   (E,E,E)-(-)- see CMN000
2,4-HEPTADIENAL see HAV450
HEPTADIENAL-2,4 see HAV450
HEPTADONE see MDO750
HEPTAFLUOROBUTYRIC ACID see HAX500
HEPTAGRAN see HAR000
HEPTALDEHYDE see HBB500
HEPTAMUL see HAR000
HEPTAN (POLISH) see HBC500
HEPTANAL see HBB500
HEPTANE see HBC500
n-HEPTANE see HBC500
1-HEPTANECARBOXYLIC ACID see OCY000
HEPTANEN (DUTCH) see HBC500
1-HEPTANETHIOL see HBD500
1-HEPTANOL see HBL500
2-HEPTANOL see HBE500
HEPTANOL-2 see HBE500
n-HEPTANOL see HBL500
n-HEPTANOL-1 (FRENCH) see HBL500
HEPTANOL, FORMATE see HBO500
HEPTANON see MDO750
HEPTAN-3-ON (DUTCH, GERMAN) see EHA600
2-HEPTANONE see MGN500
3-HEPTANONE see EHA600
HEPTAN-3-ONE see EHA600
4-HEPTANONE see DWT600
HEPTAN-4-ONE see DWT600
2,4-HEPTDIENAL see HAV450
4-HEPTENAL see HBI800
cis-4-HEPTEN-1-AL see HBI800
n-HEPTENE see HBJ000
HEPTENYL ACROLEIN see DAE450
HEPTYL ALCOHOL see HBL500
HEPTYL CARBINOL see OEI000
1-HEPTYLENE see HBJ000
HEPTYL FORMATE see HBO500
HEPTYL HYDRIDE see HBC500
HEPTYLIDENE ALDEHYDE see NNA300
HEPTYL MERCAPTAN see HBD500
n-HEPTYLMERCAPTAN see HBD500
n-HEPTYL METHANOATE see HBO500
HEPTYLMETHYLINITROSAMINE see HBP000
2-HEPTYLOXYCARBANILIC ACID-2-(1-PIPERIDINYL)
   ETHYL ESTER HYDROCHLORIDE see PIO750
(2-(HEPTYLOXY)PHENYL)CARBAMIC ACID-2-(1-PIPERIDI-
   NYL)ETHYL ESTER HYDROCHLORIDE see PIO750
N-(2-(HEPTYLOXYPHENYLCARBAMOYLOXY)ETHYL)
   PIPERIDINIUM CHLORIDE see PIO750
HEPZIDE see ENV500
HERB-ALL see MRL750
HERBAN M see MRL750
HERBATOX see DXQ500
HERBAX TECHNICAL see DGI000
HERBAZIN see BJP000
HERBAZIN 50 see BJP000
HERBEX see BJP000
HERBICIDE 273 see DXD000
HERBICIDE 326 see DGD600
HERBICIDE 976 see BMM650
HERBICIDE C-2059 see DUK800
HERBICIDE M see CIR250
HERBICIDES, MONURON see CJX750
HERBICIDES, SILVEX see TIX500
HERBICIDE TOTAL see AMY050
HERBIDAL see DAA800
HERBITOX see PCT250
HERBIZOLE see AMY050
HERBOXY see BJP000
HERCOFLAT 135 see PMP500
HERCOFLEX 260 see DVL700
HERCULES 3956 see CDV100

HERCULES 14503 see DBI099
HERCULES P6 see PBB750
HERCULES TOXAPHENE see CDV100
HERCULON see PMP500
HERKAL see DGP900
HERMAL see TFS350
HERMAT TMT see TFS350
HERMAT ZDM see BJK500
HERMAT Zn-MBT see BHA750
HERMESETAS see BCE500
HEROIEN see HBT500
HEROIIN see HBT500
HEROIN see HBT500
HEROLAN see HBT500
HEROPON see DBA800
HERPESIL see DAS000
HERPIDU see DAS000
HERPLEX see DAS000
HERPLEX LIQUIFILM see DAS000
HERYL see TFS350
HET see HCY000
HETAMIDE ML see BKE500
HETAPHENONE see DHS200
HETEROAUXIN see ICN000
HETP see HCY000
HETRAZAN see DIW200
HEV-4 see CNA750
HEXA see BBP750
HEXABALM see HCL000
HEXABARBITAL see ERD500
HEXABETALIN see PPK500
2,4,5,2',4',5'-HEXABROMOBIPHENYL see FBU509
HEXABROMOBIPHENYL (technical grade) see FBU000
HEXABUTYLDISTANNOXANE see BLL750
HEXABUTYLDITIN see BLL750
HEXACAP see CBG000
HEXACARBONYL CHROMIUM see HCB000
HEXA C.B. see HCC500
HEXACERT RED No. 3 see FAG040
HEXACHLOR see BBP750
HEXACHLOR-AETHAN (GERMAN) see HCI000
HEXACHLORAN see BBP750, BBQ500
γ-HEXACHLORAN see BBQ500
α-HEXACHLORANE see BBQ000
γ-HEXACHLORANE see BBQ500
HEXACHLORBENZOL (GERMAN) see HCC500
HEXACHLOR-1,3-BUTADIEN (CZECH) see HCD250
HEXACHLORBUTADIENE see HCD250
HEXACHLORCYCLOHEXAN (GERMAN) see BBQ000
2,3,4,4,5,6-HEXACHLORCYKLOHEXA-2,5-DIEN-1-ON
   (CZECH) see HCE000
HEXACHLORCYKLOPENTADIEN (CZECH) see HCE500
HEXACHLORFENOL (CZECH) see HCE000
HEXACHLOROACETONE (DOT) see HCL500
HEXACHLOROBENZENE see HCC500
β-HEXACHLOROBENZENE see BBR000
γ-HEXACHLOROBENZENE see BBQ500
1,2,3,4,7,7-HEXACHLOROBICYCLO(2.2.1)HEPTEN-5,6-BI-
   OXYMETHYLENESULFITE see EAQ750
α,β-1,2,3,4,7,7-HEXACHLOROBICYCLO(2.2.1)-2-HEPTENE-
   5,6-BISOXYMETHYLENE SULFITE see EAQ750
1,1,2,3,4,4-HEXACHLORO-1,3-BUTADIENE see HCD250
HEXACHLORO-1,3-BUTADIENE (MAK) see HCD250
HEXACHLORO-2,5-CYCLOHEXADIENONE see HCE000
HEXACHLORO-2,5-CYCLOHEXADIEN-1-ONE see HCE000
HEXACHLOROCYCLOHEXANE see BBP750
α-HEXACHLOROCYCLOHEXANE see BBQ000
β-HEXACHLOROCYCLOHEXANE see BBR000
1,2,3,4,5,6-HEXACHLOROCYCLOHEXANE see BBP750
1,2,3,4,5,6-HEXACHLOROCYCLOHEXANE (mixture of isomers) see BBQ750
Δ-HEXACHLOROCYCLOHEXANE see BFW500
1-α,2-α,3-α,4-β,5-α,6-β-HEXACHLOROCYCLOHEXANE see
   BFW500

1-α,2-α,3-β,4-α,5-α,6-β-HEXACHLOROCYCLOHEXANE see BBQ500

1-α,2-α,3-β,4-α,5-β,6-β-HEXACHLOROCYCLOHEXANE see BBQ000

1-α,2-β,3-α,4-β,5-α,6-β-HEXACHLOROCYCLOHEXANE see BBR000

Δ-1,2,3,4,5,6-HEXACHLOROCYCLOHEXANE see BFW500

1,2,3,4,5,6-HEXACHLOROCYCLOHEXANE, γ-ISOMER see BBQ500

γ-HEXACHLOROCYCLOHEXANE (MAK) see BBQ500

α-1,2,3,4,5,6-HEXACHLOROCYCLOHEXANE (MAK) see BBQ000

β-1,2,3,4,5,6-HEXACHLOROCYCLOHEXANE (MAK) see BBR000

HEXACHLOROCYCLOPENTADIENE see HCE500

HEXACHLOROCYCLOPENTADIENEDIMER see MQW500

1,2,3,4,5,5-HEXACHLORO-1,3-CYCLOPENTADIENE DIMER see MQW500

HEXACHLORODIBENZO-p-DIOXIN see HAJ500

1,2,3,4,7,8-HEXACHLORODIBENZO-p-DIOXIN see HCF000

1,2,3,6,7,8-HEXACHLORODIBENZO-p-DIOXIN see HAJ500

1,2,3,6,7,8-HEXACHLORODIBENZO-p-DIOXIN mixed with 1,2,3,7,8,9-HEXACHLORODIBENZO-p-DIOXIN see HCF500

1,2,3,7,8,9-HEXACHLORODIBENZO-p-DIOXIN mixed with 1,2,3,6,7,8-HEXACHLORODIBENZO-p-DIOXIN see HCF500

2,2′,3,3′,5,5′-HEXACHLORO-6,6′-DIHYDROXYDIPHENYL-METHANE see HCL000

HEXACHLORO DIPHENYL OXIDE see CDV175

HEXACHLOROEPOXYOCTAHYDRO-endo,endo-DIMETH-ANONAPHTHALENE see EAT500

HEXACHLOROEPOXYOCTAHYDRO-endo,exo-DIMETHANO-NAPHTHALENE see DHB400

HEXACHLOROETHANE see HCI000

1,1,1,2,2,2-HEXACHLOROETHANE see HCI000

HEXACHLOROETHYLENE see HCI000

1,4,5,6,77-HEXACHLORO-N-(ETHYLMERCURI)-5-NOR-BORNENE-2,3-DICARBOXIMIDE see EME050

HEXACHLOROFEN (CZECH) see HCL000

HEXACHLOROHEXAHYDRO-endo-exo-DIMETHANONA-PHTHALENE see AFK250

1,2,3,4,10,10-HEXACHLORO-1,4,4a,5,8,8a-HEXAHYDRO-1,4,5,8-DIMETHANONAPHTHALENE see AFK250

1,2,3,4,10,10-HEXACHLORO-1,4,4a,5,8,8a-HEXAHYDRO-exo-1,4,-endo-5,8-DIMETHANONAPHTHALENE see AFK250

1,2,3,4,10,10-HEXACHLORO-1,4,4a,5,8,8a-HEXAHYDRO-1,4-endo-exo-5, 8-DIMETHANONAPHTHALENE see AFK250

1,2,3,4,10,10-HEXACHLORO-1,4,4a,5,8,8a-HEXAHYDRO-1,4,5,8-endo,endo-DIMETHANONAPHTHALENE see IKO000

1,2,3,4,10,10-HEXACHLORO-1,4,4a,5,8,8a-HEXAHYDRO-1,4-endo,endo-5,8-DIMETHANONAPHTHALENE see IKO000

HEXACHLOROHEXAHYDROMETHANO 2,4,3-BENZODIOX-ATHIEPIN-3-OXIDE see EAQ750

6,7,8,9,10,10-HEXACHLORO-1,5,5a,6,9,9a-HEXAHYDRO-6,9-METHANO-2,4,3-BENZODIOXATHIEPIN-3-OXIDE see EAQ750

HEXACHLORONAPHTHALENE see HCK500

1,4,5,6,7,7-HEXACHLORO-5-NORBORNENE-2,3-DICAR-BOXYLIC ACID see CDS000

1,4,5,6,7,7-HEXACHLORO-5-NORBORNENE-2,3-DIMETHA-NOL CYCLIC SULFITE see EAQ750

3,4,5,6,9,9-HEXACHLORO-1a,2,2a,3,6,6a,7,7a-OCTAHYDRO-2,7:3,6-DIMETHANONAPHTH(2,3-b)OXIRENE see DHB400, EAT500

HEXACHLOROPHANE see HCL000

HEXACHLOROPHEN see HCL000

HEXACHLOROPHENE see HCL000

HEXACHLOROPHENE (DOT) see HCL000

HEXACHLOROPHENYL ETHER see CDV175

HEXACHLOROPLATINIC ACID see CKO750

HEXACHLOROPLATINIC(IV) ACID see CKO750

HEXACHLOROPLATININIC(4+) ACID, HYDROGEN- see CKO750

HEXACHLORO-2-PROPANONE see HCL500

1,1,1,3,3,3-HEXACHLORO-2-PROPANONE see HCL500

HEXACHLOROPROPENE see HCM000

HEXACHLOROPROPYLENE see HCM000

HEXACID 698 see HEU000

HEXACID 898 see OCY000

HEXACID 1095 see DAH400

HEXACID C-9 see NMY000

HEXACOL BRILLIANT BLUE A see FAE000

HEXACOL CARMOISINE see HJF500

HEXACOL ERYTHROSINE BS see FAG040

HEXACOL OIL ORANGE SS see TGW000

HEXACOL ORANGE GG CRYSTALS see HGC000

HEXACOL RHODAMINE B EXTRA see FAG070

HEXACOL TARTRAZINE see FAG140

HEXACYCLONAS see SHL500

HEXACYCLONATE SODIUM see SHL500

HEXADECADROL see SOW000

1-HEXADECANAMINE see HCO500

HEXADECANOIC ACID see PAE250

HEXADECANOL see HCP000

1-HEXADECANOL see HCP000

HEXADECAN-1-OL see HCP000

n-HEXADECANOL see HCP000

HEXADECENE EPOXIDE see EBX500

n-HEXADECOIC ACID see PAE250

HEXADECYL ALCOHOL see HCP000

n-HEXADECYL ALCOHOL see HCP000

N-HEXADECYLAMINE see HCO500

HEXADECYLIC ACID see PAE250

HEXADECYLPYRIDINIUM CHLORIDE see CCX000

1-HEXADECYLPYRIDINIUM CHLORIDE see CCX000

n-HEXADECYLPYRIDINIUM CHLORIDE see CCX000

1-HEXADECYLPYRIDINIUM CHLORIDE MONOHYDRATE see CDF750

HEXADECYLTRICHLOROSILANE see HCQ000

HEXADECYLTRIMETHYLAMMONIUM BROMIDE see HCQ500

(1-HEXADECYL)TRIMETHYLAMMONIUM BROMIDE see HCQ500

N-HEXADECYLTRIMETHYLAMMONIUM BROMIDE see HCQ500

N-HEXADECYL-N,N,N-TRIMETHYLAMMONIUM BROMIDE see HCQ500

1,5-HEXADIENE see HCR500

HEXA-1,5-DIENE see HCR500

HEXADIENIC ACID see SKU000

HEXADIENOIC ACID see SKU000

2,4-HEXADIENOIC ACID see SKU000

trans-trans-2,4-HEXADIENOIC ACID see SKU000

2,4-HEXADIENOIC ACID POTASSIUM SALT see PLS750

HEXADIMETHRINE BROMIDE see HCV500

HEXADIONA see DBB200

HEXADRIN see EAS500, EAT500

HEXADROL see SOW000

HEXAETHYL TETRAPHOSPHATE see HCY000

HEXAETHYL TETRAPHOSPHATE, liquid (DOT) see HCY000

HEXAETHYL TETRAPHOSPHATE, liquid, containing more than 25% hexaethyl tetraphosphate (DOT) see HCY000

HEXAFEN see HCL000

HEXAFERB see FAS000

HEXAFLUORENIUM DIBROMIDE see HEG000

HEXAFLUOROACETONE see HCZ000

HEXAFLUOROACETONE HYDRATE see HDA000

HEXAFLUOROACETONE SESQUIHYDRATE see HDE500

HEXAFLUORO ACETONE TRIHYDRATE see HDA500

HEXAFLUORODICHLOROBUTENE see HDB500

HEXAFLUORODIETHYL ETHER see HDC000

HEXAFLUOROISOBUTYRIC ACID METHYL ESTER see MKK750
HEXAFLUOROISOPROPANOL see HDC500
HEXAFLUOROKIESELSAIURE (GERMAN) see SCO500
HEXAFLUOROKIEZELZUUR (DUTCH) see SCO500
HEXAFLUOROPHOSPHORIC ACID see HDE000
1,1,1,3,3,3-HEXAFLUORO-2-PROPANOL see HDC500
HEXAFLUORO-2-PROPANONE HYDRATE see HDA000
HEXAFLUORO-2-PROPANONE SESQUIHYDRATE see HDE500
HEXAFLUOROPROPENE see HDF000
HEXAFLUOROPROPYLENE (DOT) see HDF000
HEXAFLUOROSILICATE(2-) DIHYDROGEN see SCO500
HEXAFLUOROSILICATE (2−), NICKEL see NDD000
HEXAFLUORO VANADATE (3-) TRIAMMONIUM SALT see ANI500
HEXAFLUORURE de SOUFRE (FRENCH) see SOI000
HEXAFLUOSILICIC ACID see SCO500
HEXAFLURONIUM BROMIDE see HEG000
HEXAFORM see HEI500
HEXAHYDROANILINE see CPF500
HEXAHYDROAZEPINE see HDG000
HEXAHYDRO-1H-AZEPINE see HDG000
HEXAHYDRO-2-AZEPINONE see CBF700
HEXAHYDRO-2H-AZEPIN-2-ONE see CBF700
HEXAHYDRO-2H-AZEPIN-2-ONE HOMOPOLYMER see PJY500
(2-(HEXAHYDRO-1(2H)-AZOCINYL)ETHYL) GUANIDINE HYDROGEN SULFATE see GKU000
HEXAHYDROBENZENAMINE see CPF500
HEXAHYDROBENZENE see CPB000
HEXAHYDROCRESOL see MIQ745
HEXAHYDRO-p-CYMENE HYDROPEROXIDE see IQE000
HEXAHYDRO-1,4-DIAZINE see PIJ000
1,2,3,4,5,6-HEXAHYDRO-6,11-DIMETHYL-3-(3-METHYL-2-BUTENYL)-2,6-METHANO-3-BENZAZOCINE see DOQ400
1,2,3,4,5,6-HEXAHYDRO-6-METHYLAZEPINO(4,5-b)INDOLE HYDROCHLORIDE see HDQ500
dl-1,3,4,9,10,10A-HEXAHYDRO-11-METHYL-2H-10,4A-IMINOETHANOPHENANTHREN-6-OL see MKR250
HEXAHYDROMETHYLPHENOL see MIQ745
HEXAHYDRO-1-NITROSO-1H-AZEPINE see NKI000
HEXAHYDRO-N-NITROSOPYRIDINE see NLJ500
HEXAHYDRO-3,6-endo-OXYPHTHALIC ACID see EAR000
HEXAHYDROPHENOL see CPB750
HEXAHYDROPYRAZINE see PIJ000
HEXAHYDROPYRIDINE see PIL500
HEXAHYDROTHYMOL see MCF750
HEXAHYDROTOLUENE see MIQ740
HEXAHYDRO-1,3,5-s-TRIAZINE see HDV500
HEXAHYDRO-1,3,5-TRINITROSO-s-TRIAZINE see HDV500
HEXAHYDRO-1,3,5-TRINITROSO-1,3,5-TRIAZINE see HDV500
HEXAHYDRO-1,3,5-TRINITRO-1,3,5-TRIAZIN (GERMAN) see CPR800
HEXAHYDRO-1,3,5-TRINITRO-s-TRIAZINE see CPR800
HEXAHYDRO-1,3,5-TRINITRO-1,3,5-TRIAZINE see CPR800
HEXA(HYDROXYMETHYL)MELAMINE see HDY000
HEXAKIS(μ-ACETATO-O:O'))-μ(⁴)-OXOTETRABERYLLIUM see BFT500
HEXAKIS(μ-ACETATO)-μ(⁴)-OXOTETRABERYLLIUM see BFT500
HEXAKIS(DIHYDROGEN PHOSPHATE) MYO-INOSITOL see PIB250
HEXAKIS(β,β-DIMETHYLPHENETHYL)DISTANNOXANE see BLU000
HEXAKIS(HYDROXYMETHYL)MELAMINE see HDY000
HEXAKIS(HYDROXYMETHYL)-1,3,5-TRIAZINE-2,4,6-TRI-AMINE see HDY000
HEXAKIS(2-METHYL-2-PHENYLPROPYL)DISTANNOXANE see BLU000
HEXALDEHYDE (DOT) see HEM000
HEXALIN see CPB750

HEXAMARIUM see DXG800
HEXAMETAPHOSPHATE, SODIUM SALT see SHM500
HEXAMETAPOL see HEK000
HEXAMETHIONIUM BROMIDE see HEA000
HEXAMETHONIUM BROMIDE see HEA000
HEXAMETHONIUM DIBROMIDE see HEA000
HEXAMETHONIUM DIIODIDE see HEB000
HEXAMETHONIUM IODIDE see HEB000
HEXAMETHYLBENZENE see HEC000
HEXAMETHYLDISTANNANE see HEE500
HEXAMETHYLDITIN see HEE500
HEXAMETHYLENAMINE see HEI500
HEXAMETHYLENE see CPB000
HEXAMETHYLENEAMINE see HEI500
1,1'-HEXAMETHYLENEBIS(5-(p-CHLOROPHENYL)BI-GUANIDE see BIM250
HEXAMETHYLENEBIS(DIMETHYL-9-FLUORENYLAMMO-NIUM BROMIDE) see HEG000
HEXAMETHYLENEBIS(FLUOREN-9-YLDIMETHYLAMMO-NIUM BROMIDE) see HEG000
HEXAMETHYLENE BIS(9-FLUORENYL DIMETHYLAM-MONIUM)DIBROMIDE see HEG000
HEXAMETHYLENEBIS(TRIMETHYLAMMONIUM) BRO-MIDE see HEA000
HEXAMETHYLENEBIS(TRIMETHYLAMMONIUM IODIDE) see HEB000
1,6-HEXAMETHYLENEDIAMINE see HEO000
HEXAMETHYLENE DIAMINE, solid (DOT) see HEO000
HEXAMETHYLENE DIISOCYANATE see DNJ800
HEXAMETHYLENE-1,6-DIISOCYANATE see DNJ800
HEXAMETHYLENEDIISOCYANATE (DOT) see DNJ800
1,6-HEXAMETHYLENE DIISOCYANATE (MAK) see DNJ800
HEXAMETHYLENE IMINE (DOT) see HDG000
HEXAMETHYLENEIMINE-3,5-DINITROBENZOATE see ACI250
1-(2-HEXAMETHYLENEIMINOETHYL)-2-OXOCYCLO-HEXANECARBOXYLIC ACID BENZYL ESTER HYDRO-CHLORIDE see HEI000
HEXAMETHYLENETETRAAMINE see HEI500
HEXAMETHYLENETETRAMINE see HEI500
HEXAMETHYLENETRIPEROXYDIAMINE see DCK700
HEXAMETHYLENIMINE see HDG000
2-(β-HEXAMETHYLENIMINOAETHYL)CYCLOHEXANON-2-CARBONSAUREBENZYLESTER-HYDROCHLORIDE (GERMAN) see HEI000
HEXAMETHYLENTETRAMIN (GERMAN) see HEI500
N,N,N,N',N',N'-HEXAMETHYL-1,6-HEXANEDIAMINIUM DIBROMIDE see HEA000
N,N,N,N',N',N'-HEXAMETHYL-1,6-HEXANEDIAMINIUM DIIODIDE see HEB000
2,3,3,4,4,5-HEXAMETHYL-2-HEXANETHIOL see DXT800
HEXAMETHYLMELAMINE see HEJ500
HEXAMETHYLOLMELAMIN (CZECH) see HDY000
HEXAMETHYLOLMELAMINE see HDY000
HEXAMETHYL PHOSPHORAMIDE see HEK000
HEXAMETHYLPHOSPHORIC ACID TRIAMIDE (MAK) see HEK000
HEXAMETHYLPHOSPHORIC TRIAMIDE see HEK000
N,N,N,N,N,N-HEXAMETHYLPHOSPHORIC TRIAMIDE see HEK000
HEXAMETHYLPHOSPHOROTRIAMIDE see HEK000
HEXAMETHYLPHOSPHOTRIAMIDE see HEK000
HEXAMETHYL-p-ROSANILINE HYDROCHLORIDE see AOR500
N,N,N',N',N'',N''-HEXAMETHYL-1,3,5-TRIAZINE-2,4,6-TRI-AMINE see HEJ500
HEXAMETHYL VIOLET see AOR500
HEXAMETON see HEA000
HEXAMIC ACID see CPQ625
HEXAMID see HEL500
HEXAMIDINE see DBB200
HEXAMIDINE (THE ANTISPASMODIC) see DBB200
HEXAMINE (DOT) see HEI500

HEXAMITE see TCF250
HEXAMOL SLS see SIB600
HEXANAL see HEM000
1-HEXANAL see HEM000
HEXANAMIDE see HEM500
1-HEXANAMINE see HFK000
HEXANAPHTHENE see CPB000
HEXANASTAB ORAL see ERD500
n-HEXANE see HEN000
HEXANE (DOT) see HEN000
HEXANEDIAMIDE (9CI) see AEN000
1,6-HEXANEDIAMINE see HEO000
HEXANEDINITRILE see AER250
1,6-HEXANEDIOIC ACID see AEN250
HEXANEDIOIC ACID DINITRILE see AER250
HEXANEDIOIC ACID, POLYMER with 1,4-BUTANEDIOL
    and 1,1′-METHYLENEBIS(4-ISOCYANATOBENZENE) see
    PKM250
HEXANEDIOIC ACID, POLYMER with 1,3-ETHANEDIOL and
    1,1′-METHYLENEBIS(4-ISOCYANATOBENZENE) see
    PKL750
1,2-HEXANEDIOL see HFP875
1,6-HEXANEDIOL DIISOCYANATE see DNJ800
2,5-HEXANEDIOL DIMETHYLSULFONATE see DSU000
3,3′-(1,6-HEXANEDIYLBIS-((METHYLIMINO)
    CARBONYL)OXY)BIS(1-METHYLPYRIDINIUMDIBRO-
    MIDE) see DXG800
6-HEXANELACTAM see CBF700
HEXA-NEMA see DFK600
HEXANEN (DUTCH) see HEN000
HEXANENITRILE see HER500
HEXANES (FCC) see HEN000
1-HEXANETHIOL see HES000
HEXANITRODIPHENYLSULFIDE see BLR750
HEXANITROETHANE see HET675
HEXANITROL see MAW250
HEXANOIC ACID see HEU000
n-HEXANOIC ACID see HEU000
HEXANOL see HFJ500
1-HEXANOL see HFJ500
n-HEXANOL (DOT) see HFJ500
sec-HEXANOL (DOT) see EGW000
tert-HEXANOL (9CI, DOT) see HFJ600
HEXANON see CPC000
2-HEXANONE see HEV000
HEXANONE-2 see HEV000
HEXANONE ISOXIME see CBF700
HEXANONISOXIM (GERMAN) see CBF700
1-HEXANOYLAZIRIDINE see HEW000
HEXANOYLETHYLENEIMINE see HEW000
17-α-HEXANOYLOXYPREGN-4-ENE-3,20-DIONE see
    HNT500
1,1,1,3,3,3-HEXAPHENYLDISTANNTHIANE see BLT250
HEXAPLAS M/1B see DNJ400
HEXAPLAS M/B see DEH200
HEXAPLAS M/O see ILR100
HEXASTAT see HEJ500
HEXASUL see SOD500
HEXATHANE see EIR000
HEXATHIDE see HEB000
HEXATHIR see TFS350
HEXATOX see BBQ500
HEXATYPE CARMINE B see EOJ500
HEXAVIBEX see PPK500
HEXAZANE see PIL500
HEXAZINONE see HFA300
HEXAZIR see BJK500
HEXEMAL see TDA500
HEXENAL see ERD500
2-HEXENAL see HFA500
HEX-2-ENAL see HFA500
HEX-2-EN-1-AL see HFA500
trans-2-HEXEN-1-AL see HFA525
HEXENAL (BARBITURATE) see ERD500

HEXENE see HFB000
1-HEXENE see HFB000
4-HEXENE-1-YNE-3-OL see HFF000
4-HEXENE-1-YNE-3-ONE see HFF300
4-HEXENOIC ACID, 2-ACETYL-5-HYDROXY-3-OXO, Δ-
    LACTONE, SODIUM derivative see SGD000
2-HEXENOL see HFD500
β-γ-HEXENOL see HFE000
2-HEXEN-1-OL, (E)- see HFD500
cis-3-HEXENOL see HFE000
trans-2-HEXENOL see HFD500
γ-HEXENOLACTONE see PAJ500
cis-3-HEXEN-1-OL (FCC) see HFE000
trans-2-HEXEN-1-OL (FCC) see HFD500
2-HEXEN-5,1-OLIDE see PAJ500
D″-HEXENOLLACTONE see PAJ500
cis-3-HEXENYL ISOVALERATE (FCC) see ISZ000
cis-3-HEXENYL 2-METHYLBUTYRATE see HFE550
4-HEXEN-1-YN-3-OL see HFF000
4-HEXEN-1-YN-3-ONE see HFF300
HEXERMIN see PPK500
HEXICIDE see BBQ500
HEXIDE see HCL000
HEXILMETHYLENAMINE see HEI500
HEXMETHYLPHOSPHORAMIDE see HEK000
HEXOBARBITAL see ERD500
HEXOBARBITONE see ERD500
HEXOBION see PPK500
HEXOGEEN (DUTCH) see CPR800
HEXOGEN 5W see CPR800
HEXOGEN (explosive) see CPR800
n-HEXOIC ACID see HEU000
1,6-HEXOLACTAM see CBF700
HEXOLITE see CPR800
HEXOLITE, dry or containing, by weight, less than 15% water
    (DOT) see CPR800
HEXON (CZECH) see HFG500
HEXONE see HFG500
HEXONIUM DIBROMIDE see HEA000
HEXONIUM DIIODIDE see HEB000
HEXOPHENE see HCL000
HEXOSAN see HCL000
HEXYCLAN see BBQ750
HEXYL ACETATE see HFI500
1-HEXYL ACETATE see HFI500
sec-HEXYL ACETATE see HFJ000
n-HEXYL ACETATE (FCC) see HFI500
β-HEXYLACROLEIN see NNA300
HEXYL ALCOHOL see HFJ500
n-HEXYL ALCOHOL see HFJ500
sec-HEXYL ALCOHOL see EGW000
tert-HEXYL ALCOHOL see HFJ600
HEXYL ALCOHOL, ACETATE see HFI500
HEXYLAMINE see HFK000
N-HEXYLAMINE see HFK000
HEXYLAN see BBP750
4-HEXYL-1,3-BENZENEDIOL see HFV500
HEXYL CINNAMALDEHYDE see HFO500
α-HEXYLCINNAMALDEHYDE (FCC) see HFO500
HEXYL CINNAMIC ALDEHYDE see HFO500
α-HEXYLCINNAMIC ALDEHYDE see HFO500
4-HEXYL-1,3-DIHYDROXYBENZENE see HFV500
HEXYLENE see HFB000
HEXYLENE GLYCOL see HFP875
HEXYLENIC ALDEHYDE see HFA500
HEXYL ETHANOATE see HFI500
HEXYL MERCAPTAN see HES000
HEXYL 2-METHYLBUTYRATE see HFR200
α-n-HEXYL-β-PHENYLACROLEIN see HFO500
HEXYLRESORCIN (GERMAN) see HFV500
4-HEXYLRESORCINE see HFV500
HEXYLRESORCINOL see HFV500
4-HEXYLRESORCINOL see HFV500
p-HEXYLRESORCINOL see HFV500

4-n-HEXYLRESORCINOL see HFV500
3-HEXYL-7,8,9,10-TETRAHYDRO-6,6,9-TRIMETHYL-6H-DI-
    BENZO(B,D)PYRAN-1-OL see HGK500
HEXYLTRICHLOROSILANE see HFX500
HEYDEFLON see TAI250
HF 1927 see DCW600, DCW800
HF3170 see DCS200
HFA see FIC000
H-35-F 87 (BVM) see DSQ000
HFCB see HDB500
HFE see HDC000
HFIP see HDC500
81723 HFU see TET800
Hg 532 see PGA750
H.G. BLENDING see SFT000
HGI see BBQ500
HHDN see AFK250
HI-A-VITA see VSK600
HIBERNA see DQA400
HIBESTROL see DKA600
HIBITANE see BIM250
HIBROM see NAG400
HIDACIAN see COH250
HIDACIANN see COH250
HIDACID AZO RUBINE see HJF500
HIDACID AZURE BLUE see FMU059
HIDACID DIBROMO FLUORESCEIN see BNH500
HIDACID FAST ORANGE G see HGC000
HIDACID FLUORESCEIN see FEV000
HIDACID URANINE see FEW000
HIDACO BRILLIANT GREEN see BAY750
HIDACO METHYLENE BLUE SALT FREE see BJI250
HIDACO OIL ORANGE see PEJ500
HIDACO OIL YELLOW see AIC250
HIDAN see DKQ000
HIDANTILO see DKQ000
HIDANTINA SENOSIAN see DKQ000
HIDANTINA VITORIA see DKQ000
HIDANTOMIN see DKQ000
HI-DERATOL see VSZ100
HIDRALAZIN see HGP500, PHW000
HIDRANIZIL see ILD000
HIDRASONIL see ILD000
HIDRIL see CFY000
HIDRIX see HOO500
HIDROCHLORTIAZID see CFY000
HIDROESTRON see EDP000
HIDROMEDIN see DFP600
HIDRORONOL see CFY000
HIDROTIAZIDA see CFY000
HIDRULTA see ILD000
HIDRUN see ILD000
HI-DRY see TCE250
HI-ENTEROL see CHR500
HIESTRONE see EDV000
HI-FLASH NAPHTHAETHYLEN see NAI500
HIFOL see BIO750
HIGOSAN see MEP250
HIGUERETA (CUBA, PUERTO RICO) see CCP000
HIGUERILLA (MEXICO) see CCP000
HIGUEROXYL DELABARRE see FBS000
HI-JEL see BAV750
HILDAN see EAQ750
HILDIT see DAD200
HILTHION see MAK700
HILTHION 25WDP see MAK700
HILTONIL FAST BLUE B BASE see DCJ200
HILTONIL FAST ORANGE GR BASE see NEO000
HILTONIL FAST ORANGE R BASE see NEN500
HILTONIL FAST RED KB BASE see CLK225
HILTONIL FAST SCARLET G BASE see NMP500
HILTOSAL FAST BLUE B SALT see DCJ200
HINDASOL BLUE B SALT see DCJ200
HINDASOL RED TR SALT see CLK235

HINOSAN see EIM000
HIOXYL see HIB000
HIPNAX see DLY000
HIPOFTALIN see HGP500, PHW000
HI-POINT 90 see MKA500
HIPPUZON see TEH500
HIPSAL see DLY000
HI-SEL see SCH000
HISHIREX 502 see PKQ059
HISPACID FAST ORANGE 2G see HGC000
HISPAMIN BLACK EF see AQP000
HISPAMIN BLUE 2B see CMO000
HISPAMIN BLUE 3BX see CMO250
HISPAVIC 229 see PKQ059
HISPERSE YELLOW G see AAQ250
HISPRIL see LJR000
HISTACAP see WAK000
HISTADUR see TAI500
HISTADUR DURA-TABS see TAI500
HISTADYL see TEO250
HISTALEN see TAI500
HISTALON see WAK000
HISTAMETHINE see HGC500
HISTAMETHIZINE see HGC500
HISTAMETIZINE see HGC500
HISTAMETIZYNE see HGC500
HISTAMINE see HGD000
HISTAN see WAK000
HISTANTIN see CFF500
HISTANTINE see CFF500
HISTANTINE DIHYDROCHLORIDE see CDR000
HISTAPAN see TAI500
HISTAPYRAN see WAK000
HISTARGAN see DQA400
HISTASAN see WAK000
HISTATEX see DBM800
HISTAXIN see BBV500
HISTIDINE see HGE700
l-HISTIDINE (FCC) see HGE700
HISTOCARB see CBJ000
HISTOSTAB see PDC000
HISTRYL see LJR000
HISTYN see LJR000
HISTYRENE S 6F see SMR000
HI-STYROL see SMQ500
HI-YIELD DESSICANT H-10 see ARB250
HK-141 see BAW500
H.K. FORMULA No. K. 7117 see FMU059
HL-331 see ABU500
HL 2447 see DFV400
HL 8727 see DIG400
HLS 831 see BSQ000
7-HMBA see BBH250
HMBD see HLX925
HMD see PAN100
HMDA see HEO000
HMDI see DNJ800
HMM see HEJ500
12-HM-7-MBA see HMF500
7-HM-12-MBA see HMF000
HMP see SHM500
HMPA see HEK000
HMPT see HEK000
HMT see HEI500
HMX (DOT) see CQH250
beta HMY see CQH250
HN1 see BID250, CGW000
HN2 see BIE250
HN2 AMINE OXIDE see CFA500
HN2.HCl see BIE500
HN2 HYDROCHLORIDE see BIE500
HN$_2$ OXIDE HYDROCHLORIDE see CFA750
HN$_2$ OXIDE MUSTARD see CFA500
HNT see HHD500

HNU see HKW500
HOCA see DXH250
HOCH see FMV000
HODAG GMS see OAV000
HODAG SMS see SKV150
HODAG SVO 9 see PKL100
HODOSTIN see DQY909
HODSON see DSK800
HOE 881 see FAL100
HOE 2,671 see EAQ750
HOE 2747 see CKD500
HOE 2810 see DGD600
HOE-2824 see ABX250
HOE 2872 see CLU000
HOE 2904 see ACE500
HOECHST 1082 see MDP000
HOECHST PA 190 see PJS750
HOGGAR see BNK000
HOKKO-MYCIN see SLW500
HOKMATE see FAS000
HOLBAMATE see MQU750
HOLIN see EDU500
HOLMIUM see HGF500
HOLMIUM CHLORIDE see HGG000
HOLMIUM CITRATE see HGG500
HOLMIUM(III) NITRATE, HEXAHYDRATE (1:3:6) see
    HGH000
HOLODORM see QAK000
HOLOXAN see IMH000
HOMANDREN see MPN500
HOMANDREN (amps) see TBG000
HOMBITAN see TGG760
HOMIDIUM BROMIDE see DBV400
HOMOANISIC ACID see MFE250
HOMOCHLORCYCLIZINE DIHYDROCHLORIDE see
    CJR809
HOMOCHLOROCYCLIZINE DIHYDROCHLORIDE see
    CJR809
HOMOLLE'S DIGITALIN see DKN400
HOMOOLAN see HIM000
HOMOPIPERIDINE see HDG000
HOMOSTERONE see TBF500
3-HOMOTETRA HYDRO CANNIBINOL see HGK500
HOMOTRYPTAMINE HYDROCHLORIDE see AME750
HOMOVERATRYLAMINE see DOE200
HONG KIEN see ABU500
HOOKER No. 1 CHRYSOTILE ASBESTOS see ARM268
HOPCIDE see CKF000
HOPCIN see MOV000
HORFEMINE see DKB000
HORMALE see MPN500
HORMATOX see DGB000
HORMEX ROOTING POWDER see ICP000
HORMIT see SGH500
HORMODIN see ICP000
HORMOFEMIN see DAL600
HORMOFLAVEINE see PMH500
HORMOFOLLIN see EDV000
HORMOFORT see HNT500
HORMOGYNON see EDP000
HORMOLUTON see PMH500
HORMOMED see EDU500
HORMONIN see EDU500
HORMONISENE see CLO750
HORMOTESTON see TBG000
HORMOTUHO see CIR250
HORMOVARINE see EDV000
HORSE see HBT500
HORSE HEAD A-410 see TGG760
HORTFENICOL see CDP250
HOSALON see DSK800
HOSDON GRANULE see DSK800
HOSTACORTIN see PLZ000, PMA000
HOSTACYCLIN see TBX000

HOSTAFLEX VP 150 see AAX175
HOSTAFLON see TAI250
HOSTALEN PP see PMP500
HOSTALIT see PKQ059
HOSTAQUICK see ABU500
HOSTYREN S see SMQ500
HOURBESE see DKE800
H.P. 34 see SAH000
HPA see EPI300
HPC see HNT500
β-HPN see HGP000
HPOP see HNX500
HPT see HEK000
HR 376 see CIR750
HRS 860 see CEP000
HRS 1276 see MQW500
HS see HGW500
HS-119-1 see PEE750
HT-F 76 see SMQ500
HTH see HOV500
HTP see HCY000
1-α-H,5-α-H-TROPAN-3-α-OL (±)-TROPATE (ESTER) see
    ARR000
1-α-H,5-α-H-TROPAN-3-α-OL (±)-TROPATE (ESTER), SUL-
    FATE (2:1) SALT see ARR500
HUBBUCK'S WHITE see ZKA000
HUILE d'ANILINE (FRENCH) see AOQ000
HUILE de CAMPHRE (FRENCH) see CBA750
HUILE de FUSEL (FRENCH) see FQT000
HUILE H50 see TFM250
HULS P 6500 see PMP500
HUMIFEN WT 27G see DJL000
HUNGARIAN CHAMOMILE OIL see CDH500
HUNGAZIN see ARQ725
HUNGAZIN DT see BJP000
HUNGAZIN PK see ARQ725
HUSEPT EXTRA see CLW000
HVA 2 see BKL750
HVA-2 CURING AGENT see BKL750
HW 4 see CQH250
HW 920 see DXQ500
HY 951 see TJR000
HYACINTHAL see COF000
HYACINTHIN see BBL500
HYADRINE see BBV500
HYADUR see DUD800
HYAMINE see BEN000
HYAMINE 1622 see BEN000
HYAMINE 2389 see DYA600
HYCANTHON see LIM000
HYCANTHONE see LIM000
HYCANTHONE MESYLATE see HGO500
HYCANTHONE METHANESULFONATE see HGO500
HYCANTHONE METHANESULPHONATE see HGO500
HYCANTHONE MONOMETHANESULPHONATE see
    HGO500
HYCAR see PJR000
HYCAR LX 407 see SMR000
HY-CHLOR see HOV500
HYCHOTINE see CJR909
HYCLORITE see SHU500
HYCORACE see HHR000
HYCOZID see ILD000
HYDAN see DFE200
HYDAN (ANTISEPTIC) see DFE200
HYDANTAL see DKQ000
HYDANTIN SODIUM see DNU000
HYDANTOIN see DKQ000
HYDANTOIN SODIUM see DNU000
HYDELTRA see PMA000
HYDELTRONE see PMA000
HYDOUT see DXD000, EAR000
HYDOXIN see PPK250
HYDRACRYLIC ACID β-LACTONE see PMT100

HYDRACRYLONITRILE see HGP000
HYDRAL see CDO000
HYDRALAZINE see PHW000
HYDRALAZINE CHLORIDE see HGP500
HYDRALAZINE HYDROCHLORIDE see HGP500
HYDRALAZINE MONOHYDROCHLORIDE see HGP500
HYDRAL de CHLORAL see CDO000
HYDRALIN see CPB750
HYDRALLAZINE see PHW000
HYDRALLAZINE HYDROCHLORIDE see HGP500
HYDRAPHEN see PFN000
HYDRAPRESS see HGP500
HYDRARGAPHEN see PFN000
HYDRARGYRUM BIJODATUM (GERMAN) see MDD000
HYDRATED LIME see CAT250
HYDRATROP ALDEHYDE see COF000
HYDRATROPIC ALDEHYDE see COF000
HYDRAZID see ILD000
HYDRAZIDE see ILD000
HYDRAZINE see HGS000
HYDRAZINE, anhydrous (DOT) see HGS000
HYDRAZINE, aqueous solution (DOT) see HGS000
HYDRAZINE BASE see HGS000
HYDRAZINE-BENZENE see PFI000
HYDRAZINE-BENZENE and BENZIDINE SULFATE see
   BBY300
HYDRAZINECARBOTHIOAMIDE see TFQ000
HYDRAZINECARBOXAMIDE see HGU000
HYDRAZINECARBOXAMIDE MONOHYDROCHLORIDE see
   SBW500
HYDRAZINE HYDRATE see HGU500
HYDRAZINE HYDROCHLORIDE see HGV000
HYDRAZINE HYDROGEN SULFATE see HGW500
HYDRAZINE MONOHYDRATE see HGU500
HYDRAZINE MONOSULFATE see HGW500
HYDRAZINE SULFATE (1:1) see HGW500
HYDRAZINE SULPHATE see HGW500
HYDRAZINE YELLOW see FAG140
HYDRAZINIUM SULFATE see HGW500
HYDRAZINOBENEZENE see PFI000
2-HYDRAZINOETHANOL see HHC000
2-HYDRAZINO-4-(5-NITRO-2-FURANYL)THIAZOLE see
   HHD500
2-HYDRAZINO-4-(5-NITRO-2-FURYL)THIAZOLE see
   HHD500
1-HYDRAZINO-2-PHENYLETHANE see PFC500
1-HYDRAZINO-2-PHENYLETHANE HYDROGEN SULPHATE
   see PFC750
2-HYDRAZINO-1-PHENYLPROPANE see PDN000
HYDRAZINOPHTHALAZINE see PHW000
1-HYDRAZINOPHTHALAZINE see PHW000
1-HYDRAZINOPHTHALAZINE HYDROCHLORIDE see
   HGP500
1-HYDRAZINOPHTHLAZINE MONOHYDROCHLORIDE see
   HGP500
4-HYDRAZINO-2-THIOURACIL see HHF500
HYDRAZOBENZEN (CZECH) see HHG000
HYDRAZOBENZENE see HHG000
HYDRAZODIBENZENE see HHG000
HYDRAZOETHANE see DJL400
HYDRAZOIC ACID see HHG500
HYDRAZOMETHANE see DSF600, MKN000
HYDRAZONIUM SULFATE see HGW500
2,2'-HYDRAZONODIETHANOL see HHH000
HYDRAZYNA (POLISH) see HGS000
HYDREA see HOO500
HYDRIDES see HHH500
HYDRIODIC ACID see HHI500
HYDRIODIC ACID, solution (DOT) see HHI500
HYDRIODIDE-ENTROL see CHR500
HYDRO-AQUIL see CFY000
HYDROAZOETHANE see DJL400
HYDROBROMIC ACID see HHJ000
HYDROBROMIC ACID, anhydrous (DOT) see HHJ000

HYDROBROMIC ACID MONOAMMONIATE see ANC250
HYDROCARBON GAS see HHJ500
HYDROCARBON GAS, compressed (DOT) see HHJ500
HYDROCARBON GAS, liquefied (DOT) see HHJ500
HYDROCARBON GAS, nonliquefied (DOT) see HHJ500
HYDROCERIN see CMD750
HYDROCHINON (CZECH, POLISH) see HIH000
HYDROCHLORIC ACID see HHL000
HYDROCHLORIC ACID, anhydrous (DOT) see HHL000
HYDROCHLORIC ACID, solution, inhibited (DOT) see
   HHL000
HYDROCHLORIC ACID, mixed with NITRIC ACID (3:1) see
   HHM000
HYDROCHLORIC ETHER see EHH000
HYDROCHLORIDE see HHL000
1C50 HYDROCHLORIDE see EIJ000
l-HYDROCHLORIDE ARGININE see AQW000
HYDROCHLORTHIAZID see CFY000
HYDROCINNAMALDEHYDE see HHP000
HYDROCINNAMIC ALCOHOL see HHP050
HYDROCINNAMIC ALDEHYDE see HHP000
HYDROCINNAMYL ACETATE see HHP500
HYDROCINNAMYL ALCOHOL see HHP050
HYDROCODIN see DKW800
HYDROCONQUININE see HIG500
Δ¹-HYDROCORTISONE see PMA000
HYDROCORTISONE SODIUM SUCCINATE see HHR000
HYDROCORTISONE-21-SODIUM SUCCINATE see HHR000
HYDROCOUMARIN see HHR500
HYDROCUPREINE ETHYL ESTER HYDROCHLORIDE see
   ELC500
HYDROCYANIC ACID see HHS000
HYDROCYANIC ACID, liquefied (DOT) see HHS000
HYDROCYANIC ACID, POTASSIUM SALT see PLC500
HYDROCYANIC ACID (PRUSSIC), unstabilized (DOT) see
   HHS000
HYDROCYANIC ACID, SODIUM SALT see SGA500
HYDROCYANIC ETHER see PMV750
HYDROCYCLIN see HOI000
HYDRODELTALONE see PMA000
HYDRODELTISONE see PMA000
HYDRODIURETIC see CFY000
HYDRO-DIURIL see CFY000
HYDROFLUORIC ACID see HHU500
HYDROFLUORIC ACID, anhydrous (DOT) see HHU500
HYDROFLUORIC ACID (solution) see HHV500
HYDROFLUORIC ACID, solution (DOT) see HHU500
HYDROFLUORIC ACID, SODIUM SALT (2:1) see SHQ500
HYDROFLUORIDE see HHU500
HYDROFLUORIDE-1927 WANDER see DCW800
HYDROFLUOSILICIC ACID see SCO500
HYDROFOL see PAE250
HYDROFOL ACID 1255 see LBL000
HYDROFOL ACID 1495 see MSA250
HYDROFOL ACID 1655 see SLK000
HYDROFURAMIDE see FPS000
HYDROFURAN see TCR750
HYDROGEN see HHW500
HYDROGEN (DOT) see HHW500
HYDROGEN, compressed (DOT) see HHW500
HYDROGEN, refrigerated liquid (DOT) see HHW500
HYDROGEN ANTIMONIDE see SLQ000
HYDROGEN ARSENIDE see ARK250
HYDROGENATED TERPHENYLS see HHW800
HYDROGEN AZIDE see HHG500
HYDROGEN BROMIDE (OSHA ACGIH, MAK, DOT) see
   HHJ000
HYDROGEN CARBOXYLIC ACID see FNA000
HYDROGEN CHLORIDE see HHX000
HYDROGEN CHLORIDE, anhydrous (DOT) see HHL000
HYDROGEN CHLORIDE, refrigerated liquid (DOT) see
   HHL000
HYDROGEN CHLORIDE (AEROSOL) see HHX500
HYDROGEN, CRYOGENIC LIQUID (DOT) see HHY500

HYDROGEN CYANAMIDE see COH500
HYDROGEN CYANIDE, anhydrous, stabilized (DOT) see
    HHS000
HYDROGEN CYANIDE (OSHA, ACGIH) see HHS000
HYDROGEN CYANIDE-N-OXIDE see FOS050
HYDROGEN DIOXIDE see HIB000
HYDROGENE SULFURE (FRENCH) see HIC500
HYDROGEN FLUORIDE (OSHA, ACGIH, MAK, DOT) see
    HHU500
HYDROGEN HEXACHLOROPLATINATE(4+) see CKO750
HYDROGEN HEXAFLUOROPHOSPHATE see HDE000
HYDROGEN HEXAFLUOROSILICATE see SCO500
HYDROGEN IODIDE see HHI500
HYDROGEN IODIDE, anhydrous (DOT) see HHI500
HYDROGEN IODIDE solution (DOT) see HHI500
HYDROGEN NITRATE see NED500
HYDROGEN PEROXIDE see HIB000
HYDROGEN PEROXIDE, solution (over 52% peroxide) (DOT)
    see HIB000
HYDROGEN PEROXIDE, stabilized (over 60% peroxide) (DOT)
    see HIB000
HYDROGEN PEROXIDE CARBAMIDE see HIB500
HYDROGEN PEROXIDE with UREA (1:1) see HIB500
HYDROGEN PHOSPHIDE see PGY000
HYDROGEN SELENIDE see HIC000
HYDROGEN SELENIDE, anhydrous (DOT) see HIC000
21-(HYDROGEN SUCCINATE)CORTISOL, MONOSODIUM
    SALT see HHR000
HYDROGEN SULFIDE see HIC500
HYDROGEN SULFITE SODIUM see SFE000
HYDROGEN SULFURIC ACID see HIC500
α-HYDRO-omega-HYDROXY-POLY(OXY-1,2-ETHANEDIYL)
    see PJT500
HYDROLIN see SHR500
HYDROL SW see PKF500
HYDROMAGNESITE see MAC650
HYDROMEDIN see DFP600
HYDROMIREX see MRI750
HYDROMORPHONE see DLW600
HYDROMORPHONE HYDROCHLORIDE see DNU300
HYDRONITRIC ACID see HHG500
HYDROOT see SOI500
HYDROPERIT see HIB500
HYDROPEROXIDE see HIB000
HYDROPEROXIDE, ACETYL see PCL500
1-((1-HYDROPEROXYCYCLOHEXYL)DIOXY)CYCLO-
    HEXENOL with BIS(1-HYDROXYCYCLOHEXYL)PER-
    OXIDE see CPC500
HYDROPEROXYDE de BUTYLE TERTIAIRE (FRENCH) see
    BRM250
HYDROPEROXYDE de CUMENE (FRENCH) see IOB000
HYDROPEROXYDE de CUMYLE (FRENCH) see IOB000
2-HYDROPEROXY-2-METHYLPROPANE see BRM250
HYDROPHENOL see CPB750
HYDROQUINIDINE see HIG500
HYDROQUINOL see HIH000
HYDROQUINONE see HIH000
m-HYDROQUINONE see REA000
o-HYDROQUINONE see CCP850
p-HYDROQUINONE see HIH000
α-HYDROQUINONE see HIH000
HYDROQUINONE MONOMETHYL ETHER see MFC700
HYDROQUINONE MUSTARD see BHB750
HYDRO-RAPID see CHJ750
HYDRORETROCORTIN see PMA000
HYDRORUBEANIC ACID see DXO200
HYDROSALURIC see CFY000
HYDROSCINE HYDROBROMIDE see HOT500
HYDROSILICOFLUORIC ACID see SCO500
HYDROTHAL-47 see EAR000
HYDROTHIDE see CFY000
HYDROTHOL see DXD000
HYDROTREATED HEAVY NAPHTHENIC DISTILLATE see
    MQV790

HYDROTREATED HEAVY NAPHTHENIC DISTILLATES
    (PETROLEUM) see MQV790
HYDROTREATED HEAVY PARAFFINIC DISTILLATE see
    MQV795
HYDROTREATED LIGHT NAPHTHENIC DISTILLATE see
    MQV800
HYDROTREATED LIGHT NAPHTHENIC DISTILLATES (PE-
    TROLEUM) see MQV800
HYDROTREATED LIGHT PARAFFINIC DISTILLATE see
    MQV805
HYDROXINE see CJR909
HYDROXON see HNT500
4-(4-HYDROXPHENYL)-2-BUTANONE see RBU000
N-HYDROXY-AABP see ACD000
N-HYDROXY-AAF see HIP000
N-HYDROXY-4-ACETAMIDOBIPHENYL see ACD000
N-4-(N-HYDROXYACETAMIDO)BIPHENYL see ACD000
N-HYDROXY-4-ACETAMIDODIPHENYL see ACD000
2-(N-HYDROXYACETAMIDO)FLUORENE see HIP000
N-HYDROXY-2-ACETAMIDOFLUORENE see HIP000
4-HYDROXYACETANILIDE see HIM000
p-HYDROXYACETANILIDE see HIM000
4′-HYDROXYACETANILIDE see HIM000
HYDROXYACETONE see ABC000
HYDROXYACETONITRILE see HIM500
2-HYDROXYACETONITRILE see HIM500
N-HYDROXY-4-ACETYLAMINOBIPHENYL see ACD000
N-HYDROXY-2-ACETYLAMINOFLUORENE see HIP000
N-HYDROXY-N-ACETYL-2-AMINOFLUORENE see HIP000
3-(2-HYDROXY-1-ADAMANTYL)-N-METHYLPROPLY-
    AMINE HYDROCHLORIDE see MGK750
4-HYDROXYAFLATOXIN B1 see AEW000
1-HYDROXY-4-AMINOANTHRAQUINONE see AKE250
N-HYDROXY-2-AMINONAPHTHALENE see NBI500
4-(HYDROXYAMINO)QUINOLINE-1-OXIDE see HIY500
7-β-HYDROXYANDROST-4-EN-3-ONE see TBF500
17-β-HYDROXYANDROST-4-EN-3-ONE see TBF500
17-β-HYDROXY-4-ANDROSTEN-3-ONE see TBF500
17-HYDROXY-(17-β)-ANDROST-4-EN-3-ONE see TBF500
17-β-HYDROXY-Δ⁴-ANDROSTEN-3-ONE see TBF500
2-HYDROXYANILINE see ALT000
3-HYDROXYANILINE see ALT500
o-HYDROXYANILINE see ALT000
4-HYDROXY-m-ANISALDEHYDE see VFK000
2-HYDROXYANISOLE see GKI000
o-HYDROXYANISOLE see GKI000
4-HYDROXY-1-ANTHRAQUINONYLAMINE see AKE250
4-HYDROXYAZOBENZENE see HJF000
p-HYDROXYAZOBENZENE see HJF000
4-HYDROXY-3,4′-AZODI-1-NAPHTHALENESULFONIC
    ACID, DISODIUM SALT see HJF500
4-HYDROXY-3,4′-AZODI-1-NAPHTHALENESULPHONIC
    ACID, DISODIUM SALT see HJF500
2-HYDROXYBENZAMIDE see SAH000
o-HYDROXYBENZAMIDE see SAH000
HYDROXYBENZENE see PDN750
p-HYDROXYBENZENESULFONAMIDE-O-ESTER with O,O-
    DIMETHYL PHOSPHOROTHIOATE see CQL250
3-HYDROXYBENZISOTHIAZOL-S,S-DIOXIDE see BCE500
2-HYDROXYBENZOIC ACID see SAI000
o-HYDROXYBENZOIC ACID see SAI000
p-HYDROXYBENZOIC ACID ETHYL ESTER see HJL000
2-HYDROXYBENZOIC ACID METHYL ESTER see MPI000
o-HYDROXYBENZOIC ACID, METHYL ESTER see MPI000
p-HYDROXYBENZOIC ACID METHYL ESTER see HJL500
2-HYDROXYBENZOIC ACID MONOSODIUM SALT see
    SJO000
4-HYDROXYBENZOIC ACID PROPYL ESTER see HNU500
p-HYDROXYBENZOIC ACID PROPYL ESTER see HNU500
p-HYDROXYBENZOIC ETHYL ESTER see HJL000
o-HYDROXYBENZOIC SODIUM SALT see SJO000
2-HYDROXYBENZO(a)PYRENE see BCX000
3-HYDROXYBENZO(a)PYRENE see BCX250
7-HYDROXYBENZO(a)PYRENE see BCY000

HYDROXYBENZOPYRIDINE see QPA000
N-HYDROXY-2-BENZOYLAMINOFLUORENE see FDZ000
8-HYDROXY-3,4-BENZPYRENE see BCX250
p-HYDROXYBENZYL ACETONE see RBU000
α-HYDROXYBENZYL PHENYL KETONE see BCP250
(p-HYDROXYBENZYL)TARTARIC ACID see HJO500
4-HYDROXYBIPHENYL see BGJ500
p-HYDROXYBIPHENYL see BGJ500
N-HYDROXY-N-4-BIPHENYLACETAMIDE see ACD000
3-HYDROXYBUTANAL see AAH750
1-HYDROXYBUTANE see BPW500
2-HYDROXYBUTANE see BPW750
4-HYDROXYBUTANOIC ACID LACTONE see BOV000
3-HYDROXYBUTANOIC ACID-β-LACTONE see BSX000
3-HYDROXY-2-BUTANONE see ABB500
2-HYDROXY-3-BUTENENITRILE see HJQ000
1-HYDROXY-4-tert-BUTYLBENZENE see BSE500
4-HYDROXYBUTYLBUTYLNITROSAMINE see HJQ350
3-HYDROXYBUTYRALDEHYDE see AAH750
β-HYDROXYBUTYRALDEHYDE see AAH750
γ-HYDROXYBUTYRIC ACID CYCLIC ESTER see BOV000
HYDROXYBUTYRIC ACID LACTONE see BSX000
3-HYDROXYBUTYRIC ACID LACTONE see BSX000
4-HYDROXYBUTYRIC ACID γ-LACTONE see BOV000
γ-HYDROXYBUTYROLACTONE see BOV000
2-HYDROXYCAMPHANE see BMD000
HYDROXYCARBAMINE see HOO500
8-HYDROXY-CHINOLIN (GERMAN) see QPA000
2-HYDROXY-5-CHLORO-N-(2-CHLORO-4-NITROPHENYL)
BENZAMIDE see DFV400
HYDROXYCHOLECALCIFEROL see HJV000
1-HYDROXYCHOLECALCIFEROL see HJV000
1-α-HYDROXYCHOLECALCIFEROL see HJV000
3-β-HYDROXYCHOLESTANE see DKW000
3-β-HYDROXYCHOLEST-5-ENE see CMD750
HYDROXYCINE see CJR909
o-HYDROXYCINNAMIC ACID LACTONE see CNV000
7-HYDROXYCITRONELLAL see CMS850
HYDROXYCITRONELLAL DIMETHYL ACETAL see
HJV700
HYDROXYCITRONELLAL (FCC) see CMS850
3-HYDROXYCROTONIC ACID METHYL ESTER DIMETHYL
PHOSPHATE see MQR750
3-HYDROXYCYCLOHEXADIEN-1-ONE see REA000
HYDROXYCYCLOHEXANE see CPB750
2-HYDROXY-p-CYMENE see CCM000
3-HYDROXY-p-CYMENE see TFX810
14-HYDROXYDAUNOMYCIN see AES750
14'-HYDROXYDAUNOMYCIN see AES750
14-HYDROXYDAUNORUBICINE see AES750
HYDROXYDE de POTASSIUM (FRENCH) see PLJ500
HYDROXYDE de SODIUM (FRENCH) see SHS000
HYDROXYDE de TETRAMETHYLAMMONIUM (FRENCH)
see TDK500
HYDROXYDE de TRIPHENYL-ETAIN (FRENCH) see
HON000
3-HYDROXY-4,15-DIACETOXY-8-(3-METHYLBUTY-
RYLOXY)-12,13-EPOXY-Δ⁹-TRICHOTHECENE see
FQS000
4-HYDROXY-3,5-DIBROMOBENZONITRILE see DDP000
4-HYDROXY-3,5-DI-tert-BUTYLTOLUENE see BFW750
2-(HYDROXY)-5-(2,4-DIFLUOROPHENYL)BENZOIC ACID
see DKI600
14-HYDROXYDIHYDROCODEINONE see PCG500
14-HYDROXYDIHYDROCODEINONE HYDROCHLORIDE
see DLX400
4-HYDROXY-3,5-DIIODOBENZONITRILE see HKB500
8-HYDROXY-5,7-DIIODOQUINOLINE see DNF600
β-HYDROXY-β-(2,5-DIMETHOXYPHENYL)-ISOPRO-
PYLAMINE HYDROCHLORIDE see MDW000
HYDROXYDIMETHYLARSINE OXIDE see HKC000
HYDROXYDIMETHYLARSINE OXIDE, SODIUM SALT see
HKC500
1-HYDROXY-2,4-DIMETHYLBENZENE see XKJ500

3-HYDROXYDIMETHYL CROTONAMIDE DIMETHYL
PHOSPHATE see DGQ875
3-HYDROXY-N,N-DIMETHYL-cis-CROTONAMIDE DI-
METHYL PHOSPHATE see DGQ875
3-HYDROXY-5,5-DIMETHYL-2-CYCLOHEXEN-1-ONE DI-
METHYLCARBAMATE see DRL200
2'-HYDROXY-5,9-DIMETHYL-2-(3,3-DIMETHYLALLYL)-6,7-
BENZOMORPHAN see DOQ400
dl-2'-HYDROXY-5,9-DIMETHYL-2-(3,3-DIMETHYLALLYL)-
6,7-BENZOMORPHAN see DOQ400
17-β-HYDROXY-7-α,17-DIMETHYLESTR-4-EN-3-ONE see
MQS225
(7-α,17-β)-17-HYDROXY-7,17-DIMETHYL-ESTR-4-EN-3-ONE
(9CI) see MQS225
7-HYDROXY-3,7-DIMETHYL OCTANAL see CMS850
7-HYDROXY-3,7-DIMETHYLOCTAN-1-AL see CMS850
7-HYDROXY-3,7-DIMETHYL OCTANAL:ACETAL see
HJV700
3-HYDROXY-4,5-DIMETHYLOL-α-PICOLINE see PPK250
3-HYDROXY-4,5-DIMETHYLOL-α-PICOLINE HYDROCHLO-
RIDE see PPK500
4-HYDROXY-N,N-DIMETHYLTRYPTAMINE see HKE000
5-HYDROXY-N,N-DIMETHYLTRYPTAMINE see DPG109
1-HYDROXY-2,4-DINITROBENZENE see DUZ000
β-(2-HYDROXY-3,5-DINITROPHENYL)BUTANE ACETATE
see ACE500
4-HYDROXYDIPHENYL see BGJ500
p-HYDROXYDIPHENYL see BGJ500
4,4'-HYDROXY-γ,Δ-DIPHENYL-β,Δ-HEXADIENE see
DAL600
α-HYDROXYDIPHENYLMETHANE-β-DIMETHYLAMINO-
ETHYL ETHER HYDROCHLORIDE see BAU750
2-HYDROXYDIPHENYL SODIUM see BGJ750
p-HYDROXYEPHEDRINE see HKH500
3-HYDROXY-1,2-EPOXYPROPANE see GGW500
16-α-HYDROXYESTRADIOL see EDU500
3-HYDROXYESTRA-1,3,5(10),6,8-PENATEN-17-ONE see
ECV000
3-HYDROXYESTRA-1,3,5(10),7-TETRAEN-17-ONE see
ECW000
3-HYDROXYESTRA-1,3,5(10)-TRIEN-17-ONE see EDV000
3-HYDROXYESTRA-1,3,5(10)-TRIEN-17-ONE BENZOATE
see EDV500
HYDROXYESTRIN BENZOATE see EDP000
1-HYDROXYETHANECARBOXYLIC ACID see LAG000
2-HYDROXY-1-ETHANETHIOL see MCN250
HYDROXY ETHER see EES350
17-β-HYDROXY-17-α-ETHINYL-5(10)-ESTREN-3-ONE see
EEH550
4-HYDROXY-3-ETHOXYBENZALDEHYDE see EQF000
2-(2-HYDROXYETHOXY)ETHYL ESTER STEARIC ACID see
HKJ000
1-(2-HYDROXYETHYL)-4-(3-(2-ACETYL-10-PHENOTHIA-
ZYL)PROPYL)PIPERAZINE see ABG000
2-HYDROXYETHYLAMINE see EEC600
β-HYDROXYETHYLAMINE see EEC600
2-HYDROXY-1-ETHYLAZIRIDINE see ASI000
N-(2-HYDROXYETHYL)AZIRIDINE see ASI000
N-(β-HYDROXYETHYL)AZIRIDINE see ASI000
β-HYDROXY-1-ETHYLAZIRIDINE see ASI000
1-(2-HYDROXYETHYL)-4-(3-(2-CHLORO-10-
PHENOTHIAZINYL)PROPYL)PIPERAZINE see CJM250
β-HYDROXYETHYLDIMETHYLAMINE see DOY800
2-HYDROXYETHYL ESTER STEARIC ACID see EJM500
7-(N-(β-HYDROXYETHYL)-N-ETHYL)-AMINOETHYL-8-
BENZYL-THEOPHYLLINE see BEO750
N-HYDROXYETHYL ETHYLENE IMINE see ASI000
1-(2-HYDROXYETHYL)ETHYLENIMINE see ASI000
N-(2-HYDROXYETHYL)ETHYLENIMINE see ASI000
HYDROXYETHYL HYDRAZINE see HHC000
β-HYDROXYETHYLHYDRAZINE see HHC000
N-(2-HYDROXYETHYL)HYDRAZINE see HHC000
1-HYDROXYETHYLIDENE-1,1-DIPHOSPHONIC ACID see
HKS780

3-(1-HYDROXYETHYLIDENE)-6-METHYL-2H-PYRAN-2,-4(3H)-DIONE, SODIUM SALT see SGD000

2,2'-((2-HYDROXYETHYL)IMINO BIS(N-(α,α-DIMETHYL-PHENETHYL)-N-METHYL-ACETAMIDE see DTL200

2,2'-((2-HYDROXYETHYL)IMINO)BIS(N-(1,1-DIMETHYL-2-PHENYLETHYL)-N-METHYLACETAMIDE) see DTL200

β-HYDROXYETHYL ISOPROPYL ETHER see INA500

2-HYDROXYETHYL MERCAPTAN see MCN250

1-HYDROXYETHYL METHYL KETONE see ABB500

1-HYDROXYETHYL-2-METHYL-5-NITROIMIDAZOLE see MMN250

1-(2-HYDROXYETHYL)-2-METHYL-5-NITROIMIDAZOLE see MMN250

1-(β-HYDROXYETHYL)-2-METHYL-5-NITROIMIDAZOLE see MMN250

1-(2-HYDROXY-1-ETHYL)-2-METHYL-5-NITROIMIDAZOLE see MMN250

3-(2-HYDROXYETHYL)-3-METHYL-1-PHENYLTRIAZENE see HKV000

1-(2-HYDROXYETHYL)-1-NITROSOUREA see HKW500

N-(2-HYDROXYETHYL)PHENYLAMINE see AOR750

β-HYDROXYETHYL PHENYL ETHER see PER000

5-(γ-(β-HYDROXYETHYLPIPERAZINO)PROPYL)-5H-DI-BENZO(b,f)AZEPINE DIHYDROCHLORIDE see IDF000

γ-(4-(β-HYDROXYETHYL)PIPERAZIN-1-YL)PROPYL-2-CHLOROPHENOTHIAZINE see CJM250

1-(10-(3-(4-(2-HYDROXYETHYL)-1-PIPERAZINYL)PROPYL)-10H-PHENOTHIAZIN-2-YL)ETHANONE see ABG000

10-(3-(4-(2-HYDROXYETHYL)-1-PIPERAZINYL)PROPYL)PHENOTHIAZIN-2-YL METHYL KETONE see ABG000

10-(3-(4-(2-HYDROXYETHYL)-1-PIPERAZINYL)PROPYL)PHENOTHIAZIN-2-YL METHYL KETONE DIMALEATE see ABG000

3-HYDROXY-2-ETHYL-4-PYRONE see EMA600

(2-HYDROXYETHYL)TRIMETHYLAMMONIUM CHLORIDE see CMF750

(2-HYDROXYETHYL)TRIMETHYLAMMONIUM CHLORIDE ACETATE see ABO000

(2-HYDROXYETHYL)TRIMETHYLAMMONIUM CHLORIDE SUCCINATE see HLC500

N-HYDROXY-2-FAA see HIP000

N-HYDROXY-N-(2-FLUORENYL)ACETAMIDE see HIP000

N-HYDROXY-N-2-FLUORENYLBENZAMIDE see FDZ000

3-HYDROXYGLUTACONIC ACID, DIMETHYL ESTER, DI-METHYL PHOSPHATE see SOY000

1-HYDROXYHEPTANE see HBL500

2-HYDROXYHEPTANE see HBE500

1-HYDROXYHEXANE see HFJ500

5-HYDROXY-2-HEXENOIC ACID LACTONE see PAJ500

1-HYDROXY-3-N-HEXYL-6,6,9-TRIMETHYL-7,8,9,10-TET-RAHYDRO-6-DIBENZOPYRAN see HGK500

o-HYDROXY-HYDROCINNAMIC ACID-Δ-LACTONE see HHR500

HYDROXYHYDROQUINONE see BBU250

4-HYDROXY-3-HYDROXYMETHYL-α-((tert-BUTYLAMINO)METHYL)BENZYL ALCOHOL see BQF500

17-β-HYDROXY-2-HYDROXYMETHYLENE-17-α-METHYL-3-ANDROSTANONE see PAN100

17-β-HYDROXY-2-(HYDROXYMETHYLENE)-17-METHYL-5-α-ANDROSTAN-3-ONE see PAN100

17-β-HYDROXY-2-(HYDROXYMETHYLENE)-17-α-METHYL-5-α-ANDROSTAN-3-ONE see PAN100

17-HYDROXY-2-(HYDROXYMETHYLENE)-17-METHYL-5-α-17-β-ANDROST-3-ONE see PAN100

3-HYDROXY-5-(HYDROXYMETHYL)-2-METHYLISONICO-TINALDEHYDE, HYDROCHLORIDE see VSU000

7-(3-HYDROXY-2-(3-HYDROXY-1-OCTENYL)-5-OXOCY-CLOPENTYL)-5-HEPTENOIC ACID see DVJ200

α-HYDROXY-omega-HYDROXY-POLY(OXY-1,2-ETHANE-DIYL) see PJT000

2-(HYDROXYIMINOMETHYL)-1-METHYLPYRIDINIUM CHLORIDE see FNZ000

2-HYDROXYIMINOMETHYL-1-METHYLPYRIDINIUM IO-DIDE see POS750

4-HYDROXY-3-IODO-5-NITROBENZONITRILE see HLJ500

α-HYDROXYISOBUTYRONITRILE see MLC750

4'-(1-HYDROXY-2-(ISOPROPYLAMINO)ETHYL)METH-ANESULFOANILIDE HYDROCHLORIDE see CCK250

4'-(1-HYDROXY-2-ISOPROPYLAMINO)ETHYL)METH-ANESULFONANILIDE MONOHYDROCHLORIDE see CCK250

3-α-HYDROXY-8-ISOPROPYL-1-α-H,5-α-H-TROPANIUM BROMIDE (±)-TROPATE see IGG000

(8r)-3-α-HYDROXY-8-ISOPROPYL-1-α-H,5-α-H-TROPIUM-BROMIDE-(±)-TROPATE see IGG000

3-HYDROXY-17-KETO-ESTRA-1,3,5-TRIENE see EDV000

4-HYDROXY-2-KETO-4-METHYLPENTANE see DBF750

3-HYDROXY-17-KETO-OESTRA-1,3,5-TRIENE see EDV000

HYDROXYLAMINE see HLM500

HYDROXYLAMINE NEUTRAL SULFATE see OLS000

HYDROXYLAMINE SULFATE see OLS000

HYDROXYLAMINE SULFATE (2:1) see OLS000

HYDROXYLAMMONIUM SULFATE see OLS000

HYDROXYLUREA see HOO500

N-HYDROXY-MAB see HLV000

6-HYDROXY-2-MERCAPTOPYRIMIDINE see TFR250

o-((3-HYDROXYMERCURI-2-METHOXYPROPYL)CARBAMOYL)PHENOXYACETIC ACID MONO-SODIUM SALT see SIH500

N-((3-HYDROXYMERCURI)-2-METHOXYPROPYL)-CARBA-MOYL)SUCCINAMIC ACID see MFC000

N-(γ-HYDROXYMERCURI-β-METHOXYPROPYL)SALICYL-AMIDE-o-ACETIC ACID SODIUM SALT see SIH500

4-HYDROXY-3-METHOXYALLYLBENZENE see EQR500

1-HYDROXY-2-METHOXY-4-ALLYLBENZENE see EQR500

4-HYDROXY-3-METHOXYBENZALDEHYDE see VFK000

1-HYDROXY-2-METHOXYBENZENE see GKI000

6-HYDROXY-7-METHOXY-5-BENZOFURANACRYLIC ACID Δ-LACTONE see XDJ000

o-(2-HYDROXY-4-METHOXYBENZOYL)BENZOIC ACID see HLS500

3-HYDROXY-N-METHOXY-N-METHYL-cis-CROTONAMIDE, DIMETHYL PHOSPHATE see DOL800

6-HYDROXY-3-METHOXY-N-METHYL-4,5-EPOXYMORPHI-NAN see DKW800

1-HYDROXY-2-METHOXY-4-PROPENYLBENZENE see IKQ000

4-HYDROXY-3-METHOXY-1-PROPENYLBENZENE see IKQ000

1-HYDROXY-2-METHOXY-4-PROP-2-ENYLBENZENE see EQR500

6-(HYDROXY(6-METHOXY-4-QUINOLINYL)METHYL)-1-ETHYL-3-VINYL-QUINUCLIDINIUM, IODIDE see QIS000

6-(1-HYDROXY-1-(6-METHOXY-4-QUINOLINYL)METHYL-1-ETHYL-3-VINYLQUINUCLIDINIUM, IODIDE see QIS000

2-HYDROXY-N-METHYL-1-ADAMANTANEPROPANAMINE HYDROCHLORIDE see MGK750

α-HYDROXY-β-METHYL AMINE PROPYLBENZENE see EAW000

N-HYDROXY-N-METHYL-4-AMINOAZOBENZENE see HLV000

(R)-4-(1-HYDROXY-2-(METHYLAMINO)ETHYL)-1,2-BEN-ZENEDIOL (9CI) see VGP000

1-α-HYDROXY-β-METHYLAMINO-3-HYDROXY-1-ETHYL-BENZENE see NCL500

(R)-3-HYDROXY-α-((METHYLAMINO)METHYL) BENZENEMETHANOOL see NCL500

(−)-m-HYDROXY-α-(METHYLAMINOMETHYL)BENZYL ALCOHOL see NCL500

1-m-HYDROXY-α-((METHYLAMINO)METHYL)-BENZYL ALCOHOL see NCL500

(−)-m-HYDROXY-α-((METHYLAMINO)METHYL)BENZYL-ALCOHOL HYDROCHLORIDE see SPC500

1-m-HYDROXY-α-(METHYLAMINOMETHYL)BENZYL AL-COHOL HYDROCHLORIDE see SPC500

1-HYDROXY-2-METHYLAMINO-1-PHENYLPROPANE see EAW000

2-HYDROXY-1-(3-METHYLAMINOPROPYL)ADAMANTANE HYDROCHLORIDE see MGK750

17-β-HYDROXY-17-METHYLANDROST-4-EN-3-ONE see MPN500

HYDROXY METHYL ANETHOL see IRY000

7-HYDROXYMETHYLBENZ(a)ANTHRACENE see BBH250

10-HYDROXYMETHYL-1,2-BENZANTHRACENE see BBH250

1-HYDROXY-2-METHYLBENZENE see CNX000

1-HYDROXY-3-METHYLBENZENE see CNW750

1-HYDROXY-4-METHYLBENZENE see CNX250

4-(HYDROXYMETHYL)BENZENEDIAZONIUM TETRAFLUOROBORATE see HLX925

6-HYDROXYMETHYLBENZO(a)PYRENE see BCV250

2-HYDROXY-3-METHYLCHOLANTHRENE see HMA500

7-HYDROXY-4-METHYLCOUMARIN SODIUM see HMB000

3-HYDROXY-N-METHYL-cis-CROTONAMIDE DIMETHYL PHOSPHATE see MRH209

1-(HYDROXYMETHYL)CYCLOHEXANEACETIC ACID, SODIUM SALT see SHL500

2-HYDROXY-3-METHYL-2-CYCLOPENTEN-1-ONE see HMB500

4-HYDROXYMETHYL-2,2-DIMETHYL-1,3-DIOXOLANE see DVR600

2-HYDROXYMETHYLENE-17-α-METHYL-5-α-ANDROSTAN-17-β-OL-3-ONE see PAN100

2-HYDROXYMETHYLENE-17-α-METHYL-DIHYDROTESTOSTERONE see PAN100

2-(HYDROXYMETHYLENE)-17-α-METHYLDIHYDROTESTOSTERONE see PAN100

2-HYDROXYMETHYLENE-17-α-METHYL-17-β-HYDROXY-3-ANDROSTANONE see PAN100

4-(1-HYDROXY-2-((1-METHYLETHYL)AMINO)ETHYL)-1,2-BENZENEDIOL see DMV600

(HYDROXYMETHYL)ETHYLENE ACETATE see DBF600

N-(1-(HYDROXYMETHYL)ETHYL)-d-LYSERGOMIDE see LJL000

N-(α-(HYDROXYMETHYL)ETHYL)-d-LYSERGOMIDE see LJL000

2-HYDROXYMETHYLFURAN see FPU000

3-HYDROXYMETHYL-n-HEPTAN-4-OL see EKV000

HYDROXYMETHYLINITRILE see HIM500

3-HYDROXY-1-METHYL-4-ISOPROPYLBENZENE see TFX810

HYDROXYMETHYLMERCURY see MLG000

12-HYDROXYMETHYL-7-METHYLBENZ(a)ANTHRACENE see HMF500

7-HYDROXYMETHYL-12-METHYLBENZ(a)ANTHRACENE see HMF000

1-HYDROXYMETHYL-2-METHYLDITMIDE-2-OXIDE see HMG000

17-HYDROXY-6-METHYL-16-METHYLENEPREGNA-4,6-DIENE-3,20-DIONE, ACETATE see MCB380

2-(HYDROXYMETHYL)-2-(METHYLPENTYL) BUTYLCARBAMATE CARBAMATE see MOV500

2-(HYDROXYMETHYL)-2-METHYLPENTYL ESTER, CARBAMATE, BUTYL CARBAMIC ACID see MOV500

(−)-3-HYDROXY-N-METHYLMORPHINAN see LFG000

(±)-3-HYDROXY-N-METHYLMORPHINAN see MKR250

dl-3-HYDROXY-N-METHYLMORPHINAN see MKR250

dl-3-HYDROXY-N-METHYLMORPHINAN HYDROBROMIDE see MDV250

17-β-HYDROXY-18-METHYL-19-NOR-17-α-PREGN-4-EN-20-YN-3-ONE see NNQ500

3-(HYDROXYMETHYL)-8-OXO-7-(2-(4-PYRIDYLTHIO)ACETAMIDO)-5-THIA-1-AZABICYCLO(4.2.0)OCT-2-ENE-2-CARBOXYLIC ACID, ACETATE (ESTER) see CCX500

4-HYDROXY-4-METHYL-PENTAN-2-ON (GERMAN, DUTCH) see DBF750

4-HYDROXY-4-METHYLPENTANONE-2 see DBF750

4-HYDROXY-4-METHYL-2-PENTANONE see DBF750

4-HYDROXY-4-METHYL PENTAN-2-ONE see DBF750

N-(4-((2-HYDROXY-5-METHYLPHENYL)AZO)PHENYL)ACETAMIDE see AAQ250

p-HYDROXY-α-(1-((1-METHYL-3-PHENYLPROPYL)AMINO)ETHYL)BENZYL ALCOHOL HYDROCHLORIDE see DNU200

17-HYDROXY-6-METHYLPREGNA-4,6-DIENE-3,20-DIONE ACETATE see VTF000

17-HYDROXY-6-METHYLPREGNA-4,6-DIENE-3,20-DIONE ACETATE mixed with 19-NOR-17-α-PREGNA-1,3,5(10)-TRIEN-2-YNE-3,17-DIOL see MCA500

17-α-HYDROXY-6-α-METHYLPREGN-4-ENE-3,20-DIONE ACETATE see MCA000

17-α-HYDROXY-6-α-METHYLPREGN-4-ENE-3,20-DIONE ACETATE see MCA000

17-α-HYDROXY-6-α-METHYLPROGESTERONE ACETATE see MCA000

1-HYDROXYMETHYLPROPANE see IIL000

2-HYDROXY-2-METHYLPROPIONITRILE see MLC750

1-(HYDROXYMETHYL)PROPYLAMIDE of 1-METHYL-(+)-LYSERGIC ACID HYDROGEN MALEATE see MQP500

d-N,N'-(1-HYDROXYMETHYLPROPYL)ETHYLENEDINITROSAMINE see HMQ500

3-HYDROXY-2-METHYL-4H-PYRAN-4-ONE see MAO350

5-HYDROXY-6-METHYL-3,4-PYRIDINEDICARBINOL HYDROCHLORIDE see PPK500

5-HYDROXY-6-METHYL-3,4-PYRIDINEDIMETHANOL see PPK250

5-HYDROXY-6-METHYL-3,4-PYRIDINEDIMETHANOL HYDROCHLORIDE see PPK500

3-HYDROXY-1-METHYLPYRIDINIUM BROMIDE HEXAMETHYLENEBIS(METHYLCARBAMATE) see DXG800

3-HYDROXY-2-METHYL-4-PYRONE see MAO350

3-HYDROXY-2-METHYL-γ-PYRONE see MAO350

8-HYDROXY-1-METHYLQUINOLINIUM METHYLSULFATE DIMETHYLCARBAMATE see DQY400

12-HYDROXY-4-METHYL-4,8-SECOSENECIONAN-8,11,16-TRIONE see DMX200

3-HYDROXY-α-METHYL-l-TYROSINE see DNA800

1-HYDROXYNAPHTHALENE see NAW500

2-HYDROXYNAPHTHALENE see NAX000

α-HYDROXYNAPHTHALENE see NAW500

β-HYDROXYNAPHTHALENE see NAX000

5-HYDROXY-1,4-NAPHTHALENEDIONE see WAT000

N-HYDROXYNAPHTHALIMIDE, DIETHYL PHOSPHATE see HMV000

N-HYDROXYNAPHTHALIMIDE-O,O-DIETHYL PHOSPHOROTHIOATE see NAQ500

5-HYDROXY-1,4-NAPHTHOQUINONE see WAT000

N-HYDROXY-2-NAPHTHYLAMINE see NBI500

N-HYDROXYNAPHTHYLIMIDE DIETHYL PHOSPHATE see HMV000

4-HYDROXY-3-NITROANILINE see NEM480

2-HYDROXYNITROBENZENE see NIE500

4-HYDROXYNITROBENZENE see NIF000

4-HYDROXY-3-NITROBENZENEARSONIC ACID see HMY000

4-HYDROXY-3-NITROPHENYLARSONIC ACID see HMY000

N-HYDROXY-N-NITROSO-BENZENAMINE, AMMONIUM SALT see ANO500

4-HYDROXYNONANOIC ACID, γ-LACTONE see CNF250

HYDROXYNOREPHEDRINE see HNB875

m-HYDROXY NOREPHEDRINE see HNB875

17-β-HYDROXY-19-NORPREGN-4-EN-20-YN-3-ONE see NNP500

17-HYDROXY-19-NOR-17-α-PREGN-4-EN-20-YN-3-ONE see NNP500

(17-α)-17-HYDROXY-19-NORPREGN-4-EN-20-YN-3-ONE see NNP500

17-HYDROXY(17-α)-19-NORPREGN-5(10)-EN-20-YN-3-ONE see EEH550

6-HYDROXY-3-(2H)-PYRIDAZINONE DIETHANOLAMINE
see DHF200
4-HYDROXY-2(1H)-PYRIMIDINETHIONE see TFR250
HYDROXYQUINOL see BBU250
8-HYDROXYQUINOLINE see QPA000
8-HYDROXYQUINOLINE COPPER COMPLEX see BLC250
1'-HYDROXYSAFROLE see BCJ000
14-β-HYDROXY-3-β-SCILLOBIOSIDOBUFA-4,20,22-
TRIENOLIDE see GFC000
HYDROXYSENKIRKINE see HOF000
HYDROXYSUCCINIC ACID see MAN000
4-HYDROXY-3-((4-SULFO-1-NAPHTHALENYL)AZO)-1-
NAPHTHALENESULFONIC ACID, DISODIUM SALT see
HJF500
5-HYDROXYTETRACYCLINE see HOH500
5-HYDROXYTETRACYCLINE HYDROCHLORIDE see
HOI000
7-HYDROXY-3,4-TETRAMETHYLENECOUMARIN-O,O-DI-
ETHYL THIOPHOSPHATE see DXO000
HYDROXYTOLUENE see BDX500
4-HYDROXYTOLUENE see CNX250
m-HYDROXYTOLUENE see CNW750
o-HYDROXYTOLUENE see CNX000
p-HYDROXYTOLUENE see CNX250
α-HYDROXYTOLUENE see BDX500
α-HYDROXY-α-TOLUIC ACID see MAP000
HYDROXYTOLUOLE (GERMAN) see CNW500
4'-((6-HYDROXY-m-TOLYL)AZO)ACETANILIDE see
AAQ250
(3-HYDROXY-p-TOLYL)TRIMETHYLAMMONIUM CHLO-
RIDE,METHYLCARBAMATE see HOL000
HYDROXYTRIBUTYLSTANNANE-4,4-DIMETHYLOCTA-
NOATE see TIF250
β-HYDROXYTRICARBALLYLIC ACID see CMS750
1-HYDROXY-2,2,2-TRICHLOROETHYLPHOSPHONIC ACID
DIMETHYL ESTER see TIQ250
2-HYDROXYTRIETHYLAMINE see DHO500
3-HYDROXY-4((2,4,5-TRIMETHYLPHENYL)AZO)-2,7-
NAPHTHALENEDISULFONIC ACID, DISODIUM SALT
see FAG018
3-HYDROXY-4-((2,4,5-TRIMETHYLPHENYL)AZO)-2,7-
NAPHTHALENEDISULPHONIC ACID, DISODIUM SALT
see FAG018
2-HYDROXY-1,3,5-TRINITROBENZENE see PID000
3-HYDROXY-2,4,6-TRINITROPHENOL see SMP500
HYDROXYTRIPHENYLSTANNANE see HON000
HYDROXYTRIPHENYLTIN see HON000
3-HYDROXY-l-TYROSINE see DNA200
l-o-HYDROXYTYROSINE see DNA200
3-HYDROXY-l-TYROSINE HYDROCHLORIDE see
DYC200
HYDROXYUREA see HOO500
N-HYDROXYUREA see HOO500
4-HYDROXYVALERIC ACID LACTONE see VAV000
1-α-HYDROXYVITAMIN D3 see HJV000
3-HYDROXYXANTHINE see HOP000
7-HYDROXYXANTHINE see HOP259
17-HYDROXYYOHIMBAN-16-CARBOXYLIC ACID METHYL
ESTER see YBJ000
o-HYDROXYZIMTSAURE-LACTON (GERMAN) see
CNV000
HYDROXYZINE see CJR909
HYDRURE de LITHIUM (FRENCH) see LHH000
HYDURA see HOO500
HYFLAVIN see RIK000
HYGROMIX-8 see AQB000
HYGROMYCIN B (USDA) see AQB000
HYLEMOX see EEH600
HYLENE M50 see MJP400
HYLENE-T see TGM740, TGM750
HYLENE TCPA see TGM750
HYLENE TLC see TGM750
HYLENE TM see TGM750, TGM800
HYLENE TM-65 see TGM750

HYLENE TRF see TGM750
HYLUTIN see HNT500
HYMECROMONE SODIUM see HMB000
HYMINAL see QAK000
HYMORPHAN see DLW600, DNU300
HYONIC PE-250 see PKF500
HYOSAN see MJM500
HYOSCINE see SBG000, SBG000
HYOSCINE BROMIDE see HOT500
HYOSCINE F HYDROBROMIDE see HOT500
HYOSCINE HYDROBROMIDE see HOT500
(−)-HYOSCINE HYDROBROMIDE see HOT500
1-HYOSCINE HYDROBROMIDE see HOT500
(−)-HYOSCYAMINE see HOU000
HYOSCYAMINE see HOU000
1-HYOSCYAMINE see HOU000
dl-HYOSCYAMINE see ARR000
HYOSCYINE HYDROBROMIDE see HOT500
HYOSOL see SBG000
HYOZID see ILD000
HYPCOL see QAK000
HYPERAZIN see HGP500
HYPERBUTAL see BPF500
HYPERNEPHRIN see VGP000
HYPEROL see HIB500
HYPERPAX see DNA800
HYPERTENAIN see MAW250
HY-PHI 1055 see OHU000
HY-PHI 1088 see OHU000
HY-PHI 1199 see SLK000
HY-PHI 2066 see OHU000
HY-PHI 2088 see OHU000
HY-PHI 2102 see OHU000
HYPNALETTEN see EOK000
HYPNOGEN see EOK000
HYPNOLONE see EOK000
HYPNONE see ABH000
HYPNOREX see LGZ000
HYPNOSTAN see PBT500
HYPNO-TABLINETTEN see EOK000
HYPO see SKI000, SKI500
HYPOCHLORITES see HOU500
HYPOCHLOROUS ACID, CALCIUM SALT see HOV500
HYPOCHLOROUS ACID, CALCIUM SALT (DRY MIXTURE)
see HOW000
HYPODERMACID see TIQ250
HYPONITROUS ACID see HOW500
HYPONITROUS ACID ANHYDRIDE see NGU000
HYPOPHENON see ELL500
HYPOPHTHALIN see HGP500, PHW000
HYPORENIN see VGP000
HYPOS see HGP500
HYPOTHIAZIDE see CFY000
HYPOTROL see SBM500
HYPROVAL-PA see HNT500
HYPTOR BASE see QAK000
HYRE see RIK000
HYSCO see HOT500
HYSCYLENE P see PDX000
HYSTEPS see EOK000
HYSTRENE 80 see SLK000
HYSTRENE 8016 see PAE250
HYSTRENE 9014 see MSA250
HYSTRENE 9512 see LBL000
HYTOX see MIA250
HYVAR see BMM650
HYVAREX see BMM650
HYVAR X see BMM650
HYVAR X BROMACIL see BMM650
HYVAR X WEED KILLER see BMM650

IA see IDZ000
I 337A see CDP250
IAA see ICN000

IA-BUT see BRF500
IA-PRAM see DLH630
IBA see ICP000
IBATRAN see ACE000
IBDU see IIV000
IBENZMETHYZINE see PME250
IBENZMETHYZINE HYDROCHLORIDE see PME500
IBENZMETHYZIN HYDROCHLORIDE see PME500
IBIODRAL see BBV500
IBIOFURAL see NGE500
IBIOSUC see SGC000
IBIOTON see TAI500
IBIOTYZIL see BCA000
IBN see IJD000
IBYLCAINE HYDROCHLORIDE see IAC000
IBZ see PME500
IC 6002 see PIW000
ICG see CCK000
ICI 543 see PMP500
I.C.I. 33,828 see MLJ500
ICI 38174 see INT000
ICI 45520 see ICB000
ICI 59118 see RCA375
ICI 80996 see CMX880
racemic-ICI 80,996 see CMX880
ICIG 1109 see CGV250
I.C.I. HYDROCHLORIDE see INT000
I.C.I. LTD. COMPOUND NUMBER 80996 see CMX880
ICI-PP 557 see AHJ750
ICORAL B see HNB875
ICR 10 see QDS000
ICR 377 see EHJ500
ICR 451 see CIG250
ICR 498 see BNO750
ICR-48b see DFH000
ICRF 159 see PIK250, RCA375
ICTALIS SIMPLE see DKQ000
IDALENE see MOV500
IDANTOIL see DNU000
IDANTOIN see DKQ000
IDANTOINAL see DNU000
ID 480 DIHYDROCHLORIDE see DAB800
IDEXUR see DAS000
IDOCYL NOVUM see SJO000
IDOMETHINE see IDA000
IDOXENE see DAS000
IDOXURIDIN see DAS000
IDOXURIDINE see DAS000
IDPN see BIQ500
IDRAGIN see ADA725
IDRALAZINA (ITALIAN) see PHW000
IDRAZIDE DELL'ACIDO ISONICOTINICO see ILD000
IDRAZIL see ILD000
IDRAZINA IDRATA (ITALIAN) see HGU500
IDRAZINA SOLFATO (ITALIAN) see HGW500
IDROBUTAZINA see HNI500
IDROCHINONE (ITALIAN) see HIH000
IDROESTRIL see DKA600
IDROGENO SOLFORATO (ITALIAN) see HIC500
IDROGESTENE see HNT500
IDROPEROSSIDO di CUMENE (ITALIAN) see IOB000
IDROPEROSSIDO di CUMOLO (ITALIAN) see IOB000
IDROSSIDO DI STAGNO TRIFENILE (ITALIAN) see
    HON000
1′,1-(2-IDROSSIETIL)4-(3-(2-CLORO-10-FENOTIAZIL)PRO-
    PILPIPERAZINA (ITALIAN) see CJM250
4-IDROSSI-4-METIL-PENTAN-2-ONE (ITALIAN) see
    DBF750
d-N,N′-(1-IDROSSIMETIL PROPIL)-ETILENDINITROSAMINA
    (ITALIAN) see HMQ500
4-IDROSSI-3-(3-OXO-)-FENIL-BUTIL)-CUMARINE (ITALIAN)
    see WAT200
IDROSSIZINA see CJR909
IDROTIAZIDE see CFY000

IDRYL see FDF000
IDU see DAS000
IDUCHER see DAS000
IDULEA see DAS000
IDULIAN see DLV800
IDUOCULOS see DAS000
IDUR see DAS000
IDURIDIN see DAS000
IERGIGAN see DQA400
IEROIN see HBT500
IF (fumigant) see TBV750
IFIBRIUM see LFK000
IFOSFAMID see IMH000
IFOSFAMIDE see IMH000
IGE see IPD000
IGELITE F see PKQ059
IGEPAL CA-63 see PKF500
IGEPAL CO-630 see PKF000
IGROSIN see SIH500
II-C-2 see DOQ400
IIH see ILE000
IKACLOMIN see CMX700
IKADA RHODAMINE B see FAG070
IKURIN see ANU650
ILITIA see VSZ450
ILLOXOL see DHB400
ILOPAN see PAG200
ILOTYCIN see EDH500
IM see DLH600
IMAVATE see DLH630
IMAVEROL see FPB875
IMAZALIL see FPB875
IMBRILON see IDA000
IMESONAL see SBM500
IMFERON see IGS000
IMI 115 see TGF250
IMIDA-LAB see TEH500
IMIDALINE HYDROCHLORIDE see BBJ750
IMIDAMINE see PDC000
IMIDAN see PHX250
IMIDAN (PEYTA) see TEH500
IMIDAZOL see IAL000
IMIDAZOLE see IAL000
1H-IMIDAZOLE-4-ETHANAMINE see HGD000
IMIDAZOLE-4-ETHYLAMINE see HGD000
4-IMIDAZOLEETHYLAMINE see HGD000
5-IMIDAZOLEETHYLAMINE see HGD000
IMIDAZOLE MUSTARD see IAN000
2-IMIDAZOLIDINETHIONE see IAQ000
2-IMIDAZOLIDINETHIONE mixed with SODIUM NITRITE
    see IAR000
2-(4-IMIDAZOLYL)ETHYLAMINE see HGD000
2-IMIDAZOL-4-YL-ETHYLAMINE see HGD000
β-IMIDAZOLYL-4-ETHYLAMINE see HGD000
IMIDAZO(2,1-β)THIAZOLE MONOHYDROCHLORIDE see
    LFA020
IMIDENE see TEH500
IMIDIN see NAH500
IMIDOBENZYLE see DLH600, DLH630
4,4′-(IMIDOCARBONYL)BIS(N,N-DIMETHYLAMINE)
    MONOHYDROCHLORIDE see IBA000
4,4′-(IMIDOCARBONYL)BIS(N,N-DIMETHYLANILINE) see
    IBB000
3,3′-IMIDODI-1-PROPANOL, DIMETHANESULFONATE (es-
    ter), HYDROCHLORIDE see YCJ000
IMIDOL see DLH630
IMIDOLE see PPS250
IMILANYLE see DLH630
IMINAZOLE see IAL000
2,2′-IMINOBISETHANOL see DHF000
2,2′-IMINOBISETHYLAMINE see DJG600
3,3′-IMINOBISPROPANENITRILE see BIQ500
IMINOBIS(PROPYLAMINE) see AIX250
3,3′-IMINOBIS(PROPYLAMINE) see AIX250

4,4'-((4-IMINO-2,5-CYCLOHEXADIEN-1-YLIDENE)METHY-
  LENE)DIANILINE MONOHYDROCHLORIDE-o-TOLU-
  IDINE see RMK020
2,2'-IMINODIETHANOL see DHF000
2,2'-IMINODI-ETHANOL with 1,2-DIHYDRO-3,6-PYRIDAZI-
  NEDIONE (1:1) see DHF200
2,2'-IMINODI-N-NITROSOETHANOL see NKM000
β,β-IMINODIPROPIONITRILE see BIQ500
3,3'-IMINODIPROPIONITRILE see BIQ500
IMINO-β,β'-DIPROPIONITRILE see BIQ500
β,β'-IMINODIPROPIONITRILE see BIQ500
2-IMINO-5-PHENYL-4-OXAZOLIDINONE see IBM000
IMINOPHOSPHATE see DXN600
IMINOUREA see GKW000
IMIPHOS see BGY000
IMIPRAMINA (ITALIAN) see DLH600, DLH630
IMIPRAMINE see DLH600, DLH630
IMIPRAMINEDEMETHYL HYDROCHLORIDE see DLS600
IMIPRAMINE HYDROCHLORIDE see DLH630
IMIPRAMINE MONOHYDROCHLORIDE see DLH630
IMIPRIN see DLH600, DLH630
IMIZIN see DLH600
IMIZINUM see DLH600
IMMENOCTAL see SBM500
IMMENOX see SBM500
IMOL S 140 see TNP500
IMOTRYL see BBW500
o-IMPC see PMY300
IMP DISODIUM SALT see DXE500
5'-IMP DISODIUM SALT see DXE500
IMPERIAL GREEN see COF500
IMPERON FIXER T see TND250
IMPERVOTAR see CMY800
IMPF see IPX000
IMP HYDROCHLORIDE see DLH630
IMPIRAMINE-N-OXIDE see IBP309
IMPOSIL see IGS000
IMPRAMINE see DLH600
IMPROVED WILT PRUF see PKQ059
IMPRUVOL see BFW750
IMP SODIUM SALT see DXE500
IMURAN see ASB250
IMUREK see ASB250
IMUREL see ASB250
IMUTEX see IAL000
IMVITE I.G.B.A. see BAV750
IMWITOR 191 see OAV000
IMWITOR 900K see OAV000
IN-117 see HEG000
INACID see IDA000
INACTIVE LIMONENE see MCC250
INAKOR see ARQ725
INAMYCIN see NOB000
INAPPIN see DYF200
INAPSIN see DYF200
INBUTON see BSM000
INCIDOL see BDS000
INDALCA AG see GLU000
INDALONE see BRT000
1,3-INDANDIONE see IBS000
INDAR see BPU000
INDENE see IBX000
1H-INDENE-1,3(2H)-DIONE see IBS000
INDENO(1,2,3-cd)PYRENE see IBZ000
4-(1H-INDEN-1-YLIDENEMETHYL)-N,N-DIMETHYL-
  BENZENAMINE see DOT600
INDERAL see ICB000
INDIAN BERRY see PIE500
INDIAN GUM see AQQ500, GLY000
INDIAN LICORICE SEED see AAD000
INDIAN POKE see VIZ000
INDIAN RED see IHD000, LCS000
INDIGENOUS PEANUT OIL see PAO000
INDIGO BLUE 2B see CMO000

INDIGO CARMINE see FAE100
INDIGO CARMINE (BIOLOGICAL STAIN) see FAE100
INDIGO CARMINE DISODIUM SALT see FAE100
INDIGO EXTRACT see FAE100
INDIGO-KARMIN (GERMAN) see FAE100
5,5'-INDIGOTIN DISULFONIC ACID see FAE100
INDIGOTINE see FAE100
INDIGOTINE DISODIUM SALT see FAE100
INDISULFAT (GERMAN) see ICJ000
INDIUM see ICF000
INDIUM CHLORIDE see ICK000
INDIUM CITRATE see ICH000
INDIUM NITRATE see ICI000
INDIUM SULFATE see ICJ000
INDIUM TRICHLORIDE see ICK000
INDOCID see IDA000
INDOCYBIN see PHU500
INDOKLON see HDC000
INDOL (GERMAN) see ICM000
3-INDOLACETONITRILE see ICW000
INDOLE see ICM000
3-INDOLEACETIC ACID see ICN000
β-INDOLEACETIC ACID see ICN000
β-INDOLE-3-ACETIC ACID see ICN000
1H-INDOLE-3-ACETIC ACID see ICN000
INDOLEACETONITRILE see ICW000
INDOLE-3-ACETONITRILE see ICW000
1H-INDOLE-3-ACETONITRILE see ICW000
INDOLE-3-ACRYLIC ACID see ICO000
INDOLE-3-ALANINE see TNX000
INDOLE-3-(2-AMINOBUTYL) ACETATE see AJB250
1H-INDOLE-3-BUTANOIC ACID see ICP000
INDOLE BUTYRIC see ICP000
INDOLE BUTYRIC ACID see ICP000
3-INDOLEBUTYRIC ACID see ICP000
β-INDOLEBUTYRIC ACID see ICP000
γ-(INDOLE-3)-BUTYRIC ACID see ICP000
INDOLE-3-PROPYLAMINE HYDROCHLORIDE see AME750
INDOLIN see BBW500
INDOLYACETIC ACID see ICN000
INDOLYL-3-ACETIC ACID see ICN000
3-INDOLYLACETIC ACID see ICN000
β-INDOLYLACETIC ACID see ICN000
α-INDOL-3-YL-ACETIC ACID see ICN000
INDOLYLACETONITRILE see ICW000
3-INDOLYLACETONITRILE see ICW000
3-INDOLYLACRYLIC ACID see ICO000
1-β-3-INDOLYLALANINE see TNX000
INDOLYL-3-BUTYRIC ACID see ICP000
4-(INDOLYL)BUTYRIC ACID see ICP000
3-INDOLYL-γ-BUTYRIC ACID see ICP000
4-(INDOL-3-YL)BUTYRIC ACID see ICP000
4-(3-INDOLYL)BUTYRIC ACID see ICP000
γ-(3-INDOLYL)BUTYRIC ACID see ICP000
γ-(INDOL-3-YL)BUTYRIC ACID see ICP000
3-(1-H-INDOL-3-YL)-2-PROPENOIC ACID see ICO000
γ-3-INDOLYLPROPYLAMINE HYDROCHLORIDE see
  AME750
INDOMECOL see IDA000
INDOMED see IDA000
INDOMETHACIN see IDA000
INDOMETHAZINE see IDA000
INDOMETICINA (SPANISH) see IDA000
INDONAPHTHENE see IBX000
INDOPAN see AME500
INDOPTIC see IDA000
INDO-RECTOLMIN see IDA000
INDO-TABLINEN see IDA000
INDUSTRENE 105 see OHU000
INDUSTRENE 205 see OHU000
INDUSTRENE 206 see OHU000
INDUSTRENE 4516 see PAE250
INDUSTRENE 5016 see SLK000
INERTEEN see PJL750

INETOL see INT000
INEXIT see BBQ500
INFAMIL see HNI500
INFERNO see DJA400
INFILTRINA see DUD800
INFLAMEN see BMO000
INFLAZON see IDA000
INFRON see VSZ100
INFUSORIAL EARTH see DCJ800
INGALAN see DFA400
INGALAN (RUSSIAN) see DFA400
INHALAN see DFA400
INHIBINE see HIB000
INHIBISOL see MIH275
INICARDIO see DJS200
INITIATING EXPLOSIVE DIAZODINITROPHENOL (DOT)
    see DUR800
INITIATING EXPLOSIVE FULMINATE of MERCURY (DOT)
    see MDC250
INITIATING EXPLOSIVE IMINOBISPROPYLAMINE (DOT)
    see AIX250
INITIATING EXPLOSIVE LEAD AZIDE, DEXTRINATED
    TYPE ONLY (DOT) see LCM000
INITIATING EXPLOSIVE LEAD MONONITRORESORCI-
    NATE (DOT) see LDP000
INITIATING EXPLOSIVE LEAD STYPHNATE (DOT) see
    LEE000
INITIATING EXPLOSIVE LEAD TRINITRORESORCINATE
    (DOT) see LEE000
INITIATING EXPLOSIVE NITRO MANNITE (DOT) see
    MAW250
INITIATING EXPLOSIVE NITROSOGUANIDINE (DOT) see
    NKH000
INITIATING EXPLOSIVE PENTAERYTHRITE TETRANI-
    TRATE (DOT) see PBC250
INITIATING EXPLOSIVE-TETRAZENE (DOT) see TEF500
INK ORANGE JSN see HGC000
INNOVAN see DYF200
INNOVAR see DYF200
INNOXALON see EID000
INOPSIN see DYF200
INOSINE-5'-MONOPHOSPHATE DISODIUM see DXE500
INOSIN-5'-MONOPHOSPHATE DISODIUM see DXE500
INOSITHEXAPHOSPHORSAURE (GERMAN) see PIB250
INOSITOL HEXAPHOSPHATE see PIB250
INOVAL see DYF200
INOVITAN PP see NCR000
INSARIOTOXIN see FQS000
INSECTICIDE 1,179 see MDU600
INSECTICIDE-NEMATICIDE 1410 see DSP600
INSECTICIDE No. 497 see DHB400
INSECTICIDE No. 4049 see MAK700
INSECTOPHENE see EAQ750
INSECT POWDER see POO250
6-12-INSECT REPELLENT see EKV000
INSIDON DIHYDROCHLORIDE see IDF000
INSOLUBLE SACCHARINE see BCE500
INSOM-RAPIDO see CAV000
IN-SONE see PLZ000
INSPIR see ACH000
INSULAMIN see BOM750, BQL000
INSULAMINA see DNA200
INSULTON see MKB250
INSUMIN see DAB800
INTALBUT see BRF500
INTALPRAM see DLH600, DLH630
INTEBAN SP see IDA000
INTENSE BLUE see FAE100
INTERCAIN see BQA010
INTERCHEM DIRECT BLACK Z see AQP000
INTERNATIONAL ORANGE 2221 see LCS000
INTEXAN SB-85 see DTC600
INTEXSAN CPC see CCX000
INTEXSAN LQ75 see LBW000

INTOCOSTRIN see TOA000
INTRABUTAZONE see BRF500
INTRACID FAST ORANGE G see HGC000
INTRACID PURE BLUE L see FAE000
INTRACORT see HHR000
INTRAMYCETIN see CDP250
INTRANEFRIN see VGP000
INTRASPERSE YELLOW GBA EXTRA see AAQ250
INTRATHION see PHI500
INTRATION see PHI500
INTRAVAL SODIUM see PBT500
INVENOL see BSM000
INVERSINE HYDROCHLORIDE see MQR500
INVERTON 245 see TAA100
INVISI-GARD see PMY300
IODATES see IDJ700
IODE (FRENCH) see IDM000
IODENTEROL see CHR500
IODIC ACIODIC ACID, POTASSIUM SALT see PLK250
IODIDES see IDL000
IODINE see IDM000
IODINE AZIDE see IDN000
IODINE(I) AZIDE see IDN000
IODINE CHLORIDE see IDS000
IODINE CRYSTALS see IDM000
IODINE MONOCHLORIDE see IDS000
IODINE PENTAFLUORIDE see IDT000
IODINE SUBLIMED see IDM000
IODIO (ITALIAN) see IDM000
IODOACETAMIDE see IDW000
2-IODOACETAMIDE see IDW000
α-IODOACETAMIDE see IDW000
IODOACETATE see IDZ000
IODOACETIC ACID see IDZ000
IODOAZIDE see IDN000
2-IODOBUTANE see IEH000
IODOCHLORHYDROXYQUINOL see CHR500
IODOCHLORHYDROXYQUINOLINE see CHR500
7-IODO-5-CHLORO-8-HYDROXYQUINOLINE see CHR500
7-IODO-5-CHLOROXINE see CHR500
5-IODODEOXYURIDINE see DAS000
5-IODO-2'-DEOXYURIDINE see DAS000
IODOENTEROL see CHR500
IODOFORM see IEP000
IODOMETANO (ITALIAN) see MKW200
IODOMETHANE see MKW200
3-IODOPROPIONIC ACID see IEY000
3-IODO-2-PROPYNYL-2,4,5-TRICHLOROPHENYL ETHER
    see IFA000
3-IODOTETRAHYDROTHIOPHENE-1,1-DIOXIDE see
    IFG000
IODOTRIMETHYLSTANNANE see IFN000
IODOTRIMETHYLTIN see IFN000
5-IODOURACIL DEOXYRIBOSIDE see DAS000
IODURE de MERCURE (FRENCH) see MDC750
IODURE de METHYLE (FRENCH) see MKW200
IODURIL see SHW000
IOMESAN see DFV400
IOMEZAN see DFV400
IONET S 60 see SKV150
IONOL see BFW750
IONOL (antioxidant) see BFW750
α-IONONE see IFW000
β-IONONE see IFX000
IOPEZITE see PKX250
IOTOX see HKB500
IOXYNIL see HKB500
IOXYNIL OCTANOATE see DNG200
IPA see DMV600
IPABUTONA see HNI500
IPANER see DAA800
IPD see YCJ000
IPECAC SYRUP see IGF000
IPECACUANHA see IGF000

IPHOSPHAMIDE see IMH000
IPN see ILE000, PHX550
IPNO see IBP309
IPNOFIL see QAK000
IPO 8 see TBW100
IPOGLICONE see BSQ000
IPOLINA see HGP500
IPRAL see ELX000
IPRAL SODIUM see NBU000
IPRATROPIUMBROMID (GERMAN) see IGG000
IPRATROPIUM BROMIDE see IGG000
IPRAZID see ILE000
IPRODIONE see GIA000
IPROGEN see DLH630
IPRONIAZID see ILE000
IPRONID see ILE000
IPRONIDAZOLE (USDA) see IGH000
IPRONIN see ILE000
IPROPRAN see IGH000
IPROVERATRIL see IRV000
IPSOFLAME see BRF500
IPSOTIAN see MQU750
IRADICAV see SHF500
IRAGEN RED L-U see FAG070
IRAMIL see DLH600, DLH630
IRC 453 see CIN750
IRCON see FBJ100
IRENAL see ELX000
IRENE see DLS600
IRGACHROME ORANGE OS see LCS000
IRGALITE BRONZE RED CL see BNH500
IRGALITE RED CBN see CHP500
IRGALON see EIV000
IRIDIL see HNI500
IRIDIUM see IGJ000
IRIDOCIN see EPQ000
IRIDOZIN see EPQ000
IRIFAN see TDA500
IRISH GUM see CCL250
IRISH MOSS EXTRACT see CCL250
IRISH MOSS GELOSE see CCL250
IRIUM see SIB600
IRMIN see DLH600
IROCAINE see AIT250
IROINI see HBT500
IRO-JEX see IGS000
IROMIN see FBK000
IRON see IGK800
IRON ALLOY, BASE, Fe,P (FERROPHOSPHORUS) see
   FBF000
IRON ARSENATE (DOT) see IGM000
IRON(II) ARSENATE (3:2) see IGM000
IRON(III) ARSENATE (1:1) see IGN000
IRON(III)-o-ARSENITE PENTAHYDRATE see IGO000
IRONATE see FBO000
IRON BIS(CYCLOPENTADIENE) see FBC000
IRON CARBOHYDRATE COMPLEX see IGU000
IRON CARBONYL see IHG500
IRON, CARBONYL (FCC) see IGK800
IRON CHLORIDE see FAU000
IRON(III) CHLORIDE see FAU000
IRON(II) CHLORIDE (1:2) see FBI000
IRON CHLORIDE, solid (DOT) see FAU000
IRON(III) CHLORIDE (solution) see FAW000
IRON CHOLINE CITRATE COMPLEX see FBC100
IRON CHROMITE see CMI500
IRON DEXTRAN see IGS000
IRON-DEXTRAN COMPLEX see IGS000
IRON DEXTRAN INJECTION see IGS000
IRON-DEXTRIN COMPLEX see IGU000
IRON DEXTRIN INJECTION see IGU000
IRON DICHLORIDE see FBI000
IRON DICYCLOPENTADIENYL see FBC000
IRON DIMETHYLDITHIOCARBAMATE see FAS000

IRON DISULFIDE see IGV000
IRON DUST see IGW000
IRON, ELECTROLYTIC see IGK800
IRON, ELEMENTAL see IGK800
IRON FLUORIDE see FAX000
IRON FUMARATE see FBJ100
IRON GLUCONATE see FBK000
IRON HYDROGENATED DEXTRAN see IGS000
IRON(2+) LACTATE see LAL000
IRON MONOSULFATE see FBN100
IRON NICKEL SULFIDE see NDE500
IRON ORE see HAO875
IRONORM INJECTION see IGS000
IRON OXIDE see IHD000
IRON(III) OXIDE see IHD000
IRON OXIDE, CHROMIUM OXIDE, and NICKEL OXIDE
   FUME see IHE000
IRON OXIDE FUME see IHF000
IRON OXIDE RED see IHD000
IRON OXIDE, SACCHARATED see IHG000
IRON PENTACARBONYL see IHG500
IRON PERSULFATE see FBA000
IRON-POLYSACCHARIDE COMPLEX see IHH000
IRON PROTOCHLORIDE see FBI000
IRON PROTOSULFATE see FBN100
IRON PYRITES see IGV000
IRON, REDUCED (FCC) see IGK800
IRON SACCHARATE see IHG000
IRON SESQUICHLORIDE, solid (DOT) see FAU000
IRON SESQUIOXIDE see IHD000
IRON SESQUIOXIDE HYDRATED see LFW000
IRON SESQUISULFATE see FBA000
IRON-SILICON ALLOY see FBG000
IRON SORBITEX see IHL000
IRON SORBITOL CITRATE see IHL000
IRON-SORBITOL-CITRIC ACID see IHL000
IRON SUGAR see IHG000
IRON SULFATE (2:3) see FBA000
IRON(III) SULFATE see FBA000
IRON(II) SULFATE (1:1) see FBN100
IRON(II) SULFATE (1:1), HEPTAHYDRATE see FBO000
IRON SULFIDE see IGV000
IRON TERSULFATE see FBA000
IRON TRICHLORIDE see FAU000
IRON TRIFLUORIDE see FAX000
IRON VITRIOL see FBN100
IRON VITROL see FBO000
IROSPAN see FBN100
IROSUL see FBN100, FBO000
IROX (GADOR) see FBK000
IRRADIATED ERGOSTA-5,7,22-TRIEN-3-β-OL see VSZ100
2341 I.S. see BIQ500
ISACONITINE see PIC250
ISAMIN see WAK000
ISATIDINE see RFU000
ISCEON 22 see CFX500
ISCEON 113 see FOO000
ISCEON 122 see DFA600
ISCEON 131 see TIP500
ISCOBROME see MHR200
ISCOBROME D see EIY500
ISCOTIN see ILD000
ISCOVESCO see DKA600
I-SEDRIN see EAW000
ISICAINA see DHK400
ISICAINE HYDROCHLORIDE see DHK600
ISIDRINA see ILD000
ISKIA-C see ARN125
ISMAZIDE see ILD000
ISMICETINA see CDP250
ISMIPUR see POK000
ISOACETOPHORONE see IMF400
ISOADRENALINE HYDROCHLORIDE see AMB000
ISOAMIDONE II see IKZ000

ISOAMYCIN see BBK000
ISOAMYL ACETATE see IHO850
ISOAMYL ALCOHOL see IHP000, IHP010
ISOAMYL ALKOHOL (CZECH) see IHP000
ISO-AMYLALKOHOL (GERMAN) see IHP000
ISOAMYL BENZOATE see IHP100
ISOAMYL BROMIDE see BNP250
ISOAMYL BUTYRATE see IHP400
ISOAMYL CAOPROATE see IHU100
ISOAMYL CINNAMATE see AOG600
β-ISOAMYLENE see AOI750
ISOAMYL ETHANOATE see IHO850
ISOAMYLETHYLBARBITURIC ACID see AMX750
5-ISOAMYL-5-ETHYLBARBITURIC ACID see AMX750
5-ISOAMYL-5-ETHYLBARBITURIC ACID, SODIUM DERIV-
    ATIVE see AON750
ISOAMYL FORMATE see IHS000
ISOAMYL HEXANOATE see IHU100
ISOAMYLHYDRIDE see EIK000
ISOAMYL o-HYDROXYBENZOATE see IME000
ISOAMYL ISOVALERATE (FCC) see ITB000
ISOAMYL METHANOATE see IHS000
ISOAMYL METHYL KETONE see MKW450
ISOAMYL NITRITE see IMB000
ISOAMYLOL see IHP000
ISOAMYL 3-PENTYL PROPENATE see AOG600
ISOAMYL PROPIONATE see AON350
ISOAMYL SALICYLATE (FCC) see IME000
ISOANETHOLE see AFW750
ISOBAC 20 see HCL000
ISOBAMATE see IPU000
ISOBARB see NBU000
ISOBENZAN see OAN000
1,3-ISOBENZOFURANDIONE see PHW750
ISOBICINA see ILD000
ISOBORNEOL THIOCYANATOACETATE see IHZ000
ISOBORNYL ACETATE see IHX600
ISOBORNYL THIOCYANATOACETATE see IHZ000
ISOBORNYL THIOCYANOACETATE see IHZ000
ISOBROMYL see BNP750
ISOBUTANAL see IJS000
ISOBUTANE see MOR750
ISOBUTANOL (DOT) see IIL000
ISOBUTAZINA see HNI500
ISOBUTENAL see MGA250
ISOBUTENE see IIC000
ISOBUTENYL CHLORIDE see CIU750
ISOBUTENYL METHYL KETONE see MDJ750
ISOBUTIL see HNI500
2-ISOBUTOXYETHANOL see IIP000
ISOBUTYL ACETATE see IIJ000
ISOBUTYL ACRYLATE see IIK000
ISOBUTYL ACRYLATE, INHIBITED (DOT) see IIK000
ISOBUTYL ADIPATE see DNH125
ISOBUTYL ALCOHOL see IIL000
ISOBUTYLALDEHYDE see IJS000
ISOBUTYL ALDEHYDE (DOT) see IJS000
ISOBUTYLALKOHOL (CZECH) see IIL000
ISOBUTYLAMINE see IIM000
2-ISOBUTYLAMINOETHANOL HYDROCHLORIDE ACID
    SALT, p-AMINOBENZOIC ACID ESTER see IAC000
2-(ISOBUTYLAMINO)ETHYL-p-AMINOBENZOATE HYDRO-
    CHLORIDE see IAC000
ISOBUTYLBENZENE see IIN000
ISOBUTYL BROMIDE see BNR750
ISOBUTYL BUTANOATE see BSW500
ISOBUTYL BUTYRATE (FCC) see BSW500
ISOBUTYLCARBINOL see IHP000
ISOBUTYL CELLOSOLVE see IIP000
ISOBUTYL CINNAMATE see IIQ000
ISOBUTYLDIUREA see IIV000
ISOBUTYLENE (DOT) see IIC000
ISOBUTYLENEDIUREA see IIV000
ISOBUTYL FORMATE see IIR000

ISOBUTYL-o-HYDROXYBENZOATE see IJN000
1,1'-ISOBUTYLIDENEBISUREA see IIV000
ISOBUTYLIDENEDIUREA see IIV000
ISOBUTYL ISOBUTYRATE see IIW000
ISOBUTYLISOBUTYRATE (DOT) see IIW000
ISOBUTYL KETONE see DNI800
ISOBUTYL METHACRYLATE see IIY000
ISOBUTYL-α-METHACRYLATE see IIY000
ISOBUTYL METHYL CARBINOL see MKW600
ISOBUTYL-METHYLKETON (CZECH) see HFG500
ISOBUTYL METHYL KETONE see HFG500
ISOBUTYLMETHYLMETHANOL see MKW600
ISOBUTYL NITRITE see IJD000
ISOBUTYL PHENYLACETATE see IJF400
ISOBUTYL PROPENOATE see IIK000
ISOBUTYL-2-PROPENOATE see IIK000
ISOBUTYL SALICYLATE see IJN000
ISOBUTYLTRIMETHYLETHANE see TLY500
ISOBUTYL VINYL ETHER see IJQ000
p-ISOBUTYOXYBENZOIC ACID-3-(2'-METHYLPIPERIDINO)
    PROPYL ESTER see IJR000
ISOBUTY PROPIONATE (DOT) see PMV250
ISOBUTYRALDEHYD (CZECH) see IJS000
ISOBUTYRALDEHYDE see IJS000
ISOBUTYRIC ACID see IJU000
ISOBUTYRIC ACID, BENZYL ESTER see IJV000
ISOBUTYRIC ACID, ETHYL ESTER see ELS000
ISOBUTYRIC ACID, ISOBUTYL ESTER see IIW000
ISOBUTYRIC ACID, p-TOLYL ESTER see THA250
ISOBUTYRIC ALDEHYDE see IJS000
ISOBUTYRIC ANHYDRIDE see IJW000
ISOBUTYRONITRILE see IJX000
ISOCAINE see IJZ000
ISOCAINE-ASID see AIT250
ISOCAINE BASE see PIV750
ISOCAINE-HEISLER see AIT250
ISOCARB see PMY300
ISOCARBONAZID see IKC000
ISOCARBOSSAZIDE see IKC000
ISOCARBOXAZID see IKC000
ISOCARBOXAZIDE see IKC000
ISOCARBOXYZID see IKC000
ISOCHINOL see DNX400
ISOCHLOORTHION (DUTCH) see NFT000
ISOCID see ILD000
ISOCIDENE see ILD000
ISO-CORNOX see CIR500
ISOCOTIN see ILD000
ISOCTENE see IKF000
ISOCUMENE see IKG000
ISOCYANATE de METHYLE (FRENCH) see MKX250
ISOCYANATOETHANE see ELS500
ISO-CYANATOMETHANE see MKX250
3-ISOCYANATOMETHYL-3,5,5-TRIMETHYLCYCLO-
    HEXYLISOCYANATE see IMG000
1-ISOCYANATOPROPANE see PNP000
ISOCYANIC ACID, BUTYL ESTER see BRQ500
ISOCYANIC ACID-p-CHLOROPHENYL ESTER see CKB000
ISOCYANIC ACID, CYCLOHEXYL ESTER see CPN500
ISOCYANIC ACID, DIESTER with 1,6-HEXANEDIOL see
    DNJ800
ISOCYANIC ACID, ESTER with DI-o-TOLUENEMETHANE
    see MJN750
ISOCYANIC ACID, ETHYL ESTER see ELS500
ISOCYANIC ACID, HEXAMETHYLENE ESTER see DNJ800
ISOCYANIC ACID, METHYLENEDI-p-PHENYLENE ESTER,
    POLYMER with 1,4-BUTANEDIOL see PKP000
ISOCYANIC ACID, METHYL ESTER see MKX250
ISOCYANIC ACID, METHYLPHENYLENE ESTER see
    TGM740, TGM750
ISOCYANIC ACID, 4-METHYL-m-PHENYLENE ESTER see
    TGM750
ISOCYANIC ACID-1,5-NAPHTHYLENE ESTER see
    NAM500

ISOCYANIC ACID, PHENYL ESTER see PFK250
ISOCYANIC ACID, PROPYL ESTER see PNP000
ISOCYANIDE see COI500
ISOCYANURIC CHLORIDE see TIQ750
ISODEMETON see DAP200
ISODIENESTROL see DAL600
ISODIPHENYLBENZENE see TBC620
ISODRIN see IKO000
ISODUR see IIV000
ISOENDOXAN see IMH000
ISOESTRAGOLE see PMQ750
ISOEUGENOL see IKQ000
ISOEUGENOL ACETATE see AAX750
1,3,4-ISOEUGENOL METHYL ETHER see IKR000
ISOEUGENYL ACETATE (FCC) see AAX750
ISOEUGENYL METHYL ETHER see IKR000
ISOFEDROL see EAW000, EAY500
ISOFENPHOS see IMF300
ISOFLAV see DBN400
ISOFLAV BASE see DBN600
ISOFLUOROPHATE see IRF000
ISOFLUROPHATE see IRF000
ISOFORON see IMF400
ISOFORONE (ITALIAN) see IMF400
ISOFOSFAMIDE see IMH000
ISOFTALODINITRIL (CZECH) see PHX550
ISOHEXANE see DQT400, IKS600
ISOHEXYL ALCOHOL see AOK750
ISOHOL see INJ000
ISOHOMOGENOL see IKR000
ISOL see HFP875
ISOLAN see DSK200
ISOLANE (FRENCH) see DSK200
ISOLANID see LAU000
ISOLANIDE see LAU000
ISOLEUCINE see IKX000
l-ISOLEUCINE (FCC) see IKX000
ISOL LAKE RED LCS 12527 see CHP500
ISOLYN see ILD000
ISOMEBUMAL see EJY000
ISOMEPROBAMATE see IPU000
β-ISOMER see BBR000
ISOMERIC CHLORTHION see NFT000
ISOMETASYSTOX see DAP400
ISOMETASYSTOX SULFONE see DAP600
ISOMETHADONE see IKZ000
ISOMETHEPTENE see ILK000
ISOMETHEPTENE HYDROCHLORIDE see ODY000
ISOMETHYLSYSTOX see DAP400
ISOMETHYLSYSTOX SULFONE see DAP600
ISOMETHYLSYSTOX SULFOXIDE see DAP000
ISOMIN see TEH500
ISOMYN see BBK000
ISOMYST see IQN000
ISONAL see ENB500
ISONAL (ROUSSEL) see ENB500
ISONAPHTHOL see NAX000
ISONATE see MJP400
ISONERIT see ILD000
ISONEX see ILD000
ISONIACID see ILD000
ISONIAZID see ILD000
ISONIAZIDE see ILD000
ISONICAZIDE see ILD000
ISONICID see ILD000
ISONICO see ILD000
ISONICOTAN see ILD000
ISONICOTIL see ILD000
ISONICOTINHYDRAZID see ILD000
ISONICOTINIC ACID, ETHYL ESTER see ELU000
ISONICOTINIC ACID HYDRAZIDE see ILD000
ISONICOTINIC ACID-2-ISOPROPYLHYDRAZIDE see
  ILE000
ISONICOTINOYL HYDRAZIDE see ILD000

ISONICOTINOYLHYDRAZINE see ILD000
1-ISONICOTINOYL-2-ISOPROPYLHYDRAZINE see ILE000
ISONICOTINSAEUREHYDRAZID see ILD000
ISONICOTINYL HYDRAZIDE see ILD000
1-ISONICOTINYL-2-ISOPROPYLHYDRAZINE see ILE000
ISONIDE see ILD000
ISONIDRIN see ILD000
ISONIKAZID see ILD000
ISONILEX see ILD000
ISONIN see ILD000
ISONINDON see ILD000
ISONIPECAINE see DAM600
ISONIPECAINE HYDROCHLORIDE see DAM700
ISONIRIT see ILD000
ISONITON see ILD000
ISONITOX see MIW250
ISONITROPROPANE see NIY000
ISONIZIDE see ILD000
ISONORENE see DMV600
ISONYL see ILK000
ISOOCTANE (DOT) see TLY500
ISOOCTANOL see ILL000
ISOOCTYL ALCOHOL see ILL000
ISOOCTYL ALCOHOL (2,4-DICHLOROPHENOXY)ACETATE
  see ILO000
2-ISOOCTYL AMINE see ILM000
ISOOCTYL-2,4-DICHLOROPHENOXYACETATE see ILO000
ISOOCTYL PHTHALATE see ILR100
ISOPARATHION see DJT000
ISOPENTANE see EIK000
ISOPENTANOIC ACID (DOT) see ISU000
ISOPENTANOIC ACID, PHENYLMETHYL ESTER see
  ISW000
ISOPENTANOL see IHP000
ISOPENTYL ACETATE see IHO850
ISOPENTYL ALCOHOL see IHP000
ISOPENTYL ALCOHOL ACETATE see IHO850
ISOPENTYL ALCOHOL, FORMATE see IHS000
ISOPENTYL ALCOHOL NITRITE see IMB000
ISOPENTYL BENZOATE see IHP100
ISOPENTYL BROMIDE see BNP250
ISOPENTYL FORMATE see IHS000
ISOPENTYL-2-HYDROXYPHENYL METHANOATE see
  IME000
ISOPENTYL ISOVALERATE see ITB000
ISOPENTYL METHYL KETONE see MKW450
ISOPENTYL NITRITE see IMB000
3-((ISOPENTYL)NITROSOAMINO)-2-BUTANONE see
  MHW350
ISOPENTYL SALICYLATE see IME000
ISOPESTOX see PHF750
ISOPHEN see DBA800
ISOPHENERGAN see DQA400
ISOPHENICOL see CDP250
ISOPHENPHOS see IMF300
ISOPHORONE see IMF400
ISOPHORONE DIAMINE DIISOCYANATE see IMG000
ISOPHORONE DIISOCYANATE see IMG000
ISOPHOSPHAMIDE see IMH000
ISOPHRIN see NCL500
ISOPHTHALODINITRILE see PHX550
ISOPHTHALONITRILE see PHX550
ISOPRAL see IMQ000
ISOPRENALINE see DMV600
(±)-ISOPRENALINE SULFATE see IRU000
dl-ISOPRENALINE SULFATE see IRU000
ISOPRENE see IMS000
ISOPRENE, INHIBITED (DOT) see IMS000
ISOPROCARB see MIA250
ISOPROMETHAZINE see DQA400
ISOPROPANETHIOL see IMU000
ISOPROPANOL (DOT) see INJ000
ISOPROPENE CYANIDE see MGA750
ISOPROPENIL-BENZOLO (ITALIAN) see MPK250

ISOPROPENYL-BENZEEN (DUTCH) see MPK250
ISOPROPENYLBENZENE see MPK250
ISOPROPENYL-BENZOL (GERMAN) see MPK250
ISOPROPENYL CARBINOL see IMW000
(+)-4-ISOPROPENYL-1-METHYLCYCLOHEXENE see
　LFU000
ISOPROPENYLNITRILE see MGA750
ISOPROPHYL METHYLPHOSPHONOFLUORIDATE see
　IPX000
ISOPROPILAMINA (ITALIAN) see INK000
ISOPROPILBENZENE (ITALIAN) see COE750
ISOPROPILE (ACETATO di) (ITALIAN) see INE100
(1-ISOPROPIL-3-METIL-1H-PIRAZOL-5-IL)-N,N-DIMETIL-
　CARBAMMATO (ITALIAN) see DSK200
(3)-O-2-ISOPROPOXY-CARBONYL-1-METHYLVINYL-O-
　METHYL ETHYLPHOSPHORAMIDOTHIOATE see
　MKA000
2-ISOPROPOXYETHANOL see INA500
ISOPROPOXYMETHYLPHORYL, FLUORIDE see IPX000
o-ISOPROPOXYPHENYL METHYLCARBAMATE see
　PMY300
2-ISOPROPOXYPHENYL-N-METHYLCARBAMATE see
　PMY300
o-ISOPROPOXYPHENYL-N-METHYLCARBAMATE see
　PMY300
2-ISOPROPOXYPROPANE see IOZ750
ISOPROPYDRIN see DMV600
ISOPROPYLACETAAT (DUTCH) see INE100
ISOPROPYLACETAT (GERMAN) see INE100
ISOPROPYL ACETATE see INE100
ISOPROPYL (ACETATE d') (FRENCH) see INE100
ISOPROPYLACETIC ACID see ISU000
ISOPROPYL ACETIC ACID, BENZYL ESTER see ISW000
ISOPROPYLACETONE see HFG500
ISOPROPYL ACID PHOSPHATE solid see PHE500
ISOPROPYLADRENALINE see DMV600
ISOPROPYL ALCOHOL see INJ000
ISO-PROPYLALKOHOL (GERMAN) see INJ000
ISOPROPYLAMINE see INK000
4-(2-ISOPROPYLAMINE-1-HYDROXYETHYL)METHANE-
　SULFOANILIDE HYDROCHLORIDE see CCK250
1-ISOPROPYLAMINE-3-(1-NAPHTHYLOXY)-2-PROPANOL
　see ICB000
ISOPROPYLAMINO-O-ETHYL-(4-METHYLMERCAPTO-3-
　METHYLPHENYL)PHOSPHATE see FAK000
4-(2-ISOPROPYLAMINO-1-HYDROXYAETHYL)METHANE-
　SULFONALID HYDROCHLORID (GERMAN) see CCK250
ISOPROPYLAMINOHYDROXYETHYLMETHANESUL-
　FONALIDE HYDROCHLORIDE see CCK250
ISOPROPYLAMINOMETHYL-3,4-DIHYDROXYPHENYL
　CARBINOL see DMV600
α-((ISOPROPYLAMINO)METHYL)NAPHTHALENEMETHA-
　NOL, HYDROCHLORIDE see INT000
α-(ISOPROPYLAMINOMETHYL)PROTOCATECHUYL ALCO-
　HOL see DMV600
2-ISOPROPYLAMINO-1-(2-NAPHTHYL)ETHANOL HYDRO-
　CHLORIDE see INT000
1-ISOPROPYLAMINO-3-(1-NAPHTHYLOXY)-2-PROPANOL
　see ICB000
9-((3-(ISOPROPYLAMINO)PROPYL)AMINO)-1-NITROACRI-
　DINE DIHYDROCHLORIDE see INW000
2-ISOPROPYL ANILINE see INX000
N-ISOPROPYLANILINE see INX000
o-ISOPROPYLANILINE see INX000
ISOPROPYLANTIPYRIN see INY000
ISOPROPYLANTIPYRINE see INY000
4-ISOPROPYLANTIPYRINE see INY000
ISOPROPYLARTERENOL see DMV600
4-ISOPROPYLBENZALDEHYDE see COE500
p-ISOPROPYLBENZALDEHYDE see COE500
8-ISOPROPYLBENZ(a)ANTHRACENE see INZ000
5-ISOPROPYL-1:2-BENZANTHRACENE see INZ000
ISOPROPYLBENZEEN (DUTCH) see COE750
ISOPROPYL BENZENE see COE750

p-ISOPROPYLBENZENECARBOXALDEHYDE see COE500
ISOPROPYLBENZENE HYDROPEROXIDE see IOB000
ISOPROPYLBENZENE PEROXIDE see DGR600
ISOPROPYLBENZOL see COE750
ISOPROPYL-BENZOL (GERMAN) see COE750
3-ISOPROPYL-2,1,3-BENZOTHIADIAZINON-(4)-2,2-DIOXID
　(GERMAN) see MJY500
3-ISOPROPYL-1H-2,1,3-BENZOTHIADIAZIN-4(3H)-ONE-2,2-
　DIOXIDE see MJY500
ISOPROPYL BIS(β-CHLOROETHYL)AMINE HYDROCHLO-
　RIDE see IPG000
ISOPROPYL BORATE see IOI000
ISOPROPYL CARBAMATE see IOJ000
ISOPROPYLCARBAMIC ACID, ESTER with 2-(HYDROXY-
　METHYL)-2-METHYLPENTYL CARBAMATE see IPU000
2-(p-ISOPROPYL CARBAMOYL BENZYL)-1-METHYLHY-
　DRAZINE see PME250
1-(p-ISOPROPYLCARBAMOYLBENZYL)-2-METHYLHYDRA-
　ZINE HYDROCHLORIDE see PME500
2-(p-(ISOPROPYLCARBAMOYL)BENZYL)-1-METHYLHY-
　DRAZINE HYDROCHLORIDE see PME500
1-ISOPROPYL CARBAMOYL-3-(3,5-DICHLOROPHENYL)-
　HYDANTOIN see GIA000
ISOPROPYLCARBINOL see IIL000
ISOPROPYL CELLOSOLVE see INA500
N-ISOPROPYL-2-CHLOROACETANILIDE see CHS500
N-ISOPROPYL-α-CHLOROACETANILIDE see CHS500
ISOPROPYL-3-CHLOROCARBANILATE see CKC000
ISOPROPYL-m-CHLOROCARBANILATE see CKC000
ISOPROPYL CHLOROCARBONATE see IOL000
ISOPROPYL CHLOROFORMATE see IOL000
ISOPROPYL CHLOROMETHANOATE see IOL000
ISOPROPYL-3-CHLOROPHENYLCARBAMATE see CKC000
ISOPROPYL-N-(3-CHLOROPHENYL)CARBAMATE see
　CKC000
o-ISOPROPYL-N-(3-CHLOROPHENYL)CARBAMATE see
　CKC000
ISOPROPYL-N-(3-CHLORPHENYL)-CARBAMAT (GERMAN)
　see CKC000
ISOPROPYL CRESOL see TFX810
ISOPROPYL-o-CRESOL see CCM000
6-ISOPROPYL-m-CRESOL see TFX810
ISOPROPYL CYANIDE see IJX000
ISOPROPYL-2,4-D ESTER see IOY000
ISOPROPYL DIETHYLDITHIOPHOSPHORYLACETAMIDE
　see IOT000
N-ISOPROPYL-β-DIHYDROXYPHENYL-β-HYDROXY-
　ETHYLAMINE see DMV600
ISOPROPYL DIMETHYL CARBINOL see AOK750
4-ISOPROPYL-2,3-DIMETHYL-1-PHENYL-3-PYRAZOLIN-5-
　ONE see INY000
ISOPROPYL ETHER see IOZ750
ISOPROPYL FLUOPHOSPHATE see IRF000
ISOPROPYL FORMATE see IPC000
ISOPROPYLFORMIC ACID see IJU000
ISOPROPYL GLYCIDYL ETHER see IPD000
ISOPROPYL GLYCOL see INA500
ISOPROPYLIDENEACETONE see MDJ750
4,4'-ISOPROPYLIDENEBISPHENOL see BLD500
p,p'-ISOPROPYLIDENEBISPHENOL see BLD500
2,3-ISOPROPYLIDENEDIOXYPHENYL METHYLCARBA-
　MATE see DQM600
p,p'-ISOPROPYLIDENEDIPHENOL see BLD500
4,4'-ISOPROPYLIDENEDIPHENOL DIGLYCIDYL ETHER see
　BLD750
ISOPROPYLIDENE GLYCEROL see DVR600
1,2-o-ISOPROPYLIDENE GLYCEROL see DVR600
N-ISOPROPYL ISONICOTINHYDRAZIDE see ILE000
ISOPROPYL MEPROBAMATE see IPU000
ISOPROPYL MERCAPTAN (DOT) see IMU000
N-ISOPROPYL-2-MERCAPTOACETAMIDE-S-ESTER with
　O,O-DIETHYL PHOSPHORODITHIOATE see IOT000
ISOPROPYL METHANEFLUOROPHOSPHONATE see
　IPX000

ISOVALERIC ACID see ISU000
ISOVALERIC ACID, ALLYL ESTER see ISV000
ISOVALERIC ACID, BENZYL ESTER see ISW000
ISOVALERIC ACID, BUTYL ESTER see ISX000
(E)-ISOVALERIC ACID-3,7-DIMETHYL-2,6-OCTADIENYL
    ESTER see GDK000
ISOVALERIC ACID, ETHYL ESTER see ISY000
(Z)-ISOVALERIC ACID-3-HEXENYL see ISZ000
ISOVALERIC ACID, ISOPENTYL ESTER see ITB000
ISOVALERIC ACID, METHYL ESTER see ITC000
ISOVALERONE see DNI800
ISOXANTHINE see XCA000
ISOZIDE see ILD000
ISOZYD see ILD000
ISRAVIN see DBX400
ISTONYL see DRM000
ISUPREL see DMV600
ISUPREN see DMV600
IT 40 see SMQ500
IT 931 see DLK200
ITAMID see PJY500
ITAMIDONE see DOT000
ITINEROL see HGC500
ITIOCIDE see EPQ000
ITOPAZ see EEH600
ITROP see IGG000
ITURAN see NGE000
IUDR see DAS000
5-IUDR see DAS000
IVALON see FMV000
IVAUGAN see CFY000
IVE see IJQ000
IVERMECTIN see ITD875
IVIRON see IHG000
IVORAN see DAD200
IVORIT see EJM500
IVOSIT see ACE500
IXODEX see DAD200
IXOTEN see TNT500
IXPER 25M see MAH750
IZOACRIDINA see AHS500
IZOFORON (POLISH) see IMF400

J3 see BPP750
J 400 see PMP500
JACK WILSON CHLORO 51 (oil) see CKC000
JACOBINE see JAK000, SBX500
JACUTIN see BBQ500
JADE GREEN BASE see JAT000
JAFFNA TOBACCO see BFW135
JAGUAR GUM A-20-D see GLU000
JAGUAR No. 124 see GLU000
JAGUAR PLUS see GLU000
JAIKIN see SFF000
JAMESTOWN WEED see SLV500
JANIMINE see DLH630
JANUPAP see HIM000
JAPAN AGAR see AEX250
JAPAN CAMPHOR see CBA750
JAPANESE CAMPHOR see CBB250
JAPANESE CAMPHOR OIL see CBB500
JAPANESE, OIL of CAMPHOR see CBB500
JAPAN ISINGLASS see AEX250
JASAD see ZBJ000
JASMINALDEHYDE see AOG500
JASMOLIN I or II see POO250
JAUNE AB see FAG130
JAUNE de BEURRE (FRENCH) see DOT300
JAUNE OB see FAG135
JAVA ORANGE 2G see HGC000
JAVA RUBINE N see HJF500
JAYSOL see EFU000
JAYSOL S see EFU000
JB-323 see PJA000

JB 329 see DXP800
JB 336 see MON250
JB 516 see PDN000
JB 8181 see DLS600
JEFFERSOL EB see BPJ850
JEFFERSOL EE see EES350
JEFFERSOL EM see EJH500
JEFFOX see PJT000, PJT200, PKI500
JENACAINE see AIL750
JEN-DIRIL see CFY000
JERVINE see JCS000
JESUIT'S BALSAM see CNH792
JET FUEL HEF-2 see JDA100
JEWELER'S ROUGE see IHD000
JIFFY GROW see ICP000
JIMSON WEED see SLV500
JISC 3108 see AGX000
JISC 3110 see AGX000
J-LIBERTY see MDQ250
JOD (GERMAN, POLISH) see IDM000
JODDEOXIURIDIN see DAS000
JODID SODNY see SHW000
JOD-METHAN (GERMAN) see MKW200
JOLT see EIN000
JONIT see PFA500
JON-TROL see DXE600
JOOD (DUTCH) see IDM000
JOODMETHAAN (DUTCH) see MKW200
JORCHEM 400 ML see PJT000
JOY POWDER see HBT500
J SOFT C 4 see DTC600
JUDEAN PITCH see ARO500
JULIN'S CARBON CHLORIDE see HCC500
JUMBLE BEAD see AAD000
JUNIPER BERRY OIL see JEA000
JUVASON see PLZ000
JUVOCAINE see AIT250

K-9 see MEC250
K 17 see TEH500
K 52 see OHU000
K 6451 see CJT750
K 22023 see DGD800
K62-105 see LEN000
K25 (polymer) see PKQ250
KABAT see KAJ000
KADMIUM (GERMAN) see CAD000
KADMIUMCHLORID (GERMAN) see CAE250
KADMIUMSTEARAT (GERMAN) see OAT000
KADMU TLENEK (POLISH) see CAH500
KADOL see BRF500
KADOX-25 see ZKA000
KAERGONA see MMD500
KAFAR COPPER see CNI000
KAISER CHEMICALS 11 see FOO000
KAISER CHEMICALS 12 see DFA600
KAKO BLUE B SALT see DCJ200
KAKO RED TR BASE see CLK220
KALEX see EIV000
KALGAN see PFC750
K'-ALGILINE see SEH000
KALITABS see PLA500
KALIUMARSENIT (GERMAN) see PKV500
KALIUM-BETA see PLG800
KALIUMCARBONAT (GERMAN) see PLA000
KALIUMCHLORAAT (DUTCH) see PLA250
KALIUMCHLORAT (GERMAN) see PLA250
KALIUMCYANAT (GERMAN) see PLC250
KALIUM-CYANID (GERMAN) see PLC500
KALIUMDICHROMAT (GERMAN) see PKX250
KALIUMHYDROXID (GERMAN) see PLJ500
KALIUMHYDROXYDE (DUTCH) see PLJ500
KALIUMNITRAT (GERMAN) see PLL500
KALIUMPERMANGANAAT (DUTCH) see PLP000

KALIUMPERMANGANAT (GERMAN) see CAV250, PLP000
KALLIDIN see BML500
KALLOCRYL K see PKB500
KALLODENT CLEAR see PKB500
KALMETTUMSOMNIFERUM see GFA000
KALMOCAPS see LFK000, MDQ250
KALMUS OEL (GERMAN) see OGK000
KALOMEL (GERMAN) see MCW000
KALZIUMARSENIAT (GERMAN) see ARB750
KALZIUMZYKLAMATE (GERMAN) see CAR000
KAM 1000 see SLK000
KAM 2000 see SLK000
KAM 3000 see SLK000
KAMAVER see CDP250
KAMBAMINE RED TR see CLK220
KAMFOCHLOR see CDV100
KAMPFER (GERMAN) see CBA750
KAMPOSAN see CDS125
KAMPSTOFF "LOST" see BIH250
KANAMICINA (ITALIAN) see KAL000
KANAMYCIN see KAL000
KANAMYCIN A see KAL000
KANAMYCIN SULFATE see KAM000
KANAMYTREX see KAL000
KANDISET see BCE500
KANECHLOR see PJL750
KANECHLOR 300 see PJO500
KANECHLOR 400 see PJO750
KANECHLOR 500 see PJP000
KANNASYN see KAM000
KANONE see MMD500
KANTREX see KAL000
KANTREX SULFATE see KAM000
KANZO (JAPANESE) see LFN300
KAOCHLOR see PLA500
KAOLIN see KBB600
KAON see PLG800
KAON-Cl see PLA500
KAON ELIXIR see PLG800
KAPPAXAN see MMD500
KAPROLIT see PJY500
KAPROLON see PJY500
KAPROMIN see PJY500
KAPRON see NOH000, PJY500
KAPRYLAN DI-N-BUTYLCINICITY (CZECH) see BLB250
KAPTAN see CBG000
KARAYA GUM see KBK000
KARBAM BLACK see FAS000
KARBAM WHITE see BJK500
KARBARYL (POLISH) see CBM750
KARBOFOS see MAK700
KARBORAFIN see CDI000
KARBROMAL see BNK000
KARCON see MMD500
KARDIAMID see DJS200
KARDONYL see DJS200
KAREON see MMD500
KARIDIUM see SHF500
KARIGEL see SHF500
KARION see SKV200
KARI-RINSE see SHF500
KARLAN see RMA500
KARMESIN see HJF500
KARMEX see DXQ500
KARMEX DIURON HERBICIDE see DXQ500
KARMEX DW see DXQ500
KARMEX MONURON HERBICIDE see CJX750
KARMEX W. MONURON HERBICIDE see CJX750
KARMINOMYCIN see KBU000
KARO TARTRAZINE see FAG140
KARSAN see FMV000
KARTRYL see BNK000
KASSIA OEL (GERMAN) see CCO750

KATAMINE AB see DTC600
KATCHUNG OIL see PAO000
KATHRO see CMD750
KATIV-G see MMD500
KATLEX see CHJ750
KATORIN see PLG800
KATRON see BCA000, PDN000
KATRONIAZID see PDN000
KAUTSCHIN see MCC250
KAYAFUME see MHR200
KAYAKU BLUE B BASE see DCJ200
KAYAKU DIRECT see CMO000
KAYAKU DIRECT DEEP BLACK EX see AQP000
KAYAKU SCARLET G BASE see NMP500
KAYAZINON see DCM750
KAYAZOL see DCM750
KAY CIEL see PLA500
KAYDOL see MQV750
KAYKLOT see MMD500
KAYLITE see PKQ059
KAYQUINONE see MMD500
KAYTRATE see PBC250
KAZOE see SFA000
KB (POLYMER) see SMQ500
K-BRITE see SHR500
KC-400 see PJO750
KC-500 see PJP000
K 55E see SMR000
KEDAVON see TEH500
KELACID see AFL000
KELCO GEL LV see SEH000
KELCOLOID see PNJ750
KELCOSOL see SEH000
KELENE see EHH000
KELGIN see SEH000
KELGUM see SEH000
KELOFORM see EFX000
KELSET see SEH000
KELSIZE see SEH000
KELTANE see BIO750
KELTEX see SEH000
KELTHANE (DOT) see BIO750
p,p'-KELTHANE see BIO750
KELTHANE DUST BASE see BIO750
KELTHANETHANOL see BIO750
KELTONE see SEH000
KEMESTER 105 see OHW000
KEMESTER 115 see OHW000
KEMESTER 205 see OHW000
KEMESTER 213 see OHW000
KEMICETINE see CDP250
KEMIKAL see CAT250
KEMODRIN see DBA800
KEMOLATE see PHX250
KEMPORE see ASM270
KEMPORE 125 see ASM270
KEMPORE R 125 see ASM270
KENACHROME BLUE 2R see HJF500
KENAPON see DGI400
KEPMPLEX 100 see EIV000
KEPONE see KEA000
KER 710 see EAL100
KERALYT see SAI000
KERASALICYL see SJO000
KERB see DTT600
KERECID see DAS000
KEROCAINE see AIT250
KEROSAL see SJO000
KEROSENE see KEK000
KESSCO 40 see OAV000
KESSCOFLEX MCP see DOF400
KESSCOFLEX TRA see THM500
KESSCOMIR see IQN000
KESSOBAMATE see MQU750

KESSODANTEN see DKQ000
KESSODRATE see CDO000
KESTREL (Pesticide) see AHJ750
KESTRIN see ECU750
KESTRONE see EDV000
KETAJECT see CKD750
KETALAR see CKD750
KETALGIN see MDO750
KETAMINE see CKD750
KETAMINE HYDROCHLORIDE see CKD750
KETANEST see CKD750
KETASET see CKD750
KETAVET see CKD750
KETENE see KEU000
KETENE DIMER see KFA000
KETOBEMIDONE HYDROCHLORIDE see KFK000
β-KETOBUTYRANILIDE see AAY000
KETOCYCLOPENTANE see CPW500
KETODESTRIN see EDV000
KETO-ETHYLENE see KEU000
3-KETO-l-GULOFURANOLACTONE see ARN000
KETOHEXAMETHYLENE see CPC000
2-KETOHEXAMETHYLENIMINE see CBF700
KETOHYDROXY-ESTRATRIENE see EDV000
KETOHYDROXYESTRIN see EDV000
KETOHYDROXYESTRIN BENZOATE see EDV500
KETOHYDROXYOESTRIN see EDV000
KETOLAR see CKD750
KETOLE see ICM000
γ-KETO-β-METHOXY-Δ-METHYLENE-Δα-HEXENOIC ACID
  see PAP750
15-KETO-20-METHYLCHOLANTHRENE see MIM250
KETONE METHYL PHENYL see ABH000
KETONE PROPANE see ABC750
KETONES see KGA000
KETOPENTAMETHYLENE see CPW500
β-KETOPROPANE see ABC750
l-3-KETOTHREOHEXURONIC ACID LACTONE see
  ARN000
2-KETO-1,7,7-TRIMETHYLNORCAMPHANE see CBA750
KEVADON see TEH500
6FK see HCZ000
KF-1820 see DOQ400
K-GRAN see PLA000
KH 360 see TGG760
KHAINI (INDIA) see SED400
KHAROPHEN see ABX500
KHE 0145 see MIA250
KHIMCOCCID see RLK890
KHIMCOECID see RLK890
KHIMKOKTSID see RLK890
KHIMKOKTSIDE see RLK890
KHLADON 113 see FOO000
KHLORIDIN see TGD000
KHLORTRIANIZEN see CLO750
KHP 2 see AHE250
K-IAO see PLG800
KIDOLINE see VGP000
KIEFERNADEL OEL (GERMAN) see PIH500
KIESELGUHR see DCJ800
KIESELSAURE (GERMAN) see SCL000
KIEZELFLUORWATERSTOFZUUR (DUTCH) see SCO500
K III see DUS700
KILDIP see DGB000
KILEX 3 see BSQ750
KILL-ALL see SEY500
KILLAX see TCF250
KILLEEN see CCL250
KILL KANTZ see AQN635
KILMITE 40 see TCF250
KILOSEB see BRE500
KILPROP see CIR500
KILRAT see ZLS000
KILSEM see CIR250

KILVAL see MJG500
KINAVOSYL see GGS000
KING'S GREEN see COF500
KING'S YELLOW see ARI000, LCR000
KIPCA see MMD500
KIRESUTO B see EIX500
KIRKSTIGMINE BROMIDE see DQY800
KIRKSTIGMINE METHYL SULFATE see DQY909
KIRTICOPPER see CNK500
KITON CRIMSON 2R see HJF500
KITON FAST ORANGE G see HGC000
KITON PURE BLUE L see FMU059
KITON YELLOW T see FAG140
KIWAM (INDIA) see SED400
KLAVI KORDAL see NGY000
KLEER-LOT see AMY050
KLEGECELL see PKQ059
KLIMORAL see EDU500
KLINE see BBK250
KLINIT see XPJ000
KLION see MMN250
K-LOR see PLA500
KLORAMIN see BIE500, CDP000
KLORAMINE-T see CDP000
KLOREX see SFS000
KLORT see MQU750
KLOTRIX see PLA500
KLOTTONE see MMD500
KLUCEL see HNV000
KM see KAL000
4K-2M see CIR250
KM (THE ANTIBIOTIC) see KAL000
KMTS 212 see SFO500
KNEE PINE OIL see PIH400
KNOCKMATE see FAS000
KO 7 see EAL100
KOAXIN see MMD500
KOBALT (GERMAN, POLISH) see CNA250
KOBALT CHLORID (GERMAN) see CNB599
KOBALT HISTIDIN (GERMAN) see BJY000
KOBU see PAX000
KOBUTOL see PAX000
KOCIDE see SOD500
KODAFLEX see TIG750
KODAFLEX DBS see DEH600
KODAFLEX DOP see DVL700
KODAFLEX TRIACETIN see THM500
KOFFEIN (GERMAN) see CAK500
KOHLENDIOXYD (GERMAN) see CBU250
KOHLENDISULFID (SCHWEFELKOHLENSTOFF) (GERMAN)
  see CBV500
KOHLENMONOXID (GERMAN) see CBW750
KOHLENOXYD (GERMAN) see CBW750
KOHLENSAURE (GERMAN) see CBU250
KOKOTINE see BBQ500
KOLCHAMIN see MIW500
KOLI (HAWAII) see CCP000
KOLKLOT see MMD500
KOLLIDON see PKQ250
KOLOFOG see SOD500
KOLOSPRAY see SOD500
KOLPHOS see PAK000
KOLPON see EDV000
KOMPLXON see EIV000
KONDREMUL see MQV750
KONESSIN DIHYDROBROMIDE see DOX000
KONESTA see TII250
KONLAX see DJL000
KOOLMONOXYDE (DUTCH) see CBW750
KOOLSTOFDISULFIDE (ZWAVELKOOLSTOF) (DUTCH) see
  CBV500
KOOLSTOFOXYCHLORIDE (DUTCH) see PGX000
KOPFUME see EIY500
KOP MITE see DER000

KOPOLYMER BUTADIEN STYRENOVY (CZECH) see SMR000
KOPROSTERIN (GERMAN) see DKW000
KOPSOL see DAD200
KOP-THIODAN see EAQ750
KOP-THION see MAK700
KORAD see PKB500
KORAX see CJE000
KORDIAMIN see DJS200
KOREON see NBW000
KORIUM see MJM500
KORLAN see RMA500
KORLANE see RMA500
KOROSEAL see PKQ059
KORUM see HIM000
KORUND see EAL100
KOSATE see DJL000
KOSTIL see ADY500
KOTION see DSQ000
KOTOL see BBQ750
KP 2 see PAX000
KP 140 see BPK250
K-PIN see PIB900
K-PRENDE-DOME see PLA500
K PREPARATION see BJU400
KRASTEN 1.4 see SMQ500
KRATEDYN see EAW000
KREBON see BQL000
KRECALVIN see DGP900
KREGASAN see TFS350
KRENITE (OBS.) see DUS700, DUU600
KRESAMONE see DUS700
KRESIDIN see MGO750
m-KRESOL see CNW750
p-KRESOL see CNX250
o-KRESOL (GERMAN) see CNX000
KRESOLE (GERMAN) see CNW500
KRESOLEN (DUTCH) see CNW500
o-KRESOL-GLYCERINAETHER (GERMAN) see GGS000
KRESOXYPROPANDIOL see GGS000
KREZIDINE see MGO750
KREZOL (POLISH) see CNW500
KREZONE see CIR250
KREZONITE see DUU600
KREZOTOL 50 see DUS700
KRIPTIN see WAK000
KRISTALLOSE see SJN700
KRMD 58 see MAE000
KRO 1 see SMR000
KROKYDOLITH (GERMAN) see ARM275
KROMAD see KHU000
KROMON GREEN B see CLK235
KRONITEX see TNP500
KRONITEX KP-140 see BPK250
KRONOS TITANIUM DIOXIDE see TGG760
KROTENAL see DXH250
KROTILINE see DAA800
KROTONALDEHYD (CZECH) see COB250
KROVAR II see BMM650
KRUMKIL see ABF500
KRYSID see AQN635
KRZEWOTOKS see BSQ750
KS 1675 see DBA600
KSYLEN (POLISH) see XGS000
K-THROMBYL see MMD500
KU 5-3 see EAL100
KUBARSOL see ABX500
KUEMMEL OIL (GERMAN) see CBG500
KUMADER see WAT200
KUMIAI see MIB750
KUMORAN see BJZ000
KUMULUS see SOD500
KUPFERRON (CZECH) see ANO500
KUPPERSULFAT (GERMAN) see CNP250

KUPRATSIN see EIR000
KURAN see TIX500
KURARE OM 100 see AAX250
KURON see TIX500
KUROSAL see TIX500
KUSA-TOHRU see SFS000
KUSATOL see SFS000
KUSNARIN see EID000
K-VITAN see MMD500
KW-125 see AES750
KW-4354 see DWF790, OMG000
KWAS BENZYDYNODWUKAROKSYLOWY (POLISH) see BFX250
KWAS METANIOWY (POLISH) see FNA000
KWELL see BBQ500
KWELLS see HOT500
KWIK (DUTCH) see MCW250
KWIK-KIL see SMN500
KWIKSAN see ABU500
KWIT see EEH600
KW-2-LE-T see LFA020
KYANACETHYDRAZID see COH250
KYANID SODNY (CZECH) see SGA500
KYANID STRIBRNY (CZECH) see SDP000
KYLAR see DQD400
KYOCRISTINE see LEZ000
KYPCHLOR see CDR750
KYPFOS see MAK700
KYPMAN 80 see MAS500
KYPTHION see PAK000
KYPZIN see EIR000
KYSELINA ADIPOVA (CZECH) see AEN250
KYSELINA AKRYLOVA see ADS750
KYSELINA AMIDOSULFONOVA (CZECH) see SNK500
KYSELINA BENZOOVA (CZECH) see BCL750
KYSELINA CITRONOVA (CZECH) see CMS750
KYSELINA DICHLORISOKYANUROVA (CZECH) see DGN200
KYSELINA 3,6-ENDOMETHYLEN-3,4,5,6,7,7-HEXACHLOR-Δ^4-TETRAHYDROFTALOVA (CZECH) see CDS000
KYSELINA FUMAROVA (CZECH) see FOU000
KYSELINA HET (CZECH) see CDS000
KYSELINA JABLECNA see MAN000
KYSELINA MLECNA (CZECH) see LAG000
KYSELINA NITROBENZEN-m-SULFONOVA (CZECH) see NFB500
KYSELINA STAVELOVA (CZECH) see OLA000
KYSELINA SULFAMINOVA (CZECH) see SNK500
KYSELINA-β,β'-THIODIPROPIONOVA (CZECH) see BHM000
KYSELINA TRICHLOISOKYANUROVA (CZECH) see TIQ750
KYSLICNIK DI-n-BUTYLCINICITY (CZECH) see DEF400
KYSLICNIK DIISOBUTYLCINICITY (CZECH) see DNJ000
KYSLICNIK DIISOPROPYLCINICITY (CZECH) see DNR200
KYSLICNIK DI-N-PROPYLCINICITY (CZECH) see DWV000
KYSLICNIK TRI-N-BUTYLCINICITY (CZECH) see BLL750
K-ZINC see ZKA000
KZ 3M see SCQ000

L-310 see LGK000
L343 see IOT000
L-395 see DSP400
L-561 see DRR400
L 1811 see DJT400
L-5103 see RKP000
L-01748 see DJT800
L-36352 see DUV600
LA 1 see DLY000
LA 6023 see DQR600
LA'AU-'AILA (HAWAII) see CCP000
LABDANOL see IIQ000
LABDANUM OIL see LAC000
LABICAN see MDQ250

LABOPAL see DKQ000
LACOLIN see LAM000
LACQREN 550 see SMQ500
LACQUER DILUENT see ROU000
LACQUER ORANGE V see TGW000
LACQUER ORANGE VG see PEJ500
LACQUER ORANGE VR see XRA000
LACQUER RED V3B see EOJ500
LACTATE d'ETHYLE (FRENCH) see LAJ000
LACTIC ACID see LAG000
dl-LACTIC ACID see LAG000
racemic LACTIC ACID see LAG000
LACTIC ACID, ANTIMONY SALT see AQE250
LACTIC ACID, BERYLLIUM SALT see LAH000
LACTIC ACID, BUTYL ESTER see BRR600
LACTIC ACID, BUTYL ESTER, BUTYRATE see BQP000
LACTIC ACID, CADMIUM SALT see CAG750
LACTIC ACID, ETHYL ESTER see LAJ000
LACTIC ACID, IRON(2+) SALT (2:1) see LAL000
LACTIC ACID, MONOSODIUM SALT see LAM000
LACTIC ACID SODIUM SALT see LAM000
LACTIN see LAR000
LACTOBACILLUS LACTIS DORNER FACTOR see VSZ000
LACTOBARYT see BAP000
LACTOBIOSE see LAR000
LACTOCAINE see AIT250
LACTOFLAVIN see RIK000
LACTOFLAVINE see RIK000
LACTONITRILE see LAQ000
LACTOSE see LAR000
d-LACTOSE see LAR000
LAE-32 see LJI000
LAETRILE see LAS000
LAEVORAL see LFI000
LAEVOSAN see LFI000
LAKE BLUE B BASE see DCJ200
LAKE RED C see CHP500
LAKE RED KB BASE see CLK225
LAKE SCARLET G BASE see NMP500
LAKE YELLOW see FAG140
LAMBETH see PMP500
LAMBRATEN see AJT250
LAMBROL see FIP999
LAMDIOL see EDO000
LAMITEX see SEH000
LAMORYL see GKE000
LAMP BLACK see CBT750
LANADIN see TIO750
LANATOSID C (GERMAN) see LAU000
LANATOSIDE C see LAU000
LANATOXIN see DKL800
LANAZINE see DBA800
LANDALGINE see AFL000
LANDISAN see MEO750
LANDOCAINE see BQA010
LAND PLASTER see CAX750
LANDRIN see TMD000
LANETTE WAX-S see SIB600
LANEX see DUK800
LANIAZID see ILD000
LANICOR see DKN400
LANIOZID see ILD000
LANNATE see MDU600
LANOL see CMD750
LANOPHYLLIN see TEP000
LANOXIN see DKN400
LANSTAN see CJE000
LANTHANACETAT (GERMAN) see LAW000
LANTHANUM see LAV000
LANTHANUM ACETATE see LAW000
LANTHANUM CHLORIDE see LAX000
LANTHANUM EDETATE see LAZ000
LANTHANUM NITRATE see LBA000
LANTHANUM TRIACETATE see LAW000

LAPPACONITINE see LBD000
LARD FACTOR see VSK600
LARIXIC ACID see MAO350
LARIXINIC ACID see MAO350
LARODON see INY000
LARODOPA see DNA200
LAROXIL see EAH500
LAROXYL see EAH500
LARTEN see MQU750
LARVACIDE see CKN500
LASALOCID see LBF500
LASEX see CHJ750
LASIOCARPINE see LBG000
LASIX see CHJ750
LASSO see CFX000
LATEX see PJR000
LATEXOL SCARLET R see CHP500
LATSCHENKIEFEROL see PIH400
LAUDICON see DLW600
LAUDOCAINE see BQA010
LAUDRAN DI-n-BUTYLCINICITY (CZECH) see DDV600
LAUGHING GAS see NGU000
LAURAMIDE DEA see BKE500
LAUREL CAMPHOR see CBA750
LAUREL LEAF OIL see BAT500, LBK000
LAURIC ACID see LBL000
LAURIC ACID, DIBUTYLSTANNYLENE deriv. see DDV600
LAURIC ACID, DIBUTYLSTANNYLENE SALT see DDV600
LAURIC ACID DIETHANOLAMIDE see BKE500
LAURIC ALCOHOL see DXV600
LAURIC DIETHANOLAMIDE see BKE500
LAURINE see CMS850
LAURINIC ALCOHOL see DXV600
LAUROSTEARIC ACID see LBL000
LAUROX see LBR000
LAUROYL DIETHANOLAMIDE see BKE500
LAUROYL PEROXIDE see LBR000
LAUROYL PEROXIDE, TECHNICALLY PURE (DOT) see
    LBR000
LAURYDOL see LBR000
LAURYL 24 see DXV600
LAURYL ALCOHOL (FCC) see DXV600
n-LAURYL ALCOHOL, PRIMARY see DXV600
LAURYL ALDEHYDE (FCC) see DXT000
LAURYLAMINE see DXW000
LAURYL DIETHANOLAMIDE see BKE500
LAURYL GALLATE see DXX200
LAURYLGUANIDINE ACETATE see DXX400
LAURYLISOQUINOLINIUM BROMIDE see LBW000
LAURYL MERCAPTAN see LBX000
m-LAURYL MERCAPTAN see LBX000
LAURYL SODIUM SULFATE see SIB600
LAURYL SULFATE, SODIUM SALT see SIB600
LAUSIT see IDA000
LAUXTOL see PAX250
LAUXTOL A see PAX250
LAV see CMY800
LAVANDIN OIL see LCA000
LAVATAR see CMY800
LAVENDEL OEL (GERMAN) see LCD000
LAVENDER OIL see LCD000
LAVENDER OIL, SPIKE see SLB500
LAWN-KEEP see DAA800
LAXINATE see DJL000
LAYOR CARANG see AEX250
LAZO see CFX000
LB 502 see CHJ750
LB-ROT 1 see FAG040
LC 44 see FMO129
LCR see LEY000
LDA see BKE500
LDE see BKE500
LD NORGESTREL (FRENCH) see NNQ500
LD RUBBER RED 16913 see CHP500

Le-100 see EIF000
29060 LE see VLA000
LEA-COV see SHF500
LEAD see LCF000
LEAD ACETATE see LCG000
LEAD (2+) ACETATE see LCG000
LEAD(II) ACETATE see LCG000
LEAD ACETATE, BASIC see LCH000
LEAD ACETATE TRIHYDRATE see LCJ000
LEAD ACETATE(II), TRIHYDRATE see LCJ000
LEAD ACID ARSENATE see LCK000
LEAD ARSENATE see ARC750, LCK000
LEAD ARSENATE, solid (DOT) see LCK000
LEAD ARSENATE (standard) see LCK000
LEAD(II) ARSENITE see LCL000
LEAD ARSENITE, solid (DOT) see LCL000
LEAD(II) AZIDE see LCM000
LEAD AZIDE, DRY (DOT) see LCM000
LEAD BOTTOMS see LDY000
LEAD BROWN see LCX000
LEAD CARBONATE see LCP000
LEAD(2+) CARBONATE see LCP000
LEAD CHLORIDE see LCQ000
LEAD (2+) CHLORIDE see LCQ000
LEAD (II) CHLORIDE see LCQ000
LEAD CHROMATE see LCR000
LEAD CHROMATE(VI) see LCR000
LEAD CHROMATE, BASIC see LCS000
LEAD CHROMATE OXIDE (MAK) see LCS000
LEAD CHROMATE, RED see LCS000
LEAD CHROMATE, SULPHATE and MOLYBDATE see
   LDM000
LEAD COMPOUNDS see LCT000
LEAD(II) CYANIDE see LCU000
LEAD CYANIDE (DOT) see LCU000
LEAD DIACETATE see LCG000
LEAD DIACETATE TRIHYDRATE see LCJ000
LEAD DIBASIC ACETATE see LCG000
LEAD DICHLORIDE see LCQ000
LEAD DIFLUORIDE see LDF000
LEAD DIMETHYLDITHIOCARBAMATE see LCW000
LEAD DINITRATE see LDO000
LEAD DIOXIDE see LCX000
LEAD DROSS (DOT) see LDC000, LDY000
LEAD DROSS (containing 3% or more free acid) (DOT) see
   LDC000
LEAD FLAKE see LCF000
LEAD FLUOBORATE see LDE000
LEAD(II) FLUORIDE see LDF000
LEAD FLUORIDE (DOT) see LDF000
LEAD-MOLYBDENUM CHROMATE see LDM000
LEAD MONONITRORESORCINATE (DRY) (DOT) see
   LDP000
LEAD MONOSUBACETATE see LCH000
LEAD MONOXIDE see LDN000
LEAD NAPHTHENATE see NAS500
LEAD NITRATE see LDO000
LEAD (2+) NITRATE see LDO000
LEAD(II) NITRATE see LDO000
LEAD(II) NITRATE (1:2) see LDO000
LEAD NITRORESORCINATE see LDP000
LEAD ORTHOPHOSPHATE see LDU000
LEAD ORTHOPLUMBATE see LDS000
LEAD OXIDE see LDN000
LEAD(II) OXIDE see LDN000
LEAD(IV) OXIDE see LCX000
LEAD OXIDE BROWN see LCX000
LEAD OXIDE RED see LDS000
LEAD OXIDE YELLOW see LDN000
LEAD PEROXIDE (DOT) see LCX000
LEAD PHOSPHATE see LDU000
LEAD (2+) PHOSPHATE see LDU000
LEAD PHOSPHATE (3:2) see LDU000
LEAD(II) PHOSPHATE (3:2) see LDU000

LEAD PROTOXIDE see LDN000
LEAD S2 see LCF000
LEAD SCRAP (DOT) see LDC000
LEAD STYPHNATE (DRY) (DOT) see LEE000
LEAD SUBACETATE see LCH000
LEAD(II) SULFATE (1:1) see LDY000
LEAD SULFATE, solid, containing more than 3% free acid
   (DOT) see LDY000
LEAD SULFIDE see LDZ000
LEAD SUPEROXIDE see LCX000
LEAD TETRAOXIDE see LDS000
LEAD TRINITRORESORCINATE see LEE000
LEAD TRINITRORESORCINATE (DOT) see LEE000
LEAD 2,4,6-TRINITRORESORCINOXIDE see LEE000
LEAF ALCOHOL see HFE000
LEAF ALDEHYDE see HFA500
LEAF GREEN see CMJ900
LEANDIN see COH250
LEATHER GREEN SF see FAF000
LEATHER ORANGE HR see PEK000
LEATHER PURE BLUE HB see BJI250
LEBAYCID see FAQ999
LE CAPTANE (FRENCH) see CBG000
LEDERMYCIN see MIJ500
LE DINITROCRESOL-4,6 (FRENCH) see DUS700
LEDON 11 see TIP500
LEDON 12 see DFA600
LEDON 114 see FOO509
LEDOSTEN see DJT400
LEFEBAR see EOK000
LEGUMEX DB see CIR250
LEHYDAN see DKQ000
LEINOLEIC ACID see LGG000
LEIPZIG YELLOW see LCR000
LEIVASOM see TIQ250
LEKAMIN see TNF500
LEMAC 1000 see AAX250
LEMOFLUR see SHF500
LEMON CHROME see BAK250
LEMONENE see BGE000
LEMONGRAS OEL (GERMAN) see LEG000
LEMONGRASS OIL EAST INDIAN see LEG000
LEMONGRASS OIL WEST INDIAN see LEH000
LEMON OIL see LEI000
LEMON OIL, COLDPRESSED (FCC) see LEI000
LEMON OIL, DESERT TYPE, COLDPRESSED see LEI025
LEMON OIL, DISTILLED see LEI030
LEMON OIL, EXPRESSED see LEI000
LEMONOL see DTD000
LEMON YELLOW see BAK250, LCR000
LENAMYCIN see ACJ250
LENDINE see BBQ500
LENTIZOL see EAI000
LENTOTRAN see MDQ250
LENTOX see BBQ500
LEO 72a see BHO250
LEONAL see EOK000
LEOPENTAL see PBT500
LEOSTESIN see DHK400
LEOSTESIN HYDROCHLORIDE see DHK600
LEOSTIGMINE BROMIDE see DQY800
LEOSTIGMINE METHYL SULFATE see DQY909
LEPENIL see MQU750
LEPETOWN see MQU750
LEPHEBAR see EOK000
LEPIMIDIN see DBB200
LEPINAL see EOK000
LEPINALETTEN see EOK000
LEPITOIN see DKQ000, DNU000
LEPITOIN SODIUM see DNU000
LEPSIN see DKQ000
LEPSIRAL see DBB200
LEPTAMIN see DJS200
LEPTANAL see DYF200, PDW750

LEPTOFEN see DYF200
LEPTON see DJT400
LEPTOPHOS see LEN000
LEPTRYL see LEO000
LERBEK see CMX850
LERCIGAN see DQA400
LERENOX see BDD000
LERGIGAN see DQA400
LERGOTRILE MESYLATE see LEP000
LESAN see DOU600
LESTEMP see HIM000
LETHALAIRE G-52 see TCF250
LETHALAIRE G-54 see PAK000
LETHALAIRE G-57 see SOD100
LETHALAIRE G-58 see CJT750
LETHALAIRE G-59 see OCM000
LETHANE see BPL250
LETHANE 384 see BPL250
LETHANE 384 REGULAR see BPL250
LETHANE (SPECIAL) see LEQ000
LETHELMIN see PDP250
LETHOX see TNP250
LETHURIN see TIO750
LETYL see MQU750
LEUCARSONE see CBJ000
LEUCETHANE see UVA000
LEUCIDIL see BCA000
LEUCIN (GERMAN) see LES000
LEUCINE see LES000
l-LEUCINE see LES000
dl-LEUCINE see LER000
LEUCINOCAINE see LET000
LEUCOGEN see ARN800
LEUCOHARMINE see HAI500
LEUCOL see QMJ000
LEUCOLINE see IRX000, QMJ000
LEUCOPARAFUCHSIN see THP000
LEUCOPARAFUCHSINE see THP000
LEUCOSULFAN see BOT250
LEUCOTHANE see UVA000
LEUCOVYL PA 1302 see AAX175
LEUKAEMOMYCIN C see DAC000
LEUKERAN see CDO500, POK000
LEUKERSAN see CDO500
LEUKOL see QMJ000
LEUKOMYAN see CDP250
LEUKORAN see CDO500
LEUNA M see CIR250
LEUPURIN see POK000
LEUROCRISTINE see LEY000
LEUROCRISTINE SULFATE (1:1) see LEZ000
LEVADONE see MDP250
LEVAMISOLE see LFA000, LFA020
LEVAMISOLE HYDROCHLORIDE see LFA020
LEVANIL see EHP000
LEVANOX GREEN GA see CMJ900
LEVANOX RED 130A see IHD000
LEVANOX WHITE RKB see TGG760
LEVARGIN see AQW000
LEVARTERENOL see NNO500
LEVEDRINE see BBK750
LEV HYDROCHLORIDE see LFA020
LEVIL see EHP000
LEVOARTERENOL see NNO500
LEVOGLUTAMID see GFO050
LEVOGLUTAMIDE see GFO050
LEVOMYCETIN see CDP250
LEVOMYSOL HYDROCHLORIDE see LFA020
LEVONORADRENALINE see NNO500
LEVONOREPINEPHRINE see NNO500
LEVOPHACETOPERANE HYDROCHLORIDE see LFO000
LEVOPHACETOPERAN HYDROCHLORIDE see LFO000
LEVOPHED see NNO500
LEVORENIN see VGP000

LEVORIN see CBC250, LFF000
LEVORPHAN see LFG000
LEVORPHANOL see LFG000
LEVOTHYL see MDP250
LEVUGEN see LFI000
LEVULOSE see LFI000
LEWISITE see CLV000
LEWISITE (ARSENIC COMPOUND) see CLV000
LEWISITE II see BIQ250
LEWISITE I OXIDE see DEW000
LEWIS-RED DEVIL LYE see SHS000
LEXONE see MQR275
LEYSPRAY see CIR250
LEYTOSAN see ABU500, PFP500
LFA 2043 see GIA000
L.G. 11,457 HYDROCHLORIDE see DHS200
L-GRUEN No. 1 (GERMAN) see CKN000
LGYCOSPERSE S-20 see PKL030
LH see ILE000
LIATRIS see DAJ800
LIATRIX OLEORESIN see DAJ800
LIBAVIUS FUMING SPIRIT see TGC250
LIBIOLAN see MQU750
LIBRAX see LFK000
LIBRININ see LFK000
LIBRITABS see LFK000
LIBRIUM see LFK000, MDQ250
LIBRIUM HYDROCHLORIDE see MDQ250
LICAREOL ACETATE see LFY100
LICHENIC ACID see FOU000
LICHTGRUEN (GERMAN) see FAF000
LICORICE see LFN300
LICORICE EXTRACT see LFN300
LICORICE ROOT see LFN300
LICORICE ROOT EXTRACT see LFN300
LIDA-MANTLE see DHK400
LIDAMYCIN CREME see NCG000
LIDENAL see BBQ500
LIDEPRAN HYDROCHLORIDE see LFO000
LIDOCAINE see DHK400
LIDOCAINE HYDROCHLORIDE see DHK600
LIDOL see DAM600, DAM700
LIDOTHESIN HYDROCHLORIDE see DHK600
LIFEAMPIL see AIV500, AOD125
LIGHT CAMPHOR OIL see CBB500
LIGHT GREEN FCF YELLOWISH see FAF000
LIGHT GREEN LAKE see FAF000
LIGHTHOUSE CHROME BLUE 2R see HJF500
LIGHT NAPHTHENIC DISTILLATE see MQV810
LIGHT NAPHTHENIC DISTILLATE, SOLVENT EXTRACT
  see MQV860
LIGHT NAPHTHENIC DISTILLATES (PETROLEUM) see
  MQV810
LIGHT OIL of CAMPHOR see CBB500
LIGHT ORANGE CHROME see LCS000
LIGHT PARAFFINIC DISTILLATE see MQV815
LIGHT PARAFFINIC DISTILLATE, SOLVENT EXTRACT see
  MQV862
LIGHT RED see IHD000
LIGHT SF YELLOWISH (BIOLOGICAL STAIN) see FAF000
LIGHT SPAR see CAX750
LIGNASAN FUNGICIDE see BJT250
LIGNASAN-X see BJT250
LIGNOCAINE see DHK400
LIGNOCAINE HYDROCHLORIDE see DHK600
LIGROIN see PCT250
LIKUDEN see GKE000
LILLY 1516 see DQA400
LILLY 01516 see DQA400
LILLY 22451 see MDU500
LILLY 34,314 see DRP800
LILLY 36,352 see DUV600
LILLY 37231 see LEZ000
LILYL ALDEHYDE see CMS850

LIMARSOL MALAGRIDE see ABX500
LIMAS see LGZ000
LIME see CAU500
LIME ACETATE see CAL750
LIME, BURNED see CAU500
LIME CHLORIDE see HOV500
LIMED ROSIN see CAW500
LIME-NITROGEN (DOT) see CAQ250
LIME OIL see OGO000
LIME OIL, DISTILLED (FCC) see OGO000
LIME PYROLIGNITE see CAL750
LIMESTONE (FCC) see CAO000
LIME, UNSLAKED (DOT) see CAU500
LIME WATER see CAT250
LIMIT see DKE800
LIMONENE see MCC250
1-LIMONENE see MCC500
d-LIMONENE see LFU000
(+)-R-LIMONENE see LFU000
d-(+)-LIMONENE see LFU000
dl-LIMONENE see MCC250
LIMONENE DIOXIDE see LFV000
(−)-LIMONENE (FCC) see MCC500
LIMONENE OXIDE see CAL000
LIMONITE see LFW000
LINALOL see LFX000
LINALOL ACETATE see LFY100
LINALOOL see LFX000
LINALOOL ACETATE see LFY100
LINALOOL ISOBUTYRATE see LGB000
LINALYL ACETATE see LFY100
LINALYL ALCOHOL see LFX000
LINALYL BENZOATE see LFZ000
LINALYL FORMATE see LGA050
LINALYL ISOBUTYRATE see LGB000
LINALYL PROPIONATE see LGC100
LINARODIN see MDW750
LINASEN see EOK000
LINCOCIN see LGD000
LINCOLCINA see LGD000
LINCOLNENSIN see LGD000
LINCOMYCIN see LGD000
LINCOMYCINE (FRENCH) see LGD000
LINDAGRAIN see BBQ500
LINDAN see DGP900
α-LINDANE see BBQ000
β-LINDANE see BBR000
Δ-LINDANE see BFW500
LINDANE (ACGIH, DOT, USDA) see BBQ500
LINDEROL see NCQ820
LINDOL see TNP500
LINE RIDER see TAA100
LINEX 4L see DGD600
LINFOLIZIN see CDO500
LINFOLYSIN see CDO500
LINGEL see BRF500
LINGRAINE see EDC500
LINGRAN see EDC500
LINGUSORBS see PMH500
LINOLEIC ACID see LGG000
9,12-LINOLEIC ACID see LGG000
LINORMONE see CIR250
LINOROX see DGD600
LINSEED OIL see LGK000
LINTOX see BBQ500
LINUREX see DGD600
LINURON see DGD600
LINURON (HERBICIDE) see DGD600
LIOXIN see VFK000
LIPAN see DUS700
LIPHADIONE see CJJ000
LIPO EGMS see EJM500
LIPO GMS 410 see OAV000
LIPO GMS 450 see OAV000

LIPO GMS 600 see OAV000
LIPO-HEPIN see HAQ500
LIPO-LUTIN see PMH500
LIPOPILL see DTJ400
LIPOSORB O-20 see PKL100
LIPOSORB S see SKV150
LIPOSORB S-20 see SKV150, PKL030
LIPOTRIL see CMF750
LIQUACILLIN see BDY669
LIQUAEMIN see HAQ500
LIQUAGESIC see HIM000
LIQUAMYCIN see TBX000
LIQUAMYCIN INJECTABLE see HOI000
LIQUA-TOX see WAT200
LIQUEFIED CARBON DIOXIDE see LGL000
LIQUEFIED PETROLEUM GAS see LGM000
LIQUEFIED PETROLEUM GAS (DOT) see IIC000
LIQUEMIN see HAQ500
LIQUIBARINE see BAP000
LIQUID BRIGHT PLATINUM see PJD500
LIQUID CAMPHOR see CBB500
LIQUID CARBONIC GAS see LGL000
LIQUID DERRIS see RNZ000
LIQUID ETHYENE see EIO000
LIQUIDOW see CAO750
LIQUID PITCH OIL see CMY825
LIQUID ROSIN see TAC000
LIQUIMETH see MDT750
LIQUIPHENE see ABU500
LIQUIPRIN see SAH000
LIQUI-SAN see MLH000
LIQUITAL see EOK000
LIQUOPHYLLINE see TEP000
LIRANOL see DQA600
LIRANOX see CIR500
LIROBETAREX see CJX750
LIRO CIPC see CKC000
LIROHEX see TCF250
LIROMATIN see ABX250
LIRONOX see BQZ000
LIROPON see DGI400
LIROPREM see PAX250
LIROSTANOL see ABX250
LIROTAN see EIR000
LIROTHION see PAK000
LISACORT see PLZ000
LISERGAN see ABH500
LISKONUM see LGZ000
LISSAMINE LAKE GREEN SF see FAF000
LISSENPHAN see GGS000
LISSOLAMINE see HCQ500
LISTENON see HLC500
LITAC see ADY500
LITALER see HOO500
LITEX CA see SMR000
LITHAMIDE see LGT000
LITHANE see LGZ000
LITHARGE see LDN000
LITHARGE YELLOW L-28 see LDN000
LITHICARB see LGZ000
LITHINATE see LGZ000
LITHIUM see LGO000
LITHIUM ACETYLIDE COMPLEXED with ETHYLENE-
   DIAMINE see LGQ000
LITHIUM ACETYLIDE-ETHYLENEDIAMINE COMPLEX see
   LGQ000
LITHIUM ALANATE see LHS000
LITHIUM ALUMINOHYDRIDE see LHS000
LITHIUM ALUMINUM HYDRIDE (DOT) see LHS000
LITHIUM ALUMINUM HYDRIDE, ETHEREAL (DOT) see
   LHS000
LITHIUM ALUMINUM TETRAHYDRIDE see LHS000
LITHIUM AMIDE see LGT000
LITHIUM AMIDE, POWDERED (DOT) see LGT000

LITHIUM ANTIMONIOTHIOMALATE see LGU000
LITHIUM ANTIMONY THIOMALATE see LGU000
LITHIUM BOROHYDRIDE (DOT) see LHT000
LITHIUM CARBONATE see LGZ000
LITHIUM CARBONATE (2:1) see LGZ000
LITHIUM CHLORIDE see LHB000
LITHIUM CHROMATE see LHD000
LITHIUM COMPOUNDS see LHE000
LITHIUM FLUORIDE see LHF000
LITHIUM FLUORURE (FRENCH) see LHF000
LITHIUM HYDRIDE see LHH000
LITHIUM HYPOCHLORITE see LHJ000
LITHIUM HYPOCHLORITE COMPOUND, DRY, CONTAIN-
  ING MORE THAN 39% AVAILABLE CHLORINE (DOT)
  see LHJ000
LITHIUM METAL (DOT) see LGO000
LITHIUM METAL, IN CARTRIDGES (DOT) see LGO000
LITHIUM NITRIDE see LHM000
LITHIUM PEROXIDE see LHO000
LITHIUM SILICON see LHP000
LITHIUM TETRAHYDROALUMINATE see LHS000
LITHIUM TETRAHYDROBORATE see LHT000
LITHOBID see LGZ000
LITHOGRAPHIC STONE see CAO000
LITHONATE see LGZ000
LITHOSOL ORANGE R BASE see NMP500
LITHOTABS see LGZ000
LITICON see DOQ400
LIXOPHEN see EOK000
LLD FACTOR see VSZ000
LM 91 see CJJ000
LM-2717 see CIR750
LM SEED PROTECTANT see MLH000
LOBAMINE see MDT740
LO-BAX see HOV500
LOBELINE HYDROCHLORIDE see LHZ000
(−)-LOBELINE HYDROCHLORIDE see LHZ000
LOBELIN HYDROCHLORIDE see LHZ000
LOCUST BEAN GUM see LIA000
LOFEPRAMINE see DLH630
LOISOL see TIQ250
LOMBRISTOP see TEX000
LO MICRON TALC 1 see TAB750
LO MICRON TALC, BC 1621 see TAB750
LO MICRON TALC USP, BC 2755 see TAB750
LOMIDIN see DBL800
LOMIDINE see DBL800
LOMIDINE ISOETHIONATE see DBL800
LOMUPREN see DMV600
LOMUSTINE see CGV250
LON 41 see DTM600
LONACOL see EIR000
LONAMIN see DTJ400
LONARID see HIM000
LONDOMYCIN see MDO250
LONDON PURPLE see LIC000
LONDON PURPLE, solid (DOT) see LIC000
LONGANOCT see BPF500
LONGIFENE see HGC500
β-LONGILOBINE see RFP000
LONITEN see DCB000
LONOCOL M see MAS500
LONZA G see PKQ059
LOPRESS see HGP500
LOREX see DGD600
LORINAL see CDO000
LORMIN see CBF250
LOROL see DXV600
LOROL 20 see OEI000
LOROL 22 see DAI600
LOROL 24 see HCP000
LOROL 28 see OAX000
LOROMISIN see CDP250
LOROX see DGD600

LOROXIDE see BDS000
LOROX LINURON WEED KILLER see DGD600
LORPHEN see TAI500
LORSBAN see CMA100
LOSANTIN see HOV500
LOSUNGSMITTEL APV see CBR000
LOVAGE OIL see LII000
LO-VEL see SCH000
LOVOSA see SFO500
LOVOZAL see DGA200
LOWETRATE see PBC250
LOWPSTRON see CHJ750
LOXANOL K see HCP000
LOXAPINE see DCS200
LOXON see DFH600
LOXURAN see DIW200
LOXYNIL (GERMAN) see HKB500
LPG see LGM000
L.P.G. (OSHA, ACGIH) see LGM000
LPG ETHYL MERCAPTAN 1010 see EMB100
LPT see PKB500
L.S. 3394 see BLL750
LS 4442 see CLU000
LSD see DJO000
d-LSD see DJO000
LSD-25 see DJO000
LSD-25-PYRROLIDATE see LJM000
LSD TARTRATE see LJG000
LUBERGAL see EOK000
LUBROKAL see EOK000
LUCALOX see AHE250
LUCANTHON see DHU000
LUCANTHONE see DHU000
LUCANTHONE METABOLITE see LIM000
LUCEL (polysaccharide) see SFO500
LUCEL ADA see ASM270
LUCIDOL see BDS000
LUCITE see PKB500
LUCOFLEX see PKQ059
LUCORTEUM SOL see PMH500
LUCOVYL PE see PKQ059
LUDOX see SCH000
LUETOCRIN DEPOT see HNT500
LULAMIN see TEH500, TEO250
LULLAMIN see TEO250
LUMBRICAL see PIJ000
LUMEN see EOK000
LUMESETTES see EOK000
LUMESYN see EOK000
LUMINAL see EOK000
LUMINAL SODIUM see SID000
LUMOFRIDETTEN see EOK000
LUNAR CAUSTIC see SDS000
LUNIPAK see DAB800
LUPAREEN see PMP500
LUPERCO see BDS000, DGR600
LUPEROX see DGR600
LUPEROX 500R see DGR600
LUPEROX 500T see DGR600
LUPEROX FL see BDS000
LUPERSOL see MKA500
LUPERSOL 70 see BSC250
LUPHENIL see EOK000
LURAMIN see EOK000
LURAN see ADY500
LURAZOL BLACK BA see AQP000
LURGO see DSP400
LURIDE see SHF500
LUSIL see SNM500
LUSTRAN see ADY500
LUSTREX see SMQ500
LUTALYSE see POC750
LUTATE see HNT500
LUTEAL HORMONE see PMH500

LUTEOCRIN see HNT500
LUTEOHORMONE see PMH500
LUTEOSAN see PMH500
LUTEOSKYRIN see LIV000, LIV000
LUTETIA RED CLN see CHP500
LUTETIUM CHLORIDE see LIW000
LUTETIUM CITRATE see LIX000
LUTETIUM(III) NITRATE (1:3) see LIY000
LUTEX see PMH500
3,4-LUTIDINE see LJB000
LUTINYL see CBF250
LUTOCYCLIN see PMH500
LUTOFAN see PKQ059
LUTO-METRODIOL see EQJ500
LUTOPRON see HNT500
LUTOSOL see INJ000
LUTROL see EIM000, PJT000
LUTROL-9 see EJC500
LUTROMONE see PMH500
LUVATRENE see MNN250
LUVISKOL see PKQ250
LUXISTELM see PHI500
LUXON see DFH600
LW 3170 see DCS200
LX 14-0 see CQH250
LXON see DFH600
LYCOID DR see GLU000
LYCOPERSICIN see THG250
LYDOL see DAM700
LYE see PLJ500
LYE (DOT) see SHS000
LYE, solution see SHS500
LYGOMME CDS see CCL250
LYMECYCLINE see MRV250
LYNDIOL see LJE000
LYNESTRENOL mixed with MESTRANOL see LJE000
LYNESTROL see EEH550
LYNESTROL mixed with MESTRANOL see LJE000
LYNOESTRENOL mixed with MESTRANOL see LJE000
LYOPHRIN see VGP000
LYP 97 see LBR000
LYSERGAMID see DJO000
LYSERGAURE DIETHYLAMID see DJO000
d-LYSERGIC ACID DIETHYLAMIDE see DJO000, LJF000
LYSERGIC ACID DIETHYLAMIDE-25 see DJO000
LYSERGIC ACID DIETHYLAMIDE, 1-ISOMER see LJF000
LYSERGIC ACID DIETHYLAMIDE TARTRATE see LJG000
d-LYSERGIC ACID DIETHYLAMIDE TARTRATE see
   LJG000
d-LYSERGIC ACID DIMETHYLAMIDE see LJH000
LYSERGIC ACID ETHYLAMIDE see LJI000
d-LYSERGIC ACID-1-HYDROXYMETHYLETHYLAMIDE see
   LJL000
d-LYSERGIC ACID MONOETHYLAMIDE see LJI000
LYSERGIC ACID PROPANOLAMIDE see LJL000
d-LYSERGIC ACID-1,2-PROPANOLAMIDE see LJL000
LYSERGIC ACID PYROLIDATE see LJM000
d-LYSERGIC ACID PYRROLIDIDE see LJM000
LYSERGIDE see DJO000
LYSERGSAUEREDIAETHYLAMID see DJO000
l-LYSINE HYDROCHLORIDE see LJO000
LYSINE MONOHYDROCHLORIDE see LJO000
l-LYSINE MONOHYDROCHLORIDE see LJO000
N-LYSINOMETHYLTETRACYCLINE see MRV250
LYSIVANE see DIR000
LYSOCOCCINE see SNM500
LYSOFORM see FMV000
LYSSIPOLL see LJR000
LYSTENON see HLC500
LYSTHENONE see HLC500

M 40 see CIR250
M 73 see DFV600
M-74 see DXH325

M 81 see PHI500
M 140 see CDR750
M 176 see DUD800
M 410 see CDR750
M 2060 see FIP999
M 3/158 see DAP600
M 3180 see CQH650
M-4209 see CBT250
M 5055 see CDV100
5512-M see PDC000
M-9500 see TND500
MA-110 see EHP000
MA-1214 see DXV600
MAA see AFI850
MAAC see HFJ000
MAA SODIUM SALT see DXE600
MAB see MNR500
MABLIN see BOT250
MACASIROOL see CHJ750
MACBAL see DTN200
MACE (lachrymator) see CEA750
MACE OIL see OGQ100
MACH-NIC see NDN000
MACKREAZID see COH250
MACLEYINE see FOW000
MACQUER'S SALT see ARD250
MACRABIN see VSZ000
MACRODANTIN see NGE000
MACRODIOL see EDO000
MACROGOL 1000 see PJT250
MACROGOL 4000 see PJT750
MACROGOL 400 BPC see EJC500
MACROL see EDO000
MACRONDRAY see DAA800
MACROPAQUE see BAP000
MACULOTOXIN see FOQ000
MADAGASCAR LEMONGRASS OIL see LEH000
MADHURIN see SJN700
MADIOL see MQU750
MADRIBON see SNN300
MADRIGID see SNN300
MADRINE see DBA800
MADRIQID see SNN300
MADROXIN see SNN300
MADROXINE see SNN300
MAFU see DGP900
MAG see MRF000
MAGBOND see BAV750
MAGENTA BASE see MAC500
MAGIC METHYL see MKG250
MAGMASTER see MAC650
MAGNACAT see MAP750
MAGNACIDE H see ADR000
MAGNAMYCIN see CBT250
MAGNAMYCIN A see CBT250
MAGNESIA see MAH500
MAGNESIA ALBA see MAC650
MAGNESIA MAGMA see MAG750
MAGNESIA USTA see MAH500
MAGNESIA WHITE see CAX750
MAGNESIO (ITALIAN) see MAC750
MAGNESITE see MAC650
MAGNESIUM see MAC750
MAGNESIUM ACETATE see MAD000
MAGNESIUM ALUMINUM PHOSPHIDE (DOT) see AHD250
MAGNESIUM ARSENATE see ARD000
MAGNESIUM ARSENATE PHOSPHOR see ARD000
MAGNESIUM BORINGS see MAC750
MAGNESIUM CARBONATE see MAC650
MAGNESIUM(II) CARBONATE (1:1) see MAC650
MAGNESIUM CARBONATE, PRECIPITATED see MAC650
MAGNESIUM CHLORATE see MAE000
MAGNESIUM CHLORIDE see MAE250
MAGNESIUM CLIPPINGS (DOT) see MAC750

MAGNESIUM COMPOUNDS see MAE750
MAGNESIUM DIACETATE see MAD000
MAGNESIUM DICHLORATE see MAE000
MAGNESIUM FLUOSILICATE see MAG250
MAGNESIUMFOSFIDE (DUTCH) see MAI000
MAGNESIUM GOLD PURPLE see GIS000
MAGNESIUM GRANULES COATED, PARTICLE SIZE NOT
    LESS THAN 149 MICRONS (DOT) see MAC750
MAGNESIUM HEXAFLUOROSILICATE see MAG250
MAGNESIUM HYDRATE see MAG750
MAGNESIUM HYDRIDE see MAG500
MAGNESIUM HYDROXIDE see MAG750
MAGNESIUM METAL (DOT) see MAC750
MAGNESIUM NITRATE (DOT) see MAH000
MAGNESIUM(II) NITRATE (1:2) see MAH000
MAGNESIUM OXIDE see MAH500
MAGNESIUM PELLETS see MAC750
MAGNESIUM PERCHLORATE see PCE000
MAGNESIUM PEROXIDE see MAH750
MAGNESIUM PEROXIDE, solid (DOT) see MAH750
MAGNESIUM PHOSPHIDE see MAI000
MAGNESIUM POWDER (DOT) see MAC750
MAGNESIUM RIBBONS see MAC750
MAGNESIUM SCALPINGS (DOT) see MAC750
MAGNESIUM SCRAP (DOT) see MAC750
MAGNESIUM SHAVINGS (DOT) see MAC750
MAGNESIUM SHEET see MAC750
MAGNESIUM SILICATE HYDRATE see MAJ000
MAGNESIUM SILICOFLUORIDE see MAG250
MAGNESIUM SULFATE (1:1) see MAJ250
MAGNESIUM SULFATE HEPTAHYDRATE see MAJ500
MAGNESIUM SULPHATE see MAJ250
MAGNESIUM TURNINGS (DOT) see MAC750
MAGNETIC 70,90 and 95 see SOD500
MAGNEZU TLENEK (POLISH) see MAH500
MAGRON see MAE000
MAGSALYL see SJO000
MAIPEDOPA see DNA200
MAITANSINE see MBU820
MAJOL PLX see SFO500
MAJSOLIN see DBB200
MAKAROL see DKA600
MALACHITE GREEN G see BAY750
MALACID see TGD000
MALACIDE see MAK700
MALAFOR see MAK700
MALAGRAN see MAK700
MALAKILL see MAK700
MALAMAR see MAK700
MALAMAR 50 see MAK700
MALAPHELE see MAK700
MALAPHOS see MAK700
MALASOL see MAK700
MALASPRAY see MAK700
MALATHION see MAK700
MALATHION ULV CONCENTRATE see MAK700
MALATHIOZOO see MAK700
MALATHON see MAK700
MALATHYL LV CONCENTRATE & ULV CONCENTRATE
    see MAK700
MALATION (POLISH) see MAK700
MALATOL see MAK700
MALATOX see MAK700
MALAYAN CAMPHOR see BMD000
MALDISON see MAK700
MALEATE ACIDE de l'ACETYL-3-DIMETHYLAMINO-3-
    PROPYL-10-PHENOTHIAZINE (FRENCH) see AAF750
MALEIC ACID see MAK900
MALEIC ACID ANHYDRIDE (MAK) see MAM000
MALEIC ACID, DIALLYL ESTER see DBK200
MALEIC ACID, DIBUTLY ESTER see DED600
MALEIC ACID-N-ETHYLIMIDE see MAL250
MALEIC ACID HYDRAZIDE see DMC600
MALEIC ANHYDRIDE see MAM000

MALEIC ANHYDRIDE adduct of BUTADIENE see TDB000
MALEIC HYDRAZIDE see DMC600
MALEIC HYDRAZIDE DIETHANOLAMINE SALT see
    DHF200
MALEIMIDE see MAM750
MALEINIC ACID see MAK900
MALEINIMIDE see MAM750
MALENIC ACID see MAK900
N,N-MALEOYLHYDRAZINE see DMC600
MALESTRONE see MPN500
MALESTRONE (AMPS) see TBF500
MALGESIC see BRF500
MALIC ACID see MAN000
MALIPUR see CBG000
MALIVAN see DNV000
MALIX see EAQ750
MALLOFEEN see PDC250
MALLOPHENE see PDC250, PEK250
MALLOROL see MOO250
MALMED see MAK700
MALOCID see TGD000
MALOCIDE see TGD000
MALOGEN see MPN500
MALONALDEHYDE see PMK000
MALONDIALDEHYDE see PMK000
MALONIC ACID, DIETHYL ESTER see EMA500
MALONIC ACID ETHYL ESTER NITRILE see EHP500
MALONIC ALDEHYDE see PMK000
MALONIC DIALDEHYDE see PMK000
MALONIC DINITRILE see MAO250
MALONIC ESTER see EMA500
MALONITRILE HYDRAZIDE see COH250
MALONODIALDEHYDE see PMK000
MALONONITRILE see MAO250
MALONONITRILE HYDRAZIDE see COH250
MALONYLDIALDEHYDE see PMK000
MALOPRIM see SOA500, TGD000
MALPHOS see MAK700
MALTOBIOSE see MAO500
MALTOL see MAO350
MALTOSE see MAO500
d-MALTOSE see MAO500
MALTOX see MAK700
MALTOX MLT see MAK700
MALT SUGAR see MAO500
α-MALT SUGAR see MAO500
MALYSOL see MKA250
MAM see HMG000
MAM AC see MGS750
MAM ACETATE see MGS750
MAMALLET-A see DOT000
MAMBNA see MHW350
MAMMEX see NGE500
MAMN see AAW000
MANADRIN see EAW000
MANAM see MAS500
MANCHESTER YELLOW see DUX800
MANDELIC ACID see MAP000
racemic MANDELIC ACID see MAP000
MANDELIC ACID NITRILE see MAP250
1-MANDELONITRILE-β-GLUCURONIC ACID see LAS000
MANDRIN see EAW000
MANEB see MAS500
MANEB, stabilized against self-heating (DOT) see MAS500
MANEB, with not less than 60% maneb (DOT) see MAS500
MANEBE (FRENCH) see MAS500
MANEB plus ZINC ACETATE (50:1) see EIQ500
MANEXIN see MAW250
MANGAANBIOXYDE (DUTCH) see MAS000
MANGAANDIOXYDE (DUTCH) see MAS000
MANGAAN(II)-(N,N'-ETHYLEEN-BIS(DITHIOCARBA-
    MAAT)) (DUTCH) see MAS500
MANGAN (POLISH) see MAP750

MANGAN(II)-(N,N'-AETHYLEN-BIS(DITHIOCARBAMATE)) (GERMAN) see MAS500
MANGANDIOXID (GERMAN) see MAS000
MANGANESE see MAP750
MANGANESE BINOXIDE see MAS000
MANGANESE (BIOSSIDO di) (ITALIAN) see MAS000
MANGANESE (BIOXYD de) (FRENCH) see MAS000
MANGANESE BLACK see MAS000
MANGANESE(II) CHLORIDE (1:2) see MAR000
MANGANESE COMPOUNDS see MAR500
MANGANESE CYCLOPENTADIENYL TRICARBONYL see CPV000
MANGANESE DICHLORIDE see MAR000
MANGANESE (DIOSSIDO di) (ITALIAN) see MAS000
MANGANESE DIOXIDE see MAS000
MANGANESE (DIOXYDE de) (FRENCH) see MAS000
MANGANESE ETHYLENE-1,2-BISDITHIOCARBAMATE see MAS500
MANGANESE(II) ETHYLENEBIS(DITHIOCARBAMATE) see MAS500
MANGANESE(II) ETHYLENE DI(DITHIOCARBAMATE) see MAS500
MANGANESE GREEN see MAT250
MANGANESE MANGANATE see MAT500
MANGANESE MONOXIDE see MAT250
MANGANESE OXIDE see MAS000, MAU800
MANGANESE(II) OXIDE see MAT250
MANGANESE(IV) OXIDE see MAS000
MANGANESE(III) OXIDE see MAT500
MANGANESE PEROXIDE see MAS000
MANGANESE SISQUIOXIDE see MAT500
MANGANESE(II) SULFATE (1:1) see MAU250
MANGANESE TETROXIDE see MAU800
MANGANESE TRICARBONYL METHYLCYCLOPENTA-DIENYL see MAV750
MANGANESE TRIOXIDE see MAT500
MANGANESE ZINC BERYLLIUM SILICATE see BFS750
MANGANIC OXIDE see MAT500
MANGAN NITRIDOVANY (CZECH) see MAP750
MANGANOMANGANIC OXIDE see MAU800
MANGANOUS CHLORIDE see MAR000
MANGANOUS OXIDE see MAT250
MANGANOUS SULFATE see MAU250
MANGENESE SUPEROXIDE see MAS000
MAN-GRO see MAU250
MANICOLE see MAW250
MANITE see MAW250
MANNIT-LOST (GERMAN) see MAW500
MANNIT-MUSTARD (GERMAN) see MAW500
d-MANNITOL BUSULFAN see BKM500
MANNITOL HEXANITRATE see MAW250
d-MANNITOL HEXANITRATE see MAW250
MANNITOL HEXANITRATE, containing, by weight, at least 40% water (DOT) see MAW250
MANNITOL MUSTARD DIHYDROCHLORIDE see MAW750
MANNITOL MYLERAN see BKM500
MANNITOL NITROGEN MUSTARD see MAW500
MANNOGRANOL see BKM500
MANNOMUSTINE see MAW500
MANNOMUSTINE DIHYDROCHLORIDE see MAW750
MANOXAL OT see DJL000
MAN'S MOTHERWORT see CCP000
MANTA see KAJ000
MANTHELINE see XCJ000
MANUCOL see SEH000
MANUCOL DM see SEH000
MANUFACTURED IRON OXIDES see IHD000
MANUTEX see SEH000
MANZATE see MAS500
MANZATE MANEB FUNGICIDE see MAS500
MAOA see QAK000
MAOH see MKW600
MAO-REM see PFC750
MAPLE BRILLIANT BLUE FCF see FMU059

MAPLE ERYTHROSINE see FAG040
MAPLE INDIGO CARMINE see FAE100
MAPLE LACTONE see HMB500
MAPLE PONCEAU 3R see FAG018
MAPLE TARTRAZOL YELLOW see FAG140
MAPO see TNK250
MAPP see MFX600
MAPROFIX 563 see SIB600
MAPROFIX WAC-LA see SIB600
MAQBARL see DTN200
MARALATE see MEI450
MARANHIST see WAK000
MARANYL F 114 see PJY500
MARAPLAN see IKC000
MAR BATE see MQU750
MARBLE see CAO000
MARBON 9200 see SMR000
MARCOPHANE see BGY000
MARETIN see HMV000
MAREX see HGC500
MARFOTOKS see DJI000
MARGONIL see MQU750
MARGONOVINE see LJL000
MARICAINE see DHK400
MARISILAN see AIV500
MARITUS YELLOW see DUX800
MARJORAM OIL, SPANISH see MBU500
MARKOFANE see BGY000
MARKS 4-CPA see CJN000
MARLATE see MEI450
MARLEX 9400 see PMP500
MARLOPHEN 820 see PKF500
MARMER see DXQ500
MAROXOL-50 see DUZ000
MARPLAN see IKC000
MARPLON see IKC000
MARSALID see ILE000
MARS BROWN see IHD000
MARSH GAS see MDQ750
MARSILID see ILE000
MARSIN see MNV750
MARS RED see IHD000
MARTRATE-45 see PBC250
MARVEX see DGP900
MARVINAL see PKQ059
MAS see MGQ750
MASCHITT see CFY000
MASENATE see TBG000
MASENONE see MPN500
MASEPTOL see HJL500
MASHERI (INDIA) see SED400
MASOTEN see TIQ250
MASSICOT see LDN000
MASSICOTITE see LDN000
MASTESTONA see MPN500
MASTIPHEN see CDP250
MATACIL see DOR400
MATCH see MLJ500
MATENON see MQS225
MATRICARIA CAMPHOR see CBA750
MATROMYCIN see OHO200
MATSUTAKE ALCOHOL (JAPANESE) see ODW000
MATTING ACID (DOT) see SOI500
MATULANE see PME250, PME500
MAURYLENE see PMP500
MAXATASE see BAC000
MAXIDEX see SOW000
MAXITATE see MAW250
MAXULVET see SNN300
MAYSANINE see MBU820
MAYT see MBU820
MAYTANSINE see MBU820
MAZOTEN see TIQ250
MB see MHR200

M & B 800 see DBL800
M+B 760 see TEX250
M&B 8873 see HKB500
MB 10064 see DDP000
M&B 11,461 see DNG200
MBA see BIE250
7-MBA see MGW750
MBA HYDROCHLORIDE see BIE500
MBAO see CFA500
MBAO HYDROCHLORIDE see CFA750
MBC see BAV575
MBCP see LEN000
MBDZ see MHL000
MBH see PME500
MBK see HEV000
MBNA see MHW500
MBOCA see MJM200
MBOT see MJO250
MBT see BDF000
MBTS see BDE750
MBTS RUBBER ACCELERATOR see BDE750
MBX see MHR200
6-MC see MIP750
2M-4C see CIR250
MC 474 see DJI000
MC 2303 see GGS000
MC6897 see DQM600
MCA see CEA000
3-MCA see MIJ750
MCB see CEJ125
MC DEFOLIANT see MAE000
MCE see EIF000
MCF see MIG000
MCNEIL 481 see DQU200
MCN-JR-2498 see TKK500
MCN-JR-3345 see FHG000
McN-JR 4263 see PDW750
MCN-JR-4749 see DYF200
McN-JR-16,341 see PAP250
MCN-JR-4263-49 see PDW750
M1 (COPPER) see CNI000
M2 (COPPER) see CNI000
MCP see CIR250
2M-4CP see CIR500
MCPA see CIR250
MCPP see CIR500
2-MCPP see CIR500
MCPP-D-4 see CIR500
MCPP 2,4-D see CIR500
MCPP-K-4 see CIR500
MCT see CPV000
l-(α-MD) see DNA800
MDA see MJQ000
MDAB see DUH600
3'-MDAB see DUH600
MDBA see MEL500
MDI see MJP400
pr-MDI see DPY600
MDS see DXL400
2-ME see MCN250
ME 277 see PAM000
ME-1700 see BIM500
ME 3625 see DFD000
MEA see AJT250, EEC600
MEAD JOHNSON 1999 see CCK250
MEADOW GREEN see COF500
MEASURIN see ADA725
MEB see MAS500
MEB 6447 see CJO250
MEBALLYMAL see SBM500
MEBARAL see ENB500
MEBENDAZOLE (USDA) see MHL000
MEBEREL see ENB500
MEBICHLORAMINE see BIE500

MEBR see MHR200
ME4 BROMINAL see DDP000
MEBROPHENHYDRAMINE see BMN250
MEBROPHENHYDRAMINE HYDROCHLORIDE see
    BMN250
MEBRYL see BMN250
MEBUBARBITAL see NBT500
MEBUBARBITAL SODIUM see NBU000
MEBUMAL NATRIUM see NBU000
MEBUMAL SODIUM see NBU000
MECADOX see FOI000
MECAMINE HYDROCHLORIDE see MQR500
MECAMYLAMINE HYDROCHLORIDE see MQR500
MECARBAM see DJI000
ME-CCNU see CHD250
MECHLORETHAMINE see BIE250
MECHLORETHAMINE HYDROCHLORIDE see BIE500
MECHLORETHAMINE OXIDE see CFA500
MECHLORETHAMINE OXIDE HYDROCHLORIDE see
    CFA750
MECLIZINE see HGC500
MECLIZINE HYDROCHLORIDE see MBX500
MECLOZINE see HGC500
MECLOZINE HYDROCHLORIDE see MBX500
MECODIN see MDO750
MECODRIN see BBK000
MECOMEC see CIR500
MECOPEOP see CIR500
MECOPER see CIR500
MECOPEX see CIR500
MECOPROP see CIR500
MECOTURF see CIR500
MECPROP see CIR500
MECRAMINE see AJT250
MECRLAT see MIQ075
MECS see EJH500
MeCsAc see EJJ500
MEDAMYCIN see TBX250
MEDARON see NGG500
MEDARSED see BPF000
MEDIAMID see DJS200
MEDIAMYCETINE see CDP250
MEDIBEN see MEL500
MEDICAINE see BQA010
MEDI-CALGON see SHM500
MEDIDRYL see BBV500
MEDIFLAVIN see DBX400
MEDIHALER-EPI see VGP000
MEDIHALER-TETRACAINE see BQA010
MEDINAL see BAG250
MEDIQUIL see IPU000
MEDOMET see DNA800
MEDOPREN see DNA800
MEDROL see MOR500
MEDROL DOSEPAK see MOR500
MEDRONE see MOR500
MEDROXYPROGESTERONE ACETATE see MCA000
MEE see EDP000
MEETHOBALM see BQA010
MEFEDINA see DAM700
MEFENSINA see GGS000
M.E.G. see EJC500
MEGABION see VSZ000
MEGACE see VTF000
MEGADIURIL see CFY000
MEGALOVEL see VSZ000
MEGAMYCINE see MDO250
MEGATOX see FFF000
MEGESTROL ACETATE see VTF000
MEGESTROL ACETATE + ETHINYLOESTRADIOL see
    MCA500
MEGESTROL ACETATE 4 MG., ETHINYLOESTRADIOL 50
    μg see MCA500
MEGESTRYL ACETATE see VTF000

MEGIMIDE see MKA250
6-ME-GLU-P-2 see AKS250
MEISEI TERYL DIAZO BLUE HR see DCJ200
MEK see MKA400
MEKAMIN HYDROCHLORIDE see MQR500
MEKP see MKA500
MEK PEROXIDE see MKA500
MELABON see ABG750
MELADININ see XDJ000
MELADININE see XDJ000
MELAMINE see MCB000
MELANILINE see DWC600
MELBIN see DQR600
MELDONE see CNU750
MELENGESTROL ACETATE see MCB380
MELERIL see MOO250
MELETIN see QCA000
MELILOTAL see MFW250
MELILOTIN see HHR500
MELILOTOL see HHR500
MELINITE see PID000
MELIPAN see MCB500
MELIPAX see CDV100
MELIPRAMIN see DLH600, DLH630
MELIPRAMINE see DLH600, DLH630
MELIPRAMINE HYDROCHLORIDE see DLH630
MELIPRAMIN HYDROCHLORIDE see DLH630
MELITOXIN see BJZ000
MELLARIL see MOO250
MELLARIL HYDROCHLORIDE see MOO500
MELLERETTE see MOO250
MELLERETTEN see MOO250
MELLERIL see MOO250
MELOXINE see XDJ000
MELPHALAN see PED750
MELPHALAN (RUSSIAN) see BHV000
MELPHALAN HYDROCHLORIDE see BHV250
MELPREX see DXX400
MELSEDIN BASE see QAK000
MELSOMIN see QAK000
MEMA see MEO750, MLF250
MEMC see MEP250
MEMCOZINE see SNN300
ME-MDA see MJO250
MEMMI see MLF500
MEMPA see HEK000
MENADION see MMD500
MENADIONE see MMD500
MENAGEN see EDV000
MENAPHTHON see MMD500
MENAPHTONE see MMD500
MENDEL see MQU750
MENDRIN see EAT500
MENEST see ECU750
MENFORMON see EDV000
MENHYDRINATE see DYE600
MENICHLOPHOLAN see DFD000
MENIPHOS see MQR750
MENITE see MQR750
Me₂NMOR see DTA000
MENOCIL see ALX250
MENOGEN see ECU750
MENOMYCIN see MRA250
MENOQUENS see MCA500
MENOSTILBEEN see DKA600
MENOTAB see ECU750
MENOTROL see ECU750
MENTA-BAL see ENB500
MENTHA ARVENSIS, OIL see MCB625
MENTHA ARVENSIS OIL, PARTIALLY DEMENTHOLIZED
   (FCC) see MCB625
p-MENTHA-1,3-DIENE see MLA250
p-MENTHA-1,4-DIENE see MCB750
p-MENTHA-1,5-DIENE see MCC000

p-MENTHA-1,8-DIENE see LFU000, MCC250
1,8(9)-p-MENTHADIENE see MCC250
d-p-MENTHA-1,8-DIENE see LFU000
(S)-(−)-p-MENTHA-1,8-DIENE see MCC500
(R)-(−)-p-MENTHA-6,8-DIEN-2-ONE see CCM120
1-6,8(9)-p-MENTHADIEN-2-ONE see CCM120
d-p-MENTHA-6,8,(9)-DIEN-2-ONE see CCM100
p-MENTHANE HYDROPEROXIDE see MCE000
p-MENTHANE-8-HYDROPEROXIDE see MCE000
p-MENTHANE HYDROPEROXIDE, TECHNICALLY PURE
   (DOT) see IQE000
p-MENTHAN-3-OL see MCF750
dl-3-p-MENTHANOL see MCG000
l-p-MENTHAN-3-ONE see MCG275
p-MENTHAN-3-ONE racemic see MCE250
8-p-MENTHEN-2-OL see DKV150
p-MENTH-1-EN-8-OL see TBD500
p-MENTH-8-EN-3-OL see MCE750
8(9)-p-MENTHEN-3-OL see MCE750
8-p-MENTHEN-2-ONE see DKV175
p-MENTH-8-EN-2-ONE see DKV175
MENTHEN-1-YL-8 PROPIONATE see TBE600
MENTHOL see MCF750
1-MENTHOL see MCF750
l-MENTHOL see MCG250
3-p-MENTHOL see MCG000
dl-MENTHOL see MCG000
MENTHOL racemic see MCG000
MENTHOL racemique (FRENCH) see MCG000
MENTHOL, ACETATE (8CI) see MCG500
MENTHONE see MCG275
l-MENTHONE (FCC) see MCG275
p-MENTHONE see MCG275
trans-MENTHONE see MCG275
MENTHONE, racemic see MCE250
MENTHYL ACETATE see MCG500
(−)-MENTHYL ACETATE see MCG750
l-MENTHYL ACETATE (FCC) see MCG750
dl-MENTHYL ACETATE see MCG500
1-p-MENTH-3-YL ACETATE see MCG750
l-p-MENTH-3-YL ACETATE see MCG750
MENTHYL ACETATE racemic see MCG500
(−)-MENTHYL ALCOHOL see MCG250
p-MENTH-3-YL ESTER-dl-ACETIC ACID see MCG500
MEONAL see BPF500
MEONINE see MDT740
MEP see EOS000
MEP (Pesticide) see DSQ000
MEPADIN see DAM700
MEPAMTIN see MQU750
ME-PARATHION see MNH000
MEPARFYNOL CARBAMATE see MNM500
MEPATON see MNH000
MEPAVLON see MQU750
MEPENTAMATE see MNM500
MEPENTAMATO see MNM500
MEPERIDIDE see FLV000
MEPERIDINE see DAM600
MEPERIDINE HYDROCHLORIDE see DAM700
MEPHABUTAZONE see BRF500
MEPHACYCLIN see TBX250
MEPHADRYL see BBV500
MEPHANAC see CIR250
MEPHATE see GGS000
MEPHEDAN see GGS000
MEPHEDINE see DAM700
MEPHELOR see GGS000
MEPHENAMINE HYDROCHLORIDE see OJW000
MEPHENAMIN HYDROCHLORIDE see OJW000
MEPHENSIN see GGS000
MEPHENYTOIN see MKB250
MEPHEXAMIDE see DIB500
MEPHOBARBITAL see ENB500
MEPHOBARBITONE see ENB500

MEPHOSAL see GGS000
MEPHOSFOLAN see DHH400
MEPHSON see GGS000
MEPHYTAL see ENB500
MEPIBEN see LJR000
MEPIOSINE see MQU750
MEPOSED see MQU750
MEPRANIL see MQU750
MEPRO see CIR500
MEPROBAM see MQU750
MEPROBAMAT (GERMAN) see MQU750
MEPROBAMATE see MQU750
MEPROBAMATO (ITALIAN) see MQU750
MEPROCOMPREN see MQU750
MEPROCON CMC see MQU750
MEPRODIL see MQU750
MEPRODINE (GERMAN) see NOD500
MEPROLEAF see MQU750
MEPROSAN see MQU750
MEPROSCILLARIN see MCI750
MEPROTABS see MQU750
MEPROZINE see MQU750
MEPTOX see MNH000
MEPTRAN see MQU750
MEPYRAMIN (GERMAN) see WAK000
MEPYRAMINE MALEATE see DBM800
MEPYREN see WAK000
MEQUIN see QAK000
MEQUINOL see MFC700
MER 25 see DHS000
MER-41 see CMX700
MERAKLON see PMP500
MERALLURIDE see MFC000
MERANTINE BLUE EG see FAE000
MERANTINE GREEN SF see FAF000
MERATRAN see DWK400
MERBAPHEN see CCG500
MERBENTUL see CLO750
MERBROMIN see MCV000
MERCALEUKIN see POK000
MERCAMINE see AJT250
MERCAMINE DISULFIDE see MCN500
MERCAPTAMINE see AJT250
MERCAPTAN AMYLIQUE (FRENCH) see PBM000
MERCAPTAN METHYLIQUE (FRENCH) see MLE650
MERCAPTAN METHYLIQUE PERCHLORE (FRENCH) see
  PCF300
MERCAPTANS see MCJ500
MERCAPTAZOLE see MCO500
2-MERCAPTOACETANILIDE see MCK000
α-MERCAPTOACETANILIDE see MCK000
MERCAPTOACETATE see TFJ100
MERCAPTOACETIC ACID see TFJ100
2-MERCAPTOACETIC ACID see TFJ100
α-MERCAPTOACETIC ACID see TFJ100
MERCAPTOACETIC ACID, DIESTER with DITHIO-p-
  UREIDOBENZENEARSONOUS ACID see CBI250
MERCAPTOACETIC ACID, SODIUM-BISMUTH SALT see
  BKX750
MERCAPTOACETIC ACID SODIUM SALT see SKH500
4-(MERCAPTOACETYL)MORPHOLINE O,O-DIMETHYL
  PHOSPHORODITHIOATE see MRU250
β-MERCAPTOALANINE see CQK000
o-MERCAPTOANILINE see AIF500
2-MERCAPTOBENZIMIDAZOLE see BCC500
o-MERCAPTOBENZOIC ACID, DIESTER with DITHIO-p-
  UREIDOBENZENEARSONOUS ACID see TFD750
MERCAPTOBENZOIMIDAZOLE see BCC500
2-MERCAPTOBENZOIMIDAZOLE see BCC500
MERCAPTOBENZOTHIAZOLE see BDF000
2-MERCAPTOBENZOTHIAZOLE see BDF000
2-MERCAPTOBENZOTHIAZOLEDISULFIDE see BDE750
2-MERCAPTOBENZOTHIAZOLE SODIUM DERIVATIVE see
  SIG500

2-MERCAPTOBENZOTHIAZOLE SODIUM SALT see
  SIG500
2-MERCAPTOBENZOTHIAZOLE ZINC SALT see BHA750
2-MERCAPTOBENZOTHIAZYLDISULFIDE see BDE750
(MERCAPTOBUTANEDIOATO(1-))GOLD DISODIUM SALT
  see GJC000
MERCAPTODIACETIC ACID see MCM750
MERCAPTODIMETHUR (DOT) see DST000
1-MERCAPTODODECANE see LBX000
MERCAPTOETHANOL see MCN250
2-MERCAPTOETHANOL see MCN250
β-MERCAPTOETHANOL see MCN250
(2-MERCAPTOETHYL)AMINE see AJT250
β-MERCAPTOETHYLAMINE see AJT250
β-MERCAPTOETHYLAMINE DISULFIDE see MCN500
2-MERCAPTOETHYLAMINE (OXIDIZED) see MCN500
2-MERCAPTO-4-HYDROXY-6-METHYLPYRIMIDINE see
  MPW500
2-MERCAPTO-4-HYDROXY-6-N-PROPYLPYRIMIDINE see
  PNX000
2-MERCAPTO-4-HYDROXYPYRIMIDINE see TFR250
2-MERCAPTOIMIDAZOLINE see IAQ000
MERCAPTOMERIN SODIUM see DXC000
(MERCAPTOMETHYL)BENZENE see TGO750
3-(MERCAPTOMETHYL)-1,2,3-BENZOTRIAZIN-4(3H)-ONE-
  O,O-DIMETHYL PHOSPHORODITHIOATE see ASH500
3-(MERCAPTOMETHYL)-1,2,3-BENZOTRIAZIN-4(3H)-ONE-
  O,O-DIMETHYL PHOSPHORODITHIOATE-S-ESTER see
  ASH500
2-MERCAPTO-1-METHYLIMIDAZOLE see MCO500
N-(MERCAPTOMETHYL)PHTHALIMIDE S-(O,O-DIMETHYL
  PHOSPHORODITHIOATE) see PHX250
2-MERCAPTO-6-METHYLPYRIMID-4-ONE see MPW500
2-MERCAPTO-6-METHYL-4-PYRIMIDONE see MPW500
2-MERCAPTONAPHTHALENE see NAP500
MERCAPTOPHOS see DAO600, FAQ999
2-MERCAPTOPROPANE see IMU000
3-MERCAPTOPROPANOL see PML500
2-MERCAPTO-6-PROPYL-4-PYRIMIDONE see PNX000
2-MERCAPTO-6-PROPYLPYRIMID-4-ONE see PNX000
6-MERCAPTOPURIN see POK000
MERCAPTOPURIN (GERMAN) see POK000
6-MERCAPTOPURINE see POK000
MERCAPTOPURINE RIBONUCLEOSIDE see MCQ500
6-MERCAPTOPURINE RIBOSIDE see MCQ500
2-MERCAPTO-4-PYRIMIDINOL see TFR250
2-MERCAPTO-4-PYRIMIDONE see TFR250
2-MERCAPTOPYRIMID-4-ONE see TFR250
MERCAPTOSUCCINIC ACID ANTIMONATE(III) HEXA-
  LITHIUM SALT see LGU000
MERCAPTOSUCCINIC ACID-S-ANTIMONY DERIVATIVE
  LITHIUM SALT see LGU000
MERCAPTOSUCCINIC ACID DIETHYL ESTER see
  MAK700
MERCAPTOSUCCINIC ACID, GOLD SODIUM SALT see
  GJC000
MERCAPTOSUCCINIC ACID, THIOANTHIMONATE(III),
  DILITHIUM SALT see LGU000
p-MERCAPTO SULFADIAZINE see MCR250
7-MERCAPTO-1,3,4,6-TETRAZAINDENE see POK000
MERCAPTOTHION see MAK700
MERCAPTOTION (SPANISH) see MAK700
α-MERCAPTOTOLUENE see TGO750
d-MERCAPTOVALINE see MCR750
MERCAPTURIC ACID see ACH000
(R)-MERCAPTURIC ACID see ACH000
MERCAPURIN see POK000
MERCAZOLYL see MCO500
MERCHLORATE see MEP250
MERCHLORETHANAMINE see BIE500
MERCKOGEN 6000 see AAX250
MERCOL 25 see DXW200
MERCURAM see TFS350
MERCURAMIDE see SIH500

MERCURAN see MEO750
MERCURANINE see MCV000
MERCURE (FRENCH) see MCW250
MERCURIACETATE see MCS750
MERCURIALIN see MGC250
MERCURIC ACETATE see MCS750
MERCURIC AMMONIUM CHLORIDE, solid see MCW500
MERCURIC ARSENATE see MDF350
MERCURIC BASIC SULFATE see MDG000
MERCURIC BENZOATE see MCX500
MERCURIC BENZOATE, solid (DOT) see MCX500
MERCURIC BROMIDE see MCY000
MERCURIC BROMIDE, solid (DOT) see MCY000
MERCURIC CHLORIDE (DOT) see MCY475
MERCURIC CHLORIDE, AMMONIATED see MCW500
MERCURIC CYANIDE, solid (DOT) see MDA250
MERCURIC DIACETATE see MCS750
MERCURIC IODIDE see MDD000
MERCURIC IODIDE, solid (DOT) see MDD000
MERCURIC IODIDE, solution (DOT) see MDD000
MERCURIC IODIDE, RED see MDD000
MERCURIC NITRATE see MDF000
MERCURIC OLEATE, solid (DOT) see MDF250
MERCURIC OXIDE see MCT500
MERCURIC OXIDE, solid (DOT) see MCT500
MERCURIC OXIDE, RED see MCT500
MERCURIC OXIDE, YELLOW see MCT500
MERCURIC OXYCYANIDE see MDA500
MERCURIC OXYCYANIDE, solid (desensitized) (DOT) see
  MDA500
MERCURIC POTASSIUM CYANIDE (DOT) see PLU500
MERCURIC POTASSIUM CYANIDE, solid (DOT) see
  PLU500
MERCURIC POTASSIUM IODIDE see NCP500
MERCURIC POTASSIUM IODIDE, solid (DOT) see NCP500
MERCURIC SALICYLATE see MCU000
MERCURIC SALICYLATE, solid (DOT) see MCU000
MERCURIC SUBSULFATE, solid (DOT) see MDG000
MERCURIC SULFATE see MDG500
MERCURIC SULFOCYANATE see MCU250
MERCURIC SULFOCYANIDE see MCU250
MERCURIC SULFOCYANTE, solid (DOT) see MCU250
MERCURIC THIOCYANATE see MCU250
MERCURIC THIOCYANATE, solid (DOT) see MCU250
MERCURIO (ITALIAN) see MCW250
MERCURIPHENYL ACETATE see ABU500
MERCURIPHENYL CHLORIDE see PFM500
MERCURIPHENYL NITRATE see MCU750
MERCURISALICYLIC ACID see MCU000
MERCURITAL see SIH500
MERCUROCHLORIDE (DUTCH) see MCW000
MERCUROCHROME see MCV000
MERCUROCHROME-220 SOLUBLE see MCV000
MERCUROCOL see MCV000
MERCUROL see MCV250
MERCUROME see MCV000
MERCUROPHAGE see MCV000
MERCUROTHIOLATE see MDI000
MERCUROUS ACETATE see MDE250
MERCUROUS ACETATE, solid (DOT) see MDE250
MERCUROUS AZIDE (DOT) see MCX000
MERCUROUS BROMIDE, solid (DOT) see MCX750
MERCUROUS CHLORIDE see MCW000
MERCUROUS GLUCONATE see MDC500
MERCUROUS GLUCONATE, solid (DOT) see MDC500
MERCUROUS IODIDE see MDC750
MERCUROUS IODIDE, solid (DOT) see MDC750
MERCUROUS NITRATE see MDE750
MERCUROUS OXIDE, BLACK, solid (DOT) see MDF750
MERCUROUS SULFATE, solid (DOT) see MDG250
MERCURY see MCW250
MERCURY ACETATE see MCS750, MDE250
MERCURY(2+) ACETATE see MCS750
MERCURY(II) ACETATE see MCS750

MERCURY AMIDE CHLORIDE see MCW500
MERCURY AMINE CHLORIDE see MCW500
MERCURY AMMONIATED see MCW500
MERCURY AZIDE see MCX000
MERCURY(I) AZIDE see MCX000
MERCURY(II) BENZOATE see MCX500
MERCURY BICHLORIDE see MCY475
MERCURY BINIODIDE see MDD000
MERCURY BISULFATE see MDG500
MERCURY(I) BROMIDE (1:1) see MCX750
MERCURY(II) BROMIDE (1:2) see MCY000
MERCURY(I) CHLORIDE see MCW000
MERCURY(II) CHLORIDE see MCY475
MERCURY (E)-CHLORO(2-(3-BROMOPROPIONAMIDO)CY-
  CLOHEXYL) see CET000
MERCURY COMPOUNDS, INORGANIC see MCZ000
MERCURY COMPOUNDS, ORGANIC see MDA000
MERCURY(II) CYANIDE see MDA250
MERCURY CYANIDE OXIDE see MDA500
MERCURY DIACETATE see MCS750
MERCURY DITHIOCYANATE see MCU250
MERCURY FULMINATE (DOT) see MDC000
MERCURY FULMINATE (wet) see MDC250
MERCURY(II) FULMINATE (dry) see MDC000
MERCURY(I) GLUCONATE see MDC500
MERCURY(I) IODIDE see MDC750
MERCURY(II) IODIDE see MDD000
MERCURYL ACETATE see MCS750
MERCURY, METALLIC (DOT) see MCW250
MERCURY METHYLCHLORIDE see MDD750
MERCURY MONOACETATE see MDE250
MERCURY MONOCHLORIDE see MCW000
MERCURY NITRATE see MDF000
MERCURY(I) NITRATE (1:1) see MDE750
MERCURY(II) NITRATE (1:2) see MDF000
MERCURY NITRIDE see TKW000
MERCURY NUCLEATE, solid (DOT) see MCV250
MERCURY OLEATE see MDF250
MERCURY(II) ORTHOARSENATE see MDF350
MERCURY(I) OXIDE see MDF750
MERCURY(II) OXIDE see MCT500
MERCURY OXIDE SULFATE see MDG000
MERCURY OXYCYANIDE see MDA500
MERCURY PERCHLORIDE see MCY475
MERCURY PERNITRATE see MDF000
MERCURY PERSULFATE see MDG500
MERCURY(II) POTASSIUM IODIDE see NCP500
MERCURY PROTOCHLORIDE see MCW000
MERCURY PROTOIODIDE see MDC750
MERCURY SALICYLATE see MCU000
MERCURY SUBSALICYLATE see MCU000
MERCURY(I) SULFATE see MDG250
MERCURY(II) SULFATE (1:1) see MDG500
MERCURY(II) THIOCYANATE see MCU250
MERCURY THIOCYANATE (DOT) see MCU250
MERCURY ZINC CHROMATE COMPLEX see ZJA000
MERCUSAL see SIH500
MEREX see KEA000
MERFALAN see BHT750
MERFAMIN see MDI000
MERFAZIN see PFM500
MERGE see MRL750
MERIDIL see MNQ000
MERIZONE see BRF500
MERKAPTOBENZIMIDAZOL (CZECH) see BCC500
2-MERKAPTOBENZOTIAZOL (POLISH) see BDF000
2-MERKAPTOIMIDAZOLIN (CZECH) see IAQ000
MERKAZIN see BKL250
MERMETH see SNJ000
MERN see POK000
MERONIDAL see MMN250
MEROPHAN see BHT250
o-MEROPHAN see BHT250
MERPAN see CBG000

MERPHALAN see BHT750
o-MERPHALAN see BHT750
MERPHALAN HYDROCHLORIDE see BHV000
MERPHENYL NITRATE see MCU750
MERPHOS see TIG250
MERPOL see EJN500
MERRILLITE see ZBJ000
MERSALIN see SIH500
MERSALYL see SIH500
MERSOLITE 2 see PFM500
MERSOLITE 7 see MCU750
MERTEC see TEX000
MERTESTATE see TBF500
MERTHIOLATE see MDI000
MERTHIOLATE SALT see MDI000
MERTHIOLATE SODIUM see MDI000
MERTIONIN see MDT740
MERTORGAN see MDI000
MERVAMINE see DJL000
MERZONIN SODIUM see MDI000
MESAMATE see MRL750
MESAMATE CONCENTRATE see MRL750
MESANTOIN see MKB250
MESATON see NCL500
MESCALINE see MDI500
MESCALINE HYDROCHLORIDE see MDI750
MESENTOL see ENG500
MESIDICAINE HYDROCHLORIDE see DHL800
MESIDIN (CZECH) see TLG500
MESIDINE see TLG500
MESIDINE HYDROCHLORIDE see TLH000
MESITYLAMINE see TLG500
MESITYLAMINE HYDROCHLORIDE see TLH000
MESITYLENE see TLM050
MESITYLOXID (GERMAN) see MDJ750
MESITYL OXIDE see MDJ750
MESITYLOXYDE (DUTCH) see MDJ750
MESOCAINE HYDROCHLORIDE see DHL800
MESOKAIN HYDROCHLORIDE see DHL800
MESOMILE see MDU600
MESOXALYLCARBAMIDE MONOHYDRATE see MDL500
MESOXALYLUREA MONOHYDRATE see MDL500
MESTERONE see MPN500
MESTRANOL see MKB750
MESTRANOL mixed with LYNESTRENOL see LJE000
MESTRANOL mixed with LYNESTROL see LJE000
MESTRANOL mixed with NORETHYNODREL see EAP000
MESTRENOL see MKB750
MESURAL see LFK000
MESUROL see DST000
META see TDW500
METAARSENIC ACID see ARB000
META BLACK see AQP000
METABOLITE C see SOA500
METABOLITE I see HNI500
METACE see CLO750
METACEN see IDA000
METACETALDEHYDE see TDW500
METACETONE see DJN750
METACETONIC ACID see PMU750
METACHLOR see CFX000
METACIDE see MNH000
METACIL see MPW500
METACIN see ORQ000
METACORTANDRACIN see PLZ000
METACORTANDRALONE see PMA000
METACRATE see MIB750
METADEE see VSZ100
METADELPHENE see DKC800
METADOMUS see MDO250
METAFOS see MNH000, SII500
METAFUME see MHR200
METAHYDROXYPROCAINE see DHO600
METAISOSEPTOX see DAP400

METAISOSYSTOX see DAP400
METAISOSYSTOX-SOLFON 20 315 see DAP600
METAISOSYSTOXSULFOXIDE see DAP000
METAKRYLAN METYLU (POLISH) see MLH750
METALCAPTASE see MCR750, PAP550
METALDEHYD (GERMAN) see TDW500
METALDEHYDE (DOT) see TDW500
METALDEIDE (ITALIAN) see TDW500
METALKAMATE see BTA250
METALLIBURE see MLJ500
METALLIC ARSENIC see ARA750
METALLIC OSMIUM see OKE000
METAMFETAMINA see DBA800
METAMID see PJY500
METAMIDOFOS ESTRELLA see DTQ400
METAMPHETAMIN see DBA800
METANA ALUMINUM PASTE see AGX000
METANDREN see MPN500
METANEPHRIN see VGP000
METANFETAMINA see DBA800
METANOLO (ITALIAN) see MGB150
METANTYL see XCJ000
METAOXEDRIN see NCL500
METAPHENYLENEDIAMINE see PEY000
METAPHOR see MNH000
METAPHOS see MNH000
METAPLEX NO see PKB500
METAQUALON see QAK000
METAQUEST A see EIX000
METAQUEST B see EIX500
METAQUEST C see EIV000
METARADRINE see HNB875
METARAMINOL see HNB875, HNB875
1-METARAMINOL see HNB875
METARTRIL see IDA000
METASILICIC ACID see SCL000
METASOL see MLH000
METASOL TK-100 see TEX000
METASON see TDW500
METASYMPATOL see NCL500
METASYNEPHRINE see NCL500
METASYSTEMOX see DAP000
METASYSTOX see MIW100
METASYSTOX FORTE see DAP400
METASYSTOX-R see DAP000
METASYSTOX-S see DSK600
METATHIONE see DSQ000
METATION see DSQ000
META TOLUYLENE DIAMINE see TGL750
METATOLYLENEDIAMINE DIHYDROCHLORIDE see DCE000
METATSIN see ORQ000
METAUPON see OHU000
METAXAN see XCJ000
METAXITE see ARM268
METAXON see CIR250
METAZIN see SNJ000
METAZOL see MLH000
METAZOLO see MCO500
METEPA see TNK250
METERAZIN MALEATE see PMF250
METERFER see FBJ100
METERFOLIC see FBJ100
METFORMIN see DQR600
"METH" see MDQ500
METHAANTHIOL (DUTCH) see MLE650
METHABOL see PAN100
METHACETONE see DJN750
METHACHLOR see CFX000
METHACIDE see TGK750
METHACIN see ORQ000
METHACRALDEHYDE (DOT) see MGA250
METHACROLEIN see MGA250
METHACRYLALDEHYDE (DOT) see MGA250

METHACRYLATE de BUTYLE (FRENCH) see MHU750
METHACRYLATE de METHYLE (FRENCH) see MLH750
METHACRYLIC ACID see MDN250
METHACRYLIC ACID AMIDE see MDN500
METHACRYLIC ACID, INHIBITED (DOT) see MDN250
METHACRYLIC ACID, ISOBUTYL ESTER see IIY000
METHACRYLIC ACID, METHYL ESTER (MAK) see MLH750
METHACRYLIC ACID METHYL ESTER POLYMERS see PKB500
METHACRYLIC ALDEHYDE see MGA250
METHACRYLIC AMIDE see MDN500
METHACRYLSAEUREBUTYLESTER (GERMAN) see MHU750
METHACRYLSAEUREMETHYL ESTER (GERMAN) see MLH750
METHACYCLINE HYDROCHLORIDE see MDO250
METHACYCLINE MONOHYDROCHLORIDE see MDO250
METHADONE see MDO750
METHADONE HYDROCHLORIDE see MDP000
d-METHADONE HYDROCHLORIDE see MDP500
l-METHADONE HYDROCHLORIDE see MDP250
METHADRENE see MJV000
METHAFORM see ABD000
METHALLIBURE see MLJ500
METHALLYL ALCOHOL (DOT) see IMW000
METHALLYL CHLORIDE see CIU750
α-METHALLYL CHLORIDE see CIU750
METHAMIDOPHOS see DTQ400
METHAMIN see HEI500
METHAMINODIAZEPINE HYDROCHLORIDE see MDQ250
METHAMINODIAZEPOXIDE see LFK000
METHAMINODIAZEPOXIDE HYDROCHLORIDE see MDQ250
METHAMPHETAMINE see DBB000
METHAMPHETAMINE HYDROCHLORIDE see DBA800
l-METHAMPHETAMINE HYDROCHLORIDE see MDQ500
METHANAL see FMV000
METHANAMIDE see FMY000
METHANAMINE (9CI) see MGC250
METHANE see MDQ750
METHANE, compressed (DOT) see MDQ750
METHANE, refrigerated liquid (DOT) see MDQ750
METHANEARSONIC ACID DIMERCURY SALT see DNW000
METHANE BASE see MJN000
METHANECARBONITRILE see ABE500
METHANECARBOTHIOLIC ACID see TFA500
METHANECARBOXAMIDE see AAI000
METHANECARBOXYLIC ACID see AAT250
METHANEDICARBOXYLIC ACID, DIETHYL ESTER see EMA500
METHANE DICHLORIDE see MJP450
METHANEDITHIOL-S,S-DIESTER with O,O-DIETHYL ESTER PHOSPHORODITHIOIC ACID see EEH600
METHANEPEROXOIC ACID see PCM500
METHANE, PHENYL- see TGK750
METHANESULFONIC ACID see MDR250
METHANESULFONIC ACID CHLOROETHYL ESTER see CHC750
METHANESULFONIC ACID, METHYLENE ESTER see MJQ500
METHANESULFONIC ACID TETRAMETHYLENE ESTER see BOT250
METHANESULPHONIC ACID ETHYL ESTER see EMF500
METHANESULPHONIC ACID METHYL ESTER see MLH500
METHANE TETRACHLORIDE see CBY000
METHANE TETRAMETHYLOL see PBB750
METHANETHIOL see MLE650
METHANE TRICHLORIDE see CHJ500
METHANIDE see XCJ000
METHANOIC ACID see FNA000
METHANOL see MGB150

METHANOLACETONITRILE see HGP000
METHANOL, SODIUM SALT see SIK450
METHANTHELINE BROMIDE see XCJ000
METHANTHELINIUM BROMIDE see XCJ000
METHANTHINE BROMIDE see XCJ000
METHANTHIOL (GERMAN) see MLE650
METHAPHOXIDE see TNK250
METHAPYRILENE see TEO250
METHAQUALONE see QAK000
METHAQUALONE HYDROCHLORIDE see MDT250
METHAQUALONEINONE see QAK000
METHAR see DXE600
METHARSINAT see DXE600
METHASAN see BJK500
METHAZATE see BJK500
METHAZINE see IDA000
METHCAINE HYDROCHLORIDE see IJZ000
METHDILAZINE see MPE250
METHEDRINE see DBA800, DBB000
METHEDRINE HYDROCHLORIDE see DBA800
METHELINA see XCJ000
METHENAMINE see HEI500
N,N′-METHENYL-o-PHENYLENEDIAMINE see BCB750
METHENYL TRIBROMIDE see BNL000
METHENYL TRICHLORIDE see CHJ500
METHERGINE see PAM000
METHEXENYL see ERD500
METHIACIL see MPW500
METHIAMAZOLE see MCO500
METHIDATHION see DSO000
METHILANIN see MDT740
METHIOCARB see DST000
METHIOCIL see MPW500
METHIODIDE of N-METHYLURETHANE of 3-DIETHYLAMINOPHENOL see MID250
METHIODIDE of N-METHYLURETHANE of 3-DIMETHYLAMINOPHENOL see HNO500
METHIONINE see MDT750
l-METHIONINE see MDT750
l-(−)-METHIONINE see MDT750
(±)-METHIONINE see MDT740
dl-METHIONINE see MDT740
METHOCEL HG see HNX000
METHOCILLIN-S see SLJ000
METHOCROTOPHOS see DOL800
METHOFLURANE see DFA400
METHOGAS see MHR200
METHOHEXITAL SODIUM see MDU500
METHOHEXITONE SODIUM see MDU500
METHOIN see MKB250
METHOLENE 2218 see MJW000
METHOMYL see MDU600
METHOPHENAZATE ACID FUMARATE see MDU750
METHOPHENAZINE DIFUMARATE see MDU750
METHOPLAIN see DNA800
METHOPRENE see KAJ000
METHOPROMAZINE MALEATE see MFK750
METHOPTERIN see MDV500
METHORPHINAN see MKR250
METHORPHINAN HYDROBROMIDE see MDV250
METHOTEXTRATE see MDV500
METHOTREXATE see MDV500
METHOXA-DOME see XDJ000
METHOXAMINE HYDROCHLORIDE see MDW000
METHOXANE see DFA400
METHOXCIDE see MEI450
METHOXO see MEI450
METHOXONE see CIR250, CIR500
METHOXSALEN see XDJ000
2-METHOXYACETOACETANILIDE see ABA500
o-METHOXYACETOACETANILIDE see ABA500
2′-METHOXYACETOACETANILIDE see ABA500
p-METHOXYACETOPHENONE see MDW750
4′-METHOXYACETOPHENONE see MDW750

2-METHOXY-AETHANOL (GERMAN) see EJH500
2-METHOXYAETHYLACETAT (GERMAN) see EJJ500
1-METHOXY-AETHYL-AETHYLNITROSAMIN (GERMAN)
  see MEO500
1-METHOXY-AETHYL-METHYLNITROSAMIN (GERMAN)
  see MEP500
METHOXYAETHYLQUECKSILBERCHLORID (GERMAN)
  see MEP250
p-METHOXYALLYLBENZENE see AFW750
2-METHOXY-4-ALLYLPHENOL see EQR500
2-METHOXY-1-AMINOBENZENE see AOV900
4-METHOXYAMPHETAMINE HYDROCHLORIDE see
  MEA500
4-METHOXYANILINE see AOW000
o-METHOXYANILINE see AOV900
p-METHOXYANILINE see AOW000
2-METHOXYANILINE HYDROCHLORIDE see AOX250
4-METHOXYBENZALDEHYDE see AOT500
p-METHOXYBENZALDEHYDE (FCC) see AOT500
4-METHOXYBENZENAMINE see AOW000
2-METHOXY-BENZENAMINE (9CI) see AOV900
METHOXYBENZENE see AOX750
4-METHOXYBENZENEACETIC ACID see MFE250
4-METHOXYBENZENEAMINE see AOW000
4-METHOXY-1,3-BENZENEDIAMINE see DBO000
4-METHOXY-1,3-BENZENEDIAMINE SULFATE see
  DBO400
4-METHOXY-1,3-BENZENEDIAMINE SULPHATE see
  DBO400
4-METHOXYBENZENEMETHANOL see MED500
2-METHOXY-4H-1,2,3-BENZODIOXAPHOSPHORINE-2-SUL-
  FIDE see MEC250
2-METHOXYBENZOIC ACID see MPI000
3-METHOXYBENZOIC ACID see AOU500
m-METHOXYBENZOIC ACID see AOU500
o-METHOXYBENZOIC ACID see MPI000
4-METHOXYBENZOIC ACID HYDRAZIDE see AOV500
p-METHOXYBENZOIC ACID HYDRAZIDE see AOV500
p-METHOXYBENZOIC HYDRAZIDE see AOV500
6-METHOXYBENZO(a)PYRENE see MEC500
METHOXYBENZOYL CHLORIDE see AOY250
4-METHOXYBENZOYL HYDRAZIDE see AOV500
4-METHOXYBENZOYLHYDRAZINE see AOV500
(p-METHOXYBENZOYL)HYDRAZINE see AOV500
p-METHOXYBENZYL ACETATE see AOY400
4-METHOXYBENZYL ALCOHOL see MED500
p-METHOXYBENZYL ALCOHOL see MED500
N-(p-METHOXYBENZYL)-N′,N′-DIMETHYL-N-2-PYRIDYL-
  ETHYLENEDIAMINE see WAK000
N-p-METHOXYBENZYL-N′,N′-DIMETHYL-N-α-PYRIDYL-
  ETHYLENEDIAMINE see WAK000
N-p-METHOXYBENZYL-N′-N′-DIMETHYL-N-α-PYRIDYL-
  ETHYLENEDIAMINE MALEATE see DBM800
p-METHOXYBENZYL FORMATE see MFE250
4-METHOXYBIPHENYL see PEG250
p-METHOXYBIPHENYL see PEG250
3-METHOXYBUTYL ACETATE see MHV750
2-(METHOXYCARBONYL)ANILINE see APJ250
METHOXYCARBONYL CHLORIDE see MIG000
METHOXYCARBONYLETHYLENE see MGA500
(2-METHOXYCARBONYL-1-METHYL-VINYL)-DIMETHYL-
  FOSFAAT (DUTCH) see MQR750
(2-METHOXYCARBONYL-1-METHYL-VINYL)-DIMETHYL-
  PHOSPHAT (GERMAN) see MQR750
2-METHOXYCARBONYL-1-METHYLVINYL DIMETHYL-
  PHOSPHATE see MQR750
1-METHOXYCARBONYL-1-PROPEN-2-YL DIMETHYL
  PHOSPHATE see MQR750
(1-METHOXYCARBOXYPROPEN-2-YL)PHOSPHORIC ACID,
  DIMETHYL ESTER see MQR750
METHOXYCHLOR see MEI450
p,p′-METHOXYCHLOR see MEI450
2-METHOXY-6-CHLORO-9-(4-BIS(2-CHLOROETHYL)

AMINO-1-METHYLBUTYLAMINO)ACRIDINE DIHYDRO-
  CHLORIDE see QDS000
2-METHOXY-6-CHLORO-9-DIETHYLAMINOPENTYL-
  AMINOACRIDINE see ARQ250
2-METHOXY-6-CHLORO-9-(3-(ETHYL-2-CHLOROETHYL)
  AMINOPROPYLAMINO)ACRIDINE DIHYDROCHLORIDE
  see QDS000
5-METHOXYCHRYSENE see MEJ500
6′-METHOXYCINCHONAN-9-OL see QFS000
(8-α,9R)-6′-METHOXYCINCHONAN-9-OL see QHJ000
6′-METHOXYCINCHONAN-9-OL DIHYDROCHLORIDE see
  QIJ000
6-METHOXYCINCHONINE see QHJ000
METHOXY-DDT see MEI450
2-METHOXY-3,6-DICHLOROBENZOIC ACID see MEL500
5-METHOXY-2-(DIMETHOXYPHOSPHINYLTHIO-
  METHYL)PYRONE-4 see EAS000
2-METHOXY-10-(3′-DIMETHYLAMINOPROPYL)PHENO-
  THIAZINE see MFK500
10-METHOXY-1,6-DIMETHYLERGOLINE-8-METHANOL-5-
  BROMO-3-PYRIDINECARBOXYLATE (ESTER) see
  NDM000
10-METHOXY-1,6-DIMETHYL-ERGOLIN-8-β-METHA-
  NOL-(5-BROMNICOTINAT) (GERMAN) see NDM000
1-METHOXY-2,4-DINITROBENZENE see DUP800
METHOXYDIURON see DGD600
METHOXYETHANE see EMT000
2-METHOXYETHANOL (ACGIH) see EJH500
2-METHOXYETHANOL, ACETATE see EJJ500
2-METHOXYETHANOL, ACRYLATE see MIF750
METHOXYETHENE see MQL750
METHOXY ETHER of PROPYLENE GLYCOL see PNL250
3-METHOXY-17-α-ETHINYLESTRADIOL see MKB750
3-METHOXY-17-α-ETHINYLOESTRADIOL see MKB750
2-METHOXY-ETHYL ACETAAT (DUTCH) see EJJ500
2-METHOXYETHYL ACETATE (ACGIH) see EJJ500
4-METHOXY-9-(3-(ETHYL-2-CHLOROETHYL)) see EHJ500
2-METHOXYETHYLE, ACETATE de (FRENCH) see EJJ500
1-METHOXY ETHYL ETHYLNITROSAMINE see MEO500
METHOXYETHYL MERCURIC ACETATE see MEO750
METHOXYETHYL MERCURIC CHLORIDE see MEP250
2-METHOXYETHYLMERCURIC CHLORIDE see MEP250
(β-METHOXYETHYL)MERCURIC CHLORIDE see MEP250
METHOXYETHYLMERCURY CHLORIDE see MEP250
2-METHOXYETHYLMERCURY CHLORIDE see MEP250
β-METHOXYETHYLMERCURY CHLORIDE see MEP250
1-METHOXY ETHYL METHYLNITROSAMINE see MEP500
2-METHOXYETHYL PHTHALATE see DOF400
3-METHOXY-17-α-ETHYNOESTRADIOL see MKB750
3-METHOXYETHYNYLESTRADIOL see MKB750
3-METHOXY-17-α-ETHYNYLESTRADIOL see MKB750
3-METHOXY-17-α-ETHYNYL-1,3,5(10)-ESTRATRIEN-17-β-
  OL see MKB750
3-METHOXYETHYNYLOESTRADIOL see MKB750
3-METHOXY-17-ETHYNYLOESTRADIOL-17-β see MKB750
3-METHOXY-17-α-ETHYNYL-1,3,5(10)-OESTRATRIEN-17-β-
  OL see MKB750
1-p-METHOXYFENYL-3,3-DIMETHYLTRIAZEN (CZECH)
  see DSN600
METHOXYFLUORAN see DFA400
METHOXYFLUORANE see DFA400
METHOXYFLURANE see DFA400
8-METHOXY-(FURANO-3′.2′:6.7-COUMARIN) see XDJ000
9-METHOXY-7H-FURO(3,2-g)BENZOPYRAN-7-ONE see
  XDJ000
4-METHOXY-7H-FURO(3,2-g)(1)BENZOPYRAN-7-ONE see
  MFN275
8-METHOXY-2′,3′,6,7-FUROCOUMARIN see XDJ000
8-METHOXY-4′,5′,6,7-FUROCOUMARIN see XDJ000
6-METHOXYHARMAN see HAI500
3-METHOXY-4-HYDROXYBENZALDEHYDE see VFK000
3-METHOXY-4-HYDROXYBENZOIC ACID DIETHYLAMIDE
  see DKE200

METHOXYHYDROXYETHANE see EJH500
METHOXYHYDROXYMERCURIPROPYLSUCCINYLUREA see MFC000
2-METHOXY-10-(3-(4-HYDROXYPIPERIDINO)-2-METHYL-PROPYL)PHENOTHIAZINE see LEO000
3-METHOXY-10-(3-(4-HYDROXYPIPERIDYL)-2-METHYL-PROPYL)PHENOTHIAZINE see LEO000
METHOXYMETHYL-AETHYLNITROSAMINE (GERMAN) see MEV750
(E)-(3-(METHOXYMETHYLAMINO)-1-METHYL-3-OXO-1-PROPENYL)DIMETHYL PHOSPHATE see DOL800
2-METHOXY-5-METHYLANILINE see MGO750
4-METHOXY-2-METHYLANILINE see MGO500
4-METHOXY-2-METHYLBENZENAMINE see MGO500
2-METHOXY-5-METHYL-BENZENAMINE (9CI) see MGO750
1-METHOXY-1-METHYL-3-(3,4-DICHLOROPHENYL)UREA see DGD600
METHOXYMETHYL ETHYL NITROSAMINE see MEV750
4-METHOXYMETHYL-5-HYDROXY-6-METHYL-3-PYRI-DINEMETHANOL HYDROCHLORIDE see MEY000
2-METHOXY-10-(2-METHYL-3-(4-HYDROXYPIPERIDINO) PROPYL)PHENOTHIAZINE see LEO000
METHOXYMETHYL-METHYLNITROSAMIN (GERMAN) see MEW250
METHOXYMETHYL METHYLNITROSAMINE see MEW250
3-METHOXY-5-METHYL-4-OXO-2,5-HEXADIENOIC ACID see PAP750
N-(7-METHOXY-3-METHYL-4-OXO-2-PHENYL-4H-CHROMEN-8-YL)METHYL-N,N-DIMETHYLAMINE see DNV000
4-METHOXY-4-METHYL-2-PENTANONE see MEX250
4-METHOXY-4-METHYLPENTAN-2-ONE (DOT) see MEX250
dl-p-METHOXY-α-METHYL-PHENETHYLAMINE HYDRO-CHLORIDE see MEA500
2-METHOXY-3(5)-METHYLPYRAZINE see MEX350
7-METHOXY-1-METHYL-9H-PYRIDO(3,4-b)INDOLE see HAI500
4-METHOXYMETHYLPYRIDOXINE HYDROCHLORIDE see MEY000
4-METHOXYMETHYLPYRIDOXOL HYDROCHLORIDE see MEY000
p-METHOXY-β-METHYLSTYRENE see PMQ750
METHOXYN see DBA800
2-METHOXY-5-NITROANILINE see NEQ500
2-METHOXY-5-NITROBENZENAMINE see NEQ500
2-METHOXYNITROBENZENE see NER000
4-METHOXYNITROBENZENE see NER500
p-METHOXYNITROBENZENE see NER500
1-METHOXY-2-NITROBENZENE see NER000
1-METHOXY-4-NITROBENZENE see NER500
7-METHOXY-2-NITRONAPHTHO(2,1-b)FURAN see MFB400
8-METHOXY-6-NITROPHENANTHOL-(3,4-d)-1,3-DIOXOLE-5-CARBOXYLIC ACID see AQY250
3-METHOXY-17-α-19-NORPREGNA-K,3,5(10)-TRIEN-20-YN-17-OL see MKB750
3-METHOXY-19-NOR-17-α-PREGNA-1,3,5(10)-TRIEN-10-YN-17-OL see MKB750
(17-α)-3-METHOXY-19-NORPREGN-1,3,5(10)-TRIEN-20-YN-17-OL see MKB750
d-threo-METHOXY-3-(1-OCTENYL-ONN-AZOXY)-2-BUTA-NOL see EAG000
METHOXYOXIMERCURIPROPYLSUCCINYL UREA see MFC000
S-5-METHOXY-4-OXOPYRAN-2-YLMETHYL DIMETHYL PHOSPHOROTHIOATE see EAS000
S-((5-METHOXY-2-OXO-1,3,4-THIADIAZOL-3(2H)-YL) METHYL)-O,O-DIMETHYL PHOSPHORODITHIOATE see DSO000
2-METHOXYPHENOL see GKI000
4-METHOXYPHENOL see MFC700
o-METHOXYPHENOL see GKI000
p-METHOXYPHENOL see MFC700

1-(3-(2-METHOXYPHENOTHIAZIN-10-YL)-2-METHYLPRO-PYL)-4-PIPERIDINOL see LEO000
2-(p-METHOXYPHENOXY)-N-(2-(DIETHYLAMINO)ETHYL) ACETAMIDE see DIB600
4-METHOXYPHENYLACETIC ACID see MFE250
p-METHOXYPHENYLACETIC ACID see MFE250
o-METHOXYPHENYLAMINE see AOV900
p-METHOXYPHENYLAMINE see AOW000
N-(p-METHOXYPHENYL)-1-AZIRIDINECARBOXAMIDE see MFF250
4-p-METHOXYPHENYL-2-BUTANONE see MFF580
p-METHOXYPHENYL-N-CARBAMOYLAZIRIDINE see MFF250
1-(p-METHOXYPHENYL)-3,3-DIMETHYLTRIAZENE see DSN600
4-METHOXY-m-PHENYLENEDIAMINE see DBO000
p-METHOXY-m-PHENYLENEDIAMINE see DBO000
4-METHOXY-m-PHENYLENEDIAMINE SULFATE see DBO400
4-METHOXY-m-PHENYLENEDIAMINE SULPHATE see DBO400
p-METHOXY-m-PHENYLENEDIAMINE SULPHATE see DBO400
4-METHOXYPHENYL METHYL KETONE see MDW750
p-METHOXYPHENYL METHYL KETONE see MDW750
1-(2-(p-(α-(p-METHOXYPHENYL)-β-NITROSTYRYL) PHENOXY)ETHYL)PYRROLIDINE CITRATE (1:1) see NHP500
1-(2-(p-(α-(p-METHOXYPHENYL)-β-NITROSTYRYL) PHENOXY)ETHYL)PYRROLIDINE MONOCITRATE see NHP500
1-(p-METHOXYPHENYL)PROPENE see PMQ750
2-METHOXYPROMAZINE see MFK500
METHOXYPROMAZINE MALEATE see MFK750
1-METHOXY-2-PROPANOL see PNL250
4-METHOXYPROPENYLBENZENE see PMQ750
1-METHOXY-4-PROPENYLBENZENE see PMQ750
1-METHOXY-4-(2-PROPENYL)BENZENE see AFW750
2-METHOXY-4-PROPENYLPHENOL see IKQ000
2-METHOXY-4-PROP-2-ENYLPHENOL see EQR500
2-METHOXY-4-(2-PROPENYL)PHENOL see EQR500
2-METHOXY-4-PROPENYLPHENYL ACETATE see AAX750
3-METHOXYPROPYLAMINE see MFM000
1-METHOXY-4-PROPYLBENZENE see PNE250
5-METHOXY PSORALEN see MFN275
8-METHOXYPSORALEN see XDJ000
9-METHOXYPSORALEN see XDJ000
2-METHOXYPYRAZINE see MFN285
S-((5-METHOXY-4H-PYRON-2-YL)-METHYL)-O,O-DI-METHYL-MONOTHIOFOSFAAT (DUTCH) see EAS000
S-((5-METHOXY-4H-PYRON-2-YL)-METHYL)-O,O-DI-METHYL-MONOTHIOPHOSPHAT(GERMAN) see EAS000
S-(5-METHOXY-4-PYRON-2-YLMETHYL) DIMETHYL PHOSPHOROTHIOLATE see EAS000
5-METHOXY-3-(2-PYRROLIDINOETHYL)INDOLE see MFO250
METHOXY-5-PYRROLIDINO-2'-ETHYL-3-INDOLE see MFO250
α-(6-METHOXY-4-QUINOLYL)-5-VINYL-2-QUINUCLI-DINEMETHANOL see QFS000
α-(6-METHOXY-4-QUINOYL)-5-VINYL-2-QUINCLIDINE-METHANOL see QHJ000
4-METHOXYTOLUENE see MGP000
p-METHOXYTOLUENE see MGP000
4-METHOXY-m-TOLUIDINE see MGO750
(E,E)-11-METHOXY-3,7,11-TRIMETHYL-2,4-DODECANDIE-NOATE see KAJ000
6-METHOXYTRYPTAMINE see MFS500
6-METHOXY-α-(5-VINYL-2-QUINUCLIDINYL)-4-QUINO-LINEMETHANOL see QFS000
METHVTIOLO (ITALIAN) see MLE650
12-METHYBENZ(a)ANTHRACENE-7-METHANOL see HMF000
METHYL ABIETATE see MFT500

METHYLACETAAT (DUTCH) see MFW100
METHYLACETALDEHYDE see PMT750
METHYLACETAMIDE see MFT750
N-METHYLACETAMIDE see MFT750
4-METHYLACETANILIDE see ABJ250
p-METHYLACETANILIDE see ABJ250
4'-METHYLACETANILIDE see ABJ250
METHYLACETAT (GERMAN) see MFW100
METHYL ACETATE see MFW100
METHYL ACETIC ACID see PMU750
METHYLACETIC ANHYDRIDE see PMV500
2'-METHYLACETOACETANILIDE see ABA000
METHYLACETOACETATE see MFX250
METHYL ACETONE (DOT) see MKA400
p-METHYL ACETOPHENONE see MFW250
4'-METHYL ACETOPHENONE see MFW250
METHYLACETOPYRONONE see MFW500
METHYL-β-ACETOXYETHYL-β-CHLOROETHYLAMINE see MFW750
METHYLACETOXYMALONONITRILE see MFX000
METHYL(ACETOXYMETHYL)NITROSAMINE see AAW000
6-METHYL-17-α-ACETOXYPREGNA-4,6-DIENE-3,20-DIONE see VTF000
6-α-METHYL-17-α-ACETOXYPREGN-4-ENE-3,20-DIONE see MCA000
6-α-METHYL-17-α-ACETOXYPROGESTERONE see MCA000
METHYL ACETYLACETATE see MFX250
METHYL ACETYLACETONATE see MFX250
1-METHYL-4-ACETYLBENZENE see MFW250
METHYL ACETYLENE see MFX590
METHYL ACETYLENE-PROPADIENE MIXTURE see MFX600
METHYLACETYLENE-PROPADIENE, STABILIZED (DOT) see MFX600
2-METHYLACROLEIN see MGA250
α-METHYLACROLEIN see MGA250
β-METHYL ACROLEIN see COB250, COB260
METHYLACRYLAAT (DUTCH) see MGA500
METHYLACRYLALDEHYDE see MGA250
2-METHYLACRYLAMIDE see MDN500
METHYL-ACRYLAT (GERMAN) see MGA500
METHYL ACRYLATE see MGA500
METHYL ACRYLATE, INHIBITED (DOT) see MGA500
3-METHYLACRYLIC ACID see COB500
α-METHYLACRYLIC ACID see MDN500
β-METHYLACRYLIC ACID see COB500
α-METHYL ACRYLIC AMIDE see MDN500
METHYLACRYLONITRILE see MGA750
α-METHYLACRYLONITRILE see MGA750
N-METHYLADRENALINE see MJV000
METHYLAETHYLNITROSAMIN (GERMAN) see MKB000
METHYLAL see MGA850
METHYL ALCOHOL see MGB150
METHYL ALDEHYDE see FMV000
1-METHYL-2-ALDOXIMINOPYRIDINIUM CHLORIDE see FNZ000
1-METHYL-2-ALDOXIMINOPYRIDINIUM IODIDE see POS750
METHYLALKOHOL (GERMAN) see MGB150
2-METHYL-ALLYLCHLORID (GERMAN) see CIU750
2-METHYLALLYL CHLORIDE see CIU750
β-METHYLALLYL CHLORIDE see CIU750
METHYL ALLYL CHLORIDE (DOT) see CIU750
1-METHYL-5-ALLYL-5-(1-METHYL-2-PENTYNYL)BARBI-TURIC ACID SODIUM SALT see MDU500
METHYLALLYLNITROSAMIN (GERMAN) see MMT500
METHYLALLYLNITROSAMINE see MMT500
1-α-METHYLALLYLTHIOCARBAMOYL-2-METHYLTHIO-CARBAMOYLHYDRAZINE see MLJ500
N-((1-METHYLALLYL)THIOCARBAMOYL)-N'-(METHYL-THIOCARBAMOYL)HYDRAZINE see MLJ500
METHYLAMINE see MGC250
METHYLAMINE, anhydrous (DOT) see MGC250

METHYLAMINE, aqueous solution (DOT) see MGC250
METHYLAMINEN (DUTCH) see MGC250
4-METHYL-2-AMINOANISOLE see MGO750
METHYLAMINOANTIPYRINE SODIUM METHANESULFO-NATE see AMK500
4-(METHYLAMINO)AZOBENZENE see MNR500
N-METHYL-4-AMINOAZOBENZENE see MNR500
N-METHYL-p-AMINOAZOBENZENE see MNR500
(METHYLAMINO)BENZENE see MGN750
N-METHYLAMINOBENZENE see MGN750
1-METHYL-2-AMINOBENZENE see TGQ750
2-METHYL-1-AMINOBENZENE see TGQ750
1-METHYL-2-AMINOBENZENE HYDROCHLORIDE see TGS500
2-METHYL-1-AMINOBENZENE HYDROCHLORIDE see TGS500
METHYL 2-AMINOBENZOATE see APJ250
METHYL o-AMINOBENZOATE see APJ250
METHYLAMINOCOLCHICIDE see MGF000
4-METHYLAMINO-1,5-DIMETHYL-2-PHENYL-3-PYRAZO-LONE SODIUM METHANESULFONATE see AMK500
2-METHYLAMINOETHANOL see MGG000
N-METHYLAMINOETHANOL see MGG000
β-(METHYLAMINO)ETHANOL see MGG000
m-METHYLAMINOETHANOLPHENOL see NCL500
m-METHYLAMINOETHANOLPHENOL HYDROCHLORIDE see SPC500
(R-(R*,S*))-α-(1-(METHYLAMINO)ETHYL)BENZENE-METHANOL HYDROCHLORIDE see EAY000
(−)-α-(1-METHYLAMINOETHYL)BENZYL ALCOHOL see EAW000
1-α-(1-METHYLAMINOETHYL)BENZYL ALCOHOL see EAW000
dl-α-(1-(METHYLAMINO)ETHYL) BENZYL ALCOHOL HY-DROCHLORIDE see EAX500
1-α-(1-(METHYLAMINO)ETHYL)BENZYL ALCOHOL SUL-FATE see EAY500
α-(1-METHYLAMINOETHYL)-p-HYDROXYBENZYL ALCO-HOL see HKH500
4-(2-METHYLAMINOETHYL)PYROCATECHOL HYDRO-CHLORIDE see EAZ000
2-METHYL-6-AMINOHEPTANE see ILM000
3-METHYLAMINOISOCAMPHANE HYDROCHLORIDE see MQR500
2-METHYLAMINOISOOCTANE HYDROCHLORIDE see ODY000
2-METHYLAMINO METHYL BENZOATE see MGQ250
1-METHYLAMINOMETHYLDIBENZO(b,c)BICYCLO-(2,2,2)OCTADIENE HYDROCHLORIDE see BCH750
6-METHYLAMINO-2-METHYLHEPTENE see ILK000
METHYLAMINO-METHYLHEPTENE HYDROCHLORIDE see ODY000
METHYLAMINOPHENYLDIMETHYLPYRAZOLONE METH-ANESULFONATE SODIUM see AMK500
1-2-METHYLAMINO-1-PHENYLPROPANOL see EAW000
1-(3-METHYLAMINOPROPYL)-2-ADAMANTANOL HYDRO-CHLORIDE see MGK750
5-(3-METHYLAMINOPROPYL)-5H-DIBENZO(a,d)CYCLO-HEPTENE HYDROCHLORIDE see POF250
5-(3-(METHYLAMINO)PROPYLIDENE)DIBENZO(a,e)CY-CLOHEPTA(1,5)DIENE see NNY000
5-(3-METHYLAMINOPROPYLIDENE)-10,11-DIHYDRO-5H-DIBENZO(a,d)CYCLOHEPTENE see NNY000
METHYLAMINOPROPYLIMINODIBENZYL see DSI709
N-(γ-METHYLAMINOPROPYL)IMINODIBENZYL HYDRO-CHLORIDE see DLS600
METHYLAMINOPTERIN see MDV500
1-METHYL-3-AMINO-5H-PYRIDO(4,3-b)INDOLE see ALD500
METHYLAMPHETAMINE see DBB000
N-METHYLAMPHETAMINE see DBB000
METHYLAMPHETAMINE HYDROCHLORIDE see DBA800
METHYLAMYL ACETATE see HFJ000
METHYL AMYL ACETATE (DOT) see HFJ000

METHYLAMYL ALCOHOL see AOK750
METHYL AMYL ALCOHOL see MKW600
METHYL AMYL CARBINOL see HBE500
METHYL-AMYL-CETONE (FRENCH) see MGN500
METHYL n-AMYL KETONE see MGN500
METHYL AMYL KETONE (DOT) see MGN500
METHYLAMYLNITROSAMIN (GERMAN) see AOL000
METHYLAMYLNITROSAMINE see AOL000
METHYL-N-AMYLNITROSAMINE see AOL000
METHYLANILINE see MGN750
2-METHYLANILINE see TGQ750
3-METHYLANILINE see TGQ500
4-METHYLANILINE see TGR000
m-METHYLANILINE see TGQ500
o-METHYLANILINE see TGQ750
p-METHYLANILINE see TGR000
2-METHYLANILINE HYDROCHLORIDE see TGS500
4-METHYLANILINE HYDROCHLORIDE see TGS750
o-METHYLANILINE HYDROCHLORIDE see TGS500
N-METHYL ANILINE (MAK) see MGN750
2-METHYL-p-ANISIDINE see MGO500
5-METHYL-o-ANISIDINE see MGO750
p-METHYL ANISOLE see MGP000
9-METHYLANTHRACENE see MGP750
3-METHYL-1,8,9-ANTHRACENETRIOL see CML750
3-METHYLANTHRALIN see CML750
METHYL ANTHRANILATE (FCC) see APJ250
N-METHYLANTHRANILIC ACID, METHYL ESTER see
   MGQ250
2-METHYL-1-ANTHRAQUINONYLAMINE see AKP750
METHYL APHOXIDE see TNK250
METHYLARSENIC ACID, SODIUM SALT see MRL750
METHYLARSENIC SULFIDE see MGQ750
METHYLARSINE DICHLORIDE see DFP200
METHYLARSINE SULFIDE see MGQ750
METHYLARSINIC SULFIDE see MGQ750
METHYLARSINIC SULPHIDE see MGQ750
METHYLARSONOUS DICHLORIDE see DFP200
METHYLARTERENOL see VGP000
METHYL ASPARTYLPHENYLALANATE see ARN825
1-METHYL N-l-α-ASPARTYL-l-PHENYLALANINE see
   ARN825
2-METHYLAZACYCLOPROPANE see PNL400
METHYLAZINPHOS see ASH500
2-METHYLAZIRIDINE see PNL400
METHYLAZOXYMETHANOL see HMG000
METHYLAZOXYMETHANOL ACETATE see MGS750
METHYLAZOXYMETHANOL GLUCOSIDE see COU000
METHYLAZOXYMETHANOL-β-d-GLUCOSIDE see COU000
METHYL AZOXYMETHYL ACETATE see MGS750
METHYLBEN see HJL500
10-METHYL-1,2-BENZANTHRACEN (GERMAN) see
   MGW750
4-METHYLBENZ(a)ANTHRACENE see MGV500
5-METHYLBENZ(a)ANTHRACENE see MGV750
6-METHYLBENZ(a)ANTHRACENE see MGW000
7-METHYLBENZ(a)ANTHRACENE see MGW750
8-METHYLBENZ(a)ANTHRACENE see MGW250
9-METHYLBENZ(a)ANTHRACENE see MGW500
10-METHYLBENZ(a)ANTHRACENE see MGX000
12-METHYLBENZ(a)ANTHRACENE see MGX500
3-METHYL-1,2-BENZANTHRACENE see MGV750
4-METHYL-1,2-BENZANTHRACENE see MGW000
5-METHYL-1,2-BENZANTHRACENE see MGW250
6-METHYL-1,2-BENZANTHRACENE see MGW500
7-METHYL-1,2-BENZANTHRACENE see MGX000
9-METHYL-1,2-BENZANTHRACENE see MGX500
10-METHYL-1,2-BENZANTHRACENE see MGW750
4'-METHYL-1:2-BENZANTHRACENE see MGV500
7-METHYLBENZ(a)ANTHRACENE-12-METHANOL see
   HMF500
7-METHYLBENZ(a)ANTHRACENE-5,6-OXIDE see MGZ000
N-METHYLBENZAZIMIDE, DIMETHYLDITHIOPHOS-
   PHORIC ACID ESTER see ASH500

METHYLBENZEDRIN see DBA800
2-METHYLBENZENAMINE see TGQ750
3-METHYLBENZENAMINE see TGQ500
4-METHYLBENZENAMINE see TGR000
m-METHYLBENZENAMINE see TGQ500
N-METHYLBENZENAMINE see MGN750
o-METHYLBENZENAMINE see TGQ750
p-METHYLBENZENAMINE see TGR000
2-METHYLBENZENAMINE HYDROCHLORIDE see TGS500
4-METHYLBENZENAMINE HYDROCHLORIDE see TGS750
o-METHYLBENZENAMINE HYDROCHLORIDE see TGS500
5-METHYL-1,3-BENZENDIOL see MPH500
METHYLBENZENE see TGK750
METHYL BENZENEACETATE see MHA500
METHYL BENZENECARBOXYLATE see MHA750
ar-METHYLBENZENEDIAMINE see TGL500
2-METHYL-1,4-BENZENEDIAMINE see TGM000
4-METHYL-1,3-BENZENEDIAMINE see TGL750
2-METHYL-1,4-BENZENEDIAMINE DIHYDROCHLORIDE
   see DCE200
2-METHYL-1,4-BENZENEDIAMINE SULFATE see DCE600
3-METHYL-1,2-BENZENEDIOL see DNE000
(S)-α-METHYL-BENZENEETHANAMINE SULFATE (2:1) see
   BBK500
α-METHYLBENZENEETHANEAMINE see BBK000
2-METHYLBENZENESULFONAMIDE see TGN250
o-METHYLBENZENESULFONAMIDE see TGN250
METHYL-2-BENZIMIDAZOLE see MHC250
2-METHYLBENZIMIDAZOLE see MHC250
METHYL-2-BENZIMIDAZOLE CARBAMATE and SODIUM
   NITRITE see CBN375
METHYL BENZOATE (FCC) see MHA750
4-METHYLBENZOIC ACID METHYL ESTER see MNR250
METHYLBENZOL see TGK750
4-METHYL-p-BENZOPHENONE see MHF750
6-METHYL-2H-1-BENZOPYRAN-2-ONE see MIP750
7-METHYLBENZO(a)PYRENE see MHH000
4'-METHYLBENZO(a)PYRENE see MHH000
6-METHYLBENZOPYRONE see MIP750
6-METHYL-1,2-BENZOPYRONE see MIP750
METHYL-p-BENZOQUINONE see MHI250
METHYL-1,4-BENZOQUINONE see MHI250
2-METHYL-p-BENZOQUINONE see MHI250
2-METHYLBENZOQUINONE-1,4 see MHI250
2-((2-METHYLBENZO(b)THIEN-3-YL)METHYL)-2-IMIDAZO-
   LINE HYDROCHLORIDE see MHJ500
METHYL-5-BENZOYL BENZIMIDAZOLE-2-CARBAMATE
   see MHL000
4'-METHYL-3:4-BENZPYRENE see MHH000
α-METHYLBENZYL ALCOHOL (FCC) see PDE000
1-METHYL-2-BENZYLHYDRAZINE see MHN750
N-METHYL-N-BENZYLNITROSAMINE see MHP250
METHYL-BENZYL-NITROSOAMIN (GERMAN) see
   MHP250
METHYLBIS(3-AMINOPROPYL)AMINE see BGU750
N-METHYL-BIS-CHLORAETHYLAMIN (GERMAN) see
   BIE250
METHYL-BIS-(β-CHLORAETHYL)-AMIN-N-OXYD-HYDRO-
   CHLORID (GERMAN) see CFA750
N-METHYL-BIS-β-CHLORETHYLAMINE HYDROCHLORIDE
   see BIE500
METHYLBIS(β-CHLOROETHYL)AMINE see BIE250
N-METHYL-BIS(β-CHLOROETHYL)AMINE see BIE250
METHYLBIS(2-CHLOROETHYL)AMINE HYDROCHLORIDE
   see BIE500
METHYLBIS(β-CHLOROETHYL)AMINE HYDROCHLORIDE
   see BIE500
N-METHYLBIS(2-CHLOROETHYL)AMINE HYDROCHLO-
   RIDE see BIE500
N-METHYL-BIS(2-CHLOROETHYL)AMINE (MAK) see
   BIE250
METHYL-BIS(β-CHLOROETHYL)AMINE OXIDE see
   CFA500

METHYLBIS(β-CHLOROETHYL)AMINE N-OXIDE see CFA500

METHYLBIS(β-CHLOROETHYL)AMINE-N-OXIDE HYDRO-CHLORIDE see CFA750

N-METHYLBIS(2-CHLOROETHYL)AMINE-N-OXIDE HY-DROCHLORIDE see CFA750

2-METHYL-4,5-BIS(HYDROXYMETHYL)-3-HYDROXYPYRI-DINE see PPK250

7-METHYL-2,3:9,10-BIS(METHYLENEDIOXY)-7,13a-SECO-BERBIN-13a-ONE see FOW000

β-METHYLBIVINYL see IMS000

METHYL BORATE see TLN000

METHYLBROMID (GERMAN) see MHR200

METHYL BROMIDE see MHR200

METHYL BROMOACETATE see MHR250

METHYL α-BROMOACETATE see MHR250

O-METHYL-O-(4-BROMO-2,5-DICHLOROPHENYL)PHENYL THIOPHOSPHONATE see LEN000

2-METHYLBUTADIENE see IMS000

2-METHYL-1,3-BUTADIENE (DOT) see IMS000

2-METHYLBUTANE see EIK000

METHYL n-BUTANOATE see MHY000

3-METHYLBUTANOIC ACID see ISU000

3-METHYLBUTANOIC ACID, BUTYL ESTER see ISX000

3-METHYLBUTANOIC ACID, ETHYL ESTER see ISY000

3-METHYLBUTANOIC ACID, METHYL ESTER see ITC000

3-METHYLBUTANOIC ACID, PHENYLETHYL ESTER see ISW000

3-METHYL-BUTANOIC ACID 2-PHENYLETHYL ESTER see PDF775

3-METHYLBUTANOIC ACID, 2-PROPENYL ESTER see ISV000

2-METHYLBUTANOL see MHS750

3-METHYL BUTANOL see IHP000

2-METHYL BUTANOL-1 see MHS750

2-METHYL BUTANOL-2 see PBV000

2-METHYL-2-BUTANOL see PBV000

2-METHYL-4-BUTANOL see IHP000

3-METHYLBUTAN-1-OL see IHP000

3-METHYL-1-BUTANOL (CZECH) see IHP000

3-METHYLBUTAN-3-OL see PBV000

3-METHYLBUTANOL NITRITE see IMB000

3-METHYL-2-BUTANONE see MLA750

3-METHYL BUTAN 2-ONE (DOT) see MLA750

2-METHYL-1-BUTENE see MHT000

2-METHYL-2-BUTENE see AOI750

3-METHYL-1-BUTENE see MHT250

3-METHYL-3-BUTEN-2-ON (GERMAN) see MKY500

2-METHYL-1-BUTEN-3-ONE see MKY500

3-(3-METHYL-2-BUTENYL)-1,2,3,4,5,6-HEXAHYDRO-6,11-DIMETHYL-2,6-METHANO-3-BENZAZOCIN-8-OL see DOQ400

3′-METHYLBUTTERGELB (GERMAN) see DUH600

1-METHYLBUTYL ACETATE see AOD735

3-METHYLBUTYL ACETATE see IHO850

3-METHYL-1-BUTYL ACETATE see IHO850

2-METHYL-BUTYLACRYLAAT (DUTCH) see MHU750

2-METHYL-BUTYLACRYLAT (GERMAN) see MHU750

2-METHYL BUTYLACRYLATE see MHU750

METHYLBUTYLAMINE see MHV000

N-(METHYL) BUTYL AMINE see MHV000

N-METHYL-n-BUTYLAMINE see MHV000

p-METHYL-tert-BUTYLBENZENE see BSP500

1-METHYL-4-tert-BUTYLBENZENE see BSP500

METHYL-5-BUTYL-2-BENZIMIDAZOLECARBAMATE see BQK000

1-(3-METHYL)BUTYL BENZOATE see IHP100

3-METHYLBUTYL BROMIDE see BNP250

METHYL-1-(BUTYLCARBAMOYL)-2-BENZIMIDAZOLYL-CARBAMATE see BAV575

METHYL-1,3-BUTYLENE GLYCOL ACETATE see MHV750

3-METHYLBUTYL ETHANOATE see IHO850

METHYL tert-BUTYL ETHER see MHV859

3-METHYLBUTYL FORMATE see IHS000

3-METHYLBUTYL 2-HYDROXYBENZOATE see IME000

METHYL n-BUTYL KETONE (ACGIH) see HEV000

N-3-METHYLBUTYL-N-1-METHYL ACETONYLNITROSA-MINE see MHW350

3-METHYLBUTYL NITRITE see IMB000

METHYL-BUTYL-NITROSAMIN (GERMAN) see MHW500

METHYLBUTYLNITROSAMINE see MHW500

METHYL-N-BUTYLNITROSAMINE see MHW500

5-(1-METHYLBUTYL)-5-(2-PROPENYL)-2,4,6(1H,3H,5H)-PY-RIMIDINITRIONE see SBM500

METHYL BUTYRATE see MHY000

METHYL-n-BUTYRATE see MHY000

3-METHYLBUTYRIC ACID see ISU000

β-METHYLBUTYRIC ACID see ISU000

3-METHYLBUTYRIC ACID, ALLYL ESTER see ISV000

(E)-3-METHYLBUTYRIC ACID-3,7-DIMETHYL-2,6-OCTA-DIENYL ESTER see GDK000

3-METHYLBUTYRIC ACID, ETHYL ESTER see ISY000

4-METHYL-γ-BUTYROLACTONE see VAV000

γ-METHYL-γ-BUTYROLACTONE see VAV000

8-(3-METHYLBUTYRYLOXY)-DIACETOXYSCIRPENOL see FQS000

METHYL CADMIUM AZIDE see MHY550

METHYL-CALMINAL see ENB500

METHYL CARBAMATE see MHZ000

N-METHYLCARBAMATE de 4-DIMETHYLAMINO-3-METHYL PHENYLE (FRENCH) see DOR400

METHYLCARBAMATE-1-NAPHTHALENOL see CBM750

METHYLCARBAMATE-1-NAPHTHOL see CBM750

N-METHYLCARBAMATE de 1-NAPHTYLE (FRENCH) see CBM750

METHYLCARBAMIC ACID-3-sec-BUTYL-6-CHLOROPHE-NYL ESTER see MOU750

METHYLCARBAMIC ACID-2-CHLORO-5-(1-METHYLPRO-PYL)PHENYL ESTER see MOU750

METHYLCARBAMIC ACID-o-CUMENYL ESTER see MIA250

N-METHYLCARBAMIC ACID-3-DIETHYLAMINOPHENYL ESTER, METHIODIDE see MID250

N-METHYLCARBAMIC ACID-3-(DIETHYLMETHYLAM-MONIO)PHENYL ESTER, IODIDE see MID250

METHYL CARBAMIC ACID 2,3-DIHYDRO-2,2-DIMETHYL-7-BENZOFURANYL ESTER see CBS275

METHYL-CARBAMIC ACID, ESTER with ESEROLINE see PIA500

METHYLCARBAMIC ACID, ETHYL ESTER see EMQ500

METHYLCARBAMIC ACID-2,3-(ISOPROPYLIDENEDIOXY) PHENYL ESTER see DQM600

METHYLCARBAMIC ACID-m-(1-METHYL)BUTYL)PHENYL ESTER mixed with CARBAMIC ACID, METHYL-m-(1-ETHYLPROPYL)PHENYL ESTER (3:1) see BTA250

METHYL CARBAMIC ACID-4-(METHYLTHIO)-3,5-XYLYL ESTER see DST000

METHYLCARBAMIC ACID-1-NAPHTHYL ESTER see CBM750

METHYLCARBAMIC ACID-m-TOLYL ESTER see MIB750

METHYLCARBAMIC ACID, (m-(TRIMETHYLAMMONIO) PHENYL)ESTER, IODIDE see HNO500

METHYLCARBAMIC ACID-5-(TRIMETHYLAMMONIO)-o-TOLYL ESTER, CHLORIDE see HOL000

METHYLCARBAMIC ESTER of OXYPHENYLMETHYL-DIETHYLAMMONIUM IODIDE EZsee MID250

S-METHYLCARBAMOYLMETHYL-O,O-DIMETHYL PHOS-PHORODITHIOATE see DSP400

(3-(N-METHYLCARBAMOYLOXY)PHENYL)DIETHYL-METHYL-AMMONIUM IODIDE see MID250

(3-(METHYLCARBAMOYLOXY)PHENYL)TRIMETHYL-AMMONIUM IODIDE see HNO500

1-METHYL-4-CARBETHOXY-4-PHENYLPIPERIDINE HY-DROCHLORIDE see DAM700

METHYLCARBINOL see EFU000

METHYL-4-CARBOMETHOXY BENZOATE see DUE000

METHYL CARBONATE see MIF000

METHYLCARBONYL FLUORIDE see ACM000

METHYLDI(2-CHLOROETHYL)AMINE HYDROCHLORIDE
see BIE500
METHYLDI(β-CHLOROETHYL)AMINE HYDROCHLORIDE
see BIE500
N-METHYL-DI-2-CHLOROETHYLAMINE HYDROCHLORIDE
see BIE500
N-METHYL-DI-2-CHLOROETHYLAMINE-N-OXIDE see
CFA500
METHYLDI(2-CHLOROETHYL)AMINE-N-OXIDE HYDRO-
CHLORIDE see CFA750
N-METHYLDICHLOROMALEINIMIDE see DFP800
METHYL DICHLOROSILANE (DOT) see DFS000
METHYL-DICHLORSILAN (CZECH) see DFS000
5-METHYL-7-DIETHYLAMINO-s-TRIAZOLO-(1,5-a)PYRIMI-
DINE see DIO200
3-METHYL-N,N-DIETHYLBENZAMIDE see DKC800
1-METHYL-4-DIETHYLCARBAMOYLPIPERAZINE CITRATE
see DIW200
11-METHYL-15,16-DIHYDRO-17H-CYCLOPENTA(a)PHE-
NANTHREN-17-ONE see MJE500
11-METHYL-15,16-DIHYDRO-17-OXOCYCLOPENTA(a)PHE-
NANTHRENE see MJE500
METHYLDIHYDROPYRAN see MJE750
l-α-METHYL-3,4-DIHYDROXYPHENYLALANINE see
DNA800
α-METHYL-l-3,4-DIHYDROXYPHENYLALANINE see
DNA800
l-(−)-α-METHYL-β-(3,4-DIHYDROXYPHENYL)ALANINE see
DNA800
α-METHYL-β-(3,4-DIHYDROXYPHENYL)-l-ALANINE see
DNA800
METHYL-(β-(3,4-DIHYDROXY PHENYL ETHYL) AMINE
HYDROCHLORIDE see EAZ000
METHYL-3-(DIMETHOXYPHOSPHINYLOXY)CROTONATE
see MQR750
5-METHYL-10-β-DIMETHYLAMINOAETHYL-10,11-DIHY-
DRO-11-OXO-5-DIBENZO(b,e)(1,4)DIAZEPIN see
DCW800
3'-METHYL-4-DIMETHYLAMINOAZOBENZEN (CZECH) see
DUH600
3'-METHYL-4-DIMETHYLAMINOAZOBENZENE see
DUH600
3'-METHYL-N,N-DIMETHYL-4-AMINOAZOBENZENE see
DUH600
3'-METHYLDIMETHYLAMINOAZOBENZOL (GERMAN) see
DUH600
3-METHYL-4-DIMETHYLAMINO-2,2-DIPHENYLBUTYRAM-
IDE see DOY400
METHYL-2-(DIMETHYLAMINO)-N-(((METHYLAMINO)CAR-
BONYL)OXY)-2-OXOETHANIMIDOTHIOATE see
DSP600
4-METHYL-3-DIMETHYLAMINOPHENYL ESTER-N-
METHYLCARBAMIC ACID see DQE800
METHYL-4-DIMETHYLAMINO-3,5-XYLYL CARBAMATE
see DOS000
METHYL-4-DIMETHYLAMINO-3,5-XYLYL ESTER of CAR-
BAMIC ACID see DOS000
METHYL-1-(DIMETHYLCARBAMOYL)-N-(METHYL-
CARBAMOYLOXY)THIOFORMIMIDATE see DSP600
S-METHYL-1-(DIMETHYLCARBAMOYL)-N-((METHYL-
CARBAMOYL)OXY)THIOFORMIMIDATE see DSP600
METHYL 1,1-DIMETHYLETHYL ETHER see MHV859
METHYL-N',N'-DIMETHYL-N-((METHYLCARBAMOYL)
OXY)-1-THIOOXAMIMIDATE see DSP600
N-METHYL-O,O-DIMETHYLTHIOLOPHOSPHORYL-5-THIA-
3-METHYL-2-VALERAMIDE see MJG500
2-METHYL-3,5-DINITROBENZAMIDE see DUP300
METHYLDINITROBENZENE see DVG600
1-METHYL-2,4-DINITROBENZENE see DVH000
2-METHYLDINITROPHENOL see DUS800
2-METHYL-4,6-DINITROPHENOL see DUS700
2-METHYL-4,6-DINITROPHENOL SODIUM SALT see
DUU600

N-METHYL-N,N-DIOCTYL-1-OCTANAMINIUM CHLORIDE
see MQH000
((2-METHYL-1,3-DIOXALAN-4-YL)METHYL)TRIMETHYL-
AMMONIUM IODIDE see MJH250
6-METHYL-1,11-DIOXY-2-NAPHTHACENECARBOXAMIDE
see TBX000
6-METHYL DIPYRIDO(1,2-a:3',2'-d)IMIDAZOL-2-AMINE see
AKS250
(4-METHYL-1,3-DITHIOLAN-2-YLIDENE)PHOSPHOR-
AMIDIC ACID, DIETHYL ESTER see DHH400
METHYL DODECYL BENZYL AMMONIUM CHLORIDE see
DYA600
METHYLDOPA see DNA800
l-α-METHYLDOPA see DNA800
α-METHYL-l-DOPA see DNA800
N-METHYLDOPAMINE HYDROCHLORIDE see EAZ000
METHYL DURSBAN see CMA250
METHYL-E 605 see MNH000
METHYLE (ACETATE de) (FRENCH) see MFW100
O,O-METHYLEEN-BIS(4-CHLOORFENOL) (DUTCH) see
MJM500
METHYLEEN-S,S'-BIS(O,O-DIETHYL-DITHIOFOSFAAT)
(DUTCH) see EEH600
3,3'-METHYLEEN-BIS(4-HYDROXY-CUMARINE) (DUTCH)
see BJZ000
METHYLE (FORMIATE de) (FRENCH) see MKG750
S,S'-METHYLEN-BIS(O,O-DIAETHYL-DITHIOPHOSPHAT)
(GERMAN) see EEH600
3,3'-METHYLEN-BIS(4-HYDROXY-CUMARIN) (GERMAN)
see BJZ000
3,4-METHYLENDIOXY-6-PROPYLBENZYL-n-BUTYL-DIA-
ETHYLENGLYKOLAETHER (GERMAN) see PIX250
METHYLENDIRHODANID (CZECH, GERMAN) see MJT500
METHYLENE ACETONE see BOY500
METHYLENE BICHLORIDE see MJP450
METHYLENEBIS(ANILINE) see MJQ000
2,4'-METHYLENEBIS(ANILINE) see MJP750
4,4'-METHYLENEBISANILINE see MJQ000
4,4'-METHYLENE(BIS)-CHLOROANILINE see MJM200
4,4'-METHYLENE BIS(2-CHLOROANILINE) see MJM200
METHYLENE-4,4'-BIS(o-CHLOROANILINE) see MJM200
4,4'-METHYLENEBIS(o-CHLOROANILINE) see MJM200
p,p'-METHYLENEBIS(o-CHLOROANILINE) see MJM200
p,p'-METHYLENEBIS(α-CHLOROANILINE) see MJM200
4,4'-METHYLENEBIS-2-CHLOROBENZENAMINE see
MJM200
2,2'-METHYLENEBIS(4-CHLOROPHENOL) see MJM500
METHYLENE BIS(4-CYCLOHEXYLISOCYANATE see
MJM600
METHYLENE-S,S'-BIS(O,O-DIAETHYL-DITHIOPHOSPHAT)
(GERMAN) see EEH600
4,4'-METHYLENE BIS(N,N'-DIMETHYLANILINE) see
MJN000
4,4'-METHYLENEBIS(N,N-DIMETHYL)BENZENAMINE see
MJN000
3,3'-METHYLENEBIS(4-HYDROXY-1,2-BENZOPYRONE) see
BJZ000
3,3'-METHYLENEBIS(4-HYDROXYCOUMARIN) see
BJZ000
3,3'-METHYLENE-BIS(4-HYDROXYCOUMARINE)
(FRENCH) see BJZ000
METHYLENEBIS(4-ISOCYANATOBENZENE) see MJP400
1,1-METHYLENEBIS(4-ISOCYANATOBENZENE) see
MJP400
5,5'-METHYLENEBIS(2-ISOCYANATO)TOLUENE see
MJN750
METHYLENE BIS(METHANESULFONATE) see MJQ500
4,4'-METHYLENEBIS(2-METHYLANILINE) see MJO250
4,4'-METHYLENEBIS(2-METHYLBENZENAMINE) see
MJO250
METHYLENE-BIS-ORTHOCHLOROANILINE see MJM200
1,1'-(METHYLENEBIS(OXY)BIS(2-CHLOROETHANE) see
BID750

METHYLENEBIS(4-PHENYLENE ISOCYANATE) see MJP400
METHYLENEBIS(p-PHENYLENE ISOCYANATE) see MJP400
METHYLENE BISPHENYL ISOCYANATE see MJP400
METHYLENEBIS(4-PHENYL ISOCYANATE) see MJP400
METHYLENEBIS(p-PHENYL ISOCYANATE) see MJP400
4,4'-METHYLENEBIS(PHENYL ISOCYANATE) see MJP400
p,p'-METHYLENEBIS(PHENYL ISOCYANATE) see MJP400
2,2'-METHYLENEBIS(3,4,6-TRICHLOROPHENOL) see HCL000
METHYLENE BLUE see BJI250
METHYLENE BLUE (medicinal) see BJI250
METHYLENE BLUE A see BJI250
METHYLENE BLUE BB see BJI250
METHYLENE BLUE BB ZINC FREE see BJI250
METHYLENE BLUE CHLORIDE see BJI250
METHYLENE BLUE CHLORIDE (BIOLOGICAL STAIN) see BJI250
METHYLENE BLUE D see BJI250
METHYLENE BLUE I (medicinal) see BJI250
METHYLENE BLUE NF (medicinal) see BJI250
METHYLENE BLUE POLYCHROME see BJI250
METHYLENE BLUE USP (MEDICINAL) see BJI250
METHYLENE BLUE USP XII (MEDICINAL) see BJI250
METHYLENE BROMIDE see DDP800
(4-(2-METHYLENEBUTYRYL)-2,3-DICHLOROPHENOXY)-ACETIC ACID see DFP600
METHYLENEBUTYRYL PHENOXYACETIC ACID see DFP600
METHYLENE CHLORIDE see MJP450
METHYLENE CHLOROBROMIDE see CES650
METHYLENE CYANIDE see MAO250
METHYLENEDIANILINE see MJQ000
2,4'-METHYLENEDIANILINE see MJP750
4,4'-METHYLENEDIANILINE see MJQ000
p,p'-METHYLENEDIANILINE see MJQ000
4,4'-METHYLENEDIANILINE DIHYDROCHLORIDE see MJQ100
METHYLENE DIBROMIDE see DDP800
METHYLENE DICHLORIDE see MJP450
3,4-METHYLENE-DIHYDROXYBENZALDEHYDE see PIW250
METHYLENE DIMETHANESULFONATE see MJQ500
METHYLENE DIMETHYL ETHER see MGA850
4,4'-METHYLENEDIMORPHOLINE see MJQ750
METHYLENEDINAPHTHALENESULFONIC ACID BISPHE-NYLMERCURI SALT see PFN000
3,4-METHYLENEDIOXY-ALLYBENZENE see SAD000
1,2-METHYLENEDIOXY-4-ALLYLBENZENE see SAD000
3,4-METHYLENEDIOXYBENZALDEHYDE see PIW250
3,4-METHYLENEDIOXYBENZYL ACETATE see PIX000
3,4-METHYLENEDIOXY-α-ETHYL-β-PHENYLETHYLAMINE see MJR750
1,2-METHYLENEDIOXY-4-(1-HYDROXYALLYL)BENZENE see BCJ000
3,4-METHYLENEDIOXY-α-METHYL-β-PHENYLETHYL-AMINE HYDROCHLORIDE see MJS750
1,2-(METHYLENEDIOXY)-4-(2-(OCTYLSULFINYL)PROPYL)BENZENE see ISA000
1-(3,4-METHYLENEDIOXYPHENYL)-2-AMINOPROPANE see MJS750
3,4-METHYLENEDIOXY-β-PHENYLETHYLAMINE HYDRO-CHLORIDE see MJT000
1,2-METHYLENEDIOXY-4-PROPENYLBENZENE see IRZ000
3,4-METHYLENEDIOXY-1-PROPENYL BENZENE see IRZ000
1,2-(METHYLENEDIOXY)-4-PROPYLBENZENE see DMD600
(3,4-METHYLENEDIOXY-6-PROPYLBENZYL) (BUTYL) DI-ETHYLENE GLICOL ETHER see PIX250
3,4-METHYLENEDIOXY-6-PROPYLBENZYL n-BUTYL DI-ETHYLENEGLYCOL ETHER see PIX250

4,4'-METHYLENEDIPHENYL DIISOCYANATE see MJP400
METHYLENEDI-p-PHENYLENE DIISOCYANATE see MJP400
METHYLENEDI-p-PHENYLENE ISOCYANATE see MJP400
4,4'-METHYLENEDIPHENYLENE ISOCYANATE see MJP400
METHYLENE DI(PHENYLENE ISOCYANATE) (DOT) see MJP400
4,4'-METHYLENEDIPHENYL ISOCYANATE see MJP400
METHYLENE DITHIOCYANATE see MJT500
4,4'-METHYLENE DI-o-TOLUIDINE see MJO250
METHYLENE GLYCOL see FMV000
METHYLENE GREEN see MJU250
3-METHYLENE-7-METHYL-1,6-OCTADIENE see MRZ150
3-METHYLENE-7-METHYL-1-OCTEN-7-YL ACETATE see AAW500
4-METHYLENE-2-OXETANONE see KFA000
METHYLENE OXIDE see FMV000
2-METHYLENE-3-OXO-CYCLOPENTANECARBOXYLIC ACID see SAX500
S,S'-METHYLENE O,O,O',O'-TETRAETHYL PHOSPHORO-DITHIOATE see EEH600
8-METHYLENE-4,11,11-(TRIMETHYL)BICYCLO(7.2.0) UNDEC-4-ENE see CCN000
METHYLENIUM CERULEUM see BJI250
METHYLENO-BIS-FENYLOIZOCYJANIAN (POLISH) see DNJ800
METHYLEPHEDRIN (GERMAN) see MJU750
METHYLEPHEDRINE see MJU750
N-METHYLEPINEPHRINE see MJV000
METHYL-18-EPIRESERPATE METHYL ETHER HYDRO-CHLORIDE see MQR200
METHYLERGOBASINE see PAM000
METHYLERGOBREVIN see PAM000
METHYLERGOMETRIN see PAM000
METHYLERGOMETRINE see PAM000
METHYLERGONOVIN see PAM000
METHYLERGONOVINE see PAM000
METHYL ESTER FLUOROSULFURIC ACID see MKG250
METHYL ESTER of p-HYDROXYBENZOIC ACID see HJL500
METHYLESTER KISELINY OCTOVE (CZECH) see MFW100
METHYLESTER KYSELINY ANTHRANILOVE see APJ250
METHYLESTER KYSELINY 4-FLUORMASELNE see MKE000
METHYLESTER KYSELINY ORTHOMRAVENCI (CZECH) see TLX600
METHYLESTER KYSELINY p-TOLUENSULFONOVE (CZECH) see MLL250
METHYL ESTER of METHANESULFONIC ACID see MLH500
METHYL ESTER of METHANESULPHONIC ACID see MLH500
METHYL ESTER of o-SILICIC ACID see MPI750
METHYL ESTER STEARIC ACID see MJW000
METHYL ESTER of WOOD ROSIN see MFT500
METHYL ESTER of WOOD ROSIN, partially hydrogenated (FCC) see MFT500
METHYLE (SULFATE de) (FRENCH) see DUD100
1,1'-(METHYLETHANEDILIDENEDINITRILO)BIGUANIDINE DIHYDROCHLORIDE DIHYDRATE see MKI000
4,4'-(1-METHYL-1,2-ETHANEDIYL)BIS-2,6-PIPERAZINE-DIONE see PIK250
1-METHYLETHANETHIOL see IMU000
N-METHYLETHANOANTHRACENE-9-(10H)-METHYL-AMINE HYDROCHLORIDE see BCH750
METHYL ETHANOATE see MFW100
N-METHYLETHANOLAMINE see MGG000
METHYLETHENE see PMO500
METHYL ETHER see MJW500
METHYL ETHOXOL see EJH500
2-(1-METHYLETHOXY)PHENOL METHYLCARBAMATE see PMY300

1-METHYLETHYLAMINE see INK000
4-(METHYLETHYL)AMINOAZOBENZENE see ENB000
N-METHYL-N-ETHYL-p-AMINOAZOBENZENE see ENB000
α-(((1-METHYLETHYL)AMINO)METHYL)-2-NAPHTHA-
    LENEMETHANOL, HYDROCHLORIDE see INT000
4-(1-METHYLETHYL)-BENZALDEHYDE (9CI) see COE500
2-(1-METHYLETHYL)BENZENAMINE see INX000
3-(1-METHYLETHYL)-1H-2,1,3-BENZOTHIAZAIN-4(3H)-
    ONE-2,2-DIOXIDE see MJY500
METHYLETHYLBROMOMETHANE see BMX750
(1-METHYLETHYL)CARBAMIC ACID 2-(((AMINOCARBO-
    NYL)OXY)METHYL)-2-METHYLPENTYL ESTER see
    IPU000
METHYLETHYLCARBINOL see BPW750
4-METHYL-4-ETHYL-2,6-DIOXOPIPERIDINE see MKA250
METHYLETHYLENE see PMO500
METHYLETHYLENE GLYCOL see PML000
METHYL ETHYLENE OXIDE see PNL600
METHYLETHYLENIMINE see PNL400
2-METHYLETHYLENIMINE see PNL400
1-METHYL ETHYL ESTER NITROUS ACID (9CI) see
    IQQ000
METHYL ETHYL ETHER (DOT) see EMT000
1-METHYLETHYL-2-((ETHOXY((1-METHYLETHYL)
    AMINO)PHOSPHINOTHIOYL)OXY)BENZOATE see
    IMF300
(E)-1-METHYLETHYL-3-(((ETHYLAMINO)METHOXY-
    PHOSPHINOTHIOYL)OXY-2-BUTENOATE see MKA000
1-(METHYLETHYL)-ETHYL 3-METHYL-4-(METHYLTHIO)
    PHENYL PHOSPHORAMIDATE see FAK000
3-METHYL-3-ETHYLGLUTARIMIDE see MKA250
β-METHYL-β-ETHYLGLUTARIMIDE see MKA250
2,2′-((1-METHYLETHYLIDENE)BIS(4,1-PHENYLENE-
    OXYMETHYLENE))BISOXIRANE see BLD750
METHYL ETHYL KETONE see MKA400
METHYL ETHYL KETONE PEROXIDE see MKA500
METHYL ETHYL KETONE SEMICARBAZONE see
    MKA750
METHYLETHYLKETONHYDROPEROXIDE see MKA500
METHYLETHYLMETHANE see BOR500
N-(1-METHYLETHYL)-4-((2-METHYLHYDRAZINO)
    METHYL)BENZAMIDE MONOHYDROCHLORIDE see
    PME500
N-(1-METHYLETHYL)-N′-(1-NITRO-9-ACRIDINYL)-1,3-PRO-
    PANEDIAMINE DIHYDROCHLORIDE see INW000
METHYLETHYLNITROSAMINE see MKB000
N,N-METHYLETHYLNITROSAMINE see MKB000
METHYLETHYLOLAMINE see MGG000
4-(1-METHYLETHYL)PHENOL see IQZ000
1-METHYL-5-ETHYL-5-PHENYLBARBITURIC ACID see
    ENB500
3-METHYL-5-ETHYL-5-PHENYLHYDANTOIN see MKB250
2-(1-METHYLETHYL)PHENYL METHYLCARBAMATE see
    MIA250
(1-METHYLETHYL)PHOSPHORAMIDOTHIOIC ACID O-(2,4-
    DICHLOROPHENYL)-O-METHYL ESTER see DGD800
N-(1-METHYLETHYL)-2-PROPANAMINE see DNM200
2-METHYL-5-ETHYLPYRIDINE see EOS000
6-METHYL-3-ETHYLPYRIDINE see EOS000
METHYL ETHYL PYRIDINE (DOT) see EOS000
2-METHYL-5-ETHYLPYRIDINE (DOT) see EOS000
3-METHYL-3-ETHYLPYRROLIDINE-2,5-DIONE see
    ENG500
γ-METHYL-γ-ETHYL-SUCCINIMIDE see ENG500
3-METHYLETHYNYLESTRADIOL see MKB750
18-METHYL-17-α-ETHYNYL-19-NORTESTOSTERONE see
    NNQ500
3-METHYLETHYNYLOESTRADIOL see MKB750
METHYL EUGENOL (FCC) see AGE250
METHYL FLUORIDE (DOT) see FJK000
METHYL FLUOROACETATE see MKD000
METHYL-4-FLUOROBUTYRATE see MKE000
METHYL-γ-FLUOROBUTYRATE see MKE000
METHYL-φ-FLUOROBUTYRATE see MKE000

METHYL-γ-FLUOROCROTONATE see MKE250
16-α-METHYL-9-α-FLUORO-1-DEHYDROCORTISOL see
    SOW000
16-α-METHYL-9-α-FLUORO-Δ¹-HYDROCORTISONE see
    SOW000
METHYL-γ-FLUORO-β-HYDROXYBUTYRATE see
    MKE750
METHYL-γ-FLUORO-β-HYDROXYTHIOLBUTYRATE see
    MKF000
METHYLFLUOROPHOSPHORIC ACID, ISOPROPYL ESTER
    see IPX000
16-α-METHYL-9-α-FLUOROPREDNISOLONE see SOW000
16-α-METHYL-9-α-FLUORO-1,4-PREGNADIENE-11-β,17-
    α,21-TRIOL-3,20-DIONE see SOW000
METHYL FLUOROSULFATE see MKG250
METHYL FLUOROSULFONATE see MKG250
16-α-METHYL-9-α-FLUORO-11-β,17-α,21-TRIHYDROXY-
    PREGNA-1,4-DIENE-3,20-DIONE see SOW000
METHYLFLUORPHOSPHORSAEUREISOPROPYLESTER
    (GERMAN) see IPX000
METHYLFLUORPHOSPHORSAEUREPINAKOLYLESTER
    (GERMAN) see SKS500
METHYL FLUORSULFONATE see MKG250
METHYLFLURETHER see EAT900
METHYLFORMAMIDE see MKG500
N-METHYLFORMAMIDE see MKG500
METHYL FORMATE see MKG750
METHYLFORMIAAT (DUTCH) see MKG750
METHYLFORMIAT (GERMAN) see MKG750
N-METHYL-N-FORMYLHYDRAZINE see FNW000
METHYL FORMYL see DOO600
N-METHYL-N-FORMYL HYDRAZONE of ACETALDEHYDE
    see AAH000
1-METHYL-2-FORMYLPYRIDINIUM CHLORIDE OXIME see
    FNZ000
METHYL FOSFERNO see MNH000
METHYLFURAN see MKH000
2-METHYLFURAN see MKH000
METHYL GAG see MKI000
METHYL GLYCOL see EJH500, PML000
METHYL GLYCOL ACETATE see EJJ500
METHYL GLYCOL MONOACETATE see EJJ500
METHYLGLYKOL (GERMAN) see EJH500
METHYLGLYKOLACETAT (GERMAN) see EJJ500
METHYLGUANIDIN (GERMAN) see MKI750
METHYLGUANIDINE see MKI750
METHYL GUTHION see ASH500
3-METHYL-5-HEPTANONE see EGI750
5-METHYL-3-HEPTANONE see EGI750
METHYL HEPTENONE see MKK000
6-METHYL-5-HEPTEN-2-ONE see MKK000
METHYL HEPTINE CARBONATE see MND275
2-METHYL-2-HEPTYLAMINE see ILM000
6-METHYL-2-HEPTYLAMINE see ILM000
(6-(1-METHYL-HEPTYL)-2,4-DINITRO-FENYL)-CROTO-
    NAAT (DUTCH) see AQT500
(6-(1-METHYL-HEPTYL)-2,3-DINITRO-PHENYL)-CROTO-
    NAT (GERMAN) see AQT500
2-(1-METHYLHEPTYL)-4,6-DINITROPHENYL CROTONATE
    see AQT500
METHYLHEPTYLNITROSAMIN (GERMAN) see HBP000
METHYLHEXABARBITAL see ERD500
METHYLHEXABITAL see ERD500
METHYL HEXAFLUOROISOBUTYRATE see MKK750
2-METHYL-5-HEXANONE see MKW450
5-METHYL-2-HEXANONE see MKW450
METHYL HEXYL KETONE (FCC) see ODG000
α-(N-METHYL-N-HOMOVERATRYL)-γ-AMINOPROPYL)-3,4-
    DIMETHOXYPHENYLACETONITRILE see IRV000
METHYL HYDANTOIN see MKB250
METHYL HYDRAZINE see MKN000
1-METHYL HYDRAZINE see MKN000
METHYLHYDRAZINE (DOT) see MKN000
METHYLHYDRAZINE HYDROCHLORIDE see MKN250

METHYLHYDRAZINIUM NITRATE see MRJ250
4-((2-METHYLHYDRAZINO)METHYL)-N-ISOPROPYL-
BENZAMIDE see PME250
p-(N'-METHYLHYDRAZINOMETHYL)-N-ISOPROPYL)
BENZAMIDE see PME500
p-(N'-METHYLHYDRAZINOMETHYL)-N-ISOPROPYL-
BENZAMIDE HYDROCHLORIDE see PME500
METHYL HYDRIDE see MDQ750
Δ¹-6-α-METHYLHYDROCORTISONE see MOR500
METHYL HYDROXIDE see MGB150
1-METHYL-4-HYDROXYBENZENE see CNX250
METHYL-o-HYDROXYBENZOATE see MPI000
METHYL p-HYDROXYBENZOATE see HJL500
2-METHYL-3-HYDROXY-4,5-BIS(HYDROXYMETHYL)PYRI-
DINE see PPK250
2-METHYL-3-HYDROXY-4,5-BIS(HYDROXYMETHYL)PYRI-
DINE HYDROCHLORIDE see PPK500
2-METHYL-3-HYDROXY-4,5-DIHYDROXYMETHYL-PYRI-
DIN (GERMAN) see PPK250
2-METHYL-3-HYDROXY-4,5-DI(HYDROXYMETHYL)PYRI-
DINE see PPK250
METHYL(β-HYDROXYETHYL)AMINE see MGG000
2-METHYL-1-(2-HYDROXYETHYL)-5-NITROIMIDAZOLE
see MMN250
2-METHYL-3-(2-HYDROXYETHYL)-4-NITROIMIDAZOLE
see MMN250
2-METHYL-3-HYDROXY-4-FORMYL-5-HYDROXYMETHYL-
PYRIDINE HYDROCHLORIDE see VSU000
1-METHYL-2-HYDROXYIMINOMETHYLPYRIDINIUM IO-
DIDE see POS750
1-METHYL-3-HYDROXY-4-ISOPROPYLBENZENE see
TFX810
7-METHYL-12-HYDROXYMETHYLBENZ(a)ANTHRACENE
see HMF500
17-α-METHYL-2-HYDROXYMETHYLENE-17-HYDROXY-5-
α-ANDROSTAN-3-ONE see PAN100
N-METHYL-3-HYDROXYMORPHINAN see MKR250
1-METHYL-4-(m-HYDROXYPHENYL)PIPERIDINE-4-
ETHYLKETONE HYDROCHLORIDE see KFK000
6-α-METHYL-17-α-HYDROXYPROGESTERONE ACETATE
see MCA000
6-METHYL-17-α-HYDROXY-Δ⁶-PROGESTERONE ACETATE
see VTF000
2-METHYL-3-HYDROXY-4-PYRONE see MAO350
4,4',4''-METHYLIDYNETRIANILINE see THP000
4,4',4''-METHYLIDYNETRISBENZENEAMINE see THP000
1,1',1'-(METHYLIDYNETRIS(OXY))TRIS(ETHANE) see
ENY500
1-METHYLIMIDAZOLE-2-THIOL see MCO500
2-METHYL-3-(Δ²)-IMIDAZOLINYLMETHYL)BENZO(b)-
THIOPHENE HYDROCHLORIDE see MHJ500
2-METHYL-1,3-INDANDIONE see MKV500
α-METHYL-β-INDOLAETHYLAMINE (GERMAN) see
AME500
3-METHYLINDOLE see MKV750
β-METHYLINDOLE see MKV750
3-METHYL-1H-INDOLE see MKV750
α-METHYL-β-INDOLEETHYLAMINE see AME500
METHYL IODIDE see MKW200
METHYLISOAMYL ACETATE see HFJ000
METHYL ISOAMYL KETONE see MKW450
N-METHYL-dl-ISOBORNYLAMINE HYDROCHLORIDE see
MQR500
METHYL ISOBUTENYL KETONE see MDJ750
METHYL ISOBUTYL CARBINOL see AOK750, MKW600
METHYLISOBUTYL CARBINOL see MKW600
METHYLISOBUTYLCARBINOL ACETATE see HFJ000
METHYLISOBUTYLCARBINYL ACETATE see HFJ000
METHYL-ISOBUTYL-CETONE (FRENCH) see HFG500
METHYLISOBUTYLKETON (DUTCH, GERMAN) see
HFG500
METHYL ISOBUTYL KETONE (ACGIH, DOT) see HFG500
METHYLISOCYANAAT (DUTCH) see MKX250
METHYL ISOCYANAT (GERMAN) see MKX250

METHYL ISOCYANATE see MKX250
METHYL ISOCYANATE, solutions (DOT) see MKX250
METHYL ISOEUGENOL (FCC) see IKR000
METHYLISOMIN see DBA800
METHYLISOOCTENYLAMINE see ILK000
METHYL ISOPENTANOATE see ITC000
1-METHYL-4-ISOPROPENYLCYCLOHEXAN-3-OL see
MCE750
1-METHYL-4-ISOPROPENYL-1-CYCLOHEXENE see
MCC250
1-1-METHYL-4-ISOPROPENYL-6-CYCLOHEXEN-2-ONE see
CCM120
d-1-METHYL-4-ISOPROPENYL-6-CYCLOHEXEN-2-ONE see
CCM100
METHYL ISOPROPENYL KETONE see MKY500
METHYL ISOPROPENYL KETONE INHIBITED (DOT) see
MKY500
N-METHYL-2-ISOPROPOXYPHENYLCARBAMATE see
PMY300
p-METHYLISOPROPYL BENZENE see CQI000
1-METHYL-4-ISOPROPYLBENZENE see CQI000
1-METHYL-2-p-(ISOPROPYLCARBAMOYL)BENZOHY-
DRAZINE HYDROCHLORIDE see PME500
1-METHYL-2-(-ISOPROPYLCARBAMOYL)BENZYL)
HYDRAZINE see PME250
1-METHYL-2-(p-ISOPROPYLCARBAMOYLBENZYL)
HYDRAZINE HYDROCHLORIDE see PME500
1-METHYL-4-ISOPROPYLCYCLOHEXADIENE-1,3 see
MLA250
1-METHYL-4-ISOPROPYL-1,3-CYCLOHEXADIENE see
MLA250
1-METHYL-4-ISOPROPYLCYCLOHEXADIENE-1,4 see
MCB750
2-METHYL-5-ISOPROPYL-1,3-CYCLOHEXADIENE see
MCC000
6-METHYL-3-ISOPROPYLCYCLOHEXANOL see DKV150
α-METHYL-p-ISOPROPYLHYDROCINNAMALDEHYDE see
COU500
METHYL ISOPROPYL KETONE see MLA750
2-METHYL-5-ISOPROPYLPHENOL see CCM000
5-METHYL-2-ISOPROPYL-1-PHENOL see TFX810
2-METHYL-3-(p-ISOPROPYLPHENYL)PROPIONALDEHYDE
see COU500
5-METHYL-2-ISOPROPYL-3-PYRAZOLYL DIMETHYLCAR-
BAMATE see DSK200
METHYL ISOSYSTOX see DAP400
METHYLISOTHIOCYANAAT (DUTCH) see ISE000
METHYL-ISOTHIOCYANAT (GERMAN) see ISE000
METHYL ISOTHIOCYANATE (DOT) see ISE000
METHYL ISOVALERATE see ITC000
METHYLISOVALERATE (DOT) see ITC000
5-METHYL-3-ISOXAZOLECARBOXYLIC ACID-2-BENZYL-
HYDRAZIDE see IKC000
3-METHYL-4,5-ISOXAZOLEDIONE-4-((2-CHLOROPHENYL)
HYDRAZONE) see MLC250
N'-(5-METHYL-3-ISOXAZOLE)SULFANILAMIDE see
SNK000
N'-(5-METHYL-3-ISOXAZOLYL)SULFANILAMIDE see
SNK000
N'-(5-METHYLISOXAZOL-3-YL)SULPHANILAMIDE see
SNK000
N¹-(5-METHYL-3-ISOXAZOLYL)SULPHANILAMIDE see
SNK000
METHYLJODID (GERMAN) see MKW200
METHYLJODIDE (DUTCH) see MKW200
METHYL KETONE see ABC750
2-METHYLLACTONITRILE see MLC750
METHYLLCISTOX see DAO800
METHYL LEDATE see LCW000
METHYL-LOMUSTINE see CHD250
N-METHYL-LOST see BIE250
1-METHYL-LUMILYSERGOL-8-(5-BROMONICOTINATE)-10-
METHYL ETHER see NDM000
1-METHYLLYSERGIC ACID ETHYLAMIDE see MLD500

d-1-METHYL LYSERGIC ACID MONOETHYLAMIDE see MLD500

METHYLMAGNESIUM BROMIDE (ethyl ether solution) see MLE000

METHYL MAGNESIUM BROMIDE in ETHYL ETHER (DOT) see MLE000

METHYLMERCAPTAAN (DUTCH) see MLE650

METHYL MERCAPTAN see MLE650

2-METHYLMERCAPTO-4,6-BIS(ISOPROPYLAMINO)-s-TRIAZINE see BKL250

4-METHYLMERCAPTO-3,5-DIMETHYLPHENYL N-METHYLCARBAMATE see DST000

METHYL-MERCAPTOFOS TEOLOVY see DAP400

1-METHYL-2-MERCAPTOIMIDAZOLE see MCO500

4-METHYLMERCAPTO-3-METHYLPHENYL DIMETHYL THIOPHOSPHATE see FAQ999

2-METHYLMERCAPTO-10-(2-N-METHYL-2-PIPERIDYL) ETHYL)PHENOTHIAZINE see MOO250

2-METHYLMERCAPTO-10-(2-(N-METHYL-2-PIPERIDYL) ETHYLPHENOTHIAZINE HYDROCHLORIDE see MOO500

METHYLMERCAPTOPHOS see DAO800

METHYL-MERCAPTOPHOS see MIW100

6-METHYLMERCAPTOPURINE RIBONUCLEOSIDE see MPU000

6-METHYLMERCAPTOPURINE RIBOSIDE see MPU000

4-METHYLMERCAPTO-3,5-XYLYL METHYLCARBAMATE see DST000

METHYLMERCURIC CHLORIDE see MDD750

METHYLMERCURIC CYANOGUANIDINE see MLF250

METHYLMERCURIC DICYANDIAMIDE see MLF250

METHYLMERCURICHLORENDIMIDE see MLF500

METHYLMERCURIC HYDROXIDE see MLG000

N-(METHYLMERCURI)-1,4,5,6,7,7-HEXACHLOROBICY-CLO(2.2.1)HEPT-5-ENE-2,3-DICARBOXIMIDE see MLF500

8-(METHYLMERCURIOXY)QUINOLINE see MLH000

N-METHYLMERCURI-1,2,3,6-TETRAHYDRO-3,6-ENDO-METHANO-3,4,5,6,7,7-HEXACHLOROPHTHALIMIDE see MLF500

N-METHYLMERCURI-1,2,3,6-TETRAHYDRO-3,6-METH-ANO-3,4,5,6,7,7-HEXACHLOROPHTHALIMIDE see MLF500

METHYLMERCURY see MLF550

METHYL-MERCURY(1+) (9CI) see MLF550

METHYLMERCURY(II) CATION see MLF550

METHYLMERCURY CHLORIDE see MDD750

METHYLMERCURY DICYANDIAMIDE see MLF250

METHYLMERCURY HYDROXIDE see MLG000

METHYLMERCURY β-HYDROXYQUINOLATE see MLH000

METHYLMERCURY 8-HYDROXYQUINOLINATE see MLH000

METHYLMERCURY ION see MLF550

METHYLMERCURY ION(1+) see MLF550

METHYLMERCURY OXINATE see MLH000

METHYLMERCURY OXYQUINOLINATE see MLH000

METHYLMERCURY QUINOLINOLATE see MLH000

α-METHYLMESCALINE see MLH250

METHYL MESYLATE see MLH500

METHYLMETHACRYLAAT (DUTCH) see MLH750

METHYL-METHACRYLAT (GERMAN) see MLH750

METHYL METHACRYLATE see MLH750

METHYL METHACRYLATE HOMOPOLYMER see PKB500

METHYL METHACRYLATE MONOMER, INHIBITED (DOT) see MLH750

METHYL METHACRYLATE POLYMER see PKB500

METHYL METHACRYLATE RESIN see PKB500

N-METHYLMETHANAMINE see DOQ800

N-METHYLMETHANAMINE with BORANE (1:1) see DOR200

METHYLMETHANE see EDZ000

METHYL METHANESULFONATE see MLH500

METHYL METHANESULPHONATE see MLH500

METHYL METHANOATE see MKG750

METHYLMETHANSULFONAT (GERMAN) see MLH500

METHYL METHANSULFONATE see MLH500

METHYL METHANSULPHONATE see MLH500

2-METHYL-4-METHOXYANILINE see MGO500

4-METHYL-1-METHOXYBENZENE see MGP000

1-METHYL-7-METHOXY-β-CARBOLINE see HAI500

METHYL(METHOXYMETHYL)NITROSAMINE see MEW250

METHYL-α-METHYLACRYLATE see MLH750

1-METHYL-6-(1-METHYLALLYL)DITHIOBIUREA see MLJ500

1-METHYL-6-(1-METHYLALLYL)-2,5-DITHIOBIUREA see MLJ500

N-METHYL-3′-METHYL-4-AMINOAZOBENZENE see MPY000

N-METHYL-3′-METHYL-p-AMINOAZOBENZENE see MPY000

METHYL METHYLAMINOBENZOATE see MGQ250

METHYL N-((METHYLAMINO)CARBONYL)OXY) ETHANIMIDO)THIOATE see MDU600

2-METHYL-6-METHYLAMINO-2-HEPTENE see ILK000

N-METHYL-N-(5-(N′-METHYLANILINO)-2,4-PENTA-DIENYLIDENE) ANILINIUM CHLORIDE see MLL000

METHYL-N-METHYL ANTHRANILATE see MGQ250

METHYL-4-METHYLBENZENESULFONATE see MLL250

METHYL-p-METHYLBENZENESULFONATE see MLL250

METHYL 1-(α-METHYLBENZYL)IMIDAZOLE-5-CARBOX-YLATE see MQQ500

METHYL 2-METHYLBUTANOATE see MLL600

METHYL-3-METHYLBUTANOATE see ITC000

METHYL 2-METHYLBUTYRATE see MLL600

METHYL-3-METHYLBUTYRATE see ITC000

METHYL-N-((METHYLCARBAMOYL)OXY)THIO-ACETIMIDATE see MDU600

cis-1-METHYL-2-METHYL CARBAMOYL VINYL PHOS-PHATE see MRH209

S-METHYL N-[(METHYLCARBAMOYL0OXY]THIO-ACETIMIDATE see MDU600

α-METHYL-3,4-METHYLENEDIOXYPHENETHYLAMINE HYDROCHLORIDE see MJS750

1-METHYL-2-(3,4-METHYLENEDIOXYPHENYL)ETHYL OCTYL SULFOXIDE see ISA000

d-2-METHYL-5-(1-METHYLENENYL)-CYCLOHEXANONE see DKV175

7-METHYL-3-METHYLENE-1,6-OCTADIENE see MRZ150

2-METHYL-6-METHYLENE-7-OCTEN-2-OL ACETATE see AAW500

2-METHYL-6-METHYLENE-7-OCTEN-2-YL ACETATE see AAW500

1-METHYL-4-(1-METHYLETHENYL)-(S)-CYCLOHEXENE see MCC500

(R)-1-METHYL-4-(1-METHYLETHENYL)-CYCLOHEXENE see LFU000

(S)-2-METHYL-5-(1-METHYLETHENYL)-2-CYCLOHEXEN-1-ONE see CCM100

(R)-2-METHYL-5-(1-METHYLETHENYL)-2-CYCLOHEXEN-1-ONE (9CI) see CCM120

(1S-1-α,4-α,5-α)-4-METHYL-1-(1-METHYLETHYL)-BICYCLO(3.1.0)HEXAN-3-ONE see TFW000

5-METHYL-2-(1-METHYLETHYL)CYCLOHEXANOL see MCF750

5-METHYL-2-(1-METHYLETHYL)-CYCLOHEXANOL (1-α,2-β,5-α) see MCG000

(1R-(1α,2-β,5-α))-5-METHYL-2-(1-METHYLETHYL)CYCLO-HEXANOL see MCG250

(R-(1α,2β,5α))-5-METHYL-2-(1-METHYLETHYL)-CYCLO-HEXANOL ACETATE (9CI) see MCG750

trans-5-METHYL-2-(1-METHYLETHYL)-CYCLOHEXANONE see MCG275

1-METHYL-4-(1-METHYLETHYLIDENE)CYCLOHEXENE see TBE000

N-METHYL-N-NITROSO-HARNSTOFF (GERMAN) see
    MNA750
N-METHYL-N-NITROSOHEPTYLAMINE see HBP000
N-METHYL-N-NITROSOLAURYLAMINE see NKU000
N-METHYL-N-NITROSOMETHANAMINE see NKA600
N-METHYL-N-NITROSONITROGUANIDIN (GERMAN) see
    MMP000
1-METHYL-1-NITROSO-3-NITROGUANIDINE see MMP000
N-METHYL-N-NITROSO-N'-NITROGUANIDINE see
    MMP000
5-METHYL-3-NITROSO-1,3-OXAZOLIDINE see NKU875
N-METHYL-N-NITROSO-4-OXO-4-(3-PYRIDYL)BUTYL
    AMINE see MMS500
N-METHYL-N-NITROSOPENTYLAMINE see AOL000
N-METHYL-N-NITROSOPHENETHYLAMINE see MNU250
1-METHYL-1-NITROSO-3-PHENYLUREA see MMY500
N-METHYL-N-NITROSO-N'-PHENYLUREA see MMY500
1-METHYL-4-NITROSOPIPERAZINE see NKW500
N'-METHYL-N-NITROSOPIPERAZINE see NKW500
N-METHYL-N-NITROSO-1-PROPANAMINE see MNA000
N-METHYL-N-NITROSO-2-PROPEN-1-AMINE see MMT500
METHYLNITROSO-PROPIONAMIDE see MNA250
N-METHYL-N-NITROSOPROPIONAMIDE see MNA250
METHYL-NITROSOPROPIONSAEUREAMID (GERMAN) see
    MNA250
METHYLNITROSOPROPIONYLAMIDE see MNA250
METHYLNITROSOUREA see MNA750
1-METHYL-1-NITROSOUREA see MNA750
N-METHYL-N-NITROSOUREA see MNA750
METHYLNITROSOUREE (FRENCH) see MNA750
METHYLNITROSOURETHAN (GERMAN) see MMX250
METHYLNITROSOURETHANE see MMX250
N-METHYL-N-NITROSO-URETHANE see MMX250
N-METHYL-N-NITROSOVINYLAMINE see NKY000
METHYL NONYL KETONE see UKS000
METHYL-n-NONYL KETONE see UKS000
α-METHYLNORADRENALINE HYDROCHLORIDE see
    AMB000
N-METHYLNOREPHEDRINE see EAW000
4-METHYLNORVALINE see LES000
METHYL OCTADECANOATE see MJW000
METHYL-9-OCTADECENOATE see OHW000
METHYL (Z)-9-OCTADECENOATE see OHW000
METHYL cis-9-OCTADECENOATE see OHW000
METHYLOCTENYLAMINE see ILK000
METHYL 2-OCTINATE see MND275
METHYL 2-OCTYNOATE see MND275
METHYLOL see MGB150
METHYL OLEATE see OHW000
3-METHYLOLPENTANE see EGW000
METHYLOLPROPANE see BPW500
METHYL-4-OMBELLIFERONE SODEE (FRENCH) see
    HMB000
(METHYL-ONN-AZOXY)METHANOL see HMG000
(METHYL-ONN-AZOXY)METHANOL, ACETATE (ESTER)
    see MGS750
(METHYL-ONN-AZOXY)METHYL-β-d-GLUCOPYRANOSIDE
    see COU000
METHYL ORTHOFORMATE see TLX600
METHYL ORTHOSILICATE see MPI750
9-METHYL-3-OXA-9-AZATRICYCLO(3.3.1.0^{2,4})NONAN-7-
    OL,TROPATE (ESTER) see SBG000
4-METHYL-2-OXETANONE see BSX000
METHYL OXIRANE see PNL600
METHYL OXITOL see EJH500
METHYL-3-OXOBUTYRATE see MFX250
METHYL p-OXYBENZOATE see HJL500
1-(4-METHYLOXYPHENYL)-3,3-DIMETHYLTRIAZINE see
    DSN600
2-METHYL-3-OXY-γ-PYRONE see MAO350
METHYLPARABEN (FCC) see HJL500
METHYLPARAFYNOL CARBAMATE see MNM500
METHYL PARAHYDROXYBENZOATE see HJL500
METHYL PARASEPT see HJL500

METHYL PARATHION see MNH000
METHYL PCT see DTQ600
METHYLPENTADIENE see MNH500
2-METHYLPENTANE see IKS600
3-METHYLPENTANE see MNI500
2-METHYL PENTANE-2,4-DIOL see HFP875
2-METHYL-2,4-PENTANEDIOL see HFP875
2-METHYLPENTANOL-1 see AOK750
2-METHYL-4-PENTANOL see MKW600
4-METHYLPENTANOL-2 see MKW600
4-METHYL-2-PENTANOL, ACETATE see HFJ000
4-METHYL-2-PENTANOL (MAK) see MKW600
2-METHYL-2-PENTANOL-4-ONE see DBF750
4-METHYL-2-PENTANON (CZECH) see HFG500
4-METHYL-PENTAN-2-ON (DUTCH, GERMAN) see
    HFG500
2-METHYL-4-PENTANONE see HFG500
4-METHYL-2-PENTANONE (FCC) see HFG500
4-METHYL-3-PENTENE-2-ONE see MDJ750
4-METHYL-3-PENTEN-2-ON (DUTCH, GERMAN) see
    MDJ750
2-METHYL-2-PENTEN-4-ONE see MDJ750
4-METHYL-3-PENTEN-2-ONE see MDJ750
3-METHYL-2-PENTEN-4-YN-1-OL see MNL775
3-METHYL-PENTIN-(1)-OL-(3) (GERMAN) see MNM500
4-METHYL-2-PENTYL ACETATE see HFJ000
METHYL PENTYL KETONE see MGN500
METHYL-N-PENTYLNITROSAMINE see AOL000
METHYLPENTYNOL CARBAMATE see MNM500
3-METHYL-1-PENTYN-3-OL CARBAMATE see MNM500
METHYL PERFLUOROMETHACRYLATE see MNN000
METHYLPERIDIDE see FLV000
METHYLPERIDOL see MNN250
METHYLPERIDOL HYDROCHLORIDE see MNN250
METHYLPERONE HYDROCHLORIDE see FKI000
METHYLPHENAZONIUM METHOSULFATE see MRW000
(±)-α-METHYLPHENETHYLAMINE see BBK000
dl-α-METHYLPHENETHYLAMINE see BBK000
dl-α-METHYL-PHENETHYLAMINE HYDROCHLORIDE see
    AOA750
d-α-METHYLPHENETHYLAMINE SULFATE see BBK500
dl-α-METHYLPHENETHYLAMINE SULFATE see BBK250
(α-METHYLPHENETHYL)HYDRAZINE see PDN000
3-(α-METHYLPHENETHYL)SYDONE IMINE MONOHYDRO-
    CHLORIDE see SPA000
METHYLPHENIDAN see MNQ000
METHYL PHENIDATE see MNQ000
METHYL PHENIDYL ACETATE see MNQ000
METHYLPHENOBARBITAL see ENB500
1-METHYLPHENOBARBITAL see ENB500
N-METHYLPHENOBARBITAL see ENB500
METHYLPHENOBARBITONE see ENB500
2-METHYLPHENOL see CNX000
3-METHYLPHENOL see CNW750
4-METHYLPHENOL see CNX250
m-METHYLPHENOL see CNW750
o-METHYLPHENOL see CNX000
p-METHYLPHENOL see CNX250
N-METHYLPHENOLBARBITOL see ENB500
4-METHYLPHENOL METHYL ETHER see MGP000
2-(2-(4-(2-METHYL-3-PHENOTHIAZIN-10-YLPROPYL)-1-PI-
    PERAZINYL)ETHOXY)ETHANOL DIHYDROCHLORIDE
    see DXR800
3-(2-METHYLPHENOXY)-1,2-PROPANEDIOL see GGS000
α-METHYL PHENYLACETALDEHYDE see COF000
4-METHYLPHENYL ACETATE see MNR250
p-METHYLPHENYL ACETATE see MNR250
METHYL PHENYLACETATE (FCC) see MHA500
METHYL(2-PHENYLAETHYL)NITROSAMIN (GERMAN) see
    MNU250
METHYLPHENYLAMINE see MGN750
N-METHYLPHENYLAMINE see MGN750
N-METHYL-p-(PHENYLAZO)ANILINE see MNR500

1-(2-METHYLPHENYL)AZO-2-NAPHTHALENAMINE see FAG135

1-((2-METHYLPHENYL)AZO)-2-NAPHTHALENAMINE see FAG135

1-((2-METHYLPHENYL)AZO)-2-NAPHTHALENOL see TGW000

1-(2-METHYLPHENYL)AZO-2-NAPHTHYLAMINE see FAG135

N-METHYL-N-(p-(PHENYLAZO)PHENYL)HYDROXYL-AMINE see HLV000

METHYLPHENYLBARBITURIC ACID see ENB500

METHYLPHENYLCARBAMIC ESTER OF 3-OXYPHENYL-TRIMETHYLAMMONIUM METHYLSULFATE see HNR500

N-METHYL-4-PHENYL-4-CARBETHOXYPIPERIDINE see DAM600

N-METHYL-4-PHENYL-4-CARBETHOXYPIPERIDINE HY-DROCHLORIDE see DAM700

METHYLPHENYLCARBINOL see PDE000

METHYL PHENYLCARBINYL ACETATE see MNT075

1-METHYL-4-PHENYL-4-CARBOETHOXYPIPERIDINE HY-DROCHLORIDE see DAM700

METHYLPHENYLDICHLOROSILANE (DOT) see DFQ800

1-(o-METHYLPHENYL)-3,3-DIMETHYL-TRIAZEN (GER-MAN) see MNT500

1-(2-METHYLPHENYL)-3,3-DIMETHYLTRIAZENE see MNT500

1-(3-METHYLPHENYL)-3,3-DIMETHYLTRIAZENE see DSR200

1-(m-METHYLPHENYL)-3,3-DIMETHYLTRIAZENE see DSR200

1-(o-METHYLPHENYL)-3,3-DIMETHYL-TRIAZENE see MNT500

METHYLPHENYLENEDIAMINE see TGL500

2-METHYL-p-PHENYLENEDIAMINE see TGM000

4-METHYL-m-PHENYLENEDIAMINE see TGL750

2-METHYL-p-PHENYLENEDIAMINE SULPHATE see DCE600

4-METHYL-PHENYLENE DIISOCYANATE see TGM750

METHYL-m-PHENYLENE DIISOCYANATE see TGM740

2-METHYL-m-PHENYLENE ESTER, ISOCYANIC ACID see TGM800

METHYLPHENYLENE ISOCYANATE see TGM740

4-METHYL-PHENYLENE ISOCYANATE see TGM750

2-METHYL-m-PHENYLENE ISOCYANATE see TGM800

METHYL PHENYL ETHER see AOX750

N-METHYL-5-PHENYL-5-ETHYLBARBITAL see ENB500

1-METHYL-5-PHENYL-5-ETHYLBARBITURIC ACID see ENB500

as-METHYLPHENYLETHYLENE see MPK250

3-METHYL-5,5-PHENYLETHYLHYDANTOIN see MKB250

METHYL-PHENYLETHYL-NITROSAMINE see MNU250

3-METHYL-3-PHENYLGLYCIDIC ACID ETHYL ESTER see ENC000

1-METHYL-4-PHENYLISONIPECOTIC ACID, ETHYL ESTER see DAM600

1-METHYL-4-PHENYLISONIPECOTIC ACID ETHYL ESTER HYDROCHLORIDE see DAM700

N-METHYL-β-PHENYLISOPROPYLAMIN (GERMAN) see DBB000

N-METHYL-β-PHENYLISOPROPYLAMINE see DBB000

l-N-METHYL-β-PHENYLISOPROPYLAMINE HYDROCHLO-RIDE see MDQ500

N-METHYL-β-PHENYLISOPROPYLAMINHYDROCHLORID (GERMAN) see DBA800

METHYL PHENYL KETONE see ABH000

3-METHYLPHENYL-N-METHYLCARBAMATE see MIB750

3-METHYL-2-PHENYLMORPHOLINE HYDROCHLORIDE see MNV750

METHYLPHENYLNITROSAMINE see MMU250

METHYLPHENYLNITROSOUREA see MMY500

N-METHYL-N′-PHENYL-N-NITROSOUREA see MMY500

1-METHYL-4-PHENYL-PIPERIDIN-4-CARBON-SAEURE-

AETHYLESTER-HYDROCHLORID (GERMAN) see DAM600

1-METHYL-4-PHENYLPIPERIDINE-4-CARBOXYLIC ACID ETHYL ESTER see DAM600

METHYL α-PHENYL-α-(2-PIPERIDYL)ACETATE see MNQ000

2-METHYL-1-PHENYLPROPANE see IIN000

2-METHYL-2-PHENYLPROPANE see BQJ250

2-METHYL-3-PHENYL-2-PROPENAL see MIO000

METHYL-3-PHENYLPROPENOATE see MIO500

5-METHYL-1-PHENYL-2-(PYRROLIDINYL)IMIDAZOLE see MNY750

METHYL-5 PHENYL-1 (PYRROLIDINYL-1)-2 IMIDAZOLE (FRENCH) see MNY750

3-METHYL-1-PHENYL-3-(2-SULFOETHYL)TRIAZENE SO-DIUM SALT see PEJ250

5-METHYL-1-PHENYL-3,4,5,6-TETRAHYDRO-1H-2,5-BENZOXAZOCINE HYDROCHLORIDE see NBS500

3-METHYL-2-PHENYLTETRAHYDRO-2H-1,4-OXAZINE HY-DROCHLORIDE see MNV750

1-METHYL-4-(PHENYLTHIO)PYRIDINIUM IODIDE see MOA250

4-METHYLPHENYLUREA see THG000

3-METHYL-2-PHENYLVALERIC ACID see VBK000

3-METHYL-2-PHENYLVALERIC ACID 2-DIETHYLAMINO-ETHYL ESTER METHYL BROMIDE see VBK000

METHYL PHOSPHATE see TMD250

METHYL PHOSPHITE see TMD500

METHYLPHOSPHODITHIOIC ACID-S-(((p-CHLOROPHE-NYL)THIO)METHYL)-O-METHYL ESTER see MOB250

METHYL PHOSPHONIC DICHLORIDE see MOB399

METHYLPHOSPHONOFLUORIDIC ACID, 3,3-DIMETHYL-2-BUTYL ESTER see SKS500

METHYLPHOSPHONOFLUORIDIC ACID ISOPROPYL ES-TER see IPX000

METHYLPHOSPHONOFLUORIDIC ACID-1-METHYLETHYL ESTER see IPX000

METHYLPHOSPHONOFLUORIDIC ACID 1,2,2-TRIMETHYL-PROPYL ESTER see SKS500

METHYLPHOSPHONOUS DICHLORIDE see MOC250

METHYL PHTHALATE see DTR200

METHYL PINACOLYLOXY PHOSPHORYLFLUORIDE see SKS500

METHYL PINACOLYL PHOSPHONOFLUORIDATE see SKS500

1-METHYLPIPERAZINE see MOD250

N-METHYLPIPERAZINE see MOD250

3-(4-METHYLPIPERAZINYLIMINOMETHYL)-RIFAMYCIN SV see RKP000

8-(4-METHYLPIPERAZINYLIMINOMETHYL) RIFAMYCIN SV see RKP000

8-(((4-METHYL-1-PIPERAZINYL)IMINO)METHYL) RIFAMYCIN SV see RKP000

N-(γ-(4′-METHYLPIPERAZINYL-1′)PROPYL)-3-CHLORO-PHENOTHIAZINE see PMF500

N-METHYL-PIPERAZINYL-N′-PROPYL-PHENOTHIAZIN (GERMAN) see PCK500

N-(3-(4-METHYL-1-PIPERAZINYL)PROPYL)PHENO-THIAZINE see PCK500

10-(3-(4-METHYL-1-PIPERAZINYL)PROPYL)-10H-PHENO-THIAZINE (9CI) see PCK500

N-METHYLPIPERIDINE see MOG500

1-METHYLPIPERIDINE (DOT) see MOG500

γ-(4-METHYLPIPERIDINE)-p-FLUOROBUTYROPHENONE HYDROCHLORIDE see FKI000

2-METHYL-1-PIPERIDINEPROPANOL BENZOATE HYDRO-CHLORIDE see IJZ000

2-METHYL-1-PIPERIDINOPROPANOL, BENZOATE see PIV750

METHYL-4-(3-PIPERIDINOPROPIONYLAMINO) SALICYLATE, METHIODIDE see MOK000

(2-METHYLPIPERIDINO)PROPYL BENZOATE see PIV750

3-(2-METHYLPIPERIDINO)PROPYL BENZOATE HYDRO-CHLORIDE see IJZ000

γ-(2-METHYLPIPERIDINO)PROPYL BENZOATE HYDRO-
CHLORIDE see IJZ000
dl-(2-METHYLPIPERIDINO)PROPYL BENZOATE HYDRO-
CHLORIDE see IJZ000
N-METHYLPIPERIDYL-(4)-BENZHYDRYLAETHER SALZ-
SAUREN SALZE (GERMAN) see LJR000
N-METHYL-3-PIPERIDYL BENZILATE see MON250
10-(2-(1-METHYL-2-PIPERIDYL)ETHYL)-2-(METHYLTHIO)
PHENOTHIAZINE see MOO250
10-(2-(1-METHYL-2-PIPERIDYL)ETHYL)-2-METHYLTHIO-
PHENOTHIAZINE HYDROCHLORIDE see MOO500
1-METHYL-3-PIPERIDYLIDENEDI(2-THIENYL)METHANE
see BLV000
9-(1-METHYL-4-PIPERIDYLIDENE)THIOXANTHENE see
MOO750
9-(1-METHYL-PIPERIDYL-(2)-METHYL)-CARBAZOL (GER-
MAN) see MOP500
9-(1-METHYL-2-PIPERIDYL)METHYLCARBAZOLE see
MOP500
γ-(2-METHYLPIPERIDYL)PROPYL BENZOATE see PIV750
(+−)-γ-(2-METHYLPIPERIDYL)PROPYL BENZOATE HY-
DROCHLORIDE see IJZ000
METHYL PIRIMIPHOS see DIN800
METHYLPREDNISOLONE see MOR500
6-α-METHYLPREDNISOLONE see MOR500
16-β-METHYL-1,4-PREGNADIENE-9-α-FLUORO-11-β,17-
α,21-TRIOL- 3,20-DIONE see BFV750
6-METHYL-Δ4,6-PREGNADIEN-17-α-OL-3,20-DIONE ACE-
TATE see VTF000
6-α-METHYL-4-PREGNENE-3,20-DION-17-α-OL ACETATE
see MCA000
METHYLPROPAMINE see DBA800
2-METHYLPROPANAL see IJS000
2-METHYL-1-PROPANAL see IJS000
2-METHYLPROPANE see MOR750
2-METHYL-2-PROPANETHIOL see MOS000
METHYL PROPANOATE see MOT000
2-METHYLPROPANOIC ACID see IJU000
2-METHYL PROPANOL see IIL000
2-METHYL-1-PROPANOL see IIL000
2-METHYLPROPAN-1-OL see IIL000
2-METHYL-2-PROPANOL see BPX000
N-METHYL-N-PROPARGYL-3-(2,4-DICHLOROPHENOXY)
PROPYLAMINE HYDROCHLORIDE see CMY000
2-METHYLPROPENAL (CZECH) see MGA250
2-METHYLPROPENAMIDE see MDN500
METHYL PROPENATE see MGA500
2-METHYLPROPENE see IIC000
2-METHYLPROPENENITRILE see MGA750
METHYL PROPENOATE see MGA500
METHYL-2-PROPENOATE see MGA500
2-METHYLPROPENOIC ACID see MDN250
2-METHYL-2-PROPENOIC ACID, ETHYL ESTER see
EMF000
2-METHYL-2-PROPENOIC ACID METHYL ESTER see
MLH750
2-METHYL-2-PROPENOIC ACID METHYL ESTER HOMO-
POLYMER see PKB500
2-METHYL-2-PROPENOIC ACID METHYL ESTER, POLY-
MER with TRIBUTYL(92-METHYL-1-OXO-2-PROPENYL)
OXY)STANNANE see OIY000
2-METHYL-2-PROPENOIC ACID-2-METHYLPROPYL ESTER
see IIY000
2-METHYL-2-PROPEN-1-OL see IMW000
2-METHYLPROPIONALDEHYDE see IJS000
METHYL PROPIONATE see MOT000
2-METHYLPROPIONIC ACID see IJU000
α-METHYLPROPIONIC ACID see IJU000
2-METHYLPROPIONIC ACID, ETHYL ESTER see ELS000
2-METHYLPROPIONITRILE see IJX000
2-METHYLPROPYL ACETATE see IIJ000
2-METHYL-1-PROPYL ACETATE see IIJ000
2-METHYLPROPYL ALCOHOL see IIL000
1-METHYLPROPYLAMINE see BPY000

METHYL PROPYLATE see MOT000
2-METHYLPROPYL BUTYRATE see BSW500
METHYL PROPYL CARBINOL see PBM750
METHYL-PROPYL-CETONE (FRENCH) see PBN250
3-(1-METHYLPROPYL)-6-CHLOROPHENYL METHYLCAR-
BAMATE see MOU750
6-(1-METHYL-PROPYL)-2,4-DINITROFENOL (DUTCH) see
BRE500
2-(1-METHYLPROPYL)-4,6-DINITROPHENOL see BRE500
2-(1-METHYL-N-PROPYL) 4,6-DINITROPHENOL AMMO-
NIUM SALT see BPG250
2-(1-METHYLPROPYL)-4,6-DINITROPHENYL ACETATE see
ACE500
β-METHYLPROPYL ETHANOATE see IIJ000
2-METHYL-2-PROPYLETHANOL see AOK750
N,N''-(2-METHYLPROPYLIDENE)BISUREA (9CI) see
IIV000
2-METHYLPROPYL ISOBUTYRATE see IIW000
METHYL-n-PROPYL KETONE see PBN250
METHYL PROPYL KETONE (ACGIH, DOT) see PBN250
2-METHYLPROPYL METHACRYLATE see IIY000
METHYL-N-PROPYLNITROSAMINE see MNA000
METHYLPROPYLNITROSOAMINE see MNA000
2-(1-METHYLPROPYL)PHENYL METHYLCARBAMATE see
MOV000
2-METHYL-2-PROPYL-1,3-PROPANEDIOL BUTYLCARBA-
MATE CARBAMATE see MOV500
2-METHYL-2-PROPYL-1,3-PROPANEDIOL CARBAMATE
ISOPROPYLCARBAMATE see IPU000
2-METHYL-2-N-PROPYL-1,3-PROPANEDIOL DICARBA-
MATE see MQU750
2-METHYLPROPYLPROPANOIC ACID-2-METHYLPROPYL
ESTER (9CI) see IIW000
2-METHYLPROPYL PROPIONATE see PMV250
METHYL 5-(PROPYLTHIO)-2-BENZIMIDAZOLECARBA-
MATE see VAD000
2-METHYL-2-PROPYLTRIMETHYLENE BUTYLCARBA-
MATE CARBAMATE see MOV500
2-METHYL-2-PROPYLTRIMETHYLENE CARBAMATE see
MQU750
METHYLPROTOCATECHUALDEHYDE see VFK000
2-METHYLPYRAZINE see MOW750
METHYLPYRAZOLYL DIETHYLPHOSPHATE see MOX250
3-METHYLPYRAZOLYL-5-DIETHYLPHOSPHATE see
MOX250
2-METHYLPYRIDINE see MOY000
3-METHYLPYRIDINE see PIB920
4-METHYLPYRIDINE see MOY250
α-METHYLPYRIDINE see MOY000
N-METHYLPYRIDINE-2-ALDOXIME IODIDE see POS750
1-METHYL-2-PYRIDINIUM ALDOXIME CHLORIDE see
FNZ000
N-METHYLPYRIDINIUM-2-ALDOXIME IODIDE see
POS750
N-METHYLPYRIDINIUM CHLORIDE-2-ALDOXIME see
FNZ000
1-METHYL-2-(3-PYRIDYL)PYRROLIDINE see NDN000
1-1-METHYL-2-(3-PYRIDYL)-PYRROLIDINE SULFATE see
NDR500
N¹-(4-METHYL-2-PYRIMIDINYL)SULFANILAMIDE SODIUM
SALT see SJW475
3-METHYLPYROCATECHOL see DNE000
2-METHYL PYROMECONIC ACID see MAO350
1-METHYLPYRROLIDINE see MPB250
N-METHYLPYRROLIDINONE see MPF200
1-METHYL-2-PYRROLIDINONE see MPF200
1-METHYL-5-PYRROLIDINONE see MPF200
N-METHYL-2-PYRROLIDINONE see MPF200
3-(N-METHYLPYRROLIDINO)PYRIDINE see NDN000
3-(1-METHYL-2-PYRROLIDINYL)INDOLE see MPD000
3-(1-METHYL-3-PYRROLIDINYL)INDOLE see MPD250
10-((1-METHYL-3-PYRROLIDINYL)METHYL)-PHENO-
THIAZINE see MPE250
3-(1-METHYL-2-PYRROLIDINYL)PYRIDINE see NDN000

(S)-3-(1-METHYL-2-PYRROLIDINYL)PYRIDINE (9CI) see
NDN000
(S)-3-(1-METHYL-2-PYRROLIDINYL-PYRIDINE (R-(R,R))-
2,3-DIHYDROXYBUTANEDIOATE (1:2) see NDS500
(S)-3-(1-METHYL-2-PYRROLIDINYL)PYRIDINE SULFATE
(2:1) see NDR500
METHYLPYRROLIDONE see MPF200
N-METHYLPYRROLIDONE see MPF200
1-METHYL-2-PYRROLIDONE see MPF200
N-METHYL-2-PYRROLIDONE see MPF200
(−)-3-(1-METHYL-2-PYRROLIDYL)PYRIDINE see NDN000
l-3-(1-METHYL-2-PYRROLIDYL)PYRIDINE see NDN000
1-3-(1-METHYL-2-PYRROLIDYL)PYRIDINE SULFATE see
NDR500
METHYLQUINAZOLONE HYDROCHLORIDE see MDT250
8-(METHYLQUINOLYL)-N-METHYL CARBAMATE see
MPG250
2-METHYL-1,4-QUINONE see MHI250
METHYL RED see CCE500
METHYLRESERPATE 3,4,5-TRIMETHOXYBENZOIC ACID
see RDK000
METHYL RESERPATE 3,4,5-TRIMETHOXYBENZOIC ACID
ESTER see RDK000
5-METHYLRESORCINOL see MPH500
5-METHYLRESORCINOL ORCINOL see MPH500
METHYLRHODANID (GERMAN) see MPT000
6-METHYL-9-RIBOFURANOSYLPURINE-6-THIOL see
MPU000
METHYLROSANILINE CHLORIDE see AOR500
METHYL SALICYLATE see MPI000
METHYL SELENAC see SBQ000
METHYLSENFOEL (GERMAN) see ISE000
METHYLSERGIDE BIMALEATE see MQP500
METHYL SILICATE see MPI750
METHYL SILICONE see PJR000
METHYL STEARATE see MJW000
α-METHYLSTYREEN (DUTCH) see MPK250
METHYL STYRENE see VQK650
α-METHYL STYRENE see MPK250
METHYL STYRENE (mixed isomers) see MPK500
α-METHYL-STYROL (GERMAN) see MPK250
5-METHYL-3-SULFANILAMIDOISOXAZOLE see SNK000
METHYL SULFATE (DOT) see DUD100
METHYL SULFIDE (DOT) see TFP000
METHYLSULFINYLMETHANE see DUD800
METHYL SULFOCYANATE see MPT000
METHYLSULFONIC ACID, ETHYL ESTER see EMF500
2-METHYLSULFONYL-10-(3-(4-CARBAMOYLPIPER-
IDINO)PROPYL)PHENOTHIAZINE see MQR000
1-(3-(2-(METHYLSULFONYL)PHENOTHIAZIN-10-YL)PRO-
PYL)ISONIPECOTAMIDE see MQR000
1-(3-(2-(METHYLSULFONYL)PHENOTHIAZIN-10-YL)PRO-
PYL)-4-PIPERIDINE CARBOXAMIDE see MQR000
1-(3-(2-(METHYLSULFONYL)-10H-PHENOTHIAZIN-10-YL)
PROPYL)-4-PIPERIDINE CARBOXAMIDE see MQR000
METHYL SULFOXIDE see DUD800
5-METHYL-3-SULPHANIL-AMIDOISOXAZOLE see SNK000
METHYL SULPHIDE see TFP000
METHYLSYSTOX see DAO800, MIW100
17-METHYLTESTOSTERON see MPN500
METHYLTESTOSTERONE see MPN500
17-METHYLTESTOSTERONE see MPN500
17-α-METHYLTESTOSTERONE see MPN500
METHYL-1,2,5,6-TETRAHYDRO-1-METHYLNICOTINATE
see AQT750
2-METHYL-2-(4-(1,2,3,4-TETRAHYDRO-1-NAPHTHA-
LENYL)PHENOXY)PROPANOIC ACID see MCB500
2-METHYL-2-(4-(1,2,3,4-TETRAHYDRO-1-NAPHTHYL)
PHENOXY)PROPANOIC ACID see MCB500
2-METHYL-2-(p-(1,2,3,4-TETRAHYDRO-1-NAPHTHYL)
PHENOXY)PROPIONIC ACID see MCB500
α-METHYL-α-(p-1,2,3,4-TETRAHYDRONAPHTH-1-YL-
PHENOXY)PROPIONIC ACID see MCB500

N-METHYL-Δ-TETRAHYDRONICOTINIC ACID METHYL
ESTER see AQT750
N-METHYL-1,2,3,6-TETRAHYDROPHTHALIMIDE see
MIS250
N-METHYLTETRAHYDROPYRIDINE-β-CARBOXYLIC ACID
METHYL ESTER see AQT750
N-METHYLTETRAHYDROPYRROLE see MPB250
N-METHYL-N,2,4,6-TETRANITROANILINE see TEG250
METHYLTHEOBROMIDE see CAK500
1-METHYLTHEOBROMINE see CAK500
7-METHYLTHEOPHYLLINE see CAK500
N-METHYL-3-THIA-2-METHYL-VALERAMID DER O,O-DI-
METHYLTHIOLPHOSPHORSAEURE (GERMAN) see
MJG500
2-METHYLTHIO-ACETALDEHYD-O-(METHYLCARBA-
MOYL)-OXIM (GERMAN) see MDU600
l-γ-METHYLTHIO-α-AMINOBUTYRIC ACID see MDT750
2-METHYLTHIO-4,6-BIS(ISOPROPYLAMINO)-s-TRIAZINE
see BKL250
METHYL THIOCYANATE see MPT000
4-METHYLTHIO-3,5-DIMETHYLPHENYL METHYLCARBA-
MATE see DST000
2-(METHYLTHIO)-ETHANETHIOL-O,O-DIMETHYL PHOS-
PHOROTHIOATE see MIW250
2-(METHYLTHIO)-ETHANETHIOL-S-ESTER with O,O-DI-
METHYL PHOSPHOROTHIOATE see MIW250
METHYLTHIOINOSINE see MPU000
6-METHYLTHIOINOSINE see MPU000
METHYLTHIOMETHANE see TFP000
METHYLTHIONINE CHLORIDE see BJI250
METHYLTHIONIUM CHLORIDE see BJI250
4-METHYLTHIOPHENYLDIMETHYL PHOSPHATE see
PHD250
METHYLTHIOPHOS see MNH000
2-METHYLTHIO-PROPIONALDEHYD-O-(METHYLCARBA-
MOYL)-OXIM (GERMAN) see MDU600
6-(METHYLTHIO)PURINE RIBONUCLEOSIDE see MPU000
6-METHYLTHIOPURINE RIBOSIDE see MPU000
6-METHYL-2-THIO-2,4-(1H3H)PYRIMIDINEDIONE see
MPW500
METHYLTHIOURACIL see MPW500
6-METHYLTHIOURACIL see MPW500
4-METHYL-2-THIOURACIL see MPW500
6-METHYL-2-THIOURACIL see MPW500
METHYLTHIOXOARSINE see MGQ750
4-(METHYLTHIO)-3,5-XYLENOL METHYLCARBAMATE see
DST000
4-(METHYLTHIO)-3,5-XYLYL METHYLCARBAMATE see
DST000
METHYL THIRAM see TFS350
METHYL THIURAMDISULFIDE see TFS350
METHYL-α-TOLUATE see MHA500
METHYL TOLUENE see XGS000
o-METHYLTOLUENE see XHJ000
p-METHYLTOLUENE see XHS000
METHYL TOLUENE-4-SULFONATE see MLL250
METHYL-p-TOLUENESULFONATE see MLL250
α-METHYL-α-TOLUIC ALDEHYDE see COF000
2-METHYL-p-TOLUIDINE see XMS000
4-METHYL-o-TOLUIDINE see XMS000
5-METHYL-o-TOLUIDINE see XNA000
6-METHYL-m-TOLUIDINE see XNA000
2-METHYL-p-TOLUIDINE HYDROCHLORIDE see XOJ000
4-METHYL-o-TOLUIDINE HYDROCHLORIDE see XOJ000
5-METHYL-o-TOLUIDINE HYDROCHLORIDE see XOS000
6-METHYL-m-TOLUIDINE HYDROCHLORIDE see XOS000
N-METHYL-p-(m-TOLYLAZO)ANILINE see MPY000
2-METHYL-3-o-TOLYL-4(3H)-CHINAZOLINON (GERMAN)
see QAK000
2-METHYL-3-o-TOLYL-4(3H)-CHINAZOLONE see QAK000
2-METHYL-3-TOLYLCHINAZOLON-4 HYDROCHLORIDE
(GERMAN) see MDT250
(2-METHYL-3-(o-TOLYL)-3,4-DIHYDRO-4-(QUINAZOLI-
NONE) see QAK000

2-METHYL-3-(o-TOLYL)-3,4-DIHYDRO-4-QUINAZOLINONE
see QAK000
METHYL-p-TOLYL ETHER see MGP000
METHYL-p-TOLYL KETONE see MFW250
2-METHYL-3-TOLYL-4-OXYBENSDIAZINE see QAK000
2-METHYL-3-o-TOLYL-4(3H)-QUINAZOLINONE see
QAK000
2-METHYL-3-o-TOLYL-4(3H)-QUINAZOLINONE HYDRO-
CHLORIDE see MDT250
2-METHYL-3-(2-TOLYL)QUINAZOL-4-ONE see QAK000
2-METHYL-3-o-TOLYL-4-QUINAZOLONE see QAK000
2-METHYL-3-(o-TOLYL)-4-QUINAZOLONE HYDROCHLO-
RIDE see MDT250
METHYL TOSYLATE see MLL250
METHYL-p-TOSYLATE see MLL250
METHYLTRICAPRYLYLAMMONIUMCHLORIDE see
MQH000
METHYL TRICHLORIDE see CHJ500
METHYLTRICHLOROMETHANE see MIH275
METHYLTRICHLOROSILANE see MQC500
METHYL-TRICHLORSILAN (CZECH) see MQC500
METHYL TRIFLUORIDE see CBY750
5-METHYL-2-TRIFLUOROMETHYLOXAZOLIDINE see
MQE000
METHYL-3,3,3-TRIFLUORO-2-(TRIFLUOROMETHYL)PRO-
PIONATE see MKK750
α-METHYL-3,4,5-TRIMETHOXYPHENETHYLAMINE HY-
DROCHLORIDE see TKX500
METHYLTRIMETHYLENE GLYCOL see BOS500
4-METHYL TRIMETHYLENE SULFITE see MQG500
3-METHYL-2,4,6-TRINITROPHENOL see TML500
METHYLTRIOCTYLAMMONIUM CHLORIDE see MQH000
METHYL TRITHION see MQH750
α-METHYLTRYPTAMINE see AME500
METHYL TUADS see TFS350
METHYL-4-UMBELLIFERONE SODIUM see HMB000
4-METHYLURACIL see MPW500
METHYLURETHAN see MHZ000
N-METHYL URETHAN see EMQ500
METHYLURETHANE see MHZ000
METHYL-VINYL-CETONE (FRENCH) see BOY500
METHYL VINYL ETHER see MQL750
METHYLVINYLKETON (GERMAN) see BOY500
METHYL VINYL KETONE see BOY500
METHYLVINYLNITROSAMIN (GERMAN) see NKY000
METHYLVINYLNITROSAMINE see NKY000
METHYL VIOLET 6B see MQN000
METHYLVIOLOGEN see PAJ000
METHYL VIOLOGEN (2+) see PAI990
METHYL YELLOW see DOT300
METHYL ZIMATE see BJK500
METHYL ZINEB see BJK500
METHYL ZIRAM see BJK500
METHYPHENYLMETHANOL see PDE000
METHYSERGIDE DIMALEATE see MQP500
METIAPINE see MQQ000
METICORTELONE see PMA000
METI-DERM see PMA000
METIFONATE see TIQ250
METILACRILATO (ITALIAN) see MGA500
METILAMIL ALCOHOL (ITALIAN) see MKW600
METILAMINE (ITALIAN) see MGC250
2-METIL-6-AMINO-EPTANO (ITALIAN) see ILM000
3-METIL-BUTANOLO (ITALIAN) see IHP000
METIL CELLOSOLVE (ITALIAN) see EJH500
2-METILCICLOESANONE (ITALIAN) see MIR500
METILCLOROFORMIATO (ITALIAN) see MIG000
METILCLORPINDOL see CMX850
METILE (ACETATO di) (ITALIAN) see MFW100
METILENBIOTIC see MDO250
O,O-METILEN-BIS-(4-CLOROFENOLO) (ITALIAN) see
MJM500
3,3'-METILEN-BIS(4-IDROSSI-CUMARINA) (ITALIAN) see
BJZ000

4,4-METILENE-BIS-o-CLOROANILINA (ITALIAN) see
MJM200
(6-(1-METIL-EPITL)-2,4-DINITRO-FENIL)-CROTONATO
(ITALIAN) see AQT500
METILETILCHETONE (ITALIAN) see MKA400
METIL (FORMIATO di) (ITALIAN) see MKG750
METILISOBUTILCHETONE (ITALIAN) see HFG500
METIL ISOCIANATO (ITALIAN) see MKX250
METILMERCAPTANO (ITALIAN) see MLE650
METILMERCAPTOFOSOKSID see DAP000
METIL METACRILATO (ITALIAN) see MLH750
α-METIL-β-(2-METILENE-4,5-DIIDROIMIDAZOLIL)BENZO-
TIOFANE CLORIDRATO (ITALIAN) see MHJ500
N-METIL-1-NAFTIL-CARBAMMATO (ITALIAN) see
CBM750
METILPARATION (HUNGARIAN) see MNH000
4-METILPENTAN-2-OLO (ITALIAN) see MKW600
4-METILPENTAN-2-ONE (ITALIAN) see HFG500
4-METIL-3-PENTEN-2-ONE (ITALIAN) see MDJ750
6-(1-METIL-PROPIL)-2,4-DINITRO-FENOLO (ITALIAN) see
BRE500
α-METIL-STIROLO (ITALIAN) see MPK250
2-METIL-2-TIOMETIL-PROPIONALDEID-O-(N-METIL-CAR-
BAMOIL)-OSSIMA (ITALIAN) see CBM500
6-METIL-TIOURACILE (ITALIAN) see MPW500
METILTRIAZOTION see ASH500
METINDOL see IDA000
METIONE see MDT740
METIPREGNONE see MCA000
METIZOL see MCO500
METIZOLINE HYDROCHLORIDE see MHJ500
METMERCAPTURON see DST000
METOFANE see DFA400
2-METOKSY-4-ALLILOFENOL (POLISH) see EQR500
METOKSYCHLOR (POLISH) see MEI450
METOKSYETYLOWY ALKOHOL (POLISH) see EJH500
METOLQUIZOLONE see QAK000
METOMIDATE see MQQ500
METOMIL (ITALIAN) see MDU600
METOPIMAZINE see MQR000
METOSERPATE HYDROCHLORIDE see MQR200
(2-METOSSICARBONIL-1-METIL-VINIL)-DIMETIL-FOSFATO
(ITALIAN) see MQR750
2-METOSSIETANOLO (ITALIAN) see EJH500
2-METOSSIETILACETATO(ITALIAN) see EJJ500
S-((5-METOSSI-4H-PIRON-2-IL)-METIL)-O,O-DIMETIL-
MONOTIOFOSFATO (ITALIAN) see EAS000
METOTHYRINE see MCO500
METOX see CEP000, MEI450
METOXAL see SNK000
METOXFLURAN see DFA400
METOXIDON see SNN300
METOXIFLURAN see DFA400
METOXON see CKC000
METOXYDE see HJL500
METRACTYL see MQU750
METRAMAC see DJA400
METRAMAK see DJA400
METRANIL see PBC250
METRASPRAY see BQA010
METRIBUZIN see MQR275
METRIFONATE see TIQ250
METRIPHONATE see TIQ250
METRISONE see MOR500
METRODIOL see EQJ500
METRODIOL DIACETATE see EQJ500
METROGEN RED FORMER KB SOLN see CLK225
METRON see MNH000
METRONE see MPN500
METRONIDAZ see MMN250
METRONIDAZOL see MMN250
METRONIDAZOLO see MMN250
METRO TALC 4604 see TAB750
METRO TALC 4608 see TAB750

METRO TALC 4609 see TAB750
METSO 20 see SJU000
METSO BEADS 2048 see SJU000
METSO BEADS, DRYMET see SJU000
METSO PENTABEAD 20 see SJU000
METYCAINE see PIV750
METYLAL (POLISH) see MGA850
METYLENU CHLOREK (POLISH) see MJP450
METYLESTER KYSELINY SALICYLOVE (CZECH) see
  MPI000
METYLFENEMAL see ENB500
METYLOAMINA (POLISH) see MGC250
METYLOCYKLOHEKSAN (POLISH) see MIQ740
METYLOCYKLOHEKSANOL (POLISH) see MIQ745
METYLOCYKLOHEKSANON (POLISH) see MIR250
METYLOETYLOKETON (POLISH) see MKA400
METYLOHYDRAZYNA (POLISH) see MKN000
METYLOIZOBUTYLOKETON (POLISH) see HFG500
1-METYLO-2-MERKAPTOIMIDAZOLEM (POLISH) see
  MCO500
N-METYLO-N'-NITRO-N-NITROZOGOUANIDYNY (POLISH)
  see MMP000
METYLOPARATION (POLISH) see MNH000
METYLOPROPYLOKETON (POLISH) see PBN250
METYLOWY ALKOHOL (POLISH) see MGB150
METYLPARATION (CZECH) see MNH000
METYLU BROMEK (POLISH) see MHR200
METYLU CHLOREK (POLISH) see MIF765
METYLU JODEK (POLISH) see MKW200
METYNA see ENB500
MEVASIN HYDROCHLORIDE see MQR500
MEVINFOS (DUTCH) see MQR750
MEVINPHOS see MQR750
MEXACARBATE (DOT) see DOS000
MEXENE see BJK500
MEXEPHENAMIDE see DIB600
MEXICO WEED see CCP000
MEXIDE see RNZ000
MEXIDEX see SOW000
MEXOCINE see MIJ500
MEXYL see ABX500
MEYPRALGIN R/LV see SEH000
MEZATON see NCL500
R(−)-MEZATON see NCL500
MEZCALINE see MDI500
MEZCLINE see MDI500
MEZENE see BJK500
MEZIDINE see TLG500
MEZOLIN see IDA000
MEZOTOX see DFT800
MFA see FIC000, MKD000
MFH see FNW000
MFI see IPX000
MFNA see MME809
h-MG see CGY750
MG 18037 see BQX000
MG 18370 see DTJ400
MG 18570 see DTJ400
MGA see MCB380
MGA 100 (STEROID) see MCB380
M7-GIFTKOERNER see TEM000
MGK-264 see OES000
MGK DIETHYLTOLUAMIDE see DKC800
MGK DOG AND CAT REPELLENT see UKS000
MH-30 see DHF200
MH-532 see PGA750
2M-4KH see CIR250
3-MI see MKV750
MIA see IDZ000
MIAK see MKW450
MIANESINA see GGS000
MIARSENOL see NCJ500
MIAZOLE see IAL000
MIBC see MKW600

MIBK see HFG500
MIBOLERON see MQS225
MIBOLERONE see MQS225
MIC see ISE000, MKW600
3-MIC see MKW600
MICA see MQS250
MICA SILICATE see MQS250
MICHLER'S BASE see MJN000
MICHLER'S HYDRIDE see MJN000
MICHLER'S KETONE see MQS500
p,p'-MICHLER'S KETONE see MQS500
MICHLER'S METHANE see MJN000
MICIDE see EIR000
MICOCHLORINE see CDP250
MICOFUME see DSB200
MICOFUR see NGC000
MICOL see HCQ500
MICREST see DKA600
MICROCETINA see CDP250
MICRO-CHECK 12 see CBG000
MICRO DDT 75 see DAD200
MICRODIOL see EDO000
MICROEST see DKA600
MICROFLOTOX see SOD500
MICROGRIT WCA see AHE250
MICRO-LEX GREEN 5B see BLK000
MICROLYSIN see CKN500
MICROSETILE ORANGE RA see AKP750
MICROSETILE YELLOW GR see AAQ250
MICROTEX LAKE RED CR see CHP500
MICROTHENE see PJS750
MICROZUL see CJJ000
MIDARINE see HLC500
MIDONE see DBB200
MIELUCIN see BOT250
MIEOBROMOL see DDP600
MIERENZUUR (DUTCH) see FNA000
MIGHTY 150 see NAJ500
MIH see PME250
MIH HYDROCHLORIDE see PME500
MIK see HFG500
MIKAMETAN see IDA000
MIKAMYCIN see VRF000
MIKEDIMIDE see MKA250
MIKETORIN see EAI000
MILAXEN see HEG000
MILBAM see BJK500
MILBAN see BJK500
MIL-B-4394-B see CES650
MILBEDOCE see VSZ000
MILBOL see BIO750
MILBOL 49 see BBQ500
MILCHSAURE (GERMAN) see LAG000
MIL-COL see MLC250
MILDMEN see LFK000, MDQ250
MILD MERCURY CHLORIDE see MCW000
MIL-DU-RID see BGJ750
MILEPSIN see DBB200
MILESTROL see DKA600
MILK ACID see LAG000
MILK of MAGNESIA see MAG750
MILK SUGAR see LAR000
MILK WHITE see LDY000
MILLER NU SET see TIX500
MILLER P.C. WEEDKILLER see PLC250
MILLICORTEN see SOW000
MILMER see BLC250
MILOGARD see PMN850
MILPREM see ECU750, MQU750
MILPREX see DXX400
MILTANN see MQU750
MILTON see SHU500
MILTOWN see MQU750
MILTOX see EIR000

MILTOX SPECIAL see EIR000
MIMOSA TANNIN see MQV250
MINERAL DUSTS see MQV500
MINERAL GREEN see COF500
MINERAL NAPHTHA see BBL250
MINERAL OIL see MQV750
MINERAL OIL, PETROLEUM CONDENSATES, VACUUM
  TOWER see MQV755
MINERAL OIL, PETROLEUM DISTILLATES, ACID-
  TREATED HEAVY NAPHTHENIC see MQV760
MINERAL OIL, PETROLEUM DISTILLATES, ACID-
  TREATED HEAVY PARAFFINIC see MQV765
MINERAL OIL, PETROLEUM DISTILLATES, ACID-
  TREATED LIGHT NAPHTHENIC see MQV770
MINERAL OIL, PETROLEUM DISTILLATES, ACID-
  TREATED LIGHT PARAFFINIC see MQV775
MINERAL OIL, PETROLEUM DISTILLATES, HEAVY
  NAPHTHENIC see MQV780
MINERAL OIL, PETROLEUM DISTILLATES, HEAVY
  PARAFFINIC see MQV785
MINERAL OIL, PETROLEUM DISTILLATES, HYDRO-
  TREATED HEAVY NAPHTHENIC see MQV790
MINERAL OIL, PETROLEUM DISTILLATES, HYDRO-
  TREATED HEAVY PARAFFINIC see MQV795
MINERAL OIL, PETROLEUM DISTILLATES, HYDRO-
  TREATED LIGHT NAPHTHENIC see MQV800
MINERAL OIL, PETROLEUM DISTILLATES, HYDRO-
  TREATED LIGHT PARAFFINIC see MQV805
MINERAL OIL, PETROLEUM DISTILLATES, LIGHT
  NAPHTHENIC see MQV810
MINERAL OIL, PETROLEUM DISTILLATES, LIGHT
  PARAFFINIC see MQV815
MINERAL OIL, PETROLEUM DISTILLATES, SOLVENT-DE-
  WAXED LIGHT NAPHTHENIC see MQV835
MINERAL OIL, PETROLEUM DISTILLATES, SOLVENT-DE-
  WAXED LIGHT PARAFFINIC see MQV840
MINERAL OIL, PETROLEUM DISTILLATES, SOLVENT-RE-
  FINED LIGHT NAPHTHENIC see MQV852
MINERAL OIL, PETROLEUM DISTILLATES, SOLVENT-RE-
  FINED LIGHT PARAFFINIC see MQV855
MINERAL OIL, PETROLEUM EXTRACTS, HEAVY
  NAPHTHENIC DISTILLATE SOLVENT see MQV857
MINERAL OIL, PETROLEUM EXTRACTS, HEAVY PARAF-
  FINIC DISTILLATE SOLVENT see MQV859
MINERAL OIL, PETROLEUM EXTRACTS, LIGHT
  NAPHTHENIC DISTILLATE SOLVENT see MQV860
MINERAL OIL, PETROLEUM EXTRACTS, LIGHT PARAF-
  FINIC DISTILLATE SOLVENT see MQV862
MINERAL OIL, PETROLEUM EXTRACTS, RESIDUAL OIL
  SOLVENT see MQV863
MINERAL OIL, PETROLEUM NAPHTHENIC OILS, CATA-
  LYTIC DEWAXED HEAVY see MQV865
MINERAL OIL, PETROLEUM NAPHTHENIC OILS, CATA-
  LYTIC DEWAXED LIGHT see MQV867
MINERAL OIL, PETROLEUM PARAFFIN OILS, CATALYTIC
  DEWAXED HEAVY see MQV868
MINERAL OIL, PETROLEUM PARAFFIN OILS, CATALYTIC
  DEWAXED LIGHT see MQV870
MINERAL OIL, PETROLEUM RESIDUAL OILS, ACID-
  TREATED see MQV872
MINERAL OIL, SLAB OIL see MQV875
MINERAL OIL, WHITE (FCC) see MQV750
MINERAL ORANGE see LDS000
MINERAL PITCH see ARO500
MINERAL RED see LDS000
MINERAL SPIRITS see PCT250
MINERAL THINNER see PCT250
MINERAL TURPENTINE see PCT250
MINERAL WHITE see CAX750
MINETOIN see DKQ000, DNU000
MINGIT see DFP600
MINIHIST see DBM800, WAK000
MINIUM see LDS000
MINIUM NON-SETTING RL-95 see LDS000

MINOALEUIATIN see TLP750
MINOPHAGEN A see AQW000
MINORLAR see EEH520
MINOSSIDILE (ITALIAN) see DCB000
MINOVLAR see EEH520
MINOXIDIL see DCB000
MINTACO see NIM500
MINTACOL see NIM500
MINTAL see NBU000
MINTEZOL see TEX000
MINUS see DKE800, SEH000
MINZOLUM see TEX000
MIOARTRINA see IPU000
MIOLISODAL see IPU000
MIOLISODOL see IPU000
MIORATRINA see IPU000
MIORIL see IPU000
MIORIODOL see IPU000
MIOTICOL see DNR309
MIOTISAL see NIM500
MIOTISAL A see NIM500
MIPAFOX (DOT) see PHF750
MIPAX see DTR200
MIPC see MIA250
MIPCIN see MIA250
MIPK see MLA750
MIPSIN see MIA250
MIRACIL D see DHU000
MIRACLE see DAA800
MIRAL see PDN000
MIRAMID WM 55 see PJY500
MIRAPRONT see DTJ400
MIRBANE OIL see NEX000
MIREX see MQW500
MIRLON see NOH000
MIROISTONIL see DRM000
MIRREX MCFD 1025 see PKQ059
MISHERI (INDIA) see SED400
MISHRI (INDIA) see SED400
MISODINE see DBB200
MISOLYNE see DBB200
MISPICKEL see ARJ750
MISTRON 2SC see TAB750
MISTRON FROST P see TAB750
MISTRON RCS see TAB750
MISTRON STAR see TAB750
MISTRON SUPER FROST see TAB750
MISTRON VAPOR see TAB750
MISULBAN see BOT250
MIT see ISE000
MITACIL see DOR400
MITANOLINE see PMC250
MIT-C see AHK500
MITC see ISE000
MITENON see MMD500
MITHRACIN see MQW750
MITHRAMYCIN see MQW750
MITHRAMYCIN A see MQW750
MITICIDE K-101 see CJT750
MITIGAN see BIO750
MITION see CKM000
MITIS GREEN see COF500
MITOBRONITOL see DDP600
MITO-C see AHK500
MITOCIN-C see AHK500
MITOLAC see DDJ000
MITOLACTOL see DDJ000
MITOMEN see CFA500, CFA750
MITOMIN see CFA500
MITOMYCIN see AHK500
MITOMYCIN-C see AHK500
MITOMYCINUM see AHK500
MITOSTAN see BOT250
MITOTANE see CDN000

MITOX see CEP000
MITOXAN see CQC675, EAS500
MITOXANA see IMH000
MITOXINE see BIE500
MITRAMYCIN see MQW750
MITSUI AURAMINE O see IBA000
MITSUI BLUE B BASE see DCJ200
MITSUI BRILLIANT GREEN G see BAY750
MITSUI DIRECT BLACK EX see AQP000
MITSUI DIRECT BLUE 2BN see CMO000
MITSUI METHYLENE BLUE see BJI250
MITSUI RED TR BASE see CLK220
MITSUI RHODAMINE BX see FAG070
MITSUI SCARLET G BASE see NMP500
MIXTURE of p-METHENOLS see TBD500
MIZODIN see DBB200
MIZOLIN see DBB200
MJ 1999 see CCK250
MJ 1999 HYDROCHLORIDE see CCK250
MJ 5022 see MPE250
MJF 9325 see IMH000
MK 56 see POL500
MK 125 see SOW000
MK-188 see RBF100
MK 351 see DNA800
MK 360 see TEX000
MK-595 see DFP600
MK 647 see DKI600
MK. B51 see DNA800
2M 4KHP see CIR500
MLA-74 see MLD500
MLT see MAK700
MM see BKM500
MM 4462 see AOP250
MMA see MGQ250
MMC see AHK500, MDD750
MMD see MLF250
MME see MFC700, MLH750
M'-METHYL-p-DIMETHYLAMINOAZOBENZENE see DUH600
MMH see MKN000
4-MMPD see DBO000
4-MMPD SULPHATE see DBO400
MMS see MLH500
MMT see MAV750
MNA see MMU250, NEN500
MNBK see HEV000
MNC see MMX000
MNCO see MQY325
MNE see NDM000
MNFA see MME809
MNG see MMP000
MNNG see MMP000
MNPN see MMS200
MNQ see MMD500
MNT see NMO500
MNU see MMX250, MNA750
MNU and CYCLOPHOSPHAMIDE (2:1) see CQC750
MO 709 see DRM000
MOBENOL see BSQ000
MOBILAN see IDA000
MOBILAWN see DFK600
MOBIL V-C 9-104 see EIN000
MOCA see MJM200
MOCAP see EIN000
MOCTYNOL see GGS000
MODANE SOFT see DJL000
MODOCOLL 1200 see SFO500
MODR FRALOSTANOVA 3G (CZECH) see DNE400
MODR METHYLENOVA (CZECH) see BJI250
MOENOMYCIN see MRA250
MOENOMYCIN A see MRA250
MOF see DFA400
MOGADAN see DLY000

MOHICAN RED A-8008 see CHP500
MOLASSES ALCOHOL see EFU000
MOLATOC see DJL000
MOLCER see DJL000
MOLDEX see HJL500
MOLECULAR CHLORINE see CDV750
MOLE DEATH see SMN500
MOLINAL see EOK000
MOL-IRON see FBO000
MOLLAN O see DVL700
MOLLINOX see QAK000
MOLLUSCICIDE BAYER 73 see DFV600
MOLOFAC see DJL000
MOLOL see MQV750
MOLTEN ADIPIC ACID see AEN250
MOLURAME see BJK500
MOLYBDATE see MRC250
MOLYBDENUM see MRC250
MOLYBDENUM COMPOUNDS see MRC750
MOLYBDENUM-LEAD CHROMATE see LDM000
MOLYBDENUM ORANGE see LDM000
MOLYBDENUM(VI) OXIDE see MRE000
MOLYBDENUM PENTACHLORIDE see MRD500
MOLYBDENUM TRIOXIDE see MRE000
MOLYBDIC ACID DIAMMONIUM SALT see ANM750
MOLYBDIC ACID, DISODIUM SALT see DXE800
MOLYBDIC ANHYDRIDE see MRE000
MOLYBDIC TRIOXIDE see MRE000
MOLYKOTE 522 see TAI250
MOMENTUM see HIM000
MON 0573 see PHA500
MONACRIN see AHS500
MONAGYL see MMN250
MONAMID 150-LW see BKE500
MONAQUEST see DJG800
MONARGAN see ABX500
MONASIRUP see PMQ750
MONATE see MRL750
MONAWET MD 70E see DJL000
MONDUR P see PFK250
MONDUR-TD see TGM740, TGM750
MONDUR-TD-80 see TGM740, TGM750
MONDUR TDS see TGM750
MONELAN see MRE225
MONELGIN see OAV000
MONENSIC ACID see MRE225
MONENSIN (USDA) see MRE225
MONENSIN A see MRE225
MONEX see BJL600
MONITAN see PKL100
MONITOR see DTQ400
MONKIL WP see MGQ750
MONOACETYLHYDRAZINE see ACM750
MONOAETHANOLAMIN (GERMAN) see EEC600
MONOALLYLAMINE see AFW000
MONOAMMONIUM CARBONATE see ANB250
MONOAMMONIUM GLUTAMATE see MRF000
MONOAMMONIUM l-GLUTAMATE see MRF000
MONOAMMONIUM SULFAMATE see ANU650
MONOAMMONIUM SULFIDE see ANJ750
MONOBASIC CHROMIUM SULFATE see NBW000
MONOBASIC CHROMIUM SULPHATE see NBW000
MONOBASIC LEAD ACETATE see LCH000
MONOBROMESSIGSAEURE (GERMAN) see BMR750
MONOBROMOACETIC ACID see BMR750
MONOBROMOACETONE see BNZ000
MONOBROMOBENZENE see PEO500
MONOBROMOETHANE see EGV400
MONOBROMOISOVALERYLUREA see BNP750
2-MONOBROMOISOVALERYLUREA see BNP750
MONOBROMOMETHANE see MHR200
MONO-n-BUTYLAMINE see BPX750
MONOBUTYL GLYCOL ETHER see BPJ850
MONO-tert-BUTYLHYDROQUINONE see BRM500

MONOCAINE HYDROCHLORIDE see IAC000
MONOCALCIUM ARSENITE see CAM500
MONOCHLOORAZIJNZUUR (DUTCH) see CEA000
MONOCHLOORBENZEEN (DUTCH) see CEJ125
MONOCHLORACETIC ACID see CEA000
MONOCHLORACETONE see CDN200
MONOCHLORBENZENE see CEJ125
MONOCHLORBENZOL (GERMAN) see CEJ125
MONOCHLORESSIGSAEURE (GERMAN) see CEA000
MONOCHLORETHANE see EHH000
MONOCHLORHYDRIN see CDT750
MONOCHLORHYDRINE du GLYCOL (FRENCH) see
  EIU800
MONOCHLOROACETALDEHYDE see CDY500
MONOCHLOROACETIC ACID see CEA000
MONOCHLOROACETIC ACID METHYL ESTER see MIF775
MONOCHLOROACETONE see CDN200
MONOCHLOROACETONE, inhibited (DOT) see CDN200
MONOCHLOROACETONE, stabilized (DOT) see CDN200
MONOCHLOROACETONE, unstabilized (DOT) see CDN200
MONOCHLOROACETONITRILE see CDN500
MONOCHLOROACETYL CHLORIDE see CEC250
α-MONOCHLOROANTHRAQUINONE see CEI000
MONOCHLOROBENZENE see CEJ125
MONOCHLORODIFLUOROMETHANE see CFX500
MONOCHLORODIMETHYL ETHER (MAK) see CIO250
MONOCHLORO DIPHENYL OXIDE see MRG000
MONOCHLOROETHANOIC ACID see CEA000
2-MONOCHLOROETHANOL see EIU800
MONOCHLOROETHENE see VNP000
MONOCHLOROETHYLENE (DOT) see VNP000
MONOCHLOROHYDRIN see CDT750
α-MONOCHLOROHYDRIN see CDT750
MONOCHLOROMETHANE see MIF765
MONOCHLOROMETHYL CYANIDE see CDN500
MONO-CHLORO-MONO-BROMO-METHANE see CES650
MONOCHLOROMONOFLUOROMETHANE see CHI900
MONOCHLOROPENTAFLUOROETHANE (DOT) see CJI500
MONOCHLOROPHENYLETHER see MRG000
MONOCHLOROSULFURIC ACID see CLG500
MONOCHLOROTETRAFLUOROETHANE (DOT) see
  CLH000
MONOCHLOROTRIFLUOROETHYLENE see CLQ750
MONOCHLOROTRIFLUOROMETHANE (DOT) see CLR250
MONOCHROMIUM OXIDE) see CMK000
MONOCHROMIUM TRIOXIDE see CMK000
MONOCIL 40 see MRH209
"MONOCITE" METHACRYLATE MONOMER see MLH750
MONOCLOROBENZENE (ITALIAN) see CEJ125
MONOCRATILIN see MRH000
MONOCRON see MRH209
MONOCROTALINE see MRH000
MONOCROTOPHOS see ASN000, MRH209
MONODEMETHYLIMIPRAMINE see DSI709
MONODORM see BPF500
MONOETHANOLAMINE see EEC600
N-MONOETHYLAMIDE of O,O-DIMETHYLDITHIOPHOS-
  PHORYLACETIC ACID see DNX600
MONOETHYLAMINE (DOT) see EFU400
MONOETHYLAMINE, anhydrous (DOT) see EFU400
2-N-MONOETHYLAMINOETHANOL see EGA500
MONOETHYLENE GLYCOL see EJC500
MONOETHYLENE GLYCOL DIMETHYL ETHER see
  DOE600
MONOETHYL ETHER of DIETHYLENE GLYCOL see
  CBR000
MONO(2-ETHYLHEXYL)SULFATE SODIUM SALT see
  TAV750
MONOFLUORAZIJNZUUR (DUTCH) see FIC000
MONOFLUORESSIGSAURE (GERMAN) see FIC000
MONOFLUORESSIGSAURES NATRIUM (GERMAN) see
  SHG500
MONOFLUORETHANOL see FID000
MONOFLUOROACETAMIDE see FFF000

MONOFLUOROACETATE see FIC000
MONOFLUOROACETIC ACID see FIC000
MONOFLUOROETHANE see FIB000
MONOFLUOROETHANOL see FID000
MONOFLUOROETHYLENE see VPA000
MONOFLUOROPHOSPHORIC ACID, anhydrous see PHJ250
MONOFLUOROTRICHLOROMETHANE see TIP500
MONOFURACIN see NGE500
MONOGERMANE see GEI100
MONO-GLYCOCOARD see DKL800
MONOGLYME see DOE600
MONO-N-HEXYLAMINE see HFK000
8-MONOHYDRO MIREX see MRI750
MONOHYDROXYBENZENE see PDN750
MONOHYDROXYMETHANE see MGB150
MONOIODOACETAMIDE see IDW000
MONOIODOACETATE see IDZ000
MONOIODOACETIC ACID see IDZ000
MONOIODURO di METILE (ITALIAN) see MKW200
MONOISOBUTYLAMINE see IIM000
N-MONOISOPROPYLAMIDE of O,O-DIETHYLDITHIOPHOS-
  PHORYLACETIC ACID see IOT000
MONOISOPROPYLAMINE see INK000
MONOISOPROPYL ETHER of ETHYLENE GLYCOL see
  INA500
MONOLINURON see CKD500
MONOMETHYLACETAMIDE see MFT750
N-MONOMETHYLAMIDE of O,O-DIMETHYLDITHIO-
  PHOSPHORYLACETIC ACID see DSP400
MONOMETHYLAMINE see MGC250
MONOMETHYLAMINE, anhydrous (DOT) see EDY000,
  MGC250
MONOMETHYLAMINE, aqueous solution (DOT) see
  EDY500, MGC250
MONOMETHYL-AMINOAETHANOL (GERMAN) see
  MGG000
4-MONOMETHYLAMINOAZOBENZENE see MNR500
p-MONOMETHYLAMINOAZOBENZENE see MNR500
MONOMETHYLAMINOETHANOL see MGG000
N-MONOMETHYLAMINOETHANOL see MGG000
N-MONOMETHYLANILINE see MGN750
MONOMETHYL ANILINE (OSHA) see MGN750
MONOMETHYL ETHER of ETHYLENE GLYCOL see
  EJH500
MONO METHYL ETHER HYDROQUINONE see MFC700
MONOMETHYLFORMAMIDE see MKG500
MONOMETHYL GUANIDIN (GERMAN) see MKI750
MONOMETHYLGUANIDINE see MKI750
MONOMETHYL HYDRAZINE see MKN000
MONOMETHYLHYDRAZINE NITRATE see MRJ250
MONOMETHYL MERCURY CHLORIDE see MDD750
MONONITROSOPIPERAZINE see MRJ750
MONOPENTEK see PBB750
MONOPHEN see PFC750
MONOPHENOL see PDN750
MONOPLEX DBS see DEH600
MONOPLEX DOS see BJS250
MONOPOTASSIUM ARSENATE see ARD250
MONOPOTASSIUM DIHYDROGEN ARSENATE see
  ARD250
MONOPOTASSIUM GLUTAMATE see MRK500
MONOPOTASSIUM l-GLUTAMATE (FCC) see MRK500
MONOPOTASSIUM SALT of ACETYLENEDICARBOXYLIC
  ACID see ACJ500
MONOPOTASSIUM SULFATE see PKX750
MONO-N-PROPYLAMINE see PND250
MONOPROPYLENE GLYCOL see PML000
MONOPYRROLE see PPS250
MONOSAN see DAA800
MONOSILANE see SDH575
MONOSODIOGLUTAMMATO (ITALIAN) see MRL500
MONOSODIUM ACID METHANEARSONATE see MRL750
MONOSODIUM ACID METHARSONATE see MRL750

MONOSODIUM ARSENATE see ARD600
MONOSODIUM ASCORBATE see ARN125
MONOSODIUM DIHYDROGEN PHOSPHATE see SJH100
MONOSODIUM-5-ETHYL-5-(1-METHYLBUTYL) THIOBAR-
BITURATE see PBT500
MONOSODIUM GLUCONATE see SHK800
MONOSODIUM GLUTAMATE see MRL500
α-MONOSODIUM GLUTAMATE see MRL500
MONOSODIUM-l-GLUTAMATE (FCC) see MRL500
MONOSODIUM METHANEARSONATE see MRL750
MONOSODIUM METHANEARSONIC ACID see MRL750
MONOSODIUM METHYLARSONATE see MRL750
MONOSODIUM NOVOBIOCIN see NOB000
MONOSODIUM PHOSPHATE see SJH100
MONOSORB XP-4 see SJH100
MONOSTEARIN see OAV000
MONOSULFUR DICHLORIDE see SOG500
MONOTEN see PFC750
MONOTHIOETHYLENEGLYCOL see MCN250
MONO-THIURAD see BJL600
MONOTHIURAM see BJL600
MONOTRICHLOR-AETHYLIDEN-α-GLUCOSE (GERMAN)
see GFA000
MONOVAR see NNQ500
MONOXONE see SFU500
β-MONOXYNAPHTHALENE see NAX000
MONSANTO CP 47114 see DSQ000
MONSANTO CP-49674 see DOP200
MONTANE 60 see SKV150
MONTANOX 80 see PKL100
MONTAR see PJL750
MONTECATINI L-561 see DRR400
MONTHYBASE see EJM500
MONTHYLE see EJM500
MONTMORILLONITE see BAV750
MONTREL see COD850
MONTROSE PROPANIL see DGI000
MONUREX see CJX750
MONURON see CJX750
MONUROX see CJX750
MONURUON see CJX750
MONUURON see CJX750
MOON see GGA000
MOP see NKV000
8-MOP see XDJ000
MOPA see MFE250
MOPARI see DGP900
MOPAZIN see MFK500
MOPAZINE see MFK500
MOPERONE CHLORHYDRATE see MNN250
MOPERONE HYDROCHLORIDE see MNN250
MOPLEN see PMP500
MORBOCID see FMV000
MORBUSAN see ENB500
MOREPEN see AOD125
MORFAMQUAT see BJK750
MORFINA (ITALIAN) see MRO500
MORFOTHION (DUTCH) see MRU250
MORFOXONE see BJK750
MORIN see MRN500
MORONAL see NOH500
MORPHACETIN see HBT500
MORPHANQUAT DICHLORIDE see BJK750
MORPHIA see MRO500
MORPHINA see MRO500
(−)-MORPHINE see MRO500
MORPHINE see MRO500
MORPHINE CHLORHYDRATE see MRO750
MORPHINE CHLORIDE see MRO750
MORPHINE DIACETATE see HBT500
MORPHINE HYDROCHLORIDE see MRO750
MORPHINE SULFATE see MRP250
MORPHINE SULPHATE see MRP250
MORPHINISM see MRO500

MORPHINUM see MRO500
MORPHIUM see MRO500
MORPHOLINE see MRP750
MORPHOLINE, AQUEOUS MIXTURE (DOT) see MRP750
MORPHOLINE DISULFIDE see BKU500
4-MORPHOLINENONYLIC ACID see MRQ750
MORPHOLINODISULFIDE see BKU500
l-5-(MORPHOLINOMETHYL)-3-((5-NITROFURFURYLI-
DENE)AMINO)-2-OXAZOLIDINONEHYDROCHLORIDE
see FPI150
N-MORPHOLINO NONANAMIDE see MRQ750
2-(MORPHOLINOTHIO)BENZOTHIAZOLE see BDG000
MORPHOLINYLMERCAPTOBENZOTHIAZOLE see BDG000
2-(4-MORPHOLINYLTHIO)BENZOTHIAZOLE see BDG000
MORPHOTHION see MRU250
MORSODREN see MLF250
MORTON EP332 see DSO200
MORTON SOIL DRENCH see MLF250
MORTON WP-161E see ISE000
MORTOPAL see TCF250
MOSANON see SEH000
MOSCARDA see MAK700
MOSS GREEN see COF500
MOSTEN see PMP500
MOTH BALLS (DOT) see NAJ500
MOTH FLAKES see NAJ500
MOTILYN see PAG200
MOTIORANGE R see PEJ500
MOTIROT G see XRA000
MOTOLON see QAK000
MOTOR BENZOL see BBL250
MOTOR FUEL (DOT) see GBY000
MOTOR SPIRIT (DOT) see GBY000
MOTOX see CDV100
MOTTENHEXE see HCI000
MOUNTAIN GREEN see COF500
MOUNTAIN TOBACCO see AQY500
MOUS-CON see ZLS000
MOUSE-NOTS see SMN500
MOUSE PAK see WAT200
MOUSE-RID see SMN500
MOUSE-TOX see SMN500
MOVINYL 100 see PKQ059
MOVINYL 114 see AAX250
MOXIE see MEI450
MOXONE see DAA800
MOZAMBIN see QAK000
MP see POK000
8-MP see XDJ000
MP 12-50 see TAB750
MP 25-38 see TAB750
MP 45-26 see TAB750
MP 1 (refractory) see EAL100
3-MPA see MFM000
3-MPC see MNM500
M.P. CHLORCAPS T.D. see TAI500
MPG see MRK500
MPK see PBN250
MPK 90 see PKQ250
MPN see MNA000
MPNU see MMY500
MPP see FAQ999
MRAVENCAN DI-n-BUTYLCINICITY (CZECH) see DDZ000
MRAVENCAN VAPENATY (CZECH) see CAS250
MRC 910 see GIA000
MRD 108 see DWK400
MRL 41 see CMX700
MROWCZAN ETYLU (POLISH) see EKL000
MS 33 see SKV150
MS 53 see SNK000
MS 33F see SKV150
MS 1053 see DJI000
MS 1143 see DJI000
MSG see MRL500

MSMA see MRL750
MSMED see ECU750
MSZYCOL see BBQ500
MTB 51 see XCJ000
MTBHQ see BRM500
MTD see DTQ400, TGL750
MTMC see MIB750
M.T. MUCORETTES see MPN500
MTQ see QAK000
MTU see MPW500
MTX see MDV500
MUCAESTHIN see BQA010
MUCAINE see DTL200
MUCIDRINA see VGP000
MUCOLYTICUM see ACH000
MUCOLYTICUM LAPPE see ACH000
MUCOMYCIN see MRV250
MUCOMYST see ACH000
MUCOSOLVIN see ACH000
MUCOXIN see DTL200
MUIRAMID see AAI250
MUL F 66 see PKL750
MULHOUSE WHITE see LDY000
MULSIFEROL see VSZ100
MULTAMAT see DQM600
MULTERGAN see MRW000
MULTERGAN METHYL SULFATE see MRW000
MULTEZIN see MRW000
MULTICHLOR see CDP000
MULTIN see HIM000
MURACIL see MPW500
MURATOX see DJI000
MURCIL see MDQ250
MUREX see GFA000
MURFOS see PAK000
MURFOTOX see DJI000
MURFULVIN see GKE000
MURIATE of PLATINUM see PJE000
MURIATIC ACID (DOT) see HHL000
MURIATIC ETHER see EHH000
MURIOL see CJJ000
MURITAN see DEQ000
MUROTOX see DJI000
MURPHOTOX see DJI000
MURUTOX see DJI000
MURVESCO see CJR500
MUSCARIN see MRW250
MUSCARINE see MRW250
dl-MUSCARINE see MRW250
MUSCULARON see IHH000
MUSK see MRW250
MUSK 36A see ACL750
MUSK AMBRETTE see BRU500
MUSKARIN see MRW250
N-MUSTARD (GERMAN) see BIE500
MUSTARD GAS see BIH250
MUSTARD GAS SULFONE see BIH500
MUSTARD HD see BIH250
MUSTARD OIL see AGJ250
MUSTARD SULFONE see BIH500
MUSTARD VAPOR see BIH250
MUSTARGEN see BIE250, BIE500
MUSTARGEN HYDROCHLORIDE see BIE500
MUSTINE see BIE250
MUSTINE HYDROCHLOR see BIE500
MUSTINE HYDROCHLORIDE see BIE500
MUSTRON see CFA750
MUSUET SYNTHETIC see CMS850
MUSUETTINE PRINCIPLE see CMS850
MUTAGEN see BIE250
MUTAMYCIN see AHK500
MUTAMYCIN (MITOMYCIN for INJECTION) see AHK500
MUTHESA see DTL200
MUTHMANN'S LIQUID see ACK250

MUTOXIN see DAD200
MV 119A see DLK200
MVNA see NKY000
MX 5517-02 see SMQ500
MXDA see XHS800
MY/68 see FAF000
MYACYNE see NCE000
MYANIL see GGS000
MYBASAN see ILD000
MYCAIFRADIN SULFATE see NCG000
MYCARDOL see PBC250
MYCHEL see CDP250
MYCIFRADIN see NCE000
MYCIFRADIN-N see NCG000
MYCIGIENT see NCG000
φMYCIN see TBX000
MYCINOL see CDP250
MYCOCURAN see GGS000
MYCOFARM see BFD250
MYCOPHYT see PIF750
MYCOSTATIN see NOH500
MYCOSTATIN 20 see NOH500
MYCOZOL see TEX000
MYCRONIL see BJK500
MYDRIATIN see NNM000
MYEBROL see DDP600
MYELOBROMOL see DDP600
MYELOLEUKON see BOT250
MYKOSTIN see VSZ100
MYLAXEN see HEG000
MYLEPSIN see DBB200
MYLEPSINUM see DBB200
MYLERAN see BOT250
MYLON (CZECH) see DSB200
MYLONE see DSB200
MYLONE 85 see DSB200
MYOCHRYSINE see GJC000
MYOCON see NGY000
MYOCRISIN see GJC000
MYODETENSINE see GGS000
MYODIGIN see DKL800
MYOFER 100 see IGS000
MYO-INOSISTOL HEXAKISPHOSPHATE see PIB250
MYO-INOSITOL HEXAPHOSPHATE see PIB250
MYOLAX see GGS000
MYOPAN see GGS000
MYOPLEGINE see HLC500
MYOSEROL see GGS000
MYOSTHENINE see VGP000
MYOTRATE "10" see PBC250
MYOXANE see GGS000
MYPROZINE see PIF750
MYRAFORM see PKQ059
MYRCENE see MRZ150
MYRCENYL ACETATE see AAW500
MYRCIA OIL see BAT500
MYRICIA OIL see BAT500
MYRISTIC ACID see MSA250
MYRISTICA OIL see NOG500
MYRISTYL ALCOHOL (mixed isomers) see TBY250
MYRISTYL-γ-PICOLINIUM CHLORIDE see MSB500
MYSEDON see DBB200
MYSOLINE see DBB200
MYSORITE see ARM262
MYSTOX WFA see BGJ750
MYTOMYCIN see AHK500
MYTRATE see VGP000
MYVAK see VSK900
MYVAX see VSK900

N-399 see PEM750
N 521 see DSB200
N 2790 see FMU045
N 4548 see MOB250

N 7009 see FMO129
NA see EID000, NBE500
NA-22 see IAQ000
N-1544A see PFC750
Na-AESCINAT see EDM000
NABAC see HCL000
NABAM see DXD200
NABAME (FRENCH) see DXD200
NABOLIN see MPN500
NAC see ACH000
NACARAT A EXPORT see HJF500
NACCANOL NR see DXW200
NACCONATE-100 see TGM740
NACCONATE 300 see MJP400
NACCONATE 400 see BBP000
NACCONATE H 12 see MJM600
NACCONATE 1OO see TGM750
NACELAN FAST YELLOW CG see AAQ250
NACLEX see BDE250
NACM-CELLULOSE SALT see SFO500
NAC-TB see ACH000
NACYCLYL see EDR000
NADAZONE see BRF500
NADEINE see DKW800
NADISAL see SJO000
NADISAN see BSM000
NADIZAN see BSM000
NADONE see CPC000
NADOZONE see BRF500
NAEPAINE see PBV750
NAFEEN see SHF500
NAFENOIC ACID see MCB500
NAFENOPIN see MCB500
NaFPAK see SHF500
Na FRINSE see SHF500
NAFTALEN (POLISH) see NAJ500
NAFTALOFOS see HMV000
1-NAFTILAMINA (SPANISH) see NBE000
β-NAFTILAMINA (ITALIAN) see NBE500
1-NAFTIL-TIOUREA (ITALIAN) see AQN635
2-NAFTOL (DUTCH) see NAX000
β-NAFTOL (DUTCH) see NAX000
2-NAFTOLO (ITALIAN) see NAX000
β-NAFTOLO (ITALIAN) see NAX000
α-NAFTYLAMIN (CZECH) see NBE000
β-NAFTYLAMIN (CZECH) see NBE500
1-NAFTYLAMINE (DUTCH) see NBE000
2-NAFTYLAMINE (DUTCH) see NBE500
α-NAFTYL-N-METHYLKARBAMAT (CZECH) see CBM750
β-NAFTYLOAMINA (POLISH) see NBE500
beta-NAFTYLOAMINA (POLISH) see NBE500
1-NAFTYLTHIOUREUM (DUTCH) see AQN635
NAGARSE see BAC000
NAGRAVON see VSZ000
NAH see NCQ900
NAH 80 see SHO500
NAKO H see PEY500
NAKO TEG see ALT500
NAKO TGG see REA000
NAKO TMT see TGL750
NAKO TRB see NAW500
NAKO TSA see DBO400
NAKO YELLOW EGA see ALT000
NALCO 680 see AHG000
NALCOAG see SCH000
NALCON 243 see DSB200
NALED see NAG400
NALIDIC ACID see EID000
NALIDICRON see EID000
NALIDIXIC ACID see EID000
NALIDIXIN see EID000
NALITUCSAN see EID000
NALKIL see BMM650
NALOX see MMN250

NALUTRON see PMH500
NAM see NCR000
NAMATE see DXE600
NAMEKIL see TDW500
N²-(((+)-5-AMINO-5-CARBOXYPENTYLAMINO)
  METHYL)TETRACYCLINE see MRV250
NAMURON see TDA500
5-NAN see NEJ500
NANCHOR see RMA500
NANDERVIT-N see NCR000
NANKER see RMA500
NANKOR see RMA500
NAOTIN see NCQ900
NAPA see HIM000
NAPCLOR-G see SJA000
NAPENTAL see NBU000
NAPHAZOLINE see NAH500
NAPHID see NAR000
NAPHTA (DOT) see NAI500
NAPHTAMINE BLUE 2B see CMO000, CMO250
NAPHTHA see NAI500, PCS250, ROU000
α-NAPHTHACRIDINE see BAW750
NAPHTHA DISTILLATE (DOT) see NAI500
2-NAPHTHALAMINE see NBE500
NAPHTHALANE see DAE800
2-NAPHTHALENAMINE see NBE500
NAPHTHALENE see NAJ500
NAPHTHALENE, crude or refined (DOT) see NAJ500
NAPHTHALENE, molten (DOT) see NAJ500
1,5-NAPHTHALENEDIAMINE see NAM000
1,5-NAPHTHALENE DIISOCYANATE see NAM500
1,4-NAPHTHALENEDIONE see NBA500
1,2-(1,8-NAPHTHALENEDIYL)BENZENE see FDF000
NAPHTHALENE FAST ORANGE 2GS see HGC000
NAPHTHALENE OIL see CMY825
2-NAPHTHALENESULFONIC ACID-3,3′-METHYLENEDI-
  PHENYL-MERCURY see PFM750
2-NAPHTHALENETHIOL see NAP500
NAPHTHALENE-2-THIOL see NAP500
1-NAPHTHALENOL see NAW500
2-NAPHTHALENOL see NAX000
1-(2-NAPHTHALENYL)ETHANONE see ABC500
1-NAPHTHALENYLTHIOUREA see AQN635
NAPHTHALIDINE see NBE000
NAPHTHALIN (DOT) see NAJ500
NAPHTHALINE see NAJ500
NAPHTHALOPHOS see HMV000
NAPHTHALOXIMIDE-O,O-DIETHYL PHOSPHOROTHIOATE
  see NAQ500
NAPHTHALOXIMIDODIETHYL THIOPHOSPHATE see
  NAQ500
α-NAPHTHALTHIOHARNSTOFF (GERMAN) see AQN635
NAPHTHANE see DAE800
NAPHTHANIL BLUE B BASE see DCJ200
NAPHTHANIL SCARLET G BASE see NMP500
NAPHTHANTHRACENE see BBC250
NAPHTHA PETROLEUM (DOT) see NAI500
NAPHTHA SAFETY SOLVENT see SLU500
NAPHTHA, SOLVENT (DOT) see NAI500
NAPHTHENATE de COBALT (FRENCH) see NAR500
NAPHTHENE see NAJ500
NAPHTHENIC ACID see NAR000
NAPHTHENIC ACID, COBALT SALT see NAR500
NAPHTHENIC ACID, COPPER SALT see NAS000
NAPHTHENIC ACID, LEAD SALT see NAS500
NAPHTHENIC OILS (PETROLEUM), CATALYTIC DE-
  WAXED HEAVY(9CI) see MQV865
NAPHTHENIC OILS (PETROLEUM), CATALYTIC DE-
  WAXED LIGHT (9CI) see MQV867
NAPHTHIZINE see NAH500
NAPHTHOCARD YELLOW O see FAG140
NAPHTHO(1,2,3,4-def)CHRYSENE see NAT500
NAPHTHOL see NAW000
1-NAPHTHOL see NAW500

2-NAPHTHOL see NAX000
α-NAPHTHOL see NAW500
β-NAPHTHOL see NAX000
NAPHTHOL B see NAX000
1-NAPHTHOL-N-METHYLCARBAMATE see CBM750
NAPHTHOL YELLOW see DUX800
α-NAPHTHOQUINONE see NBA500
1,4-NAPHTHOQUINONE see NBA500
NAPHTHOSOL FAST RED KB BASE see CLK225
α-NAPHTHOTHIOUREA see AQN635
β-NAPHTHYL ALCOHOL see NAX000
1-NAPHTHYLAMIN (GERMAN) see NBE000
2-NAPHTHYLAMIN (GERMAN) see NBE500
β-NAPHTHYLAMIN (GERMAN) see NBE500
1-NAPHTHYLAMINE see NBE000
2-NAPHTHYLAMINE see NBE500
6-NAPHTHYLAMINE see NBE500
α-NAPHTHYLAMINE see NBE000
β-NAPHTHYLAMINE see NBE500
NAPHTHYLAMINE BLUE see CMO250
NAPHTHYLAMINE MUSTARD see BIF250
2-NAPHTHYLAMINE MUSTARD see NBE500
N-(2-NAPHTHYL)ANILINE see PFT500
2-NAPHTHYLBIS(2-CHLOROETHYL)AMINE see BIF250
β-NAPHTHYL-BIS-(β-CHLOROETHYL)AMINE see BIF250
β-NAPHTHYL-DI-(2-CHLOROETHYL)AMINE see BIF250
1,2-(1,8-NAPHTHYLENE)BENZENE see FDF000
1,5-NAPHTHYLENEDIAMINE see NAM000
NAPHTHYLENE YELLOW see DUX800
β-NAPHTHYL HYDROXIDE see NAX000
2-NAPHTHYLHYDROXYLAMINE see NBI500
NAPHTHYLISOPROTERENOL HYDROCHLORIDE see
  INT000
2-NAPHTHYL MERCAPTAN see NAP500
β-NAPHTHYL MERCAPTAN see NAP500
1-NAPHTHYL METHYLCARBAMATE see CBM750
1-NAPHTHYL-N-METHYLCARBAMATE see CBM750
α-NAPHTHYL-N-METHYLCARBAMATE see CBM750
α-NAPHTHYLMETHYL IMIDAZOLINE see NAH500
2-(α-NAPHTHYLMETHYL)-IMIDAZOLINE see NAH500
2-(1-NAPHTHYLMETHYL)-2-IMIDAZOLINE see NAH500
2-NAPHTHYL METHYL KETONE see ABC500
β-NAPHTHYL METHYL KETONE see ABC500
1-NAPHTHYL METHYLNITROSOCARBAMATE see NBJ500
1-NAPHTHYL-N-METHYL-N-NITROSOCARBAMATE see
  NBJ500
2-NAPHTHYLPHENYLAMINE see PFT500
β-NAPHTHYLPHENYLAMINE see PFT500
2-NAPHTHYL-p-PHENYLENEDIAMINE see NBL000
α-NAPHTHYLTHIOCARBAMIDE see AQN635
1-NAPHTHYL-THIOHARNSTOFF (GERMAN) see AQN635
2-NAPHTHYL THIOL see NAP500
α-NAPHTHYLTHIOUREA see AQN635
1-(1-NAPHTHYL)-2-THIOUREA see AQN635
N-(1-NAPHTHYL)-2-THIOUREA see AQN635
α-NAPHTHYLTHIOUREA (DOT) see AQN635
1-NAPHTHYL THIOUREA (MAK) see AQN635
1-NAPHTHYL-THIOUREE (FRENCH) see AQN635
NAPHTOELAN FAST SCARLET G SALT see NMP500
NAPHTOELAN ORANGE R BASE see NEN500
NAPHTOELAN RED GG BASE see NEO500
2-NAPHTOL (FRENCH) see NAX000
β-NAPHTOL (GERMAN) see NAX000
NAPHTOL AS-KG see TGR000
NAPHTOL AS-KGLL see TGR000
NAPHTOX see AQN635
NAPOTON see LFK000, MDQ250
NAPRINOL see HIM000
NARCEOL see MNR250
NARCOGEN see TIO750
NARCOLAN see THV000
NARCOSAN see ERD500
NARCOTANE see HAG500
NARCOTANN NE-SPOFA (RUSSIAN) see HAG500

NARCOTILE see EHH000
NARCYLEN see ACI750
NARDELZINE see PFC750
NARDIL see PFC500, PFC750
NARGOLINE see NDM000
NARIGIX see EID000
NARKOLAN see THV000
NARKOSOID see TIO750
NASDOL see TBG000
NASOL see EAW000
NASS (IRAN) see SED400
NASTENON see PAN100
NASTYN see EHP000
NASWAR (PAKISTAN and AFGHANISTAN) see SED400
NATACYN see PIF750
NATAMYCIN see PIF750
NATASOL FAST ORANGE GR SALT see NEO000
NATASOL FAST RED TR SALT see CLK235
NATHULANE see PME500
NATIONAL 120-1207 see AAX250
NATIVE CALCIUM SULFATE see CAX750
NATRASCORB see ARN125
NATRASCORB INJECTABLE see ARN000
NATREEN see BCE500, SGC000
NATRI-C see ARN125
NATRINAL see BAG250
NATRIONEX see AAI250
NATRIPHENE see BGJ750
NATRIUM see SEE500
NATRIUMACETAT (GERMAN) see SEG500
NATRIUMANTIMONYLTARTRAT (GERMAN) see AQI750
NATRIUMARSENIT (GERMAN) see ARJ500
NATRIUMAZID (GERMAN) see SFA000
NATRIUMBARBITALS (GERMAN) see BAG250
NATRIUMBICHROMAAT (DUTCH) see SGI000
NATRIUMCHLORAAT (DUTCH) see SFS000
NATRIUMCHLORAT (GERMAN) see SFS000
NATRIUMCHLORID (GERMAN) see SFT000
NATRIUM CITRICUM (GERMAN) see DXC400
NATRIUM-2,4-DICHLORPHENOXYATHYLSULFAT (GER-
  MAN) see CNW000
NATRIUMDICHROMAAT (DUTCH) see SGI000
NATRIUMDICHROMAT (GERMAN) see SGI000
NATRIUMFLUORACETAAT (DUTCH) see SHG500
NATRIUMFLUORACETAT (GERMAN) see SHG500
NATRIUM FLUORIDE see SHF500
NATRIUMGLUTAMINAT (GERMAN) see MRL500
NATRIUMHYDROXID (GERMAN) see SHS000
NATRIUMHYDROXYDE (DUTCH) see SHS000
NATRIUMHYPOPHOSPHIT (GERMAN) see SHV000
NATRIUMJODID (GERMAN) see SHW000
NATRIUMMAZIDE (DUTCH) see SFA000
NATRIUMMOLYBDAT (GERMAN) see DXE800
NATRIUM NITRIT (GERMAN) see SIQ500
NATRIUMPERCHLORAAT (DUTCH) see PCE750
NATRIUMPERCHLORAT (GERMAN) see PCE750
NATRIUMPHOSPHAT (GERMAN) see SJH090
NATRIUMPROPIONAT (GERMAN) see SJL500
NATRIUMPYROPHOSPHAT see TEE500
NATRIUMRHODANID (GERMAN) see SIA500
NATRIUMSALZ DER 2,2-DICHLORPROPIONSAURE see
  DGI600
NATRIUMSELENIAT (GERMAN) see DXG000
NATRIUMSELENIT (GERMAN) see SJT500
NATRIUMSILICOFLUORID (GERMAN) see DXE000
NATRIUMSUFAT (GERMAN) see SJY000
NATRIUMSULFID (GERMAN) see SJZ000
NATRIUMTRIPOLYPHOSPHAT (GERMAN) see SKN000
NATRIUMZYKLAMATE (GERMAN) see SGC000
NATULAN see PME250, PME500
NATULANAR see PME500
NATULAN HYDROCHLORIDE see PME500
NATURAL CALCIUM CARBONATE see CAO000
NATURAL GASOLINE (DOT) see GBY000

NATURAL IRON OXIDES see IHD000
NATURAL LEAD SULFIDE see LDZ000
NATURAL RED OXIDE see IHD000
NATURAL WINTERGREEN OIL see MPI000
NAUCAINE see AIT250
NAUGARD TJB see DWI000
NAUGARD TKB see NKB500
NAUGATUCK D-014 see SOP000
NAUGATUCK DET see DKC800
NAULI "GUM" see PMQ750
NAUSEN see BBV500
NAVICALM see HGC500
NAVRON see FFF000
NAXAMIDE see IMH000
NAXOL see CPB750
NAYPER B and BO see BDS000
NB2B see CMO000
NBHA see HJQ350
N.B. MECOPROP see CIR500
NBN see BRV500
NBS 706 see SMQ500
NC5 see HER500
NC 26 see BHO250
NC 150 see PDC250, PEK250
NC-262 see DSP400
NC 5016 see DGA200
NCI-C00044 see AFK250
NCI-C00066 see ASH500
NCI-C00077 see CBG000
NCI-C00099 see CDR750
NCI-C00102 see TBQ750
NCI-C00113 see DGP900
NCI-C00124 see DHB400
NCI-C00135 see DSP400
NCI-C00157 see EAT500
NCI C00168 see TBW100
NCI-C00180 see HAR000
NCI-C00191 see KEA000
NCI-C00204 see BBQ500
NCI-C00215 see MAK700
NCI-C00226 see PAK000
NCI-C00237 see PIB900
NCI-C00259 see CDV100
NCI-C00260 see HON000
NCI-C00395 see DVQ709
NCI-C00408 see DER000
NCI-C00419 see PAX000
NCI-C00420 see DFT800
NCI-C00431 see DFV600
NCI-C00442 see DUV600
NCI-C00453 see CDO250
NCI-C00464 see DAD200
NCI-C00475 see BIM500
NCI-C00486 see BIO750
NCI-C00497 see MEI450
NCI-C00500 see DDL800
NCI-C00511 see EIY600
NCI-C00522 see EIY500
NCI-C00533 see CKN500
NCI-C00544 see DOS000
NCI-C00555 see BIM750
NCI-C00566 see EAQ750
NCI-C00920 see EJC500
NCI-C01445 see NEI000
NCI-C01478 see LBG000
NCI-C01514 see AES750
NCI-C01547 see YCJ000
NCI-C01558 see CME250
NCI-C01570 see EDR500
NCI-C01592 see BOT250
NCI-C01616 see IAN000
NCI-C01627 see PIK250
NCI-C01638 see IMH000
NCI-C01649 see TFQ750

NCI-C01661 see DDG800
NCI-C01672 see PDC250
NCI-C01683 see TGD000
NCI-C01694 see EPQ000
NCI-C01707 see TFI000
NCI-C01718 see SOA500
NCI-C01729 see TNX000
NCI-C01730 see API500
NCI-C01785 see POL500
NCI-C01810 see PME500
NCI-C01821 see DSY600
NCI-C01832 see DCE600
NCI-C01843 see NMP500
NCI-C01854 see HHG000
NCI-C01865 see DVH000
NCI-C01876 see AIB000
NCI-C01901 see AKP750
NCI-C01923 see MMG000
NCI-C01934 see NEQ500
NCI-C01956 see NHQ000
NCI-C01967 see NEJ500
NCI C01978 see NEL000
NCI-C01989 see DBO400
NCI-C01990 see MJN000
NCI-C02006 see MQS500
NCI-C02017 see PGN250
NCI-C02028 see DBF800
NCI-C02039 see CEH680
NCI-C02051 see CLK225
NCI-C02073 see EFE000
NCI-C02084 see SIQ500
NCI-C02095 see AEN000
NCI-C02108 see AAI000
NCI-C02119 see USS000
NCI-C02142 see HEM500
NCI-C02153 see THG000
NCI-C02175 see DCJ400
NCI-C02186 see TMH750
NCI-C02200 see SMQ000
NCI-C02222 see ALL750
NCI-C02244 see NKB500
NCI-C02299 see TLG250
NCI-C02302 see TGL750
NCI-C02335 see TGS500
NCI-C02368 see CLK235
NCI-C02551 see CAH500
NCI-C02653 see HCL000
NCI-C02664 see PJN000
NCI-C02686 see CHJ500
NCI-C02697 see ARP250
NCI-C02711 see CAJ750
NCI-C02722 see TGC000
NCI-C02733 see CAK500
NCI-C02766 see AMT500
NCI-C02799 see FMV000
NCI-C02813 see PIX250
NCI-C02824 see ISA000
NCI-C02835 see SGJ000
NCI-C02846 see CJX750
NCI-C02857 see EPJ000
NCI-C02868 see DJC000
NCI-C02880 see DWI000
NCI-C02891 see LCW000
NCI-C02904 see TIW000
NCI-C02915 see PFT500
NCI-C02926 see ASL250
NCI-C02937 see CAQ250
NCI-C02959 see DXH250
NCI-C02971 see MNH000
NCI-C02982 see MGO750
NCI-C02993 see MGO500
NCI-C03010 see DOU600
NCI-C03021 see NAM000
NCI-C03032 see TBR250

NCI-C03043 see AJV250
NCI-C03054 see DFE200
NCI-C03134 see EFU000
NCI-C03167 see SMD000
NCI-C03258 see ANO500
NCI-C03270 see TNC500
NCI-C03292 see CFK125
NCI-C03305 see CJY120
NCI-C03361 see BBX000
NCI-C03372 see IAQ000
NCI-C03474 see ASB250
NCI-C03485 see CDO500
NCI-C03510 see API750
NCI-C03521 see BDH250
NCI-C03554 see TBQ100
NCI-C03598 see BFW750
NCI-C03601 see PHW750
NCI-C03656 see DDA800
NCI-C03667 see DAC800
NCI-C03678 see OAJ000
NCI-C03689 see DVQ000
NCI-C03703 see HCF500
NCI-C03714 see TAI000
NCI-C03736 see AOQ000, BBL000
NCI-C03747 see AOX250
NCI-C03758 see AOX500
NCI-C03770 see CEC000
NCI-C03781 see TMD250
NCI-C03792 see ENV500
NCI-C03805 see BNK000
NCI-C03816 see DKC400
NCI-C03827 see DQD400
NCI-C03838 see CIV000
NCI-C03850 see DVR200
NCI-C03918 see DQJ200
NCI-C03963 see NEM480
NCI-C03974 see TNL250
NCI-C03985 see DGG950
NCI-C04126 see DTH000
NCI-C04159 see BKF250
NCI-C04240 see TGG760
NCI-C04251 see TGF250
NCI-C04535 see DFF809
NCI-C04546 see TIO750
NCI-C04568 see IEP000
NCI-C04579 see TIN000
NCI-C04580 see PCF275
NCI-C04591 see CBV500
NCI-C04604 see HCI000
NCI-C04615 see AGB250
NCI-C04626 see MIH275
NCI-C04637 see TIP500
NCI-C04671 see MDV500
NCI-C04693 see DAC000
NCI-C04706 see AHK500
NCI-C04717 see DAB600
NCI-C04740 see CGV250
NCI-C04762 see DDP600
NCI-C04773 see BIF750
NCI-C04784 see MPU000
NCI-C04795 see DDJ000
NCI-C04819 see DCF200
NCI-C04820 see BIA250
NCI-C04831 see HOO500
NCI-C04842 see VKZ000
NCI-C04853 see PED750
NCI-C04864 see LEY000
NCI-C04875 see DFO000
NCI-C04886 see POK000
NCI-C04897 see PLZ000
NCI-C04900 see EAS500
NCI-C04922 see CQN000
NCI-C04933 see CDN000
NCI-C04944 see BHT750

NCI-C04955 see CHD250
NCI-C05970 see REA000
NCI-C06008 see TAB750
NCI-C06111 see BDX500
NCI-C06155 see BQQ750
NCI-C06224 see EHH000
NCI-C06360 see BEE375
NCI-C06428 see MQW500
NCI-C06462 see SFA000
NCI-C06508 see BDX000
NCI-C07272 see TGK750
NCI-C08640 see CBM500
NCI-C08651 see FAQ999
NCI-C08662 see CNU750
NCI-C08673 see DCM750
NCI-C08695 see DUK800
NCI-C08991 see ARM250, ARM280
NCI C09007 see ARM275
NCI-C50011 see BCP250
NCI-C50033 see SBT000
NCI-C50044 see BII250
NCI-C50055 see DUE000
NCI-C50077 see PMO500
NCI-C50088 see EJN500
NCI-C50099 see PNL600
NCI-C50102 see MJP450
NCI-C50124 see PDN750
NCI-C50135 see EIU800
NCI-C50146 see OPM000
NCI-C50157 see RDK000
NCI-C50168 see CNU000
NCI-C50191 see SIB600
NCI-C50204 see TAV750
NCI-C50259 see HEJ500
NCI-C50317 see DCE400
NCI-C50384 see EFT000
NCI-C50395 see GLU000
NCI-C50419 see LIA000
NCI-C50442 see BJK500
NCI-C50453 see EQR500
NCI-C50464 see AGJ250
NCI-C50475 see AEX250
NCI-C50533 see TGM750
NCI-C50602 see BOP500
NCI-C50613 see AMW000
NCI-C50635 see BLD500
NCI-C50646 see CBF700
NCI-C50657 see DBL200
NCI-C50668 see MJP400
NCI-C50680 see MLH750
NCI-C50715 see MCB000
NCI-C50748 see AQQ500
NCI-C52459 see TBQ000
NCI-C52733 see DVL700
NCI-C52904 see NAJ500
NCI-C53781 see AAQ250
NCI-C53792 see CHP500
NCI-C53838 see HGC000
NCI-C53849 see HJF500
NCI-C53894 see PAW500
NCI-C53929 see PEJ500
NCI-C54262 see VPK000
NCI-C54375 see BEC500
NCI-C54546 see SBW000
NCI-C54557 see AQP000
NCI-C54568 see CMO750
NCI-C54579 see CMO000
NCI-C54604 see MJQ000, MJQ100
NCI-C54660 see DHF200
NCI-C54706 see FEW000
NCI-C54717 see ISV000
NCI-C54728 see DTD800
NCI-C54739 see RMK020
NCI-C54808 see ARN000

NCI-C54820 see CIU750
NCI-C54831 see TIQ250
NCI-C54842 see PMK000
NCI-C54853 see EES350
NCI-C54886 see CEJ125
NCI-C54922 see ALO750
NCI-C54933 see PAX250
NCI-C54944 see DEP600
NCI-C54955 see DEP800
NCI-C54966 see REF000
NCI-C54977 see EBR000
NCI-C54988 see TCF000
NCI-C54999 see CPD750
NCI-C55005 see CPC000
NCI-C55072 see CDS000
NCI-C55107 see CEA750
NCI-C55118 see CEQ600
NCI-C55130 see BNL000
NCI-C55141 see PNJ400
NCI-C55152 see AQF000
NCI-C55163 see CCP250
NCI-C55174 see DHF000
NCI-C55185 see GEO000
NCI-C55196 see NGE000
NCI-C55209 see OLA000
NCI-C55210 see RNZ000
NCI-C55221 see SHF500
NCI-C55232 see XGS000
NCI-C55243 see BND500
NCI-C55254 see CFK500
NCI-C55265 see TAI500
NCI-C55276 see BBL250
NCI-C55298 see QPA000
NCI-C55301 see POP250
NCI-C55323 see BKE500
NCI-C55345 see DFX800
NCI-C55367 see BPX000
NCI-C55378 see PAX250
NCI-C55425 see GFQ000
NCI-C55447 see MKA500
NCI-C55481 see EGV400
NCI-C55492 see PEO500
NCI-C55505 see SEM000
NCI-C55538 see EBX500
NCI-C55549 see GGW500
NCI-C55550 see TEO250
NCI-C55561 see TBX250
NCI-C55572 see LFU000
NCI-C55583 see NKM000
NCI-C55594 see MHZ000
NCI-C55607 see HCE500
NCI-C55618 see IMF400
NCI-C55652 see EAY500
NCI-C55685 see PDE000
NCI-C55696 see SNC000
NCI-C55709 see CDP250
NCI-C55721 see DNA800
NCI-C55743 see PBC250
NCI-C55765 see DKQ000
NCI-C55776 see PJD000
NCI-C55787 see HFV500
NCI-C55801 see HIM000
NCI-C55812 see MIP750
NCI-C55823 see GEM000
NCI-C55834 see HIH000
NCI-C55845 see QQS200
NCI-C55856 see CCP850
NCI-C55878 see BOV000
NCI-C55890 see HHR500
NCI-C55903 see XDJ000
NCI-C55925 see CFY000
NCI-C55936 see CHJ750
NCI-C55947 see TDY250
NCI-C55969 see AOR500

NCI-C55981 see ASM270
NCI-C55992 see NIF000
NCI-C56031 see DFI100
NCI-C56064 see NGE500
NCI-C56075 see BAU750
NCI-C56086 see AOD125
NCI-C56111 see CMP969
NCI-C56122 see RGW000
NCI-C56133 see BAY500
NCI-C56144 see IDA000
NCI-C56155 see TMN490
NCI-C56177 see FPQ875
NCI-C56188 see XNJ000
NCI-C56199 see EID000
NCI-C56202 see FPK000
NCI-C56213 see ABC250
NCI-C56224 see FPU000
NCI-C56235 see IPU000
NCI-C56246 see QFS000
NCI-C56279 see COB260
NCI-C56280 see MNQ000
NCI-C56291 see BSU250
NCI-C56326 see AAG250
NCI-C56348 see DTC800
NCI-C56360 see DBB200
NCI-C56382 see BIE500
NCI-C56393 see EGP500
NCI-C56406 see VQK650
NCI-C56417 see BMC000
NCI-C56428 see DQF800
NCI-C56439 see IPD000
NCI-C56440 see HCZ000
NCI-C56451 see PKL500
NCI-C56462 see MRH000
NCI-C56473 see HOH500
NCI-C56484 see OGQ100
NCI-C56508 see HMY000
NCI-C56519 see BDF000
NCI-C56531 see BRF500
NCI-C56553 see BRV500
NCI-C56575 see CAL000
NCI-C56586 see CHP250
NCI-C56597 see SNH000
NCI-C56600 see SNJ000
NCI-C56633 see TKV000
NCI-C56655 see PAX250
NCI-C56666 see AGH150
NCI-C60015 see COG000
NCI-C60048 see DJX000
NCI-C60071 see MRL750
NCI-C60082 see NEX000
NCI-C60093 see PFJ250
NCI-C60106 see QCA000
NCI-C60117 see TAJ000
NCI-C60128 see CGO500
NCI-C60139 see VOA000
NCI-C60173 see MCY475
NCI-C60184 see PAD500
NCI-C60195 see PEC500
NCI-C60208 see VPP000
NCI-C60219 see PGX000
NCI-C60220 see TJB600
NCI-C60231 see CEA000
NCI-C60286 see PKL100
NCI-C60297 see CNV000
NCI-C60311 see CNA250
NCI-C60344 see NDK500
NCI-C60388 see NER000
NCI-C60399 see MCW250
NCI-C60402 see EEA500
NCI-C60413 see DER000
NCI-C60537 see NMO550
NCI-C60559 see CHY250
NCI-C60560 see TCR750

NCI-C60571 see HEN000
NCI C60582 see PKQ250, PKQ500, PKQ750, PKR000,
    PKR250, PKR750, PKS000
NCI-C60639 see DYE600
NCI-C60651 see WAK000
NCI-C60662 see TMP750
NCI-C60673 see DQA400
NCI-C60695 see TMJ750
NCI-C60720 see MPE250
NCI-C60753 see DUP600
NCI-C60786 see NEO500
NCI-C60797 see PKQ059
NCI-C60811 see ZSJ000
NCI-C60822 see ABE500
NCI-C60866 see BRR900
NCI-C60899 see DIX200
NCI-C60902 see TLP500
NCI-C60913 see DSB000
NCI-C60924 see DWC600
NCI-C60935 see LBX000
NCI-C60946 see AFW750
NCI-C60968 see IJS000
NCI-C60979 see IKQ000
NCI-C60980 see BCC500
NCI-C61018 see NMW500
NCI-C61029 see PMT750
NCI-C61041 see TNP500
NCI-C61052 see IJD000
NCI-C61143 see MAU250
NCI-C61176 see ARA500
NCI-C61187 see TIV750
NCI-C61289 see CMO250
NCI-C61405 see HEO000
NCI-C61494 see BEN000
NCI-C60253A see ARM262
NCI-C61223A see ARM268
NCI-CO1763 see BSQ000
NCI-CO2131 see BSS250
NCI-CO4035 see CFK000
NCS see NBV500
NCS-14210 see BHV000
NDBA see BRY500
NDC 002-1452-01 see VKZ000
NDEA see NJW500
NDELA see NKM000
NDGA see NBR000
NDHU see NJY000
NDMA see DSY600, NKA600
NDMI see NJK500
NDPA see DWI000, NKB700
NDPhA see DWI000
NDRC-143 see AHJ750
NEA see NKD000
NEANTINE see DJX000
NEARGAL see LJR000
NEASINA see SNJ000
NEAT OIL of SWEET ORANGE see OGY000
NEAUFATIN see TEH500
NECARBOXYLIC ACID see AFR250
NECATORINA see CBY000
NECATORINE see CBY000
NECCANOL SW see DXW200
NEDCIDOL see DCM750
NEEDLE ANTIMONY see AQL500
NEFCO see NGE500
NEFOPAM see FAL000
NEFOPAM HYDROCHLORIDE see NBS500
NEFRECIL see PDC250
NEFRIX see CFY000
NEFTIN see NGG500
NEFURTHIAZOLE see NDY500
NEFUSAN see DSB200
NEGRAM see EID000
NEGUVON see TIQ250

NEGUVON A see TIQ250
NEKAL WT-27 see DJL000
NEKLACID FAST ORANGE 2G see HGC000
NEKLACID RUBINE W see HJF500
NELBON see DLY000
NELIPRAMIN see DLH600
NELLITE see PEV500
NEMA see MKB000, PCF275
NEMABROM see DDL800
NEMACIDE see DFK600
NEMACTIL see PIW000
NEMACUR see FAK000
NEMAFENE see DGG000
NEMAFOS see EPC500
NEMAFUME see DDL800
NEMAGON see DDL800
NEMAGONE see DDL800
NEMAGON SOIL FUMIGANT see DDL800
NEMANAX see DDL800
NEMAPAN see TEX000
NEMAPAZ see DDL800
NEMAPHOS see EPC500
NEMASET see DDL800
NEMATOCIDE see DDL800, EPC500
NEMATOLYT see PAG500
NEMATOX see DDL800
NEMAZENE see PDP250, PDP250
NEMAZON see DDL800
NEMBUSEN see GGS000
NEMBUTAL see NBT500
NEMBUTAL CALCIUM see CAV000
NEMBUTAL SODIUM see NBU000
NEMEROL see DAM600
NEMICIDE see LFA020
NENDRIN see EAT500
NENESIN see BNK000
NEO see TEH500
NEOANTERGAN see WAK000
NEOANTERGAN MALEATE see DBM800
NEOANTERGAN PHOSPHATE see NBV000
NEOARSOLUIN see NCJ500
NEOARSPHENAMINE see NCJ500
NEOASYCODILE see DXE600
NEOBAN see CEW500
NEOBAR see BAP000
NEOBIOTIC see NCG000
NEOBOR see SFF000
NEOBRIDAL see WAK000
NEOCAINE see AIL750, AIT250
NEO-CALMA see CJR909
NEOCARCINOSTATIN see NBV500
NEOCARZINOSTATIN see NBV500
NEOCARZINOSTATIN K see NBV500
NEOCHROMIUM see NBW000
NEOCID see DAD200
NEOCIDOL see DCM750
NEOCOCCYL see SNM500
NEO-CODEMA see CFY000
NEOCOMPENSAN see PKQ250
NEO-COROVAS see PBC250
NEOCROSEDIN see EHP000
NEO-CULTOL see MQV750
NEOCYCLINE see TBX000
NEODALIT see DCW800
NEODELPREGNIN see MCA500
NEODICOUMARIN see BKA000
NEODICOUMAROL see BKA000
NEODICUMARINUM see BKA000
NEODORM (NEW) see NBT500
NEODRENAL see DMV600
NEODRINE see DBA800
NEODYMIUM see NBX000
NEODYMIUM CHLORIDE see NBY000
NEODYMIUM(III) NITRATE (1:3) see NCB000

NEODYMIUM(III) NITRATE, HEXAHYDRATE (1:3:6) see NCB500
NEODYMIUM OXIDE see NCC000
NEO-EPININE see DMV600
NEO-ERGOTIN see EDC500
NEOESERINE BROMIDE see DQY800
NEOESERINE METHYL SULFATE see DQY909
NEO-ESTRONE see ECU750
NEO-FARMADOL see HNI500
NEO-FAT 8 see OCY000
NEO-FAT 10 see DAH400
NEO-FAT 12 see LBL000
NEO-FAT 18-61 see SLK000
NEO-FAT 90-04 see OHU000
NEO-FAT 92-04 see OHU000
NEO-FAT 18-S see SLK000
NEOFEMERGEN see LJL000
NEOFEN see HNI500
NEO-FERRUM see IHG000
NEOFLUMEN see CFY000
NEOFOLLIN see EDS100
NEO-FULCIN see GKE000
NEO-GILURYTMAL see DNB000
NEOGLAUCIT see IRF000
NEOHEXANE (DOT) see DQT200
NEO-HIBERNEX see DQA600
NEO-HOMBREOL see TBG000
NEO-HOMBREOL-M see MPN500
NEOHYDRAZID see COH250
NEO-ISTAFENE see HGC500
NEOLOID see CCP250
NEOLUTIN see HNT500
NEO-MANTLE CREME see NCG000
NEOMCIN see NCE000
NEOMIX see NCG000
NEOMYCIN see NCE000
NEOMYCINE SULFATE see NCG000
NEOMYCIN SULFATE see NCG000
NEOMYCIN SULPHATE see NCG000
NEON see NCG500
NEON, compressed (DOT) see NCG500
NEON, refrigerated liquid (DOT) see NCG500
NEONAL see BPF500
NEO-NAVIGAN see DYE600
NEO-NILOREX see DKE800
NEO-OESTRANOL 1 see DKA600
NEO-OESTRANOL II see DKB000
NEOPANTANOYL CHLORIDE see DTS400
NEOPENTANE see NCH000
NEOPENTANETETRAYL NITRATE see PBC250
NEOPENTANOIC ACID see PJA500
NEOPHARMEDRINE see DBA800
NEOPLATIN see PJD000
NEOPRENE see NCI500
NEOPROMA see MFK500
NEOPSICAINE HYDROCHLORIDE see NCJ000
NEORESTAMIN see TAI500
NEO-SALICYL see SJO000
NEOSALVARSAN see NCJ500
NEOSAR see EAS500
NEO-SCABICIDOL see BBQ500
NEOSEDYN see TEH500
NEOSEPT see HCL000
NEOSERINE BROMIDE see DQY800
NEOS-HIDANTOINA see DKQ000
NEOSTIGMETH see DQY909
NEOSTIGMINE BROMIDE see DQY800
NEOSTIGMINE METHOSULFATE see DQY909
NEOSTIGMINE METHYL BROMIDE see DQY800
NEOSTIGMINE METHYL SULFATE see DQY909
NEOSTIGMINE MONOMETHYLSULFATE see DQY909
NEOSTREPAL see SNN300
NEOSTREPSAN see TEX250
NEO-SUPRIMAL see HGC500

NEO-SUPRIMEL see HGC500
NEOSYDYN see TEH500
NEOSYNEPHRINE see NCL500
NEOTEBEN see ILD000
NEO-TESTIS see TBF500
NEOTHESIN see PIV750
NEOTHESIN HYDROCHLORIDE see IJZ000
NEO-TRAN see MQU750
NEO-TRIC see MMN250
NEOXIN see ILD000
NEOZEPAM see DLY000
NEO-ZINE see MNV750
NEO-ZOLINE see BRF500
NEOZONE D see PFT500
NEPHELOR see GGS000
NEPHENTINE see MQU750
NEPHIS see EIY500
NEPHOCARP see TNP250
NEPHRAMIDE see AAI250
NEPHRIDINE see VGP000
NEPTUNE BLUE BRA CONCENTRATION see FMU059
NEPTUNIUM see NCN500
NERACID see CBG000
NERAL see DTC800
NERIINE DIHYDRBROMIDE see DOX000
NERIOL see OHQ000
NERIOLIN see OHQ000
NERIOSTENE see OHQ000
NERKOL see DGP900
NEROL (FCC) see DTD200
NEROLI OIL, ARTIFICAL see APJ250
NERVACTON see BCA000
NERVANAID B ACID see EIX000
NERVANAID B LIQUID see EIV000
NERVANID B see EIV000
NERVATIL see BCA000
NERVOSETON see BAG250
NESDONAL SODIUM see PBT500
NESOL see MCC250
NESPOR see MAS500
NESSLER REAGENT see NCP500
NESTON see MDT740
NESTYN see EHP000
NETAGRONE 600 see DAA800
NETAZOL see CIR250
NETHALIDE HYDROCHLORIDE see INT000
NETOCYD see DJT800
NETSUSARIN see DOT000
NEU see ENV000, NKE500
NEUCHLONIC see DLY000
NEULACTIL see PIW000
NEULEPTIL see PIW000
NEUMANDIN see ILD000
NEUMOLISINA see ELC500
NEURAKTIL see BCA000
NEUROBARB see EOK000
NEUROBENZIL see BCA000
NEURODYN see TEH500
NEUROLEPTONE see BCA000
NEURONIKA see ADA725
NEUROPROCIN see EHP000
NEUROSEDIN see TEH500
NEUROZINA see CJR909
NEURYL see SHL500
NEUTRAL ACRIFLAVINE see DBX400
NEUTRAL AMMONIUM FLUORIDE see ANH250
NEUTRAL POTASSIUM CHROMATE see PLB250
NEUTRAL PROFLAVINE SULPHATE see DBN400
NEUTRAL SODIUM CHROMATE see DXC200
NEUTRAL VERDIGRIS see CNI250
NEUTRAZYME see SIB600
NEUTRONYX 600 see PKF000
NEUTRONYX 605 see PKF500
NEUTROSEL NAVY BN see DCJ200

NEUTROSEL RED TRVA see CLK235
NEUWIED GREEN see COF500
NEVANAID-B POWDER see TNL250
NEVAX see DJL000
NEVIGRAMON see EID000
NEVIN see ILD000
NEVRODYN see TEH500
NEWCOL 60 see SKV150
NEW GREEN see COF500
NEW IMPROVED CERESAN see BJT250
NEW IMPROVED GRANOSAN see BJT250
NEW PINK BLUISH GEIGY see FAG040
NEXAGAN see EGV500
NEXIT see BBQ500
NEXOVAL see CKC000
NF see NGE500
NF 246 see NDY000
NFHAA see FNO000
NFIP see NGI800
NG see NGY000
NG-180 see NGG500
NGAI CAMPHOR see NCQ820
NHA see NBI500
NH-LOST see BHN750
NHMI see OBY000
3N4HPA see HMY000
Ni 270 see NCW500
NIA 249 see AFR250
NIA 5273 see PIX250
NIA 5462 see EAQ750
NIA 5488 see CKM000
NIA-5767 see EAS000
NIA-9241 see BDJ250
NIA 10242 see CBS275
NIA 17170 see BEP500
NIA 33297 see AHJ750
NIACEVIT see NCR000
NIACIDE see FAS000
NIACIN see NCQ900
NIACINAMIDE see NCR000
NIADRIN see ILD000
NIAGARA 1240 see EEH600
NIAGARA 5,462 see EAQ750
NIAGARA 5767 see EAS000
NIAGARA 5943 see AIX000
NIAGARA 9241 see BDJ250
NIAGARA BLUE see CMO250
NIAGARA BLUE 2B see CMO000
NIAGARAMITE see SOP500
NIAGARA P.A. DUST see NDN000
NIAGARATHAL see DXD000
NIAGARATRAN see CJT750
NIAGRA 10242 see CBS275
NIALATE see EEH600
NIALK see TIO750
NIAMIDE see NCR000
NIAMINE see DJS200
NIA PROOF 08 see TAV750
NIAX CATALYST ESN see NCS000
NIAX FLAME RETARDANT 3 CF see CGO500
NIAX ISOCYANATE TDI see TGM740
NIAX TDI see TGM750, TGM800
NIAX TDI-P see TGM750
NIBREN WAX see TIT500
NIBROL see TEH500
NIBUFIN see NIM000
NICACID see NCQ900
NICAMIDE see DJS200, NCR000
NICAMIN see NCQ900
NICAMINA see NCR000
NICAMINDON see NCR000
NICANGIN see NCQ900
NICASIR see NCR000
NICAZIDE see ILD000

NICELATE see EID000
NICERGOLIN (GERMAN) see NDM000
NICERGOLINE see NDM000
NICETAL see ILD000
NICETAMIDE see DJS200
NICETHAMIDE see DJS200
NICHEL TETRACARBONILE (ITALIAN) see NCZ000
NICIZINA see ILD000
NICKEL see NCW500
NICKEL 270 see NCW500
NICKEL (ITALIAN) see NCW500
NICKEL(II) ACETATE (1:2) see NCX000
NICKEL ACETATE TETRAHYDRATE see NCX500
NICKEL ARSENIDE (As_2-Ni_5) see NDJ399
NICKEL ARSENIDE (As_8-Ni_{11}) see NDJ400
NICKEL ARSENIDE SULFIDE see NCY125
NICKEL BISCYCLOPENTADIENE see NDA500
NICKEL BISTRIPHENYLPHOSPHINE DITHIOCYANATE see BLS500
NICKEL BOROFLUORIDE see NDC000
NICKEL(II) CARBONATE (1:1) see NCY500
NICKEL CARBONYL see NCZ000
NICKEL CARBONYLE (FRENCH) see NCZ000
NICKEL CHLORIDE (DOT) see NDH000
NICKEL(II) CHLORIDE (1:2) see NDH000
NICKEL(II) CHLORIDE HEXAHYDRATE (1:2:6) see NDA000
NICKEL, COMPOUND with pi-CYCLOPENTADIENYL (1:2) see NDA500
NICKEL COMPOUNDS see NDB000
NICKEL CYANIDE (DOT) see NDB500
NICKEL CYANIDE (solid) see NDB500
NICKEL DIBUTYLDITHIOCARBAMATE see BIW750
NICKEL DIFLUORIDE see NDC500
NICKEL (DUST) see NCW500
NICKEL(II) FLUOBORATE see NDC000
NICKEL(II) FLUORIDE (1:2) see NDC500
NICKEL FLUOROBORATE see NDC000
NICKEL(II) FLUOSILICATE (1:1) see NDD000
NICKEL-GALLIUM ALLOY see NDD500
NICKEL(II) HYDROXIDE see NDE000
NICKEL HYDROXIDE (DOT) see NDE000
NICKELIC OXIDE see NDH500
NICKEL IRON SULFIDE see NDE500
NICKEL-IRON SULFIDE MATTE see NDE500
NICKEL MONOSULFATE HEXAHYDRATE see NDL000
NICKEL MONOSULFIDE see NDL100
NICKEL MONOXIDE see NDF500
NICKEL NITRATE (DOT) see NDG000
NICKEL(II) NITRATE (1:2) see NDG000
NICKEL(2+) NITRATE, HEXAHYDRATE see NDG500
NICKEL(II) NITRATE, HEXAHYDRATE (1:2:6) see NDG500
NICKELOCENE see NDA500
NICKELOUS ACETATE see NCX000
NICKELOUS CARBONATE see NCY500
NICKELOUS CHLORIDE see NDH000
NICKELOUS FLUORIDE see NDC500
NICKELOUS HYDROXIDE see NDE000
NICKELOUS OXIDE see NDF500
NICKELOUS SULFATE see NDK500
NICKELOUS SULFIDE see NDL100
NICKELOUS TETRAFLUOROBORATE see NDC000
NICKEL OXIDE see NDH500
NICKEL OXIDE (MAK) see NDF500
NICKEL(II) OXIDE (1:1) see NDF500
NICKEL OXIDE, IRON OXIDE, and CHROMIUM OXIDE FUME see IHE000
NICKEL OXIDE PEROXIDE see NDH500
NICKEL PARTICLES see NCW500
NICKEL(2+) PERCHLORATE, HEXAHYDRATE see NDJ000
NICKEL PEROXIDE see NDH500
NICKEL POTASSIUM CYANIDE see NDI000
NICKEL PROTOXIDE see NDF500

NICKEL REFINERY DUST see NDI500
NICKEL(2+) SALT PERCHLORIC ACID HEXAHYDRATE
  see NDJ000
NICKEL SELENIDE see NDJ475
NICKEL SELENIDE (3:2) CRYSTALLINE see NDJ475
NICKEL SISQUIOXIDE see NDH500
NICKEL SPONGE see NCW500
NICKEL SUBARSENIDE see NDJ399, NDJ400
NICKEL SUBSELENIDE see NDJ475
NICKEL SUBSULFIDE see NDJ500
NICKEL SUBSULPHIDE see NDJ500
NICKEL (II) SULFAMATE see NDK000
NICKEL SULFARSENIDE see NCY125
NICKEL SULFATE see NDK500
NICKEL SULFATE(1:1) see NDK500
NICKEL(II) SULFATE see NDK500
NICKEL(2+)SULFATE(1:1) see NDK500
NICKEL(II) SULFATE (1:1) see NDK500
NICKEL SULFATE HEXAHYDRATE see NDL000
NICKEL (II) SULFATE HEXAHYDRATE see NDL000
NICKEL(II) SULFATE HEXAHYDRATE (1:1:6) see NDL000
NICKEL SULFIDE see NDJ500, NDL100
NICKEL(II) SULFIDE see NDL100
α-NICKEL SULFIDE (1:1) CRYSTALLINE see NDL100
α-NICKEL SULFIDE (3:2) CRYSTALLINE see NDJ500
NICKEL SULPHATE HEXAHYDRATE see NDL000
NICKEL SULPHIDE see NDJ500
NICKEL TETRACARBONYL see NCZ000
NICKEL TETRACARBONYLE (FRENCH) see NCZ000
NICKEL(II) TETRAFLUOROBORATE see NDC000
NICKEL-TITANATE see NDL500
NICKEL TITANIUM OXIDE see NDL500
NICKEL TRIOXIDE see NDH500
NICKEL TRITADISULPHIDE see NDJ500
NICLOFEN see DFT800
NICLOFOLAN see DFD000
NICLOSAMIDE see DFV400, DFV600
NICO see NCQ900
NICO-400 see NCQ900
NICOBID see NCQ900
NICOBION see NCR000
NICOCAP see NCQ900
NICOCIDE see NDN000
NICOCIDIN see NCQ900
NICOCRISINA see NCQ900
NICODAN see NCQ900
NICODELMINE see NCQ900
NICO-DUST see NDN000
NICOFORT see NCR000
NICO-FUME see NDN000
NICOGEN see NCR000
NICOLAR see NCQ900
NICOLEN see NGG500
NICOMIDOL see NCR000
NICONACID see NCQ900
NICONAT see NCQ900
NICONAZID see NCQ900
NICONYL see ILD000
NICOR see DJS200
NICORDAMIN see DJS200
NICORINE see DJS200
NICOROL see CHJ750, NCQ900
NICORYL see DJS200
NICOSAN 2 see NCR000
NICOSIDE see NCQ900
NICO-SPAN see NCQ900
NICOSYL see NCQ900
NICOTA see NCR000
NICOTAMIDE see NCR000
NICOTAMIN see NCQ900
NICOTENE see NCQ900
NICOTERGOLINE see NDM000
NICOTIANA ATTENUATA see TGI100
NICOTIANA GLAUCA see TGI000, TGI100

NICOTIANA LONGIFLORA see TGI100
NICOTIANA RUSTICA see TGI100
NICOTIANA TABACUM see TGI100
NICOTIBINA see ILD000
NICOTIBINE see ILD000
NICOTIL see NCQ900
NICOTILAMIDE see NCR000
NICOTILILAMIDO see NCR000
NICOTINA (ITALIAN) see NDN000
NICOTINE see NDN000
(−)-NICOTINE see NDN000
l-NICOTINE see NDN000
NICOTINE, liquid (DOT) see NDN000
NICOTINE, solid (DOT) see NDN000
NICOTINE ACID see NCQ900
NICOTINE ACID AMIDE see NCR000
NICOTINE ACID TARTRATE see NDS500
NICOTINE ALKALOID see NDN000
NICOTINE BITARTRATE see NDS500
NICOTINE, COMPOUND, with NICKEL(II)-o-BENZOYL
  BENZOATE TRIHYDRATE (2:1) see BHB000
NICOTINE HYDROCHLORIDE see NDP400
NICOTINE HYDROCHLORIDE (d,l) see NDP400
NICOTINE HYDROCHLORIDE, solution (DOT) see NDP400
NICOTINE HYDROGEN TARTRATE see NDS500
(−)-NICOTINE HYDROGEN TARTRATE see NDS500
NICOTINE MONOSALICYLATE see NDR000
NICOTINE SALICYLATE (DOT) see NDR000
NICOTINE SULFATE see NDR500
NICOTINE SULFATE (2:1) see NDR500
NICOTINE SULFATE, liquid (DOT) see NDR500
NICOTINE SULFATE, solid (DOT) see NDR500
NICOTINE TARTRATE see NDS500
NICOTINE TARTRATE (1:2) see NDS500
NICOTINE TARTRATE (DOT) see NDS500
NICOTINIC ACID see NCQ900
NICOTINIC ACID AMIDE see NCR000
NICOTINIC ACID DIETHYLAMIDE see DJS200
NICOTINIC AMIDE see NCR000
NICOTINIPCA see NCQ900
NICOTINOYL HYDRAZINE see NCQ900
NICOTINSAURE (GERMAN) see NCQ900
NICOTINSAUREAMID (GERMAN) see NCR000
NICOTION see EPQ000
NICOTOL see NCR000
NICOTYLAMIDE see NCR000
NICOULINE see RNZ000
NICOVASAN see NCQ900
NICOVASEN see NCQ900
NICOVEL see NCQ900, NCR000
NICOVIT see NCR000
NICOVITOL see NCR000
NICOZIDE see ILD000
NICOZYMIN see NCR000
NICYL see NCQ900
NIDA see MMN250
NIDATON see ILD000
NIDRAZID see ILD000
NIERALINE see VGP000
NIESYMETRYCZNA DWU METYLOHYDRAZYNA (POLISH)
  see DSF400
NIFLEX see SAV000
NIFOS T see TCF250
NIFULIDONE see NGG500
NIFURADENE see NDY000
NIFURAN see NGG500
NIFUROXIME see NGC000
NIFURTHIAZOLE see NDY500
NIFUZON see NGE500
NIGLYCON see NGY000
NIH 7958 see DOQ400
NIH-5145 HYDROCHLORIDE see EIJ000
NIKARDIN see DJS200
NIKA-TEMP see PKQ059

NIKAVINYL SG 700 see PKQ059
NIKETAMID see DJS200
NIKETHAROL see DJS200
NIKETHYL see DJS200
NIKETILAMID see DJS200
NIKKELTETRACARBONYL (DUTCH) see NCZ000
NIKKOL OTP 70 see DJL000
NIKKOL SS 30 see SKV150
NIKKOL TO see PKL100
NIKORIN see DJS200
NIKO-TAMIN see NCR000
NIKOTIN (GERMAN) see NDN000
NIKOTINSAEUREAMID (GERMAN) see NCR000
NIKOTYNA (POLISH) see NDN000
NIKOZID see ILD000
NILACID see ABX500
NILODIN see DHU000
NILOX see BPF000
NILOX PBNA see PFT500
NILSTAT see NOH500
NIMCO CHOLESTEROL BASE H see CMD750
NIMERGOLINE see NDM000
NIMITEX see TAL250
NIMITOX see TAL250
NINCALUICOLFLASTINE see VKZ000
NINOL 4821 see BKE500
NINOL AA62 see BKE500
NINOL AA-62 EXTRA see BKE500, LBL000
NIOBE OIL see MHA750
NIOBIUM see NDZ000
NIOBIUM CHLORIDE see NEA000
NIOBIUM PENTACHLORIDE see NEA000
NIOCINAMIDE see NCR000
NIOFORM see CHR500
NIOMIL see DQM600
NIONATE see FBK000
NIONG see NGY000
NIOZYMIN see NCR000
NIP see DFT800
NIPA 49 see PNM750
NIPAGALLIN LA see DXX200
NIPAGALLIN P see PNM750
NIPAGIN see HJL500
NIPAGIN A see HJL000
NIPANTIOX 1-F see BQI000
NIPAR S-20 see NIY000
NIPAR S-20 SOLVENT see NIY000
NIPAR S-30 SOLVENT see NIY000
NIPASOL see HNU500
NIPAZIN A see HJL000
NIPELLEN see NCQ900
NIPEON A 21 see PKQ059
NIPERYT see PBC250
NIPERYTH see PBC250
NIPHANOID see BQA010
NIPLEN see ILD000
NIPODAL see PMF500
NIPOL 407 see SMR000
NIPOL 576 see PKQ059
NIPPON BLUE BB see CMO000
NIPPON DEEP BLACK see AQP000
NIPRIDE see SIU500
NIPSAN see DCM750
NIQUETAMIDA see DJS200
NIRAN see CDR750, PAK000
NIRAN E-4 see PAK000
NIRATIC HYDROCHLORIDE see LFA020
NIRATIC-PURON HYDROCHLORIDE see LFA020
NIRIDAZOLE see NML000
NIRVANOL see EOL000
NIRVONAL see EOK000
NIRVOTIN see PGE000
Ni 0901-S see NCW500
NISETAMIDE see DJS200

NISOTIN see EPQ000
NISSAN CATION M2-100 see TCA500
NISSAN CATION S2-100 see DTC600
NISSAN NONION SP 60 see SKV150
NISSOCAINE see AIL750
NISSOL EC see MME809
Ni 4303T see NCW500
NITADON see ILD000
NITARSONE see NIJ500
NITEBAN see ILD000
NITER see PLL500
NITHIAZID see ENV500
NITHIAZIDE see ENV500
NITICID see CHS500
NITOFEN see DFT800
NITOL see BIE500
NITOL "TAKEDA" see BIE500
NITRADOR see DUS700
NITRADOS see DLY000
NITRAFEN see DFT800
NITRALAMINE HYDROCHLORIDE see NEC000
NITRAN see DUV600
4-NITRANILINE see NEO500
m-NITRANILINE see NEN500
o-NITRANILINE see NEO000
p-NITRANILINE see NEO500
NITRAPHEN see DFT800
NITRAPYRIN (ACGIH) see CLP750
NITRATE d'AMYLE (FRENCH) see AOL250
NITRATE d'ARGENT (FRENCH) see SDS000
NITRATE de BARYUM (FRENCH) see BAN250
NITRATE MERCUREUX (FRENCH) see MDE750
NITRATE MERCURIQUE (FRENCH) see MDF000
NITRATE de PLOMB (FRENCH) see LDO000
NITRATE de PROPYLE NORMAL (FRENCH) see PNQ500
NITRATES see NED000
NITRATE de SODIUM (FRENCH) see SIO900
NITRATE de STRONTIUM (FRENCH) see SMK000
NITRATE de ZINC (FRENCH) see ZJJ000
NITRATINE see SIO900
NITRATION BENZENE see BBL250
NITRAZEPAM see DLY000
NITRAZOL CF EXTRA see NEO500
NITRE see PLL500
NITRE CAKE see SEG800
NITRENPAX see DLY000
NITRIC ACID see NED500
NITRIC ACID, over 40% (DOT) see NED500
NITRIC ACID, ALUMINUM(3+) SALT see AHD750
NITRIC ACID, ALUMIUM SALT see AHD750
NITRIC ACID, AMMONIUM SALT see ANN000
NITRIC ACID, BARIUM SALT see BAN250
NITRIC ACID, BERYLLIUM SALT see BFT000
NITRIC ACID, BISMUTH(3+) SALT see BKW250
NITRIC ACID, CADMIUM SALT see CAH000
NITRIC ACID, CADMIUM SALT, TETRAHYDRATE see
    CAH250
NITRIC ACID, CERIUM(3+) SALT, HEXAHYDRATE see
    CDB250
NITRIC ACID, CESIUM SALT see CDE250
NITRIC ACID DYSPROSIUM(3+) SALT HEXAHYDRATE
    see DYH000
NITRIC ACID, ERBIUM (3+) SALT see ECY500
NITRIC ACID, ERBIUM (3+) SALT, HEXAHYDRATE see
    ECZ000
NITRIC ACID, ETHYL ESTER see ENM500
NITRIC ACID, EUROPIUM(3+) SALT, HEXAHYDRATE see
    ERC000
NITRIC ACID, FUMING (DOT) see NEE500
NITRIC ACID, GADOLINIUM(3+) SALT see GAL000
NITRIC ACID, GALLIUM(3+) SALT see GBS000
NITRIC ACID, HOLMIUM(3+) SALT, HEXAHYDRATE see
    HGH000
NITRIC ACID, ISOPROPYL ESTER see IQP000

NITRIC ACID, LEAD (2+) SALT see LDO000
NITRIC ACID, LUTETIUM(3+) SALT see LIY000
NITRIC ACID, MAGNESIUM SALT (2:1) see MAH000
NITRIC ACID, MERCURY(I) SALT see MDE750
NITRIC ACID, MERCURY(II) SALT see MDF000
NITRIC ACID METHYL ESTER see MMF500
NITRIC ACID, NEODYMIUM SALT see NCB000
NITRIC ACID, NEODYMIUM (3+) SALT, HEXAHYDRATE see NCB500
NITRIC ACID, NICKEL(II) SALT see NDG000
NITRIC ACID, NICKEL(2+) SALT, HEXAHYDRATE see NDG500
NITRIC ACID, PHENYLMERCURY SALT see MCU750
NITRIC ACID, POTASSIUM SALT see PLL500
NITRIC ACID, PRASEODYMIUM(3+) SALT see PLY250
NITRIC ACID, PROPYL ESTER see PNQ500
NITRIC ACID (RED FUMING) see NEE500
NITRIC ACID, RED FUMING (DOT) see NEE500
NITRIC ACID, SAMARIUM(3+) SALT, HEXAHYDRATE see SAT000
NITRIC ACID, SILVER(1+) SALT see SDS000
NITRIC ACID, SODIUM SALT see SIO900
NITRIC ACID, STRONTIUM SALT see SMK000
NITRIC ACID, THALLIUM(1+) SALT see TEK750
NITRIC ACID, THORIUM(4+) SALT see TFT500
NITRIC ACID, THULIUM(3+) SALT, HEXAHYDRATE see TFX250
NITRIC ACID TRIESTER OF GLYCEROL see NGY000
NITRIC ACID (WHITE FUMING) see NEF000
NITRIC ACID, YTTERBIUM(3+) SALT, HEXAHYDRATE see YEA000
NITRIC ACID, YTTRIUM(3+) SALT see YFJ000
NITRIC ACID, ZINC SALT see ZJJ000
NITRIC ETHER (DOT) see ENM500
NITRIC OXIDE see NEG100
NITRIDAZOLE see NML000
NITRIDES see NEH000
NITRILE ACRILICO (ITALIAN) see ADX500
NITRILE ACRYLIQUE (FRENCH) see ADX500
NITRILE ADIPICO (ITALIAN) see AER250
NITRILES see NEH500
NITRIL KISELINY DIETHYLAMINOOCTOVE (CZECH) see DHJ600
NITRIL KYSELINY-o-CHLORBENZOOVE (CZECH) see CEM000
NITRIL KYSELINY ISOFTALOVE (CZECH) see PHX550
NITRIL KYSELINY MALONOVE (CZECH) see MAO250
NITRIL KYSELINY MANDLOVE (CZECH) see MAP250
NITRIL KYSELINY TEREFTALOVE (CZECH) see BBP250
NITRILOACETIC ACID TRISODIUM SALT MONOHYDRATE see NEI000
NITRILOACETONITRILE see COO000
NITRILOTRIACETIC ACID see AMT500
NITRILOTRIACETIC ACID, TRISODIUM SALT see SIP500
NITRILOTRIACETIC ACID TRISODIUM SALT MONOHY-DRATE see NEI000
NITRILO-2,2',2''-TRIETHANOL see TKP500
2,2',2''-NITRILOTRIETHANOL see TKP500
NITRINE-TDC see NGY000
NITRITES see NEJ000
NITRITE de SODIUM (FRENCH) see SIQ500
NITRITO see NGR500
5-NITROACENAPHTHENE see NEJ500
5-NITROACENAPHTHYLENE see NEJ500
5-NITROACENAPHENE see NEJ500
2-NITRO-4-ACETAMINOFENETOL (CZECH) see NEL000
3-NITRO-p-ACETOPHENETIDE see NEL000
3-NITRO-p-ACETOPHENETIDIDE see NEL000
5-NITRO-p-ACETOPHENETIDIDE see NEL000
3'-NITRO-p-ACETOPHENETIDIN see NEL000
NITRO ACID 100 percent see HMY000
NITRO ACID SULFITE see NMJ000
m-NITROAMINOBENZENE see NEN500
2-NITRO-4-AMINOPHENOL see NEM480

o-NITRO-p-AMINOPHENOL see NEM480
4-NITRO-2-AMINOTOLUENE (MAK) see NMP500
p-NITROANILINA (POLISH) see NEO500
3-NITROANILINE see NEN500
m-NITROANILINE see NEN500
o-NITROANILINE see NEO000
p-NITROANILINE see NEO500
4-NITROANILINE (MAK) see NEO500
5-NITRO-o-ANISIDINE see NEQ500
p-NITROANISOL see NER500
2-NITROANISOLE see NER000
4-NITROANISOLE see NER500
o-NITROANISOLE see NER000
p-NITROANISOLE see NER500
3-NITRO-3-AZAPENTANE-1,5-DIISOCYANATE see NHI500
NITROBENZEEN (DUTCH) see NEX000
NITROBENZEN (POLISH) see NEX000
3-NITROBENZENAMINE see NEN500
4-NITROBENZENAMINE see NEO500
NITROBENZENE see NEX000
NITROBENZENE, liquid (DOT) see NEX000
4-NITROBENZENEARSONIC ACID see NIJ500
2-NITRO-1,4-BENZENEDIAMINE see ALL750
3-NITROBENZENEDIAZONIUM PERCHLORATE see NFA500
3-NITROBENZENESULFONIC ACID see NFB500
m-NITROBENZENESULFONIC ACID see NFB500
NITROBENZOL (DOT) see NEX000
NITROBENZOL, liquid (DOT) see NEX000
5-NITROBENZOTRIAZOL (DOT) see NFJ000
5-NITROBENZOTRIAZOLE see NFJ000
5-NITRO-1H-BENZOTRIAZOLE see NFJ000
6-NITRO-1H-BENZOTRIAZOLE see NFJ000
3-NITROBENZOTRIFLUORIDE see NFJ500
m-NITROBENZOTRIFLUORIDE (DOT) see NFJ500
4-NITROBIPHENYL see NFQ000
o-NITROBIPHENYL see NFP500
NITROBROMOFORM see NMQ000
m-NITROCARBAMIDE see NMQ500
NITROCARBOL see NHM500
NITRO CARBO NITRATE see NFS500
NITROCELLULOSE see CCU250
4-NITROCHINOLIN N-OXID (SWEDISH) see NJF000
p-NITROCHLOORBENZEEN (DUTCH) see NFS525
NITROCHLOR see DFT800
m-NITROCHLOROBENZENE see CJB250
o-NITROCHLOROBENZENE see CJB750
p-NITROCHLOROBENZENE see NFS525
o-NITROCHLOROBENZENE liquid (DOT) see CJB750
m-NITROCHLOROBENZENE solid (DOT) see CJB250
p-NITROCHLOROBENZENE solid (DOT) see NFS525
p-NITROCHLOROBENZOL (GERMAN) see NFS525
3-NITRO-4-CHLOROBENZOTRIFLUORIDE see NFS700
NITROCHLOROFORM see CKN500
p-NITRO-m-CHLOROPHENYL DIMETHYL THIONOPHOS-PHATE see MIJ250
p-NITRO-o-CHLOROPHENYL DIMETHYL THIONOPHOS-PHATE see NFT000
3-NITRO-4-CHLORO-α,α,α-TRIFLUOROTOLUENE see NFS700
p-NITROCLOROBENZENE (ITALIAN) see NFS525
NITRO COMPOUNDS see NFT459
NITRO COMPOUNDS of AROMATIC HYDROCARBONS see NFT500
NITROCOTTON see CCU250
4-NITRO-m-CRESOL DIMETHYL PHOSPHATE see PHD750
2-NITRO-1,4-DIAMINOBENZENE see ALL750
4'-NITRO-2,4-DICHLORODIPHENYL ETHER see DFT800
N-NITRODIMETHYLAMINE see DSV200
NITRODIMETHYLBENZENE see NMS000
2-NITRODIPHENYL see NFP500
o-NITRODIPHENYL see NFP500
p-NITRODIPHENYL see NFQ000
N-NITRO-DMA see DSV200

NITROETAN (POLISH) see NFY500
NITROETHANE see NFY500
NITROFAN see DUS700
NITRO FAST GREEN GB see BLK000
NITROFEN see DFT800
NITROFENE (FRENCH) see DFT800
4-NITROFENOL (DUTCH) see NIF000
1-p-NITROFENYL-3,3-DIMETHYLTRIAZEN (CZECH) see
   DSX400
2-NITROFLUORENE see NGB000
NITROFORM see TMM500
5-NITRO-2-FURALDEHYDE OXIME see NGC000
5-NITROFURALDEHYDE SEMICARBAZIDE see NGE500
6-NITROFURALDEHYDE SEMICARBAZIDE see NGE500
5-NITRO-2-FURALDEHYDE SEMICARBAZONE see
   NGE500
5-NITRO-2-FURALDOXIME see NGC000
5-NITROFURAN-2-ALDEHYDE SEMICARBAZONE see
   NGE500
5-NITRO-2-FURANCARBOXALDEHYDE SEMICARBAZONE
   see NGE500
NITROFURANTOIN see NGE000
1-(((5-NITRO-2-FURANYL)METHYLENE)AMINO)-2-IMID-
   AZOLIDINONE see NDY000
3-(((5-NITRO-2-FURANYL)METHYLENE)AMINO)-2-OXAZO-
   LIDINONE see NGG500
2((5-NITRO-2-FURANYL)METHYLENE)HYDRAZINE-
   CARBOXAMIDE see NGE500
5-(5-NITRO-2-FURANYL)-1,3,4-THIADIAZOL-2-AMINE see
   NGI500
N-(4-(5-NITRO-2-FURANYL)-2-THIAZOLYL)ACETAMIDE
   see AAL750
2-(4-(5-NITRO-2-FURANYL)-2-THIAZOLYL)-HYDRAZINE-
   CARBOXALDEHYDE see NDY500
NITROFURAZOLIDONE see NGG500
NITROFURAZOLIDONUM see NGG500
NITROFURAZONE see NGE500
3-(5'-NITROFURFURALAMINO)-2-OXAZOLIDONE see
   NGG500
5-NITROFURFURAL SEMICARBAZONE see NGE500
1-((5-NITROFURFURYLIDENE)AMINO)HYDANTOIN see
   NGE000
N-(5-NITROFURFURYLIDENE)-1-AMINOHYDANTOIN see
   NGE000
N-(5-NITRO-2-FURFURYLIDENE)-1-AMINOHYDANTOIN see
   NGE000
N-(5-NITRO-2-FURFURYLIDENEAMINO)-2-IMIDAZOAI-
   DINONE see NDY000
1-((5-NITROFURFURYLIDENE)AMINO)-2-IMIDAZOLI-
   DINONE see NDY000
N-(5-NITRO-2-FURFURYLIDENE)-1-AMINO-2-IMIDAZOLI-
   DONE see NDY000
N-(5-NITRO-2-FURFURYLIDENE)-3-AMINOOXAZOLIDINE-
   2-ONE see NGG500
3-((5-NITROFURFURYLIDENE)AMINO)-2-OXAZOLIDONE
   see NGG500
N-(5-NITRO-2-FURFURYLIDENE)-3-AMINO-2-OXAZOLI-
   DONE see NGG500
(5-NITRO-2-FURFURYLIDENEAMINO)UREA see NGE500
N-(6-(5-NITROFURFURYLIDENEMETHYL)-1,2,4-TRIAZIN-3-
   YL)IMINODIMETHANOL see BKH500
NITROFUROXIME see NGC000
NITROFUROXON see NGG500
2-(5-NITRO-2-FURYL)-5-AMINO-1,3,4-THIADIAZOLE see
   NGI500
5-(5-NITRO-2-FURYL)-2-AMINO-1,3,4-THIADIAZOLE see
   NGI500
3-((5-NITROFURYLIDENE)AMINO)-2-OXAZOLIDONE see
   NGG500
3-(5-NITRO-2-FURYL)-IMIDAZO(1,2-a)PYRIDINE see
   NGI800
N-(5-(5-NITRO-2-FURYL)-1,3,4-THIADIAZOL-2-YL)ACET-
   AMIDE see FQJ000

N-(4-(5-NITRO-2-FURYL)-2-THIAZOLYL)ACETAMIDE see
   AAL750
N-(4-(5-NITRO-2-FURYL)THIAZOL-2-YL)ACETAMIDE see
   AAL750
N-(4-(5-NITRO-2-FURYL)-2-THIAZOLYL)FORMAMID (GER-
   MAN) see NGM500
N-(4-(5-NITRO-2-FURYL)-2-THIAZOLYL)FORMAMIDE see
   NGM500
N-(4-(5-NITRO-2-FURYL)-2-THIAZOLYL)-2,2,2-TRIFLUO-
   ROACETAMIDE see NGN500
6-(5-NITRO-2-FURYLVINYL)-3-(DIHYDROXYDIMETHYL-
   AMINO)-1,2,4-TRIAZENE see BKH500
((6-(2-(5-NITRO-2-FURYL)VINYL)-as-TRIAZIN-3-YL)IMINO)-
   DIMETHANOL see BKH500
N-(6-(2-(5-NITRO-2-FURYL)VINYL)-1,2,4-TRIAZIN-3-YL)
   IMINODIMETHANOL see BKH500
NITROGEN see NGP500
NITROGEN, compressed (DOT) see NGP500
NITROGEN (cryogenic liquid) see NGP510
NITROGEN, refrigerated liquid (DOT) see NGP500
NITROGEN CHLORIDE see NGQ500
NITROGEN DIOXIDE see NGR500
NITROGEN FLUORIDE see NGW000
NITROGEN FLUORIDE OXIDE see NGS500
NITROGEN GAS see NGP500
NITROGEN HALF MUSTARD see CGW000
NITROGEN IODIDE see IDN000, NGW500
NITROGEN LIME see CAQ250
NITROGEN MONOXIDE see NEG100
NITROGEN MUSTARD see BIE250
NITROGEN MUSTARD HYDROCHLORIDE see BIE500
NITROGEN MUSTARD OXIDE see CFA500, CFA750
NITROGEN MUSTARD-N-OXIDE see CFA500, CFA750
NITROGEN MUSTARD-N-OXIDE HYDROCHLORIDE see
   CFA750
NITROGEN OXIDE see NGU000
NITROGEN OXIDES mixed with OZONE (47%:53%) see
   ORY000
NITROGEN OXYCHLORIDE see NMH000
NITROGEN OXYFLUORIDE see NMH500
NITROGEN PEROXIDE, liquid (DOT) see NGR500
NITROGEN TETROXIDE see NGU500
NITROGEN TETROXIDE, liquid (DOT) see NGU500
NITROGEN TRIFLUORIDE see NGW000
NITROGEN TRIIODIDE see NGW500
NITROGLICERINA (ITALIAN) see NGY000
NITROGLICERYNA (POLISH) see NGY000
NITROGLYCERIN see NGY000
NITROGLYCERIN, liquid, desensitized (DOT) see NGY000
NITROGLYCERIN, liquid, not desensitized (DOT) see
   NGY000
NITROGLYCERINE see NGY000
NITROGLYCERIN mixed with ETHYLENE GLYCOL DINI-
   TRATE (1:1) see NGY500
NITROGLYCEROL see NGY000
NITROGLYCOL see EJG000
NITROGLYKOL (CZECH) see EJG000
NITROGLYN see NGY000
NITROGRANULOGEN see BIE500
NITROGRANULOGEN HYDROCHLORIDE see BIE500
NITROGUANIDINE see NHA500
1-NITROGUANIDINE see NHA500
2-NITROGUANIDINE see NHA500
α-NITROGUANIDINE see NHA500
NITROGUANIDINE, containing less than 20% water (DOT) see
   NHA500
NITROGUANIDINE DRY (DOT) see NHA500
3-NITRO-3-HEXENE see NHE000
NITROHYDROCHLORIC ACID (DOT) see HHM000
NITROHYDROCHLORIC ACID, diluted (DOT) see HHM000
3-NITRO-4-HYDROXYBENZENEARSONIC ACID see
   HMY000
2-NITRO-1-HYDROXYBENZENE-4-ARSONIC ACID see
   HMY000

3-NITRO-4-HYDROXYPHENYLARSONIC ACID see HMY000

1-NITRO-2-IMIDAZOLIDONE see NKL000

N-NITRO-2-IMIDAZOLIDONE see NKL000

NITROIMINODIETHYLENEDIISOCYANIC ACID see NHI500

NITROISOPROPANE see NIY000

1-NITRO-9-(3-ISOPROPYLAMINOPROPYLAMINE)-ACRIDINE DIHYDROCHLORIDE see INW000

1-NITRO-9-(3-ISOPROPYLAMINOPROPYLAMINO)-ACRIDINE DIHYDROCHLORIDE see INW000

NITRO KLEENUP see DUZ000

NITROL see NGY000

NITROLIME see CAQ250

NITROLINGUAL see NGY000

NITROLOWE see NGY000

NITROMANNITE (DOT) see MAW250

NITROMANNITE (DRY) (DOT) see MAW250

NITROMANNITOL see MAW250

NITROMETAN (POLISH) see NHM500

NITROMETHANE see NHM500

3-NITRO-6-METHOXYANILINE see NEQ500

5-NITRO-2-METHOXYANILINE see NEQ500

2-NITRO-7-METHOXYNAPHTHO(2,1-b)FURAN see MFB400

1-NITRO-2-METHYLANTHRAQUINONE see MMG000

((α-NITROMETHYL)-o-CHLOROBENZYLTHIO)ETHYLAMINE HYDROCHLORIDE see NEC000

NITROMIFENE CITRATE see NHP500

NITROMIM see CFA750

NITROMIN see CFA500

NITROMIN HYDROCHLORIDE see CFA750

NITROMURIATIC ACID (DOT) see HHM000

1-NITRONAPHTHALENE see NHQ000

2-NITRONAPHTHALENE see NHQ500

α-NITRONAPHTHALENE see NHQ000

β-NITRONAPHTHALENE see NHQ500

5-NITRONAPHTHALENE ETHYLENE see NEJ500

NITRONET see NGY000

NITRONG see NGY000

N′-NITRO-N-NITROSO-N-METHYLGUANIDINE see MMP000

5-NITRO-N-(2-OXO-3-OXAZOLIDINYL)-2-FURANMETHANIMINE see NGG500

NITROPENTA see PBC250

NITROPENTAERYTHRITE see PBC250

NITROPENTAERYTHRITOL see PBC250

1-NITROPENTANE see AOL500

NITROPHEN see DFT800

NITROPHENE see DFT800

2-NITROPHENOL see NIE500

3-NITROPHENOL see NIE000

4-NITROPHENOL see NIF000

o-NITROPHENOL see NIE500

m-NITROPHENOL (DOT) see NIE000

p-NITROPHENOL (DOT) see NIF000

p-NITROPHENOL ACETATE see ABS750

NITROPHENOLARSONIC ACID see HMY000

P-NITROPHENOL, ESTER with DIETHYL PHOSPHATE see NIM500

p-NITROPHENOL, O-ESTER with O,O-DIETHYLPHOSPHOROTHIOATE see PAK000

4-NITROPHENYL ACETATE see ABS750

p-NITROPHENYL ACETATE see ABS750

m-NITROPHENYLAMINE see NEN500

p-NITROPHENYLAMINE see NEO500

4-NITROPHENYLARSONIC ACID see NIJ500

p-NITROPHENYLARSONIC ACID see NIJ500

p-NITROPHENYLDIBUTYLPHOSPHINATE see NIM000

p-NITROPHENYLDI-N-BUTYLPHOSPHINATE see NIM000

d-(−)-threo-1-p-NITROPHENYL-2-DICHLORACETAMIDO-1,3-PROPANEDIOL see CDP250

d-threo-1-(p-NITROPHENYL)-2-(DICHLOROACETYLAMINO)-1,3-PROPANEDIOL see CDP250

p-NITROPHENYL DIETHYLPHOSPHATE see NIM500

7-NITRO-5-PHENYL-2,3-DIHYDRO-1H-1,4-BENZODIAZEPIN-2-ONE see DLY000

p-NITROPHENYLDIMETHYLTHIONOPHOSPHATE see MNH000

1-(p-NITROPHENYL-3,3-DIMETHYL-TRIAZEN (GERMAN) see DSX400

1-(4-NITROPHENYL)-3,3-DIMETHYLTRIAZENE see DSX400

1-(p-NITROPHENYL)-3,3-DIMETHYL-TRIAZENE see DSX400

NITRO-p-PHENYLENEDIAMINE see ALL750

2-NITRO-p-PHENYLENEDIAMINE see ALL750

2-NITRO-1,4-PHENYLENEDIAMINE see ALL750

o-NITRO-p-PHENYLENEDIAMINE (MAK) see ALL750

p-NITROPHENYL ETHYLBUTYLPHOSPHONATE see NIN000

o-NITROPHENYL METHYL ETHER see NER000

N-(4-NITROPHENYL)-N′-(3-PYRIDINYLMETHYL)UREA see PPP750

NITROPHOS see DSQ000

NITROPONE C see BRE500

NITROPORE see ASM270

1-NITROPROPANE see NIX500

2-NITROPROPANE see NIY000

β-NITROPROPANE see NIY000

NITROPRUSSIDNATRIUM (GERMAN) see SIU500

1-NITROPYRENE see NJA000

3-NITROPYRENE see NJA000

4-NITROPYRIDINE-1-OXIDE see NJA500

4-NITROPYRIDINE-N-OXIDE see NJA500

4-NITROQUINALDINE-N-OXIDE see MMQ250

8-NITROQUINOLINE see NJD500

4-NITROQUINOLINE-1-OXIDE see NJF000

4-NITROQUINOLINE-N-OXIDE see NJF000

NITROSAMINES see NJH000

N-NITROSAZETIDINE see NJL000

N-NITROSO-N-(1-ACETOXYMETHYL)BUTYLAMINE see BRX500

N-NITROSO-N-(ACETOXY)METHYL-N-METHYLAMINE see AAW000

N-NITROSO-N-(1-ACETOXYMETHYL)PROPYL AMINE see PNR250

N-NITROSOAETHYLAETHANOLAMIN (GERMAN) see ELG500

N-NITROSOALLYLMETHYLAMINE see MMT500

N-NITROSOAMINODIETHANOL see NKM000

4-(NITROSOAMINO-N-METHYL)-1-(3-PYRIDYL)-1-BUTANONE see MMS500

N-NITROSOAZACYCLOHEPTANE see NKI000

N-NITROSOAZACYCLOOCTANE see OBY000

1-NITROSOAZACYCLOTRIDECANE see NJK500

NITROSO-AZETIDIN (GERMAN) see NJL000

NITROSOAZETIDINE see NJL000

1-NITROSOAZETIDINE see NJL000

N-NITROSOAZETIDINE see NJL000

N-NITROSOBENZYLMETHYLAMINE see MHP250

N-NITROSOBIS(2-ACETOXYPROPYL)AMINE see NJM500

NITROSOBIS(2-CHLOROPROPYL)AMINE see DFW000

N-NITROSOBIS(2-HYDROXYETHYL)AMINE see NKM000

N-NITROSOBIS(2-HYDROXYPROPYL)AMINE see DNB200

N-NITROSOBIS(2-OXOPROPYL)AMINE see NJN000

N-NITROSO-N-BUTYL-N-(3-CARBOXYPROPYL)AMINE see BQQ250

N-NITROSO-N-BUTYLETHYLAMINE see EHC000

N-NITROSO-n-BUTYL-(4-HYDROXYBUTYL)AMINE see HJQ350

N-NITROSO-N-BUTYLMETHYLAMINE see MHW500

N-NITROSO-N-BUTYLPENTYLAMINE see BRY250

N-NITROSO-N-BUTYL-N-PENTYLAMINE see BRY250

N-NITROSOBUTYLUREA see BSA250

N-NITROSOCARBARYL see NBJ500

NITROSO COMPOUNDS see NJT500

N-NITROSO COMPOUNDS see NJT550

1′-NITROSO-1′-DEMETHYLNICOTINE see NLD500

N-NITROSODIAETHANOLAMIN (GERMAN) see NKM000
N-NITROSODIAETHYLAMINE (GERMAN) see NJW500
N-NITROSO-3,4-DIBROMOPIPERIDINE see DDQ800
N-NITROSODIBUTYLAMINE see BRY500
N-NITROSODI-n-BUTYLAMINE (MAK) see BRY500
N-NITROSO-3,4-DICHLOROPIPERIDINE see DFW200
N-NITROSODIETHANOLAMINE (MAK) see NKM000
NITROSODIETHYLAMINE see NJW500
N-NITROSODIETHYLAMINE see NJW500
NITROSO-1,1-DIETHYL-3-METHYLUREA see DJP600
N-NITROSODIFENYLAMIN (CZECH) see DWI000
p-NITROSODIFENYLAMIN (CZECH) see NKB500
1-NITROSO-5,6-DIHYDROURACIL see NJY000
N-NITROSO-N,N-DI(2-HYDROXYPROPYL)AMINE see
  DNB200
NITROSODIISOBUTYLAMINE see DRQ200
N-NITROSODIISOBUTYLAMINE see DRQ200
N-NITROSODI-ISO-BUTYLAMINE see DRQ200
N-NITROSODIISOPROPYLAMINE see NKA000
NITROSODIMETHYLAMINE see NKA600
N-NITROSODIMETHYLAMINE see NKA600
4-NITROSODIMETHYLANILINE see DSY600
p-NITROSO-N,N-DIMETHYLANILINE see DSY600
p-NITROSODIMETHYLANILINE (DOT) see DSY600
N-NITROSO-2,2'-DIMETHYLDI-n-PROPYLAMINE see
  DRQ200
NITROSO-2,6-DIMETHYLMORPHOLINE see DTA000
N-NITROSO-2,6-DIMETHYLMORPHOLINE see DTA000
NITROSODIMETHYLUREA see DTB200
N-NITROSODIMETHYLUREA see DTB200
N-NITROSO-N,N-DI(2-OXYPROPYL)AMINE see NJN000
N-NITROSODIPENTYLAMINE see DCH600
N-NITROSODI-n-PENTYLAMINE see DCH600
NITROSODIPHENYLAMINE see DWI000
4-NITROSODIPHENYLAMINE see NKB500
N-NITROSODIPHENYLAMINE see DWI000
p-NITROSODIPHENYLAMINE see NKB500
N-NITROSODIPROPYLAMINE see NKB700
N-NITROSODI-N-PROPYLAMINE see NKB700
N-NITROSO-N-DIPROPYLAMINE see NKB700
N-NITROSODI-i-PROPYLAMINE (MAK) see NKA000
N-NITROSODODECAMETHYLENEIMINE see NJK500
NITROSODODECAMETHYLENIMINE see NJK500
N-NITROSODODECAMETHYLENIMINE see NJK500
N-NITROSOEPHEDRINE see NKC000
NITROSOETHYLANILINE see NKD000
N-NITROSO-N-ETHYL ANILINE see NKD000
N-NITROSO-N-ETHYL BIURET see ENT000
N-NITROSOETHYL-N-BUTYLAMINE see EHC000
N-NITROSOETHYLENETHIOUREA see NKK500
N-NITROSOETHYLETHANOLAMINE see ELG500
N-NITROSOETHYL-2-HYDROXYETHYLAMINE see
  ELG500
N-NITROSO-N-ETHYL-N-(2-HYDROXYETHYL)AMINE see
  ELG500
N-NITROSOETHYLISOPROPYLAMINE see ELX500
N-NITROSOETHYLMETHYLAMINE see MKB000
N-NITROSOETHYLPHENYLAMINE (MAK) see NKD000
NITROSOETHYLUREA see ENV000
NITROSOETHYLURETHAN see NKE500
N-NITROSO-N-ETHYLURETHAN see NKE500
N-NITROSOETHYLVINYLAMINE see NKF000
N-NITROSO-N-ETHYLVINYLAMINE see NKF000
N-NITROSOFENYLHYDROXYLAMIN AMONNY (CZECH)
  see ANO500
2-NITROSOFLUORENE see NKF500
NITROSOGUANIDIN (GERMAN) see NKH000
NITROSOGUANIDINE see NKH000
N-NITROSOGUANIDINE see NKH000
NITROSOHEPTAMETHYLENEIMINE see OBY000
N-NITROSOHEPTAMETHYLENEIMINE see OBY000
NITROSO-HEPTAMETHYLENIMIN (GERMAN) see OBY000
N-NITROSOHEXAHYDROAZEPINE see NKI000
N-NITROSOHEXAMETHYLENEIMINE see NKI000

NITROSOHEXAMETHYLENIMINE see NKI000
NITROSO-2-HYDROXYETHYLUREA see HKW500
N-NITROSOHYDROXYETHYLUREA see HKW500
1-NITROSO-1-(2-HYDROXYETHYL)UREA see HKW500
N-NITROSOHYDROXYLAMINE see HOW500
N-NITROSO(2-HYDROXYPROPYL)(2-OXOPROPYL)AMINE
  see HNX500
N-NITROSO-2-HYDROXY-N-PROPYL-N-PROPYLAMINE see
  NLM500
NITROSO-2-HYDROXY-N-PROPYLUREA see NKO400
N-NITROSO-2-HYDROXY-N-PROPYLUREA see NKO400
N-NITROSOIMIDAZOLIDINETHIONE see NKK500
1-NITROSOIMIDAZOLIDINONE see NKL000
1-NITROSO-2-IMIDAZOLIDINONE see NKL000
N-NITROSO-IMIDAZOLIDON (GERMAN) see NKL000
N-NITROSOIMIDAZOLIDONE see NKL000
2,2'-(NITROSOIMINO)BISETHANOL see NKM000
(NITROSOIMINO)DIACETONE see NJN000
NITROSOIMINO DIETHANOL see NKM000
1,1'-NITROSOIMINODI-2-PROPANOL see DNB200
N-NITROSO-1,1'-IMINODI-2-PROPANOL see DNB200
1,1-(N-NITROSOIMINO)DI-2-PROPANOL, DIACETATE see
  NJM500
NITROSOISOPROPANOLUREA see NKO400
N-NITROSO-N-METHOXYMETHYLMETHYLAMINE see
  MEW250
N-NITROSO-N-METHYLACETAMIDE see MMT000
N-NITROSO-N-METHYL-N-ACETOXYMETHYLAMINE see
  AAW000
NITROSOMETHYLALLYLAMINE see MMT500
N-NITROSOMETHYLALLYLAMINE see MMT500
N-NITROSOMETHYLAMINACETONITRIL (GERMAN) see
  MMT750
N-NITROSOMETHYLAMINOACETONITRILE see MMT750
4-(N-NITROSO-N-METHYLAMINO)-1-(3-PYRIDYL)-1-BUTA-
  NONE see MMS500
N-NITROSO-N-METHYL-N-AMYLAMINE see AOL000
NITROSOMETHYLANILINE see MMU250
N-NITROSO-N-METHYLANILINE see MMU250
N-NITROSOMETHYLBENZYLAMINE see MHP250
N-NITROSO-N-METHYLBIURET see MMV000
N-NITROSOMETHYL-N-BUTYLAMINE see MHW500
N-NITROSO-N-METHYLCARBAMIDE see MNA750
N-NITROSOMETHYL-2-CHLOROETHYLAMINE see CIQ500
N-NITROSOMETHYLCYCLOHEXYLAMINE see NKT500
N-NITROSO-N-METHYLCYCLOHEXYLAMINE see NKT500
NITROSOMETHYLDIAETHYLHARNSTOFF see DJP600
NITROSOMETHYLDIETHYLUREA see DJP600
1-NITROSO-1-METHYL-3,3-DIETHYLUREA see DJP600
N-NITROSO-N-METHYL-N-DODECYLAMIN (GERMAN) see
  NKU000
NITROSOMETHYL-n-DODECYLAMINE see NKU000
N-NITROSO-N-METHYL-N-DODECYLAMINE see NKU000
N-NITROSOMETHYLETHYLAMINE (MAK) see MKB000
N-NITROSOMETHYLGLYCINE see NLR500
N-NITROSO-N-METHYL-HARNSTOFF (GERMAN) see
  MNA750
N-NITROSO-N-METHYLHEPTYLAMINE see HBP000
N-NITROSO-N-METHYLNITROGUANIDINE see MMP000
NITROSO-5-METHYL-1,3-OXAZOLIDINE see NKU875
N-NITROSO-5-METHYL-1,3-OXAZOLIDINE see NKU875
NITROSO-5-METHYLOXAZOLIDONE see NKU875
N-NITROSOMETHYL-2-OXOPROPYLAMINE see NKV000
NITROSOMETHYL-N-PENTYLAMINE see AOL000
N-NITROSOMETHYLPHENYLAMINE (MAK) see MMU250
N-NITROSO-N-METHYL-2-PHENYLETHYLAMINE see
  MNU250
NITROSOMETHYLPHENYLUREA see MMY500
N-NITROSO-N'-METHYLPIPERAZIN (GERMAN) see
  NKW500
1-NITROSO-4-METHYLPIPERAZINE see NKW500
N-NITROSO-N'-METHYLPIPERAZINE see NKW500
NITROSOMETHYLPROPYLAMINE see MNA000
NITROSOMETHYL-N-PROPYLAMINE see MNA000

NITROSOMETHYLUREA see MNA750
1-NITROSO-1-METHYLUREA see MNA750
N-NITROSO-N-METHYLUREA see MNA750
NITROSOMETHYLURETHAN (GERMAN) see MMX250
NITROSOMETHYLURETHANE see MMX250
N-NITROSO-N-METHYLURETHANE see MMX250
N-NITROSOMETHYLVINYLAMINE see NKY000
N-NITROSOMORPHOLIN (GERMAN) see NKZ000
NITROSOMORPHOLINE see NKZ000
4-NITROSOMORPHOLINE see NKZ000
N-NITROSOMORPHOLINE (MAK) see NKZ000
NITROSO-NAC see NBJ500
NITROSONIUM BISULFITE see NMJ000
N'-NITROSONORNICOTINE see NLD500
N-NITROSOOXAZOLIDIN (GERMAN) see NLE000
3-NITROSOOXAZOLIDINE see NLE000
N-NITROSOOXAZOLIDINE see NLE000
N-NITROSO-1,3-OXAZOLIDINE see NLE000
NITROSOOXAZOLIDONE see NLE000
N-NITROSO-N-(2-OXOBUTYL)BUTYLAMINE see BSB500
N-NITROSO-2-OXO-N-PROPYL-N-PROPYLAMINE see
    ORS000
1-NITROSO-1-PENTYLUREA see PBX500
N-NITROSOPERHYDROAZEPINE see NKI000
NITROSOPHENOL see NLF200
4-NITROSOPHENOL see NLF200
p-NITROSOPHENOL see NLF200
4-NITROSO-N-PHENYLANILINE see NKB500
N-NITROSO-N-PHENYLANILINE see DWI000
p-NITROSO-N-PHENYLANILINE see NKB500
4-NITROSO-N-PHENYLBENZENAMINE see NKB500
N-NITROSOPHENYLHYDROXYLAMIN AMMONIUM SALZ
    (GERMAN) see ANO500
N-NITROSOPHENYLHYDROXYLAMINE AMMONIUM SALT
    see ANO500
N-NITROSO-4-PICOLYLETHYLAMINE see NLH000
1-NITROSOPIPERAZINE see MRJ750
N-NITROSOPIPERAZINE see MRJ750
NITROSOPIPERIDIN (GERMAN) see NLJ500
N-NITROSO-PIPERIDIN (GERMAN) see NLJ500
1-NITROSOPIPERIDINE see NLJ500
N-NITROSOPIPERIDINE see NLJ500
1-(NITROSOPROPYLAMINO)-2-PROPANOL see NLM500
1-(NITROSOPROPYLAMINO)-2-PROPANONE see ORS000
N-NITROSO-N-PROPYLPROPANAMINE see NKB700
N-NITROSO-N-PROPYL-1-PROPANAMINE see NKB700
NITROSOPROPYLUREA see NLO500
NITROSO-N-PROPYLUREA see NLO500
N-NITROSO-N-PROPYLUREA see NLO500
1-NITROSO-2-(3-PYRIDYL)PYRROLIDINE see NLD500
N-NITROSOPYRROLIDIN (GERMAN) see NLP500
1-NITROSOPYRROLIDINE see NLP500
N-NITROSOPYRROLIDINE see NLP500
3-(1-NITROSO-2-PYRROLIDINYL)PYRIDINE see NLD500
(s)-3-(1-NITROSO-2-PYRROLIDINYL)PYRIDINE see
    NLD500
N-NITROSOSARCOSINE see NLR500
NITROSO SARKOSIN (GERMAN) see NLR500
NITROSOTRIMETHYLENEIMINE see NJL000
N-NITROSOTRIMETHYLENEIMINE see NJL000
N-NITROSO-TRIMETHYLHARNSTOFF (GERMAN) see
    TLU750
NITROSO-3,4,5-TRIMETHYLPIPERAZINE see NLY750
1-NITROSO-3,4,5-TRIMETHYLPIPERAZINE see NLY750
N-NITROSO-3,4,5-TRIMETHYLPIPERAZINE see NLY750
NITROSOTRIMETHYLUREA see TLU750
N-NITROSOTRIMETHYLUREA see TLU750
NITRO-SPAN see NGY000
NITROSTARCH see NMB000
NITROSTARCH, containing less than 20% water (DOT) see
    NMB000
NITROSTARCH, DRY (DOT) see NMB000
NITROSTAT see NGY000
NITROSTIGMIN (GERMAN) see PAK000

NITROSTIGMINE see PAK000
NITROSYL CHLORIDE see NMH000
NITROSYL ETHOXIDE see ENN000
NITROSYL FLUORIDE see NMH500
NITROSYL HYDROGEN SULFATE see NMJ000
NITROSYL HYDROXIDE see NMR000
NITROSYL PERCHLORATE see NMI000
NITROSYL SULFATE see NMJ000
NITROSYLSULFURIC ACID see NMJ000
4-NITRO-2,3,5,6-TETRACHLORANISOLE see TBR250
NITROTHIAMIDAZOL see NML000
NITROTHIAMIDAZOLE see NML000
NITROTHIAZOLE see NML000
N-(5-NITRO-2-THIAZOLYL)ACETAMIDE see ABY900
1-(5-NITRO-2-THIAZOLYL)IMIDAZOLIDIN-2-ONE see
    NML000
1-(5-NITRO-2-THIAZOLYL)-2-IMIDAZOLIDINONE see
    NML000
1-(5-NITRO-2-THIAZOLYL)-2-IMIDAZOLINONE see
    NML000
1-(5-NITRO-2-THIAZOLYL)-2-OXOTETRAHYDROIMIDAZOL
    see NML000
1-(5-NITRO-2-THIAZOLYL)-2-OXOTETRAHYDROIMID-
    AZOLE see NML000
2-NITROTOLUENE see NMO525
3-NITROTOLUENE see NMO500
4-NITROTOLUENE see NMO550
m-NITROTOLUENE see NMO500
o-NITROTOLUENE see NMO525
p-NITROTOLUENE see NMO550
mixo-NITROTOLUENE see NMO600
5-NITRO-o-TOLUIDINE see NMP500
3-NITROTOLUOL see NMO500
4-NITROTOLUOL see NMO550
NITROTRIBROMOMETHANE see NMQ000
NITROTRICHLOROMETHANE see CKN500
m-NITROTRIFLUOROTOLUENE see NFJ500
m-NITROTRIFLUORTOLUOL(GERMAN) see NFJ500
NITROUREA see NMQ500
NITROUS ACID see NMR000
NITROUS ACID-n-BUTYL ESTER see BRV500
NITROUS ACID-sec-BUTYL ESTER see BRV750
NITROUS ACID ETHYL ESTER see ENN000
NITROUS ACID, ISOBUTYL ESTER see IJD000
NITROUS ACID, ISOPROPYL ESTER see IQQ000
NITROUS ACID-3-METHYL BUTYL ESTER see IMB000
NITROUS ACID, METHYL ESTER see MMF750
NITROUS ACID-1-METHYL PROPYL ESTER see BRV750
NITROUS ACID, 2-METHYLPROPYL ESTER see IJD000
NITROUS ACID, PENTYL ESTER see AOL500
NITROUS ACID, POTASSIUM SALT see PLM500
NITROUS ACID, SODIUM SALT see SIQ500
NITROUS DIPHENYLAMIDE see DWI000
NITROUS ETHER (DOT) see ENN000
NITROUS ETHYL ETHER see ENN000
NITROUS FUMES see NEE500
NITROUS OXIDE (DOT) see NGU000
NITROUS OXIDE, compressed (DOT) see NGU000
NITROUS OXIDE, refrigerated liquid (DOT) see NGU000
NITROX see MNH000
NITROXANTHIC ACID see PID000
NITROXYL CHLORIDE see NMT000
NITROXYLENE see NMS000
NITROXYLOL (DOT) see NMS000
NITROXYNIL see HLJ500
NITROZONE see NGE500
NITRUMON see BIF750
NITRYL CHLORIDE see NMT000
NITRYL FLUORIDE see NMT500
NIUIF-100 see PAK000
NIVALENOL-4-O-ACETATE see FQR000
NIVEMYCIN see NCE000
NIVITIN see SKV200
NIX-SCALD see SAV000

NIZOTIN see EPQ000
NK 136 see DJT800
NK 171 see EAV500
NK 711 see LEN000
NK-843 see NGY000
N-LOST see BIE500
N-LOST (GERMAN) see BIE250
N-LOST-PHOSPHORSAUREDIAMID (GERMAN) see
  PHA750
NLPD see PHA750
NMA see MMU250
1-N-2-MA (RUSSIAN) see MMG000
NMBA see MHW500
NMDDA see NKU000
NMEA see MKB000
NMH see MNA750
N-6-MI see NKI000
NMO see CFA500
NMOP see NKV000
NMOR see NKZ000
NMP see MPF200
NMU see MNA750
NMUM see MMX250
NMUT see MMX250
NMVA see NKY000
NNDG see DQR600
NNK see MMS500
NNN see NLD500
N-N-PIP see NLJ500
N-N-PYR see NLP500
No. 1249 see CKN000
No. 1403 see CKN000
No. 75810 see CKN000
NOBECUTAN see TFS350
NOBEDON see HIM000
NOBEDORM see QAK000
NOBILEN see TFR250
NOBLEN see PMP500
NO BUNT LIQUID see HCC500
NOCBIN see DXH250
No. 3 CONC. SCARLET see CHP500
NOCTEC see CDO000
NOCTILENE see QAK000
NOCTOSEDIV see TEH500
NOCTOSOM see FMQ000
NOCTOVANE see ERD500
NODAPTON see GIC000
NO-DHU see NJY000
NO-DOZ see CAK500
NO-ETU see NKK500
NOFLAMOL see PJL750
NOGEDAL see DPH600
NOGEST see MCA000
NOGOS see DGP900
NOGRAM see EID000
NOHFAA see HIP000
NOLTRAN see CMA250
NOLVASAN see BIM250
NOMERSAN see TFS350
No. 907 METRO TALC see TAB750
2,6-NONADIENAL see NMV760
trans-2,cis-6-NONADIENAL see NMV760
trans,cis-2,6-NONADIENAL see NMV760
trans,cis-2,6-NONADIENAL see NMV760
γ-NONALACTONE (FCC) see CNF250
1-NONALDEHYDE see NMW500
NONALOL see NNB500
1,4-NONALOLIDE see CNF250
1-NONANAL see NMW500
NONANE see NMX000
1-NONANECARBOXYLIC ACID see DAH400
NONANOIC ACID see NMY000
NONANOIC ACID, ETHYL ESTER see ENW000
1-NONANOL see NNB500

NONAN-1-OL see NNB500
4-NONANOYLMORPHOLINE see MRQ750
2-NONENAL see NNA300
2-NONEN-1-AL see NNA300
trans-2-NONENAL (FCC) see NNA300
α-NONENYL ALDEHYDE see NNA300
NONEX 411 see HKJ000
NONION SP 60 see SKV150
NONION SP 60R see SKV150
n-NONOIC ACID see NMY000
NONOX CL see NBL000
NONOX D see PFT500
NONOX DPPD see BLE500
NONOX TBC see BFW750
NONYL ACETATE see NNB400
NONYL ALCOHOL see NNB500
n-NONYL ALCOHOL see NNB500
sec-NONYL ALCOHOL see DNH800
1-NONYL ALDEHYDE see NMW500
NONYLCARBINOL see DAI600
n-NONYLIC ACID see NMY000
NONYL METHYL KETONE see UKS000
NONYLTRICHLOROSILANE see NNE000
No. 156 ORANGE CHROME see LCS000
NOPCOCIDE see TBQ750
NO-PEST see DGP900
NO-PEST STRIP see DGP900
NOPINEN see POH750
NOPINENE see POH750
NO-PIP see NLJ500
NOPTIL see EOK000
NO-PYR see NLP500
NORACYCLINE see LJE000
(−)-NORADREC see NNO500
NORADRENALIN see NNO500
NORADRENALINA (ITALIAN) see NNO500
NORADRENALINE see NNO500
(−)-NORADRENALINE see NNO500
d-(−)-NORADRENALINE see NNO500
l-NORADRENALINE see NNO500
NORADRENLINE see NNO500
NORAL INK GRADE ALUMINUM see AGX000
NOR-AM EP 332 see DSO200
NORAMITRIPTYLINE see NNY000
NORARTRINAL see NNO500
NORBORAL see BSM000
NORCAIN see EFX000
NORDHAUSEN ACID (DOT) see SOI500
NORDICOL see EDO000
NORDICORT see HHR000
NORDIHYDROGUAIARETIC ACID see NBR000
NORDIHYDROGUAIRARETIC ACID see NBR000
NORDOPAN see BIA250
NOREPHEDRANE see BBK000
(−)-NOREPHEDRINE see NNM000
NOREPINEPHRINE see NNO500
(−)-NOREPINEPHRINE see NNO500
l-NOREPINEPHRINE see NNO500
NOREPIRENAMINE see NNO500
NORETHANDROL see EAP000
NORETHINDRONE-17-ACETATE see ABU000
NORETHINDRONE ACETATE and ETHINYLESTRADIOL see
  EEH520
NORETHINODREL see EEH550
19-NOR-ETHINYL-4,5-TESTOSTERONE see NNP500
19-NOR-ETHINYL-5,10-TESTOSTERONE see EEH550
NORETHINYNODREL see EEH550
19-NORETHISTERONE see NNP500
19-NORETHISTERONE ACETATE see ABU000
NORETHISTERONE ACETATE mixed with ETHINYL OES-
  TRADIOL see EEH520
NORETHYNODRAL see EEH550
NORETHYNODREL see EEH550
19-NORETHYNODREL see EEH550

NORETHYNODREL and ETHINYLESTRADIOL-3-METHYL ETHER (50:1) see EAP000
NORETHYNODREL mixed with MESTRANOL see EAP000
19-NOR-17-α-ETHYNYLANDROSTEN-17-β-OL-3-ONE see NNP500
19-NOR-17-α-ETHYNYL-17-β-HYDROXY-4-ANDROSTEN-3-ONE see NNP500
19-NOR-17-α-ETHYNYLTESTOSTERONE see NNP500
19-NORETHYNYLTESTOSTERONE ACETATE see ABU000
NORETHYSTERONE ACETATE see ABU000
NORFORMS see ABU500
NORGESTREL see NNQ500
d(−)-NORGESTREL see NNQ500
d-NORGESTREL see NNQ500
α-NORGESTREL see NNQ500
(±)-NORGESTREL see NNQ500
dl-NORGESTREL see NNQ500
NORGINE see AFL000
NOR-HN2 see BHO250
NOR-HN2 HYDROCHLORIDE see BHO250
NORHOMOEPINEPHRINE HYDROCHLORIDE see AMB000
NORIMIPRAMINE see DSI709
NORIMYCIN V see CDP250
NORISODRINE see DMV600
NORKOOL see EJC500
NORLESTRIN see EEH520
NOR-LOST HYDROCHLORID (GERMAN) see BHO250
NORLUTATE see ABU000
NORLUTIN see NNP500
NORLUTINE ACETATE see ABU000
NORMAL LEAD ACETATE see LCG000
NORMAL LEAD ORTHOPHOSPHATE see LDU000
NORMASTIGMIN see DQY909
NORMERSAN see TFS350
NORMETHYL EX4442 see POF250
NORMI-NOX see QAK000
NORMOCYTIN see VSZ000
NORMONSON see CHG000
NORMOSAN see CHG000
NORMOSON see CHG000
NOR-NITROGEN MUSTARD see BHN750
NORNITROGEN MUSTARD HYDROCHLORIDE see BHO250
NOROCAINE see AIL750
NORODIN see DBA800
NOROX BZP-250 see BDS000
NORPRAMIN see DLS600
19-NOR-17-α-PREGNA-1,3,5(10)-TRIEN-2-YNE-3,17-DIOL see EEH500
(17-α)-19-NORPREGNA-1,3,5(10)-TRIEN-20-YNE-3,17,DIOL see EEH500
(3-β,17-α)-19-NORPREGN-4-EN-20-YNE-3,17-DIOL DIACE-TATE see EQJ500
NOR-PRESS 25 see HGP500
NORSULFASOL see TEX250
NORSULFAZOLE see TEX250
NORTEC see CDO000
NORTIMIL see DLS600
NORTRIPTYLINE see NNY000
NORVAL see DJL000
NORVALAMINE see BPX750
NORVINYL see PKQ059
NORVINYL P 6 see AAX175
NO SCALD see DVX800
NOSPAN see MOV500
NOSPASM see IPU000
NOSTAL see EHP000
NOSTEL see CHG000
NOSTIN see EHP000
NOTENQUIL see ABH500
NOTENSIL see AAF750, ABH500
NOTESIL see ABH500
NOURALGINE see SEH000
NOURITHION see PAK000

NOVACRYSIN see GJG000
NOVADELOX see BDS000
NOVAMIDON see DOT000
NOVAMIN see DYE600, PMF500
NOVAMINE see DYE600
NOVAMONT 2030 see PMP500
NOVANTOINA see DKQ000, DNU000
NOVA-PHENO see EOK000
NOVARSAN see NCJ500
NOVARSENOBENZOL see NCJ500
NOVARSENOBILLON see NCJ500
NOVASUROL see CCG500
NOVATHION see DSQ000
NOVATONE see MDW750
NOVECYL see SAH000
NOVERIL see DCW800
NOVERYL see DCW800
NOVICODIN see DKW800
NOVID see ADA725
NOVIGAM see BBQ500
NOVIZIR see EIR000
NOVOBIOCIN MONOSODIUM see NOB000
NOVOBIOCIN, MONOSODIUM SALT see NOB000
NOVOBIOCIN, SODIUM derivative see NOB000
NOVOCAIN-CHLORHYDRAT (GERMAN) see AIT250
NOVOCAINE see AIL750
NOVOCAINE HYDROCHLORIDE see AIT250
NOVOCAIN HYDROCHLORID (GERMAN) see AIT250
NOVOCHLOROCAP see CDP250
NOVOCILLIN see BFD250
NOVOCONESTRON see ECU750
NOVODIPHENYL see DNU000
NOVODRIN see DMV600
NOVOHEPARIN see HAQ500
NOVOLEN see PMP500
NOVOMYCETIN see CDP250
NOVON 712 see PKQ059
NOVONAL see DJU200
NOVONIDAZOL see MMN250
NOVOPHENICOL see CDP250
NOVOPHENYL see BRF500
NOVOPHONE see SOA500
NOVOSCABIN see BCM000
NOVOSED see MDQ250
NOVOSIR N see EIR000
NOVOX see BSC500
NOVYDRINE see BBK000
NOXAL see DXH250
NOXFISH see RNZ000
NOXIPTILINE HYDROCHLORIDE see DPH600
NOXIPTILIN HYDROCHLORID (GERMAN) see DPH600
NOXIPTYLINE HYDROCHLORIDE see DPH600
NOXODYN see TEH500
NOXYRON see DYC800
1-NP see NIX500
2-NP see ALL750
NP 2 see NCW500
2-NP see NIY000
NP 212 see CJR909
NPA see DNB000
NPH 83 see MJG500
NPH-1091 see BDJ250
NPIP see NLJ500
2-NPPD see ALL750
NPU see NLO500
NPYR see NLP500
4-NQO see NJF000
NR.C 2294 see DNA800
NRDC 104 see BEP500
NSC 339 see DVF200
NSC 423 see DAA800
NSC 739 see AMG750
NSC-740 see MDV500
NSC-742 see ASA500

NSC 746 see UVA000
NSC-750 see BOT250
NSC 751 see EEI000
NSC 755 see POK000
NSC 759 see BCB750
NSC 762 see BIE250, BIE500
NSC-763 see DUD800
NSC 1532 see DUZ000
NSC 1895 see DCF200
NSC 2066 see TEP000
NSC-2100 see NGE500
NSC-2101 see HMY000
NSC 2107 see NGE000
NSC-2752 see FOU000
NSC 3051 see MKG500
NSC-3058 see BDH250
NSC 3060 see PKV500
NSC 3061 see TGD000
NSC-3069 see CDP250
NSC-3070 see DKA600
NSC 3072 see SFA000
NSC 3073 see FMT000
NSC-3088 see CDO500
NSC 3094 see DTP000
NSC 3096 see MIW500
NSC 3138 see DOO800
NSC-3424 see QDJ000
NSC 4170 see MMD500
NSC 4730 see EGI000
NSC 4911 see MCQ500
NSC 5354 see PEY250
NSC.5356 see DSB000
NSC-6091 see SOA500
NSC-6396 see TFQ750
NSC-6470 see NDY000
NSC-6738 see DGP900
NSC-7760 see POS750
NSC-8806 see BHV250
NSC-8806 see PED750
NSC 8819 see ADR000
NSC 9166 see TBG000
NSC 9169 see HOI000
NSC 9369 see MMP000
NSC 9659 see ILD000
NSC-9698 see MAW750
NSC-9701 see MPN500
NSC-9704 see PMH500
NSC 9706 see TND500
NSC 9717 see TND250
NSC-9895 see EDO000
NSC 10023 see PLZ000
NSC-10107 see CFA500, CFA750
NSC-10108 see CLO750
NSC 10815 see AMX750
NSC-10873 see BHN750, BHO250
NSC-12169 see EDU500
NSC 13875 see HEJ500
NSC 14083 see SLW500
NSC-14210 see BHT750
NSC 15193 see DAR600
NSC-15432 see EEH550
NSC-16498 see BIA000
NSC-16895 see LGZ000
NSC-17262 see BDC750
NSC-17592 see HNT500
NSC 17661 see GHE000
NSC 17663 see BHN500
NSC 18016 see CHC750
NSC 18321 see BHB750
NSC-19477 see HEG000
NSC-19893 see FMM000
NSC-19987 see MOR500
NSC 23519 see DCQ600
NSC-23890 see DSU000

NSC-23892 see BCC250
NSC 23909 see MNA750
NSC 24559 see MQW750
NSC-26198 see PAN100
NSC 26271 see CQC675, EAS500
NSC 26805 see EMF500
NSC 26980 see AHK500
NSC-27640 see DAR400
NSC 28693 see MRH000
NSC-29215 see TND000
NSC-29630 see DFO000
NSC-30211 see TNF500
NSC 32065 see HOO500
NSC-32606 see BOP750
NSC 32743 see ABN000
NSC.32946 see MKI000
NSC-33669 see EAL500, EAN000
NSC-34372 see DFH000
NSC-34462 see BIA250
NSC 34533 see GKE000
NSC-34652 see MKB250
NSC-35051 see BHU750
NSC 35770 see CMX700
NSC 37095 see EHV500
NSC 37448 see PDT250
NSC-37538 see BKM500
NSC 38721 see CDN000
NSC-39069 see TFU500
NSC-39084 see ASB250
NSC-39470 see BFV750
NSC 39661 see DAS000
NSC-39690 see DIG400
NSC 40774 see MPU000
NSC 44185 see EAM500
NSC-45388 see DAB600
NSC 45403 see ENV000
NSC-45624 see SKI500
NSC 47842 see VKZ000
NSC 49842 see VLA000
NSC-50256 see MLH500
NSC-50364 see MMN250
NSC-50413 see AQY250
NSC-52695 see CQN000
NSC-56410 see MLY000
NSC-57199 see BHT250
NSC-58404 see DCO800
NSC-58775 see DLY000
NSC-60195 see MQG500
NSC 60380 see CMA250
NSC-62209 see BIF250
NSC 62580 see DRO200
NSC-66847 see TEH500
NSC-67574 see LEY000, LEZ000
NSC-69536 see MLJ500
NSC 69856 see NBV500
NSC-69945 see PHA750
NSC-70731 see LGD000
NSC-71261 see TFJ250
NSC-71423 see VTF000
NSC 73438 see NKL000
NSC-77213 see PME250, PME500
NSC-78559 see DAB800
NSC-79037 see CGV250
NSC-80087 see DOT600
NSC-82151 see DAC000
NSC-82174 see EID000
NSC-82196 see IAN000
NSC-82261 see GCS000
NSC 85598 see SMD000
NSC-85998 see SMD000
NSC 87419 see CPN500
NSC 89936 see JAK000
NSC-89945 see DMX200
NSC-91523 see ICB000

NSC-92338 see CBF250
NSC-94100 see DDP600
NSC-95441 see CHD250
NSC 102627 see YCJ000
NSC 104469 see CME250
NSC-104800 see DDJ000
NSC-107430 see DOQ400
NSC-109229 see ARN800
NSC 109723 see TNT500
NSC-109724 see IMH000
NSC-110432 see DFA400
NSC 111180 see ACH000
NSC 112259 see EDR500
NSC 113926 see RKP000
NSC 114900 see DLH630
NSC-114901 see DLS600
NSC-115944 see EAT900
NSC-119875 see PJD000
NSC-122819 see EQP000
NSC-123127 see AES750
NSC-129943 see PIK250
NSC 132313 see DCI600
NSC-134434 see LIM000
NSC 138780 see FQS000
NSC-141537 see AOP250
NSC 141540 see EAV500
NSC-150014 see HGW500
NSC-153858 see MBU820
NSC-177023 see LFA020
NSC 190935 see CJJ250
NSC 190945 see DOP200
NSC 190955 see DTP800
NSC 190978 see DRR400
NSC 190981 see DVS000
NSC 190986 see DJY200
NSC 190987 see DTQ400
NSC 190997 see MPG250
NSC 191025 see DGA200
NSC 195022 see BEP500
NSC 195106 see FAK000
NSC 195154 see DOL800
NSC 247516 see PCC000
NSC-256439 see DAN000
NSC-267703 see QBS000
NSC-405124 see TIQ750
NSC-409962 see BIF750
NSC 521778 see FNO000
NSC-525334 see NDY500
NSC-528986 see AIV500
NSC-762 HYDROCHLORIDE see BIE500
NTA see SIP500
NTA SODIUM HYDRATE see NEI000
NTG see NGY000
NTM see DTR200
NTOI see NML000
NU-1932 see NOD500
NU 2017 see ABV250
NU 2206 see MDV250, MKR250
NU-BAIT II see MDU600
NUCHAR 722 see CDI000
NUCIDOL see DCM750
NUCIFERIN see DNZ000
NUCIFERINE see DNZ000, DNZ000
(−)-NUCIFERINE see NOE500
1-NUCIFERINE see DNZ000, NOE500
NUDRIN see MDU600
NUFLUOR see SHF500
NUISANCE DUSTS and AEROSOLS see NOF000
NUJOL see MQV750
NU-LAWN WEEDER see DDP000
NULLAPON B see EIV000
NULLAPON BF-78 see EIV000
NULLAPON BF ACID see EIX000
NULLAPON BFC CONC see EIV000

NU MAN see MPN500
NU-MANESE see MAT250
NUMOQUIN HYDROCHLORIDE see ELC500
NUNOL see EOK000
NUOPLAZ DOP see DVL700
NUPERCAINAL see DDT200
NUPERCAINE see DDT200
NUPERCAINE HYDROCHLORIDE see NOF500
NURELLE see TIV750
NUSYN-NOXFISH see PIX250
NUTINAL see BCA000
NUTMEG OIL see NOG500
NUTMEG OIL, EAST INDIAN see NOG500
NUTRASWEET see ARN825
NUTRIFOS STP see SJH200
NUTROSE see SFQ000
NUVA see DGP900
NUVACRON see MRH209
NUVANOL see DSQ000
NUVAPEN see AIV500
NYACOL see SCH000
NYACOL 830 see SCH000
NYACOL 1430 see SCH000
NYCOLINE see PJT200
NYCOTON see CDO000
NYCTAL see BNK000
NYDRAZID see ILD000
NYLIDRIN HYDROCHLORIDE see DNU200
NYLMERATE see ABU500
NYLOMINE ACID RED P4B see HJF500
NYLON see NOH000
NYLON-6 see PJY500
NYLOQUINONE ORANGE JR see AKP750
NYMCEL S see SFO500
NYSCAPS see WAK000
NYSCOZID see ILD000
NYSTAN see NOH500
NYSTATIN see NOH500
NYSTATINE see NOH500

O-2857 see BDE500
OAAT see AIC250
OBB see OAH000
OBELINE PICRATE see ANS500, ANS750
OBEPAR see DKE800
OBESEDRIN see BBK500
OBESONIL see BBK500
OBLIVON C see MNM500
OBLIVON CARBAMATE see MNM500
OBSTON see DJL000
OCDD see OAJ000
OCHRATOXIN A see CHP250
OCHRE see IHD000
OCI 56 see SGG500
OCIMUM BASILICUM OIL see BAR250
OCTABROMOBIPHENYL see OAH000
ar,ar,ar,ar,ar′,ar′,ar′,ar′ OCTABROMO-1,1′-BIPHENYL see
    OAH000
OCTABROMODIPHENYL see OAH000
OCTACARBONYLDICOBALT see CNB500
1,2,4,5,6,7,8,8-OCTACHLOOR-3a,4,7,7a-TETRAHYDRO-4,7-
    endo-METHANO-INDAAN (DUTCH) see CDR750
OCTACHLOR see CDR750
OCTACHLOROCAMPHENE see CDV100
OCTACHLORODIBENZODIOXIN see OAJ000 ·
OCTACHLORODIBENZO-p-DIOXIN see OAJ000
OCTACHLORODIBENZO(b,e)(1,4)DIOXIN see OAJ000
1,2,3,4,6,7,8,9-OCTACHLORODIBENZODIOXIN see
    OAJ000
OCTACHLORODIHYDRODICYCLOPENTADIENE see
    CDR750
1,2,4,5,6,7,8,8-OCTACHLORO-2,3,3a,4,7,7a-HEXAHYDRO-
    4,7-METHANOINDENE see CDR750

p-tert-OCTYLPHENOXYETHOXYETHYLDIMETHYLBENZYL
  AMMONIUM CHLORIDE see BEN000
p-tert-OCTYLPHENOXYPOLYETHOXYETHANOL see
  PKF500
OCTYL PHTHALATE see DVL600, DVL700
n-OCTYL PHTHALATE see DVL600
OCTYL SEBACATE see BJS250
OCTYLTRICHLOROSILANE see OGE000
ODA see SJN700
ODB see DEP600, EDP000
ODCB see DEP600
OE3 see EDU500
OEKOLP see DKA600
OENANTHALDEHYDE see HBB500
OENANTHOL see HBB500
OESTERGON see EDO000
OESTRADIOL see EDO000
d-OESTRADIOL see EDO000
α-OESTRADIOL see EDO000
β-OESTRADIOL see EDO000
OESTRADIOL-17-β see EDO000
cis-OESTRADIOL see EDO000
3,17-β-OESTRADIOL see EDO000
d-3,17-β-OESTRADIOL see EDO000
OESTRADIOL BENZOATE see EDP000
OESTRADIOL-3-BENZOATE see EDP000
β-OESTRADIOL BENZOATE see EDP000
β-OESTRADIOL-3-BENZOATE see EDP000
17-β-OESTRADIOL-3-BENZOATE see EDP000
OESTRADIOL DIPROPIONATE see EDR000
β-OESTRADIOL DIPROPIONATE see EDR000
17-β-OESTRADIOL DIPROPIONATE see EDR000
OESTRADIOL-3,17-DIPROPIONATE see EDR000
3,17-β-OESTRADIOL DIPROPIONATE see EDR000
OESTRADIOL MONOBENZOATE see EDP000
OESTRADIOL MUSTARD see EDR500
OESTRADIOL PHOSPHATE POLYMER see EDS000
OESTRADIOL POLYESTER with PHOSPHORIC ACID see
  EDS000
OESTRADIOL R see EDO000
OESTRAFORM (BDH) see EDP000
OESTRASID see DAL600
OESTRA-1,3,5(10)-TRIENE-3,17-β-DIOL see EDO000
17-β-OESTRA-1,3,5(10)-TRIENE-3,17-DIOL see EDO000
1,3,5(10)-OESTRATRIENE-3,17-β-DIOL 3-BENZOATE see
  EDP000
OESTRA-1,3,5(10)-TRIENE-3,16-α,17-β-TRIOL see EDU500
1,3,5-OESTRATRIENE-3-β-3,16-α,17-β-TRIOL see EDU500
(16-α,17-β)-OESTRA-1,3,5(10)-TRIENE-3,16,17-TRIOL see
  EDU500
1,3,5-OESTRATRIEN-3-OL-17-ONE see EDV000
1,3,5(10)-OESTRATRIEN-3-OL-17-ONE see EDV000
Δ-1,3,5-OESTRATRIEN-3-β-OL-17-ONE see EDV000
OESTRATRIOL see EDU500
OESTRILIN see ECU750
OESTRIN see EDV000
OESTRIOL see EDU500
16-α,17-β-OESTRIOL see EDU500
3,16-α,17-β-OESTRIOL see EDU500
OESTRODIENE see DAL600
OESTRODIENOL see DAL600
OESTRO-FEMINAL see ECU750
OESTROFORM see EDV000
OESTROGENINE see DKA600
OESTROGLANDOL see EDO000
OESTROGYNAEDRON see DKB000
OESTROGYNAL see EDO000
OESTROL VETAG see DKA600
OESTROMENIN see DKA600
OESTROMENSIL see DKA600
OESTROMENSYL see DKA600
OESTROMIENIN see DKA600
OESTROMON see DKA600
OESTRONBENZOAT (GERMAN) see EDV500

OESTRONE see EDV000
OESTROPAK MORNING see ECU750
OESTROPEROS see EDV000
OESTRORAL see DAL600
OFF see DKC800
OFFITRIL see HNI500
OFHC Cu see CNI000
OFNACK see POP000
OFNA-PERL SALT RRA see CLK235
OFTALENT see CDP250
OFTANOL see IMF300
OFUNACK see POP000
OG 1 see SEH000
OGEEN 515 see OAV000
OGEEN GRB see OAV000
OGEEN M see OAV000
OHB see SAH000
OH-BBN see HJQ350
1-α-OH-CC see HJV000
1-α-OH-D³ see HJV000
17-β-OH-ESTRADIOL see EDO000
OHIO 347 see EAT900
7-OHM-MBA see HMF000
7-OHM-12-MBA see HMF000
17-β-OH-OESTRADIOL see EDO000
OHRIC see DGF000
1-α-OH VITAMIN D3 see HJV000
OIL of ABIES ALBA see AAC250
OIL of AMERICAN WORMSEED see CDL500
OIL of ANISE see AOU250
OIL of ANISEED see PMQ750
OIL of ARBOR VITAE see CCQ500
OIL, ARTEMISIA see ARL250
OIL of BASIL see BAR250
OIL of BAY see BAT500
OIL of BERGAMOT, coldpressed see BFO000
OIL of BERGAMOT, rectified see BFO000
OIL, BITTER ALMOND see BLV500
OIL of CALAMUS, GERMAN see OGK000
OIL of CAMPHOR RECTIFIED see CBB500
OIL CAMPHOR SASSAFRASSY see CBB500
OIL of CAMPHOR WHITE see CBB500
OIL of CARAWAY see CBG500
OIL of CARDAMON see CCJ625
OIL of CASSIA see CCO750
OIL of CEDAR LEAF see CCQ500
OIL of CHENOPODIUM see CDL500
OIL of CHINESE CINNAMON see CCO750
OIL of CINNAMON see CCO750
OIL of CINNAMON, CEYLON see CCO750
OIL of CORIANDER see CNR735
OIL of EUCALYPTUS see EQQ000
OIL of FENNEL see FAP000
OIL of FUR see AAC250
OIL GAS see OGM000
OIL of GERANIUM see GDA000
OIL of GRAPEFRUIT see GJU000
OIL GREEN see CMJ900
OIL of HARTSHORN see BMA750
OIL of JUNIPER BERRY see JEA000
OIL of LABDANUM see LAC000
OIL of LAUREL LEAF see LBK000
OIL of LAVANDIN, ABRIAL TYPE see LCA000
OIL of LAVENDER see LCD000
OIL of LEMON see LEI000
OIL of LEMON, desert type, coldpressed see LEI025
OIL of LEMON, distilled see LEI030
OIL of LEMONGRASS, EAST INDIAN see LEG000
OIL of LEMONGRASS, WEST INDIAN see LEH000
OIL of LIME, distilled see OGO000
OIL of MACE see OGQ100
OIL of MARJORAM, SPANISH see MBU500
OIL of MIRBANE (DOT) see NEX000
OIL MIST, MINERAL (OSHA, ACGIH) see MQV750

OIL of MOUNTAIN PINE see PIH400
OIL of MUSTARD, artificial see AGJ250
OIL of MYRBANE see NEX000
OIL of MYRCIA see BAT500
OIL of MYRISTICA see NOG500
OIL of NIOBE see MHA750
OIL of NUTMEG see NOG500
OIL of NUTMEG, expressed see OGQ100
OIL of ONION see OJD200
OIL of ORANGE see OGY000
OIL ORANGE see PEJ500
OIL ORANGE KB see XRA000
OIL ORANGE N EXTRA see XRA000
OIL ORANGE O'PEL see TGW000
OIL ORANGE R see XRA000
OIL ORANGE 2R see XRA000
OIL ORANGE SS see TGW000
OIL ORANGE X see XRA000
OIL ORANGE XO see XRA000
OIL of ORIGANUM see OJO000
OIL of PALMA CHRISTI see CCP250
OIL of PALMAROSA see PAE000
OIL of PARSLEY see PAL750
OIL of PELARGONIUM see GDA000
OIL of PIMENTA LEAF see PIG730
OIL of PINE see PIH750
OIL RED see OHA000
OIL RED GRO see XRA000
OIL RED O see XRA000
OIL RED RO see XRA000
OIL RED XO see XRA000
OIL of ROSE GERANIUM see GDA000
OIL ROSE GERANIUM ALGERIAN see GDA000
OIL of SANDALWOOD, EAST INDIAN see OHG000
OIL of SASSAFRAS see OHI000
OIL SCARLET see XRA000, OHA000
OIL SCARLET 371 see XRA000
OIL SCARLET APYO see XRA000
OIL SCARLET BL see XRA000
OIL SCARLET 6G see XRA000
OIL SCARLET L see XRA000
OIL SCARLET YS see XRA000
OILS, CEDAR LEAF see CCQ500
OILS, CINNAMON see CCO750
OILS, CLOVE LEAF see CMY100
OILS, CORIANDER see CNR735
OILS, CUMIN see COF325
OIL of SHADDOCK see GJU000
OIL of SILVER FIR see AAC250
OIL of SILVER PINE see AAC250
OILS, LIME see OGO000
OIL SOLUBLE ANILINE YELLOW see PEI000
OIL of SPEARMINT see SKY000
OIL of SPIKE LAVENDER see SLB500
OIL of SWEET FLAG see OGK000
OIL of SWEET ORANGE see OGY000
OIL THUJA see CCQ500
OIL of THUJA see CCQ500
OIL of THYME see TFX500
OIL of TURPENTINE see TOD750
OIL of TURPENTINE, rectified see TOD750
OIL VIOLET see EOJ500
OIL of VITRIOL (DOT) see SOI500
OIL of WHITE CEDAR see CCQ500
OIL of WINTERGREEN see MPI000
OIL YELLOW see AIC250, DOT300
OIL YELLOW 21 see AIC250
OIL YELLOW 2R see AIC250
OIL YELLOW 2681 see AIC250
OIL YELLOW A see AIC250, FAG130
OIL YELLOW AAB see PEI000
OIL YELLOW AT see AIC250
OIL YELLOW C see AIC250
OIL YELLOW I see AIC250

OIL YELLOW OB see FAG135
OIL YELLOW T see AIC250
OK 622 see PAJ000
OKASA-MASCUL see TBG000
OKO see DGP900
OKTADECYLAMIN (CZECH) see OBC000
OKTAN (POLISH) see OCU000
OKTANEN (DUTCH) see OCU000
OKTATERR see CDR750
OKTOGEN see CQH250
OKULTIN M see CIR250
OLAMINE see EEC600
OLDHAMITE see CAY000
OLEAL ORANGE R see PEJ500
OLEAL ORANGE SS see TGW000
OLEAL YELLOW 2G see DOT300
OLEANDOMYCIN HYDROCHLORIDE see OHO000
OLEANDOMYCIN MONOHYDROCHLORIDE see OHO000
OLEANDOMYCIN PHOSPHATE see OHO200
OLEANDRIN see OHQ000
OLEANDRINE see OHQ000
OLEATE of MERCURY see MDF250
OLEFIANT GAS see EIO000
OLEFINS see OHS000
OLEIC ACID see OHU000
OLEIC ACID GLYCIDYL ESTER see ECJ000
cis-OLEIC ACID, METHYL ESTER see OHW000
OLEIC ACID, POTASSIUM SALT see OHY000
OLEIC ACID, SODIUM SALT see OIA000
OLEOAKARITHION see TNP250
OLEOFAC see IOT000
OLEOFOS 20 see PAK000
OLEOGESAPRIM see ARQ725
OLEOMYCETIN see CDP250
OLEOPARAPHENE see PAK000
OLEOPARATHION see PAK000
OLEOPHOSPHOTHION see MAK700
OLEORESIN TUMERIC see TOD625
OLEOSUMIFENE see DSQ000
OLEOVITAMIN A see VSK600
OLEOVITAMIN D see VSZ100
OLEOVITAMIN D3 see CMC750
OLEOVOFOTOX see MNH000
OLETAC 100 see PMP500
OLETETRIN see TBX000
OLEUM (DOT) see SOI520
OLEUM ABIETIS see PIH750
OLEUM SINAPIS VOLATILE see AGJ250
OLIBANUM GUM see OIM000
OLITREF see DUV600
OLIVE OIL see OIQ000
OLOSED see MNM500
OLOTHORB see PKL100
OLOW (POLISH) see LCF000
OLPISAN see PAX000
OMAHA see LCF000
OMAHA & GRANT see LCF000
OMAINE see MIW500
OMAIT see SOP000
OMAL see TIW000
OMCHLOR see DFE200
OMEGA CHROME BLUE FB see HJF500
OMETHOAT see DNX800
OMETHOATE see DNX800
OMF 59 see PHA750
OM-HYDANTOINE see DKQ000
OM-HYDANTOINE SODIUM see DNU000
OMIFIN see CMX700
OMITE see SOP000
OMNIBON see SNN300
OMNI-PASSIN see DJT800
OMNIPEN see AIV500
OMNIZOLE see TEX000
OMNYL see QAK000

OMP 2 see OIY000
OMPA see OCM000
OMPACIDE see OCM000
OMPATOX see OCM000
OMPAX see OCM000
OMS 2 see FAQ999
OMS 14 see DGP900
OMS-33 see PMY300
OMS 43 see DSQ000
OMS-47 see DOS000
OMS-93 see DST000
OMS 115 see DGD800
OMS 570 see EAQ750
OMS-597 see TMD000
OMS-659 see EGV500
OMS-771 see CBM500
OMS-0971 see CMA100
OMS 1075 see DRR400
OMS-1155 see CMA250
OMS-1206 see BEP500
OMS 1328 see CDS750
OMS 1804 see CJV250
OMTAN see OAN000
ONCB see CJB750
ONCO-CARBIDE see HOO500, TGN250
ONCOTEPA see TFQ750
ONCOTIOTEPA see TFQ750
ONCOVEDEX see TND000
ONCOVIN see LEY000, LEZ000
ONE-IRON see FBJ100
ONEX see CJT750
ONGROVIL S 165 see PKQ059
ONION OIL see OJD200
ONT see NMO525
ONYX see SCI500, SCJ500
ONYXOL 345 see BKE500
OOS see TAB750
OPALON see PKQ059
OPALON 400 see AAX175
OPE 30 see PKF500
OPERIDINE see DAM700
OPHTHALAMIN see VSK600
OPHTHALMADINE see DAS000
OPIPRAMOL DIHYDROCHLORIDE see IDF000
OPIUM see OJG000
OPLOSSINGEN (DUTCH) see FMV000
OPP-Na see BGJ750
OPTAL see PND000
OP-THAL-ZIN see ZNA000
OPTHOCHLOR see CDP250
OPTIMINE see DLV800
OPTIMYCIN see MDO250
OPTINOXAN see QAK000
OPTIPHYLLIN see TEP000
OPTOCHIN HYDROCHLORIDE see ELC500
OPTOQUINHYDROCHLORIDE see ELC500
8-OQ see QPA000
ORABET see BSQ000
ORACON see DNX500
ORACONAL see MCA500
ORADEXON see SOW000
ORAFURAN see NGE000
ORAGEST see MCA000
ORALCID see ABX500
ORALIN see BSQ000
ORALSONE see HHR000
ORAMID see SAH000
ORANGE #10 see HGC000
ORANGE A l'HUILE see PEJ500
ORANGE BASE CIBA II see NEO000
ORANGE BASE IRGA I see NEN500
ORANGE CHROME see LCS000
ORANGE CRYSTALS see ABC500
ORANGE G (BIOLOGICAL STAIN) see HGC000

ORANGE GC BASE see CEH675
ORANGE G DYE see HGC000
ORANGE G (INDICATOR) see HGC000
ORANGE INSOLUBLE OLG see XRA000, PEJ500
ORANGE INSOLUBLE RR see XRA000
ORANGE LEAD see LDS000
ORANGE NITRATE CHROME see LCS000
ORANGE OIL see OGY000
ORANGE OIL, coldpressed (FCC) see OGY000
ORANGE OIL KB see XRA000
ORANGE PEL see PEJ500
ORANGE RESENOLE NO. 3 see PEJ500
ORANGE 3R SOLUBLE IN GREASE see TGW000
ORANGES see BBK500
ORANGE SOLUBLE A l'HUILE see PEJ500
ORANIL see BSM000
ORANIXON see GGS000
ORANYL see BSM000
ORANZ G (POLISH) see HGC000
ORARSAN see ABX500
ORASONE see PLZ000
ORASULIN see BSM000
ORATRAST see BAP000
ORAVIRON see MPN500
ORBENIN SODIUM HYDRATE see SLJ000
ORBICIN see DCQ800
ORBON see OAV000
ORCANON see MPW500
ORCHARD BRAND ZIRAM see BJK500
ORCHIOL see TBG000
ORCHISTIN see TBG000
ORCIN see MPH500
ORCINOL see MPH500
ORDINARY AZOXYBENZENE see ASO750
ORDINARY LACTIC ACID see LAG000
OREMET see TGF250
ORESTOL see DKB000
ORETIC see CFY000
ORETON see TBG000
ORETON-F see TBF500
ORETON-M see MPN500
ORETON METHYL see MPN500
ORETON PROPIONATE see TBG000
OREZAN see BSQ000
ORGA-414 see AMY050
ORGAMIDE see PJY500
ORGANEX see CAK500
ORGANIC GLASS E 2 see PKB500
ORGANOL BORDEAUX B see EOJ500
ORGANOL FAST GREEN J see BLK000
ORGANOL ORANGE see PEJ500
ORGANOL ORANGE 2R see TGW000
ORGANOL SCARLET see OHA000
ORGANOL YELLOW see PEI000
ORGANOL YELLOW 25 see AIC250
ORGANOL YELLOW ADM see DOT300
ORGANOMETALS see OJM000
ORGASEPTINE see SNM500
ORGASTYPTIN see EDU500
ORIENTAL BERRY see PIE500
ORIENT OIL ORANGE PS see PEJ500
ORIENT OIL YELLOW GG see DOT300
ORIGANUM OIL see OJO000
ORIMON see PDP250
ORINASE see BSQ000
ORINAZ see BSQ000
ORION BLUE 3B see CMO250
ORLUTATE see ABU000
ORNID see BMV750
OROLEVOL see MQU750
ORONOL see ART250
ORPHENADRINE CITRATE see DPH000
ORPHENADRINE HYDROCHLORIDE see OJW000
ORPHENOL see BGJ750

ORPIMENT see ARI000
ORQUISTERONE see TBF500
ORSIN see PEY500
ORTEDRINE see BBK000
ORTHAMINE see PEY250
ORTHENE see DOP600
ORTHENE-755 see DOP600
ORTHESIN see EFX000
ORTHO 4355 see NAG400
ORTHO 5353 see BTA250
ORTHO-5655 see MOU750
ORTHO 5865 see CBF800
ORTHO 9006 see DTQ400
ORTHO 12420 see DOP600
ORTHOARSENIC ACID see ARB250
ORTHOARSENIC ACID HEMIHYDRATE see ARC500
ORTHOBORIC ACID see BMC000
ORTHOCAINE see AKF000
ORTHO C-1 DEFOLIANT & WEED KILLER see SFS000
ORTHOCIDE see CBG000
ORTHOCRESOL see CNX000
ORTHODERM see AKF000
ORTHODIBROM see NAG400
ORTHODIBROMO see NAG400
ORTHODICHLOROBENZENE see DEP600
ORTHODICHLOROBENZOL see DEP600
ORTHO N-4 DUST see NDN000
ORTHO N-5 DUST see NDN000
ORTHO EARWIG BAIT see DXE000
ORTHOFORM see AKF000
ORTHOFORMIC ACID, ETHYL ESTER see ENY500
ORTHOFORMIC ACID, TRIETHYL ESTER see ENY500
ORTHOFORMIC ACID, TRIMETHYL ESTER see TLX600
ORTHOHYDROXYBENZOIC ACID see SAI000
ORTHO-KLOR see CDR750
ORTHO L10 DUST see LCK000
ORTHO L40 DUST see LCK000
ORTHO-LM APPLE SPRAY see MLH000
ORTHO LM CONCENTRATE see MLH000
ORTHO LM SEED PROTECTANT see MLH000
ORTHO MALATHION see MAK700
ORTHO MC see MAE000
ORTHO-MITE see SOP500
ORTHOMRAVENCAN ETHYLNATY (CZECH) see ENY500
ORTHOMRAVENCAN METHYLNATY (CZECH) see TLX600
ORTHONAL see QAK000
ORTHONITROANILINE (DOT) see NEO000
ORTHO P-G BAIT see COF500
ORTHOPHENYLALANINE MUSTARD see BHT250
ORTHOPHOS see PAK000
ORTHO PHOSPHATE DEFOLIANT see BSH250
ORTHOPHOSPHORIC ACID see PHB250
ORTHOSAN MB see DTC600
ORTHOSIL see SJU000
ORTHOSILICATE see SCN500
ORTHO-TOLUOL-SULFONAMID (GERMAN) see TGN250
ORTHOTRAN see CJT750
ORTHO WEEVIL BAIT see DXE000
ORTIZON see HIB500
ORTOL see GGS000
ORTONAL see QAK000
ORTRAN see DOP600
ORTRIL see DOP600
ORTUDUR see PKQ059
ORVAGIL see MMN250
ORVINYLCARBINOL see AFV500
ORVUS WA PASTE see SIB600
OS 1897 see DDL800
OS 2046 see MQR750
OSAGE ORANGE see MRN500
OSAGE ORANGE CRYSTALS see MRN500
OSAGE ORANGE EXTRACT see MRN500
OSARSAL see ABX500

OSARSOLE see ABX500
OSBAC see MOV000
OSCINE see SBG000
OSCOPHEN see ARP250
OSMIC ACID see OKK000
OSMIUM see OKE000
OSMIUM(VIII) OXIDE see OKK000
OSMIUM TETROXIDE see OKK000
OSMOSOL EXTRA see PND000
OSOCIDE see CBG000
OSSALIN see SHF500
OSSIAMINA see CFA750
OSSICHLORIN see CFA750
OSSIDO di MESITILE (ITALIAN) see MDJ750
OSSIN see SHF500
OSTAMER see CDV625
OSTELIN see VSZ100
OSTEOBOND SURGICAL BONE CEMENT see PKB500
OSTREOGRYCIN see VRF000
OSVARSAN see ABX500
OTACRIL see DFP600
OTBE (FRENCH) see BLL750
OTC see HOH500
OTERBEN see BSQ000
OTETRYN see HOI000
OTOBIOTIC see NCG000
OTOFURAN see NGE500
OTOPHEN see CDP250
OTRACID see CJT750
OTS see TGN250
OTTAFACT see CFE250
OTTANI (ITALIAN) see OCU000
OTTASEPT see CLW000
OTTASEPT EXTRA see CLW000
1,2,4,5,6,7,8,8-OTTOCHLORO-3A,4,7,7A-TETRAIDRO-4,7-endo-METANO-INDANO (ITALIAN) see CDR750
OTTOMETIL-PIROFOSFORAMMIDE (ITALIAN) see OCM000
OUABAGENIN-l-RHAMNOSID (GERMAN) see OKS000
OUABAGENIN-l-RHAMNOSIDE see OKS000
OUABAIN see OKS000
OUABAINE see OKS000
OUBAIN see OKS000
OUTFLANK see AHJ750
OUTFLANK-STOCKADE see AHJ750
OVABAN see VTF000
OVADOFOS see DSQ000
OVADZIAK see BBQ500
OVAHORMON see EDO000
OVAHORMON BENZOATE see EDP000
OVANON see LJE000
OVARIOSTAT (FRENCH) see LJE000
OVASTEROL see EDO000
OVASTEROL-B see EDP000
OVASTEVOL see EDO000
OVATRAN see CJT750
OVEST see ECU750
OVESTERIN see EDU500
OVESTIN see EDU500
OVESTINON see EDU500
OVESTRION see EDU500
OVETTEN see MNM500
OVEX see CJT750, EDP000, EDV000
OVIFOLLIN see EDV000
OVIN see DNX500
OVISOT see ABO000
OVITELMIN see MHL000
OVOCHLOR see CJT750
OVOCICLINA see EDO000
OVOCYCLIN see EDO000
OVOCYCLIN BENZOATE see EDP000
OVOCYCLIN DIPROPIONATE see EDR000
OVOCYCLINE see EDO000
OVOCYCLIN M see EDP000

OVOCYCLIN-MB see EDP000
OVOCYCLIN-P see EDR000
OVOCYLIN see EDO000
OVOTOX see CJT750
OVOTRAN see CJT750
O-V STATIN see NOH500
OVULEN 50 see EQJ500
OWISPOL GF see SMQ500
OXAALZUUR (DUTCH) see OLA000
1-OXA-4-AZACYCLOHEXANE see MRP750
7-OXABICYCLO(4.1.0)HEPTANE see CPD000
7-OXABICYCLO(2.2.1)HEPTANE-2,3-DICARBOXYLIC ACID
   see EAR000
OXACYCLOPENTADIENE see FPK000
OXACYCLOPENTANE see TCR750
OXACYCLOPROPANE see EJN500
(1,2,3)OXADIAZOLO(5,4-d)PYRIMIDIN-5(4H)-ONE see
   DCQ600
OXAF see BHA750
OXAFURADENE see NDY000
3-OXA-1-HEPTANOL see BPJ850
OXAINE see DTL200
OXAL see GIK000
OXALALDEHYDE see GIK000
OXALAMIDE see OLO000
OXALATES see OKY000
OXALIC ACID see OLA000
OXALIC ACID DIAMIDE see OLO000
OXALIC ACID, DIETHYL ESTER see DJT200
OXALIC ACID DINITRILE see COO000
OXALID see HNI500
OXALONITRILE see COO000
OXALSAEURE (GERMAN) see OLA000
OXALYL CYANIDE see COO000
OXAMID (CZECH) see OLO000
OXAMIDE see OLO000
OXAMIMIDIC ACID see OLO000
OXAMMONIUM see HLM500
OXAMMONIUM SULFATE see OLS000
OXAMYL see DSP600
OXANAL YELLOW T see FAG140
OXANE see EJN500
OXANTHRENE see DDA800
3-OXAPENTANE-1,5-DIOL see DJD600
3-OXA-1,5-PENTANEDIOL see DJD600
OXAPROPANIUM IODIDE see FMX000
1,2-OXATHIOLANE-2,2-DIOXIDE see PML400
OXATIMIDE see DWF790, OMG000
OXATOMIDA see DWF790, OMG000
OXATOMIDE see DWF790, OMG000
2-H-1,3,2-OXAZAPHOSPHORINANE see EAS500
OXAZOLIDIN see HNI500
OXAZOLIDIN-GEIGY see HNI500
OX BILE EXTRACT see SFW000
m-OXEDRINE see NCL500
(−)-m-OXEDRINE see NCL500
OXETACAINE see DTL200
OXETANE see OMW000
OXETHACAINA (ITALIAN) see DTL200
OXETHAZINE see DTL200
OXIBUTOL see HNI500
OXIDATION BASE 22 see ALL750
OXIDATION BASE 25 see NEM480
OXIDATION BASE 12A see DBO400
OXIDE of CHROMIUM see CMJ900
OXIDIZED l-CYSTEINE see CQK325
OXIDOETHANE see EJN500
α,β-OXIDOETHANE see EJN500
1,8-OXIDO-p-MENTHANE see CAL000
OXI-FENIBUTOL see HNI500
OXIFENON see ORQ000
OXIFENYLBUTAZON see HNI500
OXIKON see DLX400
OXILAPINE see DCS200

OXIME COPPER see BLC250
OXIMETHOLONUM see PAN100
OXIMETOLONA see PAN100
OXINE see QPA000
OXINE COPPER see BLC250
OXINE CUIVRE see BLC250
2-OXI-PROPYL-PROPYLNITROSAMIN (GERMAN) see
   ORS000
OXIRAAN (DUTCH) see EJN500
OXIRANE see EJN500
OXIRANE-CARBOXALDEHYDE see GGW000
OXIRANEMETHANOL see GGW500
OXIRANYLMETHANOL see GGW500
OXIRANYLMETHYL ESTER of OCTADECANOIC ACID see
   SLK500
OXIRANYLMETHYL ESTER of 9-OCTADECENOIC ACID
   see ECJ000
3-OXIRANYL-7-OXABICYCLO(4.1.0)HEPTENE see
   VOA000
OXITETRACYCLIN see HOH500
OXITOL see EES350
OXITOSONA-50 see PAN100
OXLOPAR see HOI000
OXO see TAB750
2-OXO-1,2-BENZOPYRAN see CNV000
2-OXOBORNANE see CBA750
3-OXO-BUTANOIC ACID BUTYL ESTER see BPV250
3-OXOBUTANOIC ACID ETHYL ESTER see EFS000
3-OXOBUTANOIC ACID METHYL ESTER see MFX250
1-(1-OXOBUTYL)AZIRIDINE see BSY000
γ-OXO-α-BUTYLENE see BOY500
2-OXOCHROMAN see HHR500
α-OXODIPHENYLMETHANE see BCS250
3-OXO-l-GULOFURANOLACTONE see ARN000
2-OXOHEXAMETHYLENIMINE see CBF700
17-((1-OXOHEXYL)OXY)PREGN-4-ENE-3,20-DIONE see
   HNT500
2-(3-OXO-1-INDANYLIDENE)-1,3-INDANDIONE see
   ONY000
OXOLAMINE CITRATE see OOE000
OXOLANE see TCR750
OXOLE see FPK000
OXOMETHANE see FMV000
17-(1-OXOPROPOXY)-(17-β)-ANDROST-4-EN-3-ONE see
   TBG000
2-OXO-PROPYL-PROPYLNITROSAMINE see ORS000
(2-OXOPROPYL)PROPYLNITROSOAMINE see ORS000
OXOSUMITHION see PHD750
22-OXOVINCALEUKOBLASTINE see LEY000
OXSORALEN see XDJ000
OXY-5 see BDS000
OXY-10 see BDS000
OXYAMINE see CFA750
OXYBENZENE see PDN750
p-OXYBENZOESAEUREAETHYLESTER (GERMAN) see
   HJL000
p-OXYBENZOESAUREMETHYLESTER (GERMAN) see
   HJL500
p-OXYBENZOESAUREPROPYLESTER (GERMAN) see
   HNU500
OXYBENZOPYRIDINE see QPA000
OXYBIS(4-AMINOBENZENE) see OPM000
4,4′-OXYBISANILINE see OPM000
p,p′-OXYBIS(ANILINE) see OPM000
4,4′-OXYBISBENZENAMINE see OPM000
1,1′-OXYBIS(BUTANE) see BRH750
4,4′-OXYBIS(2-CHLOROANILINE) see BGT000
4,4′-OXYBIS(2-CHLORO-BENZENAMINE) see BGT000
1,1′-OXYBIS(2-CHLORO)ETHANE see DFJ050
OXYBIS(CHLOROMETHANE) see BIK000
1,1′-OXYBISETHANE see EJU000
2,2′-OXYBISETHANOL see DJD600
1,1′-OXYBISETHENE see VOP000
2,2′-(OXYBIS(ETHYLENEOXY))DIETHANOL see TCE250

OXYBISMETHANE see MJW500
2,2'-OXYBIS-6-OXABICYCLO-(3.1.0)HEXANE see BJN250
1,1-OXYBIS PENTANE see PBX000
1,1'-OXYBISPROPANE see PNM000
3,3'-OXYBIS(1-PROPENE) see DBK000
OXYBIS(TRIBUTYLTIN) see BLL750
OXYBUTANAL see AAH750
OXYBUTYRIC ALDEHYDE see AAH750
OXYCAINE see DHO600
OXYCARBON SULFIDE see CCC000
OXYCHINOLIN see QPA000
o-OXYCHINOLIN (GERMAN) see QPA000
OXYCHLORURE CHROMIQUE (FRENCH) see CML125
OXYCIL see SFS000
OXYCODEINONE see PCG500
OXYCODONE HYDROCHLORIDE see DLX400
OXYCODON HYDROCHLORIDE see DLX400
OXYCON see DLX400
OXY DBCP see DDL800
OXYDE d'ALLYLE et de GLYCIDYLE (FRENCH) see
  AGH150
OXYDE de BARYUM (FRENCH) see BAO000
OXYDE de CALCIUM (FRENCH) see CAU500
OXYDE de CARBONE (FRENCH) see CBW750
OXYDE de CHLORETHYLE (FRENCH) see DFJ050
OXYDE d'ETHYLE (FRENCH) see EJU000
OXYDE de MERCURE (FRENCH) see MCT500
OXYDE de MESITYLE (FRENCH) see MDJ750
OXYDEMETONMETHYL see DAP000
OXYDEMETON-METILE (ITALIAN) see DAP000
OXYDE NITRIQUE (FRENCH) see NEG100
OXYDEPROFOS see DSK600
OXYDE de PROPYLENE (FRENCH) see PNL600
OXYDE de TRIBUTYLETAIN see BLL750
OXYDIANILINE see OPM000
4,4'-OXYDIANILINE see OPM000
p,p'-OXYDIANILINE see OPM000
2,2'-OXYDIETHANOL see DJD600
N-(OXYDIETHYLENE)BENZOTHIAZOLE-2-SULFENAMIDE
  see BDG000
OXYDIFORMIC ACID DIETHYL ESTER see DIZ100
OXYDIMETHYLQUINAZINE see AQN000
4,4'-OXYDIPHENYLAMINE see OPM000
OXYDI-p-PHENYLENEDIAMINE see OPM000
OXYDISULFOTON see OQS000
N-OXYD-LOST see CFA500, CFA750
N-OXYD-MUSTARD see CFA500
OXYDOL see HIB000
OXYDRENE see DBA800
1-(β-OXYETHYL)-2-METHYL-5-NITROIMIDAZOLE see
  MMN250
OXYFED see DBA800
OXYFENON see ORQ000
OXYFUME see EJN500
OXYFUME 12 see EJN500
OXYFUME 20 see EJO000
OXYFUME 30 see EJO000
OXYFURADENE see NDY000
OXYGEN see OQW000
OXYGEN, compressed (DOT) see OQW000
OXYGEN, refrigerated liquid (DOT) see OQW000
OXYGEN DIFLUORIDE see ORA000
OXYGEN FLUORIDE see ORA000
OXYHYDROCHINON (GERMAN) see BBU250
OXYHYDROQUINONE see BBU250
OXYJECT 100 see HOI000
OXYKODAL see DLX400
OXYKON see DLX400
OXYLAN see DKQ000
OXYLITE see BDS000
OXYMETHALONE see PAN100
OXYMETHENOLONE see PAN100
OXYMETHOLONE see PAN100
OXYMETHYLENE see FMV000

OXYMURIATE OF POTASH see PLA250
OXYMYKOIN see HOH500
OXY-NH2 see CFA500
OXYPARATHION see NIM500
OXYPHENALON see RBU000
OXYPHENBUTAZONE see HNI500
OXYPHENIC ACID see CCP850
OXYPHENON see ORQ000
OXYPHENONIUM see ORQ000
OXYPHENONIUM BROMIDE see ORQ000
OXYPHENYLBUTAZONE see HNI500
OXYPHIONFOS see DSK600
OXYPROCAIN see DHO600
OXYPROCAINE see DHO600
p-OXYPROPIOPHENONE see ELL500
β-OXYPROPYLPROPYLNITROSAMINE see ORS000
γ-OXYPROPYLTHEOBROMIN (GERMAN) see HNZ000
γ-(γ-OXYPROPYL)-THEOBROMIN (GERMAN) see HNZ000
OXYPSORALEN see XDJ000
OXYQUINOLINE see QPA000
8-OXYQUINOLINE see QPA000
OXYQUINOLINOLEATE de CUIVRE (FRENCH) see BLC250
5-OXYRESORCINOL see PGR000
OXYSULFATOVANADIUM see VEZ000
OXYTERRACINE see HOH500
OXYTETRACYCLINE see HOH500
OXYTETRACYCLINE AMPHOTERIC see HOH500
OXYTETRACYCLINE HYDROCHLORIDE see HOI000
OXYTOCIC see EDB500
OXYTOL ACETATE see EES400
m-OXYTOLUENE see CNW750
o-OXYTOLUENE see CNX000
p-OXYTOLUENE see CNX250
OXYTRIL see HKB500
OXYTRIL M see DDP000
OXYUREA see HOO500, TGN250
OXY WASH see BDS000
OZIDE see ZKA000
OZLO see ZKA000
OZON (POLISH) see ORW000
OZONE see ORW000
OZONE mixed with NITROGEN OXIDES (53% : 47%) see
  ORY000

P07 see PKO500
P-25 see SLJ000
P-40 see DXG000
P-50 see AIV500
P-165 see ASA500
P-267 see BPR500
P 284 see DWF000
P 887 see ILE000
P 1142 see PDN000
P 1393 see BDE250
P1496 see RBF100
P 1531 see PFC750
P 2647 see BCL250
PA see PAP750
PA see PEG500
PA 144 see MQW750
PA 6 (polymer) see PJY500
PAA see BBL500
PA'AILA (HAWAII) see CCP000
PABA see AIH600
PABESTROL see DKA600, DKB000
PABS see SNM500
PAC see PAK000
PACEMO see HIM000
PACETYN see EHP000
PACIFAN see NBU000
PACITANE see BBV000
PACITRAN see MQR200
PADISAL see MRW000
PADOPHENE see PDP250

PAGANO-COR see DHS200
PAINTERS' NAPHTHA see PCT250
PAISLEY POLYMER see PMP500
PAKA (HAWAII) see TGI100
PAKHTARAN see DUK800
PAL see PEC750
PALACOS see PKB500
PALAFER see FBJ100
PALAPENT see NBU000
PALATINOL A see DJX000
PALATINOL AH see DVL700
PALATINOL BB see BEC500
PALATINOL C see DEH200
PALATINOL IC see DNJ400
PALATINOL M see DTR200
PALATONE see MAO350
PALE ORANGE CHROME see LCS000
PALESTROL see DKA600
PALINUM see TDA500
PALLADIUM see PAD250
PALLADIUM CHLORIDE see PAD500
PALLADIUM(2+) CHLORIDE see PAD500
PALLADOUS CHLORIDE see PAD500
PALLETHRINE see AFR250
PALLICID see ABX500
PALMA CHRISTI (HAITI) see CCP000
PALMAROSA OIL see PAE000
PALMITA see PGA750
PALMITIC ACID see PAE250
PALMITYL ALCOHOL see HCP000
PALMITYLAMINE see HCO500
PALOPAUSE see ECU750
PALTET see TBX250
PALUDRINE see CKB250
l-PAM see PED750
PAM (CZECH) see POS750
PAMAZONE see CJR909
2-PAM CHLORIDE see FNZ000
2-PAM IODIDE see POS750
PAMISAN see ABU500
PAMN see PNR250
PAMOLYN see OHU000
PAMOSOL 2 FORTE see EIR000
PANACELAN see DMU800
PANACIDE see MJM500
PANACUR see FAL100
PANADOL see HIM000
PANADON see PAG200
PANCAL see CAU750
PANCALMA see MQU750
PANCODINE see DLX400
PANCRIDINE see DBN400
PANDEX see PKM250
PANDRINOX see MLF250
PANDUROL see BHD250
PANESTIN see TBG000
PANETS see HIM000
PANEX see HIM000
PANFLAVIN see DBX400
PANFORMIN see BQL000
PANFURAN-S see BKH500
PANGUL see TEH500
PANITHAL see SAH000
PANMYCIN see TBX000
PANMYCIN HYDROCHLORIDE see TBX250
PANO-DRENCH 4 see MLF250
PANOFEN see HIM000
PANOGEN see MEO750, MLF250
PANOGEN TURF FUNGICIDE see MLF250
PANOGEN TURF SPRAY see MLF250
PANORAM 75 see TFS350
PANORAM D-31 see DHB400
PANOSINE see MMD500
PANOSPRAY 30 see MLF250

PANOXYL see BDS000
PANTALGINE see DAM700
PANTASOTE R 873 see PKQ059
PANTELMIN see MHL000
PANTHENOL see PAG200
d-PANTHENOL see PAG200
d(+)-PANTHENOL (FCC) see PAG200
PANTHER CREEK BENTONITE see BAV750
PANTHION see PAK000
PANTHODERM see PAG200
PANTHOJECT see CAU750
PANTHOLIN see CAU750
PANTOCAINE see BQA010
PANTOL see PAG200
PANTOLAX see HLC500
PANTOMICINA see EDH500
PANTONSILETTEN see DBX400
PANTOSEDIV see TEH500
PANTOTHENATE CALCIUM see CAU750
PANTOTHENIC ACID, CALCIUM SALT see CAU750
(+)-PANTOTHENIC ACID, CALCIUM SALT see CAU750
PANTOTHENOL see PAG200
d-PANTOTHENOL see PAG200
PANTOTHENYL ALCOHOL see PAG200
d-PANTOTHENYL ALCOHOL see PAG200
d(+)-PANTOTHENYL ALCOHOL see PAG200
PANTOVERNIL see CDP250
PAN-TRANQUIL see MQU750
PANURIN see CFY000
PAP see DRR400, PDC250
PAP-1 see AGX000
PAPAIN see PAG500
PAPANERINE see PAH000
PAPAVERINA (ITALIAN) see PAH000
PAPAVERIN CARBOXYLIC ACID, SODIUM SALT see PAG750
PAPAVERINE see PAH000
PAPAYOTIN see PAG500
PAPER BLACK BA see AQP000
PAPTHION see DRR400
PARA see PEY500
PARAAMINODIPHENYL see AJS100
PARABAR 441 see BFW750
PARABEN see HJL500, HNU500
PARACAIN see AIT250
PARACETALDEHYDE see PAI250
PARACETAMOLE see HIM000
PARACETAMOLO (ITALIAN) see HIM000
PARACETANOL see HIM000
PARACETOPHENETIDIN see ABG750
PARACHLORAMINE see HGC500
PARACHLOROCIDUM see DAD200
PARACIDE see DEP800
PARACODIN see DKW800
PARACODINE see DKW800
PARACORT see PLZ000
PARACORTOL see PMA000
PARACOTOL see PMA000
PARACRESYL ACETATE see MNR250
PARACRESYL ISOBUTYRATE see THA250
PARA CRYSTALS see DEP800
d-PARACURARINE CHLORIDE see TOA000
PARACYMENE see CQI000
PARACYMOL see CQI000
PARADERIL see RNZ000
PARADI see DEP800
PARADICHLORBENZOL (GERMAN) see DEP800
PARADICHLOROBENZENE see DEP800
PARADICHLOROBENZOL see DEP800
PARA-DIEN see DAL600
PARADONE OLIVE GREEN B see ALT000
PARADORMALENE see TEO250
PARADOW see DEP800
PARADUST see PAK000

PARAESIN see BQH250
PARAFFIN see PAH750
PARAFFIN HYDROCARBONS see PAH770
PARAFFIN OILS (PETROLEUM), CATALYTIC DEWAXED
    HEAVY (9CI) see MQV868
PARAFFIN OILS (PETROLEUM), CATALYTIC DEWAXED
    LIGHT (9CI) see MQV870
PARAFFIN WAX see PAH750
PARAFFIN WAXES and HYDROCARBON WAXES, CHLORI-
    NATED (C12, 60% CHLORINE) see PAH800
PARAFFIN WAX FUME (ACGIH) see PAH750
PARAFORM see FMV000
PARAFORMALDEHYDE see PAI000
PARAFORSN see PAI000
PARAFUCHSIN (GERMAN) see RMK020
PARAGLAS see PKB500
PARAHEXYL see HGK500
PARAL see PAI250
PARALDEHYD (GERMAN) see PAI250
PARALDEHYDE see PAI250
PARALDEIDE (ITALIAN) see PAI250
PARALEST see BBV000
PARA-MAGENTA see RMK020
PARAMAL see DBM800
PARAMANDELIC ACID see MAP000
PARAMAR see PAK000
PARAMAR 50 see PAK000
PARAMENTHANE HYDROPEROXIDE (DOT) see IQE000
PARAMETHYL PHENOL see CNX250
PARAMINE BLACK B see AQP000
PARAMINE BLUE 2B see CMO000
PARAMINE BLUE 3B see CMO250
PARAMINYL see WAK000
PARAMINYL MALEATE see DBM800
PARAMORFAN see DNU310
PARAMORPHAN see DLW600
PARAMOTH see DEP800
PARANAPHTHALENE see APG500
PARANEPHRIN see VGP000
PARANITROANILINE, solid (DOT) see NEO500
PARANITROFENOL (DUTCH) see NIF000
PARANITROFENOLO (ITALIAN) see NIF000
PARANITROPHENOL (FRENCH, GERMAN) see NIF000
PARANITROSODIMETHYLANILIDE see DSY600
PARANUGGETS see DEP800
PARAOXON see NIM500
PARAOXONE see NIM500
PARAPAN see HIM000
PARAPEST M-50 see MNH000
PARAPHENOLAZO ANILINE see PEI000
PARAPHENYLEN-DIAMINE see PEY500
PARAPHOS see PAK000
PARAPLEX P 543 see PKB500
PARAQUAT see PAI990
PARAQUAT CHLORIDE see PAJ000
PARAQUAT DICATION see PAI990
PARAQUAT DICHLORIDE see PAJ000
PARAROSANILINE see RMK020
PARAROSANILINE CHLORIDE see RMK020
PARAROSANILINE HYDROCHLORIDE see RMK020
PARASAN see BCA000
PARASCORBIC ACID see PAJ500
PARASEPT see HJL500, HNU500
PARASORBIC ACID see PAJ500
(+)-PARASORBINSAEURE (GERMAN) see PAJ500
PARASPEN see HIM000
PARASTARIN see EJM500
PARATAF see MNH000
PARATHENE see PAK000
PARATHESIN see EFX000
PARATHION see PAK000
M-PARATHION see MNH000
PARATHION, liquid (DOT) see PAK000
PARATHION (mixture, dry) see PAK250

PARATHION-ETHYL see PAK000
PARATHION METHYL see MNH000
PARATHION-METILE (ITALIAN) see MNH000
PARATOX see MNH000
PARAWET see PAK000
PARAXENOL see BGJ500
PARAXIN see CDP250
PARAZENE see DEP800
PARBENDAZOLE see BQK000
PARBOCYL-REV see SJO000
PARCIDOL see DIR000
PARCLOID see PKQ059
PARDA see DNA200
PARDIDOL see DIR000
PARDISOL see DIR000
PARDROYD see PAN100
PARENTERAL see CJR909
PARENTRACIN see BAC250
PAREST see QAK000
PAR ESTRO see ECU750
PARFENAC see BPP750
PARFENAL see BPP750
PARFEZINE see DIR000
PARFURAN see NGE000
PARGITAN see BBV000
PARIDINE RED LCL see CHP500
PARIDOL see HJL500
PARIS GREEN see COF500
PARIS RED see LDS000
PARIS YELLOW see LCR000
PARKE DAVIS CI-628 see NHP500
PARKIBLEU see CMO250
PARKIN see DIR000
PARKINSAN see BBV000
PARKIPAN see CMO250
PARKISOL see DIR000
PARKOPAN see BBV000
PARKOPHYLLIN see TEP000
PARKOSED see BNK000
PARKOTAL see EOK000
PAR KS-12 see PMC250
PARMAL see WAK000
PARMETOL see CFE250
PARMINAL see QAK000
PARMOL see HIM000
PARNATE see PET500, PET750
PAROL see CFE250, MQV750
PAROLEINE see MQV750
PAROXAN see NIM500
PAROXON see ELL500
PAROXYL see ABX500
PAROXYPROPIONE see ELL500
PARPHEZEIN see DIR000
PARPON see BCA000
PARRAFIN OIL see MQV750
PARROT GREEN see COF500
PARSIDOL see DIR000
PARSITAN see DIR000
PARSLEY HERB OIL (FCC) see PAL750
PARSLEY OIL see PAL750
PARSLEY SEED OIL (FCC) see PAL750
PARTEL see DJT800
PARTERGIN see PAM000
PARTREX see TBX250
PARVOLEX see ACH000
PARZATE see DXD200, EIR000
PARZONE see DKW800
PASCO see ZBJ000, ZKA000
PASEPTOL see HNU500
PASEXON 100T see SHK800
PASOTOMIN see PMF250
PATENT BLUE AE see FMU059
PATENT GREEN see COF500
PATRICIN see VRF000

PATROVINE see DHX800
PATTINA V 82 see PKQ059
PAVISOID see PAN100
PAXAREL see ACE000
PAXISTIL see CJR909
PAXISYN see DLY000
PAYZE see BLW750
PB see PIX250
PBB see FBU000, FBU509
PBDZ see BQK000
PBNA see PFT500
PBS see SID000
PBX(AF) 108 see CPR800
PBZ see BRF500, TMP750
PCA see MCR750, PEE750
PCB see PME250
PCB's see PJM500, PJN000
PCB (DOT, USDA) see PJL750
PCB HYDROCHLORIDE see PME500
PCBS see CJR500
PCC see CDV100
P.C. 80 CRABGRASS KILLER see PLC250
PCHO see PAI250
PCI see CJR500
PCL see HCE500
PCM see PCF300
pCMA see CIF250
PCMC see CFE250
PCMX see CLW000
PCNB see PAX000
PCP see PAX250
PCPA see CJN000
PCPBS see CJR500
PCPCBS see CJT750
PCPI see CKB000
2-N-p-PDA see ALL750
P.D.A.B. see DOT300
PDB see DEP800
PDCB see DEP800
PDD see PGT250
PDD 6040I see CJV250
PDMT see DTP000
m-PDN see PHX550
p-PDN see BBP250
PDP see PDC250
PDT see DTP000
PE see PBB750
PEACH ALDEHYDE see UJA800
PEACOCK BLUE X-1756 see FMU059
PEANUT OIL see PAO000
PEARL ASH see PLA000, PLA250
PEARLPUSS see CCL250
PEARL STEARIC see SLK000
PEARLY GATES see DJO000
PEAR OIL see AOD725, IHO850
PEARSALL see AGY750
PEB1 see DAD200
PEBC see PNF500
PEBULATE see PNF500
PECAN SHELL POWDER see PAO000
PECTALGINE see SEH000
PEDIAFLOR see SHF500
PEDIDENT see SHF500
PEDINEX (FRENCH) see CPK500
PEDRACZAK see BBQ500
PEDRIC see HIM000
PEERAMINE BLACK E see AQP000
PEG see PJT000
PEG 200 see PJT200
PEG 300 see PJT225
PEG 400 see PJT230
PEG 600 see PJT240
PEG 1000 see PJT250
PEG 1500 see PJT500

PEG 4000 see PJT750
PEG 6000 see PJU000
PEG-9 NONYL PHENYL ETHER see PKF000
PEG-9 OCTYL PHENYL ETHER see PKF500
PELADOW see CAO750
PELAGOL 3GA see ALT000
PELAGOL D see PEY500
PELAGOL DA see DBO000
PELAGOL DR see PEY500
PELAGOL EG see ALT500
PELAGOL GREY see DBO400
PELAGOL GREY C see CCP850
PELAGOL GREY D see PEY500
PELAGOL GREY GG see ALT000
PELAGOL GREY J see TGL750
PELAGOL GREY L see DBO000
PELAGOL GREY RS see REA000
PELAGOL L see DBO000
PELARGIC ACID see NMY000
PELARGIDENON 1449 see CDH250
PELARGOL see DTE600
PELARGON (RUSSIAN) see NMY000
PELARGONIC ACID see NMY000
PELARGONIC ALCOHOL see NNB500
PELARGONIC ALDEHYDE see NMW500
PELARGONIC MORPHOLIDE see MRQ750
PELARGONIUM OIL see GDA000
PELAZID see ILD000
PELENTAN see BKA000
PELIDORM see BNK000
PELLAGRAMIN see NCQ900
PELLAGRA PREVENTIVE FACTOR see NCQ900
PELLAGRIN see NCQ900
PELLCAFS see BBK500
PELLCAP see BBK500
PELLCAPS see BBK500
PELLIDOL see ACR300
PELLIDOLE see ACR300
PELLON 2506 see PMP500
PELLUGEL see CCL250
PELMIN see NCR000
PELMINE see NCR000
PELONIN see NCQ900
PELONIN AMIDE see NCR000
PELSON see DLY000
PELTOL D see PEY500
PEMAL see ENG500
PEMALIN see ENG500
PEN A see AOD125
PEN-A-BRASIVE see BFD250
d-PENAMINE see MCR750
PENBAR see NBU000
PENBRISTOL see AIV500
PENBRITIN see AIV500
PENBRITIN PAEDIATRIC see AIV500
PENBRITIN SYRUP see AIV500
PENBROCK see AIV500
PENCARD see PBC250
PENCHLOROL see PAX250
PENCIL GREEN SF see FAF000
PENCILLIC ACID see PAP750
PENCOGEL see CCL250
PENETECK see MQV750
PENFLURIDOL see PAP250
PENGITOXIN see GEW000
PENICILLAMIN see MCR750
(S)-PENICILLAMIN see MCR750
PENICILLAMINE see MCR750
d-PENICILLAMINE see MCR750
PENICILLAMINE HYDROCHLORIDE see PAP550
d-PENICILLAMINE HYDROCHLORIDE see PAP550
PENICILLIC ACID see PAP750
PENICILLIN see PAQ000
PENICILLIN G see BDY669

PENICILLIN-G, MONOSODIUM SALT see BFD250
PENICILLIN G, SODIUM see BFD250
PENICILLIN G, SODIUM SALT see BFD250
PENICLINE see AIV500
PENILARYN see BFD250
PENITE see SEY500
PENITRACIN see BAC250
PENIZILLIN (GERMAN) see PAQ000
PENNAC see SGF500
PENNAC CRA see IAQ000
PENNAC MBT POWDER see BDF000
PENNAC MS see BJL600
PENNAC ZT see BHA750
PENNAMINE see DAA800
PENNCAP-M see MNH000
PENNFLOAT M see LBX000
PENNFLOAT S see LBX000
PENNSALT TD-72 see DJI000
PENN SALT TD-183 see TBV750
PENNSALT TD 5032 see HEE500
PENNWALT C-4852 see DSQ000
PENNWHITE see SHF500
PENNYROYAL OIL see PAR500
PENNZONE E see DKC400
PENOTRANE see PFN000
PENPHENE see CDV100, TBV750
PENRECO see MQV750
PENSYN see AOD125
PENTA see PAX250
PENTAACETYLGITOXIN see GEW000
PENTA-o-ACETYLGITOXIN see GEW000
1,4,7,10,13-PENTAAZATRIDECANE see TCE500
PENTABARBITAL SODIUM see NBU000
PENTABARBITONE see NBT500
PENTABORANE(9) see PAT750
PENTABORANE(11) see PAT799
PENTABROMO PHOSPHORANE see PHR250
PENTABROMO PHOSPHORUS see PHR250
PENTACARBONYLIRON see IHG500
PENT-ACETATE see AOD725
PENTACHLOORETHAAN (DUTCH) see PAW500
PENTACHLOORFENOL (DUTCH) see PAX250
PENTACHLORAETHAN (GERMAN) see PAW500
PENTACHLORETHANE (FRENCH) see PAW500
PENTACHLORIN see DAD200
PENTACHLORNITROBENZOL (GERMAN) see PAX000
PENTACHLOROACETOPHENONE see PAV250
2′,3′,4′,5′,6′-PENTACHLOROACETOPHENONE see PAV250
PENTACHLOROANTIMONY see AQD000
PENTACHLORO DIPHENYL OXIDE see PAW250
PENTACHLOROETHANE see PAW500
PENTACHLOROFENOL see PAX250
PENTACHLORONAPHTHALENE see PAW750
PENTACHLORONITROBENZENE see PAX000
PENTACHLOROPHENATE see PAX250
PENTACHLOROPHENATE SODIUM see SJA000
PENTACHLOROPHENOL see PAX250
2,3,4,5,6-PENTACHLOROPHENOL see PAX250
PENTACHLOROPHENOL (GERMAN) see PAX250
PENTACHLOROPHENOL, DOWICIDE EC-7 see PAX250
PENTACHLOROPHENOL, DP-2 see PAX250
PENTACHLOROPHENOL, SODIUM SALT see SJA000
PENTACHLOROPHENOL, TECHNICAL see PAX250
PENTACHLOROPHENOXY SODIUM see SJA000
PENTACHLOROPHENYL CHLORIDE see HCC500
PENTACHLORURE D'ANTIMOINE (FRENCH) see AQD000
PENTACIN see CAY500
PENTACINE see CAY500
PENTACLOROETANO (ITALIAN) see PAW500
PENTACLOROFENOLO (ITALIAN) see PAX250
PENTACON see PAX250
1-PENTADECANECARBOXYLIC ACID see PAE250
1,3-PENTADIENE-1-CARBOXYLIC ACID see SKU000
PENTAERYTHRITE see PBB750

PENTAERYTHRITE TETRANITRATE see PBC250
PENTAERYTHRITE TETRANITRATE (DOT) see PBC250
PENTAERYTHRITE TETRANITRATE, DESENSITIZED, WET
  (DOT) see PBC250
PENTAERYTHRITE TETRANITRATE, DRY (DOT) see
  PBC250
PENTAERYTHRITE THERANITRATE, with not less than 7%
  wax (DOT) see PBC250
PENTAERYTHRITOL see PBB750
PENTAERYTHRITOL TETRANITRATE see PBC250
PENTAERYTHRITOL TETRANITRATE, DILUTED see
  PBC250
PENTAFIN see PBC250
PENTAFLUOROANTIMONY see AQF250
PENTAFLUOROIODINE see IDT000
PENTAGEN see PAX000
PENTAGIN see DOQ400
PENTAGIT see GEW000
2′,3,4′,5,7-PENTAHYDROXYFLAVONE see MRN500
3,5,7,3′,4′-PENTAHYDROXYFLAVONE see QCA000
PENTAHYDROXY-TIGLIADIENONE-MONOACETATE
  (C)MONOMYRISTATE(B) see PGV000
PENTA-KIL see PAX250
PENTAL see NBU000
γ-PENTALACTONE see VAV000
PENTALIN see PAW500
PENTAMETHYLBENZYL-p-ROSANILINE CHLORIDE see
  MQN000
PENTAMETHYLENE see CPV750
PENTAMETHYLENEDIAMINE see PBK500
1,5-PENTAMETHYLENEDIAMINE see PBK500
p,p′-(PENTAMETHYLENEDIOXY)DIBENZAMIDINE BIS
  (β-HYDROXYETHANESULFONATE) see DBL800
PENTAMETHYLENEDITHIOCARBAMATE see PIY500
PENTAMETHYLENEIMINE see PIL500
PENTAMETHYLENETETRAZOL see PBI500
1,5-PENTAMETHYLENETETRAZOLE see PBI500
PENTAMETHYLENE-1,5-TETRAZOLE see PBI500
PENTAMIDINE DIISETHIONATE see DBL800
PENTAMIDINE ISETHIONATE see DBL800
PENTAN (POLISH) see PBK250
PENTANAL see VAG000
n-PENTANAL see VAG000
PENTANE see PBK250
tert-PENTANE (DOT) see NCH000
3-PENTANECARBOXYLIC ACID see DHI400
1,5-PENTANEDIAL see GFQ000
1,5-PENTANEDIAMINE see PBK500
PENTANEDINITRILE see TLR500
PENTANEDIONE see ABX750
1,5-PENTANEDIONE see GFQ000
2,4-PENTANEDIONE (FCC) see ABX750
PENTANEN (DUTCH) see PBK250
1-PENTANETHIOL see PBM000
PENTANI (ITALIAN) see PBK250
PENTANOIC ACID see VAQ000
n-PENTANOIC ACID see VAQ000
tert-PENTANOIC ACID see PJA500
PENTANOL-1 see AOE000
PENTAN-1-OL see AOE000
2-PENTANOL see PBM750
PENTANOL-2 see PBM750
3-PENTANOL see IHP010
PENTANOL-3 see IHP010
PENTAN-3-OL see IHP010
N-PENTANOL see AOE000
tert-PENTANOL see PBV000
1-PENTANOL ACETATE see AOD725
2-PENTANOL, ACETATE see AOD735
4-PENTANOLIDE see VAV000
2-PENTANONE see PBN250
PENTANONE-3 see DJN750
3-PENTANONE see DJN750
PENTANTIN see DAM700

PENTANYL see PDW750
PENTAPHENATE see SJA000
PENTASODIUM TRIPHOSPHATE see SKN000
PENTASOL see AOE000, PAX250
PENTASULFURE de PHOSPHORE (FRENCH) see PHS000
PENTAZOCINE see DOQ400
PENTECH see DAD200
PENTEK see PBB750
PENTESTAN-80 see PBC250
PENTETATE TRISODIUM CALCIUM see CAY500
PENTETRATE UNICELLES see PBC250
PENTHAMIL see CAY500, DJG800
PENTHAZINE see PDP250
PENTHIOBARBITAL SODIUM see PBT500
PENTHRANE see DFA400
PENTIFORMIC ACID see HEU000
PENTIN C see MNM500
PENTINIMID see ENG500
PENTOBARBITAL see NBT500, PBS250
PENTOBARBITAL CALCIUM see CAV000
PENTOBARBITONE SODIUM see NBU000
PENTOBARBITURATE see NBT500
PENTOBARBITURIC ACID see NBT500
PENTOFRAN see DSI709
PENTOLE see CPU500
PENTONAL see NBU000
PENTOTHAL SODIUM see PBT500
PENTRAN see DFA400
PENTRANE see DFA400
PENTRATE see PBC250
PENTREX see AIV500
PENTREXL see AIV500
PENTRIOL see PBC250
PENTRYATE 80 see PBC250
PENTYL see NBU000
PENTYL ACETATE see AOD725
1-PENTYL ACETATE see AOD725
2-PENTYL ACETATE see AOD735
n-PENTYL ACETATE see AOD725
PENTYL ALCOHOL see AOE000
sec-PENTYL ALCOHOL see PBM750
tert-PENTYL ALCOHOL see PBV000
PENTYLAMINE (mixed isomers) see PBV500
2,N-PENTYLAMINOETHYL-p-AMINOBENZOATE see
    PBV750
PENTYLCARBINOL see HFJ500
3-PENTYLCARBINOL see EGW000
sec-PENTYLCARBINOL see EGW000
PENTYL CHLORIDE see PBW500
α-PENTYLCINNAMALDEHYDE see AOG500
PENTYLENETETRAZOL see PBI500
PENTYL ESTER PHOSPHORIC ACID see PBW750
PENTYL ETHER see PBX000
PENTYL FORMATE see AOJ500
m-PENTYL FORMATE see AOJ500
PENTYLFORMIC ACID see HEU000
PENTYL HEXANOATE see IHU100
PENTYL MERCAPTAN see PBM000
PENTYL NITRITE see AOL500
n-PENTYLNITROSOUREA see PBX500
PENTYL PENTYLAMINE see DCH200
p-PENTYLPHENOL see AOM250
o-(sec-PENTYL) PHENOL see AOM500
PENTYLTRICHLOROSILANE see PBY750
3-PENTYL-6,6,9-TRIMETHYL-6a,7,8,10a-TETRAHYDRO-6H-
    DIBENZO(b,d)PYRAN-1-OL see TCM250
PENWAR see PAX250
PENZYLPENICILLIN SODIUM SALT see BFD250
PEP see EDS000, PCC000
PEPPERMINT CAMPHOR see MCF750
PEPPERMINT OIL see PCB250
PEPTICHEMIO see PCC000
PERACETIC ACID, solution (DOT) see PCL500
PERACETIC ACID (MAK) see PCL500

PERAGAL ST see PKQ250
PERAGIT see BBV000
PERANDREN see TBF500, TBG000
PERATOX see PAX250
PERAWIN see PCF275
PERAZIL see CDR000
PERAZIL DIHYDROCHLORIDE see CDR000
PERAZINE see PCK500
PERBENZOATE de BUTYLE TERTIAIRE (FRENCH) see
    BSC500
PERBENZOIC ACID see PCM000
PERBUTYL H see BRM250
PERCAIN see NOF500
PERCAINE HYDROCHLORIDE see NOF500
PERCARBAMIDE see HIB500
PERCHLOORETHYLEEN, PER (DUTCH) see PCF275
PERCHLOR see PCF275
PERCHLORAETHYLEN, PER (GERMAN) see PCF275
PERCHLORATE de MAGNESIUM (FRENCH) see PCE000
PERCHLORATES see PCD000
PERCHLORATE de SODIUM (FRENCH) see PCE750
PERCHLORETHYLENE see PCF275
PERCHLORETHYLENE, PER (FRENCH) see PCF275
PERCHLORIC ACID see PCD250
PERCHLORIC ACID, AMMONIUM SALT see PCD500
PERCHLORIC ACID, BARIUM SALT·3H$_2$O see PCD750
PERCHLORIC ACID, MAGNESIUM SALT see PCE000
PERCHLORIC ACID, SODIUM SALT see PCE750
PERCHLORIDE of MERCURY see MCY475
PERCHLORMETHYLMERKAPTAN (CZECH) see PCF300
PERCHLOROBENZENE see HCC500
PERCHLOROBUTADIENE see HCD250
PERCHLORODIHOMOCUBANE see MQW500
PERCHLOROETHANE see HCI000
PERCHLOROETHYLENE see PCF275
PERCHLOROMETHANE see CBY000
PERCHLOROMETHYL MERCAPTAN see PCF300
PERCHLORON see HOV500
PERCHLOROPENTACYCLODECANE see MQW500
PERCHLOROPENTACYCLO(5.2.1.0^{2,6}.0^{3,9}.0^{5,8})DECANE see
    MQW500
PERCHLOROTHIOPHENE see TBV750
PERCHLORURE d'ANTIMOINE (FRENCH) see AQD000
PERCHLORURE de FER see FAU000
PERCHLORYL FLUORIDE see PCF750
PERCHLORYL HYPOFLUORITE see FFD000
PERCIN see ILD000
PERCIPITATED CALCIUM PHOSPHATE see CAW120
PERCLENE see PCF275
PERCLOROETILENE (ITALIAN) see PCF275
PERCOBARB see ABG750
PERCODAN see ABG750, PCG500
PERCODAN HYDROCHLORIDE see DLX400
PERCOLATE see PHX250
PERCORAL see DJS200
PERCOSOLVE see PCF275
PERCUTACRINE see PMH500
PERCUTACRINE ANDROGENIQUE see TBF500
PERCUTATRINE OESTROGENIQUE ISCOVESCO see
    DKA600
PEREMESIN see HGC500
PEREQUIL see MQU750
PERFECTA see MQV750
PERFECTHION see DSP400
PERFENAZINA (ITALIAN) see CJM250
PERFLUOROACETIC ACID see TKA250
PERFLUOROAMMONIUM OCTANOATE see ANP625
PERFLUOROBUT-2-ENE see OBO000
PERFLUORO-2-BUTENE (DOT) see OBO000
PERFLUOROCYCLOBUTANE see CPS000
PERFLUOROETHENE see TCH500
PERFLUOROETHYLENE see TCH500
PERFLUORO HYDRAZINE see TCI000
PERFLUOROMETHANE see CBY250

PERFLUOROPROPENE see HDF000
PERFLUOROPROPYLENE see HDF000
PERGACID VIOLET 2B see FAG120
PERGANTENE see SHF500
PERGITRAL see PBC250
PERGLOTTAL see NGY000
PER-GLYCERIN see LAM000
PERHYDRIT see HIB500
PERHYDROAZEPINE see HDG000
2-PERHYDROAZEPINONE see CBF700
PERHYDROGERANIOL see DTE600
PERHYDROL see HIB000
PERHYDROL-UREA see HIB500
PERHYDRONAPHTHALENE see DAE800
PERICIAZINE see PIW000
PERICYAZINE see PIW000
PERIDEX-LA see PBC250
PERILLA KETONE see PCI750
PERIMETAZINE see LEO000
PERIMETHAZINE see LEO000
O-PERIODIC ACID see PCJ250
PERIODIN see PLO500
PERIPHERMIN see ACR300
PERISTON see PKQ250
PERITRATE see PBC250
PERITYL see PBC250
PERK see PCF275
PERKE see BBK500
PERKLONE see PCF275
PERLATAN see EDV000
PERLITE see PCJ400
PERLITON ORANGE 3R see AKP750
PERLON see NOH000
PERLUTEX see MCA000
PERM-A-CHLOR see TIO750
PERMACIDE see PAX250
PERMAGARD see PAX250
PERMA KLEER see DJG800
PERMA KLEER 50 ACID see EIX000
PERMA KLEER 50 CRYSTALS see EIV000
PERMA KLEER 50 CRYSTALS DISODIUM SALT see
    EIX500
PERMA KLEER TETRA CP see EIV000
PERMA KLEER 50, TRISODIUM SALT see TNL250
PERMANENT WHITE see BAP000, ZKA000
PERMANENT YELLOW see BAK250
PERMANGANATE of POTASH (DOT) see PLP000
PERMANGANATE de POTASSIUM (FRENCH) see PLP000
PERMANGANATES see PCJ500
PERMANGANATE de SODIUM (FRENCH) see SJC000
PERMANGANIC ACID AMMONIUM SALT see PCJ750
PERMANGANIC ACID, BARIUM SALT see PCK000
PERMANGANIC ACID, SODIUM SALT see SJC000
PERMASAN see PAX250
PERMATOX DP-2 see PAX250
PERMATOX PENTA see PAX250
PERMETHRIN (USDA) see AHJ750
PERMETRINA (PORTUGUESE) see AHJ750
PERMITAL see DYE600
PERMITE see PAX250
PERNAEMON see VSZ000
PERNAEVIT see VSZ000
PERNAZINE see PCK500
PERNIPURON see VSZ000
PERNOCTON see BOR000
PERNOSTON see BOR000
PERONE see HIB000
PEROSIN see EIR000
PEROSSIDO di BENZOILE(ITALIAN) see BDS000
PEROSSIDO di BUTILE TERZIARIO (ITALIAN) see BSC750
PEROSSIDO di IDROGENO (ITALIAN) see HIB000
PEROXAN see HIB000
PEROXIDE see HIB000
PEROXIDES, INORGANIC see PCL000

PEROXIDES, ORGANIC see PCL250
1,4-PEROXIDO-p-MENTHENE-2 see ARM500
PEROXOMONOPHOSPHORIC ACID see PCN500
PEROXOMONOSULFURIC ACID see PCN750
PEROXYACETIC ACID see PCL500
PEROXYACETIC ACID, maximum concentration 43% in acetic
    acid (DOT) see PCL500
PEROXYBENZOIC ACID see PCM000
PEROXYDE de BARYUM (FRENCH) see BAO250
PEROXYDE de BENZOYLE (FRENCH) see BDS000
PEROXYDE de BUTYLE TERTIAIRE (FRENCH) see
    BSC750
PEROXYDE d'HYDROGENE (FRENCH) see HIB000
PEROXYDE de LAUROYLE (FRENCH) see LBR000
PEROXYDE de PLOMB (FRENCH) see LCX000
PEROXYDICARBONATE D'ISOPROPYLE (FRENCH) see
    DNR400
PEROXYDICARBONIC ACID, BIS(1-METHYLETHYL) ES-
    TER see DNR400
PEROXYDICARBONIC ACID DIPROPYL ESTER see
    DWV400
PEROXYDISULFURIC ACID DIPOTASSIUM SALT see
    DWQ000
PEROXYDISULFURYL DIFLUORIDE see PCM250
PEROXYFORMIC ACID see PCM500
PEROXYMONOPHOSPHORIC ACID see PCN500
PEROXYMONOSULFURIC ACID see PCN750
PEROXYNITRIC ACID see PCO000
PEROXYTRIFLUOROACETIC ACID see PCO250
PERPHENAZIN see CJM250
PERPHENAZINE see CJM250
PERSADOX see BDS000
PERSAMINE see DLH630
PERSEC see PCF275
PERSIAN RED see LCS000
PERSIA-PERAZOL see DEP800
PERSISTEN see DJT400
PERSISTOL see TND500
PERSPEX see PKB500
PERSULFATE d'AMMONIUM (FRENCH) see ANR000
PERSULFATE de SODIUM (FRENCH) see SJE000
PERSULFEN see SNN300
PERTHANE see DJC000
PERTOFRAM see DLH630
PERTOFRAN see DLS600, DSI709
PERTOFRANE see DLS600, DSI709
PERUSCABIN see BCM000
PERUVIAN BALSAM see BAE750
PERVERTIN see DBB000
PERVITIN see DBA800
PESTAN see DJI000
PESTMASTER see EIY500
PESTMASTER EDB-85 see EIY500
PESTMASTER (OBS.) see MHR200
PESTON XV see PHF750
PESTOX see OCM000
PESTOX 3 see OCM000
PESTOX 14 see BJE750
PESTOX 15 see PHF750
PESTOX 101 see NIM500
PESTOX III see OCM000
PESTOX IV see BJE750
PESTOX PLUS see PAK000
PESTOX XIV see BJE750
PESTOX XV see PHF750
PETANTIN HYDROCHLORIDE see DAM700
PETASITES JAPONICUS MAXIM see PCR000
PETERSILIENSAMEN OEL (GERMAN) see PAL750
PETHIDINE CHLORIDE see DAM700
PETHIDINETER see DAM600
PETHIDOINE see DAM600
PETHION see PAK000
PETIDIN see DAM700
PETIDION see TLP750

PETIDON see TLP750
PETILEP see TLP750
PETINIMID see ENG500
PETNAMYCETIN see CDP250
PETNIDAN see ENG500
PETROGALAR see MQV750
PETROHOL see INJ000
PETROL (DOT) see GBY000
PETROLATUM, liquid see MQV750
PETROLEUM see PCR250
PETROLEUM ASPHALT see PCR500
PETROLEUM BENZIN see NAI500
PETROLEUM CRUDE see PCR250
PETROLEUM DISTILLATE see PCS250
PETROLEUM DISTILLATES, HYDROTREATED HEAVY
  NAPHTHENIC see MQV790
PETROLEUM DISTILLATES (NAPHTHA) see NAI500
PETROLEUM ETHER (DOT) see NAI500
PETROLEUM GAS, LIQUEFIED see LGM000
PETROLEUM NAPHTHA (DOT) see NAI500
PETROLEUM PITCH see ARO500
PETROLEUM ROOFING TAR see PCR500
PETROLEUM SPIRIT (DOT) see NAI500
PETROLEUM SPIRITS see PCT250
PETROL ORANGE Y see PEJ500
PETROL YELLOW WT see DOT300
PETZINOL see TIO750
PEVIKON D 61 see PKQ059
PEVITON see NCQ900
PEYRONE'S CHLORIDE see PJD000
PF-1 see DSA800
PF-3 see IRF000
PFEFFERMINZ OEL (GERMAN) see PCB250
PFIKLOR see PLA500
PFIZER 1393 see BDE250
PFIZERPEN A see AIV500
PG 12 see PML000
PGDN see PNL000
PGE see PFH000
PGE2 see DVJ200
PGF2-α see DMU800
PGF2-α THAM see POC750
PGF2-α TRIS SALT see POC750
PGF2-α TROMETHAMINE see POC750
PH 60-40 see CJV250
PHALDRONE see CDO000
PHANAMIPHOS see FAK000
PHANANTIN see DKQ000
PHANODORM see TDA500
PHANODORN see TDA500
PHARGAN see DQA400
PHARLON see EDS100
PHARMAZOID RED KB see CLK225
PHARMETTEN see EOK000
PHAROS 100.1 see SMR000
PHASOLON see BDJ250
PHC see PMY300
M-PHDM see BKL750
PHEBUZIN see BRF500
PHEBUZINE see BRF500
α-PHELLANDRENE (FCC) see MCC000
PHEMERIDE see BEN000
PHEMEROL CHLORIDE see BEN000
PHEMETONE see ENB500
PHEMITHYN see BEN000
PHEMITON see ENB500
PHEMITONE see ENB500
PHENACETALDEHYDE DIMETHYL ACETAL see PDX000
p-PHENACETIN see ABG750
PHENACHLOR see TIW000
PHENACIDE see CDV100
PHENACITE see SCN500
PHENACYLAMINE see AHR250
PHENACYL CHLORIDE see CEA750

PHENADONE see MDO750
PHENADOR-X see BGE000
PHENAEMAL see EOK000
PHENAKITE see SCN500
PHENALCO see MCU750
PHENALGENE see AAQ500
PHENALZINE see PFC750
PHENALZINE DIHYDROGEN SULFATE see PFC750
PHENALZINE HYDROGEN SULPHATE see PFC750
PHENAMINE BLUE BB see CMO000
PHENANTHREN (GERMAN) see PCW250
PHENANTHRENE see PCW250
PHENANTOIN see MKB250
PHENANTRIN see PCW250
PHENARSAZINE CHLORIDE see PDB000
PHENARSENAMINE see SAP500
PHENASAL see DFV400
PHENATOINE see DKQ000
PHENATOX see CDV100
PHENAZINE see DKE800, PDB500
PHENAZITE see SCN500
PHENAZO see PDC250
PHENAZODINE see PDC250, PEK250
PHENAZOLINE see PDC000
PHENAZONE (PHARMACEUTICAL) see AQN000
PHENAZOPYRIDINE see PEK250
PHENAZOPYRIDINE HYDROCHLORIDE see PDC250
PHENAZOPYRIDINIUM CHLORIDE see PDC250
PHEN-BUTA-VET see BRF500
PHENBUTAZOL see BRF500
PHENCAPTON see PDC750
PHENDAL see DRR400
PHENDIMETRAZINE BITARTRATE see DKE800
PHENDIMETRAZINE HYDROCHLORIDE see DTN800
PHENE see BBL250
PHENEDRINE see BBK000
PHENEGIC see PDP250
PHENELZIN see PFC750
PHENELZINE see PFC500
PHENELZINE ACID SULFATE see PFC750
PHENELZINE BISULPHATE see PFC750
PHENELZINE SULFATE see PFC750
PHENEMALUM see SID000
(v-PHENENYLTRIS(OXYETHYLENE))TRIS(TRIETHYL-
  AMMONIUM IODIDE) see PDD300
PHENERGAN see DQA400
PHENESTERINE see CME250
PHENESTRIN see CME250
PHENETHANOL see PDD750
PHENETHIDINE see PDD500
2-PHENETHYL ACETATE see PFB250
β-PHENETHYL ACETATE see PFB250
PHENETHYL ALCOHOL see PDD750
2-PHENETHYL ALCOHOL see PDD750
α-PHENETHYL ALCOHOL see PDE000
β-PHENETHYL ALCOHOL see PDD750
β-PHENETHYLAMINE see PDE250
PHENETHYLAMINE, α-METHYL-, SULFATE (2:1) see
  BBK250
β-PHENETHYL-o-AMINOBENZOATE see APJ500
PHENETHYL ANTHRANILATE see APJ500
PHENETHYLCARBAMID (GERMAN) see EFE000
PHENETHYLENE see SMQ000
PHENETHYLENE OXIDE see EBR000
PHENETHYL ESTER ISOVALERIC ACID see PDF775
PHENETHYLHYDRAZINE see PFC500
PHENETHYLHYDRAZINE SULFATE (1:1) see PFC750
PHENETHYL ISOBUTYRATE see PDF750
PHENETHYL ISOVALERATE see PDF775
2-PHENETHYL 2-METHYLBUTYRATE see PDF790
PHENETHYL PHENYLACETATE see PDI000
N-(1-PHENETHYL-4-PIPERIDI-NYL)PROPIONANILIDE DI-
  HYDROGEN CITRATE see PDW750

N-(1-PHENETHYL-4-PIPERIDYL)PROPIONANILIDE CITRATE see PDW750
N-(1-PHENETHYL-4-PIPERIDYL)PROPIONANILIDE DIHYDROGEN CITRATE see PDW750
1-PHENETHYL-4-N-PROPIONYLANILINOPIPERIDINE see PDW500
N-PHENETHYL-4-(N-PROPIONYLANILINO)PIPERIDINE see PDW500
PHENETHYL SALICYLATE see PDK200
p-PHENETIDINE (DOT) see PDD500
p-PHENETOLCARBAMID (GERMAN) see EFE000
p-PHENETOLCARBAMIDE see EFE000
p-PHENETOLECARBAMIDE see EFE000
p-PHENETYLUREA see EFE000
PHENFLUORAMINE HYDROCHLORIDE see PDM250
PHENIBUT HYDROCHLORIDE see GAD000
PHENIC ACID see PDN750
PHENIDYLATE see MNQ000
PHENIGAMA HYDROCHLORIDE see GAD000
PHENIGAM HYDROCHLORIDE see GAD000
PHENIPRAZINE see PDN000
PHENITOL see MCU750
PHENITROTHION see DSQ000
PHENIZINE see PDN000
PHENLINE see PFC750
PHENMAD see ABU500
PHENMERZYL NITRATE see MCU750
PHENMETHYL TRIMETHYLAMMONIUM IODIDE see BFM750
PHENMETRAZINE HYDROCHLORIDE see MNV750
PHENOBAL see EOK000
PHENOBAL SODIUM see SID000
PHENOBARBITAL see EOK000
PHENOBARBITAL ELIXIR see SID000
PHENOBARBITAL Na see SID000
PHENOBARBITAL SODIUM see SID000
PHENOBARBITAL SODIUM SALT see SID000
PHENOBARBITONE see EOK000
PHENOBARBITONE SODIUM see SID000
PHENOBARBITONE SODIUM SALT see SID000
PHENOBARBITURIC ACID see EOK000
PHENO BLACK EP see AQP000
PHENO BLUE 2B see CMO000
PHENOCAINE see BQH250
PHENOCHLOR see PJL750
PHENOCLOR DP6 see PJN250
PHENODIANISYL see PDN500
PHENODIANISYL HYDROCHLORIDE see PDN500
PHENODIOXIN see DDA800
PHENODODECINIUM BROMIDE see DXX000
PHENODYNE see PFC750
PHENOHEP see HCI000
PHENOL see PDN750
PHENOL, molten (DOT) see PDN750
PHENOL ALCOHOL see PDN750
PHENOLCARBINOL see BDX500
PHENOLE (GERMAN) see PDN750
PHENOL-GLYCIDAETHER (GERMAN) see PFH000
PHENOL GLYCIDYL ETHER (MAK) see PFH000
PHENOL SODIUM SALT see SJF000
PHENOL TRINITRATE see PID000
PHENOL, 2,4,6-TRINITRO-, AMMONIUM SALT (9CI) see ANS750
PHENOLURIC see EOK000
PHENOMERCURIC ACETATE see ABU500
PHENOMET see EOK000
PHENONYL see EOK000
PHENOPLASTE ORGANOL RED B see SBC500
PHENOPROMIN see BBK500
PHENOPROPAZINE see DIR000
PHENOPROZINE see DIR000
PHENOPYRIDINE see QPA000
PHENOPYRINE see BRF500
PHENOSAN see PDP250

PHENOSANE see PEE750
PHENOTAN see ACE500, BRE500
PHENOTHIAZINE see PDP250
N-(β-(10-PHENOTHIAZINYL)PROPYL)TRIMETHYL-AMMONIUM METHYL SULFATE see MRW000
PHENOTURIC see EOK000
PHENOVERM see PDP250
PHENOVIS see PDP250
PHENOX see DAA800
PHENOXAZOLE see IBM000
PHENOXETHOL see PER000
PHENOXETOL see PER000
PHENOXUR see PDP250
PHENOXYBENZAMIDE HYDROCHLORIDE see DDG800
PHENOXYBENZAMINE see PDT250
PHENOXYBENZENE see PFA850
3-PHENOXYBENZYL (±)-3-(2,2-DICHLOROVINYL)-2,2-DIMETHYLCYCLOPROPANECARBOXYLATE see AHJ750
3-PHENOXY-1,2-EPOXYPROPANE see PFH000
PHENOXYETHANOL see PER000
2-PHENOXYETHANOL see PER000
2-PHENOXY-ETHANOL, ACRYLATE see PER250
2-(2-PHENOXYETHOXY)ETHANOL see PEQ750
PHENOXYETHYL ALCOHOL see PER000
β-PHENOXYETHYLDIMETHYLDODECYLAMMONIUM BROMIDE see DXX000
PHENOXYETHYL ISOBUTYRATE see PDS900
N-PHENOXYISOPROPYL-N-BENZYL-β-CHLOROETHYL-AMINE see PDT250
N-2-PHENOXYISOPROPYL-N-BENZYL-CHLOROETHYL-AMINE HYDROCHLORIDE see DDG800
N-PHENOXYISOPROPYL-N-BENZYL-β-CHLOROETHYL-AMINE HYDROCHLORIDE see DDG800
PHENOXYLENE SUPER see CIR250
(3-PHENOXYPHENYL)METHYL-3-(2,2-DICHLORETHENYL)-2,2-DIMETHYLCYCLOPROPANECARBOXYLATE see AHJ750
PHENOXYPROPENE OXIDE see PFH000
PHENOXYPROPYLENE OXIDE see PFH000
PHENOXYTOL see PER000
PHENSEDYL see DQA400
PHENTANYL see PDW500
PHENTANYL CITRATE see PDW750
PHENTERMINE see DTJ400
PHENTHIAZINE see PDP250
PHENTHOATE see DRR400
PHENTIN ACETATE see ABX250
PHENTINOACETATE see ABX250
PHENTYRIN see BHY500
PHENVALERATE see FAR100
PHENYBUT HYDROCHLORIDE see GAD000
PHENYGAM HYDROCHLORIDE see GAD000
PHENYLACETALDEHYDE (FCC) see BBL500
PHENYLACETALDEHYDE DIMETHYL ACETAL see PDX000
N-PHENYLACETAMIDE see AAQ500
PHENYLACETAMIDOPENICILLANIC ACID see BDY669
p-PHENYLACETANILIDE see PDY500
4'-PHENYLACETANILIDE see PDY500
PHENYLACETIC ACID see PDY850
omega-PHENYLACETIC ACID see PDY850
PHENYLACETIC ACID ALLYL ESTER see PMS500
PHENYLACETIC ACID, ETHYL ESTER see EOH000
PHENYLACETIC ACID, METHYL ESTER see MHA500
PHENYLACETIC ACID, PHENETHYL ESTER see PDI000
PHENYLACETIC ALDEHYDE see BBL500
N-PHENYLACETOACETAMIDE see AAY000
PHENYLACETONITRILE see PEA750
2-PHENYLACETONITRILE see PEA750
PHENYLACETONITRILE, liquid (DOT) see PEA750
3-(α-PHENYL-β-ACETYLETHYL)-4-HYDROXYCOUMARIN see WAT200

PHENYLCHLOROMETHYLKETONE see CEA750
2-(2-PHENYL-2-(4-CHLOROPHENYL)ACETYL)-1,3-INDAN-
DIONE see CJJ000
α-PHENYLCINNAMONITRILE see DVX600
PHENYL CYANIDE see BCQ250
1-PHENYL-1-CYCLOHEXYL-3-PIPERIDYL-1-PROPANOL
HYDROCHLORIDE see BBV000
PHENYLCYCLOPROMINE SULFATE see PET500
trans-2-PHENYLCYCLOPROPYLAMINE see PET750
trans,D,L-2-PHENYLCYCLOPROPYLAMINE SULFATE see
PET500
1-PHENYL-3,3-DIAETHYLTRIAZEN (GERMAN) see
PEU500
PHENYLDIBROMOARSINE see DDR200
PHENYL DICHLOROARSINE (DOT) see DGB600
PHENYL-5,6-DICHLORO-2-TRIFLUOROMETHYL-BENZ-
IMIDAZOLE-1-CARBOXYLATE see DGA200
β-PHENYL-o-(DIETHYLAMINOETHOXY)PROPIOPHENONE
HYDROCHLORIDE see DHS200
3-PHENYL-5-(β-(DIETHYLAMINO)ETHYL)-1,2,4-OXADI-
AZOLE CITRATE see OOE000
1-PHENYL-3,3-DIETHYLTRIAZENE see PEU500
PHENYLDIFLUOROARSINE see DKI400
1-PHENYL-2-DIMETHYLAMINOPROPANOL see MJU750
1-PHENYL-2,3-DIMETHYL-4-DIMETHYLAMINOPYRAZOL-
5-ONE see DOT000
1-PHENYL-2,3-DIMETHYL-4-DIMETHYLAMINOPYRAZO-
LONE-5 see DOT000
1-PHENYL-2,3-DIMETHYL-4-ISOPROPYL-3-PYRAZOLIN-5-
ONE see INY000
1-PHENYL-2,3-DIMETHYL-4-ISOPROPYLPYRAZOL-5-ONE
see INY000
d-2-PHENYL-3,4-DIMETHYLMORPHOLINE HYDROCHLO-
RIDE see DTN800
O-PHENYL-N,N'-DIMETHYL PHOSPHORODIAMIDATE see
PEV500
1-PHENYL-2,3-DIMETHYLPYRAZOLE-5-ONE see AQN000
1-PHENYL-2,3-DIMETHYL-5-PYRAZOLONE see AQN000
1-PHENYL-2,3-DIMETHYL-5-PYRAZOLONE-4-METHYL-
AMINOMETHANESULFONATESODIUM see AMK500
1-PHENYL-2,3-DIMETHYLPYRAZOLONE-(5)-4-METHYL-
AMINOMETHANESULFONICACID SODIUM see
AMK500
PHENYL DIMETHYL PYRAZOLON METHYL AMINOMETH-
ANE SODIUM SULFONATE see AMK500
1-PHENYL-3,3-DIMETHYLTRIAZENE see DTP000
PHENYLDIMETHYLTRIAZINE see DTP000
2-PHENYL-m-DIOXAN-5-OL see BBA000
4-PHENYLDIPHENYL see TBC750
N,N'-(m-PHENYLENE)BISMALEIMIDE see BKL750
m-PHENYLENEBIS(METHYLAMINE) see XHS800
2,2'-(1,3-PHENYLENEBIS(OXYMETHYLENE))BISOXIRANE
see REF000
1,1'-(m-PHENYLENE)BIS-1H-PYROLE-2,5-DIONE (9CI) see
BKL750
m-PHENYLENEDIAMINE see PEY000
o-PHENYLENEDIAMINE see PEY250
p-PHENYLENEDIAMINE see PEY500
1,3-PHENYLENEDIAMINE see PEY000
1,4-PHENYLENEDIAMINE see PEY500
m-PHENYLENEDIAMINE (DOT) see PEY000
1,2-PHENYLENEDIAMINE (DOT) see PEY250
1,3-PHENYLENEDIAMINE DIHYDROCHLORIDE see
PEY750
m-PHENYLENEDIAMINE HYDROCHLORIDE see PEY750
PHENYLENEDIAMINE, META, solid (DOT) see PEY000
PHENYLENEDIAMINE, PARA, solid (DOT) see PEY500
p-PHENYLENE DIAZIDE see DCL125
m-PHENYLENE DIISOCYANATE see BBP000
PHENYLENE-1,4-DIISOTHIOCYANATE see PFA500
1,4-PHENYLENEDIISOTHIOCYANIC ACID see PFA500
N,N'-(m-PHENYLENEDIMALEIMIDE) see BKL750
o-PHENYLENEDIOL see CCP850
m-PHENYLENE ISOCYANATE see BBP000

2,3-PHENYLENEPYRENE see IBZ000
2,3-o-PHENYLENEPYRENE see IBZ000
1,10-(o-PHENYLENE)PYRENE see IBZ000
1,10-(1,2-PHENYLENE)PYRENE see IBZ000
PHENYLENE THIOCYANATE see PFA500
o-PHENYLENETHIOUREA see BCC500
PHENYLEPHRINE see NCL500
(−)-PHENYLEPHRINE see NCL500
R(−)-PHENYLEPHRINE see NCL500
d-(−)-PHENYLEPHRINE HYDROCHLORIDE see SPC500
1-PHENYL-1,2-EPOXYETHANE see EBR000
PHENYL-2,3-EPOXYPROPYL ETHER see PFH000
PHENYLETHANAL see BBL500
PHENYLETHANE see EGP500
1-PHENYLETHANOL see PDE000
2-PHENYLETHANOL see PDD750
β-PHENYLETHANOL see PDD750
PHENYL ETHANOLAMINE see AOR750
N-PHENYLETHANOLAMINE see AOR750
1-PHENYLETHANONE see ABH000
PHENYLETHENE see SMQ000
PHENYL ETHER see PFA850
PHENYL ETHER-BIPHENYL MIXTURE see PFA860
PHENYL ETHER DICHLORO see DFE800
PHENYL ETHER HEXACHLORO see CDV175
PHENYL ETHER MONO-CHLORO see MRG000
PHENYL ETHER PENTACHLORO see PAW250
PHENYL ETHER TETRACHLORO see TBP250
2-PHENYLETHYL ACETATE see PFB250
α-PHENYL ETHYL ACETATE see MNT075
β-PHENYLETHYL ACETATE see PFB250
2-PHENYLETHYL ALCOHOL see PDD750
β-PHENYLETHYL ALCOHOL see PDD750
PHENYLETHYLAMINE see PDE250
2-PHENYLETHYLAMINE see PDE250
φ-PHENYLETHYLAMINE see PDE250
2-PHENYLETHYL-o-AMINOBENZOATE see APJ500
2-PHENYLETHYL ANTHRANILATE see APJ500
PHENYLETHYLBARBITURATE see EOK000
PHENYL-ETHYL-BARBITURIC ACID see EOK000
5-PHENYL-5-ETHYLBARBITURIC ACID see EOK000
PHENYLETHYLBARBITURIC ACID, SODIUM SALT see
SID000
3-PHENYL-3-ETHYL-2,6-DIKETOPIPERIDINE see DYC800
3-PHENYL-3-ETHYL-2,6-DIOXOPIPERIDINE see DYC800
PHENYLETHYLENE see SMQ000
PHENYLETHYLENE OXIDE see EBR000
2-PHENYL-2-ETHYLGLUTARIC ACID IMIDE see DYC800
α-PHENYL-α-ETHYLGLUTARIC ACID IMIDE see DYC800
α-PHENYL-α-ETHYLGLUTARIMIDE see DYC800
5-PHENYL-5-ETHYL-HEXAHYDROPYRIMIDINE-4,6-DIONE
see DBB200
2-PHENYLETHYLHYDRAZINE see PFC500
β-PHENYLETHYLHYDRAZINE see PFC500
β-PHENYLETHYLHYDRAZINE DIHYDROGEN SULFATE
see PFC750
2-PHENYLETHYLHYDRAZINE DIHYDROGEN SULPHATE
see PFC750
β-PHENYLETHYLHYDRAZINE HYDROGEN SULPHATE see
PFC750
β-PHENYLETHYLHYDRAZINE SULFATE see PFC750
PHENYLETHYLHYDRAZINE SULPHATE see PFC750
PHENYLETHYL ISOBUTYRATE see PDF750
2-PHENYLETHYL ISOBUTYRATE see PDF750
β-PHENYLETHYL ISOBUTYRATE see PDF750
PHENYLETHYL ISOVALERATE see PDF775
β-PHENYLETHYL ISOVALERATE see PDF775
PHENYLETHYLMALONYLUREA see EOK000
5-PHENYL-5-ETHYL-3-METHYLBARBITURIC ACID see
ENB500
2-PHENYLETHYL-3-METHYLBUTIRATE see PDF775
PHENYLETHYLMETHYLHYDANTOIN see MKB250
2-PHENYLETHYL-2-METHYLPROPIONATE see PDF750
2-PHENYLETHYL PHENYLACETATE see PDI000

β-PHENYLETHYL PHENYLACETATE see PDI000
2-PHENYLETHYL-α-TOLUATE see PDI000
PHENYLETHYNLCARBINOL CARBAMATE see PGE000
PHENYLETTEN see EOK000
PHENYL FLUORIDE see FGA000
PHENYLFLUOROFORM see BDH500
PHENYLFORMIC ACID see BCL750
PHENYLGAMMA HYDROCHLORIDE see GAD000
PHENYLGLYCOLIC ACID see MAP000
PHENYLGLYCOLONITRILE see MAP250
PHENYL GLYCYDYL ETHER see PFH000
PHENYLGLYOXYLONITRILE OXIME-O,O-DIETHYL PHOS-
    PHOROTHIOATE see BAT750
2-PHENYLHYDRACRYLIC ACID-3-α-TROPANYL ESTER
    see ARR000
PHENYL HYDRATE see PDN750
PHENYLHYDRAZIN (GERMAN) see PFI000
PHENYLHYDRAZINE see PFI000
PHENYLHYDRAZINE HYDROCHLORIDE see PFI250
PHENYLHYDRAZINE MONOHYDROCHLORIDE see PFI250
PHENYLHYDRAZIN HYDROCHLORID (GERMAN) see
    PFI250
PHENYLHYDRAZINIUM CHLORIDE see PFI250
1-PHENYL-2-HYDRAZINOPROPANE see PDN000
PHENYL HYDRIDE see BBL250
PHENYL HYDROXIDE see PDN750
PHENYLHYDROXYACETIC ACID see MAP000
N-PHENYLHYDROXYLAMINE see PFJ250
β-PHENYLHYDROXYLAMINE see PFJ250
1-PHENYL-2-(p-HYDROXYPHENYL)-3,5-DIOXO-4-BUTYL-
    PYRAZOLIDINE see HNI500
PHENYLIC ACID see PDN750
PHENYLIC ALCOHOL see PDN750
PHENYL-IDIUM see PDC250
PHENYL-IDIUM 200 see PDC250
5-PHENYL-2-IMINO-4-OXAZOLIDINONE see IBM000
5-PHENYL-2-IMINO-4-OXOOXAZOLIDINE see IBM000
PHENYL ISOCYANATE see PFK250
PHENYL ISOHYDANTOIN see IBM000
dl-β-PHENYLISOPROPYLAMINE HYDROCHLORIDE see
    AOA750
d-β-PHENYLISOPROPYLAMINE SULFATE see BBK500
PHENYLISOPROPYLHYDRAZINE see PDN000
β-PHENYLISOPROPYLHYDRAZINE see PDN000
3-(β-PHENYLISOPROPYL)-SIDNONIMINE HYDROCHLO-
    RIDE see SPA000
PHENYL ISOTHIOCYANATE see ISQ000
PHENYL KETONE see BCS250
PHENYL MERCAPTAN see PFL850
PHENYLMERCURIACETATE see ABU500
PHENYL MERCURIC ACETATE see ABU500
PHENYL MERCURIC CHLORIDE see PFM500
PHENYLMERCURIC DINAPHTHYLMETHANEDISULFO-
    NATE see PFM750, PFN000
PHENYL MERCURIC FIXTAN see PFN000
PHENYLMERCURIC 3,3′-METHYLENEBIS(2-NAPHTHA-
    LENESULFONATE) see PFN000
PHENYLMERCURIC NITRATE see MCU750
PHENYLMERCURIC UREA see PFP500
PHENYLMERCURIUREA see PFP500
PHENYLMERCURY ACETATE see ABU500
PHENYLMERCURY CHLORIDE see PFM500
PHENYLMERCURY
    METHYLENEDINAPHTHALENESULFONATE see
    PFN000
PHENYLMERCURY NITRATE see MCU750
PHENYL MERCURY UREA see PFP500
PHENYLMETHANE see TGK750
PHENYLMETHANETHIOL see TGO750
PHENYLMETHANOL see BDX500
PHENYLMETHYL ALCOHOL see BDX500
N-PHENYLMETHYLAMINE see MGN750
1-PHENYL-2-METHYLAMINE-PROPANOL-1-SULFATE see
    EAY500

1-PHENYL-2-METHYLAMINO-PROPAN (GERMAN) see
    DBB000
1-PHENYL-2-METHYLAMINOPROPANE see DBB000
α-PHENYL-β-METHYL AMINOPROPANE see DBB000
1-PHENYL-2-METHYLAMINOPROPANOL see EAW000
1-PHENYL-2-METHYLAMINOPROPANOL-1 see EAX500
PHENYLMETHYLCARBINOL see PDE000
1-PHENYL-5-METHYL-8-CHLORO-1,2,4,5-TETRAHYDRO-
    2,4-DIOXO-3H-1,5-BENZODIAZEPINE see CIR750
PHENYLMETHYLDICHLOROSILANE see DFQ800
N-(PHENYLMETHYL)DIMETHYLAMINE see DQP800
α-(PHENYLMETHYLENE)BENZENEACETONITRILE see
    DVX600
2-(PHENYLMETHYLENE)OCTANOL see HFO500
PHENYLMETHYL ESTER THIOCYANIC ACID (9CI) see
    BFL000
PHENYL METHYL ETHER see AOX750
1-PHENYL-3-METHYL-3-(2-HYDROXYAETHYL)-TRIAZEN
    (GERMAN) see HKV000
1-PHENYL-3-METHYL-3-(2-HYDROXYETHYL)TRIAZENE
    see HKV000
PHENYL METHYL KETONE see ABH000
PHENYLMETHYL MERCAPTAN see TGO750
PHENYLMETHYLNITROSAMINE see MMU250
3-PHENYL-1-METHYL-1-NITROSOHARNSTOFF (GERMAN)
    see MMY500
(PHENYLMETHYL) PENICILLINIC ACID see BDY669
1-PHENYL-3-METHYL-3-(2-SULFOAETHYL) NATRIUM
    SALZ (GERMAN) see PEJ250
1-PHENYL-3-METHYL-3-(2-SULFOETHYL)TRIAZENE, SO-
    DIUM SALT see PEJ250
2-PHENYL-3-METHYLTETRAHYDRO-1,4-OXAZINE HY-
    DROCHLORIDE see MNV750
PHENYLMETHYLVALERIANSAEURE-β-DIAETHYLAMI-
    NOAETHYLESTER-BROMMET HYLAT (GERMAN) see
    VBK000
PHENYL-MOBUZON see BRF500
PHENYLMONOGLYCOL ETHER see PER000
1-PHENYL-3-MONOMETHYLTRIAZENE see PFS500
PHENYL MORPHOLINE see PFS750
PHENYL MUSTARD OIL see ISQ000
PHENYL-2-NAPHTHYLAMINE see PFT500
PHENYL-β-NAPHTHYLAMINE see PFT500
N-PHENYL-2-NAPHTHYLAMINE see PFT500
N-PHENYL-β-NAPHTHYLAMINE see PFT500
4-PHENYL-NITROBENZENE see NFQ000
p-PHENYL-NITROBENZENE see NFQ000
N-PHENYL-p-NITROSOANILINE see NKB500
PHENYLOXIRANE see EBR000
1-PHENYLOXIRANE see EBR000
2-PHENYLOXIRANE see EBR000
β-PHENYL-γ-OXYPROPIONSAEURE-TROPYL-ESTER (GER-
    MAN) see ARR000
PHENYL PERCHLORYL see HCC500
4-PHENYLPHENOL see BGJ500
p-PHENYLPHENOL see BGJ500
2-PHENYLPHENOL SODIUM SALT see BGJ750
o-PHENYLPHENOL SODIUM SALT see BGJ750
α-PHENYLPHENYLACETONITRILE see DVX200
N-PHENYL-N-(1-(2-PHENYLETHYL)-4-PIPERIDINYL)-PRO-
    PANAMIDE (9CI) see PDW500
PHENYLPHOSPHINE see PFV250
PHENYLPHOSPHONOTHIOIC ACID O-(4-BROMO-2,5-
    BROMO-2,5-DICHLOROPHENYL) O-METHYL ESTER see
    LEN000
PHENYLPHOSPHONOTHIOIC ACID, O-(2,4-DICHLOROPHE-
    NYL), O-ETHYL ESTER see SCC000
PHENYL PHOSPHONYL DICHLORIDE see PFW100
PHENYLPHOSPHORODICHLORIDOTHIOUS ACID see
    PFW500
1-PHENYLPIPERAZINE see PFX000
N-PHENYLPIPERAZINE see PFX000
α-PHENYL-2-PIPERIDINEACETIC ACID METHYL ESTER
    see MNQ000

α-PHENYL-2-PIPERIDINEMETHANOL ACETATE HYDRO-
CHLORIDE see LFO000
1-PHENYL-1-(2-PIPERIDYL)-1-ACETOXYMETHANE HY-
DROCHLORIDE see LFO000
PHENYL-(2-PIPERIDYL)METHYL ACETATE HYDROCHLO-
RIDE see LFO000
2-PHENYLPROPANAL see COF000
3-PHENYLPROPANAL see HHP000
3-PHENYL-1-PROPANAL see HHP000
1-PHENYLPROPANE see IKG000
2-PHENYLPROPANE see COE750
3-PHENYLPROPANOL see HHP050
γ-PHENYLPROPANOL see HHP050
3-PHENYL-1-PROPANOL (FCC) see HHP050
3-PHENYL-1-PROPANOL ACETATE see HHP500
PHENYLPROPANOLAMINE see NNM000
3-PHENYL-1-PROPANOL CARBAMATE see PGA750
3-PHENYLPROPENAL see CMP969
3-PHENYL-2-PROPENAL see CMP969
2-PHENYLPROPENE see MPK250
β-PHENYLPROPENE see MPK250
3-PHENYLPROPENOIC ACID see CMP975
3-PHENYL-2-PROPENOIC ACID see CMP975
3-PHENYL-2-PROPENOIC ACID METHYL ESTER (9CI) see
MIO500
3-PHENYL-2-PROPENOIC ACID, 2-METHYLPROPYL ESTER
see IIQ000
3-PHENYL-2-PROPENOIC ACID PHENYLMETHYL ESTER
(9CI) see BEG750
3-PHENYL-2-PROPEN-1-OL see CMQ740
3-PHENYL-2-PROPEN-1-YL ACETATE see CMQ730
3-PHENYL-2-PROPENYLANTHRANILATE see API750
3-PHENYL-2-PROPEN-1-YL ANTHRANILATE see API750
3-PHENYL-2-PROPEN-1-YL FORMATE see CMR500
α-PHENYLPROPIONALDEHYDE see COF000
β-PHENYLPROPIONALDEHYDE see HHP000
2-PHENYLPROPIONALDEHYDE (FCC) see COF000
3-PHENYLPROPIONALDEHYDE (FCC) see HHP000
PHENYLPROPYL ACETATE see HHP500
3-PHENYL-1-PROPYL ACETATE see HHP500
3-PHENYLPROPYL ACETATE (FCC) see HHP500
PHENYLPROPYL ALCOHOL see HHP050
3-PHENYLPROPYL ALCOHOL see HHP050
γ-PHENYLPROPYL ALCOHOL see HHP050
3-PHENYLPROPYL ALDEHYDE see HHP000
α-PHENYL-α-PROPYL-BENZENEACETIC ACID-S-(DI-
ETHYLAMINO)ETHYL ESTER see DIG400
γ-PHENYLPROPYLCARBAMAT (GERMAN) see PGA750
γ-PHENYLPROPYL CARBAMATE see PGA750
2-PHENYLPROPYLENE see MPK250
β-PHENYLPROPYLENE see MPK250
1-PHENYL-2-PROPYNYL CARBAMATE see PGE000
PHENYLPSEUDOHYDANTOIN see IBM000
1-PHENYL-1-(2-PYRIDYL)-3-DIMETHYLAMINOPROPANE
see TMJ750
1-PHENYL-1-(2-PYRIDYL)-3-DIMETHYLAMINOPROPANE
HYDROCHLORIDE see PGF000
3-PHENYL-3-(2-PYRIDYL)-N,N-DIMETHYLPROPYLAMINE
see TMJ750
1-PHENYL-3-(p-2-PYRIDYLSULFAMOYLANILINO)-1,3-PRO-
PANEDISULFONIC ACID DISODIUM SALT see DXF400
PHENYLQUECKSILBERACETAT (GERMAN) see ABU500
PHENYLQUECKSILBERCHLORID (GERMAN) see PFM500
PHENYL SALICYLATE see PGG750
3-PHENYLSALICYLIC ACID see PGH000
PHENYLSENFOEL (GERMAN) see ISQ000
PHENYLSILICON TRICHLORIDE see TJA750
PHENYLSULFONIC ACID see BBS250
4-(PHENYLSULFOXYETHYL)-1,2-DIPHENYL-3,5-PYRAZO-
LIDINEDIONE see DWM000
6-PHENYL-2,3,5,6-TETRAHYDROIMIDAZO(2,1-b)THIAZOLE
see LFA000
N-PHENYL-N′-1,2,3-THIADIAZOL-5-YL-UREA see TEX600

(5-(PHENYLTHIO)-2-BENZIMIDAZOLECARBAMIC ACID,
METHYL ESTER see FAL100
PHENYLTHIOCARBAMIDE see PGN250
PHENYLTHIOPHOSPHONATE de O-ETHYLE et O-4-NITRO-
PHENYLE (FRENCH) see EBD700
1-PHENYLTHIOSEMICARBAZIDE see PGM750
1-PHENYLTHIOUREA see PGN250
N-PHENYLTHIOUREA see PGN250
1-PHENYL-2-THIOUREA see PGN250
PHENYLTOLOXAMINE HYDROCHLORIDE see DTO800
(N-PHENYL-p-TOLUENESULFONAMIDO)ETHYLMERCURY
see EME500
PHENYL-p-TOLYL KETONE see MHF750
PHENYLTRICHLOROMETHANE see BFL250
PHENYL TRICHLOROSILANE (DOT) see TJA750
PHENYL UREA-p-DI(CARBOXYMETHYL) THIOARSENITE
see CBI250
α-PHENYL-VALERATE du DIETHYLAMINO-ETHANOL
CHLORHYDRATE (FRENCH) see PGP500
2-PHENYLVALERIC ACID-2-(DIETHYLAMINO)ETHYL ES-
TER HYDROCHLORIDE see PGP500
PHENYL XYLYL KETONE see PGP750
PHENYRAL see EOK000
PHENYTOIN SODIUM see DNU000
PHERMERNITE see MCU750
PHETADEX see BBK500
PHETIDINE see DAM600
PHGABA HYDROCHLORIDE see GAD000
PHILIPS-DUPHAR PH 60-40 see CJV250
PHILODORM see TDA500
PHILOPON see DBA800
PHILOSOPHER'S WOOL see ZKA000
PHILOSTIGMIN METHYL SULFATE see DQY909
PHISODANV see HCL000
PHISOHEX see HCL000
PHIX see ABU500
PHIXIA see CMS850
PHLOROGLUCIN see PGR000
PHLOROGLUCINOL see PGR000
PHLOXINE TONER B see BNH500
PHLOX RED TONER X-1354 see BNH500
α-PHNEYLTHIOUREA see PGN250
PHOB see EOK000
PHOBEX see BCA000
PHONURIT see AAI250
PHORAT (GERMAN) see PGS000
PHORATE see PGS000
PHORATE-10G see PGS000
PHORBOL see PGS250
PHORBOL ACETATE, MYRISTATE see PGV000
PHORBOL-12,13-DIDECANOATE see PGT250
PHORBOL MONOACETATE MONOMYRISTATE see
PGV000
PHORBOL MYRISTATE ACETATE see PGV000
PHORBOLOL ACETATE MYRISTATE see PGV500
PHORBOLOL MYRISTATE ACETATE see PGV500
PHORBOL-12-o-TIGLYL-13-BUTYRATE see PGV750
PHORBYOL see CCP250
PHORON (GERMAN) see PGW250
PHORONE see PGW250
PHORTOX see TAA100
PHOSALON see BDJ250
PHOSALONE see BDJ250
PHOSCHLOR R50 see TIQ250
PHOSDRIN (OSHA) see MQR750
PHOSFENE see MQR750
PHOSFLEX 179-C see TMO600
PHOSFLEX T-BEP see BPK250
PHOS-FLUR see SHF500
PHOSFOLAN see DXN600, PGW750
PHOSGEN (GERMAN) see PGX000
PHOSGENE see PGX000
PHOSHOROTHIOIC ACID-S-(2-(ETHYLAMINO)-2-OXO-
ETHYL)-O,O-DIMETHYL ESTER see DNX600

PHOSKIL see PAK000
PHOSMET see PHX250
PHOSPHACOL see NIM500
PHOSPHAMID see DSP400
PHOSPHATE 100 see EAS000
PHOSPHATE de O,O-DIETHYLE et de O-2-CHLORO-1-(2,4-
DICHLOROPHENYL) VINYLE (FRENCH) see CDS750
PHOSPHATE de DIETHYLE et de 3-METHYL-5-PYRAZO-
LYLE (FRENCH) see MOX250
PHOSPHATE de O,O-DIMETHLE et de O-(1,2-DIBROMO-2,2-
DICHLORETHYLE) (FRENCH) see NAG400
PHOSPHATE de DIMETHYLE et de 2,2-DICHLOROVINYLE
(FRENCH) see DGP900
PHOSPHATE de DIMETHYLE et de 2-DIMETHYLCARBA-
MOYL-1-METHYL VINYLE (FRENCH) see DGQ875
PHOSPHATE de DIMETHYLE et de 2-METHOXYCARBO-
NYL-1 METHYLVINYLE (FRENCH) see MQR750
PHOSPHATE de DIMETHYLE et de 2-METHYLCARBA-
MOYL-1-METHYL VINYLE (FRENCH) see MRH209
PHOSPHATES see PGX500
PHOSPHATE, SODIUM HEXAMETA see SHM500
PHOSPHATE de TRICRESYLE (FRENCH) see TNP500
PHOSPHEMOL see PAK000
PHOSPHENE (FRENCH) see MQR750
PHOSPHENOL see PAK000
PHOSPHENYL CHLORIDE see DGE400
PHOSPHIDES see PGX750
PHOSPHINE see PGY000
1,1′,1″-PHOSPHINOTHIOYLIDYNETRISAZIRIDINE see
TFQ750
1,1′,1″-PHOSPHINYLIDYNETRISAZIRIDINE see TND250
1,1′,1″ -PHOSPHINYLIDYNETRIS(2-METHYL)AZRIDINE
see TNK250
N-(PHOSPHONOMETHYL)GLYCINE see PHA500
PHOSPHOPYRON see EAS000
PHOSPHOPYRONE see EAS000
PHOSPHORAMIDE MUSTARD CYCLOHEXYLAMINE SALT
see PHA750
PHOSPHORE BLANC (FRENCH) see PHO750
PHOSPHORE(PENTACHLORURE de) (FRENCH) see
PHR500
PHOSPHORE(TRICHLORURE de) (FRENCH) see PHT275
PHOSPHORIC ACID see PHB250
PHOSPHORIC ACID, BERYLLIUM SALT (1:1) see BFS000
PHOSPHORIC ACID, 2-CHLORO-1-(2,4,5-TRICHLOROPHE-
NYL)ETHENYL DIMETHYL ESTER see TBW100
PHOSPHORIC ACID, DIBUTYL PHENYL ESTER see
DEG600
PHOSPHORIC ACID-2,2-DICHLOROETHENYL DIMETHYL
ESTER see DGP900
PHOSPHORIC ACID, DIETHYL ESTER, with 3-CHLORO-7-
HYDROXY-4-METHYLCOUMARIN see CIK750
PHOSPHORIC ACID, DIETHYL ESTER-N-NAPHTHALIMIDE
deriv. see HMV000
PHOSPHORIC ACID, DIETHYL ESTER, NAPHTHALIMIDO
deriv. see HMV000
PHOSPHORIC ACID-DIETHYL-(3-METHYL-5-PYRAZOLYL)
ESTER see MOX250
PHOSPHORIC ACID DIETHYL 4-NITROPHENYL ESTER see
NIM500
PHOSPHORIC ACID, DIMETHYL ESTER, ESTER with DI-
METHYL 3-HYDROXYGLUTACONATE see SOY000
PHOSPHORIC ACID, DIMETHYL ESTER, ESTER with cis-3-
HYDROXY-N-METHYLCROTONAMIDE see MRH209
PHOSPHORIC ACID DIMETHYL-p-(METHYLTHIO)PHENYL
ESTER see PHD250
PHOSPHORIC ACID DIMETHYL-4-NITRO-m-TOLYL ESTER
see PHD750
PHOSPHORIC ACID, DISODIUM SALT see SJH090
PHOSPHORIC ACID, ISOPROPYL ESTER see PHE500
PHOSPHORIC ACID, LEAD (2+) SALT (2:3) see LDU000
PHOSPHORIC ACID, TRI-o-CRESYL ESTER see TMO600
PHOSPHORIC ACID TRIETHYLENE IMIDE see TND250

PHOSPHORIC ACID TRIETHYLENEIMINE (DOT) see
TND250
PHOSPHORIC ACID, TRIMETHYL ESTER see TMD250
PHOSPHORIC ACID, TRIPHENYL ESTER see TMT750
PHOSPHORIC ACID, TRIS(2,3-DIBROMOPROPYL) ESTER
see TNC500
PHOSPHORIC ACID, TRIS(2-METHYLPHENYL) ESTER see
TMO600
PHOSPHORIC ACID, TRISODIUM SALT see SJH200
PHOSPHORIC ACID, TRITOLYL ESTER see TNP500
PHOSPHORIC ANHYDRIDE see PHS250
PHOSPHORIC BROMIDE see PHR250
PHOSPHORIC CHLORIDE see PHR500
PHOSPHORIC SULFIDE see PHS000
PHOSPHORIC TRIS(DIMETHYLAMIDE) see HEK000
PHOSPHOROCHLORIDOTHIOIC ACID-O,O-DIMETHYL ES-
TER see DTQ600
PHOSPHORODICHLORIDIC ACID, ETHYL ESTER see
EOR000
PHOSPHORODIFLUORIDIC ACID see PHF250
PHOSPHORODIFLUORIDIC ACID (anhydrous) see PHF500
PHOSPHORODI(ISOPROPYLAMIDIC) FLUORIDE see
PHF750
PHOSPHORODITHIOIC ACID-S-(2-CHLORO-1-(1,3-DIHY-
DRO-1,3-DIOXO-2H-ISOINDOL-2-YL)ETHYL-O,O-DI-
ETHYL ESTER see DBI099
PHOSPHORODITHIOIC ACID-S-(2-CHLORO-1-PHTHALIMI-
DOETHYL)-O,)-DIETHYL ESTER see DBI099
PHOSPHORODITHIOIC ACID-O,O-DIETHYL ESTER-S-ES-
TER with N-ISOPROPYL-2-MERCAPTOACETAMIDE see
IOT000
PHOSPHORODITHIOIC ACID, S-((1,3-DIHYDRO-1,3-DIOXO-
ISOINDOL-2-YL)METHYL) O,O-DIMETHYL ESTER see
PHX250
PHOSPHORODITHIOIC ACID-O,O-DIETHYL ESTER-S-ES-
TER with DIETHYL MERCAPTOSUCCINATE see
MAK700
PHOSPHORODITHIOIC ACID, O,O-DIMETHYL ESTER, S-
ESTER with N-(2-MERCAPTOETHYL)ACETAMIDE see
DOP200
PHOSPHORODITHIOIC ACID, O,O-DIMETHYL-S-(2-ETHYL-
THIO)ETHYL ESTER see PHI500
PHOSPHORODITHIOIC ACID-O,O-DIMETHYL-S-(2-
(METHYLAMINO)-2-OXOETHYL) ESTER see DSP400
PHOSPHORODITHIOIC ACID, O,O-DIMETHYL-S-(2-((1-
METHYLETHYL)THIO)ETHYL) ESTER see DSK800
PHOSPHORODITHIOIC ACID, O,O-DIMETHYL S-(MORPHO-
LINOCARBONYLMETHYL) ESTER see MRU250
PHOSPHORODITHIOIC ACID-S,S'-1,4-DIOXANE-2,3-DIYL
O,O,O',O'-TETRAETHYL ESTER see DVQ709
PHOSPHOROFLUORIDIC ACID see PHJ250
PHOSPHOROFLUORIDIC ACID, DIETHYL ESTER see
DJJ400
PHOSPHOROFLUORIDIC ACID, DIISOPROPYL ESTER see
IRF000
PHOSPHOROFLUORIDIC ACID, DIMETHYL ESTER see
DSA800
PHOSPHOROTHIOIC ACID-S-(((1-CYANO-1-METHYL-
ETHYL)CARBAMOYL)METHYL)-O,O-DIETHYL ESTER
see PHK250
PHOSPHOROTHIOIC ACID, CYCLIC O,O-(METHYLENE-O-
PHENYLENE) O-METHYL ESTER see MEC250
PHOSPHOROTHIOIC ACID-O,O-DIETHYL ESTER-S-ESTER
with ETHYL MERCAPTOACETATE see DIW600
PHOSPHOROTHIOIC ACID-O,O-DIETHYL ESTER, -o-
NAPHTHALIMIDO DERIVATIVE see NAQ500
PHOSPHOROTHIOIC ACID-O,O-DIETHYL-O-(2-(ETHYL-
THIO)ETHYL) ESTER, mixed with O,O-DIETHYL S-(2-
(ETHYLTHIO)ETHYL) ESTER (7:3) see DAO600
PHOSPHOROTHIOIC ACID-O,O-DIETHYL-o-NAPHTHYL-
AMIDOESTER see NAQ500
PHOSPHOROTHIOIC ACID, O,O-DIETHYL-O-(4-NITROPHE-
NYL) ESTER see PAK000

PHOSPHOROTHIOIC ACID-O,O-DIETHYL-O-2-PYRAZINYL ESTER see EPC500
PHOSPHOROTHIOIC ACID, O,O-DIMETHYL S-(ETHYLSULFINYL-(2-ISOPROPYL)) ESTER see DSK600
PHOSPHOROTHIOIC ACID, O,O-DIMETHYL S-(2-(METHYLAMINO)-2-OXOETHYL) ESTER see DNX800
PHOSPHOROTHIOIC ACID-O,O-DIMETHYL-O-(3-METHYL-4-METHYLTHIOPHENYLE) (FRENCH) see FAQ999
PHOSPHOROTHIOIC ACID TRIETHYLENETRIAMIDE see TFQ750
PHOSPHOROTHIOIC TRICHLORIDE see TFO000
PHOSPHOROTHIONIC TRICHLORIDE see TFO000
PHOSPHOROTRITHIOUS ACID, S,S,S-TRIBUTYL ESTER see TIG250
PHOSPHOROUS ACID, BERYLLIUM SALT see BFS000
PHOSPHOROUS ACID, TRIMETHYL ESTER see TMD500
PHOSPHOROUS ACID, TRIPHENYL ESTER see TMU250
PHOSPHOROUS ACID TRIS(2-FLUOROETHYLESTER) see PHO250
PHOSPHOROUS BROMIDE (DOT) see PHT250
PHOSPHOROUS OXYBROMIDE see PHU000
PHOSPHOROUS SULFOCHLORIDE see TFO000
PHOSPHOROUS THIOCHLORIDE see TFO000
PHOSPHOROUS TRICHLORIDE SULFIDE see TFO000
PHOSPHOROUS TRIFLUORIDE see PHQ500
PHOSPHORPENTACHLORID (GERMAN) see PHR500
PHOSPHORSAEURELOESUNGEN (GERMAN) see PHB250
PHOSPHORTRICHLORID (GERMAN) see PHT275
PHOSPHORUS (red) see PHO500
PHOSPHORUS (white) see PHO750
PHOSPHORUS (white in water) see PHO740
PHOSPHORUS (yellow) see PHO750
PHOSPHORUS, AMORPHOUS, RED (DOT) see PHO500
PHOSPHORUS CHLORIDE see PHT275
PHOSPHORUS COMPOUNDS, INORGANIC see PHQ000
PHOSPHORUS FLUORIDE see PHQ500
PHOSPHORUS HEPTASULFIDE see PHQ750
PHOSPHORUS(V) OXIDE see PHS250
PHOSPHORUS OXYCHLORIDE see PHQ800
PHOSPHORUS OXYTRICHLORIDE see PHQ800
PHOSPHORUS PENTABROMIDE see PHR250
PHOSPHORUS PENTACHLORIDE see PHR500
PHOSPHORUS PENTAFLUORIDE see PHR750
PHOSPHORUS PENTASULFIDE see PHS000
PHOSPHORUS PENTOXIDE see PHS250
PHOSPHORUS PERCHLORIDE see PHR500
PHOSPHORUS PERSULFIDE see PHS000
PHOSPHORUS SESQUISULFIDE see PHS500
PHOSPHORUS(III) SULFIDE(IV) see PHS500
PHOSPHORUS TRIAZIDE see PHT000
PHOSPHORUS TRIBROMIDE see PHT250
PHOSPHORUS TRICHLORIDE see PHT275
PHOSPHORUS TRIHYDRIDE see PGY000
PHOSPHORUS TRIOXIDE see PHT500
PHOSPHORUS TRISULFIDE see PHT750
PHOSPHORWASSERSTOFF (GERMAN) see PGY000
PHOSPHORYL BROMIDE see PHU000
PHOSPHORYL CHLORIDE see PHQ800
PHOSPHORYL HEXAMETHYLTRIAMIDE see HEK000
O-PHOSPHORYL-4-HYDROXY-N,N-DIMETHYLTRYPTAMINE see PHU500
PHOSPHORYL TRIBROMIDE see PHU000
PHOSPHOSTIGMINE see PAK000
PHOSPHOTEX see TEE500
PHOSPHOTHION see MAK700
PHOSPHOTOX E see EEH600
PHOSPHURE de MAGNESIUM (FRENCH) see MAI000
PHOSPHURE de POTASSIUM (FRENCH) see PLQ500
PHOSPHURES d'ALUMIUM (FRENCH) see AHE750
PHOSPHURE de SODIUM (FRENCH) see SJI500
PHOSPHURE de ZINC (FRENCH) see ZLS000
PHOSVEL see LEN000
PHOSVIN see ZLS000
PHOSVIT see DGP900

PHOTOMIREX see MRI750
PHOXIME see BAT750
PHOXIN see BAT750
PHOZALON see BDJ250
PHP see ELL500
PHPH see BGE000
PHRENOLAN see MDU750
PHRILON see NOH000
PHTALOPHOS see HMV000
1,3-PHTHALANDIONE see PHW750
1(2H)-PHTHALAZINONE HYDRAZONE see PHW000
1(2H)-PHTHALAZINONE HYDRAZONE HYDROCHLORIDE see HGP500
PHTHALIC ACID see PHW250
PHTHALIC ACID ANHYDRIDE see PHW750
PHTHALIC ACID BIS(2-METHOXYETHYL) ESTER see DOF400
PHTHALIC ACID, DIALLYL ESTER see DBL200
o-PHTHALIC ACID, DIALLYL ESTER see DBL200
PHTHALIC ACID, DIETHYL ESTER see DJX000
PHTHALIC ACID DINITRILE see PHY000
PHTHALIC ACID DIOCTYL ESTER see DVL700
PHTHALIC ACID METHYL ESTER see DTR200
PHTHALIC ANHYDRIDE see PHW750
PHTHALIMIDIME see DNE400
PHTHALIMIDO-O,O-DIMETHYL PHOSPHORODITHIOATE see PHX250
2-PHTHALIMIDOGLUTARIMIDE see TEH500
3-PHTHALIMIDOGLUTARIMIDE see TEH500
α-PHTHALIMIDOGLUTARIMIDE see TEH500
α-(N-PHTHALIMIDO)GLUTARIMIDE see TEH500
PHTHALIMIDOMETHYL-O,O-DIMETHYL PHOSPHORODITHIOATE see PHX250
PHTHALOCYANINE BLUE 01206 see DNE400
PHTHALODINITRILE see PHY000
m-PHTHALODINITRILE see PHX550
o-PHTHALODINITRILE see PHY000
p-PHTHALODINITRILE see BBP250
PHTHALOGEN see DNE400
PHTHALOL see DJX000
PHTHALONITRILE see PHY000
PHTHALOPHOS see PHX250
N-PHTHALOYLGLUTAMIMIDE see TEH500
PHTHALSAEUREANHYDRID (GERMAN) see PHW750
PHTHALSAEUREDIAETHYLESTER (GERMAN) see DJX000
PHTHALSAEUREDIMETHYLESTER (GERMAN) see DTR200
N-PHTHALYLGLUTAMIC ACID IMIDE see TEH500
N-PHTHALYL-GLUTAMINSAEURE-IMID (GERMAN) see TEH500
α-N-PHTHALYLGLUTARAMIDE see TEH500
PHTHISEN see ILD000
1(2H)-PHTHLAZINONE, HYDRAZONE, MONOHYDROCHLORIDE see HGP500
PHYBAN see MRL750
PHYGON see DFT000
PHYGON PASTE see DFT000
PHYGON SEED PROTECTANT see DFT000
PHYGON XL see DFT000
PHYSEPTONE see MDO750
PHYSIOMYCINE see MDO250
PHYSOSTIGMINE see PIA500
PHYSOSTIGMINE SALICYLATE (1:1) see PIA750
PHYSOSTOL see PIA500
PHYSOSTOL SALICYLATE see PIA750
PHYTAR see HKC000
PHYTAR 560 see HKC500
PHYTIC ACID see PIB250
PHYTOGERMINE see VSZ450
PHYTOSOL see EPY000
PIANADALIN see BNK000
PIAPONON see PMH500
PIC-CLOR see CKN500
PICCOLASTIC see SMQ500

PICFUME see CKN500
PICLORAM see PIB900
2-PICOLINE see MOY000
3-PICOLINE see PIB920
4-PICOLINE see MOY250
α-PICOLINE see MOY000
β-PICOLINE see PIB820
γ-PICOLINE see MOY250
m-PICOLINE (DOT) see PIB920
o-PICOLINE (DOT) see MOY000
p-PICOLINE (DOT) see MOY250
PICRACONITINE see PIC250
PICRATE of AMMONIA (DOT) see ANS500
PICRATOL see ANS750
PICRIC ACID see PID000
PICRIC ACID, AMMONIUM SALT see ANS500, ANS750
PICRIDE see CKN500
PICRITE (THE EXPLOSIVE) see NHA500
PICRONITRIC ACID see PID000
PICROTIN, compounded with PICROTOXININ (1:1) see
   PIE500
PICROTOXIN see PIE500
PICROTOXINE see PIE500
PICRYLMETHYLNITRAMINE see TEG250
PICRYLNITROMETHYLAMINE see TEG250
PICRYL SULFIDE see BLR750
PID see DVV600
PIECIOCHLOREK FOSFORU (POLISH) see PHR500
PIED PIPER MOUSE SEED see SMN500
PIELIK see DAA800
PIELIK E see SGH500
PIGMENT GREEN 15 see LCR000
PIGMENT RED CD see CHP500
PIGMENT YELLOW 33 see CAP750
PIG-WRACK see CCL250
PIH see PDN000
PIKRINEZUUR (DUTCH) see PID000
PIKRINSAEURE (GERMAN) see PID000
PIKRYNOWY KWAS (POLISH) see PID000
PILLARDRIN see MRH209
PILLARON see DTQ400
PILLARZO see CFX000
PILLS (INDIA) see SED400
PILOCARPINE see PIF000
PILOCARPOL see PIF000
PILOT HD-90 see DXW200
PILOT SF-40 see DXW200
PILPOPHEN see DQA400
PIMAFUCIN see PIF750
PIMARICIN see PIF750
PIMELIC KETONE see CPC000
PIMENTA BERRIES OIL see PIG740
PIMENTA LEAF OIL see PIG730
PIMENTA OIL see PIG740
PIMENTO OIL see PIG740
PIN see EBD700
PINACOLOXYMETHYLPHOSPHORYL FLUORIDE see
   SKS500
PINACOLYL METHYLFLUOROPHOSPHONATE see SKS500
PINACOLYL METHYLPHOSPHONOFLUORIDATE see
   SKS500
PINACOLYL METHYLPHOSPHONOFLUORIDE see SKS500
PINACOLYLOXY METHYLPHOSPHORYL FLUORIDE see
   SKS500
PINAKON see HFP875
PINANG see BFW000
PINDON (DUTCH) see PIH175
PINDONE see PIH175
2-PINENE see PIH250
α-PINENE (FCC) see PIH250
β-PINENE (FCC) see POH750
2(10)-PINENE see POH750
PINE NEEDLE OIL see FBV000
PINE NEEDLE OIL, DWARF see PIH400

PINE NEEDLE OIL, SCOTCH see PIH500
PINE OIL see PIH750
PINHOLE AK 2 see ASM270
PINUS MONTANA OIL see PIH400
PINUS PUMILIO OIL see PIH400
PIPAMPERONE see FHG000
PIPANEPERONE see FHG000
PIPANOL see BBV000
PIPECURIUM BROMIDE see PII250
PIPECURONIUM BROMIDE see PII250
PIPENZOLATE BROMIDE see PJA000
PIPENZOLATE METHYLBROMIDE see PJA000
PIPERADROL HYDROCHLORIDE see PII750
PIPERAZIDINE see PIJ000
PIPERAZIN (GERMAN) see PIJ000
PIPERAZINE see PIJ000
PIPERAZINE, anhydrous see PIJ000
PIPERAZINE DIHYDROCHLORIDE see PIK000
2,6-PIPERAZINEDIONE-4,4'-PROPYLENE DIOXOPIPERA-
   ZINE see PIK250
1,4-PIPERAZINEDIYLBIS(BIS(1-AZIRIDINYL)PHOSPHINE
   OXIDE see BJC250
PIPERAZINE HYDROCHLORIDE see PIK000
PIPERIDIN (GERMAN) see PIL500
PIPERIDINE see PIL500
PIPERIDINIUM see PIY500
6-PIPERIDINO-2,4-DIAMINOPYRIMIDINE-3-OXIDE see
   DCB000
PIPERIDINOETHYL-2-HEPTOXYPHENYLCARBAMOATE
   HYDROCHLORIDE see PIO750
2-PIPERIDINOMETHYL-1,4-BENZODIOXAN HYDROCHLO-
   RIDE see BCI500
4-(3-PIPERIDINOPROPIONAMIDO) SALICYCLIC ACID
   METHYL ESTER, METHIODIDE see MOK000
2-(1-PIPERIDINO)-2-(2-THENYL)ETHYLAMINE MALEATE
   see PIT250
6-(1-PIPERIDINYL)-2,4-PYRIMIDINEDIAMINE-3-OXIDE see
   DCB000
α-(2-PIPERIDYL)BENZHYDROL see DWK400
α-(2-PIPERIDYL)BENZHYDROL HYDROCHLORIDE see
   PII750
3-(1-PIPERIDYL)-1-CYCLOHEXYL-1-PHENYL-1-PROPANOL
   HYDROCHLORIDE see BBV000
α-(2-PIPERIDYLETHYL)BENZHYDROL HYDROCHLORIDE
   see PMC250
2-(1-PIPERIDYLMETHYL)-1,4-BENZODIOXAN HYDRO-
   CHLORIDE see BCI500
PIPEROCAINE see PIV750
PIPEROCAINE HYDROCHLORIDE see IJZ000
PIPEROCAINIUM CHLORIDE see IJZ000
PIPEROCYANOMAZINE see PIW000
PIPERONAL see PIW250
PIPERONALDEHYDE see PIW250
PIPERONYL see FHG000
PIPERONYL ACETATE see PIX000
PIPERONYL ALDEHYDE see PIW250
PIPERONYL BUTOXIDE see PIX250
PIPERONYL SULFOXIDE see ISA000
PIPEROXANE HYDROCHLORIDE see BCI500
PIPERSAL see DAM600
PIPOLPHEN see DQA400
PIP-PIP see PIY500
PIPRADOL see DWK400
α-PIPRADOL see DWK400
PIPRADOL HYDROCHLORIDE see PII750
PIPRADROL HYDROCHLORIDE see PII750
PIPTAL see PJA000
PIRABUTINA see HNI500
PIRACAPS see TBX250
PIRAFLOGIN see HNI500
PIRAMIDON see DOT000
PIRARREUMOL 'B'' see BRF500
PIRAZOXON (ITALIAN) see MOX250
PIREF see DIZ100

PIRIBENZIL see TMP750
PIRID see PDC250, PEK250
PIRIDACIL see PDC250
PIRIDINA (ITALIAN) see POP250
PIRIDOL see DOT000
PIRIDOSAL see DAM600, DAM700
PIRIDROL see DWK400
PIRIDROL HYDROCHLORIDE see PII750
PIRIEX see TAI500
PIRIMECIDAN see TGD000
PIRIMETAMINA (SPANISH) see TGD000
PIRIMIFOS-METHYL see DIN800
PIRINIXIL see CLW500
PIRITON see TAI500
PIRMAZIN see SNJ000
PIROFOS see SOD100
PIROMIDINA see DOT000
PIRYDYNA (POLISH) see POP250
PISCIDEIN see HJO500
PISCIDIC ACID see HJO500
PITAYINE see QFS000
PITC see ISQ000
PITCH see CMZ100
PITCH, COAL TAR see CMZ100
PITMAL see TLP750
PITTCHLOR see HOV500
PITTCIDE see HOV500
PITTCLOR see HOV500
PITTSBURGH PX-138 see DVL700
PIVACIN see PIH175
PIVADORM see BNP750
PIVADORN see BNP750
PIVAL see PIH175
PIVALDION (ITALIAN) see PIH175
PIVALDIONE (FRENCH) see PIH175
PIVALIC ACID see PJA500
PIVALIC ACID CHLORIDE see DTS400
PIVALIC ACID LACTONE see DTH000
PIVALOLACTONE see DTH000
PIVALOLYL CHLORIDE see DTS400
PIVALOYL CHLORIDE see DTS400
2-PIVALOYL-INDAAN-1,3-DION (DUTCH) see PIH175
2-PIVALOYL-INDAN-1,3-DION (GERMAN) see PIH175
2-PIVALOYL-1,3-INDANDIONE see PIH175
2-PIVALOYLINDANE-1,3-DIONE see PIH175
PIVALYL CHLORIDE see DTS400
2-PIVALYL-1,3-INDANDIONE see PIH175
PIVALYL VALONE see PIH175
PIVALYN see PIH175
PIXALBOL see CMY800
PIX CARBONIS see CMY800
PKhNB see PAX000
PLACIDAL see MNM500
PLACIDAS see MNM500
PLACIDIL see CHG000
PLACIDOL see CJR909
PLACIDOL E see DJX000
PLACIDON see MQU750
PLACIDYL see CHG000
PLANADALIN see BNK000
PLANOCAINE see AIT250
PLANOCHROME see MCV000
PLANOMIDE see TEX250
PLANOTOX see DAA800
PLANT DITHIO AEROSOL see SOD100
PLANTDRIN see MRH209
PLANTFUME 103 SMOKE GENERATOR see SOD100
PLANTGARD see DAA800
PLANTIFOG 160M see MAS500
PLANT PROTEASE CONCENTRATE see BMO000
PLANT PROTECTION PP511 see DIN800
PLANTULIN see PMN850
PLASDONE see PKQ250
PLASKON 201 see PJY500

PLASTER of PARIS see CAX500
PLASTIBEST 20 see ARM268
PLASTORESIN ORANGE F4A see PEJ500
PLATIBLASTIN see PJD000
cis-PLATIN see PJD000
PLATIN (GERMAN) see PJD500
PLATINEX see PJD000
PLATINIC AMMONIUM CHLORIDE see ANF250
PLATINIC CHLORIDE see CKO750
PLATINOL see PJD000
PLATINOL AH see DVL700
PLATINOL DOP see DVL700
PLATINOUS CHLORIDE see PJE000
cis-PLATINOUS DIAMMINE DICHLORIDE see PJD000
PLATINOUS POTASSIUM CHLORIDE see PJD250
PLATINUM see PJD500
PLATINUM BLACK see PJD500
PLATINUM CHLORIDE see PJE000
PLATINUM(IV) CHLORIDE see PJE250
PLATINUM COMPOUNDS see PJE500
cis-PLATINUM(II) DIAMINEDICHLORIDE see PJD000
trans-PLATINUM(II)DIAMMINEDICHLORIDE see DEX000
PLATINUM ETHYLENEDIAMMINE DICHLORIDE see
　DFJ000
PLATINUM SPONGE see PJD500
PLATINUM TETRACHLORIDE see PJE250
PLAXIDOL see CJR909
PLECYAMIN see VSZ000
PLEGECYL see ABH500
PLEGICIN see ABH500
PLEGINE see DKE800
PLENASTRIL see PAN100
PLENUR see LGZ000
PLEOCIDE see ABY900
PLEXIGLAS see PKB500
PLEXIGUM M 920 see PKB500
PLICTRAN see CQH650
PLIMASINE see MNQ000
PLIOFLEX see SMR000
PLIOLITE S5 see SMR000
PLIOVAC AO see AAX175
PLIOVIC see PKQ059
PLIVA see BIE500
PLIVAPHEN see ABH500
PLOMB FLUORURE (FRENCH) see LDF000
PLOYMANNURONIC ACID see AFL000
PLUMBOPLUMBIC OXIDE see LDS000
PLUMBOUS ACETATE see LCG000, LCJ000
PLUMBOUS CHLORIDE see LCQ000
PLUMBOUS CHROMATE see LCR000
PLUMBOUS FLUORIDE see LDF000
PLUMBOUS OXIDE see LDN000
PLUMBOUS PHOSPHATE see LDU000
PLUMBOUS SULFIDE see LDZ000
PLURACOL P-410 see PJT000
PLURACOL E see PJT200
PLUTONIUM COMPOUNDS see PJI000
PLUTONIUM NITRATE (solution) see PJI500
PLYCTRAN see CQH650
PM 671 see ENG500
PMA see ABU500
PMA see PGV000
PMAC see ABU500
PMACETATE see ABU500
PMAL see ABU500
PMAS see ABU500
PMB see ECU750
PMC see PFM500
P.M.F. see PFN000
PMFP see SKS500
PMMA see PKB500
PMP see PHX250
PMP SODIUM GLUCONATE see SHK800
PMS see MRW000

PMT see PFS500
PNA see NEO500
PNB see NFQ000
PNCB see NFS525
PNOT see NMP500
PNT see NMO550
PNU see NLO500
PO-DIMETHOATE see DNX800
PODOPHYLLIN see PJJ000
PODOPHYLLUM see PJJ000
PODOPHYLLUM RESIN see PJJ000
POINT TWO see SHF500
POLAAX see OAX000
POLACARITOX see CKM000
POLAMIDON see MDP250
POLAMIDONE see MDO750
POLARAMIN see PJJ325
POLARAMINE MALEATE see PJJ325
POLARONIL (GERMAN) see TAI500
POLCOMINAL see EOK000
POLEON see EID000
POLFOSCHLOR see TIQ250
POLICAPRAN see PJY500
POLIFEN see TAI250
POLIFLOGIL see HNI500
POLIGOSTYRENE see SMQ500
POLINALIN see DOT000
POLISEPTIL see TEX250
POLISIN see BKL250
POLITEF see TAI250
POLIVAL see TEX000
POLIVINIT see PKQ059
POLLACID see HCQ500
POLONIUM see PJJ750
POLONIUM CARBONYL see PJK000
POLOPIRYNA see ADA725
POLOXAL RED 2B see HJF500
POLSTIGMINE see DQY909
POLY see SKN000
POLYAETHYLENGLYCOLE 200 (GERMAN) see PJT200
POLYAETHYLENGLYKOLE 300 (GERMAN) see PJT225
POLYAETHYLENGLYKOLE 400 (GERMAN) see PJT230
POLYAETHYLENGLYKOLE 600 (GERMAN) see PJT240
POLYAETHYLENGLYKOLE 1000 (GERMAN) see PJT250
POLYAETHYLENGLYKOLE 1500 (GERMAN) see PJT500
POLYAETHYLENGLYKOLE 4000 (GERMAN) see PJT750
POLYAETHYLENGLYKOLE 6000 (GERMAN) see PJU000
POLYAMID (GERMAN) see NOH000
POLYAMIDE 6 see PJY500
POLY(epsilon-AMINOCAPROIC ACID) see PJY500
POLYBOR see SFF000
POLYBREME see HCV500
POLYBROMINATED BIPHENYL see FBU509
POLYBROMINATED BIPHENYL (FF-1) see FBU509
POLYBROMINATED BIPHENYLS see FBU000
POLYBROMOETHYLENE see PKQ000
POLYBUTADIENE-POLYSTYRENE COPOLYMER see SMR000
POLYCAPROAMIDE see PJY500
POLY(epsilon-CAPROAMIDE) see PJY500
POLYCAPROLACTAM see PJY500
POLY(epsilon-CAPROLACTAM) see PJY500
POLYCAT 8 see DRF709
POLYCHLORCAMPHENE see CDV100
POLYCHLORINATED BIPHENYL see PJL750
POLYCHLORINATED BIPHENYL (AROCLOR 1221) see PJM000
POLYCHLORINATED BIPHENYL (AROCLOR 1232) see PJM250
POLYCHLORINATED BIPHENYL (AROCLOR 1242) see PJM500
POLYCHLORINATED BIPHENYL (AROCLOR 1248) see PJM750

POLYCHLORINATED BIPHENYL (AROCLOR 1254) see PJN000
POLYCHLORINATED BIPHENYL (AROCLOR 1260) see PJN250
POLYCHLORINATED BIPHENYL (AROCLOR 1262) see PJN500
POLYCHLORINATED BIPHENYL (AROCLOR 1268) see PJN750
POLYCHLORINATED BIPHENYL (AROCLOR 2565) see PJO000
POLYCHLORINATED BIPHENYL (AROCLOR 4465) see PJO250
POLYCHLORINATED BIPHENYL (KANECHLOR 300) see PJO500
POLYCHLORINATED BIPHENYL (KANECHLOR 400) see PJO750
POLYCHLORINATED BIPHENYL (KANECHLOR 500) see PJP000
POLYCHLORINATED BIPHENYLS see PJL750
POLYCHLORINATED CAMPHENES see CDV100
POLYCHLOROBIPHENYL see PJL750
POLYCHLOROCAMPHENE see CDV100
POLY(CHLOROETHYLENE) see PKQ059
POLYCILLIN see AIV500, AOD125
POLYCIZER DBP see DEH200
POLYCIZER DBS see DEH600
POLYCLAR L see PKQ250
POLYCLENE see DGB000
POLYCO 2410 see SMR000
POLYCRON see BNA750
POLYCYCLIC MUSK see ACL750
POLYCYCLINE see TBX000
POLYCYCLINE HYDROCHLORIDE see TBX250
POLYDIMETHYLSILOXANE see DTR850
POLYDIMETHYL SILOXANE see PJR000
POLYDIMETHYLSILOXANE RUBBER see PJR250
POLY(ESTRADIOL PHOSPHATE) see EDS000
POLYETHYLENE see PJS750
POLYETHYLENE AS see PJS750
POLYETHYLENE GLYCOL see PJT000
POLYETHYLENE GLYCOL 200 see PJT200
POLYETHYLENE GLYCOL 300 see PJT225
POLYETHYLENE GLYCOL 400 see PJT230
POLYETHYLENE GLYCOL 600 see PJT240
POLYETHYLENE GLYCOL 1000 see PJT250
POLYETHYLENE GLYCOL 1500 see PJT500
POLYETHYLENE GLYCOL 4000 see PJT750
POLYETHYLENE GLYCOL 6000 see PJU000
POLYETHYLENE GLYCOL DISTEARATE see PJU500
POLYETHYLENE GLYCOL 300 DISTEARATE see PJU500
POLYETHYLENE GLYCOL 400 (DI) STEARATE see PJU500
POLYETHYLENE GLYCOL 600 (DI) STEARATE see PJU500
POLYETHYLENE GLYCOL MONOETHER with p-tert-OCTYLPHENYL see PKF500
POLYETHYLENE GLYCOL MONO(4-OCTYLPHENYL) ETHER see PKF500
POLYETHYLENE GLYCOL MONO(4-tert-OCTYLPHENYL) ETHER see PKF500
POLYETHYLENE GLYCOL MONO(p-tert-OCTYLPHENYL) ETHER see PKF500
POLYETHYLENE GLYCOL MONOSTEARATE see PJV250
POLYETHYLENE GLYCOL MONO(p-(1,1,3,3-TETRA-METHYLBUTYL)PHENYL) ETHER see PKF500
POLYETHYLENE GLYCOL 450 NONYL PHENYL ETHER see PKF000
POLYETHYLENE GLYCOL OCTYLPHENOL ETHER see PKF500
POLYETHYLENE GLYCOL p-OCTYLPHENYL ETHER see PKF500
POLYETHYLENE GLYCOL 450 OCTYL PHENYL ETHER see PKF500
POLYETHYLENE GLYCOL p-tert-OCTYLPHENYL ETHER see PKF500

POLYETHYLENE GLYCOL p-1,1,3,3,-TETRAMETHYLBU-
TYLPHENYL ETHER see PKF500
POLY(ETHYLENE OXIDE) see PJT000
POLY(ETHYLENE TETRAFLUORIDE) see TAI250
POLYFENE see TAI250
POLYFER see IGS000
POLYFIBRON 120 see SFO500
POLYFLON see TAI250
POLYFOAM PLASTIC SPONGE see PKL500
POLYFOAM SPONGE see PKL500
POLY-G see PJT200
POLY G 400 see PJT230
POLY-GIRON see TEH500
POLYGLYCOL 1000 see PJT250
POLYGLYCOL 4000 see PJT750
POLYGLYCOL DISTEARATE see PJU500
POLYGLYCOL E see PJT200
POLYGLYCOL E1000 see PJT250
POLYGLYCOL E-4000 see PJT750
POLYGLYCOL E-4000 USP see PJT750
POLYGON see SKN000
POLYGRIPAN see TEH500
POLY-G SERIES see PJT000
POLY(IMINOCARBONYLPENTAMETHYLENE) see PJY500
POLY(IMINO(1-OXO-1,6-HEXANEDIYL)) see PJY500
POLYMERS of EPICHLOROHYDRIN and 2,2-BIS(4-HY-
DROXY PHENYL)PIPERAZINE see ECL000
POLYMERS, WATER INSOLUBLE see PKA850
POLYMERS, WATER SOLUBLE see PKA860
POLYMETHYLMETHACRYLATE see PKB500
POLYMINE D see BEN000
POLYMONE see DGB000
POLYMYXIN see PKC000
POLYMYXIN A see PKC250
POLYOESTRADIOL PHOSPHATE see EDS000
POLYOX see PJT000
POLY(1-(2-OXO-1-PYRROLIDINYL)ETHYLENE) see
PKQ250
POLY(OXY(DIMETHYLSILYLENE)) see PJR000
POLYOXYETHYLENE (75) see PJT750
POLYOXYETHYLENE 1500 see PJT500
POLYOXYETHYLENE MONO(OCTYLPHENYL) ETHER see
PKF500
POLYOXYETHYLENE-8-MONOSTEARATE see PJV250
POLYOXYETHYLENE (9) NONYL PHENYL ETHER see
PKF000
POLYOXYETHYLENE (9) OCTYLPHENYL ETHER see
PKF500
POLYOXYETHYLENE (13) OCTYLPHENYL ETHER see
PKF500
POLY(OXYETHYLENE)-p-tert-OCTYLPHENYL ETHER see
PKF500
POLYOXYETHYLENE (20) SORBITAN MONOLAURATE see
PKL000
POLYOXYETHYLENE SORBITAN MONOOLEATE see
PKL100
POLYOXYETHYLENE SORBITAN MONOSTEARATE see
PKL030
POLYOXYETHYLENE 20 SORBITAN MONOSTEARATE see
PKL030
POLYOXYETHYLENE SORBITAN OLEATE see PKL100
POLYOXYETHYLENE(8)STEARATE see PJV250
POLYOXYMETHYLENE see TMP000
POLYOXYMETHYLENE GLYCOLS see FMV000
POLYPHOS see SHM500
POLYPRO 1014 see PMP500
POLYPROPENE see PMP500
POLYPROPYLENE see PMP500
POLYPROPYLENE GLYCOL see PKI500
POLYPROPYLENE GLYCOL 750 see PKI750
POLYPROPYLENGLYKOL (CZECH) see PKI500
POLYRAM M see MAS500
POLYRAM ULTRA see TFS350
POLYRAM Z see EIR000

POLYSILICONE see PJR250
POLY-SOLV see CBR000
POLY-SOLV EB see BPJ850
POLY-SOLV EE see EES350
POLY-SOLV EE ACETATE see EES400
POLY-SOLV EM see EJH500
POLY-SOLVE MPM see PNL250
POLY-SOLV TE see EFL000
POLYSORBAN 80 see PKL100
POLYSORBATE 20 see PKL000
POLYSORBATE 60 see PKL030
POLYSORBATE 80 see PKL100
POLYSORBATE 80, U.S.P. see PKL100
POLYSTROL D see SMQ500
POLYSTYRENE see SMQ500
POLYSTYRENE-ACRYLONITRILE see ADY500
POLYSTYRENE BEADS (DOT) see SMQ500
POLYSTYRENE LATEX see SMQ500
POLYSTYROL see SMQ500
POLYTAC see PMP500
POLYTAR BATH see CMY800
POLYTEF see TAI250
POLYTETRAFLUOROETHENE see TAI250
POLYTETRAFLUOROETHYLENE see TAI250
POLY(N,N,N',N'-TETRAMETHYL-N-TRIMETHYLENE-
HEXAMETHYLENEDIAMMONIUM DIBROMIDE) see
HCV500
POLYTHERM see PKQ059
POLYTOX see DGB000
POLYURETHANE ESTER FOAM see PKL500
POLYURETHANE ETHER FOAM see PKL500
POLYURETHANE FOAM see PKL500
POLYURETHANE SPONGE see PKL500
POLYURETHANE Y-195 see PKL750
POLYURETHANE Y-217 see PKM000
POLYURETHANE Y-218 see PKM250
POLYURETHANE Y-221 see PKM500
POLYURETHANE Y-222 see PKM750
POLYURETHANE Y-223 see PKN000
POLYURETHANE Y-224 see PKN250
POLYURETHANE Y-225 see PKN500
POLYURETHANE Y-226 see PKN750
POLYURETHANE Y-227 see PKO000
POLYURETHANE Y-238 see CDV625
POLYURETHANE Y-290 see PKO500
POLYURETHANE Y-302 see PKP000
POLYURETHANE Y-304 see PKP250
POLYVIDONE see PKQ250
POLYVINYL ACETATE (FCC) see AAX250
POLYVINYL ALCOHOL see PKP750
POLY(VINYL ALCOHOL) see PKP750
POLYVINYLBROMIDE see PKQ000
POLY(n-VINYLBUTYROLACTAM) see PKQ250
POLYVINYLCHLORID (GERMAN) see PKQ059
POLYVINYL CHLORIDE see PKQ059
POLYVINYL CHLORIDE-POLYVINYL ACETATE see
AAX175
POLY(1-VINYL-2-PYRROLIDINONE) HOMOPOLYMER see
PKQ250
POLY(1-VINYL-2-PYRROLIDINONE) Hueper's polymer No. 1
see PKQ500
POLY(1-VINYL-2-PYRROLIDINONE) Hueper's polymer No. 2
see PKQ750
POLY(1-VINYL-2-PYRROLIDINONE) Hueper's polymer No. 3
see PKR000
POLY(1-VINYL-2-PYRROLIDINONE) Hueper's polymer No. 4
see PKR250
POLY(1-VINYL-2-PYRROLIDINONE) Hueper's polymer No. 5
see PKR500
POLY(1-VINYL-2-PYRROLIDINONE) Hueper's polymer No. 6
see PKR750
POLY(1-VINYL-2-PYRROLIDINONE) Hueper's polymer No. 7
see PKS000
POLYVINYLPYRROLIDONE see PKQ250

POLYVINYL SULFATE, POTASSIUM SALT see PKS250
POLYWAX 1000 see PJS750
POLY-ZOLE AZDN see ASL750
POMADEX see BBK500
POMARSOL see TFS350
POMARSOL Z FORTE see BJK500
POMASOL see TFS350
POMME EPINEUSE (FRENCH) see SLV500
PONCEAU INSOLUBLE OLG see XRA000
PONCEAU 3R see FAG018
PONCYL see GKE000
PONDERAL see PDM250
PONDERAX see PDM250
PONDIMIN see PDM250
PONECIL see AIV500
PONTACYL RUBINE R see HJF500
PONTALITE see PKB500
PONTAMINE BLACK E see AQP000
PONTAMINE BLUE BB see CMO000
PONTAMINE BLUE 3BX see CMO250
PONTAMINE DEVELOPER TN see TGL750
PONTOCAINE see BQA010
POOGIPHALAM, nut extract see BFW000
POP see ELL500
POPROLIN see PMP500
PORAMINE MALEATE see PJJ325
PORFIROMYCIN see MLY000
PORFIROMYCINE see MLY000
POROFOR 57 see ASL750
POROFOR 505 see ASM270
POROFOR ADC/R see ASM270
POROFOR ChKhZ 21 see ASM270
POROFOR ChKhZ 21R see ASM270
PORPHYROMYCIN see MLY000
PORTLAND CEMENT see PKS750
PORTLAND CEMENT SILICATE see PKS750
PORTLAND STONE see CAO000
POSTAFEN see HGC500
POSTINOR see NNQ500
PO-SYSTOX see DAP200
POTALIUM see PLG800
POTASH see PLA000
POTASH CHLORATE (DOT) see PLA250
POTASORAL see PLG800
POTASSA see PLJ500
POTASSE CAUSTIQUE (FRENCH) see PLJ500
POTASSIO (CHLORATO di) (ITALIAN) see PLA250
POTASSIO (IDROSSIDO di) (ITALIAN) see PLJ500
POTASSIO (PERMANGANATO di) (ITALIAN) see PLP000
POTASSIUM see PKT250
POTASSIUM (liquid alloy) see PKT500
POTASSIUM, metal liquid alloy (DOT) see PKT500
POTASSIUM ACID ARSENATE see ARD250
POTASSIUM ACID FLUORIDE see PKU250
POTASSIUM ACID SULFATE see PKX750
POTASSIUM ANTIMONYL TARTRATE see AQG250
POTASSIUM ANTIMONYL-d-TARTRATE see AQG250
POTASSIUM ANTIMONY TARTRATE see AQG250
POTASSIUM ARSENATE see ARD250
POTASSIUM ARSENITE see PKV500
POTASSIUM ARSENITE solution see FOM050
POTASSIUM BENZOATE see PKW760
POTASSIUM BICHROMATE see PKX250
POTASSIUM BIFLUORIDE see PKU250
POTASSIUM BIS(2-HYDROXYETHYL)DITHIOCARBAMATE
    see PKX500
POTASSIUM BISULFATE see PKX750
POTASSIUM BISULPHATE see PKX750
POTASSIUM BOROHYDRATE see PKY250
POTASSIUM BOROHYDRIDE (DOT) see PKY250
POTASSIUM BROMATE see PKY300
POTASSIUM BROMIDE see PKY500
POTASSIUM-tert-BUTOXIDE see PKY750
POTASSIUM CARBONATE (2:1) see PLA000

POTASSIUM CHLORATE see PLA250
POTASSIUM CHLORATE (DOT) see PLA250
POTASSIUM (CHLORATE de) (FRENCH) see PLA250
POTASSIUM CHLORIDE see PLA500
POTASSIUM CHLOROPLATINITE see PJD250
POTASSIUM CHROMATE(VI) see PLB250
POTASSIUM CHROMIC SULFATE see PLB500
POTASSIUM CHROMIC SULPHATE see PLB500
POTASSIUM CHROMIUM ALUM see PLB500
POTASSIUM CITRATE see PLB750
POTASSIUM CYANATE see PLC250
POTASSIUM CYANIDE see PLC500
POTASSIUM CYANIDE (solid) see PLC750
POTASSIUM CYANIDE, solution (DOT) see PLC500
POTASSIUM CYANONICKELATE HYDRATE see TBW250
POTASSIUM DICHLOROISOCYANURATE see PLD000
POTASSIUM DICHLORO-s-TRIAZINETRIONE see PLD000
POTASSIUM DICHROMATE(VI) see PKX250
POTASSIUM DICHROMATE, ZINC CHROMATE and ZINC
    HYDROXIDE (1:3:1) see ZFJ150
POTASSIUM DIHYDROGEN ARSENATE see ARD250
POTASSIUM DIOXIDE see PLE260
POTASSIUM DISULPHATOCHROMATE(III) see PLB500
POTASSIUM FLUORIDE see PLF500
POTASSIUM FLUORIDE, solution (DOT) see PLF500
POTASSIUM FLUOROACETATE see PLG000
POTASSIUM FLUORURE (FRENCH) see PLF500
POTASSIUM FLUOSILICATE see PLH750
POTASSIUM GLUCONATE see PLG800
POTASSIUM d-GLUCONATE see PLG800
POTASSIUM GLUTAMATE see MRK500
POTASSIUM GLUTAMINATE see MRK500
POTASSIUM HEXAFLUOROSILICATE see PLH750
POTASSIUM HEXAFLUOROTITANATE see PLI000
POTASSIUM HYDRATE (DOT) see PLJ500
POTASSIUM HYDRATE (solution) see PLJ750
POTASSIUM HYDRIDE see PLJ250
POTASSIUM HYDROGEN ARSENATE see ARD250
POTASSIUM HYDROGEN FLUORIDE see PKU250
POTASSIUM HYDROGEN SULFATE, solid (DOT) see
    PKX750
POTASSIUM HYDROXIDE see PLJ500
POTASSIUM HYDROXIDE, dry, solid, flake, bead, or granular
    (DOT) see PLJ500
POTASSIUM HYDROXIDE, liquid or solution (DOT) see
    PLJ500
POTASSIUM HYDROXIDE (solution) see PLJ750
POTASSIUM (HYDROXYDE de) (FRENCH) see PLJ500
POTASSIUM HYPERCHLORIDE see PLO500
POTASSIUM IODATE see PLK250
POTASSIUM IODIDE see PLK500
POTASSIUM IODOHYDRAGYRATE see NCP500
POTASSIUM ISOCYANATE see PLC250
POTASSIUM MERCURIC IODIDE see NCP500
POTASSIUM METAARSENITE see PKV500
POTASSIUM METABISULFITE (DOT, FCC) see PLR250
POTASSIUM, METAL (DOT) see PKT250
POTASSIUM MONOCHLORIDE see PLA500
POTASSIUM MONOSULFIDE see PLT250
POTASSIUM NITRATE see PLL500
POTASSIUM NITRITE (1:1) see PLM500
POTASSIUM NITRITE (DOT) see PLM500
POTASSIUM OCTACYANODICOBALTATE see PLN100
POTASSIUM cis-9-OCTADECENOIC ACID see OHY000
POTASSIUM OLEATE see OHY000
POTASSIUM OXYMURIATE see PLA250
POTASSIUM PERCHLORATE see PLO500
POTASSIUM PERMANGANATE see PLP000
POTASSIUM (PERMANGANATE de) (FRENCH) see PLP000
POTASSIUM PEROXIDE see PLP250
POTASSIUM PEROXYDISULFATE see DWQ000
POTASSIUM PEROXYDISULPHATE see DWQ000
POTASSIUM PERSULFATE (DOT) see DWQ000
POTASSIUM PHOSPHIDE see PLQ500

POTASSIUM PLATINOCHLORIDE see PJD250
POTASSIUM PYROSULFITE see PLR250
POTASSIUM RHODANATE see PLV750
POTASSIUM RHODANIDE see PLV750
POTASSIUM SALT OF POLYVINYL SULFATE see PKS250
POTASSIUM SELENATE see PLR750
POTASSIUM SILICOFLUORIDE (DOT) see PLH750
POTASSIUM SODIUM ALLOY see PLS500
POTASSIUM SORBATE see PLS750
POTASSIUM SULFATE (2:1) see PLT000
POTASSIUM SULFIDE (2:1) see PLT250
POTASSIUM SULFITE see PLT500
POTASSIUM SULFOCYANATE see PLV750
POTASSIUM TETRACHLOROPLATINATE(II) see PJD250
POTASSIUM TETRACYANOMERCURATE(II) see PLU500
POTASSIUM TETRACYANONICKELATE see NDI000
POTASSIUM TETRACYANONICKELATE(II) see NDI000
POTASSIUM TETRAIODOMERCURATE(II) see NCP500
POTASSIUM THIOCYANATE see PLV750
POTASSIUM THIOCYANIDE see PLV750
POTASSIUM ZINC CHROMATE see CMK400, PLW500
POTASSIUM ZINC CHROMATE HYDROXIDE see PLW500
POTASSURIL see PLG800
POTATO ALCOHOL see EFU000
POTAVESCENT see PLA500
POTCRATE see PLA250
POTENTIATED ACID GLUTARALDEHYDE see GFQ000
POTOMAC RED see CHP500
POUNCE see AHJ750
POVIDONE (USP XIX) see PKQ250
POWDER GREEN see COF500
POWDER and ROOT see RNZ000
POX see PHS250
PO$_x$ see PHS250
POYAMIN see VSZ000
PP511 see DIN800
PP 557 see AHJ750
PP 745 see BJK750
PP781 see MLC250
PPA see NNM000
PPD see PEY500
PPE201 see PKO500
PP FACTOR see NCQ900, NCR000
P.P. FACTOR-PELLAGRA PREVENTIVE FACTOR see NCQ900
PPZEIDAN see DAD200
PQD see DVR200
PRACARBAMIN see UVA000
PRACARBAMINE see UVA000
PRADUPEN see BDY669
PRAECIRHEUMIN see BRF500
PRAJMALINE BITARTRATE see DNB000
PRAJMALINE HYDROGEN TARTRATE see DNB000
PRALIDOXIME CHLORIDE see FNZ000
PRALIDOXIME IODIDE see POS750
PRALIDOXIME METHIODIDE see POS750
PRALUMIN see TDA500
PRAPARAT 5968 see HGP500
PRASEODYMIUM see PLX500
PRASEODYMIUM CHLORIDE see PLX750
PRASEODYMIUM(III) NITRATE (1:3) see PLY250
PRAYER BEAD see AAD000
PRAZEPINE see DLH600
PRECEPTIN see PKF500
PRECIPITATED BARIUM SULPHATE see BAP000
PRECIPITATED CALCIUM SULFATE see CAX750
PRECIPITATED SILICA see SCL000
PRECIPITATED SULFUR see SOD500
PRECIPITE BLANC see MCW000
PRECORT see PLZ000
PRECORTANCYL see PMA000
PRECORTISYL see PMA000
PREDENT see SHF500
PREDNE-DOME see PMA000

PREDNELAN see PMA000
PREDNICEN-M see PLZ000
PREDNILONGA see PLZ000
PREDNIS see PMA000
PREDNI-SEDIV see TEH500
PREDNISOLONE see PMA000
PREDNISON see PLZ000
PREDNISONE see PLZ000
PREDNIZON see PLZ000
PREDONIN see PMA000
PREDONINE see PMA000
PREEGLONE see DWX800
PREFEMIN see CHJ750
PREGICIL see AAF750
1,4-PREGNADIENE-17-α,21-DIOL-3,11,20-TRIONE see PLZ000
1,4-PREGNADIENE-3,20-DIONE-11-β,17-α,21-TRIOL see PMA000
1,4-PREGNADIENE-11-β,17-α,21-TRIOL-3,20-DIONE see PMA000
3,20-PREGNENE-4 see PMH500
PREGNENEDIONE see PMH500
PREGNENE-3,20-DIONE see PMH500
PREGN-4-ENE-3,20-DIONE see PMH500
4-PREGNENE-3,20-DIONE see PMH500
Δ^4-PREGNENE-3,20-DIONE see PMH500
PRELUDIN HYDROCHLORIDE see MNV750
PREMALIN see CKD500, DGD600
PREMARIN see ECU750, PMB000
PREMAZINE see BJP000
PREMERGE see BRE500
PREMERGE 3 see BRE500
PREMERGE PLUS see BQI000
PREMGARD see BEP500
PREMODRIN see DBA800
PRENIMON see TND000
PRENTOX see RNZ000, PIX250
PREPALIN see VSK600
PREPARATION 125 see DFT800
PREPARATION AF see HEI500
PRESAMINE see DLH630
PRESERVAL M see HJL500
PRESERVAL P see HNU500
PRESERV-O-SOTE see CMY825
PRESFERSUL see FBO000
PRESINOL see DNA800
PRESOLISIN see DNA800
PRESOMEN see ECU750
PRESPERSION, 75 UREA see USS000
PRESSOMIN HYDROCHLORIDE see MDW000
PRESSONEX see HNB875
PREVANGOR see PBC250
PREVENOL see CKC000
PREVENOL 56 see CKC000
PREVENTOL see CKC000, MJM500
PREVENTOL 56 see CKC000
PREVENTOL CMK see CFE250
PREVENTOL I see TIV750
PREVENTOL-ON see BGJ750
PREWEED see CKC000
PREZA see HBT500
PREZERVIT see DSB200
PRIADEL see LGZ000
PRIDINOL see PMC250
PRIDINOL HYDROCHLORIDE see PMC250
PRILEPSIN see DBB200
PRILTOX see PAX250
PRIMACIONE see DBB200
PRIMACLONE see DBB200
PRIMACONE see DBB200
PRIMAKTON see DBB200
PRIMARY AMYL ACETATE see AOD725
PRIMARY AMYL ALCOHOL see AOE000
PRIMARY DECYL ALCOHOL see DAI600

PRIMARY OCTYL ALCOHOL see OEI000
PRIMARY SODIUM PHOSPHATE see SJH100
PRIMATENE MIST see VGP000
PRIMATOL see ARQ725
PRIMATOL P see PMN850
PRIMATOL Q see BKL250
PRIMATOL S see BJP000
PRIMAZE see ARQ725
PRIMAZIN see SNJ000
PRIMIDON see DBB200
PRIMIDONE see DBB200
PRIMIN see DSK200
PRIMODOS see EEH520
PRIMOFOL see EDO000
PRIMOGYN B see EDP000
PRIMOGYN BOLEOSUM see EDP000
PRIMOGYN I see EDP000
PRIMOL 335 see MQV750
PRIMOLUT DEPOT see HNT500
PRIMOTEST see TBF500
PRIMROSE YELLOW see ZFJ100
PRINCEP see BJP000
PRINCILLIN see AOD125
PRINCIPEN see AIV500
PRINTEL'S see SMQ500
PRINTOP see BJP000
PRIODERM see MAK700
PRISCOL see BBJ750
PRISCOLINE HYDROCHLORIDE see BBJ750
PRIST see EJH500
PRISTACIN see CCX000
PRISTINAMYCIN see VRF000
PRIVINE see NAH500
PROADIFEN see DIG400
PROAZAIMINE see DQA400
PROAZAMINE see DQA400
PRO-BAN M see TEH500
PROBARBITAL see ELX000
PROBARBITONE see ELX000
PROBEDRYL see BBV500
PROBENECID SODIUM SALT see DWW200
PROBESE-P HYDROCHLORIDE see MNV750
PROCAINE see AIL750
PROCAINE, BASE see AIL750
PROCAINE FLUOBORATE see TCH000
PROCAINE HYDROCHLORIDE see AIT250
PROCALM see BCA000
PROCALMIDOL see MQU750
PROCARBAZIN (GERMAN) see PME500
PROCARBAZINE see PME250
PROCARBAZINE HYDROCHLORIDE see PME500
PROCARDINE see DJS200
PROCASIL see PNX000
PROCHLOROPERAZINE see PMF500
PROCHLOROPROAZINE HYDROGEN MALEATE see
    PMF250
PROCHLORPEMAZINE see PMF500
PROCHLORPERAZINE see PMF500
PROCHLORPERAZINE BIMALEATE see PMF250
PROCHLORPERAZINE DIMALEATE see PMF250
PROCHLORPERAZINE HYDROGEN MALEATE see
    PMF250
PROCHLORPERAZINE MALEATE see PMF250
PROCHLORPERIZINE MALEATE see PMF250
PROCHLORPROMAZINE see PMF500
PROCIT see DQA400
PROCORMAN see DJS200
PROCTIN see SEH000
PROCYTOX see CQC675, EAS500
PRODALUMNOL see SEY500
PRODALUMNOL DOUBLE see SEY500
PRODAN see DXE000
PRODARAM see BJK500
PRO-DEXTER see BBK500

PRODHYBASE ETHYL see EJM500
PRODICTAZIN see DIR000
PRODIERAZINE see DIR000
PRODILIDINE see DTO200
PRO-DORM see QAK000
PRODOX 133 see IQZ000
PRODUCER GAS see PMG750
PRODUCT 308 see HCP000
PRODUCT 5022 see MPE250
PRODUCT No. 161 see SIB600
PROFALVINE SULPHATE see DBN400
PROFAMINA see BBK000
PROFARMIL see TEH500
PROFAX see PMP500
PROFECUNDIN see VSZ450
PROFEMIN see CHJ750
PROFENAMINA (ITALIAN) see DIR000
PROFENAMINUM see DIR000
PROFENOFOS see BNA750
PROFENONE see ELL500
PROFERRIN see IHG000
PROFIROMYCIN see MLY000
PROFLAVIN see DBN600
PROFLAVINE see DBN600
PROFLAVINE HYDROCHLORIDE see PMH250
PROFLAVINE MONOHYDROCHLORIDE see PMH250
PROFLAVINE (SULFATE) see DBN400
PROFLAVIN SULFATE see DBN400
PROFOLIOL see DBN600, EDO000
PROFORMIPHEN see DBN600
PROFUME A see CKN500
PROFUME (OBS.) see MHR200
PROFUNDOL see DBN600
PROFURA see DBN600
PROGALLIN LA see DXX200
PROGALLIN P see PNM750
PROGARMED see DBN600
PROGEKAN see PMH500
PRO-GEN see DBN600
PROGESIC see DBN600
PROGESTEROL see PMH500
PROGESTERONE see PMH500
β-PROGESTERONE see PMH500
PROGESTERONE CAPROATE see HNT500
PROGESTERONE RETARD PHARLON see HNT500
PROGESTERONUM see PMH500
PROGESTIN see PMH500
PROGESTONE see PMH500
PRO-GIBB see GEM000
PROGUANIL see CKB250
PROGYNON see EDO000, EDS100
PROGYNON B see EDP000
PROGYNON BENZOATE see EDP000
PROGYNON-DEPOT see EDS100
PROGYNON-DH see EDO000
PROGYNON-DP see EDR000
PROGYNOVA see EDS100
PROHEPTADIENE see EAH500
PROHEPTADIEN MONOHYDROCHLORIDE see EAI000
PROKARBOL see DUS700
PROKAYVIT see MMD500
PROLATE see PHX250
PROLAX see GGS000
PROLIDON see PMH500
PROLONGAL see IGS000
PROLOXIN see GGS000
PROLUTON DEPOT see HNT500
PROMAMIDE see DTT600
PROMAR see DVV600
PROMARIT see ECU750
PROMAZINAMIDE see DQA400
PROMAZINE see DQA600
PROMAZINE HYDROCHLORIDE see PMI500
PROMETASIN see DQA400

PROMETAZIN see DQA400
PROMETHIAZINE see DQA400
PROMETHIUM see PMJ000
PROMETREX see BKL250
PROMETRIN see BKL250
PROMETRYN see BKL250
PROMETRYNE (USDA) see BKL250
PROMEZATHINE see DQA400
PROMIBEN see DLH600, DLH630
PROMIDIONE see GIA000
PROMINAL see ENB500
PROMOTESTON see TBF500
PROMPTONAL see EOK000
PROMUL 5080 see HKJ000
PROMURIT see DEQ000
PROMURITE see DEQ000
PRONAMIDE see DTT600
PRONETHALOL see INT000
PRONETHALOL HYDROCHLORIDE see INT000
PRONTALBIN see SNM500
PRONTODIN see DUO400
PRONTOSIL I see SNM500
PROPACHLOR see CHS500
PROPACHLORE see CHS500
PROPACIL see PNX000
PROPADRINE see NNM000
PROPAL see CIR500
PROPALDEHYDE see PMT750
PROPAMINE D see TDQ750
PROPANAL see PMT750
PROPANALOL see ICB000
PROPANAMINE see PND250
2-PROPANAMINE see INK000
PROPANE see PMJ750
1-PROPANECARBOXYLIC ACID see BSW000
PROPANEDIAL see PMK000
1,3-PROPANEDIAL see PMK000
1,3-PROPANEDIALDEHYDE see PMK000
1,2-PROPANEDIAMINE see PMK250
PROPANEDINITRILE see MAO250
PROPANEDINITRILE((2-CHLOROPHENYL)METHYLENE)
  see CEQ600
PROPANEDIOIC ACID, DIETHYL ESTER see EMA500
1,2-PROPANEDIOL see PML000
PROPANE-1,2-DIOL see PML000
1,2-PROPANEDIOL-1-ACRYLATE see HNT600
1,3-PROPANEDIONE see PMK000
PROPANE, 2-METHOXY-2-METHYL (9CI) see MHV859
PROPANENITRILE see PMV750
PROPANENITRILE, 3-(METHYLNITROSOAMINO)- see
  MMS200
1-PROPANESULFONIC ACID-3-HYDROXY-γ-SULTONE see
  PML400
PROPANE SULTONE see PML400
1,3-PROPANE SULTONE (MAK) see PML400
PROPANETHIOL see PML500
PROPANE-1-THIOL see PML500
2-PROPANETHIOL see IMU000
1,2,3-PROPANETRIOL see GGA000
1,2,3-PROPANETRIOL TRIACETATE see THM500
1,2,3-PROPANETRIOL, TRINITRATE see NGY000
1,2,3-PROPANETRIYL NITRATE see NGY000
PROPANEX see DGI000
PROPANID see DGI000
PROPANIDE see DGI000
PROPANIL see DGI000
PROPANOIC ACID see PJA500, PMU750
PROPANOIC ACID BUTYLESTER (9CI) see BSJ500
PROPANOIC ACID, ETHENYL ESTER see VQK000
PROPANOIC ACID, METHYL ESTER see MOT000
PROPANOIC ACID, 2-METHYLPROPYL ESTER see
  PMV250
PROPANOIC ACID, SODIUM SALT see SJL500
PROPANOIC ANHYDRIDE see PMV500

PROPANOL-1 see PND000
1-PROPANOL see PND000
PROPAN-2-OL see INJ000
2-PROPANOL see INJ000
n-PROPANOL see PND000
i-PROPANOL (GERMAN) see INJ000
PROPANOLE (GERMAN) see PND000
PROPANOLEN (DUTCH) see PND000
PROPANOLI (ITALIAN) see PND000
PROPANOLIDE see PMT100
PROPANOL NITRITE see IQQ000
PROPANONE see ABC750
2-PROPANONE see ABC750
PROPANOYL CHLORIDE see PMW500
PROPARGITE (DOT) see SOP000
PROPARGYL ALCOHOL see PMN450
PROPARGYL BROMIDE see PMN500
PROPASIN see PMN850
PROPASTE 6708 see TAV750
PROPATHENE see PMP500
PROPAZINE see PMN850
PROPELLANT 12 see DFA600
PROPELLANT 22 see CFX500
PROPELLANT 114 see FOO509
PROPELLANT C318 see CPS000
2-PROPENAL see ADR000
PROP-2-EN-1-AL see ADR000
PROPENAL (CZECH) see ADR000
PROPENAMIDE see ADS250
2-PROPENAMIDE see ADS250
2-PROPENAMINE see AFW000
2-PROPEN-1-AMINE see AFW000
PROPENE see PMO500
1-PROPENE see PMO500
PROPENE ACID see ADS750
1-PROPENE HOMOPOLYMER (9CI) see PMP500
PROPENENITRILE see ADX500
2-PROPENENITRILE see ADX500
2-PROPENENITRILE POLYMER with ETHENYLBENZENE
  see ADY500
PROPENE OXIDE see PNL600
PROPENE POLYMERS see PMP500
PROPENE TETRAMER see PMP750
2-PROPENE-1-THIOL see AGJ500
PROPENOIC ACID see ADS750
2-PROPENOIC ACID (9CI) see ADS750
2-PROPENOIC ACID-1,2-ETHANEDIYL ESTER see EIP000
2-PROPENOIC ACID-2-ETHYLBUTYL ESTER see EGZ000
2-PROPENOIC ACID, ETHYL ESTER (MAK) see EFT000
2-PROPENOIC ACID-2-HYDROXYPROPYL ESTER see
  HNT600
PROPENOIC ACID METHYL ESTER see MGA500
2-PROPENOIC ACID METHYL ESTER see MGA500
2-PROPENOIC ACID-2-METHYLPROPYL ESTER see IIK000
2-PROPENOIC ACID OXIRANYLMETHYL ESTER see
  ECH500
PROPENOL see AFV500
PROPEN-1-OL-3 see AFV500
1-PROPEN-3-OL see AFV500
2-PROPEN-1-OL see AFV500
2-PROPEN-1-ONE see ADR000
2-PROPENYLACRYLIC ACID see SKU000
PROPENYL ALCOHOL see AFV500
2-PROPENYL ALCOHOL see AFV500
4-PROPENYLANISOLE see PMQ750
p-PROPENYLANISOLE see PMQ750
p-1-PROPENYLANISOLE see PMQ750
5-(1-PROPENYL)-1,3-BENZODIOXOLE see IRZ000
5-(2-PROPENYL)-1,3-BENZODIOXOLE see SAD000
4-PROPENYLCATECHOL METHYLENE ETHER see IRZ000
PROPENYL CHLORIDE see PMR750
2-PROPENYL CHLORIDE see AGB250
PROPENYL CINNAMATE see AGC000
PROPENYL ETHER see DBK000

PROPENYLGUAETHOL (FCC) see IRY000
4-PROPENYLGUAIACOL see IKQ000
2-PROPENYL HEPTANOATE see AGH250
2-PROPENYL-N-HEXANOATE see AGA500
2-PROPENYL ISOTHIOCYANATE see AGJ250
2-PROPENYL ISOVALERATE see ISV000
3-PROPENYL METHANOATE see AGH000
2-PROPENYL 3-METHYLBUTANOATE see ISV000
4-PROPENYL-1,2-METHYLENEDIOXYBENZENE see
   IRZ000
((2-PROPENYLOXY)METHYL)OXIRANE see AGH150
2-PROPENYL PHENYLACETATE see PMS500
p-PROPENYLPHENYL METHYL ETHER see PMQ750
N-2-PROPENYL-2-PROPEN-1-AMINE see DBI600
4-PROPENYL VERATROLE see IKR000
PROPERICIAZINE see PIW000
PROPERIDOL see DYF200
PROPETAMPHOS see MKA000
PROPHENPYRIDAMINE HYDROCHLORIDE see PGF000
PROPHOS see EIN000
PROPICOL see DNR309
PROPILTHIOURACIL see PNX000
6-PROPIL-TIOURACILE (ITALIAN) see PNX000
PROPINE see MFX590
PROPIOCINE see EDH500
PROPIOKAN see TBG000
PROPIOLACTONE see PMT100
3-PROPIOLACTONE see PMT100
β-PROPIOLACTONE see PMT100
1,3-PROPIOLACTONE see PMT100
PROPIONALDEHYDE see PMT750
PROPIONATE d'ETHYLE (FRENCH) see EPB500
PROPIONATE de METHYLE (FRENCH) see MOT000
PROPIONE see DJN750
PROPIONIC ACID see PMU750
PROPIONIC ACID, solution containing not less than 80% acid
   (DOT) see PMU750
PROPIONIC ACID ANHYDRIDE see PMV500
PROPIONIC ACID CHLORIDE see PMW500
PROPIONIC ACID-3,4-DICHLOROANILIDE see DGI000
PROPIONIC ACID, ETHYL ESTER see EPB500
PROPIONIC ACID GRAIN PRESERVER see PMU750
PROPIONIC ACID, ISOBUTYL ESTER see PMV250
PROPIONIC ALDEHYDE see PMT750
PROPIONIC ANHYDRIDE see PMV500
PROPIONIC CHLORIDE see PMW500
PROPIONIC ETHER see EPB500
PROPIONIC NITRILE see PMV750
β-PROPIONOLACTONE see PMT100
PROPIONONITRILE see PMV750
PROPIONYL CHLORIDE see PMW500
PROPIONYL OXIDE see PMV500
PROPIONYL PEROXIDE (DOT) see DWQ800
PROPIONYL PEROXIDE, not more than 28% in solution see
   DWQ800
p-PROPIONYLPHENOL see ELL500
PROPISAMINE see BBK000
PROPITAN see FHG000
PROPIVANE see PGP500
PROP-JOB see DGI000
PROPOKSURU (POLISH) see PMY300
PROPOLIN see PMP500
PROPON see TIX500
PROPONEX-PLUS see CIR500
PROPOPHANE see PMP500
PROPOX see PNA500
PROPOXUR see PMY300
PROPOXYCHEL see PNA500
(+)-PROPOXYPHENE see DAB879
d-PROPOXYPHENE see DAB879
PROPOXYPHENE HYDROCHLORIDE see PNA500
(+)-PROPOXYPHENE HYDROCHLORIDE see PNA500
d-PROPOXYPHENE HYDROCHLORIDE see PNA500
α-PROPOXYPHENE HYDROCHLORIDE see PNA500

α-d-PROPOXYPHENE HYDROCHLORIDE see PNA500
d-PROPOXYPHENE MONOHYDROCHLORIDE see PNA500
PROPRANOLOL see ICB000
β-PROPRIOLACTONE (OSHA) see PMT100
β-PROPROLACTONE see PMT100
PROPROP see DGI400
PROPYCIL see PNX000
PROPYL ACETATE see PNC250
1-PROPYL ACETATE see PNC250
2-PROPYL ACETATE see INE100
n-PROPYL ACETATE see PNC250
PROPYLACETIC ACID see VAQ000
PROPYL ACETOXYMETHYLNITROSAMINE see PNR250
N-PROPYL-N-(ACETOXYMETHYL)NITROSAMINE see
   PNR250
S-PROPYL-N-AETHYL-N-BUTYL-THIOCARBAMAT (GER-
   MAN) see PNF500
N-PROPYLAJMALINE BITARTRATE see DNB000
N-PROPYLAJMALINE HYDROGEN TARTRATE see
   DNB000
N-PROPYLAJMALINIUM BITARTRATE see DNB000
N-PROPYLAJMALINIUMHYDROGENTARTRAT (GERMAN)
   see DNB000
N^4-PROPYLAJMALINIUM HYDROGEN TARTRATE see
   DNB000
PROPYL ALCOHOL see PND000
1-PROPYL ALCOHOL see PND000
n-PROPYL ALCOHOL see PND000
sec-PROPYL ALCOHOL (DOT) see INJ000
PROPYL ALDEHYDE see PMT750
i-PROPYLALKOHOL (GERMAN) see INJ000
n-PROPYL ALKOHOL (GERMAN) see PND000
PROPYLAMINE see PND250
2-PROPYLAMINE see INK000
N-PROPYLAMINE see PND250
sec-PROPYLAMINE see INK000
4-PROPYLANISOLE see PNE250
4-n-PROPYLANISOLE see PNE250
p-n-PROPYL ANISOLE see PNE250
n-PROPYLBENZENE see IKG000
PROPYL BENZENE (DOT) see IKG000
5-PROPYL-1,3-BENZODIOXOLE see DMD600
PROPYL BROMIDE see BNX750
S-PROPYL BUTYLETHYLTHIOCARBAMATE see PNF500
PROPYL CARBAMATE see PNG250
N-PROPYL CARBAMATE see PNG250
PROPYLCARBINOL see BPW500
N-PROPYLCARBINYL CHLORIDE see BQQ750
N-PROPYL CHLORIDE see CKP750
PROPYL CHLOROCARBONATE see PNH000
PROPYL CHLOROFORMATE see PNH000
n-PROPYL CHLOROFORMATE (DOT) see PNH000
PROPYL CYANIDE see BSX250
2-PROPYL-3-DIMETHYLAMINO-5,6-METHYLENEDIOXYIN-
   DENE HYDROCHLORIDE see DPY600
2-N-PROPYL-3-DIMETHYLAMINO-5,6-METHYLENEDIOXY-
   INDENE HYDROCHLORIDE see DPY600
PROPYLENE (DOT) see PMO500
PROPYLENE ALDEHYDE see ADR000, COB260
PROPYLENE CHLORIDE see PNJ400
PROPYLENECHLOROHYDRIN see CKR500
PROPYLENEDIAMINE see PMK250
PROPYLENE DIAMINE (DOT) see PMK250
PROPYLENE DICHLORIDE see PNJ400
α,β-PROPYLENE DICHLORIDE see PNJ400
4,4'-PROPYLENEDI-2,6-PIPERAZINEDIONE see RCA375
PROPYLENE EPOXIDE see PNL600
PROPYLENE GLYCOL (FCC) see PML000
α-PROPYLENEGLYCOL see PML000
1,2-PROPYLENE GLYCOL see PML000
PROPYLENE GLYCOL ALGINATE see PNJ750
PROPYLENE GLYCOL DINITRATE see PNL000
1,2-PROPYLENE GLYCOL DINITRATE see PNL000
PROPYLENE GLYCOL-1,2-DINITRATE see PNL000

PROPYLENE GLYCOL METHYL ETHER see PNL250
PROPYLENE GLYCOL MONOACRYLATE see HNT600
PROPYLENE GLYCOL MONOMETHYL ETHER see PNL250
α-PROPYLENE GLYCOL MONOMETHYL ETHER see
  PNL250
PROPYLENE GLYCOL USP see PML000
PROPYLENE IMINE see PNL400
1,2-PROPYLENEIMINE see PNL400
PROPYLENE IMINE, INHIBITED (DOT) see PNL400
PROPYLENE OXIDE see PNL600
1,2-PROPYLENE OXIDE see PNL600
1,3-PROPYLENE OXIDE see OMW000
PROPYLENE POLYMER see PMP500
PROPYLENE TETRAMER (DOT) see PMP750
PROPYLENGLYKOL-MONOMETHYLAETHER (GERMAN)
  see PNL250
n-PROPYL ESTER of 3,4,5-TRIHYDROXYBENZOIC ACID
  see PNM750
PROPYL ETHER see PNM000
β-PROPYL-α-ETHYLACROLEIN see BRI000
PROPYL-ETHYLBUTYLTHIOCARBAMATE see PNF500
PROPYLETHYL-N-BUTYLTHIOCARBAMATE see PNF500
PROPYL N-ETHYL-N-BUTYLTHIOCARBAMATE see
  PNF500
N-PROPYL-N-ETHYL-N-(N-BUTYL)THIOCARBAMATE see
  PNF500
S-(N-PROPYL)-N-ETHYL-N-N-BUTYLTHIOCARBAMATE see
  PNF500
PROPYL ETHYLBUTYLTHIOLCARBAMATE see PNF500
N-PROPYL-N-ETHYL-N-(N-BUTYL)THIOLCARBAMATE see
  PNF500
PROPYLETHYL ETHER see EPC125
PROPYL FORMATE (DOT) see PNM500
n-PROPYL FORMATE see PNM500
PROPYLFORMIC ACID see BSW000
PROPYL GALLATE see PNM750
n-PROPYL GALLATE see PNM750
PROPYL HYDRIDE see PMJ750
PROPYL p-HYDROXYBENZOATE see HNU500
n-PROPYL p-HYDROXYBENZOATE see HNU500
PROPYLIC ALCOHOL see PND000
PROPYLIC ALDEHYDE see PMT750
n-PROPYLIDENE BUTYRALDEHYDE see HBI800
PROPYL ISOCYANATE see PNP000
1-PROPYL ISOCYANATE see PNP000
m-PROPYL ISOCYANATE see PNP000
PROPYL KETONE see DWT600
PROPYL MERCAPTAN see PML500
2-PROPYL MERCAPTAN see IMU000
N-PROPYL MERCAPTAN see PML500
PROPYL METHANOATE see PNM500
PROPYLMETHANOL see BPW500
PROPYLMETHYLCARBINYLETHYL BARBITURIC ACID
  SODIUM SALT see NBU000
4-PROPYL-1,2-METHYLENEDIOXYBENZENE see DMD600
PROPYL NITRATE see PNQ500
n-PROPYL NITRATE see PNQ500
PROPYLNITROSAMINOMETHYL ACETATE see PNR250
1-(PROPYLNITROSAMINO)PROPYL ACETATE see ABT750
N-PROPYLNITROSOHARNSTOFF (GERMAN) see NLO500
1-PROPYL-1-NITROSOUREA see NLO500
N-PROPYLNITROSOUREA see NLO500
PROPYLOWY ALKOHOL (POLISH) see PND000
PROPYLPARABEN (FCC) see HNU500
PROPYLPARASEPT see HNU500
n-PROPYL PERCARBONATE see DWV400
6-(PROPYLPIPERONYL)-BUTYL CARBITYL ETHER see
  PIX250
6-PROPYLPIPERONYL BUTYL DIETHYLENE GLYCOL
  ETHER see PIX250
N-PROPYL-1-PROPANAMINE see DWR000
((PROPYLTHIO)-5-1H-BENZIMIDAZOLYL-2) CARBAMATE
  de METHYLE (FRENCH) see VAD000

(5-(PROPYLTHIO)-1H-BENZIMIDAZOL-2-YL)CARBAMIC
  ACID METHYL ESTER see VAD000
5-(PROPYLTHIO)-2-CARBOMETHOXYAMINOBENZIMID-
  AZOLE see VAD000
6-PROPYL-2-THIO-2,4(1H,3H)PYRIMIDINEDIONE see
  PNX000
PROPYL-THIORIST see PNX000
PROPYLTHIOURACIL see PNX000
4-PROPYL-2-THIOURACIL see PNX000
6-PROPYL-2-THIOURACIL see PNX000
6-N-PROPYLTHIOURACIL see PNX000
6-N-PROPYL-2-THIOURACIL see PNX000
PROPYL-THYRACIL see PNX000
n-PROPYLTRICHLOROSILANE see PNX250
PROPYLTRICHLOROSILANE (DOT) see PNX250
n-PROPYL-3,4,5-TRIHYDROXYBENZOATE see PNM750
5-PROPYL-4-(2,5,8-TRIOXA-DODECYL)-1,3-BENZODIOXOL
  (GERMAN) see PIX250
PROPYL URETHANE see PNG250
PROPYNE see MFX590
1-PROPYNE-3-OL see PMN450
PROPYNE mixed with PROPADIENE see MFX600
3-PROPYNOL see PMN450
2-PROPYN-1-OL see PMN450
2-PROPYNYL ALCOHOL see PMN450
PROPYON see PMY300
PROPYPHENAZONE see INY000
PROPYTHIOURACIL see PNX000
PROPYZAMIDE see DTT600
PROQUANIL see MQU750
PRORALONE-MOP see XDJ000
PROREX see DQA400
PROSEPTINE see SNM500
PROSEPTOL see SNM500
PROSERIN see DQY909
PROSERINE METHYL SULFATE see DQY909
PRO-SONIL see TDA500
PROSPASMIN see PGP500
PROSPASMINE see PGP500
PROSPASMINE HYDROCHLORIDE see PGP500
PROSTAGLANDIN E2 see DVJ200
(−)-PROSTAGLANDIN E2 see DVJ200
(15S)-PROSTAGLANDIN E2 see DVJ200
PROSTAGLANDIN F2-α see DMU800
PROSTAGLANDIN F2-α-THAM see POC750
PROSTAGLANDIN F2-α THAM SALT see POC750
PROSTAGLANDIN F2a TROMETHAMINE see POC750
PROSTALMON F see DMU800
PROSTAPHLIN-A see SLJ000
PROSTARMON F see DMU800
PROSTIGMINE METHYLSULFATE see DQY909
PROSTIN E2 see DVJ200
PROSTIN F2-α see DMU800
PROSTRUMYL see MPW500
PROTABEN P see HNU500
PROTACELL 8 see SEH000
PROTACHEM 630 see PKF000
PROTACHEM GMS see OAV000
PROTACTINIUM see POD500
PROTACTYL see DQA600
PROTAGENT see PKQ250
PROTANABOL see PAN100
PROTANAL see SEH000
PROTASORB O-20 see PKL100
PROTATEK see SEH000
PROTAZINE see DQA400
PROTECTONA see DKA600
PROTERNOL see DMV600
PROTEX (POLYMER) see AAX250
PROTHAZIN see DQA400
PROTHAZIN METHOSULFATE see MRW000
PROTHIUCIL see PNX000
PROTHIURONE see PNX000
PROTHOATE see IOT000

PROTHROMADIN see WAT200
PROTHYCIL see PNX000
PROTHYRAN see PNX000
PROTIURAL see PNX000
PROTOAT (HUNGARIAN) see IOT000
PROTOCATECHUIC ALDEHYDE ETHYL ETHER see EQF000
PROTOCATECHUIC ALDEHYDE METHYLENE ETHER see PIW250
PROTOCHLORURE D'IODE (FRENCH) see IDS000
PROTOPAM CHLORIDE see FNZ000
PROTOPAM IODIDE see POS750
PROTOPET see MQV750
PROTOPINE see FOW000
PROTOTYPE III SOFT see PKQ059
PROTOX TYPE 166 see ZKA000
PROTOX TYPE 167 see ZKA000
PROTOX TYPE 168 see ZKA000
PROTOX TYPE 169 see ZKA000
PROTOX TYPE 267 see ZKA000
PROTOX TYPE 268 see ZKA000
PROTRIPTYLINE HYDROCHLORIDE see POF250
PROVENTIL see BQF500
PROVIGAN see DQA400
PROVITAMIN D see CMD750
PROXAGESIC see DAB879, PNA500
PROXOL see TIQ250
PROZINEX see PMN850
PROZOIN see PMU750
PRS 640 see BML500
PRUNOLIDE see CNF250
PRURALGAN see DNX400
PRURALGIN see DNX400
PRUSSIAN BROWN see IHD000
PRUSSIC ACID (DOT) see HHS000
PRUSSIC ACID, unstabilized see HHS000
PRUSSITE see COO000
PRYSOLINE see DBB200
PRZEDZIORKOFOS (POLISH) see PDC750
PS see CKN500
PS 1 see AHE250
PSC CO-OP WEEVIL BAIT see DXE000
PSEUDOACETIC ACID see PMU750
PSEUDOACONITINE see POG250
PSEUDOBUTYLBENZENE see BQJ250
PSEUDO-BUTYLENE see BOW500
PSEUDOCUMENE see TLL750
PSEUDOCUMIDINE see TLG250
PSEUDOCUMIDINE HYDROCHLORIDE see TLG750
PSEUDOCUMOL see TLL750
PSEUDODIGITOXIN see GEU000
PSEUDOHEXYL ALCOHOL see EGW000
PSEUDOPINEN see POH750
PSEUDOPINENE see POH750
PSEUDOTHEOPHYLLINE see TEP000
PSEUDOTHIOUREA see ISR000
PSEUDOUREA see USS000
PSEUDOXANTHINE see XCA000
PSICAINE-NEU HYDROCHLORIDE see NCJ000
PSICAIN-NEW HYDROCHLORIDE see NCJ000
PSICHIAL see MDQ250
PSICOPERIDOL-R see TKK500
PSICOPLEGIL see MNM500
PSICOSAN see LFK000, MDQ250
PSICOSEDINA see MNM500
PSICOSTEN see PDN000
PSILOCINE see HKE000
PSILOCIN PHOSPHATE ESTER see PHU500
PSILOCIPIN see PHU500
PSILOTSIBIN see PHU500
PSILOTSIN see HKE000
PSL see LEN000
PSORADERM see MFN275
PSYCHAMINE A 66 HYDROCHLORIDE see MNV750

PSYCHEDRINE see BBK000
PSYCHODRINE see BBK500
PSYCHOLIQUID see TEH500
PSYCHOPERIDOL see TKK500
PSYCHOTABLETS see TEH500
PSYTOMIN see PCK500
PTC see PCC000, PGN250
PTEGLU see FMT000
PTERIDIUM AQUILINUM see BML000
PTERIS AQUALINA see BML000
PTEROYLGLUTAMIC ACID see FMT000
PTEROYL-l-GLUTAMIC ACID see FMT000
PTEROYLMONOGLUTAMIC ACID see FMT000
PTEROYL-l-MONOGLUTAMIC ACID see FMT000
PTFE see TAI250
PTG see EQP000
PTU see PGN250
PTU (THYREOSTATIC) see PNX000
PURADIN see NGG500
PURALIN see TFS350
PURAPURIDINE see POJ000
PURASAN-SC-10 see ABU500
PURATRONIC CHROMIUM CHLORIDE see CMJ250
PURATRONIC CHROMIUM TRIOXIDE see CMK000
PURATURF 10 see ABU500
PURE CHRYSOIDINE YBH see PEK000
PURE CHRYSOIDINE YD see PEK000
PURE EOSINE YY see BNH500
PURE LEMON CHROME L3GS see LCR000
PURE ORANGE CHROME M see LCS000
PURE QUARTZ see SCI500, SCJ500
PUREX see SFT000
PURE ZINC CHROME see ZFJ100
PURIFIED CHARCOAL see CBT500
PURIFIED OXGALL see SFW000
PURIMETHOL see POK000
PURINE-2,6-DIOL see XCA000
9H-PURINE-2,6-DIOL see XCA000
2,6(1,3)-PURINEDION see XCA000
PURINE-2,6-(1H,3H)-DIONE see XCA000
PURINE-6-THIOL see POK000
6-PURINETHIOL see POK000
3H-PURINE-6-THIOL see POK000
PURINETHOL see POK000
PUROCYCLINA see TBX000
PURODIGIN see DKL800
PUROSTROPHAN see OKS000
PURPLE MINT PLANT EXTRACT see PCI750
PURPURID see DKL800
PURTALC USP see TAB750
PVBR see PKQ000
PVC CORDO see AAX175
PVC (MAK) see PKQ059
PVP 1 see PKQ500
PVP 2 see PKQ750
PVP 3 see PKR000
PVP 4 see PKR250
PVP 5 see PKR500
PVP 6 see PKR750
PVP 7 see PKS000
PVP (FCC) see PKQ250
PVSK see PKS250
PX 104 see DEH200
PX-138 see DVL600
PX 404 see DEH600
PX 438 see BJS250
PYBUTHRIN see PIX250
PYCAZIDE see ILD000
PYDRIN see FAR100
PYDT see DJY000
PYKNOLEPSINUM see ENG500
PYMAFED see DBM800, WAK000
PYNACOLYL METHYLFLUOROPHOSPHONATE see SKS500

PYNAMIN see AFR250
PYNAMIN-FORTE see AFR250
PYNDT see DTV200
PYNOSECT see BEP500
PYOSTACINE see VRF000
PYRA see WAK000
PYRABUTOL see BRF500
PYRACRYL ORANGE Y see PEK000
PYRADEX see DBI200
PYRADONE see DOT000
PYRAHEXYL see HGK500
PYRALENE see PJL750
PYRALIN see CCU250
PYRAMAL see WAK000
PYRA MALEATE see DBM800
PYRAMIDON see DOT000
PYRAMIDONE see DOT000
PYRAMINE see PEE750
PYRAMIN RB see PEE750
PYRAN ALDEHYDE see ADR500
PYRANILAMINE MALEATE see DBM800
PYRANINYL see DBM800
PYRANISAMINE see WAK000
PYRANISAMINE MALEATE see DBM800
PYRANOL see PJL750
PYRANTEL TARTRATE see TCW750
PYRANTON see DBF750
PYRATHYN see TEO250
PYRAZINAMIDE see POL500
PYRAZINEAMIDE see POL500
PYRAZINECARBOXAMIDE see POL500
PYRAZINE CARBOXYLAMIDE see POL500
PYRAZINE HEXAHYDRIDE see PIJ000
PYRAZINOIC ACID AMIDE see POL500
PYRAZINOL-O-ESTER with O,O-DIETHYL PHOSPHORO-
    THIOATE see EPC500
PYRAZODINE see PDC250
PYRAZOFEN see PDC250, PEK250
PYRAZOL BLUE 3B see CMO250
PYRAZOLIDIN see BRF500
PYRAZON see PEE750
PYRAZONE see PEE750
PYRAZONL see PEE750
PYREAZID see ILD000
PYREDAL see PDC250
PYREN (GERMAN) see PON250
PYRENE see PON250
PYRENONE 606 see PIX250
PYREQUAN TARTRATE see TCW750
PYRESIN see AFR250
PYRESYN see AFR250
PYRETHERM see BEP500
PYRETHIA see DQA400
PYRETHIAZINE see DQA400
PYRETHRIN see POO000, POO100
PYRETHRIN II see POO100
PYRETHRIN I or II see POO250
PYRETHRINS see POO250
PYRETHROLONE CHRYSANTHEMUM DICARBOXLIC
    ACIDMETHYL ESTER ESTER see POO100
PYRETHROLONE ESTER of CHRYSANTHEMUMDICAR-
    BOXYLIC ACID MONOMETHYL ESTER see POO100
(+)-PYRETHRONYL (+)-PYRETHRATE see POO100
PYRETHRUM (ACGIH) see POO250
PYRETHRUM (INSECTICIDE) see POO250
PYRETRIN II see POO100
PYRIBENZAMINE see TMP750
PYRICAROYL see DJS200
PYRICIDIN see ILD000
PYRIDACIL see PDC250, PEK250
PYRIDAFENTHION see POP000
PYRIDAMAL-100 see TAI500
PYRIDAPHENTHION see POP000
PYRIDENAL see PDC250

PYRIDENE see PDC250
PYRIDIATE see PDC250
PYRIDICIN see ILD000
PYRIDIN (GERMAN) see POP250
2-PYRIDINALDOXIM METHOJODID (GERMAN) see
    POS750
PYRIDIN-2-ALDOXIN (CZECH) see POS750
3-PYRIDINAMINE see AMI250
α-PYRIDINAMINE see AMI000
PYRIDINE see POP250
2-PYRIDINEALDOXIME CHLORIDE see FNZ000
2-PYRIDINE ALDOXIME IODOMETHYLATE see POS750
PYRIDINE-2-ALDOXIME METHIODIDE see POS750
PYRIDINE-2-ALDOXIME METHOCHLORIDE see FNZ000
2-PYRIDINE ALDOXIME METHYL CHLORIDE see FNZ000
PYRIDINE-2-ALDOXIME METHYL IODIDE see POS750
PYRIDINE-3-CARBONIC ACID see NCQ900
PYRIDINE-3-CARBOXYDIETHYLAMIDE see DJS200
PYRIDINE-3-CARBOXYLIC ACID see NCQ900
3-PYRIDINECARBOXYLIC ACID see NCQ900
PYRIDINE-β-CARBOXYLIC ACID see NCQ900
PYRIDINE-3-CARBOXYLIC ACID AMIDE see NCR000
3-PYRIDINECARBOXYLIC ACID AMIDE see NCR000
PYRIDINE-3-CARBOXYLIC ACID DIETHYLAMIDE see
    DJS200
4-PYRIDINECARBOXYLIC ACID, ETHYL ESTER see
    ELU000
4-PYRIDINECARBOXYLIC ACID, HYDRAZIDE see ILD000
PYRIDINE-CARBOXYLIQUE-3 (FRENCH) see NCQ900
4-PYRIDINEETHANOL see POR500
PYRIDINE, 3-(TETRAHYDRO-1-METHYLPYRROL-2-YL) see
    NDN000
PYRIDINIUM ALDOXIME METHOCHLORIDE see FNZ000
PYRIDINIUM-2-ALDOXIME-N-METHYLIODIDE see
    POS750
PYRIDINIUM PERCHLORATE see PPC100
PYRIDIPCA see PPK500
PYRIDIUM see PDC250, PEK250
PYRIDIVITE see PDC250
PYRIDOXAL HYDROCHLORIDE see VSU000
PYRIDOXINE see PPK250
PYRIDOXINE HYDROCHLORIDE (FCC) see PPK500
PYRIDOXINIUM CHLORIDE see PPK500
PYRIDOXINUM HYDROCHLORICUM (HUNGARIAN) see
    PPK500
PYRIDOXOL see PPK250
PYRIDOXOL HYDROCHLORIDE see PPK500
PYRIDROL see DWK400, PII750
PYRIDROLE see DWK400
3-PYRIDYLAMINE see AMI250
α-PYRIDYLAMINE see AMI000
1-(PYRIDYL-3-)-3,3-DIAETHYL-TRIAZEN (GERMAN) see
    DJY000
m-PYRIDYL-DIETHYL-TRIAZENE see DJY000
1-PYRIDYL-3,3-DIETHYLTRIAZENE see DJY000
1-(PYRIDYL-3)-3,3-DIETHYLTRIAZENE see DJY000
1-(3-PYRIDYL)-3,3-DIETHYLTRIAZENE see DJY000
1-(PYRIDYL-3)-3,3-DIMETHYL-TRIAZEN (GERMAN) see
    PPL500
1-(PYRIDYL-3)-3,3-DIMETHYL TRIAZENE see PPL500
1-(m-PYRIDYL)-3,3-DIMETHYL-TRIAZENE see PPL500
N-α-PYRIDYL-N-p-METHOXYBENZYL-N′,N′-DIMETHYL-
    ETHYLENEDIAMINE PHOSPHATE see NBV000
N-3-PYRIDYLMETHYL-N′-p-NITROPHENYLUREA see
    PPP750
β-PYRIDYL-α-N-METHYLPYRROLIDINE see NDN000
1-(PYRIDYL-3-N-OXID)-3,3-DIMETHYL-TRIAZEN (GER-
    MAN) see DTV200
1-(PYRIDYL-3-N-OXIDE)-3,3-DIMETHYLTRIAZENE see
    DTV200
N-(α-PYRIDYL)-N-(α-THENYL)-N′,N′-DIMETHYLETH-
    YLENEDIAMINE see TEO250
PYRILAMINE see WAK000
PYRILAMINE MALEATE see DBM800

PYRIMIDINE PHOSPHATE see DIN800
2,4,5,6(1H,3H)-PYRIMIDINETETRONE HYDRATE see
    MDL500
2,4,6(1H,3H,5H)-PYRIMIDINETRIONE, 5-(2-BROMO-2-PRO-
    PENYL)-5-(1-METHYLPROPYL)-(9CI) see BOR000
2,4,6(1H,3H,5H)-PYRIMIDINETRIONE, 5,5-DIETHYL-,
    MONOSODIUM SALT (9CI) see BAG250
2,4,6(1H,3H,5H)-PYRIMIDINETRIONE, 5-ETHYL-5-(1-
    METHYLBUTYL)-, CALCIUM SALT (9CI) see CAV000
PYRIMIDONE MEDI-PETS see DBB200
PYRIMINYL see PPP750
PYRIMIPHOS METHYL see DIN800
PYRINAMINE BASE see TMP750
PYRINAZINE see HIM000
β-PYRINE see PON250
PYRINEX see CMA100
PYRINISTAB see TEO250
PYRINISTOL see TEO250
PYRIPYRIDIUM see PDC250, PEK250
PYRISEPT see CCX000
PYRITHEN see CHY250
PYRIZIDIN see ILD000
PYRIZIN see PDC250
PYRLEUGAN see DLH630
PYROACETIC ACID see ABC750
PYROACETIC ETHER see ABC750
PYROBENZOL see BBL250
PYROBENZOLE see BBL250
PYROCARBONATE d'ETHYLE (FRENCH) see DIZ100
PYROCARBONIC ACID, DIETHYL ESTER see DIZ100
PYROCATECHIN see CCP850
PYROCATECHINIC ACID see CCP850
PYROCATECHOL see CCP850
PYROCATECHUIC ACID see CCP850
PYROCHOL see DAQ400
PYRODONE see OES000
PYRODOXIN see PPK250
PYROGALLIC ACID see PPQ500
PYROGALLOL see PPQ500
PYROGALLOL DIMETHYLETHER see DOJ200
PYROGALLOL-1,3-DIMETHYL ETHER see DOJ200
PYROGUAIAC ACID see GKI000
PYROKOHLENSAEURE DIAETHYL ESTER (GERMAN) see
    DIZ100
M-PYROL see MPF200
PYROLUSITE BROWN see MAS000
PYROMUCIC ALDEHYDE see FPQ875
PYRONALORANGE see PEJ500
PYRONALROT R see XRA000
PYROPENTYLENE see CPU500
PYROPHOSPHATE see TEE500
PYROPHOSPHATE de TETRAETHYLE (FRENCH) see
    TCF250
PYROPHOSPHORIC ACID OCTAMETHYLTETRAAMIDE see
    OCM000
PYROPHOSPHORIC ACID, TETRAETHYL ESTER (liquid
    mixture) see TCF280
PYROPHOSPHORODITHIOIC ACID, TETRAETHYL ESTER
    see SOD100
PYROPHOSPHORODITHIOIC ACID-O,O,O,O-TETRAETHYL
    ESTER see SOD100
PYROPHOSPHORYLTETRAKISDIMETHYLAMIDE see
    OCM000
PYROSULFUROUS ACID, DIPOTASSIUM SALT see
    PLR250
PYROSULFURYL CHLORIDE see PPR500
PYRO SULFURYL CHLORIDE (DOT) see PPR500
PYROSULPHURIC ACID see SOI520
PYROSULPHURYL CHLORIDE (DOT) see PPR500
PYROTARTARIC ACID NITRILE see TLR500
PYROTROPBLAU see CMO250
PYROXYLIC SPIRIT see MGB150
PYROXYLIN see CCU250
PYROXYLIN PLASTICS (DOT) see CCU250

PYROXYLIN PLASTIC SCRAP (DOT) see CCU250
PYRROLE see PPS250
PYRROLE-2,5-DIONE see MAM750
PYRROLIDINE see PPS500
2-(PYRROLIDINYL)ETHYL-2-CHLORO-6-METHYLCARBA-
    NILATE HYDROCHLORIDE see CIK500
3-PYRROLINE-2,5-DIONE see MAM750
PYRROLYLENE see BOP500

Q-137 see DJC000
QDO see DVR200
QGH see BDD000
QIDAMP see AIV500
QIDTET see TBX250
QPB see PJA000
QSAH 7 see PKQ059
QUAALUDE see QAK000
QUADRACYCLINE see TBX250
QUAMAQUIL see PGA750
QUAMONIUM see HCQ500
QUANTRIL see BCL250
QUANTROVANIL see EQF000
QUANTRYL see BCL250
QUARTZ see SCJ500
QUARTZ GLASS see SCK600
QUATERNARIO CPC see CCX000
QUATERNOL 1 see DTC600
QUATRACHLOR see BEN000
QUATRESIN see MSB500
QUATREX see TBX250
QUAZO PURO (ITALIAN) see SCJ500
QUEBRACHIN see YBJ000
QUEBRACHINE see YBJ000
QUEBRACHO TANNIN see QBJ000
QUECKSILBER (GERMAN) see MCW250
QUECKSILBER CHLORID (GERMAN) see MCY475
QUECKSILBER(I)-CHLORID (GERMAN) see MCW000
QUECKSILBER CHLORUER (GERMAN) see MCW000
QUECKSILBEROXID (GERMAN) see MCT500, MDF750
QUELAMYCIN see QBS000
QUELETOX see FAQ999
QUELICIN see CMG250, HLC500
QUELICIN CHLORIDE see HLC500
QUELLADA see BBQ500
QUEMICETINA see CDP250
QUERCETIN see QCA000
QUERCETINE see QCA000
QUERCETOL see QCA000
QUERCITIN see QCA000
QUERTINE see QCA000
QUESTEX 4 see EIV000
QUESTEX 4H see EIX000
QUETIMID see TEH500
QUICK see CJJ000
QUICKLIME (DOT) see CAU500
QUICKSAN see ABU500
QUICKSET EXTRA see MKA500
QUICK SILVER see MCW250
QUIETIDON see MQU750
QUIETOPLEX see TEH500
QUILONUM RETARD see LGZ000
QUINACRINE see ARQ250
QUINACRINE ETHYL MUSTARD see DFH000
QUINACRINE MUSTARD see QDJ000, QDS000
QUINACRINE MUSTARD DIHYDROCHLORIDE see
    QDS000
QUINALBARBITAL see SBM500
QUINALBARBITONE see SBM500
QUINALPHOS see DJY200
QUINAMBICIDE see CHR500
QUINICARDINE see QFS000
QUINIDEX see QFS000
QUINIDINE see QFS000
(+)-QUINIDINE see QFS000

QUININE see QHJ000
β-QUININE see QFS000
QUININE BIMURIATE see QIJ000
QUININE BISULFATE see QMA000
QUININE CHLORIDE see QJS000
QUININE DIHYDROCHLORIDE see QIJ000
(−)-QUININE DIHYDROCHLORIDE see QIJ000
QUININE ETHIODIDE see QIS000
QUININE HYDROCHLORIDE see QJS000
QUININE HYDROGEN SULFATE see QMA000
QUININE MONOHYDROCHLORIDE see QJS000
QUININE MURIATE see QJS000
QUININE SULFATE see QMA000
QUINIZARIN see DMH000
QUINIZARINE GREEN BASE see BLK000
8-QUINOL see QPA000
β-QUINOL see HIH000
QUINOLINE see QMJ000
QUINOLINE-6-AZO-p-DIMETHYLANILINE see DPR000
8-QUINOLINOL see QPA000
8-(QUINOLINOLATO)METHYL MERCURY see MLH000
(8-QUINOLINOLATO)TRIBUTYLSTANNANE see TIB000
8-QUINOLINOL, MERCURY COMPLEX see MLH000
QUINOLOR COMPOUND see BDS000
N-(4-QUINOLYL)HYDROXYLAMINE-1'-OXIDE see HIY500
QUINONDO see BLC250
QUINONE see QQS200
o-QUINONE see BDC250
p-QUINONE see QQS200
QUINONE DIOXIME see DVR200
p-QUINONE DIOXIME see DVR200
QUINONE MONOXIME see NLF200
QUINONE OXIME see NLF200
p-QUINONE OXIME see DVR200
QUINONE OXIME BENZOYLHYDRAZONE see BDD000
QUINOPHENOL see QPA000
(2-QUINOXALINYLMETHYLENE)-HYDRAZINECARBOX-
    YLIC ACID METHYL ESTER-N,N'-DIOXIDE see FOI000
N-(2-QUINOXALINYL)SULFANILAMIDE see QTS000
N¹-2-QUINOXALINYLSULFANILAMIDE see QTS000
N'-2-QUINOXALYLSULFANILAMIDE see QTS000
QUINSORB 010 see DMI600
QUINTAR see DFT000
QUINTAR 540F see DFT000
QUINTESS-N see NCG000
QUINTOCENE see PAX000
QUINTOX see DHB400
QUINTOZEN see PAX000
QUINTOZENE see PAX000
QUINTRATE see PBC250
3-QUINUCLIDINOL ACETATE see AAE250
QUIRVIL see PKQ059
QUOLAC EX-UB see SIB600
QUOTANE see DNX400
QUOTANE HYDROCHLORIDE see DNX400

R 10 see CBY000
R 12 (DOT) see DFA600
R 13 see CLR250
R 14 see CBY250, PKL750
R 14 (refrigerant) see CBY250
R 20 (refrigerant) see CHJ500
R 22 (DOT) see CFX500
R 23 see CBY750
R 31 see CHI900
R 31 (refrigerant) see CHI900
R-47 see TNF500
R48 see BIF250
R50 see DAD200
R 54 see DDP600
88-R see SOP500
R 113 see FOO000
R 114 see FOO509
R161 see FIB000

R-246 see TND500
R-1303 see TNP250
R-1492 see MQH750
R 1504 see PHX250
R 1513 see EKN000
R-1608 see EIN500
R 1658 see MNN250
R 1929 see FLU000
R-2061 see PNF500
R 2170 see DAP000
R-2498 see TKK500
R 3345 see FHG000
R 40B1 see MHR200
R 4263 see PDW500, PDW750
R 4749 see DYF200
R-5,158 see DJA400
R 5240 see PDW750
R-5461 see BJD000
R 6700 see OAN000
R7000 see MFB400
R 7158 see FLJ000
R 9298 see CLS250
R 9985 see MDV500
R-12,564 see LFA020
R 16341 see PAP250
R 17635 see MHL000
R 23979 see FPB875
R 35443 see DWF790
R 35443 see OMG000
RACEMORPHAN see MKR250
RACEMORPHAN HYDROBROMIDE see MDV250
RACEPHEDRINE HYDROCHLORIDE see EAX500
RACUSAN see DSP400
RADAPON see DGI400, DGI600
RADAZIN see ARQ725
RADDLE see IHD000
RAD-E CATE 16 see HKC500
RAD-E-CATE 25 see HKC000
RADEDORM see DLY000
RADEPUR see LFK000
RADIATION see RAQ000
RADIATION, IONIZING see RAQ010
RADIOSTOL see VSZ100
RADIUM see RAV000
RADIUM F see PJJ750
RADIZINE see ARQ725
RADOCON see BJP000
RADOKOR see BJP000
RADON see RBA000
RADONIL see SNK000
RADONIN see SNN300
RADONNA see CHJ750
RADOSAN see MEO750
RADOX see CFK000
RADOXONE TL see AMY050
RADSTERIN see VSZ100
RAFEX see DUS700
RAFLUOR see SHF500
RALABOL see RBF100
RALGRO see RBF100
RALONE see RBF100
RAMETIN see HMV000
RAMIK see DVV600
RAMIZOL see AMY050
RAMOR see TEI000
R/AMP see RKP000
RAMPART see PGS000
RAMROD see CHS500
RAMUCIDE see CJJ000
RANAC see CJJ000
RANDOX see CFK000
RANEY ALLOY see NCW500
RANEY COPPER see CNI000
RANEY NICKEL see NCW500

RANKOTEX see CIR500
RANTOX T see CFK000
RAPACODIN see DKW800
RAPHATOX see DUS700
RAPHETAMINE see BBK000
RAPHONE see CIR250
RAPISOL see DJL000
RARE EARTHS see RBP000
RAS-26 see NIJ500
RASCHIT see CFE250
RASEN-ANICON see CFE250
RASIKAL see SFS000
RASPBERIN see SAH000
RASPBERRY KETONE see RBU000
RASTINON see BSQ000
RATAFIN see ABF500
RAT-A-WAY see ABF500, WAT200
RATBANE 1080 see SHG500
RAT-B-GON see WAT200
RAT-GARD see WAT200
RATINDAN 1 see DVV600
RAT & MICE BAIT see WAT200
RAT-NIP see PHO750
RAT-O-CIDE RAT BAIT see RCF000
RATOMET see CJJ000
RATOX see TEL750
RAT'S END see RCF000
RATS-NO-MORE see WAT200
RATTENGIFTKONSERVE see TEM000
RATTEX see CNV000
RATTRACK see AQN635
RAUMANON see ILD000
RAUSERPIN see RDK000
RAUWOLEAF see RDK000
RAVIAC see CJJ000
RAVINYL see PKQ059
RAVONA see CAV000
RAVONAL see PBT500
RAWETIN see HMV000
RAW SHALE OIL see COD750
RAYBAR see BAP000
RAY-GLUCIRON see FBK000
RAYOX see TGG760
RAZIDE see ILD000
RAZOL DOCK KILLER see CIR250
RAZOXANE see RCA375
RAZOXIN see RCA375, PIK250
RB see PAK000
RB 1509 see CGV250
RBA 777 see CLW000
R-C 318 see CPS000
RCA WASTE NUMBER P105 see SFA000
RCA WASTE NUMBER U203 see SAD000
RCA WASTE NUMBER U205 see SBR000
RC COMONOMER DBM see DED600
RC COMONOMER DOF see DVK600
RC 172DBM see AHE250
RC PLASTICIZER DOP see DVL700
RCRA WASTE NUMBER 7066 see DDL800
RCRA WASTE NUMBER P001 see WAT200
RCRA WASTE NUMBER P003 see ADR000
RCRA WASTE NUMBER P004 see AFK250
RCRA WASTE NUMBER P005 see AFV500
RCRA WASTE NUMBER P006 see AHE750
RCRA WASTE NUMBER P009 see ANS750
RCRA WASTE NUMBER P010 see ARB250
RCRA WASTE NUMBER P013 see BAK750
RCRA WASTE NUMBER P014 see PFL850
RCRA WASTE NUMBER P015 see BFO750
RCRA WASTE NUMBER P016 see BIK000
RCRA WASTE NUMBER P017 see BNZ000
RCRA WASTE NUMBER P018 see BOL750
RCRA WASTE NUMBER P020 see BRE500
RCRA WASTE NUMBER P021 see CAQ500

RCRA WASTE NUMBER P022 see CBV500
RCRA WASTE NUMBER P023 see CDY500
RCRA WASTE NUMBER P024 see CEH680
RCRA WASTE NUMBER P028 see BEE375
RCRA WASTE NUMBER P029 see CNL000
RCRA WASTE NUMBER P030 see COI500
RCRA WASTE NUMBER P031 see COO000
RCRA WASTE NUMBER P033 see COO750
RCRA WASTE NUMBER P034 see CPK500
RCRA WASTE NUMBER P036 see DGB600
RCRA WASTE NUMBER P037 see DHB400
RCRA WASTE NUMBER P039 see DXH325
RCRA WASTE NUMBER P042 see VGP000
RCRA WASTE NUMBER P043 see IRF000
RCRA WASTE NUMBER P044 see DSP400
RCRA WASTE NUMBER P046 see DTJ400
RCRA WASTE NUMBER P047 see DUS700
RCRA WASTE NUMBER P048 see DUZ000
RCRA WASTE NUMBER P049 see DXL800
RCRA WASTE NUMBER P050 see EAQ750
RCRA WASTE NUMBER P051 see EAT500
RCRA WASTE NUMBER P054 see EJM900
RCRA WASTE NUMBER P056 see FEZ000
RCRA WASTE NUMBER P057 see FFF000
RCRA WASTE NUMBER P058 see SHG500
RCRA WASTE NUMBER P059 see HAR000
RCRA WASTE NUMBER P060 see IKO000
RCRA WASTE NUMBER P062 see HCY000
RCRA WASTE NUMBER P063 see HHS000
RCRA WASTE NUMBER P064 see MKX250
RCRA WASTE NUMBER P065 see MDC000, MDC250
RCRA WASTE NUMBER P066 see MDU600
RCRA WASTE NUMBER P067 see PNL400
RCRA WASTE NUMBER P068 see MKN000
RCRA WASTE NUMBER P069 see MLC750
RCRA WASTE NUMBER P070 see CBM500
RCRA WASTE NUMBER P071 see MNH000
RCRA WASTE NUMBER P072 see AQN635
RCRA WASTE NUMBER P073 see NCZ000
RCRA WASTE NUMBER P074 see NDB500
RCRA WASTE NUMBER P075 see NDN000
RCRA WASTE NUMBER P076 see NEG100
RCRA WASTE NUMBER P077 see NEO500
RCRA WASTE NUMBER P078 see NGR500
RCRA WASTE NUMBER P081 see SLD000, NGY000
RCRA WASTE NUMBER P082 see NKA600
RCRA WASTE NUMBER P084 see NKY000
RCRA WASTE NUMBER P085 see OCM000
RCRA WASTE NUMBER P087 see OKK000
RCRA WASTE NUMBER P088 see DXD000, EAR000
RCRA WASTE NUMBER P089 see PAK000
RCRA WASTE NUMBER P092 see ABU500
RCRA WASTE NUMBER P093 see PGN250
RCRA WASTE NUMBER P094 see PGS000
RCRA WASTE NUMBER P095 see PGX000
RCRA WASTE NUMBER P096 see PGY000
RCRA WASTE NUMBER P097 see FAB600
RCRA WASTE NUMBER P098 see PLC500
RCRA WASTE NUMBER P101 see PMV750
RCRA WASTE NUMBER P102 see PMN450
RCRA WASTE NUMBER P104 see SDP000
RCRA WASTE NUMBER P106 see SGA500
RCRA WASTE NUMBER P108 see SMN500
RCRA WASTE NUMBER P109 see SOD100
RCRA WASTE NUMBER P110 see TCF000
RCRA WASTE NUMBER P111 see TCF250
RCRA WASTE NUMBER P112 see TDY250
RCRA WASTE NUMBER P113 see TEL050
RCRA WASTE NUMBER P115 see TEM000
RCRA WASTE NUMBER P116 see TFQ000
RCRA WASTE NUMBER P118 see PCF300
RCRA WASTE NUMBER P119 see ANY250
RCRA WASTE NUMBER P120 see VDU000
RCRA WASTE NUMBER P121 see ZGA000

RCRA WASTE NUMBER P122 see ZLS000
RCRA WASTE NUMBER P123 see CDV100
RCRA WASTE NUMBER P401 see NIM500
RCRA WASTE NUMBER P404 see EPC500
RCRA WASTE NUMBER U001 see AAG250
RCRA WASTE NUMBER U002 see ABC750
RCRA WASTE NUMBER U003 see ABE500
RCRA WASTE NUMBER U005 see FDR000
RCRA WASTE NUMBER U006 see ACF750
RCRA WASTE NUMBER U007 see ADS250
RCRA WASTE NUMBER U008 see ADS750
RCRA WASTE NUMBER U009 see ADX500
RCRA WASTE NUMBER U010 see AHK500
RCRA WASTE NUMBER U011 see AMY050
RCRA WASTE NUMBER U014 see IBB000
RCRA WASTE NUMBER U015 see ASA500
RCRA WASTE NUMBER U016 see BAW750
RCRA WASTE NUMBER U017 see BAY300
RCRA WASTE NUMBER U018 see BBC250
RCRA WASTE NUMBER U019 see BBL250
RCRA WASTE NUMBER U020 see BBS750
RCRA WASTE NUMBER U021 see BBX000
RCRA WASTE NUMBER U023 see BFL250
RCRA WASTE NUMBER U024 see BID750
RCRA WASTE NUMBER U025 see DFJ050
RCRA WASTE NUMBER U026 see BIF250
RCRA WASTE NUMBER U027 see BII250
RCRA WASTE NUMBER U028 see DVL700
RCRA WASTE NUMBER U029 see MHR200
RCRA WASTE NUMBER U031 see BPW500
RCRA WASTE NUMBER U032 see CAP500
RCRA WASTE NUMBER U033 see CCA500
RCRA WASTE NUMBER U035 see CDO500
RCRA WASTE NUMBER U036 see CDR750
RCRA WASTE NUMBER U037 see CEJ125
RCRA WASTE NUMBER U038 see DER000
RCRA WASTE NUMBER U039 see CFE250
RCRA WASTE NUMBER U041 see EAZ500
RCRA WASTE NUMBER U042 see CHI250
RCRA WASTE NUMBER U043 see VNP000
RCRA WASTE NUMBER U044 see CHJ500
RCRA WASTE NUMBER U045 see MIF765
RCRA WASTE NUMBER U046 see CIO250
RCRA WASTE NUMBER U048 see CJK250
RCRA WASTE NUMBER U049 see CLK235
RCRA WASTE NUMBER U050 see CML810
RCRA WASTE NUMBER U051 see CMY825
RCRA WASTE NUMBER U052 see CNW500, CNW750, CNX000, CNX250
RCRA WASTE NUMBER U053 see COB250, COB260, DQY800
RCRA WASTE NUMBER U055 see COE750
RCRA WASTE NUMBER U056 see CPB000
RCRA WASTE NUMBER U057 see CPC000
RCRA WASTE NUMBER U058 see EAS500
RCRA WASTE NUMBER U059 see DAC000
RCRA WASTE NUMBER U060 see BIM500
RCRA WASTE NUMBER U061 see DAD200
RCRA WASTE NUMBER U062 see DBI200
RCRA WASTE NUMBER U063 see DCT400
RCRA WASTE NUMBER U064 see BCQ500
RCRA WASTE NUMBER U067 see EIY500
RCRA WASTE NUMBER U068 see DDP800
RCRA WASTE NUMBER U069 see DEH200
RCRA WASTE NUMBER U070 see DEP600, DEP800
RCRA WASTE NUMBER U071 see DEP800
RCRA WASTE NUMBER U072 see DEP800
RCRA WASTE NUMBER U073 see DEQ600
RCRA WASTE NUMBER U074 see DEV000
RCRA WASTE NUMBER U075 see DFA600
RCRA WASTE NUMBER U076 see DFF809
RCRA WASTE NUMBER U077 see EIY600
RCRA WASTE NUMBER U078 see VPK000
RCRA WASTE NUMBER U079 see ACK000

RCRA WASTE NUMBER U080 see MJP450
RCRA WASTE NUMBER U081 see DFX800
RCRA WASTE NUMBER U082 see DFY000
RCRA WASTE NUMBER U083 see PNJ400
RCRA WASTE NUMBER U084 see DGG950
RCRA WASTE NUMBER U085 see BGA750
RCRA WASTE NUMBER U086 see DJL400
RCRA WASTE NUMBER U088 see DJX000
RCRA WASTE NUMBER U089 see DKA600
RCRA WASTE NUMBER U090 see DMD600
RCRA WASTE NUMBER U091 see DCJ200
RCRA WASTE NUMBER U092 see DOQ800
RCRA WASTE NUMBER U093 see DOT300
RCRA WASTE NUMBER U094 see DQJ200
RCRA WASTE NUMBER U095 see TGJ750
RCRA WASTE NUMBER U096 see IOB000
RCRA WASTE NUMBER U097 see DQY950
RCRA WASTE NUMBER U098 see DSF400
RCRA WASTE NUMBER U099 see DSF600
RCRA WASTE NUMBER U101 see XKJ500
RCRA WASTE NUMBER U102 see DTR200
RCRA WASTE NUMBER U103 see DUD100
RCRA WASTE NUMBER U105 see DVH000
RCRA WASTE NUMBER U107 see DVL600
RCRA WASTE NUMBER U108 see DVQ000
RCRA WASTE NUMBER U109 see HHG000
RCRA WASTE NUMBER U110 see DWR000
RCRA WASTE NUMBER U111 see NKB700
RCRA WASTE NUMBER U112 see EFR000
RCRA WASTE NUMBER U113 see EFT000
RCRA WASTE NUMBER U115 see EJN500
RCRA WASTE NUMBER U116 see IAQ000
RCRA WASTE NUMBER U117 see EJU000
RCRA WASTE NUMBER U118 see EMF000
RCRA WASTE NUMBER U119 see EMF500
RCRA WASTE NUMBER U120 see FDF000
RCRA WASTE NUMBER U121 see TIP500
RCRA WASTE NUMBER U122 see FMV000
RCRA WASTE NUMBER U123 see FNA000
RCRA WASTE NUMBER U124 see FPK000
RCRA WASTE NUMBER U125 see FPQ875
RCRA WASTE NUMBER U126 see GGW000
RCRA WASTE NUMBER U127 see HCC500
RCRA WASTE NUMBER U128 see HCD250
RCRA WASTE NUMBER U129 see BBQ500
RCRA WASTE NUMBER U130 see HCE500
RCRA WASTE NUMBER U131 see HCI000
RCRA WASTE NUMBER U132 see HCL000
RCRA WASTE NUMBER U133 see HGS000
RCRA WASTE NUMBER U134 see HHU500
RCRA WASTE NUMBER U135 see HIC500
RCRA WASTE NUMBER U136 see HKC000
RCRA WASTE NUMBER U137 see IBZ000
RCRA WASTE NUMBER U138 see MKW200
RCRA WASTE NUMBER U139 see IGS000
RCRA WASTE NUMBER U140 see IIL000
RCRA WASTE NUMBER U141 see IRZ000
RCRA WASTE NUMBER U142 see KEA000
RCRA WASTE NUMBER U143 see LBG000
RCRA WASTE NUMBER U144 see LCG000
RCRA WASTE NUMBER U146 see LCH000
RCRA WASTE NUMBER U147 see MAM000
RCRA WASTE NUMBER U149 see MAO250
RCRA WASTE NUMBER U150 see PED750
RCRA WASTE NUMBER U151 see MCW250
RCRA WASTE NUMBER U152 see MGA750
RCRA WASTE NUMBER U153 see MLE650
RCRA WASTE NUMBER U154 see MGB150
RCRA WASTE NUMBER U155 see TEO250
RCRA WASTE NUMBER U156 see MIG000
RCRA WASTE NUMBER U157 see MIJ750
RCRA WASTE NUMBER U158 see MJM200
RCRA WASTE NUMBER U159 see MKA400
RCRA WASTE NUMBER U160 see MKA500

RCRA WASTE NUMBER U161 see HFG500
RCRA WASTE NUMBER U162 see MLH750
RCRA WASTE NUMBER U163 see MMP000
RCRA WASTE NUMBER U164 see MPW500
RCRA WASTE NUMBER U165 see NAJ500
RCRA WASTE NUMBER U166 see NBA500
RCRA WASTE NUMBER U167 see NBE000
RCRA WASTE NUMBER U168 see NBE500
RCRA WASTE NUMBER U169 see NEX000
RCRA WASTE NUMBER U170 see NIF000
RCRA WASTE NUMBER U171 see NIY000
RCRA WASTE NUMBER U172 see BRY500
RCRA WASTE NUMBER U173 see NKM000
RCRA WASTE NUMBER U174 see NJW500
RCRA WASTE NUMBER U176 see ENV000
RCRA WASTE NUMBER U177 see MNA750
RCRA WASTE NUMBER U178 see MMX250
RCRA WASTE NUMBER U179 see NLJ500
RCRA WASTE NUMBER U180 see NLP500
RCRA WASTE NUMBER U181 see NMP500
RCRA WASTE NUMBER U182 see PAI250
RCRA WASTE NUMBER U184 see PAW500
RCRA WASTE NUMBER U185 see PAX000
RCRA WASTE NUMBER U187 see ABG750
RCRA WASTE NUMBER U188 see PDN750
RCRA WASTE NUMBER U189 see PHS000
RCRA WASTE NUMBER U190 see PHW750
RCRA WASTE NUMBER U191 see MOY000
RCRA WASTE NUMBER U192 see DTT600
RCRA WASTE NUMBER U193 see PML400
RCRA WASTE NUMBER U194 see PND250
RCRA WASTE NUMBER U196 see POP250
RCRA WASTE NUMBER U197 see QQS200
RCRA WASTE NUMBER U201 see REA000
RCRA WASTE NUMBER U202 see BCE500
RCRA WASTE NUMBER U206 see SMD000
RCRA WASTE NUMBER U208 see TBQ000
RCRA WASTE NUMBER U209 see TBQ100
RCRA WASTE NUMBER U210 see PCF275
RCRA WASTE NUMBER U211 see CBY000
RCRA WASTE NUMBER U212 see TBT000
RCRA WASTE NUMBER U213 see TCR750
RCRA WASTE NUMBER U214 see TEI250
RCRA WASTE NUMBER U216 see TEJ250
RCRA WASTE NUMBER U217 see TEK750
RCRA WASTE NUMBER U218 see TFA000
RCRA WASTE NUMBER U219 see ISR000
RCRA WASTE NUMBER U220 see TGK750
RCRA WASTE NUMBER U221 see TGL750
RCRA WASTE NUMBER U222 see TGS500
RCRA WASTE NUMBER U223 see TGM740, TGM750
RCRA WASTE NUMBER U225 see BNL000
RCRA WASTE NUMBER U226 see MIH275
RCRA WASTE NUMBER U227 see TIN000
RCRA WASTE NUMBER U228 see TIO750
RCRA WASTE NUMBER U230 see TIV750
RCRA WASTE NUMBER U231 see TIW000
RCRA WASTE NUMBER U232 see TAA100
RCRA WASTE NUMBER U233 see TIX500
RCRA WASTE NUMBER U234 see TMK500
RCRA WASTE NUMBER U235 see TNC500
RCRA WASTE NUMBER U236 see CMO250
RCRA WASTE NUMBER U237 see BIA250
RCRA WASTE NUMBER U238 see UVA000
RCRA WASTE NUMBER U239 see XGS000
RCRA WASTE NUMBER U240 see DAA800
RCRA WASTE NUMBER U242 see PAX250
RCRA WASTE NUMBER U244 see TFS350
RCRA WASTE NUMBER U246 see COO500
RCRA WASTE NUMBER U247 see MEI450
RD 406 see DGB000
RD 1572 see DBV400
RD 2195 see CEP000
R.D. 2786 see WAK000

RD 4593 see CIR500
RDGE see REF000
RDX see CPR800
RE-4355 see NAG400
RE 5655 see MOU750
RE 12420 see DOP600
REACID see COH250
REACTIVE BLUE 19 see BMM500
REALGAR see ARF000
REANIMIL see DNV000
REAZID see COH250
REAZIDE see COH250
REBELATE see DSP400
REBRAMIN see VSZ000
REC 7/0267 see DNV000
RECOLITE RED LAKE C see CHP500
RECONOX see PDP250
RECORDATI see BBK500
RECTHORMONE OESTRADIOL see EDP000
RECTHORMONE TESTOSTERONE see TBG000
RECTODELT see PLZ000
RECTOFASA see BRF500
RECTULES see CDO000
RED #14 see HJF500
1427 RED see FAG040
1671 RED see FAG040
1860 RED see CHP500
11411 RED see FAG070
11445 RED see BNH500
11554 RED see IHD000
11959 RED see HJF500
REDAMINA see VSZ000
REDAX see DWI000
RED B see XRA000
RED 2G BASE see NEO500
RED BASE CIBA IX see CLK220, CLK235
RED BASE IRGA IX see CLK220, CLK235
RED BASE NTR see CLK220
REDDON see TAA100
REDDOX see TAA100
RED FUMING NITRIC ACID see NEE500
REDIFAL see SNN300
REDI-FLOW see BAP000
RED IRON ORE see HAO875
RED IRON OXIDE see IHD000
REDISOL see VSZ000
RED KB BASE see CLK225
RED LEAD see LDS000
RED LEAD CHROMATE see LCS000
RED LEAD OXIDE see LDS000
RED MERCURIC IODIDE see MDD000
RED No. 5 see XRA000
RED NO 213 see FAG070
RED OCHRE see IHD000
RED OIL see OHU000
RED OXIDE of MERCURY see MCT500
RED PRECIPITATE see MCT500
RED SALT CIBA IX see CLK235
RED SALT IRGA IX see CLK235
RED SCARLET see CHP500
RED-SEAL-9 see ZKA000
REDSKIN see AGJ250
RED SQUILL see RCF000
RED TR BASE see CLK220
RED TRS SALT see CLK235
REDUCED-d-PENICILLAMINE see MCR750
REDUCTO see DKE800
REDUCTONE see SHR500
REED LV 2,4-D see ILO000
REED LV 400 2,4-D see ILO000
REED LV 600 2,4-D see ILO000
REFINED SOLVENT NAPHTHA see PCT250
REFORMIN see DJS200
REFRIGERANT 12 see DFA600

REFRIGERANT 22 see CFX500
REFRIGERANT 112 see TBP050
REFRIGERANT 112a see TBP000
REFRIGERANT 113 see FOO000
REFUSAL see DXH250
REGLON see DWX800
REGLONE see DWX800
REGONOL see GLU000
REGUTOL see DJL000
REHORMIN see DJS200
REICHSTEIN'S F see MIW500
REIN GUARIN see GLU000
REISE-ENGLETTEN see DYE600
REKAWAN see PLA500
RELA see IPU000
RELACT see DLY000
RELASOM see IPU000
RELAX see IPU000
RELAXAN see PDD300
RELAXANT see GGS000
RELAXAR see GGS000
RELDAN see CMA250
RELIBERAN see MDQ250
RELICOR see DHS200
RELICOR HYDROCHLORIDE see DHS200
RELON P see PJY500
RELUTIN see HNT500
REMALAN BRILLIANT BLUE R see BMM500
REMASAN CHLOROBLE M see MAS500
REMAZIN see HNI500
REMAZOL BRILLIANT BLUE R see BMM500
REMEFLIN see DNV000
REMICYCLIN see TBX250
RENAFUR see NDY000
RENAGLADIN see VGP000
RENAL EG see ALT500
RENALEPTINE see VGP000
RENALINA see VGP000
RENAL MD see TGL750
RENAL PF see PEY500
RENAL SLA see DBO400
RENARCOL see GGS000
RENARDIN see DMX200
RENARDINE see DMX200
RENOFORM see VGP000
RENOLBLAU 3B see CMO250
REN O-SAL see HMY000
RENOSTYPRICIN see VGP000
RENOSTYPTIN see VGP000
RENSTAMIN see DBM800
REOMAX see DFP600
REOMOL DOP see DVL700
REOMOL D 79P see DVL700
REPAIRSIN see PKB500
REPARIL SODIUM SALT see EDM000
REPEL see DKC800
REPELLENT 612 see EKV000
REPHOXITIN see CCS500
REPOCAL see CAV000
REPPER-DET see DKC800
REPRISCAL see SHL500
REPROMIX see MCA000
REPROTEROL HYDROCHLORIDE see DNA600
REPUDIN-SPECIAL see DKC800
REQUTOL see DJL000
RESARIT 4000 see PKB500
RESCUE SQUAD see SHF500
RESERPINE see RDK000
RESIDUAL OIL SOLVENT EXTRACT see MQV863
RESIDUAL OILS (PETROLEUM), ACID-TREATED (9CI) see
  MQV872
RESIN (solution) see RDP000
RESIN, solution, in flammable liquid (DOT) see RDP000
RESIN, solution (resin compound, liquid) (DOT) see RDP000

RESINOL ORANGE R see PEJ500
RESINOL YELLOW GR see DOT300
RESIN SCARLET 2R see XRA000
RESISTAMINE see TMP750
RESLOOM M 75 see HDY000
RESMETHRIN see BEP500
RESMETRINA (PORTUGUESE) see BEP500
RESOBANTIN see XCJ000
RESOFORM ORANGE G see PEJ500
RESOFORM ORANGE R see XRA000
RESOFORM YELLOW GGA see DOT300
RESORCIN see REA000
RESORCINE see REA000
RESORCINE BROWN J see XMA000
RESORCINE BROWN R see XMA000
RESORCINOL see REA000
RESORCINOL BIS(2,3-EPOXYPROPYL)ETHER see REF000
RESORCINOL DIGLYCIDYL ETHER see REF000
RESORCINOLPHTHALEIN see FEV000
RESORCINOL PHTHALEIN SODIUM see FEW000
RESORCINYL DIGLYCIDYL ETHER see REF000
RESOTROPIN see HEI500
RESPAIRE see ACH000
RESPIFRAL see DMV600
RESTAMIN see BBV500
RESTAMINE see BBV500
RESTENIL see MQU750
REST-ON see TEO250
RESTOVAR see LJE000
RESTROL see DAL600
RESTRYL see TEO250
RETALON see DAL600
RETARDER AK see PHW750
RETARDER BA see BCL750
RETARDER ESEN see PHW750
RETARDER J see DWI000
RETARDER PD see PHW750
RETARDER W see SAI000
RETARDEX see BCL750
RETENSIN see PDD300
RETINOL see VSK600
all-trans RETINOL see VSK600
RETINOL ACETATE see VSK900
RETINOL PALMITATE see VSP000
RETINYL ACETATE see VSK900
all-trans-RETINYL ACETATE see VSK900
RETINYL PALMITATE see VSP000
RETOZIDE see ILD000
cis-RETRONECIC ACID ESTER of RETRONECINE see
  RFP000
cis-RETRONECIC ACID ESTER of RETRONECINE-N-OXIDE
  see RFU000
RETRORSINE see RFP000
RETRORSINE-N-OXIDE see RFU000
RETROVITAMIN A see VSK600
REUDO see BRF500
REUDOX see BRF500
REUMACIDE see IDA000
REUMASYL see BRF500
REUMAZIN see BRF500
REUMAZOL see BRF500
REUMOX see HNI500
REUPOLAR see BRF500
REVAC see DJL000
REVENGE see DGI400
REVIDEX see BBK500
REVONAL see QAK000
REWOMID DLMS see BKE500
REWOPOL HV-9 see PKF000
REWOPOL NLS 30 see SIB600
REXALL 413S see PMP500
REXENE see PMP500
REXENE 106 see ADY500
REXOCAINE see BQA010

REXOLITE 1422 see SMQ500
REX REGULANS see GGS000
REZIFILM see TFS350
RFNA see NEE500
R-GENE see AQW000
RGH-1106 see PII250
RH-124 see BPU000
RH 315 see DTT600
RHENIUM see RGF000
RHENIUM TRICHLORIDE see RGP000
RHENOCURE CA see DWN800
RHEONINE B see FAG070
RHEOSMIN see RBU000
RHEUMIN TABLETTEN see ADA725
RHINAZINE see NAH500
RHIZOCTOL see MGQ750
RHIZOPIN see ICN000
RHODACRYST see VSZ000
RHODAMINE see FAG070
RHODAMINE 6G (biological stain) see RGW000
RHODAMINE 6GEX ETHYL ESTER see RGW000
RHODAMINE S (RUSSIAN) see FAG070
RHODAMINE 6G EXTRA BASE see RGW000
RHODANID see ANW750
RHODANIDE see ANW750
RHODIA see DAA800
RHODIA-6200 see DJA400
RHODIACHLOR see HAR000
RHODIACID see BJK500
RHODIACIDE see EEH600
RHODIA RP 11974 see BDJ250
RHODIASOL see PAK000
RHODIATOX see PAK000
RHODIATROX see PAK000
RHODINE see ADA725
RHODINOL see CMT250
RHODINOL (FCC) see DTF400
RHODINOL ACETATE see RHA000
RHODINYL ACETATE see RHA000
RHODIUM see RHF000
RHODIUM CHLORIDE see RHK000
RHODIUM(III) CHLORIDE (1:3) see RHK000
RHODIUM TRICHLORIDE see RHK000
RHODOCIDE see EEH600
RHODOLNE see SMQ500
RHODOPAS 6000 see AAX175
RHODOPAS M see AAX250
RHODULINE ORANGE see BJF000
RHOMBIC see ELC500
RHOMENE see CIR250
RHONOX see CIR250
RHOPLEX AC-33 (ROHM and HAAS) see EMF000
RHOPLEX B 85 see PKB500
RHOTHANE see BIM500
RHOTHANE D-3 see BIM500
RHYUNO OIL see SAD000
RIANIL see CLO750
RIBIPCA see RIK000
RIBODERM see RIK000
RIBOFLAVIN see RIK000
RIBOFLAVINE see RIK000
RIBOFLAVINEQUINONE see RIK000
RIBOFURANOSIDE, 9H-PURINE-6-THIOL-9 see MCQ500
β-d-RIBOSYL-6-METHYLTHIOPURINE see MPU000
RIBOSYL-6-THIOPURINE see MCQ500
RICHAMIDE 6310 see BKE500
RICHONATE 1850 see DXW200
RICHONOL C see SIB600
RICIFON see TIQ250
RICIN (HAITI) see CCP000
RICINO (PUERTO RICO) see CCP000
RICINUS COMMUNIS see CCP000
RICINUS OIL see CCP250
RICIRUS OIL see CCP250

RICKETON see CMC750
RICON 100 see SMR000
RICYCLINE see TBX250
RIFA see RKP000
RIFADINE see RKP000
RIFAGEN see RKP000
RIFALDAZINE see RKP000
RIFALDIN see RKP000
RIFAMATE see ILD000, RKP000
RIFAMPICIN see RKP000
RIFAMPICINE (FRENCH) see RKP000
RIFAMPICINUM see RKP000
RIFAMPIN see RKP000
RIFAMYCIN AMP see RKP000
RIFAPRODIN see RKP000
RIFINAH see RKP000
RIFOBAC see RKP000
RIFOLDIN see RKP000
RIFORAL see RKP000
RIGENICID see EPQ000
RIGETAMIN see EDC500
RIGIDIL see BBV500
RIGIDYL see BBV500
RIKEMAL S 250 see SKV150
RIKER 601 see TND000
RIMACTAN see RKP000
RIMACTAZID see RKP000
RIMICID see ILD000
RIMIDIN see FAK100
RIMIFON see ILD000
RIMITSID see ILD000
RIMSO-50 see DUD800
RINDEX see MDO250
RIOMITSIN see HOH500
RIP-15830 see DNG200
RIPENTHOL see DXD000
RIPERCOL-L see LFA020
RIRILIM see BBW500
RIRIPEN see BBW500
RISELECT see DGI000
RISTAT see HMY000
RITALIN see MNQ000
RITALINE see MNQ000
RITCHER WORKS see MNQ000
RITMENAL see DKQ000
RITOSEPT see HCL000
RITSIFON see TIQ250
RIVADORM see NBT500
RIVADORN see NBU000
RIVIVOL see ILE000
RIVOMYCIN see CDP250
RIVOTRIL see CMW000
RL-50 see MRL500
RMI 10,482A see MHJ500
RMI9,384A see DLS600
RO 1-5431 see MDV250, MKR250
RO 2-3599 see DLP000
RO 2-4572 see ILE000
RO 2-9757 see FMM000
RO 4-2130 see SNK000
RO 4-5360 see DLY000
RO 4-6316 see DNA200
RO 4-6467 see PME250, PME500
RO 4-6861 see CIF250
RO 4-8180 see CMW000
RO 5-0360 see DAR400
RO 5-0690 see MDQ250
RO 5-0831 see IKC000
RO 5-3059 see DLY000
RO 5-4023 see CMW000
RO 5-6901 see DAB800
RO 7-1554 see IGH000
Ro 215535 see DMJ400
Ro-5-6901/3 see FMQ000

ROACH SALT see SHF500
ROAD ASPHALT see PCR500
ROAD ASPHALT (DOT) see ARO500, ARO750
ROAD TAR (DOT) see ARO500
ROAD TAR, liquid (DOT) see ARO750
RO-AMPEN see AIV500, AOD125
ROBAMATE see MQU750
ROBANUL see GIC000
ROBENIDINE see RLK890
ROBIMYCIN see EDH500
ROBINUL see GIC000
ROBISELIN see ILD000
ROBISELLIN see ILD000
ROBITET see TBX000
ROBIZON-V see BRF500
ROBORAL see PAN100
ROCALTROL see DMJ400
ROCHIPEL see DIR000
ROCIPEL see DIR000
ROCK CANDY see SNH000
ROCK OIL see PCR250
ROCK SALT see SFT000
ROCORNAL see DIO200
RO-CYCLINE see TBX250
RODANIN S-62 (CZECH) see IAQ000
RO-DETH see WAT200
RODEX see FFF000, SMN500
RODINE see RCF000
RODINOL see CMT250
RODIPAL see DIR000
RODOCID see EEH600
ROE 101 see DBB200
ROERIDORM see CHG000
ROGODIAL see DRR400, DSP400
ROGOR see DSP400
ROGUE see DGI000
ROHYDRA see BAU750
RO-HYDRAZIDE see CFY000
RO-KO see RNZ000
ROKON see BDF000
ROLAMID CD see BKE500
ROLAZINE see HGP500
ROLL-FRUCT see CDS125
ROMACRYL see PKB500
ROMAN VITRIOL see CNP250
ROMERGAN see DQA400
ROMETIN see CHR500
ROMOSOL see ART250
ROMPARKIN see BBV000
ROMPHENIL see CDP250
ROMPUN see DMW000
ROMULGIN O see PKL100
RONDOMYCIN see MDO250
RONILAN see RMA000
RONNEL see RMA500
RONONE see RNZ000
RONTON see ENG500
ROOTONE see ICP000
ROP 500 F see GIA000
ROPTAZOL see NGG500
ROQUESSINE DIHYDROBROMIDE see DOX000
RORASUL see ASB250
RORER 148 see QAK000
ROSANIL see DGI000
ROSANILINE BASE see MAC500
p-ROSANILINE HCL see RMK020
p-ROSANILINE HYDROCHLORIDE see RMK020
ROSCOSULF see SNN300
ROSE ETHER see PER000
ROSE GERANIUM OIL ALGERIAN see GDA000
ROSEMARIE OIL see RMU000
ROSEMARY OIL see RMU000
ROSEMIDE see CHJ750
ROSEN OEL (GERMAN) see RNA000

ROSENSTHIEL see MAT250
ROSE OIL see RNA000
ROSE QUARTZ see SCI500, SCJ500
ROSMARIN OIL (GERMAN) see RMU000
RO-SULFIRAM see DXH250
ROTATE see DQM600
ROTAX see BDF000
ROT B see XRA000
ROTEFIVE see RNZ000
ROTEFOUR see RNZ000
ROTENONA (SPANISH) see RNZ000
ROTENONE see RNZ000
ROTERSEPT see BIM250
ROTESSENOL see RNZ000
ROT GG FETTLOESLICH see XRA000
ROTHANE see BIM500
ROTOCIDE see RNZ000
ROTOX see MHR200
ROUGE see IHD000
ROUGE CERASINE see OHA000
ROUGH & READY MOUSE MIX see WAT200
ROUGH & READY RAT BAIT & RAT PASTE see RCF000
ROUGOXIN see DKN400
ROUQUALONE see QAK000
ROVRAL see GIA000
ROXARSONE (USDA) see HMY000
ROXIFEN see ILD000
ROXION U.A. see DSP400
ROYAL BLUE see DJO000
ROYAL MBTS see BDE750
ROYALTAC see DAI600
ROYAL TMTD see TFS350
ROZOL see CJJ000
ROZTOZOL see CKM000
2512 R.P. see DBL800
R.P. 2512 see DBL800
R.P. 2591 see DFX400
RP 2786 see WAK000
RP 2990 see TEX250
3277 RP see DQA400
RP 3356 see DIR000
3389 R.P. see DQA400
RP 3554 see MRW000
RP 3602 see GGS000
RP 3697 see PDD300
4182 R.P. see DQA400
RP 4632 see MFK500
4753 R.P. see TGD000
RP5171 see DIG400
6140 RP see PMF500
6909 RP see PIW000
RP7293 see VRF000
RP 8167 see EEH600
RP 8228 see LFO000
8595 R.P. see DSV800
RP 8823 see MMN250
RP 8908 see PIW000
RP 9159 see LEO000
9159 RP see LEO000
RP 9965 see MQR000
RP 10192 see MIJ500
10257 R.P. see TND000
RP 13057 see DAC000
13,057 R.P. see DAC000
RP 26019 see GIA000
R-PENTINE see CPU500
2786 R.P. MALEATE see DBM800
RS 141 see CJJ250
RS 1280 see CBF250
RTEC (POLISH) see MCW250
RU-4723 see CIR750
RUBATONE see BRF500
"522" RUBBER ACCELERATOR see PIY500
RUBBER SOLVENT see ROU000

RUBEANE see DXO200
RUBEANIC ACID see DXO200
RUBENS BROWN see MAT500
RUBESOL see VSZ000
RUBIAZOL A see SNM500
RUBIDIUM see RPA000
RUBIDIUM CHLORIDE see RPF000
RUBIDIUM DICHROMATE see RPK000
RUBIDIUM FLUORIDE see RPP000
RUBIDIUM HYDROXIDE see RPZ000
RUBIDIUM HYDROXIDE, solid and solution (DOT) see
  RPZ000
RUBIDIUM METAL (DOT) see RPA000
RUBIDIUM METAL, IN CARTRIDGES (DOT) see RPA000
RUBIDOMYCIN see DAC000
RUBIDOMYCINE see DAC000
RUBIGAN see FAK100
RUBIGO see IHD000
RUBINATE 44 see MJP400
RUBINATE TDI see TGM740
RUBINATE TDI 80/20 see TGM740, TGM750
RUBITOX see BDJ250
RUBOMYCIN C see DAC000
RUBOMYCIN C 1 see DAC000
RUBRAMIN see VSZ000
RUBRIPCA see VSZ000
RUBROCITOL see VSZ000
RUBRUM SCARLATINUM see SBC500
RUCAINA see DHK400
RUCAINA HYDROCHLORIDE see DHK600
RUCON B 20 see PKQ059
RUELENE see COD850
RUELENE DRENCH see COD850
RUELENE 25E see COD850
RUKSEAM see DAD200
RULENE see COD850
RUMAPAX see HNI500
RUMESTROL 1 see DKA600
RUMESTROL 2 see DKA600
RUMETAN see ZLS000
RUN see PDN000
RUNA RH20 see TGG760
RUNCATEX see CIR500
RUSSIAN COMFREY ROOTS see RRP000
RUTGERS 612 see EKV000
RUTHENIUM see RRU000
RUTHENIUM CHLORIDE see RRZ000
RUTHENIUM COMPOUNDS see RSF000
RUTHENIUM TRICHLORIDE see RRZ000
RUTILE see TGG760
RVK see DXO200
RYANEXEL see RSZ000
RYANIA see RSZ000
RYANIA POWDER see RSZ000
RYANIA SPECIOSA see RSZ000
RYANICIDE see RSZ000
RYANODINE see RSZ000
RYOMYCIN see HOH500

S 13 see BHD250
S-33 see CIF250
S51 see BBV500
S 112A see DSQ000
S115 see NCQ900
S 140 see DAM700
S 151 see EJM500
S 154 see BPR500
S 202 see DHK600
S 276 see DXH325
S410 see DSK600
S 650 see BQH250
S 767 see FAQ800
S-805 see DCS200
S-847 see CEW500

S 940 see HMV000
S 1544 see PFC750
S 1752 see FAQ999
S 2225 see EGV500
S 2940 see DRR400
S-3151 see AHJ750
S 5602 see FAR100
S 5660 see DSQ000
S 6900 see DRR200
S 10165 see DGI000
S 65 (polymer) see PKQ059
SA see SAI000
SA 111 see SNJ000
SAATBEIZFUNGIZID (GERMAN) see HCC500
SABACIDE see VHZ000
SABADILLA see VHZ000
SABANE DUST see VHZ000
SABARI see HGC500
SACARINA see BCE500
SACCAHARIMIDE see BCE500
SACCHARATED FERRIC OXIDE see IHG000
SACCHARATED IRON see IHG000
SACCHARIN see SJN700
SACCHARINA see BCE500
SACCHARIN ACID see BCE500
SACCHARIN CALCIUM see CAM750
SACCHARINE see BCE500
SACCHARINE SOLUBLE see SJN700
SACCHARINNATRIUM see SJN700
SACCHARINOL see BCE500
SACCHARINOSE see BCE500
SACCHARIN, SODIUM see SJN700
SACCHARIN, SODIUM SALT see SJN700
SACCHARIN SOLUBLE see SJN700
SACCHAROIDUM NATRICUM see SJN700
SACCHAROL see BCE500
SACCHAROSE see SNH000
SACCHARUM see SNH000
SACCHARUM LACTIN see LAR000
SACERIL see DKQ000, DNU000
SACERNO see MKB250
SACHSISCHBLAU see FAE100
SADH see DQD400
SA 97 DIHYDROCHLORIDE see CJR809
SADOFOS see MAK700
SADOPHOS see MAK700
SADOPLON see TFS350
SADOREUM see IDA000
SAEURE FLUORIDE (GERMAN) see FEZ000
SAFFLOWER OIL see SAC000
SAFFLOWER OIL (UNHYDROGENATED) (FCC) see
  SAC000
SAFFRON YELLOW see DUX800
SAFROL see SAD000
SAFROLE see SAD000
SAFROLE MF see SAD000
SAFROTIN see MKA000
SAFSAN see DXE000
SAGATAL see NBU000
SAGE OIL see SAE500, SAE550
SAGE OIL, DALMATIAN TYPE see SAE500
SAGE OIL, SPANISH TYPE see SAE550
SAH 22 see SEM500
SAISAN see MLC250
SAKOLYSIN (GERMAN) see BHT750
SAKURAI No. 864 see YCJ000
SALACETIN see ADA725
SALACHLOR see SHJ000
SALAMID see SAH000
SALAMIDE see SAH000
SAL AMMONIA see ANE500
SAL AMMONIAC see ANE500
SALBEI OEL (GERMAN) see SAE500, SAE550
SALBUTAMOL see BQF500

SALCETOGEN see ADA725
SAL ENIXUM see PKX750
SALETIN see ADA725
SALICILAMIDE (ITALIAN) see SAH000
SALICIM see SAH000
SALICYLAMID see SAH000
SALICYLAMIDE see SAH000
SALICYL-DIAETHYL (GERMAN) see BQH250
SALICYLIC ACID see SAI000
SALICYLIC ACID, FLUOROACETATE see FFT000
SALICYLIC ACID, ISOBUTYL ESTER see IJN000
SALICYLIC ACID, ISOPENTYL ESTER see IME000
SALICYLIC ACID ISOPROPYL ESTER O-ESTER with O-
    ETHYL ISOPROPYLPHOSPHORAMIDOTHIOATE see
    IMF300
SALICYLIC ACID, METHYL ESTER see MPI000
SALICYLIC ACID with PHYSOSTIGMINE (1:1) see PIA750
SALICYLIC ACID, SODIUM SALT see SJO000
SALICYLIC ETHER see SAL000
SALICYLIC ETHYL ESTER see SAL000
SALINE see SFT000
SALIPUR see SAH000
SALISOD see SJO000
SALITHION see MEC250
SALITHION-SUMITOMO see MEC250
SALIX see CHJ750
SALIZELL see SAH000
SALOL see PGG750
SALPETERSAURE (GERMAN) see NED500
SALPETERZUUROPLOSSINGEN (DUTCH) see NED500
SALRIN see SAH000
SALSONIN see SJO000
SALT see SFT000
SALT CAKE see SJY000
SALT OF TARTER see PLA250
SALTPETER see PLL500
SALT of SATURN see LCG000
SALUFER see DXE000
SALURIN see SIH500
SALVACARD see DJS200
SALVACORIN see DJS200
SALVARSAN see SAP500
SALVO see DAA800, HKC000
SALVO LIQUID see BCL750
SALVO POWDER see BCL750
SALYMID see SAH000
SALYRGAN see SIH500
SALYZORON see BBW500
SAM see SAH000
SAMARIUM see SAQ500
SAMARIUMACETAT (GERMAN) see SAR000
SAMARIUM ACETATE see SAR000
SAMARIUM(III) CHLORIDE see SAR500
SAMARIUM NITRAT (GERMAN) see SAT000
SAMARIUM(III) NITRATE, HEXAHYDRATE (1:3:6) see
    SAT000
SAMID see SAH000
SAN 244 I see DRR200
SAN 6538 I see DJY200
SAN 6913 I see DRR200
SAN 7107 I see DRR200
SAN 230 see PHI500
SAN 52 139 I see MKA000
SANASEED see SMN500
SANATRICHOM see MMN250
SANCLOMYCINE see TBX000
SAND see SCI500, SCJ500
SAND ACID see SCO500
SANDALWOOD OIL, EAST INDIAN see OHG000
SANDIX see LDS000
SANDOCRYL BLUE BRL see BJI250
SANDOLIN see DUS700
SANDOPEL BLACK EX see AQP000
SANDORMIN see TEH500

SANDOZ 6538 see DJY200
SANDOZ 52139 see MKA000
SANEDRINE see EAW000
SANEPIL see DKQ000
SANFURAN see NGE500
SANG gamma see BBQ500
SANICLOR 30 see PAX000
SANITIZED SPG see ABU500
SANLOSE SN 20A see SFO500
SANMARTON see FAR100
SANMORIN OT 70 see DJL000
SANOCHRYSINE see GJG000
SANOCIDE see HCC500
SANOFLAVIN see DBN400
SANOHIDRAZINA see ILD000
SANOMA see IPU000
SANORIN see NAH500
SANPRENE LQX 31 see PKP000
SANQUINON see DFT000
SANREX see ADY500
SANSDOLOR see GGS000
SANSEL ORANGE G see PEJ500
SANSERT see MQP500
SANSPOR see CBF800
α-SANTALOL (FCC) see OHG000
SANTALYL ACETATE see SAU400
SANTAR see MCT500
SANTAVY'S SUBSTANCE F see MIW500
SANTHEOSE see TEO500
SANTICIZER 160 see BEC500
SANTICIZER 141 (MONSANTO) see DWB800
SANTICIZIER B-16 see BQP750
SANTOBANE see DAD200
SANTOBRITE see SJA000, PAX250
SANTOCEL see SCH000
SANTOCHLOR see DEP800
SANTOCURE MOR see BDG000
SANTOFLEX A see SAV000
SANTOFLEX AW see SAV000
SANTOFLEX IC see PEY500
SANTOMERSE 3 see DXW200
SANTONOX see TFC600
SANTOPHEN see PAX250
SANTOPHEN 20 see PAX250
SANTOQUIN see SAV000
SANTOQUINE see SAV000
SANTOTHERM see PJL750
SANTOWAX see TBC750
SANTOWAX M see TBC620
SANTOX see EBD700
SANYO FAST BLUE SALT B see DCJ200
SANYO FAST RED SALT TR see CLK235
SANYO FAST RED TR BASE see CLK220
SANYO LAKE RED C see CHP500
SAOLAN see DSK200
SAPECRON see CDS750
SAPONIN-GYPSOPHILA see GMG000
SAPPILAN see CJT750
SAPPIRAN see CJT750
SARCELL TEL see SFO500
SARCLEX see DGD600
SARCOCLORIN see BHT750
l-SARCOLYSIN see PED750
dl-SARCOLYSIN see BHT750
p-l-SARCOLYSIN see PED750
o-dl-SARCOLYSIN see BHT250
dl-SARCOLYSINE see BHT750
l-SARCOLYSINE HYDROCHLORIDE see BHV250
dl-SARCOLYSINE HYDROCHLORIDE see BHV000
SARCOLYSIN HYDROCHLORIDE see BHV000
SARCOMYCIN see SAX500
SARIN see IPX000
SARIN II see IPX000
SARKOKLORIN see BHV000

SARKOMYCIN see SAX500
SARODORMIN see DYC800
SAROLEX see DCM750
SAROTEN see EAI000
SAROTENE see EAI000
SARPIFAN HP 1 see AAX175
SASSAFRAS see SAY900
SASSAFRAS ALBIDUM see SAY900
SATECID see CHS500
SATIAGEL GS 350 see CCL250
SATIAGUM 3 see CCL250
SATIAGUM STANDARD see CCL250
SATINITE see CAX750
SATIN SPAR see CAX750
SATOX 20WSC see TIQ250
SATURN BROWN LBR see CMO750
SATURN RED see LDS000
SAURE DES PHYTINS (GERMAN) see PIB250
SAUTERALGYL see DAM700
SAUTERAZID see ILD000
SAUTERZID see ILD000
SAVENTRINE see DMV600
SAVORY OIL (SUMMER VARIETY) see SBA000
SAX see SAI000
SAXIN see BCE500, SJN700
SAXOL see MQV750
SAZZIO see AFL000
SB-23 see DQY909
S.B.A. see BPW750
SBO see DJL000
SBP-1382 see BEP500
SBP-1513 see AHJ750
S.B. PENICK 1382 see BEP500
SBS see SMR000
SC-110 see ABU500
SC-1950 see DRK600
SC 2538 see DIR000
SC 2910 see XCJ000
SC-4642 see EEH550
SC 10295 see MMN250
SC10363 see VTF000
SC 15090 see TEO500
SCABANCA see BCM000
SCALDIP see DVX800
SCANBUTAZONE see BRF500
SCANDISIL see SNN300
SCANDIUM see SBB500
SCANDIUM CHLORIDE see SBC000
SCANDIUM (3+) CHLORIDE see SBC000
SCARCLEX see DGD600
SCARLET BASE CIBA II see NMP500
SCARLET RED see SBC500
SCATOLE see MKV750
Sch 1000 see IGG000
Sch 4831 see BFV750
SCH 6673 see ABG000
SCH 7307 see IPU000
SCH 9724 see GCS000
SCH 10649 see DLV800
SCHARLACH B see PEJ500
SCH CHLORIDE see HLC500
SCHEELES GREEN see CNN500
SCHEELE'S MINERAL see CNN500
SCHEMERGIN see BRF500
SCHERING 36056 see DSO200
SCHERING 36268 see CJJ250
SCHERISOLON see PMA000
SCHINOPSIS LORENTZII TANNIN see QBJ000
SCHLAFEN see CAV000
SCHRADAN see OCM000
SCHRADANE (FRENCH) see OCM000
SCHULTENITE see LCK000
SCHULTZ No. 39 see HGC000
SCHULTZ No. 770 see FMU059

SCHULTZ No. 1038 see BJI250
SCHULTZ Nr. 208 (GERMAN) see HJF500
SCHULTZ Nr. 1309 (GERMAN) see FAE100
SCHULTZ-TAB No. 779 (GERMAN) see RMK020
SCHWEFELDIOXYD (GERMAN) see SOH500
SCHWEFELKOHLENSTOFF (GERMAN) see CBV500
SCHWEFEL-LOST see BIH250
SCHWEFELSAEURELOESUNGEN (GERMAN) see SOI500
SCHWEFELWASSERSTOFF (GERMAN) see HIC500
SCHWEFLIGE SAURE (GERMAN) see SOO500
SCHWEINFURTERGRUN see COF500
SCHWEINFURT GREEN see COF500
SCILLAGLYKOSID A (GERMAN) see GFC000
SCILLAREN A see GFC000
SCILLARENIN-3,6-DEOXY-4-o-β-d-GLUCOPYRANOSYL-α-l-
    MANNOPYRANOSIDE see GFC000
SCILLAREN & RHAMNOSE & GLUCOSE (GERMAN) see
    GFC000
SCILLIROSIDE GLYCOSIDE see RCF000
3-β-SCILLOBIOSIDO-14-β-HYDROXY-Δ-4,20,22-BUFA-
    TRIENOLID (GERMAN) see GFC000
SCINTILLAR see XHS000
SCLAVENTEROL see NGG500
SCOLINE see HLC500
SCOLINE CHLORIDE see HLC500
SCON 5300 see PKQ059
SCONATEX see AAX175, VPK000
SCOPAMIN see HOT500
SCOPARON see DRS800
SCOPARONE see DRS800
SCOPINE TROPATE see SBG000
SCOPOLAMINE see SBG000, SBG000
SCOPOLAMINE BROMIDE see HOT500
(−)-SCOPOLAMINE BROMIDE see HOT500
SCOPOLAMINE HYDROBROMIDE see HOT500
(−)-SCOPOLAMINE HYDROBROMIDE see HOT500
SCOPOLAMINIUM BROMIDE see HOT500
SCOPOLAMMONIUM BROMIDE see HOT500
SCOPOS see HOT500
SCOT see HBT500
SCOTCH PINE NEEDLE OIL see PIH500
SCURENALINE see VGP000
SCUROCAINE see AIL750, AIT250
SCYAN see SIA500
40 SD see IMF300
SD 354 see SMR000
SD 709 see DRM000
SD-1750 see DGP900
SD 1897 see DDL800
SD 3562 see DGQ875
SD 4402 see OAN000
SD 5532 see CDR750
SD 7961 see DGM600
SD 8530 see TMD000
SD 9098 see DIX600
SD 14114 see BLU000
SD 15418 see BLW750
SD 43775 see FAR100
SD ALCOHOL 23-HYDROGEN see EFU000
SDDC see SGM500
SDEH see DJL400
SDIC see SGG500
SDM see SNN300
SDM No. 5 see MAW250
SDM No 23 see PBC250
SDMH see DSF600
SDMO see SNN300
SDPH see DNU000
SEA ANEMONE TOXIN II see ARS000
SEA COAL see CMY635
SEAKEM CARRAGEENIN see CCL250
SEA-LEGS see HGC500
SEA SALT see SFT000
SEATREM see CCL250

SEAWATER MAGNESIA see MAH500
SEAZINA see SNJ000
SEBACIC ACID, DIBUTYL ESTER see DEH600
SEBACIC ACID, DIETHYL ESTER see DJY600
SEBACIL see BAT750
SECACORNIN see LJL000
SECAGYN see EDC500
SECBUBARBITAL see BPF000
SECBUTABARBITAL see BPF000
SECBUTOBARBITONE see BPF000
SECOBARBITAL see SBM500
SECOBARBITONE see SBM500
(5Z,7E)-9,10-SECOCHESTA-5.7.10(19)-TRIENE-1-α,3-β,25-
   TRIOL see DMJ400
9,10-SECOCHOLESTA-5,7,10(19)-TRIENE-1-α,3-β-DIOL see
   HJV000
(1-α,3-β,5Z,7E)-9,10-SECOCHOLESTA-5,7,10(19)-TRIENE-
   1,3,25-TRIOL see DMJ400
9,10-SECOCHOLESTA-5,7,10(19)-TRIEN-3-β-OL see
   CMC750
9,10,SECOERGOSTA-5,7,10(19),22-TETRAEN-3-β-OL see
   VSZ100
SECOMETRIN see LJL000
SECONAL see SBM500
SECONDARY AMMONIUM ARSENATE see DCG800
SECONDARY AMMONIUM PHOSPHATE see ANR500
SECONESINZ see GGS000
SECROVIN see DNX500
SECUPAN see EDC500
SECURITY see LCK000
SEDABAMATE see MQU750
SEDABAR see EOK000
SEDAFORM see ABD000
SEDALIS SEDI-LAB see TEH500
SEDAMYL see ACE000
SEDANTOINAL see MKB250
SEDAREX see EHP000
SEDA-TABLINEN see EOK000
SEDESTRAN see DKA600
SEDETINE see OAV000
SEDETOL see EJM500
SEDICAT see EOK000
SEDIMIDE see TEH500
SEDIN see TEH500
SEDISPERIL see TEH500
SEDIZORIN see EOK000
SEDLYN see EOK000
SEDMYNOL see ACE000
SEDOFEN see EOK000
SEDOMETIL see DNA800
SEDONAL see EOK000
SEDONETTES see EOK000
SEDOPHEN see EOK000
SEDOVAL see TEH500
SEDRENA see BBV000
SEDRESAN see MEP250
SEDTRAN see ACE000
SEDURAL see PDC250, PEK250
SEEDRIN see AFK250
SEEDTOX see ABU500
SEEKAY WAX see TIT500
SEGNALE RED LC see CHP500
S. EGRELTRI ATUNUN (TURKISH) see BML000
SEGURIL see CHJ750
SELECRON see BNA750
SELECTIVE see BPG250
SELEKTIN see BKL250
SELEKTON B 2 see EIX500
SELEN (POLISH) see SBO500
SELENIC ACID see SBN500
SELENIC ACID, liquid (DOT) see SBN500
SELENIC ACID, DIPOTASSIUM SALT see PLR750
SELENINYL CHLORIDE see SBT500
SELENIOUS ACID, DISODIUM SALT see SJT500

SELENIUM see SBO500
SELENIUM ALLOY see SBO500
SELENIUM BASE see SBO500
SELENIUM CHLORIDE OXIDE see SBT500
SELENIUM COMPOUNDS see SBP500
SELENIUM CYSTINE see DWY800
SELENIUM DIETHYLDITHIOCARBAMATE see DJD400
SELENIUM DIMETHYLDITHIOCARBAMATE see SBQ000
SELENIUM DIOXIDE mixed with ARSENIC TRIOXIDE (1:1)
   see ARJ000
SELENIUM(IV) DISULFIDE (1:2) see SBR000
SELENIUM(IV) DISULFIDE SHAMPOO (2.5%) see SBR500
SELENIUM DISULPHIDE (DOT) see SBR000
SELENIUM DUST see SBO500
SELENIUM ELEMENTAL see SBO500
SELENIUM FLUORIDE see SBS000
SELENIUM HEXAFLUORIDE see SBS000
SELENIUM HOMOPOLYMER see SBO500
SELENIUM HYDRIDE see HIC000
SELENIUM METAL POWDER, NON-PYROPHORIC (DOT)
   see SBO500
SELENIUM MONOSULFIDE see SBT000
SELENIUM OXYCHLORIDE see SBT500
SELENIUM SULFIDE see SBR000, SBT000
SELENIUM SULPHIDE see SBT000
SELENOCYSTINE see DWY800
SELENSULFID (GERMAN) see SBT000
SELEPHOS see PAK000
SELF ROCK MOSS see CCL250
SELINON see DUS700
SELSUN see SBW000
SELSUN BLUE see SBR000
SEL-TOX SSO2 and SS-20 see DXG000
SEM (cytostatic) see TND500
SEMAP see PAP250
SEMBRINA see DNA800
SEMDOXAN see CQC675, EAS500
SEMICARBAZIDE see HGU000
SEMICARBAZIDE HYDROCHLORIDE see SBW500
SEMICILLIN see AIV500
SEMIKON see TEO250
SEMOXYDRINE see DBA800
SEMUSTINE see CHD250
SENCOR see MQR275
SENCORAL see MQR275
SENCORER see MQR275
SENCOREX see MQR275
SENDOXAN see CQC675
SENDRAN see PMY300
SENDUXAN see CQC675, EAS500
SENECA OIL see PCR250
SENECIPHYLLIN see SBX500
SENECIPHYLLINE see SBX500
SENEGAL GUM see AQQ500
SENF OEL (GERMAN) see AGJ250
SENFOL (GERMAN) see ISK000
SENKIRKIN see DMX200
SENKIRKINE see DMX200
SENTONIL see PDW500
SENTRY see HOV500
SENTRY GRAIN PRESERVER see PMU750
SEPERIDOL see CLS250
SEPEROL see CLS250
SEPPIC MMD see CIR250
SEPTAMIDE ALBUM see SNM500
SEPTICOL see CDP250
SEPTINAL see SNM500
SEPTISOL see HCL000
SEPTOCHOL see DAQ400
SEPTOFEN see HCL000
SEPTOPLEX see SNM500
SEPTOS see HJL500
SEPTOTAN see PFN000
SEPTRA see SNK000

SEPTRAN see SNK000
SEQ 100 see EIX000
SEQUESTRENE 30A see EIV000
SEQUESTRENE AA see EIX000
SEQUESTRENE Na3 see TNL250
SEQUESTRENE Na 4 see EIV000
SEQUESTRENE SODIUM 2 see EIX500
SEQUESTRENE ST see EIV000
SEQUESTRENE TRISODIUM see TNL250
SEQUESTRENE TRISODIUM SALT see TNL250
SEQUESTRIC ACID see EIX000
SEQUESTROL see EIX000
SEREEN see HOT500
SERENIL see CHG000
SERENSIL see CHG000
SEREN VITA see MDQ250
SERIAL see MCA500
SERIL see MQU750
l-SERINE DIAZOACETATE see ASA500
l-SERINE DIAZOACETATE (ester) see ASA500
SERISOL ORANGE YL see AKP750
SERISTAN BLACK B see AQP000
SERITOX 50 see DGB000
SERMION see NDM000
SEROTONIN BENZYL ANALOG see BEM750
SERPASIL see RDK000
SERPASIL APRESOLINE see RDK000
SERPASIL APRESOLINE No. 2 see HGP500
SERPENTINE see ARM250, ARM268
SERPENTINE CHRYSOTILE see ARM268
SERRAL see DKA600
SERTAN see DBB200
SERTINON see EPQ000
SERTOFRAN see DSI709
N-SERVE NITROGEN STABILIZER see CLP750
SERVISONE see PLZ000
SES see CNW000
SESAGARD see BKL250
SESONE (ACGIH) see CNW000
SESQUISULFURE de PHOSPHORE (FRENCH) see PHS500
SET see MDI000
SETACYL DIAZO NAVY R see DCJ200
SEVENAL see EOK000
SEVICAINE see AIT250
SEVIN see CBM750
SEXADIEN see DAL600
SEXOCRETIN see DKA600
SEXTONE see CPC000
SEXTONE B see MIQ740
SF 60 see MAK700
SF6539 see HDC000
S6F HISTYRENE RESIN see SMR000
SG-67 see SCH000
SH 850 see NNQ500
SH 70850 see NNQ500
SHALE OIL (DOT) see COD750
SHAMMAH (SAUDI ARABIA) see SED400
SHAMROX see CIR250
SHARSTOP 204 see SGM500
SHED-A-LEAF see SFS000
SHED-A-LEAF "L" see SFS000
SHEEP DIP see ARE250
SHELL 40 see BQZ000
SHELL 300 see SMQ500
SHELL 4072 see CDS750
SHELL 4402 see OAN000
SHELL 5520 see PMP500
SHELL ATRAZINE HERBICIDE see ARQ725
SHELL GOLD see GIS000
SHELL MIBK see HFG500
SHELL SD-3562 see DGQ875
SHELL SD-5532 see CDR750
SHELL SD-8530 see TMD000
SHELL SD-9098 see DIX600

SHELL SD 9129 see MRH209
SHELL SD-14114 see BLU000
SHELL SILVER see SDI500
SHELLSOL 140 see NMX000
SHELL UNDRAUTTED A see AFV500
SHELL WL 1650 see OAN000
SHIGRODIN see BRF500
SHIKIMATE see SCE000
SHIKIMIC ACID see SCE000
SHIKIMOLE see SAD000
SHIKOMOL see SAD000
SHINKOLITE see PKB500
SHIN-NAITO S see TEH500
SHINNIBROL see TEH500
SHINNIPPON FAST RED GG BASE see NEO500
SHMP see SHM500
SHOALLOMER see PMP500
SHOCK-FEROL see VSZ100
SHULTZ No. 737 see FAG140
SI see LCF000
SIARKI CHLOREK (POLISH) see SON510
SIARKI DWUTLENEK (POLISH) see SOH500
SIARKOWODOR (POLISH) see HIC500
SIBOL see DKA600
SICILIAN CERISE TONER A-7127 see FAG070
SICOL 150 see DVL700
SICOL 160 see BEC500
SICO LAKE RED 2L see CHP500
SICRON see PKQ059
SIDNOFEN see SPA000
SIENNA see IHD000
SIERRA C-400 see TAB750
SIGMAMYCIN see TBX000
SIGURAN see HGC500
SILANE see SDH575
SILANE, VINYL TRICHLORO 1-150 see TIN750
SILANTIN see DKQ000
SILASTIC see PJR250
SILBER (GERMAN) see SDI500
SILBERNITRAT see SDS000
SILICA AEROGEL see SCH000, SCI000
SILICA, AMORPHOUS see SCH000
SILICA, AMORPHOUS FUMED see SCH000
SILICA, AMORPHOUS FUSED see SCK600
SILICA, AMORPHOUS HYDRATED see SCI000
SILICA (CRYSTALLINE) see SCI500
SILICA, CRYSTALLINE-CRISTOBALITE see SCJ000
SILICA, CRYSTALLINE-QUARTZ see SCJ500
SILICA, CRYSTALLINE-TRIDYMITE see SCK000
SILICA FLOUR see SCI500, SCK500
SILICA FLOUR (powdered crystalline silica) see SCJ500
SILICA, FUSED see SCK600
SILICA GEL see SCI000, SCL000
SILICA, GEL and AMORPHOUS-PRECIPITATED see
   SCL000
SILICANE see SDH575
SILICATE D'ETHYLE (FRENCH) see EPF550
SILICATES see SCM500
SILICATE SOAPSTONE see SCN000
SILICA, VITREOUS see SCK600
SILICA XEROGEL see SCI000
SILICIC ACID see SCI000, SCL000
SILICIC ACID, BERYLLIUM SALT see SCN500
SILICIC ACID TETRAETHYL ESTER see EPF550
SILICIC ANHYDRIDE see SCH000, SCJ500
SILICI-CHLOROFORME (FRENCH) see TJD500
SILICIO(TETRACLORURO di) see SCQ500
SILICIUMCHLOROFORM (GERMAN) see TJD500
SILICIUMTETRACHLORID (GERMAN) see SCQ500
SILICIUMTETRACHLORIDE (DUTCH) see SCQ500
SILICIUM(TETRACHLORURE de) (FRENCH) see SCQ500
SILICOCHLOROFORM see TJD500
SILICOETHANE see DXA000
SILICOFLUORIC ACID see SCO500

SILICON see SCP000
SILICON CARBIDE see SCQ000
SILICON CHLORIDE see SCQ500
SILICON DIOXIDE see SCI500, SCK600
SILICON DIOXIDE (FCC) see SCH000
SILICONE RUBBER see PJR250
SILICONES see SDC000
SILICON FLUORIDE see SDF650
SILICON MONOCARBIDE see SCQ000
SILICON PHENYL TRICHLORIDE see TJA750
SILICON SODIUM FLUORIDE see DXE000
SILICON TETRACHLORIDE (DOT) see SCQ500
SILICON TETRAFLUORIDE (DOT) see SDF650
SILICON TETRAHYDRIDE see SDH575
SILIKILL see SCH000
SILMURIN see RCF000
SILON see NOH000
SILOSAN see DIN800
SILOTRAS ORANGE TR see PEJ500
SILOTRAS YELLOW T2G see DOT300
SILOXANES see SDC000
SILUBIN see BQL000
SILUNDUM see SCQ000
SILVAN (CZECH) see MKH000
SILVER see SDI500
SILVER ACETYLIDE see SDJ000
SILVER ATOM see SDI500
SILVER AZIDE see SDM500
SILVER COMPOUNDS see SDO500
SILVER CYANIDE see SDP000
SILVER DIFLUORIDE see SDQ500
SILVER FIR NEEDLE OIL see AAC250
SILVER FIR OIL see AAC250
SILVER(II) FLUORIDE see SDQ500
SILVER MATT POWDER see TGB250
SILVER MONOACETYLIDE see SDR759
SILVER(1+) NITRATE see SDS000
SILVER NITRATE (DOT) see SDS000
SILVER(I) NITRATE (1:1) see SDS000
SILVER PEROXYCHROMATE see SDW000
SILVER PINE OIL see AAC250
SILVEX (USDA) see TIX500
SILVI-RHAP see TIX500
SILVISAR see HKC500
SILVISAR 510 see HKC000
SILVISAR 550 see MRL750
SIM see SNK000
SIMADEX see BJP000
SIMANEX see BJP000
SIMATIN(E) see ENG500
SIMAZIN see BJP000
SIMAZINE 80W see BJP000
SIMAZINE (USDA) see BJP000
SIMAZOL see AMY050
SIMPAMINA-D see BBK500
SIMPATEDRIN see BBK000
SIMPATOBLOCK see HEA000
SIMPLA see SGG500
SINAFID M-48 see MNH000
SINALGICO see NBS500
SINALOST see TNF500
SINAN see GGS000
SINCICLAN see DKB000
SINDESVEL see QAK000
SINDIATIL see BQL000
SINDRENINA see VGP000
SINITUHO see PAX250
SINOMIN see SNK000
SINORATOX see DSP400
SINOX see DUS700, DUU600
SINOX GENERAL see BRE500
SINOX W see BPG250
SINTESTROL see DKA600
SINTOMICETINA see CDP250

SINTOSIAN see DYF200
SINURON see DGD600
SINUTAB see ABG750
SIONIT see SKV200
SIONON see SKV200
SIPEX BOS see TAV750
SIPEX OP see SIB600
SIPLARIL see FMO129
SIPLAROL see FMO129
SIPOL L8 see OEI000
SIPOL L10 see DAI600
SIPOL L12 see DXV600
SIPOL S see OAX000
SIPOMER DAM see DBK200
SIPONOL S see OAX000
SIPON WD see SIB600
SIPTOX I see MAK700
SIRAN HYDRAZINU (CZECH) see HGW500
SIRLENE see PML000
SIRNIK AMONNY see ANJ750
SIRNIK FOSFORECNY (CZECH) see PHS000
SIROKAL see PLG800
SIRUP see GFG000
SISTOMETRENOL see LJE000
SIXTY-THREE SPECIAL E.C. INSECTICIDE see MNH000, PAK000
SK 65 see DAB879
SK-100 see TNF500
SK 555 see BHO250
SK-598 see CFA750
SK1133 see TND500
SK-3818 see TND250
SK 6882 see TFQ750
SK-15673 see PED750
SK-19849 see BIA250
SK 20501 see EAS500
SK 22591 see HOO500
SK 27702 see BIF750
SK-AMITRIPTYLINE see EAI000
SK-AMPICILLIN see AIV500
SK-Apap see HIM000
SKATOL see MKV750
SKATOLE see MKV750
φ-SKATOLE CARBOXYLIC ACID see ICN000
SK-CHORAL HYDRATE see CDO000
SK-DEXAMETHASONE see SOW000
SK-DIGOXIN see DKN400
SK-DIPHENHYDRAMINE see BAU750
SKEDULE see CBF250
SKEKhG see EAZ500
SKELLY-SOLVE-F see NAI500
SKELLY-SOLVE-L see ROU000
SKELLY-SOLVE S see PCT250
SK-ESTROGENS see ECU750
SKF 51 see ILM000
SKF 385 see PET750
SKF 1498 see DQA400
SKF 2538 see DIR000
SKF 6539 see HDC000
SKF 688A see DDG800
SKF 7988 see VRF000
SKF 20,716 see PIW000
SKF 29044 see BQK000
SKF 62979 see VAD000
SK&F 14287 see DAS000
SK&F 36914 see CLQ500
SKF-525-A see DIG400
SKI 21739 see BHV000
SKI 24464 see MNA750
SKINO #1 see BSU500
SK-LYGEN see MDQ250
SK-106N see NGY000
SK-NIACIN see NCQ900
SKOLIN see HLC500

SK-PHENOBARBITAL see EOK000
SK-PRAMINE see DLH630
SK-PRAMINE HYDROCHLORIDE see DLH630
SK-PREDNISONE see PLZ000
SKS 85 see SMR000
SK-TETRACYCLINE see TBX000, TBX250
SK-TOLBUTAMIDE see BSQ000
SLAB OIL (9CI) see MQV875
SLAKED LIME see CAT250
SLEEPAN see TEH500
SLEEPWELL see TEO250
SLIMICIDE see ADR000
SLIPRO see TEH500
S-LON see PKQ059
SLO-PHYLLIN see TEP000
S-LOST see BIH250
SLOW-FE see FBN100
SLOW-K see PLA500
SLS see SIB600
SLUDGE ACID see SEA000
SLUG-TOX see TDW500
S 75M see SFO500
SMA see SFU500
SMCA see SFU500
SMEESANA see AQN635
SMIDAN see PHX250
SMOG see SEB000
SMOKE CONDENSATE, CIGARETTE see SEC000
SMOKELESS TOBACCO see SED400
S MUSTARD see BIH250
SMUT-GO see HCC500
SN 20 see ADY500
SN 46 see CPK500
SN 36056 see DSO200
SN 36268 see CJJ250
SN 49537 see TEX600
SNATOWHITE CRYSTALS see TFC600
SNEEZING GAS see CGN000
SNIECIOTOX see HCC500
SNOMELT see CAO750
SNOW ALGIN H see SEH000
SNOWGOOSE see TAB750
SNOW WHITE see ZKA000
SNP see PAK000
SNUFF see SED400
SO see LCF000
SOAPSTONE see SCN000
SOAP YELLOW F see FEV000
SOBENATE see SFB000
SOBIODOPA see DNA200
SOBITAL see DJL000
SODA ALUM see AHG500
SODA ASH see SFO000
SODA CHLORATE (DOT) see SFS000
SODA LYE see SHS000, SHS500
SODAMIDE see SEN000
SODANIT see SEY500
SODA NITER see SIO900
SODANTON see DKQ000, DNU000
SODA PHOSPHATE see SJH090
SODAR see DXE600
SODASCORBATE see ARN125
SODESTRIN-H see ECU750
SODIO (CLORATO di) (ITALIAN) see SFS000
SODIO (DICROMATO di) (ITALIAN) see SGI000
SODIO, FLUORACETATO di (ITALIAN) see SHG500
SODIO(IDROSSIDO di) (ITALIAN) see SHS000
SODIO (PERCLORATO DI) (ITALIAN) see PCE750
SODITAL see NBU000
SODIUM see SEE500
SODIUM (dispersions) see SEF500
SODIUM (liquid alloy) see SEF600
SODIUM, metal liquid alloy (DOT) see SEF600
SODIUM ACETATE see SEG500

SODIUM ACETATE, anhydrous (FCC) see SEG500
SODIUM ACETAZOLAMIDE see AAS750
SODIUM ACID ARSENATE see ARC000
SODIUM ACID ARSENATE, HEPTAHYDRATE see ARC250
SODIUM ACID FLUORIDE see SHQ500
SODIUM ACID METHANEARSONATE see MRL750
SODIUM ACID PHOSPHATE see SJH100
SODIUM ACID PYROPHOSPHATE (FCC) see DXF800
SODIUM ACID SULFATE see SEG800
SODIUM ACID SULFATE (solid) see SEG800
SODIUM ACID SULFATE, solution (DOT) see SEG800
SODIUM ACID SULFITE see SFE000
SODIUM AESCINATE see EDM000
SODIUM ALBAMYCIN see NOB000
SODIUM ALGINATE see SEH000
SODIUM-5-ALLYL-5-ISOPROPYLBARBITURATE see
  BOQ750
SODIUM 5-ALLYL-5-(1-METHYLBUTYL)-2-THIOBARBITU-
  RATE see SOX500
SODIUM-dl-5-ALLYL-1-METHYL-5-(1-METHYL-2-PENTY-
  NYL)BARBITURATE see MDU500
SODIUM ALUMINATE, solid (DOT) see AHG000
SODIUM ALUMINOSILICATE see SEM000
SODIUM ALUMINUM HYDRIDE (DOT) see SEM500
SODIUM ALUMINUM OXIDE see AHG000
SODIUM ALUMINUM SULFATE see AHG500
SODIUM ALUMINUM TETRAHYDRIDE see SEM500
SODIUM AMIDE see SEN000
SODIUM AMINARSONATE see ARA500
SODIUM-p-AMINOBENZENEARSONATE see ARA500
SODIUM AMINOPHENOL ARSONATE see ARA500
SODIUM-p-AMINOPHENYLARSONATE see ARA500
SODIUM AMYLOBARBITONE see AON750
SODIUM ANILARSONATE see ARA500
SODIUM-ANILINE ARSONATE see ARA500
SODIUM ANTIMONYL TARTRATE see AQI750
SODIUM ANTIMONY TARTRATE see AQI750
SODIUM ARSANILATE see ARA500
SODIUM-p-ARSANILATE see ARA500
SODIUM ARSENATE see ARC000, ARD500
SODIUM ARSENATE see ARD600, ARD750
SODIUM ARSENATE DIBASIC, anhydrous see ARC000
SODIUM ARSENATE, DIBASIC, HEPTAHYDRATE see
  ARC250
SODIUM ARSENATE HEPTAHYDRATE see ARC250
SODIUM ARSENITE see DXB200, SEY500
SODIUM ARSENITE (liquid) see SEZ000
SODIUM ARSENITE, liquid (solution) (DOT) see SEY500
SODIUM ARSENITE, solid (DOT) see SEY500
SODIUM ARSONILATE see ARA500
SODIUM-l-ASCORBATE see ARN125
SODIUM ASCORBATE (FCC) see ARN125
SODIUM AUROTHIOMALATE see GJC000
SODIUM AUROTHIOSULPHATE DIHYDRATE see GJG000
SODIUM AZIDE see SFA000
SODIUM, AZOTURE de (FRENCH) see SFA000
SODIUM, AZOTURO di (ITALIAN) see SFA000
SODIUM BARBITAL see BAG250
SODIUM BARBITONE see BAG250
SODIUM 1,2 BENZISOTHIAZOLIN-3-ONE-1,1-DIOXIDE see
  SJN700
SODIUM BENZOATE see SFB000
SODIUM BENZOIC ACID see SFB000
SODIUM o-BENZOSULFIMIDE see SJN700
SODIUM BENZOSULPHIMIDE see SJN700
SODIUM 2-BENZOSULPHIMIDE see SJN700
SODIUM o-BENZOSULPHIMIDE see SJN700
SODIUM BENZYLPENICILLIN see BFD250
SODIUM BENZYLPENICILLINATE see BFD250
SODIUM BENZYLPENICILLIN G see BFD250
SODIUM BERYLLIUM MALATE see SFB500
SODIUM BERYLLIUM TARTRATE see SFC000
SODIUM BIBORATE see SFF000
SODIUM BIBORATE DECAHYDRATE see SFF000

SODIUM BICHROMATE see SGI000
SODIUM BIFLUORIDE (VAN) see SHQ500
SODIUM BIPHOSPHATE see SJH100
SODIUM BIPHOSPHATE anhydrous see SJH100
SODIUM BIS(2-ETHYLHEXYL) SULFOSUCCINATE see
   DJL000
SODIUM BISMUTH THIOGLYCOLATE see BKX750
SODIUM BISMUTH THIOGLYCOLLATE see BKX750
SODIUM BISULFATE, fused see SEG800
SODIUM BISULFATE, solid (DOT, FCC) see SEG800
SODIUM BISULFATE, solution (DOT) see SEG800
SODIUM BISULFIDE see SHR000
SODIUM BISULFITE see SFE000, SFE000
SODIUM BISULFITE (1:1) see SFE000
SODIUM BISULFITE, solid (DOT) see SFE000
SODIUM BISULFITE, solution (DOT) see SFE000
SODIUM BORATE see SFE500
SODIUM BORATE anhydrous see SFE500
SODIUM BORATE DECAHYDRATE see SFF000
SODIUM BOROHYDRIDE see SFF500
SODIUM BROMATE see SFG000
SODIUM-5-(2-BROMOALLYL)-5-sec-BUTYLBARBITURATE
   see BOR250
SODIUM BUTAZOLIDINE see BOV750
SODIUM CACODYLATE (DOT) see HKC500
SODIUM CARBOLATE see SJF000
SODIUM CARBONATE (2:1) see SFO000
SODIUM CARBOXYMETHYL CELLULOSE see SFO500
SODIUM CASEINATE see SFQ000
SODIUM CELLULOSE GLYCOLATE see SFO500
SODIUM CHLORAMINE T see CDP000
SODIUM CHLORATE see SFS000
SODIUM (CHLORATE de) (FRENCH) see SFS000
SODIUM CHLORATE, aqueous solution (DOT) see SFS000
SODIUM CHLORIDE see SFT000
SODIUM CHLORITE see SFT500
SODIUM CHLORITE (solution) see SFU000
SODIUM CHLOROACETATE see SFU500
SODIUM CHOLATE see SFW000
SODIUM CHOLIC ACID see SFW000
SODIUM CHROMATE see SGI000
SODIUM CHROMATE (VI) see DXC200
SODIUM CHROMATE (DOT) see DXC200
SODIUM CHROMATE DECAHYDRATE see SFW500
SODIUM CITRATE (FCC) see DXC400
SODIUM CLOXACILLIN MONOHYDRATE see SLJ000
SODIUM CMC see SFO500
SODIUM CM-CELLULOSE see SFO500
SODIUM COMPOUNDS see SFZ000
SODIUM CUMENEAZO-β-NAPHTHOL DISULPHONATE see
   FAG018
SODIUM CYANIDE see SGA500
SODIUM CYANIDE, solid and solution (DOT) see SGA500
SODIUM CYCLAMATE see SGC000
SODIUM CYCLOHEXANESULFAMATE see SGC000
SODIUM CYCLOHEXANESULPHAMATE see SGC000
SODIUM CYCLOHEXYL AMIDOSULPHATE see SGC000
SODIUM CYCLOHEXYL SULFAMATE see SGC000
SODIUM CYCLOHEXYL SULFAMIDATE see SGC000
SODIUM CYCLOHEXYL SULPHAMATE see SGC000
SODIUM 2,4-D see SGH500
SODIUM DALAPON see DGI600
SODIUM DBDT see SGF500
SODIUM DECYLBENZENESULFONAMIDE see DAJ000
SODIUM DECYLBENZENESULFONATE see DAJ000
SODIUM DEDT see SGJ000
SODIUM DEHYDROACETATE (FCC) see SGD000
SODIUM DEHYDROACETIC ACID see SGD000
SODIUM DELVINAL see VKP000
SODIUM DIBUTYLDITHIOCARBAMATE see SGF500
SODIUM DICHLORISOCYANURATE see SGG500
SODIUM DICHLOROACETATE see SGG000
SODIUM DICHLOROCYANURATE see SGG500
SODIUM DICHLOROISOCYANURATE see SGG500

SODIUM 2,4-DICHLOROPHENOXYACETATE see SGH500
SODIUM-2-(2,4-DICHLOROPHENOXY)ETHYL SULFATE see
   CNW000
SODIUM-2,4-DICHLOROPHENOXYETHYL SULPHATE see
   CNW000
SODIUM-2,4-DICHLOROPHENYL CELLOSOLVE SULFATE
   see CNW000
SODIUM-2,2-DICHLOROPROPIONATE see DGI600
SODIUM-α,α-DICHLOROPROPIONATE see DGI600
SODIUM 1,3-DICHLORO-1,3,5-TRIAZINE-2,4-DIONE-6-OX-
   IDE see SGG500
1-SODIUM 3,5-DICHLORO-s-TRIAZINE-2,4,6-TRIONE see
   SGG500
1-SODIUM 3,5-DICHLORO-1,3,5-TRIAZINE-2,4,6-TRIONE
   see SGG500
SODIUM DICHLORO-s-TRIAZINETRIONE, dry, containing
   more than 39% available chlorine (DOT) see SGG500
SODIUM DICHROMATE see SGI000
SODIUM DICHROMATE(VI) see SGI000
SODIUM DICHROMATE de (FRENCH) see SGI000
SODIUM DIETHYLBARBITURATE see BAG250
SODIUM-5,5-DIETHYLBARBITURATE see BAG250
SODIUM DIETHYLDITHIOCARBAMATE see SGJ000
SODIUM N,N-DIETHYLDITHIOCARBAMATE see SGJ000
SODIUM DI-(2-ETHYLHEXYL) SULFOSUCCINATE see
   DJL000
SODIUM DIHYDROGEN ARSENATE see ARD600
SODIUM DIHYDROGEN ORTHOARSENATE see ARD600
SODIUM DIHYDROGEN PHOSPHATE (1:2:1) see SJH100
SODIUM-4-(DIMETHYLAMINO)BENZENEDIAZOSUL-
   FONATE see DOU600
SODIUM-p-(DIMETHYLAMINO)BENZENEDIAZOSUL-
   FONATE see DOU600
SODIUM-4-(DIMETHYLAMINO)BENZENEDIAZOSUL-
   PHONATE see DOU600
SODIUM-p-(DIMETHYLAMINO)BENZENEDIAZOSUL-
   PHONATE see DOU600
SODIUM-(4-(DIMETHYLAMINO)PHENYL)DIAZENESUL-
   FONATE see DOU600
SODIUM DIMETHYLARSINATE see HKC500
SODIUM DIMETHYLARSONATE see HKC500
SODIUM N,N-DIMETHYLDITHIOCARBAMATE see
   SGM500
SODIUM-4,6-DINITRO-o-CRESOXIDE see DUU600
SODIUM DIOCTYL SULFOSUCCINATE see DJL000
SODIUM DIOCTYL SULPHOSUCCINATE see DJL000
SODIUM DIOXIDE see SJC500
SODIUM DIPHENYL-4,4'-BIS-AZO-2''-8''-AMINO-1''-
   NAPHTHOL-3'',6 '' DISULPHONATE see CMO000
SODIUM DIPHENYLHYDANTOIN see DNU000
SODIUM DIPHENYL HYDANTOINATE see DNU000
SODIUM-5,5-DIPHENYLHYDANTOINATE see DNU000
SODIUM-5,5-DIPHENYL-2,4-IMIDAZOLIDINEDIONE see
   DNU000
SODIUM DITHIONITE (DOT) see SHR500
SODIUM DITOLYLDIAZOBIS-8-AMINO-1-NAPHTHOL-3,6-
   DISULFONATE see CMO250
SODIUM DITOLYLDIAZOBIS-8-AMINO-1-NAPHTHOL-3,6-
   DISULPHONATE see CMO250
SODIUM DODECYLBENZENESULFONATE (DOT) see
   DXW200
SODIUM DODECYLBENZENESULFONATE, DRY see
   DXW200
SODIUM DODECYL SULFATE see SIB600
SODIUM EDETATE see EIV000
SODIUM EDTA see EIV000
SODIUM EOSINATE see BNH500
SODIUM ETASULFATE see TAV750
SODIUM ETHAMINAL see NBU000
SODIUM ETHASULFATE see TAV750
SODIUM ETHYLBARBITAL see BAG250
SODIUM ETHYLENEDIAMINETETRAACETATE see
   EIV000

SODIUM ETHYLENEDIAMINETETRAACETIC ACID see EIV000
SODIUM(2-ETHYLHEXYL)ALCOHOL SULFATE see TAV750
SODIUM 2-ETHYLHEXYL SULFATE see TAV750
SODIUM-2-ETHYLHEXYLSULFOSUCCINATE see DJL000
SODIUM ETHYLISOAMYLBARBITURATE see AON750
SODIUM ETHYLMERCURIC THIOSALICYLATE see MDI000
SODIUM-o-(ETHYLMERCURITHIO)BENZOATE see MDI000
SODIUM ETHYLMERCURITHIOSALICYLATE see MDI000
SODIUM-5-ETHYL-5-(1-METHYL-1-BUTENYL) BARBITU-RATE see VKP000
SODIUM 5-ETHYL-5-(1-METHYLBUTYL)BARBITURATE see NBU000
SODIUM-5-ETHYL-5-(1-METHYLBUTYL)-2-THIOBARBITU-RATE see PBT500
SODIUM 5-ETHYL-5-PHENYLBARBITURATE see SID000
SODIUM FLUOACETATE see SHG500
SODIUM FLUOACETIC ACID see SHG500
SODIUM FLUORACETATE de (FRENCH) see SHG500
SODIUM FLUORESCEIN see FEW000
SODIUM FLUORESCEINATE see FEW000
SODIUM FLUORIDE see SHF500
SODIUM FLUORIDE, solid and solution (DOT) see SHF500
SODIUM FLUORIDE(Na(HF₂)) see SHQ500
SODIUM FLUOROACETATE see SHG500
SODIUM FLUOROSILICATE see DXE000
SODIUM FLUORURE (FRENCH) see SHF500
SODIUM FLUOSILICATE see DXE000
SODIUM FORMATE see SHJ000
SODIUM FUMARATE see DXD800
SODIUM GLUCONATE see SHK800
SODIUM d-GLUCONATE see SHK800
SODIUM GLUTAMATE see MRL500
SODIUM l-GLUTAMATE see MRL500
l(+) SODIUM GLUTAMATE see MRL500
SODIUM GMP see GLS800
SODIUM GUANOSINE-5'-MONOPHOSPHATE see GLS800
SODIUM GUANYLATE see GLS800
SODIUM-5'-GUANYLATE see GLS800
SODIUM HEXACYCLONATE see SHL500
SODIUM HEXAFLUOROSILICATE see DXE000
SODIUM HEXAFLUOSILICATE see DXE000
SODIUM HEXAMETAPHOSPHATE see SHM500, SII500
SODIUM HYDRATE (DOT) see SHS000
SODIUM HYDRATE, solution see SHS500
SODIUM HYDRIDE see SHO500
SODIUM HYDROCORTISONE SUCCINATE see HHR000
SODIUM HYDROCORTISONE-21-SUCCINATE see HHR000
SODIUM HYDROFLUORIDE see SHF500
SODIUM HYDROGEN DIFLUORIDE see SHQ500
SODIUM HYDROGEN FLUORIDE see SHQ500
SODIUM HYDROGEN PHOSPHATE see SJH090
SODIUM HYDROGEN SULFATE, solid (DOT) see SEG800
SODIUM HYDROGEN SULFATE, solution (DOT) see SEG800
SODIUM HYDROGEN SULFIDE see SHR000
SODIUM HYDROGEN SULFITE see SFE000
SODIUM HYDROGEN SULFITE, solid (DOT) see SFE000
SODIUM HYDROGEN SULFITE, solution (DOT) see SFE000
SODIUM HYDROSULFIDE see SHR000
SODIUM HYDROSULFIDE, solution (DOT) see SHR000
SODIUM HYDROSULFITE (DOT) see SHR500
SODIUM HYDROSULPHIDE, solid (DOT) see SHR000
SODIUM HYDROSULPHIDE, with less than 25% water of crys-tallization (DOT) see SHR000
SODIUM HYDROSULPHITE see SHR500
SODIUM HYDROXIDE see SHS000
SODIUM HYDROXIDE, bead (DOT) see SHS000
SODIUM HYDROXIDE, dry (DOT) see SHS000
SODIUM HYDROXIDE, flake (DOT) see SHS000
SODIUM HYDROXIDE, granular (DOT) see SHS000

SODIUM HYDROXIDE (liquid) see SHS500
SODIUM HYDROXIDE, solid (DOT) see SHS000
SODIUM HYDROXIDE, solution (FCC) see SHS500
SODIUM o-HYDROXYBENZOATE see SJO000
SODIUM(HYDROXYDE de) (FRENCH) see SHS000
SODIUM-2-HYDROXYDIPHENYL see BGJ750
SODIUM o-((3-(HYDROXYMERCURI)-2-METHOXYPROPYL) CARBAMOYL)PHENOXY ACETATE see SIH500
SODIUM 1-(HYDROXYMETHYL)CYCLOHEXANEACETATE see SHL500
SODIUM HYPOCHLORITE see SHU500
SODIUM HYPOPHOSPHITE see SHV000
SODIUM HYPOSULFITE see SKI000, SKI500
SODIUM 5,5'-INDIGOTIDISULFONATE see FAE100
SODIUM INOSINATE see DXE500
SODIUM-5'-INOSINATE see DXE500
SODIUM IODIDE see SHW000
SODIUM IODINE see SHW000
SODIUM ISOAMYLETHYL BARBITURATE see AON750
SODIUM ISOTHIOCYANATE see SIA500
SODIUM LACTATE see LAM000
SODIUM LAURYLBENZENESULFONATE see DXW200
SODIUM-N-LAURYL SARCOSINE see DXZ000
SODIUM LAURYL SULFATE see SIB600
SODIUM LUMINAL see SID000
SODIUM MALONYLUREA see BAG250
SODIUM MERCAPTAN see SHR000
SODIUM MERCAPTIDE see SHR000
SODIUM MERCAPTOACETATE see SKH500
SODIUM 2-MERCAPTOBENZOTHIAZOLE see SIG500
SODIUM MERCAPTOMERIN see DXC000
SODIUM MERSALYL see SIH500
SODIUM MERTHIOLATE see MDI000
SODIUM METAARSENATE see ARD500
SODIUM METAARSENITE see SEY500
SODIUM METABISULFITE see SII000
SODIUM METABOSULPHITE see SII000
SODIUM METAL (DOT) see SEE500
SODIUM, METAL DISPERSION IN ORGANIC SOLVENT see SEF500
SODIUM METAPHOSPHATE see SII500
SODIUM METASILICATE see SJU000
SODIUM METASILICATE, anhydrous see SJU000
SODIUM METAVANADATE see SKP000
SODIUM METHANEARSONATE see DXE600, MRL750
4-SODIUM METHANESULFONATE METHYLAMINE-ANTI-PYRINE see AMK500
SODIUM METHARSONATE see DXE600
SODIUM METHOHEXITAL see MDU500
SODIUM METHOHEXITONE see MDU500
SODIUM METHOXIDE see SIK450
SODIUM A-dl-1-METHYL-5-ALLYL-5-(1-METHYL-2-PENTY-NYL)BARBITURATE see MDU500
SODIUM METHYLAMINOANTIPYRINE METHANESULFO-NATE see AMK500
SODIUM-4-METHYLAMINO-1,5-DIMETHYL-2-PHENYL-3-PYRAZOLONE 4-METHANESULFONATE see AMK500
SODIUM METHYLARSONATE see DXE600
SODIUM METHYLATE see SIK450
SODIUM METHYLATE, DRY (DOT) see SIK450
SODIUM MOLYBDATE see DXE800
SODIUM MOLYBDATE(VI) see DXE800
SODIUM MONOCHLORACETATE see SFU500
SODIUM MONODODECYL SULFATE see SIB600
SODIUM MONOFLUORIDE see SHF500
SODIUM MONOFLUOROACETATE see SHG500
SODIUM MONOHYDROGEN ARSENATE see ARD500
SODIUM MONOHYDROGEN PHOSPHATE (2:1:1) see SJH090
SODIUM MONOIODIDE see SHW000
SODIUM MONOSULFIDE see SJY500
SODIUM MONOXIDE see SIN500
SODIUM MONOXIDE, solid (DOT) see SIN500
SODIUM NEMBUTAL see NBU000

SODIUM-22 NEOPRENE ACCELERATOR see IAQ000
SODIUM NITRATE (DOT) see SIO900
SODIUM(I) NITRATE (1:1) see SIO900
SODIUM NITRILOTRIACETATE see SIP500
SODIUM NITRITE see SIQ500
SODIUM NITRITE and CARBENDAZIM (1:5) see CBN375
SODIUM NITRITE mixed with ETHYLENETHIOUREA see IAR000
SODIUM NITRITE and ETHYLUREA (1:2) see EQE000
SODIUM NITRITE and METHYL-2-BENZIMIDAZOLE CAR-BAMATE see CBN375
SODIUM NITRITE, mixed with POTASSIUM NITRITE see SIT500
SODIUM NITROFERRICYANIDE see SIU500
SODIUM NITROPRUSSATE see SIU500
SODIUM NITROPRUSSIDE see SIU500
SODIUM NITROSYLPENTACYANOFERRATE see SIU500
SODIUM NITROSYLPENTACYANOFERRATE(III) see SIU500
SODIUM NORAMIDOPYRINE METHANESULFONATE see AMK500
SODIUM NOVOBIOCIN see NOB000
SODIUM OCTADECANOATE see SJV500
SODIUM OLEATE see OIA000
SODIUM ORTHOARSENATE see ARD750
SODIUM ORTHOARSENITE see ARJ500
SODIUM ORTHOVANADATE see SIY250
SODIUM OXIDE see SIN500
SODIUM OXIDE (Na2-O2) see SJC500
SODIUM PCP see SJA000
SODIUM PENICILLIN see BFD250
SODIUM PENICILLIN G see BFD250
SODIUM PENICILLIN II see BFD250
SODIUM-PENT see NBU000
SODIUM PENTABARBITAL see NBU000
SODIUM PENTABARBITONE see NBU000
SODIUM PENTACHLOROPHENATE see SJA000
SODIUM PENTACHLOROPHENATE (DOT) see SJA000
SODIUM PENTACHLOROPHENOL see SJA000
SODIUM PENTACHLOROPHENOLATE see SJA000
SODIUM PENTACHLOROPHENOXIDE see SJA000
SODIUM β,β-PENTAMETHYLENE-γ-HYDROXYBUTYRATE see SHL500
SODIUM PENTHIOBARBITAL see PBT500
SODIUM PENTOBARBITAL see NBU000
SODIUM PENTOBARBITONE see NBU000
SODIUM PENTOBARBITURATE see NBU000
SODIUM PENTOTHAL see PBT500
SODIUM PENTOTHIOBARBITAL see PBT500
SODIUM PERCHLORATE see PCE750
SODIUM PERCHLORATE (DOT) see PCE750
SODIUM PERMANGANATE see SJC000
SODIUM PEROXIDE see SJC500
SODIUM PEROXYDISULFATE see SJE000
SODIUM PERSULFATE see SJE000
SODIUM PHENATE see SJF000
SODIUM PHENOBARBITAL see SID000
SODIUM PHENOBARBITONE see SID000
SODIUM PHENOLATE, solid (DOT) see SJF000
SODIUM PHENOXIDE see SJF000
SODIUM PHENYLBUTAZONE see BOV750
SODIUM-1-PHENYL-2,3-DIMETHYL-4-METHYLAMINOPY-RAZOLON-N-METHANESULFONATE see AMK500
SODIUM-1-PHENYL-2,3-DIMETHYL-5-PYRAZOLONE-4-ME-THYLAMINO METHANESULFONATE see AMK500
SODIUM PHENYLDIMETHYLPYRAZOLONMETHYL-AMINOMETHANE SULFONATE see AMK500
SODIUM PHENYLETHYLBARBITURATE see SID000
SODIUM PHENYLETHYLMALONYLUREA see SID000
SODIUM-N-PHENYLGLYCINAMIDE-p-ARSONATE see CBJ750
SODIUM-2-PHENYLPHENATE see BGJ750
SODIUM-o-PHENYLPHENATE see BGJ750
SODIUM-o-PHENYLPHENOLATE see BGJ750

SODIUM-o-PHENYLPHENOXIDE see BGJ750
SODIUM PHOSPHATE see SJH200
SODIUM PHOSPHATE, anhydrous see SJH200
SODIUM PHOSPHATE, DIBASIC see SJH090
SODIUM PHOSPHATE, MONOBASIC see SJH100
SODIUM PHOSPHATE, TRIBASIC see SJH200
SODIUM PHOSPHIDE see SJI500
SODIUM PHOSPHINATE see SHV000
SODIUM POLYACRYLATE see SJK000
SODIUM POLYALUMINATE see AHG000
SODIUM POLYMANNURONATE see SEH000
SODIUM POLYPHOSPHATES, GLASSY see SII500
SODIUM POTASSIUM ALLOY, liquid and solid (DOT) see PLS500
SODIUM PROPIONATE see SJL500
SODIUM PYROBORATE see SFF000
SODIUM PYROBORATE DECAHYDRATE see SFF000
SODIUM PYROPHOSPHATE see DXF800
SODIUM PYROPHOSPHATE (FCC) see TEE500
SODIUM PYROSULFATE see SEG800
SODIUM PYROSULFITE see SII000
SODIUM RHODANATE see SIA500
SODIUM RHODANIDE see SIA500
SODIUM SACCHARIDE see SJN700
SODIUM SACCHARIN see SJN700
SODIUM SACCHARINATE see SJN700
SODIUM SACCHARINE see SJN700
SODIUM SALICYLATE see SJO000
SODIUM SALICYL-(γ-HYDROXYMERCURI-β-METHOXY-PROPYL)AMIDE-o-ACETATE see SIH500
SODIUM SALICYLIC ACID see SJO000
SODIUM SALT of CACODYLIC ACID see HKC500
SODIUM SALT of CARBOXYMETHYLCELLULOSE see SFO500
SODIUM SALT of DICHLORO-s-TRIAZINETRIONE see SGG500
SODIUM SALT of N,N-DIETHYLDITHIOCARBAMIC ACID see SGJ000
SODIUM SALT of 4,6-DINITRO-o-CRESOL see DUU600
SODIUM SALT of ETHYLENEDIAMINETETRAACETIC ACID see EIV000
SODIUM SALT of HYDROXY-o-CARBOXY-PHENYL-FLUO-RONE see FEW000
SODIUM SALT of PHENYLBUTAZONE see BOV750
SODIUM SELENATE see DXG000
SODIUM SELENITE see SJT500
SODIUM SILICATE see SJU000
SODIUM SILICOALUMINATE see SEM000
SODIUM SILICOFLUORIDE (DOT) see DXE000
SODIUM SORBATE see SJV000
SODIUM STEARATE see SJV500
SODIUM SUCARYL see SGC000
SODIUM SULFAMERAZINE see SJW475
SODIUM SULFATE (2:1) see SJY000
SODIUM SULFATE anhydrous see SJY000
SODIUM SULFHYDRATE see SHR000
SODIUM SULFIDE (anhydrous) see SJY500
SODIUM SULFITE (2:1) see SJZ000
SODIUM SULFITE, anhydrous see SJZ000
SODIUM SULFOCYANATE see SIA500
SODIUM SULFOCYANIDE see SIA500
SODIUM SULFODI-(2-ETHYLHEXYL)6SULFOSUCCINATE see DJL000
SODIUM SULFOXYLATE see SHR500
SODIUM SULHYDRATE see SFE000
SODIUM SULPHAMERAZINE see SJW475
SODIUM SULPHATE see SJY000
SODIUM SULPHIDE see SJY500
SODIUM SULPHITE see SJZ000
SODIUM 1-(+)-TARTRATE see BLC000
SODIUM TARTRATE (FCC) see BLC000
SODIUM TCA, solution see TII250
SODIUM TELLURATE(IV) see SKC500
SODIUM TELLURITE see SKC500

SODIUM TETRABORATE see SFF000
SODIUM TETRABORATE DECAHYDRATE see SFF000
SODIUM TETRAHYDROALUMINATE(1−) see SEM500
(T-4) SODIUM, TETRAHYDROALUMINATE(1−) (9CI) see SEM500
SODIUM TETRAHYDROBORATE(1-) see SFF500
SODIUM TETRAPEROXYCHROMATE see SKF000
SODIUM TETRAPOLYPHOSPHATE see SII500
SODIUM TETRAVANADATE see SKG500
SODIUM THIAMYLAL see SOX500
SODIUM THIOCYANATE see SIA500
SODIUM THIOCYANIDE see SIA500
SODIUM THIOGLYCOLATE see SKH500
SODIUM THIOGLYCOLLATE see SKH500
SODIUM THIOPENTAL see PBT500
SODIUM THIOPENTOBARBITAL see PBT500
SODIUM THIOPENTONE see PBT500
SODIUM THIOSULFATE see SKI000
SODIUM THIOSULFATE, anhydrous see SKI000
SODIUM THIOSULFATE, PENTAHYDRATE see SKI500
SODIUM p-TOLUENESULFONYLCHLORAMIDE see CDP000
SODIUM TOSYLCHLORAMIDE see CDP000
SODIUM TRIMETAPHOSPHATE see SKM500
SODIUM TRIPHOSPHATE see SKN000
SODIUM TRIPOLYPHOSPHATE see SKN000
SODIUM TUNGSTATE see SKN500
SODIUM VANADATE see SIY250, SKP000
SODIUM VANADIUM OXIDE see SIY250
SODIUM VERONAL see BAG250
SODIUM VERSENATE see EIX500
SODIUM VINBARBITAL see VKP000
SODOTHIOL see SKI500
SO-FLO see SHF500
SOFRIL see SOD500
SOFTENIL see TEH500
SOFTENON see TEH500
SOFTIL see DJL000
SOHNHOFEN STONE see CAO000
SOILBROM-40 see EIY500
SOILBROM-85 see EIY500
SOILFUME see EIY500
SOIL STABILIZER 661 see SMR000
SOL see PKB500
SOLACEN see MOV500
SOLACIN see MOV500
SOLACTOL see LAJ000
SOLADREN see VGP000
SOLAMINE see BEN000
SOLANCARPIDINE see POJ000
SOLANIDINE-S see POJ000
SOLANTIN see DKQ000
SOLANTOIN see DNU000
SOLANTYL see DNU000
SOLAR 40 see DXW200
SOLAR BROWN PL see CMO750
SOLAR LIGHT ORANGE GX see HGC000
SOLAR RUBINE see HJF500
SOLAR VIOLET 5BN see FAG120
SOLAR WINTER BAN see PML000
SOLASKIL see LFA020
SOLASOD-5-EN-3-β-OL see POJ000
SOLASODINE see POJ000
SOLBAR see BAP000
SOLBASE see PJT200
SOLBROL A see HJL000
SOLBROL M see HJL500
SOLBUTAMOL see BQF500
SOLCAIN see DHK400
SOLDEP see TIQ250
SOLESTRO see EDP000
SOLFARIN see WAT200
SOLFO BLACK B see DUZ000
SOLFO BLACK BB see DUZ000

SOLFO BLACK G see DUZ000
SOLFO BLACK SB see DUZ000
SOLFO BLACK 2B SUPRA see DUZ000
SOLFOCRISOL see GJG000
SOLFOTON see EOK000
SOLFURO di CARBONIO (ITALIAN) see CBV500
SOLGANAL see ART250
SOLGANAL B see ART250
SOLID GREEN FCF see FAG000
SOLIWAX see DJL000
SOLKETAL see DVR600
SOLLICULIN see EDV000
SOLMETHINE see MJP450
SOLOCHROME BLUE FB see HJF500
SOLOSIN see TEP000
SOLOZONE see SJC500
SOL PHENOBARBITAL see SID000
SOL PHENOBARBITONE see SID000
SOLPRENE 300 see SMR000
SOLPYRON see ADA725
SOL SODOWA KWASU LAURYLOBENZENOSULFONO-WEGO (POLISH) see DXW200
SOLSOL NEEDLES see SIB600
SOLUBLE BARBITAL see BAG250
SOLUBLE FLUORESCEIN see FEW000
SOLUBLE GLUSIDE see SJN700
SOLUBLE GUN COTTON see CCU250
SOLUBLE INDIGO see FAE100
SOLUBLE PENTOBARBITAL see NBU000, SID000
SOLUBLE PHENOBARBITONE see SID000
SOLUBLE PHENYTOIN see DNU000
SOLUBLE SACCHARIN see SJN700
SOLUBLE SULFAMERAZINE see SJW475
SOLUBLE THIOPENTONE see PBT500
SOLUBLE VANDYKE BROWN see MAT500
SOLUBOND 0-869 see RDP000
SOLUBOND 3520 see RDP000
SOLU-CORTEF see HHR000
SOLUGLACIT see NIM500
SOLU-GLYC see HHR000
SOLUMEDINE see SJW475
SOLUPYRIDINE see DXF400
SOLUSOL-75% see DJL000
SOLUSOL-100% see DJL000
SOLUTION CONCENTREE T271 see AMY050
SOLVANOL see DJX000
SOLVANOM see DTR200
SOLVARONE see DTR200
SOLVAT 14 see BFL000
SOLVENT 111 see MIH275
SOLVENT-DEWAXED LIGHT NAPHTHENIC DISTILLATE see MQV835
SOLVENT-DEWAXED LIGHT PARAFFINIC DISTILLATE see MQV840
SOLVENT ETHER see EJU000
SOLVENT NAPHTHA see PCT250
SOLVENT ORANGE 15 see BJF000
SOLVENT RED 19 see EOJ500
SOLVENT-REFINED LIGHT NAPHTHENIC DISTILLATE see MQV852
SOLVENT-REFINED LIGHT PARAFFINIC DISTILLATE see MQV855
SOLVENT YELLOW 1 see PEI000
SOLVENT YELLOW 14 see PEJ500
SOLVIC see PKQ059
SOLVIC 523KC see AAX175
SOLVIREX see DXH325
SOLVOSOL see CBR000
SOLYACORD see DJS200
SOMA see IPU000
SOMADRIL see IPU000
SOMALGIT see IPU000
SOMALIA ORANGE 2R see XRA000
SOMALIA ORANGE A2R see XRA000

SOMALIA ORANGE I see PEJ500
SOMALIA RED III see OHA000
SOMALIA YELLOW A see DOT300
SOMALIA YELLOW R see AIC250
SOMAN see SKS500
SOMANIL see IPU000
SOMAR see DXE600
SOMBEROL see QAK000
SOMBUCAPS see ERD500
SOMBULEX see ERD500
SOMBUTOL see EOK000
SOMIO see GFA000
SOMIPRONT see DUD800
SOMLAN see DAB800
SOMNAFAC see QAK000
SOMNALERT see ERD500
SOMNASED see DLY000
SOMNIBEL see DLY000
SOMNI SED see CDO000
SOMNITE see DLY000
SOMNOLENS see EOK000
SOMNOLETTEN see EOK000
SOMNOMED see QAK000
SOMNOPENTYL see NBU000
SOMNOS see CDO000
SOMNOSAN see EOK000
SOMNUROL see BNP750
SOMONAL see EOK000
SOMONIL see DSO000
SONACIDE see GFQ000
SONAFORM see TDA500
SONAL see QAK000
SONAPAX see MOO250
SONBUTAL see BOR000
SONEBON see DLY000
SONERILE see BPF500
SONERYL see BPF500
SONISTAN see NBU000
SONNOLIN see DLY000
SONTEC see CDO000
SONTOBARBITAL NABITONE see NBU000
SOOT see SKS750
SOPENTAL see NBU000
SOPHIAMIN see MDQ250
SOPHORETIN see QCA000
SOPP see BGJ750
SOPRABEL see LCK000
SOPRACOL see MLC250
SOPRACOL 781 see MLC250
SOPRATHION see EEH600
SOPRATHION see PAK000
SOPRINAL see BAG250
SOPRINTIN see ABH500
SOPROCIDE see BBQ750
SOPRONTIN see AAF750, ABH500
SOPROTIN see ABH500
SORBA-SPRAY Mn see MAU250
SORBIC ACID see SKU000
SORBIC ACID, POTASSIUM SALT see PLS750
SORBIC ACID, SODIUM SALT see SJV000
SORBIC OIL see PAJ500
SORBICOLAN see SKV200
SORBIMACROGOL OLEATE see PKL100
SORBISTAT see SKU000
SORBISTAT-K see PLS750
SORBISTAT-POTASSIUM see PLS750
SORBITAL O 20 see PKL100
SORBITAN C see SKV150
SORBITAN MONOOCTADECANOATE see SKV150
SORBITAN, MONOOCTADECANOATE, POLY(OXY-1,2-
     ETHANEDIYL) DERIVATIVES see PKL030
SORBITAN MONOSTEARATE see SKV150
SORBITAN STEARATE see SKV150
SORBITE see SKV200

SORBITOL see SKV200
d-SORBITOL see SKV200
SORBO see SKV200
SORBO-CALCIAN see CAL750
SORBO-CALCION see CAL750
SORBOL see SKV200
SORBON S 60 see SKV150
SORBOSTYL see SKV200
SOREFLON 604 see TAI250
SORETHYTAN (20) MONOOLEATE see PKL100
SORGEN 50 see SKV150
SORLATE see PKL100
SORVILANDE see SKV200
SOSIGON see DOQ400
SOTACOR see CCK250
SOTALEX see CCK250
SOTALOL see CCK250
SOTALOL HYDROCHLORIDE see CCK250
SOTIPOX see TIQ250
SOTYL see NBU000
SOUDAN I see PEJ500
SOUDAN II see XRA000
SOUFRAMINE see PDP250
SOUP see NGY000
SOUTHERN BENTONITE see BAV750
SOVCAIN see NOF500
SOVCAINE see DDT200
SOVCAINE HYDROCHLORIDE see NOF500
SOVERIN see QAK000
SOVIET TECHNICAL HERBICIDE 2M-4C see CIR250
SOVIOL see AAX250
SOVKAIN see NOF500
SOVOCAINE HYDROCHLORIDE see NOF500
SOVOL see PJL750
SOWBUG & CUTWORM BAIT see COF500
SOXINAL PZ see BJK500
SOXINOL PZ see BJK500
75 SP see DOP600
SP 104 see TCM250
SPA see DWA600
SPAN 55 see SKV150
SPAN 60 see SKV150
SPANBOLET see SNJ000
SPANISH FLY see CBE250
SPANISH MARJORAM OIL see MBU500
SPANISH THYME OIL see TFX750
SPANON see CJJ250
SPANONE see CJJ250
SPANTOL see PGA750
SPANTRAN see MQU750
SPARIC see BRE500
SPARINE see DQA600
SPARINE HYDROCHLORIDE see PMI500
SPARTOLOXYN see GGS000
SPASEPILIN see EOK000
SPASMEDAL see DAM700
SPASMODOLIN see DAM700
SPASMOLYTIN see DHX800
SPASMOPHEN see ORQ000
SPAVIT see VSZ450
SP 60 (CHLOROCARBON) see PKQ059
SPEARMINT OIL see SKY000
SPECIAL BLUE X 2137 see EOJ500
SPECIAL TERMITE FLUID see DEP600
SPECIFEN see EID000
SPECILLINE G see BDY669
SPECTINOMYCIN DIHYDROCHLORIDE see SLI325
SPECTINOMYCIN HYDROCHLORIDE see SLI325
SPECTRACIDE see DCM750
SPECTRAR see INJ000
SPECTROLENE BLUE B see DCJ200
SPECTROLENE RED KB see CLK225
SPECULAR IRON see IHD000
SPEED see DBA800

"SPEED" see MDQ500
SPENCER 401 see PJY500
SPENCER S-6538 see DJY200
SPENCER S-6900 see DRR200
SPENT SULFURIC ACID (DOT) see SOI500
SPERLOX-S see SOD500
SPERLOX-Z see EIR000
SPERSUL see SOD500
SPERSUL THIOVIT see SOD500
SP 60 ESTER see AAX250
SPHEROIDINE see FOQ000
SPHYGMOGENIN see VGP000
SPIKE LAVENDER OIL see SLB500
SPINOCAINE see AIL750
SPIRIT of GLONOIN see SLD000
SPIRIT of GLYCERYL TRINITRATE see SLD000
SPIRIT of HARTSHORN see AMY500
SPIRIT ORANGE see PEJ500
SPIRITS of NITROGLYCERIN (DOT) see SLD000
SPIRITS of SALT (DOT) see HHL000
SPIRITS of TURPENTINE see TOD750
SPIRITS of WINE see EFU000
SPIRIT of TRINITROGLYCERIN see SLD000
SPIRIT of TURPENTINE see TOD750
SPIRIT YELLOW I see PEJ500
SPIROCID see ABX500
SPIROFULVIN see GKE000
SPIRO(ISOBENZOFURAN-1(3H),9'-(9H)XANTHENE-3-ONE,
    3',6'-DIHYDROXY-DISODIUM SALT see FEW000
SPIROZID see ABX500
SPIRT see EFU000
SPONTOX see TAA100
SPOR-KIL see ABU500
SPOROSTATIN see GKE000
SPOTRETE see TFS350
SPOTTON see FAQ999
SPRAY-DERMIS see NGE500
SPRAY-FORAL see NGE500
SPRAY-HORMITE see SGH500
SPRAYSET MEKP see MKA500
SPRAY-TROL BRANCH RODEN-TROL see WAT200
SPRENGEL EXPLOSIVES see SLG500
SPRING-BAK see DXD200
SPRITZ-HORMIN/2,4-D see DAA800
SPRITZ-HORMIT see SGH500
SPROUT NIP see CKC000
SPROUT-NIP EC see CKC000
SPUD-NIC see CKC000
SPUD-NIE see CKC000
SPURGE see BRE500
SQ 1089 see HOO500
SQ 1156 see GGS000
SQ 1489 see TFS350
SQ 9453 see DUD800
SQ 14055 see TET800
SQ 21977 see MPU000
SQ 22947 see DAR000, TET800
SQUIBB see HNT500
SQUILL see RCF000
SR 73 see DFV600
SR406 see CBG000
SRA 5172 see DTQ400
SRA 7312 see DJY200
SRA 7847 see EIM000
SRA 12869 see IMF300
SRC-II HEAVY DISTILLATE see CMY625
SRI 859 see MNA750
SRI 1720 see BIF750
SRI 1869 see NKL000
SRI 2200 see CGV250
SRI 2489 see IAN000
S-SEVEN see SCC000
S.T. 37 see HFV500
ST 155 see CBF250

ST 720 (FRENCH) see CGB000
STABILENE see BKA000
STABILISATOR C see DWN800
STABILIZATOR AR see PFT500
STABILIZED ETHYL PARATHION see PAK000
STABILIZER D-22 see DDV600
STABLE RED KB BASE see CLK225
STAFAC see VRF000
STA-FAST see TIX500
STAFLEX DBM see DED600
STAFLEX DBP see DEH200
STAFLEX DBS see DEH600
STAFLEX DOP see DVL700
STA-FRESH 615 see SGM500
STAGNO (TETRACLORURO di) (ITALIAN) see TGC250
STALFLEX DOS see BJS250
STAM see DGI000
STAM M-4 see DGI000
STAM F 34 see DGI000
STAMINE see WAK000
STAM LV 10 see DGI000
STAMPEDE see DGI000
STAMPEDE 3E see DGI000
STAM SUPERNOX see DGI000
STANDACOL CARMOISINE see HJF500
STANDACOL ORANGE G see HGC000
STANDAMIDD LD see BKE500
STANDAPOL 112 CONC see SIB600
STANDARD LEAD ARSENATE see LCK000
STANGEN see WAK000
STANGEN MALEATE see DBM800
STANILO see SLI325
STAN-MAG MAGNESIUM CARBONATE see MAC650
STANNIC BROMIDE see TGB750
STANNIC CHLORIDE, anhydrous (DOT) see TGC250
STANNIC IODIDE see TGD750
STANNIC PHOSPHIDE (DOT) see TGE500
STANNOPLUS see DAE600
STANNORAM see DAE600
STANNOUS CHLORIDE, solid (DOT) see TGC000
STANNOUS CHLORIDE (FCC) see TGC000
STANNOUS IODIDE see TGD500
STANNPLOUS see DAE600
STANOMYCETIN see CDP250
STANOZIDE see ILD000
STAPHOBRISTOL-250 see SLJ000
STAPHYBIOTIC see SLJ000
STAPHYLOMYCIN see VRF000
STAPYOCINE see VRF000
STAR see GGA000
STAR ANISE OIL see AOU250
STARCH DUST see SLJ500
STARCH GUM see DBD800
STARFOL GMS 450 see OAV000
STARFOL GMS 600 see OAV000
STARFOL GMS 900 see OAV000
STARIFEN see EOK000
STARILETTAE see EOK000
STARSOL No. 1 see AQQ500
STATHION see PAK000
STATOBEX see DKE800
STATOMIN see WAK000
STATOMIN MALEATE see DBM800
STAUFFER N-3049 see EPY000
STAUFFER N-4548 see MOB250
STAUFFER CAPTAN see CBG000
STAUFFER MV-119A see DLK200
STAUFFER N 521 see DSB200
STAUFFER N 2790 see FMU045
STAUFFER R-1,303 see TNP250
STAUFFER R-1492 see MQH750
STAUFFER R 1504 see PHX250
STAUFFER R-2061 see PNF500
STAUFFER R-3413 see DTV400

STAURODERM see FMQ000
STAY-FLO see SHF500
STAZEPIN see DCV200
STEADFAST see CCP000
STEARALKONIUM CHLORIDE see DTC600
STEAREX BEADS see SLK000
STEARIC ACID see SLK000
STEARIC ACID ALUMINUM DIHYDROXIDE SALT see
   AHA250
STEARIC ACID, CADMIUM SALT see OAT000
STEARIC ACID-2,3-EPOXYPROPYL ESTER see SLK500
STEARIC ACID, MONOESTER with ETHYLENE GLYCOL
   see EJM500
STEARIC ACID, MONOESTER with GLYCEROL see
   OAV000
STEARIC ACID, SODIUM SALT see SJV500
STEARIC ACID, ZINC SALT see ZMS000
STEARIC MONOGLYCERIDE see OAV000
STEARIX ORANGE see PEJ500
STEAROL see OAX000
STEAROPHANIC ACID see SLK000
STEAR YELLOW JB see DOT300
STEARYL ALCOHOL see OAX000
STEARYLAMINE see OBC000
STEARYLDIMETHYLBENZYLAMMONIUM CHLORIDE see
   DTC600
STEAWHITE see TAB750
STEBAC see DTC600
STECKAPFUL (GERMAN) see SLV500
STECLIN see TBX000
STECLIN HYDROCHLORIDE see TBX250
STEINAMID DL 203 S see BKE500
STEINBUHL YELLOW see BAK250, CAP750
STELADONE see CDS750
STELLAMINE see DJS200
STELLON PINK see PKB500
STEMETIL see PMF500
STEMETIL DIMALEATE see PMF250
STENOLON see MPN500
STENOSINE see DXE600
STENTAL EXTENTABS see EOK000
STEPANOL WAQ see SIB600
STERAFFINE see OAX000
STERAL see HCL000
STERANDRYL see TBG000
STERANE see PMA000
STERASKIN see HCL000
STERCORIN see DKW000
STERIDO see BIM250
STERIGMATOCYSTIN see SLP000
STERILIZING GAS ETHYLENE OXIDE 100% see EJN500
STERISEAL LIQUID #40 see SGM500
STERLING see SFT000
STERLING WAQ-COSMETIC see SIB600
STEROGYL see VSZ100
STEROLAMIDE see TKP500
STEROLONE see PMA000
STERONYL see MPN500
2,2',2''-(STIBILIDYNETRIS(THIO)TRIS-BUTANEDIOIC ACID
   HEXALITHIUM SALT see LGU000
STIBILIUM see DKA600
STIBINE see SLQ000
STIBINE, TRICHLORO- see AQC500
STIBIUM see AQB750
STICKMONOXYD (GERMAN) see NEG100
STICKSTOFFDIOXID (GERMAN) see NGR500
STICKSTOFFLOST see BIE500
STICKSTOFFWASSERSTOFFSAEURE (GERMAN) see
   HHG500
STIGMANOL BROMIDE see DQY800
STIGMANOL METHYL SULFATE see DQY909
STIGMOSAN BROMIDE see DQY800
STIGMOSAN METHYL SULFATE see DQY909
STIKSTOFDIOXYDE (DUTCH) see NGR500

STIL see DKA600
STILALGIN see GGS000
STILBEN (GERMAN) see SLR000
STILBENE see SLR000
α-STILBENECARBONITRILE see DVX600
STILBENYL-N,N-DIMETHYLAMINE see DUB800
STILBESTROL see DKA600
STILBESTROL DIETHYL DIPROPIONATE see DKB000
STILBESTROL DIMETHYL ETHER see DJB200
STILBESTROL DIPROPIONATE see DKB000
STILBESTROL PROPIONATE see DKB000
STILBESTRONATE see DKB000
STILBESTRONE see DKA600
STILBETIN see DKA600
STILBOEFRAL see DKA600
STILBOESTROFORM see DKA600
STILBOESTROL see DKA600
STILBOESTROL DIPROPIONATE see DKB000
STILBOFAX see DKB000
STILBOFOLLIN see DKA600
STILBOL see DKA600
STILKAP see DKA600
STILON see PJY500
STIL-ROL see DKA600
STILRONATE see DKB000
STIMAMIZOL HYDROCHLORIDE see LFA020
STIMINOL see DJS200
STIMULEX see DBA800, DBB000
STIMULIN see DJS200
STIMULINA see GFO050
STINERVAL see PFC500, PFC750
STINK DAMP see HIC500
STIPINE see SEH000
STIPTANON see EDU500
STIROLO (ITALIAN) see SMQ000
ST. JOHN'S BREAD see LIA000
STODDARD SOLVENT see SLU500, PCT250
STOIKON see BCA000
STOMACAIN see DTL200
STOMOLD B see BGJ750
STONE RED see IHD000
STOPAETHYL see DXH250
STOPETHYL see DXH250
STOPETYL see DXH250
STOPGERME-S see CKC000
STOP-SCALD see SAV000
STOPSPOT see PFM500
STOPTON ALBUM see SNM500
STOVAINE see AOM000
STOVARSAL see ABX500
STOVARSOL see ABX500
STOVARSOLAN see ABX500
STOXIL see DAS000
STPP see SKN000
STR see SMD000
STRAMONA (ITALIAN) see SLV500
STRAMONIUM see SLV500
STRATHION see PAK000
STRAWBERRY ALDEHYDE see ENC000
STRAZINE see ARQ725
STREL see DGI000
STREPAMIDE see SNM500
STREPCEN see SLW500
STREPTAGOL see SNM500
STREPTOCLASE see SNM500
STREPTOGRAMIN see VRF000
STREPTOL see SNM500
STREPTOMICINA (ITALIAN) see SLW500
STREPTOMYCES PEUCETIUS see DAC000
STREPTOMYCIN see SLW500
STREPTOMYCIN A see SLW500
STREPTOMYCINE see SLW500
STREPTOMYCINUM see SLW500
STREPTOMYZIN (GERMAN) see SLW500

STREPTOSIL see SNM500
STREPTOSILTHIAZOLE see TEX250
STREPTOVITACIN E 73 see ABN000
STREPTOZOCIN see SMD000
STREPTOZONE see SNM500
STREPTOZOTICIN see SMD000
STREPTROCIDE see SNM500
STRESNIL see FLU000
STREUNEX see BBQ500
STRICNINA (ITALIAN) see SMN500
STROBANE see MIH275, TBC500
STROBANE-T-90 see CDV100
STRONTIUM see SMD500
STRONTIUM ARSENITE see SME500
STRONTIUM ARSENITE, solid (DOT) see SME500
STRONTIUM CHROMATE (1:1) see SMH000
STRONTIUM CHROMATE (VI) see SMH000
STRONTIUM CHROMATE 12170 see SMH000
STRONTIUM COMPOUNDS see SMH500
STRONTIUM FLUORIDE see SMI500
STRONTIUM MONOSULFIDE see SMM000
STRONTIUM(II) NITRATE (1:2) see SMK000
STRONTIUM PEROXIDE see SMK500
STRONTIUM SULFIDE see SMM000
STRONTIUM SULPHIDE see SMM000
STRONTIUM YELLOW see SMH000
STROPHANTHIN G see OKS000
STROPHOPERM see OKS000
STRUMACIL see MPW500
STRUMAZOLE see MCO500
STRYCHININE SULFATE see SMP000
STRYCHNIDIN-10-ONE see SMN500
STRYCHNIDIN-10-ONE, SULFATE (2:1) see SMP000
STRYCHNIN (GERMAN) see SMN500
STRYCHNINE see SMN500
STRYCHNINE, solid and liquid (DOT) see SMN500
STRYCHNINE MONONITRATE see SMO000
STRYCHNINE NITRATE see SMO000
STRYCHNINE SULFATE (2:1) see SMP000
STRYCHNOS see SMN500
STRYPTIRENAL see VGP000
STRZ see SMD000
STUDAFLUOR see SHF500
STUPENONE see DLX400
STYPHNATE of LEAD (DOT) see LEE000
STYPHNIC ACID see SMP500
STYRAFOIL see SMQ500
STYRAGEL see SMQ500
STYRALLYL ALCOHOL see PDE000
STYRALYL ALCOHOL see PDE000
STYREEN (DUTCH) see SMQ000
STYREN (CZECH) see SMQ000
STYREN-ACRYLONITRILEPOLYMER see ADY500
STYRENE see SMQ000
STYRENE-ACRYLONITRILE COPOLYMER see ADY500
STYRENE-BUTADIENE COPOLYMER see SMR000
STYRENE-1,3-BUTADIENE COPOLYMER see SMR000
STYRENE-BUTADIENE POLYMER see SMR000
STYRENE EPOXIDE see EBR000
STYRENE MONOMER (ACGIH) see SMQ000
STYRENE MONOMER, inhibited (DOT) see SMQ000
STYRENE OXIDE see EBR000
STYRENE-7,8-OXIDE see EBR000
STYRENE POLYMER see SMQ500
STYRENE POLYMER with 1,3-BUTADIENE see SMR000
STYRENE POLYMERS see SMQ500
STYROFOAM see SMQ500
STYROL (GERMAN) see SMQ000
STYROLE see SMQ000
STYROLENE see SMQ000
STYROLUX see SMQ500
STYRON see SMQ000, SMQ500
STYRONE see CMQ740
STYROPOR see SMQ000

STYRYL CARBINOL see CMQ740
STYRYL OXIDE see EBR000
STZ see SMD000
SU 5879 see CFY000
SU-9064 see MQR200
SU-13437 see MCB500
SUAVITIL see BCA000
SUBACETATE LEAD see LCH000
SUBAMYCIN see TBX250
SUBARI see HGC500
SUBCHLORIDE of MERCURY see MCW000
SUBERANE see COX500
SUBICARD see PBC250
SUBITEX see BRE500
SUBLIMAT (CZECH) see MCY475
SUBLIMAZE see PDW750
SUBLIMAZE CITRATE see PDW750
SUBLIMED SULFUR see SOD500
SUBLINGULA see HAQ500
SUBTILISIN CARLSBURG see BAC000
SUBTILISIN (9CI, ACGIH) see BAC000
SUBTILISIN NOVO see BAC000
SUBTILISINS (ACGIH) see BAB750
SUBTILISINS BPN see BAB750
SUBTILOPEPTIDASE A see BAC000
SUBTILOPEPTIDASE B see BAC000
SUBTILOPEPTIDASE BPN' see BAC000
SUBTILOPEPTIDASE C see BAC000
SUBTOSAN see PKQ250
SUCARYL see CPQ625
SUCARYL ACID see CPQ625
SUCARYL CALCIUM see CAR000
SUCARYL SODIUM see SGC000
SUCCARIL see SGC000, SJN700
SUCCICURAN see HLC500
SUCCIMAL see ENG500
SUCCIMITIN see ENG500
SUCCINBROMIMIDE see BOF500
SUCCINIBROMIMIDE see BOF500
SUCCINIC ACID see SMY000
SUCCINIC ACID ANHYDRIDE see SNC000
SUCCINIC ACID BIS(β-DIMETHYLAMINOETHYL) ESTER BISMETHIODIDE see BJI000
SUCCINIC ACID BIS(β-DIMETHYLAMINOETHYL) ESTER, DIHYDROCHLORIDE see HLC500
SUCCINIC ACID BIS(β-DIMETHYLAMINOETHYL)ESTER DIMETHOCHLORIDE see HLC500
SUCCINIC ACID, CADMIUM SALT (1:1) see CAI750
SUCCINIC ACID DIESTER with CHOLINE see CMG250
SUCCINIC ACID DIESTER with CHOLINE CHLORIDE see HLC500
SUCCINIC ACID, DIESTER with CHOLINE IODIDE see BJI000
SUCCINIC ACID, DIETHYL ESTER see SNB000
SUCCINIC ACID-2,2-DIMETHYLHYDRAZIDE see DQD400
SUCCINIC ACID DINITRILE see SNE000
SUCCINIC ACID PEROXIDE (DOT) see SNC500
SUCCINIC ANHYDRIDE see SNC000
SUCCINIC-1,1-DIMETHYL HYDRAZIDE see DQD400
SUCCINIC DINITRILE see SNE000
SUCCINIC PEROXIDE see SNC500
SUCCINOCHOLINE see CMG250
SUCCINODINITRILE see SNE000
SUCCINONITRILE see SNE000
SUCCINOYLCHOLINE see CMG250
SUCCINOYLCHOLINE CHLORIDE see HLC500
SUCCINYL-ASTA see HLC500
SUCCINYLBISCHOLINE see CMG250
SUCCINYL BISCHOLINE CHLORIDE see HLC500
SUCCINYLBISCHOLINE DICHLORIDE see HLC500
SUCCINYLCHOLINE CHLORIDE see HLC500
SUCCINYLCHOLINE DICHLORIDE see HLC500
SUCCINYLCHOLINE HYDROCHLORIDE see HLC500
SUCCINYLDICHOLINE see CMG250

SUCCINYLDICHOLINE CHLORIDE see HLC500
SUCCINYLDICHOLINE IODIDE see BJI000
o,o-SUCCINYLDICHOLINE IODIDE see BJI000
SUCCINYLFORTE see HLC500
SUCCINYL OXIDE see SNC000
SUCCINYL PEROXIDE see SNC500
SUCOSTRIN see HLC500
SUCOSTRIN CHLORIDE see HLC500
SUCRA see SJN700
SUCRE EDULCOR see BCE500
SUCRETS see HFV500
SUCRETTE see BCE500
SUCROFER see IHG000
SUCROL see EFE000
SUCROSA see SGC000
SUCROSE see SNH000
SUDAN AX see XRA000
SUDAN GREEN 4B see BLK000
SUDAN III see OHA000
SUDAN ORANGE see XRA000
SUDAN ORANGE R see PEJ500
SUDAN ORANGE RPA see XRA000
SUDAN ORANGE RRA see XRA000
SUDAN RED see XRA000
SUDAN RED 7B see EOJ500
SUDANROT 7B see EOJ500
SUDAN SCARLET 6G see XRA000
SUDAN X see XRA000
SUDAN YELLOW see DOT300
SUDAN YELLOW R see PEI000
SUDAN YELLOW RRA see AIC250
SUDINE see SNN300
SUESSETTE see SGC000
SUESSTOFF see EFE000
SUESTAMIN see SGC000
SUGAI CHRYSOIDINE see PEK000
SUGAI FAST SCARLET G BASE see NMP500
SUGAI TARTRAZINE see FAG140
SUGAR see SNH000
SUGARIN see SGC000
SUGAR of LEAD see LCG000
SUGARON see SGC000
SU 8842 HYDROCHLORIDE see MQR200
SUICALM see FLU000
SUISYNCHRON see MLJ500
SULADYNE see PDC250
SUL ANILINOVA (CZECH) see BBL000
SULDIXINE see SNN300
SULEMA (RUSSIAN) see MCY475
SULFABENZPYRAZINE see QTS000
SULFACID LIGHT ORANGE J see HGC000
SULFACTIN see BAD750
SULFADENE see BDF000
SULFADIMERAZINE see SNJ000
SULFADIMETHOXIN see SNN300
SULFADIMETHOXINE see SNN300
SULFADIMETHOXYDIAZINE see SNN300
SULFADIMETHYLDIAZINE see SNJ000
SULFADIMETHYLPYRIMIDINE see SNJ000
SULFADIMETINE see SNJ000
SULFADIMETOSSINA (ITALIAN) see SNN300
SULFADIMETOXIN see SNN300
SULFADIMEZINE see SNJ000
SULFADIMIDINE see SNJ000
SULFADINE see SNJ000
SULFADSIMESINE see SNJ000
SULFA-ISODIMERAZINE see SNJ000
SULFAISODIMIDINE see SNJ000
SULFALLATE see CDO250
SULFAMATE see ANU650
SULFAMERAZINE SODIUM see SJW475
SULFAMETHALAZOLE see SNK000
SULFAMETHIAZINE see SNJ000
SULFAMETHIN see SNJ000

SULFAMETHOXAZOL see SNK000
SULFAMETHOXAZOLE see SNK000
SULFAMETHYLISOXAZOLE see SNK000
SULFAMEZATHINE see SNJ000
SULFAMIC ACID see SNK500
SULFAMIC ACID, MONOAMMONIUM SALT see ANU650
SULFAMIDIC ACID see SNK500
p-SULFAMIDOANILINE see SNM500
SULFAMIDYL see SNM500
SULFAMINSAURE (GERMAN) see ANU650
N-(5-SULFAMOYL-1,3,4-THIADIAZOL-2-YL)ACETAMIDE
    see AAI250
SULFAMUL see TEX250
SULFAN see SOR500
SULFANA see SNM500
SULFANALONE see SNM500
SULFANIL see SNM500
SULFANILAMIDE see SNM500
6-SULFANILAMIDO-2,4-DIMETHOXYPYRIMIDINE see
    SNN300
2-SULFANILAMIDO-4,6-DIMETHYLPYRIMIDINE see
    SNJ000
3-SULFANILAMIDO-5-METHYLISOXAZOLE see SNK000
2-SULFANILAMIDOQUINOXALINE see QTS000
2-SULFANILAMIDOTHIAZOLE see TEX250
2-(SULFANILYLAMINO)THIAZOLE see TEX250
N^1-SULFANILYL-N^2-BUTYLCARBAMIDE see BSM000
N^1-SULFANILYL-N^2-BUTYLUREA see BSM000
N-SULFANILYL-N'BUTYLUREE (FRENCH) see BSM000
SULFAPOL see DXW200
SULFAPOLU (POLISH) see DXW200
SULFAPYRIDINE NEUTRAL SOLUBLE see DXF400
SULFAQUINOXALINE see QTS000
SULFARSPHENAMINE BISMUTH see BKV250
SULFASAN see BKU500
SULFASAN R POWDER see BKU500
SULFASOL see SNN300
SULFASTOP see SNN300
SULFATE de CUIVRE (FRENCH) see CNP250
SULFATE D'ATROPINE (FRENCH) see ARR500
SULFATE DIMETHYLIQUE (FRENCH) see DUD100
SULFATE MERCURIQUE (FRENCH) see MDG500
SULFATE de METHYLE (FRENCH) see DUD100
SULFATE de NICOTINE (FRENCH) see NDR500
SULFATE de PLOMB (FRENCH) see LDY000
SULFATES see SNS000
SULFATE de ZINC (FRENCH) see ZNA000
SULFATHIAZOL see TEX250
SULFATHIAZOLE (USDA) see TEX250
SULFENAMIDE M see BDG000
SULFERROUS see FBN100, FBO000
SULFIDAL see SOD500
SULFIDES see SNT000
SULFIMEL DOS see DJL000
SULFINPYRAZINE see DWM000
SULFINYLBIS(METHANE) see DUD800
SULFINYL CHLORIDE see TFL000
SULFISOMEZOLE see SNK000
SULFISOMIDIN see SNJ000
SULFISOMIDINE see SNJ000
SULFITES see SNT500
o-SULFOBENZIMIDE see BCE500
o-SULFOBENZOIC ACID IMIDE see BCE500
SULFOCARBANILIDE see DWN800
SULFOCIDINE see SNM500
SULFODIMESIN see SNJ000
SULFODIMEZINE see SNJ000
SULFODOR (CZECH) see EPH000
SULFO GREEN J see FAF000
SULFONA see SOA500
SULFONAMIDE see SNM500
SULFONAMIDE P see SNM500
2-SULFONAMIDOTHIAZOLE see TEX250

2-(4-SULFO-1-NAPHTHYLAZO)-1-NAPHTHOL-4-SULFONIC
ACID, DISODIUM SALT see HJF500
SULFONATES see SNY000
o-SULFONBENZOIC ACID IMIDE SODIUM SALT see
SJN700
SULFONE-2,4,4′,5-TETRACHLORODIPHENYL see CKM000
SULFONIC ACID, MONOCHLORIDE see CLG500
SULFONIMIDE see CBF800
1,1′-SULFONYLBIS(4-AMINOBENZENE) see SOA500
4,4′-SULFONYLBISANILINE see SOA500
p,p-SULFONYLBISBENZAMINE see SOA500
4,4′-SULFONYLBISBENZAMINE see SOA500
p,p-SULFONYLBISBENZENAMINE see SOA500
SULFONYL CHLORIDE see SOT000
4,4′-SULFONYLDIANILINE see SOA500
p,p′-SULFONYLDIANILINE see SOA500
N-(SULFONYL-p-METHYLBENZENE)-N′-N-BUTYLUREA
see BSQ000
SULFOPLAN see SNN300
SULFOPON WA 1 see SIB600
SULFORON see SOD500
SULFOTEP see SOD100
SULFOTEPP see SOD100
SULFOTEX WALA see SIB600
SULFOTHIORINE see SKI500
SULFOX-CIDE see ISA000
SULFOXIDE see ISA000
SULFOXYL see BDS000, ISA000
SULFOXYPHENYLPYRAZOLIDINE see DWM000
SULFRAMIN 85 see DXW200
SULFRAMIN 40 FLAKES see DXW200
SULFRAMIN 40 GRANULAR see DXW200
SULFRAMIN 1238 SLURRY see DXW200
SULFTECH see SJZ000
SULFUR see SOD500
SULFUR CHLORIDE see SOG500, SON510
SULFUR CHLORIDE(DI) (DOT) see SON510
SULFUR CHLORIDE (MONO) (DOT) see SOG500
SULFUR CHLORIDE OXIDE see TFL000
SULFUR DECAFLUORIDE see SOQ450
SULFUR DICHLORIDE see SOG500
SULFUR DIFLUORIDE MONOXIDE see TFL250
SULFUR DIFLUORIDE OXIDE see TFL250
SULFUR DIOXIDE see SOH500
SULFUR DIOXIDE, solution see SOO500
SULFURE de 4-CHLOROBENZYLE et de 4-CHLOROPHE-
NYLE (FRENCH) see CEP000
SULFURE de METHYLE (FRENCH) see TFP000
SULFURETED HYDROGEN see HIC500
SULFUR FLOWER (DOT) see SOD500
SULFUR FLUORIDE see SOI000
SULFUR FLUORIDE OXIDE see BLD000
SULFUR HEXAFLUORIDE see SOI000
SULFUR HYDRIDE see HIC500
SULFURIC ACID see SOI500
SULFURIC ACID (mist) see SOI530
SULFURIC ACID, fuming see SOI520
SULFURIC ACID, ALUMINUM SALT (3:2) see AHG750
SULFURIC ACID, BARIUM SALT (1:1) see BAP000
SULFURIC ACID, BERYLLIUM SALT (1:1) see BFU250
SULFURIC ACID, BERYLLIUM SALT (1:1), TETRAHY-
DRATE see BFU500
SULFURIC ACID, CADMIUM(2+) SALT see CAJ000
SULFURIC ACID, CADMIUM SALT, HYDRATE see
CAJ250
SULFURIC ACID, CADMIUM SALT, TETRAHYDRATE see
CAJ500
SULFURIC ACID, CALCIUM(2+) SALT, DIHYDRATE see
CAX750
SULFURIC ACID, CHROMIUM (3+) POTASSIUM SALT
(2:1:1) see PLB500
SULFURIC ACID, CHROMIUM SALT, BASIC see NBW000
SULFURIC ACID, COBALT(2+) SALT (1:1) see CNE125
SULFURIC ACID, COPPER(2+) SALT (1:1) see CNP250

SULFURIC ACID, CYCLIC ETHYLENE ESTER see EJP000
SULFURIC ACID, DIAMMONIUM SALT see ANU750
SULFURIC ACID, DIMETHYL ESTER see DUD100
SULFURIC ACID, DIPOTASSIUM SALT see PLT000
SULFURIC ACID, DISODIUM SALT see SJY000
SULFURIC ACID, DITHALLIUM(1+) SALT (8CI, 9CI) see
TEM000
SULFURIC ACID, INDIUM SALT see ICJ000
SULFURIC ACID, IRON(2+) SALT (1:1) see FBN100
SULFURIC ACID, IRON (3+) SALT (3:2) see FBA000
SULFURIC ACID, LEAD (2+) SALT (1:1) see LDY000
SULFURIC ACID, MAGNESIUM SALT (1:1) HEPTAHY-
DRATE see MAJ500
SULFURIC ACID, MANGANESE(2+) SALT see MAU250
SULFURIC ACID, MERCURY(2+) SALT (1:1) see MDG500
SULFURIC ACID, MONOANHYDRIDE with NITROUS ACID
see NMJ000
SULFURIC ACID, MONODODECYL ESTER, SODIUM SALT
see SIB600
SULFURIC ACID, MONO(2-ETHYLHEXYL)ESTER, SODIUM
SALT (8CI) see TAV750
SULFURIC ACID, MONOPOTASSIUM SALT see PKX750
SULFURIC ACID, MONOSODIUM SALT see SEG800
SULFURIC ACID, NICKEL(2+)SALT see NDK500
SULFURIC ACID, NICKEL(2+) SALT (1:1) see NDK500
SULFURIC ACID, NICKEL(2+) SALT, HEXAHYDRATE see
NDL000
SULFURIC ACID, THALLIUM SALT see TEL750
SULFURIC ACID, THALLIUM(2+) SALT see TEM250
SULFURIC ACID, THALLIUM(1+) SALT (1:2) see TEM000
SULFURIC ACID, ZINC SALT (1:1) see ZNA000
SULFURIC ACID, ZINC SALT (1:1), HEPTAHYDRATE see
ZNJ000
SULFURIC ACID, ZIRCONIUM(4+) SALT (2:1) see ZTJ000
SULFURIC ANHYDRIDE (DOT) see SOR500
SULFURIC CHLOROHYDRIN see CLG500
SULFURIC OXIDE see SOR500
SULFURIC OXYCHLORIDE see SOT000
SULFURIC OXYFLUORIDE see SOU500
SULFUR MONOCHLORIDE see SON510
SULFUR MUSTARD see BIH250
SULFUR MUSTARD GAS see BIH250
SULFUROUS ACID see SOO500
SULFUROUS ACID ANHYDRIDE see SOH500
SULFUROUS ACID, 2-(p-tert-BUTYLPHENOXY)CYCLO-
HEXYL-2-PROPYNYL ESTER see SOP000
SULFUROUS ACID, 2-(p-tert-BUTYLPHENOXY)-1-METHYL-
ETHYL-2-CHLOROETHYL ESTER see SOP500
SULFUROUS ACID, DIPOTASSIUM SALT see PLT500
SULFUROUS ACID, cyclic ester with 1,4,5,6,7,7-HEXA-
CHLORO-5-NORBORNENE-2,3-DIMETHANOL see
EAQ750
SULFUROUS ACID, MONOSODIUM SALT see SFE000
SULFUROUS ACID, SODIUM SALT (1:2) see SJZ000
SULFUROUS ANHYDRIDE see SOH500
SULFUROUS DICHLORIDE see TFL000
SULFUROUS OXIDE see SOH500
SULFUROUS OXYCHLORIDE see TFL000
SULFUROUS OXYFLUORIDE see TFL250
SULFUR OXIDE see SOH500
SULFUR PENTAFLUORIDE see SOQ450
SULFUR PHOSPHIDE see PHS000
SULFUR SELENIDE see SBT000
SULFUR SUBCHLORIDE see SON510
SULFUR TETRAFLUORIDE see SOR000
SULFUR TRIOXIDE see SOR500
SULFUR TRIOXIDE, STABILIZED (DOT) see SOR500
SULFURYL CHLORIDE see SOT000
SULFURYL FLUORIDE see SOU500
SULKOL see SOD500
SULMET see SNJ000
SULODYNE see PDC250
SULOUREA see ISR000
SULPHABUTIN see BOT250

SULPHADIMETHOXINE see SNN300
SULPHADIMETHYLPYRIMIDINE see SNJ000
SULPHADIMIDINE see SNJ000
SULPHADIONE see SOA500
SULPHAMETHALAZOLE see SNK000
SULPHAMETHOXAZOL see SNK000
SULPHAMETHOXAZOLE see SNK000
SULPHAMETHYLISOXAZOLE see SNK000
SULPHAMIC ACID (DOT) see SNK500
SULPHANILAMIDE see SNM500
3-SULPHANILAMIDO-5-METHYLISOXAZOLE see SNK000
SULPHATHIAZOLE see TEX250
SULPHEIMIDE see CBF800
SULPHISOMEZOLE see SNK000
2-SULPHOBENZOIC IMIDE see BCE500
SULPHOBENZOIC IMIDE CALCIUM SALT see CAM750
SULPHOBENZOIC IMIDE, SODIUM SALT see SJN700
SULPHOCARBONIC ANHYDRIDE see CBV500
SULPHON-MERE see SOA500
1,1'-SULPHONYLBIS(4-AMINOBENZENE) see SOA500
p,p-SULPHONYLBISBENZAMINE see SOA500
4,4'-SULPHONYLBISBENZAMINE see SOA500
p,p-SULPHONYLBISBENZENAMINE see SOA500
4,4'-SULPHONYLBISBENZENAMINE see SOA500
SULPHONYLDIANILINE see SOA500
p,p-SULPHONYLDIANILINE see SOA500
SULPHOS see PAK000
SULPHOXIDE see ISA000
SULPHUR (DOT) see SOD500
SULPHUR, molten (DOT) see SOD500
SULPHUR, lump or power (DOT) see SOD500
SULPHUR DIOXIDE, LIQUEFIED (DOT) see SOH500
SULPHURIC ACID see SOI500
SULPHURIC ACID, CADMIUM SALT (1:1) see CAJ000
SULPHUR MUSTARD GAS see BIH250
SULPROFOS see SOU625
SULSOL see SOD500
SULTANOL see BQF500
SULXIN see SNN300
SULZOL see TEX250
SUM 3170 see DCS200
SUMATRA CAMPHOR see BMD000
SUMICIDIN see FAR100
SUMIFLY see FAR100
SUMILIT EXA 13 see PKQ059
SUMILIT PCX see AAX175
SUMINE 2015 see DQP800
SUMIOXON see PHD750
SUMIPLEX LG see PKB500
SUMIPOWER see FAR100
SUMITHIAN see DSQ000
SUMITOMO LIGHT GREEN SF YELLOWISH see FAF000
SUMITOMO PX 11 see PKQ059
SUMITOX see MAK700
SUMMETRIN see PAG500
SUNAPTIC ACID B see NAR000
SUNAPTIC ACID C see NAR000
SUNCIDE see PMY300
SUNSET YELLOW FCF see FAG150
SUPARI, nut extract see BFW000
SUPERACRYL AE see PKB500
SUPER AMIDE L-9A see BKE500
SUPERCEL 3000 see USS000
SUPER COBALT see CNA250
SUPERCOL see LIA000
SUPERCOL U POWDER see GLU000
SUPERCORTIL see PLZ000
SUPER COSAN see SOD500
SUPER-DENT see SHF500
SUPER D WEEDONE see DAA800, TAA100
SUPERFLAKE ANHYDROUS see CAO750
SUPER HARTOLAN see CMD750
SUPERLYSOFORM see FMV000
SUPERNOX see DGI000

SUPEROL see GGA000
SUPEROL RED C RT-265 see CHP500
SUPERORMONE CONCENTRE see DAA800
SUPEROX see BDS000
SUPEROXOL see HIB000
SUPERPREDNOL see SOW000
SUPER PRODAN see DXE000
SUPER RODIATOX see PAK000
SUPERSEPTIL see SNJ000
SUPERTAH see CMY800
SUPONA see CDS750
SUPONE see CDS750
SUPRA see IHD000
SUPRACAPSULIN see VGP000
SUPRACET ORANGE R see AKP750
SUPRADIN see VGP000
SUPRAMIKE see BAP000
SUPRAMYCIN see TBX250
SUPRANEPHRANE see VGP000
SUPRANEPHRINE see VGP000
SUPRANOL see VGP000
SUPRARENIN see VGP000
SUPREL see VGP000
SUPREMAL see DYE600
SUPREME DENSE see TAB750
SUP'R FLO see DXQ500
SUP'R FLO FERBAM FLOWABLE see FAS000
SUPRIFEN PSB HYDROCHLORIDE see DNU200
SUPRIMAL see HGC500
SURAMETHINIUM see HLC500
SURAUTO see IDW000
SURCHLOR see SHU500
SURCOPUR see DGI000
SUREM see DLY000
SURENINE see VGP000
SURE-SET see CJN000
SURGICAL SIMPLEX see PKB500
SURGI-CEN see HCL000
SURITAL see SOX500
SURITAL SODIUM see SOX500, SOX500
SURITAL SODIUM SALT see SOX500
SUROFENE see HCL000
SURPLIX see DLH600, DLH630
SURPRACIDE see DSO000
SURPUR see DGI000
SU SEGURO CARPIDOR see DUV600
SUSPHRINE see VGP000
SUSTANE see BFW750, BQI000, BRM500
SUSTANONE see TBF500
SUSVIN see MRH209
SUTICIDE see HCQ500
SUXAMETHONIUM CHLORIDE see HLC500
SUXAMETHONIUM see CMG250
SUXAMETHONIUM CHLORIDE see HLC500
SUXAMETHONIUM DICHLORIDE see HLC500
SUXAMETHONIUM IODIDE see BJI000
SUXCERT see HLC500
SUXEMETHONIUM see CMG250
SUXETHONIUM CHLORIDE see HLC500
SUXIL see SNE000
SUXILEP see ENG500
SUXIMAL see ENG500
SUXIN see ENG500
SUXINUTIN see ENG500
SUXINYL see HLC500
SUZORITE MICA see MQS250
SUZU see ABX250
SUZU H see HON000
SV-1522 see ABH500
SVC see ABX500
SVO 9 see PKL100
SWAT see SOY000
SWEBATE see TAL250
SWEDISH GREEN see CNN500, COF500

SWEEP see TBQ750
SWEETA see SJN700
SWEET BIRCH OIL see MPI000
SWEET DIPEPTIDE see ARN825
SWEET ORANGE OIL see OGY000
SWISS BLUE see BJI250
SYDNONE IMINE, 3-(1-METHYL-2-PHENYLETHYL)-,
 MONOHYDROCHLORIDE see SPA000
SYDNOPHENE see SPA000
SYDNOPHEN HYDROCHLORIDE see SPA000
SYKOSE see BCE500, SJN700
SYLANTOIC see DKQ000, DNU000
SYLLIT see DXX400
SYLODEX see ARM268
SYMAZINE see BJP000
SYMBIO see SNN300
SYMETRA see DKE800
SYMMETRIC DIMETHYLUREA see DUM200
SYMPAMIN see BBK500
SYMPAMINA-D see BBK500
SYMPAMINE see BBK000
SYMPATEDRINE see BBK000
SYMPATHIN E see NNO500
SYMPATHIN I see VGP000
m-SYMPATHOL see NCL500
SYMPATHOLYTIN see DCR200
m-SYMPATOL see NCL500
SYMPHYTUM OFFICINALE L see RRP000
SYMULER EOSIN TONER see BNH500
SYMULER LAKE RED C see CHP500
SYMULEX MAGENTA F see FAG070
SYMULEX PINK F see FAG070
SYMULON SCARLET G BASE see NMP500
SYNALGOS see DQA400
SYNANDRETS see MPN500
SYNANDROL see TBG000
SYNANDROL F see TBF500
SYNANDROTABS see MPN500
SYNAPAUSE see EDU500
SYNASTERON see PAN100
SYNCAINE see AIT250
SYNCAL see BCE500
SYNCURARINE see PDD300
SYNDIOL see EDO000
SYNDIOTACTIC POLYPROPYLENE see PMP500
SYNDROX see MDQ500
SYNELAUDINE see DAM700
m-SYNEPHRINE see NCL500
m-SYNEPHRINE HYDROCHLORIDE see SPC500
SYNERGIST 264 see OES000
SYNERONE see TBG000
SYNESTRIN see DKA600, DKB000
SYNESTROL see DAL600
SYNGESTERONE see PMH500
SYNGUM D 46D see GLU000
SYNGYNON see HNT500
SYNHEXYL see HGK500
SYNISTAMIN see TAI500
SYNKAY see MMD500
SYNKLOR see CDR750
SYNMIOL see DAS000
SYNOESTRON see DKB000
SYNOTOL L-60 see BKE500
SYNOVEX S see PMH500
SYNOX TBC see BSK000
SYNPENIN see AIV500
SYNPOL 1500 see SMR000
SYNPREN-FISH see PIX250
SYNSTIGMIN BROMIDE see DQY800
SYNSTIGMINE see DER600
SYNTAR see CMY800
SYNTASE 100 see DMI600
SYNTEDRIL see BBV500
SYNTES 12A see EIV000

SYNTESTRIN see DKB000
SYNTESTRINE see DKB000
SYNTEXAN see DUD800
SYNTHETIC 3956 see CDV100
SYNTHETIC AMORPHOUS SILICA see SCH000
SYNTHETIC BRADYKININ see BML500
SYNTHETIC EUGENOL see EQR500
SYNTHETIC GLYCERIN see GGA000
SYNTHETIC IRON OXIDE see IHD000
SYNTHETIC MUSTARD OIL see AGJ250
SYNTHETIC PYRETHRINS see AFR250
SYNTHETIC WINTERGREEN OIL see MPI000
SYNTHILA see DJB200
SYNTHOESTRIN see DKA600
SYNTHOFOLIN see DKA600
SYNTHOMYCINE see CDP250
SYNTHOSTIGMINE BROMIDE see DQY800
SYNTHOSTIGMINE BROMIDE see POD000
SYNTHOSTIGMINE METHYL SULFATE see DQY909
SYNTHRIN see BEP500
SYNTODRIL see BBV500
SYNTOFOLIN see DKA600
SYNTOLUTAN see PMH500
SYNTOMETRINE see LJL000
SYNTOPHEROL see VSZ450
SYNTOSTIGMIN see DER600
SYNTOSTIGMIN BROMIDE see DQY800
SYNTOSTIGMINE BROMIDE see DQY800
SYNTOSTIGMIN (TABLET) see DQY800
SYNTRON B see EIV000
SYRINGOL see DOJ200
SYRUP of IPECAC, U.S.P. see IGF000
SYSTAM see OCM000
SYSTEMOX see DAO600
SYSTOPHOS see OCM000
SYSTOX see DAO600
SYSTOX SULFONE see SPF000
SYTAM see OCM000
SYTOBEX see VSZ000

α-T see MIH275
β-T see TIN000
T-47 see PAK000
T 72 see PNX000
T 100 see TGM740
T-113 see BPG000
T-125 see TBX000
T-144 see IPX000
2,4,5-T see TAA100
T-1035 see DSA800
T-1036 see DJJ400
T-1152 see HNO500
T-1703 see IRF000
T-1835 see DEC200
T-2002 see BJE750
T-2104 see EIF000
T-2106 see IPX000
TAA see TFA000
TABAC (FRENCH) see TGI100
TABACO (SPANISH) see TGI100
TABALGIN see HIM000
TABLE SALT see SFT000
TABUN see EIF000
TAC 121 see TGG250
TAC 131 see TGG250
TACARYL see MPE250
TACAZYL see MPE250
TACE see CLO750
TACE-FN see CLO750
TACITIN see BCH750
TACOSAL see DKQ000, DNU000
TACRYL see MPE250
TAFASAN see MEP250
TAFAZINE see BJP000

TAFAZINE 50-W see BJP000
TAG see ABU500
TAG-39 see ECU750
TAGAT see SEH000
TAGATHEN see CHY250
TAG FUNGICIDE see ABU500
TAHMABON see DTQ400
TAK see MAK700
TAKAMINA see VGP000
TAKAOKA RHODAMINE B see FAG070
TAKILON see PKQ059
TALADREN see DFP600
TALARGAN see TEH500
TALBOT see LCK000
TALC, containing asbestos fibers see TAB775
TALC (powder) see TAB750
TALCORD see AHJ750
TALCUM see TAB750
TALIMOL see TEH500
TALL OIL see TAC000
TALLOL see TAC000
TALLOW BENZYL DIMETHYLAMMONIUM CHLORIDE see
   DTC600
TALMON see MAO350
TALODEX see FAQ999
TALPHENO see EOK000
TALWAN see DOQ400
TALWIN see DOQ400
TAMARON see DTQ400
TAMAS see BBW500
TAMPOVAGAN STILBOESTROL see DKA600
TAMPULES see CDP000
TANDACOTE see HNI500
TANDALGESIC see HNI500
TANDEARIL see HNI500
TANDERAL see HNI500
TANGANTANGAN OIL see CCP250
TANGERINE OIL see TAD500
TANGERINE OIL, COLDPRESSED (FCC) see TAD500
TANGERINE OIL, EXPRESSESED (FCC) see TAD500
TANIDIL see DQA400
TANNEX see IDA000
TANNIC ACID see TAD750
TANNIN see TAD750
TANNIN from CHESTNUT see CDM250
TANNIN from MIMOSA see MQV250
TANNIN from QUEBRACHO see QBJ000
TANONE see DRR400
TANTALIUM PENTAFLUORIDE see TAF250
TANTALUM see TAE750
TANTALUM-181 see TAE750
TANTALUM CHLORIDE see TAF000
TANTALUM FLUORIDE see TAF250
TANTALUM PENTACHLORIDE see TAF000
TANTUM see BBW500
TAOMYCIN see HOH500
TAOMYXIN see HOH500
TAP 85 see BBQ500
TAPAR see HIM000
TAPAZOLE see MCO500
TAPHAZINE see BJP000
TAPIOCA see DBD800
TAP 9VP see DGP900
TAR see CMY800
TAR, from tobacco see CMP800
TAR, liquid (DOT) see CMY800
TARAPACAITE see PLB250
TARAPON K 12 see SIB600
TAR CAMPHOR see NAJ500
TAR, COAL see CMY800
TARDIGAL see DKL800
TARFLEN see TAI250
TARGET MSMA see MRL750
TARICHATOXIN see FOQ000

TARIMYL see SOA500
TARLON XB see PJY500
TARNAMID T see PJY500
TARODYL see GIC000
TARODYN see GIC000
TAR OIL see CMY825
TARRAGON see AFW750
TARRAGON OIL see TAF700
TARTAN see PHK250
TARTAR EMETIC see AQG250
TARTARIC ACID see TAF750
l-TARTARIC ACID, AMMONIUM SALT see DCH000
TARTARIC ACID, DIAMMONIUM SALT see DCH000
TARTARIZED ANTIMONY see AQG250
TARTAR YELLOW see FAG140
TARTRATE ANTIMONIO-POTASSIQUE (FRENCH) see
   AQG250
TARTRATED ANTIMONY see AQG250
TARTRATE de NICOTINE (FRENCH) see NDS500
TARTRAZINE see FAG140
TARTRAZOL YELLOW see FAG140
TARZOL see DGA200
TASK see DGP900
TASK TABS see DGP900
TAT CHLOR 4 see CDR750
TATD see DXH250
TATERPEX see CKC000
TATTOO see DQM600
TAURE(o)DON see GJC000
TAXILAN see PCK500
TAYSSATAO see MEP250
TAZONE see BRF500
TB see CMO250
TBB see THX500
TBBA see BQK500
TBDZ see TEX000
TBE see ACK250
TBEP see BPK250
TBHP-70 see BRM250
TBHQ (FCC) see BRM500
TBOT see BLL750
TBP see TIA250
TBT see BSP500
TBTO see BLL750
2,4,5-TC see TIX500
TCA see TII250
T-250 CAPSULES see TBX250
T 1 (CATALYST) see DBF800
TCDBD see TAI000
TCDD see TAI000
2,3,7,8-TCDD see TAI000
TCE see TBQ100
1,1,1-TCE see MIH275
TC HYDROCHLORIDE see TBX250
TCIN see TBQ750
TCM see CHJ500
TCNA see TBR250
m-TCPN see TBQ750
2,4,5-TCPPA see TIX500
TCTH see CQH650
TCTP see TBV750
TD-183 see TBV750
TD-5032 see HEE500
TDA see TGL750
TDBP (CZECH) see TNC500
TDE see BIM500
o,p-TDE see CDN000
o,p'-TDE see CDN000
p,p'-TDE see BIM500
TDI see TGM740, TGM750
2,4-TDI see TGM750
2,6-TDI see TGM800
TDI-80 see TGM740, TGM750
TDI 80-20 see TGM740

T-82 DIFUMARATE see MDU750
TDPA see BHM000
TEA see TJN750
TEABERRY OIL see MPI000
TEBECID see ILD000
TEBENIC see ILD000
TEBERUS see EPQ000
TEBEXIN see ILD000
TEBOS see ILD000
TEBRAZID see POL500
TEC see TJP750
TECH DDT see DAD200
TECHNETIUM TC 99M SULFUR COLLOID see SOD500
TECHNICAL BHC see BBQ750
90 TECHNICAL GLYCERINE see GGA000
TECHNICAL HCH see BBQ750
TECHNOPOR see PKQ059
TECH PET F see MQV750
TECODIN see DLX400
TECODINE see DLX400
TECOFLEX HR see PKN000
TECQUINOL see HIH000
TECSOL see EFU000
TECTO see TEX000
TECZA see TJR000
TEDION see CKM000
TEDION V-18 see CKM000
TEDP (OSHA, MAK) see SOD100
TEEBACONIN see ILD000
TEF see TND250
TEFAMIN see TEP000
TEFILIN see TBX250
TEFLON see TAI250
TEFLON (various) see TAI250
TEG see TJQ000
TEGIN see OAV000
TEGIN 503 see OAV000
TEGIN 515 see OAV000
TEGOLAN see CMD750
TEGO-OLEIC 130 see OHU000
TEGOPEN see SLJ000
TEGOSEPT E see HJL000
TEGOSEPT M see HJL500
TEGOSEPT P see HNU500
TEGO-STEARATE see EJM500
TEGOSTEARIC 254 see SLK000
TEGRETAL see DCV200
TEGRETOL see DCV200
TEIB see TND000
TEKAZIN see ILD000
TEKODIN see DLX400
TEKRESOL see CNW500
TEKWAISA see MNH000
TEL see TCF000
TELAGAN see TEH500
TELARGAN see TEH500
TELARGEAN see TEH500
TELDRIN see TAI500
TELEFOS see IOT000
TELEPATHINE see HAI500
TELESMIN see DCV200
TELIDAL see HNI500
TELINE see TBX250
TELIPEX see TBG000
TELLOY see TAJ000
TELLUR (POLISH) see TAJ000
TELLURIUM see TAJ000
TELLURIUM (dust or fume) see TAJ010
TELLURIUM COMPOUNDS see TAJ500
TELLURIUM DIETHYLDITHIOCARBAMATE see EPJ000
TELLURIUM HEXAFLUORIDE see TAK250
TELLUROUS ACID, DISODIUM SALT see SKC500
TELMICID see DJT800
TELMID see DJT800

TELMIDE see DJT800
TELMIN see MHL000
TELODRIN see OAN000
TELONE see DGG000, DGG950
TELONE II SOIL FUMIGANT see DGG950
TELON FAST BLACK E see AQP000
TELOTREX see TBX250
TELTOZAN see CAL750
TELVAR see CJX750, DXQ500
TELVAR DIURON WEED KILLER see DXQ500
TELVAR MONURON WEEDKILLER see CJX750
TEMED see TDQ750
TEMEFOS see TAL250
TEMENTIL see PMF500
TEMEPHOS see TAL250
TEMIC see CBM500
TEMIK see CBM500
TEMIK G10 see CBM500
TEMLO see HIM000
TEMOPHOS see TAL250
TEMPANAL see HIM000
TEMPARIN see BJZ000
TEMPLIN OIL see AAC250
TEMPODEX see BBK500
TEMPRA see HIM000
TEMUR see TDX250
TEN see TJO000
TENAC see DGP900
TENALIN see TEO250
TENAMENE 2 see DEG200
TENDEARIL see HNI500
TENDUST see NDN000
TENIATHANE see MJM500
TENIPOSIDE see EQP000
TENITE 423 see PMP500
TENITE 800 see PJS750
TENNECETIN see PIF750
TENNECO 1742 see PKQ059
TENN-PLAS see BCL750
TENNUS 0565 see AAX175
TENOX BHA see BQI000
TENOX BHT see BFW750
TENOX HQ see HIH000
TENOX PG see PNM750
TENOX P GRAIN PRESERVATIVE see PMU750
TENOX TBHQ see BRM500
TENSINYL see MDQ250
TENSIVAL see TEH500
TENSOL 7 see PKB500
TENTON see MFK500
TENTONE see MFK500
TENTONE MALEATE see MFK750
TENTRATE-20 see PBC250
TENURID see DXH250
TENUTEX see DXH250
TEOBROMIN see TEO500
TEODRAMIN see DYE600
TEOFYLLAMIN see TEP000
TEOLAXIN see EOK000
TEONANACATL see PHU500
TEOS see EPF550
TEP see TCF280
TEPA see TND250
TEPERINE see DLH630
TEPIDONE see SGF500
TEPIDONE RUBBER ACCELERATOR see SGF500
TEPILTA see DTL200
TEPOGEN see SLJ000
TEPP see TCF250
TERABOL see MHR200
TERALUTIL see HNT500
TERAMETHYL THIURAM DISULFIDE see TFS350
TERAMYCIN HYDROCHLORIDE see HOI000
TERBENZENE see TBD000

TERBIUM see TAL750
TERBIUM CHLORIDE see TAM000
TERBIUM CITRATE see TAM500
TEREBENTHINE (FRENCH) see TOD750
TEREFTALODINITRIL (CZECH) see BBP250
TEREPHTHALIC ACID METHYL ESTER see DUE000
TEREPHTHALONITRILE see BBP250
TERETON see MFW100
TERGEMIST see TAV750
TERGIMIST see TAV750
TERGITOL 08 see TAV750
TERGITOL ANIONIC 08 see TAV750
TERGITOL TP-9 (NONIONIC) see PKF000
TERMIL see TBQ750
TERMITKIL see DEP600
TERM-I-TROL see PAX250
TERMOSOLIDO RED LCG see CHP500
TERPENE POLYCHLORINATES see TBC500
TERPENTIN OEL (GERMAN) see TOD750, PIH750
m-TERPHENYL see TBC620
o-TERPHENYL see TBC640
p-TERPHENYL see TBC750
1,3-TERPHENYL see TBC620
TERPHENYLS see TBD000
α-TERPINENE (FCC) see MLA250
γ-TERPINENE (FCC) see MCB750
TERPINEOL see TBD500
α-TERPINEOL ACETATE see TBE250
α-TERPINEOL (FCC) see TBD500
TERPINEOLS see TBD500
TERPINOLENE see TBE000
TERPINYL ACETATE see TBE250
TERPINYL PROPIONATE see TBE600
TERPINYL THIOCYANOACETATE see IHZ000
Δ-1,8-TERPODIENE see MCC250
TERRA ALBA see CAX750
TERRACHLOR see PAX000
TERRACUR P see FAQ800
TERRAFUN see PAX000
TERRAFUNGINE see HOH500
TERRAMITSIN see HOH500
TERRAMYCIN see HOH500
TERRA-SYSTAM see BJE750
TERRA-SYTAM see BJE750
TERRASYTUM see BJE750
TERR-O-GAS 100 see MHR200
TERSAN see TFS350
TERSAN 1991 see BAV575
TERSASEPTIC see HCL000
TERTRACID LIGHT ORANGE G see HGC000
TERTRACID RED CA see HJF500
TERTRAL D see PEY500
TERTRAL EG see ALT500
TERTRAL ERN see NAW500
TERTROCHROME BLUE FB see HJF500
TERTRODIRECT BLACK E see AQP000
TERTRODIRECT BLUE 2B see CMO000
TERTROGRAS ORANGE SV see PEJ500
TERTROPHENE BRILLIANT GREEN G see BAY750
TERTROPHENE BROWN CG see PEK000
TERTROSULPHUR BLACK PB see DUZ000
TERTROSULPHUR PBR see DUZ000
TERULAN KP 2540 see ADY500
o-(4-TERZ.-BUTIL-2-CLORO-FENIL)-o-METIL-FOSFORAM-
    MIDE (ITALIAN) see COD850
TESCOL see EJC500
TESERENE see DAL600
TESLEN see TBF500
TESPAMINE see TFQ750
TESTAFORM see TBG000
TESTANDRONE see TBF500
TESTAVOL see VSK600
TESTEX see TBG000
TESTHORMONE see MPN500

TESTICULOSTERONE see TBF500
TESTOBASE see TBF500
TESTODET see TBG000
TESTODRIN see TBG000
TESTOGEN see TBG000
TESTONIQUE see TBG000
TESTOPROPON see TBF500
TESTORA see MPN500
TESTORMOL see TBG000
TESTOSTEROID see TBF500
TESTOSTERONE see TBF500
trans-TESTOSTERONE see TBF500
TESTOSTERONE HYDRATE see TBF500
TESTOSTERONE PROPIONATE see TBG000
TESTOSTERONE-17-PROPIONATE see TBG000
TESTOSTERONE-17-β-PROPIONATE see TBG000
TESTOSTERON PROPIONATE see TBG000
TESTOSTOSTERONE see TBF500
TESTOVIRON see MPN500, TBG000
TESTOVIRON SCHERING see TBF500
TESTOVIRON T see TBF500
TESTOXYL see TBG000
TESTRED see MPN500
TESTREX see TBG000
TESTRONE see TBF500
TESTRYL see TBF500
TESULOID see SOD500
TETA see TJR000
TETD see DXH250
TETIDIS see DXH250
TETLEN see PCF275
TETNOR see BRF500
O,O,O',O'-TETRAETHYL-BIS(DITHIOPHOSPHAT) (GER-
    MAN) see EEH600
O,O,O,O-TETRAETHYL-DIPHOSPHAT, BIS(O,O-DIA-
    ETHYLPHOSPHORSAEURE-ANHYDRID (GERMAN) see
    TCF250
O,O,O,O-TETRAETHYL-DITHIONOPYROPHOSPHAT
    (GERMAN) see SOD100
1,3,5,7-TETRAAZAADAMANTANE see HEI500
1,4,7,10-TETRAAZADECANE see TJR000
TETRABAKAT see TBX250
TETRA-BASE see MJN000
TETRABLET see TBX250
TETRABON see TBX000
1,1,2,2-TETRABROMAETHAN (GERMAN) see ACK250
TETRABROMIDE METHANE see CBX750
TETRABROMOACETYLENE see ACK250
2,4,5,7-TETRABROMO-9-o-CARBOXYPHENYL-6-HY-
    DROXY-3-ISOXANTHONE, DISODIUM SALT see
    BNH500
1,1,2,2-TETRABROMOETANO (ITALIAN) see ACK250
S-TETRABROMOETHANE see ACK250
1,1,2,2-TETRABROMOETHANE see ACK250
2,4,5,7-TETRABROMO-3,6-FLUORANDIOL see BNH500
TETRABROMOFLUORESCEIN see BNH500
2',4',5',7'-TETRABROMOFLUORESCEIN DISODIUM SALT
    see BNH500
TETRABROMOFLUORESCEIN S see BNH500
TETRABROMOFLUORESCEIN SOLUBLE see BNH500
2-(2,4,5,7-TETRABROMO-6-HYDROXY-3-OXO-3H-XAN-
    THENE-9-YL)BENZOIC ACID, DISODIUM SALT see
    BNH500
TETRABROMOMETHANE see CBX750
1,1,2,2-TETRABROOMETHAAN (DUTCH) see ACK250
TETRA-n-BUTYLCIN (CZECH) see TBM250
TETRABUTYLSTANNANE see TBM250
TETRABUTYLTIN see TBM250
TETRABUTYLTITANATE (CZECH) see BSP250
TETRACAINE see BQA010
TETRACAP see PCF275
TETRACAPS see TBX250
TETRACEMATE DISODIUM see EIX500
TETRACEMIN see EIV000

TETRACENE see TEF500
TETRACENE EXPLOSIVE see TEF500
2,4,4′,5-TETRACHLOOR-DIFENYL-SULFON (DUTCH) see
CKM000
1,1,2,2-TETRACHLOORETHAAN (DUTCH) see TBQ100
TETRACHLOORETHEEN (DUTCH) see PCF275
TETRACHLOORKOOLSTOF (DUTCH) see CBY000
TETRACHLOORMETAAN see CBY000
1,1,2,2-TETRACHLORAETHAAN (GERMAN) see TBQ100
TETRACHLORAETHEN (GERMAN) see PCF275
N-(1,1,2,2-TETRACHLORAETHYLTHIO)CYCLOHEX-4-EN-
1,4-DIACARBOXIMID (GERMAN) see CBF800
N-(1,1,2,2-TETRACHLORAETHYLTHIO)TETRAHYDRO-
PHTHALAMID (GERMAN) see CBF800
2,4,4′,5-TETRACHLOR-DIPHENYL-SULFON (GERMAN) see
CKM000
TETRACHLORETHANE see TBQ100
1,1,2,2-TETRACHLORETHANE (FRENCH) see TBQ100
TETRACHLORKOHLENSTOFF, TETRA (GERMAN) see
CBY000
TETRACHLORMETHAN (GERMAN) see CBY000
TETRACHLOROACETONE see TBN250
TETRACHLOROBENZIDINE see TBO000
2,2′,5,5′-TETRACHLOROBENZIDINE see TBO000
3,3′,6,6′-TETRACHLOROBENZIDINE see TBO000
2,2′,5,5′-TETRACHLORO-(1,1′-BIPHENYL)-4,4′-DIAMINE,
(9CI) see TBO000
TETRACHLOROCARBON see CBY000
2,4,5,6-TETRACHLORO-3-CYANOBENZONITRILE see
TBQ750
2,2′,5,5′-TETRACHLORO-4,4′-DIAMINODIPHENYL see
TBO000
2,3,7,8-TETRACHLORODIBENZO(b,e)(1,4)DIOXAN see
TAI000
2,3,6,7-TETRACHLORODIBENZO-p-DIOXIN see TAI000
2,3,7,8-TETRACHLORODIBENZO-p-DIOXIN see TAI000
2,3,7,8-TETRACHLORODIBENZO-1,4-DIOXIN see TAI000
1,1,1,2-TETRACHLORO-2,2-DIFLUOROETHANE see
TBP000
1,1,2,2-TETRACHLORO-1,2-DIFLUOROETHANE see
TBP050
TETRACHLORODIPHENYLETHANE see BIM500
TETRACHLORODIPHENYL OXIDE see TBP250
2,4,4′,5-TETRACHLORODIPHENYL SULFONE see CKM000
2,4,5,4′-TETRACHLORODIPHENYLSULPHONE see
CKM000
TETRACHLOROETHANE see TBP750
sym-TETRACHLOROETHANE see TBQ100
1,1,1,2-TETRACHLOROETHANE see TBQ000
1,1,2,2-TETRACHLOROETHANE see TBQ100
TETRACHLOROETHENE see PCF275
TETRACHLOROETHYLENE (DOT) see PCF275
1,1,2,2-TETRACHLOROETHYLENE see PCF275
N-1,1,2,2-TETRACHLOROETHYLMERCAPTO-4-CYCLOHEX-
ENE-1,2-CARBOXIMIDE see CBF800
N-((1,1,2,2-TETRACHLOROETHYL)SULFENYL)-cis-4-CY-
CLOHEXENE-1,2-DICARBOXIMIDE see CBF800
N-(1,1,2,2-TETRACHLOROETHYLTHIO)-4-CYCLOHEXENE-
1,2-DICARBOXIMIDE see CBF800
TETRACHLOROISOPHTHALONITRILE see TBQ750
TETRACHLOROMETHANE see CBY000
1,2,4,5-TETRACHLORO-3-METHOXY-6-NITROBENZENE
(9CI) see TBR250
TETRACHLORONAPHTHALENE see TBR000
TETRACHLORONITROANISOLE see TBR250
2,3,5,6-TETRACHLORO-4-NITROANISOLE see TBR250
2,3,4,6-TETRACHLOROPHENOL see TBT000
2,4,5,6-TETRACHLOROPHENOL see TBT000
TETRACHLOROPHENYL ETHER see TBP250
m-TETRACHLOROPHTHALONITRILE see TBQ750
TETRACHLOROSILANE see SCQ500
TETRACHLOROTHIOFENE see TBV750
TETRACHLOROTHIOPHENE see TBV750
2,3,4,5-TETRACHLOROTHIOPHENE see TBV750

TETRACHLOROTHORIUM see TFT000
TETRACHLORURE d'ACETYLENE (FRENCH) see TBQ100
TETRACHLORURE de CARBONE (FRENCH) see CBY000
TETRACHLORURE de SILICIUM (FRENCH) see SCQ500
TETRACHLORURE de TITANE (FRENCH) see TGH350
TETRACHLORVINPHOS see TBW100
TETRACICLINA CLORIDRATO (ITALIAN) see TBX250
TETRACICLINA-l-METILENLISINA (ITALIAN) see
MRV250
2,4,4′,5-TETRACLORO-DIFENIL-SOLFONE (ITALIAN) see
CKM000
1,1,2,2-TETRACLOROETANO (ITALIAN) see TBQ100
TETRACLOROETENE (ITALIAN) see PCF275
TETRACLOROMETANO (ITALIAN) see CBY000
TETRACLORURO di CARBONIO (ITALIAN) see CBY000
TETRACOMPREN see TBX250
TETRACYANONICKELATE(2−) DIPOTASSIUM, HYDRATE
see TBW250
TETRACYCLINE see TBX000
TETRACYCLINE CHLORIDE see TBX250
TETRACYCLINE HYDROCHLORIDE see TBX250
TETRACYCLINE I see TBX000
TETRACYCLINE-l-METHYLENE LYSINE see MRV250
TETRACYDIN see ABG750
TETRACYN see TBX000
TETRA-D see TBX250
TETRADECANE see TBX750
TETRADECANOIC ACID see MSA250
TETRADECANOIC ACID, ISOPROPYL see IQN000
TETRADECANOL, mixed isomers see TBY250
12-TETRADECANOYLPHORBOL-13-ACETATE see PGV000
TETRADECIN see TBX000
n-TETRADECOIC ACID see MSA250
TETRADECYL ALCOHOL see TBY250
TETRADECYL DIMETHYL BENZYLAMMONIUM CHLO-
RIDE see TCA500
7,8,13,13A-TETRADEHYDRO-9,10-DIMETHOXY-2,3-
(METHYLENEDIOXY)BERBINIUM SULFATE TRIHY-
DRATE see BFN750
12-o-TETRADEKANOYLPHORBOL-13-ACETAT (GERMAN)
see PGV000
TETRADICHLONE see CKM000
TETRADIFON see CKM000
TETRADIN see DXH250
TETRADINE see DXH250
TETRADIOXIN see TAI000
TETRADIPHON see CKM000
TETRAETHOXYSILANE see EPF550
TETRAETHYLDIAMINO-o-CARBOXY-PHENYL-XANTHE-
NYL CHLORIDE see FAG070
O,O,O-TETRAETHYL-DIFOSFAAT (DUTCH) see TCF250
O,O,O,O-TETRAETHYL-DITHIO-DIFOSFAAT (DUTCH) see
SOD100
TETRAETHYL DITHIONOPYROPHOSPHATE see SOD100
TETRAETHYL DITHIOPYROPHOSPHATE see SOD100
O,O,O,O-TETRAETHYL DITHIOPYROPHOSPHATE see
SOD100
TETRAETHYL DITHIO PYROPHOSPHATE, liquid (DOT) see
SOD100
TETRAETHYLENE GLYCOL see TCE250
TETRAETHYLENEIMIDEPIPERAZINE-N,N′-DIPHOSPHORIC
ACID see BJC250
TETRAETHYLENEPENTAMINE see TCE500
TETRAETHYL LEAD see TCF000
O,O,O′,O′-TETRAETHYL S,S′-METHYLENEBISPHOSPHOR-
DITHIOATE see EEH600
O,O,O′,O′-TETRAETHYL-S,S′-METHYLENEBISPHOS-
PHORODITHIOATE see EEH600
TETRAETHYL S,S′-METHYLENE BIS(PHOSPHOROTHIOLO-
THIONATE) see EEH600
O,O,O′,O′-TETRAETHYL S,S′-METHYLENE DI(PHOSPHO-
RODITHIOATE) see EEH600
TETRAETHYL ORTHOSILICATE see EPF550
TETRAETHYL ORTHOSILICATE (DOT) see EPF550

TETRAETHYLPLUMBANE see TCF000
TETRAETHYL PYROFOSFAAT (BELGIAN) see TCF250
TETRAETHYL PYROPHOSPHATE see TCF250
TETRAETHYL PYROPHOSPHATE, liquid (DOT) see TCF250
TETRAETHYL PYROPHOSPHATE MIXTURE (liquid) see TCF280
TETRAETHYLRHODAMINE see FAG070
TETRAETHYL SILICATE see EPF550
TETRAETHYL SILICATE (DOT) see EPF550
TETRAETHYLTHIOPEROXYDICARBONIC DIAMIDE see DXH250
TETRAETHYLTHIRAM DISULPHIDE see DXH250
TETRAETHYLTHIURAM see DXH250
TETRAETHYLTHIURAM DISULFIDE see DXH250
TETRAETHYLTHIURAM DISULPHIDE see DXH250
N,N,N',N'-TETRAETHYLTHIURAM DISULPHIDE see DXH250
TETRAETIL see DXH250
O,O,O,O-TETRAETIL-DITIO-PIROFOSFATO (ITALIAN) see SOD100
O,O,O,O-TETRAETIL-PIROFOSFATO (ITALIAN) see TCF250
TETRAFIDON see CKM000
TETRAFINOL see CBY000
TETRAFLUORETHYLENE see TCH500
TETRAFLUOROBORATE(1−) compound with p-AMINOBENZOIC ACID 2-(DIETHYLAMINO)ETHYL ESTER see TCH000
TETRAFLUORO BORATE(1-) LEAD (2+) see LDE000
TETRAFLUORODICHLOROETHANE see DGL600
1,1,2,2-TETRAFLUORO-1,2-DICHLOROETHANE see FOO509
TETRAFLUOROETHENE see TCH500
TETRAFLUOROETHENE HOMOPOLYMER see TAI250
TETRAFLUOROETHENE POLYMER see TAI250
TETRAFLUOROETHYLENE see TCH500
TETRAFLUOROETHYLENE, inhibited (DOT) see TCH500
TETRAFLUOROETHYLENE HOMOPOLYMER see TAI250
TETRAFLUOROETHYLENE POLYMERS see TAI250
TETRAFLUORO HYDRAZINE see TCI000
TETRAFLUOROMETHANE (DOT) see CBY250
TETRAFLUORO-m-PHENYLENE DIAMINE DIHYDROCHLORIDE see TCI250
TETRAFLUOROSILANE see SDF650
TETRAFLUOROSULFURANE see SOR000
TETRAFORM see CBY000
TETRAFOSFOR (DUTCH) see PHO750
TETRAHELICENE see BCR750
6,7,8,9-TETRAHYDRO-5-AZEPOTETRAZOLE see PBI500
TETRAHYDROBENZALDEHYDE see TCJ100
1,2,5,6-TETRAHYDROBENZALDEHYDE see FNK025
1,2,3,6-TETRAHYDROBENZALDEHYDE (DOT) see FNK025
1,2,3,4-TETRAHYDROBENZENE see CPC579
TETRAHYDROBORATE(1−) POTASSIUM see PKY250
Δ¹-TETRAHYDROCANNABINOL see TCM250
(−)-Δ¹-3,4-trans-TETRAHYDROCANNABINOL see TCM250
(−)-Δ⁶-3,4-trans-TETRAHYDROCANNABINOL see TCM000
(−)-Δ⁸-trans-TETRAHYDROCANNABINOL see TCM000
(−)-Δ⁹-trans-TETRAHYDROCANNABINOL see TCM250
(1)-Δ¹-TETRAHYDROCANNABINOL see TCM250
1-trans-Δ⁸-TETRAHYDROCANNABINOL see TCM000
1-trans-Δ⁹-TETRAHYDROCANNABINOL see TCM250
trans-Δ⁹-TETRAHYDROCANNABINOL see TCM250
Δ⁹-TETRAHYDROCANNABINON see TCM250
(E)-1,2,3,4-TETRAHYDRO-3-α,4-β-DIHYDROXY-1-α,2-α-EPOXYBENZ(a)ANTHRACENE see DLE000
(R)-5,6,6a,7-TETRAHYDRO-1,2-DIMETHOXY-6-METHYL-4H-DIBENZO(de,g)QUINOLINE see DNZ000
TETRAHYDRO-2H-3,5-DIMETHYL-1,3,5-THIADIAZINE-2-THIONE see DSB200
TETRAHYDRO-3,5-DIMETHYL-2H-1,3,5-THIADIAZINE-2-THIONE see DSB200

TETRAHYDRO-2,4-DIMETHYLTHIOPHENE-1,1-DIOXIDE see DUD400
TETRAHYDRO-p-DIOXIN see DVQ000
TETRAHYDRO-1,4-DIOXIN see DVQ000
TETRAHYDRO-2,5-DIOXOFURAN see SNC000
1,2,3,6-TETRAHYDRO-3,6-DIOXOPYRIDAZINE see DMC600
TETRAHYDROFTALANHYDRID (CZECH) see TDB000
TETRAHYDROFURAAN (DUTCH) see TCR750
TETRAHYDROFURAN see TCR750
TETRAHYDROFURANNE (FRENCH) see TCR750
TETRAHYDRO-2-FURANONE see BOV000
TETRAHYDROFURFURYLAMINE see TCS500
TETRAHYDROGERANIOL see DTE600
TETRAHYDRO-3-IODOTHIOPHENE-1,1-DIOXIDE see IFG000
3a,4,7,7a-TETRAHYDRO-1,3-ISOBENZOFURANDIOINE see TDB000
TETRAHYDRO-p-ISOXAZINE see MRP750
TETRAHYDRO-1,4-ISOXAZINE see MRP750
TETRAHYDROLINALOOL see TCU600
3a,4,7,7a-TETRAHYDRO-4,7-METHANOINDENE see DGW000
4,6,7,14-TETRAHYDRO-5-METHYL-BIS(1,3)BENZO-DIOXOLO(4,5-c:5′,6′-g)AZECIN-13(5H)-ONE see FOW000
4a,5,7a,8-TETRAHYDRO-12-METHYL-9H-9,9c-IMINO-ETHANOPHENANTHRO(4,5-bcd)FURAN-3,5-DIOL see MRO500
1,2,5,6-TETRAHYDRO-1-METHYLNICOTINIC ACID, METHYL ESTER see AQT750
3,4,5,6,7-TETRAHYDRO-5-METHYL-1-PHENYL-1H-2,5-BENZOXAZOCINE see FAL000
(E)-4,5,6-TETRAHYDRO-1-METHYL-2-(2-(2-THIENYL)ETHENYL)PYRIMIDINE see TCW750
(E)-1,4,5,6-TETRAHYDRO-1-METHYL-2-(2-(2-THIENYL)VINYL)PYRIMIDINE TARTARATE (1:1) see TCW750
dl-TETRAHYDRONICOTYRINE see NDN000
TETRAHYDRO-N-NITROSOPYRROLE see NLP500
TETRAHYDRO-1,4-OXAZINE see MRP750
TETRAHYDRO-2H-1,4-OXAZINE see MRP750
TETRAHYDROPHENOBARBITAL see TDA500
2,3,5,6-TETRAHYDRO-6-PHENYLIMIDAZO(2,1-b)THIAZOLE see LFA000
(−)-2,3,5,6-TETRAHYDRO-6-PHENYLIMIDAZO(2,1-b)THIAZOLE HYDROCHLORIDE see LFA020
1-(−)-2,3,5,6-TETRAHYDRO-6-PHENYL-IMIDAZO(2,1-B)THIAZOLE HYDROCHLORIDE see LFA020
TETRAHYDROPHTHALIC ACID ANHYDRIDE see TDB000
TETRAHYDROPHTHALIC ANHYDRIDE see TDB000
1,2,3,6-TETRAHYDRO PHTHALIC ANHYDRIDE see TDB000
Δ⁴-TETRAHYDROPHTHALIC ANHYDRIDE see TDB000
TETRAHYDRO-6-PROPYL-2H-PYRAN-2-ONE see OCE000
TETRAHYDROPYRROLE see PPS500
1,2,3,4-TETRAHYDROSTYRENE see CPD750
6,7,8,9-TETRAHYDRO-5H-TETRAZOLOAZEPINE see PBI500
TETRAHYDROTHIOPHENE see TDC730
3a,4,7,7a-TETRAHYDRO-N-(TRICHLOROMETHANESULPHENYL)PHTHALIMIDE see CBG000
3a,4,7,7a-TETRAHYDRO-2-((TRICHLOROMETHYL)THIO)-1H-ISOINDOLE-1,3(2H)-DIONE see CBG000
1,2,3,6-TETRAHYDRO-N-(TRICHLOROMETHYLTHIO) PHTHALIMIDE see CBG000
3′,4′,5,7-TETRAHYDROXYFLAVAN-3-OL see QCA000
TETRAHYDROXYMETHYLMETANE see PBB750
TETRAIDROFURANO (ITALIAN) see TCR750
2′,4′,5′,7′-TETRAIODOFLUORESCEIN, DISODIUM SALT see FAG040
TETRAIODOFLUORESCEIN SODIUM SALT see FAG040
TETRAIODOMERCURATE(2-), DIPOTASSIUM see NCP500
TETRAKIS(DIETHYLCARBAMODITHIOATO-S,S′)TELLURIUM see EPJ000

TETRAKIS(DIETHYLDITHIOCARBAMATO)TELLURIUM see EPJ000

TETRAKISDIMETHYLAMINOPHOSPHONOUS ANHYDRIDE see OCM000

TETRAKIS(DIMETHYLCARBAMODITHIOATO-S,S') SELENIUM see SBQ000

TETRAKIS(HYDROXYMETHYL)METHANE see PBB750

TETRALENO see PCF275

TETRALEX see PCF275

TETRALISAL see MRV250

TETRALITE see TEG250

TETRALUTION see TBX250

TETRALYSAL see MRV250

TETRAM see DJA400

TETRAMEEN see TDQ750

6',7',10,11-TETRAMETHOXYEMETAN see EAL500

TETRAMETHOXY SILANE see MPI750

N,N,N'-TETRAMETHYL-3,6-ACRIDINEDIAMINE see BJF000

(R-R*,R*))-N,N,9,9-TETRAMETHYL-10(9H)-ACRIDINEPRO-PANAMINE-2,3-DIHDYROXYBUTANEDIOATE (1:1) see DRM000

TETRAMETHYLAMMONIUM CHLORIDE see TDK000

TETRAMETHYLAMMONIUM HYDROXIDE see TDK500

TETRAMETHYL AMMONIUM HYDROXIDE, liquid (DOT) see TDK500

N,N,N',N'-TETRAMETHYL-DIAMIDO-FOSFORZUUR-FLUORIDE (DUTCH) see BJE750

TETRAMETHYLDIAMIDOPHOSPHORIC FLUORIDE see BJE750

N,N,N',N'-TETRAMETHYL-DIAMIDO-PHOSPHORSAEURE-FLUORID (GERMAN) see BJE750

TETRAMETHYLDIAMINOBENZOPHENONE see MQS500

TETRAMETHYLDIAMINODIPHENYLACETIMINE see IBB000

TETRAMETHYLDIAMINODIPHENYLMETHANE see MJN000

p,p-TETRAMETHYLDIAMINODIPHENYLMETHANE see MJN000

4,4'-TETRAMETHYLDIAMINODIPHENYLMETHANE see MJN000

N,N,N',N'-TETRAMETHYL-1,2-DIAMINOETHANE see TDQ750

N,N,N',N'-TETRAMETHYLDIAMINOMETHAN (GERMAN) see TDR750

N,N,N,2-TETRAMETHYL-1,3-DIOXOLANE-4-METHANAMI-NIUM IODIDE (9CI) see MJH250

TETRAMETHYLDIURANE SULPHITE see TFS350

TETRAMETHYLENE see COW000

TETRAMETHYLENE BIS(METHANESULFONATE) see BOT250

TETRAMETHYLENE CHLOROHYDRIN see CEU500

TETRAMETHYLENE CYANIDE see AER250

TETRAMETHYLENE DIMETHANE SULFONATE see BOT250

1,4-TETRAMETHYLENE GLYCOL see BOS750

TETRAMETHYLENE OXIDE see TCR750

TETRAMETHYLENEOXIRANE see CPD000

TETRAMETHYLENETETRANITRAMINE see CQH250

TETRAMETHYLENETHIURAM DISULPHIDE see TFS350

TETRAMETHYLENIMINE see PPS500

TETRAMETHYL ETHYLENE DIAMINE see TDQ750

N,N,N',N'-TETRAMETHYLETHYLENEDIAMINE see TDQ750

N,N,N',N'-TETRAMETHYLHEXAMETHYLENEDIAMINE-1,3-DIBROMOPROPANE copolymer see HCV500

TETRAMETHYL LEAD see TDR500

N,N,N'N'-TETRAMETHYLMETHANEDIAMINE see TDR750

TETRAMETHYL METHYLENE DIAMINE (DOT) see TDR750

N,2,3,3-TETRAMETHYL-2-NORBORNANAMINE HYDRO-CHLORIDE see MQR500

p-1',1',4',4'-TETRAMETHYLOKTYLBENZENSULFONAN SODNY (CZECH) see DXW200

TETRAMETHYLOLMETHANE see PBB750

N,N,N-α-TETRAMETHYL-10H-PHENOTHIAZINE-10-ETHAN-AMINIUM METHYL SULFATE see MRW000

2-(2,3,5,6-TETRAMETHYLPHENOXY)PROPIONIC AICD see BHM000

(2,3,5,6-TETRAMETHYLPHENYL)MERCURY ACETATE see AAS500

TETRAMETHYLPHOSPHORODIAMIDIC FLUORIDE see BJE750

N,N,N,N-TETRAMETHYLPHOSPHORODIAMIDIC FLUO-RIDE see BJE750

TETRAMETHYLPLUMBANE see TDR500

TETRAMETHYLPYRAZINE see TDV725

2,3,5,6-TETRAMETHYL PYRAZINE (FCC) see TDV725

TETRAMETHYLSILICATE see MPI750

TETRAMETHYLSUCCINONITRILE see TDW250

2,4,6,8-TETRAMETHYL-1,3,5,7-TETROXOCANE see TDW500

TETRAMETHYLTHIOCARBAMOYLDISULPHIDE see TFS350

TETRAMETHYL-O,O'-THIODI-p-PHENYLENE PHOSPHO-ROTHIOATE see TAL250

O,O,O'O',-TETRAMETHYL-O,O'-THIODI-p-PHENYLENE PHOSPHOROTHIOATE see TAL250

TETRAMETHYLTHIONINE CHLORIDE see BJI250

TETRAMETHYLTHIORAMDISULFIDE (DUTCH) see TFS350

TETRAMETHYL-THIRAM DISULFID (GERMAN) see TFS350

TETRAMETHYLTHIURAM BISULFIDE see TFS350

TETRAMETHYLTHIURAM DISULFIDE see TFS350

N,N,N',N'-TETRAMETHYLTHIURAM DISULFIDE see TFS350

N,N-TETRAMETHYLTHIURAM DISULPHIDE see TFS350

TETRAMETHYLTHIURAMMONIUM SULFIDE see BJL600

TETRAMETHYLTHIURAM MONOSULFIDE see BJL600

TETRAMETHYLTHIURAMONOSULFIDE see BJL600

TETRAMETHYLTHIURAM SULFIDE see BJL600

TETRAMETHYL THIURANE DISULFIDE see TFS350

TETRAMETHYLTHIURUM DISULFIDE see TFS350

TETRAMETHYLTRITHIO CARBAMIC ANHYDRIDE see BJL600

N,N',o,o-TETRAMETHYL-(+)-TUBOCURINE see DUL800

TETRAMETHYLUREA see TDX250

1,1,3,3-TETRAMETHYLUREA see TDX250

TETRAMETHYLUREE (FRENCH) see TDX250

N,N,N',N'-TETRAMETIL-FOSFORODIAMMIDO-FLUORURO (ITALIAN) see BJE750

TETRAMINE see HOI000

TETRAMINE FAST BROWN BRS see CMO750

1-TETRAMISOLE HYDROCHLORIDE see LFA020

TETRAN see HOH500

TETRANATRIUMPYROPHOSPHAT (GERMAN) see TEE500

TETRAN HYDROCHLORIDE see HOI000

TETRANITRANILINE (FRENCH) see TDY000

TETRANITROANILINE see TDY000

2,3,4,6-TETRANITROANILINE see TDY000

N,2,4,6-TETRANITROANILINE see TDY075

TETRANITROMETHANE see TDY250

N,2,4,5-TETRANITRO-N-METHYLANILINE see TEG250

TETRANITROPENTAERYTHRITE see PBC250

1,3,5,7-TETRANITROPERHYDRO-1,3,5,7-TETRAZOCINE see CQH250

2,3,4,6-TETRANITROPHENOL see TDY600

TETRAN PTFE see TAI250

TETRA OLIVE N2G see APG500

(±)-(3,5,3',5'-TETRAOXO)-1,2-DIPIPERAZINOPROPANE see PIK250

2,4,5,6-TETRAOXOHEXAHYDROPYRIMIDINE HYDRATE see MDL500

TETRAPHENE see BBC250

TETRAPHOSPHATE HEXAETHYLIQUE (FRENCH) see HCY000

TETRAPHOSPHOR (GERMAN) see PHO750

TETRAPHOSPHORUS TRISULFIDE see PHS500
TETRAPOM see TFS350
TETRAPROPYLENE see PMP750
TETRAPROPYL LEAD see TED750
TETRASIPTON see TFS350
TETRASODIUM DIPHOSPHATE see TEE500
TETRASODIUM EDTA see EIV000
TETRASODIUM ETHYLENEDIAMINETETRAACETATE see EIV000
TETRASODIUM ETHYLENEDIAMINETETRACETATE see EIV000
TETRASODIUM (ETHYLENEDINITRILO)TETRAACETATE see EIV000
TETRASODIUM PYROPHOSPHATE see TEE500
TETRASODIUM PYROPHOSPHATE, ANHYDROUS see TEE500
TETRASODIUM SALT of EDTA see EIV000
TETRASODIUM SALT of ETHYLENEDIAMINETETRACETI-CACID see EIV000
TETRASOL see CBY000
TETRASTIGMINE see TCF250
TETRASULE see PBC250
TETRASULFIDE, BIS(ETHOXYTHIOCARBONYL) see BJU250
TETRA SYTAM see BJE750
TETRATHIURAM DISULFIDE see TFS350
TETRATHIURAM DISULPHIDE see TFS350
TETRAVEC see PCF275
TETRAVERINE see TBX000
TETRAVOS see DGP900
TETRA-WEDEL see TBX250
7,8,9,10-TETRAZABICYCLO(5.3.0)-8,10-DECADIENE see PBI500
1,2,3,3A-TETRAZACYCLOHEPTA-8A,2-CYCLOPENTA-DIENE see PBI500
TETRAZENE see TEF500
TETRAZOBENZENE-β-NAPHTHOL see OHA000
TETRAZO DEEP BLACK G see AQP000
TETRINE see EIV000
TETRINE ACID see EIX000
TETRODIRECT BLACK EFD see AQP000
TETRODONTOXIN see FOQ000
TETRODOTOXIN see FOQ000
TETRODOXIN see FOQ000
TETROGUER see PCF275
TETROLE see FPK000
TETRON see TCF250, TCF280
TETRON-100 see TCF250
TETROPIL see PCF275
TETROSAN see AFP750
TETROSOL see TBX250
TETRYL see TEG250
2,4,6-TETRYL see TEG250
TETRYL FORMATE see IIR000
TETURAM see DXH250
TETURAMIN see DXH250
TEVCOCIN see CDP250
TEVCODYNE see BRF500
TEXACO LEAD APPRECIATOR see BPV100
TEXAN RED TONER D see CHP500
TEXAPON ZHC see SIB600
TEXIN 192A see PKO500
TEXIN 445D see PKM250
TEXTILE see SFT500
TEX WET 1001 see DJL000
TFF see TNT500
T-FLUORIDE see SHF500
T-GAS see EJN500
β-TGDR see TFJ250
T-GELB BZW, GRUN 1 see QCA000
1064 TH see DUO400
TH 1314 see EPQ000
TH 346-1 see DRR400
TH 6040 see CJV250

THACAPZOL see MCO500
THALAMONAL see DYF200
THALIDOMIDE see TEH500
THALIN see TEH500
THALINETTE see TEH500
THALLIC OXIDE see TEL050
THALLIUM see TEI000
THALLIUM ACETATE see TEI250
THALLIUM(I) ACETATE see TEI250
THALLIUM(1+) ACETATE see TEI250
THALLIUM CHLORIDE see TEJ250
THALLIUM(1+) CHLORIDE see TEJ250
THALLIUM COMPOUNDS see TEJ500
THALLIUM MONOACETATE see TEI250
THALLIUM MONOCHLORIDE see TEJ250
THALLIUM MONONITRATE see TEK750
THALLIUM NITRATE see TEK750
THALLIUM OXIDE see TEL050
THALLIUM (3+) OXIDE see TEL050
THALLIUM(III) OXIDE see TEL050
THALLIUM PEROXIDE see TEL050
THALLIUM SESQUIOXIDE see TEL050
THALLIUM SULFATE see TEL750
THALLIUM(I) SULFATE (2:1) see TEM000
THALLIUM(II) SULFATE (1:1) see TEM250
THALLIUM SULFATE, solid (DOT) see TEL750
THALLOUS ACETATE see TEI250
THALLOUS CHLORIDE see TEJ250
THALLOUS NITRATE see TEK750
THALLOUS SULFATE see TEM000
THAM see POC750
THANISOL see IHZ000
THANITE see IHZ000
THBP see TKO250
THC see TCM250
Δ^1-THC see TCM250
Δ^6-THC see TCM000
Δ^8-THC see TCM000
Δ^9-THC see TCM250
THEAL TABL. see TEP000
THECODIN see DLX400
THECODINE see DLX400
THEELIN see EDV000
THEELOL see EDU500
THEIN see CAK500
THEINE see CAK500
THEKODIN see DLX400
THELESTRIN see EDV000
THELYKININ see EDV000
THENALTON see PAG200
THENARDITE see SJY000
THENARDOL see HIB500
THENOBARBITAL see EOK000
2-THENYLAMINE, 5-CHLORO-N-(2-(DIMETHYLAMINO)ETHYL)-N-2-PYRIDYL- see CHY250
THENYLENE see TEO250
THENYLPYRAMINE see TEO250
THEOBROMINE see TEO500
THEOCIN see TEP000
THEOFOL see TEP000
THEOGEN see ECU750
THEOGRAD see TEP000
THEOLAIR see TEP000
THEOLIX see TEP000
THEOLOXIN see EOK000
THEOMINAL see EOK000
THEOPHILCHOLINE see TEH500
THEOPHYL-225 see TEP000
THEOPHYLLIN see TEP000
THEOPHYLLINE see TEP000
THEOPHYLLINE, anhydrous see TEP000
THEOSALVOSE see TEO500
THEOSTENE see TEO500
THERACANZAN see SNN300

THERADERM see BDS000
THERA-FLUR-N see SHF500
THERALEPTIQUE see DUO400
THERAMINE see HGD000
THERAPOL see SNM500
THERAPTIQUE see DUO400
THERAZONE see BRF500
THERMACURE see MKA500
THERMALOX see BFT250
THERM CHEK 820 see DDV600
THERMINOL FR-1 see PJL750
"THERMIT" see TER000
THERMOASE PC-10 see BAC000
THERMOLITE 831 see BKK750
THERMOPLASTIC 125 see SMR000
THESAL see TEO500
THESODATE see TEO500
THETAMID see ENG500
THF see TCR750
1-THIA-3-AZAINDENE see BDE500
THIABEN see TEX000
THIABENDAZOLE HYDROCHLORIDE see TER500
THIABENDAZOLE (USDA) see TEX000
THIABENZOLE see TEX000
3-THIABUTAN-2-ONE, O-(METHYLCARBAMOYL)OXIME
  see MDU600
THIACETAMIDE see TFA000
THIACETIC ACID see TFA500
THIACOCCINE see TEX250
THIACYCLOPENTADIENE see TFM250
THIACYCLOPROPANE see EJP500
(N-1,2,3-THIADIAZOLYL-5)-N'-PHENYLUREA see TEX600
4-THIAHEPTANEDIOIC ACID see BHM000
THIAMAZOLE see MCO500
THIAMETON see PHI500
THIAMINE CHLORIDE HYDROCHLORIDE see TET300
THIAMINE DICHLORIDE see TET300
THIAMINE HYDROCHLORIDE see TET300
THIAMINE MONONITRATE see TET500
THIAMINE NITRATE see TET500
THIAMIN HYDROCHLORIDE see TET300
THIAMINIUM CHLORIDE HYDROCHLORIDE see TET300
THIAMIZIDE see CIP500
THIAMUTILIN see TET800
THIAMYLAL SODIUM see SOX500
THIANIDE see EPQ000
3-THIAPENTANE see EPH000
THIAPHENE see TFM250
2-THIAPROPANE see TFP000
THIARETIC see CFY000
THIATE H see DKC400
THIAZAMIDE see TEX250
THIAZINAMIUM METHYL SULFATE see MRW000
2-(THIAZOL-4-YL)BENZIMIDAZOLE see TEX000
2-(4-THIAZOLYL)BENZIMIDAZOLE see TEX000
2-(4'-THIAZOLYL)BENZIMIDAZOLE see TEX000
2-(4-THIAZOLYL)-1H-BENZIMIDAZOLE see TEX000
2-(4-THIAZOLYL)-BENZIMIDAZOLE, HYDROCHLORIDE
  see TER500
N^1-2-THIAZOLYLSULFANILAMIDE see TEX250
THIAZON see DSB200
THIAZONE see DSB200
THIBENZOLE see TEX000
THIDIAZURON see TEX600
THIERGAN see DQA400
THIFOR see EAQ750
THIIRANE see EJP500
THILLATE see TFS350
THILOPEMAL see ENG500
THILOPHENYL see DKQ000, DNU000
THIMECIL see MPW500
THIMER see TFS350
THIMEROSALATE see MDI000
THIMEROSOL see MDI000

THIMET see PGS000
THIMUL see EAQ750
THIOACETAMIDE see TFA000
THIOACETIC ACID see TFA500
THIOALKOFEN BM 4 see TFC600
THIOALLATE see CDO250
THIOAMIDE see EPQ000
THIOANILINE see TFI000
4,4'-THIOANILINE see TFI000
(THIOARSENOSO)METHANE see MGQ750
THIOBENZYL ALCOHOL see TGO750
2-THIO-2-BENZYL-PSEUDOUREA HYDROCHLORIDE see
  BEU500
4,4'-THIOBIS(ANILINE) see TFI000
4,4'-THIOBISBENZENAMINE see TFI000
4,4'-THIOBIS(6-tert-BUTYL-m-CRESOL) see TFC600
4,4'-THIOBIS(2-tert-BUTYL-5-METHYLPHENOL) see
  TFC600
4,4'-THIOBIS(6-tert-BUTYL-3-METHYLPHENOL) see
  TFC600
1,1'-THIOBIS(2-CHLOROETHANE) see BIH250
1,1'-THIOBIS(N,N-DIMETHYLTHIO)FORMAMIDE see
  BJL600
THIOBIS(DODECYL PROPIONATE) see TFD500
1,1'-THIOBISETHANE see EPH000
4,4'-THIOBIS(3-METHYL-6-tert-BUTYLPHENOL) see
  TFC600
1,1'-THIOBIS(2-METHYL-4-HYDROXY-5-tert-BUTYLBEN-
  ZENE) see TFC600
THIOBISMOL see BKX750
THIOCARB see SGJ000
THIOCARBAMATE see ISR000
THIOCARBAMIDE see ISR000
THIOCARBAMISIN see TFD750
THIOCARBAMIZINE see TFD750
THIOCARBAMYLHYDRAZINE see TFQ000
THIOCARBANIL see ISQ000
THIOCARBANILIDE see DWN800
THIOCARBARSONE see CBI250
THIOCARBONIC DICHLORIDE see TFN500
THIOCARBONYL CHLORIDE (DOT) see TFN500
THIOCARBONYL DICHLORIDE see TFN500
THIOCHRYSINE see GJG000
THIOCYANATES see TFE500
THIOCYANATE SODIUM see SIA500
THIOCYANATOACETIC ACID ISOBORNYL ESTER see
  IHZ000
THIOCYANATOETHANE see EPP000
α-THIOCYANATOTOLUENE see BFL000
THIOCYANIC ACID, AMYL ESTER see AON500
THIOCYANIC ACID, CALCIUM SALT (2:1) see CAY250
THIOCYANIC ACID, ETHYL ESTER see EPP000
THIOCYANIC ACID, MERCURY(2$^+$) SALT see MCU250
1-THIOCYANOBUTANE see BSN500
THIODAN see EAQ750
THIODEMETON see DXH325
THIODEMETRON see DXH325
p,p-THIODIANILINE see TFI000
4,4'-THIODIANILINE see TFI000
THIODIFENYLAMINE (DUTCH) see PDP250
THIODIGLYCOLIC ACID see MCM750
2,2'-THIODIGLYCOLIC ACID see MCM750
β,β'-THIODIGLYCOLIC ACID see MCM750
THIODIGLYCOLLIC ACID see MCM750
2-THIO-3,5-DIMETHYLTETRAHYDRO-1,3,5-THIADIAZINE
  see DSB200
THIODIPHENYLAMIN (GERMAN) see PDP250
THIODIPHENYLAMINE see PDP250
O,O'-(THIODI-4,1-PHENYLENE)BIS(O,O-DIMETHYL PHOS-
  PHOROTHIOATE) see TAL250
THIODI-p-PHENYLENEDIAMINE see TFI000
O,O'-(THIODI-p-PHENYLENE)-O,O,O',O'-TETRAMETHYL
  BIS(PHOSPHOROTHIOATE) see TAL250
THIODIPROPIONIC ACID see BHM000

3,3′-THIODIPROPIONIC ACID see BHM000
β,β′-THIODIPROPIONIC ACID see BHM000
THIODOW see EIR000
THIOETHANOL see EMB100
2-THIOETHANOL see MCN250
THIOETHANOLAMINE see AJT250
THIOETHYL ALCOHOL see EMB100
THIOETHYL ETHER see EPH000
THIOFACO M-50 see EEC600
THIOFACO T-35 see TKP500
THIOFIDE see BDE750
THIOFOR see EAQ750
THIOFOSGEN (CZECH) see TFN500
THIOFOZIL see TFQ750
THIOFURAM see TFM250
THIOFURAN see TFM250
THIOFURFURAN see TFM250
(1-THIO-d-GLUCOPYRANOSATO)GOLD see ART250
1-THIO-GLUCOPYRANOSE, MONOGOLD(1+) SALT see
  ART250
THIOGLUCOSE d′OR (FRENCH) see ART250
THIOGLYCOL (DOT) see MCN250
THIOGLYCOLANILIDE see MCK000
THIOGLYCOLATESODIUM see SKH500
THIOGLYCOLIC ACID see TFJ100
2-THIOGLYCOLIC ACID see TFJ100
THIOGLYCOLIC ACID ANILIDE see MCK000
THIOGLYCOLLIC ACID, AMMONIUM SALT see ANM500
THIOGLYCOLLIC ACID, SODIUM SALT see SKH500
β-THIOGUANINE DEOXYRIBOSIDE see TFJ250
2-THIO-4-HYDRAZINOURACIL see HHF500
THIOHYPOXANTHINE see POK000
THIOKARBONYLCHLORID (CZECH) see TFN500
THIOLACETIC ACID see TFA500
THIOLDEMETON see DAP200
2-THIOL-DIHYDROGLYOXALINE see IAQ000
THIOLE see TFM250
THIOL SYSTOX see DAP200
THIOLUX see SOD500
THIOMEBUMAL SODIUM see PBT500
THIOMECIL see MPW500
THIOMERIN SODIUM see DXC000
THIOMERSALATE see MDI000
THIOMETAN see DSK600
2-THIO-6-METHYL-1,3-PYRIMIDIN-4-ONE see MPW500
6-THIO-4-METHYLURACIL see MPW500
THIOMETON see PHI500
THIOMIDIL see MPW500
THIOMONOGLYCOL see MCN250
THIOMUL see EAQ750
THIOMYLAL SODIUM see SOX500
THIO-β-NAPHTHOL see NAP500
β-THIONAPHTHOL see NAP500
THIONAZIN see EPC500
THIONEMBUTAL see PBT500
THIONEX see BJL600, EAQ750
THIONEX RUBBER ACCELERATOR see BJL600
THIONIDEN see EPQ000
THIONOACETIC ACID see TFA500
THIONOBENZENEPHOSPHONIC ACID ETHYL-p-NITRO-
  PHENYL ESTER see EBD700
THIONODEMETON SULFONE see SPF000
THIONOSINE see MCQ500
THIONYLAN see TEO250
THIONYL CHLORIDE see TFL000
THIONYL DICHLORIDE see TFL000
THIONYL DIFLUORIDE see TFL250
THIONYL FLUORIDE see TFL250
2-THIO-4-OXO-6-METHYL-1,3-PYRIMIDINE see MPW500
2-THIO-4-OXO-6-PROPYL-1,3-PYRIMIDINE see PNX000
2-THIO-6-OXYPYRIMIDINE see TFR250
THIOPENTAL SODIUM see PBT500
THIOPENTAL SODIUM SALT see PBT500
THIOPENTONE SODIUM see PBT500

THIOPEROXYDICARBONIC ACID DIETHYL ESTER see
  BJU000
THIOPHEN see TFM250
THIOPHENE see TFM250
THIOPHENIT see MNH000
THIOPHENOL (DOT) see PFL850
THIOPHOS see PAK000
THIOPHOS 3422 see PAK000
THIOPHOSGENE see TFN500
THIOPHOSPHAMIDE see TFQ750
THIOPHOSPHATE de S-N-(1-CYANO-1-METHYLETHYL)
  CARBAMOYLMETHYLE et de O,O-DIETHYLE (FRENCH)
  see PHK250
THIOPHOSPHATE de O-2,4-DICHLOROPHENYLE et de O,O-
  DIETHYLE (FRENCH) see DFK600
THIOPHOSPHATE de O,O-DIETHYLE et de O-(3-CHLORO-4-
  METHYL-7-COUMARINYLE) (FRENCH) see CNU750
THIOPHOSPHATE de O,O-DIETHYLE et de O-(2,5-DI-
  CHLORO-4-BROMO) PHENYLE (FRENCH) see EGV500
THIOPHOSPHATE de O,O-DIETHYLE et de S-(2-ETHYL-
  THIO-ETHYLE) (FRENCH) see DAP200
THIOPHOSPHATE de O,O-DIETHYLE et de o-2-ISOPROPYL-
  4-METHYL-6-PYRIMIDYLE (FRENCH) see DCM750
THIOPHOSPHATE de O,O-DIETHYLE et de S-(N-METHYL-
  CARBAMOYL) METHYLE (FRENCH) see DNX800
THIOPHOSPHATE de O,O-DIETHYLE et de O-(4-NITROPHE-
  NYLE) (FRENCH) see PAK000
THIOPHOSPHATE de O,O-DIMETHYLE et de O-3-CHLORO-
  4-NITROPHENYLE (FRENCH) see MIJ250
THIOPHOSPHATE de O,O-DIMETHYLE et de O-4-CHLORO-
  3-NITROPHENYLE (FRENCH) see NFT000
THIOPHOSPHATE de O,O-DIMETHYLE ET DE S-2-(ISOPRO-
  PYLSULFINYL)-ETHYLE see DSK600
THIOPHOSPHATE de O,O-DIMETHYLE et de S-2-ETHYL-
  SULFINYLETHYLE (FRENCH) see DAP000
THIOPHOSPHATE de O,O-DIMETHYLE et de O-2-ETHYL-
  THIO-ETHYLE (FRENCH) see DAO800
THIOPHOSPHATE de O,O-DIMETHYLE et de S-2-ETHYL-
  THIOETHYLE (FRENCH) see DAP400
THIOPHOSPHATE de O,O-DIMETHYLE et de S-((5-ME-
  THOXY-4-PYRONYL)-METHYLE) (FRENCH) see EAS000
THIOPHOSPHATE de O,O-DIMETHYLE et de O-(3-METHYL-
  4-METHYLTHIOPHENYLE) (FRENCH) see FAQ999
THIOPHOSPHATE de O,O-DIMETHYLE et de O-(3-METHYL-
  4-NITROPHENYLE) (FRENCH) see DSQ000
THIOPHOSPHATE de O,O-DIMETHYLE et de O-(4-NITRO-
  PHENYLE) (FRENCH) see MNH000
THIOPHOSPHATE de O,O-DIMETHYLE et de O-(2,4,5-TRI-
  CHLOROPHENYLE) (FRENCH) see RMA500
THIOPHOSPHORIC ANHYDRIDE see PHS000
THIOPHOSPHORSAEURE-O,S-DIMETHYLESTERAMID
  (GERMAN) see DTQ400
THIOPHOSPHORYL CHLORIDE see TFO000
THIOPHOSPHORYL TRICHLORIDE see TFO000
2-THIOPROPANE see TFP000
2-THIO-6-PROPYL-1,3-PYRIMIDIN-4-ONE see PNX000
6-THIO-4-PROPYLURACIL see PNX000
β-THIOPSEUDOUREA see ISR000
6-THIOPURINE RIBONUCLEOSIDE see MCQ500
6-THIOPURINE RIBOSIDE see MCQ500
2-THIO-1,3-PYRIMIDIN-4-ONE see TFR250
THIORIDAZIN see MOO250
THIORIDAZINE see MOO250
THIORIDAZINE HYDROCHLORIDE see MOO500
THIORYL see MPW500
THIOSAN see DXH250, TFS350
THIOSCABIN see DXH250
THIOSEMICARBAZIDE see TFQ000
3-THIOSEMICARBAZIDE see TFQ000
THIOSERINE see CQK000
THIOSTOP N see SGM500
THIOSULFAN see EAQ750
THIOSULFAN TIONEL see EAQ750
THIOSULFATES see TFQ500

THIOSULFIL-A FORTE see PDC250
THIOSULFURIC ACID, DISODIUM SALT, PENTAHYDRATE
  see SKI500
THIOSULFURIC ACID, GOLD(1+) SODIUM SALT(2:1:3),
  DIHYDRATE see GJG000
THIOSULFUROUS DICHLORIDE see SON510
THIO-TEP see TFQ750
THIOTEPP see SOD100
THIOTETROLE see TFM250
THIOTEX see TFS350
THIOTHAL SODIUM see PBT500
2-THIO-1-(THIOCARBAMOYL)UREA see DXL800
THIOTHYMIN see MPW500
THIOTHYRON see MPW500
THIOTOX see TFS350
5-THIO-1H-θ-TRIAZOLO(4,5-d)PYRIMIDINE-5,7(4H,6H)-
  DIONE see MLY000
THIOTRIETHYLENEPHOSPHORAMIDE see TFQ750
THIOURACIL see TFR250
2-THIOURACIL see TFR250
6-THIOURACIL see TFR250
2-THIOUREA see ISR000
THIOUREA (DOT) see ISR000
THIOVANIC ACID see TFJ100
THIOVIT see SOD500
THIOXAMYL see DSP600
THIOXIDIL see GGS000
6-THIOXOPURINE see POK000
THIOZAMIDE see TEX250
THIPENTAL SODIUM see PBT500
THIRAM see TFS350
THIRAMAD see TFS350
THIRAME (FRENCH) see TFS350
THIRASAN see TFS350
THIRERANIDE see DXH250
THIULIX see TFS350
THIURAD see TFS350
THIURAGYL see PNX000
THIURAM see TFS350
THIURAM E see DXH250
THIURAMIN see TFS350
THIURAMYL see TFS350
THIURANIDE see DXH250
THIURETIC see CFY000
THIURYL see MPW500
THLARETIC see CFY000
THOMAPYRIN see ARP250
THOMPSON-HAYWARD TH6040 see CJV250
THOMPSON'S WOOD FIX see PAX250
THORIA see TFT750
THORIDAZINE HYDROCHLORIDE see MOO500
THORIUM see TFS750
THORIUM-232 see TFS750
THORIUM CHLORIDE see TFT000
THORIUM DIOXIDE see TFT750
THORIUM HYDRIDE see TFT250
THORIUM METAL, PYROPHORIC (DOT) see TFS750
THORIUM (4+) NITRATE see TFT500
THORIUM(IV) NITRATE see TFT500
THORIUM OXIDE see TFT750
THORIUM TETRACHLORIDE see TFT000
THORIUM TETRANITRATE see TFT500
THORN APPLE see SLV500
THOROTRAST see TFT750
THORTRAST see TFT750
THPA see TDB000
THREE ELEPHANT see BMC000
l-THREITOL-1,4-BISMETHANESULFONATE see TFU500
THREONINE see TFU750
l-THREONINE see TFU750
THRETHYLENE see TIO750
THROMBOLIQUINE see HAQ500
THU see ISR000
(1S,4R,5R)-(−)-3-THUJANONE see TFW000

THUJA OIL see CCQ500
THUJON see TFW000
THUJONE see TFW000, TFW000
l-THUJONE see TFW000
α-THUJONE see TFW000
THULIUM see TFW250
THULIUM CHLORIDE see TFW500
THULIUM(III) NITRATE, HEXAHYDRATE (1:3:6) see
  TFX250
THULOL see EDU500
THYALONE see BAG250
THYCAPSOL see MCO500
THYLATE see TFS350
THYLFAR M-50 see MNH000
THYLOGEN see WAK000
THYLOGEN MALEATE see DBM800
THYLOQUINONE see MMD500
THYME CAMPHOR see TFX810
THYME OIL see TFX500
THYME OIL RED see TFX750
THYMIAN OEL (GERMAN) see TFX500
THYMIC ACID see TFX810
THYM OIL see TFX500
THYMOL see TFX810
m-THYMOL see TFX810
o-THYMOL see CCM000
THYNESTRON see EDV000
THYNON see DLK200
THYREONORM see MPW500
THYREOSTAT see MPW500
THYREOSTAT II see PNX000
THYRIL see MPW500
TIAMIZID see CIP500
TIAMIZIDE see CIP500
TIAMULIN see TET800
TIAMULINA (ITALIAN) see TET800
TIAMUTIN see DAR000
TIAZON see DSB200
TIBAMATO see MOV500
TIBAZIDE see ILD000
TIBEMID see ILD000
TIBINIDE see ILD000
TIBISON see ILD000
TIBIVIS see ILD000
TIBIZIDE see ILD000
TIBUSAN see ILD000
TICINIL see BRF500
TIC MUSTARD see IAN000
TIDEMOL see BQL000
TIEZENE see EIR000
TIFOMYCINE see CDP250
12-o-TIGLYL-PHORBOL-13-BUTYRATE see PGV750
TIGUVON see FAQ999
TIKOFURAN see NGG500
TILCAREX see PAX000
TILDIN see BNK000
TILLAM (RUSSIAN) see PNF500
TILLAM-6-E see PNF500
TILLANTOX see BDD000
TILLRAM see DXH250
TIMAZIN see FMM000
TIMET see PGS000
TIMOLET see DLH600, DLH630
TIMONIL see DCV200
TIMOSIN see MDQ250
TIN see TGB250
TIN (α) see TGB250
TIN(IV) BROMIDE (1:4) see TGB750
TIN(II) CHLORIDE (1:2) see TGC000
TIN(IV) CHLORIDE (1:4) see TGC250
TIN CHLORIDE, fuming (DOT) see TGC250
TIN COMPOUNDS see TGC500
TINDAL see ABG000
TIN DIBUTYL DILAURATE see DDV600

TIN DICHLORIDE see TGC000
TINDURIN see TGD000
TINESTAN see ABX250
TINESTAN 60 WP see ABX250
TIN FLAKE see TGB250
TINIC see NCQ900
TIN(II) IODIDE see TGD500
TIN(IV) IODIDE (1:4) see TGD750
TINMATE see CLU000
TINNING GLUX (DOT) see ZFA000
TINOSTAT see DDV600
TINOX see MIW250
TIN PERBROMIDE see TGB750
TIN PERCHLORIDE (DOT) see TGC250
TIN (IV) PHOSPHIDE see TGE500
TIN POWDER see TGB250
TIN PROTOCHLORIDE see TGC000
TINSET see DWF790, OMG000
TIN TETRABROMIDE see TGB750
TINTETRACHLORIDE (DUTCH) see TGC250
TIN TETRACHLORIDE, anhydrous (DOT) see TGC250
TIN TETRAIODIDE see TGD750
TIN TRIAETHYLZINNACETAT (GERMAN) see ABW750
TIN TRIPHENYL ACETATE see ABX250
TIODIFENILAMINA (ITALIAN) see PDP250
TIOFINE see TGG760
TIOFOS see PAK000
TIOFOSFAMID see TFQ750
TIOFOZIL see TFQ750
TIOINOSINE see MCQ500
TIOMERACIL see MPW500
TIOPENTAL SODIUM see PBT500
TIORALE M see MPW500
TIORIDAZIN see MOO500
TIOTIRON see MPW500
TIOURACYL (POLISH) see TFR250
TIOVEL see EAQ750
TIOXANONA see TLP750
TIOXIDE see TGG760
TIPEDINE see BLV000
TIPEPIDINE see BLV000
TIPPON see TAA100
TIRAMPA see TFS350
TISIN see ILD000
TISIODRAZIDA see ILD000
TISPERSE MB-2X see NBL000
TISPERSE MB-58 see BHA750
TITAANTETRACHLORED (DUTCH) see TGH350
TITANDIOXID (SWEDEN) see TGG760
TITANE (TETRACHLORURE de) (FRENCH) see TGH350
TITANIO TETRACHLORURO di (ITALIAN) see TGH350
TITANIUM see TGF250
TITANIUM ACETONYL ACETONATE see BGQ750
TITANIUM ALLOY see TGF250
TITANIUM compounded with BERYLLIUM (1:12) see
  BFR000, BFR250
TITANIUM CHLORIDE see TGG250, TGH350
TITANIUM (III) CHLORIDE see TGG250
TITANIUM COMPOUNDS see TGG500
TITANIUM DIOXIDE see TGG760
TITANIUM METAL POWDER, DRY (DOT) see TGF250
TITANIUM NICKEL OXIDE see NDL500
TITANIUM OXIDE see TGG760
TITANIUM OXIDE BIS(ACETYLACETONATE) see BGQ750
TITANIUM, OXOBIS(2,4-PENTANEDIONATO-O,O') see
  BGQ750
TITANIUM POTASSIUM FLUORIDE see PLI000
TITANIUM SPONGE GRANULES (DOT) see TGF250
TITANIUM SPONGE POWDERS (DOT) see TGF250
TITANIUM TETRACHLORIDE see TGH350
TITANIUM TRICHLORIDE see TGG250
TITANIUM TRICHLORIDE, PYROPHORIC (DOT) see
  TGG250
TITANOUS CHLORIDE see TGG250

TITANTETRACHLORID (GERMAN) see TGH350
TITANYL BIS(ACETYLACETONATE) see BGQ750
TITRIPLEX see EIX000
TITRIPLEX III see EIX500
TIURAM see DXH250
TIURAM (POLISH) see TFS350
TIURAMYL see TFS350
TIXOTON see BAV750
TIZIDE see ILD000
TJB see DWI000
TK 1000 see PKQ059
TKB see NKB500
TL4N see DJD600
TL 69 see DGB600
TL 70 see SOQ450
TL 78 see DNJ800
TL 80 see DDL400
TL 146 see BIE250
TL 154 see CHF500
TL 161 see BHO250
TL 189 see FOY000
TL 199 see MPI750
TL 214 see DFH200
TL 262 see TFO000
TL 294 see DFP200
TL 311 see DSA800
TL 314 see ADX500
TL 329 see BID250
TL 337 see EJM900
TL 345 see DJJ400
TL 367 see FBS000
TL 373 see EOQ000
TL 389 see DQY950
TL 423 see EHK500
TL 466 see IRF000
TL 478 see BRZ000
TL 741 see FIE000
TL 792 see BJE750
TL 797 see DXR200
TL 822 see COO500
TL 833 see PHO250
TL 855 see FIM000
TL 869 see SHG500
TL 898 see MCY475
TL 944 see BEU500
TL 965 see BIA750
TL 1026 see CAG000
TL 1070 see CAG500
TL 1091 see NDC000
TL 1149 see BID250
TL 1163 see TLN250
TL 1178 see HNO500
TL 1182 see CAI000
TL 1183 see MKE750
TL 1217 see MID250
TL 1266 see DEC200
TL 1312 see FFH000
TL 1333 see MKE750
TL-1380 see PIA750
TL-1394 see DQY909
TL 1428 see MFW750
TL 1450 see MKX250
TL 1505 see ABO000
TL 1578 see EIF000
TL 1618 see IPX000
TLA see BPV100
TLD 100 see LHF000
TL 301 HYDROCHLORIDE see IPG000
TLP-607 see PAP250
TM see TDK500
TM-4049 see MAK700
TM 12008 see DJN600
TMA see TKV000, TLD500
TMAN see TKV000

7,8,12-TMBA see TLK750
TMCA see TLN750
TMEDA see TDQ750
TML see TDR500
TMP see TMD250
TMSN see TDW250
TMTD see TFS350
TMTDS see TFS350
TMTM see BJL600
TMTMS see BJL600
TMU see TDX250
T-2 MYCOTOXIN see FQS000
TNA see TDY000
TNB see TMK500
TNCS 53 see CNP250
TNG see NGY000
TNM see TDY250
TNT see TMN490
α-TNT see TMN490
TNT-TOLITE (FRENCH) see TMN490
TOABOND 40H see AAX250
TOBACCO LEAF, NICOTIANA GLAUCA see TGI000
TOBACCO PLANT see TGI100
TOBACCO REFINED TAR see CMP800
TOBACCO SMOKE CONDENSATE see SEC000
TOBACCO TAR see CMP800, SEC000
TOCE see DJT400
TOCEN see DJT400
TOCHLORINE see CDP000
α-TOCOPHEROL see VSZ450
(R,R,R)-α-TOCOPHEROL see VSZ450
(2R,4'R,8'R)-α-TOCOPHEROL see VSZ450
d-α-TOCOPHEROL (FCC) see VSZ450
dl-α-TOCOPHEROL (FCC) see VSZ450
TOCP see TMO600
TODALGIL see BRF500
TOFK see TMO600
TOFRANIL see DLH600, DLH630
TOFRANILE see DLH630
TOFURON see FQN000
N-TOIN see NGE000
TOIN UNICELLES see DKQ000
TOK see DFT800
TOK-2 see DFT800
TOKAMINA see VGP000
TOK E see DFT800
TOK E-25 see DFT800
TOK E 40 see DFT800
TOKIOCILLIN see AIV500
TOKKORN see DFT800
TOKOKIN see EDV000
TOKOPHARM see VSZ450
TOK WP-50 see DFT800
TOKYO ANILINE BRILLIANT GREEN see BAY750
TOLAMINE see CDP000
2,4-TOLAMINE see TGL750
TOLANSIN see GGS000
TOLAVAD see BBJ750
TOLAZOLINE CHLORIDE see BBJ750
TOLAZOLINE HYDROCHLORIDE see BBJ750
TOLBUSAL see BSQ000
TOLBUTAMID see BSQ000
TOLBUTAMIDE see BSQ000
TOLCIL see GGS000
TOLERON see FBJ100
TOLFERAIN see FBJ100
o-TOLIDIN see TGJ750
2-TOLIDIN (GERMAN) see TGJ750
2-TOLIDINA (ITALIAN) see TGJ750
TOLIDINE see TGJ750
2-TOLIDINE see TGJ750
o-TOLIDINE see TGJ750
3,3'-TOLIDINE see TGJ750
o,o'-TOLIDINE see TGJ750

TOLIFER see FBJ100
TOLIT see TMN490
TOLITE see TMN490
TOLL see MNH000
TOLOFREN see GGS000
TOLOMOL see AIV500
3-o-TOLOXY-1,2-PROPANEDIOL see GGS000
TOLPAL see BBJ750
TOLSEROL see GGS000
α-TOLUALDEHYDE see BBL500
TOLUEEN (DUTCH) see TGK750
TOLUEEN-DIISOCYANAAT see TGM750
TOLUEN (CZECH) see TGK750
TOLUEN-DIISOCIANATO see TGM750
TOLUENE see TGK750
TOLUENE-2-AZONAPHTHOL-2 see TGW000
o-TOLUENE-1-AZO-2-NAPHTHYLAMINE see FAG135
o-TOLUENEAZO-o-TOLUENEAZO-β-NAPHTHOL see
    SBC500
o-TOLUENEAZO-o-TOLUENE-β-NAPHTHOL see SBC500
o-TOLUENEAZO-o-TOLUIDINE see AIC250
TOLUENEDIAMINE see TGL500
m-TOLUENEDIAMINE see TGL750
p-TOLUENEDIAMINE see TGM000
TOLUENE-2,4-DIAMINE see TGL750
2,4-TOLUENEDIAMINE see TGL750
TOLUENE-2,5-DIAMINE see TGM000
p-TOLUENEDIAMINE DIHYDROCHLORIDE see DCE200
p-TOLUENEDIAMINE SULFATE see DCE600
2,5-TOLUENEDIAMINE SULFATE see DCE600
TOLUENE-2,5-DIAMINE, SULFATE (1:1) (8CI) see DCE600
p-TOLUENEDIAMINE SULPHATE see DCE600
TOLUENE-2,5-DIAMINE SULPHATE see DCE600
TOLUENE DIISOCYANATE see TGM740, TGM750
TOLUENE-1,3-DIISOCYANATE see TGM740
TOLUENE-2,4-DIISOCYANATE see TGM750
2,4-TOLUENEDIISOCYANATE see TGM750
TOLUENE-2,6-DIISOCYANATE see TGM800
2,6-TOLUENE DIISOCYANATE see TGM800
2,3-TOLUENEDIOL see DNE000
TOLUENE HEXAHYDRIDE see MIQ740
TOLUENE-2-SULFONAMIDE see TGN250
o-TOLUENESULFONAMIDE see TGN250
1-p-TOLUENESULFONYL-3-BUTYLUREA see BSQ000
α-TOLUENETHIOL see TGO750
TOLUENE TRICHLORIDE see BFL250
o-TOLUENO-AZO-β-NAPHTHOL see TGW000
α-TOLUENOL see BDX500
ar-TOLUENOL see CNW500
α-TOLUIC ACID see PDY850
m-TOLUIC ACID DIETHYLAMIDE see DKC800
α-TOLUIC ACID, ETHYL ESTER see EOH000
α-TOLUIC ALDEHYDE see BBL500
m-TOLUIDIN (CZECH) see TGQ500
o-TOLUIDIN (CZECH) see TGQ750
p-TOLUIDIN (CZECH) see TGR000
2-TOLUIDINE see TGQ750
3-TOLUIDINE see TGQ500
4-TOLUIDINE see TGR000
m-TOLUIDINE see TGQ500
o-TOLUIDINE see TGQ750
p-TOLUIDINE see TGR000
TOLUIDINE BLUE see TGS000
2-TOLUIDINE HYDROCHLORIDE see TGS500
o-TOLUIDINE HYDROCHLORIDE see TGS500
p-TOLUIDINE HYDROCHLORIDE see TGS750
p-TOLUIDINIUM CHLORIDE see TGS750
o-TOLUIDYNA (POLISH) see TGQ750
TOLUILENODWUIZOCYJANIAN see TGM750
TOLUINA see BSQ000
TOLULOX see GGS000
TOLUMID see BSQ000
α-TOLUNITRILE see PEA750
TOLUOL see TGK750

m-TOLUOL see CNW750
o-TOLUOL see CNX000
p-TOLUOL see CNX250
o-TOLUOL-AZO-o-TOLUIDIN (GERMAN) see AIC250
TOLUOLO (ITALIAN) see TGK750
p-TOLUOLSULFONSAEUREAETHYL ESTER (GERMAN) see EPW500
p-TOLUOLSULFONSAEURE METHYL ESTER (GERMAN) see MLL250
α-TOLUOLTHIOL see TGO750
p-TOLUQUINONE see MHI250
1,4-TOLUQUINONE see MHI250
TOLU-SOL see TGK750
TOLUVAN see BSQ000
m-TOLUYLENDIAMIN (CZECH) see TGL750
p-TOLUYLENEDIAMINE see TGM000
m-TOLUYLENEDIAMINE see TGL750
TOLUYLENE-2,5-DIAMINE see TGM000
2,4-TOLUYLENEDIAMINE (DOT) see TGL750
p-TOLUYLENEDIAMINE SULPHATE see DCE600
TOLUYLENE-2,5-DIAMINE SULPHATE see DCE600
TOLUYLENE-2,4-DIISOCYANATE see TGM750
p-TOLYCARBAMIDE see THG000
TOLYDRIN see GGS000
m-TOLYENEDIAMINE see TGL750
p-TOLYL ACETATE see MNR250
p-TOLYL ALCOHOL see CNX250
α-TOLYL ALDEHYDE DIMETHYL ACETAL see PDX000
TOLYLAMINE see TGR000
m-TOLYLAMINE see TGQ500
o-TOLYLAMINE see TGQ750
p-TOLYLAMINE see TGR000
o-TOLYLAMINE HYDROCHLORIDE see TGS500
5-(o-TOLYLAZO)-2-AMINOTOLUENE see AIC250
4-o-TOLYLAZO-o-DIACETOTOLUIDE see ACR300
4'-(o-TOLYLAZO)-o-DIACETOTOLUIDIDE see ACR300
1-(o-TOLYLAZO)-2-NAPHTHOL see TGW000
1-(o-TOLYLAZO)-β-NAPHTHOL see TGW000
1-(o-TOLYLAZO)-2-NAPHTHYLAMINE see FAG135
4-(o-TOLYLAZO)-o-TOLUIDINE see AIC250
o-TOLYLAZO-o-TOLYLAZO-2-NAPHTHOL see SBC500
o-TOLYLAZO-o-TOLYLAZO-β-NAPHTHOL see SBC500
1-((4-(o-TOLYLAZO)-o-TOLYL)AZO)-2-NAPHTHOL) see SBC500
TOLYL CHLORIDE see BEE375
o-TOLYL CHLORIDE see CLK100
p-TOLYL CHLORIDE see TGY075
TOLYLENEDIAMINE see TGL500
m-TOLYLENEDIAMINE see TGL750
TOLYLENE-2,4-DIAMINE see TGL750
2,4-TOLYLENEDIAMINE see TGL750
4-m-TOLYLENEDIAMINE see TGL750
p,m-TOLYLENEDIAMINE see TGM000
p-TOLYLENEDIAMINE SULPHATE see DCE600
TOLYLENE DIISOCYANATE see TGM740
m-TOLYLENE DIISOCYANATE see TGM750, TGM800
TOLYLENE-2,4-DIISOCYANATE see TGM750
2,4-TOLYLENEDIISOCYANATE see TGM750
TOLYLENE-2,6-DIISOCYANATE see TGM800
TOLYLENE ISOCYANATE see TGM740
p-TOLYL ETHANOATE see MNR250
1-o-TOLYLGLYCEROL ETHER see GGS000
α-(o-TOLYL)GLYCERYL ETHER see GGS000
p-TOLYL ISOBUTYRATE see THA250
α-TOLYL MERCAPTAN see TGO750
3-TOLYL-N-METHYLCARBAMATE see MIB750
m-TOLYL-N-METHYLCARBAMATE see MIB750
p-TOLYL METHYL ETHER see MGP000
3-(o-TOLYLOXY)PROPANE-1,2-DIOL see GGS000
o-TOLYL PHOSPHATE see TMO600
N-(p-TOLYLSULFONYL)-N'-BUTYLCARBAMIDE see BSQ000
TOLYLSULFONYLBUTYLUREA see BSQ000
3-(p-TOLYL-4-SULFONYL)-1-BUTYLUREA see BSQ000

p-TOLYLUREA see THG000
TOLYNOL see GGS000
TOLYSPAZ see GGS000
p-TOLYUREA see THG000
TOMATHREL see CDS125
A''-TOMATIDINE see THG250
TOMATIDINE GLYCOSIDE see THG250
TOMATIN see THG250
TOMATINE see THG250
α-TOMATINE see THG250
TOMATO FIX CONCENTRATE see CJN000
TOMATO HOLD see CJN000
TOMATOTONE see CJN000
TOMIL see DIR000
TONARIL see TMP750
TONARSEN see DXE600
TONCARINE see MIP750
TONEDRIN see DBA800
TONER LAKE RED C see CHP500
TONITE see CDN200
TONKA BEAN CAMPHOR see CNV000
TONOCARD see DJS200
TONOCOR see DJS200
TONOGEN see VGP000
TONOLYT ISOPROPYL MEPROBAMATE see IPU000
TONOX see MJQ000
TO NTU see BMR750
TONY RED see OHA000
TOPANE see BGJ750
TOPANEL see COB260
TOPANOL see BFW750
TOPAZONE see NGG500
TOPCAINE see EFX000
TOPEX see BDS000
TOP FLAKE see SFT000
TOP FORM WORMER see TEX000
TOPICAIN see DTL200
TOPICHLOR 20 see CDR750
TOPICLOR see CDR750
TOPICLOR 20 see CDR750
TOPICYCLINE see TBX250
TOPITOX see CJJ000
TOPITRACIN see BAC250
TOPOKAIN see AIT250
TOPOREX 855-51 see SMQ500
TOPSYM see DUD800
TOPZOL see RCF000
TORAK see DBI099
TORDON see PIB900
TORDON 10K see PIB900
TORDON 22K see PIB900
TORDON 101 MIXTURE see PIB900
TORINAL see QAK000
TORMONA see BSQ750, TAA100
TORQUE see BLU000
TORULOX see GGS000
TOSTRIN see TBG000
TOSYL see CDO000
TOSYLCHLORAMIDE SODIUM see CDP000
TOTACILLIN see AIV500
TOTALCICLINA see AIV500
TOTAPEN see AIV500
TOTOCAINE HYDROCHLORIDE see AIT750
TOTOMYCIN see TBX250
TOTP see TMO600
TOTRIL see DNG200, HKB500
TOX 47 see PAK000
TOXADUST see CDV100
TOXAFEEN (DUTCH) see CDV100
TOXAKIL see CDV100
TOXALBUMIN see AAD000
TOXAPHEN (GERMAN) see CDV100
TOXAPHENE see CDV100
TOXICHLOR see CDR750

TOXILIC ACID see MAK900
TOXILIC ANHYDRIDE see MAM000
TOXIN T2 see FQS000
TOXON 63 see CDV100
TOXYLON POMIFERUM see MRN500
TOXYPHEN see CDV100
TOYO EOSINE G see BNH500
TOYO OIL ORANGE see PEJ500
TOYO OIL YELLOW G see DOT300
TOYO ORIENTAL OIL BLUE G see BLK000
TP see TBG000
TP-21 see MOO250, MOO500
2,4,5-TP see TIX500
TPA see PGV000
TPA-3-β-OL see PGV500
TPN (pesticide) see TBQ750
TPP see TMT750
TPTA see ABX250
TPTC see CLU000
TPTH see HON000
TPU 2T see PKO500
TPU 10M see PKM250
TPZA see ABX250
TR 201 see SMR000
TRACHOSEPT see DBX400
TRAFARBIOT see AOD125
TRAGACANTH see THJ250
TRAGACANTH GUM see THJ250
TRAGAYA see SEH000
TRAKIPEAL see MDQ250
TRALGON see HIM000
TRAMADOL see THJ500
TRAMAL see THJ500
TRAMETAN see TFS350
TRAMISOL see LFA020
TRAMISOLE see LFA020
TRANCYLPROMINE SULFATE see PET500
TRANILCYPROMINE see PET750
TRANITE D-LAY see PBC250
TRAN-Q see CJR909
TRANQUIL see PGA750
TRANQUILAN see MQU750
TRANQUILLIN see BCA000
TRANS-AID see ANW750
TRANSAMINE see DAA800, PET750, TAA100
TRANSAMINE SULFATE see PET500
TRANSANNON see ECU750
TRANSENTINE see DHX800
TRANSIT see CHJ750
TRANSPARENT BRONZE SCARLET see CHP500
TRANSVAALIN see GFC000
TRANS-VERT see MRL750
TRANYLCYPRAMINE see PET750
TRANYLCYPRAMINE SULFATE see PET500
TRANYLCYPROMINE see PET750
TRANYLCYPROMINE SULFATE see PET500
TRANYLCYPROMINE SULPHATE see PET500
TRANZER see EHP000
TRANZETIL see DHX800
TRAPANAL see PBT500
TRAPANAL SODIUM see PBT500
TRAPEX see ISE000
TRAPEXIDE see ISE000
TRAPIDIL see DIO200
TRAPYMIN see DIO200
TRAQUIZINE see CJR909
TRASAN see CIR250
TRASENTIN see DHX800
TRASENTINE see DHX800
TRAUMANASE see BMO000
TRAVAD see BAP000
TRAVELIN see DYE600
TRAVELMIN see DYE600
TRAVELON see HGC500

TRAVEX see SFS000
TRAWOTOX see CDO000
TRAZENTYNA see DHX800
TRECATOR see EPQ000
TREDIONE see TLP750
TREFANOCIDE see DUV600
TREFICON see DUV600
TREFLAM see DUV600
TREFLAN see DUV600
TREFLANOCIDE ELANCOLAN see DUV600
TRELMAR see MQU750
TREMIN see BBV000
TREMOLITE ASBESTOS see ARM280
TRENIMON see TND000
TRENTADIL see THL750
TRENTADIL HYDROCHLORIDE see BEO750, THL750
TREOMICETINA see CDP250
TREOSULFAN see TFU500
TREPENOL WA see SIB600
TRESCATYL see EPQ000
TRESCAZIDE see EPQ000
TRESPAPHAN see PMP500
TRESULFAN see TFU500
TRETAMINE see TND500
TRIABARB see EOK000
TRIACETALDEHYDE (FRENCH) see PAI250
TRIACETIN (FCC) see THM500
TRIACETYL GLYCERIN see THM500
TRIAD see TIO750
TRIADIMEFON see CJO250
TRIADIMENOL see MEP250
TRIAETHANOLAMIN-NG see TKP500
TRIAETHYLAMIN (GERMAN) see TJO000
TRIAETHYLENMELAMIN (GERMAN) see TND500
TRIAETHYLENPHOSPHORSAEUREAMID (GERMAN) see
    TND250
TRIAETHYLZINNSULFAT (GERMAN) see BLN250
TRIALLYLAMINE see THN000
TRIALLYL CYANAURATE see THN500
TRIAMELIN see TND500
TRIAMIFOS (GERMAN, DUTCH, ITALIAN) see AIX000
2,4,6-TRIAMINO-s-TRIAZINE see MCB000
TRIAMINOTRIPHENYLMETHANE see THP000
4,4′,4″-TRIAMINOTRIPHENYLMETHANE see THP000
p,p′,p″-TRIAMINOTRIPHENYLMETHANE see THP000
4,4′4″-TRIAMINOTRIPHENYLMETHAN-HYDROCHLORID
    (GERMAN) see RMK020
TRIAMIPHOS see AIX000
TRIAMMINEDIPEROXOCHROMIUM(IV) see THP250
TRIAMPHOS see AIX000
TRI-p-ANISYLCHLOROETHYLENE see CLO750
TRIANOL DIRECT BLUE 3B see CMO250
TRIANTINE LIGHT BROWN 3RN see FMU059
TRIANTOIN see MKB250
TRIASOL see TIO750
TRIATOMIC OXYGEN see ORW000
1,2,3-TRIAZAINDENE see BDH250
TRIAZICHON (GERMAN) see TND000
TRIAZINE A 384 see BJP000
TRIAZINE A 1294 see ARQ725
(1,3,5-TRIAZINE-2,4,6-TRIYLTRINITRILO)HEXAKIS METH-
    ANOL see HDY000
(s-TRIAZINE-2,4,6-TRIYLTRINITRILO)HEXAMETHANOL
    see HDY000
1,1′,1″-s-TRIAZINE-2,4,6-TRIYLTRISAZIRIDINE see
    TND500
TRIAZIQUINONE see TND000
TRIAZIQUONE see TND000
TRIAZIRIDINOPHOSPHINE OXIDE see TND250
2,3,5-TRI-(1-AZIRIDINYL)-p-BENZOQUINONE see TND000
TRI(AZIRIDINYL)PHOSPHINE OXIDE see TND250
TRI-1-AZIRIDINYL)PHOSPHINE OXIDE see TND250
TRIAZIRIDINYLPHOSPHINE SULFIDE see TFQ750
TRIAZIRIDINYL TRIAZINE see TND500

TRIAZOIC ACID see HHG500
TRIAZOLAMINE see AMY050
1H-1,2,4-TRIAZOL-3-AMINE see AMY050
TRIAZOTION (RUSSIAN) see EKN000
TRIB see SNK000
TRI-BAN see PIH175
TRIBASIC SODIUM PHOSPHATE see SJH200
TRIBROMETHANOL see THV000
TRIBROMMETHAAN (DUTCH) see BNL000
TRIBROMMETHAN (GERMAN) see BNL000
TRIBROMOALUMINUM see AGX750
TRIBROMOARSINE see ARF250
TRIBROMOETHANOL see THV000
2,2,2-TRIBROMOETHANOL see ARW250
TRIBROMOETHYL ALCOHOL see THV000
2,2,2-TRIBROMOETHYL ALCOHOL see ARW250
2,4,5-TRIBROMOIMIDAZOLE CADMIUM SALT (2:1) see
    THV500
TRIBROMOMETAN (ITALIAN) see BNL000
TRIBROMOMETHANE see BNL000
TRIBROMONITROMETHANE see NMQ000
TRIBROMOPHOSPHINE see PHT250
TRIBUFON see BPG000
TRIBUTILFOSFATO (ITALIAN) see TIA250
TRIBUTON see DAA800, TAA100
TRIBUTOXYBORANE see THX750
TRI-n-BUTOXYBORANE see THX750
TRI(2-BUTOXYETHANOL PHOSPHATE) see BPK250
TRIBUTOXYETHYL PHOSPHATE see BPK250
TRI(2-BUTOXYETHYL) PHOSPHATE see BPK250
TRIBUTYLAMINE see THX250
TRI-n-BUTYLAMINE see THX250
TRI-n-BUTYL BORANE see THX500
TRIBUTYL BORATE see THX750
TRI-n-BUTYL BORATE see THX750
TRIBUTYLBORINE see THX500
TRIBUTYL CELLOSOLVE PHOSPHATE see BPK250
TRIBUTYLCHLOROSTANNANE see CLP500
TRIBUTYLE (PHOSPHATE de) (FRENCH) see TIA250
TRIBUTYLFOSFAAT (DUTCH) see TIA250
TRIBUTYL(METHACRYLOYLOXY)-STANNANE POLYMER
    with METHYL METHACRYLATE (8CI) see OIY000
TRIBUTYL(NEODECANOYLOXY)STANNANE see TIF250
(TRIBUTYL)PEROXIDE see BSC750
TRIBUTYLPHOSPHAT (GERMAN) see TIA250
TRIBUTYL PHOSPHATE see TIA250
TRI-n-BUTYL PHOSPHATE see TIA250
TRIBUTYL-PHOSPHINE compounded with NICKELCHLO-
    RIDE (2:1) see BLS250
S,S,S-TRIBUTYL PHOSPHOROTRITHIOATE see BSH250
TRIBUTYL PHOSPHOROTRITHIOITE see TIG250
S,S,S-TRIBUTYL PHOSPHOROTRITHIOITE see TIG250
TRIBUTYL(8-QUINOLINOLATO)TIN see TIB000
TRI-n-BUTYL-STANNANE OXIDE see BLL750
TRIBUTYLTIN BENZOATE see BDR750
TRI-n-BUTYLTIN CHLORIDE see CLP500
TRIBUTYLTIN NEODECANOATE see TIF250
TRIBUTYLTIN OXIDE see BLL750
S,S,S-TRIBUTYL TRITHIOPHOSPHATE see BSH250
S,S,S-TRIBUTYL TRITHIOPHOSPHITE see TIG250
TRI-N-BUTYL-ZINN BENZOATE (GERMAN) see BDR750
TRI-n-BUTYLZINN-CHLORID (GERMAN) see CLP500
TRIBUTYRIN see TIG750
TRIBUTYROIN see TIG750
TRICADMIUM DINITRIDE see TIH000
TRICALCIUMARSENAT (GERMAN) see ARB750
TRICALCIUM ARSENATE see ARB750
TRICALCIUM DIPHOSPHIDE see CAW250
TRICALCIUM PHOSPHATE see CAW120
TRICAPRYLMETHYLAMMONIUM CHLORIDE see
    MQH000
TRICAPRYLYLMETHYLAMMONIUM CHLORIDE see
    MQH000

TRICARBALLYLIC ACID-β-ACETOXYTRIBUTYL ESTER
    see ADD750
TRICARBAMIX Z see BJK500
TRICESIUM TRICHLORIDE see CDD000
TRICESIUM TRIFLUORIDE see CDD500
TRICHAZOL see MMN250
TRICHLOORAZIJNZUUR (DUTCH) see TII250
1,1,1-TRICHLOOR-2,2-BIS(4-CHLOOR FENYL)-ETHAAN
    (DUTCH) see DAD200
2,2,2-TRICHLOOR-1,1-BIS(4-CHLOOR FENYL)-ETHANOL
    (DUTCH) see BIO750
1,1,1-TRICHLOORETHAAN (DUTCH) see MIH275
TRICHLOORETHEEN (DUTCH) see TIO750
TRICHLOORETHYLEEN, TRI (DUTCH) see TIO750
(2,4,5-TRICHLOOR-FENOXY)-AZIJNZUUR (DUTCH) see
    TAA100
2-(2,4,5-TRICHLOOR-FENOXY)-PROPIONZUUR (DUTCH)
    see TIX500
O-(2,4,5-TRICHLOOR-FENYL)-O,O-DIMETHYL-MONOTHIO-
    FOSFAAT (DUTCH) see RMA500
TRICHLOORFON (DUTCH) see TIQ250
TRICHLOORMETHAAN (DUTCH) see CHJ500
TRICHLOORMETHYLBENZEEN (DUTCH) see BFL250
TRICHLOORNITROMETHAAN (DUTCH) see CKN500
TRICHLOORSILAAN (DUTCH) see TJD500
TRICHLORACETALDEHYD-HYDRAT (GERMAN) see
    CDO000
1,1,1-TRICHLORAETHAN (GERMAN) see MIH275
TRICHLORAETHEN (GERMAN) see TIO750
TRICHLORAETHYLEN, TRI (GERMAN) see TIO750
TRICHLORAMINE see NGQ500
TRICHLORAN see TIO750
1,1,1-TRICHLOR-2,2-BIS(4-CHLOR-PHENYL)-AETHAN
    (GERMAN) see DAD200
1,1,1-TRICHLOR-2,2-BIS(4-CHLORPHENYL)-AETHANOL
    (GERMAN) see BIO750
2,2,2-TRICHLOR-1,1-BIS(4-CHLOR-PHENYL)-AETHANOL
    (GERMAN) see BIO750
1,1,1-TRICHLOR-2,2-BIS(4-METHOXY-PHENYL)-AETHAN
    (GERMAN) see MEI450
TRICHLORESSIGSAEURE (GERMAN) see TII250
1,1,2-TRICHLORETHANE see TIN000
TRICHLORETHENE (FRENCH) see TIO750
TRICHLORETHYLENE, TRI (FRENCH) see TIO750
TRICHLORETHYL PHOSPHATE see CGO500
2,4,6-TRICHLORFENOL (CZECH) see TIW000
TRICHLORFENSON (OBS.) see CJT750
TRICHLORFON (USDA) see TIQ250
TRICHLORINATED ISOCYANURIC ACID see TIQ750
TRICHLORINE NITRIDE see NGQ500
TRICHLORMETHAN (CZECH) see CHJ500
TRICHLORMETHINE see TNF500
TRICHLORMETHINIUM CHLORIDE see TNF500
TRICHLORMETHYLBENZOL (GERMAN) see BFL250
N-(TRICHLOR-METHYLTHIO)-PHTHALIMID (GERMAN) see
    CBG000
TRICHLORNITROMETHAN (GERMAN) see CKN500
TRICHLOROACETALDEHYDE HYDRATE see CDO000
TRICHLOROACETALDEHYDE MONOHYDRATE see
    CDO000
TRICHLOROACETIC ACID see TII250
TRICHLOROACETIC ACID, solid (DOT) see TII250
TRICHLOROACETIC ACID, solution (DOT) see TII250
TRICHLOROACETIC ACID CHLORIDE see TIJ150
TRICHLOROACETOCHLORIDE see TIJ150
TRICHLOROACETYL CHLORIDE see TIJ150
TRICHLOROALLYLSILANE see AGU250
TRICHLOROALUMINUM see AGY750
3,5,6-TRICHLORO-4-AMINOPICOLINIC ACID see PIB900
TRICHLOROARSINE see ARF500
1,2,4-TRICHLOROBENZENE see TIK250
1,2,4-TRICHLOROBENZENE, liquid (DOT) see TIK250
unsym-TRICHLOROBENZENE see TIK250
1,1,1-TRICHLORO-2,2-BIS(p-ANISYL)ETHANE see MEI450

1,3,5-TRICHLORO-2,4,6-TRIOXOHEXAHYDRO-s-TRIAZINE
see TIQ750
TRICHLORO TRITANIUM see TGG250
TRICHLORO(VINYL)SILANE see TIN750
TRICHLOROVINYL SILICANE see TIN750
TRICHLORPHENE see TIQ250
(2,4,5-TRICHLOR-PHENOXY)-ESSIGSAEURE (GERMAN)
see TAA100
2-(2,4,5-TRICHLOR-PHENOXY)-PROPIONSAEURE (GER-
MAN) see TIX500
O-(2,4,5-TRICHLOR-PHENYL)-O,O-DIMETHYL-MONOTHIO-
PHOSPHAT (GERMAN) see RMA500
TRICHLORPHON see TIQ250
TRICHLORPHON FN see TIQ250
TRICHLORSILAN (GERMAN) see TJD500
TRICHLOR-TRIAETHYLAMIN-HYDROCHLORID (GERMAN)
see TNF500
TRICHLORURE d'ANTIMOINE see AQC500
TRICHLORURE d'ARSENIC (FRENCH) see ARF500
TRICHOCHROMOGENIC FACTOR see AIH600
TRICHOCIDE see MMN250
TRICHOFURON see NGG500
TRICHOMOL see MMN250
TRICHOMONACID "PHARMACHIM" see MMN250
TRICHOPOL see MMN250
TRICHORAD see ABY900
TRICHORAL see ABY900
TRICHOTHECIN see TJE750
TRICIONE see TLP750
TRI-CLENE see TIO750
TRI-CLOR see CKN500
TRICLORETENE (ITALIAN) see TIO750
1,1,1-TRICLORO-2,2-BIS(4-CLORO-FENIL)-ETANO (ITAL-
IAN) see DAD200
1,1,1-TRICLOROETANO (ITALIAN) see MIH275
TRICLOROETILENE (ITALIAN) see TIO750
O-(2,4,5-TRICLORO-FENIL)-O,O-DIMETIL-MONOTIOFOS-
FATO (ITALIAN) see RMA500
TRICLOROMETANO (ITALIAN) see CHJ500
TRICLOROMETILBENZENE (ITALIAN) see BFL250
TRICLORO-NITRO-METANO (ITALIAN) see CKN500
TRICLOROSILANO (ITALIAN) see TJD500
TRICLOROTOLUENE (ITALIAN) see BFL250
TRICOFURON see NGG500
TRICOM see MMN250
TRICON BW see EIX000
TRICOWAS B see MMN250
TRICRESILFOSFATI (ITALIAN) see TNP500
TRICRESYLFOSFATEN (DUTCH) see TNP500
TRICRESYL PHOSPHATE see TMO600, TNP500
TRI-o-CRESYL PHOSPHATE see TMO600
TRICRESYLPHOSPHATE, with more than 3% ortho isomer
(DOT) see TNP500
TRICURAN see PDD300
TRICYCLOHEXYLHYDROXYSTANNANE see CQH650
TRICYCLOHEXYLHYDROXYTIN see CQH650
TRICYCLOHEXYLTIN HYDROXIDE see CQH650
TRICYCLOHEXYLZINNHYDROXID (GERMAN) see
CQH650
1-TRIDECANECARBOXYLIC ACID see MSA250
TRIDESTRIN see EDU500
TRIDEZIBARBITUR see EOK000
1,2,3-TRI(β-DIETHYLAMINOETHOXY)BENZENE TRI-
ETHIODIDE see PDD300
TRI(β-DIETHYLAMINOETHOXY)-1,2,3-BENZENE TRI-
IODOETHYLATE see PDD300
TRI-DIGITOXOSIDE (GERMAN) see DKL800
TRIDILONA see TLP750
TRI(DIMETHYLAMINO)PHOSPHINEOXIDE see HEK000
TRIDIMITE (FRENCH) see SCK000
TRIDIONE see TLP750
TRIDIPAM see TFS350
TRIDONE see TLP750
TRIDYMITE see SCI500, SCK000

TRIELINA (ITALIAN) see TIO750
TRIEN see TJR000
TRI-ENDOTHAL see DXD000, EAR000
TRIENTINE see TJR000
TRI-ERVONUM see MCA500
TRIESIFENIDILE see BBV000
TRIESTE FLOWERS see POO250
TRI-ETHANE see MIH275
N,N',N''-TRI-1,2-ETHANEDIYL PHOSPHORIC TRIMIDE see
TND250
N,N',N''-TRI-1,2-ETHANEDIYLPHOSPHOROTHIOIC
TRIAMIDE see TFQ750
N,N',N''-TRI-1,2-ETHANEDIYLTHIOPHOSPHORAMIDE see
TFQ750
TRIETHANOLAMIN see TKP500
TRIETHANOLAMINE see TKP500
TRIETHANOMELAMINE see TND500
TRIETHOXYMETHANE see ENY500
TRIETHYL ACETYLCITRATE see ADD750
TRIETHYLALUMINUM see TJN750
TRIETHYLAMINE see TJO000
TRIETHYLBORANE see TJP250
TRIETHYLBORINE see TJP250
TRIETHYL CITRATE see TJP750
TRIETHYL DIALUMINUM TRICHLORIDE see TJP775
TRIETHYLENE GLYCOL see TJQ000
TRIETHYLENE GLYCOL ETHYL ETHER see EFL000
TRIETHYLENE GLYCOL MONOETHYL ETHER see EFL000
2,3,5-TRIETHYLENEIMINO-1,4-BENZOQUINONE see
TND000
TRI(ETHYLENEIMINO)THIOPHOSPHORAMIDE see
TFQ750
2,4,6-TRIETHYLENEIMINO-s-TRIAZINE see TND500
2,4,6-TRI(ETHYLENEIMINO)-1,3,5-TRIAZINE see TND500
TRIETHYLENEMELAMINE see TND500
N,N',N''-TRIETHYLENEPHOSPHOROTHIOIC TRIAMIDE see
TFQ750
TRIETHYLENEPHOSPHOROTRIAMIDE see TND250
TRIETHYLENETETRAMINE see TJR000
N,N',N''-TRIETHYLENETHIOPHOSPHAMIDE see TFQ750
N,N',N''-TRIETHYLENETHIOPHOSPHORAMIDE see
TFQ750
TRIETHYLENETHIOPHOSPHOROTRIAMIDE see TFQ750
TRIETHYLENIMINOBENZOQUINONE see TND000
2,4,6-TRIETHYLENIMINO-s-TRIAZINE see TND500
2,4,6-TRIETHYLENIMINO-1,3,5-TRIAZINE see TND500
TRIETHYLHYDROXY-STANNANE SULFATE (2:1) (8CI) see
BLN250
TRIETHYLHYDROXYTIN SULFATE see BLN250
TRIETHYL LEAD FLUOROACETATE see TJS500
TRIETHYLOLAMINE see TKP500
TRIETHYL ORTHOFORMATE see ENY500
TRIETHYLPHOSPHINEAUROUS CHLORIDE see CLQ500
TRIETHYL PHOSPHITE see TJT800
TRIETHYLTIN ACETATE see ABW750
TRIETHYLTIN SULPHATE see BLN250
TRIETILAMINA (ITALIAN) see TJO000
TRIEXIFENIDILA see BBV000
TRIFARON see CJM250
TRIFENOXYFOSFIN (CZECH) see TMU250
TRIFENSON see CJR500
TRIFENYLFOSFIT (CZECH) see TMU250
TRIFENYLTINACETAAT (DUTCH) see ABX250
TRIFENYL-TINHYDROXYDE (DUTCH) see HON000
TRIFERRIC ADRIAMYCIN see QBS000
TRIFERRIC DOXORUBICIN see QBS000
TRIFLIC ACID see TKB310
TRIFLUORACETIC ACID see TKA250
TRIFLUORALIN (USDA) see DUV600
3-(5-TRIFLUORMETHYLPHENYL)-,1-DIMETHYLHARN-
STOFF (GERMAN) see DUK800
2-(2,2,2-TRIFLUOROACETAMIDO)-4-(5-NITRO-2-FURYL)
THIAZOLE see NGN500
TRIFLUOROACETIC ACID (DOT) see TKA250

2,2,4-TRIMETHYL-1,2-DIHYDROQUINOLINE see TLP500
3,3,5-TRIMETHYL-2,4-DIKETOOXAZOLIDINE see TLP750
N,N,N-TRIMETHYL-1,3-DIOXOLANE-4-METHANAMINIUM
  IODIDE see FMX000
1,3,7-TRIMETHYL-2,6-DIOXOPURINE see CAK500
3,7,11-TRIMETHYL-2,6,10-DODECATRIEN-1-OL see
  FAB800
TRIMETHYLEENTRINITRAMINE (DUTCH) see CPR800
TRIMETHYLENE see CQD750
TRIMETHYLENE BROMIDE CHLORIDE see BNA825
TRIMETHYLENE CHLOROBROMIDE see BNA825
TRIMETHYLENE DICHLORIDE see DGF800
1,3-TRIMETHYLENEDINITRILE see TLR500
TRIMETHYLENE OXIDE see OMW000
TRIMETHYLENETRINITRAMINE see CPR800
sym-TRIMETHYLENETRINITRAMINE see CPR800
TRIMETHYLENOXID (GERMAN) see OMW000
2,2,4-TRIMETHYL-6-ETHOXY-1,2-DIHYDROQUINOLINE
  see SAV000
TRIMETHYLETHYLENE see AOI750
TRIMETHYL GLYCOL see PML000
N,N,N-TRIMETHYL-1-HEXADECANAMINIUM BROMIDE
  see HCQ500
TRIMETHYLHEXADECYLAMMONIUM BROMIDE see
  HCQ500
N-1,5-TRIMETHYL-4-HEXENYLAMINE see ILK000
TRIMETHYLHYDRAZINE HYDROCHLORIDE see TLT750
TRIMETHYL-ε-LACTONE (mixed isomers) see TLY000
TRIMETHYL((2-METHYL-1,3-DIOXOLAN-4-YL)METHYL)
  AMMONIUM IODIDE (8CI) see MJH250
TRIMETHYL (1-METHYL-2-PHENOTHIAZIN-10-YLETHYL)
  AMMONIUM METHYL SULFATE see MRW000
TRIMETHYL(1-METHYL-2-(10-PHENOTHIAZINYL)ETHYL)
  AMMONIUM METHYL SULFATE see MRW000
TRIMETHYLNITROSOHARNSTOFF (GERMAN) see
  TLU750
N-TRIMETHYL-N-NITROSOUREA see TLU750
1,1,3-TRIMETHYL-3-NITROSOUREA see TLU750
1,7,7-TRIMETHYLNORCAMPHOR see CBA750
TRIMETHYL ORTHOFORMATE see TLX600
1,3,3-TRIMETHYL-2-OXABICYCLO(2.2.2)OCTANE see
  CAL000
3,5,5-TRIMETHYL-2,4-OXAZOLIDINEDIONE see TLP750
TRIMETHYL-2-OXEPANONE (mixed isomers) see TLY000
2,2,4-TRIMETHYLPENTANE see TLY500
2,2,4-TRIMETHYL-1-PENTENE see IKF000
6,6,9-TRIMETHYL-3-PENTYL-7,8,9,10-TETRAHYDRO-6H-
  DIBENZO(B,D)PYRAN-1-OL see TCM250
N-sym-TRIMETHYLPHENYLDIETHYLAMINOACETAMIDE
  HYDROCHLORIDE see DHL800
TRIMETHYLPHENYLMETHANE see BQJ250
3,4,5-TRIMETHYLPHENYL METHYLCARBAMATE see
  TMD000
TRI 2-METHYLPHENYL PHOSPHATE see TMO600
TRIMETHYL PHOSPHATE see TMD250
O,O,O-TRIMETHYL PHOSPHATE see TMD250
TRIMETHYL PHOSPHITE see TMD500
1,2,2-TRIMETHYLPROPYL METHYLPHOSPHONOFLUORI-
  DATE see SKS500
TRIMETHYLPYRAZINE see TME270
2,3,5-TRIMETHYLPYRAZINE see TME270
TRIMETHYL SILYL-1-CHLORO-7-DIHYDRO-1,3-PHENYL-
  5,2H-BENZODIAZEPINE-1,4-ONE-2 (FRENCH) see
  CGB000
TRIMETHYLSTANNANE SULPHATE see TMI500
TRIMETHYLSTANNYL CHLORIDE see CLT000
TRIMETHYLSTANNYL IODIDE see IFN000
TRIMETHYL STIBINE see AQL000
TRIMETHYL(TETRAHYDRO-4-HYDROXY-5-METHYLFUR-
  FURYL)AMMONIUM see MRW250
TRIMETHYLTHIOUREA see TMH750
1,1,3-TRIMETHYL-2-THIOUREA see TMH750
N,N,N'-TRIMETHYLTHIOUREA see TMH750
TRIMETHYLTIN ACETATE see TMI000

TRIMETHYLTIN CHLORIDE see CLT000
TRIMETHYLTIN IODIDE see IFN000
TRIMETHYLTIN SULPHATE see TMI500
5,7,8-TRIMETHYLTOCOL see VSZ450
α,α,α'-TRIMETHYLTRIMETHYLENE GLYCOL see HFP875
2,4,6-TRIMETHYL-1,3,5-TRIOXAAN (DUTCH) see PAI250
2,4,6-TRIMETHYL-s-TRIOXANE see PAI250
2,4,6-TRIMETHYL-1,3,5-TRIOXANE see PAI250
s-TRIMETHYLTRIOXYMETHYLENE see PAI250
1,3,7-TRIMETHYLXANTHINE see CAK500
2-(TRIMETIL-ACETIL)-INDAN-1,3-DIONE (ITALIAN) see
  PIH175
3,5,5-TRIMETIL-2-CICLOESEN-1-ONE (ITALIAN) see
  IMF400
2,4,6-TRIMETIL-1,3,5-TRIOSSANO (ITALIAN) see PAI250
TRIMETIN see TLP750
TRIMETION see DSP400
TRIMETON see TMJ750, PGF000
TRIMETOPRIM-SULFA see SNK000
TRIMITAN see TNF500
TRIMSTAT see DKE800
TRIMTABS see DKE800
TRIMUSTINE see TNF500
TRIMUSTINE HYDROCHLORIDE see TNF500
TRINAGLE see CNP250
TRINATRIUMPHOSPHAT (GERMAN) see SJH200
TRINEX see TIQ250
TRINITRIN see NGY000
TRINITROACETONITRILE see TMK250
TRINITROBENZEEN (DUTCH) see TMK500
TRINITROBENZENE see TMK500
1,3,5-TRINITROBENZENE see TMK500
TRINITROBENZENE, dry (DOT) see TMK500
2,4,6-TRINITROBENZENE-1,3-DIOL see SMP500
2,4,6-TRINITRO-1,3-BENZENEDIOL see SMP500
2,4,6-TRINITROBENZOIC ACID see TML000
TRINITROBENZOIC ACID (dry) see TML000
TRINITROBENZOL (GERMAN) see TMK500
TRINITROCHLOROBENZENE see TML325
TRINITRO-m-CRESOL see TML500
2,4,6-TRINITRO-m-CRESOL see TML500
TRINITRO-m-CRESOLIC ACID see TML500
TRINITROCYCLOTRIMETHYLENE TRIAMINE see CPR800
TRINITROETHANOL (DOT) see TMM000
2,2,2-TRINITROETHANOL see TMM000
2,4,6-TRINITROFENOL (DUTCH) see PID000
2,4,6-TRINITROFENOLO (ITALIAN) see PID000
2,4,7-TRINITROFLUOREN-9-ONE see TMM250
2,4,7-TRINITRO-9-FLUORENONE see TMM250
2,4,7-TRINITROFLUORENONE (MAK) see TMM250
TRINITROGLYCERIN see NGY000
TRINITROGLYCEROL see NGY000
TRINITROMETACRESOL (DOT) see TML500
TRINITROMETHANE see TMM500
1,3,5-TRINITROPHENOL see PID000
2,4,6-TRINITROPHENOL see PID000
TRINITROPHENYLMETHYLNITRAMINE see TEG250
2,4,6-TRINITROPHENYLMETHYLNITRAMINE see TEG250
2,4,6-TRINITROPHENYL-N-METHYLNITRAMINE see
  TEG250
TRINITROPYRENE see TMN000
1,3,6-TRINITROPYRENE see TMN000
2,4,6-TRINITRORESORCINOL see SMP500
TRINITRORESORCINOL (DOT) see SMP500
TRINITRORESORCINOL, dry (DOT) see SMP500
TRINITRORESORCINOL, wetted with less than 20% water
  (DOT) see SMP500
1,3,5-TRINITROSO-1,3,5-TRIAZACYCLOHEXANE see
  HDV500
TRINITROSOTRIMETHYLENETRIAMINE see HDV500
TRINITROSOTRIMETHYLENTRIAMIN (GERMAN) see
  HDV500
2,4,6-TRINITROTOLUEEN (DUTCH) see TMN490
TRINITROTOLUENE see TMN490

s-TRINITROTOLUENE see TMN490
2,4,6-TRINITROTOLUENE see TMN490
sym-TRINITROTOLUENE see TMN490
TRINITROTOLUENE, DRY (DOT) see TMN490
s-TRINITROTOLUOL see TMN490
sym-TRINITROTOLUOL see TMN490
2,4,6-TRINITROTOLUOL (GERMAN) see TMN490
1,3,5-TRINITRO-1,3,5-TRIAZACYCLOHEXANE see CPR800
TRINOXOL see DAA800, TAA100
TRI-n-OCTYL BORATE see TMO250
TRIOCTYLMETHYLAMMONIUM CHLORIDE see MQH000
TRIODURIN see EDU500
TRIORTHOCRESYL PHOSPHATE see TMO600
TRIOSSIMETHLENE (ITALIAN) see TMP000
TRIOVEX see EDU500
TRIOXANE see TMP000
s-TRIOXANE see TMP000
1,3,5-TRIOXANE see TMP000
TRIOXIDE(S) see TGG760
TRIOXOLANES see ORY499
TRIOXON see TAA100
TRIOXONE see BSQ750, TAA100
TRIOXYMETHYLEEN (DUTCH) see TMP000
TRIOXYMETHYLEN (GERMAN) see TMP000
TRIOXYMETHYLENE see TMP000, PAI000
TRIOZANONA see TLP750
TRIPAN BLUE see CMO250
TRI-PCNB see PAX000
TRIPELENAMINE see TMP750
TRIPELENNAMINA (ITALIAN) see TMP750
TRIPELENNAMINE see TMP750
TRIPERIDOL see TKK500
TRIPHACYCLIN see TBX250
TRIPHEDINON see BBV000
TRIPHENATOL see EOK000
TRIPHENIDYL see BBV000
TRIPHENYL see TBD000
m-TRIPHENYL see TBC620
p-TRIPHENYL see TBC750
TRIPHENYLACETO STANNANE see ABX250
TRIPHENYLACRYLONITRILE see TMQ250
2,3,3-TRIPHENYLACRYLONITRILE see TMQ250
α,β,β-TRIPHENYLACRYLONITRILE see TMQ250
TRIPHENYLAMINE see TMQ250
TRIPHENYLCHLOROSTANNANE see CLU000
TRIPHENYLCHLOROTIN see CLU000
TRIPHENYLCYANOETHYLENE see TMQ250
TRIPHENYLETHYLENE see TMS250
1,1,2-TRIPHENYLETHYLENE see TMS250
TRIPHENYL PHOSPHATE see TMT750
TRIPHENYLPHOSPHINE see TMU000
TRIPHENYL PHOSPHITE see TMU250
TRIPHENYLTIN ACETATE see ABX250
TRIPHENYLTIN CHLORIDE see CLU000
TRIPHENYLTIN HYDROXIDE (USDA) see HON000
TRIPHENYLTIN OXIDE see HON000
TRIPHENYL-ZINNACETAT (GERMAN) see ABX250
TRIPHENYL-ZINNHYDROXID (GERMAN) see HON000
TRIPHOSPHORIC ACID, SODIUM SALT see SKN000
TRIPLA-ETILO see DBX400
TRIPLEX III see EIX500
TRI-PLUS see TIO750
TRIPOLI see SCI500, TMX500
TRIPOLY see SKN000
TRIPOLYPHOSPHATE see SKN000
TRIPOMOL see TFS350
TRIPOTASSIUM CITRATE MONOHYDRATE see PLB750
TRIPOTASSIUM TRICHLORIDE see PLA500
TRIPROPARGYL CYANURATE see THN500
TRIPROPYL ALUMINUM see AHH750
TRI-N-PROPYLAMINE see TMY250
TRIPROPYLAMINE (DOT) see TMY250
TRIPROPYLTIN CHLORIDE see CLU250
TRI-n-PROPYLTIN CHLORIDE see CLU250

2,4,6-TRIPROP-2-YNYLOXY-s-TRIAZINE see THN500
TRIPTIL HYDROCHLORIDE see POF250
TRIPTONE see HOT500
TRIS see TNC500
TRIS (flame retardant) see TNC500
TRISAETHYLENIMINOBENZOCHINON (GERMAN) see
  TND000
2,4,6-TRIS(ALLYLOXY)TRIAZINE see THN500
TRIS-4-AMINOFENYLMETHAN (CZECH) see THP000
TRIS(1-AZIRIDINE)PHOSPHINE OXIDE see TND250
2,3,5-TRIS(AZIRIDINO)-1,4-BENZOQUINONE see TND000
2,3,5-TRIS(1-AZIRIDINO)-p-BENZOQUINONE see TND000
TRIS(AZIRIDINYL)-p-BENZOQUINONE see TND000
TRIS(1-AZIRIDINYL)-p-BENZOQUINONE see TND000
2,3,5-TRIS(AZIRIDINYL)-1,4-BENZOQUINONE see TND000
2,3,5-TRIS(1-AZIRIDINYL)-p-BENZOQUINONE see TND000
2,3,5-TRIS(1-AZIRIDINYL)-2,5-CYLOHEXADIENE-1,4-
  DIONE see TND000
TRIS-(1-AZIRIDINYL)PHOSPHINE OXIDE see TND250
TRIS(1-AZIRIDINYL)PHOSPHINE SULFIDE see TFQ750
TRISAZIRIDINYLTRIAZINE see TND500
2,4,6-TRIS(1-AZIRIDINYL)-s-TRIAZINE see TND500
2,4,6-TRIS(1'-AZIRIDINYL)-1,3,5-TRIAZINE see TND500
2,4,6-TRIS(BIS(HYDROXYMETHYL)AMINO)-s-TRIAZINE
  see HDY000
TRIS(2-BUTOXYETHYL) ESTER PHOSPHORIC ACID see
  BPK250
TRIS(2-BUTOXYETHYL) PHOSPHATE see BPK250
TRIS-N-BUTYLAMINE see THX250
3,6,9-TRIS(CARBOXYMETHYL)-3,6,9-TRIAZAUNDECANE-
  DIOIC ACID see DJG800
N,N,N'-TRIS(2-CHLORAETHYL)-N',O-PROPYLEN-PHOS-
  PHORSAUREESTER-DIAMID (GERMAN) see TNT500
TRIS(2-CHLOROETHYL)AMINE HYDROCHLORIDE see
  TNF500
TRIS(β-CHLOROETHYL)AMINE HYDROCHLORIDE see
  TNF500
TRIS(2-CHLOROETHYL)AMINE MONOHYDROCHLORIDE
  see TNF500
TRIS(2-CHLOROETHYL)AMMONIUM CHLORIDE see
  TNF500
TRIS(2-CHLOROETHYL)ESTER PHOSPHORIC ACID see
  CGO500
N,N,N'-TRIS(2-CHLOROETHYL)-N',O-PROPYLENE PHOS-
  PHORIC ACID ESTER DIAMIDE see TNT500
TRIS(2-CHLOROETHYL) PHOSPHATE see CGO500
TRIS(β-CHLOROETHYL) PHOSPHATE see CGO500
N,N,3-TRIS(2-CHLOROETHYL)TETRAHYDRO-2H-1,3,2-
  OXAPHOSPHORIN-2-AMINE-2-OXIDE see TNT500
TRIS-1,2,3-(CHLOROMETHOXY)PROPANE see GGI000
TRIS(o-CRESYL)-PHOSPHATE see TMO600
TRIS(DIBROMOPROPYL)PHOSPHATE see TNC500
TRIS(2,3-DIBROMOPROPYL) PHOSPHATE see TNC500
TRIS(2,3-DIBROMOPROPYL) PHOSPHORIC ACID ESTER
  see TNC500
TRIS-2,3-DIBROMPROPYL ESTER KYSELINY FOS-
  FORECNE (CZECH) see TNC500
1,2,3-TRIS(2-DIETHYLAMINOETHOXY)BENZENE TRIETH-
  IODIDE see PDD300
1,2,3-TRIS(2-DIETHYLAMINOETHOXY)BENZENE TRIS
  (ETHYLIODIDE) see PDD300
2,4,6-TRIS(DI(HYDROXYMETHYL)AMINO)-1,3,5-TRIAZINE
  see HDY000
TRIS(DIMETHYLAMINO)PHOSPHINE OXIDE see HEK000
TRIS(DIMETHYLAMINO)PHOSPHORUS OXIDE see
  HEK000
2,4,6-TRIS(DIMETHYLAMINO)-s-TRIAZINE see HEJ500
2,4,6-TRIS(DIMETHYLAMINO)-1,3,5-TRIAZINE see HEJ500
TRIS(DIMETHYLCARBAMODITHIOATO-S,S')IRON see
  FAS000
TRIS(DIMETHYLDITHIOCARBAMATO)BISMUTH see
  BKW000
TRIS(DIMETHYLDITHIOCARBAMATO)IRON see FAS000

TRIS(N,N-DIMETHYLDITHIOCARBAMATO) IRON(111) see FAS000
2,3,5-TRISETHYLENEIMINOBENZOQUINONE see TND000
TRISETHYLENEIMINOQUINONE see TND000
TRIS(ETHYLENEIMINO)TRIAZINE see TND500
TRISETHYLENEIMINO-1,3,5-TRIAZINE see TND500
2,4,6-TRIS(ETHYLENEIMINO)-s-TRIAZINE see TND500
TRIS(N-ETHYLENE)PHOSPHOROTRIAMIDATE see TND250
2,3,5-TRIS(ETHYLENIMINO)BENZOQUINONE see TND000
2,3,5-TRIS(ETHYLENIMINO)-p-BENZOQUINONE see TND000
2,3,5-TRIS(ETHYLENIMINO)-1,4-BENZOQUINONE see TND000
TRIS(ETHYLENIMINO)THIOPHOSPHATE see TFQ750
2,4,6-TRIS(ETHYLENIMINO)-s-TRIAZINE see TND500
TRISFOSFAMIDE see TNT500
TRIS(2-HYDROXYETHYL)AMINE see TKP500
TRIS-N-LOST see TNF500
TRIS(p-METHOXYPHENYL)CHLOROETHYLENE see CLO750
TRIS(2-METHYL-1-AZIRIDINYL)PHOSPHINE OXIDE see TNK250
TRIS(2-METHYLAZIRIDIN-1-YL)PHOSPHINE OXIDE see TNK250
N,N',N''-TRIS(1-METHYLETHYLENE)PHOSPHORAMIDE see TNK250
TRIS(1-METHYLETHYLENE)PHOSPHORIC TRIAMIDE see TNK250
TRIS(o-METHYLPHENYL)PHOSPHATE see TMO600
TRIS(2-METHYLPROPYL)ALUMINIUM see TKR500
TRISODIUM ARSENATE, HEPTAHYDRATE see ARE000
TRISODIUM-3-CARBOXY-5-HYDROXY-1-p-SULFOPHENYL-4-p-SULFOPHENYLAZOPYRAZOLE see FAG140
TRISODIUM EDETATE see TNL250
TRISODIUM EDTA see TNL250
TRISODIUM ETHYLENEDIAMINETETRAACETATE see TNL250
TRISODIUM HYDROGEN ETHYLENEDIAMINETETRAACE-TATE see TNL250
TRISODIUM HYDROGEN (ETHYLENEDINITRILO)TETRA-ACETATE see TNL250
TRISODIUM NITRILOTRIACETATE see SIP500
TRISODIUM NITRILOTRIACETATE MONOHYDRATE see NEI000
TRISODIUM NITRILOTRIACETIC ACID see SIP500
TRISODIUM ORTHOPHOSPHATE see SJH200
TRISODIUM ORTHOVANADATE see SIY250
TRISODIUM PHOSPHATE see SJH200
TRISODIUM SALT of 3-CARBOXY-5-HYDROXY-1-SULFO-PHENYLAZOPYRAZOLE see FAG140
TRISODIUM TRIFLUORIDE see SHF500
TRISODIUM VERSENATE see TNL250
TRISOMNIN see SBM500
TRISPHOSPHAMIDE see TNT500
TRIS(TOLYLOXY)PHOSPHINE OXIDE see TNP500
TRIS(o-TOLYL)-PHOSPHATE see TMO600
1,2,3-TRIS(2-TRIETHYLAMMONIUM ETHOXY)BENZENE TRIIODIDE see PDD300
TRISULFIDE, BIS(ETHOXYTHIOCARBONYL)- see DKE400
TRISULFURATED PHOSPHORUS see PHS500
TRITHENE see CLQ750
TRITHEOM see ABY900
TRITHIOBIS(TRICHLOROMETHANE) see BLM750
TRITHION see TNP250
TRITHION MITICIDE see TNP250
TRITISAN see PAX000
TRITOFTOROL see EIR000
TRITOL see TMN490
TRITOLYL PHOSPHATE see TNP500
TRI-2-TOLYL PHOSPHATE see TMO600
TRI-o-TOLYL PHOSPHATE see TMO600
TRITON GR-5 see DJL000
TRITON X 35 see PKF500

TRITON X-40 see DTC600
TRITON X 45 see PKF500
TRITON X 100 see PKF500
TRITON X 102 see PKF500
TRITON X 165 see PKF500
TRITON X 305 see PKF500
TRITON X 405 see PKF500
TRITON X 705 see PKF500
TRIVAZOL see MMN250
TRI-VC 13 see DFK600
TRIVINYLTIN CHLORIDE see CLU500
TRIVITAN see CMC750
TRIZILIN see DFT800
TROCLOSENE see DGN200
TROCOSONE see ECU750
TRODAX see HLJ500
TROFOSFAMID see TNT500
TROFURIT see CHJ750
TROGAMID T see NOH000
TROJCHLOREK FOSFORU (POLISH) see PHT275
TROJCHLOROBENZEN (POLISH) see TIK250
TROJCHLOROETAN(1,1,2) (POLISH) see TIN000
TROJKREZYLU FOSFORAN (POLISH) see TMO600
TROJNITROTOLUEN (POLISH) see TMN490
TROLAMINE see TKP500
TROLEN see RMA500
TROLENE see RMA500
TROLITUL see SMQ500
TROLOVOL see MCR750
TROMASIN see PAG500
TROMBARIN see BKA000
TROMBIL see BKA000
TROMBOLYSAN see BKA000
TROMBOSAN see BJZ000
TROMEDONE see TLP750
TROMETE see SJH200
TROMETHAMINE PROSTAGLANDIN F2-α see POC750
TROMEXAN see BKA000
TROMEXAN ETHYL ACETATE see BKA000
TRONA see BMG400, SFO000, SJY000
TRONAMANG see MAP750
TRONOX see TGG760
TROPAEOLIN D see DOU600
dl-TROPANYL-2-HYDROXY-1-HENYLPROPIONATE SUL-FATE see ARR500
dl-TROPANYL-2-HYDROXY-1-PHENYLPROPIONATE see ARR000
TROPEOLIN see BFL000
TROPHOSPHAMID see TNT500
TROPHOSPHAMIDE see TNT500
TROPIC ACID, ESTER with SCOPINE see SBG000
TROPIC ACID, ESTER with TROPINE see ARR000
(−)-TROPIC ACID ESTER with TROPINE see HOU000
TROPIC ACID, 9-METHYL-3-OXA-9-AZATRICYCLO (3.3.1.0^{2,4})NON-7-YL ESTER see SBG000
TROPIC ACID-3-α-TROPANYL ESTER see ARR000
TROPILIDENE see COY000
TROPINE BENZOHYDRYL ETHER METHANESULFONATE see TNU000
TROPINE TROPATE see ARR000
TROPISTON see PGP500
(+,-)-TROPYL TROPATE see ARR000
dl-TROPYLTROPATE see ARR000
TROTYL see TMN490
TROTYL OIL see TMN490
TROVIDUR see VNP000, PKQ059
TROVITHERN HTL see PKQ059
TROXIDONE see TLP750
TROYKYD ANTI-SKIN BTO see BSU500
TROYSAN 142 see DSB200
TRP-P-1 see TNX275
TRP-P-2 see ALD500
TRP-P-1 (ACETATE) see AJR500
TRUCIDOR see MJG500

TRUE AMMONIUM SULFIDE see ANJ750
TRUFLEX DOP see DVL700
TRUSONO see MNM500
TRYPAFLAVINE see DBX400
TRYPANBLAU (GERMAN) see CMO250
TRYPAN BLUE see CMO250
TRYPAN BLUE SODIUM SALT see CMO250
TRYPARSAMIDE see CBJ750
TRYPTIZOL see EAH500, EAI000
TRYPTIZOL HYDROCHLORIDE see EAI000
(−)-TRYPTOPHAN see TNX000
l-TRYPTOPHAN (FCC) see TNX000
dl-TRYPTOPHAN see TNW500
dl-TRYPTOPHAN, pyrolyzate 1 see TNX275
TRYPTOPHANE see TNX000
l-TRYPTOPHANE see TNX000
TRYPTOPHAN P1 see TNX275
TRYPTOPHAN P2 see ALD500
TS-160 see TNF500
TS 219 see NIM500
TSC see TFQ000
TSD see MCR250
TSERENOX see BDD000
TSIAZID see COH250
TSIDIAL see DRR400
TSIKLODOL see BBV000
TSIKLOMITSIN see TBX000
TSIMAT see BJK500
TSINEB (RUSSIAN) see EIR000
TSIRAM (RUSSIAN) see BJK500
TSITREX see DXX400
TSIZP 34 see ISR000
TSP see SJH200
TSPA see TFQ750
TSPP see TEE500
TST see EIV000
T-STUFF see HIB000
TSUMACIDE see MIB750
TTD see DXH250, TFS350
T (²)-TRICHOTHECENE see FQS000
TTS see DXH250
TTT see HDV500
TTX see FOQ000
TU see TFR250
2-TU see TFR250
TUADS see TFS350
TUAZOLE see QAK000
TUAZOLONE see QAK000
TUBADIL see TOA000
TUBARINE see TOA000
TUBA ROOT see DBA000
TUBATOXIN see RNZ000
TUBAZID see ILD000
TUBAZIDE see ILD000
TUBECO see ILD000
TUBERCID see ILD000
TUBERIAN see ILD000
TUBERMIN see EPQ000
TUBEROID see EPQ000
TUBEROSON see EPQ000
TUBICON see ILD000
TUBOCIN see RKP000
TUBOCURARINE CHLORIDE see TOA000
(+)-TUBOCURARINE CHLORIDE see TOA000
d-TUBOCURARINE CHLORIDE see TOA000
TUBOCURARINE, CHLORIDE, HYDROCHLORIDE, (+)-
(8CI) see TOA000
d-TUBOCURARINE DICHLORIDE see TOA000
TUBOCURARINE HYDROCHLORIDE see TOA000
(+)-TUBOCURARINE HYDROCHLORIDE see TOA000
d-TUBOCURARINE HYDROCHLORIDE see TOA000
TUBOMEL see ILD000
TUBOTIN see ABX250, HON000
TUEX see TFS350

TUFF-LITE see PMP500
TUGON see TIQ250
TUGON FLIEGENKUGEL see PMY300
TUGON FLY BAIT see TIQ250
TUGON STABLE SPRAY see TIQ250
TULABASE FAST GARNET GB see AIC250
TULABASE FAST GARNET GBC see AIC250
TULABASE FAST RED TR see CLK220
TULISAN see TFS350
TULUYLENDIISOCYANAT see TGM750
TUMBLEAF see SFS000
TUMEX see QPA000
TUNG NUT MEALS see TOA500
TUNG NUT OIL see TOA510
TUNGSTEN see TOA750
TUNGSTEN COMPOUNDS see TOC500
TUNGSTIC ACID, DISODIUM SALT see SKN500
TUPHETAMINE see BBK500
TURCAM see DQM600
TURGEX see HCL000
TURISYNCHRON see MLJ500
TURMERIC see TOD625
TURMERIC OIL see COG000
TURMERIC OLEORESIN see COG000
TURPENTINE see TOD750
TURPENTINE OIL, RECTIFIER see TOD750
TURPENTINE STEAM DISTILLED see TOD750
TURPETH MINERAL see MDG000
TUSSAPAP see HIM000
TUTANE see BPY000
TUTOCAINE HYDROCHLORIDE see AIT750
TWEEN 60 see PKL030
TWEEN 80 see PKL100
TWIN LIGHT RAT AWAY see WAT200
TX 100 see PKF500
TYBAMATE see MOV500
TYBATRAN see MOV500
TYCLAROSOL see EIV000
TYDEX see BBK500
TYGON see AAX175
TYLAN see TOE600
TYLENOL see HIM000
TYLON see TOE600
TYLOSE 666 see SFO500
TYLOSIN see TOE600
TYLOSIN HYDROCHLORIDE see TOE750
TYLOSTERONE see DKA600
TYOX A see BHM000
TYPOGEN CARMINE see EOJ500
TYRANTON see DBF750
TYRIL see ADY500
TYROSINE see TOG300

U-14 see POC750
U 46 see CIR500, DAA800, TAA100
U46 see DGB000
U-1149 see FOU000
U 1363 see DVV600
U-1434 see EEI000
U-3886 see SFA000
U-4224 see DSB000
U 4513 see DRP800
U-4748 see POK000
U-4783 see ACJ500
U 4905 see HHR000
U-5043 see DAA800
U-5227 see AQN635
U-5897 see CDT750
U-5954 see DOO800
U-5965 see TBX250
U 6020 see PLZ000
U-6062 see CDP250
U-6233 see BDH250
U 6324 see PGN250

U-6421 see NGE500
U-6591 see NOB000
U 6987 see BSM000
U-8344 see BIA250
U 8771 see EHP000
U-8953 see FMM000
U-9889 see SMD000
U-10149 see LGD000
U-10,858 see DCB000
U 10997 see MQS225
U-12062 see DVJ200
U-14583 see DMU800
U-14743 see MLY000
U 15030 see DKC400
U 25,354 see DGG400
U 1 (polymer) see PKQ059
U-22394A see HDQ500
UBATOL U 2001 see SMQ500
UBRETID see DXG800
UBRITIL see DXG800
UC-21149 see CBM500
UCAR 17 see EJC500
UCAR 130 see AAX250
UCAR BUTYLPHENOL 4-T see BSE500
UCAR SOLVENT 2LM see DWT200
UCB 170 see HGC500
UCB 492 see CJR909
U.CB 4492 see CJR909
UCB 6249 see DIF600
UCC 974 see DSB200
UCET TEXTILE FINISH 11-74 (OBS.) see VOA000
U-COMPOUND see UVA000
UCON 12 see DFA600
UCON 113 see FOO000
UCON 114 see FOO509
UCON FLUOROCARBON 113 see FOO000
UCON 12/HALOCARBON 12 see DFA600
UCON 22/HALOCARBON 22 see CFX500
UCON 113/HALOCARBON 113 see FOO000
UCON 500/HALOCARBON 500 see DFB400
UCON REFRIGERANT 11 see TIP500
UDMH (DOT) see DSF400
UDOLAC see SOA500
U 46DP see DAA800
U46 DP-FLUID see DGB000
UF 1 see DMI600
UKOPEN see AOD125
U 46 KV-ESTER see CIR500
U 46 KV-FLUID see CIR500
U46KW see BSQ750
ULACORT see PMA000
ULCERFEN see TEH500
ULCINE see XCJ000
ULCUDEXTER see XCJ000
ULSTRON see PMP500
ULTRABION see AIV500
ULTRA BRILLIANT BLUE P see DSY600
ULTRABRON see AIV500
ULTRACIDE see DSO000
ULTRACORTEN see PLZ000
ULTRACORTENE-H see PMA000
ULTRAMARINE GREEN see CMJ900
ULTRAMARINE YELLOW see BAK250
ULTRAMID BMK see PJY500
ULTRAPAL CHLORIDE see HLC500
ULTRA SULFATE SL-1 see SIB600
ULTRAWET K see DXW200
ULTRON see PKQ059
ULUP see FMM000
ULV see DAO600
ULVAIR see MRH209
UMBELLATINE SULFATE TRIHYDRATE see BFN750
UMBRATHOR see TFT750
U 46 M-FLUID see CIR250

UNADS see BJL600
UNAMIDE J-56 see BKE500
1,2,3,4,5,5,6,7,9,10,10-UNDECACHLOROPENTACYCLO
(5.3.O.O$^{2.6}$.O$^{3.9}$.O^{4,8})DECANE see MRI750
γ-UNDECALACTONE see UJA800
UNDECANAL see UJJ000
1-UNDECANAL see UJJ000
n-UNDECANAL see UJJ000
UNDECANALDEHYDE see UJJ000
UNDECANE see UJS000
n-UNDECANE see UJS000
1-UNDECANECARBOXYLIC ACID see LBL000
n-UNDECANOL see UNA000
2-UNDECANONE see UKS000
10-UNDECENAL see ULJ000
1-UNDECEN-10-AL see ULJ000
UNDECYL ALCOHOL see UNA000
UNDECYL ALDEHYDE see UJJ000
N-UNDECYL ALDEHYDE see UJJ000
UNDECYLENALDEHYDE see ULJ000
10-UNDECYLENEALDEHYDE see ULJ000
UNDECYLENIC ALDEHYDE see ULJ000
UNDECYLIC ALDEHYDE see UJJ000
UNDEN see EDV000, PMY300
UNFINISHED LUBRICATING OIL see COD750
UNIBARYT see BAP000
UNICELLES see BPF000
UNICHEM see PKQ059
UNICIN see TBX250
UNICOCYDE see ILD000
UNICROP CIPC see CKC000
UNICROP DNBP see BRE500
UNICROP MANEB see MAS500
UNIDIGIN see DKL800
UNIDRON see DXQ500
UNIFLEX DOS see BJS250
UNIFOAM AZ see ASM270
UNIFORM AZ see ASM270
UNIFUME see EIY500
UNI-GUAR see GLU000
UNIMATE GMS see OAV000
UNIMOLL BB see BEC500
UNIMYCETIN see CDP250
UNIMYCIN see TBX250
UNION BLACK EM see AQP000
UNION CARBIDE A-150 see TIN750
UNIPINE see PIH750
UNIPON see DGI400, DGI600
UNIROYAL see DFT000
UNIROYAL D014 see SOP000
UNISOL RH see SFO500
UNISOMNIA see DLY000
UNISTRADIOL see EDP000
UNITANE O-110 see TGG760
UNITED CHEMICAL DEFOLIANT No. 1 see SFS000
UNITENE see MCC250
UNITERTRACID LIGHT ORANGE G see HGC000
UNITERTRACID YELLOW TE see FAG140
UNITESTON see TBG000
UNITOX see CDS750
UNIVERM see CBY000
UNIVOL U 316S see MSA250
UNLEADED GASOLINE see GCE100
UNLEADED MOTOR GASOLINE see GCE100
UNON P see TAI250
UNOXAT EPOXIDE 269 see LFV000
UNOX EPOXIDE 206 see VOA000
UP 1E see SMR000
UPIOL see BNP750
URACILLOST see BIA250
URACILMOSTAZA see BIA250
URACIL MUSTARD see BIA250
URADAL see BNK000
URAGAN see BMM650, SIH500

URAGON see BMM650
URALGIN see EID000
URAMUSTIN see BIA250
URAMUSTINE see BIA250
URANIN see FEW000
URANINE A EXTRA see FEW000
URANINE USP XII see FEW000
URANINE YELLOW see FEW000
URANIUM see UNS000
URANIUM ACETATE see UPS000
URANIUM(IV) CHLORIDE see UQJ000
URANIUM FLUORIDE (low specific activity) see UOS000
URANIUM FLUORIDE OXIDE see UQA000
URANIUM HEXAFLUORIDE, LOW SPECIFIC ACTIVITY
    (containing 0.7% or less U-235) (DOT) see UOS000
URANIUM(III) HYDRIDE see UPA000
URANIUM METAL, PYROPHORIC (DOT) see UNS000
URANIUM OXYACETATE see UPS000
URANIUM OXYFLUORIDE see UQA000
URANIUM TETRACHLORIDE see UQJ000
URANYL ACETATE see UPS000
URANYL FLUORIDE see UQA000
URANYL NITRATE (solid) see USA000
URANYL NITRATE HEXAHYDRATE see URS000
URANYL NITRATE HEXAHYDRATE, solution (DOT) see
    URS000
URAZIUM see PDC250
URBANYL see CIR750
URBASON see MOR500
URBASONE see MOR500
URBASULF see MGQ750
URBIL see MQU750
UREA see USS000
UREA DIOXIDE see HIB500
UREA HYDROGEN PEROXIDE (DOT) see HIB500
UREA HYDROGEN PEROXIDE SALT see HIB500
UREA HYDROPEROXIDE see HIB500
UREA PEROXIDE (DOT) see HIB500
UREAPHIL see USS000
UREGIT see DFP600
p-UREIDOBENZENEARSONIC ACID see CBJ000
(p-UREIDOBENZENEARSYLENEDITHIO)DI-o-BENZOIC
    ACID see TFD750
4-UREIDO-1-PHENYLARSONIC ACID see CBJ000
(p-UREIDOPHENYLARSYLENEDITHIO)DIACETIC ACID see
    CBI250
(p-UREIDOPHENYLARSYLENEDITHIO)DI-o-BENZOIC ACID
    see TFD750
UREOPHIL see USS000
URESE see BDE250
URETAN ETYLOWY (POLISH) see UVA000
URETHAN see UVA000
URETHANE see UVA000
URETHYLANE see MHZ000
UREVERT see USS000
UREX see CHJ750
URGENEA MARITIMA see RCF000
URIBEN see EID000
URIDINAL see PDC250, PEK250
URIPLEX see PDC250
URITONE see HEI500
URIZEPT see NGE000
URNER'S LIQUID see DEL000
UROBIOTIC-250 see PDC250
URODIAZIN see CFY000
URODINE see PDC250, PEK250
URODIXIN see EID000
UROFEEN see PDC250
UROMAN see EID000
UROMIDE see PDC250
UROMUCAESTHIN see BQA010
UROMYCINE see GCO000
URONEG see EID000
UROPHENYL see PDC250

UROPYRIDIN see PDC250
UROPYRINE see PDC250
UROSEMIDE see CHJ750
UROTROPIN see HEI500
UROTROPINE see HEI500
UROX B WATER SOLUBLE CONCENTRATE WEED KILLER
    see BMM650
UROX HX GRANULAR WEED KILLER see BMM650
URSOFERRAN see IGS000
URSOL BROWN RR see ALL750
URSOL D see PEY500
URSOL EG see ALT500
URSOL ERN see NAW500
URSOL OLIVE 6G see CFK125
URSOL SLA see DBO400
URTOSAL see SAH000
USACERT BLUE No. 1 see FAE000
USACERT BLUE No.2 see FAE100
USACERT RED No. 1 see FAG018
USACERT RED No. 3 see FAG040
USACERT YELLOW No. 5 see FAG140
USAF D-3 see CHU500
USAF M-7 see CEI500
USAF P-8 see CJX750
USAF A-233. see AIR250
USAF A-4600 see MAO250
USAF A-6598 see DXP200
USAF A-8564 see BIQ500
USAF A-8565 see HIM500
USAF A-9442 see SNE000
USAF A-9789 see DVX600
USAF AB-315 see DXJ800
USAF AM-5 see AAH250
USAF AM-6 see BSU500
USAF AN-7 see MFC700
USAF AN-8 see TDK000
USAF B-7 see ADC750
USAF B-15 see TFC600
USAF B-17 see BKU500
USAF B-19 see DWC600
USAF B-24 see SAV000
USAF B-30 see TFS350
USAF B-32 see BJL600
USAF B-33 see BDE750, DXH250
USAF B-35 see SGF500
USAF B-44 see DXL800
USAF B-100 see DJY800
USAF B-121 see MAL250
USAF BE-0405 see ADC750
USAF CB-2 see ILD000
USAF CB-7 see BAC250
USAF CB-13 see FMT000
USAF CB-17 see XCA000
USAF CB-19 see NCG000
USAF CB-20 see TET300
USAF CB-21 see TFA000
USAF CB-22 see NBE500
USAF CB-27 see RDK000
USAF CB-29 see ICW000
USAF CB-35 see TFJ100
USAF CB-36 see MCM750
USAF CS-6 see NNM000
USAF CY-2 see CAQ250
USAF CY-4 see NAP500
USAF CY-5 see BDE750
USAF CY-7 see BDG000
USAF CY-10 see NBA500
USAF D-1 see IDW000
USAF D-5 see BSO500
USAF DO-1 see CEB250
USAF DO-12 see HHR500
USAF DO-21 see EKR500
USAF DO-28 see DMI600
USAF DO-29 see CDY850

VAC see VLU250
VAC-10 see BRF500
VACATE see CIR250
VACUUM RESIDUUM see MQV755
VADROCID see NGE500
VAFLOL see VSK600
VAGAMIN see XCJ000
VAGANTIN see XCJ000
VAGD see AAX175
VAGESTROL see DKA600
VAGILEN see MMN250
VAGIMID see MMN250
VAGISEPT see ABX500
VAGOFLOR see ABX500
VAGOSTIGMIN see DER600
VAGOSTIGMINE BROMIDE see DQY800
VAGOSTIGMINE METHYL SULFATE see DQY909
VALADOL see HIM000
VALAMINE see IIM000
VALAN see MRW000
VALBAZEN see VAD000
VALDRENE see BAU750
VAL-DROP see SFS000
VALERAL see VAG000
n-VALERALDEHYDE see VAG000
VALERAMIDE-OM see DOY400
VALERAN DI-n-BUTYLCINICITY (CZECH) see DEA600
VALERIANIC ACID see VAQ000
VALERIANIC ALDEHYDE see VAG000
VALERIC ACID see VAQ000
n-VALERIC ACID see VAQ000
VALERIC ACID ALDEHYDE see VAG000
VALERIC ALDEHYDE see VAG000
4-VALEROLACTONE see VAV000
γ-VALEROLACTONE (FCC) see VAV000
VALERONE see DNI800
Δ-VALEROSULTONE see BOU250
VALERYLALDEHYDE see VAG000
VALERYL CHLORIDE see VBA000
VALETHAMATE see VBK000
VALETHAMATE BROMIDE see VBK000
VALEXONE see BAT750
VALFLON see TAI250
VALGESIC see HIM000
VALGIS see TEH500
VALGRAINE see TEH500
VALINE see VBP000
VALINE ALDEHYDE see IJS000
l-VALINE (FCC) see VBP000
VALIOIL see HNI500
VALKACIT CA see DWN800
VALLADAN see BCA000
VALLERGINE see DQA400
VALZIN see EFE000
VAMIDOATE see MJG500
VAMIDOTHION see MJG500
VANADIC ACID, AMMONIUM SALT see ANY250
VANADIC ACID, MONOSODIUM SALT see SKP000
VANADIC(II) ACID, TRISODIUM SALT see SIY250
VANADIC ANHYDRIDE see VDU000
VANADIC OXIDE see VEA000
VANADIO, PENTOSSIDO di (ITALIAN) see VDU000
VANADIUM see VCP000
VANADIUM CHLORIDE see VEF000
VANADIUM(III) CHLORIDE see VEP000
VANADIUM COMPOUNDS see VCZ000
VANADIUM DUST and FUME (ACGIH) see VDU000, VDZ000
VANADIUM OXIDE see VEA000
VANADIUM(V) OXIDE see VDU000
VANADIUM OXYTRICHLORIDE see VDP000
VANADIUM PENTAOXIDE see VDU000
VANADIUMPENTOXID (GERMAN) see VDU000
VANADIUM PENTOXIDE (dust) see VDU000

VANADIUM PENTOXIDE (fume) see VDZ000
VANADIUM PENTOXIDE, non-fused form (DOT) see VDU000
VANADIUMPENTOXYDE (DUTCH) see VDU000
VANADIUM, PENTOXYDE de (FRENCH) see VDU000
VANADIUM SESQUIOXIDE see VEA000
VANADIUM TETRACHLORIDE see VEF000
VANADIUM TRICHLORIDE see VEP000
VANADIUM TRICHLORIDE OXIDE see VDP000
VANADIUM TRIOXIDE see VEA000
VANADYL SULFATE see VEZ000
VANADYL TRICHLORIDE see VDP000
VANAY see THM500
VANCIDA TM-95 see TFS350
VANCIDE see MAS500
VANCIDE 89 see CBG000
VANCIDE FE95 see FAS000
VANCIDE KS see HON000
VANCIDE MZ-96 see BJK500
VANCIDE PA see BLG500
VANCIDE PA DISPERSION see BLG500
VANCIDE PB see DVI600
VANCIDE TM see TFS350
VANDEX see SBO500
VAN DYK 264 see OES000
VANDYKE BROWN see MAT500
VANGARD K see CBG000
VANGUARD N see BIW750
VANICIDE see CBG000
VANILLA see VFK000
VANILLAL see EQF000
VANILLALDEHYDE see VFK000
VANILLA PLANT see DAJ800
VANILLIC ACID DIETHYLAMIDE see DKE200
VANILLIC ACID-N,N-DIETHYLAMIDE see DKE200
VANILLIC ALDEHYDE see VFK000
VANILLIN see VFK000
p-VANILLIN see VFK000
VANILLINSAEURE-DIAETHYLAMID (GERMAN) see DKE200
VANIROM see EQF000
VANLUBE PCX see BFW750
VANOBID see CBC250, LFF000
VANOXIDE see BDS000
VANZOATE see BCM000
VAPONA see DGP900
VAPONEFRIN see VGP000
VAPO-N-ISO see DMV600
VAPONITE see DGP900
VAPOPHOS see PAK000
VAPOROLE see IMB000
VAPORPAC see ILM000
VAPOTONE see TCF250
VAPTONE see TCF280
VARAMID ML 1 see BKE500
VARIOFORM II see USS000
VARISOFT SDC see DTC600
VARITOX see TII250
VARNISH MAKERS' AND PAINTERS' NAPHTHA see PCT250
VARNISH MAKERS' NAPHTHA see PCT250
VARNOLINE see SLU500
VAROX see DRJ800
VAROX DCP-R see DGR600
VAROX DCP-T see DGR600
VARSOL see PCT250
VASALGIN see ELX000
VASAZOL see DJS200
VASCUALS see VSZ450
VASITOL see PBC250
VASOCONSTRICTINE see VGP000
VASOCONSTRICTOR see VGP000
VASODIATOL see PBC250
VASODILATATEUR 2249F see FMX000

VASODRINE see VGP000
VASOLAN see IRV000
VASOTON see VGP000
VASOTONIN see VGP000
VASO-80 UNICELIES see PBC250
VASOXINE see MDW000
VASOXINE HYDROCHLORIDE see MDW000
VASOXYL HYDROCHLORIDE see MDW000
VATERITE see CAO000
VATROLITE see SHR500
VATSOL OT see DJL000
VAZADRINE see ILD000
VAZO 64 see ASL750
V-BRITE see SHR500
VC see VNP000
V-C CHEMICAL V-C 9-104 see EIN000
VCM see VNP000
VC13 NEMACIDE see DFK600
VCR see LEY000
VCR SULFATE see LEZ000
VDC see VPK000
VDF see VPP000
VECTAL see ARQ725
VECTAL SC see ARQ725
VECTREN see CIP500
VEDERON see ILD000
VEDRIL see PKB500
VEGADEX see CDO250
VEGADEX SUPER see CDO250
VEGANTINE see DHX800
VEGETABLE GUM see DBD800
VEGETABLE PEPSIN see PAG500
VEGETABLE (SOYBEAN) OIL, brominated see BMO825
VEGFRU see PGS000
VEGFRU FOSMITE see EEH600
VEGFRU MALATOX see MAK700
VEGOLYSEN see HEA000
VEGOLYSIN see HEA000
VEHAM-SANDOZ see EQP000
VEHEM see EQP000
VEL 4283 see MKA000
VEL 4284 see DRR200
VELARDON see PAG500
VELBAN see VLA000
VELBE see VLA000
VELDOPA see DNA200
VELFLON see TAI250
VELMOL see DJL000
VELPAR see HFA300
VELPAR WEED KILLER see HFA300
VELSICOL 104 see HAR000
VELSICOL 506 see LEN000
VELSICOL 1068 see CDR750
VELSICOL COMPOUND "R" see MEL500
VELSICOL 53-CS-17 see EBW800
VELSICOL 58-CS-11 see MEL500
VELSICOL VCS 506 see LEN000
VENA see BBV500
VENCIPON see EAW000
VENDACID LIGHT ORANGE 2G see HGC000
VENDEX see BLU000
VENETIAN RED see IHD000
VENETLIN see BQF500
VENTOLIN see BQF500
VENTOX see ADX500
VENTRAMINE see DJS200
VENTUROL see DXX400
VENZONATE see BCM000
VEON 245 see TAA100
VEPESID see EAV500
VERAPAMIL see IRV000
VERATRIDINE see VHU000, VHZ000
VERATRIN (GERMAN) see VHZ000
VERATRINE see VHZ000

VERATRINE (AMORPHOUS) see VHU000
VERATROLE METHYL ETHER see AGE250
3-VERATROYLVERACEVINE see VHU000
VERATRUM VIRIDE see VIZ000
VERATRUM VIRIDE ALKALOIDS EXTRACT see VIZ000
VERAX see BBW500
VERAZINC see ZNA000
VERCIDON see DJT800
VERDICAN see DGP900
VERDIPOR see DGP900
VERDONE see CIR250
VEREDEN see MNM500
VERESENE DISODIUM SALT see EIX500
VERGEMASTER see DAA800
VERGFRU FORATOX see PGS000
VERILOID see VIZ000
VERINA see DNU200
VERMICIDE BAYER 2349 see TIQ250
VERMINUM see BQK000
VERMIRAX see MHL000
VERMITIN see DFV400, PDP250
VERMIZYM see PAG500
VERMOESTRICID see CBY000
VERMOX see MHL000
VERNAMYCIN see VRF000
VERONAL SODIUM see BAG250
VERON P 130/1 see PKQ059
VEROPHEN see DQA600
VERROL see VSZ450
VERSALIDE see ACL750
VERSAR DSMA LQ see DXE600
VERSENE 9 see TNL250
VERSENE 100 see EIV000
VERSENE ACID see EIX000
VERSENE NTA ACID see AMT500
VERSENE POWDER see EIV000
VERSENE SODIUM 2 see EIX500
VERSNELLER NL 63/10 see DQF800
VERSOMNAL see EOK000
VERSOTRANE see PFN000
VERSULIN see CDH250
VERTAC see BQI000, DGI000
VERTAC 90% see CDV100
VERTAC DINITRO WEED KILLER see BRE500
VERTAC GENERAL WEED KILLER see BRE500
VERTAC METHYL PARATHION TECHNISCH 80% see MNH000
VERTAC SELECTIVE WEED KILLER see BRE500
VERTAC TOXAPHENE 90 see CDV100
VERTAVIS see VIZ000
VERTHION see DSQ000
VERTISAL see MMN250
VERTOLAN see SNJ000
VERTON 2T see TAA100
VERTON D see DAA800
VESAKONTUHO MCPA see CIR250
VESPARAZ-WIRKSTOFF see CJR909
VESPRIN see TKL000
VESTIN see PDC250
VESTINOL AH see DVL700
VESTOLIT B 7021 see PKQ059
VESTROL see TIO750
VESTYRON see SMQ500
VESTYRON HI see SMR000
VETAFLAVIN see DBX400
VETALAR see CKD750
VETAMOX see AAI250
VETARSENOBILLON see NCJ500
VETBUTAL see NBU000
VETERINARY NITROFURAZONE see NGE500
VETICILLIN see BFD250
VETICOL see CDP250
VETIDREX see CFY000
VETIOL see MAK700

VETKALM see DYF200
VETOL see MAO350
VETQUAMYCIN-324 see TBX250
VETRANQUIL see ABH500
VETREN see HAQ500
VETSIN see MRL500
VI-ALPHA see VSK600
VIANSIN see MDQ250
VIBALT see VSZ000
VIBAZINE see HGC500
VIBISONE see VSZ000
VI-CAD see CAE250
VICCILLIN see AIV500
VICCILLIN S see AIV500
VICILLIN see AIV500
VICKNITE see PLL500
VICTORIA ORANGE see DUT600
VICTORIA YELLOW see DUT600
VICTOR TSPP see TEE500
VIDDEN D see DGG000, DGG950
VIDLON see PJY500
VIDON 638 see DAA800
VIDOPEN see AOD125
VIENNA GREEN see COF500
VIGANTOL see VSZ100
VIGORSAN see CMC750
VIKANE see SOU500
VIKANE FUMIGANT see SOU500
VILLIAUMITE see SHF500
VINAC B 7 see AAX250
VINAMAR see EQF500
VINBARBITAL SODIUM see VKP000
VINBLASTIN see VKZ000
VINBLASTINE see VKZ000
VINBLASTINE SULFATE see VLA000
VINCALEUCOBLASTIN see VKZ000
VINCALEUKOBLASTINE see VKZ000
VINCALEUKOBLASTINE SULFATE see VLA000
VINCALEUKOBLASTINE SULFATE (1:1) (SALT) see VLA000
VINCLOZOLIN (GERMAN) see RMA000
VINCOBLASTINE see VKZ000
VINCRISTINE see LEY000
VINCRISTINE SULFATE ONCORIN see LEZ000
VINCRISTINSULFAT (GERMAN) see LEZ000
VINCRISUL see LEZ000
VINCRYSTINE see LEY000
VINEGAR ACID see AAT250
VINEGAR NAPHTHA see EFR000
VINEGAR SALTS see CAL750
VINESTHENE see VOP000
VINESTHESIN see VOP000
VINETHEN see VOP000
VINETHENE see VOP000
VINETHER see VOP000
VINICIZER 80 see DVL700
VINICIZER 85 see DVL600
VI-NICOTYL see NCR000
VI-NICTYL see NCR000
VINIDYL see VOP000
VINIKA KR 600 see PKQ059
VINIKULON see PKQ059
VINILE (ACETATO di) (ITALIAN) see VLU250
VINILE (BROMURO di) (ITALIAN) see VMP000
VINILE (CLORURO didi) (ITALIAN) see VNP000
VINIPLAST see PKQ059
VINIPLEN P 73 see PKQ059
VINISIL see PKQ250
VINKEIL 100 see EIX000
VINKRISTIN see LEY000
4-VINLYCYCLOHEXENE DIOXIDE see VOA000
VINNOL E 75 see PKQ059
VINNOL H 10/60 see AAX175
VINOFLEX see PKQ059

VINOFLEX MO 400* see IJQ000
VINSTOP see SGM500
VINTHIONINE see VLU200
VINYDAN see VOP000
VINYLACETAAT (DUTCH) see VLU250
VINYLACETAT (GERMAN) see VLU250
VINYL ACETATE see VLU250
VINYL ACETATE HOMOPOLYMER see AAX250
VINYL ACETATE POLYMER see AAX250
VINYL ACETATE RESIN see AAX250
VINYL ACETATE-VINYL CHLORIDE COPOLYMER see AAX175
VINYL ACETATE-VINYL CHLORIDE POLYMER see AAX175
VINYL ACETYLENE see BPE109
VINYL ALCOHOL POLYMER see PKP750
VINYL A MONOMER see VLU250
VINYLBENZEN (CZECH) see SMQ000
VINYLBENZENE see SMQ000
VINYLBENZENE POLYMER see SMQ500
VINYLBENZOL see SMQ000
VINYLBROMID (GERMAN) see VMP000
VINYL BROMIDE see VMP000
VINYL BROMIDE, inhibited (DOT) see VMP000
VINYL BUTYL ETHER see VMZ000
VINYL-n-BUTYL ETHER see VMZ000
VINYL BUTYRATE see VNF000
VINYL BUTYRATE, INHIBITED (DOT) see VNF000
VINYLBUTYROLACTAM see EEG000
N-VINYLBUTYROLACTAM POLYMER see PKQ250
VINYL CARBAMATE see VNK000
VINYLCARBINOL see AFV500
VINYL CARBINYL CINNAMATE see AGC000
VINYLCHLON 4000LL see PKQ059
VINYLCHLORID (GERMAN) see VNP000
VINYL CHLORIDE see VNP000
VINYL CHLORIDE HOMOPOLYMER see PKQ059
VINYL CHLORIDE MONOMER see VNP000
VINYL CHLORIDE POLYMER see PKQ059
VINYL CHLORIDE-VINYL ACETATE POLYMER see AAX175
VINYL-2-CHLOROETHYL ETHER see CHI250
VINYL-β-CHLOROETHYL ETHER see CHI250
VINYL C MONOMER see VNP000
VINYL CYANIDE see ADX500
1-VINYLCYCLOHEXENE-3 see CPD750
1-VINYLCYCLOHEX-3-ENE see CPD750
4-VINYLCYCLOHEXENE-1 see CPD750
4-VINYL-1-CYCLOHEXENE see CPD750
VINYL CYCLOHEXENE DIEPOXIDE see VOA000
4-VINYLCYCLOHEXENE DIEPOXIDE see VOA000
4-VINYL-1-CYCLOHEXENE DIEPOXIDE see VOA000
4-VINYL-1,2-CYCLOHEXENE DIEPOXIDE see VOA000
VINYL CYCLOHEXENE DIOXIDE see VOA000
1-VINYL-3-CYCLOHEXENE DIOXIDE see VOA000
4-VINYL-1-CYCLOHEXENE DIOXIDE (MAK) see VOA000
VINYLE (ACETATE de) (FRENCH) see VLU250
VINYLE (BROMURE de) (FRENCH) see VMP000
VINYLE(CHLORURE de) (FRENCH) see VNP000
N,N'-VINYLENEFORMAMIDINE see IAL000
VINYL ETHER see VOP000
VINYLETHYLENE see BOP500
VINYL ETHYL ETHER see EQF500
VINYL ETHYL ETHER, INHIBITED (DOT) see EQF500
VINYLETHYLNITROSAMIN (GERMAN) see NKF000
VINYLETHYLNITROSAMINE see NKF000
VINYL FLUORIDE see VPA000
VINYLFORMIC ACID see ADS750
VINYLGLYCOLONITRILE see HJQ000
S-VINYL-dl-HOMOCYSTEINE see VLU200
VINYLIDENE CHLORIDE see VPK000
VINYLIDENE CHLORIDE (II) see VPK000
VINYLIDENE DICHLORIDE see VPK000
VINYLIDENE FLUORIDE see VPP000

VINYLIDINE CHLORIDE see VPK000
VINYL ISOBUTYL ETHER (DOT) see IJQ000
VINYL ISOBUTYL ETHER, INHIBITED (DOT) see IJQ000
VINYLITE VYDR 21 see AAX175
VINYL METHYL ETHER (DOT) see MQL750
VINYL METHYL KETONE see BOY500
VINYLOFOS see DGP900
VINYLOPHOS see DGP900
VINYLPHATE see CDS750
α-VINYLPIPERONYL ALCOHOL see BCJ000
VINYL PRODUCTS R 3612 see SMQ500
VINYL PRODUCTS R 10688 see AAX250
VINYL PROPIONATE see VQK000
1-VINYLPYRENE see EEF000
3-VINYLPYRENE see EEF000
N-VINYLPYRROLIDINONE see EEG000
1-VINYL-2-PYRROLIDINONE see EEG000
N-VINYL-2-PYRROLIDINONE see EEG000
VINYLPYRROLIDONE see EEG000
N-VINYLPYRROLIDONE see EEG000
1-VINYL-2-PYRROLIDONE see EEG000
N-VINYL-2-PYRROLIDONE see EEG000
N-VINYLPYRROLIDONE POLYMER see PKQ250
VINYLSILICON TRICHLORIDE see TIN750
VINYLSTYRENE see DXQ745
VINYL SULFONE see DXR200
VINYLTOLUENE (mixed isomers) (OSHA) see MPK500, VQK650
VINYLTOLUENE see VQK650
VINYL TOLUENES (mixed isomers), inhibited (DOT) see VQK650
VINYL TRICHLORIDE see TIN000
VINYL TRICHLOROSILANE (DOT) see TIN750
VINYL TRICHLOROSILANE, INHIBITED (DOT) see TIN750
VIOFORM see CHR500
VIOFORM N.N.R. see CHR500
VIOFURAGYN see NGG500
VIOLET 3 see XGS000
VIOLET LEAF ALDEHYDE see NMV760
VIOPSICOL see LFK000, MDQ250
VIOSTEROL see VSZ100
VIOZENE see RMA500
VI-PAR see CIR500
VI-PEX see CIR500
VIRCHEM see SHR500
VIRGIMYCIN see VRF000
VIRGINIA CAROLINA VC 9-104 see EIN000
VIRGINIAMYCIN see VRF000
VIRIDINE see PDX000
VIRORMONE see TBF500
VIROSTERONE see TBF500
VIRTEX CC see SHR500
VIRTEX D see SHR500
VIRTEX L see SHR500
VIRTEX RD see SHR500
VIRUBRA see VSZ000
VISADRON see NCL500
VISCARIN see CCL250
VISCOL 350P see PMP500
VISKING CELLOPHANE see CCT250
VISKO-RHAP see DGB000
VISKO-RHAP DRIFT HERBICIDES see DAA800
VISKO RHAP LOW VOLATILE ESTER see TAA100
VISTABAMATE see MQU750
VISUBUTINA see HNI500
VITACIN see ARN000
VITALLIUM see CNA750
VITAMIN A see VSK600
VITAMIN A1 see VSK600
VITAMIN A ACETATE see VSK900
trans-VITAMIN A ACETATE see VSK900
VITAMIN A1 ALCOHOL see VSK600
all-trans-VITAMIN A ALCOHOL see VSK600
VITAMIN A ALCOHOL ACETATE see VSK900

VITAMIN AB see HAQ500
VITAMIN A PALMITATE see VSP000
VITAMIN B^1 see TET300
VITAMIN B2 see RIK000
VITAMIN B3 see NCR000
VITAMIN B-5 see CAU750
VITAMIN B6 see PPK250
VITAMIN Bc see FMT000
VITAMIN B$_{12}$ COMPLEX see VSZ000
VITAMIN B12 (FCC) see VSZ000
VITAMIN B HYDROCHLORIDE see TET300
VITAMIN B$_6$ HYDROCHLORIDE see VSU000
VITAMIN B6-HYDROCHLORIDE see PPK500
VITAMIN B1 MONONITRATE see TET500
VITAMIN B1 NITRATE see TET500
VITAMIN C see ARN000, ARN125
VITAMIN C SODIUM see ARN125
VITAMIN D2 see VSZ100
VITAMIN D3 see CMC750
VITAMIN D^3 see HJV000
VITAMIN E see VSZ450
VITAMIN G see RIK000
VITAMIN H see AIH600
VITAMIN K3 see MMD500
VITAMIN K2(O) see MMD500
VITAMIN L see API500
VITAMIN M see FMT000
VITAMIN PP see NCR000
VITAMISIN see ARN000
VITAPLEX E see VSZ450
VITAPLEX N see NCQ900
VITARUBIN see VSZ000
VITA-RUBRA see VSZ000
VITASCORBOL see ARN000
VITAVAX see CCC500
VITAVEL-A see VSK600
VITAVEL-D see VSZ100
VITAYONON see VSZ450
VITEOLIN see VSZ450
VITINC DAN-DEE-3 see CMC750
VITON see BBQ500
VITPEX see VSK600
VITRAL see VSZ000
VITRAN see TIO750
VITREOUS QUARTZ see SCK600
VITREX see PAK000
VITRIOL BROWN OIL see SOI500
VITRIOL, OIL of (DOT) see SOI500
VITRIOL RED see IHD000
VITRUM AB see HAQ500
VI-TWEL see VSZ000
VIVACTIL see POF250
VLB see VKZ000
VLB MONOSULFATE see VLA000
VLVF see AAX175
VM-26 see EQP000
VMCC see AAX175
VM&P NAPHTHA see PCT250
VM & P NAPHTHA (ACGIH) see NAI500
VOFATOX see MNH000
VOGALENE see MQR000
VOGAN see VSK600
VOGAN-NEU see VSK600
VOGEL'S IRON RED see IHD000
VOLATILE OIL of MUSTARD see AGJ250
VOLATON see BAT750
VOLCLAY see BAV750
VOLCLAY BENTONITE BC see BAV750
VOLDYS see MCA500
VOLFARTOL see TIQ250
VOLIDAN see MCA500, VTF000
VOMEX A see DYE600
VOMISSELS see HGC500
VONAMYCIN POWDER V see NCE000

VONDACEL BLACK N see AQP000
VONDACEL BLUE 2B see CMO000
VONDACID TARTRAZINE see FAG140
VONDCAPTAN see CBG000
VONDODINE see DXX400
VONDURON see DXQ500
VONEDRINE see DBA800
VOPCOLENE 27 see OHU000
VORLEX see ISE000
VOROX see AMY050
VORTEX see ISE000
VOTEXIT see TIQ250
VP 1940 see ABX250
VP 16213 see EAV500
V-PYROL see EEG000
VUAGT-I-4 see TFS350
VULCACID D see DWC600
VULCACURE see BIX000, BJC000, BJK500, SGF500
VULCAFIX SCARLET R see CHP500
VULCAFOR BSM see BDG000
VULCAFOR TMTD see TFS350
VULCALENT A see DWI000
VULCAN RED LC see CHP500
VULCATARD see DWI000
VULCOL FAST RED L see CHP500
VULKACIT D/C see DWC600
VULKACIT DM see BDE750
VULKACIT DM/MGC see BDE750
VULKACITE L see BJK500
VULKACIT LDA see BJC000
VULKACIT LDB/C see BIX000
VULKACIT MTIC see TFS350
VULKACIT NPV/C2 see IAQ000
VULKACIT THIURAM see TFS350
VULKACIT THIURAM/C see TFS350
VULKACIT THIURAM MS/C see BJL600
VULKACIT ZM see BHA750
VULKALENT A (CZECH) see DWI000
VULKANOX 4020 see PEY500
VULKASIL see SCH000
VULKAZIT see DWC600
VULNOPOL NM see SGM500
VULTROL see DWI000
VULVAN see TBG000
VUMON see EQP000
VYAC see VLU250
VYDATE see DSP600
VYDATE L INSECTICIDE/NEMATICIDE see DSP600
VYDATE L OXAMYL INSECTICIDE/NEMATOCIDE see DSP600
VYDYNE see NOH000
VYGEN 85 see PKQ059

W 101 see PMP500
W 483 see DIR000
W 491 see CCP500
W 713 see MOV500
W 1544 see PFC500
W 1655 see PDC250, PEK250
W-2429 see DLQ400
W 6658 see BJP000
W-2946M see DNA600
WACHOLDERBEER OEL (GERMAN) see JEA000
WACKER S 14/10 see BJE750
WAIT'S GREEN MOUNTAIN ANTIHISTAMINE see WAK000
WALKO-NESIN see GGS000
WALNUT EXTRACT see WAT000
WALNUT STAIN see MAT500
WAMPOCAP see NCQ900
WANADU PIECIOTLENEK (POLISH) see VDU000
WAPNIOWY TLENEK (POLISH) see CAU500
WARBEX see FAB600
WARDAMATE see MQU750

WARECURE C see IAQ000
WARFARIN see WAT200
WARFARINE (FRENCH) see WAT200
WARKEELATE ACID see EIX000
WARKEELATE PS-43 see EIV000
WASH OIL see CMY825
WASSERSTOFFPEROXID (GERMAN) see HIB000
WATER-d2 (9CI) see HAK000
WATER GLASS see SJU000
WATER2-H2 see HAK000
WATERSTOFPEROXYDE (DUTCH) see HIB000
WATTLE GUM see AQQ500
WAXAKOL ORANGE GL see PEJ500
WAXAKOL VERMILION L see XRA000
WAXAKOL YELLOW NL see AIC250
WAXOLINE GREEN see BLK000
WAXOLINE ORANGE A see BJF000
WAXOLINE RED A see MAC500
WAXOLINE YELLOW AD see DOT300
WAXOLINE YELLOW I see PEJ500
WAXOLINE YELLOW O see IBB000
WAXSOL see DJL000
WAYNE RED X-2486 see CHP500
WEATHERBEE MUSTARD see BHB750
WEC 50 see TIQ250
WECKAMINE see BBK000
WECOLINE 1295 see LBL000
WECOLINE OO see OHU000
WEDDING BELLS see DJO000
WEED 108 see MRL750
WEED-AG-BAR see DAA800
WEEDANOL CYANOL see PLC250
WEEDAR see TAA100
WEEDAR-64 see DAA800
WEEDAR ADS see AMY050
WEEDAR MCPA CONCENTRATE see CIR250
WEEDAZIN see AMY050
WEEDAZOL see AMY050
WEEDAZOL TL see ANW750
WEEDBEADS see SJA000
WEED-B-GON see DAA800, TIX500
WEED BROOM see DXE600
WEED DRENCH see AFV500
WEED-E-RAD see DXE600, MRL750
WEEDEX A see ARQ725
WEEDEX GRANULAT see AMY050
WEEDEZ WONDER BAR see DAA800
WEED-HOE see DXE600, MRL750
WEEDOCLOR see AMY050
WEEDONE see TAA100
WEEDONE see PAX250
WEEDONE 128 see IOY000
WEEDONE 170 see DGB000
WEEDONE CRAB GRASS KILLER see PLC250
WEEDONE DP see DGB000
WEEDONE LV4 see DAA800
WEEDONE MCPA ESTER see CIR250
WEED-RHAP see CIR250
WEED TOX see DAA800
WEEDTRINE-D see DWX800
WEEDTRINE-II see ILO000
WEEDTROL see DAA800
WEEVILTOX see CBV500
WEGANTYNA see DHX800
WEGLA DWUSIARCZEK (POLISH) see CBV500
WEGLA TLENEK (POLISH) see CBW750
WEHYDRYL see BAU750
WEISS PHOSPHOR (GERMAN) see PHO750
WELDING FUMES see WBJ000
WELFURIN see NGE000
WELVIC G 2/5 see PKQ059
WEPSIN see AIX000
WEPSYN see AIX000
WEPSYN 155 see AIX000

WESCOZONE see BRF500
WESPURIL see MJM500
WEST INDIAN LEMONGRASS OIL see LEH000
WESTOCAINE see AIT250
WESTRON see TBQ100
WESTROSOL see TIO750
WETAID SR see DJL000
WET-TONE B see MSB500
WEX 1242 see PMP500
WFNA see NEF000
WH 7286 see DMW000
WH 7508 see PIW000
WHISKEY see WBS000
1700 WHITE see TGG760
WHITE ACID (DOT) see ANG250
WHITE ARSENIC see ARI750
WHITE ASBESTOS see ARM268
WHITE CAMPHOR OIL see CBB500
WHITE CAUSTIC see SHS000
WHITE CAUSTIC, solution see SHS500
WHITE CEDAR OIL see CCQ500
WHITE COPPERAS see ZNA000
WHITE CRYSTAL see SFT000
WHITE FUMING NITRIC ACID see NEF000
WHITE LEAD see LCP000
WHITE MERCURY PRECIPITATED see MCW500
WHITE MINERAL OIL see MQV750, MQV875
WHITE OIL of CAMPHOR see CBB500
WHITE PHOSPHORUS see PHO750
WHITE PRECIPITATE see MCW500
WHITE SEAL-7 see ZKA000
WHITE SPIRITS see SLU500, PCT250
WHITE STREPTOCIDE see SNM500
WHITE STUFF see HBT500
WHITE TAR see NAJ500
WHITE VITRIOL see ZNA000, ZNJ000
WIDLON see PJY500
WIE OBEN see DJY200
WIJS' CHLORIDE see IDS000
WILKINITE see BAV750
WILLBUTAMIDE see BSQ000
WILLESTROL see DKB000
WILLNESTROL see DAL600
WILLOSETTEN see SJN700
WILPO see DTJ400
WILT PRUF see PKQ059
WIN 5162 see DMV600
WIN 5606 see BCA000
WIN 12267 see DEM000
WIN 18,320 see EID000
WIN 20228 see DOQ400
WIN 24933 see LIM000
WINACET D see AAX250
WINE see WCA000
WINE ETHER see ENW000
WING STOP B see SGM500
WINIDUR see PKQ059
WINTERGREEN OIL (FCC) see MPI000
WINTERGREEN OIL, SYNTHETIC see MPI000
WINTERWASH see DUS700
WINTOMYLON see EID000
WINYLU CHLOREK (POLISH) see VNP000
WIRKSTOFF 37289 see EPY000
WITAMINA PP see NCR000
WITCIZER 300 see DEH200
WITCIZER 312 see DVL700
WITCONOL MS see OAV000
WITCONOL MST see OAV000
WITTOX C see NAS000
WL 1650 see OAN000
WL-5792 see DGM600
WL 18236 see MDU600
WL 19805 see BLW750
WL 43479 see AHJ750

WL 43775 see FAR100
WN 12 see ISE000
WNYESTRON see EDV000
WOCHEM NO. 320 see OHU000
WOFACAIN A see BQH250
WOFATOS see MNH000
WOFATOX see MNH000
WOFOTOX see MNH000
WOJTAB see PLZ000
WOLFRAM see TOA750
WOLFSBANE see AQY500
WONDER TREE see CCP000
WONUK see ARQ725
WOOD ALCOHOL (DOT) see MGB150
WOOD ETHER see MJW500
WOOD NAPHTHA see MGB150
WOOD SPIRIT see MGB150
WOOL BRILLIANT GREEN SF see FAF000
WOOLLEY'S ANTISEROTONIN see BEM750
WOOL ORANGE 2G see HGC000
WOOL VIOLET see FAG120
WOOL YELLOW see FAG140
WORM-AGEN see HFV500
WORM-CHEK see LFA020
WORM GUARD see BQK000
WOTEXIT see TIQ250
WP 155 see AIX000
WR 448 see SOA500
WR 2978 see TGD000
WR 09792 see FMH000
WRIGHTINE DIHYDROBROMIDE see DOX000
WSQ 1 see MDR250
WURM-THIONAL see PDP250
WY 509 see DQA400
WY 554 see DAM700
WY 806 see DTL200
WY 1094 see DQA600
WY-1172 see ABH500
WY 3707 see NNQ500
WY-5103 see AIV500
WY-14,643 see CLW250
WYACORT see MOR500

X 119 see DTP000
X 149 see BOT250
XA 2 see CFA750
X-AB see PKQ059
X-ALL LIQUID see AMY050
XAMAMINA see DYE600
XAN see XCA000
XANTELINE see XCJ000
XANTHACRIDINUM see DBX400
XANTHAURINE see QCA000
XANTHENE-9-CARBOXYLIC ACID, ESTER with
    DIETHYL(2-HYDROXYETHYL)METHYLAMMONIUM
    BROMIDE see XCJ000
XANTHIC OXIDE see XCA000
XANTHINE see XCA000
XANTHINE BROMIDE see XCJ000
XANTHINE-3-N-OXIDE see HOP000
XANTHINE-7-N-OXIDE see HOP259
XANTHINE-x-N-OXIDE see HOP000
XANTHOTOXIN see XDJ000
XANTHURENIC ACID see DNC200
XARIL see ABG750
XAXA see ADA725
XENENE see BGE000
XENON see XDS000
XENON, refrigerated liquid (DOT) see XDS000
XENYLAMIN (CZECH) see AJS100
XENYLAMINE see AJS100
XENYTROPIUM BROMIDE see PEM750
XERAC see BDS000
XERAL see GGS000

XILENOLI (ITALIAN) see XKA000
XILIDINE (ITALIAN) see XMA000
XILOCAINA (ITALIAN) see DHK400
XILOLI (ITALIAN) see XGS000
XITIX see ARN000
XL-50 see PDP250
XL ALL INSECTICIDE see NDN000
3,5-XMC see DTN200
XYCAINE HYDROCHLORIDE see DHK600
XYCIANE see DHK400
XYLAZINE (USDA) see DMW000
XYLENE see XGS000
m-XYLENE see XHA000
o-XYLENE see XHJ000
p-XYLENE see XHS000
1,2-XYLENE see XHJ000
1,3-XYLENE see XHA000
1,4-XYLENE see XHS000
XYLENE BLUE VSG see FMU059
m-XYLENE-α,α'-DIAMINE see XHS800
m-XYLENE-α,α'-DIISOCYANATE see XIJ000
XYLENE FAST ORANGE G see HGC000
XYLENE FAST YELLOW GT see FAG140
XYLENEN (DUTCH) see XGS000
XYLENOL see XKA000
m-XYLENOL see XKJ500
2,3-XYLENOL see XKJ000
2,4-XYLENOL see XKJ500
2,5-XYLENOL see XKS000
3,5-XYLENOL see XLS000
1,2,5-XYLENOL see XKS000
1,3,5-XYLENOL see XLS000
m-XYLENOL (DOT) see XKJ500
o-XYLENOL (DOT) see XKJ000
p-XYLENOL (DOT) see XKS000
XYLENOLEN (DUTCH) see XKA000
3,5-XYLENOL METHYLCARBAMATE see DTN200
3,5-XYLENOL-N-METHYLCARBAMATE see DTN200
XYLESTESIN see DHK400
XYLESTESIN HYDROCHLORIDE see DHK600
2,4-XYLIDENE (MAK) see XMS000
XYLIDINE see XMA000
o-XYLIDINE see XNJ000
2,3-XYLIDINE see XMJ000
2,4-XYLIDINE see XMS000
2,5-XYLIDINE see XNA000
2,6-XYLIDINE see XNJ000
3,5-XYLIDINE see XOA000
m-4-XYLIDINE see XMS000
m-XYLIDINE (DOT) see XMS000
o-XYLIDINE (DOT) see XMJ000
p-XYLIDINE (DOT) see XNA000
m-XYLIDINE HYDROCHLORIDE see XOJ000
p-XYLIDINE HYDROCHLORIDE see XOS000
2,4-XYLIDINE HYDROCHLORIDE see XOJ000
2,5-XYLIDINE HYDROCHLORIDE see XOS000
XYLIDINEN (DUTCH) see XMA000
XYLITE (SUGAR) see XPJ000
XYLITOL see XPJ000
l-XYLOASCORBIC ACID see ARN000
XYLOCAIN see DHK400
XYLOCAINE HYDROCHLORIDE see DHK600
XYLOCARD see DHK600
XYLOCITIN see DHK400
XYLOCITIN HYDROCHLORIDE see DHK600
XYLOIDIN see CCU250
XYLOL (DOT) see XGS000
m-XYLOL (DOT) see XHA000
o-XYLOL (DOT) see XHJ000
p-XYLOL (DOT) see XHS000
XYLOLE (GERMAN) see XGS000
XYLONEURAL see DHK600
XYLOTOX see DHK400
XYLOTOX HYDROCHLORIDE see DHK600

2,3-XYLYLAMINE see XMJ000
2,6-XYLYLAMINE see XNJ000
3,5-XYLYLAMINE see XOA000
1-XYLYLAZO-2-NAPHTHOL see XRA000
1-(o-XYLYLAZO)-2-NAPHTHOL see XRA000
1-(2,4-XYLYLAZO)-2-NAPHTHOL see XRA000
XYLYL BROMIDE see XRS000
m-XYLYLENDIAMIN (CZECH) see XHS800
XYLYLENDIISOKYANAT (CZECH) see XIJ000
o-XYLYLENE DICHLORIDE see DGP400
3,5-XYLYL-N-METHYLCARBAMATE see DTN200

Y 3 see CKC000
Y 195 see PKL750
Y 218 see PKM250
Y 221 see PKM500
Y-223 see PKN000
Y-238 see CDV625
Y 302 see PKP000
YAGEINE see HAI500
YAJEINE see HAI500
YALTOX see CBS275
YAMAMOTO METHYLENE BLUE B see BJI250
YANOCK see FFF000
YARMOR see PIH750
YARMOR PINE OIL see PIH750
YASOKNOCK see SHG500
YATROCIN see NGE500
YATROZIDE see ILE000
11712 YELLOW see FEV000
11824 YELLOW see FEW000
12417 YELLOW see FEW000
YELLOW AB see FAG130
YELLOW CROSS LIQUID see BIH250
YELLOW FERRIC OXIDE see IHD000
YELLOW G SOLUBLE in GREASE see DOT300
YELLOW LAKE 69 see FAG140
YELLOW LEAD OCHER see LDN000
YELLOW MERCURIC OXIDE see MCT500
YELLOW MERCURY IODIDE see MDC750
YELLOW No. 2 see FAG130
YELLOW OB see FAG135
YELLOW OXIDE of IRON see IHD000
YELLOW OXIDE of MERCURY see MCT500
1903 YELLOW PINK see BNH500
YELLOW PRECIPITATE see MCT500
YELLOW PYOCTANINE see IBB000
YELLOW ULTRAMARINE see CAP500
YELLOW Z see AAQ250
YODOMIN see TEH500
YOHIMBAN-16-CARBOXYLIC ACID DERIVATIVE of
   BENZ(G)INDOLO(2,3-A)QUINOLIZINE see RDK000
YOHIMBIC ACID METHYL ESTER see YBJ000
YOHIMBINE see YBJ000
YOMESAN see DFV400
YOSHI 864 see YCJ000
YOSHINOX S see TFC600
YPERITE see BIH250
YPERITE SULFONE see BIH500
YTTERBIUM see YDA000
YTTERBIUM CHLORIDE see YDJ000
YTTERBIUM(III) NITRATE, HEXAHYDRATE (1:3:6) see
   YEA000
YTTERBIUM TRICHLORIDE see YDJ000
YTTRIUM see YEJ000
YTTRIUM-89 see YEJ000
YTTRIUM CHLORIDE see YES000
YTTRIUM(III) NITRATE (1:3) see YFJ000
YTTRIUM OXIDE see YGA000
YTTRIUM TRICHLORIDE see YES000

Z 75 see BJK500
Z-78 see EIR000
Z 4828 see TNT500

Z 4942 see IMH000
ZACLON DISCOIDS see HHS000
ZACTIRIN COMPOUND see ABG750
ZACTRAN see DOS000
ZADINE see DLV800
ZADOLETTEN see EOK000
ZADONAL see EOK000
ZAGREB see BIE500
ZAHARINA see BCE500
ZAHLREICHE BEZEICHNUNGEN (GERMAN) see DUS700
ZAMINE see BBK500
ZANOSAR see SMD000
ZARAONDAN see ENG500
ZARDA (INDIA) see SED400
ZARLATE see BJK500
ZARODAN see ENG500
ZARONDAN-SAFT see ENG500
ZARONTIN see ENG500
ZARTALIN see ENG500
Z-C SPRAY see BJK500
ZEAPUR see BJP000
ZEARALANOL see RBF100
ZEARANOL see RBF100
ZEAZIN see ARQ725
ZEAZINE see ARQ725
ZEBENIDE see EIR000
ZEBTOX see EIR000
ZECTANE see DOS000
ZECTRAN see DOS000
ZEIDANE see DAD200
ZELAN see CIR250
ZELAZA TLENKI (POLISH) see IHF000
ZELEN OSTANTHRENOVA BRILANTNI FFB (CZECH) see JAT000
ZELIO see TEL750
ZENADRID (VETERINARY) see PLZ000
ZENALOSYN see PAN100
ZENITE see BHA750
ZENITE SPECIAL see BHA750
ZENTAL see VAD000
ZENTINIC see PAG200
ZENTRONAL see DKQ000
ZENTROPIL see DKQ000, DNU000
ZEPHROL see EAW000
ZERANOL (USDA) see RBF100
ZERDANE see DAD200
ZERLATE see BJK500
ZERTELL see CMA250
ZESET T see VLU250
ZEST see MRL500
ZESTE see ECU750
ZETAR see CMY800
ZETAX see BHA750
ZETIFEX ZN see TNC500
ZEXTRAN see DOS000
ZHENGGUANGMYCIN A2 (CHINESE) see BLY250
ZIARNIK see ABU500
ZIAVETINE see BQL000
ZIDAN see EIR000
ZIDE see CFY000
ZIMALLOY see CNA750
ZIMATE see BJK500, EIR000
ZIMATE METHYL see BJK500
ZIMCO see VFK000
ZIMTALDEHYDE see CMP969
ZIMTSAEURE (GERMAN) see CMP975
ZINADON see ILD000
ZINC see ZBJ000
ZINC ACETATE see ZBS000
ZINC ACETATE PLUS MANEB (1:50) see EIQ500
ZINC AMMONIUM NITRITE see ZDA000
ZINC ARSENATE see ZDJ000
ZINC ARSENATE, solid (DOT) see ZDJ000
ZINC ARSENATE, BASIC see ZDJ000

ZINC-m-ARSENITE see ZDS000
ZINC ARSENITE, solid (DOT) see ZDS000
ZINC-2-BENZOTHIAZOLETHIOLATE see BHA750
ZINC BENZOTHIAZOLYL MERCAPTIDE see BHA750
ZINC BENZOTHIAZOL-2-YLTHIOLATE see BHA750
ZINC BENZOTHIAZYL-2-MERCAPTIDE see BHA750
ZINC BERYLLIUM SILICATE see BFV250
ZINC-BIBUTYLDITHIOCARBAMATE see BIX000
ZINC BIS(DIMETHYLDITHIOCARBAMATE) see BJK500
ZINC BIS(DIMETHYLDITHIOCARBAMOYL)DISULPHIDE see BJK500
ZINC BIS(DIMETHYLTHIOCARBAMOYL)DISULFIDE see BJK500
ZINC CHLORATE see ZES000
ZINC CHLORIDE see ZFA000
ZINC CHLORIDE, anhydrous (DOT) see ZFA000
ZINC CHLORIDE, solid (DOT) see ZFA000
ZINC CHLORIDE, solution (DOT) see ZFA000
ZINC (CHLORURE de) (FRENCH) see ZFA000
ZINC CHROMATE see ZFJ100
ZINC CHROMATE HYDROXIDE see CMK500
ZINC CHROMATE(VI) HYDROXIDE see CMK500, ZFJ100
ZINC CHROMATE, POTASSIUM DICHROMATE and ZINC HYDROXIDE (3:1:1) see ZFJ150
ZINC CHROMATE with ZINC HYDROXIDE and CHROMIUM OXIDE (9:1) see ZFJ120
ZINC CHROME see PLW500
ZINC CHROME YELLOW see ZFJ100
ZINC CHROMIUM OXIDE see ZFJ100
ZINC COMPOUNDS see ZFS000
ZINC CYANIDE see ZGA000
ZINC DIACETATE see ZBS000
ZINC-DIBUTYLDITHIOCARBAMATE see BIX000
ZINC-N,N-DIBUTYLDITHIOCARBAMATE see BIX000
ZINC DICHLORIDE see ZFA000
ZINC DICYANIDE see ZGA000
ZINC DIETHYLDITHIOCARBAMATE see BJC000
ZINC-N,N-DIETHYLDITHIOCARBAMATE see BJC000
ZINC DIMETHYLDITHIOCARBAMATE see BJK500
ZINC N,N-DIMETHYLDITHIOCARBAMATE see BJK500
ZINC DISTERATE see ZMS000
ZINC DUST see ZBJ000
ZINC ETHIDE see DKE600
ZINC ETHYL (DOT) see DKE600
ZINC ETHYLENEBISDITHIOCARBAMATE see EIR000
ZINC ETHYLENE-1,2-BISDITHIOCARBAMATE see EIR000
ZINC FLUORIDE see ZHS000
ZINC FLUORURE (FRENCH) see ZHS000
ZINC FLUOSILICATE see ZIA000
ZINC HEXAFLUOROSILICATE see ZIA000
ZINC HYDROXYCHROMATE see CMK500, ZFJ100
ZINCITE see ZKA000
ZINC MANGANESE BERYLLIUM SILICATE see BFS750
ZINCMATE see BJK500
ZINC MERCAPTOBENZOTHIAZOLATE see BHA750
ZINC-2-MERCAPTOBENZOTHIAZOLE see BHA750
ZINC MERCAPTOBENZOTHIAZOLE SALT see BHA750
ZINC MERCURY CHROMATE COMPLEX see ZJA000
ZINC METAARSENITE see ZDS000
ZINC METHARSENITE see ZDS000
ZINC MURIATE, solution (DOT) see ZFA000
ZINC NITRATE see ZJJ000
ZINCO (CLORURO di) (ITALIAN) see ZFA000
ZINC OCTADECANOATE see ZMS000
ZINCO(FOSFURO di) (ITALIAN) see ZLS000
ZINCOID see ZKA000
ZINC OXIDE see ZKA000
ZINC OXIDE FUME (MAK) see ZKA000
ZINC PERMANGANATE see ZLA000
ZINC PEROXIDE see ZLJ000
ZINC PHOSPHIDE see ZLS000
ZINC(PHOSPHURE de) (FRENCH) see ZLS000
ZINC POTASSIUM CHROMATE see CMK400
ZINC POWDER see ZBJ000

ZINC, POWDER or DUST, non-pyrophoric (DOT) see ZBJ000
ZINC, POWDER or DUST, pyrophoric (DOT) see ZBJ000
ZINC STEARATE see ZMS000
ZINC SULFATE see ZNA000, ZNJ000
ZINC SULFATE (1:1) HEPTAHYDRATE see ZNJ000
ZINC SULFATE HEPTAHYDRATE (1:1:) see ZNJ000
ZINC SULPHATE see ZNA000
ZINC SUPEROXIDE see ZLJ000
ZINC TETRAOXYCHROMATE 76A see ZFJ100
ZINC-TOX see ZLS000
ZINC VITRIOL see ZNA000, ZNJ000
ZINC WHITE see ZKA000
ZINC YELLOW see CMK500, PLW500, ZFJ100, ZFJ120
ZINEB see EIR000
ZINK-(N,N′-AETHYLEN-BIS(DITHIOCARBAMAT)) (GER-MAN) see EIR000
ZINK-BIS(N,N-DIMETHYL-DITHIOCARBAMAAT) (DUTCH) see BJK500
ZINK-BIS(N,N-DIMETHYL-DITHIOCARBAMAT) (GERMAN) see BJK500
ZINKCARBAMATE see BJK500
ZINKCHLORID (GERMAN) see ZFA000
ZINKCHLORIDE (DUTCH) see ZFA000
ZINK-(N,N-DIMETHYL-DITHIOCARBAMAT) (GERMAN) see BJK500
ZINKFOSFIDE (DUTCH) see ZLS000
ZINKOSITE see ZNA000
ZINKPHOSPHID (GERMAN) see ZLS000
ZINN (GERMAN) see TGB250
ZINNTETRACHLORID (GERMAN) see TGC250
ZINOPHOS see EPC500
ZINOSAN see EIR000
ZINOSTATIN see NBV500
ZIRAM see BJK500
ZIRAM TECHNICAL see BJK500
ZIRAMVIS see BJK500
ZIRASAN see BJK500
ZIRBERK see BJK500
ZIRCAT see ZOA000
ZIRCONIUM see ZOA000
ZIRCONIUM CHLORIDE see ZPA000
ZIRCONIUM(IV) CHLORIDE (1:4) see ZPA000
ZIRCONIUM COMPOUNDS see ZQA000
ZIRCONIUM FLUORIDE see ZQS000
ZIRCONIUM HYDRIDE see ZRA000
ZIRCONIUM METAL see ZOA000
ZIRCONIUM METAL, dry (DOT) see ZOA000
ZIRCONIUM NITRATE see ZSA000
ZIRCONIUM OXYCHLORIDE see ZSJ000
ZIRCONIUM SHAVINGS see ZOA000
ZIRCONIUM SHEETS (DOT) see ZOA000
ZIRCONIUM(IV) SULFATE (1:2) see ZTJ000

ZIRCONIUM TETRACHLORIDE (DOT) see ZPA000
ZIRCONIUM TETRACHLORIDE, solid (DOT) see ZPA000
ZIRCONIUM TETRAFLUORIDE see ZQS000
ZIRCONIUM TURNINGS see ZOA000
ZIRCONYL CHLORIDE see ZSJ000
ZIRCONYL SULFATE see ZTJ000
ZIREX 90 see BJK500
ZIRIDE see BJK500
ZIRPON see MQU750
ZIRTHANE see BJK500
ZITEX H 662-124 see TAI250
ZITHIOL see MAK700
ZITOX see BJK500
ZITRONEN OEL (GERMAN) see LEI000
ZLUT MASELNA (CZECH) see DOT300
ZMA see ZDS000
ZMBT see BHA750
Z-METHYLPROPYL ACRYLATE see IIK000
ZnMB see BHA750
ZOALENE see DUP300
ZOAMIX see DUP300
ZOBA BLACK D see PEY500
ZOBA BROWN RR see ALL750
ZOBA EG see ALT500
ZOBA ERN see NAW500
ZOBA 3GA see ALT000
ZOBA GKE see TGL750
ZOBA SLE see DBO400
ZOGEN DEVELOPER H see TGL750
ZOLAPHEN see BRF500
ZOLIDINUM see BRF500
ZOLON see BDJ250
ZOLONE see BDJ250
ZOLONE PM see BDJ250
ZONAZIDE see ILD000
ZOOCOUMARIN (RUSSIAN) see WAT200
ZOOFURIN see NGE000
ZOOLON see BDJ250
ZOPAQUE see TGG760
ZORANE see BRF500
ZORIFLAVIN see DBX400
ZOTOX see ARB250
ZP see ZLS000
ZR 515 see KAJ000
ZUTRACIN see BAC250
ZWAVELWATERSTOF (DUTCH) see HIC500
ZWAVELZUUROPLOSSINGEN (DUTCH) see SOI500
ZYGOSPORIN A see ZUS000
ZYKLOPHOSPHAMID (GERMAN) see EAS500
ZYTEL 211 see PJY500
ZYTOX see MHR200

III.   CAS Number Cross-Index

50-00-0 see FMV000	52-49-3 see BBV000	56-81-5 see GGA000
50-02-2 see SOW000	52-51-7 see BNT250	56-85-9 see GF0050
50-06-6 see EOK000	52-53-9 see IRV000	56-86-0 see GF0000
50-07-7 see AHK500	52-67-5 see MCR750	56-89-3 see CQK325
50-10-2 see ORQ000	52-68-6 see TIQ250	57-06-7 see AGJ250
50-10-2 see DJM600	52-85-7 see FAB600	57-09-0 see HCQ500
50-12-4 see MKB250	52-89-1 see CQK250	57-10-3 see PAE250
50-13-5 see DAM700	52-90-4 see CQK000	57-11-4 see SLK000
50-14-6 see VSZ100	53-03-2 see PLZ000	57-12-5 see COI500
50-18-0 see EAS500	53-16-7 see EDV000	57-13-6 see USS000
50-21-5 see LAG000	53-19-0 see CDN000	57-14-7 see DSF400
50-24-8 see PMA000	53-46-3 see XCJ000	57-15-8 see ABD000
50-27-1 see EDU500	53-60-1 see PMI500	57-22-7 see LEY000
50-28-2 see ED0000	53-69-0 see DQI600	57-24-9 see SMN500
50-29-3 see DAD200	53-70-3 see DCT400	57-27-2 see MRO500
50-32-8 see BCS750	53-86-1 see IDA000	57-30-7 see SID000
50-33-9 see BRF500	53-95-2 see HIP000	57-33-0 see NBU000
50-35-1 see TEH500	53-96-3 see FDR000	57-37-4 see BCA000
50-37-3 see DJ0000	54-04-6 see MDI500	57-39-6 see TNK250
50-41-9 see CMX700	54-11-5 see NDN000	57-41-0 see DKQ000
50-44-2 see POK000	54-12-6 see TNW500	57-42-1 see DAM600
50-47-5 see DSI709	54-21-7 see SJ0000	57-43-2 see AMX750
50-48-6 see EAH500	54-31-9 see CHJ750	57-47-6 see PIA500
50-49-7 see DLH600	54-42-2 see DAS000	57-50-1 see SNH000
50-50-0 see EDP000	54-49-9 see HNB875	57-52-3 see BLN250
50-52-2 see MQ0250	54-62-6 see AMG750	57-53-4 see MQU750
50-55-5 see RDK000	54-64-8 see MDI000	57-55-6 see PML000
50-65-7 see DFV400	54-77-3 see DT0000	57-56-7 see HGU000
50-70-4 see SKV200	54-85-3 see ILD000	57-57-8 see PMT100
50-78-2 see ADA725	54-92-2 see ILE000	57-62-5 see CMA750
50-81-7 see ARN000	54-95-5 see PBI500	57-63-6 see EEH500
50-91-9 see DAR400	55-18-5 see NJW500	57-64-7 see PIA750
50-98-6 see EAW500	55-37-8 see DST200	57-68-1 see SNJ000
50-98-6 see EAY000	55-38-9 see FAQ999	57-74-9 see CDR750
50-99-7 see GFG000	55-43-6 see DCR200	57-83-0 see PMH500
51-02-5 see INT000	55-48-1 see ARR500	57-85-2 see TBG000
51-03-6 see PIX250	55-51-6 see BIA750	57-88-5 see CMD750
51-05-8 see AIT250	55-52-7 see PDN000	57-92-1 see SLW500
51-15-0 see FNZ000	55-56-1 see BIM250	57-94-3 see TOA000
51-17-2 see BCB750	55-63-0 see NGY000	57-96-5 see DWM000
51-18-3 see TND500	55-63-0 see SLD000	57-97-6 see DQJ200
51-21-8 see FMM000	55-68-5 see MCU750	58-08-2 see CAK500
51-28-5 see DUZ000	55-80-1 see DUH600	58-14-0 see TGD000
51-34-3 see SBG000	55-86-7 see BIE500	58-15-1 see DOT000
51-41-2 see NN0500	55-91-4 see IRF000	58-18-4 see MPN500
51-43-4 see VGP000	55-93-6 see DSU000	58-22-0 see TBF500
51-44-5 see DER600	55-97-0 see HEA000	58-25-3 see LFK000
51-45-6 see HGD000	55-98-1 see BOT250	58-27-5 see MMD500
51-52-5 see PNX000	56-04-2 see MPW500	58-28-6 see DLS600
51-55-8 see ARR000	56-18-8 see AIX250	58-34-4 see MRW000
51-60-5 see DQY909	56-23-5 see CBY000	58-38-8 see PMF500
51-61-6 see DYC400	56-29-1 see ERD500	58-39-9 see CJM250
51-62-7 see BBK750	56-35-9 see BLL750	58-40-2 see DQA600
51-63-8 see BBK500	56-38-2 see PAK000	58-54-8 see DFP600
51-71-8 see PFC500	56-38-2 see PAK250	58-55-9 see TEP000
51-75-2 see BIE250	56-40-6 see GHA000	58-56-0 see PPK500
51-79-6 see UVA000	56-49-5 see MIJ750	58-72-0 see TMS250
51-80-9 see TDR750	56-53-1 see DKA600	58-73-1 see BBV500
51-85-4 see MCN500	56-54-2 see QFS000	58-74-2 see PAH000
51-98-9 see ABU000	56-55-3 see BBC250	58-82-2 see BML500
52-24-4 see TFQ750	56-57-5 see NJF000	58-89-9 see BBQ500
52-26-6 see MR0750	56-72-4 see CNU750	58-90-2 see TBT000
52-31-3 see TDA500	56-75-7 see CDP250	58-93-5 see CFY000

59-00-7 see DNC200
59-01-8 see KAL000
59-02-9 see VSZ450
59-05-2 see MDV500
59-14-3 see BNC750
59-23-4 see GAV000
59-26-7 see DJS200
59-30-3 see FMT000
59-33-6 see DBM800
59-40-5 see QTS000
59-42-7 see NCL500
59-46-1 see AIL750
59-47-2 see GGS000
59-50-7 see CFE250
59-51-8 see MDT740
59-52-9 see BAD750
59-63-2 see IKC000
59-66-5 see AAI250
59-67-6 see NCQ900
59-85-8 see CHU500
59-87-0 see NGE500
59-88-1 see PFI250
59-89-2 see NKZ000
59-92-7 see DNA200
59-96-1 see PDT250
59-97-2 see BBJ750
60-00-4 see EIX000
60-01-5 see TIG750
60-09-3 see PEI000
60-10-6 see DWN200
60-11-7 see DOT300
60-12-8 see PDD750
60-18-4 see TOG300
60-23-1 see AJT250
60-24-2 see MCN250
60-29-7 see EJU000
60-31-1 see AB0000
60-33-3 see LGG000
60-34-4 see MKN000
60-35-5 see AAI000
60-41-3 see SMP000
60-44-6 see PBS000
60-44-6 see DJM200
60-46-8 see DOY400
60-51-5 see DSP400
60-54-8 see TBX000
60-56-0 see MC0500
60-57-1 see DHB400
60-79-7 see LJL000
60-80-0 see AQN000
60-87-7 see DQA400
60-93-5 see QIJ000
61-00-7 see ABH500
61-01-8 see MFK500
61-12-1 see NOF500
61-16-5 see MDW000
61-33-6 see BDY669
61-50-7 see DPF600
61-57-4 see NML000
61-73-4 see BJI250
61-75-6 see BMV750
61-76-7 see SPC500
61-82-5 see AMY050
61-90-5 see LES000
62-32-8 see EAZ000
62-38-4 see ABU500
62-44-2 see ABG750
62-50-0 see EMF500
62-53-3 see A0Q000
62-54-4 see CAL750
62-55-5 see TFA000
62-56-6 see ISR000
62-73-7 see DGP900
62-74-8 see SHG500

62-75-9 see NKA600
63-25-2 see CBM750
63-42-3 see LAR000
63-68-3 see MDT750
63-74-1 see SNM500
63-75-2 see AQT750
63-91-2 see PEC750
63-92-3 see DDG800
64-02-8 see EIV000
64-04-0 see PDE250
64-17-5 see EFU000
64-18-6 see FNA000
64-19-7 see AAT250
64-31-3 see MRP250
64-43-7 see AON750
64-65-3 see MKA250
64-67-5 see DKB110
64-69-7 see IDZ000
64-75-5 see TBX250
64-77-7 see BSQ000
64-95-9 see DHX800
65-22-5 see VSU000
65-23-6 see PPK250
65-29-2 see PDD300
65-30-5 see NDR500
65-31-6 see NDS500
65-45-2 see SAH000
65-85-0 see BCL750
66-25-1 see HEM000
66-27-3 see MLH500
66-32-0 see SM0000
66-56-8 see DUY800
66-75-1 see BIA250
66-76-2 see BJZ000
67-03-8 see TET300
67-20-9 see NGE000
67-21-0 see EEI000
67-42-5 see EIT000
67-43-6 see DJG800
67-45-8 see NGG500
67-48-1 see CMF750
67-56-1 see MGB150
67-63-0 see INJ000
67-64-1 see ABC750
67-66-3 see CHJ500
67-68-5 see DUD800
67-72-1 see HCI000
67-97-0 see CMC750
67-98-1 see DHS000
68-11-1 see TFJ100
68-12-2 see DSB000
68-19-9 see VSZ000
68-22-4 see NNP500
68-23-5 see EEH550
68-26-8 see VSK600
68-76-8 see TND000
68-88-2 see CJR909
68-89-3 see AMK500
69-53-4 see AIV500
69-57-8 see BFD250
69-72-7 see SAI000
69-79-4 see MA0500
69-89-6 see XCA000
70-25-7 see MMP000
70-30-4 see HCL000
70-34-8 see DUW400
70-70-2 see ELL500
71-00-1 see HGE700
71-23-8 see PND000
71-27-2 see HLC500
71-36-3 see BPW500
71-43-2 see BBL250
71-55-6 see MIH275
71-58-9 see MCA000

71-62-5 see VHU000
71-63-6 see DKL800
71-68-1 see DNU300
71-73-8 see PBT500
71-78-3 see PII750
71-79-4 see BAW500
72-14-0 see TEX250
72-17-3 see LAM000
72-18-4 see VBP000
72-19-5 see TFU750
72-20-8 see EAT500
72-33-3 see MKB750
72-43-5 see MEI450
72-44-6 see QAK000
72-54-8 see BIM500
72-55-9 see BIM750
72-56-0 see DJC000
72-57-1 see CM0250
72-69-5 see NNY000
73-22-3 see TNX000
73-32-5 see IKX000
73-78-9 see DHK600
74-31-7 see BLE500
74-82-8 see MDQ750
74-83-9 see MHR200
74-84-0 see EDZ000
74-85-1 see EIO000
74-86-2 see ACI750
74-87-3 see MIF765
74-88-4 see MKW200
74-89-5 see MGC250
74-90-8 see HHS000
74-93-1 see MLE650
74-94-2 see DOR200
74-95-3 see DDP800
74-96-4 see EGV400
74-97-5 see CES650
74-98-6 see PMJ750
74-99-7 see MFX590
75-00-3 see EHH000
75-01-4 see VNP000
75-02-5 see VPA000
75-04-7 see EDY000
75-04-7 see EDY500
75-04-7 see EFU400
75-05-8 see ABE500
75-07-0 see AAG250
75-08-1 see EMB100
75-09-2 see MJP450
75-12-7 see FMY000
75-15-0 see CBV500
75-16-1 see MLE000
75-18-3 see TFP000
75-19-4 see CQD750
75-20-7 see CAN750
75-21-8 see EJN500
75-24-1 see TLD272
75-25-2 see BNL000
75-27-4 see BND500
75-28-5 see MOR750
75-31-0 see INK000
75-33-2 see IMU000
75-34-3 see DFF809
75-35-4 see VPK000
75-36-5 see ACF750
75-37-6 see ELN500
75-38-7 see VPP000
75-39-8 see AAG500
75-43-4 see DFL000
75-44-5 see PGX000
75-45-6 see CFX500
75-46-7 see CBY750
75-47-8 see IEP000
75-50-3 see TLD500

100-00-5 see NFS525	103-36-6 see EHN000	106-27-4 see IHP400
100-01-6 see NEO500	103-37-7 see BED000	106-30-9 see EKN050
100-02-7 see NIF000	103-38-8 see ISW000	106-32-1 see ENY000
100-06-1 see MDW750	103-41-3 see BEG750	106-33-2 see ELY700
100-07-2 see AOY250	103-45-7 see PFB250	106-35-4 see EHA600
100-17-4 see NER500	103-48-0 see PDF750	106-42-3 see XHS000
100-25-4 see DUQ600	103-50-4 see BEO250	106-43-4 see TGY075
100-35-6 see CGV500	103-54-8 see CMQ730	106-44-5 see CNX250
100-37-8 see DHO500	103-65-1 see IKG000	106-46-7 see DEP800
100-38-9 see DIY600	103-69-5 see EGK000	106-47-8 see CEH680
100-39-0 see BEC000	103-71-9 see PFK250	106-48-9 see CJK750
100-40-3 see CPD750	103-72-0 see ISQ000	106-49-0 see TGR000
100-41-4 see EGP500	103-75-3 see EER500	106-50-3 see PEY500
100-42-5 see SMQ000	103-82-2 see PDY850	106-51-4 see QQS200
100-44-7 see BEE375	103-83-3 see DQP800	106-63-8 see IIK000
100-47-0 see BCQ250	103-84-4 see AAQ500	106-68-3 see EGI755
100-50-5 see FNK025	103-85-5 see PGN250	106-72-9 see DSD775
100-51-6 see BDX500	103-89-9 see ABJ250	106-87-6 see VOA000
100-52-7 see BAY500	103-90-2 see HIM000	106-89-8 see EAZ500
100-53-8 see TGO750	103-93-5 see THA250	106-90-1 see ECH500
100-56-1 see PFM500	103-95-7 see COU500	106-92-3 see AGH150
100-61-8 see MGN750	104-01-8 see MFE250	106-93-4 see EIY500
100-63-0 see PFI000	104-12-1 see CKB000	106-94-5 see BNX750
100-65-2 see PFJ250	104-45-0 see PNE250	106-95-6 see AFY000
100-66-3 see AOX750	104-46-1 see PMQ750	106-96-7 see PMN500
100-73-2 see ADR500	104-50-7 see OCE000	106-97-8 see BOR500
100-74-3 see ENL000	104-51-8 see BQI750	106-99-0 see BOP500
100-75-4 see NLJ500	104-53-0 see HHP000	107-00-6 see EFS500
100-79-8 see DVR600	104-54-1 see CMQ740	107-02-8 see ADR000
100-86-7 see DQQ200	104-55-2 see CMP969	107-03-9 see PML500
100-88-9 see CPQ625	104-57-4 see BEP250	107-05-1 see AGB250
100-97-0 see HEI500	104-61-0 see CNF250	107-06-2 see EIY600
100-99-2 see TKR500	104-65-4 see CMR500	107-07-3 see EIU800
101-02-0 see TMU250	104-68-7 see PEQ750	107-10-8 see PND250
101-14-4 see MJM200	104-75-6 see EKS500	107-11-9 see AFW000
101-21-3 see CKC000	104-78-9 see DIY800	107-12-0 see PMV750
101-27-9 see CEW500	104-90-5 see EOS000	107-13-1 see ADX500
101-31-5 see HOU000	104-91-6 see NLF200	107-14-2 see CDN500
101-37-1 see THN500	104-93-8 see MGP000	107-15-3 see EEA500
101-39-3 see MIO000	104-94-9 see AOW000	107-16-4 see HIM500
101-41-7 see MHA500	105-10-2 see DTL800	107-18-6 see AFV500
101-48-4 see PDX000	105-11-3 see DVR200	107-19-7 see PMN450
101-61-1 see MJN000	105-13-5 see MED500	107-20-0 see CDY500
101-68-8 see MJP400	105-29-3 see MNL775	107-21-1 see EJC500
101-77-9 see MJQ000	105-30-6 see AOK750	107-22-2 see GIK000
101-80-4 see OPM000	105-36-2 see EGV000	107-25-5 see MQL750
101-83-7 see DGT600	105-37-3 see EPB500	107-27-7 see CHC500
101-84-8 see PFA850	105-38-4 see VQK000	107-29-9 see AAH250
101-86-0 see HFO500	105-39-5 see EHG500	107-30-2 see CIO250
101-90-6 see REF000	105-40-8 see EMQ500	107-31-3 see MKG750
101-96-2 see DEG200	105-45-3 see MFX250	107-32-4 see PCM500
101-97-3 see EOH000	105-46-4 see BPV000	107-37-9 see AGU250
102-01-2 see AAY000	105-53-3 see EMA500	107-41-5 see HFP875
102-06-7 see DWC600	105-54-4 see EHE000	107-44-8 see IPX000
102-08-9 see DWN800	105-55-5 see DKC400	107-49-3 see TCF250
102-09-0 see DVZ000	105-56-6 see EHP500	107-49-3 see TCF280
102-20-5 see PDI000	105-57-7 see AAG000	107-66-4 see DEG700
102-50-1 see MGO500	105-58-8 see DIX200	107-70-0 see MEX250
102-54-5 see FBC000	105-60-2 see CBF700	107-71-1 see BSC250
102-62-5 see DBF600	105-64-6 see DNR400	107-72-2 see PBY750
102-69-2 see TMY250	105-67-9 see XKJ500	107-75-5 see CMS850
102-70-5 see THN000	105-74-8 see LBR000	107-81-3 see BNU500
102-71-6 see TKP500	105-75-9 see DEC600	107-82-4 see BNP250
102-76-1 see THM500	105-76-0 see DED600	107-83-5 see IKS600
102-77-2 see BDG000	105-83-9 see BGU750	107-87-9 see PBN250
102-79-4 see BQM000	105-85-1 see CMT750	107-88-0 see BOS500
102-81-8 see DDU600	105-86-2 see GCY000	107-89-1 see AAH750
102-82-9 see THX250	105-87-3 see DTD800	107-92-6 see BSW000
103-17-3 see CEP000	106-11-6 see HKJ000	107-98-2 see PNL250
103-26-4 see MIO500	106-21-8 see DTE600	107-99-3 see CGW000
103-28-6 see IJV000	106-22-9 see CMT250	108-01-0 see DOY800
103-29-7 see BFX500	106-23-0 see CMS845	108-03-2 see NIX500
103-33-3 see ASL250	106-24-1 see DTD000	108-05-4 see VLU250
103-34-4 see BKU500	106-25-2 see DTD200	108-09-8 see DQU600

116-85-8 see AKE250	122-15-6 see DRL200	124-99-2 see GFC000
117-39-5 see QCA000	122-19-0 see DTC600	125-04-2 see HHR000
117-51-1 see HGK500	122-34-9 see BJP000	125-28-0 see DKW800
117-52-2 see ABF500	122-39-4 see DVX800	125-33-7 see DBB200
117-79-3 see AIB000	122-40-7 see AOG500	125-40-6 see BPF000
117-80-6 see DFT000	122-51-0 see ENY500	125-42-8 see EMO500
117-81-7 see DVL700	122-52-1 see TJT800	125-44-0 see VKP000
117-82-8 see DOF400	122-56-5 see THX500	125-51-9 see PJA000
117-84-0 see DVL600	122-60-1 see PFH000	125-88-2 see BOQ750
118-46-7 see ALJ750	122-62-3 see BJS250	126-07-8 see GKE000
118-52-5 see DFE200	122-66-7 see HHG000	126-17-0 see POJ000
118-55-8 see PGG750	122-67-8 see IIQ000	126-22-7 see BPG000
118-58-1 see BFJ750	122-72-5 see HHP500	126-27-2 see DTL200
118-61-6 see SAL000	122-78-1 see BBL500	126-64-7 see LFZ000
118-68-3 see AJB250	122-88-3 see CJN000	126-72-7 see TNC500
118-71-8 see MAO350	122-97-4 see HHP050	126-73-8 see TIA250
118-74-1 see HCC500	122-98-5 see AOR750	126-75-0 see DAP200
118-92-3 see API500	122-99-6 see PER000	126-85-2 see CFA500
118-96-7 see TMN490	123-00-2 see AMF250	126-92-1 see TAV750
119-12-0 see POP000	123-03-5 see CCX000	126-98-7 see MGA750
119-27-7 see DUP800	123-05-7 see BRI000	126-99-8 see NCI500
119-34-6 see NEM480	123-11-5 see AOT500	127-07-1 see HOO500
119-36-8 see MPI000	123-19-3 see DWT600	127-09-3 see SEG500
119-38-0 see DSK200	123-20-6 see VNF000	127-18-4 see PCF275
119-48-2 see DUO400	123-23-9 see SNC500	127-19-5 see DOO800
119-53-9 see BCP250	123-25-1 see SNB000	127-20-8 see DGI600
119-61-9 see BCS250	123-28-4 see TFD500	127-25-3 see MFT500
119-65-3 see IRX000	123-29-5 see ENW000	127-33-3 see MIJ500
119-84-6 see HHR500	123-31-9 see HIH000	127-41-3 see IFW000
119-90-4 see DCJ200	123-32-0 see DTU600	127-47-9 see VSK900
119-93-7 see TGJ750	123-33-1 see DMC600	127-48-0 see TLP750
120-02-5 see CBI250	123-35-3 see MRZ150	127-58-2 see SJW475
120-08-1 see DRS800	123-38-6 see PMT750	127-65-1 see CDP000
120-12-7 see APG500	123-39-7 see MKG500	127-85-5 see ARA500
120-20-7 see DOE200	123-42-2 see DBF750	127-91-3 see POH750
120-36-5 see DGB000	123-51-3 see IHP000	128-04-1 see SGM500
120-40-1 see BKE500	123-54-6 see ABX750	128-08-5 see BOF500
120-47-8 see HJL000	123-61-5 see BBP000	128-37-0 see BFW750
120-51-4 see BCM000	123-62-6 see PMV500	128-44-9 see SJN700
120-57-0 see PIW250	123-63-7 see PAI250	128-46-1 see DME000
120-58-1 see IRZ000	123-66-0 see EHF000	128-53-0 see MAL250
120-61-6 see DUE000	123-68-2 see AGA500	128-58-5 see JAT000
120-62-7 see ISA000	123-72-8 see BSU250	128-80-3 see BLK000
120-71-8 see MGO750	123-73-9 see COB260	129-00-0 see PON250
120-72-9 see ICM000	123-75-1 see PPS500	129-15-7 see MMG000
120-78-5 see BDE750	123-77-3 see ASM300	129-16-8 see MCV000
120-80-9 see CCP850	123-77-3 see ASM270	129-18-0 see BOV750
120-82-1 see TIK250	123-86-4 see BPU750	129-20-4 see HNI500
120-83-2 see DFX800	123-88-6 see MEP250	129-51-1 see EDB500
120-92-3 see CPW500	123-91-1 see DVQ000	129-66-8 see TML000
120-94-5 see MPB250	123-92-2 see IHO850	129-67-9 see DXD000
121-14-2 see DVH000	123-93-3 see MCM750	129-71-5 see CDR000
121-17-5 see NFS700	124-02-7 see DBI600	129-79-3 see TMM250
121-19-7 see HMY000	124-04-9 see AEN250	130-15-4 see NBA500
121-29-9 see POO100	124-06-1 see ENL850	130-26-7 see CHR500
121-32-4 see EQF000	124-07-2 see OCY000	130-61-0 see MOO500
121-33-5 see VFK000	124-09-4 see HEO000	130-80-3 see DKB000
121-39-1 see EOK600	124-13-0 see OCO000	130-86-9 see FOW000
121-43-7 see TLN000	124-16-3 see BPL500	130-89-2 see QJS000
121-44-8 see TJO000	124-17-4 see BQP500	130-95-0 see QHJ000
121-45-9 see TMD500	124-18-5 see DAG400	131-11-3 see DTR200
121-54-0 see BEN000	124-19-6 see NMW500	131-17-9 see DBL200
121-59-5 see CBJ000	124-22-1 see DXW000	131-52-2 see SJA000
121-69-7 see DQF800	124-30-1 see OBC000	131-56-6 see DMI600
121-73-3 see CJB250	124-38-9 see CBU250	131-74-8 see ANS500
121-75-5 see MAK700	124-38-9 see CBU500	131-74-8 see ANS750
121-79-9 see PNM750	124-40-3 see DOQ800	131-79-3 see FAG135
121-82-4 see CPR800	124-40-3 see DOR000	131-89-5 see CPK500
122-00-9 see MFW250	124-41-4 see SIK450	132-17-2 see TNU000
122-03-2 see COE500	124-43-6 see HIB500	132-27-4 see BGJ750
122-09-8 see DTJ400	124-48-1 see CFK500	132-32-1 see AJV000
122-10-1 see SOY000	124-65-2 see HKC500	132-45-6 see PGP500
122-11-2 see SNN300	124-87-8 see PIE500	132-69-4 see BBW500
122-14-5 see DSQ000	124-90-3 see DLX400	133-06-2 see CBG000

300-42-5 see DBA800	352-32-9 see FMC000	467-60-7 see DWK400
300-54-9 see MRW250	352-93-2 see EPH000	469-59-0 see JCS000
300-62-9 see BBK000	353-03-7 see EKI000	469-62-5 see DAB879
300-76-5 see NAG400	353-17-3 see FHD000	469-65-8 see HJ0500
301-04-2 see LCG000	353-18-4 see CON500	470-82-6 see CAL000
301-12-2 see DAP000	353-36-6 see FIB000	470-90-6 see CDS750
302-01-2 see HGS000	353-42-4 see BMH000	471-03-4 see BIH500
302-17-0 see CD0000	353-50-4 see CCA500	471-29-4 see MKI750
302-22-7 see CBF250	353-59-3 see BNA250	471-46-5 see OLO000
302-33-0 see DIG400	357-57-3 see BOL750	471-95-4 see BON000
302-49-8 see EHV500	358-74-7 see DJJ400	474-86-2 see ECW000
302-66-9 see MNM500	359-06-8 see FFR000	475-83-2 see NOE500
302-70-5 see CFA750	359-48-8 see PCO250	475-83-2 see DNZ000
303-04-8 see DFM000	359-83-1 see DOQ400	476-32-4 see CDL000
303-33-3 see HAL500	360-54-3 see MKK750	477-30-5 see MIW500
303-34-4 see LBG000	360-68-9 see DKW000	478-84-2 see BNM250
303-45-7 see GJM000	360-89-4 see OB0000	478-99-9 see LJI000
303-47-9 see CHP250	361-09-1 see SFW000	479-45-8 see TEG250
303-54-8 see DPW600	363-24-6 see DVJ200	479-50-5 see DHU000
304-06-3 see PGH000	364-71-6 see FFT000	479-92-5 see INY000
304-20-1 see HGP500	365-26-4 see HKH500	480-16-0 see MRN500
304-28-9 see BGP250	366-70-1 see PME500	480-22-8 see APH250
304-84-7 see DKE200	366-93-8 see BHN000	480-54-6 see RFP000
305-03-3 see CD0500	367-51-1 see SKH500	480-81-9 see SBX500
305-85-1 see DNG000	368-43-4 see BBT250	481-39-0 see WAT000
305-97-5 see LGU000	368-97-8 see DKI400	483-18-1 see EAL500
306-23-0 see HNL000	371-29-9 see FIN000	484-20-8 see MFN275
306-37-6 see DSF800	371-40-4 see FFY000	487-53-6 see DH0600
306-40-1 see CMG250	371-62-0 see FIE000	487-93-4 see DPG109
309-00-2 see AFK250	371-86-8 see PHF750	488-17-5 see DNE000
309-36-4 see MDU500	373-02-4 see NCX000	488-41-5 see DDP600
311-45-5 see NIM500	375-22-4 see HAX500	491-59-8 see CML750
312-45-8 see HAQ000	378-44-9 see BFV750	492-18-2 see SIH500
313-67-7 see AQY250	379-79-3 see EDC500	492-41-1 see NNM000
314-03-4 see MOO750	389-08-2 see EID000	492-80-8 see IBB000
314-40-9 see BMM650	404-82-0 see PDM250	493-52-7 see CCE500
315-18-4 see DOS000	406-20-2 see MKE000	494-03-1 see BIF250
315-22-0 see MRH000	406-90-6 see TKB250	494-38-2 see BJF000
315-80-0 see DCW800	409-02-9 see MKK000	494-47-3 see FPS000
316-14-3 see MGW000	409-21-2 see SCQ000	495-48-7 see AS0750
316-42-7 see EAN000	420-04-2 see COH500	495-73-8 see BDD000
316-46-1 see FMN000	420-12-2 see EJP500	496-67-3 see BNP750
316-49-4 see MGV500	421-20-5 see MKG250	497-18-7 see CBS500
317-52-2 see HEG000	431-03-8 see BOT500	497-19-8 see SF0000
319-84-6 see BBQ000	434-07-1 see PAN100	497-56-3 see DUT000
319-85-7 see BBR000	437-38-7 see PDW500	499-75-2 see CCM000
319-86-8 see BFW500	438-41-5 see MDQ250	500-28-7 see MIJ250
320-72-9 see DGK200	442-51-3 see HAI500	500-38-9 see NBR000
321-54-0 see CIK750	443-30-1 see DOT600	500-92-5 see CKB250
321-55-1 see DFH600	443-48-1 see MMN250	501-53-1 see BEF500
326-61-4 see PIX000	446-86-6 see ASB250	502-39-6 see MLF250
327-98-0 see EPY000	453-18-9 see MKD000	502-55-6 see BJU000
328-38-1 see LER000	457-60-3 see NCJ500	503-01-5 see ILK000
329-65-7 see EBB500	458-24-2 see ENJ000	503-17-3 see COC500
330-54-1 see DXQ500	459-99-4 see FIM000	503-20-8 see FFJ000
330-55-2 see DGD600	460-07-1 see ACB250	503-30-0 see OMW000
330-68-7 see FFH000	460-19-5 see COO000	503-74-2 see ISU000
332-97-8 see EKK500	462-06-6 see FGA000	504-15-4 see MPH500
333-20-0 see PLV750	462-08-8 see AMI250	504-20-1 see PGW200
333-25-5 see DEW000	462-94-2 see PBK500	504-29-0 see AMI000
333-29-9 see DXN600	462-95-3 see EFT500	505-57-7 see HFA500
333-36-8 see HDC000	463-04-7 see AOL500	505-60-2 see BIH250
333-40-4 see DTV400	463-51-4 see KEU000	505-75-9 see CMN000
333-41-5 see DCM750	463-58-1 see CCC000	506-30-9 see EAF000
334-22-5 see BHN750	463-71-8 see TFN500	506-63-8 see DQR200
334-48-5 see DAH400	463-82-1 see NCH000	506-64-9 see SDP000
334-88-3 see DCP800	464-10-8 see NMQ000	506-68-3 see COO500
337-47-3 see SOX500	464-45-9 see NCQ820	506-77-4 see COO750
339-43-5 see BSM000	464-49-3 see CBB250	506-82-1 see DQW800
340-56-7 see MDT250	465-16-7 see OHQ000	506-85-4 see FOS050
341-69-5 see OJW000	465-73-6 see IK0000	506-87-6 see ANE000
342-69-8 see MPU000	466-24-0 see PIC250	506-93-6 see GLA000
342-95-0 see BHT250	466-40-0 see IKZ000	506-96-7 see ACD750
343-89-5 see FFG000	466-99-9 see DLW600	507-02-8 see AC0500

592-35-8 see BQP250	622-78-6 see BEU250	661-69-8 see HEE500
592-41-6 see HFB000	623-26-7 see BBP250	671-16-9 see PME250
592-42-7 see HCR500	623-42-7 see MHY000	671-51-2 see ASH750
592-62-1 see MGS750	623-70-1 see COB750	673-06-3 see PEC500
592-79-0 see FFV000	623-73-4 see DCN800	673-31-4 see PGA750
592-84-7 see BRK000	624-46-4 see MKA750	674-81-7 see NKH000
592-85-8 see MCU250	624-61-3 see DDJ800	674-82-8 see KFA000
593-53-3 see FJK000	624-74-8 see DNE500	676-83-5 see MOC250
593-57-7 see DQG600	624-83-9 see MKX250	676-97-1 see MOB399
593-60-2 see VMP000	624-91-9 see MMF750	680-31-9 see HEK000
593-70-4 see CHI900	624-92-0 see DRQ400	681-84-5 see MPI750
593-82-8 see DSG000	625-17-2 see DEC200	683-18-1 see DDY200
593-89-5 see DFP200	625-22-9 see DEC000	684-16-2 see HCZ000
594-42-3 see PCF300	625-55-8 see IPC000	684-93-5 see MNA750
594-71-8 see CJE250	625-58-1 see ENM500	685-09-6 see MNN000
594-72-9 see DFU000	626-17-5 see PHX550	685-91-6 see DHI200
595-33-5 see VTF000	626-38-0 see AOD735	688-74-4 see THX750
596-51-0 see GIC000	626-67-5 see MOG500	689-13-4 see FNO000
597-88-6 see DJT000	627-12-3 see PNG250	689-97-4 see BPE109
598-14-1 see DFH200	627-13-4 see PNQ500	691-88-3 see DEE200
598-31-2 see BNZ000	627-22-5 see CET250	693-07-2 see CGY750
598-55-0 see MHZ000	627-44-1 see DJO400	693-21-0 see DJE400
598-58-3 see MMF500	627-63-4 see FOY000	693-65-2 see PBX000
598-63-0 see LCP000	628-02-4 see HEM500	696-24-2 see DDR200
598-73-2 see BOJ000	628-32-0 see EPC125	696-28-6 see DGB600
599-52-0 see HCE000	628-63-7 see AOD725	702-54-5 see DJT400
600-25-9 see CJE000	628-73-9 see HER500	705-86-2 see DAF200
601-77-4 see NKA000	628-81-9 see EHA500	709-98-8 see DGI000
602-87-9 see NEJ500	628-83-1 see BSN500	712-48-1 see CGN000
602-99-3 see TML500	628-85-3 see DWU000	712-68-5 see NGI500
603-34-9 see TMQ500	628-86-4 see MDC000	720-69-4 see DBY800
603-35-0 see TMU000	628-86-4 see MDC250	723-46-6 see SNK000
605-65-2 see DPN200	628-94-4 see AEN000	732-11-6 see PHX250
605-69-6 see DUX800	628-96-6 see EJG000	738-99-8 see BJC250
605-71-0 see DUX700	629-01-0 see DCL600	749-13-3 see TKK500
606-23-5 see IBS000	629-14-1 see EJE500	753-53-7 see BMG750
606-90-6 see DWF000	629-15-2 see EJF000	757-58-4 see HCY000
607-35-2 see NJD500	629-20-9 see CPS500	758-17-8 see FNW000
607-57-8 see NGB000	629-35-6 see DEE000	759-73-9 see ENV000
607-59-0 see DSU600	629-59-4 see TBX750	759-94-4 see EIN500
608-73-1 see BBP750	630-08-0 see CBW750	760-21-4 see EGW500
609-89-2 see DFU600	630-20-6 see TBQ000	764-41-0 see DEV000
609-93-8 see DUT600	630-56-8 see HNT500	765-34-4 see GGW000
612-12-4 see DGP400	630-60-4 see OKS000	766-09-6 see EOS500
612-64-6 see NKD000	630-72-8 see TMK250	766-39-2 see DSM000
612-83-9 see DEQ800	630-93-3 see DNU000	768-52-5 see INX000
613-13-8 see APG000	631-07-2 see EOL000	771-51-7 see ICW000
613-35-4 see BFX000	631-60-7 see MDE250	776-74-9 see BNG750
613-37-6 see PEG250	631-61-8 see ANA000	777-11-7 see IFA000
613-47-8 see NBI500	632-22-4 see TDX250	779-02-2 see MGP750
613-89-8 see AHR250	633-03-4 see BAY750	780-24-5 see DDH000
613-94-5 see BBV250	635-85-8 see DOI400	781-43-1 see DQG200
614-00-6 see MMU250	636-21-5 see TGS500	786-19-6 see TNP250
614-45-9 see BSC500	636-23-7 see DCE000	794-00-3 see DRY600
614-95-9 see NKE500	638-21-1 see PFV250	794-93-4 see BKH500
615-05-4 see DB0000	638-29-9 see VBA000	800-24-8 see BDC750
615-15-6 see MHC250	638-49-3 see AOJ500	801-52-5 see MLY000
615-45-2 see DCE200	639-14-5 see GMG000	804-63-7 see QMA000
615-50-9 see DCE600	639-58-7 see CLU000	811-73-4 see IFN000
615-53-2 see MMX250	640-15-3 see PHI500	814-49-3 see DIY000
615-67-8 see CHM000	640-19-7 see FFF000	814-78-8 see MKY500
616-23-9 see DGG600	641-16-7 see TDY600	816-57-9 see NL0500
616-38-6 see MIF000	642-65-9 see DBF200	817-09-4 see TNF500
616-91-1 see ACH000	644-26-8 see AOM000	818-08-6 see DEF400
617-79-8 see EHA000	644-31-5 see ACC250	820-75-7 see DC0800
617-88-9 see CIY250	644-31-5 see ACC500	821-10-3 see DEV400
617-89-0 see FPW000	644-97-3 see DGE400	821-14-7 see ASN250
618-25-7 see CBJ750	645-05-6 see HEJ500	821-48-7 see BHO250
619-01-2 see DKV150	645-43-2 see GKU000	822-06-0 see DNJ800
621-64-7 see NKB700	645-48-7 see PGM750	822-16-2 see SJV500
621-82-9 see CMP975	657-24-9 see DQR600	824-72-6 see PFW100
621-90-9 see MNR500	657-27-2 see LJ0000	826-10-8 see MDQ500
622-45-7 see CPF000	659-70-1 see ITB000	826-39-1 see MQR500
622-51-5 see THG000	661-18-7 see FFX000	827-61-2 see AAE250

1329-86-8 see THV000	1476-53-5 see NOB000	1897-45-6 see TBQ750
1330-20-7 see XGS000	1477-55-0 see XHS800	1907-13-7 see ABW750
1330-78-5 see TNP500	1491-41-4 see HMV000	1910-36-7 see HLV000
1331-11-9 see EFH000	1492-93-9 see BHP750	1910-42-5 see PAJ000
1331-22-2 see MIR250	1493-13-6 see TKB310	1912-24-9 see ARQ725
1331-31-3 see EHH500	1498-40-4 see EOQ000	1918-00-9 see MEL500
1332-10-1 see FOM050	1498-51-7 see EOR000	1918-02-1 see PIB900
1332-21-4 see ARM250	1515-76-0 see ABM250	1918-13-4 see DGM600
1332-58-7 see KBB600	1516-32-1 see BSO500	1918-16-7 see CHS500
1332-94-1 see LAS000	1528-74-1 see DUS000	1929-82-4 see CLP750
1333-74-0 see HHW500	1544-46-3 see FFE000	1934-21-0 see FAG140
1333-74-0 see HHY500	1563-66-2 see CBS275	1936-15-8 see HGC000
1333-82-0 see CMK000	1563-67-3 see DLS800	1937-37-7 see AQP000
1333-83-1 see SHQ500	1569-69-3 see CPB625	1943-16-4 see CLT250
1333-86-4 see CBT750	1570-45-2 see ELU000	1948-33-0 see BRM500
1335-31-5 see MDA500	1582-09-8 see DUV600	1949-20-8 see OOE000
1335-32-6 see LCH000	1596-52-7 see DVE000	1955-45-9 see DTH000
1335-85-9 see DUS800	1596-84-5 see DQD400	1972-08-3 see TCM250
1335-87-1 see HCK500	1600-27-7 see MCS750	1977-10-2 see DCS200
1335-88-2 see TBR000	1609-47-8 see DIZ100	1982-36-1 see CJR809
1336-21-6 see ANK250	1615-80-1 see DJL400	1982-37-2 see MPE250
1336-36-3 see PJL750	1622-61-3 see CMW000	2016-57-1 see DAG600
1336-80-7 see FBC100	1622-79-3 see FKI000	2016-63-9 see BEO750
1338-02-9 see NAS000	1623-24-1 see PHE500	2021-21-8 see MIS250
1338-16-5 see IHL000	1624-02-8 see BLS750	2023-61-2 see DRZ000
1338-23-4 see MKA500	1632-16-2 see EKR500	2032-59-9 see DOR400
1338-24-5 see NAR000	1633-83-6 see BOU250	2032-65-7 see DST000
1338-41-6 see SKV150	1634-04-4 see MHV859	2039-87-4 see CLE750
1341-49-7 see ANJ000	1638-22-8 see BSD750	2050-92-2 see DCH200
1341-49-7 see ANJ250	1639-09-4 see HBD500	2058-46-0 see HOI000
1343-90-4 see MAJ000	1639-60-7 see PNA500	2058-52-8 see CIS750
1344-00-9 see SEM000	1642-54-2 see DIW200	2058-62-0 see MPY000
1344-28-1 see AHE250	1649-18-9 see FLU000	2058-66-4 see ENB000
1344-43-0 see MAT250	1653-64-1 see MJT000	2068-78-2 see LEZ000
1344-67-8 see CNK500	1668-19-5 see DYE409	2078-54-8 see DNR800
1344-95-2 see CAW850	1675-54-3 see BLD750	2086-83-1 see BFN500
1345-04-6 see AQL500	1689-82-3 see HJF000	2092-16-2 see CAY250
1393-62-0 see AAD000	1689-83-4 see HKB500	2104-64-5 see EBD700
1395-21-7 see BAB750	1689-84-5 see DDP000	2130-56-5 see BFX250
1397-89-3 see AOC500	1689-89-0 see HLJ500	2152-34-3 see IBM000
1399-80-0 see DYA600	1694-09-3 see FAG120	2156-56-1 see SGG000
1400-61-9 see NOH500	1698-60-8 see PEE750	2163-80-6 see MRL750
1401-55-4 see QBJ000	1705-85-7 see MIN750	2164-17-2 see DUK800
1401-55-4 see TAD750	1707-14-8 see MNV750	2179-59-1 see AGR500
1401-55-4 see CDM250	1707-95-5 see ONY000	2180-92-9 see BSI250
1401-55-4 see MQV250	1708-39-0 see BBA000	2188-67-2 see PBV750
1401-69-0 see TOE600	1709-50-8 see DIU400	2191-10-8 see CAD750
1402-68-2 see AET750	1712-64-7 see IQP000	2192-21-4 see DHS200
1403-17-4 see CBC250	1719-53-5 see DEY800	2207-85-4 see IBP309
1403-17-4 see LFF000	1738-25-6 see DPU000	2216-51-5 see MCG250
1403-66-3 see GCO000	1746-01-6 see TAI000	2219-30-9 see PAP550
1404-04-2 see NCE000	1746-77-6 see IOJ000	2223-93-0 see OAT000
1404-24-6 see PKC250	1746-81-2 see CKD500	2234-13-1 see OAP000
1405-10-3 see NCG000	1754-58-1 see PEV500	2238-07-5 see DKM200
1405-41-0 see GCS000	1757-18-2 see DIX600	2243-62-1 see NAM000
1405-87-4 see BAC250	1762-95-4 see ANW750	2244-11-3 see MDL500
1405-97-6 see GJO000	1777-84-0 see NEL000	2244-16-8 see CCM100
1406-05-9 see PAQ000	1779-25-5 see CGB500	2244-21-5 see PLD000
1406-11-7 see PKC000	1785-74-6 see DDD000	2255-17-6 see PHD750
1406-65-1 see CKN000	1789-58-8 see DFK000	2266-22-0 see FLV000
1420-04-8 see DFV600	1797-74-6 see PMS500	2270-40-8 see AOP250
1421-28-9 see DNU310	1808-12-4 see BNW500	2273-43-0 see BSL500
1421-63-2 see TKO250	1824-81-3 see ALC500	2273-45-2 see DTH400
1424-27-7 see AAS750	1836-75-5 see DFT800	2274-11-5 see EIP000
1432-75-3 see NEC000	1838-59-1 see AGH000	2275-14-1 see PDC750
1435-55-8 see HIG500	1851-71-4 see BJU250	2275-18-5 see IOT000
1455-77-2 see DCF200	1851-77-0 see DKE400	2275-23-2 see MJG500
1456-28-6 see DTA000	1866-31-5 see AGC000	2279-64-3 see PFP500
1461-22-9 see CLP250	1867-66-9 see CKD750	2279-76-7 see CLU250
1461-25-2 see TBM250	1875-92-9 see DQQ000	2294-47-5 see DCL125
1464-43-3 see DWY800	1885-14-9 see CBX109	2303-16-4 see DBI200
1464-53-5 see BGA750	1888-71-7 see HCM000	2307-55-3 see DAB020
1465-26-5 see BHV000	1892-29-1 see DXM600	2310-17-0 see BDJ250
1467-79-4 see DRF600	1893-33-0 see FHG000	2312-35-8 see SOP000

3778-73-2 see IMH000	4795-29-3 see TCS500	5910-85-0 see HAV450
3785-34-0 see BHD250	4812-22-0 see NHE000	5910-89-4 see DTU400
3792-59-4 see SCC000	4822-44-0 see MCK000	5913-82-6 see DOX000
3811-04-9 see PLA250	4824-78-6 see EGV500	5954-50-7 see DSA800
3811-49-2 see MEC250	4831-62-3 see MMQ250	5957-75-5 see TCM000
3817-11-6 see HJQ350	4891-54-7 see SPF000	5965-13-9 see DKX000
3820-53-9 see DTH800	4940-11-8 see EMA600	5965-49-1 see KFK000
3825-26-1 see ANP625	4985-15-3 see DPH600	5967-73-7 see MDP250
3844-45-9 see FAE000	4985-85-7 see AMD750	5970-32-1 see MCU000
3844-63-1 see NKL000	4988-64-1 see MCQ500	5974-19-6 see MQN000
3861-47-0 see DNG200	5034-77-5 see IAN000	5980-33-6 see HMB000
3871-82-7 see MNN250	5064-31-3 see SIP500	5985-35-3 see MDV250
3883-43-0 see DFE600	5090-37-9 see MHJ500	5989-27-5 see LFU000
3913-02-8 see BSA500	5117-17-9 see DJY800	5989-54-8 see MCC500
3913-71-1 see DAI350	5124-30-1 see MJM600	5989-77-5 see DLL000
3926-62-3 see SFU500	5131-60-2 see CJY120	6004-24-6 see CDF750
3942-54-9 see CKF000	5137-55-3 see MQH000	6012-97-1 see TBV750
3949-14-2 see MFO250	5141-20-8 see FAF000	6018-89-9 see NCX500
3953-10-4 see EGZ000	5152-30-7 see DUL800	6029-87-4 see FOT000
3963-95-9 see MDO250	5160-02-1 see CHP500	6032-29-7 see PBM750
3978-86-7 see DLV800	5169-78-8 see BLV000	6055-19-2 see CQC675
3982-91-0 see TFO000	5208-87-7 see BCJ000	6080-56-4 see LCJ000
3983-39-9 see HN0500	5221-17-0 see DDS400	6098-44-8 see ABL000
4016-14-2 see IPD000	5234-68-4 see CCC500	6104-30-9 see IIV000
4032-26-2 see DWY000	5283-66-9 see OGE000	6109-97-3 see AJV250
4044-65-9 see PFA500	5283-67-0 see NNE000	6117-91-5 see B0Y000
4075-79-0 see PDY500	5299-64-9 see MRQ750	6147-53-1 see CNA500
4098-71-9 see IMG000	5307-14-2 see ALL750	6152-43-8 see DT0800
4109-96-0 see DGK300	5329-14-6 see SNK500	6164-98-3 see CJJ250
4164-28-7 see DSV200	5332-73-0 see MFM000	6168-86-1 see ODY000
4170-30-3 see COB250	5344-27-4 see POR500	6283-24-5 see ABQ000
4194-69-8 see ICH000	5355-48-6 see ACI250	6285-05-8 see CKT500
4213-32-5 see BHU750	5377-20-8 see MQQ500	6292-91-7 see MJS750
4213-40-5 see BIA000	5392-40-5 see DTC800	6296-45-3 see CHF500
4213-45-0 see QDS000	5419-55-6 see IOI000	6304-33-2 see TMQ250
4230-97-1 see AGM500	5421-46-5 see ANM500	6305-43-7 see DDK600
4238-84-0 see LJH000	5421-48-7 see AAS250	6317-18-6 see MJT500
4242-33-5 see FIP999	5431-33-4 see ECJ000	6325-54-8 see CIG250
4245-77-6 see ENU000	5432-28-0 see NKT500	6334-11-8 see TLH000
4268-36-4 see MOV500	5459-93-8 see EHT000	6358-53-8 see D0K200
4342-03-4 see DAB600	5471-51-2 see RBU000	6365-83-9 see BPG250
4342-36-3 see BDR750	5488-45-9 see TIB000	6368-72-5 see E0J500
4368-28-9 see F0Q000	5510-99-6 see DEF800	6376-26-7 see DHP200
4386-79-2 see MJH250	5511-98-8 see ACI000	6379-46-0 see DVI600
4418-26-2 see SGD000	5522-43-0 see NJA000	6379-69-7 see TJE750
4420-79-5 see BHB750	5550-12-9 see GLS800	6414-38-6 see DLQ000
4426-51-1 see MQG500	5571-97-1 see DEM000	6423-43-4 see PNL000
4435-53-4 see MHV750	5585-67-1 see DQB800	6484-52-2 see ANN000
4439-24-1 see IIP000	5593-70-4 see BSP250	6485-34-3 see CAM750
4463-22-3 see ACD000	5598-13-0 see CMA250	6485-40-1 see CCM120
4465-94-5 see CQN000	5625-90-1 see MJQ750	6533-00-2 see NNQ500
4484-72-4 see DYA800	5626-16-4 see DBA600	6558-78-7 see BRZ000
4498-32-2 see DCW600	5632-47-3 see MRJ750	6569-51-3 see BMB500
4525-46-6 see BFM750	5675-31-0 see FLQ000	6649-23-6 see LFA000
4540-66-3 see DXL400	5688-80-2 see TKX500	6696-47-5 see OH0000
4549-40-0 see NKY000	5707-69-7 see MLC250	6708-69-6 see BJP500
4549-43-3 see MMT500	5714-00-1 see ABG000	6746-59-4 see ENK500
4549-44-4 see EHC000	5714-22-7 see S0Q450	6795-23-9 see AEW000
4553-89-3 see CMP500	5716-15-4 see DHF200	6804-07-5 see F0I000
4562-36-1 see GEU000	5743-18-0 see CAK750	6834-92-0 see SJU000
4564-87-8 see CBT250	5776-49-8 see PHA750	6842-15-5 see PMP750
4570-41-6 see AIS600	5796-14-5 see DYC200	6870-67-3 see JAK000
4584-46-7 see DRC000	5798-79-8 see BMW250	6898-43-7 see BCC250
4591-46-2 see TDY075	5800-19-1 see MQQ000	6915-15-7 see MAN000
4593-81-1 see DDY000	5809-59-6 see HJQ000	6923-22-4 see MRH209
4602-84-0 see FAB800	5827-03-2 see DKB600	6943-65-3 see EJA000
4611-02-3 see CLZ000	5831-16-3 see DLI200	6959-48-4 see CIV000
4636-83-3 see BJK750	5836-73-7 see DEQ000	6972-76-5 see DR0200
4637-56-3 see HIY500	5863-35-4 see NHP500	6974-12-5 see DDL400
4685-14-7 see PAI990	5892-15-9 see AIT000	6988-21-2 see DVS000
4691-65-0 see DXE500	5894-60-0 see HCQ000	7009-49-6 see SHL500
4724-58-7 see DPH000	5902-79-4 see MLF500	7046-61-9 see NHI500
4731-77-5 see BLB250	5903-13-9 see MME809	7047-84-9 see AHA250
4756-45-0 see HLS500	5905-52-2 see LAL000	7060-74-4 see OH0200

7068-83-9 see MHW500	7440-62-2 see VCP000	7664-41-7 see AMY500
7081-44-9 see SLJ000	7440-63-3 see XDS000	7664-93-9 see SOI500
7090-25-7 see NBJ500	7440-64-4 see YDA000	7664-93-9 see SOI530
7173-84-4 see DIX800	7440-65-5 see YEJ000	7664-98-4 see DWV000
7177-48-2 see AOD125	7440-66-6 see ZBJ000	7681-11-0 see PLK500
7187-55-5 see DXI600	7440-67-7 see ZOA000	7681-38-1 see SEG800
7203-92-1 see DSN600	7440-69-9 see BKU750	7681-49-4 see SHF500
7209-38-3 see BGV000	7440-70-2 see CAL250	7681-52-9 see SHU500
7220-81-7 see AEU750	7440-74-6 see ICF000	7681-53-0 see SHV000
7227-91-0 see DTP000	7446-09-5 see SOH500	7681-57-4 see SII000
7227-92-1 see DSX400	7446-11-9 see SOR500	7681-82-5 see SHW000
7236-83-1 see MPD000	7446-14-2 see LDY000	7681-93-8 see PIF750
7240-57-5 see MLD500	7446-18-6 see TEM000	7683-59-2 see DMV600
7241-98-7 see AEV500	7446-20-0 see ZNJ000	7693-26-7 see PLJ250
7242-04-8 see GEW000	7446-27-7 see LDU000	7697-37-2 see NED500
7287-19-6 see BKL250	7446-34-6 see SBT000	7697-37-2 see NEE500
7320-37-8 see EBX500	7446-70-0 see AGY750	7699-31-2 see DJL600
7339-53-9 see MKN250	7447-40-7 see PLA500	7699-41-4 see SCL000
7361-61-7 see DMW000	7447-41-8 see LHB000	7699-43-6 see ZSJ000
7392-96-3 see DDZ000	7452-79-1 see EMP600	7704-34-9 see SOD500
7417-67-6 see MMT000	7460-84-6 see SLK500	7704-99-6 see ZRA000
7429-90-5 see AGX000	7487-88-9 see MAJ250	7705-07-9 see TGG250
7439-88-5 see IGJ000	7487-94-7 see MCY475	7705-08-0 see FAU000
7439-89-6 see IGK800	7488-56-4 see SBR000	7705-08-0 see FAW000
7439-91-0 see LAV000	7492-70-8 see BQP000	7718-54-9 see NDH000
7439-92-1 see LCF000	7493-74-5 see AGQ750	7718-98-1 see VEP000
7439-93-2 see LGO000	7495-93-4 see CAD500	7719-09-7 see TFL000
7439-95-4 see MAC750	7521-80-4 see BSR000	7719-12-2 see PHT275
7439-96-5 see MAP750	7530-07-6 see OFI000	7720-78-7 see FBN100
7439-97-6 see MCW250	7546-30-7 see MCW000	7721-01-9 see TAF000
7439-98-7 see MRC250	7550-45-0 see TGH350	7722-64-7 see PLP000
7440-00-8 see NBX000	7553-56-2 see IDM000	7722-84-1 see HIB000
7440-01-9 see NCG500	7558-63-6 see MRF000	7722-86-3 see PCN750
7440-02-0 see NCW500	7558-79-4 see SJH090	7722-88-5 see TEE500
7440-04-2 see OKE000	7558-80-7 see SJH100	7723-14-0 see PHO500
7440-05-3 see PAD250	7563-42-0 see CAV000	7723-14-0 see PHO750
7440-06-4 see PJD500	7568-37-8 see MHY550	7723-14-0 see PHP000
7440-09-7 see PKT250	7572-29-4 see DEN600	7726-95-6 see BMP000
7440-09-7 see PKT500	7580-67-8 see LHH000	7727-15-3 see AGX750
7440-10-0 see PLX500	7601-54-9 see SJH200	7727-18-6 see VDP000
7440-15-5 see RGF000	7601-89-0 see PCE750	7727-21-1 see DWQ000
7440-16-6 see RHF000	7601-90-3 see PCD250	7727-37-9 see NGP500
7440-17-7 see RPA000	7616-94-6 see PCF750	7727-37-9 see NGR250
7440-18-8 see RRU000	7631-86-9 see SCH000	7727-43-7 see BAP000
7440-21-3 see SCP000	7631-86-9 see SCI000	7727-54-0 see ANR000
7440-22-4 see SDI500	7631-89-2 see ARD750	7733-02-0 see ZNA000
7440-23-5 see SEE500	7631-90-5 see SFE000	7738-94-5 see CMH250
7440-23-5 see SEF000	7631-95-0 see DXE800	7757-79-1 see PLL500
7440-23-5 see SEF500	7631-98-3 see DXZ000	7757-81-5 see SJV000
7440-25-7 see TAE750	7631-99-4 see SIO900	7757-82-6 see SJY000
7440-28-0 see TEI000	7632-00-0 see SIQ500	7757-83-7 see SJZ000
7440-29-1 see TFS750	7632-10-2 see DBB000	7757-93-9 see CAW100
7440-31-5 see TGB250	7632-51-1 see VEF000	7758-01-2 see PKY300
7440-32-6 see TGF250	7635-51-0 see DTN800	7758-02-3 see PKY500
7440-33-7 see TOA750	7637-07-2 see BMG700	7758-05-6 see PLK250
7440-36-0 see AQB750	7645-25-2 see ARC750	7758-09-0 see PLM500
7440-37-1 see AQW250	7646-69-7 see SHO500	7758-16-9 see DXF800
7440-38-2 see ARA750	7646-78-8 see TGC250	7758-19-2 see SFT500
7440-39-3 see BAH250	7646-79-9 see CNB599	7758-19-2 see SFU000
7440-41-7 see BFO750	7646-85-7 see ZFA000	7758-88-5 see CDA750
7440-42-8 see BMD500	7646-93-7 see PKX750	7758-94-3 see FBI000
7440-43-9 see CAD000	7647-01-0 see HHL000	7758-95-4 see LCQ000
7440-44-0 see CBT500	7647-01-0 see HHX000	7758-97-6 see LCR000
7440-45-1 see CCY250	7647-01-0 see HHX500	7758-98-7 see CNP250
7440-46-2 see CDC000	7647-10-1 see PAD500	7761-88-8 see SDS000
7440-47-3 see CMI750	7647-14-5 see SFT000	7764-50-3 see DKV175
7440-48-4 see CNA250	7647-17-8 see CDD000	7772-76-1 see ANR750
7440-50-8 see CNI000	7647-18-9 see AQD000	7772-98-7 see SKI000
7440-54-2 see GAF000	7647-19-0 see PHR750	7772-99-8 see TGC000
7440-55-3 see GBG000	7659-31-6 see SDJ000	7773-01-5 see MAR000
7440-57-5 see GIS000	7660-25-5 see LFI000	7773-06-0 see ANU650
7440-58-6 see HAC000	7664-38-2 see PHB250	7773-34-4 see DJB200
7440-59-7 see HAM500	7664-39-3 see HHU500	7774-29-0 see MDD000
7440-61-1 see UNS000	7664-39-3 see HHV500	7774-41-6 see ARC500

7775-09-9 see SFS000	7787-49-7 see BFR500	8002-09-3 see PIH750
7775-11-3 see DXC200	7787-52-2 see BFR750	8002-26-4 see TAC000
7775-14-6 see SHR500	7787-56-6 see BFU500	8002-66-2 see CDH500
7775-27-1 see SJE000	7787-62-4 see BKW750	8002-74-2 see PAH750
7778-18-9 see CAX500	7787-69-1 see CDC500	8003-03-0 see ARP250
7778-39-4 see ARB250	7787-71-5 see BMQ325	8003-19-8 see DGG000
7778-43-0 see ARC000	7789-00-6 see PLB250	8003-34-7 see P00250
7778-44-1 see ARB750	7789-06-2 see SMH000	8004-13-5 see PFA860
7778-50-9 see PKX250	7789-09-5 see ANB500	8006-39-1 see TBD500
7778-54-3 see HOV500	7789-17-5 see CDE000	8006-61-9 see GBY000
7778-74-7 see PL0500	7789-18-6 see CDE250	8006-64-2 see TOD750
7778-80-5 see PLT000	7789-20-0 see HAK000	8006-75-5 see DNU400
7779-88-6 see ZJJ000	7789-21-1 see FLZ000	8006-78-8 see LBK000
7782-39-0 see DBB800	7789-23-3 see PLF500	8006-80-2 see OHI000
7782-41-4 see FEZ000	7789-24-4 see LHF000	8006-82-4 see BLW250
7782-44-7 see OQW000	7789-25-5 see NMH500	8006-84-6 see FAP000
7782-49-2 see SB0500	7789-29-9 see PKU250	8006-90-4 see PCB250
7782-50-5 see CDV750	7789-30-2 see BMQ000	8006-99-3 see CDL500
7782-63-0 see FB0000	7789-38-0 see SFG000	8007-01-0 see RNA000
7782-65-2 see GEI100	7789-47-1 see MCY000	8007-02-1 see LEH000
7782-77-6 see NMR000	7789-59-5 see PHU000	8007-08-7 see GEQ000
7782-78-7 see NMJ000	7789-60-8 see PHT250	8007-11-2 see 0J0000
7782-79-8 see HHG500	7789-67-5 see TGB750	8007-12-3 see OGQ100
7782-89-0 see LGT000	7789-69-7 see PHR250	8007-20-3 see CCQ500
7782-92-5 see SEN000	7790-01-4 see MFX000	8007-45-2 see CMY800
7782-99-2 see S00500	7790-47-8 see TGD750	8007-46-3 see TFX500
7783-06-4 see HIC500	7790-59-2 see PLR750	8007-46-3 see TFX750
7783-07-5 see HIC000	7790-78-5 see CAE425	8007-56-5 see HHM000
7783-08-6 see SBN500	7790-79-6 see CAG250	8007-70-3 see A0U250
7783-20-2 see ANU750	7790-84-3 see CAJ250	8007-75-8 see BF0000
7783-28-0 see ANR500	7790-86-5 see CCY750	8007-80-5 see CC0750
7783-30-4 see MDC750	7790-91-2 see CDX750	8008-20-6 see KEK000
7783-33-7 see NCP500	7790-93-4 see CDU000	8008-26-2 see 0G0000
7783-35-9 see MDG500	7790-94-5 see CLG500	8008-31-9 see TAD500
7783-36-0 see MDG250	7790-98-9 see PCD500	8008-45-5 see N0G500
7783-41-7 see ORA000	7790-99-0 see IDS000	8008-51-3 see CBB500
7783-42-8 see TFL250	7791-11-9 see RPF000	8008-52-4 see CNR735
7783-46-2 see LDF000	7791-12-0 see TEJ250	8008-56-8 see LEI000
7783-48-4 see SMI500	7791-20-0 see NDA000	8008-57-9 see OGY000
7783-49-5 see ZHS000	7791-23-3 see SBT500	8008-79-5 see SKY000
7783-50-8 see FAX000	7791-25-5 see S0T000	8008-94-4 see LFN300
7783-54-2 see NGW000	7791-27-7 see PPR500	8012-54-2 see ARI500
7783-55-3 see PHQ500	7803-49-8 see HLM500	8012-74-6 see LIC000
7783-56-4 see AQE000	7803-51-2 see PGY000	8012-75-7 see CAP750
7783-60-0 see S0R000	7803-52-3 see SLQ000	8012-89-3 see BAU000
7783-61-1 see SDF650	7803-55-6 see ANY250	8012-91-7 see JEA000
7783-64-4 see ZQS000	7803-57-8 see HGU500	8012-95-1 see MQV750
7783-66-6 see IDT000	7803-62-5 see SDH575	8012-96-2 see IGF000
7783-70-2 see AQF250	8000-25-7 see RMU000	8013-75-0 see FQT000
7783-71-3 see TAF250	8000-26-8 see PIH400	8013-76-1 see BLV500
7783-79-1 see SBS000	8000-26-8 see PIH500	8014-13-9 see COF325
7783-80-4 see TAK000	8000-28-0 see LCD000	8014-19-5 see PAE000
7783-81-5 see U0S000	8000-42-8 see CBG500	8014-95-7 see S0I520
7783-95-1 see SDQ500	8000-46-2 see GDA000	8015-01-8 see MBU500
7784-21-6 see AHB500	8000-48-4 see EQQ000	8015-12-1 see EEH520
7784-33-0 see ARF250	8000-66-6 see CCJ625	8015-14-3 see LJE000
7784-34-1 see ARF500	8000-68-8 see PAL750	8015-19-8 see DNX500
7784-35-2 see ARI250	8001-15-8 see EAM500	8015-30-3 see EAP000
7784-37-4 see MDF350	8001-23-8 see SAC000	8015-54-1 see DXP800
7784-40-9 see LCK000	8001-25-0 see 0IQ000	8015-73-4 see BAR250
7784-41-0 see ARD250	8001-26-1 see LGK000	8015-79-0 see OGK000
7784-42-1 see ARK250	8001-29-4 see CNU000	8015-88-1 see CCL750
7784-44-3 see DCG800	8001-30-7 see CNS000	8015-92-7 see CDH750
7784-45-4 see ARG750	8001-31-8 see CNR000	8015-97-2 see CMY100
7784-46-5 see SEY500	8001-35-2 see CDV100	8016-20-4 see GJU000
7784-46-5 see SEZ000	8001-50-1 see TBC500	8016-26-0 see LAC000
7785-84-4 see SKM500	8001-58-9 see CMY825	8016-31-7 see LII000
7785-87-7 see MAU250	8001-61-4 see CNH792	8016-45-3 see PIG750
7786-30-3 see MAE250	8001-79-4 see CCP250	8016-68-0 see SBA000
7786-34-7 see MQR750	8001-85-2 see BMA750	8016-88-4 see TAF700
7786-67-6 see MCE750	8001-88-5 see BG0750	8021-27-0 see AAC250
7786-81-4 see NDK500	8002-03-7 see PA0000	8021-29-2 see FBV000
7787-36-2 see PCK000	8002-05-9 see PCR250	8022-00-2 see MIW100
7787-47-5 see BFQ000	8002-05-9 see PCS250	8022-15-9 see LCA000

8022-37-5 see ARL250	9004-66-4 see IGS000	10049-03-3 see FFD000
8022-56-8 see SAE500	9004-70-0 see CCU250	10049-04-4 see CDW450
8022-56-8 see SAE550	9004-70-0 see CNH000	10049-07-7 see RHK000
8023-53-8 see AFP750	9004-99-3 see PJV250	10049-08-8 see RRZ000
8024-14-4 see DAJ800	9005-08-7 see PJU500	10061-01-5 see DGH200
8024-37-1 see COG000	9005-25-8 see SLJ500	10061-02-6 see DGH000
8028-73-7 see ARE500	9005-32-7 see AFL000	10072-25-0 see DFH000
8028-75-9 see ARE250	9005-37-2 see PNJ750	10085-81-1 see BCH750
8028-89-5 see CBG125	9005-38-3 see SEH000	10087-89-5 see DWL400
8030-30-6 see NAI500	9005-46-3 see SFQ000	10099-58-8 see LAX000
8031-42-3 see DKL200	9005-49-6 see HAQ500	10099-59-9 see LBA000
8042-47-5 see MQV875	9005-64-5 see PKL000	10099-66-8 see LIW000
8047-67-4 see IHG000	9005-65-6 see PKL100	10099-67-9 see LIY000
8048-52-0 see DBX400	9005-67-8 see PKL030	10099-74-8 see LD0000
8049-17-0 see FBG000	9005-81-6 see CCT250	10101-41-4 see CAX750
8049-19-2 see FBF000	9007-13-0 see CAW500	10101-50-5 see SJC000
8050-07-5 see OIM000	9009-54-5 see PKL500	10101-97-0 see NDL000
8050-88-2 see CCU000	9011-04-5 see HCV500	10102-06-4 see USA000
8051-02-3 see VHZ000	9011-14-7 see PKB500	10102-17-7 see SKI500
8052-41-3 see SLU500	9014-01-1 see BAC000	10102-18-8 see SJT500
8052-42-4 see PCR500	9014-02-2 see NBV500	10102-20-2 see SKC500
8052-42-4 see AR0500	9015-68-3 see ARN800	10102-43-9 see NEG100
8052-42-4 see AR0750	9016-00-6 see PJR000	10102-44-0 see NGR500
8063-18-1 see SLV500	9016-45-9 see PKF000	10102-45-1 see TEK750
8064-66-2 see MCA500	9056-38-6 see NMB000	10102-49-5 see IGN000
8065-36-9 see BTA250	9076-25-9 see PCC000	10102-50-8 see IGM000
8065-48-3 see DA0600	10008-90-9 see CLU500	10102-53-1 see ARB000
8070-50-6 see EJ0000	10022-31-8 see BAN250	10103-50-1 see ARD000
9000-01-5 see AQQ500	10022-50-1 see NMT500	10103-60-3 see ARD600
9000-02-6 see AHJ000	10022-68-1 see CAH250	10108-56-2 see BQW750
9000-07-1 see CCL250	10023-25-3 see DKS800	10108-64-2 see CAE250
9000-21-9 see FPQ000	10024-93-8 see NBY000	10117-38-1 see PLT500
9000-28-6 see GLY000	10024-97-2 see NGU000	10118-76-0 see CAV250
9000-29-7 see GLY100	10025-65-7 see PJE000	10124-36-4 see CAJ000
9000-30-0 see GLU000	10025-67-9 see SON510	10124-37-5 see CAU000
9000-36-6 see KBK000	10025-73-7 see CMJ250	10124-43-3 see CNE125
9000-40-2 see LIA000	10025-74-8 see DYG600	10124-48-8 see MCW500
9000-55-9 see PJJ000	10025-76-0 see ERA500	10124-50-2 see PKV500
9000-65-1 see THJ250	10025-78-2 see TJD500	10124-56-8 see SHM500
9001-00-7 see BM0000	10025-82-8 see ICK000	10137-69-6 see CPE500
9001-33-6 see FBS000	10025-85-1 see NGQ500	10137-74-3 see CA0500
9001-73-4 see PAG500	10025-87-3 see PHQ800	10138-41-7 see ECX500
9002-18-0 see AEX250	10025-91-9 see AQC500	10138-52-0 see GAH000
9002-84-0 see TAI250	10025-99-7 see PJD250	10138-60-0 see HFF000
9002-86-2 see PKQ059	10026-04-7 see SCQ500	10138-62-2 see HGG000
9002-88-4 see PJS750	10026-08-1 see TFT000	10140-89-3 see DGH800
9002-89-5 see PKP750	10026-10-5 see UQJ000	10141-00-1 see PLB500
9002-93-1 see PKF500	10026-11-6 see ZPA000	10168-80-6 see ECY500
9003-04-7 see SJK000	10026-12-7 see NEA000	10168-81-7 see GAL000
9003-07-0 see PMP500	10026-13-8 see PHR500	10192-29-7 see ANE250
9003-20-7 see AAX250	10028-15-6 see ORW000	10210-36-3 see GJG000
9003-22-9 see AAX175	10028-18-9 see NDC500	10210-68-1 see CNB500
9003-39-8 see PKQ250	10028-22-5 see FBA000	10222-01-2 see DDM000
9003-39-8 see PKQ500	10031-13-7 see LCL000	10241-05-1 see MRD500
9003-39-8 see PKQ750	10031-18-2 see MCX750	10265-92-6 see DTQ400
9003-39-8 see PKR000	10031-53-5 see ERC000	10290-12-7 see CNN500
9003-39-8 see PKR250	10031-59-1 see TEL750	10294-33-4 see BMG400
9003-39-8 see PKR500	10034-81-8 see PCE000	10294-34-5 see BMG500
9003-39-8 see PKR750	10034-85-2 see HHI500	10294-40-3 see BAK250
9003-39-8 see PKS000	10034-93-2 see HGW500	10294-41-4 see CDB250
9003-53-6 see SMQ500	10035-10-6 see HHJ000	10294-70-9 see TGD500
9003-54-7 see ADY500	10036-47-2 see TCI000	10309-37-2 see BAD625
9003-55-8 see SMR000	10038-98-9 see GDY000	10309-79-2 see MHN750
9004-32-4 see SF0500	10039-54-0 see OLS000	10311-84-9 see DBI099
9004-34-6 see CCU150	10042-76-9 see SMK000	10318-26-0 see DDJ000
9004-51-7 see IGU000	10042-88-3 see TAM000	10325-94-7 see CAH000
9004-53-9 see DBD800	10043-01-3 see AHG750	10326-21-3 see MAE000
9004-54-0 see DBC800	10043-35-3 see BMC000	10326-24-6 see ZDS000
9004-54-0 see DBD000	10043-52-4 see CA0750	10331-57-4 see DFD000
9004-54-0 see DBD200	10045-94-0 see MDF000	10356-76-0 see FH0000
9004-54-0 see DBD400	10045-95-1 see NCB000	10361-03-2 see SII500
9004-54-0 see DBD600	10048-13-2 see SLP000	10361-44-1 see BKW250
9004-64-2 see HNV000	10048-32-5 see PAJ500	10361-79-2 see PLX750
9004-65-3 see HNX000	10048-95-0 see ARC250	10361-80-5 see PLY250

13847-65-9 see NGS500	15598-34-2 see PPC100	17088-21-0 see EEF000
13860-69-0 see MMV000	15605-28-4 see NMI000	17090-79-8 see MRE225
13863-59-7 see BMP750	15652-38-7 see DAE600	17109-49-8 see EIM000
13863-88-2 see SDM500	15662-33-6 see RSZ000	17160-71-3 see FDA880
13909-09-6 see CHD250	15663-27-1 see PJD000	17168-85-3 see THP250
13927-77-0 see BIW750	15709-62-3 see BLS500	17230-87-4 see CLS250
13952-84-6 see BPY000	15721-02-5 see TBO000	17236-22-5 see IFG000
13967-90-3 see BAI750	15721-33-2 see DNO200	17372-87-1 see BNH500
13973-88-1 see CDW000	15773-47-4 see DJN800	17381-88-3 see DBJ400
13977-28-1 see BMN250	15805-73-9 see VNK000	17406-45-0 see THG250
13980-04-6 see HDV500	15825-70-4 see MAW250	17427-00-8 see ABV250
13984-07-1 see MLL000	15829-53-5 see MDF750	17573-23-8 see DKU000
14008-44-7 see MQR000	15876-67-2 see DXG800	17575-22-3 see LAU000
14024-64-7 see BGQ750	15879-93-3 see GFA000	17605-71-9 see MJU750
14060-38-9 see ARJ500	15923-42-9 see HDQ500	17608-59-2 see NKC000
14096-51-6 see DFJ000	15930-94-6 see CMK500	17617-23-1 see FMQ000
14144-91-3 see DQE800	15954-91-3 see CAF750	17639-93-9 see CKT000
14148-99-3 see DWA600	15972-60-8 see CFX000	17673-25-5 see PGS250
14215-29-3 see CAD350	16033-21-9 see PFS500	17702-41-9 see DAE400
14220-17-8 see NDI000	16039-55-7 see CAG750	17780-75-5 see CMY000
14235-86-0 see PFN000	16066-38-9 see DWV400	17804-35-2 see BAV575
14239-51-1 see BJB750	16071-86-6 see CMO750	17822-74-1 see DHQ800
14239-68-0 see BJB500	16102-92-4 see CNI900	18186-71-5 see DXU200
14255-87-9 see BQK000	16110-13-7 see BBH250	18237-15-5 see AME750
14255-88-0 see DGA200	16111-62-9 see DJK800	18378-89-7 see MQW750
14264-16-5 see BLS250	16219-75-3 see EL0500	18433-84-6 see PAT799
14323-41-2 see TBW250	16227-10-4 see BPU000	18454-12-1 see LCS000
14324-55-1 see BJC000	16230-71-0 see DWI400	18461-55-7 see AAU250
14333-26-7 see DBC000	16291-96-6 see CDI250	18466-11-0 see MOB250
14362-31-3 see CDR250	16291-96-6 see CDI500	18559-94-9 see BQF500
14402-89-2 see SIU500	16291-96-6 see CDJ500	18662-53-8 see NEI000
14448-38-5 see HOW500	16291-96-6 see CDK000	18810-58-7 see BAI000
14452-57-4 see MAH750	16301-26-1 see ASP000	18883-66-4 see SMD000
14464-46-1 see SCJ000	16338-99-1 see HBP000	18972-56-0 see MAG250
14484-64-1 see FAS000	16339-04-1 see ELX500	19010-66-3 see LCW000
14486-19-2 see CAG000	16339-05-2 see BRY250	19049-40-2 see BFT500
14557-50-7 see DIF600	16339-07-4 see NKW500	19072-57-2 see DRK600
14628-06-9 see MPG250	16339-16-5 see CIQ500	19168-23-1 see ANF000
14644-61-2 see ZTJ000	16395-80-5 see MNA250	19287-45-7 see DDI450
14666-78-5 see DJU600	16409-45-3 see MCG500	19464-55-2 see TKT750
14667-55-1 see TME270	16423-68-0 see FAG040	19473-49-5 see MRK500
14674-72-7 see CAP000	16454-60-7 see NCB500	19525-20-3 see TER500
14684-25-4 see PFW500	16532-79-9 see BNV750	19526-81-9 see EAJ500
14696-82-3 see IDN000	16543-55-8 see NLD500	19624-22-7 see PAT750
14708-14-6 see NDC000	16561-29-8 see PGV000	19910-65-7 see BSD000
14763-77-0 see CNL250	16568-02-8 see AAH000	19982-87-7 see FMH000
14807-96-6 see TAB750	16595-80-5 see LFA020	19992-69-9 see PPL500
14807-96-6 see TAB775	16672-87-0 see CDS125	20198-77-0 see DFJ400
14808-60-7 see SCJ500	16721-80-5 see SHR000	20240-98-6 see MNT500
14816-18-3 see BAT750	16731-55-8 see PLR250	20241-03-6 see DSR200
14873-10-0 see BJY000	16752-77-5 see MDU600	20265-97-8 see AOX500
14885-29-1 see IGH000	16813-36-8 see NJY000	20296-29-1 see OCY100
14901-07-6 see IFX000	16842-03-8 see CNC230	20325-40-0 see DOA800
14901-08-7 see COU000	16853-85-3 see LHS000	20373-56-2 see DEW400
14913-29-2 see PJI500	16870-90-9 see HOP259	20548-54-3 see CAY000
14913-33-8 see DEX000	16871-71-9 see ZIA000	20627-33-2 see TLJ750
14938-35-3 see AOM250	16871-90-2 see PLH750	20738-78-7 see DWX200
14977-61-8 see CML125	16893-85-9 see DXE000	20770-41-6 see PLQ500
15120-17-9 see ARD500	16919-27-0 see PLI000	20816-12-0 see OKK000
15158-67-5 see CDA500	16919-58-7 see ANF250	20820-80-8 see EIF500
15191-85-2 see SCN500	16925-39-6 see CAX250	20830-75-5 see DKN400
15216-10-1 see NJL000	16940-66-2 see SFF500	20830-81-3 see DAC000
15284-15-8 see MDP500	16940-81-1 see HDE000	20859-73-8 see AHE750
15356-70-4 see MCG000	16941-12-1 see CKO750	20917-49-1 see OBY000
15421-84-8 see DIO200	16949-15-8 see LHT000	20941-65-5 see EPJ000
15442-77-0 see BIX500	16961-83-4 see SCO500	20977-05-3 see EDM000
15457-87-1 see TFT250	16962-07-5 see AHG875	20977-50-8 see FGV000
15468-32-3 see SCK000	16962-40-6 see ANI250	21000-42-0 see DWX000
15481-70-6 see DCE400	17010-21-8 see CAG500	21087-64-9 see MQR275
15503-86-3 see RFU000	17013-01-3 see DXD800	21109-95-5 see BAP250
15529-90-5 see CLQ500	17013-37-5 see EJY000	21243-26-5 see DLO200
15537-73-2 see BOM750	17014-71-0 see PLP250	21247-98-3 see BCV250
15546-16-4 see BHK250	17040-19-6 see DAP600	21255-83-4 see BMP500
15589-00-1 see DOG600	17068-78-9 see ARM266	21259-20-1 see FQS000

21259-76-7 see DXC000	25154-54-5 see DUQ180	27848-84-6 see NDM000
21260-46-8 see BKW000	25155-30-0 see DXW200	27858-07-7 see OAH000
21351-79-1 see CDD750	25167-67-3 see BOW250	27905-02-8 see GBO000
21416-67-1 see RCA375	25168-04-1 see NMS000	28014-46-2 see EDS000
21416-87-5 see PIK250	25168-26-7 see ILO000	28258-59-5 see XRS000
21436-96-4 see XOJ000	25201-35-8 see PAV250	28260-61-9 see TML325
21436-97-5 see TLG750	25311-71-1 see IMF300	28300-74-5 see AQG250
21450-81-7 see AAS500	25321-14-6 see DVG600	28322-02-3 see ABY000
21561-99-9 see MMY500	25322-20-7 see TBP750	28434-86-8 see BGT000
21593-23-7 see CCX500	25322-68-3 see PJT000	28675-08-3 see DFE800
21600-42-0 see DTV200	25322-68-3 see PJT200	28801-69-6 see TIF250
21600-43-1 see DJY000	25322-68-3 see PJT225	28895-91-2 see ACR400
21600-45-3 see HKV000	25322-68-3 see PJT230	28930-30-5 see BLR750
21609-90-5 see LEN000	25322-68-3 see PJT240	29232-93-7 see DIN800
21725-46-2 see BLW750	25322-68-3 see PJT250	29575-02-8 see BHK500
21736-83-4 see SLI325	25322-68-3 see PJT500	29605-96-7 see MQP500
21820-82-6 see DMB000	25322-68-3 see PJT750	29611-03-8 see AEW500
21842-58-0 see DLQ800	25322-68-3 see PJU000	29674-96-2 see MRJ250
21884-44-6 see LIV000	25322-69-4 see PKI500	29734-68-7 see DLL600
21908-53-2 see MCT500	25322-69-4 see PKI750	29767-20-2 see EQP000
22089-22-1 see TNT500	25323-30-2 see DFH800	29790-52-1 see NDR000
22144-77-0 see ZUS000	25324-56-5 see TGE500	30026-92-7 see BRR250
22224-92-6 see FAK000	25332-09-6 see PGF000	30031-64-2 see BOP750
22254-24-6 see IGG000	25340-17-4 see DIU000	30041-69-1 see DPR000
22316-47-8 see CIR750	25355-61-7 see MMW775	30223-48-4 see FDE000
22323-45-1 see ZJA000	25376-45-8 see TGL500	30525-89-4 see PAI000
22494-42-4 see DKI600	25377-73-5 see DXV000	30560-19-1 see DOP600
22670-79-7 see PEJ250	25389-94-0 see KAM000	30586-10-8 see DFX000
22750-53-4 see CAD325	25550-58-7 see DUY600	31185-56-5 see MQE000
22750-93-2 see EOD000	25551-13-7 see TLL250	31218-83-4 see MKA000
22751-24-2 see NFA500	25567-67-3 see CGL750	31242-94-1 see TBP250
22781-23-3 see DQM600	25601-84-7 see DOL800	31282-04-9 see AQB000
22839-47-0 see ARN825	25639-42-3 see MIQ745	31431-39-7 see MHL000
22960-71-0 see EEX500	25805-16-7 see PKP000	31540-62-2 see FIW000
22966-79-6 see EDR500	25843-45-2 see ASP250	31566-31-1 see OAV000
22967-92-6 see MLF550	25875-51-8 see RLK890	31959-87-2 see MKI000
23107-12-2 see DAL400	25931-01-5 see PKL750	32222-06-3 see DMJ400
23135-22-0 see DSP600	25951-54-6 see PKQ000	32238-28-1 see PKM500
23214-92-8 see AES750	25956-17-6 see FAG100	32248-37-6 see PIV750
23255-69-8 see FQR000	26043-11-8 see NDD000	32446-40-5 see AON500
23255-93-8 see HGO500	26049-68-3 see HHD500	32854-75-4 see LBD000
23257-56-9 see LFO000	26049-69-4 see DSG400	32976-88-8 see ENT000
23261-20-3 see DCI600	26134-62-3 see LHM000	33396-37-1 see MCI750
23315-05-1 see EAG000	26140-60-3 see TBC600	33401-94-4 see TCW750
23327-57-3 see NBS500	26148-68-5 see AJD750	33419-42-0 see EAV500
23414-72-4 see ZLA000	26354-18-7 see OIY000	33770-60-4 see DFC800
23422-53-9 see DSO200	26375-23-5 see PKM250	33804-48-7 see DPQ200
23435-31-6 see DON400	26401-20-7 see HFJ600	33857-26-0 see DAC800
23668-76-0 see DNR200	26401-97-8 see BKK750	33868-17-6 see MMX000
23705-25-1 see PLN100	26471-62-5 see TGM740	34099-73-5 see BMC250
23745-86-0 see PLG000	26538-44-3 see RBF100	34202-69-2 see HDA500
23746-34-1 see PKX500	26571-79-9 see CKM250	34465-46-8 see HAJ500
23757-42-8 see AMQ500	26604-66-0 see PCO000	34491-04-8 see DTP800
23795-03-1 see DWW200	26628-22-8 see SFA000	34491-12-8 see BJD000
23844-24-8 see BCL250	26645-10-3 see DJJ850	34493-98-6 see DCQ800
23950-58-5 see DTT600	26747-87-5 see EHV000	34501-24-1 see BBE750
24096-53-5 see DGF000	26782-43-4 see HOF000	34521-09-0 see AQI750
24220-18-6 see DPS200	26818-53-1 see DEB800	34562-99-7 see AJC500
24221-86-1 see EAX000	26837-42-3 see PKS250	34590-94-8 see DWT200
24426-36-6 see IPG000	26864-56-2 see PAP250	34624-48-1 see AFR500
24554-26-5 see NGM500	26952-21-6 see ILL000	35154-45-1 see ISZ000
24613-89-6 see CMI250	27083-55-2 see PKO500	35367-38-5 see CJV250
24909-09-9 see DLC000	27137-85-5 see DGF200	35400-43-2 see SOU625
24928-17-4 see PGT250	27152-57-4 see CAM500	35554-44-0 see FPB875
24961-39-5 see BNO750	27156-03-2 see DFA200	35607-66-0 see CCS500
24973-25-9 see DOJ800	27156-32-7 see MJE750	35631-27-7 see NKU875
25013-15-4 see VQK650	27193-86-8 see DXY600	35658-65-2 see CAE500
25013-15-4 see MPK500	27196-00-5 see TBY250	35725-30-5 see DYH000
25013-16-5 see BQI000	27203-92-5 see THJ500	35725-31-6 see HGH000
25036-33-3 see PKP250	27215-10-7 see DNK800	35725-33-8 see TFX250
25038-54-4 see PJY500	27262-46-0 see BON750	35846-53-8 see MBU820
25057-89-0 see MJY500	27554-26-3 see ILR100	35865-33-9 see DCI400
25103-58-6 see DXT800	27755-15-3 see DXP000	36011-19-5 see CQM250
25152-84-5 see DAE450	27774-13-6 see VEZ000	36226-64-9 see BEA500

63918-98-9 see EFG500	64742-05-8 see MQV862	68916-04-1 see BLV750
63919-01-7 see FID000	64742-10-5 see MQV863	69029-52-3 see LDC000
63937-14-4 see MDC500	64742-11-6 see MQV857	69112-96-5 see DFW000
63938-10-3 see CLH000	64742-17-2 see MQV872	69226-06-8 see DKC200
63938-24-9 see ACP500	64742-18-3 see MQV760	69226-45-5 see DVL800
63980-20-1 see DKD200	64742-19-4 see MQV770	69343-45-9 see NCJ000
63980-61-0 see PHO250	64742-20-7 see MQV765	69352-97-2 see BFN750
63981-09-9 see HHF500	64742-21-8 see MQV775	69382-20-3 see BGC250
63989-69-5 see IGO000	64742-52-5 see MQV790	70281-30-0 see DFB000
63989-82-2 see DUT200	64742-53-6 see MQV800	70288-86-7 see ITD875
63990-88-5 see BFT750	64742-54-7 see MQV795	70536-17-3 see AFI850
64036-46-0 see DJV800	64742-55-8 see MQV805	71016-15-4 see MHW350
64036-91-5 see BKD750	64742-56-9 see MQV840	71752-69-7 see NKO400
64037-50-9 see EIJ000	64742-64-9 see MQV835	72589-96-9 see CAE375
64038-56-8 see DUI400	64742-68-3 see MQV865	73419-42-8 see CAK250
64039-27-6 see TFJ250	64742-69-4 see MQV867	73622-67-0 see DBH800
64043-53-4 see CMR250	64742-70-7 see MQV868	73771-81-0 see QIS000
64046-79-3 see QDJ000	64742-71-8 see MQV870	73790-27-9 see MOK000
64046-96-4 see CAL500	65072-04-0 see VIZ000	73840-42-3 see MOA250
64047-30-9 see TLY000	65089-17-0 see CLW500	73926-87-1 see CET000
64048-13-1 see DJB600	65229-18-7 see HMQ500	73941-35-2 see THV500
64050-03-9 see HOL000	65763-32-8 see DLD800	73987-52-7 see EEP000
64050-79-9 see HNR500	65986-80-3 see ENR500	74278-22-1 see KHU000
64059-26-3 see LAH000	65996-93-2 see CMZ100	74926-97-9 see BRR500
64070-13-9 see DHA400	65997-15-1 see PKS750	75198-31-1 see NGI800
64070-83-3 see ARE000	66017-91-2 see PNR250	75321-19-6 see TMN000
64091-91-4 see MMS500	66104-24-3 see BFP500	75321-20-9 see DVD400
64092-23-5 see BHB000	66733-21-9 see EDC650	75965-74-1 see MFB400
64093-79-4 see NBW000	66788-01-0 see DKU400	76180-96-6 see AKT600
64365-11-3 see CDI000	67050-97-9 see EPC000	77500-04-0 see AJQ675
64475-85-0 see PCT250	67293-88-3 see DBB600	77536-66-4 see ARM260
64521-15-9 see DLF200	67479-03-2 see MCR250	77536-67-5 see ARM264
64598-80-7 see DLE000	67523-22-2 see DLF600	77536-68-6 see ARM280
64719-39-7 see QBS000	67730-10-3 see DWW700	77944-89-9 see CIK500
64741-49-7 see MQV755	67730-11-4 see AKS250	78109-87-2 see DII400
64741-50-0 see MQV815	67774-32-7 see FBU509	78246-54-5 see HLX925
64741-51-1 see MQV785	68006-83-7 see ALD750	83768-87-0 see VLU200
64741-52-2 see MQV810	68162-13-0 see DLD600	84837-04-7 see SLB500
64741-53-3 see MQV780	68308-34-9 see COD750	88208-15-5 see NLY750
64741-89-5 see MQV855	68476-85-7 see LGM000	91724-16-2 see SME500
64741-97-5 see MQV852	68808-54-8 see AJR500	91845-41-9 see PCR000
64742-03-6 see MQV860	68848-64-6 see LHP000	101652-07-7 see DFM800
64742-04-7 see MQV859	68855-54-9 see DCJ800	102488-99-3 see AEB750

TEMPERATURE CONVERSION TABLE

°C = (°F − 32) × 5/9 (or 0.55)   °F = °C × 9/5 (or 1.8) + 32

°F	°C	°F	°C	°F	°C
−40	−40	180	82.2	400	204.4
−35	−37.2	185	85.0	405	207.2
−30	−34.4	190	87.8	410	210.0
−25	−31.6	195	90.5	415	212.8
−20	−28.9	200	93.3	420	215.5
−15	−26.1	205	96.1	425	218.3
−10	−23.3	210	98.9	430	221.1
−5	−20.5	212	100.0	435	223.9
0	−17.78	215	101.6	440	226.6
+5	−15.0	220	104.4	445	229.4
10	−12.2	225	107.2	450	232.2
15	−9.4	230	110.0	455	235.0
20	−6.6	235	112.8	460	237.8
25	−3.9	240	115.5	465	240.5
30	−1.1	245	118.3	470	243.3
32	0	250	121.1	475	246.1
35	+1.6	255	123.9	480	248.9
40	4.4	260	126.6	485	251.6
45	7.2	265	129.4	490	254.4
50	10.0	270	132.2	495	257.2
55	12.8	275	135.0	500	260.0
60	15.5	280	137.8	550	287.8
65	18.3	285	140.5	600	315.5
70	21.1	290	143.3	650	343.3
75	23.9	295	146.1	700	371.1
80	26.6	300	148.9	750	398.9
85	29.4	305	151.6	800	426.6
90	32.2	310	154.4	850	454.4
95	35.0	315	157.2	900	482.2
100	37.8	320	160.0	950	510.0
105	40.5	325	162.8	1000	537.8
110	43.3	330	165.5	1100	593.3
115	46.1	335	168.3	1200	648.9
120	48.9	340	171.1	1300	704.4
125	51.6	345	173.9	1400	760.0
130	54.4	350	176.6	1500	815.5
135	57.2	355	179.4	1600	871.1
140	60.0	360	182.2	1700	926.6
145	62.8	365	185.0	1800	982.2
150	65.5	370	187.8	1900	1037.8
155	68.3	375	190.5	2000	1093.3
160	71.1	380	193.3		
165	73.9	385	196.1		
170	76.6	390	198.9		
175	79.4	395	201.6		